Springer Textbooks in Earth Sciences, Geography and Environment

The Springer Textbooks series publishes a broad portfolio of textbooks on Earth Sciences, Geography and Environmental Science. Springer textbooks provide comprehensive introductions as well as in-depth knowledge for advanced studies. A clear, reader-friendly layout and features such as end-of-chapter summaries, work examples, exercises, and glossaries help the reader to access the subject. Springer textbooks are essential for students, researchers and applied scientists.

Ray Cas • Guido Giordano •
John V. Wright

Volcanology

Processes, Deposits, Geology and Resources

Volume 2

Springer

Ray Cas
School of Earth, Atmosphere and Environment
Monash University
Clayton, Victoria, Australia

Ray Cas
CODES, School of Natural Sciences
University of Tasmania
Sandy Bay, Tasmania, Australia

Guido Giordano
Dipartimento di Scienze, Sezione Geologia
Università Roma Tre
Rome, Italy

John V. Wright
Consultant
Spanish Water, Curaçao

ISSN 2510-1307 ISSN 2510-1315 (electronic)
Springer Textbooks in Earth Sciences, Geography and Environment
ISBN 978-3-319-66612-9 ISBN 978-3-319-66613-6 (eBook)
https://doi.org/10.1007/978-3-319-66613-6

Cover Part of the fissure vent magma fountaining eruption at Bárðarbunga during the Holuhraun eruption, Iceland, 2014. Photo by Gro Birkefeldt Møller Pedersen

This Springer imprint is published by the registered company Springer Nature Switzerland AG
The registered company address is: Gewerbestrasse 11, 6330 Cham, Switzerland

(a)
(b)
(c)
(d)
(e)
(f)
(g)
(h)

◀ **Frontispiece** Images illustrating the diversity of volcanic activity occurring this twenty-first century—so far! **a** Lava lake at the summit crater of the Nyiragongo volcano (~3000 m a.s.l). The ultramafic composition results in eruptions with unusually fluid lava flows which, on September 2002 and May 2021, caused loss of life and destruction in the city of Goma, Democratic Republic of the Congo. (Photo by Roger Scoon, taken in March 2020). **b** Magma or fire fountaining on 4 October 2021, during the September to December eruption of Cumbre Vieja shield volcano on La Palma, Canary Islands, Spain. This built a new basanitic scoria cone, while lava flows reached the sea forming a lava delta. **c** Explosive eruption column and downwind plume from the Calbuco stratovolcano, which erupted on 22 April 2015 for the first time in four decades. The eruption lasted 90 minutes and sent a plume of volcanic ash >15 km into the atmosphere. View from Puerto Montt city, southern Chile. (Photo by Carolina Barría Kemp; *Source* Wikimedia, through Creative Commons Attribution-Share Alike 2.0 Generic licence). **d** The former village of San Miguel Los Lotes, buried by a pyroclastic flow on 3 June, 2018, erupted from the Fuego stratovolcano. It is now designated as an official national cemetery by the Government of Guatemala. (Photo by Capt. Bombero; *Source* Wikimedia, through Creative Commons Attribution-Share Alike 4.0 International licence). **e** Intra-subglacial volcanic eruption of Eyjafjallajökull volcano on 17 April, 2010, southern Iceland. The basaltic ash plume rose to a height of over 10 km causing widespread disruption of European air traffic. (Photo by Boaworm; *Source* Wikimedia, through Creative Commons Attribution 3.0 Unported licence). **f** Satellite view of the dramatically expanding, ash plume above the explosive submarine basaltic andesite eruption of Hunga Tonga–Hunga Ha'apai volcano on 15 January 2022, Tonga. The top of the main umbrella region rose to a height of at least 40 km and shows a series of concentric, outward-moving, "gravity waves" of gas, water vapour, and ash. The lower pale grey cloud is of uncertain origin but could be a radially spreading cloud of fine ash and ice particles that nucleated on fine ash particles at an altitude of ~15 km (Andrew Prata, pers. comm.) and could also be related to a buoyant ash cloud generated at sea level by near vent pyroclastic flows flowing over and into the sea. (*Source* NOAA GOES-West satellite). **g** Small, weak explosion near the summit of the submarine West Mata volcano (~1200 m water depth) ejects ash and rock, with magma glowing red. This volcanic activity was the first observed to erupt a lava of boninite composition, northeastern Lau Basin, northern end of the Tonga volcanic arc. (*Source* NSF/NOAA; ROV image taken May 2009). **h** Pristine, glassy, coherent basaltic pillow lava erupted on the seafloor at the Marianas back-arc spreading centre, western Pacific Ocean. This occurred between 2013–2015 at ~4000 m water depth. (*Source* NOAA; ROV image taken April 2016)

"We can't understand what we are looking at until we understand what we are looking for."

(Monash University Volcanics Shortcourse, 1982)

Preface

Since the publication of *Volcanic Successions* in 1987 (Cas and Wright, Allen and Unwin) there have been many advances in the field of volcanology. *Volcanic Successions* was written as a senior undergraduate, postgraduate, and industry professional-level book. Our aim here with this new book is to provide a major advance on that book at a higher level, by increasing the scope; providing much more detail on key concepts, processes, and deposit characteristics; and providing a comprehensive bibliography of each of the major topics addressed. We hope that this new book is as well received and useful as the first. In the following Introduction, we provide an overview of what we cover, the approach, and the structure of the book.

Melbourne, Australia
January 2024

Ray Cas
Guido Giordano
John V. Wright

Acknowledgements

First, we want to thank our families for their support and understanding about the time we have had to spend on the "book project", sometimes at the expense of time with them. We also thank our Institutions and Schools at Monash University, University of Tasmania, and Università Roma Tre for the opportunities, facilities, and financial support to undertake the decades of research that have led to the preparation of this book. We also thank the many corporate entities and government agencies both in Australia and from overseas and the individual course participants who have supported Monash University's annually run Volcanics Shortcourse, which since its inception in May 1982 provided the essential financial backing to complete this book. In addition, although there are too many to name specifically, we thank the many work colleagues and present and past research colleagues all over the world, in both academia and industry, for their collaborations and their encouragement to write this book. We greatly appreciate the efforts of many research students and postdoctoral research fellows we have worked with for their significant contributions to research in volcanology; some have become ongoing research collaborators. Ray Cas specifically thanks former students and postdocs Jack Simmons, Nathan Bowman, and Jozua van Otterloo for their technical assistance in drafting figures, compiling reference lists, and some proof reading of draft chapters at Monash University. Tom Orr (Orr & Associates) provided drafting facilities for several figures in Chaps. 17 and 18. In addition, we have used many figures drafted by Draga Gelt for the predecessor of this book, *Volcanic Successions* (1987). We also acknowledge the individuals who have kindly provided figures and images that we have incorporated into this book. Finally, we thank Springer for publishing our book, but which unfortunately involved multiple delays during Springer's production process, ultimately causing a significant delay in its publication.

Ray Cas
Guido Giordano
John V. Wright

Introduction: Aims, Scope, Approach, and Structure of the Book

Background

Volcanic eruptions have been one of the most fundamental influences on the Earth's evolution throughout its 4.54 Ga history. Volcanic eruptions formed the Earth's crust, were the source for the atmosphere and oceans and are arguably the basis for life on Earth. Huge eruptions have almost certainly caused climate change and influenced mass extinction events. The dynamic and varied nature of volcanic eruptions in subaerial, subaqueous, and subglacial environments and the processes by which eruption products are dispersed across the Earth's surface, the spectrum of volcanic deposit types, and their potential enormous volumes and dispersal distances make volcanic hazards among the most dangerous and threatening natural hazards on Earth. Throughout Earth's history, volcanism has produced a huge spectrum of natural resources that our society depends on, including base metals (Cu, Pb, Zn, Ni), precious metals (Au, Ag), platinum group elements (PGE), diamonds, and in modern times, it is the source for geothermal energy.

Understanding eruption styles and dispersal processes and interpreting the origins of volcanic deposits are not simple because volcanic systems and processes can be highly variable in terms of physical and chemical processes, as well as their spatial and time scales. Although we learn much about volcanic processes by witnessing modern eruptions and studying their deposits, we learn at least as much from the prehistoric volcanic rock and deposit record. Studies of the rock record also provide the database for reconstructing Earth's history. In addition, in terms of volcanic hazard evaluation, understanding the stratigraphic record of eruptions and deposits of any particular active volcano provides the best insights into the possible future eruption behaviour. However, the uncertainty of the depositional setting and its effects on eruption processes and deposit types, the huge spectrum of possible primary products and secondary deposits, and the effects of hydrothermal alteration, metamorphism, and deformation make the reconstruction of ancient volcanic successions and processes extremely challenging. The book by Cas and Wright (*Volcanic Successions* 1987) was the first to develop a facies analysis approach to the documentation and interpretation of modern and ancient volcanic successions. In this new book, we progress that approach considerably further, incorporating new approaches and developments in a total revision of the 1987 edition of *Volcanic Successions*.

Aims of the Book

The aim of this book is to help develop a better understanding of the spectrum of eruption styles in different environments and tectonic settings. In addition, we will provide a comprehensive account of the enormous range of deposit types (facies) preserved in modern and ancient volcanic terrains, their characteristics, the processes involved in their eruption and deposition, the facies and stratigraphic architecture, their depositional and tectonic setting, and their resource potential. We will also touch on the natural hazards associated with the spectrum of processes operating in volcanic terrains, although that topic is worthy of a book in its own right, and so our consideration is not comprehensive. A major aim of the book is to provide students and geoscientists with the approach and basis for making their own interpretations and reconstructions of modern, historical and ancient eruption sequences. Therefore, the book will also briefly address methods and approaches that may be useful in documenting and analysing both modern and ancient successions. The book is not an encyclopaedia, and although we present a balanced overview and review of processes and interpretations, we also present our own personal insights and assessments as research scientists with considerable experience in the range of topics discussed.

Structure, Scope, and Approach of the Book

The first-order organisation of the book is into six major parts that contain groupings of chapters related to the themes of each part: Part A: *Laying the Foundations*; Part B: *Effusive Eruptions, Lavas and Sub-Volcanic Intrusions*; Part C: *Explosive Eruptions, Processes, and Pyroclastic Deposits*; Part D: *Secondary Processes in Volcanic Settings*; Part E: *Terminology and Approaches to Understanding the Geology of Volcanoes*; and Part F: *Bringing Everything Together: The Big Picture*. This progression leads us from specific processes and deposit types to assessing the larger-scale architecture of volcanoes, their global setting, and their resources.

In Part A, many important introductory and background concepts are introduced. In Chap. 1, "Introductory Concepts", the diversity of volcanoes, eruption styles, the relative time relationships between volcanic and other processes in volcanic terrains, the facies concept, essential facies characteristics and some terminology that are important in the successful documentation and interpretation of facies are introduced. In Chap. 2, "Properties of Magmas", the variable physical and chemical properties of magmas are reviewed as the basis for understanding the nature of volcanic eruptions and the behaviour of lava flows and intrusions. In Chap. 3, "Influence of Environment on Magma Properties, Eruption Processes, and Deposits", we consider the influence that different ambient environments (atmosphere, oceans/lakes, ice, lithosphere) can have on magma properties, eruption styles, and eruption products. Finally, in Chap. 4, "Fragmentation Processes in Magmas and Volcanic Rocks", we review the diverse ways (autoclastic, explosive, hydraulic fracturing, epiclastic) by which magmas and volcanic rocks can be fragmented and the characteristics of the clast types and aggregates that characterise those different origins.

In Part B, *Effusive Eruptions, Lavas and Sub-Volcanic Intrusions*, we summarise the processes that produce largely coherent volcanic rocks, lavas, and their subvolcanic counterparts, high-level intrusions. These include Chap. 5, "Basaltic and Other Low Viscosity Lavas"; Chap. 6, "High Viscosity Silicic Lavas (Rhyolite, Dacite and Andesite"; and Chap. 7, "Komatiite Lavas and Intrusions". In each case, we consider the behaviour of these lava flow types in subaerial and subaqueous environments and the facies characteristics and facies architecture of lavas in these different settings. In Chap. 7, we also briefly address other very low-viscosity lavas, including carbonatites, and sulphur lavas. In Chap. 8, "Sub-volcanic Intrusions", we briefly consider the principles controlling the behaviour of magmas rising to the Earth's surface, the factors that determine if they will erupt or cease to rise and be emplaced as intrusions. We also assess the criteria for distinguishing intrusions, particularly sills from lavas in the geological record, which can be difficult in ancient deformed terrains but is important for understanding the chronology of events and likely resource types and settings.

Part C deals with *Explosive Eruptions, Processes, and Pyroclastic Deposits*. In Chap. 9, "Explosive Eruption Styles, Columns, and Pyroclastic Fallout Deposits", we review the characteristics and behaviour of different explosive eruption styles and columns, the dispersal processes that produce pyroclastic fallout deposits, and the deposit characteristics. We also tackle the problem of classifying or subdividing pyroclastic fallout-forming eruption styles and deposit types, which used to be simple, but is increasingly becoming more difficult, as the range of variables that affect eruption styles and deposit characteristics is better understood. As a result, we propose that existing classification schemes (D-F, VEI, Magnitude, Intensity) be recalibrated, with essential revisions proposed. In Chap. 10, "Subaerial Pyroclastic Density Currents (Pyroclastic Flows and Surges)", we assess the spectrum of pyroclastic density current types, including distinctive pyroclastic flow to pyroclastic surge types, and directed blasts. We discuss in detail the flow and sedimentation processes and the huge variety of resulting deposits and facies. We also propose a new classification scheme for ignimbrites. In Chap. 11, "Subaqueous Flows of Pyroclastic Debris and Their Deposits", we consider what happens when pyroclastic density currents enter or are erupted under water. Do they retain integrity as hot gas-supported and fluidised pyroclastic density currents, or do they ingest water and transform into water-supported transport systems? Are they still "pyroclastic", and how are such syn-eruptive deposits distinguishable from post-eruptive, resedimented deposits of pyroclastic debris and from subaerial counterparts?

This provides a segue into Part D, *Secondary Processes in Volcanic Settings*. Chapter 12, "Surface Sedimentary (Epiclastic) Processes and Deposits in Volcanic Environments", begins with an overview of the importance of surface sedimentary processes in transporting and depositing volcanic debris in both subaerial and subaqueous volcanic settings through the effects of gravity and the involvement of ice, water, or air. We consider the duration of periods of repose and their surface sedimentary processes relative to the duration of eruption episodes during the history of polygenetic volcanoes. Surface processes also pose major natural hazards in most volcanic terrains, often more so than eruption-generated primary phenomena. In Chap. 13, "Syn- to Post-emplacement Alteration Processes", we briefly address the ways in which hydrothermal, diagenetic, metamorphic alteration, and deformational processes can affect original rock textures, sometimes producing misleading pseudo-textures.

In Part E, *Terminology and Approaches*, we begin with the vexed question of "Terminology for Volcanic Deposits and Rocks" in Chap. 14, and we propose a systematic, step-by-step approach, beginning with descriptive terminology and culminating in the application of genetic terminology. In Chap. 15, "Documenting the Geology of Volcanoes, Volcanic Terrains, and Constructing Facies Models", we consider the importance of constructing facies models for all volcanoes and ancient volcanic successions, the approaches and data sets that are useful, and some of the difficulties.

In Part F, *Bringing Everything Together: The Big Picture*, we integrate concepts from preceding chapters to address some "big-picture" issues. In Chap. 16, "The Geology of Volcanoes and Their Facies Models", the facies and stratigraphic architecture and geology of different types of volcanoes and volcanic settings are considered, based on both modern and ancient examples, as well as introducing their economic significance and summarising their potential hazards. In Chap. 17, "Volcanism and Tectonic Setting", the modern plate tectonic framework is used to assess where volcanism occurs and the distinctive characteristics of volcanoes and their deposits in those settings. We then explore the tectonic regime for volcanism from the Earth's beginnings at 4.54 Ga during the Hadean and Archean and summarise how different the Earth was then (no plate tectonics until the latest Archean!) and how unique some volcanic and plutonic rocks were. Finally, in Chap. 18, "Volcanic-Hosted Resources", the spectrum of important natural resources that have formed on Earth in volcanic terrains are considered, in particular epithermal deposits, volcanic-hosted iron oxide copper–gold and uranium deposits, magmatic nickel (PGE) sulphide deposits, VMS base metal deposits, diamonds, and volcanic reservoirs hosting both hydrocarbon and geothermal energy.

We hope that you find the book to be a valuable reference volume and enjoy it as much as we have enjoyed producing it over many years.

Contents

Part D
Secondary Processes in Volcanic Settings

Aims, Scope, and Structure

In addition to primary eruption processes and deposits that are formed during active volcanism, volcanic terrains are also the sites of very active syn- to post-depositional secondary process regimes. These include surface sedimentary processes, as well as subsurface hydrothermal alteration, metamorphism and deformation. Indeed, the actual duration of eruption events at a given volcano is considerably less than of inter- or post-eruption repose periods during which vigorous surface sedimentary processes prevail. It is therefore critical for those who work in volcanic terrains to understand secondary processes, which usually produce large scale erosion at proximal sites and deposition of volcaniclastic sediment at medial and distal sites, as we discuss in Chap. 12. In addition, understanding the time scales of syn-eruptive versus post-eruptive volcanic processes is crucial to understanding the evolution of volcanoes and preserved stratigraphy. Resedimentation of volcaniclastic debris can result in wildly varying rates of sedimentation, that are very different from any other surface sedimentary environment on Earth. Volcanoes may produce huge amounts of primary volcanic debris during very short periods. This is then redistributed over time at different rates depending on volcanic sediment availability, climate, volcano-tectonics, eustatism and tectonism. By reviewing surface sedimentary or epiclastic processes and the characteristics of the resultant deposits in Chap. 12, we will highlight the features that make volcaniclastic sedimentary deposits in the rock record distinguishable from primary eruptive volcanic deposits. In Chap. 13, we then consider the range of syn- to post-emplacement alteration processes and their effects in altering the textural features of volcanic and associated sedimentary deposit types and successions, which are important in understanding ancient volcanic systems. These processes include devitrification, palagonitisation (in mafic glass), hydrothermal alteration, diagenesis, metamorphism and tectonic deformation, which may significantly modify and in some cases completely disguise original textures, which can be particularly important in ore-forming volcanic environments.

Surface Sedimentary (Epiclastic) Processes and Deposits in Volcanic Environments

12

Response to Volcanic Events, and Reconstructing Paleo-Volcanic Settings

Ray Cas, John V. Wright, and Guido Giordano

R. Cas et al., *Volcanology*, Springer Textbooks in Earth Sciences, Geography and Environment,
https://doi.org/10.1007/978-3-319-66613-6_12

(a)
(b)
(c)
(d)

Abstract

Many recent eruptions (Mt St Helens 1980, Mt Galunggung 1982, Mt Pinatubo 1991, Unzen 1991–1995, Soufrière Hills 1995–2013, Merapi 2010, and Fuego 2018) have been marked by major-impact sedimentary events both during and immediately after their respective eruptions. These secondary events have produced at least as much damage to infrastructure as the primary eruption processes, and have led to numerous deaths. In the case of the Mt Pinatubo eruption, most casualties were the result of post-eruptive hyper-concentrated flood flows that inundated the region around the volcano for at least 15 years after the eruption. Secondary sedimentary events initiated as a consequence of an eruption, or due to the instability of a high-relief volcanic edifice, therefore represent major hazards, and they must be understood as important processes in volcanic settings. In this chapter, we examine the types of surface sedimentary or epiclastic processes that operate in different volcanic terrains, including weathering and/or erosion, transportation, and deposition of volcaniclastic sediments at the Earth's surface, in both subaerial and subaqueous environments. By reviewing surface epiclastic processes and the characteristics of the resultant deposits, we will highlight the features that make volcaniclastic sedimentary deposits in the rock record distinguishable from primary eruptive volcanic deposits. These can act in concert with other pre-, syn- to post-eruption processes. Surface sedimentary processes can lead to the syn-eruptive erosion and resedimentation of unlithified pyroclastic and autoclastic deposits (epiclastic *sensu lato*), or post-eruptive erosion and reworking of older volcanic source rocks (epiclastic *sensu stricto*).

12.1 Introduction

Volcanic eruptions are short-lived and occupy only a small time interval in the total history or lifespan of a volcano, especially on large volcanoes. The principal immediate effect of eruptions is the growth or aggradation of the volcanic edifice and the surrounding volcanic landscape. This has the effect of building up the surface relief above the base level to which erosional processes work, so activating those erosional processes. Repose periods between eruptions occupy the bulk of the time interval covering the history of large polygenetic volcanoes. Given the effects of aggradation above erosional base level, the development of high slope angles regionally or locally on growing volcanic edifices, and the general abundance of loose erodible debris following volcanic explosive eruptions, it is not surprising that the long-lived repose periods are dominated by surface sedimentary processes. In these settings, erosion (Plate 12.1a, b), transportation and redeposition of volcanic debris (Plate 12.1c, d) occur at rates well above those in normal sedimentary environments.

Although the focus in understanding volcanic terrains is usually on the primary eruption processes, their deposits, and the hazards they pose, surface sedimentary processes that occur in response to an eruption, are just as perilous; moreover, they shape what is preserved in the geological record of the eruption events. For example, the 15 June 1991 explosive eruption of Mt Pinatubo stratovolcano in the Philippines (Newhall and Punongbayan 1996), erupted >7 km^3 of pyroclastic deposits ($\sim$5 km^3 DRE), much of it on the flanks of the volcano. After 5 years almost 50% had been eroded, largely by the major rivers draining the slopes of the volcano. About 250 people were killed during the eruption, during which >65,000 people were evacuated from an exclusion zone of about 15 km around the volcano. However, more than 500 people died during the monsoon-driven hyper-concentrated flood flows, which occurred in the major river valleys annually for at least 15 years after the eruption. These river valleys became choked with reworked and resedimented pyroclastic debris causing rapid aggradation and rise of the river beds (Plate 12.1d). It remains to be seen how much of the primary pyroclastic deposits of this major eruption event will be preserved in the geological record of Mt Pinatubo because of the amazingly high erosion rates that have occurred. This story has been observed at many other volcanoes, including most recently the eruption of Fuego volcano in Guatemala that began on 3 June 2018.

◀ **Plate 12.1** Illustration of surface sedimentary erosional (**a**, **b**) and depositional (**c**, **d**) processes associated with volcanic landscapes and eruptions. **a** Google Earth image of the nearly symmetrical form and deeply incised, radial drainage pattern developed around the dormant Mt Taranaki/Egmont stratovolcano, the north island of New Zealand. The last eruption was in 1854 from Fantham's Peak, the secondary cone just south of the summit. Note the deeply incised streams on the upper and middle slopes and the density of smaller streams on the lower slopes of the ring plain. The more degraded edifice to the northwest, Pouakai, may have last been active 250 kyr ago, and is therefore more degraded and incised than Taranaki volcano. (*Source* Google Earth). **b** View down one of the deeply incised box canyons on the middle slopes of Taranaki volcano, exposing a variety of slope deposits. **c** Post-eruptive choking of the North Fork Toutle River to the north of Mt St Helens 8 years after the cataclysmic 18 May 1980 eruption, USA. The once deeply incised valley was transformed into a braided river when choked by debris produced by the debris avalanche and pyroclastic flow deposits and then reworked into the river system, leading to aggradation of the river bed and damming of tributaries. **d** The choked, braided O'Donnell River draining from Mt Pinatubo, Philippines, after the 15 June 1991 eruption. The pumiceous volcanic sediment has caused aggradation and damming of the once freely flowing river system

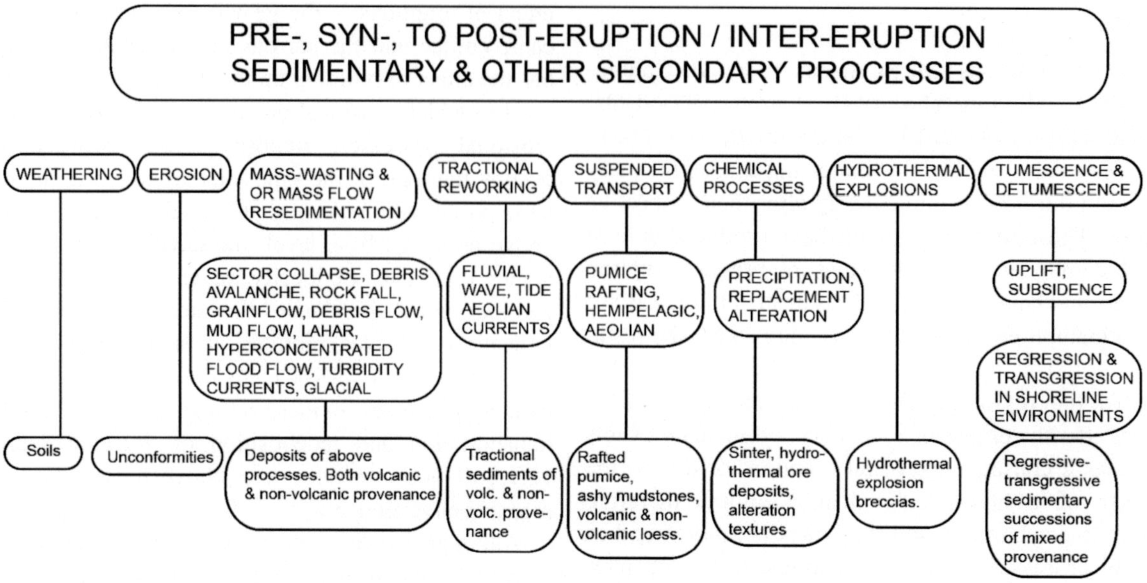

Fig. 12.1 Schematic representation of the secondary process spectrum affecting volcanic edifices before/between, during, and after primary eruption events, focusing on sedimentary surface processes to be discussed in the text

This serves to remind us that secondary surface sedimentary processes are just as important in understanding the dynamics, stratigraphy of volcanic terrains, and hazards as the primary eruption processes. It also underscores that understanding the stratigraphy of volcanoes and the frequency of eruptions is not only about what is preserved in volcanic successions but also what has been removed by surface processes, represented by erosional surfaces such as erosional scours and unconformities (Chap. 15). Given the abundance of loose debris and the overall steep slopes, the processes of erosion, transportation, and redeposition of fragmental debris constitute a major phase of activity in the majority of volcanic terrains.

In this chapter, we examine the types of surface sedimentary or epiclastic processes that operate in different volcanic terrains, as first comprehensively undertaken by Cas and Wright (1987, Chap. 10). Surface or epiclastic processes are here taken to include weathering and/or erosion, transportation and deposition of volcaniclastic sediments at the surface of the Earth (and other planets), in both subaerial and subaqueous environments (Fig. 12.1). These act in concert with other pre-, syn- to post-eruption processes, such as magmatic and hydrothermal activity, crustal movement due to magmatic tumescence and detumescence, including caldera formation, as well as tectonic processes that can trigger erosional activity and a wide spectrum of transportational processes (Fig. 12.1). The erosion can involve contemporaneous erosion of unlithified pyroclastic or autoclastic deposits (epiclastic *sensu lato*), or from contemporaneous solid rocks (e.g. lavas, welded ignimbrites), or from older volcanic source rocks (epiclastic *sensu stricto*).

12.2 Importance of Erosion and Sediment Transport in Volcanic Terrains

12.2.1 Scope of Sedimentary Processes Affecting Eroded Volcanic Debris

The surface sedimentary processes which affect eroded volcanic debris are no different from those that affect non-volcanic, terrigenous detritus. However, given the greater range of densities of the debris, due largely to varying degrees of vesiculation in volcanic eruption products, and the effects of hydraulic sorting, volcaniclastic deposits of epiclastic origin may be more poorly size-sorted, and therefore slightly different texturally, and, in terms of their sedimentary structures (Cas and Wright 1987; Manville et al. 1998, 2002) compared with terrigenous sediments. On the whole, however, the facies should reflect their surface sedimentary origins, although distinguishing them from some primary deposits is not always straightforward. By reviewing surface epiclastic processes and the characteristics of the resultant deposits, we will highlight the features that make volcaniclastic sedimentary deposits in the rock record distinguishable from primary eruptive volcanic deposits. Our review of sedimentological principles and processes cannot be comprehensive. We will summarise basic concepts and physical principles, and, by way of examples from volcanic settings, show how surface sedimentary or epiclastic processes are as relevant in volcanic settings as in non-volcanic ones.

Relatively little will be said here about the processes of weathering of volcanic material, other than, because of the metastable nature of volcanic glass and the highly weatherable, chemically reactive mineralogy of many volcanic rock types, chemical weathering and mechanical disintegration into particulate and soluble material is likely to occur at high rates under climatic conditions conducive to weathering, and given sufficient time. Louvat et al. (2008) highlight that basalts, especially basaltic glass, are more easily weathered than other rock types, and that weathering of basalts occurs more readily in tropical and temperate climates than in arctic climates. This is also clearly the case for other volcanic compositions in tropical, high-rainfall climatic settings of stratovolcanoes (Genareau et al. 2016). However, in high-slope volcanic terrains, the highly erodible, unlithified nature of many volcanic deposits ensures rapid transit of detritus through the sediment cycle to the site of deposition, often leading to rapid burial, and so, to some extent, shortcircuiting chemical weathering. In low-relief terrains, the effects of chemical weathering may be more pronounced, but may also be postponed until volcanism has ceased, because of the emplacement of new volcanic deposits associated with successive eruptive episodes, which bury previous deposits and limit the rate of short-term weathering.

Chemical weathering as we understand it requires an oxidising and hydrous atmosphere, and therefore will be limited to planetary bodies with an atmosphere and climatic conditions such as the Earth has. On other planetary bodies with other atmospheric chemistry, weathering processes are yet to be studied, but in some cases may be more akin to the effects of hydrothermal alteration. *Physical* weathering processes on Earth will be very significant in liberating volcanic fragments from source for subsequent transportation and deposition in high-relief volcanic terrains. Important physical processes include gravitational collapse or mass-wasting, and the abrasion effects of running water (rivers, rain water, and meltwater on subaerial slopes) and waves (marine and lacustrine shorelines), moving mass-flows (subaerial and subaqueous), moving ice (in high-altitude and/or high-latitude volcanic terrains), the effects of thawing and expansion, and of a permafrost (also in relatively high-altitude and/or high-latitude volcanic terrains).

12.2.2 The Duration of Eruption Events Versus Erosional Intervals in the History of Volcanoes

Since volcanism generally produces positive topographic relief edifices, this may suggest that erosion has little or no effect on volcanic terrains during the active life of a volcano. This is generally the case for short-lived monogenetic volcanoes (e.g. scoria cones, maars), although some may partially collapse during their eruption life of days, weeks, or months. For large volcanic edifices such as stratovolcanoes, explosive calderas, and oceanic shield volcanoes (Chap. 16), the actual duration of eruption events generally represents only a small fraction of the total life span of the volcano, which could be hundreds of thousands to millions of years. For stratovolcanoes, explosive calderas, and oceanic shield volcanoes, there are significant repose periods between multiple small- and large-scale eruptions (Chap. 16). During these periods of eruptive quiescence, normal surface processes operate at very high rates. In fact, in such centres, the total or cumulative duration of all the repose periods far exceeds the total or cumulative time over which eruptions take place.

In volcanic terrains with their high relief, therefore, it is not that erosion is an insignificant process during the active life of the volcanic centre, but rather that the enormous volumes of deposits that are erupted and the rates at which they are erupted and deposited during eruptive episodes, far exceeds the huge volumes removed and the very high rates of removal by surface sedimentary processes during quiescent intervals. As in normal terrigenous sedimentary successions, the time interval represented by the boundaries between depositional units (called diastems) in volcanic successions is greater than that represented by the depositional units (e.g. Sadler 1981; Martí et al. 2018; Chap. 15). However, once the active life of the volcanic centre has ceased, erosion continues to operate, and in a geologically short time period is capable of eliminating the majority of the surface relief of volcanic piles.

Vessell and Davies (1981) documented the close temporal relationships between volcanic eruption and epiclastic surface sedimentary processes in the Guatemalan volcanic arc, based on their study of Fuego volcano, which has erupted many times in historical and recent times, and most recently from mid-2018 and onwards. They suggest that the history of the Guatemalan volcanic chain in Central America is marked by cycles of activity consisting of four phases:

(a) *Phase 1*: lasts between 80 and 125 years, and is the inter-eruption phase. This is marked by relatively low rates of sediment deposition, by erosional incision of meandering rivers, and by reworking of deltas.
(b) *Phase 2*: is the eruptive phase, and is marked by eruption of lavas, air-fall deposits, and pyroclastic flows, and lasts for less than <1 year.
(c) *Phase 3*: describes the fan-building phase, and is dominated by debris flows and coarse fluvial sedimentation, which occurs in response to intense rainfall after Phase 2. Alluvial fans characteristically develop around the margins or base of stratovolcanoes during this phase. It lasts up to 2 years.

(d) *Phase 4*: or the braiding phase, results from the influx of large volumes of volcanic sediment into stream systems, transforming them from incised meandering systems to flood-prone braided systems. Rapid progradation of deltas also occurs during this stage. This phase lasts from 20 to 30 years.

Based on these observations, it is very evident the duration of surface sedimentary epiclastic processes (>99% of the ~150-year duration of a cycle) far exceeds that of the actual eruptive phases (<1%), and epiclastic surface sedimentary processes are rejuvenated in response to eruptive episodes.

12.2.3 Erosion Rates in Volcanic Terrains

There have been many assessments of the rates at which different types of volcanic terrains are eroded. Francis (1983) summarised known erosion rates, citing rates of erosional downcutting of 100–1,000 m Myr^{-1} in areas of high relief, and rates of 1–2 km Myr^{-1} in the South American Andes. He also pointed out the effects of high ratios of loose debris to lavas and the lack of vegetation cover in accelerating the erosion rate, and suggested that over a geologically significant period (1–2 Ma) the proximal, highest near-vent area of a stratovolcano would be eroded away, leaving only the core complex and part of the voluminous, epiclastic volcaniclastic dominated ring–plain sequence (Chap. 16). Karátson et al. (2012) proposed rates of down-cutting of stratovolcanoes ranging from 10 m Myr^{-1} in the extremely arid Altiplano of the Andes to 20 m Myr^{-1} in the more humid Peruvian Andes. They noted that young stratovolcanoes could experience erosional rates of down-cutting up to 100 m Myr^{-1} when covered with a lot of loose debris.

Mills (1976) calculated higher erosion rates for Mt Rainier stratovolcano in the temperate climate of the Cascades volcanic arc, USA. Rates based on the dissection of landforms were 1,100 m Myr^{-1}, and those based on river sediment load were 3,000–4,000 m Myr^{-1}, leading to a time interval of 1–3 Ma for complete erosion of a 3 km high stratovolcano. All stratovolcanoes are marked by radial systems of incising, erosional fluvial networks (Plate 12.1a). In the proximal to medial upper slopes, such erosional fluvial valleys are steep-sided, often "box canyons" (Plate 12.1b), which can create complex stratigraphic contacts and depositional geometries if they are infilled by younger eruption deposits funnelled into such valleys from the vent area (see Figs. 1.8, 1.9a, b; and Chap. 16).

These erosion rates for stratovolcanoes are higher than those inferred for scoria cones based on the studies of Wood (1980b) and Kieffer (1971; see Chap. 16), and those calculated by Walker (1982) for basaltic highlands in Iceland. Walker (1982) calculated downcutting rates of 58 m Myr^{-1} from his Icelandic study, in a terrain with an average elevation of 400 m.

Richards (1965) described the types of erosional processes, their rates, and effects in marine volcanic terrains, in discussing the erosional history of the volcano Bárcena which erupted on the island of San Benedicto, in the eastern Pacific Ocean in 1952. Bárcena is a cone of trachytic tephra and lesser lavas which reached a height of 332 m. The principal agents of erosion were gravity, wind, shoreline wave erosion, and surface runoff following several periods of heavy rain. Over 2 months 2.7×10^8 m^3 of ash, and lesser proportions of scoria lapilli, blocks and bombs were erupted. Gravitational grain flow avalanching of tephra occurred during the eruption period, producing a radiating system of downslope furrows on the cone surface. Wind erosion was prominent until the first rains resulted in a surface crust, but recurred during subsequent dry periods. Rains produced significant erosional gullies, which in time rounded the knife-like crater rim, markedly modified the initial furrow system and commenced the infilling of the crater and burial of the crater dome. Wave erosion of both tephra and a lava delta caused major changes to the position of the shoreline and removed large volumes of debris, which were then presumably redeposited subaqueously offshore. In the 4 months after eruption began, erosional retreat of tephra sea-cliffs occurred at a rate of 1.7 m day^{-1}, and about 0.71×10^6 m^3 of tephra were removed by wave erosion over a 40-day period. A lava flow was eroded at a rate of 0.12 m day^{-1} over a 158-day period in 1953. Similarly, Thorarinsson (1967) records the major impact of storm wave erosion on Surtsey island volcano in Iceland, and its satellite centres, from 1963 to 1966 (Chap. 16).

Cataclysmic erosion events, involving major mass movement events such as landslides and debris avalanches, can be instantaneous and affect all high-relief volcanic edifices, including stratovolcanoes, large shield volcanoes, and lava dome systems, in both subaerial and submarine settings (Chap. 16). They can occur syn-eruptively, triggered by an eruption, or post-eruptively, simply due to high relief, steep slopes and gravitational potential emergy. The cataclysmic eruption of Mt St Helens, also in the Cascades volcanic arc, on 18 May 1980, triggered by a magnitude 5 earthquake associated with a growing high-level cryptodome, destabilised the north flank of the stratovolcano, initiating a landslide that immediately transformed into a devastating debris avalanche directed to the north. The result was the formation of an instantaneous erosional amphitheatre-like valley on the northern slope of Mt St Helens (see Fig. 12.7d, e).

The activity levels of surface process regimes are accelerated by orders of magnitude after major, especially explosive eruptions. This is reflected initially by high rates of erosion of newly laid, unconsolidated pyroclastic deposits on the slopes of the volcano for about a 5-year period (e.g. Mt St Helens post-1980, Major et al. 2000; Mt Pinatubo, post-1991, Gran and Montgomery 2005; Gran et al. 2011). Volcaniclastic debris was funneled into major tributaries, choking them, followed by downcutting of the aggraded beds of those tributaries (Chaps. 15, 16). Sediment yields or flux in the major tributaries reflect this very high erosional capacity, rising orders of magnitude above pre-eruption levels. At Mt St Helens, pre-eruption sediment flux was $\sim$20 Mg km^{-2} yr-^{1}, whereas at the peak of erosion of new deposits after the eruption sediment flux was up to 46,000 Mg km^{-2} yr^{-1}, but dropped quickly, exponentially for the first 5 years after the eruption and then more linearly (Major et al. 2000). Even after 20 years, the sediment flux from the valleys where the debris avalanche deposits were ponded, was still 100 $\times$ that of normal, equilibrium levels. Incredibly, of the 2.5 km^{3} of debris avalanche deposit that was emplaced in minutes on May 18, 1980, 50% was remobilised as syn-eruption debris flows within hours of the start of the eruption, and a further 20% in the first year after the eruption (Major et al. 2000).

Lastly here, we emphasise, these figures are only a general guide to erosional rates, and variations will occur for a variety of reasons, including the nature of volcanic landscape (lava or pyroclastic deposit dominated), relief and slope, climatic conditions, and the various types of erosional processes.

12.2.4 Factors Affecting Accelerated Volcaniclastic Sediment Aggradation Rates During and After Explosive Eruptions

Whereas erosion rates between eruption episodes can be exceedingly high, rates of deposition and aggradation on the slopes of a volcano and the surrounding landscape can be extreme during and after eruption events because of the enormous mass and volumetric rates of eruption and lateral transport. Aggradational processes by which volcanic landscapes are built up include initial eruption-related transport and deposition on the volcanic edifice and surrounding landscape, including lava eruptions and pyroclastic events, both of which can add m to hundreds of m of deposits onto the volcanic summit and slopes during an eruption. For example, during the 15 June 1991 eruption of Mt Pinatubo, some valleys were filled with 200 m of pyroclastic flow deposits in a matter of hours (Newhall and Punongbayan 1996; Pierson and Major 2014). Secondly, there are sedimentary processes and events which can erode newly erupted and deposited volcanic deposits at rates orders of magnitude higher than pre-eruption rates, especially when the eruption products are unconsolidated deposits. These processes erode volcanic deposits from high-relief proximal parts of volcanoes and redeposit them as fans of volcanic sediment at the base and margins of the volcanic edifice.

In addition to the impact of the instantaneous availability of large volumes of unconsolidated pyroclastic debris, the impact of the eruption on vegetation plays a major role in controlling volcaniclastic sediment flux during and after explosive eruptions (Alexander et al. 2010; Pierson and Major 2014). In areas where damage to vegetation by hot pyroclastic fall deposits and pyroclastic density currents is severe, there is nothing to retard sediment movement during rain events, leading to the highest sediment flux in any environment on Earth (Gran et al. 2011; Pierson and Major 2014). Where vegetation remains in place it provides resistance to sediment movement. Finally, regional climate and ephemeral weather-controlled hydrological events also play a major role in controlling volcaniclastic sediment flux after eruptions.

Kuenzi et al. (1979) provided a valuable account of the relationship between volcanism, erosion, and sedimentation resulting from the 1902 explosive eruption of the andesite stratovolcano Santá Maria, in southern Guatemala. Santá Maria volcano, along with its near neighbour Santiaguito volcano, is drained by the headwaters of the Samalá River, which descends from an altitude of over 2,500 m around Santá Maria volcano down to the Pacific shoreline over 60 km away. The major 1902 eruption resulted in a mantle of pyroclastic deposits 20–30 m thick on the slopes of the volcano within the watershed of the Samalá River. As a result of the torrential rainfall levels (>381 cm yr^{-1}) and the associated total destruction of vegetation by the eruption, extreme erosion rates caused sediment supply to the Samalá River to increase dramatically after the eruption. Kuenzi et al. (1979) calculated that between 1902 and 1922 the bed of the Samalá rose 10–15 m. In so doing, it dammed up several major pre-1902 tributaries, producing elongate valley ponded lakes (Fig. 12.2) and re-activating older (Pleistocene) alluvial fan sedimentation. In addition, a significant delta of volcanic sediment was built at the mouth of the Samalá River, extending the shoreline nearly 7 km seaward into the Pacific Ocean, and involving 4 km^{3} of volcanic sediment (Fig. 12.2). Subsequently, reduced sediment supply has led to retreat and stabilisation of the shoreline. In addition, the active braided channel system of the Samalá River has experienced major shifts on the coastal plain (Fig. 12.2).

This pattern of extreme volcaniclastic sediment aggradation in catchment areas around volcanoes has been replicated countless times after other significant explosive eruptions since the 1902 eruption of Santá Maria volcano. The cataclysmic eruption of Mt St Helens on 18 May 1980, not only produced a sector-collapse landslide scar but also the landslide immediately transformed into a devastating

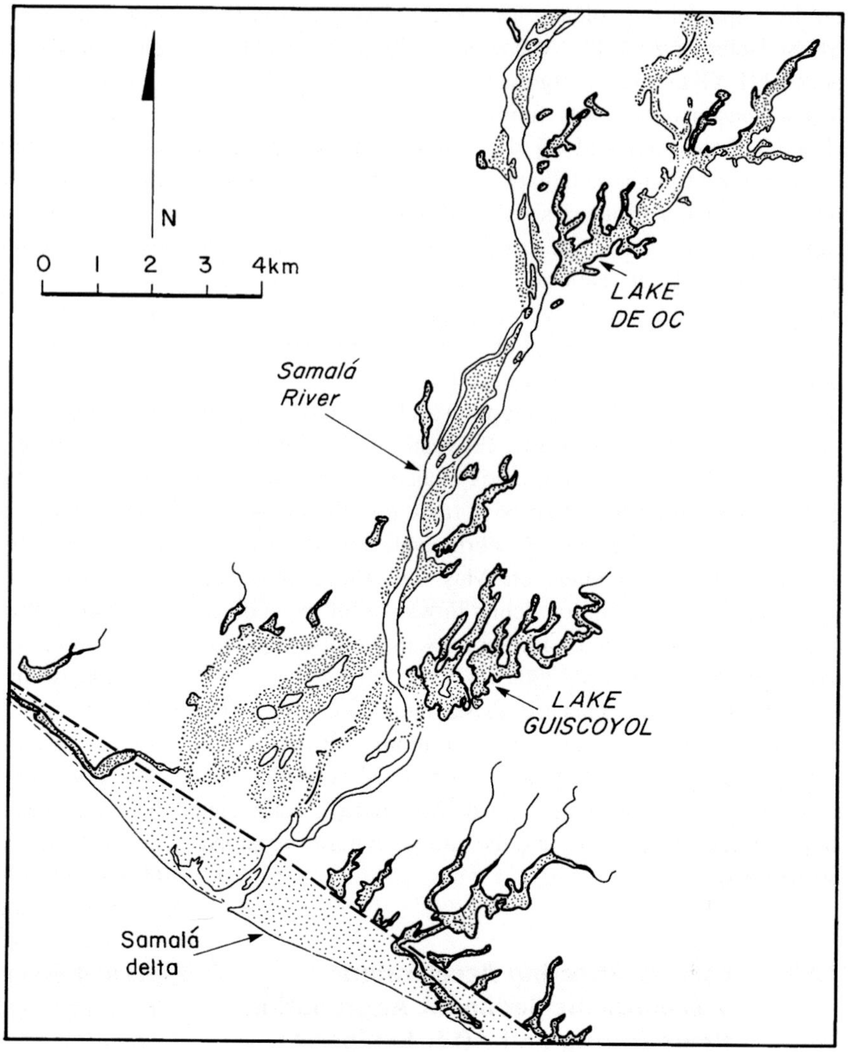

Fig. 12.2 The braidplain depositional system of the Samalá River on the Guatemalan coastal plain following the 1902 eruption of the Santá Maria volcano, which transformed the single channel, valley confined river into a wide braidplain (stippled pattern). Note the valley-ponded lakes (bold outlines) in the tributaries of the Samalá River resulting from their damming as the bed of the Samalá River rose 15 m as it was choked by volcanic sediment. Note also the major delta lobe at the mouth of the Samalá River which was built 20 years after the eruption. (From Cas and Wright 1987, After Kuenzi et al. 1979)

debris avalanche directed to the north. Much of it was deflected to the west along the North Fork Toutle River valley, burying it under tens of m of debris derived from the landslide sector of the edifice (Lipman and Mullineaux 1981, see Fig. 12.15a). The debris avalanche mixed with river and lake water from Spirit Lake to transform into debris flows and hyper-concentrated flood flows downstream, and the bed of the Toutle River aggraded instantaneously. In addition, the landslide and debris avalanche event was followed by the onset of an explosive eruption that produced a laterally directed blast flow to the north of the volcano, later that day normal pumice pyroclastic fallout and flow deposits. The landscape thus aggraded by m to tens of m and in some narrow valleys by more than 100 m in 6 h. Since the eruption, there has been erosion of loose deposits from slopes into the main watersheds, leading to further river bed aggradation. Major et al. (2000) determined that most of the loose deposits on the slopes of the edifice and surrounding landscape were eroded in 5 years after the eruption, and that most erosional effects involve incision of debris avalanche deposits ponded in major tributaries such at the Toutle River to the northwest of Mt St Helens. The Toutle River transformed from a pre-eruption, narrow incised river valley into an aggraded, wide braided river in many places, and many of its tributaries became lakes that were dammed by the aggraded bed of the Toutle River (Fig. 12.3a, Plate 12.1c).

Similarly, after the 15 June 1991 explosive eruption of Mt Pinatubo, in the Philippines >7km^3 of pyroclastic deposits were erupted onto the slopes of the stratovolcano, and in the upper reaches of major valleys draining the volcano. The eruption occurred during a major monsoonal rain event, and so the surrounding watersheds became choked with reworked volcanic sediment, which continued every year for at least 15 years afterwards during each monsoon season (Pierson et al. 1992; Newhall and Punongbayan 1996). The beds of the rivers aggraded 15–20 m, tributaries became dammed lakes (Plate 12.1d), and once fertile river banks and flood plains became inundated with volcanic sediment

Fig. 12.3 Effects of eruption triggered rapid aggradation of the river valleys and landscape around stratovolcanoes. **a** The braided river form part of the Toutle River north of Mt St Helens, USA, following the 1980 eruption. (Photo taken in 2007). **b** Aggradational and inundated river flood plain around the town of Bacalor, south of Mt Pinatubo, central Luzon, Philippines, after the 1991 eruption. Reworking of pyroclastic deposits choked the river, whose original course is shown by the line of trees, causing volcanic sediment, transported by hyper-concentrated flood flows, to spread over the once fertile agricultural flood plain and burying much of the farmlands and the town of Bacalor to depths of tens of metres in places. Note some buildings are buried to the roofs. **c** A house (red circle) buried by volcaniclastic hyper-concentrated flood flow deposits, exposed in the eroded banks of the Pasig-Potrero River, after the 1991 eruption of Mt Pinatubo. The pre-eruption flood plain level is at the contact between the lower bedded dark grey sequence with the white pyroclastic and grey upper hyper-concentrated flood flow deposits. (Photos **b** and **c** were taken in 1995)

transported by hyper-concentrated flood flow deposits, which buried towns and buildings (Fig. 12.3b, c).

Gran et al. (2011) subdivided the hydrological–sedimentological response of the river systems around Pinatubo volcano to the eruption into two phases. In Phase 1, sediment flux was at the highest levels ever recorded anywhere (3.1×10^6 m^3 km^{-2} yr^{-1}; Gran et al. 2011), which decayed exponentially for the first decade following the eruption (Sect. 16.6.5, see Fig. 16.46). In the second Phase, continued valley widening and fluvial instability maintain high sediment yields, but with sediment flux rates declining at a more linear rate for at least two decades. The stratigraphic record in these settings is clearly dominated by sedimentation over 20–30-year periods after explosive events, as proposed by Vessell and Davies (1981).

During the 1995–2013 eruption of Soufrière Hills volcano on Montserrat, major fan-deltas of volcaniclastic debris, consisting of pyroclastic density current, debris flow, sheet flood, and stream flow deposits developed radially around the volcano and island, rapidly aggrading tens of m, burying the capital city Plymouth and Bramble airport, and prograding hundreds of m into the sea (e.g. Fig. 12.4a, b; e.g. Alexander et al. 2010). Although the high aggradation and progradation rates can be attributed to abundant loose pyroclastic deposits being remobilised by rain events, a more significant factor was the destruction of vegetation during eruptions, so removing any impeding effects by vegetation to surface sedimentary transport of the pyroclastic debris (Alexander et al. 2010). Even as the eruption waned, "lahar" or hyper-concentrated flow events still occurred during every rainy season, up to the present day.

During the prolonged, largely effusive, lava dome complex forming eruption of Mt Unzen, from 1991 to 1995 in Japan (Chap. 6), multiple, gravitational, dome-margin collapse events generated about 10,000 block and ash-flows and associated ash-cloud surge events. The deposits formed fans or aprons of block and ash-flow deposits along the main pre-existing valleys (Nakada et al. 1999). During each monsoon season, as at Pinatubo, much of the block and ash-flow deposits were remobilised downstream by hot "lahars", cold debris flows, hyper-concentrated flood flows, and stream flows. The primary and secondary deposits are interstratified in the fans, the secondary deposits are texturally and compositionally very similar to the primary block and ash-flow deposits, and difficult to distinguish from them, which is an important lesson in the stratigraphic record (Uehara et al. 2015; Sect. 12.6.6). Pre-existing valleys were infilled through rapid aggradation of tens of m, hundreds of homes were destroyed or buried by pyroclastic flows, debris flows, and sheet floods (Fig. 12.5a, b), and a fan-delta prograded into the ocean.

At the very active Merapi volcano, in Java, Indonesia, the explosive and lava dome collapse events that occur every few years and produce pyroclastic density current deposits, mostly in the valleys of the upper and medial slopes, are always followed by "lahar" (debris flow, hyper-concentrated flood flow) events during rain events afterwards (Lavigne et al. 2000; Belizal et al. 2013). The lahars extend further down valleys than the primary pyroclastic density current deposits, and their deposits aggrade the river beds many metres each time, destroying bridges and hundreds of houses. Not surprisingly, Belizal et al. (2013) found that the explosive eruption of Merapi in 2010 (VEI 4/VEI*8—see Chap. 9 for explanation), which was much larger than normal, produced lahar events that extended much further downstream (into the city of Yogyakarta), than those associated with smaller eruptions in previous years, a reminder that potential "*worst-case*" hazard scenarios always need to be seriously planned for.

This message has been more recently echoed by the June 2018 eruption of Fuego volcano in Guatemala, which was also a VEI4/VEI*8 scale event. Fuego volcano is one of the most active in Central America, and is in an almost constant state of unrest. It has erupted many times in historic times. Prior to 2018, the previous large eruption was in 1974, also a VEI4/VEI*8 event. The 1974 eruption was spread over 10 days in October, marked by four major Vulcanian explosive events that produced widespread fallout, pyroclastic flows that flowed up to 7 km from the vent, and subsequent volcanic mudflows (Davies et al. 1978). Since then Fuego volcano has been regularly active with numerous small eruptions. The eruption on 5 June 2018 and subsequent days produced significant wet ash fall (because it was raining heavily during the eruption), hot pyroclastic flows, and hot lahars or mudflows which were initiated during the eruption by the heavy rainfall. Valleys aggraded rapidly due to pyroclastic flow and lahar infilling, and villages in the valleys were destroyed and buried. Sadly $\sim$200 people perished within a radius of 10 km of the volcano as a result of the impacts of wet air-fall ash loading causing roof collapses, as well as from hot pyroclastic flows, inhalation of ash, and volcanic mudflows. Unfortunately, the alert system was tardy, and ineffective. Lessons were not learned from 1974, or they were forgotten. Effective education of the communities around the volcano about the likely hazards and possible magnitude of an eruption and lahars or volcanic mudflows did not seem to have occurred, which is one of the most valuable tools in risk minimisation (Chap. 16).

Subaqueous epiclastic or sedimentary processes and rates from modern volcanoes are not as well studied or documented. However, examples of the stratigraphy and lithofacies architecture of submarine volcanic edifices from the rock record are abundant, and indicate that the aprons of submarine edifices are dominated by resedimented volcanic debris and slide-slump deposits (e.g. Jones 1967; Mitchell 1970; Cas and Wright 1987; Simpson et al. 2005). Such

Fig. 12.4 Fan deltas of Montserrat, Lesser Antilles. **a** Volcaniclastic fan-deltas on the east coast of Montserrat, prograding into the Atlantic Ocean. The southern fan (uppermost on the photo) developed at the mouth of the Tar River valley fed by block and ash and pumice and ash-flows derived from lava dome collapse and Vulcanian explosions from the Soufrière Hills volcano (also see Chap. 11, Fig. 11.8). The northern fan-delta (bottom of photo) has engulfed the former WH Bramble airport. The whole coastline has prograded hundreds of m into the sea caused by pyroclastic flows entering the sea, as well as stream flow generated progradation. (*Source* Montserrat Volcano Observatory). **b** Oblique Google Earth image of the volcaniclastic fan-delta that inundated the area where the WH Bramble airport was located, shown by airplane symbol circled in yellow. The airport was destroyed and buried by pyroclastic flows, stream floods, and their deposits. (*Source* Google Earth)

deposits are proof of the importance of sedimentary processes in submarine volcanic settings, due in large part to the relief of marine volcanic edifices above the mean depth of the surrounding sea-floor.

In Chap. 16, we describe in detail how the subaqueous parts of marine stratovolcanoes are dominated by re-deposited mass-flow volcaniclastic sediments (debris flows, turbidites, etc.). In these cases, much of this debris was probably fed from the subaerial and shallow marine parts of the volcanoes, and was eroded by subaerial and shallow marine processes such as those in the cases described above by Vessell and Davies (1981), Kuenzi et al. (1979), and Richards (1965). In addition to these sustained degradation processes, marine stratovolcanoes are particularly prone to explosively or non-explosively induced sector collapse whereby a large segment of the edifice is mobilised into a dynamic slide and/or debris avalanche. Marine geophysical and drilling investigations over the past 20 years

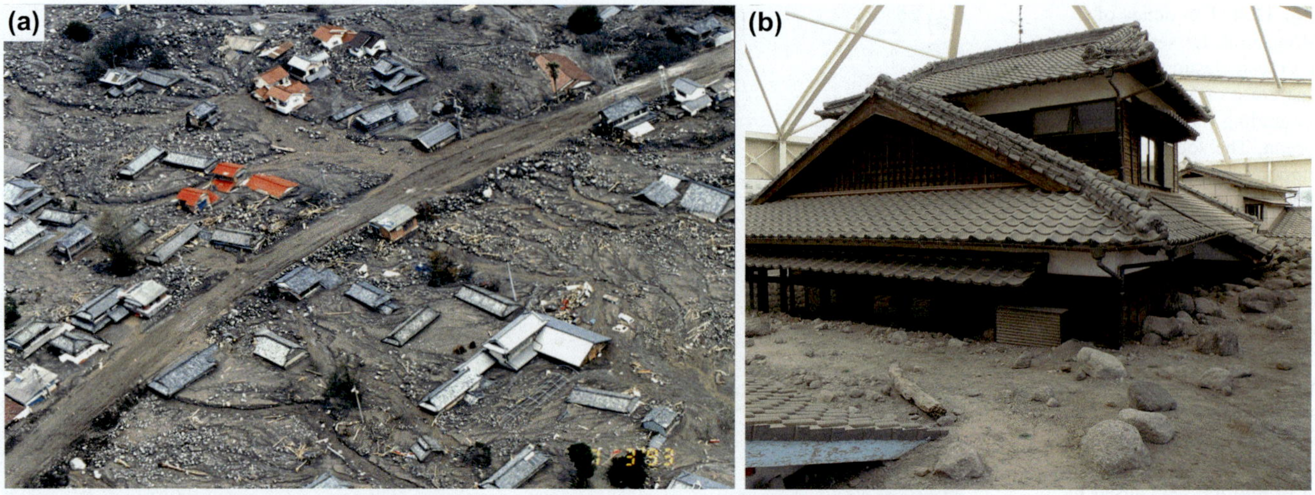

Fig. 12.5 **a** Overview of lahar, debris flow, and hyper-concentrated flood flow deposits derived from remobilisation of block and ash-flow deposits from the 1991–1993 eruption of Mt Unzen, Japan, that inundated part of the town of Shimabara at the mouth of the Mizunashi River valley near the coast (*Source* Volcanological Society of Japan). **b** Preserved houses buried to the roof by debris flow and hyper-concentrated flood flow deposits at the Shimabara Natural Disasters Museum

have highlighted two regions where sector collapse is a recurrent process and large magnitude subaqueous volcaniclastic aprons have been identified not only in and around the Canary Islands and islands of the Lesser Antilles volcanic arc (Chap. 16) but also including well-known volcanic islands such as Stromboli volcano, Italy, and during the eruption of Anak Krakatau, west of Java, on 22 December 2018. Wholly subaqueous volcanoes are likely to be less dominated by epiclastic processes and facies than are subaerially exposed volcanoes, because several of the main agents of erosion (wind, running water, and wave action) are absent. In such subaqueous settings, gravitational collapse is the main epiclastic erosional process operating, but may nevertheless be significant. However, if such volcanoes occur in restricted, small basins, then extra-formational epiclastic sediment transported from shallower basin margins will also be important.

To summarise, surface sedimentary or epiclastic processes are extremely significant in all volcanic settings, both in terms of their duration and in terms of the volumes of rock debris that may be transported. In some cases, they can initiate eruptions (see below), they are intimately associated with primary volcanic facies in the deposit record of volcanoes, and often their activity levels accelerate in response to volcanic eruptions. The hazards they pose are as dangerous and fatal as the primary eruption events. We need to learn more about them, not ignore them. In ancient successions, erosional discordances and variable proportions of epiclastic volcanic sediments should be expected, depending on the balance between erosion and sedimentation during the long-term history of the volcanic terrain (Chaps. 15, 16).

12.3 Classification of Surface Sedimentary or Epiclastic Transport Processes

Sediment transport in both volcanic and non-volcanic terrains takes place in two basic forms: in (1) *particulate* fashion, whereby each particle behaves individually, or (2) by *mass movement*, mass wasting, or mass flow, whereby a volume of sediment is moved as one, essentially instantaneously (Cas and Wright 1987, Chap. 10). In particulate sediment movement, particles have the freedom to move as individuals in response to the forces acting on them. In mass-flow movement, because a large population of fragments is moving *en masse*, as one, particle freedom is largely inhibited, and collisions and dynamic interaction between fragments are common. As well as subdividing sediment movement on the basis of particulate or mass movement, further subdivision can be made on the basis of the involvement of an interstitial medium or not, as follows (Table 12.1; Cas and Wright 1987):

- no necessary interstitial medium,
- ice as an essential interstitial medium,
- water as an essential interstitial medium, and
- air as an essential interstitial medium.

The function of an interstitial medium is to drive sediment movement, provide grain support or act as a lubricant.

In volcanic settings, the surface process phenomena we will discuss in the following sections of this chapter are fundamentally integral to the evolution of the volcano, and in some cases, to triggering volcanic eruptions. The

Table 12.1 A classification of sediment transport processes (Modified from Cas and Wright 1987)

Surface sedimentary or epiclastic process	Nearest primary volcanic analogue
1. *Sediment transport is not dependent on an interstitial medium*	
Particulate	
(a) Rock or particle free fall	Ballistics, air-fall (Chap. 9)
(b) Particle creep	
Mass movement	
(c) Rock fall	
(d) Landslides, rockslides, sector collapses	
(e) Rock or debris avalanches	Block and ash flows (Chap. 10)
(f) Grain flow	
2. *Sediment transport involving ice as an essential interstitial medium*	
Particulate	
(a) Ice rafting	
(b) Glaciers	
Mass movement	
(c) Glaciers	
(d) Permafrost creep	
3. *Sediment transport involving water as an essential interstitial medium*	
Particulate	
(a) Traction (bed-load, saltation)	
(b) Suspension	
(c) Pumice rafting	
(d) In solution	
Mass movement	
(e) Fluvial torrent flow/hyper-Lahars concentrated flood flow, sheet flood	
(f) Subaqueous granular mass flow (e.g. turbidity currents, liquefied, grain flow)	
(g) Debris flows, mud flows, Lahars	
(h) Tsunamis and storm surges	
(i) Storms, severe coastal	
(j) Slumps	
(k) Soil creep	
4. *Sediment transport involving air as an essential interstitial medium*	
Particulate	
(a) Traction (wind-blown)	Surges (Chap. 10)
(b) Suspension (wind-blown)	Surges (Chap. 10)
Mass movement	
(c) Air-lubricated rock avalanches	Pyroclastic flows (Chap. 10)

"sediment" or clasts involved could be hot or cold volcanic clasts, intraformational sediment or older basement-derived material.

12.4 Sediment Transport not Dependent on an Interstitial Medium

In this situation, an interstitial medium is not necessary to drive sediment movement, provide grain support or act as a lubricant for sediment movement. The only force causing sediment movement is gravity, and the gravitational potential of the sediment. It can occur on any planetary body where solid fragments are elevated above a topographic base level. This can occur either in particulate fashion through:

- particle or rock free fall,
- particle creep, or
- mass movement through:
  - rock fall,
  - landslides,
  - rock or debris avalanche, and
  - grain flow.

12.4.1 Rock or Particle Free Fall

In this process, a rock fragment is dislodged from an elevated position above a local base-level and free falls either through air or water, or both, and/or tumbles downslope. Although the fragment falls through air or water, neither medium contributes to the motion of the particle (Fig. 12.6a–d).

Examples include dislodgement of individual fragments or blocks from cliffs or steep slopes by earthquakes, explosive shock waves, and water or ice permeating along fractures lowering the co-efficient of friction between the block and the substrate (Fig. 12.6a, b), which can occur either subaerially or subaqueously. In addition, free rafting of debris in floating vegetation root systems in a lake or ocean, with fallout from the root system through the water column (Fig. 12.6c), or from melting of icebergs (calved from glaciers carrying eroded bedrock rock debris) as they are rafted around by currents (Fig. 12.6d), also represent examples of particle-free fall. Deposition onto a hard substrate will produce isolated fragments, and repeated events will produce an unstructured pile of fragments (Fig. 12.6b; also see Sect. 12.4.3). Where deposition occurs onto a soft, unconsolidated substrate, soft-sediment deformation and vertical, symmetrical impact structures may occur, which includes iceberg dropstones (Fig. 12.6d, e; Sect. 12.5.1). Ice berg dropstones can occur as isolated "lonestones" (Fig. 12.6e) or trails of lonestones, which form as a melting drifting iceberg progressively loses its lithic clast load as it melts. Subsequent deposition of normal ambient sediment should produce fill-in around the impact and drape structures over the top of the dropstone (Fig. 12.6d, e). Impact structures are also produced by explosively ejected ballistics (Figs. 1.15b, 9.55 g), in which case there is usually a directional asymmetry resulting from the directed trajectory of the ballistics (Chaps. 1, 9).

Individual particle rock falls, occurring over an extended period of time are significant events associated with the retreat of steep cliffs in any volcanic terrain or setting (Fig. 12.6a, b), but are also episodic events that lead to vent or crater widening (e.g. Fig. 12.7e). This can occur during active volcanism as well as during repose periods. For example, since the summit pit crater Halema'uma'u became active again in the caldera of Kilauea volcano, Hawai'i, in 2010, regular rock falls have occurred, leading to progressive widening of the crater (Chouet and Dawson 2013; Orr et al. 2013). These events have ranged from individual rocks falling to aggregates or mass-movement rock falls, and they have been particularly significant when the lava lake level has fallen and there has been deflation of the volcano. On a number of occasions such rock falls have triggered small to significant explosive degassing events, apparently caused when rocks punctured the lava crust under which gas had been accumulating (Orr et al. 2013). Some of these events produced significant ash clouds rising km above the vent. During the 2018 flank eruption and lateral fissure propagation event along the lower East Rift Zone in the Leilani Estate region (Chap. 5), the level of the lava lake in Halema'uma'u crater at the summit of Kilauea dropped dramatically as magma drained laterally through sideways propagating dykes, and this triggered increased rock fall activity and widening of Halema'uma'u crater.

Individual rock fall is also a common phenomenon from the steep margins of high aspect ratio lava domes (Chap. 6), that in time can contribute to the formation of base of dome talus aprons, and in some cases generation of block and ash flows (Chap. 10; e.g. Mt Unzen, Japan; Nakada et al. 1999; also see Sect. 12.6.6).

12.4.2 Particle Creep

Particle creep involves the slow, almost imperceptible, downslope movement of individual particles, again solely under the influence of gravity. Particles roll or slide bit by bit. Strictly, particle creep is unaided by any medium but, in practice, on exposed slopes rainfall, splash erosion, surface water sheet flow and the effects of wind will almost certainly be episodic causes of movement as well as gravity. There may also be an alternation between creep, tumbling and free fall and, where a large number of particles move together,

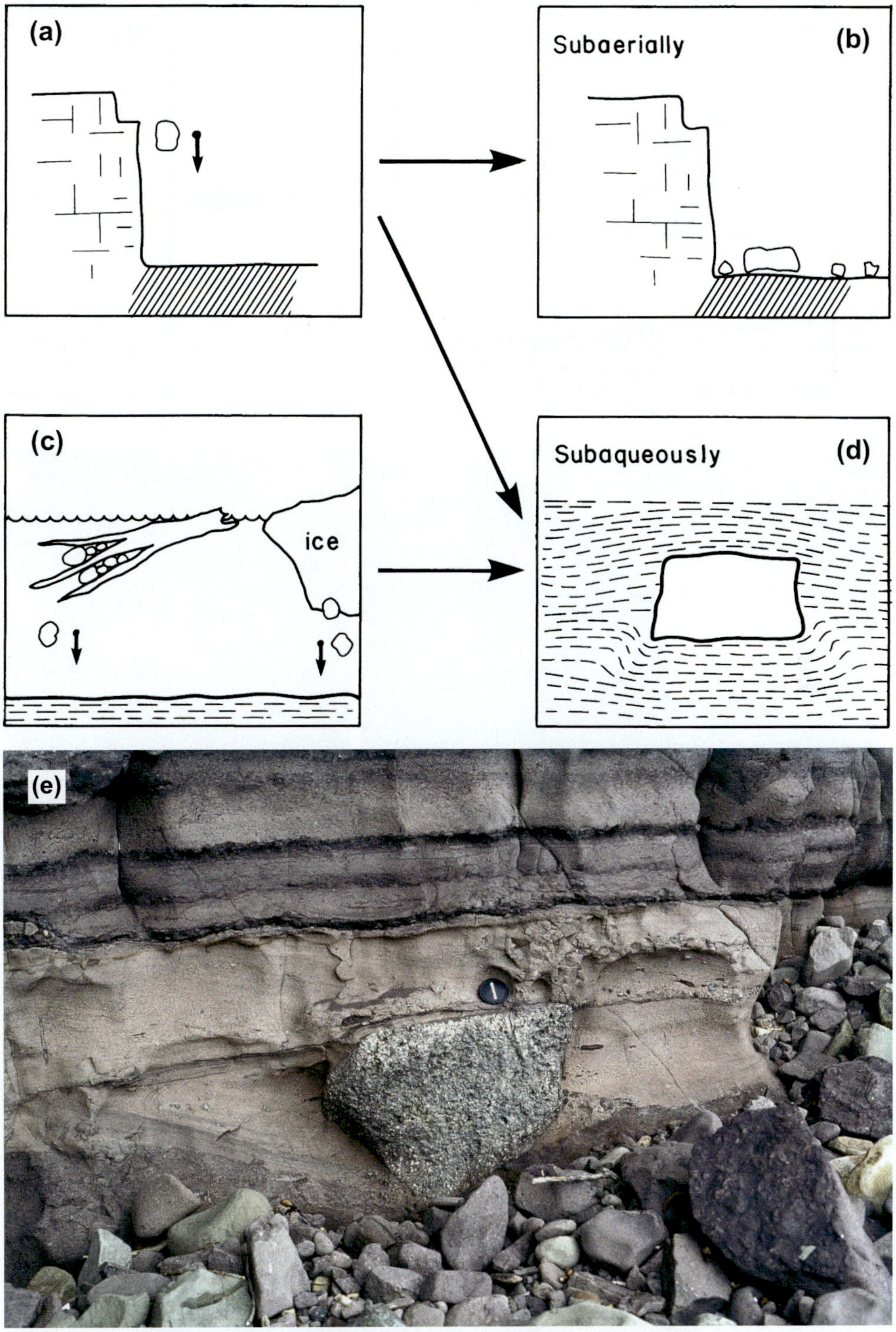

Fig. 12.6 Various situations of particulate rock free fall and resultant deposit characteristics. (**a–d** are schematic representations; From Cas and Wright 1987). **a** Individual block free fall from perched cliff. **b** Blocks accumulating on hard substrate at base of cliff. **c** Dislodgement of individual pieces of debris lodged in floating tree root systems or from melting icebergs. **d** Impact sag and sediment drape structures around a lone dropstone. **e** Impact sag structure produced by a lone Permian latite block dropstone in a continental shelf setting, impacting volcaniclastic sediment, which also infills around and slightly drapes the "lonestone". The sequence formed during volcanism in a coastal zone when southern Australia, as part of Gondwana, was in a circumpolar latitude and peri-glacial setting during the Permian ice age. The volcanic sandstones have been interpreted to be storm-generated shelf turbidites (Bull and Cas 1989). Gerringong Volcanics, Kiama, New South Wales, Australia

mass movement sliding, avalanching or grain flow will obviously occur. Particle creep can occur on any slope, whether it be subaerial or subaqueous, where a particle has gravitational potential. Particle creep will be widespread in volcanic terrains where unconsolidated debris lies on steep slopes close to or greater than the angle of repose, which varies for different-sized aggregates, but is generally a slope of $\sim$25–30°. The physical constraints for initiating movement of particles on a slow are discussed below under landslide initiation (see Fig. 12.9). Particle creep in ancient deposits may be impossible to recognise, since it produces no unique structure.

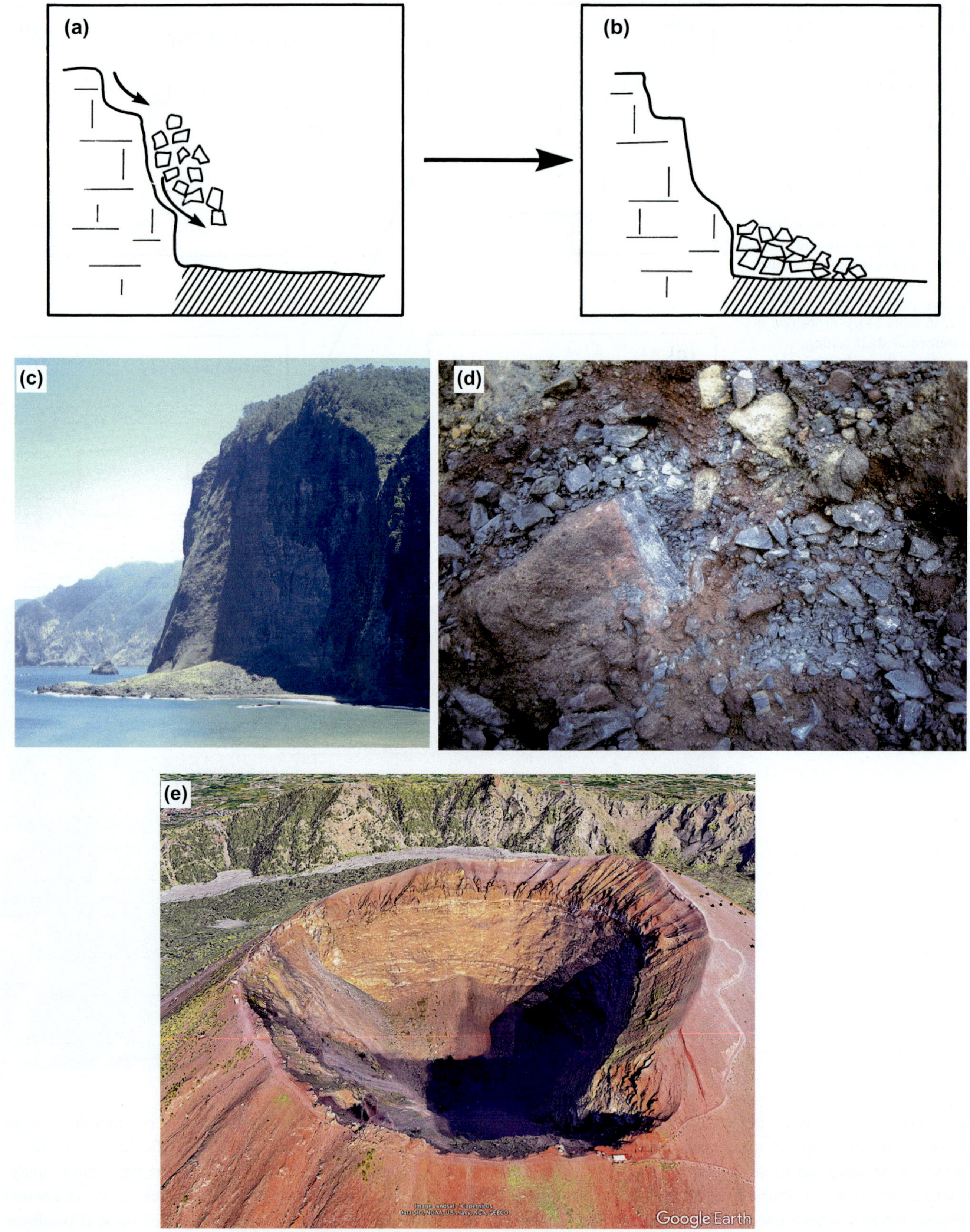

Fig. 12.7 Features of mass-movement rock-falls. **a** Schematic mass-movement rock-fall scenario, with a large volume of perched material falling from height, and **b** producing an unstructured, base of slope talus deposit (From Cas and Wright 1987). **c** Base of cliff talus lobe deposit of rock fall material sourced from an 80 m high sea cliff at Pehla d'Aguia, Madeira, Portugal. **d** Poorly sorted, fines poor, monomictic rock fall deposit, Paul do Mar, Madeira. **e** Oblique Google Earth image showing rock-fall talus fans that have been infilling the crater of the central volcanic cone 'Gran Cono' at Somma Vesuvius, Italy, since the last eruption in 1944 (*Source* Google Earth)

12.4.3 Rock Fall Mass Movement

This form of mass movement is a direct extension of individual particle or rock free fall discussed above, but involves multiple clasts in a large volume of rock debris falling *en masse* under the influence of gravity (Fig. 12.7a). This therefore has more impact in causing scarp retreat on cliffs, inner crater walls, or the margins of lava domes, than single clast rock fall events discussed above. After impact at a break in slope rock fall, rock debris may form a talus apron or fan (Fig. 12.7b–e) or transform into a lateral rock avalanche or tumble very rapidly downslope under its own momentum and the influence of gravity without the assistance of any medium.

Mass movement rock falls can take place subaerially or subaqueously, and are well documented from non-volcanic terrains (Johnson 1970; Voight 1978), but are also common in volcanic terrains (e.g. Crandell and Fahnestock 1965; Fahnestock 1978; Chouet and Dawson 2013; Orr et al. 2013). Rock fall mass movement events have occurred commonly from the walls of Halema'uma'u crater, on Kilauea volcano, Hawai'i after the crater again became an active lava lake in 2010, in some cases triggering small explosive eruptions from the lava lake crust (Orr et al. 2013). The volume of rock matter may be deposited as an unstructured pile of rubble at the base of slope (Fig. 12.7b–e), or in more open spaces may translate into another form of mass movement such as rock slide, air-lubricated rock-avalanche, or viscous debris flow; e.g. the Olokele rock fall and avalanche on the island of Kauai, Hawai'i, when it flowed into a river (Jones et al. 1984). The Olekele avalanche, originated from a 60° slope, 800 m high at the head of a canyon entrenched into the plateau-like summit of a shield volcano. The $\sim 12 \times 10^6$ m^3 rock fall from Little Tahoma Peak on the flanks of Mt Rainier stratovolcano in the Cascades volcanic arc, fell 600 m onto the Emmons glacier and transformed into a significant rock avalanche (Fahnestock 1978). The caldera margin collapse breccias of Lipman (1976) in the western San Juan Mountains, Southern Rocky Mountains Volcanic Field, Colorado (Chap. 16), are also examples of this mode of erosion and transport, and lead to progressive caldera scarp retreat and build-up of an intra-caldera margin talus deposit. After cessation of explosive eruptions, the open vents of many volcanoes fill in with crater wall collapse rock-fall deposits (e.g. crater of Vesuvius volcano, Italy after the 1944 eruption; Fig. 12.7e), which be re-excavated and ejected at the onset of the next explosive eruption. Mass movement rockfalls of hot debris from the margins of actively growing lava domes can transform into laterally flowing pyroclastic block and ash flows, as happened repeatedly during the 1990 to 1995 of Mt Unzen volcano in Japan, which is discussed in Chap. 10, Sects. 10.3.5 and 10.13.

Rock fall deposits are fines-depleted, very coarse, and poorly sorted, ranging from house-size blocks to granule size, and consist solely of coarse, angular collapse material, with little or no fine matrix material (Fig. 12.7d). The lack of a fine matrix reflects the short distance of transport and limited time for clast to clast abrasion produces some fine matrix in farther travelling rock debris avalanches.

12.4.4 Landslides, Rockslides, Sector Collapses

In landslides or rockslides the dislodged mass of rock maintains contact with and shears on a basal detachment fault over the substrate as a coherent body, under the influence of gravity (Figs. 12.8a, b; 12.9). Major landslides on the flanks of large volcanic edifices (Fig. 12.8c–g) are also called sector collapse events.

The essential requirement for a block of rock or a mass of soil or other materials to detach from the substrate is that the vertical gravitational force, $F_{(g)}$, of a mass (m) resolved parallel to the surface of the slope on which the slide occurs $F(gS)$, with a slope angle of ϕ, exceeds the frictional resistance force $F(\mu)$, which is the co-efficient of friction, μ, multiplied by the gravitational force resolved normal to the slope, $F(gN)$, trying to prevent or hinder sliding (Fig. 12.9):

$$F(gS) > F(\mu) \tag{12.1}$$

where $F(g) = mg$, $F(gS) = mg \sin \phi$, $F(gN) = mg \cos\phi$, $F(\mu) = \mu\, mg \cos\phi$, g is the acceleration due to gravity, and μ is the coefficient of friction, which is usually very high ($> \sim 0.6$); that is:

$$mg \sin \phi > \mu mg \cos \phi \tag{12.2}$$

This can be simplified and expressed as the Coulomb Friction Law (Coulomb 1776):

$$\tau = \sigma \tan \omega_s \tag{12.3}$$

where τ is the downslope shear stress, σ, is the normal force or stress, and ω_s is the static angle of friction on the slope.

In most cases slides will be initiated on relatively steep slopes (Fig. 12.8c, d), often activated along an intra-stratal, mechanically weak layer or structural surface. The sole or basal detachment (thrust) of a slide often forms through initial ductile deformation for example in a ductile clay layer or by the development of a network of brittle micro-cracks (Kilburn and Petley 2003; Petley and Petley 2006). When these connect, major brittle failure occurs along the detachment surface. In the moving landslide, internal

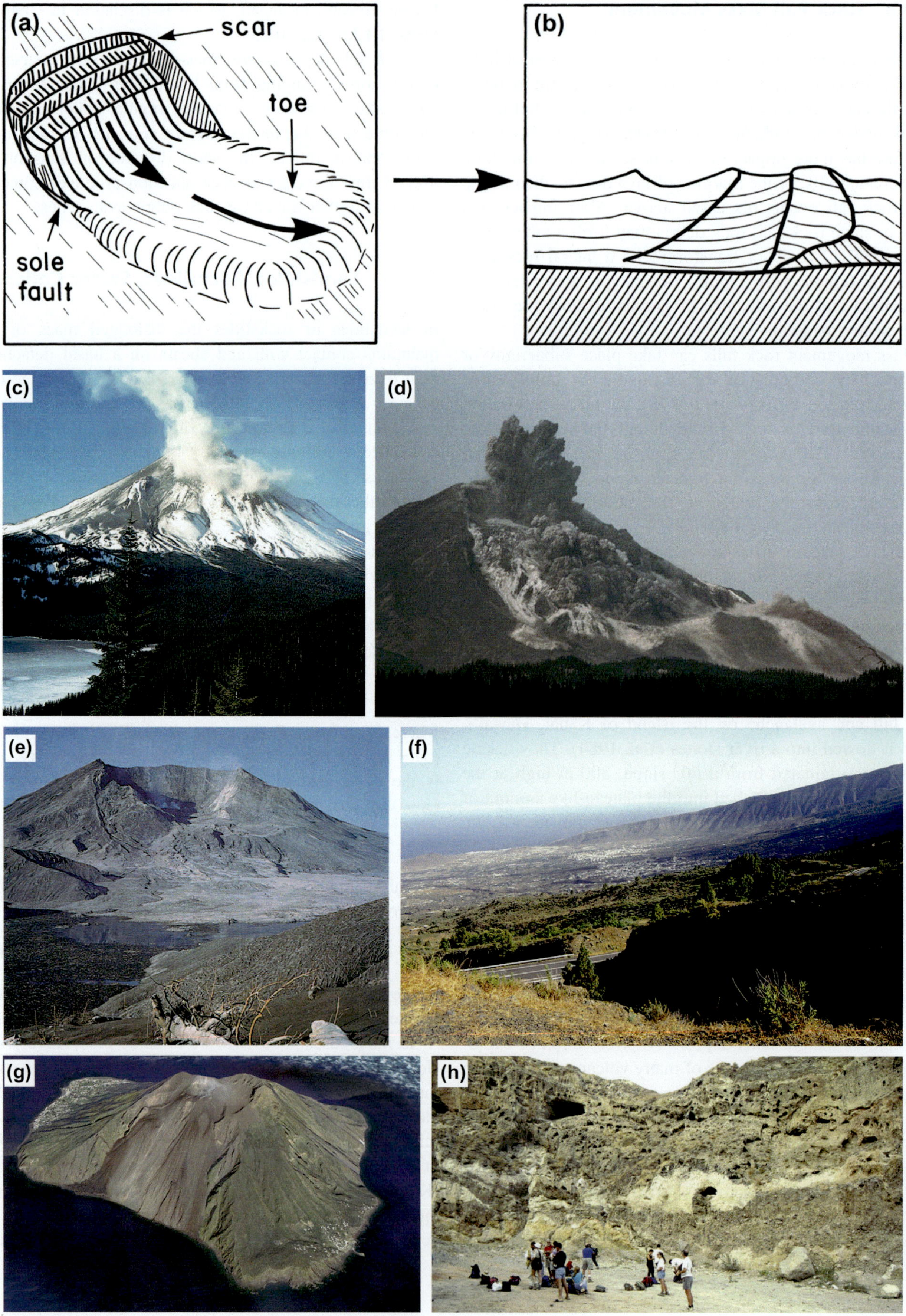
(a)
scar
toe
sole
fault
(b)
(c)
(d)
(e)
(f)
(g)
(h)

◀ **Fig. 12.8** Features of landslides. **a** Schematic representation of the initiation and morphology of a landslide and the erosion scar, and **b** a coherent slide deposit with internal brittle (fault) and ductile (folds) deformation features (From Cas and Wright 1987). **c** Mt St Helens stratovolcano before the 18 May 1980 landslide sector collapse event (From Lipman and Mullineaux 1981). **d** The initial landslide of the 18 May 1980 Mt St Helens eruption, showing discrete blocks detaching from the edifice, and transforming into a fragmental debris avalanche to the right. Also shown is the initiation of the pyroclastic blast event triggered by decompression of the magma body as the landslide moves. The removal of the slide mass forms a major concave erosion scar. Photo by G. Rosenquist taken at 08.32.57 PDT (Pacific Daylight Time), ~10 s following the first displacement of the landslide. (From Cas and Wright 1987, After Voight 1981). **e** The concave, amphitheater-like erosion scar resulting from the major slide and pyroclastic explosion event of 18 May 1980, Mt St Helens (From Lipman and Mullineaux 1981). **f** View of the very wide (10 km) ~0.85–0.57 Ma Guimar landslide valley, showing the south-western bounding side scarp fault on the oceanic island volcanic complex of Tenerife, Canary Islands, Spain. **g** Oblique Google Earth view of the Sciara del Fuoco landslide scar on the northwestern side of the marine stratovolcano Stromboli, Italy. The active summit crater system is located at the head of the landslide scar. (*Source* Google Earth). **h** Internal view of a stratified landslide body exposed at Athinios Port, Santorini, Greece, showing irregular ductile deformation of a once planar pumice fallout deposit (pale) between lithic clast-rich pyroclastic units

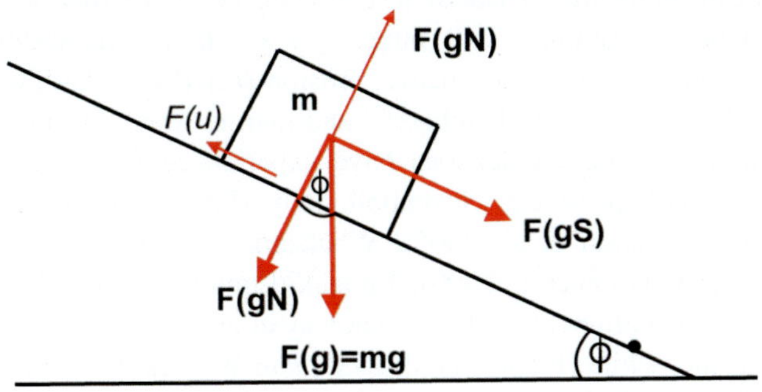

Fig. 12.9 Simple physical constraints that determine if a fragment, block of rock, landslide, or grain flow will begin to move on a slope. See text for definitions of parameters

deformation involves both ductile and brittle deformation zones or surfaces and behavior. Fluids in the fracture system help accelerate the deformations. The scale of slides may be enormously variable. Slides may be very small, local, insignificant events, or they may be major events involving millions of tonnes of rock (Fig. 12.8c–e). The source areas for large slides are marked by major topographic scars (Fig. 12.8a, c, d, e). If a slide acquires enough momentum it may even climb and slide up slopes it encounters in its path, as appears to have been the case with the catastrophic non-volcanic Vaiont landslide in northern Italy on 9 October 1963, which killed almost 2000 people (Genevois and Ghirotti 2005).

Examples of major subaerial non-volcanic slides have been documented in the volume edited by Voight (1978; also see https://en.wikipedia.org/wiki/List_of_landslides). One of the most recent, devastating examples is that of 15–17 June 2013 Kedarnath disaster, which occurred in the Himalayan state of Uttarakhand in northern India (Martha et al. 2014; Ziegler et al. 2015). Early monsoonal rains generated heavy and widespread rainfall which caused numerous landslides (as many as 3400; Martha et al. 2014), which in turn triggered a catastrophic chain of events of debris flows and floods. This was intensified where a landslide dammed the Mandakini River below the temple town of Kedarnath, which was flooded by river water backing up in the temporary lake. This eventually lead to a landslide lake outburst event in which floodwaters and glacial sediment from the breached lake above Kedarnath collided with debris from other landslides, creating a debris flow that smashed into Kedarnath from the north, devastating the town. Perhaps as many as 30,000 were killed in the Mandakini River catchment, and over 100,000 people, including many tens of thousands on the annual Hindu pilgrimage to Kedarnath, were evacuated (Ziegler et al. 2014).

Submarine slides are also common (e.g. Moore 1978; Lee 2005; Harbitz et al. 2014) given the extra lubrication provided by infiltration of water into substrate in an aqueous environment, even though water is not an essential influence in the movement of slides. One of the largest known is the 8.2 ka Storegga megaslide off the coast of mid-Norway, which moved as much as 3,500 km^3 from the submarine slide scar and generated a tsunami that hit the west coast of Norway, Scotland, and islands in between (Bondevik et al. 2003; Bryn et al. 2005). Coincidentially, the head scar also hosts one of Norway's largest gas fields (Ormen Lange, discovered in 1977).

In volcanic terrains landslides are very common, induced by steep slopes, the preponderance of seismic activity, and interbedding of lithologies of very diverse mechanical properties (van Wyk de Vries and Davies 2015). Such slides may be very local in significance, or may involve large-scale landslide or sector (flank) collapse of one side of the volcano, which may then transform into *rock* or *debris avalanche flows*, as discussed in the next section (e.g. Voight 1981; Ui 1983; van Wyk de Vries and Davies 2015; Fig. 12.8d).

Volcanic landslides on large stratovolcanoes and shield volcanoes begin by the onset of gravitational spreading of the edifice due to their gravitational potential above base level (Borgia et al. 2000). This can form horst and graben block segments as they spread, which accomodates their extension (van Wyk de Vries and Davies 2015). In particular, marine stratovolcanoes and oceanic shield volcanoes, which may rise several km above the seafloor, experience major landslide and sector collapse events because of their significant gravitational potential, and the overloading of mechanically weak substrate or interbedded sedimentary layers.

The causes of volcanic landslides (and rock avalanches) can be varied (van Wyk de Vries and Davies 2015; Hacker et al. 2018), including:

(a) Long-term gravitational instability due to the very high relief and potential energy of volcanic edifices above base level, and their steep slopes (e.g. Augustine volcano, Alaska, Begét and Kienle 1992; Mt Rainier, Cascades volcanic arc, Swanson et al. 1992).
(b) Weakening of such edifices by infiltration of meteoric water and ice into fractures, and/or hydrothermal circulation forming 'weak cored volcanoes' (e.g. 1998 Casita volcano, Nicaragua, Cecchi et al. 2004, Devoli et al. 2009; Mt Rainier, Cascades volcanic arc, Reid et al. 2001; Mt Pelée, Martinique, Le Friant et al. 2003).
(c) Intrusion of substantial volumes of magma high into an edifice, causing it to inflate and become unstable (e.g. 18 May 1980 Mt St Helens, Cascades volcanic arc, Voight et al. 1981; Kilauea volcano, Hawai'i, Swanson et al. 1976, Hacker et al. 2018).
(d) Triggered by the onset of, or during an eruption (e.g. 190 ka, Tenerife, Canary Islands, Paris et al 2017; 22 December 2018 Anak Krakatau; west of Java, Williams et al. 2018).
(e) Earthquakes and tectonic movement (e.g. Iriga volcano, Phillipines; Mt St Helens; Lagmay et al. 1999, Paguican et al. 2012).
(f) Overloading of mechanically weak substrate, which fails catastrophically under the volcano (e.g. Socompa volcano, Chile, van Wyk de Vries et al. 2001).
(g) Sea level changes leading to water saturation, or peripheral erosion and/or deposition of mechanically weak strata (e.g. Hawai'i, Moore and Clague 1992).

Hacker et al. (2018) propose that the causes of volcanic landslides and debris avalanches can be subdivided into three main types: (1) volcano edifice (or part of) collapse due to steep slopes and internal structural weaknesses; (2) laccolith (or cryptodome) intrusion high into the crust or edifice causing over-steepening of slopes and collapse; and (3) volcanic field collapse, involving large-scale shear detachment in the substrate under a significant area of a volcanic field or province.

One of the best known volcanic landslides of recent times in volcanic terrains was the spectacular one initiating the devastating 18 May 1980 eruption of Mt St Helens (Chaps. 6 and 8). The growth of a high-level magma cryptodome within the core of the stratovolcano caused gradual inflation, over-steepening and fracturing of the northern flank of the volcano. On 18 May, apparently triggered by a magnitude 5 earthquake, the oversteepened slopes collapsed, causing a huge landslide which unroofed the magma reservoir, and initiated a simultaneous phase of explosive activity (Fig. 12.8d; Christiansen and Peterson 1981; Voight et al. 1981, 1983). Several discrete initial slide blocks were identified from time-spaced photographs (Figs. 12.8d, 12.10a–d; Moore and Albee 1981; Voight et al. 1983), and under the influence of the simultaneous, lateral explosive blast, and interstitial hot water and steam, the slide masses moved away in dynamic fashion (Figs. 12.8d; 12.10b, c), transforming into a lateral fast flowing debris or rock avalanche, leaving a huge, 4 × 2 km, horseshoe amphitheatre-shaped, concave, erosion scar (Fig. 12.8d, e).

At least 50% of all subaerial and marine stratovolcanoes, and large shield volcanoes have experienced landslide or sector collapses: e.g. Stromboli, Italy (Fig. 12.8g; Tibaldi 2001; Tomassi et al. 2005), Whakaari/White Island, New Zealand (Moon et al. 2009), Taranaki/Egmont volcano, New Zealand (Palmer et al. 1991; Alloway et al. 2005), Socompa volcano, Chile (Francis et al. 1985; van Wyk de Vries et al. 2001; van Wyk de Vries and Davies 2015), Bezymianny, Kamchatka in 1956 (Gorshkov 1959; Belousov and Belousova 1998), 1964 Shuveluch volcano, Kamchatka in 1964 (Gorshkov and Dubik 1959; Belousov 1995), and Colima volcano, Mexico (Roverato et al. 2011), Tenerife (Fig. 12.8f; Watts and Masson 1995, Martí et al. 1997; Hürlimann et al. 2000; Seisdedos et al. 2012), Gran Canaria (Funck and Lykke-Andersen 1998; Schmincke and Sumita 1998) and Hawai'i (Fig. 12.11a, b; Moore 1964; Moore et al. 1989; Denlinger and Morgan 2014), to mention but a few. In Japan alone, Ui et al. (1986) identified 71 debris avalanche deposits from 54 Quaternary volcanoes, most of which would have originated as initial landslide sector collapse events. In the central Andes, Francis and Wells (1988) identified 14 major debris avalanche deposits and 28 volcanoes that have breached (collapsed) summits. They proposed that stratovolcanoes greater than 2 km high and with slopes >35° are prone to collapse. In the Lesser Antilles volcanic arc sector collapse scars, as first proposed by Roobol et al. (1983), have now been recognised on at least six of the larger volcanic islands (St Lucia, Martinique, Dominica, Guadaloupe, Montserrat and St Vincent (e.g. Le Friant et al. 2003; Boudon et al. 2007; Lebas et al. 2011), as well as the submarine Kick 'em Jenny volcano (Dondin et al. 2011). These observations suggest mutiple widespread landslides are a common feature during the long-term evolution of the volcanic islands.

Seismic interpretations of the submarine flanks of Hawai'i show multiple, stacked slides, with an architecture similar to that of fold and thrust belts, with stacked thrust

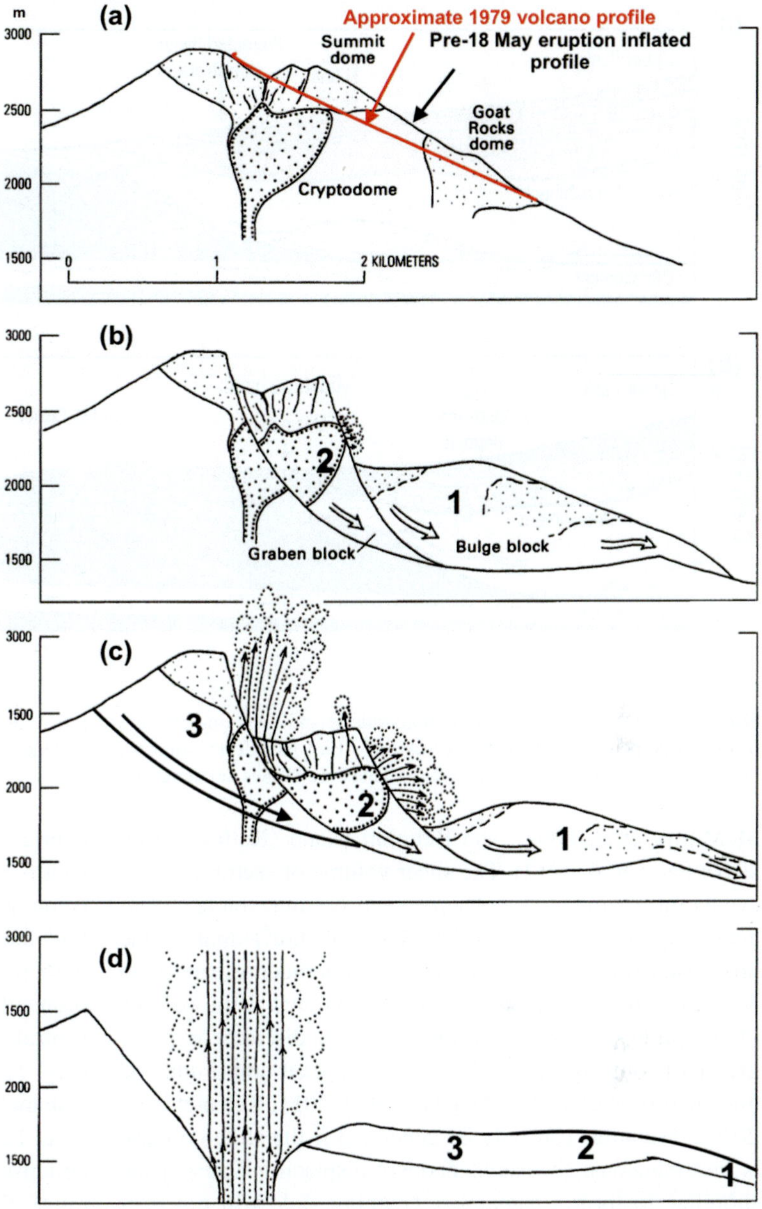

Fig. 12.10 Schematic representation of the sequence of stages in the collapse of Mt St Helens on 18 May 1980. **a** Pre-eruption geology and profile showing the degree of inflation of the north face relative to a profile measured in 1979, due to the intrusion of a cryptodome high in the edifice. **b** First and second landslide block collapses, and initial decompression of the cryptodome, triggering the start of a lateral explosion. **c** Ongoing landslide block collapse including block 3, transformation of blocks 1 and 2 into lateral rock or debris avalanche pulses, and acceleration of both the lateral and vertical blasts. **d** Onset of a buoyant plinian eruption column phase. (Modified from Moore and Albee 1981)

packages stacked on top of a basal décollement surface (Fig. 12.11a, b). Some slide masses may be as large as 150 km long and 50 km wide. The arcuate trace of Hawai'i's volcanically active East and Southwest Rift Zones onland represent the head fault of a major landslide mass (Ch. 16, see Fig. 16.27). True slide deposits that have not transformed into other mass flow types, will therefore be relatively coherent masses retaining some element of original stratigraphy (Figs. 12.8a, b, h; 12.10). Some degree of internal deformation may be apparent (faults, folds – some soft sediment ones; Figs. 12.8b, h, 12.10), and each slide mass will be separated from the substrate by a basal detachment (Figs. 12.8a, 12.11). The source area will be marked by a major scar, which is frequently concave in the direction of movement (Fig. 12.8a, e, f, g). Stranded detached slabs of the slide mass may mark the trail of the slide back to the scar.

Slides on marine stratovolcanoes and oceanic islands are also important in the triggering of very large tsunami events, which have been termed *mega-tsunamis.* For example in the Lesser Antilles volcanic arc the offshore debris avalanche fan deposits associated with onshore sector collapse scars are of a large extent and likely correspond to catastrophic events which probably triggered very large tsunami (Deplus et al. 2001). Similarly, evidence of giant tsunami waves is identified in the Canary Islands (Ward and Day 2001; Pararas-Carayannis 2002; Giachetti et al. 2011; Hunt et al. 2014; Paris et al. 2017), Hawai'i (Moore and Moore 1984;

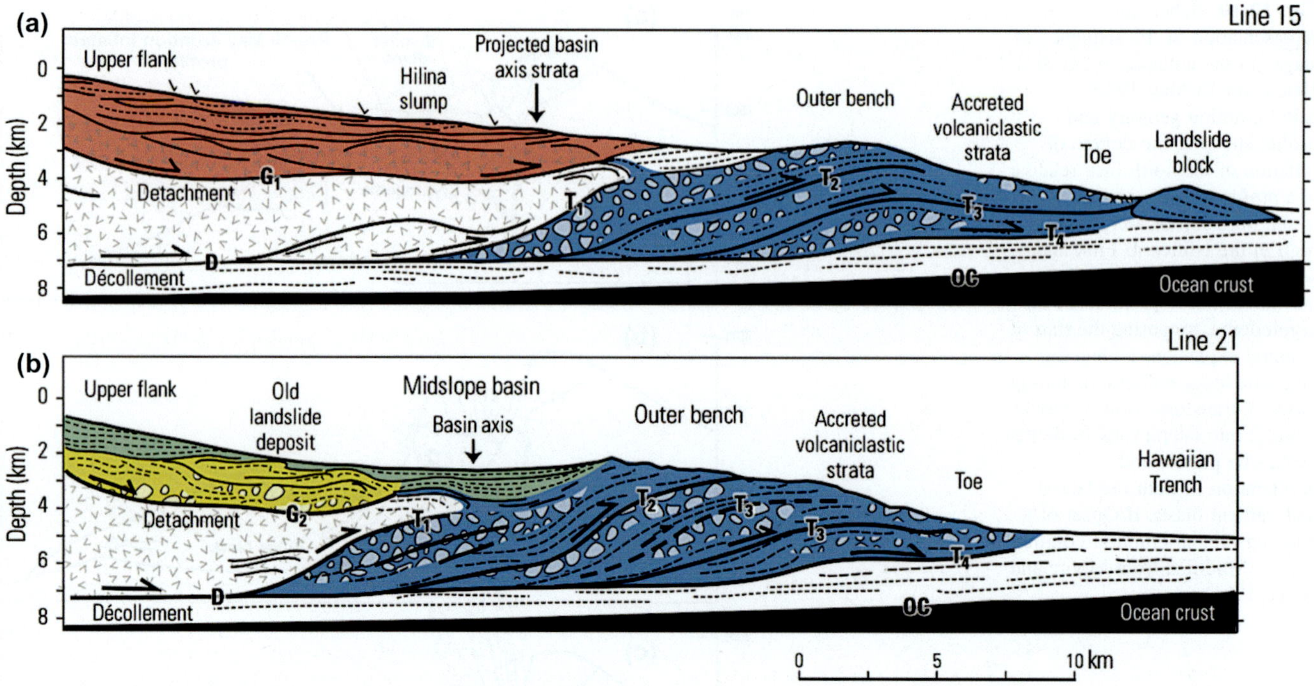

Fig. 12.11 **a**, **b** Interpretations of submarine seismic sections offshore from and perpendicular to Kilauea volcano's East Rift Zone, southern Hawai'i (Chap. 5). These show stacked landslide bodies separated by thrust detachment faults, sitting above a basal, master décollement surface. (From Denlinger and Morgan 2014)

McMurtry et al. 2004), and Réunion Island (Kelfoun et al. 2010; Paris et al. 2013). The sheer volume of sector collapse events on marine stratovolcanoes and oceanic shield volcanoes (potentially tens to hundreds of km^3) makes the propensity for generating larger tsunamis much greater than any other volcanic source mechanism. As well as the volume of the slide, the failure mechanics (e.g. discrete or retrogressive) controls the number of waves, their amplitude and thus their time of arrival (Giachetti et al. 2011; Kelfoun et al. 2010). Tsunamis can only be generated if the slide collapse is instantaneous, so instantaneously displacing water at the shoreline to form a tsunami; a creeping slide will not generate a tsunami when it moves slowly.

The largest historical volcano sector flank failure is the 1988 collapse of Ritter Island stratovolcano in Papua New Guinea (Ward and Day 2003; Silver et al. 2009); having a volume of 5 km^3 it produced a 10–15 m tsunami killing some 3,000 people along the coasts of the Bismarck Sea. More recently the landslide triggered tsunami on 22 December, 2018 at Anak Krakatau volcano in the Sunda Strait west of Java has received much attention due to the availability of multiple pre- and post-event ground-based, bathymetry and space-borne data sets (Grilli et al. 2019a, b; Williams et al. 2018; Walter et al. 2019). At approximately 20.55 local time (13.55 UTC, Coordinated Universal Time), the volcano experienced a relatively small lateral collapse during a period of eruptive activity that had begun in June (Fig. 12.12a). The collapse discharged volcaniclastic rock debris southwestwards into the 250 m deep caldera to the south of the volcano (see Fig. 11.16b). This generated a tsunami (Fig. 12.12b), that within 40 min caused at least 437 fatalities on the adjacent coasts of Java and Sumatra with run-up as high as 13 m. The principal landslide scar bisected the Anak Krakatau volcano, cutting behind the central vent and removing 50% of its subaerial extent (Fig. 12.12a). New bathymetric data has revealed a debris avalanche field containing blocks up to 70–90 m across (Fig. 12.12c; Tappin et al. 2019). Estimated submarine collapse geometries result in a primary landslide volume range of 0.22–0.30 km^3, although this subsequently has been revised down to 0.19 km^3 based on the new bathymetric data. The collapse was accompanied by explosive phreatomagmatic activity producing a series of pyroclastic density currents (see Plate 10.1b). The exact cause of the tsunamis generated by the 15th January 2022 intense explosive eruption of Hunga Tonga-Hunga Ha'apai volcano in Tonga, southwest Pacific, is still being debated (e.g. Purkis et al. 2023).

12.4.5 Rock or Debris Avalanches

Rock or debris avalanches in volcanic terrains can be initiated by vertical rock fall from steep cliffs or peaks (Fig. 12.13a; e.g. Little Tahoma rock avalanche at Mt Rainier, USA; Fahnestock 1978), or from initial sector or

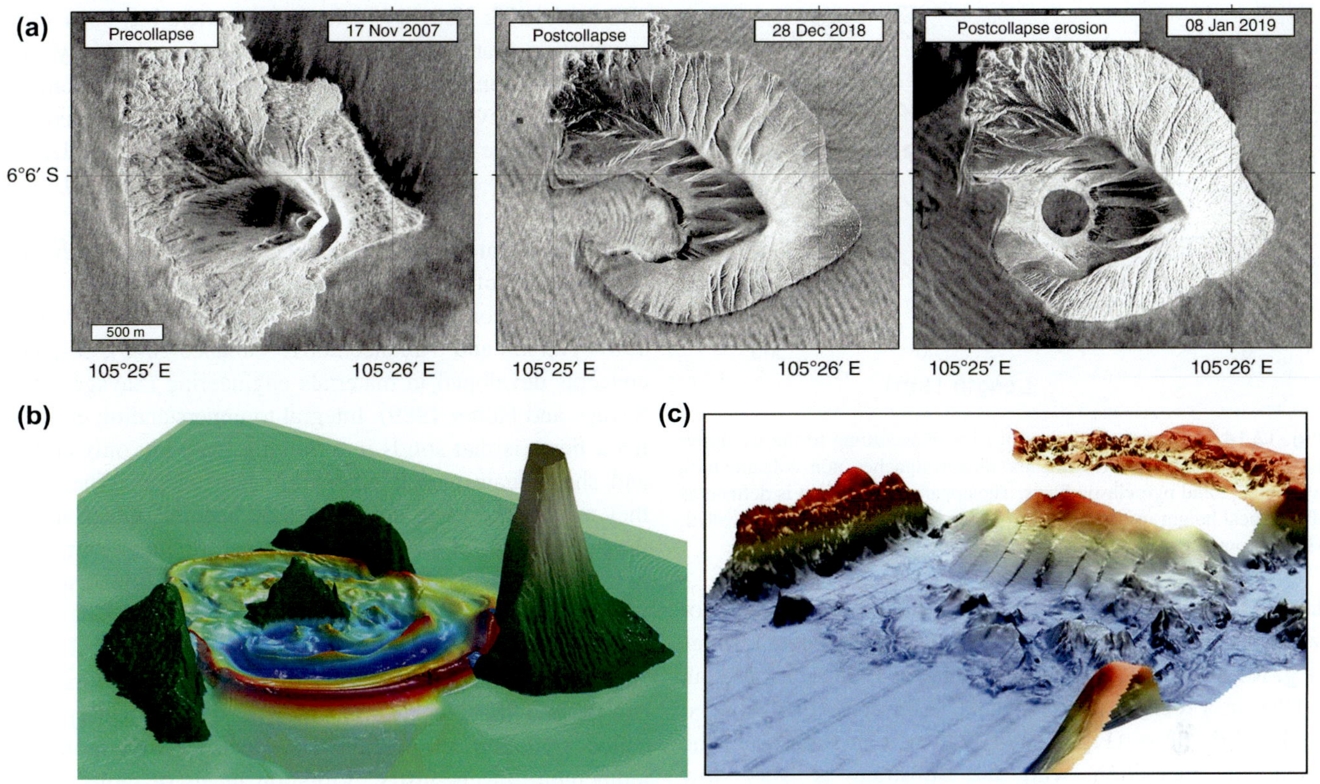

Fig. 12.12 Features of the 22 December 2018 Anak Krakatau sector collapse, Sunda Strait, west of Java, Indonesia. **a** Morphological changes by sector collapse, syn-eruptive deposition, and subsequent erosion before and after 22 December 2018. High-resolution TerraSAR-X satellite radar images show the extent of changes associated with sector collapse, and with the formation and erosion of the new coastline, tuff ring, and crater lake. (From Walter et al. 2019). **b** Computer simulation of sector collapse (after 90 s) and generation of tsunami on 22 December 2018. View looking to northeast; height of tsunami waves colour-coded from peak = darkest red (+30 m) to trough = darkest blue (−30 m). (*Source* Grilli et al. 2019b). **c** High-resolution bathymetry showing debris avalanche field which extends some 2,000 m and is littered with giant blocks up 70–90 m (*Source* Tappin et al. 2019)

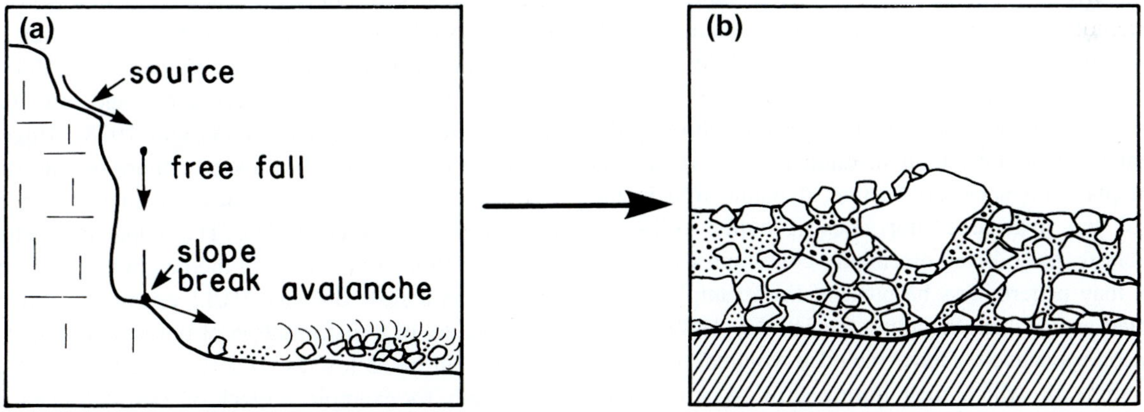

Fig. 12.13 **a** Schematic representation of the formation of a rock or debris avalanche through initial rock fall, and then deflection and transformation of the rock fall into a fast, laterally flowing rock avalanche. Volcanic rock or debris avalanches may also result from disintegration and transformation of landslides originating from sector collapse of high relief steep-sided stratovolcanoes (e.g. Mt St Helens, Fig. 12.10c). **b** Schematic representation of the internal massive texture of debris avalanche deposits, showing large mega-clasts in a very poorly sorted, very coarse matrix, with some surface hummocks or mounds, often associated with large clasts and log-jam effects around them. (From Cas and Wright 1987)

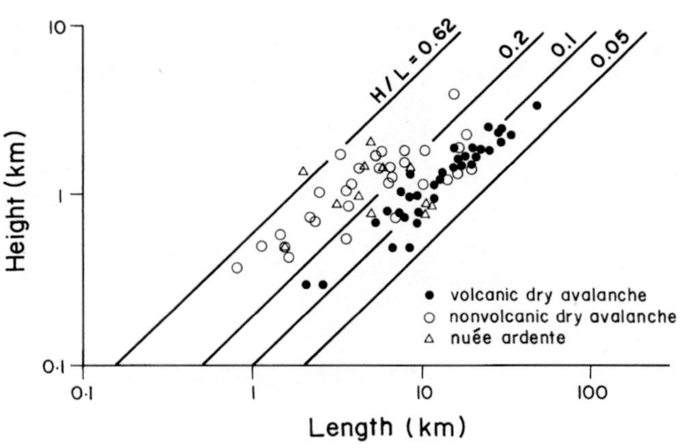

Fig. 12.14 Apparent coefficient of friction according to the Coulomb Law for sliding friction for volcanic rock avalanches, non-volcanic rock avalanches, and pyroclastic flows. The apparent coefficient is defined as the vertical height moved to the projected horizontal distance travelled, H/L. (From Cas and Wright 1987, after Ui 1983)

landslide collapse of the steep sides of high-relief stratovolcanoes, as discussed above, under volcanic landslides (e.g. Mt St Helens, USA; Glicken 1991, 1998; Hacker et al. 2018). In a rock or debris avalanche the mass of dislodged rock debris flows (Fig. 12.13a), rather than slides, and the frictional interaction with the substrate is very low (Figs. 12.13a, 12.14; Hsü 1975, 1978; Ui 1983; Siebert 1984). The term 'flow' implies a degree of relative motion and freedom between clasts, indicating that pervasive internal deformation occurs within the flow. Avalanches are therefore not coherent, and there is considerable interaction between clasts in the flow.

Hsü (1975, 1978) equated the flow mechanics of "catastrophic rock debris streams" or rock avalanches with inertial Bagnoldian grain flow. In this process, grain support, grain momentum, flow momentum and mobility are all maintained by grain-to-grain collision generated stresses leading to a progressive transfer of momentum through the flow. That is, if an interstitial medium (air, ice debris) is present it plays no necessary, active, dynamic role in the motion of the avalanche, nor does it buffer collisions between clasts, these being essentially cohesionless.

Fluids may nevertheless be present. For example, many documented avalanches (including the 18 May 1980, Mt St Helens debris avalanche) have included volumes of snow and ice, some of which melt during flow as a result of impacts. The Mt St Helens landslide and debris avalanche triggered an immediate explosive eruption, so it is possible that the debris avalanche also gained momentum from explosive thrust and volcanic gas was mixed into it as suggested by a limited amount of juvenile clast material in the deposits (Voight et al. 1981). On Earth, air is also trapped within debris avalanche flows, and fluidisation by air may also occur, although it is not essential to the mechanics of avalanches as seen from interpreted Lunar avalanches (Howard 1973; Hsü 1975). In the past there has been some speculation that rock avalanches owed their mobility to a trapped cushion of compressed air on which the avalanches rode (e.g. Shreve 1968). In this model the air is trapped at a break in slope or hydraulic jump, where the avalanche changes from a largely fall trajectory to a lateral flow trajectory.

A significant advance in understanding the flow behavior and mobility of rock or debris avalanches and other mass flows was to consider them as *granular mass flows* (Iverson 1997; Iverson and Vallance 2001; Vallance 2005), based on concepts developed in materials engineering (Savage 1984; Savage and Hutter 1989). Integral to understanding granular mass flows is that solids in a flowing mass not only collide and shear against each other, but they vibrate. The faster they vibrate the more dispersed they become, which means the bulk density of the flow, the internal frictional interaction between solids, and the bulk visocosity of the flow decrease, making the flow more mobile. Note, this is different to the dispersive pressure concept due to an interstitial fluid in Bagnoldian granular flows (Sect. 12.4.6). This vibrational energy and the level of vibration has unfortunately been called "granular temperature" by analogy with molecular temperature in chemistry. Vibration levels of molecules in chemical materials are indeed temperature dependent, but the vibrational behaviour of solids in granular flows is not related to temperature. Rather it is related to physical parameters such as momentum and internal stresses in a granular flow, and the term "granular temperature" should be renamed "*granular vibration parameter*".

Mechanically, avalanches can be distinguished from more viscous modes of transport, such as debris flows, by the greater speeds of avalanches, and by the lower frictional interaction with the substrate. Velocities of 200–300 km hr^{-1} or more are not uncommon and are based on well-constrained calculations (Voight 1978). Huge blocks, metres in diameter, hurled a km or more on ballistic trajectories by avalanches, also attest to amazing speeds (e.g. Plafker and Ericksen 1978). The apparent coefficient of friction, defined as the ratio of the height fallen or descended to the lateral flow distance (H/L; essentially reflecting the effectiveness of transformation of potential energy to kinetic energy) can be as low as 0.06 (Fig. 12.14; Ui 1983), compared with, for example, a block of rock sliding according to Coulomb's law of sliding friction, with an apparent coefficient of friction of 0.62 (Fig. 12.14; Hsü 1975, 1978).

Voight et al. (1981) analysed time lapse photographs of the 18 May 1980, Mt St Helens landslide and debris avalanche transformation and determined that the north flank of the volcano experienced at least 3 discrete progressively retrogressing landslide block collapses or phases, separated by seconds (Fig. 12.8d; 12.10c). The third was probably an amalgum of multiple smaller ones. The resulting debris avalanche was thus a succession of flow pulses sourced from different parts of the collapsing edifice (Voight et al. 1981). Voight et al. (1983) calculated that the Mt St Helens debris avalanche had a velocity of $\sim 70\ m\ s^{-1}$ ($\sim 250\ km\ hr^{-1}$) at source on the volcano, and an average velocity of $35\ m\ s^{-1}$ ($\sim 125\ km\ hr^{-1}$) over a flow distance of 23 km. The apparent co-efficient of friction was 0.1, the strength of the flow $\sim 10^4$–10^5 Pa, the viscosity $\sim 10^4$–10^5 Pa s, and the Reynolds Number between 20 and 40, indicating high-velocity laminar flow, caused by high clast concentrations in the flow. Part of the debris avalanche flow was able to climb up and over topography, surmounting a 300 to 380 m ridge, and then producing a deposit in South Coldwater Creek to the north, whereas most of the avalanche was deflected westwards along the north fork of the Toutle River (Voight et al. 1981).

Although the syn-eruptive Mt St Helens debris avalanche is now considered to be the archetypical example (e.g. Voight et al. 1981, 1983; Glicken 1991, 1998), debris avalanche events, and the full spectrum of rock falls and landslides can also occur spontaneously in volcanic terrains without a contemporaneous eruption. As previously mentioned, a significant rock fall collapse (600 m vertical fall) occurred from Little Tahoma Peak on the flanks of Mt Rainier in 1963. This produced 7 debris avalanche flow pulses, with a total volume of $\sim 12 \times 10^6\ m^3$, that flowed over the Emmons glacier, and downslope into a glacial valley for 6 km, confined by lateral moraines (Crandell and Fahnestock 1965; see Fig. 12.20b–d). The velocity was estimated to have been >100 $km\ hr^{-1}$.

Deposition of rock avalanches takes place by frictional freezing of the avalanching rock mass as they lose momentum and deflate, implying a relatively high bulk viscosity for the avalanche flow in the terminal stages, and a high degree of particle interaction. Avalanche deposits form sheets to well-defined lobes with steep margins, including levees. They can be tens of m thick, have volumes up to several tens of km^3, (e.g. Socompa, Chile which is one of the largest known and has a volume of 36 km^3; van Wyk de Vries et al. 2001), flow up to 50 km (on Earth, more on Mars), and cover areas up to several hundred km^2 (Dade and Huppert 1998). Surface pressure mounds or hummocks caused by "*log-jamming*" of large boulders are characteristic, some marked by megablocks, and others just by piles of coarse debris (Glicken 1991, 1998; Fig. 12.15a–d), suggesting a significant yield strength at the time of deposition. Because of these significant topographic variations on the surface of the flow, thicknesses are very variable.

Internally, avalanche deposits are massive and extremely poorly sorted, with fragment sizes ranging from huge megablocks (*megablock facies*) tens to hundreds of m in dimension and thousands of m^3 in volume (Fig. 12.15e–i), surrounded by a poorly sorted matrix (*matrix facies*) of boulder, cobble, pebble and minor sand and mud size debris (Fig. 12.15j; Ui 1983; Voight et al. 1983; Glicken 1991, 1998; Palmer et al. 1991; van Wyk de Vries and Davies 2015; Hacker et al. 2018). Rare individual blocks up to 500 m in diameter have been claimed (Ui 1983), although these were almost certainly slide blocks. Individual blocks are frequently highly fractured or shattered internally resulting from collisions, severe vibrations and deformational flexing in transit. Fractures form an inter-connected network, with spacing down to cm and mm, with a isotropic fabric (i.e. no preferred orientation), and they bound polygonal fragments with an *in situ*, jigsaw fit of pieces (Fig. 12.15g–i). Spalling of such fragments from big blocks has contributed to the formation of smaller matrix fragments during flow (Fig. 12.15j).

Frequently, original stratigraphy from the source is reflected by lateral relationships or an order of lithologies in the deposit from proximal to distal parts (e.g. 18 May 1980, Mt St Helens deposit; Voight et al. 1981; Glicken 1991, 1998; Hacker et al. 2018), suggesting that relatively little internal mixing of clasts has occurred, and a sequential sourcing of source stratigraphy occurred during initial slide/rock avalanche formation. Overall, grain size decreases down-flow and laterally towards the margins (Palmer et al. 1991). Megablocks are located mostly in proximal to medial areas and along the axial part of the deposit. A rock avalanche can thus be viewed as a mobile, fluidal, close packing of blocks and fragments which jostle, bump, push, collide and fragment each other in transit.

The matrix between blocks can vary greatly in character from deposit to deposit, but is very poorly sorted, and dominated by lithologies derived from the collapsing edifice and formed in the flow through abrasion of large clasts during clast collisions (Fig. 12.15j). However, it often includes material such as soil, alluvial gravel and vegetation eroded in transit. An unusual case, the very large Socompa debris avalanche deposit in Chile, is 80% dominated by lithologies of the basement underlying the volcano, which requires a special mechanism for formation. Van Wyk de Vries et al. (2001) proposed that loading of the sedimentary substrate by the very large Socompa strato-volcanic edifice caused deformational failure of the substrate leading to ejection of large volumes of liquefied debris from the substrate forming a debris avalanche.

Relatively few detailed granulometric analyses are available for avalanche deposits, and there are obvious practical problems in representing the many large blocks. Analyses of the matrix facies grain size for the 18 May 1980, Mt St Helens avalanche deposit shows that average abundances of clay, silt, sand and gravel sizes are respectively 4, 11, 42 and 43 wt% (Fig. 12.16; Voight et al. 1981). In comparison, both the 18 May 1980 blast flow and 18 May to 17 October pumice and ash flow deposits have a greater

(a)
(b)
(c)
(d)
(e)
(f)

◀ **Fig. 12.15** Features of rock or debris avalanche deposits. **a** Mound or hummock surface morphology of the 18 May 1980 debris avalanche deposit, North Fork Toutle River, Mt St Helens (Photo taken in 1989). **b** Hummocky surface of the 30 March 1956 debris avalanche deposit, Bezymianny volcano, Kamchatka, Russia (Photo courtesy of A. Belousov, taken in 1987). **c** One of the hummocks or mounds from the Mt St Helens debris avalanche deposit; note person circled for scale (From Lipman and Mullineaux 1981). **d** One of the largest hummocks or mounds from the Bezymianny debris avalanche deposit; again person circled for scale (Photo courtesy of A Belousov; taken in 1987). **e** Megablock and coarse "matrix" facies of the 1963 debris avalanche deposits from Little Tahoma Peak, Mt Rainier, USA (From Crandell and Fahnestock 1965). **f** Megablock ~50 m long in the 2,600 yr BP Kisakata debris avalanche deposit, Chokai volcano, northern Honshu, Japan. **g** Pervasive internal jigsaw-fit texture within the shattered megablock shown in (**f**). **h** Cross-section of part of a megablock (>50 m long), in the 18 May 1980 Mt St Helens debris avalanche deposit. The pale lithology is highly fractured dacite; the irregular dark body cross-cutting the dacite is the remains of a highly fractured basalt dyke that through internal vibration and fracturing during the flow of the debris avalanche has become distorted. North Fork Toutle River, Mt St Helens. **i** Close-up view of part of the contact in (**h**) between the dacite (light) and basalt dyke (dark). Note that both are pervasively fractured or shattered, the contact between them is very diffuse, and fragments of the basalt are clearly visible within the dacite near the contact, giving the impression that the whole fractured block has been shuffled, induced by vibration during transport. **j** Massive, very poorly sorted matrix facies of 18 May 1980 Mt St Helens debris avalanche deposit, dominated by angular dacite and andesite blocks and finer equivalents derived from the collapsed sector of the volcano

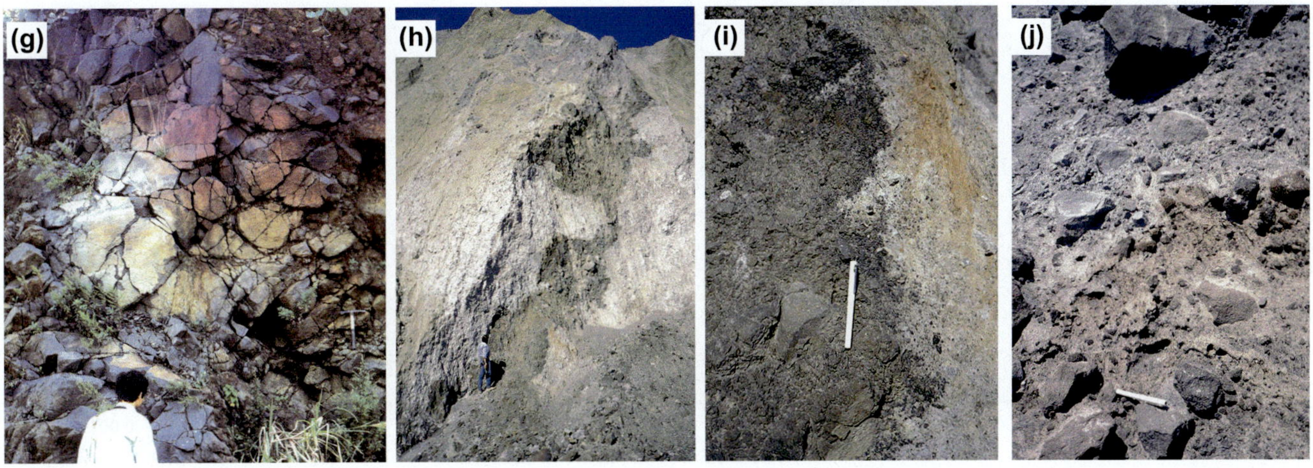

Fig. 12.15 (continued)

abundance of clay and silt sizes (equivalent to fine ash size), and much lower gravel size fragment abundances (Voight et al. 1981). Rock debris avalanche deposits are therefore very granular and fines depleted, more so than debris flow deposits, from which they may be difficult to distinguish in the field. This issue is touched on by Voight et al. (1981) in comparing the granulometry of the Mt St Helens debris avalanche deposit with other known volcanic avalanche deposits; these include the Mt Rainier ones previously referred to, and coarse deposits from both Mt St Helens and Mt Rainier which previously had been interpreted as lahars (i.e. debris flows of volcanic composition). Criteria for distinguishing debris avalanche deposits include the presence of megablocks (many m to tens of m or more in diameter) and the lack of a significant muddy matrix. Debris flows frequently develop a basal laminar, shear zone in which crude bedding develops, and grading of large clasts (coarse tail grading) may occur. Debris avalanche deposits could also be mistaken for block and ash-flow deposits and vice versa, particularly if they are dominated by a single volcanic clast type. Block and ash-flow deposits will be monmictic in composition, should show signs of having been hot and gas rich at the time of deposition (e.g. gas segregation pipes, thermal reddening, abundant charcoal; Chap. 10; Sect. 12.6.6) and they will generally lack megablocks, except very close to vent.

Although there is a significant literature on historic and geologically relatively recent debris avalanche events, there are surprisingly few documented examples from the ancient geological record, which is surprising given how common sector collapse seems to be on young volcanoes. Trofimovs et al. (2004) documented an Archean submarine example in the Yilgarn Craton in Western Australia, which consists of several megablocks of flow banded and autobrecciated dacite, pillow basalt and komatiite up to 70 m in long dimension, dispersed in a diffusely stratified sequence of massive, poorly bedded breccias consisting of clasts of the same lithologies (Fig. 12.17a–e). This is interpreted to represent collapse of a dacite lava dome complex, being intruded by komatiite, and the debris avalanche incorporating substrate, including older pillow lavas exposed in the collapse scar. Some dacite megablocks contain highly irregular

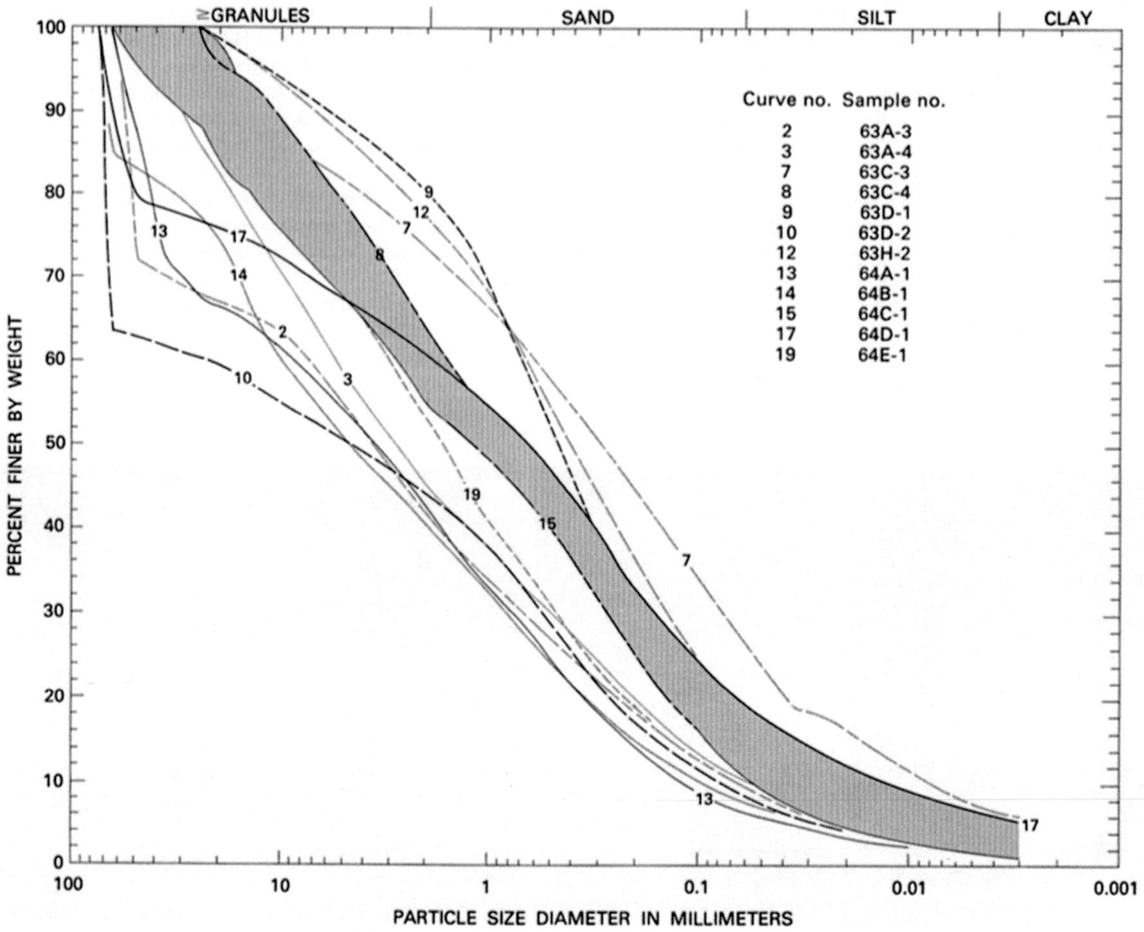

Fig. 12.16 Granulometry of multiple samples of the matrix facies of the 18 May 1980, Mt St Helens debris avalanche deposits, expressed as cumulative frequency curves. The shaded zone contains 8 grain size analyses and more or less reflects the mean grain size distribution profile. (From Voight et al. 1981)

komatiite dykes truncated by the margins of the megablocks. Simpson et al. (2005) briefly described an Ordovician example from a complex marine stratovolcano complex in central New South Wales. In an area of poor outcrop, an association of large domains of internally fractured trachyte and welded ignimbrite up to tens of m in dimension, are enclosed in a polymictic breccia matrix (Fig. 12.17f). The large domains are laterally abruptly discontinuous and cannot be correlated in diamond drill cores 50 m apart, while fragments of the large domains are mingled into the breccia matrix at the margins and mixed into it further from the contacts. The whole lithofacies can be traced along strike for about 8 km, and is part of a submarine volcaniclastic turbidite, lava, intrusion complex. The large domains were interpreted as megablocks in a succession of debris avalanche breccia deposits (Simpson et al. 2005; Chap. 15, see Fig. 15.37).

Fig. 12.17 Ancient debris avalanche deposits, Archean and Paleozoic volcanic successions, Australia. **a** Bedrock outcrop map of a ~2.7 Ga Archean debris avalanche sequence. This depicts crudely bedded polymictic breccias of dacite, basalt, and komatiite clasts, with dispersed megablocks of these lithologies up to 70 m, and the area of dacite-komatiite outcrop in the southeast representing a super-block, hundreds of metres in dimension. (From Trofimovs et al. 2004). **b** Contact between a mega-block of dacite (right) and massive matrix facies breccia, also dominated by dacite. The outcrop is extremely weathered and oxidised. Red circle around camera case for scale. **c** Internal texture of the dacite mega-block in (**b**), showing flow banded autobreccia texture. **d** Part of a 70 m long mega-block of highly weathered basaltic pillow lava, still preserving pillow facing criteria in the block, which is 90° to the strike of the crude bedding in the matrix facies breccia (see **a**), demonstrating that this is a disoriented big clast. **e** Massive, very poorly sorted, matrix breccia facies, dominated by dacite clasts in this highly weathered outcrop. **a**–**e** are all from the Alunite Locality, north of Kalgoorlie, Western Australia. **f** Drill core through part of an Ordovician debris avalanche deposit, at North Parkes, New South Wales, Australia. The hematite altered red clasts are trachyte, which form clasts up to 22 m in dimension, dispersed through a massive, dark grey matrix of small red trachyte, ignimbrite and sedimentary clasts. Drill core is ~5 cm in diameter (see also Fig. 15.37) ▶

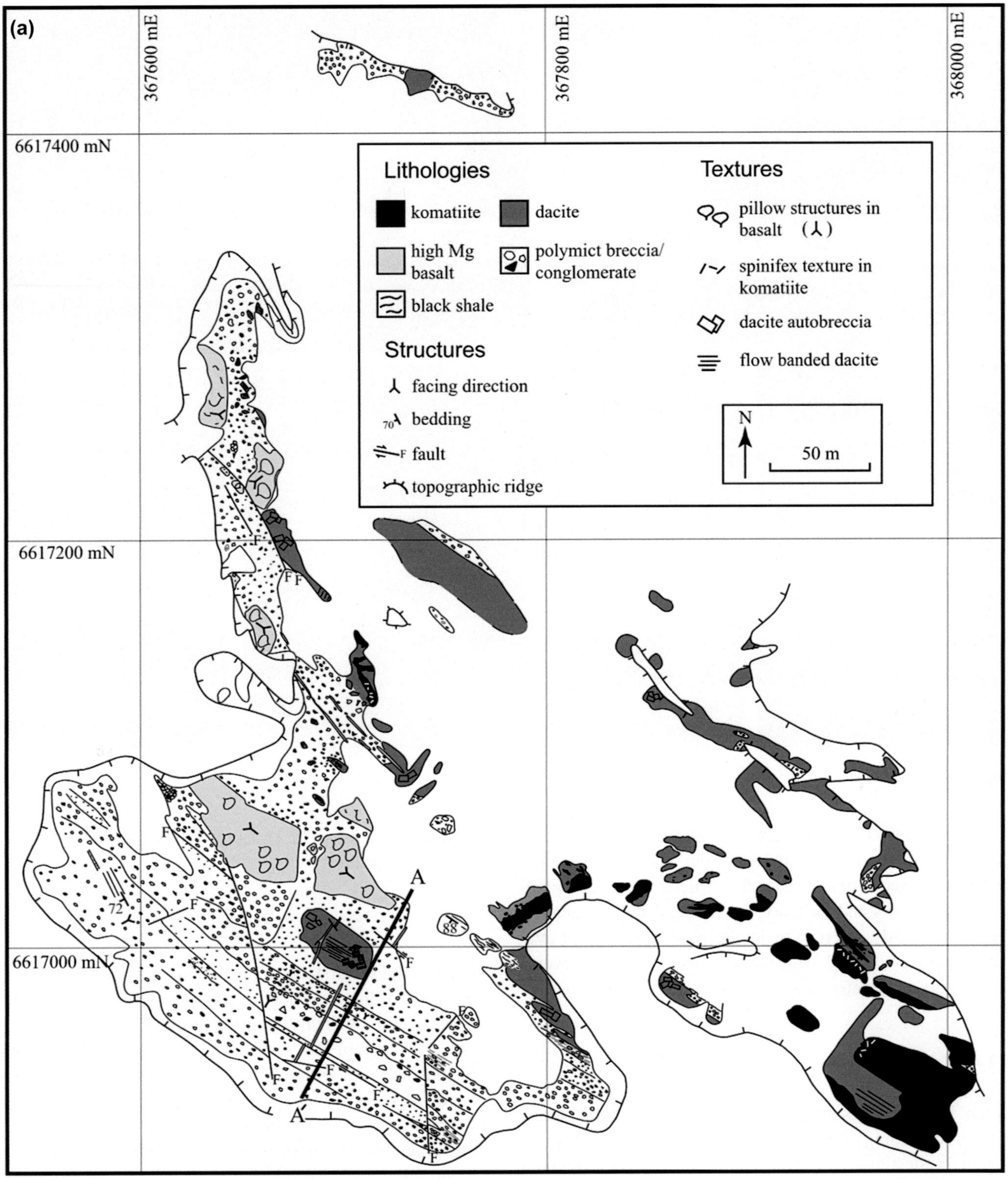
(a)
367600 mE
367800 mE
368000 mE
6617400 mN
6617200 mN
6617000 mN
Lithologies
komatiite
dacite
high Mg basalt
polymict breccia/ conglomerate
black shale
Structures
facing direction
70 bedding
F fault
topographic ridge
Textures
pillow structures in basalt
spinifex texture in komatiite
dacite autobreccia
flow banded dacite
N
50 m
A
A'
F
72
88

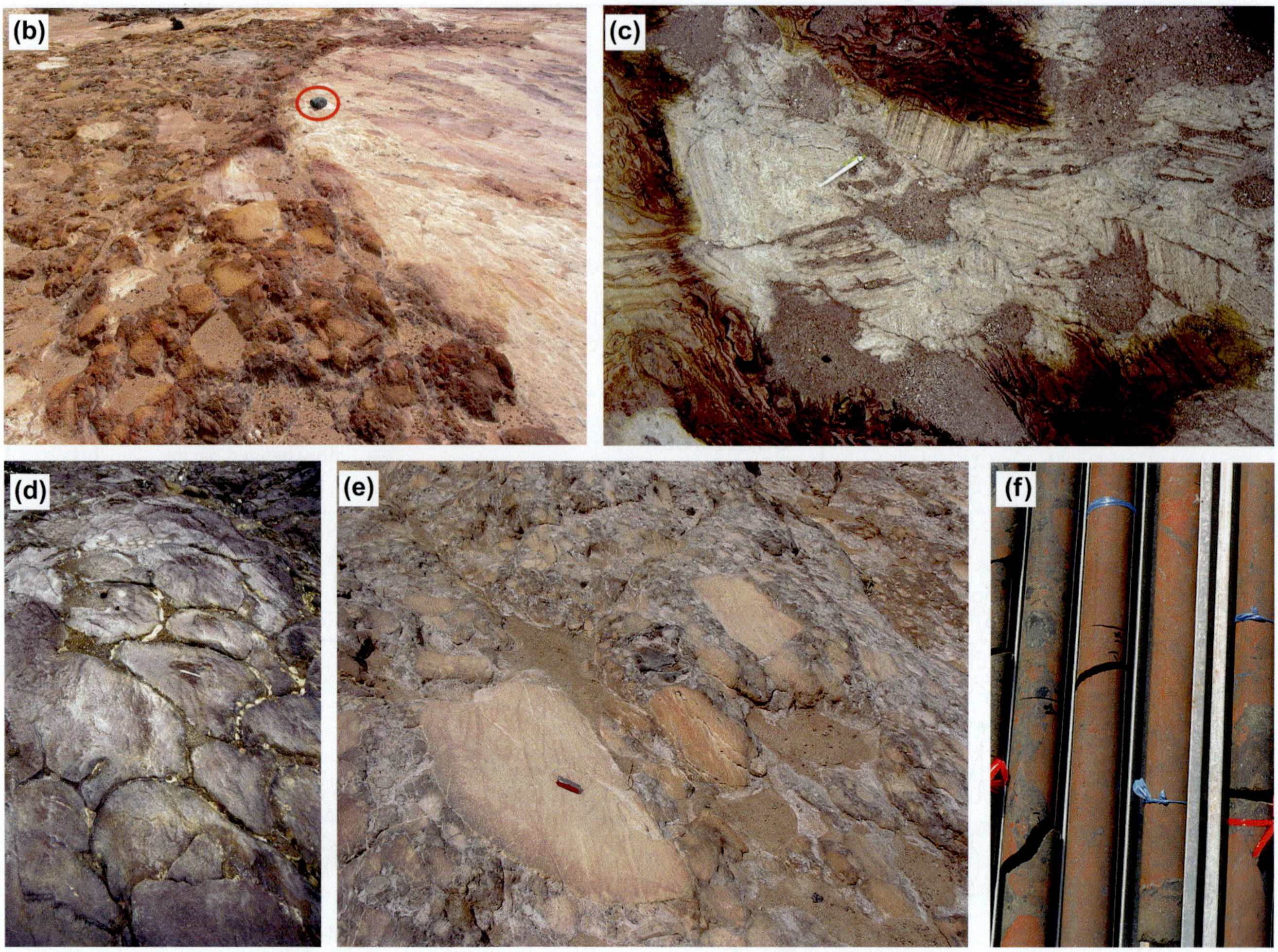

Fig. 12.17 (continued)

In an aqueous environment rock and debris avalanches probably rapidly ingest ambient water and transfom into water supported types of mass flow, including coarse granular debris flows and turbidity currents at distal margins, as we have already referred to in the case large debris avalanche fans associated with onshore sector collapse events which pass into water (Sect. 12.4.4).

12.4.6 Grain Flow

Grain flow involves the spontaneous, passive down-slope rolling and sliding of a population of cohesionless grains that together constitute a moving grain flow layer (Bagnold 1954a, b, 1966; Lowe 1976). Theoretically, grains move solely under the influence of gravity, and not under the influence of an external shear stress, induced, for example, by flowing water or wind, although in some settings those agents may provide lateral force. Although the physical constraints on the initiation of grain flows is the same as for landslides (Eqs. 12.1, 12.2), Lowe (1979) proposed that the condition for grain flow is:

$$\tau > \sigma_\mu \tan\phi + \mu \frac{du}{dy} \quad (12.4)$$

where τ is the shear stress, σ_μ is the internal normal stress, ϕ is the angle of internal friction, μ is the viscosity of the flow, and du/dy is the velocity gradient through the flow.

Shearing within the grain flow increases grain-to-grain collisions and intra-granular shearing, which produce a dispersive pressure between grains leading to transfer of momentum. These interactions maintain the mobility of the entire grain flow layer (Lowe 1976). Collisions produce a reaction force, called dispersive pressure, and induce vibration in the moving granular layer which keeps grains dispersed. Given that the grains are cohesionless and that no viscous interstitial fluid is necessary, the dispersive pressure or collisions operate in an inertial regime. The shearing grain flow layer flows in a laminar state, not turbulently. Grain flows can also occur subaqueously in (more viscous) water. In those situations the collision events are buffered by pressure gradients set up in the viscous fluids during near approaches of grains just before collision, and the buoyancy afforded by dense water. These pressure gradients also contribute to the reaction force affecting the colliding grains, so that they are subjected to viscous dispersive pressure developed in the fluid between the colliding grains, as well as the inertial dispersive pressure of the actual collision between grains.

Given these constraints, it is not surprising that true grain flow only occurs on steep slopes equal to or steeper than the maximum natural angle of stability or repose for a pile of grains. If this angle is exceeded, then downslope movement will eventually be initiated, the angle at which this occurs being called the angle of initial yield (Allen 1982). After flow, the angle of slope will be adjusted to a lower, more stable one equal to or less than the angle of repose, which Allen (1982) calls the residual angle after shearing. The natural angle of repose for an aggregate of grains depends on the grain size, sorting, packing, grain shape and grain surface texture. For most sands and gravels it is between 25° and 35°, but may be steeper for coarse, very angular aggregates.

Grain flow is different from rock avalanches discussed in the previous section, or discussed in subsequent sections on subaqueous granular mass flow, in that in grain flow there is considerable frictional contact or interaction with the substrate, which limits lateral flow on less than angle of repose slopes. In rock avalanches and other granular mass flows, the co-efficient of friction with the substrate is low, and the lateral mobility is largely due to the momentum of the whole granular mass flow.

Grain flow can occur in many natural situations and at different scales. It is responsible for the formation of angle of repose cross-stratification in aeolian and aqueous ripples and dunes by the passive cascading of sand grain flow layers down the steep, protected, angle of repose, lee-side of the ripple or dune structure (Allen 1982; Hunter 1985, Fig. 12.18a, c). It also occurs in the formation of relatively fine- gravelly to finer grained scree slopes (Fig. 12.18b). In volcanic terrains grain flow will be significant during tractional transport of volcanic sediments in both subaerial (aeolian, fluvial) and subaqueous settings where lower flow regime ripples and dunes form during tractional sediment transport (Fig. 12.18c; Sect. 12.6.1). Each cross-bed in a migrating ripple or dune represents a grain flow event from the crest of the ripple or dune as sediment is pushed up the stoss-side of the bedform by the current to the crest, which causes the upper part of the lee-side slope to oversteepen beyond the stable angle of repose, and collapse as a grain flow (Fig. 12.18a, c).

At a larger scale grain flow occurs on the surfaces of scree slopes formed along both subaerial and subaqueous steep slopes, such as cliffs (e.g. crater walls, landslide collapse scarp cliffs, ocean cliffs, where scree slopes form, fed by rock falls (Fig. 12.18b, d). In addition, the steep outer slopes of volcanic cones built around vents during pyroclastic fallout eruptions build to angle of repose slopes, and can be prone to particle creep, slides and grain flow when slopes exceed the stable angle of repose (Duffield et al. 1979). This may occur syn-eruptively and continue post eruptively. Grain flow on scoria or cinder cones often produces a radiating downslope system of furrows (e.g. Richards 1965), as well as maintaining the slopes at or about angle of repose during explosive eruptions.

Grain flow deposits will be marked by steeply inclined stratification (25° to 35°), internally diffuse, relatively thin stratification caused by the high frictional interaction between the grain flow layer and the substrate, and by the development of internal shear layering (Fig. 11.18c, e, f). Individual layers will be thin (mm to cm to a few tens of cm), they may have irregular tops due to pulsing during the grain flow event (Fig. 12.18e) and they commonly develop a reverse grading (Fig. 12.18e, f) produced by intense shearing at the contact with the substrate, which propagates upwards and forces clasts with diameters greater than the thickness of the shear layers to be forced upwards in the grain flow shear system, whilst the development of internal vibration allows fine fragments to fall downwards between coarse fragments in a process called kinetic sieving. Reverse graded pumice and scoria grain flow deposits can be distinguished from primary reverse graded pyroclastic fallout deposits by the often irregular wedge-shaped beds and the development of reverse grading through almost the entire bed in grain flow beds; whereas, fallout deposits will have a regular planar form, reverse size grading is usually confined to small thickness at the base, and the lowest part of the bed is often dominated by fine lithic fragments, resulting from efficient hydraulic sorting during fallout.

12.5 Sediment Transport Involving Ice as an Essential Interstitial Medium

In this category of transport processes, sediment particles are trapped within or on top of significant bodies of ice which acts as the transporting agent. The ice thus provides grain support in a solid state. Where the ice floats as icebergs, it and its sediment load are buoyant because the liquid water in which icebergs float is slightly denser, and movement occurs in response to currents and wind shear. Where the ice is glacial or permafrost (i.e. frozen groundwater table), the ice-sediment mass has gravitational potential and the lubrication provided by water at the base of the ice mass ensures relatively rapid downslope movement under the influence of gravity. Ice can therefore move sediment in either:

(a) Ripples and Dunes
tractive, erosive
force of current
SEPARATION
ZONE
passive
grainflow
on lee-side
(b) Scree slopes
initial angle of yield
(>angle of repose)
post-grainflow angle of
stability
(≤angle of repose)
(c)
(d)
(e)
(f)

◀ **Fig. 12.18** Grain flow processes, settings, and deposits. **a** Schematic lee-side cascading of grain flow layers on the lee-side (down-current) faces of ripples and dunes, so forming internal cross-stratification and resulting in down-current migration of the ripple or dune. A carpet of grains is eroded off the up-current side of the ripple and moved to the crest by the tractive force of the aeolian or aqueous current. Over-steepening occurs at the crest and a grain-flow cascades down the lee-side face into the space called the separation zone over which flow streamlines pass. **b** Schematic grain flow on scree slopes initiated by over-steepening to the angle of initial yield. Grain flow restores the slope to a more stable attitude equal to or less than the angle of repose. (**a**, **b** from Cas and Wright 1987). **c** Angle of repose cross-bedding in a lower flow regime dune of reworked pumice clasts, high-level terrace of Lake Taupo, New Zealand. Each cross-bed represents a grain flow deposit resulting from aqueous current flow from left to right. **d** Scree slopes in the rift vent formed by the 1886 basaltic eruption through older (white) rhyolite domes and pumice deposits resulting from the 700 yr BP Kaharoa eruption of Mt Tarawera, New Zealand. The older rhyolitic dome complex of Tarawera was mantled by basaltic tephra (dark) of the 1886 eruption and both sources are contributing to the scree slope debris. The trails of individual grain flows are represented by the different coloured streaks on the scree slope surface. **e** Explosively erupted and deposited (air fallout) debris of Pukeonake cone, New Zealand, which then moved downslope by grain flow, producing a series of four slightly irregular, reverse-graded (upward coarsening depicted by arrows) grain-supported, grain-flow layers. **f** Several reverse-graded, scoria grain flow beds, with irregular tops, forming part of a scoria cone (~75 ka), near Oia, Santorini, Greece. Dip and grain flow were towards the camera, giving an apparent shallow dip of bedding

- particulate fashion by:
  - ice rafting in icebergs, or
  - glaciers, or as
- large masses of sediment in pseudo-mass movement form by:
  - glaciers, and
  - permafrost creep.

12.5.1 Ice Rafting

The principles involved in deposition of sediment after being ice-rafted are the same as those discussed above for particle free fall (Sect. 12.4.1). An iceberg carrying sediment, calved from a glacier flowing into sea water or a lake, or from shelf-ice, begins to melt and, little by little, releases its sediment load as it drifts (Fig. 12.6c). Dropped boulders with impact structures and draped sediment layers should be expected (Figs. 12.6d, e). However, not all ice-rafted dropstones necessarily produce impact structures, especially if the dropstone still has attached to it a significant volume of ice, so reducing the total bulk density of the dropstone, and so very markedly reducing the fall velocity. Volcanic terrains in high latitudes with associated marine environments should therefore be the source for, and site of, deposition of dropstones of volcanic composition (Fig. 12.6e). Ice rafting, and subsequent deposition on melting, of air fallout ash can be also occur, transporting 'outsize' shards or other pyroclasts far from source, which can be a direct tracer of ice-flow dispersal (Shane and Frogatt 1992; Lacasse et al. 1996).

12.5.2 Glaciers (Particulate Movement)

Glaciers are relatively mobile, huge masses of ice which creep, flow or slide down slope due to their gravitational potential (Fig. 12.19a; Jiskoot 2011) at speeds that vary from the imperceptible to greater than 100 m d^{-1}. The latter occurs where frictional coupling between the glacier and the substrate has been reduced by basal melting (Raymond 1978); this lowers the basal co-efficient of friction (static angle of friction; Eq. 12.3), allowing laterally acting gravitational shear stresses to exceed frictional resistance forces, according to Coulomb's (1776) Friction Law (Eq. 12.3). Glaciers have enormous erosive capacity, as evidenced by the U-shaped valleys they gouge out of bedrock, and the huge volumes of sediment they deposit in ground, lateral and terminal moraine deposits. Glaciers transport their sediment as base load, interior load or surface load. Base load is eroded from the bedrock or substrate at the base and edges of the glacier. Interior load may originally have commenced transport as base load, but because internal flow paths often move forward and upward, it is transferred up into the interior, well above the base. Surface load results from material falling onto the surface of the glacier, for example by rockfall or rock avalanche as discussed above from Little Tahoma Peak, Mt Rainier volcano, USA, or from wind transport or pyroclastic air-fall.

Glaciers develop at relatively high latitudes and altitudes, where snowfall is common and accumulates for a substantial part of the year. This is not restricted to high latitudes, and glaciers are even known at high altitudes in the equatorial zone (e.g. Kilimanjaro volcano, East Africa Rift Zone; Cotopaxi volcano, northern Andes, Ecuador). Since volcanism occurs all over the globe independent of climate, it is not surprising that volcanoes and glaciers are frequent associates. For example, many glaciers drain off the volcanoes of Iceland (Fig. 12.20a, e), Alaska, the Cascades (Fig. 12.20b), the Andes and Antarctica, to name but a few locations. Glaciers are thus major epiclastic agents on the Atlantic mid-oceanic spreading ridge system in Iceland (Fig. 12.20a, e).

The sediment load of glaciers is dumped at the base, margins and terminus of glaciers in irregular piles or ridge-like accumulations called ground, lateral and terminal moraines (Figs. 12.19b; 12.20b–d), respectively. Moraines consist of unsorted debris varying from particles as fine as clay, resulting from the intense grinding abilities of the

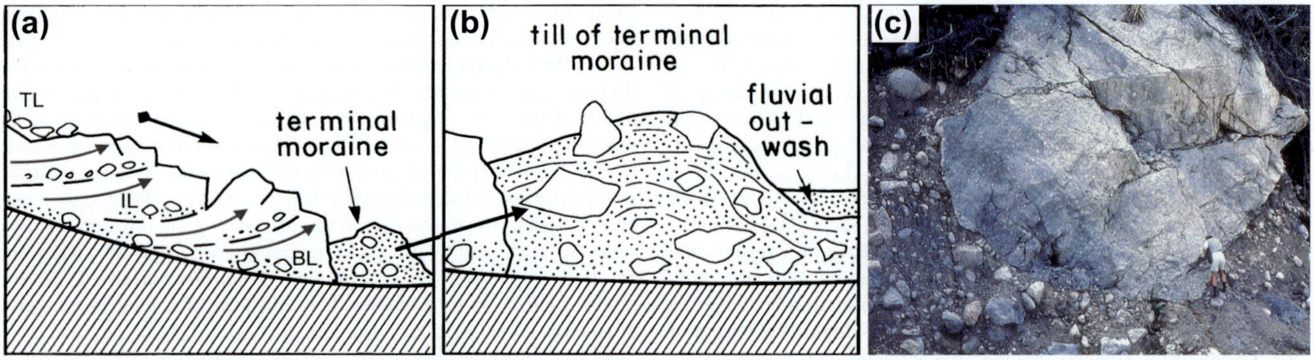

Fig. 12.19 Schematic representation of glacial transport of sediment and deposits characteristics. **a** Modes of sediment transport as bed load (BL), interior load (IL), and surface load (TL), and internal ice flow paths (curved black arrows). **b** Textural character of glacial till of the terminal moraine. which is being reworked by the fluvial outwash system. (From Cas and Wright 1987). **c** Poorly sorted nature of glacial moraine from a non-volcanic example, Pleistocene deposits, Central Highlands, Tasmania, Australia. Note the massive, bouldery texture, with grain size from huge boulders to sandy to clay size matrix

glacial ice mass, to huge blocks that have been dragged and rafted along 12.19b,c). The terms 'till', 'tillite', and 'diamictite' are used for these moraine deposits, the first being for unconsolidated moraines and the other two for lithified moraines. '*Diamictite*' is a more general, non-genetic textural term for poorly sorted aggregates such as till/tillite and debris flow deposits. The term should be used in this generic sense. The provenance of till depends on the source rocks in the drainage area of the glacial system. If these are volcanic, then volcanic till will result (e.g. Fahnestock 1963).

Sediment particles drop out one by one as melting of the ice occurs (Fig. 12.19a). If the terminus of the glacier is on land, then the moraine material may be reworked to varying degrees by the high-energy outwash river systems (Sect. 12.6.1; Fig. 12.20e) fed by the melting of ice at the glacier terminus and the base. These systems are frequently of the braided type. Given that most glacial systems have both short-term and long-term histories of advance and retreat, the upsequence facies variations through a pile of glacial sediment may be complex (Eyles et al. 1983). Facies differences also occur according to whether the glaciers are temperate, sub-polar or polar, which affects their thermal regime and the degree of basal melting. Interested readers are referred to Eyles et al. (1983) and Eyles and Miall (1984), and Menzies and van der Meer (2018).

The stratovolcano Mt Rainier and the Cascades volcanic arc have already been referred to earlier with regard to rock falls and avalanches in volcanic terrains. Mt Rainier also has a developed glacial system of about 25 glaciers, including the Emmons Glacier (Fig. 12.20b), lateral and terminal moraine deposits, and a fluvial outwash system (Fig. 12.20 b–d; Fahnestock 1963, 1978). All are sourced from the volcanic pile of Mt Rainier. Mt St Helens also had a complex of summit glaciers before 1980, much of which was destroyed by the 18 May eruption (Brugman and Meier 1981). However, a small glacier persists in the amphitheatre crater, nestled beside the intermittently active lava dome complex (Fig. 12.20f).

Apart from producing their own glacial deposits, glaciers on volcanoes are instrumental in causing major phreatomagmatic and hydrovolcanic explosive activity from subglacial eruptions, as frequently occurs in Iceland and Antarctica. The confining effects of ice produces a particular type of sub- to intra-glacial volcanic landform called tuyas, as discussed in Chap. 16. In addition to the phreatomagmatic

Fig. 12.20 Glaciers and glacial moraines in volcanic settings. **a** Oblique aerial photograph of Sólheimajökull outlet glacier, on the southern margin of Mýrdalsjökull ice sheet or glacier, which has developed on the Katla shield volcano and caldera, southern Iceland. The dark, dirty colour of the front of the glacial lobe is due to black tephra from the 1918 eruption of Katla, which is being overlain by younger snow and glacial ice. (Photo by Oddur Sigurðsson; Modified from Sigurðsson and Williams 2008). **b** The stratovolcano Mt Rainier, Cascades volcanic arc, USA, showing the Emmons Glacier, the 1963 rock avalanche burying the toe of the glacier, Little Tahoma Peak which was the source of the avalanche some 600 m above the glacier, moraine deposits of the glacier, and the White River valley containing the braided outwash river from the glacier (From Crandell and Fahnestock 1965). **c** Google Earth image of the lateral and terminal moraine from the retreated Emmons glacier and the toe of the 1963 debris avalanche deposit that originated from Little Tahoma Peak as shown in (**b**), and is mostly confined by the moraine ridges (*Source* Google Earth). **d** Oblique Google Earth image of two parallel lateral moraine ridges deposited from the retreated Emmons glacier in the White River Valley, with the 1963 debris avalanche deposit in the foreground the nearest moraine ridge is ~20 m high (*Source* Google Earth). **e** Thorsnarsk Glacier, Iceland, and its associated braided fluvial outwash system in the foreground (From Cas and Wright 1987). **f** Intra-crater Crater Glacier (left, with tensions gashes) located against the whaleback lava dome (centre) (and other domes, right) actively erupting and growing in the crater of Mt St Helens in 2005 (Photo by Steve Schilling; *Source* United States Geological Survey) ▶

(a)
Katla caldera
ÖRÆFAJÖKULL
"HÁBUNGA"
Klifurárjökull
(b)
MOUNT RAINIER
Gibraltar Rock
Little Tahoma Peak
Emmons Glacier
Mount Ruth
Avalanche debris
Goat Island Mountain
Lateral moraine
Avalanche debris
Terminal moraine
1 km
White River campground
(c)
Terminal moraine
Lateral moraine ridges
1963 debris avalanche deposits
500m
(d)
Lateral moraine ridges
1963 debris avalanche deposits
(f)

Fig. 12.20 (continued)

eruption products produced by sub- to intra-glacial eruptions, they also commonly lead to meltwater flood breakouts from the bases and margins of supra-volcano glaciers. These breakouts carry large volumes of subglacially produced hyaloclastite and pyroclastic debris into the rivers around volcanoes. In Iceland these volcanic meltwater floods are called *jökulhlaups* (Sect. 12.6.4).

12.5.3 Glaciers (Mass Movement)

Glacial transport also essentially involves mass movement as well as particulate transport. This is because in addition to carrying individual eroded fragments, glacial transport involves the carrying of aggregates of rock, perhaps picked up as base load from fluvial valley fill sedimentary piles over which the glacier is moving, or as surface load, that is deposited on the glacier surface by a rock avalanche as discussed above (Crandell and Fahnestock 1965; Fahnestock 1978). The principles of transportation, and of deposition of aggregates of rock by glacial transport and the characteristics of the resultant deposits, are the same as in the previous case of particulate movement.

12.5.4 Permafrost Creep

The term "*permafrost*" has traditionally referred to any layer of soil, sediment or rock that is consistently below 0 °C and contains a mixture of ice, water and solids. It has more recently been defined as a substrate temperature state, rather than a solid entity (Dobinski 2011). For our purposes we will consider permafrost to be a solid entity of frozen groundwater that contains enclosed solids, and on slopes is capable of creeping, so transporting the enclosed debris.

Permafrost creep involves the very slow downslope creep of a frozen surface ground layer. That layer may be soil or loose surface aggregate in which interstitial ground water has become frozen, and due to position on slope, the whole frozen layer of rock matter and ice acquires an increased gravitational potential which drives its slow passage down slope (Barsch 1992; Haeberli et al. 2006). Such motion has also been called a rock glacier. On melting, the solids deposited are contorted, perhaps preserving some original stratigraphy. Slopes of volcanoes in high latitude and/or high altitude can be subject to freezing and downslope creep, including scree slopes. The thawing of permafrost due to global climate change is a huge concern because human structures built on permafrost become unstable after thawing. In addition, permafrost in the northern circumpolar region may contain greater than a trillion tonnes of undecayed organic carbon forming vast 'carbon pools', which upon thawing release huge volumes of CO_2 and CH_4, so exacerbating the global warming trend (e.g. Zimov et al. 2006; Tarnocai et al. 2009).

12.6 Sediment Transport Involving Water as an Essential Interstitial Medium

Sediment transport involving water can occur in both particulate and mass movement forms. If the sediment movement is particulate or involves the mass movement of a cohesionless population of grains (i.e. grains which have no significant attraction or repulsion to each other) in low-viscosity, non-cohesive, relatively pure water, then the transport is said to take place in a non-viscous or inertial regime. If mass movement involves cohesive particles (e.g. electrostatically charged clay platelets) and/or a cohesive aqueous fluid (e.g. a viscous clay-water, mud mixture), then the transport is said to take place in a viscous regime. In the nonviscous or inertial regime the settling of the cohesionless particles is not significantly retarded by the interstitial fluid (water), and is essentially controlled by the momentum of the particles, their settling velocities in water and their degree of interaction (collision) with other particles.

In the different modes of sediment transport to be discussed here, water performs all of the different functions of an interstitial medium. That is, it can drive sediment movement through unidirectional or bi-directional current flow, it can provide grain-support during supsension transport, or it can act as a lubricant by reducing the co-efficent of friction between sediment particles in high sediment concentration transport processes.

Its role as a sorting agent is also extremely important in controlling the textural characteristics of the resultant deposits. In terrigenous siliciclastic sediment, in which most sediment particles have a similar density, the degree to which grain size is uniform is usually used as a measure of how well sorted a sediment is. However, as discussed in Chaps. 1 and 9, when volcanic sediments lack uniformity of grain size this does not necessarily reflect hydraulically, poorly sorted sediments. Fluids, especially water, sort clastic aggregates according to their hydraulic properties, the most important of which are density and shape. Given the variable density of volcanic detritus due to variable levels of vesicularity, hydraulically well-sorted aggregates may appear to be very poorly sorted if the grain size is highly variable in a deposit. Cashman and Fiske (1991) demonstrated that solids with different densities have markedly different settling or terminal fall velocities in water (Chap. 9; see Fig. 9.74a). Since small dense particles will fall at the same velocity as large, low-density particles (e.g. vesiculated pumice), such a deposit will appear to be poorly sorted using uniformity of size as a criterion, but will in fact be hydraulically well-sorted.

The other effect of water is to lower the effective or dynamic transportable mass of sediment particles where these are immersed in water. Given that mineral and rock fragments have densities of around 2,500 kg m^{-3}, immersion of such grains in water (density of 1,000 kg m^{-3}) lowers the effective transportable mass of such fragments by 40% relative to their mass in air, making erosion an easier prospect.

In particulate form, sediment transport takes place in the following ways (Table 12.1):

- traction,
- suspension and flotation, and
- in solution.

In mass movement sediment can be transported by:

- fluvial hyper-concentrated flood flow, torrent flow and sheet flow,
- subaqueous granular mass flow (e.g. turbidity currents),
- mudflows and debris flows,
- tsunamis and storm surges,
- severe coastal storms,
- slumps, and
- soil creep.

In volcanic terrains the Indonesian word "lahar" (or hot mudflow) has often been used somewhat indiscriminantly for a spectrum of higher particle concentration mass movement phenomena carrying volcanic debris, ranging from hyper-concentrated flow to mudflows and non-cohesive and cohesive debris flows, even when these are not hot. We believe it is important now to discriminate such deposits based on their physical textural properties and their specific transport origins, rather than lumping all these processes together, because the hazard implications could vary (Sects. 12.6.4 and 12.6.6).

12.6.1 Tractional Transport by Water (Fluvial, Shallow to Deep Bottom Currents, Waves)

Tractional sediment transport by water (Gilbert 1914) produces a very important group of sedimentary structures in the resulting deposits, called appropriately, tractional sedimentary structures (e.g. ripples, dunes, cross-stratification, horizontal or plane-bed lamination, antidunes). Tractional structures can be produced where any fluid (water, air) flows over a bed of cohesionless grains and initiates movement of the grains as bed load (Fig. 12.21a) or in saltation (Fig. 12.21b; Gilbert 1914). Tractional reworking will also winnow out most clays through suspension transport, so producing granular clastic aggregates with little or no mud matrix.

Mobilisation of sediment through current erosion usually only takes place during turbulent flow conditions when sediment particles are dense, sediment concentration is low and bed shear stress is due largely to the water mass of the current, but can also occur under laminar flow conditions when sediment particle density is low (e.g. pumice; cf. Bagnold 1955), or sediment concentration is high, and as a result, bed shear stress is high because of the higher density

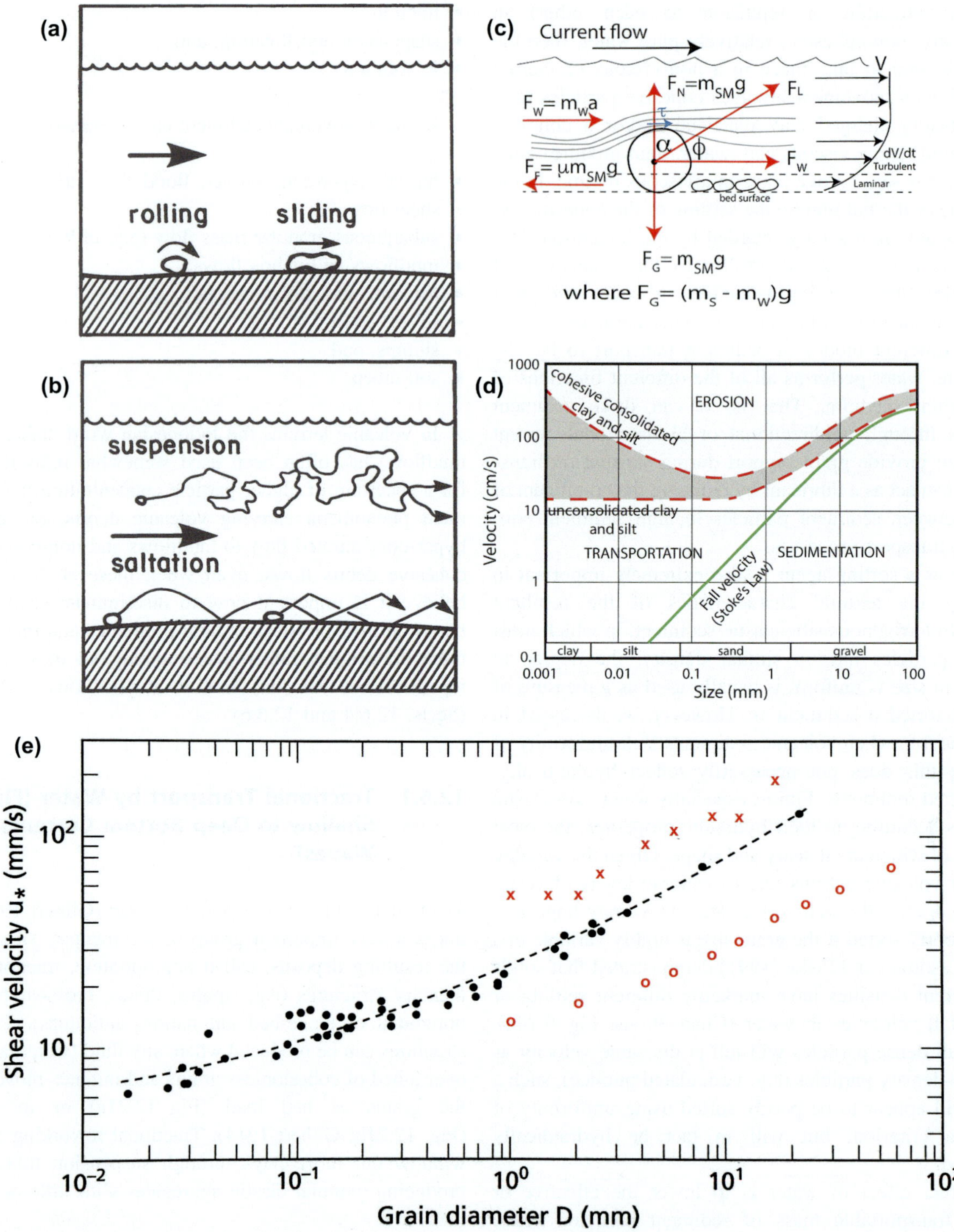

Fig. 12.21 Factors relevant to tractional sediment transport. **a**, **b** The two sediment transport modes that together constitute tractional sediment transport: (**a**) bed load (rolling and sliding), and (**b**) saltation (jumping/bouncing). Smaller grains may also be lifted into continuous or intermittent suspension after initial traction followed by uplift into the current (From Cas and Wright 1987). **c** Forces affecting sediment particles in a unidirectional aqueous current, showing the vertical velocity gradient, the basal laminar flow zone in which small particles are immersed and mostly stationary, with larger particles protruding upwards into the higher velocity and turbulent flow zone, into which they may be uplifted if the current velocity and up-lift force are high enough, as discussed in the text, where parameters are defined. **d** Simple Hjulström-Sundborg curves (red dashes) showing minimum current velocities required to erode quartz sediment and clays. The upper curve is for sediments where clay is compacted; the lower curve is for uncompacted clays. Also shown is the Stokes Settling Law curve which shows the fall velocity of different grain sizes under laminar flow conditions. **e** Aqueous current shear velocity that is required to erode different grain sizes for dense quartz grains as summarised by Miller et al. (1977) (From Leeder 2009)

of the flowing sediment–water mass. The essential criterion that determines if flow conditions in any fluid are laminar or turbulent is the Reynolds Number (*Re*):

$$Re = \frac{\rho U h}{\mu} \tag{12.5}$$

where ρ is the fluid density, U is the fluid velocity, h is the flow depth, and μ the fluid viscosity.

Bed load transport refers to the rolling or sliding of particles along the bed (Fig. 12.21a; Gilbert 1914) when a flowing fluid exerts a lateral force or shear stress at the bed surface which is great enough to overcome the weight of the particle and the frictional resistance at the bed surface (Fig. 12.21a, c). That is:

$$F_w > \mu m_{SM} g \tag{12.6}$$

where F_w is the force exerted by the flowing current on the upstream face of the particle, m_{SM} is the submerged or buoyancy effective mass of the particle, which is equivalent to mass of the particle in air minus the mass of the water that would occupy the volume of the sediment particle, μ is the co-efficient of friction between the sediment particle and the bed under water, and g is the acceleration due to gravity, which converts mass into a weight force.

However, just because a current is moving, doesn't mean that it is capable of eroding or moving sediment particles, because the shear stress in the moving current is lowest at the bed surface because of frictional drag or resistance between the current and the substrate. This produces a velocity and shear stress gradient from lowest at the bed surface to highest above the bed. If the flow velocity is low to moderate, there may even be a thin layer of water at the interface that is not moving, or is in a laminar flow or "dead" state (Re < 500), above which, where the flow velocity is higher, the flow state is turbulent (Re > 2,000), which is more suitable for causing sediment erosion. Sediment grains that have a diameter that is less than the thickness of the basal "dead" or laminar zone are unlikely to be eroded. Grains that protrude above the laminar basal layer into the overlying turbulent zone may be eroded if the applied force, *Fw*, or shear stress, τ, exceeds the minimum threshold to cause motion of that particle (Fig. 12.21c).

Another way of representing the required conditions to initiate movement of a sediment particle is to assess the shear stress, τ, the current exerts on the particle relative to the properties of the sediment particle (particle diameter, D; particle density, ρ_s; density of water, ρ_w; fluid dynamic conditions, such as Reynolds Number, *Re*, involving flow depth, flow velocity, u, fluid density and viscosity, η). These physical variables were combined by Shields (1936) into a dimensionless parameter called the Shields Number, defined as:

$$\theta_\tau = \frac{\tau}{(\rho_s - \rho_w)gD} = \frac{\rho_w u^{*2}}{(\rho_s - \rho_w)gD} \tag{12.7}$$

where θ_τ is the threshold value at which grain motion is initiated, and can be viewed as a co-efficient of friction (Miller et al. 1977), and $u*$ is the shear velocity. Shields plotted the Shields Number criterion against the grain or particle Reynolds Number, *Re**, to define the critical conditions for entrainment of particles of different size and density (Fig. 12.20d), given by:

$$Re^* = \frac{u^* D}{\upsilon} \tag{12.8}$$

The grain or particle Reynolds Number is the ratio of inertial forces to viscous forces, usually defined as the settling velocity or shear velocity of a particle times its diameter divided by the kinematic viscosity of the fluid, v. It's a measure of the effect of a particle settling through a fluid on the dynamic state of the fluid around the particle. Large dense particles settling in a fluid have a higher Re* than smaller fragments.

When the critical Shields Number criterion is achieved, the fluid has an erosive capacity, its velocity is greater than the minimum threshold velocity for erosion of the grain size in question (e.g. according to the Hjulström-Sundborg curve for dense quartz sediment particles of density ~2,650 kg m^{-3}; red curves in Fig. 12.21d), and the flow conditions in the fluid are turbulent, so that the fluid Reynolds Numbers, *Re*, are high (>2,000; Chap. 2).

The shear stress applied by a current, τ, can be converted into a critical shear velocity (Miller et al. 1977; Leeder 2009), using the relationship:

$$u^* = \sqrt{\tau \rho_w} \tag{12.9}$$

Miller et al. (1977) distilled all available data and summarised the relationship between critical shear velocity to cause entrainment of different grain sizes for dense sediment grains (Fig. 12.21e) that is more accurate than the Hjulström-Sundborg curve (Fig. 12.21d).

Saltation refers to an intermittent downstream bouncing or jumping mode of transport, involving short-term suspension of particles (Fig. 12.21b; Gilbert 1914). Saltation may be induced either by high enough velocities (and therefore bed shear stress) to lift the particles into suspension temporarily (Fig. 12.21c), or by the impact of other particles falling back to the bed. Current flow induced saltation is caused by the deflection of stream flow lines over the crest of particles, concentrating the stream lines and producing a low-pressure zone on the downcurrent side of the particle and a high-pressure gradient in the fluid above the grain where the stream flow lines are focused. The combination of these two factors cause a resultant Bernoulli effect that produces an uplift force in a downstream direction, which may be sufficient to lift the particle from the bed. The uplift force, F_L, the force exerted by the flowing water is resolved at an angle to the substrate, ϕ, and so has a vertical component, $F_N \sin\phi$, and a horizontal component, $F_w \cos\phi$ (Fig. 12.21c). The frictional resistance force, F_F (= $\mu\, m_{SM}g$; Eq. 12.6), is reduced because of the uplift force, which lowers the normal force, N (= m_{SM}g):

$$N = m_{SM}g - F_N \sin\phi \qquad (12.10)$$

so that the modifed friction force, *f*, becomes:

$$f = \mu\,(m_{SM}g - F_N \sin\phi) \qquad (12.11)$$

At the point of initiation of motion, all horizontal forces are equal to zero. That is:

$$F_W \cos\phi - f = 0 \qquad (12.12)$$

which becomes:

$$F_W \cos\phi - \mu\,(m_{SM}g - F_N \sin\phi) = 0 \qquad (12.13)$$

and:

$$F_L = \frac{\mu\, m_{SM}g}{(\cos\phi + \mu \sin\phi)} \qquad (12.14)$$

If F_L exceeds this criterion, the particle will lift off the substrate.

Hjulström (1935, modifed by Sundborg 1956) established the flow velocity at which different sediment grain-sizes can be eroded for quartz sediment spheres of known density and channel water depth (Fig. 12.21d). However, it is not accurate for particles of variable shape, and is not useful for sediment populations of highly variable density, as many volcanic sediment aggregates are. Miller et al. (1977) showed that although the Hjulström-Sundborg curves provide useful general patterns, more detailed interrogation of the data produces more accurate assessment of the relationship between minimum shear velocity required to erode different grain sizes, based on a measurement of current velocity 1 m above the sediment bed (Fig. 12.21e).

There have been very few equivalent studies specifically for pumiceous sediments. Bagnold (1955) demonstrated that very low-density particles (cf. pumice) could be mobilised by aqueous currents under laminar flow velocities and low bed shear stress. Manville et al. (2002) have provided the most comprehensive assessment of the hydrodynamic properties and behaviour of pumice sediments based on field and laboratory studies of pumice that was reworked after the 232 AD eruption from Taupo caldera, New Zealand. They determined that the settling velocity and the Shields critical boundary stress for mobilising low-density pumiceous sediments is significantly less than for dense quartz grains (Table 12.2).

The net effect of continuous bed load and saltation (i.e. tractional) transport of a large number of cohesionless grains, simultaneously under the influence of a unidirectional water flow, is to produce a range of tractional bedforms or sedimentary structures (Fig. 12.22a, b). For any particular grain size population under steadily accelerating or decelerating flow conditions, a regular succession of structures should occur with increasing or decreasing velocities. However, the succession is different for different grain sizes (Fig. 12.22b). In general, for dense particle sediments, the sequence of sedimentary structures as current velocity increases is: *ripples -> sand waves -> dunes -> plane bed or lamination -> antidunes or massive beds*, known as the flow regime sequence of sedimentary structures (Fig. 12.22a, b; Harms and Fahnestock 1965; Harms et al. 1982).

Ripples, sandwaves and dunes are all wave-like bedforms that form in sequence as current velocity is progressively increased in the lower flow regime (Fig. 12.22a, b). They are all asymmetrical structures produced by unidirectional current flow, with the crests more or less perpendicular to the direction of current flow, the steep face is on the downstream, or down-current side, and the shallow sloping face is on the up-current side (Figs. 12.18a, 12.23a). Ripples have a maximum amplitude of 4 cm, sandwaves and dunes are higher. Internally they are all marked by cross-stratification (cross-laminae if the cross-strata are <1 cm thick, cross-beds where cross-strata are >1 cm thick; Figs. 12.18a, c; 12.23b–h), which dip in the direction of flow at an angle to the base of the surface over which the current is flowing (e.g. floor of a channel). This cross-stratification is typically at or just below angle of repose (25–35°) at its upper end, and forms by the passive grain flow avalanching of packets of granular sediment from the crest of the migrating ripple or dune, down the down-current face under the influence of gravity and under lower flow regime conditions (Fig. 12.18a). Although lateral current-induced bed surface shear stress acting on the crest of the dune helps to initiate the grain flow, little or no shear stress is involved in the actual lee-side grain flow, but the initial lateral force may cause further runout of the cross bed layer at the base of the bedform compared with a static grain collapse event.

Sandwaves (or large 2D ripples or dunes) have straight to slightly sinuous, continuous crests and long wavelength crests, and produce tabular planar cross-strata (Fig. 12.23c). Ripples and dunes have diverse morphologies (Allen 1982; Leeder 2009), including straight(ish) crested and lunate forms with curved crests (e.g. Fig. 12.23a), which produce upwards concave sets of trough cross-strata (Fig. 12.23b, d–f). In exposures parallel to the current flow direction trough cross-beds dip in the direction the current is flowing and have a concave up, asymptotic geometry (Figs. 12.18a, c;

Table 12.2 Hydrodynamic properties of low-density pumice fragments compared with high-density quartz grains in water (From Manville et al. 2002)

Phi	D (mm)	Pumice							Quartz						
		(g cm^{-3})	τ_0^{crit}(N m^{-2})	U^* = (cm s^{-1})	Vs (cm s^{-1})	Re*	θ_{crit}	θ_{susp}	(g cm^{-3})	τ_0^{crit} (N m^{-2})	U^* (cm s^{-1})	V_s (cm s^{-1})	Re*	θ_{crit}	θ_{susp}
–6	64	1.26	5.1	7.14	35.4	4539	0.031	0.760	2.65	–	–	–	–	–	–
–5.5	45.3	1.26	4	6.32	33.6	2842	0.034	0.969	2.65	–	–	–	–	–	–
–5	32	1.26	3.1	5.57	29.2	1769	0.038	1.035	2.65	–	–	–	–	–	–
–4.5	22.6	1.26	2.35	4.85	25.9	1089	0.040	1.151	2.65	–	–	–	–	–	–
–4	16	1.26	1.75	4.18	19	664.7	0.043	0.876	2.65	13	0.114	66	1812	0.050	1.682
–3.5	11.3	1.26	1.2	3.46	10.8	389.2	0.041	0.400	2.65	9	0.095	56	1066	0.049	1.712
–3	8	1.26	0.75	2.74	11.5	217.6	0.036	0.642	2.65	6.5	0.081	47	640.5	0.050	1.706
–2.5	5.66	1.27	0.55	2.35	9.8	131.7	0.037	0.645	2.65	4.5	0.067	40	376.8	0.049	1.747
–2	4	1.27	0.42	2.05	8.4	81.40	0.039	0.656	2.65	3	0.055	33	217.6	0.046	1.682
–1.5	2.83	1.29	–	–	6.9	–	–	–	2.65	2	–	28	–	–	–
–1	2.00	1.30	0.30	1.73	5.6	34.40	0.051	0.537	2.65	1.2	0.035	23	68.80	0.037	1.634
–0.5	1.41	1.32	–	–	5.5	–	–	–	2.65	–	–	–	-	–	–
0	1.00	1.34	0.25	1.58	5.5	15.70	0.075	0.911	2.65	0.5	0.022	16.3	22.21	0.031	1.641
0.5	0.71	1.37	–	–	4.76	–	–	–	2.65	–	–	–	–	–	–
1	0.50	1.40	0.27	1.64	4.46	8.159	0.137	1.008	2.65	0.28	0.017	9.8	8.308	0.035	1.187
1.5	0.35	1.44	–	–	2.82	–	–	–	2.65	–	–	–	–	–	
2	0.25	1.49	0.30	1.73	2.35	4.300	0.250	0.460	2.65	0.2	0.014	4.3	3.511	0.049	0.457
2.5	0.177	1.54	–	–	2.32	-	–	–	2.65	–	–	–	–	–	–
3	0.125	1.60	0.35	1.87	1.03	2.322	0.475	0.144	2.65	0.18	0.013	1.8	1.666	0.089	0.160
3.5	0.088	1.67	–	–	0.78	–	–	–	2.65	–	-	-	-	–	–
4	0.063	1.74	0.50	2.24	0.59	1.388	1.103	0.077	2.65	0.16	0.013	0.7	0.785	0.158	0.048

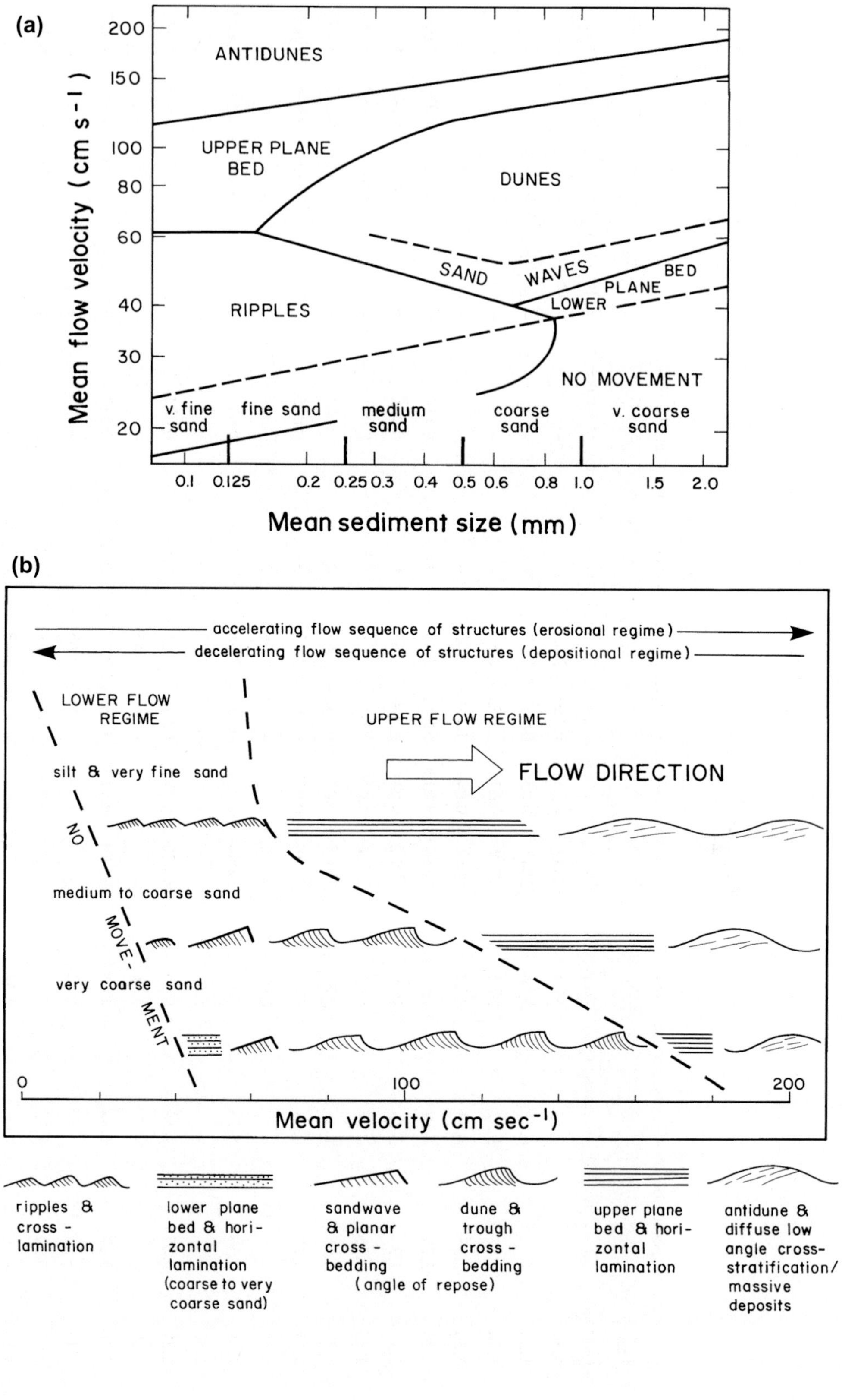

Fig. 12.22 Sedimentary structures produced under different unidirectional aqueous current flow velocities and regimes for dense siliciclastic sediment populations.
a Experimentally determined stability fields for the common tractional structures as a function of mean water flow velocity and mean grain size. Data points represent many experimental flume tank runs, using different grain sizes. Note that for different grain sizes different sequences of bedforms develop as flow velocities increase or decrease. (From Cas and Wright 1987, After Harms et al. 1982).
b Distillation of results in (**a**) showing schematically the succession of tractional bedforms and corresponding internal sedimentary structures formed under conditions of unidirectional water flow, with flow either accelerating (left to right) or decelerating (right to left). During acceleration, the succession of bedforms is eroded, whereas under deceleration the succession of bedforms and flow stages may be preserved up-sequence because net aggradation (deposition) will occur at the bed. The velocity–grain size relationships hold only for dense mineral and lithic grains; see Table 12.2 for data available for low-density, vesicular detritus. (From Cas and Wright 1987)

(a)
(b)
(c)
(d)
(e)

◀ **Fig. 12.23** Common tractional sedimentary structures reflecting different unidirectional current velocities and flow regimes. **a** Modern, unconsolidated, asymmetrical ripples in the dry floor of a modern creek bed, representing the last event when water flowed down the creek from top right to bottom left (yellow arrow). The sediment consists of volcanic fragments sourced from the Archean bedrock consisting of basalt, komatiite, dolerite, and felsic volcaniclastic rocks. Merougil Creek, Kambalda, Western Australia. **b** Ripple cross-lamination preserved in diamond drill core in Archean volcaniclastic sediments, Black Flag Group, Kambalda, Western Australia. Flow from right to left in most cross laminae sets, shown by direction of dip to left of upward concave cross-laminae. Drill core is ~5 cm wide. **c** Planar tabular cross-bedding in a sand-wave that was migrating from left to right in Early Devonian volcanic sediments, Snowy River Volcanics, W-Tree, Buchan, Victoria, Australia. Height of cross-beds 30 cm. **d** Trough cross-bedding, marked by upward concave sets of cross-beds resulting from subaqueous dune migration from left to right, probably from long-shore drift currents, Ordovician, near Penny Pass, North Wales. The cross-beds are underlain by plane horizontal lamination suggesting an upward declining current velocity and flow regime. **e** Sets of trough cross-bedding in a stack of beds, each representing a lunate fluvial dune migrating from right to left. The volcanic sandstones formed in a Cretaceous, peri-polar braided river channel succession formed during early rifting between Australia and Antarctica. Note the asymptotic relationship between the base of each cross-bed set and the underlying bed top, resulting from grain-flow avalanching of sand into the dead zone on the down-current side of the migrating dunes (cf. Fig. 12.18a). Each cross-bed represents the temporary location of the down-current face of the migrating dune. Strzelecki Group, San Remo, Gippsland Basin, Victoria, Australia. **f** Trough cross-bedding in diamond drill core in Archean volcanic sediments, Scotty Creek Formation, New Holland Mine, Agnew, Western Australia. Drill core is ~5 cm in diameter. **g** Symmetrical trough cross-bedding reworked. Pleistocene, shallow marine, pumiceous sandstones, Wanganui Basin, Rewa Hill, New Zealand. Current direction was either into or out of the outcrop. **h** 3D outcrop of fluvial Cretaceous volcanic sandstones comprising intersecting sets of large-scale trough cross-bedding. Bases of sets are concave upwards in the section transverse to the mean forset dip, and more gently curved parallel to the dip. Current direction was obliquely from right to left for the cross-bed set, and the other beds reflect channels with axes trending in the same direction. Arrow points to the prominent horizon of mudstone rip-up intra-clasts. Strzelecki Group, Gippsland Basin, Kilcunda, Victoria, Australia. **i** Detail of a lower interval of plane laminated fluvial Cretaceous volcanic sandstone, representing transition from lower to upper flow regime (Fr ~ 1) and higher energy, massive structureless sandstone, representing the upper flow regime anti-dune stage, hyper-concentrated flow conditions. Note the scouring of the massive sandstone into the laminated sandstone, and the mudstone rip-up intra-clasts at the base of the massive sandstone. Same location as (**h**)

Fig. 12.23 (continued)

12.22b; 12.23b–f, h). However, in exposures perpendicular to flow direction, they are trough shaped and upward concave in geometry (Fig. 12.23 g, h). Lower plane bed lamination forms in place of ripples in coarse sediments (Fig. 12.22a, b), under low flow velocity conditions. Ripples do not form in coarse sediment.

If the current velocity increases further, dune bedforms are smeared out by the current under very high bed shear stress conditions, and are replaced by a flat or horizontal planar bedform (Fig. 12.22a, b) which, internally, is represented by horizontal, planar, parallel lamination in sands (Fig. 12.23d, i). This bedform represents the transition from relatively low-energy conditions known as the lower flow regime, to higher energy and velocity flow conditions known as the upper flow regime, which is theoreticaly represented by a bedform called antidunes (Fig. 12.22a, b). Lower and upper flow regimes are distinguished by a parameter called the Froude Number (*Fr*), which is the ratio of inertial forces to gravitational forces:

$$Fr = \frac{U}{\sqrt{gh}} \tag{12.15}$$

where U is the velocity, g is the acceleration due to gravity and h is the flow depth. When $Fr < 1$, lower flow regime bedforms and structures occur, and this corresponds to a fluid flow state of subcritical turbulence. When $Fr \geq 1$, antidunes and standing waves form (Figs. 12.22, 12.23i), and this corresponds to supercritical turbulent flow, which means it is a very dynamic, high-energy flow system.

Antidunes are marked internally by a poorly defined, low-angle (<20°) cross-stratification that most commonly dips upstream, but may wrap over the crest of the structure, or rarely dip downcurrent. Antidunes have no clear-cut asymmetry. The low-angle cross-stratification results from the shearing of sediment over the bed surface under conditions of very high lateral bed shear stress, as discussed above. However, antidunes are rarely preserved. That flow stage is more commonly represented by massive beds of cohesionless sediment (Fig. 12.23i), if represented at all. Massive, internally structureless beds result from the rapid deposition (mass-dumping) of large volumes of sediment out of suspension. True tractional sediment transport has probably been surpassed by this stage, and the flow regime is upper, involving hyper-concentrated flow conditions (Sect. 12.6.4).

Manville et al. (2002) refer to particular depositional structures preserved in pumiceous sediments as reflecting upper or lower flow regimes. However, given the lower density of pumiceous fragments, it is unclear how the conditions (velocity *versus* grain size *versus* depositional structures) compare between dense siliciclastic and low-density pumiceous sediments until comparable experiments have been undertaken. The complete succession of structures reflecting the change in flow regime conditions during one flow cycle is only likely to be preserved from the decelerating part of the cycle, since only then will sedimentation occur. As a result, antidunes are frequently destroyed by reworking, during lower, declining flow regime conditions. Upward fining cycles and sequences of structures from fluvial channel-point bar depositional settings (especially meandering river successions) are most frequently depicted as showing the complete succession of flow regime level structures (e.g. Walker 1984). As discussed further below, subaqueous granular mass flows (Sect. 12.6.5) can also produce such successions of sedimentary structure intervals, reflecting declining flow regime levels during the decelerating or waning stages of flow.

Although many other features of tractional sediment transport and sedimentary structures could be discussed, space precludes this. However, some further comment should be made about cross-stratification and its diverse origins (see Allen 1982; Leeder 2009, for a comprehensive discussion). Cross-stratification can be subdivided into two essential types: high angle, which is at or near the angle of repose (25–35°), and low angle (usually < 20°). Only high-angle cross-stratification can be confidently interpreted as the product of lower flow regime tractional transportation and deposition. It can occur in sets of cross-laminae only centimetres thick to sets of cross-beds many metres thick. Some pyroclastic surge cross-stratification can be high angle, but there are other criteria for recognising this origin, as discussed below and in Chap. 10. Low-angle cross-stratification can have many origins, including:

(a) Tractional, representing the asymptotic toe-set portion of a once larger-scaled set of cross-beds, the upper part of which approached the angle of repose. Such a set would be associated with a migrating dune or sandwave, the top of which was truncated by contemporaneous flow erosion (see Fig. 12.23h). The downstream converging nature of the asymptotic layers should be definitive and should also distinguish such layers from upper flow regime horizontal lamination for which they could be mistaken in large-scale, broad sets.
(b) Upper flow regime anti-dune cross-stratification, as discussed above.
(c) Nested scour and fill trough shaped sets, which in sections perpendicular to flow direction (Fig. 12.23 g, h) give apparent low-angle truncations. These are end-on views of trough cross-bedded sets formed by downcurrent migration of large dunes by normal tractional processes. Compare this with the flow-parallel sections in Fig. 12.23b–f.

Fig. 12.24 Other types of cross-stratification in volcanic sediments. **a** Symmetrical wave ripples and cross-lamination in pumiceous sand (middle), overlain by well-rounded pumice conglomerate. Ancient delta sequence, Kutchan, Hokkaido, Japan. **b** Exhumed symmetrical wave ripples in sediments beneath the Ordovician Garth Tuff (ignimbrite), North Wales. **c** Hummocky cross-stratification in Permian volcanic sediments deposited by storm turbidites in a marine shelf setting. Gerringong Volcanics, New South Wales, Australia. **d** Hummocky and swaley cross-stratification in Ordovician volcanic sediments beneath the Pitts Head Tuff (ignimbrite), North Wales

(d) Two-way (bimodal) cross-stratification (Fig. 12.24a) associated with symmetrical ripples (Fig. 12.24b) formed by oscillating, near-shore wave reworking.

(e) Low-profile domal sets of cross-strata, called hummocky cross-stratification (HCS; Fig. 12.24c), and associated broad, open, low-profile trough-like sets called swaley cross-stratification (SCS; Fig. 12.24d). These are produced when the bases of storm waves surge over the sea-floor in storm-affected shelf settings, often while an off-shore directed under-flow such as a turbdiity current is triggered by water-pile up near the shoreline after a set of onshore directed storm waves. Each layer represents a short-lived wave surge event rather than passive angle of repose cross-stratification. Discrete, storm generated shelf turbidite beds may be definable (Walker 1984).

(f) Beach face low-angle cross-stratification marked by very low-angle truncations (<10°–15°) between sets which are made up of broad planar to open trough sets. The sets are produced by onshore wave surges onto the beach face, in the swash-backwash zone.

(g) Pyroclastic surge cross-stratification (Chap. 10) which is analogous to anti-dune structures in normal sediments. Layers wrap over dune crests and thicken and thin (e.g. Fig. 10.104e–g). Sorting is poor, and major grain size variations occur between successive cross-beds in set layers. Accretionary or cored lapilli may be present (Chaps. 4, 10).

The aim of this brief summary has been to highlight those surface process, current-generated tractional sedimentary structures that have hydrodynamic significance, so that some assessment of the energy regimes operating during sedimentation and of the setting can be made. However, interpretation of the paleoenvironment where tractional structures are found has to be made with care, because aqueous tractional structures can form in any setting where moving water drives sediment movement. This occurs in a subaerial landscape where wind currents can produce ripples, dunes and cross-stratification, and surface water flows in response to gravitational potential (e.g. rivers), or in standing bodies of water (lakes, oceans) where water movement is induced by winds, currents, tides, Coriolis force, or temperature or density differences. Tractional sedimentary structures can therefore form in a whole range of environments and water depths (e.g. alluvial fans, rivers, deltas, shorelines, shelf, deeper-marine) affected by bed-surface currents. They can also result from subaqueous (i.e. relatively deep water), non-viscous granular mass flows such as turbidity currents, as current velocities wane and flow conditions decline from upper to lower flow regime, as discussed further below. Tractional structures are therefore rarely indicative of any specific depositional environment, which thus has to be worked out from the overall facies characteristics of the succession.

A surprising discovery in recent decades has been the occurrence of high-amplitude and long-wavelength dune-like bedforms in deep water, including association with deeper shelf and slope contourites and submarine fan and canyon settings, as discussed further in Sect. 12.6.5.

Finally, unfortunately, there has been no research done on the quantification of flow regime current velocity and bedform behaviour and sequence of sedimentary structures in low-density pumiceous sediments, comparing the conditions, current velocities and structures with the well-documented and constrained flow regime conditions for dense fragment quartz sands. Manville et al. (2002) have documented in detail the spectrum of depositional structures developed in reworked pumiceous sediments post-dating the major 232 AD Taupo explosive eruption in New Zealand. They determined the Shields critical boundary stress for mobilising the sediments, and refer to particular depositional structures as reflecting upper or lower flow regimes. Bagnold (1955) noted that low-density ($\sim$1,004 kg m^{-3}) waxy particles could be moved by water under laminar flow conditions and velocity 3.3 cm s^{-1}, thus simulating some pumice, in flume experiments, that ripples can form and that at higher flow velocities transitional to turbulent flow, increasing suspended load had the effects of damping turbulence in the sediment laden flow system. However, he did not equate these experiments to the transportation of pumiceous sediment. However, given the lower density of pumiceous fragments, it is unclear how the conditions compare between dense siliciclastic and low-density pumiceous sediments until comparable experiments have been undertaken. This is a research project waiting to be undertaken by someone!!

12.6.2 Suspension and Flotation

Suspension involves the long-term support of grains above the bed by the fluid (Fig. 12.21b). Where the grains are silt and coarser, this takes place during turbulent flow conditions. Suspension of particulate granular sediment can occur during normal tractional sediment transport. Whereas coarser grains are moved in tractional bed-load form, finer grains can be lifted into suspended transport (Fig. 12.21c), as discussed above. During current deceleration, fallout from suspension occurs and the sediment returns to the traction mode, resulting in standard tractional sedimentary structures. High flow velocities can lead to high volumes of sediment in suspension, which then begin to verge on mass-flow conditions, which will be discussed below. Under high enough velocities, even gravel-sized material can be lifted into suspended transport, even if only intermittently.

An important parameter in understanding suspended sediment transport and sedimentation from suspension is the Rouse Number. The Rouse Number (*R#*), is defined as the ratio of the settling velocity, w_s, of sediment particles to the shear velocity, u_*, in a (turbulent) current multiplied by the von Kármán constant, *k*, which is 0.4:

$$R\# = \frac{w_s}{k\,u*} \tag{12.16}$$

When $R\# > 2.5$, the sediment is mostly being transported as bedload, when $1.2 < R\# < 2.5$, 50% of sediment is in suspension transport, and when $R\# < 0.8$, 100% of the sediment is in suspension (e.g. Yuill and Nichols 2011).

Settling velocity of sediment particles depends on their density, their size, the density of the fluid they are falling through and the viscosity of that fluid. If fallout through a fluid occurs under laminar flow conditions, particles fallout according to Stokes Settling Velocity Law (Fig. 12.21d):

$$w_s = \frac{gd^2\,(\rho_p - \rho_f)}{18\mu} \tag{12.17}$$

where w_s Is the settling velocity of a suspended sediment clast or particle, g is the acceleration due to gravity, d the clast diameter, ρ_p is the clast density, ρ_f is the fluid density, and μ the fluid viscosity.

Cashman and Fiske (1991) and Manville et al. (2002) showed that the settling velocities of low-density, vesiculated pumice through water is significantly slower (at least an order of magnitude) than of dense mineral and lithic grains (Fig. 9.74a; Table 12.2). Other factors to take into consideration are that the Stokes Law only applies to spherical particles in a stationary fluid. If particle shape varies this will affect the settling velocity, and only in a large, deep body of water are ambient fluid conditions ever likely to be static, although often a current of some type acts even at depth. In addition, the Stokes Law assumes that a single particle can fall unimpeded, without interaction with other particles, which is rarely the case. The Richardson-Zaki criterion (Richardson and Zaki 1954) accounts for the effects of multiple grains falling simultaneously, whereby the modified settling velocity of a particle, w'_s, relative to its Stokes Law solitary particle settling velocity, w_s, can be represented as:

$$w'_s = w_s\,(1 - C)^n \tag{12.18}$$

where C is the volumetric abundance of settling, suspended particles, and n is an exponent that ranges from 2.32 when the grain Reynolds number, $Re*$ (Eq. 12.7) >1 (large grains), and 4.56 for $Re* < 1$ (small grains; Leeder 2009).

Where the sediment particles are clay platelets, suspension transport can occur in even very slow laminar flow conditions. This is due to their small mass and their plate-like form which give them buoyancy, but also because of the electrostatic interaction between the water molecules and the charged clay platelets, which produces an electrostatic force support effect. Deposition of suspended clay particles will only occur under very still conditions in a standing body of water (Fig. 12.21d), such as a lake, abandoned channel, swamp, or deep ocean. This produces massive clay layers (Fig. 12.25). Such sedimentation is called hemi-pelagic when it occurs in the deep ocean and pelagic when it contains a significant planktonic biological component (see Henrich and Hüneke 2011, and Hüneke and Henrich 2011 for a thorough review). These may periodically be interrupted in fluvial settings by coarser layers introduced by fluvial flood crevasse splay events, or in relatively deep water bodies of water, subaqueous turbidity current transport, or some other short-lived, high-energy sedimentation event.

During turbidity current flow, fine clay and silt size sediment is elutriated into a fine sediment cloud that accompanies the denser and coarser surface hugging body of the flow. That sediment cloud can become buoyant and become lofted into the water mass, as ambient water mixes into the cloud (Stow and Smillie 2020). That buoyant sediment cloud can then spread at a level of neutral buoyancy in the water column, dispersing the suspended fine sediment across the host basin beyond the limits of the turbidite deposit. That fine sediment can then fallout and be deposited as an extensive clay-mud layer. Such fine layers are usually called hemipelagic, but turbidity current generated hemi-pelagic layers have been called hemi-turbidites by Stow and Smillie (2020). Where turbidity currents carry a resedimented juvenile pyroclastic grain population, such hemi-turbidite deposits will contain significant microns size shard ash particles, Such deposits could be mistaken for primary water settled ash deposits.

These surface sedimentary processes all operate in volcanic terrains in the rivers, lakes, shorelines and seas associated with different volcanic settings. The products will be suspension fallout muds, such as those illustrated in

Fig. 12.25 Grey-to-black suspension-deposited muds, and thin associated volcanic sandstone turbidites deposited in a lake, intruded elsewhere by rhyolite cryptodomes. Late Devonian Bunga Beds, Boyd Volcanic Complex, Bunga Head, southern New South Wales, Australia. (From Cas and Wright 1987)

Fig. 12.26 **a** Beached pumice on the shoreline at Magnetic Island, Queensland, Australia. The pumice is derived from an eruption in the Tongan island chain thousands of kilometres to the east. **b** Old lake deposits exposed in a 5 m terrace, shores of Lake Taupo, New Zealand. The deposits consist of muds, peat layers, phreatoplinian ashes, and pumice falls that may have settled out of suspension from a quiet water body. (From Cas and Wright 1987)

Fig. 12.25 (and also Fig. 12.26b), as well as fine sediment consisting of ash.

However, in volcanic settings material very much coarser sediment than clays, may also result from very slow settling out of the water column under quiet water conditions. Self-rafted floating pumicemay have a bulk density significantly less than that of water because of the dominant void space. It therefore has enormous flotational potential (Fig. 12.26a). There are many accounts in the literature of large volumes of pumice clasts drifting in rafts long distances, even thousands of km across the oceans before reaching continental or island shorelines after specific marine volcanic eruptions (Fig. 12.26a; Richards 1958; Bryan et al. 2012; Jutzeler et al. 2014a, b, c). Pumice can, of course, also float in lakes and can be washed onshore as a shoreline deposit (Fig. 12.26b). This includes giant pumice blocks found in caldera lakes as discussed in Chap. 16 from La Primavera volcano, Mexico (Clough et al. 1981), and Lake Taupo, New Zealand (see Fig. 16.69c, d). However, because of its porosity pumice will soak up a lot of water and may eventually become sufficiently water-logged to sink-like waterlogged wood. Studies of the hydraulic behaviour of cold and hot pumice clasts by Whitham and Sparks (1986) and Fauria et al. (2017) have shown that the water-logging of cold pumice can, predictably, take a long time, and is dependent on the degree of inter-connection between vesicles, their diameter and surface tension effects.

Surprisingly, hot pumice can draw in water very rapidly as a result of the cooling and contraction of hot gases in vesicles when the pumice falls into cold water, and the resultant suction and surface tension effects that draw water into the pumice (Whitham and Sparks 1986). The degree of "water-logging" can vary enormously, depending of vesicle sizes and the degree to which the vesicle population is linked and permeable (Fauria et al. 2017). The settling rate will vary, depending on the rate of water intake and the rate at which the bulk density of the pumice and soaked-up water increase relative to the density of water (cf. Cashman and Fiske 1991). Gravel-sized floating pumice may therefore be hydraulically equivalent to clay platelets in terms of settling velocity and, while floating, it has no equivalence! Deposits resulting from this process should therefore be essentially mud sequences containing isolated outsized pumice clasts and even concentrations of pumice in layers (e.g. Fig. 12.26b), perhaps representing a significant event (e.g. pyroclastic eruption, or tropical cyclone or hurricane) that delivered an increased volume of pumice to the water body.

12.6.3 In Solution

In all environments natural waters carry dissolved chemical components that have been weathered or leached out of ambient rock matter. This should be particularly important in primary volcanic and and associated epiclastic deposits, which may be glassy, metastable and prone to alteration. In surface and subsurface waters these leached chemical elements may flocculate as clays, or may precipitate as other mineral species as surface deposits and pore fillings, given the right chemical conditions. Flocculation of clays may occur in surface waters at the interface between fresh and saline waters.

Precipitation of siliceous sinter may occur from hydrothermal systems in subaerial settings (e.g. Fig. 12.27). Sinter deposits often contain precious metals and sulphide minerals, often precipitated under the influence of biogenic agents such as bacteria and algae, potentially forming epithermal mineral deposits (Chap. 18). In deeper water, such processes are represented by VMS suphide deposits and associated alteration assemblages. Although these processes are not strictly epiclastic, they are included here briefly to illustrate that the waters so frequently responsible for surface sedimentary transport are simultaneously or penecontemporaneously also moving considerable volumes of rock matter in dissolved form, both at the surface and in recirculating subsurface meteoric and hydrothermal systems (Chap. 18). The volumes of water and rock matter involved over a significant geological time interval should not be underestimated.

Fig. 12.27 Siliceous sinter curtain precipitated from a hot-spring, Whakarewarewa thermal area, Rotorua, New Zealand. The front-central part of the curtain is covered with a film of sulphur. (From Cas and Wright 1987)

12.6.4 Fluvial Hyper-Concentrated (Flood) Flow, Torrent Flow, and Sheet Flow

There is a spectrum of flow conditions in river valleys from normal stream flow to hyper-concentrated (flood) flows to debris flows, representing changes in sediment concentration and grain size populations. Under increasing water flux and flow velocities, progressively larger volumes of sediment may become entrained or suspended in stream flows, called "*bulking-up*", and these then transform into a transitional flow condition called hyper-concentrated (flood) flow, also sometimes called torrent flow (Smith and Lowe 1991; Major et al. 2005; Pierson 2005; Pierson and Major 2014). Further bulking-up could transform hyper-concentrated flows into non-cohesive debris flows, and then as the mass flux wanes, transformations can occur in reverse, back to hyper-concentrated flow and then stream flow (Scott et al. 1995). Where the flow system becomes more cohesive and develops a yield strength, largely due to increasing proportions of clays, the flow types become cohesive mudflows and cohesive debris flows (Sect. 12.6.6; Scott 1988; Scott et al. 1995; Pierson 2005; Pierson and Major 2014; Iverson 1997; Kataoka and Nakajo 2004; Barclay et al. 2007; Baas et al. 2009; Vallance and Iverson 2015). For current reviews on the terminology and classification of subaerial sediment–water flow types, see Germain and Ouellat (2013) and Keaton (2019).

The term "lahar", originally an Indonesian name to describe hot, subaerial aqueous flows of volcanic debris released during and immediately after volcanic eruptions, is now widely used as an umbrella term to describe the higher particle concentration flow phenomena ranging from hyper-concentrated flow to non-cohesive and cohesive debris flows, even when these are not hot. Although the term "lahar" is entrenched, it unfortunately covers a spectrum of processes, each of which has different physical properties, different flow run-out distances and therefore they each pose different hazards and risks. Hazard assessments in volcanic terrains should therefore focus on assessing different scenarios for *each* of the possible, (and sometimes combined) flow types: hyper-concentrated flow, sheet flow, non-cohesive debris flow, or cohesive mudflow and cohesive debris flow.

High sediment load (lahar) events usually occur as a consequence of the deposition of large volumes of volcanic debris, such as pyroclastic fallout and pyroclastic flow and surge deposits on the slopes and surrounding landscape around a volcano. They can occur for months to years after an eruption event, after each significant hydrological event in a volcanic region (Scott et al. 1995; Pierson 2005; Pierson and Major 2014; Vallance and Iverson 2015). For example, hyper-concentrated flow events occurred in the catchment area of rivers draining Mt Pinatubo, for 15 years during major monsoonal rain events after the major 1991 explosive eruption, which deposited ~7 km^3 of non-welded, ash rich pyroclastic deposits onto the surrounding landscape (Pierson and Major 2014).

"Lahar" events have also occurred after numerous historic eruptions on Merapi volcano in Java (Lavigne et al. 2000; De Bélizal et al. 2013). Ash-rich deposits mantling a volcanic landscape after an explosive event have low water infiltration properties during rain events, leading to high rates of surface water sheet flow run-off, which in turn causes high rates of physical erosion of the fine pyroclastic debris into the drainage watersheds around the volcano (Pierson and Major 2014; Vallance and Iverson 2015). In addition, "lahar" events can be generated by melting of snow and ice when hot pyroclastic deposits are deposited on snow and ice on volcanoes (e.g. 1985 eruption of Nevado del Ruiz stratovolcano, Colombia; Pierson et al. 1990; Sect. 12.6.6), or during subglacial eruptions, which generate subglacial breakout floods or jökulhlaups (e.g. the eruptions of Katla volcano in 1918 and Eyjafjallajökull volcano in 2010, Iceland; Gudmundsson 2015).

In addition, landslides and debris avalanches that are channeled into drainage valleys around volcanoes can also transform into water-saturated flows such as hyper-concentrated flows and debris flows, as occurred during and after the 18 May, 1980 eruption of Mt St Helens (Janda et al. 1981; Scott 1988; Hacker et al. 2018). The debris avalanche, in addition to carrying large volumes of rock debris, also contained snow, ice and groundwater, which when released as water from the debris avalanche deposits, mobilised some of the deposits into debris flows (Hacker et al. 2018). In addition, the debris avalanche and the pyroclastic flows flowed into the flowing river valleys around Mt St Helens and mixed with the river waters, transforming them into solids-laden torrents of debris (Major et al. 2000, 2005). Under these conditions large volumes of suspended granular sediment occur under conditions of high velocity, upper flow regime surface flow during peak flow or flood events, coinciding with Froude Number ≥ 1 (Eq. 12.13). River systems experiencing upper flow regime conditions, have enormous erosive capacity during the waxing flow stage, and experience very high sedimentation rates during waning flow stage. Under these conditions huge volumes of sediment are carried in suspension, supported by turbulent eddies in the water (Fig. 12.28; Smith and Lowe 1991; Pierson 2005).

Pierson (2005) proposed that in hyper-concentrated (flood) flows (HCF) the proportion of suspended fine grain sizes (clay + silt) should be high enough to develop yield strength and high fluid visocosity, which hinders the settling of sand and fine gravel. This coarser sediment load is maintained in suspension by fluid turbulence and grain interactions. According to Pierson (2005) the mean flow velocity and the sediment load transported is higher than for normal stream flow at the same water depth and slope. Pierson (2005) cautioned against imposing strict sediment concentration levels in the definition. During an HCF event in the Toutle River following an eruption from Mt St Helens on 19 to 20 March 1982 the suspended sediment load was 43%, the flow depth was 2.1 m, and the velocity was 3.5–4 m s^{-1} (Pierson 2005). During a rain triggered event at Mt Pinatubo, Philippines on 26 September 1993 in the Sacobia River valley, the suspended sediment concentration was 30%, the flow depth 0.5–0.7 m and the velocity 2.6–3.1 m s^{-1} (Pierson 2005). However, Pierson and Major (2014) propose that the distinction between hyper-concentrated flow and debris flow conditions coincides with a change from 10 to 50% solids content by volume for hyper-concentrated flow to >50% for debris flow. This change coincides with a Rouse Number (Eq. 12.14) of ~2.5.

Cronin et al. (2000) proposed that "lahars" triggered on 25 September 1995 on Ruapehu volcano, New Zealand, following phreatomagmatic explosions and spill-over of the crater lake, could be subdivided into debris flow and hyper-concentrated flow phases at different stages of the flow event, and at different distances from the source. Barclay et al. 2007 and Vallance and Iverson (2015) also emphasise the lateral, ephemeral transitions that can occur during a "lahar" event, both longitudinally down a valley channel and laterally across it. A river can transform longitudinally from normal stream flow to hyper-concentrated flow as it "bulks up" (i.e. more sediment is entrained and becomes suspended in the flow) to debris flow status (higher sediment concentration again, including clay and silt), and then wane to muddy stream flow. Across a channel, transitions can also occur, with the highest sediment concentration part of the flow event focussed in the channel (e.g. debris flow), but as the channel fills and the flow surface rises and spills over the channel margins, lower concentration flow types (e.g. hyper-concentrated flow) spill over onto the river banks or flood plain (e.g. Cronin et al. 2000).

With high suspended sediment loads, flow conditions are similar to those in subaqueous granular mass flows (Sect. 12.6.5), i.e. grains are supported by fluid turbulence and grain-to-grain collisions are common. The essential difference lies in the nature of the force driving water–sediment movement: in fluvial systems movement is initially due to the gravitational potential and momentum of the water and the sediment is entrained; in granular mass flow, the opposite is the case.

Iverson (1997) summarised this binary contrast in terms of the ratio of solid mass to fluid mass per unit volume, which can be represented by a parameter, N_{mass}:

$$N_{\text{mass}} = \frac{\rho_s \upsilon_s}{\rho_f (1 - \rho_s \upsilon_s)} \tag{12.19}$$

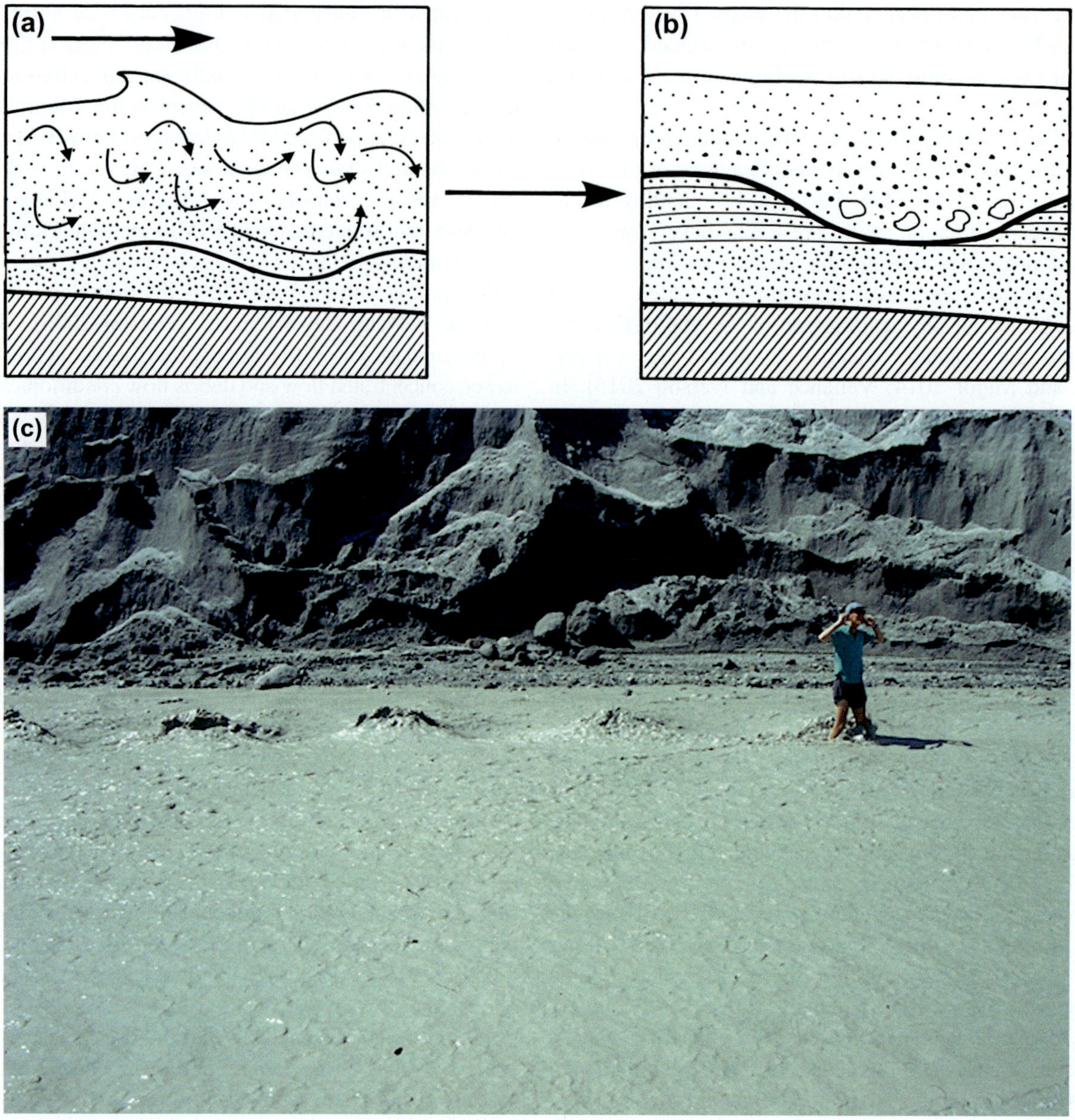

Fig. 12.28 Schematic representation of torrent flow or hyperconcentrated flow conditions in fluvial systems, and deposits. **a** Extremely high flow regime conditions corresponding with the anti-dune stage occur at the bed, whereas standing and upstream breaking waves characterise the state of the water surface. **b** Resultant deposits are frequently massive to plane bed horizontally laminated, and significant erosional scour surfaces represent new flood cycles. (Modified from Cas and Wright 1987). **c** Standing wave anti-dunes at the surface of the fast-flowing, high sediment concentration Potrero River, Mt Pinatubo, Philippines

where ρ_s and ρ_f are the (mass) densities of the solids and intergranular fluid respectively, and υ_s is the volumetric grain concentration. When $N_{\text{mass}} > 1$ it is the momentum of solid grains that dominates the dynamics of the flow system. In reality, there is again a spectrum of scenarios between these end-members.

Iverson and Vallance (2001) proposed that granular mass-flows are initiated when the intergranular shear stresses, τ, equal or exceed the normal stresses, σ, on shear surfaces in (or at the base of) an aggregate of grains times the static angle of friction, ω_s , as proposed by Coulomb (1776; Eq. 12.3).

Building on the oft cited and seminal works of Bagnold (1954b, 1966), who analysed the internal stresses between grains within a cohesionless granular flow, such as grain flows, without considering the gravitational forces acting on grains during flow, Savage (1984) proposed a parameter, called the Savage Number (Iverson and Vallance 2001). This takes into account both grain collision stresses and gravitational forces/stresses:

$$N_{\mathrm{Sav}} = \frac{\rho_s \gamma^2 \delta^2}{(\rho_s - \rho_f) g h} \tag{12.20}$$

where γ is the interstitial fluid flow strain rate, δ is the average grain diameter, g is the acceleration due to gravity and h is the depth below the flow surface. According to Savage and Hutter (1989), at $N_{\mathrm{Sav}} < 0.1$ gravitational forces dominate flow dynamics whereas at $N_{\mathrm{Sav}} > 0.1$, grain collisions dominate (Iverson and Vallance 2001).

When flow velocities wane in high-energy river systems, rapid rates of sedimentation occur, leading to the deposition of massive, structureless to diffusely stratified deposits of granular sediment (e.g. the upper layer in Figs. 12.23i, 12.29a–e), or plane bed, horizontally laminated deposits (e.g. the lower layer in Fig. 12.23i), both representing upper flow regime conditions. Such high-energy conditions are also called torrent-flow conditions and are most likely in high-energy river systems such as braided rivers during flood. For example, rivers that choked and aggraded following signficant explosive eruptions and resedimentation of pyroclastic sediment, such as at Mt St Helens 1980 (Plate 12.1c), and Pinatubo 1991 (Plate 12.1d), or entrenched highland river valleys where all runoff is concentrated into narrow steep-sided gorges and valleys.

Vessell and Davies (1981) and Davies et al. (1978c) described the fluvial dynamics and sedimentation history of the rivers draining from the Guatemalan stratovolcano Fuego following the 1974 eruption. Flow conditions were commonly in the upper flow regime. In the highlands part of the river systems, intermittent suspension transport occured for all fine-medium sand, which constitutes 25% of the sediment load. Coarser sediment was moved as bed load. However, during peak flood events the granular sediment load in suspension would have been even higher, and boulders 2–3 m in diameter were moved as bed load. In the lower reaches, on lower gradients, the proportion of suspended granular sediment transport was lower.

The outwash rivers fed by melting glaciers, perhaps caused by subglacial eruptions as commonly occurs in Iceland (e.g. Björnsson 2010; Gudmundsson 2015; Chap. 16), may at times form meltwater floods and be very dynamic, experiencing torrent flow conditions. Under these circumstances they carry very high sediment loads, which may be instantaneously dumped upon rapid deceleration, producing massive, structureless deposits. In Iceland such jökulhlaups form extensive outwash fans of volcaniclastic sediment called *sandurs* (Fig. 12.29d). Some of these glacial outbursts continued to enter the sea and are found as ash turbidites in the North Atlantic Ocean several hundred km to the south (Walker et al. 1972; Lacasse et al 1996).

Sheet floods commonly originate in areas of steep slope (>20°) and spread out onto relatively flat, lower slopes (<10°) after sudden torrential downpour. Water depths are a metre or more, and velocities reach 10 m s^{-1}. Flow conditions are therefore turbulent and highly erosive (Hogg 1982). Sheet floods dissipate quickly because they represent surges of water from a finite source. Sediment loads and sedimentation rates should be high, and upper flow regime conditions suggest that massive beds and horizontally laminated pebbly sand(stone) deposits should be common.

12.6.5 Subaqueous Granular Mass Flows—Turbidity Currents, Contourites and Others

Subaqueous granular mass flow refers to the downslope or along slope flow of a volume of cohesionless grains containing low viscosity, interstitial water, in a standing body of water (lakes, oceans). Subaqueous granular mass flows include low to high particle concentration turbidity currents, liquified and fluidised flows and grain flows. Granular mass flow processes should be common on the subaqueous slopes of volcanic edifices, whether they be wholly subaqueous or partly emergent. The debris being transported by granular mass flows could be juvenile autoclastic (e.g. hyaloclastite debris from subaqueous lavas and domes, or from lavas flowing into water from subaerial settings), pyroclastic debris derived from subaerial or subaqueous vents, or epiclastic volcanic debris. The granular mass flows could be contemporaneous with eruption, or involve post-eruption resedimentation.

12.6.5.1 Turbidity Current Spectrum and Their Deposits

Aspects of the mechanics of granular mass flow transport have been discussed by various authors including Middleton and Hampton (1976), Lowe (1979, 1982), Allen (1982), Simpson (1987), Iverson and Vallance (2001), Leeder (2009), Meiburg and Kneller (2010), and Mulder (2011). Initiation of movement of the sediment mass is maintained by its own gravitational potential, and may begin initially by creep, sliding or more dramatic gravitational collapse. However, once in motion, ambient water may become mixed into the sediment mass at its top, sides or be ingested along clefts in the flow-front.

As a consequence, the volume of the sediment and water mass expands and fluid turbulence is initiated in the interstitial spaces within the flow. If the degree of expansion is low, grain-support in the flowing mass is due to buoyancy forces, fluid vortices and turbulence, and grain-to-grain collisions. If the flow is highly expanded with grains widely dispersed in the fluid medium, then they can only be supported by the buoyancy forces and turbulence in the fluid.

Fig. 12.29 **a** Massive to diffusely layered, fluvial, hyper-concentrated flow volcanic gravel and sand deposits resulting from reworking of deposits from the 18 May 1980 debris avalanche and pyroclastic flow, and later deposits in the South Fork Toutle River, Mt St Helens, USA. **b** Massive to diffusely bedded, pumiceous volcanic sand reworked and deposited by hyper-concentrated flows following the 15 July 1991 explosive eruption of Mt Pinatubo in the Sacobia River valley, central Luzon, Philippines. **c** Massive to diffusely stratified, basaltic 1971 "lahar" or hyper-concentrated flow deposits, Villarica volcano, Chile. **d** Massive outwash fan deposits of basaltic sediment (called "sandurs") dumped by flash floods of glacial meltwater or 'jökulhlaups', near Vik, south of the Mýrdalsjökull ice sheet and Katla volcano, Iceland (see Fig. 12.20a) (From Cas and Wright 1987). **e** Fining upward succession of hyper-concentrated flow, diffusively stratified, pumiceous volcaniclastic sandstones forming Quaternary coastal apron, southwestern Bali, Indonesia. Note the presence of significant erosional scour surfaces. Height of cliff ~8 m

Although turbulence in the interstitial water provides support for the grains, the water plays a more passive role in the sense that it is initially not driving sediment motion, because its velocity and momentum is more or less the same as that of the sediment. The water is thus entrained by the downslope sediment motion. In fact, Middleton (1966) suggested that sediment movement in the body and head of such flows was faster than the speed at which the whole flow was advancing, leading to upward recirculation of the sediment in the flow-head. However, once turbulent vortices develop, the velocity of the turbulent fluid contributes to sediment entrainment and transport.

Subaqueous granular mass flows in which turbulence plays a major grain-support role, are called *turbidity currents or flows*, and their deposits are called *turbidites*. Turbidity currents are initiated by slope failure, either triggered by earthquakes, explosive shock or oversteepening of slopes beyond the stable angle of repose, due to excessive sedimentation. As described earlier in Sect. 12.4.4, onshore volcanic island sector collapses have been shown to generate large submarine debris avalanches, which in turn may disaggregate and generate debris flows and at shorelines transform into turbidity currents that run out onto adjacent deep-water abyssal plains (e.g. Hunt et al. 2014). In addtion, rivers or deltas in flood may feed sediment laden underflow currents directly into deeper water from their mouths, or storms may mobilise volumes of sediment that begin to flow downslope (Meiburg and Kneller 2010).

Based on experiments, Middleton (1966) proposed that the velocity of the turbidity current flow front, U_h, can be represented as:

$$U_h = C_h \sqrt{g' H_h} \qquad (12.21)$$

where C_h is a co-efficient of friction, (0.71 – 0.73), H_h is the average flow front thickness, and), g' is a modified gravitational constant due to the density differential, $\Delta\rho$, between the density current, ρ_f, and the interstitial water, ρ_w (Mulder 2011). g' is defined as:

$$g' = \frac{g\,\Delta\rho}{\rho_f} \qquad (12.22)$$

where $g = 9.81\ \mathrm{m\ s^{-2}}$.

The degree to which turbidity currents are expanded, which controls the sediment particle concentration, depends on the degree to which ambient water is mixed or entrained into the flow after its initiation and the level of turbulence. Entrainment and mixing are enhanced by turbulence in the flow front and at its margins. This indicates that flow Reynolds Number (Eq. 12.5) and Froude Number (Eq. 12.13) need to be high. This can also be represented by the Richardson Number (*Ri*):

$$Ri = \frac{1}{Fr\#^2} \qquad (12.23)$$

where $Fr\ \#$ is a densimetric Froude Number in which the gravitational acceleration is a modified one compared with the normal Froude Number (Eq. 12.13; Mulder 2011), as follows:

$$Fr\# = \frac{U}{\sqrt{g'\,H_h}} \qquad (12.24)$$

When Ri is < 0.25, turbulence levels are high, when $Ri \sim 1$, turbulence levels and fluid entrainment are low.

Low Sediment Concentration Turbidity Currents and Deposits

In turbidity currents with low sediment concentrations and high degrees of expansion, sediment particles are fully suspended by interstital water turbulence, particle interaction is minor, and so particles can be sorted according to their hydraulic properties, especially settling velocity. The host turbidity currents are highly turbulent with deposition occurring under waning flow behavior from supercritical (Froude Number > 1) upper flow regime conditions producing the Bouma sequence of sedimentary structures beginning with the basal massive Bouma a division transitioning into subcritical (Froude Number < 1) lower flow regime flow state leading to successive deposition of Bouma b, c, d and e divisions from the lower velocity upper parts of the current (Fig. 12.30a; Harms and Fahnestock 1965; Walker 1965; Middleton 1966). When sedimentation begins, the heaviest particles settle first. If all particles are of uniform density, the first particles to settle will also be the heaviest and largest, so producing a well-defined upward fining or normal grain size grading. Where particle densities are variable, for example due to the inclusion of variably vesicular grains, size grading of the dense components may be accompanied by a density grading unrelated to size, perhaps producing reverse grading in the pumice population. In a downflow direction, turbidite grain size also decreases as coarse grain sizes fallout out from suspension as current velocities wane. In addition, the higher flow regime Bouma divisions a and b disappear, producing thin, distal Bouma base absent deposits represented by Bouma c, d, e divisions.

High Sediment Concentration Coarse-Grained Turbidite Currents and Deposits

It was also realised that not all turbidite deposits preserve the complete Bouma sequence of structures. Many proximal turbidites (those deposited closest to the point of in initiation of the turbidity current) have top absent Bouma sequences

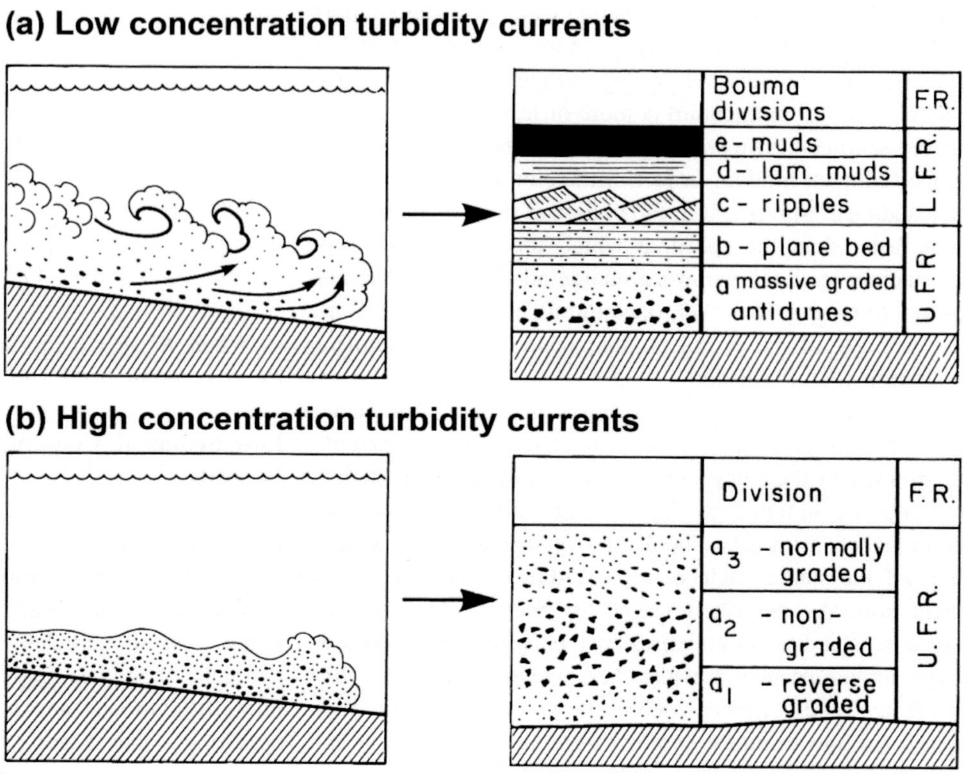

Fig. 12.30 Schematic representation of the characteristics of **a** low-concentration, and **b** high-concentration turbidity currents and their deposits. The resultant sedimentary structure divisions are interpreted in terms of the flow regime (FR) at the right (UFR = upper flow regime, Froude Number > 1, LFR = lower flow regime, Froude Number < 1). (From Cas and Wright 1987). **c** Depiction of expanded turbidite facies association depositional model for coarse gravelly to sandy turbidites related to flow state and possible depositional settings, showing new subdivisions of division b (From Postma and Cartigny 2014)

(just Bouma a or a and b divisions are present; Fig. 2.30b), resulting from high sediment concentration stages of flow. Such flows may transition laterally into low sediment concentration flows producing more distal turbidite facies deposits as outlined above, represented by finer grained beds with the normal Bouma sequence divisions and even more distally, transitioning into base absent Bouma sequence depsosits, with just Bouma c, d, e divisions represented.

In submarine turbidite fan models, proximal turbidites are deposited close to initiation point or within the main distributary channels of submarine fans. When successive granular turbidite beds lack an intervening mudstone layer (divisions d and/or e), they are said to be amalgamated. These frequently occur as channel-fill deposits in proximal to medial distances from source point, or as the deposits of channel mouth lobes in medial to distal settings, associated with submarine fan systems (e.g. Walker 1984). Distal turbidites occur in overbank settings, between subaqueous distributary channels and in the distal parts of submarine fans. They are relatively thin, often fine-grained, and/or dominated by division c ripple cross-lamination (Fig. 12.32d–g).

High particle concentration turbulent flows rapidly develop an internal in-flow stratification into a basal dense pseudolaminar granular flow layer transitioning upwards into a turbulent low-density flow layer, much as a pyroclastic flow develops a basal laminar granular flow undercurrent (the body of the flow) and an overriding turbulent ash cloud (Chap. 10). In turbidity currents with high sediment (hyper-) concentrations and low degrees of expansion, particle freedom is inhibited and size grading is only poorly developed, or is not developed at all, or is only represented by the coarsest and densest and heaviest grains (called coarse-tail grading; Lowe 1982; Kneller and Branney 1995). In such flows, sedimentation is hindered, and sedimentation follows the Richardson-Zaki Settling Law (Eq. 12.18). Sedimentation occurs in pulses or through pulsatory progressive aggradation, producing massive to diffusely stratified deposits in the "a" division (Kneller and Branney 1995). Large clasts can be supported by such a dense grain dispersion during flow. Reverse-grading may also be developed in the basal part of such deposits (Fig. 12.32b), due to intense shearing in the basal part of the highly concentrated depositing grain mass. Internal shearing may also produce diffuse layering, and escape of water during sedimentation is also hindered, leading to focussed, lateral intra-stratal flow and then vertical escape forming dish structures and pillars in the deposits immediately after deposition. Dewatering may continue after sedimentation forming further dish and pillar structures and convolute lamination in finer grained laminae higher in the deposit.

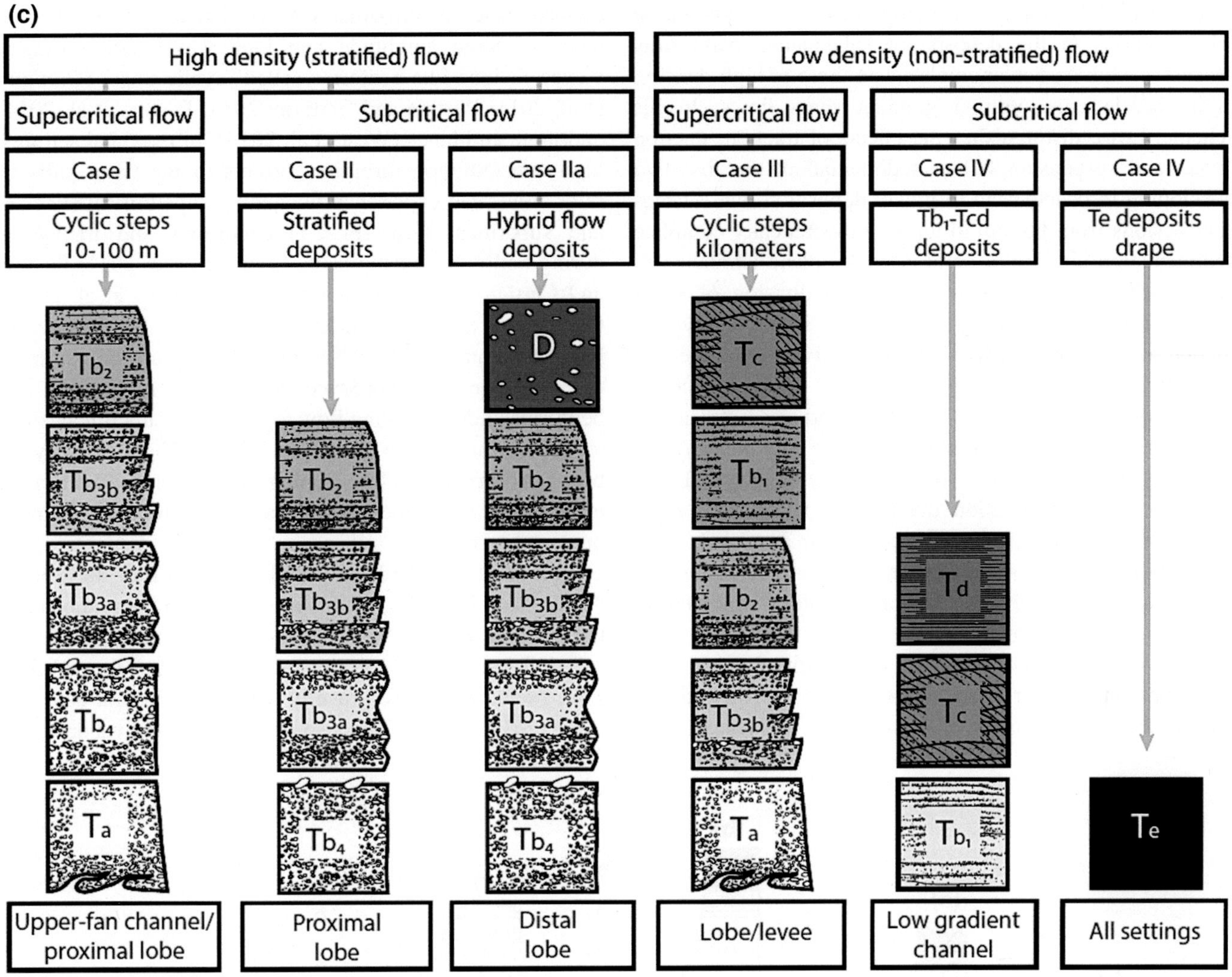

Fig. 12.30 (continued)

These variations in the nature of the basal division of the Bouma sequence (division a) led Cas (1979) to suggest that this division could be subdivided into a basal reverse graded division a_1, a middle non-graded division a_2, and an upper normally graded division a_3, which is equivalent to division a for low concentration turbidity currents (Figs. 12.30, 12.31a, b). Allen (1970) also recognised that Bouma's (1962) sequence did not take account of very coarse sand to pebbly sand turbidites, which sometimes developed a lower cross-bedded division as well as an extra division of horizontal lamination (Fig. 12.31b). All of these variations were encompassed into an extended Bouma sequence for fine- and coarse-grained turbidites by Cas (1979; Fig. 12.31a, b), based on studies of volcaniclastic megaturbidites (see Chap. 11).

Generalised, Bouma-like facies models for mass-flow conglomerate turbidites were developed by Walker (1975, 1984). The common development of reverse grading, no grading and normal grading in these suggests that these are also the deposits of highly concentrated turbidity currents, in which clast suspension, grain to grain collisions and intergranular fluid dispersive pressure play an important part during flow. However they may also show tractional features such as cross-bedding, which in isolation could be confused with fluvial gravels. Importantly, these coarse-grained conglomeratic turbidites testify to the competency of turbidity currents to transport and support gravel-sized clasts by fluid turbulence as suggested by Komar (1970). The variations in the character of conglomeratic turbidite deposits suggest that a composite facies model along the lines of the extended Bouma sequence can also be made for these (Fig. 12.31c; Walker 1984), although such a model is not as confidently constructed as for sandy turbidites. Again, in any deposit not all divisions need be present.

Lowe (1982) produced a comprehensive breakdown of the variations in the deposits of subaqueous mass flow processes, with particular emphasis on those of high-density, high particle concentration granular mass flows. In his scheme, Lowe distinguishes the effects of traction, traction carpet and suspension fallout sedimentation, the first two developing in coarse sand to gravel deposits. Lowe's facies subdivisions from bottom to top in a theoretically complete sedimentation unit are: a basal gravelly, plane laminated to crossstratified division (S_1—traction), a division of shear layers with reverse grading (S_2—traction carpet, equivalent to our a_1 division), a division of massive non-graded to normally graded, sometimes with water escape structures (S_3—suspension, equivalent to our a_2 and/or a_3 divisions), followed perhaps by normal divisions of the Bouma sequence. Where a coarse, true gravel is involved, Lowe recognised a basal traction division (R_1), followed by a reverse graded layer (R_2, equivalent to our Z_1 and Z_2 divisions) and an upper normally graded division (R_3, equivalent to our Z_2 and/or Z_3 divisions), which could be followed by the S divisions as defined above.

Relatively recent consideration of turbidite deposit characteristics on a larger lateral scale, where deposits can be traced laterally, indicates more complex associations of deposit facies in proximal settings that represent the existence of large amplitude and long wavelength sediment waves or dune-like bedforms called "cyclic steps" (Cartigny et al. 2011; Postma and Cartigny 2014; Postma et al. 2014) and large antidunes (West et al. 2019). In proximal settings, high concentration turbidity currents or the basal parts of thick flows can experience fluctuating supercritical (Fr > 1) and subcritical flow (Fr < 1) conditions and transitions during bedform development (Cartigny et al. 2011; Postma and Cartigny 2014; Postma et al. 2014; West et al. 2019). Cartigny et al. (2011) have also evaluated the changes in hydrodynamic conditions and flow state of turbidity currents that are required to produce different wave- or dune-like bedforms under changing flow regime from subcritical (Fr,1) to supercritical (Fr > 1) (Fig. 12.31d). The discovery of broad bedforms with low-sloping stoss and lee sides has led to proposals for further subdivisions in the profiles of coarse proximal turbidites.

Talling et al. (2012) proposed subdivision of the Bouma "b" division into: T_{B3}, a diffuse coarse basal interval beneath the massive Bouma "a" division; T_{B2} laminated above the "a" division gradational upwards into a finer T_{B1} subdivision. Including coarse gravelly turbidites, Postma and Cartigny (2014; Fig. 12.30c) also proposed a subdivision of the

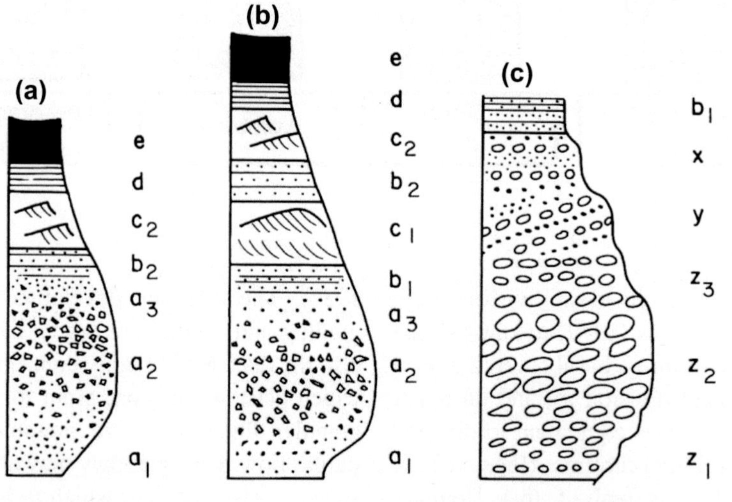

e massive mudstone
d laminated mudstone, siltstone
c_2 ripple cross - lamination
b_2 plane bed laminated sands
c_1 cross - bedded coarse sands
b_1 plane bed laminated coarse sands
a_3 normally graded sands
a_2 non - graded sands
a_1 reverse graded sands
x horizontally bedded fine gravels & sands
y cross-stratified fine gravels & sands
z_3 normally graded gravels
z_2 non - graded gravels
z_1 reverse graded gravels

Fig. 12.31 Facies models for coarse gravelly turbidity current deposits. **a–c** represent generalised schematic models for gravelly to sandy turbidites. **a** Silt to coarse sand turbidites (After Cas 1979). **b** Coarse sand to fine gravel turbidites (After Cas 1979; Allen 1970). **c** Gravel turbidites, in which imbrication may be present in the gravels (After Walker 1975). (From Cas and Wright 1987). **d** Schematic depiction of basal upper flow regime bedforms associated with proximal coarse gravelly to sandy granular turbidites, flow regimes and flow states (Modified from Cartigny et al. 2011)

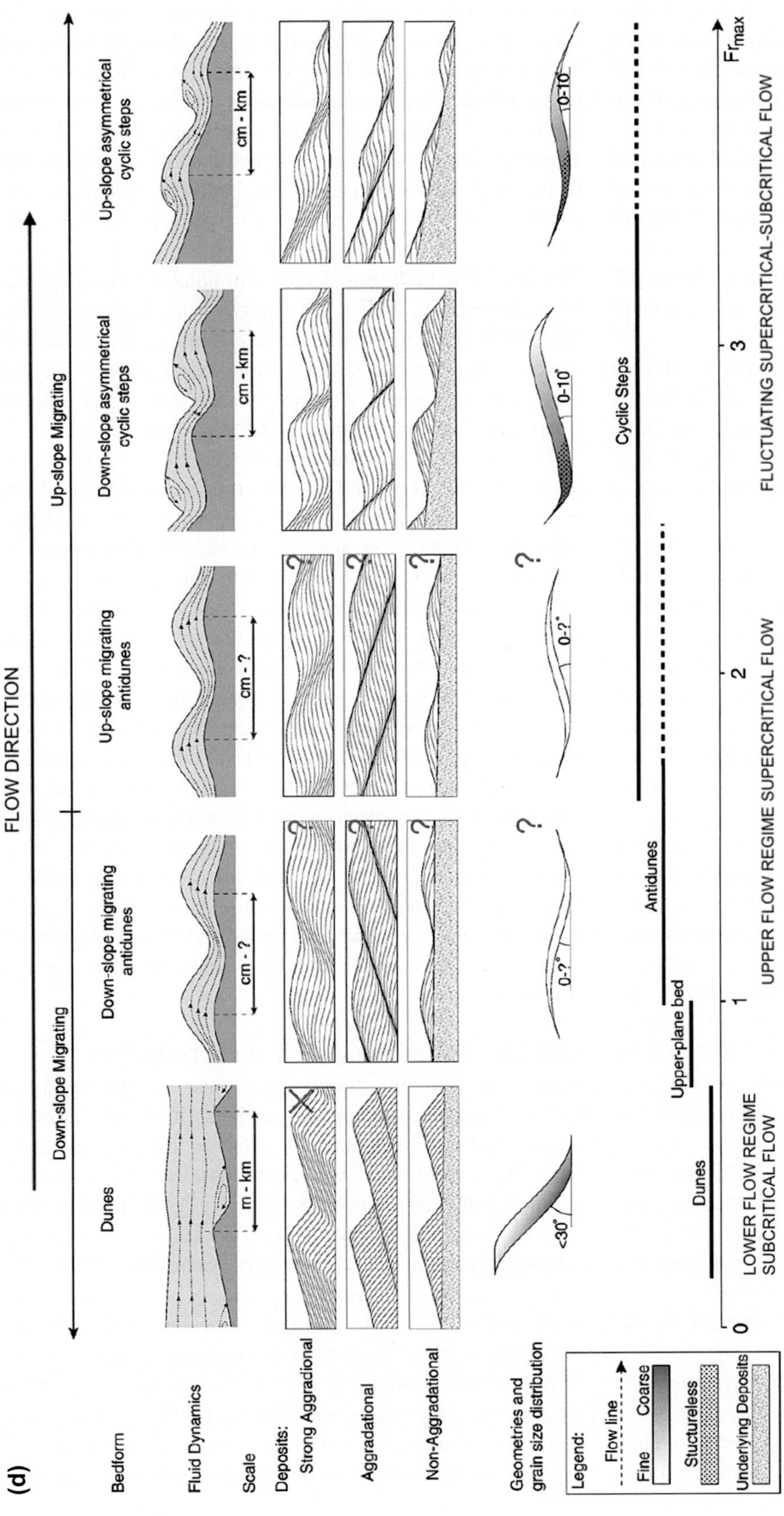

Fig. 12.31 (continued)

"b" division, with a lower Tb_4 very diffusely stratified pebbly sandy layer above a massive "a" division, transitioning upward into slightly finer and diffusely to more clearly stratified $Tb3_{a,b}$ subdivsions, then finer grained and thinly bedded to laminated Tb_2 and Tb_1 subdivisions. These transitions in subdivisions reflect transitions from supercritical to subcritical flow conditions. In particular, Postma and Cartigny (2014) and Postma et al. (2014) relate top absent turbidites with Ta-$Tb_{4,3,2}$ bed associations (Fig. 12.30c) to large wavelength (hundreds m) and high-amplitude waveform bedforms with low slopes that are called "cyclic steps" (Fig. 12.31d; Cartigny et al. 2011; Postma and Cartigny 2014).

Cyclic steps are coarse-grained bedforms with heights up to 10 m, and wavelengths up to 1 km. They are dominated by upstream shallowly dipping (<10°) bedsets, and have been interpreted to represent a series of slowly upslope migrating bedforms produced by high sediment concentration turbidity currents in their proximal parts. Cyclic steps are much larger in amplitude and wavelength than "normal" antidunes, and so have been given a different name although much of their internal structure can be reminiscent of antidunes, but scaled up significantly. Antidunes have rarely been identified in association with turbidites, but West et al. (2019) document large amplitude (2–3 m) and wavelength (75–215 m) bedforms that they interpret as large-scale antidunes in late Miocene base of slope turbidites in California, and that they feel can be distinguished from cyclic step bedforms.

Cyclic steps have been imaged in proximal settings of submarine fans, including feeder canyon floors, inner fan channels and proximal lobes at the mouths of the channels. They originate from large volume high concentration flows, which in a downflow direction transition into "normal" fully suspension supported turbulent, non-stratified turbidity current flows that deposit graded bedded traditional stratified Bouma sequence turbidite deposits in more distal mid-fan lobe and sheet settings. According to Postma and Cartigny (2014), flow state transitions occur during the formation of cyclic steps, from supercritical waxing flow in the "trough" located on the upcurrent, stoss side of the cyclic step bedform to subcritical waxing flow onto the stoss side face and over the crest of cyclic step, and then accelerating again on the lee side. Such flow state fluctuations occur because the toughs and crests represent significant hydraulic jumps. Amplitude and wavelength of cyclic steps decrease downslope and downflow.

Pope et al. (2018) have pointed out that high-amplitude sandwave structures have been known on the deep water slopes of (sub)marine volcanoes for several decades. They subdivided them into two types based on high-resolution seismic sections across such features on the slopes of two volcanoes (Macauley, Raoul) in the Kermadec (island) arc north of New Zealand—those that are wave-like ridges that represent sets of shallow thrusts forward thrust parts of submarine slides on steep slopes, and those that resulted from downslope mass flow deposition. The fields of bedforms on the slopes of Macauley volcano occur at water depths of 2000 m to 3000 m, downslope from a largely submerged summit caldera of which the only emergent part is the small Macauley island. The bedforms extend radially about 20 km from the submarine caldera rim to the southwest and north and have a combined estimated volume of 27.5 km^3. The bedforms decrease in amplitude from 140 to 10 m downslope, and in wavelength from 1500 to 250 m. The bedforms show internal layering which is mostly upstream and upslope dipping, although a topset of layers drape the morphological form. The lee side of the dipping layers (down current) appear to be truncated.

Pope et al. (2018) interpret the bedforms to be cyclic steps that were produced by syn-eruption pyroclastic flows from a subaerial eruption that transformed into a (mega) turbidity current(s) on entering the sea. The crests of the bedforms more or less coincide with subjacent ridges on a buried unconformity surface (perhaps the surface of a hummocky slope landslide deposit?), and so it is unclear whether or not the bedforms could form on a flat sloping depositional surface or whether irregularities on the depositional surface are required for the bedforms to form in phase with the substrate irregularities. Unfortunately there is little sample material and no cores through the cyclic steps to verify through componentry studies and age dating if the deposits can be related to any known eruption on Macauley island, which sits on the rim of submerged summit caldera, or from post-eruptive resedimentation of pyroclastic debris stored near the shoreline, and mobilised by earthquakes or storms as large-scale turbidity currents that flowed downslope.

12.6.5.2 Storm Generated Shelf Turbidites

Short-lived turbidity currents can also be generated on storm-dominated continental shelves. Onshore directed storm surges are balanced by offshore directed turbulent surge-like underflows. These carry shoreline and near-shore sediment offshore, depositing massive graded sands into areas of shelf mud sedimentation (Fig. 12.32h). Hummocky cross-stratification (see above) may occur (Fig. 12.24c, d), and the depositional setting should be identifiable from fossils and trace fossils, such as documented for the Permian volcaniclastic shelf succession documented by Bull and Cas (1989) in the Gerringong Volcanics, Sydney basin, New South Wales, Australia (Figs. 12.24c, 12.32h). Gravels may also be transported in this way. Storm-generated shelf-turbidites (sometimes termed *tempestites*) do not necessarily develop as much of the Bouma sequence as normal turbidites, and may be particularly deficient in divisions c and d (Fig. 12.32h). This is thought to be due to their relatively low degrees of expansion, the low slopes over which flow occurs and the low initial potential energy, all of which lead to rapid deceleration and rapid rates of sedimentation (Walker 1984).

12.6.5.3 Liquefied and Fluidised Granular Mass Flows

There has also been some debate in the literature as to the significance and likelihood of *liquefied* and *fluidised* granular mass flows (e.g. Middleton and Southard 1978, Allen 1982, Lowe 1982; Leeder 2009). Liquefaction is the process of the breaking of grain contacts in a loosely packed aggregate of cohesionless grains without there being any significant volume increase in the aggregate, so that the aggregate behaves in a fluid-like manner. This can only occur in a very inefficiently packed aggregate, and can be triggered by earthquake vibration, or by upward fluxing of water from below, which is called fluidisation when effective. After breaking of contacts, the grains are suspended in the interstitial fluid, supported by high excess pore pressures (viscous dispersive pressure), and the whole aggregate is in a 'quick', fluidal state. How far such liquefied aggregates can flow as such without evolving is uncertain. If unconfined, the interstitial water can escape, or the flow can evolve into a granular mass flow such as turbidity current.

The process of fluid escape itself causes grain support, because as the fluid streams out and up it exerts a lift force on the grains through which it streams when the upward directed fluid velocity exceeds the sediment particle settling velocity in the fluid. The streaming of a fluid through a clastic aggregate to provide grain support is called fluidisation, and has been discussed with respect to support processes in pyroclastic flows and surges in Chap. 10. The simple settling of dispersed grains causes upward displacement of fluid and very short-term self-fluidisation support, but it cannot be maintained for very long unless there is an external source for the fluidising fluid (Allen 1982), for example by ingestion of ambient water at the head of a granular mass flow. However, Allen (1970, 1982) suggested that relatively little fluidisation of this type actually occurs. Nevertheless, coarse granular mass flow deposits do show evidence of liquefaction and fluidisation processes (e.g. dish-structures, pillar structures, sand volcanoes, convolute lamination, which reflect water escape; Lowe 1975; Leeder 2009) which, if nothing else, reflect excess pore pressures at the time of sedimentation. However, it is doubtful that wholly liquefied and fluidised flows can travel very far without assistance from interstitial fluid turbulence, in which case they translate into highly concentrated turbidity currents. One extreme example of liquefaction phenomena has been described from the lowest pyroclastic flow member of the Matahina ignimbrite in New Zealand, where it apparently flowed into water (Bailey and Carr 1994; Chap. 11). In parts of the deposit that were totally submerged, liquefaction convolutions or fluid escape structures occupy its entire thickness (up to 10 m). These include tongues and diapirs (defined by pumice trains and ill-defined, finer-grained marginal zones), contorted ashy channels and subtle dish-like features (see Fig. 11.14).

12.6.5.4 Grain Flows

Grain flows have also been assessed in the subaqueous realm (Middleton and Hampton 1976), but strictly, grain flow, as discussed in Sect. 12.4.6 in subaerial settings, faces the same constraints subaqueously. Talling et al. (2012) call grain flow granular avalanching. It involves very significant frictional interaction with the substrate and will therefore only operate on slopes near or at angle of repose. In subaqueous settings, interstitial ambient water lowers the internal co-efficient of friction between grains, and grain flow is commonly represented by lee-side grain flow forming cross-stratification in migrating ripples, sandwaves and dunes and on subaqueous angle or repose scree slopes. Once initiated subaqueously on long steep slopes grain flows may evolve into a turbulent turbidity current. Sometimes the term modified grain flow is used for a process intermediate between true debris flow and grain flow, where clast interlocking is a major effect in the strength of the debris and the behavior of the flow. Modified grain flow deposits have been described from submarine pumice cones where resedimentaion of predominately water-settled fall deposits has occurred due to deposition on the steep unstable flanks (Allen and McPhie 2000; Chap. 11, see Fig. 11.36a–e). True grain flow layers will be thin and reverse-graded (Lowe 1982; Leeder 2009).

In subaqueous volcanic settings grain flow deposits should be common on aprons of resedimented hyaloclastite around subaqueously erupted lava domes, where lavas have flowed from subaerial settings into water bodies forming flow foot lava deltas, and on aprons of pumiceous debris deposited on the steep slopes of marine cones that have undergone quench fragmentation of vesicular lava domes or explosive vesiculation from relatively shallow or subaerial vents.

12.6.5.5 Slurry Flows

Lastly, *slurry flows* are essentially turbidity currents in which the interstitial fluid is not pure water, but contains a significant component of mixed clays ($\geq$5–10%), giving that interstitial fluid significant viscosity and strength. As such, grain support is also provided by the strength of the fluid and the added buoyancy caused by the lessened density contrast between the fluid and the grains (e.g. Hampton 1972, 1975). However, with increasing clay content and

(a)
(b)
(c)
(d)
(e)
(f)

◀ **Fig. 12.32** Features of turbidity current deposits (turbidites). **a** Upward finning, graded bedded sandstone turbidites with Bouma a, b d, and e divisions, interbedded with grey to black mudstones and thin "distal" turbidites. Note also the slump folded horizon near the top and the fault at the left. Late Devonian Bunga Beds, Boyd Volcanic Complex, southern New South Wales, Australia. **b** Coarse, size-graded, volcanic conglomerate turbidite, showing basal reverse grading (z1) to normal grading (z3), and coarse planar lamination (b1; see Fig. 12.30c). Tertiary, Muriwai, North Island, New Zealand. **c** Bouma a-b turbidite with grain size grading, and load casts and flame structures at the base above mudstone. Location as for (**a**). **d** Thin-graded bedded "distal" turbidites with Bouma a-d-e and Bouma b-d-e internal divisions, and spectacular load casts and flame structures. Location as for (**a**). **e** Fine-grained, grey, siltstone distal turbidite beds and laminae interbedded with black mudstones. Location as for (**a**). **f** Bouma b-c turbidite with climbing ripple cross-lamination in upper greenschist-lower amphibolite metamorphic facies. Archean Black Flag Beds, Lake Lefroy, Kambalda, Western Australia. **g** Thin, volcanic sandstone turbidites with intervening mudstone in drill core (~5 cm in diameter). Archean, Golden Grove Mine, Western Australia. **h** Permian, storm shelf, graded-bedded, volcanic sandstone turbidites with bioturbated tops. Permian Gerringong Volcanics, Broughton Formation, Kiama, New South Wales, Australia. (From Cas and Wright 1987)

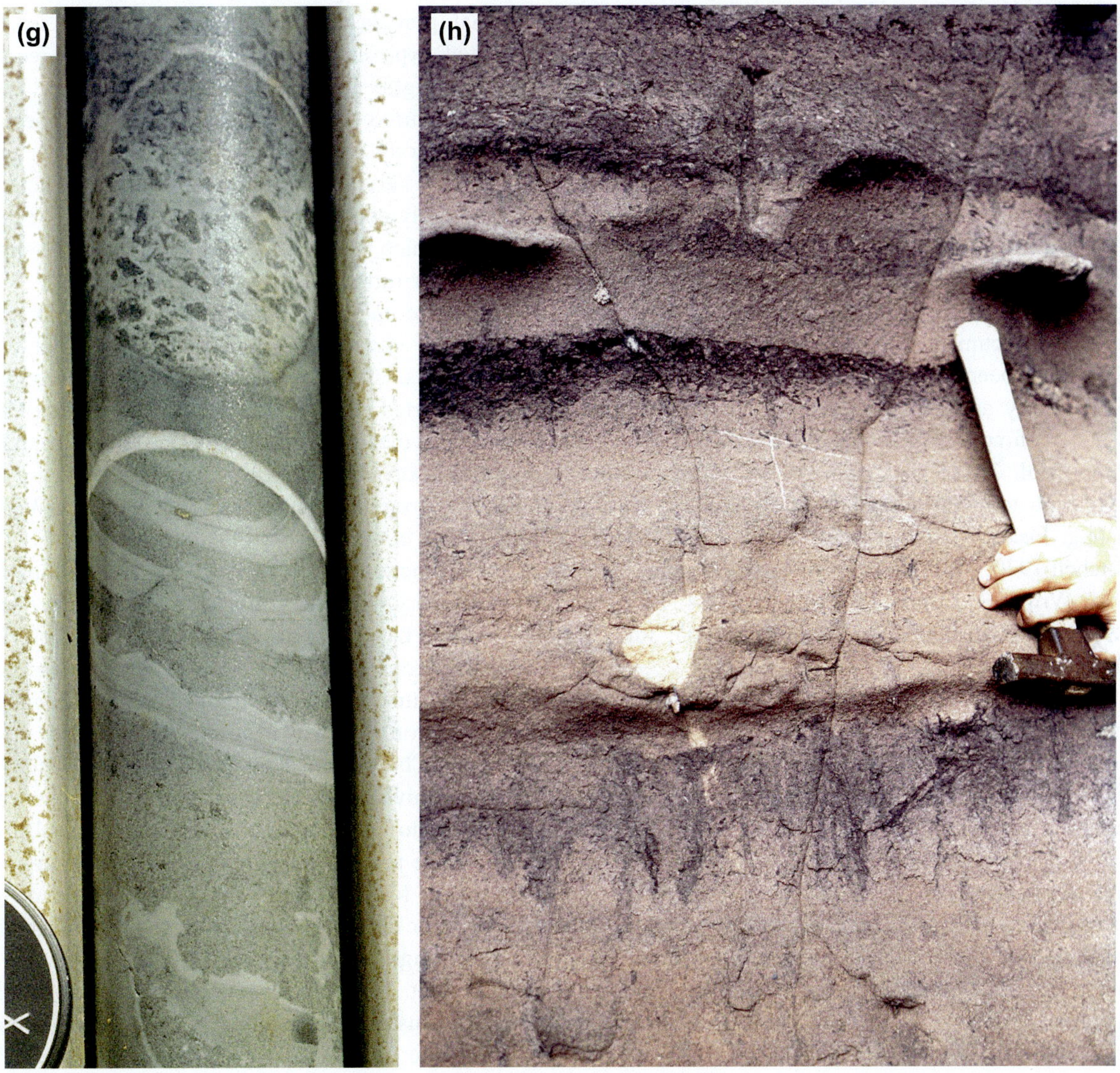

Fig. 12.32 (continued)

viscosity, turbulence levels in the fluid are damped according to the Reynolds and Froude Number criteria (Baas et al. 2009; Chap. 2). Resultant deposits should be sandy mudstones or sandstones with abundant mud matrix. Intraclasts may be important throughout the bed, and tractional structures may be notably lacking due to suppressed turbulence (Carter 1975). Slurry flows are transitional between viscous subaqueous debris flows and low-viscosity turbidity currents (e.g. Hampton 1972). Talling et al. (2012) also summarise the spectrum of mud-dominant flow types and deposits, including mud density currents or turbidites.

12.6.5.6 Volcaniclastic Granular Mass Flow Equivalents

This lengthy discussion of subaqueous granular mass-flow processes is justified by virtue of the significance of volcaniclastic mass flow deposits preserved in the geological record. They operate in relatively deep lakes, especially caldera lakes, supplied by steep, erodible margins or streams and deltas, and they operate in seas marginal to or hosting volcanic provinces. Accounts of the influences of subaqueous mass flows in marine volcanic settings are plentiful. They have been documented from modern oceanic settings, where they have been imaged and cored (e.g. Sigurdsson et al. 1980; Carey and Sigurdsson 1984; Carey and Schneider 2011; Trofimovs et al. 2008, 2012, 2013; Hunt et al. 2013; Pope et al. 2018), and amongst the successions of ancient volcanic terrains ranging from the Tertiary to the Archean (e.g. Fiske 1963; Fiske and Matsuda 1964; Jones 1967; Mitchell 1970; Tasse et al. 1978; Cas 1979, 1983; Cas et al. 1981; Wright and Mutti 1981; Lajoie 1984; Bull and Cas 1989; Hand et al. 2002; Simpson et al. 2005; Jutzeler et al. 2014a, b).

In all of these modern and ancient successions normal turbidite deposits of volcanic detritus have been commonly identified. In many the detritus is very juvenile in character, including unabraded cuspate glass shards (e.g. Cas 1979; Wright and Mutti 1981). This is important from two points of view. First, pyroclastically fragmented detritus is being redeposited by sedimentary mass-flow processes and, secondly, subaqueous mass-flow processes are capable of transporting fragile, pyroclastically fragmented detritus tens to hundreds of km without significant abrasion. So, two fundamental points raised initially in Chap. 1—that modes of fragmentation and final deposition may be quite independent and have to be evaluated independently, and that subaqueous mass-flow processes can be non-abrasive and can transport detritus long distances without significant textural modification—are reinforced.

The nature of the supply process of the volcanic detritus may be variable and is difficult to determine in some cases. Sometimes the origin can be essentially post-eruptive, epiclastic (epiclastic reworking, gravitational or sector collapse and epiclastic mass-flow redeposition; e.g. Jones 1967b; Mitchell 1970). In other cases the supply originates from syn-eruptive subaerial and subaqueous explosive eruptions (e.g. Cas 1979; Jutzeler 2014a, b; Trofimovs et al. 2008, 2012, 2013),. Pyroclastic fallout and pyroclastic flow processes may transport the debris into water and it either transforms into, or is redeposited subaqueously by sedimentary, water-supported mass-flow processes (Chap. 11). For example, Cas (1979, 1983) suggested that anomalously thick individual sedimentation units (m to tens of m thick) of crystal rich, juvenile volcanic detritus with rare preserved cuspate shards (Figs. 11.24, 11.25), were initiated when subaerial pyroclastic flows flowed into the sea, perhaps interacted explosively with it and were then reconstituted into huge, highly concentrated turbidity currents. This seems to have occurred several times, producing a coherent stratigraphic interval made up of multiple massive, graded to non-graded, very thick beds with all the characteristics of normal highly concentrated turbidites, except for their huge thickness which is explained by the nature of the supply (see Chap. 11 for further discussion).

Effusive eruptions producing hyaloclastite deposits can also lead to significant downslope transport of that autoclastic debris, either syn-eruptively or post-eruptively, by water supported granular mass flows. These contribute to the growth of volcaniclastic aprons or deltas around wholly subaqueous or partly emergent volcanic edifices, or where lava deltas prograde at shorelines (e.g. Hawai'i; Watton et al. 2013; Soule et al. 2021).

The distinction between what is pyroclastic flow and what is sedimentary mass flow in such situations may seem academic, and has been discussed in considerable detail in Chap. 11. From the point of view of establishing criteria for paleoenvironmental interpretation, it is an extremely important distinction. We will therefore reiterate here our philosophy from Chap. 11. Except in very shallow settings, it has not been convincingly demonstrated that all gas-supported pyroclastic flows can maintain their integrity subaqueously. We therefore favour water-supported flow and grain-support mechanisms for most relatively deep-water, coarse granular volcaniclastic deposits, unless they can be demonstrated to be of fallout origin, or clearly have evidence of a hot state of emplacement. Unfortunately, there are numerous references to subaqueous and submarine pyroclastic flow deposits and events in the literature, when the deposits clearly bear the imprint of sedimentary mass-flow facies characteristics (e.g. Lajoie 1984). Perhaps the intention was to identify mass flows of pyroclastic debris (i.e. debris originally fragmented pyroclastically), but redeposited by sedimentary mass-flow processes. Lajoie (1984) for example, illustrates numerous deposit types which are identical to the facies models described here in Figs. 12.30 and 12.31 (see Walker 1984), but clearly labels them to be subaqueous pyroclastic flow deposits because they contain clasts that were originally pyroclasts. In doing so Lajoie, perhaps unintentionally, implies gas-support flow mechanisms, for which there is no evidence, whereas water-supported transport is clear from deposit characteristics.

In conclusion here, we provide illustrative documentation of this spectrum of subaqueous granular mass-flow epiclastic processes and facies that are applicable to a wide variety of volcanic settings (Figs. 12.31d, 12.33). The foregoing has suggested that in any subaqueous granular mass-flow, a number of different grain-support mechanisms (Fig. 12.33a) may operate in combination, or at different times during the initiation and flow history (Fig. 12.33b). These include interstitial fluid turbulence, grain-to-grain collisions, fluidisation and liquefaction effects. Where interstitial fluids are viscous, fluid strength and buoyancy effects also operate (Fig. 12.33a). As a result, and also dependent on flow velocity and competence with distance travelled, a wide diversity of facies may result (Fig. 12.33b, c).

12.6.5.7 Contourites and Distinguishing Their Deposits from Turbidites

Sustained oceanic currents are not just confined to shallow water depths driven by surface winds. Sustained deeper water seafloor hugging currents also occur at depths from the outer shelf, continental slope and even in places on the deep ocean floor. Such currents are called *contour currents*, because of their tendency to largely move parallel to slopes rather than down them as sediment gravity currents discussed above do (Faugères and Mulder 2011). Some contour currents, such as the Gulf Stream in the western North Atlantic Ocean are driven by deep reaching major oceanic gyre circulation currents, which are initiated by global scale wind patterns and Coriolis Force modifications. Others are driven by geostrophic (large-scale) thermohaline bottom currents that move in response to density differences in the oceanic water masses caused by differences in salinity and/or temperature (Faugères and Mulder 2011). Both types of currents are sustained over long time intervals because it takes a long time for global scale currents to establish themselves and to adjust to changing conditions.

Measured contour current velocities usually range from a few cm s^{-1} to 20 cm s^{-1}, sometimes reaching 70 cm s^{-1} (Faugères and Mulder 2011). Such velocities are clearly high enough to transport mud and silt, in some cases sand and at velocities >40 cm s^{-1}, even fine gravel size sediment (Fig. 12.22a) along the sea floor producing a range of tractional sedimentary structures, as discussed above in Sect. 12.6.1, and imaged on the sea floor. More commonly, gravel and coarse sand represents lag sediment left on the seaflor by currents that have winnowed and transported finer grain sizes. The sediment that is transported by contour currents can include fine lofted terrigenous clay and from the sediment clouds that rise from turbidity currents and are then dispersed by oceanic currents, as well as deep water hemipelagic and pelagic sediments that are scoured and eroded by bottom-hugging contour currents. In addition, where contour currents encounter the slopes of marine volcanoes, volcanic sediment, particularly low-density vesiculated pumice and shards could be eroded and become incorporated into the currents.

The deposits of contour currents are called *contourite* deposits. They are typically fine sand to clay in size and consist of terrigenous, bioclastic and volcaniclastic grain types (Faugères and Mulder 2011). Fine muddy silty fine sand contourite deposits, which are the most common, are commonly cryptically laminated to cross-laminated in core, showing subtle grain size variations, but lacking the systematic grading and sequence of sedimentary structures found in deposits of short-lived turbidite currents (Faugères and Mulder 2011; Stow and Smillie 2020). Such fine deposits are commonly bioturbated, so destroying internal layering and structure. Coarse contourite deposits may preserve asymmetrical ripples, dunes, sandwaves and planar bedforms, as well as scour surfaces (Faugères and Mulder 2011; Stow and Smillie 2020).

Interestingly, where contour currents sweep large areas of flat ocean floor, extensive sheet drifts and high amplitude (10–150 m) and long wavelength (0.5–10 km) sediment wave fields are common (Faugères and Mulder 2011; Hernández-Molina et al.2018; Stow and Smillie 2020). These are mud-silt dominated and are therefore not cyclic step waves produced by coarse granular turbidity currents as described above. They show draping and inclined layering and some internal erosional surfaces in seismic sections. The dynamics of and the way in which they are constructed is unclear (Faugères and Mulder 2011). Large bedforms discovered along the Uruguayan continental slope are internally complex with depositional bedforms and erosional surfaces common, as well as significant grain size variations, suggesting involvement of not only contour currents but also downslope sediment flows (Hernández-Molina et al. 2020).

Although comment has been made about volcaniclastic components being present in contourite deposits (Faugères and Mulder 2011), there are no known studies of volcaniclastic contourites. However, given that volcanic edifices are widespread in the modern oceans, and were in ancient oceans, volcaniclastic contourite deposits must form where oceanic currents in a stratified ocean current system interact with the slopes of those volcanic edifices.

12.6.6 Mudflows and Debris Flows

These flow types differ from the previous mass movement types in being viscous flow phenomena. As the clay content

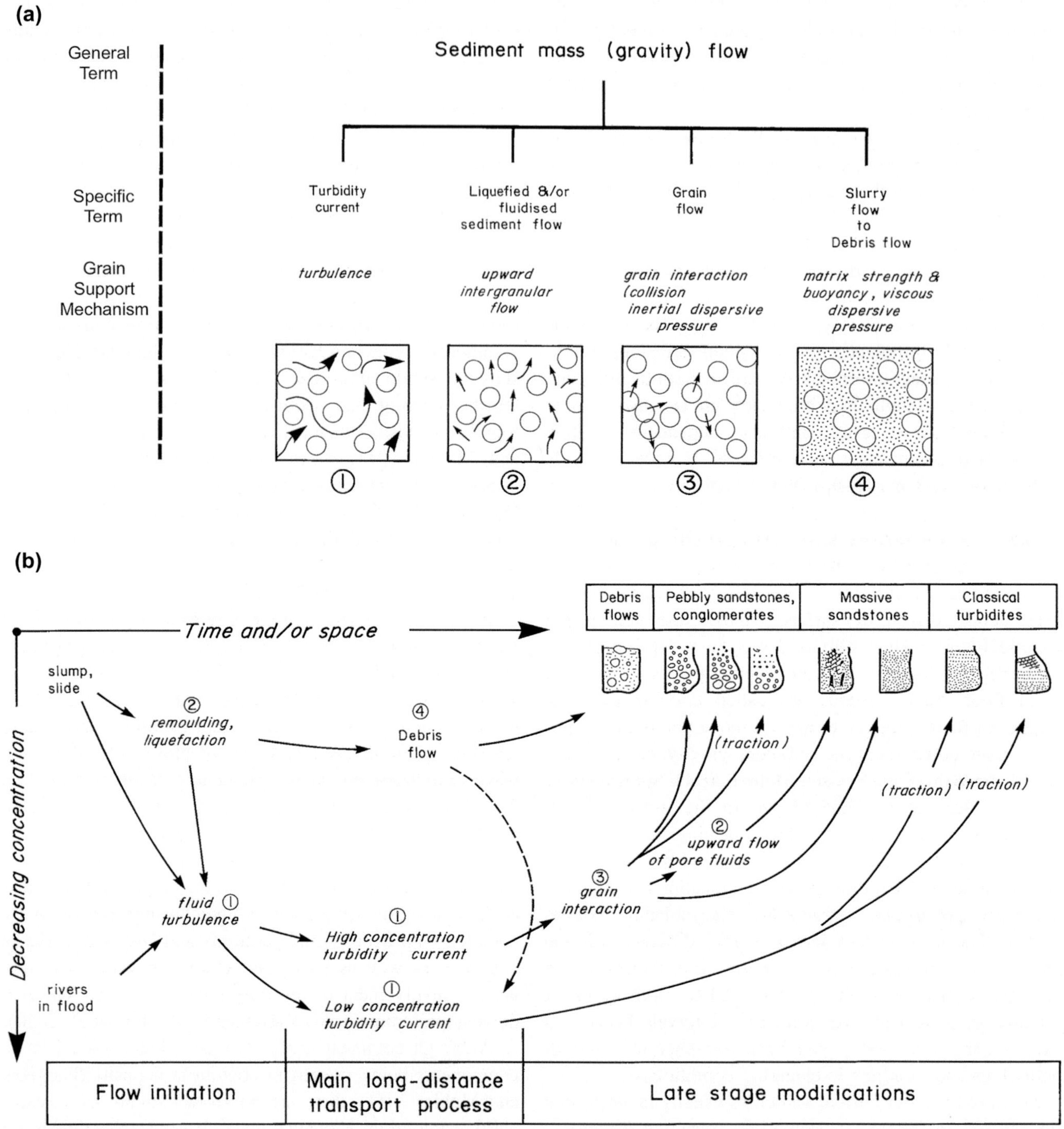

Fig. 12.33 Mass-flow process and deposit spectrum characteristics. **a** Different grain-support mechanisms and related flow types in the realms of subaqueous mass flow (After Middleton and Southard 1978). **b** Modes of initiating subaqueous mass flows, possible evolution paths, and generalised characteristics of deposits (After Walker 1978). The circled numbers correspond to the principal grain-support process operating, depicted in (**a**). **c** A generalised, process and facies spectrum for subaqueous mass flow, depicting the general facies characteristics, the flow types, the physical properties of these, the grain-support processes operating and the general physical conditions during flow (After Cas 1977). (From Cas and Wright 1987)

(c)

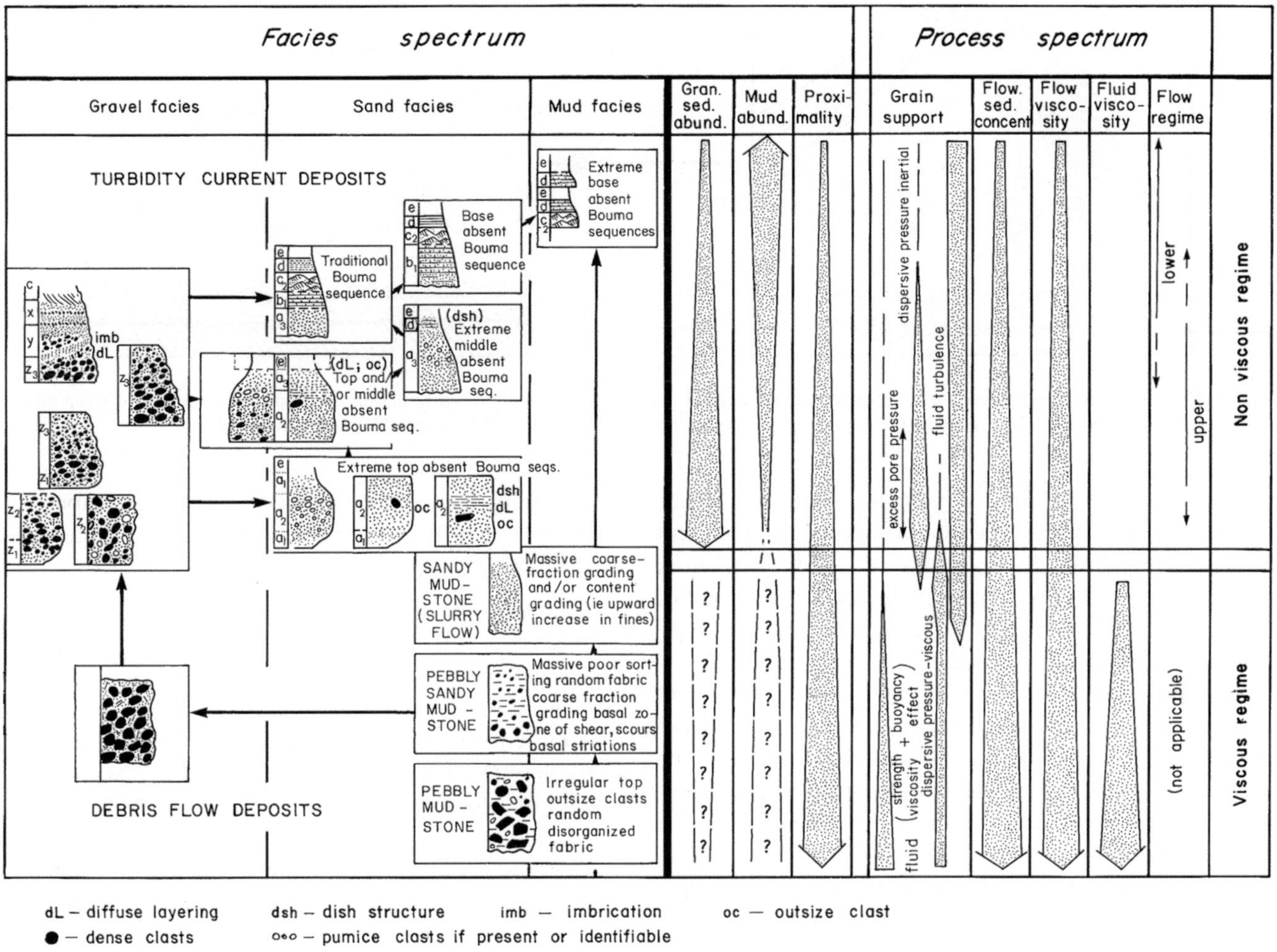

Fig. 12.33 (continued)

increases in the interstitial water in some debris flows, such flows develop a yield strength and the turbulence level is progressively suppressed (Baas et al. 2009). We separate *mudflows,* where clay and silt dominate and coarse debris is a minor component from *debris flows,* where very coarse to fine sediment is involved. As mud flows and debris flows develop a cohesive, dense mud fluid they can exert a high force on large boulders and roll them along. In addition the mud fluid develops a yield strength, capable of supporting many large clasts (Figs. 12.33a, 12.34; Hampton 1975, 1979). For large clasts there is a buoyancy effect resulting from the low-density contrast between clasts and high-density mud fluid (cf. Archimedes Principle). A third source of support comes from the viscous dispersive fluid pressure that builds up in the matrix mud fluid during near approaches of large clasts. Hampton (1979) suggested that, on this last count alone, the higher the proportion of large clasts, the greater the mobility of the viscous flow and the greater the competence to carry large clasts. Where coarse debris is abundant, clast to clast collisions cause momentum transfer, which also maintains flow mobility. According to Rodine and Johnson (1976), the support of large clasts in such debris flows occurs because of the poor sorting, and the resultant hierarchy of grain sizes. Each grain size level, together with the finer grain size levels and the interstitial fluid, provide support for the next coarser grain size level, and so on.

Debris flows can be simply initiated by failure of steep slopes of water saturated debris and soil in both subaerial and subaqueous settings and slumping-sliding downslope (Fig. 12.34a). They can also be formed by transformation from initial landslides or debris avalanches, which incorporate river water in fluvial valleys into which they flow, or seawater in submarine settings. Debris avalanche deposits may also contain ice and snow, which melt, and groundwater from the origiinal source, which as it percolates through the deposit, may mobilise some of the deposit as debris flows, as occurred on 18 May 1980 at Mt St Helens (Hacker et al.

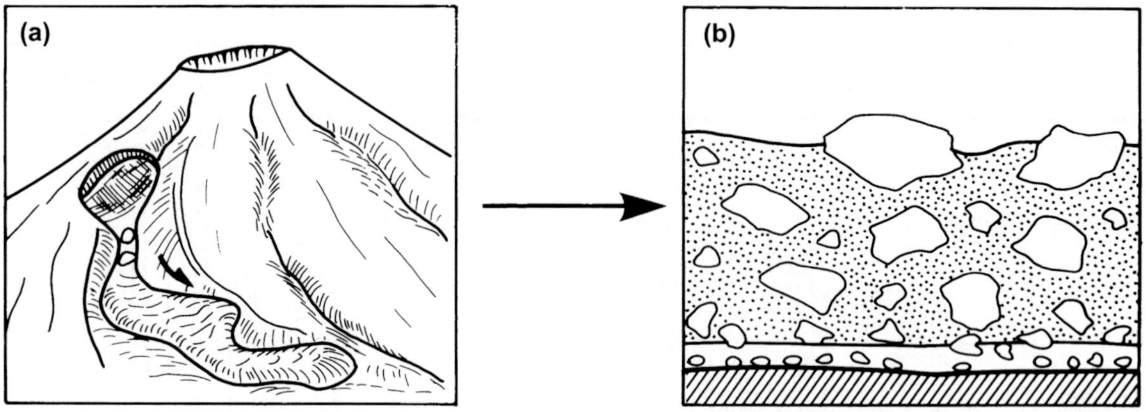

Fig. 12.34 Schematic representation of the formation and deposit textures of debris flows. **a** Initiation by failure or slumping of steep slopes. **b** General massive, poorly sorted internal texture of viscous debris flows and their deposits. (From Cas and Wright 1987)

2018). Where the detritus consists of contemporaneous volcanic material, debris flows and mudflows also represent the viscous end of the spectrum of "lahars", whereas hyper-concentrated flow represents the less viscous end of that spectrum (Sect. 12.6.4). Appreciable granular sediment can be mixed with the mud fluid without affecting its viscous properties. As long as the granular sediment particles are not in mutual contact and locking, the overall internal friction of the mass is low and viscous flow can be maintained, slopes permitting. Development of intergranular fluid pore pressure also reduces intergranular friction (Iverson 1997). Lowe (1979), Scott (1998) and Scott et al. (1995) recognised that even within the debris flow phenomenon there is a spectrum of debris flow types with varying physical properties and deposit characteristics.

Scott (1988) and Scott et al. (1995) subdivided volcanic debris flows into cohesive and non-cohesive end members. In general terms, in non-cohesive debris flows the clay content is <5 volume (vol) %, but the coarser grain population constitutes >50 vol % of the flow, with water <50 vol %. In cohesive debris flows clay constitutes >5 volume %, the grain population also represents >50 vol %, and water also <50 vol %. In comparison, in hyper-concentrated flows, clay represents <5 vol % and the coarser grain population >10–<50 vol %, with water >50% vol (Scott et al. 1995; Iverson 1997; Pierson and Major 2014). In stream flows the clay content is <5 vol %, the coarser grain content <10 vol %, and water >85% vol. Hyper-concentrated flows are therefore more watery than debris flows, and driven largely by the potential energy of the water mass in which the sediment is entrained. In debris flows, it is the potential energy of the coupled sediment and water that drives the dynamics and mobility of the flows. Because of the high sediment content of debris flows they have a higher bulk density than other water-dominated flow processes. Their bulk density, ρ_b, can be defined as (Vallance and Iverson 2015):

$$\rho_b = \rho_s m + \rho_f (1 - m) \tag{12.25}$$

where ρ_s is the average density of sediment particles, ρ_f is the density of the interstitial fluid (water + clay + silt), and m is the volume fraction of all coarser sediment particles. According to Vallance and Iverson (2015) approximate values of these parameters are: $\rho_b \sim 2{,}000$ kg m^{-3}, $\rho_s \sim 2{,}500$ kg m^{-3}, $\rho_f \sim 1{,}100$ kg m^{-3}, and $m \sim 0.6$.

This has led to the recognition of debris flows with as little as 5% interstitial mud, apparently enough to lubricate the movement of the granular sediment component (Lowe 1979, 1982, Rodine and Johnson 1976). Previously these have been called grain modified debris flows (Lowe 1979), but they are effectively hyper-concentrated flows, as discussed in Sect. 12.6.4 (Pierson 2005).

Although debris flows have previously been attributed a "Bingham" rheology, indicating that a yield stength must be overcome before flow begins, Iverson (1997, 2003) has argued against applying a "one shoe fits all" rheological model. This is because of the range of physical states of different debris flows, from some that can be very viscous or cohesive and have significant yield strength (up to 10 kPa), to those that are watery, with very low yield strength (500 Pa). According to Iverson (2003), the viscosities of debris flows can range from 0.1 to 50 Pa s. In addition, there can be significant variations in the physical state vertically and laterally, and during a debris flow event, as flow transformations can occur (e.g. Scott 1988; Scott et al. 1995; Cronin et al. 2000; Barclay et al. 2007). According to Scott et al. (1995), flow transformations are more likely to occur with non-cohesive debris flows (e.g. steam flow/flood -

> hyper-concentrated flow -> non-cohesive debris flow - > hyper-concentrated flow -> stream flow/flood as flow waxes and then wanes) than with cohesive debris flows.

Subaqueous debris flows can transform into slurry flows and perhaps turbidity currents (Fig. 12.33b) through *"flow stripping"* or detachment of the upper, lower particle concentration part (Walker 1978; Baas et al. 2009; Leeder 2009). Ambient water can be mixed into the interstitial mud-fluid, particularly along the top surface, to reduce the fluid strength and viscosity sufficiently to allow interstitial fluid turbulence to develop (Hampton 1972; Baas et al. 2009; Leeder 2009).

The dynamics of debris flows are therefore controlled by the potential energy of the mobilised sediment–water mass, the sediment concentration and vibrational energy, and the fluid content and fluid pore pressure (Iverson 1997; Vallance and Iverson 2015). Given the importance of internal fluid pore pressure in debris flows, Iverson (1997, 2003) proposed that the criterion for mobilisation of debris flows is a modified Coulomb criterion (Eq. 12.3), called the Coulomb-Terzaghi criterion, which is commonly used in soil mechanics:

$$\tau = (\sigma - P) \tan \phi + c \tag{12.26}$$

where τ is the average driving force/stress on the failure surface, σ is the gravitational stress acting on grains in the mass normal to the failure surface, P is the intergranular pore pressure, ϕ is the internal co-efficient of friction acting on the grains, and c is the cohesive or yield strength on the failure surface. During failure the cohesive/yield strength trends to 0, and the friction angle reduces in reponse to increasing pore fluid pressure. Therefore, if $P > \sigma + c$, the debris flow will remain mobile once motion is initiated.

Johnson (1970) showed that viscous, cohesive debris flows usually move by combined laminar and plug flow (Chap. 2). The marginal laminar flow zone represents the zone of maximum shear at the interface between the substrate and the body of the debris flow. In rare cross-sectional exposure through a sheet-like marine debris flow deposit resulting from the degradation of an Eocene–Oligocene surtseyan basaltic tuff cone at Bridge Point, South Island, New Zealand, Cas and Landis (1987) demonstrated that the succession preserves multiple, internal, massive plug flow channels within the debris flow deposit sheet. These show curved, upward concave, diffusely stratified laminar flow zones bounding the massive plug zones (Fig. 12.35a, b), much as in the modelled flow behaviour of Johnson (1970; see Fig. 2.17c). The exposure demonstrates that in sheet debris flows there are multiple laminar plug flow through zones or channels moving at differential velocities to the enclosing part of the deposit. Some of the diffusely stratified zones in the Bridge Point example appear to be lateral accretion layers or surfaces (cf. point bars in fluvial deposits) indicating the plug flow zones or channels could migrate laterally in their own deposit, and probably within the active debris flow (cf. multiple flow lobes in a pyroclastic flow (Chap. 10).

The mobility and competence of debris flows should not be underestimated. They are known to easily flow several km but some can flow to over 100 km from source (Pierson et al. 1990; Scott et al. 1995; Vallance and Scott 1997), carrying very large boulders well away from the source. Speeds of 40 m s^{-1} were calculated for the 1980 Mt St Helens lahars or mudflows (Janda et al. 1981). The mobility of debris flows depends on the volume of the flows, their thickness, the slope and topography of the terrain, the granulometry of the flow mass, including clay content, the volumetric concentration of solids relative to water, the interstitial fluid pore pressure and the rate at which that is dissipated (pore pressure diffusivity), which is dependent of the water content, solids concentration and granulometry (Iverson 1997; Savage and Iverson 2003; Vallance and Iverson 2015). Many documented accounts of debris flows in volcanic terrains are known, the majority being associated with stratovolcanoes; e.g. Mt St Helens (Mullineaux and Crandell 1962, Janda et al. 1981; Scott 1988; Major et al. 2005; Hacker et al. 2018), Mt Rainier (Fiske et al. 1963, Crandell 1971, Scott et al. 1995; Vallance and Scott 1997), Ruiz del Nevado (Pierson et al. 1990), and Merapi (Lavigne et al. 2000). The debris flows resulting from the 18 May, 1980 eruption of Mt St Helens, were derived directly from the landslide and avalanche event which flowed into and filled the major distributary rivers draining the volcano. The huge volumes of debris from the slide and avalanche were mixed with river water and melting snow and ice and transformed into very dynamic and destructive debris flows (Janda et al. 1981; Major et al. 2005; Hacker et al. 2018).

Debris flow deposits can be highly variable in volume, from small, local slope or valley confined to very large volume and regional in scale. The 5.6 ka, 3.8 km^3 Osceola debris flow deposit from Mt Rainier is one of the largest documented (Vallance and Scott 1997). The deposits cover and area >200 km^2, and extend at least 120 km from source. In some steep valleys the deposit is >100 m thick. It originated from a phreatomagmatic explosive eruption that initiated a debris avalanche of largely hydrothermally altered debris that then transformed into a dynamic debris flow (Vallance and Scott 1997). The debris flow was calculated to have flowed at $\sim$19 m s^{-1}, and to have had a peak discharge of 2.5×10^6 m^3 s^{-1} (Vallance and Scott 1997). The Osceola debris flow deposits show an overall vertical size grading, defined by a basal concentration of coarse boulders becoming finer upwards (normal coarse-tail grading), which Vallance and Scott (1997) attributed to progressive aggradation produced by very coarse clasts carried by the debris flow front being deposited first and then progressively buried

Fig. 12.35 Features of a submarine debris flow deposit, New Zealand. **a** Cross-sectional field sketch of the features of part of an Eocene–Oligocene sheet-like submarine basaltic volcanic debris flow deposit in the preserved stratigraphy of a Surtseyan tuff cone at Bridge Point, South Island, New Zealand. The debris flow has scoured into horizontal, planar stratified subaqueous water settled and resedimented tephra. The debris flow deposit shows not only a generally massive texture but also plug flow zones or internal channels with concentrations of large volcanic clasts, bounded by finer upwards concave, diffuse stratification that represents confining laminar shear flow zones at the margins of the plugs. In some cases, the laminar shear zone layers suggest lateral migration of the plug flow zones within the debris flow. **b** Photo of one massive, coarse clast-bearing plug flow zone (yellow boxed area, left side of cross-section leg B–C) bounded by an upward concave set of laminar shear flow layers. The left side and base of the plug zone have scoured into underlying bedded tuffaceous sediments. The coarse breccia texture inside the massive plug compared with the finer grain size of the laminar shear flow zone may reflect a type of kinetic sieving, by which clasts with diameters greater than the width of the laminar shear layers are forced away from the zone of flow shear. (From Cas and Landis 1987)

under deposits from the trailing part of the flow. Sorting improved down-flow, defined by decreasing proportions of boulders, increasing proportions of sand and gravel and depletion of clay content. In part this was explained by progressive erosional incorporation of coarse fluvial gravel debris from valley sides ("bulking-up") during passage of the flow (Vallance and Scott 1997).

Lateral levees and surface mound topography, including domains of megaclasts, are not uncommon on debris flow deposits, indicating a high yield strength (Vallance and Iverson 2015). Because of the high particle concentration, and the cohesive nature of many debris flows, tractional structures are unlikely to form in the deposits of debris flows *sensu stricto*, but they may form in overlying and later deposits during waning flow stages of an event in a river valley when the flow system transforms to hyper-concentrated flow and then stream flow. Internally debris flow deposits are very poorly sorted, with large clasts in open, matrix-supported organisation in a finer-grained matrix (Figs. 12.35a, b; 12.36; 12.37a–g) containing significant, although apparently not necessarily large, amounts of clays. Cohesive debris flows are more poorly sorted that non-cohesive debris flow, hyper-concentrated flow, and stream flow deposits, which become progressively better sorted (Fig. 12.36b), resulting from lower clast content, thus allowing the higher interstitial water fluid to effect hydraulic sorting. Hyper-concentrated flow and stream flow deposits contain progressively less clay and silt (Fig. 12.36a, b).

Debris flow deposits are mostly massive (Figs. 12.35a, b; 12.37a-g), but may be diffusely stratified around plug flow zones (Fig. 12.35a, b), and also near the top which may reflect transition to and deposition from waning hyper-concentrated flow. Thicknesses of deposits may be less than 1 m to several tens of m. A basal finer layer may be developed, representing an early, finer and watery stream flow or hyper-concentrated flow pulse before the main, bouldery debris body wave(s) or pulse(s) arrive. Longer-lived debris flow events appear to consist of multiple pulses or flow surges, which is reflected by multiple sub-units in the deposits, distinguished by grain size variations. This could reflect different debris flow pulses entering a main trunk river from tributaries, as occurred during the Armero disaster as a consequence of the 13 November, 1985 eruption of Nevado del Ruiz stratovolcano in Colombia (Pierson et al. 1990; see below).

Deposits can also show dewatering structures, including fine-grained clastic dykes (Fig. 12.37 g, h) derived from the fine matrix component of the debris flow deposit, which is entrained by vertically rising water and transported upwards through zones of weakness, which may be dyke-like (Fig. 12.37g) or pipe-like (Fig. 12.37h). These dykes have irregular geometries and margins reflecting the unconsolidated state of the debris flow deposit during dewatering, and they may contain a diffuse sub-vertical layering (Fig. 12.37g), indicating that the dewatering occurs in pulses. In addition, some debris fow deposits preserve dewatering pipes (Fig. 12.37h), in which the material in the centre of the pipe is relatively coarse compared with a late-stage annular rim that is fine-grained, and sometimes laminated as in some clastic dewatering dykes. The pipe material is overall finer grained than the surrounding debris flow deposit (Fig. 12.37g), reflecting the lower competency of the vertically streaming dewatering fluid compared with that of the original debris flow.

Because debris flows frequently accompany eruptions, they may carry hot debris (i.e. "lahars"), evidence of which may be found in charred woody remains, although such material could easily be picked up in transit by a cold debris flow. In addition, explosive eruptions that deposit hot pyroclastic deposits onto snow and ice can cause melting of the substrate snow and ice triggering volcanic debris laden water torrents, that transform into debris flows (e.g. 13 November, 1985 eruption of Nevado del Ruiz; see below). If pyroclastic flows flow over snow and ice and physically erode and incorporate snow and ice clasts into the pyroclastic flows, some may also transform into debris flows. Such "lahars" could easily have a generally similar texture to non-welded ignimbrites if much of the clast content is pumiceous, or block and ash flow deposits if the clast population is largely lithic clasts. The major distinguishing criteria to look for to identify pyroclastic flow deposits are evidence of a hot state of emplacement, including gas segregation pipes, thermal 'salmon pink" colouration, evidence of sintering and vapour-phase crystallisation, or evidence of a thermal remanent magnetisation (TRM) in pyroclastic flow deposits (Chaps. 10, 11). Debris flows that are not related to or derived from pyroclastic deposits resulting from contemporaneous explosive eruptions are more likely to be polymictic and contain variably weathered and rounded debris, although this is not an immutable guideline.

The difficulty of distinguishing primary block and ash flow (BAF) deposits from debris flow (DF) deposits sourced from such block and ash flow deposits was highlighted by Uehara et al. (2015) in a paleomagnetic study of deposits from the 1991–1995 eruption of the Mt Unzen lava dome complex, Japan. Nearly 10,000 BAFs were generated by rock falls from lava dome collapse events, which filled valleys around the volcano with a crudely bedded succession of largely monomictic lava dome-derived deposits. During rain events, especially during the monsoon seasons, these primary deposits were variably reworked and resedimented by short-lived, high-energy, non-cohesive debris flows and hyper-concentrated flows. The resultant deposits are also monomictic. In erosionally incised valleys through fans of deposits at the foot of the dome complex, it is difficult to differentiate the BAF and DF deposit types on field criteria except in some cases where gas segregation pipes sourced

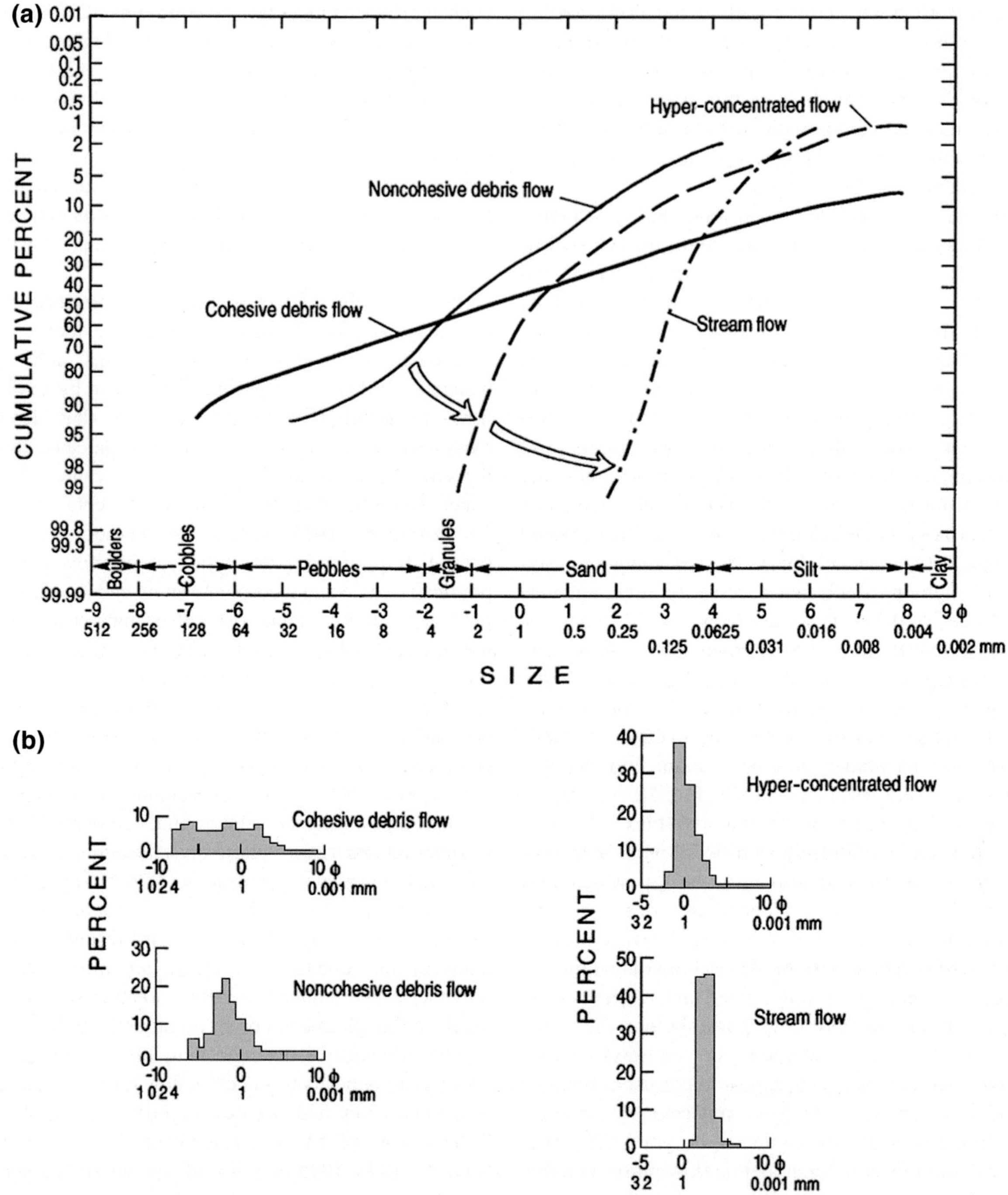

Fig. 12.36 Granulometry of debris flow, hyper-concentrated flow, and stream flow deposits. **a** Granulometry of cohesive and non-cohesive debris flow deposits compared with hyper-concentrated flow and stream flow deposits from Mt Rainier stratovolcano, USA, as represented by cumulative frequency grain size curves. **b** Granulometry of the same samples represented by histograms showing the abundance of different size classes. The spread of grain sizes and the peakedness of the histograms show the progressive better sorting characteristics from very poorly sorted cohesive debris flow to non-cohesive debris flow to hyper-concentrated flow and stream flow deposits. (From Scott et al. 1995)

from large carbonised logs, and soot coated lithic clasts near such pipes are preserved in BAF deposits. Even clast rounding was not a definitive criterion. Uehara et al. (2015) employed TRM studies of oriented clasts to determine emplacement temperatures of the deposits. Deposits in which most clasts preserve one high-temperature magnetic component that is uniformly oriented parallel to the Earth's magnetic field at the time and site of emplacement were deposited at very high temperature, between the Curie temperature of magnetite (~ 540 °C) and the glass transition

(a)
(b)
(c)
(d)
(e)
(f)
(g)
(h)

◀ **Fig. 12.37** Features of debris flow deposits. **a** Subaerial volcanic debris flow deposit mantled by air fallout and/or surge deposits, Mt Taranaki/Egmont stratovolcano, New Zealand. Note the matrix-supported texture. (From Cas and Wright 1987). **b** Subaerial, non-cohesive debris flow deposit of reworked primary block and ash flow deposits from the 1991–1995 lava dome complex forming eruption of Mt Unzen, Akamatsu River, Shimabara, Japan. Note the abundance of large angular to sub-rounded dacite boulders is a coarse matrix of the same debris. Only a few reddish boulders are sourced from an older source. **c** Cross-section through subaerial, non-cohesive debris flow deposit, Mt Unzen, Akamatsu River, Shimabara. Again note the matrix-supported texture. Height of exposure ~1.5 m. **d** Submarine debris flow deposit from the basaltic debris flow shown in Fig. 12.35, showing open, matrix-supported, clast framework texture, Eocene–Oligocene Waiareka-Deborah Volcanics, Bridge Point, Kakanui, New Zealand. The deposits include a sheet lava or pillow tube clast with black chilled margins and a spindle bomb clast. **e** Subaqueous debris flow deposit dominated by wispy scoria fragments in a matrix of black mudstone resulting from the incorporation of deep-water black muds into a subaqueous mass flow that bulked up and transformed into a viscous debris flow. Late Devonian Bunga Beds, Boyd Volcanic Complex, southern New South Wales, Australia. **f–h** Submarine basaltic debris flow deposits with dewatering clastic dykes, Eocene–Oligocene Angahook Member, Airey's Inlet, Victoria, Australia. **f** Massive, poorly sorted, matrix-supported texture with angular to irregular spindle bomb clasts and a dewatering clastic dyke (left). **g** Vertical, irregular, dewatering clastic dyke cross-cutting the massive poorly sorted debris flow deposit. Internal laminations indicating multiple flow pulses of fine ash-size material through the dewatering dyke. **h** Plan view cross-section through a large, vertical cylindrical dewatering pipe in a coarse basaltic debris flow deposit. The margins of the pipe consist of fine ash-size material whereas the centre of the pipe is coarse

temperature of the lava dome (~745 °C), and can be confidently interpreted as a BAF deposit (Chap. 10). Deposits with multiple clasts with a high- and low-temperature magnetic component, and with the high-temperature component oriented randomly in different clasts were interpreted as DF deposits (Uehara et al. 2015). Even so, there were some ambigous deposits, which could have been either BAF or hot DF deposits, which is consistent with observations and video imagery that shows that some BAF events occurred during major rain events and that hot, gas rich BAFs flowed into water swollen rivers and transformed into hot, steaming water-supported debris flows.

Debris flows can be very destructive and lethal in their impact. One of the most disastrous historic events occurred in 1985 on the Colombian stratovolcano Nevado del Ruiz. At about 9 pm on the 13 November the volcano erupted, producing hot pyroclastic flows and surges that covered ~10 km^2 of the snow and ice capped summit of the volcano (Naranjo et al. 1986). The pyroclastic density currents also physically eroded and incorporated metres of substrate snow and ice into the flows. Within minutes of the eruption beginning, these hot deposits melted the summit snow and ice and generated a torrent of water ~2×10^7 m^3 in volume that flowed into the headwaters of multiple steep sided valleys draining the top of the volcano (Pierson et al. 1990). The torrent flow carried some new pyroclastic debris, but also eroded and entrained a large volume of volcanic debris and soil from the valleys it passed through. This caused up to 4 x "bulking up" in the volume of the flows and tranformation from watery slurries to viscous debris flows, which increased to 9×10^7 m^3 in volume through sediment addition in transit. The flows flowed at velocities of up to 17 m s^{-1} and they were able to flow >100 km within 4 h, with a peak discharge of 48,000 m s^{-1} (Pierson et al. 1990). The debris flows consisted of multiple waves of gravelly muddy debris flows that were >40 m thick in narrow canyons. Deposits >10 m thick were deposited in places. The debris flows at the mouth of the Rio Laginillas valley destroyed most of the town of Armero, burying it under up to 5 m of muddy gravel (Fig. 12.38a), and killing more than 23,000 people, three quarters of the town's population. In valley axes, the debris flow deposits consist of a clast-supported massive to inverse graded bouldery-sandy-mudstone, which upwards becomes finer grained and progressively muddier in granulometry (Pierson et al. 1990).

Other volcanic debris flows may not be eruption triggered, but can also be lethal. The Christmas eve 1953 Tangiwai railway bridge disaster in New Zealand, involved collapse of a tephra dam holding water in the summit crater lake of Mt Ruapehu stratovolcano. This produced a "lahar", that flowed down the Whangaehu River valley and destroyed part of the railway bridge over the river at Tangiwai just before a passenger train came along, and plummeted into the river (Fig. 12.38b, c), killing 151 people.

During and after the 1991–1995 Mt Unzen eruption in Japan, debris flows buried and destroyed many houses and other buildings at Shimabara in the lower reaches of valleys draining the volcano slopes (Fig. 12.38d).

12.6.7 Tsunamis and Severe Storm Surges

Extreme wave conditions associated with tsunamis or severe storm surges represent major hazards along coasts worldwide, including many volcanic shorelines which can be particularly susceptible (e.g. Tappin 2007; Bourgois 2009; Peters and and Jaffe 2010; Paris et al. 2009, 2013). However, discrimination between tsunami and storm surge deposits within onshore successions can be extremely difficult, since most of their characteristics only indicate marine flooding (Fig. 12.39). During both types of event, the entrainment of shallow marine or coastal sediments and their re-deposition in terrestrial and/or transitional (e.g. back barrier, lagoonal, estuarine) sedimentary environments occurs. The

Fig. 12.38 Disastrous impact of some volcanic debris flow events. **a** The mostly buried town of Armero, Colombia, that was destroyed by debris flows triggered by an eruption of Nevado del Ruiz stratovolcano on 13 November 1985, killing over 23,000 people. The vague outline of some buried streets under the grey deposits can be seen. (*Source* United States Geological Survey). **b** The rebuilt Tangiwai railway bridge over the Whangaehu River, draining off Mt Ruapehu, New Zealand, which was destroyed on Christmas Eve, 1953, by a debris flow from the summit of the volcano, only minutes before a crowded train passed by and plummeted into the river, killing 151 people. The debris flow was triggered by the collapse of one margin of the summit crater lake of the Ruapehu stratovolcano, which can be seen on the horizon. (From Cas and Wright 1987). **c** Part of the train wreck in the Whangaehu River channel. This occurs among boulders transported by the debris flow described above. (*Source* Wikimedia, through Creative Commons Attribution-Share Alike 2.0 Generic licence; Image from New Zealand Railways Corporation collection, dated 25 December 1953, Archives New Zealand Te Rua Mahara o te Kāwanatanga). **d** Houses buried nearly to the roof by boulder debris flow deposits, in the lower reaches of the Akamatsu River, Shimabara, Japan, associated with the 1991–1995 eruption of Mt Unzen

distinctions between tsunami and storm deposits are related to differences in the hydrodynamics and sediment-sorting processes during transport (Table 12.3; Morton et al. 2007).

Tsunami deposition results from a few high-velocity, long-period waves that entrain sediment from the shoreface and beach. Tsunamis can have flow depths greater than 10 m, are generally considered to transport sediment primarily in suspension, and distribute the load over a broad region where sediment falls out of suspension when flow decelerates. In contrast, storm inundation generally is gradual and prolonged, consisting of many waves that erode beaches and dunes with no significant overland return flow until after the main flooding. Storm flow depths are commonly <3 m, sediment is transported primarily as bed load by traction and the load is deposited within a zone relatively close to the beach. While suspension and tractional mechanisms (Sects. 12.6.2, 12.6.1) are thought to be important grain support and transport mechanisms, we include onshore tsunami and storm surges in this discussion of mass movement because their deposits reflect the more or less instaneous deposition of significant amounts of sediment and, more recently, high concentration particle transport is increasingly being identified as having an potential role in their deposition (Table 12.3).

12.6.7.1 Tsunamis and Their Deposits Onshore

Tsunami flooding or inudation results from a train of long-period waves that can rapidly travel long distances from a source where they were generated. Tsunamis may be triggered by a number of non-volcanic, such as submarine

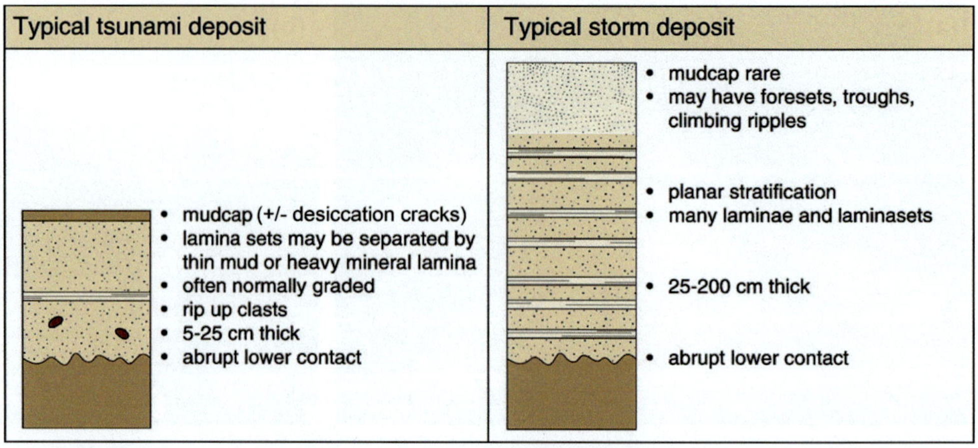

Fig. 12.39 Characteristics of typical sandy tsunami and storm surge deposits (From Morton et al. 2007)

Table 12.3 Physical characteristics of extreme waves derived from modern examples (Modified from Morton et al. 2007)

Flow characteristic	Tsunami	Coastal storm
Length of coast impacted	10–10,000 km	100–600 km
Deepwater wave height	<1 m	> 5 m
Nearshore wave height, and period	10–30 m, 100–2000 s	<10 m, 10–25 s
Potential wave run-up heights	Generally 10 s of m, but up to a few 100 s of m	A few m
No. of overland waves	Generally <10	Typically > 1000
Inundation depth	0–30 m	<5 m
Active flow duration	Minutes to hours	Hours to days
Overland flood velocity	<20 m s^{-1}	< 5 m s^{-1}
Flow direction change	Alternating run-up and backwash during an event	Return flow only at end of event
Boundary layer structure	Entire water column	Current boundary layer
Influence of wind stress	Not a factor	Increases water velocities and surge heights
Sediment transport mechanism	Suspension, some traction (hyper-concentrated to debris flow)	Mostly traction, some suspension (hyper-concentrated flow)
Phases of flooding	Repeated rapid rise and fall	Gradual initial rise, rapid medial rise, gradual fall
Event frequency	Infrequent locally, moderately frequent globally	Frequent locally and globally

earthquakes, land or seaslides and bollide impacts, as well as volcanigenic mechanisms, including pyroclastic flows entering a body of water (Chap. 11), underwater eruptions (Chaps. 9, 11), and sector or flank collapse (Sect. 12.4.4). Two main types of tsunami deposits (or *tsunamites*) have been found deposited onshore: one is composed essentially of finer-grained, usually sand-sized sediment (Fig. 12.39); the other consists of coarse clastic deposits ranging from large cobbles and boulders to megaclast-sized blocks (see Figs. 12.41, 12.42). Furthermore, it is also possible to find deposits where both grain size classes are combined. The characteristics of tsunami deposits vary greatly, but are fundamentally controlled by:

- the coastal and nearshore topography,
- the height of tsunami waves at the coast and run-up (i.e. the maximum inland elevation reached by the inundation),
- the nature and supply of existing sediment (including boulders) in any coastal setting affected by such an event, and
- the local environment and preservation of the unconsolidated sediment laid down by the tsunami inundation.

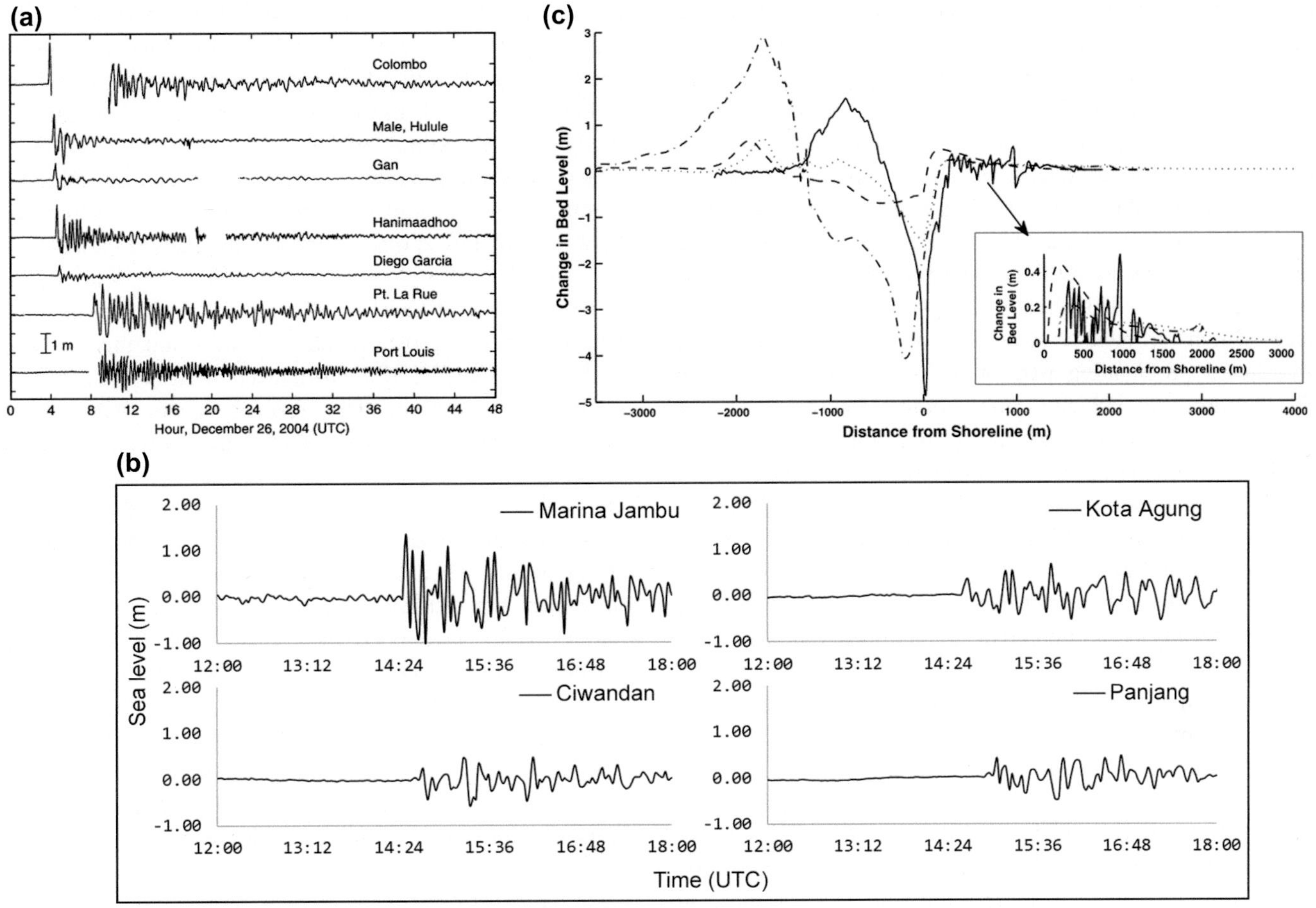

Fig. 12.40 Some features of tsunami waves. **a** Tidal gauge series from island stations in the Indian Ocean showing arrival times following the devastating tsunami of 26 December 2004. This originated from rupture of the Sunda megathrust, northern Sumatra at 00:59 UTC (Chap. 17). Note that for most locations the waves continued for hours to days after the first arrival, and the initial wave was not always the largest. Tidal contributions have been removed. (From Merrifield et al. 2005). **b** Elevation fluctuations measured at four tide-gauge stations showing arrival times around the Sunda Strait following the tsunami of 22 December 2018. This followed the relatively small (≤ 0.2 km^3) failure of the southwestern flank of Anak Krakatau which occurred at 13.55 UTC (Fig. 12.12). (From Syamsidik et al. 2020 through Creative Commons Attribution 4.0 International licence). **c** Data showing the cumulative erosion (negative values) and deposition (positive values) from test numeric simulations of the run-up (up-flow) and back-flow of various tsunami waves. Note, for all cases, the largest amount of erosion occurs near the original shoreline, while the largest amount of deposition occurs further offshore. This is where the sediment-laden back-wash collides with on-rush of the next wave forming an offshore bar near the seaward extent of the back-flow, consistent with bathymetric observations following the 26 December 2004 tsunami. (From Apotsos et al. 2011)

Maximum thicknesses of sand beds reported for modern tsunami deposits are about 1 m, however, most are generally <25 cm thick (Table 12.3; Morton et al. 2007). These usually consist of one layer or only a few layers or lamina sets, and most are normally graded (Fig. 12.39); inverse grading in natural tsunami deposits is rare (e.g. Yoshii et al. 2018). Single-layer, homogeneous and structureless tsunami deposits are indicative of extremely rapid deposition such as would occur when flow decelerates between the up-flow and backwash phases of waves. Maximum thickness of coarser-grained, conglomeratic modern tsunami deposits appears to be between 2–3 m, but most are generally relatively thin <1 m.

The catastrophic 26 December, 2004 Sumatra-Indian Ocean tsunami, which caused the death of more than 237,000 people and left in excess of one million homeless, justifiably rejuvenated research into the generation, inundation, and sediment transport and depositional mechanics of tsunami waves. From a theoretical-side, laboratory and numerical modelling has been important (e.g. Pritchard and Dickinson 2008; Apotsos et al. 2011; Johnson et al. 2016; Yoshii et al. 2017, 2018). As a tsunami wave approaches the coast (Fig. 12.40a, b), this creates onshore up-flow (or run-up), which rapidly accelerates to a maximum as the wave front passes, and then gradually decelerates before reversing direction and flowing offshore (or backwash). The resulting depth averaged flow velocities therefore have an asymmetrical or saw-tooth shape described by rapid acceleration as the wavefront passes, followed by a gradual deceleration as the wave slows and flows offshore. During the run-up and backwash, average flow velocities can exceed 10 m s^{-1}. The magnitude and timing of the saw-tooth velocity profile is partially controlled by the slope of the topography and distance from the shoreline. Thus, the onshore flow velocity decelerates more rapidly over steeper slopes; on the other hand, maximum run-up increases with increasing onshore slope. Also, steep onshore topographic features (e.g. coastal bluffs or nearshore cliffs) can affect the onshore up-flow. This creates wave reflections off the cliff face which propagate seawards, and interact with the shoreward propagating part of the wave causing a sharp deceleration in velocity. In certain circumstances the reflected wave can enhance the offshore directed flow.

Model simulations show that a significant amount of sediment is suspended and transported within the tsunami (Apotsos et al. 2011; Pritchard and Dickinson 2008). Some sediment is suspended locally by the passing wave front, however, as the turbulent energy generated by the wavefront persists for only a few hundred m, this sediment quickly settles back to the bed. Much higher concentrations of suspended sediment can occur far behind the wave front owing to transfer of sediment suspended off shore during backwash (Fig. 12.40c). The majority of sediment suspended by the tsunami is eroded during the long-duration, high-velocity backwash that occurs offshore of the shoreline. The erosive capacity of the backwash is a function of both the bed slope and the amount of water onshore as the water level recedes. On linear offshore slopes, the largest velocities and highest suspended sediment concentrations generally occur at the seaward edge of the backwash, where a water–sediment "slurry" can form (Apotsos et al. 2011). On non-linear offshore slopes the situation is more complex depending on local gradients, the slope, but in all cases the greatest erosion occurs near the original shoreline. When the sediment-laden backwash collides with the on-rush of the next wave, some of the suspended sediment is deposited locally as the flow suddenly changes direction. This may form an off-shore bar near the seaward extent of the backwash (Fig. 12.40c), as appears to have occurred following the 26 December 2004 tsunami; however, this was moving back towards the beach as a low bar within two weeks of the tsunami (Moore et al. 2008).

The sediment deposited onshore generally fines both landward and upward in the deposit, although this pattern is complicated by onshore topographic variations. These fining trends are predicted by simulations and are consistent with the majority of sediment settling from suspension. They are also consistent with field studies of the deposit of the 26 December 2004 tsunami in Banda Aceh, Indonesia, which flooded coastal northern Sumatra to a depth of over 20 m (Moore et al. 2006; Paris et al. 2007; Fig. 12.41a). This deposited a discontinuous sheet of sand up to 80 cm thick, and left mud up to 5 km inland. In most places the sand sheet is normally graded, while in some it contains a more complex internal stratigraphy which records the successive inflows and final outflow or backwash (Fig. 12.41a–c). The sand sheet contains abundant, pristine foraminifera, mostly of benthic subtidal origin, but notably has a planktonic offshore presence.

Near the shore the deposit is thin or absent, but inland it generally ranges in thickness from 5 to 20 cm but thickens where it infills low spots and stream channels (Fig. 12.41a). The dynamics of sediment suspension are essentially as described earlier in Sect. 12.6.2, and for further details and quantification see Apostas et al. (2011). In tsunamis this may result in suspended sediment-induced density stratification and hindered settling. The former occurs when large amounts of sediment are suspended in the water column, displacing less dense water, which creates a stable stratified water column. This inhibits near-bed turbulence and the mixing of sediment into the upper water column. Hindered settling occurs when suspended sediment increases to the point where the settling velocity of a particle is reduced to what it would have been in a clear fluid. Studies show suspended sediment concentrations of 20% reduce a particle's setting velocity by up to 40%, while higher concentrations can reduce the settling velocity by as much as 90%. In tsunamis the greatest reduction in settling velocity (>90%) occurs offshore of the shoreline in the backwash. Thus, the effects of hindered settling are to increase the amount of sediment suspended in the water column during inundation, as less sediment settles out offshore, and results in a thicker deposit nearer the shoreline.

As well as a discontinous sand sheet, the 26 December 2004 tsunami detached and transported large coral boulders in excess of 10 t over 500–700 m inland, which were also observed in nearshore areas on side scan sonar (Paris et al. 2010). Indeed, the fraction of boulders deposited onshore represents only a small (<10%) of the total number of boulders that apparently moved during the tsunami. This brings us on to one of the more contentious aspects in understanding the dynamics of tsunami (and storm) waves, that is, in the interpretation of coastal boulder accumulations (or *boulderites*), which may contain individual boulders weighing several hundred tonnes. These in turn have been used to calculate tsunami flow velocities (e.g. Nott 2003; Noormets et al. 2004; Paris et al. 2010; Nandasena et al. 2011). Distance travelled by a tsunami boulder is dependent on tsunami velocity, size of the boulder and material the boulder is composed of, which can be estimated from the following:

$$u_{\min} = \sqrt{\frac{2\mu m g}{C_d A_n \rho_w}} \qquad (12.27)$$

where $u_{\min}$ is the tsunami velocity at the boulders initial location, or minimum tsunami velocity (m s^{-1}), μ is the friction coefficient, which is 0.7 as suggested by Noormets

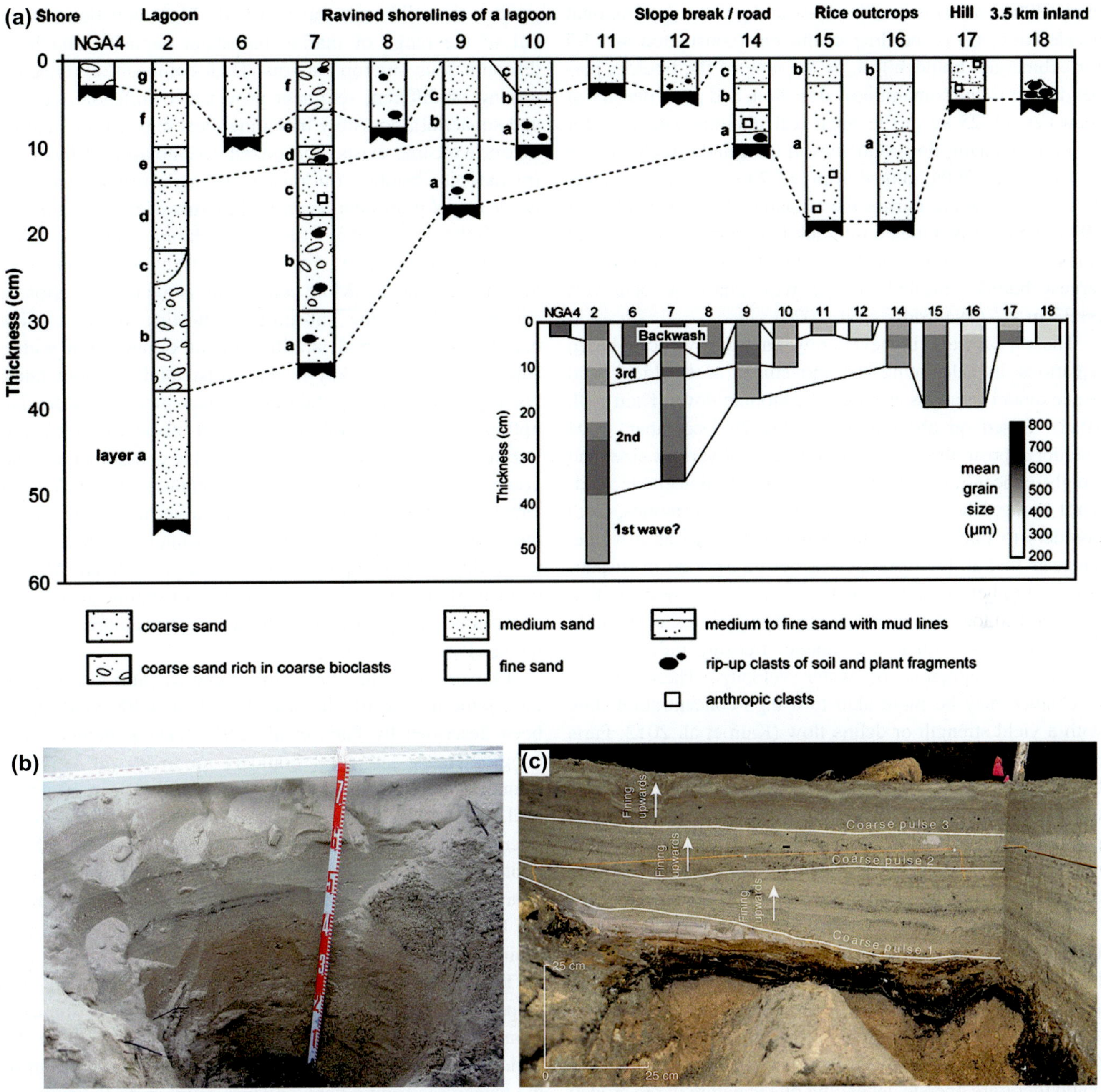

Fig. 12.41 Features of the 26 December 2004 tsunami deposit, Aceh Province, northern Sumatra, Indonesia. **a** Schematic lithostratigraphy of the tsunami deposit detected in Lhok Nga, west of Banda Aceh, showing effects of local topography. Inset shows interpretation indicating deposition by three consecutive runup waves and a final backwash (return flow or outflow). (From Paris et al. 2007). **b** Trench through tsunami deposits at the village of Lampuuk, to the north of Lhok Nga. A 73 cm thickness of sand was deposited above an erosional unconformity over soil (dark material in bottom center), in graded beds indicating multiple pulses of sand deposition. (Photo by G. Gelfenbaum; *Source* United States Geological Survey). **c** Trench through tsunami deposit found in coastal cave near the village of Lhong, 35 km south of Banda Aceh. This shows three coarse pulses and fining upwards sequences above a basal erosional unconformity above organic-rich debris. The tsunami inundated the cave and removed vegetation off the very steep limestone cliff to a height of $\sim$24 m above mean tidal level which was over 10 m above the top of the cave entrance. Note, in trenches below the 2004 tsunami deposit, a further 11 prehistoric tsunami sand layers (<7,400 yr BP) were also identified, suggesting repeated fault activity along the Sunda megathrust and inundation of the Aceh coast. (From Rubin et al. 2017)

et al. (2004), m is the boulder mass and g is the gravitational acceleration, C_d is the drag coefficient (considered as 1.95 for submerged boulders), A_n is the area of the boulder perpendicular to the tsunami flow direction, and ρ_w is density of seawater (1025 kg m^{-3}). Such calculations suggest, for example, tsunami flow velocities in the Banda Aceh area on 26 December, 2004, ranged from >3.7 to >7.5 m s^{-1} (Paris et al. 2010), while during the tsunami of 22 December of 2018, which was initiated by flank failure at Anak Krakatau (Figs. 12.12, 12.40b), the flow velocity estimated from the largest boulder located on the west coast of Java was >4.37 m s^{-1} (Syamsidik et al. 2020).

These estimates, however, rely on water-velocity-based equations in calculating the movement of boulders, and hence model tsunamis as purely Newtonian flows (Kain et al. 2012). Based on observations of the 26 December, 2004 boulder deposit, this is a mix of boulders of various sizes and weight concentrated in space, without any detectable landward fining. The coincidence of different size modes, from boulders to fine sands in the tsunami deposits as a whole, suggests that all the material was not transported in suspension, but rather through a combination of high concentration granular bed load and suspended transport modes, or possibly deposition was even by 'en masse' freezing rather than a progressive reduction of water velocity. Therefore the mechanics may be more akin to a high-concentration flow with a yield strength or debris flow (Kain et al. 2012; Paris et al. 2010). Because of the sand-beach context of many tsunami waves, their finer-grained fraction must be assessed to better understand the dynamics of tsunami waves, and the potential role of subaqueous sedimentary density flows in moving boulders (Kain et al. 2012), during in particular, their backflow. Tsunami backflow deposits interpreted as debris-flow conglomerates with hyper-concentrated flow or Bingham characteristics have been described along the Chilean continental margin (Le Roux and Vargas 2005; Le Roux et al. 2008). A spectrum of hyper-concentrated to granular and non cohesive to cohesive high-concentration flow types (as described above for other types of mass flow movement) could therefore operate depending on the nature of the eroded substrate, grain size, coastal topography and dynamics of the tsunami wave.

Whereas the 22 December, 2018 Anak Krakatau tsunami caused a significant number of human casualties and severe impacts on coastal settlements in the Sunda Strait (Syamsidik et al. 2020), this was a relatively minor geological event depositing only a thin (cm thick), discontinuous layer of fine sand and occasional megablocks of limestone. Where the deposits of prehistoric *volcanigenic tsunamis* have been identified onshore (e.g. Hawai'i, Cape Verde, Mauritius and Canary Islands) they essentially form conglomerates and gravels, sometimes containing outsize megaclasts (Paris et al. 2013, 2017). Such conglomerates are characteristically preserved at different and unusually high elevations, often out of the range of marine highstand deposits, and have discontinuous outcrop patterns. Their composition reflects a mixing of different sediment provenance (offshore, coastal and subaerial), posssibly including local exposed or shallow marine volcanic clasts, and mixed assemblages of bioclasts (planktonic, benthic, littoral and subaerial), but which are *never found in growth position*. In terms of mean grain size and thickness, landward fining and thinning may occur in tsunami conglomerates, but this is not uniquely sedimentologically diagnostic. More convincing is the identification of rip-up clasts of the substratum, attesting to an erosive emplacement. Internal stratification or organisation within subunits is neither a key-feature of tsunami conglomerates, but together with clast fabrics this can indicate successive upflows (landward imbrication) and backwash (seaward imbrication). In contrast, emerged marine conglomerates and gravels typically show great lateral continuity along elevated terraces. The elevation of these marine terraces falls within the range of sea-level changes (highstands), although the possibility of later uplift must also be considered. Marine deposits should also display a contiuity of sedimentary facies corresponding to specific shoreface to backshore environments (Paris et al. 2013).

Tsunami deposits related to a large tsunami event along the southern coast of Mauritius Island ca. 4,400 yr BP have been described by Paris et al. (2013). These include both conglomerates and coral boulders up to 90 m^3 (>100 t) found discontinuously at several locations between 10–15 m a.s.l. (Fig. 12.42). The conglomerate is 20–45 cm thick, intercalated in a lateritic soil at a depth of -50 to -80 cm depth (Fig. 12.42b). It is very poorly sorted and its composition reflects a mixing of two sediment sources: (1) marine bioclasts such as debris of corals (branching forms and brain corals), gastropods, and fragments of shells, and (2) fragments of volcanic rocks and minerals, from sand-size to pebbles. The mean thickness and grain size of the conglomerate fines landward, and the basal contact with the underlying paleosoil is erosive. The most probable origin of the tsunami is from a flank collapse of Piton de la Fournaise volcano, Réunion Island, some 200 km to the southwest (Paris et al. 2013).

On the northwestern coast of Tenerife, tsunami deposits (marine gravels with pumices) are found at altitudes up to 132 m a.s.l. The stratigraphy of the tsunami deposits and characterisation of pumice clasts found in these deposits distinguish at least two major tsunami events that occurred sequentially at about 190 ka (Paris et al 2017; Fig. 12.43a–c). Both tsunami units A and B are up to about 2 m thick, but the older A is absent at higher than 21 m a.s.l. Unit A is a very coarse-grained gravel that is clast-supported and poorly sorted (from fine pebbles to boulders); the dominant clast population comes from basaltic lavas that form nearby coastal cliffs

Fig. 12.42 Large tsunami deposit along the southern coast of Mauritius deposited ca. 4400 years ago. The most probable origin of the tsunami is a sector flank collapse of Piton de la Fournaise volcano, on Réunion Island, 200 km away. **a** Reef megaclasts having a mass >100 t. **b** Tsunami conglomerate deposit in a cutting at a depth of 70 cm and altitude of 11 m a.s.l. Note the irregular and sharp contacts with soils below and above the event unit. The marine conglomerate is composed of corals (both branching forms and brain corals) and shell fragments (bivalves and gastropods). A ^{14}C age of 4425 ± 35 yr BP was obtained on a coral branch. (From Paris et al. 2013)

and platforms, but there are also numerous marine bioclasts (bivalve and gastropod shells, foraminifera, calcareous algae, coral fragments), and rare pumices. Unit B is a coarse gravel, matrix-supported, and considerably enriched in pumices relative to unit A; clasts are composed of locally derived basalts mixed with phonolites, hydrothermally altered rocks, syenites, obsidian, and pumices, with few marine bioclasts. Field evidence and geochemical correlation of the pumices suggest the first tsunami was generated during the submarine stage of massive flank landslide failure (Icod collapse) of the northern flank of the island whereas, the second one, retrogressive sector collapse of the subaerial Las Cañadas edifice (Fig. 12.44a, b). The first collapse detached more than 100 km^3 from the northern submarine flank of Tenerife, which could have triggered significant decompression of the magma reservoir. Thus, the north-western part of Tenerife was conceivably mantled by pumice fall deposits and lithic-rich (El Abrigo) ignimbrite (Fig. 12.43a, c; Pittari et al. 2006). The second collapse (12–15 km^3) must have occurred at an advanced stage of the El Abrigo eruption. This generated a debris avalanche triggering a second, larger tsunami on the northern coasts of Tenerife which reached higher altitudes than the submarine collapse, and incorporated large amounts of pumices.

Tsunamis and their deposits should also be a common feature of caldera lake environments. Perhaps the best-described historic example comes from Karymskoye Lake, Akademii Nauk caldera, Kamchatka, Russia (Belousov and Belousova 2001; Falvard et al. 2018; Fig. 12.45). The lake is 4 km across, with steep flanks and a flat bottom at 60–65 m deep. During basaltic phreatomagmatic activity commencing 2 January 1996, a new tuff-ring formed on the northern margin of the lake. The ice covering the lake probably prevented the generation of tsunami waves during the first stages of the eruption, however, tsunamis generated by underwater explosions were observed during later activity. The limit of inundation was traced by drift wood and rounded pumice lapilli (coming from old rhyolitic pyroclastic deposits of the caldera). Tsunami run-up decreases with the distance from the eruptive source, with a maximum of 19 m on the northwestern shore (<1 km from the vent), and a minimum of 1.8 m on the southeastern shore (3 km from the vent). Boulders of volcanic tuff ≥ 1.3 m^3 were transported up to 60 m inland. Soil erosion by successive tsunamis led to the formation of talus up to 2–3 m high all around the lake, except on the southern rocky shore. Sediment reworked by tsunami waves left new beaches up to 50 m wide and sandy patches on terraces behind the talus. These tsunami deposits are often >1 m thick on the beach,

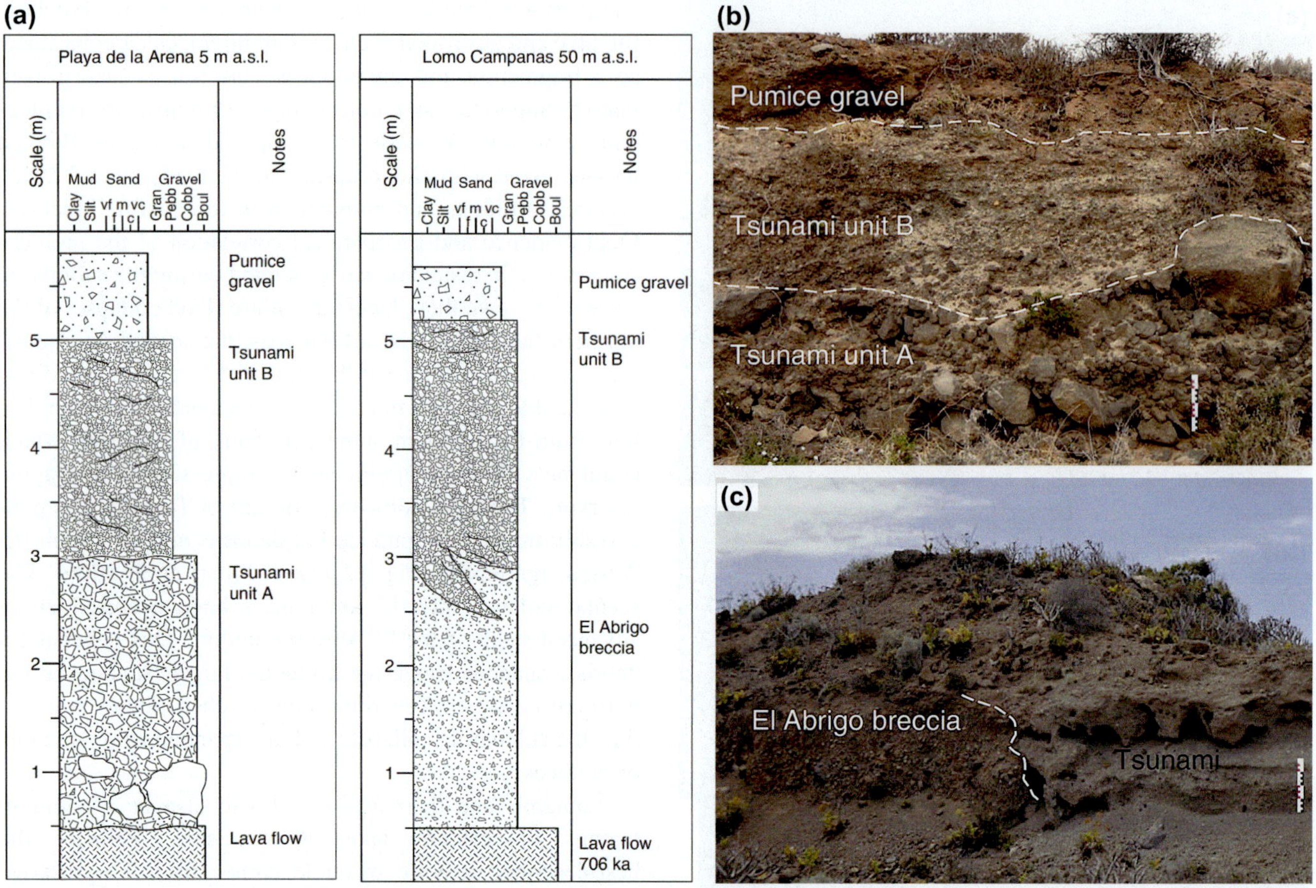

Fig. 12.43 Features of mega-tsunami deposits generated by volcano flank collapse of northern Tenerife at ca. 190 ka, Canary Islands, Spain. These deposits are found at altitudes up to 132 m a.s.l. (Modified from Paris et al. 2017). **a** Stratigraphic logs. Tsunami unit A is a very coarse-grained gravel that is clast-supported, and poorly sorted (from fine pebbles to boulders). Tsunami unit B is a coarse gravel, matrix-supported, and considerably enriched in pumices relative to unit A. Unit B is crudely structured into subunits with local variations of grain size and composition, and displays large scour-and-fill features at its base cut into a clast-supported, lithic-rich breccia. This breccia corresponds to the uppermost unit of the El Abrigo ignimbrite. Tsunami unit A is absent at elevations higher than 21 m a.s.l. **b** Succession of tsunami units A and B at Playa de la Arena dominated by locally derived basalt clasts. **c** Contact between the Abrigo breccia and tsunami unit B at Lomos de las Campanas (50 m a.s.l.). Both units have a similar locally derived clast composition but tsunami unit B is matrix-supported and cemented by carbonates

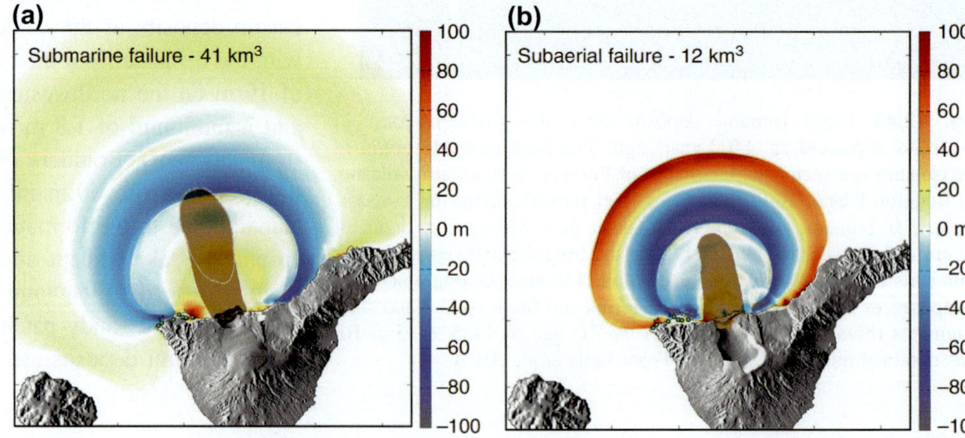

Fig. 12.44 Numerical simulations of failures on the northern flanks of Tenerife and related tsunami. Snapshots after 500 s of simulation. **a** Submarine failure followed by **b** subaerial sector collapse of the Las Cañadas edifice and El Abrigo eruption; see text for discussion. (From Paris et al. 2017)

Fig. 12.45 Section through tsunami deposits resulting from the 1996 eruption of Karymskoye Lake, Akademii Nauk caldera, Kamchatka, Russia. Trench is located on the western shore of the caldera lake 1 km from the vent, and shows a complex succession of laminated sands and silts characterised by abrupt variations of grain size. The base of the tsunami deposit is unclear, but at least 7 subunits can be distinguished. The sands contain <30% of juvenile clasts from the 1996 eruption, and are particularly rich in diatoms intercalated between coarser subunits. Inset shows backset cross-laminae in subunit B in X-CT image on a dehydrated block (left of thin section). Flow is towards the section. (From Falvard et al. 2018)

but <30 cm on the terraces and gentle slopes, and display a complex inner structure, interpreted as the result of multiple run-up and backwash events (Fig. 12.45). The structure of the tsunami deposits suggests that a major tsunami consisting of two main waves inundated the coastal terrace up to 100 m inland on the eastern shore of the lake. This largest tsunami might have occurred when underwater explosions were at a critical water depth of 40 m (corresponding to a two-third submerged explosion in the 60 m deep lake; Falvard et al. 2018).

In addition to the preceding cited examples of eruption generated tsunamis (Anak Krakatau 2018; Karymskoye Lake 1996), there are also examples of major explosive eruptions generating major tsunamis, some with significant loss of life, including the 1815 eruption of Tambora, Indonesia, the 1883 eruption of Krakatau, Indonesia, and more recently the 15th January 2022 eruption of Hunga Tonga- Hunga Ha'apai volcano, in Tonga. The 1815 Tambora eruption tsunamis resulted from the entry of pyroclastic flows (25 km^3 ?) entering the sea, which displaced the shoreline and sea seawards, creating tsunamis (Self et al. 1984). The eruption and tsunamis killed an estimated 90,000 people (Self et al. 1984). The 1883 explosive eruption of Krakatau also involved the entry of large volume pyroclastic flows into the sea, generating tsunamis 30 to 40 m high, which together with the effects of hot pyroclastic flows travelling over sea for 80 km and reaching the coastline of Sumatra killed ~30,000 people in coastal villages (Self and Rampino 1981; Simkin and Fiske 1983; Chap. 11). In addition to the pyroclastic flows generating tsunamis, the collapse of the summit of Krakatau to form the present day marine caldera (within which Anak Krakatau has subsequently been erupted and grown), is also likely to have contributed to generating tsunamis through displacing water volume during caldera collapse. Relatively little is yet known about the extremely intense explosive eruption of Hunga Tonga-Hunga Ha'apai volcano on 15th January 2022. The summit of the volcano was a pre-existing caldera that was completely submerged, except for two small islands on its rim, Hunga Tonga and Hunga Ha'apai (Cronin 2022). Following detailed submarine bathymetry surveying by NIWA, Mackay (2023) reported that the 2022 eruption produced an 850 m deep, 5 km diameter caldera involving the eruption of 9.5 km^3 of magma and rock. Much of this volume was transported by submarine mass flows of pyroclastic debris, with some flows travelling at least 80 km over the seafloor, breaking several seafloor communication cables in the process. Based on analysis of the chronology of events using several sources, Purkis et al. (2023) proposed that 3 significant tsunamis were generated during the eruption that all coincided with major explosions, the largest and most destructive tsunami coinciding with the highest intensity explosion at 4.56 UTC. That tsunami caused significant damage across the Tongan islands and involved a run-up onto cliffs as high as 45 m on some islands. Purkis et al. (2023) suggested that it seems that the tsunamis were triggered by the explosions, but recognised that other factors could also contribute. Factors that could contribute include water displacement by the shock wave of the explosions pushing the sea surface up and triggering lateral flow, caldera collapse and displacement of water by the large volume submarine mass flows of pyroclastic debris that were generated during the eruption. In addition, Matoza et al. (2022) proposed that some tsunamis could have been generated by atmospheric gravity pressure waves acting on the sea srface, which they called meteo-tsunamis. This mechanism is especially appealing for those tsunamis that formed in oceans outside the Pacific Ocean. No detailed studies documenting the characteristics of the tsunami deposits have yet been published.

12.6.7.2 Severe Coastal Storms and Their Deposits Onshore

In contrast to tsunamis, severe coastal storms are the products of coupling between the atmosphere and ocean, whereby cyclonic wind circulation and low barometric pressure combine to raise water levels and generate destructive waves. Compared to tsunamis, major coastal storms are considerably more frequent events, and hurricane Katrina in late August, 2005, is a prime example of the damage and devastation that can be caused. At least 1,500 people lost their lives, many of which occurred directly, or indirectly, as a result of the storm surge.

Extreme storms can be divided into tropical and extratropical cyclones or hurricanes (Morton et al. 2007). Tropical cyclones initially cause a gradual rise in water level, as forerunners to deep-water waves reaching the shore and the storm moving landward. There is both a very rapid rise and fall in peak storm surge as the storm crosses the coast and then a more gradual fall as flooded coastal areas and surrounding water bodies are drained. For tropical cyclones, the highest storm surges are generally restricted to a few tens of km adjacent to the eye, although elevated water levels can encompass more than 600 km of coast. Coastal flooding by extratropical cyclones (mid-latitude storms) is different in that the rise and fall of water levels are more gradual, the maximum storm surge can last for several days, and the surge can spread alongshore for more than 500 km. Within both of these flooding scenarios, the relative timing and stages of storm erosion and deposition are (Morton 2002; Morton et al. 2007):

- gradual inundation of the beach with attendant beach and dune erosion,
- overtopping of dunes or berm crest where dunes are absent,
- transfer of sand from the upper shoreface and beach to former subaerial sites, and
- deposition of perched fans or an overwash terrace.

Sand deposits from a single extreme storm may be as much as 1.5—2 m thick, and are commonly >30 cm thick (Fig. 12.39). They tend to be somewhat thicker than modern tsunami sands because overwash durations are longer, transport distances are short, flow depths are low, sediment concentrations are high, and repeated sediment transport under waves replenishes sand supply during overwash. Narrow, thick, lens-shaped deposits (terraces) that merge with the back beach are uncommon in tsunamis and may be diagnostic of storms. Both storm and tsunami deposits can be broad and thin, but this geometry is more common for tsunamis because inundation duration is short, sediment is dispersed across relatively deep flow zones, inundation distances are typically great, and volume of entrained sediment is limited by the brief onshore flow.

Sandy storm deposits typically exhibit numerous individual laminations segregated into multiple, discrete lamina sets that commonly exhibit either normal or inverse grading (Fig. 12.39). Both tsunami and storm deposits can appear to be internally massive, but most storm deposits exhibit at least some subhorizontal planar stratification. However, storm deposits commonly exhibit sedimentary structures

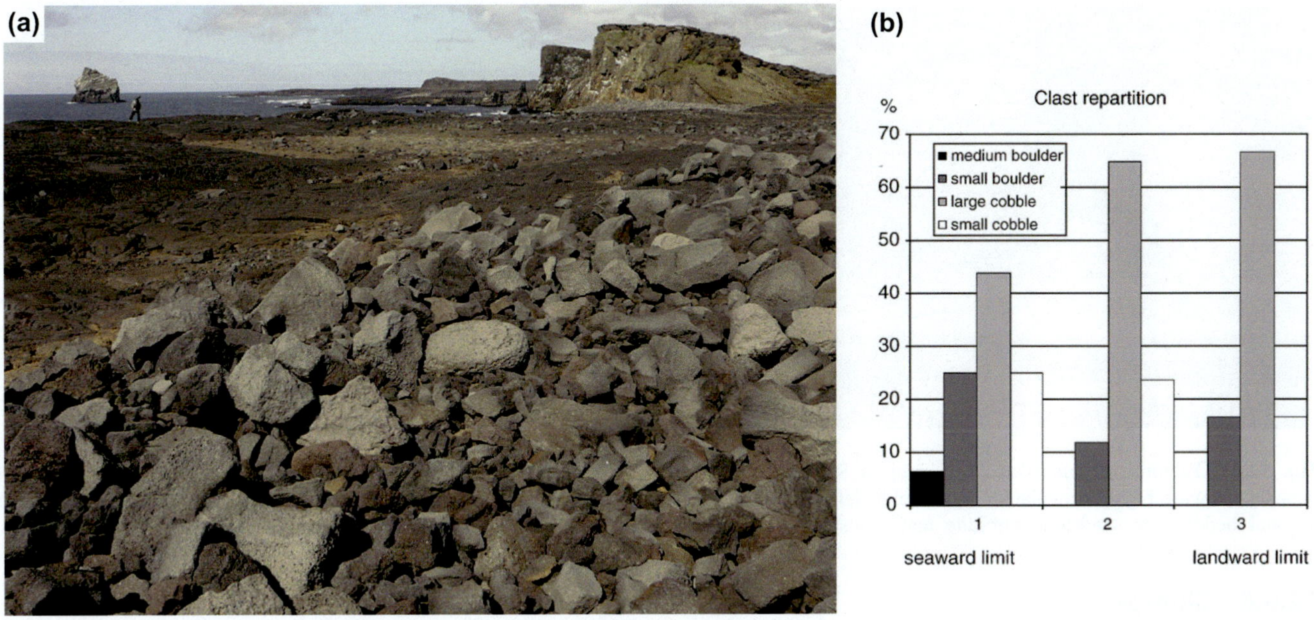

Fig. 12.46 Storm deposits. **a** Cliff-top ridge of storm deposits, located at 45–55 m from the shore at 11–12 m a.s.l., south of Valahnúkur, Reykjanes Peninsula, southern Iceland. **b** The ridge is mainly composed of basaltic pebble to medium boulder clasts, with the largest restricted to the foreshore of the accumulation, and sorting increases landward. The debris is largely fed by the upper crust of a lava flow, which typically releases large cobble-size fragments (>40% of the clasts). (From Etienne and Paris 2010)

other than planar stratification, such as foreset cross-laminae, back set (antidune) cross-laminae, and climbing ripples; although this is apparently not universal (cf. Figure 12.45). The variability of stratification types in storm deposits is probably a result of relatively shallow flow depths, variable flow velocities, and predominant bed-load transport. As discussed in Sect. 12.6.1, when antidunes form this corresponds to supercritical turbulent flow, where true tractional sediment transport has probably been surpassed by this stage by upper flow regime (Fr > 1), hyper-concentrated flow conditions (Sect. 12.6.4). Basal contacts for both storm and tsunami deposits are usually sharp and/or erosive but in contrast to tsunamis, mud rip-up clasts are rare or absent in storm deposits. This is attributed to the turbulence of the water and prolonged vigorous agitation that disaggregates and disperses incorporated mud. Although shell valves or fragments may be present in both storm and tsunami deposits, discrete lamina of whole or comminuted shells are only common in storm deposits. Mud is rare in most sandy storm deposits, while some sandy tsunami deposits contain mud laminations or plant debris that may be distributed either within or at the top of the event bed.

Cobble- to boulder-size slabs of rocks and blocks of coral may also be exhumed and deposited along coastlines during extreme storm surge events, however, the origin of many coastal boulder accumulations around the world is still under debate, especially along coastlines where both tsunamis and powerful storms have occurred in the recent past (e.g. Barbano et al. 2010; Etienne and Paris 2010; Goff et al. 2010; Dewey and Ryan 2017). Etienne and Paris (2010) describe a volcanigenic example on the south coast of the Reykjanes Peninsula in Iceland, along the emerged part of the mid-Atlantic oceanic ridge (Chap. 16). Iceland is located in the middle of the North Atlantic Ocean, and is a high-energy marine environment with frequent stormwaves, and storm days occurring at >50 per year in exposed coastal areas. The southwest coast of Iceland faces the powerful storms of the North Atlantic Ocean, with significant wave heights over 15 m. However no tsunamis have been recorded in the past 250 years, and the probability of tsunamis having hit this area of the coast is regarded as very low. Storms therefore appear as a predominant factor in the geomorphic evolution of Reykjanes coast. The boulder accumulations are composed mainly of basaltic lava and consist of cliff-top boulders, clusters and ridges, beaches, and boulder fields (Fig. 12.46a, b). Large boulders, up to 70 t in weight, have been transported and deposited up to 65 m inland and 6 m a.s.l. The maximum limit of boulder deposition and driftwood was found respectively 210 m and 550 m inland. A fundamental distinction between storms and tsunamis could be their capability of forming ridges. Indeed, the organisation of coarse clasts into ridges requires repeated reworking by waves rather than the single impact of a tsunami front wave. As far as is known, observed tsunamis did not leave boulder ridges.

Fig. 12.47 Features of slumps and soil creep. **a** Subaqueous slump deposits showing highly irregular soft-state folds defined by contorted original bedding of sandstone turbidite and mudstone layers. Late Devonian Bunga Beds, Boyd Volcanic Complex, southern New South Wales, Australia. **b** Leaning power poles indicate slow downslope creep of the soil profile, near Oamaru, South Island, New Zealand

12.6.8 Slumps

Slumps are coherent masses of aggregate that creep, slide or flow down slope leaving an erosion scar behind, to which the tail of the slump is still attached. They are therefore local in significance and represent incipient slides, with which they form a continuum. Internally soft-sediment deformation features may be apparent, but original layering is still preserved (Fig. 12.47a). The role of water in this case is to lubricate movement and lower the co-efficient of internal friction. The criterion for initiation of movement is the same as for debris flows, which many slumps evolve into (Eq. 12.26). Slumping can occur subaerially or subaqueously; in the former case slumping is not unusual after heavy rainfall.

12.6.9 Soil Creep

Imperceptible, downslope creep of soil or surface aggregate, lubricated by interstitial water is common on all landscapes with steep slopes. Soil creep zones represent incipient slides and slumps. Soil creep is increasingly becoming a problem where permafrost is melting and as global warming occurs, especially in the northern circumpolar permafrost region (Sect. 12.5.4). The effects are often visible in tilted electrical or telephone poles (Fig. 12.47b) or trees, and in damage to buildings as their foundations move.

12.7 Sediment Transport in Which Air is an Essential Interstitial Medium

The role of air in this case is either to drive sediment movement through the effects of wind, or perhaps to lubricate movement of rock avalanches. In particulate form, wind drives sediment movement by traction and suspension, whereas in mass movement it may contribute to the mobility of air-lubricated rock or debris avalanches (Sect. 12.4.5).

12.7.1 Traction and Suspension

These two modes will be discussed together since the basic principles have already been touched on in discussing the same modes subaqueously (Sect. 12.6.1). The main differences are that because particles are not immersed in water their transportable weight is greater, and because the density and viscosity of air are less, moving air or wind cannot exert as high as shear stress at equivalent velocities. As a result, wind cannot move coarse granular material (except during hurricanes, and via desert winds which can travel at speeds over 100 km hr^{-1}), the optimum sizes being fine-medium sand, although once eroded finer grain sizes are easily suspended (Allen 1982; Leeder 2009). As with all of the other modes discussed, this is based on dense grain size equivalents, and not on the hydraulic properties of low-density pumiceous material, for which no information is known.

The processes of bed load and saltation transport again constitute tractional transport, and are responsible for producing structures such as ripples and large-scale wind dune (or erg) systems, the characteristics of which are adequately described in the groundbreaking work of Bagnold (1954a), and subsequently by Allen (1982), Brookfield (1984) and Leeder (2009). According to Leeder (2009), more grain saltation occurs during wind transport of sand and silt size material than in aqueous transport, and for ripples, less lee-side avalanching of grain flow layers occurs. Leeder (2009) uses the term "ballistic ripples" to denote the role of saltation, fallout from suspension, and grain impacts in their formation. On larger dune structures, grain flow avalanching

Fig. 12.48 Windblown volcanic sediments. **a**, **b** "Shifting sands" near the Olduvai Gorge hominid site, Tanzania, are unusual barchan dunes composed of black carbonatitic ash derived from Holocene eruptions of Ol Doinyo Lengai volcano (some 100 km to the east). Most of the ash has been dispersed over the eastern Serengeti Plains, but there are still several isolated dunes of approximately 10 m high and 60 m wide that continue to migrate westward. Note large, crescent-shaped slip-face and transverse "ballistic ripples" on windward slope in (**b**). (Photos courtesy of Roger Scoon; also see Scoon 2018). **c** Coastal titanomagnetite-rich beach dunes, Bethel's Beach, north of Auckland, New Zealand. Sediment is eroded from Tertiary intermediate volcanics exposed in coastal cliffs. (From Cas and Wright 1987). **d** The Taupo ignimbrite (coarse-grained, grey, left) ponded against a paleovalley margin made of eroded Oruanui ignimbrite (grey, right), which is draped by a thick 1 m deposit of loess (pale brown, middle). Taupo Volcanic Zone, New Zealand. (From Cas and Wright 1987)

along lee-side slip-faces is important in migration of dune crests and bedforms. Aeolian dunes have more complex internal structures and can be much higher than aqueous dunes. The complexity is often reflected by the presence of small "ballistic ripple" sets on the surfaces of large aeolian dunes and in the varying ripple axis orientations compared with dune crest orientations.

In volcanic terrains, wind erosion and transport can be major influences, particularly when newly erupted pyroclastic ash mantles the surface in an unconsolidated state. Following the eruption in 1952, Richards (1965) mentions the extreme cloudiness of the atmosphere around Bárcena volcano due to wind-borne ash, and records the development of wind-generated barchan dunes. In most cases, these may only be ephemeral features, but in some circumstances, they could be preserved (Fig. 12.48a, b). Large-scale angle of repose cross-stratification is common, and cross-stratification sets may be highly irregular in geometry with well-defined bounding surfaces representing breaks in movement. Small-scale cross-lamination may slope at < angle of repose because of the high wind shear stress at the bed surface. Coastal erosion of volcanic successions could produce significant coastal beach dune successions of volcanic sands (e.g. Fig. 12.48c).

One major consequence of significant wind erosion in volcanic terrains is to loft large volumes of fine sediment and ash into the atmosphere. This material may subsequently fallout, and mantle the landscape with a blanket of ash or fine sediment, superficially being similar to air fallout ash, except that it is almost always polymictic in componentry. Figure 12.48d depicts prehistoric loess mantling the New Zealand volcanic landscape. The loess was apparently mobilised during the influence of the recent ice age and associated strong winds. The loess shows mantle bedding, similar to pyroclastic fallout deposits, is relatively thick, and to the unwary could be misinterpreted as a blanket of pyroclastic air fallout ash heralding a major eruption The deposit, however, contains mostly bedrock-derived detritus.

12.7.2 Air-Lubricated Rock/Debris Avalanches

In the discussion of rock avalanches in Sect. 12.4.5, mention was made that rock avalanches in some circumstances could glide on a trapped, compressed cushion of air, trapped at the hydraulic jump or break in slope where the rock mass changed from a fall trajectory to a lateral flow trajectory. In such a situation the air would also fluidise the flowing rock mass. Although some support this mechanism (e.g. Shreve 1968), others discount it (e.g. Hsü 1975, 1978; Howard 1978), but the concept should not be totally rejected (Allen 1982).

References

Allen JRL (1970) The sequence of sedimentary structures in turbidites, with special reference to dunes. Scot J Geol 6:146–161

Allen JRL (1982) Sedimentary structures. Their character and physical basis. Develop in sedimentol 30A, B. Elsevier, Amsterdam

Allen SR, Carey R, McPhie (2020) New insights into the eruption of submarine pumice cones: Yali revisited. Geology, in prep

Alexander J, Barclay J, Sušnik J, Loughlin SC, Herd RA, Darnell A, Crosweller S (2010) Sediment-charged flash floods on Montserrat: the influence of synchronous tephra fall and varying extent of vegetation damage. J Volcanol Geotherm Res 194(4):127–138

Alloway B, McComb P, Neall V, Vucetich C, Gibb J, Sherburn S, Stirling M (2005) Stratigraphy, age, and correlation of voluminous debris-avalanche events from an ancestral Egmont Volcano: implications for coastal plain construction and regional hazard assessmentJ Roy Soc New ZeaL 35(1–2):229–267

Apotsos A, Gelfenbaum G, Jaffe B (2011) Process-based modeling of tsunami inundation and sediment transport. J Geophys Res: Earth Surf 116(F1)

Baas JH, Best JL, Peakall J, Wang M (2009) A phase diagram for turbulent, transitional, and laminar clay suspension flows. J Sed Res 79(4):162–183

Bagnold RA (1954a) The physics of blown sand and desert dunes. Methuen and Co., Ltd., London, p 265p

Bagnold RA (1954b) Experiments on a gravity free dispersion of large solid spheres in a newtonian fluid under shear. Proc Roy Soc Lond Ser A 225:49–63

Bagnold RA (1955) Some flume experiments on large grains but little denser than the transporting fluid, and their implications. Proc Instit Civil Engin 4(2):174–205

Bagnold RA (1966) The shearing and dilation of dry sand and the 'singing' mechanism. Proc Roy Soc Lond Ser A. Mathem Phys Sci 295(1442):219–232

Barbano MS, Pirrotta C, Gerardi F (2010) Large boulders along the south-eastern Ionian coast of Sicily: Storm or tsunami deposits? Mar Geol 275(1–4):140–154

Barclay J, Alexander J, Sušnik J (2007) Rainfall-induced lahars in the Belham valley, Montserrat, West Indies. J Geol Soc Lond 164 (4):815–827

Barsch D (1992) Permafrost creep and rock glaciers. Permafrost Periglac Process 3(3):175–188

Begét JE, Kienle J (1992) Cyclic formation of debris avalanches at Mount St Augustine volcano. Nature 356(6371):701

Belousov AB (1995) The Shiveluch volcanic eruption of 12 November 1964—explosive eruption provoked by failure of the edifice. J Volcanol Geotherm Res 66(1–4):357–365

Belousov AB, Belousova MG (1998) Bezymyanni eruption on March 30, 1956 (Kamchatka): sequence of events and debris avalanche deposits. J Volcanol Seismol 20:29–47

Belousov A, Belousova M (2001) Eruptive process, effects and deposits of the 1996 and ancient basaltic phreatomagmatic eruptions in Karymskoye lake, Kamchatka, Russia. Spec Publ Int Assoc Sedimentol 30:35–60

Björnsson H (2010) JökulhlaupsinIceland: sources, release and drainage. In: Iceland in the Central Northern Atlantic: hot spot, sea currents and climatic change. May 2010, Plouzané, France. HAL-0480676

Bondevik S, Mangerud J, Dawson S, Dawson A, Lohne Ø (2003) Record-breaking height for 8000-year-old tsunami in the North Atlantic. Eos 84(31):289–293

Borgia A, Delaney PT, Denlinger RP (2000) Spreading volcanoes. Ann Rev Earth Planet Sci 28(1):539–570

Boudon G, Le Friant A, Komorowski JC, Deplus C, Semet MP (2007) Volcano flank instability in the Lesser Antilles Arc: diversity of scale, processes, and temporal recurrence. J Geophys Res: Solid Earth 112(B8)

Bouma AH (1962) Sedimentology of some Flysch deposits. A graphic approach to facies interpretation. Elsevier, Amsterdam, p 168

Bourgois J (2009) Geologic effects and records of tsunamis. In: Robinson AR, Bernard EN (eds) The sea, volume 15: Tsunamis. Harvard University Press, pp 53–91

Brookfield ME (1984) Eolian facies. In Walker RG (Ed) Facies models. Geosci Canada, 91–103

Brugman MM, Meier MF (1981) Response of glaciers to the eruptions of Mount St. Helens. In Lipman & Mullineaux (eds) The 1980 eruptions of Mount St. Helens. US Geol Surv Prof Pap 1250: 743–56

Bryan SE, Cook AG, Evans JP, Hebden K, Hurrey L, Colls P et al (2012) Rapid, long-distance dispersal by pumice rafting. PLoS One 7(7):e40583. https://doi.org/10.1371/journal.pone.0040583

Bryn P, Berg K, Forsberg CF, Solheim A, Kvalstad TJ (2005) Explaining the storegga slide. Mar Petrol Geol 22(1–2):11–19

Bull SW, Cas RAF (1989) Volcanic influences in a storm-and tide-dominated shallow marine depositional system: the late permian broughton formation, southern Sydney Basin, Kiama NSW. Aust J Earth Sci 36(4):569–584

Carey S, Sigurdsson H (1984) A model of volcanogenic sedimentation in marginal basins. Spec Publ Geol Soc Lond 16(1):37–58

Carey SN, Schneider JL (2011) Volcaniclastic processes and deposits in the deep-sea. Develop Sedimentol 63:457–515

Carter RM (1975) A discussion and classification of subaqueous mass-transport with particular application to grain-flow, slurry-flow, and fluxoturbidites. Earth Sci Rev 11(2):145–177

Cartigny MJ, Postma G, Van den Berg JH, Mastbergen DR (2011) A comparative study of sediment waves and cyclic steps based on geometries, internal structures and numerical modeling. Marine Geol 280(1–4):40–56

Cas RAF (1979) Mass-flow arenites from a Paleozoic interarc basin, New South Wales, Australia; mode and environment of emplacement. J Sed Res 49(1):29–44

Cas RAF (1983) Submarine 'crystal-tuffs': their origin using a lower devonian example from southeastern Australia. Geol Mag 120:471–486

Cas RAF, Landis CA (1987) A debris-flow deposit with multiple plug-flow channels and associated side accretion deposits. Sedimentology 34(5):901–910

Cas RAF, Powell CM, Fergusson CL, Jones JG, Roots WD, Fergusson J (1981) The lower devonian Kowmung volcaniclastics: A deep-water succession of mass-flow origin, northeastern Lachlan Fold Belt, NSW. J Geol Soc Aust 28(3–4):271–288

Cas RAF, Wright JV (1987) Volcanic successions: modern and ancient, London. Allen and Unwin, London, 526p

Cashman KV, Fiske RS (1991) Fallout of pyroclastic debris from submarine volcanic eruptions. Science 253(5017):275–280. https://doi.org/10.1126/science.253.5017.275

Cecchi E, van Wyk de Vries B, Lavest JM (2004) Flank spreading and collapse of weak-cored volcanoes. Bull Volcanol 67(1):72–91

Chouet B, Dawson P (2013) Very long period conduit oscillations induced by rock-falls at Kilauea Volcano, Hawaii. J Geophys Res: Solid Earth 118(10):5352–5371

Christiansen RL, Peterson DW (1981) Chronology of the 1980 eruptive activity. In: Lipman and Mullineaux (eds) The 1980 eruptions of Mount St. Helens. US Geol Surv Prof Pap 1250:17–30

Clough BJ, Wright JV, Walker GPL (1981) An unusual bed of giant pumice in Mexico. Nature 289(5793):49–50

Coulomb CA (1776) Sur une application des regles, de maximis et minimis a quelques problemes de statique, relatif a l'architecture. L'Imperial Royale, Paris. Memoires Mathematique Physique 7:343–384

Crandell DR, Fahnestock RK (1965) Rock-falls and avalanches from Little Tahoma Peak on Mount Rainier, Washington. US Geol Surv Bull 1221-A, 1–30

Cronin S (2022) Why the volcanic eruption in Tonga was so violent, and what to expect next. The Conversation January 16, 2022.https://mail.google.com/mail/u/0/#search/Shane+Cronin/FMfcgzGmtXCRxPtdgDXpmkstLfQGmjNv

Cronin SJ, Lecointre JA, Palmer AS, Neall VE (2000) Transformation, internal stratification, and depositional processes within a channelised, multi-peaked lahar flow. NZ J Geol Geophys 43(1): 117–128

Dade BW, Huppert HE (1998) Long-runout rock-falls. Geology 26 (9):803–806

De Bélizal E, Lavigne F, Hadmoko DS, Degeai JP, Dipayana GA, Mutaqin BW, Marfai MA, Coquet M, Le Mauff B, Robin AK, Vidal C, Cholik N, Aisyah N (2013) Rain-triggered lahars following the 2010 eruption of Merapi volcano, Indonesia: a major risk. J Volcanol Geotherm Res 261:330–347

Denlinger RP, Morgan JK (2014) Instability of Hawaiian volcanoes, chapter 4 In: Poland MP, Takahashi TJ, Landowski CM (eds) Characteristics of Hawaiian volcanoes. US Geol Surv Prof Pap 1801:149–176

Deplus C, Le Friant A, Boudon G, Komorowski JC, Villemant B, Harford C, Ségoufin J, Cheminée JL (2001) Submarine evidence for large-scale debris avalanches in the Lesser Antilles Arc. Earth Planet Sci Letts 192(2):145–157

Devoli G, Cepeda J, Kerle N (2009) The 1998 Casita volcano flank failure revisited—new insights into geological setting and failure mechanisms. Eng Geol 105(1–2):65–83

Dewey JF, Ryan PD (2017) Storm, rogue wave, or tsunami origin for megaclast deposits in western Ireland and North Island, New Zealand? Proc Natl Acad Sci 114(50):E10639–E10647

Dobinski W (2011) Permafrost. Earth Sci Rev 108(3–4):158–169

Dondin F, Lebrun JF, Kelfoun K, Fournier N, Randrianasolo A (2011) Sector collapse at Kick'em Jenny submarine volcano (Lesser Antilles): numerical simulation and landslide behaviour. Bull Volcanol 74(2):595–607

Duffield WA, Bacon CR, Roquemore GR (1979) Origin of reverse-graded bedding in air-fall pumice, Coso Range, California. J Volcanol Geotherm Res 5(1–2):35–48

Etienne S, Paris R (2010) Boulder accumulations related to storms on the south coast of the Reykjanes Peninsula (Iceland). Geomorphology 114(1–2):55–70

Eyles N, Eyles CH, Miall AD (1983) Lithofacies types and vertical profile models: an alternative approach to the description and environmental interpretation of glacial diamict and diamicitite sequences. Sedimentology 30:393–410

Eyles N, Miall AD (1984) Glacial facies. In Walker RG (ed) Facies models. Geosci Canada 15–34

Fahnestock RK (1963) Morphology and hydrology of a glacier stream —White River, Mount Rainier, Washington. US Geol Surv Prof Pap 422-A, p 70

Fahnestock RK (1965) Rockfalls and Avalanches from Little Tahoma Peak on Mount Rainier, Washington. US Geol Surv Bull 1221-A: A1–30

Fahnestock RK (1978) Little Tahoma Peak rock falls and avalanches, Mount Rainier, Washington, USA. Develop Geotech Eng 14:181–196

Falvard S, Paris R, Belousova M, Belousov A, Giachetti T, Cuven S (2018) Scenario of the 1996 volcanic tsunamis in Karymskoye Lake, Kamchatka, inferred from X-ray tomography of heavy minerals in tsunami deposits. Mar Geol 396:160–170

Faugères J-C, Mulder T (2011) Contour currents and contourite drifts. Develop Sedimentol 63:149–214

Fauria KE, Manga M, Wei Z (2017) Trapped bubbles keep pumice afloat and gas diffusion makes pumice sink. Earth Planet Sci Letts 460:50–59

Fiske RS (1963) Subaqueous pyroclastic flows in the Ohanapecosh Formation, Washington. Geol Soc Am Bull 74:391–406

Fiske RS, Matsuda T (1964) Submarine equivalents of ash flows in the Tokiwa Formation, Japan. Am J Sci 262:76–106

Francis EH (1983) Magma and sediment—II. Problems of interpreting palaeovolcanics buried in the stratigraphic column. J Geol Soc Lond 140(2):165–183

Francis PW, Gardeweg M, Ramirez CF, Rothery DA (1985) Catastrophic debris avalanche deposit of Socompa volcano, northern Chile. Geology 13(9):600–603

Francis PW, Wells GL (1988) Landsat thematic mapper observations of debris avalanche deposits in the Central Andes. Bull Volcanol 50 (4):258–278

Funck T, Lykke-Andersen H (1998) Seismic structure of the volcanic apron north of Gran Canaria. In: Weaver et al. (eds) Proc Ocean Drill Prog Sci Results 157:11–28

Genevois R, Ghirotti M (2005) The 1963 Vaiont landslide. Giornale di Geologia Applicata 1(1):41–52. https://doi.org/10.1474/GGA.2005-01.0-05.0005

Genareau K, Cronin SJ, Stewart C, Bhattacharyya S, Donahoe R (2016) Post-eruptive impacts of pyroclastic deposits from basaltic andesite stratovolcanoes on surface water composition. J Geophys Res: Biogeosci 121(5):1275–1287

Germain D, Ouellet MA (2013) Subaerial sediment-water flows on hillslopes: essential research questions and classification challenges. Prog Phys Geog 37(6):813–833

Giachetti T, Paris R, Kelfoun K, Pérez-Torrado FJ (2011) Numerical modelling of the tsunami triggered by the Güimar debris avalanche, Tenerife (Canary Islands): comparison with field-based data. Mar Geol 284(1–4):189–202

Gilbert GK (1914) The transportation of debris by running water. US Geol Surv Prof Paper 86:263p

Glicken H (1991) Sedimentary architecture of large volcanic-debris avalanches. Spec Publ Soc Econ Paleontol Mineral 45:99–106

Glicken H (1998) Rockslide-debris avalanche of May 18, 1980, Mount St. Helens Volcano, Washington. Bull Geol Surv Japan 49 (2/3):55–106

Goff J, Weiss R, Courtney C, Dominey-Howes D (2010) Testing the hypothesis for tsunami boulder deposition from suspension. Mar Geol 277(1–4):73–77

Gorshkov GS (1959) Gigantic eruption of the volcano Bezymianny. Bull Volcanol 20:77–109

Gorshkov GS, Dubik YM (1970) Gigantic directed blast at Shiveluch volcano (Kamchatka). Bull Volcanol 34(1):261–288

Gran KB, Montgomery DR (2005) Spatial and temporal patterns in fluvial recovery following volcanic eruptions: channel response to basin-wide sediment loading at Mount Pinatubo, Philippines. Geol Soc Amer Bull 117(1–2):195–211

Gran KB, Montgomery DR, Halbur JC (2011) Long-term elevated post-eruption sedimentation at Mount Pinatubo, Philippines. Geology 39(4):367–370

Grilli ST, Shambach L, Zhang C et al. (2019b) Modeling of the slide and tsunami generation from the 12/22/18 lateral collapse of Anak Krakatau volcano (Sunda Straits, Indonesia): comparison with recent field surveys of slide deposits and tsunami impact. Am Geophys Union Fall Meeting, San Francisco, Abstract NH31A-05

Grilli ST, Tappin DR, Carey S et al (2019a) Modelling of the tsunami from the December 22, 2018 lateral collapse of Anak Krakatau volcano in the Sunda Straits, Indonesia. Sci Repts 9:11946. https://doi.org/10.1038/s41598-019-48327-6

Gudmundsson MT (2015) Chapter 56 Hazards from lahars and jökulhlaups. In: Houghton B, McNutt SR, Rymer H, Stix J (eds) Sigurdsson H. The encyclopedia of volcanoes, Academic Press Elsevier, pp 971–992

Hacker DB, Rowley PD, Biek RF (2018). Catastrophic collapse features in volcanic terrains: styles and links to subvolcanic magma systems. In: Breitkreuz C, Riocchi S (eds) Physical geology of shallow magmatic systems: dykes, sills and laccoliths. Advances in volcanology, IAVCEI-Springer, 215–248

Haeberli W, Hallet B, Arenson L, Elconin R, Humlum O, Kääb A, Kaufmann V, Ladanyi B, Matsuoka N, Springman S, Mühll DV (2006) Permafrost creep and rock glacier dynamics. Permafrost Periglac Proc 17(3):189–214

Hampton M (1975) Competence of fine-grained debris flows. J Sed Res 45(4):834–444

Hampton MA (1972) The role of subaqueous debris flow in generating turbidity currents. J Sed Res 42(4):775–793

Hand JL, Cas RA, Ong L, Brown SJ, Krapez B, Barley ME (2002) Syn-and post-eruptive volcaniclastic sedimentation in Late Archaean subaqueous depositional systems of the Black Flag Group, Eight Mile Dam, Kalgoorlie, Western Australia. In: Altermann W, Corcoran PL (eds) Precambrian sedimentary environments: a modern approach to ancient depositional systems. Spec Publ Int Assoc Sedimentol 33:235–258

Harbitz CB, Løvholt F, Bungum H (2014) Submarine landslide tsunamis: how extreme and how likely? Nat Hazards 72(3):1341–1374

Harms JC, Fahnestock RK (1965) Stratification, bed forms, and flow phenomena (with an example from the Rio Grande). In Middleton GV (ed.) Primary sedimentary structures and their hydrodynamic interpretation. Soc Econ Paleontol Mineral Spec Publ 12:84–115

Harms JC, Southard JB, Walker RG (1982) Structures and sequences in clastic rocks. Lecture notes. Soc Econ Paleontol Mineral Short Course 9

Henrich R, Hüneke H (2011) Hemipelagic advection and periplatform sedimentation. Dev Sedimentol 63:353–396. Elsevier

Hernández-Molina FJ, Campbell S, Badalini G, Thompson P, Walker R, Soto M, Conti B, Preu B, Thieblemont A, Hyslop L, Miramontes E (2018) Large bedforms on contourite terraces: sedimentary and conceptual implications. Geology 46(1):27–30

Hjulström F (1935) Studies of the morphological activity of rivers as illustrated by the river Fyris. Bull Geol Inst Upsala 25:221–527

Hogg SE (1982) Sheetfloods, sheetwash, sheetflow, or...? Earth Sci Rev 18(1):59–76

Howard KA (1973) Avalanche mode of motion: Implications from lunar examples. Science 180(4090):1052–1055

Hsü KJ (1975) Catastrophic debris streams (sturzstroms) generated by rockfalls. Geol Soc Am Bull 86(1):129–140

Hsü KJ (1978) Albert Heim: observations on landslides and relevance to modern interpretations. Natural phenomena. Chapter 1. In: Voight B (ed) 1978. Rock slides and avalanches, I. Natural phenomena. Elsevier, Amsterdam, pp 70–93

Hüneke H, Henrich R (2011) Pelagic sedimentation in modern and ancient oceans. Develop Sedimentol 63:215–351. Elsevier

Hunt JE, Talling PJ, Clare MA, Jarvis I, Wynn RB (2014) Long-term (17 Ma) turbidite record of the timing and frequency of large flank collapses of the Canary Islands. Geochem Geophys Geosyst 15 (8):3322–3345

Hunter RE (1985) A kinematic model for the structure of lee side deposits. Sedimentology 32:409–422

Hürlimann M, Garcia-Piera JO, Ledesma A (2000) Causes and mobility of large volcanic landslides: application to Tenerife, Canary Islands. J Volcanol Geotherm Res 103(1–4):121–134

Iverson RM (1997) The physics of debris flows. Rev Geophys 35 (3):245–296

Iverson RM (2003) The debris-flow rheology myth. In: Rickenmann D, Chen CL (eds) Debris-flow hazards mitigation: mechanics, prediction, and assessment Millpress, Rotterdam, pp 303–314

Iverson RM, Vallance JW (2001) New views of granular mass flows. Geology 29(2):115–118

Janda RJ, Scott KM, Nolan KM, Martinson HA (1981) Lahar movement, effects, and deposits. In: Lipman PW, Mullineaux DR (eds) The 1980 Eruptions of Mount St. Helens, Washington. US Geol Surv Prof Pap 1250:461–478

Jiskoot H (2011) Dynamics of glaciers. In: Singh VP, Singh P, Haritashya UK (eds) Encyclopedia of snow, ice and glaciers. Encyclopedia of earth sciences series. Springer, Dordrecht, 245–256

Johnson AM (1970) Physical processes in geology. Freeman, Cooper, San Francisco

Johnson JP, Delbecq K, Kim W, Mohrig D (2016) Experimental tsunami deposits: linking hydrodynamics to sediment entrainment, advection lengths and downstream fining. Geomorphology 253:478–490

Jones BL, Chinn SS, Brice JC (1984) Olokele rock avalanche, island of Kauai. Hawaii. Geology 12(4):209–211

Jones JG (1967) Clastic rocks of Espiritu Santo Island, New Hebrides. Geol Soc Am Bull 78:1281–1288

Jutzeler M, Marsh R, Carey RJ, White JDL, Talling PJ, Karlstrom L (2014a) On the fate of pumice rafts formed during the 2012 Havre submarine eruption. Nat Comm 5:3660

Jutzeler M, McPhie J, Allen SR (2014b) Submarine eruption-fed and resedimented pumice-rich facies: the Dogashima Formation (Izu Peninsula, Japan). Bull Volcanol 76(10):867

Jutzeler M, McPhie J, Allen SR (2014c) Facies architecture of a continental, below-wave-base volcaniclastic basin: the Ohanapecosh formation, Ancestral Cascades arc (Washington, USA). Geol Soc Am Bull 126(3–4):352–376

Kain CL, Gomez C, Moghaddam AE (2012) Comment on 'Reassessment of hydrodynamic equations: minimum flow velocity to initiate boulder transport by high energy events (storms, tsunamis), by NAK Nandasena, R. Paris and N. Tanaka [Marine Geology 281, 70–84]. Mar Geol 319:75–76

Karátson D, Telbisz T, Wörner G (2012) Erosion rates and erosion patterns of Neogene to Quaternary stratovolcanoes in the Western Cordillera of the Central Andes: an SRTM DEM based analysis. Geomorphology 139:122–135

Kataoka K, Nakajo T (2004) Flow transformation and depositional organization of debris flow-hyperconcentrated flow-streamflow spectrum in volcanic fan-delta setting The Pleistocene Lower and Middle Formations, Yachiho Group, central Japan. J Sedimentol Soc Japan 59(59):17–26

Keaton JR (2019) Review of contemporary terminology for damaging surficial processes: stream flow, hyperconcentrated sediment flow, debris flow, mud flow, mud flood, mudslide. In: Kean JW et al (eds) Proceedings of the seventh international conference on debris-flow hazards mitigation. Golden, Colorado, USA, June 10–13, 2019. https://hdl.handle.net/11124/173051

Kelfoun K, Giachetti T, Labazuy P (2010) Landslide-generated tsunamis at Réunion Island. J Geophys Res: Earth Surf 115(F4)

Kieffer G (1971) Apercu sur la morphologie des regions volcaniques du Massif Central. In: Symposium Jean Jung: geologie, geomorphologie et structure profonde du Massif Central francais, Clermont-Ferrand, 479–510

Kilburn CR, Petley DN (2003) Forecasting giant, catastrophic slope collapse: lessons from Vajont, Northern Italy. Geomorphology 54 (1–2):21–32

Kneller BC, Branney MJ (1995) Sustained high-density turbidity currents and the deposition of thick massive sands. Sedimentology 42(4):607–616

Komar PD (1970) The competence of turbidity currents. Geol Soc Am Bull 81:1555–1562

Kuenzi WD, Horst OH, McGehee RV (1979) Effect of volcanic activity in fluvial-deltaic sedimentation in a modern arc-trench gap, southwestern Guatemala. Geol Soc Am Bull 90(9):827–838

Lacasse C, Sigurdsson H, Carey S, Paterne M, Guichard F (1996) North Atlantic deep-sea sedimentation of Late Quaternary tephra from the Iceland hotspot. Mar Geol 129(3–4):207–235

Lagmay AM, van Wyck de Vries B, Kerle N, Pyle DM (1999) Volcano instability induced by strike-slip faulting. Bull Volcanol 62(4–5):331–346

Lajoie J (1984) Volcaniclastic rocks. In Walker RG (ed) Facies models. Geosci Canada, 39–52

Lavigne F, Thouret JC, Voight B, Suwa H, Sumaryono A (2000) Lahars at Merapi volcano, Central Java: an overview. J Volcanol Geotherm Res 100(1–4):423–456

Lebas E, Le Friant A, Boudon G, Watt SF, Talling PJ, Feuillet N, Deplus C, Berndt C, Vardy ME (2011) Multiple widespread landslides during the long-term evolution of a volcanic island: insights from high-resolution seismic data, Montserrat, Lesser Antilles. Geochem Geophys Geosyst 12(5)

Lee HL (2005) Undersea landslides: extent and significance in the Pacific Ocean, an update. Nat Haz Earth Syst Sci 5(6):877–892

Leeder MR (2009) Sedimentology and sedimentary basins: from turbulence to tectonics. Wiley, p 767

Le Friant A, Boudon G, Deplus C, Villemant B (2003) Large-scale flank collapse events during the activity of Montagne Pelée, Martinique, Lesser Antilles. J Geophys Res: Solid Earth 108(B1)

Le Roux JP, Vargas G (2005) Hydraulic behavior of tsunami backflows: insights from their modern and ancient deposits. Envir Geol 49(1):65

Le Roux JP, Nielsen SN, Kemnitz H, Henriquez Á (2008) A Pliocene mega-tsunami deposit and associated features in the Ranquil formation, southern Chile. Sed Geol 203(1–2):164–180

Lipman PW (1976) Caldera-collapse breccias in the western San Juan Mountains, Colorado. Geol Soc Am Bull 87(10):1397–1410

Lipman PW, Mullineaux DR (eds) (1981) The 1980 Eruptions of Mount St. Helens, Washington. US Geol Surv Prof Pap 1250:844

Louvat P, Gislason SR, Allègre CJ (2008) Chemical and mechanical erosion rates in Iceland as deduced from river dissolved and solid material. Am J Sci 308(5):679–726

Lowe DR (1976) Grain flow and grain flow deposits. J Sed Petrol 46:188–199

Lowe DR (1982) Sediment gravity flows; II, depositional models with special reference to the deposits of high-density turbidity currents. J Sed Res 52(1):279–297

Mackay K (2023) https://www.abc.net.au/news/2022-11-21/sea-floor-map-reveals-huge-cavity-left-behind-by-tonga-volcano/101674708

McMurtry GM, Watts P, Fryer GJ, Smith JR, Imamura F (2004) Giant landslides, mega-tsunamis, and paleo-sea level in the Hawaiian Islands. Mar Geol 203(3–4):219–233

Major JJ, Pierson TC, Dinehart RL, Costa JE (2000) Sediment yield following severe volcanic disturbance—a two-decade perspective from Mount St. Helens. Geology 28(9):819–822

Major JJ, Pierson TC, Scott KM (2005) Debris flows at Mount St. Helens, Washington, USA. In: Jakob M, Hungr O, Jakob DM (eds) Debris-flow hazards and related phenomena. Springer, Berlin, Heidelberg, pp 685–731

Manville V, Segschneider B, White JDL (2002) Hydrodynamic behaviour of Taupo 1800a pumice: implications for the sedimentology of remobilised pyroclastic deposits. Sedimentology 49 (5):955–976

Manville V, White JDL, Houghton BF, Wilson CJN (1998) The saturation behaviour of pumice and some sedimentological implications. Sed Geol 119(1–2):5–16

Martha TR, Roy P, Govindharaj KB, Kumar KV, Diwakar PG, Dadhwal VK (2015) Landslides triggered by the June 2013 extreme rainfall event in parts of Uttarakhand state, India. Landslides 12 (1):135–146

Martí J, Groppelli G, da Silveira AB (2018) Volcanic stratigraphy: a review. J Volcanol Geotherm Res 357:68–91

Martí J, Hurlimann M, Ablay GJ, Gudmundsson A (1997) Vertical and lateral collapses on Tenerife (Canary Islands) and other volcanic ocean islands. Geology 25(10):879–882

Matoza RS, Fee D, Assink JD, Iezzi AM, Green DN, Kim K, Toney L, Lecocq T, Krishnamoorthy S, Lalande JM, Nishida K et al. (2022) Atmospheric waves and global seismoacoustic observations of the January 2022 Hunga eruption, Tonga. Science https://doi.org/10.1126/science.abo7063 (2022)

Meiburg E, Kneller B (2010) Turbidity currents and their deposits. Ann Rev Fluid Mech 42:135–156

Menzies J, van der Meer JJM (eds) (2018) Past glacial environments (sediments, forms and techniques). Elsevier, Amsterdam, p 858p

Merrifield MA, Firing YL, Aarup T, Agricole W, Brundrit G, Chang-Seng D, Farre R, Kilonsky B, Knight W, Kong L, Magori C (2005) Tide gauge observations of the Indian Ocean tsunami, December 26, 2004. Geophys Res Letts 32(9)

Middleton GV (1966) Experiments on density and turbidity currents: I. Motion of the head. Canad J Earth Sci 3(4):523–546

Middleton GV, Hampton MA (1976) Subaqueous sediment transport and deposition by sediment gravity flows. In: Stanley DJ, Swift DJP (eds) Marine sediment transport and environmental management. Wiley, New York, pp 197–218

Miller MC, McCave IN, Komar P (1977) Threshold of sediment motion under unidirectional currents. Sedimentology 24(4):507–527

Mills HH (1976) Estimated erosion rates on Mount Rainier. Wash Geol 4(7):401–406

Mitchell AHG (1970) Facies of an early Miocene volcanic arc, Malekula Island, New Hebrides. Sedimentology 14:201–243

Moon V, Bradshaw J, de Lange W (2009) Geomorphic development of White Island Volcano based on slope stability modelling. Eng Geol 104(1–2):16–30

Moore A, Nishimura Y, Gelfenbaum G, Kamataki T, Triyono R (2006) Sedimentary deposits of the 26 December 2004 tsunami on the northwest coast of Aceh, Indonesia. Earth Planets Space 58 (2):253–258

Moore DG (1978) Submarine slides. In Voight B (ed.) 1978. Rock slides and avalanches, I. Natural phenomena. Elsevier, Amsterdam, pp 563–604

Moore JG (1964) Giant submarine landslides on the Hawaiian Ridge. US Geol Surv Prof Pap 501–D:95–98

Moore JG, Albee WC (1981) Topographic and structural changes, March–July 1980—Photogrammetric data. In: Lipman PW, Mullineaux DR (eds) The 1980 eruptions of Mount St. Helens, Washington. US Geol Surv Prof Pap 1250:123–134

Moore JG, Clague DA (1992) Volcano growth and evolution of the island of Hawaii. Geol Soc Am Bull 104(11):1471–1484

Moore JG, Clague DA, Holcomb RT, Lipman PW, Normark WR, Torresan ME (1989) Prodigious submarine landslides on the Hawaiian Ridge. J Geophys Res 94(B12):17465–17484

Morton RA (2002) Factors controlling storm impacts on coastal barriers and beaches: a preliminary basis for near real-time forecasting. J Coast Res 1:486–501

Morton RA, Gelfenbaum G, Jaffe BE (2007) Physical criteria for distinguishing sandy tsunami and storm deposits using modern examples. Sed Geol 200(3–4):184–207

Mulder T (2011) Gravity processes and deposits on continental slope, rise and abyssal plains. Develop Sedimentol 63:25–148, Elsevier

Nakada S, Shimizu H, Ohta K (1999) Overview of the 1990–1995 eruption at Unzen Volcano. J Volcanol Geotherm Res 89(1–4):1–22

Nandasena NA, Paris R, Tanaka N (2011) Numerical assessment of boulder transport by the 2004 Indian ocean tsunami in Lhok Nga, West Banda Aceh (Sumatra, Indonesia). Comput Geosci 37 (9):1391–1399

Naranjo JL, Sigurdsson H, Carey SN, Fritz W (1986) Eruption of the Nevado del Ruiz volcano, Colombia, on 13 November 1985: tephra fall and lahars. Science 233(4767):961–963

Newhall CG, Punongbayan R (eds.) (1996) Fire and mud: eruptions and lahars of Mount Pinatubo, Philippines. Philippine Institute of Volcanology and Seismology, Quezon City and University of Washington, Seattle, p 1126

Nott J (2003) Tsunami or storm waves? Determining the origin of a spectacular field of wave emplaced boulders using numerical storm surge and wave models and hydrodynamic transport equations. J Coast Res Apr 348–356

Noormets R, Crook KAW, Felton EA (2004) Sedimentology of rocky shorelines: 3. Hydrodynamics of megaclast emplacement and transport on a shore platform, Oahu, Hawaii. Sed Geol 172 (1–2):41–65

Orr TR, Thelen WA, Patrick MR, Swanson DA, Wilson DC (2013) Explosive eruptions triggered by rockfalls at Kīlauea volcano, Hawai ʻi. Geology 41(2):207–210

Paguican EM, van Wyck de Vries B, Lagmay AM (2012) Volcano-tectonic controls and emplacement kinematics of the Iriga debris avalanches (Philippines). Bull Volcanol 74(9):2067–2081

Palmer BA, Alloway BV, Neall VE (1991) Volcanic debris avalanche deposits in New Zealand—Lithofacies organization in unconfined, wet-avalanche flows. Soc Econ Paleontol Mineral Spec Publ 45:89–98

Pararas-Carayannis G (2002) Evaluation of the threat of mega tsunami generation from postulated massive slope failures of island stratovolcanoes on La Palma, Canary Islands, and on the island of Hawaii. Sci Tsunami Haz 20(5):251–277

Paris R, Bravo JJ, González ME, Kelfoun K, Nauret F (2017) Explosive eruption, flank collapse and megatsunami at Tenerife ca. 170 ka. Nat Comm 8:15246

Paris R, Fournier J, Poizot E, Etienne S, Morin J, Lavigne F, Wassmer P (2010) Boulder and fine sediment transport and deposition by the 2004 tsunami in Lhok Nga (western Banda Aceh, Sumatra, Indonesia): a coupled offshore–onshore model. Mar Geol 268(1–4):43–54

Paris R, Kelfoun K, Giachetti T (2013) Marine conglomerate and reef megaclasts at Mauritus Island: evidences of a tsunami generated by a flank collapse of the Piton de la Fournaise volcano, Reunion Island? Sci Tsunami Haz 32(4):281–291

Paris R, Lavigne F, Wassmer P, Sartohadi J (2007) Coastal sedimentation associated with the December 26, 2004 tsunami in Lhok Nga, west Banda Aceh (Sumatra, Indonesia). Mar Geol 238(1–4):93–106

Paris R, Wassmer P, Sartohadi J, Lavigne F, Barthomeuf B, Desgages E, Grancher D, Baumert P, Vautier F, Brunstein D, Gomez C (2009) Tsunamis as geomorphic crises: lessons from the December 26, 2004 tsunami in Lhok Nga, west Banda Aceh (Sumatra, Indonesia). Geomorphology 104(1–2):59–72

Peters R, Jaffe B (2010) Identification of tsunami deposits in the geologic record; developing criteria using recent tsunami deposits. US Geol Surv Open File 210–1239:39

Petley DN, Petley DJ (2006) On the initiation of large rockslides: perspectives from a new analysis of the Vaiont movement record. In: Landslides from massive rock slope failure 2006. Springer, Dordrecht, pp 77–84

Pierson TC (2005) Hyper-concentrated flow—transitional process between water flow and debris flow. In: Jakob M, Hungr O, Jakob DM (eds) Debris-flow hazards and related phenomena. Springer, Berlin, Heidelberg, pp 159–202

Pierson TC, Janda RJ, Thouret JC, Borrero CA (1990) Perturbation and melting of snow and ice by the 13 November 1985 eruption of Nevado-del-Ruiz, Colombia, and consequent mobilization, flow and deposition of lahars. J Volcanol Geotherm Res 41:17–66

Pierson TC, Janda RJ, Umbal JV, Daag AS (1992) Immediate and long-term hazards from lahars and excess sedimentation in rivers draining Mt. Pinatubo, Philippines. US Geol Surv Water-Res Investig Rept 92:183–203

Pierson TC, Major JJ (2014) Hydrogeomorphic effects of explosive volcanic eruptions on drainage basins. Ann Rev Earth Planet Sci 42:469–507

Pittari A, Cas RA, Edgar CJ, Nichols HJ, Wolff JA, Martí J (2006) The influence of palaeotopography on facies architecture and pyroclastic flow processes of a lithic-rich ignimbrite in a high gradient setting: the Abrigo Ignimbrite, Tenerife, Canary Islands. J Volcanol Geotherm Res 152(3–4):273–315

Plafker G, Ericksen GE (1978) Nevados Huascaran avalanches, Peru. In: Voight B (ed) Rock slides and avalanches, I. Natural phenomena. Elsevier, Amsterdam, pp 277–314

Pope EL, Jutzeler M, Cartigny MJ, Shreeve J, Talling PJ, Wright IC, Wysoczanski RJ (2018) Origin of spectacular fields of submarine sediment waves around volcanic islands. Earth Planet Sci Letts 493:12–24

Postma G, Cartigny MJ (2014) Supercritical and subcritical turbidity currents and their deposits—a synthesis. Geology 42(11):987–990

Postma G, Kleverlaan K, Cartigny MJ (2014) Recognition of cyclic steps in sandy and gravelly turbidite sequences, and consequences for the Bouma facies model. Sedimentology 61(7):2268–2290

Pritchard D, Dickinson L (2008) Modelling the sedimentary signature of long waves on coasts: implications for tsunami reconstruction. Sed Geol 206(1–4):42–57

Purkis SJ, Ward SN, Fitzpatrick NM, Garvin JB, Slayback D, Cronin SJ, Palaseanu-Lovejoy M, Dempsey A (2023) The 2022Hunga-Tonga megatsunami: Near-field simulation of a once-in-a-century event. Science Adv 9(15):eadf5493

Raymond CF (1978) Mechanics of glacier movement. In Voight B (ed), Rock slides and avalanches, I. Natural phenomena. Elsevier, Amsterdam, 793–833

Reid ME, Sisson TW, Brien DL (2001) Volcano collapse promoted by hydrothermal alteration and edifice shape, Mount Rainier. Wash Geol 29(9):779–782

Richards AF (1958) Trans-Pacific distribution of floating pumice from Isla San Benedicto, Mexico. Deep-Sea Res 5:29–35

Richards AF (1965) Geology of the Islas Revillagidedo, 3. Effects of erosion on Isla San Benedicto 1952–1961 following the birth of Volcan Bárcena. Bull Volcanol 28(1):381–403

Richardson JF, Zaki WN (1954) Sedimentation and fluidisation: Part I. Trans Instit Chem Eng 32:35–53

Roobol MJ, Wright JV, Smith AL (1983) Calderas or gravity-slide structures in the Lesser Antilles island arc? J Volcanol Geotherm Res 19(1–2):121–134

Roverato M, Capra L, Sulpizio R, Norini G (2011) Stratigraphic reconstruction of two debris avalanche deposits at Colima Volcano (Mexico): insights into pre-failure conditions and climate influence. J Volcanol Geotherm Res 207(1–2):33–46

Rubin CM, Horton BP, Sieh K, Pilarczyk JE, Daly P, Ismail N, Parnell AC (2017) Highly variable recurrence of tsunamis in the 7,400 years before the 2004 Indian Ocean tsunami. Nat Comms 8:16019

Sadler PM (1981) Sediment accumulation rates and the completeness of stratigraphic sections. J Geol 89(5):569–584

Savage SB (1984) The mechanics of rapid granular flows. Adv Appl Mech 24:289–366

Savage SB, Hutter K (1989) The motion of a finite mass of granular material down a rough incline. J Fluid Mech 199:177–215

Savage SB, Iverson RM (2003) Surge dynamics coupled to pore-pressure evolution in debris flows. In: Rickenmann D, Chen CL (eds) Proceedings 3rd international conference on debris-flow hazards mitigation: mechanics, prediction, and assessment. Davos, Switzerland, Millpress, Rotterdam, pp 503–514

Schmincke HU, Sumita M (1998) Volcanic evolution of Gran Canaria reconstructed from apron sediments: synthesis of VICAP project drilling. In: Weaver et al. (eds) Proc Ocean Drill Prog Sci Res 157, 433–469

Scoon RN (2018) Geology of National Parks of Central/Southern Kenya and Northern Tanzania: Geotourism of the Gregory Rift Valley, Active Volcanism and Regional Plateaus. Springer Nature, p 221

Scott KM (1988) Origins, behavior, and sedimentology of lahars and lahar-runout flows in the Toutle-Cowlitz River system: US Geol Surv Prof Pap 1447-A:76

Scott KM, Vallance JW, Pringle PT (1995) Sedimentology, behavior, and hazards of debris flows at Mount Rainier, Washington. US Geol Surv Prof Pap 1547:56p

Seisdedos J, Ferrer M, de Vallejo LG (2012) Geological and geomechanical models of the pre-landslide volcanic edifice of Güímar and La Orotava mega-landslides (Tenerife). J Volcanol Geotherm Res 239:92–110

Self S, Rampino MR (1981) The 1883 eruption of Krakatau. Nature 294(5843):699-704

Self S, Rampino MR, Newton MS, Wolff JA (1984) Volcanological study of the great Tambora eruption of 1815. Geology12 (11):659-663

Shane PA, Froggatt PC (1992) Composition of widespread volcanic glass in deep-sea sediments of the Southern Pacific Ocean: an Antarctic source inferred. Bull Volcanol 54(7):595–601

Shreve RL (1968) The Blackhawk landslide. Geol Soc Am Spec Pap 108

Siebert L (1984) Large volcanic debris avalanches: characteristics of source areas, deposits and associated eruptions. J Volcanol Geotherm Res 22:163–197

Sigurdsson H, Sparks RS, Carey ST, Huang TC (1980) Volcanogenic sedimentation in the Lesser Antilles arc. J Geol 88(5):523–540

Sigurðsson O, Williams RS (2008). Geographic names of Iceland's glaciers: historic and modern. National energy authority Iceland. US Geol Surv Prof Pap 1746:225

Silver E, Day S, Ward S, Hoffmann G, Llanes P, Driscoll N, Appelgate B, Saunders S (2009) Volcano collapse and tsunami generation in the Bismarck volcanic arc, Papua New Guinea. J Volcanol Geotherm Res 186(3–4):210–222

Simkin T, Fiske RS (1983) Krakatau 1883—The volcanic eruption and its effects. Smithsonian Institution Press, Washington DC.464p

Simpson CJ, Cas RAF, Arundell MC (2005) Volcanic evolution of a long-lived Ordovician island-arc province in the Parkes region of the Lachlan Fold Belt, southeastern Australia. Aust J Earth Sci 52 (6):863–886

Smith GA, Lowe DR (1991) Lahars: Volcano hydrologic-events and deposition in the debris flow—hyperconcentrated flow continuum. In: Fisher RV, Smith GA (eds) Sedimentation in volcanic settings. Soc Econ Paleontol Mineral Spec Publ 45:59–70

Soule SA, Zoeller M, Parcheta C (2021) Submarine lava deltas of the 2018 eruption of Kīlauea volcano. Bull Volcanol 83(4):1–16

Stow D, Smillie Z (2020) Distinguishing between deep-water sediment facies: Turbidites, contourites and hemipelagites. Geosciences 10 (2):68

Sundborg Å (1956) The river Klarälven a study of fluvial processes. Geogr Ann 38(2–3):125–316

Syamsidik B, Luthfi M, Suppasri A, Comfort LK (2020) The 22 December 2018 Mount Anak Krakatau volcanogenic tsunami on Sunda Strait coasts, Indonesia: tsunami and damage characteristics. Nat Haz Earth Syst Sci 20(2):549–565

Swanson DA, Jackson DB, Koyanagi RY, Wright TL (1976) The February 1969 East Rift eruption of Kilauea volcano, Hawaii. US Geol Surv Prof Pap 891:33

Swanson DA, Malone SD, Samora BA (1992) Mount Rainier: a decade volcano. EOS Trans Am Geophys Union 73(16):177–186

Talling PJ, Masson DG, Sumner EJ, Malgesini G (2012) Subaqueous sediment density flows: depositional processes and deposit types. Sedimentology 59(7):1937–2003

Tappin DR (ed) (2007) Sedimentary features of tsunami deposits—their origin, recognition and discrimination: an introduction. Sed Geol 200(3–4):151–388

Tappin DR, Hunt J, Grilli ST et al. (2019) The 1883 and 2018 Krakatau tsunamis –new marine geophysical and sediment core evidence on their generation. AGU Fall Meeting, San Francisco, Abstract NH31A-04

Tarnocai C, Canadell JG, Schuur EA, Kuhry P, Mazhitova G, Zimov S (2009) Soil organic carbon pools in the northern circumpolar permafrost region. Global Biogeochem Cycles 23. https://doi.org/10.1029/2008GB003327

Tassé N, Lajoie J, Dimroth E (1978) The anatomy and interpretation of an Archean volcaniclastic sequence, Noranda region, Quebec. Canad J Earth Sci 15(6):874–888

Thorarinsson S (1967) Surtsey: the new island in the North Atlantic. The Viking Press, New York

Tibaldi A (2001) Multiple sector collapses at Stromboli volcano, Italy: how they work. Bull Volcanol 63(2–3):112–125

Tommasi P, Baldi P, Chiocci FL, Coltelli M, Marsella M, Pompilio M, Romagnoli C (2005) The landslide sequence induced by the 2002 eruption at Stromboli volcano. In: Sassa K, Fukuoka, Wang F, Wang G (eds) Landslides. Springer, Berlin, Heidelberg, 251–258

Trofimovs J, Cas RAF, Davis BK (2004) An Archaean submarine volcanic debris avalanche deposit, Yilgarn Craton, western Australia, with komatiite, basalt and dacite megablocks: The product of dome collapse. J Volcanol Geotherm Res 138(1–2):111–126

Trofimovs J, Foster C, Sparks RS, Loughlin S, Le Friant A, Deplus C, Porritt L, Christopher T, Luckett R, Talling PJ, Palmer MR (2012) Submarine pyroclastic deposits formed during the 20th May 2006 dome collapse of the Soufrière Hills Volcano, Montserrat. Bull Volcanol 74(2):391–405

Trofimovs J, Sparks RSJ, Talling PJ (2008) Anatomy of a submarine pyroclastic flow and associated turbidity current: July 2003 dome collapse, Soufrière Hills volcano, Montserrat, West Indies. Sedimentology 55(3):617–634

Trofimovs J, Talling PJ, Fisher JK, Sparks RS, Watt SF, Hart MB, Smart CW, Le Friant A, Cassidy M, Moreton SG, Leng MJ (2013) Timing, origin and emplacement dynamics of mass flows offshore of SE Montserrat in the last 110 ka: implications for landslide and tsunami hazards, eruption history, and volcanic island evolution. Geochem Geophys Geosyst 14(2):385–406

Uehara D, Cas RAF, Folkes C, Takarada S, Oda H, Porreca M (2015) Using thermal remanent magnetisation (TRM) to distinguish block and ash-flow and debris flow deposits, and to estimate their emplacement temperature: 1991–1995 lava dome eruption at Mt. Unzen Volcano, Japan. J Volcanol Geotherm Res 303:92–111

Ui T (1983) Volcanic dry avalanche deposits—identification and comparison with nonvolcanic debris stream deposits. J Volcanol Geotherm Res 18:135–50

Ui T, Yamamoto H, Suzuki-Kamata K (1986) Characterization of debris avalanche deposits in Japan. J Volcanol Geotherm Res 29(1–4):231–243

Vallance JW (2005) Volcanic debris flows. In: Jakob M, Hungr O (eds) Debris-flow hazards and related phenomena. Springer Praxis Books, Berlin, pp 247–274

Vallance JW, Iverson RM (2015) Chapter 37 Lahars and their deposits. In: Sigurdsson H, Houghton B, McNutt SR, Rymer H, Stix J (eds) The encyclopedia of Volcanoes. Academic Press Elsevier, pp 649–664

Vallance JW, Scott KM (1997) The Osceola Mudflow from Mount Rainier: sedimentology and hazard implications of a huge clay-rich debris flow. Geol Soc Am Bull 109(2):143–163

van Wyk de Vries B, Davies T (2015) Landslides, debris avalanches, and volcanic gravitational deformation. In: Sigurdsson H, Houghton B, McNutt SR, Rymer H, Stix J (eds) The encyclopedia of volcanoes. Academic Press Elsevier, 665–685

van Wyk de Vries B, Self S, Francis PW, Keszthelyi L (2001) A gravitational spreading origin for the Socompa debris avalanche. J Volcanol Geotherm Res 105(3):225–247

Vessell RK, Davies DK (1981) Non-marine sedimentation in an active fore-arc basin. In Ethridge FG, Flores RM (eds), Recent and ancient non-marine depositional environments: models for exploration. Soc Econ Paleontol Mineral Spec Publ 31:31–45

Voight B (ed) (1978) Rock slides and avalanches, I. Natural phenomena. Elsevier, Amsterdam, p 830

Voight B, Glicken H, Janda RJ, Douglass PM (1981) Catastrophic rockslide avalanche of May 18. In: Lipman PW, Mullineaux DR (eds) The 1980 eruptions of Mount St. Helens, Washington. US Geological Survey Professional Paper, vol 1250, pp 347–378

Voight B, Janda RJ, Glicken H, Douglass PM (1983) Nature and mechanics of the Mount St. Helens rockslide-avalanche of 18 May 1980. Geotechnique 33(3):243–273

Walker GPL (1982) Topographic evolution of eastern Iceland. Jökull 32:13–20

Walker GPL, Aumento F, Clarke BD, Cann JR, Ryall PJC (1972) Site 115. DSDP Initial Rep 12:361–375

Walker RG (1965) The origin and significance of the internal sedimentary structures of turbidites. Proc Yorkshire Geol Soc 35:1–32

Walker RG (1975) Generalised facies models for resedimented conglomerates of turbidite association. Geol Soc Am Bull 86:737–748

Walker RG (1978) Deep-water sandstone facies and ancient submarine fans: models for exploration for stratigraphic traps. Am Assoc Petrol Geol Bull 62(6):932–966

Walker RG (ed) (1984) Facies models, 2nd edn. Geological Association of Canada, Geosci Canada Reprint Series 1

Walter TR, Haghshenas Haghighi M, Schneider FM et al (2019) Complex hazard cascade culminating in the Anak Krakatau sector collapse. Nat Comm 10:4339. https://doi.org/10.1038/s41467-019-12284-5

Ward SN, Day S (2001) Cumbre Vieja volcano—potential collapse and tsunami at La Palma, Canary Islands 2001. Geophys Res Letts 28 (17):3397–3400

Ward SN, Day S (2003) Ritter Island volcano—lateral collapse and the tsunami of 1888. Geophys J Int 154(3):891–902

Watton TJ, Jerram DA, Thordarson T, Davies RJ (2013) Three-dimensional lithofacies variations in hyaloclastite deposits. J Volcanol Geotherm Res 250:19–33

Watts AB, Masson DG (1995) A giant landslide on the north flank of Tenerife, Canary Islands. J Geophys Res: Solid Earth 100 (B12):24487–24498

West LM, Perillo MM, Olariu C, Steel RJ (2019) Multi-event organization of deep water sediments into bedforms: long-lived, large-scale antidunes preserved in deep water slopes. Geology 47 (5):391–394

Whitham AG, Sparks RSJ (1986) Pumice. Bull Volcanol 48(4):209–223

Williams R, Rowley P, Garthwaite MC (2019) Reconstructing the Anak Krakatau flank collapse that caused the December 2018 Indonesian tsunami. Geology 47(10):973–976

Wood CA (1980) Morphometric analysis of cinder cone degradation. J Volcanol Geotherm Res 8(2–4):137–160

Wright JV, Mutti E (1981) The Dali Ash, Island of Rhodes, Greece: a problem in interpreting submarine volcanigenic sediments. Bull Volcanol 44(2):153–167

Yoshii T, Tanaka S, Matsuyama M (2017) Tsunami deposits in a super-large wave flume. Mar Geol 391:98–107

Yoshii T, Tanaka S, Matsuyama M (2018) Tsunami inundation, sediment transport, and deposition process of tsunami deposits on coastal lowland inferred from the Tsunami Sand Transport Laboratory Experiment (TSTLE). Mar Geol 400:107–118

Yuill BT, Nichols MH (2011) Patterns of grain-size dependent sediment transport in low-ordered, ephemeral channels. Earth Surf Proc Landforms 36(3):334–346

Ziegler AD, Wasson RJ, Bhardwaj A, Sundriyal YP, Sati SP, Juyal N, Nautiyal V, Srivastava P, Gillen J, Saklani U (2014) Pilgrims, progress, and the political economy of disaster preparedness–the example of the 2013 Uttarakhand flood and Kedarnath disaster. Hydrol Process 28(24):5985–5990

Zimov SA, Schuur EA, Chapin FS (2006) Permafrost and the global carbon budget. Science 312(5780):1612–1613

Syn- to Post-emplacement Alteration Processes

13

Their Effects on Original Volcanic Rock Textures

Ray Cas, John V. Wright, and Guido Giordano

R. Cas et al., *Volcanology*, Springer Textbooks in Earth Sciences, Geography and Environment,
https://doi.org/10.1007/978-3-319-66613-6_13

Plate 13.1 Apparent or "pseudo-textures" resulting from varying styles of alteration, hydraulic fracturing, and deformation. **a** *Why is this not a primary volcanic breccia?* Pseudo-breccia texture in polyphase altered coherent rhyolite, Silurian Thorkidaan Volcanics, Benambra, eastern Victoria, Australia. **b** *Why is this not a volcanic conglomerate?* Pseudo-conglomerate or pseudo-clastic texture in hydraulically fractured and chlorite altered, coherent, Archean dacite, Black Swan Mine sequence, Western Australia. **c** *Why is this not a hyaloclastite breccia?* Hydraulically fractured and chlorite–sericite altered coherent dacite, Hercules Mine sequence, Cambrian Mt Read Volcanics, western Tasmania, Australia. **d** *Why is this not a eutaxitic welded ignimbrite?* Pseudo-eutaxitic, pseudo-fiamme, or pseudo-welded ignimbrite texture in highly strained and polyphase altered, coherent Archean dacite, Perseverance Mine sequence, Leinster, and Western Australia. All drill core is ~5 cm in diameter

Abstract

We now consider the range of syn- to post-emplacement alteration processes and their effects on volcanic and associated sedimentary deposit types and successions. These processes include devitrification, palagonitisation (in mafic glass), hydrothermal alteration, diagenesis, metamorphism, and tectonic deformation, which may significantly modify and in some cases completely disguise original volcanic textures. The process that induces textural change in the first instance is devitrification of volcanic glass as it cools, which is facilitated by ion and gas diffusion through the glass. Secondly, permeation of magmatic hydrous fluids, surface waters, and meteoric water in the subsurface can begin to remobilise elements and initiate hydrothermal alteration of the volcanic glass and original minerals. In older volcanic rock successions, post-depositional processes of ongoing devitrification, hydrothermal alteration, and then metamorphism and deformation can all have the effect of overprinting original volcanic textures and mineralogy. Such processes not only have the ability to overprint original textures but also, in some instances, to obliterate them totally and produce new textures (even apparent or "*false*", pseudo-clastic, including fiamme-like textures) that can lead to misconceptions of the original rock types and their origins. This can be particularly important in ore-forming volcanic environments subjected to prolonged polyphase hydrothermal alteration and associated hydraulic fracturing, and can lead to incorrect reconstructions of their paleoenvironmental setting, eruption styles, and ore deposit types.

13.1 Introduction

One of the major problems with undertaking research or normal geological fieldwork and petrographic work on volcanic deposits is that they begin to alter almost from the moment they are emplaced, which can cause overprinting of original textures and mineralogy, and make identification and the origins of some deposits uncertain. This is because deposits that are made up of volcanic glass are very susceptible to chemical alteration and textural modification because volcanic glass is metastable at the temperature and chemical conditions existing at the Earth's surface and in the immediate subsurface (Plate 13.1).

Alteration processes can generally be grouped into those that are contemporaneous with the emplacement of volcanic deposits, and those that clearly post-date emplacement. Processes contemporaneous with deposition include devitrification, hydration-related alteration (e.g. palagonitisation in mafic glasses), hydrothermal alteration, and associated hydraulic fracturing. Processes post-dating emplacement include ongoing devitrification, prolonged hydrothermal alteration, diagenesis, metamorphism, and deformation. When ancient successions have been subjected to prolonged polyphase hydrothermal alteration, diagenesis, metamorphism, and deformation, the effects of each alteration process may be difficult to distinguish, and their cumulative effects could mask most if not all original textural features.

The process in the first instance that induces textural change is devitrification of volcanic glass as it cools, which is facilitated by ion and gas diffusion through the glass. Secondly, permeation of magmatic hydrous fluids, surface waters, and meteoric water in the subsurface can begin to remobilise elements and initiate alteration of the volcanic glass and original minerals. In some proximal volcanic settings, hydrothermal systems can be active from the time that the volcano is first formed and then continued to be active during its eruption history, and even beyond the end of active volcanism (Chap. 18). In such settings, even newly erupted volcanic deposits can be subject to alteration caused by circulating hydrothermal fluids soon after deposition.

Distinguishing original volcanic textures from superimposed secondary alteration textures is not easy, and often both are present. The existence of (relict) original textures, therefore, has to be proven rather than being assumed to be present, and unusual textures have to be explained. Each of the alteration and modification processes will now be briefly introduced. However, the coverage is not intended to be comprehensive. This topic has been addressed previously by Cas and Wright (1987), McPhie et al. (1993), and Gifkins et al. (2005).

13.2 Devitrification and Spherulite Formation

When magma is erupted, it carries with it "*crystal cargo*" consisting of magmatic phenocrysts, which have grown in a deep magma reservoir, magmatic micro-crystals or microlites, which have grown in the magma during its ascent to the Earth's surface in the conduit and during eruption and cooling, and possibly xenocrysts scavenged from country rock. Both magmatic types of crystal populations are suspended in non-crystallised groundmass magma. Upon eruption, the magma quickly cools through the glass transition temperature range, or T_g, and solidifies to form groundmass glass (Fig. 13.1a) without undergoing slow, equilibrium magmatic crystallisation.

Volcanic glass is therefore a supercooled, silicate liquid, which cools too rapidly for equilibrium groundmass crystallisation to occur completely, and consists of disordered ions and molecules. However, volcanic glass doesn't like

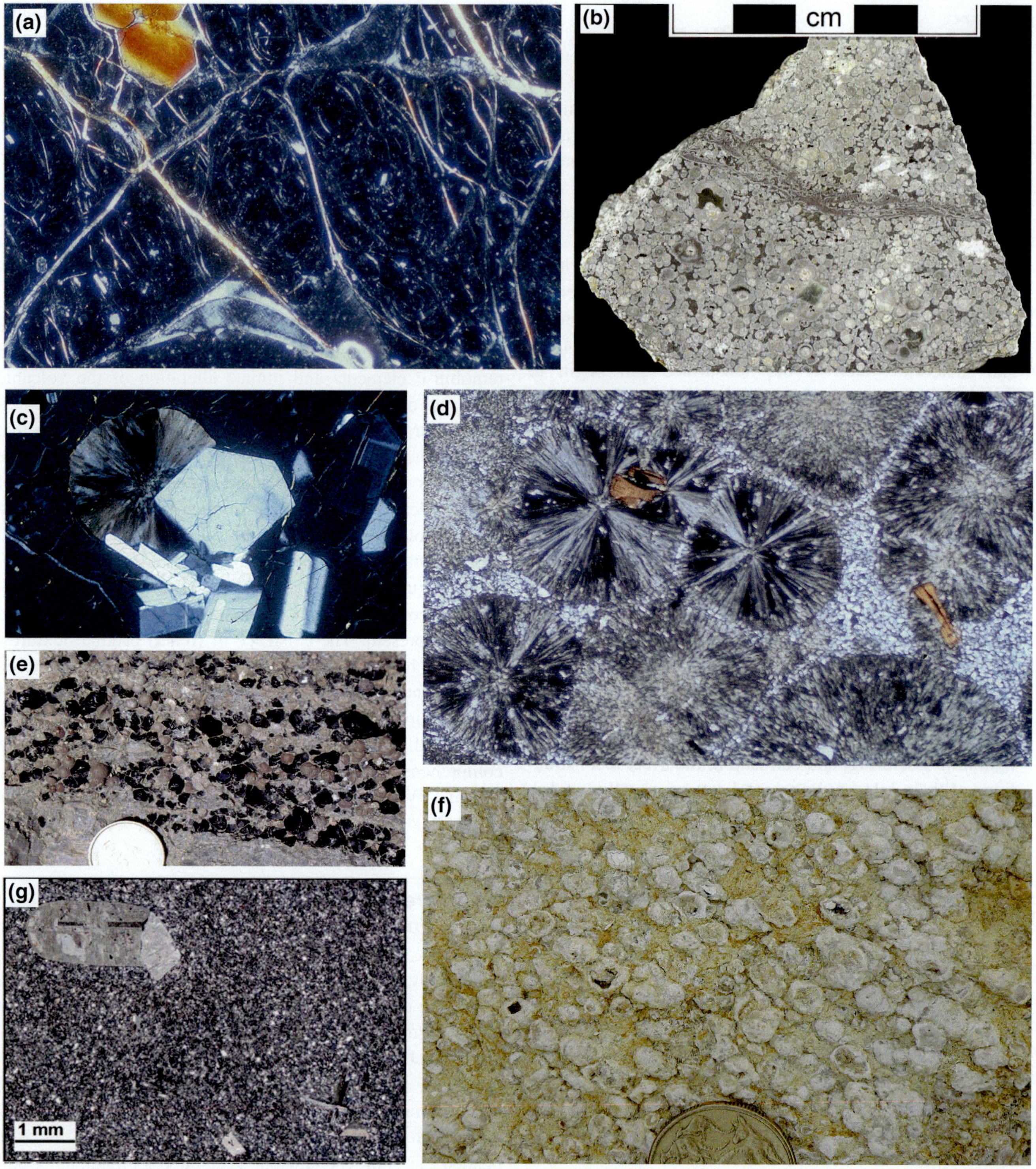

Fig. 13.1 Devitrification features of glassy rocks. **a** Porphyritic rhyolite obsidian, with a glass groundmass with suspended, tiny, tabular microlites, and cryptic curviplanar perlite fractures, without any devitrification features. Magnification 4x, x-nicols. **b** Hand-specimen view of tightly packed spherulites with dark relict glass areas that are now altered to chlorite. Miocene Shinzan Rhyolite, Japan. **c** Thin-section view of (**b**), showing a single spherulite, nucleated on the corner of a quartz phenocryst. Width of field of view 4 mm; x-nicols. **d** Thin-section view of (**b**), showing radiating, fibrous crystals of quartz and feldspar in devitrification spherulites. Width of field of view 10 mm, x-nicols. The micro-granular texture between spherulites formed after the spherulites formed. (**a**)–(**d**) are from the Miocene, submarine, Shinzan rhyolite lava, Oga Peninsula, Honshu, Japan. **e** Spherulites (pink spheres), with scattered relict black obsidian patches, Kolliedag lava dome, Cappadocia, Turkey. **f** Devitrification spherulites and lithophysae, sometimes amalgamated and arranged in trains in altered rhyolite, Late Devonian Boyd Volcanic Complex, southern New South Wales, Australia. Coin is 1.5 cm diameter. **g** Microcrystalline or microgranular groundmass texture (sometimes mistakenly called granophyric) comprising mosaics of microgranular quartz and feldspar in what was originally glassy rhyolite lava, Iberian Pyrite Belt, Portugal (From Valenzuela et al. 2011)

being glass, and would rather be a crystallised aggregate. It is thermodynamically unstable and can undergo post-eruption *devitrification,* which is the transformation from isotropic glass to a crystallised aggregate. This involves diffusion of ions through the viscous magma approaching the glass transition temperature, T_g, within the T_g temperature range, and in glass below T_g, until they meet like-minded ions and initiate the nucleation and growth of fibrous crystallites (Watkins et al. 2009). Strictly speaking, devitrification can only occur once glass has formed, but the diffusion of ions and the formation of fibrous or skeletal crystals occurs from temperatures above T_g to well below T_g (McArthur et al. 1998; Watkins et al. 2009). We use the term devitrification in the *sensu lato* sense to encompass late magmatic crystallisation just above the T_g to crystallisation in glass below T_g. This crystallisation usually occurs around a nucleus of some type, such as microlites, the corners of a phenocryst, or submicroscopic heterogeneities, leading to a radial growth of fibrous or needle-like crystallites around the nucleus. This produces growth forms such as spherulites, bow-tie aggregates, and axiolites (Lofgren 1970, 1971a, b), all of which can begin to grow above T_g (Clay et al. 2013).

13.2.1 Principal Features of Spherulites and Their Growth

Spherulites are spherical aggregates (Fig. 13.1b–f) of largely radiating needle-like or fibrous crystals (Fig. 13.1c, d), which are commonly found in the glassy carapace of rhyolitic lavas and also form in densely welded glassy ignimbrites (Lofgren 1971a, b). They commonly have diameters of 0.1–2 cm (Lofgren 1971a, b), but can be much larger and occasionally grow into *megaspherulites* tens of cm to nearly 3 m in diameter (Smith et al. 2001; Fig. 6.10f). Some spherulitic growths are called *lithophysae*, which are radiating aggregates of fibrous crystals, which have formed around an expanding vesicle or gas vugh (Fig. 13.1f). These vughs (or vugs) formed in a melt at or above the glass transition while it was still capable of deforming (McArthur et al. 1998). Given the scarcity of original volcanic glass in rocks older than Cenozoic, it must be assumed that glass either devitrifies penecontemporaneously, or is altered and undergoes (re-)crystallisation because of its metastable state, under diagenetic to low-grade metamorphic conditions.

In experimental studies, Lofgren (1970, 1971a, b) determined that the mineralogy of spherulitic growths was largely quartz (after cristobalite and rarely tridymite) and both sodium-rich and potassium-rich feldspar. Castro et al. (2008) reported abundances of 45% plagioclase, 40% quartz, 3% clinopyroxene, 1–2% magnetite, <15% interstitial glass, and <1% microvesicles in spherulites in obsidian from Krafla volcano, Iceland. In mafic rocks, spherulite-like radiating aggregates are generally rarer, consisting mainly of feathery, needle-like crystals of plagioclase and pyroxene which are called *varioles* or *variolites* and the texture, variolitic texture (see Figs. 5.47f, g and 13.16e). Varioles may in part result from devitrification, but as they have only been described in deeper water submarine glassy rocks, especially in the margins of pillows (Furnes 1973; Sandstå et al. 2011; Chap. 5), their formation may also be related to quench or chilling induced crystallisation.

Lofgren's studies of spherulite growth involved the experimental devitrification of rhyolitic obsidian glass under varying conditions of temperatures pressure, time, and fluid compositions. He was able to simulate many of the devitrification textures found in natural glasses. The stages of devitrification defined by Lofgren (1971b) are (1) an initial hydration stage, (2) a glassy stage, (3) a spherulitic stage, and (4) a granophyric (we prefer microcrystalline or microgranular) stage. The hydration stage is characterised by a polygonal mosaic of fractures in glass enclosed by a sharp curviplanar fracture, called the hydration front. In addition, strain-induced birefringence (a change in colour from pale yellow-green to pale green and pale reddish brown), and the development of micron-sized bubbles were noted. The glassy stage is marked by felsitic texture and minor spherulites (Fig. 13.1a, c), whereas the spherulitic stage is marked by abundant spherulites (Fig. 13.1b, d) and micro-poikilitic quartz. Spherulites consist of radiating clusters of fibrous crystallites and vary in size from 100 μm to several cm (Fig. 13.1b, d). When large, they are generally spherical with smooth surfaces (Fig. 13.1e), but some irregularities are found (Fig. 13.1f). Sometimes spherulites can amalgamate to produce an elongated train of connected, overlapping spherulites (Fig. 13.1f), often concentrated, and aligned along flow layering.

Lofgren (1971a) found that the morphology of spherulites varied according to the temperature of the experimental run. At runs below 400 °C, their outlines were nearly circular, between 400 and 600 °C bow-tie-shaped aggregates occurred, and, at 700 °C, lath-like fibres or open framework circular clusters of spaced fibres occurred. The microcrystalline/microgranular stage (sometimes inappropriately described as having "granophyric" texture), while not noted in Lofgren's experiments, is common in old, devitrified glassy rocks. This is dominated by fine-grained, roughly equidimensional, microcrystalline aggregates and mosaics of quartz and feldspar (i.e. microgranular texture) ± phyllosilicate minerals, rather than elongate fibres (Fig. 13.1g; also see Fig. 13.16e).

In hand specimen, devitrification can produce an apparent granularity through the development of very small spherulites and orbs, which at times may give rocks an apparent clastic character and overprint original textures (Fig. 13.1e). The aggregates of spheres can also be mistaken for accretionary lapilli and perlite orbs and cracks (Chap. 6) Patchy devitrification can also give rocks a patchy domainal texture, with the boundaries between devitrified and non-devitrified zones (glassy, altered) being called *devitrification fronts.*

Devitrification and hydration occur through the depolymerisation effects of water in breaking Si–O–Si bridges and hydrolysing tetrahedral oxygen to OH^-. The liberated SiO_4 tetrahedra are then free to reorganise and to nucleate quartz and feldspar crystal structures in the presence of alkalis (network modifiers), which diffuse through the glass network (Lofgren 1970; see also Chap. 2). Lofgren found that adding alkalis in solution increased the devitrification rate by four–five orders of magnitude, and the hydration rate by one–two orders of magnitude. In addition, devitrification may also produce a number of devitrification pore types through the cooling history of glassy volcanic rocks which may have relevance to their potential as reservoirs in hydrocarbon exploration (Zheng et al. 2018; Chap. 18).

The original textures of all glassy rocks are therefore inevitably modified by devitrification and ensuing alteration. In addition, Lofgren (1970), following the observations of Lipman (1965) and Noble (1967), found that there were significant changes in the bulk rock chemistry accompanying hydration and devitrification, especially variations in SiO_2, H_2O, Fe_2O_3/FeO ratio, and Na_2O and K_2O contents (up to 2% for individual element oxides). Therefore, ancient glassy volcanics are unlikely to reflect original major element chemistry on this count alone, not to mention the effects of alteration, diagenesis, and metamorphism.

13.2.2 Dynamics of Spherulite Growth

Castro et al. (2008) proposed that spherulites grow by advection complemented by diffusion of OH^- ions away from the radially growing fronts of spherulites, as anhydrous, fibrous quartz and feldspar crystallites grow. They calculated that growth timescales range from 1 day to 2 weeks depending on temperature of the magma/glass, with high growth rates coinciding with high temperatures (850 °C), and slow growth rates with low temperatures (650 °C). Watkins et al. (2009) further proposed that spherulite growth occurs both when the magma is above the glass transition, as well as below it in the glass state down to temperatures of about 400 °C. They propose that a 2.6 mm diameter spherulite would take about 300 days to grow at 800 °C, about 10 years at 600 °C, and about 300 years at 400 °C. Fenn (1977) demonstrated that, at low undercooling of $\Delta T < 40$ °C, small tabular feldspar crystals form, at moderate undercooling, ΔT of 75–145 °C, coarsely crystallised spherulites formed, whereas, at high undercooling, ΔT of 245–395 °C, finely crystallised, dense spherulites formed.

Based on samples collected at Tequila volcano in Mexico, Gardner et al. (2012) also determined that spherulitic growth involves diffusion of water, as well as other elements out of the spherulite as it grows, producing compositional gradients from the spherulite into the surrounding glass. They proposed that the degree to which the glass around spherulites is enriched in incompatible elements that diffuse during cristobalite-feldspar spherulite growth depends on (1) the diffusion rate of different elements, which is also temperature dependent, and (2) the size of the spherulite, which is dependent on the temperature, growth rate and timing of nucleation of spherulites during the cooling of the glass. Rapidly diffusing elements include Li, K, Na, and Cl, components with intermediate diffusion rates include H_2O, Rb, and F, and those with slow diffusion rates include Ba, Sr, Cs, and Be, which are least depleted in spherulites. They favoured most growth to occur in the T_g temperature range and below the T_g.

Clay et al. (2013) concluded that spherulite growth began at magma emplacement temperatures (~790–825 °C), continued through T_g (~750–620 °C) and into the solid state, below T_g, based on a study of spherulite growth in the rhyolitic Rocche Rosse obsidian lava of Lipari island, Italy (Chap. 6; Fig. 6.10c). They propose that at high temperatures, nucleation rates are low and crystallisation rates high, whereas at lower temperatures, nucleation rates are high and crystallisation rates low because of slow diffusion rates at low temperatures. They estimated spherulite growth times to range from 4 days at high temperatures of 800–300 °C, and further modification to take 400 years. Clay et al. (2013) collected samples, undertook detailed petrographic studies of spherulite morphologies, internal spherulite textures, relationships to groundmass textures (e.g. flow banding, microlite alignments), and between adjacent spherulites in clusters, and geochemical profiling. They found that spherulites varied from isolated, small spherical features, to polygonal forms where the obsidian was almost completely crystallised. They attributed the polygonal form to growth interference (cf. vesiculated reticulite) occurring during growth in a (nearly) solid state. They also found that some spherulites have an elongate morphology, are aligned along flow bands and aligned microlite fabrics, and some deflect such microlite alignment fabrics indicating spherulite growth during magma flow under shear flow conditions. This could suggest shear-induced crystallisation.

As a result, Clay et al. (2013) proposed a continuum of four growth stages: (1) spherulite growth at magmatic temperatures at the time of eruption and initial cooling under conditions of slow nucleation and rapid fibrous crystal growth; (2) syn-deformational spherulite growth when the increasingly viscous lava is flowing at T_g; (3) under continued flow, spherulite crystallisation occurs in high strain zones, such as along flow bands and the interface between flow bands, producing elongate spherulites and elongate clusters of anastomosing spherulites. Small spherulites may grow under lower temperature conditions favouring rapid nucleation and slow crystal growth; and, (4) under low temperatures (300 °C) solid-state modification and recrystallisation of spherulites can occur leading to polygonal forms with 120° dihedral margin angles.

Using calorimetric FTIR analysis of subaerial obsidian, von Aulock et al. (2013) found that spherulite growth occurred from temperatures above T_g (i.e. >600 °C) down to 400 °C. From about 450 °C, cracks radial to the spherulites formed, which facilitated migration of magmatic water expelled during anhydrous quartz and feldspar fibrous crystal growth in the spherulites, and these were hydrated at these temperatures.

Bullock et al. (2017) built on the work of Clay et al. (2013), undertaking further research on spherulites in the Rocche Rosse rhyolite, as well as others on Vulcano, Italy. They proposed average growth times of ~5 days and identified five different spherulite types. These represent formation in areas of low strain to high strain in the lavas, different temperatures and degrees of undercooling, and post-emplacement modifications, in a medium that ranges in physical state from viscous silicate liquid to visco-elastic to solid. Their spherulite types are: *Type I*—small spherulites with radial fabric and slight elongation formed early at high degrees of undercooling at high T_g, or temperatures above T_g; *Type II*—elongate spherulites with a brown oxidation rim, formed preferentially during lava flow shear inflow band and fold zones; *Type III*—spherulites that formed in low shear zones and at low T_g, or lower temperatures; *Type IV*—spherulites found along fractures and faults in obsidian, indicating formation after brittle deformation and deformation induced nucleation and crystallisation; *Type V*—texturally partially or completely recrystallised spherulites and lithophysae, lacking a radial fabric, and perhaps partially infilled by secondary minerals, which formed at low T_g and lower temperatures.

13.2.3 Devitrification in Ignimbrites

Devitrification in ignimbrites can be more complex. It can have a zoned distribution through ignimbrites, and be accompanied by vapour-phase crystallisation and hydrothermal or fumarolic alteration. The essential devitrification stages and textures (hydration, glassy, spherulitic, and microcrystalline/microgranular; Sect. 13.2.1) can all occur. However, because of the differences in compactional (load) welding and zonal variations that can arise in ignimbrites (Chap. 10, see Sect. 10.8.14.1 for comprehensive description), different degrees of devitrification may develop at different levels (e.g. Smith 1960; Ross and Smith 1961; Smith and Bailey 1966; Lipman and Christiansen 1964; Briggs 1976; Carr 1981; Cas and Wright 1987). Carr (1981) also suggested that devitrified pumice seemed to host spherulite growth more than the glassy welded matrix.

13.3 Vapour-Phase Alteration of Some Ignimbrites

If non-welded ignimbrites remain unlithified for a long time, they are prone to weathering and erosion, and can be lost from the geological record. However, in some porous, partially and non-welded ignimbrites, vapour-phase alteration and crystallisation can occur soon after deposition, that can partly or wholly lithify them (Fig. 13.2a–c). The Peruvian name *sillar* has often been used for such indurated rocks (Fenner 1948; Cas and Wright 1987).

Vapour-phase alteration and crystallisation results from hot volatiles in the ignimbrite deposit, which stream upwards through the porous deposits. Secondary vapour-phase minerals are precipitated in porosity between ash particles and in pumice fragments, and include tridymite, less frequently cristobalite (both of which readily invert to stable quartz), alkali feldspar (usually, sanidine), as well as iron oxides and minor metal sulphides and chlorides, and less commonly scapolite (marialite), biotite, amphibole, and clinopyroxene (Fig. 13.2d, e and g–i; Sheridan 1970; Stimac et al. 1996; Caporuscio et al. 2012). These assemblages are principally associated with high-temperature magmatic vapours (>450–700 °C) trapped during and immediately following emplacement of the ignimbrite and fumarolic activity (Sheridan 1970; Sparks et al. 1999; Stimac et al. 1996; Caporuscio et al. 2012). However, in other cases, lower temperature (50–<450 °C) mineral assemblages are important in causing cementation, in particular zeolites, such as mordenite, and clinoptilolite–heulandite series (Fig. 13.2j, k; Cas and Wright 1987; Randolph-Flagg et al. 2017; Chap. 10), and in mafic ignimbrites phillipsite, chabazite, and analcime (Fig. 13.2f, l; de'Gennaro et al. 1999; Bear et al. 2009).

We note some workers consider vapour-phase alteration to be distinct from, although transitional to, fumarolic alteration, arguing that much fumarolic activity occurs at somewhat lower temperatures, while others, that vapour-phase alteration is distinctly different from where volcanic glass reacts with condensed vapour pore fluids to form zeolites at low temperatures. The latter is sometimes regarded more strictly as hydrothermal or geothermal activity. We however take a more holistic view in that such processes (and mineral assemblages) are both spatially and time-dependent, reflecting a number of factors including an ignimbrite's eruption style, transport and emplacement processes, depositional and geomorphological setting, temperature of emplacement, composition, permeability, amount of

Fig. 13.2 Vapour-phase altered ignimbrite. **a** 1.26 Ma Upper Bandelier ignimbrite, showing non-welded relatively fresh, basal zone (pink-orange) passing upwards into vapour-phase altered and devitrified tuff, which is strongly lithified (brown-grey), crudely columnar jointed, and in which large pumice clasts have been etched out through alteration, leaving cavities. Both zones are parts of one depositional or flow unit. Outflow sheet from Valles caldera, Pajarito Plateau, New Mexico, USA. **b** Non-welded, vapour-phase altered, and columnar jointed 2.08 Ma Cerro Galán outflow ignimbrite, Paycuqui, northwestern Argentina. **c** Non-welded texture in columnar joint blocks at same locality as in (**b**). Pumice clasts show no sign of any thermal, compactional (load) flattening. **d** Photomicrograph of high-temperature vapour-phase alteration of pumice with abundant tridymite and sanidine, fumarolic pipe centre, top Upper Bandelier Tuff ignimbrite, Los Alamos, Pajarito Plateau. Width of field of view ~1 mm, x-nicols; v = vesicle. **e** As in (**d**), fumarolic pipe centre, with pumice showing acicular splays of scapolite. Width of field of view ~1 mm, ppl. (From Caporuscio et al. 2012). **f** Photomicrograph of zeolitised pumice (ppl) from the mafic (tephriphonolitic to phonolitic) Tufo Rosso a Scorie Nere ignimbrite, Vico caldera, central Italy (From Bear et al. 2009). **g**, **h** SEM images showing effects of high-temperature vapour-phase alteration, ~225 ka Mamaku ignimbrite, New Zealand. **g** Unaltered rhyolitic pumice; **h** radiating crystals of albite growing into vesicle, and **i** platy crystals of silica polymorph (probably tridymite) in which the original pumice texture has mainly been destroyed. (From Milner 2001). **j** SEM image of zeolite-cemented rim of annular column in the 1.26 Ma Upper Bandelier Tuff ignimbrite, Paliza Canyon, Pajarito Plateau. Moss-like coating of very small mordenite crystals covers pitted natural glass surfaces. (Image by Steve Self; see Fig. 10.72j for field photo). **k** High-magnification SEM image showing acicular sheaths and platy (monoclinic) zeolite crystals (clinoptilolite and heulandite series) growing into pore space and occluding microporosity. Massive coherent tuff (sillar) interconnected with "steam" pipes, 95 ka Rio Caliente ignimbrite, Rio Salado, west of La Primavera rhyolite volcano, central Mexico. (From Wright 1979; see Fig. 10.72g for field relations). **l** SEM image of rhombohedral chabazite from the, which forms the dominant zeolite species forming up to over 50% of the matrix as well as replacing pumice as shown in (**f**), and spatter clasts from the Tufo Rosso a Scorie Nere ignimbrite (From Bear et al. 2009)

local precipitation, and direction of surface and groundwater flow, which contribute to the extent and directions of vapour transport and cooling history (see Chap. 10, Sect. 10.8.14.3; also see discussion by Bear et al. 2009). For example, studies by Holt and Taylor (1998) have shown that fumarolic fluids of the 0.76 Ma Bishop Tuff ignimbrite, were not dominantly composed of magmatic H_2O as proposed by Sheridan (1970), and provide evidence suggesting a short-lived (~10 year) high-temperature (>500 °C) meteoric–hydrothermal system existed within the cooling ignimbrite. Also, in order to precipitate zeolite minerals, and notwithstanding their lower temperature stability range, a sufficient quantity of water must be available to enhance the hydration–dissolution of volcanic glass (equivalent to 10%

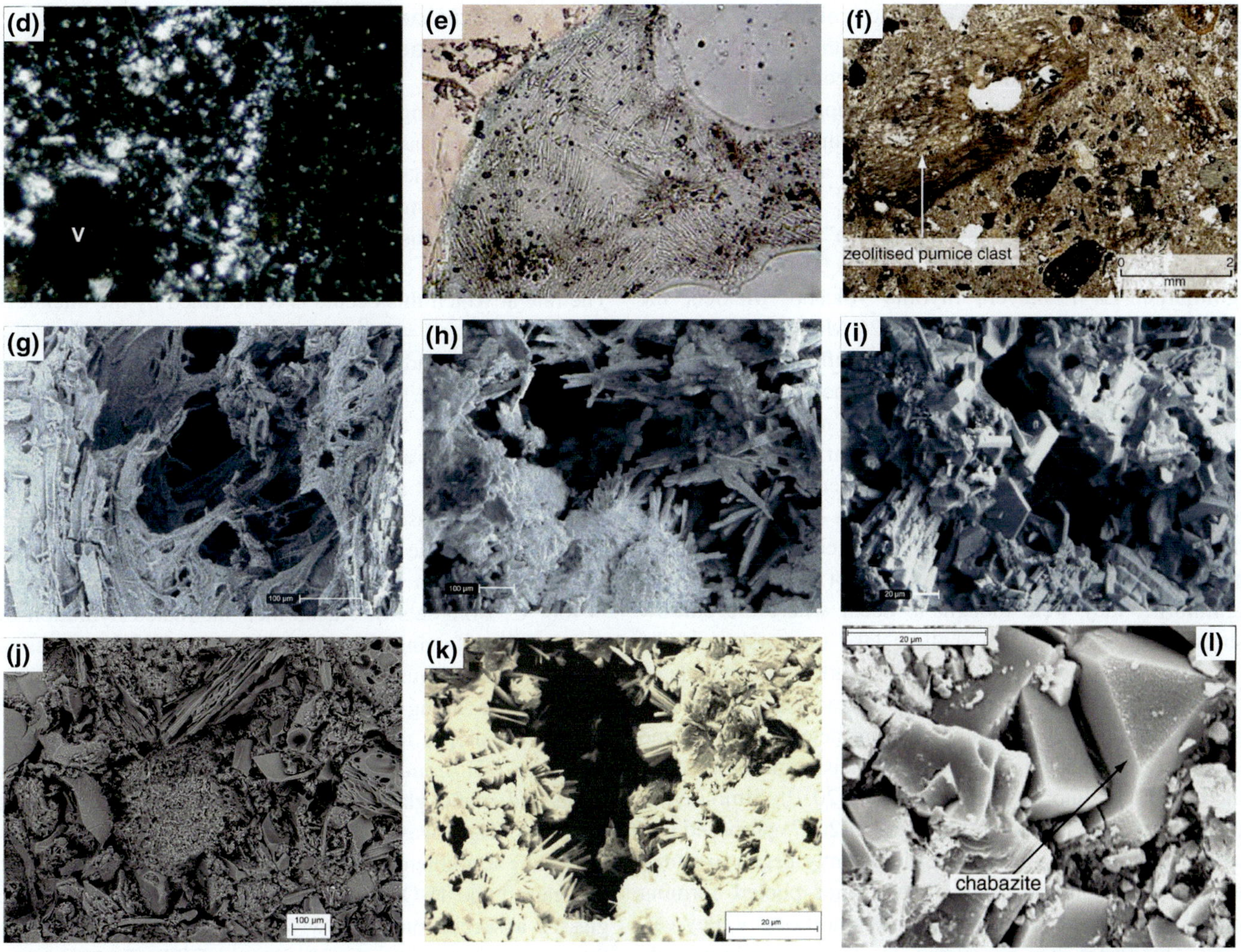

Fig. 13.2 (continued)

of its own mass; Bear et al. 2009). Magmatic H_2O is therefore solely unlikely to promote significant zeolitisation, and most of the water must be incorporated from an external source, either during eruption, upon deposition (lake, river valley or seawater), or post-depositionally (infiltration of surface or groundwater).

The volatiles that cause vapour-phase alteration include, as indicated above, magmatic volatiles, which exsolve or diffuse, or both, from juvenile glassy fragments, as well as heated ground and/or surface water (Sheridan 1970; Sparks et al. 1999; Keating 2005). Also, a significant of water vapour can be trapped water vapour within phreatomagmatic deposits, or from the interaction with geothermal fluids above the magma chamber at the time of eruption, as well as ongoing volcanic activity in near-vent settings (de'Gennaro et al. 1999, 2000; Bear et al. 2009). As secondary minerals are precipitated they infill pore space, and cement and lithify the ignimbrite (Fig. 13.2d–l). In high-temperature settings, this is commonly more intense in the upper part of the ignimbrite (see Figs. 10.65a and 10.72a). Rarely, this zone may be spatially associated with still preserved paleo-fumarolic mounds and ridges of sillar (hot gas pathways) on the surface of the ignimbrite; e.g. the 0.76 Ma Bishop Tuff, California, USA (Sheridan 1970; Fig. 10.72a); ~225 ka Mamaku ignimbrite, New Zealand (Milner 2001). Some vapour-phase altered ignimbrite is just progressively more lithified upwards (Vaniman 2006; Wright et al. 2011). In other ignimbrites more complex interrelations can occur between depositional units, vapour-phase alteration and welding zones. For example, in the 1.26 Ma Upper Bandelier ignimbrite, New Mexico, USA, vapour-phase crystallisation effectively forms an outer envelope or zone around the welded sheet (Smith and Bailey 1966; see Figs. 10.65c and 10.72b, c). Vapour phase alteration can produce the apparent appearance of two separate emplacement units within the same ignimbrite depositional unit (Cas and Wright 1987;

e.g. Fig. 13.2a). In the case of mafic ignimbrites, zeolitisation tends to be homogeneous through the thickness of the deposit, save for a thin (<2 m) non-zeolitised, unconsolidated basal zone reflecting cooler temperature conditions experienced during transport. However, thickness appears to play only a minor role and both thick and thin zeolitised deposits occur (de'Gennaro et al. 1999).

In several ignimbrites peculiar "steam" pipes or columns, have also been recognised, such as in the 95 ka Rio Caliente ignimbrite, La Primavera rhyolite volcano, Mexico, the Bandelier Tuff ignimbrites (1.8–1.26 Ma), around Valles caldera, and the 0.76 Ma Bishop Tuff (Chap. 10; see Fig. 10.72). Such pipes and columns may also occur with more massive zones (walls) up to 10 s of m high, and tabular and curvilinear bodies of similar coherent tuff, which do not obviously fit into any zonal welding scheme. These features are more resistant to erosion than the surrounding non-welded ignimbrite owing to the precipitation of zeolite minerals (Fig. 13.2j, k). In the case of the Rio Caliente ignimbrite, they were thought to occur where the outflow sheet was deposited in a shallow lake or marshy ground (Wright 1979; Cas and Wright 1987), while, in more isolated occurrences around Valles caldera, interaction between hot, non-welded ignimbrite and circulating meteoric and ground waters was associated with the rapid incision of valleys into the cooling ignimbrite (Bailey and Self 2010). Similarly, in the Bishop Tuff, Randolph-Flagg et al. (2017) proposed that as the ignimbrite was emplaced, it instantly began to slump along paleo-drainages, into which water pooled forming ponds or lakes, including in and near the Long Valley caldera. This water then percolated into the porous ignimbrite causing it to heat and boil.

Some non-welded, vapour-phase altered ignimbrites can also develop columnar jointing, which gives the misleading impression that the ignimbrite is welded (Chap. 10, Fig. 10.71). Wright et al. (2011) linked vapour-phase crystallisation of the 2.08 Ma Cerro Galán ignimbrite (northwest Argentina), and its concomitant columnar joint formation, to a paleo-topographic setting characterised by interface with external groundwater from the bottom of narrow valleys into which the ignimbrite was deposited (Fig. 13.2b, c). Here, vapourisation of water acts as a catalyst for devitrification and vapour-phase crystallisation, as well as for gas streaming through the deposit. The ignimbrite preserves non-welded texture (Fig. 13.2c), with pumice clasts showing no evidence of thermal, compactional (load) flattening (Chap. 10). The Cerro Galán ignimbrite is both welded and non-welded more distally, and in both instances, it can be columnar jointed (Cas et al. 2011b; Wright et al. 2011).

13.4 Hydration of Glass and Palagonitisation of Mafic Glass

Friedman and Long (1984) summarised the processes by which volcanic glass alters under ambient conditions. For obsidian glass, the initial most fundamental process is hydration, involving absorption of magmatic and external water through diffusion into the glass, which is counterbalanced by diffusion of Ca^{2+}, Na^{+}, and Mg^{2+} out of the glass. The rate at which this occurs is represented by the thickness of glass infiltrated = $k\,t^{0.5}$, where k is a constant and t is the time. Friedman and Long (1976) found that the hydration rate increases as a function of increasing SiO_2 content and increasing ambient temperature, and decreases as a function of increasing CaO and MgO content.

Von Aulock et al. (2013) found that in obsidian lava, crack formation associated with spherulite growth occurred at ~450 °C, while perlite crack formation resulting from cooling contraction volume changes (see below), occurred at ~400 °C. Both sets of cracks then facilitated hydration of the glass by allowing magmatic water expelled from sites of growth of anhydrous quartz and feldspar fibrous crystallites, to permeate along them. Seligman et al. (2016) used hydrogen isotopes to measure the proportions of magmatic water (as OH^-) and meteoric water (as H_2O) in vesicular felsic and mafic glasses, and found that whereas residual magmatic H_2O was as low as 0.1 wt%, after 1,500 yrs hydration at ambient atmospheric temperatures by meteoric had raised the H_2O level to 1.5–3.5 wt%, whereas mafic glasses absorbed only 1.5 wt% H_2O after 7,000 yrs. Furthermore, Giachetti et al. (2015) used FTIR (Fourier transform infrared spectroscopy) speciation of water in natural 950 yr obsidian, and experimentally hydrated and then dehydrated obsidian (at magmatic temperatures) to determine that the original magmatic water content was only 0.2–0.5 wt%, whereas meteoric water represented 1–2 wt% water, which had diffused into the glass in 950 yr. In other words, all volcanic glass will become hydrated by meteoric water after a very short (geological) time, interval, and this seems to be irrespective of climate.

When hot mafic glassy deposits are in contact with ambient water at the time of eruption, they are prone to alteration, involving hydration through a process called *palagonitisation*. As basaltic magma is chilled or quenched by contact with water it forms a pale brown to reddish brown glass called *sideromelane*, or a black glass called *tachylite* (MacDonald 1972). Tachylite was thought to be black due to very fine, dispersed grains of magnetite (Peacock and Fuller 1928; Fenner 1932; MacDonald 1972). However, Kawachi et al. (1983) established that the black grains are not magnetite but quenched crystallites of pyroxene. In pillow lavas, the two glass types may occur together in layers at the margins of pillows (Moore 1966; Kawachi et al. 1983). These glasses, especially sideromelane, are susceptible to alteration.

When sideromelane glass alters it changes to a pale yellow to yellow–brown or orange, hydrated, alteration mineral called *palagonite*, which forms rims at the interface between the glassy margins of pillows, hyaloclasts and phreatomagmatic pyroclasts and external water (Fig. 13.3; Peacock 1926; Peacock and Fuller 1928; Moore 1966; Hay and Iijima 1968; MacDonald 1972; Kawachi et al. 1983; Jakobsson and Moore 1986; Stroncik and Schmincke 2001, 2002; the last two papers present the most recent and comprehensive review of palagonite formation). Palagonite was named after the locality of Palagonia, in the Iblean Mountains of Sicily by von Waltershausen in 1845, where altered, uplifted hyaloclastite deposits occur (Stroncik and Schmincke 2002). Palagonite is either gel-like or finely fibrous (Peacock 1926; Stroncik and Schmincke 2001, 2002). Gel palagonite is clear and isotropic and occurs in contact with sideromelane glass, whereas fibrous palagonite is yellow–orange–brown, has crystalline smectite (nontronite and saponite) characteristics and represents a more advanced stage of hydration and alteration of the glass (Stroncik and Schmincke 2002).

The formation of palagonite involves hydration and ion exchange (Moore 1966). Moore found that during the formation of palagonite in submarine hawai'ian basaltic glasses, Na, Ca, and Mn were lost, whereas K, Fe, and Ti were gained relative to the unaltered glass, presumably due to exchange with seawater. He also found that the palagonite layers at the margins of pillows became thicker with age. A manganese crust precipitated from seawater was observed on the surfaces of pillows, and was also noted to be thicker as the age of the pillows increased. Moore also suggested that tachylite could be replaced by palagonite, but Kawachi et al. (1983) found that this had not happened within their Eocene pillow lava succession. Kawachi et al. (1983) found that palagonitic layers were depleted in Si, Mn, Ca, and Na, and enriched in K and H_2O. Fe was either depleted or enriched, as were Ti, Mg, and Al. Ti, contrary to its reputation as a stable element, was found to be considerably mobile.

Stroncik and Schmincke (2001, 2002) proposed that during the early gel-palagonite forming stages, Al, Mg, Ca, Na, and K are lost, H_2O is gained and Ti and Fe remain immobile. During the advanced fibrous-palagonite forming

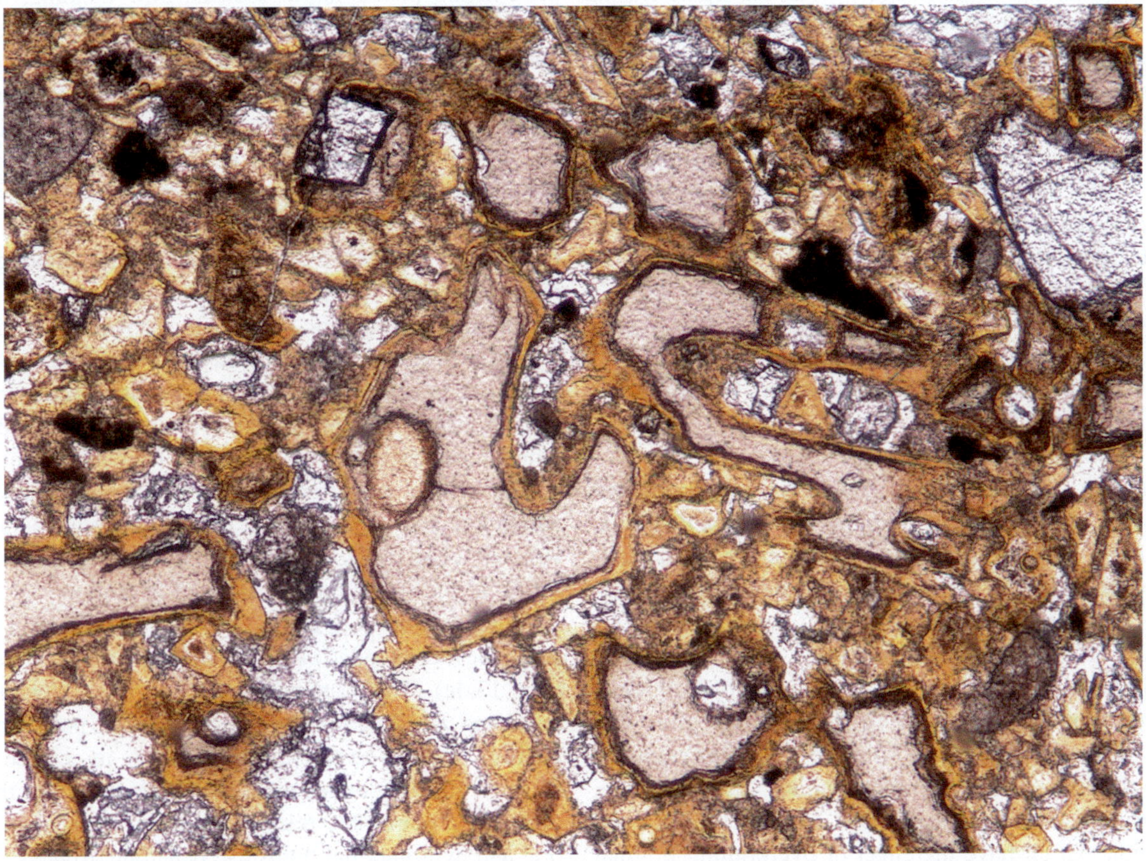

Fig. 13.3 Orange-yellow palagonite rims around the margins of phreatomagmatic sideromelane (pink) shards, and in the finer basaltic ash matrix, ~2 Ma Cape Bridgewater volcano, Newer Volcanics Province, Victoria, Australia. Magnification10x, ppl

stage, Ca, Na, H_2O, Ti and Fe are lost, and Si, Al, Mg, and K are gained, based on previous studies, their own analyses and modelling of processes. During subsequent, ongoing "diagenetic" alteration, calcite and zeolites may be precipitated in the altered glass through replacement overprinting.

Palagonitic alteration of basaltic glass is clearly pre-metamorphic, is essentially due to the interaction of ambient water with metastable basaltic glass (Moore 1966; Hay and Iijima 1968; MacDonald 1972), and produces significant compositional changes in affected rocks. It affects not only the rims of subaqueous pillows and hyaloclastite clasts but also basaltic phreatomagmatic pyroclastic ash fragments and deposits (Fig. 13.3). Jakobsson (1978) summarised his significant research on the palagonitic alteration of the tephra erupted on Surtsey during 1963. He indicates that palagonitisation can occur at normal surface temperatures and pressures over the years but is catalysed by the influence of temperature. In the area of a significant thermal anomaly, palagonitisation appears to have occurred within 1–1.5 years after eruption of deposits, under the influence of temperatures between 40 and 100 °C, and occurred in association with the contemporaneous precipitation of secondary minerals such as opaline silica, zeolites, and calcite (Jakobsson 1978; Jakobsson and Moore 1986). Palagonite rims formed at a rate of 0.1 $\mu m\ yr^{-1}$ when the tephra was at 40 °C, 1 $\mu m\ yr^{-1}$ at 80 °C, and 10 $\mu m\ yr^{-1}$ at 120 °C (Jakobsson and Moore 1986). Palagonitisation thus plays a role in stabilising and cementing Surtseyan volcanic piles (Fig. 16.103c–e) within several years after the cessation of volcanism. For an update on the hydrological and microbiological interactions with basaltic pyroclastic deposits at Surtsey over the past 50 years, see Jackson et al. (2019) and the results of the SUSTAIN drilling project.

13.5 Alteration and Formation of Pseudo-clastic Textures in Felsic Volcanic Rocks

One of the most challenging aspects of working in ancient volcanic successions, especially in the exploration and mine environments, is deciphering true or "primary" felsic volcanic textures from apparent or "*false*", *pseudo-clastic* textures. As described above, permeation of magmatic hydrous fluids, surface waters, and meteoric water in the subsurface, can begin to remobilise elements and initiate hydrothermal fluid circulation and alteration of the volcanic glass and original minerals. In older volcanic rock successions, post-depositional processes of ongoing devitrification, hydrothermal alteration, and then metamorphism and deformation can all have the effect of overprinting original volcanic textures and mineralogy, but also, in some instances, to obliterate them totally and produce new textures, which can lead to misconceptions of the original rock types and their origins, paleoenvironmental settings and hence prospectivity.

In the following discussion we examine three situations, where the effects of these processes on original felsic volcanic textures can be particularly pervasive, and where the formation of pseudo-clastic textures is common:

- perlitic cracking and alteration,
- polyphase hydrothermal alteration in coherent rocks, and
- hydraulic fracturing.

13.5.1 Perlitic Cracking and Alteration

Perlite cracks are a network of intersecting conchoidal cracks that commonly develop in isotropic volcanic glass. Perlite cracks are found in obsidian lavas (Fig. 13.4a, b; also see Fig. 6.11a–c) and in densely welded ignimbrite *vitrophyre* zones, where they cross-cut the planar welding texture (Fig. 13.4c). They are spheroidal in shape in 3D (Fig. 13.4a, b), but (sub-)concentric in 2D (e.g. in thin section, Fig. 13.4c), except where they form superimposed on a pre-existing planar fabric such as flow banding, in which case they may align along the fabric and are known as banded perlite (Allen 1988; see Fig. 13.5b). Contrary to existing models at the time, Cas and Wright (1987) proposed that as volcanic glass cools and contracts as an isotropic glass medium, conchoidal perlite cracks form through cooling contraction. The prevailing interpretation at the time and even subsequently was that perlite cracking formed later, as a consequence of hydration of glass, although it was never clear how this process worked.

Through calorimetric FTIR analysis of subaerial obsidian, von Aulock et al. (2013) showed that perlite fractures formed as the glass cooled to ~400 °C, immediately after spherulite growth and after the formation of spherulite-related cracks, which formed from T_g to ~450 °C. They also proposed that perlite cracks can form during subaqueous quench fragmentation, and are not related to late, cold-state hydration of volcanic glass. In fact, they found that cracks that formed at <200 °C showed almost no signs of hydration.

When perlitic crack texture is widely developed in volcanic glass, the rock is called *perlite*, and it has a nodular-like texture (Fig. 13.4a). That can produce some misleading textures, including *pseudo-granular* sandstone in fine perlite, potential pebbly, *pseudo-conglomerate* texture in coarse perlite (Fig. 13.4b). Of course, once perlite cracks form, they then represent a permeable fracture system that magmatic, meteoric and hydrothermal fluids can penetrate, triggering hydration and alteration of the glass bounding the cracks (Fig. 13.4c). In thin section, false shard and pyroclastic

Fig. 13.4 Features of perlite. **a** Nodular appearance of perlitically fractured obsidian, each nodule bound by concentric perlite cracks. Miocene, Kuriowa, Japan. **b** Granular, pebbly appearance of hydrated perlite. Quiron lava dome, Quevar volcano, northwestern Argentinian Andes. **c** Photomicrograph of (sub-)concentric perlite cracks in densely welded intra-caldera ignimbrite, Permian Ora ignimbrite, Bolzano, northern Italy. The brown highlighting the cracks is an oxidation effect after water. Width of field of view ~2 mm, ppl. (Image by M. Willcock)

textures can even be produced by incomplete alteration of glass bounding perlite cracks, leaving curviplanar, cuspate relict forms between cracks and where they merge, which superficially look like cuspate glass shards, but are *pseudo-shards* (Fig. 13.5; Allen 1988). The key to recognising these pseudo-clastic textures is recognising the perlite crack or fracture network through the alteration, which in some cases can be done with careful hand-lens examination of hand specimens, and/or in thin section. In the case of pseudo-shards, they are aligned parallel to the perlite cracks (Fig. 13.5).

13.5.2 Polyphase Hydrothermal Alteration in Coherent Felsic Rocks

Hydrothermal alteration is a broad term to describe the chemical reactions, and mineralogical and textural changes in rocks when subject to chemical reactions with hot water, steam, or gas, under changing thermal and chemical conditions (e.g. Henley and Ellis 1983; Chap. 18). The upper limits of this process are ill-defined and are gradational into low-grade diagenesis or burial metamorphism. According to Henley and Ellis (1983), hydrothermal alteration involves ion-exchange reactions, mineral phase transformations, mineral dissolution and new mineral growth or replacement. Jébrak (1997) has called hydrothermal alteration "*corrosive wear*". Although the original rock composition has some effect on the secondary alteration mineralogy, the main influences are the permeability of the rock pile, temperature, the composition of the fluids moving through the rock pile, and the degree of chemical disequilibrium between these fluids and the host rock and its minerals.

Hydrothermal alteration can take many forms, and occurs as a result of ion diffusion and fluid migration, ion exchange and mineral replacement. Some of the alteration minerals produced include amorphous silica, quartz, K-feldspar, albite, calcite, montmorillonite, montmorillonite–illite, illite, kaolinite, alunite, chlorite, and a wide array of zeolites and minerals that form under low-grade metamorphic conditions (Henley and Ellis 1983; Gifkins et al. 2005). Needless to say, bulk rock compositions can be very significantly altered. Stages of alteration or distinctive alteration mineral assemblages, or both, are identified by a variety of terms, which are by no means standardly defined and used, such as chloritic, sericitic, potassic, phyllic, argillic, propylitic alteration, and silicification. Hydrothermal alteration processes are an extremely important adjunct to the formation of epithermal precious and base-metal volcanic-hosted massive sulphide (VMS) mineralisation, as well as other magmatic mineralisation types (Chap. 18; Beane 1982; Barrie and

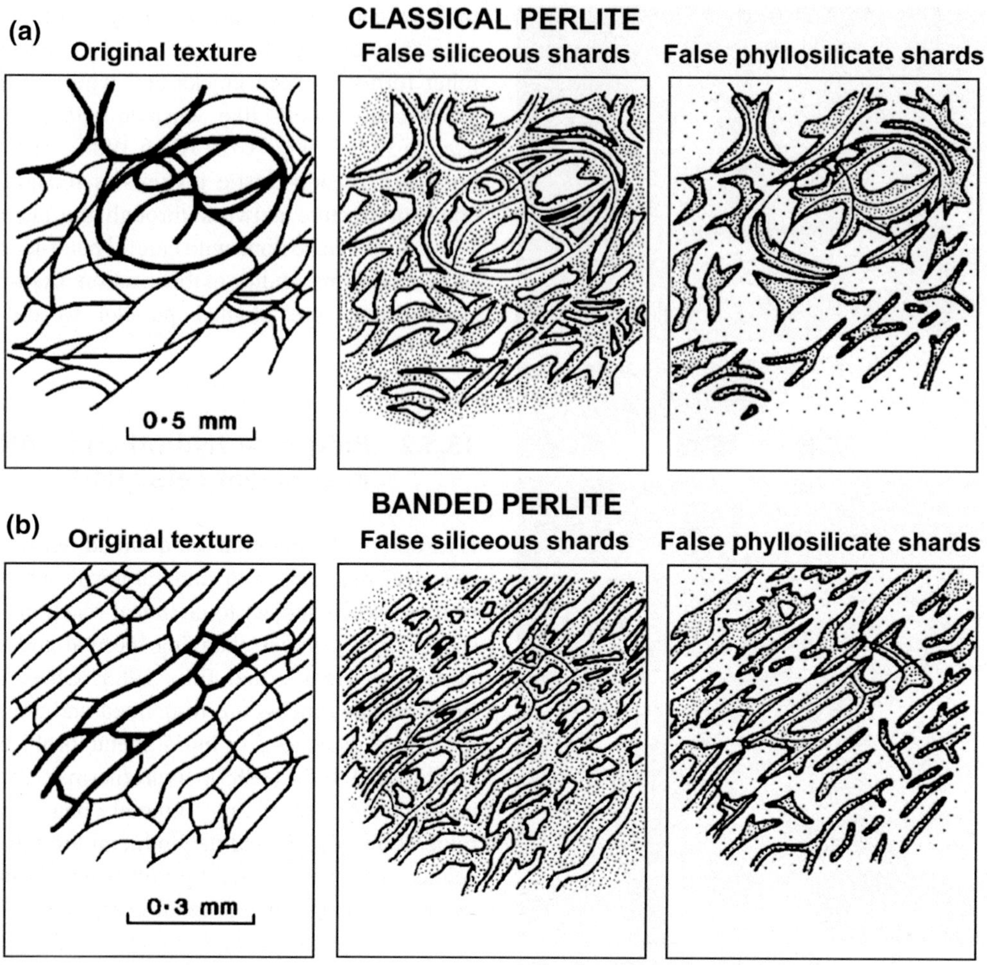

Fig. 13.5 Diagram showing the formation of pseudo-shard textures from hydration-driven alteration along original perlite cracks in coherent felsic rocks, with alteration migrating inwards from cracks isolating pseudo-shard forms. **a** Top row: standard ovoid perlite cracks. **b** Bottom row: elongate perlite crack network developed on glass with an original planar fabric such as flow banding. (From Allen 1988)

Hannington 1999; Cooke and Simmons 2000; Dubé et al. 2007; Hutchinson 1983; Lydon 1984,1988; Porritt et al. 2012; Rona 1984; Scott 1971; Titley 1982; Urabe et al. 1983). This alteration significantly affects the original host rocks to the mineralisation, with the effects being most extreme immediately around the mineralisation and the hydrothermal fluid conduits, and decreasing away from them. A detailed discussion of alteration types, processes, assemblages, and zonation patterns is beyond the scope of this chapter, and for a more detailed discussion and current references, the reader is referred to Chap. 18. However, rocks around hydrothermal systems may be altered multiple times, producing different alteration minerals. Alteration may be pervasive or partial and patchy, producing a poly-phase alteration pattern, with later alteration phases or mineral assemblages variably overprinting early alteration phases, as discussed further below.

Typical alteration minerals such as chlorite, secondary quartz and sericite, however, can also be produced by low-grade metamorphism, so that, in older, deformed and metamorphosed volcanic rock successions, deciphering the effects of hydrothermal alteration from metamorphism may be difficult. Silicification can also be produced simply from circulating meteoric waters passing through a glassy volcanic pile. It may also be difficult to distinguish the effects of devitrification from true hydrothermal silicification. Our interest here, however, is to evaluate how original volcanic deposit textures may be altered, the possible apparent or pseudo-textures that can be produced and the ambiguity and confusion that these secondary textures can pose in understanding the origins and setting of altered volcanic rock successions.

13.5.2.1 Effects on Original Volcanic Textures

According to Jébrak (1997), there are two regimes of hydrothermal alteration or corrosive wear. In the *diffusion-limited regime*, there is a marked chemical disequilibrium between the host rock and the hydrothermal fluids. The fluids pervade the exposed surface area of host rock along fractures and permeability pathways, and progressively dissolve ions and alter the rock inwards from the permeable fluid pathways. Corners where fractures intersect are more easily altered than a continuous planar surface because multiple alteration fronts are active. Originally polygonal fracture blocks begin to develop rounded cores of unaltered rock as alteration advances into the joint-bounded block (Cas et al. 2011a), as also happens with spheroidal weathering. The fresh unaltered domains become apparent clasts, whereas the altered rinds adjacent to the fractures become an apparent matrix. In the *kinetic regime,* the rate of alteration is controlled by a less vigorous or pervasive rate of chemical reaction. Alteration can be even more irregular than in the diffusion-limited regime, producing highly irregular apparent clasts, and a highly irregular distribution pattern of apparent clasts and matrix.

Hydrothermal alteration can thus be pervasive, or patchy and domainal. The latter is particularly the case when more than one alteration event or type occurs, called *polyphase alteration*, which may occur in close spatial and temporal relationship. The effects of one alteration phase may be superimposed on an older alteration phase. When either or both phases are not pervasive in their effects, complex textures due to patchy distribution of different alteration mineral phases occur that can be difficult to unravel. In particular, patchy alteration textures may give the impression of apparent or pseudo-clastic textures in coherent rocks. The relict patches of original rock appear to be clasts in a differently coloured, mineralogically altered rock that is the *apparent* or *pseudo-matrix*, or vice versa.

Allen (1988) first convincingly demonstrated that some Paleozoic VMS mineralised successions in southeastern Australia, previously interpreted as felsic pyroclastic and epiclastic volcaniclastics, are variably devitrified, polyphase altered lavas, cryptodomes and quench fragmented hyaloclastite (Cas and Wright 1987). Allen (1988) showed that the apparent clastic textures represent intensely sericite-altered patches in less sericite-altered, silica–sericite material, or intensely chlorite-altered patches in less chloritic, more siliceous material (Plate 13.1a). In both cases, original glass has been initially altered and then incompletely silicified.

When rocks are patchily, polyphase altered and there is no strong fabric in the original rock, the result can be a texture that looks like a breccia (Fig. 13.6a, b), but is more correctly called a *pseudo-breccia*, or even *pseudo-conglomerate* (Plate 13.1b; Fig. 13.6c–e). The clasts, on close inspection, tend to be highly irregular and have sharp to diffuse, gradational boundaries with the *apparent* or *pseudo-matrix* (Fig. 13.6a–f). Where a pre-existing planar flow fabric exists, alteration can superimpose and highlight such a layering or anisotropy (Fig. 13.6f).

With more intense alteration, usually associated with pervasive hydraulic fracturing (see Sect. 13.5.3, below), which provides pathways for the hydrothermal fluids, pseudo-conglomerate textures can be produced (Fig. 13.6c–e). Such textures form when intense alteration advances inwards from intersecting planar fractures that are conduits for the alteration fluids. In time, this causes corners of the rectilinear fracture blocks to be corroded and replaced, isolating rounded cores of relatively unaltered rock that constitute roundish pseudo-clasts (cf. spheroidal weathering of jointed blocks), surrounded by seams of alteration minerals, such as chlorite, which appear to be a matrix, but are a pseudo-matrix (Fig. 13.6c–e; Cas et al. 2013). Notable points about such pseudo-conglomerates are that (1) the clasts are all the same lithology (Fig. 13.6d–f), which is very unusual for high-energy, reworked sedimentary conglomerates, which are usually polymictic (2) the pseudo-clast margins are rarely smooth in detail, most showing irregular and often diffuse margins (Fig. 13.6c–e), and (3) there are gradations from pseudo-conglomerate textures into domains of in-situ jigsaw-fit (hydraulic fracture) breccia and into coherent rock (Fig. 13.6c, d). This is a textural and process impossibility in high-energy tractional sedimentary environments which are the only environments where rounding of gravel-size clasts occurs. In such settings, there should be an erosional unconformity between bedrock and any gravel or conglomerate deposits which are derived from the bedrock.

Another common effect of hydrothermal alteration is variation in the degree of alteration and even changes in the alteration mineral assemblages through a single rock unit (Fig. 13.7a). Although differently altered zones may suggest that there are different rock units, contacts between the alterations are often gradational if changes are carefully observed in continuous outcrop or drill core, not sharp, and thus indicate variable alteration of just one rock unit. In addition, hydrothermal alteration can selectively alter and mask some components, such as phenocrysts, or even specific phenocrysts, in one part of a rock unit, but not in another (Fig. 13.7b). For example, this may give the impression at first glance that one part of the unit is porphyritic, but another is not (Fig. 13.7b), or perhaps that there are two separate rock units. However, again such changes are often gradational (Fig. 13.7b), indicating that there is probably only one rock unit that is differentially altered. Close examination of the more altered part may show relicts of the original crystals or their pseudomorphs in the same abundance and of the same size as the less altered part of the rock unit.

When patchy hydrothermal alteration texture is superimposed on rocks with a pre-existing planar fabric, such as flow-layering, the apparent or pseudo-clasts may be

(a)
ASAHI
PENTAX
(b)
(c)
(d)
(e)
(f)

◀ **Fig. 13.6** Pseudo-breccia and pseudo-conglomerate alteration textures that may be confusing (*to say the least*)! **a** Gradational to sharp textural contacts between relict, early chlorite altered domains (pseudo-breccia clasts) and later quartz–sericite altered domains (pseudo-matrix) in coherent, Cambro-Ordovician altered submarine rhyolite lava. Apparent clasts are sharply to diffusely defined, irregular in shape and have the same phenocryst content as in the pseudo-matrix. Highway-Reward VMS mine, Mt Windsor Volcanics, Charters Towers, Queensland, Australia. **b** Chlorite altered (dark domains = pseudo-clasts) and extremely silicified (pinkish-grey domains = pseudo-matrix) within coherent, Cambrian altered submarine rhyolite lava. Mt Read Volcanics, Jukes Proprietary, Queenstown, western Tasmania, Australia. The highly irregular shape of the pseudo-clasts cannot be explained by any known volcanic fragmentation process! Not even pumice clasts are so irregular. The silicified domain consists of anastomosed siliceous nodules. Relict and similar feldspar phenocrysts can be seen in both domains with hand-lens. **c** Hydraulically fractured and altered Archean dacite, showing pseudo-conglomerate textures. Apparent rounded "clasts" are in gradational contact with domains of in situ hydraulic fracture breccia showing jigsaw-fit texture, and coherent dacite, which is a textural and process impossibility for true tractional sedimentary conglomerates. Black seams are chlorite alteration along original planar hydraulic fractures, which in some cases have replaced corners to produce apparent rounded clasts. **d** Close-up of pseudo-clasts in (**c**), showing a diffusely bounded round "pacman" shaped clast, and other clasts that show relict jigsaw-fit textural relationships separated by chlorite alteration seams. **e** Extreme chlorite alteration of original coherent dacite, showing some very irregular pseudo-clast margins, which could not survive the tractional, sedimentary reworking required to form rounded pebbles in a sedimentary environment. **c–e** from the Archean Black Swan nickel mine host rock succession, north of Kalgoorlie, Western Australia). **f** Close-up of sericite alteration (light) superimposed on earlier chlorite alteration preserving faint, relict flow banding in coherent submarine rhyolite lava. Note the uniform phenocryst texture throughout, and the diffuse and highly irregular boundaries of the chloritic pseudo-clasts, which because of their elongate and parallel alignment could be mistaken for fiamme in welded pyroclastic rocks. Silurian Thorkidaan Volcanics, Benambra, eastern Victoria, Australia. All drill core ~5 cm in diameter, with exception of (**f**) which is 3.5 cm (**a** and **c** from Cas and Wright 1987)

Fig. 13.7 Effects of variable alteration type or intensity, or both, on rock textures in single rock units. **a** Six core samples through the same dacite unit. In order from left to right the principal alteration phases are 1. quartz–kaolinite sericite; 2. sericite chlorite; 3. chlorite–pyrite; 4. sericite–quartz chlorite; 5. chlorite–sericite; 6. sericite (pale) chlorite (dark) resulting in an apparent clastic, breccia texture. Silurian Thorkidaan Volcanics, Benambra, eastern Victoria, Australia. **b** Variable visibility of phenocrysts in a dacite as a function of varying intensity of sericite-quartz alteration. Phenocrysts are easily visible in the core on the left, moderately visible in cores 2 and 3 and barely visible or imperceptible in the other cores. Archean Golden Grove VMS host rock succession, Yilgarn Craton, Western Australia. All drill core is ~5 cm in diameter

Fig. 13.8 Pseudo-fiamme and pseudo-eutaxitic textures. **a** Variably altered massive to flow banded Miocene dacite, Green Tuff Belt, Hokuroko Basin, Honshu, Japan. The left-hand core is relatively fresh dacite, the cores to the right are progressively more sericitically and domainally, chloritically altered. The chlorite domains give the impression of welded fiamme (i.e. collapsed pumice in welded ignimbrite), whereas the rock remains coherent dacite. The alteration changes are gradational, which is critical in identifying all cores come from the same coherent dacite body. Drill core is diameter ~5 cm. **b** Pseudo-fiamme or pseudo-welding texture in coherent rhyolite that has been polyphase domainally altered to chlorite and siliceous or quartz nodules. The wispy chlorite domains (pseudo-fiamme) are bounded by anastomosed siliceous nodules, which at the re-entrants define spikey chlorite prongs perpendicular to the sub-horizontal fabric, which real welded pumice fiamme in ignimbrites and welded fallout deposits never show. Cambrian Mt Read Volcanics, hanging-wall sequence, Hercules VMS mine, western Tasmania, Australia. **c** Pseudo-fiamme or pseudo-welding texture defined by chlorite domains in a matrix that is largely sericitically altered. The deposit is a non-welded, submarine, pumice mass-flow deposit from the foot-wall sequence, Hercules VMS mine, Cambrian Mt Read Volcanics, western Tasmania, Australia. **d** Photomicrograph from the unit shown in (**c**) showing non-collapsed, non-welded tube pumice fragments. The strong horizontal foliation in the centre is a tectonic cleavage. Width of field of view ~5 mm, ppl

elongated, parallel to the original fabric (Fig. 13.8a–c). They may even resemble collapsed welded pumice clasts or fiamme in ignimbrites (Chap. 10), and are thus called '*pseudo-fiamme*' (Fig. 13.8a–c). Although this pseudo-eutaxitic texture gives the appearance of welded ignimbrites, the original rocks could have been coherent lavas or intrusions (Fig. 13.8a, b), and are in some cases even non-welded deposits of original pyroclastic debris (Fig. 13.8c, d) that has been overprinted by a secondary misleading pseudo-eutaxitic alteration texture. In both cases, such rocks could be called '*enigmabrites*' (which thus becomes a word that can be used in the game of Scrabble!).

In summary, the principal criteria that allow pseudo-clastic textures to be recognised as alteration textures include:

(a) Diffusely defined clasts and clast margins, marked only by colour differences between apparent clasts and matrix (Fig. 13.6a, b). Colour differences may vary from stark to faint. Clast boundaries may be diffuse, and marked by a gradational colour change (Figs 13.6 and 13.8a).
(b) Clast margins and clast shapes in detail may be highly irregular in form, with inexplicable protrusions and

embayments that could not be produced by any sedimentary physical tractional process (Fig. 13.6a–f). Nevertheless, even round clast forms and pseudo-conglomerate aggregates can be generated (Fig. 13.6c–e).

(c) The texture of both pseudo-clasts and apparent matrix is the same, with the same types, same size, abundance, and distribution pattern of phenocrysts or grains in both, even though the visibility of the components in the more mineralogically different altered domains may be more cryptic (Fig. 13.6a, f).

(d) Faint, alteration-modified relict flow banding (Figs 13.6f and 13.8a), hyaloclastite and autobreccia textures (Chap. 4) also indicate the existence of original lavas or intrusions, rather than pyroclastic deposits.

(e) Features that help to distinguish true coherent rocks from fragmental or volcaniclastic rocks, in cases where the rock is strongly altered, and the origin is not immediately obvious, include (Cas et al. 2008; Hayman and Cas 2011):
   - uniform distribution of mostly euhedral–subhedral, large phenocrysts, and an apparent porphyritic texture throughout a fine medium which may have a smaller crystal population (groundmass crystallites) would be consistent with an altered coherent rock (lava, high-level intrusion);
   - the presence of fragile glomerocrysts is consistent with a coherent rock, unless they occur within a large lithic clast in a clastic rock;
   - when large crystals are dispersed in a fine medium, but the crystals are mostly broken, subhedral to angular in form, irregularly distributed in the fine medium, with the aggregate appearing to be poorly sorted, this would be consistent with a volcaniclastic texture; and

Although the emphasis here has been largely on textural changes, it is inherent from the types and intensity of processes operating that significant chemical changes should also occur, causing significant changes to the original igneous bulk–rock compositions.

13.5.3 Effects of Hydraulic Fracturing on Original Volcanic Textures

As discussed in Chap. 4, if a fluid system in the subsurface develops a pressure which exceeds the tensile strength of the enclosing rock and the minimum principal lithostatic stress component, then that fluid has the capacity to propagate a fracture or fracture system, and to open it in a tensile manner. Such fracture propagation brecciates the country rock, which is essentially shattered in situ under tensile conditions by the over-pressured fluids working through the rock (Plate 13.1c; Fig. 13.9 stages 1 and 2). The resultant fractures need not be involved in shear motion, and so the breccia fragments and the associated wall rock to the fracture need not be pervasively foliated or sheared, as would occur in a fault or shear zone. This fluid-induced fracturing and brecciation is called *hydraulic fracturing*; for further discussion of the mechanics and principles, see Secor (1969), Phillips (1972), Jébrak (1997), Cas et al. (2011a) and Chap. 4.

Hydraulic fracturing is *NOT* a primary volcanic fragmentation process, but a post-depositional subsurface process. Hydraulic fracturing propagates progressively upwards from an over-pressured, usually gaseous and/or aqueous hydrothermal source in the subsurface, in the direction of lower confining pressures higher in the crust. The major, first-order fractures in hydraulic fractures and fracture systems are therefore mostly sub-vertical in their orientation (Fig. 13.9), and the clasts bounded by hydraulic fractures are often elongate in shape with a sub-vertical orientation of clast long axes (Fig. 13.10a, b). This is not due to a tectonic strain, unless in older volcanic rock successions, the whole succession has been penetratively deformed before or after the hydraulic fracture system formed.

Hydraulic fracturing propagates several orders and scales of fractures (Fig. 13.9), producing a complex in situ breccia texture (Fig. 13.10a–d, Plate 13.1c). Second and higher order fractures are often propagated as fluid pressure increases in first-order fractures and/or as they are sealed by precipitation of hydrothermal minerals in them, leading to increased fluid pressure again, which then cracks open new, higher order fractures (Fig. 13.9, stage 3). This leads to smaller and smaller hydraulically fractured fragments being formed, most of which remain in situ, but are progressively chemically corroded and replaced, so changing their shape (Figs. 13.9 and 13.10a–d; Plate 13.1c; Cas et al. 2011a).

In volcanic systems, hydrothermal fluid systems have enormous potential for causing post-depositional hydraulic fracturing, brecciation, alteration and ore deposit-forming mineralisation (Beane 1982; Barrie and Hannington 1999; Cooke and Simmons 2000; Dubé et al. 2007; Hutchinson 1983; Lydon 1984,1988; Porritt et al. 2012; Rona 1984; Scott 1971; Titley 1982; Urabe et al. 1983). The breccia fragments are generally angular, blocky to splintery in shape and show in situ jigsaw-fit relationships (Fig. 13.10b; also see Fig. 4.33). Out of context, they could be confused with quench-fragmented jigsaw-fit hyaloclastite deposits. Where observable in outcrop, hydraulic fracture breccias should be distinguishable by their cross-cutting character and confinement to a limited fracture zone (Fig. 13.9), and associated, very strong hydrothermal alteration, represented by alteration minerals infilling fractures as veins and also variable alteration of the host rock hydraulic fracture clasts (Figs. 13.9d and 13.10a–d). The width of such hydraulic fracture zones may vary from several cm to many m.

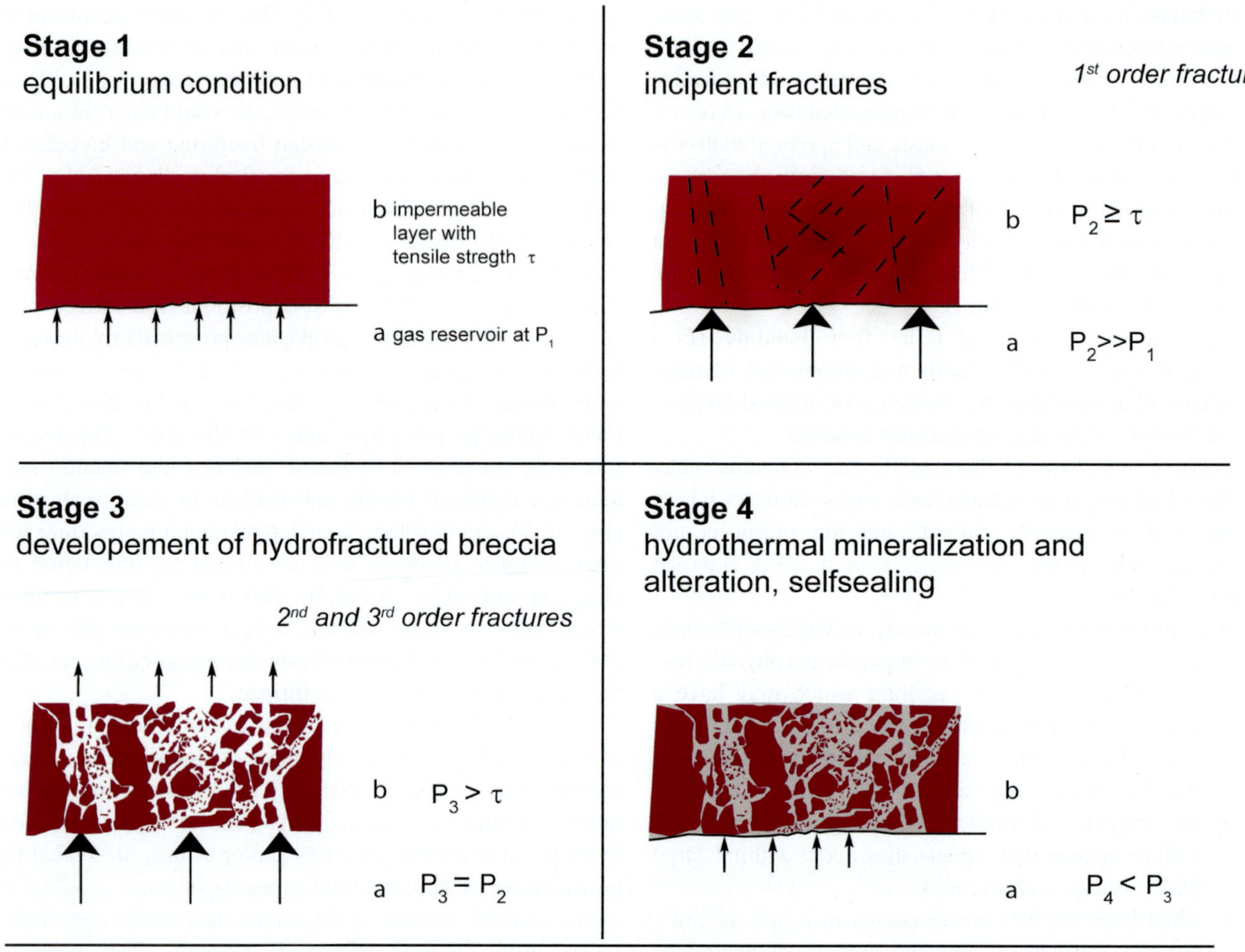

Fig. 13.9 Stages in the development of a zone of hydraulic fracture brecciation and alteration. (1) Fluid pressure build-up/equilibrium stage. (2) Incipient and first-order fracturing stage. (3) Second- and third-order cracking/fracturing stages. (4) Waning pressure stage, and final sealing of the fracture system by precipitation of alteration minerals, which could be mineralised. P_1 is the initial hydrothermal and lithostatic pressures; P_2, P_3, and P_4 are the successive fluid pressure levels; τ is the tensile strength of the country rock. Based on a study of the hierarchy of hydraulic fractures on Lisca Bianca islet, near Panarea volcanic island, Aeolian Islands, Italy. (From Cas et al. 2011a)

In narrow, incipiently brecciated zones, breccia clasts may have jigsaw puzzle-like fit relationships (Figs. 13.10a–d and 4.33). However, in well-developed breccia zones, clasts may be moved relative to each other, significantly rotated, and even transported through the system, perhaps even leading to a heterogeneous mixture of clast lithologies. Such breccia zones may also obviously be the sites for significant hydrothermal alteration and mineralisation because of the permeable network of fractures that would facilitate hydrothermal fluid flow, as discussed above under polyphase hydrothermal alteration. Also, in such zones where alteration processes have resulted in significant weakening of the rock, they may preferentially absorb strain and become highly deformed during subsequent tectonic movements.

Prolonged flow of alteration fluid through such breccia zones will lead to progressive migration of an alteration front into the blocks of rock from the bounding hydraulic fractures or joints (Figs. 13.10b–d and 13.6d–f). Just as with spheroidal weathering this will result in alteration corroding of the corners of the blocks, resulting in round kernels of fresh rock in the centres of the blocks (Figs. 13.10c, d and 13.6d–f). The resulting texture can look like an apparent conglomerate of round pseudo-pebble to boulder size kernels in a pseudo-matrix of alteration minerals (Figs. 13.10c, d and 13.6d–f). The pseudo-clasts have not been rounded by physical sedimentary abrasion, they are entirely in situ, and consist of a monomictic assemblage of pseudo-clasts. Such aggregates have gradational relationships with clusters of angular in situ, jigsaw-fit clasts and laterally into coherent rock (Figs. 13.10c, d and 13.6d). They are pseudo-conglomerates and the apparent pseudo-matrix between the clasts consists of alteration minerals, such as alunite (Fig. 13.10a, b), carbonate, and chlorite (Figs. 13.10c, d and 13.6d–f), that have been precipitated in the fractures by the

Fig. 13.10 Hydraulic fracture and alteration textures. **a** Wide hydrothermal hydraulic fracture and alteration zone, showing large relict first-order fractures bounding large blocks and second and higher order fractures bounding smaller fragments. The basaltic andesite rock is variably altered by alunite replacement, which also fills the veins sealing fractures. Quaternary Lisca Bianca islet, near Panarea volcanic island, Aeolian Islands, Italy. **b** Highly altered (white alunite) hydrothermal hydraulic fracture breccia in basaltic andesite, Quaternary Lisca Bianca islet. Note the sub-vertical first-order fracture–vein set and the elongated shape of the large clasts, whereas the second-order fractures and fragments have a more random orientation and more equant clast shape. The white vein material is alunite. (From Cas et al. 2011a, b). **c** Hydraulically fractured and altered Archean dacite from the host rock succession, Leinster nickel mine, Western Australia. The left-hand core is massive with incipient fractures, the next core to the right has numerous hydraulic fractures filled with dark chlorite. In the two middle cores, the chlorite alteration along hydraulic fractures forms wider seams, indicating inward migration alteration and replacement fronts into the hydraulic fracture clasts. Some clasts are becoming rounder in form. In the two right cores chlorite alteration seams are wider again and constituting an apparent matrix, and in the right core almost completely chlorite-replaced dacite "clasts" are ghosted and barely visible. Domains of visible relict feldspar phenocrysts from the coherent dacite protolith are visible floating in the pseudo-matrix of chlorite. This is *not* a transported volcaniclastic conglomerate or breccia, but an in situ alteration pseudo-conglomerate/breccia with pseudo-clasts and pseudo-matrix. **d** From the same succession as (**c**) with the alteration more enhanced with some impossible pseudo-clast shapes. All drill core is ~5 cm in diameter

hydrothermal fluids; the matrix is not a primary clastic matrix.

Hydrothermal hydraulic fracture breccia systems are associated with many mineralised ore systems (Phillips 1972; Laznicka 1988; Jébrak 1997; Gifkins et al. 2005), including epithermal, iron oxide copper–gold and volcanic-hosted massive sulphide (as discussed in detail Chap. 18), porphyry copper, as well as other non-volcanic generated ore systems, including orogenic gold. However, hydraulic fracturing may also occur far away from ore deposits, and also be generated during tectonic deformation (e.g. Sibson 1996).

13.6 Alteration of Kimberlites and Komatiites

Ultramafic rock types such as kimberlites and komatiites are particularly prone to alteration because of their composition and mineralogy, and most are ubiquitously altered. It is important to understand the alteration minerals and alteration textures before the original mineralogy and textures of these rocks can be reconstructed.

13.6.1 Alteration of Kimberlites

Kimberlites, and related lamproites, are ultramafic volcanic and intrusive rocks that are unique in being the principal natural elevator that brings diamonds to the Earth's surface, erupting in pipe-like conduits and vents (Chap. 18). Kimberlites range in age from Proterozoic to Holocene and are uniquely erupted through or adjacent to Archean cratons, indicating unique conditions in the sub-cratonic mantle for both generating kimberlite magmas and the xenocrystic diamonds that some kimberlites have carried to the Earth's surface. Kimberlites have very low SiO_2 content (<35 wt%), extremely low Al_2O_3 (<5 wt%), very low Na_2O (<0.5 wt%), and Na_2O/K_2O (<0.5). Primary magmatic *kimberlite* minerals may include forsteritic olivine, phlogopite, spinel, monticellite (?), and perovskite, whereas mantle-derived xenocrystic minerals include olivine, garnet, chrome diopside and phlogopite, and crustal xenoliths include everything from granite to metamorphic to sedimentary rocks, including limestone. Kimberlites and lamproites are very conducive to alteration. All primary and accessory components can contribute to the alteration phases found in kimberlites and related lamprophyres as alteration fluids mobilise elements from all components and redistribute them.

Secondary alteration minerals in kimberlite include pervasive serpentine (lizardite, antigorite), talc, magnetite, brucite (all after olivine), carbonate (calcite, dolomite), microlitic diopside, secondary phlogopite, secondary monticellite, chlorite, smectite (saponite), titanite, and andradite (Stripp et al. 2006; Cas et al. 2008; Hayman et al. 2009; Hayman and Cas 2011; Porritt et al. 2012; Pittari et al. 2015; Giuliani et al. 2014). These rocks and their original glass if present can readily be hydrated through the actions of deuteric-magmatic, hydrothermal (meteoric water heated by kimberlite) and meteoric waters. This is confirmed by detailed studies using microscopy, SEM back-scatter analyses, micro-probe and XRD analyses (Stripp et al. 2006), and oxygen, carbon, and sulphur stable isotope analyses (Giuliani et al. 2014). The alteration and replacement processes affect crystal, lithic and groundmass–matrix components, masking many original textures, and even making it difficult to distinguish between what were coherent and fragmental kimberlite rocks (Cas et al. 2008; Hayman et al. 2009; Hayman and Cas 2011; Porritt and Cas 2011; Porritt et al. 2012; Pittari et al. 2015; Chaps. 14 and 18).

Most kimberlites appear not to have not been metamorphosed during regional tectonic events because they have all been emplaced in stable cratonic settings, albeit they are often aligned along and/or above major fault structures and corridors (Chap. 8). Therefore, all the alteration has to be due to magmatic deuteric, hydrothermal and meteoric fluids, and diagenesis resulting from post-emplacement burial under younger cover sedimentary and volcanic rock sequences. This is a quite unique scenario across the spectrum of volcanic and subvolcanic rocks. Where alteration is at its lowest intensity, olivine crystal morphology, size, and distribution patterns are discernible, and can be used to distinguish coherent from fragmental kimberlite. Even when kimberlites are strongly altered, detailed textural analysis can often allow coherent and fragmental kimberlites to be distinguished (Hayman and Cas 2011). Key criteria for identifying coherent kimberlite are a uniform, evenly distributed, round, resorbed, small micro-phenocryst olivine population in the groundmass, and a scarcity of lithic clasts (Fig. 13.11a–c). Large olivine macrocrysts are often xenocrystic, not phenocrysts, they can be very irregular in shape, and give the coherent kimberlite a bi-modal olivine size population. Fragmental or volcaniclastic kimberlite will have an irregularly shaped olivine crystal fragment size population with a more normal grain size distribution, a mixture of often diverse mantle and crustal derived xenoliths, and have a poorly sorted aspect, lacking that uniform micro-phenocryst aspect of coherent kimberlite (Fig. 13.11d, e).

Typical effects of alteration on kimberlite textures include:

(a) Severe serpentine alteration and replacement of large crystals, especially olivine (Fig. 13.12a) as well as groundmass and matrix.
(b) Infilling of any original micro-porosity, usually by serpentine and/or carbonate, resulting from the breakdown of olivine, which produces a greater volume of serpentine than the volume of the original olivine crystals which is precipitated into surrounding porosity (Stripp et al. 2006).
(c) Masking of the original texture of both the groundmass of coherent rocks and the clastic matrix of volcaniclastic rocks by serpentine and carbonate, requiring close examination of crystal morphology and distribution patterns to distinguish them, as discussed above (Fig. 13.12a–c).
(d) Polyphase, patchy alteration giving the impression of large apparent clasts dominated by one alteration phase,

(a)
(b)
O
O
O
O
O
C
C
C
2 mm
(c)
(d)
(e)
cm
(f)
PM
BC
VS
DGGSC
DGG
PBG

◀ **Fig. 13.11** Features of variably altered kimberlite. **a** Relatively fresh, serpentinised coherent kimberlite, with irregular large macrocrysts of olivine in a fine non-descript groundmass. **b** Photomicrograph of (**a**), showing large macrocrysts of olivine (O) suspended in a groundmass with an even distribution of resorbed euhedral-subhedral micro-olivine crystals, themselves in a fine, non-descript, serpentinised medium which has also been replaced by carbonate (C). Both (**a**) and (**b**) are from the Eocene Koala kimberlite pipe, Ekati kimberlite field, Lac de Gras, Northwest Territories, Canada. (Image **b** is from Porritt and Cas 2011). **c** Contact between coherent and fragmental or volcaniclastic kimberlite, Jurassic Jericho kimberlite pipe, Nunavut, northwestern Canada. Note the uniform, bimodal olivine crystal population in the darkly serpentinised coherent kimberlite (left), the lithic clast-rich, differently serpentinised and carbonate-altered fragmental kimberlite (right), which contains limestone lithics (light). **d** Very crystal-rich, moderately well-sorted volcaniclastic kimberlite, with largely irregular-shaped macrocrysts of olivine, and minor pink garnet (above desiccated finger—*working with dry to wet core is tough on the skin!*). Koala kimberlite pipe, as for (**a**, **b**). **e** Heavily serpentinised almost black fragmental/volcaniclastic kimberlite, Jurassic Muskox kimberlite pipe, Nunavut, northwestern Canada. Note the poor sorting, diffuse definition of many grey olivine grains in the matrix (many are completely masked in the serpentine), and the spectacular mantle eclogite lithic clast (outlined in red; with pink garnets and pale green olivine), white limestone lithics, bright green chrome diopside, and pink garnet (in both the eclogite clast and the matrix). All diamond drill core is ~5 cm wide. **f** Different appearances of volcaniclastic kimberlite resulting largely from different forms of serpentine alteration mixed with carbonate, and in the core on the right, with magnetite alteration. BC = buff coloured carbonate; VS = vermiform serpentine; DGGSC = dark green-grey serpentine-carbonate mix with some blebs of carbonate; DGG = dark grey serpentine–carbonate mix; PBG = pale bluish green serpentine–carbonate mix; PM = patchy magnetite. All drill core is from the Cretaceous Orion kimberlite complex, Fort à la Corne, Saskatchewan, Canada. (From Pittari et al. 2015)

distributed in a domain of another alteration phase. Multiple phases of serpentine, with variable compositions, are possible (Fig. 13.12a–c; Pittari et al. 2015).

(e) Gradational contacts between domains dominated by different alteration phases, at both the macroscopic and microscopic scales (Fig. 13.12a–d).

(f) Dissolution of matrix and infilling of resultant cavities with secondary minerals such as carbonate showing inward crystal coarsening cement-like textures, which is particularly problematic in poorly sorted volcaniclastic kimberlite, which cannot have large primary cavities (Fig. 13.12d).

(g) Secondary minerals such as microlitic diopside, can be euhedral and give the impression of being primary, but close examination shows that they overprint the boundaries of primary components such as olivine and lithic fragments, with textural components such as matrix or groundmass (Fig. 13.12e), making it clear that they are secondary, overprinting alteration products, not primary magmatic crystals.

(h) Xenoliths embedded in kimberlite (e.g. Fig. 13.11e) can also be altered and replaced, and in turn release elements that can then influence the nature of the secondary alteration mineral phases in the surrounding kimberlite (Stripp et al. 2006; Hayman et al. 2009; Porritt et al. 2012). For example, chlorite, which requires a source of Al^{3+} to form, and does not occur in olivine, can be produced from the breakdown of feldspars in crust-derived xenoliths (Stripp et al. 2006; Porritt et al. 2012). Secondary carbonate (Fig. 13.12d) can be derived from the dissolution of limestone lithics (e.g. Fig. 13.11e).

Porritt et al. (2012) and Hayman et al. (2009) provided summaries of the effects of progressively increasing intensity of alteration of kimberlites and the secondary mineral phases that depict the progressive stages of alteration intensity in the Lac de Gras and Nunavut kimberlite fields in Canada. These document the transition from relatively freshly preserved olivine to progressively more serpentinised olivine and matrix or groundmass, followed by increasing clay and complete obliteration of original components and texture. Hayman et al. (2009) summarised in detail the likely origins (primary versus secondary) of the main minerals found in kimberlites, concluding that the majority are secondary, and that most could be produced through early deuteric or hydrothermal alteration, while deposits are still hot (several hundred °C). Pittari et al. (2015) undertook a very detailed analysis of the variations in alteration through the Cretaceous Orion kimberlite from the Fort a la Corne kimberlite field in Saskatchewan, Canada (Fig. 13.13). They also determined the paragenesis of alteration phases based on superposition relationships between different phases documented in thin sections and backscatter imagery, including late stage saponite formation from meteoric waters (Fig. 13.14). For a comprehensive discussion of the geology of kimberlites and lamprophyres in the context of their diamond-bearing characteristics, see Chap. 18.

13.6.2 Alteration of Komatiites

Komatiites were commonly erupted and emplaced as sub-volcanic intrusions during the Archean on many cratons. They are ultramafic in composition (<45 wt % SiO_2), with very high MgO (>18 wt %), high olivine contents, and were emplaced at temperatures up to 1650 °C (Chap. 7). Almost all komatiites are preserved in ancient submarine sedimentary and volcanic stratigraphic successions. Because of their ultramafic composition, they were very susceptible to alteration and metamorphism, and are all variably altered or metamorphosed to serpentine, talc, chlorite, carbonate, and magnetite, although some komatiites still preserve original olivine and pyroxenes. Given that the majority are Archean

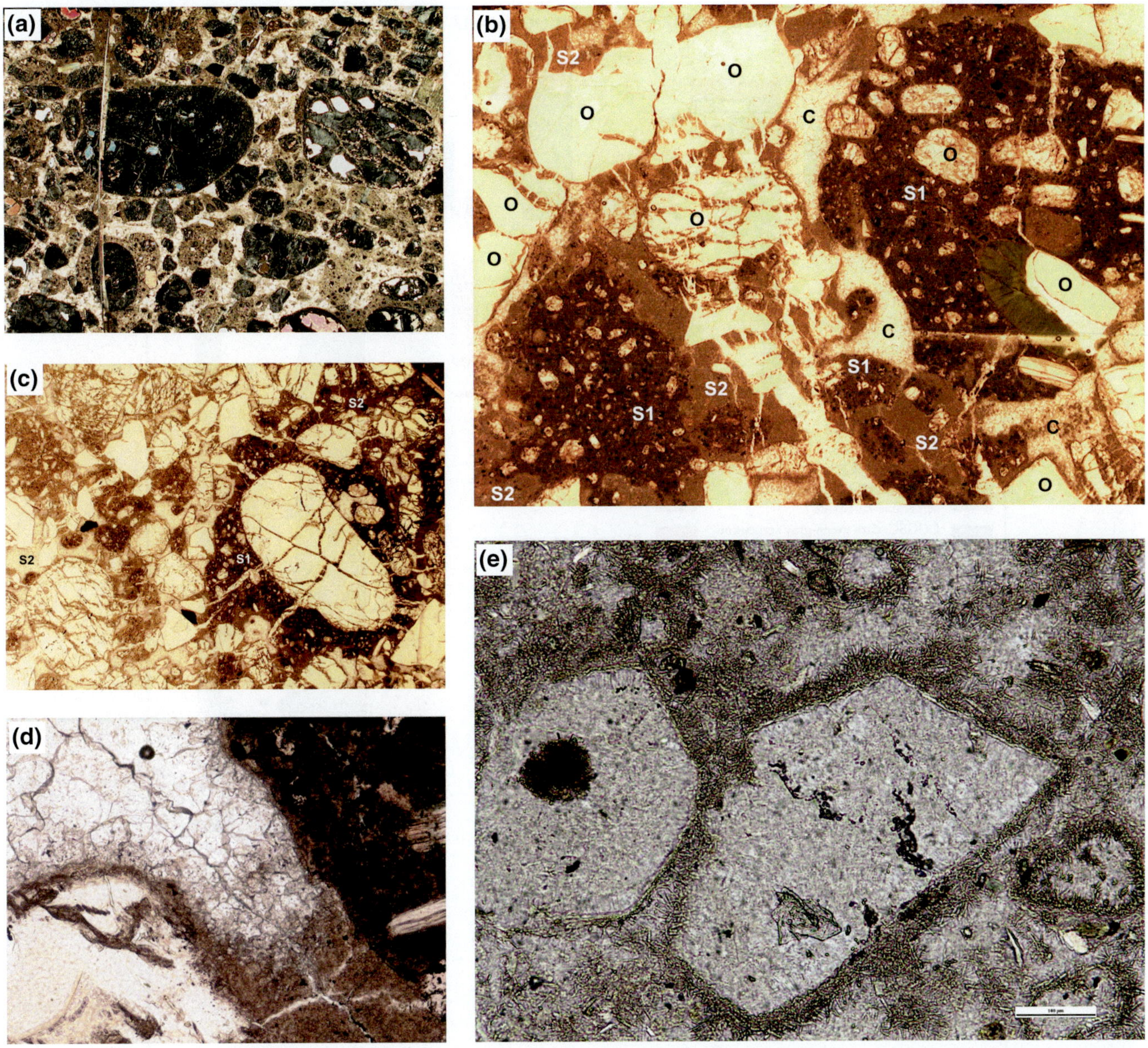

Fig. 13.12 Photomicrograph textures of altered kimberlites. **a** Pervasively serpentine altered volcaniclastic kimberlite, showing variable serpentine (shades of brown) and carbonate (white) alteration of the matrix, which give the rock a patchy texture and the appearance of large brownish clasts, which are in fact relicts of the original clastic matrix, that has also been replaced by the carbonate phase. Resorption rounded olivine grains have been almost completely replaced by dark serpentine, with a few relict fresh islands preserved in the large grains. Cretaceous Orion Central kimberlite complex, Fort à la Corne, Saskatchewan, Canada. Width of field of view 13 mm, x-nicols. (Image by Adrian Pittari). **b** Domainal alteration of the matrix of volcaniclastic kimberlite showing two phases or levels of intensity of matrix serpentine alteration, S1 serpentine and small Fe–Ti oxide blebs (dark), which represent the less altered original matrix, and then more intense S2 alteration domains (pale brown, lacking the Fe–Ti oxides), and pale granular carbonate domains (C), which appear to be infilling dissolution cavities in both matrix serpentine domains. The relict dark serpentine domains (S1) appear to be clasts, but note how irregular their margins are and the diffuse gradational contact between some S1 and S2 domains. Olivine grains (O) are almost completely replaced by pale yellow serpentine. *It's all alteration folks*! Width of field of view 7 mm, ppl. Location and source as for (**a**). **c** As for (**b**), again showing at least two phases or serpentine alteration intensity domains, with the light S2 serpentine domain lacking the small dark Fe–Ti oxide blebs present in S1 domains, suggesting they were dissolved and removed during the S2 alteration stage. Carbonate fills in amoeboid-shaped dissolution cavities in the matrix. Width of field of view 13 mm, ppl. **d** Moderately sharp to gradational relationship between carbonate replacing fine altered kimberlite matrix. Width of field of view 1 mm, ppl. Location and source as for (**a**). **e** Microlitic diopside micro-crystals rimming and overlapping olivine crystal fragments, Jurassic Muskox kimberlite, Nunavut, northwestern Canada. Width of field of view 0.85 mm, ppl. (Image by Pat Hayman)

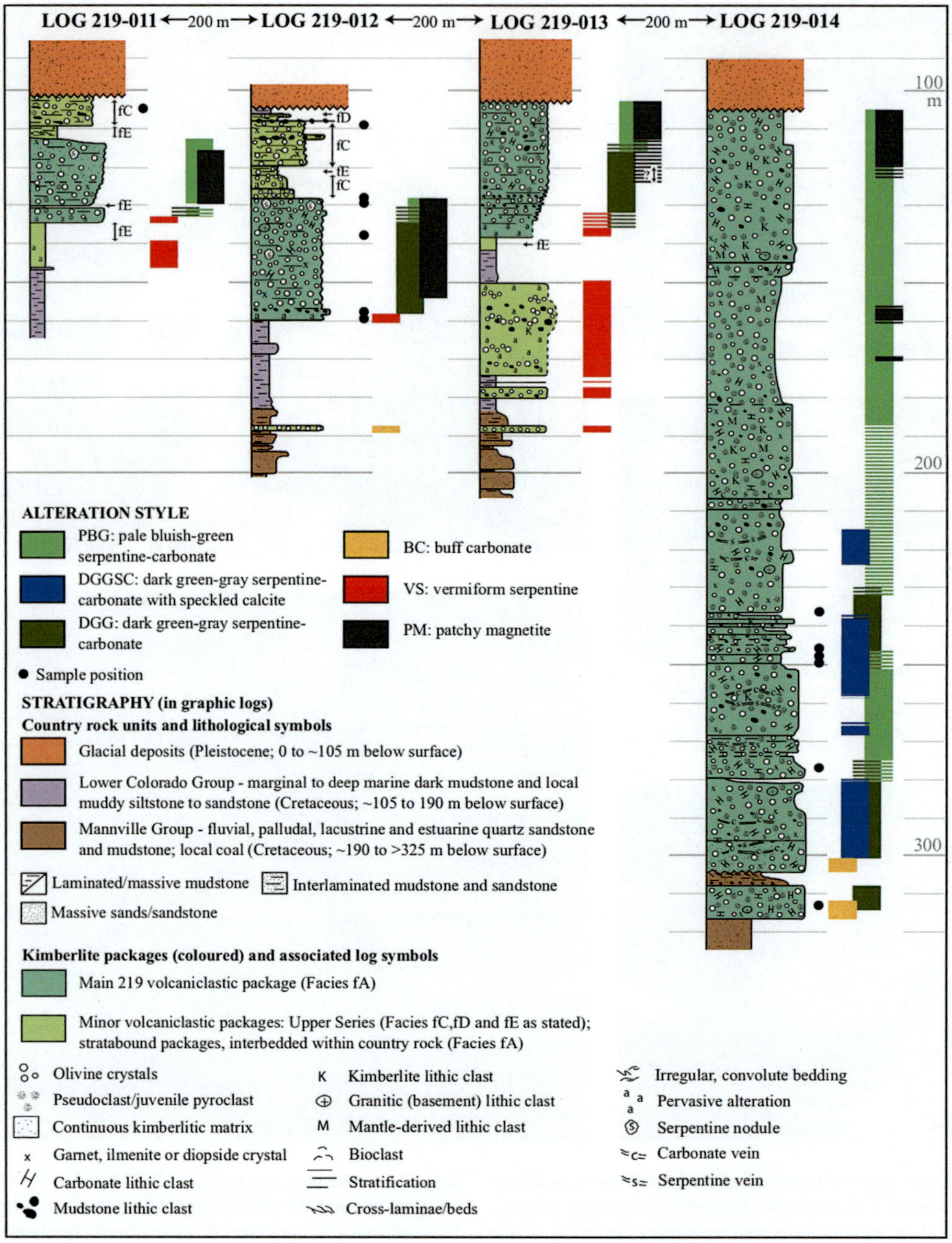

Fig. 13.13 Selected graphic logs of diamond drill cores through the Cretaceous Orion Central volcaniclastic kimberlite, Saskatchewan, Canada, showing the variations and alteration mineral assemblages through the sequence (From Pittari et al. 2015)

in age, most of the alteration is related to regional metamorphism since emplacement, but Lahaye et al. (1995) and Lahaye and Arndt (1996) showed through detailed trace element and stable oxygen isotope geochemistry that both seawater (all Archean komatiites were emplaced in submarine environments) and metamorphic fluids contributed to the alteration. In addition, for those komatiites that host magmatic nickel suphide deposits (Chap. 18), hydrothermal alteration may also be a factor. Apart from hyaloclastite breccias at the margins of some komatiite lavas, almost all komatiites are coherent lavas or intrusions, so there is less confusion about whether or not they were emplaced as coherent or fragmental rocks. However, both spinifex and cumulate textures can become pseudomorphed and overprinted by metamorphic minerals, and become obscure. In Chap. 7, common textural features that are preserved in komatiites, even when altered, are well illustrated. However, when alteration metamorphism are intense, most, if not all

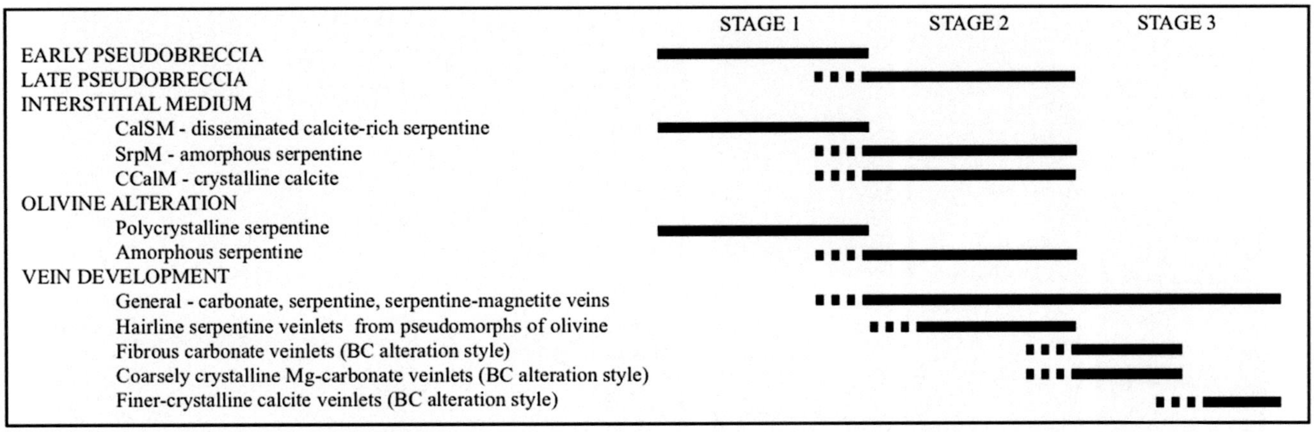

Fig. 13.14 Paragenesis of events and alteration minerals in the Cretaceous Orion kimberlite complex, Fort à la Corne kimberlite field, Saskatchewan, Canada (From Pittari et al. 2015)

original textures are overprinted and obscured. Common alteration minerals that readily obscure original textures in komatiites include serpentine, in the most intense cases forming a serpentinite rock (Fig. 13.15a, b), talc carbonate alteration (Fig. 13.15c, d), and chlorite (Fig. 13.15e). Listvenites (also spelled listwanites or listwaenites) are completely carbonated ultramafic rocks, which are generally structurally controlled, and often associated with mineralisation, particularly gold (Halls and Zhoa 1995; Qiu and Zhu 2015).

Some metamorphic minerals such as fibrous actinolite or anthophyllite may superficially be mistaken for spinifex textures in some komatiites, especially on weathered surfaces, including intrusions, making it ambiguous as to whether or not the rock units were lavas or intrusions. These minerals have pointy, almost needle-like terminations (Fig. 13.15f, g), whereas spinifex crystals tend to have skeletal stubby terminations. Although spinifex occurs most commonly in komatiite lavas, it can, however, also occur at the margins of thin intrusions, which because of their high emplacement temperature chill against crustal rocks at high crustal levels (e.g. Arndt et al. 2004).

13.7 Diagenesis

The concept of diagenesis encompasses the mineralogical and textural changes associated with lithification and the early stages of burial of any sediment or rock system (e.g. Scholle and Schluger 1979; Burley and Worden 2003; Mackenzie 2005). It is of fundamental importance in hydrocarbon exploration because it modifies original porosity, both by occluding primary porosity and generating secondary porosity, as well as causing hydrocarbon maturation. Significant textural and mineralogical changes can be produced by dissolution of original components, precipitation of and replacement by new mineral phases, and compaction. These processes occur progressively under the influence of increasing pressure, temperature, and significant fluid flux during burial under younger strata. Diagenesis can be viewed as the low grade, initial stages of metamorphism.

In volcanic rock systems, diagenesis is just as significant if not more so than in sedimentary systems (see Chap. 18, Sect. 18.9). This is due to the usual abundance of metastable glassy material and the presence of reactive mineral components, especially in basaltic to intermediate rocks, the porosity and permeability of volcanic piles, which afford access of magmatic deuteric, hydrothermal, and meteoric fluids through the pile (Surdam and Boles 1979). Alteration of volcanic glass, devitrification, palagonitisation, and micro-granular recrystallisation of glass, as discussed above, can be considered to be diagenetic processes and continue as a volcanic succession is buried and compacted under more and more volcanic and/or sedimentary overburden.

Branney and Sparks (1990) have described diagenetic fiamme-like textures in recent pumice fall deposits on São Miguel, Azores, as well as from Ordovician lacustrine volcaniclastic sedimentary rocks from the English Lake District. The modern examples are from trachytic plinian and sub-plinian pumice lapilli fall layers erupted from Sete Cidades caldera which are interlayered with soil. During soil formation, the soils developed downwards to different extents into the pumice layers, and pumice clasts were altered to halloysite and other clay minerals. The soil-and-pumice deposits were then overlain by basalt lava. Altered pumice clasts were flattened by the lithostatic load, forming aligned, bedding-parallel lenticles resembling eutaxitic texture in welded ignimbrites. Diagenetically collapsed pumice is also present in ancient submarine successions, such as in submarine Miocene pumice mass-flow deposits in the host rock successions to the Kuroko massive sulphide deposit mine sequences in Japan (Fig. 13.16a), that

(a)
(b)
(c)
(d)
(e)
(f)
(g)

◀ **Fig. 13.15** Alteration styles of komatiite. **a** Heavily serpentinised komatiite, with magnesite spots and carbonate veins, Archean, Norseman, Eastern Goldfields Province, Yilgarn Craton, Western Australia. **b** Photomicrograph of serpentine alteration of olivine crystals with carbonate alteration in komatiite adcumulate, Archean Black Swan komatiite, north of Kalgoorlie, Western Australia. Width of field of view ~8 mm, x-nicols. **c** Typical talc-carbonate alteration of komatiite, masking all original texture, Black Swan komatiite. **d** Typical talc-carbonate alteration of komatiite, masking all original texture, Archean, Agnew mine, near Leinster, Western Australia. **e** Intense chlorite alteration masking the contact between komatiite (right) and dacite (left). Black Swan komatiite. **f** Chlorite-anthophyllite(?)-serpentine-carbonate alteration of spinifex textured komatiitic basalt, Archean Dundonald komatiite sequence, Abitibi Greenstone Belt, Ontario, Canada. Note that there appear to patches of quartz replacing the large pseudomorphed spinifex crystal, which is unusual. **g** Large actinolite metamorphic crystals (not spinifex) in Archean komatiitic basalt, Welcome Well volcanic centre, east of Leonora, Western Australia. All drill core is ~5 cm in diameter

when intense, falsely give the impression of fiamme-like welded ignimbrite textures (Fig. 13.16b).

In addition to these, increasing compaction, precipitation of chemical cement in pore spaces and by overgrowth of clastic mineral grains such as volcanic quartz, the growth and overprint by low-grade burial metamorphic minerals, typically clays and zeolite facies assemblages, can be significant in causing major modification and overprinting of original rock textures and components (e.g. Coombs 1954; Raam 1968; Boles and Coombs 1977; Davies et al. 1979; Surdam and Boles 1979; Khalaf 2014).

13.8 Metamorphism

Metamorphism, whether it be regional or contact, can be thought of as a higher grade extension of diagenesis, but should not be confused with hydrothermal alteration, which is usually more local in effect. Metamorphism produces pervasive mineralogical and textural changes and, where allied with the strain effects of deformation, can cause major problems in the recognition of original textures and rock types. Nevertheless, perseverance is usually rewarded, because frequently low-strain enclaves of rock, still preserving original textures or ghosted relicts of such, can be found, particularly in greenschist or lower grade rocks, even when original components such as glass shards and pumice have been replaced by recrystallised mosaics of fine-grained quartz and feldspar, and/ or phyllosilicate minerals such as sericite, muscovite, and chlorite. For example, relict pseudomorphed shards, pumice (e.g. Fig. 13.16c, d), welded textures (Fig. 13.16e), and perlitic textures are all documented in greenschist grade volcanics (also see McPhie et al. 1993), even as old as Archean. However, where metamorphic grade (plus deformational intensity) is high, original textures are usually completely overprinted (e.g. Fig. 13.16f).

Weathering frequently etches out original components on the weathered rock surface (e.g. fiamme, flow banding). Even in high-grade metamorphic and deformed rock successions, original features may be etched out on weathered surfaces of rocks (e.g. sedimentary structures, pillow lavas, and even spinifex textures which internally in hand specimen or fresh core are totally recrystallised and contain only obvious metamorphic and/or structural features). For example, in amphibolite and granulite facies rocks, it is sometimes possible to recognise pillow lava outlines on weathered surfaces (see Fig. 13.18), whereas when the rock is broken open, it is internally metamorphically recrystallised. (Fig. 13.16f).

Metamorphic recrystallisation can change the character of original volcanic components too. For example, bi-pyramidal volcanic quartz is mono-crystalline with straight or unit extinction under crossed polars, whereas all other quartz types (plutonic, vein, metamorphic) have sweeping or undulose, strained extinction and are often poly-crystalline. So, when volcanic quartz is metamorphosed and deformed, its crystal lattice is strained, the extinction becomes undulose, and it may recrystallise into polycrystalline quartz, making its (and the host rock's) original identity (i.e. its provenance) difficult to recognise as volcanic.

Readers wanting a reminder about metamorphic facies and the typical metamorphic indicator minerals that may be found in ancient volcanic rocks are referred to books such as Philpotts and Ague (2009), and Bucher and Grapes (2011).

13.9 Deformation

Tectonic deformation can have major effects on ancient volcanic successions, firstly, from the regional scale effects in modifying original unit geometry, and secondly, on a mesoscopic and microscopic scale, where deformational fabrics and strain create major problems in recognising original textures and rock types. This is particularly a problem where pervasive deformational and metamorphic effects are both well developed. Original textures are frequently preserved in only mildly affected, low-strain enclaves, which are therefore well worth the time spent in searching for, mapping and understanding their relationships with higher strain zones. Random observations may give the impression that there are different rock types and units in a succession of variable strain. Whereas low-strain zones show largely original textures, rocks in high-strain zones may appear very different and even misleadingly give the

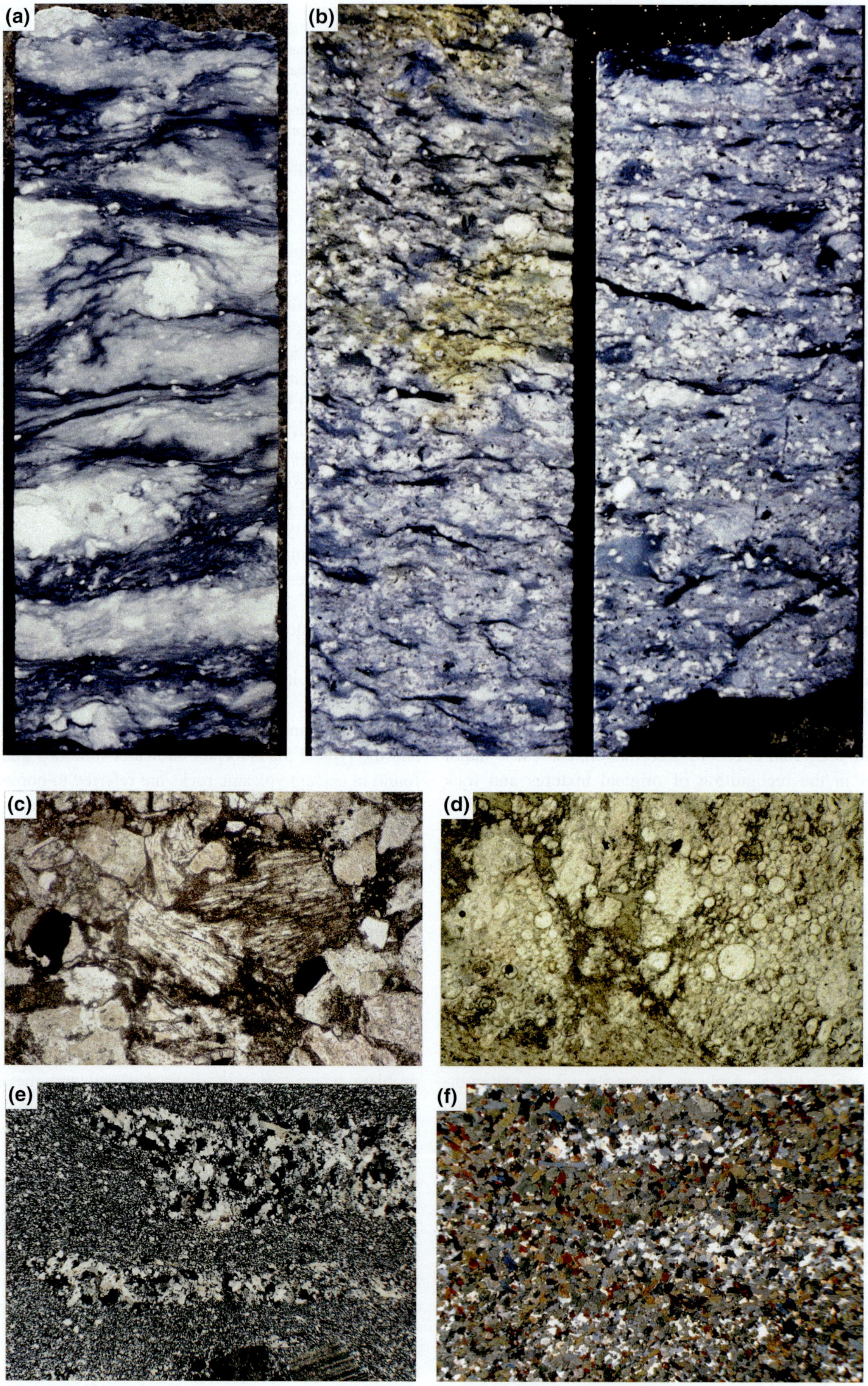
(a)
(b)
(c)
(d)
(e)
(f)

◀ **Fig. 13.16** Variable preservation of textures in ancient volcanic rocks subject to diagenesis and metamorphism. **a** Diagenetically compacted pumice in hydrothermally altered, submarine succession from diamond drill core HO44 at the base of the hangingwall sequence (upper T_3) at Fukuzawa VMS mine sequence. Miocene Green Tuff Belt, Hokuroko Basin, Honshu, Japan. **b** Diagenetically collapsed pumice in hydrothermally altered, submarine, pumice mass-flow deposits from diamond drill core HO44 in the hanging wall sequence (T_2) at the Fukuzawa mine, preserved as dark, chloritically altered flattened fiamme-like lenticle clasts. Other pumices in the same samples are not collapsed as viewed in thin section, indicating selective diagenetic collapse of some altered pumice clasts but not other less altered ones. (For graphic log of drill hole HO44, see Fig. 18.88a). **c** Photomicrograph showing two, recrystallised and sericitised tube pumice clasts with ragged ends in a crystal-rich, submarine volcaniclastic mass-flow deposit. Greenschist metamorphic facies. Cambrian Mt Read Volcanics, Anthony's Road, north of Queenstown, western Tasmania. Width of field of view ~ 4 mm, ppl. **d** Photomicrograph of perfectly pseudomorphed, highly vesicular resedimented rhyolitic pumice clasts. Lower greenschist metamorphic facies. Archean Golden Grove VMS host rock succession, Yilgarn Craton, Western Australia. Width of field of view 2.5 mm, ppl. (Image by B Clifford). **e** Photomicrograph showing thoroughly recrystallised, originally welded rhyolitic pumice clasts (fiamme) observed in x-nicols. The relict pumice lenses are seen to comprise an interlocking mosaic of quartz and feldspar, within a finer-grained quartzofeldspthic (original glassy ash-sized) matrix and scattered feldspar crystal fragments (bottom centre-right). Greenschist metamorphic facies. Ordovician Pitts Head Tuff ignimbrite, Idwal Syncline, North Wales. Width of field of view ~4 mm. (Image from McPhie et al. 1993). **f** Photomicrograph of metamorphically overprinted texture of basaltic pillow lava, at uppermost greenschist–amphibolite metamorphic facies, which has completely overprinted original basalt texture. 3.7–3.8 Ga Isua Greenstone Belt, southwest Greenland. The elongated leucocratic patches are probably deformed and recrystallised varioles. In spite of this pervasive overprint, weathering has etched out original pillow morphology (see Fig. 13.18) on surface outcrop, even at this metamorphic grade. Width of field of view ~4 mm, x-nicols

impression of different juxtaposed lithologies. Nonetheless, when the transitions between low-strain to high-strain zones are mapped in detail or where drill core is logged with care, transitions from low-strain to high-strain zones can be seen to be gradational. This can be achieved especially when some particular marker minerals or other components are visible throughout, indicating that the same rock unit prevails across the strain transition. For example, in Fig. 13.17, drill core through individual Archean dacite units in Western Australia shows such a transition from low strain to high strain through single thick coherent dacite units using plagioclase crystals as a strain indicator. This causes very different appearances, including pseudo-fiamme "enigmabrite" textures in chlorite schists (Fig. 13.17a, b; Plate 13.1b), and even pseudo-sedimentary or ash-fall-like lamination in mylonitic zones in highly deformed dacite (Fig. 13.17c).

In originally fragmental or volcaniclastic rocks, intense deformational strain can make recognition of their origins very difficult to determine (e.g. Fig. 13.17d). In Fig. 13.17d, mafic clasts are highly deformed and strained, and it is unclear if they represent a boudinaged protolith or an original, highly deformed, coarse volcaniclastic rock such as a mafic volcanic conglomerate. The answer is the latter, derived by carefully and continuously logging the diamond drill hole in which there is a gradation, from the high-strain rock in the figure to low strain, equant clast conglomerate.

On a different scale, the oldest known volcanics on Earth, within the early Archean 3.7–3.8 Ga Isua Greenstone Belt in Greenland are at upper greenschist to amphibolite facies metamorphic grade, and in outcrop show both relatively low-strain pillow lava (Fig. 13.18a) and highly strained, very stretched pillows, which look like "stripy rocks" (Fig. 13.18b). Such major strain variations occur laterally along strike and unless that transition from low-strain to high-strain rocks, which is gradational, is mapped with care, the stripy rocks in isolation could be misidentified as bedded mafic tuffs, or high-strain sheeted dykes. Microscopically, the rocks are completely metamorphically recrystallised (Fig. 13.16f), but weathering of exposed rock surfaces has etched out original pillow forms (Fig. 13.17a, b).

Many highly altered volcanic rock successions are penetratively deformed and cleaved. The question frequently arises as to whether the alteration preceded the deformation or whether deformation fabrics produced permeability for alteration fluids. There is obviously no simple answer, but it should be clear that if a rock succession is significantly altered, either in discrete zones or, at large, the affected areas will be mechanically weakened and will preferentially absorb more strain than unaltered zones will when deformation occurs. Therefore, as suggested above, there may be significant textural and broad lithological variations associated with the degree of deformation, which are not primary in origin.

13.10 High Metamorphic Grade Quartzo-feldspathic Gneisses and Metasomatic Rocks

Potentially texturally modified by all of the preceding processes at some time in their geological evolution, is the special case of high-grade quartzo-feldspathic gneisses and metasomatised rocks. Here, we are referring to upper amphibolite to granulite facies rocks which may be also extensively metasomatised or '*albitised*' by sodic (±calcic) fluids, and which are also invariably associated with high-strain (shear) zones and the thermal aureoles of syn-late metamorphic granites.

Such high-grade terrains are well represented in the Paleo-Mesoproterozoic basin successions of Australia associated with high-T/low-P metamorphism (Chap. 17). While there is no doubt that volcanic rocks occur in the lower grade

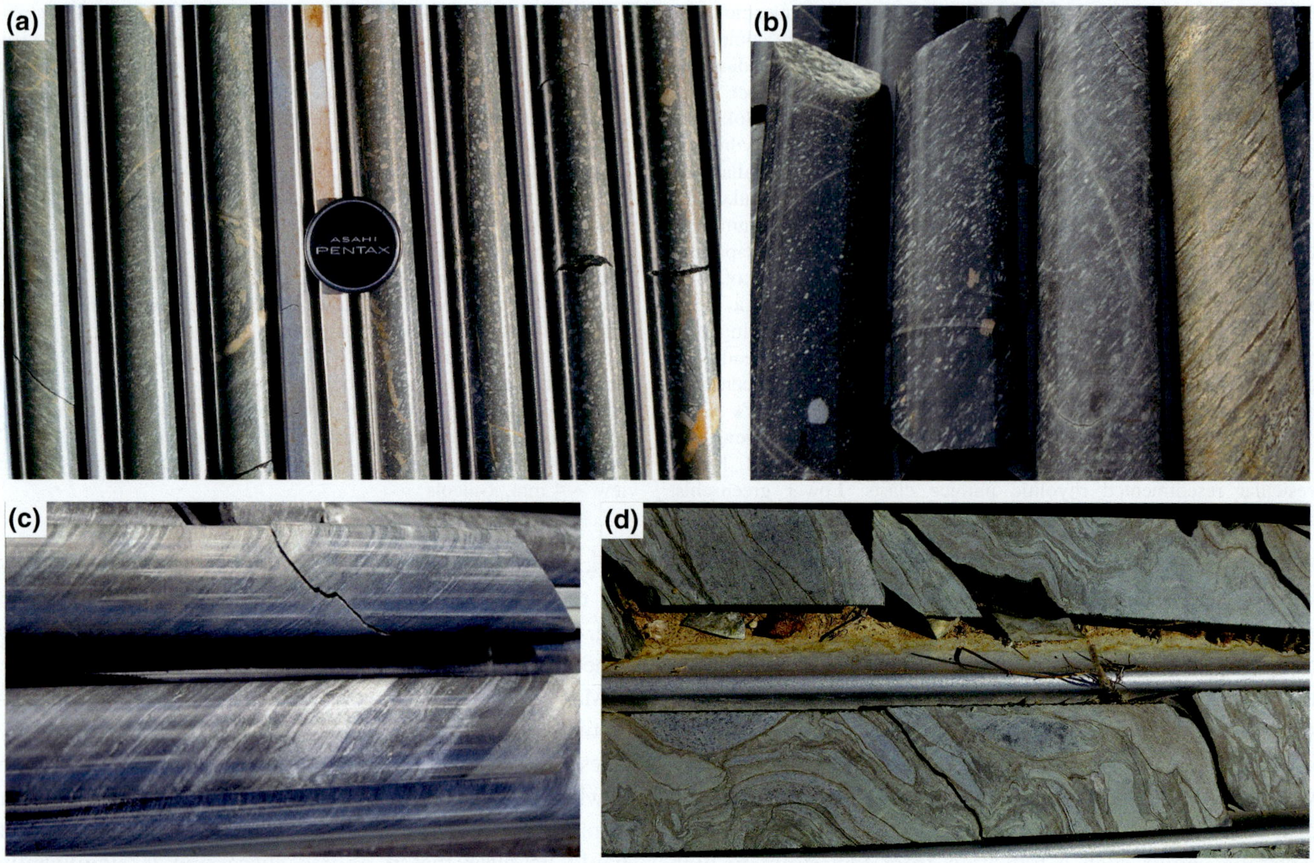

Fig. 13.17 Variable strain and its effect on original coherent porphyritic dacite. **a** At right, undeformed porphyritic dacite with well-defined feldspar phenocrysts; in centre cleaved dacite with strained, smeared feldspars; left, highly cleaved (and more pervasively altered) chloritic schist after dacite, which begins to develop a pseudo-eutaxitic fiamme texture. Archean Golden Grove VMS host rock succession, Yilgarn Craton, Western Australia. **b** Chlorite altered-metamorphosed feldspar–phyric coherent dacite (left core) that is *progressively* more tectonically strained in the cores to the right, culminating in extremely strained dacite on the right side with chlorite lithosomes that look like fiamme in welded ignimbrite, but are pseudo-fiamme in enigmabrite. The key to recognizing this is the *gradational change* in strain in the one dacite unit, without any sharp lithological changes or contacts. Archean host rock succession, Leinster nickel mine, Western Australia. **c** Extension of the strain development seen in (**b**), with the rock now a banded schist verging on mylonite in a major shear zone, that could be mistaken for laminated mudstones or laminated submarine ash fall deposits! Again, the *key* is recognising that the change from core in (**b**) to this core is gradational, reflecting progressively increasing deformational strain. **d** Highly deformed and strained Archean mafic conglomerate, Agnew mine, near Leinster, Western Australia. All drill core is ~5 cm in diameter

sections of these basins, the identification and interpretation of primary volcanic facies and textures in the high-grade terrains have been equivocal and controversial. This is no more so than at Broken Hill in western New South Wales, where the world's most iconic Pb–Zn–Ag orebody is hosted within the Willyama Supergroup. Mapping by the New South Wales Geological Survey in the early 1980s, while greatly improving the lithostratigraphic framework, also advanced the case for abundant primary pyroclastic rocks in the Willyama succession. Quartzo-feldspathic gneisses, quartz-albite "psammites" and amphibolites have been described variously as ash-flow tuffs, ignimbrites, plinian and phreatoplinian deposits, rhyolitic and scoria air-fall and ash-fall deposits (Willis et al. 1983; Stevens et al. 1988; Page and Laing 1992). However, interpretations of these volcanic rock types were rarely supported by valid volcanological criteria, much of the basis stemming from a general proposition of volcanic origin from their geochemical composition (for a full discussion, see Wright et al. 1993).

Much of the controversy at Broken Hill centres on an outlier (at Yanco Glen) of a moderate-grade metamorphic area (upper amphibolite facies) and a lithostratigraphic unit of quartz-feldspar-biotite-garnet gneiss (ca. 100 m thick). This unit, called the Potosi gneiss (which constitutes most of the Hores Gneiss stratigraphic unit) is found elsewhere in the Broken Hill Domain of the Curnamona Province, including the Mine Sequence. The apparent 'eutaxitic' texture characterising this "enigmabrite" is not typical of the texture found in other welded ignimbrites, which are

Fig. 13.18 Strain variations in 3.7–3.8 Ga pillow lava at upper greenschist to amphibolite facies metamorphic grade, Isua Greenstone Belt, Greenland. **a** Relatively low-strain pillow lava. Younging can still be determined to be up towards the top of the photo from angular re-entrants at the base of several pillows. **b** "Stripy" highly stretched and flattened to lensoidal pillow lava in a high-strain zone

macroscopically demarcated by compactionally flattened large (cm- to dm-sized) pumice clasts. Instead, this is defined by biotite aggregates generally much less than 10 mm across, which give a well-developed metamorphic schistosity to the rock (Fig. 13.19a). The size and shape of these aggregates show little variation through the unit in question, which is not what would be expected if it was composed of one or more ignimbrite flow units (pumiceous pyroclastic flows often display reverse grading of larger pumice clasts and variable pumice clast sizes), or welded cooling units (see Chap. 10). The only size variation observed is an increase in biotite size close to and around some irregular clasts, lenses or "tongues" of folded and contorted metasediment (generally up to 5 m across) within the unit (Fig. 13.18b). This is accompanied by the appearance of larger garnets, which suggests the presence and size of biotite are sensitive to metamorphic processes and host rock lithology, not primary textural variations in a pumiceous pyroclastic flow deposit. In thin section, many of the larger quartz grains (0.5–2.0 mm) do show equant, bipyramidal shapes, and the rock probably comprised a significant volcanic component (Fig. 13.18c; Stevens and Barron 2002); quartz is however more varied and includes plutonic and metamorphic populations. Microscopic evidence indicates complex metamorphic overprinting of precursor textures (whether volcanic or sedimentary), likely by contact aureole effects as the gneiss also forms the roof pendant of the Mundi Mundi granite. Notably, the Potosi gneiss also hosts many of the tin–tungsten and tourmaline occurrences in the Yanco Glen area, which are undoubtedly granite-related, and the unit is thoroughly intruded by pegmatite and apophyses of granite.

The inclusions of fine-grained metasediment are also enigmatic. In detail, these show very irregular to highly serrated margins, "ghosting" of clasts and internal domains and patches of quartz–feldspar–biotite ± garnet, suggesting metamorphic overprinting (Fig. 13.18b). Whether this has occurred to a minor degree (reaction around the margins of rip-up clasts), or represents more extensive textural replacement of a more contiguous bedded sequence is not known. What is clear is that such metasedimentary inclusions are not compatible with the interpretation of a subaerial welded ignimbrite (Wright et al. 1993; Stevens and Barron 2002). If these do represent clasts, they are more consistent with a subaqueous debris flow (sedimentary or volcaniclastic), or subaqueous syn-sedimentary slide. This would agree with sedimentological evidence from the underlying succession which suggests paleowater depths of 30–100 m (Wright et al. 1987, 1993; Stevens and Bradley 2018). One of the more imaginative models suggested to explain geochemical differences of the Potosi gneiss from a normal volcanic precursor, invokes the deposition of epiclastic, crystal-rich, ash-poor sandstones produced when pyroclastic flows flowed into and explosively interacted with a body of water (Wyborn 1998; Stevens et al. 1998; Stevens and Barron 2002; see Chap. 11). Unfortunately, by no stretch of the imagination can the Potosi gneiss be considered "crystal-enriched", nor does it show evidence of better local hydraulic sorting of crystals (e.g. layers/bands of less matrix support). Another possibility is that it represents a high-level sill emplaced into unconsolidated sediment. In the Yanco Glen domain, this might explain the association of metasedimentary "xenoclasts", euhedral quartz phenocrysts, differing populations of quartz, and irregular contact relations with metasediment, including possible larger metasediment rafts (up to 200 m wide; see Chap. 8, Fig. 8.20b). The precise origin(s) of Potosi gneiss in the

Fig. 13.19 Textures in moderate metamorphic grade Potosi gneiss in the Yanco Glen area, Broken Hill, western New South Wales, Australia. **a** Typical well-developed metamorphic schistosity defined by small biotite laths (1–2 mm long) and larger aggregates (5–20 mm long) which previously have been interpreted as fiamme and eutaxitic texture in a welded ignimbrite. This image also shows a lithic clast-like inclusion. Such clasts are generally rare and compositionally similar to the host (intraformational or cognate), and therefore are not very distinct in outcrop. **b** Irregular lenses and tongues of fine-grained, psammitic metasediment incorporated within Potosi gneiss. These metasedimentary inclusions typically have highly serrated (or ragged) to diffuse "ghosted" margins, suggesting some degree of metamorphic overprinting. In detail, the fine-grained psammitic material also includes domains of quartzo-feldspathic gneiss, suggesting complex textural interfingering. **c** Photomicrograph showing quartz-eye textures. Width of field of view 10 mm, x-nicols. Two quartz grains show distinctive equant outlines (sections parallel to c-axis of bipyramidal quartz) confirming an important volcanic phenocryst component. Detailed textural studies by Stevens and Barron (2002) indicate most quartz phenocrysts appear whole and few are broken or fragmented, which is more common with a lava or intrusion than a pyroclastic deposit. (Images courtesy of B. Stevens; From Stevens and Barron 2002)

extensive area of high granulite gneisses at Broken Hill remains equivocal (Stevens and Barron 2002).

Albitisation is a common metasomatic process whereby infiltration of Na-rich fluid (i.e. brine) leads to the alteration of granitoids and metasedimentary rocks, transforming feldspar of variable compositions into albitic feldspar (e.g. Putnis and Austrheim 2010; Condit et al. 2018). During this transformation into albite, quartz is dissolved and consumed; growth of monazite and zircon also allows the timing of high-grade Na metasomatism to be constrained geochronologically (Condit et al. 2018). The addition of Na from the brine is accompanied by the loss of Fe, K, Rb, Ba, and Ca from altered wall rocks which are enriched in the brine. These elements, when reacting with pelitic rocks, are then precipitated as metasomatic ironstones in advance of the main phase(s) of albitic alteration, producing magnetite–clinopyroxene ± biotite-actinolite alteration at high temperatures. This is now recognised as an important process in the generation of both barren and mineralised magnetite ± hematite ironstones, also known as iron oxide copper–gold deposits (Oliver et al. 2004; Chap. 18). Sodium (±calcic) alteration forms regionally extensive, networks of shear zone- and breccia-controlled hydrothermal systems that represent a massive flux of fluid and heat (300–550 °C; Oliver et al. 2004; Clark et al. 2005). Petrogenetic, geochemical and stable-isotope data indicate a significant fluid input from synchronously emplaced intrusions, which are the likely source of Cu–Au; if these elements are absent from the initial magmatic fluid, only barren ironstones form. The processes of mineral dissolution and precipitation can be extremely destructive of precursor textures, and result in pseudo-volcanic rock textures with felsic compositions that are superficially unrecognisable from their original lithologies.

The Olary Domain (western Curnamona Province) in South Australia contains albitised rocks covering many hundreds of km^2 (Kent et al. 2000; Clark et al. 2005), and known iron oxide copper–gold mineralisation. Within the so-called "Lower Albite" stratigraphic unit several volcanic or volcaniclastic units have been identified, intercalated with variably albitised, bedded finer-grained metasediment (psammopelite), banded quartz–magnetite (ironstone) units and composite gneiss interlayered with pelite and pegmatite (Ashley et al. 1996, 1997; Pepper and Ashley 1998). Such interpretations are based on the presence of aligned fiamme-like features suggesting the rocks include welded ignimbrites, and in some locations well-developed "quartz eyes". On closer examination, the so-called fiamme represent pseudo-clasts elongated in the anisotropy of the metasediment layering, which can vary with the intensity of albitisation from where more-or-less continuous relict bedding is preserved, to very pervasively metasomatised rocks with only elongated clots or schlieren (Fig. 13.20a, b; Wright and McConachy 1998). Where particularly well-developed schlieren are observed, these are highly magnetic, suggesting they result from intense albitic replacement of banded quartz-magnetite rocks (Fig. 13.20c). This is confirmed by observations of complex reaction fronts between albitite (non-magnetic) and highly magnetic metasediment (Fig. 13.20d, e). Quartz eyes are present locally in some massive albitic rocks with pseudo-fiamme textures. However, pseudo-fiamme textures always grade imperceptibly into inter-laminated metasediment, and such outcrops are thoroughly invaded by apophyses, veins, and veinlets of blue quartz pegmatite. These are preserved in various stages of dissolution or assimilation into albitite during sodium metasomatism (Condit et al. 2018), resulting in rounded and partially resorbed quartz grains. Where best developed, they occur along large shear zones, which may have been major paleofluid pathways for pegmatitic and metasomatic solutions (Clark et al. 2005).

Lastly here, and most critical, it is important to stress that all these textural types are gradational into each other (even on outcrop scale), and nowhere are any sharp lithological contacts present bounding so-called volcanic (pseudo-fiamme, quartz-eye) units. Critical to this type of analysis, as described earlier with regard to other highly hydrothermally altered successions (Sect. 13.5), is to examine gradational relationships and the total rock relations found in an outcrop. It is imperative that localised "*pseudo-volcanic*" textures are not divorced from the total rock relations present. Metasomatism is clearly recognised on the outcrop scale when a reaction interface can be seen between the protolith and a rock with a different mineralogy, often with a distinct mineral assemblage and chemistry (Fig. 18.20d, e). Such reaction fronts represent the limits of fluid infiltration, and once recognised should be a warning siren to the budding "paleo" volcanologist, not to mention geochronologist, to "*proceed with caution*"! The only unequivocal technique that will control the reliable radiometric age determination of high-grade metavolcanics, or 'apparent' volcanic rocks, is one in which a dateable mineral is analysed by electron microprobe (EPMA) in situ, and is constrained texturally and forms part of the total, including metamorphic, paragenesis of the sample being analysed. On a regional mapping scale, it is important not to confuse albitic metasomatic alteration with stratigraphy.

13.11 How Far Back into the Geological Rock Record Can We Recognise Original Volcanic Textures?

The answer to this question is, as far back as the geological record stretches, as represented in Fig. 13.18, in which recognisable pillow lava forms are preserved in the oldest

Fig. 13.20 Examples of albitic quartzo-feldspthic gneisses showing pseudo-volcanic textures from the Olary Domain, Curnamona Province, South Australia. **a** Variably albitised thinly laminated metasedimentary precursor at Mulga Bore. Observations of drill core elsewhere in the Olary Domain suggest the albitised layers are more pervasively altered siltstones, and the darker layers carbonaceous shales. **b** Felsic gneisses showing relict metasedimentary bedding defined by more-or-less continuous layers and elongated 'pseudo-clasts' giving the appearance of fiamme. Southeast of Cathedral Rock, and similarly located for remaining images. **c** Massive albitite with pseudo-fiamme defined by highly magnetic schlieren. This suggests the rock results from intense albitic replacement of banded quartz–magnetite (ironstone) rocks, with which they are spatially closely associated and show gradational relationships, as described by other workers (e.g. Pepper and Ashley 1998). **d** Reaction front between pale-coloured albitite (non-magnetic) and dark, highly magnetic (quartz–magnetite) altered metasediment. Note relict, streaky magnetic schlieren which defines pseudo-fiamme similar to those observed in (**c**). **e** Complex reaction fronts between albite and quartz–magnetite alteration phases (see text)

volcanic rocks known on Earth, in the 3.7–3.8 Ga Isua Greenstone Belt in southwest Greenland, at amphibolite metamorphic grades and high deformational strain zones.

Even though pumice is one of the most abundant components in pyroclastic deposits and epiclastic volcanic sediments in modern silicic volcanic terrains, it can be very difficult to find preserved in the ancient rock record, but it can be preserved. Not that pumice should be any less abundant in the rock record than it is today. It is, however, one of the components most susceptible to post-depositional modifications. It is porous and so readily allows passage of circulating fluids; it is glassy and metastable, so readily alters or is replaced; and it is mechanically weak, so may take up significant strain during burial compaction or deformation, or both. It also hosts vapour-phase crystallisation (in ignimbrites) and diagenetic minerals, and it readily devitrifies and may preferentially develop devitrification spherulites. In ancient volcanic successions, pumice is most readily identified in welded ignimbrites since its flattening produces the characteristic eutaxitic texture (Chap. 10). Even in metamorphosed, weathered ignimbrites a flattened pumice lenticle foliation, or fiamme foliation may be distinctive and etched out by weathering. In non-welded pyroclastic deposits and sediments, the pumice fraction could be especially susceptible to weathering and alteration. It could also be flattened by mechanical, brittle collapse during burial and compaction, or it could be "densified" by the precipitation of secondary minerals in vesicles, and so retain an original unflattened shape, perhaps making a distinction from lithic fragments difficult.

Fiske (1963) noted that pumice is preserved in rocks as old as Precambrian. We concur with this observation, but note that the form of the pumice may be preserved but the original glass has long been replaced by metamorphic minerals (Fig. 13.16c, d). Fiske (1963) suggested that a combination of both careful hand specimen and thin section examination could verify the existence of pumice and glass shards in some very ancient volcanic successions, as validated in Figs. 13.16c, d, 13.8d, and 4.22g, i.

13.12 And Finally, When All Else Fails .... Fubar-ite

Sometimes, when the rocks are so intensely deformed, altered and/or metamorphosed and that they just won't talk to you, you have to call them what they are in terms of their deformation-alteration-metamorphic state, as discussed above for high-grade metamorphic rocks and in the following chapter on terminology (Chap. 14). When there are no definitive original volcanological textures preserved, avoid using a genetic volcanological name. In addition, the interim acronym name FUBAR-ite can be applied until further information becomes available. BAR stands for "beyond all recognition"; the first two letters are self-explanatory (past tense). In some cases, with more time and exploration, outcrops may be found that reveal the rock's origin.

References

Allen RL (1988) False pyroclastic textures in altered silicic lavas, with implications for volcanic-associated mineralization. Econ Geol 83 (7):1424–1446

Arndt NT, Lesher CM, Houlé MG, Lewin E, Lacaze Y (2004) Intrusion and crystallization of a spinifex-textured komatiite sill in Dundonald Township, Ontario. J Petrol 45(12):2555–2571

Ashley PM, Cook ND, Fanning CM (1996) Geochemistry and age of metamorphosed felsic igneous rocks with A-type affinities in the Willyama Supergroup, Olary Block, South Australia, and implications for mineral exploration. Lithos 38(3–4):167–184

Ashley PM, Lawie DC, Conor CHH, Plimer IR (1997) Geology of the Olary Domain, Curnamona Province, South Australia, and field guide to 1997 excursion stops. MESA Rept Book 97(17):51p

Bailey JE, Self S (2010) The properties and formation of erosional pipe-shaped structures in ignimbrites around the VallesCaldera. Geol Soc Amer Abstr Progr 42(5):51

Barrie CT, Hannington MD (eds) (1999) Volcanic associated massive sulfide deposits: processes and examples in modern and ancient settings. Rev Econ Geol 8

Beane RE (1982) Hydrothermal alteration in silicate rocks: southwestern North America. In: Titley SR (ed) Advances in geology of the porphyry copper deposits: southwestern North America. Univ Arizona Press, Tucson, pp 117–137

Bear AN, Giordano G, Giampaolo C, Cas RAF (2009) Volcanological constraints on the post-emplacement zeolitisation of ignimbrites and geoarchaeological implications for Etruscan tomb construction (6th–3rd century BC) in the Tufo Rosso a Scorie Nere, Vico Caldera, Central Italy. J Volcanol Geoth Res 183(3–4):183–200

Boles JR, Coombs DS (1977) Zeolite facies alteration of sandstones in the Southland Syncline, New Zealand. Am J Sci 277(8):982–1012

Branney MJ, Sparks RSJ (1990) Fiamme formed by diagenesis and burial-compaction in soils and subaqeuous sediments. J Geol Soc Lond 147(6):919–922

Briggs ND (1976) Welding and crystallisation zonation in Whakamaru ignimbrite, central North Island, New Zealand. NZ J Geol Geophys 19(2):189–212

Bucher K, Grapes R (2011) Petrogenesis of metamorphic rocks. Springer, Berlin, Heidelberg, p 428p

Bullock LA, Gertisser R, O'Driscoll B (2017) Spherulite formation in obsidian lavas in the Aeolian Islands, Italy. Periodico di Mineralogia 86(1)

Burley SD, Worden R (eds) (2003) Sandstone diagenesis: recent and ancient. Int Assoc Sed Repr Ser 4:656p

Caporuscio FA, Gardner JN, Schultz-Fellenz ES, Kelley RE (2012) Fumarolic pipes in the Tshirege Member of the Bandelier Tuff on the Pajarito Plateau, Jemez Mountains, New Mexico. Bull Volcanol 74(5):1023–1037

Carr RG (1981) A scanning electron microscope study of post-depositional changes in the Matahina Ignimbrite, North Island, New Zealand. NZ J Geol Geophys 24(3):429–434

Cas RAF, Hayman P, Pittari A, Porritt L (2008) Some major problems with existing models and terminology associated with kimberlite pipes from a volcanological perspective, and some suggestions. J Volc Geotherm Res 174(1–3):209–225

Cas RAF, Giordano G, Balsamo F, Esposito A, Mastro SL (2011a) Hydrothermal breccia textures and processes: Lisca Bianca islet, Panarea volcano, Aeolian Islands, Italy. Econ Geol 106(3):437–450

Cas RAF, Wright JV (1987) Volcanic successions: modern and ancient. Allen & Unwin, London, p 528 pp

Cas RAF, Marks K, Perazzo S, Beresford SW, Trofimovs J, Rosengren N (2013) Were intercalated komatiites and dacites at the Black Swan nickel sulphide mine, Yilgarn Craton, Western Australia, emplaced as extrusive lavas or intrusive bodies? The significance of breccia textures and contact relationships. Precamb Res 229:133–149

Cas RAF, Wright HMN, Folkes CB, Lesti C, Porreca M, Giordano G, Viramonte JG (2011b) The flow dynamics of an extremely large volume pyroclastic flow, the 2.08-Ma Cerro Galán Ignimbrite, NW Argentina, and comparison with other flow types. Bull Volcanol 73 (10):1583–1609

Castro JM, Beck P, Tuffen H, Nichols AR, Dingwell DB, Martin MC (2008) Timescales of spherulite crystallization in obsidian inferred from water concentration profiles. Am Mineral 93(11–12):1816–1822

Clark C, Mumm AS, Faure K (2005) Timing and nature of fluid flow and alteration during Mesoproterozoic shear zone formation, Olary Domain, South Australia. J Met Petrol 23(3):147–164

Clay PL, O'Driscoll B, Gertisser R, Busemann H, Sherlock SC, Kelley SP (2013) Textural characterization, major and volatile element quantification and Ar–Ar systematics of spherulites in the Rocche Rosse obsidian flow, Lipari, Aeolian Islands: a temperature continuum growth model. Contribs Mineral Petrol 165(2):373–395

Condit CB, Mahan KH, Curtis KC, Möller A (2018) Dating metasomatism: monazite and zircon growth during amphibolite facies albitization. Minerals 8(5):187

Cooke DR, Simmons SF (2000) Characteristics and genesis of epithermal gold deposits. Rev Econ Geol 13:221–244

Coombs DS (1954) The nature and alteration of some Triassic sediments from Southland, New Zealand. R Soc N Z Trans 82:65–109

Davies DK, Almon WR, Bonis SB, Hunter BE (1979) Deposition and diagenesis of Tertiary-Holocene volcaniclastics, Guatemala. In: Scholle PA, Schluger PR (eds) Aspects of diagenesis. Soc Econ Paleont Mineral Spec Publ 26:281–306

de'Gennaro M, Incoronato A, Mastrolorenzo G, Adabbo M, Spina G (1999) Depositional mechanisms and alteration processes in different types of pyroclastic deposits from Campi Flegrei volcanic field (Southern Italy). J Volc Geotherm Res 91(2–4):303–320

de'Gennaro M, Cappelletti P, Langella A, Perrotta A, Scarpati C (2000) Genesis of zeolites in the Neapolitan Yellow Tuff: geological, volcanological and mineralogical evidence. Contrib Mineral Petrol 139(1):17–35

Dubé B, Gosselin PA, Mercier-Langevin P, Hannington M, Galley A (2007) Gold-rich volcanogenic massive sulphide deposits. In: Goodfellow WD (ed) Mineral deposits of Canada. Geol Assoc Canada Min Deps Div Spec Publ 5:75–94

Fenner CN (1948) Incandescent tuff flows in southern Peru. Geol Soc Am Bull 59(9):879–893

Fenner RE (1932) Concerning basaltic glass. Am Mineral 17:104–107

Fenn PM (1977) The nucleation and growth of alkali feldspars from hydrous melts. Canad Mineral 15(2):135–161

Fiske RS (1963) Subaqueous pyroclastic flows in the Ohanapecosh Formation, Washington. Geol Soc Am Bull 74:391–406

Friedman I, Long W (1976) Hydration rate of obsidian. Science 191 (4225):347–352

Friedman I, Long W (1984) Volcanic glasses, their origins and alteration processes. J Non-Cryst Solids 67(1–3):127–133

Furnes H (1973) Variolitic structure in Ordovician pillow lava and its possible significance as an environmental indicator. Geology 1 (1):27–30

Gardner JE, Befus KS, Watkins J, Hesse M, Miller N (2012) Compositional gradients surrounding spherulites in obsidian and their relationship to spherulite growth and lava cooling. Bull Volcanol 74(8):1865–1879

Giachetti T, Gonnermann HM, Gardner JE, Shea T, Gouldstone A (2015) Discriminating secondary from magmatic water in rhyolitic matrix-glass of volcanic pyroclasts using thermogravimetric analysis. Geochim Cosmochim Acta 148:457–476

Gifkins CC, Herrmann W, Large RR (2005) Altered volcanic rocks: a guide to description and interpretation. University of Tasmania, Centre for Ore Deposit Research, 274 pp

Giuliani A, Phillips D, Kamenetsky VS, Fiorentini ML, Farquhar J, Kendrick MA (2014) Stable isotope (C, O, S) compositions of volatile-rich minerals in kimberlites: a review. Chem Geol 374:61–83

Halls C, Zhao R (1995) Listvenite and related rocks: perspectives on terminology and mineralogy with reference to an occurrence at Cregganbaun, Co. Mayo, Republic of Ireland. Mineralium Deposita 30(3–4):303–313

Hay RL, Iijima A (1968) Nature and origin of palagonite tuffs of the Honolulu Group on Oahu, Hawaii. In: Coats RR, Hay RR, Anderson CA (eds) Studies in volcanology—a memoir in honor of Howel Williams. Geol Soc Am Mem 116:331–376

Hayman PC, Cas RAF (2011) Criteria for interpreting kimberlite as coherent: insights from the Muskox and Jericho kimberlites (Nunavut, Canada). Bull Volcanol 73(8):1005–1027

Hayman PC, Cas RAF, Johnson M (2009) Characteristics and alteration origins of matrix minerals in volcaniclastic kimberlite of the Muskox pipe (Nunavut, Canada). Lithos 112:473–487

Henley RW, Ellis AJ (1983) Geothermal systems ancient and modern: a geochemical review. Earth Sci Rev 19(1):1–50

Holt EW, Taylor HP (1998) $^{18}O/^{16}O$ mapping and hydrogeology of a short-lived (~ 10 years) fumarolic (>500°C) meteoric–hydrothermal event in the upper part of the 0.76 Ma Bishop Tuff outflow sheet, California. J Volcanol Geothermal Res 83(1–2):115–139

Hutchinson RW (1983) Hydrothermal concepts; the old and the new. Econ Geol 78(8):1734–1741

Jakobsson SP (1978) Environmental factors controlling the palagonitization of the Surtsey tephra, Iceland. Bull Geol Soc Denmark 27:91–105

Jakobsson SP, Moore JG (1986) Hydrothermal minerals and alteration rates at Surtsey volcano, Iceland. Geol Soc Am Bull 97:648–659

Jackson MD, Gudmundsson MT et al (2019) SUSTAIN drilling at Surtsey volcano, Iceland, tracks hydrothermal and microbiological interactions in basalt 50 years after eruption. Sci Drilling 25:35–46.e

Jébrak M (1997) Hydrothermal breccias in vein-type ore deposits: a review of mechanisms, morphology and size distribution. Ore Geol Rev 12(3):111–134

Kawachi Y, Pringle IJ, Coombs DS (1983) Pillow Lavas of the Eocene Oamaru Volcano, North Otago. Pacific Sci Congr, Dunedin, New Zealand Tour Bh 3

Keating GN (2005) The role of water in cooling ignimbrites. J Volcanol Geotherm Res 142(1–2):145–171

Kent AJ, Ashley PM, Fanning CM (2000) Metasomatic alteration associated with regional metamorphism: an example from the Willyama Supergroup, South Australia. Lithos 54(1–2):33–62

Khalaf EE (2014) Diagenetic evolution of the volcaniclastic deposits: an example from Neoproterozoic Dokhan Volcanics in Wadi Queih basin, central Eastern Desert, Egypt. Arabian J Geosci 7(7):2603–2624

Lahaye Y, Arndt N (1996) Alteration of a komatiite flow from Alexo, Ontario, Canada. J Petrol 37(6):1261–1284

Lahaye Y, Arndt N, Byerly G, Chauvel C, Fourcade S, Gruau G (1995) The influence of alteration on the trace-element and Nd isotopic compositions of komatiites. Chem Geol 126(1):43–64

Laznicka P (1988) Breccias and coarse fragmentites: petrology, environments, associations, ores. Elsevier, 832 p

Lipman PW (1965) Chemical comparison of glassy and crystalline volcanic rocks. US Geol Surv Bull 1201-D

Lipman PW, Christiansen RL (1964) Zonal features of an ash-flow sheet in the Piapi Canyon Formation, southern Nevada. US Geol Surv Prof Pap 501B:B74–B78

Lofgren G (1970) Experimental devitrification rates of rhyolitic glass. Geol Soc Am Bull 81:553–560

Lofgren G (1971a) Spherulite textures in glassy and crystalline rocks. J Geophys Res 76:5635–5648

Lofgren G (1971b) Experimentally produced devitrification textures in natural rhyolitie glass. Geol Soc Amer Bull 82:111–124

Lydon JW (1984) Ore deposit models—8. Volcanogenic massive sulphide deposits Part I: A descriptive model. Geosci Canada 12:11(4)

Lydon JW (1988) Ore deposit models—14. Volcanogenic massive sulphide deposits Part 2: Genetic models. Geosci Canada 15(1)

Macdonald GA (1972) Volcanoes. Prentice-Hall, Englewood Cliffs, New Jersey

Mackenzie FT (ed) (2005) Sediments, diagenesis, and sedimentary rocks. Treatise on geochemistry, vol 7. Elsevier, 446 pp

McArthur AN, Cas RAF, Orton GJ (1998) Distribution and significance of crystalline, perlitic and vesicular textures in the Ordovician Garth Tuff (Wales). Bull Volcanol 60(4):260–285

McPhie J, Doyle M, Allen R (1993) Volcanic textures: a guide to the interpretation of textures in volcanic rocks. University of Tasmania, CODES Key Centre, 196 pp

Milner DA (2001) The structure and eruptive history of Rotorua caldera, Taupo Volcanic Zone, New Zealand. PhD thesis, Univ. of Canterbury, Christchurch, 443 pp

Moore JG (1966) Rate of palagonitization of submarine basalt adjacent to Hawaii. US Geol Surv Prof Pap 550-D:D163–D171

Noble DC (1967) Sodium potassium and ferrous iron contents of some secondarily hydrated natural silicic glasses. Amer Mineral 52(1–2):230–252

Oliver NH, Cleverley JS, Mark G, Pollard PJ, Fu B, Marshall LJ, Rubenach MJ, Williams PJ, Baker T (2004) Modeling the role of sodic alteration in the genesis of iron oxide-copper-gold deposits, eastern Mount Isa block, Australia. Econ Geol 99(6):1145–1176

Page RW, Laing WP (1992) Felsic metavolcanic rocks related to the Broken Hill Pb-Zn-Ag orebody, Australia; geology, depositional age, and timing of high-grade metamorphism. Econ Geol 87 (8):2138–2168

Peacock MA (1926) The petrology of Iceland, Part 1. The basic tuffs. R Soc Edinburgh Trans 55:53–76

Peacock MA, Fuller RE (1928) Chlorophaeite, sideromelane and palagonite from the Columbia River Plateau. Amer Mineral 13:360–383

Pepper MA, Ashley PM (1998) Volcanic textures in quartzo-feldspathic gneiss of the Willyama Supergroup, Olary Domain, South Australia. Aust J Earth Sci 45(6):971–978

Phillips WJ (1972) Hydraulic fracturing and mineralization. J Geol Soc Lond 128(4):337–359

Philpotts A, Ague J (2009) Principles of igneous and metamorphic petrology. Cambridge Univ Press, 684 pp

Pittari A, Cas RAF, Lefebvre N, Kurszlaukis S (2015) Alteration styles in the Orion Central Volcanic Complex, Fort à la Corne Kimberlite Field, Saskatchewan, and their effects on primary volcaniclastic textures: implications for facies mapping and diamond exploration. Econ Geol 110(1):147–171

Porritt LA, Cas RAF (2011) The influence of complex intra-and extra-vent processes on facies characteristics of the Koala Kimberlite, NWT, Canada: volcanology, sedimentology and intrusive processes. Bull Volcanol 73(6):717–735

Porritt LA, Cas RAF, Schaefer B, McKnight SW (2012) Textural analysis of strongly altered kimberlite: examples from the EKATI diamond mine, Northwest Territories, Canada. Canad Mineral 50 (3):625–641

Putnis A, Austrheim H (2010) Fluid-induced processes: metasomatism and metamorphism. Geofluids 10(1–2):254–269

Qiu T, Zhu Y (2015) Geology and geochemistry of listwaenite-related gold mineralization in the Sayi gold deposit, Xinjiang, NW China. Ore Geol Rev 70:61–79

Raam A (1968) Petrology and diagenesis of Broughton Sandstone (Permian), Kiama District, New South Wales. J Sed Res 38(2):319–331

Randolph-Flagg N, Breen S, Hernandez A, Manga M, Self S (2017) Evenly spaced columns in the Bishop Tuff (California, USA) as relicts of hydrothermal cooling. Geology 45(11):1015–1018

Rona PA (1984) Hydrothermal mineralization at seafloor spreading centers. Earth Sci Rev 20(1):1–104

Ross CS, Smith RL (1961) Ash-flow tuffs: their origin, geologic relations, and identification. US Geol Surv Prof Pap 366:80p

Sandstå NR, Robins B, Furnes H, De Wit M (2011) The origin of large varioles in flow-banded pillow lava from the Hooggenoeg Complex, Barberton Greenstone Belt, South Africa. Contrib Mineral Petrol 162(2):365–377

Scholle PA, Schluger PR (eds) (1979) Aspects of diagenesis. Soc Econ Paleont Mineral Spec Publ 26:443p

Scott RB (1971) Alkali exchange during devitrification and hydration of glasses in ignimbrite cooling units. J Geol 79:100–109

Secor DT (1969) Mechanics of natural extension fracturing at depth in the Earth's crust. Research in tectonics. Geol Surv CanadaSpec Pap; 68–52

Seligman AN, Bindeman IN, Watkins JM, Ross AM (2016) Water in volcanic glass: from volcanic degassing to secondary hydration. Geochim Cosmochim Acta 191:216–238

Sheridan MF (1970) Fumarolic mounds and ridges of the Bishop Tuff, California. Geol Soc Am Bull 81(3):851–868

Sibson RH (1996) Structural permeability of fluid-driven fault-fracture meshes. J Struct Geol 18(8):1031–1042

Smith RK, Tremallo RL, Lofgren GE (2001) Growth of megaspherulites in a rhyolitic vitrophyre. Am Mineral 86(5–6):589–600

Smith RL (1960) Zones and zonal variations in welded ash-flows. US Geol Surv Prof Pap 35FF:149–159

Smith RL, Bailey RA (1966) The Bandelier Tuff: a study of ash-flow eruption cycles from zoned magma chambers. Bull Volcanol 29 (1):83–104

Sparks RSJ, Tait SR, Yanev Y (1999) Dense welding caused by volatile resorption. J Geol Soc 156(2):217–225

Stevens BP, Barron LM (2002) Volcanic textures in the Palaeoproterozoic Hores Gneiss, Broken Hill, Australia. New South Wales Geol Surv Quart Notes 113:1–22

Stevens BP, Barnes RG, Brown RE, Stroud WJ, Willis IL (1988) The Willyama Supergroup in the Broken Hill and Euriowie Blocks, New South Wales. Precamb Res 40:297–327

Stevens BP, Bradley GM (2018) Sedimentology in metamorphic rocks, the Willyama Supergroup, Broken Hill, Australia. Aust J Earth Sci 65(1):25–59

Stevens BP, Wyborn LAI, Jagodzinski EA (1998) The origin of Hores Gneiss at Broken Hill, N.S.W. and its role in mineralization. 14th Australian geological convention, Townsville. Geol Soc Aust Abstracts 49:431

Stimac J, Hickmott D, Abell R, Larocque ACL, Broxton D, Gardner JE, Chipera S, Wolff JA, Gauerke E (1996) Redistribution of Pb and other volatile trace metals during eruption, devitrification, and vapor-phase crystallization of the Bandelier Tuff, New Mexico. J Volcanol Geoth Res 73(3):245–266

Stripp GR, Field M, Schumacher JC, Sparks RS, Cressey G (2006) Post-emplacement serpentinization and related hydrothermal metamorphism in a kimberlite from Venetia, South Africa. J Met Petrol 24(6):515–534

Stroncik NA, Schmincke H-U (2001) Evolution of palagonite: crystallization, chemical changes, and element budget. Geochem Geophys Geosyst 2(7)

Stroncik NA, Schmincke H-U (2002) Palagonite—a review. Int J Earth Sci 91(4):680–697

Surdam RC, Boles JR (1979) Diagenesis of volcanic sandstones. In: Scholle PA, Schluger PR (eds) Aspects of diagenesis. Soc Econ Paleont Mineral Spec Publ 26:227–242

Titley SR (ed) (1982) Advances in geology of the porphyry copper deposits: southwestern North America. Univ Ariz Press, Tucson, pp 117–137

Urabe T, Scott SD, Hattori KE (1983) A comparison of footwall-rock alteration and geothermal systems beneath some Japanese and Canadian volcanogenic massive sulfide deposits. In: Ohmoto H, Skinner BJ (eds) The Kuroko and related volcanogenic massive sulphide deposits. Econ Geol Monogr 5:345–364

Valenzuela A, Donaire T, Pin C, Toscano M, Hamilton MA, Pascual E (2011) Geochemistry and U-Pb dating of felsic volcanic rocks in the Riotinto-Nerva unit, Iberian Pyrite Belt, Spain: crustal thinning, progressive crustal melting and massive sulphide genesis. J Geol Soc Lond 168(3):717–732

Vaniman D (2006) Tuff mineralogy. In: Heiken G (ed) Tuffs-their properties, uses, hydrology, and resources. Geol Soc Am Spec Pap 408:11–15

Von Aulock FW, Nichols ARL, Kennedy BM, Oze C (2013) Timescales of texture development in a cooling lava dome. Geochim Cosmochim Acta 114:72–80

Watkins J, Manga M, Huber C, Martin M (2009) Diffusion-controlled spherulite growth in obsidian inferred from H_2O concentration profiles. Contrib Mineral Petrol 157:163–172

Willis IL, Brown RE, Stroud WJ, Stevens BP (1983) The early Proterozoic Willyama Supergroup: stratigraphic subdivision and interpretation of high to low-grade metamorphic rocks in the Broken Hill Block, New South Wales. J Geol Soc Aust 30(1–2):195–224

Wright JV (1979) Formation, transport and deposition of ignimbrites and welded tuffs. PhD, Imperial College, University of London, 451 pp

Wright JV, Haydon RC, McConachy GW (1987) Sedimentary model for the giant Broken Hill Pb-Zn deposit, Australia. Geology 15 (7):598–602

Wright JV, Haydon RC, McConachy GW (1993) Sedimentary analysis and implications for Pb-Zn mineralisation at Broken Hill, Australia. James Cook Univ N Qld EGRU Contrib 48:91 pp

Wright HM, Lesti C, Cas RAF, Porreca M, Viramonte JG, Folkes CB, Giordano G (2011) Columnar jointing in vapor-phase-altered, non-welded Cerro Galán Ignimbrite, Paycuqui, Argentina. Bull Volcanol 73(10):1567–1582

Wright JV, McConachy GW (1998) Curnamona project generation. Rept Pasminco Explor, 64 pp (unpubl)

Wyborn L (1998) Geochemical variations in the 'Potosi gneiss': preliminary assessment of data from the Pasminco whole rock geochemical database. Aust Geol Surv Org Prof Opin 1998(08):14 pp

Zheng H, Sun X, Wang J, Zhu D, Zhang X (2018) Devitrification pores and their contribution to volcanic reservoirs: a case study in the Hailar Basin, NE China. Mar Petrol Geol 98:718–732

Part E

Terminology, and Approaches to Understanding the Geology of Volcanoes

Aims, Scope, and Structure

In this part we to consider how to describe, name and document the various deposit types and processes that are found in volcanoes and volcanic terrains. In many modern volcanic settings, where eruption and transportation processes have been witnessed and the resultant deposit types have been identified, the naming of such deposits can be quite straight forward using genetic terminology (e.g. pahoehoe lava flow, basaltic scoria fallout deposit, andesitic block and ash flow deposit, or rhyolitic pumice and ash flow deposit or ignimbrite). However, there are many eruptions that have not been witnessed, making interpretation of the deposits less straight forward, and so, genetic deposit names cannot be confidently applied. For prehistoric eruptions and sequences, and particularly for ancient successions, the origin of many deposit types may be even more uncertain, or at best, ambiguous, particularly where successions have been subject to hydrothermal alteration, metamorphism and deformation. In most cases, therefore, a cautious approach to deposit classification and nomenclature is required. Firstly, in Chap. 14, we propose an approach to systematically developing appropriate terminology from descriptive to genetic for different deposit types or facies. Flow charts summarising the key observations and steps in developing a progressive descriptive to genetic terminology are provided for both coherent (volcanic and sub-volcanic) and volcaniclastic facies. Secondly, Chap. 15 outlines the major principles and challenges in facies analysis, field survey mapping, developing an understanding of stratigraphic architecture, and documenting the geology of volcanoes and volcanic terrains, both modern and ancient. We assess the major factors that influence the preserved geometry of stratigraphic units, including the effects of eruption geometry, effects of high rates of erosion, volcano-tectonic (as opposed to regional tectonic) deformation, and provide a current review of the different approaches.

Terminology for Volcanic Deposits and Rocks

14

Descriptive to Genetic Nomenclature, with Key Observations and Steps

Ray Cas, John V. Wright, and Guido Giordano

R. Cas et al., *Volcanology*, Springer Textbooks in Earth Sciences, Geography and Environment,
https://doi.org/10.1007/978-3-319-66613-6_14

(a)
(b)
cm
(c)
(d)
(e)
(f)
(g)
(h)

Abstract

In many modern active settings where eruptions and transportation processes have been witnessed, the naming of volcanic deposits and rocks can be quite straightforward using genetic terminology. However, for prehistoric eruptions and sequences, and particularly for ancient successions, the origin of many deposit types may be uncertain or, at best, ambiguous, particularly where successions have been subject to hydrothermal alteration, metamorphism, and deformation. In most cases, therefore, a cautious approach to deposit classification and nomenclature is required. Our approach to terminology incorporates documentation of components, textures and assessing the degree and effects of alteration on both components and original emplacement textures. This allows a purely descriptive composite rock or deposit name to be constructed first, incorporating components, textural, and compositional petrological characteristics, free of any biases about emplacement settings and processes. Then important facies features such as depositional structures, contact relationships, and settings are assessed, leading to a composite descriptive and interpretive genetic name for the facies or rock unit that summarises key descriptive characteristics, emplacement processes, and settings. Colour coded flow charts summarising the key observations and steps in developing a progressive descriptive to genetic terminology are provided for both coherent (volcanic and subvolcanic) and volcaniclastic facies, including pyroclastic, autoclastic, and epiclastic, as well as hydraulic fracture deposits/rock units. These flow charts can be copied and used in the field (including drill core logging observations), and in conjunction with petrographic data, provide a useful checklist of essential observations that should be made to develop an understanding of the possible origins of any volcanic succession.

14.1 Introduction

Assigning an informative name to, or classifying a volcanic deposit or rock genetically (i.e. reflecting the process origins) is important for scientific communication, but must first be preceded by identifying the distinctive and important descriptive characteristics of a deposit that will allow its origins to be determined (Plate 14.1). Developing an informative name is the end product of a series of analytical steps undertaken to better understand the origins of a deposit or rock. It is therefore not just a mechanical process, but an essential analytical process that clarifies and facilitates the research we do.

Deciding on an appropriate name to apply to a particular deposit may seem like a straightforward matter for volcanologists who are constantly working on modern volcanoes, their eruptions, and the deposits. During such eruptions, the eruption style, the transport and depositional processes, and the immediate deposit characteristics can often be witnessed to varying degrees. However, even in such situations, the origins of some recently emplaced deposits can be ambiguous because each event and deposit cannot be viewed up close for safety reasons, and often weather or remoteness impedes observations. For example, Uehara et al. (2015) examined the deposit sequence exposed in deep erosional canyons in the apron of block and ash flow and the debris flow deposits that accumulated around Mt Unzen, Japan, during the 1991 to 1995 lava dome eruption. The eruption was witnessed from a safe distance, but details of processes in individual valleys were usually not possible because of safety issues, and often wet, cloudy, and even monsoonal weather conditions. They found that it was often very difficult to distinguish between primary block and ash flow deposits from the many debris flow deposits generated by heavy rainfall during the eruption (Plate 14.1a). This is because the componentry of the two was

◀ **Plate 14.1** A variety of coherent, fragmental, and alteration textures that need to be assessed systematically using the approach proposed in this chapter before being given a genetic rock name that reflects the origins and the textures. **a** Matrix-supported dacite breccias from the Mt Unzen 1991–1995 eruption, Kyushu, Japan. *Are these block and ash flow or debris flow deposits?* (Block and ash flow). **b** Permian latite (trachy-basalt) breccia, Gerringong Volcanics, Kiama, New South Wales, Australia. *Is this is pyroclastic agglomerate breccia or an autoclastic breccia?* (Peperite). **c** Miocene matrix-supported dacite breccia (Motoyama breccia), Hokuroko Basin, northern Honshu, Japan. *Is this a hydrothermal explosion breccia (as originally proposed), or a peperite?* (Good question!) **d** Silurian flow banded rhyolite, Rocklands, western Victoria, Australia. *Was this a lava or an intrusion?* (Lava?) **e** Crystal-rich kimberlite, Eocene Koala kimberlite pipe, Ekati kimberlite field, Lac de Gras, Northwest Territories, Canada. *Is this a coherent kimberlite or a crystal-rich volcaniclastic kimberlite, and why?* (Coherent kimberlite). **f** Archean komatiite, showing transition from random to blade spinifex to orthocumulate texture. Was this a lava or an intrusion? (Lava). **g** Obsidian-pumice-lithic clast breccia with eutaxitic texture defined by flattened black glassy fiamme, Pleistocene Guajara Formation, Tenerife, Canary Islands, Spain. *Was this a welded pyroclastic flow or fallout deposit?* (Fallout). **h** Apparent clast to matrix-supported dacite breccia, Ordovician Mt Windsor Volcanics, Charters Towers, Queensland, Australia. *Was this a volcaniclastic breccia or coherent lava?* (Hydrothermally altered lava). NOTE: In almost every case, the determination of the final origin cannot be based on the texture seen alone, but must also consider field context and contact relationships with surrounding lithofacies and stratigraphic units identical since the debris flows originated from the remobilisation of block and ash flow deposits, and both involved short-lived mass-flow transportation that produced minimal textural modification. So, even from modern eruptive sequences, it may be difficult to distinguish particular deposit types, which in turn has important implications for hazard and risk assessment for future eruptions.

There are also many eruptions which have not been witnessed, nor recorded, and for which the preserved deposit characteristics provide the only clues as to their origin. For these, investigative fieldwork and laboratory work is required to identify the diagnostic deposit characteristics that will allow an appropriate deposit name and an origin to be determined. Beyond this, for many prehistoric eruptions to the huge array of ancient volcanic successions, ranging in age from Quaternary to Archean, the deposit sequences may be only partly preserved and may be variably altered, metamorphosed, and deformed (Chap. 13). For most ancient volcanic successions the location of the vent, and even the source volcano, are unknown. For these, only detailed fieldwork, involving careful mapping, description, and documentation of deposits and their original facies characteristics will lead to a determination of their origins. For such deposits, a much more analytical approach is required, including the way that terminology is applied, beginning with descriptive properties and then progressing to genetic terms that reflect the interpreted process origins.

14.2 Previous Useful Classification Schemes for Volcanic Rocks

There have been many approaches to developing terminological schemes for volcanic rocks and deposits. In this chapter, we will focus on the most useful contributions. Fisher (1966) summarised the various approaches to developing classification schemes for deposits consisting of fragments of volcanic debris. Wentworth and Williams (1932) proposed a systematic classification scheme for pyroclasts, pyroclastic deposits, and pyroclastic rocks, subdividing them into size classes, represented by the familiar pyroclastic size terms blocks and bombs, lapilli, and ash. Fisher (1960, 1961, 1966) adopted a more comprehensive approach, by including all deposits and rocks consisting of volcanic debris, irrespective of the mode of formation. He therefore included pyroclastic, autoclastic, and epiclastic deposits (epiclastic *sensu lato*, irrespective of age of deposits or their sources, as defined in Chaps. 4, 12), and he also amended the size classes to conform exactly to the Wentworth size classes used in sedimentary geology. In addition, Fisher promoted the importance of distinguishing between descriptive and genetic terminology.

Wright et al. (1980) also focused on pyroclastic deposits, but in addition to grain size and componentry they presented a comprehensive classification of pyroclastic deposits, based on their transportation processes, and also emphasised the value of both descriptive and genetic approaches to developing terminology. Cas and Wright (1987) not only largely used the approach of Fisher, but also tried to distinguish terminological approaches for modern pyroclastic deposits (After Wright et al. 1980) and for lithified volcaniclastic deposits of the rock record. Cas and Wright (1987) proposed that to develop a genetic terminology that correctly reflects the origins of volcanic and associated sedimentary deposits, it requires a stepwise progression from initial description and documentation of deposit characteristics, from small-scale to progressively larger scale features, to the interpretation of the deposit origins, and finally the application of a genetic name. Initial deposit names used in the field should therefore be descriptive, and only after all characteristics have been considered should genetic names be applied. McPhie et al. (1993) proposed a valuable, systematic, tabulated approach to developing nomenclature for the main deposit types in volcanic terrains and rock successions, including both coherent and volcaniclastic deposits, and they provided some general guidelines as to how to construct a name.

Coherent volcanic and intrusive rocks (i.e. those not subjected to any syn- to post-eruptive fragmentation) had previously only been classified according to their petrological and geochemical characteristics, but not their emplacement origins. Cas et al. (2008a, b, 2009) reviewed the approach that has been used historically to the terminology of kimberlite deposits. They found a confusing mixture of descriptive and genetic terminology, involving many terms that are not used in modern volcanology. In an attempt to unravel the disarray, they proposed a terminological scheme involving a very systematic, step-by-step progression from descriptive to genetic terminology, which not only represents a development of the approach adopted by Fisher (1960, 1961, 1966), Wright et al. (1980), Cas and Wright (1987), and McPhie et al. (1993), but also includes consideration of both fragmental and coherent deposits. Although constructed specifically to overcome the terminological problems in kimberlite volcanology and geology, the proposed scheme (Cas et al. 2008a, b, 2009) is a generic one that can be applied to any volcanic system irrespective of age and the composition of the eruptive products.

We present that approach here, and provide guidelines for naming both coherent and volcaniclastic deposits, and including pyroclastic, autoclastic, surface sedimentary, or epiclastic, as well as hydraulic fracture deposits. The guidelines are a useful checklist of essential observations that should be made in understanding any volcanic succession. They are therefore more than just an approach to naming deposits. However, first we discuss some issues associated with using the most appropriate deposit names.

14.3 Some Problems with Developing Correct Terminology

Terminology or nomenclature can be influenced, and sometimes confused by several factors. These include grain size terminology, the perceived genetic significance of some components, the ambiguity of the multiple meanings in which some terms are used, and terminology that is driven by genetic models.

14.3.1 Understanding Grain Size Terms

Grain size and grain size distribution (e.g. sorting) are one of the first things one notices in looking at a new outcrop and deposit. How coarse or fine is it, how well sorted? Such observations immediately begin to suggest processes (energy levels of clast formation, transport competency of the transporting agents). However, grain size is not uniquely indicative of specific origins in any volcanic setting. Nonetheless, the grain size of a deposit, particularly a volcaniclastic deposit, is an important starting point to assess possible origins of deposits and developing a correct name. However, first, we need to review the origin of grain size classification schemes used in sedimentology and volcanology which we advocate.

Earth scientists and civil engineers have long used the Wentworth-Udden grain size scheme for clastic sediments based on a geometric progression of size classes using 2 as the base number (Wentworth 1922). Wentworth and Williams (1932) then first proposed a specific grain size terminology scheme for pyroclastic deposits and rocks. This was then endorsed and slightly modified by Fisher (1960, 1961, 1966), who also clearly distinguished grain size terminology for pyroclastic deposits from other fragmental deposits of volcanic debris. We agree with Fisher's rigid restriction to using terms such as bomb, block, lapilli, ash, lapilli-ash, lapillistone, tuff, and agglomerate only for deposits that were demonstrably fragmented by explosive processes and finally transported and deposited by pyroclastic processes. The question then is, "*what other grain size terms can be used for other fragmental volcanic deposits?*".

The familiar grain size terminology that is used in sedimentary geology (e.g. clay, silt, sand, pebble, and cobble boulder) is also applied in other disciplines, such as civil engineering, for loose (natural) aggregates of rock and mineral fragments. We suggest that with some qualification, it should be considered to be a generic, *descriptive* grain size terminological scheme that can be applied to any aggregate of debris, irrespective of origin, in an initial purely descriptive sense, until the origins of a deposit are determined. However, some people feel that applying sedimentary size terms in a descriptive way implies that a deposit had a sedimentary transport and depositional origin. We suggest that in the first instance, before the origin of the deposit is known, it's grain size could be described using qualifying size analogue terms, such as boulder size, cobble size, pebble size, sand size, and silt size, rather than just boulder, cobble, pebble, sand, silt, and clay, which signify a sediment *sensu stricto* size class. Once the origin of a deposit has been worked out using all deposit characteristics, then grain size terms appropriate to that origin can be applied, as discussed later in this chapter.

14.3.2 Potentially Misleading Use of Specific Components to Interpret and Name Deposits

The presence of some components may misleadingly influence thinking about the origin of particular deposits, leading to erroneous interpretations and deposit names. It is not possible to determine the origin of any deposit based on the presence of a single or several components. Several examples include:

(a) Although the presence bubble wall, glass shards in a sand size or silt size deposit may imply that the deposit is pyroclastic, but this is not necessarily so. Glass shards, even fragile ones, can be reworked by sedimentary processes and survive with their morphology intact. Such deposits should not be called "*ash*" or "*tuff*" because that implies a pyroclastic mode of transportation and deposition. An alternative may be to call them reworked or resedimented ash or tuff, or more appropriately, vitric siltstone or fine sandstone, reflecting the final sedimentary process origin.

(b) Pumice clasts, which most people assume to have a pyroclastic origin, can also form by autoclastic processes in a pumiceous carapace of a subaerial or submarine lava dome (Chaps. 3, 4, 6), and can be variably reworked by sedimentary tractional processes, or be rafted by flotation in the oceans or lakes large distances from the source, or be resedimented by sedimentary mass-flow processes long distances away from the source, as can pumice of explosive origin. Deposits that contain pumice clasts are therefore not necessarily pyroclastic deposits, and should not be given names suggesting a pyroclastic origin. Terms such as pumice conglomerate (= reworked rounded pumice), or pumiceous hyper-concentrated flood flow deposit, debris flow, or turbidite deposit are more appropriate genetic process names for those kinds of deposits, once the depositional context and processes have been evaluated from the deposit facies characteristics.

(c) Similarly, accretionary lapilli, which commonly form during phreatomagmatic and phreatic explosive eruptions (Chap. 4), do not necessarily indicate that the deposits they are found in were pyroclastic in their origin. If they were deposited in a pyroclastic deposit, which was then eroded and reworked, some accretionary lapilli could survive sedimentary transport and reworking, to be deposited in a volcanic sediment, especially if they had been hardened/lithified by heat, freezing or electrostatic forces during eruption and pyroclastic transport (Chap. 4). Accretionary lapilli occur in pyroclastic fallout, flow, and surge deposits (Chaps. 9, 10), as well as reworked and resedimented volcanic sediments. They can be found close to or distant from the source vent, and are not unique to a particular deposit or paleoenvironmental setting (e.g. Cunningham and Beard 2014).

(d) Fiamme, which are thermally and plastically, collapsed or flattened pumice fragments, which are commonly found in welded pyroclastic flow deposits or ignimbrite, can have diverse origins and do not necessarily indicate by themselves that the deposit they are found in is an ignimbrite. Welding and fiamme also occur in some near-vent, welded fallout deposits (Plate 14.1g; Sparks and Wright 1979; Houghton and Carey 2015; Chap. 9). However, pumice in non-welded pyroclastic and sedimentary deposits can also collapse diagenetically after the pumice has been weathered and buried, producing a pseudo-fiamme welding foliation (Branney and Sparks 1990; Chap. 13). Even coherent lavas and intrusions that have undergone polyphase hydrothermal alteration, as discussed in Chap. 13, may develop apparent pseudo-fiamme or pseudo-eutaxitic textures. The problem can get more acute in ancient volcanic successions, when the imposition of a deformational strain fabric, metamorphism and metasomatism can also mimic fiamme morphology, again producing misleading textures (Chap. 13).

(e) In addition, many volcanic breccia deposits, consisting of angular volcanic clasts, can be ambiguous in terms of their origin. As discussed in the next section, volcanic breccias can have multiple origins, and are not just indicative of an explosive origin (Table 14.1A, B).

In all cases, the origin of deposits containing these various components cannot be determined by the presence of those components alone, but must be determined using all the characteristics, including all components, field depositional facies characteristics, deposit contact relationships, and settings. It is only by a systematic analysis, as presented in this chapter, that a final genetic deposit name can be applied.

14.3.3 Ambiguity of the Multiple Senses in Which Some Terms Are Used

Some terms are employed in multiple senses, which can be confusing, and need to be carefully considered before using them. For example, the terms *ash* and its lithified equivalent, *tuff*, are used to describe:

- grain size in a deposit of volcanic grains < 2 mm, irrespective of the fragmentation origin,
- size class of very fine pyroclasts < 2 mm,
- aggregate of fine pyroclastic particles, irrespective of final depositional origin,
- fine-grained pyroclastic fallout, flow, or surge deposit, or
- any fine-grained deposit containing volcanic fragments, irrespective of the origin.

We suggest that the term "ash should be restricted to denote the fine grain size class of very fine pyroclasts (< 2 mm), and that "ash" and "tuff" also be used for very fine-grained deposits where it can be demonstrated that they were fragmented explosively *and* transported *and* finally deposited by pyroclastic processes, which should be added as qualifying adjectives (e.g. distal fallout ash/tuff).

The term *agglomerate* is used by some geologists to denote, either:

- any coarse breccia of volcanic debris irrespective of the fragmentation origin, or
- specifically, a coarse, near-vent, explosion breccia that will often contain fluidal bombs.

The latter is the correct use, but .... "*how do you prove that in old rock successions*"? Unless a clear, near-vent ballistic fallout origin can be demonstrated (i.e. you have to know where the vent was), any coarse aggregate of angular blocks (e.g. Plate 14.1b, c) should just be called *volcanic breccia*, which is a descriptive term without genetic connotations, until a clear origin can be demonstrated. True agglomerates are rarely preserved in the rock record.

When the term *volcanic breccia* is (erroneously) used as a synonym for agglomerate, this involves a simplistic (erroneous) assumption that coarse volcanic breccias:

- can only be produced by explosive fragmentation, and
- that they are only deposited near a volcanic vent during such an explosive event.

Both assumptions are wrong and, as Cas and Wright (1987) pointed out, there are over twenty ways to produce volcanic breccias (see below for Table 14.1A, B), the origins of which then need to be determined as discussed further below (Sect. 14.4).

Other terms may also be carelessly misused in different senses. In all cases, misuse of terminology can lead to misleading interpretations.

14.3.4 Terminology that is Driven by Genetic Models

It is not unusual to apply a genetic name to a deposit prematurely because one too hurriedly develops a preconceived idea of an eruption or depositional model. This then prejudges a deposit before the due process of carefully observing and documenting the diagnostic properties is undertaken. For example, one could decide at an outcrop that a deposit is an ignimbrite, because it is thick and poorly sorted and contains what appear to be juvenile clasts. Everything is then driven by the various facies that occur in the ignimbrite model and the eruption processes that are involved in producing ignimbrites (Chap. 10), before adequate mapping of the unit is undertaken. Such a deposit could also be a debris flow deposit, or a vent-fill lithofacies, and so misinterpretations could occur.

Kimberlite geology has had an obsession with genetic models for kimberlite pipes, and the particular deposit types that should occur in particular parts of "the kimberlite pipe model", without an understanding of the essential volcanological processes that can occur in different types of kimberlite pipes, or the types of edifices kimberlite volcanoes can build (Chap. 18). As a result, terminology that has a model-driven, genetic significance evolved over many years, and the scientific understanding of how kimberlite volcanoes behave and the volcanological significance of particular deposit types in terms of modern volcanological processes, was retarded for decades. Cas et al. (2008a, b; 2009) have comprehensively reviewed these problems and the terminological issues. Cas et al. (2008a, b; 2009), Sparks et al. (2006), and Sparks (2013) proposed that kimberlite (and including lamproite) deposit types can be interpreted in terms of modern volcanological processes, and that modern volcanological terminology can be applied after "back-stripping" the almost pervasive overprint of alteration, most of which is secondary serpentine alteration, from the preserved texture, so then giving insight into the original deposit texture and origins. The only unusual thing about kimberlite deposits is that most are preserved in intra-vent settings inside kimberlite pipes (that is, kimberlite pipes are relict vents), whereas in most other volcanic systems, vent-fill sequences are rarely preserved and exposed, and volcanologists deal mostly with supra-crustal, extra-vent deposits. The preservation of many, especially diamondiferous, kimberlite pipes is due to their rather unique setting in tectonically stable Archean cratons, whereas most other volcanoes occur in potentially more tectonically active settings, and are subject to tectonic deformation, uplift and erosion.

14.4 Clarifying Some Terms Used Here

Adopting the usage of Fisher (1966), the term *"volcaniclastic"* is a descriptive term that applies to all fragmental rocks consisting of volcanic debris, irrespective of their origins. This term thus distinguishes all fragmental volcanic deposits and rocks from coherent volcanic rocks, and from other types of sediments. It therefore includes all genetic classes of fragmental volcanic rocks, including primary fragmental deposits (pyroclastic, autoclastic), reworked and resedimented derivatives of these (epiclastic *sensu lato*), and even arguably, hydraulic fracture volcanic breccias. Unfortunately, some people use the term "volcaniclastic" to be synonymous with volcanic sediments, which is incorrect. It is a descriptive "umbrella" term for all deposits containing predominantly fragments of volcanic debris (pyroclastic, autoclastic, epiclastic) to distinguish such facies in the first place from coherent volcanic facies.

As defined by Wentworth and Williams (1932) and Fisher (1961, 1966), the term "*pyroclastic*" refers specifically to deposits and rocks that form through explosive fragmentation and are ejected explosively from a vent. We add that to be called pyroclastic, a deposit must also have been finally transported and deposited by pyroclastic processes (pyroclastic and ballistic fallout, flow, or surge), and that pyroclastic deposits can also be deposited in vents. *Pyroclasts* therefore are all fragments that are created by explosive fragmentation, irrespective of whether they are juvenile fragments (e.g. pumice, scoria, spatter, shards, and cognate lithic clasts), accessory lithic clasts, or crystal (magmatic, xenocrystic) fragments (Chap. 4). Most are ejected explosively from a vent, and are deposited outside the vent, but some never escape from the vent, or fall back into the vent and contribute to vent-fill deposits. However, some clast populations and deposits inside vents are the products of rock fall and debris collapses from unstable vent walls after eruptions have ended. Therefore, not all deposits in a vent are necessarily pyroclastic in origin, even though they consist of coarse angular volcanic debris. Angular volcanic breccias in a vent could therefore be true agglomerates, but many are rock fall deposits from the crater walls. After eruption and dispersal over the landscape many pyroclastic deposits are subsequently eroded by surface sedimentary processes (Chap. 12) and then become volcanic sediments even though the fragments may originally have been pyroclasts.

The term "*epiclastic*" was invented to describe grains, clasts, and sediment or sedimentary rock that has formed by the erosion of an older exposed source rock. Fisher (1960, pp 978–979), however, clearly included any young unconsolidated deposit, such as a pyroclastic deposit that was remobilised by, or transformed syn-eruptively into, an aqueous sedimentary transport process, as epiclastic (*sensu lato*) because its final depositional characteristics bore the imprint of sedimentary transport and deposition. Cas and Wright (1987) agreed with this application of the term epiclastic, and we do again here, because it is accidental whether or not a source material is young or old, lithified or unconsolidated at the time that surface sedimentary erosional forces and transport agents act on those materials. The key point to us is that the final mode of transport and deposition was by sedimentary processes, and NOT by primary volcanic processes. It is certainly important to know if remobilisation occurred syn-eruptively (*resedimentation,* or epiclastic *sensu lato*; Chap. 12) or post-eruptively (*reworking,* or epiclastic *sensu stricto*) in terms of understanding chronology and causes of events. If there is evidence of syn- or post-eruptive surface sedimentary or epiclastic influences in the genesis of a deposit or facies, then these terms should be used as qualifying adjectives in the final genetic deposit name. We are aware some readers may disagree with our application of the term epiclastic, as Fisher (1961, 1966) subsequently recanted, then they should at least use qualifying adjectives, such as "resedimented, or reworked pumice deposits". Nonetheless, we stress, these distinctions become increasingly difficult to differentiate further back into the geological record as context (in particularly spatial relation to a volcanic vent and direct observational data of eruptive activity) diminishes.

The term "*volcanic breccia*" is also a descriptive term to describe any deposit or rock that consists of coarse (>2 mm), angular volcanic fragments, irrespective of the mode of origin of the volcanic breccia. The term does not mean that the deposit was the result of explosive fragmentation and deposition, that is, an agglomerate, which is specifically an explosion breccia. There are in fact over 20 different ways in which volcanic breccia can be formed, including explosive, autoclastic, and epiclastic processes, as well as hydrothermal hydraulic fracturing, and even fault brecciation of volcanic rocks (see Table 14.1 below; Cas and Wright 1987, app. II). It is, therefore, a first-order descriptive textural and compositional term for a deposit of coarse, angular volcanic clasts, the origin of which then needs to be determined based on textural and depositional properties, contact relationships, and depositional settings. Only then can a more genetic name be applied.

The terms *groundmass* and *matrix* are sometimes used interchangeably in both coherent and volcaniclastic deposits. To avoid confusion, we recommend that *groundmass* be used exclusively for the fine-grained or aphanitic medium in between large phenocrysts in coherent rocks. The term *matrix* should be used exclusively for the finer-grained clastic material between large clasts or grains in a volcaniclastic deposit or sedimentary deposit. Matrix is a relative size term; it can be coarse in a breccia or conglomerate, or fine in a sand size deposit.

In kimberlite geology, there are several terms that are not used in mainstream volcanology, and are therefore confusing. Many imply a specific, poorly defined process, and/or setting in a kimberlite pipe. These have arisen because most research on kimberlites was undertaken in-house by diamond mining and exploration companies, in isolation from the volcanological community and became model driven. A strange concept arose that kimberlite volcanoes, of which kimberlite "pipes" were the vents and deeper conduits, somehow behave differently to other magma types and volcanoes and that their physical eruption and chemical processes are different and even exotic. In a series of publications, Cas et al. (2008a, b; 2009) and Sparks et al. (2006), and Sparks (2013) demonstrated that this is a nonsensical premise, and that applying normal volcanological principles to kimberlites as volcanoes and understanding the effects of alteration on original rock textures allows for realistic volcanological processes to be reconstructed.

Terms that mean almost nothing to most volcanologists include *"tuffisitic kimberlite"*, a deposit that is supposed to form specifically in one part of a kimberlite pipe only by a subterranean fluidisation process. Most volcanologists would call tuffisitic kimberlite a "lapilli-tuff" because of its poorly sorted character, and would probably explain its origin by open vent explosive eruption and depositional processes. *Hypabyssal kimberlite* has been used for coherent kimberlite, with the implication that it can only occur in the root zone, deep in a kimberlite pipe, whereas coherent kimberlite (Plate 14.1e) can actually occur at any level in the pipe, and can be erupted as lavas. The term autolith has been used for cognate kimberlite clasts. The term magmatic kimberlite has been used for coherent kimberlite, magmaclast for a juvenile pyroclast, magmaclastic for fragmented kimberlite, and so on.

Cas et al. (2008a, b; 2009) recommended that all of these "exotic" terms be abandoned, and replaced by standard volcanological terms, and process concepts. They also recommended that the approach and guidelines adopted in this chapter be adopted and used in applying terminology to any kimberlite deposit or facies. Readers specifically interested in kimberlite terminology are referred to Cas et al. (2008b, 2009) for a comprehensive review. Kimberlite volcanology is further discussed in some detail in Chap. 18, which provides a review of their geology and of their diamondiferous resources.

14.5 An Approach to Developing a Descriptive to Genetic Terminology for Volcanic Deposits and Rocks

Cas et al. (2008b; 2009) proposed that terminology for volcanic deposits and rocks should be developed sequentially in three stages:

STAGE 1: Document descriptive lithological, petrological characteristics, based on hand specimen to microscopic-scale observations:

(a) Document the descriptive lithological features seen in hand specimen in the field and in follow-up petrography and other microscopy, including components, texture (grain size, framework, matrix, cement, sorting, grain shape, rounding, vesicularity), and alteration types and effects. Selective geochemistry can also be undertaken to characterise component and whole rock compositions.
(b) Construct a composite descriptive compositional and textural facies lithological name that includes the effects of alteration.

STAGE 2: Assess and document key descriptive mesoscopic (outcrop or drill core scale) depositional/emplacement features:

(a) Document depositional structures, 2D and 3D geometry where possible (including thickness and lateral extent), and contact relationships at outcrop and drill core scale.
(b) Construct a composite descriptive facies name incorporating depositional and petrological features.

STAGE 3: Construct a genetic deposit name, if possible, by:

(a) Interpreting a process emplacement origin, using all characteristics and the interpreted (paleo)settings.
(b) Then combining key descriptive characteristics with setting and interpreted emplacement origin to produce a composite genetic deposit name.

In the following discussion, we describe the approach to developing a terminology or deposit name for any particular volcanic lithofacies, progressively from descriptive to genetic. For this, we have modified the coloured flow charts of Cas et al. (2009) for kimberlites, into a scheme that is useable for *all* volcanic successions, irrespective of age, degree of lithification, and composition (Figs. 14.1a, b). The flow charts depict two groups of deposits or rock types: coherent lithofacies (for coherent lavas or intrusions; *red flow chart*; Fig. 14.1a), and fragmental or volcaniclastic lithofacies (for autoclastic, pyroclastic, epiclastic, hydraulic fracture, and cataclastic deposits; *green flow chart*; Fig. 14.1b).

For both of these groups of lithofacies, the same three stages are represented. Within each stage, there are specific steps that require observations about specific deposit characteristics to be made and recorded. The charts therefore not only facilitate the development of an appropriate terminology, but also serve as a checklist of observations and characteristics to record in the field and in the laboratory. The observational criteria, and the range of characteristics listed, are not exhaustive or exclusive, and can be added to if needed, on a case-by-case basis. The flow charts can be copied, laminated, and taken into the field as field reference charts of observations to be made. We have field-tested these charts in undergraduate and postgraduate field classes, professional courses, and research projects undertaken by the authors.

The key question before starting to use the flow charts is: "Is the deposit/rock you are working on coherent or fragmental/volcaniclastic?" *That is, should you be using the red chart or the green chart?*

For volcanologists working on young, well-preserved, undeformed, and unaltered deposits, this will usually be a simple decision. However, spare a thought for scientists who undertake research, or work professionally in the rock record (e.g. geological mapping, mineral and geothermal energy exploration) where rocks may be hydrothermally altered, metamorphosed, and deformed, all of which can mask original rock textures (Chap. 13). For such rocks, an initial decision on whether to use the red chart or green chart may not be possible, and both should be used until a decision can be made. Even then, a final decision and appropriate rock deposit name may not be possible and an appropriate metamorphic rock name may be necessary. If all else fails, the name "*fubarite*" can be applied (Chap. 13).

Another comment needs to be made before we explain the flow charts and the steps involved. A single emplacement unit may host multiple lithofacies. For example, a rhyolite lava flow may consist of dense, crystallised (stoney) coherent rhyolite, coherent pumice domains, obsidian domains, autobreccia, and talus collapse breccias, to name but a few (Chap. 6). Each lithofacies is different and requires an appropriate name, but constitutes a part of a complex, single emplacement volcanic unit. A single, simple terminology or name can't always be applied to one emplacement unit that truly reflects diversity. Again, in the modern record that complex facies association within a single emplacement unit may be relatively easy to recognise, but in the rock record,

(a)

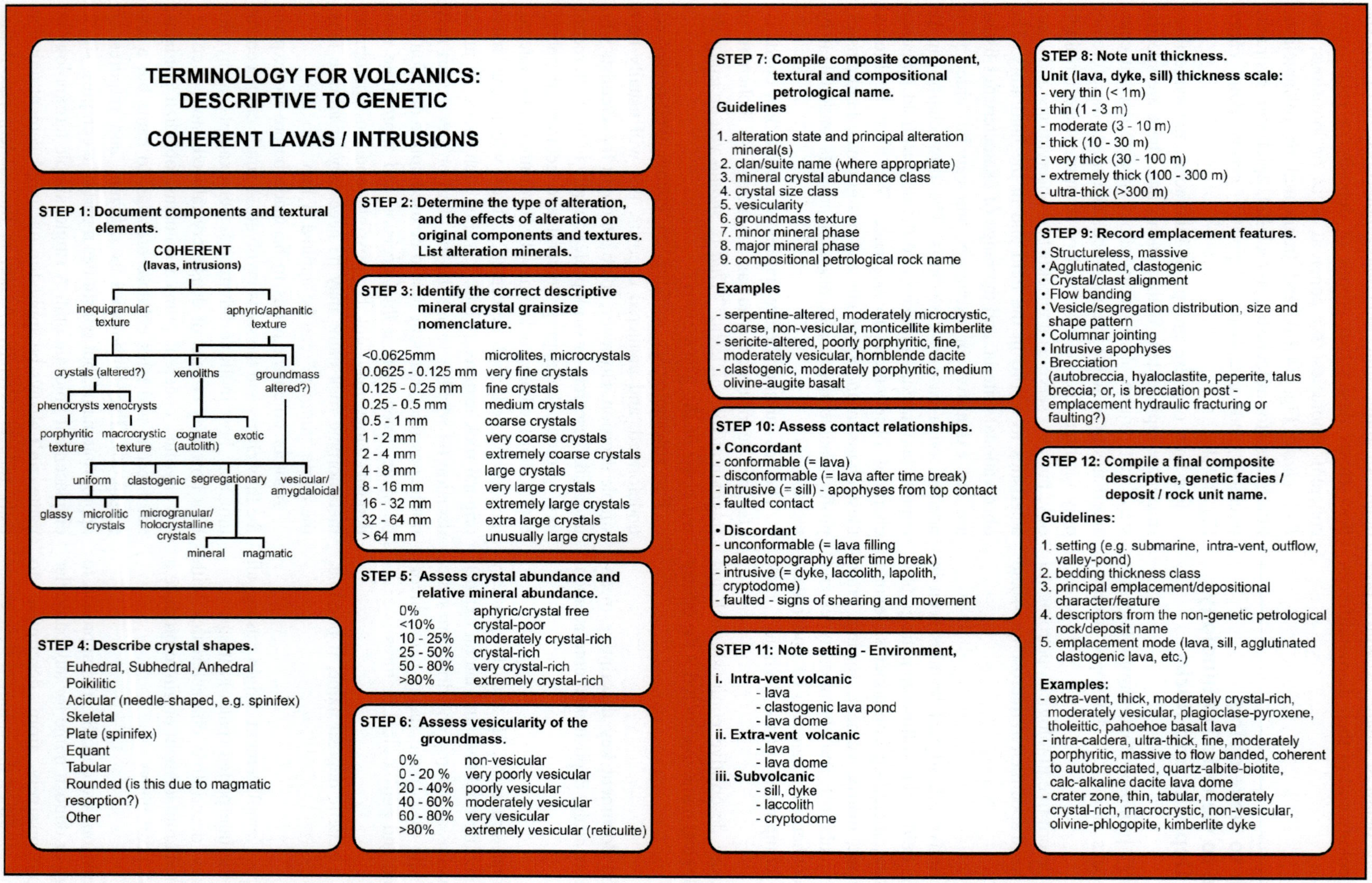

Fig. 14.1 Guidelines and observations required for developing a comprehensive deposit name for coherent volcanic and subvolcanic units (After Cas et al. 2009). As discussed in the text, the flow chart outlines a sequence of 12 steps recommended in developing a progression from initial descriptive to final genetic terminology. **a** Steps for developing a comprehensive deposit name for coherent volcanic and subvolcanic rocks. **b** Steps for developing a comprehensive deposit name for fragmental or volcaniclastic deposits

(b)

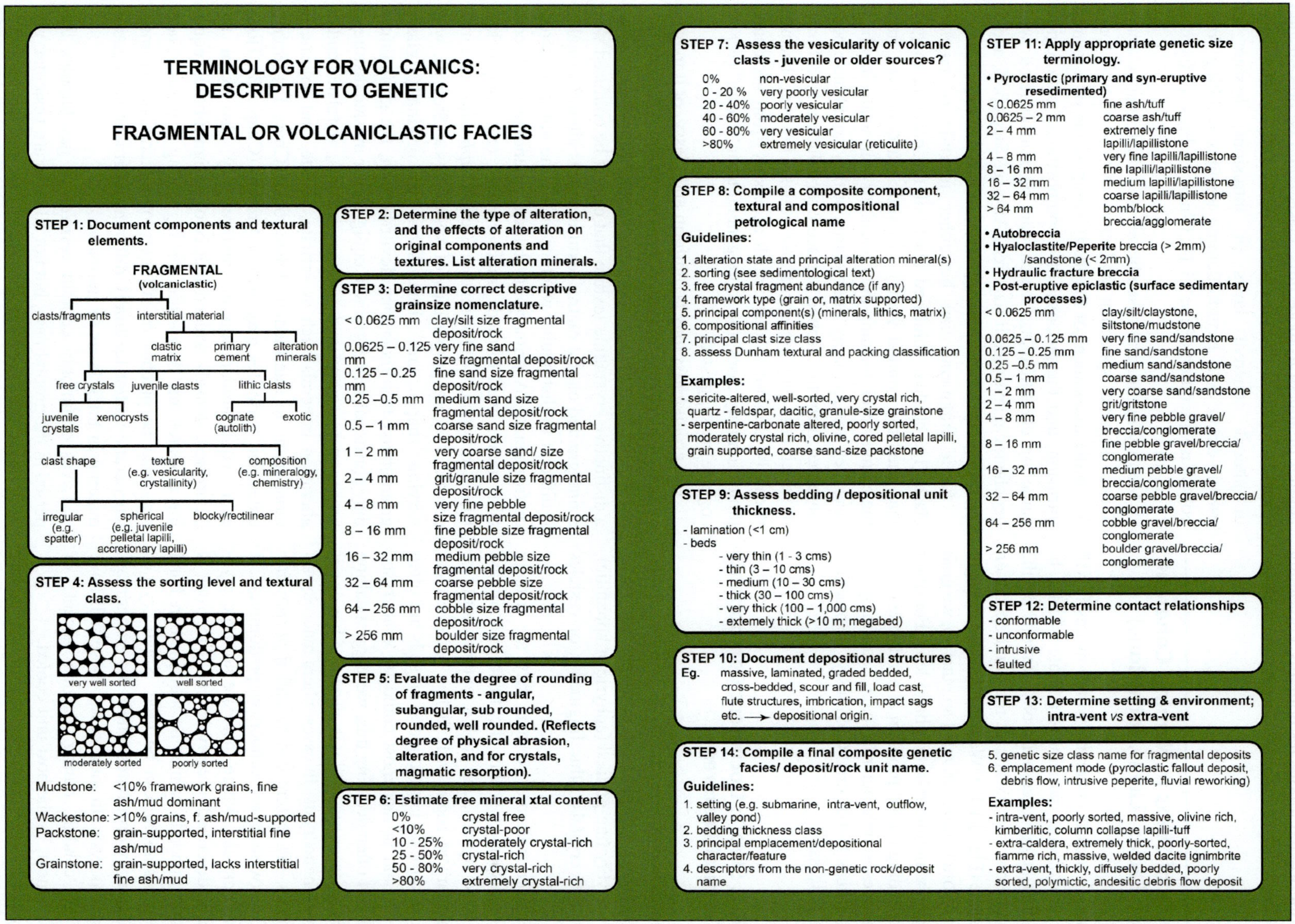

Fig. 14.1 (continued)

with discontinuous outcrop and lithology-modifying processes that overprint original lithofacies domains, combining all those facies into a single emplacement model may be challenging.

14.6 Sequence of Steps in Naming Coherent Rocks or Facies (Lavas, Intrusions)

Having decided that a rock is coherent, which means it will either be a lava or subvolcanic intrusion (e.g. Plate 14.1d–f, h; Chaps. 5–8), we first begin to compile observations on the petrological components and texture, at the hand specimen and microscopic scale (Fig. 14.1a). We have decided to begin at this scale, because often upon arrival at an outcrop, we grab a sample and begin examining it with a hand lens. For other geologists and depending on the problem at hand, one could equally start at the mesoscopic/outcrop scale, examining larger scale features (Steps 8–10) first. Obviously, in practice, this is an iterative process.

14.6.1 STAGE 1, Step 1: Document Components and Textural Features

The principal components of coherent rocks will be mineral crystals, lithic fragments, groundmass, and vesicles. However, some rocks are very fine-grained (aphyric or aphanitic), and contain no visible crystals, lithic fragments, or vesicles (Fig. 14.1a).

If large mineral *crystals* are present, they are usually reasonably evenly dispersed within the finer groundmass interstitial to the large crystals, and such a texture should be called "inequigranular" to begin with. It has become clear that in many volcanic rocks, large mineral crystals are often not all phenocrysts, and in fact most could be xenocrysts. Given the difficulty of often deciding this, we suggest calling such large crystals macrocrysts until it is clear whether or not the crystals are true phenocrysts. In cases where macrocrysts are truly phenocrysts, the texture can be called *porphyritic* (or *phaneritic*). Otherwise where xenocrysts predominate, it can be called *macrocrystic*. Where there are no large crystals, the rock can be described as *aphyric*, and if the rock is so fine-grained that minerals cannot be identified in hand specimen, the term *aphanitic* can be used. Small crystals may also occur uniformly distributed within the groundmass as microlites or microcrysts. In most cases, these will have crystallised in the conduit during magma ascent, or post-dating emplacement, as the host unit cooled.

The only fragmental rock type that could, in ancient successions, look like a porphyritic lava (massive, uniformly dispersed large crystals) is a thick massive fine-grained ignimbrite. In an ignimbrite, a significant proportion of large crystals are likely to be broken and angular or anhedral, more lithic clasts could be present, and if welded, fiamme textures should be looked for. However, in altered coherent lavas or intrusions, beware of pseudo-clastic and pseudo-fiamme textures resulting from polyphase alteration (Plate 14.1h; Chap. 13, Figs. 13.6a, b, f; 13.8a–c).

Identifying mineral assemblages and their relative abundance in hand specimen and thin section allows for a first-order assessment of the likely composition of a lava or intrusion, and the proposal of a compositional petrological name or classification, which can be corroborated with geochemical analyses if required. Also see Steps 3, 4, and 5, regarding crystal morphology, size, and abundance.

Lithic fragments can be either juvenile or *cognate* clasts (i.e. chilled, solidified clasts of the erupting magma that lined the conduit walls), or exotic clasts derived from the country rock forming the conduit or vent walls, in which case they are *accessory* lithic clasts, or they could have been entrained from the ground surface over which a lava flows, in which case they are *accidental* lithics. However, because it can be difficult to distinguish accessory from accidental lithics in many cases, the more general term "*exotic*" (*versus* cognate) is suggested.

The *groundmass* may be uniform and glassy, or it may be variably (micro)crystallised, consisting of small crystals that are smaller than the macrocrysts. Such a groundmass shows a distinctive crystalline texture that cannot be confused with a fine-grained clastic matrix texture, unless alteration makes it difficult to distinguish whether or not the unit is coherent or fragmental. A primary crystalline groundmass will contain many, well-formed small crystals (e.g. plagioclase laths), in which case the texture is called *aphanitic*, and may contain varying proportions of vesicles (see Step 6), with varying shapes and sizes. When an originally glassy groundmass has undergone devitrification, the groundmass may consist of a fine-grained, equigranular mosaic texture of tiny, equant quartz and feldspar crystals, if felsic, and/or alteration minerals such as sericite, zeolite, and chlorite in more mafic to intermediate felsic rocks (Chap. 13). Spherulites and variolites may also be present (Chap. 13). However, equigranular mosaic texture may also form the devitrified, densely welded matrix of an ignimbrite, requiring a search for diagnostic, distinguishing features. There may also be patches or stringers of mineralogically more leucocratic and/or texturally different magmatic segregation material in thick lavas or intrusions.

Lavas commonly contain *vesicles* or gas bubbles dispersed in the groundmass, but not all lavas are vesicular, and some lavas may have interior zones that are non-vesicular, but outer margins that are vesicular. Subsequent to solidifying, secondary minerals (e.g. zeolites, calcite) may precipitate in vesicles, which are then called *amygdales*. When amygdales are small, and the rock weathers, it can develop a texture with numerous dispersed spheres, ranging from a mm to several mm in diameter. Such features should not be confused with accretionary lapilli or spherulites. Sometimes in low-viscosity lavas, such as basalts, there will be late-stage vesicle segregation pipes cross-cutting otherwise non-vesicular to poorly vesicular groundmass (Chap. 5). In another variant, the groundmass may show evidence of an original clastogenic character as found in some lavas resulting from fire fountain eruptions and the fusing of spatter around the vent that then flows away as clastogenic lava that may preserve relict spatter outlines (Chap. 5).

It is always valuable to document the abundance of all the visible components of a rock, including crystals, lithic inclusions, vesicles, and groundmass. Determining primary magmatic mineral abundances can be used to assess the likely compositional character of a coherent rock. A visual comparator chart of areal/volumetric (modal) abundance of components, irrespective of what they are (e.g. Figure 14.2), can be used in the field and in the petrographic laboratory to provide a first-order estimate of mineral and other component abundances. Abundances can be quantified, if necessary, through petrographic point counting or image analysis, using ImageJ, or similar software, and rock composition can

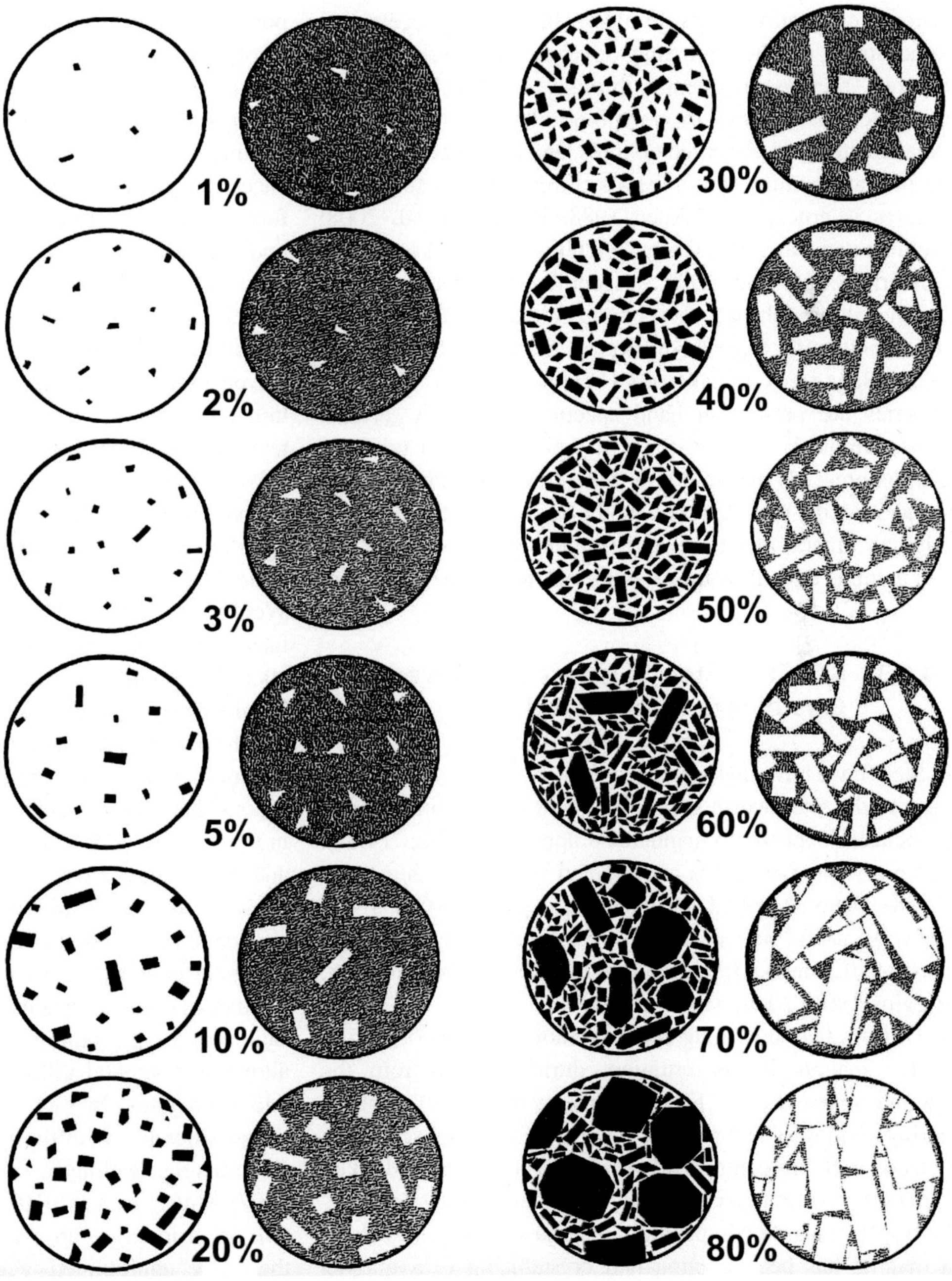

Fig. 14.2 Visual comparator chart for estimating areal/volumetric (modal) abundances of minerals in a rock, but can similarly be adapted for other components such as lithic fragments, vesicles, fossil fragments, groundmass, and matrix. Note, because dark crystals in a light background tend to be overestimated, and light crystals in a dark background underestimated, diagrams illustrating both scenarios are given for each percentage. Also, if a rock is fine-grained, dark minerals tend to be overestimated. (From Philpotts 2003)

be determined in detail using major, minor, and trace element geochemical analyses.

14.6.2 STAGE 1, Step 2: Determine the Types and Effects of Alteration (if Any) on Original Components and Textures

In prehistoric lavas and intrusions, it is essential to assess the degree to which these rock types have been altered by devitrification, hydrothermal alteration, deformation, and metamorphism (Fig. 14.1; Chap. 13). As described in Chap. 13, any of these processes can impose misleading textures onto rocks that may then become the source of major interpretation errors. In old volcanic rock successions, some styles of alteration may give the impression that a coherent rock is fragmental through the development of pseudo-clastic textures by polyphase alteration (Chap. 13). Detailed study of preserved textures, and assessment of what are primary textures and minerals compared with secondary features is crucial (e.g. Allen 1988; Hayman et al. 2008; Hayman and Cas 2011; Chap. 13). Ultramafic rocks, such as kimberlites, are especially susceptible to alteration, and distinguishing coherent facies from fragmental ones can be very challenging (e.g. Hayman et al. 2008, 2011).

The key steps to take are (1) assess if any alteration minerals are present in hand specimen and thin section; (2) determine if they overprint or replace original minerals and textures; (3) evaluate how severe this alteration is, and try to identify any domains in the rock that still preserve original mineral phases, even if pseudo-morphed, and original textures that allow the original rock type to be identified.

14.6.3 STAGE 1, Step 3: Assess the Correct Crystal Size Nomenclature

In Step 3, we have developed a crystal grain size scheme that mirrors the Wentworth grain size classes for sediments and pyroclastic deposits to standardise approaches (Fig. 14.1a). Some crystal size names are adopted from McArthur et al. (1998). Detailed crystal size distribution patterns (CSDs) can be very useful in assessing crystallisation dynamics in magmas (Cashman 1993, 2020; Cashman and Marsh 1988; Higgins 1996, 2000; Marsh 1998).

CSDs reflect how long crystallisation has been occurring in the magma before eruption, during ascent from the magma source to the Earth's surface, and after eruption, during cooling. The types and sizes of minerals present in an igneous rock are controlled by the diffusion rates of the key elements in a magma that build particular mineral frameworks, the magma composition, magma viscosity, how long a magma has been cooling and crystallising in a magma reservoir before magma ascent begins, magma ascent rate and decompression rate in the conduit, and cooling rate and cooling time before the magma cools to its solidification temperature (see Cashman 2020 for a review).

14.6.4 STAGE 1, Step 4: Describe Crystal Shapes

Describing the crystal shape may help not only to identify the mineral, but also to assist in distinguishing phenocrysts from xenocrysts (Fig. 14.1a). Crystal shapes in coherent rocks also reflect the crystallisation history, including rate (skeletal versus euhedral), which is affected by all of the factors listed with regards to CSDs above, and post-crystallisation history of crystals in magma (magmatic resorption). In Step 4, we list some common crystal shapes, but this is far from exhaustive.

Crystal morphology is influenced by many factors, but cooling rate and degree of under-cooling are major influences. In pioneering experimental studies, Lofgren (1974) found that fast rates of cooling and high degrees of under-cooling promoted the growth of skeletal plagioclase crystals under quench cooling conditions. Skeletal spinifex crystal forms are common in ultramafic Archean komatiite lavas (Plate 14.1f). Nearly all known komatiites are preserved in submarine successions, and with their high eruption temperatures, interfacing with the low ambient water temperature, their margins were subject to both high cooling rates and high degrees of under-cooling. This promoted the formation of skeletal, random, parallel plate, and dendritic spinifex textures, consisting mostly of olivine, and pyroxene (Donaldson 1976; Shore and Fowler 1999; Faure et al. 2006; Welsch et al 2012; Chap. 7). Thermal gradient is also a factor in determining whether random (high dT/dx) or parallel plate spinifex (lower dT/dx) forms (Faure et al. 2006). Spinifex texture is also known from relatively shallow and small-volume komatiite intrusions, especially into water-saturated sediments, indicating that the cooling conditions that promote crystallisation of spinifex can also exist in the subsurface (Chap. 7). Needle and microlitic pyroxene and plagioclase have also been found in the chilled glass rims of submarine pillow lavas, again representing rapid quench crystallisation (Bryan 1972).

Under conditions of slow cooling rates and low degrees of under-cooling, crystals experience equilibrium crystallisation rates, promoting growth of euhedral crystals, and allowing elemental diffusion rates to keep abreast of crystallisation rates. In such situations, crystals can equilibrate compositionally with the residual melt from which they are crystallising, leading to compositional zoning (Wiebe et al. 2007; Muller et al. 2010). Such equilibrium crystallisation will occur in large subsurface magma reservoirs, which have thermally equilibrated with the country rock, producing low-temperature gradients at the margins. Once magma begins to migrate towards the surface, cooling rates, degrees of under-cooling, and crystallisation rates will increase. Resorbed and embayed quartz and feldspar phenocrysts may also be common (e.g. Tepley et al. 1999). As revealed by CL (cathodoluminescence) studies, resorption surfaces are typically rounded dissolution patterns, which reflect reactions between crystals and the co-existing melt. The driving force behind quartz dissolution is consistent with decompression during magma ascent, resulting in a decrease in the stability field of quartz due to the shift in the position of the coetectic in the system Q–Ab–Or–H_2O. Juxtaposition of dissolving quartz phenocrysts during magma ascent may also lead to the formation of glomerocrysts thus providing a textural record of magma ascent (Hogan 1993; O'Donnell 2008).

A conspicuous form of phenocryst found in silicic volcanic rocks is equant, bipyramidal quartz. These are variously referred to as β-quartz, high quartz, or volcanic quartz, and represent the high temperature (> ~ 573 °C) polymorph of silica with a crystal structure very similar to that of quartz, but with higher symmetry (β-quartz: hexagonal; α-quartz: trigonal). Because β-quartz is only stable at temperatures that are well beyond that of hydrothermal solutions, it is usually only formed in cooling, silica-rich magmas. In most cases, quartz grains in intrusive rocks like granite are irregular, and only in rare cases do the crystals develop their ideal shape because in their final growth stages there is mutual interference with adjacent crystals. In extrusive or shallow subvolcanic rocks, well-formed β-quartz crystals (actually a pseudo-morph of α-quartz after β-quartz) are more common. Such crystals are usually grey or white, dull and translucent, and show a very simple, stubby habit. They consist of two mirrored hexagonal (bi)pyramids, so they look somewhat like ordinary quartz crystals without prism faces, and are double terminated because they grew in isolation suspended in a melt and were not attached to solid rock during growth. Only in lavas or sills are they, however, likely to remain largely intact, and not broken, whereas in pyroclastic deposits fragmented or anhedral crystals will be ubiquitous. β-quartz can also be found in well-formed crystals in gas cavities of silica-rich volcanic rocks.

Significant research has also been undertaken on the crystallisation behaviour of plagioclase under varying cooling and decompression conditions and rates, as well as magma rise history (continuous vs stepwise) (Cashman 2020). For groundmass crystallites, growth rates, crystal size, and crystal numbers are affected by cooling rate, decompression rate, and rise history, in varying degrees in different circumstances (Cashman 2020). The characteristics of phenocrysts (e.g. zonation patterns in rims, degrees of resorption) are affected by whether or not equilibrium crystallisation or magma recharge have occurred in the magma reservoir, and magma ascent and decompression rates. In lavas and subvolcanic intrusions plagioclase and potassium feldspar tend to be mostly euhedral subhedral or glomerocrystic, whereas free feldspar crystal fragments in pyroclastic deposits tend to be mostly angular to subhedral, while euhedral forms are less common and glomerocrysts are rare because they break during explosive eruptions.

Although most phenocrysts in coherent volcanic and subvolcanic rocks are euhedral to subhedral, several studies have recognised that in some lavas and shallow intrusions large magmatic crystals can be fractured and broken. This is attributed to the sudden decompression of rising magma which can cause volatiles in melt inclusions in crystals to suddenly expand causing crystals to fracture (Tait 1992; Best and Christiansen 1997; van Zalinge et al. 2018), and even due to impacts between crystals and clasts in a rising magma (Mueller et al. 2015; Jones et al. 2016).

14.6.5 STAGE 1, Step 5: Assess the Abundance of Minerals/Crystals

In Step 5, we have also provided a scheme to standardise abundance terms and the limiting percentages for each mineral/crystal abundance class and for the total crystal population (Fig. 14.1a). To facilitate estimating abundances of minerals or any other rock component, including groundmass, we have provided a visual comparator chart that can be used in the field or in the petrographic laboratory (Fig. 14.2). This will assist in comparing different units to assess if they could be related, or are different. Crystal abundance reflects the history of the magma. For example, low crystal content indicates eruption at near-liquidus temperatures, with little or no time for crystallisation, whereas high crystal content indicates prolonged residence and crystallisation before something finally triggered the eruption of magma in a highly crystallised state. Such variations lead to the question "*what determines whether or not magma erupts early or at a late stage?*". In some cases, comparing crystal abundances in coherent units, with units that are thought to be co-magmatic pyroclastic units, provides an opportunity to calculate crystal concentration processes during the formation of the pyroclastic facies. Very importantly, crystal content has a major effect on the rheology of magma (Chap. 2), affecting the bulk viscosity, and the ability of magma to migrate in the subsurface magma conduit, but also once erupted, as lava. Visible mineral crystal types and abundance also provide an initial assessment of magma composition.

14.6.6 STAGE 1, Step 6: Assess the Vesicularity of the Groundmass

Lavas and intrusions can vary from being non-vesicular to highly vesicular, even within the same unit. Lavas are often more vesicular near their tops and can even consist of coherent pumice (Chaps. 5, 6). Vesiculation occurs both in the conduit and post-eruptively, and reflects groundmass magma volatile content at the time of eruption, as well as magma rheology (Chap. 4). However, ambient pressure (e.g. atmospheric in subaerial settings, hydrostatic pressure in subaqueous environments, and lithostatic in subvolcanic settings) and decompression amount and rate, as well as the amount of magma outgassing allowed along fractures or domains of bubbles connected with the ambient environment, will also affect the level of vesiculation (Chaps 3, 4). Deep subaqueous lavas may have lower vesicularity levels than subaerial compositional counterparts, and vesiculated submarine lavas that erupt effusively may well have the capacity to erupt explosively in a subaerial setting. However, in Chaps 5 and 6, we point out that both subaerial and submarine lavas can vesiculate to very high levels (60–80% vesicularity) without undergoing explosive fragmentation. They can have scoriaceous or pumiceous carapaces. Understanding vesiculation processes, controlling factors, settings, and the effects of the environment, help to understand vesicle formation and the patterns preserved in lavas.

Detailed laboratory assessment of bubble size distribution (BSD) patterns and bubble number densities (BND; Shea et al. 2010; Toramaru 1990, 2006) can be just as useful in understanding decompression behaviour of lavas (subaerial versus subaqueous), as these have been in understanding the volatile exsolution and vesiculation history, and quantitative aspects such as decompression rates, in vesiculated pyroclasts, as discussed in Sect. 14.7.7.

The vesicle abundance classes and names we use (Fig. 14.1a) are from Houghton and Wilson (1989). Abundances can be determined in several ways. First, approximate estimates can be made using the visual comparator chart already referred to (Fig. 14.2). The second method involves determining the bulk density of a sample by immersion using the method of Houghton and Wilson (1989). Vesicularity can also be approximated petrographically using standard microscopic point counting methods. Detailed data on abundance, represented by vesicle number densities (Nv) and vesicle size distributions (VSD), can be acquired through X-ray micro-tomographic imaging, which provides 3D image reconstructions of vesicularity in small samples, or 2D methods and stereological conversion to 3D population properties, accompanied by computer processing using programs such as FOAMS. Readers are referred to the benchmark paper by Shea et al. (2010) for a summary of the methods, and application to determining and interpreting vesicularity in lavas, scoria, and pumice deposits.

14.6.7 STAGE 1, Step 7: Compose a Composite, Descriptive Component, Textural, and Compositional Petrological Name

In this step, all the observational hand specimen ± microscopic petrological characteristics can be combined to construct an informative multi-factor composite descriptive lithological name (Fig. 14.1a), for which suggested guidelines for the order of particular properties are provided in Step 7, as follows:

(1) Alteration state and principal alteration mineral(s),
(2) Compositional clan/suite name (where appropriate),
(3) Mineral crystal types and abundance class,
(4) Crystal size class,
(5) Vesicularity,
(6) Groundmass texture,
(7) Minor mineral phase,
(8) Major mineral phase, and
(9) Compositional petrological rock name.

Examples of such composite names as constructed using Step 7, include:

- serpentine-altered, moderately microcrystic, coarse, non-vesicular, monticellite kimberlite,
- sericite-altered, poorly porphyritic, fine, moderately vesicular, hornblende dacite, and
- clastogenic, moderately porphyritic, medium olivine–augite basalt.

The example names provided are more informative than simply calling the rocks kimberlite, dacite, and basalt, respectively.

14.6.8 STAGE 2, Step 8: Note the Unit Thickness

Here, we introduce a new thickness scale that reflects the range of thicknesses of lavas and intrusions and accounts for the significant viscosity variations of magma (Fig. 14.1a). For lavas, thickness is a reflection of the magma viscosity (Chaps. 2, and 5–7) and the gravitational potential and hydraulic head of the erupting magma at the vent as it feeds a lava flow. Low-viscosity magmas (basalts, komatiites; Chaps. 5 and 7) extrude and flow as thin, low aspect ratio lavas, whereas high-viscosity silicic lavas extrude and flow as thick, high aspect ratio lavas (Chap. 6). Intrusions, however, can be any thickness, irrespective of the magma viscosity because they are constrained by the strength and lithostatic pressure of the crust, allowing them to inflate to any thickness irrespective of viscosity, because they are not affected by a hydraulic head. Unit thickness allows aspect ratio to be calculated if areal extent can be determined, as well as volume calculations if sufficient thickness measurements and areal extent are available.

14.6.9 STAGE 2, Step 9: Record Emplacement Features

In this step, the important emplacement features of coherent units should be noted (Fig. 14.1a). This could include whether or not a unit is massive, flow-banded, whether there is some alignment fabric, presence of surface features such as ropy flow folds, (autoclastic) breccia, internal vesicle zoning pattern, presence or absence of apophyses, contact thermal effects, etc. See Chaps 5 to 7 to review the principal characteristics of different types of lavas, and Chap. 8 for intrusions. Such emplacement features can reflect the eruption style, fluid flow state, and emplacement processes.

14.6.9.1 STAGE 2, Step 10: Assess the Contact Relationships

This step is particularly important in stratified sequences, whether they are young or old, to distinguish whether or not coherent units are lavas or intrusions. This is not always as simple as it may seem, particularly in the older rock record. The first factor to assess is whether or not the geometry of the unit is concordant with the strata above and below, or whether it is discordant with them (Fig. 14.1a). A concordant relationship could indicate that a unit is a stratigraphically conformable lava, or a disconformable lava, or a sill, or a unit that has a bedding parallel faulted contact with the surrounding stratigraphy (Fig. 14.1a). From an event sequence point of view, it is important to distinguish these alternatives.

With a lava, we hope to find evidence of marginal autobreccia (Chaps. 4–6). With a sill, we would look for intrusive apophyses branching off the main sill into the overlying and underlying sedimentary or volcanic strata, or entrained xenoclasts or larger enclosed rafts of strata (Chap. 8). In addition, it is important to assess whether the enclosing rocks were hard or soft (and/or water-saturated) at the time of emplacement. Evidence of soft-sediment deformation and/or peperitic facies can be present around all margins of a syn-depositional intrusion (Chaps. 4, 8), but only at the base of a lava flowing over water-saturated unconsolidated sediments, and not at all if enclosing units were hard rock at the time of emplacement (Chaps 5, 6). Evidence of contact metamorphism may or may not be present, as discussed in Chap. 8.

Alternatively, if the contact relationship between a unit and its enclosing stratigraphy is discordant, then the relationship could represent an unconformable lava, a cross-cutting intrusive contact, or a faulted contact (Fig. 14.1a). Evidence for faulting includes visibly offset bedding, strong shear fabric and, within the shear fabric, slivers of the interfacing rock units.

14.6.9.2 STAGE 3, Step 11: Note the Original Setting

Understanding the origin of a unit is easier if the volcanological setting, and paleoenvironment in which it was emplaced can be established (Fig. 14.1a). This includes determining if the environment was subaerial or subaqueous if possible using both paleoenvironmentally significant associated sedimentary facies characteristics (Chap. 12) and primary volcanic features (e.g pillow lavas, abundant hyalocalstite, abundant pumiceous pyroclastic deposits), and if the setting was within a vent (intra-vent; intra-caldera), or outside of the vent (extra-vent; extra-caldera; Fig. 14.1a). In modern volcanic settings, these considerations are obvious, but in ancient successions they have to be determined by considering the overall geology of the host rock succession. This can take considerable time and effort.

14.6.9.3 STAGE 3, Step 12: Compile a Final Composite, Descriptive and Genetic Deposit or Rock Unit Name

In this final step, all of the essential characteristics of a deposit or rock unit are integrated into a composite name that reflects the essential descriptive characteristics, the mode of emplacement, and the setting. This may produce a relatively long, multi-factor name, but nonetheless this is important to reflect the essential features of the unit. Once enunciated, this name can be abbreviated for the purposes of written and oral communication. Guidelines for constructing such a name and the order of the descriptors (Fig. 14.1a) are:

(1) Setting (e.g. submarine, intra-vent, outflow, and valley-pond),
(2) Bedding thickness class,
(3) Principal emplacement/depositional characteristic/feature,
(4) Descriptors from the non-genetic rock/deposit name,
(5) Genetic size class name for fragmental deposits, and
(6) Emplacement mode (lava, sill, agglutinated clastogenic lava, etc.).

Examples of genetic deposit names include:

- extra-vent, thick, moderately crystal-rich, moderately vesicular, plagioclase–pyroxene, tholeiitic, pahoehoe basalt lava,
- intra-caldera, ultra-thick, fine, moderately porphyritic, weakly vesicular, quartz–albite–biotite, massive to flow banded, coherent (to autobrecciated), calc-alkaline dacite lava dome, and
- crater zone, thin, tabular, moderately crystal-rich, macrocrystic, non-vesicular, olivine–phlogopite, kimberlite dyke.

14.7 Sequence of Steps in Naming Fragmental or Volcaniclastic Deposits or Facies

Fragmental volcanic deposits and rocks include those fragmented by primary volcanic processes, such as pyroclastic, autoclastic and epiclastic deposits, as well as those fragmented after emplacement by secondary processes, such as hydraulic fracturing (breccias) and even tectonic processes (breccias) of volcanic source material (Chap. 4). The reason for including the latter two is that often in the rock record, it is not clear if a fragmental texture was primary or secondary in origin. In addition, we include here, alteration produced apparent or pseudo-clastic textures resulting from patchy polyphase alteration (Chap. 13), often in old coherent lavas and intrusions affected by hydrothermal alteration and/or deformation. Although they are not true primary, or even secondary fragmental rocks, we include them here to ensure that geologists are aware of their existence. Again, we propose a series of analytical stages and steps that progress from descriptive to ultimately genetic terminology, at different stages of investigation (Fig. 14.1b). These steps and stages should also sharpen our awareness of key properties that assist in working out the origin of the deposit.

14.7.1 STAGE 1, Step 1: Document Components and Textural Elements

In assessing fragmental rocks and deposits, the essential elements are the larger grains, which we call the *framework* grains or clasts, and the finer interstitial material (Fig. 14.1 b). The framework fragments could be *juvenile clasts*, *lithic clasts,* and free *crystal fragments* (Chap. 4). Interstitial material could be a fine-grained *clastic matrix*, *chemical cement* that is precipitated in pore spaces between framework grains, or post-depositional *alteration mineral assemblages*, which could variably overprint both framework and any matrix in clastic deposits, as well as creating alteration-induced apparent or pseudo-clastic textures in coherent lavas and intrusions.

Clastic matrix is recognisable through visible fine grains, which in a sand size aggregate could be silt and clay size, or in a coarser deposit, could consist of sand, silt, and clay size material. Chemically precipitated cement should be identifiable from a clear crystalline texture within pore spaces between framework grains, and may show a progressive increase in crystal size from the margins of the pore space into the centre, especially with crystalline cements such as quartz or carbonate, but not with more amorphous cements such as zeolites and iron oxides. A deposit that has silt to clay size matrix will usually not have chemical cement because there is little to no primary depositional pore space to fill. Secondary alteration materials may be patchy or pervasive in their distribution, they could selectively replace some components, but usually they overprint boundaries between framework grains and interstitial matrix and cement. Secondary alteration minerals are not chemical cement, which is only confined to pore spaces.

Free crystal fragments could be juvenile (i.e. magmatic) crystals that have been liberated from erupting magma, or they could be xenocrysts or epiclastic in origin, sourced from any crustal source rock. It is important to document exactly the origin of such crystals. If they are magmatic, the crystal assemblage is a guide to the composition of the source magma or rock, and they could be used for single crystal geochronology, or for geochemistry using microprobe analytical procedures. However, it is important to stress that crystals from the groundmass of juvenile clasts are much more reliable for both geochemical and geochronological analyses as their magmatic origin can be much better assessed. It is also important to assess the proportion of the crystals that represent juvenile magmatic crystals, to determine, in association with the juvenile clasts, if the deposit is monomictic or polymictic.

Juvenile or cognate lithic clasts should be recognisable because they represent a dominant clast population, they are (usually) texturally immature, they may be variably vesiculated, and they contain the same phenocryst population as the associated free crystal fragments. Their textures and mineral assemblages will indicate their compositional affinities. Exotic lithic clasts are derived from pre-existing sources or country rocks, and although these could be generally similar in composition to juvenile clasts, they will usually display some difference in mineral assemblage, texture, or alteration state. If the deposit is a true pyroclastic deposit, then these exotic lithic clasts will either be accessory or accidental in origin. Accessory clasts are derived from the country rock succession making up the walls of the vent and conduit, whereas accidental lithics are picked up from the ground surface by pyroclastic flows and surges. The nature of exotic lithic clasts can be determined from their mineral assemblages and textures (volcanic, plutonic, sedimentary, and/or metamorphic).

Clast shape may help to determine if clasts are juvenile or epiclastic, or if the deposit has been subject to significant surface sedimentary reworking or not. Juvenile magmatic clasts could be irregular and ragged, as for most pumice and scoria fragments, or fluidal and contorted as for spatter and spindle bombs, or angular and blocky as for lava dome autobreccia fragments, or quench fragmented hyaloclastite, or epiclasts that have not been tractionally reworked. Sedimentary mass-flow processes can, however, transport angular lithic clasts far from the source without significant rounding and abrasion (e.g. debris avalanches). In Chap. 12, we discuss the distinction between tractional reworking processes and mass-flow processes. Reworking has the effect of rounding and abrading sharp corners and edges of clasts, whereas short-lived mass flow may cause almost no modification to clast morphology.

In Chap. 13, we also alert readers to potentially misleading effects of polyphase alteration in producing misleading apparent or pseudo-clastic textures, commonly in old coherent lavas and intrusions that have been subject to hydrothermal alteration. It is important to ensure the deposits you are dealing with are true fragmental or clastic rocks and not pseudo-clastic alteration textures (Step 2).

As for coherent rocks, determining the abundance of different components of a fragmental rock, including crystal fragments, major lithic clast types, including juvenile magmatic, accessory, accidental lithics and their compositions, fossil fragments, matrix, and chemical cement, sheds light on the compositional affinities of the deposit and its fragmentation and transportation history. Again, using the visual comparator chart for estimating abundances (Fig. 14.2) provides a reasonable first-order estimate of abundances if done with care.

14.7.2 STAGE 1, Step 2: Determine the Types and Effects of Alteration (if Any) on Original Components and Textures

As for coherent deposits, assessing the level of, and impact of, alteration on the original components and textures of volcaniclastic deposits is important (Fig. 14.1b). For very recent deposits, this will be insignificant, but for any slightly older deposits this should be considered, especially in areas of hydrothermal alteration. In Step 2, it is important to document the types of alteration minerals, and assess the degree to which they have modified original components and textures. Having evaluated this aspect, it is important to reassess whether or not the components and textural elements that were identified in step one are in fact correct or if they should be changed in the light of assessing the effects of alteration. Kimberlites and other hydrothermally altered and metamorphosed rocks are susceptible to significant changes in original textures as a result of alteration. In some cases, it can be difficult distinguishing that a volcaniclastic kimberlite was fragmental (e.g. Hayman et al. 2008, 2009; Hayman and Cas 2011; Porritt et al. 2012; Pittari et al. 2015; Chap. 18).

In Chap. 13, the potential effects of alteration are reviewed, which readers are referred to for further description and discussion.

14.7.3 STAGE 1, Step 3: Determine the Correct Descriptive Grain Size Nomenclature

Grain size is one of the obvious characteristics we notice when we first look at an outcrop or a deposit. Together with sorting (next section), it provides insight into the competency and dynamics of the transport and depositional agent. However, grain size by itself does not uniquely identify the likely eruption and transportation processes.

As discussed in the introductory sections of this chapter, in assessing the grain size characteristics of a deposit, it is important to initially at least use a descriptive grain size scheme and terminology, without considering the possible origins of the deposit. At this early stage in the analysis of a volcaniclastic deposit, we recommend the adoption of the grain size scheme and terminology presented in Step 3 (Fig. 14.1b). Only after most of the other characteristics of a deposit have been documented and considered is it time to apply a genetic grain size terminology, as considered much later in Step 11. We have identified at least 70 different possible origins for deposits based on their general grain size characteristics, texture, and clast morphology (Table 14.1A–E).

The grain size of a volcaniclastic deposit is rarely completely uniform. Only extremely equigranular or well-sorted deposits will consist uniquely of just one of the size classes that are listed in the grain size scheme in Step 3. More commonly volcaniclastic deposits consist of mixtures of grain sizes. In such cases, if there is a dominant grain size (>2/3 of the clast population), then adopting the approach of Fisher (1966), that grain size class is used to name the grain size of the deposit. Where there are sub-equal proportions of two grain sizes (1/3 to 2/3), then a dual grain size terminology can be adopted.

The grain size of a deposit can be qualitatively estimated in the field, or quantified by undertaking field sieving of the coarse fractions, and/or photographing outcrops and using image processing (e.g. ImageJ) and stereological methods (e.g. Jutzeler et al. 2012, 2015; Bernard et al. 2017). For the intermediate grain sizes, laboratory sieving of samples collected in the field is used, and for the finest grain sizes

Table 14.1 Grain size and textural classes of volcaniclastic rocks, some possible origins, some diagnostic characteristics, examples, and preservation potential in modern and ancient successions (Updated from Cas and Wright 1987). (A) Deposit types that are texturally clast-supported ortho-breccias. (B) Deposit types that are texturally matrix-supported para-breccias. (C) Deposit types that are texturally clast-supported ortho-conglomerates. (D) Deposit types that are texturally clast-supported para-conglomerates. (E) Deposit types that are texturally sand size. (F) Deposit types that are texturally mud (silt and clay)-size

Grainsize-textural class	Origin	Essential characteristics	Preservation potential	Recognition potential
A. BRECCIAS -Clast-Supported (Ortho-breccia) (*Angular boulder, cobble, pebble or block, bomb, lapilli size clasts and pseudo-clasts dominant; minor or no matrix*)	1. Epiclastic mass wastage breccias, includes crater wall rock falls, talus slopes (Chap. 12)	Monomictic to polymictic; moderately sorted; massive to diffusely layered; < 1 to tens m thick (Fig. 12.7)	Poor	Moderate
	2. Epiclastic mass flow breccias, such as debris avalanche deposits (Chap. 12)	Monomictic if small to polymictic for large deposits; poorly sorted; disorganised to graded stratified; megaclasts internally fractured; local lobate geometry to more extensive sheet-like for large-scale sector collapse events; thickness up to hundreds m; associated epiclastic facies may contain tractional structures (Figs. 12.13; 12.15)	Moderate	Moderate
	3. A'a lava breccia (Chap. 5)	Monomictic (basaltic, phonolitic); very irregular spinose clast morphology; variation in the vesicularity; accidental clasts incorporated from substrate; margins brecciated and gradational into massive coherent interior; usually < 10 m thick (Plate 5.1b; Figs. 4.2a, b; 5.35)	Good for basal breccia; poor for the spinose top	Moderate
	4. Autobrecciated felsic block lavas (Chap.6)	Monomictic with angular blocks; intermediate or silicic composition; gradational into massive coherent interior; up to tens m or more thick (Figs. 4.2d, f; 6.8a)	Good	Good
	5. Autobrecciated felsic lava dome and dome margin talus deposits (Chaps 6, 12)	As for 4; intermediate or silicic composition; diffuse layering in dipping talus slope deposits marginal to core of lava dome; < 1 to tens m thick (Figs. 4.2d—f, i; 6.6b; 6.8 h)	Good	Good
	6. Hyaloclastite breccia in subaqueous lavas, domes, syndeposition intrusions (Chaps 4-6, 8)	Monomictic; any composition; coarse to fine breccia; jigsaw-fit to clast rotated textures; minor coarse matrix; blocky to splintery clasts resulting from quench fragmentation; < 1 to tens m thick; massive to diffusely stratified if resedimented at lava margins; gradational into coherent lava or intrusion; hosted by subaqueous sediments (Figs. 4.5a-g; 6.29–31)	Very good	Very good

(continued)

Table 14.1 (continued)

Grainsize-textural class	Origin	Essential characteristics	Preservation potential	Recognition potential
	7. Resedimented hyaloclastite breccia (Chaps 6, 12)	Clasts as in 6, but clast-rotated texture; diffusely stratified mass flow apron around margin of subaqueous lava or dome; lobate to sheet deposits away; < 1 to tens m thick (Fig. 6.31f, g)	Good	Good if context preserved
	8. Peperite breccia (coarse) (Chaps 4–6, 8)	As for 6, but matrix of ambient sediments injected into the fractures and spaces between blocky hyaloclasts syndepositionally; jigsaw fit to clast rotated textures; matrix may be terrigenous, volcaniclastic or carbonate sediments; < 5 m thick (Figs. 4.6;8.18d)	Very good	Very good
	9. Agglutinate spatter breccia (hawai'ian; Chaps 4, 9) (type of agglomerate)	Monomictic composition (basaltic, rarely peralkaline); moulded fluidal clast shapes and accommodation; annular geometry around vent; variable thickness up to tens of m; interbedded massive lavas (clastogenic lavas) (Fig. 9.48b, c)	Moderate	Very good
	10. Agglomerate breccias (Chaps 4, 9)	Diagnostic criterion is shaped bombs or 'hot' breadcrusted or jointed blocks that have fused together; minor coarse matrix if present; MUST have recognizable vent within tens to hundreds of m; m to tens of m thick (Figs. 4.31 g; 9.50c)	Poor	Very difficult
	11. Pumice-fallout deposits (subplinian, plinian, ultraplinian) (Chap. 9)	Monomictic; angular vesiculated felsic pumice clasts, but variable accessory lithics; no matrix; massive to diffusely layered; no cross-stratification; thickness up to 25 m, but usually < 10 m, to < 1 m distally; susceptible to weathering and alteration with breakdown of glass to clays, etc.; susceptible to tectonic deformation and layer shortening in ancient deposits; where welded, eutaxitic texture also present, and local distribution around the vent (Fig. 9.56)	Good if young covered by ignimbrites, lavas; good if old if covered by lavas, welded ignimbrites	Good if young, recognized by context; moderate for welded and old deposits
	12. Scoria fallout deposits (Chap. 9)	Mafic to intermediate angular scoria and spatter clasts; textures as for 12, but even more susceptible to weathering and alteration; tens m thick near vent in cones, < 1 m thick distally (Fig. 9.52b-e)	Moderate if young; poor if old unless buried by lavas	Very poor if old
	13. Co-ignimbrite breccias (lag, ground; Chap. 10)	Mostly polymictic lithic clast population; poorly sorted; ignimbrite matrix; 1—5 m or more thick; pass into or interlayered with ash matrix rich other ignimbrite facies (Figs. 10.50; 10.52, 10.53)	Good if capped by (welded) ignimbrite	Good
	14. Lithic rich gas segregation pipe breccias (Chap. 10)	Angular to subangular lithic fragments in pipe-like structure hosted by massive (Layer 2) ignimbrite; up to 1 m wide (Fig. 10.77c, d)	Good	Very good

(continued)

Table 14.1 (continued)

Grainsize-textural class	Origin	Essential characteristics	Preservation potential	Recognition potential
	15. Fines depleted ignimbrite (Chap. 10)	Monomictic; pumice clast rich; crystal-rich matrix if magma porphyritic; massive; thickness—several to > 10 m; succeeded by volcanic breccia (lithic-rich ground layer of ignimbrite) (Figs. 1.13; 10.57e; 10.81a, d)	Poor	Poor
	16. Pumice flow frontal lobe and marginal breccia (Chap. 10)	Monomictic; pumice clasts angular to sub-rounded; minor matrix; up to several m thick (Fig. 2.16c)	Poor	Moderate
	17. Hydraulic fracture breccias (Chaps 4, 18)	Monomictic; angular variably altered, splintery to blocky clasts; in situ 'jigsaw-fit' of clasts common; confined to crosscutting zones centimetres to metres wide; alteration minerals fill vein-like fractures between clasts (Fig. 4.33)	Very good	Very good
	18. (Proto/Ortho) Cataclastite or fault breccia in volcanic rocks (Chap. 4)	Generally monomictic; variably fractured, blocky and strained-stretched (volcanic) country rock fragments; jigsaw-fit large clasts; some (< 50%) finely fragmented matrix of same material; secondary veins; confined to fault zone; width of cataclastite zone variable; adjacent country rock fractured (Fig. 4.34b, c)	Good	Good if outcrop is good
	19. Alteration pseudo-breccia (Chap. 13)	Usually hydrothermally polyphase altered coherent lava or intrusion; patchy colour alteration (e.g. chlorite, sericite, silicia) defining pseudo- or apparent clasts in a different coloured (e.g. sericite, silica, chlorite) pseudo- or apparent matrix; pseudo-clast shapes can be very irregular; pseudo-clast boundaries diffuse, gradational into matrix; pseudo-clasts and pseudo-matrix have the same phenocrysts, with the same size, morphology, abundance, but perhaps differently altered (Plate 13.1a; Figs. 13.6a, f)	Good	Difficult

(continued)

Table 14.1 (continued)

Grainsize-textural class	Origin	Essential characteristics	Preservation potential	Recognition potential
B. BRECCIAS -Matrix-Supported) (Para-breccia) *(Angular boulder, cobble, pebble or block, bomb, lapilli size clasts and pseudo-clasts dispersed in an abundant matrix)*	20. Glacial till and moraines (diamictites) (Chap. 12)	Polymictic: poorly sorted; clast shape variable from angular to rounded; matrix includes large proportion of fine rock powder; unlikely to contain significant pumice or shards; massive to crudely bedded; up to tens m thick; associated striated pavements, clasts and fluvio-glacial facies; variable thickness (Figs. 12.19; 12.20)	Moderate	Moderate to good
	21. Glacial dropstones (Chap. 12)	Diverse lithologies; as for 19, but isolated or trails of outsize boulders to pebbles; some may be rounded; contained within bedded sub-wave base lacustrine and marine sedimentary facies; drop-stones may show impact sags; distinguished from pyroclastic bomb sags by context (Fig. 12.6e)	Good	Very good
	22. Granular (debris) mass-flow with granular matrix (Chap. 12)	Polymictc; poorly sorted; clasts angular to rounded; massive, disorganised to graded stratified facies; sand size matrix, little mud; < 1 m, to > 100 m thick where derived from entry of pyroclastic flows into water; association with other mass-flow facies; can be mistaken for block and ash flow deposit (Figs. 12.35; 12.37)	Very good	Moderate
	23. Cohesive debris flow (Chap. 12)	Mostly polymictic; very poorly sorted; clasts angular to rounded massive, disorganised; pebbly mudstone matrix; < 1 to tens m thick; association with other mass-flow facies (Fig. 12.37e)	Very good	Moderate
	24. Hyaloclastite breccia in subaqueous lavas, syn-deposition intrusions (Chaps 4, 5–8)	As for 6 and 7, but more intense quench fragmentation produces finer granular matrix to $\sim$ 1 mm size; large blocks still present; jigsaw-fit to clast rotated textures; any composition; < 1 to tens m thick (Figs. 4.5 h; 8.18 g)	Very good	Very good
	25. Clast-rotated, matrix supported peperite (Chaps. 4–6, 8)	As for 8, but less jigsaw-fit, more clast rotated textures; hyaloclasts well mixed with sedimentary matrix beneath base of lava or around margins of syn-depositional intrusion, with matrix supported texture; gradational into jigsaw-fit peperite then hyaloclastite; up to several m thick; confused with debris flow deposits? (Fig. 4.6a, f)	Very good	Moderate

(continued)

Table 14.1 (continued)

Grainsize-textural class	Origin	Essential characteristics	Preservation potential	Recognition potential
	26. Ballistic impact structure (Chaps 4, 9)	Juvenile bomb(s) or accessory block(s) impacted in a finer grained pyroclastic succession; down-sag structure in fine grained deposits; clast displacement in coarse deposits (Figs. 1.15b; 4.19h; 9.55g; 10.50e)	Very good	Very good
	27. Co-ignimbrite breccias (lag, ground; Chap. 10)	As for 13, but moderate lapilli-ash matrix (Figs. 10.50c, f; 10.52c, d; 10.53)	Very good	Good
	28. Lithic concentration zone in ignimbrite (Layer 2bL) (Chap. 10)	Polymictic; variable accessory and accidental lithics; crystal types same in pumice/scoria clasts and matrix; massive depositional units; thickness < 5 m; evidence of hot state emplacement—in welded ignimbrites, development of eutaxitic texture and columnar jointing; gas segregation pipes and pods; association with other Layer 2 ignimbrite facies and co-eruptive fallout and surge deposits (Figs. 10.40b; 10.41c; 10.47)	Good	Very good
	29. Layer 2 felsic/mafic ignimbrite (Chap. 10	Polymictic; angular to subrounded lapilli pumice/scoria clasts dispersed in dominant ash matrix; fiamme if welded; spatter if mafic; variable accessory and accidental lithic clasts; massive to diffusely stratified; extra-vent < 150 m thick; intra-caldera at least hundreds m thick; welded to non-welded; some gas segregation pipes; columnar jointing if thick (Figs. 10.35e; 10.41; 10.57; 10.58; 10.63a, b, f)	Very good	Good if welded
	30. Pumice or scoria concentration zone in ignimbrite (Chap. 10)	Angular to subrounded large pumice or scoria clasts in (Layer 2) ignimbrite; usually near top of deposit (Fig. 10.57)	Good	Good
	31. Block and ash-flow deposits (Chap. 10)	Monomictic felsic; angular, poorly to vesiculated blocks in matrix of same debris; No accessory, few accidental clasts; small volume; < 20 m thick; associated with ash cloud surge deposits above and below; some gas segregation pipes (Fig. 10.99)	Very good if young, thick or welded; poor if thin	Good if preserved
	32. Near vent base surge deposits (Chap. 10)	Monomictic to polymictic; variable vesicularity of juvenile Clasts (Chap. 4); presence of ballistics and impact sags; cored lapilli; massive, bedded and cross-bedded internal structures; single beds < 2 m; context: part of thick bedded phreatomagmatic tuff ring or cone (Figs. 1.15 g; 10.104a, b, e, h)	Poor	Good if context preserved

(continued)

Table 14.1 (continued)

Grainsize-textural class	Origin	Essential characteristics	Preservation potential	Recognition potential
	33. Ground surge deposits of ignimbrites (Chap. 10)	Monomictic to polymictic; microbreccias; stratified and low angle cross-stratified; position below and above, respectively, pyroclastic Layer 2 flow facies; thickness generally < 2 m (Fig. 10.40a)	Good if buried under thick body deposits	Moderate
	34. Hydrothermal explosion breccias (agglomerate; Chaps 4, 18)	Very poor sorting; diverse clast types and morphology; clasts variably altered; matrix of hydrothermally altered clays; may be associated with surge deposits; accretionary lapilli may occur; close to hydrothermal crater. (Fig. 4.31 g)	Poor	Good
	35. Giant pumice beds (Chap. 16)	Uniform composition of pumice blocks; enclosing matrix sediments are stratified; lacustrine (or marine) setting; radial jointing in some individual pumice blocks; chilled glassy sheath on margins of some blocks; blocks up to several m (Fig. 16.69c, d)	Rare	Very good
	36. (Meso- to Ultra-) Cataclastite or fault zone breccia (Chap. 4)	As for 18, but with > 50% matrix material the same as large clasts; greater development of foliation and strain (Figs. 4.34c; 6.27c)	Good	Good if outcrop is good
	37. Alteration pseudobreccia (Chap. 13)	As for 19, but a higher proportion of pseudo-matrix, giving an apparent matrix-supported texture; some pseudo-clasts may seem to be irregular pumice or fiamme and the rock ignimbrite (Figs. 13.6b; 13.8a-c)	Good	Very difficult
	38. Pseudo-eutaxitic fiamme texture	Patchy alteration along a pre-existing fabric, such as flow banding or more commonly a tectonic foliation in high strain deformational-metamorphic zones, giving a pseudo-welded fiamme ignimbrite texture (Figs. 13.17a, b; 13.19a; 13.20b-d); in high strain zones, gradational into low strain equivalent protolith, allowing recognition of original rock type (Fig. 13.17b)	Good in old rocks	Very difficult; trace gradation into low strain

(continued)

Table 14.1 (continued)

Grainsize-textural class	Origin	Essential characteristics	Preservation potential	Recognition potential
C. GRAVEL-SIZE or CONGLOMERATE (Clast-Supported) (Orthoconglomerate) (*Rounded boulder, cobble, pebble or block, bomb, lapilli size clasts dominant*)	39. High flow regime fluvial hyperconcentrated flow, or subaqueous mass flow (Chap. 12)	Polymictic, sometimes monomictic; massive to diffusely stratified; well-rounded pebble to boulder size clasts; coarse granular matrix context with and within volcanic(lastic) succession (Figs. 12.28; 12.29a-c; 12.32b)	Very good	Very good
	40. Tractional reworked conglomerates (Chap. 12)	Polymictic; fluvial-alluvial, shoreline to shallow marine; crude tractional sedimentary structures; pebble to boulder size clasts; coarse granular matrix, if any; context with and within (volcani)clastic succession (Figs. 12.24a; 12.26a)	Very good	Good
	41. Beached pumice raft deposits (Chap. 12)	Rounded pumice clast aggregates on beaches; distant, sometimes unknown source; encrusted marine organisms on clast surfaces; pebble to boulder size clasts (Fig. 12.26a)	Very poor	Difficult
	42. Pumice/scoria flow frontal lobe and marginal deposits (Chap. 10)	Monomictic; clast support of pumice or scoria; rough rounded pumice and scoria; moderate to low ash matrix content; crystal-enriched matrix if magma porphyritic; sheet to lensoidal geometry; intercalated with other recognisable ignimbrite facies; usually at tops of flow units (upper part of *layer 2b*); thickness < 2 m (Fig. 10.57e)	Good if ignimbrite is young	Moderate
	43. Fines-depleted ignimbrite (Chap. 10)	As for 15, but pumice clasts are rough rounded; some matrix (Figs. 1.13b; 10.57e; 10.81a, d)	Moderate if young; poor if old	Poor
	44. Accretionary lapilli filled gas segregation pipes in phreatomagmatic ignimbrite (Chap. 10)	Concentrations of accretionary lapilli in sub-vertical pipe-like structures in (Layer 2) ignimbrite; no matrix (Fig. 10.77a)	Good	Good
	45. Perlite (Chaps 4, 6, 13)	Clusters of spheroidal orbs of (mostly) felsic glass (or altered) in lavas, intrusions, welded ignimbrites; orbs bounded by subconcentric perlite cracks (Fig. 13.5a-c)	Good if young	Good
	46. Spherulites/varioles/lithophysae and variolite (Chap. 13)	Devitrification spherical features in glassy lavas, shallow intrusions, welded ignimbrites; diagnostic when spheroids are anastomosed and overlapping; distinctive uniform minerals; radial fabrics (Figs. 13.1b-f);	Good	Good

(continued)

Table 14.1 (continued)

Grainsize-textural class	Origin	Essential characteristics	Preservation potential	Recognition potential
	47. Alteration pseudo-conglomerate (Chap 13)	Hydrothermal alteration superimposed on hydraulic fracture breccia network; alteration replacement corrosion of corners of fracture blocks causing apparent rounding of blocks, invasion of block margins by alteration minerals that then form a pseudo-matrix; gradation into rectilinear hydraulic fracture breccia (Plate 13.1b; Figs. 13.6c-e; 13.10c, d)	Good in altered rocks	Very difficult
D. GRAVEL-SIZE or CONGLOMERATE (Matrix-Supported) (Paraconglomerate) *(Rounded boulder, cobble, pebble or block, bomb, lapilli size clasts and pseudo-clasts in an abundant matrix)*	48. Mass-flow redeposition (deposits with granular matrix) (Chap. 12)	Polymictic; massive to diffusely stratified; possible grading; well-rounded pebble to boulder size clasts; coarse granular matrix; context with and within volcanic(lastic) succession; < 15 m thick (Fig. 12.32b)	Moderate	Moderate
	49. Tractional reworked fluvial, alluvial, shallow marine (Chap. 12)	Polymictic; fluvial-alluvial, shallow marine; crude tractional sedimentary structures; pebble to boulder size clasts; coarse granular matrix context with and within volcanic(lastic) succession (Fig. 12.24a)	Moderate	Moderate
	50. Trails of raft transported pumice in deep water (Chap. 12)	Isolated or trails of rounded pumice clasts in deep water sedimentary successions; encrusted with marine organisms	Good	Difficult
	51. Layer 2bP pumice concentration zones in ignimbrites (Chap. 10)	As for 39, but many pumices sub-rounded to rounded due to abrasion in pyroclastic flow; context within and near top of ignimbrite is critical (Fig. 10.57a)	Moderate	Moderate
	52. Accretionary lapilli dispersed in ignimbrite (Chap. 10)	Spherical accretionary lapilli dispersed in massive ignimbrite ash matrix (Figs. 10.77a, b; 10.84f)	Good	Good
	53. Cohesive pebbly mud flows and lahars (Chap. 12)	Pebbly mudstones texturally; generally polymictic; internally massive with dispersed rounded clasts in a largely mud size matrix, some of all of which could be resedimented ash; up to a few tens of metres thick; lack evidence of hot state of emplacement (i.e. *NO* hot blocks, thermal colour alteration, thermal remanent magnetisation, gas segregation pipes) (Fig. 12.37e)	Poor	Moderate
	54. Alteration pseudo-conglomerates (Chap. 13)	As for 48, but with more alteration pseudo-matrix, giving an apparent matrix-supported texture	Good	Very difficult

(continued)

Table 14.1 (continued)

Grainsize-textural class	Origin	Essential characteristics	Preservation potential	Recognition potential
E. SAND-SIZE (*Sand or coarse ash size grain framework dominant*)	55. Epiclastic reworking in a tractional (alluvial, fluvial, shallow marine environments (Chap. 12)	Generally polymictic; abundant tractional structures; crossstratification is either high angle of repose or hummocky cross-stratification; body and trace fossils possible in marine settings depending on age (Figs. 12.23; 12.2412.24)	Very good	Very good
	56.Hyperconcentrated fluvial flood flow (Chap. 12)	Monomictic to polymictic; remobilised ash-pumice and lithic debris; massive to diffusely stratified beds; moderate to good sorting. Figures 12.23 h; 12.29b)	Moderate	Moderate
	57. Epiclastic turbidity current mass-flow redeposition; storm shelf; deep water (Chap. 12)	Turbidite facies characteristics; Bouma sequence; variably graded (Figs. 12.30; 12.32)	Very good	Very good
	58. Weathered and/or devitrified lava/dykes sills (Chap. 8)	Generally granular texture; even distribution of phenocrysts if crystallised; thick massive character; (?)relic flow banding; lithophysae, spherulites (Chap. 8); radiate fibrous to microgranular groundmass of quartz and feldspar in felsic units; absence of regular sedimentary bedding; thin to hundreds of m thick; columnar jointing if thick	Moderate	Good
	59. Resedimented distal subaqueous hyaloclastite (Chap. 12)	Monomictic felsic; bedded; angular blocky sand to pebble size hyaloclast fragments; clast-rotated; turbidite facies characteristics; small volume; within a few km of source dome; interbedded with basinal sediments (cf. Figure 12.32)	Moderate	Moderate
	60. Layer 2a of ignimbrites (Chap. 10)	Ash dominant, with increasing small pumice and scoria upwards; poorly sorted; < 1 m thick; may or may not be underlain by Layer 1 deposits; gradational into Layer 2bL lithic concentration zone upwards	Good	Good if in context
	61. Fine grained ignimbrite (Chap. 10)	Granular texture; thick massive character; (rare) shard textures in thin section; accretionary lapilli if phreatomagmatic; perhaps eutaxitic shard texture if magmatic and originally welded; some mesoscopic fiamme; gradational into other ignimbrite facies incl. lithic concentration zones; gas segregation structures; columnar jointing? (Figs. 10.77b; 10.84d)	Good if welded, poor if non-welded	Moderate

(continued)

Table 14.1 (continued)

Grainsize-textural class	Origin	Essential characteristics	Preservation potential	Recognition potential
	62. Air fall ashes and tuffs, including co-ignimbrite (Chap. 9)	Homogeneous composition; mafic or felsic; (rare) shards preserved; possible intemal diffuse lamination; thickness generally < 1 m; possible accretionary lapilli if phreatomagmatic (Fig. 9.60e)	Poor	Moderate
	63. Base surge deposits (Chap. 10)	Mafic or felsic ash or tuff; fine grained, low angle dune bedded, cross-bedded to plane laminated; < 1 m thick; accretionary lapilli; context: part of thick bedded phreatomagmatic tuff ring or cone (Fig. 10.104 g, i)	Poor	Moderate
	64. Mylonite (tectonic)	Very strong tectonic foliation, making rock look laminated; totally dynamically recrystallised rock; grain size from silt to pebble; only likely in old volcanic rocks (Fig. 13.17b, c)	Good	Difficult
F. MUD-SIZE (*Silt and clay or fine ash size dominant*)	65. Distal ground and ash cloud surge deposits (Chap. 10)	Fine grained mafic and felsic; stratified and low angle crossstratified; position below and above, respectively, pyroclastic flow Layer 2 facies; thickness generally < 0.5 m (Fig. 10.40c)	Moderate	Good
	66. Hemi-pelagic suspension settling of terrigenous fines in lake or ocean (Chap. 12)	Terrigenous clays and silts; fine grained laminated; may be (micro-) fossiliferous; black, carbonaceous if deposited in anoxic deep water (Fig. 12.25)	Very good	Very good
	67. Fine grained distal ignimbrite (Chap. 10)	Mafic or felsic; massive to diffusely stratified; occasional outsize pumice or lithic fragments; accretionary lapilli if phreatomagmatic; variable thickness (Figs. 10.77b; 10.84d)	Poor	Moderate
	68. Phreatomagmatic fallout ash/tuff (Chap. 9)	Fine grained massive to laminated; may be thick near vent (>2 m); cm to mm thick distally; accretionary lapilli (Fig. 9.60a-g)	Poor	Moderate
	69. Distal subaerial fallout ash or tuff, or subaqueous watersettled ash/tuff (Chap. 9)	Fine grained; massive to laminated; perhaps shard forms visible; older deposits may be altered, preserved as clay-like tonsteins, or subaqueously as bentonite or silicified chert-like beds; plinian or co-ignimbrite (Fig. 9.60d, e)	Poor in Subaerial sequences, good subaqueously	Good to poor, modern to ancient sequences
	70. Phreatomagmatic surge or distal co-ignimbrite surge deposits (Chap. 10)	As for 54, 55, but finer and thinner (Figs. 10.42a, b; 10.82c; 10.99 l, m; 10.104 g, i)	Poor	Moderate

laser-based granulometry techniques (if the deposit is unconsolidated), or petrographic point counting, imaging and image analysis, and further stereological analysis. Once grain size data for the coarsest size fractions have been combined with that for the finer fractions, statistical grain size parameters can be calculated, as outlined in Cas and Wright (1987, app. I) and Jutzeler et al. (2015).

Well-sorted deposits of relatively uniform grain size indicate transport processes that are efficient at separating different grain size components in transit (e.g. pyroclastic fallout deposits, tractional sedimentary transport deposits; Table 14.1A–E). In some of those deposits, there may be detectable lateral downcurrent fining of the grain size of the deposit. For example, mapping out maximum average grain size variations in pyroclastic fallout deposits allows isopleth contour patterns to be constructed, which together with thickness isopach contours allows dispersal patterns to be determined (Chap. 9). In addition, average maximum size, lithic clast, and isopleth patterns allow eruption column height to be calculated (Chap. 9).

In contrast, poorly sorted deposits, with a wide range in grain sizes, indicate processes that are so rapid or physically incapable of separating out the component grain sizes from each other in transit (e.g. pyroclastic flow deposits, debris flow, and other mass-flow deposits; Table 14.1A–E).

In Table 14.1, we have subdivided the possible range of grain sizes of all deposit types that are found in volcanic settings, modern and ancient, into the standard descriptive granulometric size classes, gravel size (> 2 mm), sand size (2 to > 0.0625 mm), and mud-size (0.0625 mm and finer). Those size class terms have no genetic connotation as to the origins of deposits. We have then subdivided the gravel size according to textural properties of the angularity or rounding of large clasts, and the overall deposit sorting.

For deposits in which the large clast population is angular and clasts are in mutual contact, that is, the deposit is clast-supported and has little or no finer matrix, the deposit class is called clast-supported breccia (Table 14.1A). There are about 20 ways of producing clast-supported breccias in volcanic settings (Table 14.1A), including autoclastic, explosive, sedimentary, and hydrothermal hydraulic fracture and alteration processes (see Table 14.1A for examples, and Chaps 4–6, 8–13 for many others).

For deposits in which the large clast population is angular and clasts are dispersed in a matrix without many touching each other, that is, the deposit is matrix-supported and has an abundant finer matrix, the deposit is called a matrix-supported breccia (Table 14.1B). There are also about 20 ways of producing matrix-supported breccias in volcanic settings (Table 14.1B), including autoclastic, explosive, sedimentary, hydrothermal hydraulic fracture and alteration, and tectonic processes (see Table 14.1B for examples, and Chaps. 4–6, 8–13 for many others).

In deposits in which the majority of large clasts are rounded and clasts are in mutual contact, that is, the deposit is clast-supported and has little or no finer matrix, the deposit class is texturally a clast-supported orthoconglomerate (Table 14.1C). There are about 10 ways of producing clast-supported conglomerates in volcanic settings (Table 14.1C), including explosive, sedimentary, and hydrothermal hydraulic fracture and alteration processes (see Table 14.1C for examples, and Chaps 4, 10, 12, 13 for many others).

In deposits in which the majority of large clasts are rounded and clasts are dispersed in an abundant matrix, the deposit is texturally a matrix-supported paraconglomerate (Table 14.1D). There are about 6 ways of producing matrix-supported conglomerates in volcanic settings (Table 14.1D), including explosive, sedimentary, and hydrothermal hydraulic fracture and alteration processes (see Table 14.1D for examples, and Chaps. 4, 10, 12, 13 for many others).

Sand and mud-size deposits (Table 14.1E, F, respectively) are dominated by those grain sizes and can be produced in over 15 ways (Table 14.1E, F). They can be produced by explosive, sedimentary, hydrothermal hydraulic fracture and alteration, and tectonic processes (e.g. Table 14.1E, F, and Chaps. 4, 9–13, for other examples).

14.7.4 STAGE 1, Step 4: Assess the Level of Sorting and Textural Class

Sorting (Fig. 14.1b) is the process by which solid particles respond to the properties and motion of a fluid when the fluid exerts a force on the particles. The fluid moves some particles but not others, causing a separation of different particles from each other, which is called sorting. It's a fluid hydraulic effect and property.

In sedimentology, sorting is determined by assessing the degree to which the grain size of a clastic deposit is uniform (e.g. Figure 14.3). Sedimentologists have long used size sorting as an indicator of energy levels and the efficiency of transport processes and the environment in being able to selectively winnow out particular grain sizes and concentrate others. For example, the best-sorted sediments occur in constantly high-energy environments such as beaches and wind-blown or aeolian environments, because energy levels are relatively high and constant. Most fluvial environments experience widely fluctuating energy levels and variable sediment transportation rates, and as a result, sediments tend to be less sorted, and also less rounded (Step 5). However, understanding sorting is not as simple as that, especially in volcanic environments, so let us first briefly consider the factors that determine how a population of solid particles is sorted (or not) by a moving fluid, which include:

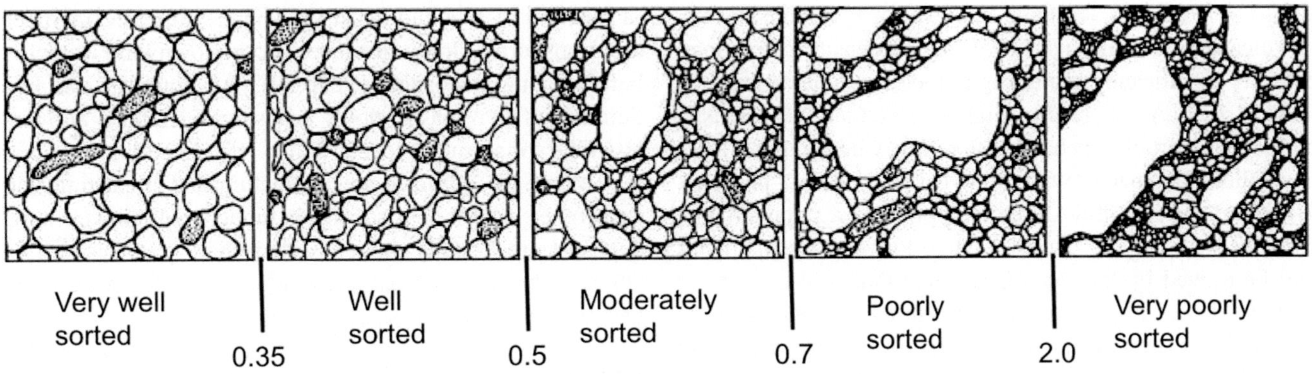

Fig. 14.3 Visual comparator chart of the main sorting classes in clastic rocks (From Prothero and Schwab 2013). The concept of sorting can only be applied to aggregates of fragments that have been dynamically moved by and segregated by the actions of a moving fluid. The term can't be applied to autoclastic, hydraulic fracture and tectonic breccias, unless in the case of autoclastic deposits they have been remobilised by surface processes. Numbers refer to the graphical standard deviation (σ_Φ) which is a measure of sorting expressed in phi units obtained by mechanical or granulometric analysis, and also typically used for examining grain size variations in unconsolidated and non-welded pyroclastic deposits (Chaps. 9, 10; for further explanation, see Cas and Wright 1987)

- the properties of the fluid, how fast it is moving, and its flow regime, which determines the fluid force,
- the properties of the solid particles, such as size, density, and shape, which determines the way in which they behave when acted on by the force of the fluid (called their *hydraulic behaviour*),
- the duration of the flow event, which determines how effective the sorting of the solids population is,
- settling velocity of solids, and
- the concentration of solids in the fluid which determines whether or not particles can respond freely to the hydraulic forces imposed on them or whether their hydraulic behaviour will be hindered.

14.7.4.1 Properties of Natural Fluids and Their Role in Sorting Clastic Aggregates

Natural fluids that transport solid particles in nature range from low-density gases (e.g. cold air in winds or hot volcanic gas in pyroclastic processes) at one end of the spectrum to water at the other end. Pure gases have a low density, and therefore to exert a significant force capable of moving a solid particle (F = ma, where F is the force, m is the mass to be moved, and a is the acceleration, which with calculus can be derived by differentiating the velocity term), they have to move at a very high velocity (e.g. a strong wind, or the upward gas velocity in an explosive eruption column capable of transporting pyroclasts high into the atmosphere). Water is much denser than gas, and therefore does not need to move as quickly as gas to exert a force that can move particles. However, in practice, the dynamic fluid that actually causes motion of some solids could also be a mixture of either of these end-member fluids together with small solid particles, such as fine ash (e.g. gas and fine ash) or clay or silt (e.g. water and clay and silt in a mud or debris flow) suspended in the pure fluids.

For example, in explosive eruption columns, pyroclastic flows, and pyroclastic surges, the fine ash fraction is almost certainly coupled with the gas (coupling is defined here as a behaviour whereby the gas molecules and fine ash particles move as one, along the same flow paths). Together, as a gas–solids mixture that is denser than pure volcanic gas, this mixture exerts a force on larger particles, and is the dynamic fluid that moves those larger particles. As particle concentrations increase in a flowing fluid different and coarser grain sizes become coupled with the gaseous and finer grain sizes, the dynamic fluid density increases and the capacity to move larger fragments increases (e.g. viscous debris flows, some high particle concentration pyroclastic flows). However, conversely, as the particle concentration and therefore the degree of particle interaction increase, the ability of the flow system to sort different size and density fragments from each other decreases, due to the effect of excessive particle interaction and hindered settling.

14.7.4.2 Role of Particle Properties in Sorting

Comparing sorting behaviour of volcanic fragments in volcaniclastic deposits compared with sedimentary epiclastic deposits requires careful consideration of the particle properties. With normal terrigenous sediments, most particles, whether mineral crystal fragments or lithic fragments, have a similar density. Therefore, they all respond similarly hydraulically when a fluid force is applied, unless they vary widely in their shape characteristics, which would then cause them to behave differently hydraulically. Because most terrigenous sediment particles behave similarly hydraulically, the degree to which the particle sizes in a deposit is uniform is a good proxy for assessing the level of hydraulic sorting. In volcaniclastic deposits, however, the density of particles can vary significantly according to the degree of vesicularity (Chap. 12). As a result, highly vesicular, low-density pumiceous particles behave differently when a fluid force is applied, compared with non-vesicular, dense particles. Therefore, when a fluid force is applied to a population of particles with varying densities, large low-density fragments will be moved by the same fluid force that is able to move smaller dense fragments. The resulting deposit will be well-sorted hydraulically, but will show a wide grain size range (Chap. 12).

Another important particle property is clast shape. An equant grain or clast is easily pushed and moved by a fluid force because it presents a reasonable upstream surface area for the fluid to impact on. By contrast, a flat, platy particle may be difficult to move initially because fluid stream paths skim over the flat surface, but once in motion and particularly if suspended, it can easily remain in motion. Even though the mass of an equant and platy but dimensionally longer fragment may be similar, the platy particle may travel further downstream and eventually be deposited together with lower mass, equant particles. The two have behaved in a similar manner hydraulically, but have very different grain sizes.

14.7.4.3 Duration of Transport in Affecting Sorting

Another factor that comes into play in determining degrees of (hydraulic) sorting is the duration of transport. Tractional sedimentary processes (Chap. 12) tend to be long-lived, whether or not they occur continuously (e.g. coastal beach), or intermittently (e.g. wind-blown or ephemeral river environments). By contrast, mass-flow events (e.g. pyroclastic flows, debris flows, turbidity currents, landslides, rock avalanches; Chaps. 10, 12) are generally short-lived events. The shorter the event, the less likely it is to produce high levels of sorting in the resultant deposits. In mass flows, the higher the particle concentration, the less likely that efficient sorting can occur due to the effects of hindered settling (Chap. 12). An exception is medial to distal turbidites, which due to a relatively expanded and turbulent state do produce reasonably high levels of sorting, as large and dense particles drop out of suspension as the turbidity current velocity wanes, producing both a vertical and lateral grain size grading (Chap.12).

In general, pyroclastic transport processes tend to be relatively short-lived, irrespective of whether they are pyroclastic fallout, pyroclastic flow, or pyroclastic surge. The transport of pyroclasts in an explosive eruption column and downwind plume produces the best sorted pyroclastic deposits because coarse and dense pyroclasts fallout closest to the vent, controlled by their terminal fall velocity (a function of size and density), whereas fine ash fragments are transported long distances in a downwind direction before being deposited (Chap. 9). By contrast, the flow duration of pyroclastic flows and surges is short, allowing less time for sorting to occur (Chap. 10). In addition, in both flow types the particles are interacting and impacting with each other, so efficient sorting is hindered. However, surge deposits are usually better sorted than pyroclastic flow deposits, because the particle load is more dispersed in the gas, which facilitates better sorting.

14.7.4.4 Settling Velocity and Sorting

As discussed in Chaps. 9 and 12, an important factor in determining when a particle is sedimented from its transport regime is its settling velocity. The factors that determine the settling velocity of a particle out of a flow of fluid and sediment particles are represented in the Stokes settling velocity equation (Chap. 12). These factors include particle density and size, as well as the fluid density and viscosity. As discussed above, low-density particles will therefore have lower settling velocities than dense particles of the same size, or another way of viewing it is that large low-density particles (e.g. pumice) will have the same settling in velocity as small dense particles (e.g. mineral crystal and lithic fragments). That is, these particles with different properties have similar hydraulic settling behaviours. The Stokes settling principle assumes that each particle can settle freely through fluid. However, if particle concentration is so high that particles impact each other, the settling behaviour is modified, or hindered (Chap. 12). Efficient hydraulic sorting cannot occur under these conditions.

In addition, the fluid flow regime affects the drag exerted by the fluid on the particle, so that the particle settling behaviour changes in a laminar versus a turbulent flow, and where gas and water are pure compared with denser gas–fine ash and water–clay and silt mixtures.

14.7.4.5 Sorting in Autoclastic Deposits?

Sorting is not a concept that can be applied to autoclastic deposits, unless they have been resedimented by surface processes. Autoclastic deposits such as autobreccias and quench fragmented hyaloclastite are fragmented in situ and are not subjected to fluid forces and transportation processes that sediments and pyroclastic deposits are subjected to. The grain size of autobreccia deposits is controlled by magma viscosity and strength, and the shear force exerted by flowing lava inside the flow on the surface crust. With hyaloclastite, the deposit grain size and uniformity of grain size are determined by the magma properties, and the degree and rate of supercooling of the magma by ambient water. Assessing sorting is therefore not appropriate for autoclastic deposits unless they are resedimented (Table 14.1A–E).

14.7.4.6 Sorting Classes or Types and Relevance to Understanding Depositional Textures and Processes

In Step 4, typical sedimentary sorting comparator images are provided as a guide to sorting classes (Figs. 14.1b; 14.3). However, although these are a useful guide for assessing sorting, it must be remembered that they do not reflect the hydraulic sorting behaviour of particles of different densities.

In general terms, fragmental or clastic deposits can be subdivided into two textural types, *clast-supported* and *matrix-supported* (Table 14.1A–D). These clearly represent different degrees of sorting. In clast-supported deposits, large framework grains are all in mutual contact, suggesting relatively high energy and efficient sorting transport processes. Even then there could be variations in that the spaces between the framework grains could be open, and perhaps subsequently in-filled with chemical cement, or they could be filled with fine-grained silt and clay-size matrix material, suggesting a lower energy and less efficient sorting transport process. In this last case, it is important to assess whether the fine-grained matrix was deposited with the coarse fraction or if it percolated later through the open voids from above. This must be assessed based on componentry and contact relationships with overlying sediments.

Carbonate sedimentologists such as Dunham (1962) devised textural sorting class schemes that allow for interpretation of the process origins of different textural configurations (Fig. 14.1b, Step 4, bottom). Dunham called clast-supported deposits without a clay and silt size matrix grainstones, and he recognised that these are the product of very high-energy environments and processes. Beach sands and aeolian sands are typical grainstones. In volcanology, coarse pyroclastic fallout deposits, which typically lack ash matrix, can be classed as grainstone equivalents. They are the product of high-energy and efficient sorting transport processes in eruption columns and downwind plumes. Dunham called clast-supported deposits, with an interstitial mud matrix, packstones. Such deposits still represent relatively high-energy transport processes, but they are less efficient in sorting grain sizes. For example, some overbank fluvial deposits and some turbidites are texturally packstones. In volcanology, some pyroclastic surge deposits are packstones, reflecting short-lived, high-energy events.

Dunham also proposed the terms wackestone and mudstone. A sedimentary wackestone is a matrix-supported deposit in which large clasts are dispersed through a fine matrix and the matrix is dominated by clay and silt size material. With primary volcanic deposits, the fine matrix component would be fine ash. The large clast framework component could vary from sand to boulder-sized material, and could be variably vesiculated, large pumice or scoria clasts, or dense lithic clasts. Typical sedimentary deposits that are wackestones are debris flow deposits and glacial till/tillite or diamictite. In volcanological realms, volcaniclastic debris flow deposits also occur, but many pyroclastic flow deposits or ignimbrites are texturally also wackestones with pumice and lithic clasts suspended in a fine ash matrix of glass shards and crystal fragments. Wackestone style deposits indicate that the transport mechanism was incapable of efficiently sorting the variable grain size components into well-sorted deposits. This is because the duration of the event is short-lived and because the particle concentration is so high that hindered settling prevents effective size sorting. In the case of viscous sedimentary debris flows, the coupling between an abundant clay size fraction and water produces a dense and viscous fluid that is capable of suspending and pushing larger clasts, and hinders sorting.

The final textural class in the Dunham (1962) scheme is called mudstone. As the name implies, this is dominated by silt and clay-sized material, with minor dispersed other size fractions. A sedimentary mudstone is usually the product of very low-energy environments involving the slow settling of very fine sediment particles from aqueous suspension to the substrate. This is particularly the case in thin and laminated mudstone deposits. Where mudstone deposits are massive and contain isolated dispersed larger clasts, they would also have originated as mud or debris flows. In volcanology, Dunham-type mudstone deposits could originate through sedimentary processes as explained, or as primary volcanic deposits that could include fine distal fallout ash, which would also show lamination, or massive extremely fine-grained, distal ash-rich pyroclastic flow deposits or ignimbrites, or surge deposits. The latter could lack large pumice and lithic clasts, but contain dispersed crystal, fine pumice, and fine lithic fragments. Such a deposit is the result a very intense fragmentation (e.g. a phreatomagmatic ignimbrite) or a distal ash-rich ignimbrite.

In old or visibly altered deposits, the effects of alteration masking the original textural configuration of a deposit can be extreme and very misleading in terms of what the original texture might have been (Chap. 13). For example, alteration can overprint and mask the existence of many framework grains as well as the original interstitial matrix or cement. Relict dispersed large fragments in a secondary matrix of an alteration mineral phase such as chlorite or sericite may give the impression of a deposit that has a matrix-supported texture, and is very poorly sorted. Because the alteration phase is secondary and fine-grained, it is what we describe as a pseudo-matrix. The relative abundance of alteration phases is no indication at all of the original abundance of a fine clastic matrix, and since it may also overprint original framework grains, their preserved apparent abundance is also misleading. Considerable petrographic work is required to reconstruct the original textural configuration and abundance of components by back stripping the secondary mineral alteration phases. In such rocks, there may be local domains that preserve parts of the original texture that should be looked for.

14.7.5 STAGE 1, Step 5: Evaluate the Degree of Rounding or Angularity of Fragments

Whether or not large clasts (> 2 mm) in a deposit are largely angular or rounded by abrasion determines if the deposit is texturally a breccia or a conglomerate (Table 14.1A–D). All large clasts are initially angular when they first form (except for accretionary lapilli) through primary volcanic processes, surface sedimentary processes, or post-volcanic hydrothermal and tectonic processes. They may then be subject to varying degrees of physical abrasion in the vent, eruption column (Chaps 4, 9), during pyroclastic flow (Chap. 10) or lava flow (Chaps. 5, 6), or during a wide range of sedimentary surface processes (Chap. 12). Depending on the intensity of fragmentation at the time of formation, and during transport, a breccia may be a clast-supported breccia or ortho-breccia (little or no matrix; Table 1.14A) or a matrix-supported breccia or para-breccia (angular clasts dispersed in a dominant matrix; Table 14.1B). There are multiple ways of producing both textural types of breccia. It should not be assumed that breccias consisting of angular volcanic clasts are explosion breccias nor that they were deposited close to the vent. Volcanic breccias can form in over 30 different ways (Table 14.1A, B).

Coarse deposits consisting of large clasts that are dominantly rounded are texturally conglomerates, irrespective of whether or not they have a sedimentary or a primary volcanic origin. It may be surprising that conglomerates can form during primary eruption processes. Lithic clasts can be milled and rounded in the vent and then be ejected explosively (Campbell et al. 2013). Delicate pumice clasts can be abraded during a pyroclastic flow event that causes significant rounding (Manga et al. 2011), particularly in frontal and marginal lobes, in fines-depleted facies in ignimbrites, and in some pumice concentration zones. Conglomerates with little or no matrix are called clast-supported or ortho-conglomerates (Table 14.1C), and those in which rounded clasts are dispersed in a dominant finer matrix are called para-conglomerates (Table 14.1D).

Rounding (Fig. 14.1b) is an important texture in clastic sediments because it reflects the degree of abrasion and smoothing of sharp corners and edges of sediment fragments during tractional sedimentary reworking (Fig. 14.4; see Chap. 12 for discussion of the processes and depositional structures produced). Efficient rounding and smoothing of sharp corners, edges, and surfaces of clasts (cf. rough rounding, see below) can only occur through prolonged abrasive interaction by sediment particles against other sediment particles under the influence of the tractional current movement. Environments in which tractional reworking of sediment particles occurs include beaches, offshore shallow water environments affected by waves and long-shore currents, aeolian wind-blown environments, and rivers. In all of these environments, sediment particles are moved through bed load rolling and sliding, and through saltation for prolonged periods of time, as discussed in Chap. 12. High-energy beach and wind-blown environments typically produce beautifully smoothly rounded sand grains, and on gravel beaches smoothly rounded and sometimes flat pebbles to boulders. In river channels, pebble to boulder size material can also become well-rounded and smoothed, but sand size material less so because current velocities can be highly variable and episodic. In other environments, the degree of smooth rounding is limited. In general, the degree of rounding of clasts increases with the duration and/or distance of tractional transport.

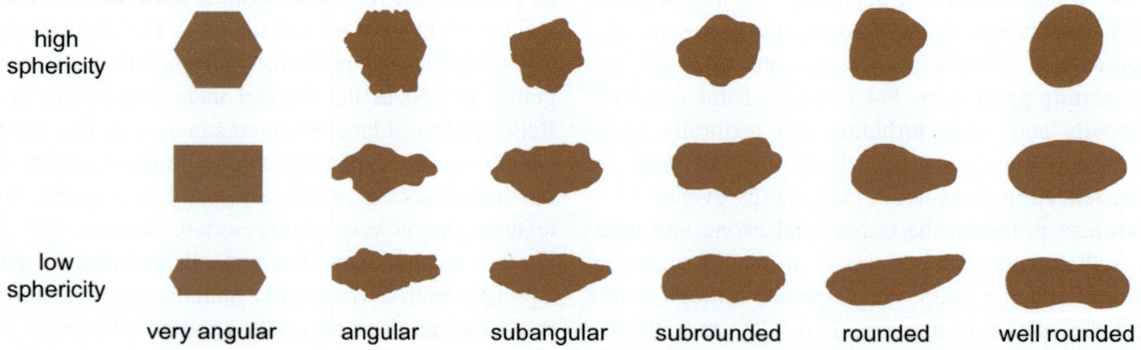

Fig. 14.4 Schematic representation of degrees of rounding of corners and edges sedimentary or volcanic clasts through physical transport and abrasion, from very angular (left) to well rounded (right). Also depicted are relative assessments of the shape factor "sphericity", which measures the degree to which an object approximates a sphere. Sphericity is inherited at source determined by whether or not a source rock has isotropic or pre-existing anisotropic fabric features such as bedding, tectonic cleavage, and jointing. (*Source* Wikimedia; Adapted from Powers 1953)

Mass-flow processes produce very limited smooth rounding because the duration of such events is short and interaction between fragments involves mostly instantaneous impacts rather than prolonged tractional grinding. At best, rough rounding may occur, marked by chipped corners and edges, and rough surfaces that begin to form a rough roundish form. If mass-flow deposits contain smooth rounded clasts, that rounding in most cases will be inherited from the environment from which the clasts were derived and incorporated into the mass flow (e.g. some submarine debris flow deposits that have sourced some of the clast population from former shoreline zone boulder beach deposits). So, in all deposits that originate through sedimentary transport and deposition, irrespective of whether they are terrigenous or volcaniclastic in composition, the presence of smooth rounding of clasts usually indicates a polycyclic history. An exception is marine pumice raft deposits that originate from marine eruptions and then travel large distances through the effects of currents, waves, and wind, during which clasts abrade against each other, and become rounded. By contrast, a mass-flow deposit with only rough rounded, or angular clasts, indicates that it is a first cycle deposit.

In the deposits of short-lived pyroclastic flows and surges, the presence of smooth, well-rounded clasts indicates that they are accidental, picked up off the ground surface from a pre-existing alluvial deposit. Rough rounded clasts, such as rounded pumice clasts in ignimbrites, or dense juvenile blocks with chipped, rough rounded corners and edges in block and ash flow deposits, will have been impact-abraded in the pyroclastic flow. Because pumice is fragile, slightly (rough) rounded pumice clasts are not unusual in ignimbrites and surge deposits, especially in pumice concentration zones (Manga et al. 2011; Chap. 10, see Figs. 10.41b, 10.57a).

Rough rounding of lithic clasts, particularly big, dense blocks can also occur in a vent, in a process called vent-milling (see Fig. 4.31 h; Campbell et al. 2013). This occurs when the gas velocity through which most pyroclasts are ejected from the vent is lower than the settling velocity for large dense lithic clasts in the vent. Such lithics may then be not only tumbled, and abraded, but also subjected to thermo-mechanical spalling of the heated margins (Campbell et al. 2013). They may remain trapped in the vent, unless a more energetic explosion and higher gas velocity are sufficient to eject the rounded blocks or boulders from the vent.

For many phenocrysts, rounding reflects the amount of magmatic resorption that has occurred in the host magma during ascent to the Earth's surface (e.g. Tepley et al. 1999), which is then preserved in primary volcanic deposits in large clasts (pumice, scoria, juvenile bombs, and blocks) and in free crystal fragments in pyroclastic deposits. In epiclastic deposits, rounding of crystal fragments may also result from tractional reworking. In lavas and sills, unbroken phenocrysts should dominate, whereas in pyroclastic deposits crystal fragments will be largely fragmented and anhedral.

14.7.6 STAGE 1, Step 6: Estimate the Free Crystal Content

Determining the types of mineral crystal fragments and their abundance in volcaniclastic deposits (Figs. 14.1b; 14.2) is important for determining the composition of the source magma or volcanic source rock from which the crystals have been derived, depending on whether the deposit is pyroclastic or epiclastic. In a pyroclastic flow deposit, comparison between the phenocryst content in juvenile clasts (on a dense rock equivalent basis, i.e. recalculating the crystal content after excluding the vesicle abundance if clasts are vesiculated) with the abundance of free crystal fragments in the matrix (after excluding the lithic and pumice/scoria content) will reflect the degree to which crystals have been concentrated in the eruption column *and* during pyroclastic flow, or in the case of some dome-collapse block and ash flows, just during flow. This occurs through the loss of the original magma groundmass fraction through the elutriation of fine vitric ash. The degree of crystal concentration will to some degree provide an indication of the height of the eruption column, with high columns creating more time for ash winnowing, which then gets dispersed downwind, compared with low columns or fountains (Walker 1972; Chap. 9). In pyroclastic fallout deposits, crystal fragments are not uniformly deposited in the deposit downwind, being least abundant in coarse, near-vent deposits, more abundant in medial coarse ash deposits, and less abundant again in distal vitric fine ash (Walker 1981).

For resedimented and reworked deposits, careful assessment of the degree to which the whole assemblage of fragments is monomictic will reflect proximity to and derivation from a nearby young source. A polymictic assemblage reflects input and mixing from multiple sources, old and young.

14.7.7 STAGE 1, Step 7: Assess the Vesicularity of Volcanic Clasts—Juvenile or from Older Sources?

Volcanic clasts in any volcaniclastic deposit could be juvenile or derived from an older source. Using the term "*juvenile clasts*" suggests that the host deposit is likely to be a primary volcaniclastic deposit, specifically pyroclastic or autoclastic. We have to be careful in making this assumption because syn-eruptive resedimentation or slight reworking of primary deposits can produce secondary deposits, which texturally and compositionally are very similar to the primary deposits because their components are the same.

As discussed in assessing the vesicularity of lavas above (Sect. 14.6.6), there are several factors that affect the level of vesiculation in an erupting magma, including magma volatile content, magma rheology, amount and rate of decompression, and ambient environment pressure. Vesiculation can occur in erupting lavas (and their derivative autoclastic breccias; Chaps 5, 6) as well as during magmatic volatile-driven explosive eruptions (Chaps 4, 9, 10). Where juvenile clasts such as pumice and scoria have formed through explosive fragmentation of an erupting magma, exsolving magmatic volatiles will have driven the magmatic explosive eruption. Volatile content by itself does not determine the level of vesiculation developed, and so the level of vesiculation cannot be used as a proxy for magma volatile content. However, the abundance and nature of vesicle population provide a lot of information about conduit processes as the magma was rising. In a seminal paper on the application of bubble number density, N_v, to understanding explosive eruption dynamics and parameters, Toramaru (2006) demonstrated that decompression rate, dP/dt, can be calculated from N_v, dP/dt is proportional to explosive eruption column height, H_T^4, H_T is proportional to magma discharge rate, $Q^{0.25}$, and Q is directly proportional to dP/dt. This is extraordinarily powerful! See Chaps. 4, 9 for more details.

Some vesiculated pyroclasts may have a simple vesicle population with normal and continuous vesicle size populations, suggesting a single, continuous, perhaps late vesiculation and decompression event. Others may have bimodal vesicle populations, indicating early vesiculation and growth (large and coalesced vesicles) and a late sudden vesiculation and decompression event (small vesicles), perhaps driven by sudden decompression and rapid acceleration of the magma up a conduit. In calculating eruption parameters resulting from the sudden decompression, the early large vesicles are ignored in the assessment of N_v. Others may have multiple populations, including coalesced large populations in the centre of pyroclasts. This could indicate ongoing vesiculation of the interior of some (large) pyroclasts after fragmentation and eruption. Where phreatomagmatic eruptions occur, the level of vesiculation is affected by the timing of magma–water interaction and quench chilling of pyroclasts, which arrests the vesiculation process. One can't assume that the preserved level of vesiculation necessarily reflects the level of volatiles, nor even the level of vesiculation at the time of magma fragmentation. Calculation of eruption phenomena using the Toramaru (2006) approach would produce minimum values of the eruption parameters because vesiculation was short-circuited at the time of magma–water interaction.

Determining the vesicularity of pyroclasts (Fig. 14.1b) is not always straightforward. If the vesicle population is large and the pyroclasts are large, first-order vesicle abundance estimates can be made using Fig. 14.2 as a visual comparator. However, vesicle populations are commonly complex and variable in size, and if detailed information is required, further techniques are required. For young unconsolidated deposits, scanning electron microscopy of vesiculated pyroclasts should be used to image the vesicle population to derive bubble number and bubble size distribution data (BND, BSD respectively) that can then be used to model eruption dynamic parameters (Chap. 9). Vesicle abundance can be determined using pyroclast immersion or gas pycnometer techniques (e.g. Shea et al. 2010; Chap. 9). For old lithified rocks, if vesicles are well enough preserved, they can be petrographically point counted, or their characteristics documented using image analysis and ImageJ. However, many very small vesicles will be overprinted by secondary diagenetic and alteration processes, and mineral phases.

Autoclastic deposits may include clasts that are non-vesicular to highly vesicular pumice clasts and deposits, the latter derived from the autobrecciation or quench fragmentation of the pumiceous carapace of lavas (Chap. 6).

In volcaniclastic sediments, the level of vesicularity of the volcanic clasts reflects that of the source material. If the sedimentary event is syn-eruptive, representing resedimentation of a primary pyroclastic or autoclastic deposit, the clasts and their levels of vesiculation will then closely reflect that of the primary source volcanic deposits. Where the source of a volcaniclastic deposit is older, and from large volcanic terrains, there will be a polymictic and diverse representation of clast types, and perhaps levels of vesicularity. The older and more distant the source is from the depositional site, the more likely that originally vesicular materials will be eliminated from the geological record because vesicular, especially glassy, materials are more prone to chemical weathering and physical abrasion breakdown than non-vesicular and crystallised volcanic sources.

14.7.8 STAGE 1, Step 8: Compile a Composite, Descriptive Component, Textural, and Compositional Petrological Name

This step combines the key observations and assessments of Steps 1 to 7 into a multi-factor descriptive name (Fig. 14.1 b). Our suggested order for compiling such a name is:

(1) Alteration state and principal alteration mineral(s),
(2) Sorting,
(3) Free mineral crystal fragment type and abundance (if any),
(4) Framework type (grain or, matrix-supported),
(5) Principal component(s) (minerals, lithics, matrix),

(6) Compositional affinities,
(7) Principal clast size class, and
(8) Dunham textural and packing classification.

Examples of petrological–textural names include:

- sericite-altered, well-sorted, very crystal-rich, quartz–feldspar, dacitic, granule-size grainstone,
- serpentine-carbonate altered, moderately sorted, moderately crystal-rich, olivine, cored pelletal lapilli, grain-supported, kimberlitic, coarse sand-size packstone, and
- poorly sorted, crystal-poor, monomictic dacite lithic, matrix-supported, boulder to silt-size wackestone.

Although long, these examples of composite names reflect the petrological make-up of the deposit, before assessing the genetic emplacement origin of the deposit by assessing and integrating meso- to macro-scale deposit characteristics that inform the transport and depositional processes, as well as the volcanological and environmental settings.

14.7.9 STAGE 2, Step 9: Assess Bedding and Depositional Unit Thickness

In Step 9 (Fig. 14.1b), we have adopted the bedding thickness scale used in sedimentary geology, but have extended it at the thickest end to accommodate decametre-plus thick mega-beds such as some ignimbrites; beds of this thickness are rare in sediments. For modern and young pyroclastic fallout deposits, mapping and measuring deposit thickness at multiple locations, then constructing isopach maps allows dispersal patterns and axis of dispersal to be determined, and volumes to be calculated (Chap. 9). For other deposits (pyroclastic flow, sedimentary volcaniclastics), volume calculations and reconstruction of transport directions and relationships to paleotopography may be possible (e.g. Chap. 10 for pyroclastic flow deposits).

14.7.9.1 STAGE 2, Step 10: Document Depositional Structures and Fabrics

Depositional structures (Fig. 14.1b) are the most important features of clastic deposits in sedimentology and pyroclastic geology for interpreting the mode of transport and deposition, as well as the depositional environment in many cases. In the geological rock record, this last aspect is extremely important in developing an understanding of volcanological settings and processes. Under depositional structures, we also include depositional fabric, such as imbrication and size grading. The importance of understanding depositional structures and fabrics and the way they can be used to understand the origins of particular facies types have been discussed in Chaps. 9 and 10 for pyroclastic deposits, and Chap. 12 for volcanic sediments.

Sedimentary volcaniclastic sediments can usually be subdivided into either tractional sedimentary deposits or mass-flow deposits (Chap. 12). The former will be marked by typical tractional depositional structures, including ripples, dunes, angle of repose cross-bedding, scour and fill structures, and are usually reasonably well sorted. Mass-flow deposits could be massive and poorly sorted when they are debris flow, landslide, and rock avalanche deposits, or continuously size graded if they are turbidite deposits (see Chap. 12 for details).

To summarise the key deposit characteristics of pyroclastic deposits, pyroclastic fallout deposits will usually show reverse size grading at the bottom, reflecting increasing eruption column height and progressively wider dispersal of coarse clasts from the beginning of the eruption, and may show normal size grading at the top of the deposit, reflecting waning of the eruption (Chap. 9). Sometimes, the top of the fallout deposit is abruptly overlain by ignimbrites, reflecting eruption column collapse. Pyroclastic flow deposits are poorly sorted, may be massive or diffusely bedded, and may be underlain by thin, low-angle, cross-stratified ground or frontal ash-cloud surge deposits, and be overlain by fine, cross-stratified trailing ash-cloud surge deposits (Chap. 10). The body deposits are usually (fine ash) matrix-supported, and may show normal coarse-tail grading of lithic clasts, and reverse coarse-tail grading of pumice clasts, although there are many variations of these themes from deposit to deposit. Base surge deposits will usually occur in a sequence of multiple low-angle cross-stratified and/or fallout deposits that form a maar rim or tuff cone sequence. As we have also discussed above, accretionary lapilli can occur in any type of pyroclastic deposit, close or far from the vent, as well as in volcanic sediments, and so by themselves, they tell us little other than they probably originated from a phreatomagmatic or hydrothermal explosion.

One of the most important questions when dealing with volcaniclastic deposits is whether they are primary or secondary (i.e. resedimented or reworked), and what was the temperature at the time of deposition? "*Were they hot or cold*?" Indications of a hot state of emplacement in primary pyroclastic deposits could include eutaxitic texture (fiamme, plastically attenuated glass shards in the matrix), columnar jointing, gas segregation (or gas escape or elutriation) pipes and structures, and steam or fumarolic pipes (Chaps. 10, 11). Although these criteria are usually reliable, some caution and due diligence are required. As discussed above, fiamme may result from diagenetic compaction, not thermally induced plastic deformation, or from polyphase hydrothermal alteration of coherent lavas and intrusions that produces pseudo-fiamme texture (Chap. 13). Plastically deformed and fused bubble wall glass shards are reliable, although false shard textures can also result from alteration along perlitic cracks in glassy lavas (e.g. Allen 1988). Columnar jointing can form in both hot welded pyroclastic flow deposits as well as vapour phase altered, non-welded ignimbrites. Although gas segregation pipes are common in pyroclastic flow deposits, steam and water escape pipes also occur in debris flow and hyper-concentrated flood flow deposits (Chap. 12); fluid escape pipes are therefore not solely indicative of pyroclastic flow deposits. Charcoal can occur in both hot pyroclastic deposits, as well as reworked and resedimented volcanic sediments. Usually, if a deposit contains only carbonised vegetation debris, it is likely to be a hot pyroclastic deposit. Where deposits contain both charcoal and non-carbonised vegetation material, the deposits are likely to be volcanic sediments. Although it varies from tree to tree, the minimum temperature required to form charcoal is about 300 °C (Chap. 10).

At the end of Step 10, one should be in a position to compile a more complete descriptive facies name, incorporating the mesoscopic or scale depositional features, together with the previous petrological name.

14.7.9.2 STAGE 2, Step 11: Apply an Appropriate Genetic Grain Size Name

At this point, all the descriptive information needed to determine the eruption and/or transportation and depositional origins should also have been assembled, and interpretation can be made. If it is decided that the deposit in question is a volcanic sediment, then the standard Wentworth sediment size terms can be adopted, without the suffix "-size" (Fig. 14.1b). If the deposit is confirmed as a pyroclastic deposit, then the Fisher (1966) size terminology for pyroclastic deposits can be adopted, as listed in Step 11 (Fig. 14.1b). For pyroclastic deposits in which sub-equal proportions of pyroclast grain size classes are present, then dual size names are recommended (e.g. block-bomb lapillistone; lapilli-ash/tuff).

For autobreccias, that is what they are called. For hyaloclastites, they will mostly be hyaloclastite or peperite breccia because of the presence of > 1/3 fragments that are > 2 mm in diameter.

14.7.9.3 STAGE 3, Step 12: Assess Contact Relationships

Unless the volcaniclastic deposit is a clastic intrusion, peperite, or in cross-cutting faulted contact, most deposits will be concordant with the enclosing stratigraphy. They may be *conformable* without a time break between underlying and overlying strata, or *disconformable*, with evidence for a time break (widespread erosional surface, soil development). Alternatively, the contact may be a bedding parallel fault, evidenced by a sheared, tectonic breccia contact. Each option has to be evaluated on its merits (Fig. 14.1b).

14.7.9.4 STAGE 3, Step 13: Determine the Setting

In Step 13 (Fig. 14.1b), the (paleo)environmental setting should be determined, as well as the volcanological setting relative to the vent system: intra-vent or extra-vent. Paleoenvironmental setting is determined using the association of facies, distinctive depositional structures, and fossils, if they are present.

14.7.9.5 STAGE 3, Step 14: Compile a Composite Genetic Name

In this final step, all the key aspects of the deposit, including petrological, facies, emplacement process, and setting are combined and represented (Fig. 14.1b). Guidelines for assembling the multi-factor name and the order of descriptors are:

(1) Setting (e.g. submarine, intra-vent, outflow, and valley pond),
(2) Bedding thickness class,
(3) Principal emplacement/depositional character/feature,
(4) Key descriptors from the non-genetic petrological rock/deposit name,
(5) Genetic size class name for fragmental deposits, and
(6) Emplacement mode (pyroclastic fallout deposit, debris flow, intrusive peperite, fluvial reworking, etc.).

Examples of final genetic deposit names include:

- intra-vent, poorly sorted, massive, olivine-rich, kimberlitic, column collapse lapilli tuff,
- extra-caldera, extremely thick, poorly sorted, fiamme-rich, massive, welded dacite ignimbrite,
- extra-vent, thickly, diffusely bedded, poorly sorted, polymictic andesitic debris flow deposit, and
- extra-vent, well-sorted, moderately thick-bedded, cross-bedded quartz-plagioclase-lithic, coarse, braided fluvial channel, volcaniclastic sandstone.

14.7.9.6 What if Rocks Are so Altered It is Impossible to Confidently Design a Name?

At times, ancient, deformed, and altered rocks have characteristics that are ambiguous as to their origins, or they are so deformed and altered that they have morphed completely into metamorphic rocks, which barely reveal their volcanological affinities (Chap.13). For rocks like that, names such as "thick, biotite–hornblende–feldspar pelitic schist", or "very thick, porphyroblastic quartz–sericite psammitic schist", or "extremely thick, quartz–feldspar–biotite gneiss with abundant 'spotty' garnets" may be appropriate. The former could originally have been a basalt lava or sill, and the second a macrocrystic felsic rock, which could have been a felsic lava or intrusion or a pyroclastic ignimbrite; an example of the latter is the enigmabritic Potosi gneiss at Broken Hill, New South Wales (Sect. 13.10). And if all else fails, "fubarite" may again be an optional name.

References

Allen RL (1988) False pyroclastic textures in altered silicic lavas, with implications for volcanic-hosted mineralization. Econ Geol 83:1424–1446

Bernard K, Thouret JC, de Vries BV (2017) Emplacement and transformations of volcanic debris avalanches-A case study at El Misti volcano. Peru J Volcanol Geotherm Res 340:68–91

Best MG, Christiansen EH (1997). Origin of broken phenocrysts in ash-flow tuffs. Geol Soc Amer Bull 109(1):63-73.

Branney MJ, Sparks RSJ (1990) Fiamme formed by diagenesis and burial-compaction in soils and subaqueous sediments. J Geol Soc Lond 147(6):919–922

Bryan WB (1972) Morphology of quench crystals in submarine basalts. J Geophys Res 77:5812–5819

Campbell ME, Russell JK, Porritt LA (2013) Thermomechanical milling of accessory lithics in volcanic conduits. Earth Planet Sci Letts 377:276–286

Cas RAF, Hayman P, Pittari A, Porritt L (2008a) Some major problems with existing models and terminology associated with kimberlite pipes from a volcanological perspective, and some suggestions. J Volcanol Geotherm Res 174:209–225

Cas RAF, Porritt L, Pittari A, Hayman P (2008b) A new approach to kimberlite facies terminology using a revised general approach to the nomenclature of all volcanic rocks and deposits: descriptive to genetic. J Volcanol Geotherm Res 174:226–240

Cas RAF, Porritt L, Pittari A, Hayman PC (2009) A practical guide to terminology for kimberlite facies: A systematic progression from descriptive to genetic, including a pocket guide. Lithos 112:183–190

Cas RAF and Wright JV (1987) Volcanic Successions: Modern and Ancient. A geological approach to processes, products and successions. Allen & Unwin, London, p 528

Cashman KV (1993) Relationship between plagioclase crystallization and cooling rate in basaltic melts. Contribs Mineral Petrol 113 (1):126–142

Cashman KV (2020) Crystal size distribution (CSD) analysis of volcanic samples: advances and challenges. Frontiers Earth Sci. 8:291. https://doi.org/10.3389/feart.2020.00291

Cashman KV, Marsh, (1988) Crystal size distribution (CSD) in rocks and the kinetics and dynamics of crystallization II. Makaopuhi Lava Lake. Contribs Mineral Petrol 99:292–305

Cunningham JK, Beard AD (2014) An unusual occurrence of mafic accretionary lapilli in deep-marine volcaniclastics on 'Eua, Tonga: palaeoenvironment and process. J Volcanol Geotherm Res 274:139–151

Donaldson CH (1976) An experimental investigation of olivine morphology. Contribs Mineral Petrol 57:187–213

Dunham RJ (1962) Classification of carbonate rocks according to depositional texture. In: Ham WE (ed) Classification of carbonate rocks. Amer Assoc Petrol Geol Mem 1: 108–121

Faure F, Arndt N, Libourel G (2006) Formation of spinifex texture in komatiites: An experimental study. J Petrol 47:1591–1610

Fowler AD, Berger B, Shore M, Jones MI, Ropchan J (2002) Supercooled rocks: development and significance of varioles, spherulites, dendrites and spinifex in Archaean volcanic rocks, Abitibi Greenstone belt. Canada. Precamb Res 115(1–4):311–328

Fisher RV (1960) Classification of volcanic breccias. Geol Soc Amer Bull 71:973–982

Fisher RV (1961) Proposed classification of volcanoclastic sediments and rocks. Geol Soc Amer Bull 72:1409–1414

Fisher RV (1966) Rocks composed of volcanic fragments and their classification. Earth Sci Rev 1(4):287–298

Hayman PC, Cas RAF (2011) Criteria for interpreting kimberlite as coherent: insights from the Muskox and Jericho kimberlites (Nunavut, Canada). Bull Volcanol 73(8):1005–1027

Hayman PC, Cas RAF, Johnson M (2008) Difficulties in distinguishing coherent from fragmental kimberlite: a case study of the Muskox pipe (Northern Slave Province, Nunavut, Canada). J Volcanol Geotherm Res 174(1–3):139–151

Hayman PC, Cas RA, Johnson M (2009) Characteristics and alteration origins of matrix minerals in volcaniclastic kimberlite of the Muskox pipe (Nunavut, Canada). In: Proceedings of the 9th international kimberlite conference, Frankfurt, Germany. Lithos 112:473–487

Higgins MD (1996) Magma dynamics beneath Kameni volcano, Greece, as revealed by crystal size and shape measurements. J Volcanol Geotherm Res 70:37–48

Higgins MD (2000) Measurement of crystal size distributions. Amer Mineral 85:1105–1116

Hogan JP (1993) Monomineralic glomerocrysts: Textural evidence for mineral resorption during crystallization of igneous rocks. J Geol 101(4):531–540

Houghton BF, Carey RJ (2015) Pyroclastic fall deposits. In: Sigurdsson H, Houghton B, McNutt SR, Rymer H, Stix J (eds) The encyclopedia of volcanoes. Academic Press Elsevier, 599–616

Houghton BF, Wilson CJN (1989) A vesicularity index for pyroclastic deposits. Bull Volcanol 51:451–462. https://doi.org/10.1007/BF01078811

Jones TJ, McNamara K, Eychenne J, Rust AC, Cashman KV, Scheu B, Edwards R (2016). Primary and secondary fragmentation of crystal-bearing intermediate magma. J Volc Geotherm Res 327:70-83.

Jutzeler M, McPhie J, Allen SR, Proussevitch AA (2015) Grain-size distribution of volcaniclastic rocks 2: Characterizing grain size and hydraulic sorting. J Volcanol Geotherm Res 301:191–203

Jutzeler M, Proussevitch AA, Allen SR (2012) Grain-size distribution of volcaniclastic rocks 1: A new technique based on functional stereology. J Volcanol Geotherm Res 239–240:1–11

McPhie J, Doyle M, Allen RL. (1993) Volcanic textures. University of Tasmania, 196p

Manga M, Patel A, Dufek J (2011) Rounding of pumice clasts during transport: field measurements and laboratory studies. Bull Volcanol 73(3):321–333

Marsh BD (1998) On the interpretation of crystal size distributions in magmatic systems. J Petrol 39:553–600

Mueller SB, Lane SJ, Kueppers U (2015) Lab-scale ash production by abrasion and collision experiments of porous volcanic samples. J Volcanol Geotherm Res 302:163-172.

Müller A, van den Kerkhof AM, Behr HJ, Kronz A, Koch-Müller M (2010) The evolution of late-Hercynian granites and rhyolites documented by quartz- A review. Earth Environ Sci Trans Roy Soc Edinburgh 100(1–2):185–204

O'Donnell SP (2008) Textural analysis of a rhyolite dike of the southern Oklahoma aulacogen at Medicine Park, Oklahoma. MSc Thesis,

Missouri Univ Sci Tech, United States, p 114 http://scholarsmine.mst.edu/masters_theses/4628
Philpotts AR (2003) Petrography of igneous and metamorphic rocks. Waveland Press, Prospect Heights, Illinois, p 178p
Pittari A, Cas RAF, Lefebvre N, Kurszlaukis S (2015) Alteration Styles in the Orion Central Volcanic Complex, Fort à la Corne Kimberlite Field, Saskatchewan, and their effects on primary volcaniclastic textures: Implications for facies mapping and diamond exploration. Econ Geol 110(1):147–171
Porritt LA, Cas RAF, Schaeffer B, McKnight SW (2012) Textural analysis of strongly altered kimberlite: Examples from the Ekati diamond mine, Northwest Territories, Canada. Canad Mineral 50:625–641. https://doi.org/10.3749/canmin.50.3.625
Powers MC (1953) A new roundness scale for sedimentary particles. J Sediment Res 23(2):117–119
Prothero DR, Schwab F (2013) Sedimentary geology: An introduction to sedimentary rocks and stratigraphy. WH Freeman and Co, New York, p 593p
Shea T, Houghton BF, Gurioli L, Cashman KV, Hammer JE, Hobden BJ (2010) Textural studies of vesicles in volcanic rocks: An integrated methodology. J Volcanol Geotherm Res 190:271–289
Shore M, Fowler AD (1999) The origin of spinifex texture in komatiites. Nature 397:691–694
Sparks RSJ (2013) Kimberlite volcanism. Ann Rev Earth Planet Sci 41:497–528
Sparks RSJ, Baker L, Brown RJ, Field M, Schumacher J, Stripp G, Walters A (2006) Dynamical constraints on kimberlite volcanism. J Volcanol Geotherm Res 155(1–2):18–48
Sparks RSJ, Wright JV 1979. Welded air-fall tuffs. In: Chapin CE, Elston WE (eds) Ash-flow tuffs. Geol Soc Amer Spec Pap 180:155–166
Tait S (1992). Selective preservation of melt inclusions in igneous phenocrysts. Amer Mineral 77(1-2):146-155.
Tepley FJ, Davidson JP, Clynne MA (1999) Magmatic interactions as recorded in plagioclase phenocrysts of Chaos Crags, Lassen Volcanic Center. California. J Petrol 40(5):787–806
Toramaru A (1990) Measurement of bubble size distributions in vesiculated rocks with implications for quantitative estimation of eruption processes. J Volcanol Geotherm Res 43:71–90
Toramaru A (2006) BND (bubble number density) decompression rate meter for explosive volcanic eruptions. J Volcanol Geotherm Res 154:303–316
Uehara D, Cas RAF, Folkes C, Takarada S, Oda H, Porreca M (2015) Using thermal remanent magnetisation (TRM) to distinguish block and ash flow and debris flow deposits, and to estimate their emplacement temperature: 1991–1995 lava dome eruption at Mt. Unzen Volcano, Japan. J Volcanol Geother Res 303: 92-111.
van Zalinge ME, Cashman KV, Sparks RS (2018). Causes of fragmented crystals in ignimbrites: a case study of the Cardones ignimbrite, Northern Chile. Bull Volcanol 80(3):22.
Walker GPL (1972) Crystal concentration in ignimbrites. Contribs Mineral Petrol 36:135–146
Walker GPL (1981) Plinian eruptions and their products. Bull Volcanol 44:223–240
Welsch B, Faure F, Famin V, Baronnet A, Bachèlery P (2012) Dendritic crystallization: A single process for all the textures of olivine in basalts? J Petrol 54(3):539–574
Wentworth CK (1922) A scale of grade and class terms for clastic sediments. J Geol 30(5):377–392
Wentworth CK, Williams H (1932) The classification and terminology of the pyroclastic rocks. Natural Res Counc Rept Comm Sediment Bull 89:19–53
Wiebe RA, Wark DA, Hawkins DP (2007) Insights from quartz cathodoluminescence zoning into crystallization of the Vinalhaven granite, coastal Maine. Contribs Mineral Petrol 154(4):439–453
Wright JV, Smith AL, Self S (1980) A working terminology of pyroclastic deposits. J Volcanol Geotherm Res 8(2–4):315–336

Documenting the Geology of Volcanoes and Volcanic Terrains 15

Mapping, Litho- to Chrono-Stratigraphic Framework and Facies Architecture, and Constructing Facies Models

John V. Wright, Guido Giordano, and Ray Cas

R. Cas et al., *Volcanology*, Springer Textbooks in Earth Sciences, Geography and Environment,
https://doi.org/10.1007/978-3-319-66613-6_15

Plate 15.1 This spectacular angular unconformity is exposed in the coastal cliffs along the west coast of Lipari Island, Aeolian Arc, Italy. But "*what is its volcanological and geological significance?*" Firstly, is the angular discordance the result of tectonic tilting? No, the lower succession of red, oxidised a'a lavas (267–150 ka) actually shows a high, primary depositional dip (foreset bedding or clinoforms), due to the steeply inclined edifice they were erupted onto. Secondly, regarding the nature of erosional surface separating the overlying succession of subhorizontal pyroclastic beds, "*is this local or more extensive?*" In fact, this represents a regionally significant erosional unconformity and associated marine terrace (ca. 125 ka), which formed as a result of glacio-eustatic sea-level rise. Correspondingly, it has larger geological and stratigraphic importance and is correlated through all the volcanoes of the Aeolian archipelago and beyond, rather than having limited, local volcanological significance. (For further explanation see text, and Figs. 15.2f and 15.20)

Abstract

This chapter outlines the major principles and challenges in facies analysis, field survey mapping, developing stratigraphic architecture, and documentation of the geology of volcanoes and volcanic terrains, both modern and ancient. The stratigraphic record of almost all volcanoes can be very incomplete, depending on the preservation potential of each eruptive unit. This is because of the prolonged inter-eruptive periods compared with the short duration of eruption events, and due to the erupted nature and volume of the deposits (e.g. large versus small volume, coherent versus clastic units). Although much of sedimentary stratigraphy is based on correlation of laterally continuous "layer-cake" type units, in volcanic terrains the stratigraphic architecture is much more complex, particularly in proximal, high relief areas on large volcanic edifices, where highly discordant contacts are common. We assess the major factors that influence the preserved geometry of stratigraphic units, including the effects of eruption geometry, effects of high rates of erosion, volcano-tectonic (as opposed to regional tectonic) deformation, and provide a current review of the different approaches (e.g. lithostratigraphic, event-sequence, and seismic sequence or architectural elements) to representing volcanic stratigraphy. Identifying event sequences, or "genetic packages", bound at their tops and bottom by unconformities is essential to understanding the history, evolution, and chrono-stratigraphy of volcanic terrains. However, unconformities are common in volcanic terrains, so it is equally significant to establish their rank or hierarchy and temporal or chronological significance. Routine methods used for the correlation of volcanic successions are briefly summarised, as well as additional tools now being employed to map modern volcanoes and volcanic terrains such as satellite and geophysical imagery. In ancient volcanic terrains facies models have been and remain the chief method to represent volcanic facies architecture and lithostratigraphy.

15.1 Introduction

Understanding the eruption behaviour of volcanoes, the hazards posed and their resource potential, is not just about understanding what is happening on the volcano today but requires an in-depth understanding of the geology of the volcanic system, including its (chrono-)stratigraphic architecture and geological history. Stratigraphy can only be established by mapping the extent of distinctive rock units or associations of them over a large region. To establish a chrono-stratigraphy, units in the stratigraphy then need to be age dated using radiometric or paleontological or geomagnetic dating. However, by establishing a stratigraphic framework, involving a largely mechanical, descriptive, classification procedural approach, only a limited understanding of geological evolution of the volcano can be developed. Establishing an accurate physical stratigraphy becomes much more powerful and valuable if detailed facies analysis, to document the key characteristics (geometry, lithology, sedimentary structures, sediment movement patterns and fossils; Chap. 1) of the lithofacies constituting stratigraphic units, is undertaken. Those facies characteristics are the basis for understanding the processes that produced each facies, and their depositional setting. Physical lithofacies stratigraphy thus has the potential to be converted into an understanding of processes and the process evolution of any rock succession, including volcanoes (Plate 15.1). Once this stage has been reached a visual process focused facies model can be constructed for any rock succession, or any stage in the evolution of that succession, that depicts the essential processes, facies distributions, and the setting.

In previous chapters, we have documented the complexity and variety of processes and deposits that occur in volcanic environments. Time- and length-scales of these processes vary enormously, resulting in multifarious architectures of rock bodies and their facies. These make it impossible to accurately document the geology of volcanoes and volcanic terrains without a full understanding of the volcanic processes associated with the related deposits. For example, the lateral facies variations of ignimbrites that occur across topography may result in significant lithofacies differences (e.g. a valley pond of thick and massive welded facies, versus a veneer of thin and stratified low-grade facies; Chap. 10), which may be very difficult to recognise as related to a single stratigraphic unit or rock formation, or event. At a broader scale, variations in eruption rates and differing spatial location of vents over time may result in very complex geometries of rock packages, controlled by the interplay of primary deposition, topography, secondary processes, and volcano-tectonics.

Mapping the geology of volcanoes in both modern and ancient terrains involves different kinds of challenges. On modern volcanoes, the surface geology may be very visible, or only variably exposed, mostly because of the effects of differing climate, extent of vegetation and anthropogenic covers from region to region around the world. By contrast, the subsurface geology is poorly exposed and other tools, such as geophysical surveys, are needed to accurately reconstruct the subsurface stratigraphy and structure of volcanic edifices. In addition, most geologists are not working in modern volcanic successions and terrains, but are exploration or geological survey geologists trying to make sense of discontinuously outcropping, variably deformed, metamorphosed, and hydrothermally altered ancient successions (Chaps. 13 and 14).

In ancient successions, all of the facies descriptors (geometry, lithology, sedimentary structures, sediment movement patterns, and fossils; Chap. 1) are subject to varying degrees of modification by processes contemporaneous with deposition or post-dating it, or both (Chap. 13). The former processes largely involve modification by erosion and hydrothermal alteration, and the latter protracted hydrothermal alteration, diagenesis, metamorphism, and tectonic deformation. These must therefore be considered before approaching facies analysis and the construction of general facies models. It is clear, on the other hand, that generalised facies models derived from modern volcanoes are necessary reference frameworks to represent significant associations of facies and, if possible, their spatial and genetic relationships, and to summarise, however schematically, the paleoenvironmental and paleogeographic context. If this can be done, then such facies models may have predictive value for ancient successions and in basin analysis studies involving volcanic successions.

To date, facies models have been the chief method of schematically representing volcanic stratigraphy and the relationships between lithostratigraphic units (defined according to the lithological characteristics), and facies analysis of volcanic deposits has been widely applied to develop those facies models (Cas and Wright 1987). The main objectives of volcanological studies until recently have often differed significantly from those of conventional stratigraphic studies on volcanic terrains, and also because volcanologists have largely ignored the possible standardisation of volcanic stratigraphy (Martí et al. 2018). However, in recent years a number of workers have attempted to introduce more formal stratigraphic principles involving the use of lithostratigraphic units. This has mainly been undertaken in studies of Quaternary volcanoes in an attempt to apply "sequence stratigraphy", as employed in the study of normal sedimentary successions (see Sect. 15.4). As applied to volcanic successions it attempts, in its simplest terms, to divide the stratigraphy into genetic packages bounded at their tops and bottom by unconformities (e.g. Plate 15.1). Such packages can often be identified as "eruption sequences", or the deposits of a major eruption phase, after which during a repose period a significant interval of erosion occurs, producing an unconformity at the top of each eruption sequence. These packages are used to provide a chrono-stratigraphic framework for the correlation and mapping of volcanic and associated sedimentary facies to synthesise the volcanic evolution of an area. As reviewed by Martí et al. (2018), so far this appears to have met with mixed degrees of success, leading in some studies to a wide range of names for volcanic units that do not necessarily imply any precise stratigraphic ranking.

In this chapter we outline the major principles and challenges of field mapping, documentation of the geology of volcanoes and volcanic terrains, both modern and ancient. We also introduce the importance of facies analysis and the development of facies models. In the following chapter (Chap. 16), we describe in detail facies models associated with the different volcanic edifices and settings.

15.2 Some Underlying Principles in Understanding Facies Analysis and Mapping of Volcanoes and Volcanic Terrains

15.2.1 Changes in Approaches to Mapping and Establishing Stratigraphy

Mapping of volcanoes and volcanic terrains, whether modern or ancient, has gone through major changes over the last 50 years. Most geological maps of volcanic areas published around the world until the 1960s (and in some cases even later) were mainly based on broad petrological distinctions of lithological or map units (e.g. basalts, dacites, rhyolites, etc.), and somewhat rudimentary lithofacies distinctions such as lavas and tuffs. For many years the "*reworked volcaniclastics*" that are commonly interbedded with lavas and major pyroclastic mappable units, received almost no attention, being either totally disregarded, or appearing on maps or legends grouped together as a single formation encompassing large parts, and even the whole history of the volcano. More recent geological maps have progressively incorporated more and more physical volcanology and geochronology, and understanding of the need to represent the lithofacies that make up stratigraphic or map units. This has allowed significant refinement of the approach to mapping volcanoes, driven by the need to show the evolution of volcanoes in terms of the facies and stratigraphic architecture of the volcanic succession that highlights the changes in time and space of eruption styles, petrology and magma compositions, and volcanic structure.

Unlike in sedimentary terrains, in which stratigraphic studies date back over two centuries, no common mapping approach has been established for volcanic areas. The criteria used in volcanic mapping have not been uniform and the results obtained differ in terms of how such maps are drawn and their ultimate purpose(s). One major challenge in mapping volcanic environments is how to apply a standard litho-stratigraphy, while at the same time being able to use all formal tools available from the *International Stratigraphic Guide* (ISG; Salvador 1994; http://www.stratigraphy.org) and *North American Stratigraphic Code* (NASC; NASCN 2005; http://nacsn.americangeosciences.

org) to also convey the complexity and specific character of volcanic terrains.

We have seen in previous chapters how dissimilar the lithofacies and emplacement mechanisms are that are associated with different eruption and transportation styles, so that the resulting deposits show wide variations in rock types, areal extent and internal changes in lithofacies and thickness. Geomorphological changes are rapid during each eruption, and the rates of change due to rapid emplacement of often very thick and voluminous eruption units do not match any other geological environment on Earth. For example, m- to hundreds of m thick ignimbrite units can be emplaced in time scales of hours or days, levelling or burying pre-existing topographies on an areal scale of hundreds to tens of thousands of km^2 (Chap. 10). For this reason, thicknesses of rock formations in volcanic environments bear no obvious relationships with average sedimentation rate (Fig. 15.1), a concept which intuitively applies, for example, to most marine and non-marine sedimentary environments. Therefore, at the scale of geological maps, we may have extensive rock formations, which represent a single eruption (e.g. one ignimbrite tens of m thick, or a single thick lava unit), and others that represent tens of thousands of years of volcanic activity and multiple eruptions (e.g. a succession of fall deposits and/or thin lava units intercalated with paleosoils and reworked material equally tens of m thick as a whole).

15.2.2 The Relationship Between Stratigraphic Units and (Litho)Facies

Litho-stratigraphic units are distinctive lithological units that can be defined and recognised at a regional or map scale, and be distinguished from the units above and below due to distinctive lithological differences that could be compositional or textural. There is a formal hierarchy of stratigraphic units, from (Super-) Group at the largest scale to Formation (the standard map unit) to Members, and the smallest scale. A Group consists of an association of two or more formations that are distinctive from the Group or Formation above and below. A Member is a distinctive element of a Formation, but not all formations are necessarily subdivided into members; only if there is something distinctive about a particular part of a formation. Formations may or may not be grouped together with other formations to form a group; they can be defined as a distinctive stand-alone map and stratigraphic unit. Some stratigraphic units take a considerable period of time to accumulate; they are thus diachronous rock units. Other rock units are emplaced instantaneously in geological time, they are valuable time markers, and are called time-rock units or isochronous rock units, if and when they can be dated. They are particularly valuable when they extend over a large area, in which case they can be used as time or isochronous correlation marker. Dating of multiple rock units in a stratigraphic succession allows a chrono-stratigraphy to be established for that succession. Stratigraphic units may be in conformable or unconformable contact with units above or below. That is, they may or may not be bound by unconformities at their tops and or bases. Stratigraphic units are named after a distinctive geographic feature (town, river, mountain name) in the region where they occur.

A (litho)facies is a distinctive deposit or part of a deposit that has distinctive, definable characteristics (lithological composition, texture [grain size, clast morphology, sorting], depositional structure, fossil assemblage) that distinguish it from adjacent facies (Chap. 1). For example, in lavas, marginal breccia (auto- and/or hyaloclastite) and the coherent core of the lava (Chaps. 4–6) represent two distinctive facies of lavas. In different situations one, or two may be present. In an ignimbrite (Chap. 10), the facies that make-up ash-cloud surge deposits, various lithic concentration deposits, the main body of the deposit, a mantling ash-cloud fallout deposit, all constitute different facies of the ignimbrite because they all have unique characteristics that can be distinguished and interpreted in process terms. Those individual facies can be laterally discontinuous, and thus are rarely all at a scale that can be represented on a geological map, whereas the ignimbrite as a whole may be representable as a map and stratigraphic unit if it is of a large enough scale and differs from the rock units above and below. The important feature about defining facies and their key properties (Chap. 1) is that they lend themselves to interpretation of the detailed physical and sometimes chemical processes and settings that formed them. Stratigraphic units on the other hand may give a general indication of their origins (e.g. sandstone formation, overlain by a limestone formation), but lack details of processes.

Facies are named after the distinctive, descriptive compositional, textural, and depositional structure features that define them (Chap. 14; Fig. 14.1a, b), not by the genetic origin, which has to be deduced from the properties. Facies of particular emplacement units can be in relatively sharp or gradational contact with the other facies of that emplacement unit. All stratigraphic units consist of facies, usually multiple ones, that help to define the process origins and setting of each stratigraphic unit.

Defining the stratigraphic architecture therefore provides the spatial and temporal (chrono-stratigraphic) framework, within which detailed facies and process analysis can be undertaken. If *all* of the stratigraphy, chrono-stratigraphy *and* the facies analysis have been undertaken conscientiously, then the geological history becomes clear. Only when a detailed chrono-stratigraphy is established for a volcanic system can trends in eruption styles, magnitude,

Fig. 15.1 The thickness and extent of mappable units in volcanic environments does not relate to average sedimentation rates in sedimentary environments. In example **a**, the lower and the upper stratified units represent prolonged periods of volcanic activity characterised by the emplacement of a sequence of multiple pumice fall deposits each separated by a period of repose and alteration to paleosoils; an equal and even greater thickness is represented by the intervening single massive unit, which is a lava flow that was emplaced during a single eruptive event. Hokkaido, Japan. In example **b**, a similar rock thickness is observed, between a lower succession of bedded fall deposits and paleosoils representing tens of thousands of years of activity, and the overlying thick brown massive Orvieto-Bagnoregio ignimbrite (OBi) representing in contrast, a few hours of eruption activity. Both the lower stratified fall succession and the ignimbrite are usually mapped at the same lithostratigraphic rank, even though their time or chrono-stratigraphic significance is vastly different. Civita di Bagnoregio, Central Italy

frequency, patterns, and rates to be evaluated in both the short and long term, which then allows all the potential hazards and their magnitudes to be evaluated. There are no short-cuts, and without detailed mapping, establishing a chrono-stratigraphy and undertaking detailed facies analysis hazard assessment cannot be comprehensive. Logging occasional or a few dispersed graphic logs of a few depositional units that are of research interest alone does not constitute mapping the essential chrono-stratigraphic and facies architecture. At best, it is "cherry picking", and won't provide the all-important understanding of the event chronology framework necessary for hazard assessment.

Stratigraphy is however more complex than simply outlined above. In addition to litho-stratigraphy and chrono-stratigraphy, an event-based approach to stratigraphy was developed to accommodate the complex stratigraphic architecture of continental margins in the form of sequence stratigraphy, which also has a defined hierarchy of units. Where seismic profiles are available, but little other outcrop control exists, a developing field of seismic stratigraphy has also been developed to address this problem. These newer approaches to stratigraphy as applied to volcanic successions and settings are discussed in detail in Sect. 15.4.

15.2.3 The Importance of Erosion and Unconformities in Volcanic Terrains

The concept of unconformities is most commonly tied to sedimentary basin settings and stratigraphic successions, representing regionally extensive periods of erosion or non-deposition, usually in response to eustatic sea-level changes, and/or tectonic movements of the crust. In the case of sea-level induced unconformities, sea level must fall, causing regression and erosion on newly exposed coastal areas. When sea-level rise and transgression occur subsequently, sedimentary strata are deposited on the unconformity, commonly producing disconformities between parallel bedded sets of strata. Tectonically induced unconformities involve uplift of the crust, which causes increased elevation above erosional base level, triggering erosion of uplifted areas. If uplift is simply block uplift, without tilting or folding, disconformities can also form, but if the tectonic uplift involves tilting or folding of strata, followed by erosion and then eventually sedimentation on the unconformity, the unconformity is an angular unconformity, separating sets of strata dipping at different angles.

Unconformities do not form in most sedimentary basins if sedimentation is more or less continuous, and if eustatic sea level and tectonic crustal movements do not occur. However, in volcanic environments sedimentation is not continuous, but is episodic, resulting from episodic eruptions, separated by long periods or repose (years to > a hundred thousand years), during which erosion and/or soil formation can occur. In addition, depending on the setting, volcanic systems can be affected by eustatic sea-level changes (Plate 15.1), and regional tectonic events. As a result, unconformities in volcanic terrains need to be analysed and subdivided into *volcanic repose unconformities*, *eustatic unconformities*, and *tectonic unconformities*. Volcanic repose unconformities usually extend over a volcanic edifice, but may also extend beyond, if dispersal of deposits and erupted volumes are large enough. Eustatic unconformities are global in extent and tectonic unconformities are regional in extent. The processes that control the formation of unconformities are divided into those that are *"autogenic"* and produced by local or internal processes (e.g. uplift, subsidence, volcanism), while those controlled by (eustacy) and regional (tectonics, climate) are refered to as *"allogenic"*. As apparent from the above, autogenic controls operate on smaller spatial and temporal scales than allogenic controls. However, distinguishing the two in practice can be especially difficult even in modern volcanic systems.

On large volcanic edifices and volcanic provinces periods of repose generally exceed the duration of periods of active volcanism, and in most cases they are much longer than eruptive periods (Chap. 12; e.g. Vessell and Davies 1981). During repose periods surface sedimentary erosion, mass wasting, volcaniclastic sediment transport, and soil formation are the dominant processes affecting the volcanic system. As discussed in Chap. 12, this is particularly the case in high relief proximal volcanic settings, particularly after the eruption of large volumes of unlithified pyroclastic deposits onto the landscape, and in rainy climatic settings (e.g. after the 1991 explosive eruption of Mt Pinatubo in the Philippines; Chap. 12). In proximal areas, high erosion rates of unconsolidated pyroclastic deposits rapidly led to the removal of large volumes of volcanic debris and formation of erosional unconformities, whereas in surrounding low gradient environments the transported volcanic sediments accumulated conformably on pre-existing strata. However, some near vent eruption products such as thick lava flows and domes are more resistant to erosion and could be largely or partially preserved in the rock record, whereas the record of contemporaneous pyroclastic sequences are often removed in proximal areas, but can be preserved as a succession of tuffaceous volcanic sediments in the surrounding sedimentary basin settings. Erosional unconformities on the steeper parts of edifices, therefore, laterally transform into, and correlate with depositional sequences in surrounding low relief areas. Unless that whole succession from proximal to distal sedimentary settings has been mapped, the selectively preserved record of eruption events in the proximal setting on the volcano is misleading, incomplete, and biased to the preservation of erosionally resistant rock units, mostly lavas, and domes. That has major consequences for hazard assessment of that volcano, giving the impression that its eruption activity is dominated by effusive activity and little or no explosive activity. The frequency of eruptions is also likely to be underestimated if the complete stratigraphy on *and around* the volcano has not been mapped and established.

The identification of *erosional unconformities* formed during periods of repose or quiescence and their extent is therefore important (Plate 15.1), as is the relationship between proximal unconformities and laterally contemporaneous and equivalent depositional sequences (e.g. Lucchi et al. 2013). This has become an essential tool in reconstructing the alternation of *constructional* and *degradational* periods (see Table 15.1; Fig. 15.2), and their interpretation in terms of interplay between volcanic activity, tectonics and volcano-tectonics, and climate changes (De Rita et al. 1998, 2000; Palladino et al. 2010; Lucchi et al. 2010; Lucchi 2013; Forni et al. 2013; Martí et al. 2018). However, not all unconformities are erosional. Periods of repose during which little or no erosion occurs, except in the channels of low gradient rivers or creeks, nor any deposition occurs, but soil formation occurs if climate is conducive, are very common in low gradient terrains, producing non-depositional unconformities such as disconformities and paraconformities.

Erosional and non-erosional unconformities are always present in volcanic successions so that in order to understand their origin it is necessary to assess their extent both spatially and chronologically, and to establish their significance, or hierarchical level in the evolution of a volcano (see

Table 15.1 Stratigraphic types, categories, and unit-terms in use to map and organise volcanic successions (Adapted from Lucchi 2013)

Type	Category	Definition	Unit-terms	Purpose of study
Time-stratigraphy (*interpretive-not mappable*)	Volcanic activity units	Periods of relatively continuous volcanic activity separated by intervals of quiescence with different duration and size (Fisher and Schmincke 1984)	Eruptive period (*kyr-Ma*), Eruptive epoch (*decades-kyr*) Eruption (*days-yr*), Eruptive phase (*min-days*), Eruptive pulse (*sec-min*)	Determination of the main steps of eruptive history
	Eruption units	Volcanic material emplaced during a single eruptive pulse, eruptive phase or eruption (Fisher and Schmincke 1984)	Lava flow, fallout, dilute and dense pyroclastic density current, lahar (etc.) eruption units	Information regarding the eruptive behaviour (e.g. processes of fragmentation, evolution of involved magmas) and emplacement mechanisms
Rock-stratigraphy (*objective or semi-objective*)	Lithostratigraphic units	Rock bodies (sedimentary or volcanic) defined and recognised on the basis of observable and distinctive lithologic properties and stratigraphic relations in a vertical succession (Salvador 1994)	Group, Formation (*primary unit*), Member	Description of rock types and reconstruction of rock successions; basic units of geological mapping
	Lithofacies	Lateral (or vertical) subdivisions of a lithostratigraphic unit based on distinctive lithologic characters (e.g. stratification, grain size, grain shape, sorting, fabric, composition) and facies associations (Cas and Wright 1987)	Near-vent, proximal, intermediate, distal facies	Description of the lateral (or vertical) variations of pyroclastic products induced by time, paleotopography and increasing distance from the source
	Unconformity-bounded stratigraphic units	Rock bodies bounded below and above by significant and demonstrable unconformities, which are surfaces of erosion and non-deposition representing a hiatus in the stratigraphic succession (Salvador 1987; 1994; NACSN 2005)	Supersynthem, Synthem (*primary unit*), Subsynthem *OR* Allogroup, Alloformation (*primary unit*), Allomember	Stratigraphic correlation and synthesis on a local to regional scale; not mappable
	Lithosomes (*informal*)	Rock bodies with a distinctive 3D geometry that may interdigitate with one or more adjacent bodies (Lucchi et al. 2010; Lucchi 2013)	Lithosome (*dimensionless units*)	Identification of independent eruptive vents and volcanic edifices, or other sedimentary bodies (e.g. slope detrital cones, marine terraces)

(continued)

Table 15.1 (continued)

Type	Category	Definition	Unit-terms	Purpose of study
Seismic-stratigraphy (*objective or semi-objective*)	Seismic sequences	As observed on seismic profiles, relatively conformable successions of genetically-related strata, bound at the top and bottom by unconformities, hiatuses and their correlative conformities (Vail et al. 1977)	Supersequence, Sequence *(primary unit)*, Parasequence, Systems tracts, Seismic facies	Stratigraphic correlation and synthesis on a local to regional scale; to date mainly applied volcanic passive margins and marginal basins where high resolution seismic data are available
	Seismic architectural elements	Set of interrelated rock units or elements observed in seismic data that form a volcanic system, comprising pre-magmatic, syn-magmatic and post-magmatic sequences separated by key surfaces (Bischoff et al. 2017; Bischoff 2018)	Sequence, Stage, Architectural element	Applicable at the volcano-scale but dependent on availability of high resolution seismic data; elements mappable in 3D using seismic, well and analogue volcanic facies models

Sect. 15.4 for further discussion; De Rita et al. 1998; Martí et al. 2018; Fig. 15.2). For example, an unconformity which is due to a small pause during an eruption, or due to changes in emplacement mechanism (e.g. the transition from a fallout to a flow phase and/or the erosion that a pyroclastic flow or a flow pulse may exert on the underlying deposits) can only extend (be visible) within the rock unit that encompasses that eruption sequence (Fig. 15.2a). Conversely, an erosional surface that is due to a more or less prolonged period of repose or quiescence, accompanied or not by climate variations and/or tectonic activity, will cut through various rock units and extends at the scale of a sector of the volcano or at the scale of the entire volcano and even beyond, depending on the intensity and duration of the geological processes that produced that erosional activity and unconformity (Fig. 15.2b–h).

In sedimentary successions, angular unconformities, especially at regional scales, can often be ascribed to a deformational event of an older succession, followed by a period of erosion before sedimentation on the sub-horizontal erosion surface (unconformity surface) begins. This is also common in volcanic terrains which can be highly tectonically active. However, in volcanic terrains, angular unconformities can also form when an older bedded succession of lavas, pyroclastic deposits, or volcanic sediments is deposited with an initial primary dip (e.g. slopes of cones and domes, talus aprons, deltas), then that succession is later eroded and a new succession is deposited on the erosional unconformity surface (e.g. Fig. 15.2b–d, f–h). Such angular unconformities tend to be more local in extent, ranging from very local to the extent of a volcanic edifice.

In addition, the erodibility and the preservation potential of lavas and volcaniclastics (with the exception of welded tuffs) are completely different and must also be taken into account. For example, effusive volcanoes are less prone to erosion than explosive volcanoes, and lavas and welded pyroclastic deposits are more resistant to erosion than non-welded pyroclastic deposits and sediments. The stratigraphic record therefore disproportionately preserves resistant volcanic units. The inherent incompleteness of the stratigraphic record of volcanic successions is more relevant for small volume and explosive eruptions than it is for large volume and effusive eruptions. Moreover, eruption style and magma viscosity affect morphology, so that low aspect ratio edifices (e.g. shield volcanoes, lava flow fields) are eroded at slower rates than are high aspect ratio edifices (e.g. stratovolcanoes; see Chap. 16).

15.2.4 The Importance of Volcanic Sediments on and Around Volcanoes to Understanding Their History

As we have already stressed above, the effects of epiclastic reworking in volcanic terrains have now been universally accepted as being important in contributing to the stratigraphy of volcanoes (e.g. Cas and Wright 1987; Fisher and Smith 1991; Manville et al. 2009; Chap. 12). The products deserve to be represented on geological maps, including for the reason that volcaniclastic sequences preserve the record of eruptions whose deposits have been eroded from the steeper parts of volcanic edifices. Epiclastic reworking of volcanic material may be prolonged during repose periods, and/or very intense during and immediately after an eruption, especially after major explosive eruptions. Large volumes of volcanic debris can be mobilised in the form of variably concentrated hyper-concentrated and debris flows (Chaps. 12 and 16). In this regard, it is important to attempt

(a)
ignimbrite
fall deposits
VSN
SVL
(b)
lower fall sequence
FogoA
(c)
(d)
(e)
Us
Sh
Ku
(f)
stratified pyroclastics
U
beach conglomerate
U
clinostratified lavas

◀ **Fig. 15.2** Erosional unconformities of different hierarchies in volcanic settings. **a** Syn-eruptive unconformity due to a change in eruption mechanism from sustained fallout to ignimbrite-forming phase in the Villa Senni Formation (VSN), Colli Albani volcano, central Italy. This unconformity surface can only be observed within the formation. By contrast, the basal unconformity, evident from the light brown paleosoil developed at the top of the lower scoria fall succession (SVL, Fontana Centogocce Formation), is due to a prolonged absence of deposition in this sector of the volcano and can be identified over a wide area not necessarily covered by the same rock formation. **b** Erosional unconformity at the base of the 4.6 ka Fogo A plinian succession, which cuts deeply through an older (≥ 8 kyr) stratified succession of plinian and sub-plinian fall deposits, all erupted from Água de Pau caldera, São Miguel, Azores, Portugal. **c** Partially preserved, degraded scoria-tuff cone (~ 2.7 Ma) buried below Upper Bandelier Tuff ignimbrite erupted from Valles Caldera at 1.26 Ma, Frijoles Canyon, Pajarito Plateau, New Mexico, USA. This unconformity marks a regionally extensive erosional surface on the southern side of the ignimbrite shield infilled by Upper Bandelier pumice flows which buried an early Pleistocene canyon of the Rio Grande cut deeply through the Cerros del Rio basaltic volcanic field. Inset in **d** shows erosional, discordant contact between scoriaceous units, well-bedded phreatomagmatic beds (Cerros del Rio cinder cone), and ignimbrite; a regolith of mafic cobbles and boulders (arrow) marks the unconformity. **e** This outcrop is located along the Kuttara volcano slopes, Hokkaido, Japan. It shows the superposition of units erupted from three different volcanoes separated by quiescence intervals: a lower aggradational stratigraphy of pyroclastic units from Kuttara (Ku) cut by a deep incision filled by an ignimbrite related to the collapse of the nearby Shikotsu caldera at 42 ka (Sh), and covered by Holocene white ash of Usu volcano (Us). **f** This marine terrace is exposed in the same area as Plate 15.1 on Lipari, and formed by relative sea level rise during Late Quaternary glacio-eustatic fluctuations. The base of the beach conglomerate (at about 30 m a.s.l.) is marked by the same erosional unconformity ~ 125 ka (U_I see Fig. 15.20; Forni et al. 2013). Away from coastal areas, the unconformity is represented by a subaerial erosion surface formed during a time-equivalent volcanic quiescence. A second unconformity (ca. 75 ka) is found at the base of the stratified pyroclastics (U_{II}); this formed as a result of subaerial erosion during sea-level fall. A period of volcanic quiescence followed prior to emplacement of the oldest part of the pyroclastic succession at 70–56 ka, or Brown Tuffs, erupted from nearby Vulcano. Both unconformities were later uplifted to their present day elevation by a combination of regional and volcano tectonics. As the processes responsible for the formation of the terrace are allogenic in nature, and of global scale (Marine Isotope Stage 5e), the resulting unconformities can be recognised well beyond the volcano and must be given the highest rank in the hierarchy of unconformities. **g** An angular unconformity (U) between the steeply dipping, bedded basaltic pyroclastic deposits of older "cinder cones of the Akrotiri Peninsula" ($\sim$400+ ka?) and the overlying, flat lying 184 ka Lower Pumice 1 eruption sequence near Cape Balos, Akrotiri, Santorini, Greece. The dips of the older pyroclastic sequence represent original cone-forming depositional dips, not deformation. **h** 300 m west of **g**, the same unconformity as in **g**, labelled U3, is shown (top left of photo). The basaltic pyroclastic cone sequence is in angular unconformable contact (U1) with subhorizontally bedded rhyodacitc "tuffs of Akrotiri" (white; >600 ka) on the left, and both successions are unconformably (U2) draped by a scoria cone sequence on the right, probably also part of the Akrotiri cinder cone sequence. There are thus three visible angular unconformities in a small area

Fig. 15.2 (continued)

to distinguish between *syn-eruptive* remobilisation, and variably long *inter-eruptive* reworking (Chap. 14).

Syn-eruptive remobilisation or *resedimentation* is here defined according to Smith and Fritz (1989), and applied to processes truly contemporaneous with, or occur immediately after the eruption. Remobilisation will continue for as long as an excess of available volcanic debris relative to normal background sediment supply levels controls the reworking processes, (Chap. 12). This syn-eruptive reworking and resedimentation phase may extend for several years after the eruption depending on climate, volume, and type of erupted deposit. For example, during the 1991 eruption at Mt Pinatubo, several km^3 of pumice and ash are deposited in a few hours by pyroclastic fallout and flows, totally inundating the pre-existing drainage network on and around the volcano. This material was then reworked and resedimented during each monsoon season for 15 years after the eruption (Pierson et al. 1992; Newhall and Punongbayan 1996;

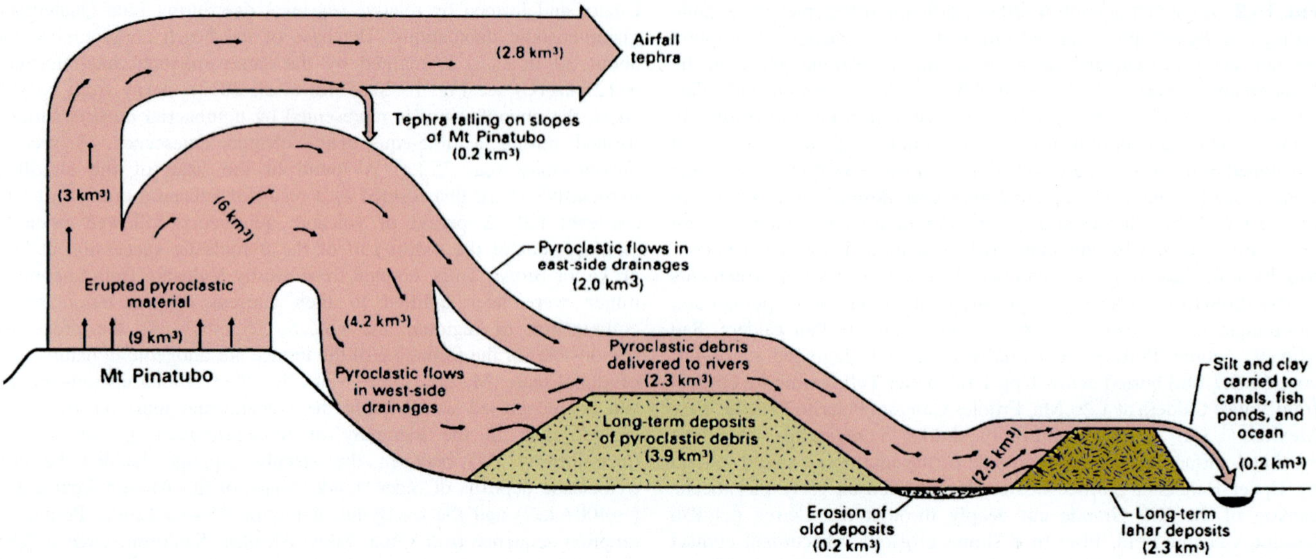

Fig. 15.3 Schematic representation of pyroclastic mass partitioning following the 1991 eruption of Mt Pinatubo, Philippines. It is shown that out of the 6.2 km^3 deposited as primary pumice and ash flow deposits, only 3.9 km^3 were in place the year after, and 2.3 km^3 were already eroded, transported away, and deposited in the lower reaches of rivers by hyperconcentrated flood flows. (After Pierson et al. 1992)

Fig. 15.3). Similar post-eruption erosion processes after other explosive eruptions (Santá Maria, Guatemala, 1902; Fuego volcano, Guatemala, 1974, 2018, which repeat similar cycles that have occurred during the past 20–30 kyr; Mt St Helens, USA, 1980; Mt Galunggung, Indonesia, 1982) have been well documented, and are described further in Chaps. 12 and 16.

As a more general rule, these studies suggest strong erosion along the upper reaches and main drainages of a volcano will occur rapidly; in the case of Mt Pinatubo this was driven by monsoonal and typhoon-generated rains during the first rainy season, and then every monsoonal season after that for 15 years. However, it takes up to several years to restore a drainage network, producing very deep gullies that only cut the new deposits which will be to the maximum depth of the pre-existing valley bottom (Fig. 15.4a), as base level has not changed during the single eruption. Resulting hyper-concentrated flows and debris flows are confined in such deep gullies until they fill up and over-spill onto the banks (Chap. 12). Their deposits are totally derived from the primary deposit (Fig. 15.4b), at least for the first few years. In distal areas, lahars transform to conglomerate and sand flash flood or hyper-concentrated flow deposits, still entirely derived from the primary deposit upslope, and they become aggradational over the pre-existing landscape (Fig. 15.4c). With time, the full restoration of the drainage network and the progressive stabilisation of the remnants of the primary deposit, induce a progressive decrease in availability of volcanic debris and the transition to normal rates of surface processes and sedimentation. This transition to full inter-eruptive conditions is testified by the re-incision of previously emplaced aggradational sequences. Deposits in this phase show cut and filling relationships and are made of polymictic fluvial conglomerate and sand deposits indicating long periods of quiescence (Fig. 15.4c, d), along with formation of *paleosoils* on topographic highs (Fig. 15.2a). The depth of re-incision illustrates the balance between average deposition and average erosion rates, and changes in base level (e.g. climate-driven and tectonic/volcanotectonic-driven). Significant hiatuses in deposition may then result in erosional unconformities which cut deeply through the volcanic succession (Plate 15.1 and Fig. 15.2b, c).

As indicated in Chap. 14 (Sect. 14.4), the ability to differentiate between syn-eruptive resedimentation (epiclastic *sensu lato*) and protracted reworking (epiclastic *sensu stricto*) becomes increasingly difficult to separate further back into the geological record, as context diminishes with regard to the spatial relation to a volcanic vent and direct observational data of eruptive activity.

15.2.5 Factors that Affect the Preserved Geometry of Emplacement and Stratigraphic Units in the Stratigraphic Record

Volcanic landscapes consist of both *constructional* and *degradational* or erosional landforms, and they usually co-exist at any time. Constructional volcanic landforms result from volcanic eruptions that build up the relief of the landscape. They can include high aspect ratio features such as lava domes, scoria cones, lava shields, ignimbrite and flood lava sheets and plateaus. The factors that affect their

Fig. 15.4 Sedimentary volcaniclastic deposits around some modern volcanoes. **a** Spectacular view south of the sloping pumice plain surrounding the northern sector of Mt St Helens, USA, which collapsed following flank failure on 18 May 1980 (see Fig. 15.9). Pumice and ash flow deposits are cut almost down to their base and the resulting deep gullies show various orders of terraces. The terrace to the right is made of debris flow deposits, while the present-day deposits at the bottom of the valley vary from flash flood to fluvial stream flow as a response to the deposits (progressive decrease in the availability of volcanic debris. **b** "Lahar" deposits, probably debris flow deposits, of the 1887 eruption of Cotopaxi volcano, Ecuador. **c** Distal resedimented pumiceous sediment deposits related to the 42 ka caldera-forming eruption at Shikotsu, Hokkaido, Japan. Note the lower planar stratified aggradational succession that is ash dominated and emplaced by flash flood, hyperconcentrated flow processes, the upward transition to a mild erosional unconformity and to coarser channelised, tractional fluvial deposits. Such a transition indicates the progressive reduction of sediment load that ultimately leads to the full re-incision of the aggradational sequence, and the restoration of the normal fluvial environment. **d** Inter-eruptive periods are dominated by deposition of polymictic fluvial deposits, for example, at Barranco del Rio, Tenerife, Canary Islands, Spain. Note the imbrication of large volcanic pebbles and cobbles

ability to build relief include the rheological properties of erupting lavas, the erupted volumes, the durations of eruptions, and the life span of the volcanic system. For pyroclastic deposits, the eruption volume is the major factor. Constructional positive relief volcanic landforms can then act as barriers to dispersal of eruptive material but they also represent confining topography against which younger deposits can pond, thus affecting the geometry of those younger deposits (e.g. Fig. 15.5a).

As discussed above and in Chap. 12, erosional surface sedimentary processes are activated or accelerated in response to most eruptions, by trying to cut newly formed, elevated relief and deposits down to pre-eruption topographic levels by various erosional processes. In high relief proximal terrains, this is most acutely expressed by the rapid incision of erosional topography such as steep-sided valleys and canyons (e.g. Fig. 15.5b), often re-incised into pre-existing valleys which were filled by say a sequence of newly erupted pyroclastic deposits. The rate of incision is mostly controlled by the erodibility of newly erupted deposits (e.g. non-welded pyroclastic deposits versus welded pyroclastic deposits and lavas).

The valleys not only act as conduits for through flow of volcanic sediment eroded from upslope and bounding valley ridges, but as sediment flux begins to wane, they become repositories or traps for deposits that can begin to fill them. These deposits can include more volcanic sediment, or new eruption products such as lavas and pyroclastic flow deposits, which selectively find the valleys to flow into. Being valley confined, the geometry of such deposits in cross-section, could be V- or U-box-shaped, and thus they are not very laterally extensive (e.g. Fig. 15.5c), except possibly down

Fig. 15.5 Examples of how constructional and degradational/erosional topography control the geometry of deposits that are subsequently emplaced onto that topography. **a** Scoria cone near Fasnia, Tenerife, Canary Islands, Spain, acting as valley confine (on left), with older eroded basalt lavas forming the right valley confine, into which multiple pyroclastic flows were funnelled, filling the paleovalley with semi-draping, on-lapping thin ignimbrites. **b** Steep sided, erosional canyon on the upper slopes of Mt Taranaki/Egmont stratovolcano, New Zealand, into which future eruptive products and sediments will be channelled, and could fill the canyon. **c** Steep-sided-box canyon filled by alluvial volcanic sediments, bounded by sub-vertical sides, Poris, Tenerife

Fig. 15.5 (continued)

slope down the valley axis, unless the valley filling units are very voluminous and they spill over the valley margins (e.g. Fig. 1.10a, b). Their lateral contacts are often very steep unconformable contacts against the valley sides. Those deposits can be re-incised multiple times producing valley-in-valley packages of deposits bounded by steep sub-vertical contacts that are not faults (Figs. 1.8 and 1.9).

In other words, the depositional surface on which new eruptive deposits are laid can be very irregular due to the existence of constructional and/or degradational/erosional topography, which will then affect the depositional geometry and across-strike continuity of newly erupted units.

15.2.6 Assessing the Contact Relationships Between Depositional/Emplacement/Stratigraphic Units

The above indicates that contacts between individual emplacement or depositional units and stratigraphic units could be conformable or unconformable. However, this involves an interpretation, that is best left until the contact relationships have been assessed based on considering the characteristics of the contact and the facies characteristics of the units that are in contact.

We suggest a descriptive approach before interpreting the contacts, using just the geometrical relationships between two units. This involves sub-dividing contacts initially into those that are concordant or discordant with subjacent strata. Then based on the details of the concordant contacts those contacts could be stratigraphic, intrusive, or faults. Possible concordant stratigraphic contacts include *conformable contacts* (Fig. 15.6a), with essentially no time break between units, *disconformable contacts*, with some erosional relief, perhaps with clasts of the lower unit at the base of the upper unit, or soil horizons on the top of the underlying unit, representing time breaks, but with all units being deposited parallel or co-planar to each other (Fig. 15.6b, c). *Paraconformities*, where two stratigraphic units are superposed parallel to each other, but the contact represents a significant time break during which no erosion has occurred, or soil horizons have formed may indicate such a time break (e.g. Fig. 15.6d). *Intrusive sills* also have largely concordant contacts along both their bottom and top contacts (Fig. 15.6e), but commonly have some cross-cutting features such as branching apophyses, and inclusions of the host rock (Chap. 8). *Bedding parallel faults* such as low angle thrusts may also juxtapose two parallel units of different ages on top of each other with essentially parallel dips, but will usually show some cataclastic textures along the contact.

(a)
(b)
(c)
(d)
P
P
P
P
(e)
(f)
(g)
(h)

◀ **Fig. 15.6** Different types of concordant (**a**–**e**) and discordant (**f**–**h**) relationships. **a** Conformable relationships between magmatic scoria fallout (dark layers) and phreatomagmatic base surge deposits (cream and grey layers) in the maar rim sequence of the 35 ka Tower Hill maar succession, showing no evidence for any time breaks (no paleosoils, nor epiclastic erosion). Newer Volcanics Province, Victoria, southeastern Australia. **b** Disconformable to paraconformable contacts between pyroclastic fallout, flow and surge depositional units (cream colour) and basalt lavas (dark grey layers) that constitute members of the <600 ka Diego Hernandez Formation, and younger lavas. Diego Hernandez caldera wall, Tenerife, Canary Islands, Spain. Each unit was emplaced instantaneously, indicating that the boundaries between each unit represent a significant time break of thousands of years in most cases, often represented by soil horizons. **c** 3D panorama of succession of pyroclastic flow deposits erupted from the Pleistocene-Holocene Ilopango caldera, separated by disconformities represented by low relief erosional paleo-channel surfaces on which soil horizons formed before being infilled by a pyroclastic flow deposit. Pan Pacific Highway, near San Salvador, El Salvador. (Photo by Steven Ralser, https://4lakesphoto.com). **d** Paraconformity (P), represented by a soil horizon without obvious erosion, separating a pyroclastic flow deposit beneath from stratified surge deposits with bomb sags above, erupted from the Pleistocene-Holocene Ilopango caldera. The succession has been normal faulted. Location as in (**c**). (Photo by Chuck DeMets, University of Wisconsin-Madison). **e** Several bedding parallel Archean komatiitic basalt sills intruding black mudstones, which in various places preserve peperite textures, indicating syn-depositional intrusion. The middle sill at the far end has apophyses branching both upwards and downwards into the mudstones (see Fig. 7.15e). Dundonald Beach, Abitibi Greenstone Belt, Ontario, Canada. **f** Sloping erosional surface or local angular unconformity, separating two oppositely dipping sequences of Late Pleistocene pyroclastic fallout deposits. The unconformity coincides with a period of glacial advance ~20 to 18 kyr ago. Chimborazo volcano, Ecuador. (Photo by Lee Siebert; *Source* Smithsonian Institute). **g** Buttress unconformity (vertical contact) between one margin of an eroded constructional rhyolite lava dome (left) and onlapping, bedded continental fluvial red beds (right), which progressively piled up against the dome and buried it. There are trails of eroded rhyolite clasts in the sedimentary rocks. Late Devonian Boyd Volcanic Complex, Eden, southern New South Wales, Australia. **h** Vertical basalt feeder dyke cross-cutting the subhorizontal stratification of the host scoria cone. Madeira, Portugal. **i** Sub-vertical normal fault scarp, against which mass-wastage collapse breccias (left) consisting of volcanic clasts, have accumulated. Pan Pacific Highway, near San Salvador, El Salvador. (Photo by Chuck DeMets, University of Wisconsin-Madison)

Fig. 15.6 (continued)

Discordant contacts include stratigraphic unconformities, which include angular unconformities, with packages of strata dipping at very different angles to each other (Plate 15.1a). In sedimentary successions, *angular unconformities* usually indicate a regionally extensive tectonic event that causes strata to be deformed (either tilted or folded), uplifted, eroded and then overlain by sub-horizontal strata, so that the two sets of strata have dips that are at an angle to each other. Angular unconformities can also be produced by global eustatic sea-level regression, leading to an erosion surface, followed by transgression and deposition of a new sequence (e.g. Plate 15.1). Such events can also occur in volcanic terrains, but the younger depositional sequence above the unconformity could be an eruption sequence prograding over the unconformity (Plate 15.1). In addition, volcano-tectonic processes at various scales (see next section), such as near surface intrusion of a thick sill, laccolith or cryptodome, or magma chamber tumescence and detumescence, may cause tilting and faulting, that could then trigger erosion, formation of an angular unconformity at the scale of the volcano or part of it, as new strata are deposited on and parallel to the erosional surface. In addition, just a period of erosion after an eruption, followed by renewed eruption activity (from a new vent location), and deposition on a sloping paleotopographic erosional surface may produce more local erosional angular unconformities (e.g. Fig. 15.6f).

In addition, *buttress unconformities* represent infilling of steep-sided box canyons with younger deposits, creating one or two sub-vertical contacts that are not faults, but are a steep erosional unconformity (Fig. 15.5c). Steep sub-vertical unconformities can also form when steep-sided constructional volcanic features such as lava domes are progressively eroded and simultaneously on-lapped by a younger sedimentary succession, forming a steep onlap unconformity (Fig. 15.6g). Discordant unconformities where the footwall sequence beneath the unconformity is an unroofed and eroded plutonic body is called a *non-conformity*, and strata above the non-conformity may contain debris eroded from the subjacent pluton. If there is relief on the unconformity on the pluton, the succeeding strata may also on-lap the unconformity, and it could be both a non-conformity and a buttress unconformity.

Steep discordant contacts could also be intrusive contacts (e.g. dykes, dills, laccoliths; Fig. 15.6h; Chap. 8), or faults (Fig. 15.6d, i). In the subsurface, faults displace unit boundaries through normal (Fig. 15.6d), reverse or strike-slip motions, but they can also have significant surface topographic relief, forming topographic constraints or traps into which post-faulting deposits are emplaced (Fig. 15.6i). It should be noted that inclined dykes or dills have the effect of displacing stratigraphy vertically on either side of the dill, much like a reverse fault would, but without significant lateral displacement (Fig. 8.8b). Steep discordant contacts may be very difficult to locate and correctly interpret in old volcanic successions in regions where there is little outcrop and only spaced diamond or chip core available to interpret the geology (see Sect. 15.8).

15.3 Volcano-Tectonics

A major challenge in volcanic terrains is recognising deformation linked to contemporaneous volcano-tectonics, and to distinguish this from deformation related to longer term, regional tectonic processes. Magma migration requires deformation of the crust at various rates and length scales (Chap. 8). For example, the emplacement of a sill or a laccolith at shallow depths (≤ 4 km below ground surface) may produce a symmetrical uplift and bulging of the overlying crust, and may result in accelerated erosion rates at the surface over a few thousand years. A striking example is the formation of the marine terrace at La Starza in the Campi Flegrei resurgent caldera, Italy (Fig. 15.7; Isaia et al. 2009; Marturano et al. 2018).

However, where present, local marine and/or fluvial terraces (controlled by autogenic processes) can only be distinguished from those related to (allogenic) global eustatic sea-level changes or regional tectonic uplifts, based on rate of uplift and the lateral extent (local *versus* regional/global). Both are only likely to be detectable in modern, active volcanic settings. Volcano-tectonic uplift rates are faster than "normal" (on average up to an order of centimetres or decimetres per year, although, with significantly higher peaks), and areas of uplift will be relatively localised (usually the wavelength is 1–10 km), though this also depends on the size and depth of the intrusion (Urbani et al. 2020). Nonetheless, shallower emplacement of cryptodomes may produce significant uplifts, faults, and fractures in very short times (days, months), though very localised in space (Fig. 15.8; Chap 6). More spectacular was the deformation associated with the 1980 Mt St Helens cryptodome that led to the flank failure and lateral blast on May 18 (Fig. 15.9; Moore and Albee 1981; Chaps. 8, 12).

At some calderas, uplifts related to shallow intrusions are impressive and eventually lead to anelastic deformation producing faults that bound km-scale resurgent blocks (Chap. 16). The extension of such resurgent faults is limited to the inside of the hosting caldera (i.e. they do not cross the caldera rim boundary) and uplifts may reach a magnitude of a thousand m or more in tens of thousands of yr (Fig. 16.68). At the Mt Epomeo resurgent dome on the Italian island of Ischia (Fig. 16.70), uplift has been >1000 m in ca. 30 ka, at an average rate of 3–4 cm yr^{-1}. Contemporaneously, such resurgence processes induce impressive geomorphological changes that generate unconformities and deposits, including reversal of drainage (i.e. from a caldera basin floor to a topographic high) and huge mass wastage processes (e.g. major landslides, sector collapses, and accelerated lahars).

Upward migration of magma produces dykes that are intimately associated with brittle deformation, as they both trigger and intrude fractures and faults (Chap. 8). Dykes may follow pre-existing structures if extension allows intrusion along favourably oriented faults and fractures. The upward propagation of dykes also triggers the neo-formation of volcano-tectonic fractures and faults, usually extensional and parallel to the extension of the dyke, to accommodate the uplift caused by the upward migration of magma (Fig. 15.10). In a generic sense, dykes are stress indicators (Chap. 8), as they are parallel to the σ_1–σ_2 plane and orthogonal to the σ_3 (i.e. the direction of minimum stress, or the direction of maximum extension). In volcanic domains

Fig. 15.7 Marine terrace of La Starza, Bay of Pozzuoli, Naples, Italy, famously described by Charles Lyell in his "*Principles of Geology*" over 150 years ago. This presently stands at about 30–55 m a.s.l., and is composed of marine sediments originally deposited in water depths of 30–50 m b.s.l. They display a maximum elevation of ~30 m a.s.l., thus the total uplift onland can be estimated to be of the order of 60–80 m. This has occurred over the last 12 kyr, with peaks in uplift related to intrusion of magma at some 3–4 km depth prior to major eruptions; e.g. as at 4 ka when a few tens of m of uplift preceded an intense period of volcanism (Isaia et al. 2009; Marturano et al. 2018). Archaeological remains found at water depths of ~10 m along the western coast of the Pozzuoli Bay indicate that the entire area underwent subsidence since Roman time. The edifice in the background is Gauro (or "majestic" in Greek) cone, located along the edge of the Campi Flegrei caldera

Fig. 15.8 Example of volcano-tectonic faults triggered by an intruding viscous, rhyolite into unconsolidated overburden, Los Humeros volcano, central Mexico. **a** Panoramic view of fault planes developed at the time of the exhumation of a rhyolite cryptodome (below, off photo) within the overlying pyroclastic succession (white, to the left). The rhyolite is also partly extrusive where it eventually shears off the overburden and sees the surface. **b** Detail of the fault plane with Riedel shear planes and dip-slip striae (arrow). Note at this location, the fault planes are more than 10 m high but they rapidly attenuate in their intensity laterally so the deformation declines to zero <1 km away; furthermore the displacement is only visible in the overburden and does not displace the rock succession at the level of the dyke feeder to the cryptodome

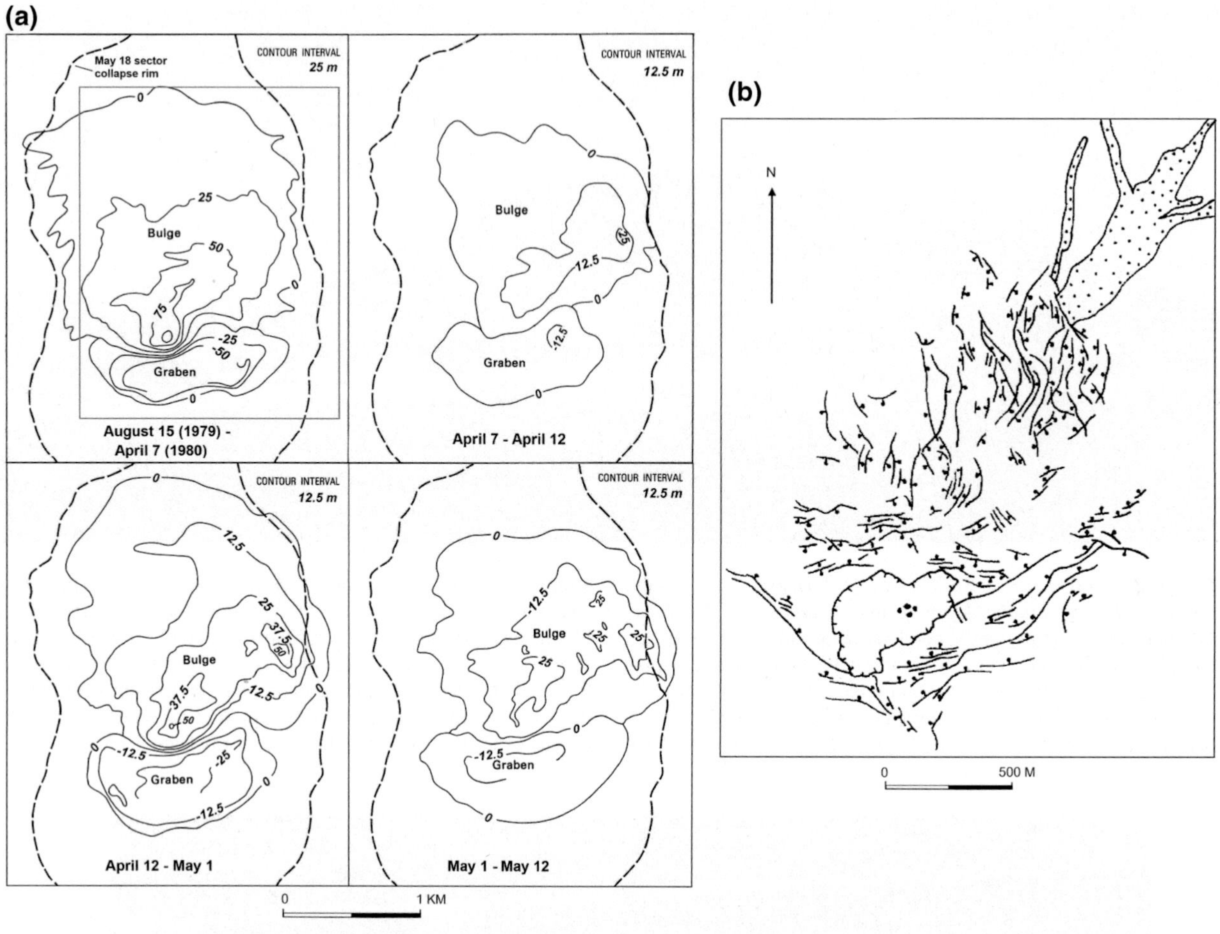

Fig. 15.9 Deformational field mapped at Mt St Helens, USA, during evolution of the cryptodome intrusion that led to the flank failure and lateral blast on 18 May, 1980. **a** Elevation changes in m between 15 August 1979–12 May 1980, and rim of the climactic 18 May sector collapse scar (bold dashed lines); box is area shown in (**b**). Note the main area of topographic upllift (or the 'bulge') compared with that of downdrop (the 'graben'). **b** Map of surface faults and fractures during final stages on 12 May within area of the future collapse scar. Dotted area is debris avalanche deposit (mainly ice and snow) caused by uplift and oversteepening of the Leschi and Forsyth Glaciers which initially avalanched on 12 May; black dots are small lakes developed in the crater which formed by phreatic explosions beginning 27 March within the graben. (Adapted from Moore and Albee 1981)

though, the stress is always a composite of several factors, including the regional tectonic stresses, stresses induced by magma pressure and those induced by crustal layering and topography (e.g. Gudmundsson 2002; Geshi et al. 2012). Regional tectonic stresses are usually far-field (Chap. 17) and are therefore dominant away from volcanic centres. They maintain their orientation at scales of hundreds to thousands of km or more, and across different structural levels. By contrast, magmatic stresses become foremost approaching the volcanic edifice, at scales of hundreds of m to km, and are dominated by radial and concentric orientations, as at the first-order magma pressure is isotropic, magma being a liquid (Figs. 8.7a–c and 15.11a; also see Fig. 16.82c). Crustal layering and contrasts in rock stiffness, as well as topography on the upper reaches of steep volcanic edifices, may become locally important, contributing to the local rotating of the stress tensor (Chap. 8).

Contrasting responses to positive deformation induced by intrusion of magma are subsidence related to caldera formation, and gravitational sector failure (Chaps. 10 and 12 and 16). These may occur in a matter of minutes or hours producing displacements of hundreds to thousands of m on volcano-tectonic faults. Such faults and fractures have the capacity to nucleate in rather different rock types, from pre-volcanic substrate to variably consolidated volcanic rocks and deposits. The resulting structures may therefore be quite complex, changing orientation, style of deformation, and branching across layers with different stiffness (e.g. Garden et al. 2017). These events and faults have the potential to change both the elevation of the water table and volume of meteoric recharge, which are principal controls on the depth of boiling and hence type of epithermal precious metal mineralisation (Chap. 18). The late-stage collapse of meteoric water into the waning stages of a porphyry copper

Fig. 15.10 Vertical rhyolite dyke (dark rock body, centre-right) cuts through extensive syn-emplacement volcano-tectonic fault-system dipping towards the dyke on the left, generated to accommodate the uplift and extension of the overburden as the dyke was propagating upward; the country rock here is hyaloclastite that is part of an uplifted early Pleistocene submarine lava and dome complex. Ponza Island, Tyrrhenian Sea, Italy

system and subsequent "*telescoping*" of epithermal alteration and mineralisation may be a common evolutionary trend (Sillitoe 1994). Caldera and sector collapses also instantaneously change the drainage system, focusing sedimentation in the newly formed depression, and excluding from deposition large areas of the volcano previously sites of intense deposition (Figs. 15.2b, c, 15.4a and 15.12a, b; see Chap. 16, Sect. 16.7). The resulting unconformities and localisation of sedimentation can be very abrupt in the volcanic record. This may be tricky to reconstruct and be erroneously interpreted as quiescence periods at older volcanoes (cf. Laurenzi 2005). At a smaller scale, explosive craters and maar-type volcanoes may also form closed, fault-bounded basins that focus sedimentation along sharp unconformities (Fig. 15.12c–e).

Another important issue that needs to be taken into account is that the magma pressure sources (e.g. magma reservoirs, dykes, cryptodomes) that create volcano-tectonic deformation are usually located in the upper crust and generally at very shallow levels, usually less than a few km. The rheology of the crust where these deformations nucleate and propagate is predominantly weak (e.g. often un- or weakly consolidated rocks), weakened (e.g. altered rocks, regions of high heat flow), or anisotropic (e.g. crustal layering of alternating weak and hard lithologies with complex 3D geometries), and in many cases water saturated (by hydrothermal and/or hydrogeologic systems in both porous and fractured domains). Furthermore, at shallow levels, confinement is poor and far from lithostatic, because of both weak mechanical properties of the rocks and possible topographic effects, if significant relief is present. Such complexity results not only in complex geometries, but also in varying mechanisms of deformation, even along the same fault. For example, discrete fault planes in hard rocks may extend into regions where rocks are non- or poorly lithified and deformation is accommodated in part or entirely by compaction and intergranular shear, defining deformation bands (Fig. 15.13a; e.g. Ramsay 1980).

In water-saturated systems such deformation bands also become pathways for fluid escape in response to compaction

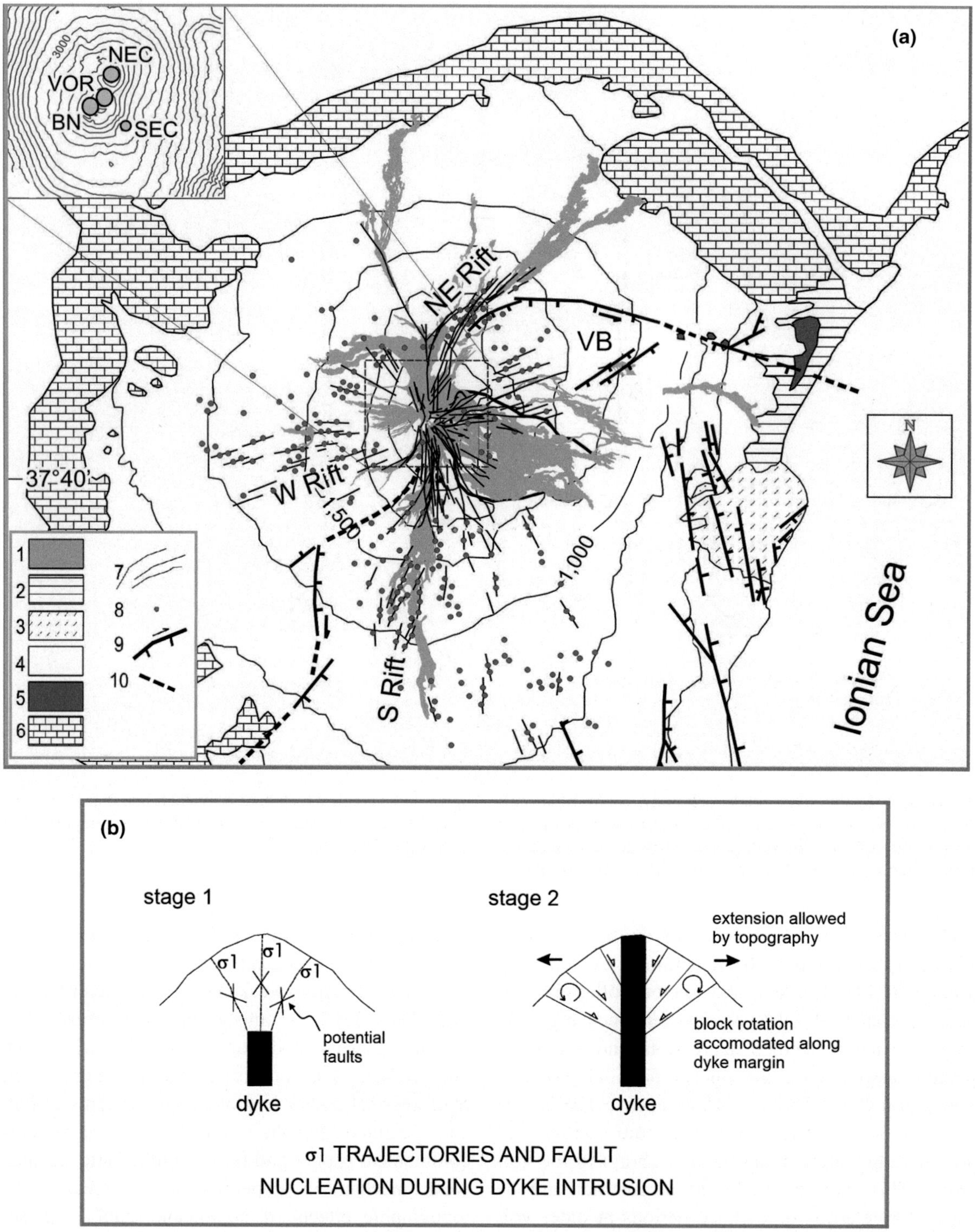

Fig. 15.11 Dyke and stress field orientations near the magma conduits of volcanoes. **a** Mt Etna in Sicily, Italy, is an excellent natural case study that shows feeder dykes are mostly radial in proximity to the summit vents (NEC = NE crater; VOR = Voragine crater; BN = Bocca Nuova crater; SEC = SE crater) where magma pressures dominate, while they change orientation progressively outward where the regional stresses dominate. Legend: 1 = Holocene lavas; 2 = Recent alluvium; 3 = Chiancone debris avalanche; 4 = pre-Holocene Etna volcanic succession; 5 = early Quaternary clays; 6 = pre-Quaternary sediments (orogenic belt); 7 = eruptive fissures; 8 = eruptive vents/pyroclastic cones; 9 = regional faults; 10 = inferred regional faults. VB = Valle del Bove; also see Fig. 8.8a. (Adapted from Lanzafame et al. 2003). **b** Schematic development of syn-dyke emplacement deformation with shallow overburden and topography. In stage 1, σ_1 stress trajectories radiate (in cross-section) away from the rising dyke; extensional faults will form at an angle, depending on the rock rheology, following Coulomb's failure criterion. In stage 2, as the dyke propagates it cuts the structures it has propagated. (Modified from De Rita et al. 2001)

Fig. 15.12 Closed basins can form rapidly in volcanic environments focusing sedimentation, while at the same time they may exclude entire sectors of the volcano from deposition. These are, in decreasing order of dimension: **a** *Calderas*, such as Quilotoa caldera, Ecuador. **b** *Sector collapses*, such as Valle del Bove at Mt Etna, Sicily (Fig. 15.11a), which formed as a result of a major lateral sector collapse during the Holocene The resulting depression is a major unconformity at the volcano scale that has focused the deposition of most lavas from the volcano's central vents. **c** *Explosive craters*, such as Vesuvius, Italy. **d**, **e** *Maar-type volcanoes*, such as found in the Newer Volcanics Province, southeastern Australia. For example, **d** Tower Hill maar volcano with a late scoria and spatter cone complex in the centre; and **e** the Red Rock volcanic complex which contains multiple isolated and coalesced maars with scalloped outlines now occupied by lakes, as well as a scoria spatter cone complex behind, to the left (northwest)

Fig. 15.12 (continued)

(Fig. 15.13b). Faults may therefore change, both along strike and along dip, from well-defined planes that define narrow cataclastic damage zones in hard rocks to diffuse shear bands in poorly consolidated and water-saturated deposits, where intergranular adjustment, compaction, and fluid escape dominate. Where fluid overpressure cannot find pathways out, conditions for hydraulic fracturing may occur (e.g. Cas et al. 2011; Vignaroli et al. 2015; Chaps. 4 and 13), which may result in in situ brecciation, or channelisation of fractured debris in clastic dykes (Fig. 15.14). Hydrothermal/hydraulic fracturing is essentially a post-lithification brecciation process, where alteration minerals may pervasively replace country rock, producing areas of pseudo-clastic and pseudo-"matrix" textures (also see Figs. 4.33 and 13.10).

Based on these foregoing considerations, the following criteria are a useful guide to distinguish volcano-tectonic structures from normal tectonic structures in the field. In particular, it is essential to establish a hierarchy of the various evidence of deformation, both in modern and in ancient successions, based on the following:

(a) *Lateral extent*: Here it is essential to document the length scale of the structure (m- to km-scale or more). This is because the shallower and smaller the pressure source is (e.g. a cryptodome intrusion versus far-field stresses), the smaller is the length scale of deformation, and vice versa. This means that volcano-tectonic faults are usually shorter (m to km and rarely tens of km in case of caldera ring-fractures) than tectonic ones (km, to hundreds of km or more). Note also, a rule of thumb is that long faults are also deep (crustal- and lithospheric-scale) faults.

(b) *Displacement rates and intensity*: Displacement across volcano-tectonic faults is usually much faster and discontinuous compared with tectonic environments. Such faults may show m to hundreds of m of displacement, which are rapidly sutured (buried) by new deposits unaffected (or very poorly affected) by such deformation (e.g. Fig. 15.6i). Displacement may also vary rapidly and abruptly; even in caldera collapses (Chaps. 10 and 16) it is rare that the caldera floor subsides equally all around the ring faults, so that the actual displacement is quite variable. This is much more so if the pressure source is shallower, as in the case of the intrusion of a cryptodome or a laccolith, or caldera resurgence, where individual blocks may experience significant uplift that rapidly attenuates laterally.

(c) *Geometry and orientation*: Volcano-tectonic deformation tends to be radial and concentric with respect to the originating pressure sources. The presence of a pre-existing structural and stratigraphic fabric of the crust may partly modify this simple symmetry, but

Fig. 15.13 Syn-volcanic faulting found in poorly- to unconsolidated volcanic deposits is accommodated by intergranular shear bands and fluid escape structures. **a** Faults in unconsolidated Holocene pumice fall deposits. Cuicuiltic Formation, Los Humeros volcano, central Mexico. **b** Syn-emplacement faults in hyaloclastite act as fluid escape pathways and are filled by fine-grained ash-size material. Ponza Island, Tyrrhenian Sea, Italy

Fig. 15.14 Hydraulic fracturing of an andesite characterised by both bulk and focused clastic domains, reflecting diffuse, pervasive and focused fluid flow and hydraulic fracturing. Lisca Bianca islet, Aeolian Islands, southern Italy

generally volcano-tectonic structures can be discriminated as they rotate in orientation at the volcano scale (or even at smaller scales). By contrast tectonic structures are persistent in orientation, usually well beyond the limit of the volcanic edifice.

(d) *Evidence of syn-volcanic faulting*: Any evidence of soft-sediment and pre-lithification deformation should be regarded as potentially associated with volcano-tectonic structures, although this is not unique to volcano-tectonics. Similarly, as indicated above in (b), syn-volcanic faulting can be recognised by the occurrence of faults that cut and are buried by volcanic deposits (even several times).

15.4 Representation and Organisation of Volcanic Units: Volcanic Stratigraphy

The representation and organisation of volcanic rock units into a volcanic stratigraphy is a fundamental component of geological mapping in volcanic areas. This is because it provides the basic criteria and essential data for identifying the spatial and temporal relationships between volcanic products and both the *intra-* and *inter-eruption* processes (earth-surface, tectonic and climatic) that provide greater understanding of the geological evolution of a region (Martí et al. 2018). Establishing precise stratigraphic relationships in volcanic successions is not only essential for understanding the past behaviour of volcanoes and for predicting how they might behave in the future, but is also critical for establishing strategies for exploration of mineral and energy resources associated with volcanic systems, or for reconstructing the evolution of sedimentary basins in which volcanism has played a significant role. In this section, we discuss some of different concepts and methods applicable in studies of volcanic stratigraphy in both modern and ancient terrains.

15.4.1 Volcanic Activity Units or Hierarchy of Eruption Events

A fundamental factor in reconstructing and mapping a volcanic succession is to represent, in some way, the evolution of volcanic activity at various scales. However, while a volcano has a birth, and ultimately an end to its activity (i.e. "*life-cycle*"), in between, it may undergo significant changes in (1) chemistry of magmas erupted (e.g. resulting from changes in the feeding and/or storage systems), (2) average output rates, (3) dominant volcanic eruption styles (e.g. effusive versus explosive), as well as (4) location of the main vents.

One of the first attempts to resolve this problem and generate maps containing more detailed information regarding eruptions was made by Fisher and Schmincke (1984). They proposed a subdivision of the cycles of volcanic activity into a hierarchy of eruption events, consisting of *eruptive periods*, *eruptive epochs*, *eruptions* and *eruptive phases* and *pulses*, where the *eruption* represents the fundamental unit, and the *eruption unit* a rock succession emplaced during a single eruption event (Table 15.1).

The eruption phases of an eruption may represent, for example, the buoyant phase of a plinian column followed by the collapsing phase (e.g. the beginning phase of the AD 79 Vesuvius eruption, Italy, on 24 August followed by the major pyroclastic flow phase, which occurred on August 25). Periods and epochs are major, or temporally longer subdivisions which include a series of eruptions with common characteristics (e.g. the 1631–1944 eruptions at Mt Vesuvius which built today's main cone of Gran Cono), distinct from the previous and following history, characterised for example by significant shifts of activity and substantial repose times in between. This method has the advantage of directly linking volcanic activity to the resulting stratigraphic record, and such volcanic stratigraphic divisions may be very useful for understanding the eruptive dynamics of a given eruption or set of eruptions. Unfortunately, the approach fails to take into account large periods of time during which inter-eruptive processes were dominant because the focus is entirely on eruptive activity. Nevertheless, it is potentially useful for providing a framework for understanding the eruption dynamics of a single eruption or sequence of eruptions. However, we stress it is critical to distinguish terminology employed to describe the activity of a volcano, from that which describes the corresponding rocks.

15.4.2 Volcanic Litho-Stratigraphy

Litho-stratigraphy is the correlation of similar lithologies or *litho-stratigraphic units*. Litho-stratigraphic units are bodies of rocks, bedded or unbedded, that are defined and characterised on the basis of their lithologic properties, and in essence have no time significance. To correlate litho-stratigraphic units, it is important to identify the key lithological properties, and look for key beds or key sequences that can be used as a datum, therefore delimiting stratigraphic (space and time) relationships. By matching litho-stratigraphic units in two or more geographically different locations based on similarity of rock or deposit type, and/or position in stratigraphy at these locations, these form the basic units of geologic mapping.

The first requirement for constructing a litho-stratigraphic framework in a volcanic terrain is to describe objectively the various lithologies found at the outcrop scale. This should follow the approach and thought processes documented in Sect. 14.5, Chap. 14, initially developing descriptive terminologies (i.e. facies descriptors) and only later applying genetic terminologies. The interpretation in terms of eruption processes and units can only be carried very late in the process, once litho-stratigraphic units are areally correlated. This can involve much more than just determining the "ash stratigraphy", because the measurements of thickness, grain size, and composition of a deposit, which are necessary for correlation, are of great volcanological value in assessing the style and scale of volcanic activity, and in understanding processes. Only then may it be appropriate to apply highly genetic terms, e.g. phreatoplinian ash fall deposit, phreatomagmatic surge pyroclastic density current, extra-caldera fines-depleted ignimbrite, intra-caldera co-ignimbrite breccia, etc. Generally, these are terms that are most readily going to be applicable to modern and relatively and young volcanic successions, where context to the source volcano is preserved. However, they may also be applicable to old volcanic rock successions in which outcrop is not continuous and the rock succession has been tectonically deformed, after a lot of hard work in mapping, establishing local stratigraphies and then correlating from location to location (e.g. Orth et al. 1989; Branney 1991; Brown et al. 2007; Willcock et al. 2013).

Different volcanic deposits and rocks can be identified from one another (see Chap. 14) by differences in:

- composition, mineralogy, and crystal content,
- grain size (for volcaniclastic units, e.g. grain size distribution, maximum grain size of juvenile and lithic components),
- componentry (for volcaniclastic units, e.g. proportion of juvenile clasts, xenoliths, crystals),
- thickness and lateral extent (or aspect ratio),
- colour,
- degree and style of welding (if present), and
- relative stratigraphic position.

It is often found that no single characteristic is indicative of a particular unit, and some are so similar that they can only be distinguished once their relative stratigraphic position, for example, to a distinctive marker or key deposit has been determined.

Once the litho-stratigraphic subdivisions of the volcanic succession have been determined to a level of satisfaction, which may take considerable time in a large project, these can be ascribed stratigraphic identities. Where there is adequate information, these are described according to conventional nomenclature of the *International Stratigraphic Guide* (ISG). Here rock formations and their subdivisions (e.g. members and lithofacies) or groupings (e.g. groups) may be used at different scales to illustrate the overall facies and rock body architecture (Table 15.1).

As an example, from a modern setting, Fig. 15.15 shows a composite stratigraphic column of Colli Albani volcano,

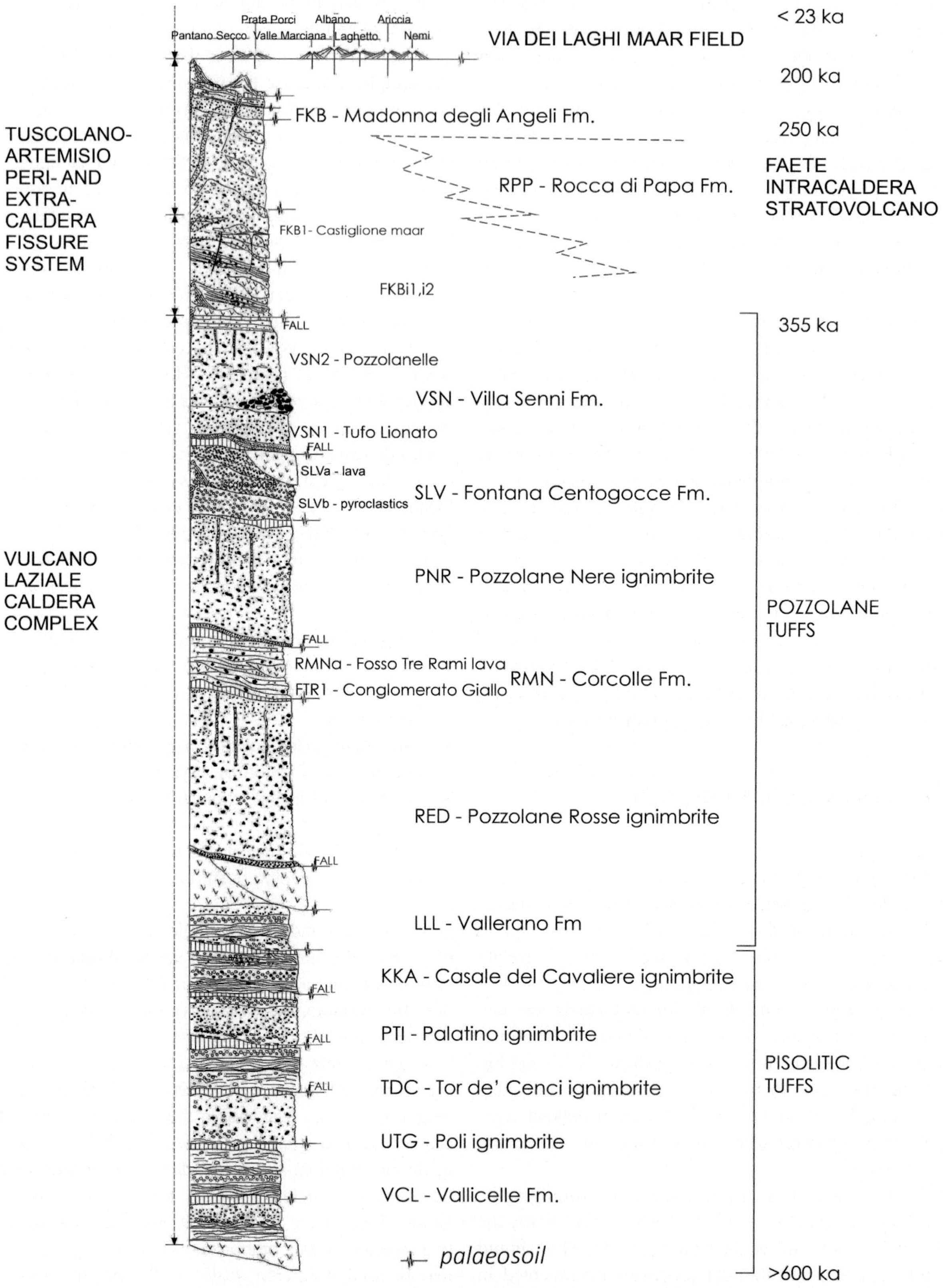

Fig. 15.15 Composite stratigraphic column of the Colli Albani volcanic succession, Roman Magmatic Province, central Italy. The quiescent Colli Albani volcano is composed of overlapping edifices, with each showing changes in volcanic style over the 600 ka history of the volcanic field. The edifices and changes in eruption style through time are indicated in the field by the major unconformities, which have been identified from surface geology and borehole stratigraphies. See text for explanation. (Modified from Giordano et al. 2010a)

Roman Magmatic Province, central Italy (From Giordano et al. 2010a). The lithostratigraphic units that comprise the volcanic succession are organised into four main "*groups*" related to the main edifices that were built over time: (1) the Vulcano Laziale caldera complex, (2) the Tuscolano-Artemisio peri- and extra-caldera fissure system, (3) the Faete intra-caldera stratovolcano, and (4) the Via dei Laghi maar field. Within the Vulcano Laziale caldera complex, at the rank of "*sub-group*", the Pisolitic Tuffs succession and the "Pozzolane Tuffs successions" indicate the lower and the upper divisions, which are dominated respectively by ash-rich (phreatomagmatic) ignimbrites, and scoria-rich ignimbrites. At the rank of "*formation*", mappable litho-stratigraphic units include deposits of either individual eruptions (e.g. ignimbrites and lava units), or deposits of a succession of eruptions and reworking events.

To illustrate the use of lower rank litho-stratigraphic units, Fig. 15.16 shows an excerpt from the legend of the Colli Albani 1:50,000 geological map (Giordano et al. 2010b). This shows subdivisions of the Villa Senni Formation (VSN), which encloses the deposits of one of the major caldera-forming events of the volcano (see Chap. 10, Sect. 10.4). In this case, the complexity of the internal stratigraphy of this formation is represented by the use of *members* and *lithofacies* (Table 15.1). For example, two lithologically distinct and superposed ignimbrites (the lower Tufo Lionato, and the upper Pozzolanelle) are represented as two members (VSN1 and VSN2 respectively). The proximal co-ignimbrite breccia of the Pozzolanelle (Colle Fumone breccia), and crystal-rich facies (Occhio di Pesce), have gradational contacts within the Pozzolanelle member so they are represented as lithofacies (VSN2b and VSN2a respectively). Syn-eruptive lahars occur in distal areas both interbedded and above the main ignimbrites so these are also treated together as lithofacies of the formation (VSNa), instead of being represented as two members.

With regard to an ancient setting, perhaps one of the most comprehensively mapped volcanic terrains worldwide is represented by the lower to middle Ordovician succession covering much of North Wales (e.g. Howells et al. 1985; Howells and Smith 1997). This region is complicated geologically because of the interdigitating nature of the deposits from local volcanic centres with the sedimentary deposits of an unstable marine basin, which then suffered significant deformation during the Caledonian Orogeny. Over several decades many geologists from the British Geological Survey have established detailed stratigraphic successions in various areas, and although each may be valid, in many cases they were not easily related to those of adjoining areas. As the detailed interpretation of the Ordovician geology of North Wales has evolved, so did the stratigraphy used to describe it, with consequent changes of classification and nomenclature. Although mapping of 1:50,000 scale geological map sheets led to many advances in the understanding of the emplacement of the volcanic sequences, especially within and the interplay with shallow marine environments (Chap. 11), it also provided the opportunity to review the litho-stratigraphic nomenclature, with emphasis on processes and events. This has led to the establishment of a rigorous litho-stratigraphic framework down to formation and member level, with some 100 defined stratal types (Rushton and Howells 1998); an excerpt, encompassing the central Snowdonia area is shown in Fig. 15.17. Not only does this give a central reference for geologists working in the region, it also provides a working manual for establishing conventional litho-stratigraphic order in other ancient volcanic terrains, which we recommend to the reader.

THE VULCANO LAZIALE IGNIMBRITE PLATEAU AND CALDERA COMPLEX

POZZOLANE TUFFS SUCCESSION

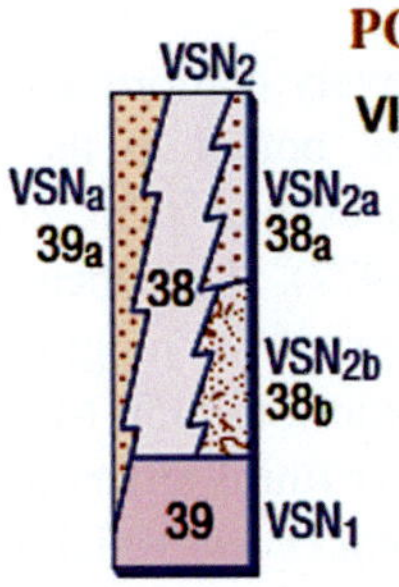

VILLA SENNI FORMATION

Pozzolanelle (VSN2). Tephri-phonolitic to phono-tephritic, massive and chaotic, unconsolidated, dark purple, coarse ash and lapilli-sized, scoria and xenolith rich ignimbrite, 20-80 m thick; (357±2 ka).

Occhio di Pesce (Fisheye) lithofacies (VSN2a). VSN2 facies with> 30% of loose leucite crystal fragments.

Colle Fumone Breccia (VSN2b). Co-ignimbrite breccia, with abundant intrusive and skarn xenoliths. K-foiditic spatter occur in the northern peri-caldera area.

Tufo Lionato (VSN1). Tephri-phonolitic lithified, massive and chaotic, red to yellow, ash-supported ignimbrite with black scoria, 2-10 m thick. To the east, the basal deposit is an up to 140 cm thick scoria fall deposit.

Pebbly-sandy lithofacies (VSNa). Syn-eruptive lahar and fluvial reworked deposits.

Fig. 15.16 Excerpt from the legend for the geological map (1:50,000) of Colli Albani volcano. This shows an example of the use of lithostratigraphic units to describe the complex internal stratigraphy of the 50 km^3, caldera-forming ignimbrite succession enclosed in the Villa Senni Formation. See text for explanation. (From Giordano et al. 2010b)

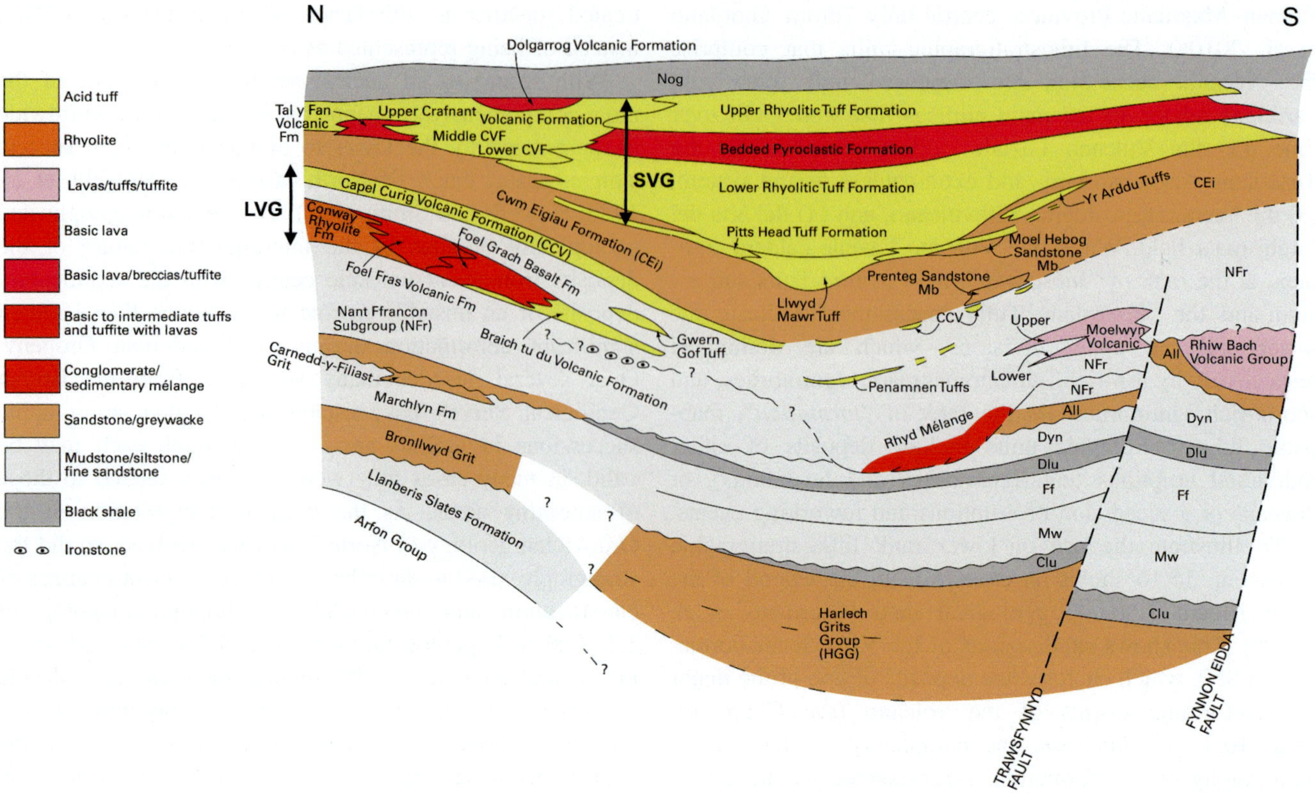

Fig. 15.17 Schematic cross-section of the lower-middle Ordovician marginal basin succession of Snowdonia, North Wales (Adapted from Rushton and Howells 1998). This illustrates the general lithostratigraphic relationships of the formations and principal members. The two main volcanic groups are also identified: LVG = Llewelyn Volcanic Group, and SVG = Snowdon Volcanic Group. These include, respectively, the Capel Curig Volcanic and Pitts Head Tuff Formations described in Chap. 11. The section is not drawn to scale, but to give some perspective represents about 50 km in an approximately N-S direction; the LVG reaches up to 1400 m in thickness, and the SVG some 1800 m. The base of the Ordovician (Tremadoc) is taken as the Dol-cyn-afon Formation (Dyn), whereas the transgressive onlap sequence of black mudstones forming the Nod Glas Formation (Nog) is latest Caradoc. This represents the order of 25 Ma of sedimentation, of which, the major part of the volcanic succession may have been deposited during an interval of about 5 Ma, although each volcanic was emplaced geologically instantaneously. Compare this with the modern volcanic marginal basin found offshore the Campanian coast, southern Italy, which has been active <1 Ma (see Figs. 15.24 and 15.25). *"What does this tell us?"*

15.4.3 The Sequence Stratigraphy Approach in Volcanic Terrains

Sequence stratigraphy is generally regarded as a relatively new concept evolving in the 1970s and 1980s as seismic sequence stratigraphy. However, its origins are much older and rooted in early geological ideas to better understand cyclic sedimentation and eustatic versus tectonic controls on sea level.

15.4.3.1 Some Basic Principles

The key element in sequence stratigraphic analysis is that the major controls on deposition in the sedimentary basins are the rates of tectonically driven subsidence and eustatic sea-level change which creates the accommodation space for sediments (Paynton 1977). Differing rates of sediment accommodation are represented by a series of surfaces about which the stacking patterns of sedimentary facies display systematic trends. Because tectonic cycles of subsidence and uplift, and eustatic cycles of sea-level change have different time spans, sequences can be classified in terms of duration (Vail et al. 1977).

Sequences are defined as relatively conformable successions of genetically-related strata, bound at the top and bottom by unconformities, hiatuses and their correlative conformities (Mitchum et al. 1977). Sequences can be divided into a number of orders or cycles ranging from in excess of 50 Myr duration (first order), to 5-50 Myr (second order, or *supersequences*), 1–5 Myr (third order) or <1 Myr (4th and fifth order, termed *parasequences*) which are 0.1–1 Myr and <100,000 years, respectively. Although debate persists concerning the principal driving mechanisms behind these cycles, at the higher scales (first and second order) tectonism is generally regarded as the main cause, whereas, glacio-eustatic changes are considered the main agents of third to fifth-order cycles, although others believe tectonism may still be significant driving forces of these. Another important concept is that of *systems tracts*, which are defined as a linkage of contemporaneous depositional systems. Particular depositional systems tracts are developed during specific phases of the relative sea-level curve: lowstand (LST), transgressive (TST), and highstand (HST). The surface between the TST and HST is termed the *maximum flooding surface,* which is coeval with the most landward transgression of the shoreline. This is marked by the finest-grained sediment or condensed sections (e.g. "hardgrounds"), and high organic content and high gamma-ray values representing anoxic or oxygen-starved bottom waters.

Sequence stratigraphy *sensu stricto* is really only applicable to marine and shoreline depositional systems, where analysis is based mainly on the interpretation of seismic sections and wire-line logs from deeper petroleum wells. Nevertheless, the basic concepts have been extended to continental sedimentary systems. For continental sequence stratigraphy, several publications appeared in the literature in the early 1990s, and later provided a general discussion of sequence stratigraphy in fluvial deposits and continental deposits (e.g. Emery and Myers 1996; Hanneman and Wideman 2010). The use of sequence stratigraphy versus *allostratigraphy* for continental deposits remains a controversial topic. Allostratigraphy refers to the use of unconformity-only bounded units, whereas sequence stratigraphy defines a sequence as "a unit bounded by unconformities or their correlative conformities". Allostratigraphy is formally defined by the North American Stratigraphic Code (NACSN 2005); its units, in decreasing rank, are *allogroup*, *alloformation*, and *allomember*. The ISG (Salvador 1994) also recognises unconformity-bounded units (UBU) but notes that these units be labelled as *synthems*. Because unconformity-bounded units are formally recognised by both the NASC and the ISG, some workers have opted to use allostratigraphic/synthem terminologies. However, the difficulty encountered in applying any unconformity-only unit stratigraphy is that it is often impossible to trace an unconformity across a basin. Generally, the situation within a basin is such that an unconformity that occurs on the basin margin is replaced by a correlative conformity near the basin's centre. Thus, an unconformity-only stratigraphy becomes a basin margin stratigraphy and in some instances may lead to restricted units and their associated nomenclature.

15.4.3.2 Volcanic Sequence Stratigraphy: Recognising Event-Sequence Packages

In recent years, to improve the capability to represent the complex stratigraphy of volcanoes and genetically related "event-sequence packages", the introduction of *unconformity-bounded units* or *stratigraphic units* (UBU or UBSU) and other stratigraphic subdivisions have been used alongside traditional lithostratigraphic units (Table 15.1).

By definition, UBSU are present at the lower or upper limits of the volcanic stratigraphic unit. However, volcanic successions commonly exhibit a wide range of stratigraphic unconformities of variable spatial and temporal extents (i.e. rank), which may not be readily apparent (De Rita et al. 1998). An excess in the identification of minor unconformities may result in a misinterpretation of the temporal variability of the volcanic succession because these unconformities may not represent important time breaks or changes in eruption dynamics or between major phases or stages in the evolution of a volcano. The same may happen with moderate unconformities that do not affect the whole volcano or volcanic field. However, the presence of minor and moderate unconformities should be indicated in a formal litho-stratigraphy as they may represent inter-eruptive events in the evolution of the volcanic system. Major unconformities caused by tectonic movements or regional non-depositional/erosion episodes may justify the use of synthems as stratigraphic units, given that they may also be interpreted in the context of conventional stratigraphy. For a thoughtful discussion of volcanic unconformities, see Martí et al. (2018).

An important feature of UBSU is that unconformities are not necessarily synchronous (i.e. they are diachronous). For example, where unconformities relate to allogenic factors external to the volcanic activity, such as for climatic and eustatic variations, the propagation of an erosional unconformity from the coast inland may span prolonged periods. Similarly, where unconformities are due to prolonged periods of quiescence, these may be diachronous across the volcano. By contrast, where unconformities are due to rapid changes of the morphology of the volcano, such as due to caldera or sector collapses, they can be considered synchronous. Another important feature of UBSU is that unconformities always correspond laterally to deposits. This is because the erosion in some areas of the volcano *must* correspond to deposition taking place somewhere else laterally, usually in the sedimentary basins that surround the volcanoes (e.g. Manville et al. 2009; Chaps. 12, 16 and 17).

In such an analysis, it is very important to be able to recognise volcaniclastic successions that correspond to increased rates of sedimentation associated with different causes. For example, as described above, the syn- and immediately post-eruptive (days to years) remobilisation of pumice and ash after a large volume explosive eruption laterally will correspond in continental settings to fast aggradation of debris flow and hyper-concentrated flood flow deposits rich in reworked monomictic juvenile material (Fig. 15.4c). This is because the large availability of light volcanic debris promotes fast removal from the upper slopes and fast deposition in distal sedimentary traps (e.g. Pierson et al. 1992; Newhall and Punongbayan 1996). Inter-eruptive periods may instead be dominated by linear re-incision with deposition of polymictic fluvial deposits in narrow channels (Fig. 15.4d), and formation of extensive paleosoils in the stabilised tops of the previously aggraded deposits. If no changes of the base level occur either due to climatic or tectonic causes, then the re-incision will bring the fluvial network down to the pre-eruptive level. Deeper re-incisions may result instead from volcano uplift or lowering of the base level. The reverse of course would be true in case of subsidence or rise of the base level.

Lithosomes are rock bodies that owe their morphology to specific geological processes and can be defined at various scales (Fig. 15.18; Lucchi 2013). In general, volcanic edifices at different scales can be considered lithosomes, as they represent the overall morphological expression of volcanic eruption styles. For example, tuff rings are low aspect ratio lithosomes that owe their morphology to the dominant

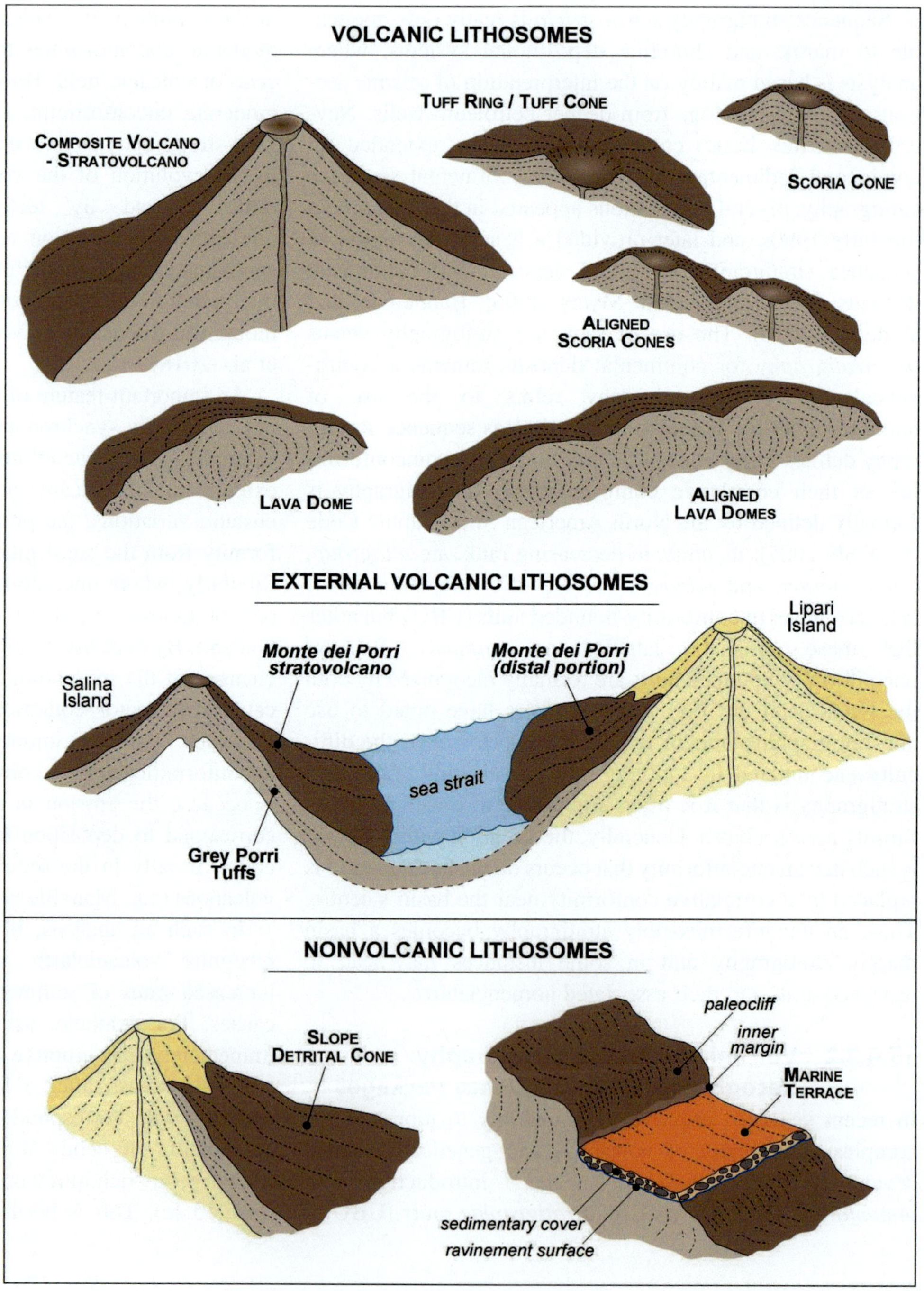

Fig. 15.18 Examples of lithosomes in volcanic environments, based on studies in the Italian Aeolian Islands (After Lucchi 2013)

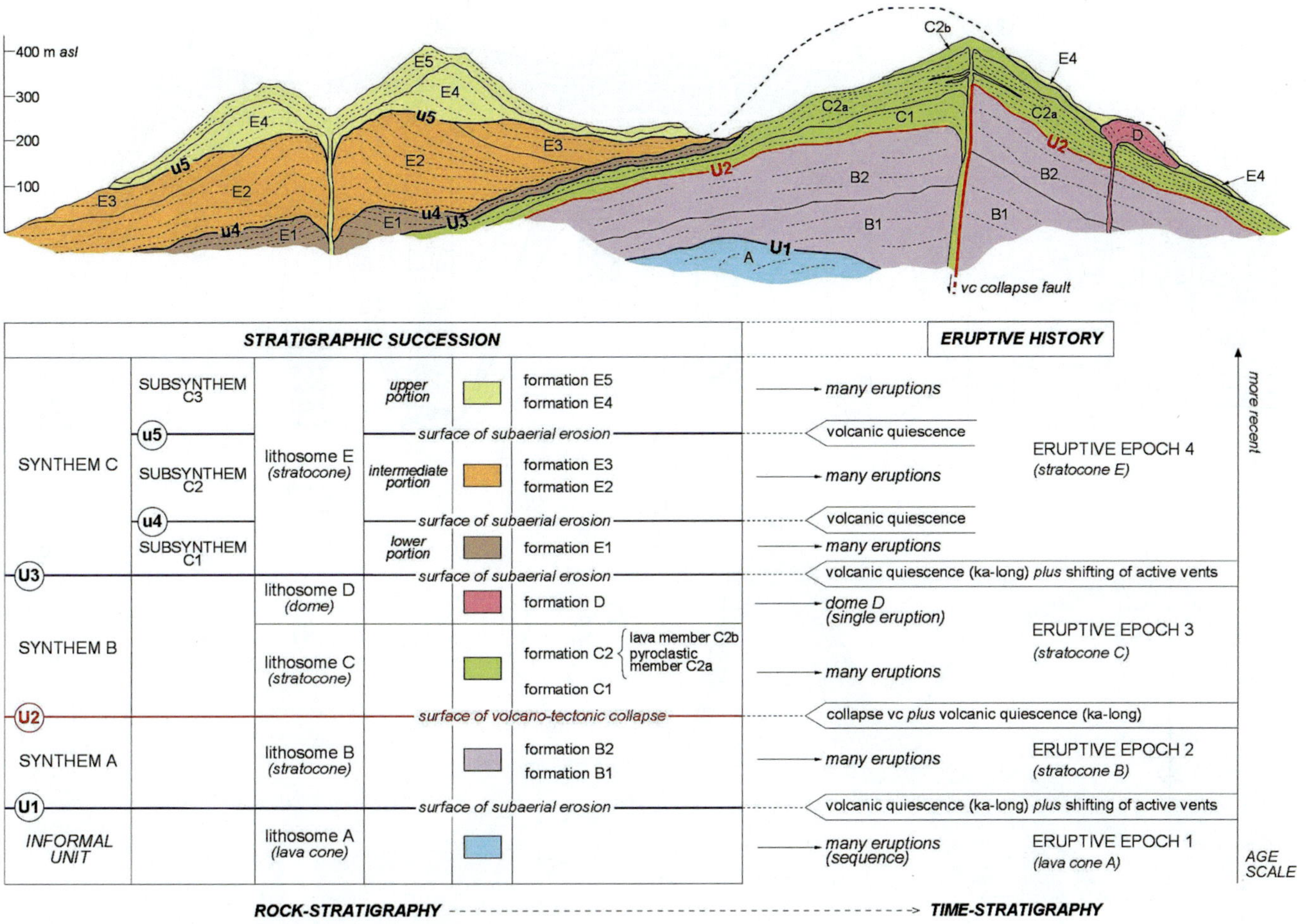

Fig. 15.19 Idealised cross-section of a volcanic area and related legend showing the relationships of unconformity bound stratigraphic units (synthems), lithosomes, and lithostratigraphic units (formations, members). The direct relationship between these rock-stratigraphic units and time-stratigraphic units (eruptive epochs and eruptions) is also displayed, with successive steps of eruptive history defined on the grounds of the reconstructed stratigraphic succession. (After Lucchi 2013)

process of base surge that emplace their deposits. Lithosomes may also be compound as for the case, for example, of the coalescence along a fissure of scoria cones, spatter rampart, and lava flows. An ignimbrite shield or a lava shield are also lithosomes. This way lithosomes may be used in a legend (*but not in a map!*) to group lithostratigraphic formations associated with distinctive spatial, structural, or eruption styles of a volcano.

Some examples of the integrated use of lithostratigraphic, UBSU, and lithosomatic subdivisions can be found in the most recent cartographic publications of the Italian volcanoes (e.g. Giordano et al. 2010a, b; Branca et al. 2011; Lucchi 2013; Forni et al. 2013). It is important to stress that geological maps should always and only represent lithostratigraphic units, whereas UBSU and lithosomes are forms to organise the stratigraphic succession and should be represented in the associated legend and ancillary schemes. Figure 15.19 summarises the possible use of the different unit types described above and the way a legend can be organised to illustrate the main stratigraphic relationships of a volcanic succession. In Fig. 15.20, an example of the level of detailed stratigraphic correlation of the volcanic succession that has been achieved on Lipari Island, which has underpinned a meticulous reconstruction of the magmatic evolution and eruptive history of the island (Forni et al. 2013).

15.4.3.3 Applied Volcanic Seismic Sequence Stratigraphy

Although concepts such UBSU can be embraced alongside lithostratigraphic terminology, the continental nature of many modern volcanic successions of interest and the general lack of access to modern seismic surveys in such areas, leaves us with significant gaps in applying sequence stratigraphy to their understanding. Nevertheless, there are two volcanic depositional environments where seismic sequence stratigraphy has been very successfully applied in recent years: (1) volcanic passive margins, and (2) marginal marine basins.

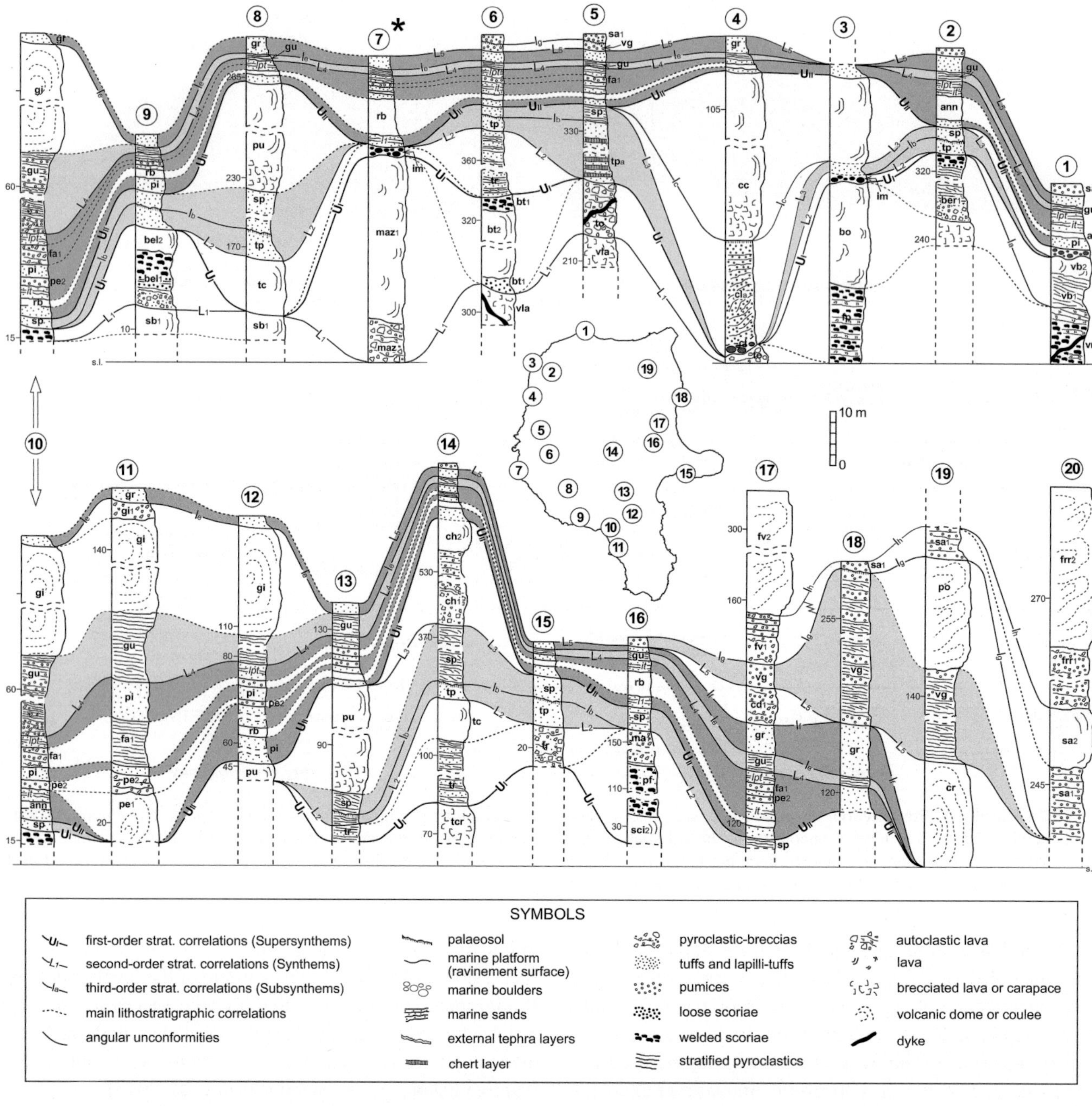

Fig. 15.20 Highly detailed, stratigraphic framework established for the island of Lipari volcanic complex, Aeolian Arc, Italy (From Forni et al. 2013). This shows measured, composite sections and the correlation of key stratigraphic units, which are described by 22 unconformity bounded stratigraphic units (synthems), 25 lithosomes, and several lithostratigraphic units. Stratigraphic correlation is based on leaf-bearing pyroclastic deposits (tp, Timpone Pataso Formation; sp, Serra Pirrera Formation); Brown Tuffs erupted from Vulcano (pi, Pianoconte Formation, gr, Piano Grottedei Rossi Formation); Grey Porri Tuffs from Salina (rb, Rocce di Barcone Formation); Lower Pollara Tuffs (lpt) from Salina; M. Guardia (gu) and Vallone del Gabellotto Formations (vg). Vertical scale is in m above sea level; Sect. 7* approximates to area shown in Plate 15.1 and Fig. 15.2f. Note the position of the major rank unconformities U_I and U_{II} which divide the stratigraphic architecture. These are allogenic unconformities associated with sea level rise and fall as discussed in Fig. 15.2f, and approximate synchronous surfaces. The other lower order unconformities are mostly autogenic. Elsewhere the main steps of eruptive history are also described in terms of eruptive epochs, sequences of eruptions and eruptions, and quiescence periods, defined according to the corresponding UBSUs and unconformities and their chrono-stratigraphy (for further description, see Forni et al. 2013)

Volcanic Passive Margins

Volcanic passive margins (VPM) record the rifting and break-up history of continents, and are associated with the eruption and intrusion of voluminous basaltic successions (Chaps. 5, 8 and 17). One of the best studied areas is the Tertiary North Atlantic Igneous Province. There the seismic volcano-stratigraphic method was pioneered in the late 1990s to study the nature, geologic history, and emplacement of extrusive volcanic rocks from 2D seismic data acquired during petroleum exploration (e.g. Planke et al. 2000; Berndt et al. 2001; Chap. 18).

Following established seismic sequence stratigraphy methodology, seismic facies units are identified and mapped, to be finally interpreted in terms of volcanic facies and emplacement processes. The volcanic sequence boundaries are defined by the top- and base-basalt reflections. Several characteristic volcanic seismic facies units have been identified, including subaerially emplaced Seaward Dipping Reflectors (SDR) and landward flows, the coastal lava delta, and the subaqueous inner flows (Fig. 15.21; Table 15.2; Chap. 17, see Fig. 17.22 and accompanying text). In this scenario, the inner flows represent an aggradational bottom set consisting of volcaniclastic sediments mixed with pillow-basalt and massive sheet flows. Lava deltas are recognised by their prograding internal architecture (clinoforms), whereas the landward flows form sheets of stacked subaerial lavas, which transition into lava delta and hyaloclastite facies at the paleo-escarpment (Fig. 15.22). Sub-basalt igneous sheets below the volcanic sequence are interpreted as sills, which are characterised by relatively high-amplitude, low-frequency saucer-shaped reflections with abrupt terminations (Planke et al. 2005; Chap. 8). These are associated with hydrothermal vent complexes which are abundant in adjacent sedimentary basins.

With the discovery of petroleum within intra-basalt reservoirs in 2004, major improvements in the seismic imaging of the basalt and sub-basalt sequences took place (Chap. 18). These improvements, especially in high resolution 3D data acquisition and fieldwork onshore on analogue seismic-scale outcrops, facilitated detailed mapping of the uppermost basalt sequence (Planke et al. 2017). High-resolution 3D imaging of the Vøring Marginal High offshore mid-Norway has provided spectacular profiles of top- and intra-basalt structures. By using 3D seismic images in map view, this has also allowed interpretation of the *seismic geomorphology* of the top-basalt horizon, revealing new insights into the late-stage development of the volcanic flow fields and the km-high coastal Vøring Escarpment (Fig. 15.23). Subaerial lava flows with compressional ridges and inflated lava lobes cover the marginal high, with a comparable structure and size to modern subaerial lava fields. The prominent Vøring Escarpment formed when eastward-flowing lava reached the coastline. The escarpment morphology is influenced by pre-existing structural highs, and these highs are locally bypassed by the lava. Volcanogenic debris flows are well-imaged on the escarpment horizon, along with large scale large slump blocks. Numerous post-volcanic extensional faults and incised channels cut into the marginal high and the escarpment, indicating the area was geologically active after the volcanism ceased.

Marginal Marine Basins

The Campanian-Latium continental margin of southern Italy comprises several marginal basins including the Bay of Pozzuoli, which represents the offshore extension of Campi Flegrei caldera volcano, and the greater Bay of Naples area. Campo Flegrei forms the emergent northern margin of a wide caldera, from which eruption of the Campanian ignimbrite occurred at about 39 ka (Chaps. 10 and 11). Another key stratigraphic marker is the Neapolitan Yellow Tuff, another major ignimbrite-forming event which was erupted at 15 ka. After the eruption of the Neapolitan Yellow Tuff the sea invaded the caldera, and the continental shelf of the Bay of Naples presently extends to water depths of 180 m below sea level. The northern sector of the Bay of Naples is made of an extensive system of banks of volcanic origin (e.g. lava domes, tuff cones, and more complex eruptive centres) interlayered with volcaniclastic and marine sediments. Seismic refection data acquired over the past 40 years, and particularly high resolution data, coupled with well-studied onshore outcrop and geothermal wells, make this region a laboratory for analysis of the interplay between continental margin volcanism and deposition of submarine (syn-eruptive and inter-eruptive) stratigraphic units (e.g. Bruno 2004; Milia 2010; Torrente and Milia 2013; Innacce et al. 2013; Aiello et al. 2011, 2012, 2016).

The Bay of Naples basin infill comprises a Lower Pleistocene basal unconformity-bounded unit (Unit A), covered by several fourth-order sequences that form a transgressive-regressive cycle deposited over the past 700 kyr (Milia 2010, and references therein). Volcanic units are interlayered in the upper part of Sequence Set C (Sequence C3) and within Sequence D (Fig. 15.24). These are subdivided into a falling/low-stand, prograding wedge (FST-LST), a transgressive system tract (TST), and a high-stand system tract (HST). Prograding wedges of volcaniclastic detritus were deposited during the fall of sea level and filled the seaways between the volcanic banks, whereas during rising sea level, deposition migrated toward these paletopographic highs until they were drowned below sea level. Thick ignimbrite units are present on the continental shelf, where it is possible to recognise (1) pre-Campania ignimbrite units, deposited between 200 and 39 ka; (2) the

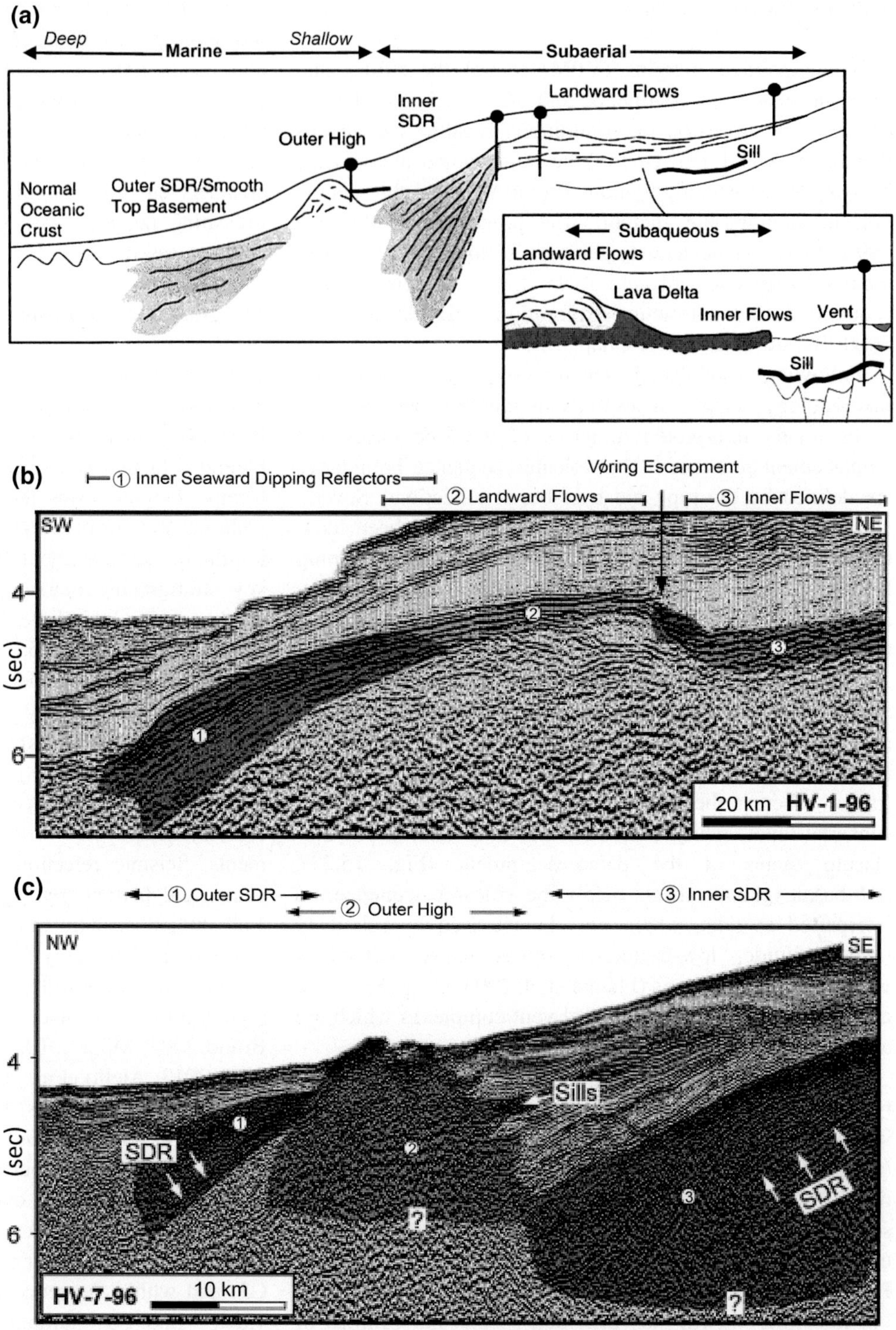

Fig. 15.21 Volcanic passive margin seismic profiles, North Atlantic Igneous Province. **a** Schematic volcanic margin profile showing characteristic seismic facies within the extrusive basaltic breakup sequence (From Planke et al. 2000). These are divided into four seismic facies units located landward of the ocean-continent boundary: outer SDR, outer high, inner SDR, and landward flows. Commonly two additional seismic facies units are identified: lava delta and inner flow. Environment of emplacement is indicated by arrows; vent refers to hydrothermal vent complexes associated with sills in adjacent sedimentary basins; wells (solid circle) schematically show where drill holes have penetrated corresponding seismic facies unit. Note, not all seismic facies units are identified on any single margin. **b**, **c** Examples of seismic profiles (in two-way travel time, TWTT) from the Vøring Margin, offshore Norway (From Berndt et al. 2001). The circled numbers within the seismic sections denote volcanic seismic facies units and correspond to the labels above the section. The Vøring Escarpment is a major paleovolcanic scarp forming the boundary between subaerial landward flows, that transition into subaqueous lava delta and inner flows (Figs. 15.22 and 15.23)

39 ka Campanian ignimbrite; (3) a thick volcanic unit made mainly of pumice, deposited between 18 and 15 ka; and (4) the 15 ka Neapolitan Yellow Tuff. In the Bay of Pozzuoli, the stratal architecture of the sediments deposited during the Sequence D transgressive and high-stand of sea level was mainly controlled by volcano-tectonics (as exemplified at La Starza, Fig. 15.7). Here, it is possible to recognise three tectono-stratigraphic units: Unit G3 (15–8 ka), Unit G2 (8–4 ka), and Unit G1 (4 ka to present), rather than transgressive and high-stand systems tracts.

Figure 15.25 presents an E-W-oriented seismic line ($\sim$ 14 km wide) collected from the Bay of Naples and along the southern margin of the Bay of Pozzuoli; the following description is taken from Milia (2010). Towards the western end the oldest seismic unit is characterised by continuous reflectors with good continuity that are sub-horizontal in the

Table 15.2 Main volcanic seismic facies characteristics identified on rifted volcanic passive margins (Adapted from Planke et al. 2000; Berndt et al. 2001; Abdelmalak et al. 2016)

Seismic facies unit	Reflection characteristics			Volcanic facies	Depositional environment
	Shape	Boundaries	Internal		
Inner flows	Sheet	Top: high amplitude, disrupted or rough, onlapped or concordant Base: negative polarity, often obscured	Chaotic or very disrupted, subparallel or hummocky	Massive and fragmented flows, volcaniclastics and hyaloclastites	Shallow marine
Lava delta	Bank	Top: high amplitude or reflection truncation Base: reflection truncation or termination	Progradational clinoforms (foreset bedding); major escarpment at paleoshoreline	Massive and fragmented basalts mainly composed of hyaloclastite debris	Coastal
Landward flows	Sheet	Top: high amplitude, smooth, onlapped, or concordant Base: low amplitude, disrupted	Parallel to subparallel; high amplitude disrupted	Flood basalts	Subaerial or shallow marine flood basalts
Inner SDR	Wedge	Top: intermediate to high amplitude, smooth or wavy, locally small escarpment Base: seldom defined	Divergent arcuate or sometimes a divergent planar pattern	Flood basalts	Subaerial flood basalts deposited in subsiding structure
Outer high	Mound	Top: high amplitude, disrupted and rough Base: not visible	Chaotic	Hyaloclastite flows and volcaniclastics	Shallow marine
Outer SDR	Wedge	Top: intermediate to high amplitude, smooth, locally small escarpment Base: seldom defined	Divergent arcuate, lower amplitude than the Inner SDR	Flood basalts mixed with pillow basalts sediments and sills	Deep marine

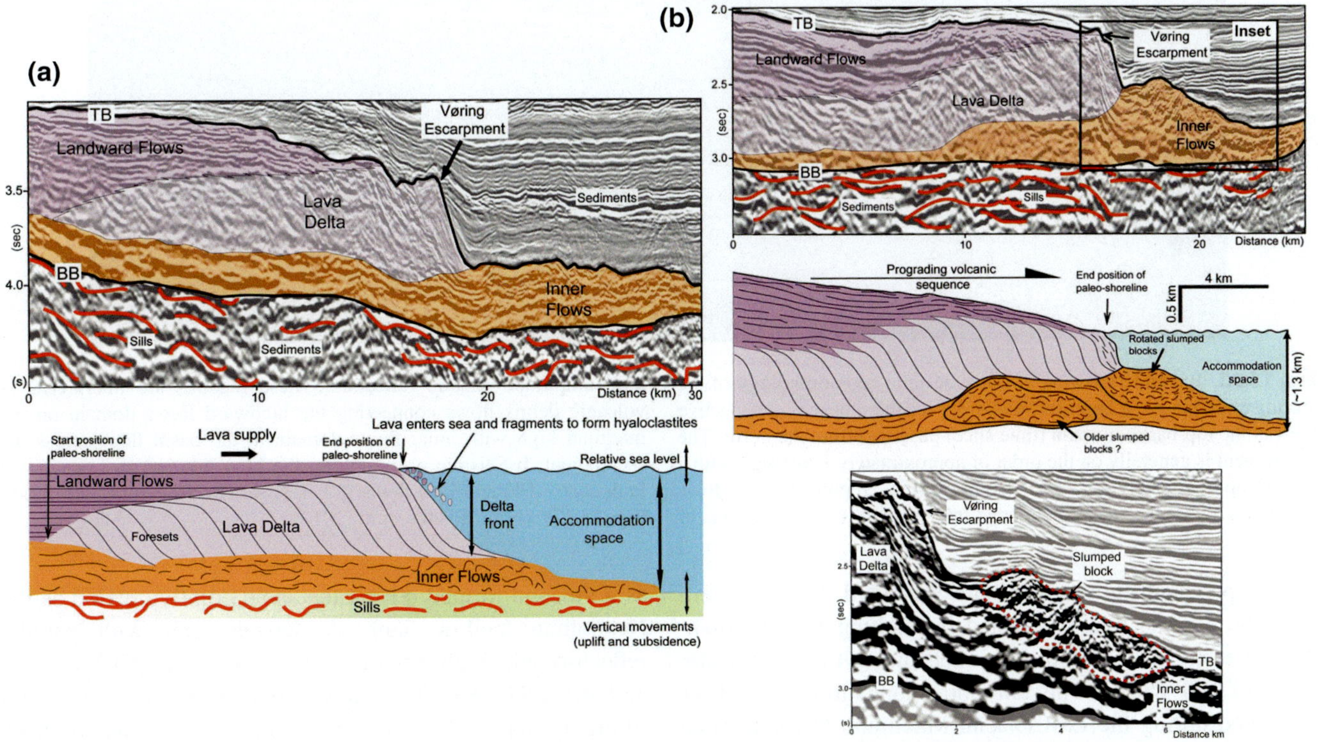

Fig. 15.22 Examples of more recent high-quality 2D seismic facies mapping along the Vøring Escarpment, offshore Norway, North Atlantic Igneous Province. **a** Southern profile (W–E) showing interpreted seismic section (in TWTT), highlighting key seismic facies units of the lava delta system and escarpment (above), and schematic representation (not to scale; below) of a developing lava-fed delta showing the accommodation space and the position of the paleo-shoreline. TB = top basalt, BB = base basalt. **b** Northern profile (NW–SE) showing interpreted seismic section (above), and schematic illustration of the depth-converted seismic section; the paleowater depth indicated is ~1.3 km. Inset shows zoom on slumped blocks along escarpment with disrupted internal reflection configuration. (Adapted from Abdelmalak et al. 2016)

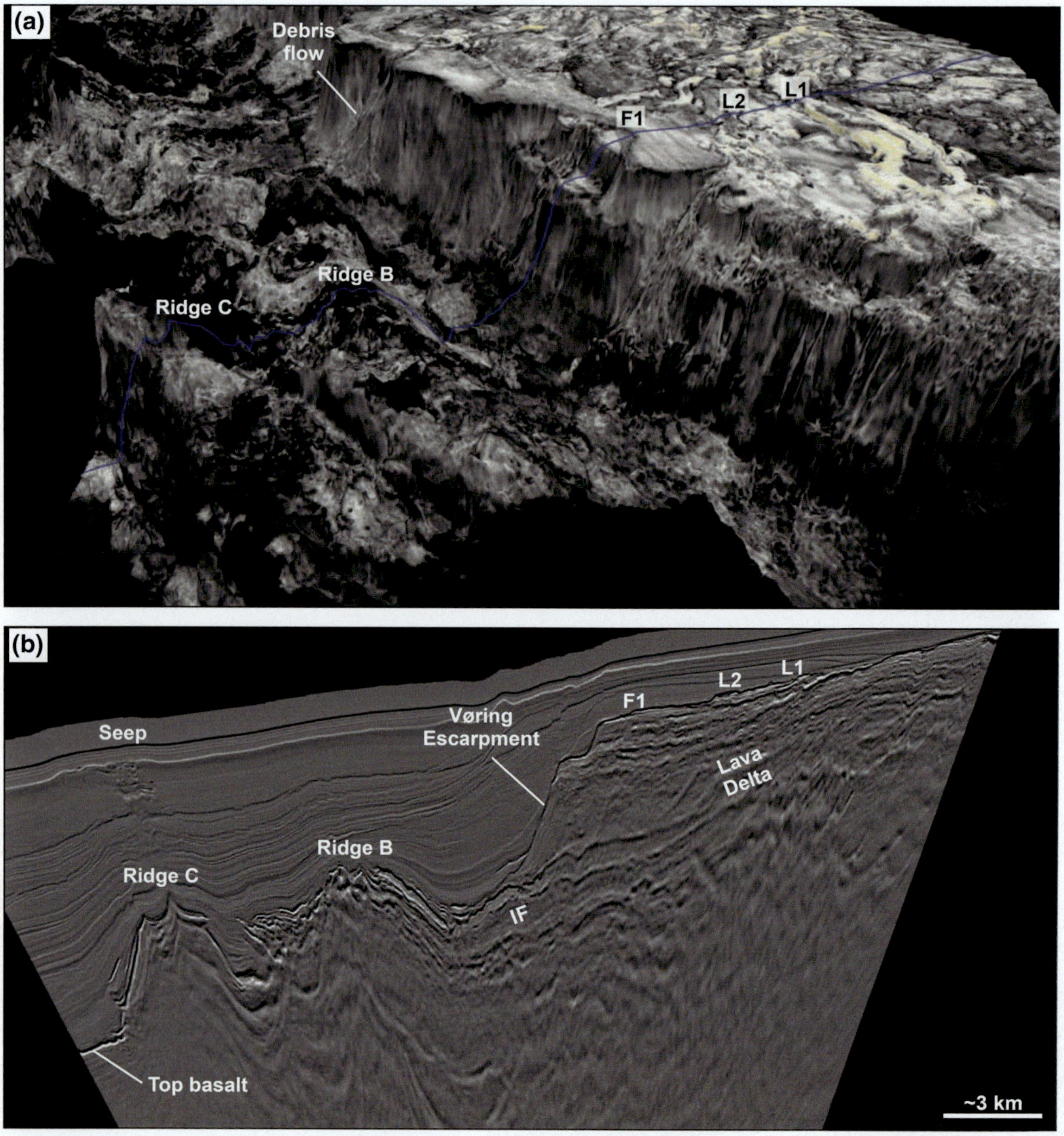

Fig. 15.23 Example showing volcanic seismic geomorphology of the Vøring Escarpment. **a** High-resolution 3D seismic image in perspective view at the top-basalt horizon (time slice) observed from the north. The escarpment is generally on the order of approximately 1 km high with a dip of approximately 25°–30°. High seismic amplitude lava flows transgress across the escarpment edge into spatially coherent aprons of high amplitude chaotic reflectors. These features are interpreted as volcanic debris flows connecting the landward flows domain on the marginal high with inner flows domain at the basin floor across the escarpment. **b** Seismic profile along the blue line in (**a**). Location of fault scarp (F1) and lava flows (L1, L2) shown; IF = inner flows (see Fig. 15.22). (From Planke et al. 2017)

middle part and dip eastward and westward. In the eastern part, the seismic unit is affected by faulting that is downthrown eastward. The top of this seismic unit lies at a water depth of 130–150 m. This unit corresponds to the FST/LST deposited during the Last Glacial Maximum (18 ka), when the sea level fell to 120 m below the present. Westward, a mounded shape (YT2) with a chaotic seismic texture or facies, ~2 km wide and 75 m high, covers the FST/LST. Based on the stratigraphic position, this mound is interpreted as a volcano formed during the same eruptive period as the Neapolitan Yellow Tuff. A seismic unit with parallel reflectors and high amplitude lies above the YT2 volcano and the FST/LST. This unit corresponds to the G3 unit filling the Bay of Pozzuoli and is interpreted as having been deposited between 15 and 8 ka. A seismic unit with reflection-free facies extends vertically across the FST/LST and G3 units. Sedimentary units are downthrown ~30 m westward. This unit is interpreted as a magmatic body

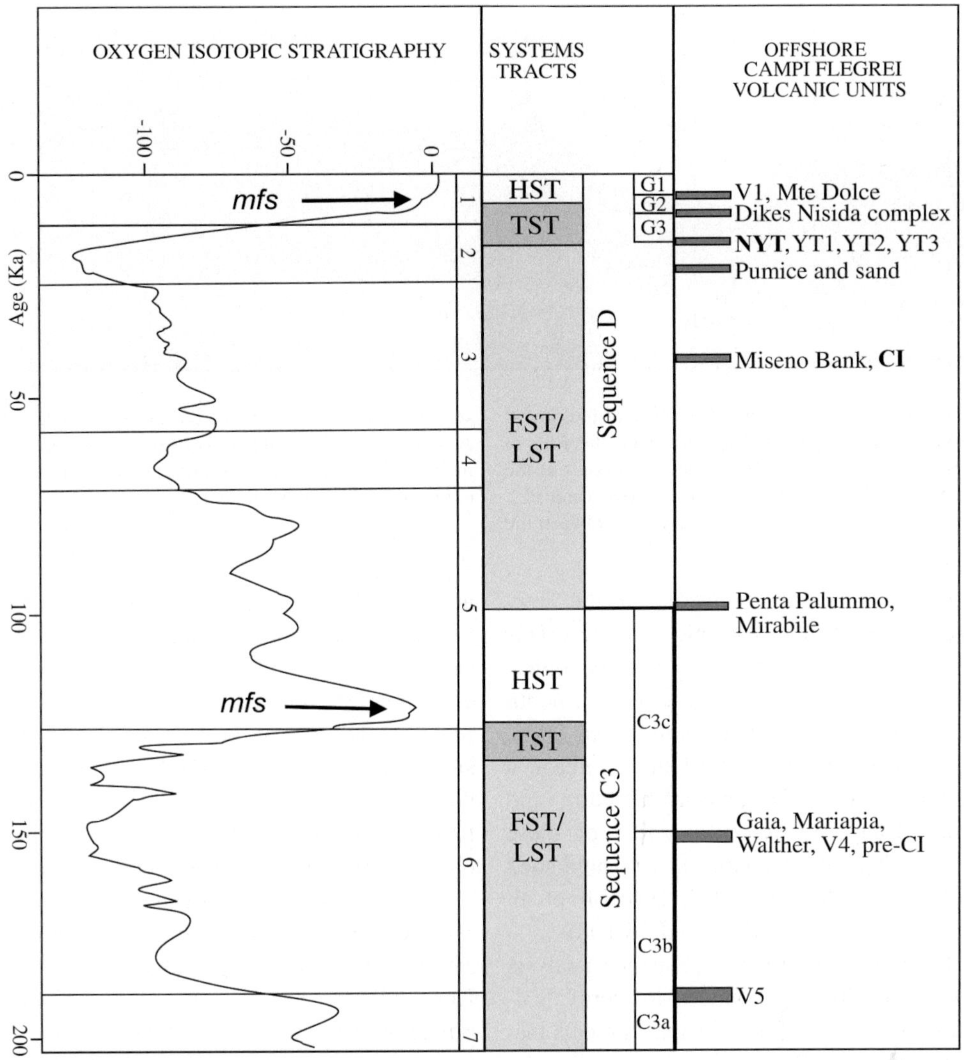

Fig. 15.24 Sequence stratigraphic scheme showing the relationship between systems tracts of the Late Quaternary depositional sequences and main volcanic eruptions, offshore Campi Flegrei, Naples, Italy. Standard marine oxygen isotope stratigraphy shows rising relative sea level to right; mfs = maximum flooding surface; HST = highstand systems tract; TST = transgressive systems tract; LST = lowstand systems tract; FST = falling systems tract. CI = Campanian ignimbrite; NYT = Neapolitan Yellow Tuff; see text for other relevant volcanic unit terminology and discussion. (From Milia 2010)

intruded along a fault; it is covered by a thin seismic unit characterised by parallel reflectors with high amplitude and good continuity, corresponding to the G2 sedimentary unit deposited in the Bay of Pozzuoli.

In the eastern part of Fig. 15.25, the seismic section crosses the Nisida Bank. This corresponds to a volcanic mound characterised by high amplitude reflectors and good continuity with convex geometries. The internal architecture of the reflectors displays a lateral slump with a concave scarp and a body with chaotic facies at the base of the scarp. The slump features are covered by minor mounds climbing upward, which represent the form of a complex tuff cone volcano (10–8 ka). Further eastward (off the coast Naples), it is possible to recognise the superposition of three thick units with chaotic seismic facies. These are interpreted to represent the superposition of thick ignimbrite units; in particular, the 39 ka Campanian ignimbrite, intercalated between the pre-Campanian ignimbrite tuff (~105–157 ka), and a chaotic unit of pumice and sand (18–15 ka).

15.4.3.4 Seismic Architectural Elements and Volcanic Systems

Kora is a submarine, Miocene andesitic polygenetic stratovolcano that is buried by 1–2 km of Pliocene-Pleistocene sediments in the Taranaki Basin, offshore New Zealand. It was discovered during petroleum exploration during the mid-1980s. Drilled in 1988, the Kora-1 well-encountered hydrocarbons in volcaniclastic reservoirs (Chap. 18,

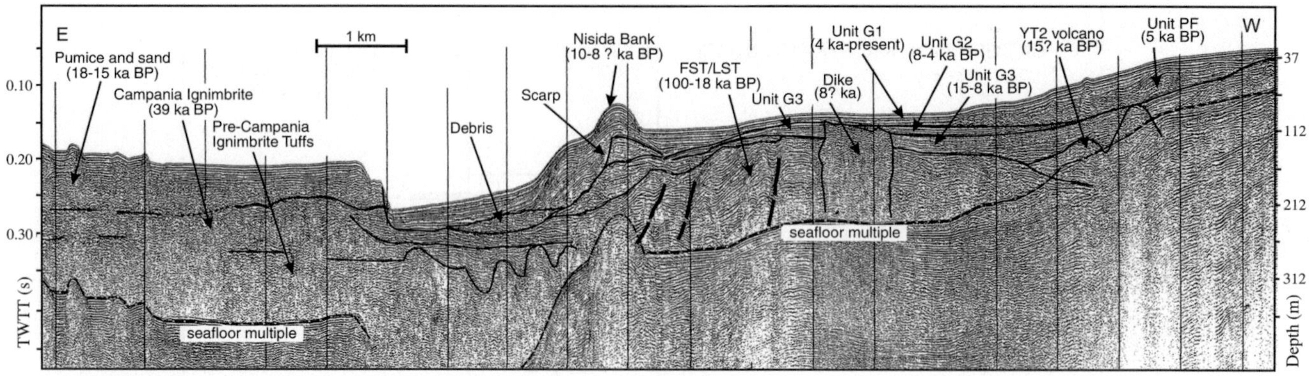

Fig. 15.25 Surf-boom seismic profile (looking south) and interpretation displaying a complex stratigraphic architecture from the Bay of Pozzuoli (W) to the Bay of Naples (E), Italy. In the eastern part of the succession the Nisida Bank tuff cone, dike, pyroclastic flows (unit PF), and a minor mounded volcanic unit (YT2) are interlayered between the sedimentary succession; FST/LST = falling/lowstand systems tract. A succession of ignimbrites is present in the eastern part of the section including the Campanian ignimbrite. See text and Fig. 15.24 for further explanation. (From Milia 2010)

Sect. 18.9). Although deemed sub-commercial at the time, Kora-1 demonstrated significant volcaniclastic reservoir properties and a viable petroleum system. To better understand Kora's complexities and better access its petroleum potential, modern 3D seismic data was acquired in 2006, thus providing a somewhat unique opportunity to image a larger volcanic edifice, its stratigraphic architecture and magmatic plumbing (Chap. 8). Recently some of these dramatic profiles have become available in the published literature (Infante-Paez and Marfurt 2017; Bischoff et al. 2017; Morley 2018; see Figs. 8.26, 8.27, and 18.108).

In a departure from a typical seismic sequence analysis approach, Bischoff et al. (2017) introduced the term "*volcanic system*" to describe the set of interrelated elements that form a complete magmatic-sedimentary complex (i.e. the pre-magmatic sedimentary strata deformed by magmatism, the intrusive and eruptive phases of the volcano, and the sedimentary strata that buried it). Within the Kora volcanic system three distinctive stratigraphic units, termed pre-magmatic, syn-magmatic, and post-magmatic sequences were identified, which are composed of 22 individual architectural elements (Fig. 15.26). Sediments deposited during the pre-magmatic sequence are undeformed by magmatic processes. The syn-magmatic sequence is characterised by rock units formed during the emplacement of intrusive bodies, eruptions, and time-equivalent epiclastic sedimentation, as well as by deformation of pre-magmatic strata due to magmatism. Sedimentary processes such as erosion and burial of the volcanic edifices dominate during the post-magmatic sequence after volcanism ceased.

The syn-magmatic sequence can be subdivided into syn-intrusive, syn-eruptive, and inter-eruptive stages (Fig. 15.27). Syn-intrusive architectural elements are characterised both by magmatic intrusions, and the resulting deformation of strata. Syn-eruptive architectural elements are controlled by volcanic eruptions and include both the volcanic and related sedimentary deposits triggered by eruptions, while inter-eruptive architectural elements are characterised by sedimentary deposition formed during periods of volcanic quiescence. The post-magmatic, erosional time interval is marked by high rates of degradation of the volcanic edifice. It is first defined by chaotic seismic facies in downlap termination over syn-magmatic reflectors on the flanks of the volcanic edifice (3a, Fig. 15.26). These seismic facies change laterally to continuous and parallel reflectors that represent sediments deposited in deep-water settings and sampled in wells tens of km distant from the Kora edifice. These wells in combination with seismic reflection data indicate that Kora volcano was progressively buried in slope to shelf environments. The base of the burial period is marked by large-scale clinoforms prograding from southeast to northwest during the Late Neogene (3b, Fig. 15.26). The 3D seismic images reveal that the volcanic edifice has had a strong influence in the development of the enclosing sedimentary systems, persisting for millions of years after the complete burial of the volcano.

15.4.3.5 Volcanic Stratigraphy—The Future

From the above discussion, it should be apparent that the systematisation of volcanic stratigraphy is a "*work in progress*". As stated by Martí et al. (2018), although the stratigraphic complexity of volcanic terrains often hinders the application of conventional methods, this should be no reason for behaving inconsistently when defining the stratigraphy to be used in a study. Otherwise, the results will not be comparable and, more importantly, considerable confusion may be created regarding the real meaning of volcanic stratigraphy. Clearly, a systematic approach needs to be adopted if we are to make volcanic stratigraphic studies more valuable as a research tool. As volcanic terrains are

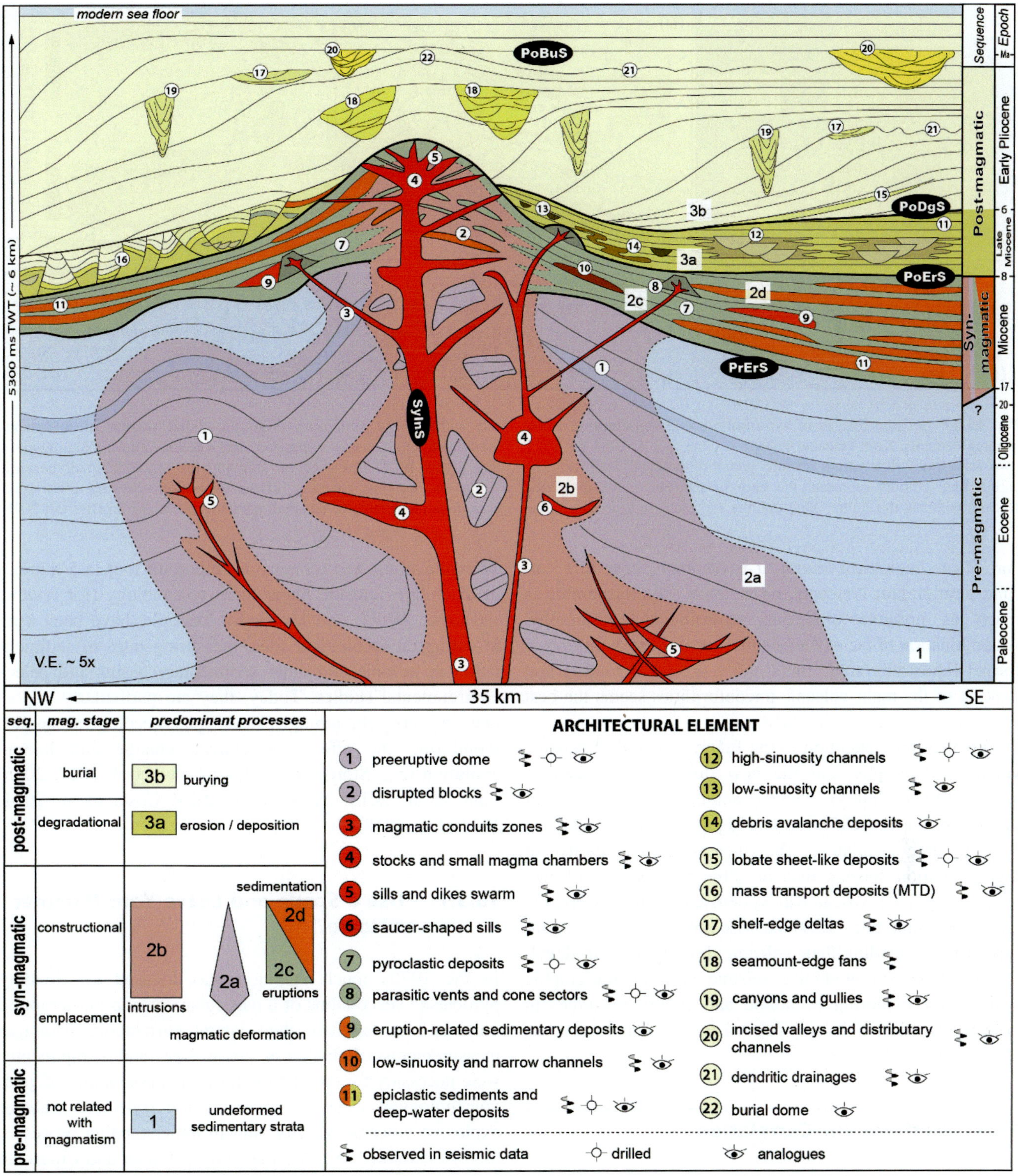

Fig. 15.26 Schematic representation of the stratigraphic sequences, magmatic stages, architectural elements, and predominant processes associated with the Kora volcanic system, offshore Taranaki Basin, New Zealand. Numbers in circles represent the location of the architectural elements indicated in the key; numbers in squares indicate predominant processes in the key. The terms "observed in seismic" represent architectural elements found in seismic data, "drilled" those sampled by wells, and "analogues" examples recognised in modern and ancient analogues. Conduit zones and dykes exaggerated for visualisation. Acronyms are SyInS = syn-intrusive surface, PrErS = pre-eruptive surface, PoErS = post-eruptive surface, PoDgS = post-degradational surface, and PoBuS = post-burial surface. (From Bischoff 2018)

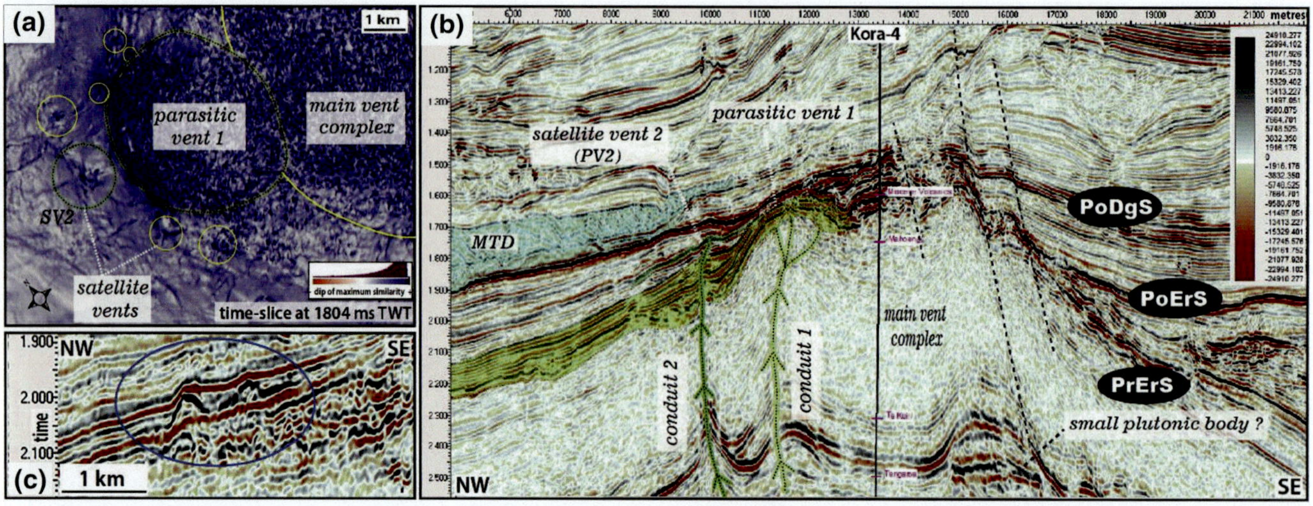

Fig. 15.27 Seismic images of syn-intrusive and syn-eruptive architectural elements, Kora volcanic system. **a** Time slice at 1084 ms TWT (two-way time) illustrating seismic geomorphological features. The darker blue colours represent the core of parasitic vents and cone sectors, whereas the lighter blue colours represent sedimentary strata that onlap these structures. Note the radial pattern of lineaments associated with the smaller parasitic and satellite vents circled in yellow and green. **b** 2D cross-section showing the relationship of parasitic vents and conduit zones. **c** 2D cross-section showing mound-like structures interpreted to be small parasitic vents. (From Bischoff 2018)

part of the geological record, it is common sense that volcanic stratigraphic concepts and methods should resemble as much as possible those of conventional stratigraphy. A combination of facies models that permits the definition of formal lithostratigraphic units and a precise definition and ranking of the main volcanic unconformities seems the best way forward (Martí et al. 2018).

We also draw attention to the future potential of seismic volcano-stratigraphy and the new discipline of volcanic seismic geomorphology in understanding the development of offshore basins especially as more high-resolution 3D seismic data become available. The increasing availability of high-resolution seismic data in some continental volcanic terrains (e.g. East African Rift System and onshore Mesozoic basins of China; Sect. 18.9) likewise holds great promise for further understanding their volcanic development. In developing robust and predictive seismic volcano-stratigraphic systems for exploring economic and energy resources in volcanic terrains, it is only logical that volcanologists are closely involved.

15.5 Methods for Correlation

The stratigraphic analysis of modern pyroclastic successions, and hence by default Quaternary volcanoes, rapidly evolved in the 1970s and 80s largely out of the study of *tephrochronology*. Tephrochronology employs the correlation and radiometric dating of discrete layers of volcanic ash (or tephra) from a single eruption to create a chronological framework. Tephrochronology, however, has a wide variety of applications, and remains an important tool in a number of disciplines outside mainstream volcanology (for review, see Lowe 2011). Ash layers, for example, have been used in dating archaeological sites, measuring rates of sedimentation in oceanic and other sedimentary basins and paleoenvironmental studies. Today, they are particularly important in paleo-climatological and paleo-ecological studies monitoring the effects of climate change and human evolution (e.g. Storey et al. 2007; Lowe et al. 2012, 2015; Lane et al. 2013; Mark et al. 2014; Costa et al. 2014; Yost et al. 2018).

15.5.1 "Use a Spade and Leave Your Hammer at Home!"

In the 1970s much attention was focused on dividing pyroclastic successions of a particular volcano into eruption units and to facilitate their lateral correlation. Although measurements of thickness, grain size, and componentry were necessary for correlation, they were potentially of great volcanological value for assessing the style and scale of explosive volcanic activity, and in understanding processes (e.g. Nakamura 1962; Aramaki 1963; Walker and Croasdale 1971; Booth et al. 1978). The usual technique is to start by constructing or measuring stratigraphic logs at all available outcrops (including digging pits), and then piece together the volcanic stratigraphy from a number of key locations by drawing correlation panels (Cas and Wright 1987; e.g. Fig. 15.28). In this way studies can be made on the whole volcanic succession, or just on the lithostratigraphic units

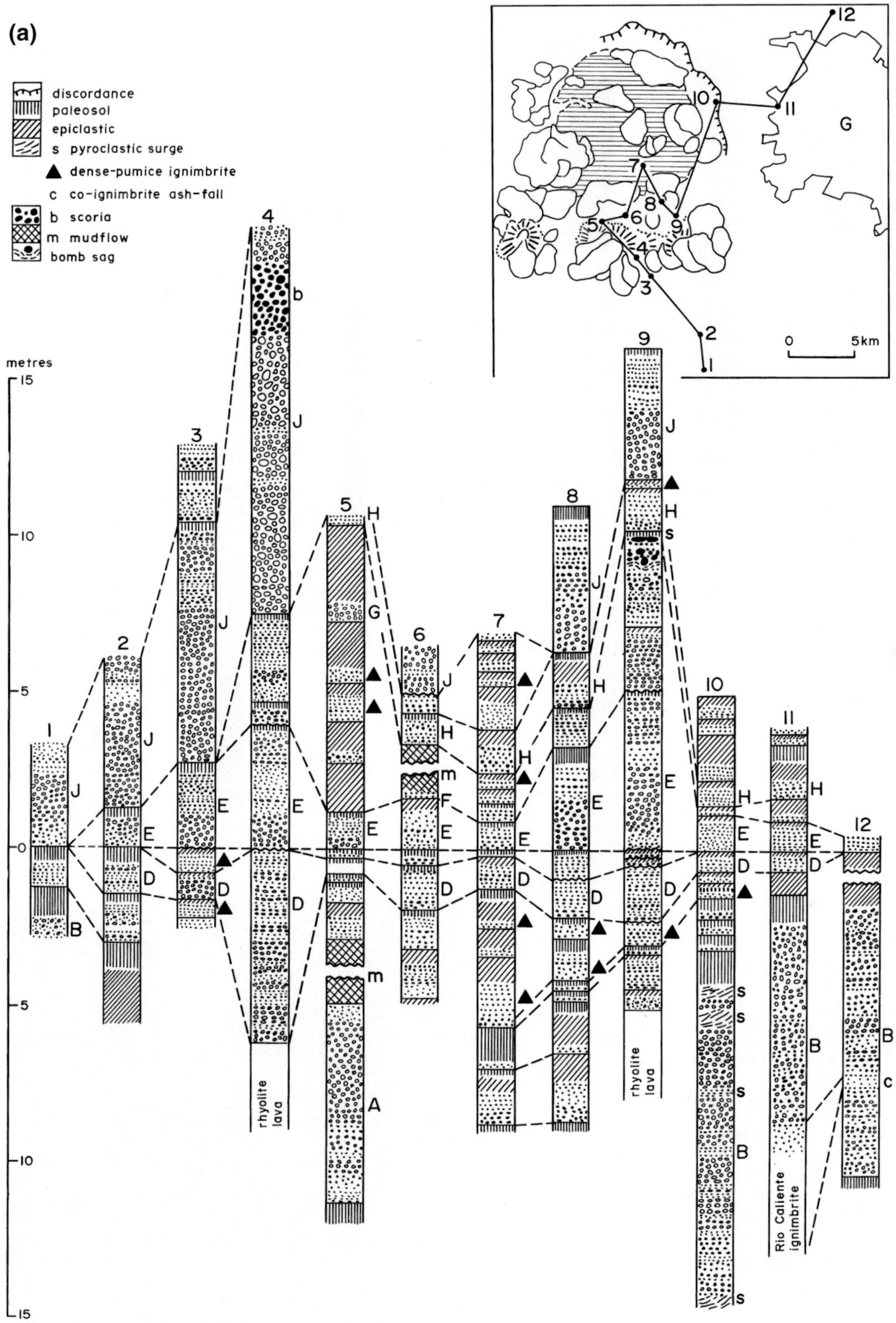

Fig. 15.28 Correlation panels illustrating the use of logged sections from a number of key locations to construct volcanic stratigraphy. **a** Measured stratigraphic sections showing lateral variations and established correlations for the rhyolitic pumice fallout succession from La Primavera volcano, Mexico (From Cas and Wright 1987, After Walker et al. 1981). **b** Measured graphic logs showing lateral variations and established correlations of the Colli Albani volcanic succession, central Italy. See Fig. 15.15 for nomenclature of the various main formations and correlation with edifices. (Modified from Giordano et al. 2010b)

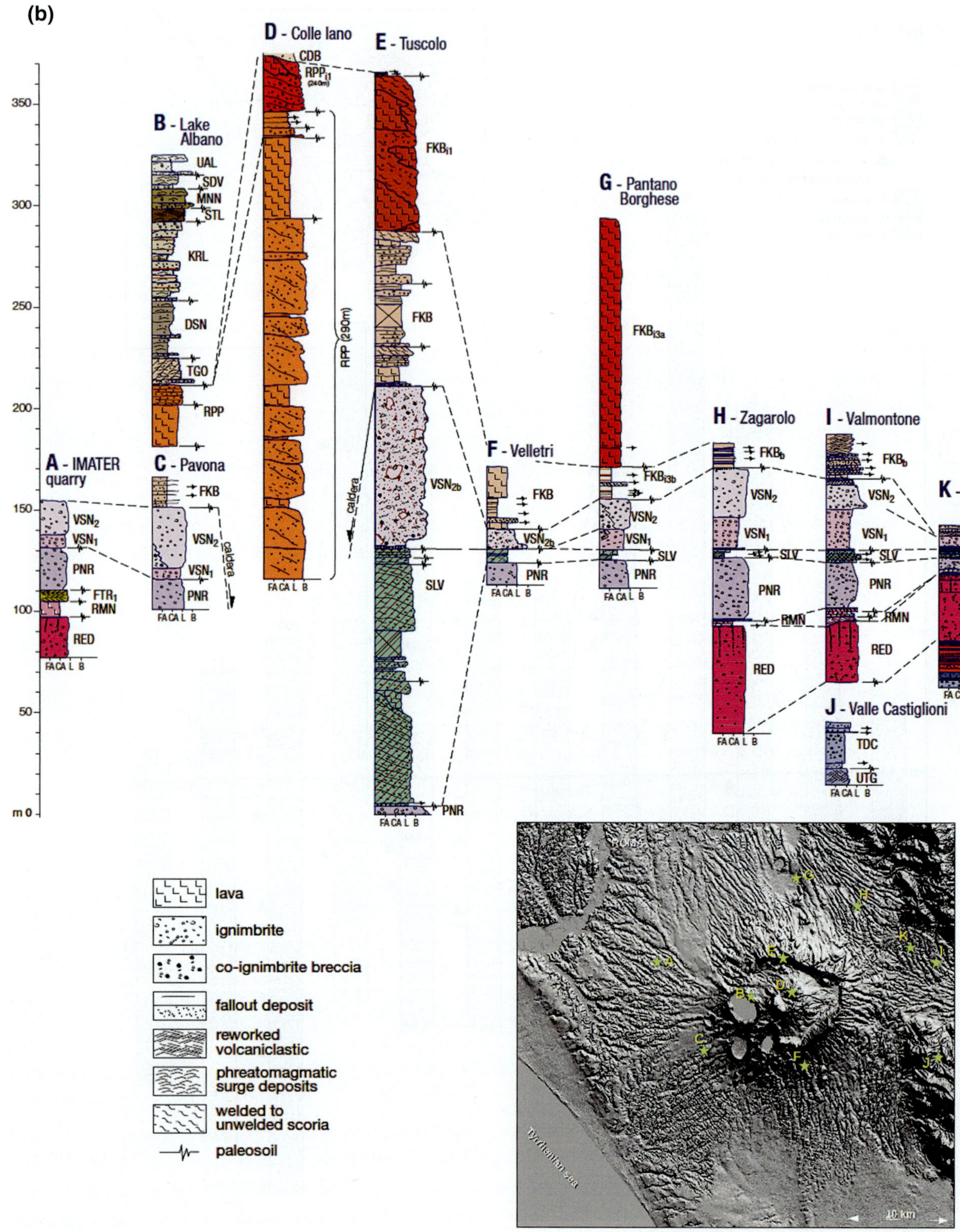

Fig. 15.28 (continued)

that relate to an individual eruption or event, or separate layers accumulated during different phases of the same eruption.

It is essential to be aware that the deposits of different eruptions are often dispersed differently around a volcano, and transport and depositional mechanisms (lava, fall, flow, or surge) will also control distribution, as well as the prevailing atmospheric and seasonal conditions (e.g. Roobol et al. 1985; Chap. 9). Deposits accumulated during the same eruption can show just as many spatial complexities in their distribution, including those deposited by syn-eruptive resedimentation. It is therefore very unlikely that any one section will show the whole stratigraphy of a volcano, or even a large part of it, or even of a single eruption. In addition, where volcanoes are in close proximity, deposits erupted from different centres will also overlap and interfinger. The whole stratigraphy is further complicated by epiclastic processes of reworking and mass-wastage.

Once individual stratigraphic logs have been measured in multiple locations across the volcano, either for an individual unit or for a package of units, correlation should be done using standard stratigraphic methods, primarily constructing cross-sections and also with the aid of correlation panels restored to more than one datum plane (i.e. flattened isochronous surfaces or synchronous beds; see Krumbein and Sloss 1963; De Rita et al. 1998; Fig. 15.29). In this method, the beds or stratigraphic surfaces that can be widely traced are used as single successive datum planes or marker horizons to correlate the stratigraphic sections, and the key beds or surfaces are "flattened" to restore the original stratigraphic relationships among the units below. In the example shown from Roccamonfina volcano, Italy (Fig. 15.29), the eruptive rock unit (synthem) is organised into an overall forestepping-backstepping sequence, which is interpreted to reflect dynamic waxing-waning eruptive and emplacement processes (De Rita et al. 1998). The internal geometry of the synthem, as well as other examples at Roccamonfina, indicate a common forestepping phase until the eruption rate peaks; during the subsequent decline in eruption energy, deposition occurs upslope (backstepping). By reconstruction of volcaniclastic rock sequences in this way it is also possible to appreciate the role of paleotopography in controlling the lithofacies association, and to reveal the presence of syn-volcanic versus post-volcanic tectonic deformation. For example, the lateral correlation of a litho-stratigraphic unit that is interpreted to be related to a single eruptive event (e.g. a caldera-forming ignimbrite succession) is only proven when its geometry and facies association are consistent with the paleotopography and with the emplacement mechanism. For further discussion see De Rita et al. (1998).

The best examples of key marker beds are individual fallout deposit layers. Examples of key stratigraphic surfaces are erosional and non-erosional unconformities and paleosoils; however, as discussed above, such surfaces are not necessarily synchronous, so when using them as a datum, an assessment could be made on their age range of formation and extent by dating the units below and above. Today, with the universal availability of a variety of radiometric dating methods and laboratories, where possible, stratigraphic sections should be geochronologically dated. These provide absolute age constraints and can be most important for correlation, and for accurate assessments of, for example, volcanic hazards and determination of magma production rates and magma dynamics. Close collaboration with geochronologists during fieldwork and laboratory analyses is essential for generating an accurate interpretation of radiometric results and establishing the chrono-stratigraphy of the volcanic succession. We also echo the words of Martí et al. (2018), that volcanic stratigraphy must be established first with field criteria, then be confirmed with geochronological data, and should never be carried out in the opposite way. In other words, it is critical that the "*tail does not wag the dog*"! The reason for this is that age dates can always be affected by anomalous factors (e.g. presence of xenocrysts, multiple growth rims on single crystals, excess argon, hydrothermal and metamorphic resetting, etc.), whereas physical stratigraphy does not lie (if done properly) based on detailed and extensive mapping and assessment of contact relationships. So be prepared to get hot and sweaty by doing the required fieldwork!

15.5.2 Geochronology

The dating methods most used for establishing a chrono-stratigraphy in volcanic successions are ^{14}C or radiocarbon dating of organic matter, which can be used for the youngest deposits (<50,000 yr BP); cosmogenic radionuclide (e.g. ^{26}Be, ^{26}Al, and ^{3}He) dating is also used for the youngest volcanic surfaces, theoretically greater than a few hundred yr BP (Darvill 2013; Brown et al. 2012); U–Th or U-series dating techniques have an older limit of ≥ 500 kyr; fission-track analysis (apatite, sphene, zircon, or volcanic glass) has a range of 100–200 Myr, although low-uranium epidote and garnet have been used to date much older samples; U–Pb dating is used for ancient volcanic rocks (>1 Myr), especially Precambrian successions; whereas $^{40}K/^{40}Ar$ and $^{40}Ar/^{39}Ar$ dating are the most versatile, with $^{40}Ar/^{39}Ar$ dating widely used on both Quaternary and ancient volcanic deposits where the method is now capable of accuracy comparable to that of the U–Pb system (Renne et al. 2010), down to as young as 4 kyr BP, especially in high K volcanics with abundant groundmass microlites (e.g. Wijbrans et al. 2011).

The usefulness of K-Ar and Ar-Ar dating depends on the material to be dated, i.e. on the amount of potassium present

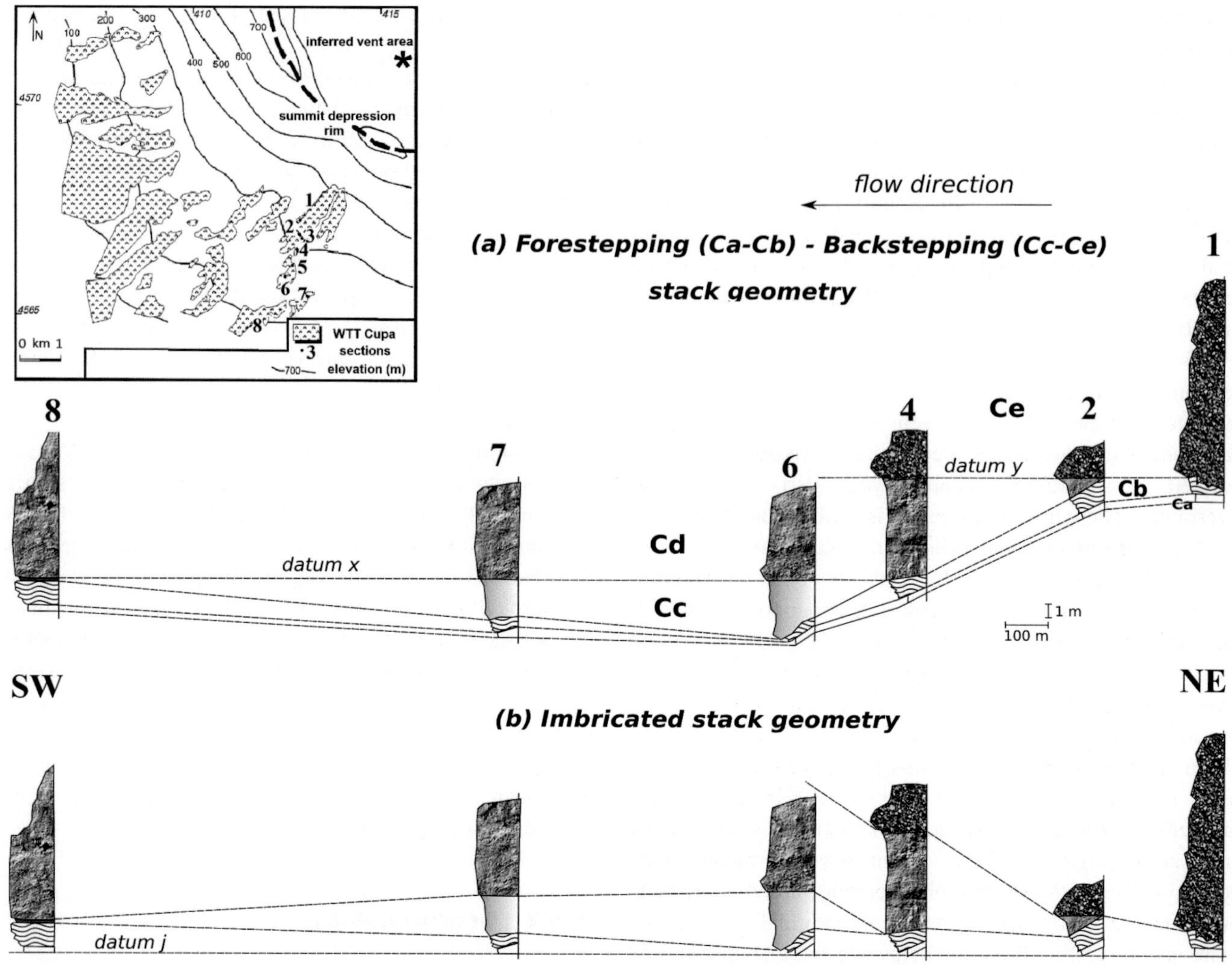

Fig. 15.29 Correlation of stratigraphic logs by flattening with respect to different datum planes within a pyroclastic eruption unit: the White Trachytic Tuff (WTT) Cupa synthem, Roccamonfina volcano, southern Italy. Ca and Cb are pyroclastic surge deposits, while Cc, Cd, and Ce are ignimbrites. (Modified from De Rita et al. 1998). **a** Datum planes "*x*" and "*y*" represent lower order unconformity surfaces bounding subsynthems; these are directly related to the eruptive activity. For this reason, their lateral extent is normally limited, but they need to be identified across the whole extent of their synthem. Unconformities that cannot be correlated laterally within the entire synthem reflect local internal discontinuities, such as caused by variations in pyroclastic flow emplacement induced by topography. This method of restoration allows one to correctly reconstruct the lenticular geometry of the units, in a fore-stepping-backstepping stack, with respect to paleotopography. By contrast, in **b**, by flattening along the base of the succession (datum "*j*"), the same logs would give an incorrect overall imbricated stack geometry with respect to flow direction, which may be misinterpreted as due to ramping of flow units on top of each other

and the age of the deposit. The older and the higher the potassium content in the volcanic deposit to be dated, the better the resolution that can be achieved. Laser ablation on single crystals provides the best results, although a lot of care must be taken in selecting true phenocrysts, not to be confused with antecrysts and xenocrysts which may give older ages. The true date of the eruption should be given by the age of the quenched groundmass; however, more generally this is variably altered and devitrified, not as rich in potassium as some crystals (such as sanidine), and difficult to separate to avoid the presence of lithics or crystals. Therefore, in practice the selection of single crystals is the best way to date volcanic deposits, provided that crystals are separated from the primary groundmass (e.g. a lava or a pumice sample, and not taken loose from the ash matrix); crystals should be carefully assessed petrographically to be in equilibrium with the groundmass. Sanidine can be routinely dated where ages are older than 70 kyr, while the practical younger limit of plagioclase is 200 kyr; in rare cases some minerals with high potassium contents (e.g. leucite), as well as whole rock dating of some basalts with abundant groundmass plagioclase microlites can be reliably dated if they are as young as a few thousands of years. Low potassium, Holocene deposits, such as Icelandic basalts for example, are difficult to date and another technique, *archeomagnetism*, has recently become a good dating approach at active basaltic volcanoes (e.g. Pinton et al. 2018; Risica et al. 2019; Magli et al. 2022).

For ancient volcanic successions U–Pb SHRIMP analysis of zircon is routinely used, together with monazite (U–Th) for silicic to intermediate rocks, and U–Pb, Pb–Pb analysis of baddeleyite (ZrO_2) for mafic and ultramafic rocks. However, with the increasing effects of deformation and metamorphism the actual volcanological or magmatic meaning of the separated population(s) of such minerals may be more problematic. For example, during certain metamorphic fluid events baddeleyite can be destroyed and even the more stable zircon and monazite are known to be reactive, can dissolve, be replaced and/or reprecipitate new crystals (Geisler et al. 2007; Condit et al. 2018). In such settings the only unequivocal technique that will control the reliable radiometric age determination of metavolcanics, is one in which a dateable mineral (zircon and/or monazite, or baddeleyite) is analysed by ion microprobe (SHRIMP) or electron microprobe (EPMA) in situ, is constrained texturally and forms part of the total (including metamorphic) paragenesis of the sample being analysed. In addition, the internal zonation needs to be assessed first using cathode luminescence imagery, to ensure that the last magmatic growth zone is being dated to assess the best-estimate eruption age. In Archean rocks, the best resolution with SHRIMP age dating provides an error of ±2–4 Myr in the most optimum situation. Development of high resolution TIMS (Thermal Ionisation Mass Spectrometry) dating of zircon or baddeleyite can produce error ranges <1 Myr in rocks of Archean age (e.g. Hayman et al. 2015). Zircon populations are carefully separated from their host rock, and then assessed microscopically before dating, to identify the least altered crystals for dating. In volcaniclastic rocks of Archean age, mixed zircon populations are commonly found. Rapid LA-ICPMS (Laser ablation inductively coupled plasma mass spectrometry) age dating of large populations of zircon crystal separates can be undertaken. Each age determination generates an average age for each mineral grain, and then statistically similar populations can be distinguished. The youngest coherent population then gives a maximum depositional age for the sediment (e.g. Squire et al. 2010). Older populations usually reflect xenocrystic or older source provenance. If a volcaniclastic sediment overwhelmingly contains one age population it can be interpreted to have been deposited almost directly, penecontemporaneously from a single volcanic source, and even a major eruption event (e.g. Squire et al. 2010).

15.5.3 Biostratigraphy

Traditional biostratigraphy may also be very useful in establishing a chronostratigraphic framework in some volcanic settings, in particular marine environments. Fossil contents provide the relative ages of sediments which are intercalated with volcanic units. A good example where biostratigraphy has been employed to provide a detailed chronostratigraphic reference for lithostratigraphic correlation is the Ordovician volcanic succession found in North Wales (Fig. 15.17; Rushton and Howells 1998). The criteria for recognising stages and substages are based largely on the shelly faunal succession (mainly brachiopods and trilobites). There are also important graptolitic faunas present at some levels, which are assigned to established graptolite zones, as well as microfossils (acritarchs and chitinozoa).

15.5.4 Geochemistry

Geochemistry has many uses in volcanology. First, it helps to characterise the composition of rock units/deposits, which is a key aspect of discriminating different stratigraphic units. Secondly, it can be used in correlation and tephrochronology, and third it is fundamental in understanding the petrogenesis of magmas from each volcano. However, only where a detailed stratigraphy has been established, can petrogenetic trends within that stratigraphy and during the history of the volcano be identified.

Geochemical analysis involves first-order major and minor element analyses, and more detailed trace element, rare earth element and isotopic analyses, depending on the specific aims of the study. Before geochemistry is undertaken on any sample, detailed petrography should be undertaken to assess to what degree the sample is contaminated by xenocrystic or accessory lithic components, and to see if they can be separated from the juvenile magmatic components (phenocrysts, groundmass). In addition, petrography will give an indication of the degree of alteration of a sample (see Chap. 13). Highly contaminated and altered samples do not produce accurate original magmatic compositions.

Samples of lavas and subvolcanic intrusions should be as fresh and free of antecrystic, xenocrystic, and xenolithic material as possible. In pyroclastic deposits, only fresh, uncontaminated juvenile clasts of pumice, scoria, and juvenile bombs should be analysed, again after undertaking petrographic examination of juvenile clasts. Where mineral chemistry analyses are to be undertaken, again, only juvenile magmatic crystals should be analysed, which could involve painstaking separation of crystal fragments from crushed samples, or very careful targeting of magmatic crystals in in situ (e.g. microprobe, ICP-MS laser ablation) analyses. For old volcanic rock samples, the likelihood of some alteration or metamorphism is greater than for younger rocks, and so analytical strategies need to be carefully considered. Again, detailed petrography before analysis is essential. Even so, in old volcanic successions, many rocks are so altered that only so-called stable or immobile minor and trace elements (e.g. Zr, Ti, Y, Nb) can be used to decipher geochemical affinities and origins. In the end, some rocks are so altered that chemical analysis does not truly reflect original composition, but the alteration assemblages and history of the rock. The isocon method (not to be confused with isochron) can be used to statistically assess the degree to which particular elements in a rock deviate from the norm, using ratios of elements to immobile elements (Grant 1986, 2005).

As well as geochemical analysis for routine and advanced petrogenetic investigations, in some studies of pyroclastic deposits *geochemical fingerprinting* can be essential for correlation and tephrochronology. This has been very successively used, specifically, in correlation of deep-sea and distal ash layers (Chap. 9; e.g. Ninkovich et al. 1978; Song et al. 2000; Smith et al. 2011, 2016). Rapid and routine electron microprobe analysis of individual glass shards now provides a particularly powerful correlative tool (Lowe 2011; Lowe et al. 2015; Ponomareva et al. 2015, and references therein).

Geochemistry, of course, is also used to model the evolution of magmas and their petrogenesis. Unfortunately, such studies frequently fall short of achieving their full potential because of poor sampling strategies in the field on volcanoes. When suites of samples are collected over a volcano they will of course show variation trends when the samples are analysed and analytical data are plotted graphically, because magmas will show variations from one batch to another, or even within a single batch. However, variation trends from analyses of such bulk sample suites will only show crude first or second-order trends.

On the other hand, if samples are collected within a well-established chrono-stratigraphy, any variations can be related first to the unconformity-bounded stratigraphic sequences that they are contained within. Comparisons can then be made with compositional patterns and trends in overlying and underlying sequences, that by definition are separated by a significant period of time, to assess if differences from one eruption sequence to another are significant. Do differences indicate injection of a new magma batch and source, or evolution of a pre-existing magma source? Such a detailed interrogation strategy then allows identification of third and fourth-order differentiation or fractionation trends, if they exist. For example, monogenetic volcanoes have commonly been considered to be mono-magmatic, based on historic bulk sampling at just one or two random localities within the volcano sequence. Such a sampling strategy assumed that the magmatic composition is uniform because such volcanoes involve very small erupted volumes. However, detailed mapping, definition of a detailed eruption stratigraphy and sampling within such a stratigraphy, has now identified several cases where it is clear that more than one magma batch was erupted during the history of the small monogenetic volcanoes (e.g. Németh et al. 2003; Van Otterloo et al. 2014; Boyce et al. 2015). If such is the case with small volcanoes, larger ones have the potential to hold a treasure trove of yet unrecognised geochemical complexity, if they are sampled within a detailed chrono-stratigraphic framework.

In summary, if you want to optimise petrogenetic studies, you must first map and establish a chrono-stratigraphy and sequence of events, or work with a team of researchers who have done that. Sampling should only be undertaken in such an established stratigraphic framework.

For those interested in how to use and interrogate geochemical data, the book by Rollinson (2014) remains invaluable. For those interested in petrogenetic modelling the widely used MELTs suite of spreadsheet-based petrogenetic calculators (MELTs, Ghiorso and Sacks 1995; pMELTs, Ghiorso et al. 2001; and Rhyolite MELTs, Gualda et al. 2012), and a computerised approach (Janoušek et al. 2016) may be useful.

15.6 Additional Tools: Remote Sensing, Geophysics, and Drill Holes

Mapping both modern and ancient volcanic terrains may require additional tools in combination with routine field work. This is particularly true where outcrops are poor or scarce, logistics are difficult, and where dissection of the volcanic pile is not enough to allow the reconstruction of the 3D geometry of the volcanic succession. Remote sensing (satellite or aerial imagery) is one of the best tools to be used in poorly accessible areas, where little geology is known, and/or in hazardous terrains (Teeuw 2007). This has evolved from early Landsat-based studies in the late 1970s of some of the world's most remote regions (e.g. Francis and Baker 1978; Francis and de Silva 1989), to detailed studies of morphology and morphometric analysis of volcanoes and volcanic features (e.g. Flynn et al. 2001; Inbar et al. 2011; Grosse et al. 2012, 2014), to monitoring active volcanoes (Ramsey and Flynn 2004; Reath et al. 2016; Mia et al. 2017). For topical reviews see Kervyn et al. (2007), van der Meer et al. (2012), and Ramsey and Harris (2013).

Integrated with geophysical data sets remote sensing imagery now provide powerful tools for mapping the extent and distribution of volcanic products, vent locations and for understanding the regional context, including the thermal state, of larger volcanic provinces (e.g. Cas et al. 2017; see below).

15.6.1 Remote Sensing Imagery

A large variety of satellite and aerial imagery is now available in some form for all parts of the Earth's surface, including the ocean basins. This includes multispectral images (e.g. Landsat and Enhanced Thematic Mapper Plus, MODIS, ASTER), and digital elevation models or DEMs (e.g. SRTM, LiDAR), all of which can be used to identify different types of terrains, landforms, and volcanic features. Morphology and colours can be used to identify the extent of geological units, such as individual lava flows and pyroclastic flow fans. By working on individual and combinations of spectral bands, it is possible to study the reflectance spectral signature of different rocks, and refine the use of satellite images to identify for example the role of alteration which may then be related to ages of similar lithotypes (for example, lava of similar composition but of different ages). In this way, it is possible to both produce preliminary maps that can be then validated by ground truthing, but also to extend areally spot information that is acquired in the field.

Airborne light detection and ranging (LiDAR) is one of the most rapidly-advancing remote sensing technologies and is found to be particularly useful in surveying younger volcanic terrains (e.g. Mazzarini et al. 2007; Deardorff and Cashman 2012; Whelley et al. 2014; Nomikou et al. 2014, Bisson et al. 2016), even where there is dense vegetation. This collects very high-resolution elevation data using an aircraft-mounted or drone-mounted oscillating laser scanner that emits many pulses per second; these pulses are reflected by ground features and distance reflectance data are recorded.

LiDAR data are particularly useful in vegetated regions, where post-processing of the data provides a close approximation to "bare earth topography", often with <1 m horizontal resolution and a vertical accuracy of <20 cm. This compares favourably with the 20 m resolution for conventional regional DEM, and 30 and 90 m resolutions for the global SRTM, and is able to resolve detailed textural and topographic features that cannot be resolved in the regional and global scale DEM datasets (e.g. Fig. 15.30). LiDAR surface mapping allows rapid, high-precision 3D models to be created with relative ease, which can be very useful in identifying the superposition relationships of volcanic features both at a regional and a local volcano scale. Patterns in surface roughness can be characterised using co-occurrence texture statistics to define various lithologies, e.g. pristine-pyroclastic flow, reworked-pyroclastic, lahar, boulder beds, braided streams, and fresh lava versus eroded lava flows (Whelley et al. 2014). This allows the relative eruption order for each volcanic complex to be determined, which can then be followed up with targeted ground mapping, sampling, and geochronology to establish a chrono-stratigraphy. The same method applies when determining the superposition relationships between discrete volcanic centres.

LiDAR is also proving to be very effective as a hazard analysis tool, as detailed topography information can be used to analyse the potential flow paths of any topography confined lavas, pyroclastic flows, and sedimentary mass flows (Hofton et al. 2006; Joyce et al. 2009; Jessop et al. 2012). This then allows rapid analysis and prediction of the real-time hazards associated with pyroclastic flows or debris flows/"lahars" from a stratovolcano, or eruption of lava flows from a monogenetic basaltic volcanic field, where an eruption may derive from any location at any time within the field. LiDar has also been shown to be effective for monitoring the dispersal of volcanic ash plumes that pose a threat to jet air traffic, such as during the 2010 Eyjafjallajökull eruption in southern Iceland (Ansmann et al. 2010; Sicard et al. 2012; Wiegner et al. 2012). In new developments, ground-based LiDAR (using DIAL or Differential Absorption LiDAR) is being used to monitor volcanic CO_2 flux in volcanic plumes for the early detection and warning of eruptions (e.g. Aiuppa et al. 2017).

15.6.2 Geophysical Data Sets

Depending on the location of the study or project area, a number of regional and/or high resolution digital geophysical data sets may be available. These include heat flow, seismic tomography, regional magneto-telluric, aeromagnetic, gravity, and radiometric image processed data, which in many countries are now routinely collected and published by various Governmental agencies.

Regional heat flow, seismic tomography, and magnetotelluric data are used to locate and interpret the presence of heat flow anomalies, seismic velocity attenuation and magnetotelluric conductivity anomalies in the crust and lithospheric mantle. Thus, in active volcanic terrains they can be used to locate the presence high-temperature zones of melt or partial melt, whereas in less active regions (e.g. no historic eruptions) they may confirm that these are potentially still active. Regional aeromagnetic, gravity, and radiometric imagery have now been publicly available in some countries (especially Australia) since the late 1980s, and therefore have been used routinely in both industry and academic mapping projects of volcanic terrains for many years (see Sect. 17.7).

Regional aeromagnetic images can also be utilised very effectively to map the first-order extent of younger volcanic units where they are in contact with units with much lower

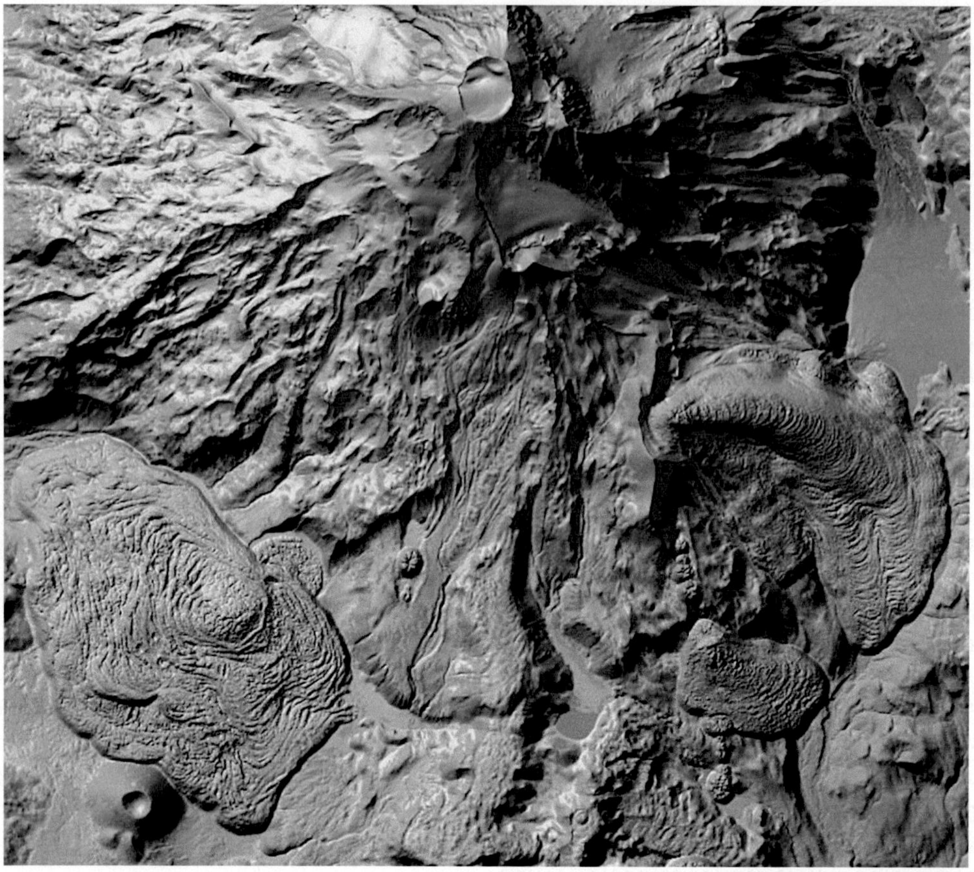

Fig. 15.30 Example of LiDAR DEM captured in a glaciated, active volcanic terrain, Three Sisters volcanic chain, Cascades Range, Oregon, USA. The scene shows the summit crater and southern flank of South Sister (or "Charity") stratovolcano. Activity began between 50 and 45 ka, with eruption of numerous rhyolite lava flows and domes. From 37–30 ka a broad cone was built of rhyolite to andesite domes and flows, which was followed by intense activity between 30 and 22 ka centred on the summit. This fed the lava flows that built up the steep radial apron of the cone and formed an 800 m-wide summit crater. After a quiescent period of some 20 kyr two episodes of rhyolite eruptions occurred; the only postglacial eruptions. The first, about 2.2 ka, produced several small domes including the Rock Mesa coulée at the southwest toe of the volcano. At about 2.0 ka two linear zones opened on the southeast and north flanks. These eruptions formed the Devils Chain, a 5 km-long chain of 20 domes and couleés to the southeast; most of the magma released during this episode was issued from a vent at 2,400 m elevation, which formed the large Newberry coulée. (*Source* Department of Oregon Geology and Mineral Industries, DOGAMI)

magnetic signature. For example, the late Miocene to Holocene Newer Volcanics Province (NVP) of southeastern Australia is extremely well-delineated in regional aeromagnetic data (Fig. 15.31; Cas et al. 2017; Chap. 16). Basaltic lava plains cover a large area of the NVP and have a distinctive stippled magnetic texture and a high magnetic anomaly response. This response is superimposed on the smooth, non-magnetic response of the Otway Basin sedimentary sequence in the south of the province, making the extent of the lavas easily traceable using the magnetic anomaly map (Fig. 15.31a, b), including those areas where lavas are covered by younger regolith or are underwater. The distinctive magnetic signature of the lava flow fields of the NVP immediately allows a first-order map of the extent of the NVP to be drawn, which shows it is more extensive than previously thought because lava flows buried under soil and under water were previously not detectable.

In the northern parts of the province the geophysical response of the basement rocks (Paleozoic metasediment and granites of the Lachlan and Delamerian orogens) is more complex, consisting of large circular-to-ovoid shaped anomalies (magnetic highs or lows) that define granite bodies, and a roughly NNW-SSE-trending magnetic fabric that represents folded volcanic and turbidite sequences. Although the basement response is more complex, the NVP lavas are still distinguishable by their stippled texture, especially if this texture is further enhanced through data- and image-processing techniques. Some lava flows, confined by pre-existing valleys, appear as strong magnetic highs (or lows where they have reverse remanent magnetisation).

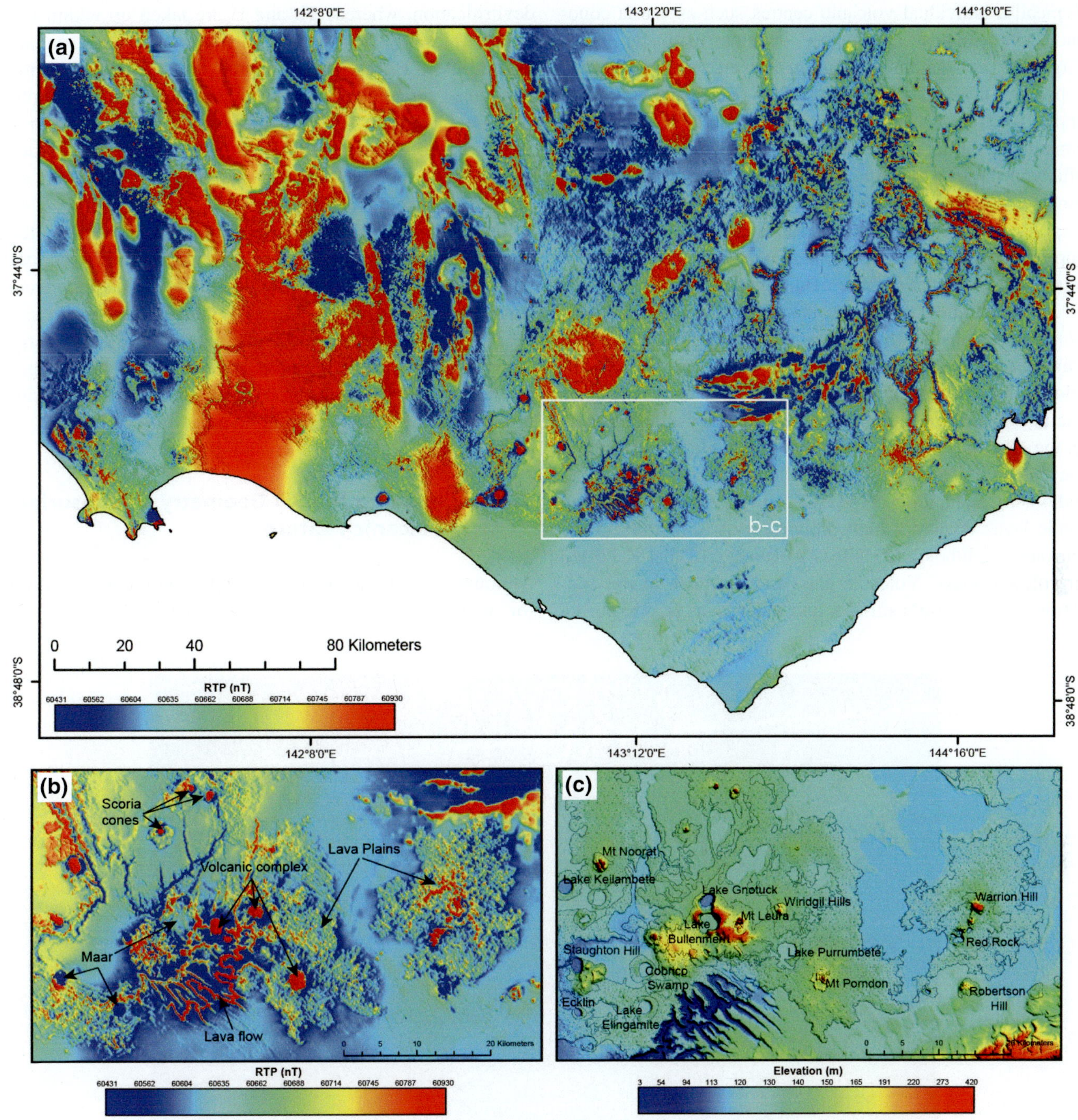

Fig. 15.31 Aeromagnetic response of the basaltic Newer Volcanics Province lavas and volcanoes, Victoria, southeastern Australia. **a** Colour Reduced-to-Pole (RTP) aeromagnetic anomaly (Total Magnetic Intensity, TMI) map of the Newer Volcanics Province (NVP). This shows the extent of the NVP (defined by stippled texture) superimposed on the low magnetic response of the Otway Basin, and the NNE-SSW-trending fabric of the Paleozoic basement rocks in the north. **b** Colour RTP TMI image of part of the Western Plains sub-province of the NVP, which highlights the different volcanic features and their associated magnetic response. (*Source* **a**, **b** Geological Survey of Victoria, through Creative Commons Attribution 4.0 International licence). **c** 90 m SRTM DEM data for the same area as (**b**) showing the mapped outline of lavas and major volcanic centres (From Cas et al. 2017)

Typically, individual volcanic centres such as scoria cones and lava shields appear as circular magnetic highs (Fig. 15.31a, b), although some are reverse magnetised and appear as deep magnetic lows. The maars and tuff rings appear as circular magnetic lows (Fig. 15.31b, c), sometimes containing subtle magnetic highs within their centre, and often cut through the stippled texture of the underlying lava flows. Volcanic complexes have varied geophysical responses depending on their eruptive history, but usually appear as a strong, near-circular magnetic high, which is related to a magmatic phase of the eruption.

Aerial gamma-ray survey data or radiometric data for volcanic terrains highlight variations in potassium (K), thorium (Th), and uranium (U) in the eruption products. Variation in the abundance of these three radioactive elements is useful for identifying changes in the lithology at the surface. In the case of the NVP, this technique is ideal for mapping the extent of some young lava flows, which have strong radiometric signatures (Fig. 15.32). In each case, the young mafic lavas display a high proportion of K relative to the surrounding lithologies, with a percentage of that K hosted in volcanic glass. Volcanic glass is metastable in the conditions at the Earth's surface, and older lavas lose K during devitrification, whereas Th and U are taken up within iron oxides during pedogenesis (Marshall 1961; Dickson and Scott 1997). Hence, eroded mafic lavas eventually show radiogenic signatures that are similar to that of local granites (Dickson and Scott 1997). Young lava flows also tend to be well exposed with little-to-no regolith cover, which is ideal as airborne radiometric surveys sample gamma rays that are derived from the near surface (30–40 cm depth).

In certain cases, used in combination with other remotely sensed data, aeromagnetic and radiometric data can be used to identify individual lava flow lobes from a single volcano (Fig. 15.33). The correlation of boundaries mapped using high resolution digital elevation models with boundaries mapped using magnetic anomaly and radiometric images, provides a quick and rigorous tool for mapping superimposed thin flows, and the total lava flow extent.

15.6.3 Imaging the 3D Geometry of Subsurface Volcanic Features

Although regional aeromagnetic and gravity data are useful for mapping on a province-wide scale, they lack the

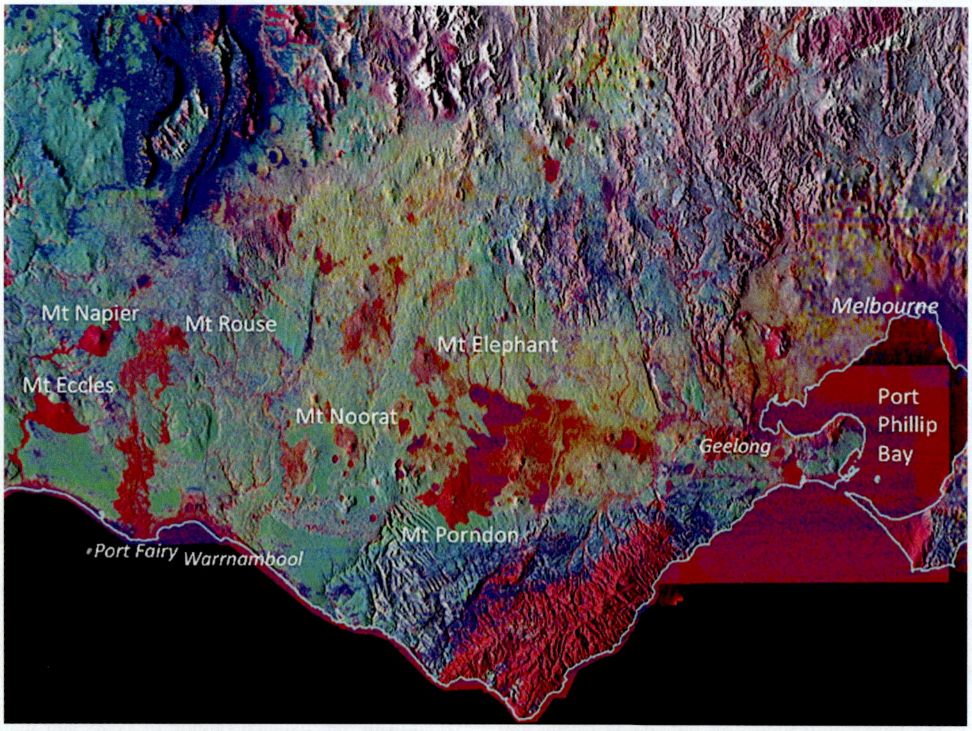

Fig. 15.32 Digitally enhanced radiometric image from LANDSAT data for western Victoria. The processed radiometric data are overlain on a DEM. The NVP is represented by the muted, relatively smooth green domain, with areas of enhanced radiometric response within this domain representing radiogenically enhanced lavas and other eruptive products, shown in red. Several volcanoes and lavas are particularly responsive, including Mt Porndon lava shield volcano in the east and Mt Rouse, Mt Napier and Mt Eccles (and their lavas) in the west. Seawater and saline lakes, including small circular maar crater lakes, also give an enhanced response because of elevated K contents, as does the outcrop belt of the volcaniclastic Cretaceous Otway Group in the south. (*Source* Geological Survey of Victoria VIMP data release, through Creative Commons Attribution 4.0 International licence; From Cas et al. 2017)

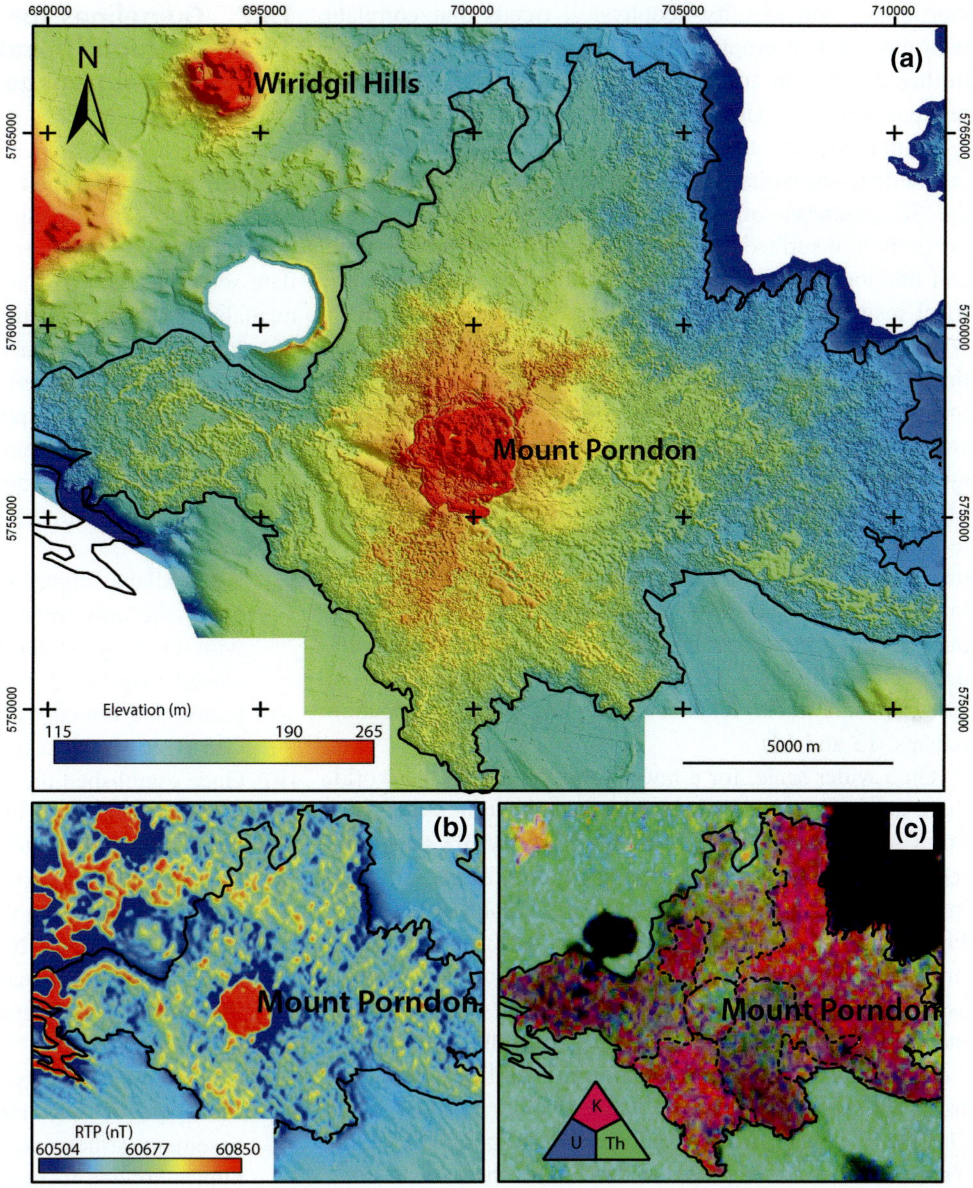

Fig. 15.33 Geophysical images over the Mt Porndon shield volcano showing the extent of its associated lava flows, as defined by different textures, intensities and ratios attributed to the lavas (From Cas et al. 2017). **a** LiDAR DEM gridded at 5 m (*Source* Corangamite Shire Catchment Authority, Victoria). **b** RTP magnetic anomaly TMI image. **c** Ternary aerial radiometric image showing geochemical zones related to discrete superimposed lava flow lobes (dashed lines). (*Source* **b**, **c** Geological Survey of Victoria VIMP data release, through Creative Commons Attribution 4.0 International licence)

resolution needed to model subsurface volcanic features, which require detailed high resolution (15–20 m station spacing) ground surveys. At the scale of individual eruption centres, very high-resolution ground-based gravity and magnetic surveys and forward and inversion data processing have been effective in the NVP in modelling the subsurface structures and geometry of maar volcanoes and volcanic complexes. These allow the generation of detailed 3D models of these volcanic centres from interpretations of the gridded gravity and magnetic data, as well as 2D forward and 3D inverse modelling (for a description of the modelling technique, see Blaikie et al. 2014a). The models are constrained with petrophysical property data, such as the density and magnetic susceptibility of different lithofacies and their components, measured for each eruptive centre, together with geological data (e.g. bedrock stratigraphy, crater morphology, eruption stratigraphy and componentry (including accessory crustal lithic fragment types) and the interpreted eruption style. Where the subsurface litho-stratigraphy is well established the stratigraphic provenance of the crustal accessory lithics can also be established, providing insight into the depth at which the explosive activity was occurring. This can then be used to constrain the minimum depth of the diatreme (maar volcanoes), and identify geophysical trends that correlate with the different eruptive styles (Blaikie et al. 2014b). For further description and discussion see Cas et al. (2017 and references therein; also see Sect. 16.3, Fig. 16.16).

Particularly in the exploitation of hydrocarbon and geothermal reservoirs (Chap. 18), accurate reconstruction of the 3D geometry of volcanic successions also needs the integration of surface data with geophysics, and stratigraphic and lithological information from drill holes. Seismic, electrical resistivity, and magnetotelluric surveys are the most widely geophysical methods used, providing good data on the depth and geometry of the main contrast. Electrical and magnetotelluric data are highly affected by the presence of water, especially hydrothermal, and alteration, so that they are mostly used in reconstruction of underground pathways of volcanic fluids in prospecting for geothermal resources (e.g. Heise et al. 2008; Sect. 18.10). The acquisition of lithological and stratigraphic data from drill holes is very important but, especially with regard to geothermal systems, pervasive hydrothermal alteration is likely to have a severely detrimental effect on the preservation of primary volcanic textures and rock relations, and hamper classification (Chaps. 13 and 14).

On a wider scale, for a few volcanoes around the world, there are sufficient drill hole data (mainly from water bore holes in populated areas) available to allow 3D geometric reconstruction and accurate deposit volume calculations; e.g. in the Rome hinterland in central Italy (Diano et al. 2010; Fig. 15.34). While these subsurface data are extremely important both for hydrological and volcanic hazard assessment, they can also provide a much more accurate picture of rock volumes erupted than generally achieved. At Colli Albani, the major unconformity surfaces identified from surface geology and a large number of bore holes (1,503) have been mapped geospatially (e.g. Fig. 15.32b). These surfaces are then used to reconstruct the relevant rock packages, which allow computation of the precise volume of the preserved deposits for each of the volcanic intervals analysed, as well as determination of the paleotopographic evolution of the volcano. As most of the collected data are from boreholes drilled in the ignimbrite plateau surrounding the caldera, the reconstructions refer to the extra-caldera deposits. The total minimum deposit volume of the Colli Albani volcano (see Fig. 15.15) calculated by this method is 208.1 km^3; to this should be added an intra-caldera ignimbrite volume of 80 km^3 to give a total volume of 290 km^3 (Diano et al. 2010). The minimum extra-caldera volume of the ignimbrite-forming succession of the Villa Senni Formation (Fig. 15.16) is calculated to be 30 km^3, to which can be added an intra-caldera volume of 20 km^3 to give a total of 50 km^3.

15.7 Guidelines for Mapping Modern Volcanic Terrains Using Remote Sensing Imagery and Geophysical Data Sets

The techniques outlined above are widely applicable to mapping modern volcanic provinces and individual volcanoes employing remotely sensed images and geophysical data sets; the main caveat being, in the first instance, their availability. Aeromagnetic, gravity, and radiometric data coupled with satellite/aerial images and moderate- to high-resolution DEMs derived from SRTM and LiDAR data are excellent tools to apply in first-order mapping when approaching a new volcanic field or centre.

The following guidelines outline a systematic approach for mapping and studying modern volcanic terrains:

(a) All available digital data sets should be accessed that cover the area of interest. These should be compiled with existing (if any) digital geological data (either spatial map-based vector data, or georeferenced raster scans) and linked within an appropriate data base to a GIS (Geographical Information Systems) platform.
(b) Once established the GIS database allows multiple data sets to be analysed in parallel, and new layers or thematic maps to be generated. Initially these should be used to define the extent of a volcanic field or province. In a simple example this may only be a case of refining existing geological linework using aeromagnetic data, while in others it may entail using multiple data sets (depending on vegetation, topography, etc.) to define new boundaries.
(c) Once an initial GIS-based digital map is generated, aeromagnetic and DEM data can be used to map basement structures, and spatial data analysis techniques applied to investigate the relationships between eruption points and these basement structures. This may lead to identifying previously unrecognised volcanic centres.
(d) Airborne radiometric data and, where vegetation cover is minimal, ASTER (Advanced Spaceborne Thermal Emission and Reflection Radiometer) can then be used to define major volcanic units such as lava flows, lava domes, pyroclastic flow deposit lobes, and fans. Spatial analysis can again be used to classify the volcanic landforms.
(e) Then at more local levels LiDAR imagery, other DEM imagery or traditional stereographic aerial photogrammetric interpretation can be used to develop local, relatively detailed maps and stratigraphic relationships that are based on identifying the superposition relationships

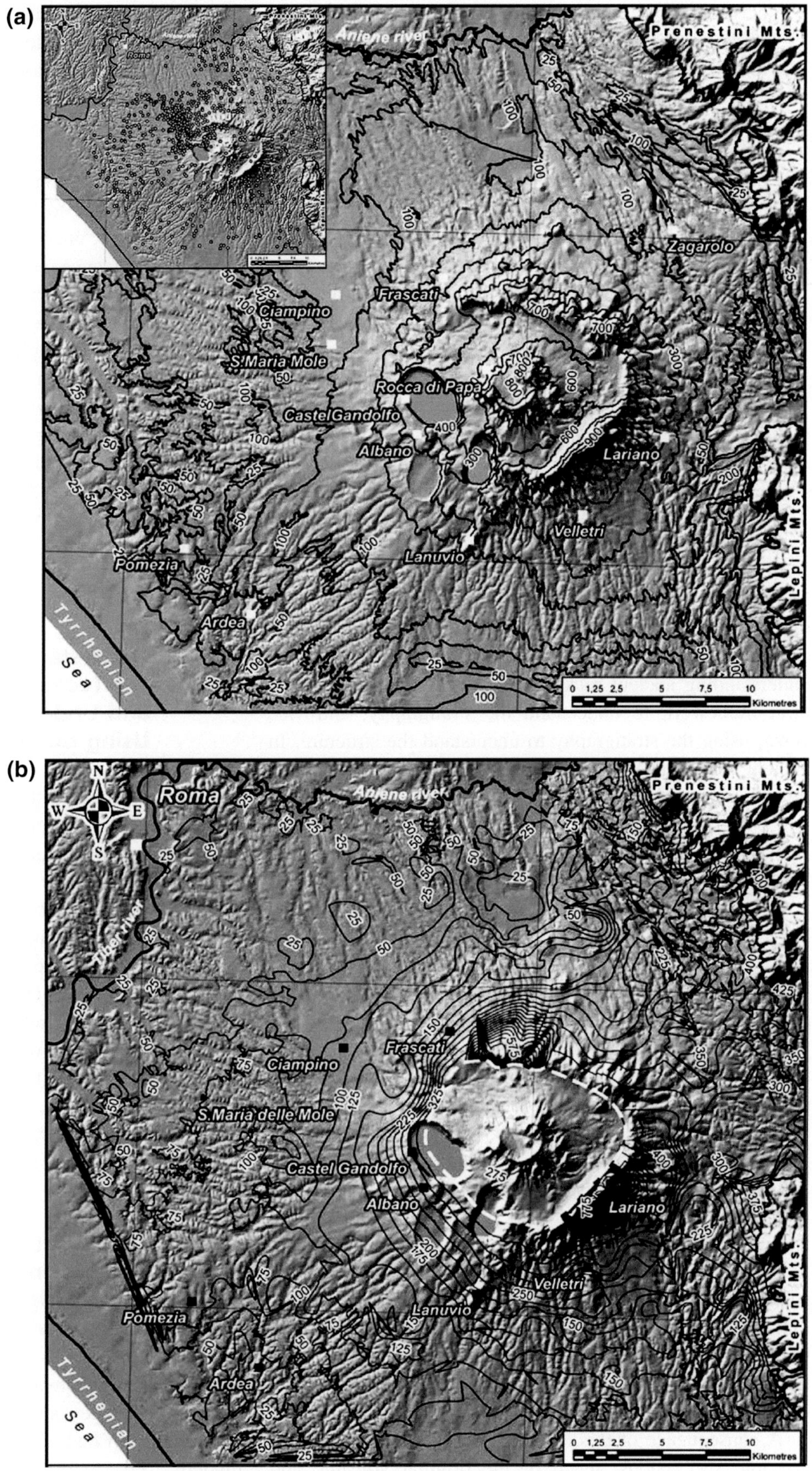

Fig. 15.34 3D reconstruction and deposit volume calculations, Colli Albani volcano, central Italy. **a** Total isopach map (in m) of the Colli Albani volcanic products erupted over the past 600 kyr (See Fig. 15.15 for stratigraphy), derived by subtracting the elevation of the reconstructed base of the volcanics from the DEM elevation of the present-day topography. Inset shows areal distribution of the 1,503 water bore hole stratigraphies used in deposit volume calculations; to provide accurate data these were first validated based on location, elevation and reliability of descriptions, and then coded corresponding to rock formations used in the (1:50,000) geological map. Note, most of the data points are located in the extracaldera ignimbrite plateau. **b** Example showing the altitude (in m) of the unconformity surface below the Villa Senni Formation, taken as the base of the Tufo Lionato ignimbrite at 355 ka (VSN_1; see Figs. 15.2a, 15.15); white dashed line encircles the caldera. (From Diano et al. 2010)

between deposit units, landform features, and roughness analysis to allow a chronology of events to be established.

(f) The resulting local maps can then be followed up with fieldwork to support the stratigraphic relationships and collect samples for geochronology, geochemistry, and petrography. This will allow a chronostratigraphic map and framework to be completed, as well as provide a detailed lithostratigraphic characterisation of the succession of map units. Such maps and stratigraphic reconstructions (including any drill-hole data, or seismic and geological cross-sections) then provide the basis for detailed volcanological studies to be undertaken. All these kinds of point data and spatial 3D data can be managed within the GIS database.

15.8 Documenting Ancient Volcanic Successions

Mapping, establishing a stratigraphy and facies analysis still remain the most important tools for describing and studying ancient volcanic successions. A significant approach in understanding ancient rock successions is to use the structural framework to understand the stratigraphy, and vice versa, using the stratigraphy to understand the structure. In ancient rock successions, you cannot do one without the other.

15.8.1 Factors Affecting Original Geometrical and Lithological Characteristics of Volcanic Facies

Ancient volcanic successions are, on the whole, the erosional and deformed relics of complex volcanic centres and their lateral deposit sequences, neither of which will look the same as the deposits on presently active volcanoes. Previously, we have identified that the preservation of the constituent facies and their geometry is dependent on the interplay between construction or aggradation (deposition), and degradation (erosion; Chap. 12). Subsequent deposits therefore develop irregular facies geometry. Because of the relatively high slopes and the generally high availability of loose debris in proximal and near-vent settings (especially subaerially), erosion rates are usually high. Unless a particular unit is resistant (e.g. lavas, welded tuffs) or is rapidly buried under a resistant unit, or both, it will be eroded, and the record of such rock units may be lost or their original extent and geometry severely modified.

A consequence of this is that, in many ancient subaerial volcanic successions, much of the original volcanic and sedimentation record has been lost because of post-depositional erosion. Resistant units (e.g. lavas, welded units) may be disproportionately preserved relative to original abundance, and original depositional geometries are frequently not conserved, because of erosional modification and/or tectonic strain. Lateral facies and age relationships may therefore be complex, change abruptly which is not necessarily related to fault contacts (e.g. Figs. 1.8 and 15.6 g). Furthermore, unconformities may be common, relatively local in significance, not related to regional deformation, and the erosion that unconformities represent may have been initiated by an aggradational (eruptive or sedimentary) event rather than tectonic uplift. For example, so-called *buttress unconformities*, which represent near-vertical contacts with say the margin of a lava dome and an onlapping sedimentary succession (Fig. 15.6g).

In contrast to proximal settings, in volcanic successions in deep-water settings the preservation of original facies and stratigraphy is better because erosional processes are not as common and intense, and so lateral stratigraphic continuity is also better.

15.8.2 Understanding Stratigraphy Using the Tectonic Structures, and Understanding Tectonic Structures Using the Stratigraphy

Earlier we discussed the role of volcano-tectonic deformation (Sect. 15.3) and presented some guidelines on how the effects may be separated from tectonic deformation. Nonetheless, volcanic terrains are also the sites of contemporaneous crustal movements and, given that many are associated with orogenic belts, they may also be the sites of subsequent penetrative deformation, including both folding and faulting. Deformation may therefore severely change original stratigraphic relationships and, where strain is high, the original geometry of rock units (Chap. 13). Both folding and faulting may obliterate 3D perspective of lateral facies geometry and extent, and make assessment of dispersal patterns and processes, that are taken for granted in modern terrains, very difficult to impossible.

15.8.3 The Difficulties of Using Diamond Drill Core to Understand Regional Structure and Stratigraphy in Areas of Poor Outcrop of Ancient Volcanics

Difficulties in reconstructing stratigraphy and structure in ancient volcanic successions are compounded when deformed, uplifted volcanic successions are subject to modern, post-deformational erosion, leaving a very fragmented preserved record, which is even more difficult to decipher if poorly exposed.

In areas of poor outcrop due to soil and/or vegetation cover, geophysics must first be used to determine bedrock regional structure and geology, as discussed above. in addition to limited outcrop, if there is diamond drill core available that has penetrated through the soil cover into the bedrock, for example from ore deposit exploration programs, that is also used, however sparse the distribution is. Diamond drill core is a variably fractured cylindrical rod of rock that is usually between ~5 and 8 cm in diameter. Single drill holes can be drilled and oriented in any direction and plunge, depending on the aim of the drill hole and what is already known about the regional geology and structure, although with increasing depth the diamond drill bit may begin to diverge in direction. Depending on the bedrock geology, diamond drill holes that produce continuous core can be drilled up to ~3,000 m long, although they are usually shorter given how expensive they are to drill. Diamond drill core is thus, essentially 1D continuous outcrop. However, if a series of drill holes are drilled that are tens to hundreds of m apart, often a good 2D, and sometimes a 3D reconstruction of the local geology can be achieved. It is slow and painstaking work and involves very carefully logging the facies and structural characteristics, to produce a detailed graphic log for each hole before correlations between holes can be attempted.

If diamond drill core is oriented as it is drilled (called "ori-core") by frequently marking the top edge of intervals of core with a straight line and the down or uphole direction, as they are extracted from the drill stem, which is very time consuming and expensive, it can be used to measure dip and strike of bedding and the orientation of structural features such as cleavage and lineations. Without oriented drill core the core may not contain enough information to decipher the structure of the succession. If excellent younging or facing indicators in sedimentary or volcanic facies are present, much of the structure, and thus stratigraphy, can be determined even in one drill core, and even more so where multiple ori-cores are available in a particular area. Without ori-core ambiguous alternative interpretations of the structure and stratigraphy are possible even if younging directions are available (Fig. 15.35a–d).

If ori-core is not available, but some outcrops are present nearby that display key in situ structural-bedding relationships (e.g. bedding-cleavage relationships and lineations, younging directions) then stratigraphic and structural orientations can be interpolated from bedding-cleavage relationships in the drill core that are schematically oriented in the regional 3D space to be parallel to those in outcrop. Ideally, however, multiple drill holes, and many younging indicators, preferably in 3D oriented drill core, are required to resolve both the structure and stratigraphy in such situations.

As an example, in an area where there is sparse outcrop with limited information such as younging in sedimentary strata interleaved with volcanic units such as lavas (e.g. basalt and komatiite lavas or sills in the Archean Kalgoorlie Terrane, Yilgarn Craton, Western Australia; see below), and isolated drill core exists nearby that preserves alternating core intervals of coherent igneous and sedimentary intervals, the stratigraphy and structure in the drill core could be interpreted in multiple ambiguous ways, unless multiple, good younging indicators are present in the drill core (Fig. 15.35a–d). The alternative scenarios may be distinguished using facies contact relationships between lavas (Chaps. 5–7) and intrusions (Chap. 8) with sediments, geochemistry, and geochronology (Fig. 15.35).

Another example of the ambiguous interpretations of drill core intersections is represented in Fig. 15.36, also involving an alternation of coherent rock intervals and sedimentary strata that appear to be concordant. Such a scenario could be interpreted in terms or either a succession of an association of contemporaneous sills intruding a stratified sedimentary succession (Fig. 15.36a), or lavas and sediments (Fig. 15.36b). These alternatives could be distinguished by careful assessment of the contact relationships between the coherent units through contact facies analysis (lavas, Chaps. 5–7; intrusions Chap. 8), and geochemistry and perhaps geochronology.

A final example of the difficulty of interpreting stratigraphy in areas of poor outcrop where drill core is the principal data source comes from an Ordovician volcanic succession in western New South Wales, Australia, where drill holes as little as 25 and 50 m apart could not be correlated (Fig. 15.37). The drill holes appeared to consist mostly of massive, coarse polymictic volcanic breccia, in which red oxidised trachyte clasts are prominent, interleaved with intervals of trachyte from several to tens of m thick, which were originally interpreted as lavas. However, those thick intervals of trachyte could not be correlated from one drill hole to another (Fig. 15.37). Further examination of the randomness of the occurrence of the trachyte intervals, indicated that the intervals were in fact megaclasts in a very thick debris avalanche deposit, tens of m thick (Simpson et al. 2005; Chap. 12). On close examination, the megaclasts are internally pervasively micro-fractured and clasts are seen to spall off their margins into the matrix breccia, which is consistent with a debris avalanche deposit (Chap. 12).

Another problem with folded ancient volcanic successions is that although deformation may cause steeply dipping beds, in volcanic terrains steeply dipping beds may reflect primary depositional dips orientations and are not necessarily the consequence of tectonic deformation. Volcanoes may have steep initial slopes upon which deposition occurs. Fall deposits may mantle an irregular, steep topography, producing steep depositional attitudes (Chap. 9), which may

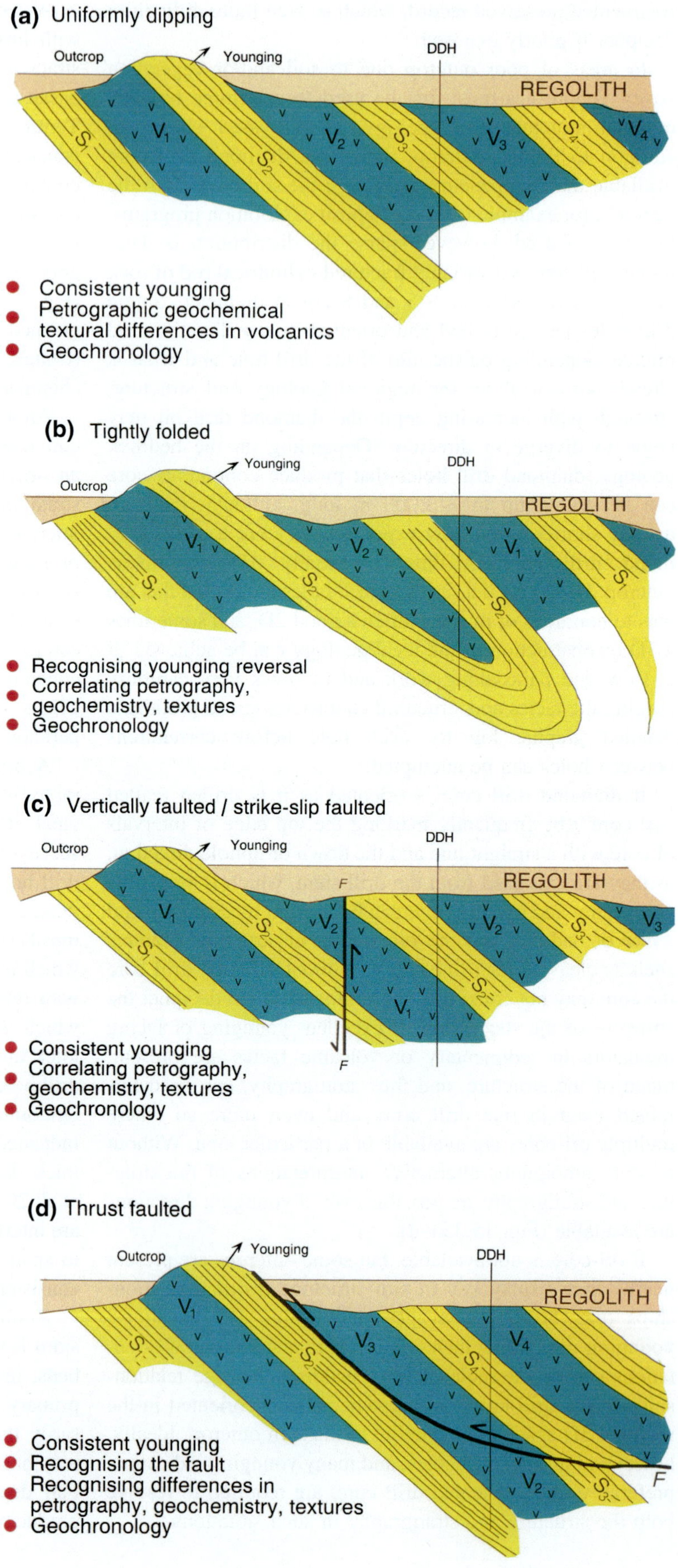

Fig. 15.35 Hypothetical alternative interpretations of the poorly outcropping geology of a deformed bedrock association of alternating lavas and sedimentary strata with only sparse outcrop, limited younging information, and one nearby vertical diamond drill core that intersects two intervals of lava and sediments. Ways of perhaps resolving the alternatives are proposed. **a** Simple conformable uniformly dipping sequence. **b** Conformable, isoclinally folded succession. **c** Vertically up-faulted stratigraphy through a blind fault (i.e. not visible at the surface). **d** Younger stratigraphy thrust faulted into juxtaposition with older stratigraphy

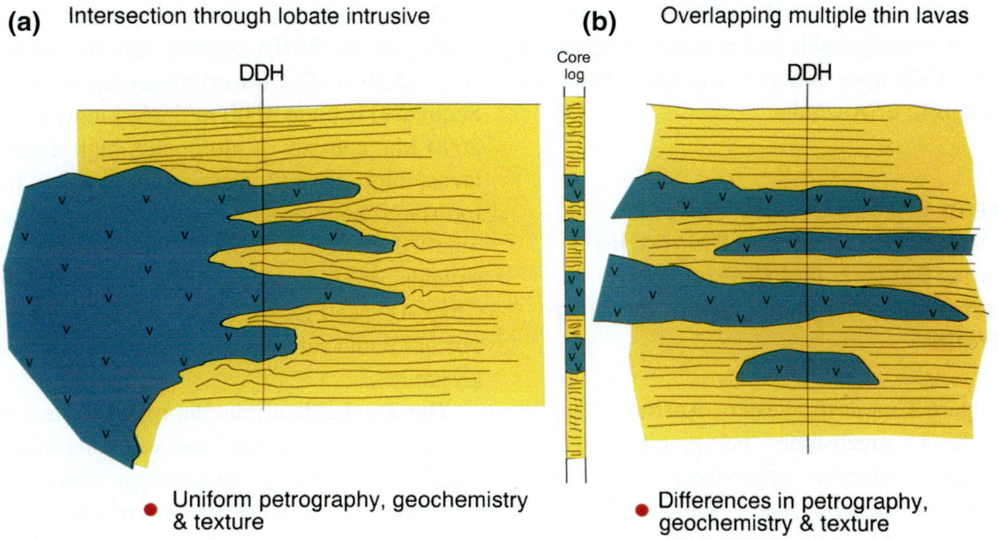

Fig. 15.36 Hypothetical scenarios interpreting a similar succession of coherent rock units and sedimentary strata. **a** Succession of lavas and sediments. **b** Multiple sills intruding a sedimentary succession, perhaps originating from a major intrusive body

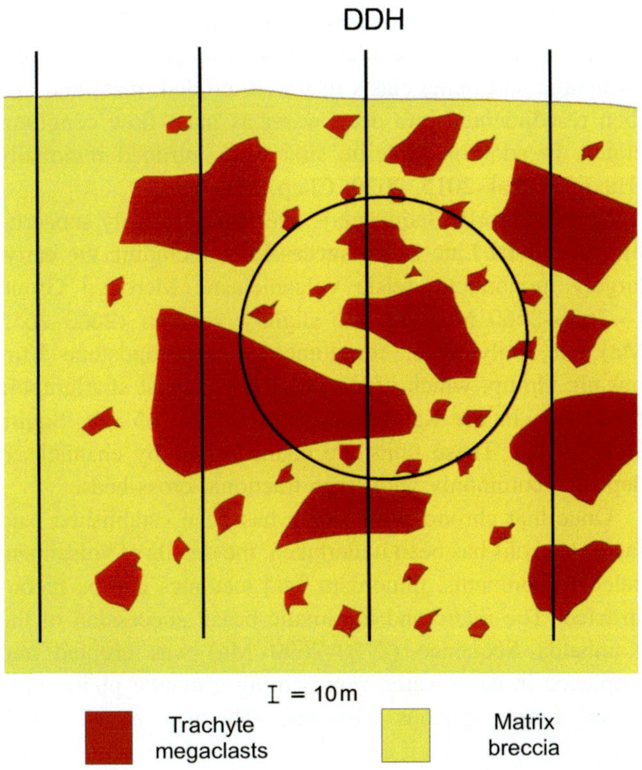

Fig. 15.37 Schematic representation of the difficulty of interpreting intersections of coherent trachyte and intervening breccia intervals when correlation between diamond drill holes are only tens of m apart. Looking only at a single drill hole, the question that should be asked is: *What is the circled area—a succession of irregular lavas and/or intrusions and mass-flow breccias, or megaclasts in a debris avalanche megabreccia?* Based on an Ordovician submarine deposit in the Ordovician Goonumbla Volcanics, New South Wales, Australia (see Simpson et al. 2005)

be at least partly preserved in the rock record (see Fig. 9.3a). Not all dips preserved in the rock record are therefore necessarily the result of deformation.

15.8.4 Mapping and Establishing a Chrono-Stratigraphic Framework in Ancient Volcanic Successions

Ancient volcanic successions have to be mapped in detail regionally to understand their internal stratigraphy and facies architecture, with the superposed effects of deformation back-stripped, if possible, to restore original relationships and facies geometry as best as possible. Again, geophysics can be invaluable as a first-order mapping tool, especially where outcrop is poor. Aeromagnetics highlights bedrock stratigraphic units with high magnetic susceptibility, from those with low susceptibility, and can also highlight regional scale folds and faults when magnetic stratigraphic units are good stratigraphic markers and are deformed. Regional gravity distinguishes units dominated by dense rocks from those with low density. Radiometrics highlights units with a high radiogenic response. After analysing geophysical data maps and identifying possible bedrock units that have positive and negative anomalies, ground mapping and logging of diamond drill core can then define the lithological characteristics of those rock units and allow measurement of basic structural data such as dips and strikes, to facilitate understanding of the regional structure. Measurement of detailed stratigraphic and graphic logs in outcrop and from drill core at numerous localities along strike, accompanied by sampling for petrography, componentry, geochemistry, and geochronology then allow correlations to be constructed and a detailed chrono-stratigraphic and structural framework to be established. Following are two examples from the Archean and Paleozoic of Australia.

15.8.5 Archean Chrono-Stratigraphy and Volcanology of the Kalgoorlie Terrane, Eastern Goldfields Superterrane, Yilgarn Craton, Western Australia

The Archean Yilgarn Craton of Western Australia consists of typical Archean greenstone belt-granite tectono-stratigraphic terranes. Archean greenstone belts consist mostly of submarine ultramafic, mafic, and felsic volcanic rock units interleaved with sedimentary strata, subvolcanic intrusions, metamorphosed to greenschist to amphibolite metamorphic facies (hence called "greenstones"), and intruded by contemporaneous and post-stratigraphic granitoid intrusions (Chap. 17). In the Yilgarn Craton, the best-known terrane is the Kalgoorlie Terrane in the eastern Yilgarn Craton, which has been extensively mapped by the Western Australia Geological Survey (e.g. Swager et al. 1995; Chap. 17, see Sect. 17.7.8), and mining companies exploring for the many ore deposits in the terrane, including komatiite hosted nickel sulphide deposits and orogenic gold deposits (Chap. 18). The climate is temperate-arid, the craton has been tectonically stable for hundreds of millions of years and has been subject to long term weathering. Outcrop is thus poor and weathered, covered by sand, soil, arid vegetation. As a result, regional geophysics is essential to understanding the first-order bedrock structure and stratigraphic architecture. Regional aeromagnetic maps represent an essential first-order regional mapping tool (Fig. 15.38a). A combination of regional geophysics, ground mapping, logging of tens of thousands of diamond drill cores through the stratigraphy, sampling, petrography, geochemistry and U–Pb SHRIMP, high resolution TIMS and LA-ICPMS age dating have produced a good understanding of the regional chrono-stratigraphy and chronology of events (Figs. 15.38a, b and 15.39a, b). Detailed local studies and facies analysis of both the volcanics and the sedimentary succession have produced a good understanding of the paleoenvironmental, volcanological, and the tectonic processes and setting and their evolution (e.g. Squire et al. 2010; Hayman et al. 2015, 2020; Chap. 17).

The Kalgoorlie Terrane stratigraphic succession of volcanics and sedimentary rocks (Fig. 15.39a, b) is highly folded and faulted, metamorphosed from greenschist to amphibolite facies, ranges in age from 2720 to ~2655 Ma and is intruded by granitoids that range in age from ~2700 to 2630 Ma. The stratigraphic succession has been subdivided into three major sequences (Krapez et al. 2000; Squire et al. 2010), which can be correlated regionally (Fig. 15.39b). The lowest stratigraphic package (Kambalda Sequence) is up to 4000 m thick, ranges in age from 2720 to 2690 Ma, consists of ultramafic, high magnesium komatiite lavas, high to low magnesium basalts, their subvolcanic intrusive counterparts (dunites, dolerites respectively), interbedded black anoxic hemipelagic mudstones, and extends >600 km along strike (Figs. 15.38b and 15.39a, b; Chap. 18, see Fig. 18.34). In the northern half, some dacitic volcanics and subvolcanic intrusions are interleaved with the komatiites.

The Kambalda ultramafic-mafic succession is overlain by a largely felsic volcanic and volcaniclastic and interbedded largely black anoxic hemipelagic mudstone succession. It constitutes the Black Flag Group, which is at least 3000 m thick in places, ranges in age from 2690 to ~2670 Ma (Fig. 15.39a, b; Squire et al. 2010), and is also called the Kalgoorlie Sequence. The volcaniclastics include monomictic volcaniclastic megaturbidites (Chap. 11) and mass flow resedimented volcanic conglomerates with a totally intra-basinal provenance derived from intra-basinal felsic volcanic island edifices, as well as granite clasts and ultramafic and mafic clasts that were eroded, reworked and then resedimented into deep water as mass flow conglomerates. In addition, dolerite sills were intruded regionally (Hayman et al. 2015, 2020; Chap. 17).

The Kalgoorlie Sequence is overlain by largely subaerial fluvial-alluvial Late Basin successions, including the early, largely monomictic felsic volcaniclastic Merougil Group (~2670–2660 Ma) and the slightly younger (2660–2655 Ma) late, polymictic, conglomeratic and sandstone Kurrawang Group, which represents the youngest stratigraphic succession in the Kalgoorlie Terrane (Fig. 15.39b; Squire et al. 2010). These units are characterised by channelised deposits, commonly with large tractional cross-beds.

Once that chrono-stratigraphy has been established, and facies analysis has been undertaken, the details of volcanism, paleoenvironments, plutonism, and tectonics can be reconstructed. The mafic and ultramafic basalt succession of the Kambalda Sequence (2720–2690 Ma) was erupted and emplaced in deep water, representing a mantle plume generated Large Igneous Province (Hayman et al. 2015; Chap. 17). Although it extends at least 600 km along strike, localised subvolcanic intrusions indicate that there were many eruption centres. Both the anoxic black mudstones and the absence of any pyroclastic deposits indicate water depths were probably >1 km (Chaps. 3 and 4). The overlying felsic volcanic succession of the Kalgoorlie Sequence (2690–2670 Ma) was also deposited in deep water, as attested by anoxic black mudstone interbeds. The regional change from ultramafic-mafic to felsic magmatism is attributed to the waning of mantle plume activity but onset of mantle anatexis

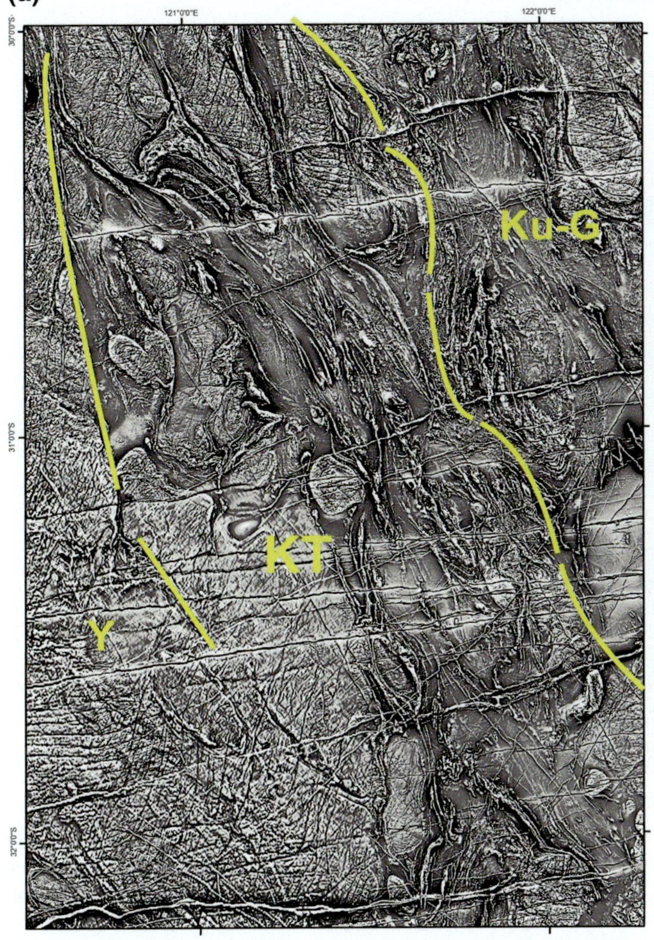

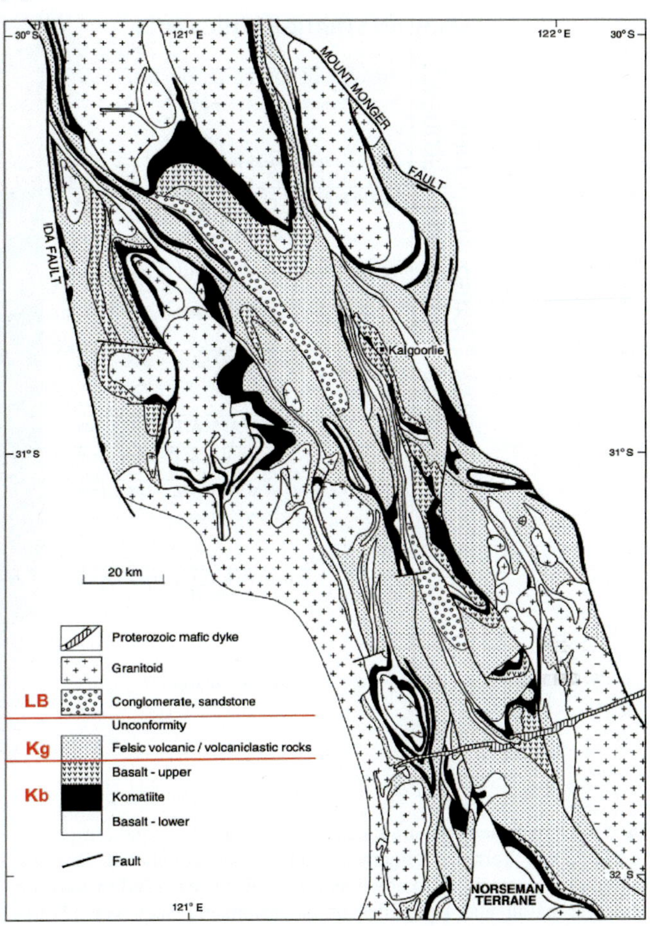

Fig. 15.38 Regional maps of the Archean Kalgoorlie Terrane, Yilgarn Craton, Western Australia, showing the distribution of greenstone belts consisting of volcanic, sub-volcanic intrusion and interleaved sedimentary and volcaniclastic strata intruded by and deformed around contemporaneous and later granitoid bodies. **a** Regional aeromagnetic anomaly map (TMI First Vertical Derivative) in grey scale of the southern half of the Kalgoorlie Terrane (KT) showing dark grey lineated, deformed greenstone belt strata (volcanics, sediments) with high magnetic susceptibility, and "squeezed" between weakly magnetic granitoid bodies in light grey with cross-hatched joint patterns. Ku-G is the adjacent Kurnalpi and Gindalbie terranes; Y is the Youanmi Terrane. (*Source* Western Australia Department of Mines, Industry Regulation and Safety, through Creative Commons Attribution-Share Alike 4.0 International licence). **b** Simplified geological map of the southern half of the Kalgoorlie Terrane. Kb = Kambalda Sequence (2720–2690 Ma); Kg = Kalgoorlie Sequence or Black Flag Group (2690–2670 Ma); LB = Late Basin Sequence (2670–2655 Ma). (From Swager et al. 1995)

of metasomatised upper mantle, and crustal anatexis of a deep mafic crust induced by regional heating of the crust by the mantle plume. This triggered not only TTG (tonalite-trondhjemite-granodiorite) composition felsic volcanism, but also regional TTG granitoid plutonism, with many plutons intruding high into the greenstone belt stratigraphic succession (Fig. 15.39b). Some plutons domed the crust upwards exposing komatiites and basalts to erosion and resedimentation of the debris, as well as being unroofed and eroded themselves (Hayman et al. 2020). Some plutons were the magma reservoirs for the felsic volcanism, including shallow water to subaerial explosive volcanism which generated monomictic volcaniclastic megaturbidites as pyroclastic flows entered the sea. There is no evidence for plate tectonic processes or settings up to this time.

From 2670 Ma onwards there is evidence for regional uplift to almost wholly subaerial environments, initially driven by pluton updoming, still involving some explosive volcanism, represented by monomictic fluvially reworked tuffaceous sediment in fluvial braidplain and alluvial fan environments. The later polymictic conglomeratic fluvial-alluvial Kurrawang deposits include a far-field provenance as well as deep crustal metamorphic and plutonic clasts, indicating a major regional compressional and

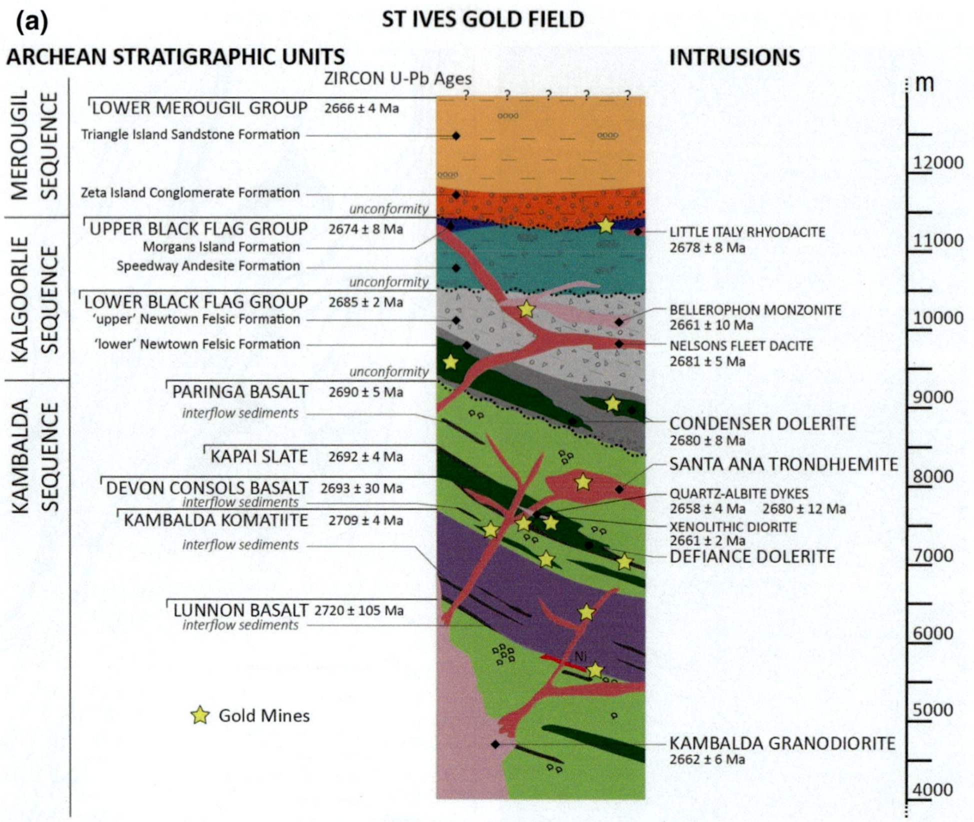

Fig. 15.39 **a** Schematic stratigraphic column for the southern part of the Kalgoorlie Terrane, Yilgarn Craton, around Kambalda, showing stratigraphic contacts and known age dates. Principal subdivisions are the lower basalt and komatiites of the Kambalda Sequence (2720–2690 Ma), the felsic Kalgoorlie Sequence (2690–2670 Ma) and the youngest tuffaceous felsic volcaniclastic, early Late Basin Merougil Group (2670–2660 Ma). (From Doutch 2019, incorporating data from Connors et al. 2005). **b** Regional correlative chrono-stratigraphy for the Kalgoorlie Terrane, including the same major subdivisions as in (**a**) but also including the youngest, polymictic sedimentary package, the late Late Basin Kurrawang Group, and correlatives (From Squire et al. 2010). Ages in both (**a**) and (**b**) are derived from multiple sources

uplift tectonic event, and perhaps representing the first evidence of plate subduction generated crustal convergence and uplift, at about 2660 Ma.

So, the paleogeography changed from totally open, deep-water ocean without any landmasses, to deep ocean marked by dispersed emergent felsic volcanic edifices, to wholly subaerial fluvial-alluvial, over a 60 Ma period. The volcanic products reflect the effect of paleoenvironment and water depth in controlling whether or not explosive activity occurred, as happens today, but the unique plume influenced geodynamic regime of the Archean controlled the composition of the volcanic products and their distribution. No simple linear arc-like magmatic belt is discernible, and so the magmatism is more akin to intraplate plume driven activity. Similar stratigraphies, with slightly different ages and timing of events, are common on other Archean cratons around the world.

The important message from this example is that unravelling the volcanology of the rock record is much more time consuming and complicated, requiring a full geological approach, compared with working on young modern volcanoes. Secondly, we can't assume that all magmatism in Earth history occurred within the sphere of the plate tectonic framework. It is now widely being recognised that the most primitive form of plate tectonics did not develop until the late Archean, between 3000 and 2500 Ma, and perhaps from ~2700 Ma onwards (e.g. Cawood et al. 2018; Smithies et al. 2019; Hawkesworth et al. 2020; Chap. 17).

15.8.6 The Lower Devonian Merrions (Tuff) Formation, Hill End Trough, New South Wales, Australia

The Lower Devonian Merrions (Tuff) Formation (MTF) is a regionally extensive important stratigraphic and structural marker in the Siluro-Devonian, deep marine Hill End Trough (HET), Lachlan Orogen, New South Wales, Australia (Cas 1977; Fig. 15.40a). The MTF and the HET were originally mapped on foot and using aerial photos.

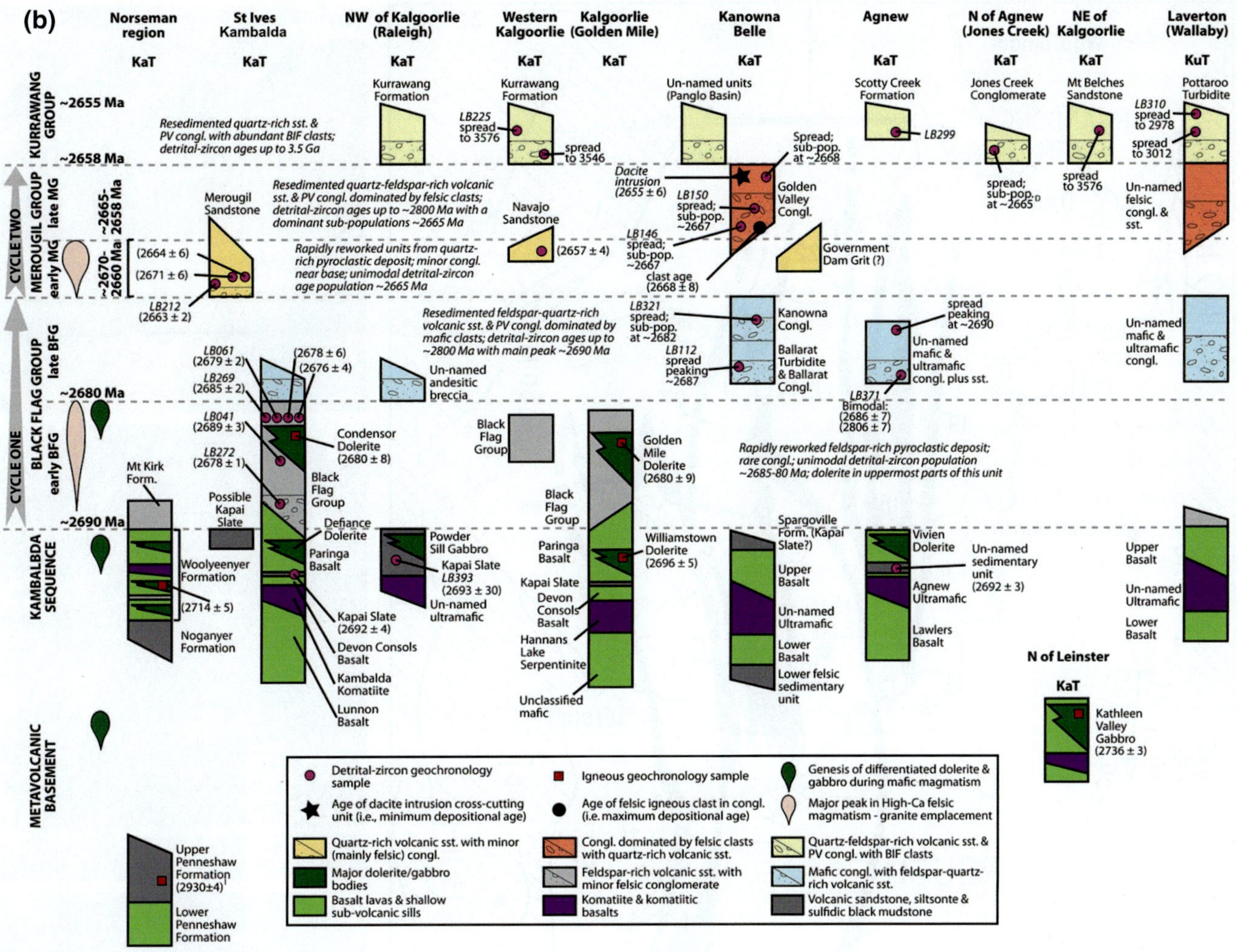

Fig. 15.39 (continued)

Availability of modern regional geophysical data sets (Fig. 15.40a, b), would have made that mapping program much easier! Geophysical datasets including aeromagnetics (Fig. 15.40b) and radiometrics (Fig. 15.40c) show that the HET succession is tightly folded into regional scale north-south trending shallow plunging folds, as well as faults, and that the MTF shows up both magnetically and radiometrically as a regional stratigraphic and structural marker unit. The HET succession has been metamorphosed to greenschist facies.

The MTF in its currently deformed state outcrops over 1800 km^2, which when the succession is unfolded palinspastically the minimum outcrop area is 3400 km^2, but originally would have been much more extensive. Its average thickness is $\sim$0.5 km, indicating that the minimum original volume was 1700 km^3. Detailed graphic logging of the MTF in the field at strategic spaced localities regionally, shows that the MTF consists of multiple monomictic dacitic submarine tuffaceous sandstone megaturbidite depositional units (Chap. 11) and associated far flowing very thick dacite lava flows (Chap. 6), organised into stratigraphic members (A–P) hosted within a very thick, basinal terrigenous turbidite succession. The MTF succession is up to 1000 m thick, and volcaniclastic depositional units range from a metre to >100 m thick, dominated by angular volcanic plagioclase and minor quartz crystal fragments, although one member, Member D is unique in containing abundant orthoclase (probably after sanidine) and volcanic quartz, allowing it to be used as a datum for correlation of members within the MTF. Three dacite lavas are interstratified with the crystal-rich volcaniclastic units, the thickest being up to 390 m thick (Fig. 15.41a).

In its deformed state, it is not possible to reconstruct what the paleo-depositional, paleotopographic surface was like, although because the host succession was a deep-water turbidite basin, the paleo-depositional surface was probably

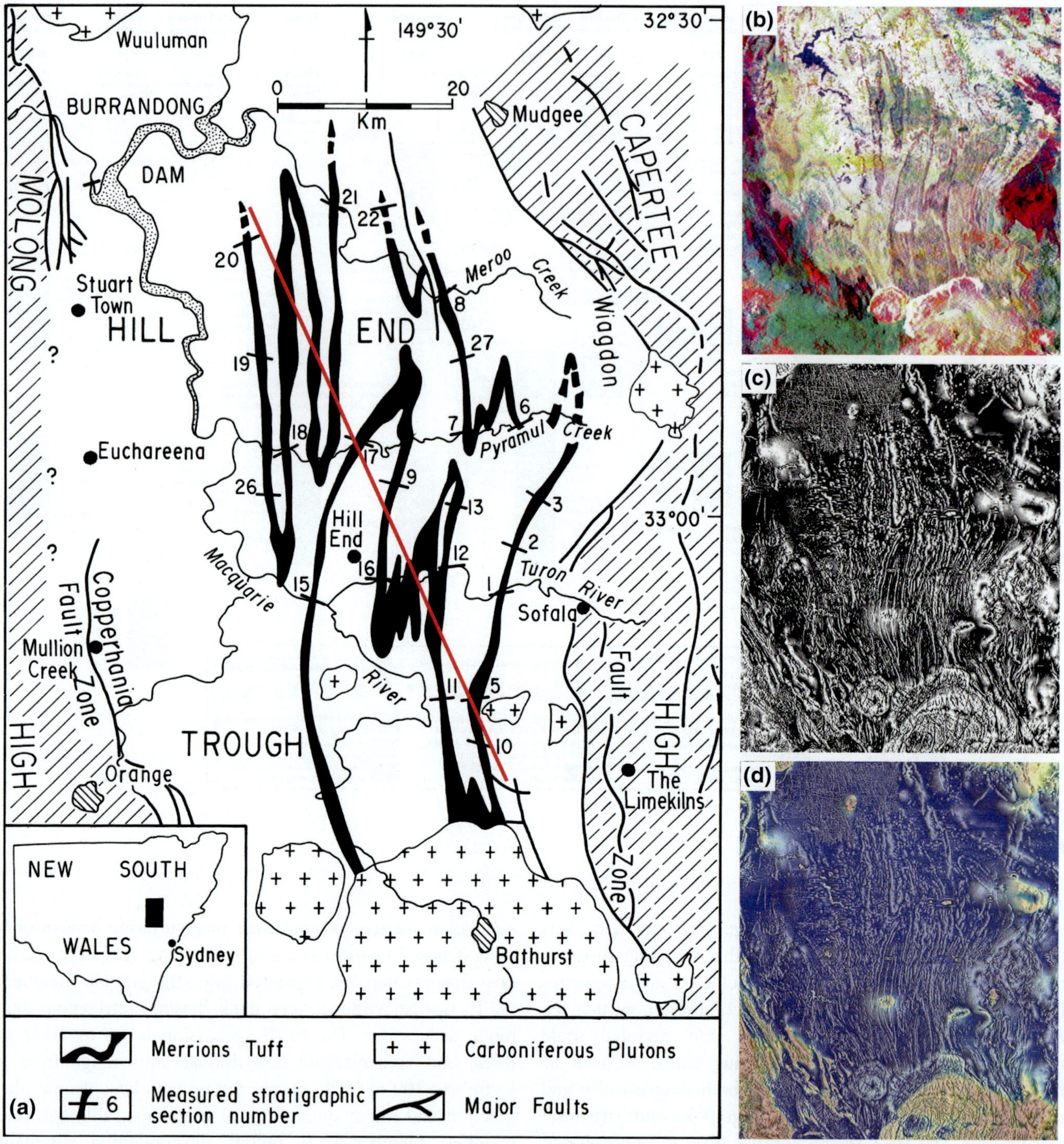

Fig. 15.40 Geological and geophysical maps of Siluro-Devonian Hill End Trough (HET), New South Wales, southeastern Australia, highlighting the regional stratigraphic and structural marker, the Lower Devonian Merrions (Tuff) Formation (MT). **a** Simplified geological map of the HET region, highlighting in black the regionally extensive MT. Numbers with cross-cutting dashes show the locations where detailed graphic logs of the MT were measured (e.g. Fig. 15.41a, b). The red line shows the trace of the correlated graphic logs in Fig. 15.41a. The regional mapping was done before regional airborne geophysical surveys were undertaken. (From Cas 1977). **b** Regional aerial radiometrics map of the same area as in (**a**). The MT shows up as a distinctive folded pink-mauve unit. **c** Regional RTP TMI with tilt grey scale map for the HET, as in (**a**). Note how much stratigraphic and structural data exists due to the shadow effect of tilt perspective. **d** Regional RTP TMI with colour filter and tilt map, as in (**a**) again showing stratigraphic and structural detail, including the trace of the MT that can be compared with its trace in (**a**). (*Source* **b–d** State of New South Wales, Department of Regional NSW (Mining, Exploration, and Geoscience), through Creative Commons Attribution 4.0 International licence)

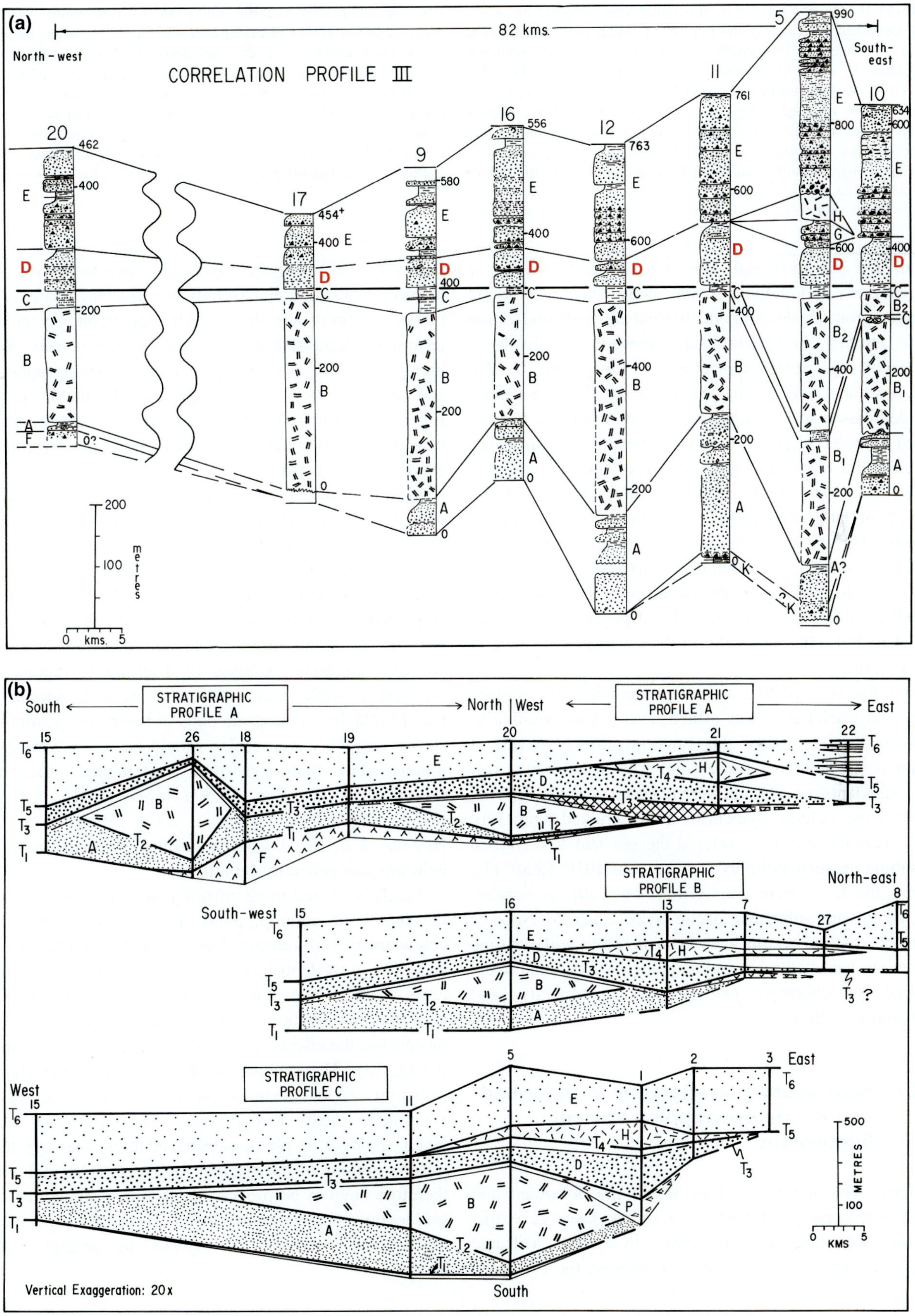
(a)
82 kms.
North-west
South-east
CORRELATION PROFILE III
20
17
9
16
12
11
5
10
E
D
C
B
A
F
H
G
B2
B1
K
metres
0 kms. 5
(b)
STRATIGRAPHIC PROFILE A
South
North
West
East
STRATIGRAPHIC PROFILE B
South-west
North-east
STRATIGRAPHIC PROFILE C
Vertical Exaggeration: 20x
METRES
KMS

◀ **Fig. 15.41** Facies characteristics and stratigraphic architecture of the Lower Devonian Merrions (Tuff) Formation (MT), Hill End Trough, New South Wales. **a** North-northwest trending section line showing correlations of units in multiple graphic logs through the MT. Section 10 is in the southeast, Section 20 is in the northwest, 82 km away; locations of graphic logs are shown on Fig. 15.40a. Stippled units are crystal rich megaturbidity depositional units, hachured units are dacite lavas. Member D is highlighted in red D's because it is unique in being the only volcaniclastic stratigraphic unit to contain abundant angular crystal fragments ($\sim$30% modal abundance each) of volcanic quartz and orthoclase (after sanidine), as well as plagioclase, whereas other volcaniclastic units consist dominantly of angular plagioclase ($\sim$90%) and minor volcanic quartz ($\sim$10%). Interbedded pelitic units (dashes) indicate considerable time breaks between major explosive eruptions that produced the crystal-rich megaturbidite units. **b** Schematic reconstruction of the stratigraphic architecture of the MT using correlated detailed graphic logs (locations shown in Fig. 15.40a). Note that the thickness of volcaniclastic Member D thins over the crest of Lava Member B, and then Lava Member H is displaced to the northeast of B, indicating it still exerted a topographic effect on the flow path of H. (From Cas 1977)

smooth and sub-horizontal. However, detailed analysis of the internal stratigraphic architecture made possible through regional scale componentry that allowed subdivision of the internal stratigraphy has allowed assessment of the evolving paleo-depositional topography. The pre-depositional topography seems to have had a central axial depression striking northwest. Unit A appears to have formed a fan-like deposit thinning northwards. In particular, it is clear the high relief morphology of the thickest, first erupted lava, Member B, controlled the thickness and distribution of later depositional units, including Member D megaturbidite unit that thins over maximum thickness axis of B lavas, and thickens along the northeast margin of the lava, which formed a shoulder depocentre. The succeeding lava Member H was emplaced in that depocentre depression parallel to but adjacent to the northeast margin of lava B (Fig. 15.41b).

After the stratigraphic and facies architecture, componentry and provenance was established, it was possible to reconstruct the origin of this huge, regionally extensive volcaniclastic and lava succession. The voluminous volcaniclastic depositional units were interpreted as the deposits of very large volume pyroclastic flows erupted from subaerial caldera(s) that then entered the sea and transformed into submarine megaturbidity currents (Cas 1979; Chap. 11), whereas the lavas were erupted intra-basinally as voluminous, far flowing submarine dacite lavas (Chap. 6). None of this reconstruction would have been possible without time consuming, hard fieldwork and mapping on a regional scale, followed by laboratory petrography, componentry, and geochemical studies.

15.8.7 Representing Stratigraphy in Deformed and Poorly Outcropping Ancient Successions

In ancient volcanic successions which are deformed and poorly outcropping, establishing a simple layer cake like (chrono-)stratigraphy is not simple. In modern volcanic settings, we know that in near-vent settings, the effects of high relief and irregular topography due to the interaction between constructional (eruptions) and degradational processes (erosion), and the diversity and frequency of events (eruptive, depositional, erosional, tectonic) can produce highly irregular preserved geometries of stratigraphic units and contacts (e.g. high angle unconformable contacts; Fig. 15.6g). Unconformable contacts may be common, and need not represent major, regionally extensive deformational events or time breaks.

However, in addition to unconformable contacts, discordant stratigraphic relationships may also be due to the impact of tectonic deformation, including folding and faulting, and intrusions in both modern and ancient settings, making it difficult to accurately depict spatial and temporal relationships between units in an area. In these situations, we suggest it is more appropriate to construct *interactive stratigraphic-facies columns* or diagrams to represent and interpret possible rock relations (Cas and Wright 1987; Fig. 15.42a, b). In the distal parts of basins away from vents, slopes may be lower, facies should be more laterally continuous and more-consistent stratigraphic relationships should occur (e.g. Figs. 15.39a, b and 15.41a, b; Cas and Wright 1987; Manville et al. 2009; Squire et al. 2010; Hayman et al. 2015). In these settings, standard stratigraphic columns can generally be utilised.

Lastly here, and more critically, we have emphasised that the preserved outcrop, hand specimen, and petrographic characteristics of volcanic facies may be significantly influenced by post-depositional processes. Where ancient successions have been subjected to the protracted effects of hydrothermal alteration, diagenesis, deformation, and metamorphism, the effects of each may be difficult to distinguish (Chap. 13). These processes not only have the ability to overprint original textures, but in some instances to obliterate them totally and produce new, misleading textures (even apparent or pseudo-clastic textures) that can lead to misconceptions of the original rock types. Original textures therefore have to be proven rather than assumed, and unusual textures have to be explained. These effects also test the efficacy of geochemical analysis in ancient volcanic successions.

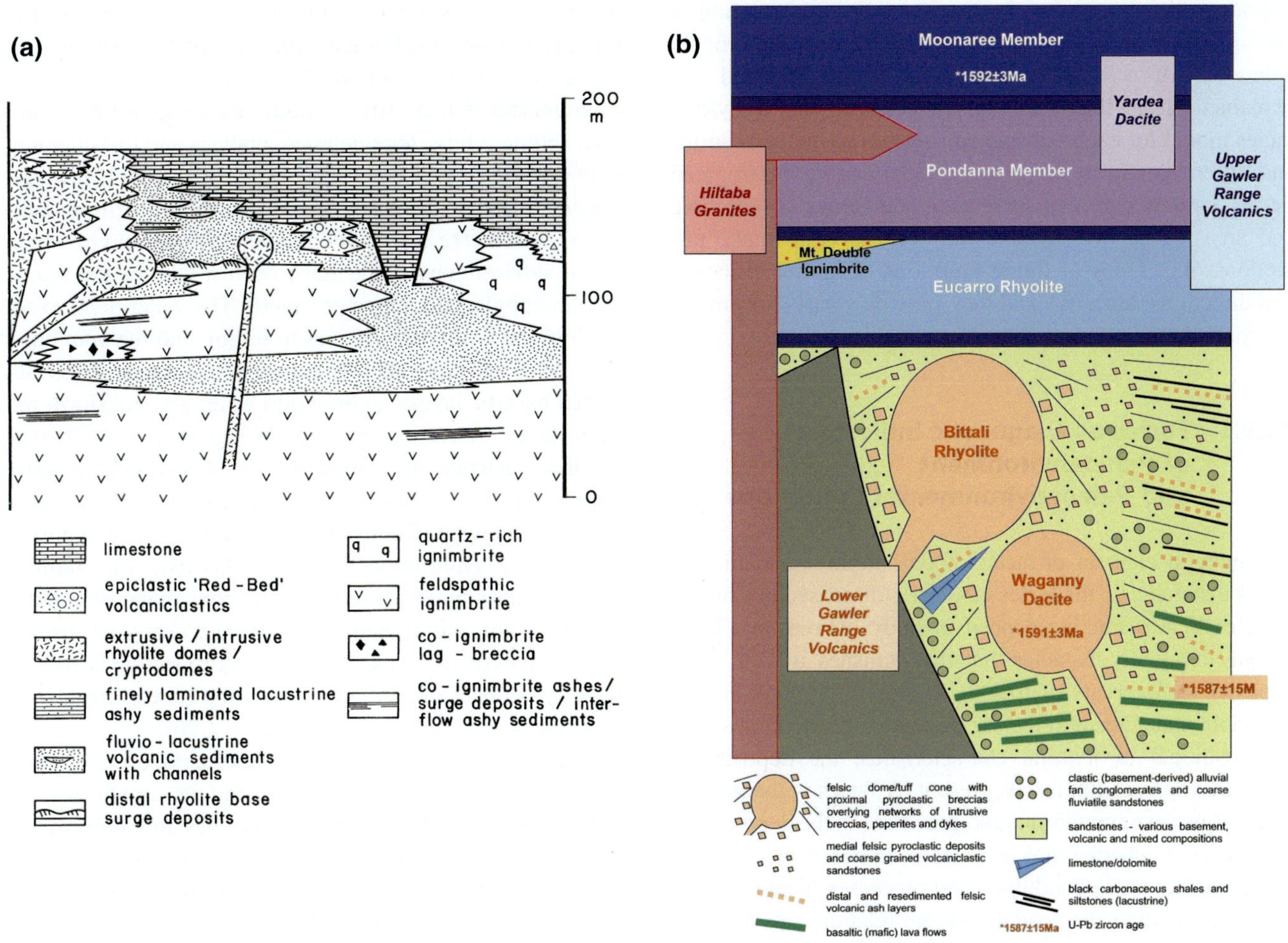

Fig. 15.42 Examples of interactive stratigraphic-facies columns. **a** Based on the mapped geology of uppermost Early Devonian Snowy River Volcanics, Murrindal-W-Tree area, eastern Victoria, Australia (From Cas and Wright 1987). **b** Mesoproterozoic Gawler Range Volcanics, coeval intrusive Hiltaba Suite Granites and sedimentary successions (~ 1592 Ma), central and southern Gawler Ranges, South Australia. The facies relations illustrated could represent a cross-section through either an early rift margin or a caldera fault. (From Wright 2008; based on several published sources, including Allen et al. 2003)

15.9 Developing Facies Models for Modern and Ancient Volcanoes and Volcanic Terrains

In Chap. 16, we describe the geology of the main types of volcanoes and volcanic systems and depict the geology of many of them using facies models. *Facies models* are schematic reconstructions of spatial and age relationships of facies, depicting the facies and stratigraphic architecture, (paleo)environments and processes in a sedimentary and/or volcanic system. The concept originated in sedimentary geology, and facies models were developed to distinguish the facies associations, architecture, and the inferred sedimentary processes of one sedimentary environment from another (e.g. Walker 1984; Reading 1996; Selley 2004). However, in volcanology, facies models can also be usefully constructed to summarise the general characteristics of each volcano type and system, and to distinguish them all from each other (Cas and Wright 1987). Such models have to be based on understanding of both modern and ancient volcanoes and their products because modern volcanoes reveal morphological features and deposits on the surface, whereas ancient successions provide a window as to what's inside a volcano and more widely dispersed surrounding settings.

Generic facies models for particular types of volcanoes and volcanic settings can be used as a guide for what to expect in the geology of a new example to be studied. They can therefore be used for comparisons between the generic model and what is actually found in the new example. Features that are common between the generic model and the new model provide rapid understanding of the new example. However, any differences that can be recognised between the two models are equally important because they reflect important differences (e.g. in magma compositions, eruption and dispersal processes, and possible environment) in the behaviour of a volcano that need to be assessed and explained. It is thus essential to construct an individual facies model for each volcano and system so that similarities and differences from the generic model can be interrogated. Facies models can also serve to guide future observations and areas to be investigated, they can serve to make predictions (e.g. types of future eruptions, dispersal processes and directions, hazards), and they provide a basis for process interpretations and modelling them.

15.9.1 Facies as Diagnostic Indicators of Paleoenvironment and Paleoenvironmental Conditions

The integral element of facies models is the association of facies that make up the facies model. As discussed in Chap. 1, a facies is a distinctive deposit or body of rock that has a unique definable character that distinguishes it from other deposits, or intervals of rock (Cas and Wright 1987). The definable character may be the geometry of the deposit, its compositional or textural characteristics, the depositional structures or even the fossils present. A facies is the product of a unique set of conditions and processes in the depositional environment, whether they be volcanic or not. A facies model should depict those conditions and processes (Chap. 1).

Understanding the spatial and age relationships between facies and associations of facies is crucial to understanding whether or not spatially related facies are genetically related, and in deciphering the facies and stratigraphic architecture, and chronology of events. By doing this, models of the deposits produced by particular events can be developed, eruptive histories can be reconstructed, clearer pictures of the emplacement processes and environments emerge, and hazards and resource potential can be evaluated (Chaps. 1, 16, and 18).

The main purpose of carefully documenting the characteristics of facies is to identify those characteristics which give clues to their origin, mode of deposition or environmental conditions, or both, at the time of emplacement (Chap. 1). For example, angle of repose cross-bedding in epiclastic deposits suggests tractional sediment transport and deposition; structureless to faintly laminated epiclastic claystones or mudstones suggest settling out of suspension and very quiet aqueous conditions; well-defined eutaxitic texture in a thick, massive, poorly sorted, undeformed volcaniclastic deposit suggests welding and deposition by pyroclastic flow.

However, there are few facies or facies characteristics which by themselves unequivocally identify the host depositional environment or the spatial context of a facies within a broader paleogeography. This applies to both sedimentary and volcanic terrains. For example, eutaxitic texture by itself may indicate a near vent, welded air-fall deposit (Chap. 9), or ignimbrite (Chap. 10), in both cases suggesting a subaerial setting. Only more detailed analysis and consideration of related facies will determine which it is and, if it is ignimbrite, whether it is proximal or distal. Pillow lavas (Chap. 5) indicate a subaqueous depositional environment, but the environment could have been a lake or the sea, or intra- or subglacial meltwater cavern. The water depth could be anywhere from only several m to km, and the vent could have been subaqueous or subaerial, because subaerially erupted basaltic lavas can flow into water and transform into pillow lavas. Accretionary lapilli (Chap. 4; Fig. 4.27) indicate the existence of a subaerial or atmospheric eruption column, but final deposition may have occurred close to or far away from vent (Chaps. 9 and 14), and even be deposited in deep-sea environments (Fig. 16.100d, e). Ballistic bomb-sag structures are one of the few diagnostic facies features of volcanic terrains, suggesting a near-vent (usually within hundreds of m and up to a few km) setting (Chap. 9). However, impact sag structures (or dropstones) can also be produced by epiclastic processes (Fig. 12.6).

Therefore, to arrive at a sound interpretation and reconstruction of the paleoenvironment and its conditions, the geologist working in ancient terrains has to establish significant associations of facies. The context of any single facies is most likely to be revealed by its relationship to and association with other facies. For example, lacustrine facies deposited in a subaerial volcanic terrain could be part of an eruptive centre (crater or larger caldera lake), or they could have formed in the surrounding landscape far from an eruptive centre. The facies by itself does not uniquely indicate which setting, but the associated facies could help to identify a silicic caldera lake paleoenvironment (e.g. perhaps thick, aerially restricted rhyolite flows and cryptodomes, proximal thick ignimbrite facies, giant pumice deposits, critical epiclastic facies, and evidence of hydrothermal alteration and hot spring activity; Chap. 16).

Similarly, a succession of ignimbrites could represent an intra-caldera fill succession, or a series of outflow ignimbrites deposited tens of km from the vent. For example, McPhie (1983) identified a succession of Carboniferous ignimbrites (both welded and non-welded) in New South Wales, Australia, as an outflow succession. This was based on the association of moderately thick ignimbrites (cf. very thick intra-caldera fill ignimbrites, Chap. 10) with valley fill morphologies, fluvial volcaniclastic sedimentary rocks, and the absence of the expected association of caldera-related facies as indicated above. In contrast, Willcock et al. (2013) have described the Permian Ora Formation, in northern Italy, as a well-preserved intra-caldera ignimbrite succession. The ignimbrites of the Ora Formation are voluminous (>1200 km^3), crystal-rich ($\sim$25 to 55%), ubiquitously welded and the basal member shows the conspicuous development of co-ignimbrite breccias. The stratigraphic architecture provides evidence for caldera collapse at the onset of the eruption, a multiple discharge point, fissure eruption style, and progressive, incremental caldera in-filling by numerous pyroclastic flow pulses within the caldera.

So, in summary, there are few facies in volcanic terrains that will be unique to a particular setting or environment, so that interpretation should be based on the association of facies. Associated sedimentary facies will be the most valuable indicators of immediate sedimentary environments and the sedimentary processes and conditions operating.

15.9.2 Constructing Facies Models

Constructing a facies model for any modern or ancient volcano or volcanic system will inevitably involve several stages, beginning at the map scale and proceeding to a thorough, careful analysis of facies and associations of facies. The final facies models must therefore be fact-based; the stages involved are:

(a) Where surface exposure or subsurface mine exposure occurs, produce an outcrop map, which is a map showing where each outcrop occurs, its size, its composition and any relevant structural information. Outcrop maps, being "fact maps", are important because they constrain the degree of confidence that can be put on final interpretations, and serve as a basis for critically evaluating the possible spatial and age relationships between different facies. Various mediums are now available to use as base maps, including conventional topographic maps as well as remotely sensed imagery compiled onto various GIS data base programs (See Chap. 17, Sect. 17.7).

(b) Determine the structure of the succession. In particular, identify any major faults and high strain zones where volcanic textures may potentially be highly modified. If available, aeromagnetic and other geophysical data sets should be used in the field to help identify and extend structures under cover in non-exposed areas. To understand regional scale structure the stratigraphy must be known, and conversely to understand the stratigraphy, especially lateral changes in stratigraphy, the regional and local structure must be known.

(c) Where surface outcrop is sufficient stratigraphic sections and detailed graphic logs of facies should be measured (e.g. Figs. 10.85b, 15.28a, b, and 15.41a, b); when subsurface drill core is available, detailed logging will be necessary, and two- or 3D correlation fence diagrams should be attempted.

(d) Identify and describe all facies present based on field outcrop, core characteristics, hand-specimen characteristics, and personal thin-section observations, using the facies descriptors outlined in Chap. 14 (Fig. 14.1a, b). In successions that have been variably altered and deformed, facies analysis should first be done in the least altered areas or cores, and only when confidence in recognition of original facies has been developed, work back into more altered areas and cores. Recognition of significant associations of facies, and an understanding of the depositional context, may be very important guides in interpreting problematic rocks. Petrographic work should be done personally by geologists; petrographic descriptions made by consultants who do not see the rocks in context should be avoided (if possible).

(e) Work out the spatial and relative age relationships of facies (sharp, gradational, conformable, unconformable, faulted, or intrusive) from outcrop patterns and cross-sections. Only at this stage should samples be taken for geochronological and/or geochemical analysis (i.e. when there is reasonable knowledge of what is actually being sampled and for what purpose). It is important that sampling is chronostratigraphically controlled otherwise variation diagram trends do not represent detailed petrogenetic trends through time, just generalised patterns.

(f) Unless critical contacts are well exposed, it may be difficult to ascertain the age relationships between facies and therefore unrealistic to construct conventional "layer-cake" stratigraphies or columns. In many instances, interactive stratigraphic-facies columns or diagrams are preferable as a means of schematically representing spatial facies-stratigraphic relationships, the possible age relationships, and the preserved geometries (Fig. 15.42).

(g) Consider the possible genetic relationships between associated facies, and then make genetic interpretations of each facies and the total association of facies in terms of origins, environment of deposition, and environmental conditions.

(h) If a broader paleogeographic context is required, then look at the identified significant associations of facies and their relationships to each other. Where outcrop permits, this could be supplemented by collecting data on transport and source directions such as paleocurrent measurement, contouring of maximum clast sizes, and thickness variations.

(i) Finally, assemble all the facies, facies association(s), spatial stratigraphic and geochronological data and construct a *fact-based facies model* or 2D cross-sections in key directions and schematically reconstruct the facies and stratigraphic architecture and relationships in the subsurface part of the cross-section, and on the surface depict the environments (e.g. Fig. 15.43a).

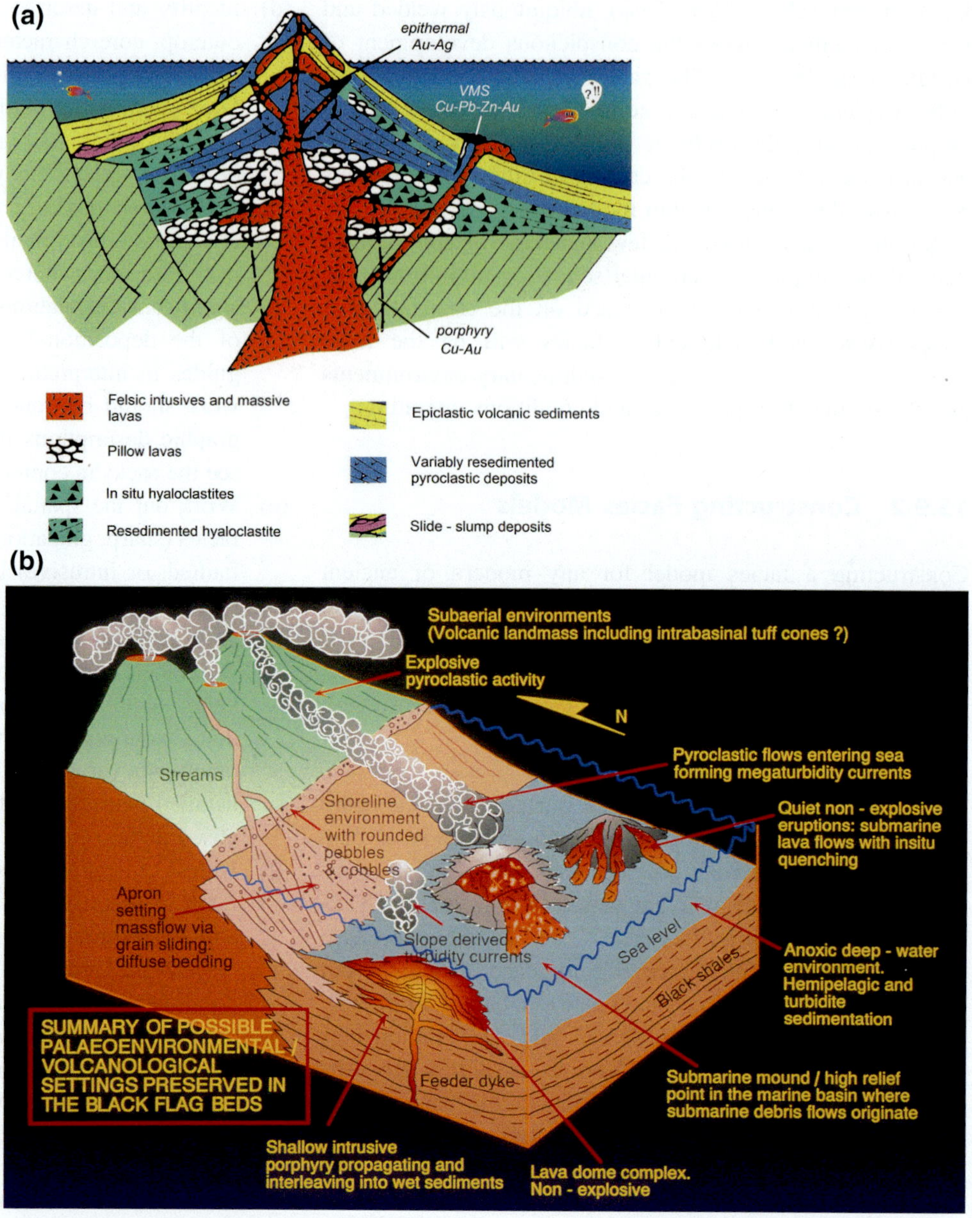

Fig. 15.43 Fact-based and dynamic-fact facies models. **a** Facies model for a marine stratovolcano, combining observations from the subaerial parts of modern marine examples, and the uplifted and eroded remnants of ancient examples (Modified from Cas 1992). **b** Schematic dynamic facies model for part of the 2690 to 2670 Ma Black Flag Group volcanic, intrusive and sedimentary succession around Kalgoorlie, in the Kalgoorlie Terrane, Yilgarn Craton, Western Australia (*Source* Hand 1999)

In Fig. 15.43a, a fact-based facies model for marine stratovolcanoes is presented as an example. This is based on many studies of modern and ancient marine stratovolcanoes. Well-studied modern marine stratovolcanoes include Stromboli volcano, Italy, where the subaerial edifice is dominated by lavas and some pyroclastic deposits, but some pillow lava is exposed at sea level, and Augustine volcano, Alaska, USA, where both lavas and voluminous pyroclastic deposits occur above sea level. The submarine part of the model is based on studies of the uplifted remnants of the submarine parts of ancient marine stratovolcanoes (e.g. Roobol and Hackett 1987; Houghton and Landis 1989; Cas 1992). The potential location of various mineral occurrences of economic importance can also be added to the model (Cas 1992, Chap. 18). For a *dynamic facies model*, processes can be schematically added to the model that are representative and reflect the origins of the facies when the association of facies formed, which sometimes can be presented as a 3D block diagram (Fig. 15.43b). Figure 15.43b depicts the association of facies, the paleoenvironmental settings, and volcanological, intrusive, and sedimentary processes, reconstructed for part of the ~2690 to 2670 Ma Black Flag Group, in the Kalgoorlie Terrane, Yilgarn Craton, Western Australia (Hand 1999), as discussed above (Sect. 15.8.5).

In Chap. 17 (Sect. 17.8), we build on this methodology for mapping and documenting ancient successions in the field and outline additional GIS data base and interrogation techniques for evaluating the tectonic and paleogeographic context of ancient and mineralised volcanic terrains.

15.9.3 Distinguishing Volcanic Systems According to Their Complexity as a Guide to Constructing Facies Models

In Chap. 16, we summarise the geology of the many different types of volcanoes and settings, and present facies models for most. As an introduction, we can subdivide volcanoes into *simple volcanoes* and complex or *multiplex volcanoes* (see Chap. 16). Simple volcanoes undergo a uniform eruption behaviour throughout their life. Many simple volcanoes are monogenetic (e.g. scoria cones, maar volcanoes), that is, they experience essentially just one eruption phase, and although eruption styles can change (e.g. from explosive to effusive) magma composition remains uniform, and no significant time breaks can be identified in the deposits. Some simple volcanoes however can be polygenetic, experiencing multiple eruption phases, but the styles and magma composition remain relatively uniform (e.g. Mt Fuji stratovolcano, Japan).

On the other hand, multiplex volcanoes are all polygenetic and can experience many changes in eruption styles, magma compositions, locations of vents, and even volcano edifice style. For example, Santorini island volcano in Greece has fluctuated between being a basaltic shield, an andesitic stratovolcano, and explosive rhyolitic to dacitic caldera volcano multiple times during its 0.6 Ma history (Fig. 16.78; Druitt et al. 1999). Currently, it has a lava flow-dome edifice growing in the middle of its marine caldera crater, and parasitic basaltic scoria cones on its outer subaerial slopes. This is a truly complex volcano, as well as being polygenetic. Similarly, the Las Cañadas volcano forming the main edifice on the Island of Tenerife, Spain, has had an incredibly complex history (Martí 2019; Cas et al. 2022), fluxing from being the second largest marine basaltic shield volcano from >16 Ma to present, to an explosive phonolitic caldera volcano from ~4 Ma to present, interspersed with being a phonolitic-basaltic stratovolcano (Plate 16.1; Fig. 16.79). It can be argued that currently Tenerife is all three: shield volcano, explosive caldera, and stratovolcano, given the timescales of eruptive activity. Historic shield building basaltic volcanism is well documented, with the most recent in 1909, Teide stratovolcano last erupted in 1492, and explosive caldera-forming eruptions occur cyclically, ~every 200 kyr, with the last occurring ~190 ka.

Finally, a little crystal ball gazing, with regard to understanding large complex volcanoes. Figure 15.44a shows a photograph from the crest of the Fossa crater on Vulcano island volcano in the Aeolian arc, southern Italy, looking northwards to Lipari volcanic island just 2 km to the north of Vulcano, and Salina, the twin volcanic island which lies ~6 km north of Lipari. Figure 15.44b shows a Google Earth view of these island volcanoes as well as eastern Sicily to the

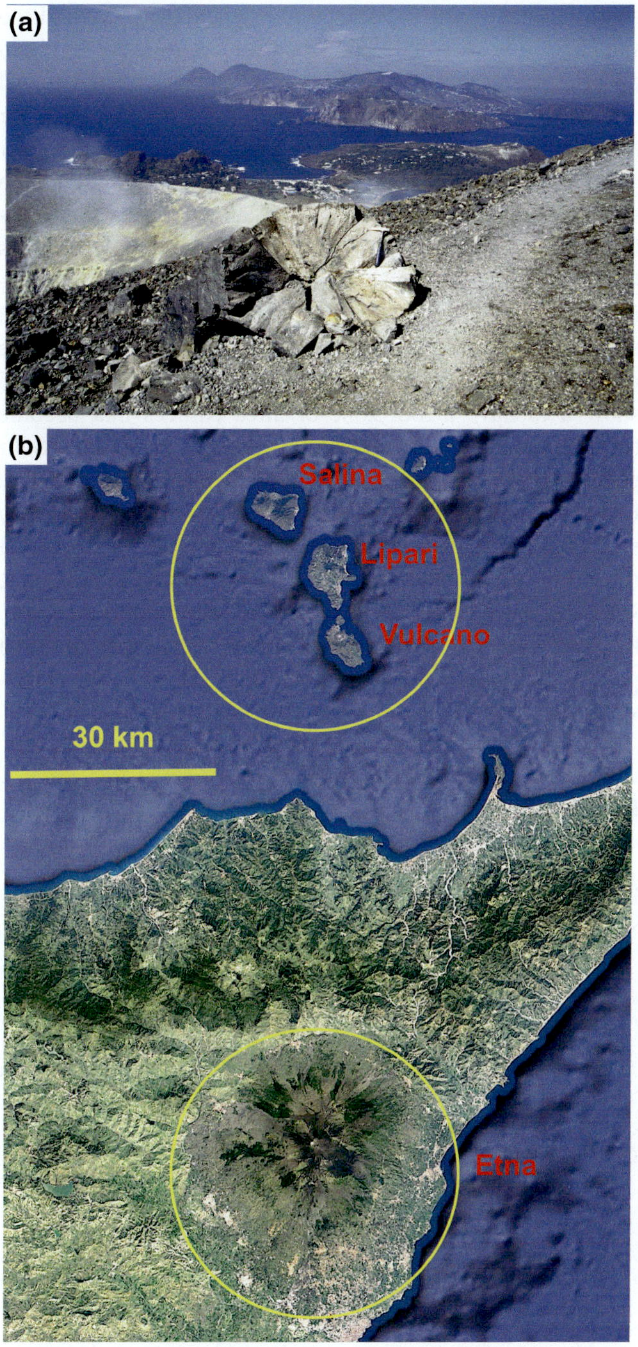

Fig. 15.44 Crystal ball gazing about how large volcanic edifices may originate and evolve (see text). **a** Photo of view from the time of the Fossa crater, Vulcano, looking across to Lipari and in the distance Salina volcano, Aeolian Islands, Italy. **b** Comparison of the scale of Mt Etna on eastern Sicily, with a circle the same diameter (~50 km) encircling Salina, Lipari and Vulcano island volcanoes (*Source* Google Earth)

south with the >50 km diameter Mt Etna shield-stratovolcano circled. The distance from the southern tip of Vulcano to the northern tip of Salina is ~30 km. All are still considered to be active and have probably been for at least 200–300 kyr, whereas Etna has been active for ~500 kyr. If they remain active they will merge into a huge subaerial, complex volcanic edifice about the same size as Etna. However, the submerged part of these three volcanoes has already merged and forms an elongated composite edifice 2 to 3 km high, of which the three islands only represent the emergent tops (Pensa et al. 2019). A facies model of the new merged volcanic edifice (let's call it "Salipano") will show it to have a root zone of three edifices that become fused by overlapping products of the new combined "Salipano" edifice, perhaps with multiple active vents, much as Etna now has. *Is that how Etna evolved, from a cluster of multiple small edifices that merged?* It is also likely that other big volcanoes experience similar complex histories, which in the rock record could be a nightmare to decipher, let alone depict in a facies model. Now read on to the final part of the book (Section F), and Chap. 16 to better understand the geology of our main volcano types.

References

Abdelmalak MM, Planke S, Faleide JI, Jerram DA, Zastrozhnov D, Eide S, Myklebust R (2016) The development of volcanic sequences at rifted margins: new insights from the structure and morphology of the Vøring Escarpment, mid-Norwegian Margin. J Geophys Res Solid Earth 121(7):5212–5236

Aiello G, Cicchella AG, Di Fiore V, Marsella E (2011) New seismo-stratigraphic data of the Volturno Basin (northern Campania, Tyrrhenian margin, southern Italy): implications for tectono-stratigraphy of the Campania and Latium sedimentary basins. Ann Geophys 54(3):265–283

Aiello G, Giordano L, Giordano F (2016) High-resolution seismic stratigraphy of the Gulf of Pozzuoli (Naples Bay) and relationships with submarine volcanic setting of the Phlegrean Fields volcanic complex. Rendiconti Lincei 27(4):775–801

Aiello G, Giordano L, Marsella E, Passaro S (2012) Seismic stratigraphy and marine magnetics of the Naples Bay (Southern Tyrrhenian sea, Italy): the onset of new technologies in marine data acquisition, processing and interpretation. Mar Geophys Res. https://doi.org/10.1007/s11001-012-9150-8

Aiuppa A, Fiorani L, Santoro S, Parracino S, D'Aleo R, Liuzzo M, Maio G, Nuvoli M (2017) New advances in Dial-LiDAR-based remote sensing of the volcanic CO2 flux. Front Earth Sci 5:15

Allen SR, Simpson CJ, McPhie J, Daly S (2003) Stratigraphy, distribution and geochemistry of widespread felsic volcanic units in the Mesoproterozoic Gawler Range Volcanics, South Australia. Aust J Earth Sci 50:97–112

Ansmann A, Tesche M, Groß S, Freudenthaler V, Seifert P, Hiebsch A, Schmidt J, Wandinger U, Mattis I, Müller D, Wiegner M (2010) The 16 April 2010 major volcanic ash plume over central Europe: EARLINET lidar and AERONET photometer observations at Leipzig and Munich, Germany. Geophys Res Letts 37(13)

Aramaki S (1963) Geology of Asama volcano. J Fac Sci Tokyo Univ Soc II 14:229–443

Berndt C, Planke S, Alvestad E, Tsikalas F, Rasmussen T (2001) Seismic volcanostratigraphy of the Norwegian Margin: constraints on tectonomagmatic break-up processes. J Geol Soc Lond 158 (3):413–426

Bischoff AP (2018) Architectural elements of buried volcanic systems and their impact on geoenergy resources. PhD thesis, University of Canterbury, Christchurch, New Zealand, 226p

Bischoff AP, Nicol A, Beggs M (2017) Stratigraphy of architectural elements in a buried volcanic systems and implications for hydrocarbon exploration. Interpretation 5(3):141–159

Bisson M, Spinetti C, Neri M, Bonforte A (2016) Mt. Etna volcano high-resolution topography: airborne LiDAR modelling validated by GPS data. Int J Digit Earth 9(7):710–732

Blaikie TN, Ailleres L, Betts PG, Cas RAF (2014a) Interpreting subsurface volcanic structures using geologically constrained 3-D gravity inversions: examples of maar-diatremes, Newer Volcanics Province, southeastern Australia. J Geophys Res Solid Earth 119 (4):3857–3878

Blaikie TN, Ailleres L, Betts PG, Cas RAF (2014b) A geophysical comparison of the diatremes of simple and complex maar volcanoes, Newer Volcanics Province, south-eastern Australia. J Volcanol Geotherm Res 276:64–81

Booth B, Croasdale R, Walker GPL (1978) A quantitative study of five thousand years of volcanism on Sao Miguel, Azores. Philos Trans R Soc Lond Ser A Math Phys Sci 288(1352):271–319

Boyce JA, Nicholls IA, Keays RR, Hayman PC (2015) Variation in parental magmas of Mt Rouse, a complex polymagmatic monogenetic volcano in the basaltic intraplate Newer Volcanics Province, southeast Australia. Contrib Mineral Petrol 169(2):11

Branca S, Coltelli M, Groppelli G, Lentini F (2011) Geological map of Etna volcano 1:50,000 scale. Ital J Geosci 130(3):265–291

Branney MJ (1991) Eruption and depositional facies of the Whorneyside Tuff Formation, English Lake District: an exceptionally large-magnitude phreatoplinian eruption. Geol Soc Am Bull 103 (7):886–897

Brown RJ, Kokelaar BP, Branney MJ (2007) Widespread transport of pyroclastic density currents from a large silicic tuff ring: the Glaramara tuff, Scafell caldera, English Lake District, UK. Sedimentology 54(5):1163–1190

Brown RJ, Manya S, Buisman I, Fontana G, Field M, Mac C, Sparks RSJ, Stuart FM (2012) Eruption of kimberlite magmas: physical volcanology, geomorphology and age of the youngest kimberlitic volcanoes known on earth (the Upper Pleistocene/Holocene Igwisi Hills volcanoes, Tanzania). Bull Volcanol 74(7):1621–1643

Bruno PP (2004) Structure and evolution of the Bay of Pozzuoli (Italy) using marine seismic reflection data: implications for collapse of the Campi Flegrei caldera. Bull Volcanol 66(4):342–355

Cas RAF (1977) Basin characteristics of the early Devonian part of the Hill End Trough, New South Wales, based on a stratigraphic analysis of the Merrions Tuff. J Geol Soc Aust 24(7–8):381–401

Cas RAF (1979) Mass-flow arenites from a Palaeozoic interarc basin, New South Wales, Australia: mode and environment of emplacement. J Sediment Petrol 49:29–44

Cas RAF (1992) Submarine volcanism; eruption styles, products, and relevance to understanding the host-rock successions to volcanic-hosted massive sulfide deposits. Econ Geol 87(3):511–541

Cas R, Giordano G, Balsamo F, Esposito A, Mastro SL (2011) Hydrothermal breccia textures and processes: Lisca bianca islet, Panarea volcano, Aeolian Islands, Italy. Econ Geol 106(3):437–450

Cas RAF, Van Otterloo J, Blaikie TN, Van Den Hove J (2017) The dynamics of a very large intra-plate continental basaltic volcanic province, the Newer Volcanics Province, SE Australia, and implications for other provinces. In: Németh K et al (eds) Monogenetic volcanism, vol 446(1). Geological Society of London, Special Publications, pp 123–172

Cas RAF, Wright JV (1987) Volcanic successions: modern and ancient. Allen & Unwin, London, p 528p

Cas RAF, Wolff JA, Marti J, Olin P, Edgar C, Pittari A, Simmons JM (2022) Tenerife, a complex oceanic island volcanic system of evolving basaltic oceanic shield, explosive polygenetic phonolitic calderas, and phonolitic-basaltic stratovolcanoes. Earth Sci Rev 230:103990.

Cawood PA, Hawkesworth CJ, Pisarevsky SA, Dhuime B, Capitanio FA, Nebel O (2018) Geological archive of the onset of plate tectonics. Philos Trans R Soc Ser A 376:20170405. https://doi.org/10.1098/rsta.2017.0405

Condit CB, Mahan KH, Curtis KC, Möller A (2018) Dating metasomatism: monazite and zircon growth during amphibolite facies albitization. Minerals 8(5):187

Connors K, Donaldson J, Morrison B, Davy's C, Neumayr P (2005) The stratigraphy of the Kambalda-St Ives district: workshop notes. St Ives Gold Mine, Gold Fields Pty. Ltd., Kambalda

Costa A, Smith VC, Macedonio G, Matthews NE (2014) The magnitude and impact of the Youngest Toba Tuff super-eruption. Front Earth Sci 2(16):1–8

Darvill CM (2013) Cosmogenic nuclide analysis. In: Cook SJ, Clarke LE, Nield JM (eds) Geomorphological techniques. British Society for Geomorphology, London

De Rita D, Fabbri M, Giordano G, Rodani S, Trigari A (2000) L'utilizzo delle unita stratigrafiche a limiti inconformi in ambiente vulcanico e vulcanoclastico; proposta di una metodologia di terreno e di organizzazione informatica dei dati. Boll Soc Geol Ital 119 (3):749–760

De Rita D, Giordano G, Cecili A (2001) A model for submarine rhyolite dome growth: Ponza Island (central Italy). J Volcanol Geother Res 107(4): 221-239.

De Rita D, Giordano G, Milli S (1998) Forestepping-backstepping stacking pattern of volcaniclastic successions: Roccamonfina volcano, Italy. J Volcanol Geotherm Res 78(3):267–288

Deardorff ND, Cashman KV (2012) Emplacement conditions of the c. 1,600-year BP Collier Cone lava flow, Oregon: a LiDAR investigation. Bull Volcanol 74(9):2051–2066

Diano G, Bonamico A, Giordano G (2010) Three-dimensional reconstruction of the main unconformities of the Colli Albani stratigraphy and deposit volume calculations. In: Funiciello R, Giordano G (eds) The Colli Albani volcano, vol 3. Geological Society of London Special Publications and IAVCEI, pp 177–188

Dickson BL, Scott KM (1997) Interpretation of aerial gamma-ray surveys-adding the geochemical factors. AGSO J Aust Geol Geophys 17:187–200

Doutch D (2019) Origin, geochemistry, stratigraphic and structural setting of the Archean Invincible gold deposit, St Ives gold camp, Yilgarn Craton, Western Australia. PhD thesis, University of Tasmania, Australia

Druitt TH, Edwards L, Mellors RM, Pyle DM, Sparks RS, Lanphere M, Davies M, Barreirio B (1999) Santorini volcano, vol 19. Geological Society of London, Memoirs, 165p

Emery D, Myers KJ (1996) Sequence stratigraphy. Blackwell Science, Oxford, 297p

Fisher RV, Schmincke H-U (1984) Pyroclastic rocks. Springer, Berlin Heidelberg

Fisher RV, Smith GA (eds) (1991) Sedimentation in volcanic settings, vol 45. Soc Econ Pal Mineral Spec Pub, 257p

Flynn LP, Harris AJ, Wright R (2001) Improved identification of volcanic features using Landsat 7 ETM+. Remote Sens Environ 78 (1–2):180–193

Forni F, Lucchi F, Peccerillo A, Tranne CA, Rossi PL, Frezzotti ML (2013) Stratigraphy and geological evolution of the Lipari volcanic complex (central Aeolian archipelago). In: Lucchi F et al (eds) The Aeolian islands volcanoes, vol 37(1). Geological Society of London Memoirs, pp 213–279

Francis PW, Baker MC (1978) Sources of two large ignimbrites in the central Andes: some Landsat evidence. J Volcanol Geotherm Res 4 (1–2):81–87

Francis PW, de Silva SL (1989) Application of the landsat thematic mapper to the identification of potentially active volcanoes in the Central Andes. Remote Sens Environ 28:245–255

Garden TO, Gravley DM, Kennedy BM, Deering C, Chambefort I (2017) Controls on hydrothermal fluid flow in caldera-hosted settings: evidence from Lake City caldera, USA. Geosphere 13 (6):1993–2016

Geisler T, Schaltegger U, Tomaschek F (2007) Re-equilibration of zircon in aqueous fluids and melts. Elements 3(1):43–50

Geshi N, Kusumoto S, Gudmundsson A (2012) Effects of mechanical layering of host rocks on dike growth and arrest. J Volcanol Geotherm Res 223:74–82

Ghiorso MS, Hirschmann MM, Reiners PW, Kress VC (2001) The pMELTS: a revision of MELTS for improved calculation of phaserelations and major element partitioning related to partial melting of themantle to 3 GPa. Geochem Geophys Geosyst 3 (5):1030. https://doi.org/10.1029/2001GC000217

Ghiorso MS, Sack R (1995) Chemical mass transfer in magmatic pro-cesses IV. A revised and internally consistent thermodynamic model forthe interpolation and extrapolation of liquid–solid equilibria in magmaticsystems at elevated temperatures and pressures. Contrib Mineral Petrol 119(2–3):197–212. https://doi.org/10.1007/bf00307281

Giordano G, De Benedetti AA, Diana A, Diano G, Esposito A, Fabbri M, Gaudioso F, Marasco F, Mazzini I, Miceli M, Mincione V (2010a) Stratigraphy, volcano tectonics and evolution of the Colli Albani volcanic field. Geol Soc Lond IAVCEI Spec Publ 1(3):43–97

Giordano G, Mattei M, Funiciello R (2010b) Geological map 1:50,000 scale of the Colli Albani volcano. In: Funiciello R, Giordano G (eds) The Colli Albani Volcano, vol 1(3). Geological Society of London and IAVCEI Special Publication

Grant JA (1986) The Isocon diagram—a simple solution to Gresens' equation for metasomatic alteration. Econ Geol 81:1976–1982

Grant JA (2005) Isocon analysis: a brief review of the method and applications. Phys Chem Earth 30:997–1004

Grosse P, Euillades PA, Euillades LD, van Wyk de Vries B (2014) A global database of composite volcano morphometry. Bull Volcanol 76(1):784

Grosse et al., 2012Grosse P, van Wyk de Vries B, Euillades PA, Kervyn M, Petrinovic IA (2012) Systematic morphometric characterization of volcanic edifices using digital elevation models. Geomorphology 136(1):114–131

Gualda GAR, Ghiorso MS, Lemons RV, Carley TL (2012) Rhyolite-MELTS: a modified calibration of MELTS optimized forsilica-rich, fluid-bearing magmatic systems. J Petrol 53(5):875–890. https://doi.org/10.1093/petrology/egr080

Gudmundsson A (2002) Emplacement and arrest of sheets and dykes in central volcanoes. J Volcanol Geotherm Res 116(3):279–298

Hand JL (1999) Sedimentology, volcanology and tectonostratigraphic significance of the Archaean Black Flag Beds, Kalgoorlie region, Western Australia. PhD thesis, Monash University, Australia

Hanneman DL, Wideman CJ (2010) Continental sequence stratigraphy and continental carbonates. Dev Sedimentol 62:215–223

Hawkesworth CJ, Cawood PA, Dhuime B (2020) The evolution of the continental crust and the onset of plate tectonics. Front Earth Sci 8:326. https://doi.org/10.3389/feart.2020.00326

Hayman PC, Fullard R, Cas RAF, Squire RJ (2020) Clasts in Archean conglomerates and implications for uplift: evidence from the 2.7 Ga

Agnew Greenstone Belt (Western Australia). Precambrian Res 344:105758

Hayman PC, Thébaud N, Pawley MJ, Barnes SJ, Cas RAF, Amelin Y, Sapkota J, Squire RJ, Campbell IH, Pegg I (2015) Evolution of a ~ 2.7 Ga large igneous province: a volcanological, geochemical and geochronological study of the Agnew Greenstone Belt, and new regional correlations for the Kalgoorlie Terrane (Yilgarn Craton, Western Australia). Precambrian Res 270:334–368

Heise W, Caldwell TG, Bibby HM, Bannister SC (2008) Three-dimensional modelling of magnetotelluric data from the Rotokawa Geothal field, Taupo Volcanic Zone, New Zealand. Geophys J Int 173(2):740–750

Hofton MA, Malavassi E, Blair JB (2006) Quantifying recent pyroclastic and lava flows at Arenal Volcano, Costa Rica, using medium-footprint lidar. Geophys Res Lett 33(21)

Houghton BF, Landis CA (1989) Sedimentation and volcanism in a Permian arc-related basin, southern New Zealand. Bull Volcanol 51 (6):433–450

Howells MF, Reedman AJ, Leveridge BE (1985) Geology of the country around Bangor. Explanation of 1:50 000 sheet sheet 106 (England and Wales). British Geological Survey, London, 34p

Howells MF, Smith M (1997) The geology of the country around Snowdon. Memoir of 1:50 000 sheet 119 (England and Wales). British Geological Survey, London, 104p

Inbar M, Gilichinsky M, Melekestsev I, Melnikov D, Zaretskaya N (2011) Morphometric and morphological development of Holocene cinder cones: a field and remote sensing study in the Tolbachik volcanic field, Kamchatka. J Volcanol Geotherm Res 201(1–4):301–311

Infante-Paez L, Marfurt KJ (2017) Seismic expression and geomorphology of igneous bodies: a Taranaki Basin, New Zealand, case study. Interpretation 5(3):121–140

Innacce P, Milia A, Torrente MM (2013) 4D geologic evolution in Gaeta Bay sedimentary infill (Eastern Tyrrhenian Sea). GeoActa 12:201–212

Isaia R, Marianelli P, Sbrana A (2009) Caldera unrest prior to intense volcanism in Campi Flegrei (Italy) at 4.0 ka BP: implications for caldera dynamics and future eruptive scenarios. Geophys Res Letts 36(21)

Janoušek V, Moyen JF, Martin H, Erban V, Farrow C (2016) Geochemical modelling of igneous processes–principles and recipes in R language. Springer, Berlin, 2016;10:978-3

Jessop DE, Kelfoun K, Labazuy P, Mangeney A, Roche O, Tillier JL, Trouillet M, Thibault G (2012) LiDAR derived morphology of the 1993 Lascar pyroclastic flow deposits, and implication for flow dynamics and rheology. J Volcanol Geotherm Res 245:81–97

Joyce KE, Samsonov S, Manville V, Jongens R, Graettinger A, Cronin SJ (2009) Remote sensing data types and techniques for lahar path detection: a case study at Mt Ruapehu, New Zealand. Remote Sens Environ 113(8):1778–1786

Kervyn M, Kervyn F, Goossens R, Rowland SK, Ernst GG (2007) Mapping volcanic terrain using high-resolution and 3D satellite remote sensing. In Teeuw RM (ed) Mapping hazardous terrain using remote sensing, vol 283(1). Geological Society of London Special Publications, pp 5–30

Krapez B, Brown SJ, Hand J, Barley ME, Cas RAF (2000) Age constraints on recycled crustal and supracrustal sources of Archaean metasedimentary sequences, Eastern Goldfields Province, Western Australia: evidence from SHRIMP zircon dating. Tectonophysics 322(1–2):89–133

Krumbein WC, Sloss LL (1963) Stratigraphy and sedimentation. Freeman, San Francisco, p 660p

Lane CS, Chorn BT, Johnson TC (2013) Ash from the Toba supereruption in Lake Malawi shows no volcanic winter in East Africa at 75 ka. Proc Natl Acad Sci 110(20):8025–8029

Lanzafame G, Neri M, Acocella V, Billi A, Funiciello R, Giordano G (2003) Structural features of the July–August 2001 Mount Etna eruption: evidence for a complex magma supply system. J Geol Soc 160(4): 531-544.

Laurenzi MA, Marra F, Taddeucci J, Freda C, Marzocchi W, Scarlato P (2005) Recurrence of volcanic activity along the Roman Comagmatic Province (Tyrrhenian margin of Italy) and its tectonic significance. Tectonics 24(5)

Lowe DJ (2011) Tephrochronology and its application: a review. Quat Geochronol 6(2):107–153

Lowe JJ, Barton N, Blockley S, Ramsey CB, Cullen VL, Davies W, Gamble C, Grant K, Hardiman M, Housley R, Lane CS (2012) Volcanic ash layers illuminate the resilience of Neanderthals and early modern humans to natural hazards. Proc Natl Acad Sci 109 (34):13532–13537

Lowe et al, 2015Lowe JJ, Ramsey CB, Housley RA, Lane CS, Tomlinson EL, RESET Team, RESET Associates (2015) The RESET project: constructing a European tephra lattice for refined synchronisation of environmental and archaeological events during the last c. 100 ka. Quat Sci Rev 118:1–7

Lucchi F (2013) Stratigraphic methodology for the geological mapping of volcanic areas: insights from the Aeolian archipelago (southern Italy). In: Lucchi F et al (eds) The Aeolian Islands volcanoes, vol 37 (1). Geological Society of London Memoirs, pp 37–53

Lucchi F, Keller J, Tranne CA (2013) Regional stratigraphic correlations across the Aeolian archipelago (southern Italy). In: Lucchi F et al (eds) The Aeolian Islands volcanoes, vol 37(1). Geological Society of London Memoirs, pp 55–81

Lucchi F, Tranne CA, Rossi PL (2010) Stratigraphic approach to geological mapping of the late-Quaternary volcanic island of Lipari (Aeolian archipelago, Southern Italy). In: Groppelli G, Viereck-Goette L (eds) Stratigraphy and geology of volcanic areas, vol 464. Geological Society of America Special Papers, pp 1–32

Magli A, Branca S, Speranza F, Risica G, Siravo G, Giordano G (2022) Paleomagnetic dating of prehistoric lava flows from the urban district of Catania (Etna volcano, Italy). Geol Soc Am Bulletin 134 (3-4): 616-628.

Manville V, Németh K, Kano K (2009) Source to sink: a review of three decades of progress in the understanding of volcaniclastic processes, deposits, and hazards. Sediment Geol 220(3):136–161

Mark DF, Petraglia M, Smith VC, Morgan LE, Barfod DN, Ellis BS, Pearce NJ, Pal JN, Korisettar R (2014) A high-precision $^{40}Ar/^{39}Ar$ age for the Young Toba Tuff and dating of ultra-distal tephra: forcing of Quaternary climate and implications for hominin occupation of India. Quat Geochronol 21:90–103

Marshall RR (1961) Devitrification of natural glass. Geol Soc Am Bull 72(10):1493–1520

Martí J (2019) Las Cañadas caldera, Tenerife, Canary Islands: a review, or the end of a long volcanological controversy. Earth Sci Rev 196:102889

Martí J, Groppelli G, da Silveira AB (2018) Volcanic stratigraphy: a review. J Volcanol Geotherm Res 357:68–91

Marturano A, Isaia R, Aiello G, Barra D (2018) Complex dome growth at Campi Flegrei caldera (Italy) in the last 15 ka. J Geophys Res Solid Earth 123(9):8180–8197

Mazzarini F, Pareschi MT, Favalli M, Isola I, Tarquini S, Boschi E (2007) Lava flow identification and aging by means of lidar intensity: Mount Etna case. J Geophys Res Solid Earth 112 (B0220). https://doi.org/10.1029/2005JB004166

McPhie J (1983) Outflow ignimbrite sheets from Late Carboniferous calderas, Currabubula Formation, New South Wales, Australia. Geol Mag 120(5):487–503

Mia M, Fujimitsu Y, Nishijima J (2017) Thermal activity monitoring of an active volcano using Landsat 8/OLI-TIRS sensor images: a case study at the Aso volcanic area in southwest Japan. Geosciences 7 (4):118

Milia A (2010) The stratigraphic signature of volcanism off Campi Flegrei (Bay of Naples, Italy). In: Groppelli G, Viereck-Goette L (eds) Stratigraphy and geology of volcanic areas, vol 464. Geological Society of American Special Papers, pp 155–170

Mitchum RM, Vail PR, Thompson S (1977) Seismic stratigraphy and global changes of sea level: Part 2. The depositional sequence as a basic unit for stratigraphic analysis. In: Paynton CE (ed) Seismic stratigraphy: applications to hydrocarbon exploration, vol 26. American Association of Petroleum Geologists Memoir, pp 53–62

Moore JG, Albee WC (1981) Topographic and structural changes, March-July 1980- Photogrammetric data. US Geol Surv Prof Pap 1250:123–134

Morley CK (2018) 3-D seismic imaging of the plumbing system of the Kora Volcano, Taranaki Basin, New Zealand: the influence of syn-rift structure on shallow igneous intrusion architecture. Geosphere 14(6):2533–2584

NACSN, 2005NACSN (North American Commission on Stratigraphic Nomenclature) (2005) North American stratigraphic code. Am Assoc Pet Geol Bull 89(11):1547–1591

Nakamura K (1962) Volcano-stratigraphic study of Oshima volcano, Izu. Bull Earthq Res Inst Univ Tokyo 42:649–728

Németh K, White JD, Reay A, Martin U (2003) Compositional variation during monogenetic volcano growth and its implications for magma supply to continental volcanic fields. J Geol Soc Lond 160(4):523–530

Newhall CG, Punongbayan R (eds) (1996) Fire and mud: eruptions and lahars of Mount Pinatubo, Philippines. Philippine Institute of Volcanology and Seismology, Quezon City and University of Washinton, Seattle, 1126p

Ninkovich D, Sparks RS, Ledbetter MT (1978) The exceptional magnitude and intensity of the Toba eruption, Sumatra: an example of the use of deep-sea tephra layers as a geological tool. Bull Volcanol 41(3):286–298

Nomikou P, Parks MM, Papanikolaou D, Pyle DM, Mather TA, Carey S, Watts AB, Paulatto M, Kalnins ML, Livanos I, Bejelou K (2014) The emergence and growth of a submarine volcano: the Kameni islands, Santorini (Greece). Geo Res J 1:8–18

Orth K, Cas RAF, Wright JV (1989) Facies analysis and facies associations in the recognition of volcanic centres in silicic terranes: an example from the Early Devonian of Australia. Aust J Earth Sci 36(2):167–188

Palladino DM, Simei S, Sottili G, Trigila R, Groppelli G (2010) Integrated approach for the reconstruction of stratigraphy and geology of quaternary volcanic terrains: an application to the Vulsini Volcanoes (central Italy). In: Groppelli G, Viereck-Goette L (eds) Stratigraphy and geology of volcanic areas, vol 464. Geological Society of America Special Papers, pp 63–84

Paynton CE (ed) (1977) Seismic stratigraphy: applications to hydrocarbon exploration. American Association of Petroleum Geologists Memoir 26

Pensa A, Pinton A, Vita L, Bonamico A, De Benedetti AA, Giordano G (2019) ATLAS of Italian submarine volcanic structures. Mem Descr Carta Geol D'it 104:77–183

Pierson TC, Janda RJ, Umbal JV, Daag AS (1992) Immediate and long-term hazards from lahars and excess sedimentation in rivers draining Mt. Pinatubo, Philippines, vol 92. US Geological Survey, Water-Resources Investigations Report, pp 183–203

Pinton A, Giordano G, Speranza F, Thordarson T (2018) Paleomagnetism of Holocene lava flows from the Reykjanes Peninsula and the Tungnaá lava sequence (Iceland): implications for flow correlation and ages. Bull Volcanol 80(1):10

Planke S, Millett JM, Maharjan D, Jerram DA, Abdelmalak MM, Groth A, Hoffmann J, Berndt C, Myklebust R (2017) Igneous seismic geomorphology of buried lava fields and coastal escarpments on the Vøring volcanic rifted margin. Interpretation 5(3):1–17

Planke S, Rasmussen T, Rey SS, Myklebust R (2005) Seismic characteristics and distribution of volcanic intrusions and hydrothermal vent complexes in the Vøring and Møre basins. Geol Soc Lond Pet Geol Conf Ser 6(1):833–844

Planke S, Symonds PA, Alvestad E, Skogseid J (2000) Seismic volcanostratigraphy of large-volume basaltic extrusive complexes on rifted margins. J Geophys Res Solid Earth 105(B8):19335–19351

Ponomareva V, Portnyagin M, Davies SM (2015) Tephra without borders: far-reaching clues into past explosive eruptions. Front Earth Sci 3(83):1–16

Ramsay JG (1980) Shear zone geometry: a review. J Struct Geol 2(1–2):83–99

Ramsey MS, Flynn LP (2004) Strategies, insights, and the recent advances in volcanic monitoring and mapping with data from NASA's earth observing system. J Volcanol Geotherm Res 135(1–2):1–11

Ramsey MS, Harris AJ (2013) Volcanology 2020: how will thermal remote sensing of volcanic surface activity evolve over the next decade? J Volcanol Geotherm Res 249:217–233

Reading HG (1996) Sedimentary environments: processes, facies and stratigraphy. Blackwell Science Ltd., Oxford, p 704p

Reath KA, Ramsey MS, Dehn J, Webley PW (2016) Predicting eruptions from precursory activity using remote sensing data hybridization. J Volcanol Geotherm Res 321:18–30

Renne PR, Mundil R, Balco G, Min K, Ludwig KR (2010) Joint determination of ^{40}K decay constants and ^{40}Ar∗/^{40}K for the Fish Canyon sanidine standard, and improved accuracy for ^{40}Ar/^{39}Ar geochronology. Geochim Cosmochim Acta 74(18):5349–5367

Risica G, Speranza F, Giordano G, De Astis G, Lucchi F (2019) Paleomagnetic dating of the Neostromboli succession. J Volcanol Geother Res 371: 229-244.

Rollinson HR (2014) Using geochemical data: evaluation, presentation, interpretation. Routledge, 384p

Roobol MJ, Hackett D (1987) Paleovolcanic facies and exhalite geochemistry; guides for selecting exploration areas in volcano-sedimentary complexes. Econ Geol 82(3):691–705

Roobol MJ, Smith AL, Wright JV (1985) Dispersal and characteristics of pyroclastic fall deposits from Mt. Misery volcano, West Indies. Geol Rundsch 74(2):321–335

Rushton AWA, Howells MF (1998) Stratigraphical framework for the Ordovician of Snowdonia and the Lleyn Peninsula. Br Geol Surv Res Rept RR/99/08:41p

Salvador A (1987) Unconformity-bounded stratigraphic units. Geol Soc Am Bull 98(2):232–237

Salvador A (ed) (1994) International stratigraphic guide: a guide to stratigraphic classification, terminology, and procedure. Int Union Geol Sci Geol Soc Am 214p

Selley RC (2004) Ancient sedimentary environments: and their sub-surface diagnosis, 4th edn. Routledge, London, p 320p

Sicard M, Guerrero-Rascado JL, Navas-Guzmán F, Preißler J, Molero F, Tomás S, Bravo-Aranda JA, Comerón A, Rocadenbosch F, Wagner F, Pujadas M (2012) Monitoring of the Eyjafjallajökull volcanic aerosol plume over the Atmos. Chem Phys 12(6):3115–3130

Sillitoe RH (1994) Erosion and collapse of volcanoes: causes of telescoping in intrusion-centered ore deposits. Geology 22(10):945–948

Simpson CJ, Cas RAF, Arundell MC (2005) Volcanic evolution of a long-lived Ordovician island-arc province in the Parkes region of the Lachlan Fold Belt, southeastern Australia. Aust J Earth Sci 52 (6):863–886

Smith GA, Fritz WJ (1989) Volcanic influences on terrestrial sedimentation. Geology 17(4): 375-376.

Smith VC, Isaia R, Engwell SL, Albert PG (2016) Tephra dispersal during the Campanian Ignimbrite (Italy) eruption: implications for ultra-distal ash transport during the large caldera-forming eruption. Bull Volcanol 78(6):45

Smith VC, Pearce NJ, Matthews NE, Westgate JA, Petraglia MD, Haslam M, Lane CS, Korisettar R, Pal JN (2011) Geochemical fingerprinting of the widespread Toba tephra using biotite compositions. Quat Int 246(1–2):97–104

Smithies RH, Lu Y, Johnson TE, Kirkland CL, Cassidy KF, Champion DC, Mole DR, Zibra I, Gessner K, Sapkota J, De Paoli MC (2019) No evidence for high-pressure melting of Earth's crust in the Archean. Nat Commun 10(1):1–2

Song SR, Chen CH, Lee MY, Yang TF, Iizuka Y, Wei KY (2000) Newly discovered eastern dispersal of the youngest Toba Tuff. Mar Geol 167(3–4):303–312

Squire RJ, Allen CM, Cas RA, Campbell IH, Blewett RS, Nemchin AA (2010) Two cycles of voluminous pyroclastic volcanism and sedimentation related to episodic granite emplacement during the late Archean: Eastern Yilgarn Craton, Western Australia. Precambrian Res 183(2):251–274

Storey M, Duncan RA, Swisher CC (2007) Paleocene-Eocene thermal maximum and the opening of the northeast Atlantic. Science 316 (5824):587–589

Swager CP, Griffin TJ, Witt WK, Wyche S, Ahmat AL, Hunter WM, McGoldrick PJ (1995) Geology of the Archaean Kalgoorlie Terrane–an explanatory note. West Aust Geol Surv GSWA Rep 48:1–26

Teeuw RM (ed) (2007) Mapping hazardous terrain using remote sensing, vol 283. Geological Society of London Special Publications, 179p

Torrente MM, Milia A (2013) Volcanism and faulting of the Campania margin (Eastern Tyrrhenian Sea, Italy): a three-dimensional visualization of a new volcanic field off Campi Flegrei. Bull Volcanol 75(6):719

Urbani S, Giordano G, Lucci F, Rossetti F, Acocella V, Carrasco-Núñez G (2020) Estimating the depth and evolution of intrusions at resurgent calderas: Los Humeros (Mexico). Solid Earth 11(2):527–545

Vail PR, Mitchum RM, Thompson S (1977) Seismic stratigraphy and global changes of sea level: Part 4. Global cycles of relative changes of sea level. In: Paynton CE (ed) Seismic stratigraphy: applications to hydrocarbon exploration. Am Assoc Petrol Geol Mem 26:83–98

van der Meer FD, van der Werff HM, van Ruitenbeek FJ, Hecker CA, Bakker WH, Noomen MF, Van Der Meijde M, Carranza EJ, De Smeth JB, Woldai T (2012) Multi- and hyperspectral geologic remote sensing: a review. Int J Appl Earth Obs Geoinformation 14 (1):112–128

Van Otterloo J, Raveggi M, Cas RAF, Maas R (2014) Polymagmatic activity at the monogenetic Mt Gambier volcanic complex in the Newer Volcanics Province, SE Australia: new insights into the occurrence of intraplate volcanic activity in Australia. J Petrol 55 (7):1317–1351

Vessell RK, Davies DK (1981) Non-marine sedimentation in an active fore-arc basin. In: Ethridge FG, Flores RM (eds) Recent and ancient non-marine depositional environments: models for exploration, vol 31. Society of Economic Paleontologists and Mineralogists Special Publication, pp 31–45

Vignaroli G, Aldega L, Balsamo F, Billi A, De Benedetti AA, De Filippis L, Giordano G, Rossetti F (2015) A way to hydrothermal paroxysm, Colli Albani volcano, Italy. Geol Soc Am Bull 127(5–6):672–687

Walker GPL, Croasdale R (1971) Two Plinian-type eruptions in the Azores. J Geol Soc Lond 127(1):17–55

Walker GPL, Wright JV, Clough BJ, Booth B (1981) Pyroclastic geology of the rhyolitic volcano of La Primavera, Mexico. Geol Rundsch 70(3):1100–1118

Walker RG (ed) (1984) Facies models. Geosci Canada, Geol Assoc Canada. Reprint Series 1:317p

Whelley PL, Glaze LS, Calder ES, Harding DJ (2014) LiDAR-derived surface roughness texture mapping: application to Mount St. Helens pumice plain deposit analysis. IEEE Trans Geosci Remote Sens 52 (1):426–438

Wiegner M, Gasteiger J, Groß S, Schnell F, Freudenthaler V, Forkel R (2012) Characterization of the Eyjafjallajökull ash-plume: potential of lidar remote sensing. Phys Chem Earth Parts A/B/C 45:79–86

Wijbrans J, Schneider B, Kuiper K, Calvari S, Branca S, De Beni E, Norini G, Corsaro RA, Miraglia L (2011) 40Ar/39Ar geochronology of Holocene basalts; examples from Stromboli, Italy. Quat Geochronol 6(2):223–232

Willcock MA, Cas RAF, Giordano G, Morelli C (2013) The eruption, pyroclastic flow behaviour, and caldera in-filling processes of the extremely large volume (> 1290 km3), intra-to extra-caldera, Permian Ora (Ignimbrite) Formation, Southern Alps, Italy. J Volcanol Geotherm Res 265:102–126

Wright JV (2008) Uranium prospectivity of the Gawler Craton: final report. Report to Quasar Resources, Adelaide, Australia, 77p (unpubl.)

Yost CL, Jackson LJ, Stone JR, Cohen AS (2018) Subdecadal phytolith and charcoal records from Lake Malawi, East Africa imply minimal effects on human evolution from the ~74 ka Toba supereruption. J Hum Evol 116:75–94

Part F
Bringing Everything Together: The Big Picture

Aims, Scope, and Structure

In this final part we attempt to piece together the 'bigger picture', thereby providing a global overview of the larger-scale aspects of volcanic systems and terrains. So far we have examined the principal processes and mechanisms that generate the enormous spectrum of volcanic deposits and rock types, ranging from ultramafic to silicic compositions, from subvolcanic to effusive to pyroclastic to epiclastic facies, and from those deposited in subaerial to subglacial and subaqueous environments, while in the preceding Part E, we have provided a systematic approach to describing their terminology, and recording and mapping their characteristic features. Part F is divided into three substantive chapters. Chap. 16 first documents the geology and the important morphological, physical, facies and stratigraphic characteristics found in and around modern volcanoes and volcanic centres. In addition, we again assess the duration of primary volcanic versus syn- to post-eruptive sedimentary or epiclastic processes. This leads to the development of general facies models for the main volcano and volcanic terrain types, which are relevant for modern settings, and are valuable guides for making meaningful geological reconstructions of ancient volcanic successions, including the economic significance of potentially prospective volcano types. In Chap. 17, we provide an updated review of the known tectonic settings in which volcanism occurs today and in the recent past, and the overall geological characteristics of these settings using the plate tectonic model as a global framework for Phanerozoic and Proterozoic times. However, there is little evidence that plate tectonics operated during the Hadean and Archean, which represent almost 50% of geological time, and so we consider what the tectonic paradigms were for volcanism during the Earth's early, fascinating history. Volcanoes and volcanic rocks are host to economically very significant quantities of natural resources, which are described in Chap. 18. Here we focus predominantly on the geology of the principal ore deposit types that characterise volcanic successions and some high-level subvolcanic intrusive settings, but also consider volcanic rock reservoirs in hydrocarbon exploration, and geothermal energy.

16 The Geology of Volcanoes and Their Facies Models

Morphology, Dynamics, Evolution, Successions, Economic Significance and Hazards

Guido Giordano, Ray Cas, and John V. Wright

R. Cas et al., *Volcanology*, Springer Textbooks in Earth Sciences, Geography and Environment,
https://doi.org/10.1007/978-3-319-66613-6_16

Plate 16.1 Extremes of volcano/volcanic system types and complexity. **a** The oceanic island of Tenerife in the Canary Islands, Spain, represents one of the most complex volcanic systems on Earth (basaltic shield-explosive phonolitic multi-caldera-stratovolcano complex), and rises nearly 8 km from the ocean floor to its subaerial summit. This oblique Google Earth image (looking west) shows the three overlapping nested explosive calderas of the summit Las Cañadas caldera complex, and the steep-sided stratovolcanoes of El Teide (centre right), Pico Viejo with its well-defined crater (top centre), and the fissure vent system of Montaña Blanca (lower right), located along the northern ring fracture of the caldera complex. The Las Cañadas calderas formed during three, major ignimbrite-forming eruptions from at least 1.6 Ma to ~190 ka. The last eruption of El Teide occurred ~1,150 BP (dark lavas or "Lavas Negras" which cover much of the flanks of the volcano), of Pico Viejo in 1798 (black lava, top centre on its southwestern flank), and Montaña Blanca ~2,000 years ago. The last eruption on Tenerife, a scoria cone and lava forming eruption, occurred in 1909. (*Source* Google Earth). **b** In contrast, a cross-section through a small scoria cone which became exposed during formation of the AD 2000 caldera collapse vent at Myjakejima volcano, Myjakejima, Izu Islands, Japan.This shows a simple architecture of the scoria cone with inward and outward dipping scoria beds, and a central zone of agglutinate filling the crater/vent above a narrow tapering lava lake fed by a narrow dyke that has propagated irregularly through the substrate of lavas (centre left)

Abstract

Modern volcanoes and volcanic centres encompass a wide variety of morphological, physical, facies and stratigraphic characteristics, and duration of primary volcanic versus sedimentary or epiclastic processes. This leads to the development of general facies models, which are relevant for modern settings, and valuable guides for making meaningful geological reconstructions of ancient volcanic successions. Our discussion covers the full scale of volcanic landforms, ranging from small individual monogenetic scoria and pumice cones and phreatomagmatic maar-type volcanoes, to larger polygenetic volcanoes. These include marine basaltic shields that from their base on the seafloor represent some of the largest volcanoes on the planet, as well as eruptive vents for flood plateau and plains basalt provinces, stratovolcanoes both found in both continental and marine settings, and the largest silicic explosive caldera volcanoes. The latter are sometimes called "*supervolcanoes*" which are responsible for some of the biggest explosive eruptions recorded in geological history, resulting from structural caldera collapse and in some settings contributing to regional ignimbrite "*flare-up*" events. Moreover, the special case of marine silicic calderas is also considered. We also now recognise a number of more complex intermediate to silicic volcanic systems; these include "*multiplex volcanoes*" which have changed in character through their evolution, and other multi-vent centres that appear to occur in the absence of a large central cone structure, or conversely are not controlled by caldera collapse depressions. The remaining two sections are grouped into intra- or subglacial volcanoes (formed under ice and meltwater lakes), and mafic oceanic volcanic centres, namely, spreading ridges, plateaus, seamounts, and surtseyan tuff cones. We also introduce the reader to the economic significance of potentially prospective volcano types and their successions prior to their detailed description in Chap. 18, as well as concluding with a summary of the potential hazards posed by different types of volcanoes.

16.1 Introduction

In this chapter, we document the important morphological, physical, facies, and stratigraphic characteristics of modern volcanoes, which range from being very complex (Plate 16.1a) to simple (Plate 16.1b) systems. By emphasising the processes, products, and the duration of primary volcanic versus surface sedimentary or epiclastic processes found in and around modern volcanic centres, we bring these together to construct general but workable facies models, which have predictive value for ancient successions and in basin analysis studies involving volcanic successions. As introduced in Chap. 1, there are many different types of volcanoes, each with distinctive features. Our attention is focused here more on the main individual volcano types (Fig. 16.1a) rather than provinces, with the exception of monogenetic volcanic provinces. This is an important distinction, given the relatively shorter life-span of individual volcanoes compared with the longer lived, larger volcanic-sedimentary provinces of which they are part. The latter will be discussed further in Chap. 17, where we will provide an overview on volcanism in different geodynamic settings. We also comment briefly on the economic significance of potentially prospective volcano types and their successions as a prelude to the detailed description of different types of volcanic-hosted natural resources in Chap. 18. In the concluding section of this chapter we summarise the important natural hazards that can potentially affect modern volcanoes and their environments.

Volcanoes can be conveniently subdivided primarily into two types: *monogenetic* and *polygenetic*. Monogenetic volcanoes are built up by the products of one eruption or one eruptive phase (e.g. Plate 16.1b), whereas polygenetic volcanoes are those resulting from many eruptions, separated by relatively long repose periods and often involving different magmas (e.g. Cas and Wright 1987; Thouret 1999; Plate 16.1a; Fig. 16.1b). An eruption from a monogenetic volcano could last from days to several years but is essentially one prolonged eruption involving a single magma type, usually basaltic (e.g. Parícutin 1943–1952, Capelinhos 1957–1958, Surtsey 1963–1967, Heimaey 1973, La Palma 2021). These all represent the activity of monogenetic centres, even though the activity varied from months, in the case of the 1973 activity on Heimaey, to 9 years in the case of the Parícutin activity. By contrast, large polygenetic volcanoes such as stratovolcanoes, explosive calderas and volcanic island shield volcanoes frequently involve not only repeated eruptive phases separated by significant repose periods from the one vent, but are also often marked by multiple vents or eruptive centres, and changes in magma composition through their history (Plate 16.1).

A significant distinction between monogenetic and polygenetic volcanoes appears to be that monogenetic volcanoes have a simple magma conduit system used only during one eruption, or one prolonged eruptive phase (Fig. 16.1b). By contrast, polygenetic volcanoes may have complex "*plumbing systems*", with an intricate, complicated conduit structure (Fig. 16.1b) that is used many times, or different parts of which are used, during spaced eruptive phases. This distinction also usually corresponds to the complexity and longevity of the associated plumbing system, where polygenetic volcanoes reflect the presence at depth of persistent magmatic sources, which are likely

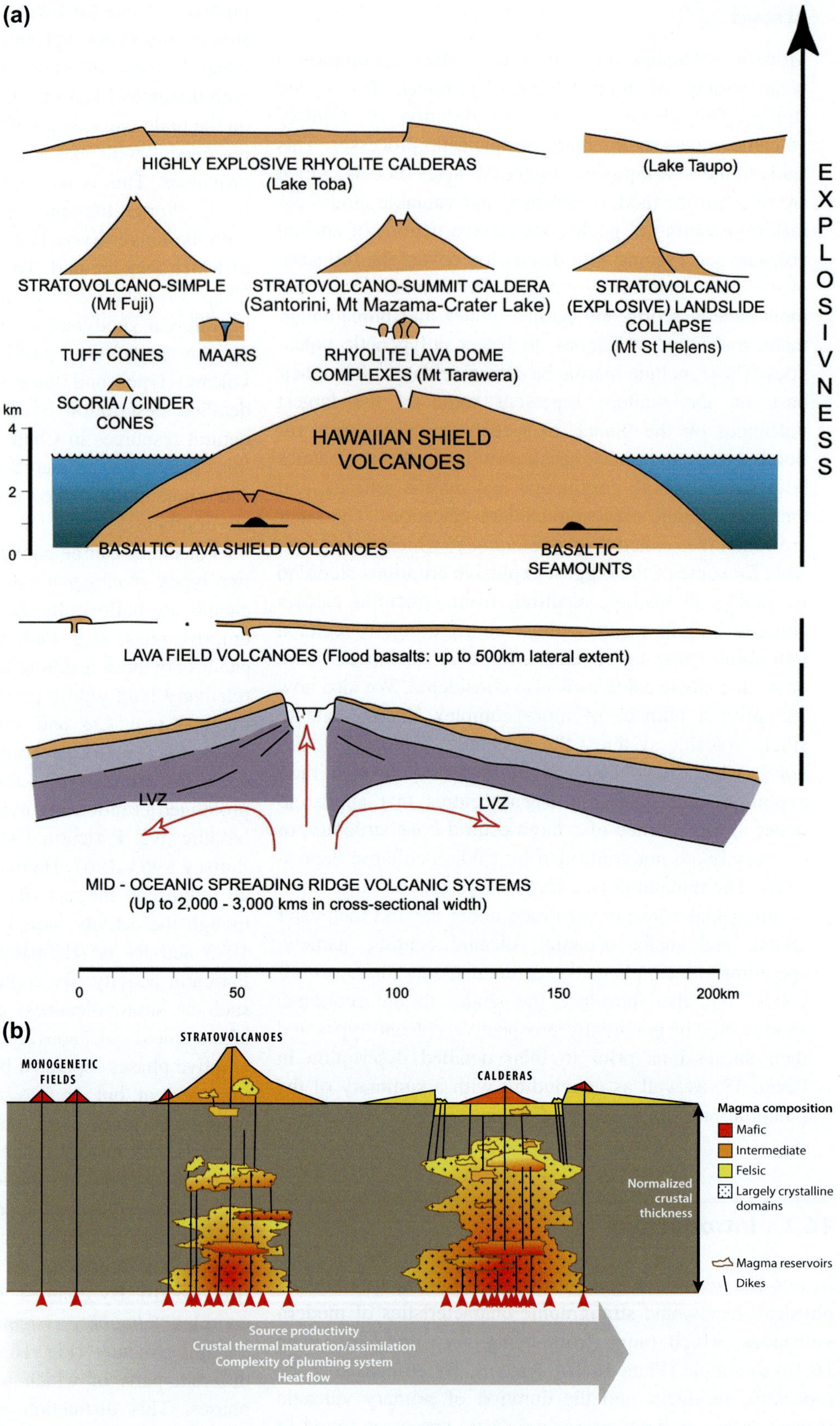

Fig. 16.1 Volcano types and their magma plumbing systems. **a** Schematic representation of the principal types of volcano, with an indication of their dimensions and relative explosiveness. **b** Schematic comparison of monogenetic versus polygenetic volcanoes and related plumbing systems (From Giordano and Caricchi 2022). The crustal thickness is normalised to represent variable geodynamic settings with variable Moho depths. Mafic monogenetic volcanoes at intraplate provinces are envisaged as underlain by simple dyke/conduits connecting the mantle source to the surface. Polygenetic volcanoes at the convergent arc and extensional settings (e.g. stratovolcanoes and calderas) are associated with increasingly higher magma production rates (greater number of red arrows at crust base), building over time complex multi-level, trans-crustal magmatic systems. Magma composition and crystallinity: red = mafic; orange = intermediate; yellow = felsic; cross pattern = largely crystalline domains

multi-level and compositionally, variably differentiated (Fig. 16.1b). In addition, monogenetic volcanoes tend to be smaller, whereas polygenetic volcanoes tend to be large (Fig. 16.1a), which also reflects their longevity and the large magnitude of their magma sources. However, the story is not so simple. Monogenetic basaltic edifices, for example, may both characterise volcanic provinces made only or dominantly of many single volcanoes or eruptive points, as well as occurring on the flanks and surroundings of polygenetic volcanoes. In the first case, basalts effectively erupt directly from a mantle source (e.g. Cas et al. 2017), whereas in the second, scattered monogenetic edifices may be fed by shallower and more differentiated magma reservoirs. Interestingly, it is also becoming apparent that even monogenetic volcanoes, such as scoria cones and maar volcanoes and multi-vent complexes may in fact be poly-magmatic, and erupt compositionally different, but co-genetic and contemporaneous magma batches (e.g. Németh et al. 2003; Strong and Wolff 2003; Smith et al. 2008; van Otterloo et al. 2014; Boyce et al. 2015; Jordan et al. 2015; Báez et al. 2017).

The morphology and dimensions of different types of volcanoes (Fig. 16.1a) are influenced by magma composition and its properties (Chap. 2), the resulting eruption styles and dispersal processes (explosive—Chaps. 4, 9, 10 and 11; effusive—Chaps. 5, 6 and 7), the environment in which eruptions occur (Chaps. 3, 5, 6, 7, 9, 10 and 11), and the surface processes that are active in different environments and on each volcano type (Chap. 12).

In the following sections, we first focus on the volcano types commonly built during monogenetic activity, principally scoria and pumice cones, and phreatomagmatic volcanoes, including maars, tuff rings and tuff cones. We then describe polygenetic eruptive centres, commencing with subaerial-to-emergent edifices built by the eruption of mafic magmas: basaltic shields, and source vents for flood basalt plateau and plains basalt provinces. We then describe stratovolcanoes, which are the most abundant type of large volcano found on Earth in continental and island arc settings. These are built from repeated eruptions of either calc-alkaline or alkaline magmas, and in marine settings evolve from basaltic shield volcanoes or seamounts. We also highlight flank or sector collapse events especially in large marine basaltic (Hawai'ian and Canarian) shields and stratovolcanoes (Roverato et al. 2021). We then describe caldera volcanoes, which include very large, multi-vent calderas and volcano-tectonic depressions. Such "*supervolcanoes*" are dominated by evolved, intermediate to rhyolitic and alkaline magmas, and are responsible for some of the largest eruptions recorded in geological history now known as ignimbrite "*flare-ups*". Less well known are the characteristics of marine, especially deep water, silicic calderas. We follow this with a discussion of some more complex volcanic systems, including multiplex volcanoes (Plate 16.1), and other multi-vent centres that appear to occur in the absence of a large central cone structure, or conversely are not controlled by caldera collapse depressions. Intra- or subglacial volcanoes (formed under ice and meltwater lakes) are then described, and lastly mafic oceanic volcanic centres, namely: spreading ridges, plateaus, seamounts, and surtseyan tuff cones.

There are presently many databases that are freely searchable online where a large deal of information can be found on active and modern volcanoes. In particular here, we mention the Global Volcanism Program which is run by the Smithsonian Institution at http://volcano.si.edu/.

16.2 Scoria, Spatter and Pumice Cones

Scoria (or cinder) cones (Plate 16.1b; Fig. 16.2a–e) are small volcanic landforms (Fig. 16.1a) built typically during relatively low intensity subaerial explosive eruptions involving fallout from a reasonably sustained low eruption column, from which fallout of coarse pyroclasts builds a cone around the vent (Kereszturi and Németh 2013). The eruption style can vary from strombolian *sensu stricto* to violent strombolian and microplinian (e.g. Fig. 16.2a), and sometimes even some minor phases of hawai'ian fire fountaining and even phreatomagmatic activity (see Chap. 9; Valentine and Gregg 2008; Németh 2010; Kereszturi and Németh 2013). Spatter cones are dominated by hawai'ian fire fountaining and strombolian *sensu stricto* eruption styles, leading to the formation of variably agglutinated spatter deposits and in some cases lavas. Magma compositions erupted by scoria and spatter cones are mostly of basaltic and basaltic andesite composition but can also include other low viscosity magma types such as phonolite. Pumice cones are dominated by weak explosive eruptions of low volatile content silicic magmas. An older but excellent review of the characteristics and evolution of scoria cones is given by Wood (1980a), who indicated they are the most common type of volcano on Earth. Many scoria cone-building eruptions have been observed and recorded through historic times and we have previously mentioned the eruptions of Parícutin in 1944 and Heimaey in 1973, and a slightly older example would be Sunset Crater erupted around AD 1085 in the San Francisco Volcanic Field, Arizona (Alfano et al. 2019). The most recent scoria

Fig. 16.2 Young scoria cones. **a** 1973 Eldfell scoria cone-forming eruption, Heimaey, Vestmannaeyjar Islands, Iceland, showing a microplinian eruption column deflected downwind, with fallout building a cone around the vent and also more widely dispersed fallout deposits (Photo by George Carman). **b** The youthful De La Laguna scoria cone rising above a complex lava flow field or shield characterised by the typical very low-angle profile, Antofagasta de la Sierra Basin, southern Puna Plateau, Argentina (see Fig. 16.4a). **c** Mt Elephant scoria cone, Newer Volcanics Province, Victoria, southeastern Australia; at 300 m high, the largest scoria cone in the province (Photo by H. Handley; interpretation by R'Casso demonstrates indeed that it is an elephant. *How obvious!*). **d** The new southeast crater at Mt Etna, Italy, is marked by a large scoria cone built between 2011 and 2012 (see Behncke et al. 2014); note the very steep slopes and people climbing the zig-zag track up the middle of the front face of the cone. **e** Low hawai'ian fire fountains on 21 September 2021 building a new basanitic scoria cone on the La Palma (Canary Islands, Spain), while feeding lava flows from a breached sector. The eruption started on 19 September and finished on 13 December 2021. (*Source* Smithsonian Institution Global Volcanism Program, provided by Cabildo La Palma)

cone-forming eruption, from September 19 to 13 December 2021 on La Palma, Canary Islands built a multi-vent edifice more than 150 m high, through sustained fire fountains as high as 500 m, forming scoria and ash plumes that rose up to 3–4 km above the vent (Carracedo et al. 2022). Contemporaneously, eruption of extensive and sustained basanitic lava flows occurred and these severely impacted some urbanised areas of the island destroying houses and infrastructure (Di Fiore et al. 2023; Fig. 16.2e).

16.2.1 Morphometry

Scoria, spatter and pumice cones are often approximately circular in plan view. Elongate forms are built when eruptions continue along a large part of a fissure which does not localise to a single point source vent. Scoria cones have central bowl-shaped craters, although these tend to become infilled by subsequent mass-wastage grain flow and slide deposits. In a morphometric analysis of 910 scoria cones

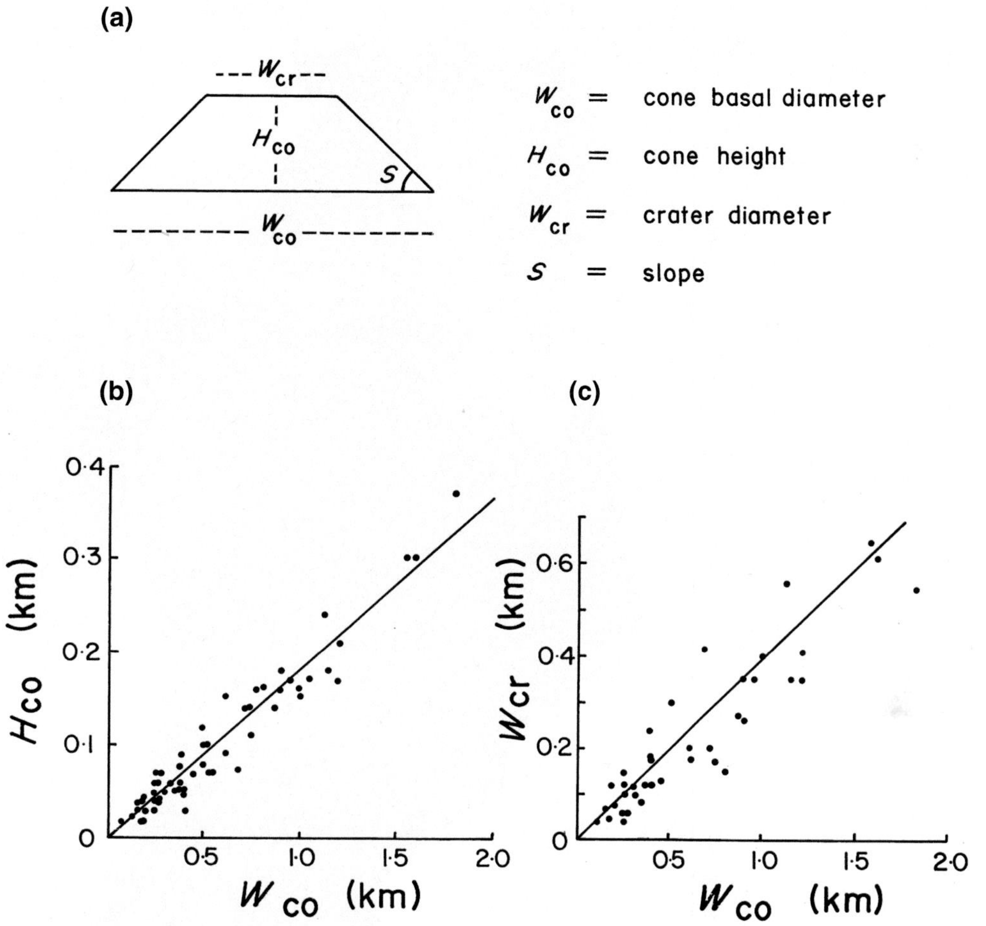

Fig. 16.3 Cone dimensions for relatively fresh basaltic scoria cones from various parts of the world (n = 83). W_{co} = cone diameter at the base; W_{cr} = cone crater diameter; H_{co} = cone height. (From Cas and Wright 1987, After Wood 1980a)

Wood (1980a) found that the basal diameter W_{co}, ranges between 0.25 and 2.5 km (mean 0.9 km), and for relatively fresh cinder or scoria cones (e.g. Fig. 16.2a–e), cone height $H_{co} = 0.18\ W_{co}$ (Fig. 16.3a, b), up to 400 m high. These volcanoes were also found to be characterised by relatively wide craters for any given basal diameter of the cone, and crater width $W_{cr} = 0.40\ W_{co}$ (Fig. 16.3c). This is because most of the cone is built of larger fragments, most of which are ejected on short ballistic trajectories and so fall only around the vent. A very common characteristic of scoria cones is the presence of breaching, sector collapses and scars, from which late stage lava flows issue to produce variably large lava fields around them (Fig. 16.4). Cone breaching may be influenced by feeder dyke orientation, by the slope and tectonic influences on the underlying ground surface, and by transport of part of the early formed cone as rafts on top of lava flows issuing from vents at the base of the cones (e.g. Corazzato and Tibaldi 2006; Valentine and Gregg 2008; Németh 2010; Kereszturi et al. 2012).

Scoria cones are very susceptible to weathering and erosion (including by the wind), which changes their morphology. In the Massif Central in France, Kieffer (1971) first showed that 10,000 yr old cones have slopes several degrees shallower than the most recent cones, and that slopes of cones 1 Ma old are as low as 15° (Fig. 16.5). Cones ≥4 Ma old are marked only by residual volcanic necks, the scoria having been completely eroded. Wood (1980b) and Hooper and Sheridan (1998) studied the degradation of scoria cones and modelled their changes in shape with time. With time and increasing degradation, cones show decreases in H_{co}, H_{co}/W_{co} ratio and slope, but the ratio of W_{cr}/W_{co} does not appear to change (Wood 1980b; Fig. 16.6). Variation in H_{co} and W_{co} for scoria cones of different ages from the San Francisco Volcanic Field, Arizona are shown in Fig. 16.6a (Wood 1980b). In Fig. 16.6b the future degradation history of Sunset Crater, which is the youngest cone in the San Francisco field, is predicted. This has been derived from past degradation trends of other cones in this field. A summary of some relevant measurable quantities associated with scoria cones is shown in Table 16.1 (McKnight and Williams 1997; Báez et al. 2017). For a detailed review of the morphometry and degradation of scoria cones, see Kereszturi and Németh 2012).

Given the large variability of climate conditions worldwide and in time it is not possible to propose a unique pattern of degradation with time for scoria cones.

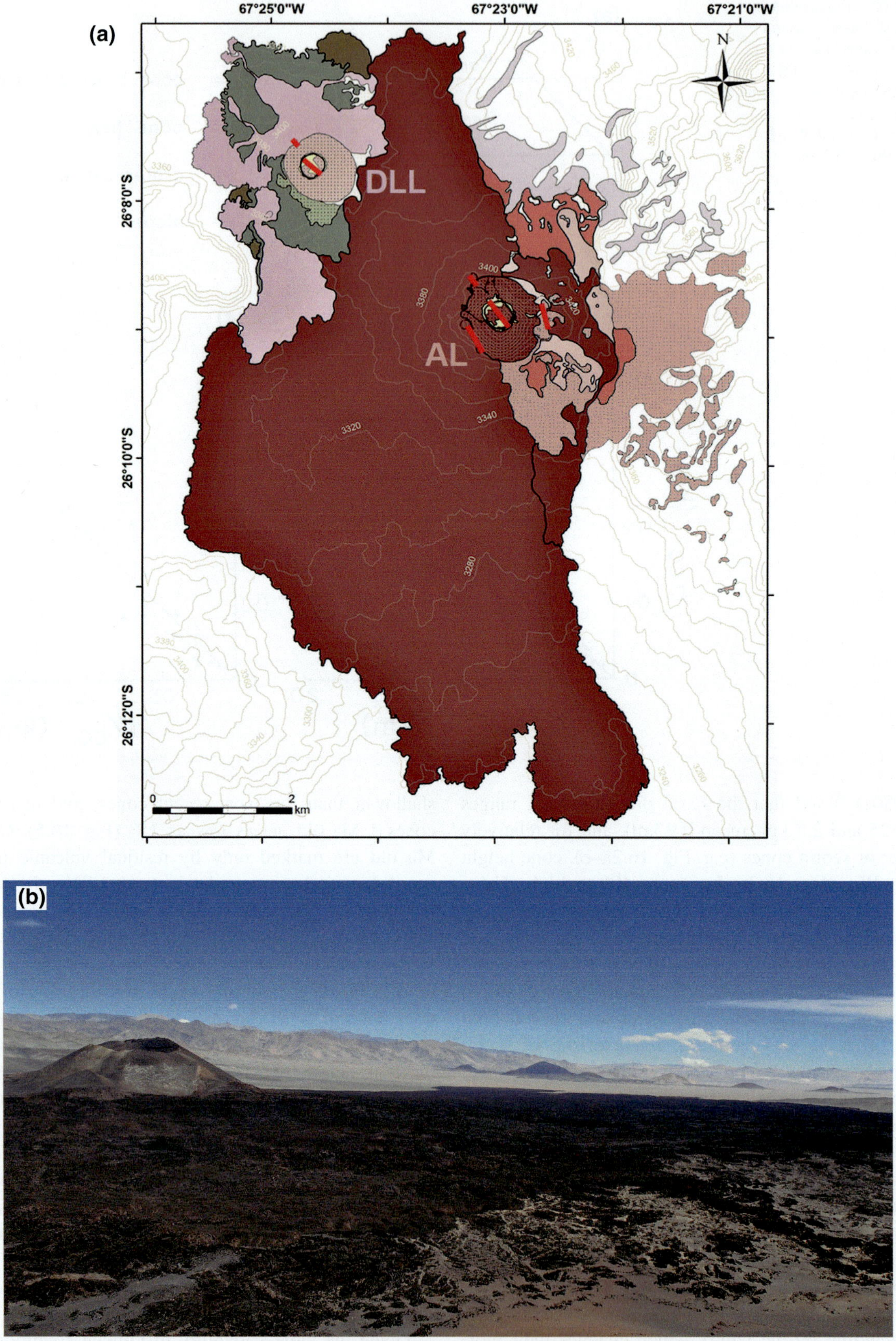

Fig. 16.4 Scoria cones with associated lava flows. **a** Geological map and stratigraphy of the freshly preserved Quaternary Alumbrera (AL) and De La Laguna (DLL; see Fig. 16.2b) scoria cones in the Antofagasta de la Sierra Basin, southern Puna Plateau, Argentina. The scoria cones are associated with large lava flow fields or shields mainly fed by late stage effusive activity; red lines = NW–SE eruptive fissures. The Alumbrera cone is breached towards the northwest from which emanates the youngest, most extensive lava field (dark brown). Detailed mapping of ventifacts and the identification of temporal hiatuses shows that each volcano was built by more than one eruption. (Modified from Báez et al. 2017; see Table 16.1). **b** View from the north of Alumbrera cone and the youngest lava flow field mapped in (**a**). In the background a series of aligned and older scoria cones show smoother topographic profiles

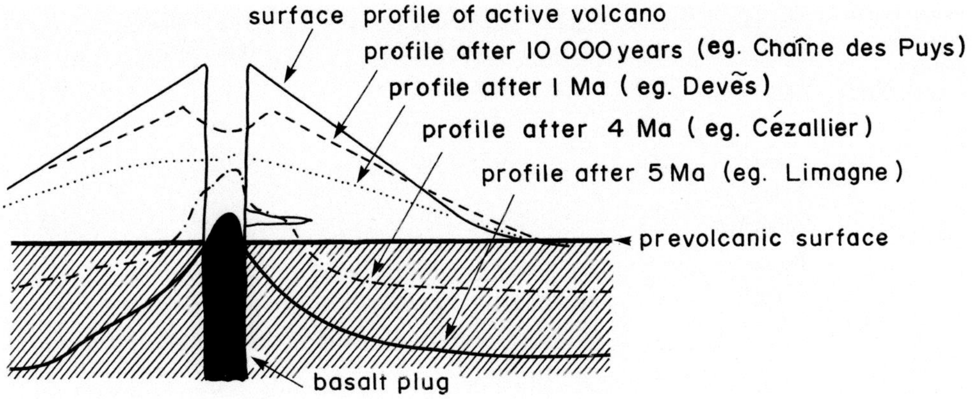

Fig. 16.5 Changes in cone morphology with increasing age in the Massif Central, France (From Cas and Wright 1987, After Kieffer 1971)

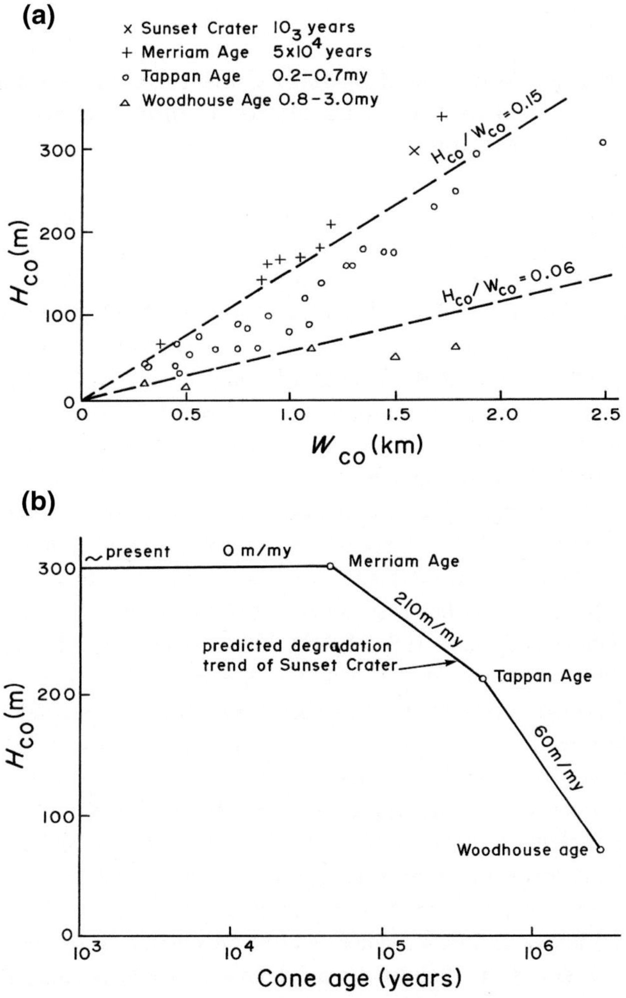

Fig. 16.6 **a** Cone height versus basal diameter for cones (n = 38) of different stratigraphic ages in the San Francisco Volcanic Field, Arizona, USA. **b** Possible future degradation history for the 340 m high Sunset Crater erupted ca. AD 1085, derived from past degradation trends in the San Francisco Volcanic Field. (From Cas and Wright 1987, After Wood 1980b)

Nonetheless, provided that climatic conditions are similar, comparisons can be made based on different degrees of erosion (e.g. Hasenaka and Carmichael 1985; Dohrenwend et al. 1986; Fornaciai et al. 2012; Kereszturi and Németh 2013; Báez et al. 2017). However, other workers have observed that denudation trends and changes in morphometric parameters over time do not support a simple degradation model for scoria cones in the same volcanic field (e.g. Bandas del Sur, Tenerife; Kereszturi et al. 2013).

16.2.2 Eruption Duration, Repose Periods or Eruption Frequency

Even though scoria cones are the most common volcanoes on Earth, surprisingly few have been observed erupting in historical times (Valentine and Connor 2015; e.g. 1973 Eldfell eruption, Heimaey, Iceland, Fig. 16.2a). Scoria and pumice cone-forming eruptions range in duration from a few days to a few years (e.g. the 9-year eruption of Parícutin in Mexico; see Chap. 9). Of observed scoria cone eruptions, 50% have lasted for less than 30 days, and 95% stopped within 1 year (Wood 1980a). Later eruptions produce new cones at separation distances of a few kilometres, and it has been speculated (McGetchin and Settle 1975) that these distances may be approximately equal to the depths of the magma reservoirs. However, as indicated above, it is now recognised that some scoria cones may actually involve multiple eruptions separated by tens to hundreds of thousands of years with geochemical variations (Németh and Kereszturi 2015). In the southern Puna Plateau, Baez et al. (2016) suggest the polycyclic nature of the Alumbrera and De La Laguna scoria cones (Figs. 16.2b and 16.4) is related to a combination of a high magma input, a favourable regional stress field and the interaction of rising magma with pre-existing faults (Chap. 17). Although not witnessed, the

Table 16.1 Some morphometric parameters of selected scoria cones (After McKnight and Williams 1997; Báez et al. 2017)

Volcano	Cone height (H_{co} m)	Crater width/cone width (W_{cr}/W_{co})	Total volume (km^3 DRE)	Pyroclastics volume/total volume (%)
Parícutin	420	0.31	2.1	67
Jorullo	350	0.25	~2.0	~60
Cerro Negro	250	0.18–0.36	0.2	77
Alumbrera	163	0.31	0.4	30
De La Laguna	153	0.31	0.2	42

duration of rhyolitic pumice cone-forming eruptions is likely to be similar to scoria cone-forming eruptions since the explosive intensity and fallout dispersal distances would be similar to produce cones at the vent.

Repose periods in monogenetic basaltic fields or provinces are in the order of 10^2–10^3 yr, and such fields may be active for 10^6 yr, and even several Ma (e.g. Condit and Connor 1996; Valentine and Connor 2015; Cas et al. 2017). There are many examples of this type of basaltic volcanic field which are potentially active today. One very well-described example is the Lunar Crater volcanic field in central Nevada which has been active over the past 6 Ma (Valentine et al. 2017). Another example is the late Cenozoic Newer Volcanics Province in western Victoria and South Australia, southeastern Australia (Fig. 16.7), where volcanic centres are dated from <8 Ma to as young as ~5 ka (Matchan and Phillips 2011; Cas et al. 2017; Oostingh et al. 2017). The average eruption frequency in this monogenetic province is in the order of ~15–20 kyr (Cas et al. 2017) although dating of the >400 volcanoes is limited, but as Condit and Connor (1996) and van den Hove (2017a) pointed out, clusters of volcanoes in a province are likely to have a shorter eruption frequency than the province-wide general average. Basaltic volcanism may be active for much longer periods (5×10^7 yr) in the same region when a number of separate magmatic provinces (10^6 yr) are relatively closely related in space and time. The Older Volcanics Province of eastern Victoria is an example of this, and basaltic volcanism persisted spasmodically from 95 to 17 Ma (Wellman 1974; Day 1983), as well as along the eastern seaboard of Australia following the break-up of Gondwana.

Lesti et al. (2008) recognised the occurrence of clusters of alkali-basaltic scoria cones in the Newer Volcanics Province, with spacing between clusters interpreted as related to the structure and thickness of the crust, and often located in proximity to major crustal faults (e.g. Boyce 2013; Cas et al. 2017; Chap. 17). Van den Hove et al. (2017b) propose that the relationship between the location of vents and faults is clearest where volcanoes overlie the Paleozoic basement, where dykes are likely to be the main magma plumbing conduit. However, the pattern is much less clear where volcanoes overlie weakly lithified, flat lying, Tertiary sedimentary basin strata, in which sills may be more prominent as part of the magma plumbing system.

16.2.3 Scoria and Spatter Cone Eruptions, Cone Characteristics, Deposits and Facies Models

Scoria cones are built during relatively sustained eruptions involving variations from strombolian *sensu stricto* to hawai'ian to violent strombolian and microplinian (Fig. 16.2a, e), although minor phases of phreatomagmatic activity may also occur (see Chap. 9; Valentine and Gregg 2008; Németh 2010; Kereszturi and Németh 2013). Their deposits are therefore dominated by coarse, massive to diffusely stratified scoria and bombs (e.g. Plate 16.1b; Figs. 16.8a and 16.9a–e).

Although the primitive nature of many cone-forming mafic magmas, as well as the presence of mantle xenoliths, indicate rapid rise rates from mantle sources of up to ≤ 1 m s^{-1}, and implied dykes, often associated with fissure vent systems (e.g. Wilson and Head 1981; Parfitt 2004; Valentine and Gregg 2008), relatively recent work in sedimentary basins indicates that inter-connected saucer-shaped sills may also be conduits that feed scoria cones (e.g. Hansen 2006; Jackson 2012; Reynolds et al. 2018; Chap. 8). These can form where the near-surface crustal sequences are only moderately lithified sedimentary strata, which have low bulk density and relatively weak mechanical strength, in which situations dense mafic magmas exceed their neutral buoyancy levels. This is conducive to forming saucer-shaped sills which, as discussed in Chap. 8, at shallow crustal levels can then become the feeders to cone-forming eruptions. The more convoluted the magma plumbing system is to the Earth's surface, the more likely the magma will become contaminated by crustal assimilation and be subjected to crystal fractionation.

Fresh scoria cones have steep angle of repose slopes of about 33° (e.g. Figs. 16.2a–e, 16.8a and 9.52a). The principal depositional facies making up most of the cone consists

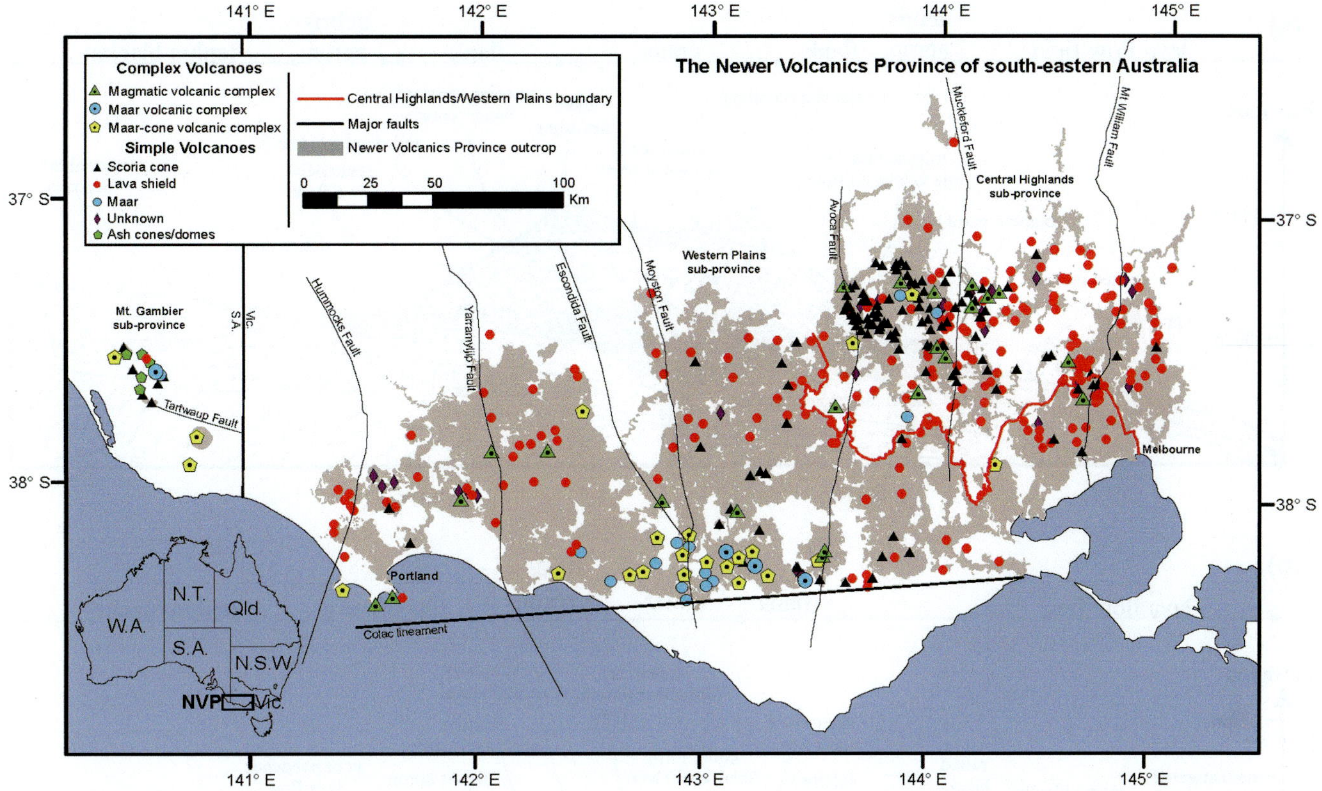

Fig. 16.7 Map showing distribution of scoria cones and other volcanic centres within the late Cenozoic, intraplate Newer Volcanics Province, southeastern Australia (From Boyce 2013). The volcanic centres are surrounded by extensive plains basaltic lava fields (grey shading; see Sect. 16.6.2). Note that most eruption centres in the south are dominantly phreatomagmatic, whereas in the north, they are dominantly magmatic, which reflects the nature of the subjacent crust. In the south the crust is marked by several Tertiary aquifers in the Otway Basin, whereas to the north the crust is "dry" Paleozoic crust (see Fig. 16.13)

of a very thick, massive to diffusely stratified pile of scoria and bombs (Plate 16.1b; Figs. 16.8a and 16.9a–c; Keresztui and Németh 2013). Ballistic bombs include spatter bombs dispersed in scoria (Fig. 16.9c), or concentrated in distinct horizons, which sometimes form agglutinated layers (Chaps. 4 and 9). The cone sequence may be cut by one (Plate 16.1b; Fig. 16.9a) or more feeder dykes (Plate 8.1a; Fig. 16.8a). Avalanching and rolling of coarse near-vent, mainly ballistic and fallout scoria down the outer slopes, as well as back into the crater, are important depositional processes occurring on the cone during an eruption (Fig. 16.9d, e). Many of the scoria layers found in the cones are therefore of mass-flow origin and more specifically, inertial grain-flows which can develop reverse size grading during flow (Fig. 16.9e; Chap. 12).

Frequently, the gas content of magma will decrease towards the end of a scoria cone-forming eruption, so the proportion of lava spatter ejected increases. Many cones are therefore capped by a collar or rim of spatter (Németh 2010). Sometimes dissected scoria cones show welded interiors with columnar jointing and compaction profiles broadly similar to welded tuffs (Németh 2010; Valentine and Connor 2015). The different deposits found in the interiors of scoria cones have been fully discussed and illustrated in Fig. 16.9 and in Chap. 9 (e.g. Fig. 9.52). A large proportion of the explosively ejected scoria is transported downwind to form a mantling fall deposit of usually limited dispersal, although more widely dispersed (Fig. 9.52c; sub-plinian and plinian) examples are known (Chap. 9).

Most scoria cones, and even many pumice cones, have accompanying lava flows which form during cone-building, or when degassed magma continues to be extruded after explosive activity has ceased (Figs. 16.4, 16.8, 16.7, 16.9d and 16.10a; Németh 2010; Kereszturi and Németh 2013; Valentine and Connor 2015; Báez et al. 2017). Scoria cones generally have smaller volumes than their associated lavas, and therefore the cone only represents a small part of the total magma erupted. At La Palma, the ratio of explosive versus effusive products during the 2021 eruption was estimated at ca. 1:5 (Civico et al. 2022). Scoria cones are typically monogenetic. They most commonly occur as isolated cones in large basaltic volcanic fields (e.g. Boyce 2013; Cas et al. 2017; Figs. 16.2b, 16.4 and 16.7), but they may also show more complex, polycyclic activity which possibly evolved over thousands of years (e.g. Alumbrera and De La Laguna scoria cones, Fig. 16.4; Báez et al. 2017). Scoria

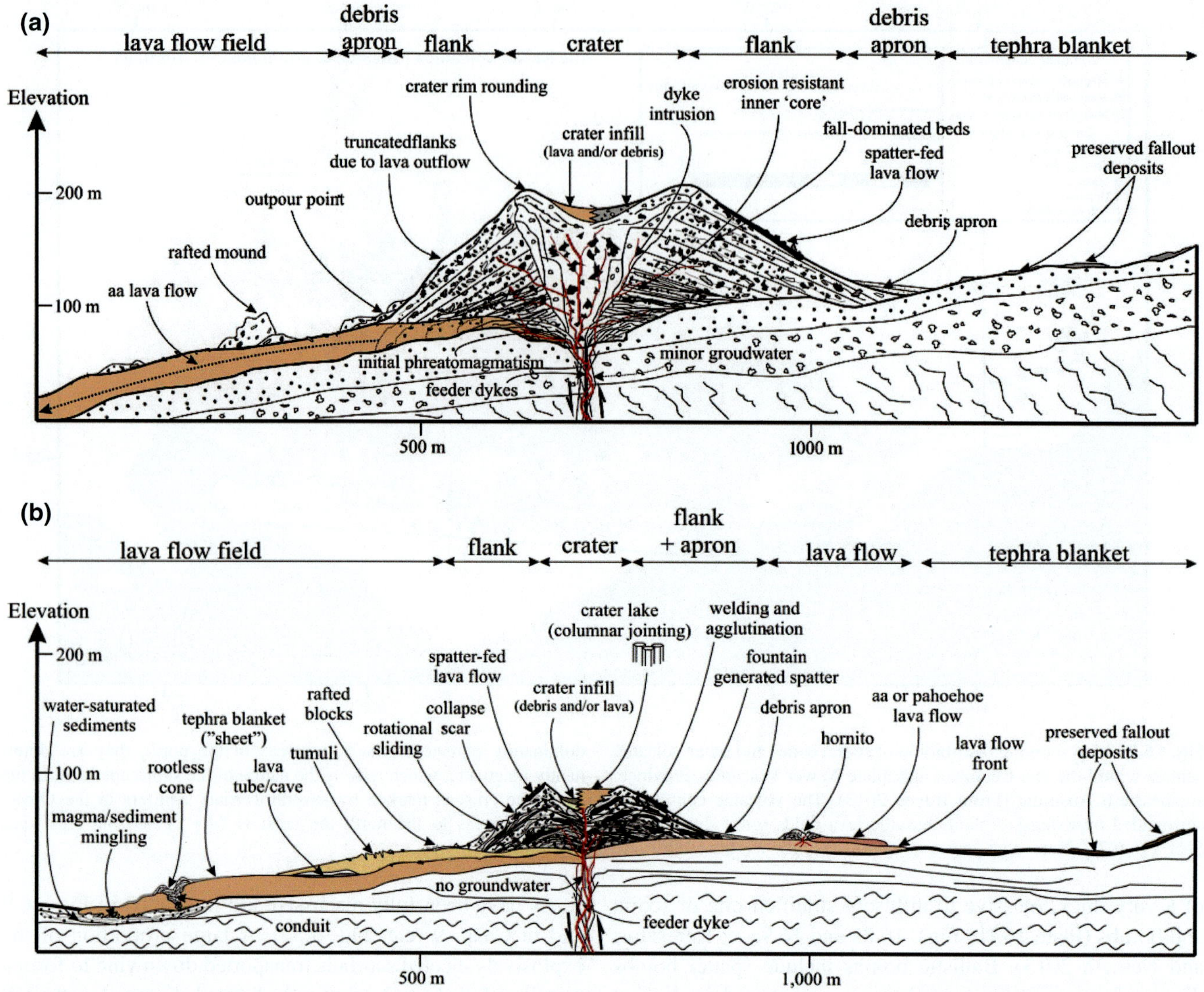

Fig. 16.8 Facies models for scoria and spatter cones. **a** Schematic cross-section and facies model through a typical scoria cone showing distribution of various facies types, dominated by massive to diffusely stratified scoria, spatter beds and in some, lava flows. **b** Schematic cross-section and facies model through a typical spatter cone showing distribution of various facies types, dominated by massive to diffusely stratified spatter agglutinate beds, with lesser scoria and commonly (clastogenic) lava flows. (Adapted from Kereszturi and Németh 2013; *Source* InTechOpen through Creative Commons Attribution 3.0 Unported licence)

cones may also occur in nested clusters within complex tuff ring and maar centres (Sect. 16.3, see Fig. 16.12c–f), and as parasitic cones on polygenetic shield and stratovolcanoes, on which they are also essentially monogenetic elements (Fig. 16.2d; e.g. McGuire and Pullen 1989).

More silicic pumice-forming eruptions can also build steep-sided cones at vent, although pumice cones are not well documented in the literature. An example is the 729 AD Monte Pilato pumice cone, which also fed the Rocche Rosse obsidian flow described in Chap. 6 (Fig. 16.10a; Dellino and La Volpe 1995; Davì et al. 2011) along its northern breached margin. Such pumice cones are the product of silicic magmas that have a low magmatic H_2O content, enough to drive a weak micro- to sub-plinian explosive eruption (Chap. 9), but well short of producing a high intensity plinian scale eruption. In terms of the facies (massive, diffusely stratified) and texturally (sorting, vesicularity), pumice cones are similar to scoria cones, although there will generally be a lower proportion of fluidal bombs unless magma compositions are phonolitic or trachytic. Deposits include dominantly fallout, pyroclastic flow and surge deposits (Fig. 16.10b, c; Dellino and La Volpe 1995; Davì et al. 2011). Pumice cones can also form shallow water to emergent edifices, on top of emerging lava domes (Reynolds and Best 1976; Reynolds et al. 1980; Cas et al. 1990; Allen and McPhie 2000; Figs. 9.78 and 9.79), in which case evidence

(a)
(b)
(c)
(d)
(e)
(f)
(g)

◀ **Fig. 16.9** Features of scoria and spatter cones. **a** Cross-section through a nearly complete scoria cone exposed on the coast of Madeira, Portugal. Note the gentle outward dips of massive to diffusely stratified scoria either side of the vertical, central, cross-cutting feeder dyke. **b** Massive, to diffusely stratified scoria and ballistic bomb deposits making up Mt Leura scoria cone, Newer Volcanics Province, Victoria, southeastern Australia. Person circled for scale. **c** Close-up view of massive scoria with spatter bombs, Kizil Hill scoria cone, Turkey. **d** Crater of the Alumbrera scoria cone in the foreground showing both agglutinated and loose scoria, and bomb layers draping the inner walls, Antofagasta de la Sierra Basin, Argentina. In the middle background the fresh morphology of the De La Laguna scoria cone and surrounding lava field, contrasts with older more denuded cones in the distance (see also Fig. 16.4, Table 16.1). **e** Reverse graded scoria beds, of probable grain-flow origin, scoria cone near Oia, Santorini, Greece. **f** Wedging bed set of agglutinated spatter layers, making up a low spatter cone, Cape Bridgewater volcano, Newer Volcanics Province, Portland, Victoria. Also see Figs. 9.7a–e and 9.48a–c for spatter spray explosions, and other spatter cones and deposits. **g** Close-up view of **f** showing the gradational contacts between the densely welded layers and intervening non-welded scoria and spatter horizons. Height of exposure ~2 m

Fig. 16.10 Features of felsic pumice cones. **a** The Monte Pilato rhyolitic pumice cone (MP) erupted in 729 AD on Lipari, Italy. This breached seaward allowing the Rocche Rosse obsidian lava flow (RR) to flow seawards (i.e. the vegetated tongue visible right of centre in the picture. (*Source* NOAA through Google Earth). **b** Crudely bedded pumice and ash horizons in the thick Mt Pilato pumice cone sequence. The poorly sorted thick wedging unit at head height is a pyroclastic flow deposit, the brown ash units are surge deposits and the other coarse units are fallout deposits. **c** Diffusely stratified, surge-modified pumice fallout deposit, between ash-rich surge deposits, Mt Pilato pumice cone sequence. Also see Figs. 9.78 and 9.79 for features of dome-top pumice cones and their deposits in aqueous environments

of interaction with water will be found (e.g. blocky and bread crusted, quench fragmented clasts, surge dune forms and cross-stratification, subaqueous mass flow resedimentation; Chaps. 9 and 11).

Spatter cones, resulting from sustained fire fountaining and/or strombolian spattering, are dominated by agglutinated spatter deposits, which form low mounds or ramparts up to a few tens of m high (Figs. 16.8b and 16.9f, g, 9.7d, g, h and 9.48; Kereszturi and Németh 2013). Clastogenic lava layers can form from the rapid accumulation of spatter, which fuses together forming lava like layers (Fig. 5.15), that may be able to flow away from the vent (e.g. Mattsson and Höskuldsson 2005; Carracedo-Sánchez et al. 2012; Chaps. 5 and 9; Fig. 9.7g–i).

16.2.4 Resource Potential and Economic Significance

Scoria and spatter cones are most commonly exploited for industrial aggregate, but can also host gemstones, in particular sapphires and zircon; e.g. Eastern and Southeastern Asia where they are related to Cenozoic continental plains basalt provinces (Sect. 16.6.2). Most commercial exploitation is from alluvial placer deposits some of which are confined to particular scoria cones (Izokh et al. 2010).

16.3 Maars, Tuff Rings and Tuff Cones

These volcanic landforms and craters (Figa. 16.11a–g and 16.12a–f) are usually monogenetic and are produced by phreatomagmatic, phreatic and sometimes hydrothermal eruptions (Chaps. 4 and 9). They are the second most abundant volcano type after scoria cones, and all of these volcanic landforms are often associated with each other in large basaltic volcanic fields (e.g. Fort Rock, Christmas Lake Valley Basin, Oregon, USA; Heiken 1971; Brand and Clarke 2009; and the Newer Volcanics Province, southeastern Australia; Joyce 1975; Boyce 2013; Cas et al. 2017; Fig. 16.7). However, maars, tuff rings and tuff cones can also form on the flanks of or within the craters of larger central volcanoes and stratovolcanoes (Fig. 16.11a, f), in emergent or coastal edifices (Fig. 16.11e) and where conspicuous, multi-layered both pressurised and phreatic/hydrothermal aquifers occur in the substrate (e.g. in the Roman Magmatic Province, Italy; Giordano et al. 2014).

There is usually a geological and geographical control on the distribution of these different volcano types, determined essentially by hydrological and hydrogeological factors. Along the Nejapa-Miraflores fault fissure-system in Nicaragua, phreatomagmatic craters are found close to the Nicaragua Lake, whereas scoria cones occur only outside the lake basin (Pardo et al. 2009). In the Newer Volcanics Province, phreatomagmatic craters are concentrated in the southern half of the province where a cover of aquifer-bearing Cenozoic sediments overlies the Paleozoic basement (Figs. 16.7 and 16.13; Joyce 1975; Cas et al. 2011a, 2017; Boyce 2013). As well as mafic compositions, examples of other magmas involved in forming these phreatomagmatic volcanoes are also common, such as rhyolitic (Sheridan and Updike 1975; Austin-Erickson et al. 2008; Davies et al. 2008), dacitic (Sillitoe et al. 1984), phonolitic (Schmincke et al. 1973), foiditic (De Benedetti et al. 2008; Cross et al. 2014), and carbonatitic (Mattsson and Tripoli 2011). Ross et al. (2017) have reviewed the known occurrences of maar-diatreme systems of felsic compositions (andesite, dacite, rhyolite), many of which host epithermal gold(-silver) mineralisation.

16.3.1 Classification and Terminology

The term "*maar*", describing a shallow crater lake, originates from the Eifel district of Germany. Although it has been used in a general sense for any broad, low rimmed volcanic craters with or without lakes (including tuff rings and tuff cones), it also has a more restricted meaning. *Maars sensu stricto* are phreatomagmatic craters with steep to vertical inner slopes, floor levels below the level of the surrounding terrain, and with a wide, gently sloping apron (or rim sequence) of mostly phreatomagmatic deposits, which dip radially away from the crater at shallow angles (Figs. 16.11a, 16.12a–f and 16.14a). Important reviews describing these volcanoes are given by Ollier (1967), Wohletz and Sheridan (1983), Lorenz (2003), White and Ross (2011), Kereszturi and Németh 2013, and Németh and Kereszturi (2015).

Maars, when new and little eroded, have craters which are cut into the surrounding country rock (Fig. 16.14a; Kereszturi and Németh 2013). They generally lack or have only minor inward dipping beds, and frequently exhibit near-vertical crater wall scarps below rim beds (Figs. 11.3a, b; 16.14a), reflecting collapse of maar rim deposits adjacent to the vent back into the vent as slide blocks. Maar craters range in diameter from a few hundred m (Fig. 11.3a) to ~3 km (Figs. 16.12a–f, 16.14a and 16.15). W_{cr} values for a sample of maars from around the world measured by Wood (1980c) are given in Fig. 16.15. They form when rising magma explosively interacts with ground water (or less commonly surface-derived water) below the original topographic surface, and because a crater or diatreme is explosively excavated into the substrate (Fig. 16.16a–f), their deposits contain a lot of accidental country rock xenoliths, resulting from the explosive excavation of the subsurface diatreme into the country rock succession (see Chap. 4 for discussion of mechanisms of phreatomagmatic explosive activity). The deposits may contain relatively little or, in the case of phreatic eruptions, no juvenile magmatic pyroclasts.

The youngest maarsexcavated in weakly lithified substrate have crater depth-to-diameter ratios of 1:5, but this ratio increases (i.e. 1:4, 1:3, 1:2) with age as the craters become infilled with epiclastic sediment, often deposited in ponded shallow lakes and as erosion increases their diameter. The term *diatreme* is used to describe a pipe- or inverted cone-shaped excavation into the country rocks filled at least in part with coarse-grained volcanic debris, and are found below maar craters (Figs. 16.14a and 16.16; Sect. 16.3.5), although, these are rarely exposed in cross-section (Fig. 16.16d–f). The diatreme not only becomes a depocentre for pyroclastic debris but also the conduit through which dykes of magma are injected.

The traditional model for the geometry of a simple maar diatreme (e.g. Lorenz (2003) suggested that the ratio of the

Fig. 16.11 Maars, tuff rings and tuff cones. All three morphologies may form during phreatomagmatic eruptions, but maar craters may also result from phreatic and hydrothermal explosive activity. **a** Viti maar, Krafla volcanic complex, Iceland, erupted in 1724. It is ~350 m diameter, and has the typical maar steep inner crater wall and gently outward sloping maar rim surface (Photo by Simisa, image enhanced; *Source* Wikimedia through Creative Commons Attribution 3.0 Unported licence). **b** Hverfjall tuff ring, Mývatn, Iceland, has a wide crater that is ~1 km diameter. **c** Inner gently sloping crater wall of Hverfjall tuff ring, Mývatn, Iceland. (Photo by Martin Falbisoner, image enhanced; *Source* Wikimedia through Creative Commons Attribution 4.0 International licence). **d** Monte di Procida tuff cone, Campi Flegrei, Naples, Italy. **e** Tuff cone formed during shallow marine eruption between 19 December 2014 to 26 January 2015, Hunga Tonga-Hunga Ha'apai islands, Tonga. Note the network of rapidly incised drainage gullies. (Photographed September 2015 by Shane Cronin, University of Auckland). **f** Phreatic craters at the top of the active dome at Guagua Pichincha, Ecuador. **g** Hydrothermal explosion crater, surrounded by hydrothermally altered ground and substrate. Nisyros caldera, southeastern Aegean Sea, Greece

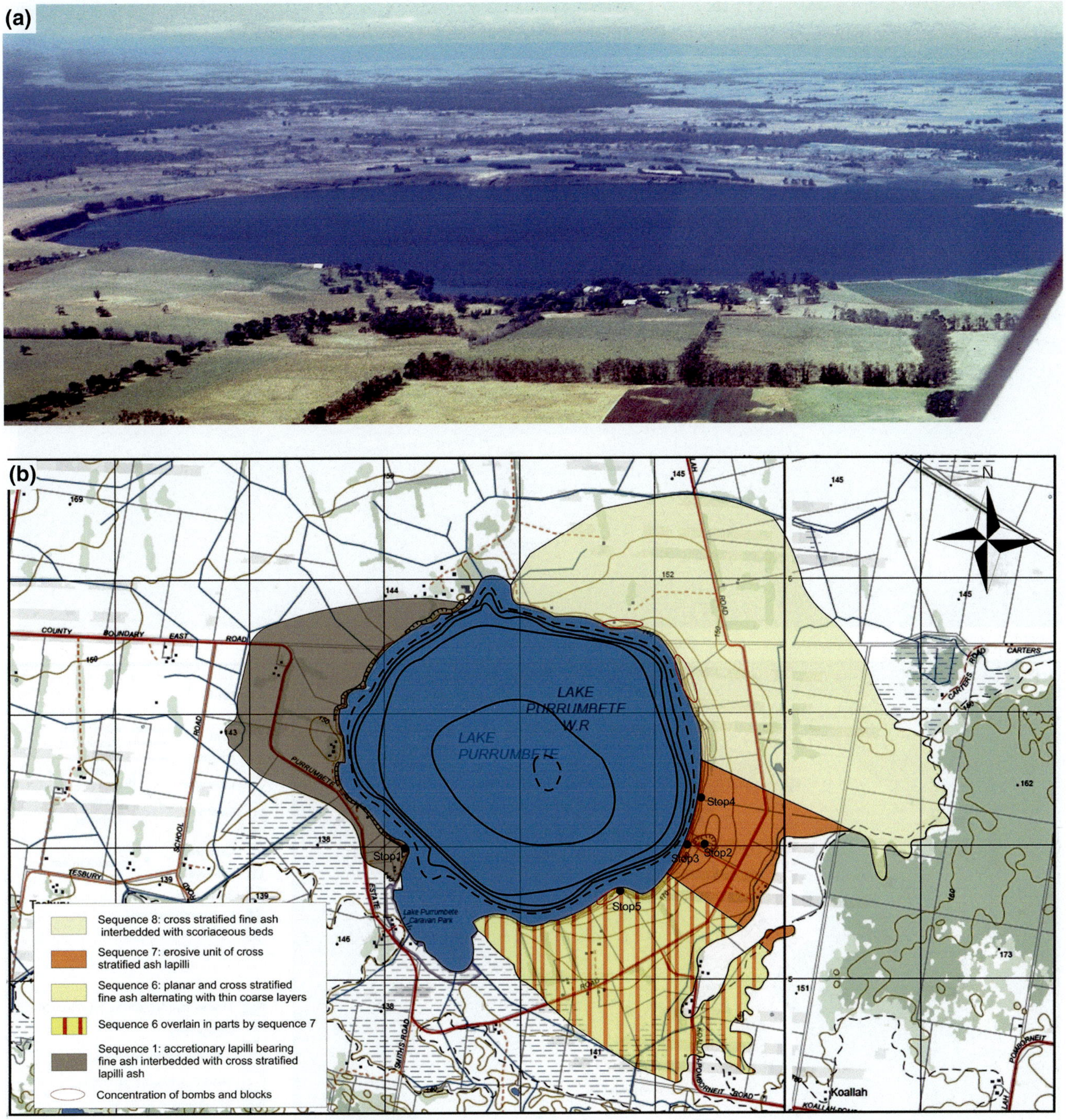

Fig. 16.12 Aerial photos and maps of three selected maar volcanoes in the late Cenozoic, Newer Volcanics Province, Victoria and South Australia. **a** Oblique aerial photo of Lake Purrumbete maar; looking south. **b** Map of Lake Purrumbete maar. **c** Oblique aerial photo of the Tower Hill volcanic complex; looking south. See Fig. 16.16a for schematic cross-section A-B. **d** Map of the Tower Hill volcanic complex, which is a maar >3 km diameter, and has steep crater wall and gently sloping outer rim with a late interior scoria cone complex 35 ka. **e** Oblique aerial photo of Mt Gambier, which is a very large nested maar-scoria cone complex ~5 ka, and one of the youngest volcanoes in the province; looking west-northwest. **f** Map of Mt Gambier maar-scoria cone complex. (**b**, **d**, and **f** from Cas et al. 2011a)

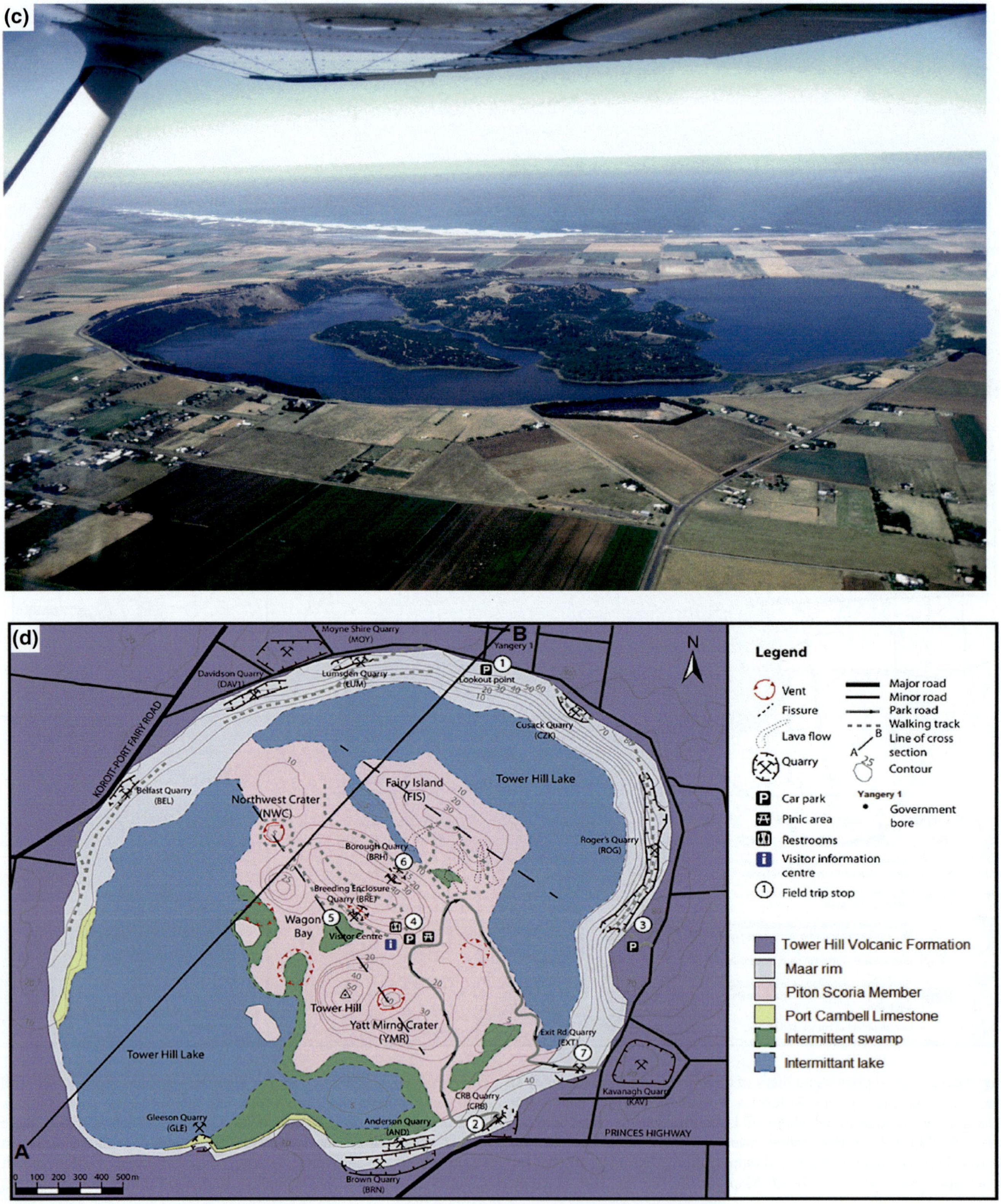

Fig. 16.12 (continued)

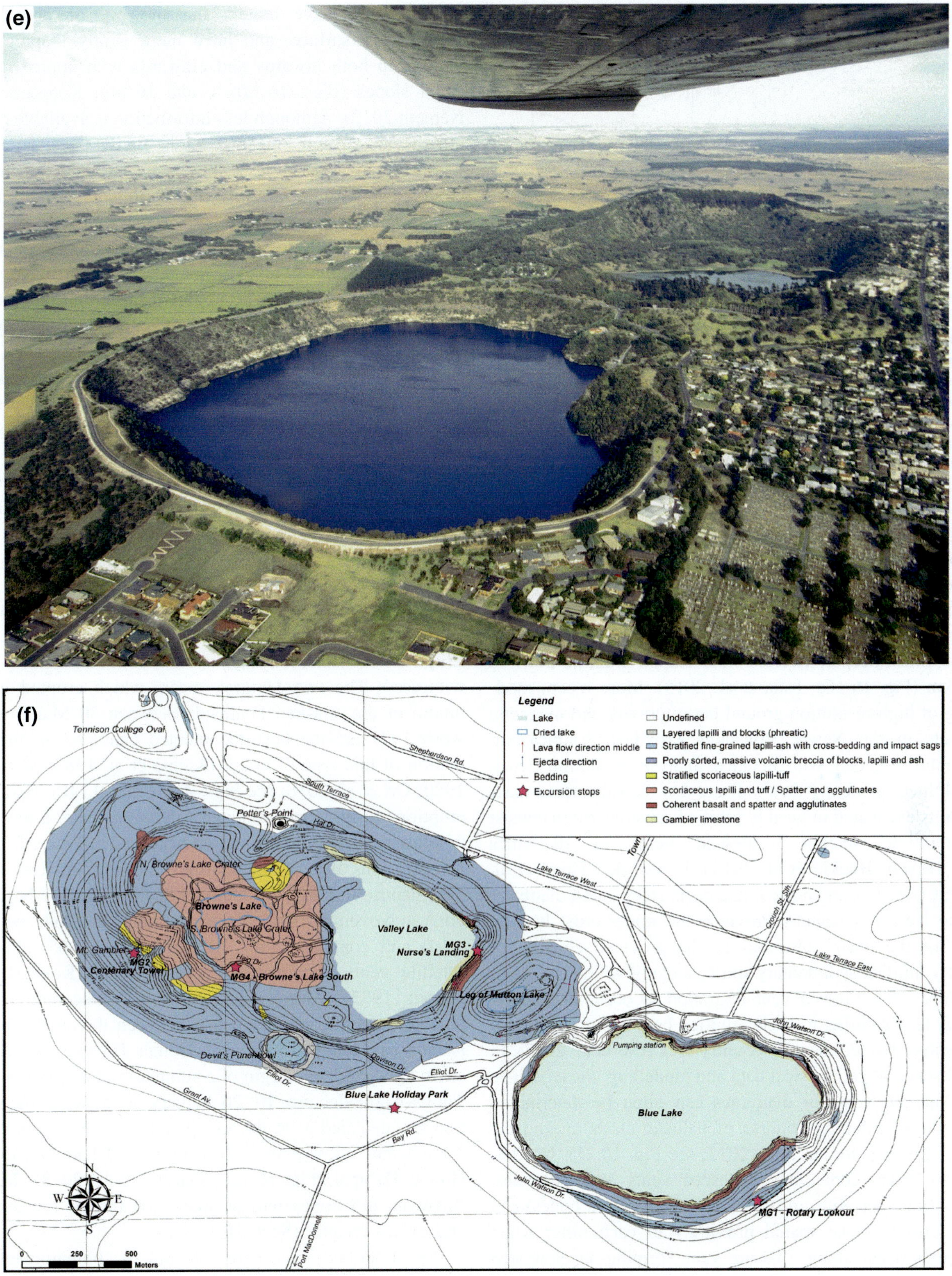

Fig. 16.12 (continued)

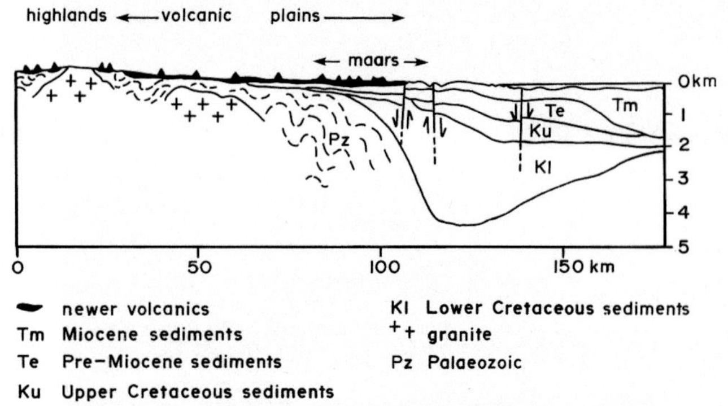

Fig. 16.13 North–south schematic cross-section through the late Cenozoic Newer Volcanics Province in western Victoria, southeastern Australia (see Fig. 16.7). This shows the distribution of maar-type craters and other volcanoes (scoria cones and lava shields) in relation to the basement of Paleozoic rocks overlain by a trough-like succession of Cretaceous and Tertiary sediments forming the Otway Basin, a continental margin break-up basin that formed during the separation of Australia from Antarctica during the break-up of Gondwana. (From Cas and Wright 1987, After Joyce 1975)

diameter of the maar rim (crest to crest) to the depth of the diatreme was about 1: 1. This was based on maars hosted by lithified or hard country rock. Geophysical data of the famous Solfatara maar crater, Campi Flegrei, Italy suggest a single, much deeper, inverted cone-shaped structure down to 2–3 km (Fig. 16.16c; Isaia et al. 2015). More recent modelling of high-resolution ground based gravity and magnetic datasets in the Newer Volcanics Province, southeastern Australia, together with detailed analysis of country rock xenolith componentry data indicate that the subsurface maar conduits excavated in weakly lithified substrate often consist of multiple shallow coalesced vents or diatremes (Fig. 16.16b; Blaikie et al. 2014a, b; van den Hove et al. 2015; Chap. 17), and not a single deep, steep-sided diatreme. In the Newer Volcanics Province the upper crust hosting most of the maar volcanoes in the province consists of weakly lithified sedimentary strata, some of which are major aquifers. This allowed vent migration to occur at shallow crustal levels during a single eruption because of the lack of strength of the sedimentary country rock (Cas et al. 2017). In the absence of geophysical data and modelling, the existence of coalesced vents or diatremes can often be determined from the distribution patterns of ballistic clasts around the crater rim (e.g. Jordan et al. 2013; see Fig. 16.21). In addition to maars containing coalesced vents and diatremes, maars (together with tuff cones and even scoria cones) can also form in close spatial proximity to each other when eruptions occur along a fissure vent, forming an elongated cluster or complex of vents (e.g. the ~5 ka Mt Gambier volcanic complex, Newer Volcanics Province, South Australia; Fig. 16.12e, f; van Otterloo et al. 2013, 2014).

Tuff rings have craters that lie on, or above, the pre-eruption surface, and have more gently sloping rims which dip both inwards and outwards with approximately equal slopes (Figs. 16.11b, c and 16.14b; Kereszturi and Németh 2013). Although less information is available on the dimensions of tuff rings (possibly partly because of the confusion in terminology), they seem to have dimensions comparable with maars. Tuff rings form when rising magma interacts explosively with abundant water close to or at the ground surface. As a result, they usually contain a much higher proportion of juvenile material than do maars. Tuff rings are also usually mafic in composition, but more silicic examples are also common (e.g. Campi Flegrei, Italy; Di Vito et al. 1999).

Tuff cones differ from tuff rings by generally having smaller craters and having larger height to width ratios and steeper outer slopes (Figs. 16.11d, e and 16.14c; Kereszturi and Németh 2013). They are commonly formed in areas where surface water (lake or shallow marine) is located above the vent. They show bedding angles of at least 20–25° near rim crests.

16.3.2 Eruptions, Characteristics and Deposits

Eruptions of maars, tuff rings and tuff cones have rarely been witnessed. The two Ukinrek maars which erupted in the tundra of the Alaskan Peninsula between 30 March and 9 April 1977 are the best-documented examples of the formation of this type of crater (Kienle et al. 1980; Self et al. 1980; Ort et al. 2018; see Figs. 4.18c and 9.11c). Historical eruptions forming maars suggest durations from a few days to a few weeks. The best-known historical tuff cone-forming eruption was that of the marine volcano, Surtsey, in the Vestmannaeyjar Islands south of Iceland, which began to erupt in November 1963 (Thorarinsson et al. 1964; see Plate 4.1d) and continued into the summer of 1965, although satellite vents remained active until 1967 (Kokelaar and Durant 1983). More recently, on 19 December 2014, phreatomagmatic activity began between the small Tongan islands of Hunga Tonga and Hunga Ha'apai, above the submerged caldera margin of a submarine volcano along the Tonga-Kermadec arc. By 26 January 2015, when the eruption was declared over, the emergent tuff cone was some 1 to 2 km wide, 2 km long, 140 m high, and had joined with Hunga Ha'apai (Fig. 16.11e; Garvin et al. 2018). Marine surtseyan tuff cones will be discussed further towards the end of this chapter (Sect. 16.11.4).

As described in Chaps. 4 and 9, phreatomagmatic explosive eruptions are marked by closely spaced explosions, representing a distinctive pulsing style of explosive eruption (e.g. Fig. 4.18a–d). Pyroclasts are ejected from the

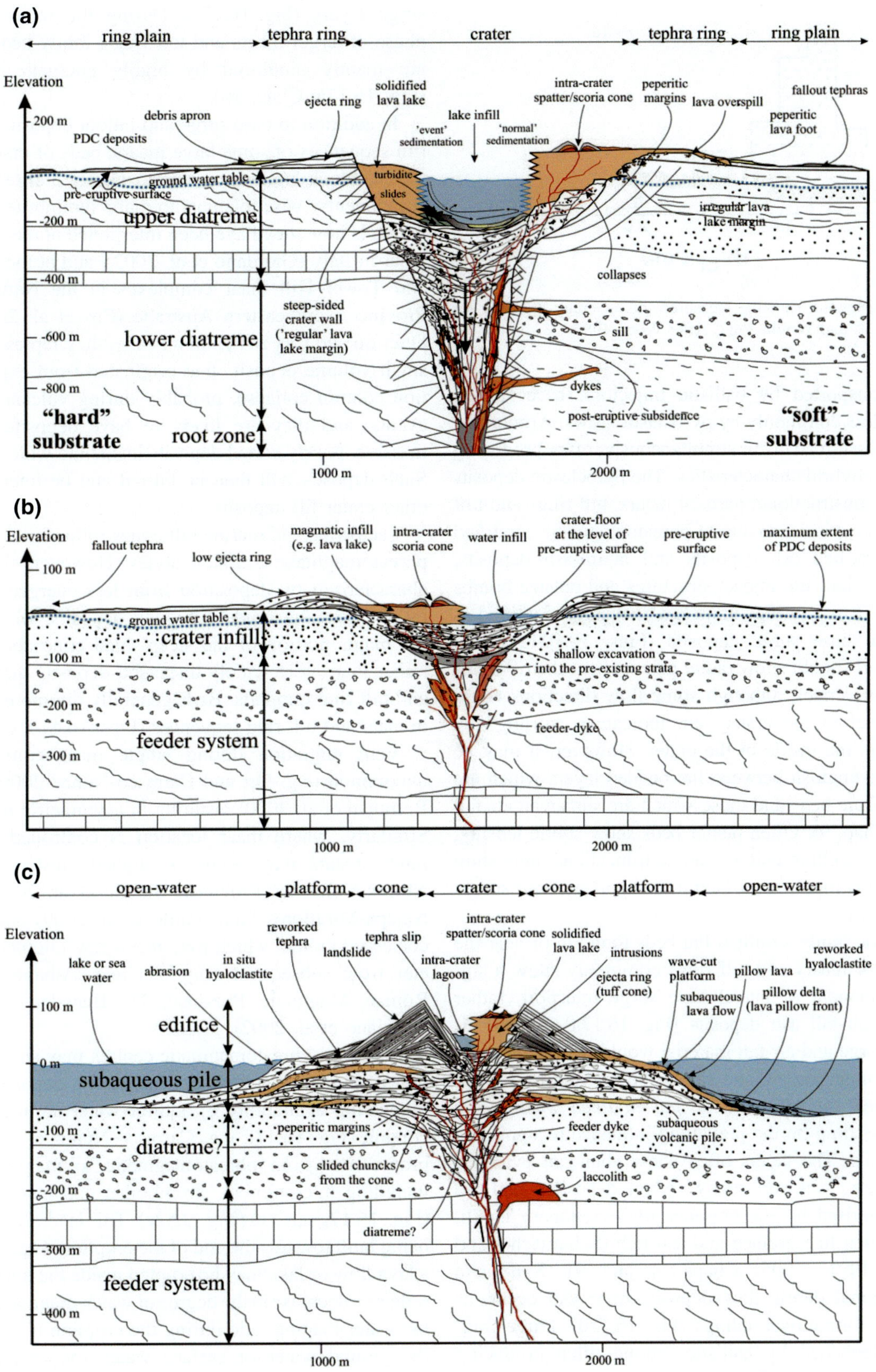

Fig. 16.14 Schematic cross-sections through **a** maars, **b** tuff rings and **c** tuff cones. These show differences between three types of crater formed by phreatomagmatic eruptions (and for maars also by phreatic activity), and the distribution of various facies types. Below maar craters, the vent fill, comprising volcanic breccia and slide blocks of collapsed rim pyroclastic deposits, is termed a diatreme. (Adapted from Kereszturi and Németh 2013; *Source* InTechOpen through Creative Commons Attribution 3.0 Unported licence)

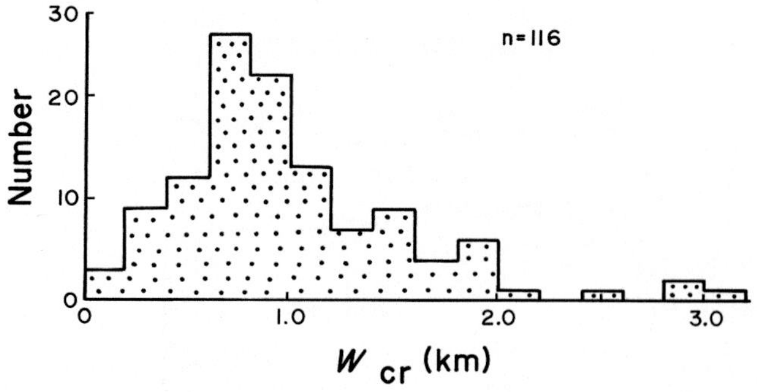

Fig. 16.15 Histogram showing measurements of crater diameter for maars (From Cas and Wright 1987, After Wood 1980c)

vent and transported by ballistic trajectory, base surges, fallout, and less commonly by pyroclastic flows. Around the vent all processes can occur simultaneously, often producing deposits with hybrid characteristics. The pyroclastic deposits forming the constructional parts of maars, tuff rings and tuff cones therefore consist mostly of phreatomagmatic, stratified and cross-stratified ash deposits and lapilli-ash deposits, punctuated by ballistic impact structures and outsize bombs and blocks. Cross-stratified units can be readily identified as dune bedded base surge deposits (Figs. 10.104e–g and 16.17a, b) and should show radial paleocurrent distributions away from the crater, although some flow directions could be sub-parallel to the crater rim, indicating "surfing" by surges around the inside of the crater. However, it may be difficult to distinguish between flat-bedded layers which are base surge plane beds and those which are surtseyan air-fall layers (cf. Chap. 9). Often planar beds show subtle lensing, pinching and swelling and low-angle truncations, indicating at least some component of lateral transport by base surges (Fig. 16.17a–c).

Most of the thinly stratified flat beds found in or near the crater walls of maars and tuff rings commonly show a significant component of lateral base surge transport, rather than simple air-fall ash deposits (Fig. 16.17a–c). Realistically, base surge and air-fall material would often have been deposited simultaneously and should be referred to as *base surge-modified fall deposits* or *fall-modified base surge deposits* depending on the prevailing influence recognisable in the deposits. The layers often show abundant evidence that they were "wet and sticky" when deposited and are often characterised by the presence of accretionary lapilli, especially when fine grained and ash-rich (Schumacher and Schmincke 1991, 1995; Chaps. 4 and 9). Syn- and post-depositional debris flow deposits may also occur, as well as intra-formational slumps (Fig. 16.17d). Large fragments are deposited by ballistic fallout, often producing impact sag structures (Fig. 16.17c, e), but these may be interbedded with less-coarse fall deposits and fine-grained surge layers (Fig. 16.17f). During the main construction phase of larger maars and tuff rings, thinly bedded deposits are mainly emplaced by highly energetic base surges (Fig. 16.17a; Chap. 10).

In addition to base surge and fallout deposits, in the maar rim sequences of some large maars, beds of massive, poorly sorted block-lapilli-tuff occur. These are texturally and componentry wise identical to mafic ignimbrites (Chap. 10), which is how they have been interpreted at the Lago Albano maar in Italy (Giordano et al. 2002), and at the Mt Gambier and Tower Hill maar complexes in the Newer Volcanic Province, southeastern Australia (Cas et al. 2011a, b; van Otterloo and Cas 2016). In all cases they represent relatively small volume deposits that originated from explosive eruption column collapse, probably during vulcanian explosive events, and they are likely to have deposited equivalent massive, poorly sorted deposits inside the wide maar craters. Such deposits will then be buried and be interbedded with other crater fill deposits.

Lacustrine and marine tuff cones follow similar, pulsating phreatomagmatic eruption styles. However, these may be characterised by deposition from less energetic surges and fallout, forming crudely bedded deposits, which construct the major portion of the steep-sided cone above the vent, rather than more widely dispersed deposits that build maar and tuff ring ramparts. Beds generally dip more steeply than in maar and tuff ring sequences (Figs. 16.11d, e and 16.14c).

Vent migration during single maar eruptions is not uncommon (e.g. Ort and Carrasco-Núñez 2009; De León--Barragán et al. 2020) resulting in asymmetric maar edifices. Similarly, where maar location is controlled by regional faults, maars may occur as aligned clusters (e.g. in the Newer Volcanics Province, Blaikie et al. 2012; along the Nejapa-Miraflores fault, Pardo et al. 2009) and even form coalesced craters which may reach few kilometres in diameter from subsequent eruptions (e.g. Albano maar in the Roman Magmatic Province, De Benedetti et al. 2008; Giordano et al. 2002).

Subaerial phreatomagmatic centres may show an alternation of phreatomagmatic and magmatic deposits due to the fluctuating availability of aquifer water around the magma conduit (Fig. 16.17f; e.g. Carrasco-Núñez et al. 2007), or a progression from phreatomagmatic to magmatic activity, including microplinian, strombolian and hawai'ian activity (Fig. 16.17g; also see Fig. 9.52c). This leads to scoria cone(s) being built towards the end of the eruption (Fig. 16.12c–f), or a lava flow or lake may be erupted inside the maar crater, and reflects a decrease in the degree of water-magma interaction or the water-magma ratio during the eruption (Chap. 4). Either the ground water or surface water source was gradually exhausted, or the aquifer was damaged during subsurface explosions, so reducing permeability and water flow, or rising

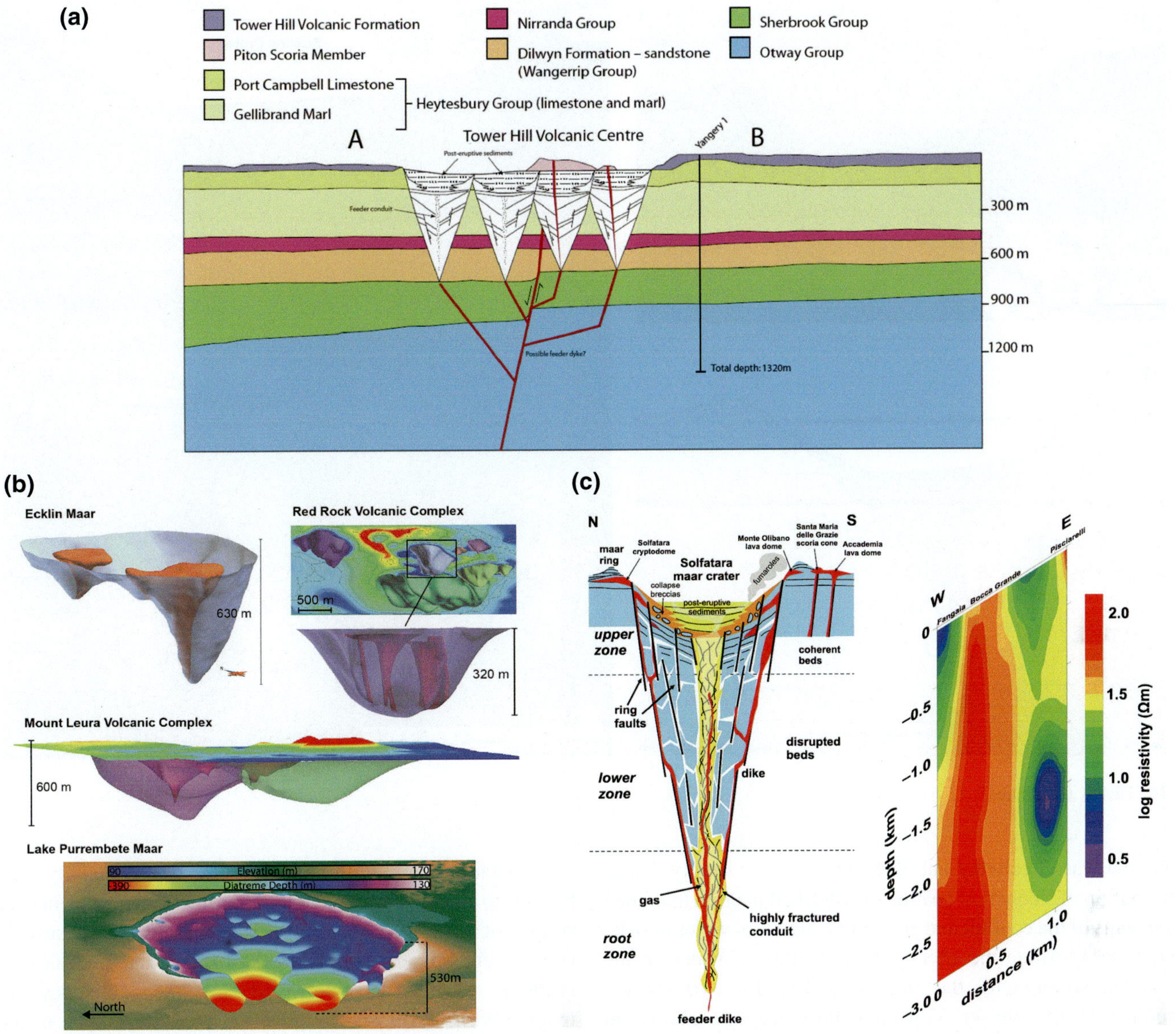

Fig. 16.16 Subsurface structure of maar diatremes. **a** Schematic cross-section showing the setting of the 35 ka Tower Hill maar-scoria cone complex, Newer Volcanics Province, Victoria, southeastern Australia (From Prata 2012, Modified after Orth 1988). Line of section A-B is located on Fig. 16.12d. Subsurface geology of the Otway Basin is inferred from xenoliths and geophysics. The ground water contained in the sandstones of the Dilwyn Formation controlled the predominantly phreatomagmatic course of the eruption. **b** Reconstructed subsurface geometry of maar diatremes in the Newer Volcanics Province, based on forward modelling and inversions of high resolution, ground gravity and magnetic data sets. The models highlight the multiple coalesced shallow diatreme roots, resulting from excavation of weakly lithified sedimentary substrate with little mechanical strength, rather than one very large and deep one as is commonly the case if the substrate are lithified (cf. Fig. 16.23a, b). (From Cas et al. 2017). **c** Internal structure of the Solfatara alkaline maar diatreme, Campi Flegrei, Naples, Italy, inferred by integration of (left) surface and (right) magnetotelluric data (From Isaia et al. 2015). **d**, **e**, **f** Views through the Suoana maar-diatreme exposed along the wall of the Myiakejima caldera which collapsed in July–August 2000, Izu Islands, Japan. The maar formed in the seventh century. Exposed in this exceptional cross-section are a 400 m wide maar crater, an associated tuff apron with a maximum thickness of 20 m, a diatreme extending vertically to a depth of about 220 m, and a feeder dike connected to the base of the diatreme. The conical central diatreme pipe is filled by very coarse, breccia-size volcaniclastic debris, which is divided into 6 units based on lithological and structural characteristics. The upper half of the diatreme comprises landslide deposits, mainly derived from the surrounding crater wall, while the bottom is occupied by massive explosion breccia. Some coherent blocks (unit F) were detached and partially rotated from the wall of the diatreme, to be preserved in the diatreme fill. Area detailed in image (**e**) is located in (**f**) by orange box. (From Geshi et al. 2011)

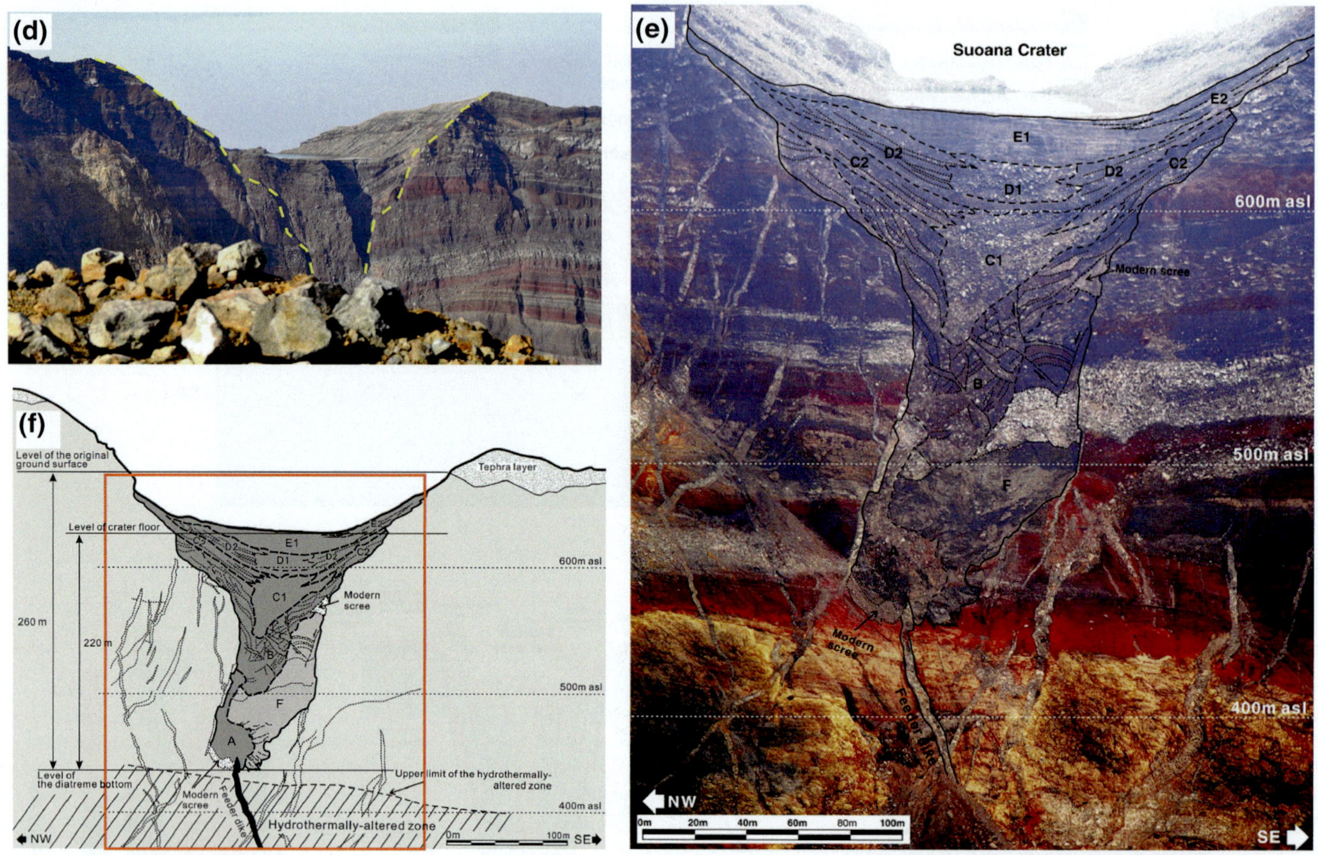

Fig. 16.16 (continued)

magma was effectively cut off from the water supply due to formation of a coherent coating of solidified magma that lined the walls of the conduit and prevented ingress of water into the vent. To illustrate the effects of availability of water on the volcanic stratigraphy, the successions found in two selected centres in the Newer Volcanics Province, are shown in Fig. 16.18 (see also Cas and Wright 1987). At Lake Purrumbete maar (Figs. 16.12a, b, 16.17a and 16.18a; Jordan et al. 2013, 2015) abundant ground water was available throughout the eruption which was almost wholly phreatomagmatic (Jordan et al. 2013). The 35 ka Tower Hill maar (Figs. 16.12c, b and 16.16a; Prata, 2012) is a major "*nested*", tuff ring- or maar-scoria cone complex, over 3 km in diameter. The ground water availability must have been intermittent, because the maar crater wall shows a succession of interbedded phreatomagmatic and magmatic deposits (Figs. 16.17f and 16.18b, c; Cas et al. 2011a, 2017). The late stage of the eruption became magmatic, building a considerable scoria cone complex in the centre of the maar (Fig. 16.12b, c), with isolated minor lavas associated with these. Eruption points were clearly aligned along NNW-trending fissures (Figs. 16.12d and 16.16a).

Constantly alternating phases in the style of activity may be due to a very complex interrelationship between rate of supply of water, from a ruptured and fracturing subsurface aquifer, and magma supply rate. In particular, pulsation in eruptive energy is controlled by the efficiency in magma-water explosive interaction that depends on the complex interplay between (1) magma ascent rate; (2) state of the multi-phase magma suspension at interaction with water, and (3) hydrogeology of the aquifers, including the flow rate or permeability, which may fluctuate if the aquifer is being damaged at the point of explosive interaction in the subsurface.

In a review of six felsic maar-tuff ring-diatreme centres around the world, Ross et al. (2017) summarised the deposit characteristics and eruption processes. Generally, the vesicularity of juvenile clasts was low, although in some cases increasing upwards, and together with dune bedded base surge deposits and abundant country rock xenoliths, this suggests a phreatomagmatic style of volcanism. In general, they found that the deposits of the maar rim and tuff ring sequences are similar to those of mafic maar and tuff ring volcanoes, i.e. bedded, up to ≥ 50 m thick, and include

(a)
(b)
(c)
(d)
(e)
(f)

◀ **Fig. 16.17** Deposits of maar and tuff ring sequences. **a** Cross-stratified base surge beds and gently undulating planar strata forming the rim sequence of the basaltic Lake Purrumbete maar, Newer Volcanics Province, Victoria, Australia. Base surge flow direction was from right to left based on the asymmetry of the dune bed forms. The quarry face is about 10 m high. **b** Remains of the 8.7 ka Vallone del Gabelloto maar/tuff ring/tuff cone sequence, Lipari, Italy, showing well-formed base surge dune bed forms, cross-stratification and gently undulating planar strata. **c** Rhyolitic trough shaped base surge deposits in the middle, in between more planar beds above and below that show subtle thickness changes, which are interpreted as surge-modified fallout deposits. Miocene Ramadas tuff ring, Argentina. **d** Intra-formational slumping of wet base surge beds, Citara Formation, Ischia, Italy. **e** Impact sags of ballistics, Carapacho tuff ring, Llancanelo Volcanic Field, Argentina. **f** Interbedded "wet" phreatomagmatic base surge beds (light, cream coloured) "dry" phreatomagmatic surge and surge-modified fallout (grey) and magmatic scoria fallout beds (dark), exposed in the maar rim sequence of the Tower Hill volcanic complex (see Fig. 16.12c, d), Victoria, southeastern Australia, reflecting variations in availability of water in the magma conduit. **g** Transition from tuff ring forming phreatomagmatic base surge deposits (fawn coloured) at the base of the exposure through a tuff ring, overlain by black scoria fallout deposits, which close to the vent (off to the right of the field of view) built thick scoria cone deposits (see Fig. 16.9b). Pleistocene Mt Leura volcanic complex, Victoria

Fig. 16.17 (continued)

low-angle cross-bedded base surge deposits, and ballistic impact bombs and blocks of cognate and country rock lithologies. In the diatremes bedded sequences also occur, often with steep dips, sometimes at considerable depth in the diatreme. In addition, megaclasts of bedded rim and crater floor deposits were found to have subsided into the diatreme, as is found with kimberlite pipes. Another common facies with kimberlite pipes is a massive poorly sorted thick lapilli tuff facies, which have been interpreted in different ways. Cas et al. (2008a) and Porritt et al. (2008) interpreted this massive facies as fall back deposits collapsing into the diatreme from an explosive eruption column much as occurs during column collapse that produces pyroclastic flows because of the similar poorly sorted granulometry and textures as found in ignimbrites. Other interpretations for the massive facies, particularly where they occur as units with sub-vertical contacts in the diatreme succession, is that they result from suppressed tephra jets emitted from explosion deep in the diatreme through deposits already in the diatreme, so homogenising them Ross and White 2006; White and Ross 2011). The two alternative interpretations are not mutually exclusive; both can and almost certainly do occur.

There is one additional facies in felsic maar-diatreme systems that do not occur in mafic maars, and that is felsic

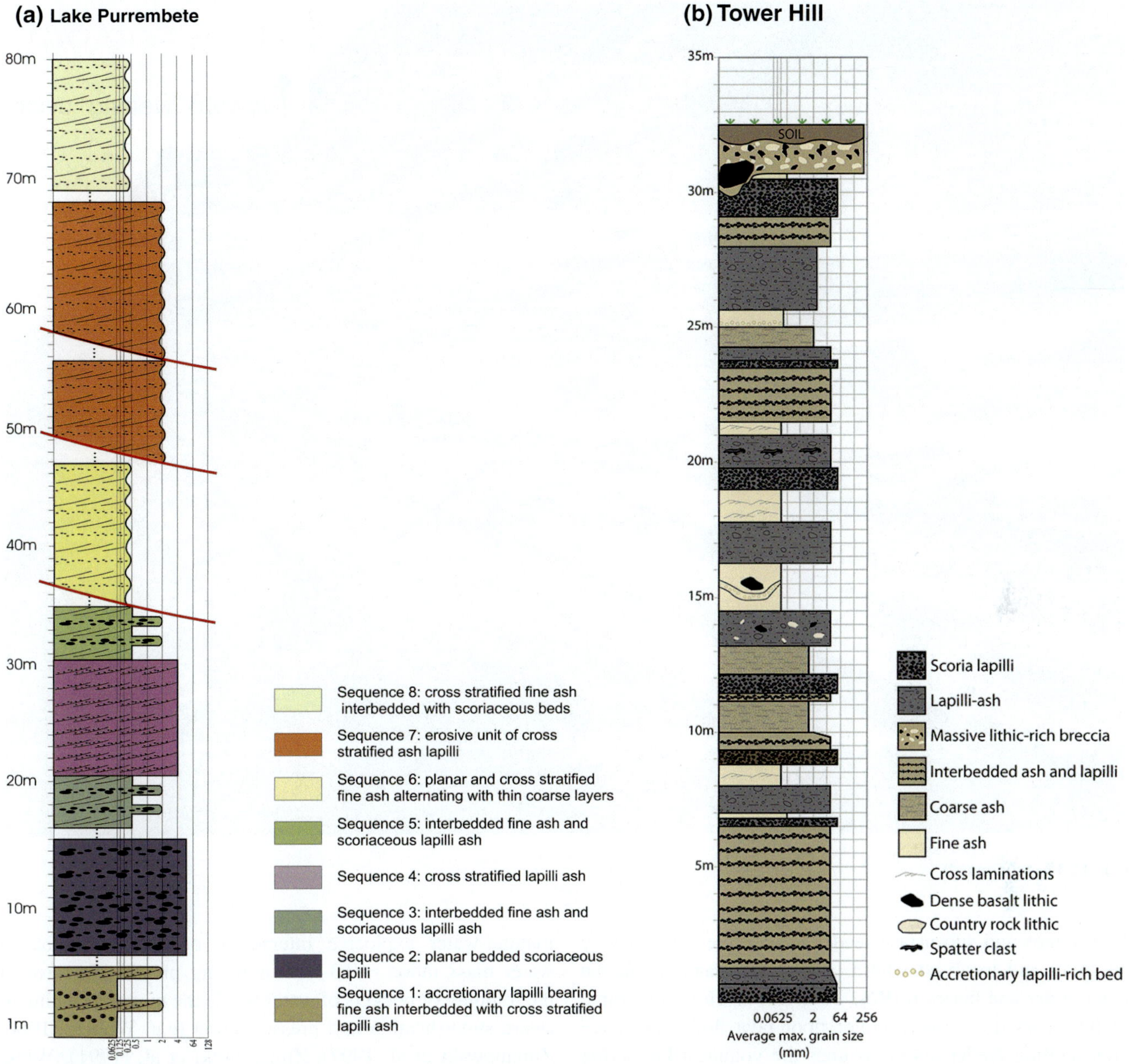

Fig. 16.18 Deposit features found in basaltic maar rim sequences of two selected centres in the late Cenozoic Newer Volcanic Province, Victoria, southeastern Australia. **a** Composite graphic log of Lake Purrumbete maar rim sequence (see Fig. 16.12a, b), showing mostly "wet" to "dry" pheatomagmatic deposits and only minor scoria deposits (From Cas et al. 2011a). Red lines = syn-eruptive faults and low-angle slides formed when segments of maar rim collapsed into the crater; also see Fig. 10.109d, e. **b** Graphic log of Tower Hill maar-scoria complex (see Figs. 16.12c, d and 16.17f), showing interbedded magmatic scoria, and coarse and fine phreatomagmatic deposits, indicating fluctuating degrees of explosive magma-water interaction during the eruptions (From Cas et al. 2011a). At both Purrumbete and Tower Hill volcanoes all sections are essentially free of major weathering and erosional breaks, implying a short eruption and accumulation time. **c** Interbedded phreatomagmatic (light colour) and scoria fall deposits (dark layers) in the upper maar crater wall at the basaltic Tower Hill maar-scoria complex, as graphically logged in (**b**). The changes from fine ash to coarse lapilli to scoria layers (and vice versa) can be interpreted in terms of "drying" and "wetting" cycles, reflecting fluctuating degrees of magma-water interaction. (From Cas et al. 2017)

Fig. 16.18 (continued)

lava domes and subsurface feeder felsic dykes, which may be small and localised within part of the maar, or fill the maar crater and buries it (Ross et al. 2017). However, some mafic maars do preserve crater floor lava flows and have cross-cutting feeder dykes, as does the youngest kimberlite volcano, the ~10 ka Igwisi Hills volcanic system in Tanzania (Brown et al. 2012; Chap. 18).

16.3.3 The Role of Aquifers

White and Houghton (2000) summarised ascent rates inferred for observed phreatomagmatic eruptions and recognised that in many cases the magma was ponding, slowly rising or even draining back into the conduit. Furthermore, other styles observed during the phreatomagmatic eruptions varied from effusive to strombolian/hawai'ian, indicating that maximum magma fluxes are at the lower end of intensity (see Chap. 9). This is a reasonable scenario, as it has been demonstrated experimentally (Chap. 4) that efficient magma-water explosive interaction occurs at magma to water mass ratios ~0.3 and involving pre-mixing through the trapping of pockets of water inside the magma, leading to steam super-heating and pressurisation (e.g. Wohletz 1986; Zimanowski et al. 1997). Zimanowski et al. (1991), White (1996), and Schipper et al. (2011) proposed that efficient phreatomagmatic explosive activity could occur at lower water:magma mass ratios, perhaps even an order of magnitude lower, if a mixture of fragments and water (or "impure coolant", Chap. 4) exists in the conduit at the point of explosive magma-water interaction.

As a consequence, to achieve efficient phreatomagmatic explosive interaction, at increasing magma fluxes through the conduit a comparable increase of not just of water availability, but also of conditions promoting efficient pre-mixing with water, are required. Such a scenario suggests that for underground phreatomagmatism, where the ground water flow rate is limited by the aquifer hydrostatic head and water transmissivity, slow magma ascent rates, such as those associated with strombolian eruptions (Chap. 9), favour conditions for efficient interaction. This explains the frequent association of phreatomagmatic deposits and strombolian to effusive deposits just described above (Fig. 16.18). Faster ascending magma columns would require progressively more transmissive aquifers and large hydrostatic heads to trigger phreatomagmatism, such as may be the case of deep confined aquifers, including hydrothermal systems.

Identifying the aquifer(s) involved can be done by studying the nature of the xenoliths present within the phreatomagmatic deposits (e.g. Barberi et al. 1989; Jordan et al. 2013; van Otterloo et al. 2013). For example, in both the maars of the Newer Volcanics and Roman Magmatic Provinces the large presence of limestone xenoliths indicate that the main aquifers involved are the pressurised aquifers hosted in fractured limestones at several hundreds to thousands of m below ground (Figs. 16.19 and 16.20; De Benedetti et al. 2010; Danese and Mattei 2010). Based on deposit observations, it appears that the involvement of shallow aquifers may result in highly pulsatory style, with repetitious wet and dry phases, whereas the involvement of deeper pressurised and fractured aquifers may result in more sustained conditions of efficient phreatomagmatism. Changes in eruptive energy during the course of the eruption may also result in impulsive phases of vent/conduit enlargement, testified by the presence of interbedded shallow to deep derived breccias (Fig. 16.20). Detailed study of the xenolith componentry upsequence is the key to understanding the fluctuations of the depth of the fragmentation level in the conduit (see Chap. 9).

Lorenz and Kurszlaukis (2007) proposed that the progressive depletion of the aquifer involved in maar-diatreme formation should drive a progressive deepening of the locus of magma-water interactions. Other authors have suggested instead that many fluctuations in physical factors controlling explosive activity may occur, and not just in a simple fashion (e.g. Valentine and White 2012). We emphasise that not only the water flux towards the conduit may change in time based on the local hydrogeology, but so will the changing physical state of the aquifer around the conduit as

Fig. 16.19 Large ballistic blocks of country rock at the Pleistocene Albano maar, Colli Albani volcano, Roman Magmatic Province, Italy. The abundant white coloured clasts, both in the lower stratified base surge and surge-modified fallout succession and in the upper massive ignimbrite unit, are Mesozoic limestone clasts. The local subsurface stratigraphy indicates that these fragments of limestone were brought to the surface from a depth of >1 km where a pressurised hydrothermal aquifer is present, which triggered the hydrovolcanic eruption. The dark clasts are mafic

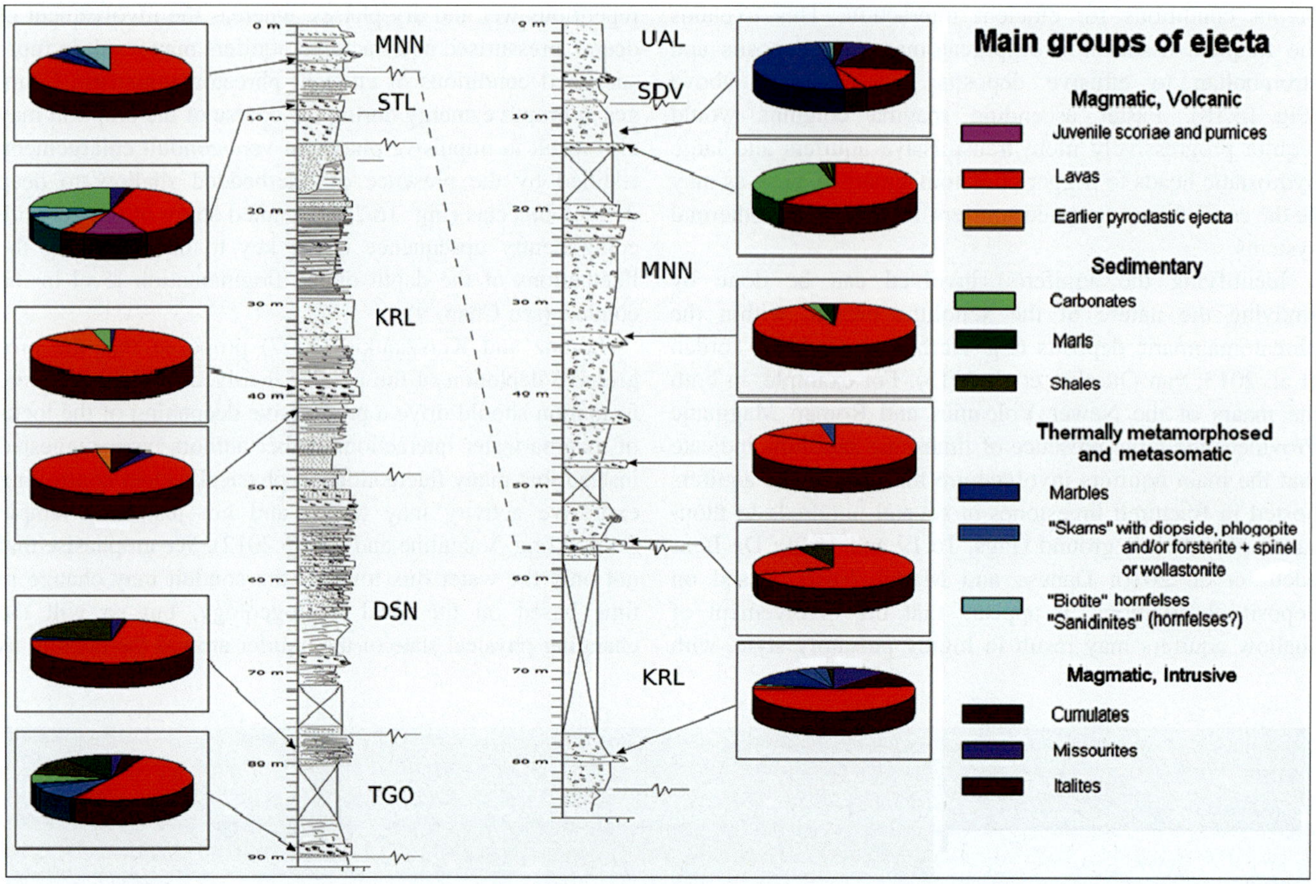

Fig. 16.20 Composite stratigraphic logs of the mafic Albano maar succession. Letters refer to different eruption units, as the maar is composite and its activity spanned over at least seven eruptions between ca. 70 and ca. 25 ka. The pie diagrams show the lithic componentry of breccia beds and show variable proportions of volcanic versus deep-seated lithics, indicating fluctuations in the fragmentation level

explosions pulverise it, as well as the state of the magma, which affects the extent and efficiency of interaction. For example, magma may cross a productive aquifer in a coherent, vesicular or even already fragmented state, changing substantially the chances for ground water to penetrate into the magma conduit and for pre-mixing and explosive interaction. At the same time it is important to assess the initial porosity (i.e. primary versus secondary), the permeability and transmissivity characteristics of the aquifers involved with phreatomagmatism (e.g. Ort et al. 2000). Even though specific studies that analyse this topic have not been performed yet, we believe that a vesicular or even already fragmented magma (e.g. due to magmatic explosive fragmentation deeper in the conduit) is more likely to produce explosive phreatomagmatism because the magma surface area available for heat transfer to the external water is higher. This is even more so for surtseyan activity where interaction occurs with unconfined open and shallow water bodies, where conditions for efficient pre-mixing and steam super-heating are very limited if a coherent lava enters the water (see Chaps. 5 and 6). This has some support in the relatively high levels of vesicularity of the juvenile pyroclasts from the 1963 eruption of Surtsey volcano, Iceland (Thorarinsson et al. 1964), and the 2009 and 2014–15 eruptions of Hunga Tonga-Hunga Ha'apai volcano, Tonga (Colombier et al. 2018).

16.3.4 Palagonitisation and Hydrothermal Alteration

Another important consideration is that the original basaltic glass (sideromelane) in the phreatomagmatic deposits of some craters is quickly altered by hydration to palagonite (Hay and Iijima 1968; Stroncik and Schmincke 2002; Chap. 13). In alkali-rich phreatomagmatic deposits, the main alteration phases are zeolites such as chabazite and phillipsite (de'Gennaro et al. 2000). This produces indurated and lithified tuffs, and primary textures become obscured, especially in layers composed of fine ash. Palagonitisation seems to be more commonly found in tuff cones rather than the other types of craters, although many tuff rings are altered. Wohletz and Sheridan (1983) suggest that this is because tuff cones are largely constructed by less-inflated surges (Chap. 9) containing a high proportion of water which is then trapped in the ash after emplacement. Alteration of the glass is also undoubtedly caused by steam moving through the deposits with mild hydrothermal activity, as has occurred at Surtsey and Hunga Tonga-Hunga Ha'apai (Fig. 16.11e). Indeed, the basaltic pyroclastic deposits at Surtsey were almost completely consolidated within 15 years of the eruption (Jakobsson and Moore 1982, 1986; Jackson et al. 2019, and results of SUSTAIN drilling project at Surtsey). More generally, lithification of wet or moist phreatomagmatic structures is likely to occur geologically rapidly within tens to thousands of years (Thordarson 2004).

16.3.5 Diatremes and Maar Lakes

Diatremes are only exposed by deeper erosion and are therefore generally only seen in older or rapidly uplifted volcanic successions, although, remarkable geological coincidences can always occur (Fig. 16.16d). Recent literature now often refers to maar-diatreme systems (Lorenz 2003; Lorenz and Kurszlaukis 2007; White and Ross 2011; Ross et al. 2017), and the term has become somewhat synonymous with the conduit or vent resulting from relatively shallow to moderate level phreatomagmatic, phreatic or hydrothermal explosive activity (e.g. Figs. 16.11a, f, g and 16.14a), although this is not always so. Diatremes are typically composed of volcanic breccias generated by one or more of these processes and, internally, breccia facies may show complex geometrical relations (e.g. Fig. 16.16d; Geshi et al. 2011; Lefebvre et al. 2013; Davies et al. 2008; Ross et al. 2017; also see Chap. 18).

Traditionally the term "*diatreme*" has, however, been employed to describe kimberlitic pipes. Traditional kimberlite pipe models depict steep tapering pipes which range in size from a few tens of m to over 400 m in diameter and may extend to depths of 2 km (Chap. 18). While explosive interaction with water and phreatomagmatic activity may have been important during the eruptive phases of some kimberlitic pipes, our recent studies suggest their evolution can be significantly more multifarious, and involvement of external water is not a necessary driving mechanism for their formation (e.g. Cas et al. 2008a; Porritt et al. 2008; Sect. 18.7.3). For views to the contrary, see Lorenz and Kurszlaukis (2007), and Kurszlaukis and Lorenz (2008). Understanding kimberlite diatremes specifically, and kimberlitic rocks more generally, has also been hindered by inconsistent and often highly genetic terminology as highlighted previously in Chap. 14 and by Cas et al. (2008b, 2009).

Geophysical imaging of modern maar diatremes also reveals complex structures, with simple to multiple vent-pipes (Blaikie et al. 2012, 2014a, b; Isaia et al. 2015; Cas et al. 2017). Blocks of the crater wall may collapse along inward dipping faults, susceptible to gravitational readjustment, causing the enlargement of the crater by maar rim retreat (Figs. 16.14a, 16.18a and 16.21). Such ring fault systems can be highly prospective sites for the development of epithermal vein systems in evolved magma systems (Chap. 18).

Bathymetric studies of maar lakes have also become important in identifying the spatial evolution of the vent system over time (e.g. Anzidei et al. 2008; Pardo et al. 2009; van den Hove et al. 2015; Fig. 16.22). Deep lakes in active maars may also be the site of intense degassing which may cause storage at depth of significant amounts of dangerous gas species such as CO_2. Severe natural accidents caused by gas release (so-called "limnic eruptions") have occurred for example in 1984 and 1986 at Lakes Monoun and Nyos, in

Fig. 16.21 Collapse fault, shown by dashed yellow line, in the Albano maar wall displaces the massive Peperino Albano phreatomagmatic ignimbrite downwards and crater-wards (i.e. to the right)

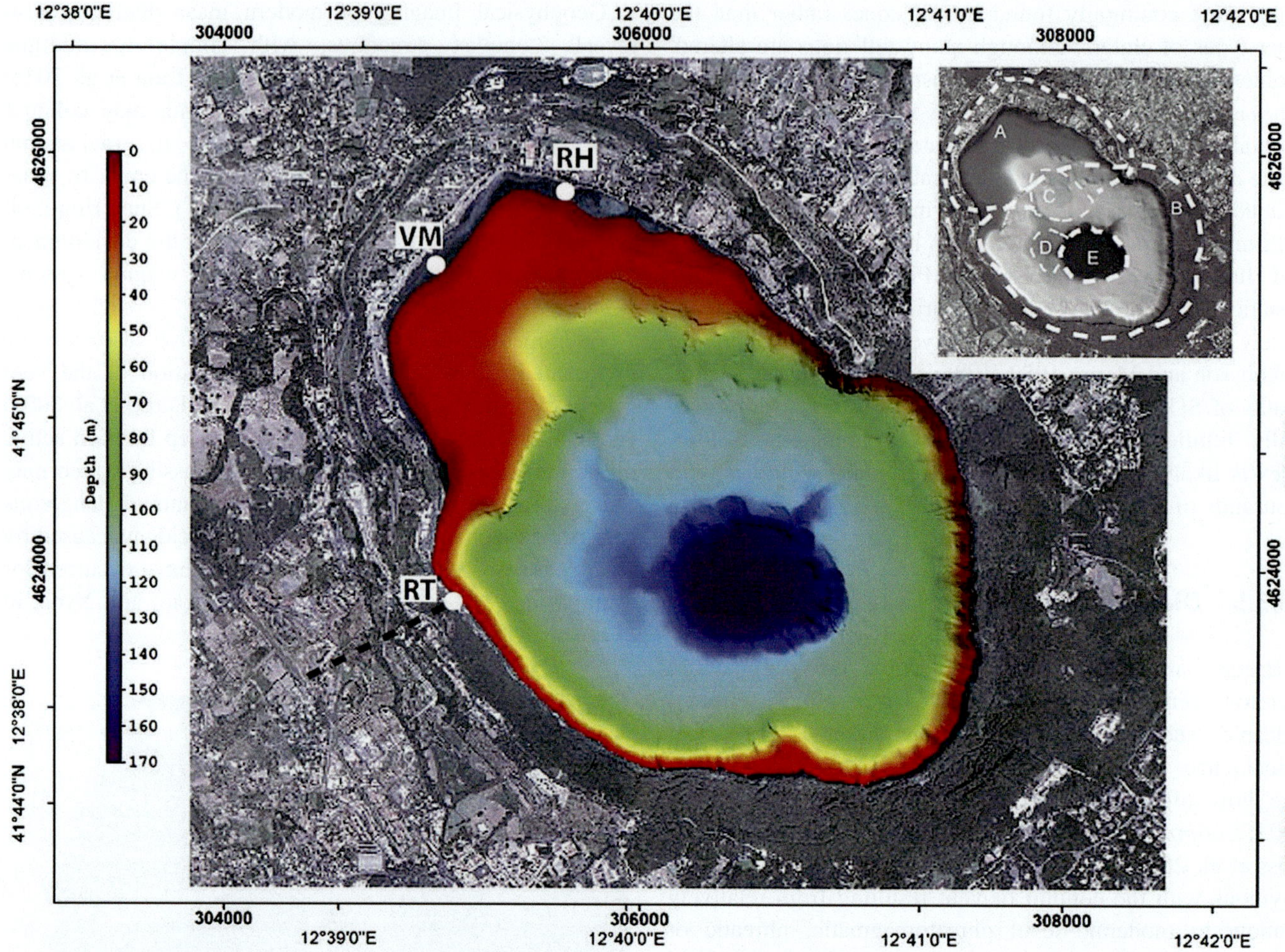

Fig. 16.22 Bathymetry of the Albano maar. Lake Albano is 170 m deep and CO_2 degassing at the bottom, coupled with a temperature and density stratification of the water. This poses hazards in terms of gas storage and possible release in cases of lake roll-over. Triggers may include landslides, earthquakes, shallow intrusions (hot water injection) and phreatic eruptions. VM indicates a Middle Bronze Age village that was submerged by the lake; RH and RT indicate the location of the Roman Harbour and the Roman Tunnel drain both affected by lake level oscillations. Letters A-E in the inset indicate the outlines of the coalesced craters. (From Anzidei et al. 2008)

Cameroon, respectively (Sigurdsson et al. 1987; Kling et al. 1987; Kusakabe et al. 1989; Kusakabe 2017), and at Lake Albano, Italy, during Ancient Roman times (Funiciello et al. 2003; De Benedetti et al. 2008; Fig. 16.22). Destabilisation of the deep gas-rich waters may occur as a result of earthquakes and/or landslides which induce lake tsunamis and overflows of water and CO_2 (such as responsible for over 1700 deaths in the case of Lake Nyos in 1986; Sect. 16.12.2).

16.3.6 Facies Models for Maars, Tuff Rings and Tuff Cones

In Fig. 16.14 we have presented schematic cross-sections (After Kereszturi and Németh 2013) of the surface morphological elements, subsurface geometry, and deposit architecture of maar (Fig. 16.14a), tuff ring (Fig. 16.14b) and tuff cone volcanoes (Fig. 16.14c), and the relationship to upper crustal substrate. These serve as general guides to the morpho-facies architecture of these related but slightly different, monogenetic phreatomagmatic volcano types, especially for mafic examples.

However, we also need to distinguish differences between mafic and felsic types, and we will focus on maar and tuff ring volcanoes, and differences between the compositional end members and substrate type (Fig. 16.23; e.g. Ross et al. 2017). In both cases the morphology of the two compositional types is the same (Figs. 16.14a and 16.23a, b), with a shallow sloping outward dipping apron of deposits for maars (Figs. 16.11a, 16.12a–f and 16.14a), or dipping both inwards and outwards for tuff rings and cones (Figs. 16.11b–e and 16.14b, c), and relatively thin, bedded maar rim or tuff ring sequences up to ~ 50 m thick. These are dominated by base

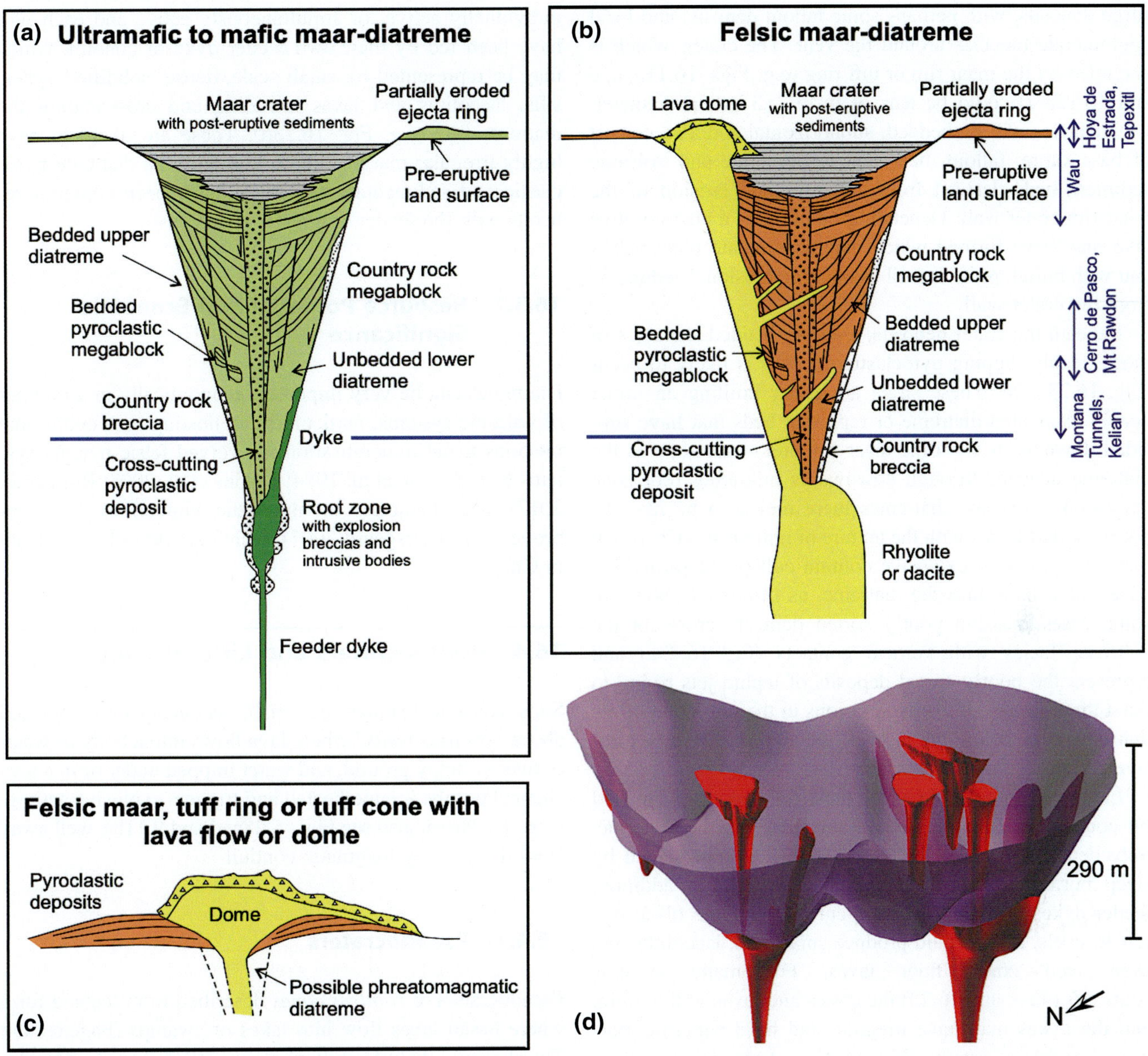

Fig. 16.23 Facies models for mafic to felsic maar volcanoes. Models in (**a**–**c**) are drawn for a hard lithified substrate, leading to excavation of steep-sided diatremes. In cases where the substrate consists of weakly lithified sediments (**d**), the subsurface diatreme shape will be highly irregular and probably consist of several coalesced shallower diatremes, as also shown in Fig. 16.16b. **a** Mafic (-ultramafic) maars, showing surface morphology, with bedded maar rim deposits, shallow-dipping bedded maar crater deposits, steeper dipping bedded pyroclastic deposits that have filled and/or subsided into the diatreme, deeper massive diatreme facies, cross-cutting vertical massive tephra jet deposits, subsided country and maar rim megablocks, intra-maar subsidence faults, and intrusive dykes into diatreme fill. The blue line shows the possible top of the aquifer that caused the phreatomagmatic explosive eruption. **b** Felsic maars. As for **a** but note, the wider feeder felsic dyke and surface lava dome. **c** Felsic maar, tuff ring and tuff cone buried by a relatively voluminous crater filling lava dome and flow. (**a**–**c** from Ross et al. 2017). **d** Multi-diatreme subsurface structure of the late Pleistocene Coragulac maar, Red Rock volcanic complex, Newer Volcanics Province, Victoria, southeastern Australia. This was excavated into a weakly lithified sedimentary substrate, reconstructed from forward modelling of detailed ground gravity and magnetic survey data. The red core of each diatreme represents a dense body of rock, probably solidified basalt feeder. (From Blaikie et al. 2012)

surge deposits, with perhaps some fallout deposits and local agglomerate breccias around the vent. The crater, which is encircled by the maar rim or tuff ring (e.g. Figs. 16.11a, c, e and 16.12a–f) could be tens of m to ≥ 3 km in diameter. Crater facies will be bedded, subhorizontal and also consist of base surge, fallout, near-vent agglomerate and volcanic sediment beds derived from collapses and erosion of the maar rim crater wall. Depending on climate, a post-eruptive lake may have formed which may accumulate plant debris and with burial, peat, as well as ongoing erosional sediments from the crater wall.

Beneath the crater deposits, a deeper bedded sequence of more steeply dipping pyroclastic deposits is likely to occur (Fig. 16.23a, b). These strata are either infilling an initial deeply excavated diatreme or represent beds that have subsided downwards from the maar rim and crater zone as the diatreme deepens through downwards migrating root zone explosions. In some diatremes there may also be massive poorly sorted facies with the texture of ignimbrite (Chap. 10) that may represent explosive column collapse deposits that have fallen back into the diatreme, as discussed above. In some cases massive poorly sorted deposits cross-cut the diatreme facies with vertical contacts (Fig. 16.23a) and represent the poorly sorted deposits of tephra jets trying to burst through the fill from explosions in the root zone of the diatreme at or below the level of the aquifer driving explosive activity.

Large megablocks of bedded maar rim deposits or spalled off country rock may also have subsided into the diatreme. Subsidence of segments of strata should be represented by steep normal faults within the diatreme fill. In addition, feeder dykes are likely to cross-cut the diatreme fill at different levels. Some could produce small sills, laccoliths and even feed crater floor lavas. For mafic systems (Figs. 16.14a–c and 16.23) the dykes and lavas will be thin, and the dykes may have irregular and have peperitic margins. For felsic systems (Fig. 16.23b, c) the dykes will be wide because of the higher viscosity of the magma, and the lavas will be high aspect domes and lavas, that could fill in parts or even bury the crater.

The simplest geometry and facies architecture will be preserved in the diatremes of relatively narrow maars and tuff rings, up to several hundred m in diameter, excavated in solid bedrock (Fig. 16.23a–c). These will tend to be relatively steep-sided and tapering narrow inverted cones. However, in larger diatremes from several hundred m to several km diameter, and where the host stratigraphy is relatively weak or poorly lithified, the subsurface geometry will be highly irregular, and will commonly consist of a cluster of coalesced diatremes, each with its own complex infill facies sequence that interfaces irregularly with neighbouring diatremes, Each diatreme will have been individually active, or simultaneously active and each will have been fed by their own feeder dyke of conduit, which may be represented by small scale, dense, solidified dykes, sills, laccoliths and lavas intruding and cross-cutting the diatreme fills (e.g. Fig. 16.23d). These are likely to have highly irregular margins given that both the diatreme pyroclastic and sedimentary fill would have been contemporaneous with the feeder magmatic intrusions.

16.3.7 Resource Potential and Economic Significance

Diatremes can be very important economically in a number of volcanic systems, particularly, in hosting both epithermal precious metal mineralisation in evolved felsic magma systems (e.g. Sillitoe et al. 1984; Davies et al. 2008; Ross et al. 2017), and diamonds in ultramafic kimberlites and lamproites (e.g. Kjarsgaard 2007), which are described in detail in Chap. 18.

16.4 Pseudocraters and Littoral Cones

Such volcanic features are small, secondary cones formed above "*rootless vents*" where lava flows into a body of water or over swampy ground, and water trapped under or in a lava channel or tube is superheated and triggers explosive activity (Fig. 16.24a–d; also see Figs. 5.28c, d and 9.11b), well away from the primary magmatic conduit.

16.4.1 Pseudocraters

Pseudocraters or *rootless cones* described from Iceland form where basalt lavas flow into lakes or swamps (Fagents and Thordarson 2007; Hamilton et al. 2010; Fig. 16.24a, b). Phreatic or phreatomagmatic (Chap. 4) explosions are then caused by the interaction of external water with magma which erupt to build small volcaniclastic cones. These may form large fields of scattered small cones on flat pahoehoe lava fields. Cone deposits may vary from a diffusely stratified association of spatter agglutinate to scoria and ash beds and there is the ubiquitous lack of xenoliths, especially ballistics, or deep-seated mantle or crustal ones, with the exception of soft sediment or soil that may be derived from the underlying wet environment. Some rare cross-sectional exposures of the interior of lava fields cut through by rootless vents show very irregular pattern of columnar jointing and upper transition to fragmental facies. Similar rootless cone fields have been identified on Mars (Fagents and Thordarson 2007).

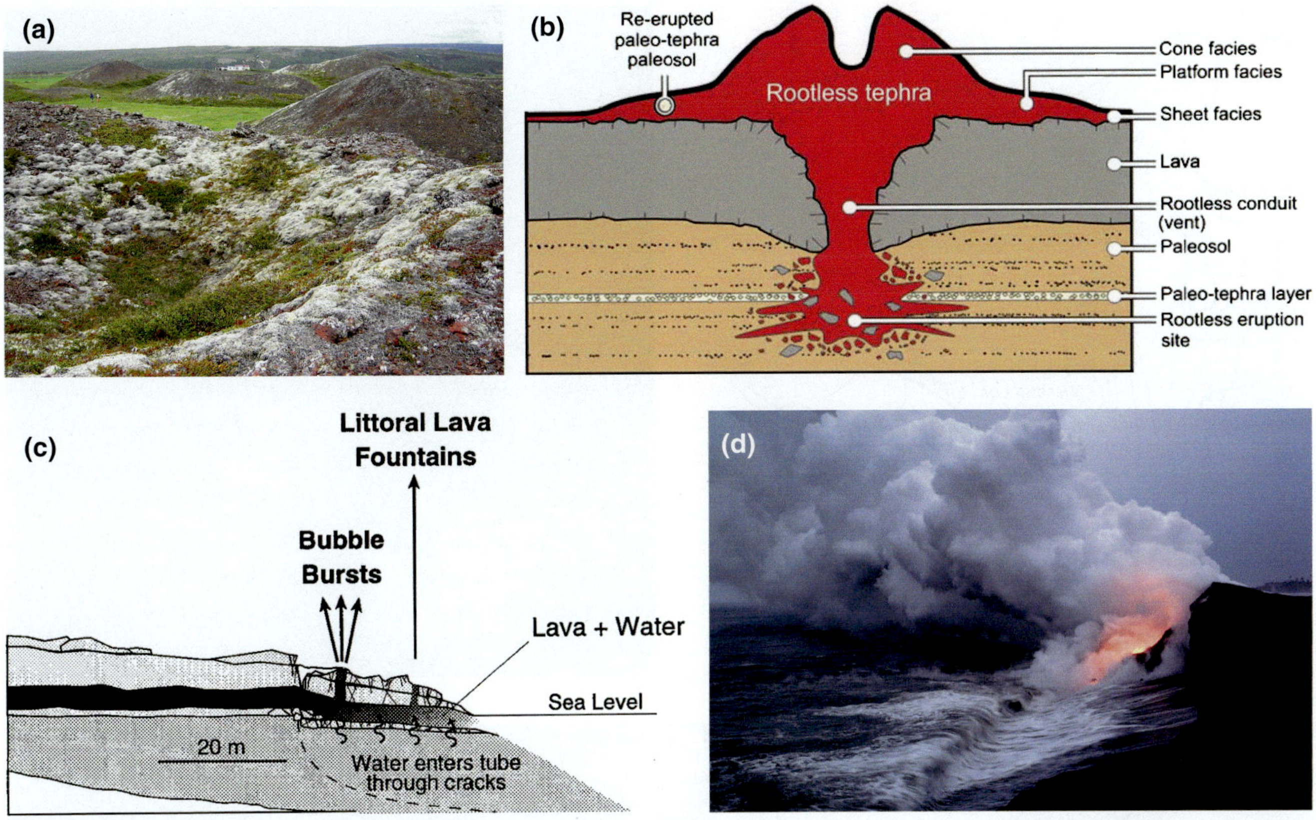

Fig. 16.24 Pseudocraters and littoral cones. **a** Cluster of pseudocraters or rootless cones, punctuating the top of one of the large-volume basaltic flows within the active Icelandic rift zones. **b** Model of pseudocrater facies. These typically include three morphological subdivisions: distal sheet-like fall deposits, an intermediate tephra platform, and a proximal cone-shaped accumulation of tephra and agglutinated lava spatter with a central crater above a rootless vent. (Based on the 1783–84 Laki lava flow; Modified from Hamilton et al. 2010). **c** Model for littoral explosions following partial collapse of lava bench, which submerges and fractures a portion of an active lava tube. Explosions rupture the roof of the tube producing circular mounds of welded spatter. (Based on observations of lava flows entering the ocean from 1992–94, Kilauea volcano, Hawai'i; From Mattox and Mangan 1997). **d** Formation of littoral cone which sits on top of the lava cliff (to the right), east of Kupapa 'u Point, Kilauea volcano, Hawai'i (Photo by T. Takahashi; *Source* United Sates Geological Survey)

16.4.2 Littoral or Rootless Cones

Littoral cones form in a similar way when basaltic lavas enter the sea and are best described from Hawai'i (Fisher 1968; Jurado-Chichay et al. 1996; Mattox and Mangan 1997; Fig. 16.24c, d). The 1868 Mauna Loa eruption formed the Pu'u Ho'u littoral cones in a five-day period when two lava channels from the flow entered the sea (Fisher 1968). A cone complex consisting of overlapping half-cones was constructed on land. They are built of poorly bedded hyaloclastite fragments and layers of agglutinated spatter which show crude welding profiles and form clastogenic lavas. Older littoral cones along this stretch of coastline at Pu'u Ki show identical features (Fig. 5.28d; Cas and Wright 1987). In more recent eruptions of Kilauea volcano's East Rift Zone, littoral cones have been observed forming on several occasions when lavas reached the sea (Fig. 16.24c, d; see below). These types of volcanic cone are important paleoenvironmental indicators, and it is important not to misinterpret examples as primary centres or vice versa (e.g. Walker 1992). The distinctive association of facies to look for is the combination of: (1) pillow lavas, (2) palagonitised hyaloclastite breccias which may show crude foreset bedding (flow-foot breccias) and define a lava delta (Chap. 5), and (3) shallow marine and lacustrine sediments (and marine or freshwater fossils to distinguish between marine and lacustrine shorelines).

16.5 Basaltic Shield Volcanoes

In plan view, basaltic shield volcanoes are symmetrical and circular to elliptical in shape. In profile, they are convex-up piles of basaltic lava flows, with gentle slopes of <10°. They are built up by frequent single eruption events or multiple lava-forming pulses during more prolonged eruptions of fluidal basaltic lava from central vents or flank eruptions, or both. The island of Hawai'i hosts several coalescing major

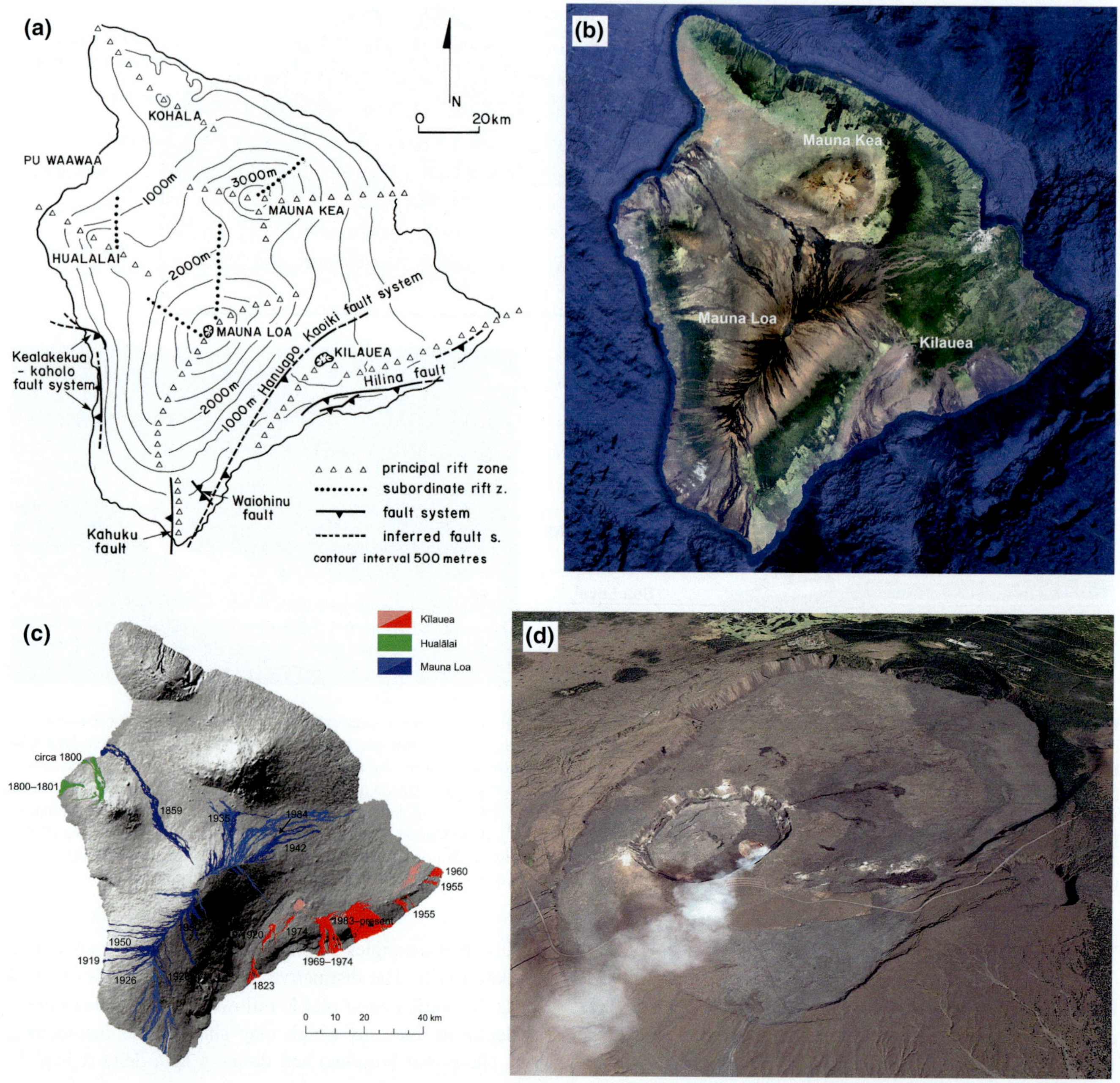

Fig. 16.25 Volcanic architecture of Hawai'i. **a** Map showing the location of the main shield volcanoes, the associated rift zones and faults (From Cas and Wright 1987). **b** Google Earth image of Hawai'i, highlighting Mauna Loa, Kilauea, and their lava flows and fields, and Mauna Kea shield volcanoes (*Source* Google Earth). **c** Map of historic basaltic lava flows on Hawai'i, colour-coded according to their source volcano (From Cashman and Mangan 2014). **d** Oblique Google Earth image of Kilauea shield caldera, about 5 km long, showing Halemaumau crater, with an active lava lake, in the foreground; lava flows can be seen in the floors of both the caldera and crater (*Source* Google Earth)

shield volcanoes in their own right: Mauna Loa, Mauna Kea, Kohala, Hualalai and Kilauea (Fig. 16.25a–c), not to mention numerous other smaller ones (e.g. Mauna Ulu). The whole island is clearly a major polygenetic complex. The smallest shields have basal diameters (W_s) of a few km (e.g. Mauna Ulu), whereas for the largest shields found in the Hawai'ian Islands, W_s is over 100 km (e.g. Mauna Loa and Mauna Kea; Figs. 16.25 and 16.26; Pike and Clow 1981). Shield heights (H_s) are on average = 0.05 W_s. The flanks of basaltic shields are almost entirely composed of overlapping lava flows (Fig. 16.25). Both pahoehoe and a'a lava types occur; most lavas are pahoehoe near the vent and may change to a'a with distance and along steep slopes (Chap. 5).

Other than basalt lavas in the pile, there may be minor (<5%) pyroclastic deposits, including scoria fall deposits of limited dispersal (hawai'ian, strombolian), and deposits produced by phreatomagmatic and phreatic explosions. Oxidised soil horizons and epiclastic layers which may be of diverse origins (e.g. glacial as on Mauna Kea; Porter 1979) may also be interbedded. In detail, the lava pile will have a complex stratigraphy, as exemplified in Fig. 16.25b, c, by lavas spreading radially from vent as tongue-like lobes. The stratigraphy is certainly not of a simple layer-cake system because lavas are not widespread sheets; their flow paths are controlled by axes of low topography. Such low topography is not necessarily erosional valleys but could be lows between adjacent lava lobes. Irregular lobate geometry and stratigraphy may occur and deciphering the relative timing of lava units may be almost impossible in the subsurface stratigraphy where overlapping superposition relationships are not determinable.

In the literature, basaltic shield volcanoes are sometimes divided into different types, based principally on geographical location:

- Hawai'ian or large shields,
- Icelandic or small shields,
- Galapagos shields, and
- Canarian shields.

Although Hawai'i and Hawai'ian shields are commonly considered to be typical of oceanic shield volcanic system, in fact Hawai'i represents only the mafic end member of oceanic island shield volcanic systems. Hawai'ian oceanic shield volcanoes are mostly effusive. Other oceanic shield volcanoes can develop more evolved magma systems (phonolitic, trachytic) late in their evolution, and can become quite explosive (e.g. Tenerife and Gran Canaria in the Canary Islands; Cas et al. 2022; Schmincke and Sumita 2010; Troll and Carracedo, 2016; the western volcanic islands in the Galapagos Islands, Pacific Ocean, Harpp and Geist 2018); São Miguel and Faial islands in the Azores, Atlantic Ocean, Zanon 2015; and, Reunion Island in the Indian Ocean, Lénat et al. 2001).

16.5.1 Hawai'ian Shields

The Hawai'ian archipelago is formed by a chain of shield volcanoes that are interpreted as the result of overpass of the Pacific plate over a stationary mantle hot-spot, with younging direction of the line of islands opposite to the direction of plate movement (Fig. 16.26). The rapid passage of the Pacific Plate over the Hawai'ian hot spot ensures that the active life-span of Hawai'ian shield volcanic systems is relatively short, mostly <1–2 Myr, and all <6 Myr. Magma eruption rates are high ($\sim$90 $km^3 kyr^{-1}$; Dvorak and Dzurisin 1993; Robinson and Eakins, 2006). The Hawai'ian volcanoes are the largest polygenetic basaltic shields on Earth. They have summit calderas (Fig. 16.25d) and major rift zones (Figs. 16.25a and 16.27a) marked by spatter cones, spatter ramparts, collapse craters (pit craters), scoria cones and smaller superimposed (Icelandic-type) monogenetic shields, (e.g. Mauna Ulu, Pu'u O'o, Kupaianaha on Kilauea (Clague and Sherrod, 2014). A valuable summary of what is known about all facets of the volcanically active island of Hawai'i has been compiled by Poland et al. (2014).

The two most active shields on Hawai'i are Kilauea and Mauna Loa (Fig. 16.25); Kilauea has been erupting more or less continuously from a series of vents since January 1983 (Chap. 3), while Mauna Loa most recently erupted in November 2022, after a quiescence of 38 years (see Plate 3.1 a). Mauna Loa is the world's largest active volcano, rising nearly 9 km from the Pacific Ocean floor to its summit at 4169 m above sea level. The total volume of Mauna Loa is around 100,000 km^3 (Clague and Sherrod 2014 and references therein). The composite shield complex of Hawai'i has a basal diameter of $\sim$200 km on the seafloor at a depth of some 5 km. Together with an isostatically subsided root of 8 km, the volcanic pile is thus $\sim$17 km thick. At Kilauea, the long-term eruption rate is $\sim$0.1 $km^3\ yr^{-1}$ (Swanson 1972; Clague and Sherrod 2014) and its age is reported as 300,000 yr (Calvert and Lanphere 2006). Stratigraphic accumulation rates obtained in Hawai'ian shields from detailed chronostratigraphy range between 1 and 16 m kyr^{-1} (Clague and Sherrod 2014). Much of Mauna Loa and Kilauea is likely to consist of pillow lava and hyaloclastite formed by submarine extrusions as the volcanoes rose from the seafloor (Figs. 16.26 and 16.27; Chap. 5), initially in the same way as seamounts grow (Sect. 16.8). These shields therefore closely interface with the marine environment.

The progressive availability of drill-hole data from the oldest and deepest parts of Hawai'ian shields has allowed different constructional phases to be distinguished, as summarised by Clague and Sherrod (2014). The pre-shield phase is entirely submarine and consists of tholeiitic, transitional, alkalic, and strongly alkalic lavas. The transition to the main shield phase at all hawai'ian volcanoes is marked by the onset of voluminous eruptions of tholeiitic basalt interpreted as the establishment of a mature plumbing system connected to the underlying mantle hot spot. Although rare, explosive and effusive eruption of highly differentiated trachytic magma has occurred during the shield phase. Restoration of dominant alkali basalt to hawai'ite and trachyte eruptions marks the transition to the post-shield phase. This phase is related to the displacement of the volcano from the axis of its deep hot-spot source as the Pacific plate overrides it.

Fig. 16.26 Digital elevation model of the Hawai'ian archipelago. Recent lava flows on the southernmost island of Hawai'i are indicated in red and issue from rifts, flanks and summit; K = Kilauea, ML = Mauna Loa, MK = Mauna Kea. Māhukona is a submerged shield volcano on the northwestern flank of Hawai'i. The summit caldera of this shield volcano subsided below sea level ca. 435–365 kyr ago making it the oldest volcano in the building of Hawai'i. The youngest volcano is the active Loihi Seamount on the submerged southeast flank of Mauna Loa. The green arrow shows the approximate direction of the Pacific plate motion, which is opposite to the younging ages of the volcanic island chain, as a result of the persistent hot-spot position with respect to the overriding plate. (Modified from Clague and Sherrod 2014)

Important structural features of the shields are summit calderas and radial or "triple star" rifts (Fig. 16.25a), which can also be traced offshore. Eruptions within the calderas, such as the recent eruption in Halema'uma'u crater in Kilauea caldera (e.g. Fig. 16.25d), occur slightly more frequently than on the rifts, but it is the eruptions from the lateral rifts (Figs. 16.25a, c and 16.27a) that give these shields their elongate form. Since 1983, eruptions have occurred almost continuously from flank vents of Kilauea, such as Pu'u O'o and Kupaianaha, along the East Rift Zone, lateral to Kilauea caldera (Fig. 16.27a). Mauna Loa and Kilauea have the largest calderas on Hawai'i, both about 5 km in diameter (e.g. Fig. 16.25d), but older shields in the Hawai'ian island chain have calderas of 10–20 km in diameter. Hawai'ian calderas are collapse structures, related to the lateral withdrawal of magma from under the summit magma reservoir and conduit system, which is then erupted from lateral rift zone vents, or intruded at depth during the shield-forming stage (Michon et al. 2011; Figs. 16.27a–c and 16.28). Such lateral magma withdrawal facilitates collapse or subsidence of the roof of a central magma reservoir (Fig. 16.25c). This occurs when the magma pressure in the deep reservoirs becomes broadly equal to or less then the lithostatic load reached at the same level, which increases over time due to the growth of the volcano. Rather than erupt effusively from the summit of the volcano magma therefore preferentially stagnates in shallow reservoirs. Under these pressure conditions, lateral withdrawal of magma, either effusively or intrusively, is more likely and the roof of these shallow reservoirs may collapse (Fig. 16.28). These calderas are not related to major explosive eruptions.

An essential feature of Hawai'ian calderas is the mismatch between the volume of effusion and volume of the collapse, which clearly relates to a significant intrusive volume that is directed towards rift zones, as well as "open-system" magma recharge from the mantle. Vents for eruptions within the calderas are generally fissures that cut across their floors. The Halema'uma'u crater within Kilauea caldera (Fig. 16.25d) is somewhat of an exception. It may have a pipe-like vent, although eruptions still tend to occur through fissures at the bottom of the crater, and it has been the site of a lava lake many times, including the time of

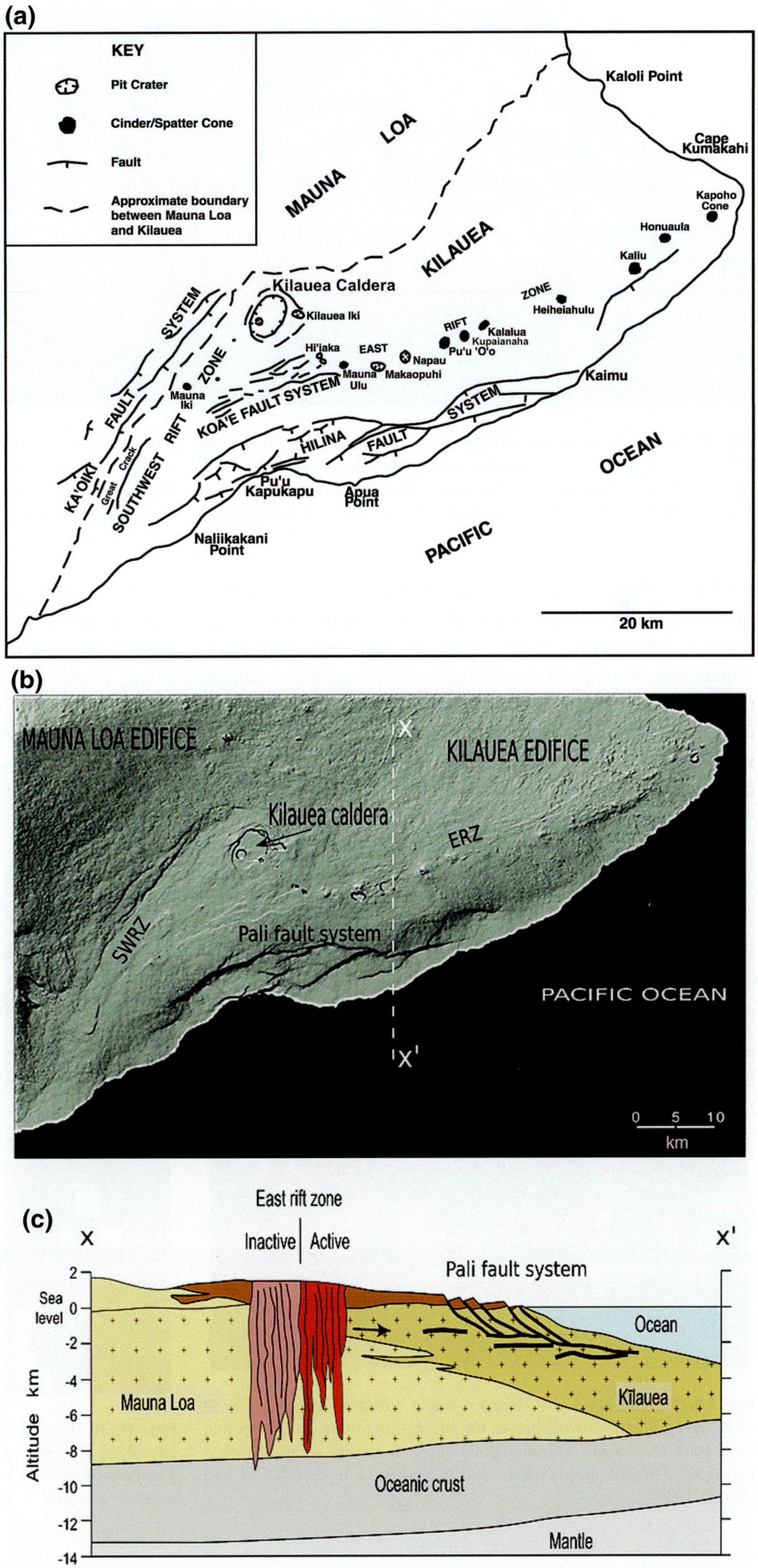

Fig. 16.27 **a** Map showing the major faults, the structural architecture of the East Rift Zone, the location of Kilauea caldera and of other eruption points along the rift (From Parfitt and Peacock 2001). **b** Digital elevation map of the main structural features of Kilauea volcano. SWRZ = Southwestern Rift Zone; ERZ = East Rift Zone; X and X' show line of cross-sectional illustrated below. (Modified from Poland et al. 2014). **c** North–south cross-section showing the interfingering of the Mauna Loa and Kilauea edifices. The lower areas (overprinted by crosses) indicate the dominant presence of pillow lava and hyaloclastite formed during the subaqueous stage of edifice building. The low mechanical strength of palagonitised basalts and interbedded abyssal sediments is mainly responsible for the nucleation and detachment of gravitationally-driven faults (Pali fault system), which accommodate the lateral spreading of the edifice under its load. (From Clague and Sherrod 2014)

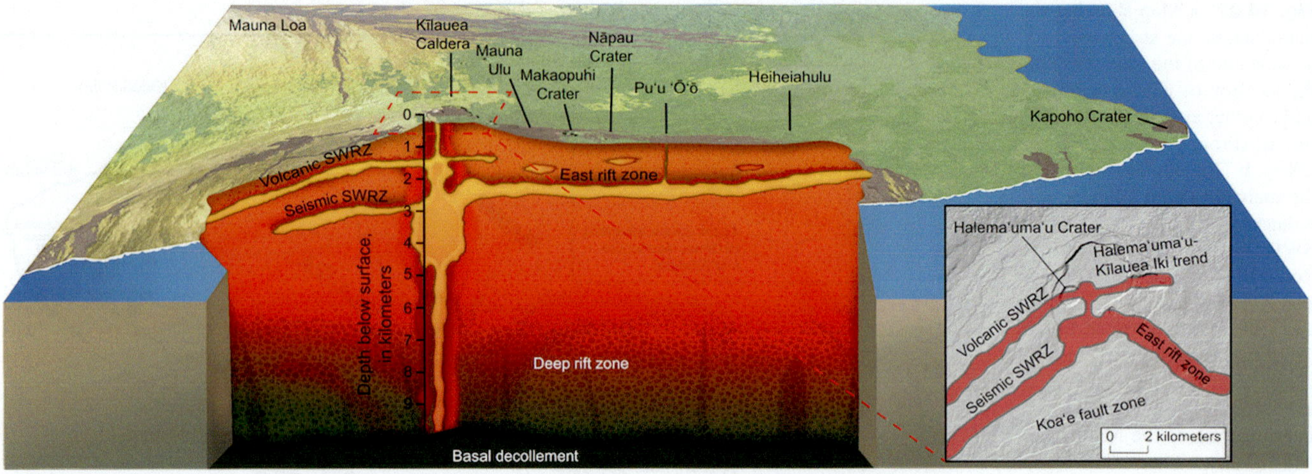

Fig. 16.28 Schematic representation of the geometry of the Hawai'ian plumbing system. This offers insights on possible mechanisms of effusive caldera formation (the right inset highlights the Kilauea summit caldera). These include magma intrusion and apical extension of the edifice, as well as the lateral withdrawal of magma from shallow reservoirs, either by lateral intrusions and/or by lateral eruptions. This leads to decompression within the reservoir inducing gravitational instability of the roof and consequential collapse. (From Poland et al 2014)

writing. As well as subaerial pahoehoe and a'a lavas, marine pillow and sheet lavas and hyaloclastite, and minor pyroclastics, the shields are also built by high-level intrusions which may be present in the walls of the summit calderas (Fig. 16.29a, b). Radial rifts usually extend from summit calderas (Fig. 16.25a–c). Their origins and roles are debated in terms of radial spreading of the volcano, potentially contributing to lateral flank or sector collapse, and lateral injection of magma as a result of the laterally unconfined volcano growth (Fig. 16.27c; e.g. Poland et al. 2014), or forceful intrusion of magma from deep sources (Swanson et al. 1976). Scoria cones are concentrated along the rift zones and reach a density of 8 per km^2 (Fig. 16.30).

Gravity sliding and slumping of shield flanks into the ocean basin are also major structural features related to the gravitational instability induced by the rising height and

Fig. 16.29 High-level sills, Kilauea shield volcano, Hawai'i. **a** Wall of Kilauea caldera showing part of the succession of thin lava flows that have built the shield. Irregular body (bottom centre) is an intrusion called the Uwekahuna laccolith, which is seen in **b**. (From Cas and Wright 1987)

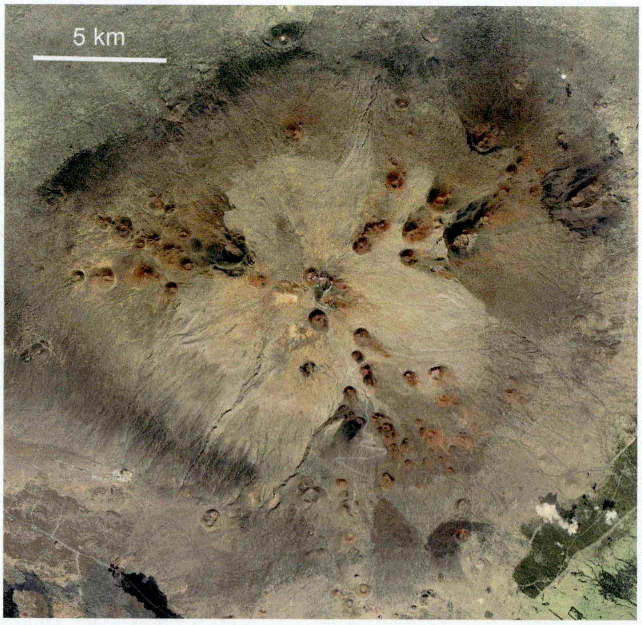

Fig. 16.30 Google Earth image showing distribution and density of late stage alkaline scoria cones on the upper slopes of Mauna Kea. Their distribution clearly defines the principal rift zones on Mauna Kea; also see Cas and Wright (1987). (*Source* Google Earth)

increasing load of Hawai'ian volcanoes (Duffield et al. 1982; Moore et al. 1989, 1994). These gravity collapse features are among the largest known globally; some individual debris avalanches are more than 200 km in length and 5000 km^3 in volume, forming a vast basaltic volcaniclastic apron (some 10^5 km^2) around the archipelago. The southern flank of Kilauea, for example, is constantly being displaced towards the sea along normal faults ("*pali*") as a result of load caused by addition of lava flows and the intrusion of magma into the summit and rift zones (Fig. 16.27). Seismic imaging of the volcano interior shows that such normal faults become low-angle at depth (Fig. 16.27), possibly detaching on the low strength pre-volcanic abyssal sediments (e.g. Fig. 12.11 a, b; Denlinger et al. 2014) or within the palagonitised pillow lava succession (Swanson et al. 1976; Lipman et al. 1985).

16.5.2 Icelandic Shields

Iceland formed astride the Atlantic spreading ridge where the rate of extension and magma production are so high that large volumes of basaltic magmas are emplaced during active rift formation (Fig. 16.31). These characteristically erupt as large (>1 km^3), high discharge (>1000 m^3s^{-1}) fissure-fed flood basalt lavas exemplified by the historic eruptions of Laki in 1783–84 and Eldgjá 939–40 AD (Chap. 5), and the prehistoric Thjorsa lava flow (see Fig. 16.38). At lower discharge rates (≤ 300 m^3s^{-1}) these eruptions produce lava shields surrounded by pahoehoe flow fields fed from lava lakes residing in summit craters (Thordarson and Larsen 2007). The main lava shields are Holocene in age and, except for the lava shield on top of Surtsey in the Vestmannaeyjar Islands, all are located in the North and West Volcanic Zones, their highest density being on the Reykjanes Peninsula (Fig. 16.31a; Andrew and Gudmundsson 2007).

Icelandic shields are smaller (W_s < 12 km) than Hawai'ian shields, almost symmetrical and entirely, or nearly entirely, built up by effusive eruptions from a central summit vent (Rossi 1996). The summit craters of Icelandic shields are usually <1 km across, and often have raised rims of spatter. There are generally few radial fissures, or lines of parasitic cones, although pseudocraters may be common, and as indicated above they are composed of large numbers of thin pahoehoe flows. Two of the best known are Theistareykjabunga in the North Volcanic Zone and Skjaldbreidur in the West Volcanic Zone (Fig. 16.31a). Theistareykjabunga has a W_s of 11.2 km, H_s of 140 m and a slightly convex shapes with mean slopes of 1.6°; Skjaldbreidur has a W_s of 8.7 km, H_s of 520 m and an overall concave-convex shape with steeper ($\sim 10°$) slopes (Fig. 16.32a, b).

Most Icelandic shield volcanoes are probably monogenetic, and even the larger ones such as Skjaldbreidur may have been constructed in less than 10 yr. Although no historic eruptions producing Icelandic shield volcanoes are known, the eruptive activity of Surtsey in the period 1963–7 is perhaps an example of the duration of the eruptive event required to produce Icelandic shield volcanoes. The Surtsey eruption involved the eruption of >2 km^3 of basalt over the 4 years of activity and produced a lava shield during its effusive phase. Iceland also has a number of larger modern and potentially active polygenic central volcanoes (e.g. Hekla, Askja, Krafla, and Katla). Their form is irregular, variable and has been influenced by ice-volcano interactions to varying degrees. Only rarely do these conform morphologically to hawai'ian shields, while several have steeper slopes matching the profiles of stratovolcanoes (Thordarson and Larsen 2007). However, older Icelandic central vent volcanoes of Neogene age, bear stronger relationships to Hawai'ian shields in terms of their morphometry and lithofacies associations.

Climatic conditions also play a part in determining the form of Icelandic volcanoes. When eruptions occur under significant ice caps, or during glacial stages, widespread ice sheets, ice-constrained, steep-sided subglacial table or tuya volcanoes (Sect. 16.9; Figs. 16.86b, 16.87 and 16.91a) form, whereas where and when (inter-glacial stage) there is no significant ice cover, low-profile lava shield volcanoes form.

16.5.3 Galapagos Shields

The Galapagos Islands of Isabela and Fernandina comprise several major polygenetic shield volcanoes (Fig. 16.33), thought to coincide with the current location of the Galapagos hot spot. These shields are very regularly spaced, occurring about 35–40 km apart at the intersections of orthogonal tectonic lineaments. W_s at sea level varies from 45 to 80 km, and H_s is between 3,000 and 3,500 m above sea level (Pike and Clow 1981). These shields have gentle lower slopes which rise to steeper central slopes that flatten off around spectacular summit calderas (Munro and Rowland 1996; Amelung et al. 2000). The summit calderas are between 3.0 and 8.7 km in diameter (the largest being Sierra Negra on Isabela) and are surrounded by well-defined concentric fissures which are marked by small scoria and spatter cones (Chadwick and Dieterich 1995). Lavas have been erupted from these, some pouring back into calderas, and lavas have also issued from radial fissures on the flanks of the shields (Michon et al. 2011; Harpp and Geist 2018). Alkaline basalt and evolved magmas (such as rhyolite) are more important, especially in the eastern volcanic islands (Harpp and Geist 2018), than in the Hawai'ian shields. It has been suggested the high-level intrusion of dykes and sills may account for their characteristic shape.

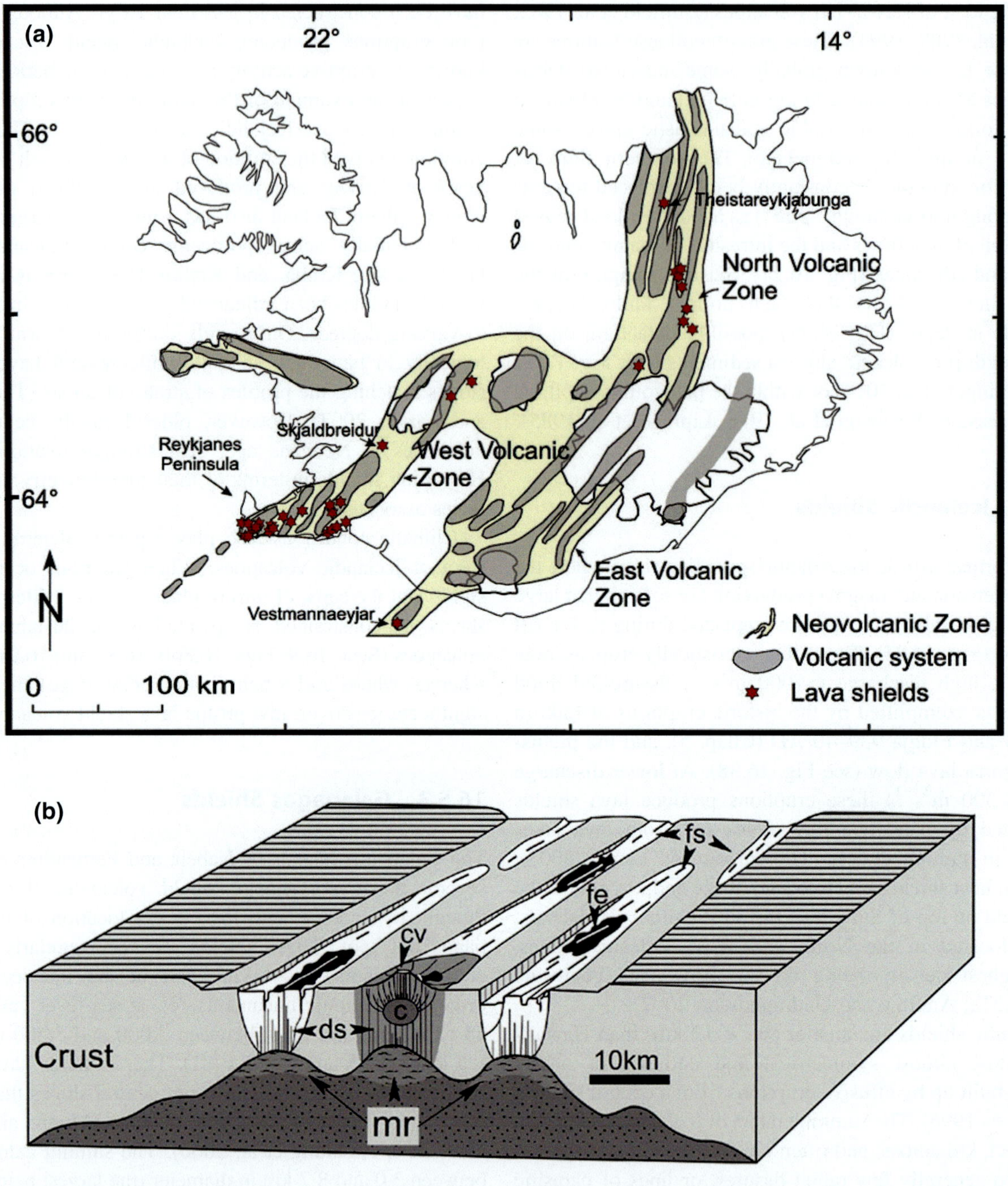

Fig. 16.31 Setting of Icelandic shield volcanoes. **a** Distribution of active Icelandic rift systems and location of Holocene lava shields (From Andrew and Gundmundsson 2007). **b** Key elements of rift systems. This shows relative location of magma reservoirs (mr), crustal magma chambers (c), dyke swarms (ds), central polygenetic volcanos (cv), fissure swarms (fs) and fissure eruptions (fe). At high magma discharge fissure eruptions produce flood basalt lavas, but at low discharge rates, monogenetic shield volcanoes. (From Thordarson and Larsen 2007)

16.5.4 Canarian Shields

A hot spot origin for the Canary Islands, located off northwest Africa, has been proposed and this different style of plume behaviour is thought to have been caused by edge-driven convection through the interaction of the Canaries plume with the continental lithosphere of Africa (Carracedo and Perez-Torrado 2013). Taylor et al. (2020) suggest that the plume was inclined with its head lying under the youngest western islands, and with an outflow tail continuing eastwards towards Africa, so accounting for ongoing volcanism in the oldest part of the island chain (Cas et al. 2022). Importantly, sustained plume magmatism over >10 Myr has allowed evolved felsic magma reservoirs to form

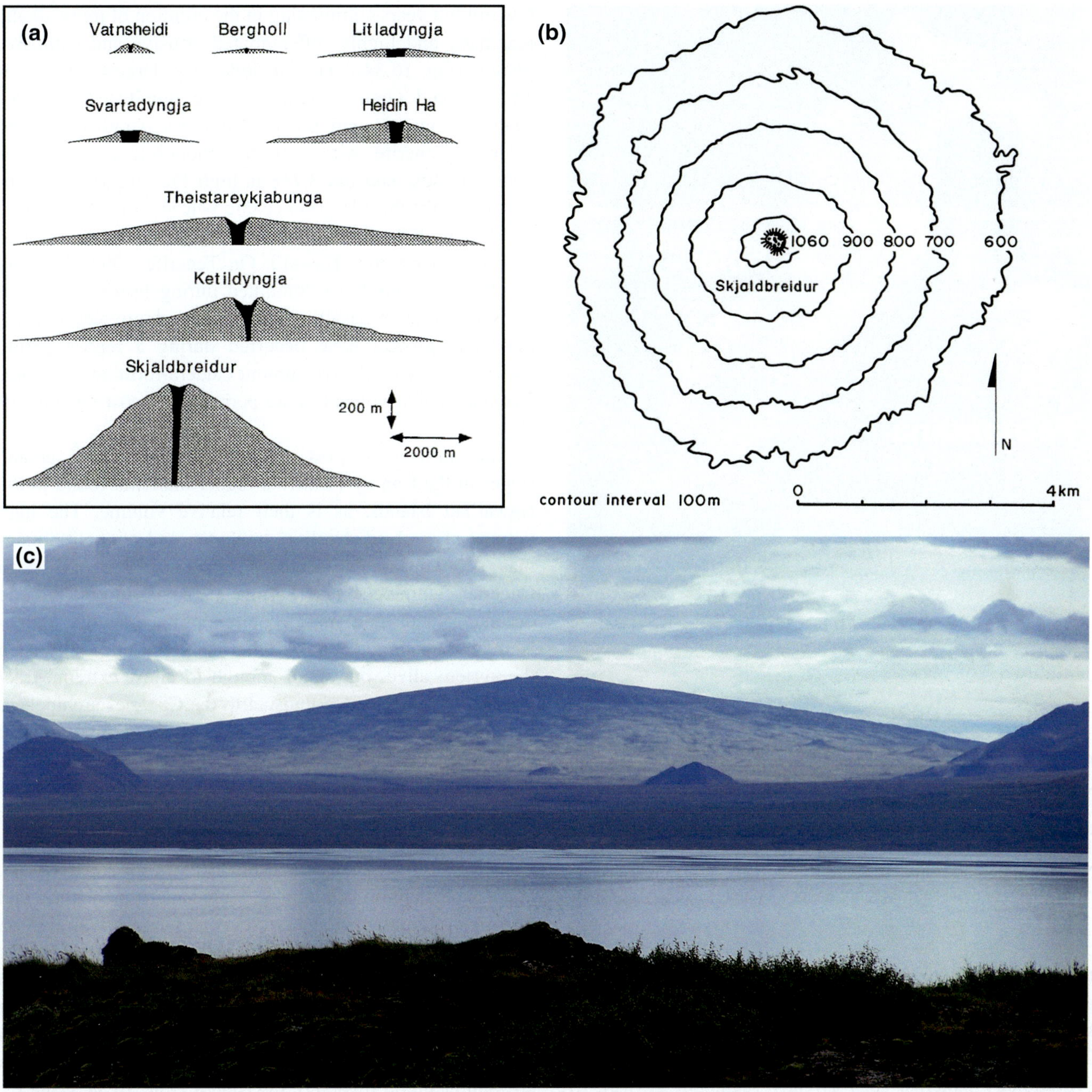

Fig. 16.32 Morphology of Icelandic shield volcanoes. **a** Measured profiles of selected Holocene shield volcanoes, Iceland, showing different sizes and slope forms (From Rossi 1996). **b** The nearly perfectly symmetrical shield of Skjaldbreidur, southwestern Iceland, depicted by evenly spaced topographic contours, and formed by lavas erupted from a central summit vent (From Cas and Wright 1987). **c** Photograph of Skjaldbreidur shield volcano; note the scoria cones in the foreground and on the slopes of the volcano (Photo by Thor Thordarson)

under Tenerife and Gran Canaria, leading to both explosive calderas and effusive-explosive stratovolcanoes to form on both (Schmincke and Sumita 2010; Troll and Carracedo 2016). However, Canarian shields have low eruption rates (<1 km^3 kyr^{-1} for Tenerife; Ancochea et al. 1990) compared with Hawai'ian shields.

It is now well established that the main stage in island growth involved the construction of one or more large basaltic shield volcanoes (e.g. Schmincke and Sumita 1998; Carracedo et al. 1998; Carracedo 1999; Carracedo and Perez-Torrado 2013; Fig. 16.34). These may make up 90% or more by volume of any given island, although most of

Fig. 16.33 The islands of Isabela and Fernandina (furthest west) in the Galapagos Archipelago, Ecuador. The large shield volcanoes appear regularly spaced and occur at the intersections of NNW-SSW with SW-NE structures, probably representing deep-seated fracture systems. All shield volcanoes have a summit caldera, the largest being Sierra Negra, forming the central southern part of Isabella. (Landsat 7 shaded relief colour composite image; *Source* NASA and University of Miami)

their evolution took place subaqueously. Still, these shields appear smaller in volume than those of Hawai'i, with W_s at sea level perhaps between 20 and <50 km and H_s 1000–2500 m. While they show common constructional and evolutionary patterns, such as early seamount growth, rift zones, progressive volcano instability and multiple gravitational collapses, there are significant differences between Canarian shield volcano systems and the other types discussed above.

Notably, the Canarian shields are dominated by alkali basalts which were probably erupted at lower eruptive rates and involved cooler magmas. Construction of several shields along the length of the island chain (490 km-long) occurred from about 22 to <2 Ma. The youngest and most westerly islands of La Palma and El Hierro, whose last eruptions occurred in 2021 and 2011 respectively, are still in their shield phase, whereas the other islands have evolved, with this shield phase followed by erosional intervals lasting a few million years culminating in the eruption of more alkalic magmatic and highly differentiated effusive and explosive phases (Fig. 16.34). This includes the largest islands of Tenerife and Gran Canaria that are dominated by large central phonolitic and trachytic edifices (stratocones) and calderas. Tenerife, with a summit phonolitic explosive caldera complex, and the 3,718 m-high Pico de Teide stratovolcano is the third largest oceanic volcano after Mauna Loa and Mauna Kea, and the second largest, intraplate, oceanic volcanic island after Hawai'i. On Tenerife, shield building basaltic eruptions have continued during historic times, at the same time as basaltic to phonolitic stratovolcano eruptions and growth have occurred during a repose period between major explosive summit caldera-forming eruptions. Canarian shield volcanoes are perhaps the most complex of all.

The number of post-erosional or rejuvenated alkalic stages in the Canary Islands is higher than that found in the Hawai'ian Islands, as is their relative volume. The high volume of evolved rhyolitic, trachytic, and phonolitic magmas on the largest islands (e.g. >1,000 km^3 on Gran Canaria; Schmincke and Sumita 1998) is unusual among oceanic islands and must require optimal magma replenishment rates large enough to thermally and materially keep the system alive, but small enough to allow extreme fractionation to occur. Also unusual to the Canaries archipelago if it is sourced by hot spot activity, is the ongoing volcanic activity on the oldest islands in the east contemporaneous with eruption activity on the youngest islands in the west. Subaerial activity on Fuerteventura and Lanzarote islands has occurred intermittently for about 20 Ma, on Gran Canaria for about 15 Ma, and on Tenerife for some 12 Ma. With the exception of Gomera, Holocene to prehistoric and historic eruptions have occurred on all islands, with that on Lanzarote in 1730–1736 (3–5 km^3) being one of largest basaltic eruptions (After Eldgjá and Laki in Iceland; Chap. 5). The most recent basaltic eruption in the west of the Canary Islands on La Palma island (see Fig. 16.110a, b), lasted from 19 September–13 December 2021. Sector collapse features are significant on several of the islands (Fig. 16.34, stage E), and have generated vast submarine volcaniclastic aprons and turbidite fans (see Fig. 16.46; Hunt et al. 2014).

Tenerife is one of the most complex volcanic systems on Earth (Cas et al. 2022; Fig. 16.34a–d; Sect. 16.8.1). Basaltic shield volcanism began >12 Ma, and basaltic volcanism still occurs from 3 diffuse "rift" zones, with the most recent eruption being in 1909, when a scoria cone and lava flow formed. At ~3.5 Ma phonolitic volcanism began mostly from the summit region, initiating the formation of the Las Cañadas caldera and stratovolcano complex (Plate 16.1, Fig. 16.34e). Over the last 1.8 Ma it has experienced at least three, major explosive caldera-forming cycles, each lasting 200 to 300 kyr and separated by similar periods of

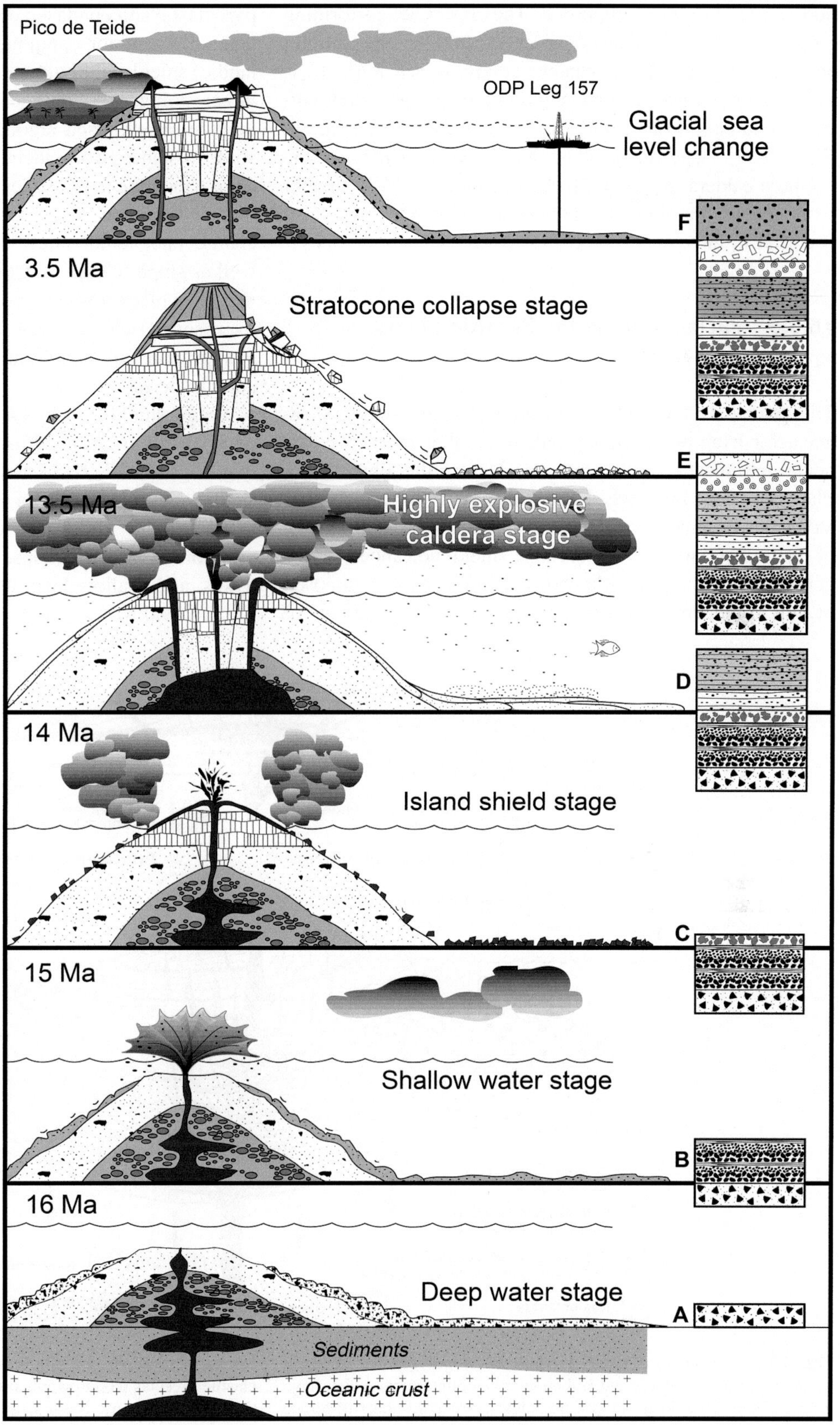

Fig. 16.34 Schematic evolutionary stages illustrating the submarine growth and emergence of Canarian basaltic shield volcanoes, and their subsequent caldera and stratocone development. (Based on studies of Gran Canaria; From Schmincke and Sumita 1998). Right of figure shows main types of volcaniclastic apron rocks drilled during ODP Leg 157. Volcaniclastic sediments produced during submarine stage A are chiefly hyaloclastite debris flow deposits, dominated by increasingly vesicular sideromelane shards during stage B as the edifice grows up into shallower water. Shoreline derived lava delta breccias and offshore epiclastic turbidites are common in the sediments of emergent stage C. Epiclastic rhyolitic, trachytic and offshore phonolitic turbidites are produced during ash flow eruptions and erosional intervals during stage D, with widespread debris avalanche deposits resulting from stratocone collapse in stage E. Glacially driven sea-level changes result in episodic beach sand transport to the depositional basin during Quaternary stage F, which is also characterised by abundant distal ash layers erupted from Tenerife

non-caldera-forming eruptions. The last caldera-forming eruption was ~190 ka (Edgar et al. 2007). The summit area is now a coalesced caldera complex (see Plate 16.1, Fig. 16.79e). Since 190 ka a basaltic-phonolitic stratovolcano complex, El Teide-Pico Viejo (Plate 16.1a, b; Fig. 16.34, stage F, 16.48f), has been growing on the northern caldera rim, with the last eruption in 1798. Smaller monogenetic basaltic and phonolitic eruptions have also occurred in the caldera, producing lavas and scoria cones.

16.5.5 Facies Models for Basaltic Shield Volcanoes

The principal elements of subaerial basaltic shield volcanoes, for example as in Iceland, or smaller ones in continental intraplate basaltic provinces such as the Snake River Plain Volcanic Province, USA (Greeley 1982) and Newer Volcanics Province, southeastern Australia (Cas et al. 2017), or the subaerial parts of oceanic shield volcanoes, are a gently dipping basaltic pahoehoe and a'a lava pile composed of thin to moderate thick lava flows, with lesser amounts of pyroclastic and epiclastic deposits (Fig. 16.35a). The latter is generally a minor component of smaller monogenetic (Icelandic) shields, although it could be somewhat greater within larger polygenetic edifices. Such larger shields may also include red oxidised regolith surfaces and soil horizons. At some distance from their eruption point, lavas transition from pahoehoe to a'a facies, but in any given individual case this depends on many physical factors (e.g. magma discharge rate, temperature, vesicularity and composition) as well as slope (Chap. 5). Coupled with the fact that eruptions may take place from many vents (e.g. along the same fissure) over short time intervals, this can give rise to complex stratigraphies of alternating and interdigitating coherent and autoclastic facies.

Large shield volcanoes also often develop summit calderas resulting from effusive eruptions (effusive calderas) (e.g. Figs. 16.27, 16.33 and 16.35a). Collapse of the roof of shallow magma reservoirs results from the lateral drainage of magma along dyke conduits that feed flank vents and eruptions. Caldera collapse exposes the proximal lava pile, dykes, and shallow level intrusions in the caldera wall (e.g. Fig. 16.29a, b). Larger shields may develop parasitic scoria

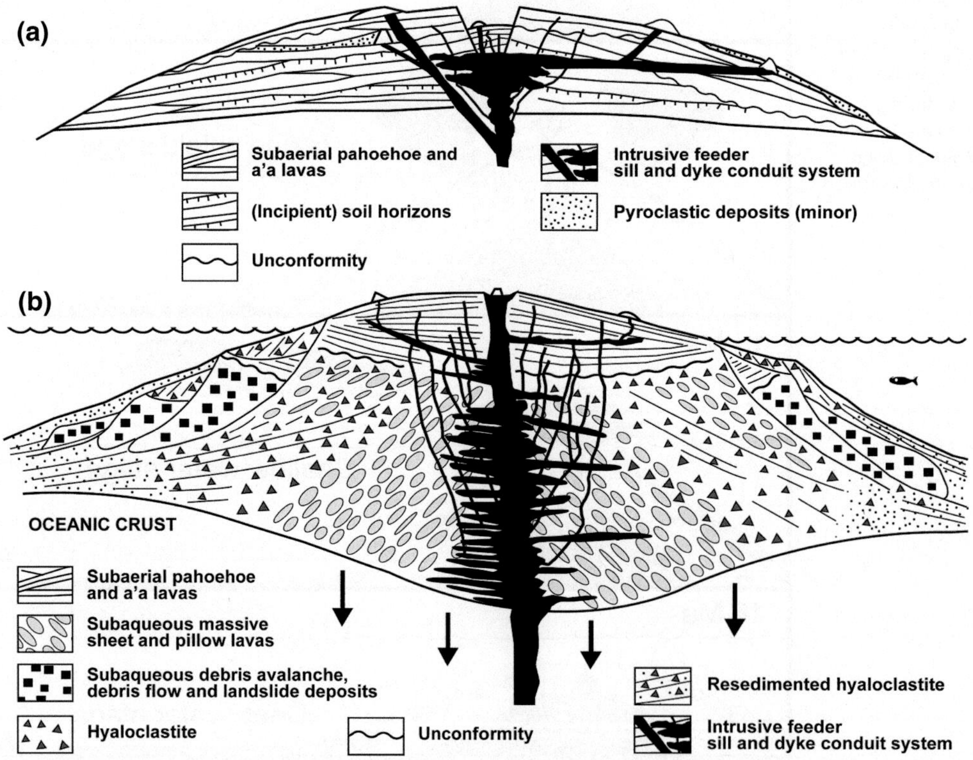

Fig. 16.35 Generalised facies models for **a** polygenetic subaerial basaltic shield volcanoes, and **b** their marine counterparts which have emerged above sea level. In (**a**) the shield consists of sheet to lobate pahoehoe and a'a lavas. Time breaks are represented by some erosional discordances and (incipient) soil formation. An intrusive feeder dyke and sill complex supplies magma to both the central vent system and flank eruptions that facilitate summit caldera collapse. In (**b**) a basal submarine complex of pillow and sheet lavas and associated hyaloclastite breccias pass upwards into a subaerial shield complex similar to that depicted in (**a**). The transition will be marked by an unconformity and passage beds feeding prograding lava and hyaloclastite deltas. The submarine flanks will be marked by sector collapse landslide, debris avalanche, debris flow and volcaniclastic turbidite aprons. The feeder intrusive sill and dyke complex may be very pervasive

cones, small shields and/or phreatomagmatic/phreatic craters oriented along major rift zone axes and fault lines.

In emergent marine shield volcanoes, distally, lava flows may enter the sea causing phreatic explosions and the formation of littoral cones (e.g. Fig. 5.28d), as well as feeding lava deltas comprising subaqueous pillow lava and hyaloclastite facies (e.g. Figs. 4.3 and 5.38). These are subsequently reworked into shoreline depositional systems characterised by "black" and "green" (olivine-rich) sands (e.g. Fig. 4.3a, b).

However, as we have described above, in many cases this only represents the exposed subaerial tops of major oceanic volcanoes. Similar to marine stratovolcanoes, marine oceanic island shield volcanoes (e.g. Hawai'ian, Galapagos, and Canarian) have foundations built in a submarine environment which initially develops in the same manner as seamounts (see below). Their lower submarine parts are therefore constructed of a thick succession of the pillow and massive sheet lavas and associated hyaloclastite breccias, built around a central intrusive core (Fig. 16.35b). Where these submarine edifices emerge above sea level, the basal marine part of the edifice transitions upwards into a transitional marine-subaerial succession and then into a subaerial volcanic succession, similar in facies and architecture to subaerial shield edifices (Fig. 16.35b). Even at this stage, however, marine facies may form where lavas flow into the sea from subaerial vents and where submarine eruptions occur from lateral parasitic submarine vents. These volcanoes are in most cases associated (although not exclusively) with intraplate hot spot or mid-oceanic rift and transform systems (Chap. 17). They are therefore built on oceanic crust, although in some cases (e.g. Canary Islands) this may be in part overlain by continental margin sediments.

Such settings are associated with significant amounts of epiclastic erosion and transport which during the submarine and emergent stages deposit flank facies consisting of redeposited basaltic volcaniclastic (mainly hyaloclastite) facies, which as the subaerial part of the shield grows may include large slides, debris flows, debris avalanches caused by sector collapse (Chap. 12; Fig. 16.35b). Magmas may evolve from tholeiitic basalts to more alkaline varieties, including phonolites and trachytes, which may lead to the development of central explosive calderas and/or explosive-effusive stratovolcanoes (e.g. Tenerife and Gran Canaria, Canary Islands archipelago). Such evolved central eruption centres could generate large volume and widespread pyroclastic deposits which can be dispersed offshore and also produce derivative epiclastic trachytic and phonolitic sediments, including far flowing turbidites that form offshore volcaniclastic aprons (see Fig. 16.46a, b). Together with widespread debris avalanche deposits resulting from stratovolcano and shield flank collapse these greatly expand the submarine volcanic apron. These are discussed further in the more general context of marine stratovolcanoes below (Sect. 16.6.4).

16.5.6 Resource Potential and Economic Significance

Although none of the shield volcanoes described appear to have yielded significant mineral discoveries, similar large oceanic mafic to alkaline volcanoes in the Pacific region do host some very large epithermal gold deposits. Most significant are the Emperor gold-telluride deposit located on Viti Levu (Fiji), and the supergiant Ladolam gold deposit on Lihir Island (Papua New Guinea). Perhaps a critical difference is that in both these cases alkaline volcanism appears to be subduction-related (Chaps. 17 and 18). Because of its special location Iceland has been a pioneer in geothermal energy and heating. There has also been some minor development of geothermal resources at Kilauea, while pre-feasibility studies have identified prospects on Isabela (Galapagos) and the Canary Islands.

16.6 Basaltic Volcanic Provinces and Their Source Vents: Flood Basalt Plateau Provinces and Plains Basalt Provinces

So far, we have examined mostly basaltic volcanic centres dominated by effusive, lava-forming eruptions and small explosive centres. In this section, we look at the nature of regionally extensive basaltic volcanic systems that may host multiple and even regionally extensive vent systems with high magma discharge rates. The two end member provinces we will consider are extensive flood basalt plateau provinces and plains basalt provinces, as discussed in Chap. 5. This is done to show that not all vents are necessary parts of central or point source volcanoes. In a strict sense, these vent systems are source vents to entire provinces (or mafic supervolcano systems) rather than single localised volcanoes, and because of their importance, we consider them here.

16.6.1 Continental Flood Basalt Plateau Provinces

Continental flood basalts are some of the largest single eruptive units known and their lavas are believed to have flowed great distances (hundreds of km) from the source vents (Hooper 2000; Self et al. 2015; Chap. 5). Repeated eruptions build up a flood basalt lava pile, forming vast lava plateaus many of which cover areas of $>10^6$ km^2

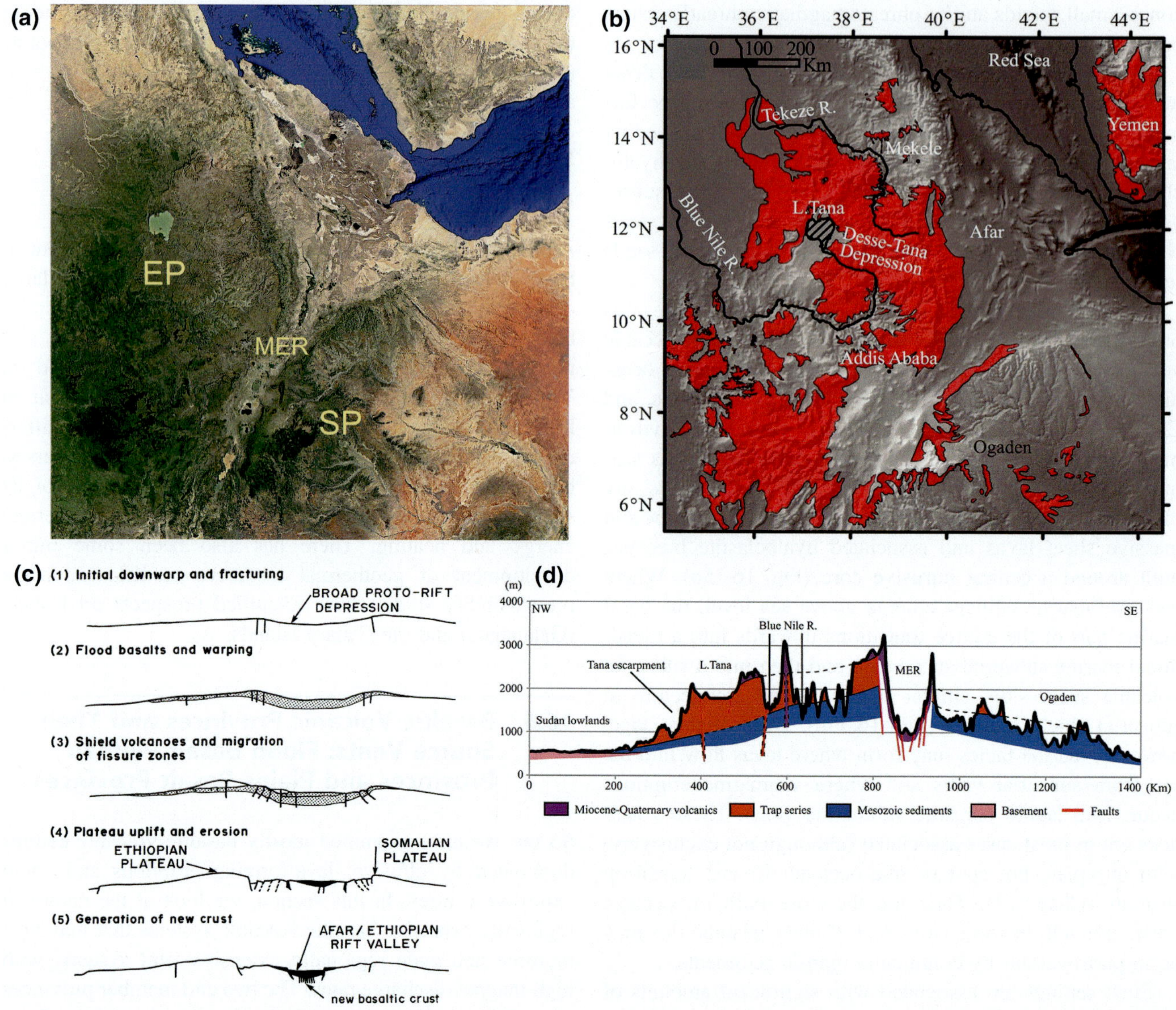

Fig. 16.36 Ethiopian Rift and Ethiopian-Yemen Flood Basalt Province. **a** Google Earth overview showing location; EP = Ethiopian Plateau, MER = Main Eithiopian Rift, SP = Somalian (Harar) Plateau (*Source* Google Earth). **b** Digital elevation model showing the preserved extent of the Miocene Ethiopian-Yemen Flood Basalt Province (From Sembroni et al. 2016). **c** Schematic evolution of the Ethiopian flood basalt plateau during the past 40 Ma and development of the Afar mantle plume(s). Length of sections is about 1000 km. Note, at time (2) extensive tholeiitic flood basalts were erupted in the northern sectors of the province at ca. 30 Ma with the development of the Afar continental margin and the initiation of rifting in the southern Red Sea and Gulf of Aden. In the southern sectors further into continental Africa coeval flood basalts were mildly alkaline. At time (3), immediately following peak activity, eruptions of tholeiitic and then alkaline basalts built shield volcanoes (30–10 Ma) over the present Ethiopian Plateau surface, which are discernable in (**a**). (From Cas and Wright 1987, After Mohr 1983). **d** Schematic NW–SE geological and topographic cross-section showing late stage updoming, possibly the result of long-term, deep mantle processes (From Sembroni et al. 2016)

(Hooper 2000) and generally have slopes <2–3°. These are typified by the Columbia River Plateau in Washington, Oregon and Idaho which is of mid-Miocene age. The Columbia River Plateau basalts cover an area of about 220,000 km^2 (Figs. 5.1, 5.14 and 5.36a–c) have an estimated volume of 195,000 km^3 and were erupted during a very short interval of only 2 or 3 Ma (Thordarson and Self 1998; Self et al. 2015 and references therein). Other young examples occur in Iceland, the Tertiary North Atlantic Igneous Province, and the mid-Cenozoic Ethiopian-Yemen Plateau (Fig. 16.36; Mohr and Zanettin 1988; Hooper 2000; Corti 2009; Sembroni et al. 2016). Older examples include the Cretaceous Deccan Traps in northwestern India, which cover an area of more than 500,000 km^2 (Figs. 5.1a, 5.36d and 5.37) and have a volume of more than 1 million km^3 (Hooper 2000; Jerram and Widdowson 2005; Self et al.

2008), the Cretaceous Parana Plateau of southern Brazil-Uruguay, the Jurassic Karoo basalts in South Africa, the Siberian Traps of the central sector of the Siberian platform which are early Mesozoic in age, and the Proterozoic Keweenawan district of mid-continental North America. Magma compositions range from predominantly tholeiitic to alkali basalts, but can also include more differentiated compositions such as basaltic andesites, and in some cases even high-temperature rhyolites (Self et al. 2015).

Many flood basalt plateau provinces are closely associated with mantle plume activity, and the initiation and early development of rifted continental margins (Courtillot et al. 1999; Geoffroy 2005; Self et al. 2015; Chap. 17). Some of the above examples became active during the formation of new continental margins born with the Mesozoic-Cenozoic fragmentation of Pangea. The Eocene-Miocene (40–20 Ma) flood basalts in Ethiopia were mainly erupted at about 30 Ma, in a short period of ~ 1 Ma duration (Fig. 16.36a, b; Hofmann et al. 1997; Corti 2009; Sembroni et al. 2016). These form a lava succession between 0.5 and 2 km thick that covered an area of some 1 million km^2, before plateau uplift and ensuing erosion occurred during Pleistocene time (Fig. 16.36c; Sembroni et al. 2016). The flood basalt eruptions occurred through fissures mostly controlled by Precambrian and reactivated Mesozoic zones of weakness (Chap. 17). After this peak of activity, a number of large, tholeiitic to mainly alkaline, shield volcanoes developed on the surface of the volcanic plateau (Fig. 16.36a; Kieffer et al. 2004), after which subsequent volcanism was largely confined to regions of rifting (Corti 2009). This culminated with opening along the Red Sea and Gulf of Aden and initiation of seafloor spreading.

The characteristics of flood basalt lavas and their eruptions are described in Chap. 5. Plateaus are built mostly of simple sheet pahoehoe lava flows which cover vast areas. They used to be thought to erupt at discharge rates as high as 10^6 m^3 s^{-1}, but more recently estimates are 4 or 5 $\times$ 10^3 m^3 s^{-1} (Self et al. 2015; Chap. 5). Although flood basalts were generally believed to be erupted from fissures, recognising their vent systems has been difficult, and Swanson et al. (1975) were probably the first to document vent systems for a large flood basalt plateau. Part of the problem is that these vents tend to be covered by younger flood basalts. Also, features that may mark the vents, such as spatter cones and ramparts, are ephemeral features which are easily buried and eroded. Swanson et al. (1975) documented vent systems for the Roza Member (Fig. 16.37; Chap. 5) and the Ice Harbour flows in the Columbia River Plateau. They are both associated with one linear dyke system at least 20 km wide and 450 km long. The vent systems are confined to linear zones within this dyke system, are tens of km long, a few km wide, and have the same trend. The vent and near-vent areas (Fig. 16.37) are defined by remnants of spatter cones and ramparts, agglutinates and welded scoria, which are the typical near-vent products of lava-fountaining. Lava flow units are also more abundant in these areas. For an updated description of the Roza member vent system, also see Thordarson and Self 1998).

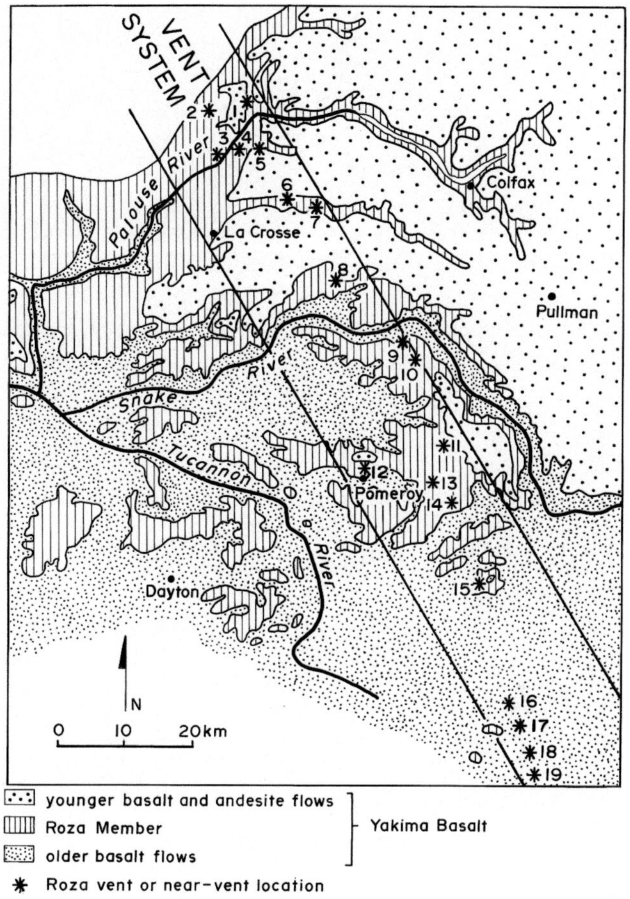

Fig. 16.37 Map locating linear-vent system for the Roza Member flood basalt flows in the Columbia River Plateau, USA. Total areal extent of the Roza Member is shown in Fig. 5.14; also see Cas and Wright (1987) for detailed documentation of vent and near-vent facies. (From Cas and Wright 1987, After Swanson et al. 1975)

Vent systems within flood basalt plateaus are therefore likely to be large and to have complex stratigraphies, with overlapping spatter cones and interdigitating agglutinated layers, clastogenic lavas and welded scoria fall deposits, in addition to the lavas (Chaps. 5 and 9). The most similar modern analogue to vent systems for flood basalts can be found in southern Iceland. The northeast-trending Bárdarbunga-Veidivötn system extends for nearly 80 km along the Eastern Volcanic Zone in the active rift and has produced some 57 km^3 of lavas from subaerial eruptions and 10 km^3

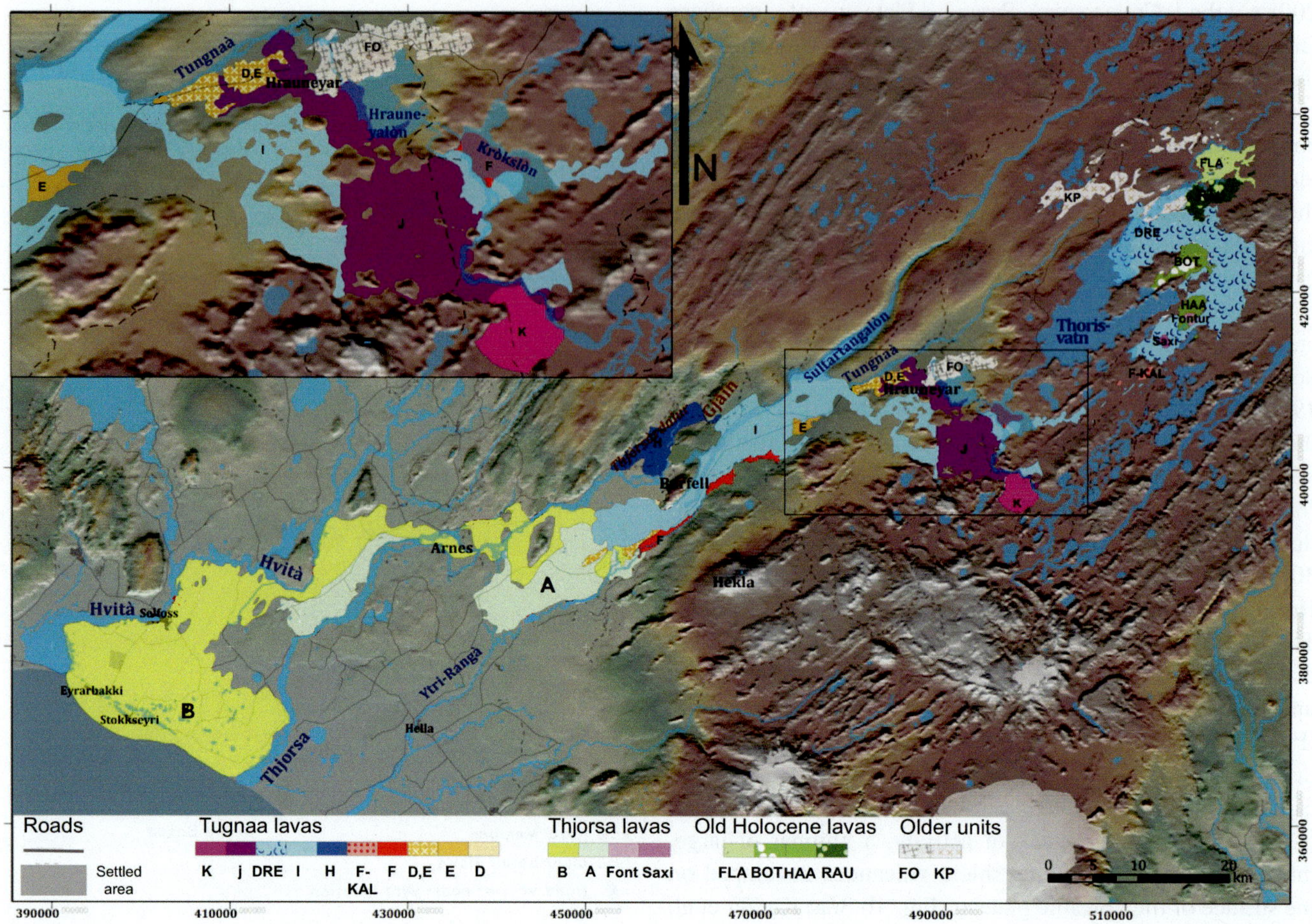

Fig. 16.38 Geological map of the Holocene Tugnaà flood basalt lava sequence in the Thjorsa valley, Eastern Volcanic Zone, southeastern Iceland. Upstream, the younger Tugnaà lavas form a complex overlapping lava pile close to the likely fissure source vents, while downstream only the largest and oldest Thjorsa (A, B) flows reached the ocean, being buried by younger flows in proximal-to-vent areas (e.g. inset). Numbers refer to sites that have been used to paleomagnetically date the lava succession (Pinton 2015; Pinton et al. 2018). These suggest both Thjorsa lava flows A and B represent the same eruption at about 6000–5300 yr BP, and with an estimated volume of ≥ 25 km^3 make this the largest Holocene eruption known. The youngest Tugnaà lava flow (Thj) dated paleomagnetically is 400–500 yr BP. (From Pinton 2015)

from subglacial eruptions (Thordarson et al. 2003). Here, several overlapping fissure systems have fed large-volume lava flows dating back to ca. 8 ka; these are referred to as Tugnaà lava flows and constitute the most voluminous lava sequence in Iceland (Fig. 16.38). The oldest are called the Thjorsa lava flow(s), which have an estimated volume of 25 km^3, and considered to be the largest known eruption during the Holocene. Both the Thjorsa and the younger Tugnaà lavas developed along the valley that follows the Thjorsa river, from the Thorisvatn lake down to the lowlands between the Hvità and Thjorsa rivers and to the ocean (Fig. 16.38). The fissure source vents were buried and covered by successively younger lavas which in proximal areas form a thick lava pile, so the older Thjorsa lavas only generally outcrop downstream. The Thjorsa lava eruption has been recently dated at 6000–5300 yr BP (Pinton 2015; Pinton et al. 2018).

16.6.2 Intraplate (Monogenetic or Distributed) Continental Plains Basalt Provinces

Continental plains basalt volcanic provinces consist of multiple small, usually monogenetic volcanoes (lava shields, scoria cones, maars, tuff cones) distributed over a significant area that can range from >10^2 to >10^4 km^2 (Connor and Conwasy 2000; Le Corvec et al. 2013a; Valentine and Connor 2015; Cas et al. 2017). The area between the volcanoes can be variably occupied by lava flows originating from the monogenetic volcanoes and are known to flow tens of km from the source vents. Often the volcanoes show a distinct clustering around crustal fractures, indicating that rising magmas have intersected pre-existing faults and joint sets and utilised them as magma conduits (Boyce 2013; Le Corvec et al. 2013a, b; van den Hove et al. 2017b). Crustal structures have acted as fissures to small volumes of magma, often repeatedly along the length of regional structures that may extend tens to hundreds of km. Often individual volcanoes form visible lines above those structures, and even within individual volcanoes there are alignments of vents.

So, as discussed in Chap. 5, the vents in plains basalt provinces are both of the central vent type, largely within low-profile shields and scoria cones, as well as fissure systems (Greeley 1982). As such, these provinces share characteristics of both flood basalt plateaus and shield volcanoes. According to Greeley (1982), in describing the Snake River Plain Volcanic Province, USA, some lava fields consist of coalescing very low-profile shields, others of moderate-sized, fissure-fed lavas. The lavas are frequently multi-lobate or compound, and compared to flood basalts, their volumes are small to moderate and the inferred magma discharge rates are considered to be considerably less. The lava shields have slopes less than 0.5°, except at the summit where slopes may rise to 5°. The summit region may also be marked by one or more irregular craters or topped by scoria and spatter cones (e.g. Fig. 16.2b). The shields can be up to 15 km in diameter, but many are smaller, and have volumes of less than 7 km^3, similar to low-profile Icelandic shields. In the Snake River Plain, many contemporaneous sets of shields appear to be aligned in distinctive rift zones, representing fissures along which localised eruption points formed (Greeley 1982).

The late Cenozoic Newer Volcanics Province of south-eastern Australia is one of the largest intraplate continental basaltic plains or monogenetic provinces on Earth. It comprises over 400 individual lava shield, scoria cone and maar volcanoes, which are distributed over an area of >23,000 km^2 (Fig. 16.7; Joyce 1975; Cas et al. 2011a, 2017; Boyce 2013; Van den Hove et al. 2017b) marked by multiple overlapping relatively small-volume lava flows, the longest of which is $\sim$60 km long (Cas et al. 2017). Magma compositions range from primitive picrites to tholeiitic and alkali basalts to trachytes (Cas et al. 2017). Many volcanoes are located above or near known crustal faults, and many individual volcanoes are elongate indicating their formation above crustal fissure magma conduits (Fig. 16.12c, d; Cas et al. 2017). However, the widely distributed volcanic centres clearly indicate that, unlike flood basalt provinces, the vents are widely distributed over a large region, not focused on a single major fissure vent and dyke system. Hence, these volcanic systems have also been called distributed volcanic provinces.

The longevity of plains basalt provinces can range from 10^5 kyr to several Myr (Valentine and Connor 2015), the magma compositions can be wide ranging, eruption volumes from each centre can be small, and across each province, there is no necessary progression of volcanism or systematic age trend, although eruptions can occur in clusters or domainally at different times (e.g. Cas et al. 2017). Although these provinces are dominated by monogenetic volcanoes, the provinces are clearly long-lived, "leaky" polygenetic provinces. Given that they are intraplate, and long-lived they are sourced by locally anomalously hot mantle, but the origin of such hot mantle regions is not clear in each case (Cas et al. 2017). It is also unclear as to why they generate multiple small batches of magma of such diverse compositions for such a long time over a large area.

16.6.3 Facies Models for Continental Basalt Terrains and Basins

The principal volcanic elements of continental basaltic terrains and the basins in which they occur will be flood and plains lava flows (depending on the extent and volume of basalt erupted), together with shield volcanoes, scoria cones, maars or tuff rings and cones, as well as dykes and sills (Fig. 16.39a, b). The difference between the two volcanic systems is scale. Flood basalt provinces consist of and are dominated by voluminous, very extensive sheet lavas (Fig. 16.39a), whereas in plains basalt provinces the lavas are smaller and contain small distributed monogenetic volcanoes, including explosive and effusive centres (Fig. 16.39b). Important accessory sedimentary environments will include fluvial channels, which will be constantly relocated as flood lavas pond in valleys. The fluvial systems will be incisive and erosional, acting against the aggradational effects of the basaltic volcanism in building up the topography. Fluvial successions will therefore not be thick or extensive and will only be preserved in the rock record through the capping effects of younger valley-fill lavas. Lacustrine and swamp deposits and environments may also be relatively short-lived. They could originate either within maars or tuff rings, or craters, or by the damming of valleys by lavas or ponding within the flat landscape dominated by lavas. They may be short-lived because contemporaneous erosion could breach their physical margins or because they become buried under a new lava flow.

The basaltic succession may be underlain by alluvial-fluvial deposits and basement from which it may be separated by a regional unconformity, and upwards may pass again into alluvial-fluvial successions as volcanism wanes (Fig. 16.39b). Locally, where lavas have flowed over swampy areas, or into lakes or rivers, hyaloclastites, and even pillow lavas, may be found. If regional volcanism occurred in a continental edge setting, it may have been accompanied by regional updoming or tumescence, which is compensated by regional subsidence after volcanism ceases, perhaps leading to marine transgression. Throughout the section paleosoils should be found, reflecting the long erosion-weathering intervals between short-lived volcanic episodes. Magmas in these provinces may be tholeiitic to alkaline in composition.

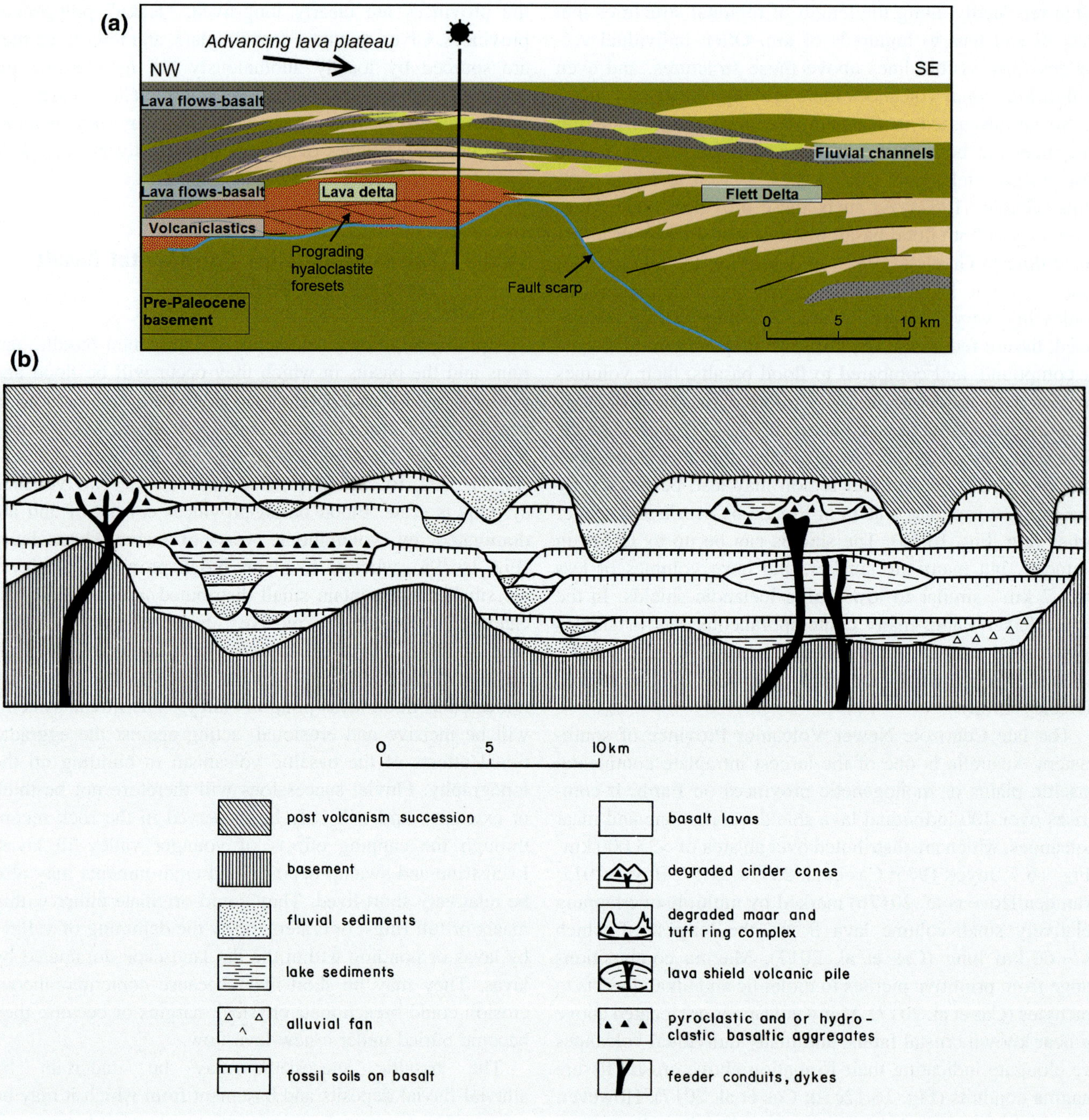

Fig. 16.39 Facies models for continental basalt successions. **a** Flood basalt succession associated with continental break-up based on seismic exploration data from the Paleogene North Atlantic Igneous Province, Faroes-Shetland Basin. This shows an easterly advancing succession of rift fed lavas and volcaniclastics (several km thick) which interface with contemporaneous continental-derived fluvio-deltaic systems prograding to the northwest. A succession of basalt lava flows, hyaloclastites and volcaniclastic sediments separated by a series of intra-basaltic fluvial to marginal marine clastic sequences results. Well shows position of Rosebank oil and gas discovery. (Modified from Duncan et al. 2009). **b** Plains basalt facies model association based on the late Cenozoic Newer Volcanics Province, southeastern Australia (see text) (From Cas and Wright 1987)

16.6.4 Resource Potential and Economic Significance

Continental flood basalts represent major magmatic events and some are associated with important copper-nickel-PGE (platinum group element) sulphide ore deposits (e.g. at Norils'k in the Siberian Traps, the Duluth Complex in the Keweenawan mid-continental rift, and Insizwa in the Karoo). These deposits are generally thought to be found in subvolcanic intrusions and formed by orthomagmatic processes (Chap. 18). At Norils'k several giant deposits are hosted within sill-like feeder systems to the overlying flood basalt succession. Most continental flood basalts are inferred to have mantle plume origins (Chap. 17), and therefore there may be potential for association with other mantle plume-related mineral deposits, perhaps diamondiferous kimberlitic pipes.

Continental plains basalt volcanic provinces may also host concentrations of alluvial or placer gold (derived from basement lodes) and precious stones in river channel successions (called "deep lead" deposits in eastern Australia) buried by valley confined basalt lavas. Primary sapphires and other gem-quality minerals (zircon, garnet and spinel) may also be found in some scoria cones, maars and tuff rings but these are generally sub-economic. Volcanic rifted continental margins have become increasingly attractive petroleum exploration targets since discovery of the Rosebank oil and gas field in the North Atlantic Igneous Province in 2004, for which the reservoir is intra- and sub-basalt fluvial and shallow marine siliciclastic sands (Fig. 16.39a; Chap. 18).

16.7 Stratovolcanoes

Stratovolcanoes or composite volcanoes are the characteristic volcanic landform found at subduction plate margins and are the most abundant type of large volcano on the Earth's surface in continental and island arc settings (Davidson and de Silva 2000; Fig. 16.40). At continental margin arcs and mature island arcs, stratovolcanoes are wholly subaerial in setting (Fig. 16.40a–h, but in less evolved island arcs they form significant volcanic edifices that rise from the deep seafloor. Some of these emerge above sea level (Fig. 16.40i; also see book Frontispiece d), and so have both submarine and subaerial parts, but others are totally submarine edifices (Fig. 16.40j, k).

Stratovolcano morphology results from multiple eruptions of relatively short lava flows and pyroclastic deposits from a central vent over an extended period of tens to hundreds of thousands of yr, leading to the building of a large cone around the vent at rates that exceed the erosional down-cutting rates of surface sedimentary processes. The textbook diagram usually shows a symmetrical cone with steeply dipping, "layer-cake" pyroclastic deposits interbedded with lava flows. However, in detail the shapes of these volcanoes are more variable (e.g. there may be several eruptive points, or the presence of a summit caldera; Fig. 16.41a–g), as are their individual deposits, and so they rarely have a simple layer-cake geology. This is because eruption volumes during any eruption may be small, deposits from a single eruption are dispersed in variable directions, and topography, which controls dispersal of lavas, pyroclastic flows and sedimentary mass flows, changes. It is also important to stress that mass wastage is just as important as, if not more important than, volcanic processes in determining their morphology and deposits. This is because repose periods are usually much longer than eruptive phases (Chap. 12; Vessell and Davies 1981; Cas and Wright 1987), although an aggradational landform would not be present without the volcanic activity.

Volcaniclastic deposits (pyroclastic and epiclastic) are usually very important volumetrically, and on some stratovolcanoes can make up to 80% or more of the succession. However, the variations in the proportions of lavas to volcaniclastics can vary enormously from one stratovolcano to another, and some may be dominated by effusive lava eruptions (e.g. Etna volcano, Italy; Fuji volcano, Japan; Licancabur volcano, Chile, Fig. 16.40b; and Tuzgle volcano, northwestern Argentina, Fig. 16.40c), whereas others, such as Ngauruhoe volcano, New Zealand (Fig. 16.40a), Agung volcano, Indonesia (Fig. 16.40e), and Vesuvius volcano, Italy (Fig. 16.40f), mostly experience explosive eruptions. However, even on volcanoes such as Ruapehu, New Zealand, where lavas are important edifice builders, they are surrounded by extensive ring plains of volcaniclastic debris, most of it being resedimented volcaniclastic sediment (Fig. 16.42f–i).

In subduction settings at continental margins, stratovolcanoes are constructed by eruptions of calc-alkaline magmas (Table 16.2; Chap. 17). They are usually broadly andesitic or basaltic andesite in composition, but their products may include basalts, dacites and rhyolites, and some are basaltic centres (e.g. Fuji in Japan, Klyuchevskoy in Kamchatka, Stromboli in Italy). Alkaline magmas also generate stratovolcanoes, which are on average larger than their calc-alkaline counterparts (Table 16.2). Examples built up of trachytic and phonolitic products are common on the intra-oceanic Atlantic islands of the Azores and Canary Islands, which are not associated with subduction margins (Chap. 6). Vesuvius is an example of a K-alkaline stratovolcano (Fig. 16.40f). Island arc stratovolcanoes are basaltic to basaltic andesite in composition, with more (island arc) tholeiitic affinities than calc-alkaline. Again, it should be clear that chemical composition cannot be the sole control of volcano type and shape as testified to by the basaltic Fuji, the "*classical*" stratovolcano, which has much lower average magma discharge rates than basaltic shield-forming volcanoes.

(a)
(b)
(c)
(d)
(e)
(f)
Google Earth
(g)

◀ **Fig. 16.40** Examples of stratovolcanoes. **a** Line of Ruapehu (snow covered, obscured, left), Ngauruhoe (dark prominent cone middle ground) and the heavily degraded Tongariro (foreground, including Red Crater) stratovolcano cluster, central North Island, New Zealand. **b** Licancabur volcano, Chile. Note the narrow lava lobes extending beyond the base of the cone. (*Source* Google Earth). **c** Tuzgle volcano in the Andes, northwestern Argentina, with young block lavas draping the eroded older part of the cone. **d** Villarrica volcano, Chile, with a gas plume rising above the lava-filled, summit crater. **e** Agung volcano, Bali, Indonesia, with ash plume, November 2017 (Photo by Michael Ishak; *Source* Wikimedia through Creative Commons Attribution 4.0 International licence). **f** Vesuvius volcano, Italy. The rim to the left is of the Somma caldera, which resulted from multiple plinian eruptions and caldera collapse events, the most recent being the famous 79 AD Pompeii-Herculaneum eruption, and was last active in 1944. Inside the rim to the right stands the prominent Gran Cono (right summit) formed after the 79 AD eruption. (*Source* Google Earth). **g** Mt St Helens, USA. The summit is truncated by the amphitheatre that formed as a result of the sector collapse that triggered the devastating 18 May 1980 eruption. The prominent mounds at the centre of the amphitheatre are the lava domes that were extruded from 1980–84, and 2004–08. The sloping incised fan emerging from the amphitheatre is the "pumice plain", consisting of debris avalanche deposits overlain by pyroclastic flow and debris flow deposits. **h** Relief map showing Mt Pelée (top), and the eroded Pitons du Carbet and Morne Jacob volcanic centres to the southeast and east respectively, Martinique, Lesser Antilles. Dashed white lines outline major sector collapse scars. (SRTM image courtesy of Alan L. Smith). **i** Stromboli volcano, Aeolian Islands, southern Italy. Note summit fumarole activity, and the spectacular Sciara del Fuoco sector collapse scar, which is now the locus of eruptive activity. (*Source* Google Earth). **j** The fully submarine Haungaroa stratovolcano in the Kermadec island arc, north of New Zealand, shown as a digital elevation model draped with bathymetric contours. The cone is relatively symmetrical, but the flanks are marked by several sector collapse scars. The base is at >3,000 m water depth, and the summit marked by a flat plateau (lava lake?) with several summit cones, is at <1,000 m water depth. (From Wright et al. 2006). **k** Digital elevation model (top) and contoured bathymetry (bottom) of the wholly submarine Brothers stratovolcano, Kermadec arc, showing a summit caldera and older caldera rim, and younger cones growing in the caldera. Dashed lines are regional faults, coloured lines and dots represent tracks and sites from measuring chemical anomalies associated with the hydrothermal field in the summit region. (From de Ronde et al. 2005)

16.7.1 Morphometry

Pike and Clow (1981) collated morphometric data for over 200 stratovolcanoes (Table 16.2). They divided stratovolcanoes into a number of classes according to whether the magma type is calc-alkaline or alkaline, and according to the presence of a summit crater (e.g. Ngauruhoe, New Zealand, Fig. 16.40a; Ruapehu, New Zealand, Fig. 16.40a; Tuzgle volcano, Argentina, Fig. 16.40c), sector collapse scar (e.g. Mt St Helens, USA, Fig. 16.40g; Mt Pelée, Martinique, Fig. 16.40h; Stromboli, Italy, 16.40i), or a summit caldera (e.g. Mt Pinatubo, Luzon, Philippines, Fig. 16.41d; Batur, Bali, Indonesia, see Fig. 16.48d; Tambora, Sumbawa, Indonesia, Fig. 16.41e; Fogo, São Miguel, Azores, Fig. 16.41f. Those stratovolcanoes with summit calderas were further subdivided according to those with greater proportions of silicic and pyroclastic products (e.g. Crater Lake, USA, Fig. 16.41a, b; Santorini, Greece, see Fig. 16.78e, f; Somma caldera, Vesuvius, Italy, Figs. 16.40f and 16.41g; Las Cañadas, Tenerife, Plate 16.1, and see Fig. 16.79e, f). Data for these different classes are summarised in Table 16.2. For a global update employing analysis of SRTM DEM data, see Grosse et al. (2014).

Average upper slope angles of stratovolcanoes range from about 15° to 33°, levelling out to a few degrees on the surrounding ring plains. This defines their classical hyperbolic profile (Fig. 16.40), which represents the equilibrium between aggradational and degradational processes along the stratovolcano slopes. Some stratovolcanoes are topographically very impressive, rising steeply as high as 5 km above their bases. The tallest above sea level seems to be Queen Mary's Peak on Tristan da Cunha, while the largest generally quoted is Teide on Tenerife which represents a massive, multi-vent stratovolcano and caldera (Las Cañadas) complex rising nearly 8 km base-to-peak from the ocean floor (Plate 16.1; see Sect. 16.8.1), but sitting atop a basaltic shield volcano which itself is emergent above sea level. For most stratovolcanoes, however, summit height is between 1000 and 3000 m above base level, and the basal diameter 20–40 km. With these dimensions, volumes of stratovolcanoes are usually in the order of $>10^1$–10^2 km^3 (Table 16.2), although significantly larger edifices exist (Table 16.3).

The maximum height of a volcano is mostly controlled both by: (1) the limit of its gravitational stability that favours mass-wastage (Chap. 12), as volcano growth increases its potential energy and thus mass wasting potential, and (2) by the excess pressure within the shallow magma chamber that is progressively counteracted by the increase of lithostatic load as the volcano grows (Davidson and de Silva 2000). The latter process explains the common observation that early products of stratovolcanoes are more voluminous and less evolved. This can be understood by taking into account the simple relationship that controls the propagation of a dyke-conduit from a magma reservoir:

$$Pm + dPm = Pl + c \tag{16.1}$$

where Pm is the magmastatic pressure and Pl is the lithostatic pressure. For dyke formation to occur the excess magmatic pressure, dPm, must overcome the tensile strength, c, of the country rocks (see Chaps. 3 and 8).

As the volcano grows Pl increases the pressure confinement of the magma reservoir, therefore, progressively limits the tendency for the intrusion of dykes and eruptions. This leads to longer repose periods between eruptions and magma differentiation, which in turn allows larger magmatic overpressures to be achieved (Davidson and de Silva 2000). Some spectacular examples of stratovolcanoes that develop

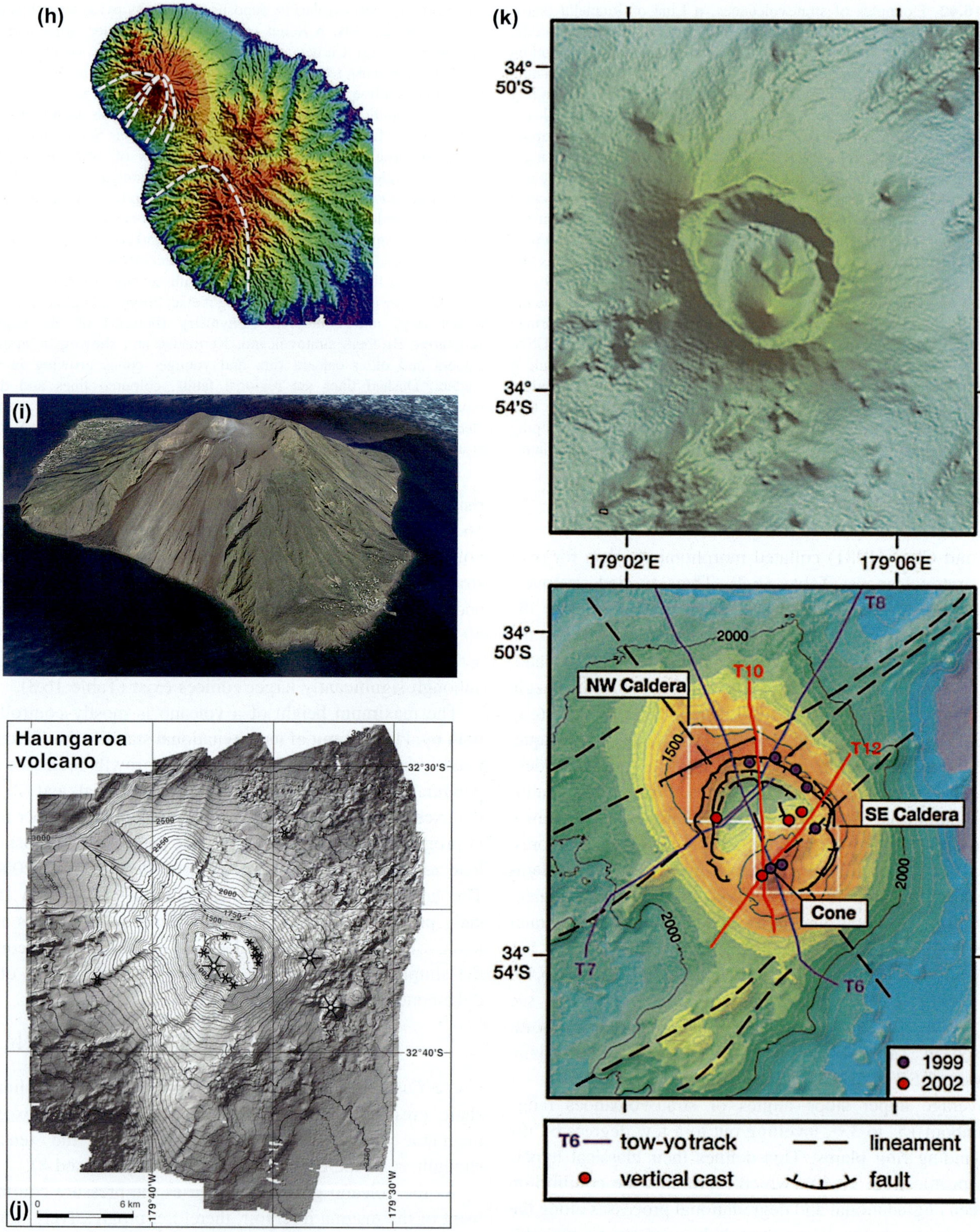

Fig. 16.40 (continued)

(a)
(b)
(c)
(d)
(e)
(f)
(g)

◀ **Fig. 16.41** Calderas at the summit of stratovolcanoes. **a** Panorama view of Crater Lake caldera, Mt Mazama stratovolcano, Oregon, USA, with the young volcanic cone of Wizard Island growing in the caldera. Crater Lake is ~10 km in diameter. (Photo by Jack Simmons). **b** Oblique Google Earth image of Crater Lake caldera. Mt Mazama lost about 1,600 m of its top during the final caldera-forming eruption about 6.8 ka. (*Source* Google Earth). **c** Changbaishan volcano summit caldera, China-North Korea border (Photo by Martin Menzies). **d** The 15 June 1991 eruption of Mt Pinatubo, Luzon, Philippines, removed part of the summit, leaving a small caldera 2.5 km in diameter (Photo by Chris Tomnong; *Source* Wikipedia through Creative Commons Attribution 3.0 Unported licence). **e** Astronaut's view of the summit caldera of Tambora volcano, Sumbawa, Indonesia. The caldera is 6 km in diameter and 1,100 m deep, and formed during the largest eruption in recent historic times on 10–11 April 1815, when the top of the volcano's 4,300 m high peak collapsed. The summit edge of the caldera now stands at 2,850 m high. (Photo by NASA Expedition 20 crew; *Source* NASA). **f** Summit caldera lake of Fogo volcano, São Miguel, Azores, Portugal. **g** Inner wall of Somma caldera, Vesuvius, Italy (See Fig. 16.40f), cut through by several dykes

from earlier less evolved and highly productive lava effusion periods to more periodic explosive eruptions of differentiated magmas are found in Italy, such as Vesuvius, Roccamonfina and Vico volcanoes (Cioni et al. 1999; De Rita and Giordano 1996; Conticelli et al. 2013). Mt Etna and Mt Fuji are examples of basaltic stratovolcanoes where their huge dimensions have favoured the increase in explosivity of eruptions at increasing height of the edifice (Branca et al. 2011; Miyaji 1988).

Gravitational instability induced by the volcano growth can be accommodated either by continuous and low intensity degradation of the upper and steeper slopes of the stratovolcano, or by impulsive events of sector collapse and related catastrophic debris avalanches (Figs. 10.4f, 12.8c–e and 16.40g; Chaps. 10 and 12). At least 50% of the Earth's stratovolcanoes have experienced one or more sector collapse events. These may be frequently associated with magmatic intrusion or eruptive episodes, as observed during the 1980 eruption of Mt St Helens when a cryptodome was emplaced prior to failure (Voight et al. 1983; Fig. 8.19d). However, sector collapse can potentially be triggered seismically or by extreme weather at any time when a volcano becomes unable to support its own load (Chap. 12). The longer term causes of this can be many (e.g. van Wyk de Vries and Davies 2015; Roverato et al. 2021; Chap. 12), including basement (especially strike-slip) faulting, gravitationally induced spreading of basement substrata, hydrothermal (hypogene) alteration within the edifice (Chap. 18), and climatic, including the location of paleo-water tables and supergene alteration. Sector collapses appear to be especially common on stratovolcanic islands and are a potential trigger for large tsunamis (Chap. 12); e.g. Stromboli (Tinti et al. 2000; Tibaldi 2001), Réunion (Kelfoun et al. 2010), Canary Islands (Seisdedos et al. 2012; Paris et al. 2017), and Lesser Antilles (Deplus et al. 2001; Le Friant et al. 2003; Poisson and Pedreros 2010).

It is important to note that while the classic picture of a stratovolcano shows a simple cone with one main central vent, vent migration is also a rather common occurrence (e.g. the Doña Juana volcanic complex, Pardo et al. 2019, 2023). Summit areas may be much more complex and contain several main vents, e.g. at Stromboli and Etna. Another effect of the increased lithostatic load and the reduction of summit eruption frequencies is the increased possibility for a tall stratovolcano to give rise to flank eruptions (fissures or craters). Such was the case with the last and most explosive plinian eruption at Mt Fuji in 1707, which occurred entirely from three flank craters (Miyaji et al. 2011). Larger and more complex edifices also occur, e.g. the Mt Tongariro Volcanic Massif, New Zealand (Fig. 16.40a; Miller and Williams-Jones 2016). The Tongariro complex is constructed of at least 17 overlapping vents built during 6 main cone-building episodes, indicating a complex plumbing system covering an area of 65 km^2. The massif has been extensively modified by glaciation since the first eruptions around 275 ka, so surface exposures of early vents are obscured by later eruptions or have been removed by erosion. Since 7 ka the area has been dominated by the growth of Mt Ngauruhoe (2280 m; Fig. 16.40a). However, two complex volcanic systems, Santorini and Tenerife, standout as having changed their character and form repeatedly during their history, including more than one explosive caldera collapse event, and we propose they deserve to be treated as a new class of larger volcano named "multiplex volcanoes" (Sect. 16.9.1).

It has generally been assumed that the morphology of marine stratovolcanoes is similar to subaerial ones. However, many marine stratovolcanoes, which are mostly elements of subduction-related island arcs, are in fact still fully submerged, and so relatively little has been known about them. Relatively recently, very detailed marine surveys of such submerged volcanoes using very high-resolution multi-beam bathymetry surveys, deep-tow photography, submersibles and sampling, are rapidly changing our understanding of these systems (e.g. Wright et al. 2002, 2006; de Ronde et al. 2005, 2007; Carey et al. 2018). These do indeed show that marine stratovolcanoes appear to be similar morphologically to subaerial counterparts and can rise several hundred m to 3 km above the surrounding seafloor (Fig. 16.40j, k). The summits can be marked by single or multiple summit craters, and quite a number have summit calderas; e.g. in the Tonga-Kermadec arc, north of New Zealand: Brothers volcano (Fig. 16.40k; Wright et al. 2006; de Ronde et al. 2005, 2007), and Hunga Tonga-Hunga Ha'apai volcano, which displayed emergent phreatomagmatic activity in 2009 and again in 2014–15 (Fig. 16.11e,

(a)
(b)
(c)
(d)
(e)
(f)
(g)
(h)

◀ **Fig. 16.42** Facies and facies architecture of stratovolcanoes. **a** Cluster of lobate lavas and domes near the summit of Colima volcano, Mexico, indicative of the core and proximal zones of stratovolcanoes. **b** Stacked, variably eroded columnar jointed lavas forming the upper slopes of Mt Taranaki/Egmont, New Zealand. **c** Core zone of an old edifice of Ruapehu, New Zealand showing stacked columnar jointed lavas, and a brown hydrothermally altered zone. **d** Irregular relationships and geometry of the Fogo A pumice fallout deposit (white unit) mantling an orange paleosoil in an eroded ridge and valley paleotopography, overlain by a valley filling, ponded, massive to diffusely stratified, pyroclastic flow deposit (grey). Fogo volcano, São Miguel, Azores, Portugal. **e** Thin sheet and lobate a'a lavas infilling the Sciara del Fuoco sector collapse landslide scar on Stromboli volcano, Italy (see Fig. 16.40i). These layers dip at angle of repose and are very discordant to the stratified older sequence on the right. Vertical field of view is several hundred metres. **f** Walls of a box canyon, middle slopes of Mt Taranaki/Egmont volcano, showing very coarse, massive, resedimented, slope breccia deposits of volcanic debris flow (massive) and hyper-concentrated flood flow (diffusely stratified) origin. **g** Bedded sequence of hyper-concentrated flood flow, debris flow and stream flow deposits, Mt Taranaki/Egmont. Height of outcrop ~15 m. **h** Fining upward and variably channelised succession of hyper-concentrated flood flow, diffusively stratified, pumiceous volcaniclastic sandstones, distal Quaternary coastal apron, southwestern Bali, Indonesia. Height of cliff ~8 m. **i** Distal debris avalanche breccia deposit in the ring plain succession to the Tongariro and Ruapehu stratovolcano complex. **j** Bedded fallout sequence in the ring plain sequence to the Tongariro and Ruapehu stratovolcano complex. **k** Lobate pillow (PL) to massive (ML) lava with flat base above plane bedded volcaniclastic turbidites and overlain by a thick sequence of conglomeratic volcaniclastic mass flow beds that are onlapping the high relief lobate morphology of the pillow lava. The sequence forms part of the submarine apron of a Miocene marine stratovolcano, Maori Bay, Muriwai, New Zealand. **l** Bedded massive normal to reverse graded Miocene volcaniclastic turbidites. Height of outcrop ~3 m. Setting and location as for (**k**)

Fig. 16.42 (continued)

Table 16.2 Summary of dimensions of different classes of stratovolcano (Updated after Pike and Clow 1981)

Class	Basal diameter (km)	Height* (m)	Summit crater, collapse scar or caldera diameter	Average volume (km^3)	Examples
1. calc-alkaline with summit crater	6–55	625–4650	0.1–1.75	60	Fuego, Popocatepetl, Mt Adams, Tuzgle, Fuji, Mayon, Agung, Ngauruhoe, Mt Egmont Klyuchevskoy
2. alkaline with summit crater	1.5–60.5	425–4900	0.15–0.9	80	Etna, Gran Cono Vesuvius, Hekla, Queen Mary's Peak, Teide, Mt Erebus
3. calc-alkaline with sector collapse scar	11–30	1020–2800	1.2–4.5	100	Bezymianny, Shiveluch, Mt Pelée, Lassen Peak, Mt St Helens, White Island, Stromboli
4. calc-alkaline with caldera	7.5–68.5	500–3350	1.75–11	160	Cosegüina, Toluca, Aniakchak, Batur, Ruapehu, Nysiros, Mt Pinatubo
5. calc-alkaline with caldera, more silicic and more pyroclastics than (4)	8–59	200–1700	2.2–18	80	Crater Lake, Santorini#, Krakatau, Rabaul, Deception Island
6. alkaline with caldera	6.5–87	350–4700	0.75–9	250	Sete Cidades, Fogo, Furnas, Kilimanjaro, Tambora, Mt Erebus,
7. alkaline with caldera, more silicic and more pyroclastics than (6)	7.5–64.5	100–2200	1.8–18.5	200	Roccamonfina, Vico, Somma Vesuvius, Las Cañadas#, Pantelleria, Ngorongoro, Fantale

*Average height of crater or caldera rim above pre-volcano topographic datum. #We also describe these as examples of "multiplex volcanoes" which have changed their character and form repeatedly during their history (see text, Sect. 16.9.1). Based on a sample size of approximately 250 volcanoes

Table 16.3 Average lifetime output rates for some stratovolcanoes, including peak rates for some well-studied volcanoes (After Wadge 1982; Mattei et al. 2010; Branca et al. 2011; Francalanci et al. 2013; Tibaldi 2001)

Volcano	Volume (km^3)	Age (ka)	Average output rate (km^3 kyr^{-1})	Peak output rates (km^3 kyr^{-1})
Sakurajima	40	13–15	2.8	
Shikotsu	140	32	4.4	
Asama	60	30	2.0	
Fuji	400	80	5.0	
Hakone	200	400	0.5	
Aso	45	25	1.8	
Kirishima	50	17	2.9	
Klyuchevskoy	250–300	10	27.5	
Shiveluch	1000	200	5.0	
Avachinsky	100	60	1.7	
Mt Adams	200	940	0.2	1.6–5.0
Fuego	50	100	0.5	
Santa Maria	20	30	0.7	
Tatara-San Pedro	55	930	0.06	0.2–0.3
Mt Tongariro	50	275	0.2	0.9–1.0
Vesuvius	100	22	4.5	
Roccamonfina	130	580	0.2	0.4
Vico	65	300	0.2	0.3
Etna	500	150	3.3	
Stromboli	9*	85	0.1	

*Includes sector collapse volumes

and see Figs. 9.11b and 10.100c; Vaughan and Webley 2010; Garvin et al. 2018; Colombier et al. 2018) before the catastrophic, new caldera forming explosive eruption of 15th January 2022. Those that had or have summit calderas within several hundred m of sea level could have formed calderas through explosive events (see Sects. 16.7.4 and 16.8.9). Calderas in deeper water (>1 km) could have formed through lateral lava-forming magma withdrawal through flank eruptions, summit lava dome eruptions, and caldera collapse during explosive eruptions when the summits were at shallower water depths (<1 km). Many volcanoes also show evidence of lateral sector collapse, sometimes multiple times, marked by collapse scars on the slopes and depositional lobes or fans at the base onto the submarine floor (e.g. Fig. 16.40j).

16.7.2 Output Rates, Repose Periods and Life Expectancy

Data on magma output rates at stratovolcanoes rely on availability of strong stratigraphic and geochronological datasets that allow the integration of volumes erupted over long time scales. Such detailed datasets are only available for a limited number of volcanoes. In Table 16.3 a compilation is given of estimates of the volumes of some large active stratovolcanoes and the time taken to build them. Average lifetime output rates can be calculated from these and usually range between about 1 and 10 km^3 kyr^{-1} but note that the volumes in general represent less than the total volume of magma erupted at each volcano because dispersed pyroclastic deposits are not accounted for.

Klyuchevskoy, which is the largest and highest stratovolcano in the Kamchatka Peninsula arc, Russia, appears to have the highest output rate and has been repeatedly active for over the past three centuries since records began in 1697. Periods of dormancy for stratovolcanoes may be much longer than periods of activity (e.g. Vessell and Davies 1981), during which peak output rates may be much higher than average eruption rates. Repose periods between eruptions seem to vary considerably. At Agung volcano, Bali, Indonesia (Fig. 16.40e), Fontijn et al. (2015) recorded 52 fallout deposits in the last 5.2 kyr, giving an explosive eruption frequency of at least one every 100 yr, whereas the island volcano Stromboli, Italy (Fig. 16.40i), seems to be almost persistently active (Fig. 9.7a–f; Rosi et al. 2000). Other stratovolcanoes seem to have repose periods as long as 10^3 and 10^4 yr, and these are sometimes erroneously termed dormant, and even extinct, although they may be very much alive (e.g. Nevado de Toluca, Mexico, Garcia-Palomo et al. 2002; Panarea volcano, Aeolian Arc, Pensa et al. 2019).

Most active stratovolcanoes appear to be less than 10^6 yr old (Table 16.3). However, it is perhaps more important to know the length of time over which overlapping and closely spaced stratovolcanoes of a polygenetic complex can be active in an area. It is unlikely that within ancient terrains (Paleozoic and older), all or many individual centres can be distinguished, so what time periods could we be dealing with for coalesced complexes of stratovolcanoes? Usually volcanic provinces appear to be active on timescales of 10^6–10^7 yr or more, in many cases reflecting the age of the volcanic arc of which stratovolcanoes are characteristic elements (Cas and Wright 1987; Smith et al. 2013).

16.7.3 Eruptions, Characteristics and Deposits

Eruptions of stratovolcanoes are very variable in style, duration and frequency. They can experience a variety of explosive and effusive eruption styles, their setting can be subaerial or submarine, or partially both, eruptions can be of variable magnitudes, and as a result, even modern subaerial stratovolcanoes present a myriad of problems in the analysis of their facies and stratigraphic architecture.

Some modern stratovolcanoes experience nearly continuous, mild strombolian explosive activity (Chap. 9), the type example being Stromboli, Italy (Figs. 9.7a–e and 16.40i). Vulcanian eruptions (Fig. 9.8a–g, Chap. 9; Fig. 10.4a, b, Chap. 10) are common and these may generate small scoria/spatter and ash flows (e.g. the eruption of Ngauruhoe, New Zealand, 1975, Figs. 10.4b and 10.88, Nairn and Self 1978; Stromboli, 2019, Figs. 10.89 and 11.9, Plate 11e, Giordano and De Astis 2021; Mt Fuji, Yamamoto et al. 2005), or pumice and ash flows (e.g. Soufrière Hills, Montserrat, 1997, Fig. 10.4a, Druitt and Kokelaar 2002). Phreatomagmatic eruptions are also common where volcanoes have, or had, a crater lake, e.g. Ruapehu, New Zealand (Kilgour et al. 2010). Block and ash flows (Chap. 10) are commonly associated with the explosive and/or gravitational collapse of viscous or "pelean" lava domes. Such eruptions may also generate more catastrophic lateral blasts (Chap. 10), as exemplified by the 8 May 1902 eruption of Mt Pelée, on Martinique, or occasionally during the 1991–1995 eruption of Mt Unzen, Japan, and at Soufrière Hills volcano, Montserrat, on 26 December 1997. Less frequently, much larger (>0.1–10 km^3; Cioni et al. 2015) felsic sub-plinian, plinian and ignimbrite-forming eruptions occur from these same cones, including Mt Pelée, Martinique (Smith and Roobol 1990; Wright et al. 2016), Mt Scenery, Saba and the Quill, St. Eustatius (Roobol and Smith 2004), and Mt Mazama, USA (Fig. 16.41a, b; Bacon et al. 1997) Indeed, these two contrasting eruption styles have repeatedly occurred throughout the history of Mt Pelée, which has undergone 13 such alternations, composed of eight pelean and five plinian (pumice-dominated) sequences, erupted during the past 3000 yr.

Many other stratovolcanoes are known to have frequently produced large-volume pumice fallout and flow deposits. These include, for example, Vesuvius, Italy (Fig. 16.40f); Crater Lake, Mt Mazama, USA (Fig. 16.41a, b); Fogo, São Miguel, Azores (Figs. 16.41f and 16.42d), and Mt Pinatubo in the Philippines (Fig. 16.41d) (e.g. Cas and Wright 1987; Bacon et al. 1997; Cioni et al. 2008; Pensa et al. 2015; Newhall and Punongbayan 1996). Welded tuffs of both air-fall and pyroclastic flow origin can be formed during this type of activity (Chaps. 9 and 10). Such eruptions also cause the more important changes in cone morphology, such as happened at Mt St Helens in 1980 (Fig. 16.40g) and Mt Pinatubo in 1991 (Fig. 16.41d). While cataclysmic caldera collapse can occur during or after larger ignimbrite-forming events (Fig. 16.41), such as during the eruptions of Vesuvius at 18.3 ka (Fig. 16.40f), Mt Mazama about 6.8 ka (Fig. 16.41a, b), Fogo a 4.6 ka (Fig. 16.41f), Changbaishan about 1000 yr BP (Fig. 16.41c; Sun et al. 2014a), Tambora in 1815 (Fig. 16.41e), and Krakatau in 1883 (Fig. 16.48b). We treat the mechanics of caldera formation separately in Sect. 16.7; although this is in the context of larger caldera volcanoes, the same principles are applicable to the summit calderas of stratovolcanoes.

As well as explosive eruptions, effusive eruptions are frequently observed, and lavas, domes and dome complexes may be important in the stratigraphy of many stratovolcanoes (Figs. 16.42a–c, 5.19 and 6.1e, f; Chaps. 5 and 6), although these may vary considerably in composition, even within the same volcanic edifice. Lava domes with their surrounding block and ash flow aprons may form juxtaposed complexes of differing stratigraphic age (e.g. Mt Scenery, Saba; Roobol and Smith 2004). Other eruptions on stratovolcanoes can occur from parasitic centres, and scoria cones, spatter cones and phreatomagmatic craters may form, perhaps aligned along a fissure. Parasitic cones and dykes may also have radial distributions (Chap. 8). Magma is also intruded into the piles, and dykes and shallow intrusive bodies are seen in the crater or caldera walls of stratovolcanoes (Fig. 16.41g).

The volcanic successions of modern subaerial stratovolcanoes therefore can show complex and rapid changes in lithologies, both vertically and laterally, and which may be distributed quite differently around each volcanic centre (Fig. 16.42). Viscous lava domes and flows will occur mostly near-vent in core or proximal zones (Fig. 16.42a–c). The dispersal of pyroclastic fallout deposits will be controlled by the height of the eruption column and the prevailing wind direction at the time of the eruption (Chap. 9). This means that on many stratovolcanoes the distribution of fallout deposits may be quite asymmetrical (Sigurdsson et al. 1980; Cas and Wright 1987). Fallout deposits will mantle topography, even irregular topography (Fig. 16.42d), and may be welded near-vent (Chap. 9). Pyroclastic flows will be channelled down canyons and valleys (Fig. 16.42d), that are often repeatedly re-excavated and infilled (Chap. 12) and may be sectorally or radially distributed. In addition, secondary mass-wastage and sedimentary surface processes (sector collapse, landslide, debris avalanche, hyperconcentrated flood flow, stream flow; Chap. 12) are very dynamic and important features of subaerial stratovolcanoes. These processes can be highly erosional, forming erosional surfaces, upon which later deposition occurs, creating stratigraphic discordances (e.g. Fig. 16.42e). Sedimentary surface process volcaniclastic deposits form major parts of the successions (Fig. 16.42f–h; Chap. 12). These are commonly funnelled down deeply incised "box" canyons on middle slopes (Fig. 16.42f) and spread at the base of slope to form volcaniclastic ring plains, dominated by hyperconcentrated flood flow, stream flow deposits (Fig. 16.42g, h), distal debris avalanche deposits (Fig. 16.42i, as well as some primary pyroclastic fallout sequences (Fig. 16.42j).

Chemical composition of magmas may also show complex variations, and products range from mafic to evolved silicic compositions. Obviously, the generation, evolution and type of magma erupted from these volcanoes can be complex. This could involve magma chambers at different levels and a complex plumbing system, as well as different batches of primary magma periodically rising into the system (Fig. 16.1).

16.7.4 Marine Stratovolcanoes and Their Emergence

Marine stratovolcanoes (Fig. 16.40i–k) have many similarities to their wholly subaerial counterparts (e.g. Wright et al. 2006). For example, they: (1) have significant relief above their base, (2) are large and steep-sided, (3) may have summit calderas, (4) can have diverse magmatic compositions, and (5) show very diverse deposit associations. However, they are significantly different in that they have a foundation built in submarine environments and a summit region which may (or not) be subaerial. If the initial eruptions are basaltic to basaltic andesite, then the lower submarine parts of the stratovolcanoes may develop similar stratigraphy and characteristics to volcanic seamounts, being dominated by submarine lavas including pillow and sheet lavas, and hyaloclastite (e.g. Figs. 5.45 and 5.46), and if they are more evolved, viscous lava domes become more prominent. However, stratovolcanoes may be longer lived than individual oceanic seamount volcanoes and can build substantial subaerial cones above sea level showing similar eruption styles, characteristics and deposits to those just described for subaerial stratovolcanoes. The Aeolian Arc is made of a cluster of stratovolcanoes up to 3000 m high, of which more than two-thirds are subaqueous, with summits that may reach almost 1000 m a.s.l. (e.g. Stromboli volcano, Italy; Figs. 5.19; and 16.40i; Pensa et al. 2019).

Detailed marine surveys of such submarine stratovolcanoes in the Kermadec volcanic arc (e.g. Wright et al. 2002 2006; de Ronde et al. 2005, 2007; Fig. 16.40j, k) show that they are dominated by lavas and hyaloclastite at depths >1,000 m, but if they become shallower, there is an increase in fragmental volcanic debris, including pumice. Wright et al. (2002, 2006) interpreted these empirical observations as reflecting a transition from high hydrostatic pressures at $> \sim 1{,}000$ m that suppressed explosive exsolution of magmatic volatiles to lower hydrostatic pressures at shallower depths, that in some cases permitted explosive eruptions. Wright et al. (2006) also recognised that in parts of the Kermadec arc where the arc crust was thin, water depths are high, and volcanoes tend to be basaltic to basaltic andesite, whereas where the arc crust is thicker and more evolved, anatexis of arc crust has led to more evolved magma compositions. This includes rhyolites, with higher magmatic water contents, which leads to eruptions more prone to be explosive, especially where the volcanoes grew to shallow water depths (Sect. 16.8.9). In addition to primary eruptive products, the subaerial and submarine slopes of marine stratovolcanoes are also marked by sector collapse scars, and lobate debris avalanche and base of slope fan deposits (Fig. 16.40i). Debris flow and turbidity current deposits of resedimented volcanic debris will be common (Fig. 16.42k, l). These all contribute to building volcaniclastic aprons, as for subaerial counterparts, but the deposits are distinguishable by their subaqueous mass flow facies characteristics, as discussed in Chap. 12, whereas the ring plain successions of subaerial stratovolcanoes are of fluvial-alluvial origin.

A number of stratovolcanoes in the Lesser Antilles island arc show a common evolutionary pattern from submarine to subaerial edifice (Smith et al. 2013). Successions pass upwards from tholeiitic submarine hyaloclastites, breccias, pillow and massive lava flows and interbedded calcareous sediments into emergent shallow subaqueous lavas and phreatomagmatic centres, pyroclastics and their reworked equivalents associated with coralline limestones. These are then capped by calc-alkaline subaerial facies and edifices dominated by basaltic andesitic to dacitic eruptions.

One well-documented succession is exposed on the southern flank of the Quill stratovolcano, St Eustatius. This is composed of about 200 m of limestones, volcaniclastic and pyroclastic debris that has been uptilted up to 40° (Roobol and Smith 2004; Fig. 16.43a). The limestones (known as the White Wall-Sugar Loaf) are very rich in fossils of Recent shallow water corals, mollusc shells and echinoid fragments. Most of the corals are in upright positions of original growth perpendicular to the stratification and lived in water depths of <10 m. These are interstratified with volcaniclastic horizons including beds of rhyolitic ash and pumice breccia, epiclastic material, and mixed ash and claystone beds (Fig. 16.43a). The beds of pumice breccia contain angular blocks up to 75 cm in diameter and are repeated several times throughout the succession (including beneath that depicted in Fig. 16.43a). They are thought to represent submarine rhyolitic eruptions that formed floating pumice rafts, as first observed during the submarine rhyolite eruption at Tuluman volcano in 1953–57, Papua New Guinea (Reynolds and Best 1976; Reynolds et al. 1980), and more recently described during the 2012 submarine eruption of Havre volcano (e.g. Jutzeler et al. 2014; Carey et al. 2018; Chap. 6).

It is likely these eruptions were accompanied by shallow water phreatomagmatic activity as the volcano grew close to sea level, and emergence of tuff rings or cones occurred, accounting for the many beds dominated by fine-grained rhyolitic ash. Beds with well-rounded pumice indicate significant epiclastic reworking in a shoreline environment, and redeposition during storm degradation of any emergent edifice. The upper most part of the section (present beach level) is composed of internally chaotic, massive to poorly stratified, ash-rich deposits with large andesite blocks and abundant coralline limestone debris. These include blocks of dark basaltic andesite, which show thick chilled margins with many radial cooling joints, or ropey fluidal textures, while some appear preserved within impact craters. These are interpreted as juvenile blocks and ballistic bombs formed during phreatomagmatic explosive activity when basaltic andesite magma was erupted into shallow sea water through a partially lithified carbonate bank.

The White Wall-Sugar Loaf succession provides documentation of shallow water, emergent volcanic activity over an extended period (Fig. 16.43b, c; Roobol and Smith 2004). The rhyolitic ashes and giant pumice beds record shallow water subaqueous rhyolitic eruptions at times of high sea level when emergent tuff rings were built above sea level and accompanying phreatomagmatic explosions produced very fine-grained ashes, which were often reworked by currents. The exact site of the vent for these deposits remains unknown but may well underlie a shallow water bank immediately west of the Quill. The basaltic andesite explosive breccias at the top of the succession are seen as forming during phreatomagmatic activity at a time of high sea level. This activity took place, around 40–50 ka, with the vent probably under the present cone of the Quill. Another very well-described emergent succession is that of the Miocene Cabo de Gata volcanic zone in southeast Spain (Soriano et al. 2012, 2013, 2016; Cas and Giordano 2014).

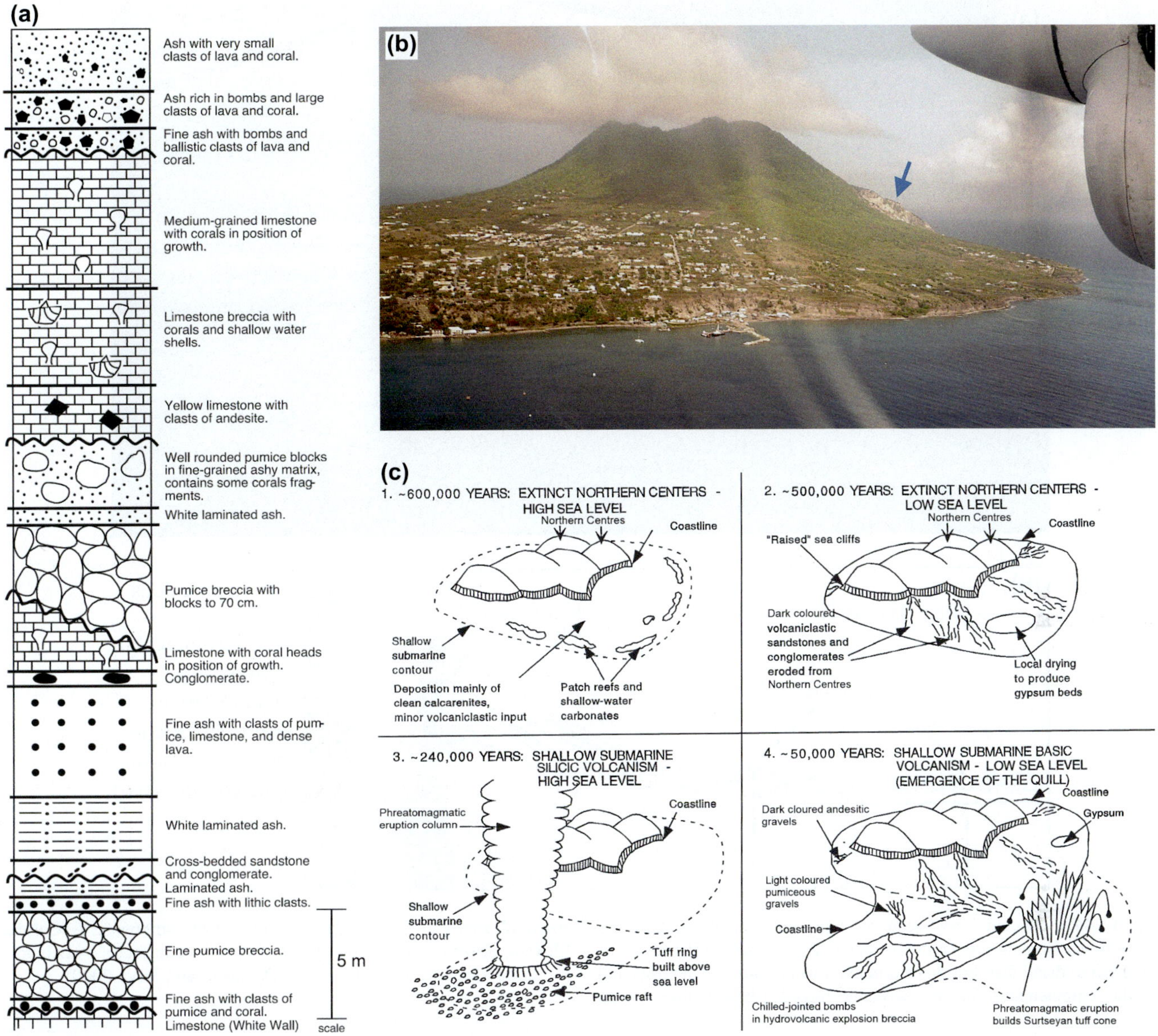

Fig. 16.43 Emergent succession exposed on the southern flank of the Quill volcano, St Eustatius, northern Lesser Antilles. **a** Detailed stratigraphy of the upper 50 m exposed in the Sugar Loaf Ridge and beach, which represents the most reliably dated part of the approximately 200 m thick section in its entirety. U-Th and U-Th series dating of corals indicate the start of this upper part is some 300 ka, while corals near the top, unconformably below the basaltic andesite explosion breccia described in the text, are 68 ka. (Modified from Roobol and Smith 2004). **b** View of the Quill and uplifted limestones forming the White Wall-Sugar Loaf (arrow). **c** Schematic model showing the main types of emergent facies that make up the White Wall-Sugar-Loaf succession. The Northern Centres are an older, eroded andesitic complex of pelean domes, lava flows and explosion craters (≤ 1 Ma). (From Roobol and Smith 2004)

16.7.5 Mass-Wastage and Surface/Epiclastic Processes

The influence of sedimentary processes on modern volcanoes has been increasingly recognised as fundamental to the understanding of their evolution (Cas and Wright 1987). If we observe modern volcanic regions, it is quite evident that reworked and resedimented volcaniclastic deposits are volumetrically important and must be significant in the geological record, a feature affirmed in Chap. 12.

Stratovolcanoes are particularly prone to mass wastage because they are high topographic features, although all modern volcanoes are continually being degraded (e.g. Karátson et al. 2012). Large-scale sedimentary repositories formed around stratovolcanoes are called "*volcaniclastic ring plains*" because they are dominated by volcaniclastic

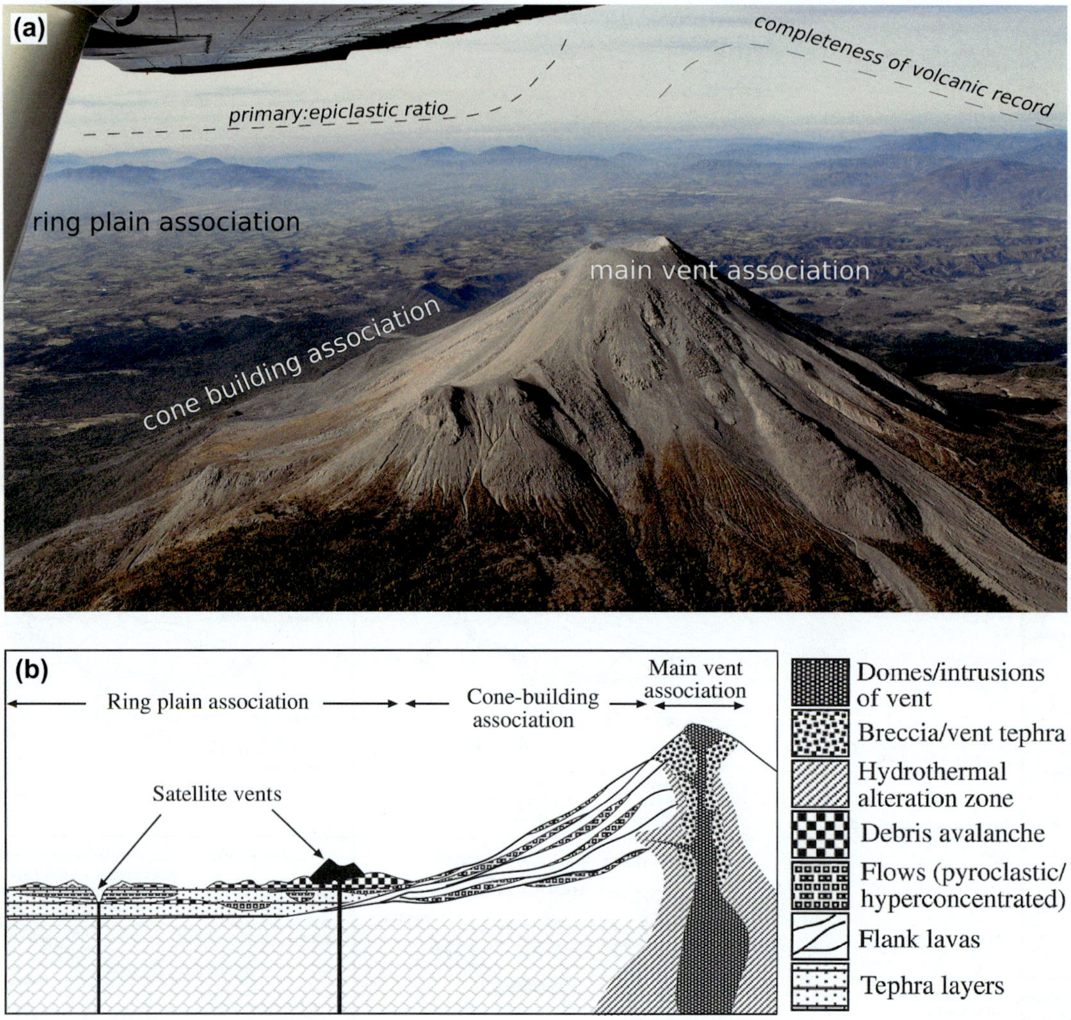

Fig. 16.44 Schematic cross-sectional representation of eruption and depositional domains of stratovolcanoes. **a** Schematic illustration of proximal to distal distribution of primary versus reworked deposits and of the completeness of the volcanic record (dashed lines) from stratovolcanoes to surrounding ring plains; the stratocone is Colima, Mexico, also imaged in Fig. 16.42a. **b** Three main facies associations for stratovolcanoes (Modified from Davidson and Silva 2000)

detritus (Fig. 16.44a, b). Following Davidson and de Silva (2000), the deposits of stratovolcanoes can be divided into three main facies associations (Fig. 16.44b):

(a) *Main vent association*: Made mostly of very proximal lava and volcaniclastic breccia deposits and intrusions (Fig. 16.42a–c); this has also been called the core zone facies association by Vessell and Davies (1981).
(b) *Cone-building association*: This forms the main cone and comprises mostly proximal and medial primary and secondary volcaniclastic facies of volcanic breccias (block and ash flow deposits, pumice flow deposits, colluvium breccias), air-fall deposits, and highly channelised debris flow, hyper-concentrated flood and stream flow deposits (collectively sometimes called lahars, Chap. 12; Fig. 16.42d–f). Vessell and Davies (1981) called this the proximal zone facies association.
(c) *Ring plain association*: This often forms vast, low gradient basins around stratovolcanoes and fills them with abundant volcaniclastic sediment. It comprises distal volcaniclastic facies dominated by fluvial-alluvial sands, breccias and conglomerates, including more laterally persistent hyper-concentrated flood flow deposits (Fig. 16.42g, h), with more rarely interbedded pyroclastic fallout (Fig. 16.42j) or pyroclastic flow and debris avalanche (Fig. 16.42i) deposits resulting from the most intense eruptive events, as well as local deposits from satellite or parasitic vents. This constitutes the medial and distal facies associations or zones of Vessell and Davies (1981).

The intensity and style of erosion during repose periods between eruptions are largely controlled by topography, climate and type of volcanic debris present on the slopes of the volcano. Explosive eruptions provide much more erodible volcanic debris than effusive eruptions, so erosion is accelerated immediately during and/or following major explosive events. This declines progressively through time, in accordance with the local climate (e.g. wet, versus dry). Seminal studies on the temporal relationships between volcanic and epiclastic processes were undertaken by Vessell and Davies (1981) and Kuenzi et al. (1979) in the Guatemalan volcanic chain, and more recently following the 1991 eruption of Pinatubo stratovolcano in the Philippines (Pierson et al. 1992).

Vessell and Davies (1981) examined a succession deposited in the past 20–30 ka around Fuego volcano in Guatemala. On Fuego, only eruptions producing >6 × 10^7 m^3 of ejecta were found to be capable of triggering a larger scale sedimentary response, with repose periods of some 100–150 years. Eruptions, which typically lasted 1 year, were followed by periods dominated by debris flows (2 yr), braided fluvial transport and delta construction along the coastline (20–30 yr) and stream incision and delta reworking (80–125 yr). Kuenzi et al (1979) described fluvial and deltaic sedimentation on the Guatemalan coastal plain in front of Santá Maria volcano following a plinian eruption in 1902. The dramatic increase in volcaniclastic sediment supply raised the bed of the Samalá River 10–15 m, and from 1902–1922 a deltaic platform prograded nearly 7 km seaward, depositing a prism of sediment of some 4 km^3. With waning sediment supply, the delta was gradually destroyed over the next 25 years and sands redistributed laterally into shoreface and beach environments. Such sporadic episodes of delta formation and subsequent reworking of deltaic sands along sedimentary strike would produce a complex sand body showing a variety of vertical sedimentary sequences generated by a number of shoreline processes. These would overlie deeper water facies (e.g. prodelta turbidites) and be overlain by an alluvial complex which would inter-tongue landward with proximal volcanic facies. For more detailed discussion of these two examples see Cas and Wright (1987) and Chap. 12.

Studies of more recent eruptions at Mt St Helens 1980, Mt Galunggung 1982 and Mt Pinatubo 1991 have demonstrated exceedingly high values of post-eruption accumulation of volcaniclastic sediment in downstream channels and other environments in the first year, particularly in the first few months (Pierson et al. 1992; Fig. 16.45). Mt Pinatubo provides an excellent and very well-studied example of post-eruption erosion processes (Pierson et al. 1992; Newhall and Punongbayan 1996). During the June 1991 plinian eruption pyroclastic flow deposits filled and buried narrow, deeply incised radial valleys with up to 220 m of highly erodible pumice and ash. Monsoonal and typhonic rains during the first rainy season drove very heavy removal of fresh pyroclastic detritus as "hot" lahars from the upper reaches of especially the eastern side of the volcano, with up to several tens of lahar events per individual drainage system in August. This led to some of the highest sediment yields ever recorded (some $10^6 m^3$ km^2 yr^{-1}; Fig. 16.45a), which decayed exponentially for the first decade following the eruption (Hayes et al. 2002; Gran and Montgomery 2005). However, since 2001 alluvial fans have been aggrading faster than the decay model would predict, with rates that have levelled off and are no longer declining exponentially (Fig. 16.45b; Gran et al. 2011). Patterns of long-term sedimentation are similar to those at Mt St. Helens, with two phases of erosion and sediment export (Fig. 16.45c). In the first, erosion of hill slope tephra and carving of the initial valley network lead to high sediment yields that decline exponentially over 5–10 yr. In the second phase, continued valley widening and fluvial instability maintain high sediment yields for at least two decades. A comparison of post-eruption deposition with pre-eruption yields indicates that century- to millennial-scale sedimentation is dominated by sediment associated with discrete eruptive events.

Similarly, marine stratovolcanoes are also subject to major mass-wastage, involving landslides, debris flows, and high energy, fluvial gullying, and torrent flow, on subaerial flanks, as well as seaslides, debris flows and turbidity current processes on their submarine flanks (Chap. 12). In addition to these sustained degradation processes, marine stratovolcanoes are particularly prone to explosively or non-explosively induced sector collapse whereby a large segment of the edifice is mobilised into a dynamic slide and/or debris avalanche (Chaps. 10, 11 and 12). Their submarine flanks could therefore grow considerably by the radial or sectoral growth of a large submarine volcaniclastic apron fed by such mass-flow currents. Marine geophysical and drilling investigations over the past 20 years have highlighted two regions where sector collapse is a recurrent process and large-magnitude submarine volcaniclastic aprons have been identified, namely, in and around the Canary Islands and islands of Lesser Antilles volcanic arc which are highlighted below, but also including well-known volcanic islands such as Stromboli (Bosman et al. 2009; Casalbore et al. 2011), and Réunion (Kelfoun et al. 2010).

As previously indicated, the Canary Islands represent some of the largest oceanic volcanoes on Earth, which are surrounded by thick subaqueous volcaniclastic aprons extending several tens of km seawards to depths of >4 km (e.g. Schmincke and Sumita 1998; Funck and Lykke-Andersen 1998; Krastel and Schmincke 2002; Seisdedos et al. 2012; Fig. 16.46). These aprons closely correlate with a number of onshore and proximal marine landslides.

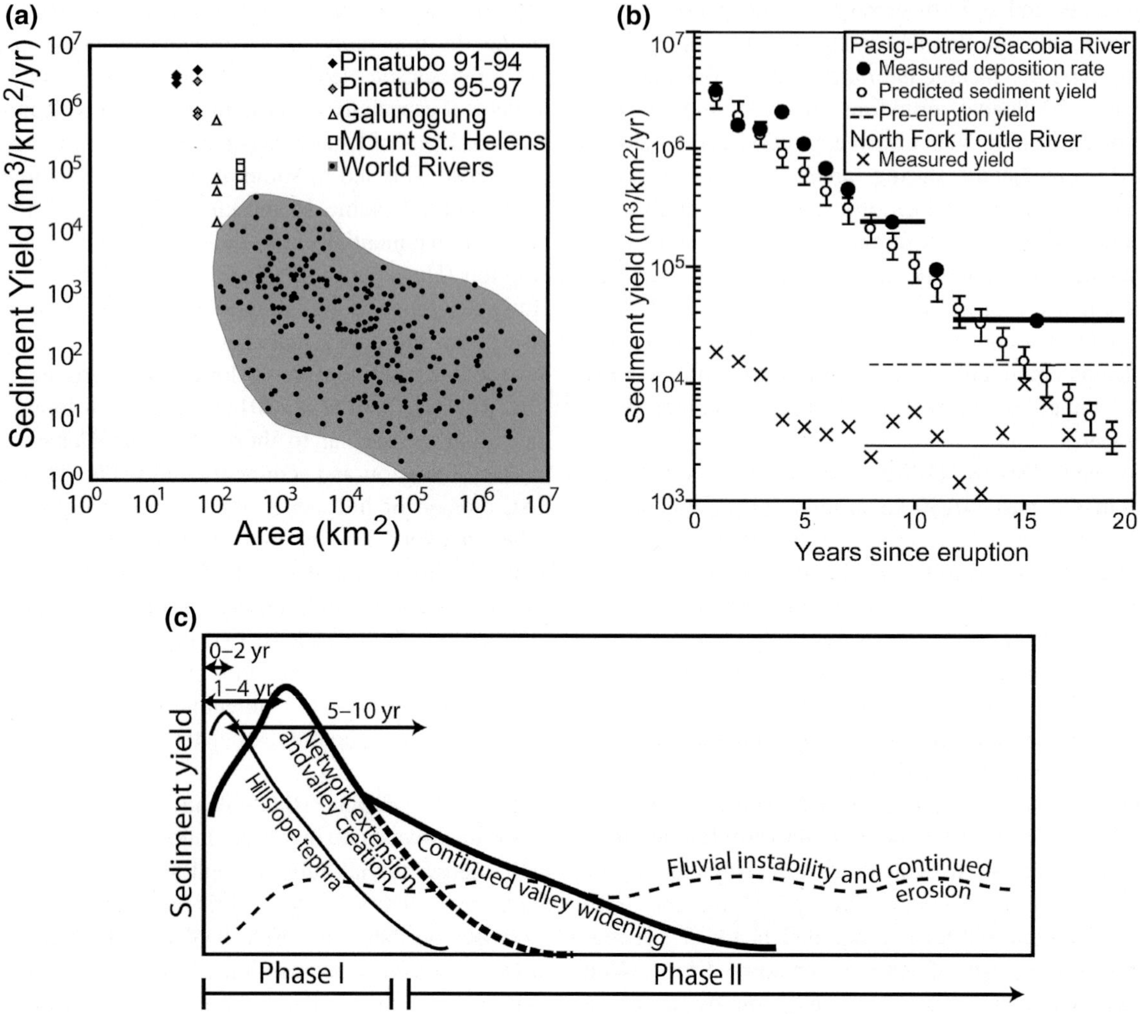

Fig. 16.45 Effects of three recent eruptions on sediment budget. **a** Sediment yields on volcanically disturbed rivers as compared to normal world rivers. Shown are sediment yields from the 4 years following eruption, and in the case of Pinatubo those for 1995–1997. (From Gran and Montgomery 2005). **b** Longer term sediment yields at Mt Pinatubo (Pasig-Potrero and Sacobia Rivers) and Mt St. Helens (North Fork Toutle River). Bold lines indicate periods where measured sedimentation data are averaged over multiple years. Pre-eruption yields from Mt Pinatubo show that in the first 19 yr the equivalent of >750 yr of sediment was transported at pre-eruption rates (dashed horizontal line); excluding a major monsoonal flood in 1972 (solid line) this is equivalent to 4000 yr at pre-eruption rates. The last previous eruption at Mt Pinatubo was ca. 500 yr ago. Other eruptions have occurred several thousand years apart, which were even larger than 1991. Post-eruption sediment export thus dominates century- to millennial-scale sediment budgets. (From Gran et al. 2011). **c** Model illustrating change of dominant sediment sources through time. Magnitudes of each source vary greatly depending on the eruption. In phase I, hill slope tephra and valley creation lead to high sediment yields early that decline exponentially through time. In phase II, continued valley widening and fluvial instability maintain high sediment yields for decades. (From Gran et al. 2011)

The onshore landslide scars represent some of the biggest known landslides containing 50–200 km^3 mobilised rock (e.g. Carracedo et al. 1999; Seisdedos et al. 2012). Offshore, these correlate with a number of distal volcanic island-sourced turbidites which extend >1000 km onto the surrounding abyssal plains (Hunt et al. 2014). These volcaniclastic turbidites (comprising some 125 beds or events) on the Madeira Abyssal Plain provide a record of multiple, large-volume volcanic island sector collapses over the past 17 Ma. The turbidite records demonstrate that landslides often coincide with protracted periods of volcanic edifice growth, suggesting that loading of the volcanic edifices may be a key preconditioning factor for landslide triggers. Furthermore, the last large-volume failures from Tenerife coincide with explosive volcanism at the end of caldera-forming eruptive cycles (Paris et al. 2017; Hunt et al. 2011, 2014; Cas et al. 2022). Many large-volume Canarian landslides also appear to have occurred during periods of warmer and wetter climates associated with sea level rise and subsequent high-stand conditions.

In the Lesser Antilles, as first proposed by Roobol et al (1983), sector collapse has now been recognised on several

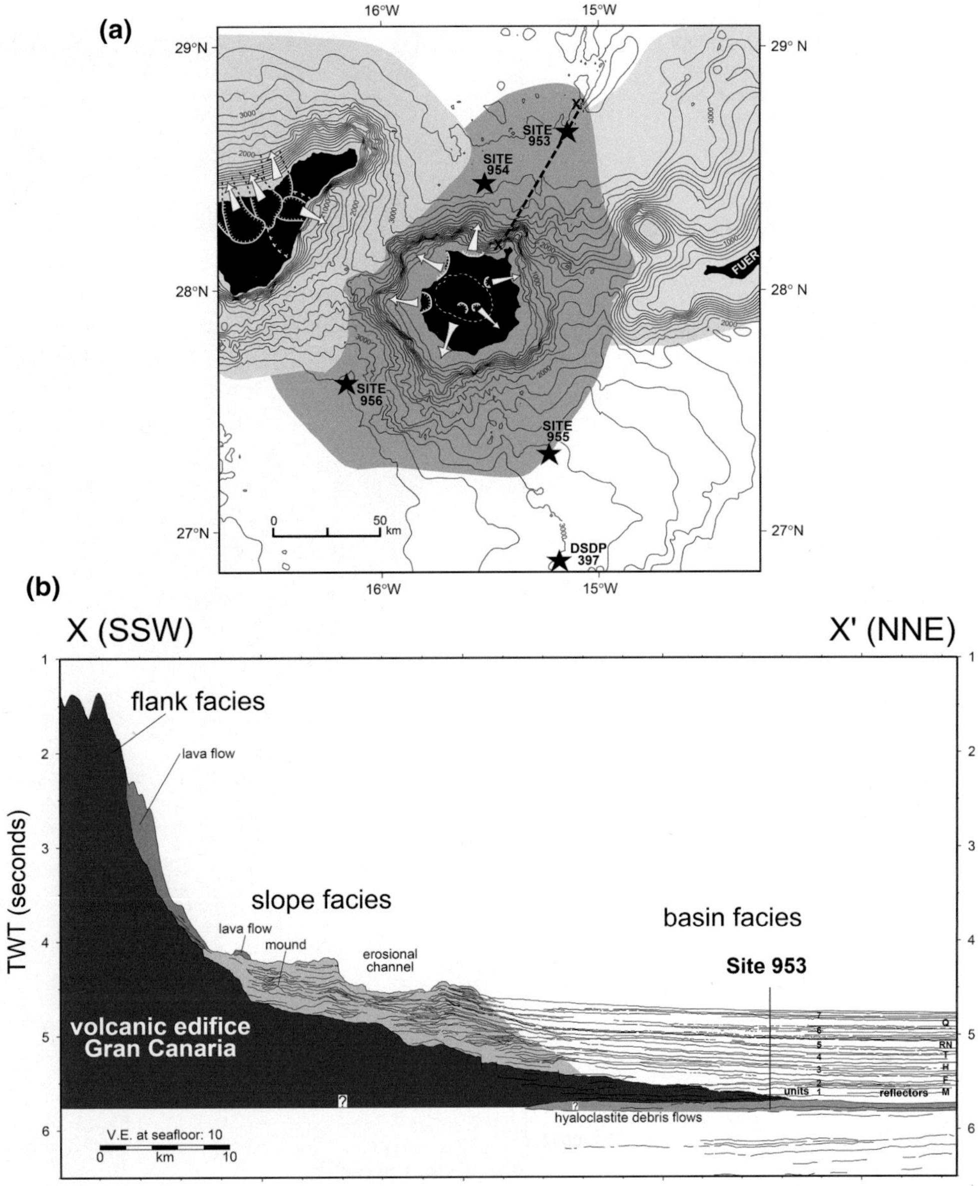

Fig. 16.46 Sedimentation around Gran Canaria, Canary Islands, Spain. **a** Contour map of Gran Canaria and parts of Fuerteventura and Tenerife showing distributions of submarine volcaniclastic aprons which decrease in age from east to west. Amphitheatres of major landslides on Tenerife and Gran Canaria (scars) and main pathways of debris avalanches (white arrows). ODP Leg 157 drilling sites are shown, X-X′ is line of seismic section below. (Modified from Schmincke and Sumita 1998). **b** Interpreted seismic line through marine volcaniclastic apron north of Gran Canaria subdivided into slope and basin facies (From Funck and Lykke-Andersen 1998)

volcanoes and confirmed by the identification of large-magnitude debris flow avalanche deposits off the west coasts of Dominica, Martinique and St. Lucia (e.g. Deplus et al. 2001; Boudon et al. 2007). Seafloor mapping shows that these form several subaqueous volcaniclastic fans, which are characterised by hummocky topography and megablocks, and extend >90 km into the Granada Basin with associated volumes up to 20 km^3 (Fig. 16.47). Each fan was produced by a sector collapse, that can be traced to "horse-shoe-shaped" collapse scars on active volcanoes onshore, which then flowed westwards down into the Caribbean Sea. These fans are therefore restricted to the back-arc sectors of the arc volcanoes, where both aerial and submarine slopes are steeper than on the eastern slopes. This asymmetry probably results from progressive loading by accumulation of volcanic products on the western slopes and development of long-term gravitational instabilities. Meteoric and hydrothermal fluid circulation also creates weakened areas favouring the recurrence of flank collapses (Le Friant et al. 2003). Recurrent episodes of mass wasting during the cyclic growth of the submarine Los Frailes volcanic complex (Miocene Cabo de Gata volcanic arc, Spain) are well described in Soriano et al. (2012), made of shattered lava megablocks with contractional jointing embedded within a chaotic hyaloclastite matrix.

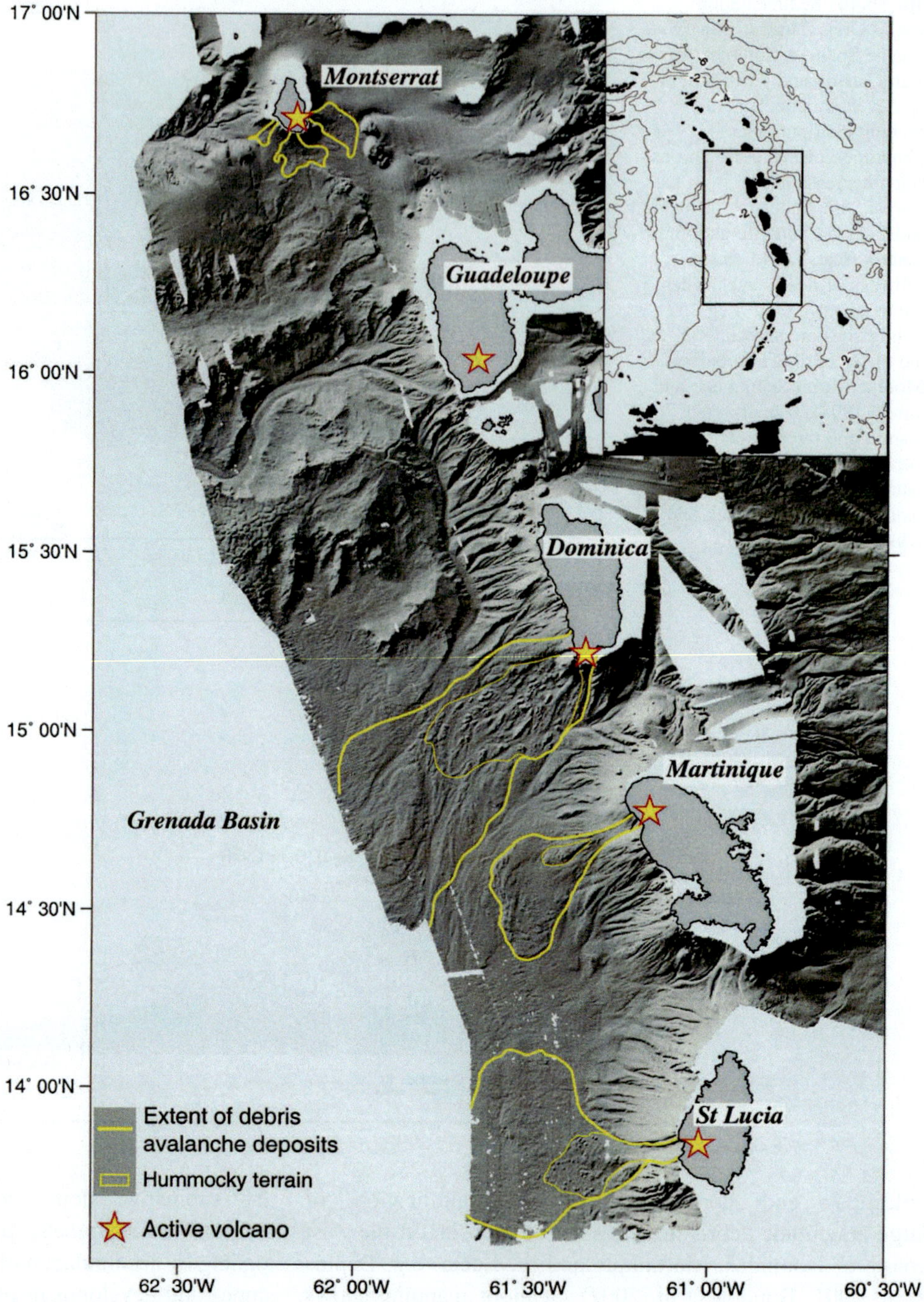

Fig. 16.47 Extent of debris avalanche deposits on the seafloor surrounding the Lesser Antilles island arc, eastern Caribbean Sea. Volcanoes active in the last 3000 yr are shown by stars. Limits of the deposits are superimposed on swath bathymetry. Thick yellow lines show the whole extent of debris avalanche deposits deduced from the integrated geophysical analysis; thin yellow lines are the limits of hummocky terrain covered by megablocks. The inset shows the regional bathymetry; contour interval = 2 km. Note in the northern part of the arc, several smaller debris avalanche deposits are also present on the east, south and west submarine flanks of Soufrière Hills volcano, Monserrat, which erupted from 1995 to recently. Flank collapse may be a recurrent process in the recent history of this volcano; the eastern deposit is probably partly related to the formation of English's Crater interpreted as a collapse structure formed 4000 yr ago. (From Deplus et al. 2001)

16.7.6 Post-caldera and Post-sector Collapse Stratovolcano Building Volcanism

In some cases, stratovolcanoes can grow on the floor or along the ring fractures of calderas after their formation, either in the summit calderas of stratovolcanoes, or in larger caldera volcanoes (Sect. 16.8.7). This is the case at Santorini in the southern Aegean Sea, which is a very complex volcanic system (Sect. 16.9.1.1). Since the eruption of the Minoan ignimbrite, the summit of a dacitic lava shield named Nea Kameni (or "new island") has emerged during the past 3.5 kyr from the submerged caldera floor and is currently active (Figs. 16.48a and 16.78e, f; Fytikas et al. 1990; Druitt et al. 1999). A similar new and active stratovolcano called Anak Krakatau ("child of Krakatau"), emerged at the centre of the 1883 Krakatau caldera (Fig. 16.48b, c; Dahren et al. 2012); this itself suffered flank collapse on 22 December 2018, generating a devastating tsunami which inundated the west Java and southern Sumatra coastlines (e.g. Williams et al. 2019; Chap. 12).

Fig. 16.48 Examples of post-caldera collapse stratovolcanoes and post-sector collapse volcanism. **a** Nea Kameni, Santorini, Greece, is a dacite lava shield that could develop into a stratovolcano (Photo by Olaf Tausch; *Source* Wikimedia through Creative Common Attribution 3.0 Unported licence). **b** Krakatoa caldera, 50 km west of Java, Indonesia; the active central cone Anak Krakatau first emerged in 1927 (Landsat 7 satellite ca. 2001; *Source* Landsat programme NASA). **c** Anak Krakatoa has most recently been active since 2017, culminating in flank collapse and phreatomagmatic eruption on 22 December 2018, dramatically transforming its morphology (Chap. 12) (Photo by Tyke, taken January 2016; *Source* Wikimedia through Creative Commons Attribution 4.0 International licence). **d** Batur caldera and young stratovolcano, Bali, Indonesia; note caldera wall and lake in background (Photo by Tropical Living; *Source* Wikimedia through Creative Commons Attribution 3.0 Unported licence). **e** Rinjani caldera with its active cone Barujari which last erupted in 2015, Lombok, Indonesia; the caldera-forming eruption is thought to have taken place in the thirteenth century AD (Photo courtesy of J-C Komorowski). **f** Pico El Teide (large cone, centre) and Pico Viejo (smaller cone, left) stratovolcanoes, which developed on the southern rim of the Las Cañadas coalesced caldera complex after the last caldera-forming eruption ~190 ka, Tenerife, Canary Islands, Spain. The black lava from a flank vent on Pico Viejo erupted in 1798, the dark lavas on Teide in ~850 AD (*Source* Google Earth; also see Plate 16.1a, b). **g** Petit and Gros Piton dacite lava "spires" erupted inside the Qualibou depression (shown schematically by yellow dashed line); small town of Soufrière is nestled along the northern margin of the structure, St Lucia, eastern Caribbean Sea (Photo by Aneil Lutchman, looking south; *Source* Wikimedia through Creative Commons Attribution-Share Alike 2.0 Generic licence)

Fig. 16.48 (continued)

Elsewhere in Indonesia, the Batur caldera contains a 700 m high active stratovolcano (Mt Batur) rising above the surface of Lake Batur (Fig. 16.48d). The first documented eruption was in 1804 and it has been frequently active, as recently as 2000. Located just 18 km to the northwest stands Agung stratovolcano, and the Batur caldera stands as an omen to the likely future evolutionary trajectory of Agung volcano (Fig. 16.40e). On the neighbouring island of Lombok, Rinjani caldera contains the cone of Barujari (Fig. 16.48e), which has been frequently active since 1847.

At Vesuvius, the central "Gran Cono" has grown since the calamitous eruption of 79 AD, within a caldera formed by the collapse of the original Somma edifice, a polyphase caldera, which first collapsed at 18.3 ka (see Fig. 16.40f). This has given rise to the term "*somma*" used to describe a high, circular or crescent-shaped ridge with steep inner walls including the rim of an older crater or caldera surrounding a central volcanic cone, and sometimes for sector collapse scars. Mt Kilimanjaro volcano (Tanzania) comprises a deeply eroded remnant caldera formed about 1.9 Ma, with two younger cones, the youngest being Kibo, which has a basal diameter of some 24 km and forms the highest peak in Africa (5,895 m a.s.l.). However, perhaps the most spectacular example is the double stratovolcanoes of El Teide and Pico Viejo, which both formed over the past 200 ka within the Las Cañadas caldera, on Tenerife, Canary Islands, Spain (Plate 16.1; Fig. 16.48f).

Post-sector collapse volcanism is also a common feature, as found within several of the collapse scars in the Lesser Antilles volcanoes, such as on the islands of Martinique and St. Lucia (Boudon et al 2013; see Fig. 16.47), Stromboli (Fig. 16.40i) and Mt St Helens (Figs. 6.1e and 16.40g). At Mt Pelée, following a 32 ka flank collapse, magma compositions changed from silicic to mafic generating abundant scoria and ash flows. Rapid building of a new cone increased the load on magma bodies at depth and their density threshold. Magma production rate decreased while composition of the erupted products changed to more silicic compared to the preceding period of activity. These younger, low-density magmas generated plinian and dome-forming eruptions up to the Present. In contrast, in the Pitons du Carbet centre to the south of Mt Pelée and Qualibou depression on St Lucia (Fig. 16.48h), sector collapse had an opposite effect. In both cases, silicic magmas, erupted immediately after the flank-collapses, are highly porphyritic (up to 60% phenocrysts) and much more viscous than the magmas erupted before the flank-collapses. They have been generally emplaced as voluminous and very high aspect ratio lava domes (i.e. "*pitons*"). Such magmas could not ascend without a significant decrease of the density threshold effect produced by the volcanic edifice loading before the flank collapse.

Geothermal activity is also a common feature of some stratovolcanoes which have suffered caldera or sector collapse, including the Qualibou depression (Sulphur Springs). Other examples are Nysiros volcano, Hellenic arc, Greece (Marini et al. 1993; Ambrosio et al. 2010) and Whakaari-White Island, New Zealand (Mayer 2015). A hydrothermal explosion at Whakaari Island on 9 December 2019, killed 21 and injured a further 26 tourists visiting the geothermal area at the time (Gibson 2020).

16.7.7 Facies Models for Stratovolcanoes

Given that environment (subaerial, subaqueous) affects volcanic eruption styles and the resultant types of deposits, there will be clear differences in the facies associations between subaerial and marine stratovolcanoes.

16.7.7.1 Continental Stratovolcanoes—Facies Model

Continental stratovolcanoes show complex primary variations in erupted products in time and space, and epiclastic processes and mass-wastage of the steep-sided cones further complicate the facies architecture and stratigraphy of the pile (Fig. 16.49). In the preceding discussion, we systematically described the principal volcanic elements of stratovolcanoes. Short lavas, domes and shallow intrusions (commonly basaltic andesite to dacite in composition) are the main cone-building elements, which together with radial dykes armour the cone against erosion. These are flanked by, and interdigitate with, various types of pyroclastic and epiclastic deposits. Pyroclastic deposits may show a wide range in composition and eruptive style. However, it will only be the deposits of large plinian and ignimbrite eruptions, especially welded deposits, that tend to be preserved in the geological record. Non-welded pyroclastic deposits will, for the most part, be eroded from the volcano slopes and be redeposited as mass-flow deposits and reworked epiclastic sediments in basin areas marginal to the volcanic centre, in the surrounding ring plain setting. Much of this probably occurs very shortly after their eruption (1–10 yr) with much of the deposition taking place on the developing ring plain of these volcanoes.

Perhaps the one element that characterises stratovolcanoes is their variability. As concluded by Cas and Wright (1987; Fig. 16.49), the facies architecture and stratigraphy of stratovolcanoes demonstrate:

(a) Rapid inconsistent lithological and compositional changes in the vertical succession.
(b) Rapid lateral lithofacies changes which, moving away from the source, may fit into a systematic facies pattern.

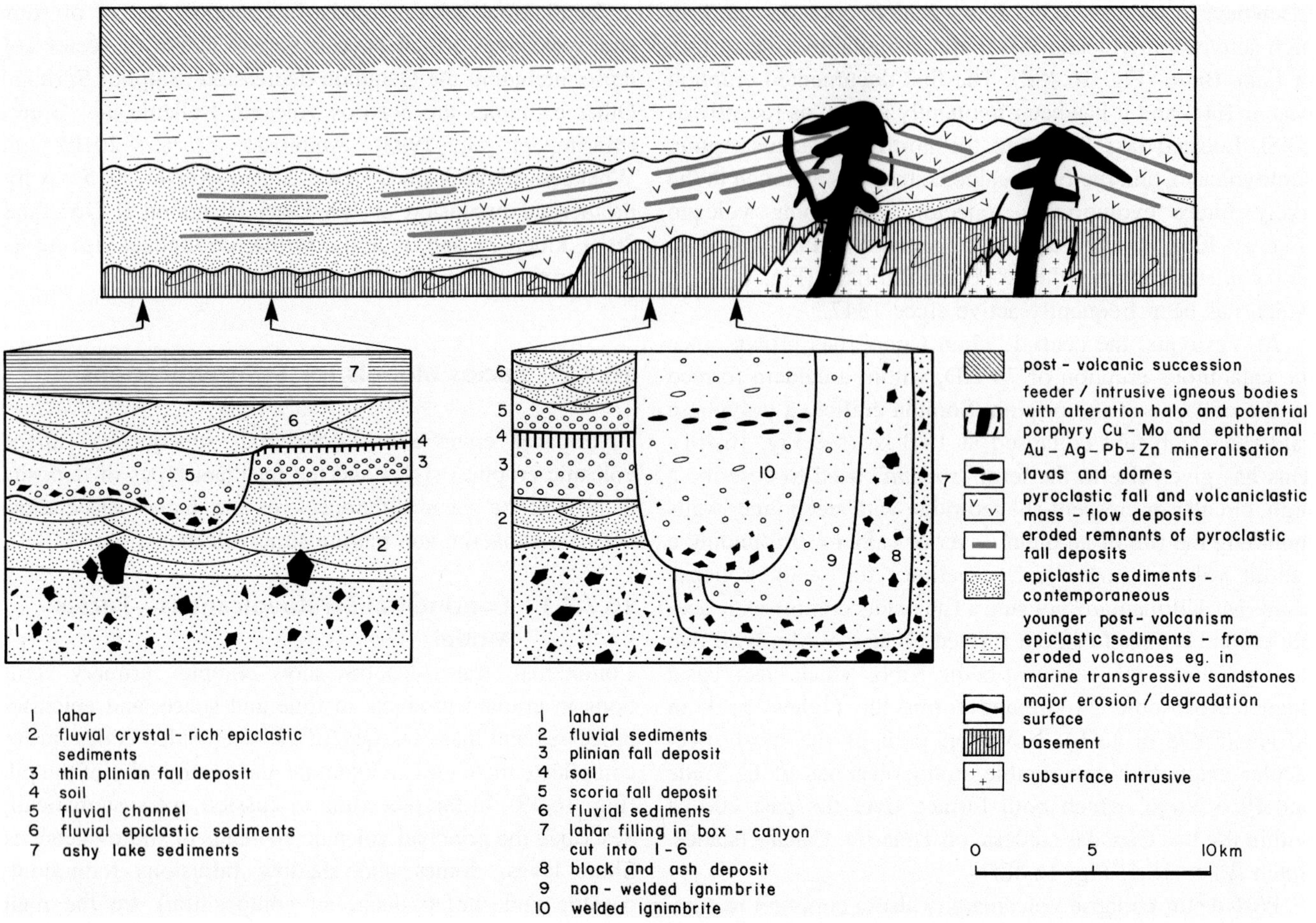

Fig. 16.49 Generalised facies model for continental stratovolcanoes. Significant and geometrically complex facies variations occur between the proximal near-vent cone facies and the distal ring plain environment where epiclastic volcaniclastics are dominant. Note that this model is highly generalised, depicting a relatively large proportion of volcaniclastic deposits. As indicated in the text many stratovolcanoes consist of a much higher proportion of lavas and intrusions than shown here, especially in the near-vent "core" zone. (From Cas and Wright 1987)

Around the centre of the cone, facies will be discontinuous, controlled by deep valleys, with channelling of lavas, coarse pyroclastic flow deposits, and other types of volcaniclastic flows (e.g. 2018 eruption of Mayon volcano, Philippines). Lavas, domes and shallow intrusions together constitute the "core" of the volcano, which will also have active and extinct hydrothermal systems.

(c) Away from the vent on the middle and lower slopes, there will be an increasing proportion and variety of volcaniclastic breccias, which may be pyroclastic flow deposits of various types and facies (Chap. 10) and secondary epiclastic flows due to mass-wastage (Chap. 12).

(d) The lower slopes and surrounding ring plain will be marked by thick successions of immature volcanic detritus that have accumulated in alluvial-fluvial, or marine settings, or both.

16.7.7.2 Marine Stratovolcanoes—Facies Model

Marine stratovolcanoes are generally similar to their solely subaerial counterparts: they have significant relief above their base, they are large and steep-sided, they may have summit calderas, and they have diverse magmatic compositions. However, they are significantly different in that they have a base built of submarine facies dominated by pillow and sheet lavas, lava domes and variably resedimented hyaloclastite (Fig. 16.50), and a summit region which may be built of subaerial facies, if the edifice has emerged above sea level (see Fig. 15.43a). Sandwiched between these submarine and subaerial facies associations, relatively thinner successions of shallow marine facies, including increased proportions of pyroclastic deposits, may be preserved; these may form weak or ductile layers along which landslides may be prone. Such volcanoes are almost invariably associated with island arc systems (Chap. 17) built on prominent arc crustal blocks which consist of older,

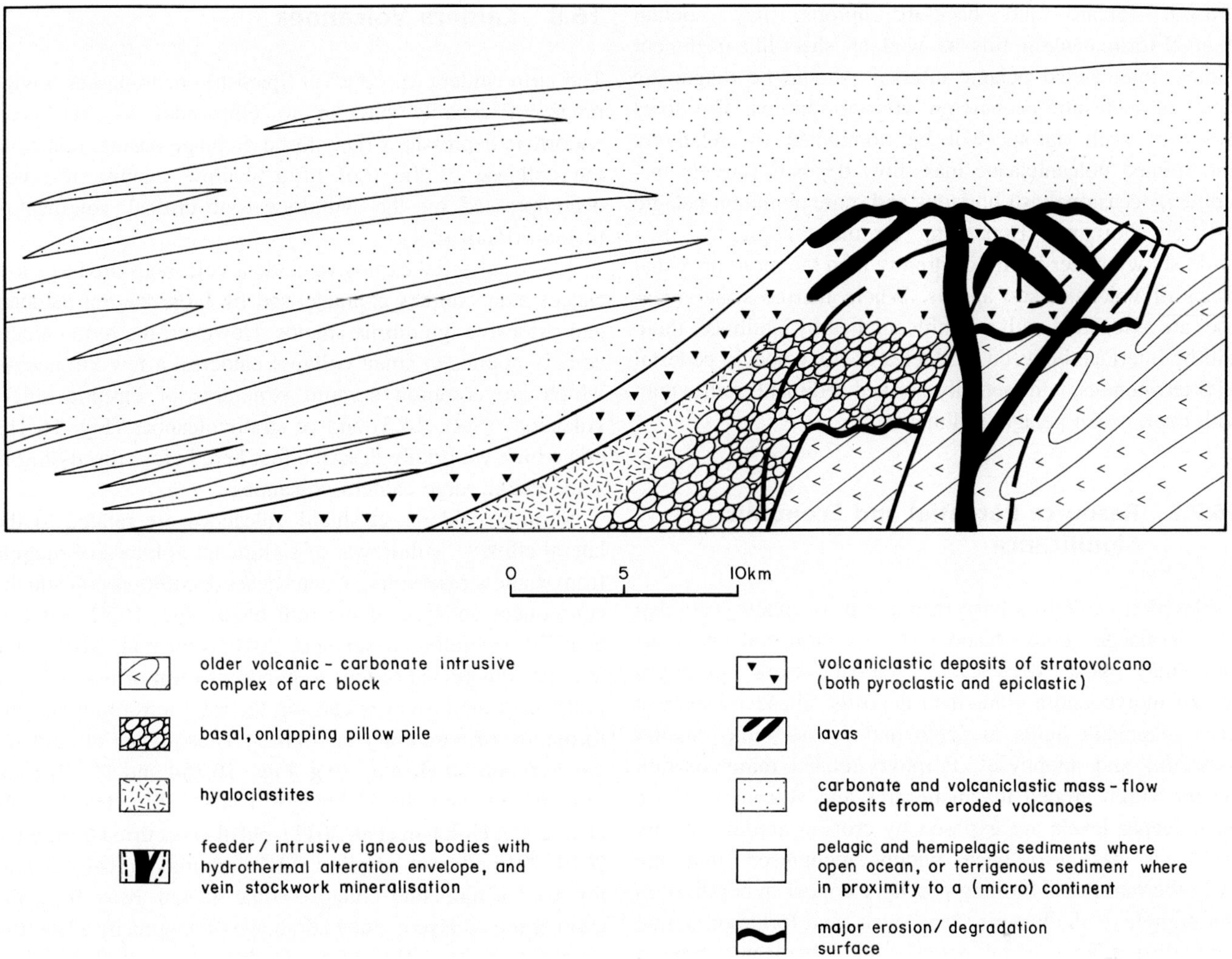

Fig. 16.50 Generalised facies model for marine stratovolcanoes. Compared with continental stratovolcanoes, the principal differences are the presence of submarine pillow lavas and hyaloclastites forming the sub-structure of the edifice, and thick successions of submarine mass-flow deposits potentially forming very extensive volcaniclastic aprons (From Cas and Wright 1987; also see Fig. 15.43)

deformed arc volcanic piles, associated intrusions and younger carbonate cappings, upon a foundation of submerged oceanic, arc or continental crust (Fig. 16.50). Such settings are associated with significant epiclastic erosional, transportational and depositional processes (Chap. 12), which continue after the volcano has ceased to be active, leading to erosional removal of most of the subaerial top of the volcano if it was emergent, and formation of a wave planed, submarine platform.

High rates of subaerial mass-wastage and shoreline erosion will be accompanied by major submarine mass-flow processes, shedding debris into associated basins. These produce coarse, proximal forearc and back-arc sediment aprons built up largely by mass-flow deposits (debris flows, debris avalanches, turbidity currents, slumps and slides). In this regard, fully emergent stratovolcanoes will be different from seamounts in that their initial lava-dominated foundation will be buried by, and flanked by a volumetrically, equally or even more significant, volcaniclastic apron (Fig. 16.50). The back-arc apron, which is under the direct influence of the active volcanic arc, will contain much juvenile, contemporaneous volcanic debris, as well as older epiclastic, volcanic detritus.) In the Lesser Antilles arc the back-arc apron is depleted in primary pyroclastic fall material because of the preferential transport of ashes eastwards by strong prevailing westerly wind systems at higher altitudes (Sigurdsson et al. 1980; Cas and Wright 1987). Most of the back-arc apron therefore consists of mass-flow deposits including major debris avalanche fans as a result of sector collapse.

Both forearc and back-arc aprons may contain channel-form canyon fills as well as sheet-like sediment bodies. Intercalated pelagic intervals should be diagnostic facies of such intra-oceanic arc volcano terrains. The distal facies of such aprons will be dominated by relatively fine-grained volcaniclastic mass-flow deposits (largely turbidites), occasional ash horizons and more abundant pelagic sediment horizons. Depending on the topography, they could show greater lateral continuity than the more proximal facies of volcaniclastic aprons. Where marine stratovolcanoes are associated with a significant, nearby landmass, there will be interface between the facies of the volcanic pedestal with terrigenous, continentally derived sediment, rather than with open ocean pelagic sediments.

16.7.8 Resource Potential and Economic Significance

Exploration geologists have recognised for many years that andesite-dacite stratovolcanoes (both continental and marine) may host deep-level, vein-stockwork porphyry copper-molybdenum-gold-silver deposits. These are sourced from magmatic fluids at depth and focused into smaller intrusions and apophyses. Porphyry-related mineralisation occurs within eroded magmatic arcs and volcanoes where their deeper levels are exposed by erosion and/or tectonic uplift. It has also been widely recognised that the hydrothermal fluids forming porphyry copper mineralisation can migrate to shallower crustal levels, and form epithermal gold-silver $\pm$ base metal deposits in subvolcanic breccia pipes, diatremes and vein systems in overlying volcanic successions, and in surface hydrothermal systems, as discussed more fully in Chap. 18. Both active continental and marine stratovolcanoes that are emergent above sea level also have excellent geothermal energy resource potential.

In addition to porphyry copper and epithermal systems, where a substantial part of the stratovolcano is submerged, seafloor hydrothermal springs may have potential for producing massive, exhalative sulphides or replacement bodies (e.g. de Ronde et al. 2005, 2007; Chap. 18). These could occur in two sites, either: on the flanks, associated with lateral hydrothermal springs, or where the summit is a collapsed caldera, in a summit basin fed directly by the vent hydrothermal system (e.g. Santorini, Krakatau and Deception Island, Antarctica). The steep flanks may be subject to considerable epiclastic erosional and mass-flow depositional processes long after eruption has ceased, so that ore precipitation could be interrupted or could also be redeposited downslope by mass-flow processes, or both.

16.8 Caldera Volcanoes

The term caldera (or "pot" in Spanish) encompasses a wide range of broadly circular to ellipsoidal to rectilinear, endorheic depressions (i.e. closed drainage basins) related to the collapse of the roof of a shallow magma reservoir decompressed by the withdrawal of critical volumes of magma (Plate 16.1).

Large subaerial explosive caldera volcanoes are huge low aspect ratio edifices characterised by large central calderas and extensive ignimbrite shields. However, the term caldera equally applies to small collapse calderas, a few km across, which are common summit structures of basaltic shield volcanoes (Sect. 16.5) and of stratovolcanoes (Fig. 16.41), and which we briefly describe first here before discussing in some detail, large caldera volcanoes.

Summit calderas at shield volcanoes are related to the lateral effusive withdrawal of significant volumes of magma from shallow reservoirs, which causes decompression and the consequent collapse of the roof block (Fig. 16.51; Michon et al. 2011; Gudmundsson et al. 2016; Neal et al. 2019). This may be constructed of tens to hundreds of m of basaltic lava units, in some instances causing the total unroofing and the formation of temporary lava lakes. These types of calderas are common on Hawai'i (e.g. Figs. 16.25d and 16.29) Reunion Island, and the Galapagos Islands (Fig. 16.33; Cole et al. 2005; Holohan et al. 2011), but also occurred during the 2014–2015 effusive eruption at Bárdarbunga in Iceland. This involved a moderate collapse some 48 km away from the main eruptive fissure, due to drainage of magma by a laterally propagating dyke (Fig. 16.51; Gudmundsson et al. 2016).

Explosive collapse calderas can vary enormously in their dimensions (Fig. 16.52), and can form distinctive edifices in their own right, or occur at the summits of some stratovolcanoes, or form parts of more complex volcanic systems. Summit calderas at stratovolcanoes are related to the explosive ejection of significant volumes of magma, usually in the order of 1–10 km^3 DRE (Dense Rock Equivalent), and VEI 5–6/VEI* 9–10 magnitude eruptions (Volcanic Explosive Index, Newhall and Self 1982; and modified as discussed in Chap. 9, see Sects. 9.2 and 9.10.5). Historically this has happened at many volcanoes including Vesuvius in 79 AD, Krakatau in 1883, and Mt Pinatubo in 1991. The structural diameter of such calderas is usually ≤ 10 km. The eruption of Mt Pinatubo in 1991 removed a small part of the summit, leaving a small caldera 2.5 km in diameter (Fig. 16.52; Stix and Kobayashi 2008). Pre-historic examples include the various caldera-forming eruptions of Vesuvius (Fig. 16.40f; Cioni et al. 1999), the Cerro Blanco volcano (Argentina; Báez et al. 2020) and Mt Mazama, USA

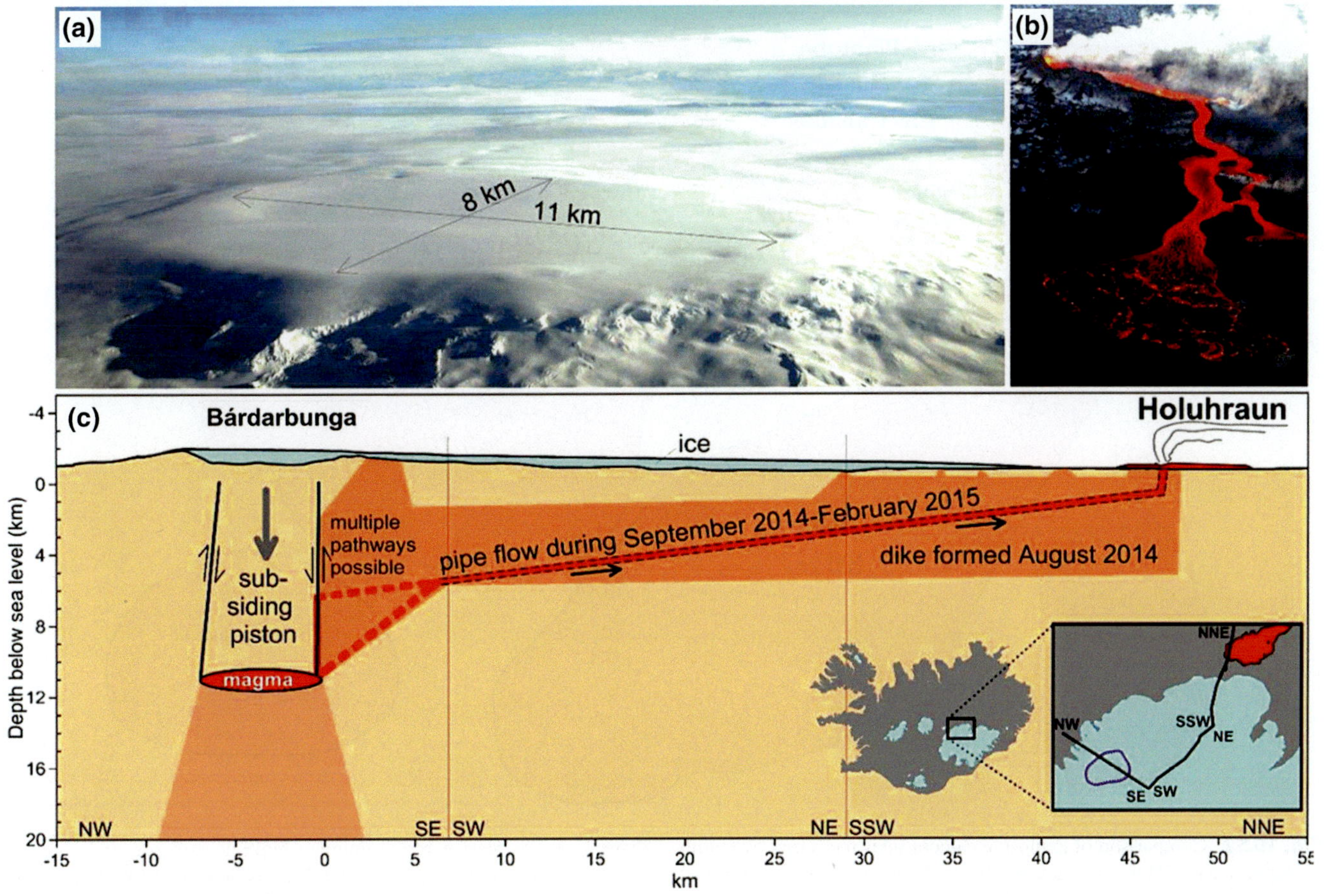

Fig. 16.51 Bárdarbunga lateral magma withdrawal and caldera collapse, Eastern Volcanic Zone, Iceland (From Gudmundsson et al. 2016). **a** Ice-covered Bárdarbunga caldera below the Vatnajökull ice cap. **b** 2014–2015 Holuhraun fissure and active lava field, which erupted 2 km beyond the northern limit of the ice cap. **c** Interpretation of eruptive, seismic and geophysical data, in terms of lateral intrusion of a dyke which extended for 48 km and fed the eruption by withdrawing magma laterally from the central magma reservoir

(Figs. 16.41a, b and 16.52; Bacon, 1983). During the 7.7 ka eruption of Mt Mazama, which produced the spectacular Crater Lake summit caldera, ~1,600 m of the top of the pre-existing stratovolcano collapsed, for the most part into the magma chamber. Such types of major explosive eruption and their resulting ignimbrite deposits are described in Chap. 10.

Young, discrete calderas often appear to record just one major caldera-forming explosive eruption; e.g. the 1257 AD Samalas (Rinjani) caldera, Lombok (Lavigne et al. 2013; Fig. 16.48e), the 1815 Tambora caldera (Self et al. 1984; Fig. 16.41e), the 1883 Krakatau caldera (Self and Rampino 1981; Fig. 16.48b), or the ~1000 AD or Millennium eruption of Changbaishan volcano, China-North Korea border (Sun et al. 2014; Fig. 16.41c). However, in most cases, larger calderas result from multiple caldera-forming eruptions, so that they are actually coalesced clusters of calderas or "nested" caldera complexes formed during multiple, individual major explosive eruptions. For example, Rabaul caldera on New Britain, Papua New Guinea, comprises a larger 8 × 14 km diameter caldera (Fig. 16.53a), within which the most recent phase of caldera collapse (6 × 10 km diameter) occurred during a major VEI 6/VEI* 10 (Chap. 9) explosive eruption about 1,400 yr BP (~680 AD; McKee et al. 2015; McKee and Duncan 2016). Major explosive activity at Rabaul, however, commenced at about 200 ka and has produced a sequence of at least ten dacitic ignimbrites, and calderas may have been formed as a consequence of the large volumes of tephra produced during some of these eruptions (McKee and Duncan 2016). The 14 km × 14 km Campi Flegrei caldera was formed at 39 ka by the 265 km^3 (DRE) Campanian Ignimbrite eruption (Silleni et al. 2020), later followed by the collapse of the smaller and nested Neapolitan Yellow Tuff caldera at 15 ka (Orsi et al. 1996). The iconic Santorini caldera (Figs. 16.48a and 16.53b; also see Fig. 16.78a–f), which is ~12 × 7 km in diameter, is the result of at least 4 caldera collapse forming explosive eruptions, the last being the 3.6 ka (1600 BC) Minoan eruption (VEI 6–7/VEI* 10–11 [Chap. 9]; Druitt et al. 1999). The Santorini Volcanic

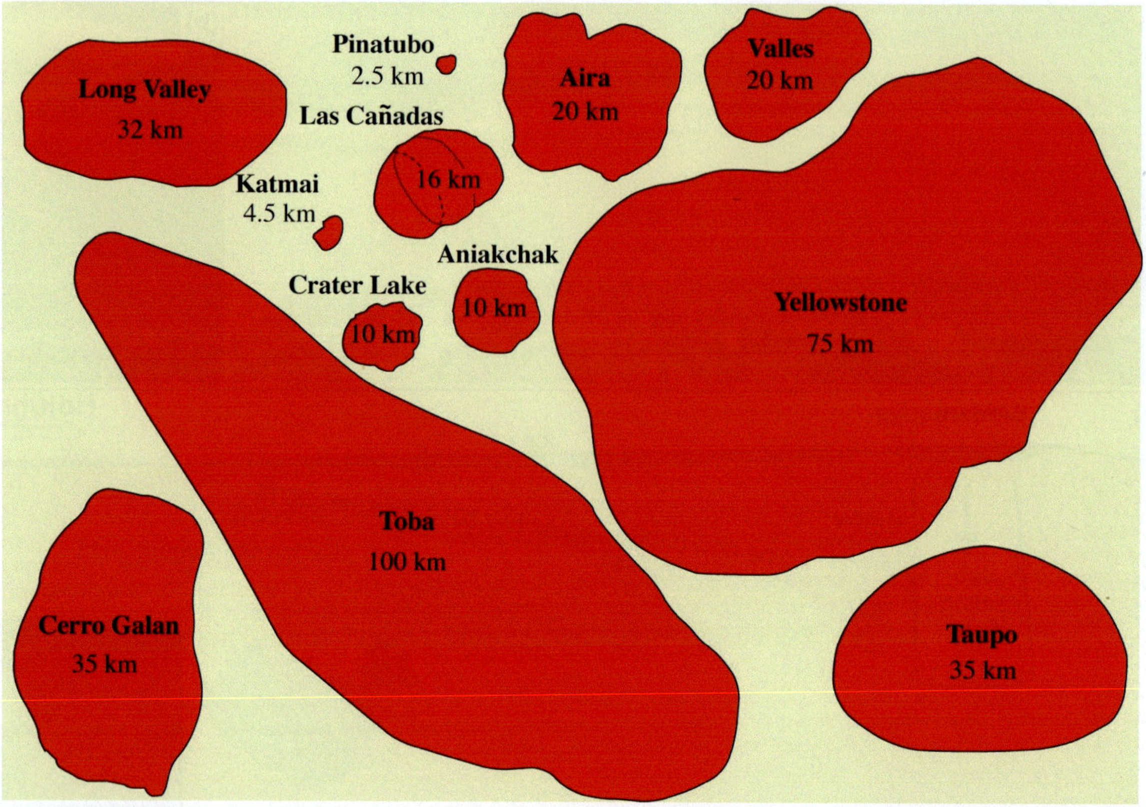

Fig. 16.52 Comparison of explosive caldera sizes measured by length of longest axis. This encompasses the full range of dimensions and shapes known on Earth. Las Cañadas (Tenerife, Canary Islands, Spain; Plate 16.1a) is shown representing a multiple collapse or polyphase caldera, which is also true of Valles and Yellowstone (western United States). (Image courtesy of Adrian Pittari)

Complex will be discussed more in Sect. 16.8.1. Krakatau volcano is well known for its caldera-forming eruption of 1883 (Fig. 16.48b; see Chap. 11, Figs. 11.16 and 11.17), however, historical observations and modern investigations suggest this was the site of a previous collapse possibly taking place at ca. 535 AD (Escher 1931; van Bemmelen 1949; Wohletz 2000).

Superimposed, larger explosive eruptions, in the order of 10 to >1000 km^3 (VEI 6–8/VEI* 10–12, Chap. 9), can result in much larger calderas, up to several tens of km in diameter. The largest known include Lake Toba in Sumatra Indonesia, which is almost 100 km long (Figs. 16.52 and 16.53c; Chap. 17), Yellowstone, USA (75 km long; Figs. 16.52 and 16.53d, e), Lake Taupo, North Island of New Zealand (35 km long; Figs. 16.52 and 16.53f, g; Chap. 10), and Cerro Galan, Argentina (30 km long; Fig. 10.61, Chap. 10). These are called "*supervolcanoes*" and they are some of the largest volcanic landforms on the Earth's surface (Figs. 16.52 and 16.53c–g). In the remainder of this section we therefore largely focus attention on this special family of volcanoes.

16.8.1 Identification of Caldera Volcanoes

Caldera volcanoes are polygenetic and invariably consist of multiple eruption points or vents. For example, as indicated above, they may consist of several coalesced individual calderas, and other volcano types described earlier (scoria cones, lava domes, stratovolcanoes), may grow within them, along their margin or on their flanks. They are found in extensional and trans-tensional tectonic settings in rifts, grabens and marginal basins of continents and microcontinents (Chap. 17). Their morphometric form is in marked contrast to stratovolcanoes. Typically, they lack a topographically impressive cone, and in extreme cases they form large, broad volcano-tectonic depressions, the type example being Lake Taupo and other calderas which constitute the Taupo Volcanic Zone (Fig. 16.53f, g; Cas and Wright 1987). Such calderas have very low aspect ratios in association with large ignimbrite-forming eruptions which form extensive *outflow ignimbrite shields*, often containing more than one ignimbrite (Figs. 10.26 and 10.28). Large caldera volcanoes, especially older eroded ones, can, therefore, be difficult to

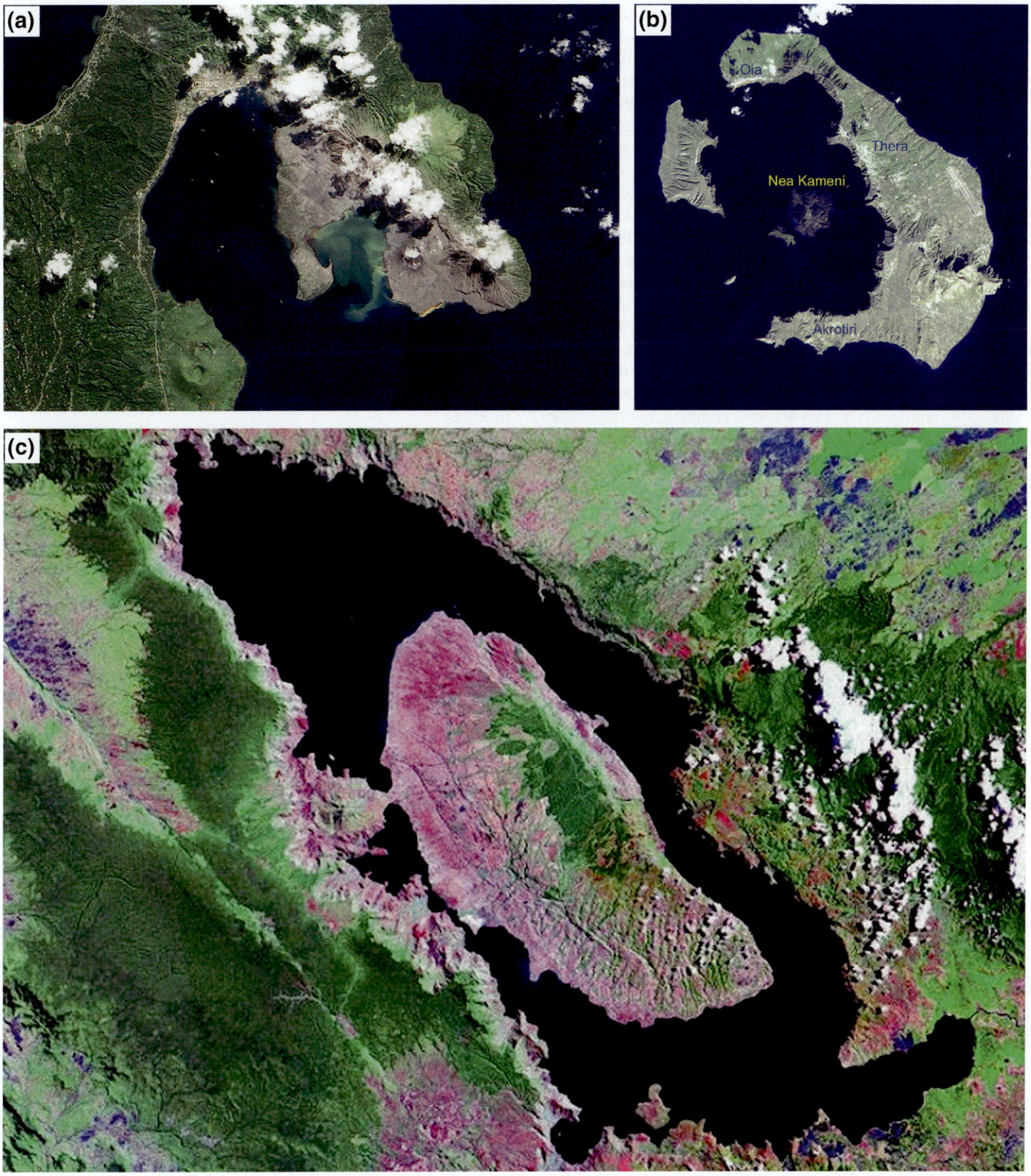
(a)
(b)
Oia
Thera
Nea Kameni
Akrotiri
(c)

◀ **Fig. 16.53** Examples of caldera volcanoes. **a** Rabaul caldera, New Britain, Papua New Guinea, which has dimensions of ~14 × 8 km. The prominent cone near the end of the southeastern rim of the caldera is Tavurvur, which is the most frequently active of several satellite vents around the perimeter, including Vulcan, to the southwest which was last active in 1994. (*Source* NASA Earth Observatory, image created from EO-1 ALI data). **b** Google Earth image of Santorini caldera, Aegean Sea, Greece, which has dimensions of 12 × 7 km. Nea Kameni island in the middle is a developing dacite lava shield or stratovolcano (see Fig. 16.48a). (*Source* Google Earth). **c** Landsat image of Toba caldera, Sumatra, Indonesia. The length of the caldera is almost 100 km. The prominent elongate island in the centre id the resurgent dome, Samosir Island. The prominent NNW-trending lineament 10–20 km to the SW of the caldera rim is the strike-slip Sumatra Fault Zone, which in large part influences the elongate form of the caldera. The artificial pink colour represents areas of high mid-wavelength reflectivity in Band 7. (*Source* NASA). **d** Simplified geological map and earthquake activity of the Yellowstone supervolcano caldera, Wyoming, USA. This shows a complex, multiple caldera structure resulting from three major ignimbrite-forming eruptions over the past 2.1 Ma. The youngest took place at 0.64 Ma forming the structure known as the Yellowstone caldera, while, earlier eruptions at 1.3 Ma (smaller caldera to west) and 2.1 Ma (the largest structure) created the older calderas. (*Source* United States Geological Survey). **e** View of Yellowstone caldera complex showing the many active hot spring areas and fumaroles (Photo by Ken McGee; *Source* United States Geological Survey). **f** Geologic setting of Lake Taupo caldera and other rhyolitic calderas within the Taupo Volcanic Zone, New Zeand. Inset shows tectonic context of northern New Zealand (CA = Colville Ridge, CB = continental boundary, KA = Kermadec arc). Distribution of major caldera-forming ignimbrites colour-coded according to age in table. Calderas: TA = Lake Taupo, KA = Kapenga, MO = Mangakino, OH = Ohakuri, OK = Okataina, RE = Reporoa, RO = Rotorua, WH = Whakamaru. (Modified from Rowland and Simmons 2012). **g** View of the Lake Taupo caldera depression which, because of its very gentle inward dipping slopes, has also been termed an "*inverse*" volcano (*Source* New Zealand Geological and Nuclear Sciences, GNS; under Creative Commons Attribution 3.0 New Zealand licence)

recognise because of the wide aerial extent of their "slope" deposits, and lack of a prominent positive morphology.

Occasionally many large calderas are found to cluster together within the same region. This phenomenon has been related to large magmatic events referred to as ignimbrite "*flare-up*" (e.g. de Silva and Gosnold 2007; Best et al. 2013; Murray et al. 2013; Gravley et al. 2016; see Sect. 16.7.5, Table 16.4). One of the best known is the mid-Cenozoic (36–18 Ma) flare-up in the southwestern United States focused on central Nevada (Fig. 16.54). However, this only represents the small part of an area of intense explosive volcanic activity extending southwards into the Sierra Madre Occidental and elsewhere in Mexico, which possibly erupted 10^6 km^3 of silicic magma (Aguirre-Díaz and Labarthe-Hernández 2003; Aguirre-Díaz et al. 2008). The Taupo Volcanic Zone in New Zealand experienced such a flare-up in the late Pleistocene, with at least 8 identified calderas active in a relatively small region (Gravley et al. 2016). Many such systems have only been identified since satellite imagery has become available, especially in remote areas such as the Miocene-Quaternary (11–1 Ma) ignimbrite flare-up of the Altiplano Puna (de Silva 1989).

Even more challenging to identify are ignimbrite linear fissure vent systems and so-called *graben calderas*. Although the concept of fissure eruptions for ignimbrites is not new, supportive field evidence has generally been meagre or controversial (Cas and Wright 1987). However, studies in the Sierra Madre Occidental have convincingly shown the eruption of several large-volume silicic ignimbrites are controlled by major rift or graben faults, usually through several vents aligned along normal faults which form the caldera walls (Aguirre-Díaz and Labarthe-Hernández 2003; Aguirre-Díaz et al. 2008). Such calderas occur in areas of regional extension where faulting and subsidence likely continues, displacing the intra-graben caldera products downward into and hidden within the tectonic depression.

Although many, if not most, Cenozoic calderas around the world have been identified, there is probably still scope for new discoveries as the source origins of many ignimbrites are still debated (e.g. Aguirre-Díaz and Labarthe-Hernández 2003; Petrinovic et al. 2010), and especially further back into the older rock record.

16.8.2 Composition

The very existence of caldera volcanoes implies the presence of a long-lived, complex and large trans-crustal magma plumbing system, with their volcanic products, therefore, dominated by evolved, intermediate to rhyolitic compositions (e.g. Fig. 16.55; Hildreth 2004; Cashman et al. 2017; Chap. 8). This complexity may be reflected not only in the magmas erupted during the caldera-forming event (e.g. Cashman and Giordano 2014), but also in both pre- and post-caldera volcanism. During these earlier and later periods, smaller eruptions of variable chemical composition may form lava domes and flows, scoria and pumice cones, and even stratovolcanoes within the collapsed caldera but also outside (Fig. 16.55a). Large calderas are therefore always multi-vent centres, and globally span the full spectrum of andesite, alkaline and higher silica compositions.

Andesitic examples include the Laguna de Bay and Taal volcano-tectonic depressions in the Philippines (Wolfe and Self 1983; Cas and Wright 1987). Laguna de Bay has the appearance of a caldera with dimensions of 25 km × 12 km (300 km^2), nested in a large collapse structure of uncertain origin. This latter structure is 35 km × 45 km in diameter and 1500 km^2 in area. Major Pleistocene eruptions of alkali-rich andesitic ignimbrites from Laguna de Bay and

(d)

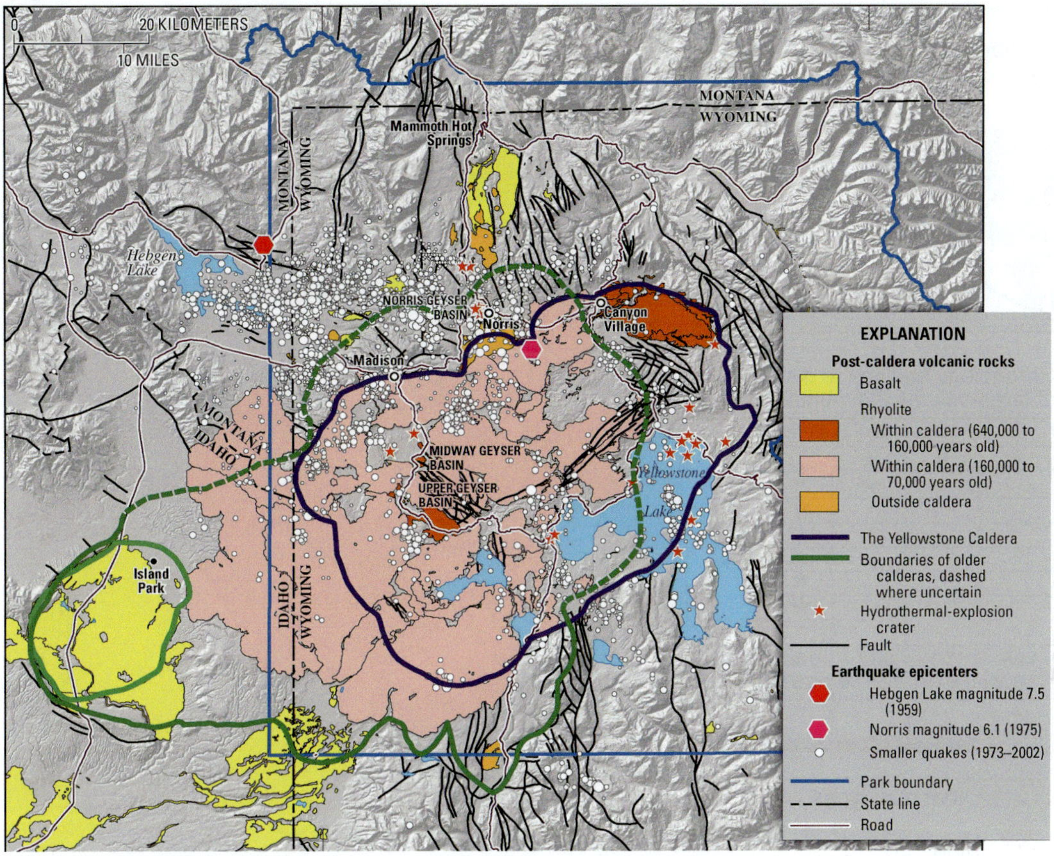

Fig. 16.53 (continued)

(f)

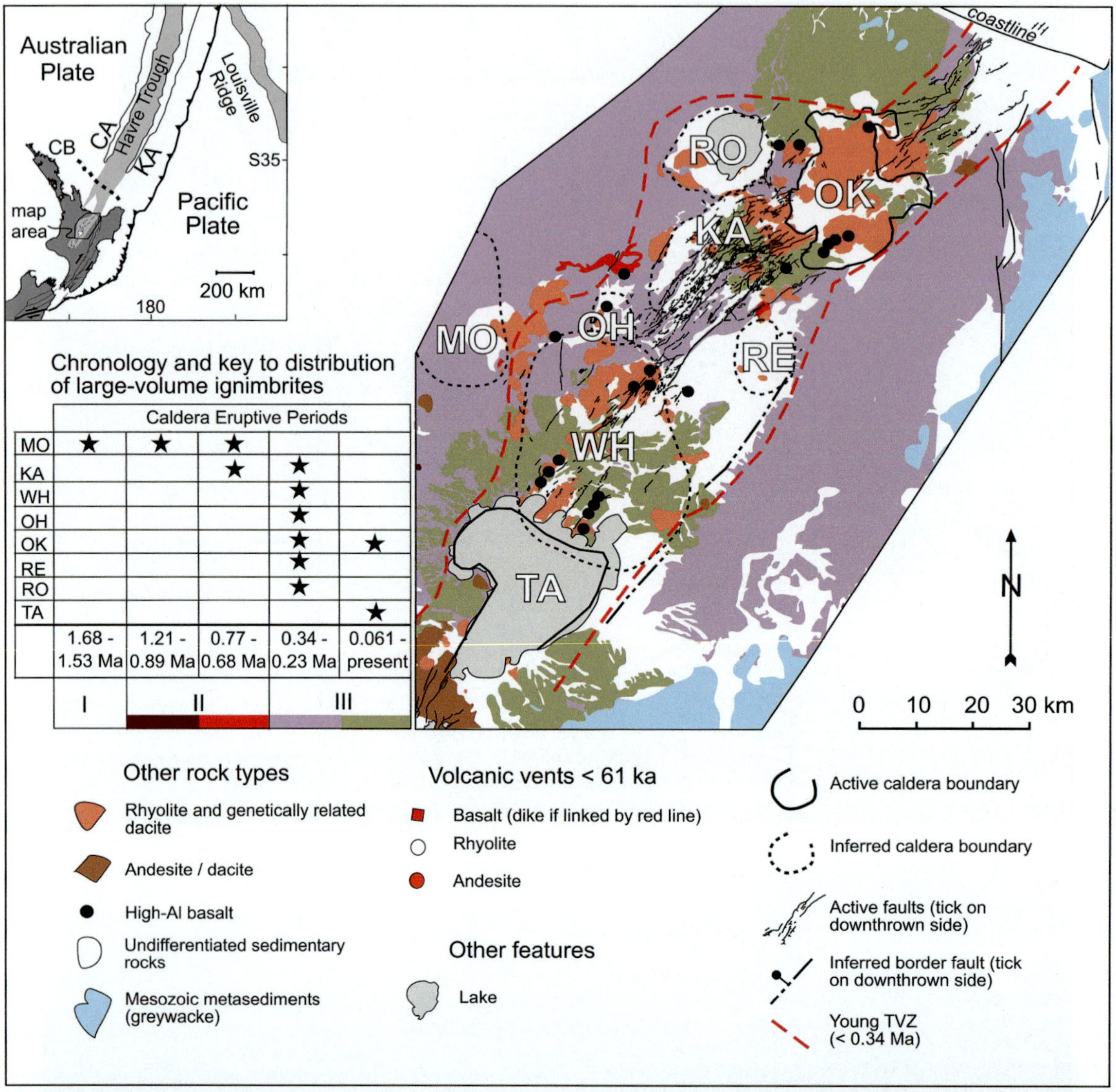

(g)

Fig. 16.53 (continued)

Table 16.4 Time scales and categories of ignimbrite flare-ups (After Gravley et al. 2016)

Time-scale	Ignimbrite flare up	Duration (Ma)	Erupted volume (km^3)	Average output rate ($km^3\ kyr^{-1}$)	CFE[a]	CFE/10 kyr
Category 1 (≥ 10 Ma)	Sierra Madre Occidental (main[1])	38–20	393,000	21.7	15[2]	0.01
	Southern Rocky Mountain	37–23	17,500	1.2	30	0.02
	Altiplano Puna	11–1	15,000	1	21	0.02
	Central Nevada	36–18	25,000	1.4	17	0.01
Category 2 (~2 Ma)	Southern Rocky Mountain	28.7–26.85	10,000	5.4	9	0.05
	Southwestern Nevada	13.4–11.4	4300	2.2	4	0.02
	Yellowstone	2.1–present	6300	3	3	0.01
	Taupo Volcanic Zone	1.6–present	6000	3.8	25	0.16
	Central Nevada	27.6–25.7	13,500	7.1	11	0.06
Category 3 (10 s–100 s kyr)	Taupo Volcanic Zone	0.35–0.28	3000	38	7	1.00
	Southern Rocky Mountain	26.9–26.85	1450	29	4	0.8
	Central Nevada	26.98–26.36	5600	90	4	0.6
	Southwestern Nevada	11.6–11.4	2100	10.5	2	0.1

[a]CFE = Number of caldera-forming events. [1]Main outcrop area; mid-Tertiary ignimbrite-forming activity however covers a more extensive area of Mexico and into the southwestern United States, possibly erupting in excess of 10^6 km^3 of silicic magma. [2]Represents a minimum as so far identified, and does not include an unknown number of potential graben calderas (Aguirre-Díaz and Labarthe-Hernández 2003; Aguirre-Díaz et al. 2008)

Taal have joined together several Cenozoic island volcanoes to form the present-day landmass of southwestern Luzon.

The Quaternary Roman Magmatic Province in central Italy is an unusual potassic alkaline ignimbrite flare-up which erupted tens of 10–100 km^3 volume caldera-forming ignimbrites associated with a cluster of seven large caldera volcanoes (Fig. 16.56a, b; see Sect. 16.8.5). Compositions range from tephritic to phonolitic, and trachytic (Mattei et al. 2010), and include large volume, caldera-forming scoria and ash flow deposits, e.g. those from the alkaline Colli Albani mafic caldera volcano, on the outskirts of Rome (Fig. 16.56c; Giordano et al. 2010; Sect. 16.8.5). To the south, the Roman Magmatic Province incorporates the active Neapolitan volcanoes and the Campi Flegrei caldera (Fig. 16.56). This is a large multi-vent alkaline centre surrounded by the 265 km^3 (DRE) Campanian ignimbrite, which is compositionally a potassic trachyte (Fedele et al. 2003; Di Salvo et al. 2020; Silleni et al. 2020; Chaps. 10 and 11). Similar alkali-trachytic compositions characterise the Ischia multi-vent caldera, which hosts probably the most impressive and fastest growing resurgent dome on Earth (De Vita et al. 2010; Selva et al. 2019; see Figs. 16.68d and 16.70).

As discussed in Chap. 10, we now differentiate relatively large scoria and ash flow deposits, that may result from the explosive eruption of mafic magmas or other low viscosity magmas (e.g. basaltic andesites, phonolites, some trachytes), as *mafic ignimbrites*. These are rarer than those associated with felsic magmas, although in recent years they have been more frequently recognised in the literature. Mafic pyroclastic flow deposits of intermediate volume ignimbrites (up to VEI 7), may be common products of caldera-forming eruptions at explosive mafic calderas as in the Roman Magmatic Province (Chap. 10), but no examples are known of >100 km^3 DRE.

More silicic examples include the very large dacitic ignimbrite-forming centres such as Cerro Galan caldera (Sparks et al. 1985; Cas et al. 2011b; Folkes et al. 2011; Chap. 10; Fig. 10.61). The large calderas of Kyushu: Aso, Aira, Ata, and further offshore, Kikai all of which are surrounded by extensive andesitic to rhyodacitic ignimbrite sheets, fall into this group. The Aira and Ata calderas have collapsed beneath Kagoshima Bay at the southern end of Kyushu, which is a large volcano-tectonic graben open to the sea (Aramaki 1984; see Fig. 16.72). Calderas related to large-volume ignimbrite flare-ups are mostly associated with rhyolitic compositions, such as the Southern Rocky Mountains Volcanic Field (Lipman 2007; Lipman and Bachmann 2015), the Taupo Volcanic Zone (Wilson et al. 2009; Gravley et al. 2016), the Sierra Madre Occidental (Ferrari et al. 2007; Aguirre-Díaz et al. 2008), and the Altiplano Puna (de Silva and Gosnold 2007).

Current thinking suggests the magma systems that occur below large silicic caldera volcanoes involve crystallisation of mafic to intermediate forerunners in the middle to upper crust, yielding an intermediate to silicic crystal mush (e.g. Allan et al. 2013; Giordano and Caricchi 2022; Chap. 8). The *magmatic mush* model was initially developed to explain

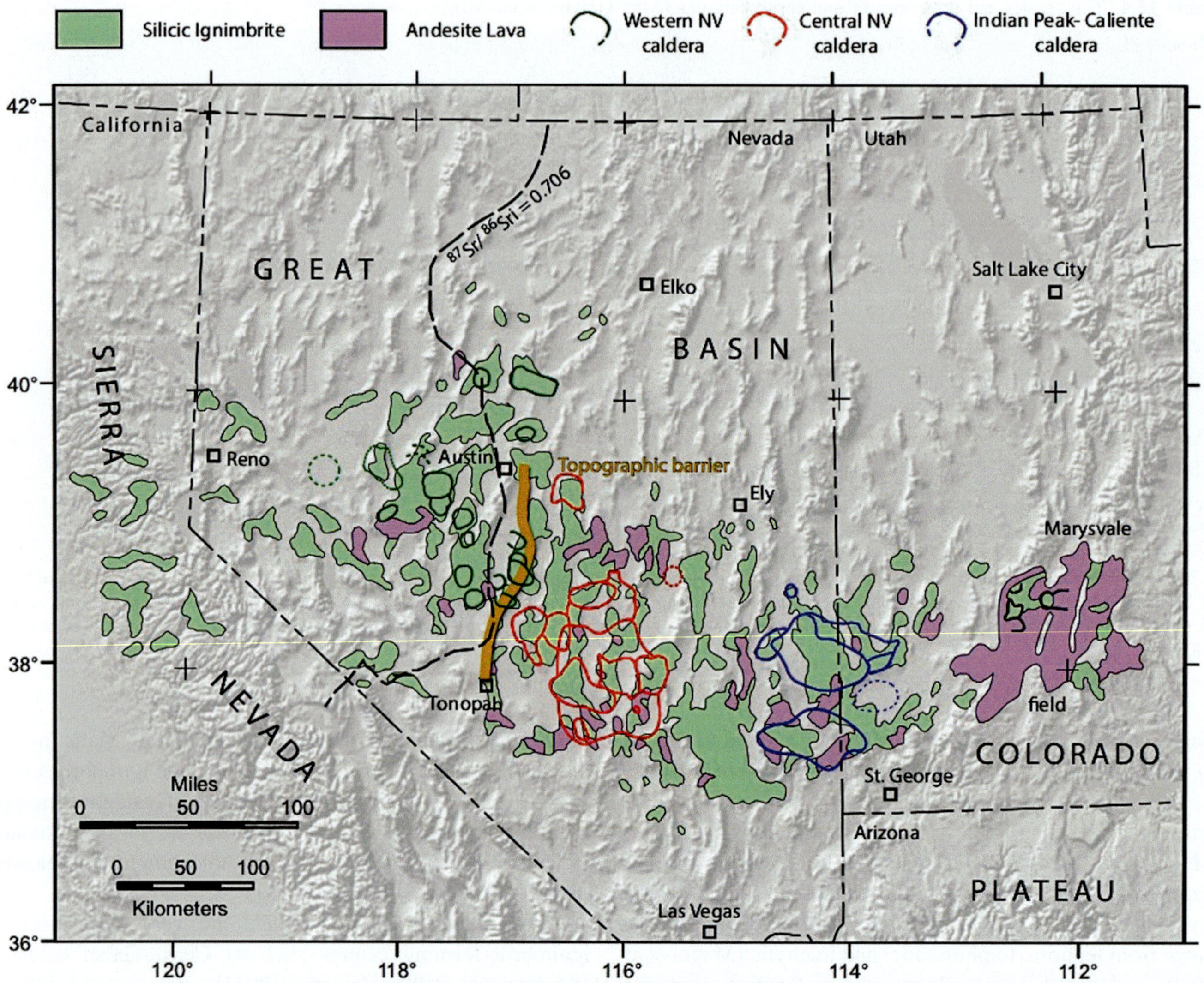

Fig. 16.54 The 36–18 Ma central Nevada (western United States) ignimbrite (green and purple areas) and caldera flare-up accounts for an impressive number of calderas (in heavy outlines in various colours) which are partly coalesced (From Best et al. 2013)

the Long Valley magma reservoir from which the 0.76 Ma Bishop Tuff (650 km^3) was erupted (Fig. 16.55; Hildreth 2004; Hildreth and Wilson 2007). In this model, the magma system comprises a relatively thin boundary layer of granitoid solidified against the country rocks, inboard of which is a rigid "sponge" of mainly crystals with minor interstitial melt (Fig. 16.55b). Within this rigid zone is an extensive reservoir of crystal-rich (>60% solid) mush that is maintained in a partially molten state via fluxing of heat and mafic magma through the basal parts of the crustal reservoir. The amalgamation of melt-rich lenses near the roof creates a low-density barrier that blocks the ascent of mantle-derived mafic magma to the surface. The crystal-poor, melt-dominated zone near the roof of such a system is a magma chamber that may be tapped to feed rhyolitic eruptions. More recent work explains the compositional variability in the Bishop Tuff in terms of syn-eruptive activation of multiple magma batches laterally juxtaposed (Gualda and Ghiorso 2013). Cashman and Giordano (2014) propose that crystal mushes may be internally much more complex than previously thought and eruptible magma can be scattered in a plexus of scattered pockets, sills and dykes which can become inter-connected at the time of the eruption, defining the size and chemical gradient of the resulting ignimbrite, as well as the style of caldera collapse (see Fig. 16.62a–d). Deciphering such subtleties in the magma architecture of such large volcanoes can only happen when detailed geochemical analyses and sampling are undertaken within a very carefully determined, detailed volcanic stratigraphy (Chap. 15).

Occasionally basaltic dykes reach the surface, but generally only outside the rhyolitic centre (defined by the caldera ring fracture zone) which may be a "shadow zone"

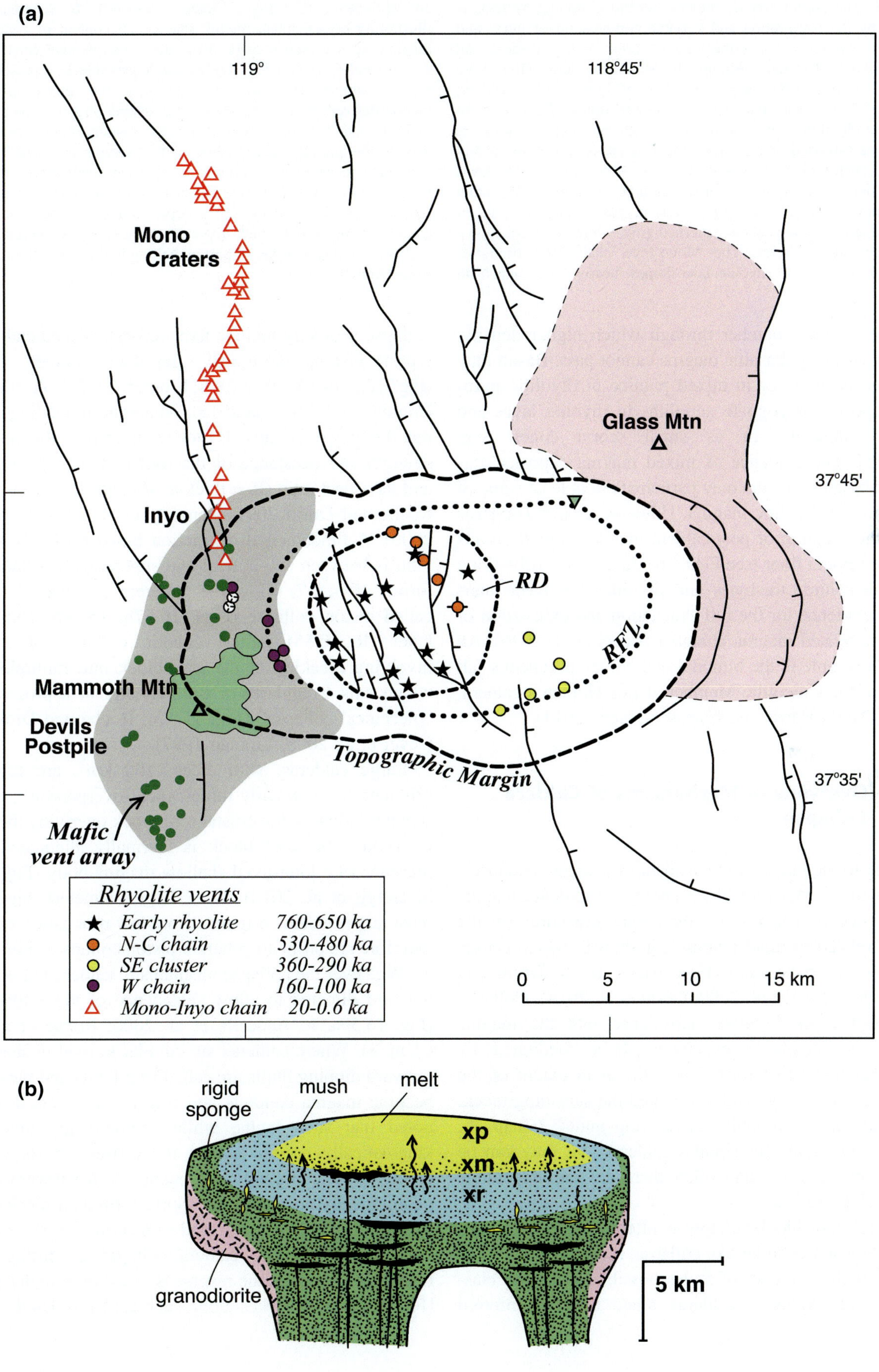
(a)
119°
118°45'
37°45'
37°35'
Mono Craters
Glass Mtn
Inyo
RD
RFZ
Mammoth Mtn
Devils Postpile
Topographic Margin
Mafic vent array
Rhyolite vents
Early rhyolite 760-650 ka
N-C chain 530-480 ka
SE cluster 360-290 ka
W chain 160-100 ka
Mono-Inyo chain 20-0.6 ka
0 5 10 15 km
(b)
rigid sponge
mush
melt
xp
xm
xr
granodiorite
5 km

◀ **Fig. 16.55** The complexity of caldera volcano plumbing systems is illustrated by the distribution and variable composition of intra- and extra-caldera products, as in the case of Long Valley caldera and adjacent Glass Mountain, Mammoth Mountain, and Mono-Inyo systems, California, USA. **a** Outline map of Long Valley caldera; RFZ = ring fault zone, and RD = resurgent dome. Four sets of intra-caldera rhyolite vents (clustered by both age and location) are shown; dome (inverted triangle in northeast moat) is early post-caldera rhyodacite. Distribution of pre-caldera Glass Mountain rhyolite lavas and thick pyroclastic apron is shown in pink. Mammoth Mountain trachydacite-rhyodacite dome complex is in green, and the zone of contemporaneous mafic vents is patterned grey, while vents are shown in as filled green circles. The Mono-Inyo chain (red triangles) comprises more than 30 rhyolite lava domes, nearly all of which all are Holocene; IC = Inyo Craters (phreatic). **b** Schematic figure illustrating crystal mush model. Phenocryst content of zoned mobile magma: xp = crystal-poor (0–5%), xm = intermediate crystal content; xr = crystal-rich (15–55%). Dykes and lenses (black) represent the few mafic magmas able to penetrate into the upper crust; most mantle-derived basalt lodges in the deeper crust, inducing partial melting and hybridism and thermally sustaining the system. This depicts the pre-climactic Bishop Tuff reservoir ca. 0.76 Ma, which continued to segregate eruptible crystal-poor melt lenses throughout Early rhyolite time (760–650 ka) and as late as 300 ka but has since largely crystallised. Root zones depict mush columns consisting of quartz + feldspar-rich (melt-depleted) cumulates and migmatised protoliths extending to zones of partial melting in the lower crust. (Adapted from Hildreth 2004)

above the magma chamber through which higher density, but lower viscosity, basaltic magma cannot pass. Basalt may be present as scoria or in mixed pumice in rhyolitic pyroclastic deposits, as cognate xenoliths in rhyolite lavas and ignimbrite deposits, or as small scoria cones (e.g. Fig. 16.55). The evidence of mixed magma deposits suggests that basaltic magma may participate in at least some of the eruption of rhyolitic magma. However, at Los Humeros caldera, the eruption of post-caldera mantle-sourced basalts across the caldera floor (see Figs. 16.63a and 18.110b) along with small-volume trachytes and rhyolites has been interpreted as evidence for the full evacuation and exhaustion of the caldera-related magma chamber (Lucci et al. 2020). An excellent example is the bimodal and contemporaneous fall deposits of the Cuicuiltic Member at Los Humeros caldera, Mexico (Dávila-Harris and Carrasco-Núñez 2014).

16.8.3 Overview of Mechanisms of Caldera Collapse

According to the classical mechanism of caldera "collapse", failure of the crustal roof block above a high-level magma reservoir occurs because of the depressurisation of the reservoir related to rapid magma withdrawal, which causes progressive under-pressure (Figs. 16.57, 16.58; Druitt and Sparks 1984; Martí et al. 2009; Geshi et al. 2014). Collapse of the roof block involves subsidence into the magma reservoir as magma is evacuated. It is facilitated by sub-vertical faults that form inside the areal extent of the underlying magma reservoir and bound the subsiding block. These faults are commonly called "ring faults" (Chap. 8), which is appropriate for circular calderas, but they can be complex in geometry, and often there are multiple faults involved (Fig. 16.57a).

Although roof block collapse is often depicted as a simple, single piston cylinder like collapse, it can be much more complex, with a number of styles possible. Common classifications of explosive calderas relate to their inferred collapse geometry include five well-established end member types: piston, piecemeal, trap-door, funnel, downsag (Fig. 16.57b-f; Walker 1984; Lipman 1997; Cole et al. 2005; Branney and Acocella 2015). Analogue modelling suggests that the type of collapse is mostly affected by the mechanical strength and resistance of the roof of the magma chamber and its aspect ratio (R = thickness/width; Roche et al. 2000; Roche and Druitt 2001; Kennedy et al. 2004; Geyer et al. 2006). A fundamental distinction based on the roof aspect ratio is between sub-critical and supercritical collapses. The former ($R < 0.7$) allow the coherent collapse of the roof, called piston collapse (Figs. 16.57b, 16.58b, 16.59b); the latter ($R > 0.85$) induce chaotic collapse of the roof, involving breaking of the roof block into multiple blocks, which sink irregularly or differentially in a process called "piecemeal collapse" (Fig. 15.57c; Roche and Druitt 2001; Geyer et al. 2006; Lipman 1997).

Large calderas, with areas >100 km^2, are thought to originate from an early tumescence (overpressure) and subsequent failure of the crustal roof block overlying the magma reservoir. This roof block is thermally attenuated by the presence of a long-lived shallow magma body (Fig. 16.59a, b; Gregg et al. 2012). Their large diameters suggest that most large calderas originate from low roof aspect ratios and therefore are prone to coherent piston collapses (Fig. 16.60a, b). Whatever the collapse mechanism, caldera faults must be outward directed to allow subsidence of the caldera block (Fig. 16.59a, b; Acocella et al. 2000; Roche et al. 2000; Chap. 8). Where calderas are circular to oval in shape, such outward dipping faults are called ring faults and they usually become magma conduits and ring fissure vents. It must be noted that such faults, while geometrically inverse, are kinematically normal as the main stress σ_1 is vertically oriented. An important consequence is that inward dipping, geometrically normal faults would form as a result of pressurisation of the reservoir and tumescence of the overlying crust, either prior to caldera collapse or during caldera resurgence, whereas the reverse is true during under-pressure (Fig. 16.61a, b). Dykes along such caldera-related faults are

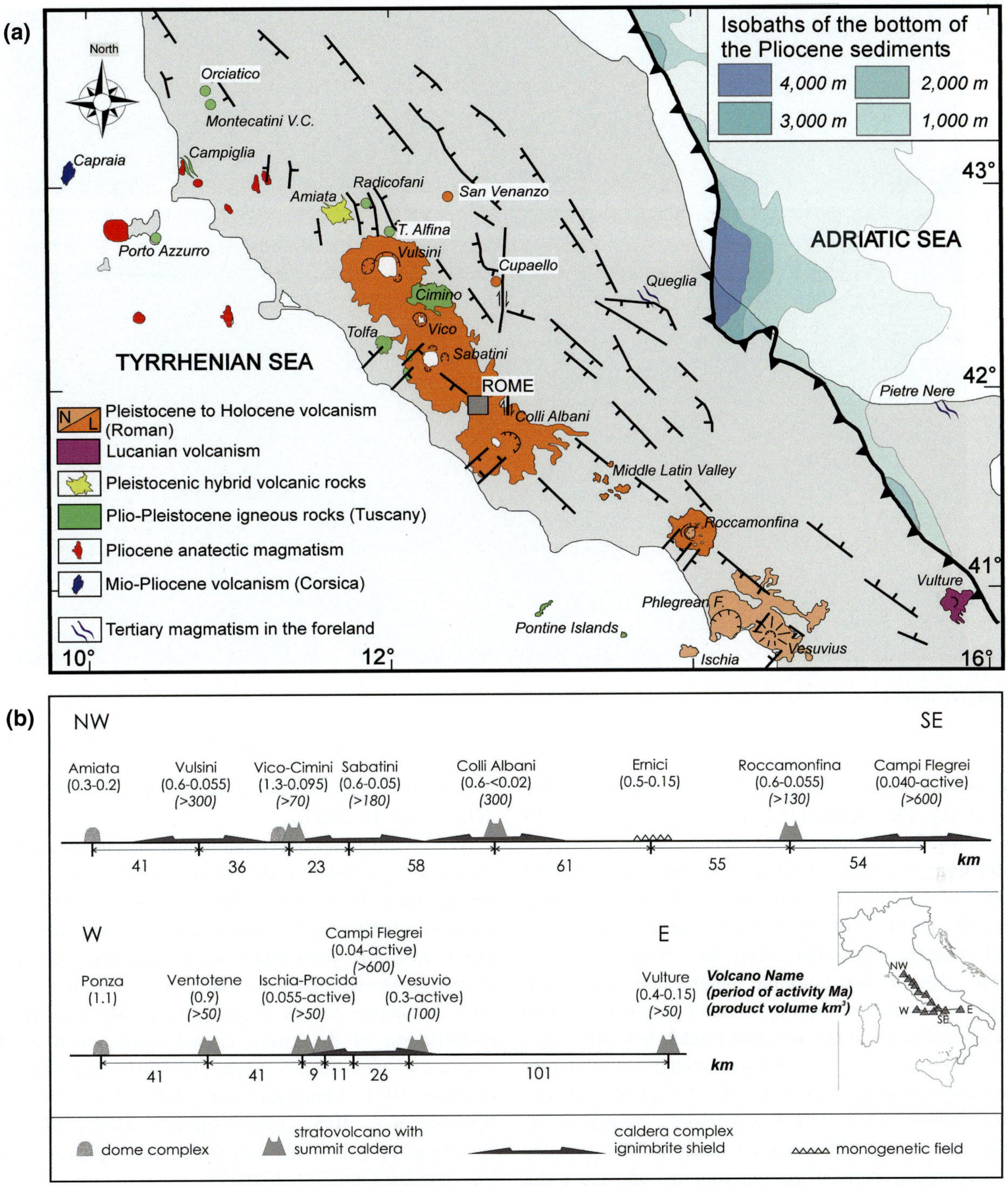

Fig. 16.56 Calderas of the alkalic Quaternary Roman Magmatic Province in central Italy. **a** Map showing the location of calderas (From Conticelli et al. 2010). **b** Spacing between calderas and other stratovolcanoes and volcanic edifices along the main northwest-southeast Roman trend (upper profile), and east–west Neapolitan trend (lower profile) (From Mattei et al. 2010). **c** Copernicus Sentinel-2 true colour image of the Colli Albani mafic caldera, southeast of Rome. The caldera (or more correctly Vulcano Laziale caldera complex) is 10 km in diameter and formed during seven major explosive eruptions between 600 and 355 kyr ago, producing at least 255 km^3 of ejecta. The floor is partly filled by the post-caldera scoria cone/developing stratovolcano of Mt della Faete (see Fig. 16.65c) in the centre of the caldera, and partly by the products of several maar volcanoes, including two presently occupied by the small Lake Nemi and the larger, oval-shaped Lake Albano (Sect. 16.3.5, see Figs. 16.22 and 16.109). The town of Castel Gandolfo overlooks Lake Albano and is known for its apal summer residence. (*Source* European Space Agency through Creative Commons Attribution-ShareAlike 3.0 IGO (CC BY-SA 3.0 IGO) licence)

(c)

Fig. 16.56 (continued)

defined as cone-sheets when intruding the inward dipping faults, and ring-dikes when intruding outward dipping faults (Anderson 1937a, b; Chap. 8).

Cashman and Giordano (2014) suggested that the timing of caldera collapse and geometry of the vent system reflect not only the crustal roof aspect ratios but dominantly the style of magma extraction from the reservoir. Their hypothesis suggests that eruptible magma does not necessarily only accumulate at the top of a reservoir of crystal-rich magma as envisaged in the standard model, but rather in a plexus of dykes, sills, lenses, and pods within a much larger and thermally stable crystal mush (Fig. 16.62; Chap. 8). The conduit therefore may extend deeper into the magma reservoir and tap (or not) different melt batches. This explains common complex chemical zoning patterns, and the usual very large depth ranges from which magma is drained (drainage height) recognised from melt inclusions (e.g. Hildreth and Wilson 2007). It is noticeable that whereas drainage heights known for ignimbrites are usually in the order of several km, the associated caldera collapse heights are much smaller, generally <1 km (Geyer and Martí 2008). This also challenges the common models for caldera collapse which assume equal melt drainage height and caldera collapse height (e.g. Scandone 1990; Branney 1995), and that these are solely a function of the roof mechanical strength and aspect ratio (e.g. Roche et al. 2000; Geyer et al. 2006; Acocella 2007).

Within this new conceptual model, it is the strength of the reservoir crystalline framework or mush that also dictates the timing and style of collapse. Thus, early mobilisation of only volatile-rich and crystal-poor inter-connected melt network results in single point eruptions. If, and when the reservoir strength is overcome by the load of the overlying roof block rocks, collapse occurs (Fig. 16.62c). Alternatively, early mobilisation of the entire crystal mush, such as during eruption of very large volume, crystal-rich ignimbrites (e.g. Hildreth 1981; Maughan et al. 2002; Cas et al. 2011b; Willcock et al. 2015), could result in a caldera collapse fully synchronous with the eruption (i.e. piston collapse, Fig. 16.62d; Gregg et al. 2012).

16.8.4 Morphometry

Detailed morphometric data for caldera volcanoes can be found in the Collapse Caldera Data Base freely accessible online at http://www.gvb-csic.es/CCDB/. W_s, which is the diameter of the ignimbrite shield, varies from about 50 to >200 km, and W_{vc}, which is the diameter of the volcanic centre, defined by the morphological rim of the caldera (see

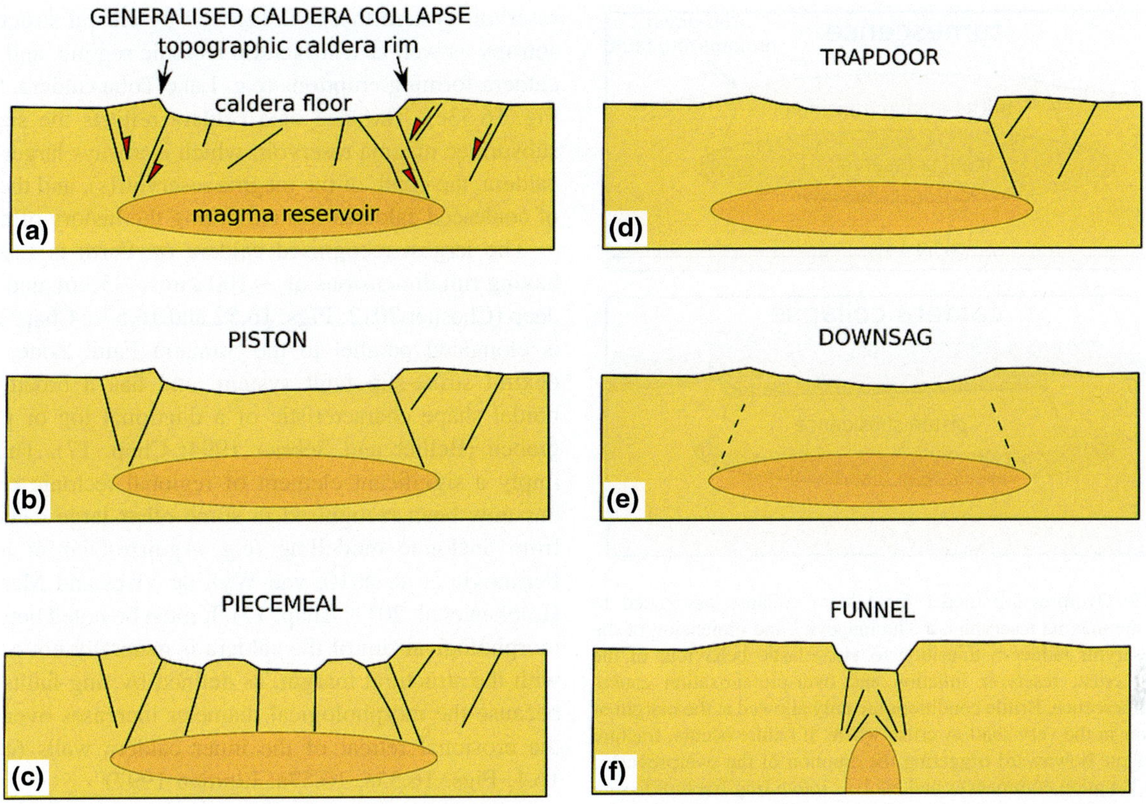

Fig. 16.57 Principal mechanisms of caldera collapse (After Lipman 1997). **a** Generalised collapse mechanism and main features. **b** Piston collapse involves the coherent subsidence of the caldera floor and activation of annular ring vents during the eruption; this kind is likely for large-volume eruptions, and oblate and shallow magma reservoirs. **c** Piecemeal collapse involves the differential collapse of the floor and activation of several fissure vents. **d** Trapdoor collapse involves the presence of a hinge and an asymmetric collapse with only a section of ring vents active during the eruption. **e** Down-sag is the early stage of caldera subsidence and only involves incipient faults and/or lateral migration of magma at depth. **f** Funnel calderas are small features probably related to chaotic scavenging of the roof of a prolate and deep magma reservoir, and/or sector collapses of the walls of a large vent into the vent

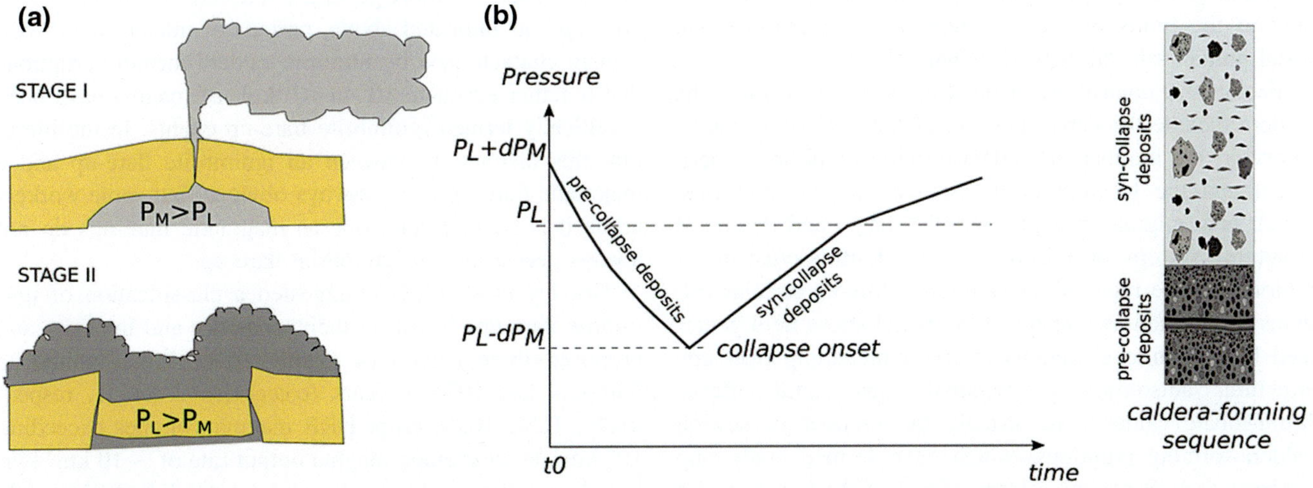

Fig. 16.58 Classical under-pressure model for caldera collapse. **a** Stage I is associated with the withdrawal of critical amount of magma at magmatic pressure (P_m) larger than lithostatic (P_l) prior to collapse. Stage II refers to the mechanical destabilisation and consequent collapse of the roof of the magma reservoir at $P_l > P_m$. (Modified from Druitt and Sparks 1984). **b** Same concept visualised on a time-pressure diagram; the volumetric percentage (dense rock equivalent) of pre-collapse products compared to the total erupted volume is termed *f* (see Fig. 16.60). Pre-collapse deposits are dominated by fallout deposits, whereas syn-collapse deposits are dominated by voluminous ignimbrites. (Modified from Geyer and Martí 2008)

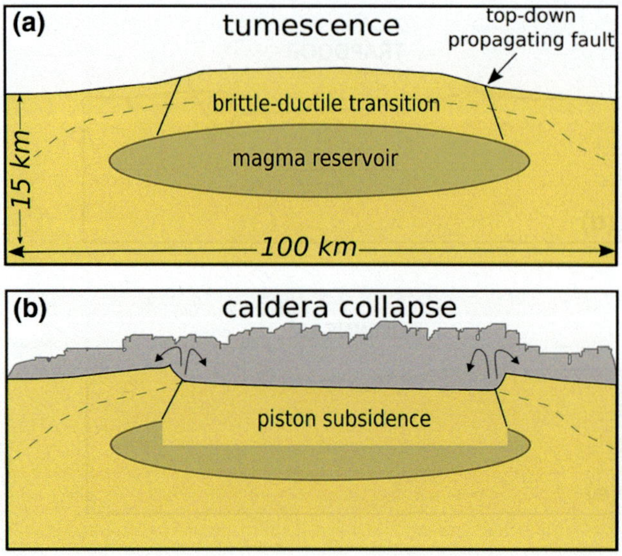

Fig. 16.59 Overpressure model for caldera collapse associated to large-volume magma reservoirs. **a** The longevity and dimension of the magma reservoir induce a transition to viscoelastic behaviour of the embedding crust; reservoir inflation and over-pressurisation causes regional tumescence. Brittle conditions are only allowed at the margin of the reservoir in the very shallow crust, where, if failure occurs, fracture may propagate downward triggering the eruption of the overpressured magma. **b** Eruption commences and evolves into a ring fracture fissure vent system, while deflation accommodates crustal detumescence. In this type of model collapse may be synchronous with the initiation of the eruption. (Adapted from an original concept of Gregg et al. 2012)

Fig. 16.52), is usually 10–60 km. Slope of the ignimbrite shield varies from <1° to 5° and H_s (height of shield) would generally be a few hundred m. The total volume of these shields varies from 10^2 to $>10^3$ km^3. However, these volume estimates generally ignore the large amounts of ash lost far beyond the limits of the volcanoes as co-ignimbrite and distal plinian ash-fall deposits (Chap. 10).

Factors that control the shape of morphological rim of the caldera include the depth and size of the subjacent magma reservoir, the number of caldera-forming eruptions experienced and the location of the source magma reservoirs, which may migrate through time, the nature and degree of structural isotropy or anisotropy of the Earth's crust in the region, and the regional stress regime. Simple circular calderas are formed as a result of the radial stress field generated by a high-level magma reservoir prevailing over any structural anisotropy or regional stress field effects. Multi-lobate calderas are usually the product of several caldera-forming eruptions, which have formed coalescing calderas (e.g. Santorini caldera, Fig. 16.53b; Las Cañadas caldera, Tenerife, Plate 16.1). Calderas which are rectilinear are usually controlled by a well-structured (i.e. multiple faults) crust and/or an active extensional to trans-tensional regional stress regime (e.g. Okataina caldera, New Zealand, Fig. 16.53f). Very elongate calderas reflect a magma reservoir system that is influenced by crustal structural anisotropy as well as a strike-slip tectonic regime, and multiple caldera-forming eruptions (e.g. Lake Toba caldera, Sumatra, Fig. 16.53c). The size of a caldera reflects the size of the subsurface magma reservoir, which is always larger than the caldera, the depth of the magma reservoir(s), and the number of coalesced calderas formed during the history of a caldera.

The largest recognised caldera on Earth is Lake Toba, having rim dimensions of ~ 100 km $\times$ 35 km, and ≥ 2 km deep (Chesner 2012; Figs. 16.52 and 16.53c; Chap. 10). This is elongated parallel to the Sumatra Fault Zone, a major dextral strike-slip fault system, and has a broadly rhomboidal shape characteristic of a dilational jog or pull-apart graben (Bellier and Sébrier 1994; Chap. 17). This would imply a significant element of regional tectonic control as has now been recognised in some other large calderas and from analogue modelling (e.g. Aguirre-Díaz et al. 2008; Petrinovic et al. 2010; van Wyk de Vries and Merle 1998; Holohan et al. 2011; Chap. 17). It must be noted here that the morphological rim of the caldera is generally not coincident with the structural margin, as defined by ring faults. That is because the morphological diameter increases over time by the erosional retreat of the inner caldera walls (e.g. Plate 16.1; Figs. 16.55a, 16.57a; Lipman 1997).

16.8.5 Output Rates, Repose Periods and Life Expectancy

Caldera volcanoes are usually clustered both in space and time and reflect periods of very high magma productivity with accumulation of large volumes of magma in the crust. Whereas these are more generally referred to as a "*magmatic flare-up*", as indicated above, periods of intense silicic volcanism characterised by multiple caldera-forming eruptions that together evacuate 10^3 to 10^4 km^3 of magma have been specifically termed ignimbrite flare-up events. In the literature this distinction between an ignimbrite flare-up and a magmatic flare-up is not always obvious, but some workers argue it is an important one as magmatic flare-ups do not always necessitate an ignimbrite flare-up.

Gravley et al. (2016) suggested a classification of ignimbrite flare-ups based on their longevity and intensity and proposed three categories (Table 16.4). Most ignimbrite flare-ups last 10^6–10^7 years (categories 2 and 1, respectively, Table 16.4), erupt DRE magma volumes exceeding 10^4 km^3, at an average magma output rate of ~ 10 km^3 kyr^{-1}, and are interpreted to be driven from depth by high mantle flux (i.e. magmatic flare-ups). So, for example, the Sierra Madre Occidental, Mexico, and Southern Rocky Mountains Volcanic Fields, USA, represent category 1, and the Altiplano Puna, Argentina and Chile, and the Taupo Volcanic Zone, New Zealand, represent category 2. However, much

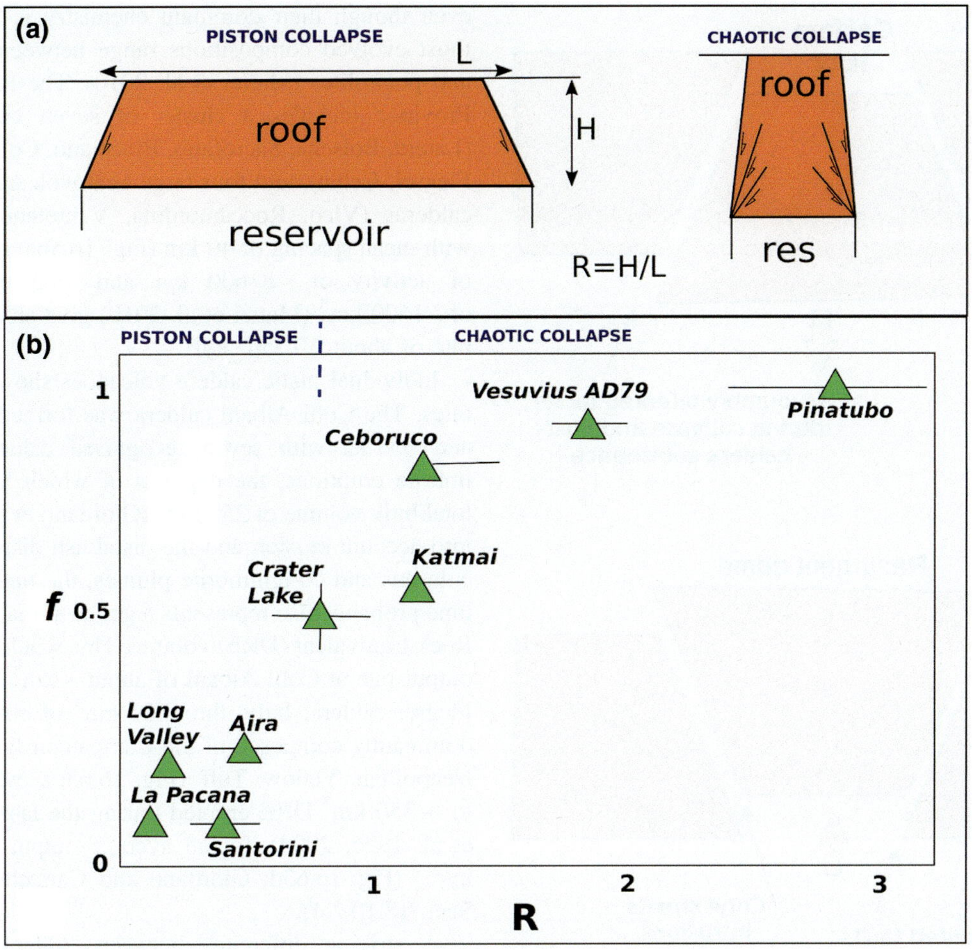

Fig. 16.60 Experimental types of caldera collapse from sandbox analogue models. **a** The aspect ratio (R = thickness/horizontal extent) of the roof of the magma chamber controls the style of collapse: coherent, piston-like collapses (left in the panel) with early subsidence are achieved for low aspect ratios that may correspond in nature to laterally extensive and shallow magma reservoirs; chaotic collapses (right in the panel) involving progressive upward migrating fault excavation and late stage subsidence are related to high aspect ratios corresponding in nature to small and relatively deep, vertically extensive magma reservoirs. (After Roche and Druitt 2001). **b** Relationship between roof aspect ratio (R) and the percentage of pre-collapse deposits (*f*), for several caldera-forming eruptions. While mainly from stratovolcanoes, these data suggest that piston collapses of shallow and extensive magma reservoirs (typical of larger caldera volcanoes, e.g. Long Valley, Aira and La Pacana) require progressively lesser amount of magma withdrawal for collapse. This possibly also reconciles, as end members of a spectrum, the under-pressure model for the chaotic collapse and the overpressure model for the piston collapses. (Modified from Cashman and Giordano 2014)

of the present-day geological footprint of the active part of the Taupo Volcanic Zone was established in a remarkable ignimbrite flare-up event between 350 and 280 ka. During this time, eight caldera and ignimbrite-forming eruptions occurred, evacuating some 3000 km^3 of magma, with calderas occurring in a 3600 km^2 area (Wilson et al. 2009). These much shorter timescale flare-up events (10^4–10^5 yr) represent category 3 and appear to reflect more intense periods of magmatic output (e.g. caused by accelerated rifting and/or mantle input) within the longer timescale categories. Output rates vary between 1 and 100 km^3 kyr^{-1}, with larger values typical of faster flare-ups (Table 16.4).

Individual silicic caldera volcanoes show variable output rates in these regional flare-up regions. The largest output rates known from the most productive rhyolitic volcanoes, such as Lake Taupo and Okataina calderas in the Taupo Volcanic Zone, New Zealand, with an average of about 4 km^3 kyr^{-1} DRE for each. For La Primavera volcano in Mexico, which has been active for the past 100 kyr, the average output is 0.6 km^3 kyr^{-1} DRE (Walker et al. 1981; Cas and Wright 1987; Fig. 13.37). Los Humeros Volcanic Complex, the largest caldera in the Trans-Mexican Volcanic Belt, experienced two caldera-forming explosive eruptions at 164 and 69 ka, following an extended period of rhyolite lava

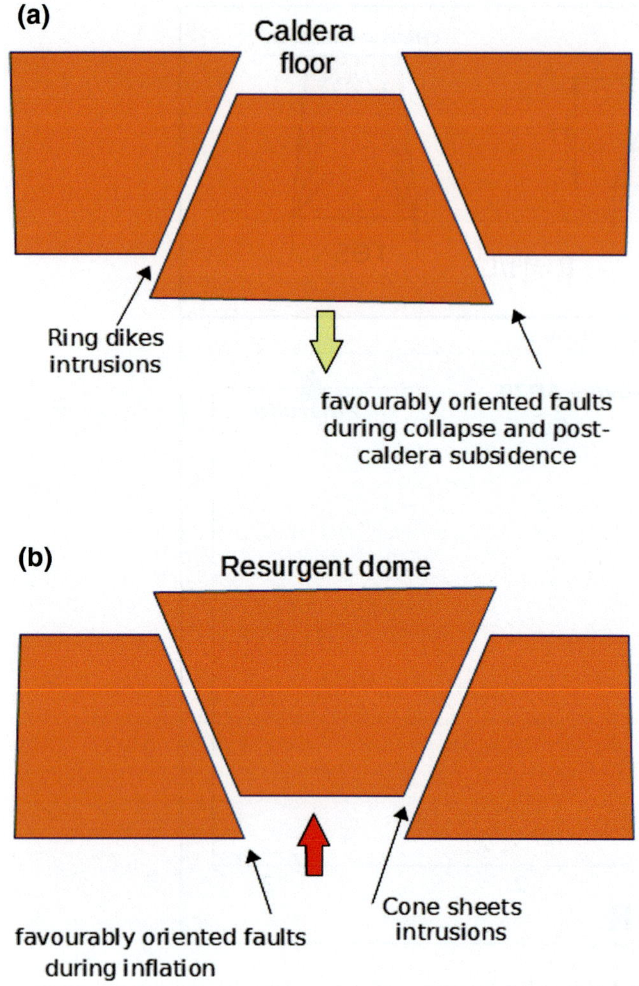

Fig. 16.61 The geometry of collapse faults controls the post-caldera subsidence and intrusive history. **a** In the case of post-caldera subsidence, outward directed faults can further dilate and accommodate dyke intrusions that form *ring dyke* complexes. **b** In case of resurgence, outward directed faults lock, while new or reactivated inward directed faults accommodate uplift; intrusions along such faults form *cone sheet* complexes

dome eruptions from ~700 ka to 270 ka and followed by post-caldera explosive and lava dome forming volcanism (Fig. 16.63a, b; Carrasco-Núñez et al. 2018; also see Sect. 18.10.2.3, 'Case Study'). The first caldera-forming phase (164 ka) erupted 115 km^3 DRE of magma, whereas the second caldera phase was preceded by a 10 km^3 DRE plinian stage (70 ka) followed by the 69 ka caldera stage during which 15 km^3 of magma was erupted (Fig. 16.63b; Carrasco-Núñez et al. 2018). The average magma output rate calculated by Giordano and Caricchi (2022) over 700 kyr was ~4.1 km^3 DRE kyr^{-1} (Fig. 16.63b).

Ignimbrite flare-ups have been attributed so far only to silicic provinces. However, the mafic alkaline calderas of the Roman Magmatic Province also fit, by volume and time-scale, within the broad definition of ignimbrite flare-ups, even though their dominant chemistry is alkaline and the most evolved compositions range between alkali-trachytes and phonolites (Mattei et al. 2010). The Roman Magmatic Province includes a cluster of seven caldera volcanoes (Latera, Bolsena, Sacrofano, Bracciano, Colli Albani, Campi Flegrei, Ischia), and four large stratovolcanoes with summit calderas (Vico, Roccamonfina, Ventotene and Vesuvius) with mean spacing of 40 km (Fig. 16.56a, b). The age range of activity of ~8–600 kyr and total erupted volumes of >1600 km^3 (Mattei et al. 2010) give an average eruption rate of about 2.7 km^3 kyr^{-1}.

Individual mafic caldera volcanoes show variable output rates. The Colli Albani caldera was formed between ~600 and 355 ka with seven recognised caldera-forming ignimbrite eruptions, the deposits of which have a preserved total bulk volume of 255 km^3 (Giordano et al. 2010). Taking into account erosion and the distal ash dispersed by plinian columns and co-ignimbrite plumes, the preserved bulk volume probably also represents a good estimation of the Dense Rock Equivalent (DRE) volume. This would give an average output rate at Colli Albani of about 1 km^3 kyr^{-1}. At Campi Flegrei caldera, Italy, the >600 km^3 of erupted material is dominantly composed of the Campanian Ignimbrite and the Neapolitan Yellow Tuff (Fig. 16.63c), which correspond to ~350 km^3 DRE erupted during the last ca 80 kyr (Orsi et al. 1996, 2004), for an average output rate of 4.1 km^3 kyr^{-1} (Fig. 16.63d; Giordano and Caricchi 2022; also see Sect. 18.10.2.4).

A striking difference between caldera volcanoes and stratovolcanoes and their types of activity is the repose periods between eruptions. For Taupo caldera, the average repose period over the past 50,000 yr is ~3000 yr, but this has varied from a few hundred yr to 16,000 yr, and for Okataina caldera over the same period it is ~2500 yr (Wilson et al. 2009). Repose periods between eruptions of very large ignimbrites may be of the order of 10^5 yr, and perhaps even 10^6 yr (e.g. Cerro Galan caldera, Argentina; Folkes et al. 2011). Such intervals suggest that some calderas, such as Long Valley, Valles and Yellowstone, are probably capable of erupting another large-volume ignimbrite (e.g. Wolff and Gardner 1995); although, detailed petrological models for the Long Valley caldera (Fig. 16.55) suggest the main magma reservoir may be moribund (Hildreth 2004). Recent studies show that magma plumbing systems evolve significantly across their life time, so that average and peak magma output rates vary (Fig. 16.63b, d), reflecting the variable interplay between magma input and thermal maturation of the crust over time (Giordano and Caricchi 2022). The life expectancy of some individual rhyolitic centres could therefore be greater than 10^6 yr. The Valles caldera formed during two major rhyolitic ignimbrite-forming eruptions (the Lower and Upper Bandelier Tuffs) at 1.6 and 1.26 Ma (Smith and Bailey 1966;

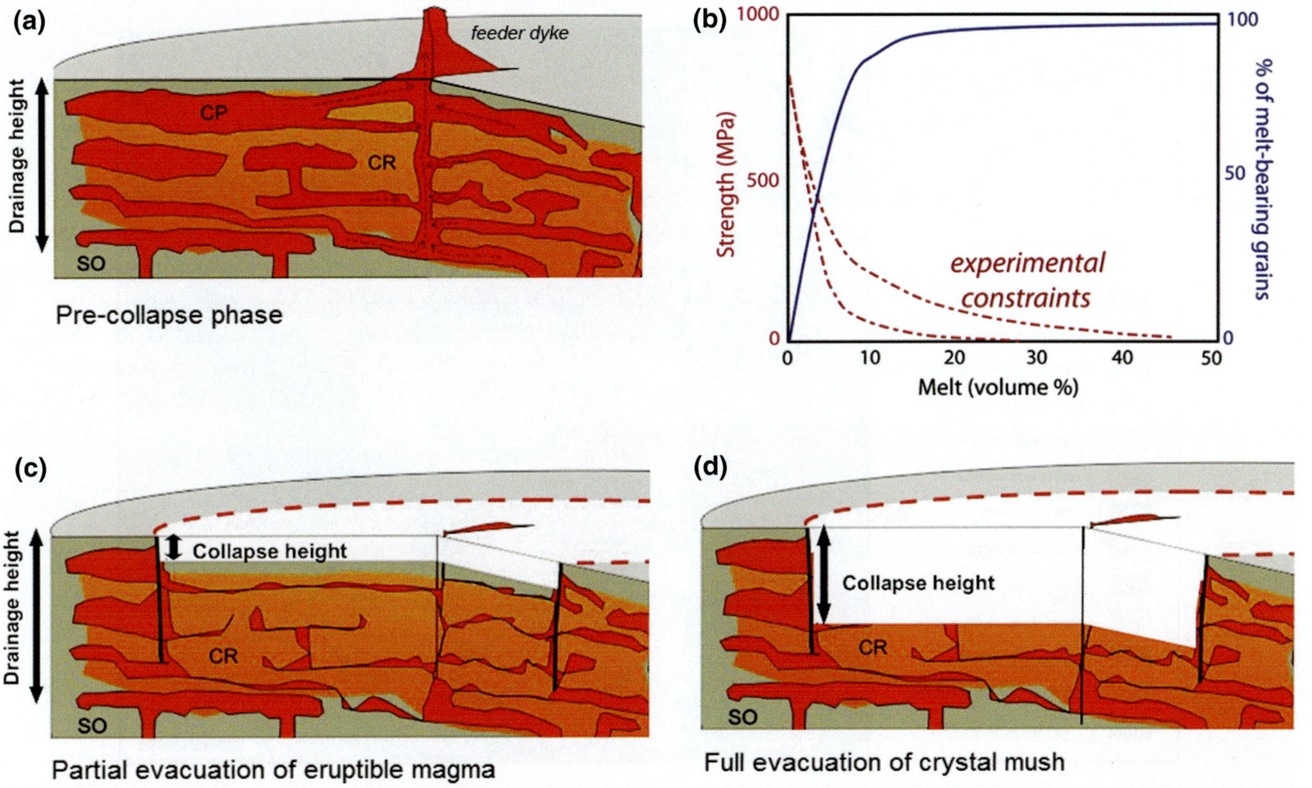

Fig. 16.62 An alternative model for caldera collapse modulated by the strength of the magma reservoir (Modified from Cashman and Giordano 2014). **a** Conceptual model for magma reservoirs made of variably inter-connected and inter-connectable pods, lenses, dykes and sills of eruptible (usually) crystal-poor magma (CP) set in a framework of progressively crystal-richer mush (CR) and intrusive rocks (SO). **b** Variation of strength as a function of percentage of melt (dashed lines) versus crystal framework (solid line) illustrates that driving pressures needed to mobilise magma are progressively larger at higher crystal contents (After Rosenberg et al. 2007). **c** For withdrawals that leave a substantial part of the crystal-rich framework, the strength of the framework contributes to modulating the timing of collapse along with the strength and aspect ratio of the roof (compare with Fig. 16.60); in this case, unlike classic models, the collapse height is different from the drainage height, i.e. the depth range from which the magma is mobilised. **d** For withdrawals that involve the wholesale mobilisation of the magma reservoir including its crystal-rich framework, the model coincides with classic models where collapse height and drainage heights are the same; this would be the case for eruption of crystal-rich ignimbrites and possibly justifies very early collapses

Self et al. 1986, 1996; see Fig. 16.66), while the Yellowstone centre is as at least as old as 2.1 Ma, involving four major explosive eruptions (Christiansen 1979; Christiansen et al. 2007; Fig. 16.53d, e). The Cerro Galan caldera in northwestern Argentina first erupted a caldera-forming ignimbrite at ~6 Ma and has a record of six such magnitude eruptions from then to the last one at ~2.1 Ma (Folkes et al. 2011), but it still has an active geothermal field in the caldera. Lake Atitlán caldera in southwestern Guatemala, results from the youngest of at least three overlapping caldera-forming events occurring over the past 14 Myr, culminating in the very large-volume Los Chocoyos Ash at 75 ka (Cisneros de León et al. 2021a; Chap. 9). The large San Juan calderas (Southern Rocky Mountain Volcanic Field) and associated ignimbrites, were all formed within a period of ~7 Ma, between 29 and 22 Ma (Lipman 2007). The Taupo Volcanic Zone (Fig. 16.53f) has been active for 1.6 Ma (Wilson et al. 2009; Gravley et al. 2016).

16.8.6 Eruptions, Characteristics and Deposits

From the discussion above, it is apparent the primary eruptive products which characterise caldera volcanoes are related to large explosive eruptions of pumice and ash as plinian fallout deposits and voluminous ignimbrites (Chaps. 9 and 10). These kinds of large-volume explosive eruptions are relatively infrequent and separated by relatively long repose periods (Fig. 16.63b, d). Such periods may be characterised by total quiescence, but more frequently, by post-caldera, smaller volume, effusive lava and dome forming eruptions, and mild explosive eruptions, forming scoria or pumice cones, issued from around or within the collapsed caldera (Fig. 16.63a, c).

We distinguish two main environments at caldera volcanoes: (1) the extra-caldera outflow ignimbrite shield, which slopes gently away from caldera rim, and (2) the intra-caldera fill (Figs. 10.25, 10.26, 10.27, 10.28 and 10.29).

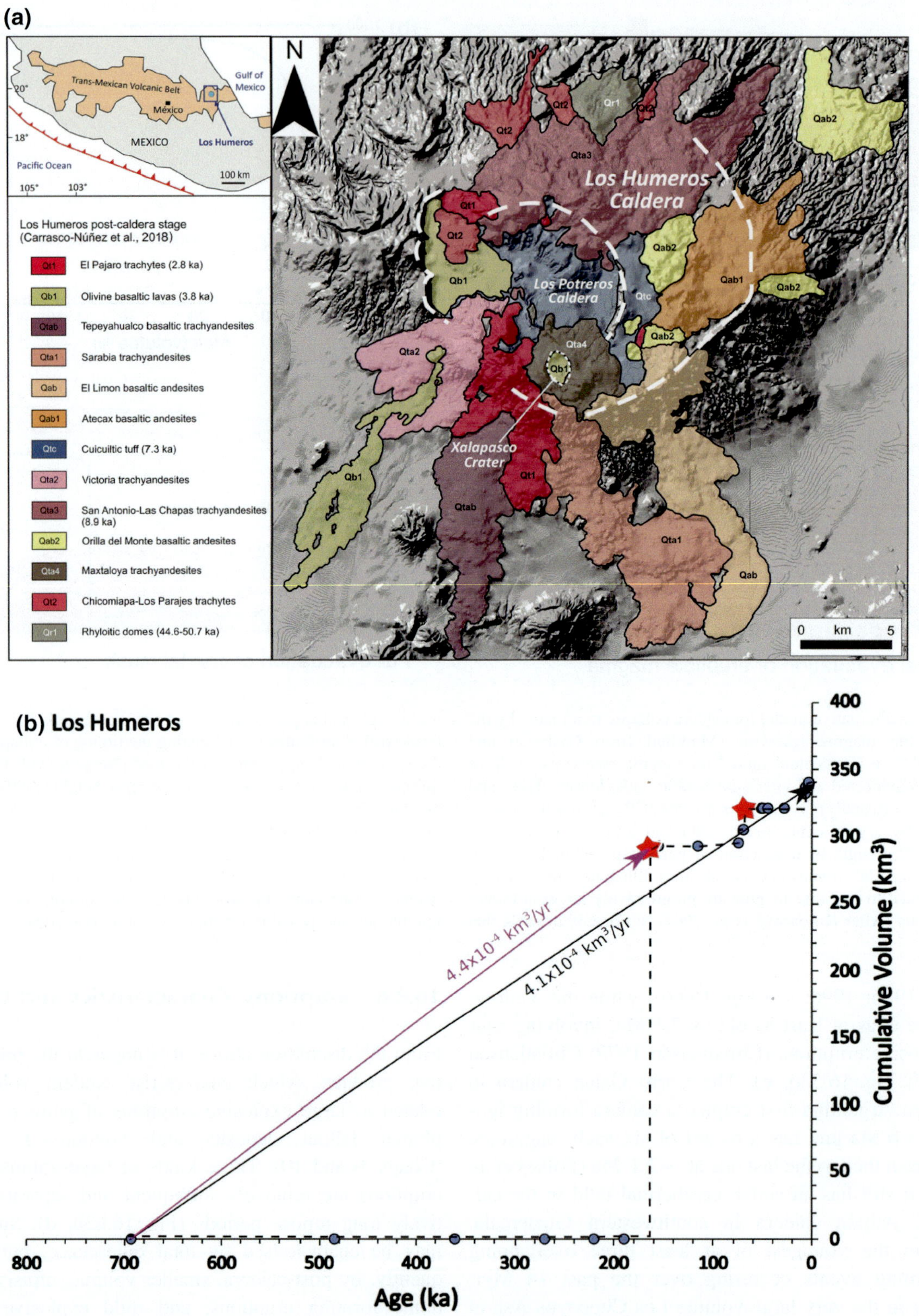

Fig. 16.63 Magma output rates for individual caldera volcanoes in magma flare-up regions. **a** Geological map of the Holocene post-caldera products from Los Humeros caldera volcano, central Mexico. Both lavas and explosive deposits range from basaltic to trachyte and rhyolite in composition. (From Lucci et al. 2020). **b** Cumulative DRE magma volume erupted versus age for Los Humeros caldera. Red stars indicate caldera-forming events; the grey arrow and related value is the long-term average eruption rate, whereas the pink arrow and value indicate peak values at caldera-forming events. (From Giordano and Caricchi 2022). **c** Geological map of the caldera-forming and post-caldera products and vents at Campi Flegrei caldera, Italy (From Vitale and Isaia 2014). **d** As for (**b**) for Campi Flegrei caldera (From Giordano and Caricchi 2022)

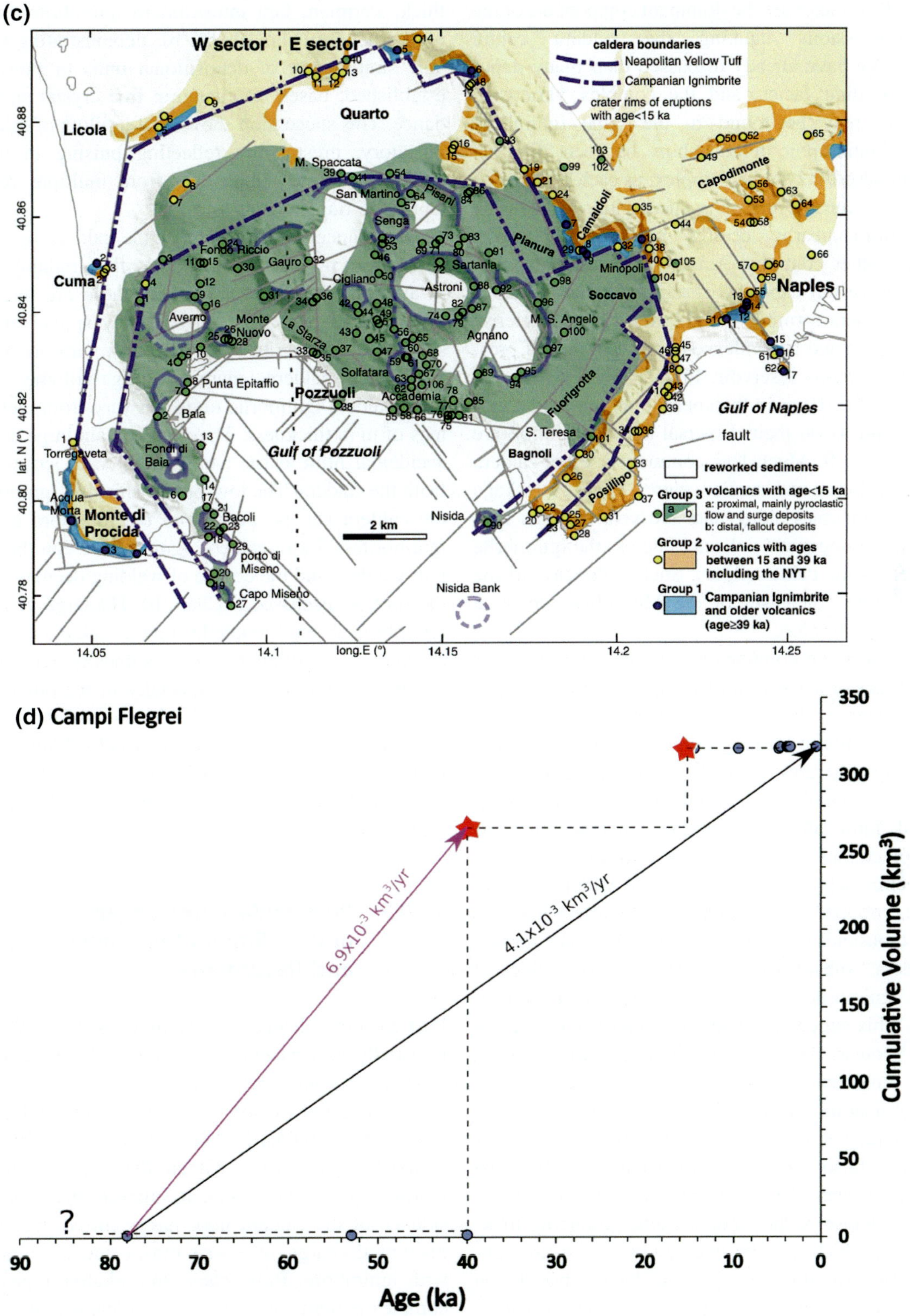

Fig. 16.63 (continued)

Ignimbrite outflow facies are the dominant components of the low-angle successions flanking the caldera centre (Fig. 16.64a). We have already discussed their many depositional and welding facies, and the complex variety of lithologies that may be found in the same ignimbrite (Chap. 10). Stratigraphic relationships between different ignimbrites are also likely to be complex, as each one fills an erosional surface cut into the older succession. In addition, ignimbrites from nearby centres may often overlap (Fig. 16.64b). Other common primary pyroclastic deposits are plinian pumice fall deposits. These may either be associated with caldera-forming ignimbrites or characterise pre-caldera periods and be evidence of the build-up of pressure within the magma reservoir. Plinian fall deposits will vary from <1 m to >10 m in thickness in and around a caldera centre; controls on their dispersal and distribution are discussed in Chap. 9. Much less common in extra-caldera settings are lava domes and smaller centres, although leakage of less evolved magmas along parasitic vent systems may form chains of monogenetic volcanoes across the ignimbrite shield (as well as the caldera floor). Such is the case of the Holocene rhyolitic volcanism of the Mono-Inyo chain at Long Valley (Fig. 16.55a).

The caldera-forming eruptions are associated with major structural changes of the volcano leading to caldera collapse (Chap. 10). Caldera collapse may occur from the beginning or during the ignimbrite-forming eruption, or late, around a system of ring fractures formed above the drained or draining magma chamber. The caldera is therefore filling with ponded intra-caldera ignimbrite simultaneously as subsidence or collapse of the caldera is occurring, which forms a specific depocentre or basin into which erupting ignimbrite is trapped. Intra-caldera ignimbrites may be hundreds to thousands of m thick filling in the space formed during caldera subsidence (Fig. 10.29). The thickest intra-caldera ignimbrites may be >3 km thick and are usually welded. This suggests collapse commenced very early during the course of the eruption, possibly during pre-eruption tumescence of the viscoelastic crust above an inflating large magma reservoir (Fig. 16.59; Jellinek and De Paolo 1981; Gregg et al. 2012). At this time, *caldera collapse breccias* (also called meso-breccias) of the type envisaged by Lipman (1976) may form by the caving or gravitational collapse of the caldera walls, producing rockslides and falls, which may be interbedded within the ignimbrite. These should be distinguishable based on componentry and distribution from *co-ignimbrite breccias* (Chap. 10).

There have been surprisingly few studies that have documented the internal stratigraphic and facies architecture of intra-caldera ignimbrite successions. Willcock et al. (2013) described the detailed internal architecture of the 1200 m thick, Permian, Ora intra-caldera ignimbrite succession in northern Italy (Fig. 10.85a, b), demonstrating that a definable stratigraphy of depositional units or pulses could be established, based on changing free crystal size and abundance. This succession shows that infilling of calderas can be pulsatory, presumably reflecting pulsing in the eruption and/or feed into the caldera from multiple vents, mostly along the ring fracture.

Simultaneous to infilling of the caldera by intra-caldera ignimbrite, pyroclastic flows also flow radially outside and away from the caldera. Because these are unconfined (cf. intra-caldera pyroclastic flows), they spread forming more sheet-like ignimbrites that can flow tens to over 100 km from the caldera margin. Large-volume outflow or extra-caldera ignimbrite units can vary from metres to many tens of m in thickness. They can contain large accessory and accidental lithic clasts and lithic breccia lenses tens of km from the caldera. The low-profile shield of ignimbrite around the caldera can be built from one or more eruptions and ignimbrites. Extra-caldera ignimbrites can be welded or non-welded, and the degree of welding decreases away from the vent, as discussed in Chap. 10. The large-volume outflow ignimbrite sheet(s) erupted from one caldera can overlap and complexly interdigitate stratigraphically with the outflow sheets of other calderas especially in regions of ignimbrite flare-up, which can make identification of the source caldera problematical; e.g. the Southern Rocky Mountain Volcanic Field (Stevens and Lipman 1976; Cas and Wright 1987), and Mexican Volcanic Belt (Aguirre-Díaz and López-Martínez 2009).

16.8.7 Post-caldera Volcanism and Sedimentation: Subsidence and Resurgence

Post-caldera volcanic activity may last for $>10^5$–10^6 yr and is usually accompanied or followed by phases of general subsidence and/or resurgence of the caldera floor, either evenly distributed (homogeneous) or focused on structural blocks (Fig. 16.65a, b; Marsh 1984; Kennedy et al. 2012; Wilcock et al. 2013). Given that most calderas have a morphological expression, it suggests that significant subsidence of the caldera floor has occurred after the peak of magma discharge, otherwise most calderas would be infilled with ignimbrite. Both silicic and alkaline types may show complex patterns of post-caldera volcanism, with domes and cones distributed in various geometrical patterns. These are primarily controlled by the caldera-forming ring and other magma-tapping, extensional or trans-tensional faults, but are also dependent on whether the caldera is in a deflating or resurgent stage and relationships with pre-existing regional

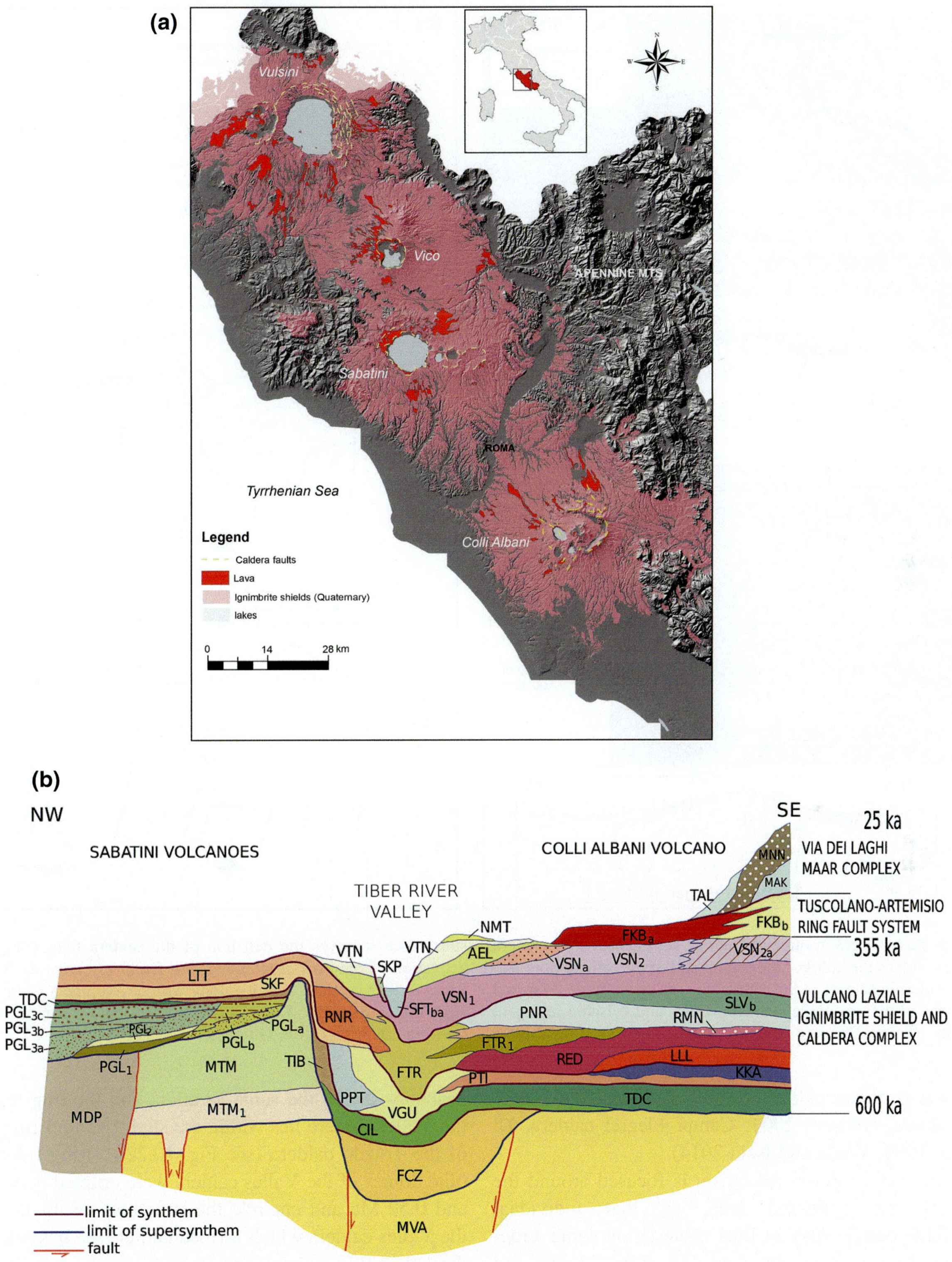

Fig. 16.64 Location of the main caldera volcanoes and stratigraphy around Rome, central Italy. **a** Digital elevation map showing the calderas and outflow ignimbrite sheets (also see Fig. 16.56). **b** Schematic representation of the stratigraphic relationships across Rome. The right side (southeast) shows the ignimbrites of the Colli Albani volcano (TDC, PTI, KKA, RED, PNR, VSN, which were erupted from Vulcano Laziale caldera complex) extending across the plateau and into the Tiber river valley, where they interfinger with sedimentary terraces (CIL, VGU, FTR, AEL). Younger terraces host volcaniclastic sedimentation from post-caldera activity (Tuscolano-Artemisio ring fault system, and Via dei Laghi maar complex in VTN and SKP). Here, Colli Albani ignimbrites interfinger with products coming from the Sabatini calderas to the northwest (TIB, PPT, SKF, RNR, LTT). The volcanic succession rests above a marine to continental sedimentary substrate (MVA, MTM, MDP, FCZ, PGL). (Excerpt from Funiciello et al. 2008, Geological Map of Rome, 1:50,000 scale)

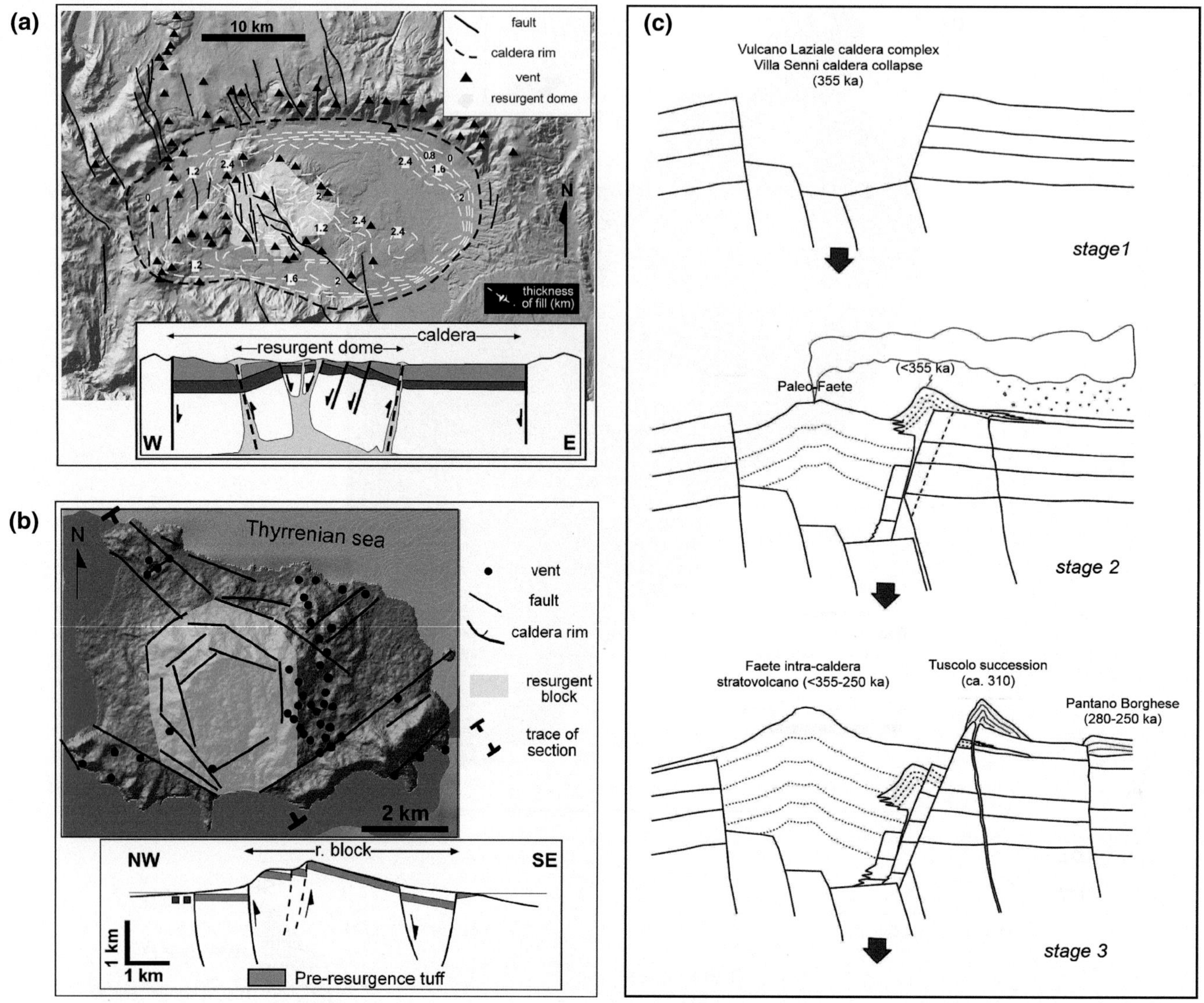

Fig. 16.65 Post-caldera resurgence versus subsidence. Post-caldera resurgence involves the uplift of the central part of the caldera floor and may focus volcanic activity at the edge of the resurgent dome. Map and cross-sections of **a** Long Valley caldera, USA, and **b** Ischia resurgent block, Italy (From Branney and Acocella 2015). **c** Post-caldera subsidence involves the deflation of the caldera floor (*stage 1*), and activation of outward progressing ring fractures (*stage 2*). These may focus further volcanic activity along with development of intra-caldera stratovolcanoes *(stage 3),* as found at Colli Albani volcano, central Italy (see Fig. 16.56c). (From Giordano et al. 2010)

structures (e.g. Valles caldera, Self et al. 1986, 1996; Long Valley caldera, Hildreth 2004; Campi Flegrei caldera, Di Vito et al. 1999, Vitale and Isaia 2014).

Generally, post-caldera volcanism is focused around the margin of the collapsed area, so that individual small-volume centres may in time coalesce on some structures to produce a nearly complete ring of lava domes and other edifices. For example, at Valles caldera, most rhyolite lavas are located on the ring fracture of the Valles caldera (Figs. 16.66 and 10.22d, e), or part of an incomplete ring related to the earlier Toledo caldera (Smith and Bailey 1968; Self et al. 1986, 1996; Heiken et al. 1990; Phillips et al. 2007; Chap. 10). The Valles caldera and the Toledo calderas probably shared the same location, and the ring fracture of the Valles caldera lies within the slightly larger ring fracture of the Toledo caldera (see Fig. 10.22d). Ring fracture rhyolite domes of the Valles caldera were erupted between 1.24 and 0.52 Ma and encircle the structural resurgent dome of the Valles caldera which formed earlier (see below). Each of these domes is probably monogenetic, with repose periods of 10^3–10^4 yr between eruptions (Self et al. 1996). La Primavera caldera is somewhat more complicated (Walker et al. 1981; Cas and Wright 1987). Some of the rhyolite lavas are located on an almost complete ring, and some lavas on a transverse line (probably a deep normal fault) which crosses it. Most of these domes are again monogenetic. However, there are three larger, polygenetic structures which appear to lie on an outer ring structure and could be called independent volcanoes in their own right. Caldera volcanoes without a well-defined ring structure also occur. The rhyolite lavas at Rotorua caldera have a more linear distribution, and many of the vents seem to have been located along fissures where regional extension seems likely to have been a much more important structural control on the distribution of eruptive points than magmatic processes (Ashwell et al. 2013).

At the alkaline Colli Albani caldera, the geometry of its ring faults has controlled the pattern of post-caldera volcanism (Fig. 16.65c). The first stage of post-caldera volcanism, lasting some 10^5 years, was a prolonged period of subsidence and down-sagging, which progressed over time outside of the original structural caldera margin. During this period, effusive to strombolian edifices were erupted, but only along the northern and eastern sections of the caldera, where deep ring faults were outward dipping and therefore favourably oriented for magma intrusion during caldera subsidence. Contemporaneous intra-caldera volcanism progressively filled the caldera floor with growth of the Faete stratovolcano, from which effusive and strombolian to plinian eruptions occurred. Between 180 and 20 kyr BP a reversal of caldera floor long-term deformation to resurgence produced a marked shift in the location of monogenetic centres (maars and scoria cones), which are aligned only along the western ring faults (Giordano et al. 2010). This shift in activity is attributed to the activation of inward dipping faults, reconstructed by gravity and seismic data, favourable to dilation during resurgence, and also mirroring closure of the previously active outward dipping structures on the eastern side of the caldera.

When the central parts of some calderas are up-domed they give rise to resurgent domes, typically composed of uplifted or up-arched welded intra-caldera ignimbrite, but also including post-caldera sediments (Fig. 16.66; Smith and Bailey 1968; Marsh 1984; Cole et al. 2005; de Silva et al. 2015; Mucek et al. 2017). At Valles caldera, post-collapse (early) rhyolite lavas were first associated with the resurgent uplift, as faults developed on the resurgent dome and were utilised as conduits for intra-caldera magmas (Fig. 16.66; Phillips et al. 2007; Wilcock et al. 2013). These faults are aligned NE-SW and parallel the deep, trans-crustal Jemez lineament (Self et al. 1986). The time window between the eruption of the Upper Bandelier Tuff ignimbrite and the oldest ring fracture rhyolite dome, suggests resurgence likely took place within a period of 30–50 kyr (Phillips et al. 2007). Today the summit of the resurgent dome at Redondo Peak is $\sim$840 m above the undomed caldera floor, although originally the height was probably significantly greater (Fig. 16.66). More generally, resurgence appears to occur on time scales of 10^3–10^5 yr and may involve 1000 m or more of uplift (e.g. Kennedy et al. 2012; de Silva et al. 2015; Mucek et al. 2017). Figure 16.67 presents some conceptual models to explain the observed resurgence of the floor of caldera volcanoes.

The first two models relate to readjustment of the plumbing system after the withdrawal of the ignimbrite-forming magma (Fig. 16.67a, b). In comparison, the third model involves an open plumbing system where the magma reservoir is progressively replenished to overpressure conditions (Fig. 16.67c); the overlying caldera block remains essentially impermeable to magma and accommodates the increased pressure by uplifting through flexing. This process appears somewhat similar to pre-caldera tumescence, observed at many large calderas (Smith and Bailey 1968). We interpret this to be a result of progressive weakening of the crust that becomes largely viscoelastic due to high heat-flux associated with the presence of a large magma body at shallow depths and is able to accommodate large degrees of inflationary deformation (Jellinek and DePaolo 1981; Gregg et al. 2012). The fourth model (Fig. 16.67d) attributes updoming to inflation of a very shallow laccolith intruded into the caldera fill succession. Galetto et al. (2017) propose that eruption in this last scenario is constrained by a hardened, altered intra-caldera ignimbrite succession that blocks dykes and promotes high-level sill formation. This, together with cooling and crystallisation of the upper magma in the sill causes a rheological boundary to form that limits magma rise to the surface and causes inflation of the sill into a laccolith, causing resurgent updoming. Each of these mechanisms may occur in different settings. Time scales of post-caldera resurgence also indicate that in this case deformation is at least partly accommodated by viscoelastic deformation, although, in comparison with pre-caldera tumescence, the presence of ring faults focusses shear and produces anelastic behaviour.

Impressive resurgent domes with 1000 m or more of uplift have been described at Cerro Galan (Folkes et al. 2011; Fig. 16.68a, b) and Aguas Caliente (Petrinovic et al. 2010) calderas in northwestern Argentina; Lake Toba, Indonesia (van Bemmelen 1939; Chesner 2012; de Silva et al. 2015; Mucek et al. 2017; Figs. 16.53c and 16.68c); Creede (Fig. 16.68d), Valles (Figs. 10.22d and 16.66), Lake City, San Juan, Timber Mountain and Long Valley calderas to name a few in the western United States (Smith and Bailey 1968; Phillips et al. 2007; Kennedy et al. 2012). An intriguing feature is that although such impressive levels of resurgence call for remarkable pressure increases in the related magma reservoirs, many resurgent domes have ended their "life-cycles" without giving rise to a further large-volume ignimbrite eruption, although lava domes have been erupted during resurgence. This indicates the magma

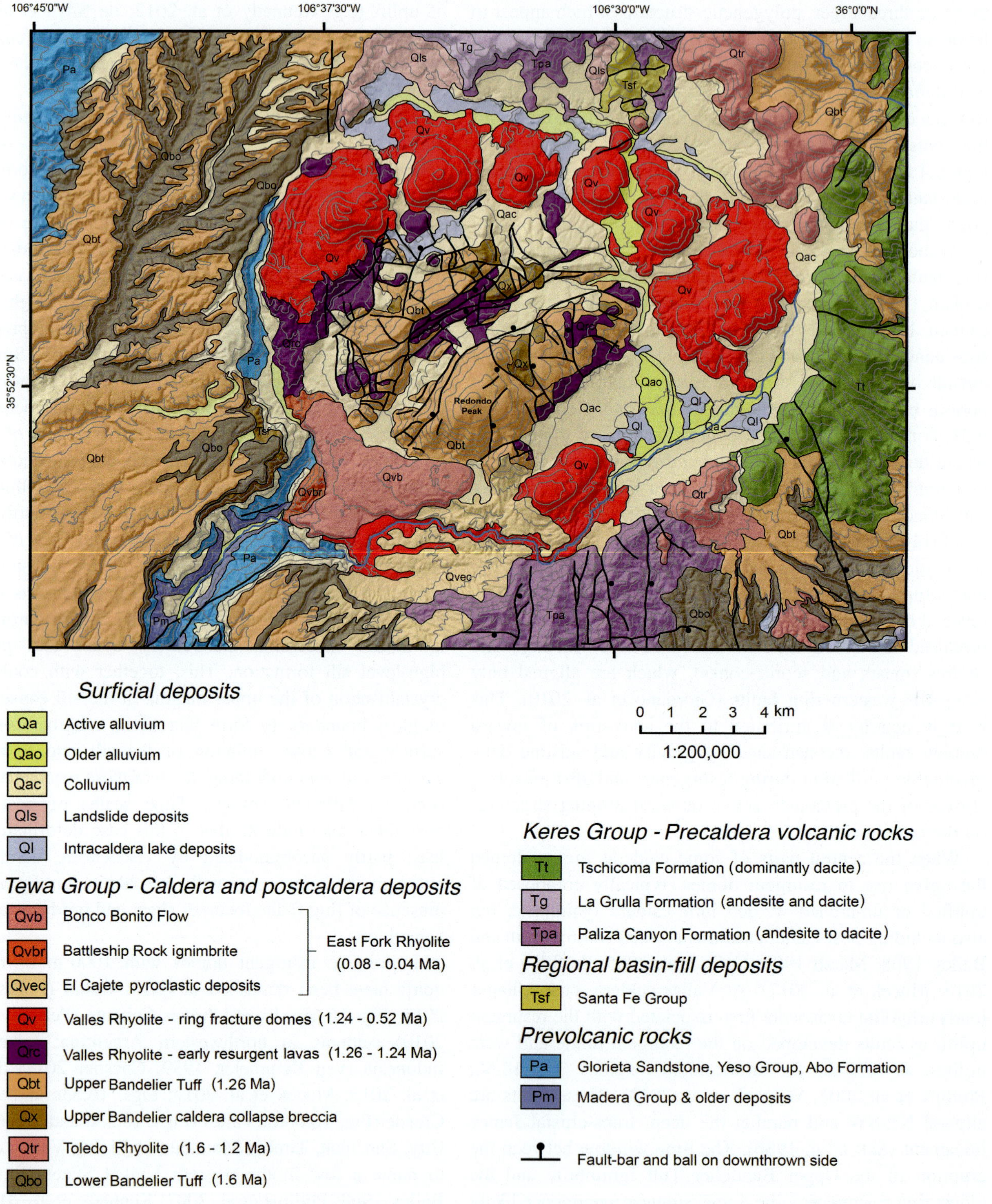

Fig. 16.66 Geological map of the 1.26 Valles and 1.6 Ma Toledo calderas, Jemez Mountains, New Mexico, USA. This shows an almost perfect arcuate alignment of post-caldera collapse rhyolite lava domes inside the topographic rim of the Valles caldera, so defining the caldera collapse ring fracture. Also shown is the resurgent dome of the Valles caldera, the top of which forms Redondo Peak (3430 m), consisting of uplifted welded Upper Bandelier Tuff intra-caldera ignimbrite, caldera collapse breccias, caldera lake sediments, and early post-caldera rhyolite lavas associated with resurgence. The Valles and Toledo calderas largely overlap, with the ring fracture of the Valles caldera lying or nested within the slightly larger ring fracture of the older Toledo caldera, so that little post-Toledo collapse rhyolite domes eruptions between the two caldera-forming eruptions are preserved; although these include the Cerro Toledo Formation group of domes and sub-plinian fallout deposits (Chap. 10, see Figs. 10.28b and 10.40a). The youngest volcanic eruptions at Valles caldera are represented by the East Fork Rhyolite which is petrologically different from the high silica, ringfracture domes, and thought to represent a new cycle of magmatic activity (Wolff and Gardner 1995). Generalised ages are from Self et al. (1996), Phillips et al (2007), Wolff et al. (2011), Zimmerer et al. (2016). (Modified from Wolff and Thompson 2017, After Goff et al. 2011; also see Fig. 10.22d, e for regional map and Landsat image)

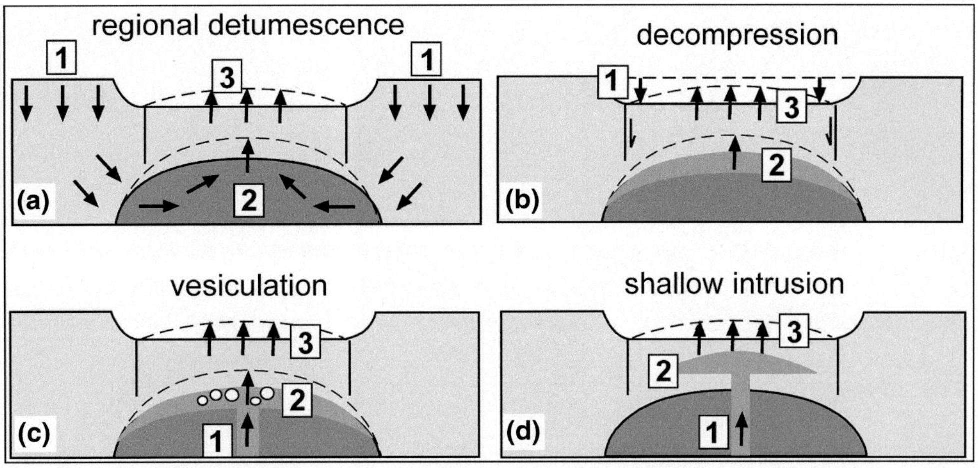

Fig. 16.67 Main processes that may induce resurgence during post-caldera activity according to Marsh (1984). **a** Regional tumescence (1) induces compression over the magma reservoir (2) and then uplift (3). **b** Decompression (1) due to magma withdrawal induces magma buoyancy (2) and uplift (3). **c** Magma recharge and vesiculation (1) induce volumetric changes and buoyancy (2) and uplift (3). **d** Intrusion (1) and growth of a magma body at shallow levels (2) induce uplift (3). (Adapted from Branney and Acocella 2015)

causing resurgence is probably largely degassed, representing in part residual crystal mush (e.g. Kennedy et al. 2012; Wilcock et al. 2013).

Based on the "*type example*" of Valles caldera, Smith and Bailey (1968) proposed a life cycle for caldera volcanoes, which is initiated with pre-caldera regional tumescence leading to eruption and caldera collapse. This is followed by post-caldera sedimentation, resurgent doming and ring fracture volcanism, and terminal hydrothermal activity. However, this cycle is not seen at many calderas. For example, the calderas in the Taupo Volcanic Zone and in the Roman Magmatic Province have largely not been affected by resurgent doming. Only some of the calderas associated with the Altiplano Puna flare-up in the Andes have been affected by caldera resurgence, whereas others have not. Nor does the chemistry of the magma appear to be a discriminating factor. While it is true that most resurgent calderas are rhyolitic or rhyodacitic in composition, the most recent and fastest rising example is the K-trachytic dome of Mt Epomeo (Ischia). There about 1100 m of uplift has occurred in less than 33 kyr, or on average ≥ 30 mm yr^{-1} (Fig. 16.68e; Tibaldi and Vezzoli 1998; Selva et al. 2019). This compares well with calculated initial uprise rates for the Lake Toba resurgent dome of ~ 5 cm yr^{-1}, declining to 1 cm yr^{-1} (de Silva et al. 2015; Mucek et al. 2017), and 1.9–3.7 cm yr^{-1} at Valles caldera (Phillips et al. 2007). However, these represent average resurgence rates, as the nature of resurgence as a continuous versus an intermittent process is unknown. We are therefore far from fully understanding the factors that promote resurgent doming at calderas and suggest that tectonic stresses can also act on time- and length-scales that may interplay with processes that enhance resurgence (Acocella et al. 2000; Chap. 17, see Sect. 17.6).

In addition, most calderas experienced more than one caldera-forming cycle, with many showing increasing duration of quiescence periods and erupted volumes over time (e.g. Cerro Galan; Folkes et al. 2011; Toba; Liu et al. 2021). Caricchi and Giordano (2022) suggest that this pattern relates to the changing thermal and chemical structure of the trans-crustal plumbing system, which over time is able to receive and incubate progressively larger volumes of eruptible magma, promoting longer quiescence periods, eventually resulting in larger volume eruptions. This view highlights the significance of post-caldera formation quiescence periods and reassesses the common use of average recurrence time of eruption (which embeds the notion of a static system) to define the state of activity of volcanoes. A prolonged quiescence period may equally relate to a true waning of the magma input or formation of larger accumulations of magma at depth. Such first-order uncertainty calls for innovative quantitative investigation methods to assess the amount and availability of eruptible magma below quiescent calderas across the entire crust, something at present far from being within reach of current geophysical, geochemical and petrological approaches.

Along with post-caldera volcanism, subsidence and resurgence, surface or epiclastic processes and deposits are also important in caldera volcanoes, and it is important to stress the repose periods between eruptions, and the possible effect on the hydrologic and geomorphic system of a large ignimbrite eruption. Caldera volcanoes are characterised by the rapid production of vast volumes of loose, easily erodible pyroclastic deposits, which can be easily remobilised as sediment after emplacement over irregular intervals. This pumiceous and tuffaceous sediment can be transported out of the volcano by rivers cutting through the ignimbrite shield,

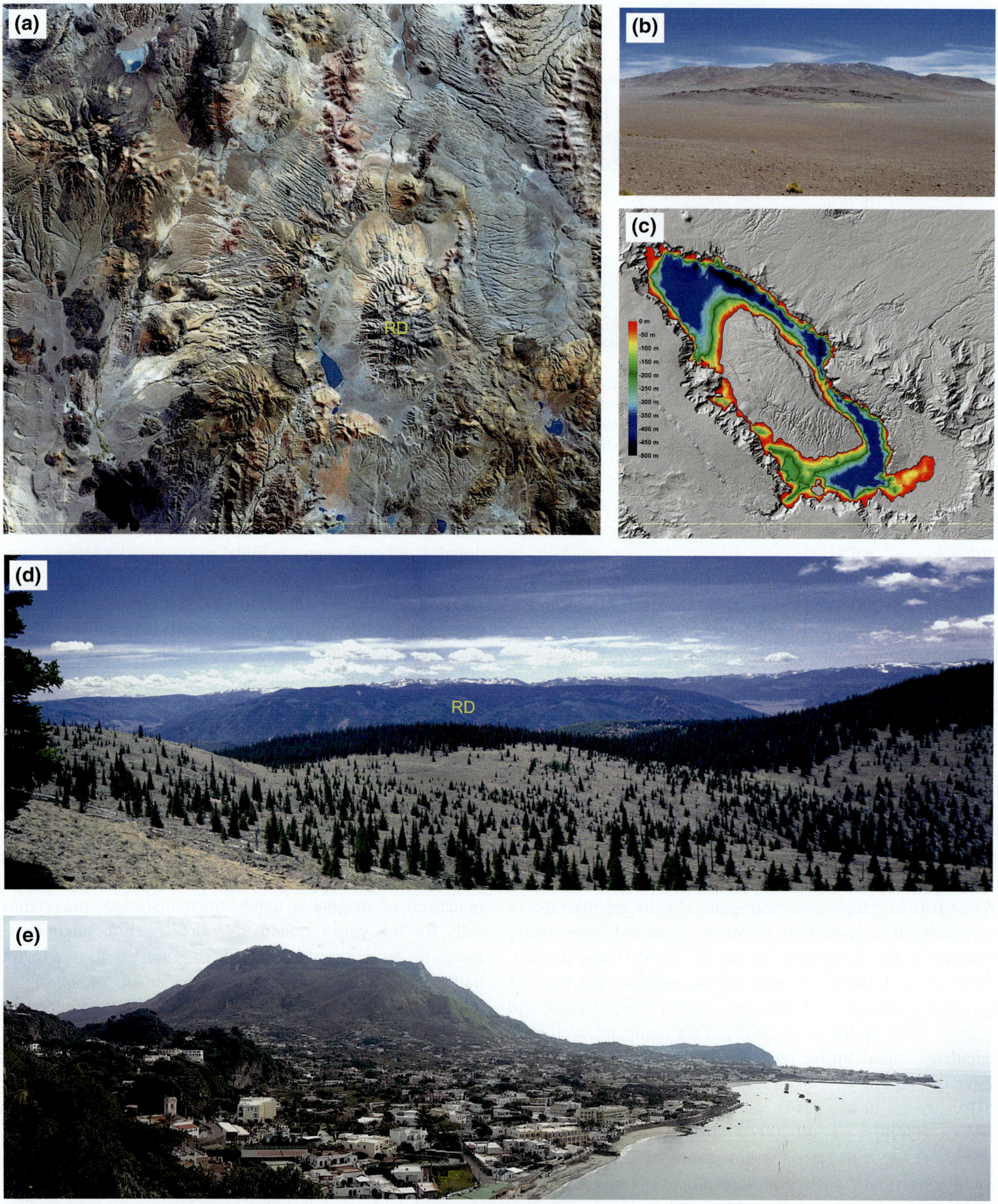

Fig. 16.68 Examples of resurgent domes in calderas. **a** Resurgent dome (RD) in the centre of the Cerro Galan caldera, Argentina. Both the caldera and resurgent dome are elongate in shape and are surrounded by an outward dipping, erosionally incised outflow ignimbrite shield. (Landsat Thematic Mapper image from Earth Explorer; *Source* United States Geological Survey). **b** Eastward view of the resurgent Cerro Galan dome from the caldera floor. **c** Digital elevation model of and bathymetric contours of Lake Toba caldera showing the resurgent dome of Samosir island, Sumatra, Indonesia. Both are elongated parallel to the Sumatran Fault Zone. (From Chesner 2012). **d** Symmetrical resurgent dome (RD) in the centre of Creede Caldera, Southern Rocky Mountains Volcanic Field, San Juan Mountains, USA. **e** Mt Epomeo on the Italian island of Ischia is the fastest actively growing resurgent dome known

or back into the system to be deposited inside the caldera. Many caldera volcanoes are partly or totally submerged either by lakes (e.g. Bolsena, Taupo, Toba, and Toya, Japan), or the sea (e.g. Taal, Philippines; Campi Flegrei, Italy; Aira, Ata and Kikai, Japan; Santorini and Kos-Yali, Greece; see Sect. 16.8.9 for discussion of marine silicic calderas). Caldera fill sequences are therefore complex associations of fluvial and lacustrine and/or marine epiclastic sediments, pyroclastic deposits and lavas (Fig. 16.69a–f). The up-domed caldera fill sequence at La Primavera contains an unusual bed of giant pumice with blocks of pumice up to 8 m in diameter (Fig. 16.69c; Clough et al. 1981; Cas and Wright 1987; Figs. 13.45 and 13.46). These are thought to represent vesiculated blocks spalled from the pumiceous carapace of a rhyolitic lava during the subaqueous eruption (Chap. 6), which then floated away, eventually becoming waterlogged and deposited within lacustrine sediment. A similar giant pumice bed has been described from Lake Taupo (Fig. 16.69d; Wilson and Walker 1985). In addition, at rapidly rising resurgent domes, such as Mt Epomeo on Ischia, impulsive and massive sector collapses and debris avalanches may occur, possibly triggered by decompression of the hydrothermal and magmatic system, as well as other hazards such as tsunamis if near the coastline (Fig. 16.70; Della Seta et al. 2012; Chap. 12).

16.8.8 Geothermal Activity and Hydrothermal Alteration

The near-surface intrusion of domes and/or magma at depth below long-lived caldera centres can promote active geothermal systems which may extend well beyond the caldera margin (e.g. the Roman Geothermal Province, Giordano et al. 2014). These are responsible for hydrothermal alteration of volcanic rocks and facies over wide areas, and the formation of epithermal precious metal and other mineral deposits in a wide variety of volcanic settings and facies (Chap. 18). Such caldera-related geothermal systems have huge potential (e.g. the Roman Geothermal Province, Giordano et al. 2014) and are increasingly being used to generate geothermal energy or power in caldera volcano provinces such as the Taupo Volcanic Zone (e.g. Wairakei geothermal field, Rosenberg et al. 2009; Chap. 18).

Important volcanic manifestations of geothermal activity are hydrothermal pools, geysers, sinter deposits and hydrothermal or phreatic explosions (Chaps. 4 and 18). These occur when superheated water at shallow depths flashes to steam, disrupting the confining rocks and ejecting solid debris, water and steam. In these explosions, magma is not directly involved (cf. phreatomagmatic eruptions described earlier). Hydrothermal explosion craters are small (W_{cr} < 1 km; Fig. 16.11g) but often occur as clusters of multiple craters, often aligned along faults. They are closely associated with larger volcanic edifices characterised by shallow, up-flowing geothermal systems (Henley and Ellis 1983; Chap. 18). Most commonly, therefore, they are found within large rhyolitic calderas and volcano-tectonic depressions (e.g. Taupo Volcanic Zone; Bogie et al. 2005).

Hydrothermal explosions may produce both ballistic ejecta and pyroclastic surges. Deposits are often crudely bedded and consist of angular to subrounded fragments of pervasively altered, older lava and other volcanic, epiclastic and local country rocks within a matrix of more-finely granulated fragments and hydrothermal clays (Fig. 4.31g; and Chap. 18). Some of the finer-grained layers may show evidence of being deposited by base surges. Hydrothermal explosions have occurred in most of the active geothermal fields of the Taupo Volcanic Zone during the late Quaternary, including several historic events. As well as hydrothermal explosion craters these fields are characterised by hot springs, geysers and pools, steaming ground, fumaroles, and sinter deposits and terraces (Chap. 18). These include the famous Pink and White Terraces of Lake Rotomahana, considered at the time one of the wonders of the world, which were blasted away by the explosion of the hydrothermal system associated with the 1886 eruption of Mt Tarawera (de Ronde et al. 2016). Good descriptions of these rock types in the Quaternary Kawerau geothermal field are also given by Nairn and Wiradiradja (1980), Browne and Lawless (2001) and Rowland and Simmons (2012).

Hydrothermal explosions result from boiling of an up-flowing geothermal fluid, under a localised silica cap (Hedenquist and Henley 1985). They are initiated by fault rupture, which causes a large pressure decrease, resulting in violent steam flashing and intense fracturing (Hedenquist and Henley 1985; Cas et al. 2011c; Rowland and Simmons 2012; Vignaroli et al. 2015). As the geothermal fluid in contact with a large surface area of fractured rock boils, precipitation of gold and other precious metals may occur (Chap. 18). Gold is focused in the up-flow zone within diatremes, breccia pipes and surrounding fault-fracture networks, formed below such hydrothermal craters. Many repeated hydrothermal explosion events could lead to formation of large, near-surface epithermal vein deposits (Rowland and Simmons 2012). Because of their important exploration potential, we leave more detailed description and illustration of hydrothermal exploration craters and their geothermal setting to Chap. 18.

16.8.9 The Special Case of Marine Silicic Calderas

Although our preceding description and discussion of caldera volcanoes has largely focussed on their continental settings,

Fig. 16.69 Examples of sedimentary caldera fills. **a** Horizontal, tuffaceous "moat" sediments deposited in an annular lake formed between the resurgent dome and caldera wall at Creede caldera, Southern Rocky Mountains Volcanic Field, San Juan Mountains, USA. **b** Lacustrine deposits overlying the Permian intra-caldera Ora ignimbrite, Italy. **c** Giant pumice blocks enclosed in lacustrine sediments, La Primavera rhyolite volcano, Mexico. **d** Giant pumice block (approximately 5 m across) embedded within Taupo caldera lake sediments, New Zealand; this block and those at La Primavera are characterised by cooling joints normal to their margins and quenched (breadcrusted) surfaces. **e** Ancestral (pre-232 AD) Taupo caldera lake sediments; this section displays symmetrical ripples formed by near shore wave action (Chap. 12). **f** Thinly laminated lacustrine sediments of the Huka Falls Formation, Taupo Volcanic Zone; these represent an extended period of ancestral lacustrine sediments (>200 kyr) deposited within extensional basins hosting the paleo-Lake Huka. The long-lived lacustrine setting is recognised to have been continually modified by episodic volcanism and gradual tectonism. The Huka Falls Formation is an important regional seal to geothermal and hydrothermal activity (Chap. 18)

silicic calderas should also occur in at least shallow marine environments on submerged continental and micro-continental arc crust, and at the shallow summits of some marine arc volcanoes. Such volcanoes may similarly be highly explosive, and produce ignimbrites, that may be deposited subaerially, in shallow to deep waters, or on exposed islands (Chap. 11), together with abundant water-lain fallout or resedimented deposits (Chaps. 6 and 11), abundant redeposited pyroclastic debris (Chap. 12), and lava domes (Chap. 6). Deep-water calderas will be more lava dominated

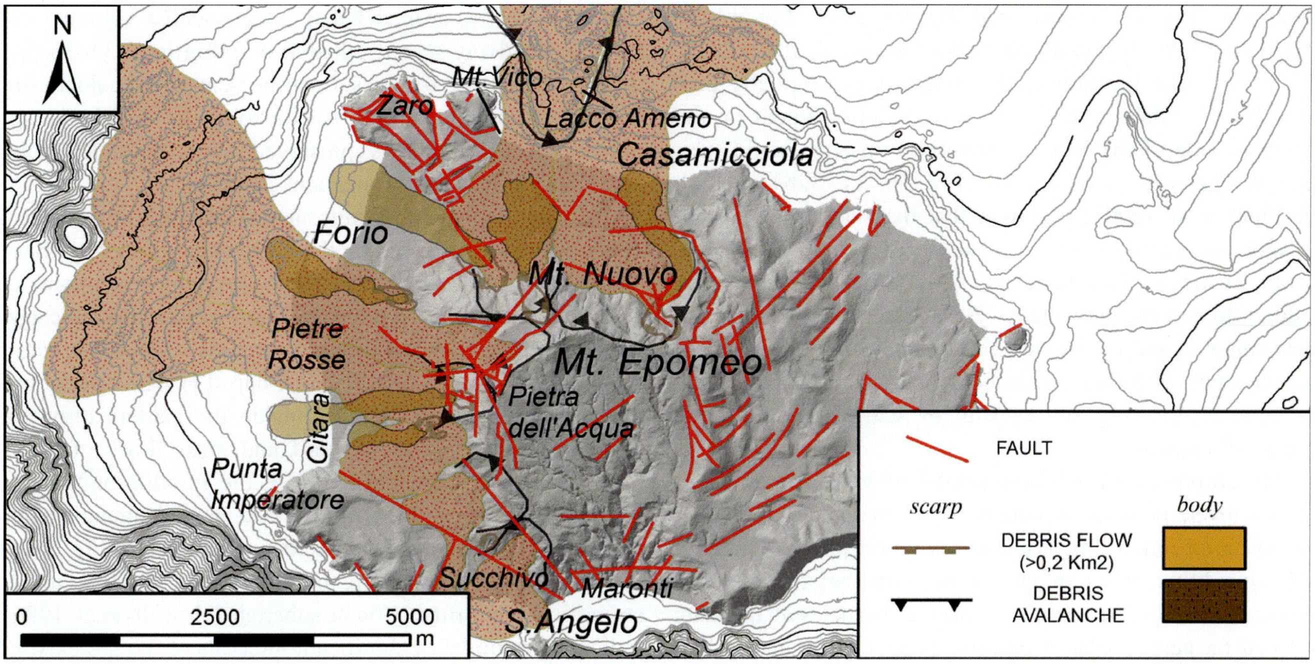

Fig. 16.70 Source areas and recent extension of debris avalanches and debris flows offshore from the Mt Epomeo resurgent dome, Ischia. The dome is actively rising at the impressive rate of uplift >1000 m in <33 ka. The grey shaded area represents the island terrain a.s.l.; the contour lines indicate bathymetry at 10 m intervals. (thick lines at 50 m intervals). (From Della Seta et al. 2015)

due to the suppressing effects of hydrostatic pressure on explosive eruptions in deep water (Chaps. 3 and 4).

16.8.9.1 Significance of Depth of Caldera Floor?

An important caveat, however, is that centres that are now subaqueous may have originated from subaerial vent systems followed by caldera collapse to submarine water depths. That is, just because the caldera floor is under water now, doesn't mean that the caldera-forming eruption necessarily occurred from submarine vents; instead, it only represents the post-eruption subsidence depth (see Johnston et al. 2014). For both modern and ancient marine explosive calderas understanding the water depth at which the caldera-forming eruptions occurred is therefore critical, especially given there could be hundreds of m to 2 km of caldera subsidence. The maximum possible depth of an eruption is constrained by the depth of the rim of the caldera, which represents the maximum pre-subsidence water depth of the summit of the volcanic edifice. However, the actual eruption depth may have been even shallower if there was a significant part of the edifice at shallower pre-eruption water depths that has been removed during the eruption through collapse downwards into the caldera, or outside of it as a sector collapse event, or was destroyed explosively. In ancient successions, it is almost impossible to assess the absolute water depth of the eruption let alone determine the paleoenvironmental water depth of the seafloor.

Understanding the eruption water depth is important, because as discussed in Chaps. 3, 4, and 9, the intensity of subaqueous explosions will be increasingly suppressed with increasing water depth and hydrostatic pressure. Volatile exsolution and explosive eruptions at water depths greater than 1,000 m will be significantly less intense than subaerial counterparts and may be completely suppressed at water depths greater than 1,500–2,000 m. Even though volatile exsolution is still possible at such depths it occurs more slowly, and exsolved volatiles will be supercritical or highly compressed volatile phases (Cas and Simmons 2018). With eruptions becoming increasingly effusive and lava forming with increasing water depth, lavas may be highly vesicular and could be prone to fire fountaining, compared with subaerial counterparts. The carapaces of submarine lavas could be more pumiceous than subaerial counterparts, and quench fragmentation of such coherent pumice carapaces can form voluminous deposits of pumice hyaloclastite by non-explosive processes (Cas and Wright 1987; Cas 1992; Cas and Giordano 2014; Allen et al. 2010; Manga et al. 2018). The problem then is to discriminate autoclastic pumice from pyroclastic pumice (Carey et al. 2018; Manga et al. 2018), and from resedimented pumice.

16.8.9.2 Shallow-Intermediate Depth Calderas

The inferred 161 ka Kos-Yali caldera, Greece, from which the ~90 km^3, Kos Plateau Tuff was erupted, appears to be wholly submerged but parts of its rim may have been emergent (Fig. 16.71; Cas and Wright 1987, Fig. 13.44; Allen 2001; Pe-Piper et al. 2005; Nomikou et al. 2013, 2018; Chap. 11). The location of the caldera was originally proposed to lie offshore, encompassing the embayment on the southern side of the Kefalos Peninsula on Kos Island (Cas and Wright 1987; Allen 2001; Fig. 16.71). Subsequently, Pe-Piper et al. (2005) proposed that the caldera lay further offshore along the northwest margin of Nisyros stratovolcano. However, Nomikou et al. (2013) report a deep caldera-like depression that is 700 m deep, just to the east, between Kos and the islands of Yali and Strogyli; this most recent study is based on detailed bathymetry surveys and underwater imaging of deposits in scarps bounding the proposed caldera.

The eruption probably originated from vents just above or at sea level, perhaps associated with an andesitic edifice, based on an initial phreatoplinian eruption phase, abundant andesitic lithic clasts in ignimbrite, and that the exposed products are almost entirely subaerial pyroclastic deposits located on nearby islands and the Turkish mainland (Allen 2001; Chap. 11). They include minor phreatoplinian fall and surge deposits, and the low aspect ratio Kos Plateau ignimbrite, which resulted from flow over both land and the sea (see Figs. 11.18 and 11.19). Other volcanic facies on Kos include pre- and post-caldera-forming fall-surge tuff cones, rhyolite lava dome complexes and associated reworked volcanic sediment (Cas and Wright 1987; Allen 2001; Allen and Cas 2001). On the island of Yali, an uplifted dome and quenched pumiceous cone complex gives some indication of the products of post-caldera-forming submarine volcanism in moderately shallow marine calderas (Allen and McPhie 2000; Chaps. 9 and 11). Similarly, the catastrophic 1883 explosive eruption of Krakatau in Indonesia began at or above sea level on the former emergent edifice of Krakatau, but resulted in a submerged caldera, of which the remaining islands, apart from post-eruptive Krakatau, represent emergent parts of the caldera rim (Simkin and Fiske 1983).

Other relatively shallow water modern caldera centres are the Aira, Ata and Kikai calderas, Kyushu, Japan (Fig. 16.72a, b). The 7.3 ka eruption that produced the 17 × 20 km offshore Kikai caldera probably occurred from wholly subaerial vents on an offshore island. This produced widely dispersed fallout and ignimbrite deposits (∼150 km^3; VEI 7), including the low aspect ratio Koya pyroclastic flow that was able to flow over the sea for >60 km to the mainland (Maeno and Taniguchi 2007; Chap. 11). Pyroclastic deposits exposed on islands along the caldera rim and on the mainland are subaerial in character, and include plinian pumice fallout, ignimbrite and welded surge deposits (∼ 5 km^3; see Fig. 11.21c). Shimizu et al. (2024) mapped an equivalent megaturbidite deposit on the seafloor spread radially ∼40 km around Kikai caldera with a volume ∼71 km^3. A relatively recent effusive eruption in 1934–35 produced a lava dome from a submarine vent on the caldera ring fracture. Although initially subaqueous, the dome emerged above sea level and is preserved as an island, Showa Iwo-jima (Maeno and Tanaguchi 2005).

In ancient volcanic settings, an example of a very shallow water caldera is provided by the Ordovician Lower Rhyolitic Tuff Formation (LRFT) in North Wales, which is dominated by up to >500 m of welded and non-welded rhyolitic ignimbrite ponded in a contemporaneous caldera structure (Fig. 16.73; Howells et al. 1986; Chap. 11). The caldera was at least 15 km in diameter. Outside the caldera only one non-welded ignimbrite-like unit and mass flows of remobilised tuff occur intercalated with the fossiliferous marine succession (see Fig. 11.41c–g). These are substantially thinner than the massive intra-caldera succession (see Fig. 11.41 a, b). Over most of its extent, the LRTF lies disconformably to strongly unconformably on its substrate (Howells et al. 1986), indicating removal of significant intervals of the pre-existing marine sedimentary substrate and significant parts of the older welded Pitts Head Tuff. This indicates an extended period of significant erosion, probably under subaerial conditions, prior to the emplacement of the LRTF, suggesting that much of the caldera may have been emergent during the initial parts of the eruption. A likely cause of pre-eruption regional updoming from shallow marine settings to above sea level may have been the intrusion of a large volume of magma into shallow crustal levels prior to eruption causing regional tumescence (Chap. 11).

Busby-Spera (1984, 1986) described two very thick Triassic units consisting of pumiceous debris preserved within a deeper water sedimentary succession in the Mineral King Caldera Complex, Sierra Nevada, California. The Vandever Mountain Ash Flow Tuff is up to 480 m thick, enclosed in largely pelitic and turbiditic ambient sedimentary facies, indicative of water depths below storm wave base (>100–300 m). The Vandever Mountain Tuff is considered to fill in a 7 km diameter caldera subsidence structure, defined by syn-emplacement normal faults, and associated fault scarp breccias and megabreccias, including slide blocks up to 0.5 km in diameter. The main tuff unit is massive and consists of lapilli tuff. Deformation and metamorphism have largely destroyed original textures, although, Kokelaar and Busby (1992) demonstrated plastic shard and pumice welding textures in the matrix of the ignimbrite using back-scatter images (see Chap. 11). The Monarch Rhyolite Ash Flow Tuff is considered to be a trap-door caldera fill deposit because of its thickness of 600 m and its geometry (Busby-Spera, 1986). Over an 8 km exposed strike length it changes from a massive, unsorted pumice lapilli tuff in

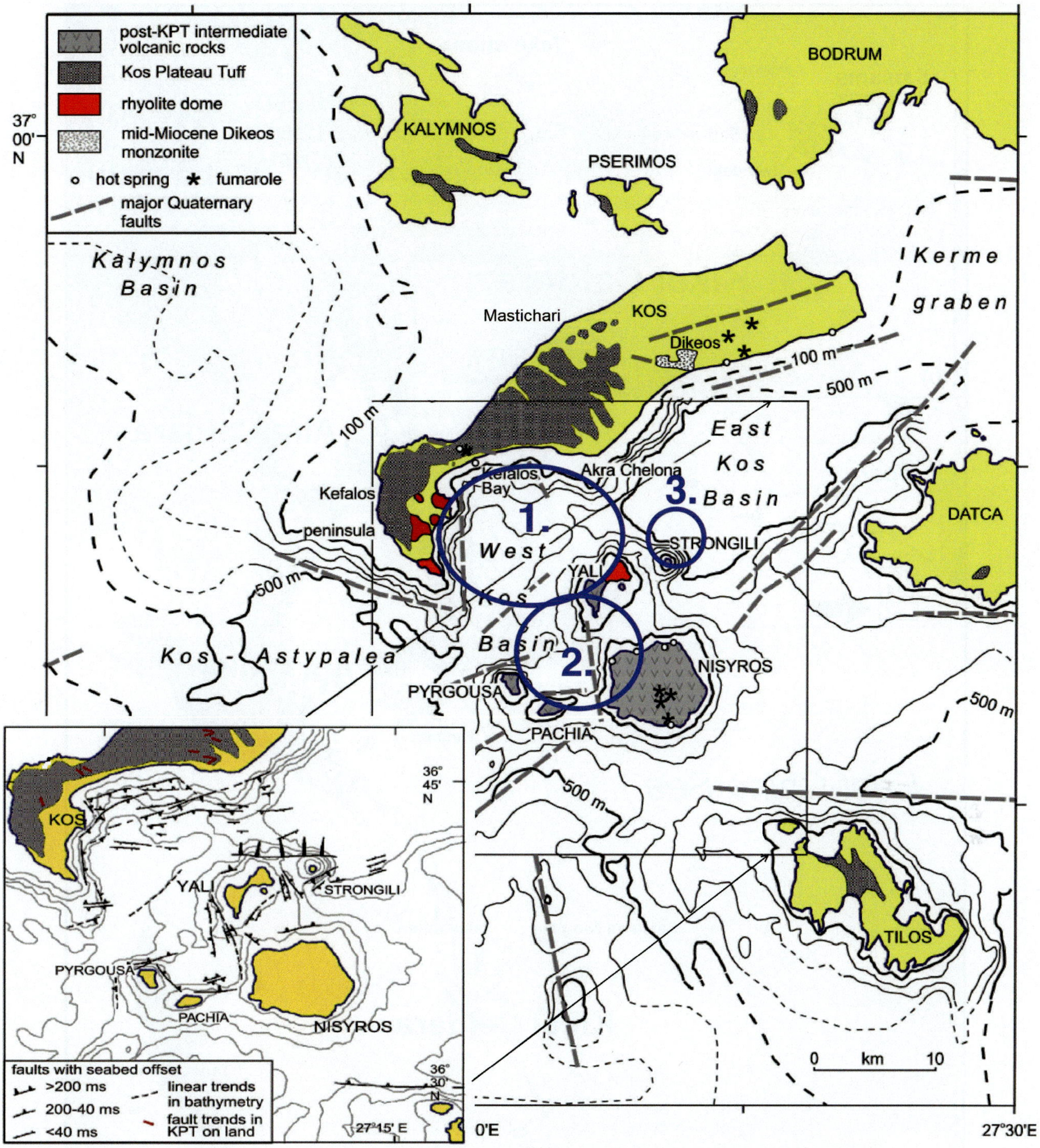

Fig. 16.71 Location of outcrops of the 161 ka Kos Plateau Tuff (grey shade) on the island of Kos and surrounding islands in the southeast Aegean Sea, and the Datca and Bodrum peninsulas of Turkey. Various proposed locations of the offshore caldera are: 1 = Cas and Wright (1987), Allen (2001); 2 = Pe-Piper et al. (2009); 3 = Nomikou et al. (2013). Inset (bottom left) shows major sea floor faults identified from seismic sections. (Modified from Pe-Piper et al. 2005)

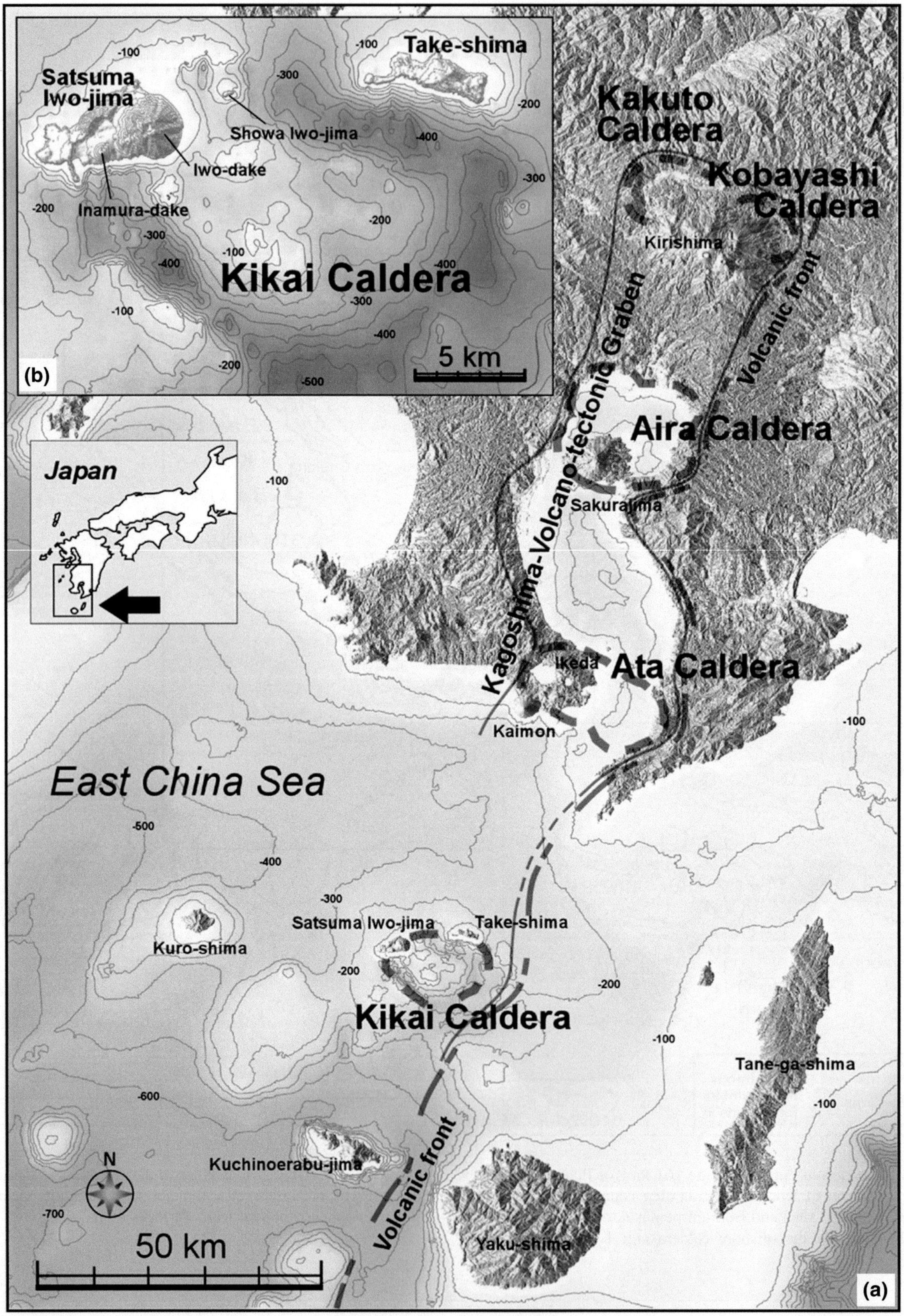

Fig. 16.72 Location and setting of the Kikai caldera, Kyushu, Japan. **a** Map showing the location of the Aira, Ata and Kikai relatively shallow marine calderas along the coastline and offshore of southwest Kyushu. **b** Inset showing the bathymetry of the offshore Kikai caldera. (From Maeno and Taniguchi 2007)

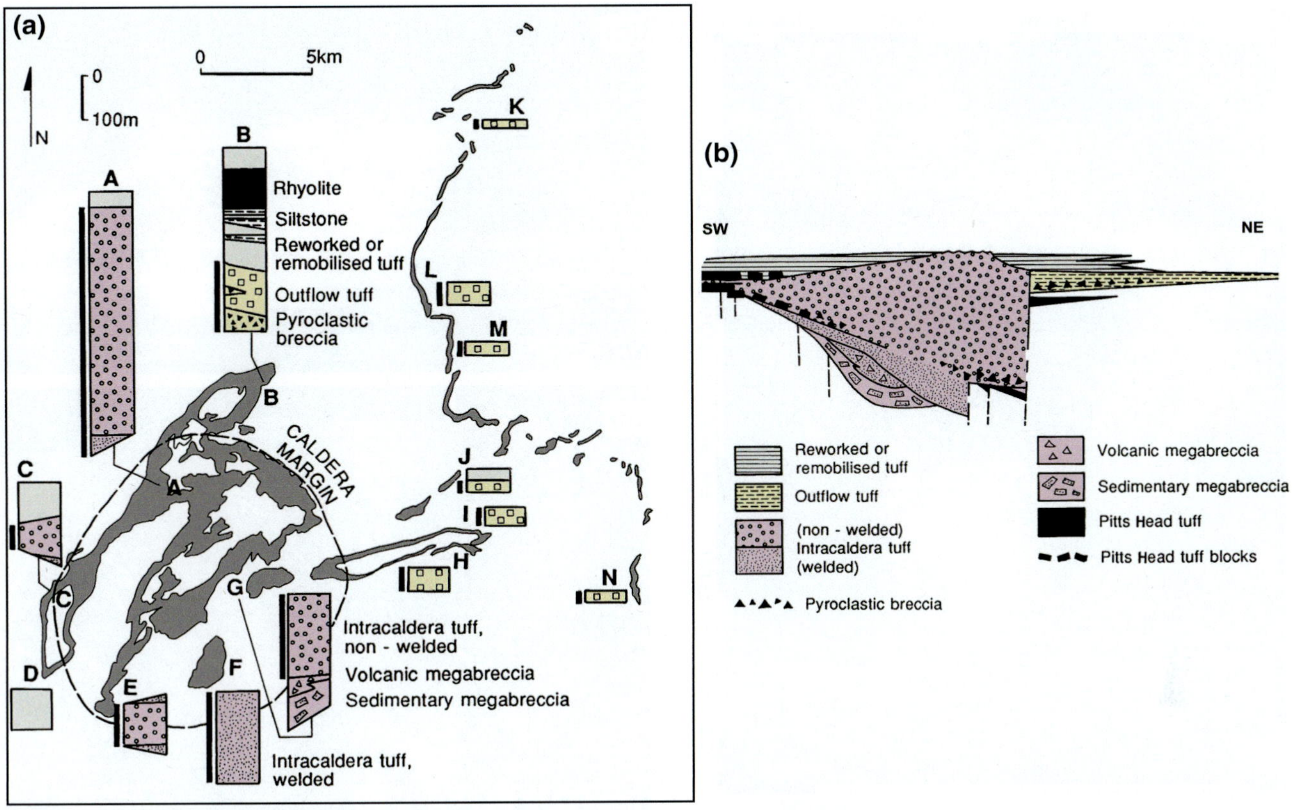

Fig. 16.73 Features of the Ordovician Lower Rhyolitic Tuff Formation caldera and outflow ignimbrite sheet, North Wales. **a** Simplified geological map and facies distribution, and **b** cross-section of the shallow marine caldera and outflow succession. Vertical bars to the left of the stratigraphic sections in (**a**) indicate thickness of primary ignimbrite present. The base of the Lower Rhyolitic Tuff Formation is unconformable to disconformable on substrate, indicating an extended period of significant erosion, probably under subaerial conditions, prior to the emplacement. Much of its top is also reworked, suggesting that the whole ignimbrite was likely emergent when deposited (Chap. 11). (After Cas 1992; Modified from Howells et al. 1986)

proximal areas, gradationally into finer, bedded, sorted tuffs to the south in distal areas. In both cases, the high proportion of pumiceous deposits and their thickness is consistent with ponded caldera fills (Chap. 11). Given that caldera subsidence of several hundred m probably occurred, the explosive eruption would have commenced in relatively shallow water, but below wave base, possibly <300 m deep, before caldera collapse occurred.

16.8.9.3 Wholly Submarine Calderas

Although totally submarine calderas have been recognised in the modern oceans (e.g. Halbachet al. 1989; Wright et al. 2006; de Ronde et al. 2011; Carey et al. 2018), their characteristics and deposits remain poorly described because of access difficulties. Remote sensing technology, such as very high-resolution side-scan sonar and deep-tow bathymetry, are now capable of producing high-resolution bathymetry maps (resolution <5 m) that can be transformed into unprecedented, photo-quality digital elevation models (e.g. Havre volcano summit caldera, Kermadec arc, Fig. 16.74; Carey et al. 2018; Chap. 6). Together with deep-tow and submersible photography, these are allowing superposition relationships to be established in surficial deposits on the seafloor. However, because of the limitations of coring depths, stratigraphies are constrained to assessing superposition relationships in surficial sea floor deposits (e.g. Carey et al. 2018).

The 2012 wholly submarine eruption of the Havre submarine volcano, Kermadec arc north of New Zealand, which has a summit caldera (Fig. 16.74a, b), produced a lot of floating pumice, that was initially interpreted as originating from submarine explosive eruptions. Subsequent detailed sea floor bathymetry surveys and sampling of the wholly submarine eruption products demonstrated that the primary eruption products were submarine rhyolitic lava domes and lava flows erupted from the southern ring fractures on the caldera rim at water depths of 900 m (Fig. 16.74c; Carey et al. 2018; Chap. 6). These had highly pumiceous (40–70% vesicles) carapaces, which through quench fragmentation produced large volumes of pumice (Carey et al. 2018;

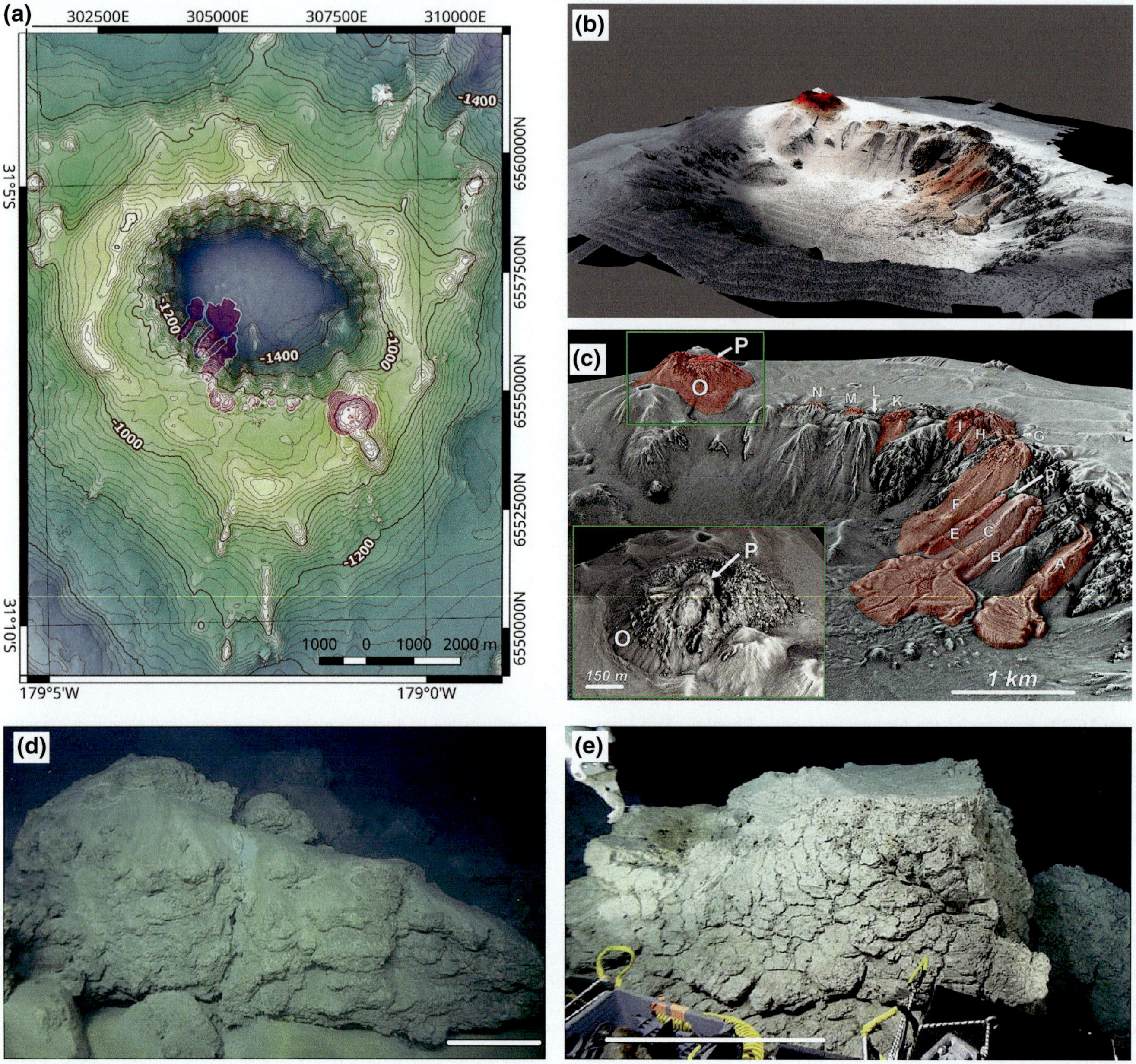

Fig. 16.74 Features of the wholly submarine Havre volcano, Kermadec arc, north of New Zealand, and its 2012 eruption products. **a** General bathymetry map of the wholly submarine Havre caldera with its rim at 900–1,000 m b.s.l., and caldera floor at 1,400 m b.s.l. Also outlined are the rhyolitic lavas erupted during the 2012 eruption from the southern caldera rim ring fracture. (From Ikegami et al. 2018). **b** Oblique digital elevation model constructed from very high-resolution bathymetry, spotlighting the southern caldera rim and its 2012 lava flows and domes, some of which flowed down the caldera wall to the caldera floor (From Carey et al. 2018). **c** Detail of the southern caldera rim. Mapping of stratigraphic relationships and sampling demonstrate that the vent responsible for the 2012 pumice raft is overlain by a 250 m high, 0.11 km^3 pair of domes (O and P) also erupted in 2012. (From Manga et al. 2018). **d** and **e** ROV images of submerged giant pumice megaclasts up to 6 m in diameter, and showing curviplanar surfaces and quenched margins with breadcrusting and cooling joints; white line in each image is 1 m (From Carey et al. 2018)

Manga et al. 2018), some of which rose to the sea surface and was then rafted away and was widely dispersed (Jutzeler et al. 2014; Chap. 12). Much also became waterlogged and sank as giant pumice megaclasts (Fig. 16.74d, e) and lapilli size pumice and ash. Minor explosive activity also produced a dispersed ash and fine pumice deposit (Murch et al. 2019).

Because of the difficulties of accessing the deposits and stratigraphy of modern deep-water submarine caldera eruptions, much of our current understanding comes from the uplifted rock record, and especially from closely studied mining districts which host volcanic-hosted massive suphide (VMS) deposits; e.g. the Miocene Kuroko successions of Japan, the Cambrian Mt Read Volcanics, Tasmania and those of the Archean Abitibi greenstone belt, Canada (Chap. 18). This in itself creates some difficulties because recognising ancient submarine calderas in the rock record can be very challenging, especially when regional tectonic events have variably deformed and dismembered the ancient successions, when outcrop is poor, and original water depth constraints are poor, relative only, ambiguous, or non-existent. For example, early interpretations of the setting for the Miocene Kuroko VMS deposits in Japan was a cluster of submarine, deep-water calderas (Ohmoto and Skinner 1983). However, there is no structural or facies evidence (ring faults, unusually thick, intra-caldera facies) of their existence (Chap. 18). The principal volcanic elements of the Kuroko successions are subaqueous dome-like lavas, cryptodomes and associated volcaniclastics, including resedimented pumice deposits derived from shallow water vents, together with deep water mudstones, massive sulphides and other associated chemical sedimentary facies (Cas 1992).

A better candidate for an ancient deep-marine caldera setting is the Ordovician rhyolitic Bald Mountain Caldera, in Maine (Busby et al. 2003; Kessel and Busby 2003; Chap. 11). VMS mineralisation is hosted within a thick sequence of ponded (intra-caldera) pumiceous deposits ($\sim$500 m thick), which subdivide the succession into a nearly aphyric footwall and a more crystal-rich hanging wall, having different geochemistry. The sulphidic horizon therefore appears to represent a significant boundary and possibly a hiatus in volcanism. The 150–200 m intervals of footwall deposits lack fine ash matrix, are dominated by blocky pumice clasts, and are interpreted as ignimbrites produced by subaqueous phreatomagmatic explosive activity. The $\sim$300 m thick hanging wall sequence occurs in depositional units up to 60 m thick, some of which are graded, and in which the pumice clasts are not blocky. These are intercalated with volcanic breccias, interpreted as caldera wall collapse breccias, and overlain by up to 1,500 m of coherent rhyolite lavas, intrusions into mudstones producing peperitic margins and subaqueous lava dome collapse mass-flow deposits. Busby et al. (2003) interpret the Bald Mountain succession to have resulted from a piecemeal caldera collapse event in an arc rift setting. They propose that the water depth of the explosive eruption was $\sim$1,400 m, based on fluid inclusion paleo-barometry data from the sulphide deposits. However, if sulphide mineralisation formed after caldera collapse associated with the eruption of the footwall succession, the water depth at which the sulphide deposits formed could be several hundred m greater than that at which the explosive caldera-forming eruption began.

16.8.10 Facies Models for Explosive Caldera Volcanoes

As described with continental and marine stratovolcanoes (Sect. 16.7.7), because environment affects volcanic eruption styles and the resultant types of deposits, there will be clear differences in the facies associations between subaerial and marine caldera volcanoes, as well as between shallow-moderate depth and deep submarine silicic calderas.

16.8.10.1 Continental Explosive Calderas—Facies Models

The principal volcanic elements of continental explosive caldera volcanoes are topographically subdued clusters of hills rising above gently sloping ignimbrite shields. The main focal element of this association of primary volcanic facies is the caldera structure, which will contain multiple lava and ignimbrite eruption points (Fig. 16.75). The margins of the caldera may be abrupt, steep erosional scarps (Fig. 16.53c), or gently inward dipping slopes (Fig. 16.53g) and may have associated caldera margin collapse breccias. Silicic domes and thick lavas will lie within or at/around the margins of the caldera structure, often along the trace of the original caldera collapse ring fracture fault (Fig. 16.66), although some may be erupted outside the caldera, and may be associated with near-vent pyroclastic falls and other volcaniclastic deposits. The extra-caldera shields contain various ignimbrite facies (Figs. 10.26, 10.27 and 10.28) and rhyolitic fall deposits (Fig. 9.55) erupted from the caldera. Aside more evolved parasitic centres, basaltic scoria cones may occur around the margins or beyond the limits of caldera structures, but rarely within the caldera confines (cf. Lucci et al. 2020). Hydrothermal systems including sinter deposits and geothermal pools will be common within and outside the caldera, and these may be associated with hydrothermal explosion craters.

The caldera subsidence basin will host not only lavas and domes (or cryptodomes), but also intra-caldera ignimbrites. These can be extremely thick (Fig. 10.29), much thicker than ignimbrites of the outflow shield, and will often have associated near-vent co-ignimbrite breccias (Figs. 10.50,

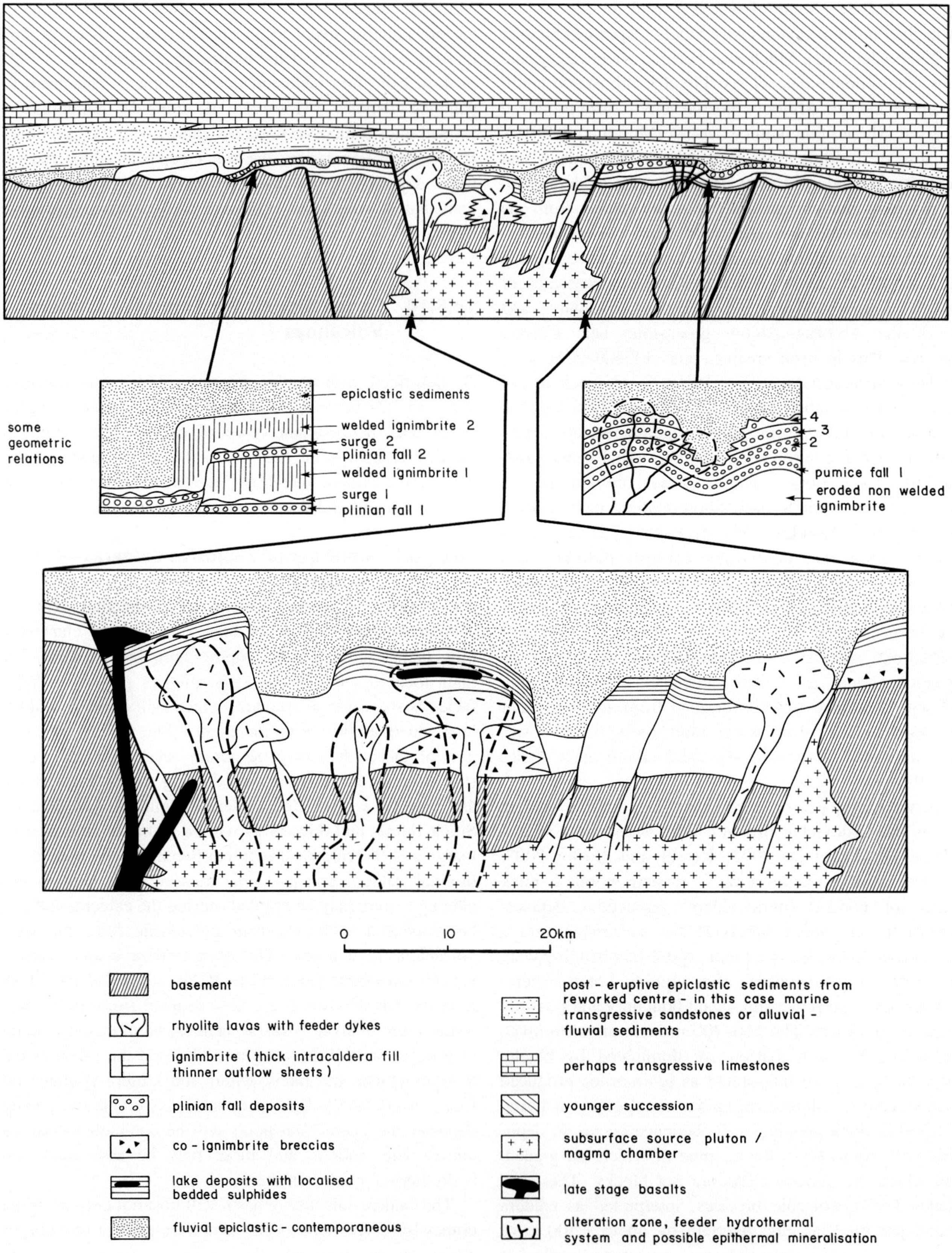

Fig. 16.75 Facies model for continental silicic explosive caldera volcanoes. This depicts the variations in facies associations from the focal caldera complex to the ignimbrite shield outside the caldera, including thick intra-caldera fill ignimbrites, caldera lake sediments, intra-caldera lava domes and intrusions, compared with the thinner sheet-like ignimbrites and pumice fallout deposits in the outflow ignimbrite shield, with varying proportions of intercalated reworked fluvial-alluvial volcanic sediments. (From Cas and Wright 1987)

10.51, 10.52 and 10.53). This caldera fill succession will also contain intercalated epiclastic sediments. Such deposits could include fluvial sediments, subaerial mass flows, and caldera and valley-ponded lake deposits, subaqueous debris flows and turbidites of pumiceous debris, lacustrine and diatomaceous muds, and horizons of floated giant pumice (Fig. 16.69). Where caldera centres are resurgent, caldera fill strata could be significantly tilted, faulted, and folded.

Within the extra-caldera shield, erosion will be important between ignimbrite eruptions (Fig. 16.75). Valleys, typically steep-sided box canyons, are quickly incised into freshly erupted ignimbrite as the drainage pattern tries to re-establish itself. If welded, the non-welded top will be stripped off in time (e.g. 10^2–10^5 yr, depending on size and climate), otherwise rapid degradation of the whole sheet will continue unless buried by another ignimbrite, or other deposits. Epiclastic processes within the shield are largely erosional, and depositional processes are restricted to alluvial settings and to local, shallow lakes formed in valleys, or perhaps developed on the low slopes of the shield. The eroded volcaniclastic debris is largely removed from the volcano and is deposited in basins marginal to the centre, and the shields are, therefore, composed dominantly of ignimbrite with volumetrically small amounts of epiclastic interbeds. Large volumes of pumice, ash, and crystals will be removed to be deposited in large fluvial outwash plains flanking the ignimbrites (perhaps >50% of most ignimbrites ends up in such epiclastic sequences) or are even transported into the marine environment. Many, especially small-volume, non-welded ignimbrites have a low probability of even being preserved in the geological record. When a number of silicic centres are associated together in time, ignimbrite shields will overlap. Any one stratigraphic section could therefore contain a number of ignimbrites erupted from different centres, which may even infill the older calderas of other centres (e.g. caldera cluster in the San Juan Mountains, USA). Facies associations will be diverse, lateral facies relationships will be abrupt and stratigraphic relationships will be complex.

16.8.10.2 Marine Silicic Calderas-Facies Model(s)

As a rule of thumb, moderate depth to shallow water silicic calderas (<1,000 m, and especially <500 m) will contain high proportions of in situ pyroclastic deposits whereas deeper water calderas might be expected to host increasing proportions of autoclastic hyaloclastite deposits derived from quench fragmentation of submarine lavas. The Kos-Yali and Kikai calderas provide the basis, albeit largely predictive, for the facies model of a shallow-intermediate depth rhyolitic caldera (Fig. 16.76a). From left-to-right facies pass from emergent, with primary pyroclastic deposits and incised fluvial valleys, to shoreline at the caldera margin with silicic domes associated with autoclastic breccias and fans, and into deeper marine environments, including the opposite margin, with silicic domes and cryptodomes associated with hyaloclastites and quench fragmented breccias. This also schematically shows a welded ignimbrite which was erupted subaerially but now resides deep within the submarine caldera structure, well below where it initially was emplaced. Localised pyroclastic successions such as shallow marine dome-top rhyolitic tuff and pumice cones may also be present (Fig. 16.76b).

Kuroko successions are also likely to be equally valid models for deeper water silicic calderas with submarine lavas and hyaloclastites, and shallow cryptodomes intruding water-saturated sediments form the principal elements (Fig. 16.77a, b). The domes may be covered by hyaloclastite and marginal aprons of hyaloclastite and they may be pumiceous including giant pumice clasts, while cryptodomes will have marginal peperite deposits (Fig. 16.77b). Lavas may occur in overlapping clusters, associated with mass flow redeposited volcaniclastics, including perhaps some pyroclastic pumice derived from shallower eruption points, and deep-marine pelagic-hemipelagic and turbidite sediments (Fig. 16.77a–c), which represent hiatus in eruptive activity. The presence of unusually thick pumiceous deposits ($\pm$ evidence of welding) may imply ponded intra-caldera deposits (Chap. 11). Depending on tectonic setting, variable proportions of resedimented terrigenous sediments may occur, if in proximity to land masses (Fig. 16.77c).

16.8.11 Resource Potential and Economic Significance

Precious and other epithermal metal deposits, and geothermal energy are the most important resources associated with continental silicic calderas. Epithermal deposits of arsenic, antimony, gold, silver, lead and zinc may be found associated with shallow level hydrothermal systems. Characteristically these are of the low suphidation type and controlled by regional scale geothermal circulation so there is potential for mineralisation to occur in a wide range of volcanic settings and facies, both inside and outside the caldera margins (Chap. 18). In addition, porphyry copper although more generally porphyry molybdenum systems may be associated with the deeper-level intrusions of large silicic calderas. Bedded sulphides $\pm$ gold-silver may occur in caldera lakes, but generally they are sub-economic. However, these include the super-large San Cristobal disseminated silver-zinc deposit, Bolivia. Mineralisation occurs mainly as disseminations formed within lacustrine facies formed by either replacement and/or from hot springs within a small rhyolitic caldera (or large crater) lake (Chap. 18).

Massive stratiform and stratabound copper-lead–zinc sulphides ($\pm$ gold and silver) are the most significant

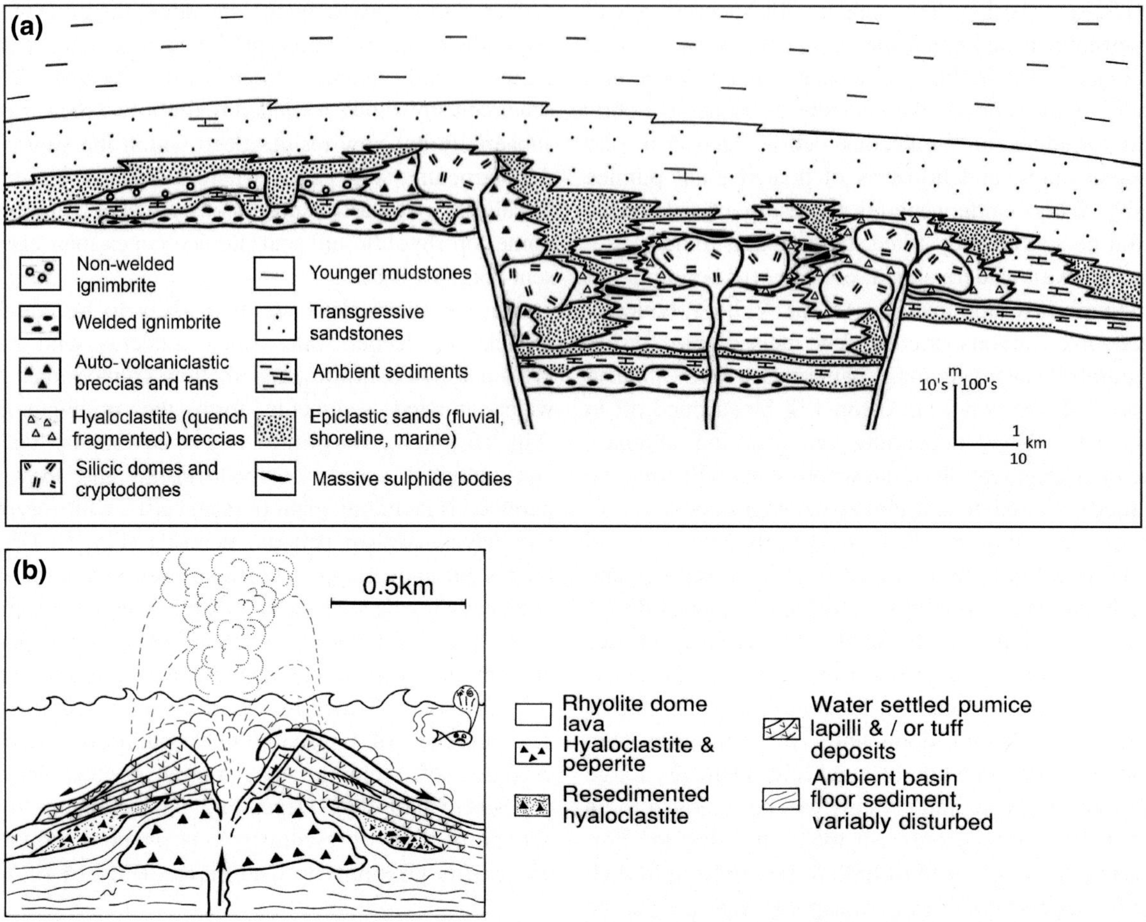

Fig. 16.76 Facies models for moderate to shallow marine water silicic caldera volcanoes. **a** Intermediate depth submarine rhyolitic caldera based on the setting of the Aegean islands of Kos and Yali, and Kikai caldera, offshore Kyushu, Japan (see Figs. 16.71 and 16.72). From left-to-right facies pass from emergent, showing incised fluvial valleys, to shoreline at the caldera margin and into deeper marine environments, including the opposite caldera margin. This also schematically shows a welded ignimbrite which was erupted subaerially but now resides deep within the submarine caldera structure, well below where it initially was emplaced. Also see Fig. 16.73. **b** Shallow marine dome-top rhyolitic tuff and pumice cone as might be found on the submerged caldera ring fracture (From Cas 1992)

resource likely in submarine caldera settings. Traditionally these were thought to accumulate from hydrothermal springs (or "black smokers") which were trapped in local sub-basins. This leads to the precipitation of delicately laminated deposits of sulphides interbedded with pelagic-hemipelagic sediments. However, research over the past three decades has demonstrated many VMS deposits formed as sulphide mounds or lenses on and just below the seafloor where mineralisation infills pore-space and replacement is important (Chap. 18). It is also apparent that in some examples, precipitation of sulphides occurred quite deep (tens of m) beneath the seafloor. This is typically found where host facies are very porous, permeable and/or reactive, such as glassy, pumiceous units or perlitised domes, lavas and shallow intrusions. In this case, mineralisation may occur as multiple lenses which may be discordant and have gradational contacts with the host facies (i.e. stratabound).

Although Kuroko successions were interpreted as submarine calderas more recent studies have indicated that volcanism more generally took place in deep-water extensional rift basins, presumably from multi-vent sources. Nevertheless, the two ideas are not mutually exclusive (Fig. 16.77c; also see discussion Sect. 18.8.7), and calderas cannot be disqualified as potential ore-forming centres. Many summit calderas of submarine stratovolcanoes of the Kermadec arc north of New Zealand have active sea floor hydrothermal vents, including sulphide deposit building black smoker vents (e.g. de Ronde et al. 2005, 2007). Several ancient VMS districts and deposits have also been attributed to submarine caldera environments, in particular,

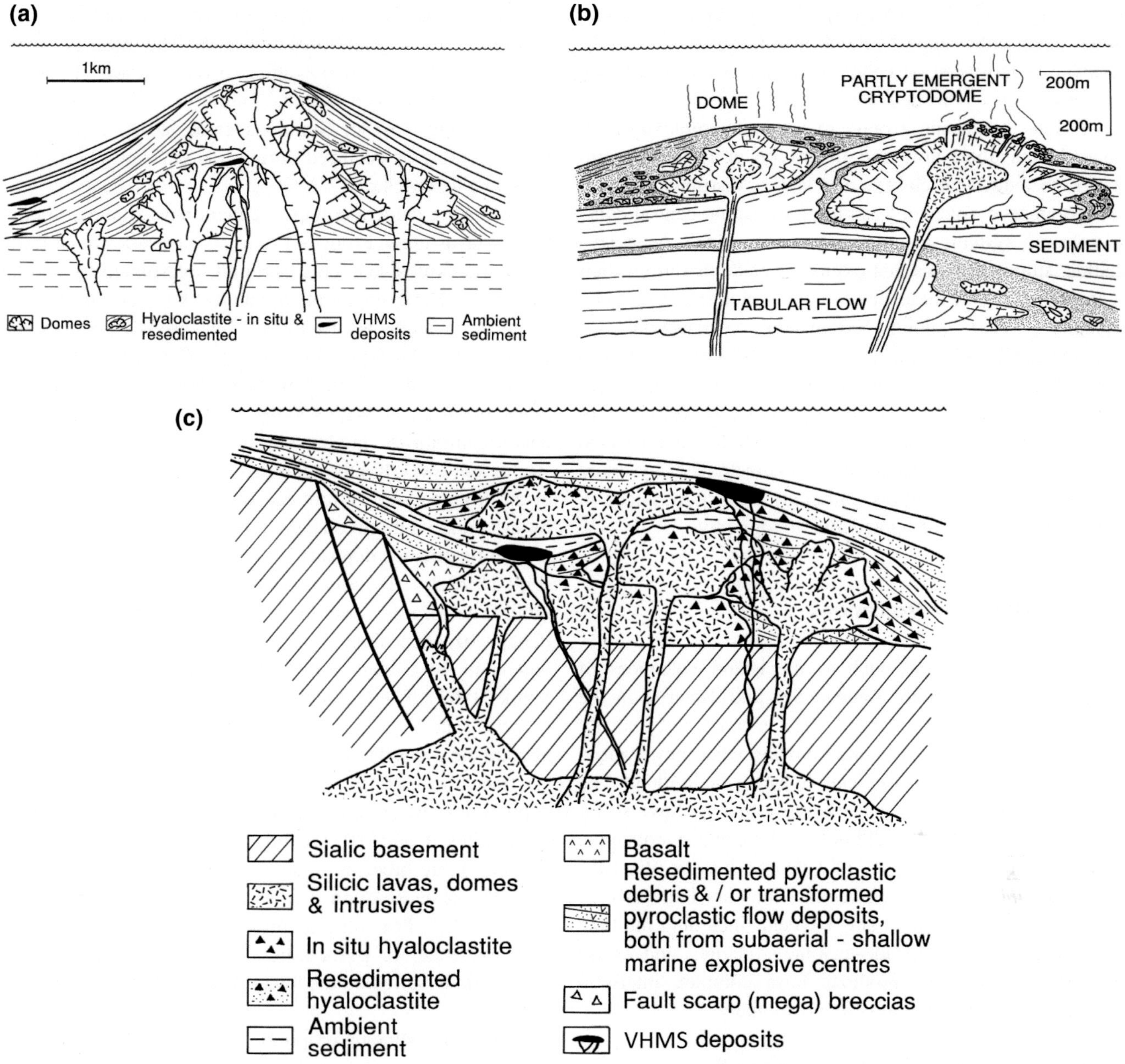

Fig. 16.77 Facies models for deep-marine silicic volcanic centres (From Cas 1992). **a** Submarine silicic dome complex. **b** Various forms of silicic submarine lavas, intrusive cryptodomes with margins of hyaloclastite and peperite (Adapted from Allen 1988) **c** Kuroko-type succession dominated by deep-water silicic lava extrusions rather than pyroclastic eruptions; this can be viewed as a section through the margin of a deep-water extensional rift basin (see Sect. 18.8.7) or, alternatively, a submarine subsidence caldera

those found within Archean greenstone belts of the Superior Province, Canada (e.g. Sturgeon Lake, Hunter Mine and Noranda; Chap. 18). However, studies that use fluid inclusion paleo-barometry from massive sulphides from calderas to assess eruption water depths (e.g. Kessel and Busby 2003) may actually record sulphide formation after caldera collapse, or even deep, sub-seafloor sulphide replacement (Chap. 18), and fluid inclusions can be reset during metamorphism and alteration, requiring a cautious approach when using them to reconstruct paleo-water depths.

16.9 More Complex Volcanic Systems: Multiplex Volcanoes, Intermediate-Silicic Multi-vent Centres, Clusters and Fields

There are other centres of mostly intermediate to silicic volcanic and subvolcanic intrusive activity, although mafic and alkaline magmas may occur, that are best described as hybrids of the major volcano types discussed above. These may have changed in character through their evolution, while others either appear to lack major constructional edifices (i.e. a large central cone structure), or conversely are not obviously controlled by caldera or sector collapse depressions. Such centres range in size and complexity from isolated lava dome-flow and maar-type or tuff ring complexes, to multi-vent clusters and larger "distributed" fields of monogenetic volcanoes lacking a dominant central cone, to volcanic systems that have at various times been lava shields, stratovolcanoes and explosive calderas during their evolution. Many of these show strong structural control and typically occur along continental extensional or trans-tensional and strike-slip fault systems or are aligned along regional tectonic lineaments. They therefore commonly occur within extensional rift basins as well as transcurrent pull-apart basins and structures. Some larger island arc volcanoes also appear to be characterised by more complex multi-vent centres, although these may occur in and/or form regions of appreciable topography.

16.9.1 Multiplex Volcanoes

Some volcanic systems, such as the island volcanoes of Santorini in the southern Aegean Sea and Tenerife in the Canary Islands, have changed their character repeatedly during their evolution, not quite knowing what they want to be. Their various personas can however individually be represented by the major volcano types that we have so far described, although it is difficult to generate a collective facies model for these types of volcanic system. We term these "*multiplex volcanoes*", because they are large to very large edifices that contain many elements including multiple volcanic centres and volcano-tectonic structures, and whose eruptive and erosional products form complex interrelationships in space, time and compositionally.

16.9.1.1 Santorini Volcano, Aegean Sea

Santorini volcano in Greece (Fig. 16.53b) is one of the most iconic island volcanoes on Earth and has, for example, a history of having been a lava dome, mafic lava shield volcano, stratovolcano, and highly explosive caldera volcano, with some of these recurring several times throughout its 0.65 Ma history (Druitt et al. 1999). It has had a cyclical volcanic history, beginning with eruption of marine rhyodacite lavas, hyaloclastite, and tuffaceous deposits from ~650 to 550 ka (Fig. 16.78a). From 530 to 430 ka a significant andesitic stratovolcano complex, the Peristeria complex (Fig. 16.78b) existed, represented by lavas, associated breccias, and tuffaceous deposits. From 450 to 340 ka mafic scoria cones and lavas are preserved. The first major explosive cycle occurred from 360 to 180 ka. This begins with the Cape Therma sequence of andesitic to rhyodacitic fallout deposits, ignimbrites and lavas. The cycle is capped by the extensive rhyodacitic Lower Pumice 1 and 2 sequences of pumice fallout, ignimbrite and co-ignimbrite breccias (Fig. 16.78c; Simmons et al. 2016, 2017). Two caldera collapses occurred during this latter eruption phase, at ~172 to 184 ka.

Following caldera collapse, a mafic lava shield edifice of mostly lavas grew. From ~170 to 50 ka a series of andesitic to dacitic explosive events occurred, interrupted by another caldera collapse event at about 100 ka, but poorly constrained, which is followed by growth of another mafic-dacitic, lava shield edifice, the Skaros shield at ~68 ka (Fig. 16.78d). Another major caldera-forming explosive event occurred at ~20 ka, represented by the Cape Riva dacitic ignimbrites, lag breccias and pumice fallout deposits. At 3.6 ka (1600 BC) the Late Bronze Age Minoan explosive event produced the current caldera and harbour, represented by the stark, white rhyodacitic plinian fallout, ignimbrite and phreatomagmatic deposits at the top of the sequence (Fig. 16.78e). Pyroclastic flows of at least six ignimbrite-forming eruptions, both felsic and mafic types (Chap. 10), deposited proximal lag deposits now exposed widely in the caldera wall (Druitt et al. 1989; Mellors and Sparks 1991). The lag deposits include coarse-grained lithic breccias (andesitic to rhyodacitic eruptions) and spatter-rich agglutinate deposits (andesitic eruptions only). The latter can be traced around most of the caldera wall, thicken in depressions and are interpreted as a proximal facies of mafic ignimbrites (Chap. 10, see Fig. 10.87). Facies associations between lithic breccias, spatter agglutinates, and ignimbrite from the same eruption can be very complex (Mellors and Sparks 1991). For some eruptions, lag deposits provide the only evidence for pyroclastic flows, because most of the ignimbrite is buried on the lower flanks of Santorini or under the sea.

Since the Minoan eruption, a new dacitic lava shield edifice has been growing in the middle of the caldera harbour at Nea Kameni (Fig. 16.78f) and Palaea Kameni islands, with submarine and shallow marine phreatomagmatic and hyaloclastite formation during the early stages, followed (since ~726 AD) by numerous subaerial effusive and weakly explosive events, the last in 1950 (Johnston et al. 2014). The volcanic history of Santorini cannot be represented by a single, standard or "type" volcano facies model. Its history has been cyclical, very changeable, and very complex, which can only be represented by multiple models depicting each stage in its history. Reconstructing the eruption chronology and eruption scenarios of such a volcanic complex preserved in the geological rock record would be very challenging, given that on Santorini it has all occurred in $\ll$1 Ma.

Fig. 16.78 Outcrops representing some of the developmental stages of the Santorini volcanic complex, Aegean Sea, Greece. **a** White bedded, probably submarine rhyodacitic tuffaceous deposits of the early Akrotiri centres (~650–550 ka) unconformably draped by basaltic scoria and tuff cone deposits (450–350 ka), Balos Bay, Akrotiri Peninsula. **b** View of Mikro Profitis Ilias (main peak), which is the remains of the andesitic Peristeria3 stratovolcano (530–430 ka), truncated to the right by the modern caldera wall (see Fig. 16.53b). **c** View of the deposits of the first two rhyodacitic caldera-forming pyroclastic deposits of the 184 ka Lower Pumice 1 and 172 ka Lower Pumice 2 eruption sequences, West Cape Balos, Akrotiri. The grey sequence above the orange paleosoil is the LP1 Sequence of pumice fallout, surge, ignimbrite and co-ignimbrite lag breccias. The upper pale grey to white sequence is the LP2 sequence of pumice fallout, co-ignimbrite lag breccias and ignimbrite. Person for scale. **d** Sequence of basaltic-andesitic lavas of the Skaros lava shield edifice (~68 ka), looking north from Thera township to Oia, in the distance. **e** View from Nea Kameni of the caldera wall resulting from the second rhyodacitic, caldera-forming explosive eruption, the Late Bronze Age 1600 BC (3.6 ka) Minoan eruption, which produced the white pyroclastic sequence top right of the caldera wall. The other unit in the middle of the cliff on the right is the Lower Pumice 2. It is truncated and mantled by a thin black unit, which is the welded pumice fallout facies (Thera welded tuff) of the 144.6 ka Middle Pumice. The white buildings are part of the town of Thera. **f** View from the top of the caldera wall at Thera looking over the very young Nea Kameni dacite lava shield edifice growing in the middle of the caldera (also see Fig. 16.48a)

16.9.1.2 Tenerife, Canary Islands

Tenerife, largest of the Canary Islands (Fig. 16.79a, b), is located off northwest Africa and is the second largest oceanic shield volcano on Earth, after Hawai'i. It has also had an extremely complex volcanic history (Martí et al. 1994; Martí 2019; Cas et al. 2022; Fig. 16.79b). For >12 Ma Tenerife experienced largely lava-forming mafic volcanism, that built stacked lava shield edifices, the Teno, Anaga and Roque del Conde shields (Fig. 16.79b, c), which were initially submarine and then subaerial. Basaltic shield-forming volcanism has continued along the Dorsal and Santiago ridges, and the Southern Volcanic Zone (Fig. 16.79b), represented by scoria cones and lava flow fields (Fig. 16.79d), with the last subaerial basaltic eruption occurring in 1909. However, before 3.5 Ma, a phonolitic magma reservoir system developed under the summit region of Tenerife, as a result of fractionation and differentiation of basaltic magma sources at depths of >8 km beneath the surface. From 3.5 to 2 Ma the nature of the phonolitic volcanism (the so-called Lower Group of the Las Cañadas edifice; Fig. 16.79b) is not well known, however, some exposures include mainly lavas, suggesting perhaps a stratovolcano complex existed (Martí et al. 1994).

More is known about the phonolitic Upper Group (<1.6 Ma), which represents the deposits of at least three explosive caldera-forming volcanic cycles, each lasting about 200 kyr, and separated by ~200 kyr period of non-caldera-forming activity (Martí et al. 1994; Bryan et al. 1998; Edgar et al. 2007; Martí 2019; Cas et al. 2022). The three calderas are coalesced at the summit of Tenerife, forming a caldera complex called the Las Cañadas caldera complex (Fig. 16.79e; also see Plate 16.1a; Martí et al. 1994; Martí and Gudmundsson 2000; Cas et al. 2022). Each caldera-forming cycle is represented by multiple, widely dispersed pyroclastic deposits, each ranging from <1 km^3 to >15 km^3 in volume, suggesting likely incremental caldera collapse or subsidence during each cycle (Edgar et al. 2007; Cas et al. 2022). Since the end of the last caldera-forming cycle at ~190 ka, a stratovolcano complex has been growing along the projected trend of the northern caldera wall or ring fracture (Plate 16.1a; Fig. 16.79e). The largest cone, El Teide (Plate 16.1a, Fig. 16.42g) has experienced both basaltic and phonolitic effusive and explosive activity (Ablay and Martí 2000; García et al. 2011; Martí et al. 2012; Carracedo and Troll 2013). The caldera walls expose welded and non-welded phonolitic pyroclastic fallout and ignimbrite deposits (Fig. 16.79f, g).

The outer slopes of the Las Cañadas edifice, fully down to the coast, comprise a complexly stratified succession of pyroclastic fallout, flow and associated surge deposits (e.g. Fig. 16.79h). The pyroclastic deposits of the caldera-forming cycles contain many dense accessory mafic, phonolitic and syenitic (representing old, solidified, subsurface magma reservoirs) lithic clasts (Pittari et al. 2000). These suggest the existence of significant edifices (stratovolcanoes?) that may have grown, like Teide, between caldera cycles, and were then progressively destroyed during the next explosive caldera-forming cycles. Coincident with the explosive caldera cycles, Tenerife also experienced several major sector collapse landslide and tsunami events including some thought to be explosively induced during caldera-forming eruptions (Hunt et al. 2011, 2018; Paris et al. 2017; Chap. 12). These are represented in the present-day morphology by huge landslide valley scars, that are being variably infilled by sedimentation and monogenetic basaltic volcanoes and lava flows (e.g. Fig. 16.79i), as well as offshore by extensive submarine aprons and distal turbidite fans (Hunt et al. 2014).

Again, it is difficult to construct a simple facies model for such a multifarious volcanic system, because simultaneously it is an oceanic basaltic island shield volcano, an explosive caldera system and a stratovolcano system. Alternatively, different models can be proposed to reflect each stage in the evolution of the system.

16.9.2 Extensional Rift Basin Systems and Graben Calderas

Within large volcano-tectonic grabens and broader regions of high extensional strain, rhyolitic rift volcanism often leads to caldera formation as described in the preceding section. As we have indicated, large calderas are essentially multi-vent structures and dome complexes and clusters of domes appear to characterise the post-caldera periods of many silicic and alkaline examples. However, even in these settings not all silicic and alkaline volcanism is necessarily the result of ring fracture collapse and ignimbrite-forming activity.

The Basin and Range Province in the western United States is a wide region of late Cenozoic large-magnitude extension (100–250%) and magmatism, now interpreted in terms of the extensional collapse of the North American Cordilleran foreland fold and thrust belt (Chap. 17). Since the mid-Miocene (~16.5 Ma; i.e. succeeding the central Nevada caldera and ignimbrite flare-up event described in Fig. 16.54), this has produced alternating basins and ranges about 20–50 km apart (Eaton 1982; John 2001). This period was characterised by bimodal basalt-rhyolite volcanism and subvolcanic intrusive activity (Fig. 16.80), closely associated with the development of abundant and often isolated basins across the region. Continental extension allowed rapid ascent of mafic magmas through the crust forming monogenetic basaltic fields aligned on major fissures. However, underplating of basalt at the base of the crust resulted in partial melting forming rhyolitic melts. These rhyolite magmas erupted as domes and lava flows with

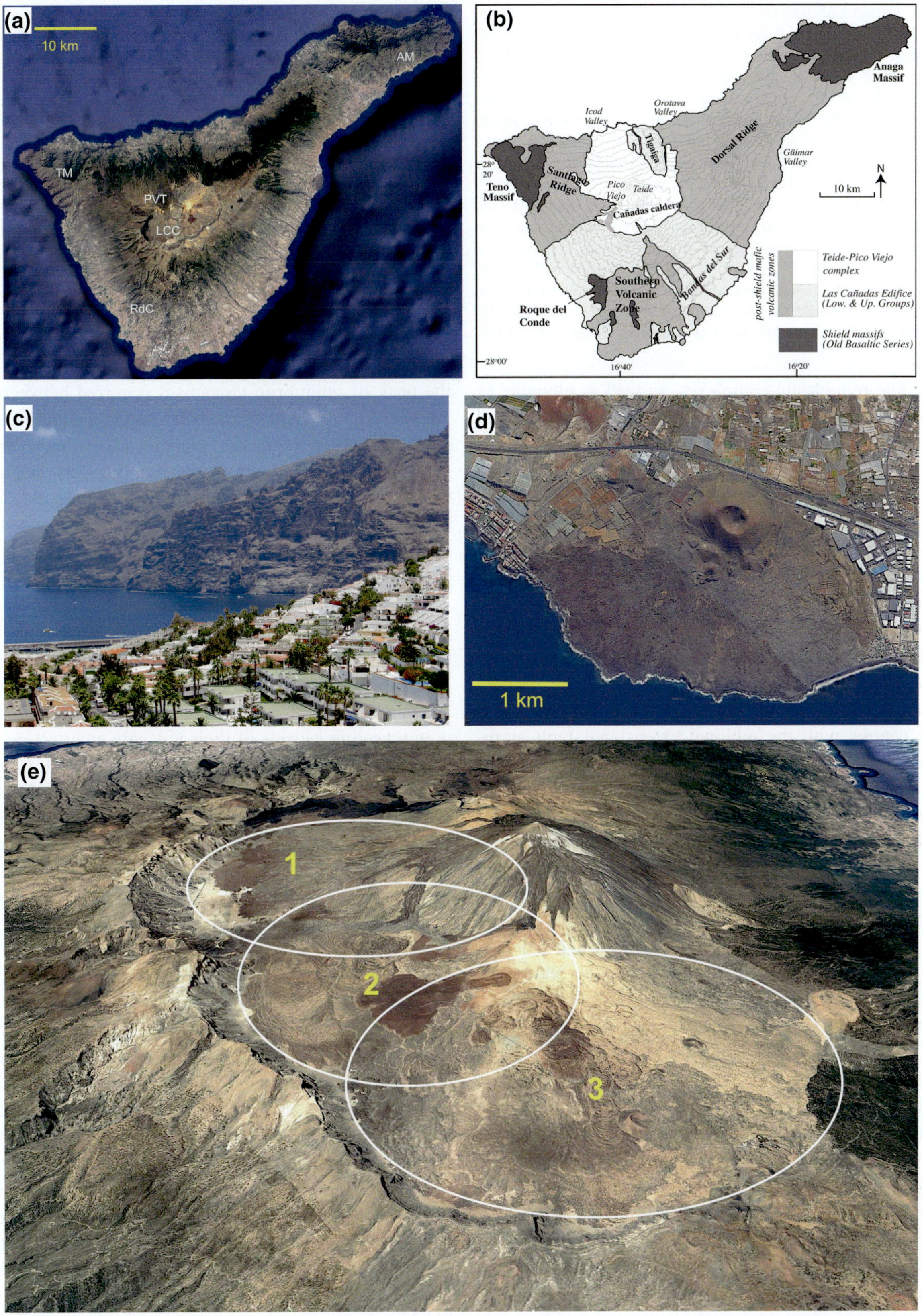
(a)
10 km
AM
TM
PVT
LCC
RdC
(b)
Anaga Massif
Orotava Valley
Icod Valley
Tigaiga
Dorsal Ridge
Güimar Valley
28° 20'
Teno Massif
Santiago Ridge
Pico Viejo
Teide
Cañadas caldera
10 km
N
Southern Volcanic Zone
Bandas del Sur
Roque del Conde
post-shield mafic volcanic zones
Teide-Pico Viejo complex
Las Cañadas Edifice (Low. & Up. Groups)
Shield massifs (Old Basaltic Series)
28°00'
16°40'
16°20'
(c)
(d)
1 km
(e)
1
2
3

◀ **Fig. 16.79** Features of the Tenerife volcanic complex, Canary Islands, Spain. **a** Overhead Google Earth image of Tenerife, highlighting the Las Cañadas volcanic complex (LCC) at the summit, marked by a coalesced caldera complex, along the northern margin of which the Pico-Viejo-Teide (PVT) stratovolcano complex has grown. The triangular apices of the island preserve the pre- 4 Ma basaltic shield volcanic complexes, the Teno Massif (TM) to the northwest, the Anaga Massif (AM) to the northeast, and the Roque del Conde (RdC) Zone to the south. (*Source* Google Earth). **b** Simple map of the distribution of the main litho-stratigraphic sequences representing the main volcanic morphological and evolutionary stages of Tenerife (Modified from Pittari et al. 2008). **c** Dissected shield volcano system consisting of sheet basalt lavas making up the Teno Massif in northwestern Tenerife, at Los Gigantos. **d** Malpais de Guimar scoria cone and lava flow field, Guimar, southern Tenerife (*Source* Google Earth). **e** Oblique Google Earth image looking west of the Las Cañadas coalesced caldera complex. 1 = the oldest Ucanca caldera, 2 = the Guajara caldera, and 3 = the youngest, Diego Hernandez caldera. The oldest caldera has the most eroded wall, and the youngest, the least eroded. Also shown are El Teide (large cone), Pico Viejo (smaller cone behind) stratovolcanoes, and the Montagna Blanca lava fissure-system foreground, which developed on the northern rim of the caldera after the most recent caldera-forming eruption ~190 ka. (*Source* Google Earth; also see Plate 16.1a). **f** Part of the Guajara caldera wall, marked by prominent, thick welded units. **g** Profile through a very thick fallout deposit on the Guajara caldera wall rim, showing transition from non-welded (lower 80%) to welded at the top, Passege de Giujiron. Person circled for scale. **h** Multiple pyroclastic fallout and flow (ignimbrite) deposits on the southern slopes of the volcano, Barranco del Rio. Person circled for scale. **i** Guimar sector collapse landslide valley on the southeast slopes of Tenerife. At the coast, the Malpais de Guimar scoria cone and lava flow field is just visible, as seen in (**d**)

small-volume pyroclastic complexes, as well as elongate or elliptical subvolcanic intrusions within the basins. Local accumulation of such melts in moderate depth magma chambers led to eruptions of ignimbrites, and the formation of calderas, often peralkaline in composition (e.g. Sawyer et al. 1994). In the north of the province these include several catalysed by the early Yellowstone hot spot, e.g. the large McDermitt caldera (Henry et al. 2017), as well as Yellowstone itself (Huang et al. 2015).

While caldera-derived eruptive products may be present within Basin and Range rift basins many show limited evidence of caldera development (e.g. Yellowstone caldera trail), but a closer correlation of smaller, volcanic and intrusive centres with the regional extensional fault architecture, in particular basin margin faults and transfer faults or accommodation zones (Faulds and Stewart 1998; Chap. 17). Transfer faults accommodate segmentation and displacement along rift basins and serve as important magma and fluid pathways. Typically, these basins show a complex interplay of silicic effusive, pyroclastic and other volcaniclastic rocks with sediments dominated by alluvial to lacustrine facies. The silicic volcanics typically form relatively small volcanic fields ≤ 1000 km^2, perhaps volumetrically ≤ 200 km^3, which were erupted over short durations of 2–3 Ma (John 2001). These mid-Miocene and younger rift basins and calderas (although to a lesser extent) are economically important and may host significant epithermal precious metal and other mineral deposits (Chap. 18).

The Basin and Range Province extends southwards into Mexico into the Sierra Madre Occidental volcanic province, which represents the largest, continuous ignimbrite province on Earth, with an estimated, erupted volume of volcanic rocks of 400,000 km^3 (Aguirre-Díaz et al. 2008). The province is marked by a graben and horst tectonic fabric, and many grabens were eruption centres for voluminous ignimbrite eruptions, giving rise to the concept of graben calderas (Aguirre-Díaz et al. 2008; Aguirre-Díaz and López-Martínez 2009; Sect. 16.8.1). Graben formation and the peak of ignimbrite volcanism occurred between 38 and 23 Ma (Aguirre-Díaz et al. 2008). In the southern part of the province, where the largest graben calderas exist, the major graben bounding faults were magma conduits, and some are filled with welded tuff. Magma chambers are considered to have been batholith size, and grabens are interpreted to have formed during major eruptions when regional scale magma chambers were evacuated (Aguirre-Díaz et al. 2008).

Elsewhere many younger rift basins contain intermediate-silicic volcanic clusters exposed along inverted half-graben margins (e.g. Inci 1998; Murray and Busby 2015) or buried by subsequent post-rift subsidence and sedimentation. Oil exploration in some of these basins is now revealing aspects of their volcanic architecture, such as in the Cretaceous Songliao Basin, northeast China (e.g. Wang and Chen 2015; Chap. 18). A 3D map of the central Xujiaweizi graben based on seismic mapping and constrained by numerous petroleum wells is shown in Fig. 16.81. Most of the buried volcanoes appear to be domal features in profile and are confirmed by proximal rhyolitic facies associations in cored intervals, coupled with seismic data. These form clusters aligned along major north-northwesterly basin-forming faults, and easterly transfer faults. Some domal features could be original structures, but they are often also accompanied by widespread truncations implying significant post-eruption erosion and reworking. These paleo-volcanic highs represent favourable locations for gas migration, and accumulation in well-developed reservoir porosity in proximal volcanic facies associations (Chap. 18). A good example of an alkaline multi-vent complex originating in a tectonically controlled graben structure with no evident central edifice is the Quaternary Monte Cimino volcanic complex in central Italy (Conticelli et al. 2013). This forms as a cluster or small field (covering some 300 km^2) of high-K alkaline domes with block and ash flows and ignimbrite aprons.

Other examples of regional scale, Basin and Range type, extensional continental rift basin systems with extensive

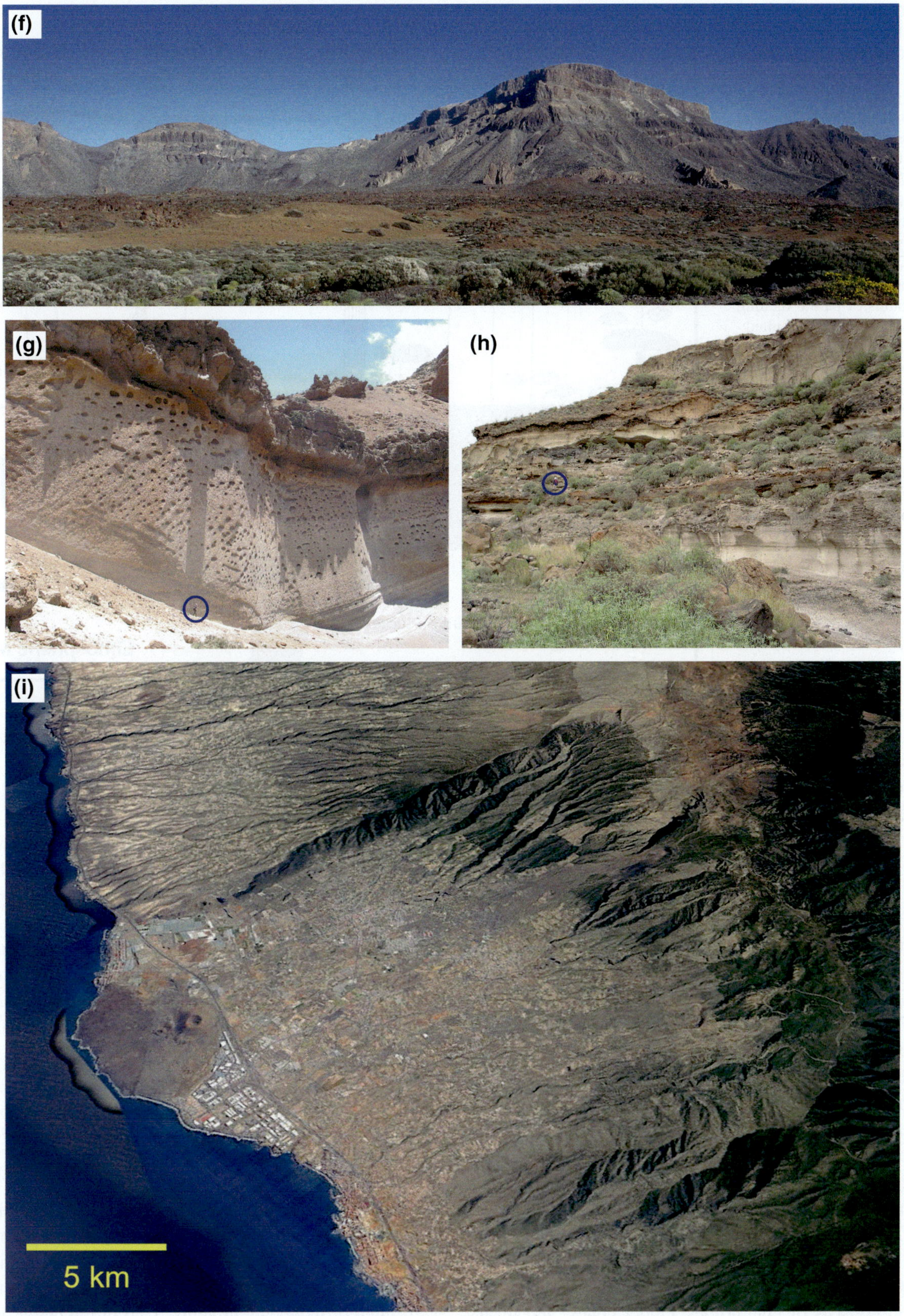

Fig. 16.79 (continued)

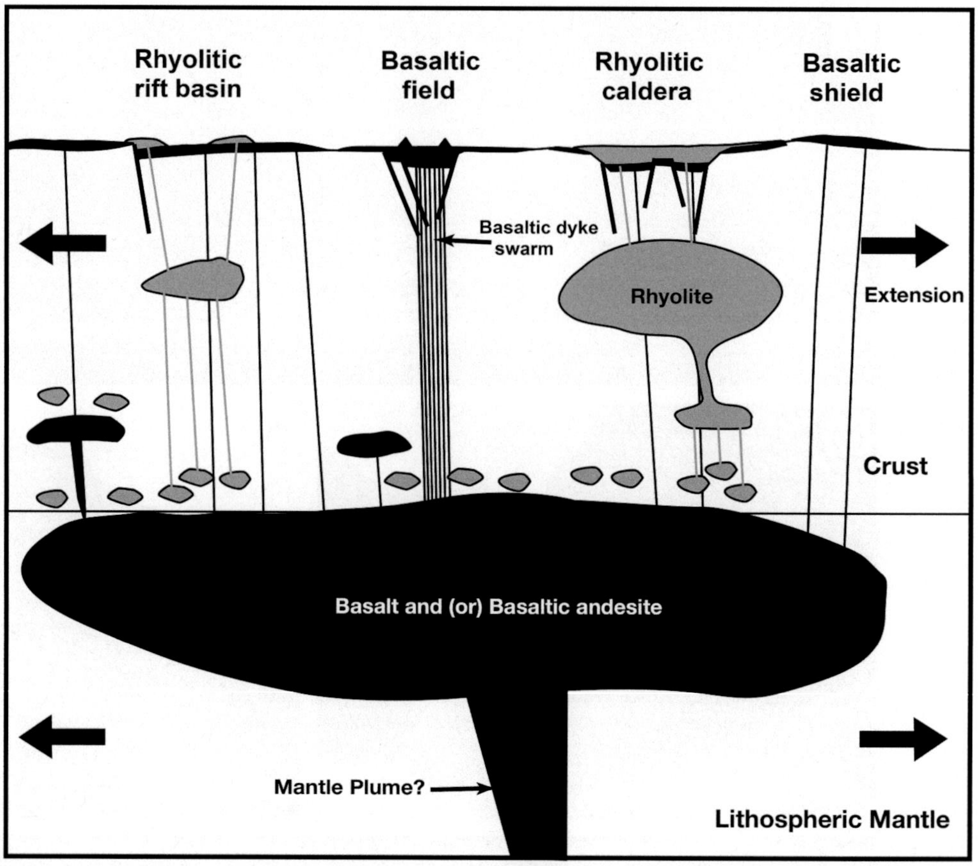

Fig. 16.80 Volcano-tectonic setting of the bimodal basalt-rhyolite assemblage in the Basin and Range Province, western United States. Upwelling of athenospheric mantle into the lithosphere likely related to a mantle plume led to partial melting. Continental extension allowed rapid ascent of mafic magmas through the crust and eruption as thin lava flows, shield volcanoes, and dyke swarms with little interaction with the crust. Small amounts of partial melting of the base of the crust resulted from basalt underplating, and formed evolved rhyolitic melts. The rhyolite magmas erupted as domes and lava flows in extensional basins and rifts, whereas local accumulation of these melts in moderate depth magma chambers led to eruptions of ignimbrites and the collapse of calderas. (From John 2001)

magmatism, include the late Cenozoic Pannonian Basin in the Carpathian tectonic system (Fodor et al. 1999), and the Silurian to Early Devonian, the Late Devonian rock systems of the Lachlan Orogen, southeastern Australia (Cas 1983).

16.9.3 Transcurrent "Pull-Apart" Basins

The Salton Sea Trough represents a transform/strike-slip basin at the southern end of the San Andreas fault system, California (McKibben et al. 1988; Brothers et al. 2009; Chap. 17). This is an active continental rift zone forming the landward extension of the Gulf of California-East Pacific Rise spreading system, but strike-slip movement controls its development. The sedimentary fill comprises Pliocene to Pleistocene deltaic, lacustrine and evaporite sediments, with an active geothermal field associated with a volcanic centre comprising an arc of five Quaternary rhyolite domes. Similarly controlled along the San Andreas transform fault system in California's northern Coast Ranges, the Clear Lake volcanic field is the location of late Pliocene to early Holocene activity (Donnelly-Nolan et al. 1993). The volcanic field consists of lava domes, cinder cones, and maars with eruptive products varying from basalt to rhyolite. Rhyolitic and dacite domes (1.0–1.1 Ma) of Cobb Mountain form the highest peak in the volcanic field (1,440 m). The field is also the site of the Geysers geothermal field, and the Clear Lake volcanics are also thought to have been the heat source for the hot springs and hydrothermal activity that formed gold ore at the famous McLaughlin mine (Chap. 18).

Several transcurrent, trans-Andean, fault systems also focus the development of volcanism (Chap. 17; Fig. 17.45), such as the 80 km^2, Pleistocene Tocomar pull-apart basin, along the Calama-Olacapato-El Toro fault system, which hosts a monogenetic rhyolitic explosive volcanic field, currently site of an active medium enthalpy geothermal system (Giordano et al. 2013; Filipovich et al. 2020).

16.9.4 Multivent Intermediate-Silicic Volcanic Fields of the "High" Andes

Geological investigations in the Andean continental margin over the past 20 years have identified several Quaternary bimodal basalt-rhyolite volcanic fields which contain few large contemporaneous edifices or calderas, but numerous distributed or independent vents. Two of the better described are the Puelche and Laguna del Maule volcanic fields,

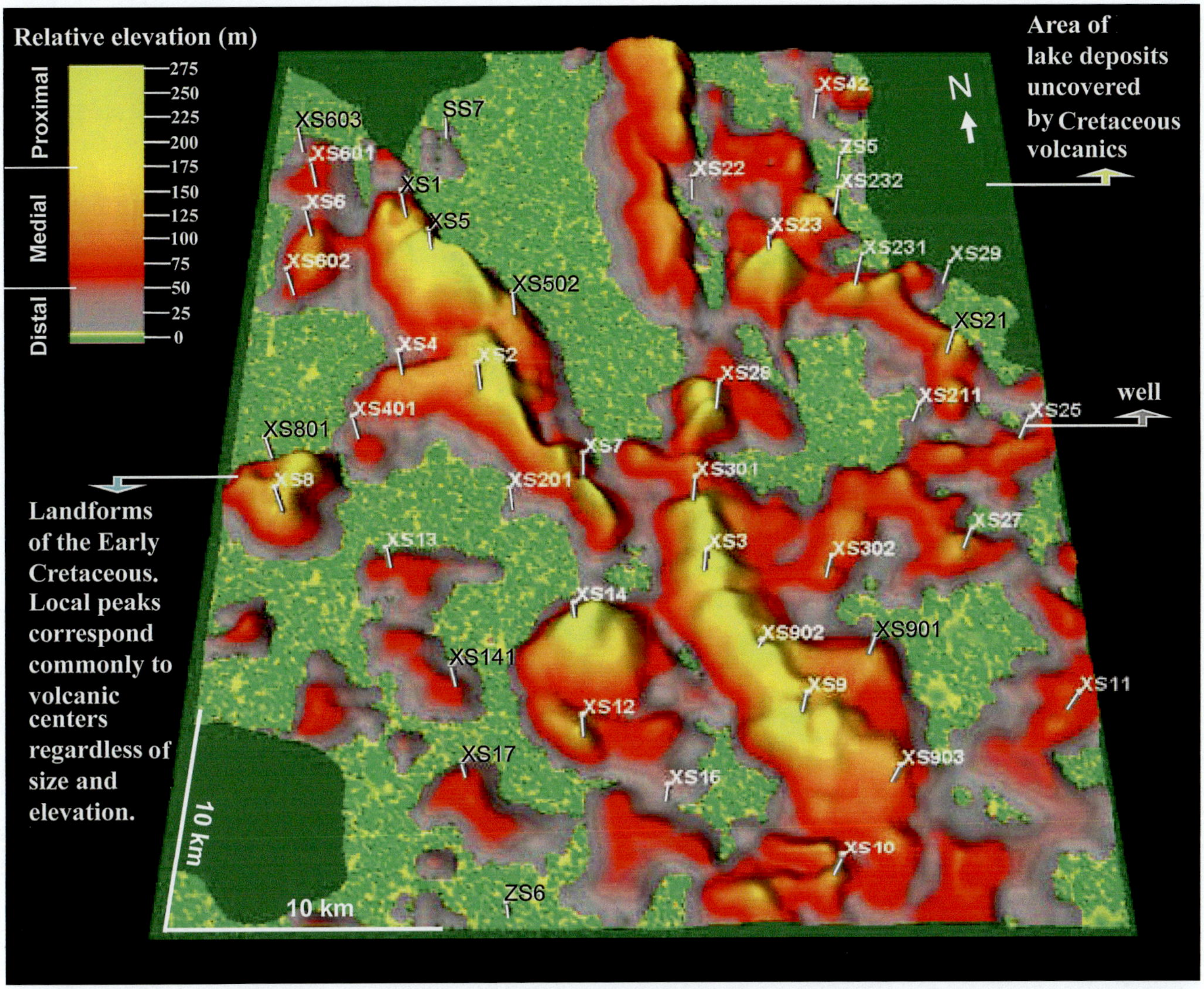

Fig. 16.81 Three-dimensional map, based on petroleum well and seismic data, showing buried volcanic facies in the Xujiaweizi Graben, within the Mesozoic Songliao Basin, China. The proximal (domes), medial, and distal facies associations are transitional and interfingering with each other, commonly without clear boundaries. In plan view, they approximately cover approximately 20%, 30%–50%, and 30%–50% of the area, respectively. (From Wang and Chen 2014)

straddling the Andean Cordillera range crest along the Chile-Argentina border (Hildreth et al. 1999, 2010; Fig. 16.82).

The larger Laguna del Maule volcanic field surrounds a lacustrine basin and covers an area of some 500 km^2 from which >350 km^3 of volcanic products have been erupted in the past 1.5 Ma. On its western edge, it is contiguous with a large, polygenetic stratovolcano (Volcán Tatara-Pedro) which defines the volcanic front of the active Andean arc. The field extends as far as 40 km beyond the volcanic front and includes 14 glaciated (early and middle Pleistocene) medium-sized, polygenetic mafic-intermediate stratovolcanoes and shields, with some 115 monogenetic cones, domes and lava flows of post- and pre-glacial age. These lesser vents are divided into basaltic andesite (43%), dacitic (11%), and rhyodacite and rhyolite (46%).

A large welded ignimbrite was erupted at 1.5 Ma followed by another at about 950 ka, producing a modest 12 km × 8 km-wide caldera underlying the northern half of the lake basin (Fig. 16.82b). At least 50 post-glacial (<26 ka) eruptions have occurred from 24 vents surrounding the wider lake basin producing rhyolite and rhyodacitic domes and coulées (Fig. 16.82d), most of them accompanied by pumice falls. The large number of post-glacial silicic vents around the lake basin and the general progression from rhyodacite to rhyolite (as well as lack of mafic enclaves) suggest growth of a latest Pleistocene to Holocene magma chamber some 20 km in diameter beneath the basin (Singer

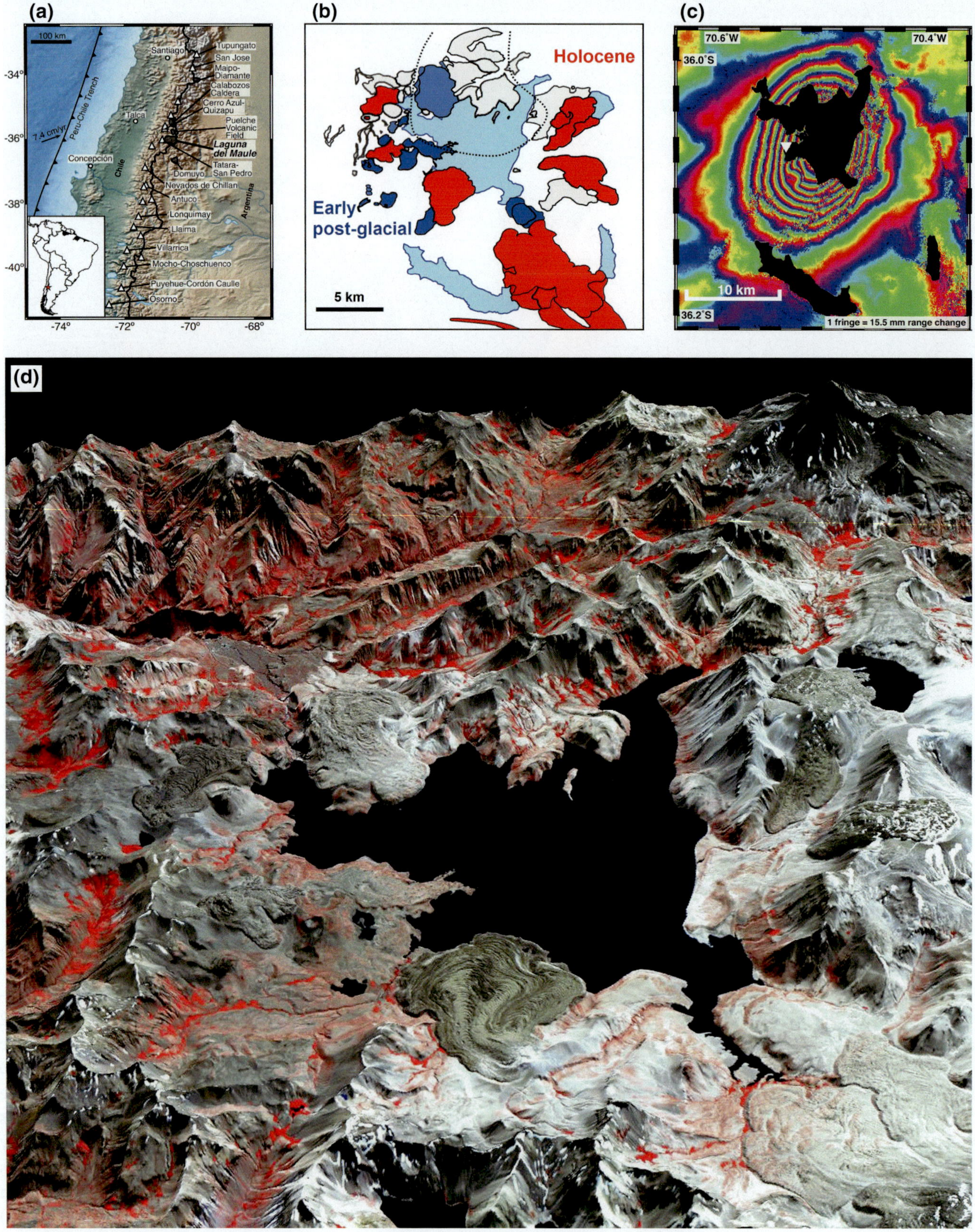
(a)
100 km
Tupungato
San Jose
Santiago
Maipo-Diamante
Calabozos Caldera
Cerro Azul-Quizapu
Talca
Puelche Volcanic Field
Laguna del Maule
Concepción
Tatara-San Pedro
Domuyo
Nevados de Chillan
Antuco
Lonquimay
Llaima
Villarrica
Mocho-Choschuenco
Puyehue-Cordón Caulle
Osorno
Peru-Chile Trench
7.4 cm/yr
Chile
Argentina
-34°
-36°
-38°
-40°
-74°
-72°
-70°
-68°
(b)
Holocene
Early post-glacial
5 km
(c)
70.6°W
70.4°W
36.0°S
36.2°S
10 km
1 fringe = 15.5 mm range change
(d)

◀ **Fig. 16.82** Laguna del Maule volcanic field, southern Andes volcanic zone. **a** Regional location map showing selected frontal arc stratovolcanoes (triangles), and multi-vent silicic volcanic fields (dark grey) some of which include calderas (From Andersen et al. 2017). **b** Map showing the distribution of silicic lavas erupted during the early post-glacial (<26–19 ka) and latest Pleistocene to Holocene (ca. 8–2 ka); dotted line is ca. 950 ka caldera (From Andersen et al. 2017) **c** Interferometric analysis of synthetic aperture radar (InSAR) data spanning the period 14–04-2013 to 25–01-2014. One fringe denotes 15.5 mm of range change between satellite and ground. These equate to uplift rates exceeding 200 mm yr^{-1}, which are among the highest ever measured at a volcano that is not actively erupting. (From Singer et al. 2014). **d** ASTER satellite image showing Lake Maule surrounded by a complex volcanic landscape, including small stratovolcanoes, lava domes, and scoria cones. Some of the most prominent features surrounding the lake are rhyolite coulées, which protrude into the lake. ASTER produces images using infrared, red, and green wavelengths of light: vegetated areas range in colour from red to pink, snow is white, water is black, and bare rock is earth-toned. (*Source* NASA Earth Observatory)

et al. 2014; Andersen et al. 2017). This area has also been inflating (tumescence) at an astounding rate of >200 mm yr^{-1} since 2007 making it perhaps the today's most active rhyolitic centre (Fig. 16.82c).

The most recent hypothesis suggests that the silicic vents have tapped an extensive layer of crystal-poor, rhyolitic melt which began to form atop a magmatic mush zone established by ca. 20 ka, causing a renewed phase of rhyolite eruptions during the Holocene (Andersen et al. 2017). Modelling of surface deformation and geophysical changes suggests that magma is currently intruding at a depth of $\sim$5 km at a rate of 0.03 km^3 kyr^{-1}. Extrapolating the volume change for the inflation yields a rate of mass addition consistent with that which produced the rhyolitic Long Valley caldera-forming eruption. However, whether this unrest is a prelude to a caldera event or an imminent eruption of any particular style remains unknown.

Similar multi-vent, intermediate to silicic volcanic fields are not restricted to Plio-Pleistocene activity but are also common through the central and northern Andes within the middle to late Miocene (17–5 Ma). These are highly significant because they host some of the larger epithermal gold deposits so far discovered on the planet (e.g. Bissig et al. 2015; Chap. 18). The majority of these Andean epithermal Au–Ag deposits occur at high elevations (3500–5200 m a.s.l.), typically 200–500 m below low relief volcanic landforms, which are mostly located near the water shed along the range crest of the orogen (akin to the Quaternary fields). Probably the best described is the Yanacocha volcanic field in northern Peru which hosts the largest of these, a super-giant deposit containing >1,400 tonnes of gold (Longo et al. 2010). The Yanacocha volcanic field comprises a co-genetic series of middle-late Miocene andesitic and dacitic lavas, domes, ignimbrites, and minor subvolcanic intrusions, dykes and breccias covering an area of some 350 km^2. These are controlled by a northeast-southwest structural corridor parallel to major regional Andean lineaments. The youngest, the Coriwachay Dacite (10.8–8.4 Ma), represents the most silica rich volcanic rocks, forming small volumes of dacite-rhyolite domes, porphyry dikes and plugs spatially closely coincident with the main corridor of hydrothermal alteration and gold. Yanacocha and similar Andean epithermal precious metal deposits and their volcanic host facies are documented in detail in Chap. 18.

16.9.5 Large Island Arc Multi-vent Volcanoes

Some larger island arc volcanoes also appear to be multi-vent complexes lacking a central cone. Cas and Wright (1987) referenced St Lucia and Dominica in the Lesser Antilles island arc. The tropically forested central highlands of St Lucia form an older volcanic core from which radiate Quaternary valley-filling dacitic pumice flow fans (Cas and Wright 1987). Studies of this pumice flow succession suggested that eruptions occurred from a number of vents located over a wide area, involving different dacitic magma batches at various times (Wright et al. 1984), rather than the so-called Qualibou depression on the southwest side of island. The Qualibou depression was initially described as a caldera but reinterpreted as a gravity collapse or sector collapse scarp (Roobol et al. 1983; Fig. 16.42f). Wohletz et al. (1986) revived the caldera origin for a smaller structure within the depression. Subsequent studies have since clearly demonstrated that two massive sector collapses occurred from this site, generating submarine debris avalanche deposits extending at least 60 km across the floor of the Caribbean Sea (e.g. Deplus et al. 2001; Fig. 16.48). While post-sector collapse volcanism has been important inside the collapse scarp (Sect. 16.7.6), the significance of circular faults and structures within it remains equivocal. However, any young source vents plugged by dacitic pelean domes also remain undetected in the densely vegetated interior (Lindsay 2005).

Dominica, in the centre of the island arc (Fig. 16.48), is similarly dominated by a rugged central range of peaks with elevations in excess of 1,000 m. Unlike the other islands of the Lesser Antilles, the youngest volcanic centres of Dominica (late Pleistocene to Holocene) are located throughout the entire length of the island (Smith et al. 2013). The centres are dominated by pelean-style andesite and dacite volcanism characterised by the formation of lava

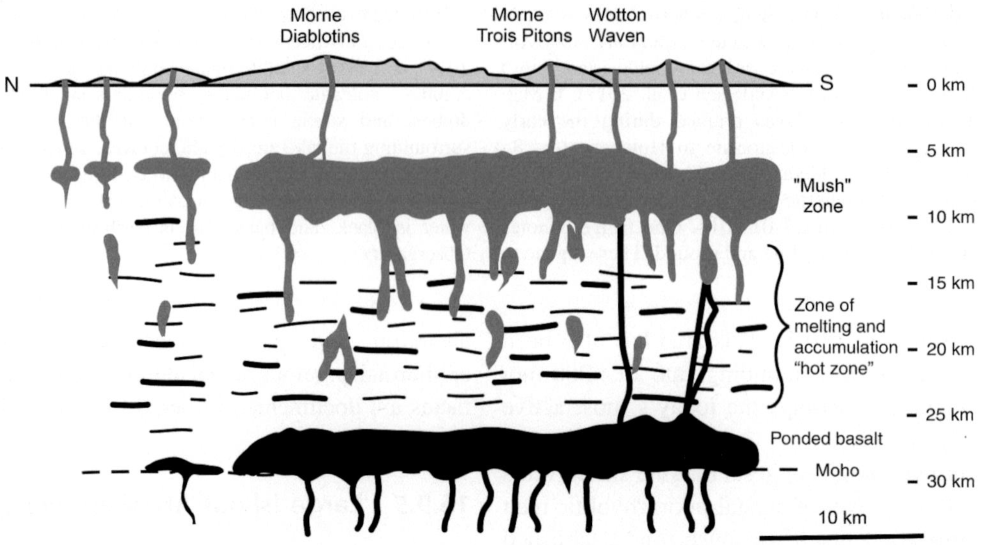

Fig. 16.83 Quaternary magmatic plumbing system, Dominica, Lesser Antilles. Increase in mantle flux continues injection of sills and/or dykes beneath most of southern and central Dominica, producing a "hot zone" under the island and generation of intermediate hybrid magmas. Greatest flux is under southern Dominica, as witnessed by the presence of most of the potentially active centres and the caldera-forming eruptions of Morne Trois Pitons and Wotten Waven. Mid-crustal magma chambers expand and eventually merge to form mid-crustal batholith under most of island. Expansion of magmatic system initiates new volcanic zone of activity below northern Dominica. (From Smith et al. 2013)

domes, and block and ash flows. The mountainous topography of the island is in part due to the presence of >40 pelean domes and dome complexes scattered throughout the island (some 750 km^2). Six centres have also undergone plinian-style eruptions, with three of these centres Morne Diablotins in the north, the Morne Trois Pitons and Wotten Waven producing widespread ignimbrite flow fans, some of which are welded. These eruptions formed two calderas at Morne Trois Pitons and Wotten Waven, within a central graben structure. Three centres have also experienced sector collapses. The widely distributed nature of these young volcanic centres and their deposits, together with micro-seismic swarms and geothermal phenomena, led Smith et al. (2013) to propose most of the island is underlain by a batholithic-size magma chamber formed in the late Pleistocene (Fig. 16.83). This produced numerous volcanic centres dominated almost exclusively by andesites and dacites of such similar compositions that the different centres cannot be geochemically distinguished from each other. The amalgamation of individual magma chambers beneath these different centres appears to have produced a mid-crustal batholith, resulting in uplift of the island, formation of a central half-graben, and the current island-wide distribution of volcanic-related phenomena.

Recent tracing of the extent of the submarine portions of the Aeolian Arc volcanic islands, Italy, shows that the cluster of Vulcano-Lipari-Salina (cf. "*Salipano*" in Chap. 15, see Fig. 15.44a) is in fact a single merged NE-elongated structure, with its emerged top punctuated by several explosive and effusive centres, domes and calderas, whose dimension and multi-vent association is very similar to other fissure-controlled complex volcanic complexes such as the submarine Marsili and Palinuro volcanoes (Pensa et al. 2019; Fig. 16.84).

16.9.6 Facies Models for (Caldera-Deficient) Multi-vent Intermediate-Silicic Volcanic Centres

Continental bimodal rift centres are in many respects very similar to those of rhyolitic caldera volcanoes, however, rather than the ring fault structure, in this case the focal element of primary volcanic facies is the rift basin fault architecture. This is principally controlled by normal and transfer faults in extensional rifts and grabens, and trans-tensional fault arrays in strike-slip settings (Fig. 16.85a–c; also see Busby 2012, for a valuable review from a perspective of the cordilleran evolution of the southwestern United States).

Rhyolitic domes and short lavas erupted from multiple eruption points will lie along and/or intersections of these structures, which act as the principal magma conduits and are associated with near-vent pyroclastic and other volcaniclastic and terrigenous alluvial-fluvial fan deposits (Fig. 16.85a–c). However as indicated above, in broader extensional settings these basins are not mutually exclusive from calderas. Large-volume welded and non-welded

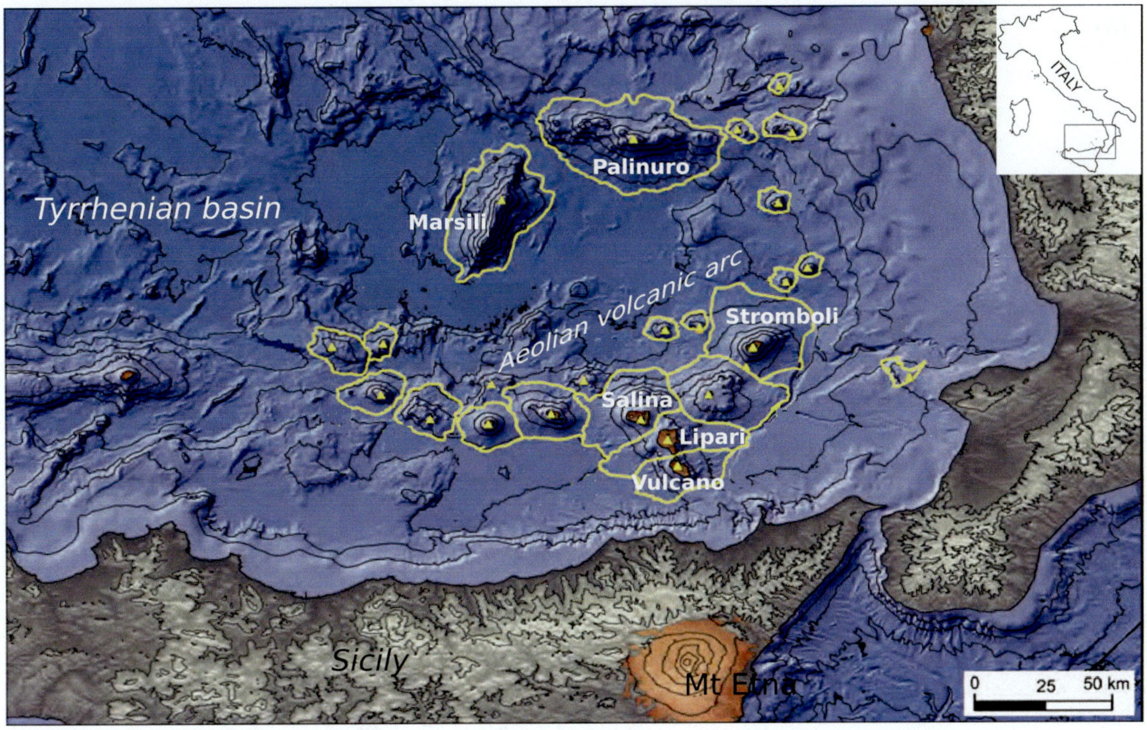

Fig. 16.84 Bathymetric map of the southeastern Tyrrhenian Sea basin and Aeolian Arc, Italy. Bathymetric contours are each 500 m, so maximum depth of the bathyal plain is 3000 m b.s.l. Yellow lines indicate the extent of both submarine and emergent volcanoes. Marsili and Palinuro are recognised as fissural, multi-vent submarine volcanic complexes. Salina, Lipari and Vulcano are the emergent parts of similar coalesced multi-vent volcanic complex. (Modified from Pensa et al. 2019)

ignimbrites originating from calderas and graben calderas and building ignimbrite shields may extend widely along and outside the rift system (Fig. 16.85c). In addition, small-volume non-welded ignimbrites which do not extend large distances from the rift system may be prominent. Contemporaneous basaltic volcanism commonly occurs within the basin structure forming dykes, sills, basaltic lava shield and half-graben ponded lava piles, scoria cones and associated lava flow fields. Where these erupt into, or flow into, lacustrine basins in the extensional landscape, peperites, and hyaloclastites will form. Dome top tuff cones may also occur in lacustrine environments. Several examples represent the basis for the facies models presented in Fig. 16.85a–c.

The late Devonian Boyd Volcanic Complex of southeastern Australia was part of a regionally extensive, bimodal volcanic Basin and Range type rift zone terrain in the Paleozoic Lachlan Fold Belt or Orogen (Cas 1983). The preserved remains of the Boyd Volcanic Complex are contained within a north–south trending rift graben, and the succession is mostly a subaerial association of basalts, rhyolites, including ignimbrites, fluvial-alluvial sediments, and local deep-water lacustrine subaqueous volcanics and sediments (Fig. 16.85a, b; Fergusson et al. 1979; Cas et al. 1993; see below). The rift basin system was subject to trans-tension (Giordano and Cas 2001) producing pull-apart depocentres along the rift zone. Volcanism is thought to have been controlled by basin faults, which acted as conduits for rising magmas (Fig. 16.85a, b).

The early Jurassic, Mt Wrightson Formation, Arizona (Fig. 16.85c; Riggs and Busby-Spera 1990; Busby 2012), provides an ancient example of a multi-vent intermediate-silicic volcanic complex, broadly comparable to those found in the “high” Andes. While many of the facies assemblages are also characteristic of stratovolcano complexes, the paucity of debris flow deposits and the burial of near-vent facies by ignimbrites suggest that the Wrightson Formation was deposited in a low relief multi-vent complex, with no high standing central edifice. Intra-arc graben subsidence resulted in a complex juxtaposition of vent, proximal, medial and distal facies as vent areas were repeatedly buried by volcanic debris derived from adjacent vents within the multi-vent field. Although, some of the thickest and mostly densely welded tuffs probably represent outflow ignimbrite sheets erupted from a caldera or calderas outside the field. This resulted in the accumulation and preservation of a thick succession (4.5 km thick) of terrestrial volcanic, subvolcanic and sedimentary rocks. Irregular hypabyssal intrusions of andesite to rhyolitic compositions occur throughout the succession and are interpreted as vent related.

(a)

(b)

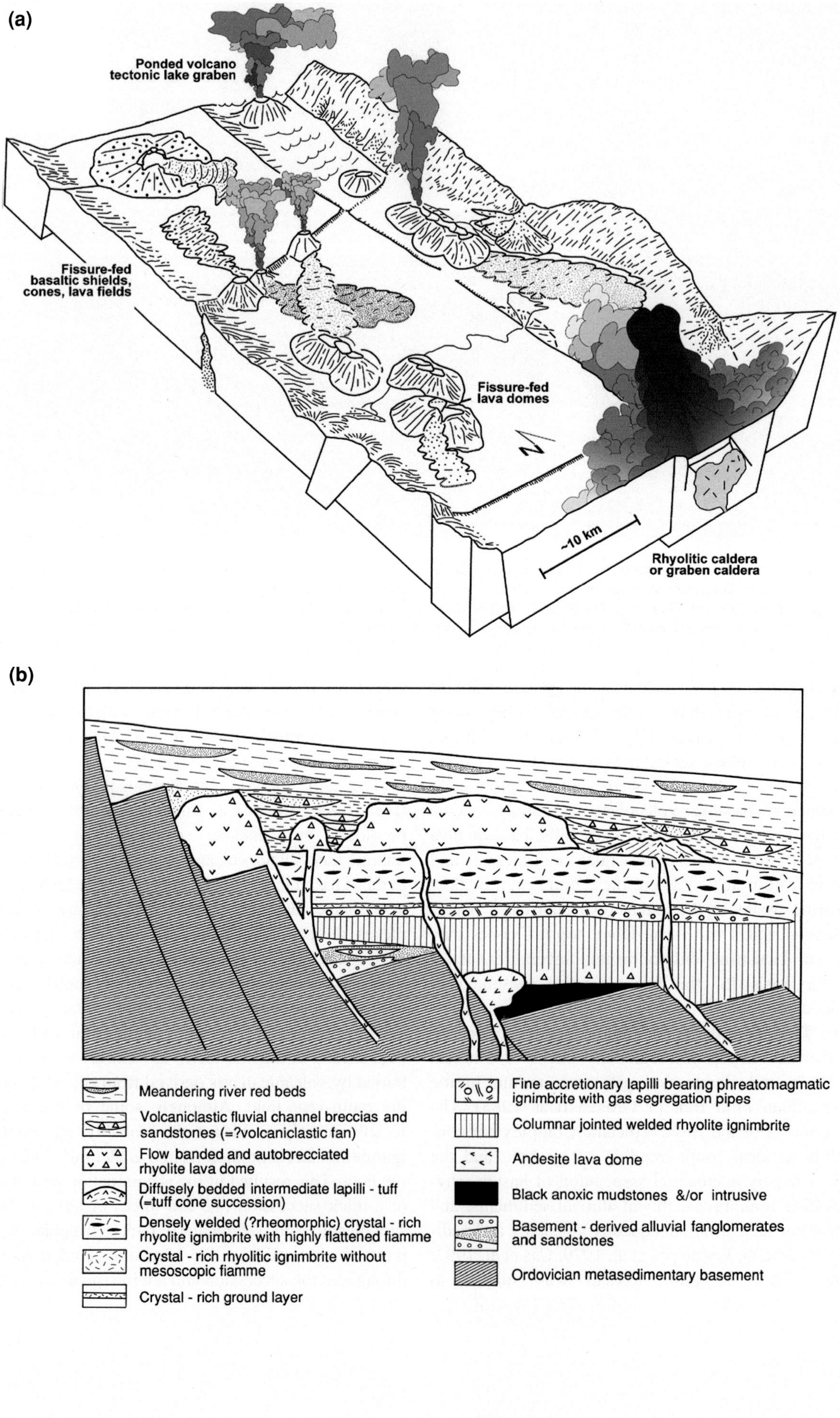
Ponded volcano tectonic lake graben
Fissure-fed basaltic shields, cones, lava fields
Fissure-fed lava domes
N
~10 km
Rhyolitic caldera or graben caldera
Meandering river red beds
Volcaniclastic fluvial channel breccias and sandstones (=?volcaniclastic fan)
Flow banded and autobrecciated rhyolite lava dome
Diffusely bedded intermediate lapilli - tuff (=tuff cone succession)
Densely welded (?rheomorphic) crystal - rich rhyolite ignimbrite with highly flattened fiamme
Crystal - rich rhyolitic ignimbrite without mesoscopic fiamme
Crystal - rich ground layer
Fine accretionary lapilli bearing phreatomagmatic ignimbrite with gas segregation pipes
Columnar jointed welded rhyolite ignimbrite
Andesite lava dome
Black anoxic mudstones &/or intrusive
Basement - derived alluvial fanglomerates and sandstones
Ordovician metasedimentary basement

◀ **Fig. 16.85** Features and facies associations of continental rift basin volcanic systems. **a** Paleogeographic reconstruction for subaerial, multi-vent, bimodal extensional rift basins and grabens, based on the Late Devonian Boyd Volcanic Complex, southern New South Wales, Australia (After Fergusson et al. 1979). **b** Cross-sectional representation of the stratigraphic and facies architecture and associations of the continental, Late Devonian Boyd Volcanic Complex around the town of Eden. **c** Facies model illustrating the paleogeography and volcanic setting of the early Jurassic Mt Wrightson Formation, Arizona, USA. Eolian dunes derived from the northwest are interstratified with the products of active and effusive vents, and more distal caldera complexes within an intra-arc graben. These vent areas include abundant peperite breccias suggesting active subsidence and deposition of wet sediments during emplacement of this multi-vent complex. (From Busby 2012, After Riggs and Busby-Spera 1990)

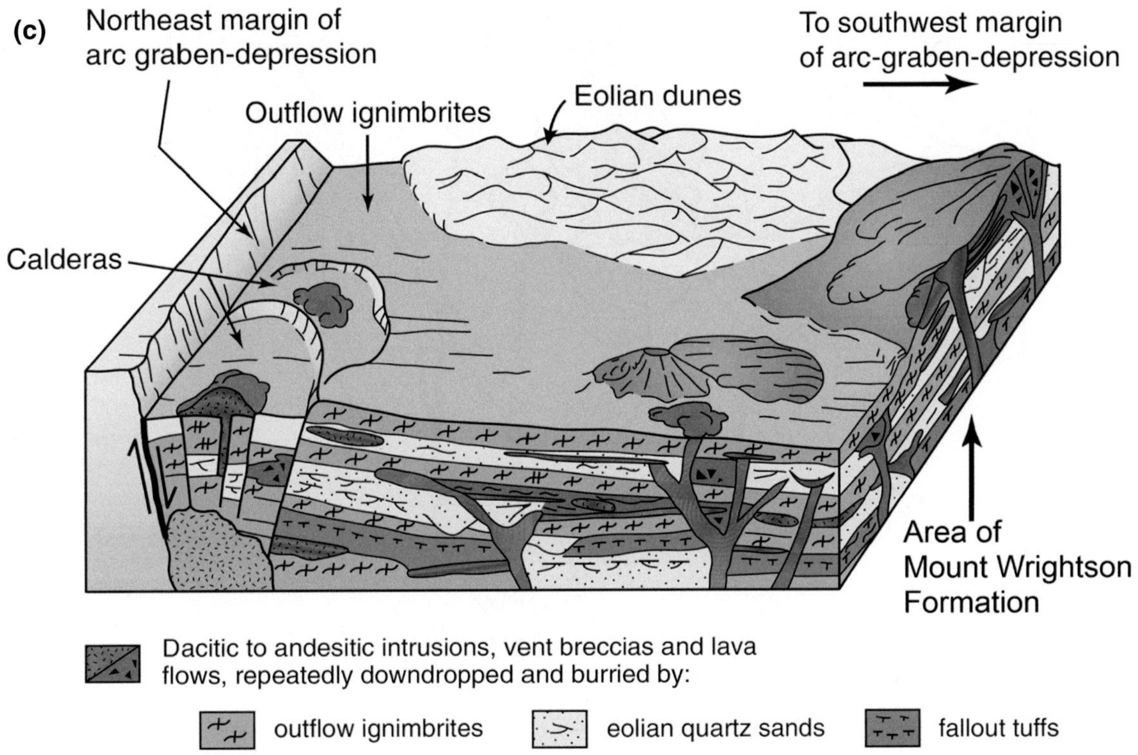

Fig. 16.85 (continued)

These commonly show peperitic margins, formed when magma or lava intrudes or mixes with wet unconsolidated sediment (Chap. 4). Peperites indicate sediments were wet and unlithified and their deposition penecontemporaneous with shallow intrusive activity. Similarly, at the world class Baja de la Alumbrera porphyry copper deposit, peperites have been identified throughout the volcanic succession (Harris et al. 2006; see Chap. 18). Notwithstanding the exploration implications (Chap. 18), this may be a key facies indicator separating the successions of low relief multi-vent fields (with standing bodies of water and/or water-saturated sediment) from stratovolcanoes.

Parts of such extensional to trans-tensional rift settings will trap water and can form substantial, deep, volcanically active volcano-tectonic graben basins (Fig. 16.86a, b; e.g. Lake Tanganyika, East Africa Rift Zone, which is up to 1,500 m deep; Chap. 17). Bimodal rift basins also occur in deeper water environments, where major continental rift systems transition into marine settings, forming marine volcano-tectonic graben basins, as well as during back-arc rifting of mature arcs (e.g. Miocene rifting that produced the Kuroko basins in northwestern Japan; Chap. 18).

A well-documented wholly subaqueous, bimodal, volcano-tectonic, trans-tensional rift basin also occurs in the late Devonian Boyd Volcanic Complex, southeastern Australia, in the Bunga Beds basin, (Fig. 16.86a, b; Cas et al. 2001; Giordano and Cas 2001; Chap. 8), which was part of a broader regional bimodal extensional terrain (Cas 1983). Sedimentary environments include basin margin fan deltas and deep-water basin centre settings marked by both terrigenous and volcaniclastic mass flow turbidite facies associations and intercalated hemipelagic sediments. Although there are resedimented scoria deposits derived from basin margin centres, and relatively shallow water dome-top pumice and tuff cones, deep water had the effect of limiting explosive eruptions, leading to a higher proportion of

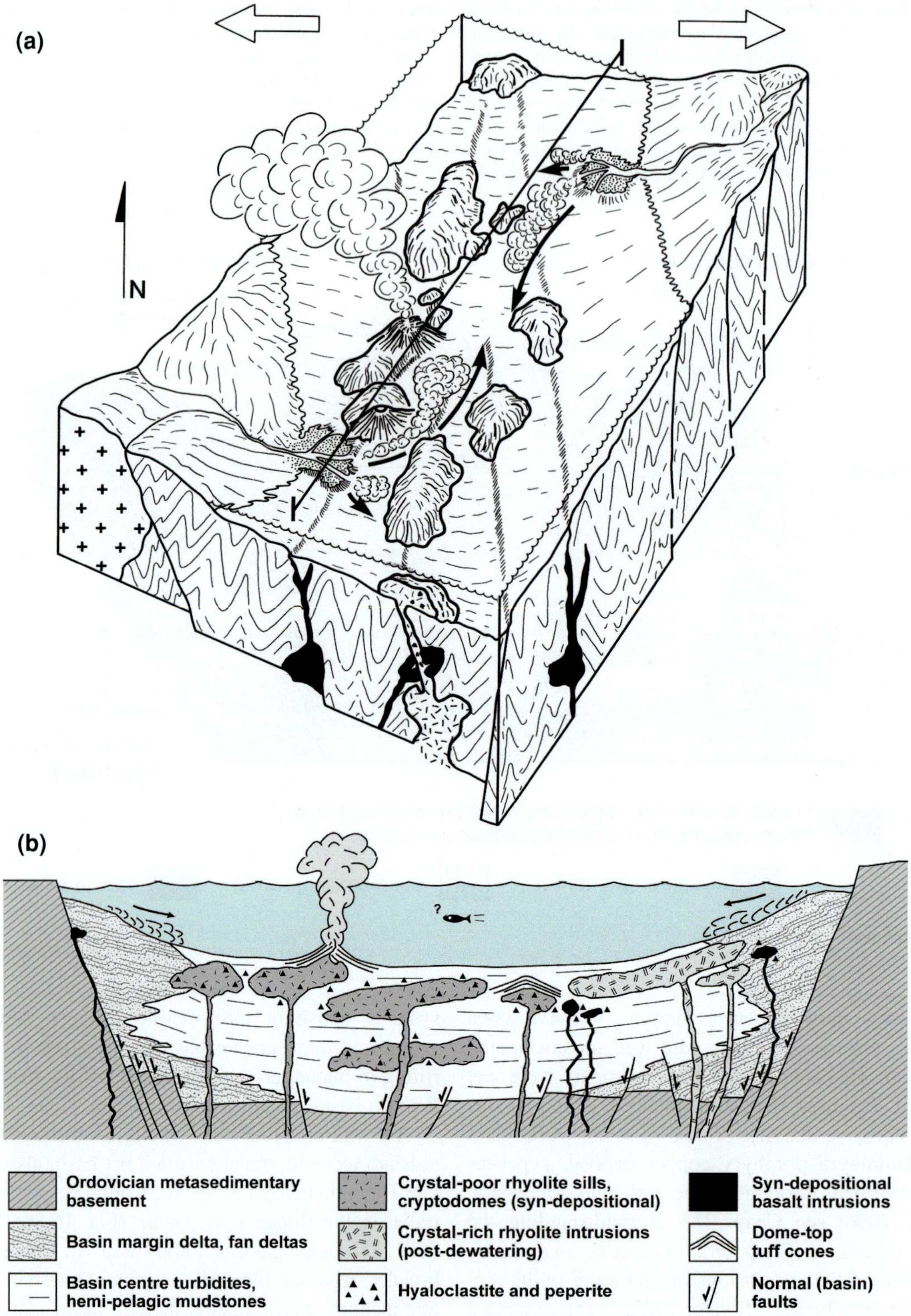

Fig. 16.86 Features and facies associations of subaqueous rift basin volcanic systems. **a** Paleogeographic reconstruction for subaqueous, multi-vent, bimodal extensional rift basins and grabens. This model is based on our studies of the Late Devonian Bunga Beds outlier, Boyd Volcanic Complex, southern New South Wales, Australia. This shows basin margin (fan-)deltas, subaqueous high relief lava domes and cryptodomes along normal faults, localised dome-top tuff cones, and regional extensional context of the lacustrine basin. (From Cas et al. 2001). **b** Cross-sectional facies model

lavas, and in the sedimentary fills, syn-depositional sills, dykes, cryptodomes and associated hyaloclastite and peperitic breccias, of both silicic and mafic compositions (Fig. 16.86a, b; see Figs. 8.18b–d and 8.37a–c for textures).

It is important to stress such rift and trans-tensional volcanic basins, both continental and subaqueous, are principally sedimentary depocentres. Modern studies on rift basins show that they can be complexly segmented structurally, especially by transfer faults, and so the volcanic and sedimentary facies architecture along the axis of the basin could change abruptly and vertically (Chap. 17). Volcanic activity is generally associated with the earliest phase(s) of extension and rifting, although multi-stage rifting events are not unknown, which is followed by post-rift ("sag" or thermal relaxation) sedimentation, dominated by fine-grained facies which extend outside the main basin margin faults (Chap. 17). In trans-tensional settings smaller volcanic centres (rhyolite domes) are clustered along releasing bends of strike-slip faults, while major calderas (such as Lake Toba) are created by dilational jogs or pull-apart basins at larger releasing stepovers (Chap. 17). In these settings, even more complex and abrupt interfacing of volcanic and sedimentary facies can be anticipated laterally and vertically.

16.9.7 Resource Potential and Economic Significance

While some may argue, perhaps correctly, that multi-vent intermediate-silicic volcanic centres represent a disparate category rather than a family of volcano types, there is no doubt as to their economic importance. Singularly they account for more than 20% of the world's epithermal gold (Chap. 18), including the premier deposit at Yanacocha, in Peru. From the available data, rift basins contain substantially more epithermal gold than silicic caldera volcanoes (Chap. 18). However, the amount of gold associated with calderas may be underestimated due to the difficulty in recognising such structures in the older record, although, the large majority of epithermal deposits discovered are post-Mesozoic in age. We have also stressed rift basins and calderas are not mutually exclusive and, it may be calderas, or more precisely their underlying magma chambers, that drive large-scale geothermal flow. Some multi-vent centres clearly overlie large, shallow magma bodies, and this may be the critical factor in driving magmatic and hydrothermal fluids and producing supergiant gold deposits (Chap. 18).

Many volcanic-hosted massive sulphide deposits (including the Kuroko-style deposits of Japan) are now believed to be hosted within deep submarine rift basins rather than submarine calderas. As with bimodal continental rift centres described above (Sect. 16.9.6), these similarly are not necessarily mutually exclusive volcanologically or in their host rock facies (Sect. 16.8.9). Again, rather than the caldera ring fault, the focal element of primary volcanic facies becomes the rift basin fault architecture, which will also control the migration of hydrothermal fluids (Chap. 18). There is also some indication that porphyry copper systems are not solely confined to stratovolcanoes but may also be associated with topographically subdued multi-vent centres. Lastly, intermediate-silicic volcanic centres (paleo-highs) and facies in younger continental rift basins are increasingly being recognised as a valid petroleum play (Chap. 18).

16.10 Intra- or Subglacial Volcanoes

Volcanoes commonly erupt within and beneath ice sheets and glaciers (Figs. 16.87a and 16.51). More recently this has been termed *glacio-volcanism*, which is described as the interaction of magma with ice in all its forms, including snow, firn (granular, partially compacted snow), and any meltwater (e.g. Edwards et al. 2015; Smellie 2018). Typically, these form distinctive steep-sided ridges called *tindas*, or steep circular table mountains called *tuyas* (Fig. 16.87b, c), which were first described from British Columbia by Mathews (1947). Examples are well known from currently and formerly glaciated regions of the world, particularly Iceland, British Columbia and Antarctica (Russell et al. 2014). Interest in this topic has expanded dramatically since the April–May 2010 eruption of Eyjafjallajökull volcano in Iceland through a summit glacier (Fig. 16.87a). This highlighted two of the major hazards that are exceptional to glacio-volcanism: local massive flooding and enhanced production of fine-grained, airborne, phreatomagmatic ash, leading to disruption of aviation. Also, research on the geology of Mars has required an improved understanding of terrestrial analogues of glacio-volcanic eruptions and deposits, while, glacio-volcanic deposits help map distributions of ice sheets in space and time and, thus, may be critical in paleo-climatic reconstructions.

16.10.1 Pioneering Icelandic Studies

Possibly still the best-documented intra- or subglacial volcanoes are described from Iceland, following pioneering work beginning in the early 1900s, and a renaissance in the late 1960s (e.g. Jones 1969). Compositionally, all types occur: basaltic, andesitic, dacitic and rhyolitic (Furnes et al. 1980). Basaltic subglacial volcanoes consist principally of masses of pillow lavas, palagonitised hyaloclastite breccias

Fig. 16.87 Intra-subglacial volcanoes. **a** View of the Eyjafjallajökull volcano, southern Iceland, towards the end of the first explosive phase on 17 April 2010. View is from north looking south across the summit glacier; in the foreground is the main outlet point for jökulhlaups which were triggered by repeated meltwater outbursts on 14–15 April (see Fig. 16.90). (Photo by Oddur Sigurdsson, Iceland Meteorological Office). **b** Tuya mountain named Bláfjall near Mývatn lake, northern Iceland. **c** Northern Victoria Land, Antarctica, showing in the foreground a deglaciated tuya called Shield Nunatak which comprises a base made of hyaloclastite and the top an emergent scoria cone. The stratovolcano in the background is Mt Melbourne

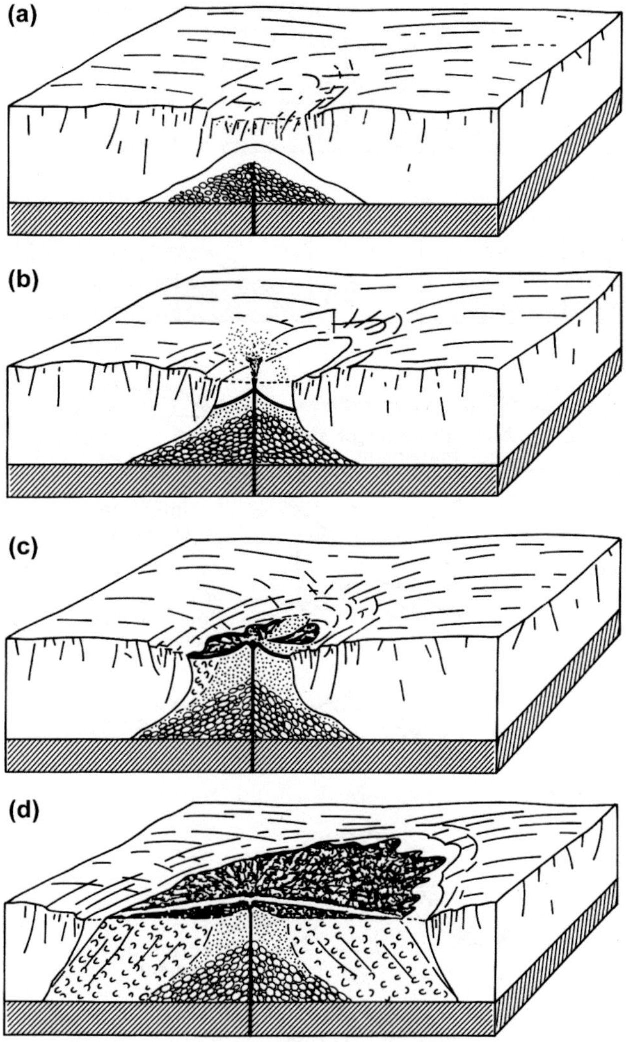

Fig. 16.88 Growth stages of a tuya. **a** Magmatic heat causes melting of ice sheet above eruptive fissure forming meltwater vault. Within this vault erupting lava builds a steep-sided pillow lava pile. **b** As pile builds, roof of vault collapses forming an ice cauldron infilled with an intraglacial lake. Effusion gives way to explosive phase on emergence and resulting volcaniclastic deposits accumulate between walls of ice on top of pillow lava pile. **c** Emergent explosive phase give sway to aerial effusive phase. Lava issues from vent and pushes out into meltwater lake on deltas of flow-foot breccia. **d** Advanced stage of aerial effusive phase. Products of earlier eruptive phases overwhelmed and buried by flow-foot breccia. (From Cas and Wright 1987, After Jones 1969)

and sideromelane fragments. Jones (1969) envisaged that a subglacial basaltic eruption first builds up a steep-sided pillow pile, as lava from a fissure is extruded into a meltwater vault created under the ice sheet (Fig. 16.88). Much of the hyaloclastites appear to form later as lava deltas build out into a meltwater lake as the growing pile emerges. Flow-foot breccias develop as the deltas capped by lava grow out into the meltwater, eventually forming an equidimensional table mountain. Intra-glacial basaltic eruptions in Iceland have formed very substantial deposits, often referred to as "*móberg*", which are deposits of hyaloclastite volcanic breccia formed during subglacial eruptions. Overlapping centres form compounded deposits extending tens of km and forming mountains up to 1,500 m in height and resulted from eruptions during the latter parts of the late Pleistocene. Older móberg masses (late Pliocene-early Pleistocene) are also known and are buried by younger volcanic rocks.

Although discussion of tuyas mostly revolves around subglacial basaltic eruptions, subglacial rhyolite eruptions can also build tuyas (Tuffen et al. 2002; McGarvie 2009; Russell et al. 2014). The study of Tuffen et al. (2002; Fig. 16.89a) documents a relatively large (~ 1 km^3) subglacial to emergent rhyolite tuya at Torfajökull central volcano, southern Iceland, which was emplaced in late Pleistocene time beneath a glacier >350 m thick. Torfajökull is the largest silicic complex in Iceland and contains an outer ring of several flat-topped, rhyolitic tuyas some of which are coalesced into larger structures (or complex tuya), as the one described here. The eruption began initially with a sequence of subglacial phreatomagmatic explosive activity in a subglacial vault from which most water has drained through factures in the ice of channels at the bedrock-ice interface, building a succession of lapilli tuff deposits up to 300 m thick. This sequence is succeeded and overlain by rhyolite lava once the glacier above the vent melts, forming a meltwater lake; however, there is now confirmatory evidence that this was preceded by phase of significant subaerial explosive activity not preserved, or covered by Holocene fallout deposits and scree, in proximal locations (see below). The lower part of the lava succession includes a gradational passage zone of irregular lobes and hyaloclastite breccia with the underlying lapilli tuffs as the erupting lava progrades in the lake over the underlying succession. More lava eruption then builds a subaerial lava pile above the lake level, displacing the lake water. The lava varies from highly perlitic coherent to pumiceous obsidian to marginal quench fragmented breccias, including peperite at the base. The vesicularity of eruptive products increases upwards as glaciostatic and hydrostatic pressures decrease (McGarvie 2009). McGarvie (2009) has produced a comprehensive compilation of known occurrences of rhyolite-ice interaction in Iceland, documenting effusion of lava, tuya formation and magma-ice interaction on stratovolcanoes. The lava-forming eruptions produced coherent lava and associated hyaloclastite, whereas tuya formation is along the lines proposed by Tuffen et al. (2002).

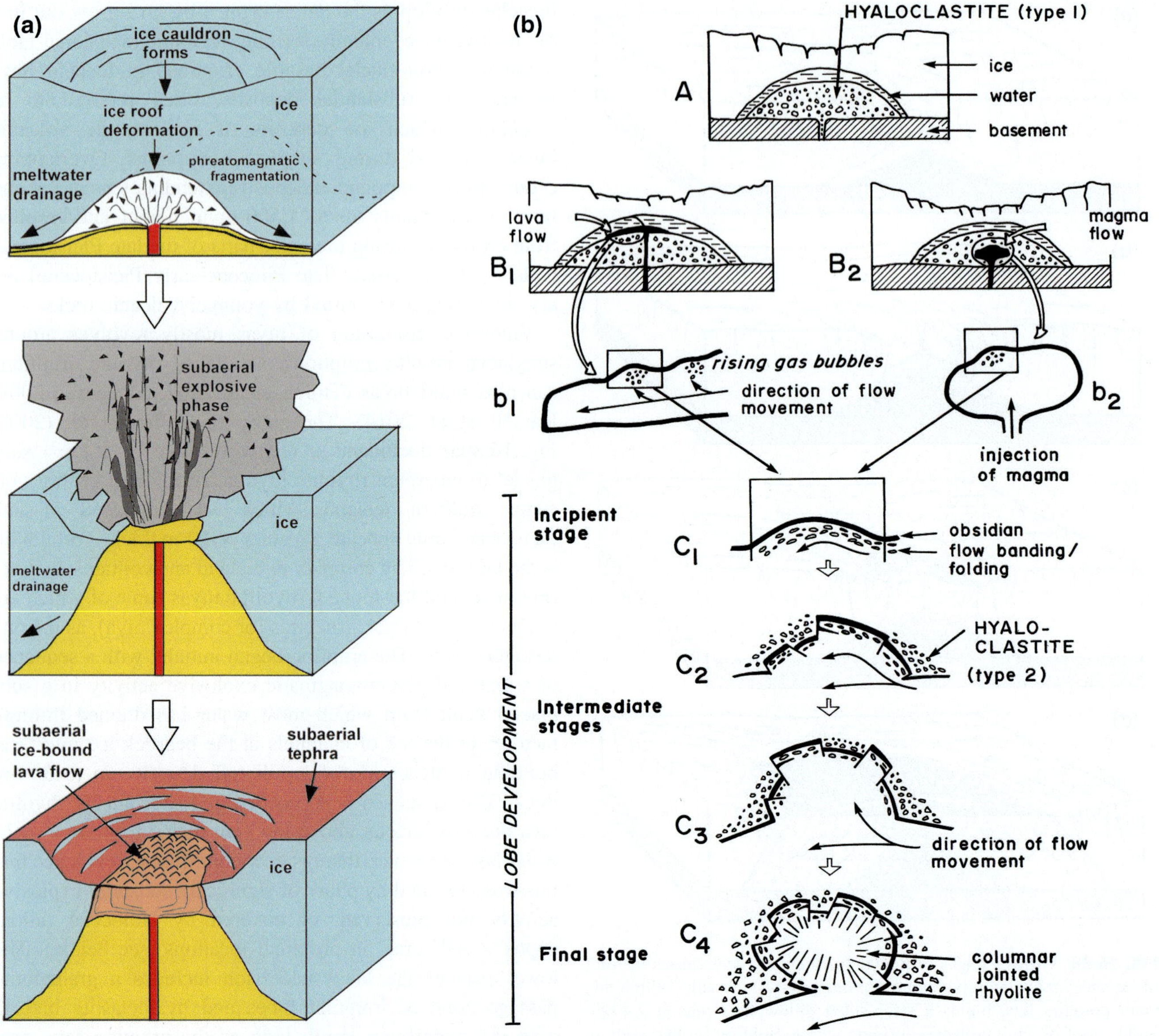

Fig. 16.89 Evolutionary models for rhyolitic subglacial volcanic eruptions, Iceland. **a** Relatively large (≥ 1 km^3), rhyolitic tuya based on studies of the Torfajökull central volcano, southern Iceland. The eruption commenced with vigorous phreatomagmatic phase within a well-drained ice vault, possibly followed by a subaerial explosive phase (not identified in proximal deposits on-tuya), and then a subaerial effusive phase. Basal diameter of tuya edifice represented is $\sim$2–5 km. (Modified from Tuffen et al. 2002). **b** Formation of two types of silicic hyaloclastite and lava lobes described from small volume (0.01–0.1 km^3), rhyolitic subglacial eruptions (From Cas and Wright 1987, After Furnes et al. 1980)

Several rhyolitic subglacial accumulations of smaller volume (0.01–0.1 km^3) also occur in Iceland as described by Furnes et al. (1980), which still stands as one of the few published accounts of facies and emplacement mechanisms of subglacial rhyolite. Facies consist of two main components (Fig. 16.89b): (1) hyaloclastites and (2) lava lobes, averaging a few metres (<10 m) in diameter. Two types of rhyolitic breccia are found, and these are believed to represent major changes in the style of activity during subglacial eruptions. *Type 1*, by far the most important type, is brecciated pumice and glass and is thought to result from explosive activity (stage A, Fig. 16.89b). By analogy, with subaerial silicic eruptions the initial phase of the eruption of rhyolite magma beneath ice is also likely to have been explosive (plinian or phreatoplinian) if the confining pressure is not too high. *Type 2* is characterised by fragments of obsidian and stony (microcrystalline) rhyolite and is genetically related to the lava lobes (stages B and C). The lobes generally consist of (from the margin inwards), a chilled obsidian rind, a zone of flow-banded and flow-folded pumiceous rhyolite, and a central zone of radially joined rhyolite. These lobes are thought to be the remnants of disintegrated subglacial lava flows or lava intrusions into the waterlogged pumiceous hyaloclastite. *Type 2* breccias are hyaloclastites formed by the brecciation and quench fragmentation of the lobes. The lava flows or intrusive bodies are emplaced by a continuous process of alternating construction and fragmentation of lobes. Rhyolite lavas which have been emplaced into wet sediments show similar lava lobes at their margins and a similar suite of quench fragmented facies and textures (Chaps. 4 and 6).

Tuffen et al. (2001) have recognised the complex nature of the interactions between sub-glacially erupting rhyolite and the ice, leading to the formation of multiple thermal erosive channels at the interface, followed by extrusion of coherent lobes and associated hyaloclastite breccias, some of which could be resedimented sub-glacially within meltwater vaults. While, detailed correlations by Moles et al. (2019), describe the formation of pyroclastic density currents (pdcs) associated with a subaerial explosive phase during formation of the rhyolitic tuya at Torfajökull volcano (Fig. 16.89a; Tuffen et al. 2002). These deposited the variably welded Thórsmörk ignimbrite (1.5–2.0 km^3 DRE) beyond the ice margin and $\sim$30 km from source, as well as widely dispersed ash fallout marker across the North Atlantic region and the Greenland ice sheet.

Huge meltwater floods called "*jökulhlaups*" breaking out from the margins of glaciers into surrounding river systems and resulting from subglacial eruptions were also first recognised in Iceland (Chap. 12). These have deposited large debris fans or sandurs (see Figs. 12.29d and 16.114b), and some of the larger flows are believed to have continued into the North Atlantic Ocean and deposited volcaniclastic turbidites (Laughton et al. 1972; Lacasse et al. 1996).

16.10.2 Antarctica

Antarctica is the largest glacio-volcanic province in the world. There are many volcanoes extending from the sub-Antarctic South Sandwich Islands (including Deception Island, which erupted sub-glacially in 1969), through the Antarctic Peninsula and Marie Byrd Land, and into East Antarctica, a distance of some 5,000 km (Smellie et al. 2008, 2009; Davies et al. 2012).

Eruptions coincided with the development of the Antarctic Ice Sheet. The volcanoes are mostly basaltic, but there are few examples of more evolved magma compositions. They range from very large stratovolcanoes with summit elevations up to 4,000 m above sea level and basal diameters of 40–60 km (e.g. Mt Melbourne in the background of Fig. 16.87c), to volcanic fields composed of multiple small centres (e.g. Shield Nunatak in the foreground of Fig. 16.87c). Volcanic outcrops in Antarctica are characteristically very well exposed, although, extensive snow and ice cover, and the remote locations can make accessing them challenging. Cliff sections up to 2 km high may extend 10 or 20 km laterally, although, many volcanoes have minimal exposure or have been extensively removed by multiple overriding ice sheets, particularly in the Antarctic Peninsula. Much larger, sub-glacially erupted edifices are formed of alternating thick sections of lava and fragmental rocks and are therefore technically stratovolcanoes. However, the frequent development of lava-fed deltas has resulted in volcano profiles with slopes less than 15°, which are described as shield volcanoes. Both terms have been used to describe Antarctic volcanoes. The Antarctic Peninsula hosts numerous mainly small volcanic edifices with ages extending between 7.5 Ma and present, although, several much larger stratovolcanoes are present in the northern part of the region.

Two generic types of glacio-volcanic sequence have been identified in the Antarctic Peninsula (Fig. 16.90), described as (1) thick sequences dominated by multiple lava-fed deltas, and (2) thinner, sheet-like sequences. The lava-fed delta sequences conform very well with the facies model of Jones (1969), displaying subaerial capping lava top-sets, overlying subaqueous foreset (or flow-foot) *hyaloclastite* breccias. These are thought to have been erupted under much thicker ice, hundreds of m thick, rather than 10 s of m in the case of sheet-like deposits. Documentation of such facies differences has enabled a detailed history of ice sheet thicknesses to be deduced. This shows that the Antarctic Peninsula Ice Sheet has varied in thickness from up to ca. 850 m, but to typically much less (<400 m) for the period 6.25 Ma to present. Thus, the ice sheet appears to have simply draped rather than drowned the landscape, although it was probably persistent even through ice-poor periods corresponding to inter-glacial stages.

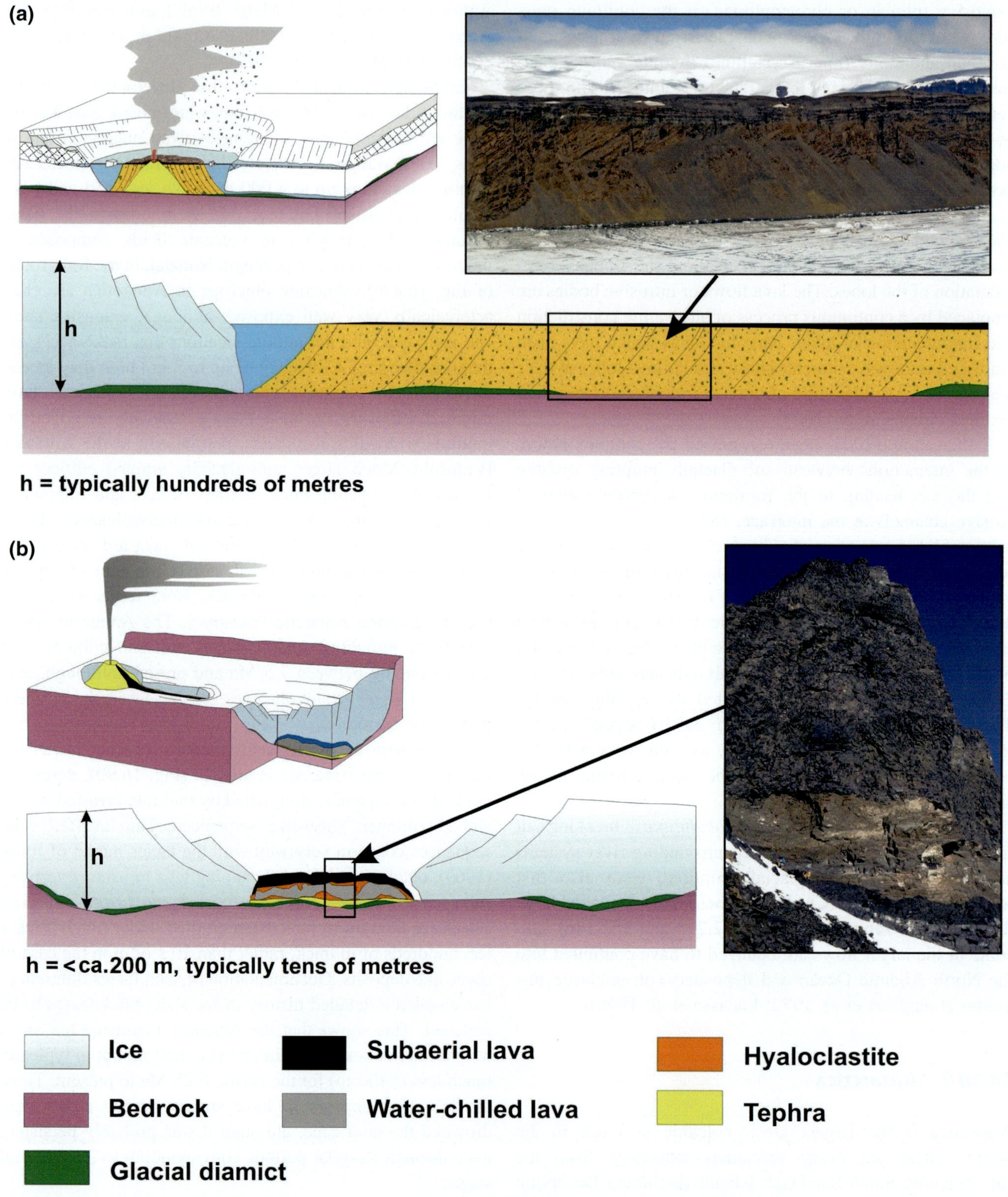

Fig. 16.90 Two generic types of basaltic glacio-volcanic sequence described in the Antarctic Peninsula. These show facies dominated by **a** multiple pahoehoe-fed lava deltas with thick flow-foot breccias, and **b** thinner, sheet-like sequences. (From Smellie et al. 2009)

16.10.3 April–May 2010 Subglacial Eruption of Eyjafjallajökull Volcano, Iceland

Eyjafjallajökull (Fig. 16.87a) is a central volcano in Iceland completely covered by an ice cap and is located at the southern end of Iceland's East Volcanic Zone (Figs. 16.91a and 16.61). The ice cap covers an area of about 100 km^2 over Eyjafjallajökull volcano, which lies 25 km west of another subglacial volcano, Katla that historically been much more active and known for powerful subglacial eruptions (see Sect. 16.12.8, Fig. 16.114b).

Intense seismicity and uplift in January–March 2010 were followed by a small basaltic flank eruption on 20 March–12 April, forming a 1.3 km^2 lava flow field, about 8 km east of the summit caldera of the volcano (Fig. 16.91a; Gudmundsson et al. 2010; Magnússon et al. 2012). This first eruption, in the form of a fissure vent, did not occur under the glacier and was relatively small in scale. After a brief pause the summit eruption started at ca. 15.30 UTC on 14 April and was preceded by an intense three-hour-long earthquake swarm. This phase was characterised by the discharge of more silicic (benmoreite-rhyolitic, 60–69 wt% SiO_2) magma into a water-filled vent as the eruption melted its way through the 200 m thick glacier in the summit caldera. This supported a 5–10 km high plume that alternated between shades of white (steam-rich) and dark grey (loaded with ash). In the first two days, the eruption formed ice cauldrons over the active vents (Fig. 16.91a, b). The main activity on 14 April was in the south ice cauldron but moved to the west cauldron on 15 April where it remained until the end of the eruption (Fig. 16.91b). Ice-melting caused repeated jökulhlaups to flood down the nearby rivers on 14–15 April, requiring 800 people to be evacuated. This eruption was explosive (phreatomagmatic), due to meltwater getting into the volcanic vent. It was estimated to be ten to twenty times larger than the opening phase. This explosive eruption threw volcanic ash several km up in the atmosphere, which led to air travel disruption in northwest Europe for six days from 15 to 21 April 2010 and again, in May 2010, including the closure of airspace over many parts of Europe (Gudmundsson et al. 2010, 2012; see Fig. 16.114a).

16.10.4 Facies Models for Intra- or Subglacial Volcanoes

The principal elements of mainly basaltic glacio-volcanic edifices are thick successions of pillow lavas, surrounded by even thicker successions of hyaloclastite breccias (Fig. 16.92). Volcanic activity commences with an early, effusive phase erupting pillow lavas with, or without, minor volumes of hyaloclastites and volcanic ash. Where ice thicknesses are sufficient (>500 m) to suppress volatile exsolution, explosive fragmentation is restricted to phreatomagmatic processes, although this may also be suppressed until a shallow meltwater lake has formed on the glacier surface (Chap. 4). The initial pressure on the magma is the pressure at the base of the glacier, which is the combined load of the overlying ice (glaciostatic) and any accumulated meltwater (hydrostatic). Evidence suggests that the volume of melted ice invariably exceeds the volume of erupted material, therefore, a meltwater lens or intraglacial lake rapidly forms at the base of the ice. This may leak at any time along the base of the ice sheet or by lake overflows (Fig. 16.92a), generating jökulhlaups.

The largely coherent facies are superseded vertically and laterally by predominantly fragmental facies. This change is linked to the growth in edifice height within the intraglacial lake. This is because, firstly, as the growing pillow mound enlarges and steepens, gravitational collapse increasingly generates deposits of volcaniclastic material (pillow breccia, hyaloclastite). Secondly, as the eruption proceeds, the effective water pressure on the erupting magma decreases, causing greater magmatic and phreatomagmatic explosivity. This is due to the combined effects of (1) the increasing height of the edifice within the lake, and (2) deformation and subsidence of the ice above it, which as it melts, its surface subsides creating an open cauldron above the growing edifice. When the bulk of the edifice breaches the lake surface, lava moves laterally away from the vent until it reaches the shoreline. Here it builds a delta formed by pillow lavas and hyaloclastites. As the delta continues to prograde, the subaerial top of the edifice is covered by subaerial lava flows. This boundary between the subaqueous and subaerial eruption products is referred to as the "*passage zone*" (Smellie 2006). Variations in passage zone elevations may record changes to the lake level, which are controlled by the glacial hydrology, or subglacial and/or supraglacial drainage (Fig. 16.92b; Smellie 2006).

While this facies model serves as an important useful starting point, we emphasise that many aspects and controls on glacio-volcanic eruptions are far from fully understood, e.g. effusive versus explosive initiation, recognition of magmatic versus phreatomagmatic components, tuyas as monogenetic versus polygenetic edifices (e.g. Edwards et al. 2015). Russell et al. (2014) propose to encompass the large variability of subglacial edifices or glacio-volcanoes by simplifying the terminology indicating them as tuyas sided by a morphological descriptor such as flat-topped tuya, conical tuya, linear tuya (or tindar), and complex tuya. This

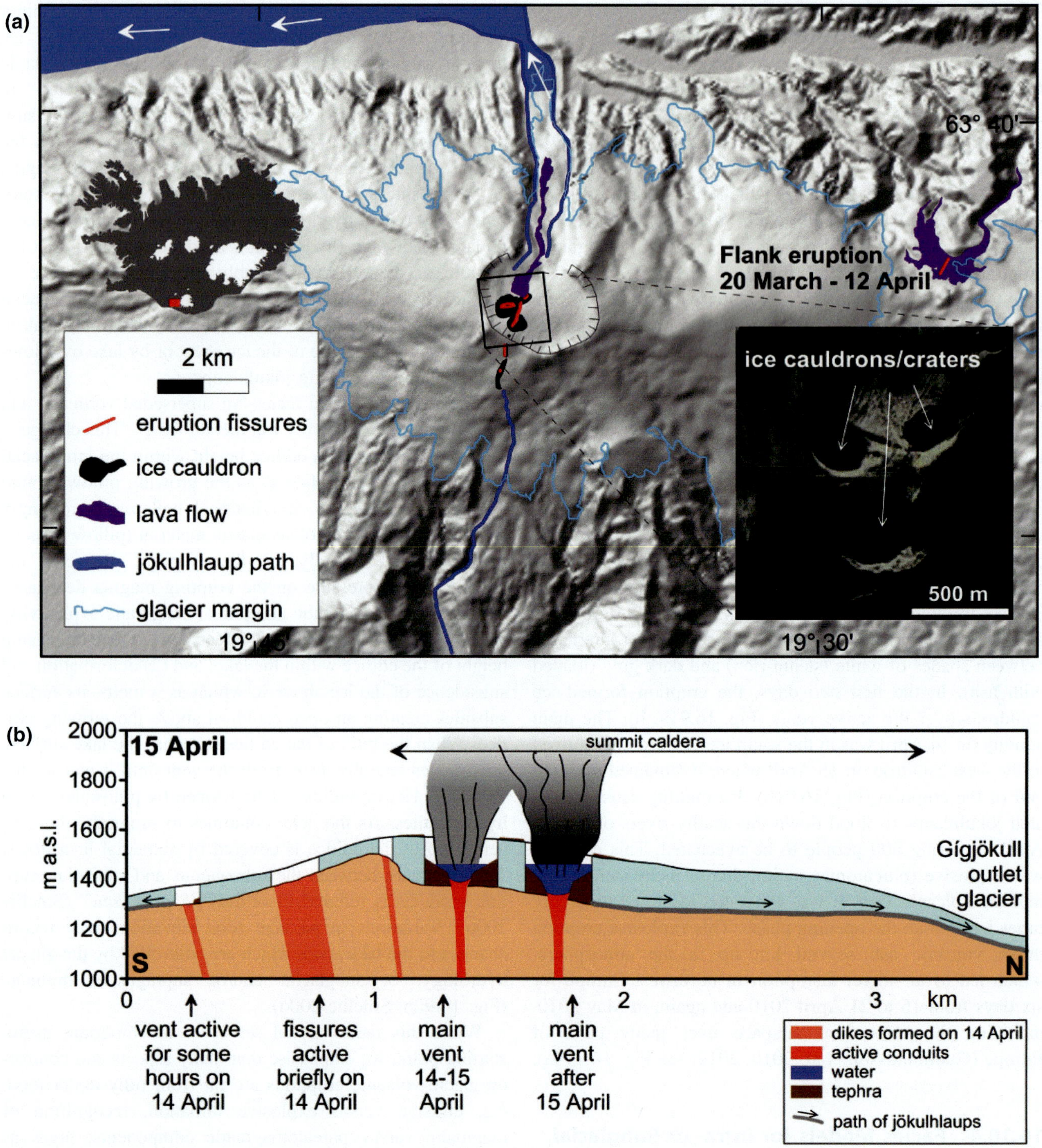

Fig. 16.91 Setting of the April–May 2010 explosive Eyjafjallajökull subglacial eruption, southern Iceland. **a** Location of eruption sites, ice cauldrons and pathways of jökulhlaups. Inset on right: an airborne synthetic aperture radar image from 15 April; inset on left: location of Eyjafjallajökull volcano (red). **b** Schematic north–south cross-section through the main craters within the summit caldera. This shows the two main ice cauldrons with the vents (south cauldron active 14–15 April and west cauldron active from 15 April), and meltwater flow on 15 April. (Modified from Gudmundsson et al. 2012)

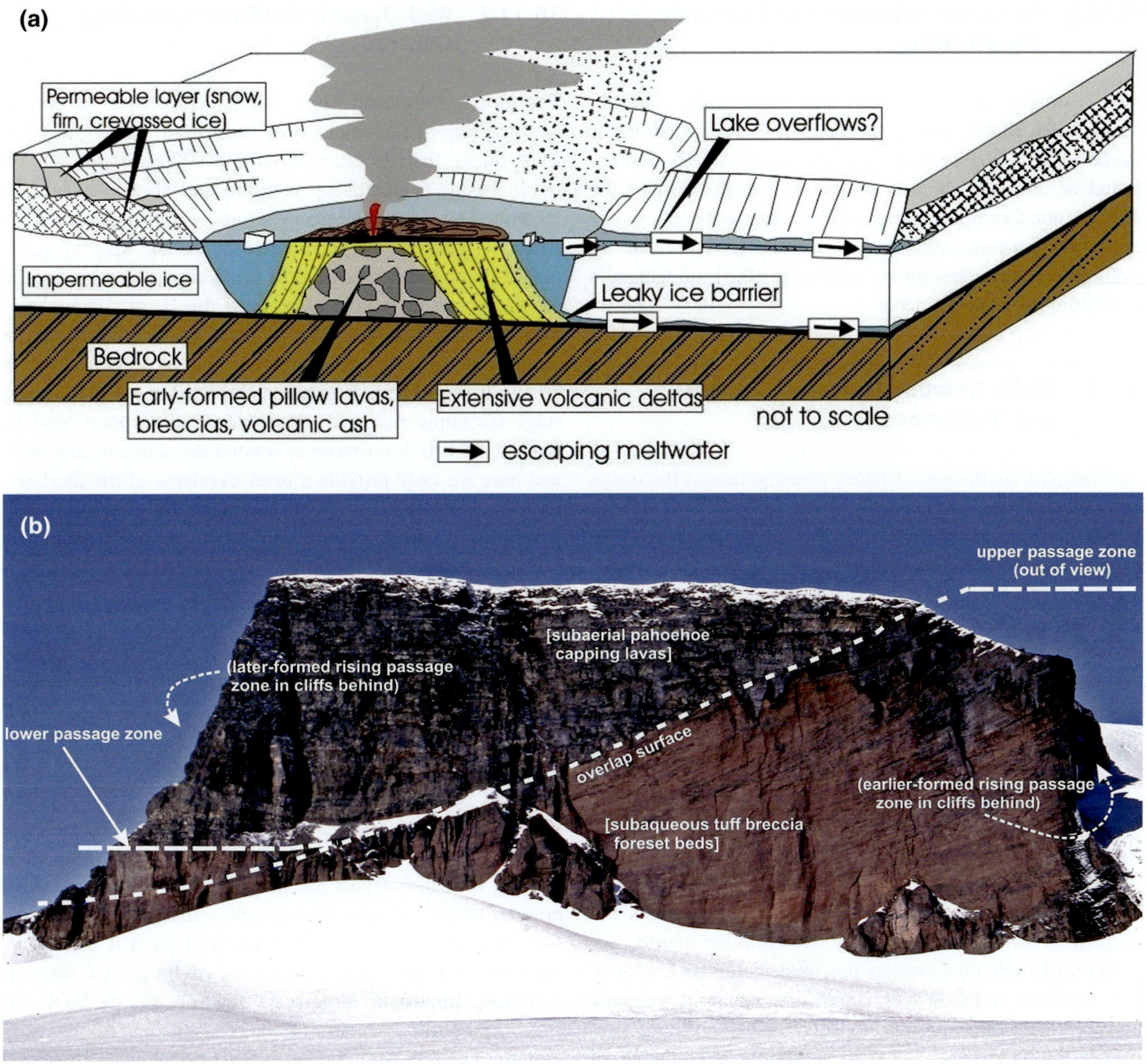

Fig. 16.92 **a** Generalised facies model for intra-subglacial basaltic eruption and tuya formation (From Smellie 2013). **b** Passage zone (i.e. fossil water level) geometry is one of the most diagnostic ways of identifying a glacial setting for a volcanic eruption; there are several types. This example shows a pahoehoe-fed delta in which the passage zone has dropped during catastrophic drainage during eruption. (Photo courtesy of John Smellie)

classification scheme can be further addressed by adding to each edifice type the description of the main lithofacies association, such as for example flat top lava-dominated tuya or pillow-dominated conical tuya. For comprehensive description and discussion of the many different types of glacio-volcanic landforms and facies models, for both mafic and silicic centres, see Smellie (2013, 2018), Russell et al. (2014) and Edwards et al. (2015).

16.10.5 Resource Potential and Economic Significance

Since subglacial volcanoes are essentially mostly ice-confined shield volcanoes, their resource potential is similar to shield volcanoes as discussed above. As demonstrated in Iceland, which is unique in being a part of the Mid-Atlantic Ocean spreading ridge system that is above sea level, the surrounding terrain is highly endowed in geothermal energy resources, supplying 100% of Iceland's non-automotive energy needs.

16.11 Mafic Oceanic Volcanic Centres and Volcanoes

Over the past six decades detailed investigation of the ocean floor has provided a remarkable wealth of information about submarine basaltic volcanic processes at constructive plate margins and in the ocean basins (Chap. 5). Since the 1950s this work has advanced from underwater photography and early seismic studies to deep-sea drilling, the use of submersibles, deep-tow instrumentation, side-scan sonar and high-resolution bathymetric and seismic surveying. Although fissure-fed basaltic lavas create most of the newly accreted upper oceanic crust, large-scale volcanic cones that are the products of central volcanism are also important components in both axial rift systems as well as oceanic plate interiors.

Submarine oceanic plate volcanism is important because it produces the largest volume of volcanic rocks on Earth, especially at active spreading centres. However, because of subduction, the vast majority of these volcanic rocks are destined to be recycled by gravity back into the mantle. So, in terms of their preservation potential in the rock record, they represent a much less significant group of volcanic products than their present extent suggests. Where preserved, the volcanics occur as stratigraphic elements of oceanic crustal slices. These are usually tectonically emplaced as *ophiolite* bodies in forearc, subduction-related accretionary prisms, or obducted onto continental margins, and as the basement to island arc complexes (Chap. 17).

Although much has been written on the stratigraphy of oceanic crust, only more recently has insight been gained into the nature of the volcanic centres and volcanoes that produce oceanic crustal volcanic rocks. For our discussion we can divide these into four main types, which broadly reflect submarine to emergent mafic oceanic volcanic activity:

- oceanic seafloor spreading centres,
- oceanic plateaus,
- seamounts (and other submarine lava piles), and
- surtseyan tuff cones.

16.11.1 Mid-Oceanic Seafloor Spreading Centres

The global mid-ocean ridge (MOR) system can be considered the largest magmatic system on Earth (see Plate 17.1). This linear feature extends over 60,000 km through the ocean basins at depths up to 5 km and above sea level in Iceland. The ridge crest itself is only about 5–30 km wide, which is relatively small compared to the size of ocean plates, that can be thousands of km across. Nonetheless, this produces greater than two-thirds of the Earth's annual volcanic output. As such, the topic is vast and includes an understanding of global magmatic and volcanic processes arising from magma generation, through ascent and storage, ridge spreading rate and geometries, and eruption onto the seafloor. A full description is beyond the scope of this book, and here we only provide a brief overview of the literature. For more substantive reviews, the older study by Macdonald (1982) is still important, and for more up-to-date analysis, see Macdonald (2001), Searle (2013), Soule (2015), Rubin (2014), and Michibayashi et al. (2019).

16.11.1.1 Mid-Oceanic Spreading Ridge Systems

Volcanic activity occurs predominantly on the axis of the mid-ocean seafloor spreading ridges (MOR), issuing from fissures within a median rift valley (Chap. 17, see Fig. 17.2). Rift valleys are dominated by ridge-parallel fractures and faults, producing a very rough terrain with horsts and grabens (Fig. 16.93). Within this graben zone, small volcanic mounds with a topography built of overlapping lava flow-fronts are prominent. Lava types that are erupted to form Layer 2 of oceanic crust include both pillow lavas as well as non-pillowed sheet lavas (Chap. 5). At the foot of fault scarps, talus fans of brecciated lava are common. Important large-scale features of all MOR are transform faults, which link large segments of the ridge system. These accommodate relative horizontal slip as new seafloor is being injected, and the older seafloor slowly migrates sideways away from the ridge crests. Hydrothermal vents and fissures are also very important features, and in particular, the direct observation of active sulphide-depositing vents, or black "smokers" has been a highlight of submarine exploration of MORs (Fig. 16.94). These were first discovered on the East Pacific Rise (e.g. Ballard et al. 1984) but have now been widely observed at many spreading ridges, including marginal sea or back-arc basin spreading ridges. Diffuse venting of hydrothermal fluids has been detected to continue through the carapace of newly erupted lava flows well after the eruption wanes. However, focused-flow hydrothermal activity (i.e. from well-defined chimneys and mounds) can persist at a site over multiple eruptions.

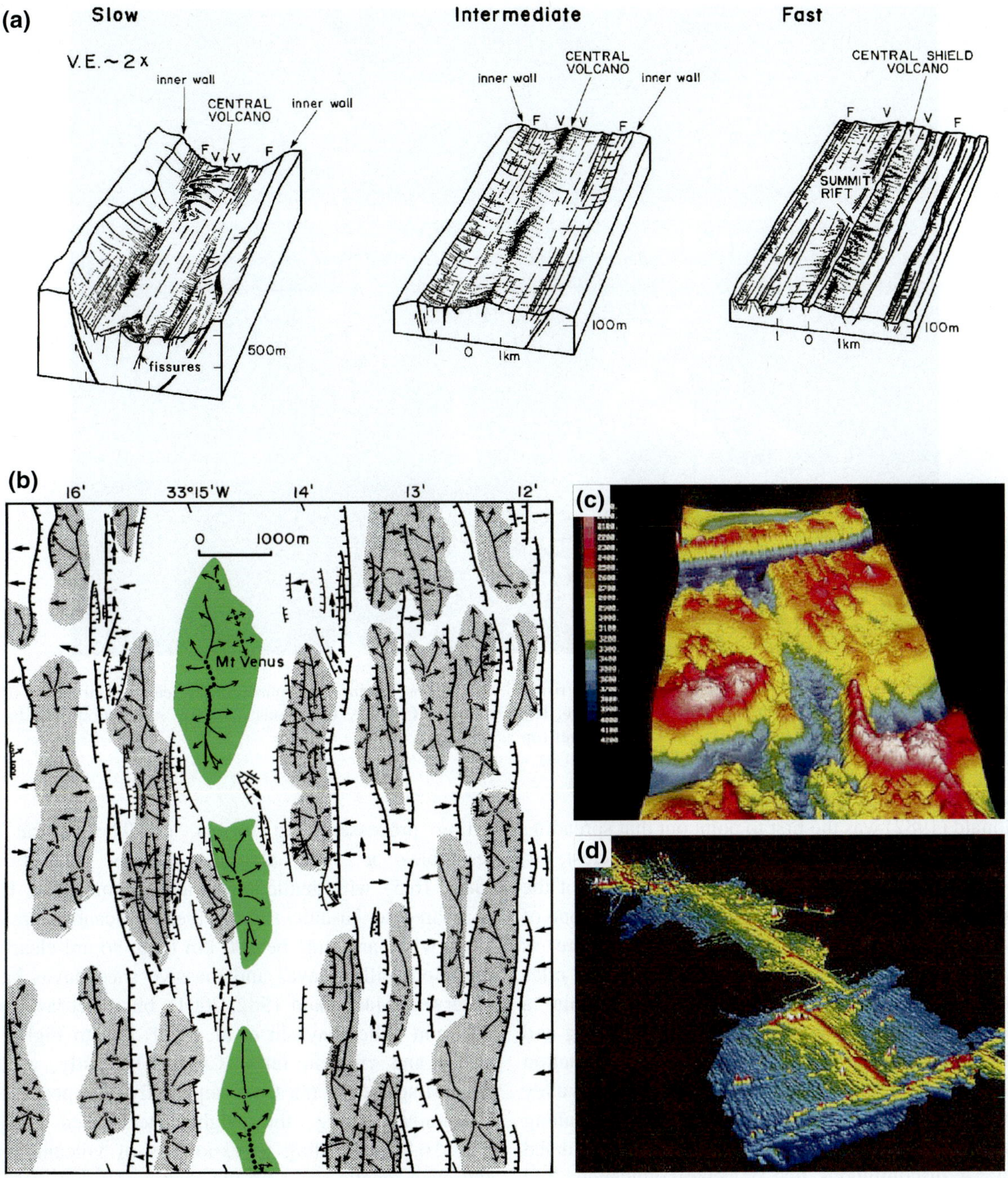

Fig. 16.93 Characteristics of mafic mid-oceanic spreading ridges. **a** Schematic cross-sections and 3D block diagrams through the neovolcanic zones of typical slow-, intermediate- and fast spreading ridges (From Cas and Wright 1987, After Macdonald et al. 1982). **b** Map of fault patterns (hachured lines) and volcanic mounds (stippled areas in the inner rift valley of part of the Mid-Atlantic Ridge. Green shading indicates youngest basaltic volcanism on present rift axis (ie. neovolcanic zone). Heavy fault lines indicate wall faults; light fault lines, inner-floor faults. Black dots (most recent) and open circles (older) indicate vents and crest lines of volcanoes; thinner arrows show flow lobes. Dip and plunge of fault blocks on walls shown with thick short arrows. Fault lines are placed near the base of fault scarps and throw on faults varies from <0.1 to 25 m. Mount Venus rises about 200 m from its base. Data are from bathymetric map, dive observations, bottom photography and deep-towed geophysical studies. (From Cas and Wright 1987, After Ballard and van Andel 1977). **c** Shaded relief map of the Mid-Atlantic Ridge. Here, the ridge is the plate boundary between the South American and African plates, which are moving apart at about 35 mm yr^{-1}. The axis of the ridge is marked by a 1–2 km deep rift valley, which is typical of most slow spreading ridges. The map reveals a major discontinuity called the Cox transform fault truncating the rift valley, and a smaller 12 km structural offset or jog within it. Colours indicated depths of from 1900 m (pink) to 4200 m (dark blue). **d** Shaded relief map of a 1000 km stretch of the East Pacific Rise extending from 8° to 17° N. Here, the East Pacific Rise is the boundary between the Pacific and Cocos plates, which are separating at a fast rate of 110 mm yr^{-1}. The map reveals two kinds of discontinuities: large offsets or transform faults about 100 km long, and smaller offsets about 10 km long, called overlapping spreading centres. Colours indicate depths from 2400 m (pink) to 3500 m (dark blue). (**c**, **d** from Macdonald 2001)

Fig. 16.94 Black smoker seafloor hydrothermal vent activity rising above sulphide mound at Turtle Pits, southern Mid-Atlantic Ridge. The Turtle Pits vent field has some of the highest recorded temperatures (up to 407 °C) for emanating hydrothermal fluids. (*Source* MARUM - Centre for Marine Environmental Research, University of Bremen; through Creative Commons Licence Attribution 4.0 International licence)

Macdonald (1982) was the first to point out that spreading rates not only affect the morphology of the ridge crest, but also the volcanoes from which the basalts of layer 2 of the oceanic crust are erupted. The *neovolcanic zone* (the zone of recent and ongoing volcanism at the spreading centre) is only 1–2 km wide (Fig. 16.93a). At slow spreading rates (10–40 mm yr^{-1}) the neovolcanic zone lies within a well-defined rift valley, where tectonic forces (faulting and plate separation) dominate over volcanic construction (Fig. 16.93a, c). This leads to the common axial valley profile, with volcanic accretion commonly localised along smaller ridges within the valley and fault block-dominated rift valley. A discontinuous line of central volcanoes and lava fields form, which are elongate parallel to the spreading axis (Fig. 16.93b), and typically have dimensions of 1–4 km in basal diameter and heights of about 250 m. They are predominantly constructed of pillow lava (Macdonald 1982, 2001). At intermediate spreading rates the medial rift valley is less well defined (Fig. 16.93a), and the volcanoes are more elongate along strike, appear to be fissure fed, and may be interrupted by en echelon offsets. Heights are only a maximum of 50 m and pahoehoe-like sheet flow lavas are more common. At high spreading rates (>90 mm yr^{-1}) there is no definable axial rift valley, and bathymetric relief is lower because volcanic construction dominates over tectonic rifting processes (Fig. 16.93a, c). The central volcano resembles a very elongate basaltic shield volcano (Sect. 16.5), with gentle slopes and a summit rift. However, compared to basaltic shields these volcanoes are only 1–2 km wide, and may be 100 km long, so are clearly fissure fed. Both pillow lavas and massive sheet lavas have been observed (Macdonald 1982, 2001), but sheet lavas are more common (Michibayashi et al. 2019) due to higher magma uprise and effusion rates (Chap. 5). Lastly, at the very slowest-spreading rates, volcanic activity appears to be discontinuous along the highly tectonised ridge axis (Fig. 16.93b), leading to exposures of volcanic, plutonic, and even mantle rocks on the seafloor (Rubin 2014).

By far the predominant eruption style represented at mid-ocean ridges is effusive lava flows. Lava flow thickness, run out, and surface morphology reflect variations in magma viscosity, effusion rate, local slope, topographic obstructions, and the sequence of emplacement events during individual eruptions (Rubin et al. 2012; Chap. 5, and references therein). Submarine lava flow morphologies are typically classified by the length scale of the quenched-crust units that collectively make up a lava flow, which themselves form as a function of flow rate and cooling rate. Common different types of lava deposits are known as pillow, lobate, and sheet lavas. Methods of detection and

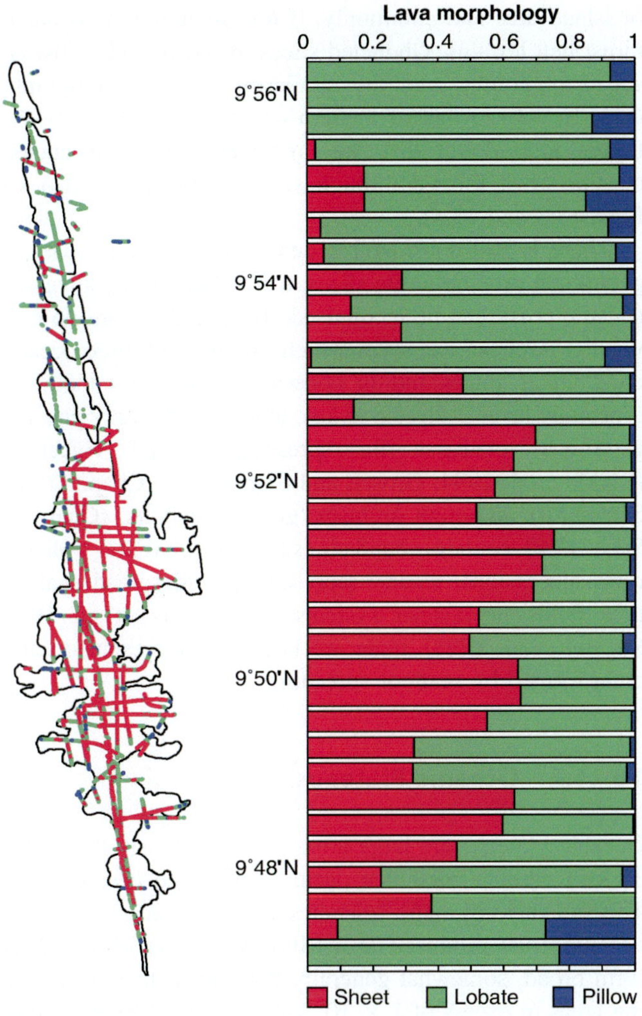

Fig. 16.95 Map of the 2005–2006 eruption of the East Pacific Rise from seafloor photography, submersible dives, and acoustic seafloor imaging. Flow morphology was characterised by over 10,000 seafloor photographs along towed camera tracks. The morphology along the eruptive fissure shows a high proportion of basaltic sheet flows (red) towards the centre of the fissure, and an increase in the proportion of lobate (green) and pillow flows (blue) towards its distal ends. Lower eruption rates at the ends of the fissures are consistent with decreased magma driving pressures that would have accompanied a cessation of fissure opening. (From Soule 2015; Modified from Fundis et al. 2010)

mapping of these flow types involve direct or remote visual observation from submersibles and towed camera systems (e.g. Fig. 16.95). High-resolution sonar data from autonomous underwater vehicles have recently been employed to generate detailed maps of lava flow fields, their extent and morphology over relatively large areas of seabed (e.g. Chadwick et al. 2013; Clague et al. 2013; McClinton et al. 2013). Volumetrically minor deposits of lapilli- to ash-sized fragments also occur throughout the depth range of observations in the deep sea. Although it is often assumed these have explosive origins, these may be generated and incorporated in several different ways including submarine fire fountaining, quench fragmentation or implosion of pillow lavas (Chaps. 4 and 5).

Similar to terrestrial volcanoes, the frequency of MOR volcanism reflects the rate of magma input to the volcano, conditions within the crustal magma reservoir and tectonic stress regime. Marine scientists currently lack the information to determine eruption frequency, size, and duration for all but a very small number of submarine mid-ocean ridge volcanoes (Rubin et al. 2012; Tolstoy and Tan 2019). Individual eruptions can last just days to perhaps a year. Many mid-ocean ridge volcanoes likely operate at a quasi-steady state in which eruption volume and repose (the time between eruptions) correlate positively as the result of a relatively constant magma flux. Using such a steady-state assumption and calculating magma supply rate from the spreading rate and an assumption of constant crustal thickness, the shortest average eruption repose interval (~ 10 yr) should occur at the fastest-spreading rates (e.g. the Southern East Pacific Rise; Rubin 2014). An average repose interval of roughly 1,000 yr should occur on slower-spreading ridges like the Mid-Atlantic Ridge.

The only sediment deposited is from pelagic sources. A pelagic sediment mantle only becomes prominent at the edges of the median rift valleys, and on the ridge flanks and abyssal plains as the lithosphere spreads with time away from the MORs. The pelagic sediment layer and the underlying basaltic crustal layer make up layers 1 and 2, respectively, of the oceanic crust. These should be closely associated with an underlying mafic sheeted dyke complex, and gabbroic and mafic cumulate igneous rocks (layer 3). Such crustal profiles coincide with ophiolite profiles, as discussed below and more extensively in Chap. 17.

16.11.1.2 Marginal Sea or Back-Arc Basin Spreading Centres

Marginal sea or back-arc basin seafloor spreading centres (Chap. 17, see Fig. 17.5) also produce a similar crustal profile, but in addition the basaltic crust could be underlain by remnant island arc or sialic crust, while overlain by a mixed volcaniclastic-carbonate clastic mass-flow sediment apron derived from associated island arcs, or even siliciclastic sediment derived from continental margins. It remains uncertain, however, what the nature of the volcanic centres is. Although axial spreading systems are usually topographically elevated ridge regions of high heat flow, the axial rift morphology that is relatively easily definable for mid-oceanic ridge spreading systems is often not so easily observed in marginal basins. Evidence from the Manus spreading centre in the Bismarck Sea (Papua New Guinea) is that these are characterised by neovolcanic zones comprising overlapping en echelon rifts (or pull-apart structures), which are underlain by thinned arc crust and linked by major strike-slip transform faults (Martinez and Taylor 1996;

Chap. 17). Within these rifts, volcanic activity, including evolved andesitic to dacitic magmas, and hydrothermal venting, are focused, forming en echelon volcanic ridges and lava fields with sulphide chimneys (Chap. 18). In addition, magnetic anomaly patterns are also difficult to define clearly, suggesting that the seafloor spreading process may be more diffuse or irregular process.

In the Tyrrhenian back-arc basin, oceanic seafloor spreading started in early Pleistocene and led to the formation, between ca. 1 Ma and the Present, of the Marsili volcanic complex, which rises from −3500 m below sea level for a total height of 2700 m (Fig. 16.84). The Marsili is a 60 km long, NNE-SSW elongated volcanic ridge with an average width of 30 km (Pensa et al. 2019). The edifice is made of a lower more voluminous succession of medium to low-K basalts overlain by high-K calc-alkaline products during late Pleistocene and Holocene, currently hosting an active hydrothermal system (Ventura et al. 2013).

16.11.1.3 Facies Model for Oceanic Crust and Ophiolites

Seafloor spreading ridge systems are highly faulted and fractured parallel to the trend of the ridge axes due to extensional normal faulting of the newly formed crust as spreading occurs. As a result, the submarine eruptions occur from fissure fracture vents producing seafloor lavas fed by subjacent vertical, parallel sheeted dykes. These centres thus produce the ocean crustal layer 2 pillow and massive basalt lavas and associated hyaloclastite breccias, and the subjacent layer 3 sheeted dolerite dyke complex and deeper plutonic complex composed of gabbroic and mafic–ultramafic cumulate rocks. As the newly formed crust spreads away for the spreading ridge axis, it becomes mantled with pelagic and hemipelagic sediments, which constitute Layer 1 in the classic oceanic crust profile and found in many ophiolites (Fig. 16.96a; Chap. 17).

In spreading ridge systems, fissure vents are generally located in the axial rifts (Fig. 16.93a–d) and extend for 10 s to 100 s of km and more along the trend of the ridge (Fig. 16.93d). The fissure vents lead to the eruption of overlapping lava fields (variously sheet, lobate and pillow; Fig. 16.95), and to the formation of very elongate shield-like volcanoes or point source lava mounds and cones along the fissure (Fig. 16.93b). The layer of pelagic sediment (layer 1), thickens away from the spreading axis with time.

The ridge-parallel fractures and fault systems, which form in the environs of their MORs due to tectono-isostatic adjustments as the lithosphere ruptures and spreads away from the ridge crest, will dismember many original volcanic centres. Most oceanic crust and lithosphere is destined to be subducted later in its geological life cycle. Those segments that are preserved in the geological rock record have been accreted or obducted tectonically along continental margins or island arcs and commonly, if not ubiquitously, occur in thrust belt terrains. Obducted slices of old oceanic crust are called ophiolites, and the similarity of their stratigraphy (Fig. 16.96b) to that of modern oceanic crust (Fig. 16.96a) leaves little doubt that they represent pieces of ancient oceanic crust formed through seafloor spreading (also see Figs. 17.3 and 17.4).

There is still considerable debate as to how far back in geological time ophiolite slices, truly representing slivers of oceanic crust, can be recognised. To date there are no convincing Archean examples which preserve all three oceanic crust layers intact and in correct juxtaposition, including especially layer 3 sheeted dykes, which are the only physical geological evidence for spreading related volcanism (Bickle et al. 1994; Grosch and Slama 2017; Chap. 17, see Sect. 17.8). Although Archean "greenstone belts" (Chap. 17, Sect. 17.8) contain abundant, submarine basaltic volcanic rock successions, clearly representing widespread seafloor lava fields or plateau systems (Desrochers et al. 1993; Puchtel et al. 1998), none are known to represent oceanic crust formed by seafloor spreading.

16.11.2 Oceanic Plateaus

Oceanic plateaus are large areas of anomalously thick (10–35 km) oceanic crust that occur in the deep ocean basins. This thickness of mafic volcanics is well in excess of normal oceanic crust produced at a MOR spreading centre. They form broad, somewhat generally flat-topped features covering areas in excess of 1×10^6 km^2, rising 2,000 m or more above the surrounding seafloor. They represent the oceanic equivalent of continental flood basalts and are one of the main categories of large mafic igneous provinces (oLIP; Chaps. 5 and 17). It is now generally recognised that oceanic plateaus were formed as a result of mantle plume activity, marking the position of initial plume-head stage of hot-spot development (e.g. Mann and Taira 2004; Fitton et al. 2004a, b; Kerr 2014). Hotspot tracks (e.g. seamount chain) represent continuing magmatism associated with the remaining plume conduit or "tail".

The plume-head model predicts that plateaus were formed in massive eruptive outpourings of basaltic magma lasting only a few million years or less. Nearly all examples found in the today's ocean basins were formed in the Cretaceous (see Fig. 5.54). Such eruptions are thought to have had major climatic, oceanographic and biospheric impacts, including generation of some of the planet's most prolific petroleum source rocks (e.g. Kerr 2005). Today, thick blankets (100 s of m to >1 km) of pelagic sediments effectively cover them; however, in detail their surfaces may show complex seafloor geometries with intricate systems of channels, atolls and seamounts.

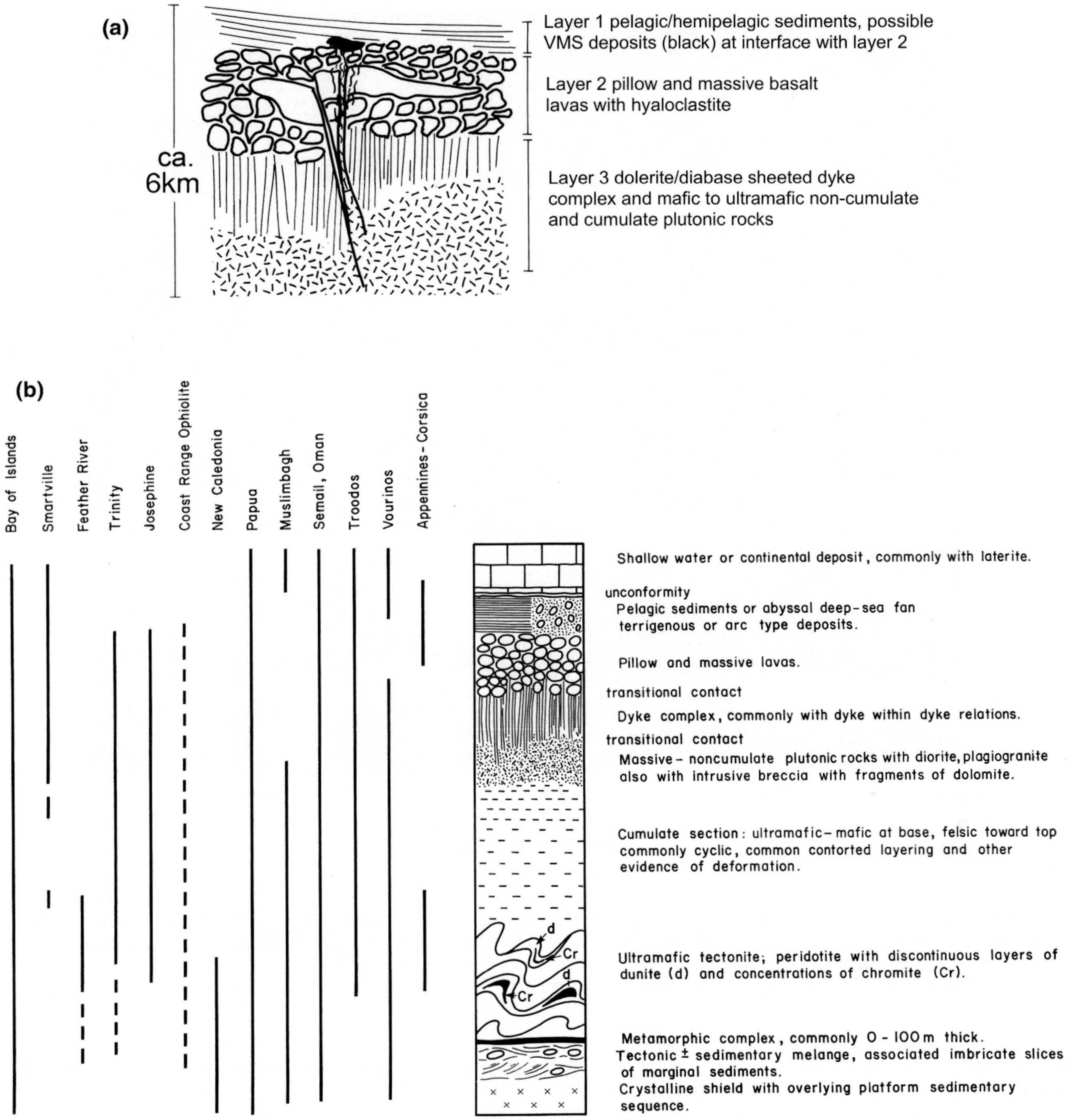

Fig. 16.96 Stratigraphy of modern ocean crust and ophiolites. **a** Schematic representation of the stratigraphy of ocean crust generated at mid-oceanic spreading ridges based on many Deep-Sea Drilling Programme (DSDP) and Ocean Drilling Programme (ODP) drill holes (From Cas 1992). **b** Representation of the stratigraphy of many ophiolites (obducted slices of ancient oceanic crust) around the world, with the base of the ophiolite being a highly tectonised thrust contact with continental margin or island arc crust (From Cas and Wright 1987, After Moores 1982)

16.11.2.1 Size, Age and Characteristics of Modern Oceanic Plateau Examples

The Ontong Java Plateau (OJP) near the Solomon Islands in the western Pacific is the largest igneous province in the world (Fig. 16.97). Erupted at ca. 122 Ma, the OJP has a surface area of 2.0×10^6 km^2 (roughly the size of Western Europe), and a volume as high as 6×10^7 km^3 (Fitton et al. 2004a, b). If, as currently understood, most of the OJP was formed essentially during a single volcanic episode lasting only a few million years, then the magma production rate

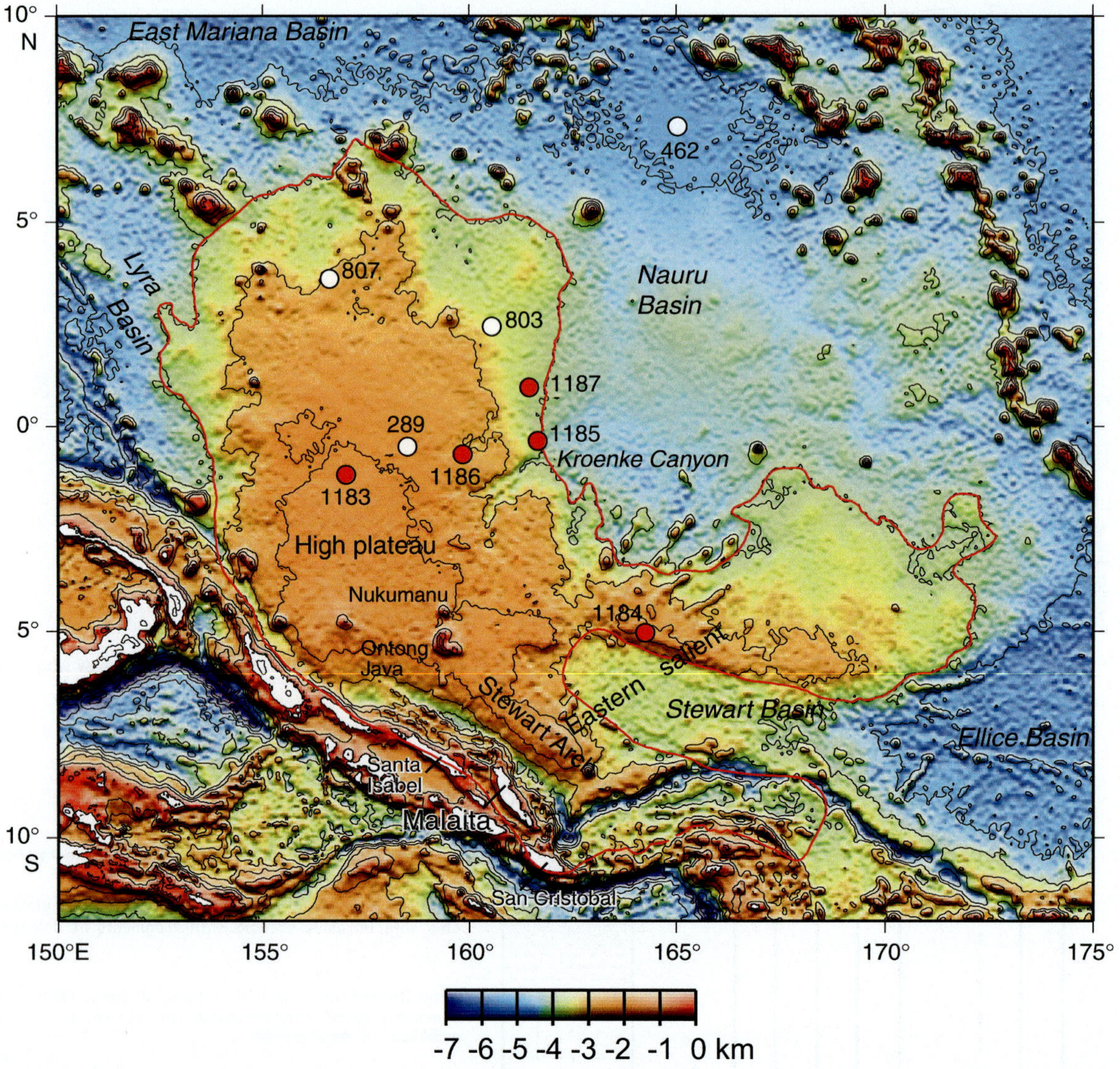

Fig. 16.97 Predicted bathymetry of the Ontong Java Plateau and surrounding areas showing the location of DSDP and ODP igneous basement drill sites. Red circles = ODP Leg 192 drill sites, white circles = pre-Leg 192 drill sites. The edge of the oceanic plateau is defined by the −4000 m contour except in the southeastern part, where it has been uplifted through collision with the Solomon Islands volcanic arc. (From Fitton et al. 2004b)

required for its formation would have rivalled that of the entire global mid-ocean ridge system at the time. The OJP may therefore represent the largest igneous event of the last 200 Ma. The OJP is also geologically significant because it is currently converging on an island arc (Solomon Islands; Fig. 16.97) and is a key area for understanding the fate of large, thick and therefore far more buoyant plateaus on reaching subduction zones. Oceanic plateaus are therefore more difficult to subduct, which plays an important role in "*arc polarity reversal*" (Mann and Taira 2004; Chap. 17). This is a process in which subduction below an island arc ceases, the arc is accreted to the formerly consuming plate (here, the Pacific plate) and subduction reverses direction to consume the formerly overriding plate (the Australian plate or OJP).

The next largest oceanic plateaus in order of decreasing volume are the Kerguelen Plateau (KP) of the Indian Ocean and the Caribbean-Colombian Plateau (CCP). A few plateaus like the KP appear to have had sufficient magma flux to build above sea level although many, like the OJP, appear to have remained submerged well below sea level throughout most of their development. The CCP represents the only example where arc polarity reversal, intrusion of arc-related plutons and tectonic imbrication around its margins has exposed what may be the deeper levels of the oceanic plateau. Of all the oceanic plateaus, the KP has the widest range in ages and magma compositions reflecting formation in a variety of tectonic settings over a lifespan beginning with the break-up of Gondwanaland (i.e. separation of India, Australia and Antarctica). Within the young widening ocean, the first massive pulse of plume magmatism created the southern KP (118–110 Ma), which was followed by several later plume melting episodes (Elan Bank, central KP, Broken Ridge, Ninety-east Ridge; see Fig. 5.54) persisting to about 30 Ma (Wallace et al. 2002; Kerr 2014). Volcanism in an oceanic intraplate setting continues to the present-day (Kerguelen Archipelago, and Heard and McDonald Islands). The Elan Bank is noteworthy because evolved felsic (trachytic to rhyolitic) lavas have been drilled.

As a result of their deeply submerged location, enormous size and thick sedimentary cover, the internal volcanic structure and facies architecture of today's oceanic plateaus remain poorly understood, especially the form of their eruptive centres. Nevertheless, the basaltic basement of several ocean plateaus has been sampled in the course of Deep-Sea Drilling Project (DSDP) and Ocean Drilling Programme (ODP) legs, in particular the OJP (ODP Leg 192; Fitton et al. 2004a, b; Figs. 16.97 and 16.98). Collision of the OJP with the old Solomon arc also resulted in obduction of the OJP's southeastern margin to create some on-land exposures of basaltic basement. The thickest exposures are found on the remote island of Malaita which reveal a monotonous succession of Early Cretaceous tholeiitic pillow basalt, sheet flows, and sills (the Malaita Volcanic Group) 3–4 km thick. Rare and very thin interbeds composed of laminated pelagic chert or limestone suggest high eruption frequency and emplacement into deep water. Similar pillowed and massive flows were cored on the submerged plateau (Fig. 16.98), while multichannel seismic reflection data shows its uppermost igneous crust is characterised by intra-basaltic basement reflectors (Inoue et al. 2008). These are interpreted to comprise many lava flows, including hyaloclastites, several tens of m thick erupted at varying effusion rates. High effusion rates produced thicker massive sheet flows perhaps extending several hundred km from their vent or fissure sources, and low effusion rates produced thinner pillow flows of lesser extent (Fig. 16.99; Chap. 5). While the data did not image any vents or fissures, or larger volcanic structures, they did reveal multiple volcanic events erupted at both high and low effusion rates. Inoue et al. (2008) concluded significant vent and/or fissure eruptions during the main eruptive phase emanated probably from the crest of the OJP.

More recently, integrated drilling and seismic investigation (IODP Expedition 324) has identified an immense basaltic shield volcano as part of the Shatsky Rise, in the northwestern Pacific Ocean (Sager et al. 2011, 2013; see Fig. 5.54). The Tamu Massif appears to form the oldest and largest edifice (with an average basal diameter >500 km; Fig. 16.100) within the oceanic plateau and is a single supervolcano larger than Hawai'i which is the Earth's largest active volcano, and comparable in size with Olympus Mons on Mars, often considered the largest volcano in the solar system. The volcano is believed to be constructed predominantly of massive sheet lava flows (up to >20 m thick) that emanated from the volcano centre to form a broad, shield-like shape (Figs. 16.98 and 16.100). Its anomalously low slopes (<1–1.5°) are interpreted to result from very high effusion rates of the erupting lavas. Seismic lines crossing the summit of Tamu Massif show intra-basement reflectors extend downhill from the summit apex, implying that this

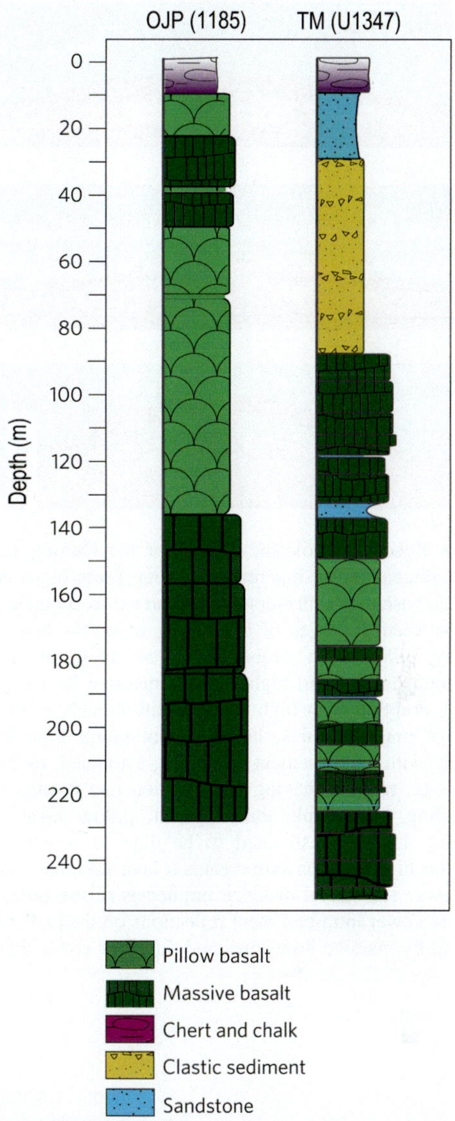

Fig. 16.98 Generalised logs of drill core from the top igneous section of the Ontong Java Plateau (left), and comparison with the Tamu Massif, Shatsky Rise (right). Ontong Java Plateau site 1185 is from ODP Leg 192, and Tamu Massif site U1347 is from IODP Expedition 324 (see Figs. 16.97 and 16.100 for locations). (Modified from Sager et al. 2013)

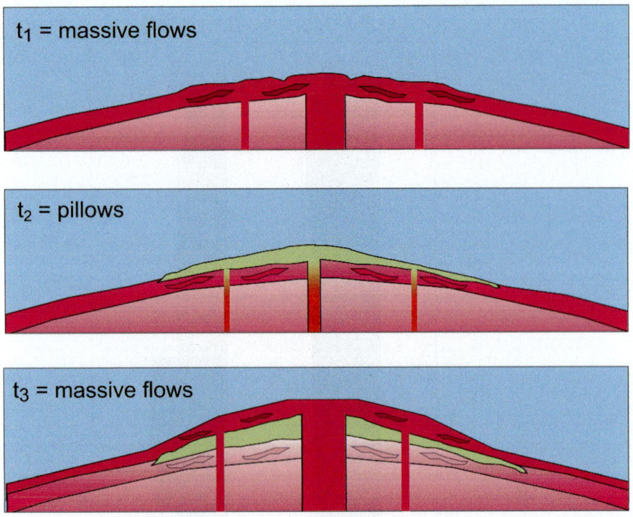

Fig. 16.99 Schematic volcanic model for the Ontong Java Plateau based on multichannel seismic reflection data (From Inoue et al. 2008). Intra-basaltic basement reflectors are interpreted as acoustic impedance contrasts between packages of alternating relatively low-density and low-velocity pillow lava sequences 30 m thick, alternating with relatively high density and high-velocity massive lava sequences. At time 1 (t1), eruption with high effusion rate creates a lava flow unit consisting of multiple, thick, laterally continuous, tube fed massive flows, each with emplacement durations estimated to be years to decades. At t2, eruption having low effusion rate creates a lava flow unit consisting of multiple thin, layered, pillow lavas, each with emplacement durations estimated to be days to months. At t3, an eruption with high effusion rate creates a lava flow unit consisting of massive flows. The lateral distance implied is of the order of several hundred km. Fewer intra-basement reflections on the OJP's flanks may be explained by massive flow cores and thin flow crusts that are below resolution

feature is the flow source (Fig. 16.100). Along the apex at least three depressions, ranging up to 170 m in depth and 5–15 km across and 100 km apart are identified. These may represent collapse calderas or pit craters above a rift zone where shallow magma chambers are drained, as observed on large subaerial Hawai'ian shield volcanoes (Sect. 16.5.1). The seismic data also reveal a number of tall (up to a km high) secondary peaks, which probably are late stage volcanic features, analogous to parasitic cones observed on shield volcanoes (see Fig. 16.30). Tamu Massif may be the largest single, central volcano on Earth and other oceanic plateaus having similar morphologies possibly formed in the same way. Although the volcanic and vent structures of the larger oceanic plateaus (OJP, KP and CCP) are less well constrained, they may have evolved as similar composite or multiple shield-like features erupted from many sources.

16.11.2.2 Some Cryptic Pyroclastic Occurrences and Remaining Controversies

One of the most unexpected results of ODP Leg 192 was the intersection of a thick succession (338 m) of basaltic volcaniclastic rocks at Site 1184, in the eastern lobe of the OJP (Fitton et al. 2004a, b; Fig. 16.97). This is described as a succession predominantly of massive coarse lithic vitric tuffs, lapilli tuffs and lapillistones, with abundant accretionary and armoured lapilli (Fig. 16.101a–c; Mahoney et al. 2001; Thordarson 2004). Based on sedimentological and geochemical characteristics these were divided into six subunits, which where complete, are between 60 and 80 m thick. Accretionary lapilli occur throughout the succession

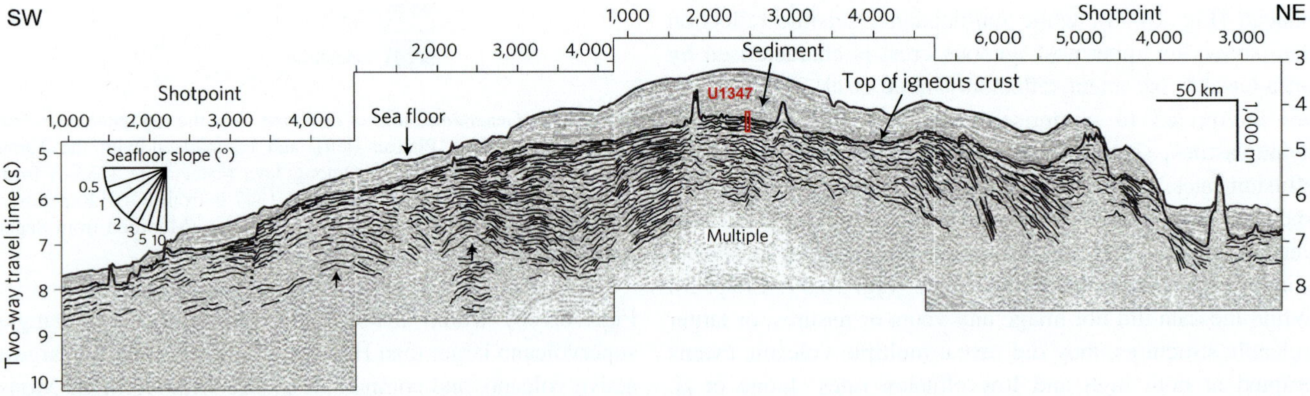

Fig. 16.100 Interpreted multichannel seismic reflection line along the main, NE-SW axis of the Tamu Massif, which forms southern end of the Shatsky Rise oceanic plateau (From Sager et al. 2013). Dark lines highlight the intra-basaltic basement reflectors. In this case they are mainly interpreted thick, dense, massive flows, thick packages (groups) of lava flows with similar properties, as well as the contrast of lava flows with occasional thick layers of volcaniclastics (hyaloclastites) or interflow sediments. Location of IODP drill site U1347 is shown (see Fig. 16.97)

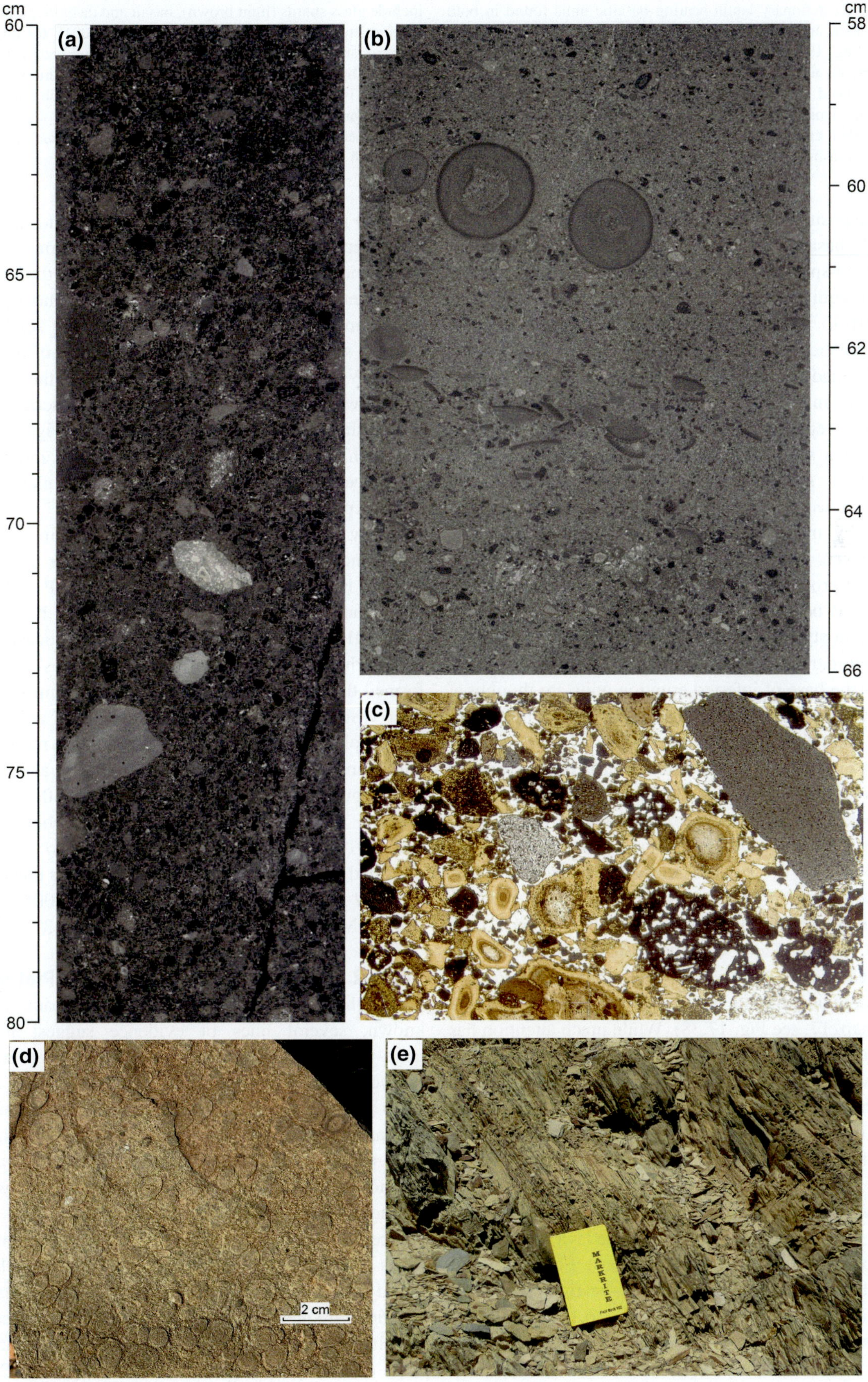
cm
60
65
70
75
80
(a)
(b)
cm
58
60
62
64
66
(c)
(d)
2 cm
(e)
MARKRITE

◀ **Fig. 16.101** Accretionary lapilli-bearing basaltic units found in both the Ontong Java and Caribbean-Colombia Plateaus, correspondingly from ODP Leg 192, drill site 1184 (**a–c**), and the Aruba Lava Formation (**d**, **e**). **a** Lapilli tuff comprising coarse (≤ 2 cm) angular, poorly vesiculated clasts; the large pale clast (centre) is dolerite. **b** Accretionary lapilli, both whole (round) and broken fragments, in diffusely bedded grey lithic vitric tuff. **c** Photomicrograph showing wide range of ash-sized basaltic lithic and vitric fragments. These include glass shards (light brown), basalt and tachylite (black); the glass is completely altered. (From Mahoney et al. 2001). **d** Concentration of accretionary lapilli from a 30 cm thick massive bed, which in turn is overlain by 15 cm of diffusely-wavy laminated, fine-grained basaltic debris. **e** Thinly bedded, basaltic ash turbidites which contain dispersed accretionary lapilli; these occur in a package some 20 m thick sandwiched between coarse-grained, basaltic volcaniclastic debris flow units (3–4 m thick)

found in depositional packages <0.5–20 m thick. Most of the juvenile basaltic component consists of blocky, subangular, non-to-sparsely vesicular sideromelane fragments (or altered equivalents) with larger basaltic lithic clasts (Fig. 16.101c). This implies the eruptions were phreatomagmatic, with accretionary lapilli indicating that the eruptions formed subaerial or atmospheric eruption columns.

Shipboard investigators determined the volcaniclastic deposits were redeposited, perhaps by turbidity currents, on the flanks of a surtseyan volcano at, or below wave base (Sect. 16.11.4, see Fig. 16.108). In contrast, Thordarson (2004) concluded they were primary phreatomagmatic tuffs dominated by medial fallout and subordinate base surge layers, that were deposited on land. They were interpreted to represent the products of between 6 to 10 phreatomagmatic eruptions, and because of their relative thickness, perhaps comparable in style to the 1783–84 Laki fissure eruption in Iceland (Thordarson and Self 1993; Chap. 5). Wood fragments near the base of inferred eruption units are thought at represent the remains of standing trees buried by tephra fallout. At least four subunits have a distinct glass composition with no intermixing confirming each is the result of one eruptive phase, and that the voluminous subaerially erupted rocks belong to the same magmatic event as that responsible for the construction of the main plateau.

Around the circum-Caribbean region, the Caribbean-Colombian Plateau (CCP) has been partially obducted making large segments available for study (e.g. Kerr et al. 1998). The CCP comprises thickened (up to 20 km) oceanic crust believed to have formed in the Pacific, approximately over the site of the Galapagos hot spot around 90 Ma (Chap. 17, see Fig. 17.38). While much attention has been given to geochemical studies and petrological modelling, few studies applying modern volcanological facies analysis have occurred. Most successions are simply described in terms of proportions of basaltic pillowed and massive flows, and diabase (dolerite) sills. Intervals of komatiites (Gorgona Island, Colombia; Kerr et al. 1996a) and more abundant picrites (e.g. Curaçao; Kerr et al. 1996b) have been located in the lowest parts of the lava successions. A facies-based study of the Aruba Lava Formation has recently described accretionary lapilli-bearing sequences comparable in thickness to the OJP (Fig. 16.101d, e). These occur in repeated packages from 5 to ≥ 10 m thick, within vertically continuous sequences of basaltic detritus up to >100 m thick, but in this case, appear to have all been deposited by water-supported density currents. This and other evidence indicate they are more likely the syn-eruptive, resedimented products of surtseyan volcanoes formed by phreatomagmatic explosive activity on the shallow marine top of a seamount, then deposited at depths of probably >2–3 km (Wright et al. in prep). Together with the identification of pillow fragment breccias previously described as fluvial and/or shallow marine conglomerates (see Figs. 5.55e–g and 5.56), this has important implications for interpreting the paleotectonic significance of similar volcaniclastic successions exposed elsewhere in the CCP.

Many aspects of oceanic plateaus remain poorly understood and controversial. For example, their uplift and subsidence histories as represented by the OJP and elsewhere (Mann and Taira 2004; Roberge et al. 2005; Korenaga 2005; Kerr 2014). Secondly, the paradigm that oceanic plateaus and perhaps more generally large igneous provinces are formed through more or less instantaneous magmatic events over vast areas within a few million years at the initiation mantle plumes. Clearly the KP has a much more protracted and complex volcanic history. The OJP also includes a minor but well-established pulse some 30 million years younger than the main plume event at about 90 Ma (e.g. Kerr 2014). Furthermore, in the CCP, $^{40}Ar/^{39}Ar$ ages have extended the role of basaltic volcanism by as much as 70 Ma (139–69 Ma) in Costa Rica (Hoernle et al. 2004), and 30 Ma (92–63 Ma) in Curaçao (Loewen et al. 2013).

16.11.2.3 Facies Model for Oceanic Plateaus

Although the vent systems for oceanic plateaus are less well known, several lines of evidence suggest fissure fed, large-volume eruptions are likely to have been important in their construction. As we have described earlier, ocean plateaus are characterised by very thick, km-scale successions of many lava flows, including hyaloclastite breccias, several 10 s of m thick erupted at varying effusion rates. High effusion rates produced thicker massive sheet flows perhaps extending several 100 s of km from their vent or fissure sources, and low effusion rates produced thinner pillow flows of lesser extent. Rare and very thin interbeds composed of laminated pelagic chert or limestone suggest high eruption frequency and emplacement into deep water. A schematic representation of the evolution and facies of large oceanic plateau is illustrated in Fig. 16.102 (also see

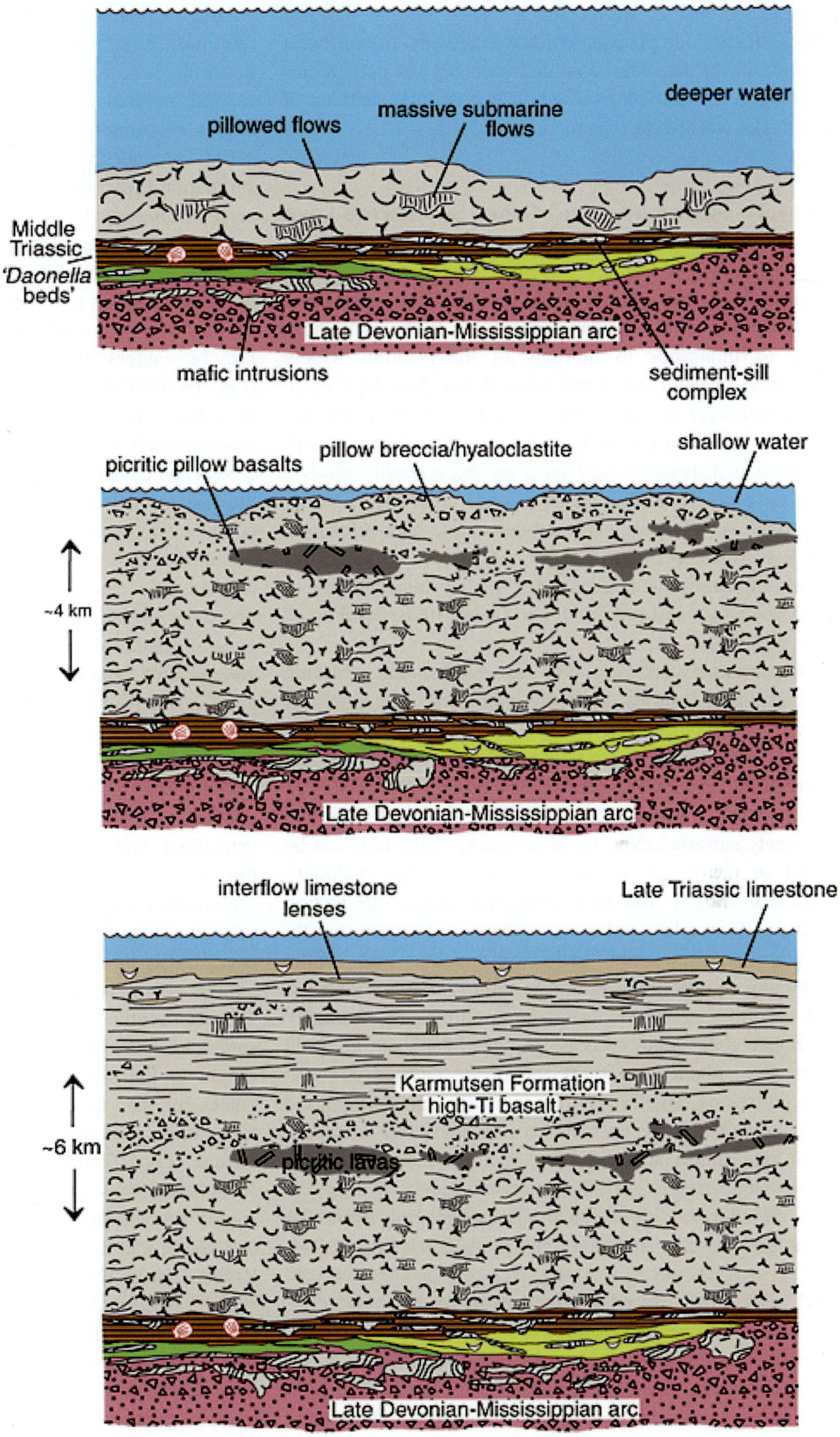

Fig. 16.102 Evolution of an oceanic plateau showing generalised facies architecture. This is based on the reconstruction of the Triassic Wrangellia oceanic plateau, which extends as thrust-dissected slices some 2300 km in the Pacific Northwest, from Alaska to Vancouver Island, British Columbia. This image represents Vancouver Island only. Volcanic activity was initiated with submarine flows emplaced onto pelagic sediment overlying submerged mid-Paleozoic island arc crust. The plateau was built during a single phase of tholeiitic volcanism ca. 230–225 Ma, and possibly within as few as 2 Ma. Wrangellia probably was accreted to North America in the late Jurassic-early Cretaceous. (From Greene et al. 2010)

Figs. 16.99 and 16.100). This is based on the Triassic Wrangellia oceanic plateau, which was accreted to northwest America in the late Mesozoic and probably (in part) represents some of the best exposed, if not unique, sections of such rocks worldwide (Greene et al. 2010).

16.11.3 Seamounts (and Other Submarine Lava Piles)

Seamounts, perhaps surprisingly, represent the most common physiographic volcanic features on Earth, with hundreds of thousands to millions of them found widely distributed over all the ocean basin floors (Staudigel et al. 2010; Staudigel and Koppers 2015; Fig. 16.103). Despite this they remain a poorly known group of submarine volcanoes which, because they have not been readily observable, have received little attention until relatively recently. Nevertheless, advances with submersibles, high-resolution side-scan sonar and narrow-beam bathymetric surveys have greatly increased our understanding of the morphologies of these volcanoes and provide a good basic understanding of their geology.

Seamounts, with very few exceptions, are volcanic aggradational features built by extrusive and intrusive activity, and volcano-tectonic processes such as crater and caldera formation or sector collapse. Larger seamounts may temporarily emerge as volcanic islands, and be subject to surface erosion, provide the foundation for coral reefs and ultimately subside below sea level again. Seamounts can be found in nearly all seafloor tectonic settings, including mid-ocean ridges and their transform faults, subduction zones, and intraplate and oceanic plateaus. They are mostly basaltic in composition, including tholeiitic and alkalic seamounts at mid-ocean ridges or within intraplate settings, and calc-alkaline seamounts in arc settings. Occasionally seamount rocks and successions are found on land, in accreted or obducted terrains and in the uplifted cores of oceanic islands. The Tyrrhenian sea is a young poly-phased Pliocene to Holocene back-arc basin which hosts in a very small area almost all kinds of seamount types, from small monogenetic centres, to large stratovolcanoes, to huge fissural complexes, with chemical affinities ranging from OIB to tholeiitic to calc-alkaline (Pensa et al. 2019; Fig. 16.84).

16.11.3.1 Results of Modern Marine Investigations

Marine investigations in the early 1980s (e.g. Searle 1983, 1984; Batiza and Vanko 1983) added considerable insight to the nature of submarine seamounts and guyots (a flat-topped seamount) documented by earlier workers such as Menard (1964, 1969). These central volcanoes are highly variable in size and shape (Fig. 16.103a–d). Seamounts that are near circular in plan view have basal diameters ranging from the minimum limits of resolution (<1 km) to 25 km, although generally less than 10 km, while their heights are usually several hundred m to <2 km but may exceed 5 km. Slope angles of seamounts tend to be steeper than those of subaerial volcanoes, partly due to the more rapid cooling of submarine lavas, and partly due to the common presence of steep, sub-vertical fault scarps. They have dominantly concave-up or sometimes convex-up slopes, which at their maximum are inclined at up to 40°, and on average about 20°. The summits are often flat-topped with diameters up to 10 km, and frequently these have well-defined craters <1 km, or caldera collapse structures >1–6 km in diameter (Fig. 16.103b, d; Staudigel et al. 2010; Staudigel and Koppers 2015). Given that such volcanoes are (at these depths) non-explosive in eruptive style (Chaps. 4 and 5), summit craters and calderas reflect post-eruption magma withdrawal, drawdown and subsidence above a magma chamber or conduit following eruptions of lavas from flank vents. Although seamounts are probably mostly relatively small monogenetic centres (erupted over timescales of years to decades), some may evolve into very large, complex polygenetic oceanic shield centres, which can grow upwards and emerge above sea level, such as Hawai'i, and the Galapagos and Canary Islands. In these cases, the subaerial part of the volcano has evolved into a major shield volcano (Sect. 16.5; see also Schmidt and Schmincke 2002).

Variations in shape and form stem from variations in conduit geometry (Batiza and Vanko 1983), magma supply (Staudigel and Koppers 2015), and sediment (layer 1) thickness above oceanic crust (Medialdea et al. 2017). As seamounts grow, they may develop rift zones, mostly two in opposing directions, giving them an elongate shape, but occasionally three or four, resulting in a star-pattern defining a radial geometry (Fig. 16.103c). However, seamounts at hot spots or oceanic spreading ridges with rather prolific magma supply systems often form on the flanks of a previous seamount in the same seamount trail, buttressing against or coalescing with the pre-existing volcanic edifice. This results in complex composite seamounts with multiple eruption centres (Fig. 16.103b). Seamounts may form conspicuous seamount chains extending over thousands of km across significant portions of an ocean basin (e.g. the Hawai'ian-Emperor and Louisville seamount chains, and Ninety-East Ridge; Staudigel and Koppers 2015; Chap. 17, see Fig. 17.29). Many shorter seamount chains occur in all the major ocean basins, but none of them is as voluminous and most tend to be discontinuous. Discontinuous chains may have extended sections missing, or still to be identified due to their small size or complex interference with other chains. Alternatively, seamounts may be solitary features or form groups or clusters, in near-ridge settings or in intraplate settings.

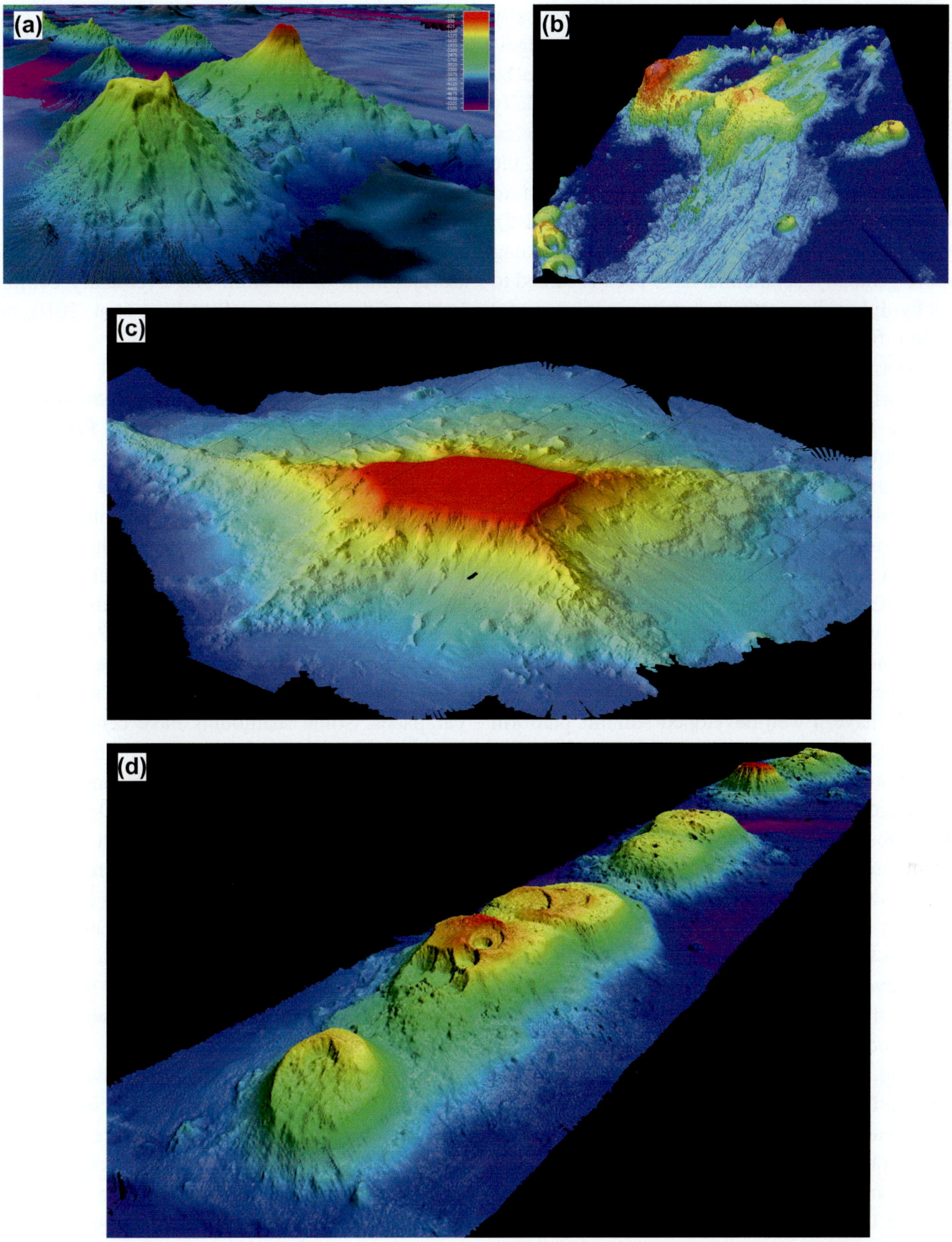

Fig. 16.103 Examples of oceanic seamounts. **a** Pao Pao seamount (background) and an unnamed guyot (foreground) is one example of nearby seamounts with very different geomorphology, Tokelau seamount chain, southwest Pacific. Pao Pao is 4776 m high, coming to a very sharp peak at around −300 m and shows steep flanks, while the unnamed feature has a distinct flat top. **b** Axial seamount located on the Juan de Fuca Ridge, northeast Pacific Ocean. This seamount features an unusual "rectangular-shaped" caldera, and its flanks are pockmarked by fissures, vents, sheet flows, and pit craters. Its geology is further complicated by its intersection with several smaller seamounts surrounding it. Note: perspective image shows exaggerated bathymetry, which highlights the ridge-parallel faults in the foreground and top right. **c** Ladd seamount is an irregular, star-shaped guyot, controlled by at least three intersecting rift axes, Hawai'ian-Emperor seamount chain. The flat top appears to have been a lava lake. **d** Flat topped seamounts (basal diameters of ≤ 10 km) forming the President Jackson seamount chain, west of the Gorda Ridge, northeast Pacific Ocean. The summits of all but the second from the northwestern end (top right) are flat, and most are modified by nested or cross-cutting calderas and smaller pit craters, some of which scallop the flanks of the volcanoes. At $\sim$1500–1800 m below sea level, these are unrelated to explosive activity, but magma withdrawal, perhaps associated with a flank fissure lava eruption. Several of the volcanoes have remnants of centrally located summit shields, suggesting that their flat tops formed by the infilling and overflowing of earlier large calderas (Clague et al. 2000b). (*Source* Swath Bathymetry: **a–c** from NOAA Ocean Explorer; **d** from MBARI —Monterey Bay Aquarium Research Institute)

Oceanic seamounts are cones and domes that typically grow upwards from the abyssal plains near spreading ridges, or away from them, in which case they are thought to be derived from mantle hot spots. They are therefore essentially basaltic, and pillow lavas appear to be prominent in the basal parts where, due to the great water depth of the vent, the hydrostatic pressure of the water column is sufficient to prevent explosive eruption of the basalt (Chaps. 4 and 5), although hyaloclastites could also form. Near spreading ridges, and more generally, seamounts have been assumed to have been fed by dykes. Recent high-resolution seismic and bathymetric data from the Canarian Basin, an intraplate region in the eastern Central Atlantic, shows volcanic cones are associated with the intrusion of sill complexes (Fig. 16.104a, b; Medialdea et al. 2017). The sills are intruded into about 1600 m of Cretaceous to Quaternary sediments and occur as stratified or branched configurations, surrounded by peripheral zones of hydrothermally altered sediment. The former comprise a vertical chimney with lateral migration of strata-concordant sills and topped with a single volcanic cone, while the latter, an inverted tree-shaped plumbing system with a volcanic cone at their top surrounded by a branched divergent system of inclined sills topped by hydrothermal domes (Fig. 16.104b).

Seamounts that have a well-developed summit platform are called *guyots* (Fig. 16.103). Generally, such seamount forms are thought to be the result of continued wave erosion during their emergent phase of development, followed by later isostatic subsidence (see Fig. 16.106b, c). Such flat-topped table-mounts also may show characteristic wave-cut terraces due to changes in sea level, or drowned valleys formed subaerially by erosion of rivers (Fig. 16.106d). However, some seamounts with flat tops appear never to have reached sea level (Searle 1983). Searle (1983) suggested that the steep slopes are built of pillow breccias and hyaloclastites tumbling over the edge of the flat-topped summit, which is predominantly built of sheet flows with high effusion rates. Clague et al. (2000a) proposed that deep water, flat-topped seamounts originated as point source lava mounds that enlarged as lavas continued to erupt and spread laterally, and then formed summit lava lakes (e.g. Fig. 16.103c), which episodically over-spilled to form steep prograding pillow and hyaloclastite flanks. Flat tops may also represent the infilling and overflowing of earlier large calderas (Clague et al. 2000b; Fig. 16.103d). Flat tops may, therefore, not necessarily solely be produced by wave erosion and truncation, although most tall guyots (e.g. Fig. 16.103a) are generally thought most likely to have been flattened near sea level, as evidenced by summit carbonate reef systems.

Older seamounts are stabilised by precipitation of volcanogenic ferromanganese crusts and elements. Shallow water seamounts may become colonised by fringing, framework-building biota, which will then contribute carbonate clastic debris to the submarine flanks of the volcanic pedestal. Prolonged growth of the seamount could produce an apron of coarse volcaniclastic and carbonate clastic detritus, transported and deposited by mass-flow processes fed through a well-developed canyon system. The facies of such a seamount should merge laterally into the pelagic sediment facies in deep water. Isostatic subsidence of the seamount and cessation of volcanism could lead to the development of an atoll, and eventually complete submergence of the seamount (Staudigel et al. 2010; Staudigel and Koppers 2015).

16.11.3.2 Facies Models for Seamounts and Submarine Lava Piles

Deep-marine basaltic seamounts vary in size from <1 km to 10 s of km in diameter, and from 100 s of m to as much as 2 km in height. Because of their depth of formation, these are effectively lava volcanoes and although pillow lavas are characteristic, sheet lavas and hyaloclastite will be significant components (Fig. 16.105a). Submarine fire fountain spatter deposits as well as limu o Pele (Chaps. 4 and 5) might also occur. The submarine facies are also likely to be dominated by complex feeder dykes and their hyaloclastite carapaces. Some seamounts have centrally located summit shields or their remnants where younger calderas have destroyed these, and their flanks and tops may contain numerous small "parasitic" cones. Deep-water craters and calderas (Fig. 16.103b, d), are almost certainly due to collapse of the summit area following eruptions of lavas from flank vents, including from the base of the edifice. The flanks of older seamounts may be fluted or embayed by erosional gullies and seaslide sector collapses and surrounded by hummocky aprons of volcaniclastic debris avalanche and debris flow deposits which interface with deep-water pelagic sediment facies (Fig. 16.103d).

For seamounts that grow to sea level, the growing basal (pillow) lava and hyaloclastite breccia pile (Fig. 16.106a) will be subject to lower hydrostatic pressure. Both explosive exsolution of magmatic volatiles as well as phreatomagmatic explosive eruptions, perhaps building surtseyan cones, and ongoing quench fragmentation of erupting magma will occur, leading to a mantle of pyroclastic and hyaloclastite deposits (Fig. 16.106b). If the seamount emerges above sea level and the vent is no longer intruded by seawater, more-passive lava eruptions will produce a lava cap over the volcaniclastic layer (Fig. 16.106b). Lavas flowing into the sea may transform into pillow lavas or, more commonly, will quench-fragment and build out coarse hyaloclastite deltas and flow-foot breccias (identical to those observed in intra-subglacial volcanoes, e.g. Fig. 16.88). Complex relative sea level changes associated with large tidal ranges or tectonic movements could produce complex stratigraphic-structural relationships

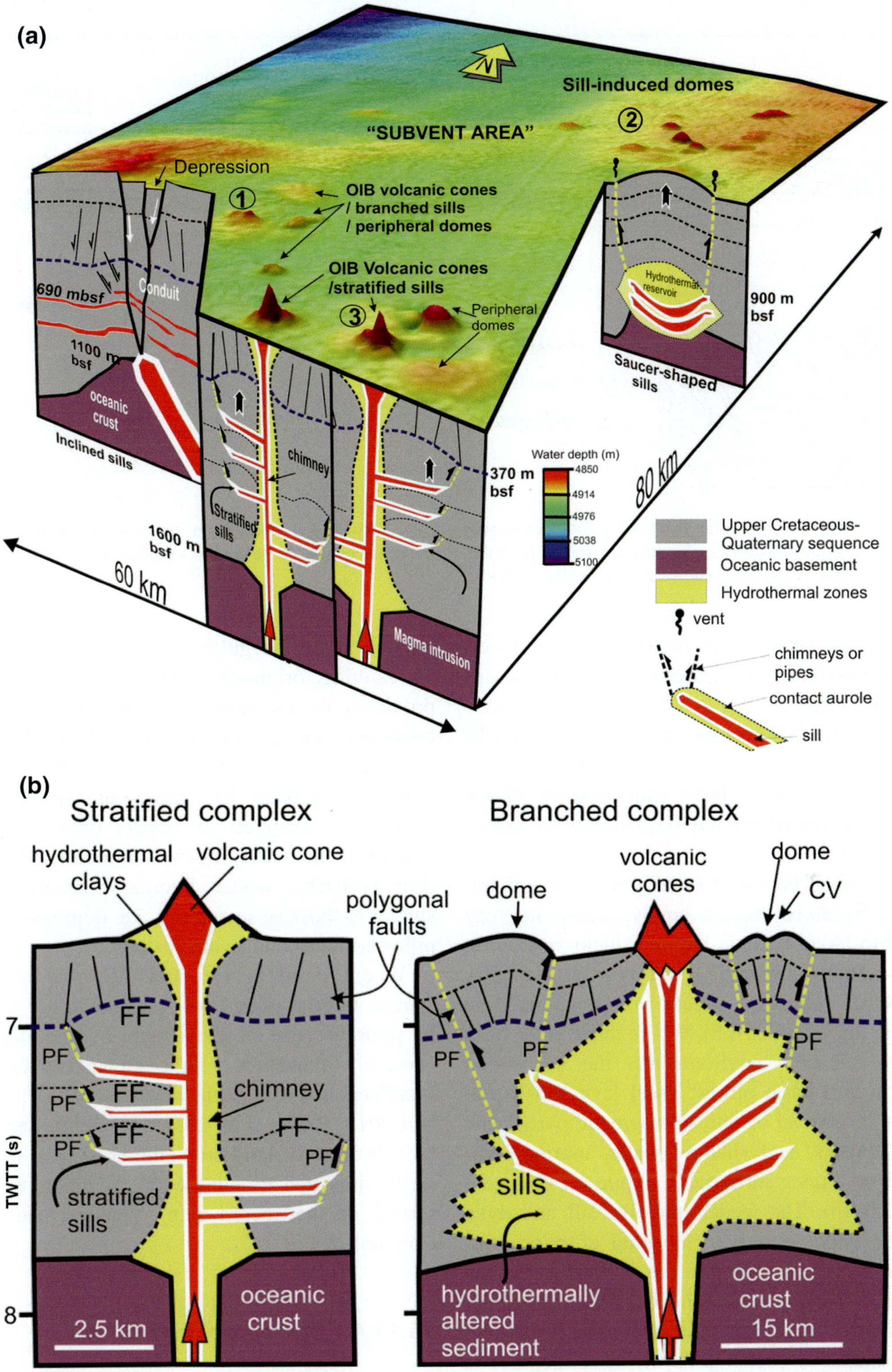

Fig. 16.104 Internal sub-structure of young (1.6 Ma–25 ka) seamounts and associated sills and hydrothermal vents, Canary Basin intraplate basaltic field, eastern Central Atlantic (~ 500 km west of the Canary Islands) (Modified from Medialdea et al. 2017). **a** 3D plan showing the multi-beam bathymetry and the architecture of the three main types of magmatic-hydrothermal structures: (1) mound with a hydrothermal vent depression related to inclined sills, (2) cluster of seabed domes mainly induced by saucer-shaped sills (Chap. 8), and (3) volcanic cones (seamounts) composed of OIB-type basanites sometimes surrounded by rims of hydrothermal domes. **b** Schematic cross-sections (from high-resolution seismic interpretation) of two main types of volcanic cone-sill complex. Stratified complexes are about 100–250 m high, 4 km in diameter, and have slopes of up to 23°; branched complexes are about 100 m high, 1.5–3 km in diameter, slopes have of 10–12°, and have cones surrounded by hydrothermal domes. PF = peripheral faults developed at the edge of the saucer-shaped sills; FF = forced folds; CV = central vent associated with hydrothermal dome

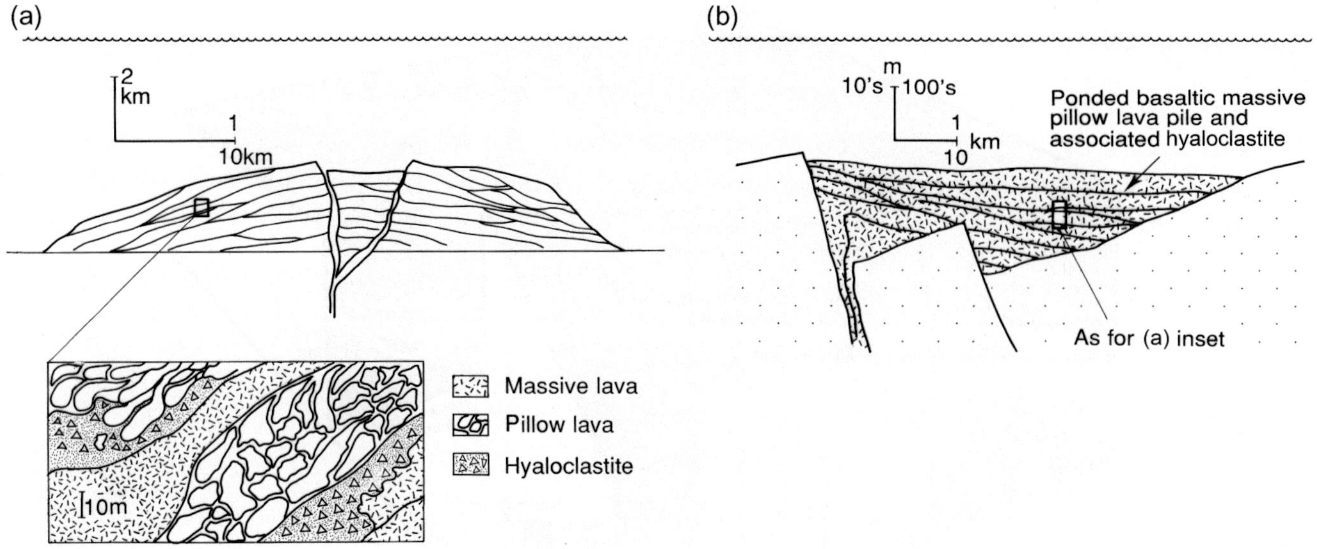

Fig. 16.105 Facies models for basaltic seamounts and tectonically ponded lava sequences. **a** Deep-marine seamounts dominated by lavas, including pillow and sheet lavas; inset shows typical facies. **b** Basalt lava successions ponded in tectonic-topographic half-graben basin traps; facies should be similar to those shown in inset to (**a**). (From Cas 1992)

in the upper part of the stratigraphy, including several levels of coherent lava-flow-foot breccia transition (as first described by Jones and Nelson 1970; Fig. 16.106c, d). After cessation of volcanism, the shallow tops of some centres may be sites of erosion (McPhie 1995), or for atoll and limestone cap formation. Both coarse volcaniclastic and carbonate clastic detritus could be transported and redeposited in thick aprons around the seamount.

On the other hand, if the sea floor eruptions are voluminous and have a sustained magma supply, seamounts may evolve further into larger oceanic basaltic shield volcanoes, built of piles or mounds of basaltic lava flows (Fig. 16.35b). The successions of submarine shield volcanoes may be 100 s of m to km thick (Sect. 16.5). Where these grow from subaqueous into subaerial environments, the succession should change from a basal pillow, sheet lava and hyaloclastite pile into an interval of phreatomagmatic pyroclastic and epiclastic deposits, and shallow water, marine sediments, followed by a thick subaerial lava pile (or emergent facies; Fig. 16.35a, b). The feeder intrusive sill and dyke complex may be very pervasive. Subaerial eruptions frequently produce lavas that flow into the sea and cause the edifice to enlarge laterally through the progradation of shoreline and offshore lava deltas. The submarine flanks will be marked by extensive sector collapse landslide, debris avalanche, debris flow, and volcaniclastic turbidite aprons (Fig. 16.35b).

In ancient deformed successions, seamount lava sequences may also be difficult to distinguish from a thick succession of lavas ponded in a subsiding (half) graben depression (Fig. 16.105b). The topography of the seafloor may be highly irregular, which to a large part is controlled by faulting originating either from tectonic activity in proximity to spreading ridges, or in association with rifted continental margins or arc crustal blocks. Prior to smoothing by long-term sedimentation, the sea floor topography may thus be marked by numerous small fault-controlled sub-basins. Eruption of basalts onto such a topography will result in locally thick, submarine ponded basalt piles (Fig. 16.105b), whose internal characteristics should be similar to those of seamounts, i.e. dominated by massive and pillowed lavas and hyaloclastites.

Where basaltic seamounts develop on oceanic lithosphere, their preservation potential in the rock record must be poor because of the effects of ongoing subduction of oceanic lithosphere, although exceptions occur (e.g. MacPherson 1983; Ortiz-Hernández 2003; Schnur and Gilbert 2012; Buchs et al. 2011). If basaltic seamounts develop on submerged island arc sialic crust in a setting not necessarily subjected to subduction tectonics, they may be preserved in the record (e.g. Watkins and Browne 1989; Corcoran 2000).

16.11.4 Surtseyan Tuff Cones

Surtseyan volcanoes are examples of basaltic seamounts, or the tops of seamounts or sub-emergent submarine volcanoes on which explosive eruptions were initiated in relatively shallow seas. As a result, the initial products are phreatomagmatic pyroclastic deposits, and in some cases the usual early pillow lava and hyaloclastite stage of deep-water

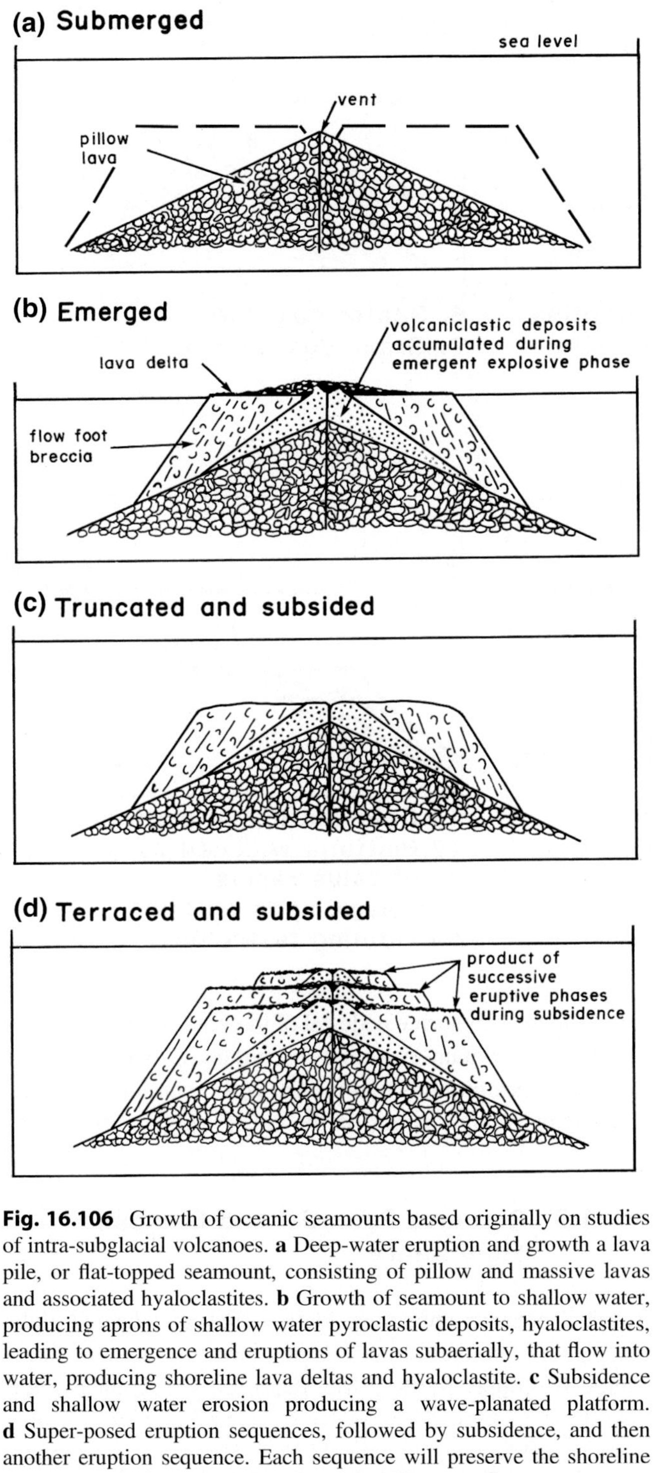

Fig. 16.106 Growth of oceanic seamounts based originally on studies of intra-subglacial volcanoes. **a** Deep-water eruption and growth a lava pile, or flat-topped seamount, consisting of pillow and massive lavas and associated hyaloclastites. **b** Growth of seamount to shallow water, producing aprons of shallow water pyroclastic deposits, hyaloclastites, leading to emergence and eruptions of lavas subaerially, that flow into water, producing shoreline lava deltas and hyaloclastite. **c** Subsidence and shallow water erosion producing a wave-planated platform. **d** Super-posed eruption sequences, followed by subsidence, and then another eruption sequence. Each sequence will preserve the shoreline passage zone of lava and shoreline delta. The most flat-topped guyots (especially tall ones) are probably the result of erosion near sea level, but in other cases their morphology may result from cappings of lava associated with summit lakes, which could develop even in deep water during stage (**a**), as represented with dashed outline of flat top. (From Cas and Wright 1987, After Jones 1966)

seamounts may be by-passed. Tuff cones may also form in lacustrine environments when basaltic magma erupts onto a shallow basin floor (Fig. 16.14c).

16.11.4.1 Size, Eruptions and Characteristics

Marine surtseyan tuff cones are largely phreatomagmatic explosive centres that typically form in water depths up to 200–300 m (Figs. 16.107a–d and 16.108a–c). Their explosiveness is due to phreatomagmatic explosive interaction between seawater and rising basaltic magma (Chaps. 4 and 9; Sects. 4.6.9, 9.5.3; Sect. 16.3.2). The eruption style is typically a pulsing succession of phreatomagmatic blasts that produce ballistic blasts or tephra jets, outward radiating base surges and near-vent fallout (see Plate 4.1d, Fig. 4.18a), which together build a bedded tuff cone of deposits (Fig. 16.14c). The best-known surtseyan tuff cone is the namesake, Surtsey, which began to erupt in early November 1963 off the southern coast of Iceland at a water depth of 130 m, building a cone above sea level within 3–7 days that eventually rose to 150 m high (Thorarinsson et al. 1964; Jakobsson and Moore 1982; Sect. 16.3.2). Other well-documented examples include those of the Tongan islands of Hunga Tonga and Hunga Ha'apai in 2009 and 2014–2015, which occurred on the submerged caldera margin of a submarine volcano (Vaughan and Webley 2010; Garvin et al. 2018; Colombier et al. 2018). The larger eruption in 2014 commenced on 19 December 2014 and, at the end of activity on 26 January 2015, the emergent tuff cone was 1–2 km wide, 2 km long, 140 m high (Fig. 16.11e).

Surtseyan tuff cones, more generally, in marine settings may be several hundred m up to several km in diameter, and as high as 100–300 m, as well as overlap or cluster together. Marine tuff cones are commonly lithified due to the rapid (mild hydrothermal) alteration of basaltic glass called palagonitisation (Sect. 16.3.4). They may also evolve into larger island structures, increasingly dominated by subaerial hawai'ian and strombolian activity, and lava flow emplacement (e.g. Heimaey, ~20 km to the northeast of Surtsey; Mattsson and Höskuldsson 2003, 2005).

16.11.4.2 Facies Models for Surtseyan Tuff Cones

The submarine part of the cone succession consists of diffusely bedded, water settled tuffs and lapilli tuffs (see Fig. 9.82a, b), ballistic breccias, and resedimented tuffaceous turbidites (Fig. 9.82c), debris flow (see Fig. 12.35a, b), and gravitational slide deposits (Cas et al. 1989; Fig. 16.107a-d). For cones that ultimately emerge above sea level (e.g. Fig. 16.11e), the subaerial part of the succession comprises a bedded sequence of phreatomagmatic base surge,

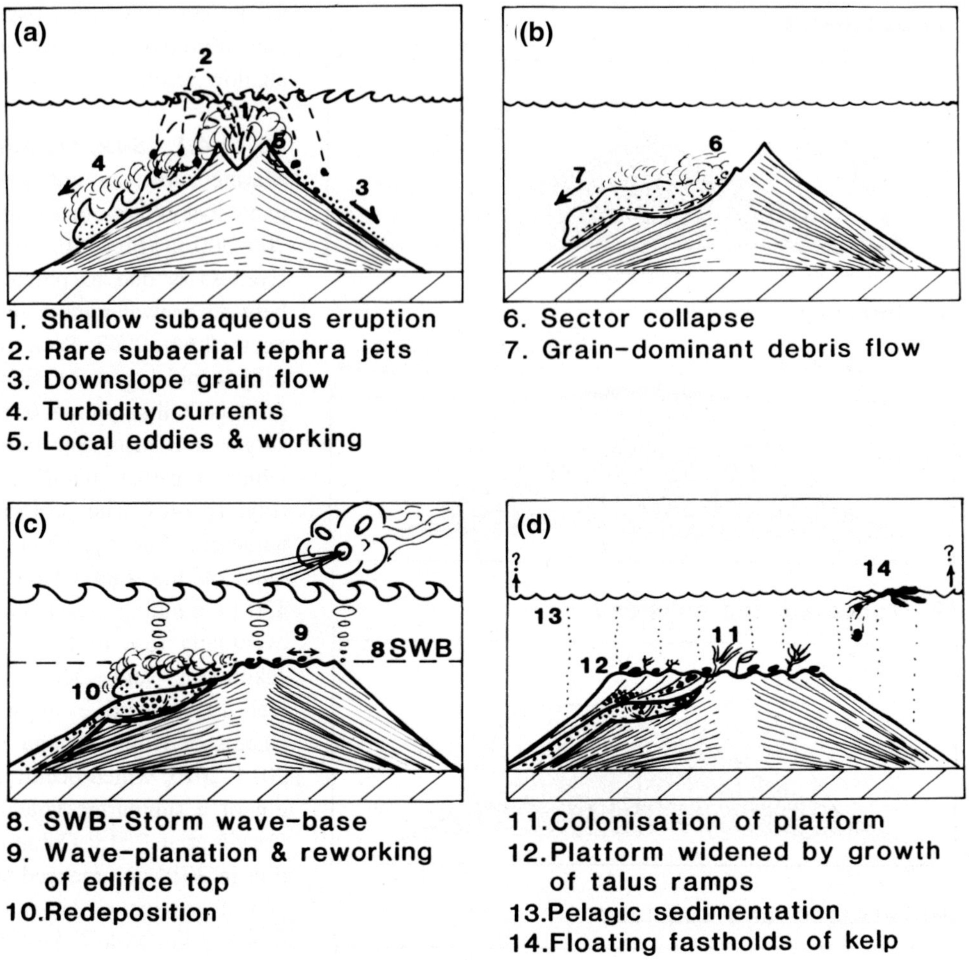

Fig. 16.107 Sequence of stages in the formation of the Eocene–Oligocene Bridge Point submarine tuff cone, Otago, New Zealand. **a** Initial submarine phreatomagmatic explosive eruption leading to growth of a submarine tuff cone, accompanied by syn-eruptive mass flow resedimentation. **b** Submarine sector collapse of part of the cone, producing a large-volume debris flow. **c** Post-eruptive wave planation of the top of the wholly submarine edifice. **d** Colonisation of the wave-planated platform by fauna and flora. (Modified from Cas et al. 1989)

agglomerate breccias and air fallout deposits (Fig. 16.108a–c; Sect. 16.3.2). Accretionary lapilli may be preserved, particularly in subaerial deposits of the cone, but also in distal submarine turbidite deposits (Fig. 16.101d, e).

If tuff cones rise above sea level and develop a stable cone in which the vent is protected from ingress by the sea, eruption activity can change to magmatic, building scoria cones above the tuff cones (Fig. 16.108c) and even erupting lavas (Fig. 16.108a) that armour the edifice against wave erosion, as happened during late stages of the Surtsey eruption. Such lavas may build deltas of lava (e.g. Fig. 5.45b, g) and hyaloclastite (Fig. 5.50d) into the sea (Fig. 16.108a). Tuff cones can form in isolation (e.g. Cas et al. 1989; Plate 4.1d; Fig. 16.107a–d), or in overlapping clusters, creating complex internal architecture (Fig. 16.108a–c; Coombs et al. 1986; Moorhouse et al. 2015). Both solitary and clustered cone complexes in marine settings are subject to marine wave and shallow water current erosion, leading to wave planated, submerged edifice tops, which can be colonised by shallow marine fauna and flora. Mixed bioclastic and pyroclastic debris is resedimented down the slopes by mass flows, enlarging the submarine pedestal of the volcano. Cone cluster edifices will contain erosional discordances, which become colonised hard grounds that in turn will be on-lapped by the deposits of younger cones on the cluster (Fig. 16.108a, c).

16.11.5 Resource Potential and Economic Significance

Basaltic oceanic spreading centres fed by sheeted dyke complexes and their hydrothermal systems are the likely source of copper-pyrite VMS deposits, which have traditionally been termed Cyprus-type, named after where they were mined in the Troodos Massif. However, in resource terms these are of limited grade and tonnage and remain more an academic curiosity than a modern commercial exploration target. Although modern seafloor hydrothermal systems are known to contain significant accumulations of metals from hydrothermal vents (Hannington et al. 2011), these remain mainly of scientific interest. Nonetheless, plans to commence deep-sea exploitation of gold-rich VMS

Fig. 16.108 Features of shallow marine and lacustrine nested surtseyan tuff cone complexes. **a** Facies model based on the geology of the Eocene-Oligocene Oamaru tuff cone complex, New Zealand, depicting multiple, overlapping eruption packages, cross-cutting feeder dykes, partially collapsed edifices, with internally discordant relationships, late stage emergence of the edifice above sea-level, stabilisation of the cone, and effusion of lava from a subaerial vent, creating a prograding lava and hyaloclastite delta into the sea. (For further description, see Coombs et al. 1986; Moorhouse et al. 2015). **b** Steeply dipping part of the tuff cone succession in (a), marked by bedded, and wedging bed sets of phreatomagmatic fallout and surge-modified fallout deposits. **c** Drammatic coastal cliff cross-sectional exposure (looking northeast) through the eroded surtseyan tuff cone and scoria cone complex produced during the 1957–1958 shallow marine to subaerial trachytic Capelhinos explosive eruption, Faial, Azores, Portugal. The early shallow marine phreatomagmatic phase of the eruption produced the bedded, pale grey tuff cone, part of which is preserved on the right. Internal discordances and channels attest to the role of base surges and breaks in the eruption and edifice collapses. The second phase of the eruption produced a superposed scoria cone complex (red-black) with a complex, cross-cutting internal branching feeder magma plumbing system (grey), which formed once the tuff cone complex had built above sea level allowing magmatic explosive activity to occur. Sea cliffs are 100–130 m high. (Photo by Lisa Buckley, taken in 2019)

mineralisation along the Manus spreading centre in the Bismarck Sea back-arc basin (Papua New Guinea) have advanced to feasibility stage (Chap. 18). The mining of seafloor polymetallic nodules (containing nickel, copper, cobalt, and manganese) is also actively being assessed, and seamounts may hold some future potential for the exploitation of their Mn-rich encrustations (containing vanadium, molybdenum and platinum).

In an interesting development, deep-sea muds (layer 1) highly enriched in rare-earth elements and yttrium (or REY) have now been discovered on the western North Pacific Ocean floor, east of the Mariana Trench (Takaya et al. 2018). These contain over 5,000 ppm total REY and are possibly an enormous future resource because of the large aerial extent of the pelagic muds and advantageous mineralogical features. Preliminary estimates suggest a world class resource of rare-earth oxide for the most promising area, accounting for about 50 years of average global demand of Y, Eu, Tb, and Dy.

Thick sequences of oceanic mafic and ultramafic volcanics located in Archean and Paleoproterozoic greenstone belts may be prospective both for VMS and komatiite-hosted nickel sulphide mineralisation (Chaps. 7 and 18). However, in these settings basaltic protoliths also play a very different role in ore genesis, behaving as source rocks, as well as potential host rocks for orogenic lode gold mineralisation hosted by veins and shear zones (e.g. Phillips and Powell 1993; Chap. 18). Metamorphic devolatilisation of mafic and/or greywacke successions at the greenschist-amphibolite transition produce low salinity fluids capable of transporting large quantities of gold, which are then precipitated by reaction with Fe-rich host rocks (i.e. mafic lavas and sills) and/or structural trap sites.

16.12 Potential Hazards Associated with Modern Volcanoes: Summary

In this final Section of Chap. 16 we briefly summarise the important natural hazards as understood from modern volcanoes and their settings. In terms of volcanic hazard evaluation, understanding the stratigraphic record of eruptions and deposits of any particular active or potentially volcano provides the best insights into the possible future eruption behaviour (e.g. Mendoza-Rosas and De la Cruz-Reyna, 2008; Scandone et al. 2016; Giordano and Caricchi 2022). However, hazard evaluation and risk mitigation is a major and important applied branch of volcanology worthy of a book in its own right, and so our consideration is not comprehensive.

It has been estimated that today, over half a billion people live in regions of the world directly subject to volcanic risk (Auker et al. 2013; Loughlin et al. 2015a, b). Volcanoes directly threaten major population centres (e.g. Rome and Naples, Italy; Mexico City and Guadalajara, Mexico; Manila, Philippines; Yogyakarta in Java, Indonesia;, and the island of Kyushu, Japan), and can have an important influence on the socio-economic development of extensive regions, especially in the developing world. They can cause serious environmental and economic impacts at the global level in the form of climate change. The assessment and mitigation of risk in volcanology is extremely complex, since it encompasses several different hazardous natural phenomena which may occur in concert (Thierry et al. 2008; Selva et al. 2019; Hayes et al. 2020; Nieto-Torres et al. 2021). *Volcanic hazards* are volcanic or related geophysical phenomena or events that could pose risks to communities around a volcano. *Risk* is an assessment of the likely impact of a hazard or volcanic event, including loss of life, destruction of assets including infrastructure, vulnerability which is the likely proportion of the value to be lost, and hazard area is probability a certain area is likely to be affected by an event (Nieto-Torres et al. 2021) A volcanic hazard assessment determines the probability that a volcanic eruption, phenomenon or related geophysical event will occur in a given geographic area and within a specified window of time, while the *risk* that can be associated with a volcanic hazard depends on the proximity and vulnerability of a population of people, or an asset near to where a volcanic event might occur (Martí 2018; Nieto-Torres et al. 2021).

To evaluate and manage volcanic risk we need first to assess the volcanic hazard(s), that is, identify how an active volcano or volcanic area has behaved in the past and then use this information to infer how it may behave in the future. This task requires a compilation of all existing geological, geochemical, geochronological and geophysical information concerning the eruption style of the volcanic system in question, its eruptive recurrence, structural constraints (e.g. on the likely opening of new vents), and the characteristics and potential extent of its main hazards. All this information can be used to draw up eruption scenarios and hazard zonation maps (e.g. Lindsay and Robertson 2018) that will constitute the basis for civil authorities designing risk management programmes, as well as an essential material to develop the educational and communication programmes that should also form part of a risk reduction process. For example, in contrast, during the deadly 2018 eruption of Fuego volcano, Guatamala, as well as the alert system being tardy and ineffective, lessons appear not to have been learned from the volcano's major eruption in 1974, or they were forgotten. Effective education of the communities around the volcano about the likely hazards and possible magnitude of an eruption and lahars or volcanic mudflows did not seem to have occurred, which is one of the most valuable tools in risk minimisation.

Today modern volcanic risk assessment and mitigation involve the development of complex volcano-monitoring arrays (mostly seismological, geochemical, and deformational networks), Geographical Information Systems (GIS), computer models, vulnerability databases, and probabilistic risk assessment protocols. Such probabilistic volcanic hazard assessment (PVHA) methods have become important tools for risk mitigation of volcanic events (e.g. Selva et al. 2015; Takarada 2017). Several platforms, which provide a number of numerical simulations and volcanic hazards assessment tools, are now available online. These include the Volcanic Hazard Assessment Support System (VHASS: http://volcano.g-ever1.org), which aims to implement a user-friendly, WebGIS-based, open-access online system for potential hazards assessment and risk mitigation of Quaternary volcanoes in the world (for further details, see Takarada 2017). Here users can get up-to-date information such as eruption chronology and geophysical monitoring data of a specific volcano using the direct link system to major volcano databases on the system. Nevertheless, as we have indicated in Chap. 1, for only about 20% of the 1,551 volcanoes that are classified as having been active in the Holocene, is there adequate geological information that can be used to assess the history and frequency of volcanic activity in risk and hazard assessment (Loughlin et al. 2015a, b). In addition, many other volcanoes around the world could potentially be active even if their last eruption is older than the Holocene, so the number of active volcanoes is likely to be much larger (Giordano and Caricchi 2022). This means that as geoscientists we still have a long way to go to not only adequately document the remaining 80% of the world's active volcanoes, but identify older, potentially active centres!

As indicated initially, our consideration of volcanic hazards is not wide ranging. We will not discuss details of hazards evaluation during volcanic unrest (typically dealt with by event tree probabilistic approaches; e.g. Neri et al. 2008; Sobradelo et al. 2014), nor about the expected duration and size of eruptions, which remain the main unknowns prior to the eruption itself. We will instead summarise the spectrum of the main hazards associated with the eruptive styles and related phenomena which can be expected at different volcano types. In step with the foregoing sections of this chapter, potential hazards can be subdivided into the following categories of volcanoes and environments:

- scoria and spatter and (pumice) cones,
- maars, tuff rings and tuff cones,
- littoral and rootless cones,
- basaltic shield volcanoes,
- continental basaltic provinces,
- continental and marine stratovolcanoes,
- continental and marine caldera volcanoes,
- intra- or subglacial volcanoes,
- mafic oceanic volcanic centres, and
- submarine to emergent seamounts and surtseyan tuff cones.

We will say little here about the potential hazards from more complex or multiplex volcanoes as described in Sect. 16.9.1, except to refer readers to what we will list for shield volcanoes, stratovolcanoes and explosive caldera volcanoes.

For major, current compilations on the subject we refer the reader to Sigurdsson et al. (2015); Loughlin et al. (2015a, b); Mackie et al. (2016); Fearnley et al. (2017). For a more dated, but excellent text, see Blong (1984). We also refer the reader to the work of the IAVCEI Task Group on Crisis Protocols (Giordano et al. 2016).

16.12.1 Scoria and Spatter (and Pumice) Cones

Mafic scoria and spatter cone-forming eruptions, although small in volume, can produce significant hazards for communities in close proximity:

(a) Hawai'ian to violent strombolian weak plumes are usually <10 km high, are easily affected by local winds, and may affect civil aviation operations with short-term disruption to air traffic and nearby airports in the direction of the downwind ash plume.
(b) Scoria fallout can be several m thick within 1 km, thinning to cm thick downwind (3–4 km; Chap. 9) from the vent. This can impact nearby buildings and infrastructure through roof loading and collapse.
(c) Ballistic bombs and blocks can be ejected up to 1 km from vent during violent eruption phases.
(d) Downwind dispersal of fine scoria and ash can contaminate water supplies and infiltrate power generation infrastructure, causing it to malfunction.
(e) Spatter cones and some scoria cones can also erupt lava flows that can travel several tens of km (Chap. 5) cutting roads and railway tracks, as well as damaging and destroying buildings.

Scoria cones may be solitary in monogenetic volcanic fields or parasitic to larger central edifices. In the first case the risk is largely due to the potential proximity to urbanised areas, such is the case of Auckland, New Zealand, which is built within such a basaltic monogenetic field. In the second case, scoria cones may form at any elevation from the summit (i.e. very far from human settlements), to the low slopes, where urban settlements are much closer, on a large volcanic edifice such as the case of Catania at the toe of Mt Etna, Sicily. The recent 19 September–13 December 2021 eruption at La Palma, Canary Islands, occurred along the flank of the Cumbre Vieja edifice and severely impacted on human settlements and activities both with lava flows destroying vast urbanised and farm areas, as well as with tephra which have covered most of the southern part of the island. There have been relatively few hazard assessments documented of monogenetic scoria cone fields (e.g. Auckland, Lindsay et al. 2011; Mexico City, Siebe and Macías 2006; La Palma, Marrero et al. 2019; see Fig. 16.110a, b). These present some significant differences compared to strategies developed for stratovolcanoes (Sect. 16.7), one of the main problems in such fields being predicting the unknown "new" vent location, and the fact that they commonly are built of multiple eruptive events, with deposits spread over the next possible vent opening area. Another problem is the recognition now that some scoria cones are polycyclic in nature (e.g. Alumbrera and De La Laguna cones, southern Puna, Argentina; Baez et al. 2016; Fig. 16.4). The occurrence of more than one eruption in a single scoria cone and changes in the eruptive style during long-lasting eruptions may need to be considered in some monogenetic fields. Although no rhyolitic cone-forming eruptions have been observed, their risk to local communities is likely to be similar to scoria cones.

16.12.2 Maars, Tuff Rings and Tuff Cones

Eruptions that form these types of volcano occur in similar settings as scoria cones, i.e. associated with monogenetic fields or with larger central volcanic edifices, where magma interacts explosively with water. Although small in volume, maars tuff rings and tuff cones produce more significant hazards than scoria and pumice cone eruptions because of the large proportion of fine ash that is produced during hydrovolcanic eruptions, especially phreatomagmatic activity (Chap. 4). Potential hazards include:

(a) Base surges (±higher concentration pdcs) can flow radially up to 10 km from vent (Chap. 10), causing physical injuries to people and livestock, and damage to buildings, water and power installations, and to crops.
(b) Phreatomagmatic fallout deposits can be several m thick within 1 km, to cm thick tens of km downwind (Chap. 9) which can impact nearby buildings through building roof loading and collapse, and crop loading and collapse.
(c) Base surge and fallout deposits can build a rampart or ring of deposits tens of m thick around the maar crater edge, thinning radially for at least 1 km in all directions
(d) Ballistic blasts during phreatomagmatic blasts, and especially during vulcanian blasts when the vent becomes blocked, can eject ballistic blocks up to 2 km or more from the maar crater.
(e) Downwind dispersal of fine ash will contaminate water supplies and infiltrate power generation infrastructure, causing it to malfunction.
(f) Fine phreatomagmatic ash lofted into the atmosphere can be dispersed hundreds to thousands of km (Chap. 9), causing chaos and shutting down air traffic and airports for days to weeks because of the clogging effects of fine ash on aircraft engines (cf. 2010 eruption of Eyjafjallajökull volcano, Iceland; Sect. 16.10.3).
(g) Fine ash can also cause severe respiratory and other health problems, especially for people with pre-existing health conditions, even when the ash concentration is low.
(h) Some maar volcanoes can release large volumes of magmatic CO_2, especially from poorly circulating crater lakes, sometimes termed "limnic eruptions". Carbon dioxide being a heavy gas, hugs the ground, and can even flow downslope, causing asphyxiation of human and animal populations (e.g. during the evening of 21 August 1986 at Lake Nyos, Cameroon which killed over 1700 people and some 8000 in livestock; Kusakabe 2017). CO_2 is mostly odourless and is therefore an extremely dangerous volcanic gas because of this. Destabilisation of the deep CO_2-rich waters may occur as a result of earthquakes and/or landslides which induce lake tsunamis and overflows of water and CO_2.
(i) The emplacement of thick and wet ash may trigger lahars which may flow farther away than primary deposits even in absence of significant rainfall or mixing with other kinds of external water (Chap. 12).
(j) Secondary hazards arise from extinct maar craters which have been lakes or swamps after eruptive activity ceased, leading to build-up of decayed vegetation. If this vegetation becomes thick, and buried, and converts into peat, it poses a fire hazard during hot summer fire seasons. This occurred in southeastern Australia, in the late Cenozoic Newer Volcanics Province, in the summer of 2018, when fires in two extinct maars burned in buried peat for 2 months and required huge volumes of water to be used to extinguish, at a time when water was scarce and precious due to extended drought. The constant smoke near ground level was a considerable health hazard for local communities.

Maar-type volcanoes often occur in monogenetic fields and in close proximity to scoria cones (e.g. Newer Volcanics Province in southeastern Australia; Cas et al. 2017); indeed maar-forming phreatomagmatic eruptions may evolve into scoria cone-building and even effusive events both vertically throughout a single eruption (e.g. Hopi Buttes Volcanic Field; White 1991) or laterally across a linear fissure (e.g. Nejapa-Miraflores volcanic fault, Managua, Nicaragua; Pardo et al. 2009). The major risk mitigation problem is identifying where the next vent opening may occur (Marrero et al. 2019). However phreatomagmatic eruptions producing maars, tuff rings and tuff cones may occur at any volcano where either surface water or substantial subsurface aquifers occur. These may be associated with caldera settings (e.g. Astroni in Campi Flegrei caldera; Isaia et al. 2004), where volcano slopes approach coastal areas (e.g. Capelinhos, Azores; Cole et al. 2001), or in conjunction with active geothermal systems (Wohletz and Heiken 1992). One of their most destructive syn-eruptive hazards is the discharge of base surges and associated pdcs. For example, this has occurred several times over the last tens of thousands of years on the southeastern outskirts of Rome, from eruptions from maar craters in the Lake Albano area (Fig. 16.109; also see Fig. 16.22). Lake Albano has also been the site of CO_2 degassing events in Ancient Roman times (Funiciello et al. 2003; De Benedetti et al. 2008), as well as in historical times, which presently represents the major volcanic threat along with local seismicity for this densely populated area.

For a thorough review of the syn- and post-eruptive products and hazards of maar volcanoes, see Lorenz (2007), White and Ross (2011), and Valentine and White (2012).

16.12.3 Littoral and Rootless Cones

The main potential hazards associated with the entry of lavas into the sea or lakes, or flow over marshy/snowy/iced ground, and the formation of littoral cones and rootless cones (pseudocraters) are:

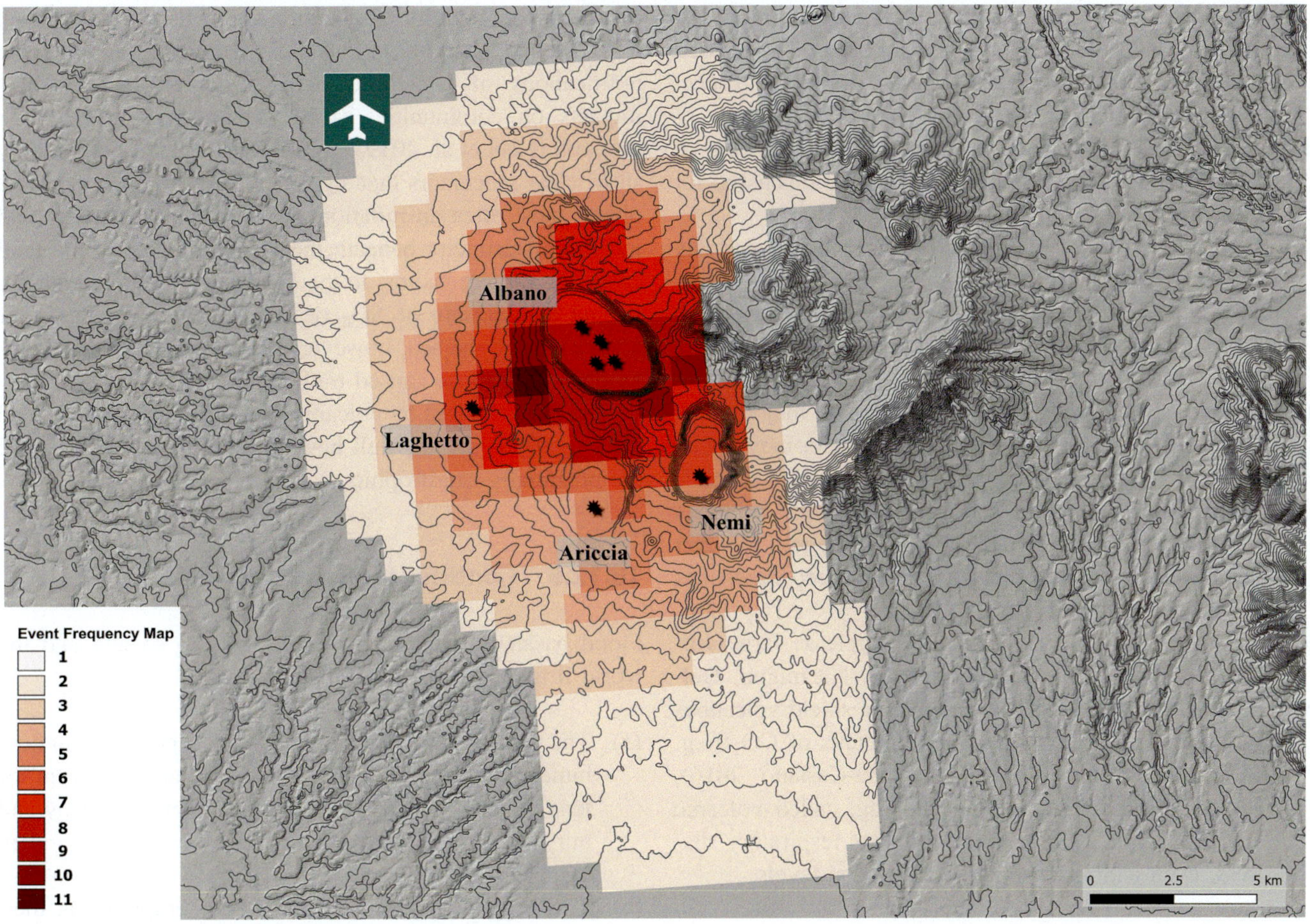

Fig. 16.109 Boolean map of base surge and pdc deposits infilling of the Colli Albani caldera area, some 10–15 km southeast of Rome, central Italy. The reference period is the last 100 ka in relation to the maar-type activity from Lake Albano (7 eruptions), Laghetto (2 eruptions), Ariccia (1 eruption), and Lake Nemi (2 eruptions). The numbers in the legend refer to the number of times a certain area has been covered by pdc deposits from the geological record. The outer extent of each eruptive unit has been extended to the inferred max run-out based on thickness decay patterns of the related deposits. Also shown is location of Ciampino International Airport. (After Pensa et al. 2020)

(a) Phreatic (or molten fuel–coolant) explosions can occur unexpectedly due to the interaction of external water with magma (Chap. 4) in a lava channel or tube.
(b) Asphyxiation can result from the formation of large volumes of steam at lava entry points into the sea and potential release of CO_2 from lava tubes.
(c) Collapse of parts of the lava benches and littoral cones into the sea where lavas are entering the sea are extremely hazardous, and
(d) They can cause ejection of ballistic blocks and widespread dispersion of fine ash.

Many people have been killed in Hawai'i from the above phenomena because they came to close to the point of entry of lavas into the sea. On land they can form large fields of rootless cones (e.g. Laki eruption of 1783–84) and are potential hazards because of their sudden initiation.

16.12.4 Basaltic Shield Volcanoes

There are many potential hazards associated with shield volcano eruptions:

(a) Low viscosity basaltic lava flows (Chap. 5) can be fast moving and far reaching, and may issue from fractures that propagate along rifts down to inhabited areas along the volcano slopes, where they can disrupt and destroy buildings, towns, roads, railway tracks, recreational facilities tens of km from the vent (e.g. the 2002 eruption at Nyiragongo volcano, Democratic Republic of the Congo; the eruption at Kilauea, Hawai'i in 2018; Las Palmas, Canary Islands in 2021).
(b) Explosive eruptions may occur intermittently both at the summit and along the gently sloping flanks, including magmatic scoria (cone) forming eruptions

and phreatomagmatic eruptions that generate base surges and widespread dispersal of fine ash, as discussed for scoria cones and maar-forming eruptions respectively above (Chaps. 9 and 10).

(c) Marine shield volcanoes also experience passage of lavas into the sea, with attendant hazards discussed for littoral cones above (Chap. 5).
(d) Lateral withdrawal of lava along flank eruptions may trigger caldera and pit collapses in summit regions (e.g. Miyakejima, Izu Islands, Japan, in 2000, Geshi et al. 2002; along Kilauea's lower East Rift Zone in 2018, Neal et al. 2019, Chap. 5).
(e) Marine shield volcanoes are prone to major flank landslides or sector collapses, which may or may not be eruptively triggered (Chap. 12). If they occur at or in the marine interface, they can generate major tsunamis (mega-tsunamis) if the rate of slip and the rate of water displacement are instantaneous. Slow creep landslides do not cause tsunamis.
(f) Some basaltic shield volcanoes can release large volumes of SO_2 (e.g. Masaya volcano, Nicaragua; Delmelle et al. 2002), while some marine shield volcanoes are known to release significant amounts of CO_2 (e.g. El Hierro, Canary Islands; Santana-Casiano 2016). Mixed volcanic SO_2 rich air and fog, called volcanic smog or "*vog*", can cause significant respiratory distress, even if SO_2 levels are not at lethal levels.

Probably the deadliest eruption of a shield volcano was in 2002 at Mt Nyiragongo at the border between the Democratic Republic of the Congo and Rwanda, when fractures opened along the low slopes of the volcano and fast, ultra-low viscosity ($\sim$30% wt % SiO_2) foidite basalt lava flows (Chap. 2) invaded and partially destroyed the town of Goma leaving 170 people dead and 350,000 homeless (Tedesco et al. 2007). More recently, lava and scoria and ash fallout from the 19 September–13 December 2021 eruption of the Spanish island of La Palma damaged more than 50 km of roads and many tens of hectares of farmland, and engulfed over 3,000 homes and buildings causing some 7,000 people to be evacuated (Fig. 16.110a, b). The total damage caused by the eruption is estimated to amount up to €900 M. The lava field has an area of >10 km^2 and has formed prograding lava deltas at the coast forming a new peninsula. Reports on the eruption are available at the Smithsonian Institution Global Volcanism Program La Palma website: https://volcano.si.edu/volcano.cfm?vn=383010.

Kilauea and the other active Hawai'ian shield volcanoes represent some the most intensely monitored volcanoes in the world. For more than 100 years geoscientists at the Hawai'ian Volcano Observatory (HVO) have developed new volcano-monitoring instruments and networks to record and document activity. These monitoring data are employed to better understand eruptions and earthquakes and to issue notifications of hazardous activity, which helps reduce impacts on communities, such as occurred in the 2018 eruption (Neal et al. 2019). Hawai'ian volcanoes commonly show clear precursors that signal the start of a new eruptive phase; most often an inflation of the volcano's summit and increased seismicity beneath the volcano. Most of the fatal casualties among visitors to the Volcanoes National Park are related to exposure of SO_2 or a combination of gases, or lavas entering the sea, however, these are significantly outnumbered by those caused traffic incidents.

The hazard posed by volcanic mega-tsunamis may extend 100s to 1000s of km, well beyond the source volcano, and are potentially catastrophic.

16.12.5 Continental Basaltic Provinces

Flood basalt provinces pose huge risks to society and the world if they were to be initiated today:

(a) The volumes of individual lavas are huge, they can flow hundreds of km and cover huge areas and can erupt for tens of years (Chap. 5). They can therefore destroy and bury all towns, infrastructure and agricultural land in their path.
(b) The biggest impact of flood basalt eruptions is their cooling effect on global climate in the long term because they release huge volumes of SO_2 (e.g. 12,000 Mt during the Miocene Roza Member flood basalt eruption, Columbia River Flood Basalt Province, USA; Chap. 5), which is converted into H_2SO_4 aerosol gas in the atmosphere. The impact of flood basalt eruptions on global climate during geological history has been discussed in Chap. 5, and it has also been demonstrated that they coincide with major global mass extinction events.
(c) Continental plains basalt provinces present smaller scale hazards, which can however be significant for communities within their confines, including:
(d) Lavas can flow tens of km from the vent (Chap. 5) cutting roads, railway tracks, destroying building and other infrastructure.
(e) Explosive eruptions can produce fallout deposits (Chap. 9), as discussed for scoria cones and maar volcanoes above, damaging and destroying buildings, impacting water and power installations, destroying crops, and causing health problems.
(f) Base surges can cause significant damage and cause injuries for up to 10 km radially from vent (Chap. 10).

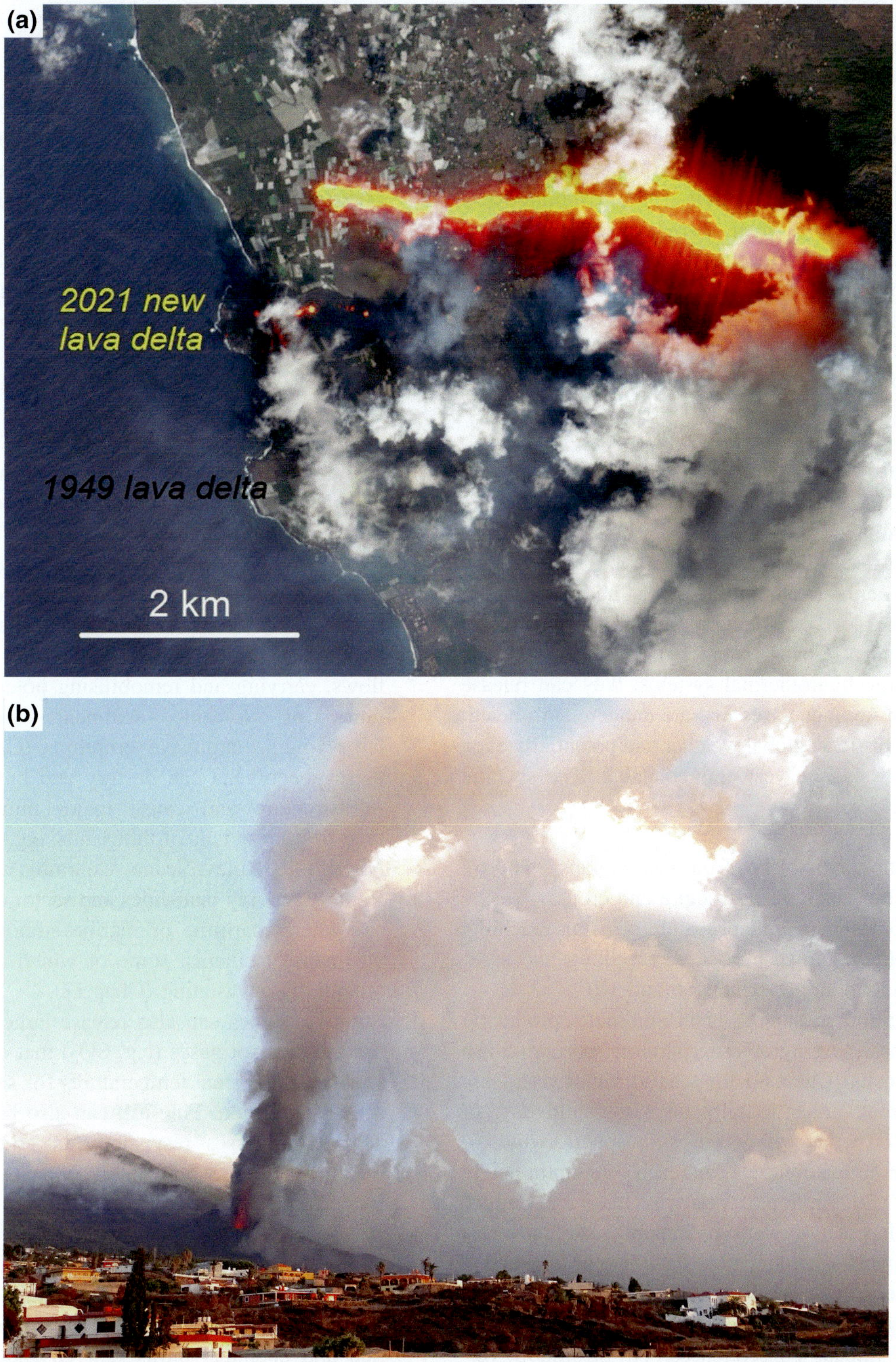

Fig. 16.110 19 September–13 December 2021 eruption of Cumbre Vieja shield volcano on La Palma, Canary Islands, Spain. **a** Copernicus Sentinel-2 true colour image taken on 10 October. The eruption commenced on 19 September and lava reached the sea 10 days later after forcing more than 6000 people to leave the affected area and destroying hundreds of homes in the Todoque neighbourhood. The lava flow field reached up to 600 m wide in places, and crossed the main coast road, dividing communities on either side and burned through plantation greenhouses and fertiliser, giving off toxic fumes. This image captures a new flow of lava (enhanced using shortwave infrared channel) following collapse of parts of the northern face of the volcano's cone on 9 October. This new flow destroyed the few remaining buildings still standing north of Todoque. (*Source* European Space Agency through Creative Commons Attribution-ShareAlike 3.0 IGO (CC BY-SA 3.0 IGO) licence). **b** View looking south taken on 4 October showing the vent area and neighbourhood of Todoque. While only a mild explosive and effusive eruption, the activity has been a local disaster which will disrupt communities for many years

The few large-volume historical basaltic flood events, involving the eruption of several cubic km of lava, had occurred in Iceland, i.e. at Eldgjá in 939–40 AD and the 1783–84 Laki eruption, but other similar events also occurred in prehistoric times, all candidates for severe impacts on climate with serious consequences for human communities (Thordarson and Self 1993; Gudmundsson et al. 2008; Oppenheimer et al. 2018). Larger events are fortunately very rare. For recent discussions on the relationships and possible causal links between flood basalt events and mass extinctions, see Rampino and Self (2015), Clapham and Renne (2019) and Adatte et al. (2020), and references therein.

16.12.6 Stratovolcanoes

Both continental and marine stratovolcanoes pose many potential hazards because they are so large, high above base level, they can erupt voluminous lavas and pyroclastic deposits, with explosive styles ranging from phreatic to vulcanian, to plinian (Chaps. 9 and 10). They can generate vigorous and large hydrothermal systems, they can release large volumes of noxious gases that are dangerous to health and can affect short-term global weather and potentially even climate, and they can be very unpredictable. The specific hazards include:

(a) Lavas can flow tens of km from vent (Chaps 5 and 6), destroying buildings, infrastructure and crops.
(b) Summit domes may collapse and form far reaching, usually valley confined block and ash flows and associated ash-cloud surges (Chaps. 6 and 10)
(c) Explosive eruptions can be very voluminous, producing landscape mantling pyroclastic pumice, scoria and ash fallout deposits (Chap. 9) for tens to hundreds of km from the vent, either radially on a no wind day, or downwind on a day with strong stratospheric winds.
(d) Explosive eruption columns can reach >30 km in height (e.g. the AD 79 eruption of Vesuvius, Italy) and pumice and ash dispersal are at continental scale, severely affecting aircraft operations for days, weeks or more.
(e) Ballistic blocks can be ejected up to 4–6 km from the vent in case of Plinian intensities with fatal and very destructive impact consequences. At Stromboli volcano, Italy, ballistics have fallen onto the towns of Stromboli and Ginostra, both located on the shoreline on opposite sides of the island, during often unexpected, violent, vulcanian blasts from the summit vents.
(f) Fallout deposits will load and cause roof collapse (especially if composed of wet ash; e.g. 2018 eruption of Fuego volcano, Guatamala), bury buildings, destroy crops and other vegetation, destroy food options for animals, leading to significant deaths, and cause respiratory and other health problems.
(g) Pyroclastic flows and surges (pdcs), especially associated with vulcanian and plinian eruptions, or caldera collapse for larger scale eruptions, can travel many kilometres and up to tens to over 100 km, sometimes over high mountains (Chap. 10), and sometimes over the sea (e.g. the eruption Krakatau 1883, Sunda Strait; Chap. 11), destroying all buildings and infrastructure in their path, setting fire to forests, and leading to many human and animal fatalities.
(h) Particularly devastating are lateral blasts, the deposits of which may have very low preservation potential in the stratigraphic record (Chap. 10).
(i) The entry of pyroclastic flows into the sea or a large lake can cause tsunamis that radiate from the entry point into the water (Chap. 11).
(j) The steep slopes of volcanoes generate intense surface process phenomena, including landslides, sector collapses, lahars of various types, i.e. debris flows, hyper-concentrated flood flows and intense stream flows, carrying and remobilising potentially large volumes of volcanic sediment, particularly after large-volume explosive eruptions (Chap. 12). All of these phenomena can destroy and bury buildings and infrastructure and cause major human and animal fatalities (up to tens of thousands; e.g. 1985 eruption of Nevado del Ruiz volcano, Colombia).
(k) Rapidly moving landslides and sector collapses into the sea where maritime or marine stratovolcanoes occur, will cause tsunamis, some of which can be huge and potentially devastating (Chap.12).
(l) Stratovolcanoes can also release large volumes of fine ash and aerosol gases (e.g. SO_2) that can at least affect weather and global temperatures for several years after a major eruption. Fine ash can also be remobilised by aeolian processes (especially in arid environments), which can compound primary impacts on short to even longer timescales (Jarvis et al. 2020).

(m) Hydrothermal systems on stratovolcanoes can produce sometimes unexpected hydrothermal explosions that can produce numerous fatalities and very severe thermal and chemical burns (e.g. Whakaari-White Island stratovolcano on 9 December 2019).

Many stratovolcanoes pose a high risk for populated areas and as such volcanic hazard assessments have been applied to a large number of stratovolcanoes. Hazard information can be combined to create integrated hazard zonation maps, which define zones of high to low risk; e.g. the Lesser

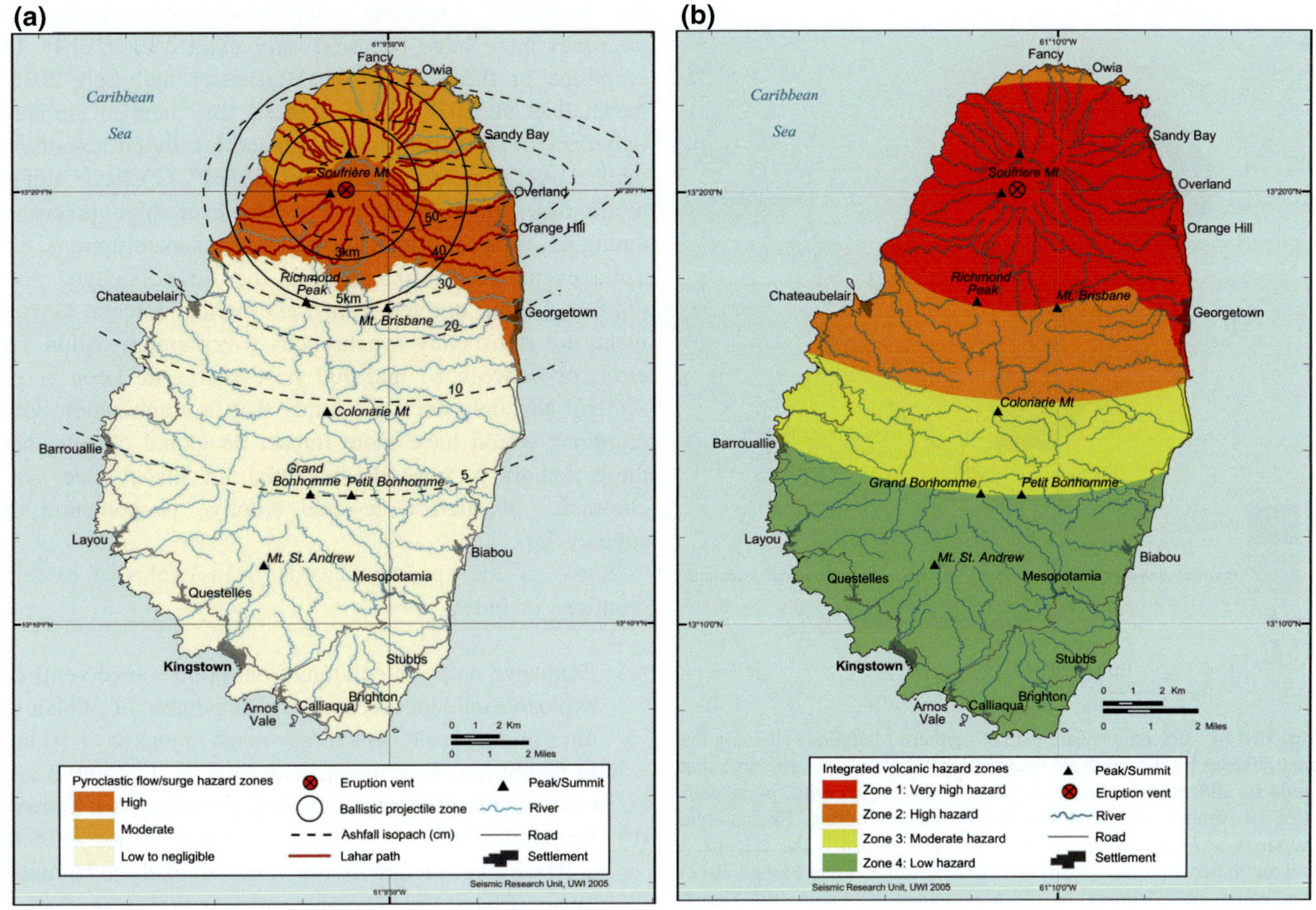

Fig. 16.111 Volcanic hazard map **a**, and integrated volcanic hazard zones **b** for the most likely scenario for a future eruption of Soufrière volcano, St. Vincent, Lesser Antilles, involving both effusive dome-forming and explosive activity (From Lindsay and Robertson 2018, reproduced from Lindsay et al. 2005; through Creative Commons Attribution 4.0 International licence)

Antilles volcanic arc (Lindsay et al. 2005; Lindsay and Robertson 2018), El Chichón and Nevado de Toluca, Mexico (Macías et al. 2008; Capra et al. 2008). This approach is illustrated for the Soufrière, St Vincent stratovolcano (Fig. 16.111), which during the past 4000 yr has had on average one explosive eruption every 100 yr, the latest being in 2020–21. However, imperative to this approach is that "boots on-the ground" studies correctly identify all the critical volcanic, especially pyroclastic facies. As an example, on Martinique, present-day volcanic hazard zones essentially correspond to the risk from pelean-style activity from Mt Pelée, and have ignored the potential effects of low aspect ratio, pumiceous pdcs (misidentified as distal pumice fallout deposits), formed during prehistoric plinian-style activity ca. 2000 yr BP (Fig. 16.112; Wright et al. 2016). These passed over topography 500 m high 10 km from vent, and the zone of potential devastation from such an eruption extends south to the main population centre of Fort-de-France, presently regarded as low to very low risk.

16.12.7 Caldera Volcanoes

Caldera volcanoes in any continental to shallow marine environment pose huge hazards because they are potentially highly explosive and can erupt huge volumes of pumice, scoria, fine ash and gas, as well as much smaller eruptions at any time, which include almost all of the above-mentioned eruptions and related hazards. Since calderas range in diameter (Fig. 16.52) from < 3 km (Mt Pinatubo; Fig. 16.41d) to ~100 km (Toba; Fig. 16.53c), their impacts range from significant to catastrophic. For example, during the 1991 eruption of Mt Pinatubo in the Philippines, erupted ~5 km^3 dense rock equivalent of magma, dumping some 8–10 km^3 of pumiceous deposits and ash on the landscape and releasing ~20 Mt of SO_2 into the atmosphere, causing a −0.5 °C cooling of the atmosphere for 2 years after the eruption. At the other end of the spectrum, the 74 ka (VEI 8/VEI* 12; see discussion Chap. 9, Table 9.1) eruption of Toba caldera on Sumatra, Indonesia, erupted >2,000 km^3 DRE of magma, a deposit volume

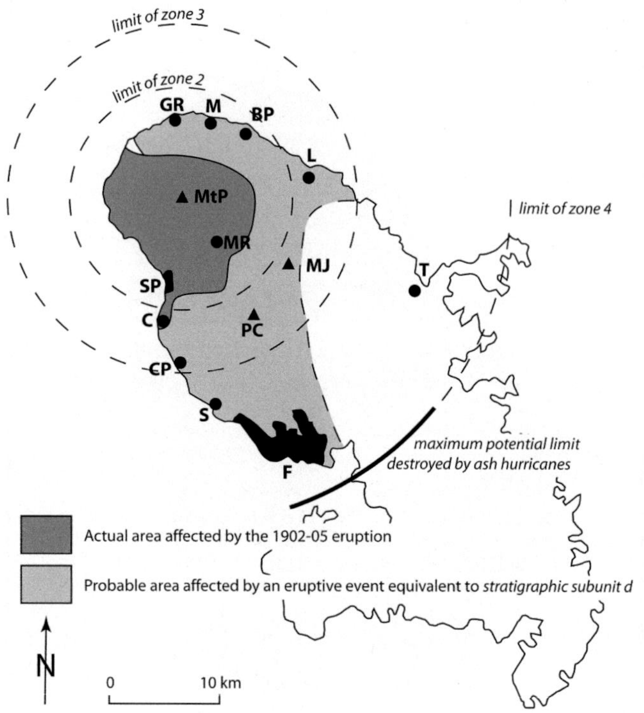

Fig. 16.112 Volcanic hazard map of northern Martinique showing the area affected by the 1902–05 eruption of Mt Pelée, and the area that could be affected by a future eruption of low aspect ratio, pumiceous pdcs of similar magnitude to those erupted during Plinian-style eruptions of Mt Pelée ca. 2000 yr BP. Also shown is the calculated maximum potential run-out distance (After Andrews and Manga 2011) that could be achieved by pdcs surmounting Pitons du Carbet (=28 km). Circles are volcanic hazard zonation contours from the Volcanic Atlas of the Lesser Antilles (Lindsay et al. 2005); within zones 1–4 occurs the area of moderate to highest hazard (closely corresponding to the major risk from Pelean-style activity), zone 5 is the area regarded as having low risk, and zone 6 low to very low risk. This study indicates the risk posed in these later two zones from future Plinian-style activity requires reassessment; i.e. zone 5 should be upgraded to high (equating to the minimum run-out distance from pdcs capable of surmounting Pitons du Carbet, = 15 km), and zone 6 to moderate. SP = St. Pierre; C = Carbet; CP = Case Pilote; S = Schoelcher; F = Fort-de-France; T = Trinité: L = Le Lorrain; BP = Basse Pointe; M = Macouba; GR = Grand'Rivière; MR = Morne Rouge; MtP = Mt Pelée; PC = Pitons du Carbet; MJ = Morne Jacob. (From Wright et al. 2016)

of $\sim$3,000 km^3, released $\sim$5,000 Mt of SO_2, and may have caused a global cooling of as much as −5 to −10 °C, although the effects on early human societies remain highly debated. Nonetheless, the 74 ka event was a truly supereruption, that if it occurred with current global population of almost 8 billion people, would be by any measure disastrous! While the 74 ka Toba event represents the largest Quaternary eruption, certainly much larger eruptions occurred in the geological record.

As eruption frequency and magnitude are inversely related (such as for earthquakes), the frequency of large volume, caldera-forming eruptions is fortunately low. According to the LaMEVE Database (http://globalvolcanomodel.org/vogripa/) there have occurred only eight VEI 7/VEI* 11 eruptions in the past 10 kyr (Gertisser and Self 2015; Newhall et al. 2018). This suggests they happen globally only once or twice per millennium and locally far less often, while larger supereruptions (VEI 8/VEI* 12) have global return times of 10 s–100 s kyr. However, high precision dating has now confirmed that two recent supereruptions, the Toba event (73.8 ± 0.3 ka) and the Los Chocoyos Ash (74.8 ± 1.8 ka) erupted from Lake Atitlán caldera, Guatamala, are statistically synchronous (overlapping within 1σ error, or between 87 and 400 yr; Cisneros de León et al. 2021b). Not only does this imply that in combination both eruptions would have more impact on global climate, but these authors propose the potential for planet-wide synchronicity of such large-scale eruptive phenomenon or supercycles.

Some of the specific hazards from explosive caldera eruptions include:

(a) Explosive eruption columns from point source vents of explosive caldera volcanoes are inevitably of plinian to ultra-plinian scale, which may reach in excess of 50 km in height and may merge with huge co-ignimbrite ash columns producing the largest dispersals of ash known.
(b) Eruption columns or fountains from syn- to post-caldera collapse ring fracture fissure vents are likely to be much lower (Chap. 10; Giordano and Cas 2021), producing catastrophic pyroclastic flows and voluminous "co-ignimbrite) ash fallout deposits (Chaps. 9 and 10).
(c) Point source vent generated high eruption columns can produce large volumes of fallout deposits that can be dispersed hundreds to thousands of km downwind. Close to vent they can be >20 m thick, thinning to >1 m thick 30 km from vent, and then thinning further. Such thick deposits of even low-density pumice overload roofs, causing collapse and other damage to buildings and infrastructure.
(d) Fallout deposits infiltrate and contaminate water and power facilities for hundreds or more km from vent, rendering them inoperable.
(e) Widespread fallout deposits destroy crops for hundreds of km in the area of dispersal, could lead to food shortages, famine and pestilence (Newhall et al. 2018).
(f) The fine ash causes respiratory problems for humans and animals for hundreds of km from vent and beyond, leading to many fatalities.
(g) Pyroclastic flows can devastate the landscape in all directions for up to 100 km, and sometimes more if they are large enough, destroying and burying all buildings, infrastructure, forests, and causing huge human and animal fatalities.

(h) The finest ash can circumnavigate the globe many times in the stratosphere which, together with large volumes of aerosol gases (e.g. 20,000 Mt of SO_2 from the 74 ka Toba caldera eruption), can increase global albedo and cause global cooling (e.g. perhaps −10 °C after the 74 ka Toba eruption).
(i) Surface processes will also represent huge hazards after a significant explosive eruption because huge volumes of unlithified pyroclastic debris can be remobilised into powerful debris flows, hyper-concentrated flood flows and stream flows in temperate to tropical climates where significant annual surface stream flow occurs.
(j) Fine ash, as indicated above, can also be redistributed by aeolian processes, causing the formation of resuspension clouds which could impact the health of humans and livestock, and aviation for centuries after an eruption (Jarvis et al. 2020).
(k) Caldera volcanoes for most of their life do not erupt but are in unrest, such as at Campi Flegrei (Orsi et al. 1996), where smaller scale, non-caldera-forming eruption may occur at any time, of almost any size, from small phreatic to plinian scale, and also impressive caldera resurgence (or subsidence) may pose serious hazards, such as at Ischia volcano (Selva et al. 2019).

Caldera volcanoes are therefore particularly dangerous close to and even within densely populated metropolitan areas. One of the world's best examples of a high volcanic risk area is Naples, which has been growing within the Campi Flegrei caldera and along the slopes of Somma Vesuvius (e.g. Marzocchi and Woo 2009; Alberico et al. 2011; Selva et al. 2012). The volcanic hazard of the caldera is extremely high because its magmatic system is active, with the last eruption occurred in 1538 AD (Monte Nuovo), widespread fumaroles and hot spring activity, and uplift in the Pozzuoli area of ≥ 4 m cumulatively during the past 50 years which has triggered mass evacuations during episodes of unrest (Orsi et al. 2004; Charleton et al. 2020). The volcano has a history of highly explosive activity, including the Campanian ignimbrite (39 ka) and Neapolitan Yellow Tuff (15 ka), which produced two major ignimbrite-forming eruptions resulting in at least two major caldera collapses. This is all compounded by the fact that about 1.5 million people live within and around the caldera. A volcanic hazard map for the Campi Flegrei area is shown in Fig. 16.113. This takes into account the eruptive behaviour of the volcano over the past 5 ka years, when the present-day stress regime was established. From 2005, a new inflation phase has initiated, still ongoing at the time of writing, with a total uplift of $\sim$0.7 m as of 2020 (Bevilacqua et al. 2022).

16.12.8 Intra- or Subglacial Volcanoes

Numerous subglacial eruptions in Iceland have provided many case studies of the potential hazards associated with this type of volcanic activity:

(a) Even small eruptions that produce a meltwater pond can have a huge socio-economic impact when they facilitate phreatomagmatic eruptions. The April–May 2010 eruption of Eyjafjallajökull volcano in Iceland produced a fine ash-laden eruption plume that was blown over Europe and halted air traffic for almost a week (Fig. 16.114a; Sect. 16.10.3), causing an estimated €3.5 B of economic damage.
(b) The other significant hazard is meltwater breakout floods called jökulhlaups, which emanate from the base of the glacier hosting the subglacial eruption (Chap. 12). The meltwaters are generated by melting of the base of the glacier as magma is erupted and then find their way along the base of the glacier and through fractures in the ice, forming devastating floods that can destroy infrastructure such as bridges and power installations.
(c) More proximal ash fallout can have significant effects on buildings, crops and the health of humans and animals, and more rarely, pdcs are known to occur during emergent rhyolitic subglacial eruptions and can travel tens of km across and beyond the ice mass (Moles et al. 2019).

As an example, the hazard map of the civil protection system in southern Iceland is shown in Fig. 16.114b (Elíasson 2014), where a subglacial eruption at Katla volcano could flood any of three lower lying sectors of the Mýrdalsjökull ice sheet with torrents of meltwater. A jökulhlaup will be generated at each location, Kötlu in the east, Enta in the west and Sólheimar in the south. The major potential hazards are the jökulhlaup floods that can overflow three different floodplains (sandurs), an associated tsunami that could inundate the south coast of Iceland, and a volcanic ash that endangers civil aviation on an even larger scale than the Eyjafjallajökull eruption in 2010 (Elíasson 2014).

16.12.9 Mafic Oceanic Volcanic Centres

Little is known about the potential hazards of submarine mafic volcanic systems, but we can make a few comments:

(a) Mid-oceanic ridges release a constant, sometimes pulsing (depending on spreading rates) rate of CO_2 into the oceans, some of which will be transferred into the

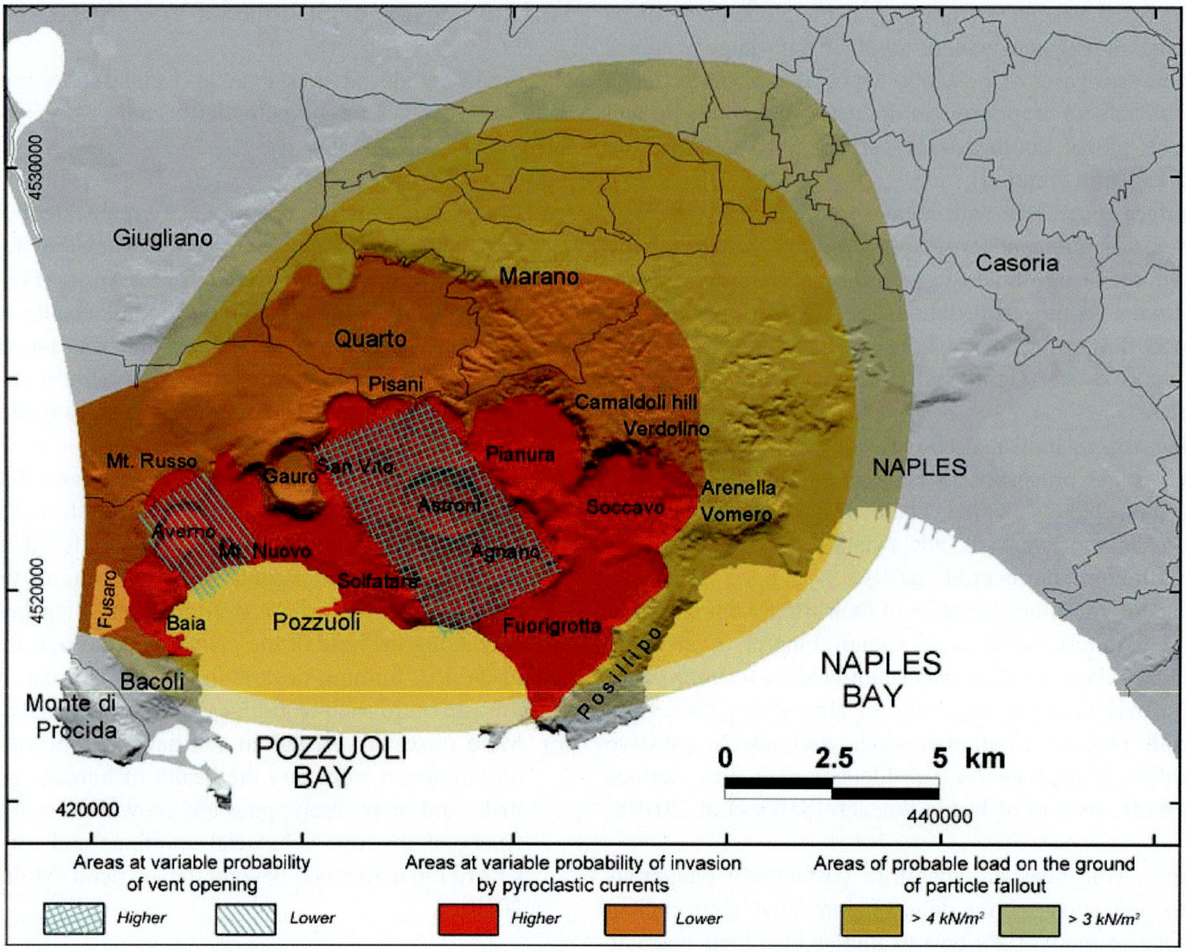

Fig. 16.113 Volcanic hazard map of the Campi Flegrei caldera, Naples, Italy. This has been constructed based on the eruptive history over the past 5 kyr (From Orsi et al. 2004). A future vent most likely will open in the area between Agnano and San Vito, which has been the site of the large majority of the eruptions during the past 5 kyr, corresponding to the portion of a resurgent caldera block under tensile stress regime. Another possible location of the vent could be in the Monte Nuovo-Averno area at the junction of two fault systems. While the largest eruption of the past 5 kyr, could be considered the maximum expected, the most probable maximum eruption might be larger. Very energetic pdcs could surmount morphological barriers and in favourable conditions, exit the lowland towards the north

atmosphere, contributing to the greenhouse gas budget of the Earth's atmosphere. However, it has been doing this for at least 2 billion years, and so is not impacting climate change at the rate that the Earth has been experiencing since the start of the industrial revolution, ~150 years ago.

(b) Oceanic plateau volcanism would have raised sea level because the volumes of magma erupted as lavas on the seafloor is huge, in some cases much higher than for continental flood basalts. Some of that sea level would then have subsided as the plateaus began to isostatically subside.

(c) Oceanic plateau volcanism would have increased the supply of CO_2 into the oceans and then into the atmosphere above normal mid-oceanic ridge background supply rates, so enhancing global greenhouse warming effects (e.g. Kerr 2005).

(d) As for continental flood basalt event oceanic plateau volcanism would have released significant volumes of SO_2, Cl, F, and H_2S. However, these gases would have been absorbed into the oceans, contributing to acidification of the oceans, and impacting life in the oceans.

Much remains unknown about seafloor volcanism and its potential effects on long-term climate variability, but seismo-acoustic coverage of much of the ocean is a potential future development (Tolstoy 2015; Tolstoy and Tan 2019). This would allow seafloor eruptions to be identified with some confidence without the need for on-site confirmation. Such monitoring would better quantify present-day eruption rates in the ocean environment and help understand the contribution of mid-ocean ridges to present-day volcanism, and the effects on the global carbon cycle and flow of greenhouse gases.

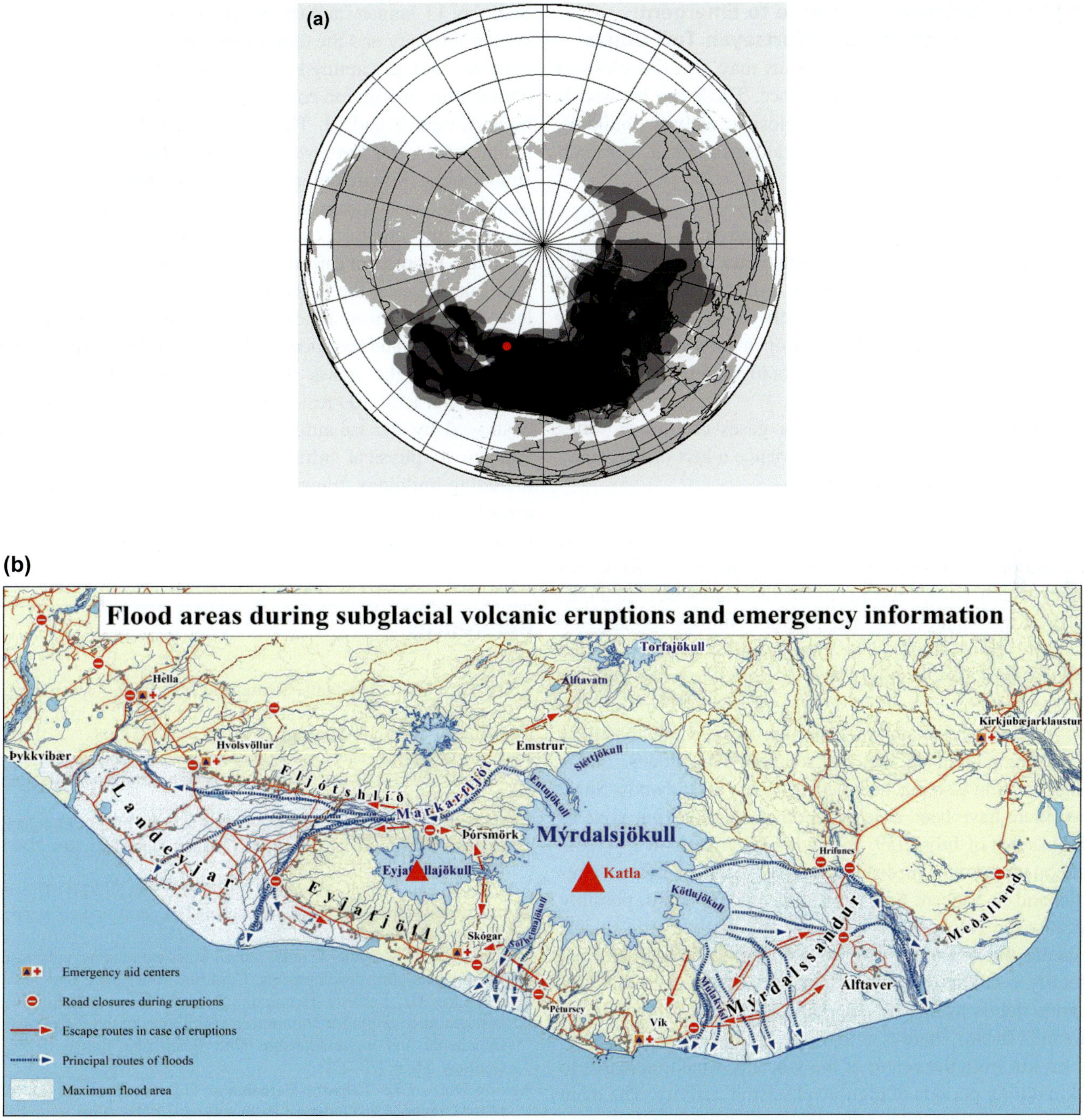

Fig. 16.114 Glacio-volcanic hazards in southern Iceland. **a** Cumulative map showing the extent of the phreatomagmatic ash-cloud erupted between 14 and 25 April 2010 from Eyjafjallajökull volcano (red circle). This closed European airspace on several days. (From Volcanic Ash Advisory Centre, UK—VAACUK; *Source* Wikimedia through Creative Commons Attribution-Share Alike 3.0 Unported licence). **b** Hazard map of the civil protection system showing high risk areas from flooding during subglacial eruptions and jökulhlaup events from Katla volcano. Katla features a 110 km^2 ice-filled summit caldera that is up to 700 m deep; some 25 km to the west is the summit caldera of the smaller Eyjafjallajökull volcano. (After Elíasson 2014; through Creative Commons Attribution-Share Alike 3.0 Unported licence)

16.12.9.1 Shallow Submarine to Emergent Seamounts and Surtseyan Tuff Cones

Submarine volcanoes and seamounts may also rise close to and even breach the seawater surface. These can be associated with phreatomagmatic explosive activity building surtseyan tuff cones. The major hazards associated with such settings are:

(a) Airborne fine ash may impact aviation.
(b) Explosive tephra jets that evolve into base surges could flow over water for perhaps 1–2 km, and sector collapses could generate relatively small tsunamis (with amplitudes of up to 10 m in open waters), that nevertheless could still pose dangers to shipping and coastal areas.
(c) Bubbles of escaping volcanic gases can lower water density, and boats could experience a loss of buoyancy creating a sinking hazard.

An example of an active shallow submarine-emergent volcano or seamount that is closely monitored is Kick 'em Jenny which rises 1300 m above the Caribbean Seafloor, north of the island of Grenada, Lesser Antilles volcanic arc (Lindsay et al. 2005). The first recorded activity of the volcano was 23–24 July 1939 an eruption broke the sea surface, sending a cloud of steam and debris 275 m into the air. A series of tsunamis $\sim$2 m high reached the coastlines of northern Grenada and the southern Grenadines, and a small tsunami also reached the west coast of Barbados. Since then there have been at least 13 other eruptions but none as large as that of July 1939. Following an eruption in 1943, the schooner "Island Queen" mysteriously sank between Grenada and St Vincent with over 60 people on board, possibly because the volcano was still degassing. Bathymetric studies indicate the summit is 185 m below the sea surface, and identify a collapse feature which was the source of a submarine debris avalanche extending 15 km west towards the Grenada Basin. There is a maritime exclusion zone normally at 1.5 km from the centre of the volcano, which increases to 5 km during periods of high-level seismic activity. The main future risk is that of a fully emergent surtseyan tuff cone eruption with column heights >1–5 km.

Other famous surtseyan eruptions in historical times have been witnessed in 1831 at Isola Ferdinandea (or Graham Bank) in the Sicily channel, Italy (Coltelli et al. 2016), in 1957–58 at Capelinhos, Azores (Cole et al. 2001), in 1963–67 at Surtsey, Iceland (Thorarinsson et al 1964), and in 2009, 2014–15 and 2021–22 Hunga Tonga-Hunga Ha'apai islands, Tonga (Vaughan and Webley 2010; Garvin et al. 2018; Cronin 2022; Brenna et al. 2022). The eruption of 20 December 2021 to 14 January 2022 at Hunga Tonga-Hunga Ha'apai was typically of surtseyan explosive style, but on Saturday 15 January at 4.08 pm (UTC) it erupted with much greater intensity and the eruption column grew spectacularly over an hour as captured in satellite imagery, producing a phreatoplinian eruption column at least 40 to possibly 57 km high (Chaps. 3 and 9). This dispersed fallout ash deposits radially for >100 km over the Tonga Islands, and also a stratospheric downwind plume of very fine ash and SO_2 westwards over Australia and beyond. The phreatoplinian phase of the eruption caused tsunamis in Tonga, Fiji and along the Pacific rim causing damage in New Zealand, Japan, the United States, the Russian Far East, Chile, and Peru. At least four people were killed in Tonga from tsunami waves up to 15 m high, and two people drowned in Peru when a 2 m wave struck the coast. In Tonga some 600 structures including at least 300 residential buildings were damaged or destroyed by the tsunami waves, with the total estimated damage to physical infrastructure estimated at US$90 M including buildings, transport, power and water, sea and air infrastructure, repairs to submarine telecommunication cables, agriculture, forestry, fishing, and ash fallout clean up.

References

Ablay GJ, Martí J (2000) Stratigraphy, structure, and volcanic evolution of the Pico Teide-Pico Viejo formation, Tenerife, Canary Islands. J Volcanol Geotherm Res 103(1–4):175–208

Acocella V (2007) Understanding caldera structure and development: an overview of analogue models compared to natural calderas. Earth Sci Rev 85(3–4):125–160

Acocella V, Cifelli F, Funiciello R (2000) Analogue models of collapse calderas and resurgent domes. J Volcanol Geotherm Res 104(1–4):81–96

Acocella V, Funiciello R, Marotta E, Orsi G, De Vita S (2004) The role of extensional structures on experimental calderas and resurgence. J Volcanol Geotherm Res 129(1–3):199–217

Adatte T, Bond DP, Keller G (eds) (2020) Mass extinctions, volcanism, and impacts: new developments. Geol Soc Amer Spec Pap 544

Aguirre-Díaz GJ, Labarthe-Hernández G (2003) Fissure ignimbrites: fissure-source origin for voluminous ignimbrites of the Sierra Madre Occidental and its relationship with Basin and Range faulting. Geology 31(9):773–776

Aguirre-Díaz GJ, Labarthe-Hernández G, Tristán-González M, Nieto-Obregón J, Gutiérrez-Palomares I (2008) The ignimbrite flare-up and graben calderas of the Sierra Madre Occidental, Mexico. Developm in Volcanol 10:143–180

Aguirre-Díaz GJ, López-Martínez M (2009) Geologic evolution of the Donguinyó-Huichapan caldera complex, central Mexican Volcanic Belt, Mexico. J Volcanol Geoth Res 179(1–2):133–148

Alberico I, Petrosino P, Lirer L (2011) Volcanic hazard and risk assessment in a multi-source volcanic area: the example of Napoli city (Southern Italy). Nat Haz Earth Sys Sci 11(4):1057–1070

Alfano F, Ort MH, Pioli L, Self S, Hanson SL, Roggensack K, ... Clarke AB (2019) Subplinian monogenetic basaltic eruption of Sunset Crater, Arizona, USA. Geol Soc Am Bull 131(3–4):661–674

Allan ASR, Morgan DJ, Wilson CJN, Millet M-A (2013) From mush to eruption in centuries: assembly of the super-sized Oruanui magma body. Contribs Mineral Petrol 166(1):143–164

Allen RL (1988) False pyroclastic textures in altered silicic lavas, with implications for volcanic-associated mineralization. Econ Geol 83 (7):1424–1446
Allen SR (2001) Reconstruction of a major caldera-forming eruption from pyroclastic deposit characteristics: Kos Plateau Tuff, eastern Aegean Sea. J Volcanol Geotherm Res 105(1–2):141–162
Allen SR, Cas RAF (2001) Transport of pyroclastic flows across the sea during the explosive, rhyolitic eruption of the Kos Plateau Tuff, Greece. Bull Volcanol 62(6–7):441–456
Allen SR, Fiske RS, Tamura Y (2010) Effects of water depth on pumice formation in submarine domes at Sumisu, Izu-Bonin arc, western Pacific. Geology 38(5):391–394
Allen SR, Mcphie J (2000) Water-settling and resedimentation of submarine rhyolitic pumice at Yali, eastern Aegean, Greece. J Volcanol Geotherm Res 95(1–4):285–307
Ambrosio M, Doveri M, Fagioli MT, Marini L, Principe C, Raco B (2010) Water–rock interaction in the magmatic-hydrothermal system of Nisyros Island (Greece). J Volcanol Geotherm Res 192 (1–2):57–68
Amelung F, Jónsson S, Zebker H, Segall P (2000) Widespread uplift and 'trapdoor'faulting on Galapagos volcanoes observed with radar interferometry. Nature 407(6807):993
Ancochea E, Fúster JM, Ibarrola E, Cendrero A, Coello J, Hernán F, Cantagrel JM, Jamond C (1990) Volcanic evolution of the island of Tenerife (Canary Islands) in the light of new K-Ar data. J Volcanol Geotherm Res 44:231–249
Anderson EM (1937a) Cone-sheets and ring-dykes: the dynamical explanation. Bull Volcanol 1(1):35–40
Anderson EM (1937b) The dynamics of the formation of cone-sheets, ring-dykes, and caldron-subsidences. Proc R Soc Edinburgh 56:128–157
Andersen NL, Singer BS, Jicha BR, Beard BL, Johnson CM, Licciardi JM (2017) Pleistocene to Holocene growth of a large upper crustal rhyolitic magma reservoir beneath the active Laguna del Maule volcanic field, central Chile. J Petrol 58(1):85–114
Andrew RE, Gudmundsson A (2007) Distribution, structure, and formation of Holocene lava shields in Iceland. J Volcanol Geotherm Res 168(1–4):137–154
Anzidei M, Carapezza ML, Esposito A, Giordano G, Lelli M, Tarchini L (2008) The Albano Maar Lake high resolution bathymetry and dissolved CO2 budget (Colli Albani volcano, Italy): constrains to hazard evaluation. J Volcanol Geotherm Res 171(3–4):258–268
Aramaki S (1984) Formation of the Aira Caldera, southern Kyushu, ~22,000 years ago. J Geophys Res Solid Earth 89(B10):8485–8501
Ashwell PA, Kennedy BM, Gravley DM, Von Aulock FW, Cole JW (2013) Insights into caldera and regional structures and magma body distribution from lava domes at Rotorua Caldera, New Zealand. J Volcanol Geotherm Res 258:187–202
Auker MR, Sparks RS, Siebert L, Crosweller HS, Ewert J (2013) A statistical analysis of the global historical volcanic fatalities record. J Appl Volcanol 2(1):2
Austin-Erickson A, Büttner R, Dellino P, Ort MH, Zimanowski B (2008) Phreatomagmatic explosions of rhyolitic magma: experimental and field evidence. J Geophys Res Solid Earth 113(B11)
Bacon CR (1983) Eruptive history of Mount Mazama and Crater Lake caldera, Cascade Range, USA. J Volcanol Geother Res 18(1–4):57–115
Bacon CR, Mastin LG, Scott KM, Nathenson M (1997) Volcano and earthquake hazards in the Crater Lake region, Oregon. US Geol Surv Open-File Rept 97(487):32
Báez W, Bustos E, Chiodi A, Reckziegel F, Arnosio M, de Silva S, ... Peña-Monné JL (2020) Eruptive style and flow dynamics of the pyroclastic density currents related to the Holocene Cerro Blanco eruption (Southern Puna plateau, Argentina). J South Am Earth Sci 98:102482
Báez W, Nuñez GC, Giordano G, Viramonte JG, Chiodi A (2017) Polycyclic scoria cones of the Antofagasta de la Sierra basin, Southern Puna plateau, Argentina. Geol Soc Lond Spec Publ 446 (SP446):443
Ballard RD, Hekinian R, Francheteau J (1984) Geological setting of hydrothermal activity at 12 50′ N on the East Pacific Rise: a submersible study. Earth Planet Sci Letts 69(1):176–186
Ballard RD, van Andel TH (1977) Morphology and tectonics of the inner rift valley at lat 36 50′ N on the Mid-Atlantic Ridge. Geol Soc Amer Bull 88(4):507–530
Barberi F, Cioni R, Rosi M, Santacroce R, Sbrana A, Vecci R (1989) Magmatic and phreatomagmatic phases in explosive eruptions of Vesuvius as deduced by grain-size and component analysis of the pyroclastic deposits. J Volcanol Geotherm Res 38(3–4):287–307
Batiza R, Vanko D (1983) Volcanic development of small oceanic central volcanoes on the flanks of the East Pacific Rise inferred from narrow-beam echo-sounder surveys. Marine Geol 54(1):53–90
Behncke B, Branca S, Corsaro RA, De Beni E, Miraglia L, Proietti C (2014) The 2011–2012 summit activity of Mount Etna: Birth, growth and products of the new SE crater. J Volcanol Geotherm Res 270:10–21
Bellier O, Sébrier M (1994) Relationship between tectonism and volcanism along the Great Sumatran fault zone deduced by SPOT image analyses. Tectonophys 233(3–4):215–231
Best MG, Christiansen EH, Gromme S (2013) Introduction: The 36–18 Ma southern Great Basin, USA, ignimbrite province and flareup: Swarms of subduction-related supervolcanoes. Geosphere 9 (2):260–274
Bevilacqua A, De Martino P, Giudicepietro F, Ricciolino P, Patra A, Pitman EB, ... Neri A (2022) Data analysis of the unsteadily accelerating GPS and seismic records at Campi Flegrei caldera from 2000 to 2020. Sci Rep 12(1):19175
Bickle MJ, Nisbet EG, Martín A (1994) Archean greenstone belts are not oceanic crust. The J Geol 102(2):121–137
Bissig T, Clark AH, Rainbow A, Montgomery A (2015) Physiographic and tectonic settings of high-sulfidation epithermal gold–silver deposits of the Andes and their controls on mineralizing processes. Ore Geol Rev 65:327–364
Blaikie TN, Ailleres L, Betts PG, Cas RAF (2014a) A geophysical comparison of the diatremes of simple and complex maar volcanoes, Newer Volcanics Province, south-eastern Australia. J Volcanol Geotherm Res 276:64–81
Blaikie TN, Ailleres L, Betts PG, Cas RAF (2014b) Interpreting subsurface volcanic structures using geologically constrained 3-D gravity inversions: examples of maar-diatremes, Newer Volcanics Province, southeastern Australia. J Geophys Res Solid Earth 119 (4):3857–3878
Blaikie TN, Ailleres L, Cas RAF, Betts PG (2012) Three-dimensional potential field modelling of a multi-vent maar-diatreme—the Lake Coragulac maar, Newer Volcanics Province, south-eastern Australia. J Volcanol Geotherm Res 235:70–83
Blong RJ (1984) Volcanic hazards: a sourcebook on the effects of eruptions. Academic Press, Elsevier, 424 pp. https://doi.org/10.1016/C2009-0-21853-8
Bogie I, Lawless JV, Rychagov S, Belousov V (2005) Magmatic-related hydrothermal systems: classification of the types of geothermal systems and their ore mineralization. In: Rychagov S (ed) Geothermal and mineral resources of modern volcanism areas. In: Proceedings of international Kuril-Kamchatka field workshop
Bosman A, Chiocci FL, Romagnoli C (2009) Morpho-structural setting of Stromboli volcano revealed by high-resolution bathymetry and backscatter data of its submarine portions. Bull Volcanol 71 (9):1007

Boudon G, Le Friant A, Komorowski JC, Deplus C, Semet MP (2007) Volcano flank instability in the Lesser Antilles Arc: diversity of scale, processes, and temporal recurrence. J Geophys Res: Solid Earth 112(B8)

Boudon G, Villemant B, Le Friant A, Paterne M, Cortijo E (2013) Role of large flank-collapse events on magma evolution of volcanoes. Insights from the Lesser Antilles Arc. J Volcanol Geotherm Res 263:224–237

Boyce J (2013) The Newer Volcanics Province of southeastern Australia: a new classification scheme and distribution map for eruption centres. Aust J Earth Sci 60(4):449–462

Boyce JA, Nicholls IA, Keays RR, Hayman PC (2015) Variation in parental magmas of Mt Rouse, a complex polymagmatic monogenetic volcano in the basaltic intraplate Newer Volcanics Province, southeast Australia. Contribs Mineral Petrol 169(2):11

Branca S, Coltelli M, Groppelli G, Lentini F (2011) Geological map of Etna volcano, 1: 50,000 scale. Ital J Geosci 130(3):265–291

Brand BD, Clarke AB (2009) The architecture, eruptive history, and evolution of the Table Rock Complex, Oregon: from a Surtseyan to an energetic maar eruption. J Volcanol Geotherm Res 180(2–4):203–224

Branney M, Acocella V (2015) Calderas. In: Sigurdsson H, Houghton B, Mcnutt SR, Rymer H, Stix J (eds) The encyclopedia of volcanoes. Academic Press Elsevier, pp 299–315

Branney MJ (1995) Downsag and extension at calderas: new perspectives on collapse geometries from ice-melt, mining, and volcanic subsidence. Bull Volcanol 57(5):303–318

Brenna M, Cronin SJ, Smith IE, Pontesilli A, Tost M, Barker S, ... Vaiomounga R (2022) Post-caldera volcanism reveals shallow priming of an intra-ocean arc andesitic caldera: Hunga volcano, Tonga, SW Pacific. Lithos 412:106614

Brothers DS, Driscoll NW, Kent GM, Harding AJ, Babcock JM, Baskin RL (2009) Tectonic evolution of the Salton Sea inferred from seismic reflection data. Nature Geosci 2(8):581

Brown RJ, Manya S, Buisman I, Fontana G, Field M, Mac Niocaill C, Sparks RSJ, Stuart FM (2012) Eruption of kimberlite magmas: physical volcanology, geomorphology and age of the youngest kimberlitic volcanoes known on earth (the Upper Pleistocene/Holocene Igwisi Hills volcanoes, Tanzania). Bull Volcanol 74 (7):1621–1643

Browne PRL, Lawless JV (2001) Characteristics of hydrothermal eruptions, with examples from New Zealand and elsewhere. Earth Sci Rev 52(4):299–331

Bryan SE, Martí J, Cas RAF (1998) Stratigraphy of the Bandas del Sur Formation: an extracaldera record of Quaternary phonolitic explosive eruptions from the Las Cañadas edifice, Tenerife (Canary Islands). Geol Mag 135(5):605–636

Buchs DM, Arculus RJ, Baumgartner PO, Ulianov A (2011) Oceanic intraplate volcanoes exposed: example from seamounts accreted in Panama. Geology 39(4):335–338

Busby CJ (2012) Extensional and transtensional continental arc basins: case studies from the southwestern United States. In: Busby CJ, Azor (eds) Tectonics of sedimentary basins: recent advances. Blackwell Publishing, pp 382–404

Busby CJ, Kessel LO, Schulz KJ, Foose MP, Slack JF (2003) Volcanic setting of the Ordovician Bald Mountain massive sulfide deposit, northern Maine. Econ Geol Monogr 11:219–244

Busby-Spera CJ (1984) Large-volume rhyolite ash flow eruptions and submarine caldera collapse in the Lower Mesozoic Sierra Nevada, California. J Geophys Res Solid Earth 89(B10):8417–8427

Busby-Spera CJ (1986) Depositional features of rhyolitic and andesitic volcaniclastic rocks of the Mineral King submarine caldera complex, Sierra Nevada, California. J Volcanol Geotherm Res 27 (1–2):43–76

Ca-L J (2012) Seismic reflection imaging and controls on the preservation of ancient sill-fed magmatic vents. J Geol Soc Lond 169(5):503–506

Calvert AT, Lanphere MA (2006) Argon geochronology of Kilauea's early submarine history. J Volcanol Geotherm Res 151(1–3):1–18

Capra L, Norini G, Groppelli G, Macías JL, Arce JL (2008) Volcanic hazard zonation of the Nevado de Toluca volcano, México. J Volcanol Geother Res 176(4): 469-484.

Carey R, Soule SA, Manga M, White JDL, Mcphie J, Wysoczanski R, Jutzeler M, Tani K, Yoerger D, Fornari D (2018) The largest deep-ocean silicic volcanic eruption of the past century. Sci Adv 4 (1):e1701121

Carracedo JC (1999) Growth, structure, instability and collapse of Canarian volcanoes and comparisons with Hawaiian volcanoes. J Volcanol Geotherm Res 94(1–4):1–19

Carracedo JC, Day S, Guillou H, Badiola ER, Canas JA, Torrado FJP (1998) Hotspot volcanism close to a passive continental margin: the Canary Islands. Geol Mag 135(5):591–604

Carracedo JC, Day SJ, Guillou H, Torrado FJP (1999) Giant quaternary landslides in the evolution of La Palma and El Hierro, Canary Islands. J Volcanol Geotherm Res 94(1–4):169–190

Carracedo JC, Perez-Torrado FJ (2013) Geological and geodynamic context of the teide volcanic complex. In: Carracedo JC, Troll VR (eds) Teide Volcano, active volcanoes of the world. Springer, Heidelberg, Berlin, pp 23–36

Carracedo JC, Troll VR (eds) (2013) Teide volcano: Geology and eruptions of a highly differentiated oceanic stratovolcano. Springer-Verlag, p 279

Carracedo JC, Troll VR, Day JM, Geiger H, Aulinas M, Soler V, ... Albert H (2022) The 2021 eruption of the Cumbre Vieja Volcanic Ridge on La Palma, Canary Islands. Geology Today 38(3):94–107

Carracedo-Sánchez MC, Sarrionandia F, Arostegui J, Eguiluz L, Ibarguchi JG (2012) The transition of spatter to lava-like body in lava fountain deposits: features and examples from the Cabezo Segura volcano (Calatrava, Spain). J Volcanol Geotherm Res 227:1–14

Carrasco-Núñez G, Bernal JP, Dávila P, Jicha B, Giordano G, Hernández J (2018) Reappraisal of Los Humeros volcanic complex by new U/Th zircon and 40Ar/39Ar dating: implications for greater geothermal potential. Geochem Geophys Geosyst 19(1):132–149

Carrasco-Núñez G, Ort MH, Romero C (2007) Evolution and hydrological conditions of a maar volcano (Atexcac crater, Eastern Mexico). J Volcanol Geother Res 159(1–3):179–197

Cas RAF (1983) A review of the palaeogeographic and tectonic development of the Palaeozoic Lachlan Fold Belt of southeastern Australia. Geol Soc Aust Spec Paper 10, 104 p.

Cas RAF (1992) Submarine volcanism; eruption styles, products, and relevance to understanding the host-rock successions to volcanic-hosted massive sulfide deposits. Econ Geol 87(3):511–541

Cas RAF, Allen RL, Bull SW, Clifford BA, Wright JV (1990) Subaqueous, rhyolitic dome-top tuff cones: a model based on the Devonian Bunga Beds, southeastern Australia and a modern analogue. Bull Volcanol 52(3):159–174

Cas RAF, Blaikie T, Boyce J, Hayman P, Jordan S, Piganis F, Prata G (2011a) Factors that influence varying eruption styles (from magmatic to phreato-magmatic) in intra-plate continental basaltic volcanic provinces. IUGG2011a General Assembly, July 2011a, Melbourne, Australia, 117 pp

Cas RAF, Bull SW, Allen RL, Clifford BA, Wright JV (1993) Influence of environment on contrasting styles of volcanism, volcaniclastics, and sedimentation: the Devonian Boyd Volcanic Complex, southeastern Australia. A2 Field Trip Guide: IAVCEI Scientific Assembly, Canberra. Aust Geol Surv Bull 1993/64, 68 pp

Cas RAF, Edgar C, Allen RL, Bull S, Clifford BA, Giordano G, Wright JV (2001) Influence of magmatism and tectonics on

sedimentation in an extensional lake basin: the upper Devonian Bunga Beds, Boyd Volcanic Complex, South-Eastern Australia. In: Riggs N, White JDL (eds) Int Assoc Sed Spec Publ, pp 81–108
Cas RAF, Giordano G (2014) Submarine volcanism: a review of the constraints, processes and products, and relevance to the Cabo de Gata volcanic succession. Ital J Geosci 133(3):362–377
Cas R, Giordano G, Balsamo F, Eposito A, Mastro SL (2011c) Hydrothermal breccia textures and processes; Lisca Bianca islet, Panarea volcano, Aeolian Islands, Italy. Econ Geol 106(3):437–450
Cas RAF, Hayman P, Pittari A, Porritt L (2008a) Some major problems with existing models and terminology associated with kimberlite pipes from a volcanological perspective, and some suggestions. J Volcanol Geotherm Res 174(1–3):209–225
Cas RAF, Landis CA, Fordyce RE (1989) A monogenetic, surtla-type, Surtseyan volcano from the Eocene-Oligocene Waiareka-Deborah volcanics, Otago, New Zealand: a model. Bull Volcanol 51(4):281–298
Cas RAF, Porritt L, Pittari A, Hayman PC (2009) A practical guide to terminology for kimberlite facies: a systematic progression from descriptive to genetic, including a pocket guide. Lithos 112:183–190
Cas RAF, Porritt LA, Pittari A, Hayman P (2008b) A new approach to kimberlite facies terminology using a revised general approach to the nomenclature of all volcanic rocks and deposits: descriptive to genetic. J Volcanol Geotherm Res 174(1–3):226–240
Cas RAF, Simmons J (2018) Why deep-water eruptions are so different from subaerial eruptions. Front Earth Sci 6:198. https://doi.org/10.3389/feart.2018.00198
Cas RAF, Van Otterloo J, Blaikie TN, Van den Hove J (2017) The dynamics of a very large intra-plate continental basaltic volcanic province, the Newer Volcanics Province, SE Australia, and implications for other provinces. Geol Soc Lond Spec Publ 446 (SP446):448
Cas RAF, Wright HMN, Folkes CB, Lesti C, Porreca M, Giordano G, Viramonte JG (2011b) The flow dynamics of an extremely large volume pyroclastic flow, the 2.08-Ma Cerro Galán Ignimbrite, NW Argentina, and comparison with other flow types. Bull Volcanol 73 (10):1583–1609
Cas RAF, Wolff JA, Martí J, Olin PH, Edgar CJ, Pittari A, Simmons JM (2022) Tenerife, a complex end member of basaltic oceanic island volcanoes, with explosive polygenetic phonolitic calderas, and phonolitic-basaltic stratovolcanoes. Earth-Sci Rev 230:103990
Cas RAF, Wright JV (1987) Volcanic successions: modern and ancient, London. Allen and Unwin, London
Casalbore D, Romagnoli C, Bosman A, Chiocci FL (2011) Potential tsunamigenic landslides at Stromboli Volcano (Italy): Insight from marine DEM analysis. Geomorphology 126(1–2):42–50
Cashman KV, Giordano G (2014) Calderas and magma reservoirs. J Volcanol Geotherm Res 288:28–45
Cashman KV, Mangan MT (2014) A century of studying effusive eruption in Hawaii. In: Poland MP, Takahashi TJ, Landowski CM (eds) Characteristics of Hawaiian volcanoes. US Geol Surv Prof Pap 1801:357–396
Cashman KV, Sparks RSJ, Blundy JD (2017) Vertically extensive and unstable magmatic systems: a unified view of igneous processes. Science 355(6331):eaag3055
Chadwick WW, Clague DA, Embley RW, Perfit MR, Butterfield DA, Caress DW, Paduan JB, Martin JF, Sasnett P, Merle SG (2013) The 1998 eruption of Axial Seamount: new insights on submarine lava flow emplacement from high-resolution mapping. Geochem Geophys 14(10):3939–3968
Chadwick WW Jr, Dieterich JH (1995) Mechanical modeling of circumferential and radial dike intrusion on Galapagos volcanoes. J Volcanol Geotherm Res 66(1–4):37–52
Chesner CA (2012) The Toba caldera complex. Quat Int 258:5–18
Charlton D, Kilburn C, Edwards S (2019) Volcanic unrest scenarios and impact assessment at Campi Flegrei caldera, Southern Italy. J Appl Volcanol 9(1):1–26
Christiansen RL (1979) Cooling units and composite sheets in relation to caldera structure. Geol Soc Amer Spec Pap 180:29–42
Christiansen RL, Lowenstern JB, Smith RB, Heasler H, Morgan LA, Nathenson M, Mastin LG, Muffler LJP, Robinson JE (2007) Preliminary assessment of volcanic and hydrothermal hazards in Yellowstone National Park and vicinity. US Geol Surv Prof Pap 1071:94
Cioni R, Bertagnini A, Santacroce R, Andronico D (2008) Explosive activity and eruption scenarios at Somma-Vesuvius (Italy): towards a new classification scheme. J Volcanol Geotherm Res 178(3):331–346
Cioni R, Pistolesi M, Rosi M (2015) Plinian and Subplinian Eruptions. In: Sigurdsson H, Houghton B, Mcnutt SR, Rymer H, Stix J (eds) The encyclopedia of volcanoes, 2nd edn. Academic Press Elsevier, pp 520–534
Cioni R, Santacroce R, Sbrana A (1999) Pyroclastic deposits as a guide for reconstructing the multi-stage evolution of the Somma-Vesuvius Caldera. Bull Volcanol 61(4):207–222
Cisneros de León A, Mittal T, de Silva SL, Self S, Schmitt AK, Kutterolf S (2021a) On synchronous supereruptions. Essoar. https://doi.org/10.1002/essoar.10507418.2
Cisneros de León A, Schindlbeck-Belo JC, Kutterolf S, Danišík M, Schmitt AK, Freundt A, Pérez W, Harvey JC, Wang KL, Lee HY (2021b) A history of violence: magma incubation, timing and tephra distribution of the Los Chocoyos supereruption (Atitlán Caldera, Guatemala). J Quat Sci 36(2):169–179
Civico R, Ricci T, Scarlato P et al. (2022) High-resolution Digital Surface Model of the 2021 eruption deposit of Cumbre Vieja volcano, La Palma, Spain. Sci Data 9:435
Clague DA, Dreyer BM, Paduan JB, Martin JF, Chadwick WW, Caress DW, Portner RA, Guilderson TP, McGann ML, Thomas H (2013) Geologic history of the summit of Axial Seamount, Juan De Fuca Ridge. Geochem Geophys 14(10):4403–4443
Clague DA, Moore JG, Reynolds JR (2000a) Formation of submarine flat-topped volcanic cones in Hawai'i. Bull Volcanol 62(3):214–233
Clague DA, Reynolds JR, Davis AS (2000b) Near-ridge seamount chains in the northeastern Pacific Ocean. J Geophys Res Solid Earth 105(B7):16541–16561
Clague DA, Sherrod DR (2014) Growth and degradation of Hawaiian volcanoes. In: Poland MP, Takahashi TJ, Landowski CM (eds) Characteristics of Hawaiian volcanoes. US Geol Surv Prof Pap 1801:97–146
Clapham ME, Renne PR (2019) Flood basalts and mass extinctions. Ann Rev Earth Planet Sci 47:275–303
Clough BJ, Wright JV, Walker GPL (1981) An unusual bed of giant pumice in Mexico. Nature 289:49–50
Cole JW, Milner DM, Spinks KD (2005) Calderas and caldera structures: a review. Earth Sci Rev 69(1–2):1–26
Cole PD, Guest JE, Duncan AM, Pacheco JM (2001) Capelinhos 1957–1958, Faial, Azores: deposits formed by an emergent surtseyan eruption. Bull Volcanol 63:204–220
Colombier M, Scheu B, Wadsworth FB, Cronin S, Vasseur J, Dobson KJ, Hess KU, Tost M, Yilmaz TI, Cimarelli C, Brenna M (2018) Vesiculation and quenching during surtseyan eruptions at Hunga Tonga-Hunga Ha'apai volcano, Tonga. J Geophys Res Solid Earth 123(5):3762–3779
Coltelli M, Cavallaro D, D'Anna G, D'Alessandro A, Grassa F, Mangano G, ... Gresta S (2016) Exploring the submarine Graham Bank in the Sicily Channel. Ann Geophys
Condit CD, Connor CB (1996) Recurrence rates of volcanism in basaltic volcanic fields: an example from the Springerville volcanic field, Arizona. Geol Soc Amer Bull 108(10):1225–1241

Connor CB, Conwasy FM (2000) Basaltic volcanic fields. In In: Sigurdsson H, Stix J, Houghton BF, Mcnutt SR, Rymer H (eds) Encyclopedia of volcanoes. Academic, San Diego, pp 331–343

Conticelli S, Avanzinelli R, Poli G, Braschi E, Giordano G (2013) Shift from lamproite-like to leucititic rocks: Sr–Nd–Pb isotope data from the Monte Cimino volcanic complex vs. the Vico stratovolcano, Central Italy. Chem Geol 353:246–266

Conticelli S, Laurenzi MA, Giordano G, Mattei M, Avanzinelli R, Melluso L, Tommasini S, Boari E, Cifelli F, Perini G (2010) Leucite-bearing (kamafugitic/leucititic) and-free (lamproitic) ultra-potassic rocks and associated shoshonites from Italy: constraints on petrogenesis and geodynamics. J Virtual Expl 36(20)

Coombs DS, Cas RAF, Kawachi Y, Landis CA, Mcdonough WF, Reay A (1986) Cenozoic volcanism in north, east and central Otago. R Soc New Zeal Bull 23:278–312

Corazzato C, Tibaldi A (2006) Fracture control on type, morphology and distribution of parasitic volcanic cones: an example from Mt. Etna, Italy. J Volcanol Geotherm Res 158(1–2):177–194

Corcoran PL (2000) Recognizing distinct portions of seamounts using volcanic facies analysis: examples from the Archean Slave Province, NWT, Canada. Precamb Res 101(2–4):237–261

Corti G (2009) Continental rift evolution: from rift initiation to incipient break-up in the Main Ethiopian Rift, East Africa. Earth Sci Rev 96(1–2):1–53

Courtillot V, Jaupart C, Manighetti I, Tapponnier P, Besse J (1999) On causal links between flood basalts and continental breakup. Earth Planet Sci Letts 166(3):177–195

Cronin S (2022) Why the volcanic eruption in Tonga was so violent, and what to expect next. The Conversation January 16, 2022. https://mail.google.com/mail/u/0/#search/Shane+Cronin/FMfcgzGmtXCRxPtdgDXpmkstLfQGmjNv

Cross JK, Tomlinson EL, Giordano G, Smith VC, De Benedetti AA, Roberge J, Manning CJ, Wulf S, Menzies MA (2014) High level triggers for explosive mafic volcanism: Albano Maar, Italy. Lithos 190:137–153

Dahren B, Troll VR, Andersson UB, Chadwick JP, Gardner MF, Jaxybulatov K, Koulakov I (2012) Magma plumbing beneath Anak Krakatau volcano, Indonesia: evidence for multiple magma storage regions. Contribs Mineral Petrol 163(4):631–651

Danese E, Mattei M (2010) The sedimentary substrate of the Colli Albani. In: Funiciello R, Giordano G (eds) The Colli Albani Volcano. Spec Publ IAVCEI and Geol Soc Lond, London, pp 141–152

Davì M, De Rosa R, Donato P, Sulpizio R (2011) The Lami pyroclastic succession (Lipari, Aeolian Islands): a clue for unravelling the eruptive dynamics of the Monte Pilato rhyolitic pumice cone. J Volcanol Geotherm Res 201(1–4):285–300

Davidson J, De Silva S (2000) Composite volcanoes. In: Sigurdsson H, Stix J, Houghton BF, Mcnutt SR, Rymer H (eds) Encyclopedia of volcanoes. San Diego: Academic Press, pp 663–681

Davies AGS, Cooke DR, Gemmell JB, Simpson KA (2008) Diatreme breccias at the Kelian gold mine, Kalimantan, Indonesia: precursors to epithermal gold mineralization. Econ Geol 103(4):689–716

Davies BJ, Hambrey MJ, Smellie JL, Carrivick JL, Glasser NF (2012) Antarctic Peninsula ice sheet evolution during the Cenozoic Era. Quat Sci Rev 31:30–66

Dávila-Harris P, Carrasco-Núñez G (2014) An unusual syn-eruptive bimodal eruption: the Holocene Cuicuiltic Member at Los Humeros caldera, Mexico. J Volcanol Geotherm Res 271:24–42

Day RA (1983) Petrology and geochemistry of the Older Volcanics, Victoria (distribution, character sation and petrogenesis). Unpubl PhD thesis, Monash Univ, Australia

De'Gennaro M, Cappelletti P, Langella A, Perrotta A, Scarpati C (2000) Genesis of zeolites in the Neapolitan Yellow Tuff: geological, volcanological and mineralogical evidence. Contribs Mineral Petrol 139(1):17–35

De Benedetti AA, Caprilli E, Rossetti F, Giordano G (2010) Metamorphic, metasomatic and intrusive xenoliths of the Colli Albani volcano and their significance for the reconstruction of the volcano plumbing system. In: Funiciello R, Giordano G (eds) The Colli Albani Volcano. IAVCEI Special Publication 3. Geological Society London, pp 153–176

De Benedetti AA, Funiciello R, Giordano G, Diano G, Caprilli E, Paterne M (2008) Volcanology, history and myths of the Lake Albano maar (Colli Albani volcano, Italy). J Volcanol Geotherm Res 176(3):387–406

De León-Barragán L, Carrasco-Núñez G, Ort MH (2020) Stratigraphy and evolution of the Holocene Aljojuca Maar volcano (Serdán-Oriental basin, Eastern Trans-Mexican Volcanic Belt), and implications for hazard assessment. J Volcanol Geother Res 392:106789

De Rita D, Giordano G (1996) Volcanological and structural evolution of Roccamonfina volcano (Italy): origin of the summit caldera. Geol Soc Lond Spec Publ 110(1):209–224

de Ronde CE, Massoth GJ, Butterfield DA, Christenson BW, Ishibashi J, Ditchburn RG, ... Takai K (2011) Submarine hydrothermal activity and gold-rich mineralization at Brothers Volcano, Kermadec Arc, New Zealand. Mineral Deposita 46:541–584

de Ronde CEJ, Baker ET, Massoth GJ, Lupton JE, Wright IC, Sparks RJ, Bannister SC, Reyners ME, Walker SL, Greene RR (2007) Submarine hydrothermal activity along the mid-Kermadec arc, New Zealand: large-scale effects on venting. Geochem Geophys Geosyst 8(7)

de Ronde CEJ, Fornari DJ, Ferrini VL, Walker SL, Davy BW, Leblanc C, Tontini FC, Kukulya AL, Littlefield RH (2016) The Pink and White Terraces of Lake Rotomahana: what was their fate after the 1886 Tarawera Rift eruption? J Volcanol Geotherm Res 314:126–141

de Ronde CEJ, Hannington MD, Stoffers P, Wright IC, Ditchburn RG, Reyes AG, Baker ET, Massoth GJ, Lupton JE, Walker SL (2005) Evolution of a submarine magmatic-hydrothermal system: brothers volcano, southern Kermadec arc, New Zealand. Econ Geol 100 (6):1097–1133

de Silva SL (1989) Altiplano-Puna volcanic complex of the central Andes. Geology 17(12):1102–1106

de Silva SL, Gosnold WD (2007) Episodic construction of batholiths: insights from the spatiotemporal development of an ignimbrite flare-up. J Volcanol Geotherm Res 167(1–4):320–335

de Silva SL, Mucek AE, Gregg PM, Pratomo I (2015) Resurgent Toba—field, chronologic, and model constraints on time scales and mechanisms of resurgence at large calderas. Front Earth Sci 3:25

De Vita S, Sansivero F, Orsi G, Marotta E, Piochi M (2010) Volcanological and structural evolution of the Ischia resurgent caldera (Italy) over the past 10 ky. Stratigraphy and Geology in Volcanic Areas. Geol Soc Amer Spec Pap 464:193–239

Della Seta M, Esposito C, Marmoni GM, Martíno S, Paciello A, Perinelli C, Sottili G (2015) Geological constraints for a conceptual evolutionary model of the slope deformations affecting Mt. Nuovo at Ischia (Italy). Ital J Eng Geol Environ 15(2):15–28

Della Seta M, Marotta E, Orsi G, De Vita S, Sansivero F, Fredi P (2012) Slope instability induced by volcano-tectonics as an additional source of hazard in active volcanic areas: the case of Ischia island (Italy). Bull Volcanol 74(1):79–106

Dellino P, Volpe LL (1995) Fragmentation versus transportation mechanisms in the pyroclastic sequence of Monte Pilato-Rocche Rosse (Lipari, Italy). J Volcanol Geotherm Res 64(3):211–231

Delmelle P, Stix J, Baxter P, Garcia-Alvarez J, Barquero J (2002) Atmospheric dispersion, environmental effects and potential health hazard associated with the low-altitude gas plume of Masaya volcano, Nicaragua. Bull Volcanol 64(6):423–434

Denlinger RP, Morgan JK (2014) Instability of Hawaiian volcanoes, Chapter 4 In: Poland MP, Takahashi TJ, Landowski CM (eds) Characteristics of Hawaiian volcanoes: US Geol Surv Prof Pap 1801:149–176

Deplus C, Le Friant A, Boudon G, Komorowski J-C, Villemant B, Harford CL, Ségoufin J, Cheminée J-L (2001) Submarine evidence for large-scale debris avalanches in the Lesser Antilles Arc. Earth Planet Sci Letts 192(2):145–157

Desrochers J-P, Hubert C, Ludden JN, Pilote P (1993) Accretion of Archean oceanic plateau fragments in the Abitibi, greenstone belt, Canada. Geology 21(5):451–454

Di Fiore F, Vona A, Scarani A, Giordano G, Romano C, Giordano D, ... Pankhurst M (2023) Experimental constraints on the rheology of lavas from 2021 Cumbre Vieja eruption (La Palma, Spain). Geophys Res Lett 50(4):e2022GL100970

Di Salvo S, Avanzinelli R, Isaia R, Zanetti A, Druitt T, Francalanci L (2020) Crystal-mush reactivation by magma recharge: evidence from the Campanian Ignimbrite activity, Campi Flegrei volcanic field, Italy. Lithos 376:105780

Di Vito MA, Isaia R, Orsi G, Southon J, De Vita S, D'antonio M, Pappalardo L, Piochi M (1999) Volcanism and deformation since 12,000 years at the Campi Flegrei caldera (Italy). J Volcanol Geotherm Res 91(2–4):221–246

Dohrenwend JC, Wells SG, Turrin BD (1986) Degradation of Quaternary cinder cones in the Cima volcanic field, Mojave Desert, California. Geol Soc Amer Bull 97(4):421–427

Donnelly-Nolan JM, Burns MG, Goff FE, Peters EK, Thompson JM (1993) The Geysers-Clear Lake area, California; thermal waters, mineralization, volcanism, and geothermal potential. Econ Geol 88 (2):301–316

Doyle MG, Allen RL (2003) Subsea-floor replacement in volcanic-hosted massive sulfide deposits. Ore Geol Rev 23(3–4):183–222

Druitt TH, Edwards L, Mellors RM, Pyle DM, Sparks RSJ, Lanphere M, Davies M, Barriero M (1999) Santorini volcano. Geol Soc Lond Mem 19:1–165

Druitt TH, Kokelaar BP (eds) (2002) The eruption of Soufrière Hills Volcano, Montserrat, from 1995 to 1999. Geol Soc London Mem 21:645pp

Druitt TH, Sparks RSJ (1984) On the formation of calderas during ignimbrite eruptions. Nature 310(5979):679–681

Duffield WA, Christiansen RL, Koyanagi RY, Peterson DW (1982) Storage, migration, and eruption of magma at Kilauea volcano, Hawaii, 1971–1972. J Volcanol Geotherm Res 13(3–4):273–307

Duncan l, Helland-Hansen D, Dennehy C (2009) The Rosebank discovery: a new play type in intra basalt reservoirs of the North Atlantic Volcanic Province. In: Abstracts 6th European Production and Development Conference and Exhibition (DEVEX), Aberdeen, Scotland. http://www.devex-conference.org/pdf/Presentations_2009/

Dvorak JJ, Dzurisin D (1993) Variations in magma supply rate at Kilauea Volcano, Hawaii. J Geophys Res Solid Earth 98 (B12):22,255–22,268

Eaton GP (1982) The Basin and Range province: origin and tectonic significance. Ann Rev Earth Planet Sci 10(1):409–440

Edgar CJ, Wolff JA, Olin PH, Nichols HJ, Pittari A, Cas RAF, Reiners PW, Spell TL, Martí J (2007) The late Quaternary Diego Hernandez Formation, Tenerife: volcanology of a complex cycle of voluminous explosive phonolitic eruptions. J Volcanol Geotherm Res 160(1):59–85

Edwards BR, Gudmundsson MT, Russell JK (2015) Glaciovolcanism. In: Sigurdsson H, Houghton B, Mcnutt SR, Rymer H, Stix J (eds) The encyclopedia of volcanoes. Academic Press Elsevier, pp 377–394

Elíasson J. Katla volcano in Iceland, potential hazards and risk assessment. Nat Sci 6(3):43224

Escher BC (1931) Gloedwolken en lahars. Vulkanische Katastrophes in Nederlandsch-Indië. Tropisch Nederl 3(19–20):291–320

Fagents SA, Thordarson T (2007) Rootless volcanic cones in Iceland and on Mars. In: Chapman M (ed) The geology of Mars: evidence from earth-based analogs. Cambridge Univ Press, Cambridge, pp 151–177

Faulds JE, Stewart JH (eds) (1998) Accommodation zones and transfer zones: the regional segmentation of the Basin and Range Province. Geol Soc Amer Spec Pap 323:Pages

Fearnley CJ, Bird DK, Haynes K, McGuire WJ, Jolly G (eds) (2017) Observing the volcano world: volcano crisis communication. In: Advances in Volcanology, IACEI. Springer Nature, 773 pp. https://doi.org/10.1007/11157_2017_27

Fedele FG, Giaccio B, Isaia R, Orsi G (2003) The Campanian Ignimbrite Eruption, Heinrich Event 4, and Palaeolithic Change in Europe: a high-resolution investigation. Volcanism and the earth's atmosphere. Amer Geophys Union Monogr 139:301–328

Fergusson CL, Cas RAF, Collins WJ, Craig GY, Crook KAW, Powell CM, Scott PA, Young GC (1979) The Upper Devonian Boyd Volcanic Complex, Eden, New South Wales. J Geol Soc Aust 26(1–2):87–105

Ferrari L, Valencia-Moreno M, Bryan SE (2007) Magmatism and tectonics of the Sierra Madre Occidental and its relation with the evolution of the western margin of North America. Geol Soc Amer Spec Pap 422:1–39

Filipovich R, Báez W, Groppelli G, Ahumada F, Aldega L, Becchio R, ... Giordano G (2020) Geological map of the Tocomar basin (Puna Plateau, NW Argentina). Implication for the geothermal system investigation. Energies 13(20):5492

Fisher RV (1968) Puu Hou littoral cones, Hawaii. Geol Rundsch 57 (3):837–864

Fitton JG, Mahoney JJ, Wallace PJ, Saunders AD (2004a) Origin and evolution of the Ontong Java Plateau: introduction. Geol Soc Lond Spec Publ 229(1):1–8

Fitton JG, Mahoney JJ, Wallace PJ, Saunders AD (2004b) Leg 192 synthesis: origin and evolution of the Ontong Java Plateau. In: Fitton et al (eds) Proceedings ocean drilling program, scientific results, vol 192, pp 1–18. http://www-odp.tamu.edu/publications/192_SR/VOLUME/SYNTH/SYNTH.PDF

Fodor L, Csontos L, Bada G, Györfi I, Benkovics L (1999) Tertiary tectonic evolution of the Pannonian Basin system and neighbouring orogens: a new synthesis of palaeostress data. Geol Soc Lond Spec Publ 156(1):295–334

Folkes CB, Wright HM, Cas RAF, De Silva SL, Lesti C, Viramonte JG (2011) A re-appraisal of the stratigraphy and volcanology of the Cerro Galán volcanic system, NW Argentina. Bull Volcanol 73 (10):1427–1454

Fontijn K, Costa F, Sutawidjaja I, Newhall CG, Herrin JS (2015) A 5000-year record of multiple highly explosive mafic eruptions from Gunung Agung (Bali, Indonesia): implications for eruption frequency and volcanic hazards. Bull Volcanol 77(7):59

Fornaciai A, Favalli M, Karátson D, Tarquini S, Boschi E (2012) Morphometry of scoria cones, and their relation to geodynamic setting: a DEM-based analysis. J Volcanol Geotherm Res 217:56–72

Francalanci L, Lucchi F, Keller J, De Astis G, Tranne CA (2013) Eruptive, volcano-tectonic and magmatic history of the Stromboli volcano (north-eastern Aeolian archipelago). Geol Soc Lond Mem 37(1):397–471

Funck T, Lykke-Andersen H (1998) Seismic structure of the volcanic apron north of Gran Canaria. In: Weaver (ed) Proceedings of the ocean drilling program, scientific results, vol 157, pp 11–28

Fundis AT, Soule SA, Fornari DJ, Perfit MR (2010) Paving the seafloor: volcanic emplacement processes during the 2005–2006 eruptions at the fast spreading East Pacific Rise, 9 50′ N. Geochem Geophys Geosyst 11(8)

Funiciello R, Giordano G, De Rita D (2003) The Albano maar lake (Colli Albani Volcano, Italy): recent volcanic activity and evidence of pre-Roman Age catastrophic lahar events. J Volcanol Geotherm Res 123(1–2):43–61

Funiciello R, Giordano G, Mattei M (2008) Carta geologica di Roma. Memorie Descrittive della Carta Geologica d'Italia, 80

Furnes H, Fridleifsson IB, Atkins FB (1980) Subglacial volcanics—on the formation of acid hyaloclastites. J Volcanol Geotherm Res 8 (1):95–110

Fytikas M, Kolios N, Vougioukalakis G (1990) Post-Minoan volcanic activity of the Santorini volcano. Volcanic hazard and risk, forecasting possibilities. In: Doumas C (ed) Thera and the Aegean world III, vol 2, pp 183–198

Galetto F, Acocella V, Caricchi L (2017) Caldera resurgence driven by magma viscosity contrasts. Nat Commun 8(1):1750

García O, Martí J, Aguirre G, Geyer A, Iribarren I (2011) Pyroclastic density currents from Teide-Pico Viejo (Tenerife, Canary Islands): implications for hazard assessment. Terra Nova 23(3):220–224

Garvin JB, Slayback DA, Ferrini V, Frawley J, Giguere C, Asrar GR, Andersen K (2018) Monitoring and modeling the rapid evolution of earth's newest volcanic island: Hunga Tonga Hunga Ha'apai (Tonga) using high spatial resolution satellite observations. Geophys Res Lett 45(8):3445–3452

Geoffroy L (2005) Volcanic passive margins. CR Geosci 337 (16):1395–1408

Gertisser R, Self S (2015) The great 1815 eruption of Tambora and future risks from large-scale volcanism. Geol Today 31(4):132–136

Geshi N, Németh K, Oikawa T (2011) Growth of phreatomagmatic explosion craters: a model inferred from Suoana crater in Miyake-jima Volcano, Japan. J Volcanol Geotherm Res 201(1):30–38

Geshi N, Ruch J, Acocella V (2014) Evaluating volumes for magma chambers and magma withdrawn for caldera collapse. Earth Planet Sci Letts 396:107–115

Geyer A, Folch A, Martí J (2006) Relationship between caldera collapse and magma chamber withdrawal: an experimental approach. J Volcanol Geotherm Res 157(4):375–386

Geyer A, Martí J (2008) The new worldwide collapse caldera database (CCDB): a tool for studying and understanding caldera processes. J Volcanol Geotherm Res 175(3):334–354

Gibson P (2020) New Zealand Whakaari volcano eruption and occupational hygiene responses. Occup Health South Afr 26(1):29

Giordano G, Carg-Team (2010) Stratigraphy, volcano tectonics and evolution of the Colli Albani volcanic field. In: Funiciello R, Giordano G (eds) The Colli Albani Volcano. Geol Soc Lond Spec Publ of IAVCEI, London, vol 3, pp 43–98

Giordano G, Caricchi L (2022) Determining the state of activity of transcrustal magmatic systems and their volcanoes. Ann Rev Earth Planet Sci 50:231–259

Giordano G, Cas RAF (2001) Structure of the Upper Devonian Boyd Volcanic Complex, south coast New South Wales: implications for the Devonian-Carboniferous evolution of the Lachlan Fold Belt. Aust J Earth Sci 48(1):49–61

Giordano G, Cas RAF (2021) Classification of ignimbrites and their eruptions. Earth Sci Rev 220:103697

Giordano G, De Astis G (2021) The summer 2019 basaltic Vulcanian eruptions (paroxysms) of Stromboli. Bull Volcanol 83:1–27

Giordano G, De Benedetti AA, Bonamico A, Ramazzotti P, Mattei M (2014) Incorporating surface indicators of reservoir permeability into reservoir volume calculations: application to the Colli Albani caldera and the Central Italy Geothermal Province. Earth Sci Rev 128:75–92

Giordano G, Bretton R, Calder, E, Cas R, Gottsmann J, Lindsay J, Newhall, C, Pallister J, Papale P, Rodriguez I (2016) Toward IAVCEI guidelines on the roles and responsibilities of scientists involved in volcanic hazard evaluation, risk mitigation and crisis response. Bull Volcanol 78(31). https://doi.org/10.1007/s00445-016-1021-8

Giordano G, De Rita D, Cas R, Rodani S (2002) Valley pond and ignimbrite veneer deposits in the small-volume phreatomagmatic 'Peperino Albano' basic ignimbrite, Lago Albano maar, Colli Albani volcano, Italy: influence of topography. J Volcanol Geoth Res 118(1–2):131–144

Giordano G, Pinton A, Cianfarra P, Baez W, Chiodi A, Viramonte J, ... Groppelli G (2013) Structural control on geothermal circulation in the Cerro Tuzgle–Tocomar geothermal volcanic area (Puna plateau, Argentina). J Volcanol Geother Res 249:77–94

Goff F, Gardner JN, Reneau SL, Kelley SA, Kempter KA, Lawrence JR (2011) Geologic map of the Valles caldera, Jemez Mountains, New Mexico. New Mexico Bur Mines Min Res GM 79. Scale 1:50,000. ISBN: 9781883905293

Gran KB, Montgomery DR (2005) Spatial and temporal patterns in fluvial recovery following volcanic eruptions: Channel response to basin-wide sediment loading at Mount Pinatubo, Philippines. Geol Soc Amer Bull 117(1–2):195–211

Gran KB, Montgomery DR, Halbur JC (2011) Long-term elevated post-eruption sedimentation at Mount Pinatubo, Philippines. Geology 39(4):367–370

Gravley DM, Deering CD, Leonard GS, Rowland JV (2016) Ignimbrite flare-ups and their drivers: a New Zealand perspective. Earth Sci Rev 162:65–82

Greeley R (1982) The Snake River Plain, Idaho: representative of a new category of volcanism. J Geophys Res Solid Earth 87(B4):2705–2712

Greene AR, Scoates JS, Weis D, Katvala EC, Israel S, Nixon GT (2010) The architecture of oceanic plateaus revealed by the volcanic stratigraphy of the accreted Wrangellia oceanic plateau. Geosphere 6(1):47–73

Gregg PM, De Silva SL, Grosfils EB, Parmigiani JP (2012) Catastrophic caldera-forming eruptions: Thermomechanics and implications for eruption triggering and maximum caldera dimensions on Earth. J Volcanol Geotherm Res 241–242:1–12

Grosch EG, Slama J (2017) Evidence for 3.3-billion-year-old oceanic crust in the Barberton greenstone belt, South Africa. Geology 45 (8):695–698

Grosse P, Euillades PA, Euillades LD, van Wyk de Vries B (2014) A global database of composite volcano morphometry. Bull Volcanol 76(1):784

Gudmundsson MT, Jónsdóttir K, Hooper A, Holohan EP, Halldórsson SA, Ófeigsson BG, Cesca S, Vogfjörd KS, Sigmundsson F, Högnadóttir T (2016) Gradual caldera collapse at Bárdarbunga volcano, Iceland, regulated by lateral magma outflow. Science 353 (6296):aaf8988

Gudmundsson MT, Pedersen R, Vogfjörd K, Thorbjarnardóttir B, Jakobsdóttir S, Roberts MJ (2010) Eruptions of Eyjafjallajökull Volcano, Iceland. Eos 91(21):190–191

Gudmundsson MT, Thordarson T, Höskuldsson Á, Larsen G, Björnsson H, Prata FJ, Oddsson B, Magnússon E, Högnadóttir T, Petersen GN, Hayward CL (2012) Ash generation and distribution from the April-May 2010 eruption of Eyjafjallajökull, Iceland. Sci Rept 2(1):1–12

Guillou H, Carracedo JC, Paris R, Torrado FPJ (2004) Implications for the early shield-stage evolution of Tenerife from K/Ar ages and magnetic stratigraphy. Earth Planet Sci Letts 222(2):599–614

Halbach P, Nakamura K-I, Wahsner M, Lange J, Sakai H, Käselitz L, Hansen R-D, Yamano M, Post J, Prause B (1989) Probable modern analogue of Kuroko-type massive sulphide deposits in the Okinawa Trough back-arc basin. Nature 338(6215):496

Hamilton CW, Thordarson T, Fagents SA (2010) Explosive lava–water interactions I: architecture and emplacement chronology of volcanic rootless cone groups in the 1783–1784 Laki lava flow, Iceland. Bull Volcanol 72(4):449–467

Hannington M, Jamieson J, Monecke T, Petersen S, Beaulieu S (2011) The abundance of seafloor massive sulfide deposits. Geology 39 (12):1155–1158

Hansen DM (2006) The morphology of intrusion-related vent structures and their implications for constraining the timing of intrusive events along the NE Atlantic margin. J Geol Soc Lond 163(5):789–800

Harpp KS, Geist DJ (2018) The evolution of Galápagos volcanoes: an alternative perspective. Front Earth Sci 6:50

Harris AC, Bryan SE, Holcombe RJ (2006) Volcanic setting of the Bajo de la Alumbrera porphyry Cu-Au deposit, Farallón Negro volcanics, northwest Argentina. Econ Geol 101(1):71–94

Hasenaka T, Carmichael ISE (1985) The cinder cones of Michoacán—Guanajuato, central Mexico: their age, volume and distribution, and magma discharge rate. J Volcanol Geotherm Res 25(1–2):105–124

Hay RL, Iijima A (1968) Petrology of palagonite tuffs of Koko craters, Oahu, Hawaii. Contribs Mineral Petrol 17(2):141–154

Hayes JL, Wilson TM, Deligne NI, Lindsay JM, Leonard GS, Tsang SW, Fitzgerald RH (2020) Developing a suite of multi-hazard volcanic eruption scenarios using an interdisciplinary approach. J Volcanol Geother Res 392:106763

Hayes SK, Montgomery DR, Newhall CG (2002) Fluvial sediment transport and deposition following the 1991 eruption of Mount Pinatubo. Geomorphology 45(3–4):211–224

Hayman PC, Thébaud N, Pawley MJ, Barnes SJ, Cas RAF, Amelin Y, Sapkota J, Squire RJ, Campbell IH, Pegg I (2015) Evolution of a ~2.7 Ga large igneous province: a volcanological, geochemical and geochronological study of the Agnew Greenstone Belt, and new regional correlations for the Kalgoorlie Terrane (Yilgarn Craton, Western Australia). Precamb Res 270:334–368

Hedenquist JW, Henley RW (1985) Hydrothermal eruptions in the Waiotapu geothermal system, New Zealand; their origin, associated breccias, and relation to precious metal mineralization. Econ Geol 80(6):1640–1668

Heiken G, Goff F, Gardner JN, Baldridge WS, Hulen JB, Nielson DL, Vaniman D (1990) The Valles/Toledo Caldera Complex, Jemez Volcanic Field, New Mexico. Ann Rev Earth Planet Sci 18(1):27–53

Heiken GH (1971) Tuff rings: examples from the Fort Rock-Christmas Lake Valley Basin, south-central Oregon. J Geophys Res 76 (23):5615–5626

Henley RW, Ellis AJ (1983) Geothermal systems ancient and modern: a geochemical review. Earth Sci Rev 19(1):1–50

Henry CD, Castor SB, Starkel WA, Ellis BS, Wolff JA, Laravie JA, Mcintosh WC, Heizler MT (2017) Geology and evolution of the McDermitt caldera, northern Nevada and southeastern Oregon, western USA. Geosphere 13(4):1066–1112

Hildreth W (1981) Gradients in silicic magma chambers: implications for lithospheric magmatism. J Geophys Res Solid Earth 86 (B11):10153–10192

Hildreth W (2004) Volcanological perspectives on Long Valley, Mammoth Mountain, and Mono Craters: several contiguous but discrete systems. J Volcanol Geotherm Res 136(3–4):169–198

Hildreth W (2010) Laguna del Maule Volcanic Field: Eruptive History of a Quaternary basalt-to-rhyolite distributed volcanic field on the Andean rangecrest in central Chile, Servicio Nacional de Geología y Minería-Chile. 63:143

Hildreth W, Fierstein J, Godoy E, Drake RE, Singer (1999) The Puelche Volcanic Field: extensive Pleistocene rhyolite lava flows in the Andes of central Chile. Revista Geològica de Chile 26(2):275–309

Hildreth W, Wilson CJN (2007) Compositional Zoning of the Bishop Tuff. J Petrol 48(5):951–999

Hoernle K, Hauff F, Van Den Bogaard P (2004) 70 my history (139–69 Ma) for the Caribbean large igneous province. Geology 32(8):697–700

Hofmann C, Courtillot V, Feraud G, Rochette P, Yirgu G, Ketefo E, Pik R (1997) Timing of the Ethiopian flood basalt event and implications for plume birth and global change. Nature 389 (6653):838–841

Holohan EP, Schöpfer MPJ, Walsh JJ (2011) Mechanical and geometric controls on the structural evolution of pit crater and caldera subsidence. J Geophys Res Solid Earth 116(B7)

Hooper DM, Sheridan MF (1998) Computer-simulation models of scoria cone degradation. J Volcanol Geotherm Res 83(3–4):241–267

Hooper PR (2000) Flood basalt provinces. In: Sigurdsson H, Stix J, Houghton BF, Mcnutt SR, Rymer H (eds) Encyclopedia of volcanoes. Academic, San Diego, pp 345–359

Howells MF, Reedman AJ, Campbell SDG (1986) The submarine eruption and emplacement of the Lower Rhyolitic tuff Formation (Ordovician), N Wales. J Geol Soc Lond 143(3):411–423

Huang HH, Lin FC, Schmandt B, Farrell J, Smith RB, Tsai VC (2015) The Yellowstone magmatic system from the mantle plume to the upper crust. Science 348(6236):773–776

Hunt JE, Cassidy M, Talling PJ (2018) Multi-stage volcanic island flank collapses with coeval explosive caldera-forming eruptions. Sci Rept 8(1):1–1

Hunt JE, Talling PJ, Clare MA, Jarvis I, Wynn RB (2014) Long-term (17 Ma) turbidite record of the timing and frequency of large flank collapses of the Canary Islands. Geochem Geophys Geosyst 15 (8):3322–3345

Hunt JE, Wynn RB, Masson DG, Talling PJ, Teagle DA (2011) Sedimentological and geochemical evidence for multistage failure of volcanic island landslides: a case study from Icod landslide on north Tenerife, Canary Islands. Geochem Geophys Geosyst 12(12)

Ikegami F, McPhie J, Carey R, Mundana R, Soule SA, Jutzeler M (2018b) The eruption of submarine rhyolite lavas and domes in the deep ocean–Havre 2012, Kermadec Arc. Front Earth Sci 6:147. https://doi.org/10.3389/feart.2018.00147

Inci U (1999) Miocene synvolcanic alluvial sedimentation in lignite-bearing Soma Basin, Western Turkey. Turk J Earth Sci 7 (2):63–78

Inoue H, Coffin MF, Nakamura Y, Mochizuki K, Kroenke LW (2008) Intrabasement reflections of the Ontong Java Plateau: implications for plateau construction. Geochem Geophys Geosyst 9(4)

Isaia R, D'Antonio M, Dell'Erba F, Di Vito M, Orsi G (2004) The Astroni volcano: the only example of closely spaced eruptions in the same vent area during the recent history of the Campi Flegrei caldera (Italy). J Volcanol Geother Res 133(1–4):171–192

Isaia R, Vitale S, Di Giuseppe MG, Iannuzzi E, Tramparulo FD, Troiano A (2015) Stratigraphy, structure, and volcano-tectonic evolution of Solfatara maar-diatreme (Campi Flegrei, Italy). Geol Soc Amer Bull 127(9–10):1485–1504

Izokh AE, Smirnov SZ, Egorova VV, Anh TT, Kovyazin SV, Phuong NT, Kalinina VV (2010) The conditions of formation of sapphire and zircon in the areas of alkali-basaltoid volcanism in Central Vietnam. Russian Geol and Geophys 51(7):719–733

Jackson MD, Gudmundsson MT et al (2019) SUSTAIN drilling at Surtsey volcano, Iceland, tracks hydrothermal and microbiological interactions in basalt 50 years after eruption. Sci Drilling 25:35–46

Jakobsson SP, Moore JG (1982) The Surtsey research drilling project of 1979. Surtsey Res Prog Rep 9:76–93

Jakobsson SP, Moore JG (1986) Hydrothermal minerals and alteration rates at Surtsey volcano, Iceland. Geol Soc Amer Bull 97(5):648–659

Jarvis PA, Bonadonna C, Domínguez L et al (2020) Aeolian remobilisation of volcanic ash: outcomes of a workshop in the Argentianian Patagonia. Front Earth Sci 8:575184

Jellinek AM, DePaolo DJ (2003) A model for the origin of large silicic magma chambers: precursors of caldera-forming eruptions. Bull Volcanol 65:363–381

Jerram DA, Widdowson M (2005) The anatomy of Continental Flood Basalt Provinces: geological constraints on the processes and products of flood volcanism. Lithos 79(3):385–405

John DA (2001) Miocene and early Pliocene epithermal gold-silver deposits in the northern Great Basin, western United States: characteristics, distribution, and relationship to magmatism. Econ Geol 96(8):1827–1853

Johnston EN, Sparks RSJ, Nomikou P, Livanos I, Carey S, Phillips JC, Sigurdsson H (2014) Stratigraphic relations of Santorini's intra-caldera fill and implications for the rate of post-caldera volcanism. J Geol Soc Lond 172(3):323–335

Jones JG (1966) Intraglacial volcanoes of south-west Iceland and their significance in the interpretation of the form of the marine basaltic volcanoes. Nature 212:586–588

Jones JG (1969) Pillow lavas as depth indicators. Amer J Sci 267 (2):181–195

Jones JG, Nelson PHH (1970) The flow of basalt lava from air into water–its structural expression and stratigraphic significance. Geol Mag 107(1):13–19

Jordan SC, Cas RAF, Hayman PC (2013) The origin of a large (>3 km) maar volcano by coalescence of multiple shallow craters: Lake Purrumbete maar, southeastern Australia. J Volcanol Geotherm Res 254:5–22

Jordan SC, Jowitt SM, Cas RAF (2015) Origin of temporal-compositional variations during the eruption of Lake Purrumbete Maar, Newer Volcanics Province, southeastern Australia. Bull Volcanol 77(1):883

Joyce EB (1975) Quaternary volcanism and tectonics in southeastern Australia. In: Suggate RP, Cresswell MM (eds) Quaternary studies. Bull R Soc N Z 13:169–176

Jurado-Chichay Z, Rowland SK, Walker GPL (1996) The formation of circular littoral cones from tube-fed pāhoehoe: Mauna Loa, Hawai'i. Bull Volcanol 57(7):471–482

Jutzeler M, Marsh R, Carey RJ, White JDL, Talling PJ, Karlstrom L (2014) On the fate of pumice rafts formed during the 2012 Havre submarine eruption. Nature Coms 5

Karátson D, Telbisz T, Wörner G (2012) Erosion rates and erosion patterns of Neogene to Quaternary stratovolcanoes in the Western Cordillera of the Central Andes: an SRTM DEM based analysis. Geomorphology 139:122–135

Kelfoun K, Giachetti T, Labazuy P (2010) Landslide-generated tsunamis at Réunion Island. J Geophys Res Earth Surf 115(F4)

Kennedy B, Stix J, Vallance JW, Lavallée Y, Longpré M-A (2004) Controls on caldera structure: results from analogue sandbox modeling. Geol Soc Amer Bull 116(5–6):515–524

Kennedy B, Wilcock J, Stix J (2012) Caldera resurgence during magma replenishment and rejuvenation at Valles and Lake City calderas. Bull Volcanol 74(8):1833–1847

Kereszturi G, Geyer A, Martí J, Németh K, Dóniz-Páez FJ (2013) Evaluation of morphometry-based dating of monogenetic volcanoes—a case study from Bandas del Sur, Tenerife (Canary Islands). Bull Volcanol 75(7):734

Kereszturi G, Jordan G, Németh K, Dóniz-Páez JF (2012) Syn-eruptive morphometric variability of monogenetic scoria cones. Bull Volcanol 74(9):2171–2185

Kereszturi G, Németh K (2013) Monogenetic basaltic volcanoes: genetic classification, growth, geomorphology and degradation. In: Updates in volcanology—new advances in understanding volcanic systems. IntechOpen. https://doi.org/10.5772/51387

Kerr AC (2005) Oceanic LIPs: the kiss of death. Elements 1(5):289–292

Kerr AC (2014) Oceanic plateaus. In: Turekian KK, Holland HD (eds) Treatise on geochemistry: volume 4: the crust, 2nd edn. Elsevier, vol 4(18), pp 631–667

Kerr AC, Marriner GF, Arndt NT, Tarney J, Nivia A, Saunders AD, Duncan RA (1996a) The petrogenesis of Gorgona komatiites, picrites and basalts: new field, petrographic and geochemical constraints. Lithos 37(2–3):245–260

Kerr AC, Tarney J, Marriner GF, Klaver GT, Saunders AD, Thirlwall MF (1996b) The geochemistry and petrogenesis of the late-Cretaceous picrites and basalts of Curaçao, Netherlands Antilles: a remnant of an oceanic plateau. Contribs Mineral Petrol 124(1):29–43

Kerr AC, Tarney J, Nivia A, Marriner GF, Saunders AD (1998) The internal structure of oceanic plateaus: inferences from obducted Cretaceous terranes in western Colombia and the Caribbean. Tectonophysics 292(3–4):173–188

Kessel LG, Busby CJ (2003) Analysis of VHMS-hosting ignimbrites erupted at bathyal water depths (Ordovican Bald Mountain sequence, Northern Maine). In: White DL, Smellie JL, Clague DA (eds) Explosive subaqueous volcanism. Amer Geophys Union Monogr 140:361–379

Kieffer B, Arndt N, Lapierre H, Bastien F, Bosch D, Pecher A, Yirgu G, Ayalew D, Weis D, Jerram DA, Keller F (2004) Flood and shield basalts from Ethiopia: magmas from the African superswell. J Petrol 45(4):793–834

Kieffer G (1971) Aperçu sur la morphologie des régions volcaniques du Massif Central. In: Symposium Jean Jung: Géologic, géomorphologie et structure profounde du Massif Central français. Clermont-Ferrand 479–510

Kienle J, Kyle PR, Self S, Motyka RJ, Lorenz V (1980) Ukinrek Maars, Alaska, I. April 1977 eruption sequence, petrology and tectonic setting. J Volcanol Geotherm Res 7(1):11–37

Kilgour G, Manville V, Della Pasqua F, Graettinger AH, Hodgson KA, Jolly GE (2010) The 25 September 2007 eruption of Mount Ruapehu, New Zealand: directed ballistics, surtseyan jets, and ice-slurry lahars. J Volcanol Geotherm Res 191(1):1–14

Kjarsgaard BA (2007) Kimberlite diamond deposits. In: Goodfellow WD (ed) Mineral deposits of Canada: a synthesis of major deposit types, district metallogeny, the evolution of geological provinces, and exploration methods. Geol Assoc Canada Min Deps Div Spec Publ 5:245–272

Kling GW, Clark MA, Wagner GN, Compton HR, Humphrey AM, Devine JD, Evans WC, Lockwood JP, Tuttle ML, Koenigsberg EJ (1987) The 1986 Lake Nyos gas disaster in cameroon, west Africa. Science 236(4798):169–175

Kokelaar BP, Durant GP (1983) The submarine eruption and erosion of Surtla (Surtsey), Iceland. J Volcanol Geotherm Res 19(3–4):239–246

Kokelaar P, Busby C (1992) Subaqueous explosive eruption and welding of pyroclastic deposits. Science 257(5067):196–201

Korenaga J (2005) Why did not the Ontong Java Plateau form subaerially? Earth Planet Sci Letts 234(3–4):385–399

Krastel S, Schmincke H-U (2002) The channel between Gran Canaria and Tenerife: constructive processes and destructive events during the evolution of volcanic islands. Int J Earth Sci 91(4):629–641

Kuenzi WD, Horst OH, Mcgehee R (1979) Effect of volcanic activity on fluvial-deltaic sedimentation in a modern arc-trench gap, southwestern Guatemala. Geol Soc Amer Bull 90(9):827–838

Kurszlaukis S, Lorenz V (2008) Formation of "Tuffisitic Kimberlites" by phreatomagmatic processes. J Volcanol Geotherm Res 174(1–3):68–80

Kusakabe M (2017) Lake Nyos and Monoun gas disasters (Cameroon)—limnic eruptions caused by excessive accumulation of magmatic CO_2 in crater lakes. Geochem Monogr Ser 1:1–50

Kusakabe M, Ohsumi T, Aramaki S (1989) The Lake Nyos gas disaster: chemical and isotopic evidence in waters and dissolved gases from three Cameroonian crater lakes, Nyos, Monoun and Wum. J Volcanol Geotherm Res 39(2–3):167–185

Lacasse C, Sigurdsson H, Carey S, Paterne M, Guichard F (1996) North Atlantic deep-sea sedimentation of Late Quaternary tephra from the Iceland hotspot. Marine Geol 129(3–4):207–235

Laughton AS, Berggren WA et al (1972) Initial reports of the Deep Drilling Project Volume XII, Washington DC, US Government Printing Office. Washington DC

Lavigne F, Degeai JP, Komorowski JC, Guillet S, Robert V, Lahitte P, Oppenheimer C, Stoffel M, Vidal CM, Pratomo I, Wassmer P (2013) Source of the great AD 1257 mystery eruption unveiled, Samalas volcano, Rinjani Volcanic Complex, Indonesia. Proc Nat Acad Sci 110(42):16742–16747

Le Corvec N, Menand T, Lindsay J (2013b) Interaction of ascending magma with pre-existing crustal fractures in monogenetic basaltic volcanism: an experimental approach. J Geophys Res Solid Earth 118(3):968–984

Le Corvec N, Spörli KB, Rowland J, Lindsay J (2013a) Spatial distribution and alignments of volcanic centers: clues to the formation of monogenetic volcanic fields. Earth-Sci Rev 124:96–114

Le Friant A, Boudon G, Deplus C, Villemant B (2003) Large-scale flank collapse events during the activity of Montagne Pelée, Martínique, Lesser Antilles. J Geophys Res Solid Earth 108(B1)

Lefebvre NS, White JDL, Kjarsgaard BA (2013) Unbedded diatreme deposits reveal maar-diatreme-forming eruptive processes: Standing Rocks West, Hopi Buttes, Navajo Nation, USA. Bull Volcanol 75 (8):739

Lénat JF, Gibert-Malengreau B, Galdéano A (2001) A new model for the evolution of the volcanic island of Réunion (Indian Ocean). J Geophys Res Solid Earth 106(B5):8645–8663

Lesti C, Giordano G, Salvini F, Cas RAF (2008) Volcano tectonic setting of the intraplate, Pliocene-Holocene, Newer Volcanic Province (southeast Australia): role of crustal fracture zones. J Geophys Res Solid Earth 113(B7)

Lindsay JM (2005) Saint Lucia. In: Lindsay JM, Robertson REA, Shepherd JB, Ali S (eds) Volcanic hazard atlas of the Lesser Antilles. Univ West Indies, Trinidad and Tobago, pp 218–238

Lindsay JM, Robertson RE (2018) Integrating volcanic hazard data in a systematic approach to develop volcanic hazard maps in the Lesser Antilles. Front Earth Sci 6:42

Lindsay JM, Robertson REA, Shepherd JB, Ali S (eds) (2005) Volcanic hazard atlas of the Lesser Antilles. Univ West Indies, Trinidad and Tobago, p 279

Lipman PW (1976) Caldera-collapse breccias in the western San Juan Mountains, Colorado. Geol Soc Amer Bull 87(10):1397–1410

Lipman PW (1997) Subsidence of ash-flow calderas: relation to caldera size and magma-chamber geometry. Bull Volcanol 59(3):198–218

Lipman PW (2007) Incremental assembly and prolonged consolidation of Cordilleran magma chambers: evidence from the Southern Rocky Mountain volcanic field. Geosphere 3(1):42–70

Lipman PW, Bachmann O (2015) Ignimbrites to batholiths: integrating perspectives from geological, geophysical, and geochronological data. Geosphere 11(3):705–743

Lipman PW, Lockwood JP, Okamura RT, Swanson DA, Yamashita KM (1985) Ground deformation associated with the 1975 magnitude-7.2 earthquake and resulting changes in activity of Kilauea volcano, Hawaii. US Geol Surv Prof Pap 1276:45

Liu PP, Caricchi L, Chung SL, Li XH, Li QL, Zhou MF, Lai YM et al (2021) Growth and thermal maturation of the Toba magma reservoir. PNAS 118: 45, e2101695118. https://doi.org/10.1073/pnas.2101695118

Loewen MW, Duncan RA, Kent AJR, Krawl K (2013) Prolonged plume volcanism in the Caribbean Large Igneous Province: new insights from Curaçao and Haiti. Geochem Geophys Geosyst 14 (10):4241–4259

Longo AA, Dilles JH, Grunder AL, Duncan RA (2010) Evolution of calc-alkaline volcanism and associated hydrothermal gold deposits at Yanacocha, Peru. Econ Geol 105(7):1191–1241

Lorenz V (2003) Maar-diatreme volcanoes, their formation, and their setting in hard-rock or soft-rock environments. Geolines 15:72–83

Lorenz V (2007) Syn-and posteruptive hazards of maar–diatreme volcanoes. J Volcanol Geotherm Res 159(1–3):285–312

Lorenz V, Kurszlaukis S (2007) Root zone processes in the phreatomagmatic pipe emplacement model and consequences for the evolution of maar–diatreme volcanoes. J Volcanol Geotherm Res 159(1–3):4–32

Loughlin SC, Sparks RSJ, Brown SK, Jenkins SF, Vye-Brown C (eds) (2015a) Global volcanic hazards and risk. Cambridge University Press. https://doi.org/10.1017/CBO9781316276273

Loughlin SC, Sparks RSJ, Brown SK, Jenkins SF, Vye-Brown C (eds) (2015b) Global volcanic hazards and risk. Cambridge University Press, 1208 pp. https://doi.org/10.1017/CBO9781316276273

Lucci F, Carrasco-Núñez G, Rossetti F, Theye T, White JC, Urbani S, Azizi H, Asahara Y, Giordano G (2020) Anatomy of the magmatic plumbing system of Los Humeros Caldera (Mexico): implications for geothermal systems. Solid Earth 11(1):125–159

Macdonald KC (1982) Mid-ocean ridges: fine scale tectonic, volcanic and hydrothermal processes within the plate boundary zone. Ann Rev Earth Planet Sci 10(1):155–190

Macdonald KC (2001) Mid-Ocean ridge tectonics, volcanism, and geomorphology. Encyclopedia of ocean sciences. Academic Press, Oxford, pp 852–866

Macías JL, Capra L, Arce JL, Espíndola JM, García-Palomo A, Sheridan MF (2008) Hazard map of El Chichón volcano, Chiapas, México: Constraints posed by eruptive history and computer simulations. J Volcanol Geother Res 175(4): 444-458.

Mackie S, Cashman K, Ricketts H, Rust A, Watson M (eds) (2016) Volcanic ash: hazard observation. Elsevier, 300 pp. https://doi.org/10.1016/C2014-0-03381-3

Macpherson GJ (1983) The Snow Mountain volcanic complex: an on-land seamount in the Franciscan terrain, California. J Geol 91 (1):73–92

Maeno F, Taniguchi H (2006) Silicic lava dome growth in the 1934–1935 Showa Iwo-jima eruption, Kikai caldera, south of Kyushu, Japan. Bull Volcanol 68(7–8):673–688

Maeno F, Taniguchi H (2007) Spatiotemporal evolution of a marine caldera-forming eruption, generating a low-aspect ratio pyroclastic flow, 7.3 ka, Kikai caldera, Japan: implication from near-vent eruptive deposits. J Volcanol Geotherm Res 167(1):212–238

Magnússon E, Gudmundsson MT, Roberts MJ, Sigurðsson G, Höskuldsson F, Oddsson B (2012) Ice-volcano interactions during the 2010 Eyjafjallajökull eruption, as revealed by airborne imaging radar. J Geophys Res Solid Earth 117(B7)

Mahoney JJ, Fitton JG, Wallace PJ et al. (2001) Leg 192 preliminary report: basement drilling of the Ontong Java Plateau Sites 1183–1187. In: Proceedings ocean drilling program, initial reports, vol 192. http://www-odp.tamu.edu/publications/192_IR/192ir.htm

Manga M, Fauria KE, Lin C, Mitchell SJ, Jones M, Conway CE, Degruyter W, Hosseini B, Carey R, Cahalan R, Houghton BF, White JDL, Jutzeler M, Soule SA, Tani K (2018) The pumice raft-forming 2012 Havre submarine eruption was effusive. Earth Planet Sci Letts 489:49–58

Mann P, Taira A (2004) Global tectonic significance of the Solomon Islands and Ontong Java Plateau convergent zone. Tectonophysics 389(3–4):137–190

Marini L, Principe C, Chiodini G, Cioni R, Fytikas M, Marinelli G (1993) Hydrothermal eruptions of Nisyros (Dodecanese, Greece). Past events and present hazard. J Volcanol Geotherm Res 56(1–2):71–94

Marsh BD (1984) On the mechanics of caldera resurgence. J Geophys Res Solid Earth 89(B10):8245–8251

Marrero JM, García A, Berrocoso M, Llinares Á, Rodríguez-Losada A, Ortiz R (2019) Strategies for the development of volcanic hazard maps in monogenetic volcanic fields: the example of La Palma (Canary Islands). J Appl Volcanol 8(1):6

Martí J (2018) Assessing volcanic hazard: a review. Oxford Handbooks Online, 73 pp. https://doi.org/10.1093/oxfordhb/9780190699420.013.32

Martí J (2019) Las Cañadas caldera, Tenerife, Canary Islands: a review, or the end of a long volcanological controversy. Earth-Sci Rev 196:102889

Martí J, Geyer A, Folch A (2009) A genetic classification of collapse calderas based on field studies, and analogue and theoretical modelling. Volcanology: the Legacy of GPL Walker. IAVCEI-Geological Society of London, London, pp 249–266

Martí J, Gudmundsson A (2000) The Las Cañadas caldera (Tenerife, Canary Islands): an overlapping collapse caldera generated by magma-chamber migration. J Volcanol Geotherm Res 103(1–4):161–173

Martí J, Hurlimann M, Ablay GJ, Gudmundsson A (1997) Vertical and lateral collapses on Tenerife (Canary Islands) and other volcanic ocean islands. Geology 25(10):879–882

Martí J, Mitjavila J, Araña V (1994) Stratigraphy, structure and geochronology of the Las Cañadas caldera (Tenerife, Canary Islands). Geol Mag 131(6):715–727

Martí J, Sobradelo R, Felpeto A, García (2012) Eruptive scenarios of phonolitic volcanism at Teide–Pico Viejo volcanic complex (Tenerife, Canary Islands). Bull Volcanol 74(3):767–782

Martinez F, Taylor B (1996) Backarc spreading, rifting, and microplate rotation, between transform faults in the Manus Basin. Mar Geophys Res 18(2–4):203–224

Marzocchi W, Woo G (2009) Principles of volcanic risk metrics: Theory and the case study of Mount Vesuvius and Campi Flegrei, Italy. J Geophys Res: Solid Earth 114(B3)

Matchan E, Phillips D (2011) New 40Ar/39Ar ages for selected young (< 1 Ma) basalt flows of the Newer Volcanic Province, southeastern Australia. Quat Geochron 6(3–4):356–368

Mathews WH (1947) "Tuyas," flat-topped volcanoes in northern British Columbia. Amer J Sci 245(9):560–570

Mattei M, Conticelli S, Giordano G (2010) The Tyrrhenian margin geological setting: from the Apennine orogeny to the K-rich volcanism. The Colli Albani Volcano. Spec Publ of IAVCEI and Geol Soc Lond 3:7–28

Mattox TN, Mangan MT (1997) Littoral hydrovolcanic explosions: a case study of lava–seawater interaction at Kilauea Volcano. J Volcanol Geotherm Res 75(1):1–17

Mattsson H, Höskuldsson Á (2003) Geology of the Heimaey volcanic centre, south Iceland: early evolution of a central volcano in a propagating rift? J Volcanol Geotherm Res 127(1–2):55–71

Mattsson HB, Höskuldsson Á (2005) Eruption reconstruction, formation of flow-lobe tumuli and eruption duration in the 5900 BP Helgafell lava field (Heimaey), south Iceland. J Volcanol Geotherm Res 147(1–2):157–172

Mattsson HB, Tripoli BA (2011) Depositional characteristics and volcanic landforms in the Lake Natron-Engaruka monogenetic field, northern Tanzania. J Volcanol Geotherm Res 203(1–2):23–34

Maughan LL, Christiansen EH, Best MG, Grommé CS, Deino AL, Tingey DG (2002) The Oligocene Lund Tuff, Great Basin, USA: a very large volume monotonous intermediate. J Volcanol Geotherm Res 113(1–2):129–157

Mayer K, Scheu B, Gilg HA, Heap MJ, Kennedy BM, Lavallée Y, Letham-Brake M, Dingwell DB (2015) Experimental constraints on phreatic eruption processes at Whakaari (White Island volcano). J Volcanol Geotherm Res 302:150–162

Mcclinton T, White SM, Colman A, Sinton JM (2013) Reconstructing lava flow emplacement processes at the hot spot-affected Galápagos Spreading Center, 95 W and 92 W. Geochem Geophys Geosyst 14 (8):2731–2756

McGarvie D (2009) Rhyolitic volcano–ice interactions in Iceland. J Volcanol Geotherm Res 185(4):367–389

McGetchin TR, Settle M (1975) Cinder cone separation distances-implications for depth of formation of gabbroic xenoliths. Eos 56 (12):1070

McGuire WJ, Pullen AD (1989) Location and orientation of eruptive fissures and feederdykes at Mount Etna; influence of gravitational and regional tectonic stress regimes. J Volcanol Geotherm Res 38 (3–4):325–344

McKee CO, Baillie MG, Reimer PJ (2015) A revised age of AD 667–699 for the latest major eruption at Rabaul. Bull Volcanol 77(7):65

McKee CO, Duncan RA (2016) Early volcanic history of the Rabaul area. Bull Volcanol 78(4):24

McKibben MA, Andes JP, Williams AE (1988) Active ore formation at a brine interface in metamorphosed deltaic lacustrine sediments; the Salton Sea geothermal system, California. Econ Geol 83(3):511–523

McKnight SB, Williams SN (1997) Old cinder cone or young composite volcano?: the nature of Cerro Negro, Nicaragua. Geology 25(4):339–342

McPhie J (1995) A Pliocene shoaling basaltic seamount: Ba volcanic group at Rakiraki, Fiji. J Volcanol Geotherm Res 64(3–4):193–210

Medialdea T, Somoza L, González FJ, Vázquez JT, de Ignacio C, Sumino H, Sánchez-Guillamón O, Orihashi Y, León R, Palomino D (2017) Evidence of a modern deep water magmatic hydrothermal system in the Canary Basin (eastern central Atlantic Ocean). Geochem Geophys Geosyst 18(8):3138–3164

Menard HW (1964) Marine Geol of the Pacific. Mc Graw-Hill, New York, p 217

Menard HW (1969) Growth of drifting volcanoes. J Geophys Res 74 (20):4827–4837

Mendoza-Rosas AT, De la Cruz-Reyna S (2008) A statistical method linking geological and historical eruption time series for volcanic hazard estimations: applications to active polygenetic volcanoes. J Volcanol Geother Res 176(2):277–290

Michibayashi K, Tominaga M, Ildefonse B, Teagle DA (2019) What lies beneath. The formation and evolution of oceanic lithosphere. Oceanography 32(1):138–149

Michon L, Massin F, Famin V, Ferrazzini V, Roult G. (2011) Basaltic calderas: collapse dynamics, edifice deformation, and variations of magma withdrawal. J Geophys Res: Solid Earth 116:B03209

Miller CA, Williams-Jones G (2016) Internal structure and volcanic hazard potential of Mt Tongariro, New Zealand, from 3D gravity and magnetic models. J Volcanol Geotherm Res 319:12–28

Miyaji N (1988) History of younger Fuji volcano. J Geol Soc Jpn 94:433–452

Miyaji N, Kan'no A, Kanamaru T, Mannen K (2011) High-resolution reconstruction of the Hoei eruption (AD 1707) of Fuji volcano, Japan. J Volcanol Geotherm Res 207(3):113–129

Mohr PA (1983) Ethiopian flood basalt province. Nature 303:577–584

Mohr PA, Zanettin B (1988) The Ethiopian flood basalt province. In: Macdougall JD (ed) Continental flood basalts. Springer, Dordrecht, pp 63–110

Moles JD, McGarvie D, Stevenson JA, Sherlock SC, Abbott PM, Jenner FE, Halton AM (2019) Widespread tephra dispersal and ignimbrite emplacement from a subglacial volcano (Torfajökull, Iceland). Geology 47(6):577–580

Moore JG, Clague DA, Holcomb RT, Lipman PW, Normark WR, Torresan ME (1989) Prodigious submarine landslides on the Hawaiian Ridge. J Geophys Res Solid Earth 94(B12):17465–17484

Moores EM (1982) Origin and emplacement of ophiolites. Rev Geophys 20(4):735–760

Moorhouse BL, White JDL, Scott JM (2015) Cape Wanbrow: a stack of Surtseyan-style volcanoes built over millions of years in the Waiareka-Deborah volcanic field, New Zealand. J Volcanol Geotherm Res 298:27–46

Mucek AE, Danišík M, De Silva SL, Schmitt AK, Pratomo I, Coble MA (2017) Post-supereruption recovery at Toba Caldera. Nature Comms 8:15248

Mueller WU, Mortensen JK (2002) Age constraints and characteristics of subaqueous volcanic construction, the Archean Hunter Mine Group, Abitibi Greenstone Belt. Precamb Res 115(1–4):119–152

Munro DC, Rowland SK (1996) Caldera morphology in the western Galapagos and implications for volcano eruptive behavior and mechanisms of caldera formation. J Volcanol Geotherm Res 72(1–2):85–100

Murch AP, White JD, Carey RJ (2019) Characteristics and deposit stratigraphy of submarine-erupted silicic ash, Havre volcano, Kermadec Arc, New Zealand. Front Earth Sci 7:1

Murray BP, Busby CJ (2015) Epithermal mineralization controlled by synextensional magmatism in the Guazapares Mining District of the Sierra Madre Occidental silicic large igneous province, Mexico. J South Amer Earth Sci 58:54–71

Murray BP, Busby CJ, Ferrari L, Solari LA (2013) Synvolcanic crustal extension during the mid-Cenozoic ignimbrite flare-up in the northern Sierra Madre Occidental, Mexico: evidence from the Guazapares Mining District region, western Chihuahua. Geosphere 9(5):1201–1235

Nairn IA, Self S (1978) Explosive eruptions and pyroclastic avalanches from Ngauruhoe in February 1975. J Volcanol Geother Res 3(1–2):39–60

Nairn IA, Wiradiradja S (1980) Late Quaternary hydrothermal explosion breccias at Kawerau geothermal field, New Zealand. Bull Volcanol 43(1):1–13

Neal CA, Brantley SR, Antolik L, Babb JL, Burgess M, Calles K, ... Damby D (2019) The 2018 rift eruption and summit collapse of Kīlauea Volcano. Science 363(6425):367–374

Németh K (2010) Monogenetic volcanic fields: origin, sedimentary record, and relationship with polygenetic volcanism. Geol Soc Amer Spec Pap 470:43–66

Németh K, Kereszturi G (2015) Monogenetic volcanism: personal views and discussion. Int J Earth Sci 104(8):2131–2146

Németh K, White JDL, Reay A, Martín U (2003) Compositional variation during monogenetic volcano growth and its implications for magma supply to continental volcanic fields. J Geol Soc Lond 160(4):523–530

Neri A, Aspinall WP, Cioni R, Bertagnini A, Baxter PJ, Zuccaro G, Andronico D, Barsotti S, Cole PD, Ongaro TE, Hincks TK (2008) Developing an event tree for probabilistic hazard and risk assessment at Vesuvius. J Volcanol Geotherm Res 178(3):397–415

Newhall C, Self S, Robock A (2018) Anticipating future Volcanic Explosivity Index (VEI) 7 eruptions and their chilling impacts. Geosphere 14(2):572–603

Newhall CG, Punongbayan R (eds) (1996) Fire and mud: eruptions and lahars of Mount Pinatubo, Philippines: Philippine Instit Volcanol Seismol and Univ Washinton, 1126 pp

Newhall CG, Self S (1982) The volcanic explosivity index (VEI) an estimate of explosive magnitude for historical volcanism. J Geophys Res Oceans 87(C2):1231–1238

Nieto-Torres A, Guimarães LF, Bonadonna C, Frischknecht C (2021) A new inclusive volcanic risk ranking, part 1: methodology. Front Earth Sci 672

Nomikou P, Papanikolaou D (2018) Geodynamics and volcanism in the Kos-Yali-Nisyros volcanic field. Nisyros Volcano. Springer, pp 13–55

Nomikou P, Papanikolaou D, Alexandri M, Sakellariou D, Rousakis G (2013) Submarine volcanoes along the Aegean volcanic arc. Tectonophysics 597–598:123–146

Norini G, Groppelli G, Sulpizio R, Carrasco-Núñez G, Dávila-Harris P, Pellicioli C, Zucca F, De Franco R (2015) Structural analysis and thermal remote sensing of the Los Humeros Volcanic Complex: implications for volcano structure and geothermal exploration. J Volcanol Geotherm Res 301:221–237

Ohmoto H, Skinner BJ (eds) (1983) The Kuroko and related volcanogenic massive sulfide deposits. Econ Geol Monogr 5

Ollier CD (1967) Maars their characteristics, varieties and definition. Bull Volcanol 31(1):45–73

Oostingh KF, Jourdan F, Matchan EL, Phillips D (2017) 40Ar/39Ar geochronology reveals rapid change from plume-assisted to stress-dependent volcanism in the Newer Volcanic Province, SE Australia. Geochem Geophys Geosyst 18(3):1065–1089

Oppenheimer C, Orchard A, Stoffel M, Newfield TP, Guillet S, Corona C, Sigl M, Di Cosmo N, Büntgen U (2018) The Eldgjá eruption: timing, long-range impacts and influence on the Christianisation of Iceland. Clim Change 147(3–4):369–381

Orsi G, De Vita S, Di Vito MA (1996) The restless, resurgent Campi Flegrei nested caldera (Italy): constraints on its evolution and configuration. J Volcanol Geotherm Res 74(3–4):179–214

Orsi G, Di Vito MA, Isaia R (2004) Volcanic hazard assessment at the restless Campi Flegrei caldera. Bull Volcanol 66(6):514–530

Ort MH, Carrasco-Núñez G (2009) Lateral vent migration during phreatomagmatic and magmatic eruptions at Tecuitlapa Maar, east-central Mexico. J Volcanol Geother Res 181(1-2): 67-77

Ort MH, Lefebvre NS, Neal CA, McConnell VS, Wohletz KH (2018) Linking the Ukinrek 1977 maar-eruption observations to the tephra deposits: new insights into maar depositional processes. J Volcanol Geother Res 360:36–60

Ort MH, Wohletz K, Hooten JA, Neal CA, McConnel VS (2000) The Ukinrek maars eruption, Alaska, 1977: a natural laboratory for the study of phreatomagmatic processes at maars. Terra Nostra 6:396–400

Ortiz-Hernández LE, Acevedo-Sandoval OA, Flores-Castro K (2003) Early Cretaceous intraplate seamounts from Guanajuato, central Mexico: geochemical andmineralogical data. Revista Mexicana De Ciencias Geológicas 20(1):27–40

Orth K (1988) Geology of the Warrnambool 1:50 000 map. Geol Surv Victoria Rept 86

Pardo N, Macias JL, Giordano G, Cianfarra P, Avellán DR, Bellatreccia F (2009) The ~ 1245 yr BP Asososca maar eruption: the youngest event along the Nejapa–MiRAFlores volcanic fault, Western Managua, Nicaragua. J Volcanol Geotherm Res 184 (3):292–312

Pardo N, Pulgarín B, Betancourt V, Lucchi F, Valencia LJ (2019) Facing geological mapping at low-latitude volcanoes: The Doña

Juana Volcanic Complex study-case, SW-Colombia. J Volcanol Geother Res 385:46–67

Pardo N, Sulpizio R, Lucchi F, Giordano G, Cronin S, Pulgarín BA, ... Cabrera MA (2023) Late Holocene volcanic stratigraphy and eruption chronology of the dacitic Young Doña Juana volcano, Colombia. Geol Soc Am Bulletin 2023. https://doi.org/10.1130/B36557.1

Parfitt EA (2004) A discussion of the mechanisms of explosive basaltic eruptions. J Volcanol Geotherm Res 134(1):77–107

Parfitt EA, Peacock DCP (2001) Faulting in the south flank of Kilauea Volcano, Hawai'i. J Volcanol Geotherm Res 106(3–4):265–284

Paris R, Bravo JJC, González MEM, Kelfoun K, Nauret F (2017) Explosive eruption, flank collapse and megatsunami at Tenerife ca. 170 ka. Nature Comms 8:15246

Pe-Piper G, Piper DJW, Perissoratis C (2005) Neotectonics and the Kos Plateau Tuff eruption of 161 ka, South Aegean arc. J Volcanol Geotherm Res 139(3–4):315–338

Pensa A, Bisson M, Nave R, Bonamico A, Isaia R, Giordano G (2020) Long dormant volcano hazard assessment based on geological data: the Colli Albani case study (Roma, Italy). Final report—INGV FISR 2016 project "Colli Albani"

Pensa A, Cas RAF, Giordano G, Porreca M, Wallenstein N (2015) Transition from steady to unsteady Plinian eruption column: the VEI 5, 4.6 ka Fogo A Plinian eruption, São Miguel, Azores. J Volcanol Geotherm Res 305:1–18

Pensa A, Pinton A, Vita L, Bonamico A, De Benedetti AA, Giordano G (2019) Atlas of the Italian submarine volcanic structures. Mem Descr Carta Geol It 104:77–183. ISBN 978-88-9311-074-7

Petrinovic IA, Martí J, Aguirre-Diaz GJ, Guzmán S, Geyer A, Paz NS (2010) The Cerro Aguas Calientes caldera, NW Argentina: an example of a tectonically controlled, polygenetic collapse caldera, and its regional significance. J Volcanol Geotherm Res 194(1–3):15–26

Phillips EH, Goff F, Kyle PR, McIntosh WC, Dunbar NW, Gardner JN (2007). The $^{40}Ar/^{39}Ar$ age constraints on the duration of resurgence at the Valles caldera, New Mexico. J Geophys Res Solid Earth 112 (B8). https://doi.org/10.1029/2006JB004511

Phillips GN, Powell R (1993) Link between gold provinces. Econ Geol 88(5):1084–1098

Pierson TC, Janda RJ, Umbal JV, Daag AS (1992) Immediate and long-term hazards from lahars and excess sedimentation in rivers draining Mt. Pinatubo, Philippines. US Geol Surv Water-Res Investig Rept 92:183–203

Pike RJ, Clow GD (1981) Revised classification of terrestrial volcanoes and catalog of topographic dimensions, with new results of edifice volume. US Geol Surv Open file Rept OF 81-1038

Pinton A (2015) Reconstruction of the post-glacial volcanic activity in iceland through paleomagnetic survey and remote sensing. Unpubl PhD thesis, Università Roma Tre, Italy

Pinton A, Giordano G, Speranza F, Þórðarson Þ (2018) Paleomagnetism of Holocene lava flows from the Reykjanes Peninsula and the Tungnaá lava sequence (Iceland): implications for flow correlation and ages. Bull Volcanol 80(1):10

Pittari A (2004) Eruption dynamics and emplacement processes for the climactic Abrigo Member, Tenerife, Canary Islands. Unpubl PhD, Monash Univ, Australia

Pittari A, Cas RAF, Wolff JA, Nichols HJ, Larson PB, Martí J (2008) The use of lithic clast distributions in pyroclastic deposits to understand pre-and syn-caldera collapse processes: a case study of the Abrigo Ignimbrite, Tenerife, Canary Islands. Develop Volcanol 10:97–142

Poisson B, Pedreros R (2010) Numerical modelling of historical landslide-generated tsunamis in the French Lesser Antilles. Nat Haz Earth Syst Sci 10(6):1281–1292

Poland MP, Miklius A, Montgomery-Brown EK (2014a) Magma supply, storage, and transport at shield-stage Hawaiian volcanoes. In: Poland MP, Takahashi TJ, Landowski CM (eds) Characteristics of Hawaiian volcanoes. US Geol Surv Prof Pap 1801:179–236

Poland MP, Takahashi TJ, Landowski CM (eds) (2014b) Characteristics of Hawaiian volcanoes: US Geol Surv Prof Pap 1801

Porritt LA, Cas RAF, Crawford BB (2008) In-vent column collapse as an alternative model for massive volcaniclastic kimberlite emplacement: an example from the Fox kimberlite, Ekati Diamond Mine, NWT, Canada. J Volcanol Geotherm Res 174(1–3):90–102

Porter SC (1972) Distribution, morphology, and size frequency of cinder cones on Mauna Kea volcano, Hawaii. Geol Soc Amer Bull 83(12):3607–3612

Porter SC (1979) Quaternary stratigraphy and chronology of Mauna Kea, Hawaii: a 380,000-yr record of mid-Pacific volcanism and ice-cap glaciation. Geol Soc Amer Bull 90(7_Part_II):980–1093

Prata G (2012) Understanding cyclicity in the deposits of maar volcanoes, Tower Hill Volcanic Complex, Victoria. Unpubl MSc, Monash Univ, Australia

Puchtel IS, Hofmann AW, Mezger K, Jochum KP, Shchipansky AA, Samsonov AV (1998) Oceanic plateau model for continental crustal growth in the Archaean: a case study from the Kostomuksha greenstone belt, NW Baltic Shield. Earth Planet Sci Letts 155 (1):57–74

Rampino MR, Self S (2015) Large igneous provinces and biotic extinctions. In: Sigurdsson H, Stix J, Houghton B, Mcnutt S, Rymer H (eds) Encyclopedia of volcanoes. Academic Press, San Diego, pp 1049–1058

Reynolds MA, Best JG (1976) Summary of the 1953–57 eruption of Tuluman volcano, Papua New Guinea. In: Johnson RW (ed) Volcanism in Australasia, pp 287–296

Reynolds MA, Best JG, Johnson RW (1980) 1953–57 eruption of Tuluman volcano: rhyolitic volcanic activity in the northern Bismarck Sea. Geol Surv Papua New Guinea Mem 7

Reynolds P, Schofield N, Brown RJ, Holford SP (2018) The architecture of submarine monogenetic volcanoes–insights from 3D seismic data. Basin Res 30(S1):437–451

Riggs NR, Busby-Spera CJ (1990) Evolution of a multi-vent volcanic complex within a subsiding arc graben depression: Mount Wrightson Formation, Arizona. Geol Soc Amer Bull 102(8):1114–1135

Roberge J, Wallace PJ, White RV, Coffin MF (2005) Anomalous uplift and subsidence of the Ontong Java Plateau inferred from CO2 contents of submarine basaltic glasses. Geology 33(6):501–504

Robinson JE, Eakins BW (2006) Calculated volumes of individual shield volcanoes at the young end of the Hawaiian Ridge. J Volc Geotherm Res 151:309–317

Roche O, Druitt TH (2001) Onset of caldera collapse during ignimbrite eruptions. Earth Planet Sci Letts 191(3–4):191–202

Roche O, Druitt TH, Merle O (2000) Experimental study of caldera formation. J Geophys Res Solid Earth 105(B1):395–416

Roobol MJ, Smith AL (2004) Volcanology of Saba and St. Eustatius, Northern Lesser Antilles. R Netherlands Acad Arts Sciences, Amsterdam, 320 pp

Roobol MJ, Wright JV, Smith AL (1983) Calderas or gravity-slide structures in the Lesser Antilles island arc? J Volcanol Geotherm Res 19(1–2):121–134

Rosenberg CL, Medvedev S, Handy M (2007) On the effects of melting on continental deformation and faulting. Tectonic Faults: Agents of Change on a Dynamic Earth. Dahlem Workshop Rep 95:357–402

Rosenberg MD, Bignall G, Rae AJ (2009) The geological framework of the Wairakei–Tauhara geothermal system, New Zealand. Geothermics 38(1):72–84

Rosi M, Bertagnini A, Landi P (2000) Onset of the persistent activity at Stromboli volcano (Italy). Bull Volcanol 62(4–5):294–300

Ross PS, Carrasco-Núñez G, Hayman P (2017) Felsic maar-diatreme volcanoes: a review. Bull Volcanol 79(2):20

Ross P-S, White JDL (2006) Debris jets in continental phreatomagmatic volcanoes: a field study of their subterranean deposits in the Coombs Hills vent complex, Antarctica. J Volcanol Geotherm Res 149:62–84

Rossi MJ (1996) Morphology and mechanism of eruption of postglacial shield volcanoes in Iceland. Bull Volcanol 57(7):530–540

Roverato M, Dufresne A, Procter J (Eds) (2021) Volcanic Debris Avalanches: From Collapse to Hazard. Springer Nature

Rowland JV, Simmons SF (2012) Hydrologic, magmatic, and tectonic controls on hydrothermal flow, Taupo Volcanic Zone, New Zealand: implications for the formation of epithermal vein deposits. Econ Geol 107(3):427–457

Rubin KH (2014) Mid-ocean ridge magmatism and volcanism. In: Harff J et al (eds) Encyclopedia of marine geosciences. Springer, Dordrecht. https://doi.org/10.1007/978-94-007-6238-1_28

Rubin KH, Soule SA, Chadwick WW Jr, Fornari DJ, Clague DA, Embley RW, Baker ET, Perfit MR, Caress DW, Dziak RP (2012) Volcanic eruptions in the deep sea. Oceanography 25(1):142–157

Russell JK, Edwards BR, Porritt L, Ryane C (2014) Tuyas: a descriptive genetic classification. Quat Sci Rev 87:70–81

Russell JK, Porritt LA, Lavallée Y, Dingwell DB (2012) Kimberlite ascent by assimilation-fuelled buoyancy. Nature 481(7381):352

Sager WW, Sano T, Geldmacher J (2011) IODP Expedition 324: ocean drilling at Shatsky Rise gives clues about oceanic plateau formation. Sci Drilling 12:24–31

Sager WW, Zhang J, Korenaga J, Sano T, AaP K, Widdowson M, Mahoney JJ (2013) An immense shield volcano within the Shatsky Rise oceanic plateau, northwest Pacific Ocean. Nature Geosci 6 (11):976

Santana-Casiano JM, Fraile-Nuez E, González-Dávila M, Baker ET, Resing JA, Walker SL (2016) Significant discharge of CO_2 from hydrothermalism associated with the submarine volcano of El Hierro Island. Nature Sci Repts 6:25686

Sawyer DA, Fleck RJ, Lanphere MA, Warren RG, Broxton DE, Hudson MR (1994) Episodic caldera volcanism in the Miocene southwestern Nevada volcanic field: revised stratigraphic framework, 40Ar/39Ar geochronology, and implications for magmatism and extension. Geol Soc Amer Bull 106(10):1304–1318

Scandone R (1990) Chaotic collapse of calderas. J Volcanol Geotherm Res 42(3):285–302

Scandone R, Bartolini S, Martí J (2016) A scale for ranking volcanoes by risk. Bull Volcanol 78:1–8

Schipper CI, White JDL, Zimanowski B, Büttner R, Sonder I, Schmid A (2011) Experimental interaction of magma and "dirty" coolants. Earth Planet Sci Letts 303(3):323–336

Schmidt R, Schmincke H-U (2002) From seamount to oceanic island, Porto Santo, central East-Atlantic. Int J Earth Sci 91(4):594–614

Schmincke HU, Sumita M (1998) Volcanic evolution of Gran Canaria reconstructed from apron sediments: synthesis of VICAP project drilling. In: Weaver (ed) Proc Ocean Drilling Prog Sci Results 157:433–469

Schmincke H-U, Fisher RV, Waters AC (1973) Antidune and chute and pool structures in the base surge deposits of the Laacher See area, Germany. Sedimentology 20(4):553–574

Schmincke H-U, Sumita M (2010) Geological evolution of the Canary Islands. Gorres-Verlag, Koblenz, 188 pp

Schnur SR, Gilbert LA (2012) Detailed volcanostratigraphy of an accreted seamount: implications for intraplate seamount formation. Geochem Geophys Geosyst 13(12)

Schroder JF, Papale P eds (2015) Volcanic hazards, risks and disasters. Elsevier, 532 pp. https://doi.org/10.1016/C2011-0-07012-6

Schumacher R, Schmincke H-U (1991) Internal structure and occurrence of accretionary lapilli—a case study at Laacher See Volcano. Bull Volcanol 53(8):612–634

Schumacher R, Schmincke H-U (1995) Models for the origin of accretionary lapilli. Bull Volcanol 56(8):626–639

Searle MP, Waters DJ, Garber JM, Rioux M, Cherry AG, Ambrose TK (2015) Structure and metamorphism beneath the obducting Oman ophiolite: evidence from the Bani Hamid granulites, northern Oman mountains. Geosphere 11(6):1812–1836

Searle R (1984) GLORIA survey of the East Pacific Rise near 3.5 S: tectonic and volcanic characteristics of a fast spreading mid-ocean rise. Tectonophysics 101(3–4):319–344

Searle R (2013) Mid-ocean ridges. Cambridge Univ Press, 330 pp

Searle RC (1983) Submarine central volcanoes on the Nazca plate—high-resolution sonar observations. Marine Geol 53(1–2):77–102

Seisdedos J, Ferrer M, De Vallejo LG (2012) Geological and geomechanical models of the pre-landslide volcanic edifice of Güímar and La Orotava mega-landslides (Tenerife). J Volcanol Geotherm Res 239:92–110

Self S, Coffin MF, Rampino MR, Wolff JA (2015) Large igneous provinces and flood basalt volcanism. In: Sigurdsson H, Stix J, Houghton B, Mcnutt S, Rymer H (eds) Encyclopedia of volcanoes. Academic Press, San Diego, pp 441–458

Self S, Goff F, Gardner JN, Wright JV, Kite WM (1986) Explosive rhyolitic volcanism in the Jemez Mountains: vent locations, caldera development and relation to regional structure. J Geophys Res 91 (B2):1779–1798

Self S, Heiken G, Sykes ML, Wohletz K, Fisher RV, Dethier DP (1996) Part I: field guide to the Bandelier Tuff and Valles caldera. Field excursions to the Jemez Mountains, New Mexico. New Mexico Bur Mines Min Res Bull 134:7–55

Self S, Jay AE, Widdowson M, Keszthelyi LP (2008) Correlation of the Deccan and Rajahmundry Trap lavas: are these the longest and largest lava flows on Earth? J Volcanol Geotherm Res 172(1):3–19

Self S, Kienle J, Huot J-P (1980) Ukinrek Maars, Alaska, II. Deposits and formation of the 1977 craters. J Volcanol Geotherm Res 7 (1):39–65

Self S, Rampino MR (1981) The 1883 eruption of Krakatau. Nature 294:699–704

Selva J, Acocella V, Bisson M, Caliro S, Costa A, Della Seta M, De Martino S, Di Martino P, Giordano G, ... Cardaci C (2019) Multiple natural hazards at volcanic islands: a review for the Ischia volcano (Italy). J Appl Volcanol 8(1):1–43

Selva J, Orsi G, Di Vito MA, Marzocchi W, Sandri L (2012) Probability hazard map for future vent opening at the Campi Flegrei caldera, Italy. Bull Volcanol 74:497–510

Sembroni A, Faccenna C, Becker TW, Molin P, Abebe B (2016) Long-term, deep-mantle support of the Ethiopia-Yemen Plateau. Tectonics 35(2):469–488

Sheridan MF, Updike RG (1975) Sugarloaf Mountain Tephra—A Pleistocene rhyolitic deposit of base-surge origin in northern Arizona. Geol Soc Amer Bull 86(4):571–581

Shimizu S, Nakaoka R, Seama N, Suzuki-Kamata K, Kaneko K, Kiyosugi K, Iwamaru H, Sano M, Matsuno T, Sugioka H, Tatsumi Y (2024) Submarine pyroclastic deposits from 7.3 ka caldera-forming Kikai-Akahoya eruption. J Volcanol Geotherm Research 448:108017

Siebe C, Macías JL (2006). Volcanic hazards in the Mexico City metropolitan area from eruptions at Popocatépetl, Nevado de Toluca, and Jocotitlán stratovolcanoes and monogenetic scoria cones in the Sierra Chichinautzin volcanic field. Geol Soc Amer Spec Pap 402. Penrose Conf Ser 253–239

Sigurdsson H, Devine JD, Tchua FM, Presser FM, Pringle MK, Evans WC (1987) Origin of the lethal gas burst from Lake Monoun, Cameroun. J Volcano Geotherm Res 31(1–2):1–6

Sigurdsson H, Houghton B, McNutt SR, Rymer H, Stix J (eds) (2015) The encyclopedia of volcanoes. Academic Press, Elsevier

Sigurdsson H, Sparks RSJ, Carey SN, Huang TC (1980) Volcanogenic sedimentation in the Lesser Antilles arc. J Geol 523–540

Silleni A, Giordano G, Isaia R, Ort MH (2020) The magnitude of the 39.8 ka Campanian Ignimbrite eruption, Italy: method, uncertainties and errors. Front Earth Sci 8:543399

Sillitoe RH, Baker EM, Brook WA (1984) Gold deposits and hydrothermal eruption breccias associated with a maar volcano at Wau, Papua New Guinea. Econ Geol 79(4):638–655

Simkin T, Fiske RS (1983) Krakatau 1883—The volcanic eruption and its effects. Smithsonian Institution Press, Washington DC. 464p

Simmons JM, Cas RAF, Druitt TH, Carey RJ (2017) The initiation and development of a caldera-forming Plinian eruption (172ka Lower Pumice 2 eruption, Santorini, Greece). J Volcanol Geotherm Res 341:332–350

Simmons JM, Cas RAF, Druitt TH, Folkes CB (2016) Complex variations during a caldera-forming Plinian eruption, including precursor deposits, thick pumice fallout, co-ignimbrite breccias and climactic lag breccias: the 184ka Lower Pumice 1 eruption sequence, Santorini, Greece. J Volcanol Geotherm Res 324:200–219

Singer BS, Andersen NL, Le Mével H, Feigl KL, Demets C, Tikoff B, Thurber CH, Jicha BR, Cardona C, Córdova L (2014) Dynamics of a large, restless, rhyolitic magma system at Laguna del Maule, southern Andes, Chile. Geol Soc Amer Today 24(12):4–10

Smellie JL (2006) The relative importance of supraglacial versus subglacial meltwater escape in basaltic subglacial tuya eruptions: an important unresolved conundrum. Earth-Sci Rev 74(3–4):241–268

Smellie JL (2013) Quaternary vulcanism: subglacial landforms. In: Elias SA (ed) The encyclopedia of quaternary science, vol 1. Elsevier, Amsterdam, pp 780–802

Smellie JL (2018) Glaciovolcanism—a 21st century proxy for palaeo-ice. In: Menzies J, Van Der Meer JJM (eds) Past glacial environments (sediments, forms and techniques). Elsevier, Amsterdam, pp 335–375

Smellie JL, Haywood AM, Hillenbrand C-D, Lunt DJ, Valdes PJ (2009) Nature of the Antarctic Peninsula Ice Sheet during the Pliocene: geological evidence and modelling results compared. Earth Sci Rev 94(1–4):79–94

Smellie JL, Johnson JS, Mcintosh WC, Esser R, Gudmundsson MT, Hambrey MJ, van Wyk de Vries B (2008) Six million years of glacial history recorded in volcanic lithofacies of the James Ross Island Volcanic Group, Antarctic Peninsula. Palaeogeog Palaeoclim Palaeoecol 260(1–2):122–148

Smith AL, Roobol MJ (1990) Mt. Pelée, Martínique: a study of an active island-arc volcano. Geol Soc Amer Mem 175

Smith AL, Roobol MJ, Mattioli GS, Fryxell JE, Daly GE, Fernandez LA (2013) The volcanic geology of the mid-arc Island of Dominica. Geol Soc Amer Spec Pap 496:249p

Smith IEM, Blake S, Wilson CJN, Houghton BF (2008) Deep-seated fractionation during the rise of a small-volume basalt magma batch: Crater Hill, Auckland, New Zealand. Contribs Mineral Petrol 155 (4):511–527

Smith RL, Bailey RA (1966) The Bandelier Tuff: a study of ash-flow eruption cycles from zoned magma chambers. Bull Volcanol 29 (1):83–104

Smith RL, Bailey RA (1968) Resurgent cauldrons. In: Coats RR, Hay RL, Anderson CA (eds) Studies in volcanology (Howell Williams volume). Geol Soc Amer Mem 116:153–210

Sobradelo R, Bartolini S, Martí J (2014) HASSET: a probability event tree tool to evaluate future volcanic scenarios using Bayesian inference. Bull Volcanol 76(1):1–5

Soriano C, Cas RAF, Riggs NR, Giordano G (2016) Submarine volcanism of the Cabo de Gata magmatic arc in the Betic-Rif Orogen, SE Spain: processes and products. INTECH. https://doi.org/10.5772/63579

Soriano C, Giordano G, Cas RAF, Riggs N, Porreca M (2013) Facies architecture, emplacement mechanisms and eruption style of the submarine andesite El Barronal complex, Cabo de Gata, SE Spain. J Volcanol Geotherm Res 264:210–222

Soriano C, Riggs N, Giordano G, Porreca M, Conticelli S (2012) Cyclic growth and mass wasting of submarine Los Frailes lava flow and dome complex in Cabo de Gata, SE Spain. J Volcanol Geotherm Res 231:72–86

Soule SA (2015) Mid-ocean ridge volcanism. In: Sigurdsson H, Houghton B, Mcnutt SR, Rymer H, Stix J (eds) The encyclopedia of volcanoes. Academic Press, Elsevier, pp 395–403

Sparks RSJ, Francis PW, Hamer RD, Pankhurst RJ, O'callaghan LO, Thorpe RS, Page R (1985) Ignimbrites of the Cerro Galan Caldera, NW Argentina. J Volcanol Geotherm Res 24(3–4):205–248

Staudigel H, Koppers AA (2015) Seamounts and island building. In: Sigurdsson H, Houghton B, Mcnutt SR, Rymer H, Stix J (eds) The encyclopedia of volcanoes. Academic Press, Elsevier, pp 405–421

Staudigel H, Koppers AA, Lavelle JW, Pitcher TJ, Shank TM (2010) Mountains in the sea. Oceanography 23(1):16–213

Steven TA, Lipman PW (1976) Calderas of the San Juan volcanic field, south-western Colorado. US Geol Surv Prof Pap 958

Stix J, Kobayashi T (2008) Magma dynamics and collapse mechanisms during four historic caldera-forming events. J Geophys Res: Solid Earth 113(B9)

Stroncik NA, Schmincke H-U (2002) Palagonite–a review. Int J Earth Sci 91(4):680–697

Strong M, Wolff J (2003) Compositional variations within scoria cones. Geology 31(2):143–146

Sun C, You H, Liu J, Li X, Gao J, Chen S (2014) Distribution, geochemistry and age of the Millennium eruptives of Changbaishan volcano, Northeast China—a review. Front Earth Sci 8(2):216–230

Swanson DA (1972) Magma supply rate at Kilauea Volcano, 1952–1971. Science 175(4018):169–170

Swanson DA, Duffield WA, Fiske RS (1976) Displacement of the south flank of Kilauea Volcano; the result of forceful intrusion of magma into the rift zones. US Geol Surv Prof Pap 963

Swanson DA, Wright TL, Helz RT (1975) Linear vent systems and estimated rates of magma production and eruption for the Yakima Basalt on the Columbia Plateau. Amer J Sci 275(8):877–905

Takaya Y, Yasukawa K et al (2018) The tremendous potential of deep-sea mud as a source of rare-earth elements. Nature Sci Repts 8 (1):5763

Takarada S (2017) The Volcanic Hazards Assessment Support System for the online hazard assessment and risk mitigation of Quaternary volcanoes in the world. Front Earth Sci 5:102

Taylor RN, Davila-Harris P, Branney MJ, Farley ER, Gernon TM, Palmer MR (2020) Dynamics of a chemically pulsing mantle plume. Earth Planet Sci Letts 537:116182

Tedesco D, Vaselli O, Papale P, Carn SA, Voltaggio M, Sawyer GM, Durieux J, Kasereka M, Tassi F (2007) January 2002 volcano-tectonic eruption of Nyiragongo volcano, Democratic Republic of Congo. J Geophys Res Solid Earth 112(B9)

Thierry P, Stieltjes L, Kouokam E, Nguéya P, Salley PM (2008) Multi-hazard risk mapping and assessment on an active volcano: the GRINP project at Mount Cameroon. Nat Haz 45:429-456

Thorarinsson S, Einarsson T, Sigvaldason G, Elisson G (1964) The submarine eruption off the Vestmann Islands 1963–64. Bull Volcanol 27(1):435–445

Thordarson T (2004) Accretionary-lapilli-bearing pyroclastic rocks at ODP Leg 192 Site 1184: a record of subaerial phreatomagmatic eruptions on the Ontong Java Plateau. Geol Soc Lond Spec Publ 229(1):275–306

Thordarson T, Larsen G (2007) Volcanism in Iceland in historical time: volcano types, eruption styles and eruptive history. J Geodyn 43 (1):118–152

Thordarson T, Self S (1993) The Laki (Skaftár Fires) and Grímsvötn eruptions in 1783–1785. Bull Volcanol 55(4):233–263

Thordarson T, Self S (1998) The Roza Member, Columbia River Basalt Group: a gigantic pahoehoe lava flow field formed by endogenous processes? J Geophys Re Solid Earth 103(B11):27411–27445

Thordarson T, Self S, Miller DJ, Larsen G, Vilmundardóttir EG (2003) Sulphur release from flood lava eruptions in the Veidivötn, Grímsvötn and Katla volcanic systems, Iceland. In: Oppenheimer C, Pyle DM, Barclay J (eds) Volcanic degassing. Geol Soc Lond Spec Publ London, vol 213, pp 103–121

Thouret J (1999) Volcanic geomorphology—an overview. Earth Sci Rev 47(1–2):95–131

Tibaldi A (2001) Multiple sector collapses at Stromboli volcano, Italy: how they work. Bull Volcanol 63(2–3):112–125

Tibaldi A, Vezzoli L (1998) The space problem of caldera resurgence: an example from Ischia Island, Italy. Geol Rundsch 87(1):53–66

Tinti S, Bortolucci E, Romagnoli C (2000) Computer simulations of tsunamis due to sector collapse at Stromboli, Italy. J Volcanol Geotherm Res 96(1–2):103–128

Tolstoy M (2015) Mid-ocean ridge eruptions as a climate valve. Geophys Res Letts 42(5):1346–1351

Tolstoy M, Tan YJ (2019) Mid-ocean ridge eruption rates, long-term climate change, and the importance of marine seismoacoustics. J Acoust Soc Amer 146(4):3067

Troll VR, Carracedo JC (2016) The geology of the Canary Islands. Elsevier, 636 pp

Tuffen H, Gilbert J, McGarvie D (2001) Products of an effusive subglacial rhyolite eruption: Bláhnúkur, Torfajökull, Iceland. Bull Volcanol 63(2–3):179–190

Tuffen H, McGarvie DW, Gilbert JS, Pinkerton H (2002) Physical volcanology of a subglacial-to-emergent rhyolitic tuya at Rauðufossafjöll, Torfajökull, Iceland. Geol Soc Lond Spec Publ 202(1):213–236

Valentine GA, Connor CB (2015) Basaltic volcanic fields. In: Sigurdsson H, Houghton B, Mcnutt S, Rymer H, Stix J (eds) The encyclopedia of volcanoes, 2nd edn. Elsevier, pp 423–439

Valentine GA, Cortés JA, Widom E, Smith EI, Rasoazanamparany C, Johnsen R, Briner JP, Harp AG, Turrin B (2017) Lunar Crater volcanic field (Reveille and Pancake Ranges, Basin and Range Province, Nevada, USA). Geosphere 13(2):391–438

Valentine GA, Gregg TKP (2008) Continental basaltic volcanoes—processes and problems. J Volcanol Geotherm Res 177(4):857–873

Valentine GA, White JDL (2012) Revised conceptual model for maar-diatremes: subsurface processes, energetics, and eruptive products. Geology 40(12):1111–1114

Van Bemmelen RW (1939) The volcano-tectonic origin of Lake Toba (north Sumatra). De Ingenieur in Nederlandsch Indie 6–9:126–140

Van Bemmelen RW (1949) The geology of Indonesia. Gov Print Office, The Hague, Netherlands, 732 pp

Van Den Hove J, Grose L, Betts PG, Ailleres L, Van Otterloo J, Cas RAF (2017a) Spatial analysis of an intra-plate basaltic volcanic field in a compressional tectonic setting: South-eastern Australia. J Volcanol Geotherm Res 335:35–53

Van den Hove JC, Ailleres L, Betts PG, Cas RAF (2015) Subsurface structure of a large basaltic maar volcano examined using geologically constrained potential field modelling, Lake Purrumbete Maar, Newer Volcanics Province, southeastern Australia. J Volcanol Geotherm Res 304:142–159

Van den Hove JC, Van Otterloo J, Betts PG, Ailleres L, Cas RAF (2017b) Controls on volcanism at intraplate basaltic volcanic fields. Earth Planet Sci Letts 459:36–47

Van Otterloo J, Cas RAF (2016) Low-temperature emplacement of phreatomagmatic pyroclastic flow deposits at the monogenetic Mt Gambier Volcanic Complex, South Australia, and their relevance for understanding some deposits in diatremes. J Geol Soc Lond 173 (4):701–710

Van Otterloo J, Cas RAF, Sheard MJ (2013) Eruption processes and deposit characteristics at the monogenetic Mt. Gambier Volcanic Complex, SE Australia: implications for alternating magmatic and phreatomagmatic activity. Bull Volcanol 75(8):737–758

Van Otterloo J, Raveggi M, Cas RAF, Maas R (2014) Polymagmatic activity at the monogenetic Mt Gambier volcanic complex in the Newer Volcanics Province, SE Australia: new insights into the occurrence of intraplate volcanic activity in Australia. J Petrol 55 (7):1317–1351

van Wyk de Vries B, Davies T (2015) Landslides, debris avalanches, and volcanic gravitational deformation. In: Sigurdsson H, Houghton B, McNutt SR, Rymer H, Stix J (eds) The encyclopedia of volcanoes. Academic Press, Elsevier, pp 665–685

van Wyk de Vries B, Merle O (1998) Extension induced by volcanic loading in regional strike-slip zones. Geology 26(11):983–986

Vaughan RG, Webley PW (2010) Satellite observations of a surtseyan eruption: Hunga Ha'apai, Tonga. J Volcanol Geotherm Res 198(1–2):177–186

Ventura G, Milano G, Passaro S, Sprovieri M (2013) The Marsili Ridge (Southern Tyrrhenian Sea, Italy): an island-arc volcanic complex emplaced on a 'relict'back-arc basin. Earth-Sci Rev 116:85–94

Vessell RK, Davies DK (1981) Non-marine sedimentation in an active fore arc basin. In: Ethridge FG, Flores RM (eds) Recent and ancient non-marine depositional environments: models for exploration. Soc Econ Pal Mineral Spec Publ 31:31–45

Vignaroli G, Aldega L, Balsamo F, Billi A, De Benedetti AA, De Filippis L, Giordano G, Rossetti F (2015) A way to hydrothermal paroxysm, Colli Albani volcano, Italy. Geol Soc Am Bull 127(5–6):672–687

Vitale S, Isaia R (2014) Fractures and faults in volcanic rocks (Campi Flegrei, southern Italy): insight into volcano-tectonic processes. Int J Earth Sci 103(3):801–819

Voight B, Janda RJ, Glicken H, Douglass PM (1983) Nature and mechanics of the Mount St. Helens rockslide-avalanche of 18 May 1980. Geotechnique 33(3):243–273

Wadge G (1982) Steady state volcanism: evidence from eruption histories of polygenetic volcanoes. J Geophys Res Solid Earth 87 (B5):4035–4049

Walker GPL (1984) Downsag calderas, ring faults, caldera sizes, and incremental caldera growth. J Geophys Res Solid Earth 89 (B10):8407–8416

Walker GPL (1992) Puu Mahana near South Point in Hawaii is a primary Surtseyan ash ring, not a sandhills-type littoral cone. Pacific Sci 46(1):1–10

Walker GPL, Wright JV, Clough BJ, Booth B (1981) Pyroclastic geology of the rhyolitic volcano of La Primavera, Mexico. Geol Rundsch 70(3):1100–1118

Wallace PJ, Frey FA, Weis D, Coffin MF (2002) Origin and evolution of the Kerguelen Plateau, Broken Ridge and Kerguelen archipelago. J Petrol 43:1105–1413

Wang P, Chen S (2015) Cretaceous volcanic reservoirs and their exploration in the Songliao Basin, northeast China. Amer Assoc Petrol Geol Bull 99(3):499–523

Watkins R, Browne QJ (1989) An Ordovician continental-margin sequence of turbidite and seamount deposits in the Roberts Mountains allochthon, Independence Range, Nevada. Geol Soc Amer Bull 101(5):731–741

Wellman P (1974) Potassium-argon ages on the Cainozoic volcanic rocks of eastern Victoria, Australia. J Geol Soc Aust 21(4):359–376

White JDL (1996) Impure coolants and interaction dynamics of phreatomagmatic eruptions. J Volcanol Geotherm Res 74(3–4):155–170

White JDL, Houghton B (2000) Surtseyan and related phreatomagmatic eruptions. In: Sigurdsson H, Stix J, Houghton BF, Mcnutt SR, Rymer H (eds.) Encyclopedia of volcanoes. Academic, San Diego, pp 495–513

White JDL, Ross P-S (2011) Maar-diatreme volcanoes: a review. J Volcanol Geotherm Res 201(1):1–29

Wilcock J, Goff F, Minarik WG, Stix J (2013) Magmatic recharge during the formation and resurgence of the Valles Caldera, New Mexico, USA: evidence from quartz compositional zoning and geothermometry. J Petrol 54(4):635–664

Willcock MAW, Bargossi GM, Weinberg RF, Gasparotto G, Cas RAF, Giordano G, Marocchi M (2015) A complex magma reservoir system for a large volume intra-to extra-caldera ignimbrite: mineralogical and chemical architecture of the VEI8, Permian Ora ignimbrite (Italy). J Volcanol Geotherm Res 306:17–40

Willcock MAW, Cas RAF, Giordano G, Morelli C (2013) The eruption, pyroclastic flow behaviour, and caldera in-filling processes of the extremely large volume (>1290 km3), intra- to extra-caldera, Permian Ora (Ignimbrite) Formation, Southern Alps, Italy. J Volcanol Geotherm Res 265:102–126

Williams R, Rowley P, Garthwaite MC (2019) Reconstructing the Anak Krakatau flank collapse that caused the December 2018 Indonesian tsunami. Geology 47(10):973–976

Wilson CJN, Gravley DM, Leonard GS, Rowland JV (2009) Volcanism in the central Taupo Volcanic Zone, New Zealand: tempo, styles and controls. Studies in Volcanology: The Legacy of George Walker. Spec Publ IAVCEI and Geol Soc Lond 2:225–247

Wilson CJN, Walker GPL (1985) The Taupo eruption, New Zealand I. General aspects. Phil Trans R Soc Lond A Math Phys Engin Sci 314 (1529):199–228

Wilson L, Head JW III (1981) Ascent and eruption of basaltic magma on the Earth and Moon. J Geophys Res Solid Earth 86(B4):2971–3001

Wohletz K (2000) Were the dark ages triggered by volcano-related climate changes in the 6th century? Eos 48(81):F1305

Wohletz K, Heiken G, Ander M, Goff F, Vuataz F-D, Wadge G (1986) The Qualibou Caldera, St. Lucia, West Indies. J Volcanol Geotherm Res 27(1–2):77–115

Wohletz KH (1986) Explosive magma-water interactions: thermodynamics, explosion mechanisms, and field studies. Bull Volcanol 48:245–264

Wohletz KH, Sheridan MF (1983) Hydrovolcanic explosions; II, Evolution of basaltic tuff rings and tuff cones. Amer J Sci 283 (5):385–413

Wolfe JA, Self S (1983) Structural lineaments and Neogene volcanism in southwestern Luzon. The Tectonic and Geologic Evolution of Southeast Asian Seas and Islands. Amer Geophys Union Monogr 27:157–172

Wolff JA, Brunstad KA, Gardner JN (2011) Reconstruction of the most recent volcanic eruptions from the Valles caldera. New Mexico. J Volcanol Geotherm Res 199(1–2):53–68

Wolff JA, Gardner JN (1995) Is the Valles caldera entering a new cycle of activity? Geology 23(5):411–414

Wolff JA, Thompson RA (2017) The Jemez Mountains and Valles caldera. In: Thompson RA et al (eds) Field guides to the Rio Grande rift, Jemez Mountains, and San Juan volcanic field, New Mexico and Colorado. U.S. Geol Surv Sci Invest Rept 2017-5022-R. In press

Wood CA (1980a) Morphometric evolution of cinder cones. J Volcanol Geotherm Res 7(3–4):387–413

Wood CA (1980b) Morphometric analysis of cinder cone degradation. J Volcanol Geotherm Res 8(2–4):137–160

Wood CA (1980c) Maars. In: Green (ed) Encyclopedia of volcanology

Wright IC, Stoffers P, Hannington M, De Ronde CEJ, Herzig P, Smith IEM, Browne PRL (2002) Towed-camera investigations of shallow–intermediate water-depth submarine stratovolcanoes of the southern Kermadec arc, New Zealand. Marine Geol 185(3):207–218

Wright IC, Worthington TJ, Gamble JA (2006) New multibeam mapping and geochemistry of the 30–35 S sector, and overview, of southern Kermadec arc volcanism. J Volcanol Geotherm Res 149 (3):263–296

Wright JV, Roobol MJ, Smith AL, Sparks RSJ, Brazier SA, Rose WI, Sigurdsson H (1984) Late quaternary explosive silicic volcanism on St Lucia, West Indies. Geol Mag 121(1):1–15

Wright JV, Sastrohardjo F, Curet EA (in prep) Volcanic facies analysis of the Aruba Lava Formation, Leeward Antilles: eruption and depositional processes on the distal submarine flank of an emergent oceanic plateau volcano

Wright JV, Smith AL, Roobol MJ, Mattioli GS, Fryxell JE (2016) Distal ash hurricane (pyroclastic density current) deposits from a ca. 2000 yr BP Plinian-style eruption of Mount Pelée, Martínique: distribution, grain-size characteristics, and implications for future hazard. Geol Soc Amer Bull 128(5–6):777–791

Yamamoto T, Takada A, Ishizuka Y, Miyaji N, Tajima Y (2005) Basaltic pyroclastic flows of Fuji volcano, Japan: characteristics of the deposits and their origin. Bull Volcanol 67:622–633

Zanon V (2015) Chapter 5 The magmatism of the Azores islands. Geol Soc Lond Memoirs 44:51–64

Zimanowski B, Büttner R, Lorenz V (1997) Premixing of magma and water in MFCI experiments. Bull Volcanol 58(6):491–495

Zimanowski B, Fröhlich G, Lorenz V (1991) Quantitative experiments on phreatomagmatic explosions. J Volcanol Geotherm Res 48 (3):341–358

Zimmerer MJ, Lafferty J, Coble MA (2016) The eruptive and magmatic history of the youngest pulse of volcanism at the Valles caldera: implications for successfully dating late Quaternary eruptions. J Volcanol Geotherm Res 310:50–57

Volcanism and Tectonic Setting

17

The Modern Global Tectonic Framework, and an Approach for Evaluating Ancient Tectonic Settings, including the Hadean and Archean

John V. Wright, Guido Giordano, and Ray Cas

R. Cas et al., *Volcanology*, Springer Textbooks in Earth Sciences, Geography and Environment,
https://doi.org/10.1007/978-3-319-66613-6_17

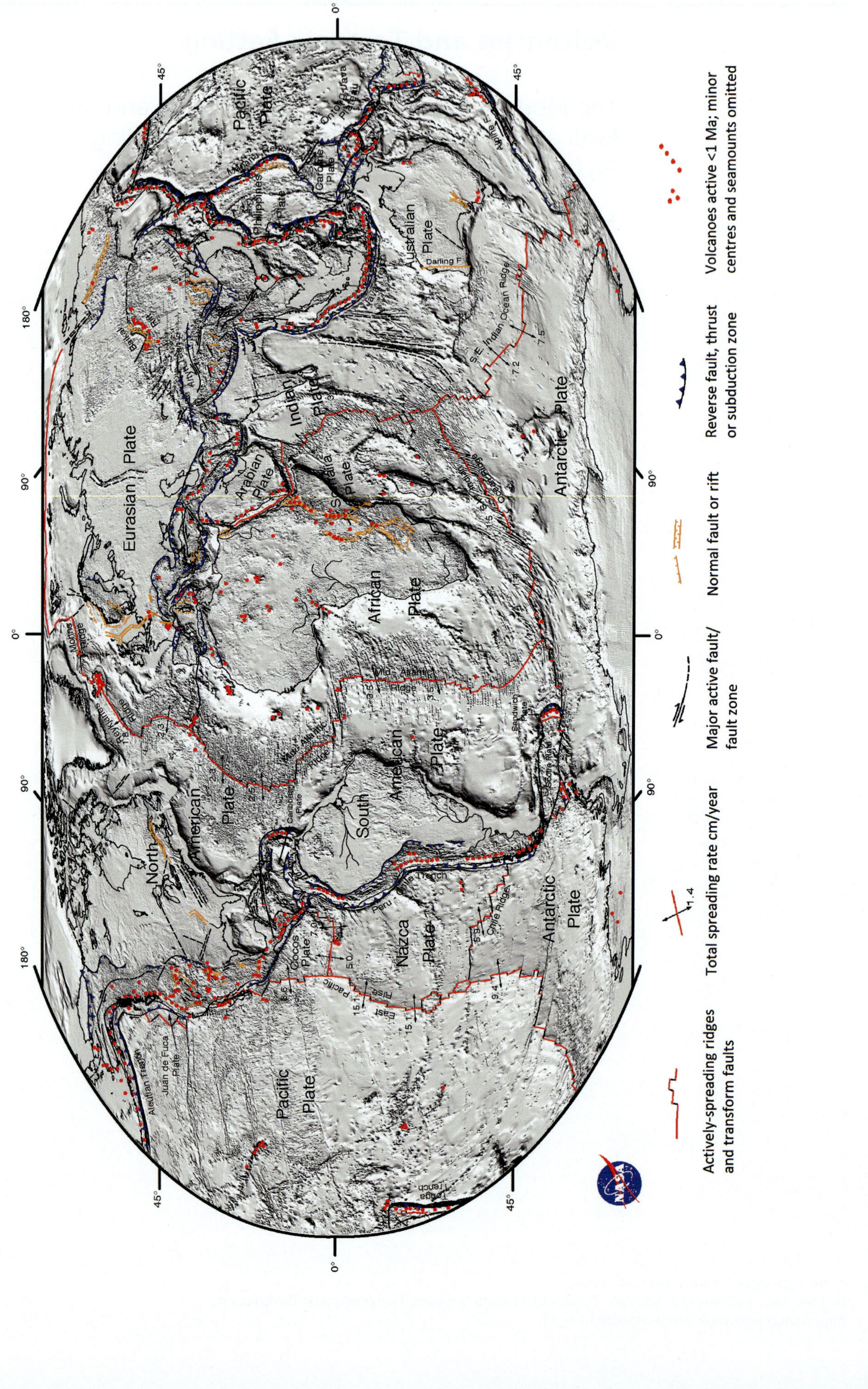
Eurasian Plate
African Plate
North American Plate
South American Plate
Pacific Plate
Antarctic Plate
Nazca Plate
Australian Plate
Indian Plate
Arabian Plate
Philippines Plate
Caroline Plate
Cocos Plate
Caribbean Plate
Juan de Fuca Plate
Sandwich Plate
Mid-Atlantic Ridge
East Pacific Rise
Peru Chile Trench
Chile Ridge
S.E. Indian Ocean Ridge
Darling F.
Tonga Trench
Aleutian Trench
Reykjanes Ridge
Mohns Ridge
Ontong Java Plateau
Actively-spreading ridges and transform faults
Total spreading rate cm/year
Major active fault/ fault zone
Normal fault or rift
Reverse fault, thrust or subduction zone
Volcanoes active <1 Ma; minor centres and seamounts omitted

Abstract

Since publication of Cas and Wright (1987), the relationship between volcanism and tectonic setting has expanded into an enormous and multidisciplinary topic. In this chapter, we provide an updated review of the known tectonic settings in which volcanism occurs today and the relatively recent past, and the overall geological characteristics of these settings using the plate tectonic model as a global framework for Phanerozoic and Proterozoic times. However, there is little evidence that plate tectonics operated earlier during the Hadean and Archean, which represents almost 50% of geological time, and so we will also consider what the tectonic paradigm for volcanism was during the early history of the Earth. Initially, during the two billion years prior to when plate tectonics developed, the Earth was a magma ocean. This evolved into a single mafic–ultramafic crustal shell (stagnant lid) as the Earth cooled and was affected by mantle plume activity that influenced most volcanism. Buoyant diapiric granitoid plutonism resulting from anatexis of mafic lower crust allowed the Earth's crust to evolve and "felsify", leading to cratonic nuclei. Localised, limited subduction (mobile lid) commenced in the late Archean, but global plate tectonics probably didn't develop as we know it until the Proterozoic (<2,500 Ma). In the post-Proterozoic plate tectonic framework, the first-order division of settings for volcanism is based on the overall prevailing regional tectonic regime: divergent, oblique/strike-slip, "passive or hot spot" (i.e. plume-related), or convergent settings. We describe the styles of volcanism typically found in the various plate margin and intraplate tectonic settings, incorporating information on crustal setting, magma generation, as well as the potential for the formation of Large Igneous Provinces (LIPs), providing examples from ancient terrains. In this discussion we will also introduce some of the more recent geodynamic concepts and their relationship to magmatic and volcanic activity. In particular, seismic tomography has evolved as one of the most powerful tools for studying the structure and dynamics of subduction zones and convergent plate margins. Finally, guidelines are given for the evaluation of the tectonic context in ancient volcanic and mineralised terrains. This chapter provides a dynamic framework to accompany our concluding chapter (Chap. 18) on volcanic-hosted resources.

17.1 Introduction

One of the original aims in Cas and Wright (1987) was to discuss volcanism in a modern global tectonic framework as a guide to determining the tectonic settings of ancient volcanic successions. For many years, one approach to this subject had been from a geochemical point of view, the methodology being to correlate particular geochemical properties, fingerprints or trends in modern volcanic successions with their specific plate tectonic setting, and to use the patterns derived from modern successions in trying to identify the tectonic affinities of ancient volcanic successions.

There is no doubt that recent advances in major, trace element and isotopic geochemistry, modern data analysis and geochemical modelling have allowed identification of similarities and differences between the rock-forming conditions, and geological processes that are prevalent in different plate tectonic settings. These include the pressure and temperature conditions during partial melting, phase relations during melting (i.e. degree of partial melting and residual phase assemblages), and source compositions (e.g. anhydrous versus hydrous, variations in the amount of recycled material, and source fertility). Traditionally, two-dimensional binary or ternary diagrams plotting only a few elements have been widely used for the geochemical discrimination of the tectonic setting of magmatism. Although simple to construct and visually easy to understand, recent research has suggested that even amongst proponents of this approach, the use of only a limited number of elements means that such diagrams cannot identify several tectonic settings, while more generally casting doubt on the reliability of the results they generate (e.g. Li et al. 2015; Ueki et al. 2017). These authors conclude that effective tectonic discrimination requires numerous elements and recommend the use of multi-element variation diagrams for qualitative tectonic discrimination. For quantitative discrimination, in addition to such multi-element diagrams, more advanced statistical and computational (including machine learning-based) techniques may be necessary.

Nevertheless, in spite of these advances over three decades we still contend that the geochemical approach offers little hope of unambiguously clarifying the original tectonic context by itself, providing at best a probability level of a particular suite of samples having affinity with one tectonic setting or another (e.g. Snow 2006; Li et al. 2015). This is problematic because ambiguity remains, and even if the

◀ **Plate 17.1** Digital volcano-tectonic activity map of the Earth. This shows the current tectonic plates and their boundaries, which represent where most active tectonism and volcanism have occurred over the last one million years. (*Source* NASA/Goddard Space Flight Centre; image and data processing by Paul Lowman and Jacob Yates. Download at https://visibleearth.nasa.gov/images/88415/digital-tectonic-activity-map/88415t)

likelihood of affinity with a particular tectonic setting is only 5%, that could be correct, whereas a 90% probability with another setting could be wrong. Relying on just geochemical discrimination diagrams to assign tectonic affinity, which is specific to the plate tectonic regime, is just a poor approach. The fact that plate tectonics did not occur for almost 50% of Earth history during the Hadean and Archean (see Sect. 17.7), further reinforces this point, especially but not exclusively, for volcanic successions of Hadean-Archean age.

Tectonic setting can only be confidently evaluated by considering the overall regional geological framework of the volcanic rock succession in question, combined with a consideration of the original tectonic controls that allow volcanism to occur in the first place. More importantly, ancient volcanic successions not only suffer from the effects of deformation in masking the original tectonic context, but also from the chemical overprints of metamorphism and alteration (Chaps. 13, 15). A more holistic approach is, therefore, necessary to determine the original tectonic settings of ancient volcanic successions (see Sect. 17.8). Since publication of Cas and Wright (1987) the relationship between volcanism and tectonic setting has expanded into an enormous and multidisciplinary topic. In particular, advances in geodynamics such as seismology, the availability of high-resolution seismic reflection data, potential field and satellite imagery, and numerical modelling have all greatly improved our understanding of today's tectonic setting of global volcanism. Seismic tomography has evolved as one of the most powerful tools for studying the structure and dynamics of subduction zones. Tomographic imaging combines abundant information from the various criss-crossing seismic (body or P- and S-, and surface) waves triggered by earthquakes or artificial seismic sources (explosions, vibrosis). These data are used to create 2D and 3D images of velocity anomalies, which may be interpreted as structural, thermal, or compositional variations.

In this chapter, we review the known tectonic settings in which volcanism occurs and the overall geological characteristics of those settings. The plate tectonic model, as a modern global tectonic framework (Plate 17.1; Fig. 17.1), has produced a reasonably clear picture of the patterns and conditions of volcanism associated with different tectonic settings during the Phanerozoic and Proterozoic. Building on Cas and Wright (1987), this still remains a useful starting place for developing an approach to interpreting the tectonic setting of ancient volcanic successions (Table 17.1) from the Proterozoic to the present. We will discuss understanding of the tectonic regime for volcanism in the Hadean-Archean, representing almost 50% of Earth history, at the end of this chapter (Sect. 17.7).

The first-order division of plate tectonic settings is based on the overall prevailing regional tectonic kinematic stress regime: divergent, oblique/strike-slip, "passive" or hot spot (i.e. principally plume-related), or convergent settings. From here, we describe the styles of volcanism typically found in the various plate settings, incorporating information on crustal setting, magma generation, the potential for the generation of Large Igneous Provinces (LIPs), and provide examples from ancient terrains. In these discussions, we will also highlight some of the more recent geodynamic and tectonic concepts, and how they relate to understanding and interpreting the regional evolution of magmatic and volcanic activity. This review is followed by a brief discussion on the relation of volcanism to stress fields and magma pathways. The foregoing analysis serves both to provide, at the end of this chapter, some updated guidelines for the evaluation of tectonic context in ancient volcanic (and mineralised) terrains, and a dynamic framework to accompany our final chapter on volcanic-hosted resources (Chap. 18).

17.2 Divergent Tectonic Settings

A *divergent* plate boundary (also known as a *constructive* plate boundary) is a broadly linear, fault-controlled rupture or "*rift*" created by separation of crustal rocks (oceanic or continental), when two tectonic plates are moving away from each other driven by upwelling and diverging limbs of mantle convection cells. Most of today's active divergent plate boundaries occur between oceanic plates, and exist as a system of submerged, mid-oceanic spreading ridges or *MORs* that encircle the Earth (Plate 17.1; Fig. 17.1a). Spreading ridges are not perfectly straight and continuous but are offset by orthogonal major transform faults, giving it a distinctive "jig-saw" outline. Spreading is generally not uniform and spreading rates at adjacent offset ridge segments an be different.

Historically, it was at mid-ocean spreading ridges that one of the key pieces of evidence leading to acceptance of plate tectonic theory was discovered. The symmetrical pattern of strips of the oceanic crust showing a stripe-like pattern of alternating magnetic normal and reversed polarity linear anomalies on opposite sides of the ridges, which corresponded with the Earth's geomagnetic pole reversal timescale, provided proof of continuous, relatively symmetrical seafloor spreading. Smaller spreading ridges also occur in back-arc and interarc settings, or marginal seas. Divergent boundaries within continents initially produce narrow extensional to trans-tensional basins that may eventually grow to become semi-contiguous, continental-scale rift valleys, as best represented today by the East African Rift System (Plate 17.1). Here crustal separation may ultimately lead to development of a passive margin and a new mid-oceanic spreading ridge.

In divergent settings magmas are generated by partial melting of the asthenospheric mantle wedge that wells

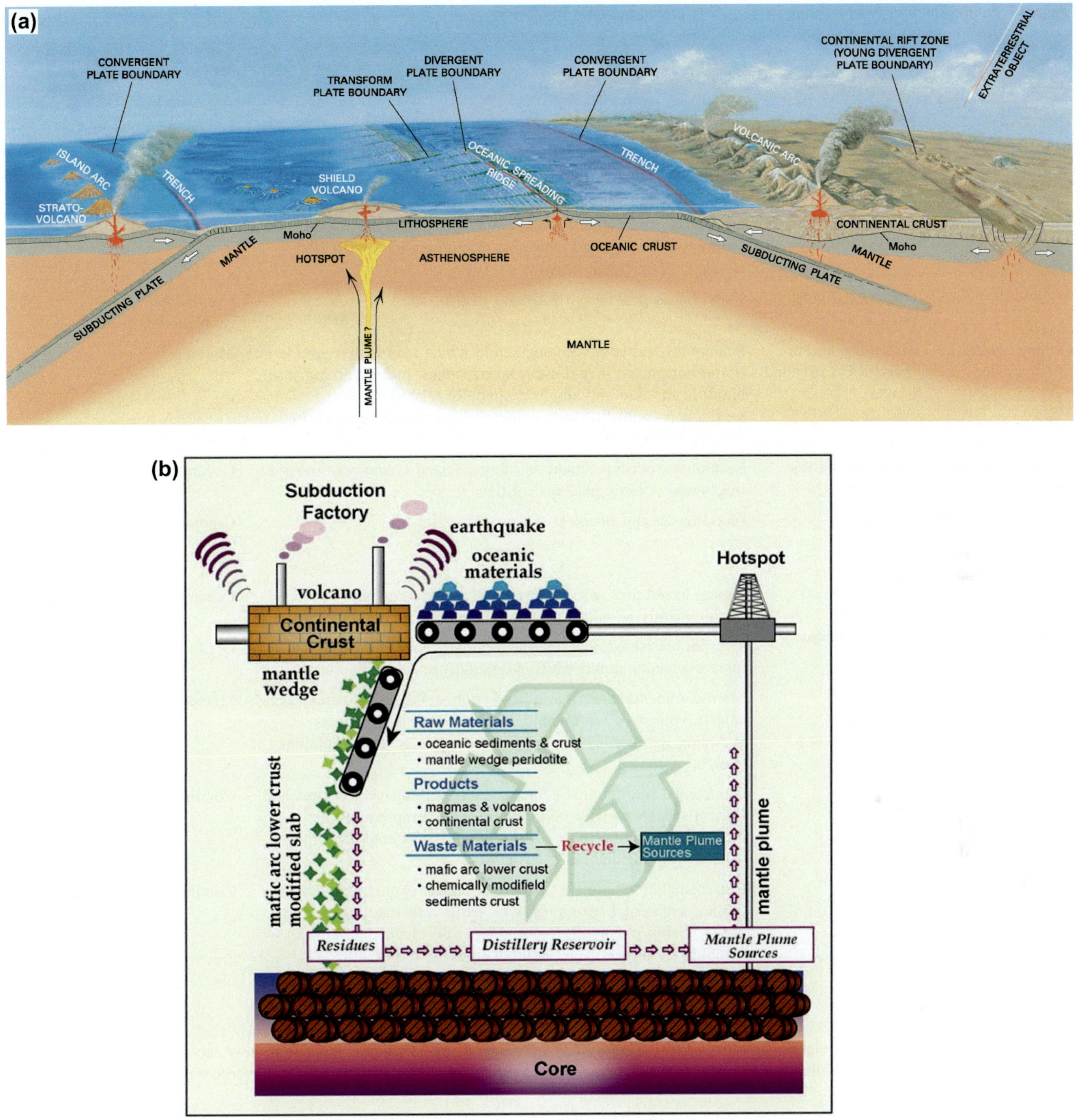

Fig. 17.1 The global tectonic framework of volcanism. **a** Schematic cross-sectional perspective of the modern plate tectonic model as well established by the late 1970s, showing the main types of plate boundaries and associated magmatism. As illustrated, very rarely, massive asteroid or comet impact can also cause melting of the Earth's crust and form magma bodies. The best-known example of this type of superheated melt sheet is the Sudbury Igneous Complex in Canada (ca. 1.85 Ga; Zieg and Marsh 2005), which has hosted the world's largest nickel sulphide resource. (*Source* United States Geological Survey). **b** Cartoon illustrating the plate tectonic cycle. The "subduction factory" processes raw materials, such as oceanic sediments, oceanic crust and mantle lithosphere and manufactures them into arc magmas and continental crust. Aqueous fluids, extracted from oceanic raw materials via dehydration reactions during subduction, dissolve particular chemical elements and generate magmas in the overlying mantle wedge. The waste materials or residues, such as chemically modified slab components (oceanic crust and sediments) and delaminated mafic lower arc crust are transported and stored in the deep mantle and recycled as raw materials for mantle plume-related magmatism. (From Tatsumi and Stern 2006, under Creative Commons Attribution-Share Alike 4.0 International licence; After Tatsumi 2005)

Table 17.1 Introduction to volcanism in a modern global tectonic framework

Prevailing regional tectonic regime	Plate tectonic setting	Type of volcanism[1]	Crustal setting	Dominant melt generation process
Divergent	Mid-oceanic ridges	Sea floor spreading; mid-ocean ridge fissures and shield volcanoes	Oceanic	Decompression
	Marginal or back-arc basins	Sea floor spreading (smaller scale); fissures and more evolved effusive and volcanic products	Mixed	
	Continental extensional rifts and basins	Bimodal: basaltic fissures and scoria cones, silicic dome-tuff ring complexes, and caldera volcanoes characterised by voluminous, ignimbrite flare-up and extensive granite batholiths (SLIP). Basaltic: flood basalt provinces (cLIP)	Continental	
	Continental volcanic passive margins	Flood basalt provinces (cLIP and oLIP)	Continental (transitional)	
Oblique/strike-slip	Transform and/or transcurrent rifts and basins	Seamounts and some oceanic shield which may evolve into large island volcanoes; lava domes, scoria cones, maars varying from basalt to rhyolite and alkaline; spatially related to subduction-controlled arcs, caldera volcanoes and ignimbrite flare-up (SLIP)	Mixed	
"Passive" or hot spot	Intraplate oceanic volcanism	Seamounts, oceanic shield and larger island volcanoes, aseismic ridges and oceanic plateaus (oLIP)	Oceanic	
	Intraplate continental volcanism[2]	Flood basalt and plains basalt provinces (cLIP)	Continental	
Convergent	Subduction controlled volcano-plutonic arcs	Young island arcs: marine stratovolcanoes composed predominantly of island arc tholeiites; may include boninites	Oceanic	Flux
		Transitional island arcs: marine stratovolcanoes characterised by more evolved magmas which are broadly andesitic in composition	Oceanic	
		Microcontinental arcs: marine and continental stratovolcanoes and caldera volcanoes with abundant silicic eruptive products including ignimbrite flare-up and co-magmatic granite plutons (SLIP)	Mixed	
		Continental margin arcs: stratovolcanoes, multi-vent centres and caldera volcanoes with voluminous silicic eruptive products, including ignimbrite flare-up and extensive granite batholiths (SLIP); may include adakitic melts	Continental	
		Continental/arc-continent collisional zones: magmatism occurs during continental margin growth, early collision, collision and post-collision potentially incorporating all of the above in a complex tectonic collage; compositions span tholeiitic, calk-alkaline to potassic and ultrapotassic overlapping in time, adakites, leucogranites; voluminous post-collisional "A-type" magmatism (SLIP)	Continental	

[1]Potential for Large Igneous Provinces: basaltic, *oLIP* oceanic, *cLIP* continental; *SLIP* silicic. [2]*Note* it is now recognised some continental intraplate basaltic volcanism (including flood basalts and cLIPS) may have subduction-related origins within convergent regimes (see Sect. 17.5.6)

up beneath the rift width as a consequence of crustal thinning (e.g. McKenzie 1978; McKenzie and Bickle 1988; Grove and Till 2015). As asthenospheric mantle rises its pressure is lowered, leading to *decompressional melting*. Decompressional melting occurs when the adiabat for the convecting mantle crosses the solidus, which has a shallower slope than the adiabat in the upper mantle. The solidus is encountered at different depths in the mantle in different tectonic settings, and this location depends on the ambient mantle temperature and composition of the upwelling mantle. The latent heat of fusion for mantle peridotite, the heat capacity, the ambient mantle temperature, and the pressure and compositional dependence of the mantle solidus control the rate of melt production.

At MORs, magma production is dominated by ascending hotter, deeper mantle driven by convection (Fig. 17.2; Langmuir et al. 1992; Langmuir and Forsyth 2007; Grove and Till 2015). As the mantle buoyantly ascends and crosses its solidus at a depth determined by the ambient mantle temperature, fractional adiabatic decompression melting begins (Fig. 17.2a). With time, the newly formed oceanic lithosphere cools, thickens and becomes denser, and is

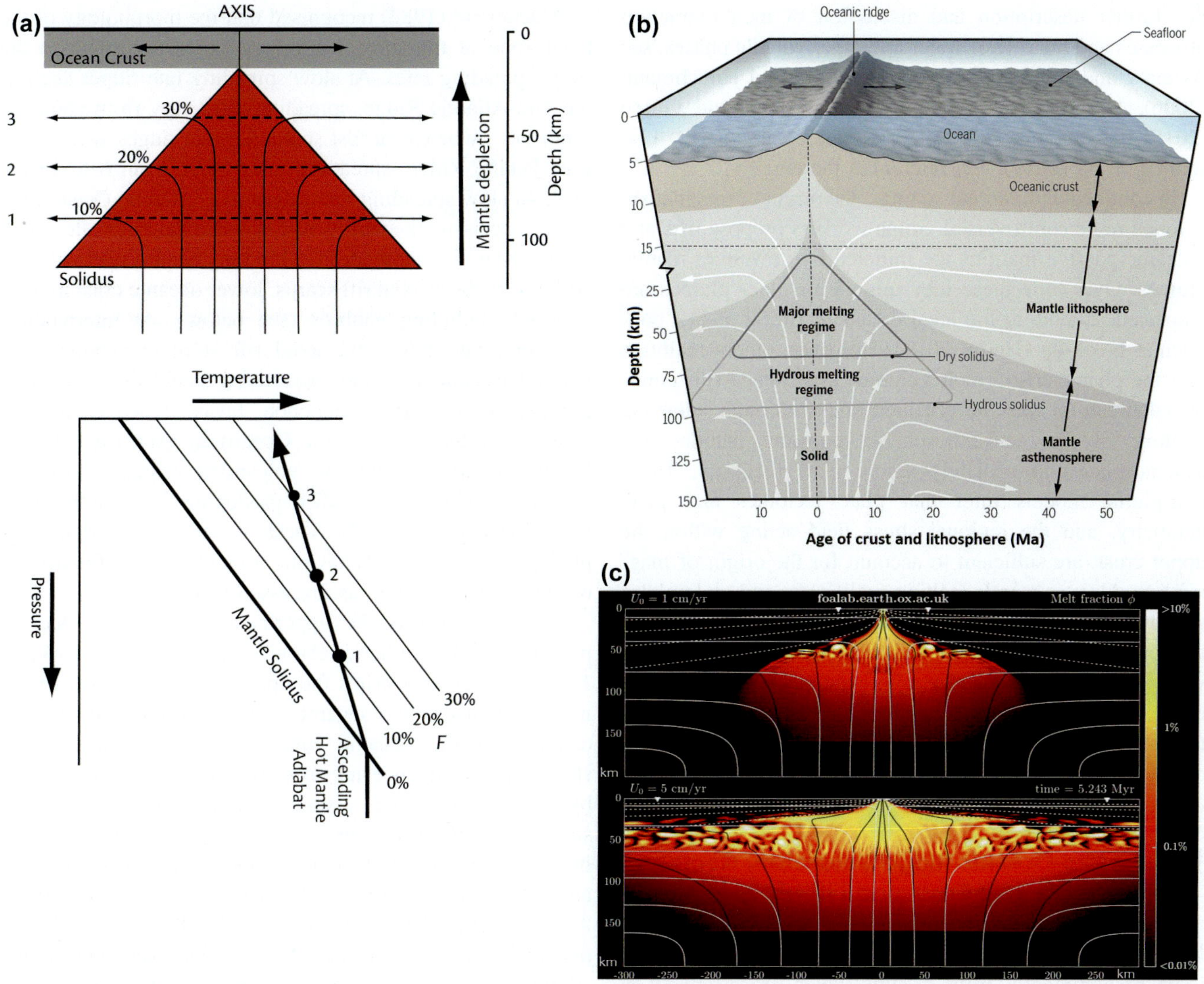

Fig. 17.2 Adiabatic decompression melting beneath mid-oceanic ridge systems. **a** Partial melting occurs when the adiabat for ascending mantle crosses the mantle solidus, which has a shallower slope than the adiabat in the upper mantle. As the solidus is crossed, melting progressively increases adiabatically, whereby, the latent heat of fusion only continues by decrease with pressure decrease. The numbers on the bottom figure correspond to the pressures where melting stops for the melt fraction (*F*%) indicated on the upper diagram. As a consequence, hotter mantle crosses the solidus at greater depths, leading to a larger melting regime, greater degrees of melting and depletion of the residual mantle column (as shown here), faster spreading centres and thicker crust than that produced by ascending cooler mantle. (From Langmuir and Forsyth 2007, After Langmuir et al. 1992). **b** Current view of the structure of a mid-ocean ridge. Solid-state flow induced by plate spreading brings hot asthenospheric mantle material upward into the triangular melt-region outlined, which in 3D forms a triangular prism. Upwelling mantle first crosses the hydrous solidus, where melting begins as a result of trace water in the mantle, and subsequently the dry solidus, where decompressional melting accelerates as illustrated in (**a**). Melts segregate from the residue, rising to form oceanic crust. (From Asimow 2017). **c** 2D simulation of magma-mantle dynamics beneath mid-ocean ridges. This illustrates the degree of partial melting (right column) for slower/cooler (top = 1 cm yr^{-1}) versus fast/hotter (below = 5 cm yr^{-1}) spreading centres; snapshot is after 5 Ma. (Simulation performed by Foalab group, University of Oxford; *Source* Tobias Keller, Youtube)

"pulled" by gravity (termed "*ridge push*") in opposite directions away from the topographically elevated ridge (Fig. 17.2b, c). This leads to the formation of a three-dimensional, triangular prism-shaped melting region, where melting ends at greater depths as a particular part of a plate moves away from the ridge. The melts aggregate at shallow depths and erupt as a mixture of partial melts produced over a range of depths from the mantle that has experienced progressively higher degrees of melt depletion as it rises to shallower depths (Fig. 17.2b). This process is known as *polybaric near fractional melting*, which explains the basalt compositions that have been sampled globally at spreading ridges. The basalt of layer 2 of oceanic crust is typically N-type MORB. A number of other factors, including the presence of volatiles, can influence the initiation of melting and melt productivity as a function of depth.

For further description and discussion of melt extraction processes and the role of water and other volatile phases, see Asimow and Langmuir (2003), Dasgupta and Hirschmann (2006), Langmuir and Forsyth (2007), Benn and Grove (2015), Grove and Till (2015), Keller et al. (2017), Lissenberg et al. (2019), and references therein.

In continental divergent settings, the origin of magmatism is much more controversial. Some workers propose rifts and volcanic passive margins are initiated by upwelling mantle plumes. Over time these may migrate or other plumes are emplaced, extending the areal extent of rifting and/or continental break-up (Heron 2019). However, others maintain that the characteristic melt volumes associated with continental break-up may be obtained solely through adiabatic melting of the sub-lithospheric buoyant mantle (i.e. asthenosphere) immediately before plate break-up. These non-plume models infer that plate tectonics and plate geometry, and the regional stress field acting within the upper crust, are sufficient to account for the origin of magmatism. Also, as mantle melting proliferates and underplates the lower crust, there is increasing potential for the generation of more evolved silicic and felsic melts. In this section we will therefore commence with examination of spreading ridges at MORs and in marginal basins, before embarking on the more complex picture of continental volcanism in divergent settings, even though non-sequential geologically.

17.2.1 Mid-Oceanic Ridge Volcanism, Geology of Oceanic Crust and Lithosphere

Mid-oceanic spreading ridge (MOR) volcanism appears to issue predominantly from fissures which are arranged *en echelon* and/or small, elongate, central volcanoes within a median valley (e.g. Macdonald 1982, 2001; Sect. 16.10.1). Small volcanic mounds and a topography of overlapping flow-fronts are prominent within this median valley. The volcanic eruption products are overwhelmingly basaltic N-type MORB, and both pillow and sheet lavas and associated hyaloclastite breccias occur based on the results of the Deep-Sea Drilling Project (DSDP) and subsequent programmes (ODP and IODP), and direct or remote observation from submersibles and sonar imaging (e.g. see Fig. 16.94). Hydrothermal vents and fissures may also occur (Fig. 16.93). MORs are dominated by ridge-parallel fractures and faults, producing a very rough ridge and chasm terrain (Fig. 16.92a–d). In the open ocean the only source of sediment is through pelagic sources, whereas in proximity to continents and oceanic arcs, hemipelagic and even terrigenous sediment may impinge. A pelagic sediment layer only becomes prominent at the edges of the median valleys and on the ridge flanks and abyssal plains as the lithospheric plates spread away from the MORs (for more detailed description, see Sect. 16.11.1).

Macdonald (1982) recognised that the morphology of the axial zone of the spreading ridges varies according to the ridge spreading rates. At slow spreading rate ridges such as the mid-Atlantic Ridge, spreading rates vary from about 2–4 cm yr^{-1}, whereas at fast spreading rate ridges such as the East Pacific Ridge, spreading rates range from 6 to 18 cm yr^{-1}. At slow spreading rate ridges (e.g. Atlantic Ocean), the axial zone is marked by well-defined axial rift valley with steep scarps, and elongate 1–4 km long central vent volcanic edifices. In these axial rift scarps, lower oceanic crust may be exposed, including gabbros (see below). At intermediate spreading rate ridges the axial rift is more subdued and marked by low elongate, fissure-fed shield-like volcanoes dominated by sheet pahoehoe lavas, whereas at fast spreading ridges (e.g. Pacific Ocean) an axial rift valley is barely discernible and the volcanism originates from a continuous shield-like edifice producing dominantly sheet lavas (Macdonald 1982). These variations in ridge morphology and volcanic behaviour coincide with differences in oceanic crustal profile, as discussed below.

Oceanic crust is the uppermost layer of the oceanic portion of a lithospheric tectonic plate (Juteau and Maury 1999). The crust overlies the uppermost, solid layer of the mantle, with the two separated by the Moho and together constituting the oceanic lithosphere. Although the full stratigraphy of the oceanic crust has still not been penetrated by deep ocean drilling, seismic, gravity and magnetic studies suggest that it closely resembles the stratigraphy of ophiolites, which are subducted or obducted thrust slices of ancient oceanic crust (Coleman 1977; Moores 1982; Anderson et al. 1982; Newmark et al. 1985; Fig. 17.3a), as discussed in the next Sect. 17.2.2. Understanding of the stratigraphy of oceanic crust has emerged through a combination of geological studies of ophiolites, dredging and sampling of modern seafloor escarpments, deep-sea drilling through the DSDP, ODP and IODP programmes, and geophysical surveys of the modern seafloor. Sonar surveys using remotely operated submersible vehicles (ROVs) have produced detailed maps of lava flow fields, their extent and morphology in ridge crest areas (e.g. Juan de Fuca Spreading Ridge, northeast Pacific Ocean; Chadwick et al. 2013; Clague et al. 2013; Chaps. 5, 16, Figs. 5.39 and 5.40).

Seismic studies of modern oceanic crust have defined a generally comparable stratigraphy to ophiolites based on seismic properties. The principal layers identified based on compressional wave seismic velocities (V_P) are seismic layer 1, layers 2A, 2B and 2C, layer 3 and layer 4 (Fig. 17.3a; Anderson et al. 1982; Juteau and Maury 1999). A complete section of oceanic crust has a thickness of 5–7 km (Anderson et al. 1982; Juteau and Maury 1999; Soule 2015; Michibayashi et al. 2019; see Fig. 17.21). One of the most successful deep-sea drilling penetrations through this stratigraphy has been in DSDP/ODP drill-hole 504B, on the

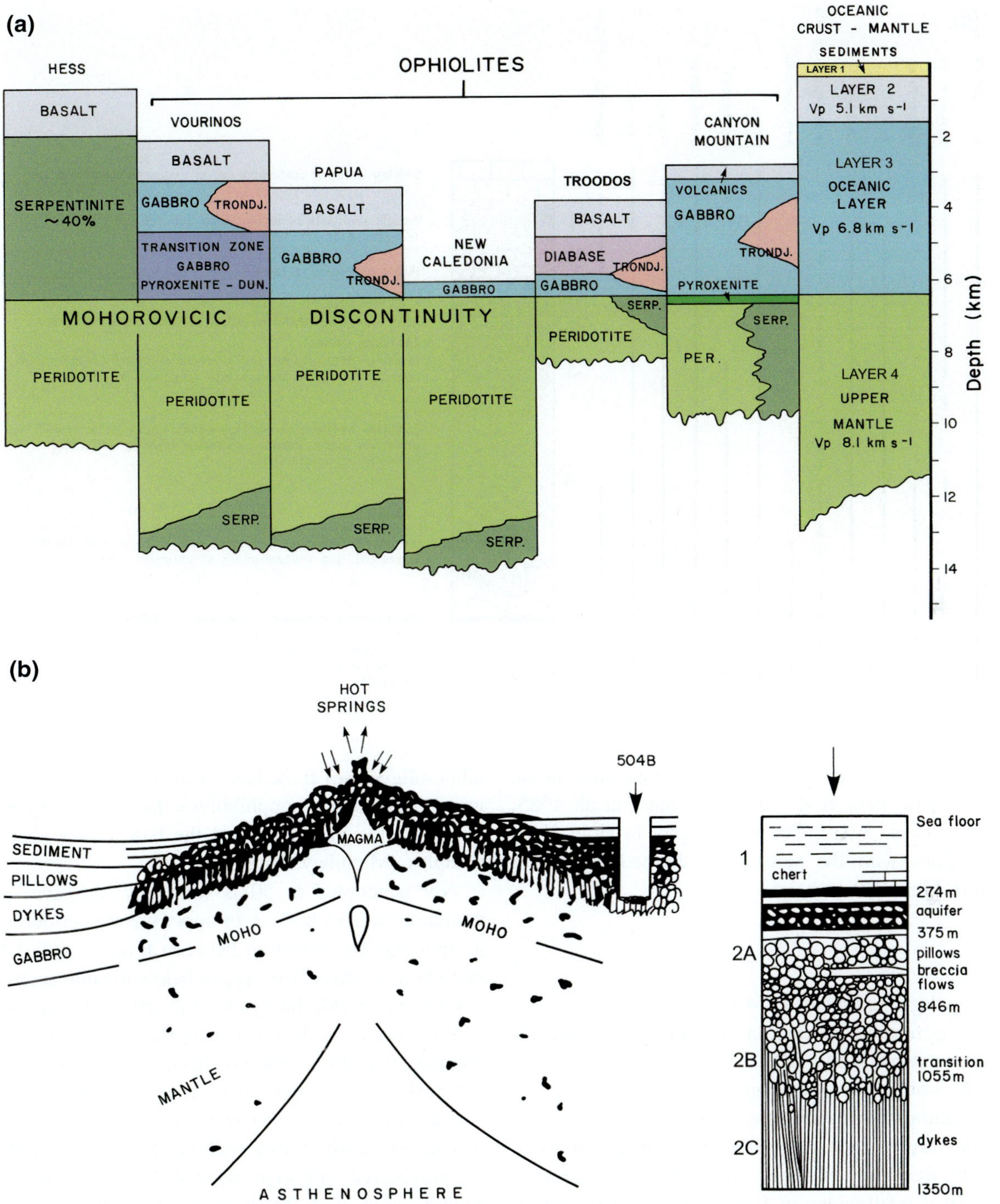

Fig. 17.3 Stratigraphy of oceanic crust and ophiolites. **a** Comparison of stratigraphy of several known ophiolites with the seismic stratigraphy of oceanic crust and mantle on the right (From Cas and Wright 1987, After Moores 1982). **b** Schematic representation of the section intersected in Deep-Sea Drilling Project/Ocean Drilling Programme (DSDP/ODP) Hole 504B, drilled to 2,111 m below seafloor on the southern flanks of the Galapagos spreading centre, south of the Costa Rica Rift. Hole 504B still represents the deepest drill-hole ever undertaken by scientific ocean drilling programmes; this was carried out in 8 legs between 1978 and 1993, in 3,945 m of water. (From Cas and Wright 1987, After Anderson et al. 1982). **c** Expanded definition of a complete ophiolite (From Cas and Wright 1987, After Moores 1982; see text for description). Numbers in (**b**) and (**c**) refer to oceanic crust layers as defined in the text

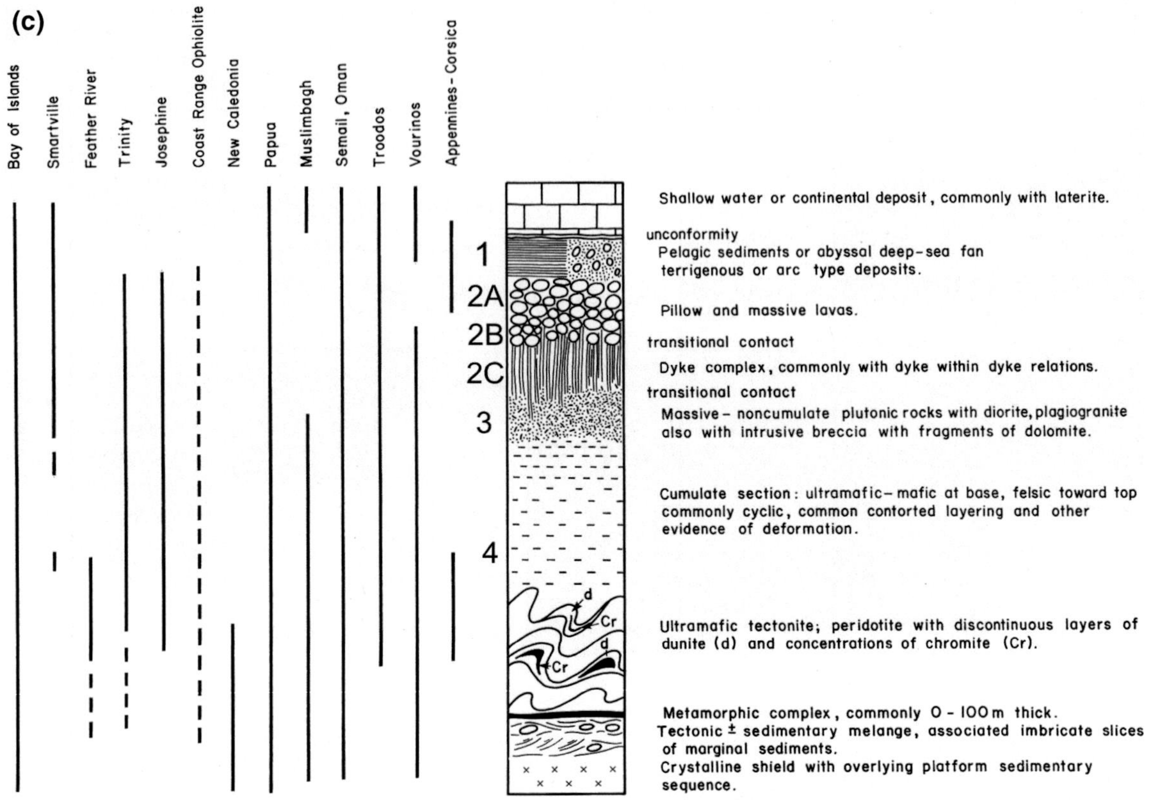

Fig. 17.3 (continued)

southern flanks of the Galapagos fast spreading ridge in the Nazca Plate (Anderson et al. 1982; Newmark et al. 1985; Michibayashi et al. 2019). The stratigraphy in Hole 504B and correlations with the seismically defined layers are illustrated in Fig. 17.3b, where:

(*Top*).

- *layer 1*: pelagic sediments (274 m),
- *layer 2A*: permeable succession of pillow lavas, breccias and some massive flows (100 m),
- *layer 2B*: more pillow lavas, breccias and minor massive flows; fractures filled with smectite (475 m),
- *transition layer*: decreasing frequency of pillow lava, increasing frequency of dykes, and includes a stockwork of Fe, Zn, Cu-sulphides in laumontite, chlorite, calcite and quartz veins in a fractured and brecciated pillow lava sequence (210 m),
- *layer 2C*: sheeted dyke complex (1,050+ m), and
- (*layer 3*: gabbros; not intersected by drilling but imaged geophysically).

(*Bottom*)

The succession was progressively more metamorphosed or hydrothermally altered with increasing depth in the hole, reaching greenschist facies (chlorite, actinolite, albite-oligoclase) at the base of the layer 1 basalts, and upper greenschist(-lower amphibolite) facies (hornblende, secondary calcic plagioclase) at the base of the sheeted dyke complex (Michibayashi et al. 2019).

Not penetrated in 504B were any intrusive (gabbroic) complex rocks of layer 3, or upper mantle peridotite of layer 4, although the transition between layers 2 and 3 has been identified seismically to lie just below the intersected sheeted dyke complex (Michibayashi et al. 2019). However, in Hole 1256D, located in the Guatamala Basin northwest of 504B, a complex transition between sheeted dykes and several gabbro horizons was intersected. In Hole 1256D an interval of 810 m of basaltic lavas passes down into 346 m of sheeted dykes, below which two gabbroic horizons 60 and 20 thick are interleaved with more sheeted dykes (Teagle et al. 2006; Michibayashi et al. 2019), representing the uppermost transition from layer 2 to layer 3. Holes 504B and 1256D intersected oceanic crust from fast spreading ridges in the Pacific Ocean. In addition, sampling and drilling in the Hess Deep, an extensional rift in the Cocos-Nazca spreading ridge, sampled lower oceanic crustal primitive olivine gabbros and troctolites of layer 3 (Michibayashi et al. 2019). Drilling and seafloor sampling in oceanic crust generated by slow spreading ridges in the Indian and Atlantic oceans indicate some significant differences in stratigraphy and depth of important lithologies. In these studies, lower crustal gabbros and serpentinised peridotites were commonly encountered, lateral continuity of rock units was less apparent than from fast spreading centre crust, and small plutons are common (Michibayashi et al. 2019). It has also been found that at slow spreading ridges, low-angle detachment faults are common, leading to denudation of the mantle, exposing serpentinised peridotite in submarine scarps and promoting uprise of seafloor gabbroic core complexes which are not found at fast speading ridges (Michibayashi et al. 2019).

The thickness and character of the layer 1 sediments is highly variable and depends on the age of the crust (thin for young crust, thick for old crust), biogenic productivity and proximity to areas of influx of continentally derived terrigenous clastic sediments. Calcareous or siliceous pelagic sediment, or both will mantle true open ocean oceanic crust. Where oceanic crust comes under the influence of continental margin sedimentation it will also be marked by significant thicknesses of terrigenous sediments. Oceanic crust formed near island arcs will be mantled by pyroclastic and redeposited volcanic sediments (see Sect. 17.2.2).

The areal extent of oceanic crust during active ocean basin life is huge and measurable in terms of millions of km^2. However, because of the development of subduction systems, the preservation potential of oceanic crust is low, as most will be subducted (Fig. 17.1a, b). Oceanic crust is only likely to be found in older orogenic belts, as thrust slices of ophiolite caught up in accretionary prisms of subduction complexes, or slabs obducted over the leading edge of the overriding plate and its volcanic arc, whether it be oceanic or continental (Cas and Wright 1987; Dilek and Robinson 2003; Fig. 17.4). Oceanic crust is pervasively normally faulted. This faulting occurs during the initial stages of rifting, as the oceanic crust moves from the initial volcanic

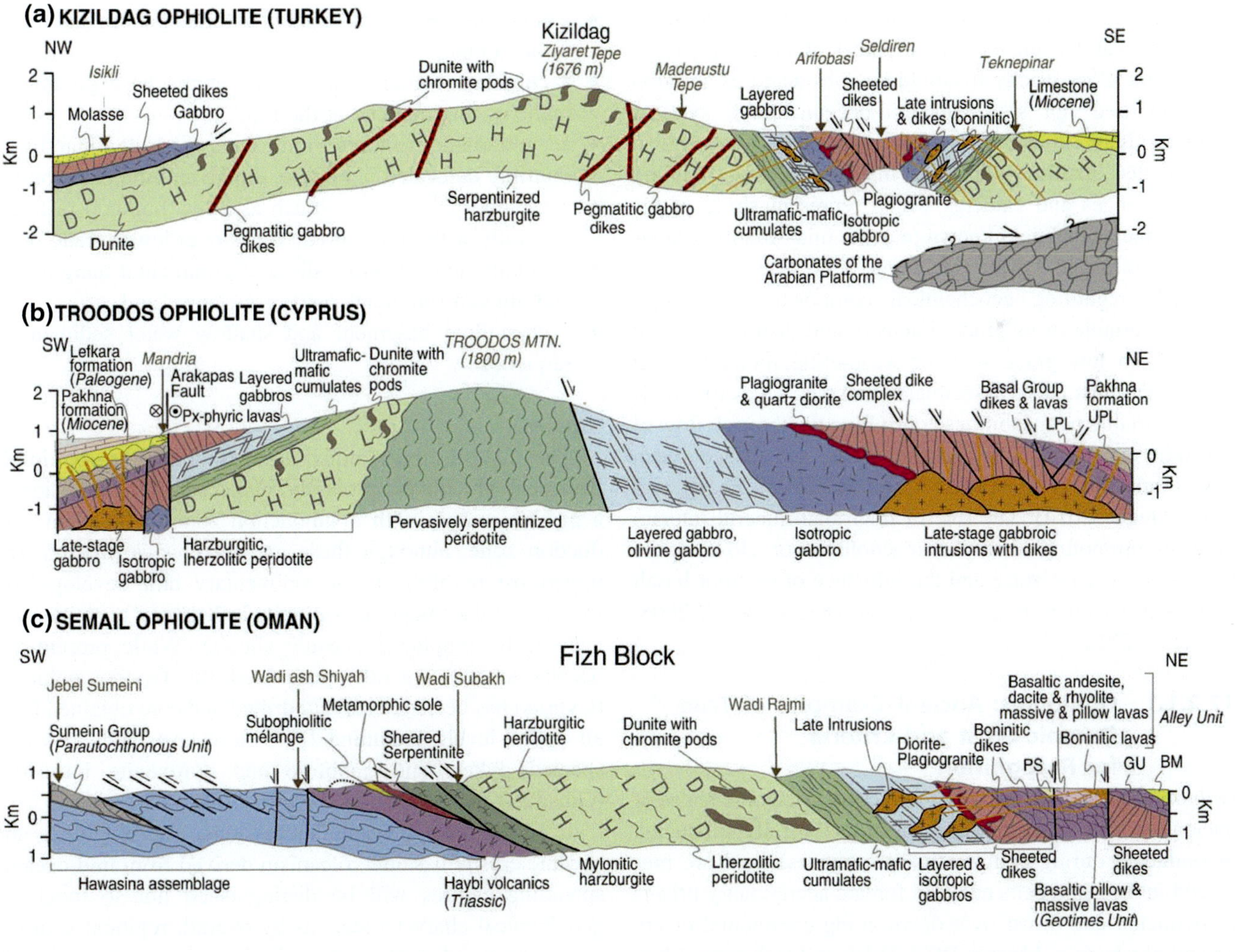

Fig. 17.4 Lithological and structural cross-sections illustrating complex tectonic relationships associated with obduction of three well-studied Tethyan ophiolites: **a** Kizildag, Turkey; **b** Troodos, Cyprus; and **c** Semail, Oman. These are all located within the Mesozoic-Cenozoic Alpine-Mediterranean continental collisional zone (Sect. 1.5.5). No vertical exaggeration; D = dunite, H = harzburgite, L = lherzolite; in **c**, BM = Batinah mélange, GU = Geotimes Unit, PS = metalliferous umbers, pelagic chalk, and radiolarite. (From Dilek and Furnes 2009)

rift valley beyond the flanks of the ridge and need not be entirely the result of subsequent deformation.

The geochemistry of the basalts of layer 2 has been extensively documented as being oceanic tholeiite (most commonly "N-type MORB" or normal) in character, and their geochemical characteristics indicate derivation from poorly evolved magmas (e.g. Klein and Langmuir 1987; Hémond et al. 2006; Gale et al. 2013). However, although the magmatism is clearly mantle-derived, it need not be exclusively mafic. Silicic differentiates and products of melting of already formed crust, perhaps altered by permeation of seawater, forming lavas and pyroclastic deposits, including small-scale ignimbrites, are known on the rare subaerial exposures of mid-oceanic ridges such as in Iceland (Walker 1962, 1966; Thordarson and Larsen 2007; Gurenko et al. 2015). The major active rifts also include central volcanoes, which, although typified by effusive basalt eruptions, are also characterised by a range of explosive styles that include andesitic, dacitic and rhyolitic eruptions of sub-plinian and plinian styles. It should be noted here, however, that Iceland is very atypical of spreading ridge volcanism; more strictly it should be considered as a region of juxtaposed hot spot and mid-oceanic ridge systems (Gudmundsson 2000). Elsewhere, especially along transform faults, alkaline volcanism may also occur, building stratovolcanoes and calderas which represent some of the largest volcanic edifices known (e.g. the mid-Atlantic islands of the Azores and Tristan da Cuhna; Sect. 17.3).

Lastly regarding geochemical composition, alteration may be variable in its grade, intensity and distribution, and varies from low grade seafloor weathering products to at least greenschist facies assemblage minerals. The variation is due to variable permeabilities controlling the through flow of hydrothermal fluids into and out of the subsurface ocean (e.g. Anderson et al. 1982; Alt et al. 1992; Zhang and Smith-Duque 2014). The role of microbial-mineral interactions in promoting alteration reactions at or close to the basalt-seawater interface and the influence of seafloor basalt biomes is an interesting development (Kruber et al. 2008; Lee et al. 2015).

17.2.1.1 Ophiolites: Ancient Examples of True Oceanic Crust and Criteria for Recognition

Ophiolites are nearly all, if not universally, tectonic slices of ancient oceanic crust that have been tectonically subducted or obducted during compressional orogenesis and are preserved in orogenic belts either in forearc accretionary prisms or obducted and thrust over down-going continental or arc crustal blocks (e.g. Moores 1982; Dilek and Robinson 2003; Fig. 17.4). Although some ophiolites may be slices of open ocean (MOR) crust, most are probably slices of back-arc basin (BAB) crust that have been sliced and diced between colliding island arc crustal blocks or island arc blocks and continental masses. Ophiolite stratigraphy was originally defined in 1972 by a Penrose Conference and later expanded by Moores (1982; Fig. 17.3a, c) which still remains, with some minor modifications, the standard stratigraphy (sometimes termed the "*Penrose-type pseudo-stratigraphy*") recognised today. This ideally includes, from bottom to top:

(*Top*)

- post-emplacement deposits of laterite, reef limestone, or shallow marine or subaerial sediments,
- an abyssal or bathyal sediment sequence which may include radiolarian chert, red pelagic limestone, metalliferous sediments, breccias and/or pyroclastic deposits,
- an extensive section of massive and pillowed flows and intercalated sediments,
- a sheeted dyke complex,
- a non-cumulus unit of variably textured gabbro and minor trondhjemite,
- a cumulate complex, ultramafic at the base grading to mafic or intermediate at the top,
- an ultramafic tectonite unit composed dominantly of multiply deformed peridotite, dunite and minor chromite,
- a metamorphic unit, as much as a few hundred m thick, generally with higher grade rocks over lower grade ones,
- a tectonic unit of thrust slices of continental margin, rise and abyssal sediments, and/or mélange, and
- a crystalline basement and shallow water sedimentary sequence.

(*Bottom*)

The bottom three units are essentially pre-ophiolitic basement which were probably emplaced during collision of a passive margin with a subduction zone or incipient subduction zone, although these are not always present. The uppermost regolith and/or sedimentary unit develops following emplacement. Those units in between constitute the tectonically emplaced oceanic slice(s). While present-day oceanic crust is not fully penetrated, the Troodos ophiolite (Cyprus) has been extensively drilled and core obtained from all crustal levels, including 1,175 m of lavas and 2.3 km of sheeted dykes, and gabbroic and pyroxenite intrusions (Gibson et al. 1991).

In older mafic volcanic successions, true ophiolites and oceanic crustal basaltic volcanism derived from mid-oceanic spreading centres will be distinguished not so much by geochemical characteristics as by overall, regional geological context and stratigraphic similarity to modern oceanic crust profiles. This is because the original geochemical signatures are usually, and pervasively overprinted by alteration. The basement to the volcanic succession should be clearly ultramafic, representing uppermost mantle; the layer 2 basalts should interface directly downwards with a sheeted dyke complex, which is the principal physical evidence of the tectonic extension associated with seafloor spreading; and, an essentially full oceanic lithosphere stratigraphy should be developed (layers 1–4, Fig. 17.3a). The relict oceanic lithospheric slices should form an integral part of a complex orogenic terrain, irrespective of whether the final tectonic emplacement of the slices was by subduction-related accretionary prism off-scraping and thrusting, or by obduction (Fig. 17.4; Dilek and Furnes 2014). Subduction-related ophiolites should be spatially associated with blueschist-bearing tectono-stratigraphic units and mélanges. The scale of the whole complex should be large enough to be compatible with an original large oceanic basin setting, and a subsequent major subduction setting. Probably the best described examples of ophiolites are represented by the Tethyan ophiolites within the Alpine-Himalayan collisional orogenic system (Fig. 17.4), while the oldest that can be confidently ascribed (i.e. evidence of Penrose-type stratigraphy) are Neoproterozoic ophiolites of the Arabian Shield (Dilek and Robinson 2003).

Archean greenstone belts are also characterised by thick sequences of mafic and ultramafic lavas and intrusive rocks, which has attracted considerable discussion over the years as to the degree they represent true oceanic crust (i.e. ophiolites), erupted at MOR spreading ridges. However, there are no convincing examples of Archean ophiolites presently known (Bickle et al. 1994; Dilek and Robinson 2003). Recently, Grosch and Slama (2017) have described, mainly from a geochemical approach, evidence for 3.3 Ga oceanic crust in the Barberton greenstone belt (Onverwacht Group) in South Africa. Nonetheless, without clear evidence of a sheeted dyke complex juxtaposed beneath the chert-pillow association, and above gabbros and then mantle peridotites in a semi-intact crustal slice several km thick, the Kromberg sequence remains equivocal. Sheeted dykes are the only physical geological evidence of oceanic seafloor spreading, and these have not been recognised in Archean greenstone belts. In other Archean greenstone belts mantle plumes have been shown to be important, such as in the Yilgarn Craton in Western Australia (Campbell and Hill 1988; Campbell et al. 1989; Barnes et al. 2012; Sect. 17.7), which appears to have an older continental basement even though the mafic volcanic sequences are predominantly, if not entirely, submarine (Mole et al. 2014; Chap. 16). While mantle plumes may have been very important in the Archean, there is considerable debate concerning whether or not island arcs and Phanerozoic-style subduction existed. We will discuss the tectonic regime in which volcanism occurred during the Archean, later in the chapter (Sect. 17.7).

17.2.2 Oceanic Back-Arc Basin, Inter-Arc Basin and Marginal Sea Spreading Volcanism

With the development of plate tectonic theory, geologists initially thought convergent plate margins were zones entirely dominated by compression, and so wide areas of strong extension above subduction zones were not expected. Karig (1970, 1971, 1974) first proposed that *back-arc* basins were initiated by crustal extension producing first rifts, and then new ocean crust by seafloor spreading. This type of basin is formed behind or within a continental margin or intra-oceanic arcs, in which case they are also termed *interarc basins*. Because they occur behind volcanic island arcs and commonly along continental margins, as well as convergent plate margins they are also referred to as "*marginal basins*". Presently most (75%) of these are found in the circum-Pacific region, with most concentrated on the western side of the Pacific Ocean (Fig. 17.5; Plate 17.1; Tamaki and Honza 1991; Marsaglia 1995).

In the western Pacific many back-arc or marginal basins are the product of the successive splitting and rifting of subduction-related volcanic arc blocks (Karig 1970, 1971, 1974; Tamaki and Honza 1991). This is followed by oceanward migration of the frontal half of the rifted arc block, allowed by migration of the hinge of the subducting plate seawards, called "roll back", driven by gravitational sinking of the down-going plate due its increasing density. This extension triggers the opening of a new back-arc or interarc basin through the formation of a small-scale seafloor spreading centre (Fig. 17.6a). The part of the rifted block that stays behind is called a remnant arc (Fig. 17.6a–c). Evidence for the extensional origin of these basins is indicated by: (1) the irregular topography and presence of normal faults, (2) little or no sediment cover, (3) high heat flow (active basins), and (4) limited and/or chaotic magnetic anomaly patterns over elevated basin centre areas (Sclater et al. 1972; Lawver and Hawkins 1978; Weissel 1981; Taylor et al. 1996). The successive splitting of the arc block and the seaward migration of the frontal arc should produce a sequence of progressively younger back-arc basins oceanwards (Fig. 17.7a), by successive "small-scale" seafloor spreading in these basins, which stops when each new basin forms.

The basins and ridges or rises east of Australia and east of the Chinese mainland have ages predicted by this model and are floored by oceanic crust (Cas and Wright 1987 and references therein). The currently active and youngest of these are the Mariana Trough and Lau Basin-Havre Trough (Fig. 17.7a, b; Stern et al. 2004; Baker et al. 2008; Ruellan et al. 2003; Smith and Price 2006; Stern 2010). Seafloor spreading in the Marianas Trough commenced at about 3–4 Ma, and presently varies from 1.5 to 4.5 cm yr^{-1} along the

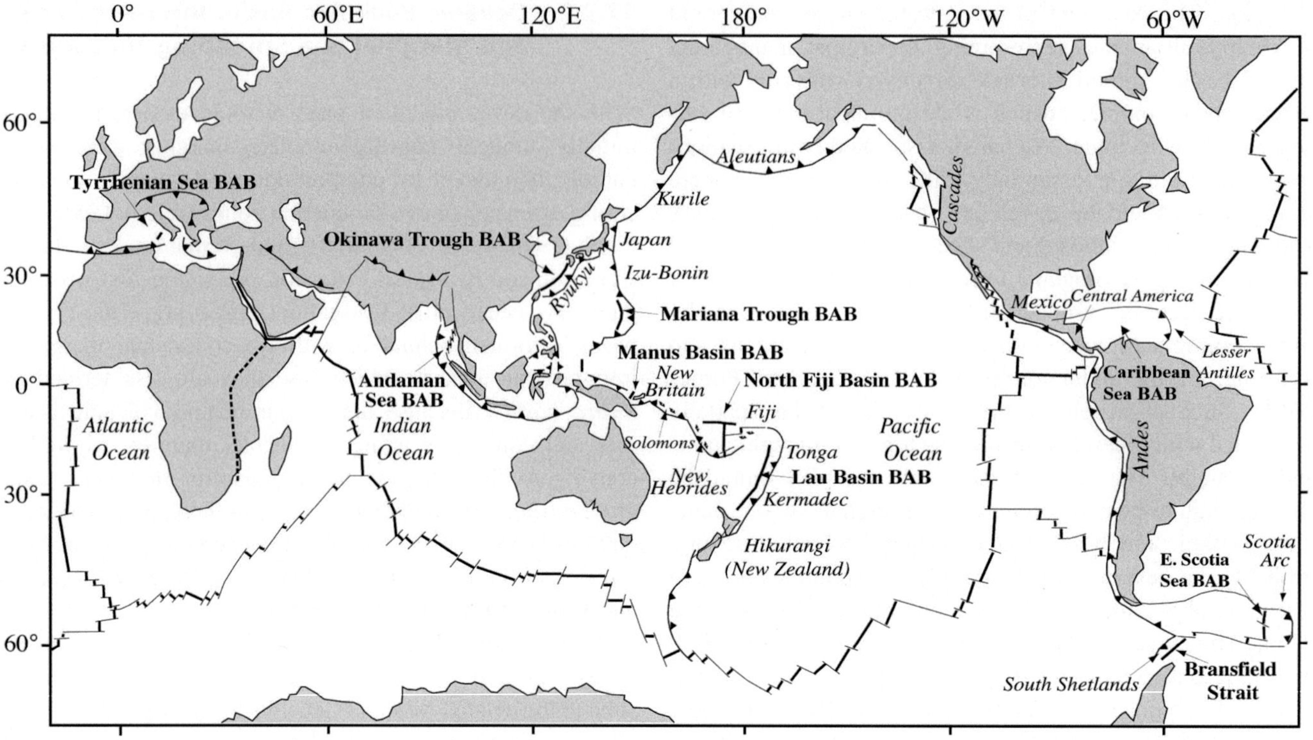

Fig. 17.5 Present-day distribution of active back-arc or marginal sea basins (*Source* Wikipedia through Creative Commons Attribution-Share Alike 3.0 Unported licence)

north-to-south trending spreading ridge as the basin extends in an east-west direction (Kato et al. 2003). In the Lau Basin-Havre Trough (Fig. 17.7c, d), seafloor spreading began <5.5 Ma and ranges from 15.9 cm yr^{-1} in the northern Lau Basin, to 1.5 cm yr^{-1} in the southern Havre Trough (Ruellan et al. 2003). These regions are marked by high axial heat flow and have a sufficiently linear, though not always symmetrical or well defined, magnetic anomaly patterns to suggest axial, if diffuse, spreading from a basin axial, irregular topographic rise. These can be offset along transform faults, or alternatively, by overlapping spreading centres, as observed on mid-oceanic ridges. The crustal thickness is thin, 5–15 km (Brooks et al. 1984; Takahashi et al. 2008; Fig. 17.6b, c) and oceanic in nature, and the crust could also be expected to have an ophiolitic profile consisting of layers 1–3, overlying a layer four upper mantle basement (as described above).

However, not all arc systems necessarily have back-arc basins originating from the rifting of an arc block. For example, the Aleutians arc appears to have formed upon old open ocean lithosphere and to have "trapped" part of this in the associated back-arc basin. Tamaki (1985) has suggested that some associations of marginal basins may have opened simultaneously. He cites the two areas of the Sea of Japan that are floored by oceanic crust but separated by the submerged block of continental crust of the Yamoto Rise, suggesting that these back-arc basins opened simultaneously as the micro-continental arc of Japan separated from the Asian mainland in the early Tertiary. He further suggested that this dual basin system resulted because of a shallow-dipping subduction zone, while single back-arc basins form where subduction zones are steep. It is also possible that the refractory and thick nature of the continental basement under the Japan arc will have contributed to the opening of multiple basins. In both cases, however, seaward roll back of the hinge of the subducting plate or slab must occur to allow extensional spreading in the back-arc basins. The back-arc basin of the Lesser Antilles (Fig. 17.5) was initiated during the late Cretaceous when an oceanic plateau formed in the Pacific collided with a west-facing volcanic arc (see Fig. 17.38). This triggered the formation of the east-facing ancestral Lesser Antilles arc (also known as the "Great Arc of the Caribbean"), which led to slab rollback and eastward migration of the proto-Caribbean seafloor between the Americas since about 85 Ma (Sect. 17.5.2).

The geodynamic origins of back-arc basin formation have been much debated (e.g. Marsaglia 1995; Xu et al. 2014). Most recent thinking and analysis, as just described in the Caribbean, suggests back-arc basins form as a result of slab (or trench or hinge) rollback. This is the backward or seaward motion of the subduction zone relative to the motion of the plate which is being subducted (see Sect. 17.5). As the subduction zone and its associated trench pull backward, the overriding plate is stretched, thinning the crust which is

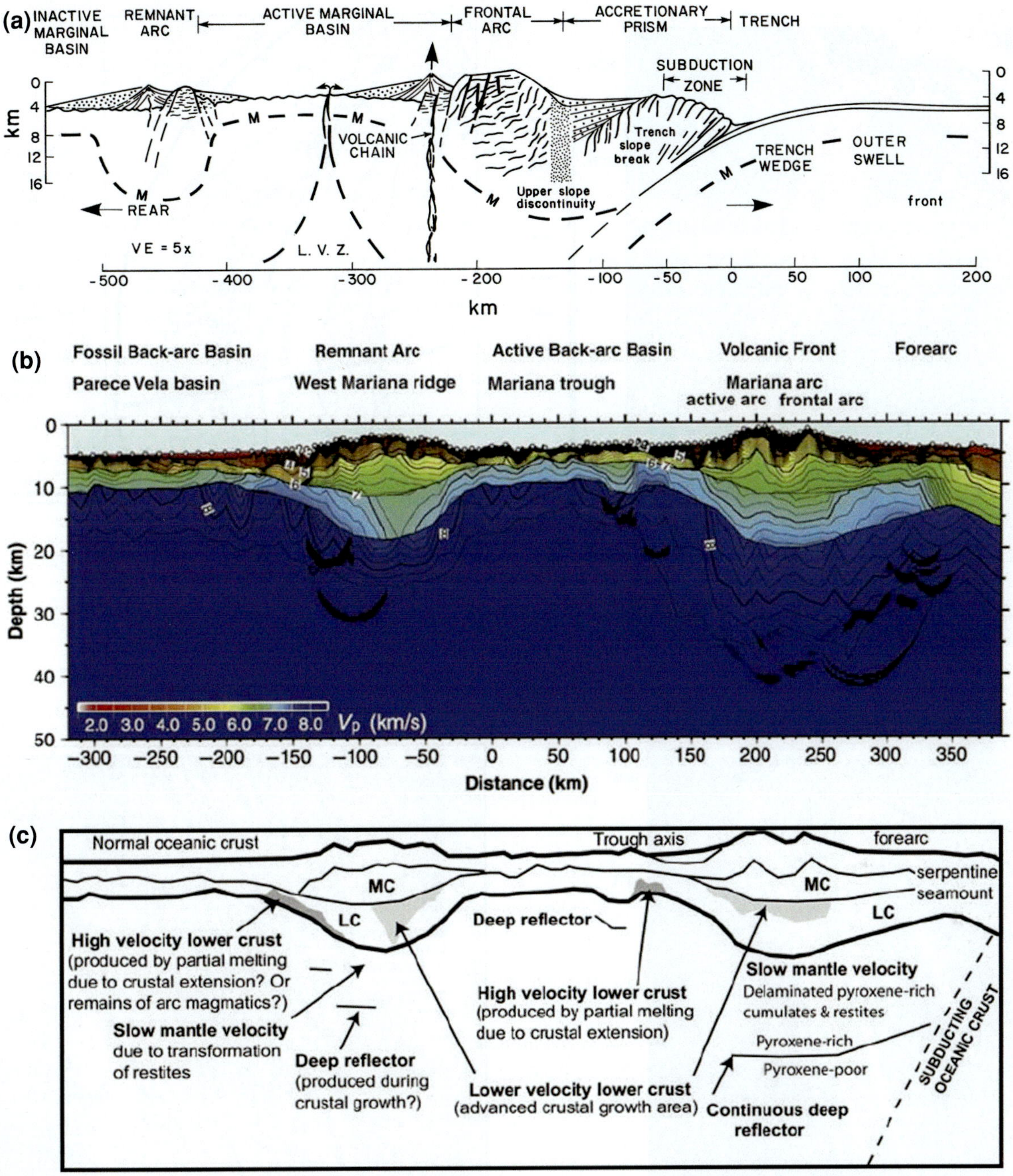

Fig. 17.6 Features of back-arc basins and their crustal and lithospheric structure. **a** Essential elements of an active back-arc—interarc—marginal basin system and its associated arc—subduction complex based on the western Pacific region (From Cas and Wright 1987, After Karig 1974). **b** Seismic velocity structure (top), and **c** interpretation (below) of the Mariana arc—back-arc—interarc—marginal basin system. Numbers indicate P-wave velocity (km s^{-1}); MC = middle crust; LC = lower crust; dark blue = inferred mantle; thick black lines in mantle indicate positions of deep reflectors, which possibly mark boundaries in the uppermost mantle separating upper regions with abundant pyroxene and pyroxenite, from those below dominated by peridotite. Note serpentinite diapirs/mud volcanoes in forearc. (From Stern 2010, After Takahashi et al. 2008; see Fig. 17.7b for location of profile)

manifest in the back-arc basin. Time–space plate reconstructions and kinematic modelling (Sdrolias and Müller 2006) indicate that back-arc basins develop when: (1) the age of subducting normal oceanic lithosphere is >55 Ma; (2) the intermediate (first 100 km) dip of the subducting slab during back-arc spreading is >30°; and (3) they are preceded by an absolute motion of the overriding plate away from the subduction hinge, thereby creating accommodation space between the overriding and subducting plates. Seaward migration of the overriding plate as a precursor to back-arc

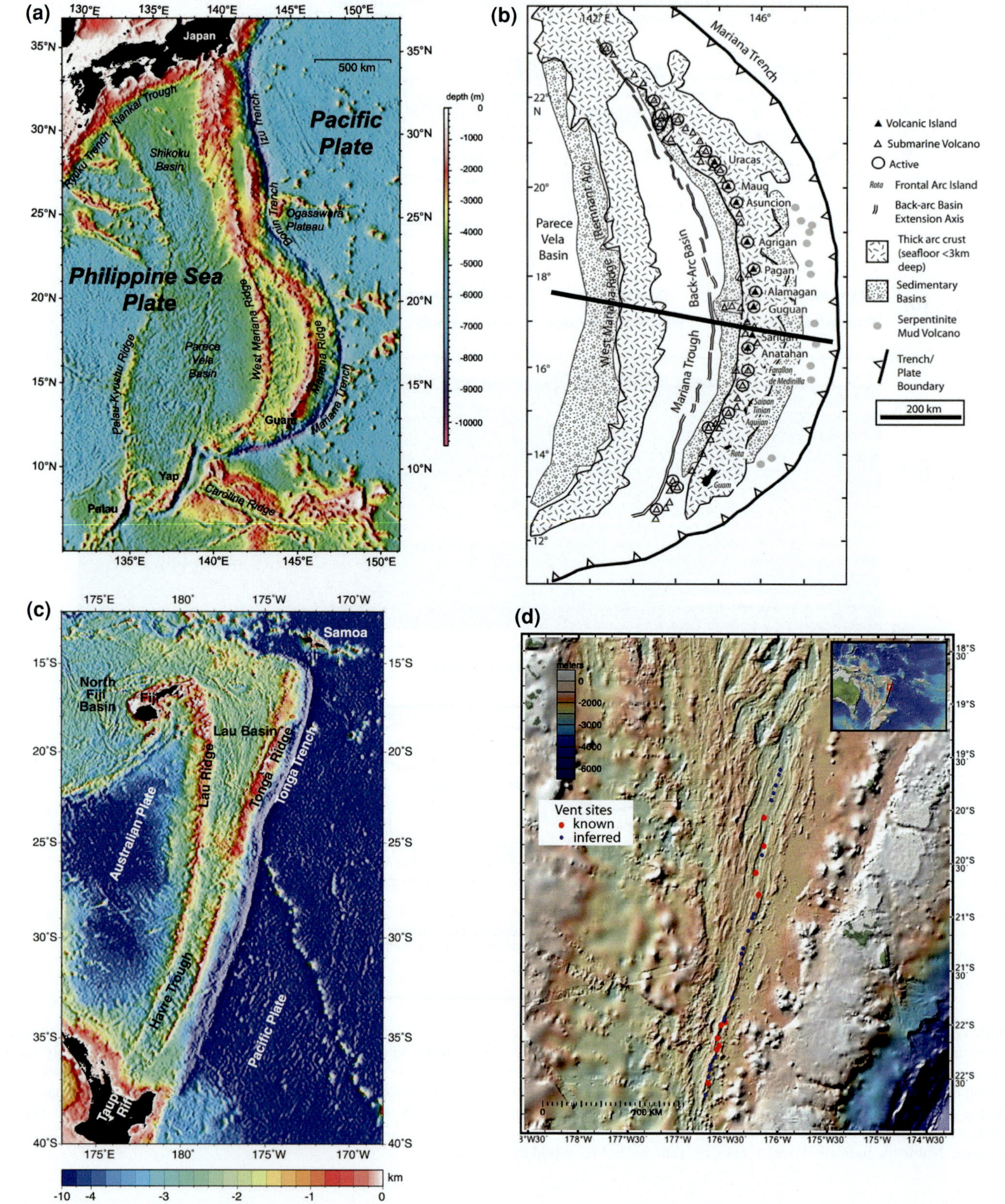

Fig. 17.7 Tectonic setting of the Mariana and Lau-Havre back-arc basins. **a** Mariana Basin regional bathymetry and plate tectonic setting (*Source* NOAA). **b** Mariana Basin simplified tectonic map showing the back-arc basin spreading ridge, active arc volcanoes, forearc uplifted islands, remnant arc, major sedimentary basins, and serpentinite mud volcanoes; black line locates position of seismic line shown in Fig. 17.6b (From Stern 2010). **c** Lau Basin-Havre Trough regional bathymetry and plate tectonic setting (*Source* NOAA). **d** Detailed bathymetry of eastern Lau Basin showing axial spreading centre and positions of hydrothermal vents, which are restricted on the seafloor to the neovolcanic zone within a narrow ($\sim \pm 1.5$ km) axial ribbon (*Source* Ridge 2000, http://ridge2000.marine-geo.org/)

extension may indicate that extension on the overriding plate is influenced by the oceanward lateral flow of the mantle. However, once back-arc extension is established, rollback of the trench hinge appears to be the primary force responsible for the continued creation of accommodation space.

Geological and geophysical observations and constraints on the dynamic evolution of the central Mediterranean subduction zone system (i.e. Tyrrhenian back-arc basin, Fig. 17.5) similarly support such a time-dependent scenario. This commenced with: (1) slow initiation of subduction (between 80 and 35–30 Ma), associated with shallow dip of the subducting slab; (2) progressive increase of the subduction speed and dip, related to the slab pull with increasing length of the subducted slab (from 35–30 to 15 Ma) resulting in arc migration and fast back-arc spreading; and (3) a pulsatory stage (from 15 Ma to present day) related to the interaction of the subducting slab with the 660 km mantle discontinuity, resulting in pulses of back-arc spreading (Faccenna et al. 2001 (Sect. 17.5.5). The driving mechanism for back-arc extension is therefore a combination of surface kinematics, properties of the down-going slab, the effect of lateral mantle flow on the slab, and mantle wedge dynamics (Sdrolias and Müller 2006). For a radically different view proposing a large-scale, linked dextral pull-apart marginal basin system in the northwest Pacific (formed in response to the Indo-Eurasia collision), see Xu et al. (2014). This does not however change the seafloor spreading origin of these marginal basins, but potentially adds another regional tectonic control on their initiation.

Spreading gives rise to new oceanic crust. However, spreading axes may be spatially closely associated with convergent margins, and down-going slabs can influence the pattern of mantle flow and prevent the kind of simple upwelling observed at open ocean ridges (Fig. 17.2). Back-arc spreading centres are also influenced by the flux of water and other elements that come from the slab as it subducts (Taylor and Martinez 2003; Langmuir et al. 2007). Whereas the magmas erupted along the major ocean spreading ridges are predominantly N-type MORB, many back-arc basins are floored by basalts (sometimes termed BABB) transitional between N-type MORB and island arc or even calc-alkaline basalts (e.g. Taylor and Karner 1983; Saunders and Tarney 1984; Tian et al. 2005; Masuda and Freyer 2015; Peccerillo 2017). On a broader scale, it is possible to relate basalt composition, tectonic setting of the basin, and maturity of the adjacent subduction zone. Thus, the Parece Vela Basin, formed during the earliest stages of the Mariana subduction system (Fig. 17.7a), is floored by basalts indistinguishable from N-type MORB, whereas the younger Mariana Trough back-arc basin (Fig. 17.7a, b) is erupting MORB and basalts with calc-alkaline characteristics, commonly in close spatial proximity.

A calc-alkaline component is best developed in narrow, ensialic basins, where the extension is adjacent to mature, and continent-based magmatic arcs. For example, the Bransfield Strait is a narrow back-arc basin separating the South Shetland Islands and the Antarctic Peninsula (Weaver et al. 1979; Keller et al. 2002; Fig. 17.5), and the oceanic Havre-Tough is a narrow back-arc basin developing behind the Tonga-Kermadec island arc system, extending into the micro-continental mass of New Zealand, that has been described as an ensialic marginal basin (Cole 1984; Fig. 17.7c). This range of compositions (N-type MORB to calk-alkaline) can only be explained by invoking chemical variations in the composition of the mantle material supplying the back-arc basin crust, implying the transfer of material from the subducted slab (Sect. 17.5, see Fig. 17.42).

In the central Mediterranean region, the fast back-arc spreading and its spatial migration resulted in the coexistence of different magma types from intraplate (MORB-tholeiites, Na-transitional and alkaline) to arc products (arc tholeiites, calc-alkaline and shoshonitic) with strong evidence for a very heterogeneous mantle source. Such heterogeneity has been explained in part due to focussed upwelling of the upper mantle in response to slab fragmentation (Rosenbaum et al. 2008; Faccenna et al. 2010) and in part due to metasomatism of the mantle upper wedge by recycling of subducted sediments (Avanzinelli et al. 2009).

Sedimentation in back-arc basins is strongly asymmetric, with most of the sediment supplied from the active magmatic arc, which regresses in step with the rollback of the trench (Fig. 17.6a). Layer 1 sediments are largely pelagic where they are formed well away from the influence of sedimentation from the rifted arc blocks bounding the developing marginal sea. However, even within a hundred km of the frontal arc and remnant arc blocks, redeposited volcaniclastic sediments transported by submarine mass-flow processes will form a significant pile of sediments above the layer 2 basalts (Carey and Sigurdsson 1984; Cas and Wright 1987; Marsaglia 1995). Various amounts of redeposited carbonate debris may also be included. Ash-fall layers derived from the active frontal arc may also occur. The volcaniclastic character of this sedimentary succession may help to distinguish marginal sea crust and lithosphere from large ocean basin oceanic crust. The latter is mantled by only pelagic sediments or terrigenous, continentally derived clastics, or both, except where it is being subducted at an island arc plate margin, in which case it may also contain an arc derived volcaniclastic element in layer 1. Continentally derived clastics are likely to be more compositionally mature (quartzrich) and diverse in their provenance than the sediment fill of an interarc basin. However, marginal basins directly adjacent to continental masses may contain both provenance elements.

17.2.2.1 Ancient Examples of Oceanic Back-Arc Basin or Marginal Sea Crust

In terms of recognising ancient back-arc basin or marginal sea crust volcanism in the rock record, the most compelling criteria are the ophiolitic stratigraphy outlined above, the distinctive volcaniclastic character of layer 1 sediments, and the association with a regionally extensive, well-documented contemporaneous arc succession and its forearc subduction complex, which will all be incorporated into a deformed ancient orogenic belt (e.g. Yule et al. 2006). The scale of the whole system should be compatible with suitable modern analogues (e.g. Cas et al. 1980; Cas 1983). Modern marginal basins are hundreds of km long and similarly wide at their widest parts, but may narrow right down to converging apices, such as in the Mariana system (Fig. 17.7a, b). Deformational style of the preserved marginal sea succession could be variable depending on its tectonic history. As first indicated by Karig (1972), the active life of marginal seas may be terminated by the initiation of subduction zones within the marginal seas, during *arc polarity reversal* of the system (see Sect. 17.5.1). Evidence of this should occur in the form of an intra-basinal subduction accretionary prism complex. Simpler deformational styles involving regional compressional deformation, perhaps associated with the jamming up of the subduction zone in front of the arc, could also occur.

17.2.3 Continental Rift Volcanism

Active extension or tectonic stretching of continental lithospheric plates induces surface deformation, high heat flow and volcanism due to the effects of normal faulting, and changes in crustal and mantle thickness, structure and state (e.g. McKenzie 1978; Allen and Allen 1990; Leeder 1995). Stretching of a continental plate takes place when external forces, ultimately controlled by plate motions and mantle convection, pull it in opposed or near opposed directions. As the lithosphere is stretched horizontally, it thins vertically through listric fault extension (Fig. 17.8). Hot, lower mantle (asthenosphere), which is plastic and can flow, rises below the thinned crust creating a thermal anomaly in the base of the lithosphere. This causes the overlying crust to weaken, in turn leading to brittle failure and normal faulting at the surface, and subsidence. This is referred to as the *rift phase*. A number of researchers have since argued that the rifting forces available are insufficient to rupture thick, cold lithosphere, however, the presence of melt in the form of intruded of dykes and sills could weaken the lithosphere. This process has been called "*magma-assisted*" rifting (e.g. Buck 2004; Kendall and Lithgow-Bertelloni 2016; see Fig. 17.14).

As the rift grows, slip on faults drops blocks of crust downwards producing low trough-like rift basin areas

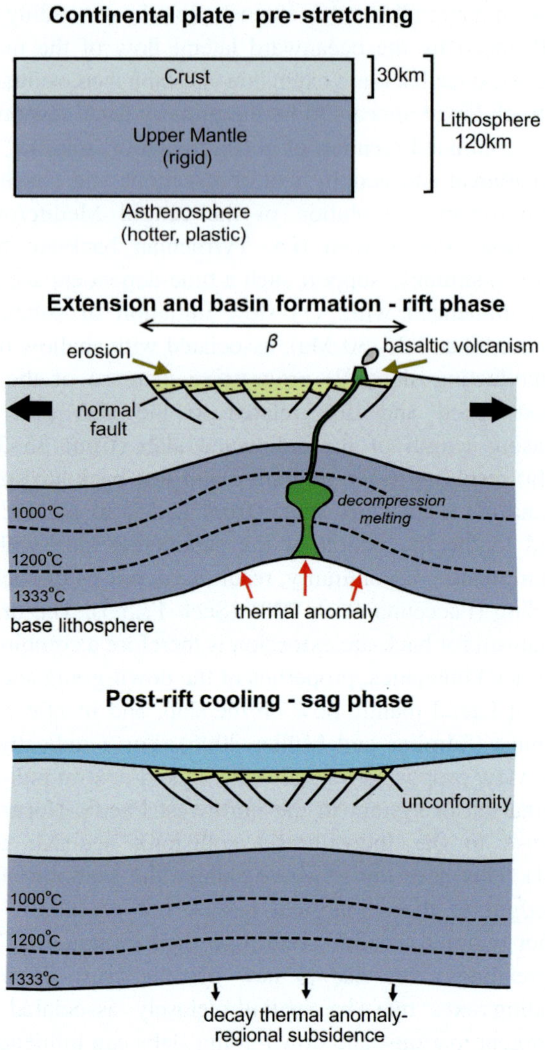

Fig. 17.8 McKenzie model for the development of extensional sedimentary basins. Both volume and geochemical characteristics of magma can be related to the amount of stretching, or β factor (e.g. McKenzie and Bickle 1988; Brown and White 1995). The rift stage shows simultaneous necking of the lithosphere with uprise of an asthenospheric diapir. Decompression associated with the latter causes melting of the mantle to give alkaline basaltic magmas

bordered by narrow, uplifted mountain ridges or shoulders. Erosion of basement along basin margins causes significant flux of coarse clastic sediments into the troughs. These form thick fluvial-alluvial wedges of sandstone, gravels and conglomerates. This phase is often associated with decompressional basaltic or bimodal volcanic activity which, depending on regional geodynamic setting (e.g. amount of extension or *beta (β) factor*, distance from subducting plate margin, mantle flow), can be of limited or copious extent and more varied compositionally. Generally, where $\beta \leq 2$ a rift section devoid of volcanics develops (McKenzie and Bickle 1988). With increasing β factors, alkali basalt melts are generated and change to tholeiites as the amount of melting increases which underplate the lower crust, increasing the potential for the generation of more silicic and felsic melts.

Once stretching has ceased the continental plate cools, which causes further subsidence, called the *post-rift* or "*sag*" *phase*. This leads to the deposition of a succession of overlapping sedimentary layers, mainly composed of finer-grained mudstones and shales, limestones and evaporites, generally deposited in lacustrine and marine environments. Ideally an unconformity is found between the rift and sag phase sediments.

Continental or intra-continental rifts can be divided into two end members (e.g. Cas and Wright 1987; Leeder 1995):

- narrow, linear rift zones or grabens (which are 50–100 km wide, and >1000 km long); and
- broad (>1000 km wide) rift provinces, which include many individual narrow (15–30 km) rift basins, often half grabens, separated by range forming, often listric faults.

The classic example of the former is the East African Rift System that hosted the earliest hominid evolution, while of the latter, the Basin and Range Province in the western United States is a widespread rift terrain. Narrow, linear rifts may form as a prelude to full continental break-up (complete continental lithosphere separation), and creation of new oceanic crust and large ocean basins by seafloor spreading. These have been described in the past as Atlantic-type margins, but are now more generally referred to as volcanic passive margins (Sect. 17.2.4). The full extensional spectrum, from continental rifting to oceanic crustal generation, can be observed in the Ethiopian Rift-Afar-Red Sea region (Figs. 16.36a–d and 17.9a, b).

17.2.3.1 Narrow Linear Continental Rift Zones

Evidence from the East African Rift System (EARS) and elsewhere (e.g. the Rhine Graben, and Baikal and Rio Grande Rifts; Plate 17.1), suggests these sit on top of major regional surface upwarps or domal features covering several thousands of km^2 (Fig. 17.9a, b; Morley et al. 1999; Chorowicz 2005; Whichura et al. 2011; Sembroni et al. 2016). Sometimes these uplifts are referred to as an intra-continental ridge system (Chorowicz 2005) and are interpreted as the surface effects of rising mantle plume(s) and asthenospheric intrusion at depth. Over time these may migrate, or other plumes are emplaced, probably the result of long-term, deep mantle processes (Sembroni et al. 2016; Civiero et al. 2016; Kendall and Lithgow-Bertelloni 2016; Rooney 2017).

The East African Rift System extends for nearly 3000 km from its triple junction with the Red Sea and Gulf of Aden (Indian) spreading ridges in the Afar region in Ethiopia in the north to Mozambique in the south (Figs. 17.9a, b and 17.10a). From a singular rift system in the north it splits into a complex dual rift system to the south. The basement within which the EARS is developing is Precambrian continental crust. In the EARS, several uplift-updoming phases in the Cenozoic were accompanied by widespread continental flood basalt volcanism (e.g. Cas and Wright 1987; Morley et al. 1999; Chorowicz 2005; Corti 2009; Macgregor 2015; Chap. 16; Figs. 16.36b and 17.9b). In the late Cenozoic, the areas of principal updoming were then affected by rifting and extension, reflected by the formation of an axial graben system (Figs. 16.36c and 17.10a–e) and more aerially restricted fault-controlled volcanism. Crustal failure preferentially occurred along pre-existing weak zones in the Precambrian basement, particularly in orogenic suture zones and terrane boundaries. Igneous products are petrologically and geochemically diverse, ranging from mafic to silicic in character, and are alkaline to peralkaline rocks. Calderas (Fig. 17.10c) and ignimbrites are common, as are lavas, pyroclastic and epiclastic deposits. These were initially and, with the exception of the Afar region, are today deposited in subaerial settings dominated by fluvial, alluvial fan and lake (ranging from freshwater to alkaline) environments, within a normal fault-controlled topography (Figs. 17.10 and 17.11a). Nonetheless, these settings were accompanied by major climatic and paleo-vegetation shifts, with more pronounced dry conditions at the expense of rainforests at 16–14 Ma, and the expansion of grasslands at 8–7 Ma (Whichura et al. 2011).

Continental linear rifts (Figs. 17.10a-e and 17.11a) are dominated by half-graben geometries (Figs. 17.10b, d and 17.11b). Petroleum industry-driven advances in extensional tectonics and evolution of sedimentary basins made in the mid-late 1980s showed extensional rifts and basins are rarely symmetrical (as inferred by Fig. 17.8). Instead they are commonly highly asymmetric, accommodated by dip-slip and rotation (or roll-over) along *listric normal faults* which sole-out into a shallow-dipping *detachment fault zone*, upon which the basin rides in the hanging wall (e.g. Gibbs 1984; Lister et al. 1986; Coward et al. 1987; Etheridge et al. 1987; Lambiase 1995; Fig. 17.11b). These form series of tilted half grabens having opposite dip polarities, which are linked through moderately-steeply dipping high-angle, transverse structures that control the along strike segmentation or compartmentalisation of extensional rifts. These structures have been variously termed *transfer* or *accommodation zones and faults* (e.g. Morley et al. 1990; Bosworth 1987, 1994), and *relay ramps* (e.g. Gawthorpe and Hurst 1993). Accommodation and transfer zones are commonly used to describe larger (or interbasin) structures that link rift segments or individual half grabens (e.g. Fig. 17.11b). These may be associated with changes in structural geometry (e.g. dip-sense of normal and detachment faults) either side of the

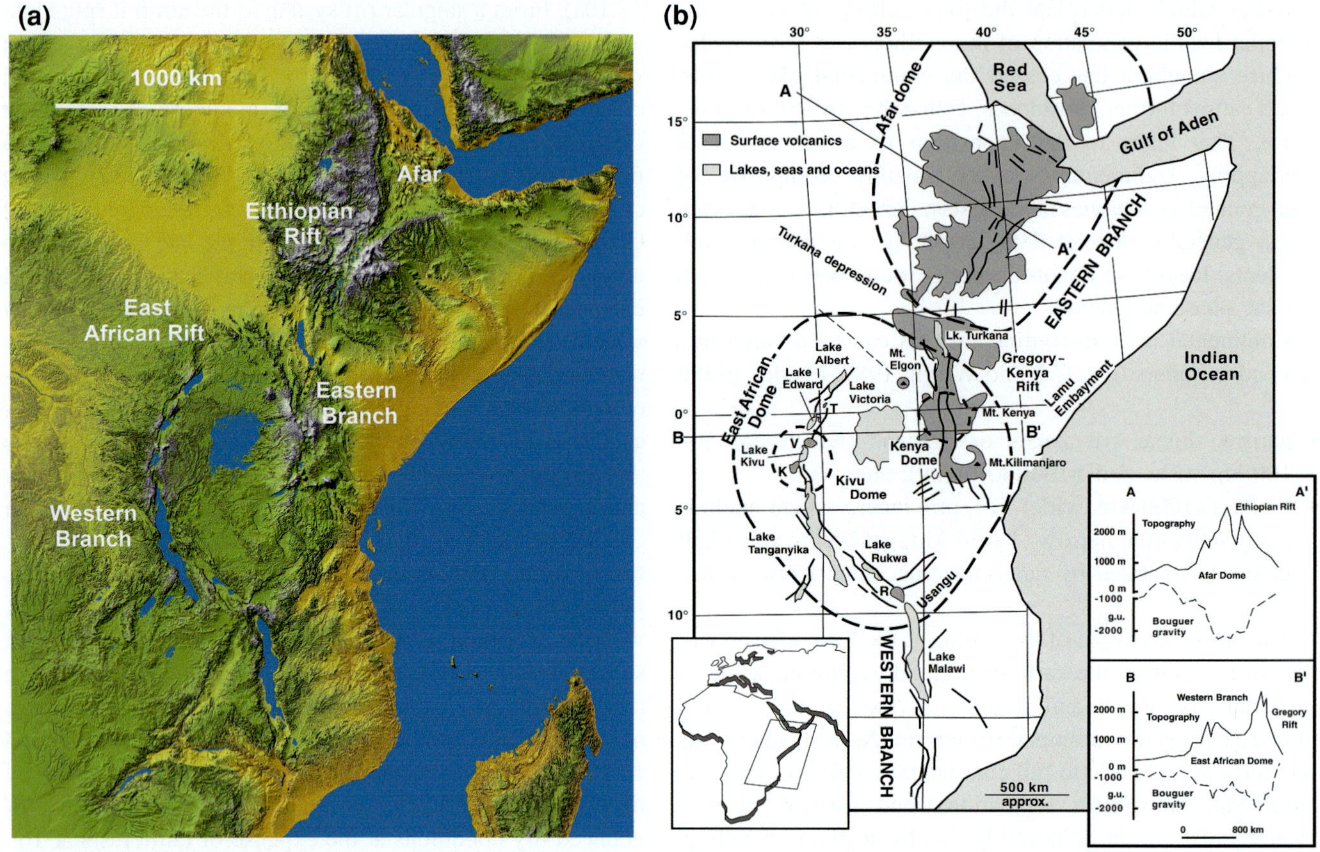

Fig. 17.9 The East African Rift System. **a** Space Shuttle radar topography image showing main rift segments. The oldest and best-defined rift occurs in the Afar region of Ethiopia. Further to the south a series of rifts occur which include a largely amagmatic Western branch, containing the Lake Albert or Albertine Rift and the East African Great Lakes, and a volcanically active Eastern branch that roughly bisects Kenya and northern Tanzania north-to-south. These two branches together have been termed the East African Rift (EAR), while the Eastern branch has been variously called the Gregory Rift, Kenya Rift or Great Rift Valley. The two EAR branches are grouped with the Ethiopian Rift to define the East Africa Rift System (EARS). (*Source* NASA-JPL Caltech). **b** Distribution of surface volcanics (dark grey; mostly representing the area covered by flood basalts) and large, pre-rift topographic domes in relation to rift structure. Isolated Western branch volcanic centres: R = Rungwe, T = Toro-Ankole, V = Virunga, K = South Kivu. (From Morley et al. 1999)

structure. Intrabasin transfer faults accommodate changes in structural style and displacement between faults within individual half grabens. Relay ramps link displacement between two *en echelon* faults, forming a zone dipping into the hanging wall; these are now recognised as the most common type of intrabasin transfer structure (Fossan and Rotevatn 2016).

As well as accommodating extension, transfer zones including relay ramps are important in controlling sedimentation (see below). They also provide "*short-circuit*" pathways for both the lateral and longitudinal flow of buoyant basinal (i.e. hydrocarbon), geothermal and magmatic fluids (e.g. Cartwright 1987, 1992; Morley et al. 1990; Wright 1990, 1994; Gawthorpe and Hurst 1993; Rowland and Simmons 2012; Fossan and Rotevatn 2016; Williams et al. 2020; Chap. 18).

Although it had been recognised for many years, that rifts are highly segmented and, as in the case of the EARS, may even bifurcate into separate branches (Figs. 17.9 and 17.10), it was only in the late-1980s that the nature and extent of this became fully appreciated (Bosworth 1987, 1994; Rosendahl 1987; Ebinger 1989a, b; Morley 1989, 1995). For example, Bosworth (1987) interpreted part of the EARS in Kenya (Gregory Rift) as a series of detachment systems compartmentalised by wrench-dominated accommodation (or transfer) zones (Fig. 17.12). These detachment systems are manifest at the surface as a series of breakaway zones and regional bounding faults (e.g. Elgeyo escarpment) that produce sub-basins with half-graben form (Fig. 17.12a, b). The asymmetry of sub-basins alternates along the rift axis, indicating that the polarity of the underlying active detachment systems also reverses. In this model, the

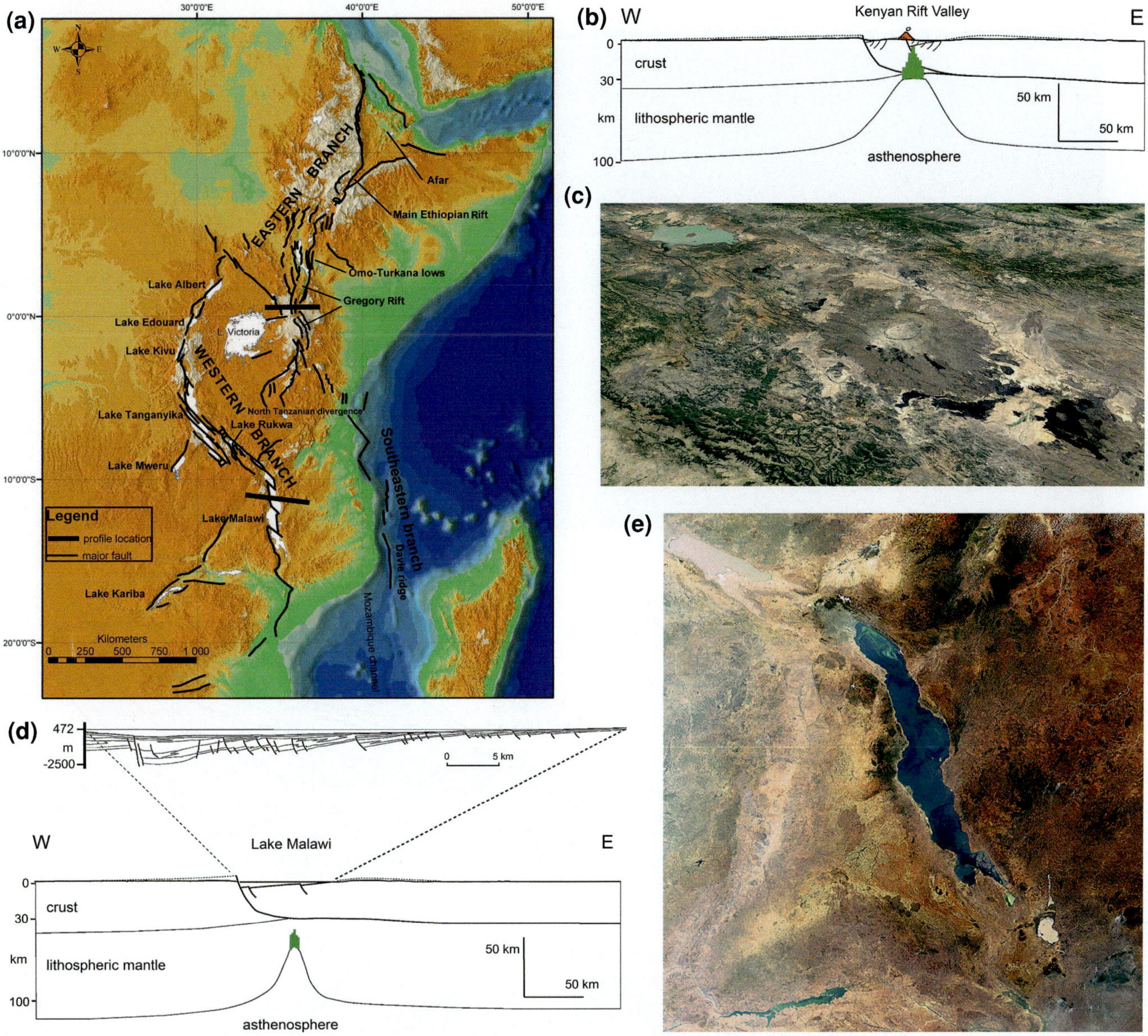

Fig. 17.10 Tectonic segmentation and deep structure of the East African Rift System. The rift system comprises a series of aligned, adjacent individual tectonic basins (rift valleys) separated from each other by fault-bounded shoals (transverse or accommodation zones, or transfer faults), and generally bordered by uplifted shoulders. **a** Digital elevation model; black lines = main faults; white surfaces = lakes; E-W dotted black lines = locations of cross-sections (From Chorowicz 2005). **b** Lithospheric cross-section of the Kenya Rift Valley, Gregory Rift; green = inferred dyke intrusions (After Chorowicz 2005). **c** Oblique Google Earth view south along the volcanically active, central rift axis, Kenya. This shows the prominent Mt Silali caldera (slightly above and right of centre), Mt Emuruangogolak volcano to its north, and Lake Baringo in the distance (see Fig. 17.12a). (*Source* Google Earth). **d** Line drawing of seismic reflection profile (above) and lithospheric cross-section (below) of Lake Malawi, Western branch, showing predominance of asymmetric, half-graben geometries (After Chorowicz 2005). **e** Envisat image of Lake Malawi and Lake Rukwa. Note dramatic change in orientation of rift depocentres (i.e. lakes) between the right-lateral NW–SE Tanganyika-Rukwa-Malawi transverse zone and the N–S Malawi Rift (see Fig. 17.13a). The Malawi Rift is also compartmentalised into a number of en echelon sub-basins separated by NW-trending transfer faults. Mt Rungwe volcano (R) sits in a right-stepping dilational releasing bend. (*Source* European Space Agency through Creative Commons Attribution-ShareAlike 3.0 IGO (CC BY-SA 3.0 IGO) licence)

(a)

(b)

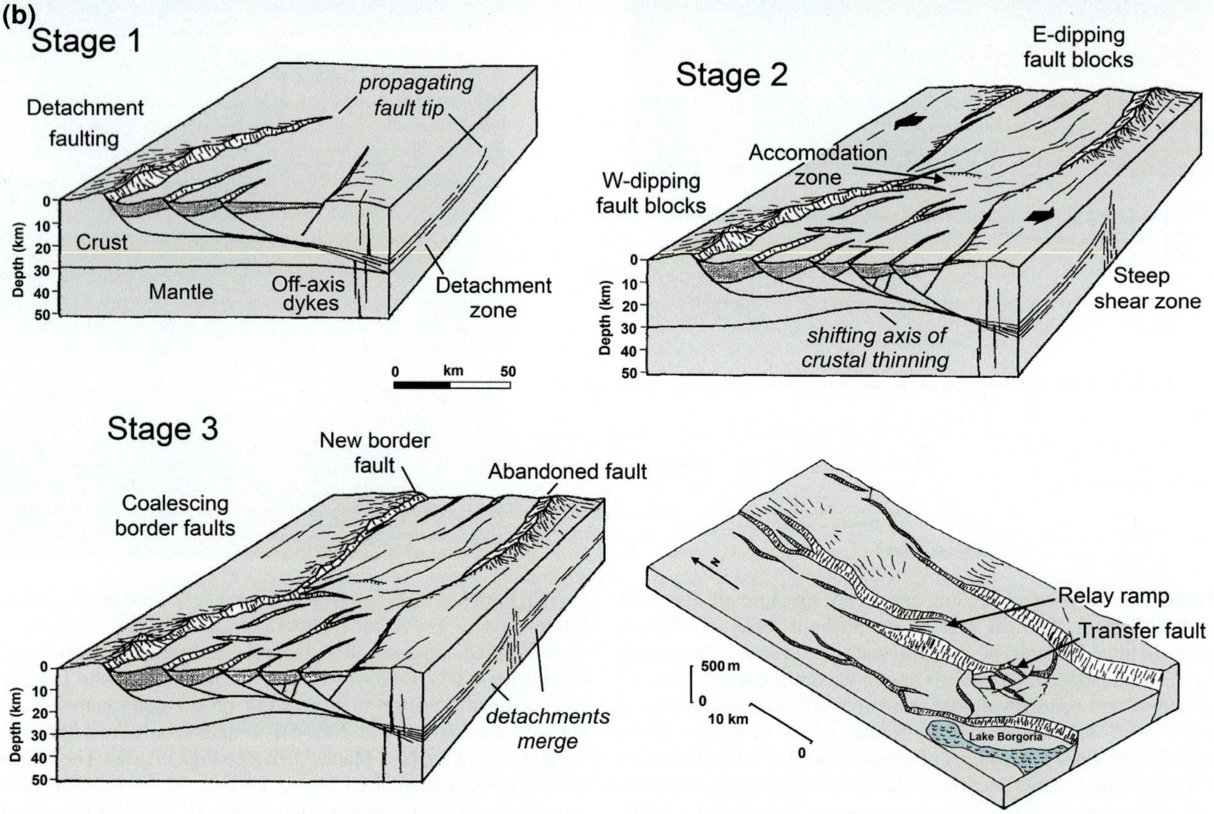

Fig. 17.11 Features of the East Africa Rift System. **a** Panoramic image of Albertine Rift, Western branch of the East African Rift System. From foreground to background this shows Lake Tanganyika (northern part), Lake Kivu, Lake Edward and Lake Albert. (Computer rendered image by Christoph Hormann; *Source* Wikimedia through Creative Commons Attribution-Share Alike 3.0 Unported licence). **b** Structural evolution of continental rift basins exemplified by the East African Rift System. Stage 1: Young rift with half-graben geometry developed above single detachment fault. Stage 2: Half-grabens linked along strike by accommodation zones as new detachments form. Polarity reversal may arise due to random selection or reflect pre-existing basement structure. Stage 3: At higher strains, alternating detachments reorganise, eliminating faults that dip in one direction and linking border faults of the same polarity. (From Bosworth 1994). Final diagram shows line drawing of extensional fault organisation with relay ramp and diffusely fault-dominated transfer zone on the eastern margin of the Gregory Rift, near Lake Bogoria, Kenya (see Fig. 17.12a) (From Morley 1990)

detachments are separated laterally by regional oblique-slip accommodation zones, while off-axis from the rift, the detachments are inferred to merge along strike as they cut to the base of the lithosphere (Fig. 17.12c, d). Above these zones, the largest volcanoes of the rift (Mts Eldon, Marsabit, Kenya and Kilimanjaro) occur in two belts 100–150 km to the west and east of the central axis of the rift valley (i.e. the Gregory Rift valley proper), and the zone of maximum subsidence and crustal thinning. The development of major volcanic cones was seen as places where leaky accommodation structures interact with the zones of thinned lithosphere. While imaginative, and one of the first contributions applying modern extensional concepts to the EARS, Bosworth's model does not adequately explain the complex distribution and evolution of volcanic sequences and extensional structures spatially and through time. Other workers have emphasised the role of continental-scale transverse (rift-oblique) structures or lineaments in controlling rifting events and location of volcanism (e.g. Katz 1987; Daly et al. 1989; Smith and Mosley 1993; Morley 1995; Chorowicz 2005; Robertson et al. 2016; Corti et al. 2018). In particular, Chorowicz (2005) proposed NW–SE divergence (or oblique extension) between the African (Nubian) and Somalian Plates was accommodated on large northwest-trending faults, which he regarded as intra-continental transform fault zones, linking rift-parallel normal faults and basins (Fig. 17.13a).

Such lineaments are related to Proterozoic basement mobile belts and suture zones (terrain boundaries), Elsewhere in the African continent inherited basement features control the location of sedimentary basins, aligned igneous complexes, in particularly kimberlites and carbonatites, and mineral deposits (Daly et al. 1989; Chap. 18). In particular, crustal-scale steep basement shear zones (ancient lateral and oblique ramps of thrusts) appear to have exerted profound influence on rift basin formation and volcanism.

These structures are difficult to reactivate by dip-slip movement, and thus tend to operate as major strike-slip transfer zones during later extensional events (Daly et al. 1989). In the EARS, the most prominent oblique structure is the Aswa transverse zone, which is a major (>1000 km in length, and as much as 100 km wide), long-lived Proterozoic mylonitic shear zone(s) (Fig. 17.13a; Katumwehe et al. 2016; Saalmann et al. 2016), and fundamentally controls segmentation and architecture of the Gregory Rift, as discussed above. This wide diffuse fault zone comprises several large sinistral NW-striking faults, which are responsible for some of the strongest earthquakes recorded in Africa (Gaulon et al. 1992; Mulwa and Kimata 2014). These include the strong 20 May 1990 South Sudan earthquake (magnitude 7.2) which was purely strike-slip with a strong left-lateral (sinistral) component. Field observations and satellite imagery also indicate sinistral reactivation movement along this large structural feature during Cenozoic times. The transverse zone incorporates the Mt Elgon volcano probably in left-stepping, fault pull-apart structure (Fig. 17.13a). At the intersection of the Aswa shear zone with the Eastern branch, the Gregory Rift turns and aligns sub-parallel with transverse zone, which constitutes the boundary with the central and southern Gregory Rift. The transverse zone continues southeast-ward in part controlling volcanic and fault belt in northern Tanzania or North Tanzania Divergence zone (NTD). This links the Eastern branch with the and the Southeastern branch of the EARS, which has been thought to represent the early stages of a dextral transform fault extending offshore into the Indian Ocean (Fig. 17.10a; Chorowicz 2005). The Tanganyika–Rukwa–Malawi (TRM) transverse zone is another major northwest-trending feature, which connects the two main segments of the Western branch (Fig. 17.13a). This forms a complex right-stepping, dextral (right lateral) en echelon extensional fault zone, including a releasing bend opening responsible for the siting of the Rungwe volcanic centre (Figs. 17.9b and 17.10e).

Reactivation of Precambrian zones of weakness along oblique transverse faults, and the opening of dilational fractures, is now the more generally accepted explanation for off-axis volcanism in the EARS (Korme et al. 1997; Chorowicz 2005; Corti 2009; Le Gall et al. 2008; Isola et al. 2014; Corti et al. 2018; although other models have been forwarded, e.g. Maccaferri et al. 2014). This also agrees with evidence from elsewhere that rift volcanism is controlled by more fundamental and deeper transverse structures, e.g. late Carboniferous-early Permian rifting in northwest Europe and the Mesozoic North Sea Rift (Cartwright 1987, 1992), and Cenozoic Rio Grande Rift (Self et al. 1986). A schematic reconstruction of a rift segment-bounding transverse structure is illustrated in Fig. 17.13d. Mid-lithospheric magma chambers comprising evolved melts are connected to the surface by narrow and sub-vertical, lower crustal shear zones. These splay in the upper crust into a braided, discontinuous pattern of normal and strike-slip faults, and provide the main conduits for magma to reach the surface. Sedimentation, volcanism and rift propagation are viewed as the products of differential movement along successive transverse lineaments.

More recent present-day plate-kinematic analysis of the EARS and wider region employing satellite imagery, GPS and fault plane slip vector data, coupled with detailed field structural analysis, are now suggesting a more complicated tectonic picture (Le Gall et al. 2008; Morley 2010; Delvaux et al. 2012; Laó-Dávila et al. 2015; Muirhead and Kattenhorn 2018; Stamps et al. 2018; Glerum et al. 2020). Overall, present-day plate-kinematic vectors in the region are

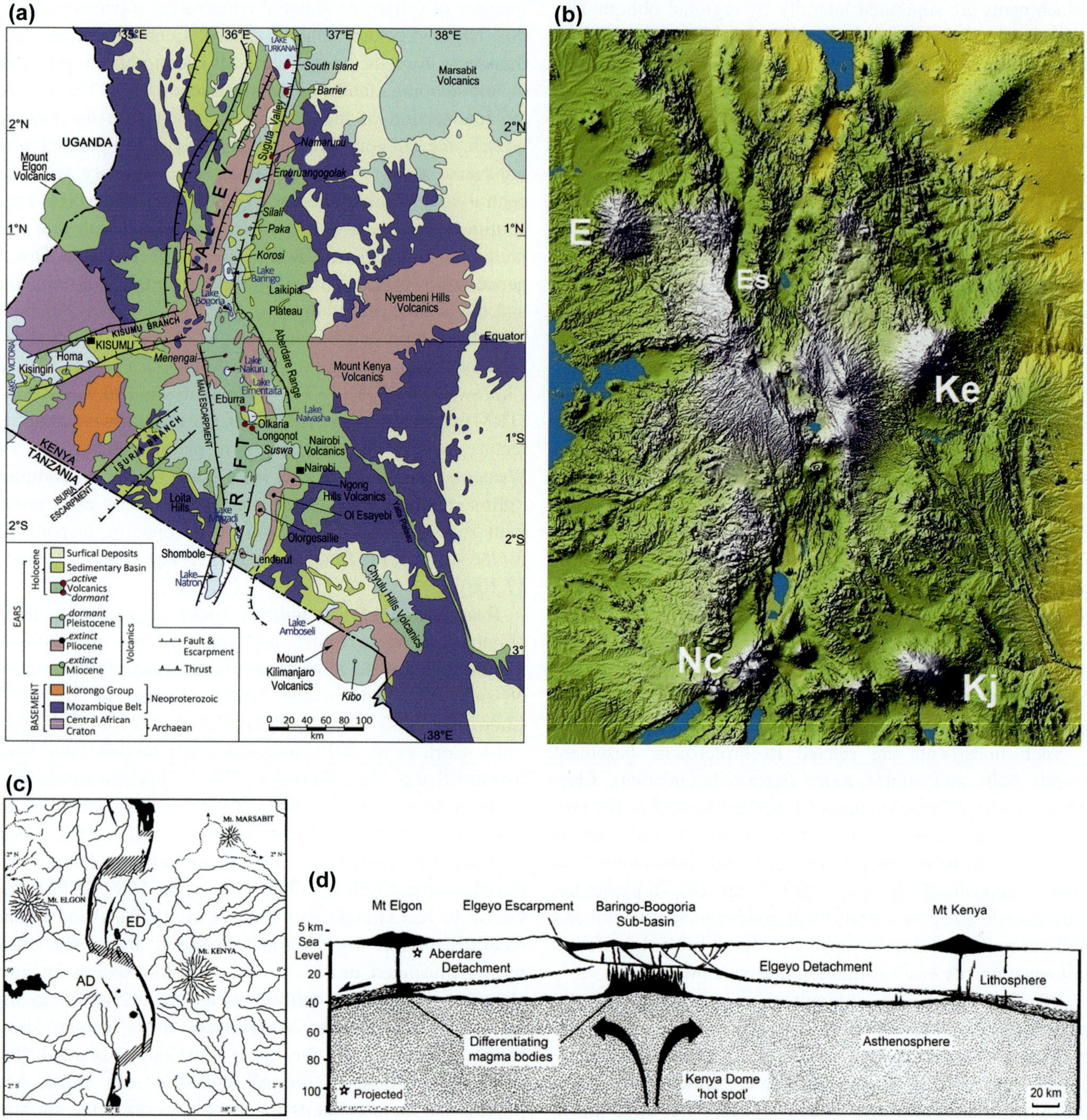

Fig. 17.12 Features of the Gregory Rift. **a** Simplified geological map of the Gregory Rift, central and southern Kenya (From Scoon 2018). **b** Space Shuttle radar topography image of the area covered in (**a**). Reference points: E = Mt Elgon, Es = Elgeyo fault scarp, M = Mt Marsabit, Ke = Mt Kenya, Kj = Mt Kilimanjaro; and Nc = Ngorongoro caldera, Od = Ol Doinyo Lengai in Tanzania, not shown in (**a**). (*Source* NASA-JPL Caltech). **c** Rift half-graben bounding faults and accommodation zone (diagonal rule) structure as interpreted by Bosworth (1987). Map view showing drainage patterns; ED = Elgon detachment, AD = Aberdare detachment; lakes are in black. (Adapted from Lambiase and Bosworth, 1995, After Bosworth 1987). **d** Schematic cross-section from Mt Elgon to Mt Kenya. Main active detachment beneath Baringo-Bogoria sub-basin (Elgeyo detachment) dips to east and produces lithospheric thinning beneath region of Mt Kenya. Zones of major lithospheric thinning and accompanying off-axis volcanism (Elgon, Kenya, Marsabit and Kilimanjaro) are offset from zone of crustal thinning (i.e. rift valley proper). Oppositely directed detachment systems interact in wrench-dominated accommodation zones. (From Bosworth 1987)

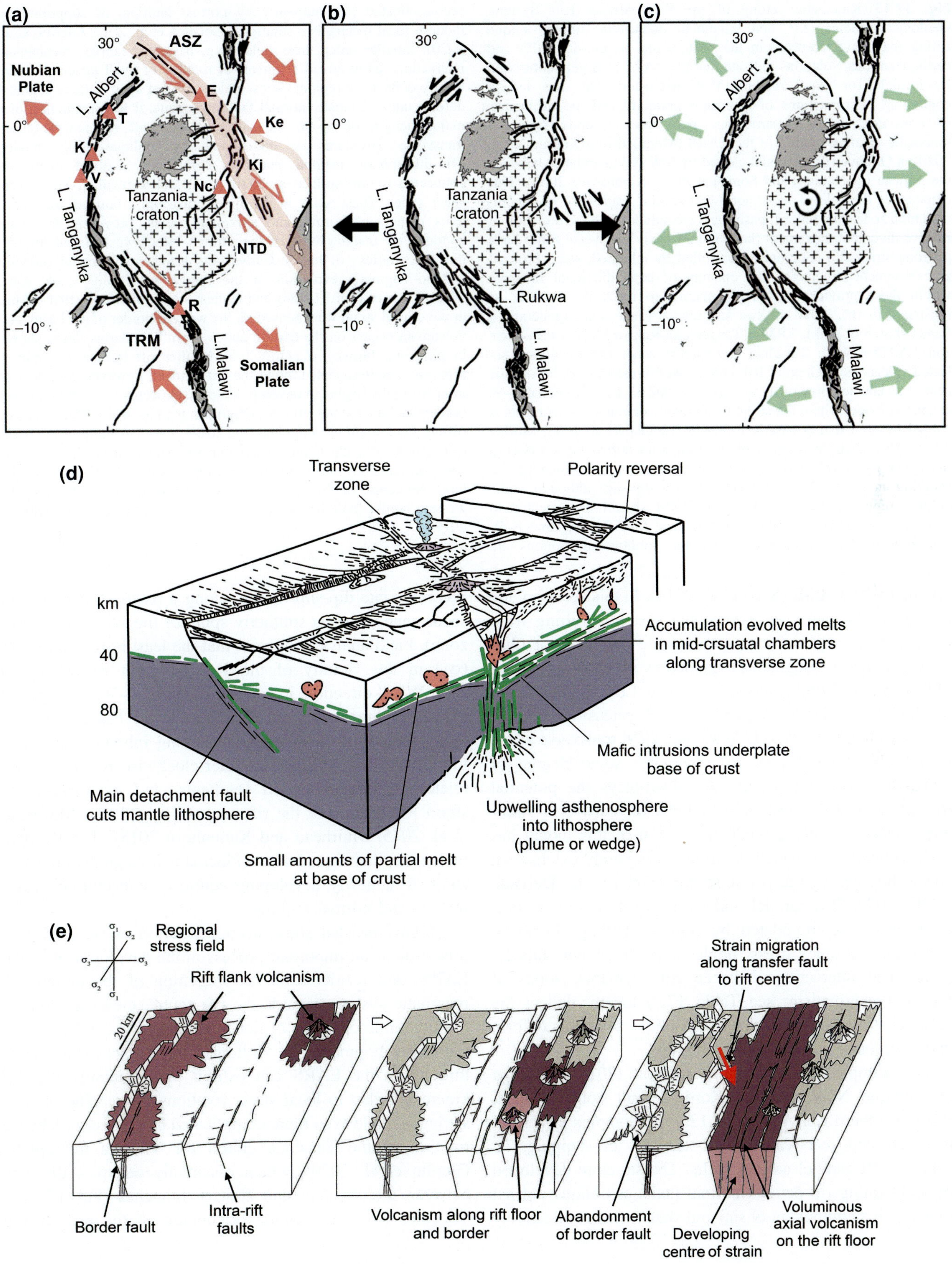
(a)
30°
ASZ
Nubian Plate
L. Albert
E
T
0°
Ke
K
V
Kj
Nc
Tanzania craton
L. Tanganyika
NTD
R
−10°
TRM
L.Malawi
Somalian Plate
(b)
30°
0°
L. Tanganyika
Tanzania craton
L. Rukwa
−10°
L. Malawi
(c)
30°
L. Albert
0°
L. Tanganyika
−10°
L. Malawi
(d)
Transverse zone
Polarity reversal
km
40
80
Accumulation evolved melts in mid-crsuatal chambers along transverse zone
Mafic intrusions underplate base of crust
Main detachment fault cuts mantle lithosphere
Upwelling asthenosphere into lithosphere (plume or wedge)
Small amounts of partial melt at base of crust
(e)
Regional stress field
Rift flank volcanism
20 km
Strain migration along transfer fault to rift centre
Border fault
Intra-rift faults
Volcanism along rift floor and border
Abandonment of border fault
Developing centre of strain
Voluminous axial volcanism on the rift floor

◀ **Fig. 17.13** Kinematic setting of the East African Rift System. Geological studies (e.g. field kinematic indicators) imply oblique rifting may have occurred in the past, while present-day GPS and earthquake focal solutions indicate that the EARS is largely controlled by far-field (or plate) regional E-W extension. This has led to conflicting interpretations of extension obliquity and the mechanics of transverse (transfer) structures. **a–c** Three recently proposed kinematic opening models of the EARS (Modified from Glerum et al. 2020). **a** Oblique rifting model caused by NW–SE divergence between the African (or Nubian) and Somalian Plates (Chorowicz 2005). This emphasises the role of large northwest-striking transverse or accommodation zones that link rift basins and are parallel with the direction of relative divergent movement (large red arrows). Transverse zones have a strong strike-slip component controlled by inherited, steep Proterozoic basement structures. The opening of large dilational fractures within these provides routes for magma to reach the surface, as illustrated in (**b**). ASW = Aswa and related shear zones (generalised trends shown in pink), TRM = Tanganyika-Rukwa-Malawi transverse zone, NTD = North Tanzania Divergence zone; red triangles = selected volcanoes in Gregory Rift and isolated Western branch volcanic centres as labelled previously in Figs. 17.9 and 17.12. **b** Regional E–W extension model where the strike of obliquely oriented rift segments is guided by pre-existing basement fabrics (Le Gall et al. 2008; Morley 2010). This leads to rotation of the minimum horizontal stress (σ_3) from the regional (plate tectonic) extension direction by oblique fabrics, resulting in faults that are predominantly dip-slip, although some oblique segments like the ASZ and TRM should be expected to experience a degree of strike–slip motion. **c** Modified regional E–W extension model, where rift-parallel and -orthogonal motions are accommodated predominantly by normal faulting, as supported by modern focal mechanism solutions (Delvaux et al. 2012; Glerum et al. 2020). Smaller order stress fields (e.g. basement fabrics, mechanical interactions between rift segments or rotation of the Tanzania Craton) locally deflect the regional stress field such that even oblique segment deform under a mostly normal faulting regime. **d** Reconstruction of a major (length of up to thousands of km) rift segment-bounding transverse structure as envisioned by oblique-rifting model. Mid-lithospheric magma chambers comprising evolved melts are connected to the surface by narrow and sub-vertical, lower crustal shear zones. These splay in the upper crust into a braided, discontinuous pattern of normal and strike-slip faults, providing the main conduits for magma to reach the surface. (Redrawn from an original diagram courtesy of Joe A. Cartwright). **e** Schematic model showing tectono-magmatic evolution of the transverse Kordjya fault zone, Magadi Basin. (Left) During the earliest stages of rifting (first 2 Ma), upper crustal strain was focused on the western border of the basin, and volcanism occurs on the eastern and western boundaries (lavas shown in pink and brown). (Centre) 2–5 Ma after rift initiation. Strain is primarily accommodated along the rift borders; however, a system of intra-rift faults begins to develop. Volcanism occurs along both the rift borders and the central rift axis. (Right) In the past few million years of rifting, fault strain has migrated (red arrow) into the centre of the basin in response to a developing magmatic segment beneath the rift. This was facilitated by the transverse Kordjya fault, which kinematically links the centre of the basin with the older locus of strain along the western border fault system. (Adapted from Muirhead and Kattenhorn 2018)

oriented WNW–ESE (Saria et al. 2014). In this scenario, the EARS is considered an extensional system resulting from relative E–W regional extension, where the strike of obliquely oriented rift segments is guided by pre-existing fabrics of Proterozoic mobile belts (Fig. 17.13b; Le Gall et al. 2008; Morley 2010). Under regional ~E–W extension, oblique structures like the NW–SE Aswa and TRM transverse zones would also be expected to experience some degree of strike-slip motion; however, this downplays the potential role of large-scale strike-slip motions on transverse structures. More recently a modified version of this model has been forwarded, in which deformation is purely extensional and orthogonal to the rift segments (Fig. 17.13c; Delvaux et al. 2012). Rift-parallel and -orthogonal motion is predominantly accommodated by normal faulting. Local or non-plate tectonic stress fields (e.g. basement fabrics, mechanical interactions between rift segments, zones of focussed magmatism; see Table 17.2) locally deflect the regional stress field such that even oblique segments deform under a mostly normal faulting regime.

For example, Muirhead and Kattenhorn (2018) show that the oblique NNW-trending Kordjya fault zone (Magadi Basin, southern Kenya; Fig. 17.12a) is a complex system of inherited and reactivated NNE-striking, left-stepping, en echelon, rift-parallel normal faults. The structure developed by reactivating rift-parallel normal faults as oblique-normal faults with a component of sinistral shear, which collectively accommodate dip-slip motion along a reactivated basement shear zone (likely a southerly splay of the Aswa transverse zone). Field analysis suggests that the least horizontal stress (σ_3) in the vicinity of the Kordjya fault rotated to an ENE-WSW direction, in contrast to the ESE–WNW regional extension direction found elsewhere in the basin. This was likely the result of an interplay among inherited basement fabrics and the localised counter-clockwise rotation of σ_3 related to interacting rift segments, and magma-assisted strain localisation at the rift centre (Fig. 17.13e; Muirhead et al. 2015; Muirhead and Kattenhorn 2018). The Kordjya fault zone, therefore, forms a kinematic linkage that transfers strain to a newly developing centre of concentrated magmatism and normal faulting.

Magma-assisted strain accommodation or weakening is now seen as an important process in the development of the EARS, and results from the migration of the underlying magmatic system towards the rift centre, creating a potentially voluminous flow of hot fluids that assist in transverse fault development. It follows that transverse faults throughout the EARS can exhibit predominantly dip-slip kinematics due to local stress rotations of σ_3. The general setting is made even more complicated by counter-clockwise rotation of the Tanzania craton (or Victoria microplate; Glerum et al. 2020), which essentially occupies the area between the Western and Eastern branches (Fig. 17.13c). Together with the oblique orientation of the pre-existing

Table 17.2 Summary of crustal stress field hierarchy

Stress field[1]	Cause	Length-scale (km)	Strain rates	Time scales	Examples	Control on magmatism/volcanism
Regional (*First-order*)	Inter-plate; far-field plate boundary forces	>>1000	mm yr^{-1}– cm yr^{-1}	Long term (100–10 Myr)	Mid-oceanic ridge "push"; subducting slab pull and rollback; shear traction at the base of the lithosphere; continental collisional resistance forces; orogenic gravitational collapse and escape	Localisation and orientation of magmatic/volcanic provinces; type of mantle magmatic sources/geochemical signatures (e.g. rift, arc and intraplate magmatism)
Regional (*Second-order*)	Intraplate; lithospheric and mantle processes	>1000–100	mm yr^{-1}– cm yr^{-1}	Medium term (>1 Myr–100 kyr)	Lithospheric flexure; mantle (thermal) buoyancy forces; rotation of cratonic block or microplate	Migration of magmatic/volcanic loci and changes of magmatic sources/geochemical signatures within existing volcanic provinces
Local	Crustal and magmatic processes	<100	mm yr^{-1}– m s^{-1}	Short term (>100 kyr–hr/min)	Pre-existing weaknesses and anisotropies; mechanical interactions between fault segments (e.g. transfer zones in rift basins); overpressures by magma and other fluids; volcano topography and gravitational loading	Localisation of individual volcanic edifices; intra-edifice volcanic structure and deformation; localisation of vents; geometry of magma reservoirs and caldera or other volcano-tectonic collapse structures; magma-external fluid interactions during storage and transport

[1]s-order regional and local stress fields have the ability to drive stress perturbations causing rotations of the larger scale regional stress field imposed by plate boundary forces

basement weaknesses, the rotation leads to predominantly normal faulting oblique to the regional and local extension direction. As also mentioned by many authors, the stress field may have changed over time, complicating the interpretation of fault–slip measurements in terms of regional extension direction.

Regarding magmatic typology, as an example, we may briefly examine volcanism in the vicinity of the Gregory Rift (Fig. 17.12). Flood basalt volcanism (Afar plume) commenced in the northern EARS at about 40 Ma, erupting thick sequences of predominantly tholeiitic basalt centred on the Ethiopian Plateau and Yemen (Chap. 16; Figs. 16.36a–d and 17.9a, b). Initiation of magmatism then migrated southwards, reaching the Gregory Rift in the early Miocene ca. 24–22 Ma (Fig. 17.12a; Bosworth 1987; Macdonald 2002; Macgregor 2015; Rooney 2020a). The trend of volcanic activity within the nascent rift valley was generally one of early alkali basalts, followed by more heterogeneous accumulations of phonolites, trachytes and basalts. On the margins of the rift, fissure-fed flood basalts and large central-type volcanic eruptions occurred. The initiation of off-axis volcanism on the western side of the rift was marked by eruption of nepheline-phonolite sequences with alkaline intrusions and carbonatites (e.g. Mt Elgon and Kisingiri volcanoes; Fig. 17.12a). This changed in the late Miocene with the eruption of extensive (flood) phonolitic lavas (13–11 Ma) along both the western and eastern rift shoulders associated with rifting and half-graben formation; (?) concomitant with intrusion of an asthenospheric plume into the lithospheric upper mantle (Smith and Mosley 1993). Volcanism has continued through to the Holocene with eruption of basalts (including extensive flood lavas) to phonolites and trachytes. This now appears to have been associated with at least two further pulses of basaltic volcanism followed by silicic volcanism (Rooney 2020a). This includes today where mildly peralkaline (comenditic) high silica rhyolitic dome fields (Eburru and Olkaria volcanic centres), and pantelleritic trachyte calderas dominate the central axis (e.g. Mts Suswa, Longonot and Menengai; Fig. 17.12a). During this period, off-axis volcanism shifted east with the eruption of basaltic to phonolitic-trachytic complexes (3–<1 Ma) forming Mts Kenya and Marsabit to the north, while further south in northern Tanzania an abrupt widening of the rift (NTD) occurred where it extends from iconic caldera of Ngorongoro in the west to Kilimanjaro in the east (Fig. 17.12b). Volcanoes forming the NTD (Fig. 17.13a) are broadly distributed across the rift valley floor and rift margins, while the locus of volcanism over this period progressed from the southwest to the north and east, with a main pulse occurring at about 2.3 Ma, possibly marking the inception of a main rifting event (Mana et al. 2015).

Magmatism in the Gregory Rift is closely linked with asthenospheric ascent (e.g. Macdonald 2002), where decompressional partial melting may occur at all levels within the plume, producing magmas that are tholeiitic at shallow level (as found in the Ethiopian Plateau), transitional at intermediate levels and alkaline at deeper levels. As pointed out by Huppert and Sparks (1988), the time scale for melting and crystallisation in basalt-induced crustal melting (10^2–10^3 yr) is very short compared to the lifespan of large silicic magma chambers (>10^6 yr), or the time scale of thermal relaxation of the continental crust (>10^7 yr). If basalt is repeatedly intruded into the same region during an episode of mantle plume activity, later intrusions are likely to encounter more silicic or felsic differentiated products of earlier intrusions. Repeated intrusion of mafic dykes and mixing, assimilation or partial remelting of such lithologies followed by fractional crystallisation could generate a whole range of magmas such as trachytes, phonolites, and peralkaline rhyolites from mafic magmas of alkaline to transitional affinity. This complexity may reflect the open-system behaviour of the central rift volcanoes, where mafic dikes periodically intrude compositionally evolved shallow magma chambers (e.g. Mt Suswa caldera; White et al. 2012). A further petrological highlight of the Gregory Rift is the active carbonatitic volcano Ol Doinyo Lengai in northern Tanzania. While best known for the eruption of natro-carbonatite lavas and pyroclastics (Fig. 7.25a, b; Table 7.2), the volcano is dominated by peralkaline silicate lavas (Dawson et al. 1994). For more detailed discussion of magmatic aspects of the Gregory (Kenya) Rift, see Macdonald (2002), Mana et al. (2015), Rooney (2020a, b).

In advanced stages of rifting, continued extension and axial graben subsidence should produce marked marine incursions and, in time, an overall (post-rift thermal subsidence) transgressive succession should develop (Fig. 17.8c). The basal (rift phase) parts of the succession will contain volcanics or thick volcaniclastics, or both, intercalated with non-marine successions dominated by alluvial fans and braided alluvial plain systems, with associated fan deltas where these systems interface with bodies of water, either lakes (as common in the EARS) or marine incursions. Rivers supplying debris from active basin-bounding faults and internal master fault scarps often feed into trunk river systems which then move sediment parallel with the central rift axis between fault blocks (Leeder and Gawthorpe 1987; Lambiase and Bosworth 1995; Leeder 1995; Gawthorpe and Leeder 2000). In marginal lacustrine or marine environments, clastic sediments may be reworked into various beach and shoreline facies and/or be transported into deeper water turbidite fan systems. Like with river systems, turbidite fans may be oriented parallel with the rift axis. Transfer faults can have pronounced effects on sedimentation and may be zones of severe erosion, as well as conduits guiding rivers and or subaqueous mass flows (Gawthorpe and Hurst 1993; Williams et al. 2020).

The basement in the initial stages of rift formation is clearly continental explaining, in some cases, the abundance of silicic volcanics, derived from sub-crustal melting heating of the lower crust. As the extensional tectonic regime develops, giving access to more mantle magmas, the thinning continental crust becomes injected with mafic magmas, becoming transitional in character and, finally, when spreading proper commences, it becomes oceanic in character. The proportion of volcanics (including sills) will slowly disappear upsequence as the rift widens into a narrow sea with a MOR, such as occurs with the Afar Depression at the northern end of the EARS (Fig. 16.36a, b). During this, volcanism should become more oceanic tholeiitic in character. The upsequence change from subaerial fluvial-alluvial and lacustrine successions, to marine sedimentation should be accompanied by fundamental lithological changes. Mixed basaltic to evolved silicic volcanics and volcaniclastics or basement-derived, immature, lithic sediments, or both, should become increasingly mature, more quartz-rich, and may give way to marine carbonates. A current model for the evolution of continental rifts based on the EARS is described in Fig. 17.14a–d (Muirhead et al. 2016). Failed rifts or aulacogenes could cease development at any stage (e.g. Burke 1978), meaning that they never progress to the break-up and passive margin stage.

For a series of time reconstruction (paleogeographic) maps of the EARS illustrating the progressive development of fault trends, subsidence, volcanism and topography, as well as the scale of the system, see Macgregor (2015), and Purcell (2018). For a current and comprehensive review of magmatism and magmatic evolution of the EARS, see the series of papers by Rooney (2017, 2020a, b, c and d, and references therein). For further discussion of the mechanics of rifting at the lithospheric-scale and the role of magma-assisted rifting in the EARS, see Kendall and Lithgow-Bertelloni (2016). For more general descriptions of the Gregory Rift and its active volcanoes, see Dawson (2008) and Scoon (2018).

Ancient Examples of Narrow Continental Rift Zones

In the rock record narrow continental rifts should be identified as a series of relatively confined (across strike), linked or semi-continuous fault-bounded depocentres, containing km-thick accumulations of sediment with variable volcanic components. Basal syn-rift sediments will be dominated by non-marine coarse clastic, basin margin fanglomerates, especially close to basin-bounding faults and consequently comprise high amounts of basement-derived clast populations. These may pass laterally and upsequence into coarse-grained, braid-dominated fluvial facies and lacustrine facies. Contemporaneous basaltic rocks should be commonly intercalated within the syn-rift successions, including

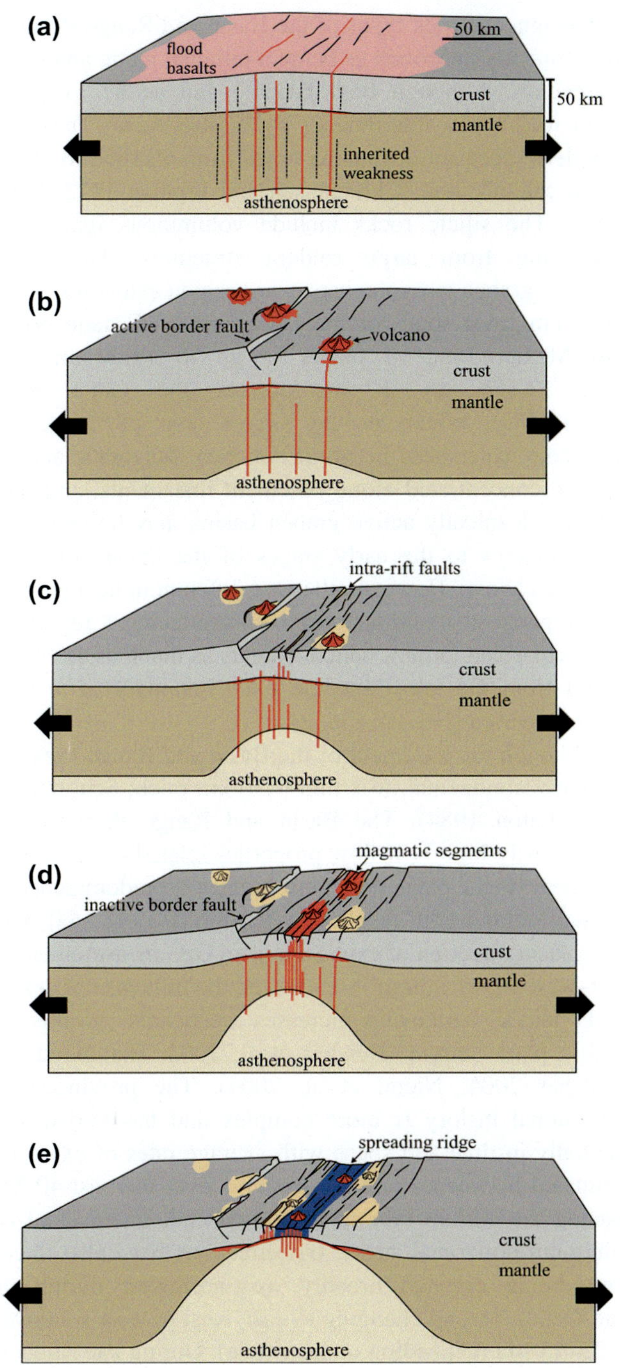

◀ **Fig. 17.14** "Magma-assisted" rifting model illustrating tectonic evolution of a continental rift based upon the East African Rift System (Adapted from Muirhead et al. 2016; also see Corti 2009, and Muirhead et al. 2015). **a** Extensive flood-basalt volcanism associated with mantle plume event initiates plateau uplift contemporaneously or shortly after. Note presence of zone of inherited weakness that localises deformation in subsequent rifting phases. **b** Within 10 Ma, the lithosphere thins through brittle faulting and ductile deformation. This gives rise to high-angle, deep crustal-penetrating, normal faults leading to the development of large half grabens, such as those found today in the Western branch of the EARS. Asthenosphere begins to rise in response to the thinning lithosphere, producing magmas that intrude lithosphere as dykes, or accrete along the crust-mantle boundary (see Fig. 17.13b). Magmatism that reaches the surface primarily forms volcanoes along the rift margins and transverse zones initiated on pre-existing, deep crustal shear zones. **c** After about 10 Ma, brittle strain begins to focus on smaller intra-rift faults along the rift axis. Volcanic activity is controlled by stresses related to the mechanical interaction between rift segments as well as inherited crustal structures, which together also control the orientation of oblique faults within transfer zones. Ongoing lithospheric thinning leads to further upwelling of asthenosphere and generation of greater melt volumes (which may or may not erupt at the surface). Dyke intrusions fed from these melt sources heat and weaken the lithosphere reducing its yield stress by an order of magnitude (Buck 2004). **d** Continued injection of magma along the rift axis results in the development of narrow zones of intense magmatism within the rift centre, termed magmatic segments, as now observed in the Gregory Rift. Strain is localised into these magmatic segments, reducing mechanical interactions between basin segments in transfer zones, and concentrating volcanic activity. Subsurface dyke emplacement accommodates further extension, rendering border faults inactive. As soon as volcano-tectonic activity is localised within magmatic segments, feedback between deformation and magmatism develops: the thinned lithosphere is strongly modified by the extensive magma intrusion and extension is facilitated and accommodated by a combination of magmatic intrusion, dyking and faulting. **e** After 5–10 Ma, following further thinning, heating and magma intrusion, the tectonically and magmatically thinned lithosphere may rupture in the heavily intruded zones. Nascent spreading ridges, as seen in the Afar Depression, form as magma injection along the rift axis creates new oceanic lithosphere (dark blue) as continental crust is replaced by mafic igneous material

subvolcanic sills and dykes. As extension continues (increasing β factors), extensional faulting will become more pervasive, and accompany the eruption of more evolved silicic magmas and increased mafic dyke intrusion. This should occur together with an increasingly complex provenance of the basin-fill succession, as more and more reworked volcanic detritus enters into the depositional cycle.

In high-strain rifts extensive crustal thinning and mantle underplating leads to the development of mid-lithospheric zones of partial melting and remelting, leading to highly variable evolved magmas and volcanic centres (Fig. 17.13d).

From a broader regional perspective, the central rift depocentre and main volcanic axis may be expected to be associated or interface with slightly younger or broadly contemporaneous flood basalt lava successions, as well as the successions of large distinctly alkaline central vent or stratovolcanoes as in the EARS. As also in the EARS, there is the possibility that the central rift depocentre may interface with a rifted continental or passive margin as rifting progresses towards break-up, as evidenced by marine incursions and seafloor spreading. It is lastly important to point out that in ancient terrains, major tectonic basin inversion can cause thrusting and tectonic extrusion of the rift fill. Listric normal faults and detachment faults are ideal weaknesses available for reactivation as reverse faults and thrusts during compressional deformation.

One of the most studied older, narrow rifts systems, as a result of its prolific hydrocarbon production, is the North Sea Basin. Rifting activity in the North Sea commenced during the earliest Triassic, peaked during the late Jurassic-earliest Cretaceous, and terminated during the Paleocene. During the early-middle Jurassic volcanism and a rift system developed in the central North Sea area where flood basaltic lavas were extruded. These were coincident with mantle doming in the middle of the North Sea where the Iapetus Suture intersected the Tornquist-Teisseyre Fault System (Cartwright 1987, 1992). Doming was centred at the junction of the Viking Graben, Central Graben and Moray Firth Basin, where the Forties basaltic volcanic centre erupted. Discovered in 1970, the Forties field (5 Bbbl of oil in place) was the first major and largest oil find in the UK sector of the North Sea (Wills 1991). Hydrocarbons are reservoired in extensive Paleocene submarine fan sandstones, while the trap is a four-way dip-closed anticline, formed by drape and compaction over the structural high of middle Jurassic basalts.

From a geophysical perspective, the one of the most prominent features on gravity and magnetic maps of North America is the Keweenawan or Midcontinent Rift, a 3,000 km long band of buried igneous rocks and associated sediments extending from Lake Superior (Stein et al. 2015). Gravity and magnetic anomalies and modelling of seismic reflection data indicating thick successions of mafic rocks, mostly flood basalts but also major subvolcanic intrusions exemplified by the Duluth Complex, which is one of the largest gabbroic sills on Earth (some 16 km thick and covering nearly 5,000 km^2). The Midcontinent Rift is commonly viewed as a type example of a failed rift that formed and expired within a continental interior, far from its margins. However, recent studies propose that it was associated with a successful seafloor spreading event (Stein et al. 2014).

17.2.3.2 Broad Continental Rift Provinces

The Basin and Range Province is dominated by an expansive terrain of graben formation (Fig. 17.15a-c), normal faulting (Fig. 17.15c) and volcanism (Fig. 17.15c) with a great diversity of petrological characteristics. The Basin and Range Province is located in an extraordinarily complex tectonic setting, lying east of the western North American subduction zones that are separated by the transcurrent San Andreas Fault Zone, which links separated oceanic spreading ridges in the northeast Pacific Ocean (Juan de Fuca spreading ridge) and the northern end of the East Pacific spreading ridge in the Gulf of California (Fig. 17.15d, e; Plate 17.1).

The igneous rock types of the Basin and Range Province are frequently described as being a bimodal mafic and silicic association. Although both basaltic and acidic rocks are major rock types, intermediate rocks also occur. There is a complete spectrum of magma types from alkaline to tholeiitic to calc-alkaline (Christiansen and Lipman 1972; Eaton 1982). The silicic rocks include voluminous ignimbrites originating from large caldera structures. The whole volcanic-graben terrain covers a vast region extending nearly 4,000 km from south of the United States-Canada border into Mexico, and in places is up to 1,000 km wide (Fig. 17.15d, e). It incorporates the Great Basin (Fig. 17.15e), which is the largest area of contiguous endorheic watersheds in North America. Magmatic activity may be concentrated along basement lineaments, and individual volcanically active graben basins may be similar in characteristics to the early stages of the linear rift zones described above. However, they are different in having many counterparts on a regional scale, so producing a regionally broad rift-basin terrain, which extends as much as 1,500 km inland from the late Cenozoic plate boundary of western North America (Fig. 17.15d, e).

Although the basement of the Basin and Range Province is clearly continental, it is thin (<30 km thick; Scholz et al. 1971; Eaton 1984). The Basin and Range Province has anomalously high heat flow properties related to its thinned basement. Fault plane solutions suggest a predominance of normal faulting with resultant extension (Fig. 17.16a), with the present direction of extension from GPS measurements is east–west at 3–4 mm yr^{-1}. However, the influence of dextral (right lateral) strike-slip increases westwards towards the Pacific plate margin (Bennett et al. 2003; Hammond and Thatcher 2004; Niemi et al. 2004). The province-wide extensional history is more complex and has varied substantially in time and space with average rates of extension estimated between 4 and 20 mm yr^{-1} over the past 40 Ma, placing the current rate at the lowest end. Extensional (or minimum horizontal stress, σ_3) directions have also fluctuated, being oriented broadly west-northwest during the mid-Cenozoic, and changing to east–west to west-southwest at 25–15 Ma (depending on longitude). During late Miocene times, least principal stress rotated back clockwise to west-northwest (Wernicke 1992; McQuarrie and Wernicke 2005).

Concepts as to origins of the Basin and Range Province have evolved over many years and include subduction of an active MOR under the western North American continent (Menard 1964) and the development of a back-arc continental basin system analogous to oceanic back-arc basins (Scholz et al. 1971; Eaton 1984). Zoback et al. (1981) and Eaton (1984) expanded on the latter hypothesis, suggesting

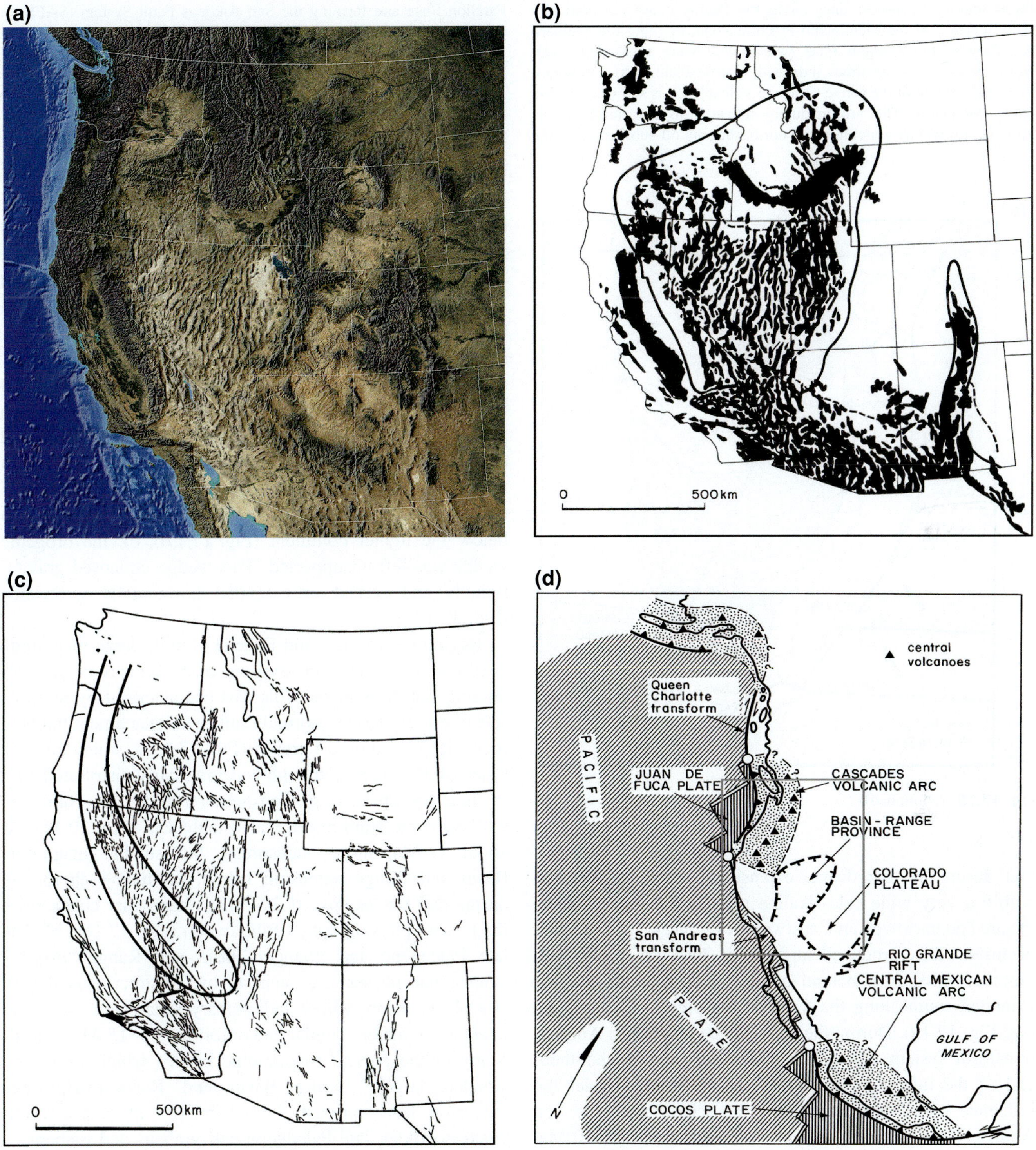

Fig. 17.15 Tectonic setting of the Basin and Range Province, western United States. **a** Relief image of the western United States at same size as (**b**) and (**c**) (Face of the Earth and Relief image; *Source* UNAVO). **b** Distribution of grabens and other structural basins containing Miocene, Pliocene and Quaternary sedimentary as well as Quaternary volcanic rocks. Bold line shows full extent of Basin and Range Province, extending to the Rio Grande Rift in the southeast. Note the wide region of graben development in the central Great Basin region as also highlighted in the relief image. (From Cas and Wright 1987, After Eaton 1982). **c** Distribution of Miocene and younger normal faults with known or suspected Quaternary movement. The extent of the Miocene and younger Cascades continental margin volcanic arc is shown by the bold line, the modern arc being confined to the northern third of the area shown. Strike-slip faults in southern and coastal California are also shown. (From Cas and Wright 1987, After Eaton 1984). **d** Broader regional plate tectonic context; box shows area illustrated in (**e**) (From Cas and Wright 1987, After Dickinson 1979). **e** Geotectonic map showing the main volcanic and tectonic features adjoining the Great

(continued)

Basin region. East–west lines along the Pacific coast schematically show location of the Mendocino Fracture Zone, marking the southern boundary of the Farallon Plate (which was being subducted under North America) at the times indicated; Pacific Plate occurs to south. Maximum extent of the active Cascades volcanic arc is shown at 21 Ma (early Miocene). The arc gradually retreated to the north as the Mendocino fracture zone migrated north, shutting off subduction of the Farallon Plate and forming the San Andreas Fault System (SAFS) or transform boundary to the southwest (see also Fig. 17.19b). The position of the Yellowstone "hot spot" is shown at about 2 Ma. MRJ = Mendocino triple junction; JFP = Juan de Fuca Plate; CRB = Columbia River Basalts; MC = McDermitt Caldera; NNR = North Nevada Rift. (From John 2001)

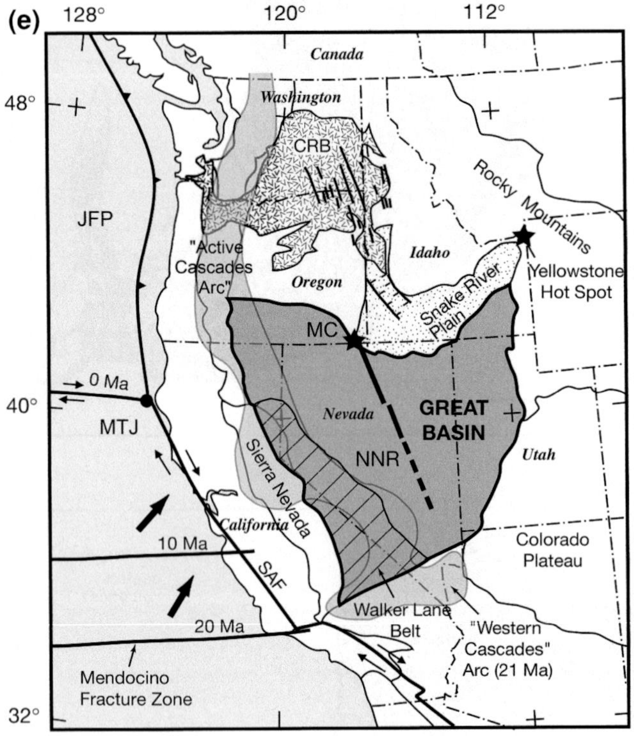

Fig. 17.15 (continued)

that during the Oligocene, extensional tectonics occurred within a very wide calc-alkaline arc. By the mid-Miocene the arc had narrowed and, eastwards of it, back-arc extension occurred. This produced an ensialic back-arc basin with bimodal volcanism, behind a linear calc-alkaline arc that extended south along the trend of the present Cascade arc (see Fig. 17.19). During the early Miocene, the San Andreas transform system propagated northwards and the southern end of the magmatic arc retreated northwards. Under the influence of transcurrent faulting associated with the San Andreas system, and as indicated above at about 12–10 Ma, extension became oriented west-northwest (Zoback et al. 1981). This extension was described as "back-transform extension" (Eaton 1984).

More recently the Basin and Range Province has been interpreted in terms of extensional collapse of the North American Cordilleran foreland fold and thrust belt (e.g. Coney 1987; Wernicke 1992; Constenius 1996; Sonder and Jones 1999; Fig. 17.17). As described above, late Mesozoic-early Cenozoic convergence resulted in eastward subduction of the Farallon oceanic plate at a shallow angle below the continental margin of North America (Fig. 17.17a). Intraplate shortening during Laramide compression created an orogenic wedge (i.e. folds and thrust belt). A rapid decrease in the rate of Farallon-North America plate convergence and/or steepening of the subducted oceanic slab at ca. 50 Ma resulted in a large reduction in east–west horizontal compressive stress in the Cordillera. As the dense oceanic plate sank into the mantle to form a slab below the lighter continent (Fig. 17.17b, c), the orogenic wedge was left unsupported. This wedge collapsed gravitationally and spread or extended horizontally westwards (Fig. 17.17a).

Beginning in the mid-Eocene, rapid, large-magnitude (>100–400%) extension resulting from rollback of the subducting slab took place. This was characterised by multiple sets of closely spaced normal faults and detachment faults in more deeply exposed terranes (Fig. 17.16a), which affected much of the region, although was irregularly distributed both in time and space (e.g. Constenius 1996). This continued until dynamic equilibrium was established at ca. 20 Ma. At about 17 Ma, more widespread, but smaller magnitude Basin and Range extension began, producing alternating basins and ranges spaced 20–50 km apart that characterise the present physiography of the region (Fig. 17.16a, b). The late Paleogene and younger Basin and Range "*proper*" extensional phases are seen as two distinct episodes of crustal extension, whose sedimentary successions are separated by a major angular unconformity (3–15 Ma hiatus). Many of the present topographic basins, which were constructed largely during Basin and Range-to-Holocene extension, have variously buried, broken up, and/or reactivated the older, late Paleogene half grabens and grabens.

The stimulus for this "new thinking" in the region was the discovery of large-scale, shallow-dipping normal faults (or detachment faults), that often juxtapose upper crustal rocks on metamorphic tectonites from the middle crust forming metamorphic core complexes (e.g. Wernicke and Burchfiel

(a)

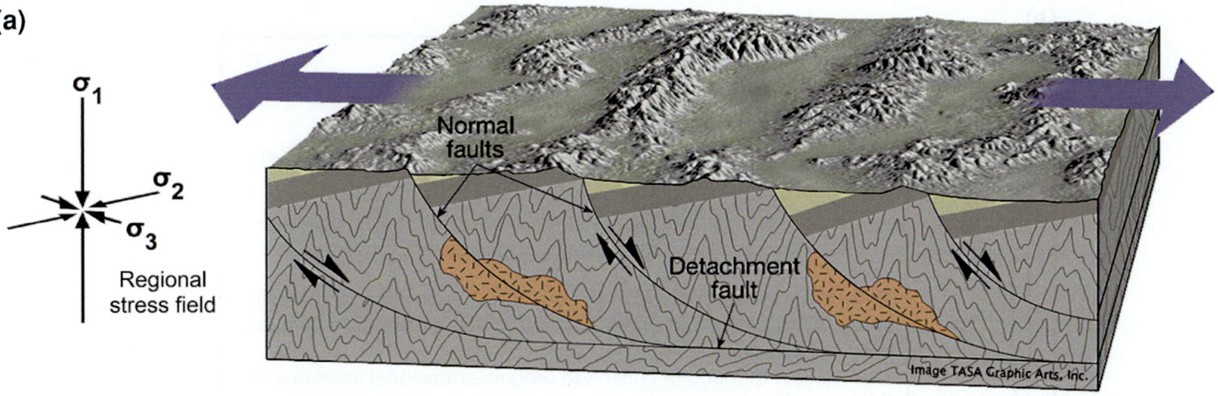

(b)

Fig. 17.16 Basin and Range topography, and general structural model. **a** Basin and Range extension takes place on large, gently dipping detachment faults on which normal faulted blocks, tilt or slide along. Instead of the whole block subsiding, only one side if the block slips along the detachment fault, tilting towards the fault plane so creating a series of mountains (ranges) and valleys (basins). Blocks are rotated in one direction at their tops due to the motion of their bottoms along the main detachment fault, with the resulting basin (half-graben) having one steep side and the other more gradual. (*Source* TASA Graphic Arts, Idaho State University). **b** Typical basin and range topography consisting of parallel mountains and valleys (*Source* Stanford University Department of Earth Sciences, https://tectonics.stanford.edu/extensional-tectonics-basin-and-range)

1982; Wernicke 1995; Fig. 17.18a–d). The detachment model allows for large-magnitude intra-continental extension to take place along a major shear zone (or main detachment). This consists of a flat lying upper crustal segment, and a steeper lower crustal part segment, part of which continues into the mantle. The upper part undergoes brittle deformation and results in rifting of the upper crust and basin formation. As depth increases the deformation mechanism alters into brittle-ductile and ductile processes, and faults pass into shear zones and mylonites. Because crustal thinning is asymmetric, upwelling of the asthenosphere occurs where the main detachment cuts down into the mantle and not the centre of basin. This has the effect of uplifting and even "breaking" the main detachment at surface, culminating in exposure of a metamorphic core complex (Fig. 17.18c). Transverse, accommodation or transfer zones and structures, some of which clearly reactivate basement structures, accompanied rift segmentation, as well as controlled igneous and hydrothermal activity (e.g. Faulds and Stewart 1998).

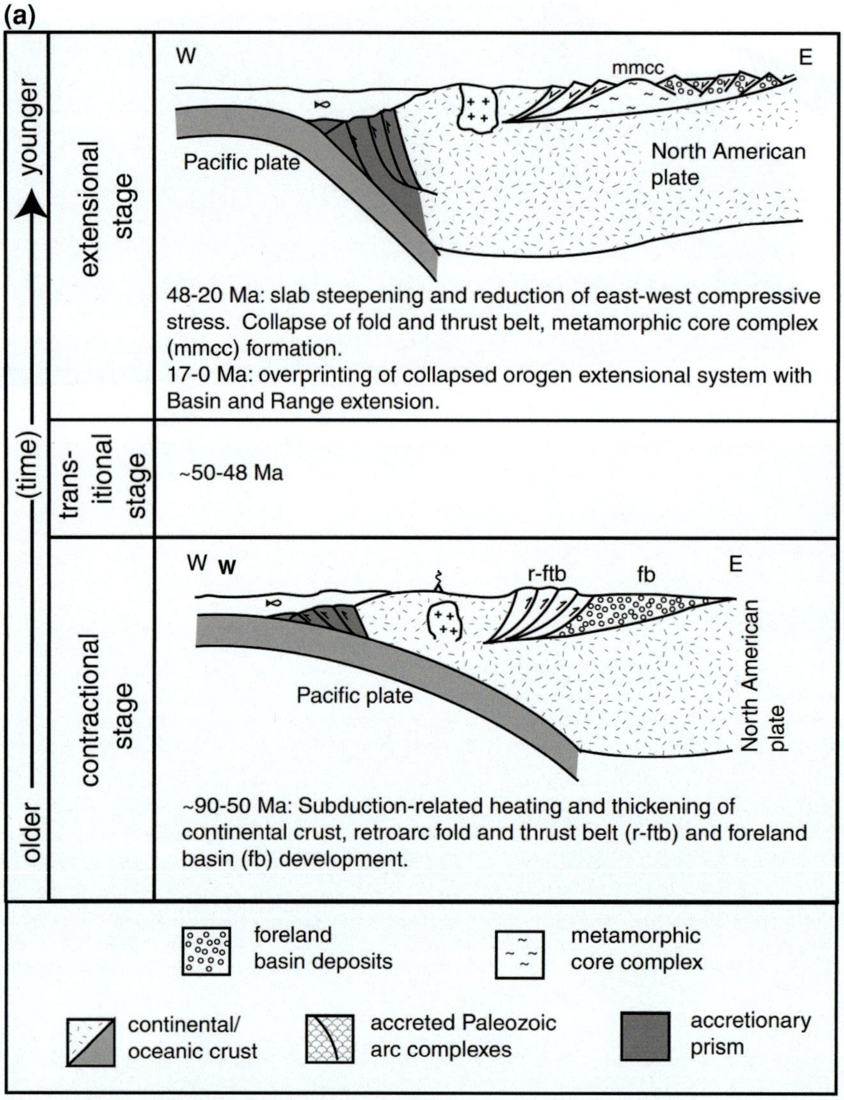
(a)
younger
(time)
older
extensional stage
W
E
mmcc
Pacific plate
North American plate
48-20 Ma: slab steepening and reduction of east-west compressive stress. Collapse of fold and thrust belt, metamorphic core complex (mmcc) formation.
17-0 Ma: overprinting of collapsed orogen extensional system with Basin and Range extension.
trans-itional stage
~50-48 Ma
contractional stage
W
E
r-ftb
fb
Pacific plate
North American plate
~90-50 Ma: Subduction-related heating and thickening of continental crust, retroarc fold and thrust belt (r-ftb) and foreland basin (fb) development.
foreland basin deposits
metamorphic core complex
continental/ oceanic crust
accreted Paleozoic arc complexes
accretionary prism

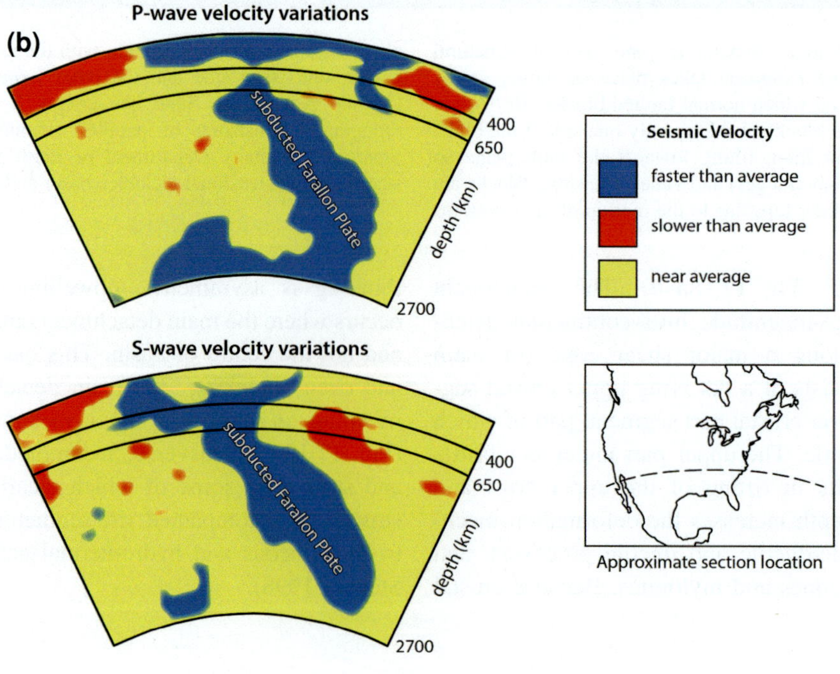
P-wave velocity variations
(b)
subducted Farallon Plate
400
650
depth (km)
2700
Seismic Velocity
faster than average
slower than average
near average
S-wave velocity variations
subducted Farallon Plate
400
650
depth (km)
2700
Approximate section location

◄ **Fig. 17.17** Features of the Western Cordillera, USA. **a** Simplified geological model illustrating extensional orogenic collapse of the western Cordilleran margin of the United States accompanying slab roll back of the Pacific (Farallon) Plate through the Cenozoic. As well as hyperextension forming metamorphic complexes, this was accompanied by copious volcanic and intrusive activity. As described in the text, this took place in two main extensional episodes, which were somewhat different in space, magnitude and magmatism. Only the younger of which is strictly regarded as Basin and Range which was centred on the Great Basin. (Adapted from Johnson et al. 2001). **b** Seismic tomographic cross-section showing P- and S-wave velocity variations in the mantle across southern North America. The images show variations in seismic velocity relative to the global mean at depths from the surface to the core-mantle boundary. Blues indicate faster than average and reds slower than average seismic velocity. The large tabular blue anomaly that crosses the entire lower mantle is interpreted as the descending Farallon Plate. (Adapted from Grand et al. 1997; *Source* Wikimedia through Creative Commons Attribution-Share Alike 4.0 International licence). **c** 3D visualisation of the remnants of the Farallon Plate in the mantle below North America. Note the intricate geometry reflecting ca. 100 Ma of subduction, involving roll back or steepening, tearing, buckling and folding of the slab, while more current imaging suggest remnants of at least two additional oceanic plates may be present (Sigloch and Mihalynuk 2013). (Image by Stuart A. Snodgrass, Hans-Peter Bunge; *Source* NASA/Goddard Space Flight Centre Scientific Visualisation Studio)

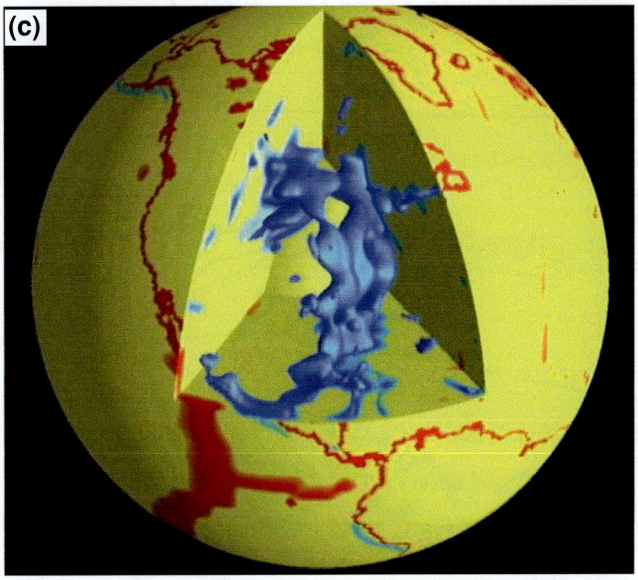

Fig. 17.17 (continued)

Extensional basin and metamorphic core complex formation were also broadly coeval with regional intermediate to silicic magmatism that swept through the Cordillera in Cenozoic time (Coney 1987; Wernicke et al. 1987; Wernicke 1992; Constenius 1996; John 2001; Dickinson 2006; Fig. 17.19a, b). While in general this correlation exists, the migratory pattern of volcanism is complex and there is still considerable debate on the precise nature of such relationships (e.g. Wernicke 1992; Sonder and Jones 1999; Dickinson 2006; Busby 2012; Putirka and Platt 2012). Overall, high-K calc-alkaline magmatism during the mid-Cenozoic extension (Eocene-early Miocene) appears to have commenced at the northern and southern (Sierra Madre Occidental) extremes of the Basin and Range, migrating towards one another in "waves" along the axis of the province. In the Pacific Northwest and Great Basin southerly migrating volcanic fronts mirrored slab rollback (Fig. 17.19a). Magmatism was associated with major episodes of ignimbrite flare up and the formation of large caldera volcanoes, particularly in the southwestern United States and Sierra Madre Occidental (Best et al. 2013; Chap. 16). The eruption of several large-volume silicic ignimbrites in the Sierra Madre Occidental has now been shown to be controlled by major rift or graben faults, which must have extended and cut batholith-sized magma chambers at depth (Aguirre-Diaz and Labarthe-Hernández 2003; Aguirre-Diaz et al. 2008).

Copious andesitic-silicic magmatic activity during this first extensional phase is generally linked to partial melting of overthickened, post-compressional (Laramide) lithosphere (Sonder et al. 1987). More recent studies suggest these magmas were formed by partial mixing of mantle-derived basalt with crustal melts. One model that explains these observations is removal of the subducted Farallon Plate by tearing of the slab near both the Canadian and Mexican borders (Fig. 17.19a; Humphreys 1995; Liu and Stegman 2011). This led to upwelling of asthenospheric

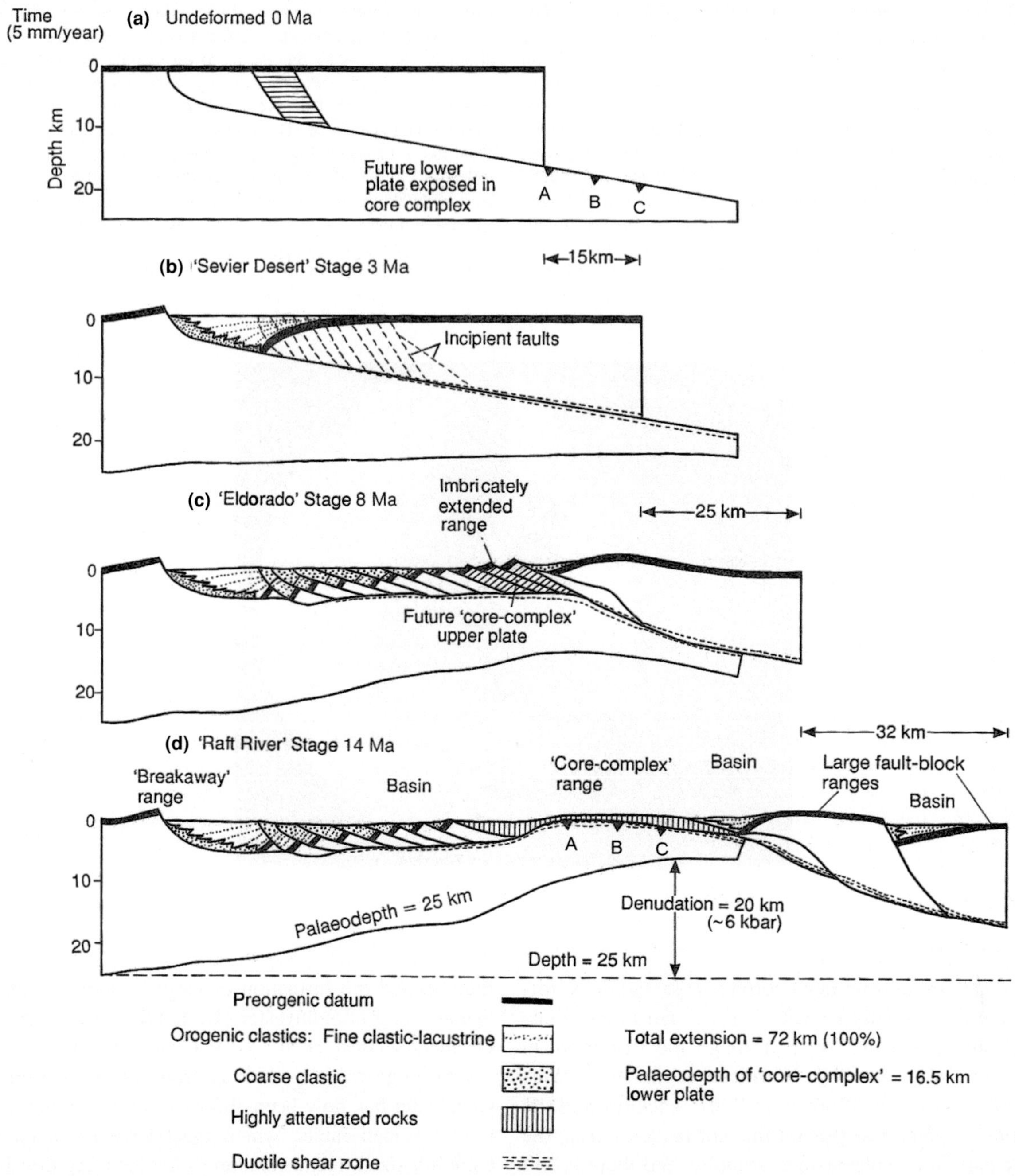

Fig. 17.18 Wernicke-model of detachment extensional shear deformation, and origin of metamorphic complexes in broad continental rift provinces, showing **a**, **b** initial failure of thick crustal plates, to **c**, **d** long-term extensional segmentation and crustal thinning. Mid-crustal rocks in the hanging wall may initially pass through greenschist or amphibolite facies in the ductile shear zone. As a result of uplift and tectonic erosion the detachment zone is progressively reworked at lower temperatures and brittle deformation overprints the ductile deformation. (From Allen and Allen 1990, After Wernicke 1995)

mantle behind the trailing edge of the delaminating lithospheric slab, decompressional melting, and formation of basaltic melts that partially mixed with lower crustal melts (Colgan et al. 2011). The scenario here is similar to that which has been described during flat slab subduction along the South American convergent plate margin, allowing for rapid subduction of slab tears (Schepers et al. 2011; Sect. 17.5.4). Concurrent with initiation of high-K calc-alkaline magmatism was formation of large sediment-hosted (Carlin-type) gold deposits that occurred

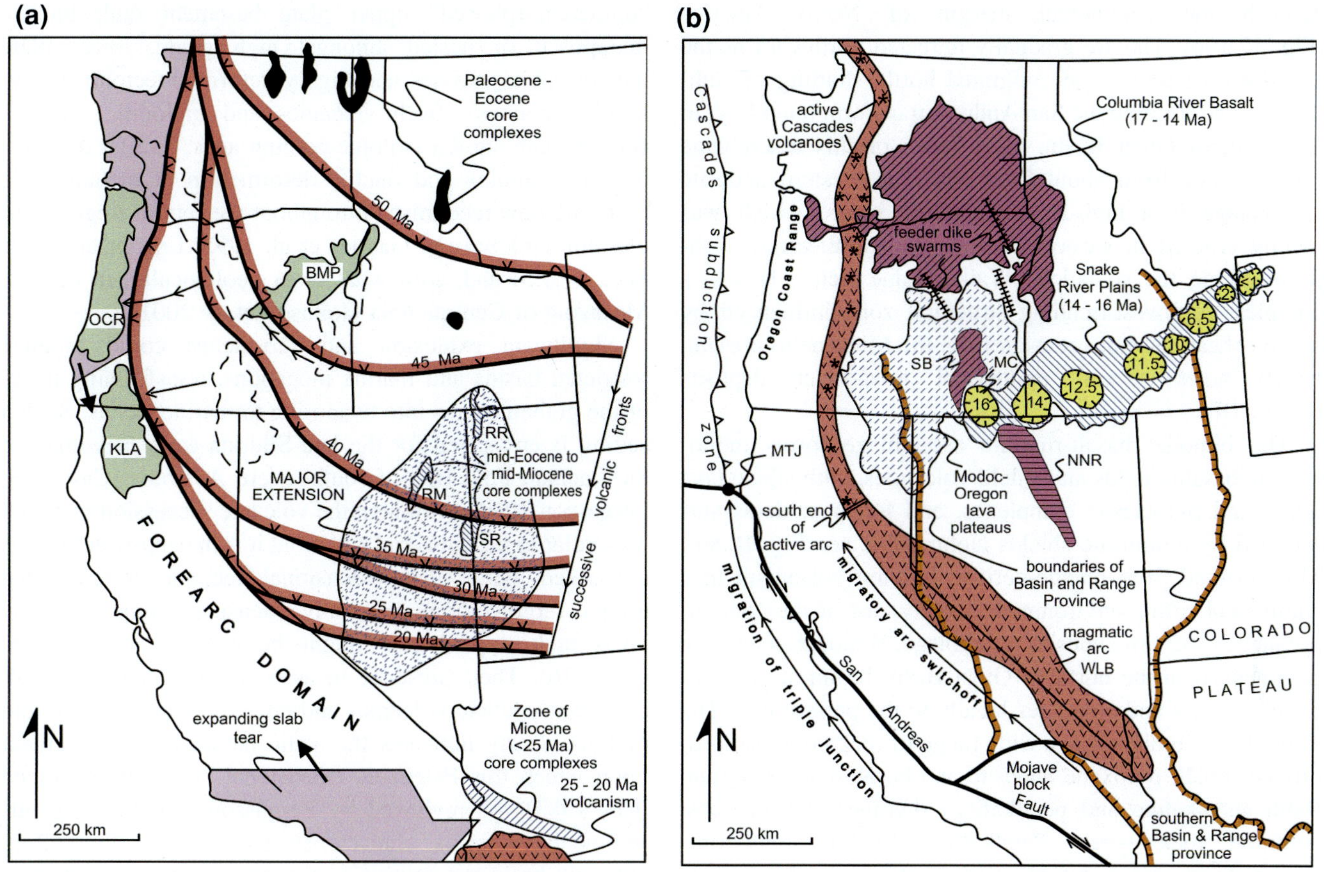

Fig. 17.19 Cenozoic paleotectonic development of the western United States during extensional collapse of the Cordilleran Orogen (From Dickinson 2006). **a** Mid-Cenozoic (Eocene-early Miocene) showing volcanic fronts and metamorphic core complex ages. Curved arrows in Pacific Northwest indicate Paleogene rotational displacements of Oregon-Washington Coast Range (OCR) and Blue Mountains Province (BMP), and translational displacement of Klamath Mountains block (KLA), with original positions indicated by dashed outlines. Selected Great Basin core complexes are: RM = Ruby Mountain; RR = Raft River; SR = Snake Range. Extent of slab tear is adapted from Humphreys (1995). **b** Middle Miocene-present (post-17 Ma) showing approximate area of Basin and Range extension and magmatism (within brown hachured line). Also indicated is extent of Miocene and younger coastal Cascades magmatic arc. Post-arc volcanic centres active near California-Nevada state line (Walker Lane Belt = WLB) are not separated from within span of earlier arc volcanism; these formed as Mendocino triple junction (MTJ) migrated northwards. Chain of Snake River Plain calderas indicating track of Yellowstone (Y) hot spot was initiated with formation of the McDermitt caldera (MC) at 16.5 Ma. Mid-Miocene Columbia River dyke swarms (west-to-east) Monument, Cornucopia and Chief Joseph). Modoc-Oregon lava plateaus are mainly Pliocene age at surface. Other features: SB = Steens Basalt (coeval with the Columbia River Flood Basalt Province); NNR = Northern Nevada rift (geophysical extent)

between 40 and 36 Ma; although strong temporal and spatial relationships exist, the role of magmatism in the formation of Carlin-type deposits is still much debated (Chap. 18).

By the time magmatism arrived in the central Basin and Range (southern edge of the Great Basin) in the middle Miocene (ca. 20 Ma; Fig. 17.19a), major changes in the extensional architecture took place. This developed into a broad region of tilted fault blocks accommodating moderate amounts of extension (Fig. 17.19b). During this period, coinciding with growth of the North America transform boundary, a significant portion of northwest-southeast right-lateral shear developed closer to the Pacific margin, which persists today (Hammond and Thatcher 2004; McQuarrie and Wernicke 2005). By contrast mid-Miocene volcanism was also accompanied by the widespread appearance of basaltic volcanics, consistent with significant crustal thinning during the earlier extension, including the Columbia River flood basalt (17–14 Ma) and Snake River (16–14 Ma) plains basalt plateaus (Fig. 17.19b).

Magmatically the Great Basin, however, is more complicated, and comprises two distinct assemblages (John 2001), termed: (1) western andesite and (2) bimodal basalt-rhyolite. The western andesite assemblage is composed predominantly of lava flows of mostly andesite-dacite (with minor rhyolite and basalt) composition erupted as stratovolcanoes, dome fields and subvolcanic plutons. This assemblage is part of the remnant continental margin arc that formed in response to subduction of the Farallon plate beneath the continental margin of North America (Fig. 17.19b). The arc gradually retreated northward as the Mendocino fracture zone migrated north, shutting off subduction and forming the San Andreas transform boundary. In the northern Great Basin, most rocks of this assemblage range in age from about 22–4 Ma. The western andesite assemblage is a high-K calc-alkaline series, which was mainly erupted in western Nevada and northeastern California along the so-called "Walker Lane Belt". This is a complex northwest-trending structural zone influenced by dextral shear and trans-tension from the Miocene to present, which hosts several epithermal gold-silver deposits (Chap. 18).

The bimodal basalt-rhyolite assemblage forms monogenetic basaltic fields and shield volcanoes, with silicic lava dome and pyroclastic complexes, and larger calderas surrounded by ignimbrite shields (John 2001; see Fig. 16.80). This resulted from mantle-derived basaltic volcanism in a continental rifting environment, largely east of the contemporaneous western andesite assemblage. Bimodal volcanism commenced in the northern Great Basin began at approximately 17 Ma and continues locally to the present day. The bimodal assemblage has a wide range of rock compositions; olivine basalt, pyroxene andesite and basaltic andesite, and (both subalkaline and peralkaline) rhyolites are the most common rock types. Peralkaline rhyolite ignimbrites (comendites), are a notable part of this assemblage (e.g. McDermitt caldera, Henry et al. 2017). Another important complication has been impingement of the Yellowstone hot spot (mantle plume) on the crust which occurred simultaneously at about 16.5 Ma along the Nevada-Oregon border, near McDermitt caldera (Fig. 17.19b). This has since tracked east-northeast to its present position below the Yellowstone volcanic plateau (e.g. Parsons et al. 1998; Pierce and Morgan 2009). Continental extension over a mantle plume may have accelerated normal faulting and caldera formation, as well the development of shallow geothermal systems and epithermal gold mineralisation in the region (Hames et al. 2009; Chap. 18).

Ancient Examples of Broad Continental Rift Provinces

In broad, highly extended terrains such as the Basin and Range Province, extensive sediment successions are associated and interstratified with volcanic successions. These are typical continental sediments (fluvial, alluvial fan and lacustrine), frequently developed within (half) graben basins, and often associated with volcanic landforms and successions. In the rock record, the principal criteria to look for are the widespread (hundreds of km width) occurrence of bimodal volcanics, associated co-magmatic intrusions and multiple, localised, structurally controlled sedimentary basins. These basins should ideally comprise highly rotated "unmetamorphosed" upper plate basement fault blocks, juxtaposed or nested amongst high grade, lower plate metamorphic gneisses and highly deformed tectonites (mylonitic gneisses). Both volcanics and mylonitic gneisses should share similar isotopic cooling ages. Similar domains of core complex and ductile deformation of metamorphic rocks are now recognised throughout the central Aegean and western Turkey (e.g. Gautier et al. 1993, Lips et al. 2001; Sect. 17.3.5) and, somewhat older geologically, in the late Mesozoic of Central Asia (Johnson et al. 2001).

Significant extension and subsidence could produce restricted basins and marine incursions, transforming many of the graben basins into evaporitic to marine basins. Such a setting is envisaged for the late Silurian-early Devonian of the Lachlan Fold Belt of southeastern Australia (Cas 1983; Fergusson 2010). Although the volcanic succession may still be similar to the subaerial situation, it is in deep-water basins associated with such extensional tectonic terrains that unique, large volume, regionally extensive, submarine silicic lavas and cryptodomes might be expected (Cas 1978; Chap. 16). Their presence in association with widespread, dominantly silicic volcanics and contemporaneous granitoid plutons clearly indicates the sialic nature of the basement upon which this Paleozoic extensional province developed (Cas 1983). As reviewed by Fergusson (2010), numerous sedimentary basins characterise the Lachlan Fold Belt, although basin recognition is constrained by later compressional deformations which have obscured original basin geometries, with uplift having caused significant erosion. This is now generally interpreted as having a continental back-arc setting with a convergent margin to the east, with extensional and magmatic activity driven by rollback of the subduction slab. Another clearly bimodal Late Devonian extensional to trans-tensional province then formed after a mid-Devonian compressional event (Cas et al. 2001; Giordano and Cas 2001; Chaps. 8, 16). To date, coeval metamorphic core complexes and associated tectonites have not been identified, in these Paleozoic terrains, indicating extension was probably more moderate compared to the Basin and Range Province, although the Omeo metamorphic complex is a possible contender.

The intra-continental Mesoproterozoic "superbasins" of northern Australia were also characterised by protracted periods of rift phase extension and elevated heat flow (ca. 1800–1670 Ma), with eruption of voluminous flood basalts and mafic sills, and widespread bimodal volcanics and granite intrusions (Southgate et al. 2000; Betts et al. 2006; Betts and Giles 2006). These are intercalated with predominantly continental-shallow marine sediments deposited within demonstrable half-graben structures, with evidence for large, extensional detachment shear zones at mid-crustal levels. The basins are inferred to have evolved by protracted lithospheric extension caused by the retreat of a north-dipping subduction zone located along the southern margin of the continent. Eventually these basins were terminated by orogenesis which caused lesser to greater degrees of inversion and metamorphism.

17.2.4 Continental Volcanic Passive Margins

Volcanically active continental passive margins (VPMs) represent a type of Large Igneous Province or LIP (Table 17.1; Chap. 5). They evolve following continental break-up and develop sequential to the extrusion of flood basalt plateaus (or traps) and/or narrow continental rift settings with more evolved magmas (White and McKenzie 1989; Geoffroy 2005). As indicated above, if a continental extensional regime fully develops, the thinning continental crust is injected with more and more mafic magma, becoming transitional in character, and welding ruptured continental crust to nascent oceanic crust (Fig. 17.20). Repeated injection of mafic dyke swarms and the eruption of lavas cover the transition from thinned continental crust to oceanic crust. Also occurring during this phase, mafic to ultramafic rocks underplate the transitional crust, forming high-velocity seismic zones below the thinned and heavily dyked, continental or transitional crust (Geoffroy 2005; Geoffroy et al. 2007). Asthenospheric upwelling leads to the formation of a mid-ocean ridge, and new oceanic crust progressively separates the once-conjoined continental rift halves. Continued volcanic eruptions spread lava flows across transitional crust and onto oceanic crust. Due to the high rate of magmatic activity the new oceanic crust forms much thicker than typical oceanic crust (Fig. 17.21).

Compared to non-volcanic examples (i.e. sedimentary passive margins), VPMs mark continental break-up over a hotter mantle. Much of our present knowledge of passive margin systems has advanced dramatically over the past two decades from the increasingly available wide angle and high-resolution offshore and onshore seismic surveys, potential field and well data as a result of exploration interest in their petroleum prospectivity. Today there is evidence that a majority of passive margins belong to the VPM type. Compared to sedimentary passive margins, VPMs are

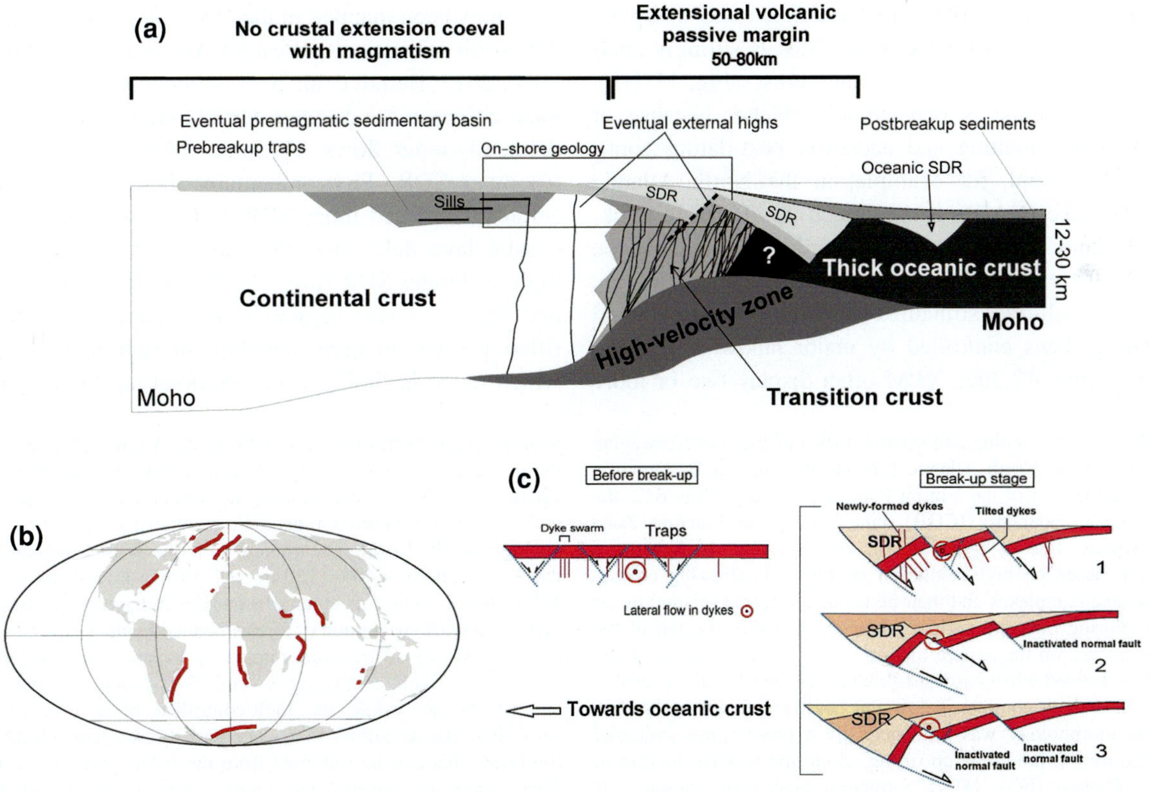

Fig. 17.20 Architecture, location and formation of volcanic passive margins. **a** Key elements and architecture of volcanic passive margins (From Geoffroy et al. 2007). Note the emplacement of continental flood basalts (traps) before the syn-magmatic stretching and thinning of the continental crust. SDR = seaward-dipping reflectors. **b** Worldwide distribution of volcanic passive margins (From Geoffroy 2005). **c** Model for formation of seaward-dipping reflectors; SDR are syn-magmatic roll-over flexures accommodated by continentward-dipping normal faults (From Geoffroy 2005)

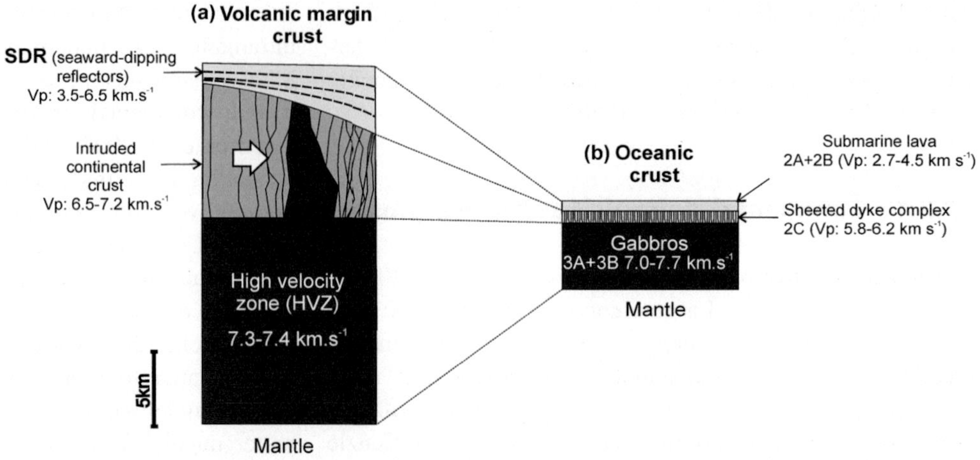

Fig. 17.21 Comparison of **a** volcanic passive margin igneous crust versus, **b** normal oceanic crust. This shows comparison of layering and compressional wave seismic velocities (Vp). (From Geoffroy et al. 2007)

narrower and associated with a thick magmatic crust, which extends along strike thousands of km; Fig. 17.20). From the surface down to the Moho (Fig. 17.21a), this is composed of: (1) wedges of seaward-dipping volcanic rocks (also known as seismic "*seaward-dipping reflectors*" or *SDR*), (2) intensively intruded and stretched transitional crust, and (3) large volumes of material of high compressional wave velocity, generally interpreted as underplated high-Mg gabbros. This top-to-down trilogy of basalts, sheeted complex, and gabbros postulated for the VPM crust is strongly analogous with the structure of oceanic crust (Fig. 17.21b). However, VPM crust is considerably thicker because of enhanced mantle melting and accretion post-dating continental break-up and, for example, in the North Atlantic Igneous Province (NAIP), it may reach up 40 km in thickness in Iceland and along the Greenland-Iceland-Faeroe ridge (Geoffroy et al. 2007; Plate 17.1). SDR are syn-magmatic roll-over structures formed by the growth and fill of half grabens controlled by major landward dipping listric faults (Fig. 17.20c). VPM often display two or more accreted wedges of SDR several km-thick, the inner one usually located immediately above the stretched and intruded continental crust. Breakup of the continental crust is believed to take place at the climax of the SDR and, for example, more than half the length of the South Atlantic continental margins appears to be dominated by basaltic rocks marked by them (Gladczenko et al. 1997).

A number of characteristic volcanic seismic facies units have now been mapped in the NAIP and other VPM, such as the North West Shelf, Western Australia (e.g. Planke et al. 2000, 2017; Berndt et al. 2001; Abdelmalak et al. 2016). The most common are divided into (1) landward flows, (2) lava delta, (3) inner flows, (4) inner SDR, (5) outer high, and (6) outer SDR. Environmentally these represent the subaerially emplaced inner SDR and landward lava flows, the coastal lava delta, and the subaqueous inner flows, outer high, and outer SDR (Fig. 17.22; also see Figs. 15.21, 15.22 and 15.23). These suggest that volcanic constructions on rifted passive margins develop through five main phases which range in timing from syn-break-up to post-break-up:

Fig. 17.22 Volcanic setting and seismic facies of the mid-Norwegian rifted margin of the North Atlantic Igneous Province. **a** Bathymetric map showing location of the Vøring Plateau (VP), ODP Site 642, the continental-ocean boundary (COB), and Jan Mayan Fracture Zone (JMFZ); yellow line is position of cross-section. **b** Schematic cross-section based on high-resolution seismic data showing characteristic volcanic complexes and their associated volcanic seismic facies units. The Vøring Escarpment is a major scarp within the top of the volcanic sequence on the eastern margin of the Vøring Marginal High (VMH). This formed when eastward-flowing lava reached the coastline creating a lava delta complex below the landward flows facies unit. Escarpment morphology was influenced by a pre-existing structural high formed during pre-break-up rifting, along the western margin of the Fenris Graben (FG). HTVC = hydrothermal vent complex; K (K-reflection) = base of seaward-dipping seismic reflectors (SDR). (**a**, **b** from Planke et al. 2017). **c** Dramatic image generated on the top-basalt seismic horizon revealing the sub-seafloor paleogeomorphology of the Vøring Escarpment (looking southwest). This shows a well-developed subaerial lava field on the Vøring Marginal High and the prominent, approximately 1 km high coastal escarpment. Volcaniclastic debris flows are well-imaged on the escarpment and spread out as lobes into the Fenris Graben. (From Planke et al. 2017). **d** On-land analogue of the Vøring Escarpment from the West Greenland sector of the NAIP; photo ~400–1150 m a.s.l. Brown lava flows of the Nuuk Killeq Basalt (414 La) entered a marine embayment to forming lava delta, and thick succession of foreset bedded hyaloclastites, or flow-foot breccias (414 hy). Three minor upper hyaloclastite horizons merge with the main horizon, indicating several flow groups separated in time entered the sea while the fault-controlled basin was continuously subsiding. Arrow points to a very dark brown hyaloclastite horizon (outlined) that can be followed from the paleoshore into to the basin. These units are covered by younger grey picrite basalt flows and hyaloclastites. (From Pedersen et al. 2017 through Creative Commons Attribution-Share Alike 4.0 International licence) ▶

(a)
70°
642
VP
JMFZ
65°
COB
0°
(b)
642
VMH
Landward Flows
FG
Top basalt
Vøring Escarpment
Sediments
Lava Delta
Inner Flows
HTVC
Lower Series Flows
SDR
K
Base basalt
COB
km
2
4
Basalt flows
Hyaloclastites
Volcaniclastic sediments
Sills
Sediments
(c)
Escarpment
Lava flow
Debris flow
(d)
417 La
414 La
414 La
414 hy
417 hy
414 hy
414 hy

(a) Initially, small-volume volcanism in a "wet" basin or shallow marine environment; the deposits of this phase form the base of the volcanic complex.
(b) Emplacement of large-volume basaltic effusive volcanics at seaward propagating volcanic centres and fissures, creating the landward directed flows. Coeval hydrovolcanic fragmentation and coastal erosion of their landward termination generates a lava delta complex between the bounding escarpment, and thin basalt flows and volcaniclastic debris (inner flows) of a former rift valley or half-graben. The lava deltas are analogous to those described from subglacial and seamount environments and characterised by progradational foresets or clinoforms termed flow-foot breccias (Chaps. 5, 16).
(c) Effusive basaltic volcanism fills the wedge shape accommodation space caused by continued subsidence at a seaward propagating, fault-controlled half-graben, creating the inner SDR.
(d) As subsidence continues, vent areas and eruption points eventually submerge below sea level, with phreatomagmatic (surtseyan) activity and later hyaloclastite and pillow lava build-ups and mounds. These create the outer high, which rapidly erodes and creates basins dominated by redeposited mass flows of hyaloclastite and pillow breccia debris.
(e) Increasing water depth and pressure finally prohibits explosive volcanism, and the last volcanic facies emplaced is an outer or oceanic SDR (Fig. 17.20a), consisting of pillow basalts and sheet flows, before volcanism continues as normal oceanic seafloor spreading.

Also, numerous igneous sheet intrusions, mainly sill complexes, and associated hydrothermal vent complexes are abundant in adjacent sedimentary basins, and are important for determining their thermal history, and therefore petroleum potential (Planke et al. 2005; Rateau et al. 2013; Chaps. 8, 18).

The mechanism of continental break-up has more recently also been reassessed in the case of the NAIP (Geoffroy 2005; Geoffroy et al. 2007). As we have so far described (such as in the EARS), the consensus has been that continental break-up and VPM volcanism takes place over a hot plume head or residual tail (Heron 2019). An alternative approach is to establish the pattern of magma flow, which can be achieved by mapping the orientations of dykes, and flow direction via their magnetic susceptibilities (AMS techniques; Chaps. 5, 8). Calculated flow vectors within dyke swarms feeding both early flood basalts and subsequent SDR lavas suggest most of the mafic lavas forming the NAIP transited through a relatively small number of volcanic centres (Geoffroy et al. 2007). The magma was intruded centrifugally in dyke swarms at some distance away from individual centres (up to 400 km) along the trend of maximum horizontal stress (σ_H) acting in the crust, feeding lava piles where dykes intersected the ground surface (Fig. 17.23). The distribution of mafic magmas at the surface is therefore primarily controlled by the regional stress field acting within the upper crust, but also by magma viscosity (Chap. 8). Geoffroy et al. (2007) conclude igneous centres feeding LIPs most likely represent the crustal expression of small-scale convective cells of buoyant mantle, naturally located beneath a thermal boundary layer below the Moho separating larger scale convecting mantle (Fig. 17.23c), rather than *per se* mantle plumes. In this model the location of igneous (accretion) centres is key to understanding the distribution of melting in the mantle underlying LIPs; in the case of the flood basalt stage in the NAIP this is estimated to be a minimum of about 40 Ma. Their distribution appears partly independent of pre-magmatic rift zones in the North Atlantic, but clearly inherited Caledonian basement features also have influence.

17.2.4.1 Ancient Examples of Continental Volcanic Passive Margins

The NAIP and Western Australian rifted continental margin and others provide detailed accounts of Mesozoic and younger rift margin history (e.g. Eldholm et al. 2000; Geoffroy 2005 and references therein). However, although there is strong evidence of repeated continental break-up and creation of new oceanic crust in the rock record, there appear to be very few volcanic passive margins that have been positively identified, in particular, by means of ancient seaward-dipping reflectors. One of the best described examples, including thick basaltic sequences resembling SRDs, formed at ca. 600 Ma during fragmentation of the eastern margin of proto-Gondwanaland (Direen and Crawford 2003). Rocks forming this latest Neoproterozoic passive margin occur as disrupted and discontinuous basalt-dominated packages that can be traced in potential field data over a total strike length >3000 km in southeastern Australia, usually associated with shallow marine sedimentary rocks. In many places, they were overthrust at ca. 510 Ma by boninite-dominated allochthons derived from the forearc region of a colliding Early Cambrian intra-oceanic arc.

These mafic volcanic successions occur in three elongate belts in the Delamerian Orogen, and are best exposed in western New South Wales and Tasmania where they are brought to surface by listric thrust (inverted extensional)

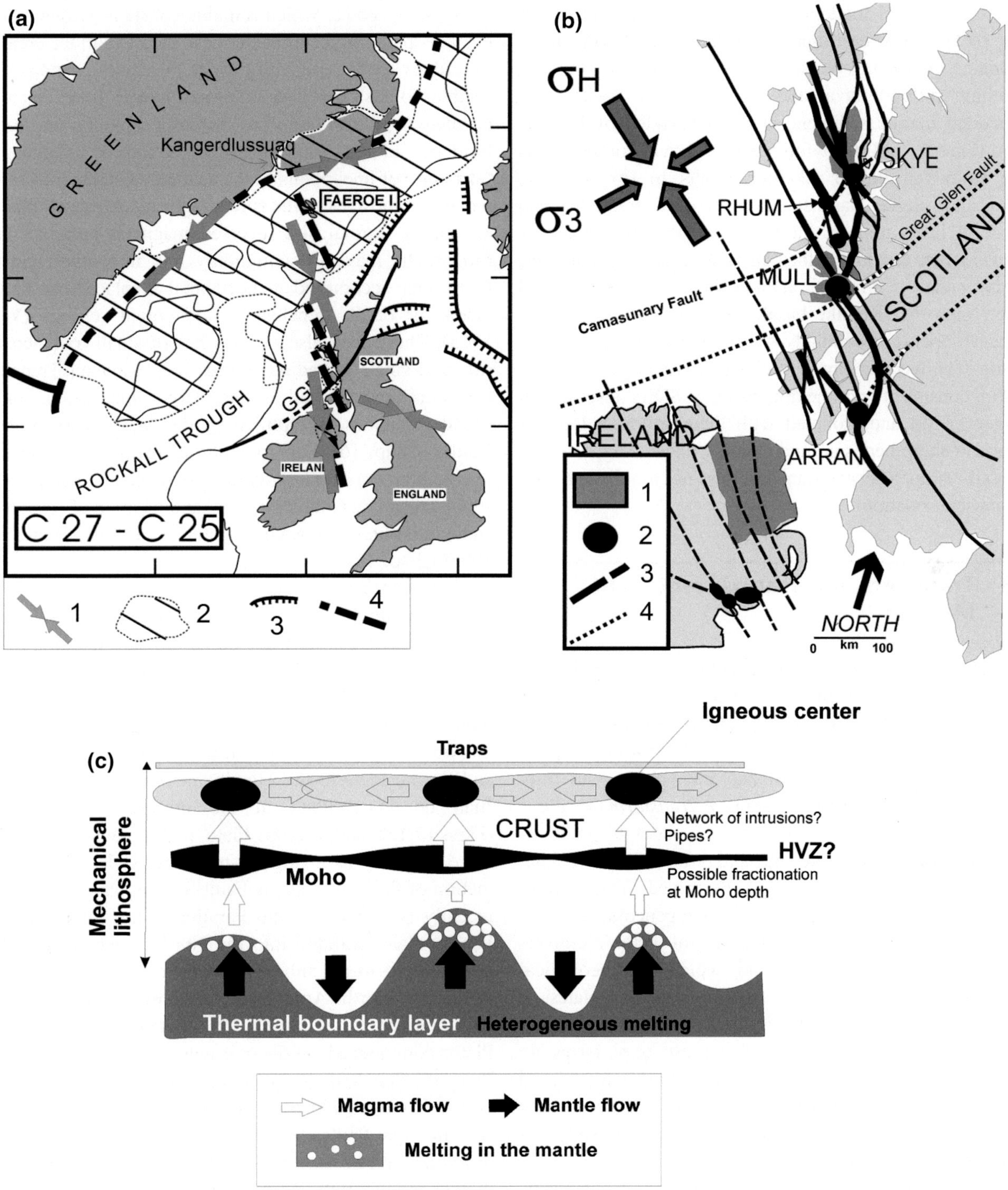

Fig. 17.23 Regional stress fields, magma flow and distribution of major igneous centres in the North Atlantic Igneous Province. The distribution of mafic magmas at the surface is primarily controlled by the regional stress field acting within the upper crust, although inherited basement features also have an influence. In this model, the location of igneous (accretion) centres is key to understanding the distribution of melting in the mantle underlying LIPs (see text). (From Geoffroy et al. 2007). **a** NAIP during the C27-C25 flood basalt (trap) stage. GG = Great Glen fault; 1 = trend of the maximum horizontal stress inferred from dyke swarms and fault tectonics; 2 = distribution of Paleocene lavas and sills; 3 = faults; 4 = dike swarms. **b** The British Tertiary Igneous Province; σ_H = maximum horizontal stress; σ_3 = minimum stress; 1 = "traps" (lavas); 2 = igneous centre; 3 = dyke swarms; 4 = main crustal faults. **c** Hypothetical relationship between mantle thermal boundary layer below the Moho and crustal distribution of igneous centres (see text); HVZ = high-velocity zone

faults. Geophysical modelling suggests the subsurface volcanic rocks form east-dipping wedges 1.5–6 km thick, comprising thickly layered sequences of dominantly basaltic (including pillow) lava and volcaniclastic rocks, and interbedded marine sedimentary rocks intruded by basaltic dykes (Direen and Crawford 2003). Lava compositions range from transitional alkaline basalts to rift tholeiitic basalts and picrite basalts; more evolved trachytic and comenditic lavas are rare. Although undoubtedly affected by early Paleozoic tectonism, the consistent geometries of these volcanic suites, with more alkaline lavas in the west and thicker, more MORB-like basalt piles in the east, suggest that ocean opening eventually occurred east of their present location. The wedges have individual strike lengths of 200–240 km, consistent with the dimension of SDRs described in younger VPMs and, coupled with their orthogonal distribution and sharp bends around cratonic nuclei, suggests that the fossil continent-ocean transition was not too distant from their present location.

17.3 Transform and Transcurrent Fault Margins

In a plate tectonic setting strike-slip faults may be divided into *transform* and *transcurrent faults* (Allen and Allen 1990; Nilsen and Sylvester 1995). The term transform fault should strictly be used for strike-slip faults that define plate boundaries, penetrate the crust and lithosphere, and fully accommodate motion between plates. A *transform margin* is a boundary where the motion between plates is predominantly horizontal or strike-slip and where, in principle, lithosphere is neither created nor destroyed. Transcurrent faults are restricted to intraplate faults that penetrate only the brittle crust. These may occur in zones of tectonic convergence and/or orogenic escape and, which over geological time may evolve to separate micro-continental plates or distinct regional tectonic domains (i.e. terrane boundaries). However, a review of the literature suggests these terms are used more loosely and, for instance, transform fault is used to describe intra-continental faults that do not necessarily separate plates, and some transcurrent faults probably cut through the entire lithosphere.

Most transform faults are hidden in the deep ocean basins where they offset divergent spreading ridge boundaries in short to larger "zig-zags" resulting from seafloor spreading as described in Sect. 17.2.1 (Plate 17.1; Fig. 17.1a; and see Fig. 16.92c). Transform faults accommodate the lateral offset between segments of ocean crust with similar spreading vectors, which results from rotation and stretching of the mid-oceanic ridge. Transform faults move mechanically differently from simple strike-slip faults, as slip along transform faults do not increase the distance between the ridges it separates, which remains constant because the ridges are spreading centres. As new ocean seafloor created at the ridges moves gradually away, it causes portions of the seafloor to push past each other in opposite directions, which is where transform faults are currently active on the seafloor. Evidence of this motion can be found in paleomagnetic "striping" of linear belts of the seafloor basaltic crust that are parallel to the axes of the spreading ridge segments that have alternately normal and reversed magnetic polarity. Transform faults may also commonly link into other transform faults that accommodate horizontal displacement between plates, or ocean trenches in convergent settings (Wilson 1965). These may also be the sites of small trans-tensional rifts and small spreading centres (e.g. Cayman Trench and Bismarck Sea), as well as some larger mid-oceanic alkaline island volcanoes exemplified by the Azores and Tristan da Cunha (Chap. 16). Superposition of melting anomalies (i.e. mantle plumes) on old structures may also be common (e.g. at the Tristan da Cunha hot spot, and location of carbonatites and kimberlites along continental extensions; White et al. 1995; Chap. 18).

Transform faults are therefore not limited to oceanic crust and spreading centres, and indeed some of the best known and most destructive occur on-land at the margins of continental plates. Perhaps the most studied example is the San Andreas Fault System on the Pacific coast of the United States. The San Andreas Fault links the East Pacific Rise off the west coast of Mexico (Gulf of California; Fig. 17.24a) and Pacific Plate to the Mendocino Triple Junction and the Juan de Fuca Plate in the northeast Pacific Ocean (Figs. 17.15b and 17.19b; Plate 17.1), making it described more precisely as a ridge-to-transform-style fault. The formation of the San Andreas Fault System has occurred fairly recently (<30 Ma), as the Farallon Plate, followed by the Pacific Plate, collided into the North American Plate. The collision led to the subduction of the Farallon Plate underneath the North American Plate (Fig. 17.17). Once the spreading centre separating the Pacific and the Farallon Plates commenced subducting beneath the North American Plate, the San Andreas transform fault was initiated. In more detail, this is a wide zone of dextral strike-slip deformation with multiple faults, fault strands and inter-fault basins often developed at releasing bends or stepovers (variously termed dilational jogs, pull-apart basins or rhombochasms). These are characterised by complex patterns of sedimentary and structural evolution as depositional settings may change from trans-tensional to transpressional, or vice versa (Mann et al. 1983; Nilsen and Sylvester 1995; Busby 2012).

Within this system there are also high heat flow, or volcanically and geothermally active basins. The best documented is the Salton Sea Trough, which occurs at the southern end of the San Andreas Fault system (Fig. 17.24a–d). This is in effect an active continental rift or pull-apart basin forming the landward extension of the Gulf of California-East Pacific Rise spreading system, but clearly strike-slip motion controls its development (Brothers et al. 2009). Within this is the world class Salton Sea geothermal field, which is an area of active hydrothermal mineralisation in Pliocene–Pleistocene deltaic, lacustrine and evaporite sediments (McKibben et al. 1988). These thermal anomalies are allied with oblique, proto-seafloor spreading and opening of the Gulf of California. The geothermal field is spatially associated with an arc of five Quaternary rhyolite domes located within a trans-tensional stepover between the San Andreas Fault and Imperial Fault (Fig. 17.24b). The projected location of maximum subsidence is near the southern shoreline, approximately coincident with the locus of Quaternary volcanism and a northeast-trending band of seismicity and very high heat flow. Farther south, the rapidly subsiding Mesquite Basin (Brawley geothermal field) appears to define a separate pull-apart basin within the larger San Andreas-Imperial stepover. Similarly controlled along the San Andreas transform fault system in California's northern Coast Ranges is the Clear Lake volcanic field. The volcanic field comprises late Pliocene to early Holocene lava domes, cinder cones, and maars with eruptive products varying from basalt to rhyolite and is also the site of the Geysers geothermal field (Donnelly-Nolan et al. 1993; Chap. 16). The Clear Lake volcanics are also thought to have been the heat source for the hot springs and hydrothermal activity that formed gold ore at the McLaughlin mine (Chap. 18).

Another region where strike-slip tectonics controls sedimentary basin development, and magmatic and geothermal activity is on the large Indonesian island of Sumatra. The Sumatra (also known as Semangko) Fault Zone (SFZ) runs the entire length of the island (>1500 km). Although high heat flow and volcanicity in this case are ultimately controlled by subduction along the Sunda Trench, this major dextral fault system is located some 300 km onshore to the east (e.g. Aldiss and Gazhale 1984; McCaffrey 2009; Bradley et al. 2017; Fig. 17.25). This fault zone accommodates most of the strike-slip motion associated with convergence between the Indo-Australian and Sunda (Eurasian) Plates (Plate 17.1). Convergence is not perpendicular to the plate boundary in this region, but instead the two plates move at an oblique angle. Most of the convergent strain is accommodated by motion at the plate boundary "*megathrust*" fault that defines the Sunda Trench, whereas the oblique component of plate motion sub-parallel to the plate boundary is largely accommodated by the SFZ. The area between the main plate boundary thrust fault and the SFZ forms what has been termed a "*sliver plate*", which includes the entire offshore forearc, forearc islands (accretionary ridge complex), and the continental portion of Sumatra west of the SFZ. Another major fault, the Mentawai Fault (and site of the 9.15 magnitude 26 December 2004 earthquake and resultant tsunami; Chap. 12), separates the forearc basin from the forearc islands. In Sumatra dextral displacements of some 130–150 km have taken place through the Cenozoic.

On land, eleven active volcanoes, five calderas, five degassing volcanoes and profuse geothermal activity, all of which constitute the Sunda Arc, are focussed along this strike-slip fault system (Fig. 17.25). The most striking example is Lake Toba, which is the largest caldera recognised in the Quaternary having rim dimensions of 100 km × 35 km (Chap. 16, Fig. 16.53c). The 2 km deep, steep-walled caldera erupted at least 2800 km^3 of magma in a single event 74,000 yr ago, ranking it as the largest volcanic eruption of the Quaternary (VEI 8.8; Chesner 2012; Chap. 10). A vast portion of northern Sumatra was drowned in ignimbrite that extended from coast to coast, an extensive ash deposit blanketed large portions of Southeast Asia including the entire subcontinent of India, and an aerosol cloud and dust cloud probably encompassed the Earth (Chap. 9). The caldera is elongated parallel with the fault system and is rhomboidal rather than circular in shape, and appears to be related to a wide, right-stepping pull-apart structure, suggesting a high degree of structural control during its formation (Bellier and Sebrier 1994; Bellier et al. 1997; cf. Sieh and Natawidjaja 2000). At least three older ignimbrite-forming eruptions are recognised to have occurred over the past 1.2 Ma and, although originally thought to have been the result of a single catastrophic eruption, Lake Toba is now regarded as a nested caldera complex consisting of overlapping, broadly co-located, caldera collapses. Today a resurgent dome (Samosir Island), with an uplifted thick veneer of diatomaceous lake sediments, dominates the caldera (van Bemmelen 1939; Chesner 2012; de Silva et al. 2015; see Figs. 16.53c and 16.68c).

The crust ranges from 30 to 40 km thick beneath Lake Toba caldera, and the depth to the Benioff Zone is approximately 125 km indicating a subduction angle of about 30°. Dextral displacement along the SFZ near Toba has annual slip rates ranging from 1.4 to 2.3 cm yr^{-1} (Chesner 2012; Bradley et al. 2017). A conspicuous geodynamic feature on the Indian-Australian Plate, subducting almost directly beneath Toba, is a 2 km high aseismic ridge known as the Investigator Fracture Zone (IFZ; Fig. 17.26a). Earthquake and seismic tomography data suggest Toba coincides with a bend in the subducting slab, while subduction of the IFZ is associated with high seismicity. It has been suggested this may represent a possible area of basaltic underplating which focuses volatile release into the mantle wedge beneath Toba (Chesner 2012 and references therein). Low velocity regions extend below the caldera all the way to the subducted slab, and are suggestive of magma chambers at 15 km and possibly shallowing to 5 km. More recent tomographic modelling now reveals a more complex multi-level plumbing

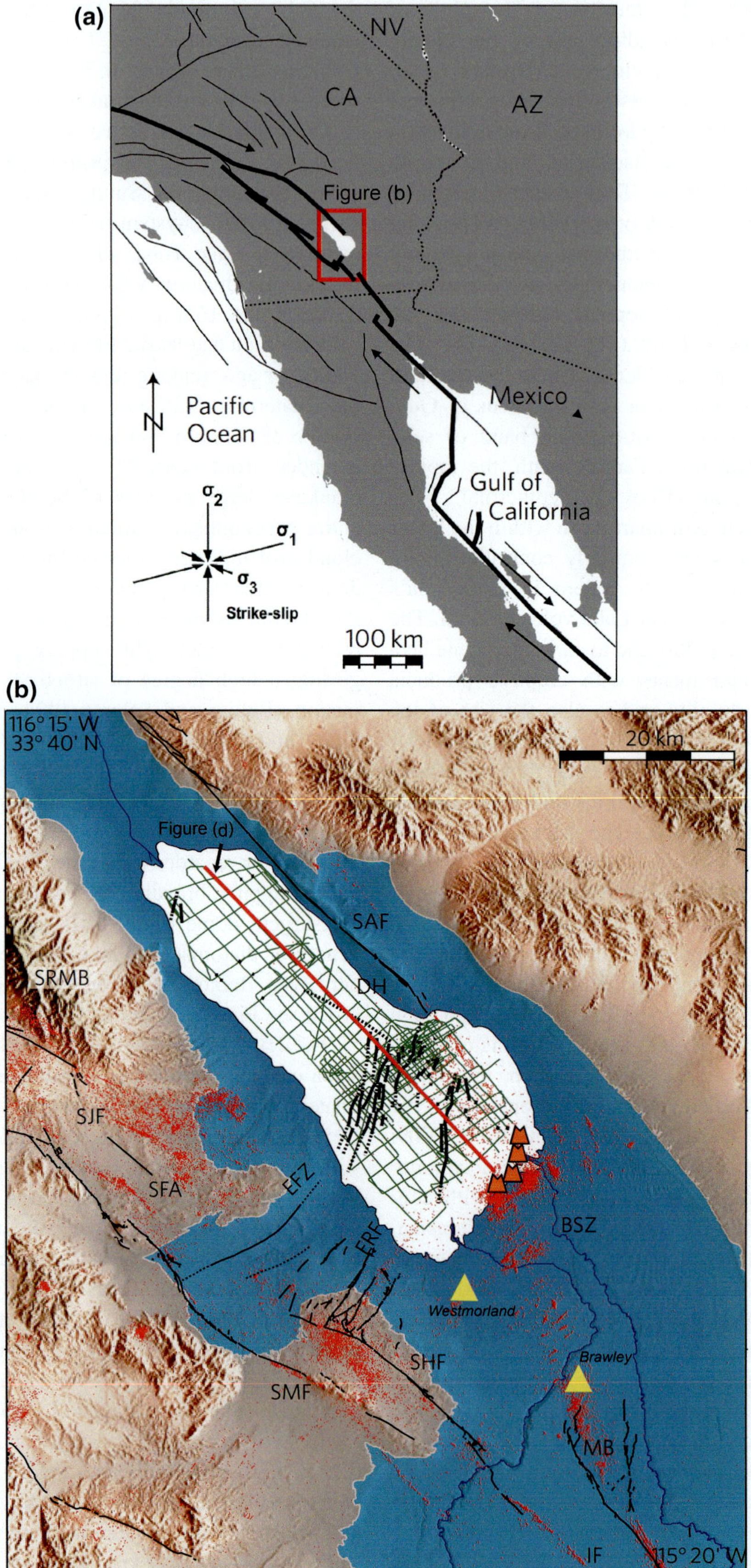
(a)
NV
CA
AZ
Figure (b)
Pacific
Ocean
Mexico
Gulf of
California
σ2
σ1
σ3
Strike-slip
100 km
(b)
116° 15' W
33° 40' N
20 km
Figure (d)
SAF
DH
SRMB
SJF
SFA
EFZ
ERF
BSZ
Westmorland
Brawley
SHF
SMF
MB
IF
115° 20' W
32° 45' N

◀ **Fig. 17.24** Oblique extension across strike-slip faults causes subsidence and leads to the formation of pull-apart basins such as the Salton Sea, in southern California, USA. **a** Regional tectonic location map showing position of San Andreas Fault Zone and Gulf of California spreading centre. **b** Map of the Salton Sea geothermal area and region, including the Westmorland and Brawley fields (yellow triangles). Green lines represent seismic reflection profiles outlining geothermal exploration area; red line is location of seismic reflection profile shown in **d**. Black lines = Holocene faults; red dots = earthquakes; orange triangles = rhyolite domes; shaded blue = extent of prehistoric Lake Cahuilla; MB = Mesquite Basin. **c** Rose diagram of fault strikes. The average strike of oblique-normal faults in the Salton Sea geothermal field is N15° E; other trends are the San Andreas Fault (SAF), Imperial Fault (IF), and the North America-Pacific (NA-PA) plate motion vector at this latitude. **d** Long-axis seismic reflection profile and interpretation (inset). A truncation surface (dashed line) separates the Holocene (CFm, Cahuilla Formation) from the underlying Pleistocene (BFm, Brawley Formation); coloured lines represent time horizons in the CFm. North of the EFZ (Extra Fault Zone) reflectors in the CFm are concordant and record little deformation, while those in the BFm show evidence of compressional deformation. South of the EFZ layers in the CFM diverge and thicken indicating maximum subsidence occurs near the southern shoreline, and that extensional faults may provide fluid migration pathways for young rhyolitic volcanism. (From Brothers et al. 2009)

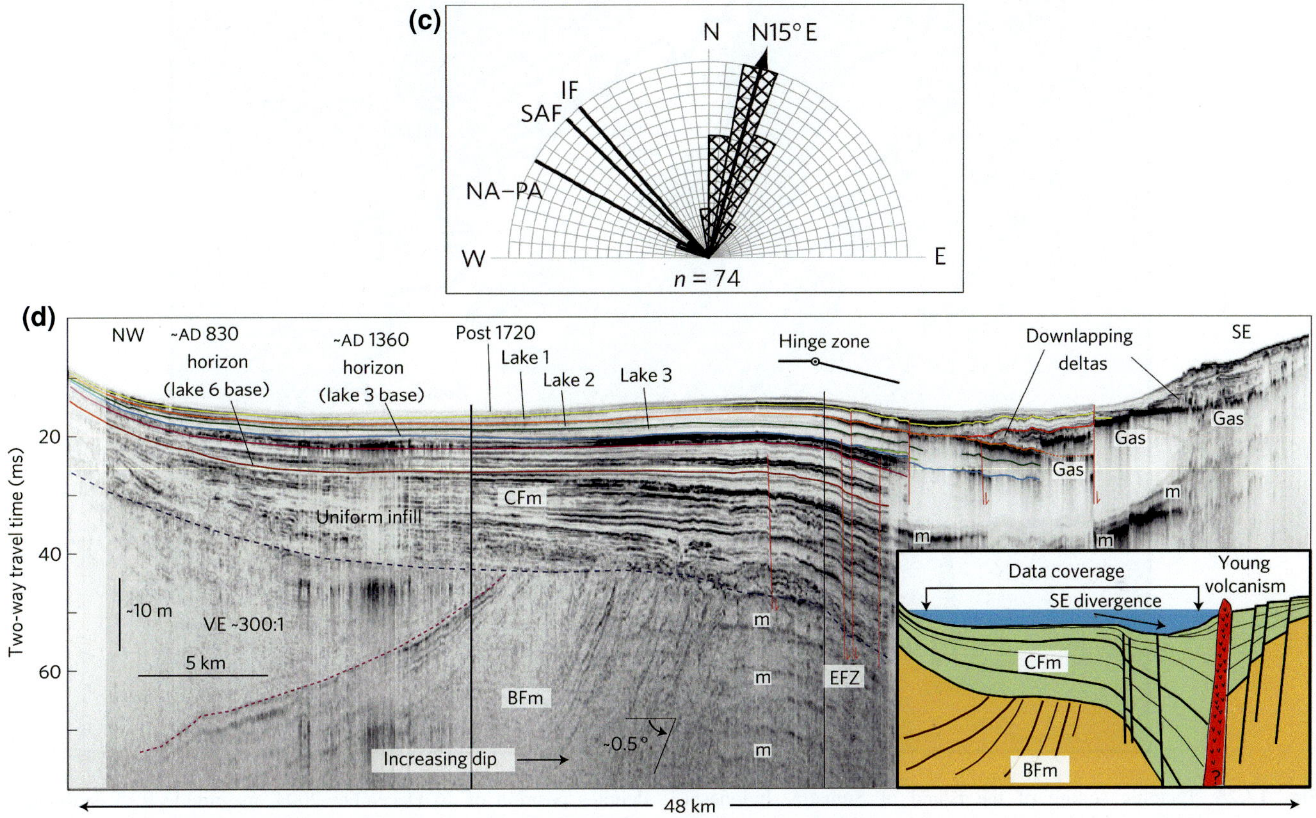

Fig. 17.24 (continued)

system beneath Toba (Koulakov et al. 2016, Fig. 17.26b). Anomalously large amounts of fluids and volatiles are thought to originate in the subducting slab of IFZ at about 150 km depth. Possibly also aided by a tear in the slab (see Fig. 17.46c), the volatiles migrate upward and cause active melting in the mantle wedge. The volatile-rich basic magmas accumulate at the base of the crust in a large ($\sim 5 \times 10^4$ km^3) reservoir which serves as a powerful heat source. The overheated volatiles continue ascending through the middle and upper crust and cause melting of silicic rocks. This leads to the formation of shallow crustal reservoir strongly enriched in volatiles and potentially highly explosive. Koulakov et al. (2016) conclude the Toba magma-generating engine continues to be active at present, despite its current period of inactivity.

From a wider perspective it is important to recognise the SFZ is only one of several major strike-slip systems in Asia (Fig. 17.27). These have been described as "*escape shears*" formed during collision and the indentation of India into Asia as crustal segments were extruded laterally away from the advancing Indian Plate (Tapponnier et al. 1982, 1986, 2000). Thrusting, especially in continent–continent

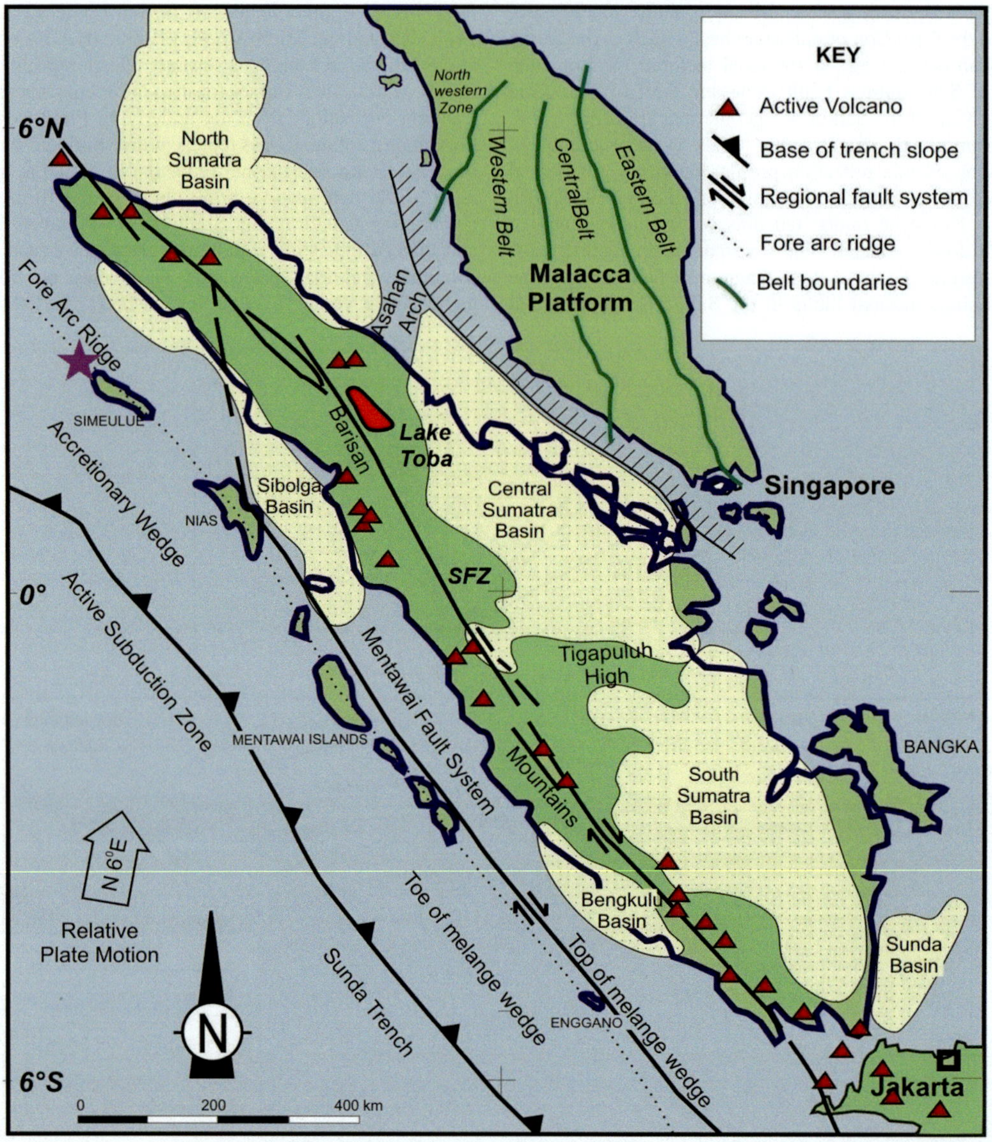

Fig. 17.25 Tectonic setting of the island of Sumatra, Indonesia, showing the location of active volcanoes of the Sunda Arc, Lake Toba caldera, the Sumatra Fault Zone (SFZ) and Cenozoic basins. Purple star denotes position of 26 December 2004, magnitude 9.15 earthquake. (Modified after Herman Darman; *Source* Wikimedia through Creative Commons Attribution-Share Alike 4.0 International licence)

collisional belts, may result in extensive crustal thickening, which in the Tibetan Plateau is anomalously thick ($\sim$50–90 km; Rai et al. 2006; Mechie and Kind 2013), and the highest and largest topographic feature on Earth. The amount of thickening is however limited by thermal (convective) and gravitational instabilities developed in the root zone of thickened lithosphere, and gravitational instabilities caused by sharp elevation increases (e.g. Dewey 1988; England and Houseman 1989; Rey et al. 2001; Fournier et al. 2004). The effect of the loss of the root of an orogenic belt leads to thinning of the lithosphere, upwelling and decompression melting of the asthenosphere, a rapid increase in geothermal gradient, high-T/low-P metamorphism and crustal anatexis. As the lower crust loses competency and cannot support further thickening, it begins to spread "sideways" so escaping the overburden.

Analogue experiments have indicated deformation is driven by two processes (Fournier et al. 2004): (1) collision which produces thickening accommodated by folding and thrust faulting, which is followed by strike-slip movements on steepened thrust faults, laterally expelling material from the thrust pile; and (2) gravitational spreading which produces thinning accommodated by normal faulting. However, it not easy to discriminate in the finite strain the part related to collision and the part related to gravitational spreading. Major strike-slip fault zones should, therefore, be extensively developed where continental blocks indent into one another, as during Himalayan mountain building. The general tectonic framework of the Himalayas is illustrated in Fig. 17.27, showing transcontinental-scale thrusts and strike-slip zones. Major strike-slip faults are sinistral on the western side of the Indian indentor, and dextral on the eastern side; north of the indentor polarities are opposite. Major strike-slip motion on the eastern side of the Indian indentor extends south through Myanmar (Sagaing Fault), linking the Andaman Sea marginal basin and spreading centres into the SFZ (e.g. Tun and Watkinson 2017). The framework that emerges is one in which large-scale, collision induced deformation has encompassed most of the crust and lithosphere of Asia since the Eocene. This has transformed convergence at the head of the "head" of the indentor, into extension at its "tails" via major strike-slip faulting.

More recent regional tectonic syntheses and analogue modelling have also now emphasised the potential role of subduction rollback along the East Asian continental margin in causing widespread strike-slip and extensional deformation (e.g. Schellart and Lister 2005; Yin 2010; Schellart et al. 2019). These studies suggest that extrusion tectonics is not the (only) driving mechanism for extensional collapse of the Himalayan orogen, and that eastward migration of the western Pacific trench system through most of the Cenozoic may have played an important role. With increasingly refined data sets (e.g. GPS and seismic monitoring, as well as high precision dating), the relative near versus far-field roles of extrusion tectonics, gravitational collapse *sensu stricto*, subduction rollback and peripheral opening of marginal sea basins remains a hotly contested topic in understanding Himalayan and Asian geology (e.g. Searle et al. 2011; Yin 2010; Xu et al. 2014; Tian 2015; Schellart et al. 2019; see also Sect. 17.5.6).

Strike-slip faults within orogenic belts may also reflect the expulsion of melts at depth (Yin 2010). As strike-slip movements induce local regions of extension (e.g. releasing bends and various pull-apart structures) into an otherwise compressive environment, these are liable to focus magmatic and geothermal activity. From a metals point of view, and in particular epithermal gold exploration, they are therefore also likely to be important, and indeed these major strike-slip systems of Asia are extensively mineralised (e.g. Sumatra, Myanmar and the Philippines), as well as those transecting the Andean continental margin of South America (Chap. 18).

17.3.1 Ancient Examples of Transform and Transcurrent Fault Margins

The recognition of ancient transform and transcurrent basins, without or comprising, a significant volcanic component can be challenging, as their structural and depositional histories are complex (Mann et al. 1983; Nilsen and Sylvester 1995; Busby 2012). Most long-lived basins, and therefore those more likely having preservation potential, are polycyclic and undergo repeated episodes of trans-tensional and trans-pressive subsidence and uplift. These movements can locally remove or "cannibalise" major parts of the sedimentary and structural record, making unconformities frequent and stratigraphic relations difficult to reconstruct. Lateral movements on major bounding faults and other faults commonly detach, rotate and translate basins or parts of basins from their places of origin, making along strike correlations tenuous. In ancient terrains, strike-slip faults are commonly reactivated as normal, oblique, thrust and more complex zones of faulting, which may result in only the preservation of disparate parts of the original strike-slip basins. Few oceanic transform basins are likely to survive subsequent episodes of subduction, or obduction to provide sufficient evidence of their structural and depositional evolution.

If there is any common theme, therefore, it is their complexity. Laterally displaced source areas (identified by provenance studies), facies and depocentres and kinematic (e.g. Riedel shear arrays) evidence provide general criteria for the recognition of strike-slip faulting. Complex and abrupt facies changes should occur within narrow, generally elongate basins characterised by high sedimentation rates, very thick stratigraphic successions and abundant evidence of paleo-seismic activity. Depo-centres will be strongly asymmetric and half-graben patterns common in transverse sections. Coarse-grained debris flow aprons characterise active strike-slip margins and fluvio-deltaic facies the opposite margins, while submarine or lacustrine turbidites, fluvio-deltaic, playa lake-evaporite and eolian facies typify axial depo-systems. Depo-systems migrate parallel to the most active strike-slip margin, which can result in the deposition of extraordinary thick sedimentary sections in small basins. Complex, and inconsistent cycles of coarsening- and fining-upward may reflect repeated tectonic movements and changes in base level along different fault segments.

A number of Late Jurassic strike-slip basins developed on continental crust along the Sawmill Canyon Fault, Arizona

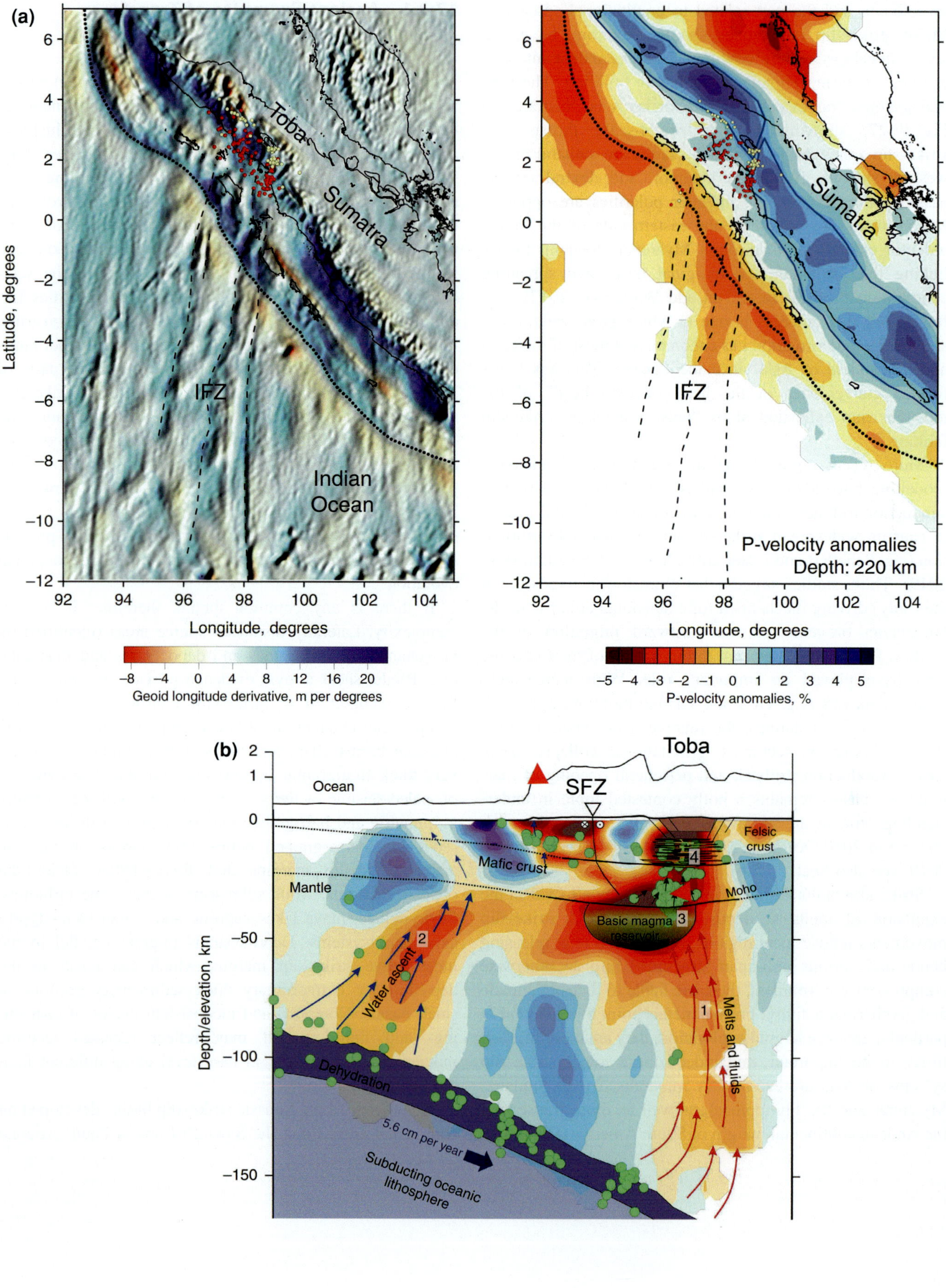
(a)
Toba
Sumatra
IFZ
Indian Ocean
Latitude, degrees
Longitude, degrees
Geoid longitude derivative, m per degrees
P-velocity anomalies
Depth: 220 km
P-velocity anomalies, %
(b)
Toba
Ocean
SFZ
Felsic crust
Mafic crust
Mantle
Moho
Basic magma reservoir
Water ascent
Melts and fluids
Dehydration
5.6 cm per year
Subducting oceanic lithosphere
Depth/elevation, km

◀ **Fig. 17.26** Location of the Investigator Fracture Zone (IFZ) and spatial relationship to Lake Toba caldera. In addition to bringing an anomalous amount of fluids into the mantle, subduction of the IFZ could trigger slab tear also intensifying the process of magma generation (see Sect. 17.5). **a** Setting of the IFZ: geoid model image; dots indicate the slab-related seismicity at two depth-intervals (red = 50–120 km, yellow = 120–170 km). **b** P-velocity anomalies at 220 km depth. The dark blue line highlights the possible boundaries of the slab. **c** Model of the magmatic plumbing system beneath Toba based on interpretation of tomographic S-wave velocity model (anomalies 1–4); exaggerated topography is shown above the section; Sumatra Fault Zone (SFZ) with possible depth of propagation indicated; green dots = earthquakes. Blue arrows (2) = ascent of water due serpentinization and/or high fluid release from the slab at 80–90 km depth beneath the forearc; the temperature of this pathway is not high enough to melt rocks and generate considerable volcanic activity. Red arrows (1) = path of ascending fluids and melts originated from the slab at ~150 km depth; this leads to the growth of ascending diapirs containing high-temperature, volatile-rich basic melts produced by partial melting of the mantle wedge. At depths of 30–50 km, these ascending magma diapirs form a large basic magma reservoir (3), which is too dense to continue ascending through the lower density continental crust. On reaching the middle and upper crust, overheated volatiles cause melting of rocks to form a shallow, silicic magma reservoir (4). Red triangle schematically indicates position of strato-volcanoes along the coastal region which are characterised by mafic melts (and their derivatives) generated from the mantle wedge, whereas, Toba consists almost exclusively of melts derived from continental crust. (From Koulakov et al. 2016)

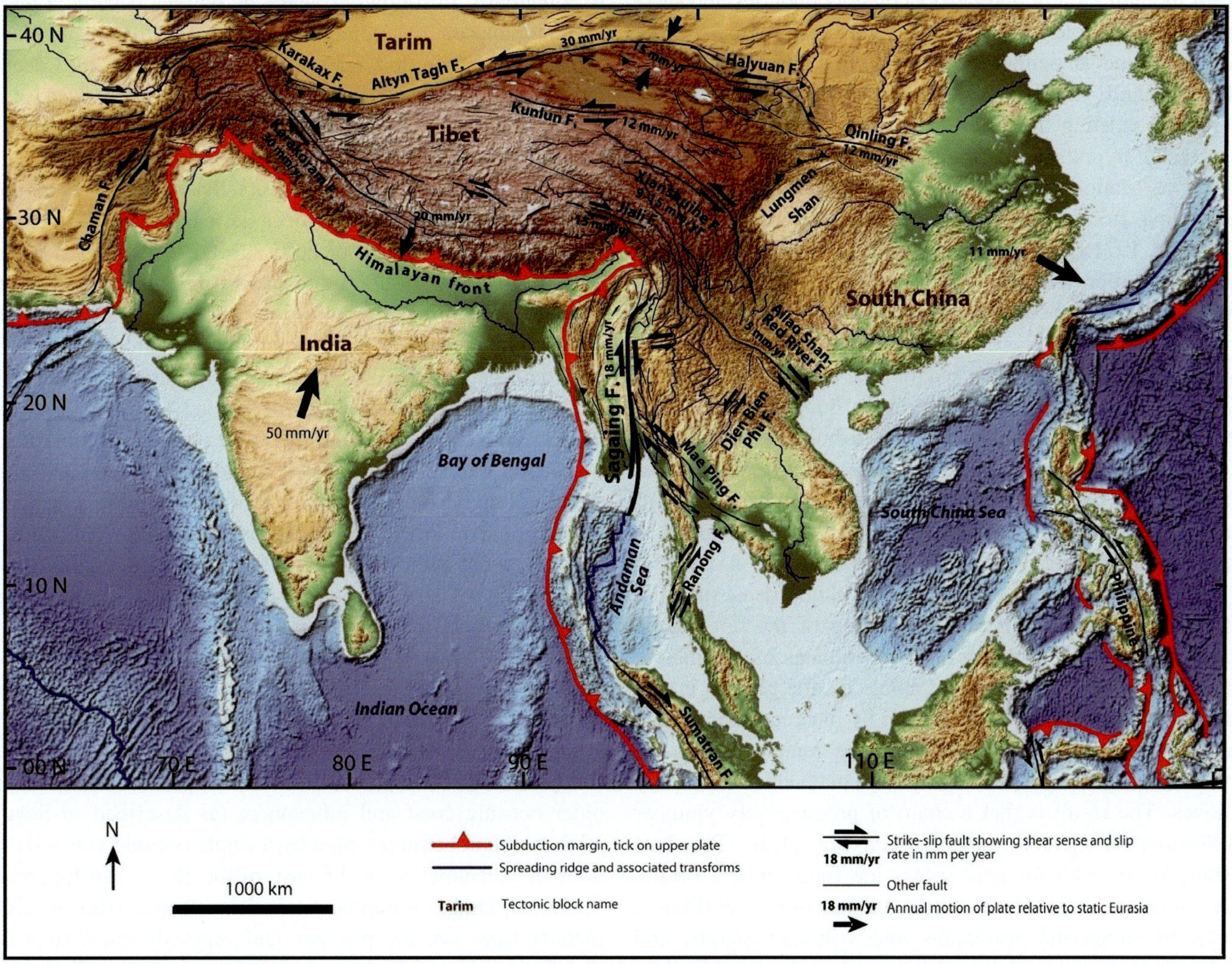

Fig. 17.27 Topography-bathymetry map with tectonic framework of the Cenozoic India-Asia collisional orogen (*Source* Ian Watkinson, After NASA and base digital elevation model from NOAA; http://www.sagaingfault.info/index.html). This illustrates active gravitational, extensional collapse or "extrusion" of thickened crust in the Tibetan Plateau along major lithospheric-scale escape shears (Tapponnier et al. 1982, 1986). These include, on the eastern side of the Indian "indentor", the dextral Sagaing Fault Zone which represents one of the longest and most active strike-slip systems in the world (Tun and Watkinson 2017), and links the Andaman Sea marginal basin and spreading centres into the Sumatra Fault Zone; while on its western side, the sinistral Chaman Fault Zone. As discussed in the text, more recent studies have also indicated the potential role of subduction rollback along the East Asian continental margin in causing widespread strike-slip and extensional deformation

(Busby et al. 2005; Busby and Bassett 2007), provides an ancient analogue of the Sunda arc in Sumatra. The sinistral Sawmill Canyon Fault possibly represents an inboard strand of the so-called "Mojave-Sonora megashear", thought to have controlled tectonic transport along the paleo-Pacific Jurassic subduction margin.

17.4 "Passive" or "Hot Spot" Intraplate Volcanism

The next first-order tectonic regime to consider is termed "passive" or *hot-spot* volcanism, which describes volcanic activity that occurs within tectonic plates and has generally been thought largely unrelated to plate boundaries and plate movements (Table 17.1). Passive is used in the sense that the stress regime is neither extensional nor compressive (i.e. neutral), and volcanism is supposedly, chiefly associated with pre-existing fractures (Cas and Wright 1987). As originally hypothesised, intraplate volcanoes form as the result of "stationary", vertical mantle upwellings or plumes, extending from the core-mantle boundary to the surface (Morgan 1971; Heron 2019). Mantle plumes are primarily thought to represent hot gravitational instabilities formed at the core-mantle boundary due to a core-mantle thermal boundary layer. They are therefore bodies of hotter, and thus lighter, material within the mantle that rises by convection. Conceptually plumes are thought to buoyantly rise as a plastically deforming mass that has a bulbous plume-head fed by a long, narrow plume tail (Fig. 17.28). As the head impinges on the base of the lithosphere, it spreads outward into a mushroom shape. Such plume heads are imagined to have diameters that could reach up to 2,000 km in diameter (White and McKenzie 1989). As the plume rises, pressure drops causing decompressional melting of the hot mantle material and the generation of large volumes of magma.

A compelling feature of intraplate mantle plumes is that they appear to be largely unaffected by plate motions. While a plume that feeds hot-spot volcanoes remains stationary relative to the mantle, the lithospheric plate above it usually moves. The result is that a chain of progressively younger volcanoes is created on the overlying plate. The best examples of such *hot-spot tracks* are found in the Pacific Ocean (Plate 17.1). The Pacific plate contains several linear belts of submarine seamounts and volcanic islands, and formation of at least some of these can be attributed to volcanism above a mantle hot spot. As the Pacific plate move across stationary hotspots, volcanism generates volcanoes that are only active as long as they are above the mantle plume. As the plate moves them away from their source, their magma supply is eventually cut off and they become extinct, while new volcanoes are formed above the hot spot. Thus, a linear chain of seamounts and island volcanoes is formed, the Hawai'ian and Emperor seamount chains being the classic case (Fig. 17.29a; Chap. 16). However, similar hot-spot tracks can be identified within intraplate continental settings, one of the best known being the Yellowstone "track" (Fig. 17.19b). In eastern Australia, there are several Cenozoic hot-spot tracks which include the Cosgrove track now recognised as the longest on Earth (Fig. 17.29b; Davies et al. 2015), and offshore in the post-Gondwana break-up marginal sea basins of the Tasman and Coral Seas there are parallel submarine seamount trails (Fig. 17.30a; Sutherland 2003; Sutherland et al. 2012).

From Morgan's (1971) initial estimate of some 20 plumes in the Earth's mantle, their proposed number expanded exponentially in following years, as plumes became a geological panacea to explain all intraplate volcanism. Now most lists contain about 50 plumes. Subdivision into deep, intermediate, and shallow hotspots has been proposed, and the more recent estimates for the number of plumes ascending from the core-mantle boundary are below 10 (e.g. Fig. 17.28a). In more recent years the role and even the existence of mantle plumes became a highly controversial topic (e.g. Anderson 2000, 2005a, 2006; Ernst and Buchan 2001; Foulger et al. 2005 and papers therein; Fig. 17.28b), and debate continues today as to whether or not they can form spontaneously or are generated by the sinking of plates into the deep mantle (Anderson 2013; Heron 2019). Certainly not every hot spot can be related to a deep mantle plume, and there are a number of examples of intraplate volcanism that have never been suggested to be related to a mantle plume. The mechanism of origin for these intraplate volcanic events often remains poorly understood; however, even the most enthusiastic plume proponents generally now accept that not all hotspots are underlain by a deep mantle plume (see www.MantlePlumes.org).

17.4.1 Intraplate Oceanic Volcanism

Intraplate oceanic volcanisms occurs upon a basement of older oceanic crust and lithosphere (as described in Sect. 17.2.1). It can be represented by a single oceanic volcano, or lines of volcanoes as in the case of the Hawai'ian-Emperor seamount chain. Intraplate oceanic volcanic edifices also include large oceanic plateaus and supervolcanoes such as the Ontong Java Plateau and Tamu Massif, which are thought to be the submarine equivalents of flood basalt eruptions or oLIPS. These are described in Chap. 16, and not considered further below.

The Hawai'ian volcanoes are the most studied examples of hot-spot volcanoes. The five volcanoes that form the island of Hawai'i (Chap. 16, see Figs. 16.25 and 16.26) at

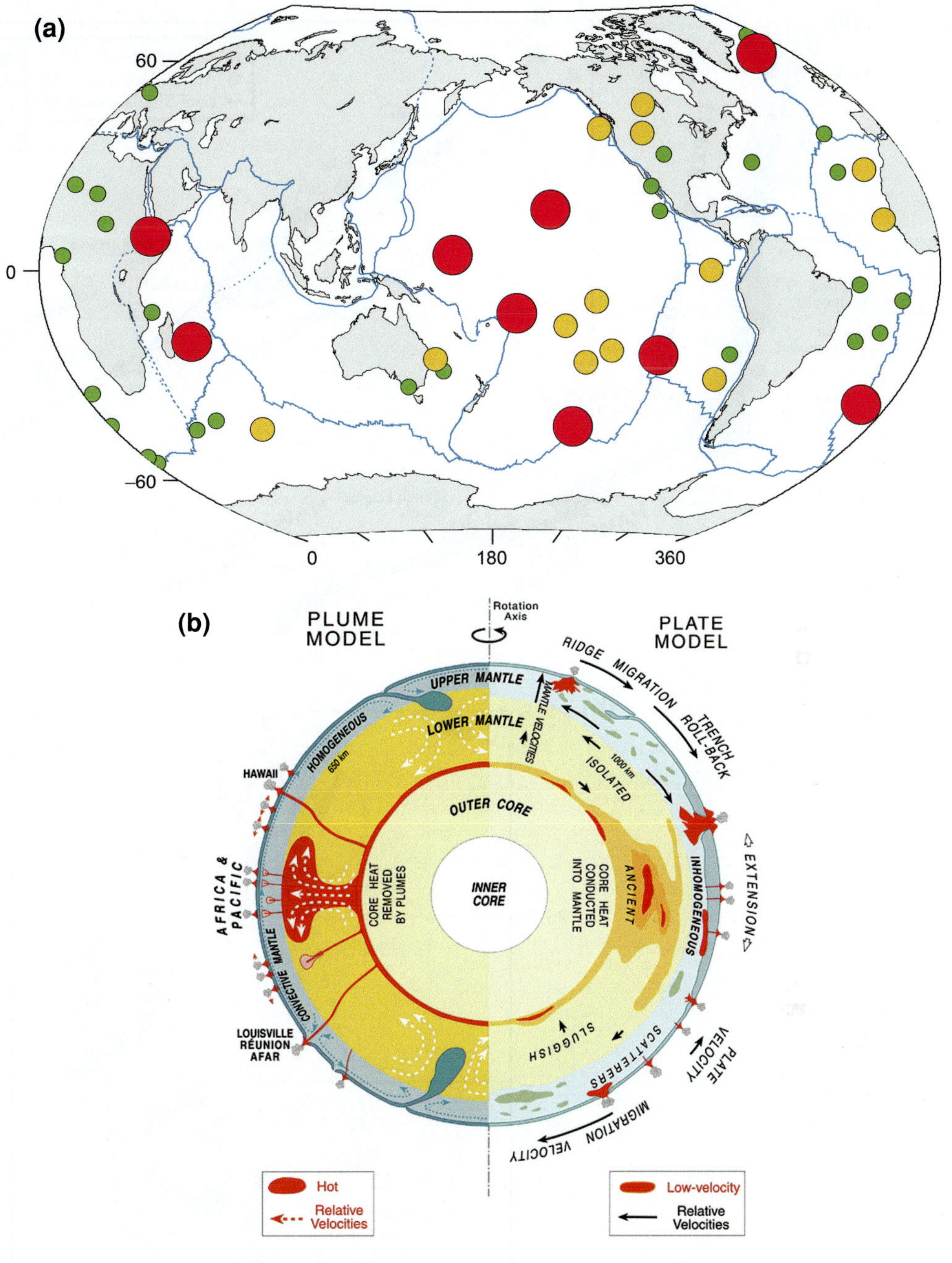

Fig. 17.28 The Earth's present-day "hot spots" and plumes. **a** Courtillot et al. (2003) divided hotspots into three categories, corresponding to proposed origins at the core-mantle boundary (red), at the base of the upper mantle (yellow), and in the lithosphere (green) (From Foulger and Natland 2003). **b** Schematic cross-section of the Earth showing the plume versus plate hypotheses. The plume model shows the different categories of hot spots, where the deep mantle provides the material and deep mantle or core the heat. Red regions are assumed to be hot and buoyant, while blue cold and dense. In the deep mantle both narrow "feeder tubes" and giant upwellings exist. As an alternative to the plume model, a growing number of models treat anomalous intraplate volcanism as a consequence of plate tectonics. The plate model shows its important attributes: variable depth of recycling of material, migrating ocean ridges and trenches, a concentration of volcanism in tensile regions of the plates, an inhomogeneous and inactive upper mantle, an isolated and sluggish lower mantle, and pressure-broadened ancient features in the deep mantle. (From Anderson 2005b)

the southeast end of the Hawaiian chain are all <1 Ma. Two of these, Kilauea and Mauna Loa, are two of the most active volcanoes in the world, while the youngest volcano is the seamount named Loihi. Northwestward along the Hawai'ian chain each island is progressively older (Fig. 17.29a). The extinct volcano or volcanoes that formed the island of Kauai are about 5 Ma. Bathymetric images show a major submarine continuation of the Hawai'ian Ridge to the northwest of the Hawai'ian Islands and then a dogleg bend into the Emperor Seamounts, which comprise an entirely submarine ridge continuing northward to the edge of the Pacific Plate (Plate 17.1). Ages of rocks obtained by dredging and drilling the Emperor Seamounts indicate that they are a continuation of the Hawai'ian chain, and that the Hawai'ian hot spot has been active for at least 80 Ma. Generally, the interpretation has been the Pacific Plate has moved over this centre of volcanism, first northward and later northwestward, at a rate of approximately 8–10 cm yr^{-1}. The bend between the Emperor Seamounts and the Hawai'ian Ridge occurs at about 45 Ma, indicating a significant shift in the direction of movement of the Pacific Plate. However, paleomagnetic evidence suggests Hawai'i has not remained stationary, but changed direction radically at the time of the Hawaiian-Emperor bend, while the Pacific Plate did not

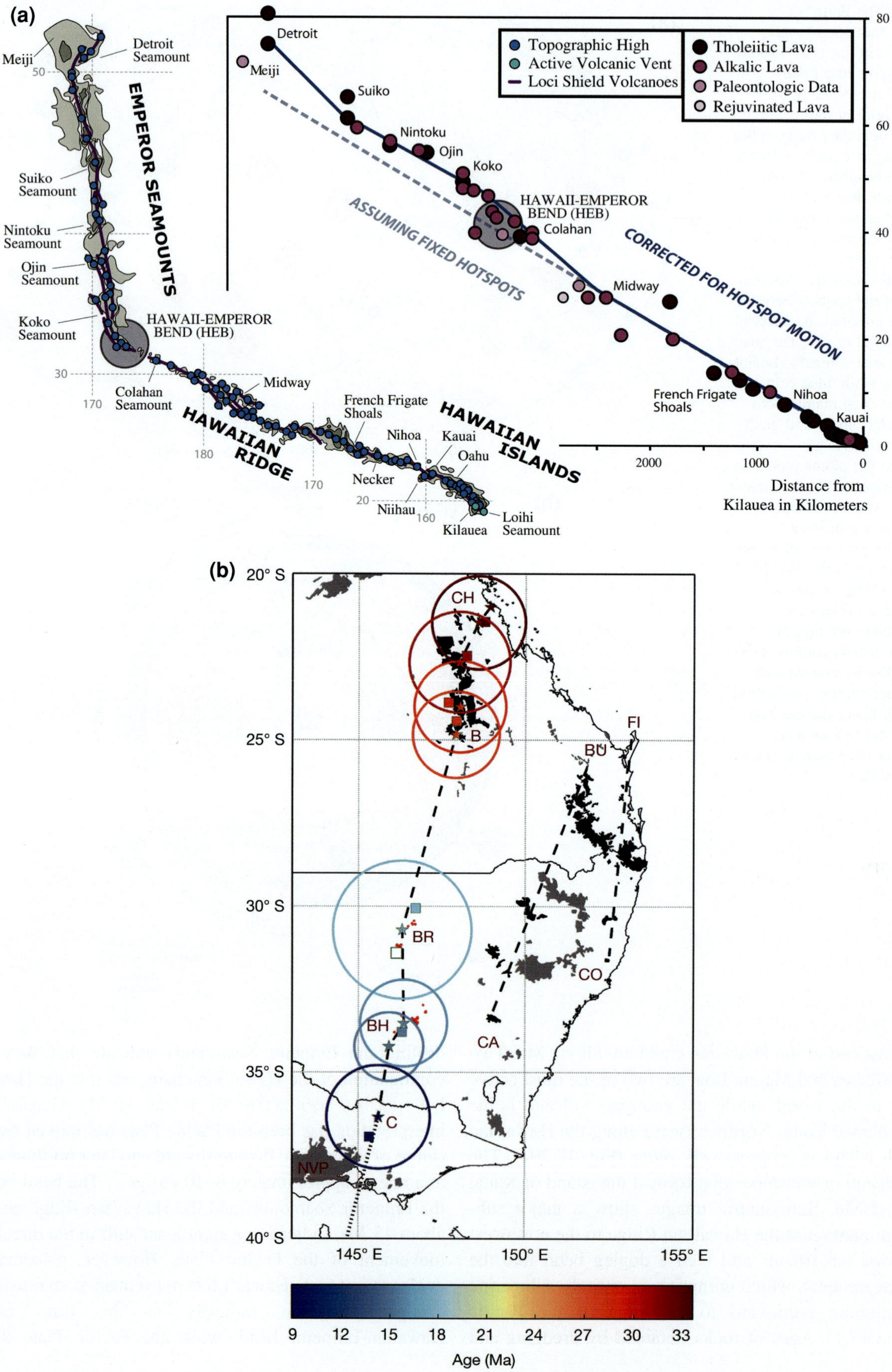
(a)
Meiji
Detroit Seamount
EMPEROR SEAMOUNTS
Suiko Seamount
Nintoku Seamount
Ojin Seamount
Koko Seamount
HAWAII-EMPEROR BEND (HEB)
Colahan Seamount
Midway
HAWAIIAN RIDGE
French Frigate Shoals
Necker
Nihoa
Kauai
Oahu
HAWAIIAN ISLANDS
Niihau
Kilauea
Loihi Seamount
50
30
20
170
180
170
160
Topographic High
Active Volcanic Vent
Loci Shield Volcanoes
Tholeiitic Lava
Alkalic Lava
Paleontologic Data
Rejuvinated Lava
Detroit
Meiji
Suiko
Nintoku
Ojin
Koko
HAWAII-EMPEROR BEND (HEB)
Colahan
Midway
French Frigate Shoals
Nihoa
Kauai
ASSUMING FIXED HOTSPOTS
CORRECTED FOR HOTSPOT MOTION
Age (Ma)
80
60
40
20
0
2000
1000
0
Distance from Kilauea in Kilometers
(b)
20° S
25° S
30° S
35° S
40° S
145° E
150° E
155° E
CH
B
FI
BU
BR
CO
BH
CA
C
NVP
9
12
15
18
21
24
27
30
33
Age (Ma)

◀ **Fig. 17.29** Hot spot volcano tracks. **a** Age-distance relationships and location of dated samples or the Hawai'ian-Emperor seamount chain (From Staudigel and Koppers 2015). The age progression along the Hawai'i-Emperor seamount track validates the hot spot concept and is corroborated by GPS measurement of Pacific Plate motion. However, there is considerable discussion about the interpretation of age discussions along the Emperor seamounts (see text). **b** As proposed for eastern Australian Cenozoic volcanic centres; which are coloured as follows: black = central volcanoes, grey = basaltic lava fields, and red = leucitite-bearing volcanoes. Age progressive hot spot tracks are denoted by black dashed lines, and include the Earth's longest continental hot spot track, the 2,000 km long Cosgrove track (33–9 Ma). Stars denote a selection of known volcanic centres along this track that are coloured by age (in Ma), while squares mark predicted volcanic locations, based on reconstructed absolute motions of the Australian Plate. Circles surrounding each dated volcanic centre are measure of the uncertainty in predicted locations, such as in the underlying mantle plume diameter and the associated melt region's lateral extent. The dotted line to the south of Cosgrove marks the predicted extension of the Cosgrove track towards its present-day location. B = Buckland; BH = Begargo Hill; BR = Bokhara River; BU = Bunya; C = Cosgrove; CA = Canobolas; CH = Cape Hillsborough; CO, Comboyne; FI = Fraser Island; NVP = Newer Volcanics Province. (From Davies et al. 2015; see Sect. 17.5.6, Fig. 17.59 for further discussion)

change direction (Koppers et al. 2001). This invalidates the assumption of hot-spot "fixity" making the interpretation of plate motion non-unique and remains a contentious topic (Staudigel and Koppers 2015).

Hawai'i remains an excellent case study for these types of volcanoes (Chap. 16 and references therein). The thickness of the lithosphere under Hawai'i is ca. 90 km, and earthquakes associated with magma ascent and ponding suggest the active magma migration zones under Hawai'i are at depths of 30–45 km, which is well within the lithospheric mantle. Most lavas erupted from Hawai'ian shield volcanoes have experienced extensive fractional crystallisation in shallow crustal reservoirs. From studies of the most primitive magmas, at least three types of mantle melting can be identified (Grove and Till 2015). The most voluminous lavas in the Hawai'ian systems are those that erupt on the shields. These are related to a primary magma that is very high in MgO (picrite). Experimental studies indicate these melts are in equilibrium with a depleted harzburgite source at 30–45 km. Another set of primary magmas (alkali basalt) is found to be in equilibrium with garnet lherzolite, at depths near the asthenosphere-lithosphere boundary ($\sim$90 km). These are found as eruptive products that form during the early stages of Hawai'ian volcano growth, and are present at Loihi Seamount. A possible common link between these two distinct primary magmas is when asthenospheric magmas produced at a 90 km depth (i.e. alkali basalts) ascend into the mantle lithosphere they react with the colder solid lithosphere (Wagner and Grove 1998). Here, alkali basalts heat the lithosphere above its melting point and assimilate orthopyroxene (high MgO), and crystallise olivine altering the melt composition to picrite. A third type of primary magma at Hawai'i is generated when CO_2 stored in the lithosphere lowers the melting point and produces melts saturated in this volatile component. These melts are rare, but ubiquitous at Hawaii and other hot spots and form the third mantle-derived melt associated with intraplate volcanism.

Individual seamounts may be <1 km to tens of km in diameter and will have a volcanic pedestal stratigraphy up to several km thick (Chap. 16). This will consist of a basal pile of predominantly basaltic pillow lavas, hyaloclastites, and if the seamount approaches sea level, a veneer of pyroclastic-hyaloclastic layers, capped by coherent submarine lavas, perhaps all fringed and ultimately surmounted by limestones and biogenic colonies as atolls form. The basaltic rocks may be alkaline or tholeiitic and may be geochemically variable (Staudigel and Koppers 2015). Associated sediments are likely to be pelagic and volcaniclastic, and carbonate mass-flow deposits are likely to form aprons. We also note not all seamounts are necessarily produced by mantle hot spots, but rather may be the result of deep fractures within the plates (e.g. transform faults) that allow pockets of magma to leak to the surface (Presnall and Gudfinasson 2011). These pockets originate in the low-velocity zone asthenosphere where a small percentage may be molten. Once a sufficient volume of magma forms in the subsurface it will rise buoyantly, and any fracture/fault system within plates, or at their margins, could facilitate this process.

17.4.1.1 Ancient Intraplate Oceanic Volcanism: Characteristics

As for volcanism associated with mid-oceanic ridges and ocean crust formation, the preservation potential of intraplate volcanic islands and seamounts in the rock record is low and they, too, should mostly be subducted along with the oceanic lithosphere. Similarly, they could conceivably be preserved as parts of ocean crust slices in convergent margin accretionary complexes, or as discrete off-scraped, sheared-off blocks and tectonic slices in these settings as has happened along the northern California Coastal Range (e.g. MacPherson 1983; Schnur and Gilbert 2012) and Central America (Ortiz-Hernández et al. 2003; Buchs et al. 2011), in which case they may be highly deformed and probably dismembered.

17.4.2 Intraplate Continental Volcanism

Intraplate continental volcanism is also essentially basaltic, although rarer, more silicic centres are also known. This builds basaltic provinces that are characterised by plains of valley-fill lavas (plains basalts) dotted with cinder cones and tuff rings (Cas and Wright 1987; Németh 2010; Valentine and Connor 2015; Sect. 16.5). Plains basalt provinces are exemplified by the Newer Volcanics Province (NVP) of southeastern Australia (e.g. Joyce 1975; Cas et al. 2017) and the Snake River Plain, Idaho, USA (Greeley 1982; Hughes et al. 1999). However, more copious volumes of magma generation may erupt, regionally extensive flood lava sheets, forming plateau basalts or traps, such as the Ethiopian Plateau and Yemen (Sect. 17.2.3) and the Columbia River Basalt Plateau (Chaps. 5, 16). Widespread continental basaltic volcanism probably suggests that the lithosphere is in an extensional or generally "neutral" state, allowing large volumes of mantle magmas to reach the surface. As we have previously described in this chapter, plumes can lead to continental rifting, the eruption of massive flood basalt plateaus (cLIPS), the development of volcanic passive margins, and ultimately lithospheric separation and seafloor spreading, although alternative geodynamic models exist (e.g. Anderson 2005a; Geoffroy et al. 2007; Heron 2019; Figs. 17.23 and 17.28b).

The NVP occurs at the southern end of a 4,000 km long belt of intraplate volcanic provinces that run parallel to the eastern seaboard of Australia (Fig. 17.30a). It is perhaps the largest representative on Earth of basaltic plains volcanism, with an area of >23,000 km^2 and a dense rock equivalent erupted deposit volume of <900 km^3 (Fig. 17.30b, c; Cas et al. 2017). The vast majority of the >400 volcanoes (Joyce 1975; Lesti et al. 2008; Boyce 2013; Cas et al. 2017) are monogenetic although some are polygenetic, and the province has been active since ca. 8 Ma. Lava fields, shields, scoria cones are common, and there are >40 maars and volcanic complexes. Maars occur dominantly in the south where magmas erupted through Tertiary sedimentary aquifers, whereas in the north, over Paleozoic crust, there are few (see Figs. 16.10 and 16.13). Complex interactions of the magma volatile content, magma ascent rates, conduit characteristics and the availability and depth of aquifers caused diverse eruption styles (van Otterloo and Cas 2013, 2016; van Otterloo et al. 2013; Blaikie et al. 2015; Cas et al. 2017; Chaps. 4, 9 and 16). Volcanoes commonly occur close to major crustal faults (Fig. 17.30c), which acted as the magma conduits. The available age data (which is relatively sparse) show no obvious migration pattern of volcanism across the province, although do suggest that volcanism occurred in particular parts of the province at different times (Cas et al. 2017). Most rock compositions range from picritic to basaltic andesitic; however, there is a wider range of compositions present from nephelinite to andesite. Although some compositions occur more commonly in certain regions, alkali basalts are widespread throughout the province.

The availability of regional potential field and other geophysical imagery have helped clarify many of the structural aspects of the NVP (Cas et al. 2017). The eruptive centres are positioned on or in close proximity to basement faults (Fig. 17.30b, c), although the precise nature of how these localise volcanism remains a matter of debate (cf. Lesti et al. 2008; Cas and van Otterloo 2011; Boyce 2013; van Otterloo et al. 2014). In the northern half of the province there is a good correlation of volcanoes with the north–south to NNW-trending faults in the Paleozoic basement, and in the southern half with east–west-trending faults in the Mesozoic and Tertiary sedimentary rocks of the Otway Basin. Many of the volcanic centres contain multiple vents and fissures that are aligned parallel to nearby fault structures. Examples include the Mt Gambier Volcanic Complex, having 14 vents which are aligned parallel with the nearby Tartwaup Fault System, a principle basin-margin fault zone (van Otterloo et al. 2013; see Fig. 16.12d, e). There also appears to be good spatial correlation between the location of larger volcanic complexes and the crustal-scale faults. Large-volume, multi-vent complexes occur along or in very close proximity (<2 km) to crustal-scale faults, and a good example is again the Mt Gambier Volcanic Complex. Many occurrences of mantle xenoliths in the volcanic deposits (Fig. 17.31a, b) are also close to the crustal-scale faults, implying faster rates of ascent for the magmas that carried the xenoliths. This leads to the conclusion that the larger crustal-scale structures associated with more voluminous volcanic centres and faster rising magmas were those that aided the ascent of magma through the crust (Fig. 17.31c; van Otterloo et al. 2014).

On a broader scale, as indicated above, the NVP forms part of a more extensive, 4,000 km long belt of isolated pockets of intraplate volcanism that has affected eastern Australia through the late Mesozoic-Cenozoic (ca. 100 Ma; Fig. 17.30a). The geochemical character can range from alkaline to tholeiitic and more differentiated rocks, including trachytic rocks, which are also common and form "central volcanoes" mainly in southern Queensland and northern New South Wales (Sutherland et al. 2012). Linear age trends in the eastern Australia have been assigned to the passage of a lithospheric plate over a mantle hot spot (Wellman and McDougall 1974; Sutherland 2003, 2012; Davies et al. 2015). Plume-like trace element and isotopic components (East Australia Plume signature) were recognised in both basalts associated with central volcanoes, and larger basaltic fields and provinces. Plate motion projections and presumed past plume line positions have been inferred to indicate that their original asthenospheric sources were in the Bass Strait-west Tasman Sea region (the Eastern Australian and Tasmantid Hotspots; Fig. 17.30a). This postulated plume zone shows a high heat flow anomaly and a deep, slow seismic-wave structure (Montelli et al. 2006) and is closest

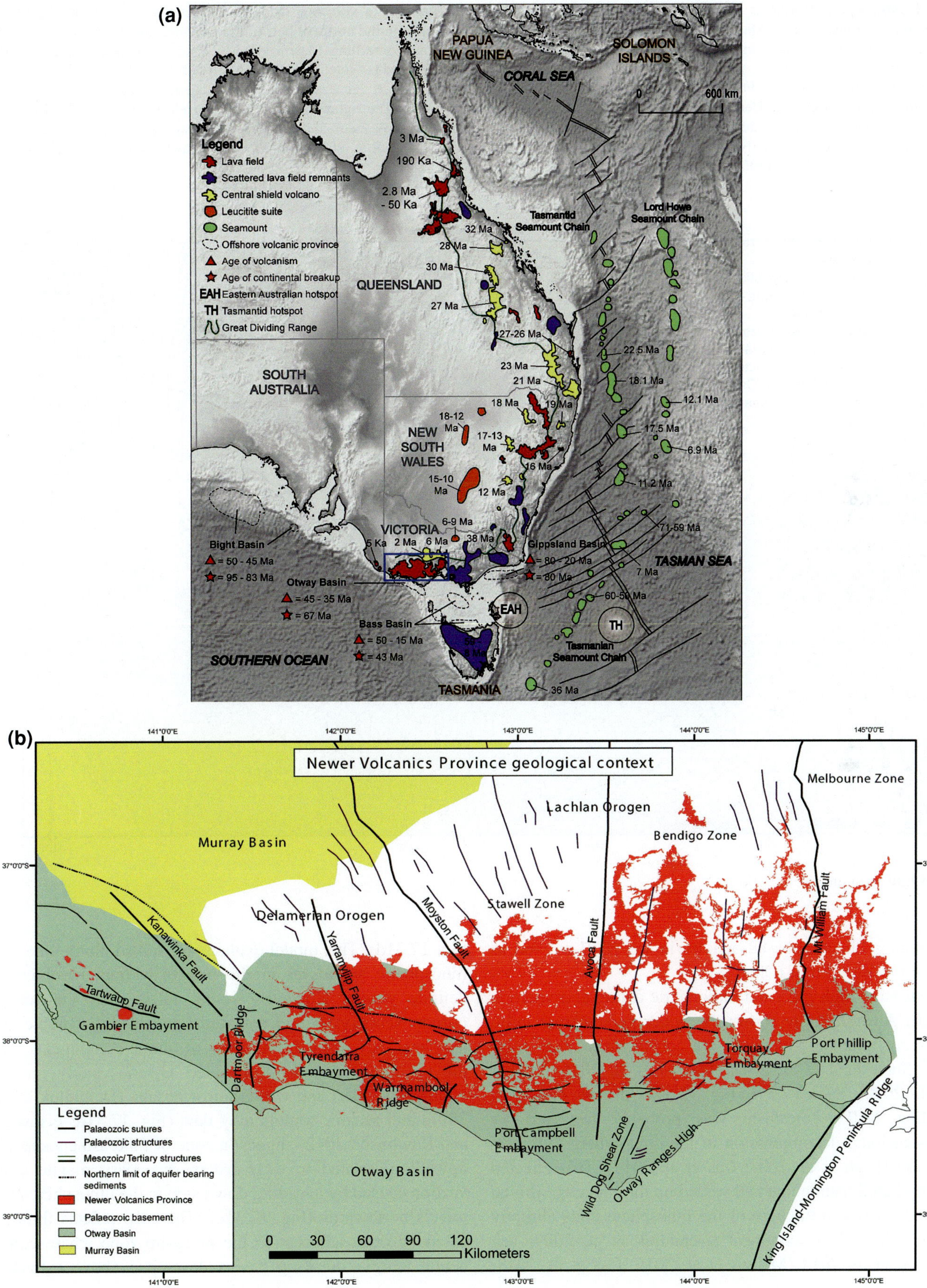
(a)
PAPUA NEW GUINEA
CORAL SEA
SOLOMON ISLANDS
0 600 km
Legend
Lava field
Scattered lava field remnants
Central shield volcano
Leucitite suite
Seamount
Offshore volcanic province
Age of volcanism
Age of continental breakup
EAH Eastern Australian hotspot
TH Tasmantid hotspot
Great Dividing Range
3 Ma
190 Ka
2.8 Ma - 50 Ka
32 Ma
28 Ma
30 Ma
27 Ma
27-26 Ma
23 Ma
21 Ma
18 Ma
19 Ma
18-12 Ma
17-13 Ma
15-10 Ma
12 Ma
16 Ma
6-9 Ma
38 Ma
5 Ka
2 Ma
6 Ma
Tasmantid Seamount Chain
Lord Howe Seamount Chain
22.5 Ma
18.1 Ma
12.1 Ma
17.5 Ma
6.9 Ma
11.2 Ma
71-59 Ma
7 Ma
60-50 Ma
36 Ma
QUEENSLAND
SOUTH AUSTRALIA
NEW SOUTH WALES
VICTORIA
TASMANIA
TASMAN SEA
SOUTHERN OCEAN
Bight Basin
= 50 - 45 Ma
= 95 - 83 Ma
Otway Basin
= 45 - 35 Ma
= 67 Ma
Bass Basin
= 50 - 15 Ma
= 43 Ma
Gippsland Basin
= 80 - 20 Ma
= 80 Ma
EAH
TH
Tasmanian Seamount Chain
(b)
Newer Volcanics Province geological context
141°0'0"E
142°0'0"E
143°0'0"E
144°0'0"E
145°0'0"E
37°0'0"S
38°0'0"S
39°0'0"S
Murray Basin
Lachlan Orogen
Melbourne Zone
Bendigo Zone
Stawell Zone
Delamerian Orogen
Moyston Fault
Yarramyljup Fault
Avoca Fault
Mt William Fault
Kanawinka Fault
Tartwaup Fault
Gambier Embayment
Dartmoor Ridge
Tyrendarra Embayment
Warrnambool Ridge
Port Campbell Embayment
Torquay Embayment
Port Phillip Embayment
Wild Dog Shear Zone
Otway Ranges High
King Island-Mornington Peninsula Ridge
Otway Basin
Legend
Palaeozoic sutures
Palaeozoic structures
Mesozoic/ Tertiary structures
Northern limit of aquifer bearing sediments
Newer Volcanics Province
Palaeozoic basement
Otway Basin
Murray Basin
0 30 60 90 120 Kilometers

◄ **Fig. 17.30** Newer Volcanics Province of southeastern Australia. **a** Regional setting of late Mesozoic-Cenozoic volcanic regions of eastern Australia; blue box in south = Newer Volcanics Province (From Meeuws et al. 2016; see also Fig. 17.59). **b** Relationship between the Newer Volcanics Province and the Paleozoic basement to the north, and late Mesozoic-Cenozoic Otway Basin to the south. The Newer Volcanics Province trends E–W, structural trends in the Paleozoic trend north–south to NNW, whereas structures in the Otway Basin trend EW to WNW. **c** Detailed crustal structural map. This shows the location of volcanoes (dots) in relation to faults in the Paleozoic basement in the northern half of the province and, in the southern half, to those controlling Mesozoic and Tertiary sedimentary rocks of the Otway Basin. Note also the ENE-trending faults, which are upper crustal faults generated during Neogene compression and basin inversion. The figure shows a good correlation between the locations of many volcanoes and clusters of volcanoes, and known faults. (**b**, **c** from Cas et al. 2017; compiled and drawn by J. van Otterloo)

(c)

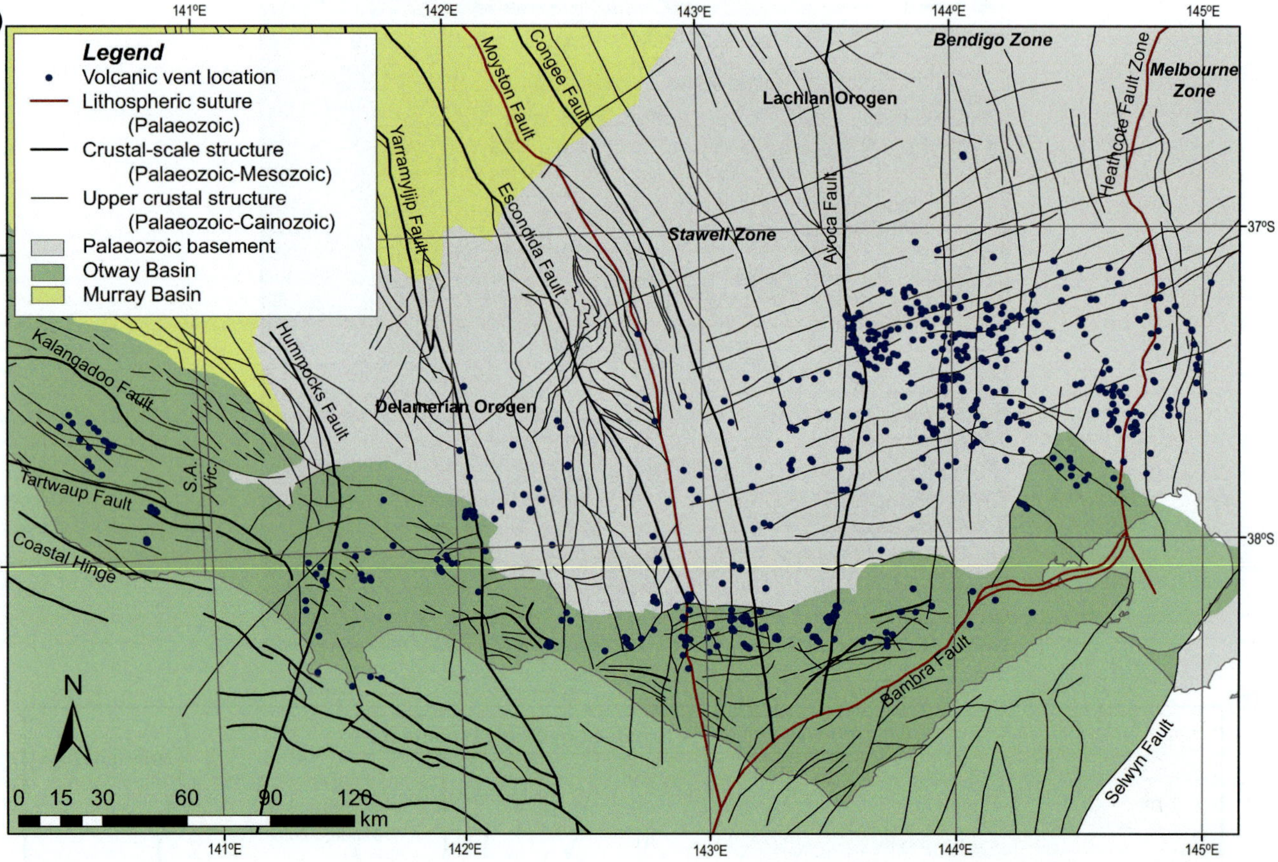

Fig. 17.30 (continued)

to fringing Pliocene basalt activity in the NVP as well as within the Bass Strait (Chap. 8).

Extensive high precision radiometric dating elsewhere in eastern Australia now shows the wider geodynamic influences on intraplate volcanism are considerably more complex (Sutherland et al. 2012; Meuws et al. 2016; Mather et al. 2020; Sect. 17.5.6, Fig. 17.59). In the east–west trending NVP, volcanism was probably triggered by trans-tensional decompression in the crust where fault sets intersect (Lesti et al. 2008; Cas et al. 2017). This affected hot, hydrated mantle that had welled up through edge-driven convection where the base of the lithosphere thins abruptly at the edge of the continent (Demidjuk et al. 2007; van Otterloo et al. 2014; Meuws et al. 2016; Cas et al. 2017; Fig. 17.31d). This model explains the mixed plume-like and lithospheric signatures in the basalts without involving a definite plume. However, young felsic melt indications (zircon xenocrysts) and signatures in xenoliths (noble-gas isotopes) may result from inputs of plume components into the upwelling system. A detailed U-series study on the youngest (55 ka) and westernmost basalts (Demidjuk et al. 2007) and seismic tomography data (Fig. 17.31d; Fishwick and Rawlinson 2012) suggest a zone of mild asthenospheric upwelling, related to edge-driven secondary mantle convection caused by a leading down step in the lithosphere as it travels northwards (Fig. 17.31e). This step reflects thinner, extended crust adjacent to the break-up zone where continental margin break-up basins developed above extended

Paleozoic crust, whereas to the north of the break-up basins, the Paleozoic crust is thicker. In addition, Mather et al. (2020) propose that interaction between lithosphere subducted at the New Zealand subduction system with the mantle transition zone beneath Australia generated intraplate magmatism in Eastern Australia, as appears to have occurred in eastern China, associated with the subducting slab flattening out in the mantle transition zone, passing under Australia, delivering fluids and sediments into the mantle transition zone and triggering partial melting that generates magmas for the intraplate provinces.

An interesting interpretation of the NVP was proposed by Lesti et al. (2008) who proposed that volcanism has developed at the tip of the northward projection of Tasman Fracture Zone into the Australian continental plate (Fig. 17.32a). The Tasman Fracture Zone in the Southern Ocean is one of the major transform faults that accommodate considerable offset of the Southeast Indian mid-oceanic spreading ridge, which was characterised by significant pulses of spreading during the Cenozoic (Fig. 17.32b). The southern projection of the Tasman Fracture Zone cuts through Northern Victoria Land in Antarctica where it connects to the intense intraplate Miocene-Holocene alkali basalt-trachyte volcanism of the McMurdo Volcanic Group (Rocchi et al. 2003; Vignaroli et al. 2015) broadly contemporaneous to the NVP. From this larger perspective it is possible to see how "push" associated with spreading of the Southern Ocean ridge system cannot be fully accommodated by Antarctic plate motion, which is in a stable polar position almost fully encircled by spreading ridges. At the same time, the northward motion of the Australian plate is modulated by the rate of subduction below the Pacific and Asian plates. Under these boundary conditions it is suggested both northward and southward transcurrent movement is propagated well beyond the ocean transform zone into their respective intra-continental settings (Salvini et al. 1997).

As described above, the clustering of the more than 400 eruption points in the NVP suggests their strong tectonic control on magma emplacement (Fig. 17.30c). Structural data indicate the Otway basin has a complex post-rift tectonic history, characterised by intermittent extensional, compressional and transcurrent phases (Lesti et al. 2008). In particular, inversion during the Pliocene and Quaternary has been associated with NW–SE oriented regional compression, which parallels the present-day stress field (Sandiford 2003; Fig. 17.32a). The major importance of the NW–SE alignment as magma pathways would reflect the opening (or reopening) of extensional fractures favourably oriented parallel to the orientation of the maximum principal stress (σ_1). This would be best accommodated in a trans-tensional-transpressional environment controlled by plate boundary forces, rather than in a typical continental passive margin (Hillis et al. 2008). In this framework, both the NVP (along with its precursory late Cretaceous-Miocene Older Volcanics Province; Wellman and McDougall 1974; Price et al. 2014) in southeast Australia and the McMurdo Volcanic Group in Northern Victoria Land, though several thousands of km apart and on two different plates, are potentially linked geodynamically. It is suggested that transform-related, far-field stresses have interfered with the "normal" evolution of the continental passive margins, inducing decompression and coeval extraction of intraplate magma at these two specific sites (Lesti et al. 2008; Cas et al. 2017).

It is also important to note, the Tasman Fracture Zone extends across the major Paleozoic suture zone that occurs in southeastern Australia between the double-vergent Tasman Fold Belt and its equivalent terranes in Antarctica (e.g. Miller et al. 2005; Boger and Miller 2004). This pre-existing zone of weakness likely controlled the geometry of the Jurassic continental break-up, subsequent evolution of the Southern Ocean Ridge, the loci where transcurrent faulting propagated beyond the transform zone, and the fertility of the tectonically-decompressed mantle in this particular zone during the Miocene-Holocene. The existence of coeval twin intraplate, transform-related volcanic provinces, such as the NVP and McMurdo Volcanic Group, envisages a new geodynamic scenario, where genetically related volcanism can occur on two different plates. Other potential candidates of such types of volcanism can be recognised across the Atlantic, where chains of seamounts and emergent islands, like the Canary archipelago (Chap. 16), are located along the projection of transform zones into continental passive margin intraplate settings.

For further discussion on the wider geodynamic influences of subduction-related processes on intraplate volcanism, in particularly along the East Asia- and eastern Australian-Pacific plate margins, see Sect. 17.5.6. Here we describe novel concepts that argue some active intraplate volcanoes are not hot spot-related to deep mantle plumes but are caused by convergent plate tectonic processes in the upper mantle.

17.4.2.1 Ancient Intraplate Continental Volcanism: Characteristics

Intraplate continental volcanic provinces should have moderate preservation potential in the rock record. The major counteracting influence is long-term post-eruptive erosion in a tectonically, relatively stable continental setting; indeed, we are not currently aware of any well-described ancient examples (other than larger flood basalt systems associated with rifting). Volcanic successions of this type, erupted upon a stable continental rock record, should be recognisable by virtue of a regionally extensive basaltic suite erupted upon a demonstrable continental (sialic) basement. The effect of the outpouring of fluidal basaltic lavas into valley topography is

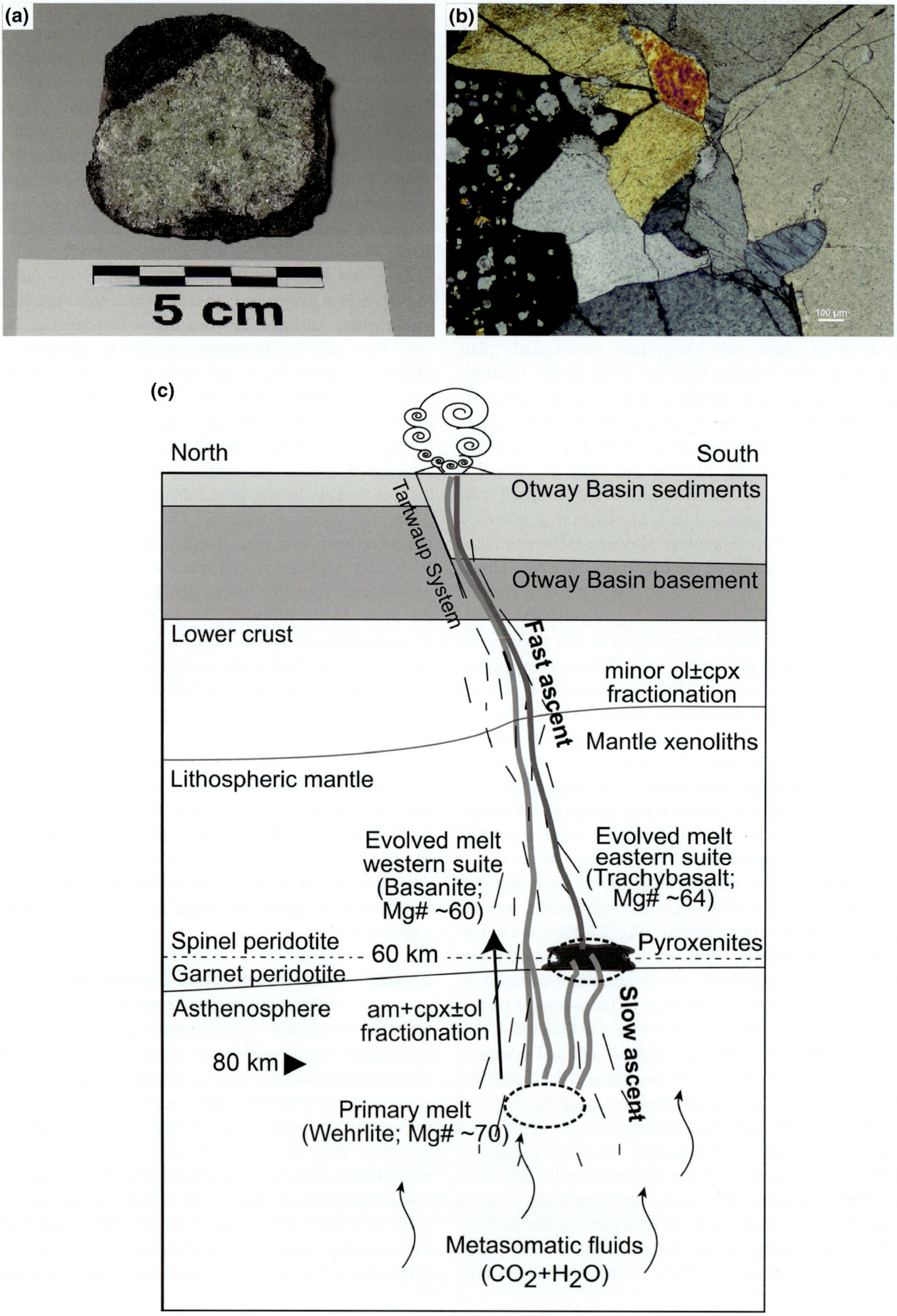
(a)
5 cm
(b)
100 μm
(c)
North
South
Otway Basin sediments
Tartwaup System
Otway Basin basement
Lower crust
Fast ascent
minor ol±cpx fractionation
Mantle xenoliths
Lithospheric mantle
Evolved melt western suite (Basanite; Mg# ~60)
Evolved melt eastern suite (Trachybasalt; Mg# ~64)
Spinel peridotite
60 km
Pyroxenites
Garnet peridotite
Asthenosphere
am+cpx±ol fractionation
Slow ascent
80 km
Primary melt (Wehrlite; Mg# ~70)
Metasomatic fluids (CO_2+H_2O)

◀ **Fig. 17.31** Geodynamic characterisation of the Newer Volcanics Province, southeastern Australia. **a**, **b** Photograph of a volcanic bomb with a lherzolite xenolithic core, and photomicrograph of the rim of a spinel lherzolite from the Mt Gambier volcanic complex (From Cas et al. 2017). **c** Petrogenic model for the magmatic source and evolution of the Mt Gambier basalts; ol = olivine; cpx = clinopyroxene; Mg# = magnesium number. Depths are approximate based on the results of melting models realtive the southeastern Australian geotherm (From van Otterloo et al. 2014). **d** Depth section at 50 km from body wave tomography with superimposed topography, bathymetry and the locations of Cenozoic volcanics (From Fishwick and Rawlinson 2012). The fact that almost all volcanic centres and lava fields overlie low-velocity regions is likely to be related to the presence of a thin lithosphere. In the case of the NVP, which does not overlie significantly elevated topography, the lower velocities that can clearly be observed in the underlying mantle are most likely due to the elevated temperatures associated with a hot-spot source. The fact that the lower velocity anomalies cannot be observed below 150 km depth could indicate that the hot spot is localised in the uppermost mantle. If so, the presence of a thermal anomaly of this magnitude may be explained by edge-driven convection. **e** Schematic representation of the edge-driven convection model proposed to explain volcanism in the NVP (From Demidjuk et al. 2007; also see text)

Fig. 17.31 (continued)

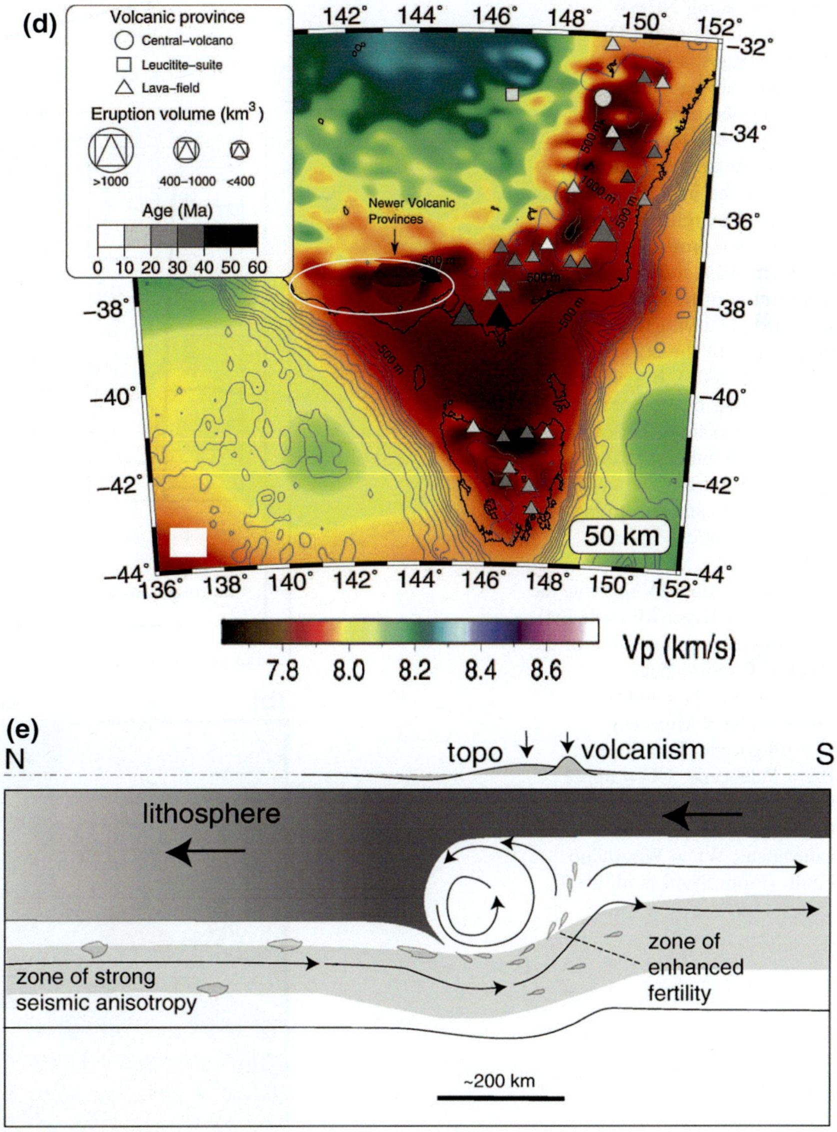

to reorganise the drainage, in some cases damming up valleys, producing swamps and lakes. Fluvial, swamp, and lake sediments should be interspersed with the basaltic volcanics (see Fig. 16.39a, b) but are probably best developed below and above the relatively short-lived volcanic succession. Intraplate basaltic volcanic provinces along Australia's eastern seaboard extend back to the Cretaceous in age. Kimberlites could also be expected in those settings where intraplate volcanism occurs in proximity to Archean-Paleo-Proterozoic cratonic crust.

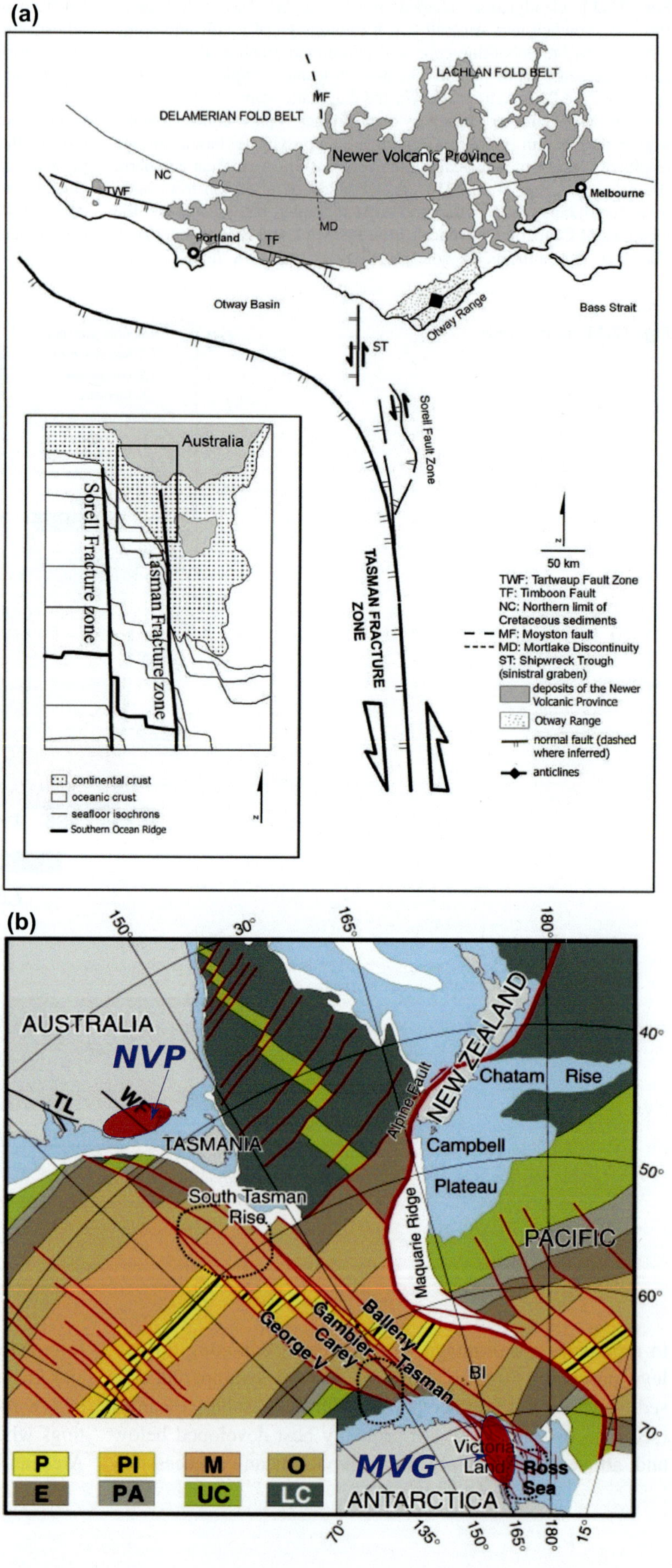

Fig. 17.32 Relationships between intraplate volcanic provinces in southeastern Australia and Antarctica to the Tasman Fracture Zone, Southern Ocean. **a** Tectonic setting of the Newer Volcanic Province in relation to the Tasman Fracture Zone. The major volcanic features and present-day stress field orientation are compatible with the orientation of the main tectonic structures as described in Fig. 17.30c. Intermittent basin inversions along NE-SW structures (such as the Otway Range) suggest a geodynamic scenario where a major role is played by the propagation of major N–S trending sinistral faults associated the Tasman Fracture Zone into the Australian continental margin. This interpretation promotes the role of the Tasman Fracture Zone on the reactivation of trans-tensional structures, and on triggering the magmatism of the NVP. (From Lesti et al. 2008). **b** Seafloor anomaly map of the Southern Ocean. This shows the location across the Southeast Indian spreading centre of the coeval, intraplate volcanism forming the Newer Volcanics Province (NVP) and the McMurdo Volcanic Group (MVG). Note the significantly larger Miocene [M] band of oceanic crust west of the Tasman Fracture Zone. P = Pliocene; PI = lower Pliocene; M = Miocene; O = Oligocene; E = Eocene; PA = Paleocene; UC = upper Cretaceous; LC = lower Cretaceous; TL = Torrens Lineament; WF = Woorndoo Fault. (From Storti et al. 2007)

17.5 Subduction-Controlled Volcano-Plutonic Arcs

Subduction zone magmatism is a direct response to tectonic and chemical processes operating at convergent margins (Plate 17.1). Early models (ca. 1970) emphasised magmatism was the result of frictional heating along the Benioff zone, melting of the subducting slab and compressional shortening of the overriding plate in the arc-trench gap. In contrast, modern thinking proposes most magmas originate from the *mantle wedge* above the subduction zone. This is by a process where hydrous fluids are released from minerals that break down during metamorphism of the subducting basalt slab and Layer 1 sediments, rise into the mantle and initiate partial melting (e.g. Wallace 2005; Grove et al. 2012; Grove and Till 2015; Xiao et al. 2016; Fig. 17.33). This process has been termed *flux melting* and produces mafic magma which rises and differentiates, forming the volcanic and intrusive rocks of the overlying arc. It is distinctively different from adiabatic decompression melting, because the melt continually reacts with the hotter shallower overlying mantle. If the melting conditions were to be adiabatic, no melting would occur. However, under special arc situations and geometries (e.g. tear of the subducting slab) decompression melting can also occur at subduction zones.

It is also now recognised that the process of subduction should be viewed in a dynamic way (e.g. Hamilton 1995), rather than as a "static", single instant in time view that portrays the subducting slab rolling over a stationary hinge and sliding down a slot fixed in the mantle. This approach has been revolutionised with the increasing availability of tomographic images of seismic velocity anomalies, which may be interpreted as thermal or compositional variations, as well as structural. Viewed in a dynamic way most subduction zones are characterised by retreat of the hinge with time (or *rollback*), and that the mantle wedge above the subduction zone is constantly replenished by an inflow of hot, undepleted mantle (Fig. 17.33a). This counteracts melt extraction in the depleting wedge, maintaining "steady-state" arc magmatism.

Two particularly distinctive "end-member" subduction-related magmatic suites, which we will discuss because of their apparent paleotectonic significance in Phanerozoic arcs, are *boninites* and *adakites*. Both have been attributed to anomalously high temperatures in the mantle wedge and/or subducted slab, either during arc initiation or during hot subduction (Crawford et al. 1989; Defant and

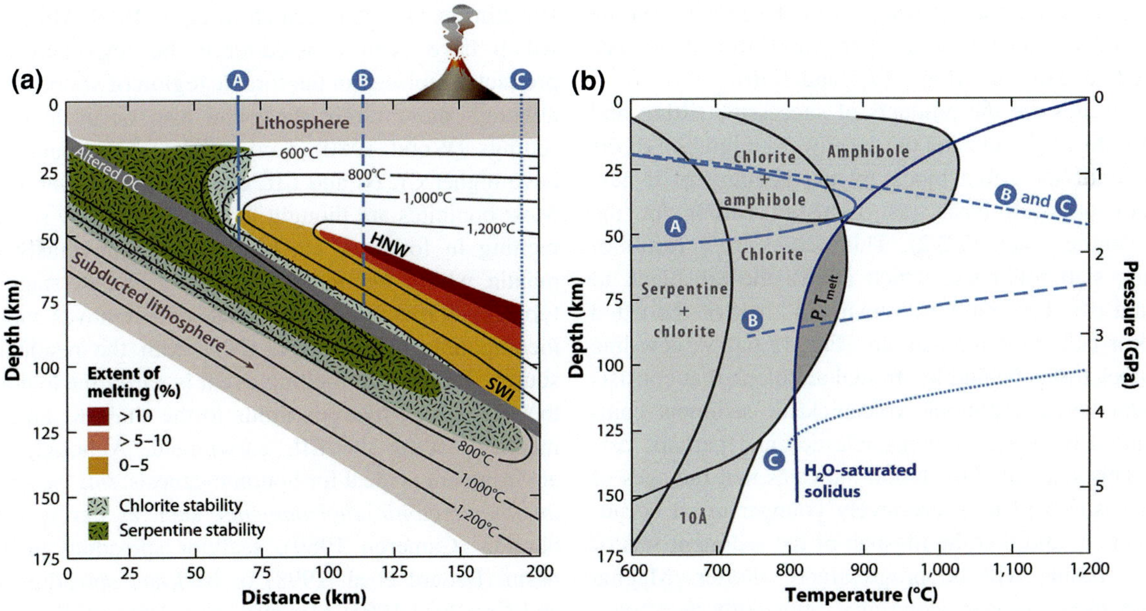

Fig. 17.33 Conditions limiting hydrous flux melting in subduction zones (From Grove and Till 2015). **a** Typical cross-section through a subduction zone using a thermal model for a slab dip of 30° and a convergence rate of 40 km Myr^{-1}. Also shown are the stability limits for high-H_2O (>10 wt%) minerals. The melting model predicts the extent of hydrous flux melting in the bottom half of the mantle wedge. HNW (hot nose of the wedge) denotes the hottest, shallowest nose of the mantle wedge where the maximum amount of melting occurs. The part of the mantle wedge directly above the interface between subducted oceanic lithosphere and the mantle is indicated as SWI (slab-wedge interface) and extends from path A to path C; OC = oceanic crust. **b** Phase diagram with the temperature-depth paths from **a** superimposed. The dark grey region labelled P,T_{melt} shows the region of pressure–temperature space where the stability of hydrous phases in the base of the mantle wedge allows H_2O-saturated melting to begin

Drummond 1990). Changes in the nature of recycled material, and lithologies of the overriding plate may also influence the composition of the volcanic products erupted, and intrusive rocks emplaced into the arc.

17.5.1 Young Oceanic Island Arcs

The locus of volcanism is typically a narrow line or chain of volcanoes, some submerged and some emergent above sea level, lying parallel to a deep-sea trench (Plate 17.1; Fig. 17.34a). The trench is the site of subduction or underplating of ocean crust including seamounts and aseismic ridges (sometimes termed buoyant bathymetric impactors). The trench and line of volcanoes (see Fig. 17.7) may have a straight trace (e.g. Tonga-Kermadec Islands), or more generally be curved or arc-shaped (e.g. Mariana arc). The arc system has a regular structure: trench; inner trench wall; forearc basin slope (or arc-trench gap), perhaps above an accretionary prism; arc block, including an outer non-volcanic line of islands; volcanic arc-back-arc/marginal sea/interarc basin; and, in some systems, a remnant or third arc, all developed above an arc-ward-dipping seismic (or Benioff) zone (Fig. 17.6). Arcs are at least several hundred km long, the line of active volcanoes is less than 50 km wide, the distance between the active arc and trench may be tens of km or more, and the width of the basin between the active frontal and inactive remnant arc varies from a few tens to hundreds of km wide (Figs. 17.5 and 17.6).

Remnant arcs are the products of successive rifting and splitting of the arc block, and the oceanward migration of the frontal half of the rifted block by the opening of a new interarc or back-arc basin (as discussed earlier for the western Pacific; Sect. 17.2.2). This occurs as a result of subducting slab roll back, which allows the arc block to extend and rift. The other or "remaining" half of the rifted arc block is called the remnant arc (Fig. 17.6a). A new line of volcanoes forms behind the frontal arc block that consists of an older volcanic-plutonic-volcaniclastic sediment complex mantled by younger volcaniclastics and perhaps carbonates. This sequential evolution indicates that the ages of marginal basins will be successively younger in an oceanward direction, and that the lifespan of the sediment fills of successive basins will be progressively shorter. Magma production rates can vary along arcs, apparently as a function of variable along-arc subduction and back-arc spreading rates (Sample and Karig 1982).

In these settings, island marine stratovolcanoes generally have a definable stratigraphy (Chap. 16; see Fig. 16.50), with a volcanic-intrusive core, flanked by volcaniclastic aprons. The sediment fills of the associated basins will be volcaniclastic debris avalanche and mass-flow deposits laid down as aprons on the flanks of the arc volcanoes, as thinning mass-flow sheets in the near reaches of associated basins, and as wedges and sheets in the forearc basin (Fig. 17.6a). Carbonate debris may also be included in these deposits. Pelagic sediments will also be important. During the active life of the volcanic arc, pelagic sediments will be best represented furthest from the active arc apron but are thin or non-existent in the basin centre. In volcanically inactive basins pelagic sediments will also mantle the previously active volcanic apron and will be interspersed with a declining proportion of epiclastic volcaniclastic sediments derived from the remnant arcs. Whereas some arcs are wholly intra-oceanic, others develop in relatively close proximity to continental masses. In ancient successions terrigenous clastics in basinal sequences can be used as an indication of proximity to continental crustal masses (e.g. Cas et al. 1980).

Most young island arc rocks are basalts or basaltic andesites of "island arc tholeiite" character derived from the subducting lithosphere and/or the overlying mantle and crust (Gill 1981, 1987; Hawkins et al. 1984; Gill et al. 1984; Hawkesworth et al. 1993; Murphy 2007 and references therein). On the other hand, *boninites* are a distinctive but generally volumetrically small group of rocks, named for their occurrence in the Izu-Bonin-Mariana arc south of Japan. Boninites are high-MgO (up to 15% MgO) *primitive* andesites containing 56–60% SiO_2 (Cameron et al. 1983; Hawkins et al. 1984; Crawford et al. 1989; Murphy 2007), which have been considered to be important magmatic products emplaced in the forearc region of young island arcs, although more rarely these can also occur in continental settings (Wood 1980). Compared to other andesites they have higher Cr, Ni and LIL elements, but lower Ti and Nb. Most boninites are thought to have formed by second stage melting in forearcs, via hydration of previously depleted mantle within the mantle wedge. The first stage melting typically forms island arc basalt and removes most of the incompatible elements (Ti, Nb) from the residual mantle source. The second melting event is partly made possible by the addition of hydrous fluids to the shallow, hot, depleted mantle, leading the LIL enrichment. Although a forearc environment is ideal for boninite genesis, *this may take place in other tectonic environments*, such as during ridge subduction (Cameron 1989), seafloor spreading in a forearc basin (Bedard et al. 1998) or back-arc spreading (Falloon and Crawford 1991; Monzier et al. 1993). Silicic rocks can also be found as early stages of oceanic arc development (Karig 1970; Gill 1984; Dobson et al. 2006). Dobson et al. (2006) describe an Eocene submarine boninite volcanic centre exposed on the island of Chichijima, Bonin Islands, which includes andesite, dacite, quartz dacite, and rhyolite. Trends for both major and trace element data suggest that the more evolved lavas were derived primarily through high-level fractional crystallisation of boninite. Boninites

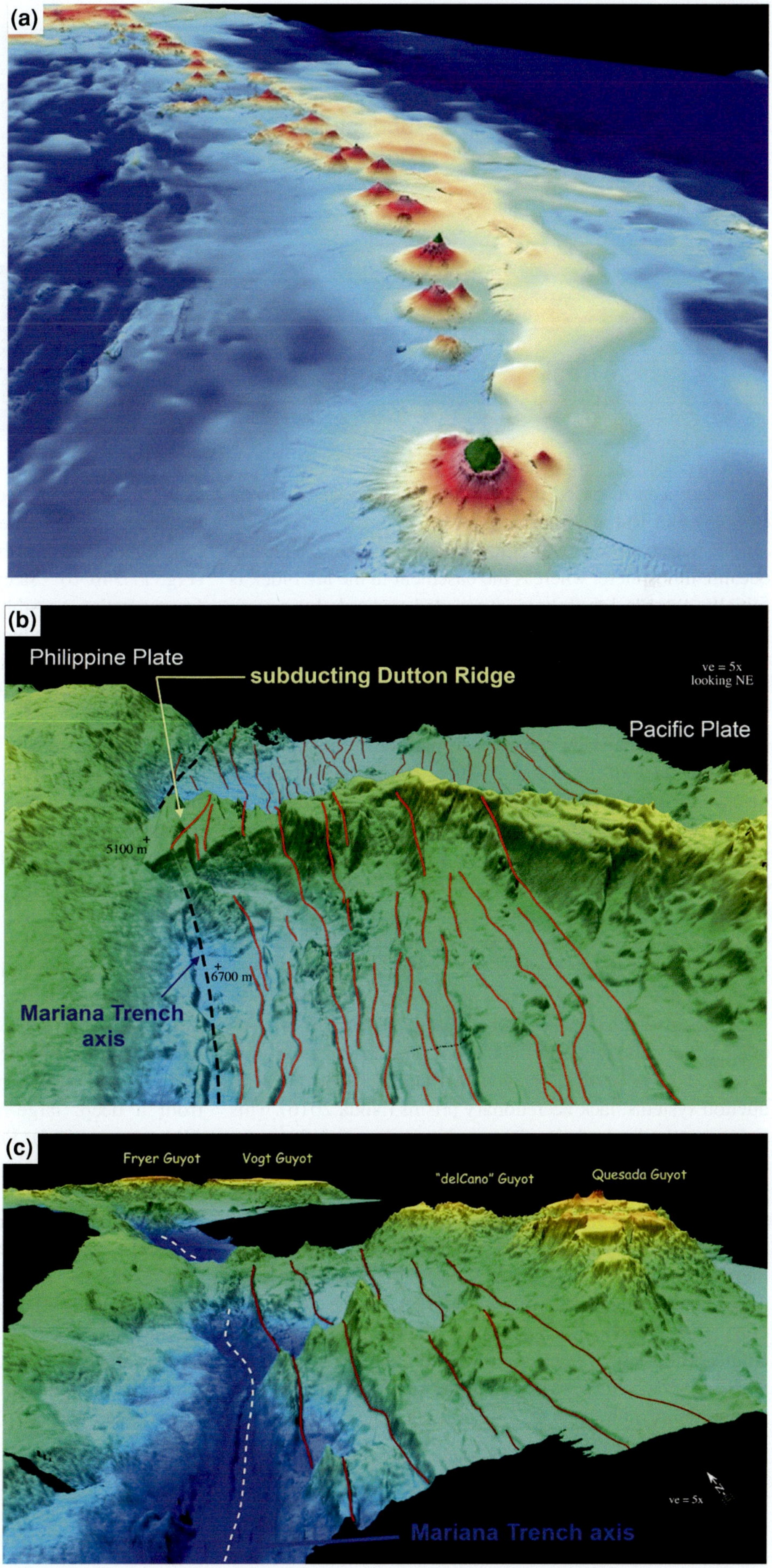

Fig. 17.34 Some selected features of young, oceanic island arcs and trenches, as exemplified by the Mariana Arc, western Pacific (see also Figs. 17.6 and 17.7a). **a** Seafloor bathymetry showing chain of both submerged and emergent (green) volcanoes lying parallel to deep-sea trench (right). Note the extent of the forearc region and the Mariana Trough back-arc basin to the west (left), characterised by extensional fault topography. Vertical exaggeration is 3x. (*Source* NOAA Ocean Explorer; Image created by Susan Merle). **b** Perspective view of bathymetry showing Dutton Ridge being subducted into the Mariana Trench. Numerous extensional faults (red lines) disrupt the seafloor and the ridge proper. (*Source* CCOMJHC, University of New Hampshire). **c** Perspective view of the bathymetry looking NNE along the Mariana Trench (white dashed lines), this shows several guyots (or flat-topped seamounts) that are being carried into the trench on the Pacific Plate. The red lines again mark major extensional faults that disrupt both the Pacific Plate and the seamounts. (*Source* CCOMJHC, University of New Hampshire). **d** Schematic representation of fragmentation of seamount entering subduction zone (From Staudigel and Koppers 2015)

Fig. 17.34 (continued)

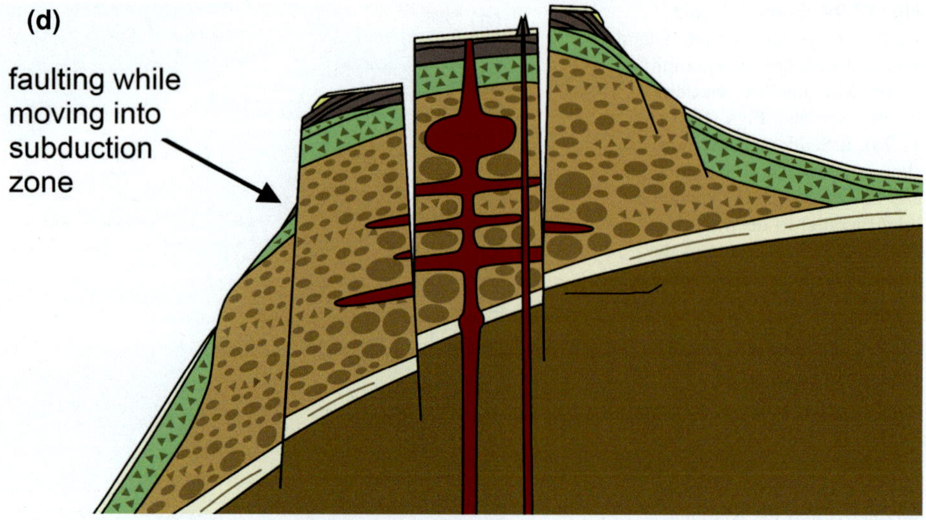

have also been recorded in the northeast Lau back-arc basin (Embley and Rubin 2018).

The basement or crust in the surrounding basins is clearly oceanic lithosphere, whereas the basement in the arc block is initially oceanic lithosphere, succeeded upwards by a variably deformed volcanic-volcaniclastic-pelagic sediment-mafic intrusive complex, about 15 km thick (Karig 1970, 1974; Fig. 17.6a). The arc block itself is often characterised by arrays of normal faults, particularly towards the rear (Fig. 17.6a). The inner trench wall may be made up of an imbricate stack of thrust faulted slices or off-scraped slivers of oceanic crust (ophiolite) and deep-water sediments, both pelagic and redeposited volcaniclastic in origin, in which the faults dip towards the arc (Fig. 17.6a). This deformed, off-scraped sediment wedge, or accretionary prism may be subducted some depth and subjected to high-P/low-T *blueschist* metamorphism (Miyashiro 1972; Ernst 2010). Tectonic *mélanges*, sometimes remobilised as diapiric masses (e.g. Timm et al. 2014), may also be produced at this time.

Although accretionary prisms should be sought-after as evidence of subduction in the rock record, it is apparent that several, and in particularly young, modern island arc subduction systems, lack accretionary prisms (Noda 2016). This is apparently due to their relatively low sediment influx rates and to a low degree of frictional coupling between the plates. At the beginning of subduction initiation between two plates the forearc setting may be extensional until the sinking plate retreats the hinge to obtain a sufficient down-dip motion (e.g. Stern and Bloomer 1992; Stern 2004). This extension leads to subsidence associated with normal faults in the overriding crust (Fig. 17.34b–d). Pelagic and terrigenous materials simply fill these extensional accommodation spaces, so allowing what sediment there is to be subducted in step-like graben sediment traps in the down-going plate (Uyeda 1983; von Huene 1984; Noda 2016, and references therein). The transport of bathymetric impactors to subduction zones typically ends with their descent into the trench (Fig. 17.34b–d). Steeper subduction likely involves their fragmentation possibly leading to slab tears/holes at deeper depths, while subduction into a shallower sloping accretionary prism typically causes them to "plough" into the sediment deforming the front of this wedge; nevertheless, seamounts are known to have accreted onto a continent (e.g. MacPherson 1983).

Tomographic imaging has emerged as one of the most instructive methods for analysis of the structure and dynamics of subduction zones (e.g. Spakman et al. 1988, 1993; Zhao et al. 2001, Zhao 2012). These studies have revealed down-going slabs do not have simple geometries and may have complex histories involving roll back or steepening, tearing, buckling and folding and polarity reversal. High-resolution data is particularly available from dense seismic networks in Japan and East China (Zhao 2012). This has enabled the subduction zone geometry and mantle tomography below the Izu-Bonin-Mariana island arc system to be determined (Fig. 17.35a, b), including results from a deep, large-magnitude (7.8) earthquake in 2015 (Zhao et al. 2017). Prominent dipping high-velocity (blue) zones are imaged in the upper mantle down to a depth of ~410 km (top of the mantle transition zone, MTZ). These correspond to the subducting Pacific and Philippine Sea slabs in which intermediate-depth and deep earthquake activity occurs. The active arc volcanoes are located above low-velocity (red) zones in the mantle wedge above the subducting slabs. The subducting Philippine Sea slab is also imaged clearly to the west of the Ryukyu Trench, indicating that the Philippine Sea slab has also reached the MTZ depth. North of latitude ~28° N the Pacific slab becomes flat in the

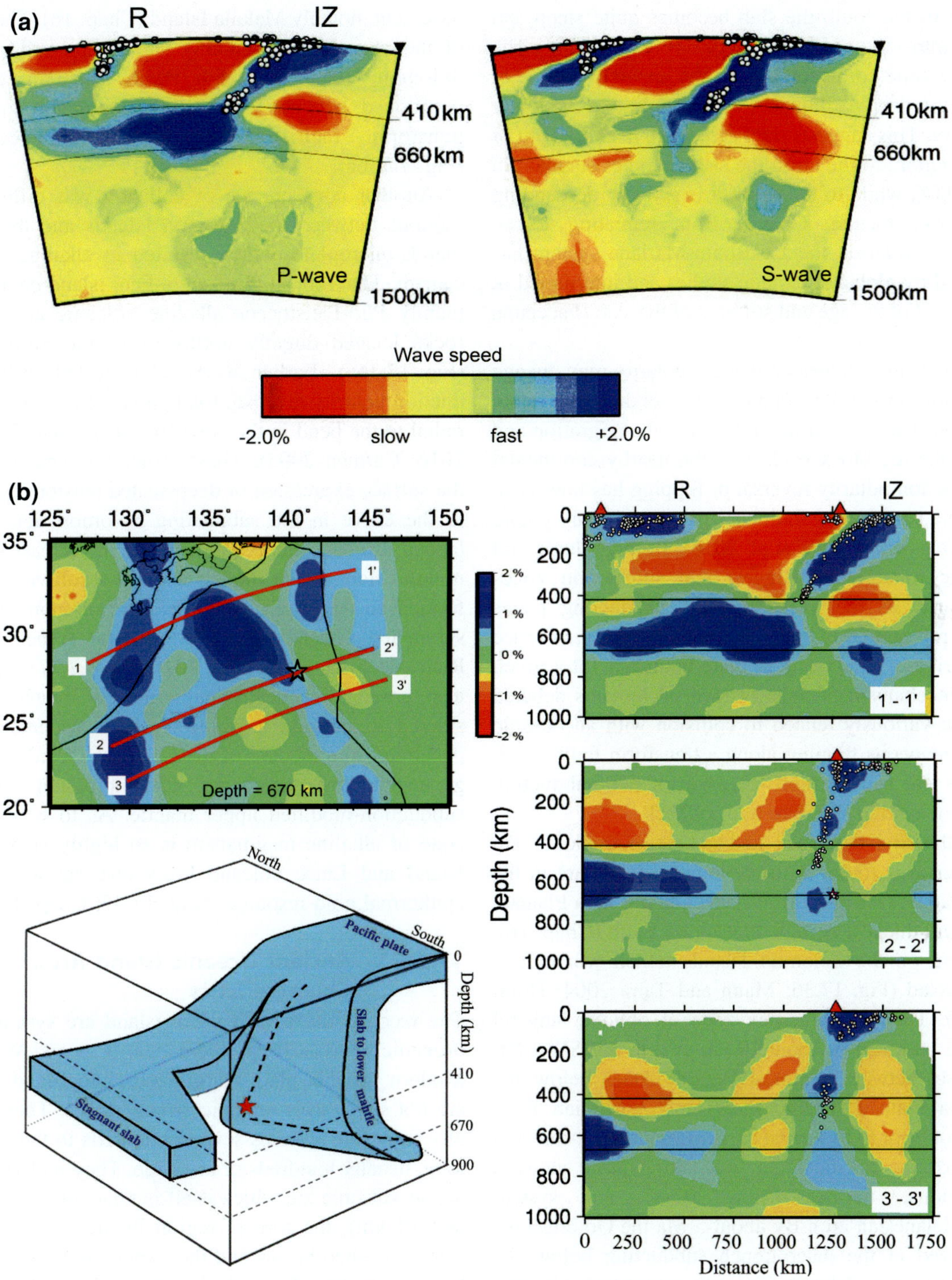

Fig. 17.35 Vertical cross-sections of seismic tomographic images through the Izu-Bonin and Ryukyu trenches, northwest Pacific. High and low-velocity perturbations are shown in blue and red colours respectively; scale (in %) as indicated. **a** Older study showing P- (left) and S-wave (right) data which reveal lateral variations in rigidity of deep slabs and mantle; profile at ~31° N latitude; IZ = Izu-Bonin trench (Pacific slab), R = Ryukyu trench (Philippines Sea slab); open dots depict earthquake hypocentres of magnitude >5.5; lines represent the 410 and 660 km discontinuities which define the Mantle Transition Zone (MTZ). (From Widiyantoro et al. 1999). **b** Tomographic results and interpretation of P-wave data from 2015 Bonin deep, large (magnitude 7.9) earthquake. (Top left) Inset map and location of cross-sections; map view depth = 670 km which is at the MTZ-lower mantle boundary; star is epicentre of the 2015 Bonin deep earthquake. (Right) Vertical P-wave cross-sections along the profiles shown in inset map; active volcanoes are shown in red triangles; white circles denote seismicity within a 30-km width of each profile; two lines represent the MTZ; red star is hypocentre 2015 Bonin earthquake. (Bottom left) Interpretation showing the subducting Pacific slab and position of slab tear relative to hypocentre of 2015 Bonin earthquake (red star). (From Zhao et al. 2017)

MTZ, while to the south the slab becomes quite steep and penetrates into the lower mantle (Fig. 17.35b). The high-velocity zone corresponds with the apparent source of the 2015 Bonin deep earthquake, which is close to the lower slab boundary. This suggests the Pacific slab is likely split or torn in this region; to the north the slab becomes stagnant in the in the MTZ, while to the south it is actively descending into the lower mantle. Change from retreating to an advancing trench-arc at the Izu-Bonin-Mariana system has been revealed by global kinematic studies and interpreted in terms of control of the age and stiffness of the slab (Faccenna et al. 2009).

Karig (1972) first suggested that arc polarity may change in time, leading to the development of a subduction system in the previous back-arc basin, and then to the migration and collision of the arc block back onto the nearby continental mass. Subduction polarity reversal or flipping has now been shown to be a common evolutionary pattern of many young island arcs (e.g. von Hagke et al. 2016). Both geologic and now seismic tomography suggest that subduction zones change polarity as the overriding and subducting plates switch their roles either in time (at a given location), or space along the strike of a convergent plate boundary (at a given time). The mechanism of polarity reversal remains debated but has been variously related to collision with an oceanic plateau, spontaneous flipping along a transform fault, propagating slab tear and breakoff, collision of two subduction zones, or propagating slab tear and rollback.

One of the first described and a type example of arc polarity reversal is found in the Solomon Islands arc in the southwest Pacific Ocean. Currently the Ontong Java Plateau, the largest and thickest oceanic plateau on Earth (Chap. 16), is converging on the Solomon Islands trench where it is being subducted (Fig. 17.36; Mann and Taira 2004; Holm et al. 2016). This is a complex zone of oblique sinistral convergence (at a rate of about 10 cm yr^{-1} in a WSW-ENE direction), set between opposed Benioff zones below the Ontong Java Plateau (Pacific Plate) and Australian Plate (Fig. 17.36a). Prior to its arrival ca. 6 Ma southwest-directed subduction of the Pacific Plate was taking place along a semi-continuous, northeast-facing island arc system (Northern Melanesian arc). By about 5 Ma the Ontong Java Plateau arrived at the paleo-trench subducting below the present-day Solomon Islands. This was associated with initiation of subduction and polarity reversal creating the San Cristobal trench (Fig. 17.36b). Whereas, along the paleo-trench, arrival of the plateau "choked" the subduction zone causing it to step seaward, forming the present-day North Solomon trench. Subduction of the Ontong Java Plateau continues today along the North Solomon trench with about 20% of the upper 30 km-thick plateau being transferred by subduction to the overriding accretionary prism, creating some on-land exposures of basaltic basement, notably Malaita Island (Chap. 16). The lower 80% of the plateau is subducting southwest-wards beneath the Solomon Island arc, with Pacific-Australia convergence partitioned between two active subduction zones, and a transform fault marking the pre-Pliocene trench (Fig. 17.36a).

Another consequence of collision was differential plate motions between the Solomon Islands and the North Bismarck microplate, which resulted in alkaline oceanic volcanism. The Tabar-Lihir-Tanga-Feni island chain comprises mainly Plio-Pleistocene alkaline volcanic and subvolcanic rocks located directly northwest of the Solomon Islands (Fig. 17.36a). Earlier workers suggested, following subduction polarity reversal, rotation caused extension on faults radial to the bend in the New Britain Trench (Lindley 1988, 2016; Carmen 2003). These structures were interpreted as the surface expression of deep-seated tensional gashes radial to the curve in the subducting Solomon Sea slab, which tapped asthenospheric partial melts in the east New Britain region. More recent models suggest a sub-vertical slab tear formed on the nose of the curve in Solomon slab as it was subducted beneath the New Britain Trench (Holm and Richards 2013; Holm et al. 2016; Fig. 17.37). The tip of the tear terminates beneath southern New Ireland and is considered to provide a window where asthenosphere could penetrate from the rear of the slab to the front, with magmas generated by adiabatic decompression melting of subduction-modified upper mantle. As to why this limited zone of alkaline magmatism is so highly significant, Lihir Island and Luise volcano hosts one the world's premier epithermal gold resources (Fig. 17.37; Chap. 18).

17.5.1.1 Ancient Oceanic Island Arcs: Characteristics

The recognition of an ancient island arc system within an orogenic belt can be problematical (Cas and Wright 1987). Modern oceanic plate margin, volcanic island arc systems are not small-scale tectonic arrangements. They are regionally extensive systems several hundreds to thousands of km long, usually hundreds of km wide. They include at least the active volcanic arc which itself is relatively narrower (a few tens of km), the forearc region including the accretionary prism, perched basins and the trench, and they may include back-arc basins. The volcanic arc consists of a line of stratovolcanoes with basal diameters of at least 20 km. The spacing between individual stratovolcanoes in a modern plate margin-island arc system is in the order of 10^1–10^2 km or more. Therefore, in an ancient orogenic belt a linear volcanic belt only tens to a hundred km long and a few to tens of km wide does not necessarily constitute a plate margin-island arc system. This is especially so without the proven coexistence of a contemporaneous forearc accretionary prism system, which is the physical evidence of the

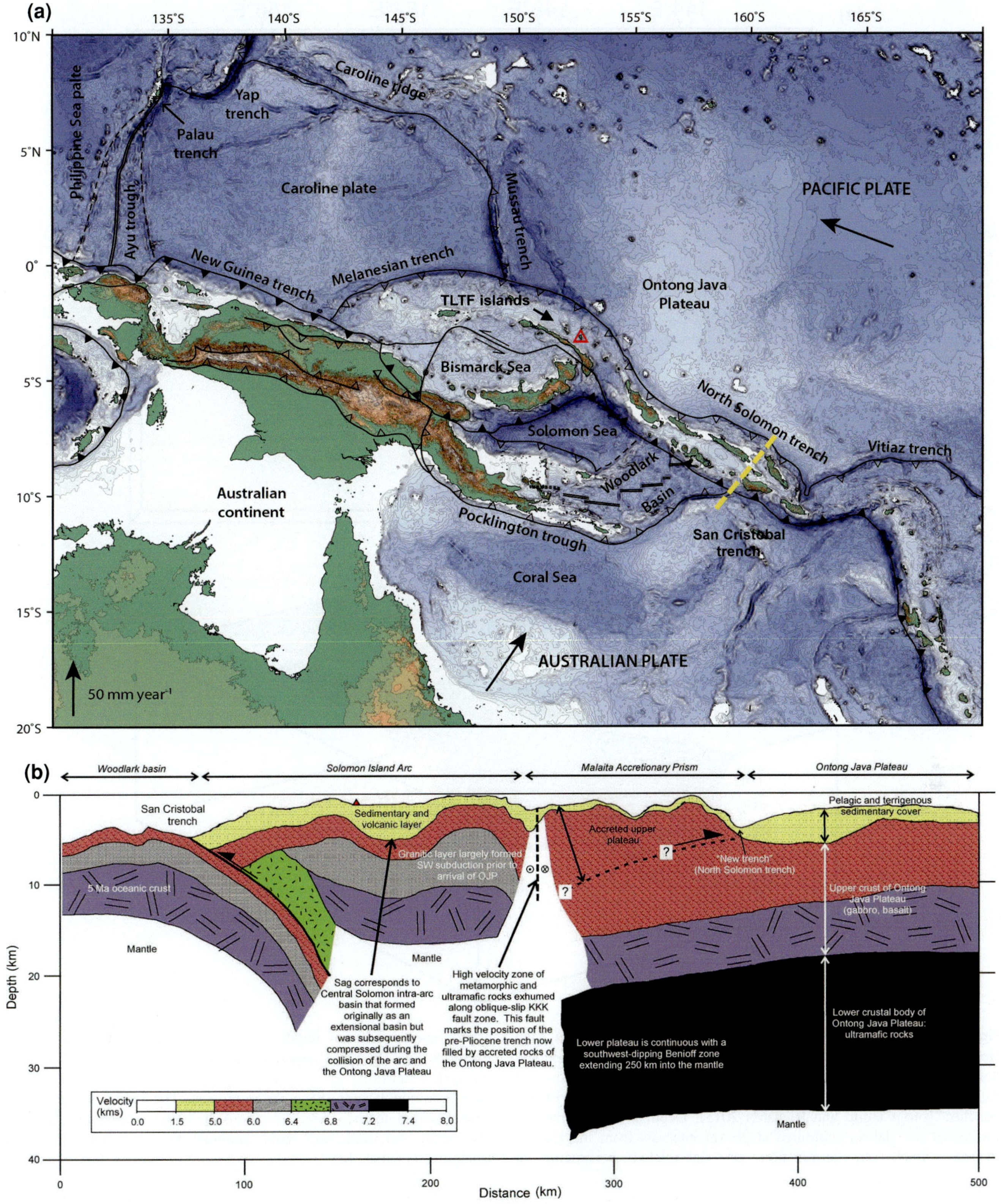

Fig. 17.36 Solomon Islands arc and trench system. **a** Topography, bathymetry and wider plate tectonic setting of the Solomon Islands convergent zone. Arrows indicate rate and direction of plate motion of the Australian and Pacific Plates; dashed yellow line is interpreted seismic transect shown below; TLTF is the alkalic Tabar-Lihir-Tanga-Feni island chain; red triangle locates Lihir Island hosting the supergiant Ladolam epithermal gold deposit. (From Holm et al. 2016). **b** Geophysical and geologic interpretation of combined reflection and deep refraction data in a cross-arc transect of the Solomon Islands. Note the location of the pre-Pliocene paleo-trench (northeast-facing North Melanesian Arc) now located along the Kia-Kaipito-Korigole (KKK) transform fault. (From Mann and Taira 2004)

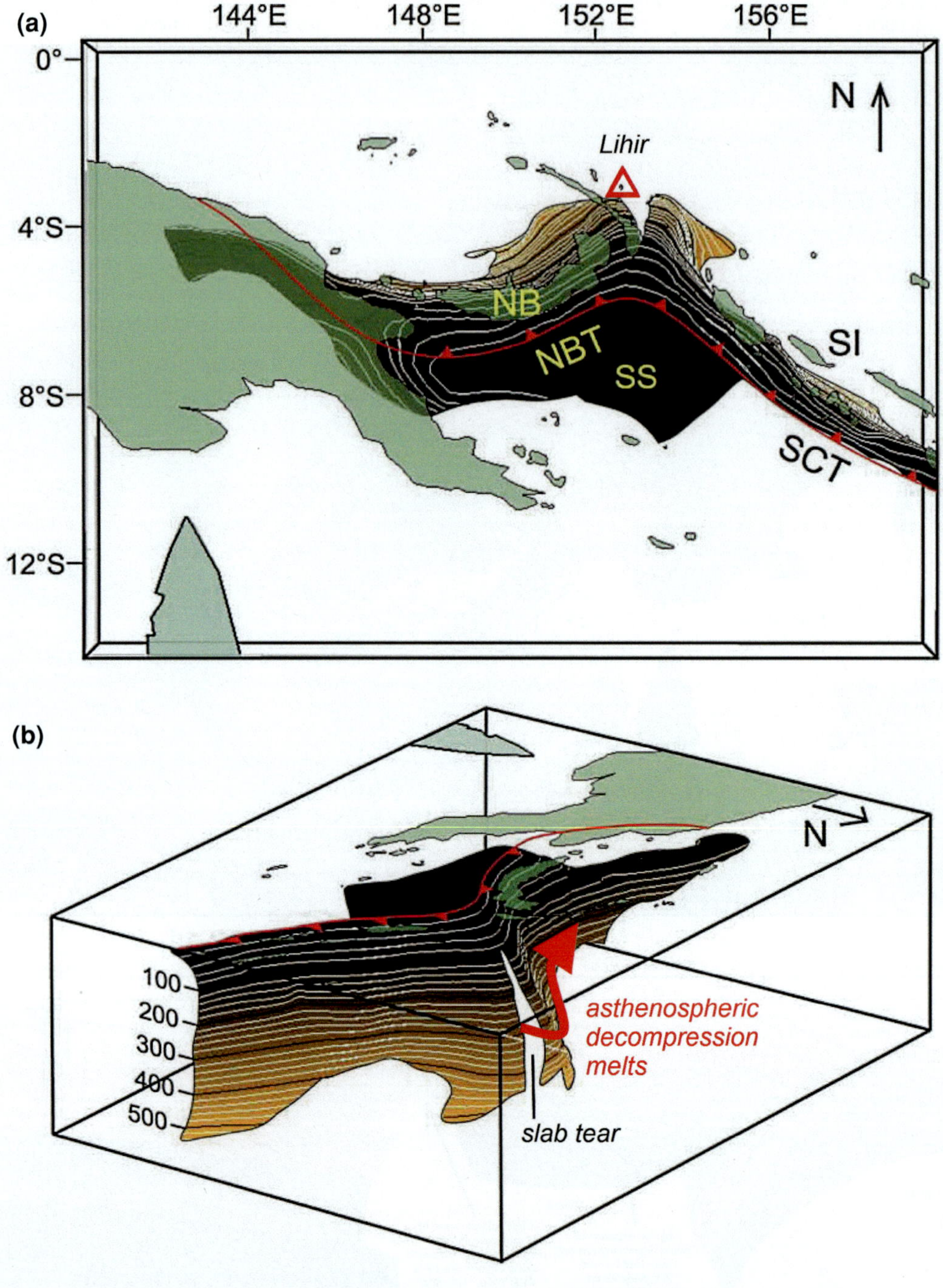

Fig. 17.37 Geometry of the subducted Solomon Sea Plate. **a** Map view and **b** 3D model of the Solomon slab comprising the subducted Solomon Sea Plate, and associated crust of the Woodlark Basin and Australian Plate subducted at the New Britain and San Cristobal trenches. (From Holm and Richards 2013). Depth is in km; the top surface of the slab is contoured at 20 km intervals from the Earth's surface (black) to termination of slab-related seismicity at approximately 550 km depth (light brown). Red line indicates the locations San Cristobal trench (SCT)-New Britain trench (NBT)-and on-land extensions in Papua New Guinea. NB = New Britain, SI = Solomon Islands, SS = Solomon Sea. The model highlights the spatial and potential genetic relationship between slab tear, alkaline volcanism and epithermal gold mineralisation at Lihir Island (red triangle; see Chap. 18)

subduction process (e.g. blueschists intimately mixed with ophiolites). Boninites are also found almost exclusively, but not totally (see above), in the forearc of primitive island arcs, and in ophiolite complexes thought to represent former forearc settings or at least formed above a subduction zone (e.g. Pe-Piper et al. 2004; Viruete et al. 2006).

17.5.2 Transitional Ocean Island Arcs

We use the term "transitional" for some oceanic island arcs in the sense that they presently are producing with chemical compositions close to average continental crust (Tatsumi and Stern 2006; Gazel et al. 2015), which is andesitic in composition and enriched in elements incompatible in the mantle (e.g. K, Rb, U, Th, LREE). As proposed in the plate tectonic cycle, raw materials such as oceanic sediments, oceanic crust, mantle lithosphere and wedge materials, are fed into the subduction factory and manufactured into arc magmas and continental crust (see Fig. 17.1b; Tatsumi and Stern 2006). The significant contamination of magmas by incompatible elements that do not fit into mantle minerals, and rocks higher in feldspar and silica differentiates transitional island arcs from young oceanic island arcs described above, which show the least continent-like (or depleted) geochemical signatures. It also distinguished them from micro-continental arcs (below), which form within older continental crust and show inherited signatures from pre-existing sialic material. In transitional oceanic island arcs, a higher continental signature may result from melts from special conditions of the subducting oceanic crust (Gazel et al. 2015), such as an enriched mantle plume or "hot spot" (e.g. Central American land bridge arc, and western Aleutians island arc), or may reflect recycling of a component derived from subducted continental sediments (e.g. Lesser Antilles and eastern Aleutians island arcs).

The Lesser Antilles arc also appears to be the oldest, currently active intra-oceanic island arc (Plate 17.1; Fig. 17.5), separating the Caribbean from the Atlantic Plates. The geological history and evolution of the Caribbean Plate and its fringing volcanic arc systems is both complex and controversial (e.g. Pindell 1991; Kerr et al. 2003; Pindell et al. 2006; James 2009; Escalona and Mann 2011; Wright and Wyld 2011; Loewen et al. 2013; van Benthem et al. 2013; Boschman et al. 2014). However, there is now general agreement that the plate interior is composed of Pacific oceanic crust, thickened by oceanic plateau magmatism. The Caribbean-Colombian Plateau (CCP; Chap. 16) is composed of late Cretaceous (95–90 Ma) oceanic crust (some 8–20 km thick). This is believed to have been erupted in the eastern Pacific Ocean, near the present-day Galapagos Islands and site of a long-lived mantle plume (Fig. 17.38). Subduction of the eastern proto-Caribbean seafloor led to insertion of the Caribbean Plate between the Americas at about 90 Ma. Here it first collided with a volcanic arc above an eastward-dipping subduction zone approximately in the same position as the present-day Central American trench (or land bridge arc). Collision of the arc with the relatively thick ocean plateau crust resulted in polarity reversal, whereby, a new volcanic arc formed above a westward-dipping subduction zone on the plate's eastern margin (Fig. 17.38). Evidence for such younger subduction-related magmatic activity (89–86 Ma) is found on Aruba, which is underlain by a large tonalitic batholith, and Curaçao where quartz diorite dykes, are intruded into the slightly older submarine basaltic lava succession.

This new volcanic arc was, however, only the first or oldest in a series of volcanic arcs that gradually migrated eastwards by hinge rollback retreat of the west-dipping subducting slab over the past 85 Ma (Fig. 17.39; also see Fig. 17.82). This island arc system has been labelled the Great Arc of the Caribbean (GAC). Whether this represents

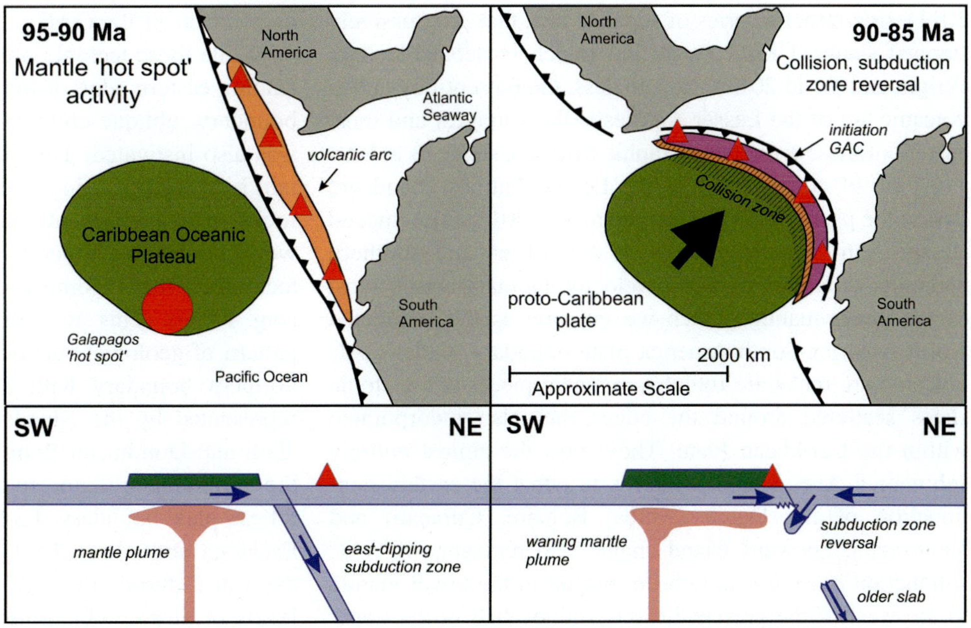

Fig. 17.38 Schematic evolution of the proto-Caribbean Plate, Caribbean-Colombian Plateau (CCP), and initiation of the east-facing Great Arc of the Caribbean (GAC) in the late Cretaceous. The model shows two time slices at 95–90 Ma (left), and 90–85 Ma (right) when the CCP collided with the GAC. (Adapted from Wright and Wyld 2011; Loewen et al. 2013)

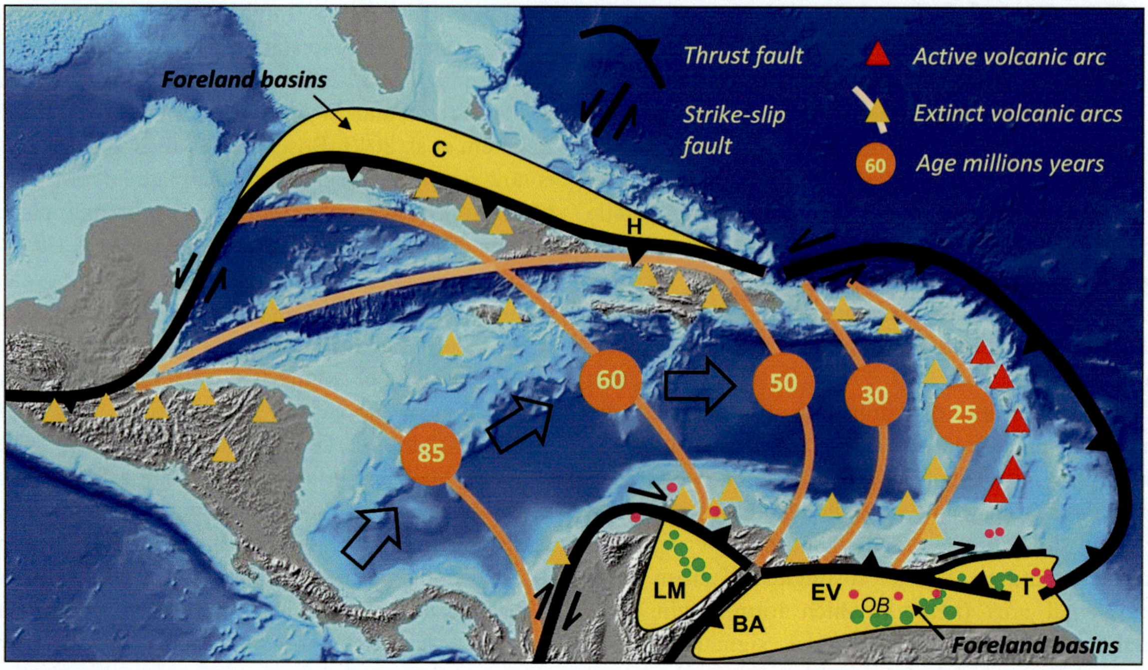

Fig. 17.39 Synoptic view showing eastwards migration of the Caribbean Plate and position of the Great Arc at its leading edge over the past 85 million years. The arc(s) gradually migrated ca. 1500 km eastwards by hinge rollback retreat of the west-dipping subduction zone to its present location along the active Lesser Antilles island arc. Oblique collision associated with eastward displacement of the Caribbean Plate relative to the North and South American Plates formed a sequence of foreland basins which young eastwards in tandem with the age of arc volcanism; namely: LM = Lake Maracaibo, BA = Barinas-Apure, EV = East Venezuelan, T = Trinidad (Columbus), C = Cuban, H = Hispaniola. The foreland basins on the southern Caribbean plate margin were filled by clastic sediments shed both from the colliding arc and cratonic areas to the south. These are economically highly significant as they presently contain the world's largest petroleum reserves. For simplicity only major oil (green) and gas/condensate (pink) fields are highlighted; smaller circles = giant fields (containing ≥ 500 MMbbl of recoverable oil, or ≥ 3 TCF of gas equivalent), while the larger = supergiant oil fields (≥ 5 Bbbl). The majority of reserves are contained within the ginormous Orinoco heavy oil and tar belt (OB), which is probably the largest hydrocarbon accumulation on the planet, with total original oil and natural gas in place of 1400 Bbbl and 180 TCF respectively; technically recoverable reserves have been recently estimated as > 500 Bbbl oil and 135 TCF gas. (Based on an original concept of Pindell 1981, and Escalona and Mann 2011; for paleogeographic reconstructions see Fig. 17.82)

rollback of a single or unique arc (e.g. Escalona and Mann 2011), or a rather a series of separate arcs that progressively stepped seaward from the active trench is debatable (e.g. Wright and Wyld 2011). Nonetheless, the present-day active volcanic arc of the Lesser Antilles is the youngest and most eastern of these arcs. Tomographic evidence suggests at least 1100 km of convergence at the Lesser Antilles island arc during the past ~45 Myr (Benthem et al. 2013). The imaged Lesser Antilles slab consists of a northern and southern anomaly, separated by a low-velocity anomaly across most of the upper mantle, which we interpret as the subducted North America-South America plate boundary. Older GAC volcanic arc rocks are found as remnant pieces and tectonic slices scattered around the edges, but also incorporated within the Caribbean Plate. They form the almost entirely submerged Aves Ridge, which is in effect the eastern continuation of the Dutch (Aruba, Bonaire, Curaçao) and Venezuelan Leeward Island chain. Slab remnants of GAC subduction have also now been imaged in the lower mantle to the west of the present Lesser Antilles slab in the upper mantle (Benthem et al. 2013). These suggest Eocene slab detachment of the GAC.

While these tectonic processes were commencing on the leading eastern edge of the Caribbean Plate, on its southern boundary, oblique collision with the South American Plate was also instigated. This produced a fold and thrust mountain belt, incorporating obducted slices of the CCP and arc rocks, and a series of foreland basins which possess the world's largest petroleum reserves, followed by strike-slip tectonics that dominate the present plate margin (Fig. 17.39). This was mirrored by a somewhat similar pattern of geological development on the Caribbean Plate's northern boundary with the North American Plate, now represented by the islands of the Greater Antilles (Cuba, Haiti and Dominican Republic). It is important to note that the prolific petroleum source rocks of the southern Caribbean plate boundary (La Luna Formation and equivalents; Escalona and Mann 2011) are the same age as the CCP, deposited during the "Turonian anoxic event". Increased levels of global submarine basaltic (mantle plume) activity during this period, including the CCOP, gave rise to emission of large quantities of volcanic gases (in particularly CO_2) into the oceans and atmosphere, leading to deposition of organic-rich black shales in ocean basins worldwide (e.g. Kerr 1998, 2005).

17.5.2.1 Ancient Transitional Island Arcs: Characteristics

As described for young oceanic arcs the same geological constraints apply. In these somewhat more mature island arc systems volcanic products should typically be generally andesitic and include a higher proportion of evolved silicic magmas as typified by the volcanoes of the Lesser Antilles, and evidence of larger plutonic systems and mid-crustal magma chambers (Smith et al. 2013; Chap. 16). Also, there is potential for wider and thicker forearc prisms and the products of mass-wastage and sector collapse avalanches in back-arc settings (see Fig. 16.47). Figure 17.39 serves to illustrate the scale, time framework and complexity of structural and stratigraphic relations that potentially may be encountered in the older rock record (see also Fig. 17.82). Eastward displacement of the Caribbean Plate relative to the North and South American Plates, shows how the arc has migrated diachronously over a distance of some 1500 km from western Venezuela in the Paleocene ($\sim$60 Ma), to a zone of presently active deformation east of Trinidad. Each diachronous stage follows a similar pattern of pre-, syn- and post-collisional basin formation (Escalona and Mann 2011). Initial collision is characterised by overthrusting of arc and forearc terranes onto the Mesozoic sedimentary passive margin (including petroleum source rocks) of northern South America. This was accompanied by foreland basin formation, and as collision extended towards Trinidad, progressively younger foreland basins formed to the east. From an economic perspective, the complex distribution of basins and oil and gas fields in the region requires a complete understanding of the relation between large-scale plate interactions between the Caribbean and South American Plates and volcanic arc terrains.

As a further interesting analogue, the generation of young continental crust in transitional oceanic arcs by subduction of "hot-spot" enriched crust has significant implications for the discussion of the generation of Archean continental crust which is geochemically similar (specifically the tonalite-trondhjemite-granodiorite or TTG suite; Gazel et al. 2015; see Sects. 17.7.2.6 and 17.7.4).

17.5.3 Microcontinental Arc Volcanism

The arc block structure is much the same as for young and transitional island arc systems, but with further development of continental crust, and is exemplified by islands of Japan, Indonesia and New Zealand which are arc systems incorporating older continental crust (Plate 17.1). The main differences are that the arc block is substantially thicker (30–$\leq$ 40 km), there are considerable proportions of recycled, mature supracrustal sediments, and the arc block is much more sialic in character (Cas and Wright 1987; Gazel et al. 2015).

Tomographic images illustrating the lithospheric structure below the northeast Japan arc are shown in Fig. 17.40 (Zhao 2012). Two profiles pass through three active volcanoes on Hokkaido (Komagatake, Taisetsuzan, and Meakandake). The subducting Pacific slab is imaged clearly as a high-velocity zone with seismic velocities 4–6% higher than that of the average mantle. Intermediate depth earthquakes occur actively in the subducting Pacific slab and they form a clear double seismic zone. Prominent low-velocity zones are visible in the crust and uppermost mantle beneath the active arc volcanoes, extending to a depth of 150–200 km in the central portion of the mantle wedge. The mantle wedge low-velocity zone is roughly parallel with the subducting slab, and is close to, or connected with, the slab at varying depths. Normal crustal earthquakes occur in the upper crust down to about 20-km depths, but some earthquakes occur in the lower crust and uppermost mantle around the Moho discontinuity and usually are of lower frequency than normal crustal earthquakes. Such low-frequency events are considered to be related to magmatic activity under the active arc volcanoes.

Similarly, beneath southern Japan the Philippine Sea Plate low-velocity anomalies are clearly visible in the crust and upper mantle wedge under the volcanic front in Kyushu (Fig. 17.41a). A prominent dipping low-velocity zone is visible in the crust and mantle wedge under the recently active Unzen volcano, connecting the surface edifice with the Philippine Sea slab at about 100 km depth. This infers that the Unzen magmatism is likely related to the dehydration process of the Philippine Sea slab at about 100 km depth, even though the volcano is located in a back-arc position (see Fig. 17.35). This is examined in more detail in Fig. 17.41b, which displays a high-resolution crustal tomography image beneath Unzen. This shows that a well-defined cone-shaped, low-velocity zone is visible in the crust, thought to represent high-temperature anomalies containing a magma reservoir (Zhao 2012). The cut off-depth of crustal seismicity shallows towards the crater of the volcano, which is in good agreement with the upper boundary of the low-velocity zone (Fig. 17.41b). These features indicate a thinning of the brittle seismogenic layer beneath the volcano. A summary model to explain across-arc tomographic imagery of Japanese subduction processes is shown in Fig. 17.42.

Thicker continental crust has profound effects on the magmatic products which are mainly silicic and are in large

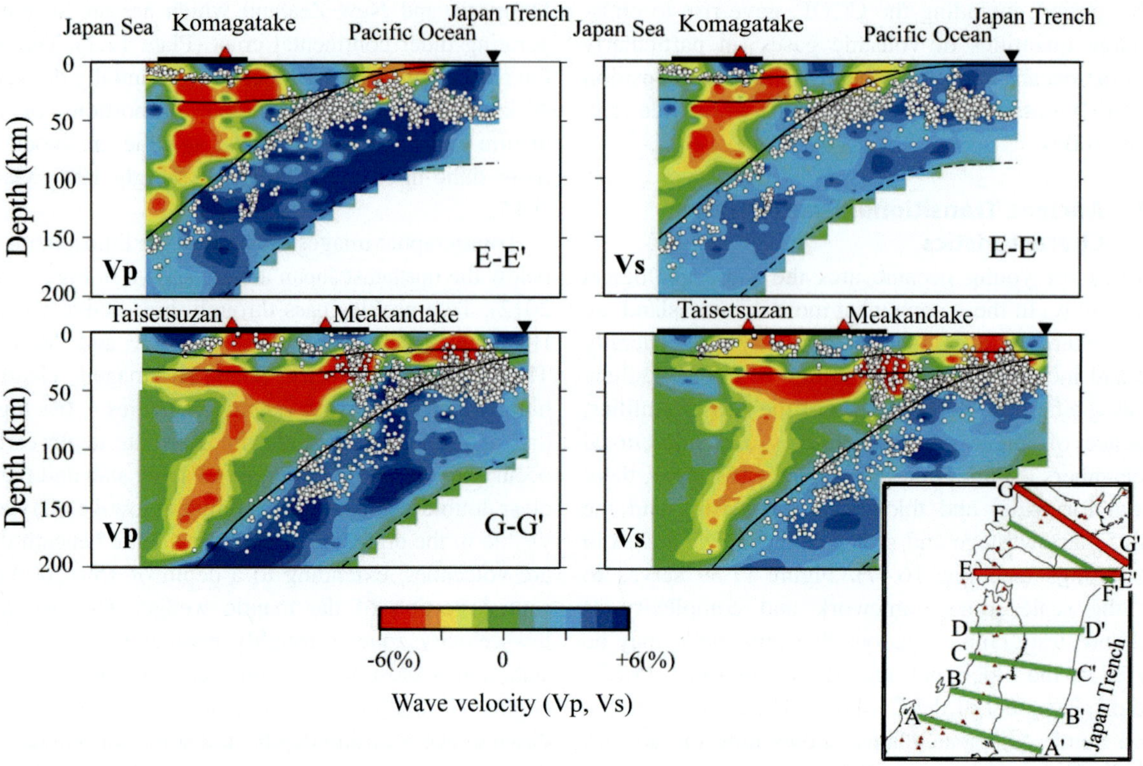

Fig. 17.40 Seismic tomography images illustrating lithospheric structure, northern Japan volcanic arc. Vertical cross-sections of P- and S-wave velocity images along the profiles E-E′ (above) and G-G′ (below) as located on the inset map in Hokkaido. Low and high-velocity perturbations are shown in red and blue colours respectively; scale (in %) as indicated. Red triangles denote active arc volcanoes, while inverted triangles show location of the Japan oceanic trench. Small white dots show earthquakes which occurred within a 15 km width from the profile. The three curved lines show the Conrad and Moho discontinuities and the upper boundary of the subducting Pacific slab; dashed lines denote the estimated lower boundary of the Pacific slab. No vertical exaggeration of cross-sections. (From Zhao 2012, After Wang and Zhao 2005)

part derived from the lower crust (Cole 1979, 1984; Graham et al. 1995). However, rock types fully range from mafic to intermediate to silicic in character (Ewart et al. 1977; Whitford et al. 1979; Cole 1984; Murphy 2007). Calc-alkaline rocks are very prominent, but as in Japan and Indonesia, an across-arc trend from tholeiitic nearest the trench to calc-alkaline to alkaline and even shoshonitic volcanoes furthest from the trench is known (e.g. Miyashiro 1972, 1974; Whitford et al. 1979; Wheller et al. 1987; Sendjaja et al. 2009; Yoshida et al. 2014). The active volcanic belt is again relatively narrow (several tens of km wide) but may migrate towards or away from the trench with time (e.g. Sugimura and Uyeda 1973) or be widened by extensive outflow ignimbrite sheets, as in the Sunda arc in Sumatra (Sect. 17.3).

Along-arc variations also occur. The more oceanic, juvenile parts of arc systems (e.g. east Indonesian or Banda arc, and Tonga-Kermadec arc) being more mafic in character, with the more mature parts (e.g. Sunda arc in Java and Sumatra, and New Zealand) having more silicic products (Hall and Smyth 2008; Chesner 2012; Ewart et al. 1977, 1998; Cole 1979, 1984; Wilson et al. 2009). Large silicic ignimbrites, ignimbrite "flare-up", caldera volcanoes, some submarine in their setting (e.g. Krakatau), and supervolcanoes (e.g. Lake Toba, Lake Taupo) are common, which occur interspersed among aligned stratovolcanoes (e.g. Figs. 17.25, 17.26; Chap. 16). Co-magmatic granitic intrusions may also be prominent (Polve et al. 2001; Hall and Smyth 2008), and testify to the existence of a sialic crust at least 20 km thick, that is thick enough for low-P/high-T metamorphism and anatexis of the lower crust to occur, as well as very fast cooling and rapid exhumation (e.g. Hennig et al. 2001). Basaltic volcanics and dykes testify to mantle influences in magma genesis in Indonesia (Wheller et al. 1987; Carn and Pyle 2001), Japan (Shuto et al. 2015), and New Zealand (Cole 1979, 1984; Smith et al. 1993; Huang et al. 2000; Cook et al. 2004; Linnell et al. 2016). The sialic crust is not only thick, but regionally extensive. For example, Sumatra, west Java, Borneo and the shallow marine shelf between, is underlain by sialic crust, formed by amalgamation of continental blocks during the Triassic-early Cretaceous (Fig. 17.43a). This includes Proterozoic

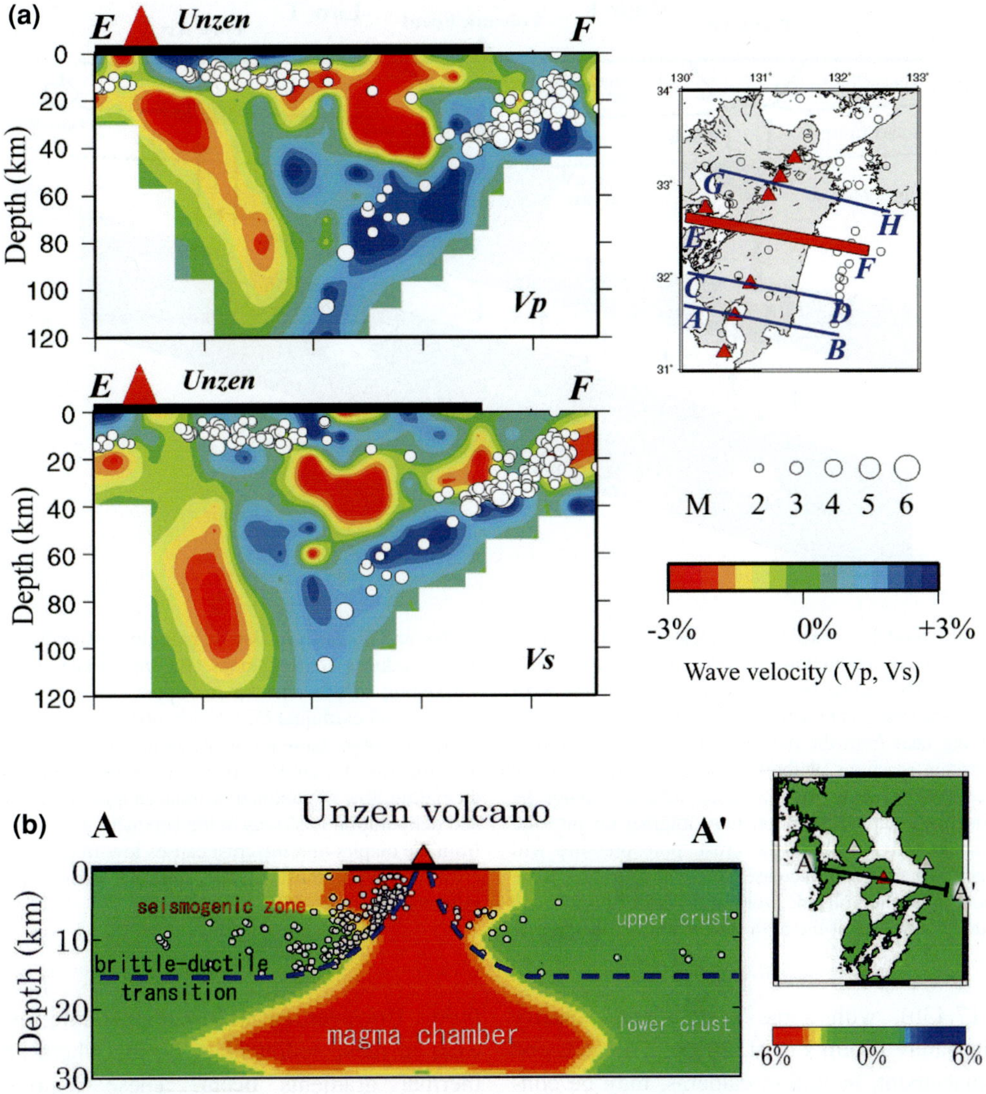

Fig. 17.41 Seismic tomography images illustrating lithospheric structure, southern Japan volcanic arc. **a** Vertical cross-sections of P- and S-wave velocity images along the profiles as shown on the inset map, Kyushu. The subducting Philippine Sea slab is clearly imaged as a high-velocity (blue) zone, while prominent low-velocity (red) zones are visible in the crust and upper mantle wedge beneath the active arc volcanoes, in particularly Unzen. White dots show earthquakes and magnitude occurring within a 15 km width from the profile. No vertical exaggeration of cross-sections. **b** High-resolution P-wave image along profile A-A′ directly below Unzen, showing cone-shaped low-velocity zone (high-temperature anomaly) interpreted as magma chamber below this historically active volcano, which last erupted 1991–95. (From Zhao 2012, After Wang and Zhao 2006 and Zhao et al. 2002)

basement intruded by Permo-Triassic granites and records a long record of subduction-related magmatism extending back into the early Mesozoic or late Paleozoic (Hall and Smyth 2008; Hall 2009). Surrounding New Zealand, the submarine Lord Howe Rise and the New Caledonia-Norfolk Ridge to the northwest and, Chatham Rise and Campbell Plateau to the southeast are also composed of sialic crust, which includes Mesozoic granites and Paleozoic schists (Fig. 17.43b). This has more recently been termed the "Zealandia microcontinent" and represents a mostly submerged (>90%), mass of extended continental crust that separated from and sank isostatically after breaking away from Australia 85–60 Myr ago (Mortimer et al. 2017; also see discussions Sects. 17.5.6, 17.7.4).

Within the active volcanic belt a significant structural feature is the association of grabens, extensional basins and normal faulting, e.g. the Beppu-Shimabara and Kagoshima Bay grabens, southern Kyushu where there is a strong correlation between extensional faulting and the location of active volcanism (Matsumoto et al. 2015; Chaps. 16, 18). In Indonesia (especially Sumatra) and New Zealand's Taupo Volcanic Zone there is a significant strike-slip component of movement associated with the extensional terrains (e.g. Hamilton 1979; Spörli 1980; Cas and Wright 1987,

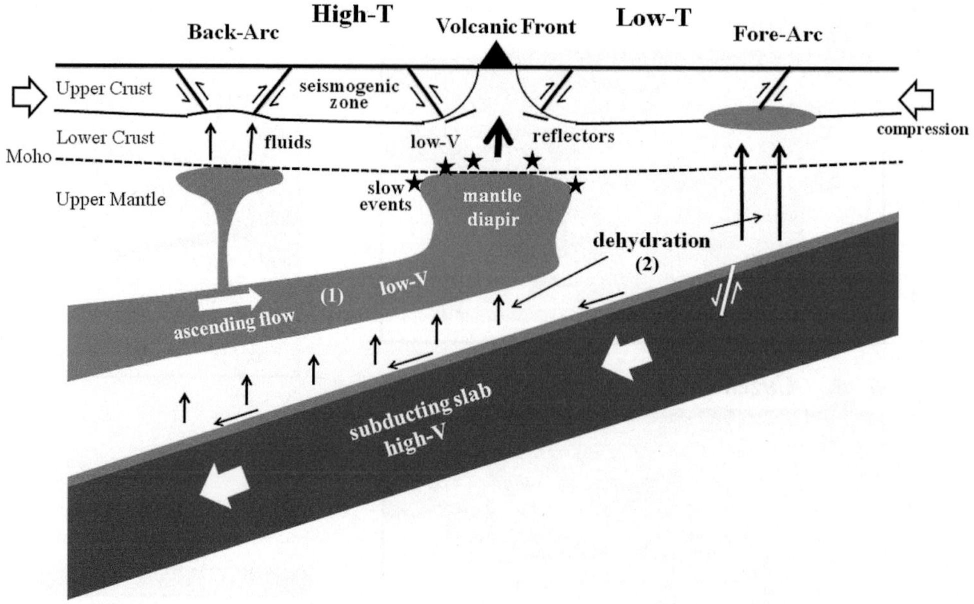

Fig. 17.42 Schematic model to explain seismogenesis and tomographic imagery of Japanese subduction processes (From Zhao 2012, After Zhao et al. 2010). Two processes are considered to be the most important in a subduction zone: (1) ascending ("corner") flow into the mantle wedge bringing heat from the deeper mantle through convection, and (2) dehydration reactions of the subducting oceanic slab. In the forearc region temperature is lower, hence magma cannot be produced, but fluids from slab dehydration may migrate up into the crust. If fluids enter an active fault in the crust, pore pressure will increase and fault zone friction will decrease, which may trigger large crustal earthquakes. Below the volcanic front and back-arc areas, arc magma can be produced because of the high temperature in the mantle wedge and fluids from slab dehydration. Migration of magma up to the crust produces arc volcanoes and causes lateral heterogeneities and a weakening of the seismogenic upper brittle crust, which can lead to large crustal earthquakes. Dehydration reactions of the subducting slab may also trigger large intraplate earthquakes. The low-velocity zones in the uppermost mantle represent mantle diapirs associated with an ascending flow of subduction-induced convection in the mantle wedge, and dehydration reactions in the subducting slab. Magma rising further from the diapirs into the crust causes low-frequency events in the lower crust and uppermost mantle, making their appearance as S-wave reflectors

Fig. 15.8; Fig. 17.43b), with some basins being clearly trans-tensional in nature (Spinks et al. 2005). Volcanic and subvolcanic emplacement, including calderas, may be controlled by smaller pull-apart and dilational structures (releasing bends and stepovers; Sect. 17.3). These fault systems can facilitate heat transfer from widespread shallow crustal intrusions, emplaced over wide areas to the surface. Active and degassing volcanoes are associated with volcanic geothermal reservoirs containing magmatic gases and acid fluids. However, large, low temperature resources may also exist in these sedimentary basins where anomalously high thermal gradients occur. These regions are therefore prospective for epithermal gold mineralisation in both these settings. In New Zealand the Taupo Volcanic Zone can be viewed as an on-land extension of the Havre interarc basin (Karig 1970) and has been described by Cole (1984) as a back-arc ensialic marginal basin.

Fig. 17.43 Principal tectonic elements of micro-continental arc systems. **a** Southeast Asia. The yellow shaded area is the zone of collision between the Eurasian, Indian-Australian, and Pacific-Philippine Sea Plates, which can be considered as a wide suture zone, within which there are smaller sutures. The area commonly considered to form the Mesozoic Sundaland continental core is outlined, which was surrounded by subduction zones (i.e. sutures) for much of the Mesozoic and Cenozoic. Dashed lines show incremental growth of the Indonesian region (or accretionary orogen) at different stages by addition of fragments of continental, volcanic arc and ophiolitic rocks. This occurred in an episodic way, related primarily to arrival of continental fragments at subduction margins, after which subduction resumed in new locations. (From Hall 2009; also see Hall and Smyth 2008). **b** New Zealand is located within a much more extensive, nearly submerged mass of extended continental crust that sank after breaking away from Australia at 60–85 Ma, having separated from Antarctica between 85–130 Ma. This has been referred to as the "Zealandia microcontinent" and, with an area of 4.9×10^6 km^2, currently represents the Earth's largest microcontinent. Zealandia (outlined in pink) comprises thinned continental crust detached from Gondwana and probably is composed of Mesozoic granites and Paleozoic basement (e.g. Lord Howe Rise, Chatham Rise and Campbell Plateau) overlain by rift basins with up to 4 km of Late Cretaceous-Cenozoic sediments; NC-N = New Caledonian-Norfolk Ridge, interpreted as a thin, continental ribbon. TVZ = Taupo Volcanic Zone; AF = Alpine Fault; HB = Havre back-arc basin; L-C = Lau-Colville remnant arc; T-K = Tonga-Kermadec volcanic arc. (Digital base relief model from NOAA) ▶

(a)

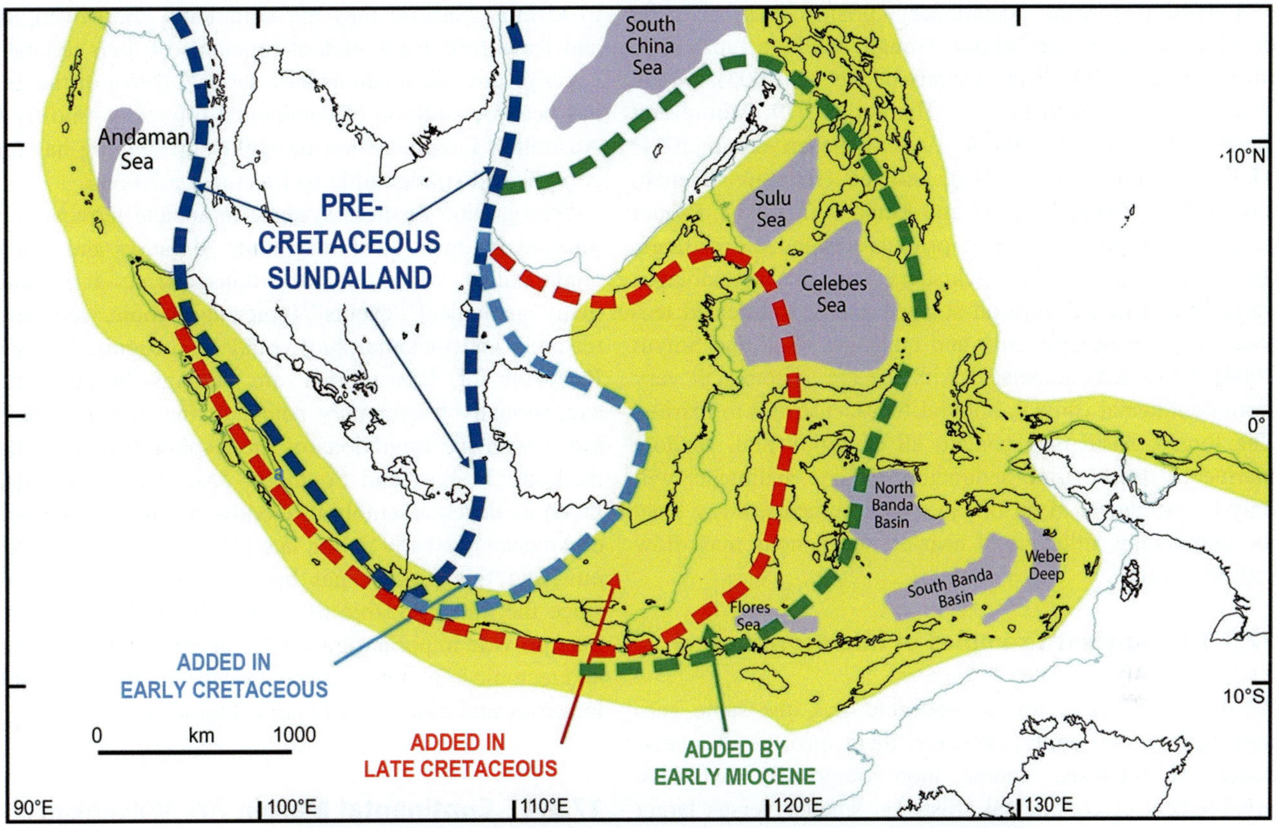
South China Sea
Andaman Sea
10°N
PRE-CRETACEOUS SUNDALAND
Sulu Sea
Celebes Sea
0°
North Banda Basin
Weber Deep
South Banda Basin
Flores Sea
ADDED IN EARLY CRETACEOUS
10°S
0 km 1000
ADDED IN LATE CRETACEOUS
ADDED BY EARLY MIOCENE
90°E
100°E
110°E
120°E
130°E

(b)

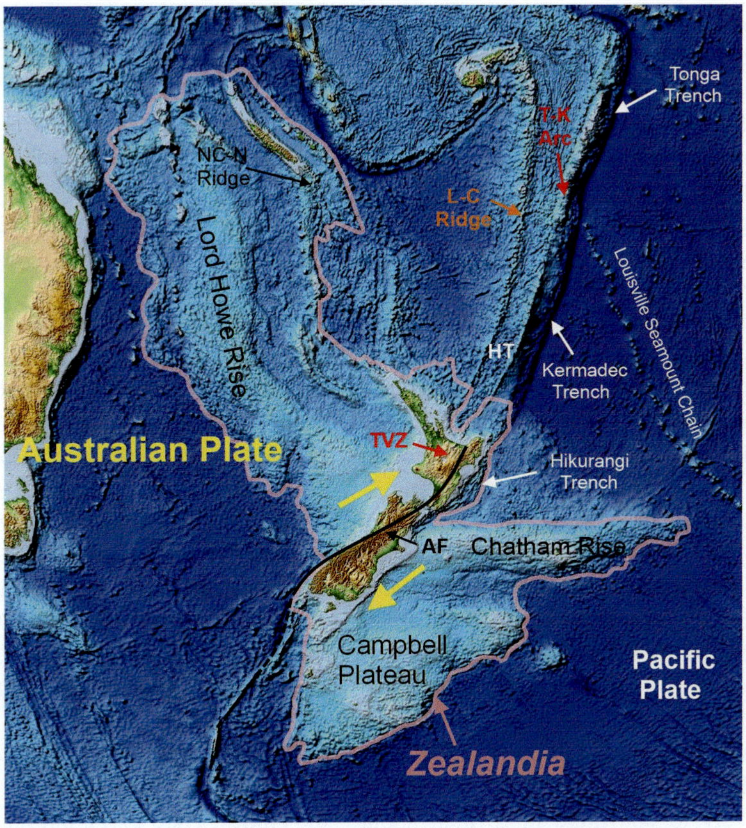
Tonga Trench
T-K Arc
NC-N Ridge
L-C Ridge
Lord Howe Rise
Louisville Seamount Chain
HT
Kermadec Trench
Australian Plate
TVZ
Hikurangi Trench
AF
Chatham Rise
Campbell Plateau
Pacific Plate
Zealandia

Extensive forearc accretionary prisms may also be developed, such as the Nankai Trough, southern Japan (e.g. Moore et al. 1990, 2015; Underwood et al. 2003), Sunda Trench, western Sumatra (e.g. Hamilton 1979; Moore and Karig 1980; van der Werff 1996; Lüschen et al. 2011; McNeill and Henstock 2014), and the Hikurangi Trench, eastern New Zealand (Bland and Kemp 2006, 2014; Paquet et al. 2011). As, with more primitive island arc settings, these should include blue schist metamorphic assemblages variably intermixed with off-scraped and/or imbricated tectonic slices of oceanic crust and mélanges (Hall and Smyth 2008). Sediments associated with the arc system will vary from continental (fluvial, alluvial fan, lacustrine) to marine. The forearc basin may contain all of continental, shallow marine to deep-marine sediments. Pelagic sediments will only be prominent in far offshore environments, away from the dominating influence of near-arc terrigenous mass-flow sedimentation.

17.5.3.1 Ancient Examples of Microcontinental Arcs

As discussed above for intra-oceanic arcs, the same geological principles apply; however, there must be clear evidence of localised and/or more regionally extensive involvement of continental crust (i.e. smaller versus larger sialic fragments), a higher proportion of evolved silicic magmas and evidence of low-P/high-T metamorphism. This should be parallel high-P/low-T metamorphism found in rocks of the forearc region (i.e. paired metamorphic belts; Miyashiro 1972; Ernst 2010). In total, the framework of the geological framework and tectonic evolution of the system should be considerably more complex. The Cenozoic history of the Southeast Asia region shows that an arc can be formed and destroyed in a short time and may ultimately leave relatively little trace in the stratigraphic record. The complexity of tectonic evolution in the region shows that a great deal can happen in a 1–10 Ma time span, and much of this complexity could be missed in the study of ancient arcs.

The Southeast Asia region has been described as an *accretionary orogen* (Hall 2009). It is ultimately a mosaic of different orogens, and multiple sutures accreted progressively around an older continental core (Fig. 17.43a). There has, however, been no simple story of addition of large pieces of crust by collision, and then termination of orogenic development (Hall and Smyth 2008; Hall 2009). Subduction has resumed and new collisions have followed. There has been considerable growth of the orogen, but little material has been added by subduction accretion, or by accretion of oceanic plateaus. Volcanic arcs do not seem to have contributed much additional crust in terms of relative width, and plutonic rocks are unlikely to constitute more than about 10% of the entire orogen in its widest parts. Intra-orogen ocean basins can completely disappear; they are also difficult to identify because they are commonly short-lived, small, and leave little trace such as remnants of their sedimentary cover or arc volcanism as they subduct. Most of the growth has been by addition of continental fragments (mostly from Australia). Fragmentation by strike-slip faulting has further complicated stratigraphic and tectonic relations.

Recognition of these systems in ancient terrains is complicated by their areal magnitude, duration and structural complexities, which are time transgressive. It is unlikely many geological "events" (volcano-plutonic, sedimentary, tectonic, and metamorphic) could be recognised in similar Paleozoic or Precambrian arc settings because of the exceptionally fragmentary nature of the rock record, combined with the limitations of the resolution of most dating methods. In East and Southeast Asia, and China there is much evidence assembly of Gondwanaland-derived terrains commenced as early as the late Devonian-early Carboniferous, and this process has been repeated numerous times since the mid-Paleozoic (e.g. Metcalfe 1996, 2013). These may provide important guides for deciphering the magmatic and tectonic evolution (crustal growth) in other geographical locations and older micro-continental arcs.

17.5.4 Continental Margin Arc Volcanism

Magmatism in these settings takes place upon a wholly sialic, continental-type crust, which in the Andes is very thick, up to 60–80 km (Fig. 17.44; Wigger et al. 1994; Beck and Zandt 2002; Chulick et al. 2013). Crustal involvement in magma genesis is evident from the high proportion of silicic volcanic rocks, including very large ignimbrites (Chaps. 10, 16), and contemporaneous granitoids. Intermediate volcanic successions may also be common (e.g. Andes, Cascades), and they appear to be largely calc-alkaline. Forearc basin(s) and accretionary prisms should be expected (Kulm et al. 1982; Moberly et al. 1982; von Huene et al. 1996), although, significant along-strike variations controlled by sediment flux, basal friction and subduction erosion of the outer wedge, and pore fluid pressure are documented (e.g. von Huene et al. 1999; von Huene and Ranero 2003; Noda 2016). As with micro-continental arc settings, extensional basins may be associated with the volcanic arc and back arc, although these may evolve into foreland basins in compressional settings (Wilson 1991; Horton et al. 2016). Sediments associated with the arc and back arc will be continental: fluvial, alluvial fan and lacustrine. Sediments in the forearc area will range from continental to shallow marine to deep-marine. The volcanic arc will again be several 100 s to thousands of km long (8,000 km in the case of the Andes; Plate 17.1), but relatively narrow at any one time (tens of km). However, the arc may migrate due to slab roll-back and changing slab subduction angle, producing a time-transgressive, wider belt of volcanic rocks and associated granitoids. In margins with protracted subduction durations, magmatic arcs may overlap or form magmatic belts parallel to the trench and coastline. During periods of shallow subduction, magmatic activity can migrate significantly inland compared with average distance from the trench. In the Andes magmatic belts have migrated in general eastwards over the past 200 Ma (Coira et al. 1982; Scheuber and Reutter 1992).

Continental margin magmatic arcs were once considered geodynamically relatively simple, where, subduction of oceanic crust generates the formation of a magmatic arc, which gives rise to thermal weakening of the crust and its subsequent thickening in a compressive regime. Recent research in the Andes has however emphasised that compression and extension alternate, leading to repeated orogenic cycles of mountain building and their subsequent extensional collapse (Ramos 2009; Ramos and Folguera 2009; Ramos et al. 2014). These are principally controlled by the absolute motion of the South American continent, which in turn is controlled by two competing factors: (1) negative trench rollback velocity in the upper plate, which generates extension, and (2) positive roll back, when the upper plate overrides the trench and compression dominates through the margin. Over an extended period of subduction, trench retreat below the Andes has been controlled by opening of the South Atlantic Ocean. Conversely, westwards motion of the upper plate and overriding of the trench began after the break-up of western Gondwana and separation of Africa at ca. 126 Ma (Ramos et al. 2014). Nonetheless, detailed examination of different Andean segments has shown alternating tectonic regimes superimposed on these first-order controls (Ramos and Ghiglione 2008).

Such studies have recognised several orogenic cycles associated with periods of shallowing and steepening of the subducting slab (Ramos 2009; Ramos and Folguera 2009; Ramos et al. 2014). Cycles commence with a normal-to-thick crust associated with calc-alkaline stratovolcanoes along the main axis of the Andes. This is followed by magmatic expansion and migration of arc volcanism and deformation towards the foreland region (flat slab), which in turn is followed by crustal thickening and mountain uplift, leading to extensional collapse and normal faulting (slab roll back and steepening). Most cycles are characterised by large rhyolitic calderas and ignimbrite flare-up (e.g. the Miocene-Quaternary Altiplano Puna flare-up; de Silva et al. 2006; Chap. 16), thought to be produced by the injection of hot asthenosphere in the subduction wedge. In these cases, initially thick crust favours delamination of the lower crust and the lithospheric mantle during steepening of the subduction zone. Crustal delamination is concentrated only in a narrow central belt along the cordilleran axis (Kay et al. 1994). In areas of normal-to-thinner crust (≤ 40 km), steepening of the oceanic subducted slab and extension may be also linked to eruption of flood basalts in retro-arc foreland settings (Ramos et al. 2014). These eruptions are also linked to strain partitioning which has taken place over space and time to accommodate subduction and crustal thickening, whereby the Andes are cross-cut by strike-slip fault systems that focus off-axis magmatism and volcanism. Examples are the Calama-Olacapato-El Toro lineament (Matteini et al. 2002; Norini et al. 2013) and the Archibarca lineament (Fig. 17.45). Back-arc mafic magmas are erupted along such structures (e.g. Kay et al. 1994; Acocella et al. 2011; Risse et al. 2008) indicating relatively fast uprise of poorly differentiated magmas across the Andean crust, with only minor shallow reservoirs (Bàez et al. 2017). Such volcanism may be followed by a new cycle of arc magmatism which initially constructs a series of stratovolcanoes on normal, thick crust (Ramos et al. 2014).

Today the Andean subducting margin contains two prominent "flat slab" sections located in Peru and central Chilean (termed the "Peruvian" and "Pampean" flat slabs; Fig. 17.44b), which are both some 200–300 km long (Ramos and Folguera 2009; Hu and Liu 2016; Schepers et al. 2011). Not only does the presence of these flat slabs strongly affect present seismic and volcanic activity (which are quiet regions), evidence is emerging they may also control ore generation (Bissig et al. 2015; Chap. 18). As well as these two flat slabs that formed at about 11 and 12 Ma respectively, several older flat slab subduction episodes have been interpreted to have occurred in segments along the Andean subduction zone since 42 Ma. It was formerly proposed that formation of the Peruvian and Pampean flat slabs was the result of the positive buoyancy of the subducting Nazca Plate, induced by volcanic-related thickening over an oceanic plateau (Inca Plateau) and along two aseismic ridges (Nazca and Juan Fernandez Ridges; Fig. 17.46a). For pre-Miocene flat slab episodes, however, the correlation between flat slab subduction and arrival of such positively buoyant bathymetric impactors at the trench are not supported by plate-kinematic reconstructions, implying that other mechanisms likely also contribute to triggering flat slab subduction (Schepers et al. 2011). The Peruvian and Pampean segments also differ significantly in that landward arc migration has been largely absent, suggesting that the flattening process of the slabs (incorporating aseismic ridge subduction) may be more complex than reduction of local slab dip angle.

Analogue modelling suggested that slab flattening in the Andes is a long-term process (ca. 10 Ma) which can only occur if hundreds of km of a buoyant section of slab are forced into the subduction zone (Espurt et al. 2008). Numerical modelling and kinematic reconstruction of shortening have now been able to quantify the relationship between the flat slab length, rollback of the slab bend, and

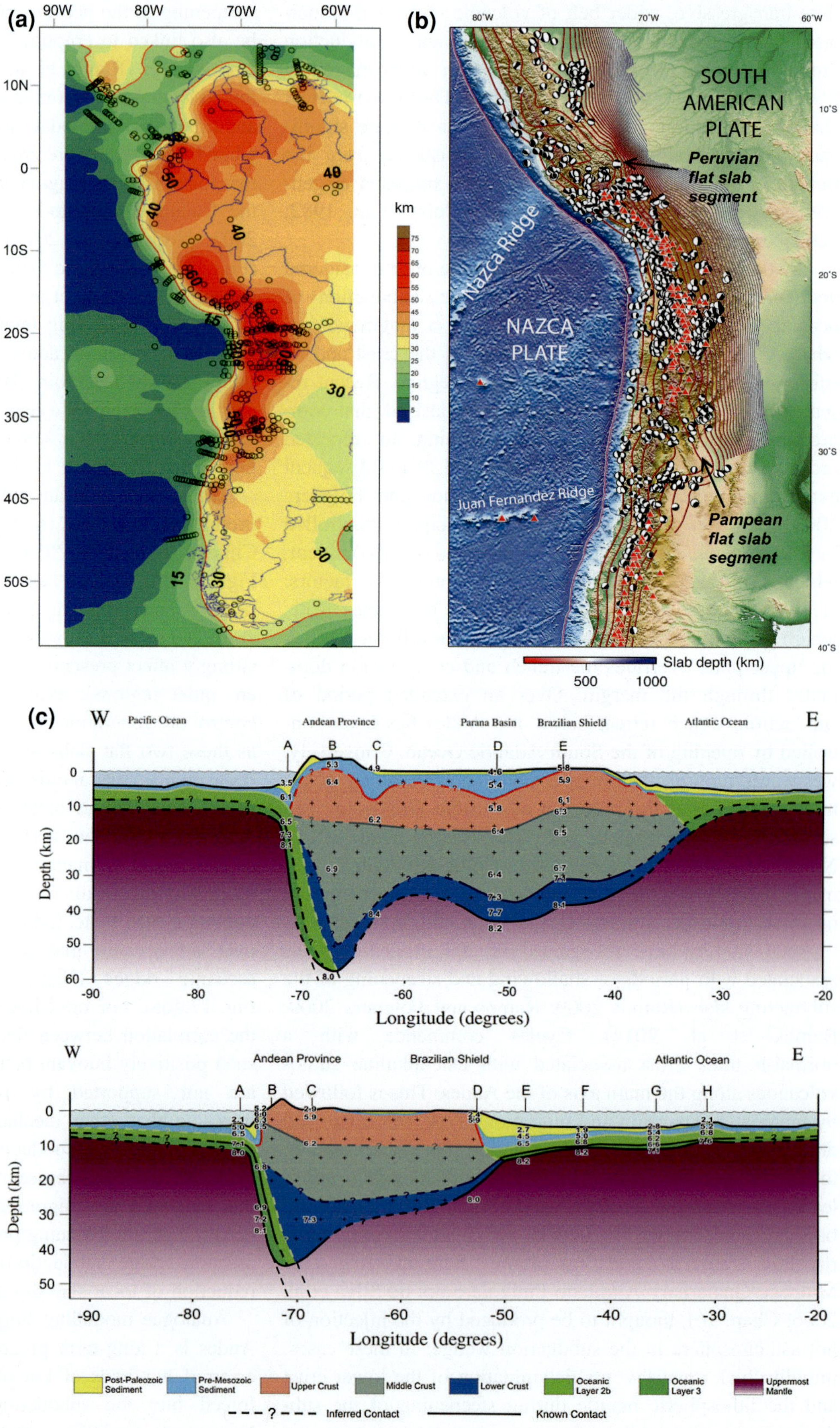

Fig. 17.44 Crustal structure of the Andean plate margin. **a** Contour map of crustal thickness of Andes adjacent areas derived from seismic data. The map shows the crust along much of the length of the Andes is very thick (>50 km), but gradually decreases southwards (From ~50 km under central Chile to ~30 km beneath Tierra del Fuego). The red line highlights the 25 km contour which delineates the landward edge of the South American continental margin. (Adapted from Chulick et al. 2013). **b** Topography overlain by depth contours of the interpolated Benioff zone, with distribution of modern volcanoes (red triangles) and seismicity (white-black circles). Labelled are the Peruvian and Pampean flat slab segments which are seismically and volcanically quiet regions. (From Hu and Liu 2016). **c** Cross-sections showing the lithospheric structure through the South American continent at latitudes 20° S (top), and 33.5° S (bottom). Letters represent P-wave velocity profiles on or near the cross-sectional line; numbers show velocities as indicated. Note the thick crustal root observed beneath the Andes. (From Chulick et al. 2013)

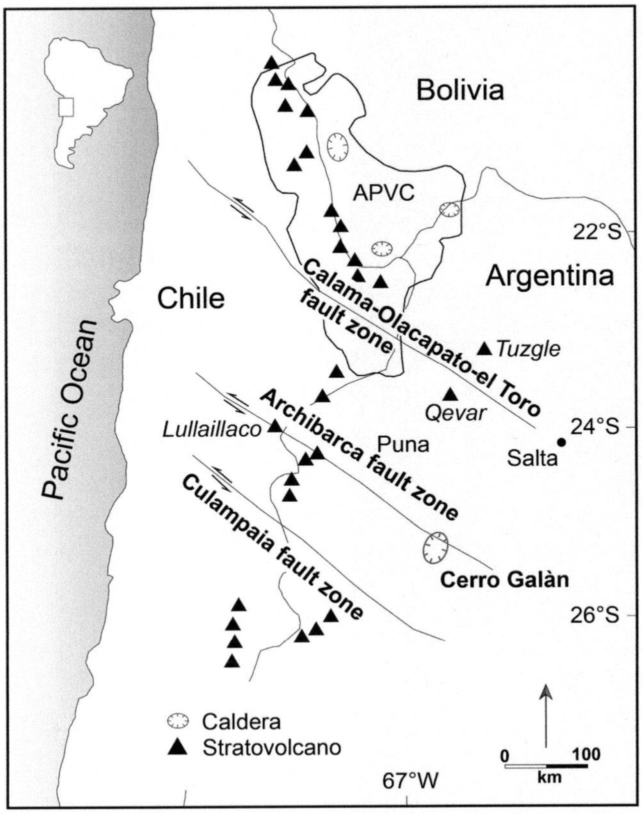

Fig. 17.45 Schematic map of the main transcurrent fault systems that cut across the central Andes allowing the formation of offset, transversal volcanoes respect to arc and back-arc locations. Area indicated with APVC, across the Bolivian, Chilean and Argentinian territories, is the Altiplano Puna Volcanic Complex. Note the location of the major Cerro Galan caldera (Chaps. 10, 16). (Modified from Lesti et al. 2011)

the amount of subduction since the onset of flat slab formation (Schepers et al. 2011). These results suggest that the Nazca slab has retreated at about 2 cm yr^{-1} since 50 Ma. In the flat slab portions, no rollback has occurred since their formation at about 12 Ma, generating "horse-shoe" slab geometries. Schepers et al. (2011) propose that, in concert with other drivers (i.e. subduction of an aseismic ridge, combined with a delayed basalt-to-eclogite transition in the subducted crust), creation of overpressured sub-slab mantle would assist hydrodynamic slab suction above the slab. This would tend to lift the slab, in effect supporting the weight of the slab in an advancing upper plate motion setting and, therefore, locally impeding rollback. This could be maintained until slab tearing releases the overpressure.

Slab tearing at the edges of the flat slab, or formation of slab tears/holes in the centre of the flat slab allows escape of the trapped mantle (Fig. 17.46; Funiciello et al. 2003, 2004; Hu and Liu 2016; Schepers et al. 2011). This releases the sub-slab overpressure, while the flat slab segment steepens and becomes part of renewed slab rollback. Tear subduction therefore re-establishes a continuous slab and allows the process to recur, providing a mechanism for the transient nature of flat slabs. Reconstructed Cenozoic subduction rates of the Nazca Plate (80–90 km Myr^{-1}) suggest slab tears that terminated previous periods of flat subduction would have rapidly disappeared in the deeper upper mantle. Consequently, the renewed build-up of sub-slab pressure sets the stage for a new phase of flat subduction.

The upwelling asthenosphere from beneath the tearing flat slabs also provides an ideal condition for the slab (especially its basaltic crust) to melt, generating adakitic magmas as found in Peru and central Chile (Fig. 17.46c). In Peru, strong localised upwelling first occurred at 9 Ma along the eastern edge of the subducting Inca Plateau and at 4 Ma along the western edge as it further subducted (Fig. 17.46a; Hu and Liu 2016). Another phase of prominent upwelling started at 7 Ma along the down-going Nazca Ridge (Fig. 17.47a). As subduction continued, the slab gaps and associated asthenosphere upwelling in Peru migrated southeast-ward and are sustained to the present. In central Chile, a similar yet weaker upwelling started around 5 Ma when the slab tear initiated and subsequently waned towards the present-day (Fig. 17.47b).

Adakites are volcanic rocks of intermediate to silicic composition ($\geq 56\%$ SiO_2) with $Al_2O_3 > 15$ wt%, MgO < 3 wt%, $Na_2O \leq 7.5$ to ≥ 3.5 wt%, low Y (<18 ppm) and heavy rare-earth elements, Sr typically > 400 ppm, $^{87}Sr/^{86}Sr$ values typically <0.7040, high Sr/Y (>40) ratios and have geochemical characteristics of magma thought to have formed by partial melting of altered basalt subducted below volcanic arcs at the amphibolite to eclogite transition at depths between 75 and 85 km (Defant and Drummond 1990; Drummond et al. 1996; Hastie 2020; Zhang et al. 2021). Typically, this is thought to occur when young "warmer" oceanic crust (in particular spreading or aseismic ridges) is subducted, which enables partial melting of the metamorphosed subducted basalt (eclogite) rather than the mantle above. Tearing of the flat slab also causes local temperature increase which, together with the upwelling asthenosphere, likely would melt the much-thinned slab and its basaltic crust. Given that: (1) basaltic oceanic crust has copper concentrations about two times greater than those in the continental crust, and (2) high oxygen fugacity in the subduction environment promotes release of copper during partial melting, slab melts should contain higher copper (and also gold) concentrations (e.g. Thiéblemont et al. 1997; Mungall 2002; Sun et al. 2011, 2012; cf. Richards 2011 for an alternative view). Identification of slab melt adakites is therefore potentially important for predicting exploration targets for copper- and gold-porphyry ore deposits (Hastie 2020; Zhang et al. 2021), and perhaps explains the close association of the Andes margin with some of the world's largest porphyry copper and epithermal gold deposits (Chap. 18).

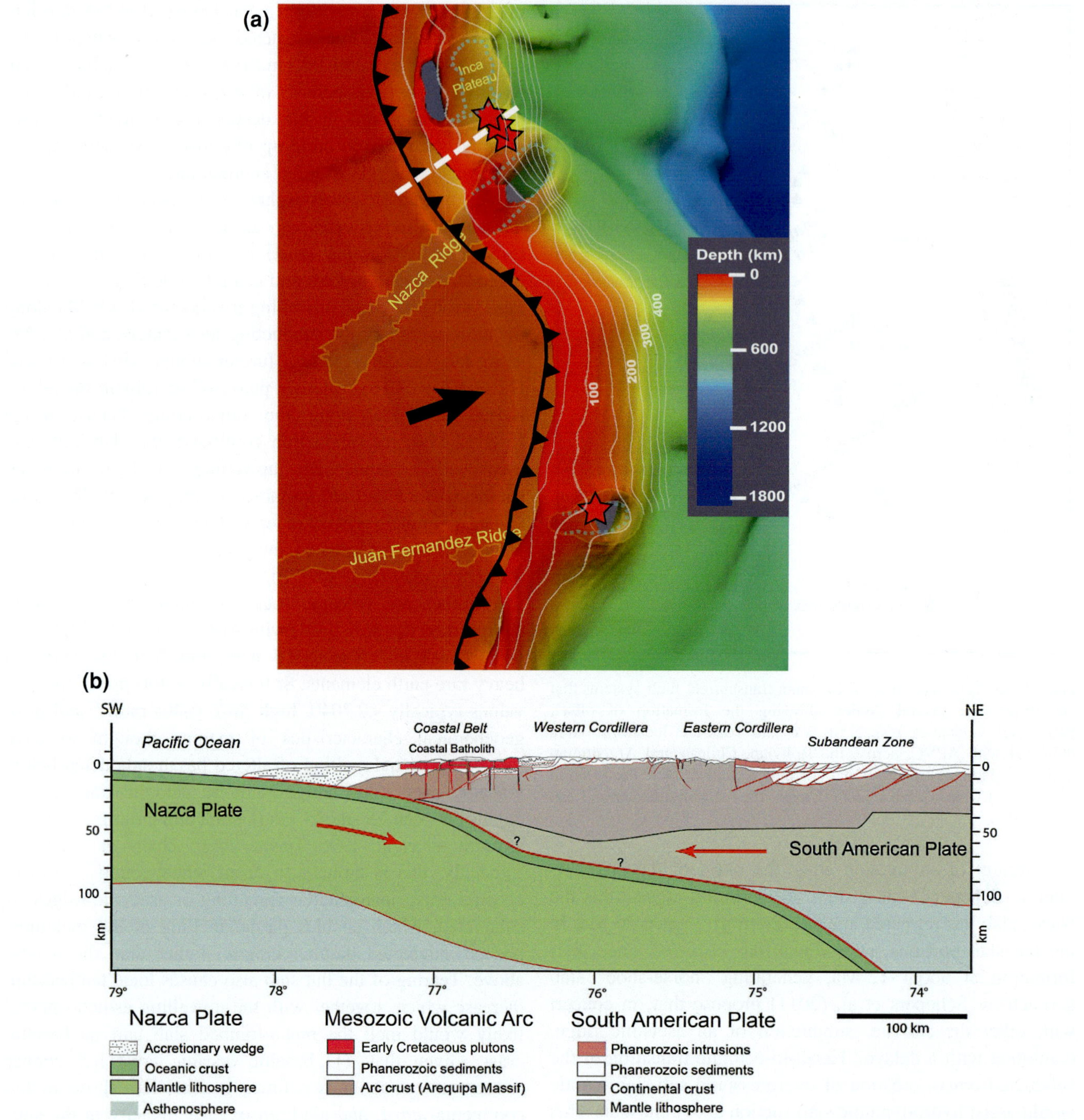

Fig. 17.46 Geometry and geology of the predicted present-day flat slab segments and slap tears beneath South America, and that of the Nazca Plate west of the trench. **a** 3D aerial view of the subducting Nazca Plate outlined using an isotherm of 300 °C cooler than the ambient mantle. Colours represent the depth of the slab's upper surface on the right side of the trench, and depth of the plate's lower surface on the left side. Slab tears are illustrated by both temperature isosurface and the evolution of buoyancy features (grey indentations); dotted lines within the subducting buoyancy features outline their original intact geometry; red stars indicate the locations of adakitic eruptions; dashed white line is profile shown in (**b**). (From Hu and Liu 2016). **b** Geological cross-section showing the geometry of the plate boundary and the crustal structure in the Peruvian flat slab segment (From Pfiffner and Gonzalez 2013). **c** Schematic view of flat slab and tear formation. The key feature is a down-dip oriented slab tear within the subducting, positively buoyant, aseismic ridge. Formation of the slab tear induces asthenosphere upwelling from beneath the slab. Upwelling asthenosphere warms up and melts the severely stretched oceanic crust, leading to adakitic volcanism during the initial stage of slab tearing. Slab tearing events should represent a common consequence of flat slab formation due to subduction of small- to medium-sized buoyant, bathymetric impactors. (From Hu and Liu 2016)

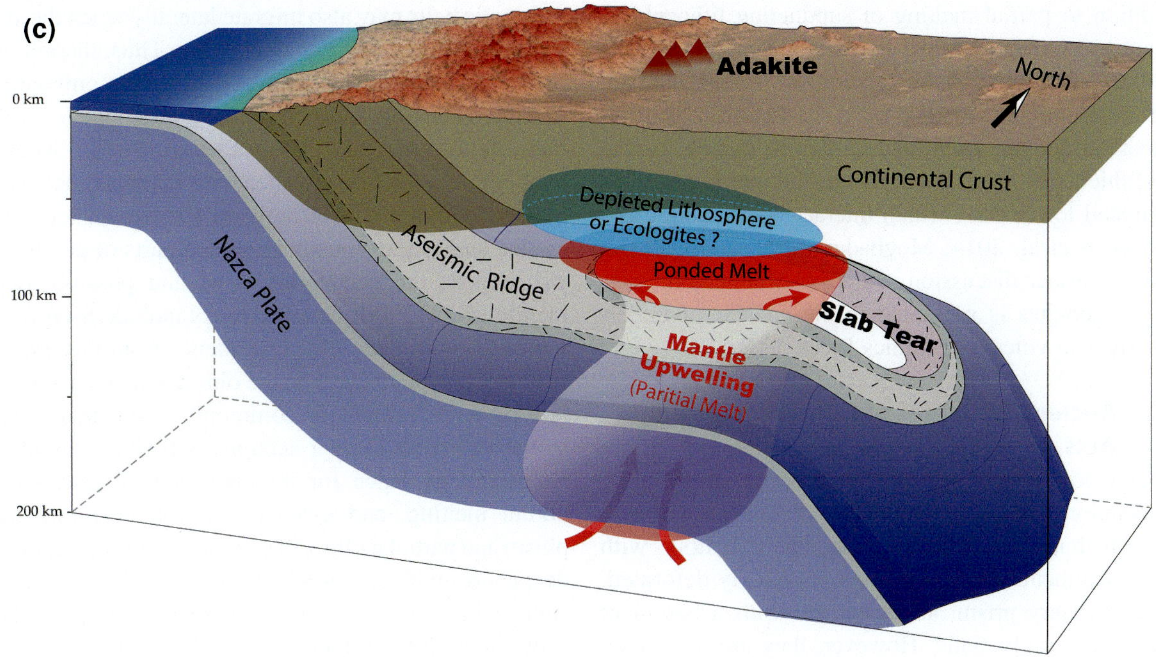

Fig. 17.46 (continued)

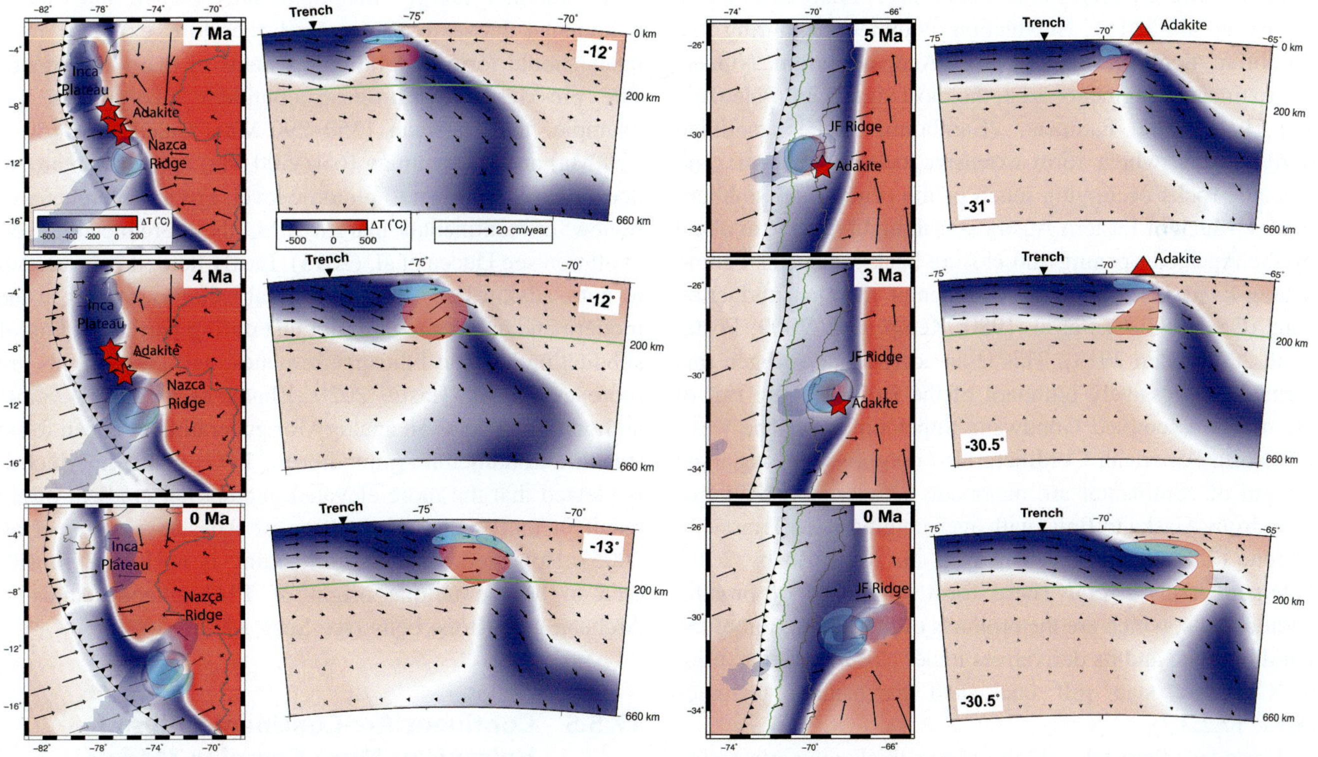

Fig. 17.47 Thermal and velocity models showing evolution of the **a** Peruvian and **b** Pampean, central Chilean slab tears. In each panel, both the map (left) and cross-sectional (right) views are shown, corresponding to different times. All maps are at 100 km depth, and cross-sections are east–west at the given latitude. The background colour is for temperature (note the different colour scales for the maps and cross-sections) and arrows for velocity fields. The slab tear induced upwelling likely generates partial melting (red pattern), eclogitised crustal fragments at shallower depths (cyan patterns) and adakitic eruptions (red stars in map views and triangles in cross-sectional views). (From Hu and Liu 2016)

In addition to partial melting of subducting lithospheric plate, there are indications that some adakites can also be generated by fractional crystallisation of mantle-derived magma that assimilates country rock, fractional crystallisation of magma derived from metasomatized mantle, partial melting of thickened lower crust, and partial melting of slabs of delaminated lower crust which interact with mantle peridotite (Rossetti et al. 2014; Moghadam et al. 2016; Zhang et al. 2019). Further discussion of the possible tectonic significance of adakites is presented in the context of understanding Archean volcano-tectonics (Sect. 17.7.5.7).

17.5.4.1 Ancient Examples of Continental Margin Arcs

In the rock record such a setting should be recognisable by the association of a regionally extensive, linear, silicic-intermediate calc-alkaline magmatic belt(s) with continental sediments, paralleled by a complexly deformed, forearc accretionary prism, all associated with a basement that was demonstrably sialic. However, they are long-lived (10–100 Ma) and, therefore, likely to be spatially and tectonically complex and should include evidence of paired metamorphic belts. For example, the Andes have actually been the site of continental margin magmatic activity for some 500 Ma (Lucassen et al. 2000, 2011; Franz et al. 2006; Ducea et al. 2015). Subduction commenced during the Cambrian along the western proto-Andean margin. Numerous extinct arc segments exist, the most significant Paleozoic arcs being the Pampean (Cambrian) and Famatinian (Ordovician), which form extensive belts of plutonic and volcanic rocks occupying much of the central and southern Andes. Much of Eastern Australia is thought to have evolved by the repeated opening and closure back-arc basins behind a long-lived (515–230 Ma) continental margin subduction zone (Cas 1983; Cas et al. 1980; Kemp et al. 2009; Foster et al. 2009; Glen 2013). This is an accretionary orogen that formed progressively outboard of the Pacific-facing rim of the newly assembled Gondwana Supercontinent. The western North American Cordillera also comprises a long-lived margin of continental arc magmatic activity (mostly intrusive) from Alaska to Baja, with ages ranging from Triassic to Eocene (250–50 Ma; e.g. Busby et al. 2005; Busby 2012; Paterson et al. 2011). These form a number of broadly parallel belts, which are the products of the subduction of the Farallon Plate and its derivatives under North America, prior to Cenozoic slab roll back and extensional tectonics (Sect. 17.2.3.2).

Episodes of crustal and lithospheric thickening alternating with periods of extension (slab roll back, or root foundering) produce cyclic vertical changes in arcs including, for example, alternating periods of foreland and extensional basin development (e.g. Rossetti et al. 2014). The typical width of active continental arc magmatism is about 20–30 km, however, magmatic belts may also migrate laterally when the slab dip is shallow or under extreme thickening. Thus, theoretically, the frontal arc magmatism can sweep inboard or outboard relative to a fixed location; most commonly, arcs appear to migrate at rates of 1–5 mm yr^{-1} (Ducea et al. 2015). Additionally, back-arc regions that also generate magmas may extend the magmatic footprint well beyond the frontal arc. The more eroded and/or tectonically exhumed parts of continental arcs will expose the deeper arc crust and possibly uppermost mantle (Fig. 17.48). Granitic rocks include S-types, metaluminous I-types, and A-type; evidence from Eastern Australia suggests S-types tend to be the oldest in a given area, whereas A-type emplacement is commonly associated with extensional tectonics. Deeper sections will show abundant mafic rocks and evidence for incorporation of crustal rocks via partial melting, and extensive high-temperature metamorphism and partial melting to generate granulitic residues. Even deeper assemblages of arc roots may be found in xenolith suites which are either granulitic or eclogitic; e.g. the Cretaceous Sierra Nevada batholith, California, USA (Saleeby et al. 2003; Ducea et al. 2015; Fig. 17.48). The deepest part of the Cretaceous Fiordland magmatic arc in South Island, New Zealand, tectonically exposes similar eclogitic rocks (De Paoli et al. 2009).

Continental margin magmatic arcs evolve into stable continental masses over time and, therefore, exhumed remnants should form integral components of many continents. The pre-Mesozoic European continent incorporates, for example, the Variscan (350–290 Ma), Caledonian (500–400 Ma) and Cadominian (650–500 Ma) arcs, which are all deeply exposed down to plutonic level. For a comprehensive review of continental margin arcs, their architecture and evolution see Ducea et al. (2015). Lastly here, it is important to note, adakites can form in several tectonic settings in the modern and ancient record, not just during partial melting of subducting basaltic lithospheric plates, and their origins are more equivocal (Hastie 2020; Zhang et al. 2021). Some authors use them as an analogy for generating Archean felsic (tonalite-trondhjemite-granodiorite; TTG) melts, in which it is viewed that the more elevated geothermal gradient might result in more common melting of the slab (Martin et al. 2005). However, a subduction setting is no longer considered to be the unique origin of TTG igneous suites in Archean greenstone belts (see Sect. 17.7).

17.5.5 Continent/Arc-Continent Collision Volcanism: More Complex Models

Continental collision can be viewed as an extreme "end member" variant of subduction, whereby an ocean basin and the subduction zone is destroyed, mountains are produced, and two continents ± intervening island arcs are

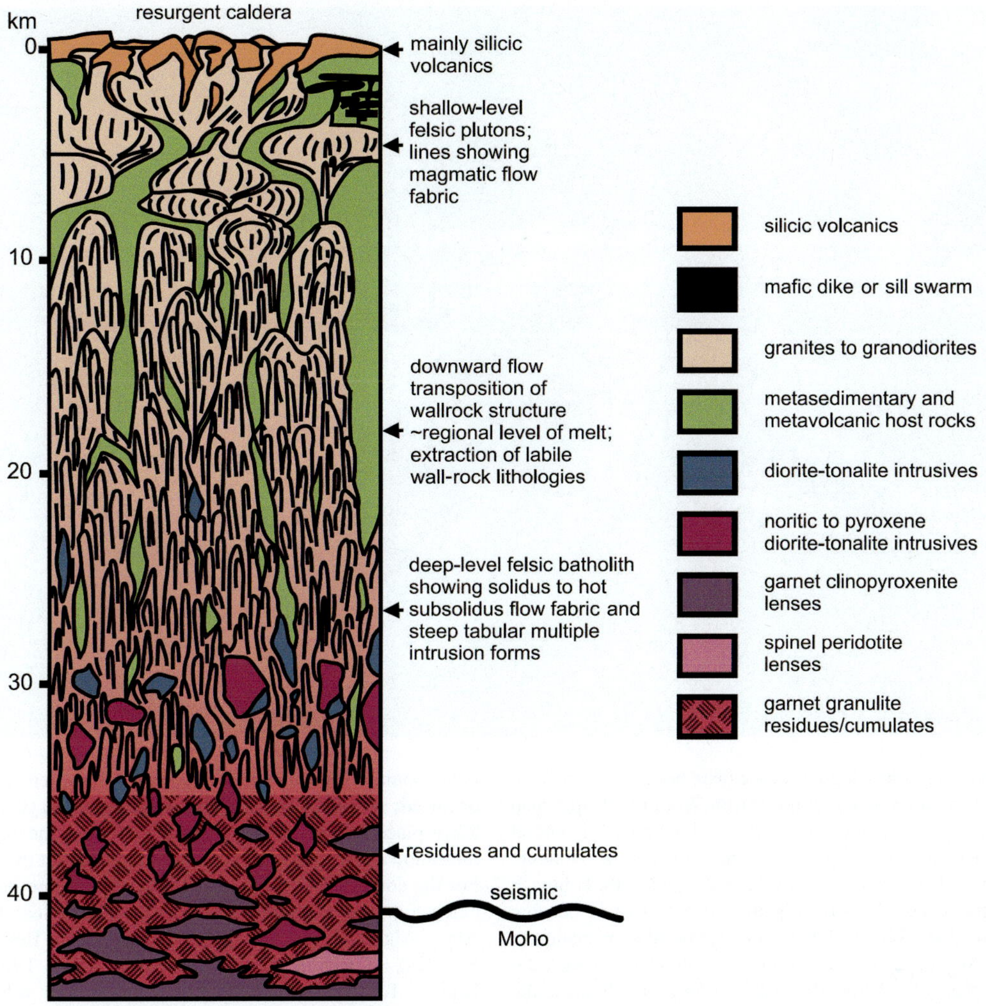

Fig. 17.48 Schematic cross-section through a continental margin magmatic arc from the volcanic cover to the top of the arc root. The vertical continuation of the root is represented by abundant garnet- and/or amphibole-bearing pyroxenites, which gradually give way to typical upper mantle peridotites below 60–100 km depth. The depicted section is based on structural relations of an actual crustal section from ~5 to 35 km depth in the Mesozoic Sierra Nevada magmatic arc, California, USA. Although subduction beneath such arcs may be active for hundreds of millions of years, most magmatism is emplaced during magmatic "flare-ups" of 10–30 Myr duration (Paterson et al. 2011; Ducea et al. 2015). Individual plutons and batholiths can grow in <0.5–10 Myr. Pulse sizes moving through these magma plumbing systems vary from small dike-like to large diapir-like pulses, both of which may form from earlier amalgamation of poorly defined pulses. (Adapted from Paterson et al. 2011, After Saleeby et al. 2003)

sutured together. Continental collision is not an instantaneous event, but may take several tens of millions of years before faulting and folding caused by collisions stops. The collision between India and Asia (Plate 17.1) has been taking place for ca. 50 Ma already and shows no signs of waning. One of the best studied and complex continental convergent plate boundaries is between Eurasia and Africa, which is termed the Alpine collisional zone. This comprises a "Rubik's cube" of micro-plates, subduction zones, mountains, thrust belts, volcano-plutonic arcs, suture zones (ophiolites, high-P/low-T metamorphic rocks), as well as back-arc basins and extended zones of lithosphere along the convergent plate boundary (e.g. Jolivet and Faccenna 2000; Ring et al. 2010; Jolivet and Brun 2010; Jolivet et al. 2013; Plate 17.1). All are still active today!

The geological history of the Alpine-Mediterranean region records the creation of a Mesozoic Neotethys ocean and its subsequent Cenozoic destruction between Eurasia and Africa. The sites of former subduction zones have been variously inferred by the location of thrust belts, ophiolites and calk-alkaline volcano-plutonic arcs. The Hellenic and Calabrian trenches appear to be the only presently active subduction zones in the region (Fig. 17.49). Associated with retreating plate boundaries (i.e. slab retreat) are wide zones of extension in the upper plate (e.g. the Aegean Sea, Tyrrhenian Sea, Pannonian Basin, Liguro-Provençal Basin, and Balearic,

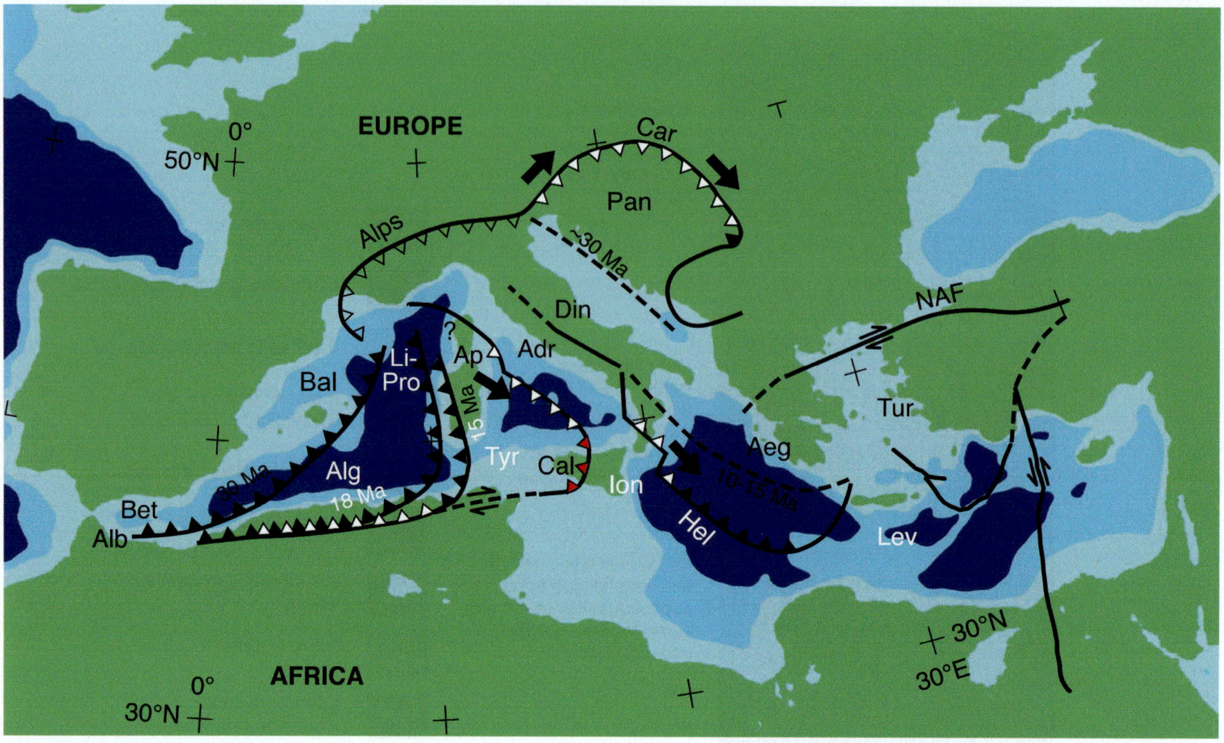

Fig. 17.49 Synoptic view showing Cenozoic plate boundary evolution of the wider Alpine-Mediterranean region (From Wortel and Spakman 2000). Large black arrows indicate inferred directions of lateral migration of slab detachment along the Apennines-Calabria, Hellenic and Carpathian Arcs. Blue colours indicate bathymetry; dark blue = deeper than 2.5 km (generally corresponding to presence of oceanic lithosphere); intermediate blue = 2.5–1.0 km; light blue = shallower than 1.0 km. The Aegean Sea is underlain by (extended) continental lithosphere, as is probably below the Adriatic Sea. In all arcs, the outermost sawtooth curves indicate the present location of the convergent boundary; black sawtooth indicate where the subducting slab is/or was considered to be continuous; white sawtooth indicate plate boundary segments where slab detachment is assumed to have occurred; red sawtooth (Calabria) indicate slab detachment may have taken place recently. For the western Mediterranean, three stages in the migration of the plate boundary are displayed at the times as indicated. For the southern Aegean, 10–15 Ma refers to the most recent phase of extension. Adr = Adriatic Sea; Aeg = Aegean Sea; Alb = Alboran Sea; Alg = Algerian Sea; Ap = Apennines; Bet = Betics; Bal = Balearic Sea; Cal = Calabria; Car = Carpathians; Din = Dinarides; Hel = Hellenic Trench; Ion = Ionian Sea; Lev = Levantine Basin; Li-Pro = Liguro-Provençal Basin; NAF = North Anatolian Fault; Pan = Pannonian Basin; Tur = Turkey; Tyr = Tyrrhenian Sea

Algerian and Alboran Seas; Fig. 17.49). A large strike-slip component of later motion on delimiting faults, principally the North Anatolian Fault Zone is also now observed.

In Turkey and the Aegean, closing of the northern leading edge of Neotethys commenced in the late Cretaceous, although details of this remain controversial (e.g. Eyuboglu et al. 2011, 2016). The widely held view is that closure of Neotethys took place along a northward-dipping subduction zone (although others have emphasised the opposite direction), with formation of the Pontides volcanic arc. In the Eastern Pontides this is a several km thick submarine volcanic succession dominated by calc-alkaline basaltic andesite to dacitic and rhyolitic hyaloclastite successions intruded by subvolcanic dykes forming a volcanic core complex, which are intercalated with turbiditic sediments (Wright et al. 1995; Wright 1996). Initial collision may have commenced in Paleocene-early Eocene times ($\sim$60–50 Ma), and the suture is believed to be along the Izmir-Ankara-Erzincan, and Vadar ophiolite belts in Turkey and Greece, respectively (Fig. 17.50). These include large bodies of ophiolite including well-preserved sections of oceanic lithosphere and accretionary mélanges with local blueschist assemblages.

Following collision between the Pontide and Tauride microcontinents, andesitic-alkaline volcanism (including adakites) produced a submarine to subaerial volcanic succession overlapping the older arc, while intra-continental convergence, north–south shortening and thrusting continued after this collision. Origins of the adakitic magmatism are enigmatic. It has been correlated with partial melting of thickened continental crust in a post-collisional setting by either delamination during onset of extensional collapse, or slab-break off induced asthenospheric upwelling (Karsli et al. 2010, 2013), or evidence of slab roll back of a southerly-dipping subduction zone (Eyuboglu et al. 2011, 2016).

In the Aegean region, compression caused folding, thrusting and crustal thickening with high-P/low-T

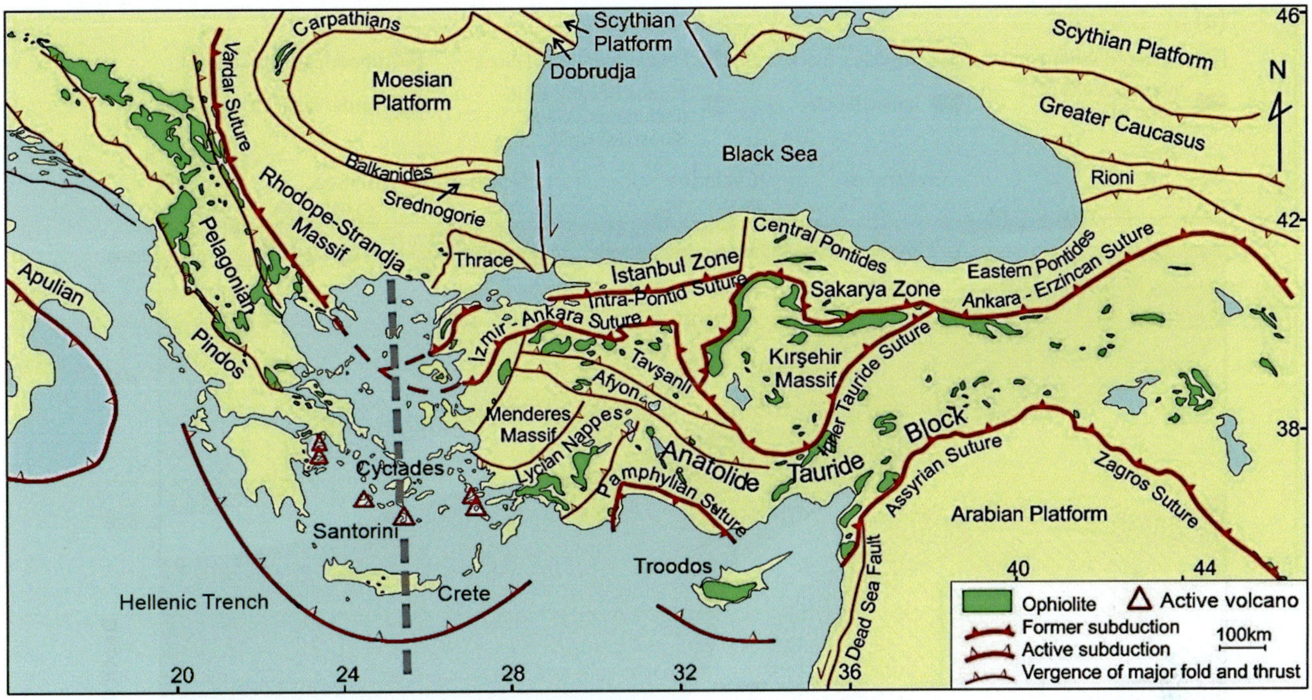

Fig. 17.50 Regional location map showing continental blocks, major sutures and related ophiolite belts of the Mesozoic-Cenozoic, eastern Mediterranean continental collisional orogen. The Aegean volcanic arc is located by line of active and potentially active centres; north of the arc volcanism has migrated southward since the late Eocene-early Oligocene with rollback of the Hellenic trench. Thick dashed line shows crustal and tomographic profile of the subducted slab illustrated in Fig. 17.51a and described in text. (Modified from Tavlan et al. 2011)

blueschist metamorphism at about 50–40 Ma (mid-Eocene). This was succeeded by a geodynamic regime which appears to have been dominated by calc-alkaline (with high-K to shoshonitic) magmatism and extensional tectonics persisting to the present day (Jolivet and Brun 2010; Jolivet et al. 2013, 2015). This includes development of high extensional strain zones and metamorphic core complexes in the eastern Mediterranean, e.g. Naxos (Cyclades), and the Rhodope and Menderes Massifs (Fig. 17.50; Lister et al. 1984; Gautier et al. 1993; Sokoutis et al. 1993; Brun and Sokoutis 2007; Lips et al. 2001; Ring et al. 2010; Jolivet et al. 2013, 2015). Volcanism and granitoid magmatism show progressive younging trends from the late Eocene-early Oligocene (ca. 40 Ma) southwards towards the presently active arc. Earlier studies postulated two subduction zones to explain earlier phases of Oligocene-mid-Miocene activity, separate from the younger active volcanic front. However, the results of seismic tomographic imaging structure did not corroborate this model (Spakman et al. 1988, 1993; Bijwaard et al. 1998).

Rather, from south of Crete, these profiles show a steep, northward subducting downward buckled slab (high-velocity anomaly) now known to descend ~1800 km into the lower mantle, and >2000 km long (Ring et al. 2010; Fig. 17.51). The geometry of the slab changes distinctly, draping the 660 km discontinuity (lower mantle boundary) sub-horizontally for some 300 km. Directly above this, the slab is thicker than in its upper part, possibly due to folding. The slab in the lower mantle is ~800 km long, but approximately twice as thick as the upper mantle slab. The important question of debate today is whether the upper and lower mantle slabs are directly connected (Ring et al. 2010). Most workers agree the lower mantle high-velocity anomaly probably represents ~1100–1500 km of material subducted in the Mesozoic, whereas the upper mantle anomaly represents some 1000 km of subduction in the Cenozoic. Do they indeed reflect continuous subduction of a single slab possibly for as long as 200 Ma? More recent, higher resolution tomographic solutions suggest both may be true (Wei et al. 2019). These confirm previous results that in the south Aegean the slab has subducted to a depth of at least 1100 km. However, in addition, there is evidence that the subducting slab is breaking, and has already broken off at the 660 km discontinuity. In contrast, under the northern Hellenides region, at least four high-velocity anomalies representing a succession of slabs have been imaged, suggesting step-wise subduction rather than continuous subduction.

The central and western Mediterranean regions have experienced more complex Cenozoic to Quaternary convergence histories (Fig. 17.49). Northward convergence of Africa commenced at ca. 80 Ma and occurred mostly due to subduction of the Ionian-Adriatic lithosphere below the European Plate. The position of the convergent margin has

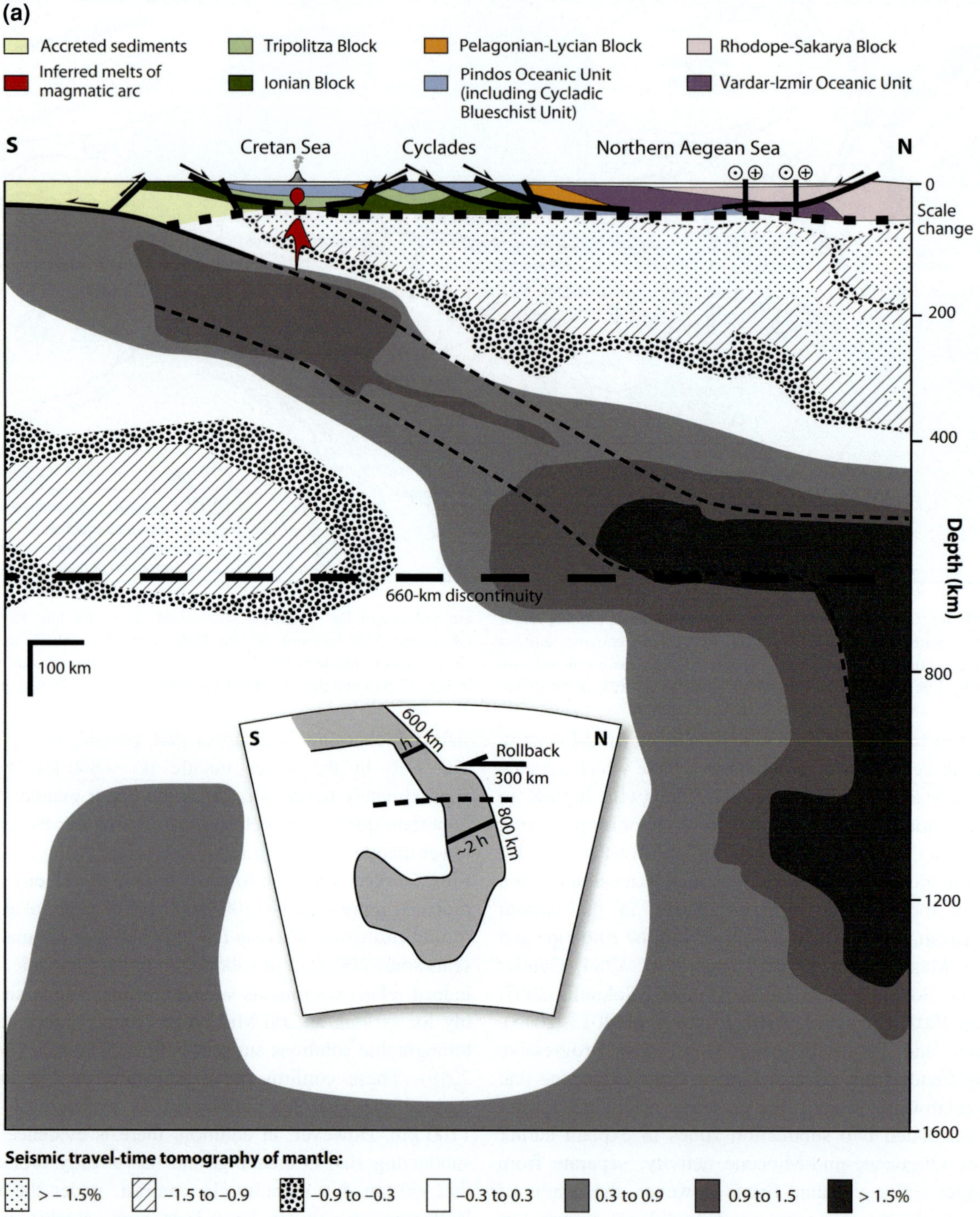

Fig. 17.51 Rollback and geometry of the subducting Aegean slab, eastern Mediterranean. **a** Schematic geodynamic cross-section through the Aegean Sea, from south of Crete to its northern margin The thick dotted line marks the crust-mantle boundary; a northward-dipping Benioff plane can be followed down to a depth of 180 km, while tomographic image shows a high-velocity anomaly of the descending slab down to ~1800 km. The geometry of the high-velocity anomaly changes distinctly across the upper-lower mantle discontinuity at 660 km (see inset); dashed lines represent the Cenozoic portion of the down-going slab, which drapes backwardly over the 660 km discontinuity. The Aegean extensional region is underlain by low-velocity upper mantle. (From Ring et al. 2010). **b** Model showing interaction of a subducting slab with the 660 km discontinuity. The discontinuity slows down slab descent into the lower mantle, causing it to fold (t1). At t2 the slab starts to drape over the 660-km discontinuity, and seaward migration of the retreating slab relative to the trench accelerates extension in the overriding plate. Draping of the slab along the discontinuity allows the slab to maintain a similar geometry, which results in pronounced retreat and a phase hyperextension through the back arc (t3). (Modified from Ring et al. 2010, After Schellart and Lister 2005)

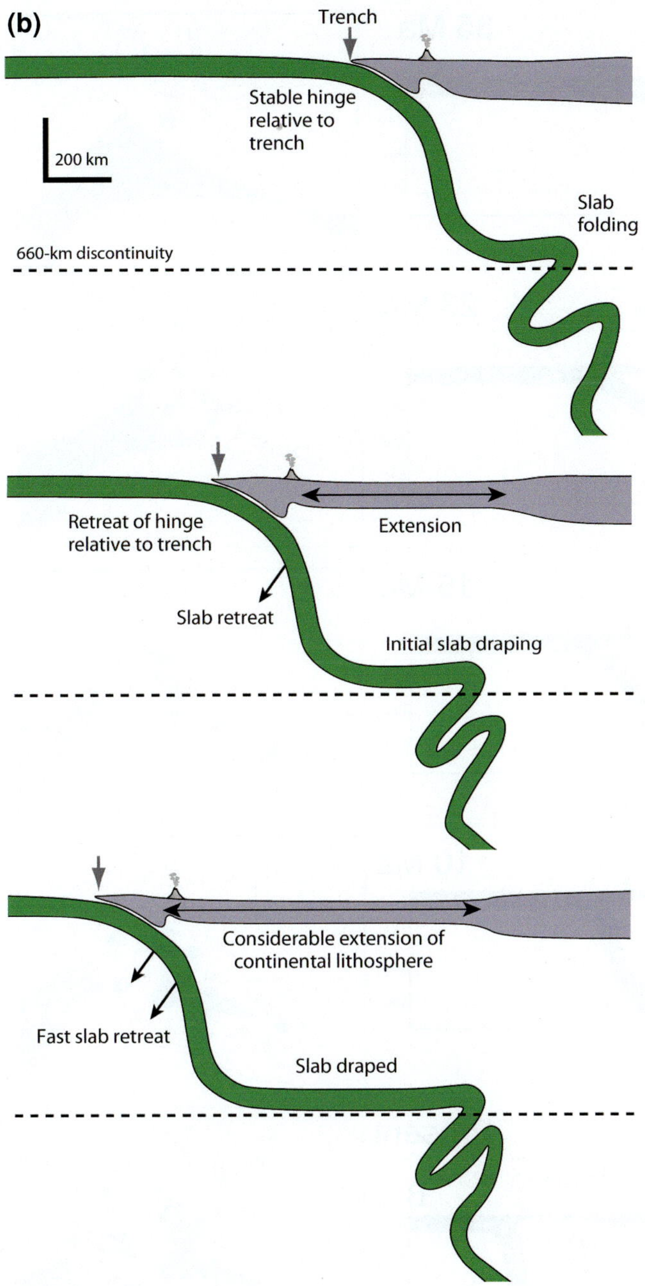

Fig. 17.51 (continued)

migrated in pulses, driven by fragmentation and "roll back" of the subducting slab (Faccenna et al. 2001, 2004, 2010). The initial configuration of the arc was stable for about 20–30 Ma, trending approximately E-W, and associated with low convergence rates of about 1 cm yr^{-1}. A subsequent progressive increase in subduction rate is thought to be related to the beginning of slab pull at ca. 65 Ma and related initiation of the slab roll back.

From 35 to 30 Ma, collision of the Adria continental microplate changed the configuration of the trench-arc system, leading to localised faster retreat along the Ionian oceanic lithosphere corridor and fragmentation of the slab (Fig. 17.52). At this stage, extension started behind the accretionary wedge and was coeval with arc volcanism in Sardinia and Provence (34–32 Ma). From 21 to 16 Ma extension accelerated to form the Liguro-Provençal oceanic back-arc basin along with counter-clockwise rotation of the Sardinia-Corsica continental block by 25–30° (Speranza et al. 2002; Fig. 17.49). At around 12–10 Ma rifting shifted into the Tyrrhenian region, leading to the fast, eastward migration and spreading of two oceanic back-arc basins during the Pliocene–Pleistocene: the Vavilov Basin at 5–4 Ma, and the Marsili Basin at 2 Ma. Concomitant decrease in width of the subducting slab resulted in tight arching of the surrounding orogenic belts in the Apennines, Calabrian arc and Sicily (Cifelli et al. 2007).

Present-day subduction of the Ionian Plate below the Calabrian Arc is imaged in the southern Tyrrhenian Sea by the presence of a steep (70°), NW-dipping Benioff zone and seismicity distributed down to 450 km (Selvaggi and Chiarabba 1995).The Moho in the Tyrrhenian back-arc basin is thinned to around 20 km in thickness. Tomographic images show the subducting slab as a well-defined and steeply inclined high-velocity zone that flattens at about 500 km depth. 3D images show that the slab is largely detached or torn across the eastern part of the Calabrian Arc and similarly in the Carpathians (Wortel and Spakman 2000; Fig. 17.49). Tomographic imagery has had important implications for tectonic reconstructions in the region and elsewhere. In the past there has been a tendency for geologists to place a new subduction zone (and a new ocean to close) when the locus of volcanism shifts. This is an impulse that has also hindered understanding of other ancient orogenic zones. Instead we effectively appear to be witnessing the evolution of a single subduction zone during continental collision, with roll back of the slab, and parts of the slab tearing and detaching into the mantle at various times (e.g. Rosenbaum et al. 2008; Figs. 17.53 and 17.54).

The dynamics of slab roll back have important consequences on the nature and evolution in time and space of volcanism (Fig. 17.53). The present-day active arc volcanism is located in a very narrow and short belt of volcanoes in the southern Tyrrhenian Sea region, extending from the Aeolian arc to the Neapolitan volcanoes. This belt should also be extended to include the quiescent Upper Pleistocene calderas of the Roman volcanoes (Fig. 17.54), whose composition overlap with that of the Neapolitan volcanoes. The age of last preserved eruption is also as recent as 25 ka at Colli Albani (Mattei et al. 2010; Conticelli et al. 2010a). Magma compositions span from typical calc-alkaline to potassic and ultrapotassic overlapping in time at many volcanoes (e.g. in the active Aeolian volcanoes; Francalanci et al. 1993; De Astis et al. 2000; Conticelli et al. 2009; Peccerillo et al. 2013). Differentiated magmas prevail over

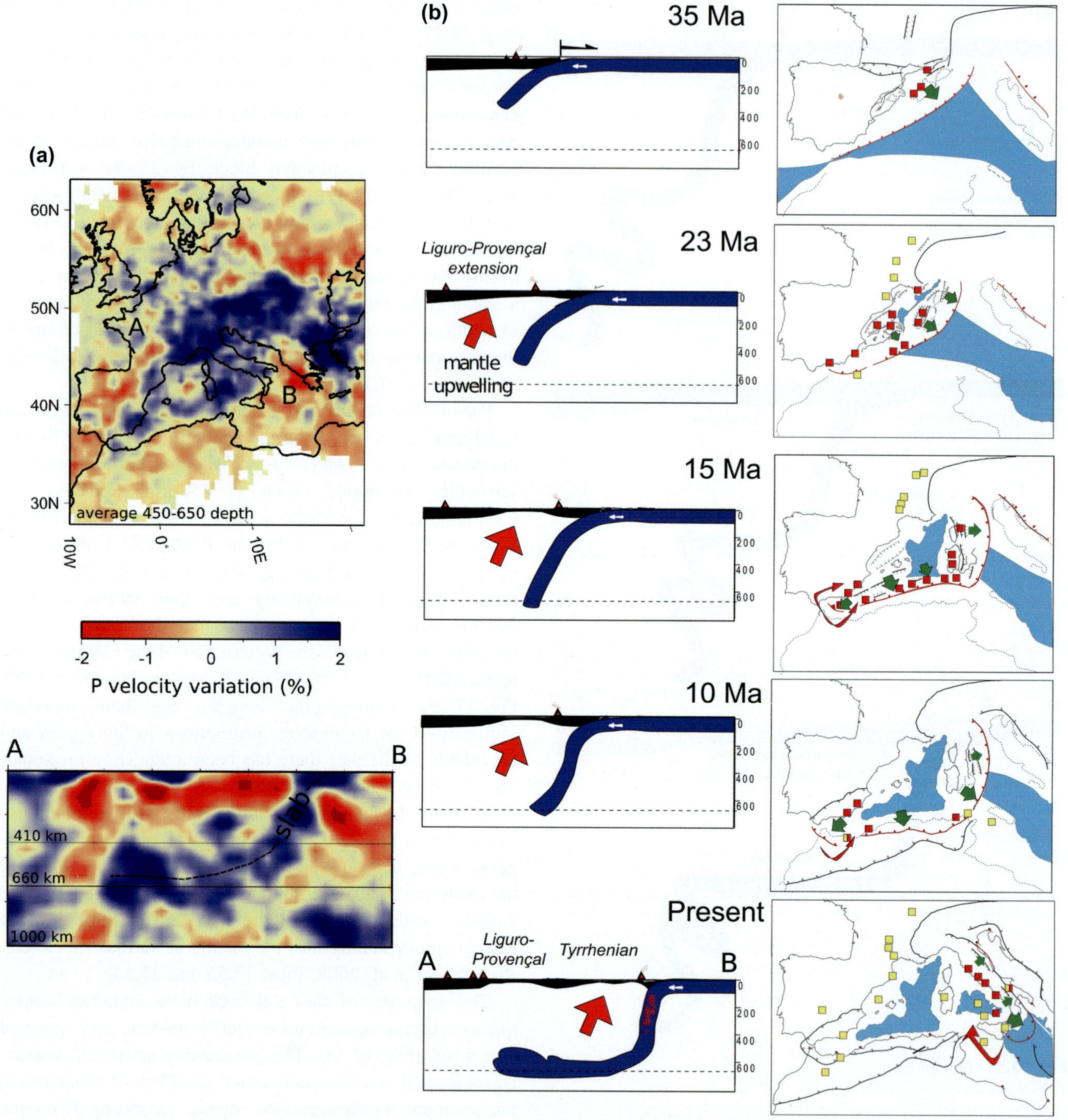

Fig. 17.52 Seismic tomographic image of the present-day southwestern European mantle and dynamic interpretation of its evolution (Modified from Faccenna et al. 2010). **a** Tomographic image and cross-section along the NW–SE transect A–B; map view averages the P-wave velocity variations between 450 and 650 km depth. The high-velocity anomaly (blue) shows the steeply subducted African slab that flattens along the 660 km mantle discontinuity. **b** Reconstruction over the last 35 Ma of subduction and progressive roll back of the African lithosphere below the European lithosphere in the central Mediterranean. Left panels show a schematic cross-section oriented NW–SE as A–B in **a**. These illustrate progressive roll back was accompanied by SE-ward migration of the trench and opening of the Liguro-Provençal and Tyrrhenian back-arc basins with asthenospheric mantle upwelling (red arrows); the present-day panel is a reconstruction of today's configuration based on seismic tomography data (red dots are locations of deep earthquakes along the Benioff zone). Right panels indicate the same in plan view, with migration and arching of the arc-trench system (green arrows); red and yellow squares are representative of subduction-related and anorogenic volcanism, respectively; red arrows indicate inferred return flow triggered by retreating slabs

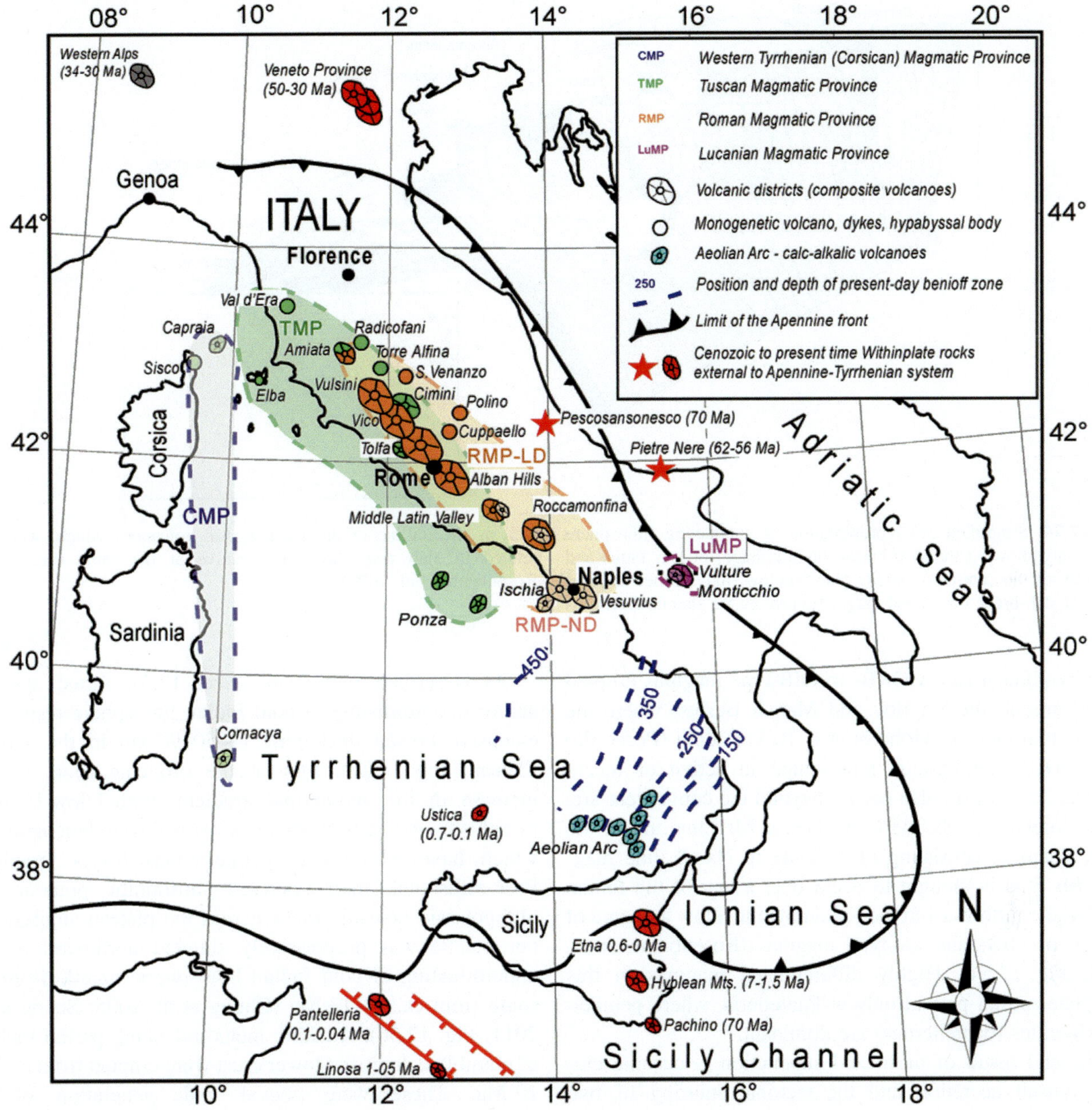

Fig. 17.53 Main volcanic provinces in the Tyrrhenian region, Italy, showing the complexity and superposition of many geodynamic processes. The diachronous western Tyrrhenian, Tuscan, Roman and Lucanian provinces show progressive and increasing evidence of mixing between subduction-related and anorogenic magmas. (From Conticelli et al. 2010b)

primitive and most evolved compositions span from rhyolite (e.g. Vulcano and Lipari), to K-trachyte (e.g. Campi Flegrei) and K-phonolites (e.g. Vesuvius), controlling the dominant explosive character of the active volcanism.

The potassic character of these orogenic magmas is thought to be aided by recycling of pelitic sediments in the subduction zone, being responsible for the upward migration of K-rich fluids and K-metasomatic enrichment in veins of the upper plate peridotitic lithosphere (Conticelli et al. 2002; Avanzinelli et al. 2009). Partial pressure of CO_2 at the mantle source, likely related to the decarbonation of calcareous sediments and limestones subducted along with the pelitic component, has been experimentally correlated with the degree of silica saturation of the primary potassic magmas (Foley 1992). This explains the simultaneous occurrence of ultrapotassic silica-saturated lamproites and silica under-saturated kamafugites and leucitites (Conticelli et al. 2010b, 2013).

At the same time, back-arc extension in the Tyrrhenian Sea facilitates asthenospheric upwelling, controlling

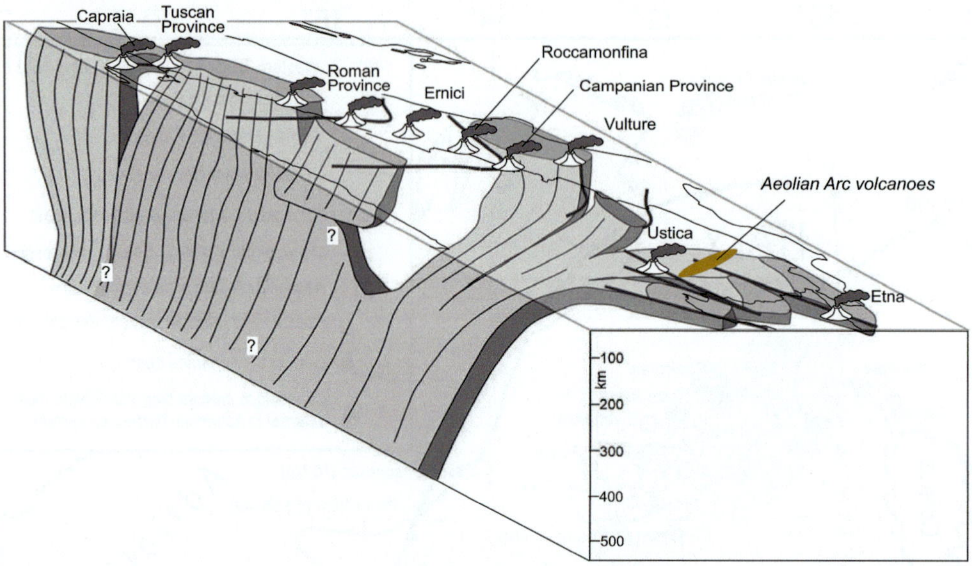

Fig. 17.54 Simplified 3D representation of subducting lithosphere below Italy showing the distribution (approximate) of tear faults and geometry of the central Apennine slab window, Italy. Other selected tear- and slab-break off related magmatic centres are also indicated, as well as the outline of the Aeolian Arc volcanoes which, while not related to slab tear, have been active at the same time. (After Rosenbaum et al. 2008)

Pliocene–Quaternary MORB- to OIB-type magmas erupted in and around the Vavilov and Marsili Basins, where the crust is thin and the Moho is at 6–10 km depth (Peccerillo 2017). OIB-type basalts, represented in active or recent intraplate volcanism also occurs beyond the convergent arc, in the foreland of the African plate at Mt Etna and in the Sicily channel, consisting of tholeiite to Na-alkaline magmas. Mt Etna is thought to occur over a tear of the Ionian slab, where inflows from the Ionian mantle allow a degree of mixing of OIB- and arc-type magmas (Rosenbaum et al. 2008; Fig. 17.54). Highly differentiated magmas in this intraplate setting prevail only at Pantelleria, where peralkaline rhyolites (pantellerites) are dominant.

The end result of protracted subduction is oceanic closure, crustal accretion and the tectonic suturing of two continents and intervening arcs. Evidence from the Himalayas suggests subduction pulled the Tethys oceanic plate down into the mantle until the Indian continent finally collided with Eurasia at ca. 45 Ma (e.g. Husson et al. 2014). Pulled from below and pushed from behind by the underlying mantle flow, the Indian continent moved northward. It gradually stepped over its own slab, flipped it, and folded it under itself in the mantle, several hundreds of km to the south of the Himalayan belt (e.g. van der Voo et al. 1999; Replumaz et al. 2010; Capitanio and Replumaz 2013; Fig. 17.55). Entrained by its oceanic precursor, the continental slab partially entered the subduction zone and also actively contributed to its own subduction (e.g. Capitanio et al. 2010).

As described previously (Sect. 17.3), thrust stacking and/or underthrusting of cold Indian lithosphere resulted in extensive crustal thickening ($\sim$50–90 km in the Tibetan Plateau; Fig. 17.27), loss of the orogenic root, a rapid increase in the geothermal gradient, high-T/low-P metamorphism and crustal anatexis (Himalayan leucogranites, which have affinities to peraluminous S-type granites). Post-collisional, mantle-derived shoshonitic potassic and ultrapotassic volcanic rocks across the plateau suggest that hot mantle was progressively shunted northward by the underthrusting of cold Indian lithospheric mantle from the south from ca. 50–13 Ma (Chung et al. 2005; Searle et al. 2011; Fig. 17.56). Adakitic melts indicating partial melting of possibly eclogitised lower crust were erupted from ca. 26–10 Ma. These were coeval with generation of the leucogranites (23–15 Ma; Phillips et al. 2013) resulting from crustal anatexis. These are thought to represent a partially molten, middle crustal layer intruded southwards from beneath southern Tibet to the Greater Himalaya (Searle et al. 2011). The so-called Himalayan "*channel flow model*" involves rocks wholly of Indian crustal origin that have been underthrust to the north, thickened, heated, metamorphosed and returned to the surface along a mid-crustal channel bounded by low-angle ductile shear zones below (Main Central Thrust) and above (South Tibetan Detachment).

By about 13–10 Ma underthrusting of the Indian lithospheric plate closed off heat-supply from the asthenosphere, terminating igneous activity in the southern Tibetan Plateau. The youngest post-collisional volcanic rocks (<15 Ma) are

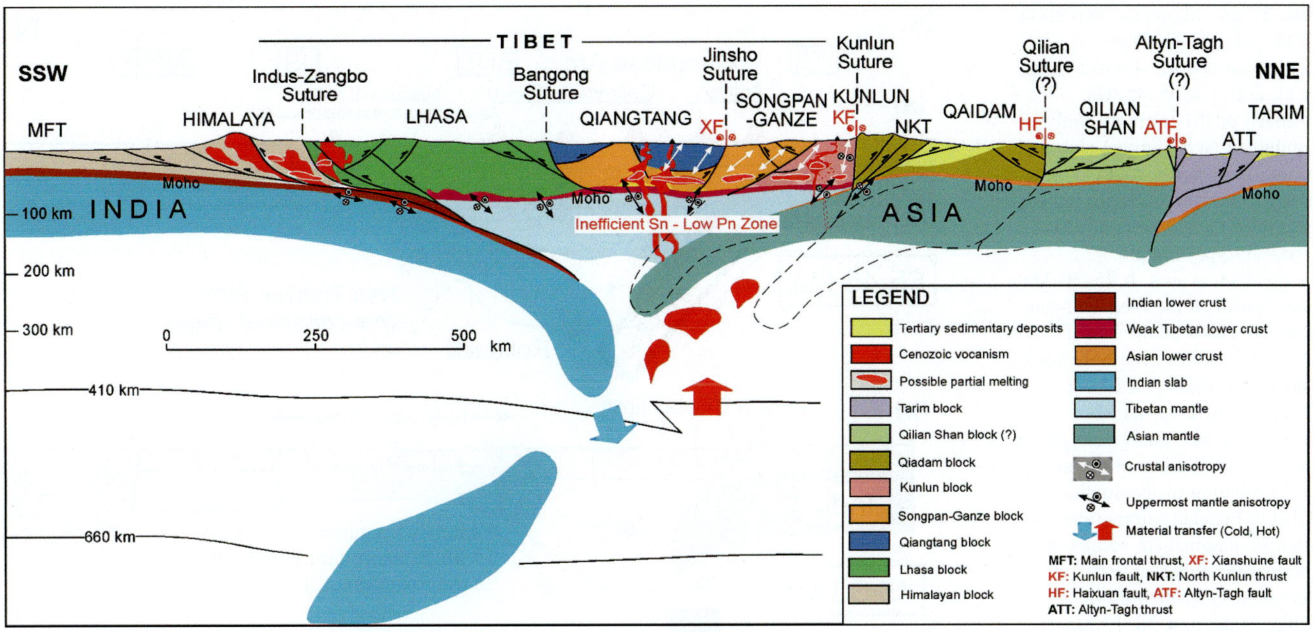

Fig. 17.55 Synoptic lithospheric cross-section illustrating the Cenozoic evolution of the Himalaya-Tibet collisional orogen between the Indian and Asian continents. Seismic data identify a zone of inefficient Sn-slow Pn waves below the Tibetan Plateau, which suggests the presence of warm(er) bodies, supporting the interpretation that this part of the cold lithosphere has been removed and replaced by asthenosphere. Low V_p/V_s, Poisson's ratios and high conductivity at shallower levels, indicate presence of fluid/melt in the crust. The cross-section also schematically illustrates crustal and uppermost mantle seismic anisotropy below Tibet. Seismic anisotropy beneath suture and younger strike-slip faults indicates the reactivation of old structures during Cenozoic restructuring of Tibet. Seismic fabric dipping steeply southward and trending WNW-ESE support lithospheric-scale, northward thrusting. A fabric dipping gently northward is also present, that might be related to a near-horizontal rock fabric induced by crustal flow within a low-velocity layer (Ozacar and Zandt 2004; Wang et al. 2016); this broadly correlates with the zone of partial melting as illustrated. (Image courtesy of the author Arda Ozacar, METU, Ankara, Turkey)

shoshonitic, requiring a deep and hot mantle source, and are only located in northernmost Tibet (Fig. 17.56). While this was taking place along the main zone of continental collision, extrusion and collapse of the orogen, possibly amplified by subduction rollback on the western Pacific margin, occurred via strike-slip and extensional fault systems over an enormous area of East Asia (Sect. 17.3).

17.5.5.1 Ancient Examples of Continent/Arc-Continent Collision

Loss of subducted slabs, collision, huge crustal thickening, high-T/low-P metamorphism, anatexis and extensional collapse are some of the main geodynamic processes observed in modern collisional systems, and by analogy, should also occur in ancient continental collisional orogens. Regarding scale, from what we understand of the Himalayan-Alpine (Tethyan) orogenic system, collision occurred along a deformation front >10,000 km long, and in places up to 2,000 km wide, evolving over a period of >100 Ma. Magmatism taking place during the processes of initial subduction and continental margin arc growth, early collision, collision and post-collision will likely play a key role in deciphering such examples in the rock record. As found in the Himalayas (ca. 120–50 Ma), magmatism commenced along an Andean-type margin forming the southern edge of Asia (Chung et al. 2005; Searle et al. 2011; Fig. 17.56). This formed a 2,500 km long, over 100 km wide, subduction-related I-type granite batholith complex (Gangdese granites) comprising abundant granodiorites and granites, with an extensive calc-alkaline volcanic supra-structure (Linzizong volcanics). Early collisional compositions (ca. 50 Ma) vary from low-K tholeiitic through calc-alkaline to shoshonitic magma suites, suggesting more involvement of a mantle component. These were followed by eruption of post-collisional potassic magmatism across Tibet as described above (Fig. 17.56), associated with direct involvement of asthenospheric melting in magma generation.

The Himalayan post-collisional magmatic suite also includes a conspicuous component of leucogranites, which are also common in ancient collisional orogens (Nabelek and Liu 2004). Although there is general agreement that leucogranites are anatectic products of mainly metapelitic source rocks, the heat sources necessary for their production has remained controversial (e.g. Crawford and Windley 1990; Nabelek and Liu 2004; Searle et al. 2011; Phillips et al. 2013; Weinberg 2016). These include overthrusting of

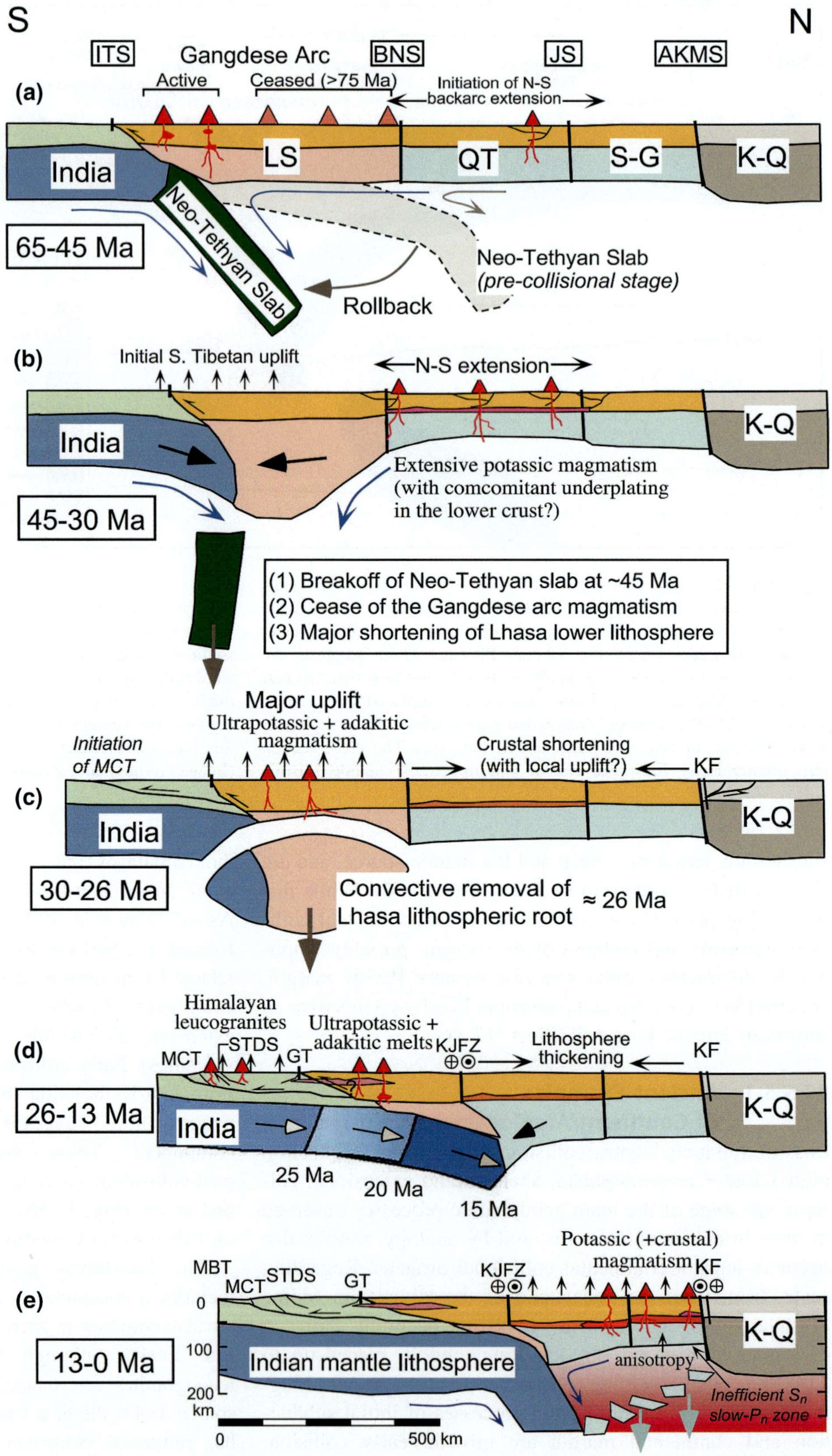

Fig. 17.56 Magmatic evolution of the Himalayan-Tibet collisional orogen (From Chung et al. 2005). **a** 65–45 Ma: rollback of the formerly flatly subducted Neo-Tethyan slab and initiation of north–south extension in the back-arc region; extensive Gangdese continental (Andean-type) margin magmatism ceases. **b** 45–30 Ma: detachment of the Neo-Tethyan slab and onset of the "*hard collision*" particularly in the lower part of the India and Lhasa lithospheres. **c** 30–26 Ma: removal of the Lhasa lithospheric mantle root leading to the uplift and formation of the southern Tibetan Plateau; ultrapotassic and adakitic magmatism results from thermal perturbation and decompression melting of asthenosphere. **d** 26–13 Ma: Northward underthrusting of the Indian mantle lithosphere and associated Himalayan post-orogenic magmatism; leucogranites generated by crustal anatexis forming a partially molten middle crust layer extruded south from beneath southern Tibet; underthrusting closes off post-orogenic magmatic activity by about 13 Ma. **e** 13 Ma to present: "diffused" loss of the lower part of the lithospheric mantle beneath the Qiangtang and Songpan-Ganze terranes; rejuvenation of potassic magmatism in the northern part of the plateau. LS = Lhasa Block; QT = Qiangtang Block; S-G = Songpan-Ganze Block; K-Q = Kunlun-Qaidam Block; ITS = Indus-Tsangpo Suture; BNS = Bangong-Nujiang Suture; JS = Jinsha Suture; AKMS = Ayimaqi-Kunlun-Muztagh Suture; MCT = Main Central Thrust; STDS = South Tibet Detachment System; GT = Gandese Thrust; KJFZ = Karakorum-Jiali Fault Zone; KF = Kunlun Fault

low grade sediments by high grade gneisses and subsequent fluid fluxing, the partial melting of decompressing crustal blocks during rapid exhumation, deformation-induced heating along shear zones, or lower crustal partial melting during/following crustal thickening and slab loss. For the Miocene leucogranites of the Himalayas, these now appear contemporaneous with loss of the lithospheric root, allowing upwelling and decompression melting of the asthenosphere, which then intruded the base of the underplated Indian crust promoting melting (Fig. 17.56). A-type (alkaline or "anorogenic") granites, widely believed to be partial melts of F and/or Cl enriched "dry" granulitic residue at high temperatures (>830–1000 °C), are compatible with such a source region, and the likely final magmatic events of continental collision. These are ultimately thought to give rise to a post-orogenic A-type granite suite and extrusion of extensive silicic (ignimbrite or lava) volcanics (e.g. Whalen et al. 1987; Crawford and Windley 1990; Eby 1992; Windley 1993; and for more current reviews of their classification see Bonin 2007; Grebennikov 2014). These are expected to intrude and/or be spatially associated with a tectonic collage comprising exhumed high-T/low-P metamorphic gneisses (bounded either by contractional or extensional ductile shear zones; e.g. Lee et al. 2000), fragments of lower crustal rocks possibly underplated by large mafic bodies, subduction zone remnants (ophiolites and blue schist rocks), and more exotic high-P/low-T metamorphic rocks such as eclogites, as well as contemporaneous basin-fill sediments.

Possibly the best described example of such a setting in the older rock record is represented by Carboniferous-Permian magmatism in Europe, which is a remarkable, continental scale, rift-related period of extensive silicic volcanism (Wilson et al. 2004). The Permo-Carboniferous collision of Gondwana and Laurasia, or Variscan orogeny, led to their amalgamation and formation of the late Paleozoic supercontinent of Pangaea (Wilson et al. 2004; McCann et al. 2006). In addition, collapse of the thickened Variscan orogenic crust resulted in late orogenic crustal extension (Henk 1997). The post-Variscan period was one of intense crustal re-equilibration and reorganisation under an alternating trans-tensional and transpressional tectonic regime. This tectonic activity controlled the kinematic patterns and subsidence of approximately 70 basins, all of which are characterised by a major strike-slip component associated with escape tectonics in their evolution. Most of the Carboniferous-Permian rift basins trace long-lived Variscan fault systems and involved extensive, predominantly clastic sedimentation as well as were loci for volcanic activity (McCann et al. 2006). Magmatism and volcanism affected almost the entire continental crust, characterised by different histories and thicknesses, producing widespread high-T/low-P metamorphism. Ignimbrite flare-ups and large calderas are exhumed at various localities (Martí 1991; Willcock et al. 2013) associated with calc-alkaline to alkaline, mafic to variably differentiated rocks (Quick et al. 2009; Marocchi et al. 2008).

Evidence from the Paleo-Mesoproterozoic basins and terrains of Australia also suggests orogenic collapse may have been important in their evolution. Metamorphism of basin-fills was widespread with the development of high-T/low-P facies (e.g. Scrimgeour and Sandiford 1993; Forbes et al. 2005; Foster and Rubenach 2006). This was accompanied by extensive silicic volcanic activity (e.g. between ca. 1880–1840 Ma and 1600–1500 Ma) implying large-scale collapse of the orogenic belts. Focussing on the younger event, referred to as the Carpentarian Orogen, this was characterised by a period of orogenesis involving episodic crustal thickening, metamorphism and magmatism extending across the central, southern and eastern Proterozoic terrains of the continent. Although there are differences in the detailed deformation and thermal histories between some terrains the overall pattern of crustal shortening is consistent with the evolution of a major orogenic belt. Age data suggest peak metamorphism and syn- to post-tectonic magmatic activity were diachronous throughout the belt. Late granitic plutons are common in most of the higher grade terrains; however, copious anorogenic magmatic activity appears concentrated in the southern parts of the orogenic belt (referred to as the "Hiltaba Event"). In the Gawler Craton extrusion of extensive bimodal Gawler Range Volcanics (ca. 1592 Ma), including the Olympic Dam volcano-plutonic complex, was accompanied by emplacement of co-magmatic A-type granites forming the Hiltaba Suite (1600–1580 Ma; Budd et al. 2001). The Olympic Dam volcano-plutonic complex hosts one of the largest known accumulations of metals, in particularly uranium, on Earth and is regarded as the "gangbuster" of the iron-oxide-copper–gold (IOCG) deposit class (Chap. 18). Less clear are the geodynamic drivers of crustal extension. Some workers have speculated that the Hiltaba Event was a magmatic flare-up caused by retreat and roll back of a northward-dipping subducting slab over a mantle plume (Betts and Giles 2006; Betts et al. 2009), whereas others have suggested delamination and loss of the lithospheric root (Skirrow et al. 2009; see Fig. 18.40).

17.5.6 The Subduction-Related Origin of Some Continental Intraplate Volcanism

As a footnote to Sect. 17.5 on subduction-controlled volcano-plutonic arcs, we briefly describe novel concepts on the potential role of deep subduction, stagnant slabs and the mantle transition zone (MTZ) in controlling some continental intraplate volcanism. Previously in Sect. 17.4.2 we

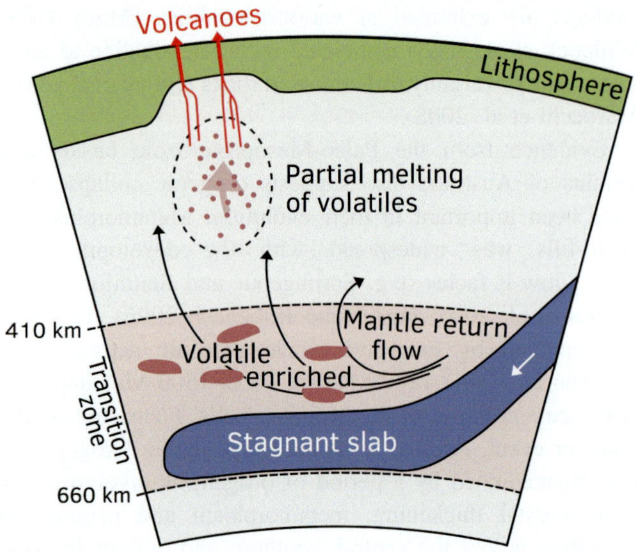

Fig. 17.57 Conceptual subduction-related model of intraplate continental volcanism. The subducting slab meets strong resistance when it encounters the 660 km discontinuity, bending horizontally and where it may stagnate in the mantle transition zone (MTZ) for a long time (ca. 100–140 Ma). This mobilises mantle from a H_2O-enriched MTZ that rises to the surface. Following decompression, the surrounding mantle will destabilise and melt during ascent to generate non-age progressive volcanism. Increases in slab flux promote more vigorous mantle upwelling, which enhances the frequency of volcanic eruptions. Finally the slab collapses to fall down to the core-mantle boundary because of the very large gravitational instability from mineral phase transitions that take place (see text). This has also been called the *big mantle wedge* (BMW) model as first proposed to explain the origin of the intraplate Changbaishan volcano (Zhao et al. 2009; Zhao 2012); the BMW is defined as a broad region in the upper mantle and upper part of the MTZ overlying the long stagnant Pacific slab (see Fig. 17.58). (From Mather et al. 2020)

have discussed continental intraplate volcanism from a general tectonic perspective in terms of "hot spots" related to high heat-flux, lower mantle-rooted, convective upwellings or plumes (e.g. Fig. 17.28) and, as in the case of the well-studied Newer Volcanics Province of southeastern Australia, perhaps more complex models invoking edge-driven convection (Fig. 17.31e), and/or transform-related far-field stresses (Fig. 17.32). More recently, global seismic tomography has suggested that some continental intraplate basaltic volcanism (including flood basalts) may be related to the upwelling of hotter and wet asthenospheric materials associated with the deep subduction, and stagnation of the down-going plate where it penetrates along the MTZ (Fig. 17.57). This occurs at depths between 410 and 660 km (also see Fig. 17.35). Examples include the Pacific slab under East Asia (e.g. Zhao 2012; Wang et al. 2015), and the Pacific Plate under eastern Australia (Mather et al. 2020).

In the East Asian continent, there are many late Cenozoic intraplate basaltic fields and one larger, polygenetic central volcano which is Changbaishan caldera (see Fig. 16.41c) on the China-North Korean border (Fig. 17.58; Plate 17.1). Changbaishan occurs some 1300 km inboard of the active, consuming Pacific plate margin beneath the Japan Trench (Fig. 17.58a, b), and is the largest and most active volcano in northeast Asia, erupting several times historically, including the major AD 1050 Millennium eruption. Compositionally it is bimodal ranging from basaltic to rhyolitic, with few andesitic products. The basaltic magmas are primarily alkaline, but some tholeiitic (subalkaline) basalts and basaltic andesites also occur (Wang et al. 2015). Tomographic images across the region show the subducting Pacific slab as a sub-horizontally deflected high-velocity zone or flattened *"stagnant slab"* above the 660 km discontinuity at the base of the upper mantle (Fig. 17.58a, b). As well as the stagnant slab in the MTZ beneath the Korean Peninsula and northeast China, a localised low-velocity anomaly with a plume-like geometry is revealed in the crust directly beneath the active Changbaishan, extending down to the 410 km discontinuity (Fig. 17.58c; Zhao et al. 2004; Zhao 2012; Tian et al. 2016; Kuritani et al. 2019). This suggests intraplate volcanism is caused by an upper mantle plume sourced from the stagnant slab stored in hydrous MTZ. Rather than having significant excess heat, it is proposed the plume upwelled largely because of buoyancy resulting from water supplied from the Pacific slab in the MTZ (Kuritani et al. 2019; see below). Geochemical studies of the basalts also show involvement of the stagnant slab and metasomatised peridotite materials from the MTZ in their mantle source.

As well as Changbaishan volcano, the Cenozoic intraplate volcanic province of central-east Asia includes a number of basaltic fields and larger plateau (Fig. 17.58a; Wang et al. 2015). One of the better studied are the Chifeng continental flood basalts, which were again erupted far from any plate boundaries, mainly between 24 and 6 Ma. Tomographic data indicates they were erupted directly above the inferred edge of the Pacific stagnant slab (red dashed line in Fig. 17.58a), extending about 200 km further to the west and producing 100–450 m thick sheet-like lava flows that cover a total area of 3,000 km^2. Compositionally they include quartz tholeiites, olivine tholeiites, and alkali basalts. Detailed petrochemical studies (Wang et al. 2015) suggest the source of the basalts is dominated by two end member reservoirs: (1) the stagnant Pacific slab, and (2) fertile peridotite. Mixing of melts from these two end members explains the major element, incompatible trace element, and oxygen isotope compositions, while the two end member mixing calculation shows some lavas may also contain minor subducted sediments. Wang et al. (2015) propose that slab-triggered wet upwelling and upward percolation/hydration, originating from the MTZ (which contains up to 2.5 wt% H_2O in water-enriched minerals), produced large volumes of melt (Fig. 17.57).

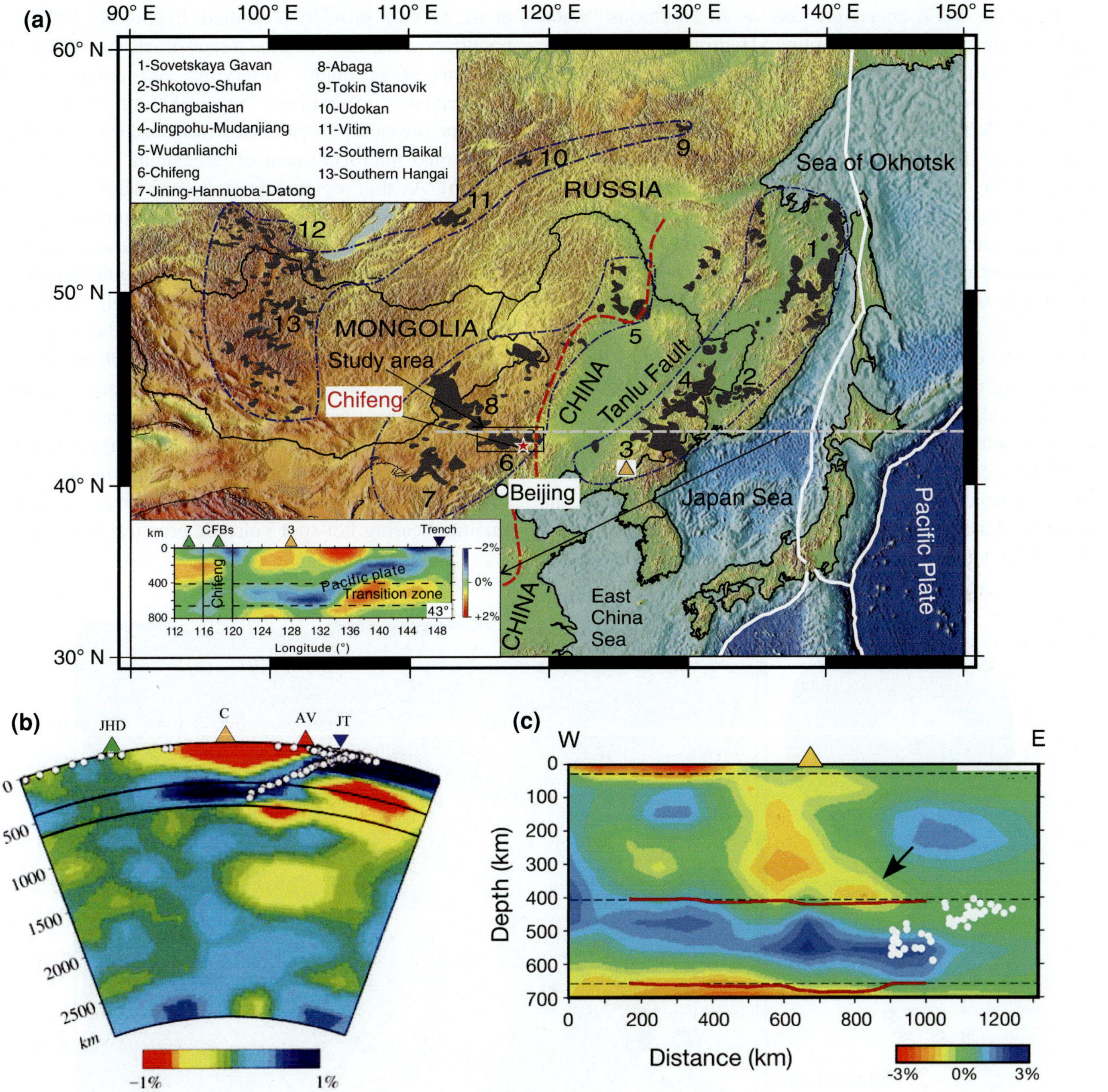

Fig. 17.58 Late Cenozoic intraplate volcanic province in Central-East Asia. **a** Map on digital topographic base showing location of late Cenozoic volcanism; dark grey areas = basaltic lava fields; orange triangle = Changbaishan polygenetic central volcano. Thick white lines indicate plate boundaries. The red dashed line indicates the western edge of the present-day stagnant Pacific slab. Location of Chifeng flood basalt field and study area described in text (box and star). Inset shows tomographic cross-section (along grey dashed horizontal line) of the inferred stagnant paleo-Pacific slab within the mantle transition zone between 410–660 km depth. High and low-velocity perturbations are shown in blue and red colours respectively; scale (in %) as indicated. (From Wang et al. 2015). **b** Vertical mantle profile through the same area from the Earth's surface to >2,500 km showing P-wave velocity images determined by a global tomographic inversion. Lines denote the 410 and 660 km discontinuities. In the upper mantle (above the 660 km discontinuity), the subducting Pacific slab is clearly imaged as a high-velocity (blue) zone, while earthquakes (white dots) occur down to about 600 km depth within the slab. Under northeast China, the slab becomes stagnant in the mantle transition zone. In the lower mantle, pieces of fast anomalies are visible under the stagnant slab in the transition zone. JT = Japan Trench; AV = arc volcanoes; C = Changbaishan caldera (basalt-rhyolite); D = Jining-Hannuoba-Datong basaltic lava fields. (From Zhao et al. 2004). **c** Detailed vertical cross-section of P-wave tomography passing (approximately E–W) through the Changbaishan volcano. The three dashed lines denote the Moho, 410 and 660 km discontinuities. The red lines represent geometries of the 410 and 660 km discontinuities estimated from teleseismic data. Arrow points to the prominent low-velocity (warm colours) anomaly with a plume-like shape observed rising out of the mantle transition zone. (From Kuritani et al. 2019)

Thus, the MTZ effectively acts as an enormous "water tank" in deep-Earth fluid cycling. Hydrous phases in the MTZ eventually decompose into anhydrous phases by releasing fluids to that produce wet upwellings that pool within the MTZ. The long residence time (>100 Myr) of the Pacific slab in this zone promotes the loss of water by dehydration reactions, while the declining rate of Pacific-Eurasia convergence from a late Cretaceous convergence rate of 120–140 mm yr^{-1} year to a minimum in the Eocene of 30–40 mm yr^{-1} year may also have contributed to slab stagnation. The stagnant Pacific slab is likely destined to eventually fall into the lower mantle (as rocks undergo pressure-induced phase transitions and become denser, e.g. pyroxene-garnet transformation; Frost 2008; van Mierlo et al. 2013), and release most of its water by dehydration melting (e.g. Tian et al. 2016).

Similar concepts are now being applied to understanding the Cenozoic intraplate volcanism which dominates the eastern Australian seaboard, as discussed earlier in Sect. 17.4.2 (see Fig. 17.30a; Mather et al. 2020). Mather et al. (2020) provide a unified hypothesis linking plate boundary and slab volume reconstructions of the Pacific Plate to upper mantle reservoirs and intraplate volcano geochemistry to explain the ~5000 km wide intraplate volcanic province which occurs from eastern Australia to the submerged microcontinent of "Zealandia" (Figs. 17.59 and 17.43b). As this region migrated northward, it passed over mantle upwellings which produced a number of hot-spot volcanoes. Mather et al. (2020) distil these down to three distinct chains (Fig. 17.59): (1) the leucitite-bearing Cosgrove track, (2) the Tasmantid seamount chain, and (3) the Lord Howe seamount chain. In between these central-type volcanoes, there are abundant basaltic plains lava fields (notably the Newer Volcanics Province) and volcanism is pervasive across continental and oceanic regions. Contrary to earlier studies, new statistical analysis suggests these lava field volcanoes do not show age progression, and considerably outnumber the age progressive central and leucitite-bearing hot-spot volcanic eruption products.

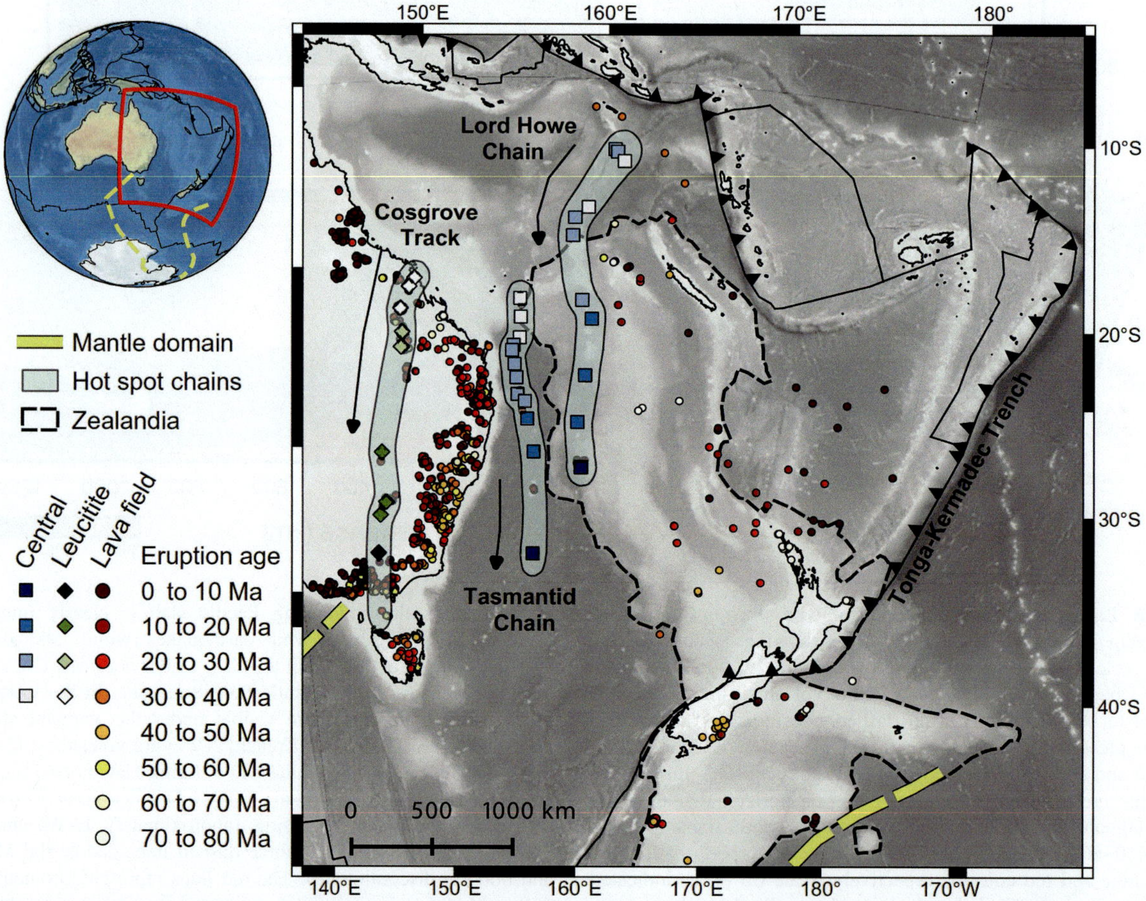

Fig. 17.59 Map showing the distribution of Cenozoic volcanism in the eastern Australia and Zealandia micro-continental region, southwestern Pacific. This illustrates the major plate boundaries and submarine continental lithosphere overlain with volcanoes coloured by eruption age. These are divided into age progressive central and leucitite-bearing volcanoes (="hot spot" or lower mantle plume related), and non-age progressive basaltic lava field volcanoes (=subduction slab or mantle transition zone related). The age progressive volcanoes occur in three distinct chains or hot-spot tracks both onshore (Cosgrove track) and offshore (Tasmantid and Lord Howe seamount chains). (Modified from Mather et al. 2020; digital base relief model from NOAA)

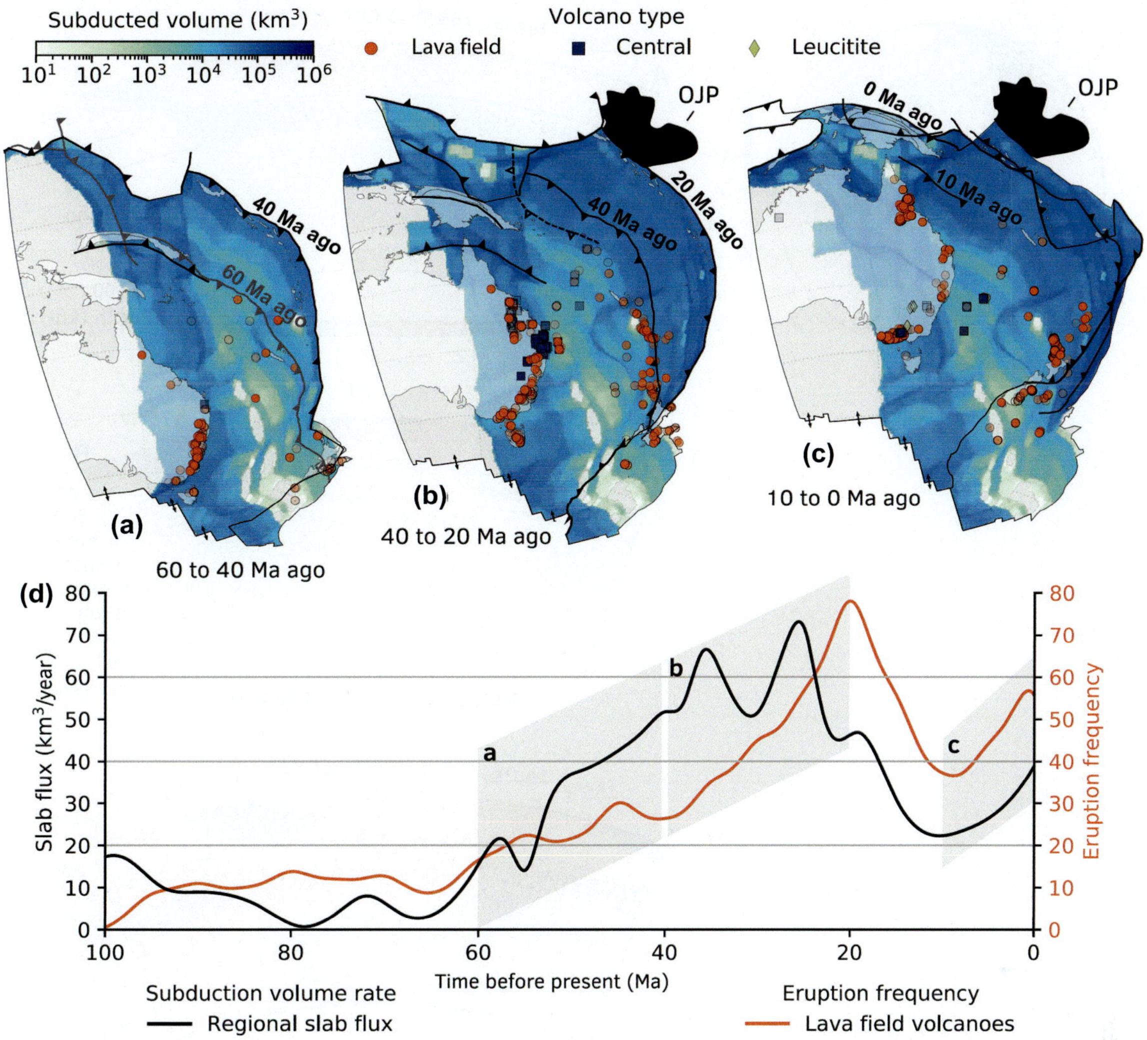

Fig. 17.60 Paleogeographic snapshots based on tectonic reconstructions showing cumulative volume of subducted oceanic lithosphere in the southwestern Pacific during the Cenozoic (From Mather et al. 2020). **a** 60–40 Ma as subduction rollback progresses along the western Pacific margin. **b** 40–20 Ma as Papua New Guinea amalgamated and subduction rollback along the western Pacific continues until the collision of the Ontong Java Plateau with the Melanesian Trench, which causes subduction to re-establish along the Trobriand Trough. **c** 20–0 Ma as Australia and Zealandia migrate northward to its present-day location and the reversal of subduction polarity along the northern margin of Australia. Eruptions are grouped into basaltic lava field (red circles), central volcanoes (blue squares), and leucitite-bearing volcanoes (green diamonds). **d** Regional slab flux plotted against time overlain with the eruption frequency of basaltic lava field volcanoes. (Refer to Mather et al. 2020 for further details of materials and methods, and video of animated reconstruction)

Mather et al. (2020) modelled the subducted slab volume over 100 Myr (Fig. 17.60) and found that slab flux drives volcanic eruption frequency, indicating stimulation of an enriched mantle transition zone reservoir. Volcanic isotope geochemistry distinguishes a high $^{238}U/^{204}Pb$ ratio (=high μ or HIMU) mantle reservoir derived from recycled MORB (>1 Ga old) in the slab-poor south, from a northern EM1/EM2 (geochemically enriched mantle) reservoir, reflecting a more recent voluminous influx of oceanic lithosphere into the mantle transition zone. This compares well with P-wave tomographic cross-sections across the region shown in Fig. 17.61. Two gradual increases in volcanism occurred between 60 and 21 Ma and between 10 and 2 Ma were separated by a brief lull in eruption frequency (Fig. 17.60). The first peak is correlated with an increase in regional slab flux from 20 to 70 $km^3\ yr^{-1}$ accommodated by eastward rollback of the subduction boundary (Fig. 17.60a, b, d). This 35 Myr period of increasing slab flux was terminated by the collision of the Ontong Java Plateau (OJP) with the Melanesian trench at ~26 Ma ago (Figs. 17.60b and 17.36a, b; Chap. 16, see Fig. 16.97). Following a rapid slowdown of the Australian Plate there was a 50% drop in the eruption frequency of lava field volcanoes after 21 Ma, as northward migration of the Australian Plate stabilised at 6.8 $cm\ yr^{-1}$. Increase in eruption frequency between 10 and 2 Ma corresponds to an increase

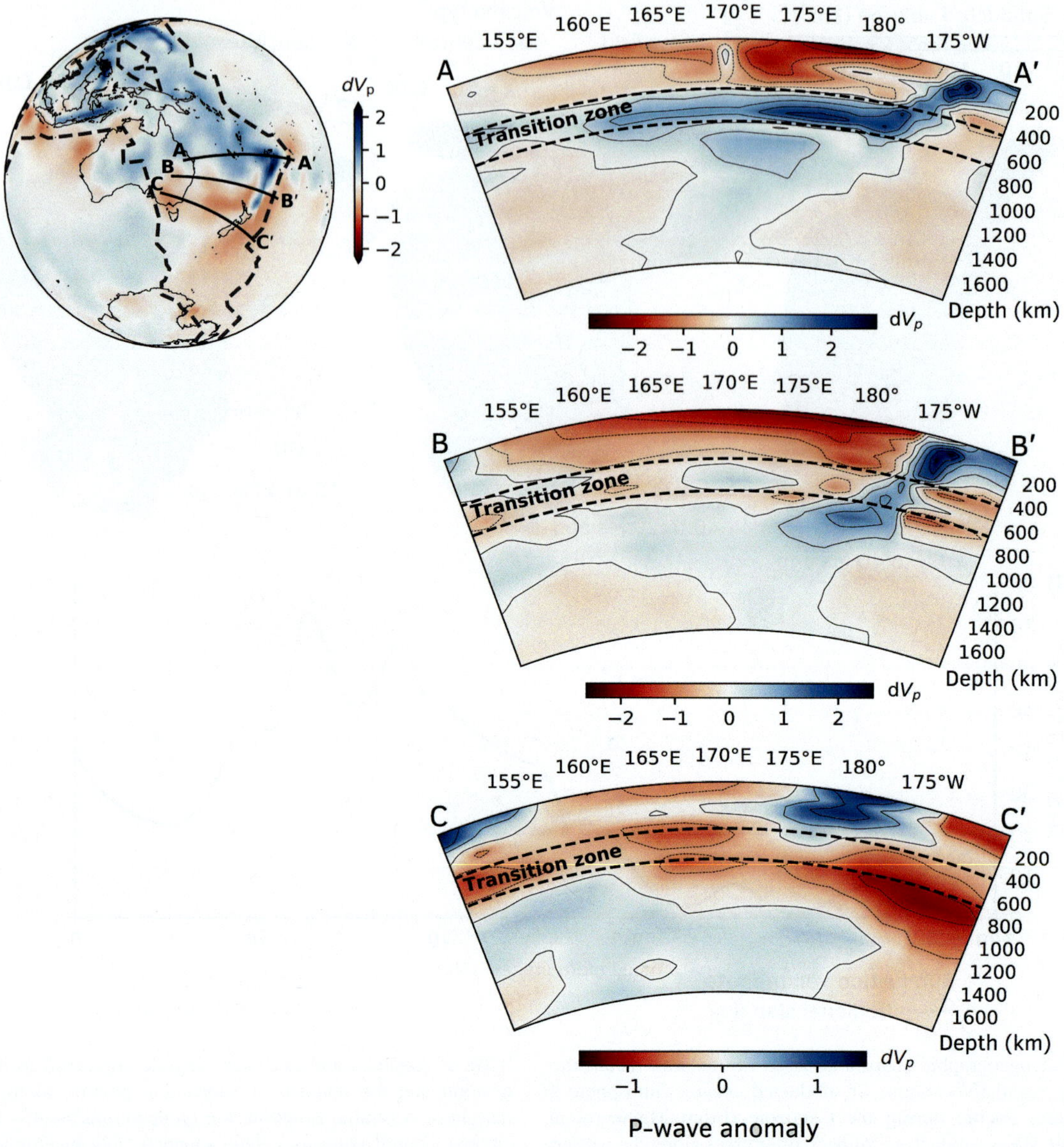

Fig. 17.61 P-wave velocity tomographic variations in the mantle below eastern Australia and Zealandia, southwesten Pacific. Inset: globe shows average velocity variations in the mantle transition zone (MTZ; 410–660 km depth), and location of three accompanying vertical cross-sections. Dashed line on globe indicates the extent of subducted slab material integrated from 120 Ma to the present day. The entrainment of the high-velocity (blue) Pacific slab within the MTZ zone can be clearly observed in northernmost profile (A-A') where the MTZ is entirely occupied by the stagnant slab across the whole region. In the central profile (B-B') slab dip is steeper, while some high-velocity anomalies are resolved west of the subduction zone, suggesting remnants of slab reside. In the most southerly profile (C–C') the position of the subduction zone is less evident, and no high-velocity anomalies are present within the MTZ. Here, subduction has likely been congested by the Hikurangi Plateau, thus limiting the volume of oceanic lithosphere subducted and having entered the mantle over the last 60 Ma compared to more northerly sections of the subduction zone. Note, the offshore Australia-Zealandia region suffers from poor earthquake station coverage, and therefore tomographic models are not as well constrained as for East Asia. (From Mather et al. 2020)

in slab flux associated with acceleration in the northward migration of Australia (Fig. 17.60c, d). Much of the volcanism observed during this period occurs in the Newer Volcanics Province and northern Queensland. While volcanism in southeastern Australia may be partially sourced from the deep mantle plume associated with the Cosgrove track, because of their proximity in the late Miocene (see Fig. 17.29b), it is also synchronous with an increase in slab flux of 16 $km^3\ yr^{-1}$, and a northward migration of the Australian Plate at 7.5 $cm\ yr^{-1}$. This younger peak has been alternatively attributed to edge-driven convection to produce the Newer Volcanics Province along the southeastern margin of Australia (Demidjuk et al. 2007; van Otterloo et al. 2014; Meuws et al. 2016; Cas et al. 2017). Nonetheless, Mather et al. (2020) argue the majority of volcanism occurs far beyond the boundaries of lithospheric step changes, and high rates of asthenospheric shear are not predicted across the entire 5000 km wide volcanic province.

In both East Asia and eastern Australia, it is now proposed the generation of intraplate volcanism mainly occurs during increases in slab flux as a large volume of material enters the subduction zone and displaces the mantle to generate return flow (Figs. 17.57 and 17.60d). These upwellings mobilise mantle from a water-enriched MTZ that rises to the surface. Following decompression, the surrounding mantle will destabilise and melt during ascent to generate non-age progressive volcanism. In both regions a similar mantle return flow has been proposed as the vehicle to drive volcanism. It remains to be tested how this model may be applied to other intraplate regions universally. However, if the entrainment of ancient volatiles in upwelling mantle is linked to flat subduction where there is high slab flux, then the mechanism driving volcanism should be relevant to other areas of intraplate volcanism. Perhaps those, in particular, associated with the assembly of supercontinents, which are surrounded by circum-supercontinent subduction zones (i.e. accretionary orogens; Sect. 17.5.3.1). The assembly of supercontinents therefore potentially leads to accumulation of a large volume of hydrated oceanic slabs within the MTZ, which may possibly change the mode of mantle convection. Accumulated melts can escape to the surface through weak zones, such as pre-existing suture zones, within the lithosphere. This may explain why continental flood basalt provinces (cLIPS) are commonly located over suture zones, as accumulated hydrous melts at the boundary between the lithosphere and asthenosphere have the potential to behave as a lubricant to continent rifting and thus enhance the break-up of supercontinents.

17.6 Volcanism, Stress Regimes and Magma Pathways

Migration of magma upward through the lithosphere occurs in dykes and conduits formed when magma overpressure overcomes the tensile strength of the country rocks, largely related to their rheology and the lithostatic pressure but also to regional stresses. Detailed discussion of how magma is transported and magma-related fractures form is beyond the scope of this book, but many of the salient features have been introduced in Chap. 8. So far in this chapter, we have described the various geodynamic environments where magmatism and volcanism can occur within the well understood Proterozoic-Phanerozoic global plate tectonic framework, each characterised by different magma production rates, rates of intrusive versus extrusive products and related proportion of evolved versus primitive magmas which in turn affect the viscosity and the mobility through the crust. In terms of plate tectonics, these can be broadly related to divergent, oblique/strike-slip, and convergent prevailing regional tectonic regimes (Fig. 17.1a; Table 17.1). In the case of "passive" intraplate settings, volcanism may be triggered in part locally by one or more ambient stress regimes (extensional, trans-tensional/transpressional, or compressional).

Volcanoes seen at different scales, show many collective and individual features that illustrate the interplay between such far-field regional tectonic stresses, and local (non-plate tectonic) variations of the stress fields determined by magma reservoir pressure, crustal heterogeneities and anisotropies, and volcano growth (topography) and gravitational loading by the edifice (van Wyk de Vries and van Wyk de Vries 2018). For example, volcanoes at the regional scale usually form chains or volcanic arcs parallel to the main regional tectonic features (e.g. Nakamura 1977; Acocella 2014) and have recognisable average spacing related to the crustal thickness and rheology (e.g. Vogt 1974; Mohr and Wood 1976; Brink 1991). At the volcano scale, we may recognise preferential elongations of edifices, alignment of vents, and orientation of dykes and parasitic cones, as well as the coalescence of diachronous edifices (e.g. Isola et al. 2014; Wadge et al. 2016). Modern satellite radar (InSAR) images of actively deforming volcanoes show a simple bulls eye pattern of motion typically attributed to varying magmatic pressure (inflation or deflation) within a magma reservoir (Fig. 17.62; Biggs et al. 2009, 2011). However, volcano-tectonic features rarely appear solely influenced by magmatic pressures, so that many calderas and sector collapses have preferential orientations in response to the interplay of local stress fields and pre-existing mechanical weaknesses and anisotropies (e.g. Robertson et al. 2016; Cecchi et al. 2004). For a summary of the hierarchy of stress fields that interplay in controlling magmatism and volcanism, see Table 17.2.

Although this discussion about stress fields (magma generated, regional) and the resulting magma pathways and settings is predicated largely within the plate tectonic framework, most of the principles and magma migration mechanisms would also have been valid during pre-plate tectonic regime times during the Hadean and Archean, as discussed in Sect. 17.7.

17.6.1 Competing Stress Regimes, Rates and Time Scales

In order to correctly understand the role of far-field regional stresses on volcanism we first need to frame the time scales for the various processes that are at play (Table 17.2). Strain rates associated with plate tectonics are in the order of mm yr^{-1} to cm yr^{-1}. By contrast, magma ascent rates, and the related fracture propagation rates that allow magma to migrate through solid rocks, are many orders of magnitude larger. Basaltic magma rise rates in "cold" intraplate settings

Fig. 17.62 Overlay of satellite radar (InSAR) image showing bulls eye patterns of magmatic deformation over the Kenya Rift topography, with Silali caldera (foreground) and Paka shield volcano to the south, which are about 25 km apart; also see Figs. 17.10c and 17.12a. The magma sources are located at depths of 2–5 km, and the strain patterns are equivalent to the deformation produced by a point or spherical/ellipsoidal pressure source typically attributed to changes in magmatic pressure caused by magma and/or volatile fluxes in a magma reservoir. While such deformation patterns are an indicator of an active magmatic system and can be shown to have a statistical link to the likelihood of eruption (Biggs et al. 2014), the interplay of the different stress fields that are associated with volcanic events is poorly understood and potentially complex (see Fig. 17.64). (*Source* Juliet Biggs)

from depths exceeding 60–70 km can be calculated by the presence of heavier mantle xenoliths, and may reach a few m s^{-1} (Wilson and Head 1981; Demouchy et al. 2006; Rutherford 2008). More viscous magmas in the upper crust may move much slower (e.g. 10^{-3}–10^{-2} m s^{-1} at Soufrière Hills volcano, Montserrat 1995–1997; Devine et al. 1998) but still several orders of magnitude higher than average tectonic strain rates.

The main driving overpressure for magma rise is provided by its buoyancy at the reservoir(s) level(s); rise rates are modulated by magma density, viscosity and volatile contents. Overpressure is achieved by volumetric growth of the reservoir(s), which are estimated at rates of 10^{-3}–10^{-2} km^3 yr^{-1} (Annen 2009 and references therein). Dry magmatic liquids (i.e. volatile poor) should in principle rarely erupt as buoyancy overpressure is lost upwards along the ascent path, compensated by magmastatic pressure (see Chap. 8), and most of the rocks in the shallowest crust have densities much lower than any of the known magmas. The presence of volatiles is fundamental with this respect, especially in the upper crust, where magma ascent rates may become much faster if driven by volatile exsolution (e.g. Humphrey et al. 2008; see also Chaps. 9, 10). Indeed, global figures for rates of intrusive versus extrusive magmatism propose variable ratios, on average of 10:1 (e.g. Crisp 1984; Shaw 1985;White et al. 2006), suggesting most magmas never make it to the surface, with ratios that vary according to the thermal and chemical evolution of the crust in response to magma flux (Giordano and Caricchi 2022).

Work on Tenerife and Iceland (Gudmundsson et al. 1999; Gudmundsson and Philipp 2006) also indicates that the majority of dykes are arrested in the shallow crust, at both stratigraphic and structural rheological boundaries. Crustal rheology is also not only dependent on the local stratigraphy and structure, but also is a transient function of the longevity and rate of magma injection and storage. Large silicic systems associated with ignimbrite flare-up are, for example, characterised by heat flows that may exceed 400 mW m^{-2}, where the crust surrounding the main magma reservoirs has a ductile behaviour (see Chap. 16). This largely reduces tectonic faulting (e.g. Metcalf and Kapp 2015), as well as the upward propagation of dykes so favouring growth of large magma reservoirs and plutonic complexes (Jellinek and DePaolo 2003; Annen 2009; Gregg et al. 2012; Giordano and Caricchi, 2022). Thermal effects on the crust are progressively less important as the magma intrusion rates decrease, and the dimensions of the magma reservoirs are smaller.

This brief overview emphasises that once magma is formed, its buoyant overpressure promotes deformation rates orders of magnitude higher than any tectonic strain rate. The tendency of magma to propagate upward along dykes and conduits is therefore almost inevitable (e.g. Cañón-Tapia and Walker 2004) and is mostly controlled by the rheology of the rising magma and of the crust (e.g. Galetto et al. 2017). However, tectonic stresses, and tectonically loaded pre-existing faults and fractures in critical state (i.e. close to rupture) may favour or oppose the upward propagation of vertical or near-vertical fractures as much as the rheological layering of the crust (Maccaferri et al. 2010), especially where magma overpressure is reduced. On modern volcanoes, the stress field associated with topographic and structural anomalies is also important, so that dyke orientations may vary significantly. The 2001 eruption at Mt Etna is a good example showing the interplay of lithostatic, tectonic and magmatic pressures on dyke orientation (see Sect. 8.6.2).

17.6.2 Controls of Magma Pathways

In general terms, vertical or sub-vertical fractures are most likely to form when σ_1, the maximum principal stress, is orientated vertically (or subvertically) or horizontally, and σ_3, the minimum principal stress, is orientated horizontally (Cas and Wright 1987; Fig. 17.63). Given that magma is a liquid, the related overpressures are near-isotropic; in the case of buoyancy, the induced σ_1 is vertically oriented. The orientation of the incipient fracture-dyke will depend on the orientation of σ_3, if overpressure is high enough to exceed the tensile strength of the rock. In theory, under lithostatic conditions and in the absence of any pre-existing weakness (e.g. inherited structures) or deviatoric stresses (e.g. tectonic), magma-related fractures may be orientated randomly, producing radially and concentric oriented dyke complexes (Chap. 8).

In reality, country rocks always have a pre-existing structure and mechanical anisotropy, and far-field tectonic stresses occur. Fractures will then propagate parallel or sub-parallel to σ_1 and orthogonal to σ_3, with failure

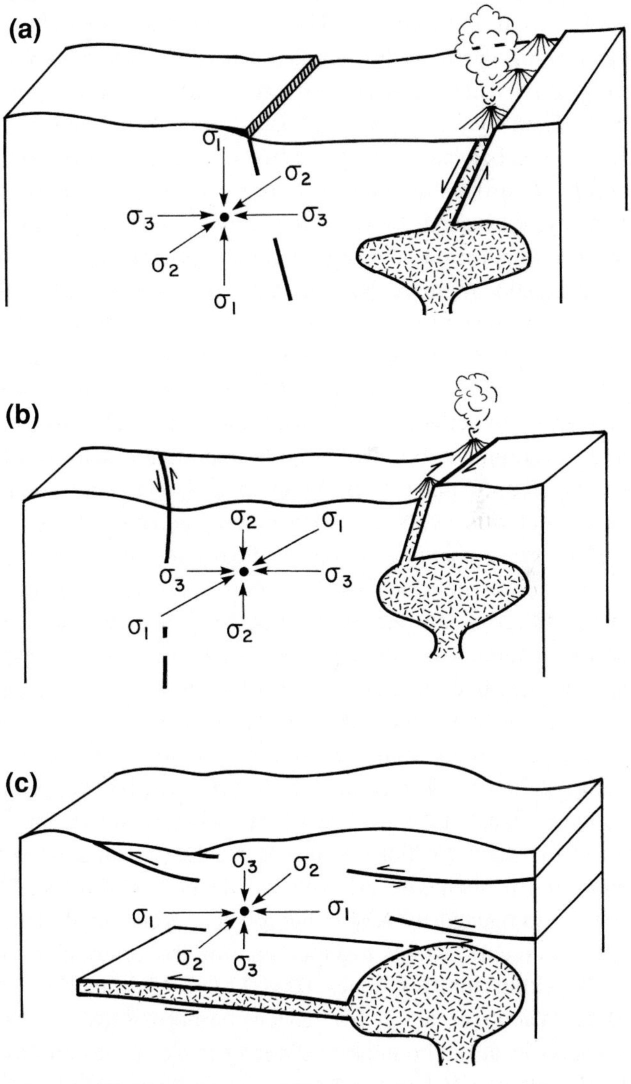

Fig. 17.63 Schematic representation of regional crustal lithospheric stress field configurations, orientations of resultant fractures and the controls of these on the passage of magma through the crust and lithosphere. **a** Normal or extensional faulting, **b** strike-slip faulting, and **c** thrust or compressional faulting. (From Cas and Wright 1987)

mechanics (Shaw 1980) that range from pure extensional fracture (rare) to extensional shear failure (more common). Once the rock has fractured, magma will move along these fractures only if the magma pressure remains greater than σ_3 (Shaw 1980). In a situation where the far-field tectonic stress tensor has a vertical σ_1 and a horizontal σ_3 (equivalent to normal faulting in extensional settings), the magma will easily move through sub-vertical fractures parallel or near-parallel to the regional σ_1–σ_2 plane, or at fracture intersections (Fig. 17.63a). If the regional σ_1 is horizontal and σ_3 is also horizontal, vertical strike-slip fractures and faults will be produced (Fig. 17.63b). In this case magma will also preferentially open fractures parallel to the regional σ_1-σ_2 plane. Furthermore, strike-slip regimes are always associated with local trans-tensional systems, such as dilational jogs, releasing bend and stepover basins (Mann et al. 1983; Nilsen and Sylvester 1995), where repeated faulting and magma uprise is favoured, potentially leading to volcanic activity and formation of major caldera complexes as exemplified by Lake Toba, Sumatra (Sect. 17.3). Stepover basins form between the ends of two parallel to sub-parallel, *en echelon* strike-slip faults or strands that are not connected; typically, they occur between left-stepping strands in sinistral faults, and right stepping strands in dextral faults.

Strike-slip regimes also include the special case of transfer or accommodation zones and faults in continental rift basins, for example, the East African Rift System (EARS), as described in Sect. 17.2.3. Transfer zones not only reveal the effects of inherited basement weaknesses on fault development, but also provide important clues regarding developing magmatic and tectonic systems as young continental rifts evolve. Such structures activate/deactivate in response to changes in both far-field plate boundary forces, as well as more localised rotations of σ_3 as segmented basins progressively merge and interact, and new zones of focused magmatism develop (see Fig. 17.13). The orientation of magmatic activity in the EARS is also a good example of how competing stress regimes may control volcanism. Young cone lineament and dyke trends are highly variable. This results from the interplay between: (1) the regional stress field; (2) local magma-induced stress fields; (3) stress rotations related to mechanical interactions between rift segments; as well as (4) pre-existing weaknesses (dominated by NW-trending shear zones in Proterozoic basement), which together control the orientation of faults within transfer zones (Fig. 17.64; Muirhead et al. 2015; Wadge et al. 2016). Extension-normal dykes and cones occur at the boundary between transfer zones and the distal ends of rift basin segments, where their orientations are controlled by the regional extensional stress field. Zones of shallow magma storage, situated below larger volcanoes, create a local magmatic stress field responsible for the formation of radial dikes. Extension-oblique dykes and cones are focused along transfer zones and intrude inherited basement structures and/or form by local rotations of σ_3 from rift segment interactions. This also suggests that dyke intrusion, rather than tectonic faulting, may be the preferred mode of upper crustal strain accommodation within these structural settings (e.g. Rowland et al. 2010). As a consequence, the dyke direction may not be controlled directly by tectonic stresses, but rather by a reduced tensile strength of country rock induced from previous intrusive activity.

In compressional regimes, where the regional σ_1 is horizontal, the resultant tectonic fractures will be subhorizontal if σ_3 is vertical, which could lead, if there is little magma overpressure, to the emplacement of subhorizontal sills (Fig. 17.63c; Cas and Wright 1987; Menand et al. 2010). However, extensional, strike-slip and compressional stress regimes are homogeneous only at a regional scale. In reality, strain partitioning always occurs and allows the development of local favourable pathways for magma uprise in any stress regime. So, although it is intuitive that in divergent (both intra-continental rifts and continental passive margins) and oblique/strike-slip geodynamic settings, magma uprise is helped by the presence of a favourably oriented stress field, faults and fractures, the understanding of the structural and tectonic stress field controls at convergent plate margins is more complex. The first important issue is that convergence is very rarely perfectly orthogonal to the plate boundaries, so that arcs may develop in a range of regional stress regimes from purely compressional to largely extensional (Acocella and Funiciello 2010; Fig. 17.65). Extension in arc environments is favoured by slab rollback and inception of back-arc extension (as described extensively in Sect. 17.5; e.g. the Taupo Volcanic Zone, Fig. 17.65a; the Alpine-Mediterranean region, Fig. 17.49). Trans-tensional/ transpressional stress regimes occur in oblique settings where some of the strain is partitioned in strike-slip motions (Fig. 17.65b), exemplified by the Sunda Arc in Sumatra (Sect. 17.3). Even where compression is dominant, such as the eastern Japan Sea margin and northern Honshu (Fig. 17.65c), the thickening of the wedge eventually reaches a condition by which σ_2 and σ_3 are relatively similar in magnitude and their "*flipping*" may promote transition from thrusting to strike-slip tectonics (e.g. Norini et al. 2013) Thickened arc wedges may also undergo delamination of their roots and collapse if the supporting compressional stress wanes (e.g. Teng 1996).

It is now clear, from both field observations and analogue modelling, that in convergent geodynamic settings, where the overall stress field is compressional and the associated regional σ_1 is horizontal, there are many ways to facilitate the rise of magma through the crust (e.g. van Wyk de Vries and Merle 1998; Melnick et al. 2006; Galland et al. 2007a, b; Tibaldi 2008; Mathieu et al. 2011; Tibaldi et al. 2010; Norini et al. 2013; van Wyk de Vries and van Wyk de Vries 2018).

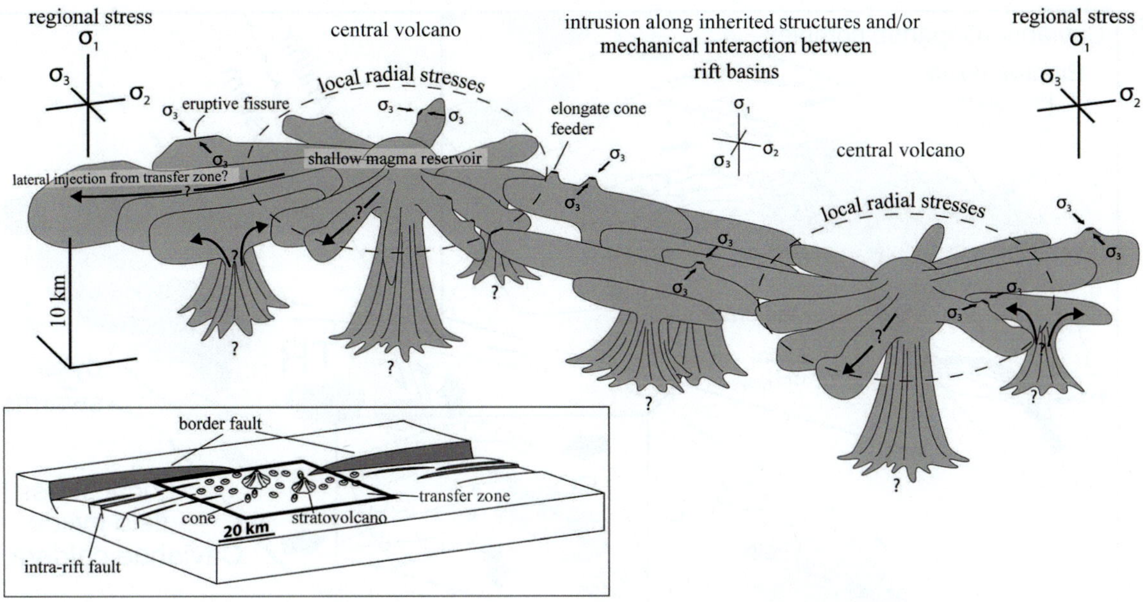

Fig. 17.64 Shallow magmatic architecture of transfer zones within extensional rift basins, as exemplified by the East African Rift System (After Muirhead et al. 2015). The strikes of young dykes and volcanic cones are complexly controlled by variations in the stress state within the transfer zone. Extension-oblique dykes are focused along transfer zone and intrude inherited basement structures and/or form under a rotated stress field driven by the mechanical interaction between adjacent rift basin segments. Extension-normal dykes, with orientations controlled by the regional stress state, intrude the distal ends of rift segments. Zones of shallow magma storage situated below active volcanoes are large enough to create a local magmatic stress field (see Fig. 17.62) responsible for the formation of radial dykes. Dykes are either fed laterally from shallow magma reservoirs or ascend from mantle depths. Inset shows structural location of the volcanic cones and larger volcanoes fed by the schematic subvolcanic dyke and reservoir network

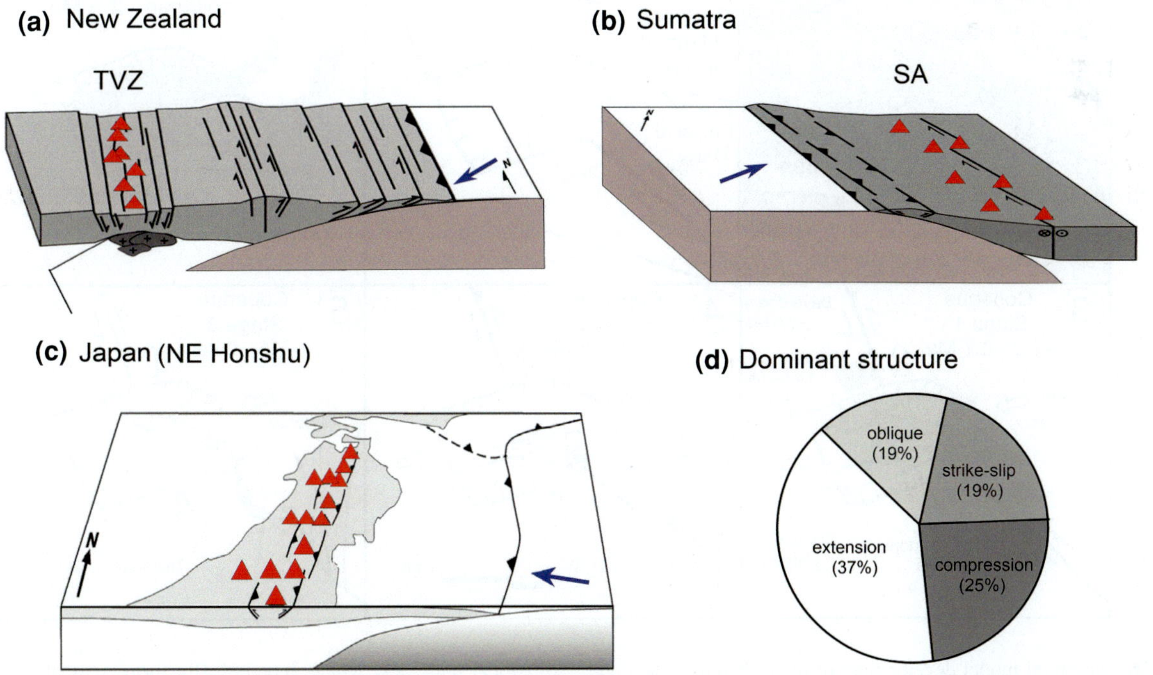

Fig. 17.65 Main tectonic settings for volcanic arcs. **a** Extensional regime represented by the Taupo Volcanic Zone (TVZ), New Zealand. **b** Oblique/strike-slip represented by the Sunda Arc (SA), Sumatra, Indonesia. **c** Compression represented by the Japan Arc in northeast Honshu. **d** Frequency distribution of the dominant stress field along the volcanic arc (for 16 arcs worldwide). The blue arrow on the subducting plate indicates the direction of plate motion; red triangles are schematic volcano locations. (Adapted from Acocella and Funciello 2010)

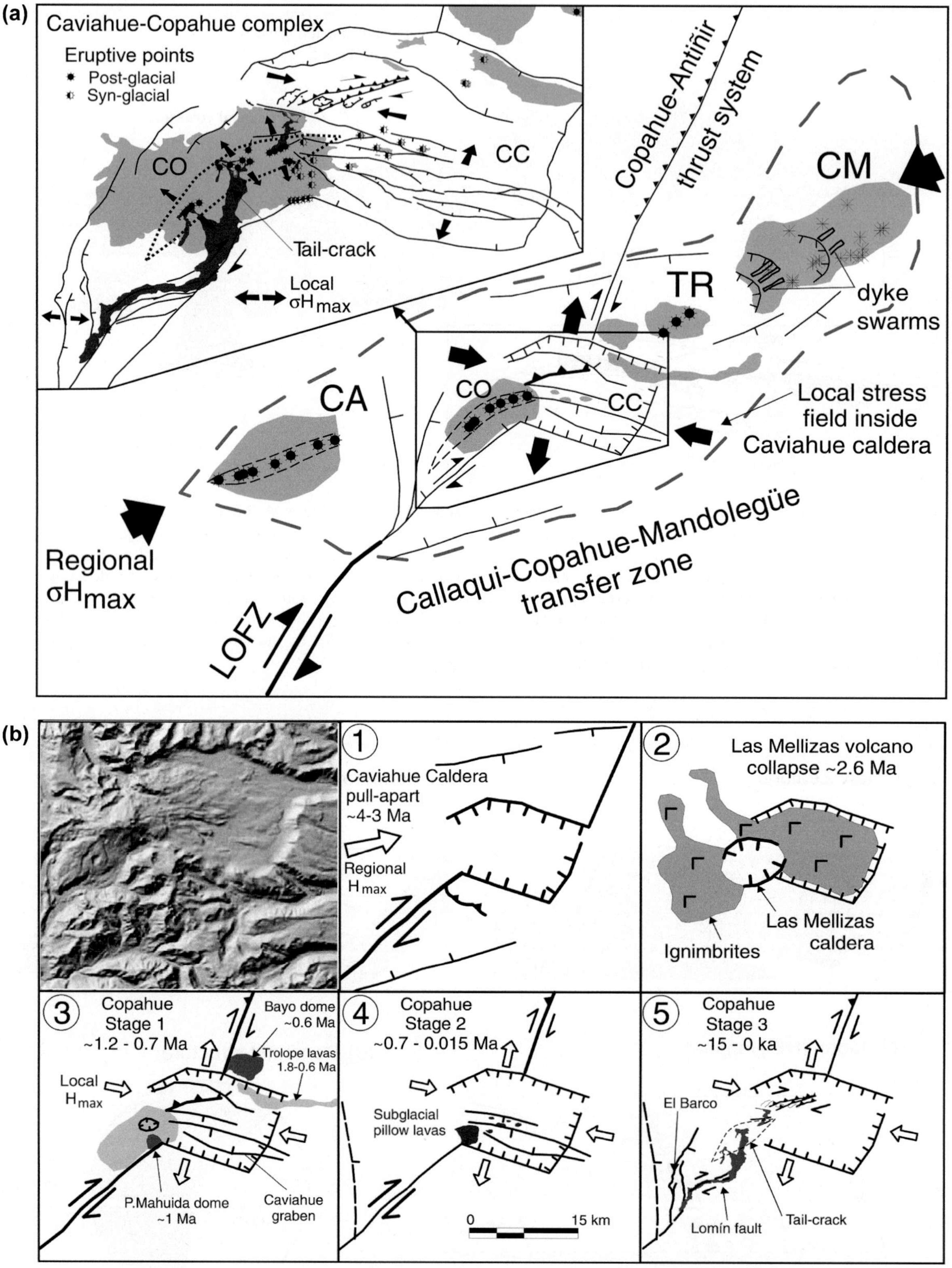

Fig. 17.66 Structural model development of the Callaqui caldera and Copahue volcano complex, southern border between Chile and Argentina. **a** Interpretive map; black arrows indicate the direction of σH_{max}. The local stress field inside the Caviahue caldera (CC) is formed by the interaction between the Liquiñe-Ofqui strike-slip system, the Copahue-Antiñir thrust system, and Callaqui-Copahue-Mandolegüe transfer zone. The Copahue volcano complex (CO) developed within a dilational tail-crack, which geometrically forms part of a "*horse tail*" structure at the termination of the dextral Liquiñe-Ofqui fault zone. Note the continuous clockwise rotation in the Copahue volcano and inside the Caviahue caldera. CA = Callaqui volcano; TR = Trolo′n eruptive centre; CM = Cordillera de Mandolegüe. **b** Structural evolution; (upper left) shaded relief elevation model; stages 1–5 as described in text. (From Melnick et al. 2006)

For example, the Caviahue caldera and Copahue volcano complex located on the border between Chile and Argentina in the southern Andes is a striking example of the tectonic control on volcanic activity in an oblique subduction setting (Melnick et al. 2006; Fig. 17.66a, b). Its evolution commenced at ca. 4–3 Ma with the opening of the 20 km × 15 km Caviahue pull-apart caldera. Las Mellizas volcano formed inside the caldera, and then collapsed at approximately 2.6 Ma (stages 1 and 2, Fig. 17.66b). Since then the Copahue volcano has evolved in a further three stages: (3) 1.2–0.7 Ma formed the approximately 1 km thick andesitic edifice, (4) 0.7–0.01 Ma erupted andesitic–dacitic subglacial pillow lavas, and (5) 0.01–0 Ma erupted basaltic–andesites and pyroclastic flows from fissures, aligned cones, and summit crater. Volcanism observed in the Caviahue-Copahue complex occurs at the intersection of three regional structural systems: the Liquiñe-Ofqui strike-slip system, the Copahue-Antiñir thrust system, and Callaqui–Copahue–Mandolegüe volcano-tectonic lineament. The Liquiñe-Ofqui fault zone is a > 1000 km long, dextral strike-slip system that accommodates part of the margin-parallel component of oblique subduction in the southern Andes and is a major control of volcanism. The Callaqui-Copahue-Mandolegüe structure evolved as a basin-forming transfer zone during Cenozoic rifting and inversion, and probably represents a crustal-scale discontinuity. This links the strike-slip Liquiñe-Ofqui fault zone and the compressive Copahue-Antiñir thrust system (Fig. 17.66a), accommodating the step configuration of these two regional fault systems and decoupling their contrasting kinematics.

Within zones of crustal thickening (thrusting), deep crustal ramps and associated transfer faults also may localise magmatic fluids. Examples of such structures control volcanism and active geothermal systems in northern Argentina in the Tocomar Basin (Giordano et al. 2013; Chiodi et al. 2019; Filipovich et al. 2020; Chap. 18), as well as major exhumed intrusion-related copper–gold and gold deposits in New Guinea (Fig. 17.67a; Wright and van der Veen 1993; Hill 1991; Hill et al. 2002; White et al. 2014). These, including the supergiant Grasberg-Ertsberg gold-rich porphyry and skarn deposits which host the largest gold resource in the world (>200 Moz or 5,000 tonnes; Leys et al. 2012), are located on long-lived basin-forming transfer zones. Later these behaved as lateral ramps during contraction of the New Guinea fold and thrust belt, focussing dioritic magmas through basement-involved tear faults (or dilational jogs) into overlying Mesozoic-Miocene host rocks (Fig. 17.67b, c). Nonetheless, it is important to conclude that compressional regimes, where the crust is thickened, thermally attenuated and structurally layered, will generally favour intrusion of large plutons and batholiths that may grow by the sequential injection and vertical stacking of granitic sills (Glazner et al. 2004; Miller 2008; Annen 2009). That there are also large sections of collisional zones where volcanism is relatively minor to absent, both in continent–continent settings such as along the Himalaya (Fig. 17.56), but also in continent-ocean settings where the subducting slab is flat or shallowly inclined as found in the Andes (Fig. 17.44).

17.6.3 Correlation Between Earthquakes and Eruptions?

A final remark, appropriate here, concerns the potential relationships between active tectonics and, in particular, seismic energy released during earthquakes and eruptions (e.g. Manga and Brodsky 2006). This issue has been much debated in the literature, but no unequivocal outcomes have been obtained from statistical analyses of the occurrence of volcanic eruptions even after major earthquakes such as those of Sumatra, Indonesia in 2004 and Tohoku, Japan, in 2011 (Bebbington and Marzocchi 2011; Yukutake et al. 2011). The main problem appears to be understanding the scale at which seismic energisation of magmatic fluids in the crust may be effectively promoting their upward movement. For example, in the literature, a link is often proposed between the large earthquake that struck Pompeii in AD 62 and the eruption of Vesuvius that occurred in AD 79. Whatever, the correlation, anecdotal or causative, and the time lag between seismic activity and eruption, remain scientifically unexplored. Is it possible that large earthquakes can "stimulate" the deep magma source, triggering the recharge of shallow magma reservoirs that may then erupt years or decades later? In general, for magma to be affected by far-field seismic energisation, it is reasonable to believe that magma should be in a critical state, something that fortunately occurs rarely even in very active magmatic systems. This may explain why there is no clear statistical relationship between eruptions and earthquake occurrence.

17.7 Volcanism and Plutonism During the Hadean—Archean and Trying to Reconstruct Volcano-Tectonic Regimes: Oh, What a Challenge!

To begin, it is important to remind us that the Hadean (4.54–4.0 Ga) and Archean (4.0–2.5 Ga) eons represent ~45% of Earth history. Many people may assume that the well-known present-day plate tectonic paradigm existed throughout Earth history. Although this was considered to be the case for several decades from the late 1970s to about 2000, it has since become apparent that many of the key geodynamic elements of modern plate tectonics are not represented in the Archean geological record.

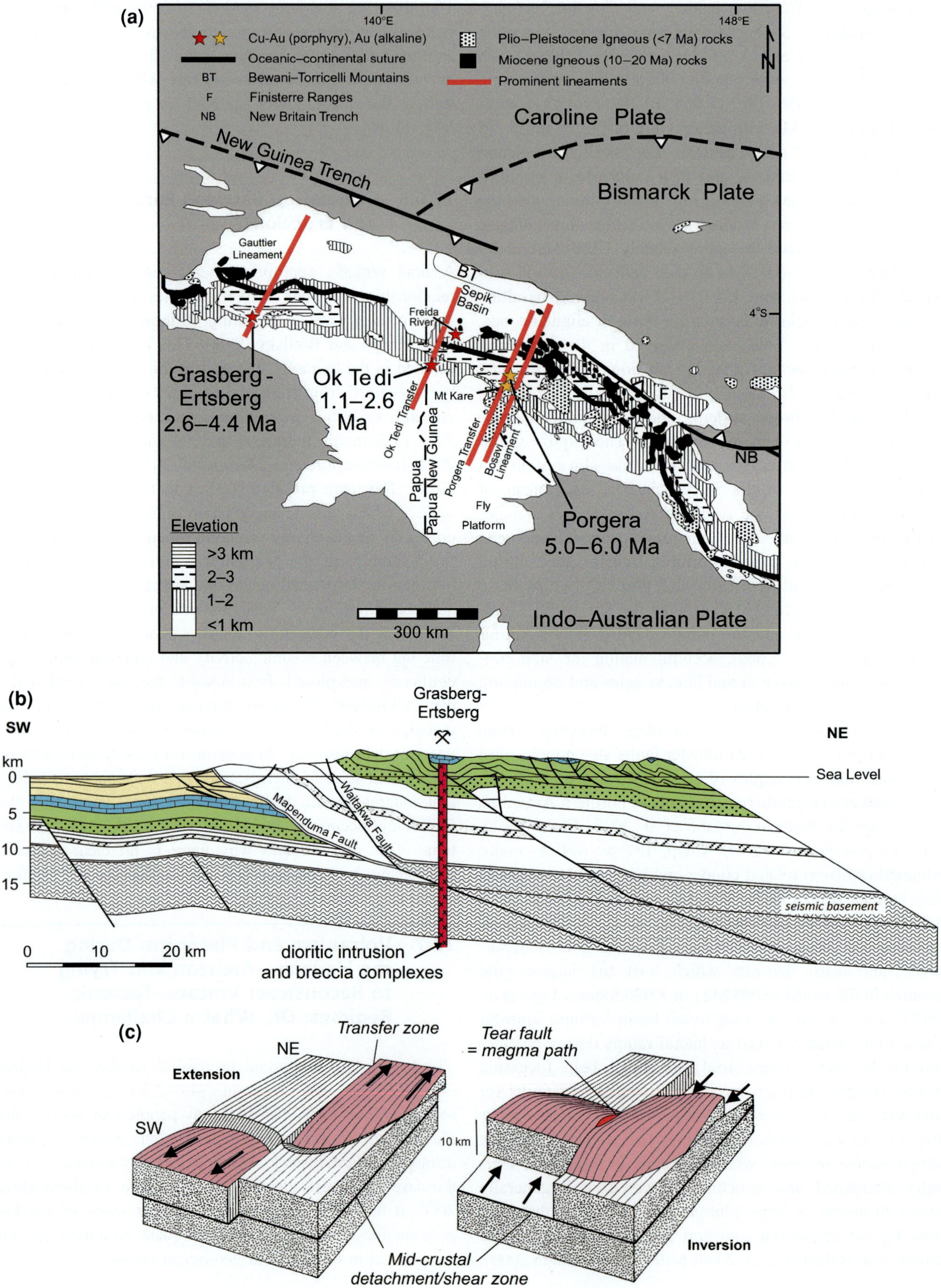
(a)
140°E
148°E
Cu-Au (porphyry), Au (alkaline)
Oceanic–continental suture
BT Bewani–Torricelli Mountains
F Finisterre Ranges
NB New Britain Trench
Plio–Pleistocene Igneous (<7 Ma) rocks
Miocene Igneous (10–20 Ma) rocks
Prominent lineaments
N
Caroline Plate
New Guinea Trench
Bismarck Plate
Gauttier Lineament
BT
Sepik Basin
Freida River
4°S
Grasberg - Ertsberg 2.6–4.4 Ma
Ok Tedi 1.1–2.6 Ma
Mt Kare
F
Ok Tedi Transfer
Papua New Guinea
Porgera Transfer
Bosavi Lineament
NB
Fly Platform
Porgera 5.0–6.0 Ma
Elevation
>3 km
2–3
1–2
<1 km
300 km
Indo–Australian Plate
(b)
Grasberg-Ertsberg
SW
NE
km
0
5
10
15
Sea Level
Mapenduma Fault
Waitakwa Fault
seismic basement
0 10 20 km
dioritic intrusion and breccia complexes
(c)
Transfer zone
NE
Extension
SW
Tear fault = magma path
10 km
Inversion
Mid-crustal detachment/shear zone

◀ **Fig. 17.67** Tectonic framework of Indonesian Papua (formerly Irian Jaya) and Papua New Guinea **a** Elevation, key geodynamic and structural elements, and main orebodies (Modified from Hill et al. 2002). **b** Regional balanced cross-section for the Grasberg-Ertsberg gold-copper mine area. The orebody (nearly 5000 m a.s.l.) was emplaced ~20 km above an inverted crustal-scale extensional fault. Apatite fission track ages record cooling below ~100 °C and illustrate development of the upper fold and thrust belt in the Late Miocene and thrusting on the Mapenduma Fault associated with basement thrusting in the Pliocene, coeval with dioritic intrusion and mineralisation. (After Hill et al. 2002). **c** Model for intrusion-related mineralisation. During Mesozoic basin formation, extensional normal faults soled-out on a mid-crustal detachment, compartmentalised by transfer zones. During Late Miocene to Early Pliocene compression, extensional faults were inverted to form basement cored anticlines, while transfer faults were reactivated as steep, lateral ramps. These steep structures are favourable pathways which allow magma and deep fluid migration through basement-involved tear faults and dilational structures (e.g. breccia pipes) into thin-skinned Mesozoic-Miocene host lithologies. At Grasberg-Ertsberg these have reacted to form extensive skarn mineralisation. Green = Mesozoic; blue = Paleocene-middle Miocene limestones (New Guinea Limestone Group); yellow = late Miocene-Recent. (Modified from Hill 1989)

Why is it important to understand the tectonic regimes within which volcanism occurred during the Archean? The reasons are many and include:

(a) Given the long duration of the Hadean-Archean (4.54–2.5 Ga), the rock record provides major insight into the early history and evolution of our planet Earth and allows us to interrogate how much the Earth's dynamics have changed or remained constant.
(b) This will then provide understanding of what might have happened on other planets during their early history.
(c) Since much of the Earth's atmosphere and oceans originated from the release of volcanic gases during eruptions after the formation of the original atmosphere during planetesimal collisions and accretion (Kasting 1993), understanding volcanic eruption processes and setting during the Archean will improve understanding of the development of the atmosphere and oceans during this time.
(d) It is also considered likely that the earliest life forms originated in submarine volcanic hydrothermal vent systems during the Archean (Davies 1998; Dodd et al. 2017).

17.7.1 Geological Architecture of Archean Cratons

Archean cratons are the remnants of Archean crust. They consist of associations of granitoids and supracrustal volcanic-sedimentary "greenstone belts" (e.g. Figure 17.68). In many cratons these associations occur in distinctive crustal belts that are bound by major crustal, perhaps lithospheric-scale faults. For better or worse these have been called tectono-stratigraphic terranes and super-terranes, resulting from belief that plate tectonics did occur in the Archean, despite the considerations of Bickle et al. (1994) that there is little or no geological evidence for it. It is possible that the development of the stratigraphy of the preserved crust and lithosphere that make up these structural terranes and super-terranes did not involve plate tectonics (Mole et al. 2014, 2019; Johnson et al. 2017), but that the terranes were formed and juxtaposed during the transition to a plate tectonic regime at the end of the Archean, as will be discussed later for the Yilgarn Craton, Western Australia, below (Sect. 17.7.8).

The surface area of most Archean cratons is dominated by clusters huge granitoid plutons that intrude supracrustal greenstone belts, which are deformed volcanic and sedimentary successions that wrap around or are squeezed between the plutons (e.g. Yilgarn Craton, Western Australia; Fig. 17.68a, c; e.g. Jones et al. 2020; Witt et al. 2020; Pilbara Craton, Western Australia, Collins et al. 1998; Van Kranendonk et al. 2004). This architecture is frequently described as an overall regionally widespread dome and keel architecture and does not reflect the largely linear asymmetrical architecture of subduction generated volcanic arcs and trench-subduction zones (Bédard 2006). The emplacement ages of the granitoid plutons are contemporaneous with to post-dating the age of the greenstone stratigraphic units. The important question that faces us is: "*do granitoid-greenstone belt associations preserve clear evidence about their tectonic origins despite their great age?*".

17.7.2 Principal Volcanic and Plutonic Rock Types in Hadean-Archean Cratons and Greenstone Belts

Archean greenstone and granite cratons can be very complex in their lithologies and geology, and although it is tempting to attribute their origins to processes similar to modern plate tectonic processes and settings, this is far from certain, and almost certainly not the case at various times during their evolution. Distinctive rock types include:

- komatiite lavas and subvolcanic dunitic sills,
- basalt lavas and subvolcanic dolerite (diabase) sills,
- andesites,
- rare boninites,

- dacitic and less dominant rhyolite volcanics and subvolcanic intrusions, and
- granitoids (in particular the suite of tonalite-trondhjemite-granodiorite [TTG] rocks).

17.7.2.1 Komatiites

Komatiites (Chap. 7) are ultramafic, very low SiO_2 (<45 wt %), very high-MgO (>18 wt%) lavas and intrusions. They are mostly confined to the Archean, are generally agreed to be the eruption and intrusive products of mantle plumes (Arndt 2003; Barnes et al. 2012; Mole et al. 2014) and were erupted at temperatures up to 1650 °C, reflecting the high temperature of the Earth at the time. There is some debate as to whether or not the plumes originated from deep in the mantle near the core-mantle boundary (Campbell et al. 1989; cf. Morgan 1971; French and Romanowicz 2015; Heron 2019), or from the shallow mantle transition zone (Wyman 2020; Sect. 17.5.6), or in fact probably both. Komatiite lavas are distinctive for their zoned olivine spinifex and cumulate textures, and are commonly associated with co-magmatic, subvolcanic peridotite and dunite sills (see Chap. 7). Almost all preserved komatiite lavas were emplaced in (deep) submarine environments because there was very little exposed landmass until the end of the Archean. Komatiites are commonly found in association with voluminous basalt lavas and subvolcanic dolerites, which are also considered to be plume generated, compositionally related, and co-genetic, as well as intercalated deep-water anoxic black shales and minor turbidites. Together they formed extensive submarine lava flow fields defining Archean Large Igneous Provinces (Barnes et al. 2012; Hayman et al. 2015). Komatiites are major sources for nickel sulphide (Ni-Cu-PGE) ore deposits (Chap. 18).

17.7.2.2 Basalts

Basalts (Chap. 5) result from partial melting of the mantle, and the numerous compositional variants can be explained by varying degrees of partial melting (high for tholeiite basalts, low for alkali basalt), different temperatures of partial melting (high for tholeiites, low for alkali basalts), depths and pressures of partial melting (low for tholeiites, high for alkali basalts), varying degrees of reaction with wall rocks, and varying degrees of fractional crystallisation of melts (Green 1969). These physical constraints would have been just as valid in the Archean as today, although the temperature gradient inside the Earth was higher than today, and the lithostatic pressure gradient in the upper mantle and crust was also higher than today because the Archean crust was largely mafic and very thick ($\sim$30 km) and the uppermost mantle was less depleted, more primitive, less evolved and not as reworked as today. Some basalts, especially high magnesium basalts are clearly related to and are a by-product of plumes rising through the mantle (Campbell and Jarvis 1984; Campbell et al. 1989; Arndt 2003; Barnes et al. 2012). Archean tholeiites have higher Ni, Co, Cr, and Fe compared with modern tholeiites, which Arndt (1991) attributes in part to mixing of basaltic and komatiite magmas associated with a rising plume derived from near the core-mantle boundary. It may not be straightforward to simply use geochemical fingerprinting of basalts to try to match Archean basalts with those from the modern world with known tectonic settings and assume that the same tectonic settings and conditions were involved. Li et al. (2015) and Condie (2015) have demonstrated that there is enormous overlap in the compositional attributes of even modern magmas from different tectonic settings using so-called "discrimination" diagrams that at best provide only a possible statistical probability of a match with a specific tectonic setting being valid, let alone attempting matches with Archean examples. Li et al. (2015) and Condie (2015) refer to such diagrams for the tectono-magmatic fingerprinting of ancient volcanic successions as "*indiscrimination*" diagrams.

Barnes et al. (2012) summarised in detail the characteristics of the basalts associated with komatiites in the Eastern Goldfields Super-Terrane in the Yilgarn Craton, Western Australia. They subdivided basalts into three types: high-, medium-, and low thorium (Th) basalts, which are all notable in having higher Ni and Co contents than all modern basalts but are most like the plume generated Large Igneous Province (LIP) basalts of Eastern Greenland and offshore, North Atlantic Igneous Province (see Fig. 17.23). The low Th basalts are typical Archean tholeiites, also found on other cratons such as the Superior Craton in Canada. Their chemistry includes: SiO_2 content of 50–53 wt%; Al_2O_3 $\sim$15 wt%; high Cr and Ni; MgO between 5.5 and 8 wt%; "flat REE–HFSE [rare-earth elements; high field strength] patterns with slight depletion in Th and very minor positive and negative Nb anomalies" (Barnes et al. 2012). In contrast, the high Th (siliceous) basalts have high SiO_2, (>54 wt%); MgO from 6 to 9 wt%; relatively low Fe and Ti; "lower Ni for given Mg#, not matched by depletion in Cr; strongly enriched LREE and Th; strongly negative Nb anomalies and mantle-like Zr/Nb, Nb/Y and HREE slope" (Barnes et al. 2012). The medium Th basalts have chemistry intermediate between the two end members. Barnes et al. (2012) interpret the low Th basalts as being plume head in origin (cf. Campbell et al. 1989), and the high Th (which are also high Mg) siliceous basalts as resulting from crustal contamination of komatiite magma, followed by homogenisation and crystal fractionation of olivine in a crustal magma reservoir before eruption. In addition, there are very high-MgO basalts, called komatiitic basalts, worldwide in Archean cratons, which could also form through such processes, but undergoing less fractionation of olivine. None of the basalts are like modern N-MORB, E-MORB, or OIB basalts (Barnes et al. 2012).

17.7.2.3 Andesites

Andesites in modern arcs are rarely primary magmas derived from partial melting of down-going basaltic plates or slabs, but rather the result of mixing or mingling of basaltic and silicic magmas (Reubi and Blundy 2009; Sect. 17.5). They can also result from fractional crystallisation of basaltic magmas, or partial melting of mafic crustal material. They can thus be formed in many tectonic settings, not just volcanic arcs. For example, andesites have been documented in the extensional continental Basin and Range Province in the western United States (Christiansen and Lipman 1972; Glazner and Ussler 1989; Sect. 17.2.3.2). Andesites occur in Iceland where they are known as icelandite (Walker 1963). Andesites also occur in Archean greenstone belts but are very subordinate to other igneous rock types. Although their incompatible trace element characteristics are similar to those from modern arc systems, they have unusually high MgO, Ni and Co (Barnes and Van Kranendonk 2014) and different Th/Yb vs Nb/Yb plot trends (Bédard 2018). Barnes and Van Kranendonk (2014) have modelled the formation of andesites in the Eastern Goldfields Super-Terrane (EGST) of the Yilgarn Craton, Western Australia, as being the product of the mixing or contamination of plume derived tholeiitic magmas with contemporaneous felsic (TTG) partial melts of the pre-existing crust. The stratigraphic record of the EGST includes a widespread komatiite-basalt association interpreted as a plume derived LIP by Barnes et al. (2012) and Hayman et al. (2015). The andesites succeed most of the basalts and komatiites, but they can clearly have originated through interaction between residual komatiite and more felsic crustal contaminants.

17.7.2.4 Boninites

As discussed in Sect. 17.5, boninites in the modern plate tectonic regime are mostly diagnostic of island arc forearc settings and magmatism, although they also occur in oceanic back-arc settings (e.g. Northeast Lau Basin, Embley and Rubin 2018 and references therein). They have also been reported in Archean cratons, and according to Bédard (2018) Archean boninites are much less depleted in incompatible elements than Phanerozoic examples, which owe their depletion to an extended period of fluxing by volatiles in the subduction environment. This then indicates that the source Archean mantle was not subject to long-term "wet-melting" in subduction settings (Bédard 2018). Bédard (2018) instead proposes that they, and some calc-alkaline magmas formed from limited "subcretion" of slabs of oceanic crust against and under continental nuclei leading to eclogitisation of the leading edge of the slabs and metasomatic induced melting of that region.

Lowrey et al. (2019) reported the occurrence of two $\sim$2.8 Ga "boninitic-like" sequences in the Youanmi Terrane, in the western Yilgarn Craton, Western Australia, which are also sourced from less depleted mantle. They are associated 100 km along strike with coeval high-Mg andesites and sanukitoid and hydrous mafic intrusions (Lowrey et al. 2019). Although they are attributed as boninitic, they can also be classified as high-Mg komatiitic basalts. Lowrey et al. (2019) concluded the boninite and associated rocks indicate a metasomatised mantle source and subduction operating in the Yilgarn between 2.82 and 2.73 Ga. However, the mode of Archean subduction may have been very different to the styles that dominate in the Phanerozoic, or another process operated that might replicate key features of subduction. Unfortunately, no location for the subduction-like zone has been identified, and so the interpretation relies purely on geochemical fingerprinting, which can be highly ambiguous (Li et al. 2015; Condie 2015). It would be fair to say, that there is some uncertainty about the tectonic setting and petrogenetic processes associated with Archean boninitic rocks.

17.7.2.5 Dacitic Volcanic and Sub-volcanic Intrusions

Dacitic volcanic rocks are the most common felsic volcanic composition associated with Archean greenstone belts, although some rhyolites also occur. They occur as coherent submarine lava domes, associated autoclastic deposits and resedimented mass flow aprons (e.g. Trofimovs et al. 2004; Rosengren et al. 2008), and monomictic juvenile crystal-rich volcaniclastic mass-flow (megaturbidite, Chap. 11; or normal turbidite, Chap. 12) reflecting some explosive eruptions in shallow water or slightly emergent volcanic island settings, and the passage of pyroclastic flows into the sea (e.g. Hand et al. 2002; Brown et al. 2002; Chap. 11). Although such juvenile volcaniclastic deposits can be voluminous, testifying to major explosive events, definite welded ignimbrites are very rare, indicating insignificant subaerially exposed volcanic edifices at the time. However, variably reworked polymictic volcanic sediments, including volcanic conglomerates and sandstones that range from felsic to tri-modal (felsic-mafic–ultramafic) provenance do occur and signify erosion and reworking in tractional subaerial fluvial and shoreline settings of some volcanic islands (McGoldrick et al. 2013; Hayman et al. 2020). In addition, greenstone belts are commonly intruded by porphyritic quartz-feldspar dacite sills and dykes.

These felsic dacite suites often increase in importance upwards in the greenstone stratigraphy. They can also occur contemporaneous with early komatiite and basalt sequences (Trofimovs et al. 2004; Rosengren et al. 2008), so constituting a tri-modal magmatic association, and then higher in the stratigraphy they become the dominant compositional lithofacies type (e.g. eastern Yilgarn Craton, Western Australia; Krapež et al. 2000; Squire et al. 2010; Kositcin et al. 2008).

In the eastern Yilgarn Craton, these dacitic volcanics were erupted contemporaneously with and were co-magmatic with the widespread Tonalite-Trondhjemite-Granodiorite (TTG) suite of plutons that were intruded into high crustal levels from 2.69 to 2.665 Ga (Smithies et al. 2019). Most are compositionally similar to TTG suite and sanukitoid plutons, and some of the TTG plutons were the magma sources for the volcanics (Smithies et al. 2019). Some TTG plutons intruded so high into the greenstone crustal sequence that they emerged above the seafloor and even sea level were unroofed and shed porphyritic and granitoid clasts into the volcaniclastic aprons around the volcanic islands (Squire et al. 2010; McGoldrick et al. 2013; Hayman et al. 2020).

17.7.2.6 The Archean Tonalite-Trondhjemite-Granodiorite (TTG) Suite—The Elephant in the Room

Archean cratons are generally dominated by granitoid batholiths of diverse compositions, distributed over huge areas, not in relatively linear belts. For example, in the Yilgarn Craton, Western Australia, granitoids represent ~60% of the surface area of the craton (Fig. 17.63a). The granitoids are represented by >60% high-Ca (granodiorite, granite, trondhjemite), >20% low-Ca (granodiorite, granite), 5–10% high-HFSE granite, minor granodiorite), 5–10% mafic granitoid, and <5% syenite bodies (Champion and Sheraton 1997).

Associations of *tonalite-trondhjemite-granodiorite* plutonic rock bodies (TTG) were recognised as being archetypical of Archean cratons in the 1970s, although not all Archean granitoids have TTG characteristics (Moyen and Martin 2012). All TTG rocks are sodic and have >20% modal quartz content. Granodiorite has a higher plagioclase than K-feldspar (orthoclase) content, tonalite is a very plagioclase rich (oligoclase to andesine; see Fig. 2.4b) variety of granodiorite with both amphibole and pyroxene as minor mafic phases, and trondhjemite is a variety of tonalite that is dominated by oligoclase, and has minor biotite as the only mafic mineral phase. Granodiorite is a minor component of the TTG suite, which is dominated by tonalite and trondhjemite. The volcanic equivalent of these three TTG plutonic rock types are all varieties of dacite, with the equivalent mineralogy.

Geochemically, TTG's are silica rich (>64 wt%, but commonly >70 wt%), have high Na_2O (3–7 wt%), $K_2O/Na_2O < 0.6$, low ferromagnesian element content ($Fe_2O_3 + MgO + MnO = TiO_2 < 5$ wt%), average Mg# ~ 0.43, Ni ~ 18 ppm, and Cr ~ 40 ppm (Moyen and Martin 2012). They have high LREE and low HREE contents with $(Nb/Ta)_{average}$ being 8.7, $(Lu/Hf)_{average}$ is 0.033 and $(Zr/Sm)_{average}$ is 44.3 (Moyen and Martin 2012; Smithies et al. 2019). According to Smithies et al. (2019), their Na rich nature together with Sr/Y and La/Yb > 40 indicate melting of a hydrated or metasomatised mafic source at pressures >1 GPa or depths >30 km. TTGs are more enriched in Na than modern calc-alkaline systems and have low HREE abundances compared with Proterozoic and Phanerozic crust produced under plate tectonic and subduction-related processes (Moyen and Martin 2012).

In addition to the overall petrological and geochemical characteristics summarised here, Johnson et al. (2019) identified changes and trends in some elemental abundances and ratios in TTGs from the Hadean and through the Archean. These include increasing Mg#, K_2O, Ni, Ba, Sr abundances, and K_2O/Na_2O, Sr/Y ratios, which they interpret in terms of a transition between different tectonic regimes, especially in the interval from 3.3 Ga to 3.0 Ga.

Moyen (2011) recognised that there is a spectrum of compositional attributes that reflect the different pressure conditions under which different TTG rock types form. He distinguished high (HP), medium (MP) and low-pressure (LP) TTG rocks. The high-pressure group is characterised by elevated values in Al_2O_3, Na_2O, Sr; and low values in Y, Yb, Ta and Nb. It is also somewhat more leucocratic (lower MgO, higher SiO_2 contents, but higher Mg#) than the other TTG groups. The low-pressure rocks show opposite features, whereas the medium pressure group is intermediate between the two. The high-pressure group is clearly identified whereas the low and medium pressure groups are more difficult to separate, the more discriminating elements being Sr and LREE (Moyen 2011). Johnson et al. (2017), Mole et al. (2019) and Smithies et al. (2019) have highlighted other geochemical parameters that reflect different pressure regimes under which partial melting of source rocks occurred. HP-TTG have high La/Yb and Sr/Y ratios and were formed at depths ~50 km, LP and MP pressure TTG have lower La/Yb and Sr/Y ratios and were formed at depths ~30–45 km (Johnson et al. 2017; Moyen and Laurent 2018).

It is now commonly accepted that TTG magmas are derived from partial melting of a metasomatised or hydrated mafic source (Moyen and Martin 2012; Johnson et al. 2017, 2019). However, there is considerable debate as to the nature of the tectonic regimes that produced TTG rocks. Many authors considered the TTG suite to be equivalent to calc-alkaline suites that are considered to be typical of modern subduction generated volcanic arcs, even though there are compositional differences (e.g. alkalis, REE patterns). Today, the production of young continental crust with compositions similar to Archean TTG rocks in transitional island arcs is an unusual process; e.g. limited to locations where there are especially voluminous partial melts of oceanic crust due to the subduction of a young hot plate, formed above a mantle plume or "hot spot", as in the Central American land bridge arc (Gazel et al. 2015; Sect. 17.5.2).

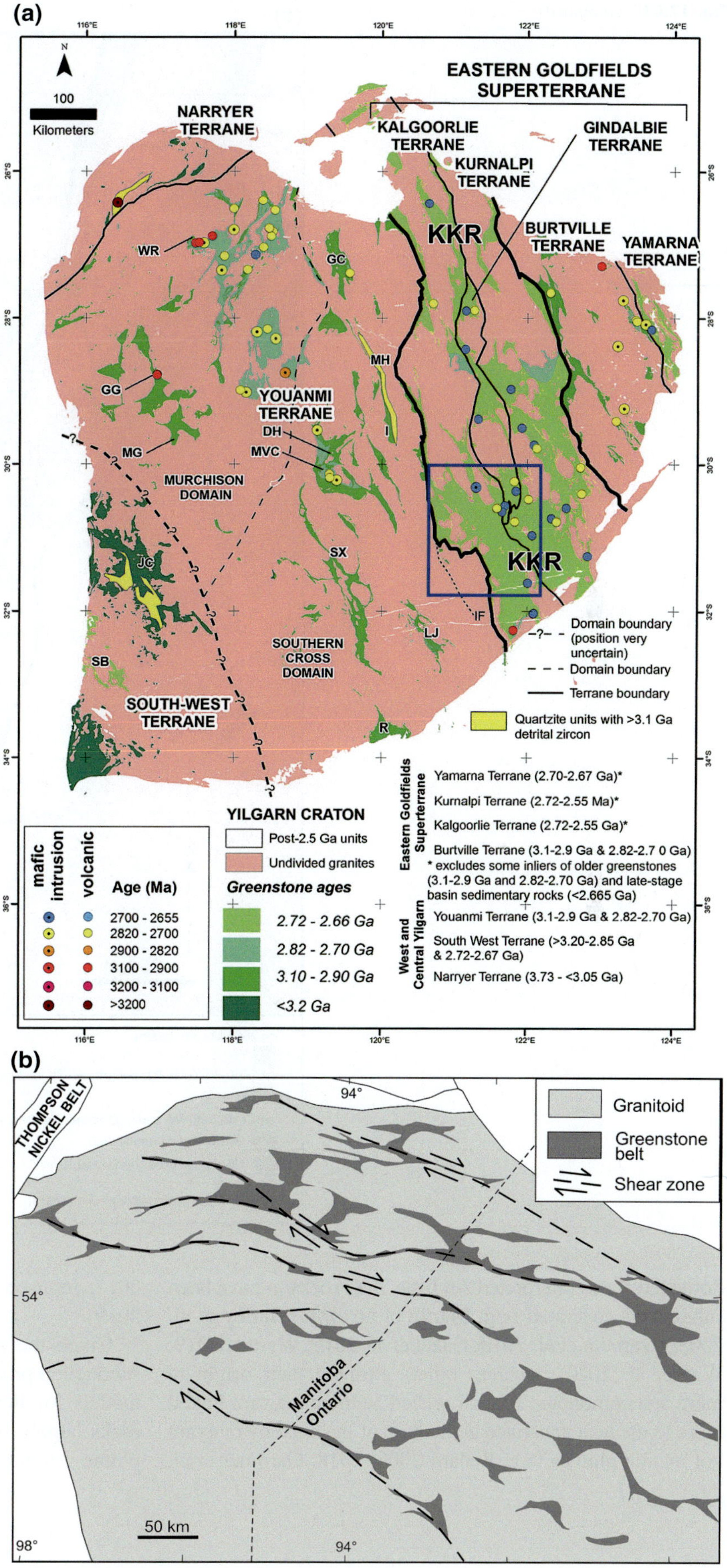

Fig. 17.68 Geological maps of the Archean cratons. **a** Map of the entire Yilgarn Craton, Western Australia showing in pink the areal extent of granitoids and in various shades of green, deformed greenstone belts of komatiite, basalt, felsic volcanics and sedimentary rock types. The various shades of green depict greenstone belts of different ages. Blue box shows area shown in more detail in Figs. 17.68c and 17.74d. (From Witt et al. 2020). **b** Simplified geological map of the northwestern Superior Craton, Canada, again showing the overwhelming areal extent of granitoid plutons (pale grey) relative to greenstone belts (dark grey) (From Parmenter et al. 2006). **c** Detailed geological map of the southern part of the Archean Kalgoorlie and Kurnalpi terranes, eastern Yilgarn Craton, Western Australia (blue box in **a**) distinguishing the major greenstone lithofacies and stratigraphic units, and showing their deformed relationship to the contemporaneous to younger granitoids. Granitoids = pink; komatiites = purple; basalts = green; felsic volcanics and volcaniclastics—cream; late basin sediments = brown. Note how stratigraphy mostly wraps around the margins of granitoids, rather than being truncated, indicating uplift of the stratigraphy by granitoid plutons in a dome and keel manner. (From Jones et al. 2020)

Fig. 17.68 (continued)

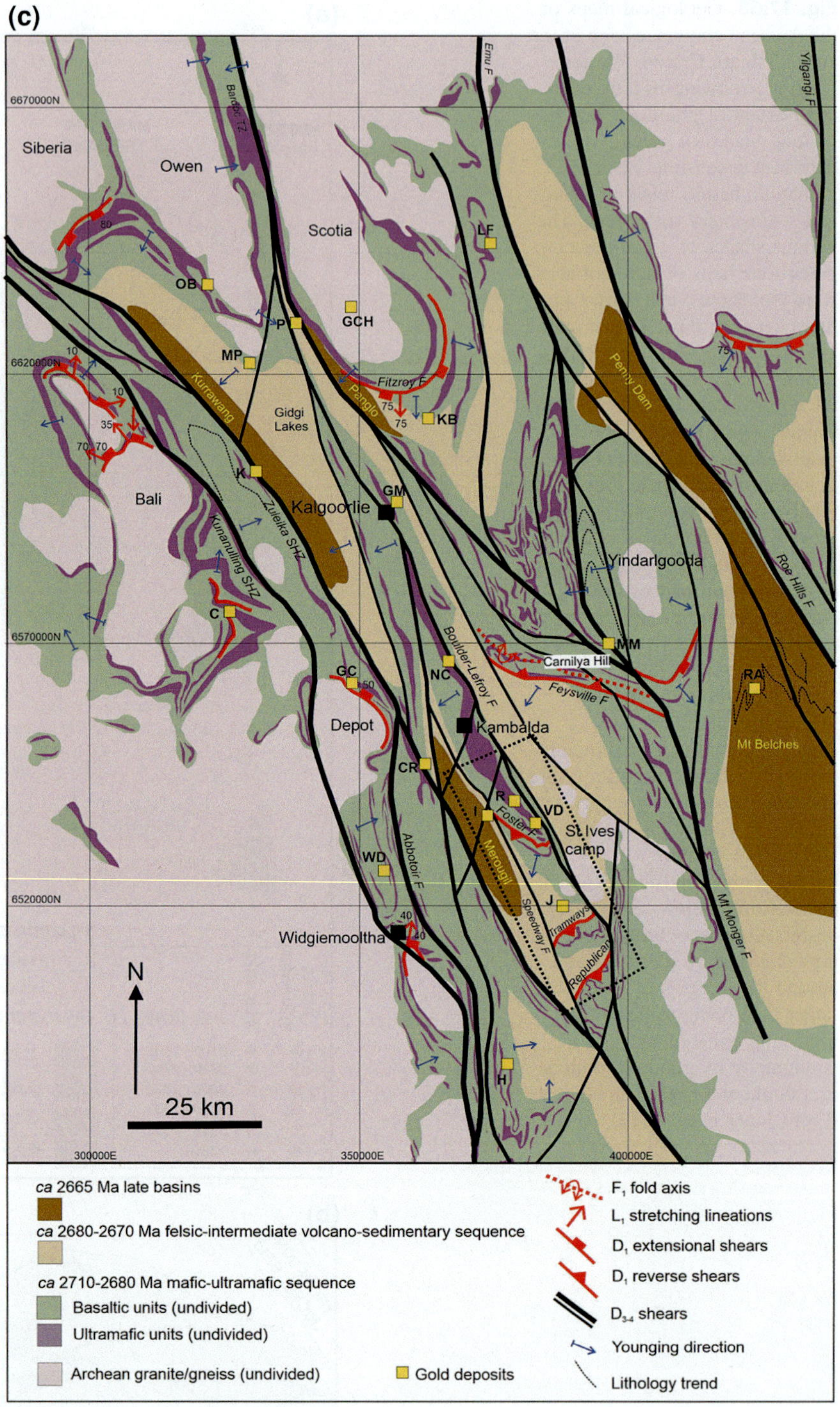

Some have also interpreted Archean TTG rocks to have been subduction generated (e.g. Martin et al. 2005; Barley et al. 2008; Czarnota et al. 2010; Gazel et al. 2015; Wyman 2019; Witt et al. 2020), whereas others attribute their origin to more autochthonous, vertical related tectonic regimes linked more to the heat generated at the base of the crust by buoyant hot mantle plumes (e.g. Bédard 2006, 2018; Gardiner et al. 2017; Johnson et al. 2017; Smithies et al. 2018; Mole et al. 2019).

Given the absence of physical geological evidence for subduction processes during most of the Archean, we favour models for the origin of TTG suite plutons and volcanic rocks largely related to the influence of mantle plumes and plume generated heat causing partial melting of a thickened

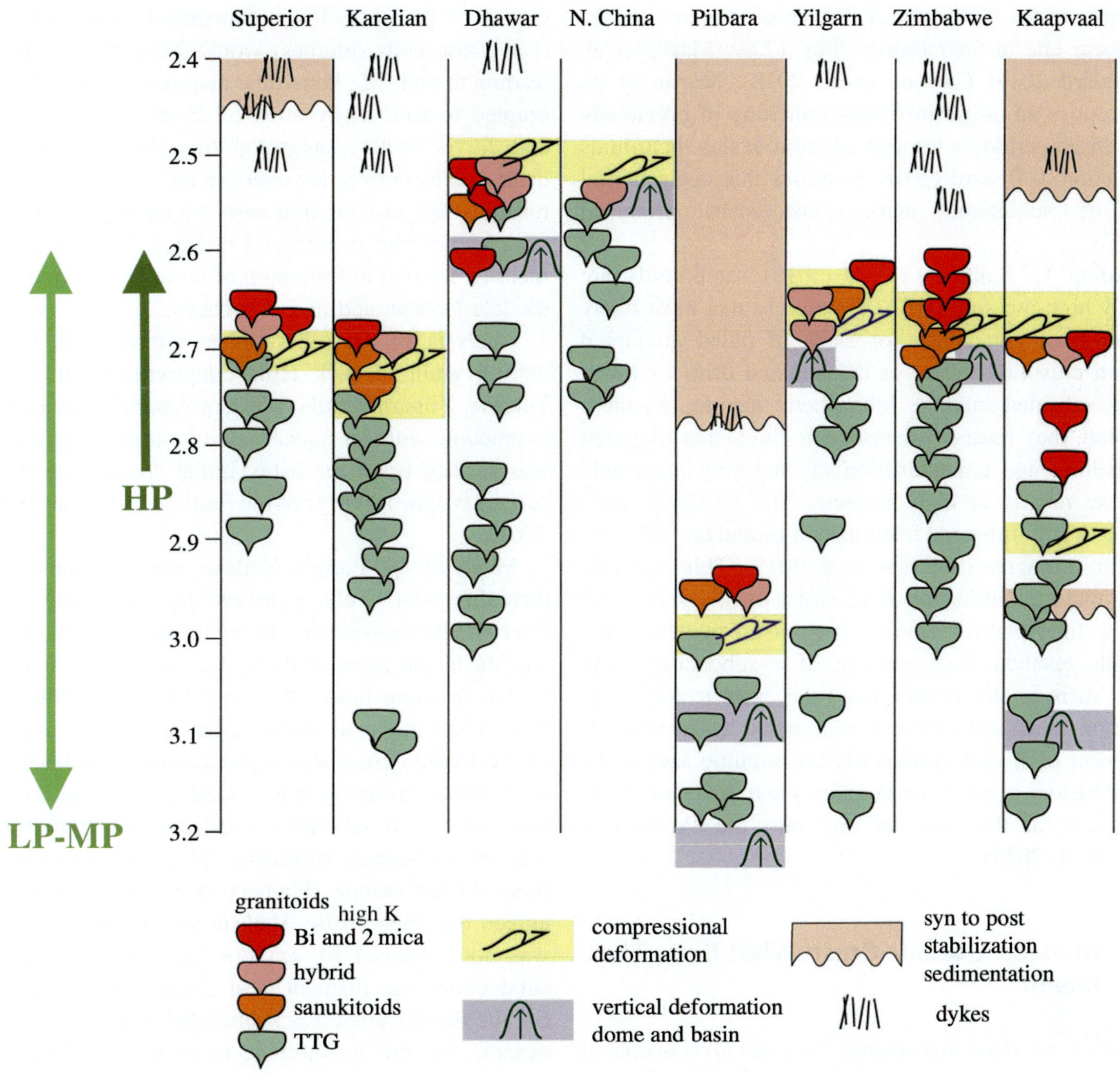

Fig. 17.69 Timing and changing nature of TTG magmatism in Archean cratons around the world. Note the change from standard TTG early to more potassic types including sanukitoid, hybrid and syenitic types late in the history of each craton. (From Cawood et al. 2018, through Creative Commons Attribution-Share Alike 4.0 International licence)

metasomatised or hydrated mafic crust. Low and medium pressure TTG rocks can be explained by partial melting of in situ thick mafic crust, at depths of 30–35 km, whereas high-pressure TTG rocks require partial melting of very thick or mafic crust, and/or perhaps delaminated slabs or blocks from the base of the crust sinking into deeper, high lithostatic pressure regions of the lithospheric mantle at depths of ~50–60 km (Johnson et al. 2017; Nebel et al. 2018).

Moyen and Laurent (2018) pointed out that through the Archean there is an increase in the proportion of HP-TTG relative to LP and MP-TTG plutons which they relate to an increasing thickness of the evolving Archean crust. There is also a change from TTG to potassic and peralkaline felsic plutons into the latest Archean, which is also consistent with partial melting of thickened crust (Fig. 17.69; Nebel et al. 2018; Cawood et al. 2018; Cawood and Hawkesworth 2019).

Sanukitoids are a specific type of granitoid with a calc-alkaline profile and high MgO, Mg# > 60, Cr, Ni and K_2O values, which distinguish them from rest of the TTG plutonic suite, whereas other key element discriminants are similar (LREE, Sr, Ba and high Sr/Y and La/Yb ratios; Martin et al. 2005; Smithies et al. 2019). They are an Na rich end member of the TTG suite. On Archean cratons, sanukitoids appear late in their history (Fig. 17.69; Martin et al. 2005; Bédard 2018; Cawood et al. 2018). Martin et al. (2005) propose an origin involving remelting of previously metasomatised peridotite through addition of slab melt/fluids (= subduction?). Bédard (2018) proposes that again partial melting of sub-creted mafic crust slabs produced sanukitoids.

According to Smithies et al. (2019) sanukitoids are essentially high-pressure (HP-)TTG plutons that most likely originated from detachment of slabs of failed attempted subduction crust/lithosphere or delaminated drips of lower crust that subsided into the lithospheric mantle. As these were heated, they released metasomatic fluids that triggered partial melting and contamination of enclosing peridotitic lithospheric mantle at high pressure. The resulting melts were subject to hornblende fractionation producing HP-TTG or sanukitoid magma (Smithies et al. 2019). This probably occurred under the influence of an earlier mantle plume that heated the lithospheric mantle. There is no evidence that sanukitoids resulted from melting of a subducting slab. Although there is still debate about the exact tectonic significance of sanukitoids, there is widespread agreement that emplacement of K-rich sanukitoids and syenites late in the magmatic history of each craton reflect the transition to plate tectonics late in the Archean and into the Proterozoic (Cawood et al. 2018).

17.7.3 Archean Oceanic Crust: What Does That Mean?

The term oceanic crust immediately conjures up concepts of Phanerozoic layered basaltic oceanic crust produced by seafloor spreading, including the dynamically diagnostic sheeted dyke complex (Sect. 17.2.1, see Fig. 17.3). However, this element is not present in preserved Archean crust; the story is more complex than this, and we have to go back to the very beginning.

When the Earth first formed through accretion of planetesimals at 4.54 Ga, because of the frictional heat generated by impact coalescence, it was molten. This allowed iron to segregate rapidly to the core, enclosed in a silicate molten envelope. As the Earth cooled its initial primitive basalt crust (Amelin et al. 1999) is thought to have formed through crystallisation from a magma ocean, above a convecting mantle (Schaefer and Elkins-Tanton 2018). That is, the Earth's initial basaltic crust was not formed through volcanism *per se*. Nevertheless, once a crust was formed, convection generated fissures in the crust would have allowed magma to erupt on the crust. Campbell and Jarvis (1984) proposed that due to the higher mantle temperature, mantle convection would have been vigorous, and the degree of partial melting of mantle in upwelling limbs of convection cells (plumes) would have been high ($\sim 10\%$) leading to high-Mg komatiitic magmas, and less basalt being erupted to form early crust on Earth. However, Campbell and Jarvis (1984) suggested that this earliest crust was destroyed by convective overturn and ingestion back into the mantle. Heat lost through early vigorous convection would have cooled the mantle and reduced the degree of partial melting, leading to formation of more basaltic magmas from the late Hadean and in the Archean.

Analysis of the oldest dated zircon grain (4.4 Ga), a detrital grain in Jack Hills conglomerate in the Narryer Terrane, Yilgarn Craton, Western Australia, are indicative of interaction with an aqueous environment suggesting there was surface water, probably initial "ocean" on the Earth's surface within 100 Myr of the Earth's formation (Wilde et al. 2001).

Much of the Earth's Hadean and Archean mafic crust thereafter was likely generated by submarine volcanism, often producing oceanic plateaus (Chaps. 5, 16), and in time leading to the accumulation of a mafic crust that was 30–35 km or more thick (Bédard 2006, 2018; Johnson et al. 2017; Capitanio et al. 2020). During the Archean only 2–3% of the Earth's crust was exposed above sea level (Flament et al. 2008), meaning that at least 97% of the Earth's crust was submerged submarine crust during the Archean. This was due to isostatic subsidence of a dense crust in a hotter, more ductile mantle (Flament et al. 2008). It is generally agreed that most of the Hadean and Archean basaltic crust was not produced by seafloor spreading accompanied by subduction (e.g. Flament et al. 2008; Johnson et al. 2017). So, the Earth's crust was submerged hence justifiably called oceanic, but this has nothing to do with seafloor spreading and the way that modern oceanic crust is formed.

17.7.4 Archean Continental Crust: What Does That Mean?

The answer to this question depends of how continental crust is defined. In general, most people would agree that continental crust is more felsic than oceanic crust. Modern upper continental crust has an intermediate bulk composition (cf. granodiorite, Taylor and McLennan 1995; andesitic, Gazel et al. 2015; Cawood et al. 2018). However, the early crust on Earth was primitive basaltic (Amelin et al. 1999; O'Neil and Carlson 2017; Bédard 2018; Piccolo et al. 2019), and progressively evolved and thickened through partial melting of uppermost mantle, by adding more submarine basalt and komatiite lavas as sheets onto pre-exisiting submarine mafic crust, the development of mafic plateaus and by addition of intrusions leading to progressive geochemical depletion of the uppermost mantle (Johnson et al. 2017; Cawood et al. 2018; Capitanio et al. 2020).

The thickening mafic crust was subject to partial melting at its hot base, which was subject to Rayleigh–Taylor instabilities, that led to "drips" of dense metamorphosed lower crust sagging downwards, and detachment or delamination of these. Crossing the solidus, these drips were subject to partial melting, releasing hydrous fluids that triggered more mantle melting and mafic mantle generation, but also producing progressively more evolved intermediate to silicic plutons, including TTG suite plutons, which rose buoyantly and intruded the mafic crust (Johnson et al. 2017; Bédard 2006, 2018; Piccolo et al. 2019; Cawood et al. 2018).

In addition, compositionally evolved plutons formed through meteorite impact generated melting of the developing mafic crust during the Hadean period of impact bombardment (Johnson et al. 2018). Cawood et al. (2018) also point out that Archean cratons generally lack high-grade metamorphic rocks (e.g. granulite facies), which is suggestive of continental crust that was thinner than during the Proterozic and Phanerozic.

The "felsification" of the Earth's early crust thus represents the initiation of localised "continental-type crust" during Eo-Archean to Meso-Archean times, which was probably hybrid mafic-felsic "continental crust" (Johnson et al. 2017), becoming progressively more felsic through much of the Archean. Less than 3% of the Earth's surface was emergent above sea level (Flament et al. 2008), and there is little evidence for the existence of large emergent continents until the Proterozoic (Taylor and McLennan 1995). The lack of significant continental freeboard (i.e. emergent above sea level) can be attributed to the largely dense mafic crust and the hotter more plastic upper mantle during the Hadean and most of the Archean (Flament et al. 2008). Even today large areas of continental crust can remain submerged beneath sea level if the crust is not isostatically buoyant enough, due to a dense lower crust or keel.

For example, the huge 4.9×10^6 km^2 area of continental crust called the "Zealandia microcontinent", which is situated east of Australia (see Figs. 17.43b and 17.59). With the exception of New Zealand which is the largest part of Zealandia above sea level, followed by New Caledonia, most of the landmass is totally submerged beneath the Pacific Ocean but for a few rocky volcanic islands rising above sea level (e.g. Lord Howe Island, Norfolk Island, and the Chatham Islands cluster east of New Zealand). Its crust is generally 10–30 km thick, consisting of inferred Precambrian basement and overlying Phanerozic sedimentary basin successions, which were extended and thinned during Gondwana break-up (Mortimer et al. 2017; Fig. 17.43b), and has P-wave seismic velocity of $\sim$7 km s^{-1} This is intermediate between continental and oceanic crust (Mortimer et al. 2017), but is marked by a large area of underplated, mafic (Pacific Plate) lower crust (Grobys et al. 2009; Fig. 17.61), associated with intraplate basaltic volcanism represented on the emergent volcanic islands (e.g. Sorrentino et al. 2011; Mather et al. 2020; Sect. 17.5.6). Such a huge area of submerged continental or hybrid crust is an excellent analogue for the scenario of Archean cratons, with their very limited area of emergent crust (Flament et al. 2008).

Taylor and McLennan (1995) used REE patterns of sediments as a proxy for assessing the abundance of "continental crust" through time. They found that Archean shales showed less enrichment in light REE and no Eu anomaly compared with post-Archean sediments, which they attribute to sources dominated by mantle-derived rocks, whereas post-Archean shales reflect a dominant intra-crust source derivation. They estimated that only 10% of the Earth's continental crust had formed by 4 Ga, but 50% by 2.6 Ga. Cawood et al. (2019) propose that increasing Rb/Sr ratios of Archean rocks through time are a good proxy for increasing crustal thickness and SiO_2 content, which they also relate to increasing surface area and volume. They propose that the surface area of continental crust had reached about that of the present-day Earth by the end of the Archean. Interestingly, they also propose that both continental crustal volume and thickness have been declining since $\sim$0.8 Ga. This was probably due to cooling of the Earth, transition to cold and more rigid plate subduction, which then caused more frictional coupling and tectonic erosion of the overriding plates by the down-going plates at subduction zones.

Tang et al. (2016) used Ni/Co and Cr/Zn ratios as proxies to assess the bulk MgO content of upper Archean continental crust and found that the upper crust changed from a very mafic composition before 3 Ga to a very felsic one by 2.5 Ga, which they relate to the onset of plate tectonics at $\sim$3 Ga, with subduction generated magmas contributing to the felsification of the upper continental crust. Rollinson (2017) also proposed a rapid growth of continental crust volume between 3 and 2.5 Ga. He suggests that bursts of apparent continental crust generation during the Eo- and Meso-Hadean proposed by other workers based on peaks of zircon ages and the interpretation of Hf isotope and Lu/Hf data to calculate model ages of their source rocks, may be incorrect because some of the assumptions those calculations are based on may not be correct.

Eriksson et al. (2013) reviewed the evolution of environments and associated sedimentary facies through Earth history. They noted that although isolated examples of continental sediments are found in the Meso-Archean (3.2–2.8 Ga) rock record, they become more abundant during the Neo-Archean (2.8–2.5 Ga) and are then very widespread from the beginning of the Proterozoic. This again suggests the progressive emergence of continental areas above sea level during the late Archean and more rapidly at the Archean-Proterozoic transition, when most researchers agree that transition into a plate tectonic regime was well advanced.

The development of continental-type crust also seems to be fundamentally related to the activity of mantle plumes and how they contributed to the formation of both oceanic and continental-type crust (Bédard 2006, 2018; Bédard et al. 2013), as discussed further in Sect. 17.7.5.3 below.

17.7.5 Proposed Archean Tectonic Models

There have been a number of tectonic models or paradigms proposed for the origin of Archean granite-greenstone terrains that constitute Archean Cratons. The use of the term and spelling of "terrain" is intentional to indicate a geographic region of Archean rocks and not a tectono-stratigraphic terrane in the plate tectonic sense of a dismembered slice of lithosphere that is bounded by major crustal faults and has been displaced from its source area and juxtaposed adjacent to other (exotic/suspect/exogenous) terranes, generally by far-travelled transcurrent faulting (e.g. terranes of northwestern North America; Coney 1989; Colpron et al. 2017) associated with plate tectonic movements and dynamics. Exogenous origins must be demonstrated by geological evidence (e.g. significant mismatches in rock types, stratigraphy, ages, deformational and metamorphic events between adjacent terranes). Unless that can be demonstrated fault-bounded terrains should be considered to be endogenous. The various tectonic models or regimes that have been proposed for the Archean or at different time periods during the Archean include:

- magma ocean,
- stagnant lid,
- plumes,
- drips and delamination,
- mobile lid subcretion or embryonic subduction,
- pluton diapirism and greenstone sagduction, and
- plate tectonics.

17.7.5.1 Magma Ocean—The First "Tectonic" Regime (Eo-Hadean)

As discussed above, when the Earth first formed through accretion of planetesimals at 4.54 Ga it was molten (cf. Fig. 17.70a) because of the frictional heat generated. As the Earth cooled the primitive basalt crust crystallised from a magma ocean, above a convecting mantle (Schaefer and Elkins-Tanton 2018). Although the Earth's initial basaltic/ultramafic crust was not formed through volcanism *sensu stricto*, at some time a soft, skin-like crust formed above an underlying convection system in the magma ocean that may have been several hundred km deep/thick (Schaefer and Elkins-Tanton 2018). An isotropic network of fissures would have formed in the crust, forming squishy small platelets drifting above a convecting magma sea, much as occurs in large summit lava lakes of basaltic magma on large basaltic volcanoes (e.g. Nyiragongo volcano lava lake, Democratic Republic of the Congo; Fig. 17.70b; Valade et al. 2018). At points of vigorous upwelling almost random eruption points would have formed together with points of simultaneous downwelling causing recycling of very ductile ephemeral crust back into the lava pool (Fig. 17.70c, d; Valade et al. 2018).

This ductile, ephemeral regime was very unlike the modern tectonic regime in that there was no lithosphere as defined today (a rigid crust and the uppermost mantle above a plastic asthenosphere). There was a very ductile crust underlain immediately by a hot, plastic, convecting asthenospheric mantle. Recycling of crust driven by underlying convection would have occurred in a very plastic deformational regime, and perhaps, as in modern lava lakes, could also have involved leading edge roll-over of the advancing upper crustal "plate" and forced sinking of the over-ridden lower crustal "plate". The crust alone was therefore the Earth's earliest lithosphere in the Eo-Hadean. Although there is no known preserved Eo-Hadean mafic crust/lithosphere, isotopic and trace element data of preserved younger Archean igneous rocks indicate their existence and suggest derivation from primitive chondritic sources (Debaille et al. 2013; O'Neil and Carlson 2017; Johnson et al. 2018).

17.7.5.2 Stagnant Lid—Tectonic Regime and the Evolution of a Crust/Lithosphere (Hadean-Meso-Neo-Archean

There is general agreement that after the initial magma ocean formed, the Earth progressively formed an essentially continuous mafic crust above a convecting mantle (Fig. 17.71a) without any clearly defined plate tectonic recycling of lithosphere until at least 3 Ga. In fact, the mafic crust was buoyant on the underlying mantle, which had not yet differentiated, and so the crust was in effect "unsubductable" (Bédard 2018; Piccolo et al. 2020). This regime has been called a single or "stagnant" lid" by many authors (e.g. Solomatov and Moresi 2000; Bédard et al. 2013; Bédard 2018; Debaille et al. 2013; O'Neill et al. 2013; Cawood et al. 2018; Nebel et al. 2018; Johnson et al. 2019; Smithies et al. 2019; Capitanio et al. 2020).

It is envisaged that this lid was deformable under conditions of an upper mantle temperature between 150 and 250 °C hotter than today's ~1350 °C temperature (Piccolo

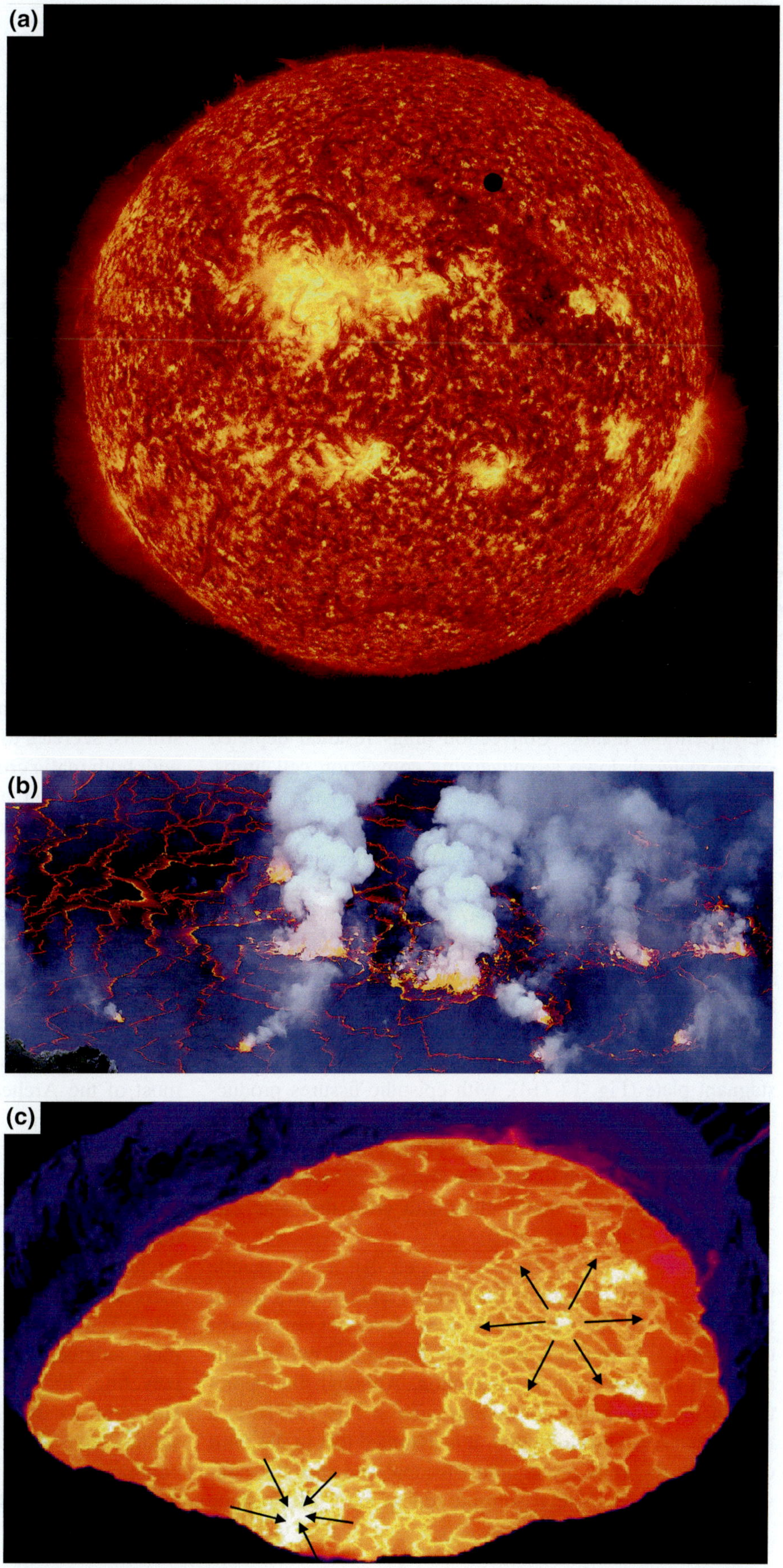

Fig. 17.70 Representation of the early Earth during the Eo-Hadean. **a** Earth as a molten planet with a vigorously convecting magma ocean at the surface. The image is actually one of the gaseous Sun taken in 2012 from the Solar Dynamics Observatory showing the planet Venus passing in front of the sun as a black ball in the top right of the image. (*Source* NASA). **b** The Earth's earliest crust would have consisted of a very plastic basalt crust that due to vigourous convection underneath would have broken into deformable platelets, as seen here from the lava lake of Nyiragongo volcano, Democratic Republic of the Congo (Photo by Cai Tjeenk Willink; *Source* Wikimedia through Creative Commons Attribution-Share Alike 3.0 Unported licence). **c** Infrared image showing convective upwelling point, indicated by diverging black arrows, and downwelling point, indicated by converging black arrows, Nyiragongo volcano (Screenshot from Valade et al. 2018, Appendix A Video 2). **d** Schematic representation of the scenario in (**c**), showing a convective upwelling zone in the centre and downwelling zone on the left (From Valade et al. 2018)

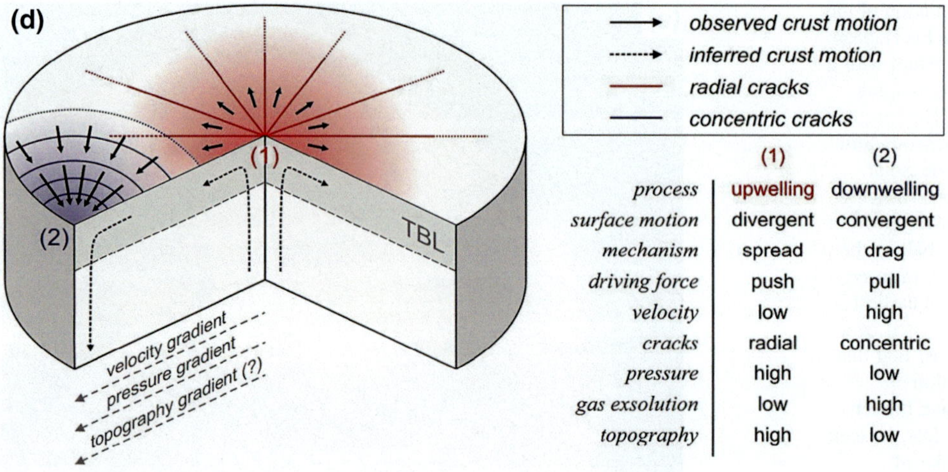

Fig. 17.70 (continued)

et al. 2020). Under these conditions, modelling and experiments have demonstrated that the sub-crustal/lithospheric mantle beneath the stagnant lid was marked by a structure of multiple, closely spaced ($\sim$several hundred km), relatively vigorous convection cells (Fig. 17.71b; Solomatov and Moresi 2000; Capitanio et al. 2020). Melting occurs at the interface between the crust, to which the melt is added, and the uppermost mantle. Convection drags residual depleted source material down at converging convection cell limbs, allowing this material to mix back into the mantle (Fig. 17.71b), and so leading to progressive depletion of the uppermost mantle (Capitanio et al. 2019). This stage of crust development is perhaps similar to phases of activity in modern large volcanic basalt lava lakes, beginning with a thin ductile lava crust that breaks into platelets that move in response to underlying convection currents, places crumpling in ductile fashion (Fig. 17.71c). As the lava lake activity wanes and convection turnover declines, the crust cools, becomes thicker and platelets coalesce into a single stagnant plate (Fig. 17.71d), with episodic fissures producing eruptions of lava onto the hardened crust, so thickening it further. It is considered that the Earth's stagnant lid crust also thickened through time by the eruption of lavas onto the seafloor leading to a mafic–ultramafic crust that was 20–30 km thick (Palin et al. 2020).

Capitanio et al. (2019) demonstrated that as the mantle temperature decreases the crust thickens and becomes stiffer, the diameter of the convection cells decreases and fewer drips form. These drips, discussed further below, perhaps represent the precursor to subduction. Although the depiction of the sub-crustal convection cells in Fig. 17.71b is a two-dimensional cross-section, it is assumed that in the third dimension the cells are equidimensional and that the convecting mantle is largely isotropic unless the Coriolis force induced some element of anisotropy on the convection system in the hotter, less viscous mantle.

Capitanio et al. (2020) develop this model of crust formation and upper mantle depletion further. They propose that as upwelling mantle in sub-crustal convection cells undergoes decompressional melting, the upper mantle progressively dehydrates, becomes more depleted, cools and stiffens, accretes to the base of the crust, forming lithosphere with both mafic crust and uppermost mantle. Continuation of this process then leads to top down accretion of more mantle to the lithosphere, thickening of the lithosphere, and increasing rheological rigidity, paving the way to ultimate plate tectonic subduction in the late Archean.

Debaille et al. (2013) used ^{147}Sm-^{143}Nd and ^{146}Sm-^{142}Nd systematics of a 2.7 Ga tholeiitic lava flow in the Abitibi Greenstone Belt, Superior Craton, Canada, to show that it preserves a ^{142}Nd anomaly. They proposed that the preservation of this anomaly is due to the extended preservation of a stagnant lid crust/lithosphere throughout the Hadean and most of the Archean to at least 2.7 Ga. They argue that if plate tectonic-like lithospheric recycling ("mobile lid") had occurred for significant durations during the Archean there would have been greater elemental and isotopic homogenisation of the upper mantle, on a time scale of several hundred Myr, so obliterating any potential ^{142}Nd anomaly in late Archean volcanics. Debaille et al. (2013) acknowledge that there are peaks in ages of dated zircons in Archean rocks at 4.1, 3.7, 3.3 and 3.1 Ga, suggesting short bursts of localised plate subduction or mobile lid tectonics leading to plate recycling, arc creation and continental crust generation. However, as already discussed above, other and subsequent studies have shown that felsification of the crust leading to continental nuclei probably occurred through TTG generation by partial melting of metasomatised mafic lower crust

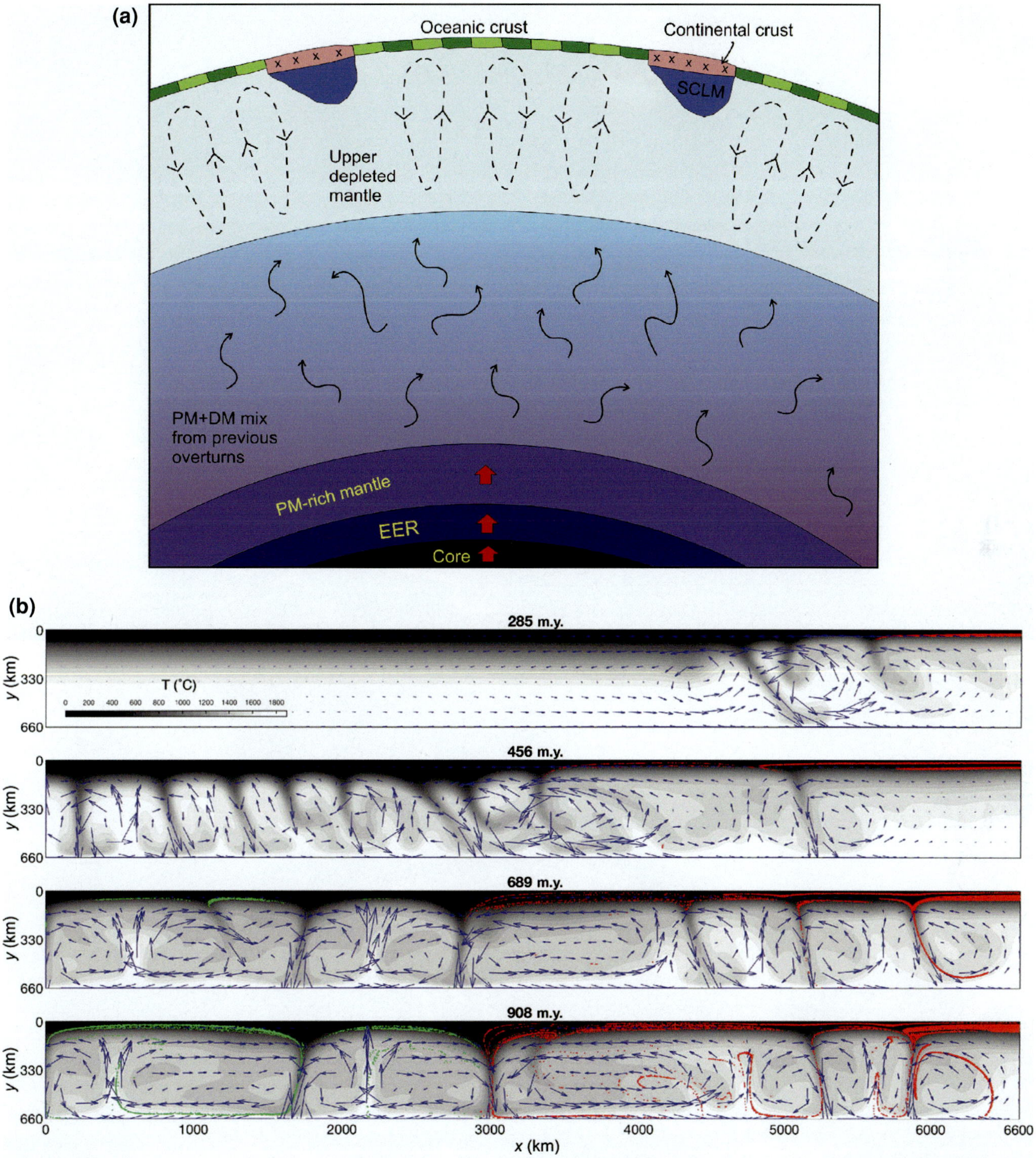

Fig. 17.71 Features of a stagnant lid Earth. **a** Representation of a stagnant lead crust on earth during the Archean showing widespread mafic crust and localised areas of developing "continental "crust above a vigorously convecting mantle (From Bédard 2018). **b** Results of modelling depicting and evolving complex system of upper mantle convection cells beneath a continuous stagnant lead crust. Red and green trails depict areas of melt that are formed at the base of the crust and are dragged down where downwelling limbs of convection cells converge, leading to drips of crust mixing downwards into the mantle. (From Capitanio et al. 2019). **c** Ductile platelets of basalt forming a crust on the surface of the Halema'uma'u crater lake, Hawai'i. Note the ductile folding of the lava lake crust at the left and right. **d** Continuous lava lake crust, with minor incipient fissures, Halema'uma'u crater lake, Hawai'i, 2018. The different platelets that have fused together to form the continuous crust are identifiable by variations in their colour. (Both **c** and **d** screenshots from video *"Kilauea summit eruption—lava returns to Halema'uma'u"*; *Source* United States Geological Survey)

Fig. 17.71 (continued)

(LP-, MP-TTG) and detached, delaminated blocks of such crust (HP-TTG; Bédard 2006, 2018; Gardiner et al. 2017; Johnson et al. 2017; Smithies et al. 2018; Mole et al. 2019; Piccolo et al. 2019).

Debaille et al. (2013) also ascribe the lack of evidence for plate tectonic processes on Mars and Venus to the presence of a stagnant lid tectonic regime throughout their formative history. In contrast, on Earth, the tectonic regime changed from a largely stagnant lid regime to a plate tectonic regime during a transition from the Late Archean into the Proterozoic. On some cratons that transition may have commenced at ~3 Ga, whereas on others the onset may have been ~2.7 Ga (Cawood et al. 2018).

Finally, although the stagnant lid prevailed until at least 3 Ga, during that time there were disruptions to the continuous stagnant lid, through the activity of intermittent plume activity, as discussed next.

17.7.5.3 Plume Tectonics (Hadean–Archean)

There has been considerable discussion of the role of mantle plumes throughout Earth history in the evolution and dynamics of the Earth (e.g. Campbell and Griffiths 1990; Condie 2001; Campbell 2005; Ernst and Buchan 2001; Foulger et al. 2005; Heron 2019; Sect. 17.4, see Fig. 17.28), but also specifically during the Hadean and Archean (e.g. Campbell and Jarvis 1984; Campbell and Hill 1988; Campbell et al. 1989; Barnes et al. 2012; Bédard 2006, 2018; Mole et al. 2014; Piccolo et al. 2020). In the modern plate tectonic world mantle plumes are precursors to (super)continental break-up, causing regional crustal updoming and then onset of seafloor spreading (Sect. 17.2.3.1, Fig. 17.14. They generate central volcano, age progressive hot-spot tracks of volcanoes both in the oceans (e.g. Hawai'ian-Emperor seamount chain, Fig. 17.29a; the Tasmantid and Lord Howe seamount chains; Figs. 17.30a and 17.59), and on-land (e.g. Cosgrove hot spot track, Figs. 17.29b, 17.30a; 17.59; Yellowstone hot spot track, Fig. 17.19b), and are also the source of LIPS (e.g. flood basalt provinces, oceanic plateaus; Table 17.1; Chaps. 5, 16). During the Hadean and Archean mantle plumes were probably even more active, given the higher internal temperature of the Earth, and they were vital in contributing to the formation of both oceanic and continental crust (Fig. 17.72a, b; Bédard 2006, 2018).

According to Campbell and Jarvis (1984) the higher mantle temperature in the Hadean and Archean, would have caused vigorous mantle convection, and a high degree of partial melting of mantle (>10%; others suggest up to 30%, Bédard 2006, 2018; Bédard et al. 2013; Mole et al. 2014) in upwelling limbs of convection cells (plumes) producing komatiitic magmas, and less basalt being erupted. As heat was lost through vigorous convection cooling of the mantle, this would have reduced the degree of partial melting forming more basaltic magmas, which dominate over komatiite in the Archean.

Nonetheless, significant periods of komatiite volcanism accompanied by high-Mg and tholeiitic basalt volcanism at different times in the Archean (~3.5, 3.1, 2.9, 2.7 Ga; Chap. 7, see Fig. 7.1), attest to major periods of plume activity through much of the Archean. Most of these events produced widespread submarine lava flow fields and crust, oceanic plateaus and associated subvolcanic komatiite and basalt intrusions, equivalent to younger LIPs (Bédard 2006; Barnes et al. 2012; Hayman et al. 2015).

Although modern mantle plumes can cause uplift of continental crust prior to continental break-up (e.g. East African Rift Zone, Sect. 17.2.3.1), the rheology of the Archean crust and mantle probably limited this effect. Because the mantle was significantly hotter, its viscosity was lower, and because the crust, even "continental crust" was largely mafic, thick and denser than today, it was not as buoyant as modern crust. In the early hot, weak, and less viscous crust and mantle thermal buoyancy of the crust was probably compensated by lateral spreading or extension of the up-buoying crust (Fig. 17.72a; Bédard et al. 2013; Rey et al. 2014; Piccolo et al. 2020). Uplift was thus limited, and this is verified by the minor areas (~3%) of emergent crust throughout most of the Archean (Flament et al. 2008). Only in the Late Archean, when the Earth had cooled significantly, some of the crust had felsified, become less dense and more rigid, is there evidence for widespread uplift (Fig. 17.72b) and on most cratons, emergence above sea level.

Rising plumes did however trigger anatexis of lower mafic crust (Fig. 17.72b–d) by raising the geotherm in the crust above them as they rose, inducing the solidus to rise and triggering the formation of TTG suite granitoids as well as other types (Campbell and Hill 1988; Smithies et al. 2019; Piccolo et al. 2020). In weak, hotter less viscous early Archean crust, rising plutons did not cause any uplift, but potential thermo-buoyancy uplift was probably transformed into lateral thermo-gravitational spreading or extension of the crust above plumes (Bédard et al. 2013; Rey et al. 2014). However, in the late Archean of the Yilgarn Craton of Western Australia, as will be discussed below, the rise of buoyant diapiric granitoid plutons through the cooler, more rigid, stronger late Archean crust coincided with progressive regional uplift, deformation and continental emergence above sea level and caused updoming of the supracrustal greenstone sequences they were intruding (Campbell and Hill 1988; Hayman et al. 2020).

Piccolo et al. (2019, 2020) modelled the impact of hot mantle plumes on Archean crust thermo-dynamically/mechanically. They propose that the effect depends on how much higher the temperature of the upper mantle (asthenosphere) was at the time compared with today

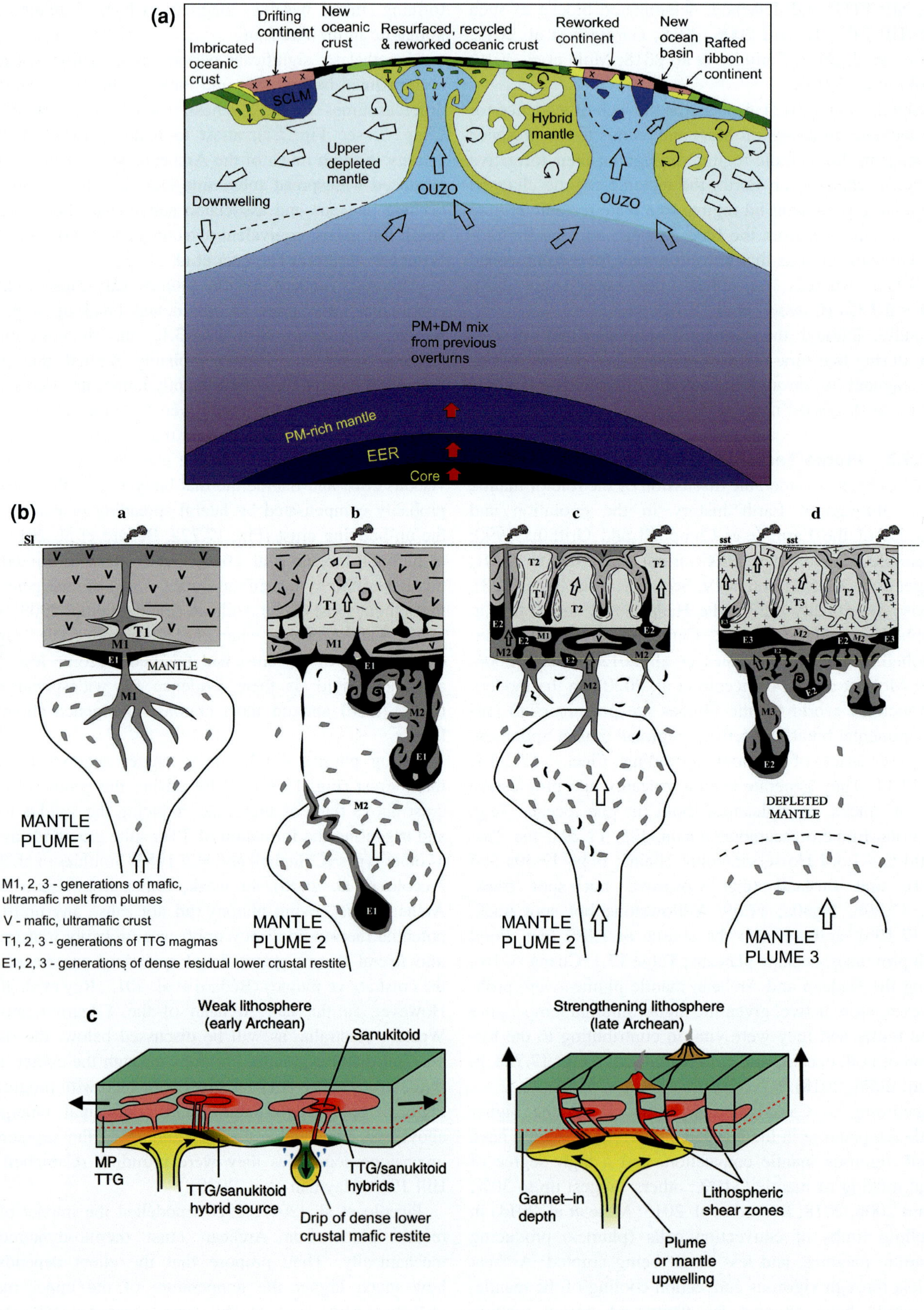
(a)
Drifting continent
New oceanic crust
Resurfaced, recycled & reworked oceanic crust
Reworked continent
New ocean basin
Rafted ribbon continent
Imbricated oceanic crust
SCLM
Hybrid mantle
Upper depleted mantle
Downwelling
OUZO
OUZO
PM+DM mix from previous overturns
PM-rich mantle
EER
Core
(b)
a
b
c
d
Sl
sst
sst
T1
M1
E1
MANTLE
M1
MANTLE PLUME 1
T1
M1
E1
M2
E1
M2
E1
MANTLE PLUME 2
T2
T1
T2
T2
E2
M1
E2
M2
M2
E2
M2
V
MANTLE PLUME 2
T2
T2
T3
T3
E3
M2
E3
E2
M2
E3
E2
E2
M3
E2
DEPLETED MANTLE
MANTLE PLUME 3
M1, 2, 3 - generations of mafic, ultramafic melt from plumes
V - mafic, ultramafic crust
T1, 2, 3 - generations of TTG magmas
E1, 2, 3 - generations of dense residual lower crustal restite
c
Weak lithosphere (early Archean)
Sanukitoid
MP TTG
TTG/sanukitoid hybrid source
TTG/sanukitoid hybrids
Drip of dense lower crustal mafic restite
Strengthening lithosphere (late Archean)
Garnet–in depth
Lithospheric shear zones
Plume or mantle upwelling

◀ **Fig. 17.72** Effects of mantle plumes during the Hadean and Archean. **a** Cartoon representation of plume activity during the late Archean, showing plumes (or overturn upwelling zones = OUZO, sky blue) under both oceanic (centre) and continental crust (right) where they spread laterally. Under continental crust they cause the continental crust to extend, as well as causing ultramafic, mafic and felsic (TTG) intrusions into the crust, eruptions onto the crust, and detachment of lower crustal "drips" or slabs into the upper mantle where they release hydrous fluids, triggering partial melting and generation of high-pressure (HP-)TTG. Plumes under the oceanic crust lead to eruptions of ultramafic and mafic lavas, and as the plumes spread the mantle convection currents impact the keels of continents and drive their drift. (From Bédard 2018). **b** Various scenarios showing the impacts of three mantle plumes (1, 2, 3) on the evolution of the crust and upper mantle. The plumes release and generate new mafic and ultramafic magmas (M1, 2,3) and felsic TTG magmas (T1, 2, 3) by causing partial melting of both upper mantle and the lower mafic crust respectively. Mafic and ultramafic magmas from the mantle are both intruded and extruded (V) causing thickening of the crust in time (scenario A). TTG magmas from partial melting of the lower mafic crust also intrude into and erupt onto the crust, and emplacement of plutons causes deformation and sagduction of supracrustal volcanic and sedimentary sequences, forming the typical intrusive dome and supracrustal stratigraphy keel architecture of cratons (scenarios B-D). Dense metamorphosed mafic residues (restites) after partial melting of the lower mafic crust (E1, 2, 3) may detach or delaminate from the base of the crust as drips, sinking into the upper mantle, where they mix, partially melt, release hydrous fluids which trigger partial melting of the upper mantle, and may produce some HP-TTG magmas which rise buoyantly into the crust. The crust thus felsifies and the upper mantle becomes depleted. (From Bédard 2006). **c** Impact of plume on hot, ductile early Archean crust. Processes as for (**b**), but potential buoyant uplift is dissipated by lateral gravitational extension (horizontal black arrows). **d** Impact of plume on colder, more rigid late Archean crust. Processes as for (**b**), but buoyant uplift of the crust (vertical black arrows) and brittle deformation (faulting) begin to occur. (**c**, **d** modified from Smithies et al. 2019)

(~1350 °C). Their modelling suggests that where upper mantle temperature was >1500 °C and the plume temperature was 1600 °C, the plume begins to spread radially at the base of the hot ductile lithosphere triggering mafic magma generation in that area. Eruption and intrusion of komatiites, high-Mg basalts, and tholeiitic basalts produce an embryonic mafic "continental" crust and thicken it. Rayleigh–Taylor instabilities at the base of the crust begin to trigger generation of felsic/TTG melts which rise into the crust. The developing mafic-felsic crust spreads laterally and initiates subsidence of the crust being over-ridden as slabs and drips that generate further felsic melts, while thinning and decompression of the lithospheric mantle above the plume continue to cause mafic magmas to form.

At lower upper mantle temperatures of <1450 °C, the lithosphere is more rigid, more TTG felsic crust is produced and as the crust above the plume spreads radially, it more definitely overthrusts a more rigid plate-like slab around the margins. This is more akin to plate subduction but occurs peripheral to the plume margins, and the depth of the "subduction" is limited. Piccolo et al. (2020) describe this process as "sub-accretion". The spreading inner crust is replaced by newly erupted mafic and felsic crust, producing an annular structure. These contrasting results may thus reflect the progressive evolution of magma-tectonic regime during the Archean, as the Earth cooled.

Bédard (2006, 2018) analysed the role of mantle plumes or overturn upwelling zones (OUZOs), as he called them, in forming the crust and sub-cratonic mantle through the Archean (Fig. 17.72a). He proposed that mantle plumes were a consequence of a stagnant lid crust, which would reduce mantle heat escape causing heat build-up in the lower mantle from radioactive decay and core heat flux. Intermittently this would trigger a major mantle convective overturn, and plume upwelling. Plumes or OUZOs were intermittent in their timing and they disrupted the stagnant lid crust/lithosphere that was restored between major OUZO events. Plumes generated eruptions of basalts and komatiite above the plume, thickening the crust, and also leading to partial melting of lower crust, as well as delamination (dripping) of dense mafic granulitic lower crust blobs or slabs, and their interaction with the subjacent mantle, producing TTG suite plutons as previously discussed (Fig. 17.72b). Thus, continental crust nuclei were formed, during which depleted sub-cratonic upper mantle also developed (Bédard 2018). Pre-existing continents above active mantle plumes would be subject to extension and be invaded by mafic, ultramafic and felsic magmas (Bédard et al. 2013). Barnes et al. (2012) and Mole et al. (2014) suggested this cratonisation process drove the hottest and most voluminous komatiite eruptions to the edge of established continental blocks. The dynamic evolution of the early continents thus directly influenced the addition of deep mantle material to the Archean crust, oceans and atmosphere.

Continents distant from plume heads would be rafted away by upper mantle convection currents flowing away from the plume head impacting the sub-cratonic keel and dragging them along (Fig. 17.72a). Interestingly, this was the basis for the concept of continental drift, prior to the discovery that continents are passive passengers being rafted on rigid lithospheric plates in the plate tectonic regime. Bédard (2018) proposes that OUZOs were larger than modern plumes, and that resulting magma compositions are different, but these differences seem to reflect secular variations including the Earth being hotter in the Hadean-Archean and due to the onset of plate tectonics in the latter half of Earth history. Hadean-Archean plumes were intermittent in timing, apparently occurring at about 4.34, 3.9–3.85, 3.4, 3.1–2.9 and 2.7 Ga, which coincide with bursts of crust formation, including continental crust, based on zircon age peaks (Bédard 2018).

17.7.5.4 Dripping and Delamination of Lower Hadean-Archean Mafic Crust

We have previously discussed the effect of both a hotter, thicker crust and hotter mantle in causing gravitational detachment or delamination of "drips" of dense, metamorphosed lower mafic crust into the upper mantle (Fig. 17.67b, c; Bédard 2006, 2018; Kamber 2015; Chowdhury et al. 2017; Nebel et al. 2018; Johnson et al. 2017; Smithies et al. 2019; Piccolo et al. 2019, 2020; Capitanio et al. 2019, 2020). This is *not subduction* because it only involves irregular ductile masses or parts of the lowermost crust detaching downwards from the base of the crust. There is no evidence that the drips were extensive slabs in three dimensions. Lower crustal drips are a geochemical construct based on the need to generate high-pressure TTG suite plutons by partial melting of mafic lower crust and upper mantle at depths and pressures below the base of the thick Archean mafic crust, and thus as a mechanism to generate depleted upper mantle. This has been discussed further in Sect. 17.7.2.6.

17.7.5.5 Mobile Lid-Duction and Sub-cretion Tectonics—Embryonic Local Plate Tectonics (Meso-Neo-Archean)

Between the hot, ductile crust/lithosphere stagnant lid regime of the Hadean and at least Meso-Archean and the rigid cold(er) plate tectonic regime of the Proterozoic and Phanerozoic, there was probably a transition period of localised "*mobile or squishy lid*", or embryonic attempted plate subduction (Bédard 2018; Cawood et al. 2018). Opinions on when plate tectonics commenced on Earth vary. Some suggest the period from 3.2 to 3.0 Ga based on increasing felsification of the cratonic crust (i.e. a change from mafic crustal composition to intermediate crustal composition), accompanied by increasing Rb/Sr ratios of the crust, increasing volume of continental crust, increasingly more potassic nature of TTG suites through the Archean, increasing preservation of shallow marine to subaerial sedimentary successions (e.g. Cawood et al. 2018; Hawkesworth et al. 2020; Piccolo et al. 2020). Shirey and Richardson (2011) and Van Kranendonk (2011) report that dated sub-cratonic mantle mineral inclusions in diamonds are peridotitic before 3 Ga and eclogitic after 3 Ga, which they interpret as the first evidence of deep subduction of mafic crust to generate high-pressure eclogite. Although this is plausible, it is also possible the as the Earth and mantle cooled the solidus dropped and became deeper, so limiting the amount of (partial) melting of slabs of detached, delaminated lower mafic crust descending downwards into the mantle. By 3 Ga they could then metamorphose to eclogite. Other suggestions for the onset of subduction are the end of the Archean at ~2.5 Ga, based on Archean shales showing less enrichment in light REE and no Eu anomaly compared with post-Archean sediments, attributed to sources dominated by mantle-derived rocks in the Archean and continental cratonic sources after the Archean (Taylor and McLennan 1995), and geological evidence of the lack of typical subduction-related geotectonic elements in Archean cratons (Bédard 2006, 2018; Bédard et al. 2013).

Bédard et al. (2013) and Bédard (2018) suggested that drifting continents, driven by mantle convection currents sub-(ac)creted inclined or imbricated slabs of oceanic plateaus or crust at their leading edge without leading to deep subduction (Fig. 17.73) because the crustal slabs were too buoyant and not rigid enough in the hotter Archean Earth. Nonetheless, the deeper parts of accreting slabs may undergo

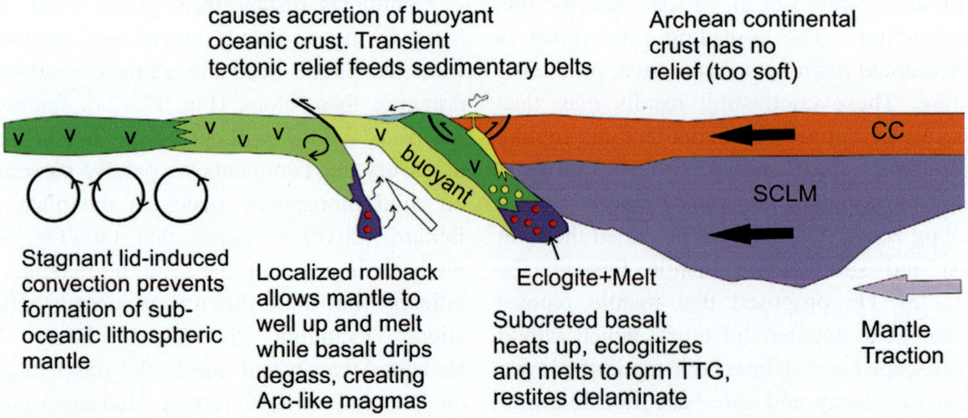

Fig. 17.73 Model of sub-(ac)cretion of slabs of mafic oceanic crust (*not* formed by modern seafloor spreading processes) against the leading edge of drifting continental crust. Because of the buoyancy of oceanic crust in hotter mantle, slabs do not descend far into the mantle, but they may undergo partial melting to generate TTG magmas. (From Bédard 2018)

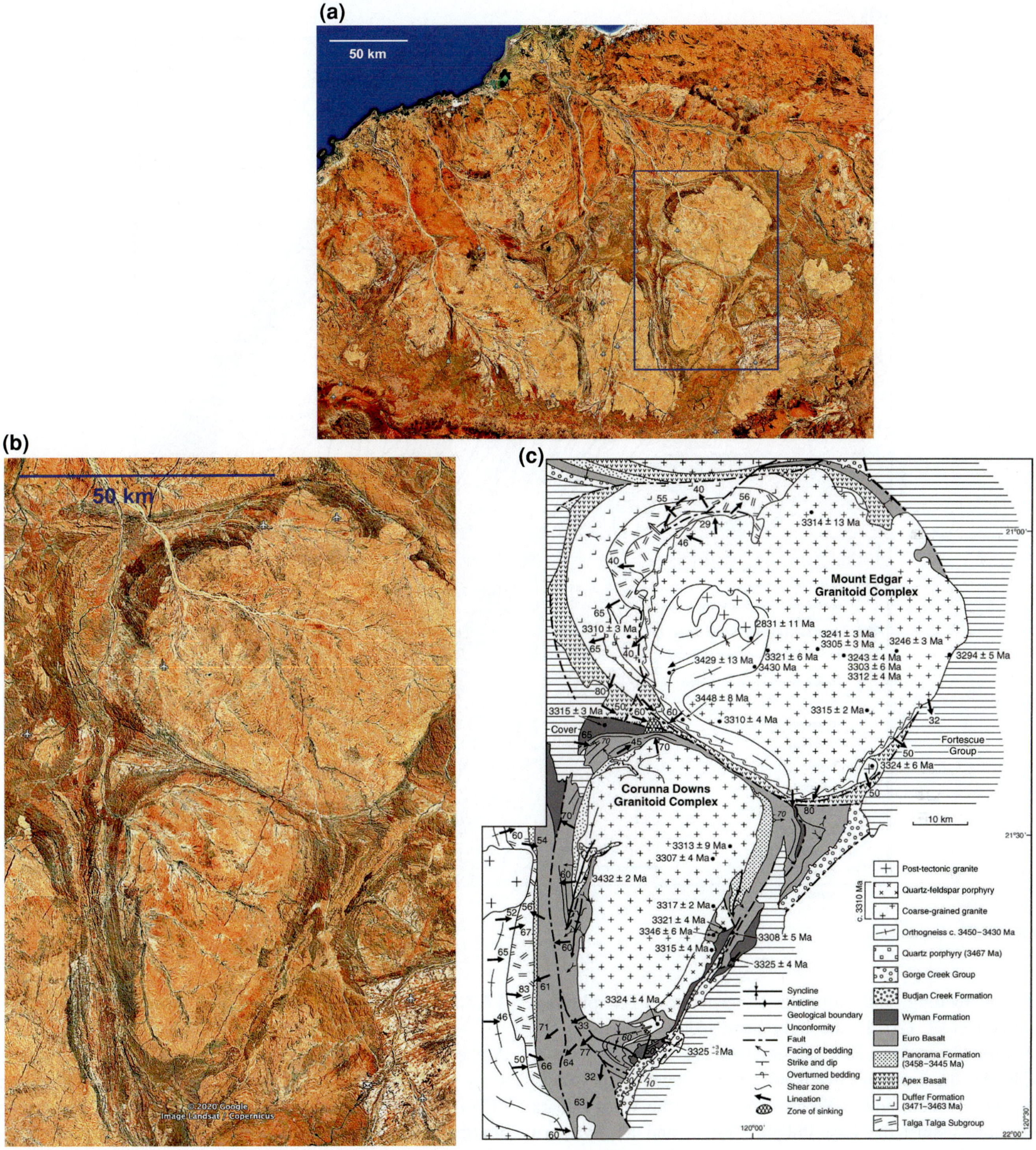

Fig. 17.74 Relationships between granitoid plutons and greenstone belts in Archean cratons. **a** Google Earth image of the Pilbara Craton, northwestern Western Australia. This highlights the pale orange granitoid domes and their close spacing, separated and surrounded by narrow dark grey-brown greenstone belts. (*Source* Google Earth). **b** Close-up Google Earth view of the area in blue box in (**a**), showing two of the plutons in the eastern Pilbara Craton and the stratified nature of the greenstone belts, with strata wrapping around the margins of the plutons. **c** Interpretative geology map of the area in (**b**). Bedding and younging directions (pale grey, thin arrows) radiate away from the pluton margins. Similarly, mineral lineations (bold black arrows) pitch/plunge radially away from and around the margins of the plutons. (From Van Kranendonk et al. 2004). **d** Structural map of the southern part of the Kalgoorlie and Kurnalpi terranes, eastern Yilgarn Craton, Western Australia, shown in Fig. 17.68c. Note that stratigraphic marker ultramafic units, and S1 foliation (red dotted lines) generally wrap around the margins of granitoid plutons (pink), and bedding/younging (blue arrows) and L_1 lineation (red arrows) generally radiate away from the margins of the plutons. The regional compressional D_2–D_3 deformation S_2 cleavage trends NNW and cuts across stratigraphic trends and pluton margins, as do $D_{3\text{-}4}$ shears. (From Jones et al. 2020)

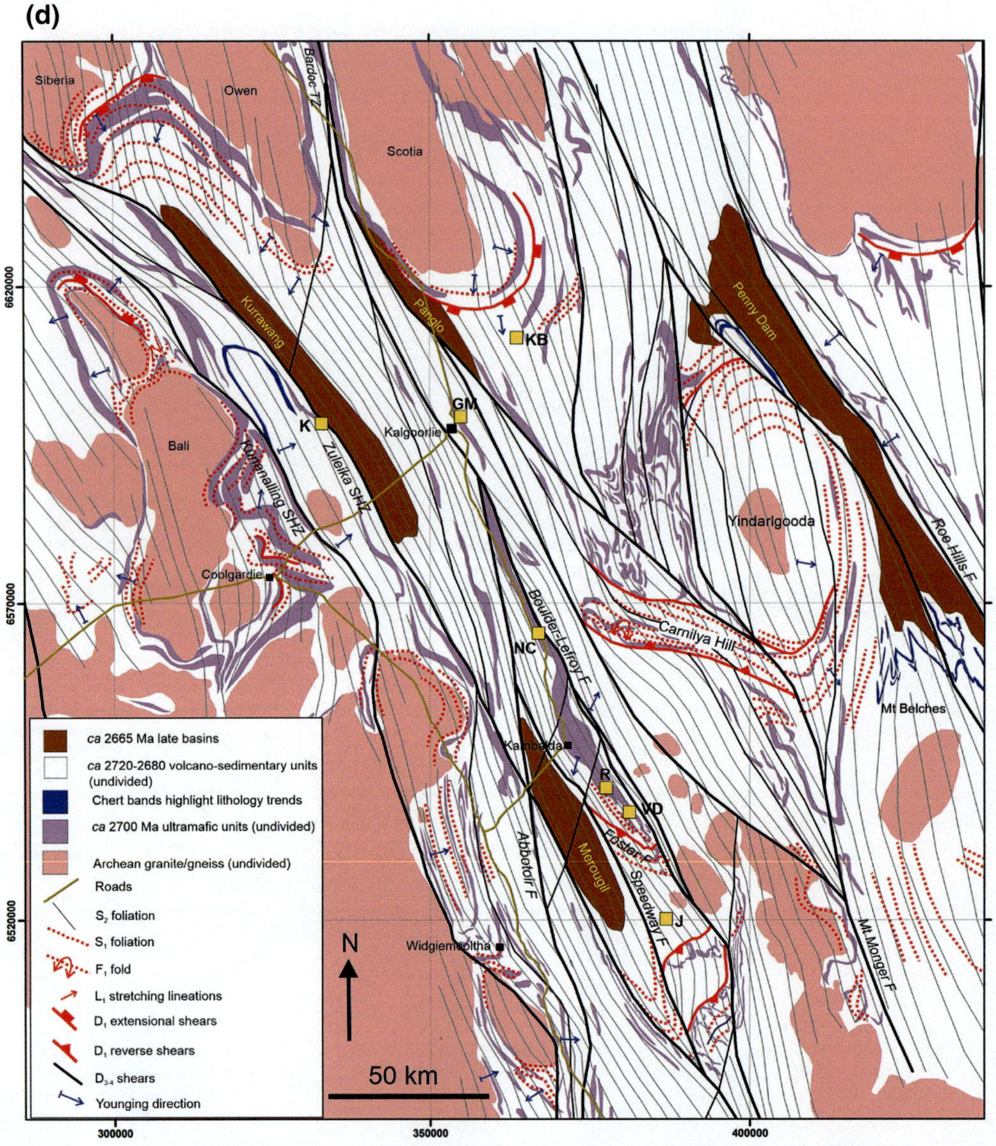

Fig. 17.74 (continued)

partial melting and dripping to generate TTG plutons. Bédard et al. (2013) cite imbricated tectonic slices in the Abitibi Greenstone Belt marginal to the Superior Craton, Canada, the western Greenland terranes, against the North Atlantic Craton, and in Western Australia, the West Pilbara terranes against the East Pilbara Craton, and the Western Gneiss Terrane against the rest of the Yilgarn Craton as examples of possible sub-creted tectonic domains.

On the other hand, O'Neill et al. (2013) proposed that plate(let) subduction could have occurred episodically throughout the Archean. Piccolo et al. (2020) propose that a crude form of subduction is likely to have commenced around the margins of extending continents above plumes once the Earth and its mantle and crust had cooled to the point where a rigid subductable crust had formed.

17.7.5.6 Pluton Diapirism and Greenstone Belt Sagduction Tectonics

Archean cratons are marked by granitoid dome and greenstone keel architecture, consisting of voluminous granitoid (TTG) plutons enclosed by folded supracrustal greenstone stratigraphic successions (Fig. 17.74a–d; Anhaeusser et al. 1969; Archibald et al. 1978; Gorman et al. 1978; Platt 1980; Anhaeusser 1984, 2014; Campbell and Hill 1988; Swager and Griffin 1990; Shackleton 1995; Collins et al. 1998; Bleeker 2002; Dirks et al. 2002; Van Kranendonk et al. 2004; Bédard 2006; Bédard et al. 2013; Kamber 2015; Thurston 2015; Jones et al. 2020; Zibra et al. 2020). Granitoid domes cover ~60–70% of the surface area of cratons (Figs. 17.68a, b and 17.74a) and often intrude supracrustal sequences of the same age (Fig. 17.75). That is, the granitoid

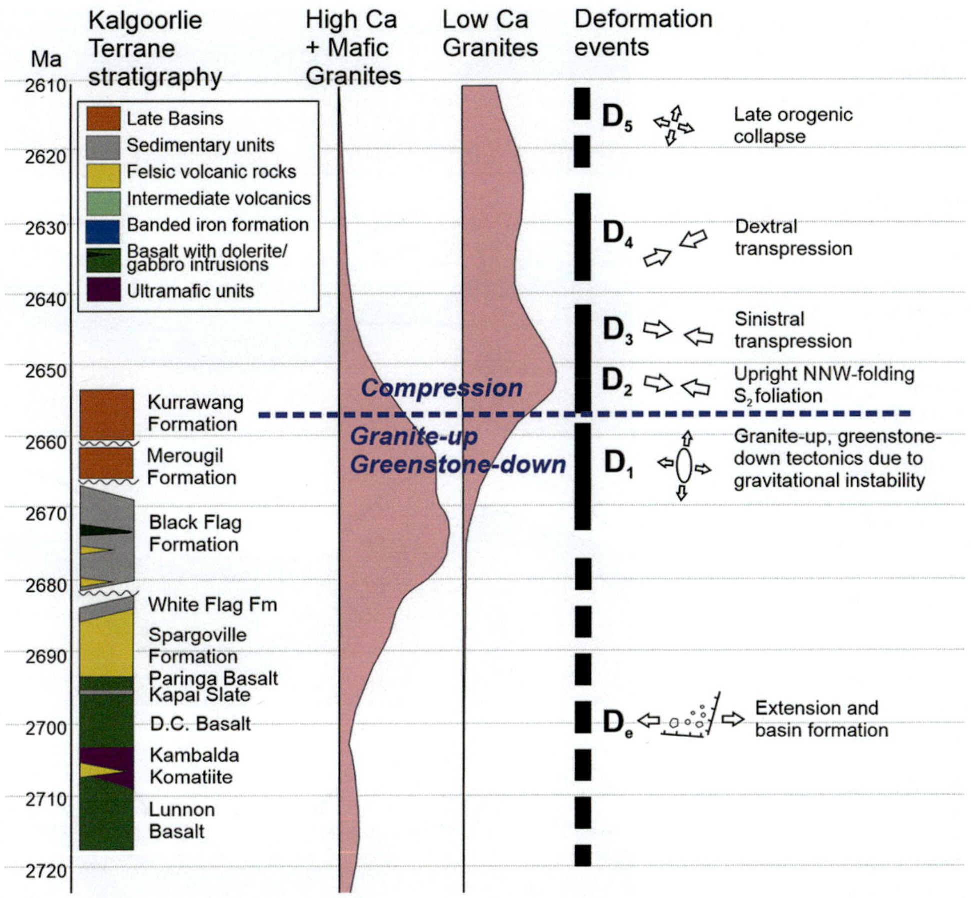

Fig. 17.75 Temporal relationships between the stratigraphic units, granitoids and deformational events of the Kalgoorlie Terrane, Eastern Goldfields Province, Western Australia. These show that plutonism was occurring contemporaneously with volcanism and sedimentation during a regional extensional or atectonic regime, and that regional compressional events did not begin until the deposition of the Late Basin Kurrawang Formation, which is a subaerial, fluvial-alluvial polymictic conglomerate and sandstone succession, resulting from regional compression and uplift, which may represent the onset of plate tectonic subduction and collision in the Eastern Yilgarn Craton. All other stratigraphic units, except the tuffaceous Merougil Formation, were emplaced in deep-marine environments. The Merougil Formation was deposited in a braid-plain environment and is the product of a major explosive event related to emergence of a granitoid pluton. (From Jones et al. 2020, After Czarnota et al. 2008)

plutons are coeval with some of the supracrustal volcanic rocks they are intruding, and in some cases, they are co-magmatic with them (e.g. Smithies et al. 2019). However, a long-standing debate over the decades has been the extent to which vertical or autochthonous tectonics (i.e. granite diapirism) versus horizontal allochthonous tectonics (i.e. thrusting) has contributed to the evolution and final tectonic architecture of granite dome-greenstone synclinal terrains. That is, in a "nut shell", were greenstone belts "squeezed" between granitoid plutons during later regional compressional deformation, or did a significant part of that deformation occur early, contemporaneous with vertical pluton emplacement and accommodating marginal greenstone downsag deformation, which was subsequently accentuated during regional compressional/transpressional deformation (as per references above; for recent reviews, see Anhaeusser 2014; Kamber 2015; Thurston 2015)?

Deep seismic reflection data which have become more readily available, especially from the Kalgoorlie Terrane of the eastern Yilgarn Craton, have suggested more complicated crustal geometries characterised by domes and antiformal structures (Blewett and Hitchman 2006; Blewett et al. 2010a). These domal structures have been interpreted as thrust controlled duplexes or antiformal thrust stacks (Drummond et al. 2000; Goleby et al. 2002), but more recently are thought to represent core complexes formed during regional extension (Blewett and Hitchman 2006; Blewett et al. 2010a). Core complexes have also been reported from other granite-greenstone terrains, such as the southern Barberton Mountain Land, Kaapvaal Craton, South Africa (Kisters et al. 2003).

Nonetheless, detailed field-based structural studies of the margins of some such granitoid domes have also demonstrated that there are relations and fabrics which precede the

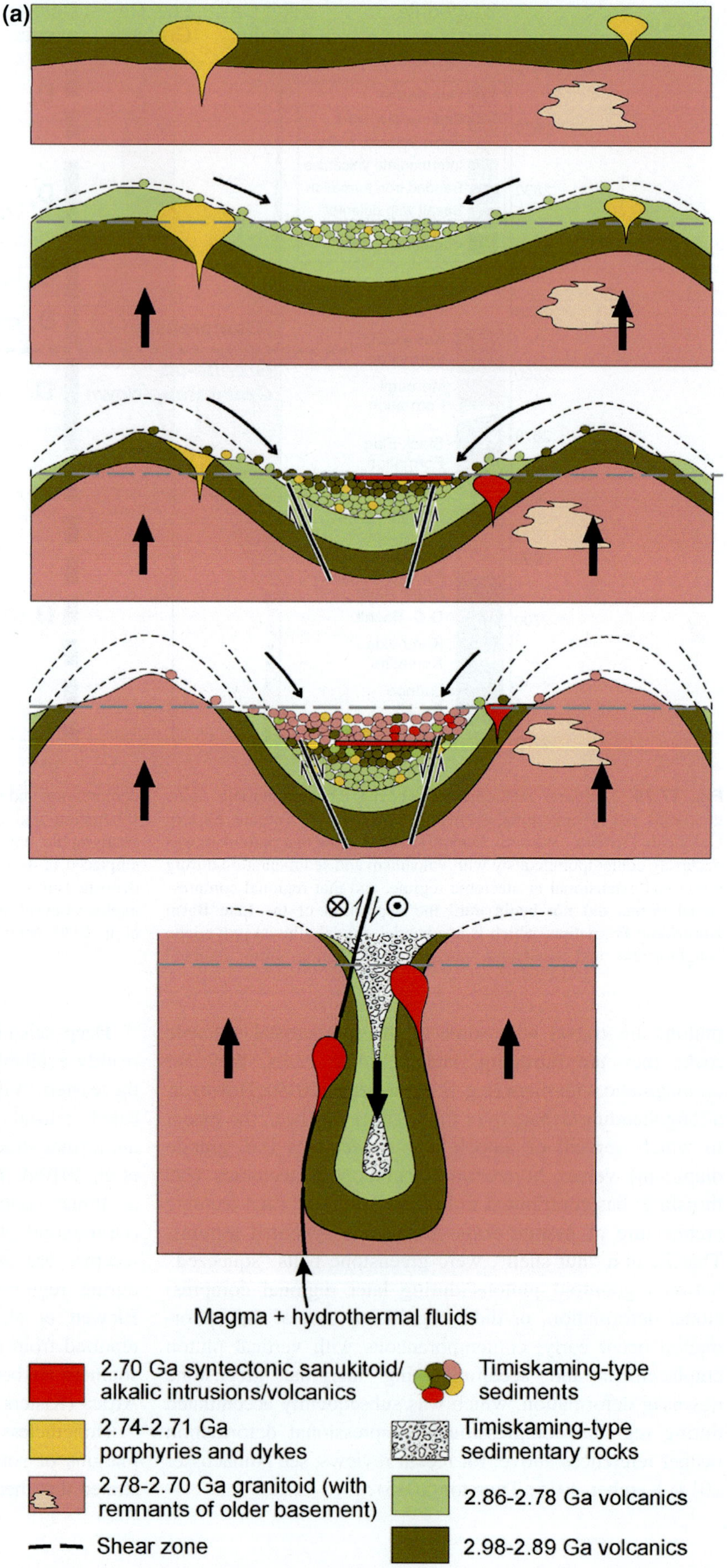

Fig. 17.76 Granitoid diapirism and associated sagduction. **a** Model of granitoid pluton diapirism and contemporaneous sagduction of supracrustal volcanic and sedimentary successions. This leads (From top to bottom) to emergence of the pluton top, erosion and sedimentation of pluton derived detritus in adjacent sedimentary late basins which progressively sag, deform and are sagducted. Based on the 2.78–2.7 Ga, Island Lake Group, northwestern Superior Craton, Canada. (From Lin et al. 2013). **b** Diapiric and sagduction model for plutons and greenstone belts showing simultaneous pluton uprise and greenstone downsag during a time of extensional or no tectonic stress field (early D_1). This leads to sedimentation in a supra-greenstone sedimentary Late Basins (late D_1 subsidence and uplift). Followed by the onset of regional compressional stress causing regional penetrative deformation ($D_{2\text{-}3}$). Based on the 2.69–2.65 Ga, Kalgoorlie and Kurnalpi terranes, Yilgarn Craton, Western Australia. (From Jones et al. 2020). **c** Progressive diapiric pluton uplift and derived volcanism, simultaneous greenstone uplift and unroofing, and downsag deformation (Vivien Formation). This leads to progressive erosion of uplifting greenstone stratigraphy (Maria Mine Formation), as well as unroofing and erosion of the pluton system (Scotty Creek Formation). Based on the 2.69–2.65 Ga, Agnew area, Yilgarn Craton, Western Australia. (From Hayman et al. 2020)

Fig. 17.76 (continued)

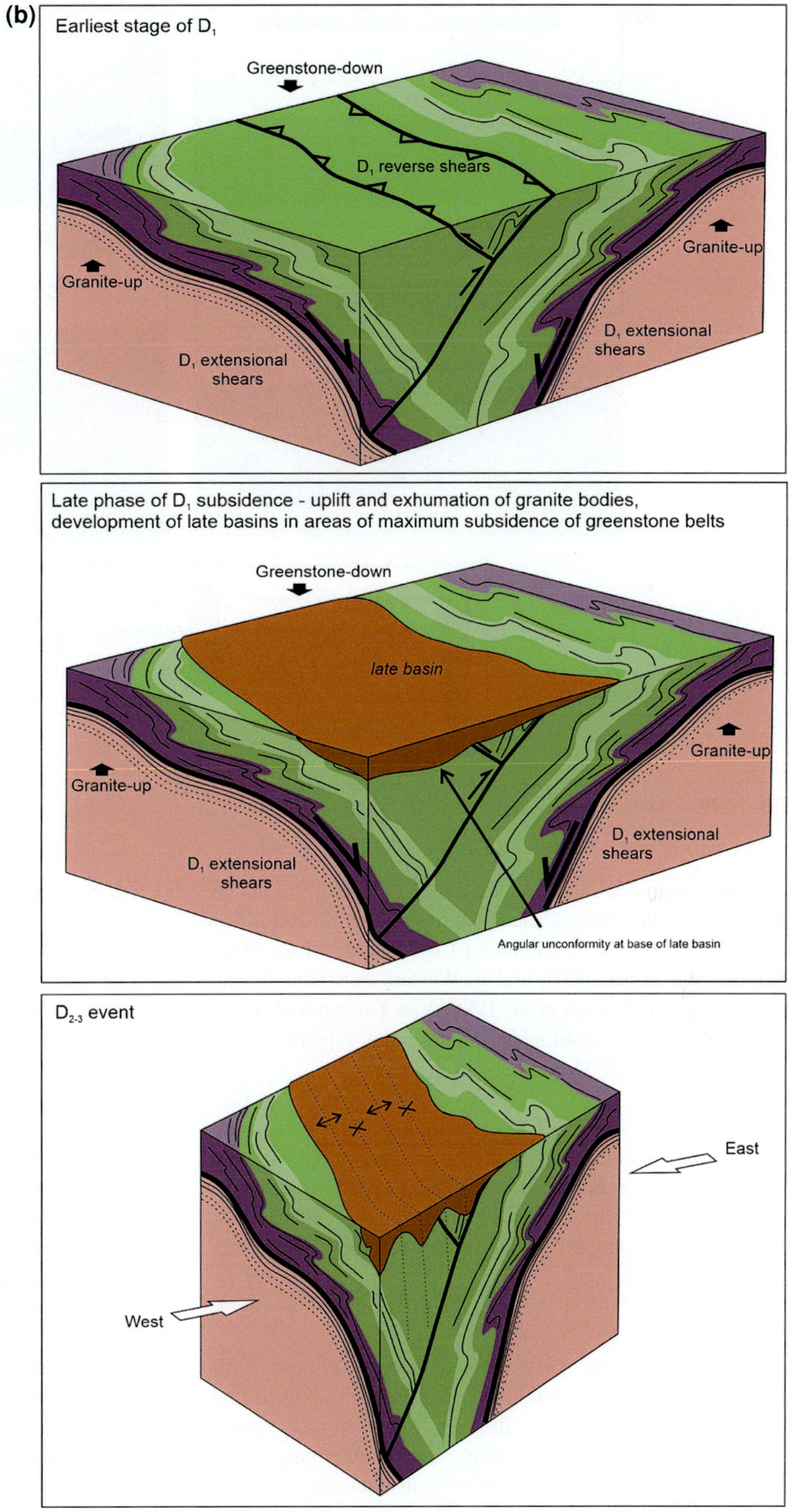

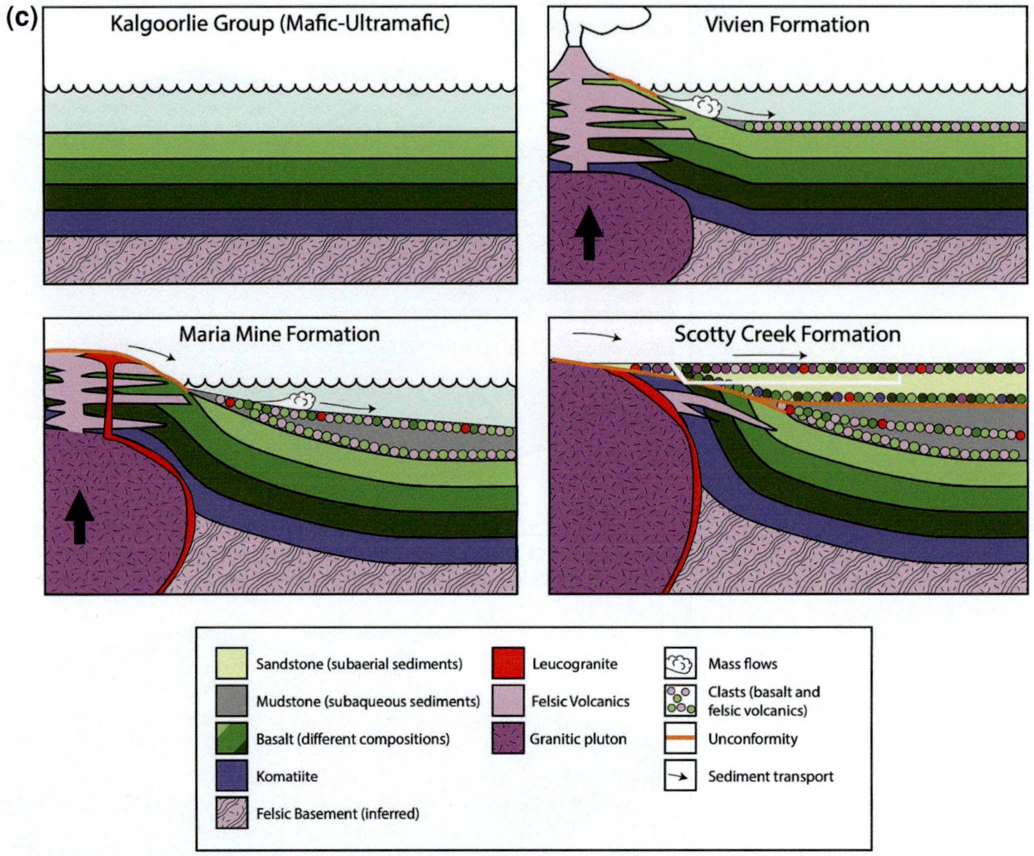

Fig. 17.76 (continued)

regional compressional/transpressional structural events in the Superior Province, Canada (Parmenter et al. 2006; Lin et al. 2013), Barberton Mountain Land, South Africa (Van Kranendonk et al. 2009), and in Western Australia, in the Pilbara Craton (Collins et al. 1998; Van Kranendonk et al. 2004) and Eastern Goldfields Superterrane, Yilgarn Craton (Jones et al. 2020). These include:

(a) Greenstone rocks are often tightly folded between plutons into upright, faulted synclinal keels, and are typically composed of a coherent, upward-younging stratigraphy which faces away from the granite domes, with the youngest rocks in the centre of the keels (Figs. 17.74c and 17.76a, b; Van Kranendonk et al. 2004; Parmenter et al. 2006; Lin et al. 2013; Jones et al. 2020).

(b) Chaotic internal geometries of granite plutons with otherwise simple outlines, ring faults along dome margins and in flanking greenstone belts, mushroom-shaped fold flaps around the margins of some plutons, and sedimentation in inter-diapir synclines (Fig. 17.74c; Van Kranendonk et al. 2004).

(c) An S_1 foliation (axial planar to F_1 folds) occurs at the margins of the granite plutons and in the adjacent greenstones, is concentric around the pluton margin and dips outwards, as does the vergence of F_1 recumbent isoclinal folds (Figs. 17.74d and 17.76b; Parmenter et al. 2006; Van Kranendonk et al. 2004; Jones et al. 2020).

(d) Where D1 extensional shear zones (Sh_1) occur associated with pluton margins they are also concentric, outwards dipping, and kinematically indicate pluton up, greenstone-down relative motions (Fig. 17.74d; Jones et al. 2020).

(e) Steeply plunging L_1 metamorphic mineral stretching lineations are oriented radially axisymmetric around the margins of the plutons (Fig. 17.74c, d; Van Kranendonk et al. 2004; Jones et al. 2020).

(f) S_0 bedding is concentric around the pluton; dips and younging are radially axisymmetric outwards from the pluton margins (Fig. 17.74c, d; Van Kranendonk et al. 2004; Jones et al. 2020).

(g) Later regional S_2 compressive foliation and Sh_2 shear zones cross-cut all D_1 fabrics and are regionally co-planar with regional D_2–D_3 folds (Fig. 17.74d; Parmenter et al. 2006; Jones et al. 2020).

These studies contend this type of deformation results from crustal overturn due to density inversions between buoyant (partially molten) granitoid intrusions and cooler, denser greenstones. Mechanically, the diapiric intrusive emplacement of domal granitoid plutons into supracrustal sequences (Fig. 17.76a–c) occurs as they create their own stress fields and structural fabrics, unrelated to any regional deformational stress field (e.g. Van Kranendonk et al. 2004; Jones et al. 2020). While occurring broadly contemporaneously over a significant geographical area, they are not *per se* a regional tectonic or deformational event. Thus, if multiple plutons are emplaced at different times, they will all generate their own structural fabrics that are similar to those of others but are each of different ages. This raises concerns about the danger of traditional structural methodologies of simply establishing chronologies of fabrics in different locations and correlating them regionally as the same regional tectonic deformational event (e.g. eastern Yilgarn Craton, Jones et al. 2020; see Sect. 17.7.8). These fabrics need to be clearly distinguished from truly regional deformational tectonic events, and age dating at each location is fundamental to that.

In addition, Squire et al. (2010), Lin et al. (2013), McGoldrick et al. (2013), and Hayman et al. (2020) have documented the provenance of many volcaniclastic sediment sequences contemporaneous with pluton emplacement and peripheral to them. They demonstrated that the detritus, such as conglomerate clasts and mineral fragments (including zircons), are derived directly from the erosion of plutons emerging above sea level and the landscape and were deposited directly into surrounding "late" basins (Fig. 17.76a, c). Squire et al. (2010), Smithies et al. (2019), and Hayman et al. (2020) have also shown that some of the emerging diapiric plutons sourced major felsic explosive eruptions that shed huge volumes of pyroclastic debris into surrounding basins.

The contemporaneous vertical downsag of supracrustal sequences around the margins of simultaneously, diapirically uprising plutons (Fig. 17.76a–c) was first recognised decades ago (Anhaeusser et al. 1969) in the Kaapvaal and Zimbabwe cratons in southern Africa, at a time when the significance of horizontal tectonics associated with plate tectonics was gathering steam. The term "sagduction" was coined by Goodwin and Smith (1980) for the downsag of greenstone belts sandwiched between rising granitoid plutons in the northwestern Superior Craton. The concept has also been applied in the Dharwar Craton in India (Bouhallier et al. 1995; Bhaskar Rao et al. 2020), and the Pilbara (Collins et al. 1998; Van Kranendonk et al. 2004) and eastern Yilgarn cratons (Jones et al. 2020; Zibra et al. 2020) cratons in Western Australia.

Kinematically, did greenstones sink down triggering complementary upwards rising pluton diapirism, as suggested in the Pilbara Craton? There a 5–10 km thick pile of dense basalts and komatiites were erupted onto the lower density felsic crust, insulating it from normal heat loss, accentuated by mantle plume heat from beneath causing heating of the crust, and triggering gravitational instability (e.g. Collins et al. 1998; Van Kranendonk et al. 2004). This could also be the case in the Yilgarn Craton and the Superior Province. Alternatively, is it simply the buoyancy of granitic magma that drives low-density magma to rise through denser crust, which in turn sinks around the rising diapirs (Bleeker 2002; Squire et al. 2010; Lin et al. 2013; McGoldrick et al. 2013; Hayman et al. 2020), or both?

17.7.5.7 And then There is …. Plate Tectonics

The majority of researchers agree that plate tectonics as we know it today did not fully develop until at least the late Archean or more likely the Proterozoic (e.g. Bédard 2006, 2018; Chowdury et al. 2017; Cawood et al. 2018, Condie 2018; Hawkesworth et al. 2020, Brown and Johnson 2019; Brown et al. 2020; Palin et al. 2020; Piccolo et al. 2020). Its onset required the Earth to have cooled enough so that the evolving crust and lithosphere were: (1) sufficiently cold to behave relatively rigidly over large areas of the Earth's surface, (2) for the crust and uppermost or lithospheric mantle to evolve and differentiate, and (3) for the mantle solidus to become deeper as the Earth's mantle cooled. A colder more rigid lithosphere and crust and a deeper solidus would allow high-pressure metamorphism of the subducting solid lithosphere to eclogite facies at depth, so causing the down-going plate to become denser at depth and then exert gravitational slab pull. This in turn would produce extensional forces elsewhere in the lithosphere so that rifting, pulling apart and mantle upwelling could initiate extensive seafloor spreading ridge systems to form. Simultaneously mantle plumes have remained fundamentally important in affecting what happens to the lithosphere.

The fundamental tenets of the modern plate tectonic paradigm are seafloor spreading and subduction (Sects. 17.2, 17.5; Plate 17.1, Figs. 17.1 and 17.2). Seafloor spreading that generates new oceanic crust and lithosphere (plates) represented dynamically and geologically by the sheeted dyke complex (layer 2C) in the stratigraphy of oceanic crust and lithosphere, which represent continuous lateral accretion through vertical dyke injections of magma at the mid-oceanic ridges as plates spread away from ridges under the extensional stress regime at ridges (Sect. 17.2; Figs. 17.2 and 17.3).

Subduction of plates involves recycling plates back into the mantle and is represented by the trench, forearc off-scraping of layer 1 sediments from the down-going plate producing accretionary prisms and tectonic mélange wedges, high-pressure blueschist metamorphic assemblages at depth, and forearc and arc magmatism (Sect. 17.5.1; e.g. Fig. 17.6a). Typical forearc volcanism in modern subduction generated oceanic volcanic arcs is represented by boninites, although boninites can also form in back-arc settings (Embley and Rubin 2018), whereas arc volcanism is represented by island arc tholeiites in juvenile arcs, and by adakites in more evolved arc settings. However, as mentioned in Sect. 17.5.4.1 adakites are also no longer considered to be necessarily diagnostic of arc magmatism, and can form through fractional crystallisation of mantle-derived magma that assimilates country rock, fractional crystallisation of magma derived from metasomatised mantle, partial melting of thickened lower crust, and partial melting of slabs of delaminated lower crust which interact with mantle peridotite (Hastie 2020; Zhang et al. 2021) . Also, some workers clearly cast doubt on using tectono-magmatic geochemical discrimination diagrams to fingerprint the tectonic setting of ancient volcanic rocks, calling such diagrams "indiscrimination" plots (Li et al. 2015), or questioning their indiscriminate use (Condie 2015; Zhang et al. 2021). Such approaches, in isolation, are not diagnostic because there is too much overlap between the compositional fields for magmas from different tectonic settings. Given the uncertainty that exists about the tectonic regimes in the Archean, tectono-magmatic geochemical affinity diagrams based on Cenozoic plate tectonic compositions are not likely to be very useful for Archean rocks (Condie 2015; Sect. 17.8.3).

In addition, mantle plumes, producing oceanic plateaus, flood basalt provinces, hot spot tracks or lines of oceanic islands and of continental volcanoes, are also a fundamental part of the modern plate tectonic framework. In many cases plume volcanism is a precursor to and heralds the break-up of supercontinents, in other cases it occurs in intraplate settings, both oceanic and continental.

17.7.6 Plate Tectonic Elements Missing in Archean Cratons

To evaluate if plate tectonics operated during the Hadean and Archean, it is reasonable to seek geological evidence in the ancient rock record that supports the fundamental processes associated with plate tectonics.

Bickle et al. (1994) recognised that although Archean greenstone belts contain abundant mafic and ultramafic volcanic and co-genetic intrusive associations, these do not preserve the full stratigraphy and therefore, the dynamic elements of modern oceanic lithosphere. In particular, sheeted dykes are absent, and although ophiolites, which are obducted slices of complete oceanic lithosphere are found in Phanerozoic and Proterozoic orogenic belts, they are not unequivocally known in Archean greenstone belts. Palin et al. (2020) also point out that the only unequivocal ophiolites are <2 Ga old, raising the fundamental question: "*did lithospheric plates and plate tectonic processes only begin in the Proterozoic?*".

For the lithosphere and lithospheric plates to exist there has to be an asthenosphere that separates the lithosphere from the deeper mantle and facilitates the detachment gliding of the lithosphere over the deeper mantle. So does the absence of Archean ophiolites imply a lack of a lithosphere and asthenosphere, or was the lithosphere much thicker and the asthenosphere much deeper, so that it was rheologically impossible for the lithosphere to break into plates and for plates to collapse gravitationally, so inducing plate subduction? Subduction and slab pull are necessary to then induce extensional rifting of the lithosphere elsewhere, that in turn would induce upwelling of mantle magma, injection of sheeted dykes into the rifting lithosphere, forming spreading ridges.

In addition, for plate tectonic processes to have occurred, Archean terrains need to have a geometrical asymmetry or polarity, with an identifiable subduction accretionary complex, and a parallel, linear, relict volcanic arc magmatic belt, and a back-arc region (Bédard et al. 2013). The superimposed deformational fabric may or may not be parallel to the original arc trend, and so the only way to reconstruct the original spatial relationships of geological elements is to palinspastically undo major fault displacements, including thrusts normal listric faults, strike-slip or transcurrent faults, as well as regional scale folds (Sect. 17.8). We are not aware of any study that has attempted this.

Palin et al. (2020) have evaluated all the geological requirements of the plate tectonic paradigm and tried to evaluate how they would have been affected by the different physical regime of the early Earth. Some of the factors they have assessed are now discussed.

17.7.6.1 Blueschist Metamorphic Assemblages

Bickle et al. (1994) pointed out the absence of high-pressure blueschist domains in Archean terrains, and in addition that no convincing associated accretionary prism and tectonic mélange zones have been documented. According to Palin et al. (2020) the oldest dated high pressure (<2.3 GPa; = ~75 km)–low temperature (<500 °C; ~70 km) blueschist rocks are NeoProterozoic (~0.8 Ga) in age. This could be real, suggesting there was no deep subduction prior to this, or it could be due to a lack of preservation/exposure of older blueschist rocks, or due to higher geotherms in earlier subduction settings, so preventing the formation of key indicator minerals such as glaucophane, or sodic amphibole in mafic protoliths. However: (1) the density of mafic crust and mantle would have been less in a hotter Earth, so the lithostatic pressure gradient would have been less, requiring a greater depth than at present to generate the high pressure required, and (2) the geothermal gradient would have been higher, producing higher temperatures at shallower levels in the Earth than at present. Alternatively, there may have been a compositional control on blueschist formation in the Archean and older Proterozoic, whereby high-MgO basalts typical of Precambrian ocean crust would not have been able to stabilise sodic amphibole (Palin and White 2016). Instead, the subduction of MgO-rich oceanic crust would have created greenschist-like rocks formed today at low temperatures and pressures. Jadeitite and lawsonite, other blueschist marker minerals have only be found in Phanerozoic blueschist rocks (Palin et al. 2020).

17.7.6.2 High Pressure to Ultra-High-Pressure Minerals and Rocks

The transition of quartz to coesite under ultra-high-pressure conditions of ~2.7 Ga (>100 km depth) and 500–900 °C has been cited as evidence for deep subduction, but the oldest rocks with coesite are 620 Ma (Palin et al. 2020). There is also now some doubt that such conditions to produce coesite are unique, with evidence of formation at <40 km in deformed continental crust (Palin et al. 2020). Similarly, eclogite can form at pressures of ~1.6 Ga, temperatures of ~750 °C, and has been found in rocks as old as Meso-Archean (Palin et al. 2020). However, given our discussion above of delamination or dripping of lower mafic crust during the Archean, slabs of such mafic crust could have sunk deep enough into the mantle to metamorphose to eclogite facies.

17.7.6.3 Accretionary Prisms and Tectonic Mélanges

Accretionary prisms (Sects. 17.5.1–17.5.4) are *en echelon*, stacked thrust slices of oceanic sediments and sometimes layer 2 basalts scraped off the down-going lithospheric plate in the subduction zone and accreted against the leading edge of the overriding plate. These prisms can sometimes include slices or lenses of highly foliated, chaotically tectonically mixed lithosomes of lithologies associated with subduction settings, including meta-sediments, volcanics, ophiolite, and metamorphic rocks discussed above, called tectonic mélanges. According to Palin et al. (2020), unequivocal accretionary prisms older than ~0.9 Ga have not been found, although several Archean rock associations have been suggested as examples.

17.7.7 Summary of the Evolution of Tectono-Magmatic Processes and Regimes During the Hadean and Archean

The changes from the beginning of Earth history are numerous, but not surprising given that we started with a molten Earth at 4.54 Ga (Fig. 17.77). Earth was initially a molten magma ocean in the Eo-Hadean, with convective overturn of the cooling plastic, probably komatiitic proto-crust (Figs. 17.70 and 17.77), much like in a large lava lake. Then the Earth formed a more or less stagnant lid mafic–ultramafic crust through the Neo-Hadean and most of the Archean (Figs. 17.71 and 17.77), initially underlain

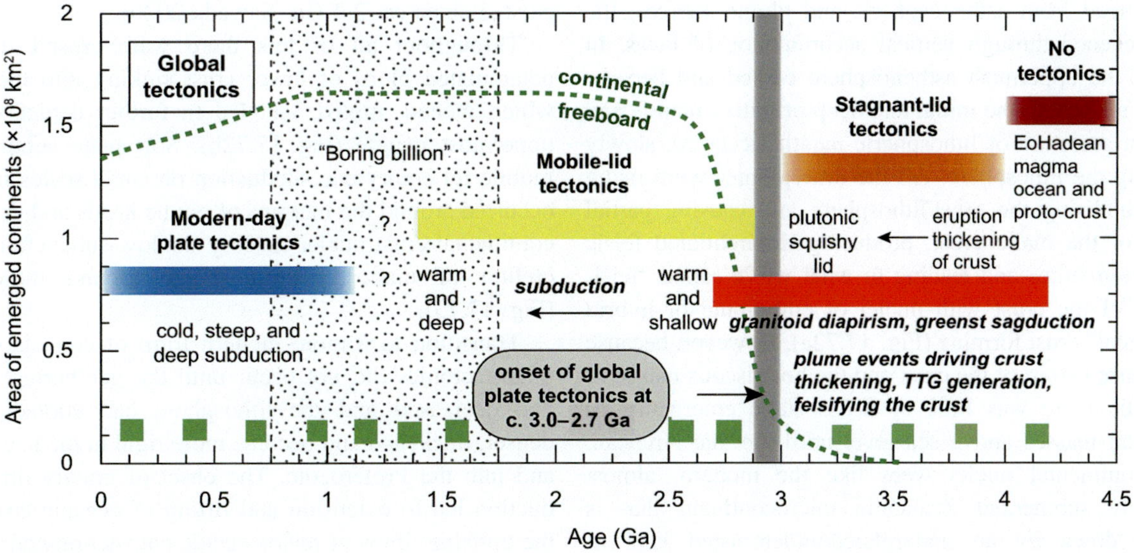

Fig. 17.77 Synthesis of tectonic evolution of the Earth, highlighting the major tectonic regimes and their approximate age ranges (Modified from Palin et al. 2020)

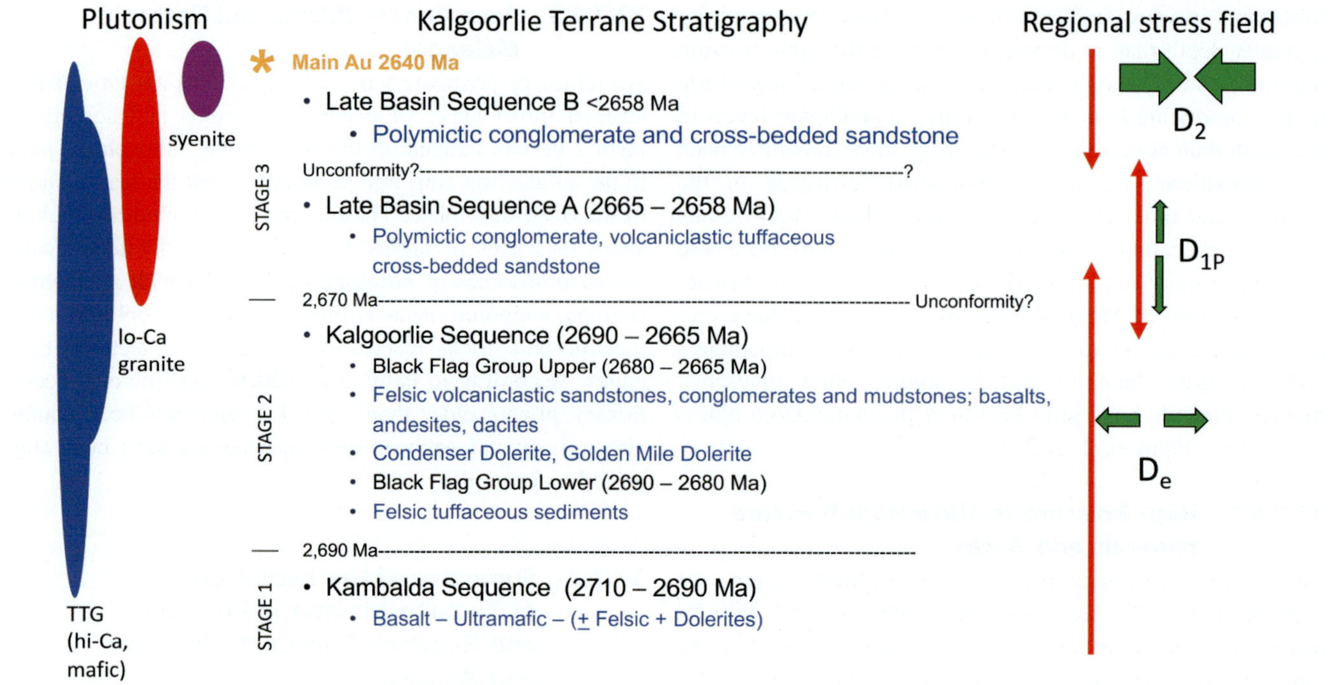

Fig. 17.78 Summary of the stratigraphy of the Kalgoorlie Terrane (>2.7–<2.658 Ga), Yilgarn Craton, Western Australia, also showing the timing of major granitoid pluton groups and regional stress regimes. D_e represents a possible weakly extensional or passive regime, probably generated by mantle plume spreading, or an atectonic regime; D_{1P} represents a vertical stress regime associated with diapiric uprise of granitoid plutons, and late downward sagduction of surrounding greenstone belts; and D_2 represents the first of several regional compressional tectonic events. Main gold mineralisation event (e.g. Golden Mile supergiant deposit) occurred over a relative short period of time at about 2.64 Ga and is spatially associated with a late phase of feldspar-phyric dykes which have similar orientations to the lode rocks (Vielreicher et al. 2016). The Au-bearing fluids were derived from low salinity, H_2O-CO_2 fluids of deeper, probable metamorphic origin (e.g. devolatilisation of mafic and ultramafic volcanic source rocks of the Kambalda Group), ± higher salinity fluids derived from the late more potassic (syenitic) plutons (Witt et al. 2020)

directly by the convecting asthenosphere, and punctuated by mantle plume events that erupted lavas onto the crust and thickened it. Progressively as eruptions occurred onto this crust sourced from asthenospheric and plume sources, the crust thickened through vertical accumulation of lavas. In addition, the uppermost asthenosphere cooled and became depleted relative to the initial and deep primitive mantle and top down accretion of lithospheric mantle occurred, slowly thickening the lithosphere. All the time plumes were rising and destabilising the crust/lithosphere and causing partial melting of the mafic crust, producing differentiated felsic TTG pluton suites and leading to what we've called "felsification" of the crust with nuclei of embryonic or hybrid "continental" crust forming (Fig. 17.72a). However, because of the dense nature of the crust and the hot viscous nature of the mantle there was little or no isostatic emergence of continental masses above sea level until the late Archean. These continental nuclei were like the modern, almost completely submerged Zealandia microcontinent that is weighed down by an underplated/undercrusted keel of basalt, although more on the scale of just the Campbell Plateau, southeast of New Zealand (Fig. 17.43b). During the Archean continents emerged with widths and lengths greater than 1,000 km, and only through further amalgamation have continents with widths and lengths of nearly 10,000 km existed since ca. 2.4 Ga (Sawada 2020).

Throughout all of this there were crustal drips and delaminated blobs of lower crust sinking into the mantle, which melted, mixed, and led to further depletion of the uppermost mantle (Fig. 17.72b). Soft plate subduction or mobile lid embryonic subduction on small scales may have occurred around the margins of plume heads and developing continental crust nuclei causing shallow subduction or subcretion of mafic crust/lithosphere around these nuclei (Fig. 17.73).

Transition to a more modern form of cold, larger scale plate tectonics did not occur until the geothermal gradient had decreased and the lithosphere had cooled, become denser, thickened and become more rigid in the late Archean and into the Proterozoic. The onset of gravity driven subduction led to extension and rifting of oceanic crust above the uprising limbs of major mantle convection cells, leading to the onset of seafloor spreading and the lateral accretion of oceanic crust at spreading ridges. It was only when a rigid crust had formed that brittle crustal deformation, thickening and uplift caused large areas of continental crust to become emergent above sea level. And then there's the story of the diapric uprise of TTG plutons into coeval supracrustal greenstone volcanic and sedimentary successions causing sagduction, uplift and erosional unroofing of these plutons without a regional tectonic compressional event in the Neo-Hadean and through the Archean (Figs 17.74 and 17.77).

Because many physical geological pieces of evidence of the various proposed tectonic regimes are not preserved, many of these regimes are based on geochemical numerical modelling, with assumptions about how much hotter the Earth's upper mantle may have been during the Hadean and Archean. Given that the Earth's interior had not differentiated very much, and certainly not to the degree reached today, comparison of geochemical signatures of Archean volcanic and plutonic rocks with Phanerozoic volcanic and plutonic rocks generated during cold, global plate tectonics, is fraught with problems.

17.7.8 A Case Study—The Neo-Archean Kalgoorlie Terrane, Eastern Goldfields Superterrane, Yilgarn Craton, Western Australia

The well-defined north-northwest structural grain of the late Archean Eastern Goldfields Superterrane (EGST) of the eastern Yilgarn Craton is made up of NNW-trending greenstone belts and granitoid domes separated by an anastomosing network of major shear zones (Figs. 17.68a, c, 17.75 and 17.78). Strain has been partitioned heterogeneously, with large areas of relatively intact greenstone stratigraphy (with weak fabric development), which envelop and dip and young away from broad, elongate, gently NNW–SSE-plunging granitoid domes. These areas contrast with intervening localised zones of shear-dominated high-strains up to 5 km in width with intense foliation, steep dips and dismembered stratigraphy. These high-strain zones are commonly areas of significant tectonic reworking and were subject to intense extensional and thrust and strike-slip (both sinistral and dextral) contractional deformation.

The regional tectonic history has been described and interpreted by numerous authors, and various competing deformational schemes have evolved (e.g. Swager 1989, 1997; Blewett et al. 2010b; Jones et al. 2020). Here we adopt that of Jones et al. (2020; Fig. 17.75), a slight variation of that of Swager (1989, 1997), as this corresponds most closely with our studies over the past 30 years and represents regional scale events best. The regional structural architecture is typically attributed to intense east–west compression/transpression during D_2 and D_3 events. The early pre-D_2 deformation history of the EGST is less well understood but incorporates the period of diapiric granitoid intrusion into supracrustal sequences (granite-up, greenstone-down D_1 tectonics) discussed in Sect. 17.7.5.6. Local extensional deformation thought to be related to early (greenstone) basin development is assigned to the D_e event. D_2 is characterised by upright north-trending folds and a well-developed axial planar S_2 fabric is widely recognised and is a key structural marker across the EGST.

Stratigraphically the Kalgoorlie Terrane (Fig. 17.73; >2.720–2.658 Ga) preserves a succession describing an evolution of magmas, eruption processes, paleoenvironments, sediment provenance, and deformation that are inconsistent with a plate tectonic setting, except perhaps at the very end. The succession represents a common cycle or trend in Archean greenstone belts, from mafic–ultramafic (Stage 1) to felsic magmatism (Stage 2), to late basin clastic sedimentation (Stage 3), each stage accompanied by granitoid plutonism, including TTG, which are now briefly described.

17.7.8.1 Deep Marine Volcanic Basin(s)

An early, deep submarine succession ($\sim$2.72–2.69 Ga) of komatiite (Fig. 17.79a) and pillow to massive tholeiitic and komatiitic basalt lavas (Fig. 17.79b), hyaloclastite breccias (Fig. 17.79c), co-magmatic subvolcanic intrusions, with intercalated chert, black mudstones (Fig. 17.79d) and minor felsic volcanic rock, represents *Stage 1*, and is thought to overlie an older poorly exposed greenstone belt sequence (Krapež et al. 2000; Squire et al. 2010; Jones et al. 2020; Austin et al. 2022; Figs. 17.75 and 17.78). The succession is 2–3 km thick and extends >600 km along strike (e.g. Figure 17.68c). It represents regionally widespread mantle plume volcanism that produced an oLIP (Barnes et al. 2012; Hayman et al. 2015; Austin et al. 2022), during a weak regional extensional (D_e) or passive tectonic stress regime that may have been initiated by the mantle plume at $\sim$2.80 Ga. Austin et al. (2022) propose that this succession is part of a regionally extensive mafic–ultramafic association ranging in age from about 2810–2710 Ma, and furthermore propose that this same tholeiitic association of the same age range is also recognisable on other cratons, including the Zimbabwe and Slave cratons. They suggest that this associations represents magmatism associated with a globally important rifting event.

The intercalated hemipelagic sediments (Fig. 17.79d, e) indicate that volcanism occurred in an open, anoxic, deep-water setting with no evidence of nearby emergent continents. The lack of pyroclastic deposits indicate that water depths were great enough (>1,000 m) such that hydrostatic pressure suppressed all explosive activity (Cas and Simmons 2018; Chaps. 4, 6, 9, and 11). All eruptive activity was therefore deep-water effusive. The

(a)
(b)
(c)
(d)

◀ **Fig. 17.79** Stratigraphic lithofacies of the *Stage 1*, >2.7–2.69 Ga Kambalda Sequence, Kalgoorlie Terrane, Yilgarn Craton, Western Australia. **a** Textural profile through a 10 m thick komatiite lava, from top to base, showing random spinifex (left), blade spinifex (middle), and orthocumulate (right) textures, Kambalda. **b** Pillow basalt in drill core, Jaguar Mine sequence. **c** High MgO basalt hyaloclastite breccia, Kambalda. All drill core is ~5 cm wide. **d** Interflow silicified, laminated, hemipelagic mudstone, Kambalda. **e** Silver Lake Peridotite Member, from the slopes of Red Hill, Kambalda. The poorly outcropping orange outcrops, numbered 1–3, represent silicified interflow sediments (as shown in **d** and long, time breaks between successive komatiite lavas (grey ground, not outcropping). The two komatiite lavas are each about 20 m thick, whereas sediment interval 1 sits above the basal komatiite, which is about 40 m thick at this locality. The sequence young away from the viewer. **f** Imaginary internal architecture within an Archean oLIP such as the thick, Kambalda Sequence sheet-like lavas as represented today by the stacked, laterally continuous sheet flood basalt lavas and thin, impersistent interflow sedimentary layers (arrows point to known horizons) of the Columbia River Flood Basalt Province. The sheet flow lobes observed in the cliff are ~20–40 m thick, and therefore comparable in scale to the komatiite lavas in (**e**). Grande Ronde lavas at Heller Bar, Washington, USA. (Photo and annotation by Steve Self)

paleogeography was therefore like a submerged flood basalt province (Fig. 17.79f). Some TTG magmatism (high-Ca granitoids, mafic granitoids) was beginning to occur contemporaneously with ultramafic and mafic volcanism.

17.7.8.2 Felsic TTG Magmatism

From ~2.69 to ~2.665 Ga (*Stage 2*) felsic TTG magmatism and volcanism was dominant (Figs. 17.75 and 17.78). Deposition of black to grey hemi-pelagic mudstones (Fig. 17.80a) continued, indicating a deep water, anoxic open ocean setting. Felsic volcanism is represented by submarine lavas, hyaloclastite, monomictic tuffaceous megaturbidites (Fig. 17.80b), polymictic volcanogenic turbidite and conglomeratic mass flow deposits (Fig. 17.80c), and contemporaneous high-level granitoid intrusions (Fig. 17.78). A significant mafic event at ~2.72–2.69 Ga indicates the ongoing influence of the mantle plume and heat that caused regionally extensive felsic crustal anatexis and magmatism. There is no clearly defined linear magmatic arc-like pattern. The felsic volcanics and volcaniclastics represent marine intra-basinal felsic volcanoes (Fig. 17.80d, e; e.g. lava domes, stratovolcanoes?). Some of these grew into shallow water, became emergent, and were explosive under low ambient pressures, producing the submarine tuffaceous megaturbidites. The *Stage 2* volcanic conglomerates, including some granitoid clasts (Fig. 17.80c), were not generated by orogenic deformation and uplift, but were derived exclusively from the intra-basinal volcanic centres and subvolcanic plutons, and from *Stage 1* komatiite and mafic stratigraphy that was up-domed and exposed by diapiric plutons. Ongoing plume buoyancy and contemporaneous granitoid diapirism caused regional and local uplift (D_{1P}), producing steep lineations, concentric foliations and shear zones around plutons (Jones et al. 2020), and shallowing of basins during ongoing regional extension (D_e).

17.7.8.3 Late Subaerial Basins and Continental Collision

From ~2.665 Ga a transition into more widespread subaerial paleoenvironments is represented by *Stage 3A*, or "early Late Basin" cross-bedded and massive felsic tuffaceous volcanic sediments (Fig. 17.81a), dominated by well-preserved bipyramidal volcanic quartz and angular feldspar crystal fragments, deposited in fluvial braid-plain, alluvial fan, shallow marine, and local deep-water settings. These deposits represent emergence of a regional landmass driven by ongoing regional diapiric plutonism, and volcanoes above sea level and on-land, triggering intense and voluminous explosive activity. The voluminous pyroclastic deposits were quickly reworked under lower and upper flow regime flow conditions (Chap. 12) into surrounding fluvial braid-plain environments. This period was also marked by a change in granite compositions from high-Ca to low-Ca granites (Fig. 17.78).

At <2.658 Ga, *Stage 3B* or "late Late Basin" polymictic sandstones (Fig. 17.81b) and conglomerates (Fig. 17.81c) represent major, regional crustal compression (D_2), uplift and emergence of large landmasses, leading to far-field sediment provenance. Deposition occurred in a high relief alluvial-fluvial landscape (Fig. 17.81d). By this stage TTG plutonism was waning, and volcanism had ceased, consistent with a significant east–west regional compressional stress field. TTG magmatism was progressively replaced by more potassic granite and syenitic plutonism (Fig. 17.78), which also coincides with the onset of regioanll orogenic gold mineralisation. *Stage 3* thus perhaps represents the first indirect evidence for the onset of large-scale crustal compression, regional uplift, probably associated with subcretion peripheral to the craton. This was probably a *transition* into plate tectonic subduction into the Proterozoic, but in the absence of evidence of subduction (no ophiolites, accretionary prisms, or blueschist metamorphism) plate tectonics does not appear to have played a role.

This stage corresponds with the main phase of greenstone-hosted orogenic lode gold (Chap. 18) mineralisation. The Kalgoorlie Terrane contains past and present resources amounting to some 6,000 t (or nearly 200 Moz) of gold, making it arguably the richest auriferous region on Earth. Continuous mining at the world-famous Kalgoorlie mines since 1893 has produced ca. 2,000 t (>60 Moz) of gold, which is mainly hosted by the Golden Mile Dolerite. Most gold mineralisation is shear-hosted and/or associated

Fig. 17.79 (continued)

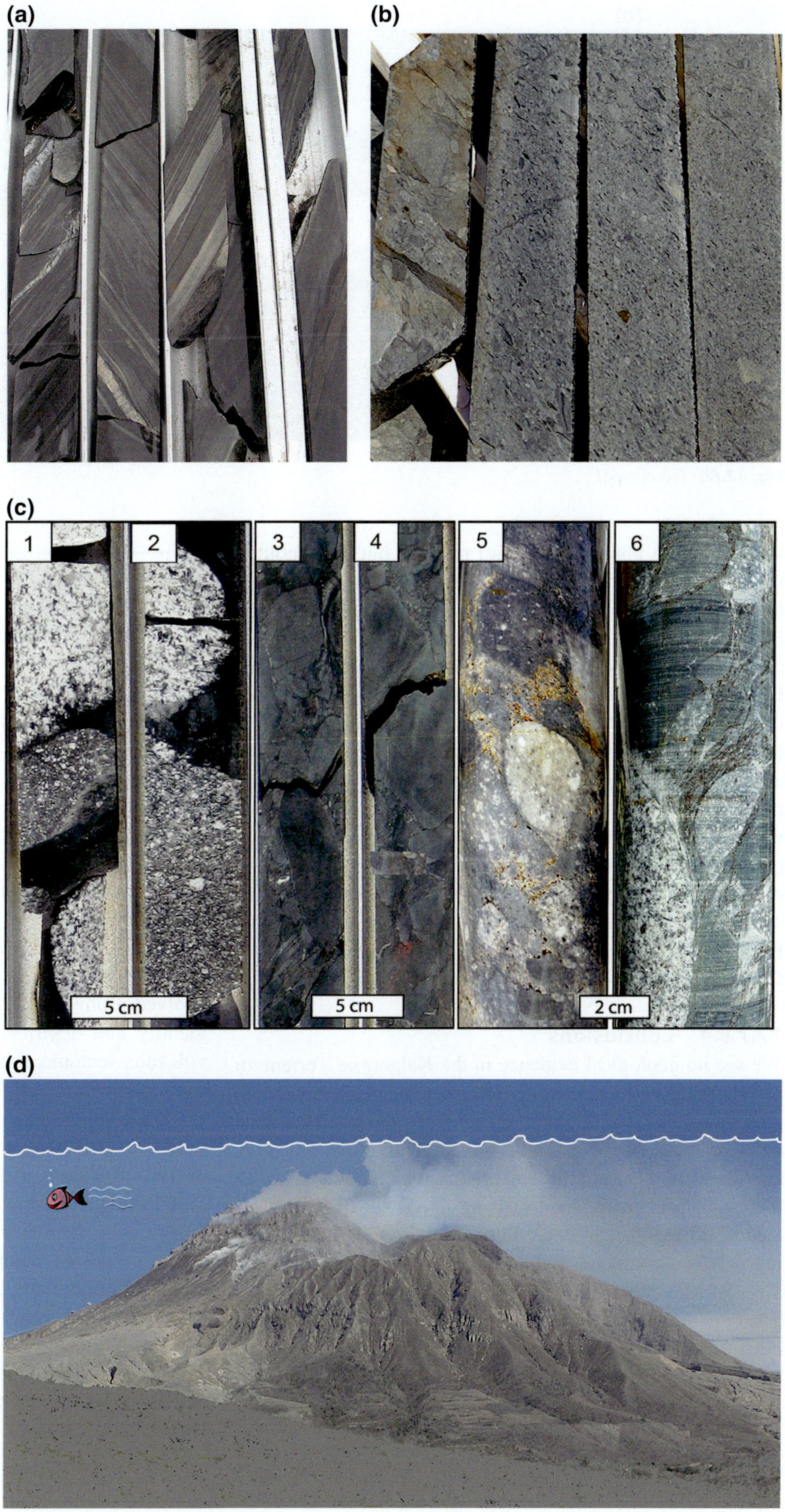

Fig. 17.80 Stratigraphic lithofacies of the *Stage 2*, 2.9–2.65 Ga Kalgoorlie Sequence, Kalgoorlie Terrane, Yilgarn Craton, Western Australia. **a** Laminated black mudstone, representing ambient deep-water anoxic conditions. Kambalda. **b** Core showing the transition from a basal conglomeratic facies (left) grading upwards into finer and more crystal-rich facies (right) in a 30 m thick volcaniclastic megaturbidite deposit. Kambalda. **c** Intra-basinally derived conglomerates from left to right: (rows 1 and 2) mixed clast conglomerate including granite clasts; (rows 3 and 4) mafic–ultramafic clast conglomerate; (rows 5 and 6) mixed felsic-mafic clast conglomerates. Agnew. All drill core in (**a**-**c**) is ~5 cm wide. **d** Imaginary setting of an intra-basinal lava dome-stratovolcano erupting in moderately deep water. (*Actually* Soufrière Hills volcano, Montserrat; Photo by Patrick Smith taken June 2007; *Source* Wikimedia). **e** Emergent intra-basinal stratovolcano showing pyroclastic flow deposits (light coloured slopes) reaching the shoreline, which would have generated volcaniclastic megaturbidite deposits offshore, as envisaged during Kalgoorlie Sequence time. (*Actually* Augustine volcano, Alaska, USA; *Source* United States Geological Survey)

(e)

Fig. 17.80 (continued)

with the more potassic (syenitic) plutons, but most gold is shear-hosted. In addition to the shear zones which provide the fluid conduits, the nature of host rocks (Fe-rich mafic lavas and sills) that are cross-cut are also important. Many of the very big deposits are dolerite-hosted, including the Golden Mile supergiant deposit, and the Junction, Condenser and Cave Rocks dolerites at Kambalda-St Ives. It is usually the thicker, more evolved dolerites with differentiated, very coarse granophyre horizons/lenses (almost pegmatitic in form) of dacite composition in their middle to upper parts that act as both a chemical trap (magnetite is an oxidising catalyst for the reduced mineralising fluids), and coarsely crystalline brittle textures that provide inter-crystal surfaces and hydraulic fracture porosity for the gold and gangue phases to precipitate within (Hayman et al. 2020; Chap. 8).

17.7.8.4 Conclusions

We see no geological evidence in the Kalgoorlie Terrane or eastern Yilgarn Craton of a plate tectonic regime prior to 2.658 Ga (i.e. no ophiolites, accretionary prisms, blueschist belts, uplifted orogenic belts, or mountainous continents). A long-lived mantle plume caused partial melting of thick metasomatised mafic crust causing uprise of TTG plutons, leading to regional plutonic diapiric updoming, and changing paleoenvironments and eruption styles from *Stages 1 to 3* ($\sim$50 Ma). Widespread TTG magmatism does not signify a subduction setting, but widespread anatexis of a lid-like crust/lithosphere. The change to regional horizontal tectonics appears to coincide with the development of the late Late Basins, and onset of one of the principal gold-forming events in the Earth's history from ca. 2.64 Ga.

17.8 Evaluating the Regional Tectonic Context of Ancient Volcanic (and Mineralised) Terrains

Previously in Chap. 15 we have described the essential techniques necessary for mapping and documenting ancient successions in the field. Building on that, here we will briefly outline the approach required for regional tectonic and paleogeographic analysis. Since publication of Cas and Wright (1987), major advances in computing power, the widespread availability of digital data sets, and GIS data base management and software interrogation techniques have not simply transformed, but revolutionised, how to approach evaluation of the tectonic context of ancient volcanic terrains. With these advances, such studies have evolved from mainly academic endeavours attempting to identify and describe ancient terrains in terms of modern volcanic, tectonic and paleoenvironmental analogues (e.g. Cas 1983), to exploration industry-based or academic-supported, and government project generation initiatives involving spatial analysis of large multi-relational datasets to assess their mineral prospectivity. Such exploration initiatives were largely pioneered in the late 1980s in Australia by the major mineral exploration companies, when image processed potential field (aeromagnetic and gravity) data, and radiometric data became publicly available for big swaths of the country. By the late-1990s such studies had evolved into being completely computer based, and capable of incorporating large, disparate legacy data sets from extensive geographical areas and different countries (e.g. Barnes et al. 1998; Wright et al. 1999).

Fig. 17.81 Stratigraphic lithofacies of the *Stage 3*, 2.65–2.58 Ga Late Basin Sequence, Kalgoorlie Terrane, Yilgarn Craton, Western Australia. **a** *Stage 3A* early Late Basin cross-bedded tuffaceous fluvial quartzo-feldspathic sandstone, Merougil Group, Lake Lefroy. **b** *Stage 3B* late Late Basin cross-bedded fluvial-alluvial polymictic sandstone, Kurrawang Group, Kalgoorlie. **c** Polymictic fluvial-alluvial conglomerate, Kurrawang Group, Kalgoorlie. **d** Envisaged subaerial braided fluvial-alluvial setting (without vegetation and snow!) during the *Stage 3B* Late Basin Sequence time. The view is of the upper Rangitata River, South Island, New Zealand (Photo by B. Edwards; *Source* New Zealand Department of Conservation)

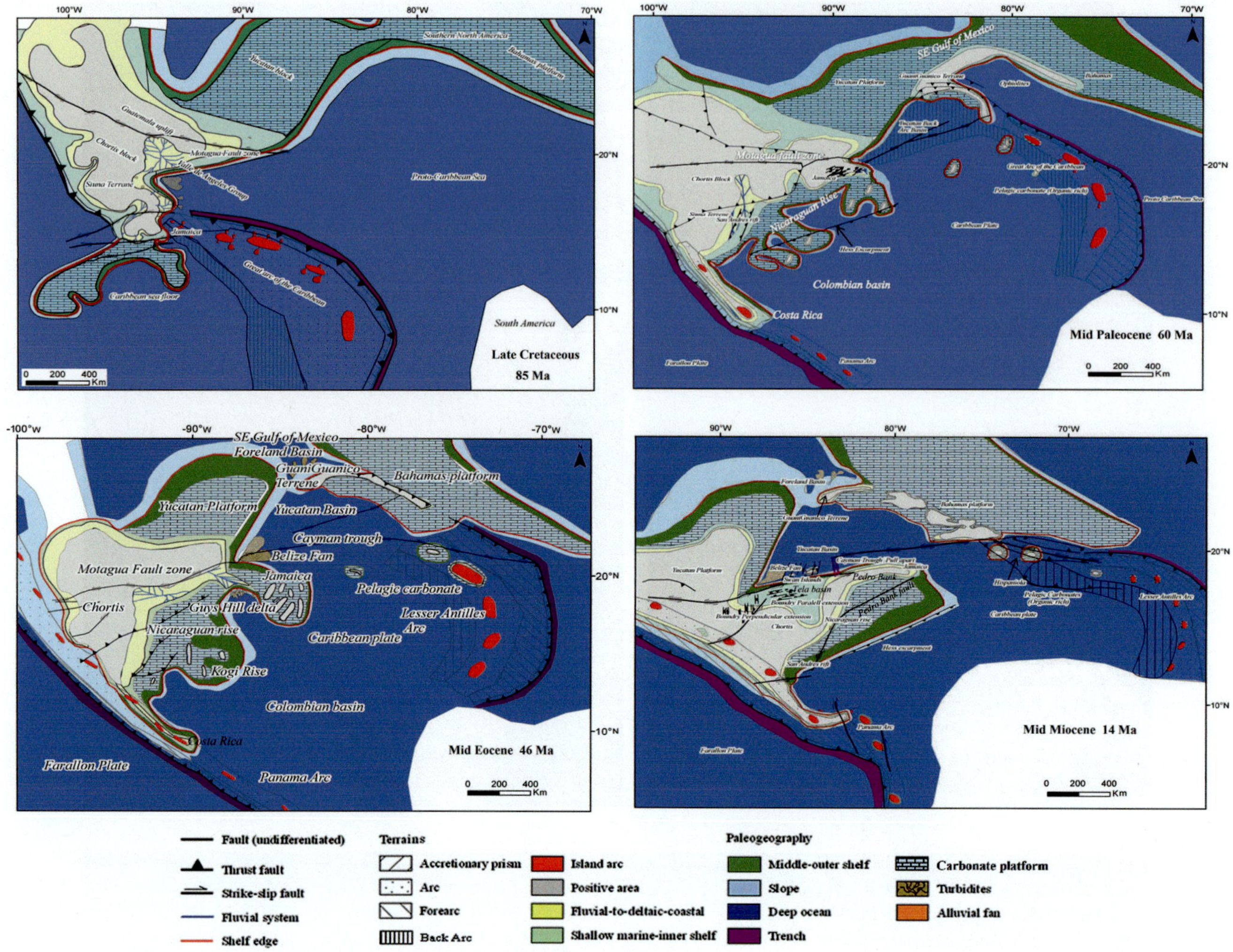

Fig. 17.82 Example of a paleogeographic reconstruction employing contemporary GIS-based techniques. This shows evolution of the northern Caribbean region and Lesser Antilles island arc system for four selected time slices from initiation of the Great Arc of the Caribbean at about 85 Ma. For detailed description see Sect. 17.5.2, Figs. 17.38, 17.39. (Adapted from Ahmad 2014; generated using PaleoGIS software)

17.8.1 Spatial Data and Relational Data Bases

Over the past 30 years increasing amounts of spatial digital data have become available variously from government, industry and academic sources. These include spatial map-based vector (e.g. geology, structure, tectonic elements) and raster (geophysical and satellite imagery) data, and point data (various geochemistry, geochronology, mineral deposit, drill-hole), which are nowadays GIS referenced, and can be superimposed in interactive layers. Many academic studies can rely entirely on publicly available spatial data, which in some countries, especially Australia, are routinely issued by both State and the National geological authorities. In many regions of the world, however, original hard-copy map data have to be digitised, georeferenced and attributed which, depending on the size of the project and number countries involved, may involve an extensive reclassification process. The aim is to provide seamless GIS layers for geology, structures (faults, lineaments, folds) and tectonic units, which generally entails systematically working through the region of interest on a state by state, or country by country basis. Depending on scale, for a large study, this could involve between as many as 40–50 separate maps often with differing projections and sometimes several languages (Barnes et al. 1998).

In practice, this also means the construction of a stratigraphic lexicon. This library contains stratigraphic information for all geological units defined by the project, such as age, lithological association, class (volcanic, intrusive, sediment or metamorphic) and stratigraphic parentage (e.g. sequence, megasequence, supersequence, etc.). This should also be linked to a geological timescale library, which enables a data base view to be constructed approximating to a true stratigraphic column. Depending on the vintage of the available maps from which the various GIS coverages are derived, a literature search will be essential to incorporate new data that have since been published, which can be used to supplement or supersede the attributes supplied with the original data. Even where digital data is available, to enable geological interpretation and eventually GIS-based prospectivity modelling, it is often necessary to edit to varying degrees existing data or construct new map layers. For example, and critically, this often requires the generation of a new structural map(s) redrawn using contemporary aeromagnetic and gravity data, satellite imagery, and any available seismic data.

Various platforms (including online geoinformation services) are now accessible for the design and application of template database tables and forms, and data capture procedures, to facilitate data entry and update. It is important system integration is established to facilitate the sharing of information between the various software products. A typical database may consist of several components (or sub-systems), for example: (1) regional geology, structure and tectonics, (2) whole rock geochemistry, (3) geochronology and, which in an exploration context may further include, (4) mineral deposit occurrences, (5) mineral deposit analytical data (e.g. isotope, fluid inclusion analyses), (6) exploration geochemistry (e.g. stream sediment, soil, calcrete analysis and indicator minerals), (7) drill-hole data (e.g. lithological and geophysical logs, assays), and (8) references. The first is related spatially to regions (polygons) or lines, which are in GIS format, the next six are point datasets and the final is non-spatial. Each spatial sub-system has a corresponding GIS layer associated with it. Attributes and metadata for GIS layers are held within the database and are linked by common key fields.

17.8.2 Methodology

Despite these technological advances, many of the fundamental principles however remain unchanged, none more so than that any evaluation should be an appraisal of the total geologic framework, not simply the geochemistry or any other single factor (Cas and Wright 1987). Based on the above GIS approach, the following are some suggested guidelines:

(a) The most fundamental element in any geoscience GIS database is the creation of a geological map base. This should cover the whole terrain or larger region of interest, and be compiled at the highest resolution possible available source data allows. Cartographic variations and projection issues are routine problems which have to be dealt with on a case by case basis. As are "edge-map" issues when regionally correlating specific stratigraphic units across source maps; these may have to be remapped (if geophysical and/or satellite data are available), reassigned and related to the data base.

(b) At the same time, it is critical to design and commence building code libraries, in particularly the stratigraphic lexicon. The stratigraphic lexicon provides the source for all the stratigraphic and age related information within the data base. This will control the length of time intervals or "slices" of interest that can be reasonably subdivided from the database (point f, below).

(c) Once map and stratigraphic data are visually "up and running", preliminary investigation and verification of the data can commence. From both volcano-tectonic and exploration perspectives it is important to separate the volcanic and plutonic phases, differentiating their compositions and, in the case of volcanic rock successions, their sedimentary facies associations, even in just simple terms, such as continental versus marine. This should be supplemented with the tabulation and spatial integration of available geochronological data, which may require redefinition and reassignment of volcanic and plutonic units.

(d) While this preliminary assessment is taking place, it is important to compile the best structural map justifiable by the available source data. This can include mapped faults on original geological maps, an assessment of the geometry and kinematic style of faults, representing the fold styles, scales and chronology of fold generations, as well as incorporating other published information. However, as indicated earlier, this will often involve creating a new structural map if aeromagnetic and other geophysical data are available. At this time, information from seismic data should be assessed and incorporated. Structures can then be variously described and coded such as magnetic trend (lineament) or mapped fault, and if deemed to form a segment of a regional fault named as such, so building a hierarchy of structures.

(e) Once a map of the current structural framework is available, it is then essential to attempt to reconstruct the original geological framework and spatial relationships by palinspastically undoing major fault displacements, and unfolding regional folds so original spatial relationships, distancing between key geological elements (e.g. plutons) and areal extent of these can be restored. In practice this is done by first constructing balanced structural cross-sections that can then be palinspastically reconstructed by restoring package bed lengths; latitudes and longitudes from the structure section can then be moved with the bed lengths as reference points. Only then can the approximate original spatial relationships between the original lithofacies and the paleogeographic elements and paleoenvironments they represent be assessed. This is key to then reconstructing paleotectonic scenarios. With the advent of deep crustal seismic imaging this can be done with first and even higher order precision, even for metamorphosed Proterozoic terrains and Archean cratons.

(f) The next step is to subdivide the stratigraphic record into significant time slices that are representative of the evolution of the ancient rock system. These should be based on available biostratigraphy and geochronology. Even for many Proterozoic and late Archean successions, with modern geochronological methodologies, time slices of 10–20 Ma can now be achieved, and for Paleozoic successions shorter duration. For each time slice the key components of the stratigraphic sequence, including volcanic, sedimentary and plutonic units the ages of which fall within each time slice should be plotted on a palinspastically undeformed map base. Each time slice map base will then show the original spatial distribution, composition of volcanic and associated sedimentary facies, coeval plutonic and metamorphic rocks and potentially active structures.

(g) From these data, thematic maps are now generated for all time slices of potential interest. These can be saved as individual workspaces sharing their common components that can be viewed, zoomed or panned, and layers can be turned on or off using standard GIS controls. These time slice geology snapshots at original spatial distances can thus trace the evolution of the stratigraphic succession, which can then be interrogated for distinctive trends and the onset of new phases of magmatic, sedimentary or tectonic evolution. Changes in magma compositions herald significant influences in the evolution of the region, as do changes in the provenance, textures and paleoenvironmental settings of the sedimentary facies, which can often be linked to tectonic influences.

(h) For volcano-tectonic evaluation of Phanerozoic to? Neo-Archean successions (i.e. the time interval of recognisable plate tectonics), there are now two options available. First, find the best available modern paleogeographic analogue for the facies configuration plotted. The analogue must be of the same scale and should have an equivalent facies distribution pattern, and evidence for equivalent dynamic elements (Cas and Wright 1987). Alternatively, GIS data can now be imported into a paleogeographic and/or basement terrane map base, which have the advantage of being able to directly assess the dynamic evolution over a selected time period (e.g. Fig. 17.82). Digital paleogeographic map bases are now readily available for Phanerozoic time and, although with decreasing accuracy, well into the Proterozoic, where at some stage the first option may be the better approach. For older, Archean successions ($\geq$ 2.5 Ga) there are still many uncertainties with regard to the relevance of plate tectonic models (as discussed in Sects. 17.7.5 and 17.7.6, and noted below).

(i) GIS-based prospectivity modelling techniques have been developed and are increasingly being employed to assist mineral exploration targeting, in an attempt to reduce risk and increase exploration effectiveness. To undertake this, additional data sets to be included to thematic maps could include mineral deposit information, exploration geochemistry and ore deposit geochemistry. Employing these data, prospective parts of the stratigraphy and/or structures are identified for various commodities (e.g. Au–Ag, Cu–Au, U, Zn–Pb), and/or styles or "*plays*" (e.g. epithermal gold, Archean greenstone-hosted lode gold, iron-oxide copper–gold, volcanic-hosted uranium and volcanic-hosted massive sulphide).

17.8.3 A Final Comment on Geochemical Fingerprinting

Although a lot of emphasis is placed on the geochemical profiles or "fingerprints" of igneous rocks in ancient terrains to reconstruct their original tectonic origins in terms of modern tectonic settings for magmatism, this is often over-done. It relies too much on the assumption that magma generation processes in the past were subject to the same controls and tectonic settings as in the Phanerozoic plate tectonic framework. Multiple factors affect magma generation and compositions, including the nature of the source region, including its composition, especially water content, prior history of melt extraction, and history of metasomatism and metamorphism. In addition, the pressure regime and temperature gradient are pivotal. It cannot be assumed that the same pressure regime as occurs at modern subduction settings existed in the past, and certainly not during the Archean (Sect. 17.7).

Since the Earth was hotter in its early history (Flament et al. 2008; Herzberg et al. 2010), it also had a higher temperature gradient than now, which would have affected the depth at which the solidus temperature of the mantle and thick crust was reached (Johnson et al. 2017). If the solidus was crossed at shallower depths than now for particular rock types, the prevailing pressures would have been lower at those shallower depths, affecting the degree of partial melting. In addition, because the mantle in the Archean had not been subject to as much "reworking" (convective mixing, melt extraction, fractionation) as the mantle has by today, the products of melting and their chemical signatures would have been different than those of today, which is likely reflected by the Archean TTG suite of igneous rocks, as discussed in Sect. 17.7.2.6. However, this all suggests that using geochemical fingerprinting to interpret tectonic settings, especially in the Archean by comparison with Phanerozoic compositions related to modern plate tectonic magma forming processes, has numerous problems (Li et al. 2015; Condie 2015).

References

Abdelmalak MM, Planke S, Faleide JI, Jerram DA, Zastrozhnov D, Eide S, Myklebust R (2016) The development of volcanic sequences at rifted margins: New insights from the structure and morphology of the Vøring Escarpment, mid-Norwegian Margin. J Geophys Res: Solid Earth 121(7):5212–5236

Acocella V (2014) Structural control on magmatism along divergent and convergent plate boundaries: overview, model, problems. Earth Sci Rev 136:226–288

Acocella V, Funiciello F (2010) Kinematic setting and structural control of arc volcanism. Earth Planet Sci Letts 289(1–2):43–53

Acocella V, Gioncada A, Omarini R, Riller U, Mazzuoli R, Vezzoli L (2011) Tectonomagmatic characteristics of the back-arc portion of the Calama-Olacapato-El Toro Fault Zone, Central Andes. Tectonics 30(3)

Aguirre-Díaz GJ, Labarthe-Hernández G (2003) Fissure ignimbrites: fissure-source origin for voluminous ignimbrites of the Sierra Madre Occidental and its relationship with Basin and Range faulting. Geology 31(9):773–776

Aguirre-Díaz GJ, Labarthe-Hernández G, Tristán-González M, Nieto-Obregón J, Gutiérrez-Palomares I (2008) The ignimbrite flare-up and graben calderas of the Sierra Madre Occidental, Mexico. Devel Volcanol 10:143–180

Ahmad SS (2014) Paleogeographic reconstruction of the northern Caribbean region from Late Cretaceous to the Recent. MSc Thesis, University of Stravanger, Norway, 116p

Aldiss DT, Ghazali SA (1984) The regional geology and evolution of the Toba volcano-tectonic depression, Indonesia. J Geol Soc Lond 141(3):487–500

Allen PA, Allen JR (1990) Basin analysis: principles and application. Blackwell Scientific, Oxford, London, 451p

Alt JC, France-Lanord C, Floyd PA, Castillo P, Galy A (1992) low-temperature hydrothermal alteration of Jurassic ocean crust, Site 801. In: Larsen et al. (eds) Proc Ocean Drill Prog Sci Results 129, 415–427

Amelin Y, Lee DC, Halliday AN, Pidgeon RT (1999) Nature of the Earth's earliest crust from hafnium isotopes in single detrital zircons. Nature 399(6733):252–255

Anderson DL (2000) Thermal state of the upper mantle; no role for mantle plumes. Geophys Res Letts 27(22):3623–3626

Anderson DL (2005a) Large igneous provinces, delamination, and fertile mantle. Elements 1(5):271–275

Anderson DL (2005b) Scoring hotspots: the plume and plate paradigms. In: Foulger GR, Natland JH, Presnall DC, Anderson DL (eds) Plates, plumes, and paradigms. Geol Soc Am Spec Pap 388:31–54

Anderson DL (2006) Speculations on the nature and cause of mantle heterogeneity. Tectonophysics 416(1–4):7–22

Anderson DL (2013) The persistent mantle plume myth. Aust J Earth Sci 60(6–7):657–673

Anderson RN, Honnorez J, Becker K, Adamson AC, Alt JC, Emmermann R, Kempton PD, Kinoshita H, Laverne C, Mottl MJ, Newmark RL (1982) DSDP Hole 504B, the first reference section over 1 km through Layer 2 of the oceanic crust. Nature 300 (5893):589

Anhaeusser CR (1984) Structural elements of Archaen granite-greenstone terranes as examplified by the Baberton Mountain Land, Southern Africa. In: Kröner A, Greiling R (eds) Precambrian tectonics illustrated. E. Schweizerbart'sche Verlagsbuchhandlung, Stuttgart, pp 57–58

Anhaeusser CR (2014) Archaean greenstone belts and associated granitic rocks—a review. J Afr Earth Sci 100:684–732

Anhaeusser CR, Mason R, Viljoen MJ, Viljoen RP (1969) A reappraisal of some aspects of Precambrian shield geology. Geol Soc Am Bull 80(11):2175–2200

Annen C (2009) From plutons to magma chambers: Thermal constraints on the accumulation of eruptible silicic magma in the upper crust. Earth Planet Sci Letts 284(3–4):409–416

Archibald NJ, Bettenay LF, Binns RA, Groves DI, Gunthorpe RJ (1978) The evolution of Archaean greenstone terrains, Eastern Goldfields Province, Western Australia. Precamb Res 6(2):103–131

Arndt NT (1991) High Ni in Archean tholeiites. Tectonophysics 187 (4):411–419

Arndt NT (2003) Komatiites, kimberlites, and boninites. J Geophys Res: Solid Earth 108(B6)

Asimow PD (2017) A measure of mantle melting. Science 355 (6328):908–909

Asimow PD, Langmuir AC (2003) The importance of water to oceanic mantle melting regimes. Nature 421(6925):815–820

Austin JM, Hayman PC, Murphy DT, Wingate MT, Lu Y, Lowrey J, Rose K (2022) The voluminous 2.81–2.71 Ga Goldfields Tholeiitic Super Event: implications for basin architecture in the Yilgarn Craton and global correlations. Precamb Res 369:106528

Avanzinelli R, Lustrino M, Mattei M, Melluso L, Conticelli S (2009) Potassic and ultrapotassic magmatism in the circum-Tyrrhenian region: significance of carbonated pelitic versus pelitic sediment recycling at destructive plate margins. Lithos 113(1–2):213–227

Báez W, Carrasco-Nuñez GC, Giordano G, Viramonte JG, Chiodi A (2017) Polycyclic scoria cones of the Antofagasta de la Sierra basin, Southern Puna plateau, Argentina. In: Németh K et al (eds) Monogenetic volcanism. Geol Soc Lond Spec Publ 446(1):311–336

Baker ET, Embley RW, Walker SL, Resing JA, Lupton JE, Nakamura KI, de Ronde CE, Massoth GJ (2008) Hydrothermal activity and volcano distribution along the Mariana arc. J Geophys Res: Solid Earth 113(B8)

Barley ME, Brown SJ, Krapež B, Kositcin N (2008) Physical volcanology and geochemistry of a Late Archaean volcanic arc: Kurnalpi and Gindalbie terranes, Eastern Goldfields superterrane, Western Australia. Precamb Res 161(1–2):53–76

Barnes GJ, Wright JV, Robinson D (1998) Asia database project: documentation. Report to Normandy Mining Asia, vols I-III:54p (unpubl)

Barnes SJ, Van Kranendonk MJ (2014) Archean andesites in the east Yilgarn craton, Australia: products of plume-crust interaction. Lithosphere 6(2):80–92

Barnes SJ, Van Kranendonk MJ, Sonntag I (2012) Geochemistry and tectonic setting of basalts from the Eastern Goldfields Superterrane. Aust J Earth Sci 59(5):707–735

Bebbington MS, Marzocchi W (2011) Stochastic models for earthquake triggering of volcanic eruptions. J Geophys Res: Solid Earth 116 (B5)

Beck SL, Zandt G (2002) The nature of orogenic crust in the central Andes. J Geophys Res: Solid Earth 107(B10)

Bédard JH (2006) A catalytic delamination-driven model for coupled genesis of Archaean crust and sub-continental lithospheric mantle. Geochim Cosmochim Acta 70(5):1188–1214

Bédard JH (2018) Stagnant lids and mantle overturns: Implications for Archaean tectonics, magmagenesis, crustal growth, mantle evolution, and the start of plate tectonics. Geosci Front 9(1):19–49

Bédard JH, Harris LB, Thurston PC (2013) The hunting of the snArc. Precamb Res 229:20–48

Bédard JH, Lauziere K, Tremblay A, Sangster A (1998) Evidence for forearc seafloor-spreading from the Betts Cove ophiolite, Newfoundland: oceanic crust of boninitic affinity. Tectonophysics 284 (3–4):233–245

Behn MD, Grove TL (2015) Melting systematics in mid-ocean ridge basalts: application of a plagioclase-spinel melting model to global variations in major element chemistry and crustal thickness. J Geophys Res: Solid Earth 120(7):4863–4886

Bellier O, Sébrier M (1994) Relationship between tectonism and volcanism along the Great Sumatran fault zone deduced by SPOT image analyses. Tectonophysics 233(3–4):215–231

Bellier O, Sebrier M, Pramumijoyo S, Beaudouin T, Harjono H, Bahar I, Forni O (1997) Paleoseismicity and seismic hazard along the Great Sumatran Fault (Indonesia). J Geodynam 24(1–4):169–183

Bennett RA, Wernicke BP, Niemi NA, Friedrich AM, Davis JL (2003) Contemporary strain rates in the northern Basin and Range province from GPS data. Tectonics 22(2)

Berndt C, Planke S, Alvestad E, Tsikalas F, Rasmussen T (2001) Seismic volcanostratigraphy of the Norwegian Margin: constraints on tectonomagmatic break-up processes. J Geol Soc Lond 158 (3):413–426

Best MG, Christiansen EH, Gromme S (2013) Introduction: the 36–18 Ma southern Great Basin, USA, ignimbrite province and flareup: swarms of subduction-related supervolcanoes. Geosphere 9(2):260–274

Betts PG, Giles D (2006) The 1800–1100 Ma tectonic evolution of Australia. Precamb Res 144(1–2):92–125

Betts PG, Giles D, Foden J, Schaefer BF, Mark G, Pankhurst MJ, Forbes CJ, Williams HA, Chalmers NC, Hills Q (2009) Mesoproterozoic plume-modified orogenesis in eastern Precambrian Australia. Tectonics 28:TC3006

Betts PG, Giles D, Mark G, Lister GS, Goleby BR, Ailleres L (2006) Synthesis of the Proterozoic evolution of the Mt Isa Inlier. Aust J Earth Sci 53(1):187–211

Bhaskar Rao YB, Kumar TV, Sreenivas B, Babu EV (2020) A Review of Paleo-to Neoarchean crustal evolution in the Dharwar craton, southern India and the transition towards a plate tectonic regime. Episodes 43(1):51–68

Bickle MJ, Nisbet EG, Martin A (1994) Archean greenstone belts are not oceanic crust. J Geol 102(2):121–137

Biggs J, Anthony EY, Ebinger CJ (2009) Multiple inflation and deflation events at Kenyan volcanoes, East African Rift. Geology 37 (11):979–982

Biggs J, Bastow ID, Keir D, Lewi E (2011) Pulses of deformation reveal frequently recurring shallow magmatic activity beneath the Main Ethiopian Rift. Geochem Geophys Geosyst 12(9)

Biggs J, Ebmeier SK, Aspinall WP, Lu Z, Pritchard ME, Sparks RS, Mather TA (2014) Global link between deformation and volcanic eruption quantified by satellite imagery. Nat Comms 5(1):1–7

Bijwaard H, Spakman W, Engdahl ER (1998) Closing the gap between regional and global travel time tomography. J Geophys Res: Solid Earth 103(B12):30055–30078

Bissig T, Clark AH, Rainbow A, Montgomery A (2015) Physiographic and tectonic settings of high-sulfidation epithermal gold–silver deposits of the Andes and their controls on mineralizing processes. Ore Geol Rev 65:327–364

Blaikie TN, van Otterloo J, Ailleres L, Betts PG, Cas RAF (2015) The erupted volumes of tephra from maar volcanoes and estimates of their VEI magnitude: examples from the late Cenozoic Newer Volcanics Province, south-eastern Australia. J Volcanol Geotherm Res 301:81–89

Bland KJ, Kamp PJ (2006) Geological structure of the forearc basin in central Hawke's Bay, eastern North Island. In: New Zealand petroleum conference proceedings, pp 1–13

Bland KJ, Kamp PJ (2014) Hawke's Bay forearc basin (eastern North Island, New Zealand): stratigraphy, biostratigraphy, chronology, geological maps and paleogeography. Ministry of Business, Innovation and Employment, New Zealand Unpub Rept. PR4883:66p

Bleeker W (2002) Archaean tectonics: a review, with illustrations from the Slave craton. Geol Soc Lond Spec Pub 199(1):151–181

Blewett RS, Czarnota K, Henson PA (2010b) Structural-event framework for the eastern Yilgarn Craton, Western Australia, and its implications for orogenic gold. Precamb Res 183(2):203–229

Blewett RS, Henson PA, Roy IG, Champion DC, Cassidy KF (2010a) Scale-integrated architecture of a world-class gold mineral system: the Archaean eastern Yilgarn Craton, Western Australia. Precamb Res 183(2):230–250

Blewett RS, Hitchman AP (2006) Final Report, 3D Geological models of the eastern Yilgarn Craton Project Y2. Geosci Aust Rec 2006 (05):276p

Boger SD, Miller JM (2004) Terminal suturing of Gondwana and the onset of the Ross-Delamerian Orogeny: the cause and effect of an Early Cambrian reconfiguration of plate motions. Earth Planet Sci Letts 219(1–2):35–48

Bonin B (2007) A-type granites and related rocks: evolution of a concept, problems and prospects. Lithos 97(1–2):1–29

Boschman LM, van Hinsbergen DJ, Torsvik TH, Spakman W, Pindell JL (2014) Kinematic reconstruction of the Caribbean region since the Early Jurassic. Earth-Sci Rev 138:102–136

Bosworth W (1987) Off-axis volcanism in the Gregory Rift, east Africa: implications for models of continental rifting. Geology 15(5):397–400

Bosworth W (1994) A model for the three-dimensional evolution of continental rift basins, north-east Africa. Geol Rundsch 83(4):671–688

Bouhallier H, Chardon D, Choukroune P (1995) Strain patterns in Archaean dome-and-basin structures: the Dharwar craton (Karnataka, South India). Earth Planet Sci Letts 135(1–4):57–75

Boyce J (2013) The Newer Volcanics Province of southeastern Australia: a new classification scheme and distribution map for eruption centres. Aust J Earth Sci 60(4):449–462

Bradley KE, Feng L, Hill EM, Natawidjaja DH, Sieh K (2017) Implications of the diffuse deformation of the Indian Ocean lithosphere for slip partitioning of oblique plate convergence in Sumatra. J Geophys Res: Solid Earth 122(1):572–591

Brink UT (1991) Volcano spacing and plate rigidity. Geology 19 (4):397–400

Brooks DA, Carlson RL, Harry DL, Melia PJ, Moore RP, Rayhorn JE, Tube SG (1984) Characteristics of back-arc regions. Tectonophysics 102(1–4):1–6

Brothers DS, Driscoll NW, Kent GM, Harding AJ, Babcock JM, Baskin RL (2009) Tectonic evolution of the Salton Sea inferred from seismic reflection data. Nat Geosci 2(8):581

Brown JW, White RS (1995) The effect of finite extension rate on melt generation at continental rifts. J Geophys Res: Solid Earth 100 (B9):18011–18029

Brown M, Johnson T (2019) Metamorphism and the evolution of subduction on Earth. Amer Mineral 104(8):1065–1082

Brown M, Johnson T, Gardiner MJ (2020) Plate tectonics and the Archean Earth. Ann Rev Earth and Planet Sci 48:291–320

Brown SJ, Barley ME, Krapež B, Cas RAF (2002) The Late Archaean Melita Complex, Eastern Goldfields, Western Australia: shallow submarine bimodal volcanism in a rifted arc environment. J Volcanol Geotherm Res 115(3–4):303–327

Brun JP, Sokoutis D (2007) Kinematics of the southern Rhodope core complex (North Greece). Int J Earth Sci 96(6):1079–1099

Buchs DM, Arculus RJ, Baumgartner PO, Ulianov A (2011) Oceanic intraplate volcanoes exposed: example from seamounts accreted in Panama. Geology 39(4):335–338

Buck WR (2004) Consequences of asthenospheric variability on continental rifting. In: Karner et al (eds) Rheology and deformation of the lithosphere at continental margins. Columbia University Press, pp 1–30

Budd AR, Wyborn AI, Bastrakova IV (2001) The metallogenic potential of Australian Proterozoic granites. Geosci Aust Rec 2001/12:108 p

Burke K (1978) Evolution of continental rift systems in the light of plate tectonics. In: Tectonics and geophysics of continental rifts. Springer, Dordrecht, pp 1–9

Busby CJ (2012) Extensional and transtensional continental arc basins: case studies from the southwestern United States. In: Busby CJ, Azor (eds) Tectonics of sedimentary basins: recent advances. Blackwell Publishing Limited, pp 382–404

Busby CJ, Bassett KN (2007) Volcanic facies architecture of an intra-arc strike-slip basin, Santa Rita Mountains, Southern Arizona. Bull Volcanol 70(1):85–103

Busby CJ, Bassett KN, Steiner MB, Riggs NR, Anderson TH, Nourse JA, McKee JW (2005) Climatic and tectonic controls on Jurassic intra-arc basins related to northward drift of North America. In: Anderson TH et al (eds) The Mojave-Sonora megashear hypothesis: development, assessment, and alternatives. Geol Soc Amer Spec Pap 393:359–376

Cameron WE, McCulloch MT, Walker DA (1983) Boninite petrogenesis: chemical and Nd-Sr isotopic constraints. Earth Planet Sci Letts 65(1):75–89

Cameron WE (1989) Contrasting boninite-tholeiite associations from New Caledonia. In: Crawford AJ (ed) Boninites and Related Rocks, Unwin-Hyman, London, pp 314–338

Campbell IH (2005) Large igneous provinces and the mantle plume hypothesis. Elements 1(5):265–269

Campbell IH, Griffiths RW (1990) Implications of mantle plume structure for the evolution of flood basalts. Earth Planet Sci Letts 99 (1–2):79–93

Campbell IH, Griffiths RW, Hill RI (1989) Melting in an Archaean mantle plume: heads it's basalts, tails it's komatiites. Nature 339 (6227):697–699

Campbell IH, Hill RI (1988) A two-stage model for the formation of the granite-greenstone terrains of the Kalgoorlie-Norseman area, Western Australia. Earth Planet Sci Letts 90(1):11–25

Campbell IH, Jarvis GT (1984) Mantle convection and early crustal evolution. Precamb Res 26(1):15–56

Cañón-Tapia E, Walker GP (2004) Global aspects of volcanism: the perspectives of "plate tectonics" and "volcanic systems." Earth Sci Rev 66(1–2):163–182

Capitanio FA, Morra G, Goes S, Weinberg RF, Moresi L (2010) India-Asia convergence driven by the subduction of the Greater Indian continent. Nat Geosci 3(2):136

Capitanio FA, Nebel O, Cawood PA (2020) Thermochemical lithosphere differentiation and the origin of cratonic mantle. Nature 588 (7836):89–94

Capitanio FA, Nebel O, Cawood PA, Weinberg RF, Chowdhury P (2019) Reconciling thermal regimes and tectonics of the early Earth. Geology 47(10):923–927

Capitanio FA, Replumaz A (2013) Subduction and slab breakoff controls on Asian indentation tectonics and Himalayan western syntaxis formation. Geochem Geophys Geosyst 14(9):3515–3531

Carey S, Sigurdsson H (1984) A model of volcanogenic sedimentation in marginal basins. In: Kokelaar P, Howells MF (eds) Marginal basin geology. Geol Soc Lond Spec Publ 16(1):37–58

Carman GD (2003) Geology, mineralisation, and hydrothermal evolution of the Ladolam gold deposit, Lihir Island, Papua New Guinea. Soc Econ Geol Spec Publ 10:247–284

Carn SA, Pyle DM (2001) Petrology and geochemistry of the Lamongan Volcanic Field, East Java, Indonesia: primitive Sunda Arc magmas in an extensional tectonic setting? J Petrol 42(9):1643–1683

Cartwright JA (1987) Transverse structural zones in continental rifts- an example from the Danish sector of the North Sea. Petrol Geol North West Eur 1:441–452

Cartwright JA (1992) Fundamental crustal lineaments and transverse structural zones in continental rifts, Basement Tectonics 7. Springer, Dordrecht, pp 209–217

Cas RAF (1978) Silicic lavas in Paleozoic flyschlike deposits in New South Wales, Australia: behavior of deep subaqueous silicic flows. Geol Soc Am Bull 89(12):1708–1714

Cas RAF (1983) A review of the palaeogeographic and tectonic development of the Lachlan Fold Belt of southeastern Australia. Geol Soc Aust Spec Publ 10:104p

Cas RAF, Edgar C, Allen RL, Bull S, Clifford BA, Giordano G, Wright JV (2001) Influence of magmatism and tectonics on sedimentation in an extensional lake basin: the Upper Devonian Bunga Beds, Boyd Volcanic Complex, south-eastern Australia. In: White JDL, Riggs NR (eds) Volcaniclastic sedimentation in lacustrine settings. Int Assoc Sed Spec Publ 30:81–108

Cas RAF, Powell CM, Crook KA (1980) Ordovician palaeogeography of the Lachlan Fold Belt: a modern analogue and tectonic constraints. J Geol Soc Aust 27(1–2):19–31

Cas RAF, Simmons JM (2018) Why deep-water eruptions are so different from subaerial eruptions. Front Earth Sci 6:198

Cas RAF, Van Otterloo J (2011) Introduction to the IUGG excursion guide. Factors that influence varying eruption styles (from magmatic to phreatomagmatic) in intraplate basaltic volcanic provinces: the Newer Volcanics Province of south-eastern Australia. Field trip guide. XXV IUGG General Assembly. Melbourne, IAVCEI, 7–31

Cas RAF, Wright JV (1987) Volcanic successions: modern and ancient. Allen & Unwin, London, p 528p

Cas RAF, Van Otterloo J, Blaikie TN, Van Den Hove J (2017) The dynamics of a very large intra-plate continental basaltic volcanic province, the Newer Volcanics Province, SE Australia, and implications for other provinces. In: Németh K et al (eds) Monogenetic Volcanism. Geol Soc Lond Spec Publ 446(1):123–172

Cawood PA, Hawkesworth CJ (2019) Continental crustal volume, thickness and area, and their geodynamic implications. Gondwana Res 66:116–125

Cawood PA, Hawkesworth CJ, Pisarevsky SA, Dhuime B, Capitanio FA, Nebel O (2018) Geological archive of the onset of plate tectonics. Phil Trans Roy Soc Lond Ser A: Math Phys Engin Sci 376(2132):20170405

Cecchi E, van Wyk de Vries B, Lavest JM, (2004) Flank spreading and collapse of weak-cored volcanoes. Bull Volcanol 67(1):72–91

Chadwick WW, Clague DA, Embley RW, Perfit MR, Butterfield DA, Caress DW, Paduan JB, Martin JF, Sasnett P, Merle SG (2013) The 1998 eruption of Axial Seamount: new insights on submarine lava flow emplacement from high-resolution mapping. Geochem Geophys 14(10):3939–3968

Champion DC, Sheraton JW (1997) Geochemistry and Nd isotope systematics of Archaean granites of the Eastern Goldfields, Yilgarn

Craton, Australia: implications for crustal growth processes. Precamb Res 83(1–3):109–132
Chesner CA (2012) The Toba caldera complex. Q Int 258:5–18
Chiodi A, Tassi F, Báez W, Filipovich R, Bustos E, Galli MG, Suzaño N, Ahumada MF, Viramonte JG, Giordano G, Pecoraino G (2019) Preliminary conceptual model of the Cerro Blanco caldera-hosted geothermal system (Southern Puna, Argentina): inferences from geochemical investigations. J South Am Earth Sci 94:102213
Chorowicz J (2005) The east African rift system. J Afr Earth Sci 43(1–3):379–410
Chowdhury P, Gerya T, Chakraborty S (2017) Emergence of silicic continents as the lower crust peels off on a hot plate-tectonic Earth. Nat Geosci 10(9):698–703
Christiansen RL, Lipman PW (1972) A Discussion on volcanism and the structure of the Earth-Cenozoic volcanism and plate-tectonic evolution of the Western United States. II. Late Cenozoic. Phil Trans R Soc Lond A 271(1213):249–284
Chulick GS, Detweiler S, Mooney WD (2013) Seismic structure of the crust and uppermost mantle of South America and surrounding oceanic basins. J South Am Earth Sci 42:260–276
Chung SL, Chu MF, Zhang Y, Xie Y, Lo CH, Lee TY, Lan CY, Li X, Zhang Q, Wang Y (2005) Tibetan tectonic evolution inferred from spatial and temporal variations in post-collisional magmatism. Earth-Sci Rev 68(3–4):173–196
Cifelli F, Mattei M, Rossetti F (2007) Tectonic evolution of arcuate mountain belts on top of a retreating subduction slab: the example of the Calabrian Arc. J Geophys Res: Solid Earth 112(B9)
Civiero C, Goes S, Hammond JO, Fishwick S, Ahmed A, Ayele A, Doubre C, Goitom B, Keir D, Kendall JM, Leroy S (2016) Small-scale thermal upwellings under the northern East African Rift from S travel time tomography. J Geophys Res: Solid Earth 121 (10):7395–7408
Clague DA, Dreyer BM, Paduan JB, Martin JF, Chadwick WW, Caress DW, Portner RA, Guilderson TP, McGann ML, Thomas H (2013) Geologic history of the summit of Axial Seamount. Juan De Fuca Ridge. Geochem Geophys 14(10):4403–4443
Coira B, Davidson J, Mpodozis C, Ramos V (1982) Tectonic and magmatic evolution of the Andes of northern Argentina and Chile. Earth Sci Rev 18(3–4):303–332
Cole JW (1979) Structure, petrology, and genesis of Cenozoic volcanism, Taupo Volcanic Zone, New Zealand-a review. NZ J Geol Geophys 22(6):631–657
Cole JW (1984) Taupo-Rotorua depression: an ensialic marginal basin of North Island, New Zealand. In: Kokelaar P, Howells MF (eds) Marginal basin geology. Geol Soc Lond Spec Publ 16(1):109–120
Coleman RG (1977) Ophiolites: ancient oceanic lithosphere? Springer, Berlin, Heidelberg
Colgan JP, Egger AE, John DA, Cousens B, Fleck RJ, Henry CD (2011) Oligocene and Miocene arc volcanism in northeastern California: evidence for post-Eocene segmentation of the subducting Farallon plate. Geosphere 7(3):733–755
Collins WJ, Van Kranendonk AM, Teyssier C (1998) Partial convective overturn of Archaean crust in the east Pilbara Craton, Western Australia: driving mechanisms and tectonic implications. J Struct Geol 20(9–10):1405–1424
Colpron M, Nelson JL, Murphy DC (2017) Northern Cordilleran terranes and their interactions through time. GSA Today 17(4/5):4. https://doi.org/10.1130/GSAT01704-5A.1
Condie KC (2001) Mantle plumes and their record in Earth history. Cambridge University Press, 306 p
Condie KC (2015) Changing tectonic settings through time: indiscriminate use of geochemical discriminant diagrams. Precamb Res 266:587–591
Condie KC (2018) A planet in transition: the onset of plate tectonics on Earth between 3 and 2 Ga? Geosci Front 9(1):51–60
Coney PJ (1987) The regional tectonic setting and possible causes of Cenozoic extension in the North American Cordillera. In: Coward MP, Dewey JF, Hancock PL (eds) 1987. Continental extensional tectonics. Geol Soc Lond Spec Publ 28(1):177–218
Coney PJ (1989) Structural aspects of suspect terranes and accretionary tectonics in western North America. J Struct Geol 11(1–2):107–125
Constenius KN (1996) Late Paleogene extensional collapse of the Cordilleran foreland fold and thrust belt. Geol Soc Am Bull 108 (1):20–39
Conticelli S, Avanzinelli R, Poli G, Braschi E, Giordano G (2013) Shift from lamproite-like to leucititic rocks: Sr-Nd-Pb isotope data from the Monte Cimino volcanic complex versus the Vico stratovolcano, Central Italy. Chem Geol 353:246–266
Conticelli S, Boari E, Avanzinelli R, De Benedetti AA, Giordano G, Mattei M, Melluso L, Morra V, Funiciello R (2010a) Geochemistry, isotopes and mineral chemistry of the Colli Albani volcanic rocks: constraints on magma genesis and evolution. In: Funiciello R, Giordano G (eds) The Colli Albani Volcano. Spec Publ IAVCEI and Geol Soc Lond 3:107–139
Conticelli S, D'antonio M, Pinarelli L, Civetta L (2002) Source contamination and mantle heterogeneity in the genesis of Italian potassic and ultrapotassic volcanic rocks: Sr–Nd–Pb isotope data from Roman Province and Southern Tuscany. Miner Petrol 74(2–4):189–222
Conticelli S, Guarnieri L, Farinelli A, Mattei M, Avanzinelli R, Bianchini G, Boari E, Tommasini S, Tiepolo M, Prelević D, Venturelli G (2009) Trace elements and Sr-Nd-Pb isotopes of K-rich, shoshonitic, and calc-alkaline magmatism of the Western Mediterranean Region: genesis of ultrapotassic to calc-alkaline magmatic associations in a post-collisional geodynamic setting. Lithos 107(1):68–92
Conticelli S, Laurenzi MA, Giordano G, Mattei M, Avanzinelli R, Melluso L, Tommasini S, Boari E, Cifelli F, Perini G, Beltrando M (2010b) Leucite-bearing (kamafugitic/leucititic) and-free (lamproitic) ultrapotassic rocks and associated shoshonites from Italy: constraints on petrogenesis and geodynamics. J Virt Explor 36(20)
Cook C, Briggs RM, Smith IE, Maas R (2004) Petrology and geochemistry of intraplate basalts in the South Auckland volcanic field, New Zealand: evidence for two coeval magma suites from distinct sources. J Petrol 46(3):473–503
Corti G (2009) Continental rift evolution: from rift initiation to incipient break-up in the Main Ethiopian Rift, East Africa. Earth Sci Rev 96(1–2):1–53
Corti G, Sani F, Agostini S, Philippon M, Sokoutis D, Willingshofer E (2018) Off-axis volcano-tectonic activity during continental rifting: insights from the transversal Goba-Bonga lineament, Main Ethiopian Rift (East Africa). Tectonophysics 728:75–91
Courtillot V, Davaille A, Besse J, Stock J (2003) Three distinct types of hotspots in the Earth's mantle. Earth Planet Sci Letts 205(3–4):295–308
Coward MP, Dewey JF, Hancock PL (1987) Continental extensional tectonics. Geol Soc Lond Spec Publ 28(1):637p
Crawford AJ, Falloon TJ, Green DH (1989) Classification, petrogenesis and tectonic setting of boninites. In: Crawford AJ (ed) Boninites and related rocks. Unwin Hyman, London, pp 1–49
Crawford MB, Windley BF (1990) Leucogranites of the Himalaya/Karakoram: implications for magmatic evolution within collisional belts and the study of collision-related leucogranite petrogenesis. J Volcanol Geotherm Res 44(1–2):1–9
Crisp JA (1984) Rates of magma emplacement and volcanic output. J Volcanol Geotherm Res 20(3–4):177–211
Czarnota K, Blewett RS, Goscombe B (2008) Structural and metamorphic controls on gold through time and space in the central Eastern Goldfields Superterrane—a field guide. Geol Surv Western Aus Rec 2008(9):66p

Czarnota K, Champion DC, Goscombe B, Blewett RS, Cassidy KF, Henson PA, Groenewald PB (2010) Geodynamics of the eastern Yilgarn Craton. Precamb Res 183(2):175–202

Daly MC, Chorowicz J, Fairhead JD (1989) Rift basin evolution in Africa: the influence of reactivated steep basement shear zones. In: Cooper Ma, Williams GD (eds) Inversion tectonics. Geol Soc Lond Spec Publ 44(1):309–334

Dasgupta R, Hirschmann MM (2006) Melting in the Earth's deep upper mantle caused by carbon dioxide. Nature 440(7084):659–662

Davies P (1998) The Fifth Miracle. The Penguin Press, The Search for the Origin of Life, p 320

Davies DR, Rawlinson N, Iaffaldano G, Campbell IH (2015) Lithospheric controls on magma composition along Earth's longest continental hotspot track. Nature 525(7570):511

Dawson JB (2008) The Gregory rift valley and Neogene-recent volcanoes of northern Tanzania. Geol Soc Lond Mem 33

Dawson JB, Pinkerton H, Pyle DM, Nyamweru C (1994) June 1993 eruption of Oldoinyo Lengai, Tanzania: exceptionally viscous and large carbonatite lava flows and evidence for coexisting silicate and carbonate magmas. Geology 22(9):799–802

De Astis G, Peccerillo A, Kempton PD, La Volpe L, Wu TW (2000) Transition from calc-alkaline to potassium-rich magmatism in subduction environments: geochemical and Sr, Nd, Pb isotopic constraints from the island of Vulcano (Aeolian arc). Contribs Miner Petrol 139(6):684–703

Debaille V, O'Neill C, Brandon AD, Haenecour P, Yin QZ, Mattielli N, Treiman AH (2013) Stagnant-lid tectonics in early Earth revealed by 142Nd variations in late Archean rocks. Earth Planet Sci Letts 373:83–92

Defant MJ, Drummond MS (1990) Derivation of some modern arc magmas by melting of young subducted lithosphere. Nature 347 (6294):662

Delvaux D, Kervyn F, Macheyeki AS, Temu EB (2012) Geodynamic significance of the TRM segment in the East African Rift (W-Tanzania): active tectonics and paleostress in the Ufipa plateau and Rukwa basin. J Struct Geol 37:161–180

Demidjuk Z, Turner S, Sandiford M, George R, Foden J, Etheridge M (2007) U-series isotope and geodynamic constraints on mantle melting processes beneath the Newer Volcanic Province in South Australia. Earth Planet Sci Letts 261(3):517–533

Demouchy S, Jacobsen SD, Gaillard F, Stern CR (2006) Rapid magma ascent recorded by water diffusion profiles in mantle olivine. Geology 34(6):429–432

De Paoli MC, Clarke GL, Klepeis KA, Allibone AH, Turnbull IM (2009) The eclogite-granulite transition: mafic and intermediate assemblages at Breaksea Sound. New Zealand. J Petrol 50 (12):2307–2343

de Silva SL, Mucek AE, Gregg PM, Pratomo I (2015) Resurgent Toba —Field, chronologic, and model constraints on time scales and mechanisms of resurgence at large calderas. Front Earth Sci 3:25

de Silva S, Zandt G, Trumbull R, Viramonte JG, Salas G, Jiménez N (2006) Large ignimbrite eruptions and volcano-tectonic depressions in the Central Andes: a thermomechanical perspective. Geol Soc, London, Spec Publ 269(1):47–63

Devine JD, Rutherford MJ, Gardner JE (1998) Petrologic determination of ascent rates for the 1995–1997 Soufriere Hills Volcano andesitic magma. Geophys Res Letts 25(19):3673–3676

Dewey JF (1988) Extensional collapse of orogens. Tectonics 7 (6):1123–1139

Dickinson WR (1979) Cenozoic plate tectonic setting of the Cordilleran region in the United States. In: Armentrout JM, Cole MR, Terbest H (eds) Cenozoic palaeogeography of the western United States, Pacific Coast Palaeogeography Symptoms, vol 3, pp 1–13

Dickinson WR (2006) Geotectonic evolution of the Great Basin. Geosphere 2(7):353–368

Dilek Y, Robinson PT (2003) Ophiolites in earth history. Geol Soc Lond Spec Publ 218(1):700p

Dilek Y, Furnes H (2009) Structure and geochemistry of Tethyan ophiolites and their petrogenesis in subduction rollback systems. Lithos 113(1–2):1–20

Dilek Y, Furnes H (2014) Ophiolites and their origins. Elements 10 (2):93–100

Direen NG, Crawford AJ (2003) Fossil seaward-dipping reflector sequences preserved in southeastern Australia: a 600 Ma volcanic passive margin in eastern Gondwanaland. J Geol Soc Lond 160 (6):985–990

Dirks PH, Jelsma HA, Hofmann A (2002) Thrust-related accretion of an Archaean greenstone belt in the Midlands of Zimbabwe. J Struct Geol 24(11):1707–1727

Dobson PF, Blank JG, Maruyama S, Liou JG (2006) Petrology and geochemistry of boninite-series volcanic rocks, Chichi-Jima, Bonin Islands, Japan. Int Geol Rev 48(8):669–701

Dodd MS, Papineau D, Grenne T, Slack JF, Rittner M, Pirajno F, O'Neil J, Little CT (2017) Evidence for early life in Earth's oldest hydrothermal vent precipitates. Nature 543(7643):60–64

Donnelly-Nolan JM, Burns MG, Goff FE, Peters EK, Thompson JM (1993) The Geysers-Clear Lake area, California; thermal waters, mineralization, volcanism, and geothermal potential. Econ Geol 88 (2):301–316

Drummond BJ, Goleby BR, Swager CP (2000) Crustal signature of Late Archaean tectonic episodes in the Yilgarn craton, Western Australia: evidence from deep seismic sounding. Tectonophysics 329(1–4):193–221

Drummond MS, Defant MJ, Kepezhinskas PK (1996) Petrogenesis of slab-derived trondjemite-tonalite-dacite/adakite magmas. Earth Environ Science Trans Roy Soc Edinburgh 87(1–2):205–215

Ducea MN, Saleeby JB, Bergantz G (2015) The architecture, chemistry, and evolution of continental magmatic arcs. Ann Rev Earth Planet Sci 43:299–331

Eaton GP (1982) The Basin and Range province: origin and tectonic significance. Ann Rev Earth Planet Sci 10(1):409–440

Eaton GP (1984) The Miocene Great Basin of western North America as an extending back-arc region. Tectonophysics 102(1–4):275–295

Ebinger CJ (1989a) Tectonic development on the western branch of the East African rift system. Geol Soc Am Bull 101:885–903

Ebinger CJ (1989b) Geometric and kinematic development of border faults and accommodation zones, Kivu-Rusizi Rift, Africa. Tectonics 8(1):17–133

Eby GN (1992) Chemical subdivision of the A-type granitoids: petrogenetic and tectonic implications. Geology 20(7):641–644

Eldholm O, Gladczenko TP, Skogseid J, Planke S (2000) Atlantic volcanic margins: a comparative study. In: Nøttved A et al. (eds) Dynamics of the Norwegian Margin. Geol Soc Lond Spec Publ 167(1):411–428

Embley RW, Rubin KH (2018) Extensive young silicic volcanism produces large deep submarine lava flows in the NE Lau Basin. Bull Volcanol 80(4):36

England P, Houseman G (1989) Extension during continental convergence, with application to the Tibetan Plateau. J Geophys Res: Solid Earth 94(B12):17561–17579

Eriksson PG, Banerjee S, Catuneanu O, Corcoran PL, Eriksson KA, Hiatt EE, Laflamme M, Lenhardt N, Long DG, Miall AD, Mints MV (2013) Secular changes in sedimentation systems and sequence stratigraphy. Gondwana Res 24(2):468–489

Ernst RE, Buchan KL (2001) Mantle plumes: their identification through time. Geol Soc Am Spec Pap 352:595p

Ernst WG (2010) Subduction-zone metamorphism, calc-alkaline magmatism, and convergent-margin crustal evolution. Gondwana Res 18(1):8–16

Escalona A, Mann P (2011) Tectonics, basin subsidence mechanisms, and paleogeography of the Caribbean-South American plate boundary zone. Marine Petrol Geol 28(1):8–39

Espurt N, Funiciello F, Martinod J, Guillaume B, Regard V, Faccenna C, Brusset S (2008) Flat subduction dynamics and deformation of the South American plate: insights from analog modeling. Tectonics 27(3)

Etheridge MA, Branson JC, Stuart-Smith PG (1987) The Bass, Gippsland and Otway basins, southeast Australia: a branched rift system formed by continental extension. In: Sedimentary basins and basin-forming mechanisms. Amer Assoc Petrol Geol Mem 12: 147–162

Ewart A, Brothers RN, Mateen A (1977) An outline of the geology and geochemistry, and the possible petrogenetic evolution of the volcanic rocks of the Tonga-Kermadec-New Zealand island arc. J Volcanol Geotherm Res 2(3):205–250

Ewart A, Collerson KD, Regelous M, Wendt JI, Niu Y (1998) Geochemical evolution within the Tonga–Kermadec–Lau arc–back-arc systems: the role of varying mantle wedge composition in space and time. J Petrol 39(3):331–368

Eyuboglu Y, Dudas FO, Santosh M, Zhu DC, Yi K, Chatterjee N, Jeong YJ, Akaryalı E, Liu Z (2016) Cenozoic forearc gabbros from the northern zone of the Eastern Pontides Orogenic Belt, NE Turkey: implications for slab window magmatism and convergent margin tectonics. Gondwana Res 33:160–189

Eyuboglu Y, Santosh M, Bektas O, Ayhan S (2011) Arc magmatism as a window to plate kinematics and subduction polarity: example from the eastern Pontides belt. NE Turkey. Geosci Front 2(1):49–56

Faccenna C, Becker TW, Lallemand S, Lagabrielle Y, Funiciello F, Piromallo C (2010) Subduction-triggered magmatic pulses: a new class of plumes? Earth Planet Sci Letts 299(1–2):54–68

Faccenna C, Becker TW, Lucente FP, Jolivet L, Rossetti F (2001) History of subduction and back arc extension in the central Mediterranean. Geophys J Int 145(3):809–820

Faccenna C, Di Giuseppe E, Funiciello F, Lallemand S, Van Hunen J (2009) Control of seafloor aging on the migration of the Izu-Bonin-Mariana trench. Earth Planet Sci Letts 288(3–4):386–398

Faccenna C, Piromallo C, Crespo-Blanc A, Jolivet L, Rossetti F (2004) Lateral slab deformation and the origin of the western Mediterranean arcs. Tectonics 23(1)

Falloon TJ, Crawford AJ (1991) The petrogenesis of high-calcium boninite lavas dredged from the northern Tonga ridge. Earth Planet Sci Letts 102(3–4):375–394

Faulds JE, Stewart JH (1998) Accommodation zones and transfer zones: the regional segmentation of the Basin and Range Province. Geol Soc Amer Spec Pap 323:257p

Fergusson CL (2010) Plate-driven extension and convergence along the East Gondwana active margin: Late Silurian-Middle Devonian tectonics of the Lachlan Fold Belt, southeastern Australia. Aust J Earth Sci 57(5):627–649

Filipovich R, Báez W, Groppelli G, Ahumada F, Aldega L, Becchio R, Giordano G (2020) Geological map of the Tocomar basin (Puna Plateau, NW Argentina). Implication for the geothermal system investigation. Energies 13(20):5492

Fishwick S, Rawlinson N (2012) 3-D structure of the Australian lithosphere from evolving seismic datasets. Aust J Earth Sci 59 (6):809–826

Flament N, Coltice N, Rey PF (2008) A case for late-Archaean continental emergence from thermal evolution models and hypsometry. Earth Planet Sci Letts 275(3–4):326–336

Foley S (1992) Vein-plus-wall-rock melting mechanisms in the lithosphere and the origin of potassic alkaline magmas. Lithos 28 (3–6):435–453

Forbes CJ, Betts PG, Weinberg R, Buick IS (2005) A structural metamorphic study of the Broken Hill Block, NSW, Australia. J Met Petrol 23(8):745–770

Fossen H, Rotevatn A (2016) Fault linkage and relay structures in extensional settings—a review. Earth Sci Rev 154:14–28

Foster DA, Gray DR, Spaggiari C, Kamenov G, Bierlein FP (2009) Palaeozoic Lachlan orogen, Australia; accretion and construction of continental crust in a marginal ocean setting: isotopic evidence from Cambrian metavolcanic rocks. In: Cawood PA, Kröner A (eds) Earth accretionary systems in space and time. Geol Soc Lond Spec Publ 318(1):329–349

Foster DR, Rubenach MJ (2006) Isograd pattern and regional low-pressure, high-temperature metamorphism of pelitic, mafic and calc-silicate rocks along an east–west section through the Mt Isa Inlier. Aust J Earth Sci 53(1):167–186

Foulger GR, Natland JH (2003) Is "hotspot" volcanism a consequence of plate tectonics? Science 300(5621):921–922

Foulger GR, Natland JH, Presnall DC, Anderson (eds) (2005) Plates, plumes, and paradigms. Geol Soc Am Spec Pap 388

Fournier M, Jolivet L, Davy P, Thomas JC (2004) Backarc extension and collision: an experimental approach to the tectonics of Asia. Geophys J Int 157(2):871–889

Francalanci L, Taylor SR, McCulloch MT, Woodhead JD (1993) Geochemical and isotopic variations in the calc-alkaline rocks of Aeolian arc, southern Tyrrhenian Sea, Italy: constraints on magma genesis. Contribs Miner Petrol 113(3):300–313

Franz G, Lucassen F, Kramer W, Trumbull RB, Romer RL, Wilke HG, Siebel W (2006) Crustal evolution at the Central Andean continental margin: a geochemical record of crustal growth, recycling and destruction. The Andes: Active Subduction Orogeny, 45–64

Funiciello F, Faccenna C, Giardini D (2004) Role of lateral mantle flow in the evolution of subduction systems: insights from laboratory experiments. Geophys J Int 157(3):1393–1406

Funiciello F, Faccenna C, Giardini D, Regenauer-Lieb K (2003) Dynamics of retreating slabs: 2. Insights from three-dimensional laboratory experiments. J Geophys Res: Solid Earth 108(B4)

Gale A, Dalton CA, Langmuir CH, Su Y, Schilling JG (2013) The mean composition of ocean ridge basalts. Geochem Geophys Geosyst 14(3):489–518

Galetto F, Acocella V, Caricchi L (2017) Caldera resurgence driven by magma viscosity contrasts. Nature Comms 8(1):1750

Galland O, Cobbold PR, de Bremond d'Ars J, Hallot E (2007a) Rise and emplacement of magma during horizontal shortening of the brittle crust: insights from experimental modeling. J Geophys Res: Solid Earth 112(B6)

Galland O, Hallot E, Cobbold PR, Ruffet G, de Bremond D'Ars J (2007b) Volcanism in a compressional Andean setting: a structural and geochronological study of Tromen volcano (Neuquén province, Argentina). Tectonics 26(4)

Gardiner NJ, Hickman AH, Kirkland CL, Lu Y, Johnson T, Zhao JX (2017) Processes of crust formation in the early Earth imaged through Hf isotopes from the East Pilbara Terrane. Precamb Res 297:56–76

Gaulon R, Chorowicz J, Vidal G, Romanowicz B, Rouit G (1992) Regional geodynamic implications of the May–July 1990 earthquake sequence in southern Sudan. Tectonophysics 209(1–4): 87–103

Gazel E, Hayes JL, Hoernle K, Kelemen P, Everson E, Holbrook WS, Hauff F, Van Den Bogaard P, Vance EA, Chu S, Calvert AJ (2015) Continental crust generated in oceanic arcs. Nat Geosci 8(4):321–327

Gautier P, Brun JP, Jolivet L (1993) Structure and kinematics of upper Cenozoic extensional detachment on Naxos and Paros (Cyclades Islands, Greece). Tectonics 12(5):1180–1194

Gawthorpe RL, Hurst JM (1993) Transfer zones in extensional basins: their structural style and influence on drainage development and stratigraphy. J Geol Soc Lond 150(6):1137–1152

Gawthorpe RL, Leeder MR (2000) Tectono-sedimentary evolution of active extensional basins. Basin Res 12(3–4):195–218

Geoffroy L (2005) Volcanic passive margins. Comptes Rendus Geosci 337(16):1395–1408

Geoffroy L, Aubourg C, Callot J, Barrat J (2007) Mechanisms of crustal growth in large igneous provinces: the north Atlantic province as a case study. In: Foulger GR, Jurdy DM (eds) Plates, plumes, and planetary processes. Geol Soc Am Spec Pap 430:747–774

Gibbs AD (1984) Structural evolution of extensional basin margins. J Geol Soc Lond 141(4):609–620

Gibson ILJ, Malps LJ, Robinson PT, Xenophontos C (1991) Cyprus crustal study project: initial report, Holes CY-1 and 1a. Geol Surv Canada Pap 90–20:393p

Gill JB (1981) Orogenic andesites and plate tectonics. Springer Science & Business Media

Gill JB (1984) Sr-Pb-Nd isotopic evidence that both MORB and OIB sources contribute to oceanic island arc magmas in Fiji. Earth Planet Sci Letts 68(3):443–458

Gill JB (1987) Early geochemical evolution of an oceanic island arc and backarc: Fiji and the South Fiji Basin. J Geol 95(5):589–615

Gill JB, Stork AL, Whelan PM (1984) Volcanism accompanying back-arc basin development in the southwest Pacific. Tectonophysics 102(1–4):207–224

Giordano G, Caricchi L (2022) Determining the state of activity of transcrustal magmatic systems and their volcanoes. Ann Rev Earth Planet Sci. 50:231–259

Giordano G, Cas RAF (2001) Structure of the Upper Devonian Boyd Volcanic Complex, south coast New South Wales: implications for the Devonian-Carboniferous evolution of the Lachlan Fold Belt. Aust J Earth Sci 48(1):49–61

Giordano G, Pinton A, Cianfarra P, Baez W, Chiodi A, Viramonte J, Norini G, Groppelli G (2013) Structural control on geothermal circulation in the Cerro Tuzgle-Tocomar geothermal volcanic area (Puna plateau, Argentina). J Volcanol Geotherm Res 249:77–94

Gladczenko TP, Hinz K, Eldholm O, Meyer H, Neben S, Skogseid J (1997) South Atlantic volcanic margins. J Geol Soc Lond 154 (3):465–470

Graham IJ, Cole JW, Briggs RM, Gamble JA, Smith IE (1995) Petrology and petrogenesis of volcanic rocks from the Taupo Volcanic Zone: a review. J Volcanol Geotherm Res 68(1–3):59–87

Glazner AF, Bartley JM, Coleman DS, Gray W, Taylor RZ (2004) Are plutons assembled over millions of years by amalgamation from small magma chambers? GSA Today 14(4/5):4–12

Glazner AF, Ussler W (1989) Crustal extension, crustal density, and the evolution of Cenozoic magmatism in the Basin and Range of the western United States. J Geophys Res: Solid Earth 94(B6):7952–7960

Glen RA (2013) Refining accretionary orogen models for the Tasmanides of eastern Australia. Aust J Earth Sci 60(3):315–370

Glerum A, Brune S, Stamps DS, Strecker MR (2020) Victoria continental microplate dynamics controlled by the lithospheric strength distribution of the East African Rift. Nat Comms 11(1):1–5

Goleby BR, Korsch RJ, Fomin T, Bell B, Nicoll MG, Drummond BJ, Owen AJ (2002) Preliminary 3–D geological model of the Kalgoorlie region, Yilgarn Craton, Western Australia, based on deep seismic-reflection and potential-field data. Aust J Earth Sci 49(6):917–933

Goodwin AM, Smith IE (1980) Chemical discontinuities in Archean metavolcanic terrains and the development of Archean crust. Precamb Res 10(3–4):301–311

Gorman BE, Pearce TH, Birkett TC (1978) On the structure of Archean greenstone belts. Precamb Res 6(1):23–41

Grand SP, Van der Hilst RD, Widiyantoro S (1997) High resolution global tomography: a snapshot of convection in the Earth. Geol Soc Am Today 7(4):1–7

Grebennikov AV (2014) A-type granites and related rocks: petrogenesis and classification. Russian Geol Geophys 55(11):1353–1366

Green DH (1969) The origin of basaltic and nephelinitic magmas in the Earth's mantle. Tectonophysics 7(5–6):409–422

Gregg PM, De Silva SL, Grosfils EB, Parmigiani JP (2012) Catastrophic caldera-forming eruptions: thermomechanics and implications for eruption triggering and maximum caldera dimensions on Earth. J Volcanol Geotherm Res 241:1–12

Grobys JW, Gohl K, Uenzelmann-Neben G, Davy B, Barker D (2009) Extensional and magmatic nature of the Campbell Plateau and Great South Basin from deep crustal studies. Tectonophysics 472(1–4):213–225

Grosch EG, Slama J (2017) Evidence for 3.3-billion-year-old oceanic crust in the Barberton greenstone belt, South Africa. Geology 45 (8):695–698

Grove TL, Till CB (2015) Melting the Earth's upper mantle. In: Sigurdsson H, Houghton B, McNutt SR, Rymer H, Stix J (eds) The encyclopedia of volcanoes. Academic Press Elsevier, pp 35–47

Grove TL, Till CB, Krawczynski MJ (2012) The role of H_2O in subduction zone magmatism. Ann Rev Earth Planet Sci 40:413–439

Gudmundsson A (2000) Dynamics of volcanic systems in Iceland: example of tectonism and volcanism at juxtaposed hot spot and mid-ocean ridge systems. Ann Rev Earth Planet Sci 28(1): 107–140

Gudmundsson A, Marinoni LB, Marti J (1999) Injection and arrest of dykes: implications for volcanic hazards. J Volcanol Geotherm Res 88(1–2):1–13

Gudmundsson A, Philipp SL (2006) How local stress fields prevent volcanic eruptions. J Volcanol Geotherm Res 158(3–4):257–268

Gurenko AA, Bindeman IN, Sigurdsson IA (2015) To the origin of Icelandic rhyolites: insights from partially melted leucocratic xenoliths. Contribs Miner Petrol 169(5):1–21

Hall R (2009) The Eurasian SE Asian margin as a modern example of an accretionary orogen. In: Cawood PA, Kröner A (eds) Earth accretionary systems in space and time. Geol Soc Lond Spec Publ 318(1):351–372

Hall R, Smyth HR (2008) Cenozoic arc processes in Indonesia: Identification of the key influences on the stratigraphic record in active volcanic arcs. In: Draut AE, Clift PD, Scholl DW (eds) Formation and applications of the sedimentary record in arc collision zones. Geol Soc Amer Spec Pap 436:27–54

Hames W, Unger D, Saunders J, Kamenov G (2009) Early Yellowstone hotspot magmatism and gold metallogeny. J Volcanol Geotherm Res 188(1–3):214–222

Hamilton WB (1979) Tectonics of the Indonesian region. US Geol Surv Prof Pap 1078

Hamilton WB (1995) Subduction systems and magmatism. In: Smellie JL (ed) Volcanism associated with extension at consuming plate margins. Geol Soc Lond Spec Publ 81(1):3–28

Hammond WC, Thatcher W (2004) Contemporary tectonic deformation of the Basin and Range province, western United States: 10 years of observation with the Global Positioning System. J Geophys Res: Solid Earth 109(B8)

Hand JL, Cas RA, Ong L, Brown SJ, Krapez B, Barley ME (2002) Syn-and Post-Eruptive Volcaniclastic Sedimentation in Late Archaean Subaqueous Depositional Systems of the Black Flag Group, Eight Mile Dam, Kalgoorlie, Western Australia. Int Assoc Sediment Spec Pub 33:235–258

Hastie ARA (2020) Adakites. Chapter in reference module in Earth systems and environmental sciences Elsevier. https://doi.org/10.1016/B978-0-12-409548-9.12496-0

Hawkesworth C, Cawood PA, Dhuime B (2020) The evolution of the continental crust and the onset of plate tectonics. Front Earth Sci 8:326. https://doi.org/10.3389/feart.2020.00326
Hawkesworth CJ, Gallagher K, Hergt JM, McDermott F (1993) Mantle and slab contributions in arc magmas. Ann Rev Earth Planet Sci 21 (1):175–204
Hayman PC, Fullard R, Cas RAF, Squire RJ (2020) Clasts in Archean conglomerates and implications for uplift: evidence from the 2.7 Ga Agnew Greenstone Belt (Western Australia). Precamb Res 23:105758
Hayman PC, Thébaud N, Pawley MJ, Barnes SJ, Cas RAF, Amelin Y, Sapkota J, Squire RJ, Campbell IH, Pegg I (2015) Evolution of a ~2.7 Ga large igneous province: a volcanological, geochemical and geochronological study of the Agnew Greenstone Belt, and new regional correlations for the Kalgoorlie Terrane (Yilgarn Craton, Western Australia). Precamb Res 270:334–368
Hawkins JW, Bloomer SH, Evans CA, Melchior JT (1984) Evolution of intra-oceanic arc-trench systems. Tectonophysics 102(1–4): 175–205
Hémond C, Hofmann AW, Vlastelic I, Nauret F (2006) Origin of MORB enrichment and relative trace element compatibilities along the mid-Atlantic Ridge between 10 and 24 N 2006. Geochem Geophys Geosyst 7(12):22
Henk A (1997) Gravitational orogenic collapse vs plate-boundary stresses: a numerical modelling approach to the Permo-Carboniferous evolution of Central Europe. Geol Rundsch 86(1): 39–55
Hennig J, Hall R, Forster MA, Kohn BP, Lister GS (2001) Rapid cooling and exhumation as a consequence of extension and crustal thinning: inferences from the Late Miocene to Pliocene Palu Metamorphic Complex, Sulawesi, Indonesia. Tectonophysics 712:600–622
Henry CD, Castor SB, Starkel WA, Ellis BS, Wolff JA, Laravie JA, McIntosh WC, Heizler MT (2017) Geology and evolution of the McDermitt caldera, northern Nevada and southeastern Oregon, western USA. Geosphere 13(4):1066–1112
Heron PJ (2019) Mantle plumes and mantle dynamics in the Wilson cycle. Geol Soc Lond Spec Pub 470(1):87–103
Herzberg C, Condie K, Korenaga J (2010) Thermal history of the Earth and its petrological expression. Earth Planet Sci Letts 292(1–2): 79–88
Hill KC (1989) The Muller anticline, Papua New Guinea; basement-cored, inverted extensional fault structures with opposite vergence. Tectonophysics 158(1–4):227–245
Hill KC, Kendrick RD, Crowhurst PV, Gow PA (2002) Copper-gold mineralisation in New Guinea: tectonics, lineaments, thermochronology and structure. Aust J Earth Sci 49(4):737–752
Hillis RR, Sandiford M, Reynolds SD, Quigley MC (2008) Present-day stresses, seismicity and Neogene-to-Recent tectonics of Australia's 'passive' margins: intraplate deformation controlled by plate boundary forces. In: Johnson HD et al (eds) The nature and origin of compressive margins. Geol Soc Lond Spec Publ 306(1):71–90
Holm RJ, Richards SW (2013) A re-evaluation of arc–continent collision and along-arc variation in the Bismarck Sea region, Papua New Guinea. Aust J Earth Sci 60(5):605–619
Holm RJ, Rosenbaum G, Richards SW (2016) Post 8 Ma reconstruction of Papua New Guinea and Solomon Islands: microplate tectonics in a convergent plate boundary setting. Earth Sci Rev 156:66–81
Horton BK, Fuentes F, Boll A, Starck D, Ramirez SG, Stockli DF (2016) Andean stratigraphic record of the transition from backarc extension to orogenic shortening: a case study from the northern Neuquén Basin, Argentina. J South Am Earth Sci 71:17–40
Hu J, Liu L (2016) Abnormal seismological and magmatic processes controlled by the tearing South American flat slabs. Earth Planet Sci Letts 450:40–51
Huang Y, Hawkesworth C, Smith I, van Calsteren P, Black P (2000) Geochemistry of late Cenozoic basaltic volcanism in Northland and Coromandel, New Zealand: implications for mantle enrichment processes. Chem Geol 164(3):219–238
Hughes SS, Smith RP, Hackett WR, Anderson SR (1999) Mafic volcanism and environmental geology of the eastern Snake River Plain, Idaho. In: Guidebook to the geology of eastern Idaho. Idaho Mus Nat Hist 143–168
Humphreys E (1995) Post-Laramide removal of the Farallon slab, western United States. Geology 23:987–990
Humphreys MC, Menand T, Blundy JD, Klimm K (2008) Magma ascent rates in explosive eruptions: Constraints from H2O diffusion in melt inclusions. Earth Planet Sci Letts 270(1–2):25–40
Huppert HE, Sparks RSJ (1988) The fluid dynamics of crustal melting by injection of basaltic sills. Earth Environ Sci Trans R Soc Edinburgh 79(2–3):237–243
Husson L, Bernet M, Guillot S, Huyghe P, Mugnier JL, Replumaz A, Robert X, Van der Beek P (2014) Dynamic ups and downs of the Himalaya. Geology 42(10):839–842
Isola I, Mazzarini F, Bonini M, Corti G (2014) Spatial variability of volcanic features in early-stage rift settings: the case of the Tanzania Divergence, East African Rift System. Terra Nova 26(6):461–468
James KH (2009) In situ origin of the Caribbean: discussion of data. In: James KH, Lorente Ma, Pindell JL (eds) The Origin and Evolution of the Caribbean Plate. Geol Soc Lond Spec Publ 328(1):77–125
Jellinek AM, DePaolo DJ (2003) A model for the origin of large silicic magma chambers: precursors of caldera-forming eruptions. Bull Volcanol 65(5):363–381
Juteau T, Maury RC (1999) The oceanic crust, from accretion to mantle recycling. Springer, London
John DA (2001) Miocene and early Pliocene epithermal gold-silver deposits in the northern Great Basin, western United States: characteristics, distribution, and relationship to magmatism. Econ Geol 96(8):1827–1853
Johnson CL, Webb LE, Graham SA, Hendrix MS, Badarch G (2001) Sedimentary and structural records of late Mesozoic high-strain extension and strain partitioning, East Gobi basin, southern Mongolia. Geol Soc Am Mem 194:413–434
Johnson TE, Brown M, Gardiner NJ, Kirkland CL, Smithies RH (2017) Earth's first stable continents did not form by subduction. Nat 543 (7644):239–242
Johnson TE, Gardiner NJ, Miljković K, Spencer CJ, Kirkland CL, Bland PA, Smithies H (2018) An impact melt origin for Earth's oldest known evolved rocks. Nat Geosci 11(10):795–799
Johnson TE, Kirkland CL, Gardiner NJ, Brown M, Smithies RH, Santosh M (2019) Secular change in TTG compositions: implications for the evolution of Archaean geodynamics. Earth Planet Sci Letts 505:65–75
Jolivet L, Brun JP (2010) Cenozoic geodynamic evolution of the Aegean. Int J Earth Sci 99(1):109–138
Jolivet L, Faccenna C (2000) Mediterranean extension and the Africa-Eurasia collision. Tectonics 19(6):1095–1106
Jolivet L, Faccenna C, Huet B, Labrousse L, Le Pourhiet L, Lacombe O, Lecomte E, Burov E, Denèle Y, Brun JP, Philippon M (2013) Aegean tectonics: strain localisation, slab tearing and trench retreat. Tectonophysics 597:1–33
Jolivet L, Menant A, Sternai P, Rabillard A, Arbaret L, Augier R, Laurent V, Beaudoin A, Grasemann B, Huet B, Labrousse L (2015) The geological signature of a slab tear below the Aegean. Tectonophysics 659:166–182
Jones SA, Cassidy KF, Davis BK (2020) Unravelling the D1 event: evidence for early granite-up, greenstone-down tectonics in the Eastern Goldfields. Western Australia. Aust J Earth Sci 1–35. https://doi.org/10.1080/08120099.2020.1755364

Joyce EB (1975) Quaternary volcanism and tectonics in southeastern Australia. Bull R Soc New Zealand 13:169–176

Kamber BS (2015) The evolving nature of terrestrial crust from the Hadean, through the Archaean, into the Proterozoic. Precamb Res 258:48–82

Karig DE (1970) Ridges and basins of the Tonga-Kermadec island arc system. J Geophys Res 75(2):239–254

Karig DE (1971) Structural history of the Mariana island arc system. Geol Soc Am Bull 82(2):323–344

Karig DE (1972) Remnant Arcs. Geol Soc Am Bull 83(4):1057–1068

Karig DE (1974) Evolution of arc systems in the western Pacific. Ann Rev Earth Planet Sci 2(1):51–75

Karsli O, Dokuz A, Uysal İ, Aydin F, Kandemir R, Wijbrans J (2010) Generation of the Early Cenozoic adakitic volcanism by partial melting of mafic lower crust, Eastern Turkey: implications for crustal thickening to delamination. Lithos 114(1):109–120

Karsli O, Uysal I, Dilek Y, Aydin F, Kandemir R (2013) Geochemical modelling of early Eocene adakitic magmatism in the Eastern Pontides, NE Anatolia: continental crust or subducted oceanic slab origin? Int Geol Rev 55(16):2083–2095

Kasting JF (1993) Earth's early atmosphere. Science 259(5097):920–926

Kato T, Beavan J, Matsushima T, Kotake Y, Camacho JT, Nakao S (2003) Geodetic evidence of back-arc spreading in the Mariana Trough. Geophys Res Letts 30(12)

Katumwehe AB, Abdelsalam MG, Atekwana EA, Laó-Dávila DA (2016) Extent, kinematics and tectonic origin of the Precambrian Aswa Shear Zone in eastern Africa. Gondwana Res 34:241–253

Katz MB (1987) East African rift and northeast lineaments: continental spreading-transform system? J Afr Earth Sci 6(1):103–107

Kay SM, Coira B, Viramonte J (1994) Young mafic back arc volcanic rocks as indicators of continental lithospheric delamination beneath the Argentine Puna plateau, central Andes. J Geophys Res: Solid Earth 99(B12):24323–24339

Keller RA, Fisk MR, Smellie JL, Strelin JA, Lawver LA (2002) Geochemistry of back arc basin volcanism in Bransfield Strait, Antarctica: subducted contributions and along-axis variations. J Geophys Res: Solid Earth 107(B8)

Keller T, Katz RF, Hirschmann MM (2017) Volatiles beneath mid-ocean ridges: deep melting, channelised transport, focusing, and metasomatism. Earth Planet Sci Letts 464:55–68

Kemp AI, Hawkesworth CJ, Collins WJ, Gray CM, Blevin PL (2009) Isotopic evidence for rapid continental growth in an extensional accretionary orogen: The Tasmanides, eastern Australia. Earth Planet Sci Letts 284(3–4):455–466

Kendall JM, Lithgow-Bertelloni C (2016) Why is Africa rifting? In: Wright TJ et al (eds) Magmatic rifting and active volcanism. Geol Soc Lond Spec Publ 420(1):11–30

Kerr AC (1998) Oceanic plateau formation: a cause of mass extinction and black shale deposition around the Cenomanian-Turonian boundary? J Geol Soc Lond 155(4):619–626

Kerr AC (2005) Oceanic LIPs: the kiss of death. Elements 1(5):289–292

Kerr AC, White RV, Thompson PM, Tarney J, Saunders AD (2003) No oceanic plateau-no Caribbean plate? The seminal role of an oceanic plateau in Caribbean plate evolution. In: Bartolini C, Buffler RT, Blickwede J (eds) The circum-Gulf of Mexico and Carribean. Assoc Amer Petrol Geol Mem 79:126–168

Kisters AF, Stevens G, Dziggel A, Armstrong RA (2003). Extensional detachment faulting and core-complex formation in the southern Barberton granite–greenstone terrain, South Africa: evidence for a 3.2 Ga orogenic collapse. Precamb Res 127(4):355–378

Klein EM, Langmuir CH (1987) Global correlations of ocean ridge basalt chemistry with axial depth and crustal thickness. J Geophys Res: Solid Earth 92(B8):8089–8115

Koppers AA, Morgan JP, Morgan JW, Staudigel H (2001) Testing the fixed hotspot hypothesis using $^{40}Ar/^{39}Ar$ age progressions along seamount trails. Earth Planet Sci Letts 185(3):237–252

Korme T, Chorowicz J, Collet B, Bonavia FF (1997) Volcanic vents rooted on extension fractures and their geodynamic implications in the Ethiopian Rift. J Volcanol Geotherm Res 79(3–4):205–222

Kositcin N, Brown SJ, Barley ME, Krapež B, Cassidy KF, Champion DC (2008) SHRIMP U-Pb zircon age constraints on the Late Archaean tectonostratigraphic architecture of the Eastern Goldfields superterrane, Yilgarn craton. Western Australia. Precamb Res 161 (1–2):5–33

Koulakov I, Kasatkina E, Shapiro NM, Jaupart C, Vasilevsky A, El Khrepy S, Al-Arifi N, Smirnov S (2016) The feeder system of the Toba supervolcano from the slab to the shallow reservoir. Nat Comms 7:12228

Krapež B, Brown SJ, Hand J, Barley ME, Cas RAF (2000) Age constraints on recycled crustal and supracrustal sources of Archaean metasedimentary sequences, Eastern Goldfields Province, Western Australia: evidence from SHRIMP zircon dating. Tectonophysics 322(1–2):89–133

Kruber C, Thorseth IH, Pedersen RB (2008) Seafloor alteration of basaltic glass: textures, geochemistry, and endolithic microorganisms. Geochem Geophys Geosyst 9(12)

Kulm LD, Thornburg TM, Schrader HJ, Resig JM (1982) Cenozoic structure, stratigraphy and tectonics of the central Peru forearc. In: Legget JK (ed) Trench-forearc geology: sedimentation and tectonics in modern and ancient active plate margins. Geol Soc Lond Spec Publ 10(1):151–169

Kuritani T, Xia QK, Kimura JI, Liu J, Shimizu K, Ushikubo T, Zhao D, Nakagawa M, Yoshimura S (2019) Buoyant hydrous mantle plume from the mantle transition zone. Sci Rep 9(1):1–7

Lambiase JJ (1995) Hydrocarbon habitat in rift basins. Geol Soc Lond Spec Publ 80:381p

Lambiase JJ, Bosworth W (1995) Structural controls on sedimentation in continental rifts. In: Lambiase JJ (ed) Hydrocarbon habitat in rift basins. Geol Soc Lond Spec Publ 80(1):117–144

Lawver LA, Hawkins JW (1978) Diffuse magnetic anomalies in marginal basins: their possible tectonic and petrologic significance. Tectonophysics 45(4):323–339

Langmuir CH, Forsyth DW (2007) Mantle melting beneath mid-ocean ridges. Oceanography 20(1):78–89

Langmuir CH, Klein EM, Plank T (1992) Petrological systematics of mid-ocean ridge basalts: Constraints on melt generation beneath ocean ridges. In: Phipps et al (eds) Mantle flow and melt generation at mid-ocean ridges. Am Geophys Union Monogr 71:183–280

Laó-Dávila DA, Al-Salmi HS, Abdelsalam MG, Atekwana EA (2015) Hierarchical segmentation of the Malawi Rift: the influence of inherited lithospheric heterogeneity and kinematics in the evolution of continental rifts. Tectonics 34(12):2399–2417

Lee J, Hacker BR, Dinklage WS, Wang Y, Gans P, Calvert A, Wan J, Chen W, Blythe AE, McClelland W (2000) Evolution of the Kangmar Dome, southern Tibet: structural, petrologic, and thermochronologic constraints. Tectonics 19(5):872–895

Lee MD, Walworth NG, Sylvan JB, Edwards KJ, Orcutt BN (2015) Microbial communities on seafloor basalts at Dorado Outcrop reflect level of alteration and highlight global lithic clades. Front Microbiol 6:1470

Leeder MR (1995) Continental rifts and proto-oceanic rift troughs. In: Busby CJ, Ingersoll RV (eds) Tectonics of sedimentary basins. Blackwell Science, Cambridge, pp 119–148

Leeder MR, Gawthorpe RL (1987) Sedimentary models for extensional tilt-block/half-graben basins. In: Coward MP, Dewey JF, Hancock PL (eds) Continental extensional tectonics. Geol Soc Lond Spec Publ 28(1):139–152

Le Gall B, Nonnotte P, Rolet J, Benoit M, Guillou H, Mousseau-Nonnotte M, Albaric J, Déverchère J (2008) Rift propagation at craton margin: distribution of faulting and volcanism in the North Tanzanian Divergence (East Africa) during Neogene times. Tectonophysics 448(1–4):1–9

Lesti C, Giordano G, Salvini F, Cas R (2008) Volcano tectonic setting of the intraplate, Pliocene-Holocene, Newer Volcanic Province (southeast Australia): role of crustal fracture zones. J Geophys Res: Solid Earth 113(B7)

Lesti C, Porreca M, Giordano G, Mattei M, Cas RA, Wright HM, Folkes CB, Viramonte J (2011) High-temperature emplacement of the Cerro Galán and Toconquis Group ignimbrites (Puna plateau, NW Argentina) determined by TRM analyses. Bull Volcanol 73 (10):1535–1565

Leys CA, Cloos M, New BT, MacDonald GD (2012) Copper-gold ±molybdenum deposits of the Ertsberg-Grasberg District. Papua, Indonesia: In: Hedenquist JW, Harris M, Camus F (eds) Geology and genesis of major copper deposits and districts of the world. Soc Econ Geol Spec Publ 16:215–235

Li C, Arndt NT, Tang Q, Ripley EM (2015) Trace element indiscrimination diagrams. Lithos 232:76–83

Lin S, Parks J, Heaman LM, Simonetti A, Corkery MT (2013) Diapirism and sagduction as a mechanism for deposition and burial of "Timiskaming-type" sedimentary sequences, Superior Province: Evidence from detrital zircon geochronology and implications for the Borden Lake conglomerate in the exposed middle to lower crust in the Kapuskasing uplift. Precamb Res 238:148–157

Lindley ID (1988) Early Cainozoic stratigraphy and structure of the Gazelle Peninsula, east New Britain: an example of extensional tectonics in the New Britain arc-trench complex. Aust J Earth Sci 35:231–244

Lindley ID (2016) Plate flexure and volcanism: Late Cenozoic tectonics of the Tabar-Lihir-Tanga-Feni alkalic province, New Ireland Basin, Papua New Guinea. Tectonophysics 677:312–323

Linnell T, Shane P, Smith I, Augustinus P, Cronin S, Lindsay J, Maas R (2016) Long-lived shield volcanism within a monogenetic basaltic field: The conundrum of Rangitoto volcano. New Zealand. Geol Soc Amer Bull 128(7–8):1160–1172

Lips AL, Cassard D, Sözbilir H, Yilmaz H, Wijbrans JR (2001) Multistage exhumation of the Menderes massif, western Anatolia (Turkey). Int J Earth Sci 89(4):781–792

Lissenberg CJ, MacLeod CJ, Bennett EN (2019) Consequences of a crystal mush-dominated magma plumbing system: a mid-ocean ridge perspective. Phil Trans R Soc A 377(2139):20180014

Lister GS, Banga G, Feenstra A (1984) Metamorphic core complexes of Cordilleran type in the Cyclades, Aegean Sea, Greece. Geology 12(4):221–225

Lister GS, Etheridge MA, Symonds PA (1986) Detachment faulting and the evolution of passive continental margins. Geology 14 (3):246–250

Liu L, Stegman DR (2011) Segmentation of the Farallon slab. Earth Planet Sci Letts 311(1–2):1–10

Loewen MW, Duncan RA, Kent AJ, Krawl K (2013) Prolonged plume volcanism in the Caribbean Large Igneous Province: new insights from Curaçao and Haiti. Geochem Geophys Geosyst 14(10):4241–4259

Lowrey JR, Wyman DA, Ivanic TJ, Smithies RH, Maas R (2019) Archean boninite-like rocks of the northwestern Youanmi Terrane, Yilgarn Craton: Geochemistry and genesis. J Petrol 60(11):2131–2168

Lucassen F, Becchio R, Wilke HG, Franz G, Thirlwall MF, Viramonte J, Wemmer K (2000) Proterozoic-Paleozoic development of the basement of the Central Andes (18–26 S)—a mobile belt of the South American craton. J South Amer Earth Sci 13(8):697–715

Lucassen F, Becchio R, Franz G (2011) The Early Palaeozoic high-grade metamorphism at the active continental margin of West Gondwana in the Andes (NW Argentina/N Chile). Int J Earth Sci 100(2–3):445–463

Lüschen E, Müller C, Kopp H, Engels M, Lutz R, Planert L, Shulgin A, Djajadihardja YS (2011) Structure, evolution and tectonic activity of the eastern Sunda forearc, Indonesia, from marine seismic investigations. Tectonophysics 508(1–4):6–21

Maccaferri F, Bonafede M, Rivalta E (2010) A numerical model of dyke propagation in layered elastic media. Geophys J Int 180 (3):1107–1123

Maccaferri F, Rivalta E, Keir D, Acocella V (2014) Unloading-driven off-axis volcanism in rift zones. Nat Geosci 7:297–300

Macdonald KC (1982) Mid-ocean ridges: fine scale tectonic, volcanic and hydrothermal processes within the plate boundary zone. Ann Rev Earth Planet Sci 10(1):155–190

Macdonald KC (2001) Mid-Ocean ridge tectonics, volcanism, and geomorphology. Encyclopedia of Ocean Sciences. Academic Press, Oxford, pp 852–866

Macdonald R (2002) Magmatism of the Kenya Rift Valley: a review. Earth Environ Sci Trans Royal Soc Edinburgh 93(3):239–253

Macgregor D (2015) History of the development of the East African Rift System: a series of interpreted maps through time. J Afr Earth Sci 101:232–252

MacPherson GJ (1983) The Snow Mountain volcanic complex: an on-land seamount in the Franciscan terrain, California. J Geol 91 (1):73–92

Mana S, Furman T, Turrin BD, Feigenson MD, Swisher III CC (2015) Magmatic activity across the East African North Tanzanian divergence zone. J Geol Soc 172(3):368–389

Manga M, Brodsky E (2006) Seismic triggering of eruptions in the far field: volcanoes and geysers. Ann Rev Earth Planet Sci 34:263–291

Mann P, Hempton MR, Bradley DC, Burke K (1983) Development of pull-apart basins. J Geol 91(5):529–554

Mann P, Taira A (2004) Global tectonic significance of the Solomon Islands and Ontong Java Plateau convergent zone. Tectonophysics 389(3):137–190

Marocchi M, Morelli C, Mair V, Klötzli U, Bargossi GM (2008) Evolution of large silicic magma systems: new U-Pb zircon data on the NW Permian Athesian Volcanic Group (Southern Alps, Italy). J Geol 116(5):480–498

Marsaglia KM (1995) Interarc and backarc basins. In: Busby CJ, Ingersoll RV (eds) Tectonics of sedimentary basins. Blackwell Science, Cambridge, pp 299–329

Martí J (1991) Caldera-like structures related to Permo-Carboniferous volcanism of the Catalan Pyrenees (NE Spain). J Volcanol Geotherm Res 45(3–4):173–186

Martin H, Smithies RH, Rapp R, Moyen JF, Champion D (2005) An overview of adakite, tonalite–trondhjemite–granodiorite (TTG), and sanukitoid: relationships and some implications for crustal evolution. Lithos 79(1–2):1–24

Masuda H, Fryer P (2015) Geochemical characteristics of active backarc basin volcanism at the southern end of the Mariana Trough. In: Subseafloor biosphere linked to hydrothermal systems. Springer, Tokyo, pp 261–273

Mather B, Müller RD, Seton M, Ruttor S, Nebel O, Mortimer N (2020) Intraplate volcanism triggered by bursts in slab flux. Sci Adv 6: eabdo0953

Mathieu L, de Vries BV, Pilato M, Troll VR (2011) The interaction between volcanoes and strike-slip, transtensional and transpressional fault zones: analogue models and natural examples. J Struct Geol 33(5):898–906

Matsumoto S, Nakao S, Ohkura T, Miyazaki M, Shimizu H, Abe Y, Yamashita Y (2015). Spatial heterogeneities in tectonic stress in

Kyushu, Japan and their relation to a major shear zone. Earth Planets Space 67(1):1–9
Mattei M, Conticelli S, Giordano G (2010) The Tyrrhenian margin geological setting: from the Apennine orogeny to the K-rich volcanism. In: Funiciello R, Giordano G (eds) The Colli Albani Volcano. Spec Publ IAVCEI and Geol Soc Lond 3:7–28
Matteini M, Mazzuoli R, Omarini R, Cas R, Maas R (2002) Geodynamical evolution of Central Andes at 24 S as inferred by magma composition along the Calama–Olacapato–El Toro transversal volcanic belt. J Volcanol Geotherm Res 118(1):205–228
McCaffrey R (2009) The tectonic framework of the Sumatran subduction zone. Ann Rev Earth Planet Sci 37:345–366
McCann T, Pascal C, Timmerman MJ, Krzywiec P, López-Gómez J, Wetzel L, Krawczyk CM, Rieke H, Lamarche J (2006) Post-Variscan (end Carboniferous-Early Permian) basin evolution in western and central Europe. In: Gee DG, Stephenson RA (eds) European lithosphere dynamics. Geol Soc Lond Mem 32 (1):355–388
McGoldrick KL, Squire RJ, Cas RAF, Briggs M, Tunjic J, Allen CM, Campbell IH, Hayman PC (2013) The largest Au deposits in the St Ives Goldfield (Yilgarn Craton, Western Australia) may be located in a major Neoarchean volcano-sedimentary depo-centre. Miner Deposita 48(7):861–881
McKenzie D (1978) Some remarks on the development of sedimentary basins. Earth Planet Sci Letts 40(1):25–32
McKenzie D, Bickle MJ (1988) The volume and composition of melt generated by extension of the lithosphere. J Petrol 29(3):625–679
McKibben MA, Andes JP, Williams AE (1988) Active ore formation at a brine interface in metamorphosed deltaic lacustrine sediments; the Salton Sea geothermal system, California. Econ Geol 83(3):511–523
McNeill LC, Henstock TJ (2014) Forearc structure and morphology along the Sumatra-Andaman subduction zone. Tectonics 33 (2):112–134
McQuarrie N, Wernicke BP (2005) An animated tectonic reconstruction of southwestern North America since 36 Ma. Geosphere 1 (3):147–172
Mechie J, Kind R (2013) A model of the crust and mantle structure down to 700 km depth beneath the Lhasa to Golmud transect across the Tibetan plateau as derived from seismological data. Tectonophysics 606:187–197
Meeuws FJE, Holford SP, Foden JD, Schofield N (2016) Distribution, chronology and causes of Cretaceous-Cenozoic magmatism along the magma-poor rifted southern Australian margin: links between mantle melting and basin formation. Marine Petrol Geol 73:271–298
Melnick D, Folguera A, Ramos VA (2006) Structural control on arc volcanism: The Caviahue-Copahue complex, Central to Patagonian Andes transition (38° S). J South Am Earth Sci 22(1–2):66–88
Menand T, Daniels KA, Benghiat P (2010) Dyke propagation and sill formation in a compressive tectonic environment. J Geophys Res: Solid Earth 115(B8)
Menard HW (1964) Marine geology of the Pacific (No. QE39. M46 1964). McGraw Hill, New York
Metcalfe I (1996) Pre-Cretaceous evolution of SE Asian terranes. In: Hall R, BlundellD (eds) Tectonic evolution of Southeast Asia. Geol Soc Lond Spec Publ 106(1):97–122
Metcalfe I (2013) Gondwana dispersion and Asian accretion: tectonic and palaeogeographic evolution of eastern Tethys. J Asian Earth Sci 66:1–33
Metcalf K, Kapp P (2015) Along-strike variations in crustal seismicity and modern lithospheric structure of the central Andean forearc. In: DeCelles PG, Ducea MN, Carrapa B, Kapp PA (eds) Geodynamics of a Cordilleran orogenic system: the Central Andes of Argentina and Northern Chile. Geol Soc Am Mem 212:61–78
Michibayashi K, Tominaga M, Ildefonse B, Teagle DA (2019) What lies beneath. The formation and evolution of oceanic lithosphere. Oceanography 32(1):138–149
Miller JM, Phillips D, Wilson CJ, Dugdale LJ (2005) Evolution of a reworked orogenic zone: the boundary between the Delamerian and Lachlan Fold Belts, southeastern Australia. Aust J Earth Sci 52 (6):921–940
Miller JS (2008) Assembling a pluton… one increment at a time. Geology 36(6):511–512
Miyashiro A (1972) Metamorphism and related magmatism in plate tectonics. Am J Sci 272(7):629–656
Miyashiro A (1974) Volcanic rock series in island arcs and active continental margins. Am J Sci 274:321–355
Moberly R, Shepherd GL, Coulbourn WT (1982) Forearc and other basins, continental margin of northern and southern Peru and adjacent Ecuador and Chile. In: Legget JK (ed) Trench-forearc geology: sedimentation and tectonics in modern and ancient active plate margins. Geol Soc Lond Spec Publ 10(1):171–189
Moghadam HS, Rossetti F, Lucci F, Chiaradia M, Gerdes A, Martinez ML, Nasrabady M (2016) The calc-alkaline and adakitic volcanism of the Sabzevar structural zone (NE Iran): implications for the Eocene magmatic flare-up in Central Iran. Lithos 248:517–535
Mohr PA, Wood CA (1976) Volcano spacings and lithospheric attenuation in the Eastern Rift of Africa. Earth Planet Sci Letts 33 (1):126–144
Mole DR, Fiorentini ML, Thebaud N, Cassidy KF, McCuaig TC, Kirkland CL, Romano SS, Doublier MP, Belousova EA, Barnes SJ, Miller J (2014) Archean komatiite volcanism controlled by the evolution of early continents. Proc Nat Acad Sci 111(28):10083–10088
Mole DR, Kirkland CL, Fiorentini ML, Barnes SJ, Cassidy KF, Isaac C, Belousova EA, Hartnady M, Thebaud N (2019) Time-space evolution of an Archean craton: A Hf-isotope window into continent formation. Earth Sci Rev 196:102831
Montelli R, Nolet G, Dahlen FA, Masters G (2006) A catalogue of deep mantle plumes: new results from finite-frequency tomography. Geochem Geophys Geosyst 7(11)
Monzier M, Danyushevsky LV, Crawford AJ, Bellon H, Cotten J (1993) High-Mg andesites from the southern termination of the New Hebrides island arc (SW Pacific). J Volcanol Geotherm Res 57 (3–4):193–217
Moore GF, Karig DE (1980) Structural geology of Nias Island, Indonesia; implications for subduction zone tectonics. Am J Sci 280 (3):193–223
Moore GF, Shipley TH, Stoffa PL, Karig DE, Taira A, Kuramoto S, Tokuyama H, Suyehiro K (1990) Structure of the Nankai Trough accretionary zone from multichannel seismic reflection data. J Geophys Res: Solid Earth 95(B6):8753–8765
Moore GF, Boston BB, Strasser M, Underwood MB, Ratliff RA (2015) Evolution of tectono-sedimentary systems in the Kumano Basin, Nankai Trough forearc. Marine Petrol Geol 67:604–616
Moores EM (1982) Origin and emplacement of ophiolites. Rev Geophys 20(4):735–760
Morgan WJ (1971) Convection plumes in the lower mantle. Nature 230 (5288):42
Morley CK (1989) Extension, detachments, and sedimentation in continental rifts (with particular reference to East Africa): Tectonics 8 1175–1192
Morley CK (1995) Developments in the structural geology of rifts over the last decade and their impact on hydrocarbon exploration. In: Lambiase JJ (ed) Hydrocarbon habitat in rift basins. Geol Soc Lond Spec Publ 80(1):1–32

Morley CK (2010) Stress re-orientation along zones of weak fabrics in rifts: an explanation for pure extension in 'oblique' rift segments? Earth Planet Sci Letts 297(3–4):667–673

Morley CK, Nelson RA, Patton TL, Munn SG (1990) Transfer zones in the East African rift system and their relevance to hydrocarbon exploration in rifts. Amer Assoc Petrol Geol 74(8):1234–1253

Morley CK, Ngenoh DK, Ego JK (1999) Introduction to the East African Rift System. In: Morley CK (ed) Geoscience of rift systems: evolution of East Africa. Am Assoc Petrol Geol Stud Geol 44:1–18

Mortimer N, Campbell HJ, Tulloch AJ, King PR, Stagpoole VM, Wood RA, Rattenbury MS, Sutherland R, Adams CJ, Collot J, Seton M (2017) Zealandia: Earth's hidden continent. GSA Today 27(3):27–33

Moyen J-F, Laurent O (2018) Archaean tectonic systems: a view from igneous rocks. Lithos 302–303:99–125

Moyen J-F, Martin H (2012) Forty years of TTG research. Lithos 148:312–336

Muirhead JD, Kattenhorn SA (2018) Activation of preexisting transverse structures in an evolving magmatic rift in East Africa. J Struct Geol 106:1–8

Muirhead JD, Kattenhorn SA, Le Corvec N (2015) Varying styles of magmatic strain accommodation across the East African Rift. Geochem Geophys Geosyst 16(8):2775–2795

Muirhead JD, Kattenhorn SA, Lee H, Mana, S, Turrin BD, Fischer TP, Kianji G, Dindi E, Stamps DS (2016) Evolution of upper crustal faulting assisted by magmatic volatile release during early-stage continental rift development in the East African Rift. Geosphere 12 (6):1670–1700. https://doi.org/10.1130/GES01375.1

Mulwa JK, Kimata F (2014) Tectonic structures across the East African Rift based on the source parameters of the 20 May 1990 M7. 2 Sudan earthquake. Nat Haz 73:493–506

Mungall JE (2002) Roasting the mantle: slab melting and the genesis of major Au and Au-rich Cu deposits. Geology 30(10):915–918

Murphy JB (2007) Igneous rock associations 8. Arc magmatism II: geochemical and isotopic characteristics. Geoscience Canada 34(1)

Nabelek PI, Liu M (2004) Petrologic and thermal constraints on the origin of leucogranites in collisional orogens. Earth Environ Sci Trans R Soc Edinburgh 95(1–2):73–85

Nakamura K (1977) Volcanoes as possible indicators of tectonic stress orientation: principle and proposal. J Volcanol Geotherm Res 2 (1):1–6

Nebel O, Capitanio FA, Moyen JF, Weinberg RF, Clos F, Nebel-Jacobsen YJ, Cawood PA (2018) When crust comes of age: on the chemical evolution of Archaean, felsic continental crust by crustal drip tectonics. Phil Trans Roy Soc Lond Ser a: Math Phys Engin Sci 376(2132):20180103

Németh K (2010) Monogenetic volcanic fields: origin, sedimentary record, and relationship with polygenetic volcanism. In: Canon-Tapia E, Szakacs A (eds) What is a Volcano? Geol Soc Am, Special Papers, 470, 43–66

Newmark RL, Anderson RN, Moos D, Zoback MD (1985) Structure, porosity and stress regime of the upper oceanic crust: Sonic and ultrasonic logging of DSDP hole 504B. Tectonophysics 118(1–2):1–42

Niemi NA, Wernicke BP, Friedrich AM, Simons M, Bennett RA, Davis JL (2004) BARGEN continuous GPS data across the eastern Basin and Range province, and implications for fault system dynamics. Geophys J Int 159(3):842–862

Nilsen TH, Sylvester AG (1995) Strike-slip basins. In: Busby CJ, Ingersoll RV (eds) Tectonics of sedimentary basins. Blackwell Science, Cambridge, pp 425–457

Noda A (2016) Forearc basins: types, geometries, and relationships to subduction zone dynamics. Geol Soc Am Bull 128(5–6):879–895

Norini G, Baez W, Becchio R, Viramonte J, Giordano G, Arnosio M, Pinton A, Groppelli G (2013) The Calama-Olacapato-El Toro fault system in the Puna Plateau, Central Andes: geodynamic implications and stratovolcanoes emplacement. Tectonophysics 608: 1280–1297

O'Neill C, Debaille V, Griffin W (2013) Deep earth recycling in the Hadean and constraints on surface tectonics. Am J Sci 313(9): 912–932

O'Neil J, Carlson RW (2017) Building Archean cratons from Hadean mafic crust. Science 355(6330):1199–1202

Ortiz-Hernández LE, Sandoval OAA, Castro KF (2003) Early Cretaceous intraplate seamounts from Guanajuato, central Mexico: geochemical and mineralogical data. Rev Mex Cien Geol 20 (1):27–40

Ozacar AA, Zandt G (2004) Crustal seismic anisotropy in central Tibet: implications for deformational style and flow in the crust. Geophys Res Letts 31(23)

Palin RM, Santosh M, Cao W, Li SS, Hernández-Uribe D, Parsons A (2020) Secular change and the onset of plate tectonics on Earth. Earth Sci Rev 207:103172

Palin RM, White RW (2016) Emergence of blueschists on Earth linked to secular changes in oceanic crust composition. Nat Geosci 9 (1):60–64

Paquet F, Proust JN, Barnes PM, Pettinga JR (2011) Controls on active forearc basin stratigraphy and sediment fluxes: the Pleistocene of Hawke Bay, New Zealand. Geol Soc Amer Bull 123(5–6):1074–1096

Parmenter AC, Lin S, Timothy Corkery M (2006) Structural evolution of the Cross Lake greenstone belt in the northwestern Superior Province, Manitoba: implications for relationship between vertical and horizontal tectonism. Canad J Earth Sci 43(7):767–787

Parsons T, Thompson GA, Smith RP (1998) More than one way to stretch: a tectonic model for extension along the plume track of the Yellowstone hotspot and adjacent Basin and Range Province. Tectonics 17(2):221–234

Paterson SR, Okaya D, Memeti V, Economos R, Miller RB (2011) Magma addition and flux calculations of incrementally constructed magma chambers in continental margin arcs: combined field, geochronologic, and thermal modeling studies. Geosphere 7 (6):1439–1468

Pe-Piper G, Tsikouras B, Hatzipanagiotou K (2004) Evolution of boninites and island-arc tholeiites in the Pindos Ophiolite, Greece. Geol Mag 141(4):455–469

Peccerillo A (2017) The Aeolian arc. In: Cenozoic Volcanism in the Tyrrhenian Sea Region. Advance in Volcanol. Springer, pp 217–263

Peccerillo A, De Astis G, Faraone D, Forni F, Frezzotti ML (2013) Compositional variations of magmas in the Aeolian arc: implications for petrogenesis and geodynamics. In: Lucchi FP et al (eds) The Aeolian Islands Volcanoes. Geol Soc Lond Mem 37 (1):491–510

Pedersen AK, Larsen LM, Pedersen GK (2017) Lithostratigraphy, geology and geochemistry of the volcanic rocks of the Vaigat Formation on Disko and Nuusuag, Paleocene of West Greenland. Geol Surv Den Greenland Bull 39:244

Pfiffner OA, Gonzalez L (2013) Mesozoic-Cenozoic evolution of the western margin of South America: case study of the Peruvian Andes. Geosciences 3(2):262–310

Phillips RJ, Searle MP, Parrish RR (2013) The geochemical and temporal evolution of the continental lithosphere and its relationship to continental-scale faulting: The Karakoram Fault, eastern Karakoram, NW Himalayas. Geochem Geophys Geosyst 14(3):583–603

Piccolo A, Kaus BJ, White RW, Palin RM, Reuber GS (2020) Plume—Lid interactions during the Archean and implications for the generation of early continental terranes. Gondwana Res 88:150–168

Piccolo A, Palin RM, Kaus BJ, White RW (2019) Generation of Earth's early continents from a relatively cool Archean mantle. Geochem Geophys Geosyst 20(4):1679–1697

Pierce KL, Morgan LA (2009) Is the track of the Yellowstone hotspot driven by a deep mantle plume?-Review of volcanism, faulting, and uplift in light of new data. J Volcanol Geotherm Res 188(1–3):1–25

Pindell J, Kennan L, Stanek KP, Maresch WV, Draper G (2006) Foundations of Gulf of Mexico and Caribbean evolution: eight controversies resolved. Geol Acta 4(1–2):303–341

Pindell JL (1991) Geologic rationale for hydrocarbon exploration in the Caribbean and adjacent regions. J Petrol Geol 14(2):237–258

Planke S, Millett JM, Maharjan D, Jerram DA, Abdelmalak MM, Groth A, Hoffmann J, Berndt C, Myklebust R (2017) Igneous seismic geomorphology of buried lava fields and coastal escarpments on the Vøring volcanic rifted margin. Interpretation 5(3):1–17

Planke S, Rasmussen T, Rey SS, Myklebust R (2005) Seismic characteristics and distribution of volcanic intrusions and hydrothermal vent complexes in the Vøring and Møre basins. In: Petroleum geology conference series. Geol Soc Lond 6(1):833–844

Planke S, Symonds PA, Alvestad E, Skogseid J (2000) Seismic volcanostratigraphy of large-volume basaltic extrusive complexes on rifted margins. J Geophys Res: Solid Earth 105(B8):19335–19351

Platt JP (1980) Archaean greenstone belts: a structaral test of tectonic hypothesis. Tectonophysics 65(1–2):127–150

Polve M, Maury RC, Vidal P, Priadi B, Bellon H, Soeria-Atmadja R, Joron JL, Cotten J (2001) Melting of lower continental crust in a young post-collision setting; a geochemical study of Plio-Quaternary acidic magmatism from central Sulawesi (Indonesia). Bull De La Soc Géol France 172(3):333–342

Presnall DC, Gudfinnsson GH (2011) Oceanic volcanism from the low-velocity zone: without mantle plumes. J Petrol 52(7–8):1533–1546

Price RC, Nicholls IA, Day A (2014) Lithospheric influences on magma compositions of late Mesozoic and Cenozoic intraplate basalts (the Older Volcanics) of Victoria, south-eastern Australia. Lithos 206:179–200

Purcell PG (2018) Re-imagining and re-imaging the development of the East African Rift. Petrol Geosci 24(1):21–40

Putirka K, Platt B (2012) Basin and Range volcanism as a passive response to extensional tectonics. Geosphere 8(6):1274–1285

Quick JE, Sinigoi S, Peressini G, Demarchi G, Wooden JL, Sbisà A (2009) Magmatic plumbing of a large Permian caldera exposed to a depth of 25 km. Geology 37(7):603–606

Ramos VA (2009) Anatomy and global context of the Andes: main geologic features and the Andean orogenic cycle. In: Kay nSM, Ramos VA, Dickinson WR (eds) Backbone of the Americas: shallow subduction, plateau uplift, and ridge and terrane collision. Geol Soc Amer Mem 204:31–65

Ramos VA, Folguera A (2009) Andean flat-slab subduction through time. In: Murphy JB, Keppie JD, HynesAJ (eds) Ancient orogens and modern analogues. Geol Soc Lond Spec Publ 327(1):31–54

Ramos VA, Ghiglione MC (2008) Tectonic evolution of the Patagonian Andes. Develop Quat Sci 11:57–71

Ramos VA, Litvak VD, Folguera A, Spagnuolo M (2014) An Andean tectonic cycle: from crustal thickening to extension in a thin crust (34°-37° SL). Geosci Front 5(3):351–367

Rateau R, Schofield N, Smith M (2013) The potential role of igneous intrusions on hydrocarbon migration, west of Shetland. Petrol Geosci 19(3):259–272

Replumaz A, Negredo AM, Villasenor A, Guillot S (2010) Indian continental subduction and slab break-off during Tertiary collision. Terra Nova 22(4):290–296

Reubi O, Blundy J (2009) A dearth of intermediate melts at subduction zone volcanoes and the petrogenesis of arc andesites. Nature 461 (7268):1269–1273

Rey P, Vanderhaeghe O, Teyssier C (2001) Gravitational collapse of the continental crust: definition, regimes and modes. Tectonophysics 342(3–4):435–449

Rey PF, Coltice N, Flament N (2014) Spreading continents kick-started plate tectonics. Nature 513(7518):405–408

Richards JP (2011) High Sr/Y arc magmas and porphyry Cu±Mo±Au deposits: just add water. Econ Geol 106(7):1075–1081

Ring U, Glodny J, Will T, Thomson S (2010) The Hellenic subduction system: High-pressure metamorphism, exhumation, normal faulting, and large-scale extension. Ann Rev Earth Planet Sci 38:45–76

Risse A, Trumbull RB, Coira B, Kay SM, van den Bogaard P (2008) $^{40}Ar/^{39}Ar$ geochronology of mafic volcanism in the back-arc region of the southern Puna plateau, Argentina. J South Amer Earth Sci 26 (1):1–5

Robertson EA, Biggs J, Cashman KV, Floyd MA, Vye-Brown C (2016) Influence of regional tectonics and pre-existing structures on the formation of elliptical calderas in the Kenyan Rift. In: Wright TJ et al (eds) Magmatic rifting and active volcanism. Geol Soc Lond Spec Publ 420(1):43–67

Rocchi S, Storti F, Di Vincenzo G, Rossetti F (2003) Intraplate strike-slip tectonics as an alternative to mantle plume activity for the Cenozoic rift magmatism in the Ross Sea region, Antarctica. In: Storti F, Holdsworth RE, Salvini F (eds) Intraplate strike-slip deformation belts. Geol Soc Lond Spec Publ 210(1):145–158

Rollinson H (2017) There were no large volumes of felsic continental crust in the early Earth. Geosphere 13(2):235–246

Rooney TO (2017) The Cenozoic magmatism of East-Africa: Part I—flood basalts and pulsed magmatism. Lithos 286:264–301

Rooney TO (2020a) The Cenozoic magmatism of East Africa: Part II–rifting of the mobile belt. Lithos 360:105291

Rooney TO (2020b) The Cenozoic magmatism of East Africa: Part III–rifting of the craton. Lithos 360:105390

Rooney TO (2020c) The Cenozoic magmatism of East Africa: Part IV–the terminal stages of rifting preserved in the Northern East African Rift System. Lithos 360:105381

Rooney TO (2020d) The Cenozoic magmatism of East Africa: part V–magma sources and processes in the East African Rift. Lithos 360: 105296

Rosenbaum G, Gasparon M, Lucente FP, Peccerillo A, Miller MS (2008) Kinematics of slab tear faults during subduction segmentation and implications for Italian magmatism. Tectonics 27(2)

Rosendahl BR (1987) Architecture of continental rifts with special reference to East Africa. Ann Rev Earth Planet Sci 15(1):445–503

Rosengren NM, Cas RAF, Beresford SW, Palich BM (2008) Reconstruction of an extensive Archaean dacitic submarine volcanic complex associated with the komatiite-hosted Mt Keith nickel deposit, Agnew-Wiluna greenstone belt, Yilgarn Craton, Western Australia. Precamb Res 161(1–2):34–52

Rossetti F, Nasrabady M, Theye T, Gerdes A, Monié P, Lucci F, Vignaroli G (2014) Adakite differentiation and emplacement in a subduction channel: The late Paleocene Sabzevar magmatism (NE Iran). Geol Soc Am Bull 126(3–4):317–343

Rowland JV, Simmons SF (2012) Hydrologic, magmatic, and tectonic controls on hydrothermal flow, Taupo Volcanic Zone, New Zealand: implications for the formation of epithermal vein deposits. Econ Geol 107(3):427–457

Rowland JV, Wilson CJ, Gravley DM (2010) Spatial and temporal variations in magma-assisted rifting, Taupo Volcanic Zone, New Zealand. J Volcanol Geotherm Res 190(1–2):89–108

Ruellan E, Delteil J, Wright I, Matsumoto T (2003) From rifting to active spreading in the Lau Basin-Havre Trough backarc system (SW Pacific): locking/unlocking induced by seamount chain subduction. Geochem Geophys Geosyst 4(5)

Rutherford MJ (2008) Magma ascent rates. Rev Miner Geochem 69 (1):241–271

Saalmann K, Mänttäri I, Nyakecho C, Isabirye E (2016) Age, tectonic evolution and origin of the Aswa Shear Zone in Uganda: activation of an oblique ramp during convergence in the East African Orogen. J Afr Earth Sci 117:303–330

Saleeby J, Ducea M, Clemens-Knott D (2003) Production and loss of high-density batholithic root, southern Sierra Nevada, California. Tectonics 22(6):1064

Salvini F, Brancolini G, Busetti M, Storti F, Mazzarini F, Coren F (1997) Cenozoic geodynamics of the Ross Sea region, Antarctica: crustal extension, intraplate strike-slip faulting, and tectonic inheritance. J Geophys Res: Solid Earth 102(B11):24669–24696

Sample JC, Karig DE (1982) A volcanic production rate for the Mariana island arc. J Volcanol Geotherm Res 13(1–2):73–82

Sandiford MIK (2003) Neotectonics of southeastern Australia: linking the Quaternary faulting record with seismicity and in situ stress. Spec Pap Geol Soc Am 107–120

Saria E, Calais E, Stamps DS, Delvaux D, Hartnady CJ (2014) Present-day kinematics of the East African Rift. J Geophys Res: Solid Earth 119(4):3584–3600

Saunders AD, Tarney J (1984) Geochemical characteristics of basaltic volcanism within back-arc basins. In: Kokelaar P, Howells MF (eds) Marginal basin geology. Geol Soc Lond Spec Publ 16:59–76

Sawada H (2020) Estimation of secular change in the size of continents for understanding early crustal development. Front Earth Sci 8:541094. https://doi.org/10.3389/feart.2020.541094

Schaefer L, Elkins-Tanton LT (2018) Magma oceans as a critical stage in the tectonic development of rocky planets. Phil Trans Roy Soc Lond Ser A: Math Phys Engin Sci 376(2132):20180109

Schellart WP, Chen Z, Strak V, Duarte JC, Rosas FM (2019) Pacific subduction control on Asian continental deformation including Tibetan extension and eastward extrusion tectonics. Nat Commun 10(1):1–5

Schellart WP, Lister GS (2005) The role of the East Asian active margin in widespread extensional and strike-slip deformation in East Asia. J Geol Soc Lond 162(6):959–972

Schepers G, Van Hinsbergen DJ, Spakman W, Kosters ME, Boschman LM, McQuarrie N (2011) South-American plate advance and forced Andean trench retreat as drivers for transient flat subduction episodes. Nature Comms 8:15249

Scheuber E, Reutter KJ (1992) Magmatic arc tectonics in the central Andes between 21° and 25° S. Tectonophysics 205(1–3):127–140

Schnur SR, Gilbert LA (2012) Detailed volcanostratigraphy of an accreted seamount: implications for intraplate seamount formation. Geochem Geophys Geosys 13(12)

Scholz CH, Barazangi M, Sbar ML (1971) Late Cenozoic evolution of the Great Basin, western United States, as an ensialic interarc basin. Geol Soc Amer Bull 82(11):2979–2990

Sclater JG, Hawkins JW, Mammerickx J, Chase CG (1972) Crustal extension between the Tonga and Lau ridges: petrologic and geophysical evidence. Geol Soc Amer Bull 83(2):505–518

Scoon RN (2018) Geology of National Parks of Central/Southern Kenya and Northern Tanzania: Geotourism of the Gregory Rift Valley, active volcanism and regional plateaus. Springer Nature, 221p

Scrimgeour I, Sandiford M (1993) Early Proterozoic metamorphism at The Granites gold mine, Northern Territory: implications for the timing of fluid production in high-temperature, low-pressure terranes. Econ Geol 88(5):1099–1113

Sdrolias M, Müller RD (2006) Controls on back-arc basin formation. Geochem Geophys Geosyst 7(4)

Searle MP, Elliott JR, Phillips RJ, Chung SL (2011) Crustal–lithospheric structure and continental extrusion of Tibet. J Geol Soc Lond 168(3):633–672

Self S, Goff F, Gardner JN, Wright JV, Kite WM (1986) Explosive rhyolitic volcanism in the Jemez Mountains: vent locations, caldera development and relation to regional structure. J Geophys Res: Solid Earth 91(B2):1779–1798

Selvaggi G, Chiarabba C (1995) Seismicity and P-wave velocity image of the southern Tyrrhenian subduction zone. Geophys J Int 121 (3):818–826

Sembroni A, Faccenna C, Becker TW, Molin P, Abebe B (2016) Long-term, deep-mantle support of the Ethiopia-Yemen Plateau. Tectonics 35(2):469–488

Sendjaja YA, Kimura JI, Sunardi E (2009) Across-arc geochemical variation of Quaternary lavas in West Java, Indonesia: mass-balance elucidation using arc basalt simulator model. Island Arc 18(1): 201–224

Shackleton RM (1995) Tectonic evolution of greenstone belts. In: Coward MP, Reis, AC (eds) Early Precambrian processes. Geol Soc Lond Spec Publ 95(1):53–65

Shaw HR (1980) The fracture mechanisms of magma transport from the mantle to the surface. In Hargraves RB (ed) Physics of magmatic processes. Princeton Universiy Press, pp 201–64

Shaw HR (1985) Links between magma-tectonic rate balances, plutonism, and volcanism. J Geophys Res: Solid Earth 90 (B13):11275–11288

Shirey SB, Richardson SH (2011) Start of the Wilson cycle at 3 Ga shown by diamonds from subcontinental mantle. Science 333 (6041):434–436

Shuto K, Nohara-Imanaka R, Sato M, Takahashi T, Takazawa E, Kawabata H, Takanashi K, Ban M, Watanabe N, Fujibayashi N (2015) Across-arc variations in geochemistry of Oligocene to quaternary basalts from the NE Japan arc: constraints on source composition, mantle melting and slab input composition. J Petrol 56 (11):2257–2297

Sieh K, Natawidjaja D (2000) Neotectonics of the Sumatran fault, Indonesia. J Geophys Res: Solid Earth 105(B12):28295–28326

Sigloch K, Mihalynuk MG (2013) Intra-oceanic subduction shaped the assembly of Cordilleran North America. Nature 496(7443):50–56

Skirrow RG, Jaireth S, Huston DL, Bastrakov EN, Schofield A, Van der Wielen SE, Barnicoat AC (2009) Uranium mineral systems: processes, exploration criteria and a new deposit framework. Geosci Aust Rec 20009(20):44p

Smith AL, Roobol MJ, Mattioli GS, Fryxell JE, Daly GE, Fernandez LA (2013) The volcanic geology of the mid-arc Island of Dominica. Geol Soc Am Spec Pap 496:249p

Smith IE, Price RC (2006) The Tonga-Kermadec arc and Havre-Lau back-arc system: their role in the development of tectonic and magmatic models for the western Pacific. J Volcanol Geotherm Res 156(3–4):315–331

Smith IE, Okada T, Itaya T, Black PM (1993) Age relationships and tectonic implications of late Cenozoic basaltic volcanism in Northland, New Zealand. NZ J Geol Geophys 36(3):385–393

Smith M, Mosley P (1993) Crustal heterogeneity and basement influence on the development of the Kenya Rift, East Africa. Tectonics 12(2):591–606

Smithies RH, Ivanic TJ, Lowrey JR, Morris PA, Barnes SJ, Wyche S, Lu YJ (2018) Two distinct origins for Archean greenstone belts. Earth Planet Sci Letts 487:106–116

Smithies RH, Lu Y, Johnson TE, Kirkland CL, Cassidy KF, Champion DC, Mole DR, Zibra I, Gessner K, Sapkota J, De Paoli MC (2019) No evidence for high-pressure melting of Earth's crust in the Archean. Nat Comms 10(1):1–2

Snow CA (2006) A re-evaluation of tectonic discrimination diagrams and a new probabilistic approach using large geochemical databases: Moving beyond binary and ternary plots. J Geophys Res: Solid Earth 111(B6).

Sonder LJ, England PC, Wernicke BP, Christiansen RL (1987) A physical model for Cenozoic extension of western North America. In: Coward MP, Dewey JF, Hancock PL (eds) 1987. Continental extensional tectonics. Geol Soc Lond Spec Publ 28(1):187–201
Sonder LJ, Jones CH (1999) Western United States extension: how the west was widened. Ann Rev Earth Planet Sci 27(1):417–462
Sokoutis D, Brun JP, Van Den Driessche J, Pavlides S (1993) A major Oligo-Miocene detachment in southern Rhodope controlling north Aegean extension. J Geol Soc Lond 150(2):243–246
Solomatov VS, Moresi LN (2000) Scaling of time-dependent stagnant lid convection: application to small-scale convection on Earth and other terrestrial planets. J Geophys Res Solid Earth 105(B9): 21795–21817
Sorrentino L, Cas RAF, Stilwell JD (2011) Evolution and facies architecture of Paleogene Surtseyan volcanoes on Chatham Islands, New Zealand, Southwest Pacific Ocean. J Volcanol Geotherm Res 202(1–2):1–21
Soule SA (2015) Mid-ocean ridge volcanism. In: Sigurdsson H, Houghton B, McNutt SR, Rymer H, Stix J (eds) The encyclopedia of volcanoes. Academic Press Elsevier, 395–403
Southgate PN, Bradshaw BE, Domagala J, Jackson MJ, Idnurm M, Krassay AA, Page RW, Sami TT, Scott DL, Lindsay JF, McConachie BA (2000) Chronostratigraphic basin framework for Palaeoproterozoic rocks (1730–1575 Ma) in northern Australia and implications for base-metal mineralisation. Aust J Earth Sci 47 (3):461–483
Spakman W, Wortel MJ, Vlaar NJ (1988) The Hellenic subduction zone: a tomographic image and its geodynamic implications. Geophys Res Letts 15(1):60–63
Spakman W, van der Lee S, van der Hilst R (1993) Travel-time tomography of the European-Mediterranean mantle down to 1400 km. Phys Earth Planet Int 79(1–2):3–74
Speranza F, Villa IM, Sagnotti L, Florindo F, Cosentino D, Cipollari P, Mattei M (2002) Age of the Corsica-Sardinia rotation and Liguro-Provençal Basin spreading: new paleomagnetic and Ar/Ar evidence. Tectonophysics 347(4):231–251
Spinks KD, Acocella V, Cole JW, Bassett KN (2005) Structural control of volcanism and caldera development in the transtensional Taupo Volcanic Zone, New Zealand. J Volcanol Geotherm Res 144(1–4): 7–22
Spörli KB (1980) New Zealand and oblique-slip margins: tectonic development up to and during the Cainozoic. In: Balance PF, Reading HG (eds) Sedimentation in oblique-slip mobile zones 147–170
Squire RJ, Allen CM, Cas RAF, Campbell IH, Blewett RS, Nemchin AA (2010) Two cycles of voluminous pyroclastic volcanism and sedimentation related to episodic granite emplacement during the late Archean: Eastern Yilgarn Craton, Western Australia. Precamb Res 183(2):251–274
Stamps DS, Saria E, Kreemer C (2018) A geodetic strain rate model for the East African Rift System. Sci Repts 8(1):1–8
Staudigel H, Koppers AA (2015) Seamounts and island building. In: Sigurdsson H, Houghton B, McNutt SR, Rymer H, Stix J (eds) The encyclopedia of volcanoes. Academic Press Elsevier, pp 405–421
Stein CA, Kley J, Stein S, Hindle D, Keller GR (2015) North America's Midcontinent Rift: when rift met LIP. Geosphere 11(5):1607–1616
Stein CA, Stein S, Merino M, Randy Keller G, Flesch LM, Jurdy DM (2014) Was the Midcontinent Rift part of a successful seafloor-spreading episode? Geophys Res Letts 41(5):1465–1470
Stern RJ (2004) Subduction initiation: spontaneous and induced. Earth Planet Sci Letts 226(3–4):275–292
Stern RJ (2010) The anatomy and ontogeny of modern intra-oceanic arc systems. In: Kusky TM, Zhai MG, Xiao W (eds). The evolving continents: understanding processes of continental growth. Geol Soc Lond Spec Publ 338:7–34
Stern RJ, Bloomer SH (1992) Subduction zone infancy: examples from the Eocene Izu-Bonin-Mariana and Jurassic California arcs. Geol Soc Amer Bull 104(12):1621–1636
Stern RJ, Fouch MJ, Klemperer SL (2004) An overview of the Izu-Bonin-Mariana subduction factory. In: Eiler J, Hirschmann M (eds) Inside the subduction factory. Am Geophys Union Monogr 138:175–222
Storti F, Salvini F, Rossetti F, Morgan JP (2007) Intraplate termination of transform faulting within the Antarctic continent. Earth Planet Sci Letts 260(1–2):115–126
Sugimura A, Uyeda S (1973) Island arcs, Japan and its environments. In: Developments in tectonics, vol 3. Elsevier, Amsterdam
Sun W, Zhang H, Ling MX, Ding X, Chung SL, Zhou J, Yang XY, Fan W (2011) The genetic association of adakites and Cu–Au ore deposits. Int Geol Rev 53(5–6):691–703
Sun WD, Ling MX, Chung SL, Ding X, Yang XY, Liang HY, Fan WM, Goldfarb R, Yin QZ (2012) Geochemical constraints on adakites of different origins and copper mineralization. J Geol 120 (1):105–120
Sutherland FL (2003) 'Boomerang' migratory intraplate Cenozoic volcanism, eastern Australian rift margins and the Indian–Pacific mantle boundary. In: Hillis R, Müller RD (eds) Evolution and dynamics of the Australian Plate. Geol Soc Aust Spec Publ 22 and Geol Soc Am Spec Pap 372:203–221
Sutherland FL, Graham IT, Meffre S, Zwingmann H, Pogson RE (2012) Passive-margin prolonged volcanism, East Australian Plate: outbursts, progressions, plate controls and suggested causes. Aust J Earth Sci 59(7):983–1005
Swager C (1989) Structure of Kalgoorlie greenstones: regional deformation history and implications for the structural setting of the Golden Mile gold deposits. Geol Surv Western Aust Rept 25:59–84
Swager C, Griffin TJ (1990) An early thrust duplex in the Kalgoorlie-Kambalda greenstone belt, eastern Goldfields Province. Western Australia. Precamb Res 48(1–2):63–73
Swager CP (1997) Tectono-stratigraphy of late Archaean greenstone terranes in the southern Eastern Goldfields, Western Australia. Precamb Res 83(1–3):11–42
Takahashi N, Kodaira S, Tatsumi Y, Kaneda Y, Suyehiro K (2008) Structure and growth of the Izu-Bonin-Mariana arc crust: 1. Seismic constraint on crust and mantle structure of the Mariana arc-back-arc system. J Geophys Res: Solid Earth 113(B1)
Tamaki K (1985) Two modes of back-arc spreading. Geology 13 (7):475–478
Tamaki K, Honza E (1991) Global tectonics and formation of arginal basins: role of the western Pacific. Episodes 14:224–230
Tang M, Chen K, Rudnick RL (2016) Archean upper crust transition from mafic to felsic marks the onset of plate tectonics. Science 351 (6271):372–375
Tapponnier P, Peltzer G, Armijo R (1986) On the mechanics of the collision between India and Asia. In: Coward MP, Ries AC (eds) Collision tectonics. Geol Soc Lond Spec Publ 19(1):113–157
Tapponnier P, Peltzer GL, Le Dain AY, Armijo R, Cobbold P (1982) Propagating extrusion tectonics in Asia: new insights from simple experiments with plasticine. Geology 10(12):611–616
Tapponnier P, Zhiqin X, Roger F, Meyer B, Arnaud N, Wittlinger G, Jingsui Y (2000) Oblique stepwise rise and growth of the Tibet Plateau. Science 294(5547):1671–1677
Tatsumi Y (2005) The subduction factory: how it operates in the evolving Earth. Geol Soc Am Today 15(7):4
Tatsumi Y, Stern RJ (2006) Manufacturing continental crust in the subduction factory. Oceanography 19(4):104–112
Tavlan M, Thorne R, Herrington RJ (2011) Uplift and lateritisation history of the Çaldağ ophiolite in the context of Neo-Tethyan ophiolite obduction and uplift: implications for the Cenozoic

weathering history of western Anatolia. J Geol Soc Lond 168 (4):927–940
Taylor B, Karner GD (1983) On the evolution of marginal basins. Rev Geophys 21(8):1727–1741
Taylor B, Martinez F (2003) Back-arc basin basalt systematics. Earth Planet Sci Letts 210(3–4):481–497
Taylor B, Zellmer K, Martinez F, Goodliffe A (1996) Sea floor spreading in the Lau back-arc basin. Earth Planet Sci Letts 144(1–2):35–40
Taylor SR, McLennan SM (1995) The geochemical evolution of the continental crust. Rev Geophys 33(2):241–265
Teagle DAH, Alt JC, Umino S, Miyashita S, Banerjee NR, Wilson DS, and the Expedition 309/312 Scientists (2006) Proceedings of the integrated ocean drilling program 309/312:1–127, U.S. Govt. Print. Off., Washington, D.C.
Teng LS (1996) Extensional collapse of the northern Taiwan mountain belt. Geology 24:949–952
Thiéblemont D, Stein G, Lescuyer JL (1997) Epithermal and porphyry deposits: the adakite connection. Comptes Rendus De L'academie Des Sci Ser IIA Earth Planet Sci 2(325):103–109
Thordarson T, Larsen G (2007) Volcanism in Iceland in historical time: volcano types, eruption styles and eruptive history. J Geodynam 43 (1):118–152
Thurston PC 2015. Igneous Rock Associations 19. Greenstone belts and granite−greenstone terranes: constraints on the nature of the Archean world. Geosci Canada 437–484
Tian X, Chen Y, Tseng TL, Klemperer SL, Thybo H, Liu Z, Xu T, Liang X, Bai Z, Zhang X, Si S (2015) Weakly coupled lithospheric extension in southern Tibet. Earth Planet Sci Letts 430:171–177
Tian L, Zhao G, Zhao G, Shi X, Lv H (2005) Geochemistry of basaltic lavas from the Mariana Trough: evidence for influence of subduction component on the generation of backarc basin magmas. Int Geol Rev 47(4):387–397
Tian Y, Zhu H, Zhao D, Liu C, Feng X, Liu T, Ma J (2016) Mantle transition zone structure beneath the Changbai volcano: Insight into deep slab dehydration and hot upwelling near the 410 km discontinuity. J Geophys Res: Solid Earth 121(8):5794–5808
Tibaldi A (2008) Contractional tectonics and magma p, aths in volcanoes. J Volcanol Geotherm Res 176(2):291–301
Tibaldi A, Pasquarè F, Tormey D (2010) Volcanism in recerse and strike-slip fault settings. In: Cloetingh S, Negendank J (eds) New frontiers in integrated solid earth sciences. Springer Science and Business Media B.V., 315–348
Timm C, Davy B, Haase K, Hoernle KA, Graham IJ, De Ronde CE, Woodhead J, Bassett D, Hauff F, Mortimer N, Seebeck HC (2014) Subduction of the oceanic Hikurangi Plateau and its impact on the Kermadec arc. Nature Comms 5:4923
Trofimovs J, Cas RAF, Davis BK (2004) An Archaean submarine volcanic debris avalanche deposit, Yilgarn Craton, western Australia, with komatiite, basalt and dacite megablocks: The product of dome collapse. J Volc Geotherm Res 138(1–2):111–126
Tun ST, Watkinson IM (2017) The Sagaing Fault, Myanmar. In: Barber AJ, Kin Zaw, Crow MJ (eds) Myanmar; geology, resources and tectonics. Geol Soc Lond Mem 48(1):413–441
Ueki K, Hino H, Kuwatani T (2017) Geochemical discrimination and characteristics of magmatic tectonic settings; a machine learning-based approach. arXiv preprint arXiv:1712.09016
Underwood MB, Moore GF, Taira A, Klaus A, Wilson ME, Fergusson CL, Hirano S, Steurer J (2003) Sedimentary and tectonic evolution of a trench-slope basin in the Nankai subduction zone of southwest Japan. J Sed Res 73(4):589–602
Uyeda S (1983) Comparative subductology. Episodes 2:19–24
Valade S, Ripepe M, Giuffrida G, Karume K, Tedesco D (2018) Dynamics of Mount Nyiragongo lava lake inferred from thermal imaging and infrasound array. Earth Planet Sci Letts 500:192–204
Valentine GA, Connor CB (2015) Basaltic volcanic fields. In: Sigurdsson H, Houghton B, McNutt SR, Rymer H, Stix J (eds) The encyclopedia of volcanoes. Academic Press Elsevier, 423–439
van Bemmelen RW (1939) The volcano-tectonic origin of Lake Toba (north Sumatra). De Ingenieur in Nederlandsch Indie 6–9:126–140
van Benthem S, Govers R, Spakman W, Wortel R (2013) Tectonic evolution and mantle structure of the Caribbean. J Geophys Res: Solid Earth 118(6):3019–3036
Van Kranendonk MJ (2011) Onset of plate tectonics. Science 333 (6041):413–414
Van Kranendonk MJ, Collins WJ, Hickman A, Pawley MJ (2004) Critical tests of vertical vs. horizontal tectonic models for the Archaean East Pilbara granite–greenstone terrane, Pilbara craton, western Australia. Precamb Res 131(3–4):173–211
Van Kranendonk MJ, Kröner A, Hegner E, Connelly J (2009) Age, lithology and structural evolution of the c. 3.53 Ga Theespruit Formation in the Tjakastad area, southwestern Barberton Greenstone Belt, South Africa, with implications for Archaean tectonics. Chem Geol 261(1–2):115–139
van Otterloo J, Cas RAF (2013) Reconstructing the eruption magnitude and energy budgets for the pre-historic eruption of the monogenetic ~5 ka Mt. Gambier Volcanic Complex, south-eastern Australia. Bull Volcanol 75(12):769–787
van Otterloo J, Cas RAF (2016) Low-temperature emplacement of phreatomagmatic pyroclastic flow deposits at the monogenetic Mt Gambier Volcanic Complex, South Australia, and their relevance for understanding some deposits in diatremes. J Geol Soc Lond 173 (4):701–710
van Otterloo J, Cas RA, Sheard MJ (2013) Eruption processes and deposit characteristics at the monogenetic Mt. Gambier Volcanic Complex, SE Australia: implications for alternating magmatic and phreatomagmatic activity. Bull Volcanol 75(8):737–758
van Otterloo J, Raveggi M, Cas RAF, Maas R (2014) Polymagmatic activity at the monogenetic Mt Gambier volcanic complex in the Newer Volcanics Province, SE Australia: new insights into the occurrence of intraplate volcanic activity in Australia. J Petrol 55 (7):1317–1351
van Wyk de Vries B, Merle O (1998) Extension induced by volcanic loading in regional strike-slip zones. Geology 26(11):983–986
van Wyk de Vries B, van Wyk de Vries M (2018) Tectonics and volcanic and igneous plumbing systems. In: Volcanic and igneous plumbing systems. Elsevier, pp 167–189. https://doi.org/10.1016/B978-0-12-809749-6.00007-8
van der Voo R, Spakman W, Bijwaard H (1999) Tethyan subducted slabs under India. Earth Planet Sci Letts 171(1):7–20
van der Werff W (1996) Variation in forearc basin development along the Sunda Arc, Indonesia. J SE Asian Earth Sci 14(5):331–349
Vielreicher NM, Groves DI, McNaughton NJ (2016) The giant Kalgoorlie gold field revisited. Geosci Front 7(3):359–374
Vignaroli G, Balsamo F, Giordano G, Rossetti F, Storti F (2015) Miocene-to-Quaternary oblique rifting signature in the Western Ross Sea from fault patterns in the McMurdo Volcanic Group, north Victoria Land, Antarctica. Tectonophysics 656:74–90
Viruete JE, de Neira AD, Huerta PH, Monthel J, Senz JG, Joubert M, Lopera E, Ullrich T, Friedman R, Mortensen J, Pérez-Estaún A (2006) Magmatic relationships and ages of Caribbean Island arc tholeiites, boninites and related felsic rocks, Dominican Republic. Lithos 90(3):161–186
Vogt PR (1974) Volcano spacing, fractures, and thickness of the lithosphere. Earth Planet Sci Letts 21(3):235–252
von Hagke C, Philippon M, Avouac JP, Gurnis M (2016) Origin and time evolution of subduction polarity reversal from plate kinematics of Southeast Asia. Geology 44(8):659–662

von Huene R, Pecher IA, Gutscher MA (1996) Development of the accretionary prism along Peru and material flux after subduction of Nazca Ridge. Tectonics 15(1):19–33

von Huene R, Weinrebe W, Heeren F (1999) Subduction erosion along the North Chile margin. J Geodynam 27(3):345–358

von Huene R, Ranero CR (2003) Subduction erosion and basal friction along the sediment-starved convergent margin off Antofagasta, Chile. J Geophys Res: Solid Earth 108(B2)

von Huene RV (1984) Tectonic processes along the front of modern convergent margins-research of the past decade. Ann Rev Earth Planet Sci 12(1):359–381

Wadge G, Biggs J, Lloyd R, Kendall JM (2016) Historical volcanism and the state of stress in the East African Rift System. Front Earth Sci 4:86. https://doi.org/10.3389/feart.2016.00086

Wagner TP, Grove TL (1998) Melt/harzburgite reaction in the petrogenesis of tholeiitic magma from Kilauea volcano. Hawaii. Contribs Miner Petrol 131(1):1–12

Walker GPL (1962) Tertiary welded tuffs in eastern Iceland. Quarterly J Geol Soc Lond, London 118(1–4):275–290

Walker GPL (1963) The Breiddalur central volcano, eastern Iceland. Quat J Geol Soc Lond 119(1–4):29–63

Walker GPL (1966) Acid volcanic rocks in Iceland. Bull Volcanologique 29(1):375–402

Wallace PJ (2005) Volatiles in subduction zone magmas: concentrations and fluxes based on melt inclusion and volcanic gas data. J Volcanol Geotherm Res 140(1–3):217–240

Wang Q, Hawkesworth CJ, Wyman D, Chung SL, Wu FY, Li XH, Li ZX, Gou GN, Zhang XZ, Tang GJ, Dan W (2016) Pliocene-Quaternary crustal melting in central and northern Tibet and insights into crustal flow. Nature Comms 7:11888

Wang XC, Wilde SA, Li QL, Yang YN (2015) Continental flood basalts derived from the hydrous mantle transition zone. Nature Comms 6(1):1–9

Wang Z, Zhao D (2005) Seismic imaging of the entire arc of Tohoku and Hokkaido in Japan using P-wave, S-wave and sP depth-phase data. Phys Earth Planet Interiors 152(3):144–162

Wang Z, Zhao D (2006) Vp and Vs tomography of Kyushu, Japan: New insight into arc magmatism and forearc seismotectonics. Phys Earth Planet Interiors 157(3–4):269–285

Weaver SD, Saunders AD, Pankhurst RJ, Tarney J (1979) A geochemical study of magmatism associated with the initial stages of back-arc spreading. Contribs Miner Petrol 68(2):151–169

Wei W, Zhao D, Wei F, Bai X, Xu J (2019) Mantle dynamics of the eastern Mediterranean and Middle East: Constraints from P-wave anisotropic tomography. Geochem Geophys Geosyst 20(10):4505–4530

Weinberg RF (2016) Himalayan leucogranites and migmatites: nature, timing and duration of anatexis. J Met Petrol 34(8):821–843

Weissel JK (1981) Magnetic lineations in marginal basins of the western Pacific. Phil Trans R Soc Lond A 300(1454):223–247

Wellman P, McDougall I (1974) Cainozoic igneous activity in eastern Australia. Tectonophysics 23(1–2):49–65

Wernicke B (1992) Cenozoic extensional tectonics of the US Cordillera. In: Burchfiel BC, Lipman PW, Zoback ML (eds) The Cordilleran orogen: conterminous US. Geol Soc Am, Geol N Am G-3:553–558

Wernicke B (1995) Low-angle normal faults and seismicity: a review. J Geophys Res: Solid Earth 100(B10):20159–20174

Wernicke B, Burchfiel BC (1982) Modes of extensional tectonics. J Struct Geol 4(2):105–115

Wernicke BP, England PC, Sonder LJ, Christiansen RL, (1987) Tectonomagmatic evolution of Cenozoic extension in the North American Cordillera. In: Coward MP, Dewey JF, Hancock PL (1987) Continental extensional tectonics. Geol Soc Lond Spec Publ 28(1):203–221

Whalen JB, Currie KL, Chappell BW (1987) A-type granites: geochemical characteristics, discrimination and petrogenesis. Contribs Miner Petrol 95(4):407–419

Wheller GE, Varne R, Foden JD, Abbott MJ (1987) Geochemistry of Quaternary volcanism in the Sunda-Banda arc, Indonesia, and three-component genesis of island-arc basaltic magmas. J Volcanol Geotherm Res 32(1–3):137–160

White JC, Espejel-García VV, Anthony EY, Omenda P (2012) Open system evolution of peralkaline trachyte and phonolite from the Suswa volcano, Kenya rift. Lithos 152:84–104

White LT, Morse MP, Lister GS (2014) Lithospheric-scale structures in New Guinea and their control on the location of gold and copper deposits. Solid Earth 5(1):163–179

White R, McKenzie D (1989) Magmatism at rift zones: the generation of volcanic continental margins and flood basalts. J Geophys Res: Solid Earth 94(B6):7685–7729

White SH, De Boorder H, Smith CB (1995) Structural controls of kimberlite and lamproite emplacement. J Geochem Explor 53(1–3):245–264

White SM, Crisp JA, Spera FJ (2006) Long-term volumetric eruption rates and magma budgets. Geochem Geophys Geosys 7(3)

Whitford DJ, Nicholls IA, Taylor SR (1979) Spatial variations in the geochemistry of Quaternary lavas across the Sunda arc in Java and Bali. Contribs Miner Petrol 70(3):341–356

Wichura H, Bousquet R, Oberhänsli R, Strecker MR, Trauth MH (2011) The Mid-Miocene East African Plateau: a pre-rift topographic model inferred from the emplacement of the phonolitic Yatta lava flow, Kenya. In: van Hinsbergen et al (eds) The formation and evolution of Africa: a Synopsis of 3.8 Ga of earth history. Geol Soc Lond Spec Publ 357(1):285–300

Widiyantoro S, Kennett BL, Van der Hilst RD (1999) Seismic tomography with P and S data reveals lateral variations in the rigidity of deep slabs. Earth Planet Sci Letts 173(1–2):91–100

Wigger PJ, Schmitz M, Araneda M, Asch G, Baldzuhn S, Giese P, Heinsohn WD, Martínez E, Ricaldi E, Röwer P, Viramonte J (1994) Variation in the crustal structure of the southern Central Andes deduced from seismic refraction investigations. In: Tectonics of the southern Central Andes. Springer, Berlin, Heidelberg, pp 23–48

Wilde SA, Valley JW, Peck WH, Graham CM (2001) Evidence from detrital zircons for the existence of continental crust and oceans on the Earth 4.4 Gyr ago. Nature 409(6817):175–178

Willcock MA, Cas RA, Giordano G, Morelli C (2013) The eruption, pyroclastic flow behaviour, and caldera in-filling processes of the extremely large volume (>1290 km^3), intra-to extra-caldera, Permian Ora (Ignimbrite) Formation, Southern Alps, Italy. J Volcanol Geotherm Res 265:102–126

Williams RM, Underhill JR, Jamieson RJ (2020) The role of relay ramp evolution in governing sediment dispersal and petroleum prospectivity of syn-rift stratigraphic plays in the Northern North Sea. Petrol Geosci 26(2):232–246

Wills JM (1991) The Forties Field, Block 21/10, 22/6a, UK North Sea. Geol Soc, London, Memoirs 14(1):301–308

Wilson CJ, Gravley DM, Leonard GS, Rowland JV (2009) Volcanism in the central Taupo Volcanic Zone, New Zealand: tempo, styles and controls. In: Thordarson T et al (eds) Studies in volcanology: the legacy of George Walker. Spec Publ IAVCEI and Geol Soc Lond 2:225–247

Wilson JT (1965) A new class of faults and their bearing on continental drift. Nature 207(4995):343–347

Wilson L, Head JW (1981) Ascent and eruption of basaltic magma on the Earth and Moon. J Geophys Res: Solid Earth 86(B4):2971–3001

Wilson M, Neumann ER, Davies GR, Timmerman MJ, Heeremans M, Larsen BT (2004) Permo-Carboniferous magmatism and rifting in Europe. Geol Soc Lond Spec Publ 223(1):490p

Windley BF (1993) Proterozoic anorogenic magmatism and its orogenic connections. J Geol Soc Lond 150(1):39–50

Witt WK, Cassidy KF, Lu YJ, Hagemann SG (2020) The tectonic setting and evolution of the 2.7 Ga Kalgoorlie–Kurnalpi Rift, a world-class Archean gold province. Miner Deposita 55(4):601–631

Wood CP (1980) Boninite at a continental margin. Nature 288 (5792):692

Wortel MJ, Spakman W (2000) Subduction and slab detachment in the Mediterranean-Carpathian region. Science 290(5498):1910–1917

Wright JE, Wyld SJ (2011) Late Cretaceous subduction initiation on the eastern margin of the Caribbean-Colombian Oceanic Plateau: One Great Arc of the Caribbean (?). Geosphere 7(2):468–493

Wright JV (1990) Basins and styles of basin, epithermal and other high crustal level mineralisation. James Cook University of North Queensland, EGRU Contribution 35:158p

Wright JV (1994) Guides to metals exploration in basins. Short Course presented to Normandy Exploration Ltd (Darwin, Australia; Trabzon, Turkey), 360p (unpubl)

Wright JV (1996) Mineralisation associated with volcanic terrains: the Mesozoic-Cenozoic Alpine collisional zone. Normandy Mining Group Geol Conf, 18–25 November, Darwin, Northern Territory, Australia. Rept no: CR21049:12p (unpubl)

Wright JV, Barnes GJ, Laes DYM, Presnell RD, Smith PC, Wallace CA (1999) Belt-Purcell terrane project: Final report. Report to Kennecott Exploration Co, 52p (unpubl)

Wright JV, Piantone H, Cassard D, Bahçler, Erkül F (1995) Results of 1995 field program, Eastern Pontides, Turkey. LaSource Project Generation Group, Orleans, France. Rept. no. 10.014: 41p (unpubl.)

Wright JV, van der Veen A (1993) Preliminary research investigations Irian Jaya exploration programme. Report to Normandy Anglo Asia, 24p (unpubl.)

Wyman DA (2019) 2.8 Ga subduction-related magmatism in the Youanmi Terrane and a revised geodynamic model for the Yilgarn Craton. Precamb Res 327:14–33

Wyman DA (2020) Komatiites from mantle transition zone plumes. Front Earth Sci 8:540744. https://doi.org/10.3389/feart.2020.540744383

Xiao Y, Niu Y, Wang KL, Lee DC, Iizuka Y (2016) Geochemical behaviours of chemical elements during subduction-zone metamorphism and geodynamic significance. Int Geol Rev 58(10):1253–1277

Xu J, Ben-Avraham Z, Kelty T, Yu HS (2014) Origin of marginal basins of the NW Pacific and their plate tectonic reconstructions. Earth Sci Rev 130:154–196

Yin A (2010) Cenozoic tectonic evolution of Asia: a preliminary synthesis. Tectonophysics 488(1–4):293–325

Yoshida T et al. (2014) Evolution of late Cenozoic magmatism and the crust–mantle structure in the NE Japan Arc. In: Gómez-Tuena A, Straub SM, Zellmar GF (eds) Orogenic andesites and crustal growth. Geol Soc Lond Spec Publ 385(1):335–387

Yukutake Y, Honda R, Harada M, Aketagawa T, Ito H, Yoshida A (2011) Remotely-triggered seismicity in the Hakone volcano following the 2011 off the Pacific coast of Tohoku Earthquake. Earth Planets Space 63(7):41

Yule JD, Saleeby JB, Barnes CG (2006) A rift-edge facies of the Late Jurassic Rogue-Chetco arc and Josephine ophiolite, Klamath Mountains, Oregon. In: Snoke AW, Barnes CG (eds) Geological studies in the Klamath Mountains province, California and Oregon. Geol Soc Amer Spec Pap 410:53–76

Zhang GL, Smith-Duque C (2014) Seafloor basalt alteration and chemical change in the ultra-thinly sedimented South Pacific. Geochem Geophys Geosyst 15(7):3066–3080

Zhang L, Li S, Zhao Q (2021) A review of research on adakites. Int Geol Rev 15:1–8

Zhao D (2012) Tomography and dynamics of western-Pacific subduction zones. Environmental, Earth Planets Monograph 1(1):1–70

Zhao D, Fujisawa M, Toyokuni G (2017) Tomography of the subducting Pacific slab and the 2015 Bonin deepest earthquake (Mw 7.9). Nature Sci Repts 7:44487

Zhao D, Mishra OP, Sanda R (2002) Influence of fluids and magma on earthquakes: seismological evidence. Phys Earth Planet Interiors 132(4):249–267

Zhao D, Lei J, Tang R (2004) Origin of the Changbai intraplate volcanism in Northeast China: evidence from seismic tomography. Chinese Sci Bull 49(13):1401–1408

Zhao D, Santosh M, Yamada A (2010) Dissecting large earthquakes in Japan: Role of arc magma and fluids. Island Arc 19(1):4–16

Zhao D, Tian Y, Lei J, Liu L, Zheng S (2009) Seismic image and origin of the Changbai intraplate volcano in East Asia: role of big mantle wedge above the stagnant Pacific slab. Phys Earth Planet Inter 173 (3–4):197–206

Zibra I, Lu Y, Clos F, Weinberg RF, Peternell M, Wingate MT, Prause M, Schiller M, Tilhac R (2020) Regional-scale polydiapirism predating the Neoarchean Yilgarn Orogeny. Tectonophysics 779:228375

Zieg MJ, Marsh BD (2005) The Sudbury Igneous Complex: viscous emulsion differentiation of a superheated impact melt sheet. Geol Soc Amer Bull 117(11–12):1427–1450

Zoback ML, Thompson GA, Anderson RE (1981) Cainozoic evolution of the state of stress and style of tectonism of the Basin and Range province of the western United States. Phil Trans R Soc Lond A 300 (1454):407–434

Volcanic-Hosted Resources

18

Characteristics and Controls on Formation of Metallic Ore Deposits, Diamonds, Hydrocarbon Reservoirs, and Geothermal Energy

John V. Wright, Ray Cas, and Guido Giordano

R. Cas et al., *Volcanology*, Springer Textbooks in Earth Sciences, Geography and Environment,
https://doi.org/10.1007/978-3-319-66613-6_18

Plate 18.1 Snapshots of the major classes of volcanic-hosted resources described in this chapter. **a** Yanacocha gold mine, northern Peru. Since mining began in 1993 the world's most productive epithermal precious metal deposit. (*Source* Wikimedia through Creative Commons Attribution 3.0 Unported licence). **b** Drill core showing high-grade, disseminated chalcopyrite mineralisation within hematite-rich breccia from the supergiant Olympic Dam iron oxide copper–gold deposit, South Australia (From Reynolds 2001; *Source* Department for Energy and Mining, Government of South Australia, through Creative Commons Attribution-Share Alike 4.0 International licence). **c** Uranophane ore in brecciated and altered welded ignimbrite, the Peña Blanca uranium deposit, northern Mexico (From Murphy and Pickett 2005). **d** Komatiite-hosted nickel–copper massive sulphide ore, Kambalda, Western Australia (*Source* ABC News, Australia). **e** Open pit at the Koala diamond pipe discovered in 1991, Ekati kimberlite field, Lac de Gras, Northwest Territories, Canada. **f** Subsea sampling of mineralised chimney ($\sim$1,550 m below sea level), Solwara seafloor volcanic-hosted massive sulphide project, Bismarck Sea, Papua New Guinea (*Source* Nautilus Minerals). **g** Natural gas flare, Rosebank appraisal well. Oil was discovered in intra-basalt reservoirs of the North Atlantic Igneous Province in 2004, 130 km northwest of the Shetland Islands, Scotland. (*Source* Chevron North Sea Ltd.). **h** The iconic Great Geysir in southwest Iceland is located on the landward extension of the Mid-Atlantic (Reykjanes) Ridge, which is actively being drilled to test the feasibility, technology, and economics of producing very high-temperature or supercritical geothermal resources

Abstract

Volcanoes and volcanic rocks are hosts to economically very significant quantities of natural resources. Initially, we provide a brief overview of volcanic-hosted ore deposits and their typology. We then focus predominantly on the geology of the principal ore deposit types that characterise volcanic successions and some high-level subvolcanic intrusive settings. These are divided into epithermal precious metal, iron oxide copper–gold (IOCG), volcanic-hosted uranium, komatiitic-hosted nickel sulphide, kimberlitic-hosted diamonds, and volcanic-hosted massive sulphide (VMS) deposits. Epithermal gold deposits represent, in terms of exploration expenditure, the most important type of volcanic-hosted resource thus warranting the greatest attention. We discuss these in terms of today's active geothermal systems that occupy the same volcano-tectonic environments as the fossil hydrothermal systems which host them and further assess these in more detail by dividing them by volcanic landform or feature and facies associations. IOCG deposits are a relatively new family of hydrothermal ore deposit types, only some of which have a clear spatial and genetic relationship to subvolcanic and volcanic processes and rocks. Most volcanic-hosted uranium deposits are relatively small and low grade; however, some larger and higher-grade resources are known. Both komatiite lavas and komatiite intrusions in Archean (and some Proterozoic) greenstone terrains locally host significant orthomagmatic nickel sulphide (Ni, Cu, PGE) deposits and are one of the world's major sources of Ni production. Kimberlites and lamproites transport diamond xenocrysts in their "mantle cargo" to the Earth's surface, where they erupt explosively to form pipe-like bodies representing the remains of volcanoes and their vent systems. Polymetallic VMS deposits represent a distinctive, although highly variable, submarine magmatic–hydrothermal ore deposit type having direct genetic links to both a submarine environment and volcanism, including subvolcanic intrusions. Consideration is also given to volcanic rock reservoirs in hydrocarbon exploration which, with significant exploration successes over the past fifteen years, represent a new frontier field of petroleum exploration globally. Finally, we discuss geothermal energy, which is likely to become a more significant energy resource in the future.

18.1 Introduction

Volcanoes and volcanic rocks are hosts to economically very significant quantities of natural resources (Sillitoe and Bonham 1984; Cas and Wright 1987; McPhie and Cas 2015; Chap. 16). Since obsidian tools were first crafted in Paleolithic times, over the millennia volcanic materials have supplied essential requirements to human development (Chap. 1). For our purposes, we can divide these principally into (1) *volcanic-hosted ore* deposits, (2) *volcanic rock reservoirs*, hosting oil and gas accumulations, and (3) *geothermal energy*. By far the most economically important are the "*traditional*" volcanic-hosted ore deposits, comprising both metallic and non-metallic ores. The former includes multiple metal species, while the latter many industrial minerals, such as perlite, clays, zeolites, borate, and sulphur, but also diamonds and especially gems (Plate 18.1). In this chapter, we focus predominantly on the principal metallic ores including uranium which currently provides over 10% of world electricity demand, in addition to primary diamond deposits that characterise volcanic successions and some high-level subvolcanic intrusive settings (Fig. 18.1).

Attention will be also given to volcanic rock reservoirs in hydrocarbon exploration. Although such reservoirs have been known for many years, they were largely regarded as a curiosity, in general largely ignored and even avoided as valid exploration plays because of their perceived lack of reservoir quality. Exploration successes over the past fifteen years, particularly in the Mesozoic–Cenozoic basins of eastern China, the North Atlantic Tertiary Igneous Province, and East African rift basins have, however, changed that perception, and the accumulation of oil and gas in volcanic facies is a new frontier field of hydrocarbon exploration globally.

Many of today's active geothermal systems occupy the same volcano-tectonic environments as fossil hydrothermal systems which host important classes of precious metal deposits. Their geology and hydrogeology are therefore first considered within this context (Sect. 18.3), prior to discussing geothermal energy as a resource in the final section of this chapter. Geothermal power presently contributes <1% to world energy consumption (both direct use and electricity generation); however, with the increasing need to decarbonise the energy sector, the pace of development is likely to accelerate in the future.

18.2 Volcanic-Hosted Ore Deposit Types: A Brief Overview and Typology

Economic geologists have recorded many different types of ore deposits hosted within volcanic and subvolcanic rocks and their successions. These are usually divided firstly into (1) those that have a direct relation to magma and its crystallisation, termed *orthomagmatic*, and (2) those formed by precipitation and/or interaction with metal-bearing solutions of hot water which are called *hydrothermal* (Fig. 18.1).

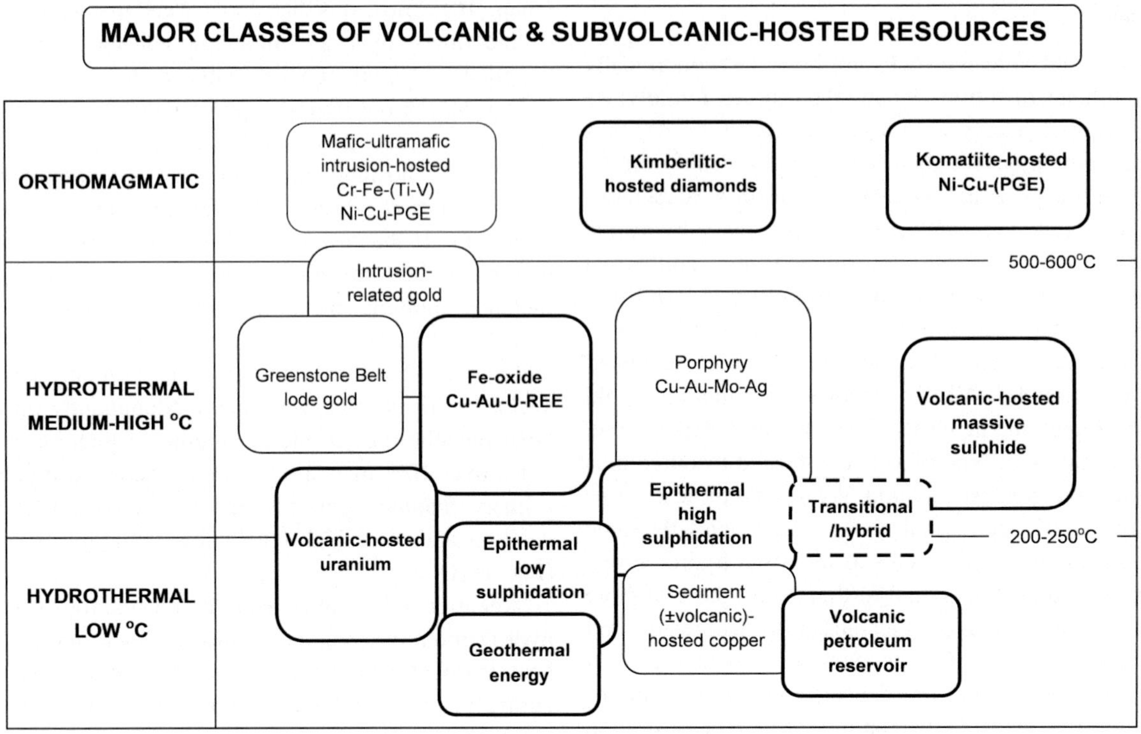

Fig. 18.1 Scheme showing major classes of ore deposits and natural resources found hosted within volcanic and subvolcanic rocks and their successions. Overlap indicates some level of synergy. Those shown in bold are featured in this chapter. Note diamonds are not strictly orthomagmatic because they are xenocrysts resulting from ultra-metamorphism in the mantle, which are then scavenged and carried by kimberlite magmas. Transitional/hybrid refers to an ill-defined group of deposits showing features of both VMS and epithermal deposits that have recently been recognised in shallower marine environments (Sect. 18.8). For an overall review of major gold deposit types, particularly those not addressed in this chapter, see Robert et al. (2007). For current reviews of porphyry and other types of copper deposits, see Hedenquist et al. (2012), and for mafic–ultramafic intrusion-related ore deposits, see Mondal and Griffin (2018)

18.2.1 Orthomagmatic Ore Deposits

Orthomagmatic ore deposits primarily form from mantle-derived, mafic–ultramafic magmas, which may occur as larger intrusions, subvolcanic bodies (sills, laccoliths, or lopoliths) as well as being erupted as lava extrusions (continental flood basalts, and the special case of *komatiites*; Chap. 7). Key processes in forming ore deposits of this type are fractional crystallisation, liquid immiscibility, and magma (i.e. wall rock) contamination, in particular, the input of sulphur from crustal rocks (Lightfoot and Hawkesworth 1997; Arndt et al. 2005; Keays and Lightfoot 2010; Song et al. 2011; Naldrett 2013). Fractional crystallisation separates ore and non-ore minerals according to their crystallisation temperature. As early crystallising minerals form they incorporate metal elements, which, as they settle onto the floor of the intrusion or lava, concentrate ore minerals there. Chromite and magnetite are oxide ore minerals that form in this way. As magma cools, it can become density stratified, and parts may separate from the main body of magma leading to liquid immiscibility. In this process, sulphides may separate and sink below less dense, more silicate-rich parts of the magma-forming layers or be dynamically injected into surrounding magma and wall rocks creating massive sulphide ores containing nickel and copper $\pm$ platinum group elements. Larger mafic and ultramafic intrusions form both sulphide and oxide orebodies, while komatiites are solely sulphidic orebodies.

Diamonds are also generally regarded as an orthomagmatic type of ore deposit, despite the fact diamonds are not generated by the host kimberlitic magmas, or processes that produced those magmas, but are mantle-derived xenocrysts entrained by kimberlite magmas as they pass through the sub-cratonic mantle (Fig. 18.1).

18.2.2 Hydrothermal Ore Deposits

Hydrothermal ore deposits incorporate a very wide spectrum of mineralisation styles formed from hot aqueous solutions. Such solutions were either originally rich in magmatic ore-forming elements, or these were added by fluid–rock interactions during transport, or some combination of both. Hydrothermal deposits are further subdivided on the basis of temperature into low and medium-to-high-temperature types (Fig. 18.1). Note we have purposely avoided the term "*mesothermal*" for such higher temperature fluids to avoid confusion with those traditionally allied with orogenic (or metamorphic) lode gold deposits. While many orogenic lode gold deposits are found hosted by volcanic rocks (particularly Precambrian greenstone belt successions; Chap. 17, see Sect. 17.7.8), gold mineralisation is generally significantly younger and unrelated to volcanism. The ore fluids mainly have distinctive metamorphic origins (Phillips and Powell 1993, 2009), although in some terrains these are divided into orogenic and intrusion-related (± a magmatic fluid) sub-types, even though both formed during the latest stages of orogenesis (Groves et al. 2003; Witt et al. 2020). The volcanic host rocks provide a reactive (Fe-rich) chemical-textural trap (e.g. mafic volcanics and sills), and/or competency contrasts that channel gold-bearing fluids into shear zones and other dilatant features (Robert et al. 2007; Hayman et al. 2019).

Epithermal precious metal deposits represent, in terms of exploration expenditure, the most important type of volcanic-hosted resource. The term epithermal was defined by Waldemar Lindgren over 80 years ago to include a range of precious metal (gold and silver), base metal, mercury, and stibnite deposits which he believed formed from aqueous solutions charged with igneous emanations at low temperatures (<200 °C) and moderate pressures. Today epithermal is used in a much broader sense; it has since been recognised, for example, that in many deposits, mineralisation may have occurred at more elevated temperatures (200–300 °C), or at low pressures (less than a few hundred bars). Many studies of large geothermal systems associated with silicic volcanic terrains (calderas and/or extensional rifts and basins) have stressed the importance of meteoric water and convection. These form a distinctive class of epithermal vein deposit termed *low suphidation* (Fig. 18.1). While magmas at depth may provide the heat energy to drive hydrothermal fluid circulations, the precise role of magmatic processes, even in deposits hosted or spatially close to volcanic rocks, can often be unclear or complex.

Many higher temperature hydrothermal deposits show more direct genetic, as well as spatial relations to magmatic and intrusive bodies. These are exemplified by a number of types of ore deposits associated with subvolcanic intrusive complexes, generally termed "*porphyry-related*" (Fig. 18.1). Porphyry-related ore deposits contain varying amounts of copper, gold, molybdenum, and silver, derived from magmatic sources at depth, subsequently focused into smaller intrusions and apophyses at higher crustal levels. Since the early 1970s, it has been appreciated that porphyry-related mineralisation is found within eroded magmatic arcs and volcanoes (e.g. Sillitoe 1973; Chap. 16). Although we do not specifically describe such deeper-level porphyry systems in this chapter, it has also been widely recognised that the hydrothermal fluids forming copper mineralisation, if not trapped, will continue to migrate to shallower crustal levels (≤ 2 km). These may form *high sulphidation* epithermal gold–silver deposits (Fig. 18.1) that occur in permeable volcanic and brecciated host rocks, which are incorporated within extensive zones of hydrothermal alteration caused by magmatically charged fluids.

Iron oxide copper–gold (or IOCG) deposits form a second, very economically important class of medium- to high-temperature hydrothermal deposits (Fig. 18.1). As their name implies, these are linked together by an abundance of iron oxides (hematite ± magnetite) with the ore, which is mainly sulphidic. The discovery of the Olympic Dam Cu–Au–U–REE (South Australia) in 1975 attracted significant attention from the world's exploration industry because of its exceptionally large tonnage and metal content (in particular uranium) and its association with abundant iron-oxide alteration and host granite ± volcanic breccias. Initially, the Proterozoic host rocks were regarded as having a sedimentary origin and not until 1983, when underground drives were made into the orebody, was it realised these represented some type of magmatic and/or hydrothermal breccia complex (Porter 2000). IOCG deposits are now known to occur throughout geological time, be distributed worldwide, and are principally located either in intra-cratonic, anorogenic rift settings or along continental margins above subduction zones. Some clearly have close spatial as well as genetic relations to subvolcanic and volcanic processes and host rocks, while in others this is not so obvious or not apparent at all. The iron oxides may be an early evolutionary phase of a large and continuing magmatically driven hydrothermal system as possibly at Olympic Dam, or alternatively are little related. There is also debate about the source of the metals, whether these are magmatic or have been leached by fluid wall rock reactions with surrounding country rocks, whereas a number of workers have invoked the importance of evaporites in the formation of the ore fluids.

Volcanic uranium deposits are hosted within similar and, in some cases, the same continental successions as epithermal gold systems associated with subaerial silicic volcanism in highly extensional tectonic terrains (e.g. Basin and Range Province, western United States). Alkaline rhyolite magmas (within a broader bimodal setting) drive shallow hydrothermal fluid circulation, as well as the volatile component transporting metals in solution. Uranium mineralisation develops within or adjacent to so-called "topaz" or "rare metal" rhyolite domes and lavas. Most volcanic uranium deposits of this type are relatively small and low grade, especially when compared to classic styles of unconformity-related and sandstone-hosted uranium mineralisation. However, significantly larger and higher-grade volcanic-hosted resources are known in the Transbaikal uranium province, which includes Streltsovska caldera (Russia), second only to Olympic Dam in size as a natural repository of uranium. The association of uranium minerals with, in particular, fluorite and other lithophile elements is well known from anorogenic magmatic and rift basin settings undergoing extensional orogenic collapse.

As well as terrestrial settings, magmatic and volcanic-related geothermal activity is also found in submarine environments. The discovery of hydrothermal vents or "black smokers" on modern oceanic spreading centres in the Pacific and Atlantic Oceans (Fig. 16.92) and deep marine exploration in the western Pacific back-arc basins has confirmed the importance of medium-to-high-temperature hydrothermal fluids in the formation of *volcanic-hosted massive sulphide (or VMS)* deposits (Fig. 18.1). These modern examples occur at an average depth of about 2000 m below the ocean's surface, although, the deepest known are located at depths of some 5000 m in the Cayman Trough, Caribbean Sea. Sulphides are precipitated at or close to the seafloor where metal-rich hydrothermal fluids vent and mix with cold seawater. The hydrothermal fluid originates from seawater requiring heat energy to drive its deep circulation, and seawater is also the major source of sulphur. Such deposits, therefore, have direct genetic links to both a submarine environment and volcanism, including subvolcanic intrusions. The potential for deep-sea mining of some modern VMS deposits is also now being taken seriously with advanced projects in the Bismarck Sea, Papua New Guinea (Lipton et al. 2018).

Ancient VMS ore deposits are a major source globally of zinc, lead, and copper, while some examples also produce significant silver and gold. They occur in submarine volcanic or volcano-sedimentary settings. Host rocks can be both compositionally (rhyolite to basalt) and lithologically diverse and are intercalated with deep marine sedimentary sequences dominated by turbidites. For ancient VMS deposits, the depth of seawater is typically inferred to be deeper than 1000 m, where the higher confining pressure allows hydrothermal fluids to discharge at temperatures in the range of 400–200 °C without boiling. In the older literature, many VMS deposits were described as exhalative deposits or called "*exhalites*", implying a high degree of lateral dispersal and fallout from down-vent plumes. However, it is quite clear from modern analogues that hydrothermal vent activity builds aggradational features (e.g. mounds), and it is within these that significant tonnages of massive sulphides are likely to be located. Detailed analysis of ancient volcanic facies has also now shown, in some cases, ore deposition occurred somewhat deeper (tens of m) beneath the seafloor. In the near vicinity of orebodies, alteration and deformation can be intense, making interpretation of primary volcanic textures and stratigraphic relations very difficult (Chap. 13).

18.3 Epithermal Precious Metal Deposits

Epithermal deposits are exploited principally as sources of gold and silver, and in some cases associated base metals (lead–zinc). Gold and silver may occur as a native element, electrum, in sulphides (especially pyrite), or in tellurides, and a variety of sulphide and sulphosalt minerals are commonly present. Epithermal deposits form at shallow depths, generally in the range of hundreds of m to ≤ 2 km below the Earth's surface. Geological and geochemical studies, and exploration successes in active volcanic regions over the past 30 years, in particular the Pacific Rim, have shown that many of today's geothermal systems are located in the same volcano-tectonic settings as fossil hydrothermal systems, which host epithermal deposits in Cenozoic and older successions. A range of types of active geothermal systems can form due to variations in depth, size, and age of the magmatic body (or heat engine) driving the system, the proportion of fluid components derived from solidifying magma, the magnitude of structural or seismic activity, and variations in the near-surface hydrological and geomorphic environment.

18.3.1 Modern Geothermal Systems

Studies of modern, larger geothermal systems show hot, deeply circulating convective groundwaters moving through fault-fracture networks usually heated and driven by shallow igneous intrusions, which provide the heat energy (Henley and Ellis 1983; Henley 1985). Henley and Ellis (1983) divided modern volcanically controlled geothermal systems into two end-member types: silicic and andesitic. In more recent years, these have become more generally described as either *basinal* or *stratovolcano* systems (Bogie et al. 2005; Fig. 18.2). As the name implies, the former are located in topographic basins or depressions, mainly extensional basins, rifts, and rhyolitic calderas but also strike-slip or pull-apart basins within magmatic arcs. These can be subdivided further into low- and high-salinity types. Stratovolcanoes are high-relief structures which, while often occurring in volcanic arc chains, are commonly sufficiently isolated from each other that they do not form a continuous region of high elevation.

Low salinity types are the most common basinal geothermal systems, exemplified by the Taupo Volcanic Zone, New Zealand. Meteoric recharge of waters that feed these systems is derived from equal or higher elevations than

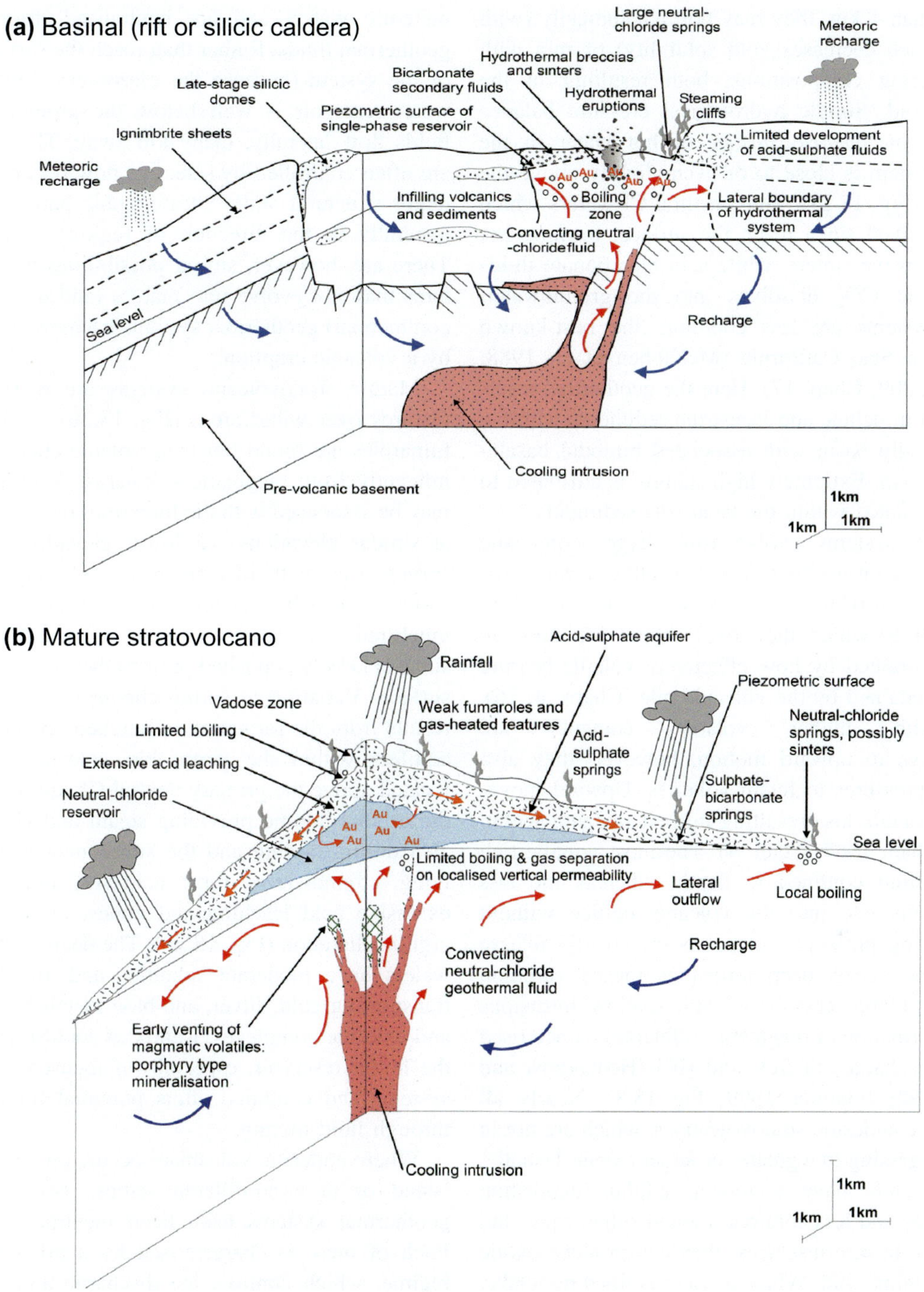

Fig. 18.2 Two end-member types of modern geothermal systems and potential sites of precious metal precipitation. **a** Basinal geothermal system developed in a structural depression (extensional basin or rift, pull-apart basin, or caldera) associated with silicic volcanism. **b** Stratovolcano geothermal system developed in a large composite cone associated with andesitic volcanism. Such a geothermal system evolves from a degassing (or immature) stratovolcano by the influx of meteoric waters, which over time are heated and convect. (Adapted from Bogie and Lawless 1987)

the ground surface over the geothermal field (Fig. 18.2a). The geothermal reservoirs contain low salinity, reduced neutral-Cl waters evolved by water–rock ± magma interaction at depths of 5–10 km. Magmas at depth also provide volatile compounds that play a role in transporting metals in solution (primarily H_2S, CO_2, HCl, SO_2). As these waters are heated and convectively circulate, gold is taken into solution as bisulphide complexes. When solutions rise to depths of less than 1 km, they may boil adiabatically (with loss of H_2S which decreases gold solubility) or mix with waters of differing compositions, both resulting in the deposition of gold. Simple hydrological pressure balance shows that the *piezometric surface* of hot water in the hydrothermal system is close to or even slightly above the ground surface (Fig. 18.2a). Thus, neutral-Cl waters which are rich in dissolved silica reach the surface, forming hot springs and extensive sinters, while acid bicarbonate fluids may form where CO_2 dissolves into the groundwater. High-salinity systems are less common, the best-known being the Salton Sea, California (McKibben et al. 1988; Brothers et al. 2009; Chap. 17). Here the geothermal reservoirs are hosted in deltaic and lacustrine sediments within a transform/strike-slip basin with associated bimodal basalt–rhyolite magmatism. Extremely high salinity is attributed to evaporites intercalated within the lacustrine sediments.

Stratovolcano systems evolve from large, composite cones which allow intrusions to rise to shallow levels, possibly within one km of the ground surface (Fig. 18.3). The minimum depth to which they rise without initiating an eruption is constrained by how effectively volatile-bearing melts can be contained by the volcanic pile (Chaps. 4, 16). Intrusions with higher volatile contents are density-conducive to upward motion; however, they also require higher pressures to retain volatiles. Upward movement triggers volatile loss resulting in melt freezing within the crust (Burnham 1967; Chap. 4). The most volatile-rich melts are therefore confined to depth, whereas the less volatile-rich melts rise into the volcanic edifice without erupting. Differing pressure constraints on volatile release between shallow versus deep intrusions means variation exists in their volatile chemistry, with shallow intrusions favouring the formation of magmatic solfataras characterised by fumarolic discharges of SO_2 and HCl (Hedenquist and Lowenstern 1994; Fournier 1999; Fig. 18.3). Nearly all potentially active andesitic stratovolcanoes, which are not in eruption, are degassing to a greater or lesser extent. Usually, magmatic solfataras have abundant sulphur deposition around fumaroles and less common molten sulphur pits and flows (Chap. 7). In rare instances, they can produce exotic sublimates including gold. Where a crater is filled by a lake, magmatic volatiles may condense to produce a potentially more hazardous, hot acid-Cl lake.

The point at which a degassing (or immature) stratovolcano develops an active or convecting geothermal system (mature stratovolcano) is dependent on the duration meteoric waters within the surrounding volcanic pile are heated and begin to convect (Bogie et al. 2005; Fig. 18.2b). Geothermal systems developed in such high-relief settings are largely recharged from low-elevation areas (i.e. ring plain; Chap. 16) surrounding the edifice. Once a convective system is initiated, rising magmatic vapours and waters interact with meteoric waters and are neutralised, forming neutral-Cl geothermal fluids. Rather than reach the surface in the centre of the system (because the piezometric surface of the hot water reservoir is well below the ground surface), these fluids flow laterally, many km away. These outflow zones are often controlled by selective flow occurring along a few major structural and/or stratigraphic pathways or conduits, generally in the direction of regional groundwater flow. There are, however, strong possibilities that immature systems may not evolve into mature (and therefore potentially commercial) geothermal systems because they are destroyed by a volcanic eruption.

Mature stratovolcano systems are marked by thermal features over wider areas (Fig. 18.2b). At high elevations, fumaroles are found but their volatile chemistry differs significantly from magmatic solfataras. Acid-SO_4 hot springs may be associated with the fumaroles or occur by themselves at similar elevations. At lower elevations, neutral-Cl hot springs occur with, in some instances, hydrothermal eruption craters and silica sinters, but typically these are on a much-reduced scale than found in basinal settings because silica is largely deposited prior to thermal water reaching the surface. Variation in spring chemistry relative to elevation results from the formation of perched secondary geothermal aquifers within the permeable volcanic pile. These are located above the primary neutral-Cl aquifer, which boils in the upflow regime providing steam and H_2S. Oxidation of H_2S during boiling and the subsequent mixing with overlying groundwaters form acid-SO_4 fluids, which cause extensive acid leaching and zones, or caps, of advanced argillic alteration (Fig. 18.2b). The deep neutral-Cl reservoir waters have moderate salinities and are thus capable of transporting gold, silver, and base metals by both bisulphide and chloride complexes. Reflux of acid-SO_4 fluids back into the hotter reservoir, especially if focused via fault-fracture systems and conduits, offers potential for gold deposition through fluid mixing.

Where andesitic volcanism occurs on an oceanic alkaline island or in a cordilleran setting, two further types of geothermal systems have been identified (Fig. 18.4a, b). Each of these is characterised by a different hydrological regime, which controls the discharge and recharge of the geothermal system, the distribution of fluid conduits and outlets, the type and distribution of hydrothermal alteration products (as discussed in the next section), and the potential sites for ore deposition.

18.3.2 Genetic Types of Epithermal Deposit

Within this geological and hydrological framework derived from modern geothermal systems, two fundamental genetic types of epithermal precious metal deposits have been

Fig. 18.3 Degassing or immature stratovolcano and hydrothermal system. **a** Hydrologic model showing potential sites and types of copper and gold mineralisation deposited from degassing magmatic volatiles and fluids (From Bogie et al. 2005). **b** and **c** High-temperature fumaroles or solfataras charged with magmatic vapours at Whakaari (White) Island, New Zealand. The outlet temperatures of fumaroles vary between 100 and >850 °C in response to variations in the rates of magma degassing and eruptive cycle. The hydrothermal system has been active for approximately 10,000 yr, in which time some 10^6 t of Cu and 45 t of Au may have discharged from the magma (Hedenquist et al. 1993). Although these amounts are mainly released as aerosols into the atmosphere, this illustrates the potential to form moderate to large orebodies in the subsurface. As a cautionary note, on 9 December 2019, a hydrothermal explosion killed 22 and injured a further 25 tourists and tour guides visiting the geothermal area inside the active crater at the time (Gibson 2020)

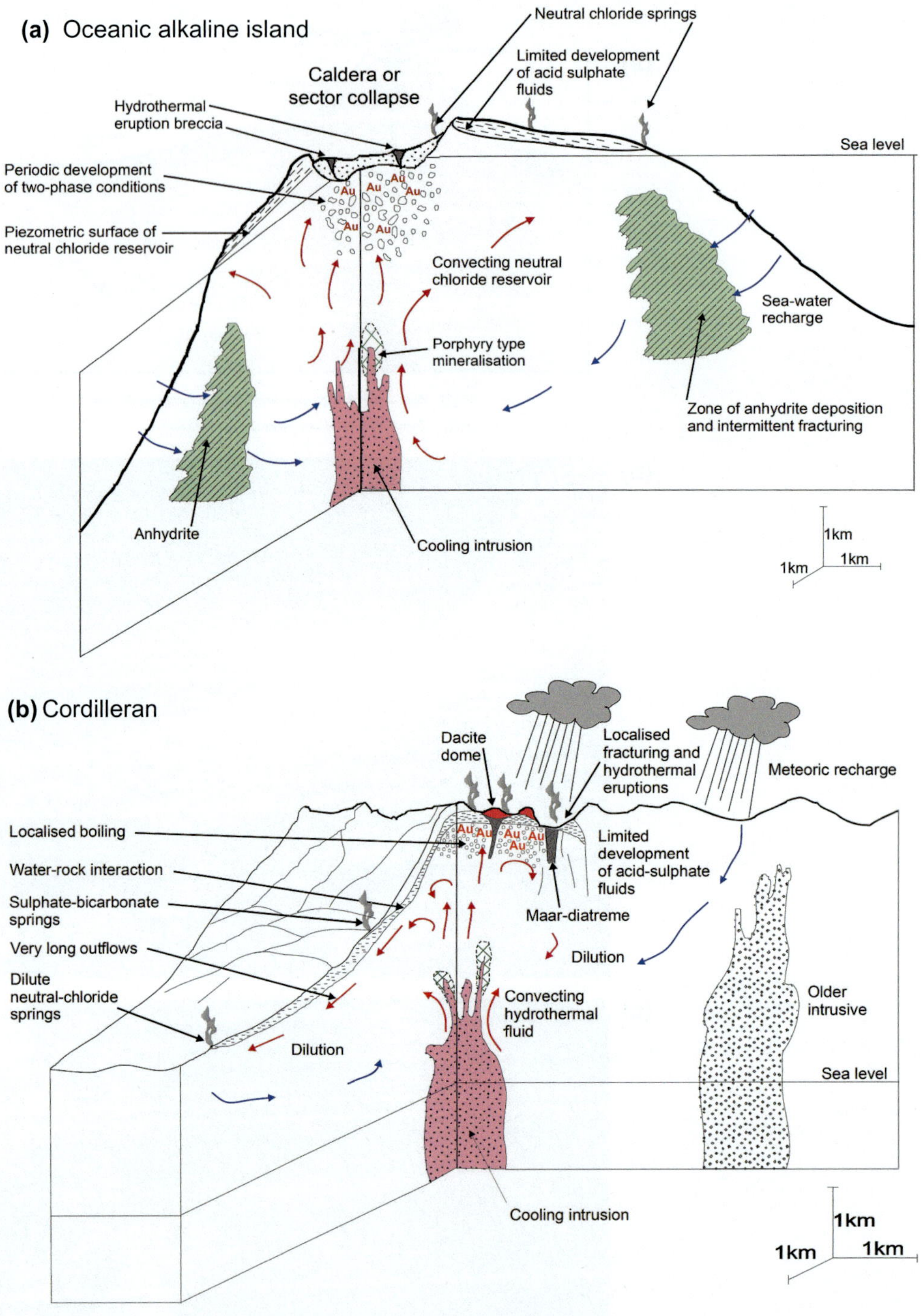

Fig. 18.4 Additional types of modern geothermal system associated with stratovolcanoes and andesitic volcanic terrains, and their potential sites of precious metal precipitation. **a** Oceanic alkaline islands, which are susceptible to caldera and/or sector collapse. **b** Cordilleran setting; here volcanism is coeval with compressional tectonics and mountain building (i.e. fold and thrust belt development), such as at a persistently subducting plate margin. The combination of repeated volcanism and shallow groundwater can lead to the formation of maar-diatremes, which can be significant locations of ore deposition. (Adapted from Bogie and Lawless 1987)

established: *adularia–sericite* and *acid-sulphate* (Hayba 1985; Heald et al. 1987). These are based on differences in the alteration assemblages and mineralogy produced by two contrasting hydrothermal fluids (Fig. 18.5), having, respectively, near-neutral pH (and reducing) or acid pH (and oxidising). Based on the sulphidation states of their hypogene (below the level of supergene oxidation) sulphide assemblages, adularia–sericite-type are now more generally termed *low sulphidation* (or *LS*) whereas acid-sulphate-type, *high sulphidation* (or *HS*) deposits (Sillitoe 1993; Arribas 1995; Hedenquist et al. 2000; Cooke and Simmons 2000; Sillitoe and Hedenquist 2003; Simmons et al. 2005; Corbett 2002, 2013; Fig. 18.6).

LS deposits are characterised by relatively narrow (m- to 10 s of m-scale) zones of quartz–adularia–sericite (illite) alteration in or adjacent to gold ± silver mineralised quartz veins in the ore environment (Fig. 18.5b). This vein-related alteration assemblage is surrounded by an envelope of

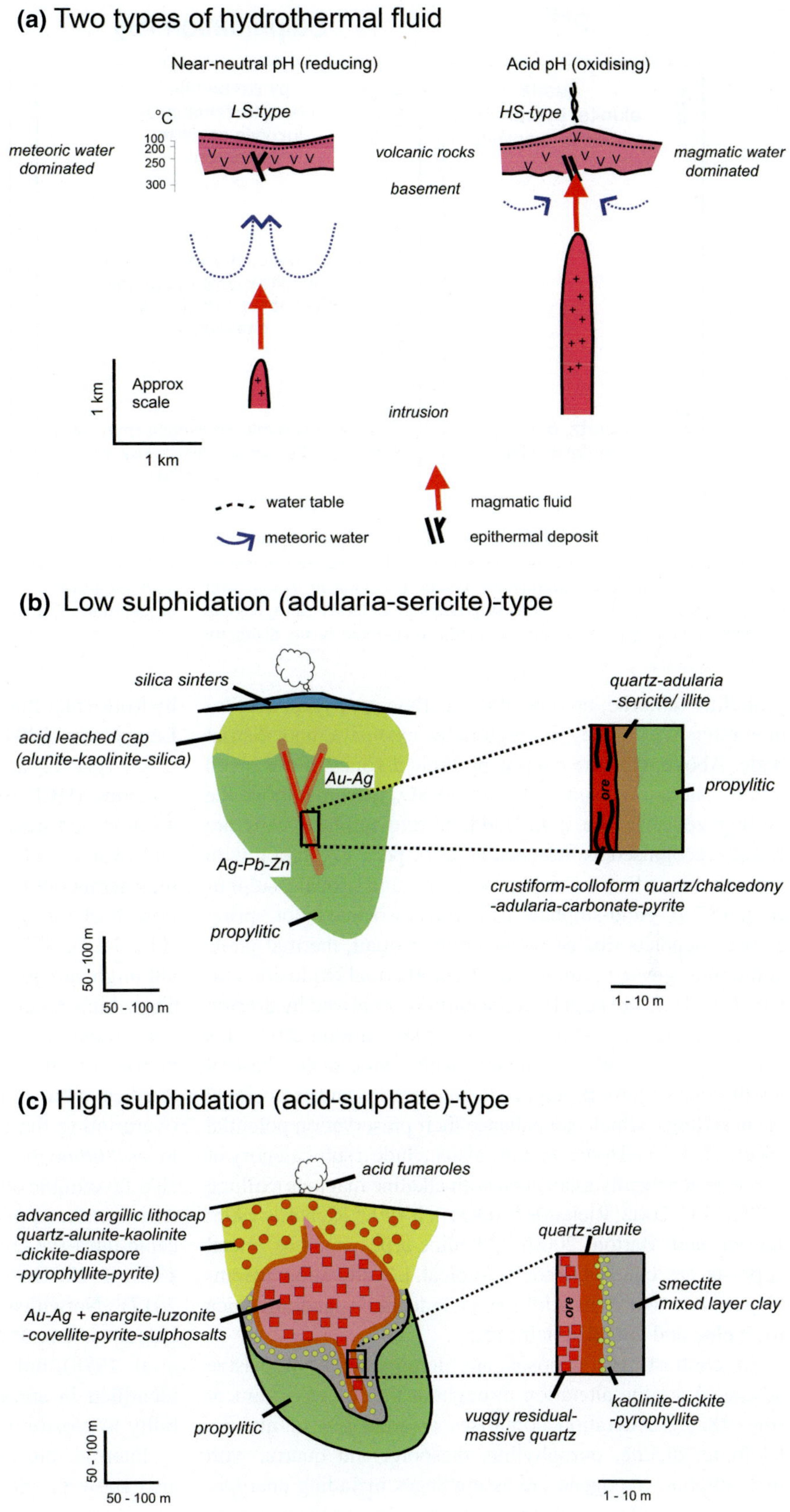

Fig. 18.5 Hydrothermal fluid and alteration types. **a** Two contrasting types of hydrothermal fluid, which lead to two fundamentally different genetic types of epithermal precious metal deposit termed low sulphidation (or adularia-sericite) illustrated in (**b**), and high sulphidation (or acid-sulphate) in (**c**). The different zonal patterns of wall rock alteration in (**b**) and (**c**) are used as vectors towards mineralisation; the diagrams on the left show the large-scale pattern, and the rectangle area outlined is magnified on the right to show alteration zonation patterns in the vicinity of ore. Alteration minerals form under specific pH and temperature conditions and therefore detailed mineralogical analysis is a valuable tool in the analysis of hydrothermal flow paths and as a guide to ore. (Adapted from Simmons et al. 2005)

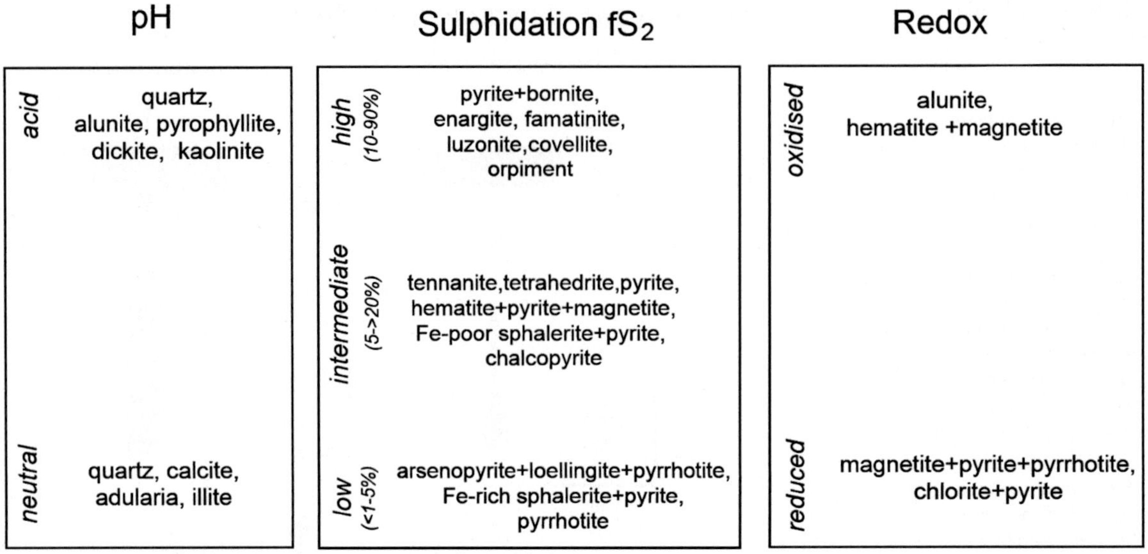

Fig. 18.6 Key indicator minerals of various states of pH, sulphidation, and oxidation used to distinguish epithermal ore-forming environments. The term "*intermediate*" sulphidation has been employed to represent the transition from high to low sulphidation in terms of hypogene sulphide mineralogy. As discussed in the text, there is no distinctly different intermediate sulphidation hydrothermal fluid, and intermediate and low sulphidation alteration is similar, in marked contrast to that of high sulphidation alteration. Percentages indicate a relative abundance of sulphides. (Adapted from Simmons et al. 2005)

propylitic (chlorite, epidote albite, carbonate, pyrite, zeolite) alteration, which may be regionally extensive and district wide. Above, they are capped by limited zones of advanced argillic alteration resulting from acid-SO_4 leaching above the boiling zone. In the field, modern acid-sulphate caps are readily recognised as bleached areas of poor vegetation with amorphous silica, alunite, kaolin, and local sulphur (Fig. 18.7a). Acid-sulphate caps and associated hot spring activity deposits (including steaming ground, thermal pools and ponds, geysers, sinters, and hydrothermal explosion craters; Fig. 18.7) are largely, if not entirely, removed by erosion in most ancient (pre-Miocene) examples (Sillitoe 2015). LS deposits principally correlate with wide-scale basinal geothermal systems in silicic volcanic terrains and extensional basin settings, which can enhance their preservation potential (Sect. 18.3.4.1). However, they also include a sub-category of deposits genetically associated with alkaline magmas (Sillitoe 1989, 1993, 2002; Richards and Kerrich 1993; Richards 1995; Jensen and Barton 2000). Alkaline-type precious metal deposits are characterised by typical LS alteration assemblages, but which in addition include the vanadium mica roscoelite and abundant tellurides.

In contrast, HS deposits are dominated by pervasive advanced argillic alteration throughout the ore environment (Fig. 18.5c). Alteration comprises assemblages of alunite, kaolinite, dickite, pyrophyllite, diaspore, and quartz, with high sulphur, hypogene ore assemblages including enargite, luzonite, covellite, and pyrite hosted by residual quartz rocks. Mineralogical and isotopic characteristics indicate the direct involvement of a magmatic component in the hydrothermal fluids (Hedenquist et al. 1993; Hedenquist and Lowenstern 1994; Arribas 1995). Such fluids are strongly acidic (pH <2) and composed of high-temperature magmatic volatiles (HCl, SO_2, and HF) which reflect only limited dilution and neutralisation by mixing with meteoric waters and water–rock interaction. These are comparable with high-temperature solfataras and vapours and acid springs now discharging from degassing (immature) stratovolcanoes (Fig. 18.3). HS systems are, therefore, characteristic of, although not restricted to, large andesitic stratovolcanoes. Their high relief provides a thick section of permeable volcanic rocks susceptible to leaching by aggressive fluids. This blanket of intense hypogene alteration shares many characteristics with zones of advanced argillic alteration found overprinting the tops of porphyry copper deposits, referred to as "*lithocaps*" (Sillitoe 1989; Fig. 18.8). Lithocaps are also favourable sites for epithermal gold deposition, and the term is now applied to the entire area between the subvolcanic intrusive environment and paleosurface (Sillitoe 1995, 1999, 2010; Hedenquist and Taran 2013; Hedenquist et al. 2017). Notwithstanding the discovery of HS deposits of Paleozoic and Proterozoic age (e.g. Hallberg 1994; Dube et al. 1998), hot spring and surface features have not been identified in ancient successions because of their susceptibility to erosion (Sect. 18.3.4.2).

Intermediate and andesitic volcanoes and terrains may also show the development of LS epithermal ore deposits. These may form from neutral-chloride fluids in the distal outflow regime of mature stratovolcanoes (Fig. 18.2b) or marine stratovolcanoes where the surrounding sea has a

Fig. 18.7 Features of active geothermal areas associated with low suphidation epithermal precious metal systems in basinal geothermal settings, as exemplified by the central Taupo Volcanic Zone, North Island, New Zealand (see Fig. 18.13 for locations). **a** Steaming cliff, Waimangu geothermal field. **b** Champagne Pool, whose name is derived from the abundant efflux of carbon dioxide (CO_2) similar to a glass of bubbling champagne. The hot spring was formed 900 years ago by a hydrothermal eruption in the Waiotapu geothermal field, forming a crater 65 m in diameter and up to 60 m deep. The orange colour originates from precipitates of arsenic and antimony sulphide which contrast with the grey-white silica sinter surrounding the pool. **c** Siliceous sinter curtain precipitated from the inactive hot-spring pond, Whakarewarewa, Rotorua geothermal field (From Cas and Wright 1987). **d** Geyser near Champagne Pool. **e** Sinter falls, Waimangu geothermal field; note these occur near the site of the historic Pink and White Terraces of Lake Rotomahana, once called the eighth wonder of the world, which were blasted away by the explosion of the hydrothermal system associated with the 1886 eruption of Mt Tarawera (de Ronde et al. 2016). **f** Karapiti hydrothermal explosion crater, Crater of the Moons, Wairakei geothermal field (Photo by Michael Rosenberg, GNS Science, New Zealand)

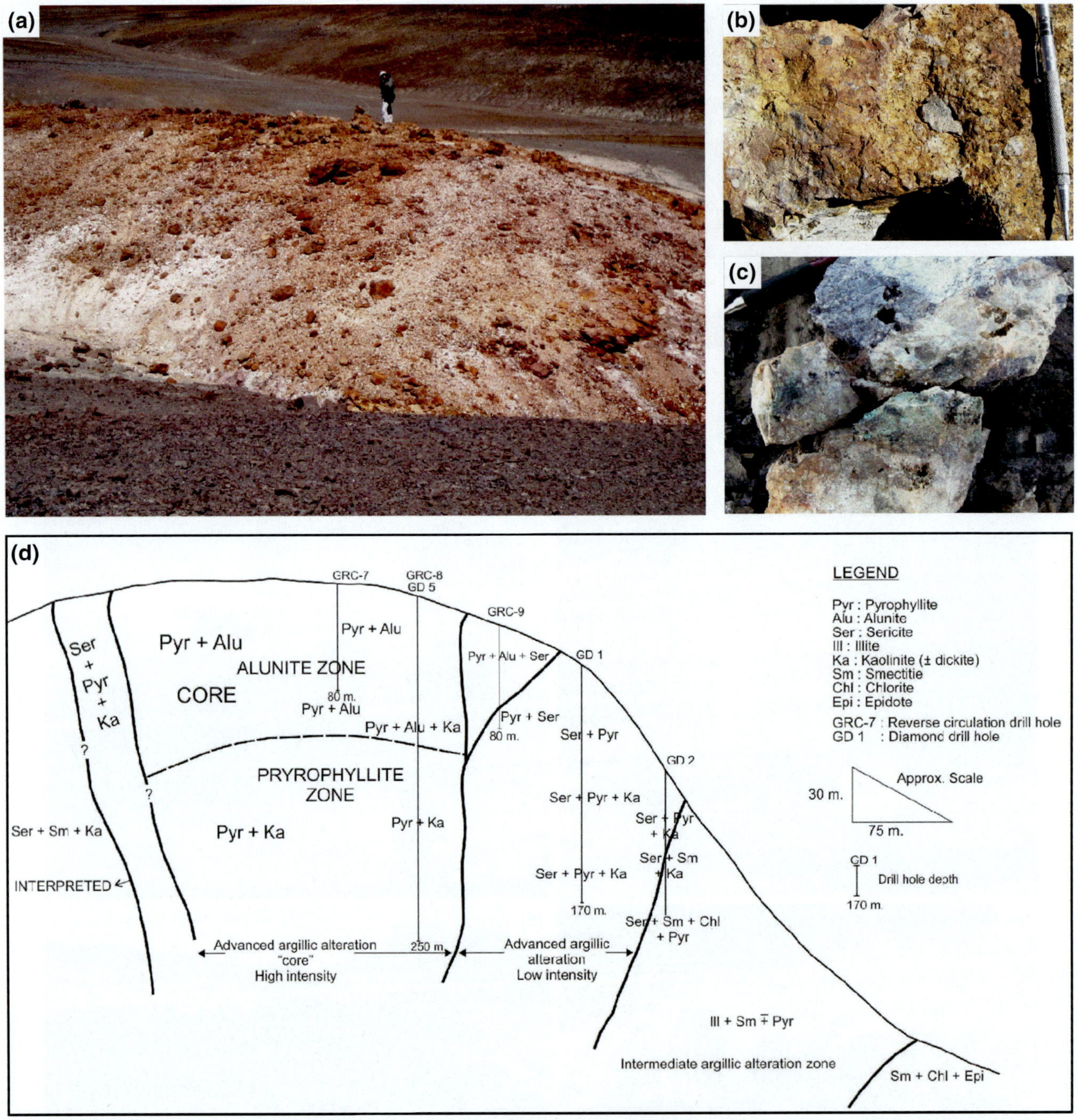

Fig. 18.8 Features of high sulphidation lithocaps. **a** An outcropping lithocap area of intense silica, alunite, dickite, and localised pyrophyllite alteration surrounded by gravel deposits; local host rock is andesite lava. **b** Close-up showing brecciated silica clasts within a matrix rich in limonite, and **c** silicified hydrothermal breccia with copper oxides. (*Source* Mirasol Resources, copper–molybdenum–gold prospect, central Chile). **d** Lateral and vertical zonation of alteration assemblages in higher-level lithocap at the Kuscayiri gold prospect, northwestern Turkey. Local volcanic host rocks are andesite lavas and related volcaniclastics, dacitic ignimbrites, and a quartz porphyry lava dome/stock which have lost most of their original volcanic textures. An advanced argillic assemblage of quartz, pyrophyllite, alunite, and kaolinite (±dickite) occurs at the centre of the system and shows a gradual change outward into sericite and pyrophyllite, then to sericite, illite/smectite mixed layer, and smectite. This halo surrounds and encloses zones of vertical to subvertical silica–quartz structures, including breccia pipe and stockworks with pyrite, pyrite disseminations, and Au–(Cu) mineralisation. Locally these silica bodies are strongly leached and vuggy (see text). (From Yilmaz 2003)

major control on the recharge and hydrology of the geothermal system (e.g. Nisyros, Greece; Ambrosio et al. 2010; Fig. 18.4a). More specifically with regard to oceanic alkaline island volcanoes, the deep submerged substructure of individual volcanoes may be very large compared to the small island(s) emergent above the sea. These systems are characterised by major precipitation of anhydrite from the recharge of heated seawater. This may dramatically affect permeability and hydrodynamic equilibrium and lead to cyclical fluctuations in the water table, boiling, and gold deposition (Bogie and Lawless 1987). Such volcanoes are also susceptible to caldera or sector collapse causing considerable disturbances to the hydrothermal system, perhaps creating the conditions to form very large epithermal gold deposits (e.g. Lihir Island, Papua New Guinea; Sect. 18.3.4.5). Widespread development of LS systems may also occur in cordilleran settings (Fig. 18.4b), where repeated volcanism on a persistently tectonically active plate margin constructs a mountain belt (e.g. Andes, Papua New Guinea, and Philippines). Here andesitic volcanic centres are surrounded by areas of sufficiently higher relief to provide meteoric recharge, such that water levels in the geothermal reservoir relative to the ground surface are also high, potentially creating very long outflows of neutral-Cl waters.

From the above discussion, it should be apparent that considerable overlap between both LS and HS genetic types could occur. Indeed, in the economic geology literature, much effort has been made to further categorise these into *intermediate sulphidation* or *IS* deposits, as well as other subtypes based on igneous association and metal content (Fig. 18.6; Sillitoe 1993; Hedenquist et al. 2000; Sillitoe and Hedenquist 2003). Rather than "pigeon-hole" similar attributes of the mineralisation, for our purposes, it is more preferable to regard these as a process, representing extremes in a possible continuum between meteoric-dominated (LS) and magmatic-dominated (HS) hydrothermal fluids. There is also potential for their juxtaposition or overprinting, a process which has been termed "*telescoping*", in particular, when applied to the superposition of epithermal mineralisation on the porphyry copper environment (Sillitoe 1994). However, this could hypothetically occur in any geothermal setting and be caused by a number of factors: (1) waning magmatic activity and availability of magmatic fluids; (2) tectonic uplift and erosion; (3) climactic caldera or sector collapse, or caldera resurgence; and/or (4) climatic changes which affect both the elevation of the water table and volume of meteoric recharge (which are principal controls on the depth of boiling and hence type of epithermal mineralisation). The late-stage "collapse" of meteoric water into the waning stages of an HS system and subsequent LS alteration and mineralisation may be a common evolutionary trend. Renewed magmatic activity recharging hydrothermal fluids with magmatic vapours could superimpose HS assemblages on LS deposits. For further typological description and discussion, also see Corbett (2002, 2013).

Low sulphidation deposits are easily the most abundant systems (Table 18.1; Fig. 18.9a). Historically most of the volcanic-hosted epithermal gold production in the western United States (Berger and Henley 1989) and 70% of the principal deposits in the western Pacific are from hydrothermal systems of this type (Sillitoe 1989). Irrespective of the epithermal deposit being low or high sulphidation, the range of volcanic host rock settings and facies is broadly similar, and these occur in a wide range of tectonic settings (Sillitoe and Hedenquist 2003; Fig. 18.9). Epithermal gold has accounted for about 6% of the world's gold production and 16% of its silver (Simmons et al. 2005). Regarding the size of epithermal deposits, we have divided them here on the basis of contained gold into superlarge (>200 t), giant (>400 t), and supergiant (>800 t).

18.3.3 Styles of Mineralisation

Many diverse styles of precious metal mineralisation are found in epithermal deposits. These are generally described in terms of vein, stockwork, and disseminated mineralisation, which essentially reflect host rock permeability (e.g. fracture versus primary and/or secondary porosity) during the mineralisation process. Many deposits combine in variable proportions characterising these three end members. All styles may be intimately associated with various types of brecciated host rocks, and determination of the fragmentation process(es), that is, volcanic, magmatic, and/or hydrothermal, is critical to fully elucidate the setting of mineralisation (e.g. Sillitoe 1985; Cas and Wright 1987). Essentially steep normal, oblique-slip, or strike-slip faults provide the principal fluid conduits, with dilational fault jogs and releasing bends on strike-slip or transfer faults providing particularly favourable trap sites (Sibson 1987; Sillitoe 1993; Rowland and Simmons 2012; Chap. 17). Feeder faults commonly give rise to conventional veins up to several m wide, while prolonged, repeated dilational activity and mineralisation may result in veins tens of m wide and more complicated geometries. Stockwork and disseminated gold mineralisation are amenable to bulk mining.

Within LS systems, hydrothermal fluids progressively become dilute and neutralised, and silica dissolves. Silica is later precipitated in veins as quartz followed by gold and silver. This results in banded veins, each band representing a repeated phase or separate event during the mineralising cycle (Fig. 18.10). Banded *crustiform* and *colloform* textures are the most characteristic primary depositional feature of LS quartz–adularia veins (Fig. 18.10a–e); their essential morphological difference being crystal size. Crustiform texture shows recurrent bands of crystals, with many pointed

Table 18.1 Selected features of epithermal precious metal deposits referred to in the text

Location	Volcanic setting/facies association	Age mineralisation (Ma)	Tectonic setting	Major host rock	Alteration type	Mineralisation style	Gold/ *Silver* content Moz/ (tonnes)
Round Mountain, Nevada	Silicic caldera (intra-caldera)	26	Extensional continental rift (Basin and Range)	Rhyolitic ignimbrite	LS	Veins, fracture fills, and disseminations	15 (425)
Cerro Vanguardia, Argentina	Silicic caldera (extra-caldera)	155–153	Continental margin back arc	Rhyolitic ignimbrite	LS	Veins	4.5 (125)
McLaughlin, California	Basaltic andesite lava/intrusions	2.2–0.75	Continental transform fault (dilational jog on San Andreas Fault)	Jurassic mudstone, metabasalt, and serpentinite	LS	Sheeted veins	4 (110)
Sleeper, Nevada	Silicic dome (cryptodome)	16.1–15.5	Extensional continental rift (Basin and Range)	Porphyritic rhyolite lava	LS	High-grade banded quartz veins, breccias, stockworks, and disseminations	2.7 (75)
DeLamar, Idaho	Silicic dome	15.7	Extensional continental rift (Basin and Range)	Quartz latite–rhyolite lava, domes, and breccias	LS	Veins and stockworks	1 (28)
Bodie, California	Silicic dome	8.4–8.2	Extensional continental rift (Basin and Range)	Dacite lava, domes, and breccias	LS	Veins	1.6 (45) *7.3 (200)*
Goldfield, Nevada	Silicic dome (on caldera ring fracture)	21–20	Extensional continental rift (Basin and Range)	Rhyodacitic domes	HS	Veins and breccias	4.2 (120)
Summitville, Colorado	Silicic dome (on caldera ring fracture)	22.5	Continental margin back arc	Quartz latite dome, flows, and pyroclastics	HS	Breccia pipes	<1 (23)
Cerro de Potosí, Bolivia	Silicic dome	13.8–13.5	Continental margin back arc	Rhyodacitic porphyry dome	HS	Veins, stockworks in massive and vuggy quartz	*3,500 (100,000)*
San Cristobal, Bolivia	Silicic dome	7.9–7.5	Continental margin back arc	Porhyritic rhyodacitic domes and crater lake sediments	LS	Barite veins and veinlets, mainly disseminations (lake sediments)	*450 (12,750)*
Ovaçik, Turkey	Intermediate dome	18.2	Extensional continental rift (Anatolian)	Porhyritic andesite lava	LS	Banded quartz veins, stockwork, and sheeted veinlets	>1 (>28)
Wau, Papua New Guinea	Maar-diatreme	4–2.4	Island arc (cordilleran)	Phreatomagmatic breccias intruded by andesitic–dacitic domes	LS	Veins, stockworks, and disseminations	1 (28)
Kelian, Indonesia	Maar-diatreme	19.8–19.5	Island arc (cordilleran)	Phreatomagmatic (carbonaceous matrix-rich) breccias intruded by rhyolite dykes	LS	Hydrothermal breccias, veins, and disseminations	>8.5 (>240)
Montana Tunnels, Montana	Maar-diatreme	50–45	Continental margin back arc	Matrix-rich phreatomagmatic breccias	LS	Sulphide disseminations and veinlets	2.1 (60) *38 (1,000)*
Cripple Creek, Colorado	Maar-diatreme	31–28	Continental margin back arc	Phonolitic to alkali basalt diatreme complex	LS (alkaline)	Veins and disseminations	30 (835)

(continued)

Table 18.1 (continued)

Location	Volcanic setting/facies association	Age mineralisation (Ma)	Tectonic setting	Major host rock	Alteration type	Mineralisation style	Gold/ *Silver* content Moz/ (tonnes)
Pueblo Viejo, Dominican Republic	Maar- diatreme	109.6–108.7	Primitive Intra-oceanic island arc	Carbonaceous sediments and andesitic–dacitic volcaniclastics, intruded dacite dome, and breccias	HS	Veins, disseminations (conglomerates), replacements (carbonaceous sand-silt layers)	28 (800)
Lepanto, Luzon, Philippines	Maar-diatreme	1.4–1.3	Island arc (cordilleran)	Dacitic maar pyroclastics intruded dacite dome	HS	Vuggy quartz, massive sulphide, hydrothermal breccia bodies; replacement ore along basal contact rim pyroclastics	4.3 (123)
Mount Leyshon, Queensland	Subvolcanic breccia pipe	280	Continental margin back arc (transtensional pull-apart on regional lineament)	Late breccia pipe and tuffisite breccia dykes associated andesite and rhyolitic stocks	LS	Breccia cavity infills, veins, infilling fracture networks	3.5 (100)
Kidston, Queensland	Subvolcanic breccia pipe	332	Continental margin back arc (transtensional pull-apart on regional lineament)	Breccia pipe associated rhyolitic stocks and minor andesite	LS	Breccia cavity infills, sheeted veins around periphery of pipe	5 (140)
Choquelimpie, Chile	Andesitic stratovolcano (minor erosion of central cone)	7	Compressive continental margin arc	Summit andesitic porphyry dome	HS	Veins, hydrothermal breccia bodies	<1 (25)
Nansatsu, Kyushu, Japan	Andesitic cones in (?) caldera (shallow erosion)	5	Island arc (cordilleran)	Andesitic volcanics	HS	Vuggy quartz, massive sulphide bodies	<0.5 (12)
Nena, Papua New Guinea	Andesitic stratovolcano (moderate erosion)	13.1–11.2	Island arc (cordilleran)	Andesitic volcano-sedimentary succession	HS	Massive sulphide bodies, vuggy quartz, hydrothermal breccias	2 (56)
Nalesbitan, Luzon, Philippines	Andesitic stratovolcano (shallow erosion)	Pliocene	Island arc (cordilleran)	Andesitic volcanics	HS	Massive quartz, hydrothermal breccia bodies	1 (28)
Tampakan, Philippines	Andesitic stratovolcano (deep erosion)	3.34–3.23	Island arc (cordilleran)	Andesitic lava flows	HS	Massive-vuggy quartz, hydrothermal breccias along intra-volcanic unconformity	10 (287)
Bor, Serbia	Andesitic stratovolcano (deep erosion)	86	Continental margin collisional arc	Andesitic lava facies (coherent, autoclastic)	HS	Massive sulphide bodies, hydrothermal breccias and stockworks; sediment-hosted disseminations	1.8 (50)
Chelopech, Bulgaria	Maar-diatreme (?shallow marine caldera) margin)	91	Continental margin collisional arc	Andesitic to dacitic breccias	HS	Veins, hydrothermal breccias and massive sulphide replacements and disseminations	5.5 (156) *14.6* *(415)*
Štiavnica-Hodruša, Slovakia	Andesitic stratovolcano (resurgent caldera)	12.2–10.7	Continental margin collisional arc	Resurgent horst (pre- and post-caldera andesite lavas, volcaniclastics, and epiclastics)	LS	Veins and stockworks	3 (80) *120* *(3400)*

(continued)

Table 18.1 (continued)

Location	Volcanic setting/facies association	Age mineralisation (Ma)	Tectonic setting	Major host rock	Alteration type	Mineralisation style	Gold/ *Silver* content Moz/ (tonnes)
Hishikari, Japan	Eroded andesitic terrain	0.9	Island arc (cordilleran)	Metasedimentary basement, andesitic volcanics	LS	Banded high-grade quartz veins	10 (300)
Baguio, Luzon, Philippines	Eroded andesitic terrain	0.65	Island arc (cordilleran)	Dacitic domes and diatreme (Acupan); conglomerates	LS	Veins, stockworks	35 (1000)
Baguio, Luzon, Philippines	Eroded andesitic terrain	0.65	Island arc (cordilleran)	Dacitic domes and diatreme (Acupan); conglomerates			
Paracale, Luzon, Philippines	Eroded andesitic terrain	2.4	Island arc (cordilleran)	Basement trondhjemite dome intruded dacite porphyry dikes	LS	Veins	6 (160)
Gunung Pongkor, Indonesia	Eroded andesitic terrain	2	Island arc (cordilleran)	Andesitic volcanics and sediments	LS	Veins	3.7 (105) *36 (1030)*
Yanacocha, Peru	Intermediate multivent	11.0–8.2 (main stages)	Compressive continental margin arc	Dacitic–rhyolitic domes, ignimbrite	HS	Massive-vuggy quartz bodies as disseminations and replacements	>50 (>1400)
El Indio, Chile	Non apparent	8–5	Compressive continental margin arc	Oligocene dacitic volcaniclastics, early Miocene andesite lavas (no coeval igneous rocks)	HS	Veins, bonanza veins	>10 (>300)
Veladaro, Argentina	Intermediate multivent	12.7–10.7	Compressive continental margin arc	Volcano-sedimentary breccias, subvolcanic domes	HS	Hydrothermal breccia bodies	12.2 (350)
La Coipa, Chile	Intermediate multivent	20–17	Compressive continental margin arc	Dacitic domes and pyroclastics, diatreme breccias	HS	Vuggy quartz massive sulphide bodies, breccias	3.5 (100)
Pascua-Lama, Chile–Argentina	Subvolcanic breccia pipe	9.1–8	Compressive continental margin arc	Pipe-like bodies of brecciated Permo-Triassic to Jurassic granitoids	HS	Breccia and stockwork, disseminated bodies	17.6 (500) *20 (585)*
Lagunas Norte, Peru	Maar-diatreme	17.4–16.5	Compressive continental margin arc	Brecciated Cretaceous quartzites (75%), early Miocene dacitic diatreme breccias (25%)	HS	Vuggy quartz, disseminated, and stockwork	13 (370)
Emperor, Vitu Levu, Fiji	Alkaline oceanic volcano (with caldera)	3.9	Island arc (post subduction zone reversal and arc rotation)	Mafic lavas, minor air-fall tuffs	LS (alkaline)	Veins, flat-makes, shatter zones	10 (280)
Ladolam, Lihir Island, Papua New Guinea	Alkaline oceanic volcano (with sector collapse)	0.35-?present	Island arc (post subduction zone reversal and arc rotation and/or slab tear)	Mafic-intermediate volcano-sedimentary succession	LS (alkaline)	Breccias, stockworks	46 (1300)
Betze-Post, Nevada	No clear or direct relation to volcanism; coeval rhyolitic–andesitic dykes	40–39	Continental margin back arc	Fine-grained calcareous and carbonaceous Paleozoic metasediment	LS	Disseminated and breccias in tabular bodies, and "seams" along dark decalcified fractures	35 (1000)

This summary is based on multiple sources including text references, current global industry reviews, and company reports

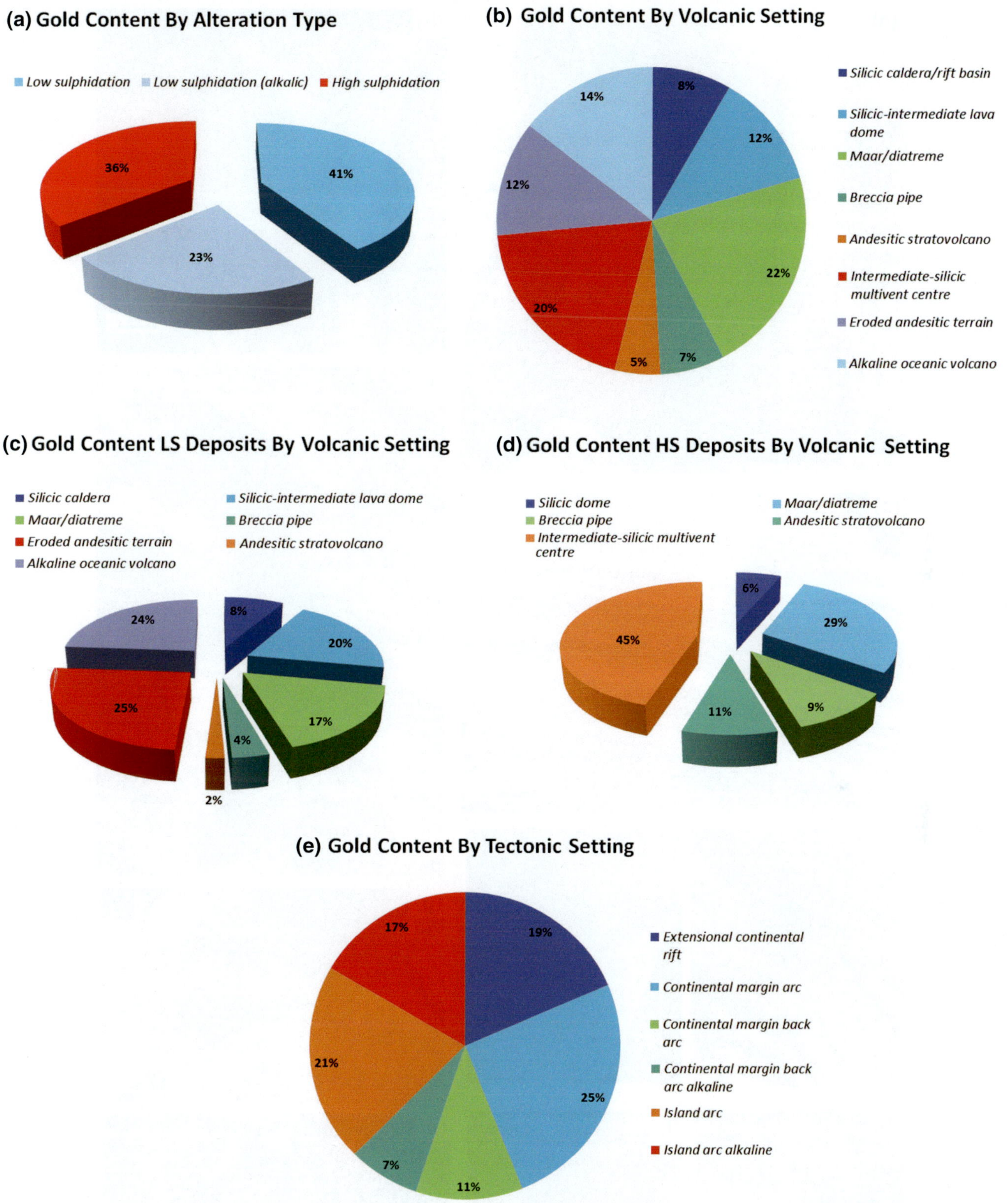

Fig. 18.9 Summary of contained gold in epithermal gold deposits by alteration type, volcanic setting, and tectonic setting. This includes most of the larger known deposits and represents a global resource of approximately 15,000 tonnes or 500 Moz of gold. Selected features of examples referred to in the text are described in Table 18.1

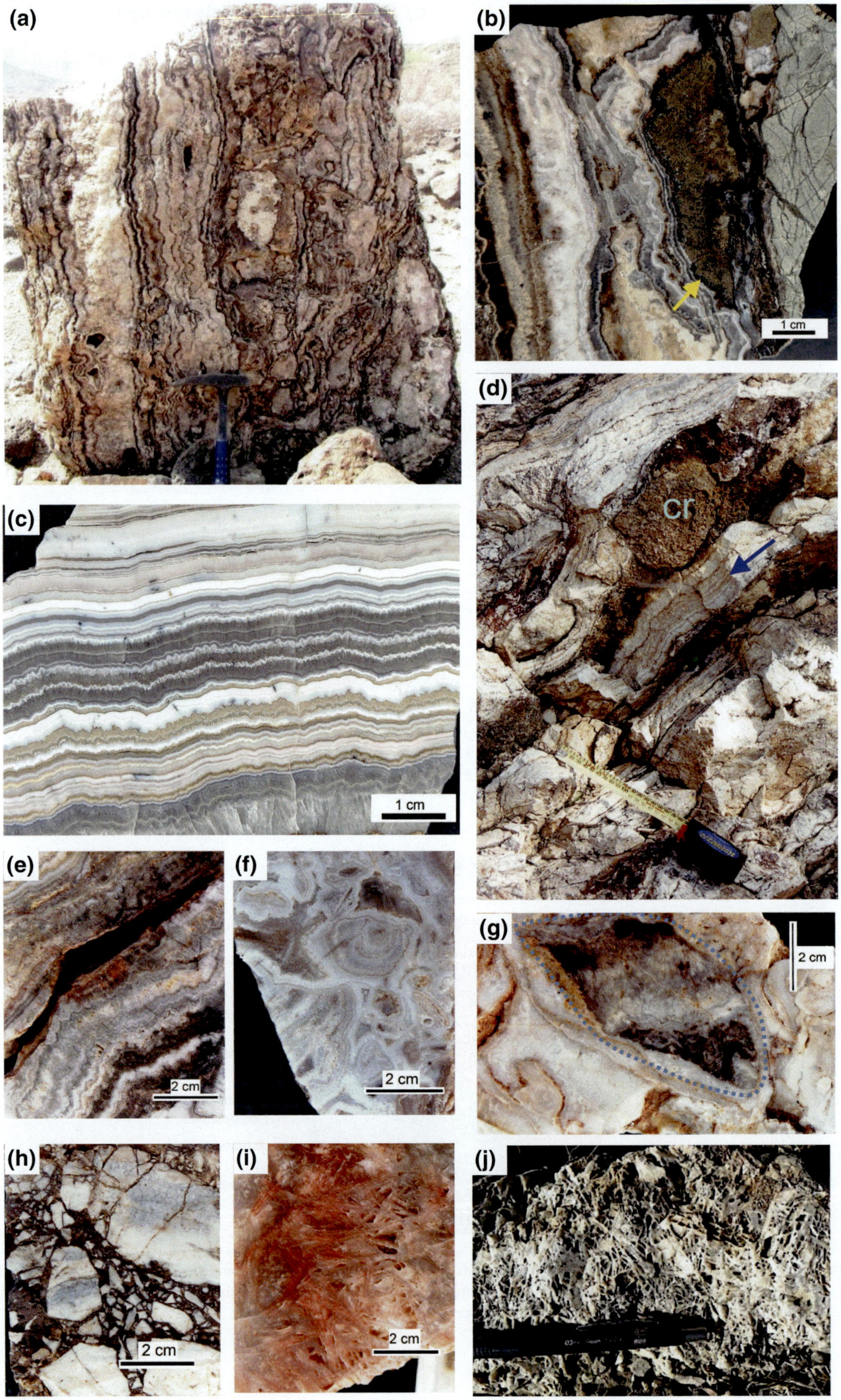
(a)
(b)
1 cm
(c)
1 cm
(d)
cr
(e)
2 cm
(f)
2 cm
(g)
2 cm
(h)
2 cm
(i)
2 cm
(j)

◀ **Fig. 18.10** Characteristic primary epithermal quartz vein features and textures of low suphidation vein systems. **a** Prospector's "reward"! Classic crustiform and chalcedonic colloform banded epithermal boulder from discovery outcrop of the Pandora vein system, Djibouti. Hosted by basalt lavas (2–2.7 Ma), this was discovered in 2012, although, the vein had been previously mapped as rhyolite. (*Source* Thani Stratex Resources). **b**, **c** McLaughlin "hot spring" gold deposit, northern California, USA. **b** Gold-bearing crustiform vein from the South Pit. The brown and tan coloration in the quartz–chalcedony bands is due to finely disseminated hydrocarbon residue, which often indicated the presence of higher-grade ore. The yellow arrow points to the outer fringe of a dendritic electrum (gold–silver alloy) mass; the dendrites may have formed by scavenging particulate Au by electrostatic attraction to the hydrocarbon–silica phase (Sherlock and Lehrman 1995). The grey-green wall rock at the right of the image is Jurassic metabasalt. (Photo and description by Dean Enderlin). **c** Fine crustiform and colloform banded chalcedonic vein which is typical of the McLaughlin deposit and carried 3–6 g/t Au (Photo by Catherine Koehler; *Source* University of California, Davis-McLaughlin Reserve). **d**–**i** Ovaçik gold deposit, northwestern Turkey (Photos by Hüsseyin Yilmaz; for further description, see Yilmaz 2002 and text). **d** Field view showing one of the two main quartz–adularia veins ("S" vein) with crustiform and chalcedonic banding (blue arrow) with abundant carbonate replacement textures (cr), which contains 23.1 g/t Au over 3 m. **e** Crustiform banded quartz–adularia vein textures containing 15.7 g/t Au over 1 m. **f** Cockade banded quartz breccia, with late-stage vug infills. **g** High-grade quartz breccia (containing 91.0 g/t Au over 1 m), with rotated clast (blue outline) showing crustiform banding and plumose intergrowth textures (after adularia). **h** Late-stage hydraulic breccia showing abundant jigsaw-fits and mobilised splintery fragments of quartz–adularia vein and re-fragmented quartz-cemented breccia with hematite ± silica ± gold matrix; this contains 20.8 g/t Au over 1 m. **i** Bladed quartz after calcite replacement texture indicative of hydrothermal boiling. **j** Quartz after calcite lattice texture, Goya vein, northeast Nevada, USA; assays from this outcrop range up to 25 g/t Au over 3 m (*Source* Logan Resources). **k** Quartz–adularia vein displaying prominent "ginguro" bands of dark grey silver sulphides, Konomai gold–silver deposit, Hokkaido, Japan (*Source* Japan Gold). **l**, **m** Sheeted epithermal vein and stockwork, McLaughlin gold deposit. Formation of sheeted vein and stockwork was characterised by syntectonic multistage vein-filling of extensional and transtensional fractures. **l** Specimen with at least five generations of veining (Photo by Dean Enderlin). **m** The most impressive sheeted vein complex formed the root of the San Quentin sinter and accounted for approximately 25% of the total gold production at the McLaughlin mine (Photo by Catherine Koehler; *Source* University of California, Davis-McLaughlin Reserve; for detailed description, see Sherlock et al. 1995). **n** Bonanza-grade gold ore in quartz–adularia vein from the Sleeper mine, northern Nevada. Native gold occurs in this rock as colloform bands, partially replaces breccia clasts, and also is disseminated in the matrix. Field of view 10.5 cm across. (Photo by James St John; *Source* Wikimedia through Creative Commons Attribution 2.0 Generic licence). **o** Bonanza zone ore, Koryu gold–silver deposit, Hokkaido. Coarse chalcopyrite (cpy) and some Au–Ag minerals precipitated during late comb-quartz formation, which were likely dissolved and remobilised from the earlier ginguro stage. (Adapted from Shimizu 2014)

Fig. 18.10 (continued)

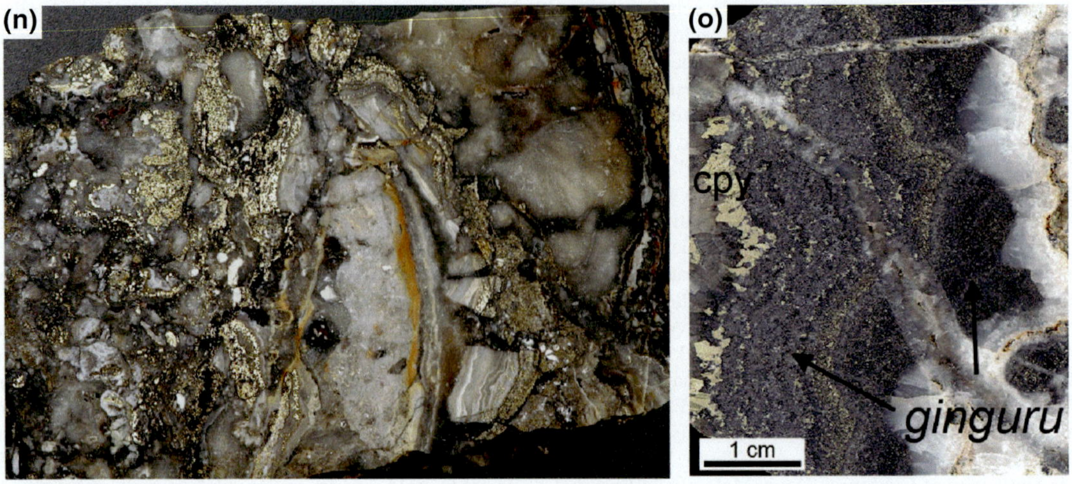

Fig. 18.10 (continued)

terminations and oriented parallel to vein walls, which are defined by differences in mineralogy and/or colour, while colloform texture reveals rhythmic or oscillatory bands with reniform surfaces of chalcedonic silica. Such textures are generally attributed to rapid deposition which would occur during rapid boiling, in open spaces such as associated with sudden dilation of a fracture (Henley and Hughes 2000; Moncada et al. 2012; Shimizu 2014; see Fig. 18.15). *Cockade* texture is confined to brecciated zones, where rock or vein fragments are surrounded by concentric bands of later chalcedonic silica (Fig. 18.10f). Other common textural types are *platy* or *lattice-bladed* quartz, where silica pseudo-morphs early calcite cements (Fig. 18.10d, i, j); this is interpreted to result from the rapid growth of calcite as CO_2 is lost to the vapour phase during boiling geothermal systems (Simmons and Christenson 1994; Etoh et al. 2002; Moncada et al. 2012). Another term sometimes used is *chalcedony–ginguro* mineralisation, which comprises low-temperature microfine chalcedony as the main form or quartz combined with black sulphidic "*ginguro*" (Corbett 2002, 2013; Fig. 18.10k, o). The term ginguro is derived from nineteenth-century Japanese miners who recognised the association of elevated precious metal grades with this material as bands, breccia clasts, and matrix. Ginguro comprises fine pyrite, silver sulphosalts, electrum, native gold, and minor chalcopyrite.

As indicated earlier, alteration of marginal wall rocks to LS veins (or other geometrical forms) grades laterally outwards from illite to smectite and commonly into an outer propylitic zone (Fig. 18.5b). Vein systems are similarly vertically zoned from barren surficial acid-sulphate caps, sinter, and hydrothermal eruption breccias (if preserved), downwards into veinlets and stockworks emplaced into the base of hydrothermal breccias with low-grade mineralisation and into fissure veins hosting most economic gold–silver mineralisation. These may show multiple episodes of hydraulic fracturing, brecciation, and rebrecciation (Fig. 18.10l, m). They often terminate downwards in quartz veins with elevated base metal contents and sub-economic gold–silver (Fig. 18.5b). Most fissure veins occur in dilatant fractures which host wider and higher gold-grade vein portions within ore shoots. Zones of exceptionally high gold grade (several hundreds to thousands of g/t Au) are referred to as *bonanza* veins (Fig. 18.10n, o).

On the other hand, HS deposits occur in permeable volcanic and brecciated host rocks, linked by permeable structures to the magmatic source. Within these feeder faults, magmatic vapour-enriched fluids react with groundwater to form strong acids (Hedenquist and Taran 2013; Fig. 18.11a). Such acids leach and dissolve surrounding rock leaving only residual silica behind, often as a "sponge-like" rock with abundant voids or large pores known as *vuggy* (or *vughy*) *silica* or *quartz* within which volcanic textures are often obliterated beyond recognition (Sillitoe 1993; Corbett 2002). Vugs (or vughs) represent sites of leached phenocrysts, pumice, and lithic fragments, which provide secondary porosity for later mineralisation (Fig. 18.11d–f). Zoning develops outwards (and downward) from cores of vuggy silica, through advanced argillic lithocap assemblages containing alunite (Fig. 18.11b, c) to distal argillic alteration. The shape of mineralisation is generally determined by the distribution of vuggy silica, which ranges from steeper dipping, tabular bodies aligned along feeder faults to shallower dipping or "*mushroom*- or *christmas tree*-shaped" bodies reflecting fluid migration along permeable beds and horizons. Mineralisation occurs as massive and semi-massive sulphides (>50% sulphide minerals; Fig. 18.11g), vuggy quartz infills (Fig. 18.11h), and replacements, veins, hydrothermal breccias, and disseminations. Most HS ore systems display elements of structural,

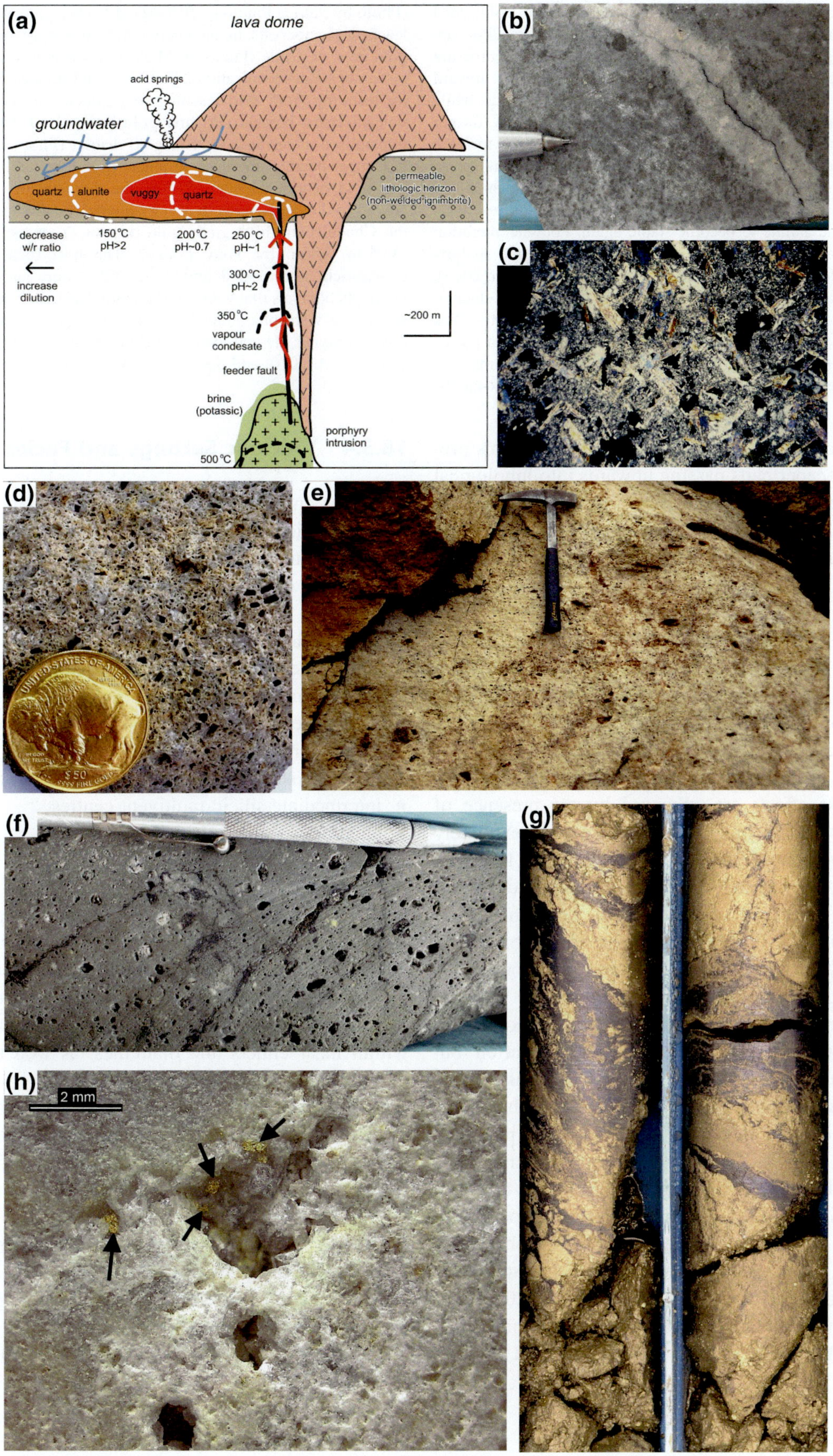
(a)
lava dome
acid springs
groundwater
quartz
alunite
vuggy
quartz
permeable lithologic horizon (non-welded ignimbrite)
decrease w/r ratio
150 °C pH>2
200 °C pH~0.7
250 °C pH~1
increase dilution
300 °C pH~2
350 °C
vapour condesate
~200 m
feeder fault
brine (potassic)
porphyry intrusion
500 °C
(b)
(c)
(d)
(e)
(f)
(g)
(h)
2 mm

◀ **Fig. 18.11** Characteristic features of high sulphidation epithermal systems. **a** Schematic development of advanced argillic lithocap, resulting from vapour condensate ascending along a feeder structure which intersects a permeable horizon. The alteration widens upward due to an increase in reactivity (lower pH) at lower temperatures, while lateral flow within the permeable horizon is accompanied by further cooling (dashed isotherms) and increased reactivity (lower pH) forming a lithocap of residual vuggy quartz with a halo of quartz–alunite. In distal positions, the water/rock ratio decreases, accompanied by increased dilution by groundwater, both of which cause the reactivity to decrease and the alunite to become stable again. Potassic alteration associated with the porphyry forms at the same time as the shallow advanced argillic alteration. (Modelled by Hedenquist and Taran 2013). **b** Quartz–alunite alteration halo and veinlet from the Bor gold district, Serbia (Photo by Dejan Kozelj). **c** Thin section from alunite–silica lithocap showing (primary) hypogene euhedral alunite laths in places forming "star" patterns in a vuggy microcrystalline quartz, Ergama gold prospect, northwest Turkey. Field of view ~3 mm wide, x-nicols. (Photo by Patrice Piantone, BRGM). **d** Vuggy porosity developed from leached phenocrysts in an intermediate porphyry intrusive, El Indio gold deposit, Chile (Photo by Michael Thomsen; *Source* https://www.researchgate.net/publication/282704937). **e** Lithocap with vuggy porosity resulting from acid leaching of fragments in a matrix-supported, volcaniclastic breccia, Ergama gold prospect. **f** Vuggy porosity in mineralised andesitic lava, Bor gold district (Photo by Dejan Kozelj). **g** Copper-bearing massive sulphide (pyrite-covellite) mineralisation, Bor gold district. Depth = 574 m, the diameter of drill core ~6 cm (Photo by Dejan Kozelj). **h** Vuggy quartz containing 300 g/t Au from the Chaquicocha high-grade gold deposit, Yanacocha, Peru; depth = 33.5 m (Photo by Tony Longo). The host rock began life as a crystal-rich, non-welded ignimbrite. Vugs were sites of feldspar and mafic phenocrysts that were subsequently leached. Native gold (arrows) and barite were precipitated late in the vugs after the fluids evolved towards higher pH and lower temperature. (For further description, see Longo et al. 2010, and Sect. 18.3.4.5)

breccia, or lithological control, with structural controls predominating in deeper parts passing upwards into lithological at shallower levels. The intersection of dilatant structures and permeable horizons or with the margins of diatremes and breccia pipes represent important sites for ore formation (Corbett 2002). HS deposits may also locally contain bonanza zones (Corbett 2002; Longo et al. 2010).

Lastly here regarding lithocaps, it is imperative during exploration to distinguish those associated with the hypogene alteration from those formed by supergene oxidation (weathering) of pyritic rocks (Sillitoe 1993; Fig. 18.12). Hydrothermal and supergene alteration often have complementary effects on rocks, increasing porosity, and decreasing cohesion. These effects are enhanced by the presence of disseminated sulphide mineralisation, which is extremely sensitive to hydrolysis and gives rise to acid leaching. Lithocaps formed by hypogene alteration result, as described above, either from the circulation of deep magmatic volatiles in the HS mineralising environment or at the paleosurface by condensation of steam and oxidation of H_2S in acid-leached zones capping both HS and LS deposits (Fig. 18.12a, b). In the first case, these are potentially ore-forming, while in the second, these are barren but potentially occur above ore. Supergene alteration takes place close to the water table and may or may not have any relation, depending on the origin of the oxidised pyrite, to mineralisation (Fig. 18.12c). While some criteria have developed to distinguish these different reactions and processes in the field (e.g. alunite crystallinity; Sillitoe 1993), only detailed mineralogical, radio-crystallographic, and stable isotopic analysis coupled with radiometric age determination provide definitive data (e.g. Deyell et al. 2005). For a current review of the multiple formation environments of advanced argillic minerals, see Hedenquist and Arribas (2021).

18.3.4 Volcanic Settings and Facies

To assess epithermal deposits in more detail with regard to their volcanic host rocks and successions, a convenient way is to divide them by volcanic landform or feature and facies associations (Table 18.1; Fig. 18.9):

- silicic caldera and rift basin successions,
- silicic-intermediate lava domes (and intrusive cryptodomes),
- maar-type volcanic centres (including diatremes),
- breccia pipes,
- andesitic stratovolcanoes and eroded terrains,
- intermediate-silicic multivent centres,
- alkaline oceanic volcanoes, and
- sediment-hosted.

Such categories are, however, not mutually exclusive. Lava domes and maar-type volcanoes may be important epithermal hosts in both larger-scale silicic caldera and intermediate stratovolcano and multivent settings, while lava domes and cryptodomes may be important facies found within maar craters and diatremes, as well as subvolcanic breccia pipes. Additionally, some workers view sediment-hosted disseminated gold deposits, frequently referred to as *Carlin-type*, to represent distal volcanic or magmatic settings. These are therefore briefly included as the final association described here.

18.3.4.1 Silicic Calderas and Rift Basin Successions

This association is exemplified by the large geothermal and hot spring systems active in the central Taupo Zone in North Island of New Zealand, studies of which formed the basis for

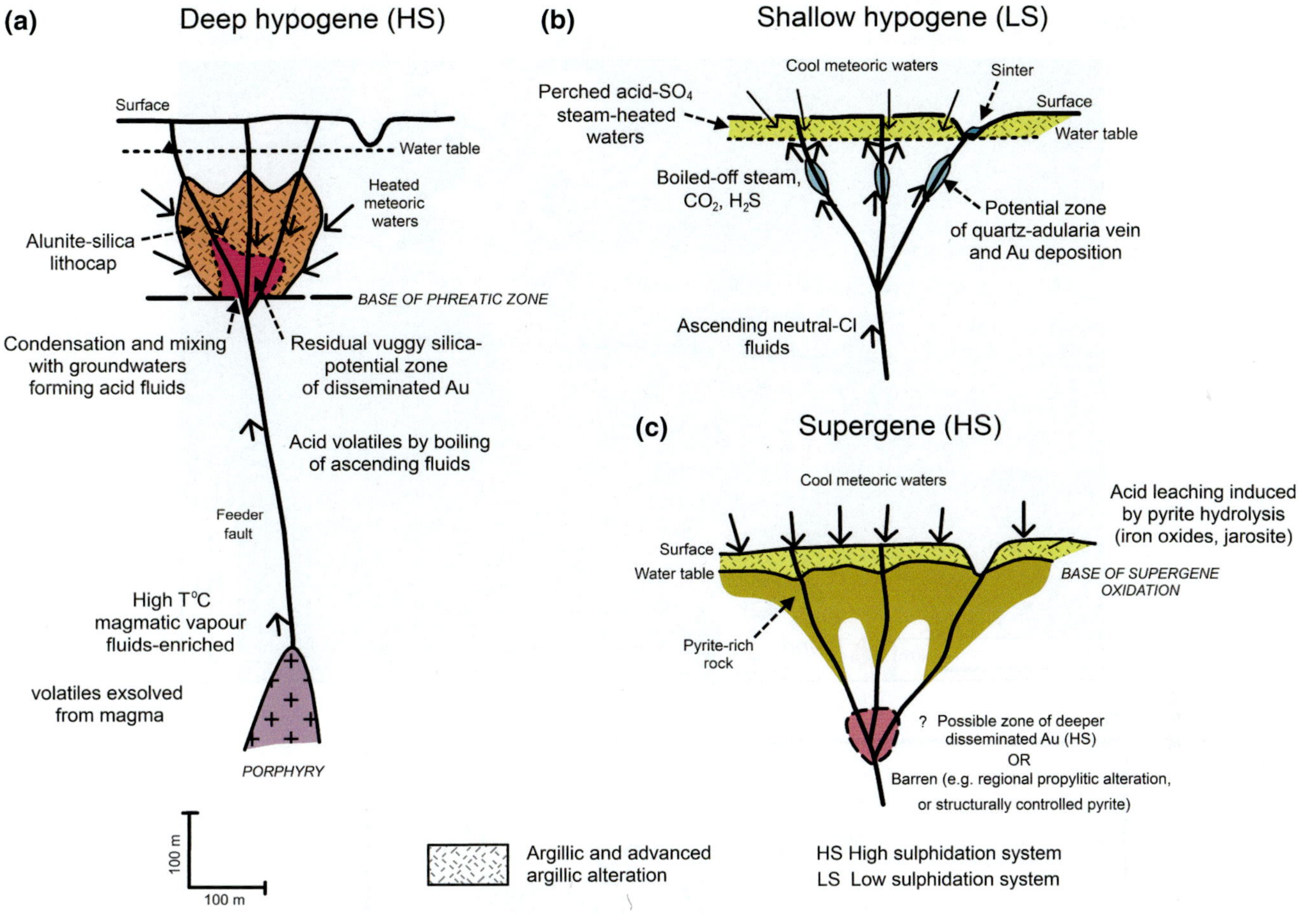

Fig. 18.12 Characterisation of hydrothermal and supergene alteration in epithermal systems. Distinguishing the three types of advanced argillic alteration is important because of their relationships to gold and silver mineralisation. Only detailed mineralogical, radio-crystallographic, and isotopic analysis of the differing clay assemblages can discriminate these definitively. Regarding alunite, only euhedral laths forming "pseudo-stars" in a quartz matrix are evidence of deep, hypogene alteration involving magmatic fluids (Fig. 18.11c). (After Sillitoe 1993)

understanding LS hydrothermal systems worldwide (Figs. 18.2a, 18.13). The discovery of gold-rich sulphide scales in geothermal wells at the Broadlands-Ohaaki field (Brown 1986) was an important step in confirming the precious metal ore-forming capabilities of these active systems, although, all known mineralisation is presently sub-economic (Simmons et al. 2016). The geothermal systems are intimately associated with major silicic calderas occurring within the Taupo Volcanic Zone (TVZ), a back-arc volcano-tectonic graben dominated by continental extension and copious silicic volcanism commencing 1.6 Ma ago (Chap. 16; Wilson et al. 1995, 2009; Rowland and Simmons 2012; Wilson and Rowland 2016).

Nearly all geothermal fields are located on caldera ring faults and/or major extensional and transfer faults (also known as accommodation or transverse zones, and relay ramps; Chap. 17). Transfer faults accommodate segmentation and displacement along rift basins and serve as important fluid pathways (Rowland and Sibson 2001, 2004; Rowland and Simmons 2012; McLellan et al. 2010). Host facies are a thick volcanic succession (1 to >3 km) dominated by voluminous pumice fall deposits and ignimbrites (welded and non-welded), which accompanied the formation of the calderas and volcano-tectonic depressions (e.g. Heap et al. 2020). These explosive eruptions were followed by the extrusion of rhyolitic or dacitic lava domes and flows or high-level intrusion of cryptodomes. Calderas may also contain thick successions of intercalated lacustrine and fluvial (both epiclastic- and Mesozoic greywacke basement-derived) sediments. Lacustrine sediments are particularly important in forming impermeable seals or cap rocks to the geothermal fields (e.g. the Huka Falls Formation; Fig. 18.14).

A schematic representation showing likely fluid flow conditions below a geothermal field is given in Fig. 18.14a (Rowland and Simmons 2012). The fields themselves are characterised by hot springs, geysers and pools, steaming ground, fumaroles, sinter deposits and terraces, and

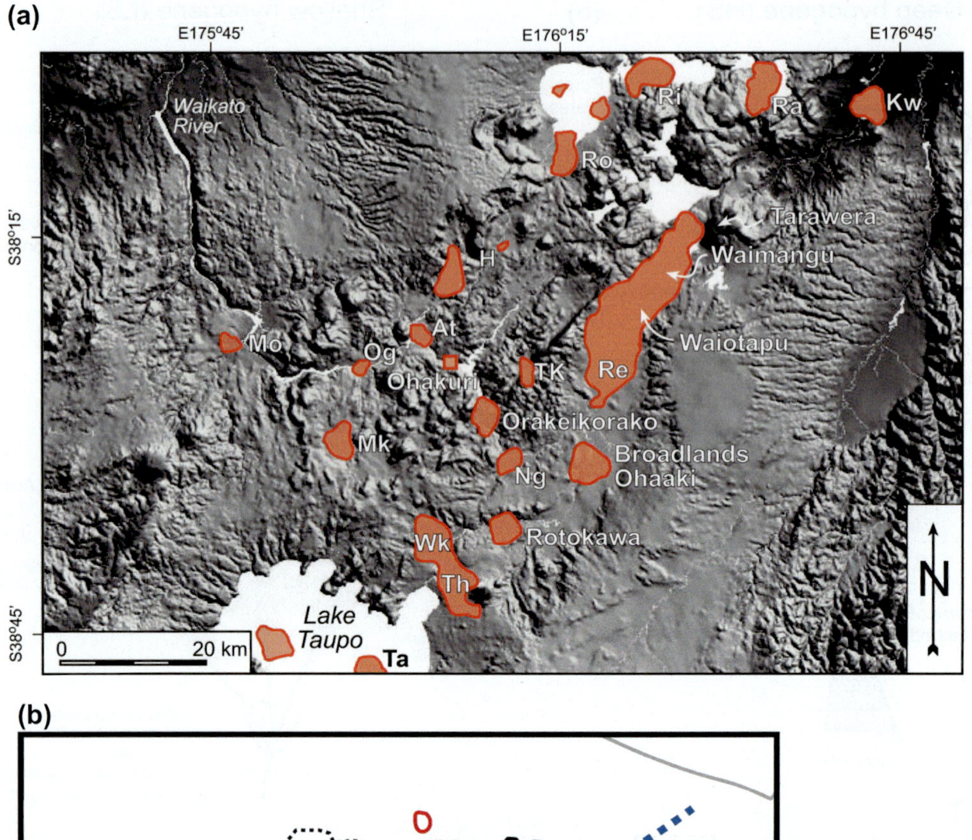

Fig. 18.13 Distribution of geothermal fields within the central Taupo Volcanic Zone. **a** Digital elevation model showing positions of active fields (orange) delimited by 30 Ωm resistivity boundary, and the extinct Ohakuri geothermal system (box). Some larger fields are labelled; abbreviations are names of other geothermal fields: At = Atiamuri, H = Horohoro, Kw = Kawerau, Mk = Mokai, Mo = Mangakino, Ng = Ngatamariki, Og = Ongaroto, Ra = Rotoma, Re = Reporoa, Ri = Rotoiti, Ro = Rotorua, Ta = Taupo, Th = Tauhara, TK = Te Kopia, and Wk = Wairakei. Mount Tarawera rhyolite dome complex is also located. **b** Average precious metal flux for selected geothermal fields (After Simmons and Brown 2007). Active (solid black line) and inferred (dashed black line) caldera boundary faults: Ta = Taupo, OK = Okataina, Mo = Mangakino, Wh = Whakamaru, Oh = Ohakuri, Re = Repora, Ka = Kapenga, and Ro = Rotorua. Distribution of young (<61 ka) volcanic vents shown by circles (white = rhyolitic, orange = andesitic) and red squares = basalt (dike as linked by the red line). Modern rift axes indicated by the blue dashed lines. (**a**, **b** from Rowland and Simmons 2012)

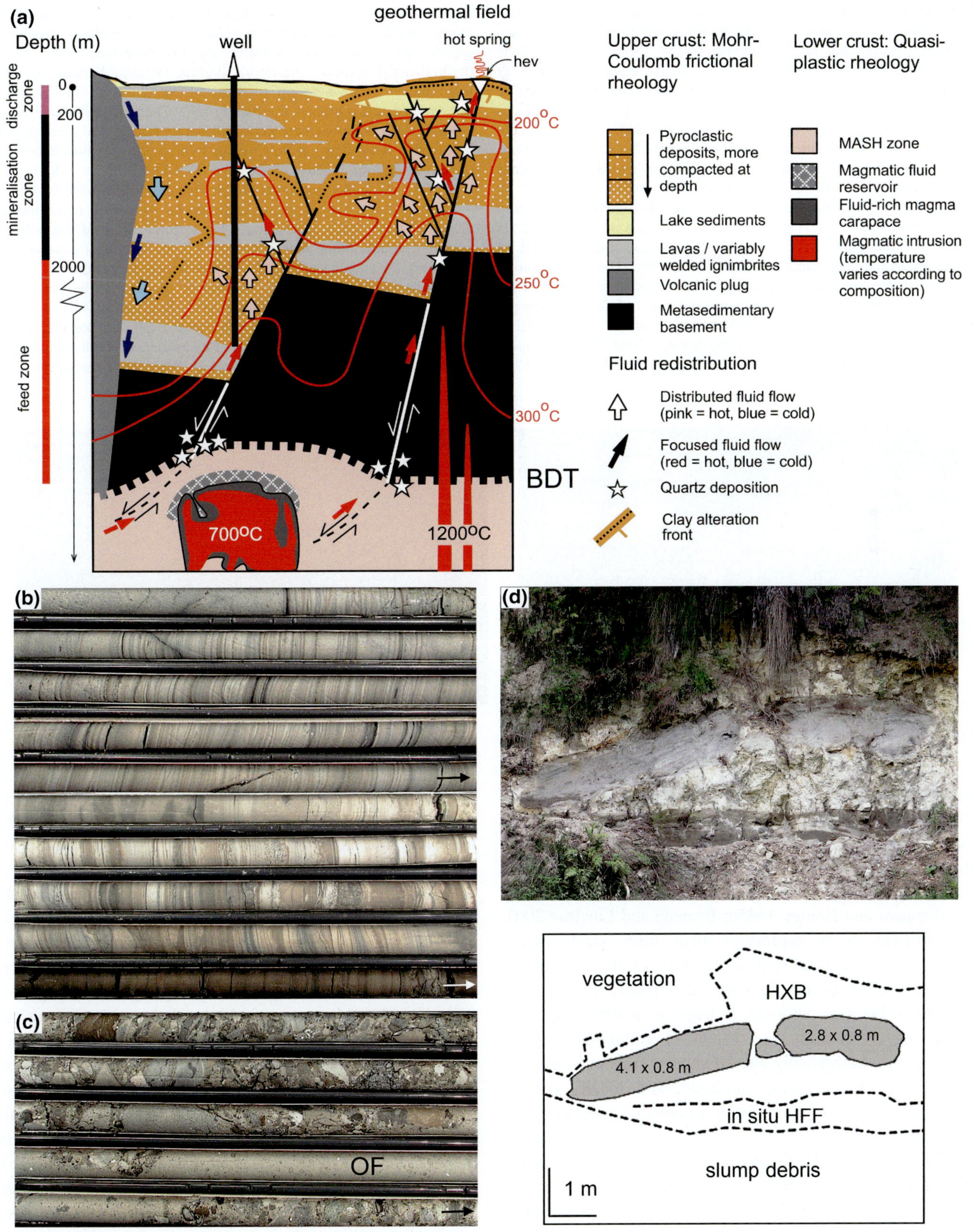
(a)
geothermal field
Depth (m)
well
hot spring
hev
discharge zone
mineralisation zone
feed zone
0
200
2000
200°C
250°C
300°C
BDT
700°C
1200°C
Upper crust: Mohr-Coulomb frictional rheology
Lower crust: Quasi-plastic rheology
Pyroclastic deposits, more compacted at depth
Lake sediments
Lavas / variably welded ignimbrites
Volcanic plug
Metasedimentary basement
MASH zone
Magmatic fluid reservoir
Fluid-rich magma carapace
Magmatic intrusion (temperature varies according to composition)
Fluid redistribution
Distributed fluid flow (pink = hot, blue = cold)
Focused fluid flow (red = hot, blue = cold)
Quartz deposition
Clay alteration front
(b)
(c)
OF
(d)
vegetation
HXB
2.8 x 0.8 m
4.1 x 0.8 m
in situ HFF
slump debris
1 m

◀ **Fig. 18.14** Features of basinal geothermal fields and deposits. **a** Schematic section showing likely fluid flow conditions within typical a basinal geothermal field developed in the Taupo Volcanic Zone (From Rowland and Simmons 2012). Fluids transect a relatively impermeable basement, and an overlying cover sequence comprising weak and porous lithologies intercalated with and punctured in places by more competent and less permeable layers. BDT = brittle–ductile transition, hev = hydrothermal explosion vent. Note depth intervals representing the feed, epithermal mineralisation, and discharge zones for geothermal fluids. **b** The top of the system is capped by impermeable lacustrine sediments of the Huka Falls Formation (HFF). The upper tray (167–172 m, in direction of the arrow) is typical of the upper HFF composed of planar, thinly bedded to laminated silt to fine sand and shows weak hydrothermal alteration dominated by smectite-group clay. The lower tray (296–301 m) illustrates some diversity in sediment type and bedding, including a brown organic-rich siltstone in the last core typical of the lower HFF. Tauhara geothermal field. (*Source* GNS Science, New Zealand/Contact Energy Ltd.; description by Michael Rosenberg, GNS Science). **c** Core through post-Oruanui (~12 ka) hydrothermal explosion breccia informally named the Crown Breccia (124–129 m). The clasts are brown organic-rich siltstone, grey siltstone, and fine sandstone (HFF), green (hydrothermal chlorite) tuffaceous breccia (Waiora Formation), and large blocks (elsewhere up to 4 m) of accretionary lapilli-bearing fine-grained tuff (Oruanui Formation, OF). Tauhara geothermal field. (*Source* GNS Science, New Zealand/Contact Energy Ltd.; description by Michael Rosenberg). **d** Pre-Oruanui hydrothermal explosion breccia (HXB) overlying in situ HFF, close to Craters of the Moon (see Fig. 18.7f), Wairakei geothermal field. This is somewhat remarkable in containing some exceptionally large blocks of poorly consolidated, unaltered HFF which appear too weak to have survived the ballistic impact. The bulk of the breccia host rock is massive (unstructured) and highly argillic altered sandy clay with fine to very coarse gravel. Mostly the clasts are highly altered (variously illite, smectite, kaolinite, chlorite) pyroclastic rock. (Photo and field sketch by Michael Rosenberg). **e** Coarse-grained hydrothermal explosion breccia, Waiotapu geothermal field. **f** Finer-grained hydrothermal explosion deposit showing diffuse "base surge" stratification, Waimangu geothermal field

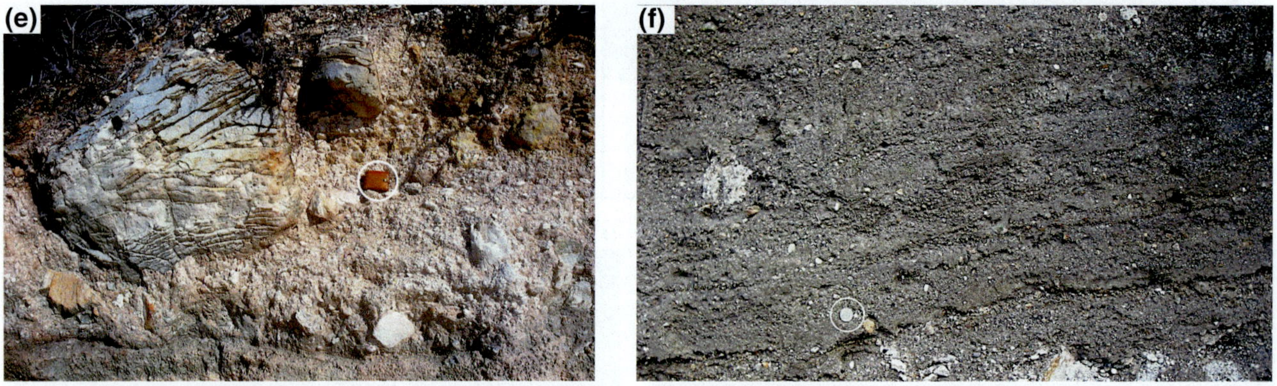

Fig. 18.14 (continued)

hydrothermal explosion craters (see Fig. 18.7). Many of these meso-scale features are also aligned on extensional or caldera margin faults. Hydrothermal explosions result from the accumulation of volatiles separated by boiling of the upflowing geothermal fluid, under a localised silica cap (Hedenquist and Henley 1985a; Browne and Lawless 2001; Fig. 18.14a). Such explosions may form small craters (usually less than 1 km in diameter; Fig. 18.7f), but multiple structures often occur. Deposits are crudely bedded volcaniclastic breccias, although in larger examples base surge layers may be present (Fig. 18.14c–f). Lithologically they consist of hydrothermally altered angular fragments of older ignimbrite, rhyolite lava, epiclastic, and locally derived rocks in a matrix of more-finely granulated fragments and hydrothermal clays.

Temperature and pressure conditions prevailing in the underlying low sulphidation quartz vein system are further illustrated in Fig. 18.15a. The epithermal mineralisation zone (>200–2,000 m depth) is the depth interval in which boiling conditions develop, and high-flux conduits form in narrow vertical fractures or pipes. Such high-flux conditions are likely short-lived and transient, concomitant with tectonic fault activation (or reactivation), as well as dyke intrusion. The main controlling factors on hydrothermal flow (Rowland and Simmons 2012) are (1) presence and/or development of a vertically extensive fault-fracture network; (2) primary and secondary porosity–permeability of host lithologies; (3) heterogeneities in tensile strength of the host rocks, and their ongoing transformations due to hydrothermal alteration (potentially both weakening and strengthening); (4) role of basement faults localising growth faults in the cover layer; and (5) complexity arising from the interplay of lithological and structural controls on fluid movement. Boiling creates additional fluid buoyancy and two-phase fluid flow, which in an open conduit results in continuous isenthalpic boiling in which gases remain in contact with liquid to the surface (Spycher and Reed 1989; Fig. 18.15b).

Epithermal vein deposits are, however, typically located in dilatant zones or fault intersections. These are also potential locations where gases may physically separate from boiling water, ascending due to buoyancy along secondary near-vertical structures (Fig. 18.15c). Loss of the gas

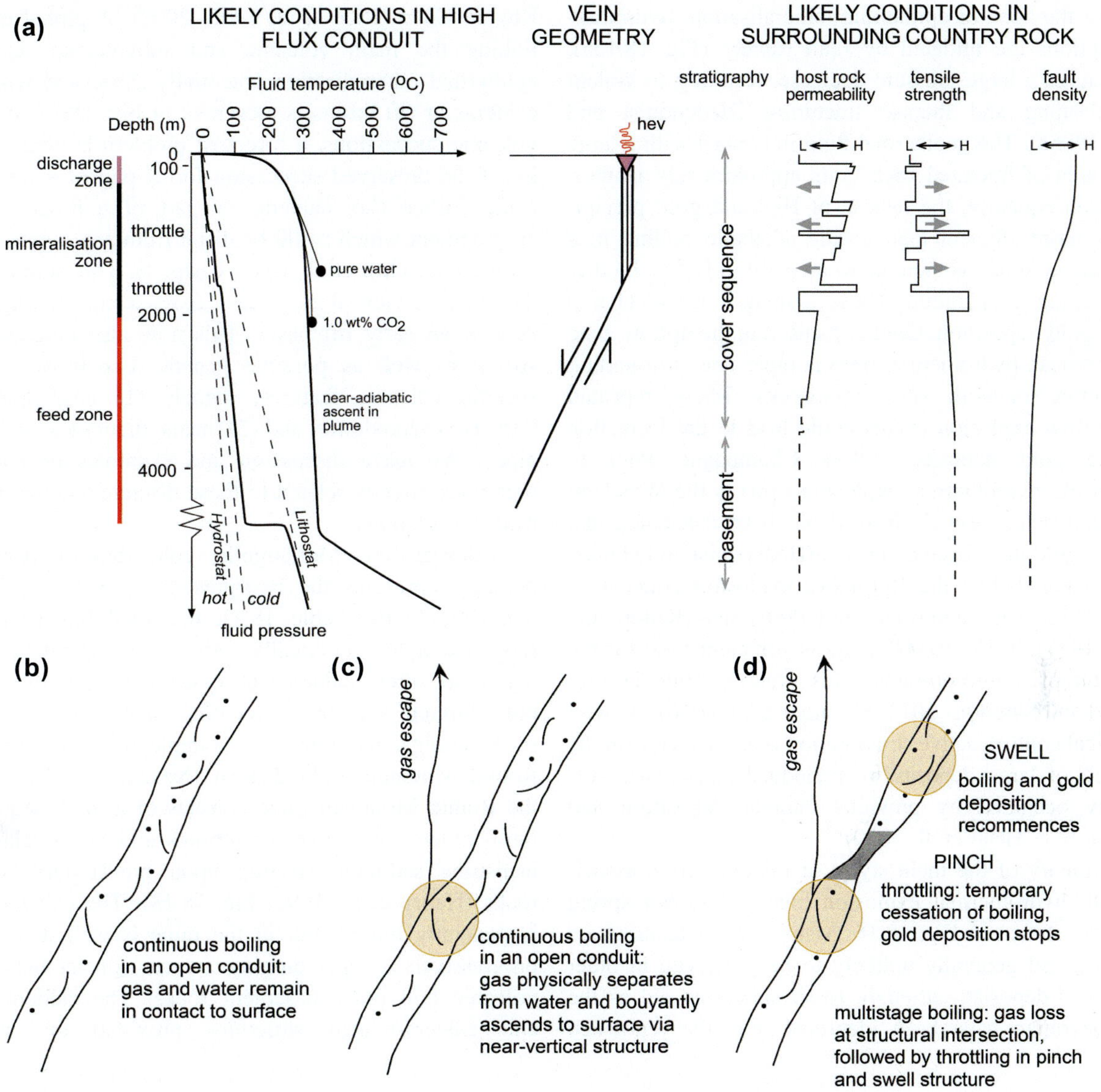

Fig. 18.15 **a** Predicted temperature and pressure conditions in an epithermal vein system with the geometry as shown. Mineralisation potentially only occurs during high-flux transient events, i.e. tectonic fault activation (or reactivation), as well as dyke intrusion. Variations in the physical properties of the surrounding country rocks with depth are schematically indicated. L = low, H = high, while grey arrows indicate the effects of mineral precipitation (reduced permeability/increased tensile strength) on the country rocks which, when high, can result in a series of "throttles". Above the epithermal mineralising zone, hydrothermal explosions (hev) and breccia formation may be triggered by the pressure drop associated with fault rupture. (From Rowland and Simmons 2012). **b–d** Some possible boiling scenarios for low suphidation epithermal veins. **b** Continuous isenthalpic boiling of neutral-chloride waters in an open conduit, where gases remain in contact with liquid. **c** Continuous isenthalpic boiling in an open conduit, but gases separate from boiling fluid at a structural intersection and are lost along secondary near-vertical structures. **d** Multistage isenthalpic boiling showing a gas loss at a structural intersection followed by throttling in a pinch and swell feature. Boiling and gold deposition temporarily cease as the fluid passes through the constriction (pinch) and hydrodynamic pressure is elevated; as the fluid passes into the overlying dilatant zone (swell), boiling and gold deposition recommence. Yellow circles identify areas with potential for high-grade and bonanza ore shoots. (Modified from Cooke and McPhail 2001)

phase could have dramatic chemical effects promoting precious metal deposition at such sites (Cooke and McPhail 2001). The geometry of the fault-fracture network, or pore space (possibly choked by mineral precipitation), may also result in "throttling" as the fluid passes through narrow constrictions raising hydrodynamic pressures, followed by localised pressure drops in overlying dilatant zones, boiling, and gold deposition (Fig. 18.15a, d). This process is likely to be episodic as fluids continue to ascend through the crust and may explain the erratic nature of bonanza ore shoots away from structural intersections (Cooke and McPhail 2001).

Above the zone of epithermal mineralisation, hydrothermal eruptions are initiated by fault rupture (Fig. 18.15a), which causes a large pressure decrease, resulting in violent steam flashing and intense fracturing (Hedenquist and Henley 1985a). The geothermal fluid in contact with a large surface area of fractured rock boils approximately isothermally. Consequently, the release of H_2S and gold precipitation is more efficient than during adiabatic boiling in a convecting column, as there is no increase in H_2S solubility due to falling temperature. These systems form ideal trap sites for gold deposition. Gold is focused in the upflow zone within vertical, hydrothermal breccia pipes and surrounding fault-fracture network (i.e. stockwork). Many repeated hydrothermal explosion events could lead to the formation of large gold deposits. Below Champagne Pool (a 900-year-old hydrothermal explosion vent) in the Waiotapu geothermal field, New Zealand, it has been calculated that 10^5 oz of gold have been deposited (Hedenquist and Henley 1985a), while below the Rotokawa explosion crater 1.7–3.3 × 10^6 oz were deposited in 6,000 years (Krupp and Seward 1987). Thus, 10,000 years is sufficient time for the deposition of a million-ounce-plus orebody (Fig. 18.13b; Rowland and Simmons 2012; Simmons et al. 2016). As well as a critical control, active structuration is a key factor for the formation of large deposits by periodically renewing permeability occluded by previous mineral deposition and alteration (e.g. Heap et al. 2020).

A summary of the main styles of mineralisation associated with hydrothermal explosion breccia and hot spring systems is shown in Fig. 18.16. Many surface features are ephemeral and generally unlikely to be preserved in older and ancient deposits, although, fossil sinters as old as the Early Devonian have been identified (e.g. the renowned Rhynie Chert; Rice et al. 1995, 2002). A global overview linking the main volcanic and subvolcanic settings of epithermal mineralisation potentially associated with silicic calderas or rift basins is presented in Fig. 18.17. In ancient volcanic successions, a foremost problem is often identifying if the preserved succession forms part of a former volcanic edifice (i.e. caldera) or part of a broader basinal environment which could be distal from any larger volcanic source (e.g. Orth et al. 1989; Chap. 1). This summary may therefore be viewed as either a cross-section through a caldera or an early rift basin. This now also emphasises the spatial as well as possible genetic associations to other specific volcanic features, namely (1) lava domes and intrusive cryptodomes and (2) maars, diatremes, and breccia pipes. We leave discussion and examples of epithermal orebodies directly related to these discrete features until the next two sections.

Although the McLaughlin gold deposit (California) perhaps represents the best-preserved LS hot spring system (Fig. 18.10; Table 18.1), the TVZ has become the type example generically. However, specifically with regard to silicic caldera successions, exploration has had only limited success in locating large gold deposits in such settings, the foremost example being the superlarge Round Mountain gold deposit (Nevada). The geology of the Round Mountain mine consists of a thick sequence of intra-caldera Oligocene ignimbrites and volcaniclastic and lacustrine sediments resting upon pre-Tertiary basement rocks (Henry et al. 1997; Fig. 18.18). The caldera margin is generally buried, but in the mine is well defined by a progressively steeper dipping arcuate contact between the volcanic and older basement rocks. The caldera margin and caldera-related structures provided the structural

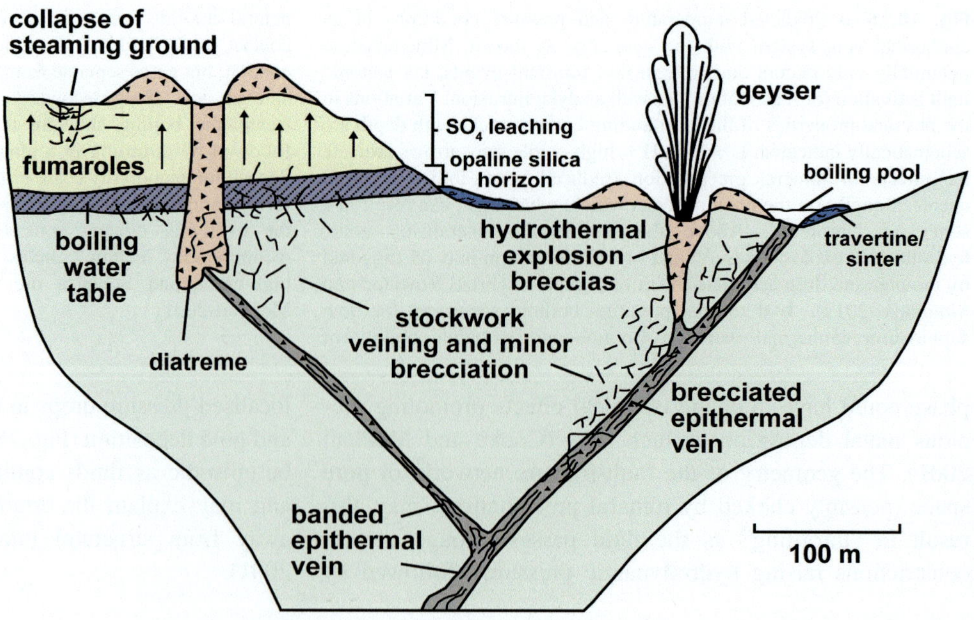

Fig. 18.16 Summary of hydrothermal explosion breccia and hot spring epithermal environments showing main styles of precious metal mineralisation (Adapted from Baker et al. 1986)

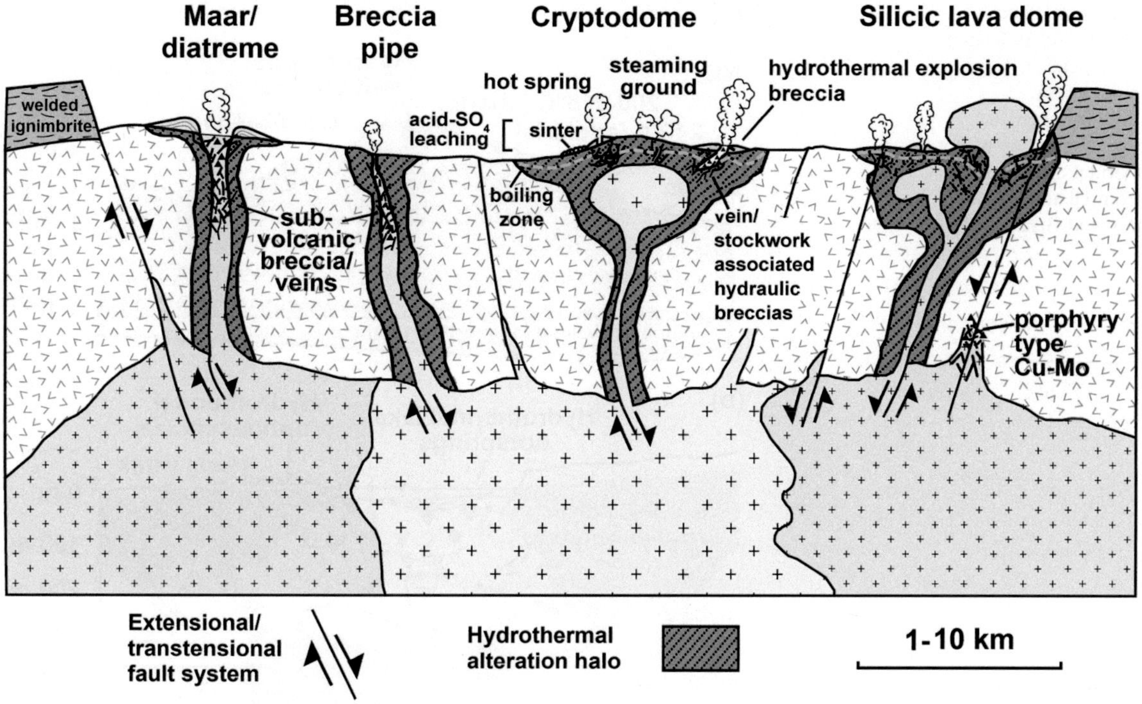

Fig. 18.17 Global overview illustrating the main volcanic and subvolcanic settings of epithermal precious metal mineralisation potentially associated with either silicic calderas or rift basins

ground preparation for the hydrothermal system, with gold mineralisation hosted by a variably welded rhyolitic ignimbrite. Welding zonation is an important control of gold style and content, with porous, now intensely altered and mineralised, non-to-poorly welded tuff forming the main host (100–210 m thick), bounded by impermeable densely to moderately welded tuffs (aquicludes), or basement. Significant hot spring gold and silver deposits hosted by extra-caldera or outflow ignimbrite successions are known in Patagonia of the Jurassic age, including well-preserved fossil sinters (e.g. Cerro Vanguardia; Schalamuk et al. 1997; Guido et al. 2002).

18.3.4.2 Silicic to Intermediate Lava Domes

Numerous epithermal deposits are known to be spatially associated with silicic to intermediate subaerial lava domes and their high-level intrusive equivalents (or cryptodomes). Although in many cases a genetic link between contemporaneous magmatism and mineralisation may occur, this is not necessarily always true, and lava domes significantly older than the ore-forming hydrothermal activity, especially with regard to LS geothermal systems, may potentially form host rocks (e.g. Fig. 18.14a).

Subaerial andesitic to rhyolitic domes and lavas may occur as isolated bodies or form complexes, which are commonly located on faults. As described above, they may form part of the intra-caldera associations of silicic caldera successions marking the position of the ring fault and/or other extensional faults (Fig. 18.17) or the successions of stratovolcanoes and multivent centres (Sect. 18.3.4.5). They also occur independently, especially in regions of high extension, along basin-forming normal and transfer faults. Such lavas are characterised by a high aspect ratio, reflecting the higher viscosities of the magmas (Chaps. 2, 6). Single emplacement units are typically very thick (up to several hundred m) and have generally short runout or lateral extents (from a few hundred m to ≤ 5 km), although some intermediate flows may flow further. High viscosity also results in the production of a thick carapace of autobreccia (Chap. 4). These effusive eruptions may be preceded, accompanied, and/or followed by small volume (phreatic, phreatomagmatic, and/or magmatic) explosive eruptions, so pyroclastic deposits may be intercalated with or form maars and tuff rings surrounding domes and dome complexes (Chap. 16).

High-grade epithermal gold–silver mineralisation at Sleeper (Basin and Range Province, Nevada) is an example hosted by more silicic, rhyolitic lava. The Sleeper rhyolite forms part of a bimodal volcanic sequence, interbedded with volcaniclastic rocks deposited in a Miocene extensional basin. Since mining commenced in 1986, the host rocks have been interpreted as a subaerial rhyolitic flow dome complex (Nash et al. 1989). More recent analysis suggests the Sleeper rhyolite may rather represent an intrusive laccolith-like body or cryptodome emplaced into tuffaceous

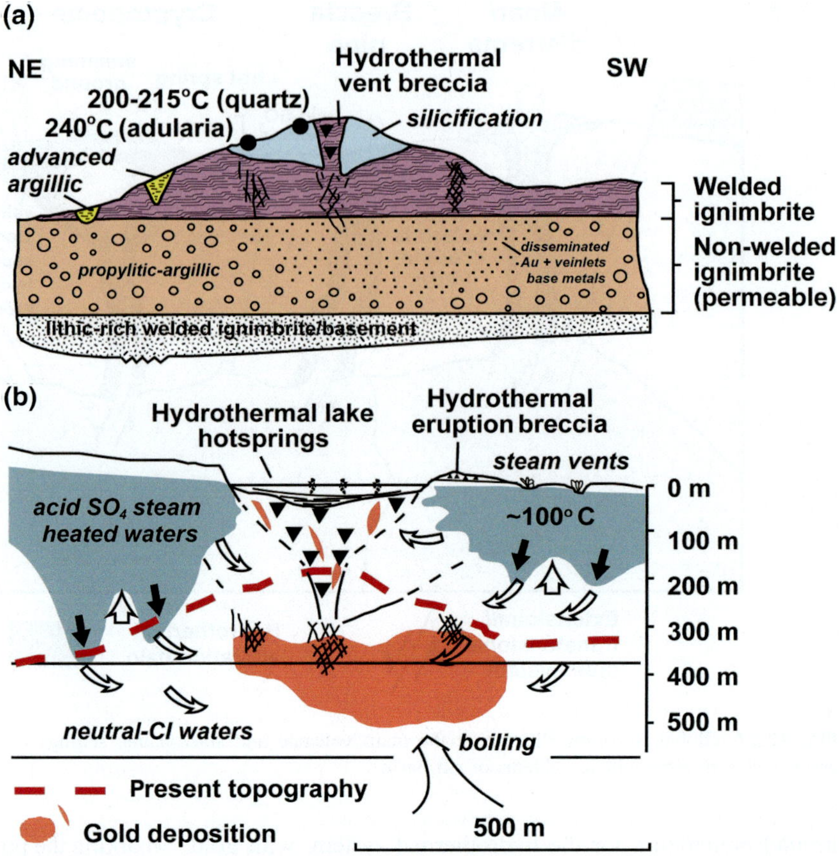

Fig. 18.18 Setting of the superlarge, intra-caldera Round Mountain gold deposit, Nevada, USA. **a** Geology and mineralisation. **b** Schematic hydrologic reconstruction of hot spring epithermal environment. (Adapted from Henley 1985, After Tingley and Berger 1985)

(pumice-bearing) rhyolitic rocks (Jackson and Chevillon 2007). The Sleeper deposit is a LS bonanza gold–silver vein ore body (1.7 Moz Au, 2.3 Moz Ag, now exploited), with an envelope of bulk minable, disseminated low-grade gold ore (some 1 Moz). Silicified porphyritic rhyolite is the major host rock. High-grade banded quartz veins comprised >60% of the mined orebody together with breccia and stockwork mineralisation. The veins were semi-continuous for more than 1,200 m along the strike, extending more than 500 m down dip. The average gold grade was about 10 g/t, while bonanza grades locally averaged 600 g/t but ranged as high as 5,000 g/t (Fig. 18.10n). Gold vein, breccia, and disseminated mineralisation occurred along reactivated, basin-controlling normal faults and cross faults.

Detailed radiometric dating suggests mineralisation at Sleeper occurred during the final stages of igneous hydrothermal activity (Conrad et al. 1993). The Sleeper rhyolite was emplaced at about 16.5 Ma with volcanic and intrusive activity possibly persisting until 16.1 Ma. Formation of high-grade veins, representing the main period mineralisation, began at about 16.2 Ma and lasted approximately 1 Ma. Brecciated vein ore and cross-cutting quartz–adularia veins represent waning stages of vein formation and mineralisation that may have continued until about 14 Ma. Thus, the lifespan of the entire igneous–hydrothermal system is well constrained to be less than 3 Ma. Younger volcanic rocks, including peralkaline ignimbrites erupted between 16 and 15 Ma from McDermitt caldera, are not related to any mineralisation or alteration at the Sleeper deposit, which appears to have formed solely within the localised flow dome or cryptodome complex during its magmatic–hydrothermal evolution.

The Sleeper deposit exemplifies a number of silicic to intermediate lava dome-hosted precious metal deposits of the Miocene age in the Basin and Range Province and Colorado Plateau. These include DeLamar, Comstock Lode, Bodie, Goldfield, and Summitville (Table 18.1). Although some epithermal deposits are related to larger silicic calderas and their ring faults, others, like Sleeper, are found in isolated volcanic complexes, and both low and high sulphidation types occur. Examples with well-dated geochronologies again suggest epithermal mineralisation forms part of individual magmatic–hydrothermal system generally lasting 2–3 Ma, in which, hydrothermal activity commences at or near the end of volcanic activity (e.g. Berger and Henley 1989). Such systems are now envisaged to be part of a larger episode of regional tectonic and magmatic activity. In the late Cenozoic, this region was dominated by extensional tectonics and bimodal volcanism, although intermediate compositions are common (Chap. 17). A number of workers have indicated correlations between continental margin arc subduction beneath the west coast of North America, Basin and Range extensional tectonics, magmatic heat sources, including migration of the ancestral Yellowstone hotspot, and mineralising events (John 2001; Hames et al. 2009; see Fig. 17.19b). It is important to note that the Basin and Range Province is characterised by many styles of epithermal, both volcanic and sediment-hosted Carlin-type, as well as deeper-level gold-rich mineralisation (e.g. Bonham 1989).

Within this category, we may also include some other metallogenic associations, namely polymetallic silver-rich deposits, and "topaz" rhyolites hosting beryllium, lithium, and uranium mineralisation (Sect. 18.4). Many tin–silver deposits of the central Andean highlands of Bolivia are associated with middle Miocene silicic lava dome and flow dome complexes (Cunningham et al. 1991; Table 18.1). Mineralisation is generally located in fractures and faults, including regional faults that may have controlled dome emplacement, or permeable zones. Permeable zones may include the deposits of pyroclastic phases that erupted prior to the effusion of lava. The presence of advanced argillic lithocaps and vuggy residual quartz are consistent with HS epithermal deposits (Sillitoe et al. 1998). These include some of the world's largest silver deposits (e.g. Cerro de Potosí, mined since the mid-1500s). Potosí is a bonanza polymetallic silver–tin vein system and represents the world's largest silver deposit (>100,000 t Ag). This formed between about 13.8–13.5 Ma and subsequently underwent several million years of supergene oxidation (Rice et al. 2005). Silver-rich base metal deposits include the superlarge San Cristobal disseminated silver–zinc deposit hosted by an andesitic–rhyodacitic dome and crater complex (Jacobsen et al. 1969; Phillipson and Romberger 2004). Mineralisation appears genetically associated with late rhyodacitic domes emplaced around the crater margin and occurs as barite veins and veinlets cutting lava, although mainly as disseminations formed within lacustrine facies formed by either replacement and/or from hot springs within a crater lake. Mineralisation is dominantly of LS type, but may have evolved into a late-stage HS system (Phillipson and Romberger 2004).

The Ovaçik gold–silver deposit, Turkey (1.12 Moz Au), is an example of an LS vein system hosted by a subaerial lava dome at the more intermediate end of the compositional spectrum. This and several other epithermal vein deposits are located within the western Anatolian volcanic and extensional province, a region of widespread Cenozoic calk-alkaline volcanism (Yilmaz 2002; Yilmaz et al. 2007). Gold mineralisation occurs closely associated with systems of early Miocene northeast–southwest-trending grabens and rift basins. The host rock succession consists of subaerial andesitic–dacitic lava dome facies comprising coherent lava, autobreccias, volcaniclastic debris flows, and fluviatile epiclastics (Fig. 18.19). At Ovaçik, lithologies consist of massive, porhyritic andesite lava with minor andesite

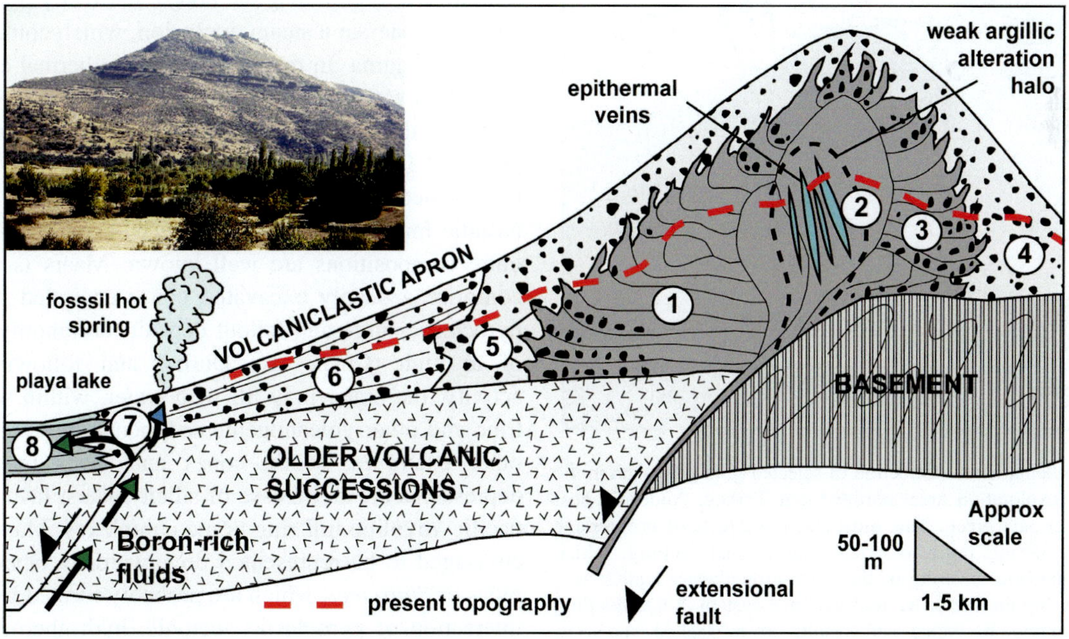

Fig. 18.19 Facies model for subaerial intermediate lava dome hosting epithermal vein mineralisation based on the Ovaçik gold deposit, northwestern Turkey. The inset shows a view of the early Miocene andesitic lava dome complex forming the margin of Bergama Graben. Facies: (1) coherent, variably porphyritic, and flow-banded andesitic–dacitic lava; (2) coarsely porphyritic "intrusive" core; (3) autobreccias; (4) talus slope breccias; (5) lava lobes ("*pseudo-pillows*"); (6) volcaniclastic debris flows; (7) fluviatile epiclastics; (8) lacustrine and volcaniclastic turbidites. It is also important to note such fault-controlled basins and playa lakes are host to some of the world's largest borate deposits. (Adapted from Wright 1996; for further description, see Yilmaz 2002, Yilmaz et al. 2007)

autobreccia. Two major lens-shaped quartz veins form the main economic deposit. These have a length of about 1 km, average about 10 m in width, and extend down dip some 200 m. The margins of the veins have transitional lateral boundaries (up to 25 m on either side) of the main quartz lodes, through stockwork and sheeted quartz veinlets to the argillic alteration within competent andesite wall rocks. Textural and alteration assemblages are similar to typical banded LS veins (Fig. 18.10). Alteration minerals are dominated by smectite, mixed-layer illite/smectite, chalcedonic quartz, and adularia, which indicate mineralisation formed mainly at temperatures of about 170–200 °C.

Age data suggest that early Miocene extension-related mineralisation took place in the broader Ovaçik area at 18 ± 0.2 Ma, within volcanic rocks having an age of 19.7 Ma (Yilmaz et al. 2007). This includes a number of other gold prospects, including Narliça where gold was mined in pre-Roman times (about 600 years BC), as well as several levels or ledges of barren silicification within the andesitic host rock succession. The latter represent the episodic descent of the silica-saturated groundwater aquifer from the upper levels of the hydrothermal system (Fig. 18.20). This was probably associated with zones of descending steam-heated, acid-SO_4 waters coincident with

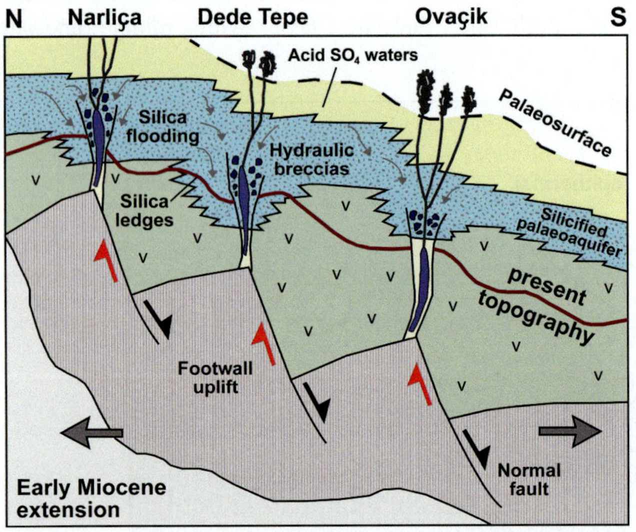

Fig. 18.20 Schematic reconstruction of selected gold prospects in the wider Ovaçik exploration area, northwestern Turkey. Andesitic host facies shown in pale green. This illustrates the effects of progressive footwall uplift during repeated extensional normal faulting, which causes the systematic lowering of the paleo-groundwater aquifer (i.e. water table) during the life of the hydrothermal system. Silica flooding causes overprinting by silica and results in a stacked series of syn-hydrothermal barren silica ledges marking the descent of the aquifer; this potentially may cause dilution of gold deposited earlier. The distance between Ovaçik and Narliça is approximately 2.5 km, and the difference in elevation is 200 m. Today fossil sinters occur along the footwall of basin-controlling faults at lower elevations (as shown in Fig. 18.19). (Modified from Wright 1996)

steep fault structures. Progressive footwall uplift during extension along these faults would cause the systematic lowering of the groundwater aquifer (or water table) during the life of the hydrothermal system and overprinting by silica. Silicification can be detrimental to prospectivity when occurring late in the epithermal cycle, causing dilution of earlier deposited gold. At Ovaçik, these effects were less intense, and present topography preserves most of the vein system. Silica flooding occurred deeper into the other vein systems, with what is preserved generally barren. In such prospects, gold will generally only be located in hydraulically and or structurally rebrecciated zones associated with renewed phases of hydrothermal activity. This may have been the case at Narliça, which is located at a higher elevation (270 m) than Ovaçik (70 m). Paleotemperatures of gold mineralisation at Narliça (220–300 °C) are significantly higher, consistent with deeper levels of erosion and lower gold potential at depth. Groundwater aquifer descent of as much as 200–400 m has been described from other LS vein systems (Sillitoe 2015).

18.3.4.3 Maar-Type Volcanic Centres

Maars, tuff rings, and tuff cones are varieties of small volcanic craters (generally ≤ 2 km in diameter) formed by phreatomagmatic and phreatic eruptions (Chap. 16). The main driving force of these eruptions is the interaction of magma with surface or groundwater or a shallow aquifer. A phreatic eruption is one where hot wall rock (e.g. surrounding a rising magma body) indirectly interacts with water to generate a steam explosion, which contains no new juvenile magma. In one sense, a hydrothermal eruption is a type of phreatic explosion, but it is important to differentiate these by the intensely altered nature of the erupted clasts (Sect. 18.3.4.1; Chap. 4). Generally, these types of the monogenetic volcanic centre are formed by the eruption of basaltic magmas; however, those with intermediate to rhyolitic compositions are well known. Maars (*sensu stricto*) comprise a deeply excavated crater encircled by a rim of pyroclastic surge and fallout deposits commonly a few 10 s of m high (Chap. 16). During and following eruptive activity, the crater may fill with water, within which lacustrine and other volcanogenic sediments may accumulate. In other examples, scoria cones and lavas may infill the depressions or, in the case of silicic maars, lava domes, and dacitic–rhyolitic tuff ring dome complexes. Maar craters are envisaged to be represented downward by tapering breccia pipes or *diatremes*, which are probably formed by a complex interaction of pyroclastic, tectonic, hydrothermal, and surface processes (see Fig. 16.16).

Intermediate-silicic maar-type craters and the shallower levels of diatremes are known to be important hosts to epithermal gold–silver mineralisation (Fig. 18.21; Table 18.1; for a current review, see Ross et al. 2017). Two

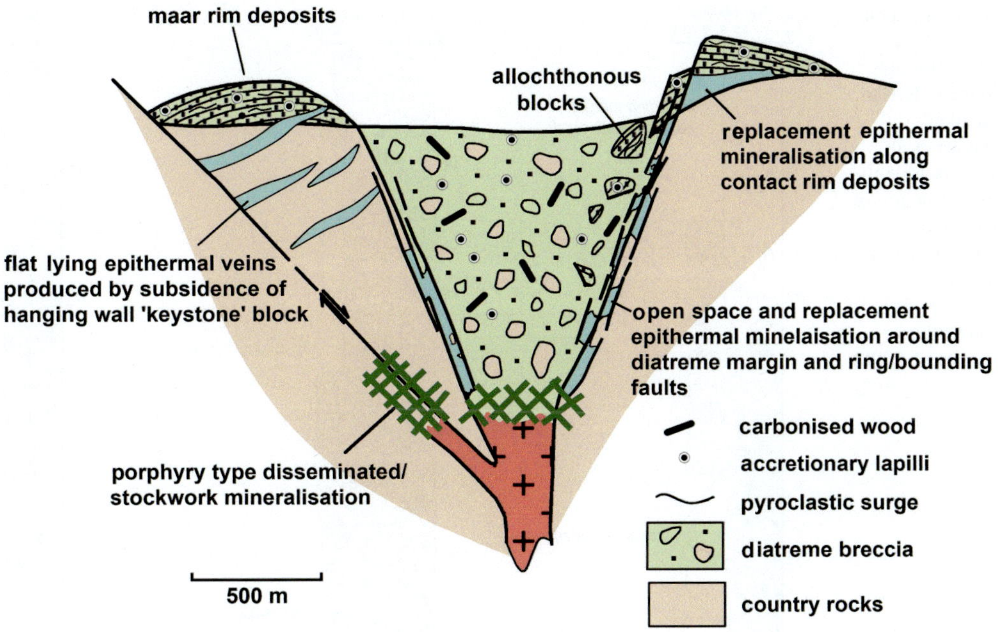

Fig. 18.21 Schematic setting of intermediate-silicic maar-diatremes showing main controls of epithermal precious metal mineralisation, based in part on the Wau gold deposit, Papua New Guinea. Coal and carbonised wood fragments derived from the paleosurface or carbonaceous wall rocks may work their way downwards deep into diatreme breccias. (Adapted from Baker et al. 1986)

well-described examples, from a volcanic facies aspect, are Wau in Papua New Guinea (Sillitoe et al. 1984) and Kelian in Kalimantan, Indonesia (Davies et al. 2008). At Wau, the maar crater is located close to a major fault scarp, which likely controlled dacitic magmatic activity. The maar deposits are a complex sequence of breccias, but include pyroclastic surge deposits, accretionary lapilli, and lava domes, the identification of accretionary lapilli being critical in the prognosis. Following initial maar formation, hydrothermal brecciation and gold deposition were controlled by low-angle extensional faults which developed during the subsidence of a “keystone” block between the regional fault zone and the ring fault controlling the maar (Fig. 18.21). The margins of the diatreme around the “ring” or bounding fault system and along the contact between rim pyroclastics with country rocks also represent important sites for ore formation.

At Kelian, the setting appears more complex, comprising three “nested” diatreme breccia bodies, which internally are themselves composite in nature, emplaced into a small (ca. 1.5 km wide), strike-slip pull-apart basin. Prior to brecciation, andesite was intruded into a package of felsic volcaniclastic rocks and overlying carbonaceous sandstones and mudstones, and a weakly mineralised geothermal system was established. The later intrusion of porphyritic (quartz–phyric and later quartz–feldspar–phyric) rhyolite into the active geothermal system triggered subsurface phreatomagmatic and phreatic explosions which disrupted the pre-existing hydrothermal system producing diatreme breccia bodies (the Tepu, Runcing, and Burung Breccias). Each body consists of polymict breccias and sandstones that contain abundant carbonaceous matrix and variable amounts of early andesite and coeval rhyolitic clasts (e.g. Tepu Breccia; Fig. 18.22). Surficial products of maar-diatreme volcanic activity are preserved as large blocks of well-stratified breccias. These contain accretionary lapilli and were deposited by a combination of wet pyroclastic base surge and co-surge fallout processes (Chaps. 10, 16). Evidence for syn-eruptive resedimentation of the pyroclastic deposits is preserved in poorly stratified breccia beds. Allochthonous megablocks of phreatomagmatic deposits subsided several hundred metres down from the maar environment into the underlying diatremes. The carbonaceous matrix-rich diatreme breccias acted as aquicludes during subsequent hydrothermal activity, focusing fluids into the wall rocks adjacent to the diatremes, where hydrothermal brecciation and epithermal gold–silver mineralisation occurred.

Focusing on the three main breccia bodies, these are characterised by abundant carbonaceous components (clasts and/or matrix) derived from the surrounding sedimentary country rocks (Fig. 18.22). Geometrically each breccia body has a complex internal arrangement of breccia facies. Contacts between facies may vary from sharp (and erosional) to gradational and are generally steep–subvertical and/or cross-cutting suggesting they consist of a number of breccia pipe-like bodies and dykes, formed by multiple brecciation episodes. Typically, most internal and country rock

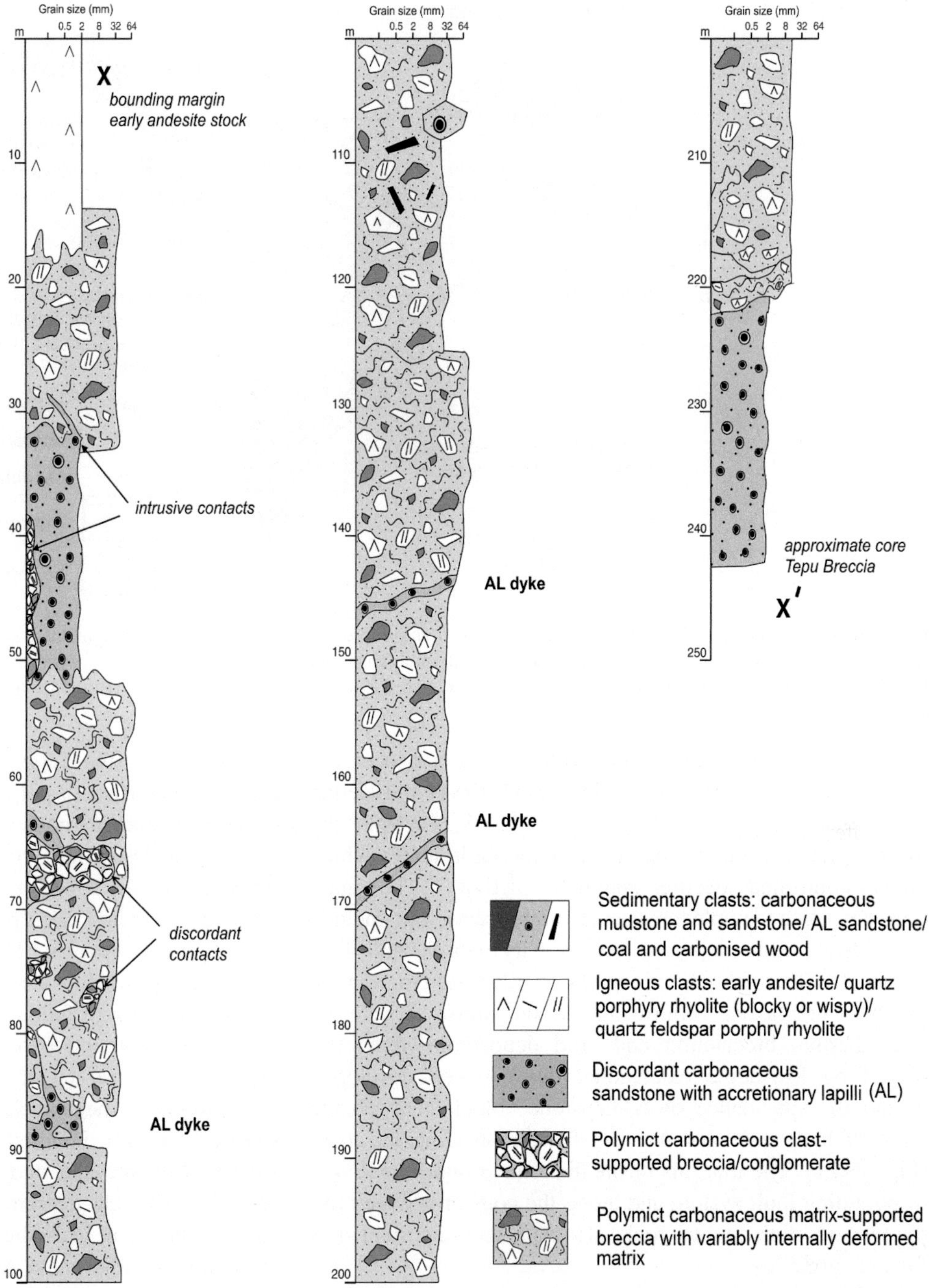

Fig. 18.22 Pseudo-stratigraphic graphic log (X–X[1]) based on horizontal pit mapping through the core of Teru Breccia, Kelian gold deposit, Kalimantan, Indonesia (From Davies et al. 2008). Note the steep and highly intrusive and discordant nature of many of the internal facies contacts which are described in the text. (For a detailed description and accompanying images of breccia and other lithofacies, see Davies et al. 2008)

contacts are gradational over metres to tens of m, especially where earlier breccias have been cross-cut and stopped or removed by emplacement of younger breccias. Some internal facies occur as subvertical pipes that are smaller in volume than, but have similar gross geometry to, the main enclosing breccia bodies. This type of complex interfingering within breccia pipe bodies has previously been termed "*antler geometries*" (Wormald et al. 1991; Wormald 1993; see Fig. 18.24). An essential criterion again in elucidating the volcanic setting at Kelian was the recognition of accretionary lapilli (Fig. 18.22). Not only are accretionary lapilli commonly associated with surface-deposited pyroclastic surge and fallout facies, but have also been found within subsurface diatreme breccia facies. These appear to occur in finer-grained units (higher degree of fragmentation), which again have discordant geometries forming bodies up to 20 m in diameter or smaller dykes. While possibly forming surficially, being then reworked into the subsurface environment, evidence elsewhere described below (Sect. 18.3.4.4) shows accretionary lapilli may also form at depth within subvolcanic breccia pipes. The evolution of the Kelian maar-diatreme complex is summarised in Fig. 18.23.

Other examples of maar-hosted epithermal gold–silver deposits are Montana Tunnels, Montana (Sillitoe et al. 1985), and the supergiants Cripple Creek, Colorado (Thompson et al. 1985; Kelley et al. 1998), and Pueblo Viejo HS deposit, Dominican Republic (Russell et al. 1986; Russell and Kesler 1991; Nelson 2000). Pueblo Viejo has more recently been at the centre of critical debate, as to whether it represents an early Cretaceous maar-diatreme ± dome complex or is related to younger late Cretaceous–Tertiary intrusive activity, occurring after the region was tectonically buried under a thick sequence of ophiolitic rocks (Sillitoe et al. 2006, 2007; Muntean et al. 2007). Zircon U–Pb ages now better constrain the timing of epithermal mineralisation to between 109.6 and 108.7 Ma, consistent with an Albian (late Cretaceous) age, just prior to ophiolite emplacement which preserved the deposit from erosion (Mueller et al. 2008). At deeper levels, some maar-diatreme HS deposits are closely genetically associated with porphyry-related styles of copper–gold mineralisation, for example, Lepanto and the Far Southeast porphyry system (Luzon, Philippines), which share overlapping lithocaps with common evolutionary trends (Hedenquist et al. 1998, 2017).

18.3.4.4 Breccia Pipes

We differentiate breccia pipes from diatremes in that the former may never have reached the surface or show no evidence of having explosively vented (i.e. pyroclastic fragmentation) at the paleosurface. The largest examples include porphyry copper and some iron oxide copper–gold deposits (Sect. 18.3) associated with major subvolcanic intrusive complexes. Here however we focus on smaller hypabyssal, porphyry-related breccia bodies (≤ 5 km^2 in plan), which typify a number of gold deposits such as Kidston and Mt Leyshon, northeast Queensland, Australia (Baker and Andrew 1991; Wormald 2017). Such host rocks may contain a wide spectrum of breccia types, and both recognition and their classification have been problematical. These difficulties are compounded because many involve multiple breccia-forming, intrusive, and hydrothermal events.

Kidston (Carboniferous) and Mt Leyshon (Permo-Carboniferous) are similarly located on regional northeast–southwest-trending lineaments, which have protracted histories of ductile to brittle faulting and igneous intrusion. Both are broadly rhomboid or trapezoid in shape, located on major lithological boundaries (between granitoids and high-grade metamorphic rocks at Kidston, or granitoids and metasedimentary rocks at Mt Leyshon), and have highly irregular margins suggesting they represent pull-apart structures (or rhombochasms) formed along transcurrent faults (Wright 1990; Chap. 17). In both examples, brecciation is envisaged at shallow crustal levels ≥ 1–4 km, with gold mineralisation late, deposited from fluids similar to LS epithermal systems. Neither show evidence of connectivity to the paleosurface (i.e. allochthonous blocks of rim pyroclastics, coeval lacustrine sediments or contemporaneous sedimentary cover, or carbonised wood).

From an internal facies perspective, the Mt Leyshon gold deposit has been documented in most detail. The effects of variably pervasive hydrothermal alteration and misidentification of false pyroclastic textures (Chap. 13) hampered initial studies made in the mid-late 1980s. Fragmental units and porphyritic igneous rocks were interpreted as a near-surface diatreme and maar-type volcanic centre (Paull et al. 1990). Subsequent lithological and structural mapping, combined with extensive investigation and graphic logging of the drill core, showed mineralisation was hosted by an intrusive subvolcanic breccia complex (Wormald et al. 1991; Wormald 1993, 2017). Detailed facies analysis charting the progressive fragmentation and transport (expansion) of brecciated lithologies and their geometrical relations has demonstrated that fault-derived brittle fracturing coupled to magma decompression and gas discharge at depth, rather than hydrovolcanic processes, drove fragmentation.

The breccia complex (2.0 × 1.5 km) comprises three principal elements (Fig. 18.24): (1) basement, (2) Main Pipe breccia, and (3) an interactive magma-breccia sequence, including the Mt Leyshon breccia (principal orebody host rock). Strike-slip fault motions accompanied all stages of breccia complex evolution and likely controlled pre-breccia basement fragmentation, magma recharge, vapour discharge into dilated host rock, hydraulic fracturing, and brecciation, which were synchronous or stepwise interactive processes. The facies characteristics of the breccia units suggest processes of gas/solids mobilisation and fluidisation controlled their formation. Jigsaw-fits are ubiquitous on all scales.

Pre-breccia basement (metasediments, granite, rhyolite, and dolerite dykes) both defines the margins of the complex and occurs as partially brecciated, somewhat rotated blocks and larger more or less in situ masses within the complex (Fig. 18.25a). The first and most extensive breccia-forming event is referred to as the Main Pipe breccia, which was complex-wide defining the irregular contact of the complex with the basement (Fig. 18.25b). Intense pre- to syn-breccia brittle deformation in the basement continues outside the breccia envelope for 10–100 m. Main Pipe breccia is a polymict matrix-supported breccia consisting of sub-angular to sub-rounded clasts in a matrix of finely fragmented basement lithologies (Fig. 18.26a). Breccia contacts with host rocks are either sharp or more diffuse where expanded but not transported blocks and masses of the basement (1–600 m in size) are injected by fine clastic, rock flour matrix. As a result, the Main Pipe breccia has intrusive, interfingering contacts with both the margins of the complex and

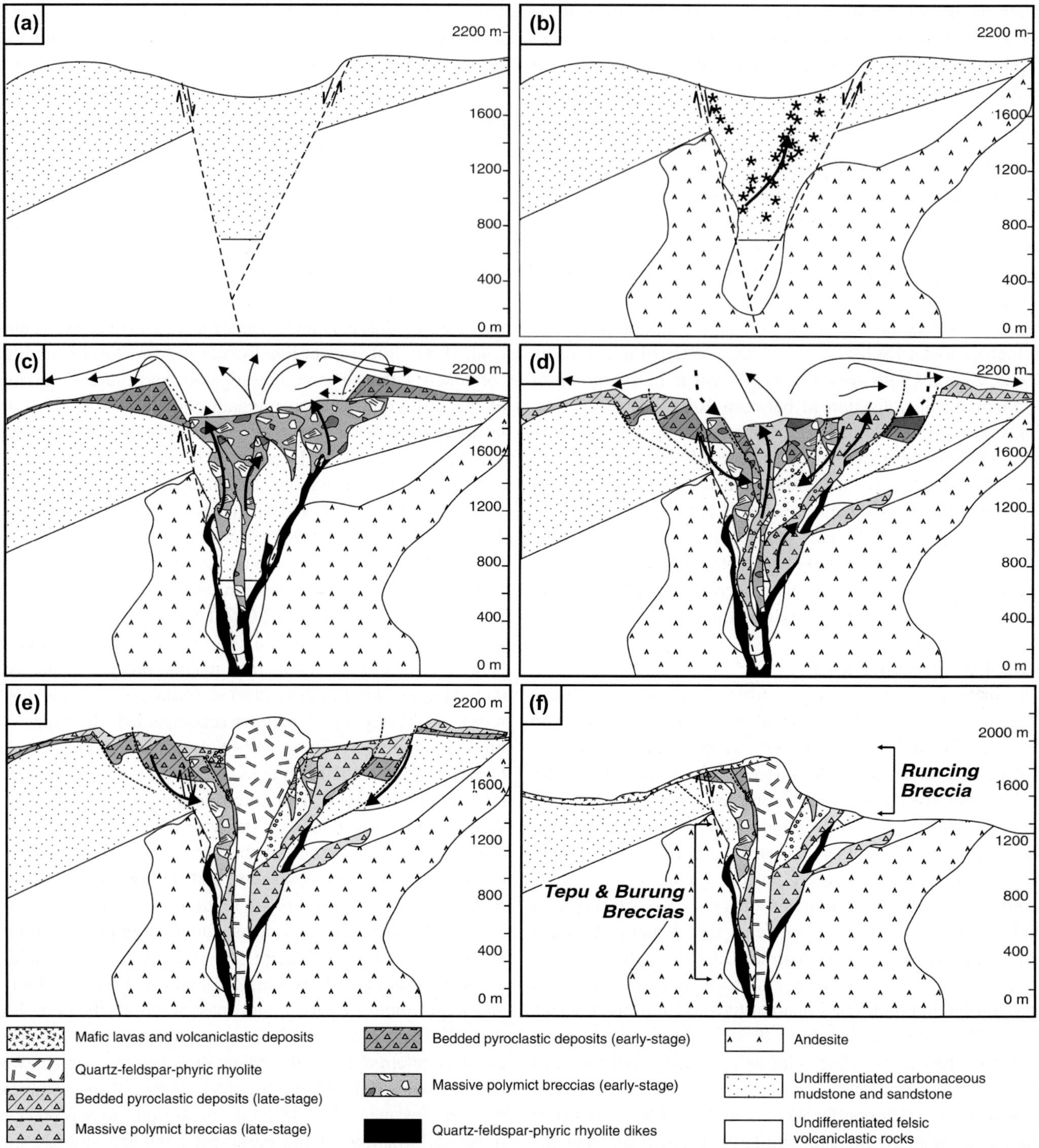

Fig. 18.23 Schematic model for the formation of the Kelian diatremes (From Davies et al. 2008). **a** Ground preparation: downfaulting juxtaposes Eocene–Oligocene carbonaceous sediments against late Cretaceous volcaniclastic sediments (basement). **b** Intrusion of Miocene early andesite stocks along the same fault system; arrows and stars indicate possible minor phreatic explosions. **c** Quartz–phyric rhyolite dykes (19.8 Ma) similarly intruded into the same fault system encountered groundwater initiating phreatomagmatic explosive activity. This produced unstratified polymict carbonaceous breccias in the subsurface (thick arrows) base surge and fallout deposits at the surface (thin arrows) building a maar rim. Synvolcanic resedimentation of volcaniclastic material (dashed arrows) into the maar also occurred. **d** Continued phreatomagmatic explosions continually widened and increased the depth of the diatreme producing multiple cross-cutting breccia bodies. This was accompanied by the collapse of the maar rim along the ring and bounding faults, thus blocks of pyroclastic rim deposits, earlier diatreme phases, and other wall rocks slumped several hundreds of m into the diatreme. **e** Late-stage stocks and plugs of quartz–feldspar–phyric rhyolite (19.5 Ma) were intruded into the Burung and Runcing Breccias accompanied by minor phreatomagmatic and widespread phreatic eruptions. This caused fluidisation of maar sediments which were intruded through older breccia phases along active faults, as the gravitational collapse of the maar and diatreme continued. **f** Miocene gold and base metal mineralisation, hydrothermal brecciation, and veining, followed by uplift and erosion to the present position. Mafic lava and volcaniclastic sediments overlie a Plio-Pleistocene paleosurface

basement blocks within the main pipe, displaying an overall antler-like geometry (Fig. 18.24).

Two distinct Main Pipe breccia sub-facies are present: a coarse variant (typically 10–30% matrix), which is the most extensive, and a fine variant (typically 70–95% matrix), originally interpreted as vent breccia and lapilli tuff, respectively (Paull et al. 1990). However, contacts between the two variants are gradational, steeply dipping, and intrusive (Fig. 18.24). Gradual changes occur over tens to hundreds of m, but locally oscillation between fine and coarse varieties is recognised over shorter distances. The fine sub-facies typically form steeply dipping, slot- or pipe-like zones within the more voluminous coarse variant. Accretionary lapilli (or pellets) also occur within these steeply dipping features (Fig. 18.26b) and are known to occur over 400 m of vertical section. In the thin section, they show nuclei of fine rock fragments identical to the breccia matrix outside the structures, coated by finer-grained, clay-size material. These features are consistent with fine particle accretion during upward transport in gas-rich particulate systems (McCallum 1985). Rare andesite to rhyolite fragments are texturally and mineralogically very similar to younger (cognate) igneous units within the later interactive magma-breccia sequence, strongly supporting the presence of an established magma system at depth. Later hydrothermal alteration re-lithified the complex, setting the stage for fragmentation of the Main Pipe breccia and basement during subsequent events.

The interactive magma-breccia sequence is represented by multiple smaller-scale breccia pipes/dykes, *tuffisite* breccia dykes, and igneous bodies intruded primarily into the western half of the complex (Fig. 18.25c–e). During this later stage, intrusive phases were focused higher in the crust, resulting in a closer spatial and temporal link between igneous emplacement and brecciation. Igneous stocks and dykes have a high-K, calk-alkaline character, and porphyritic textures with an aphanitic groundmass and can be subdivided into andesite and quartz andesite, and rhyodacitic and rhyolite types. The largest breccia body and main ore host is the Mt Leyshon breccia pipe (Figs. 18.25d, 18.26g). This has margins bound by strike-slip faults, which therefore are generally sharp and curvilinear, except locally where dyke-like apophyses extend. Pre-mine surface mapping suggests the Mt Leyshon breccia pipe narrowed upwards. The pipe has a known vertical extent of at least 800 m, but deep drilling and other evidence suggest prior to erosion this may have been as much as 2 km (Fig. 18.27).

The Mt Leyshon breccia is generally clast supported, with angular to sub-rounded fragments between 2 and 10 cm in size (Fig. 18.26d, e). Where polymictic, the breccia can contain fragments of all previous lithologies (i.e. basement, lithified Main Pipe breccia, early dykes, and porphyry phases). In contrast to Main Pipe breccia, the proportion of rock flour matrix is generally small (1–5%). It is composed of two principal parts: (1) largely unmoved, 1–10 m-sized blocks of fault-derived, hydraulically fractured, and weakly expanded basement and Main Pipe breccia, and (2) expanded and transported, cm- to m-scale anastomosing zones of rock flour matrix-bearing intrusive monomict to polymict breccia. The pipe can be further subdivided into several zones or sub-facies based on variations in fragmentation characteristics, which reflect differences in intensity of host rock fragmentation, rock type (rheology), and degree of fragment transport or expansion (Fig. 18.27).

Tuffisite breccia dykes form shallow to steeply dipping bodies (0.01 m to >15 m thick) which intrude and locally rework the Mt Leyshon breccia pipe (Figs. 18.25e, 18.26c). They are a major host of high-grade gold mineralisation when cut by later structures and hydrothermal leaching that generated secondary porosity. Fine-, medium-, and coarse-grained types are defined according to their modal fragment size, although the fine variant is dominant (Fig. 18.26a, c, f). These three types are well sorted and can occur independently or as composite bodies, with gradational to sharp contacts between each other. They commonly show grading or fine layering in bands (1–10 mm thick) and the fine variant occasionally displays convoluted flow textures in areas where the fluidised material passed around obstructions (Fig. 18.26a). Tuffisite breccia dykes contain fragments of all previous described rock units and felsic (cognate) igneous fragments. In the fine variant petrography indicates fragments may consist of >99% (<2 mm-sized) igneous material, implying a juvenile magmatic origin. Significantly, the mineralogy, microtexture, and trace element geochemistry of fragments are identical to the rhyodacite igneous dykes that immediately followed them in the intrusive sequence (Fig. 18.25e), indicating a close genetic relationship.

Taken together such features strongly suggest the tuffisite breccia dykes formed by the injection of juvenile magma particles into structures during explosively driven hydrofracture. Coupled with prolonged gas discharge (and dilation), this led to milling (of cogenetic igneous units/host rocks), fluidisation (gas and magma-solids mobilisation), and density stratification in bodies stretching for up to several hundred m long. They show comparable facies characteristics to tuffisites (grading and layering) in well-sorted magma-derived units injected through penetration of gases into host rock as first described by Cloos (1941).

Hydrothermal alteration did not accompany the fragmentation process, but was superimposed during succeeding cycles of fluid generation and successive rupture of fault/vein structures. Hydrothermal pulses occurred either post-Main Pipe breccia or post the interactive magma-breccia sequence (Wormald 1993, 2017). In both cases, the breccia-forming events were followed first by low-temperature alteration

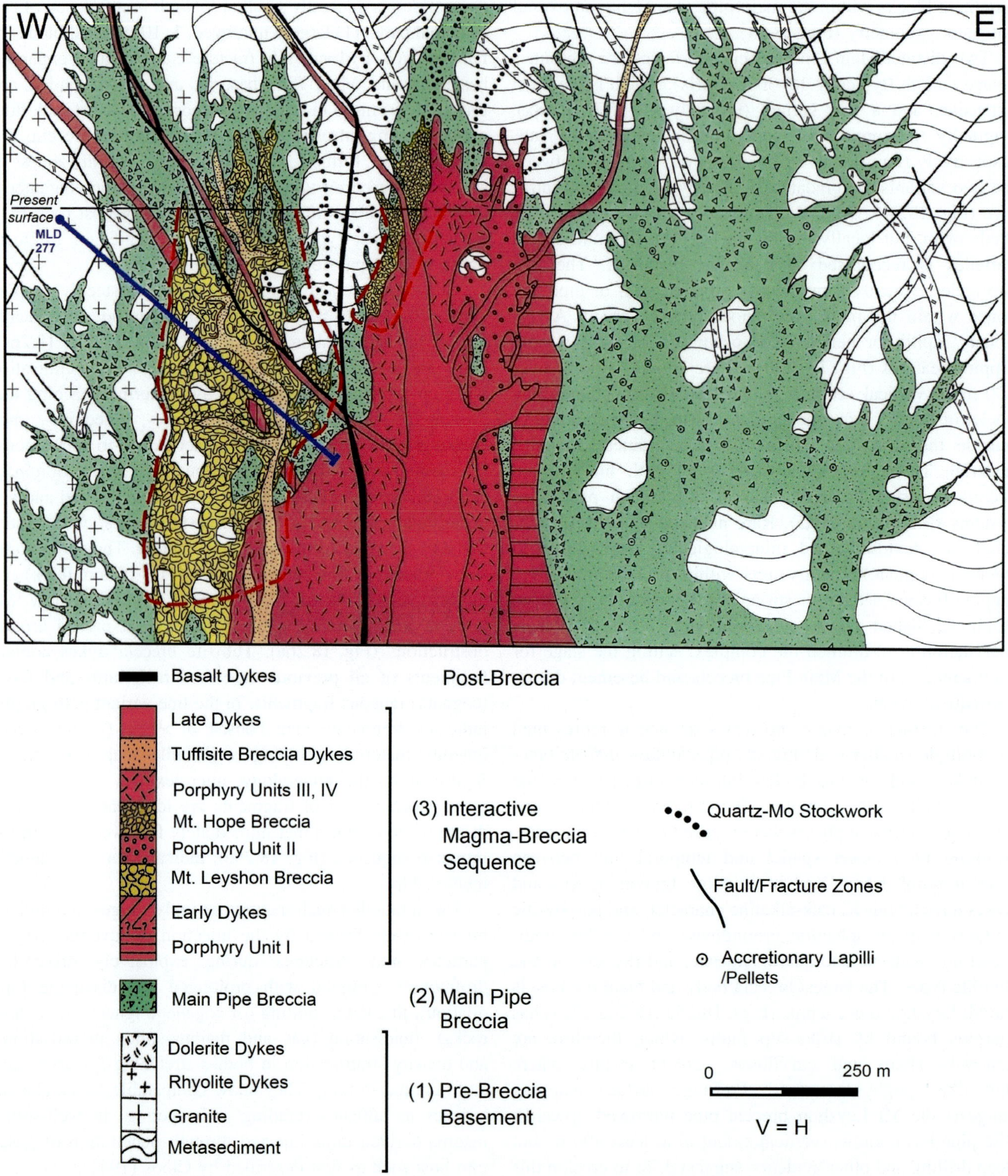

Fig. 18.24 Schematic reconstruction showing facies architecture of the Permo-Carboniferous Mt Leyshon breccia complex, northeastern Queensland, Australia (After Wormald 1993). The area illustrated is approximately 1.5 × 1 km. For clarity, syn- to post-complex faults are not shown and some late igneous dykes have been omitted. Gold mineralisation is highlighted within the red dashed lines. One drill hole used to constrain the interpretation is shown for reference; this contains 160 m @ 3.8 g/t Au, including 10 m @ 20.1 g/t Au. *Note* (1) the largely in situ or little rotated blocks of the Lower Paleozoic basement inside the breccia complex; (2) the steeply dipping, chaotic antler-like geometry of the Main Pipe breccia; (3) intricate variations and gradation between Main Pipe breccia (coarse and fine) sub-facies, including accretionary lapilli (pellets); (4) steeply dipping and interfingering porphyry contacts; (5) gross textural differences between the Main Pipe, Mt Leyshon, and Mt Hope breccia pipes; and (6) the interfingering and intrusive nature of the tuffisite breccia dykes (a major host of high-grade gold mineralisation) inside the main orebody

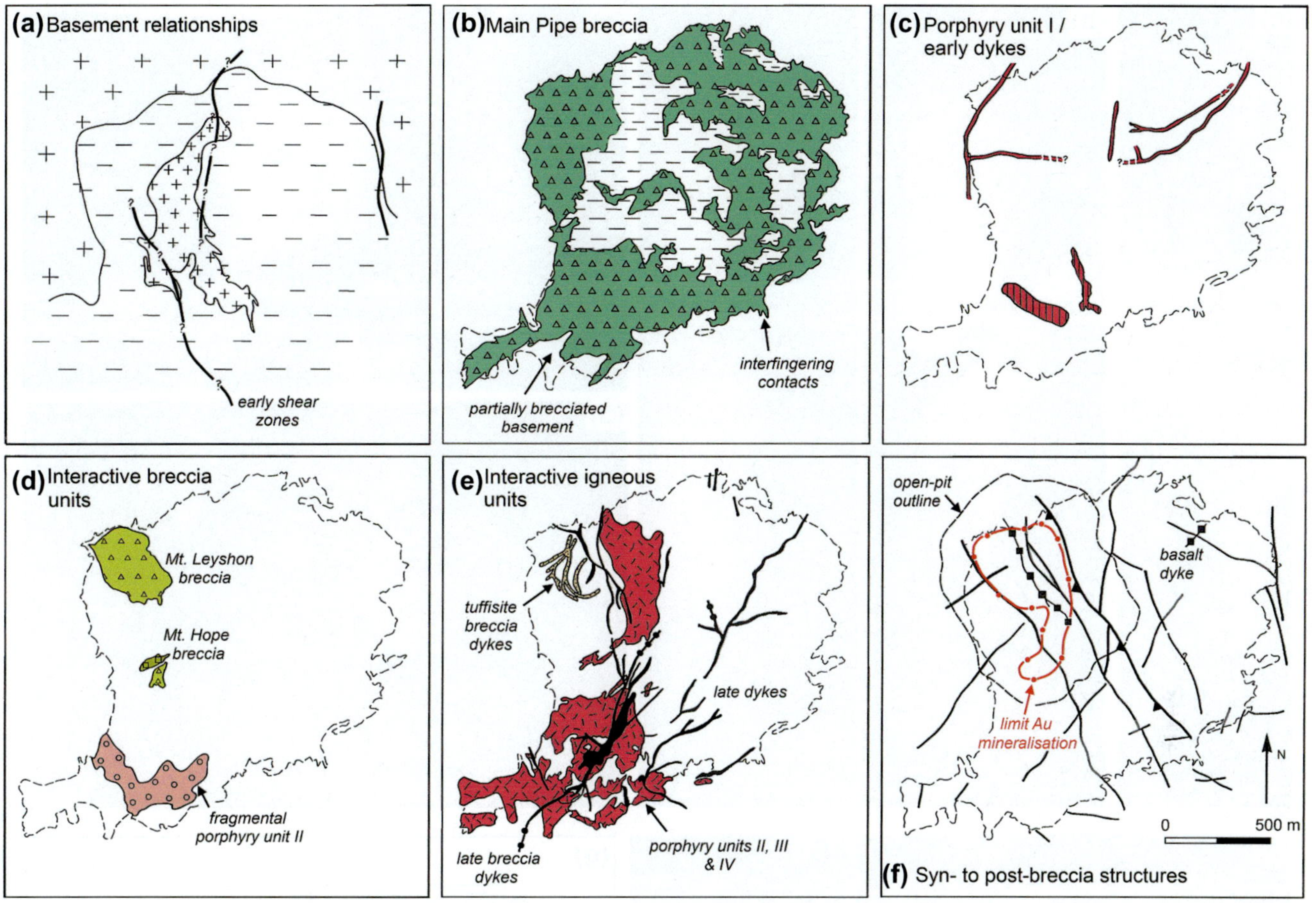

Fig. 18.25 Evolution of the Mt Leyshon breccia complex illustrating the sequential development of the key facies components (see legend to Fig. 18.24). **a** Basement relationships; note the embayed nature of Cambrian metasediment (Puddler Creek Formation) and Ordovician granite contact and the position (reconstructed) of the large Silurian rhyolite dyke body. **b** Main Pipe breccia generation; closed spaced dashes indicate little rotated blocks of the undifferentiated basement, which preserve tectonically driven fracturing, formed prior to the intrusive, interfingering breccia–basement contacts. **c** Pre-Mt Leyshon breccia igneous units which generally trend N, E, or NNE. **d** Major interactive breccia units; the known expression of these units is confined to the western half of the complex. **e** Major interactive igneous units; porphyry units II, III, and IV show close spatial and temporal relations to the interactive breccia bodies; also shown are the tuffisite breccia dykes, late igneous dykes, and late breccia dykes. Note the change in igneous unit trend to predominantly NE. **f** Syn- to post-breccia structures; also shown are post-breccia and post-mineral basalt dykes, the approximate limit of gold mineralisation, and the outline of the open pit. (After Wormald 1993)

phases (illite–smectite, calcite), which suggest meteoric fluid quickly ingressed the complex. The lithified rock flour matrix restored competence, but in turn restricted further meteoric fluid access, favouring the flow of magmatic-sourced fluids in a network of fault-vein structures. Following lithification of the Main Pipe breccia, prograde fluids resulted in an overprint of potassic alteration and laminated quartz–molybdenum veins, similar to those associated with porphyry copper–molybdenum systems. Similarly, the interplay between tectonic, magmatic, and fluid phase processes that formed the Mt Leyshon breccia pipe (and other members of the interactive sequence) was immediately followed by low-temperature alteration, before fluids of mixed origin cemented the matrix and resulted in successive cycles of gangue and ore mineral infill (Wormald 1993).

The bulk of gold mineralisation is paragenetically very late and post-dates all coeval igneous emplacements, and a considerable sequence of prograde to retrograde hydrothermal alteration and infill (including base metals; Wormald 1993, 2017). Gold principally occurs in fractured sulphides (sphalerite, chalcopyrite, galena ± pyrite) or in association with bismuth-sulphides and sulphosalts, within reactivated fault/vein structures, breccia voids, or interlinked diffuse fracture networks cutting the Mt Leyshon breccia and tuffisite breccia dykes (Fig. 18.26e, g), as well as the Mt Hope breccia pipe (Fig. 18.25d–f). Igneous units do not host ore unless cut by these later structures (typically along or near

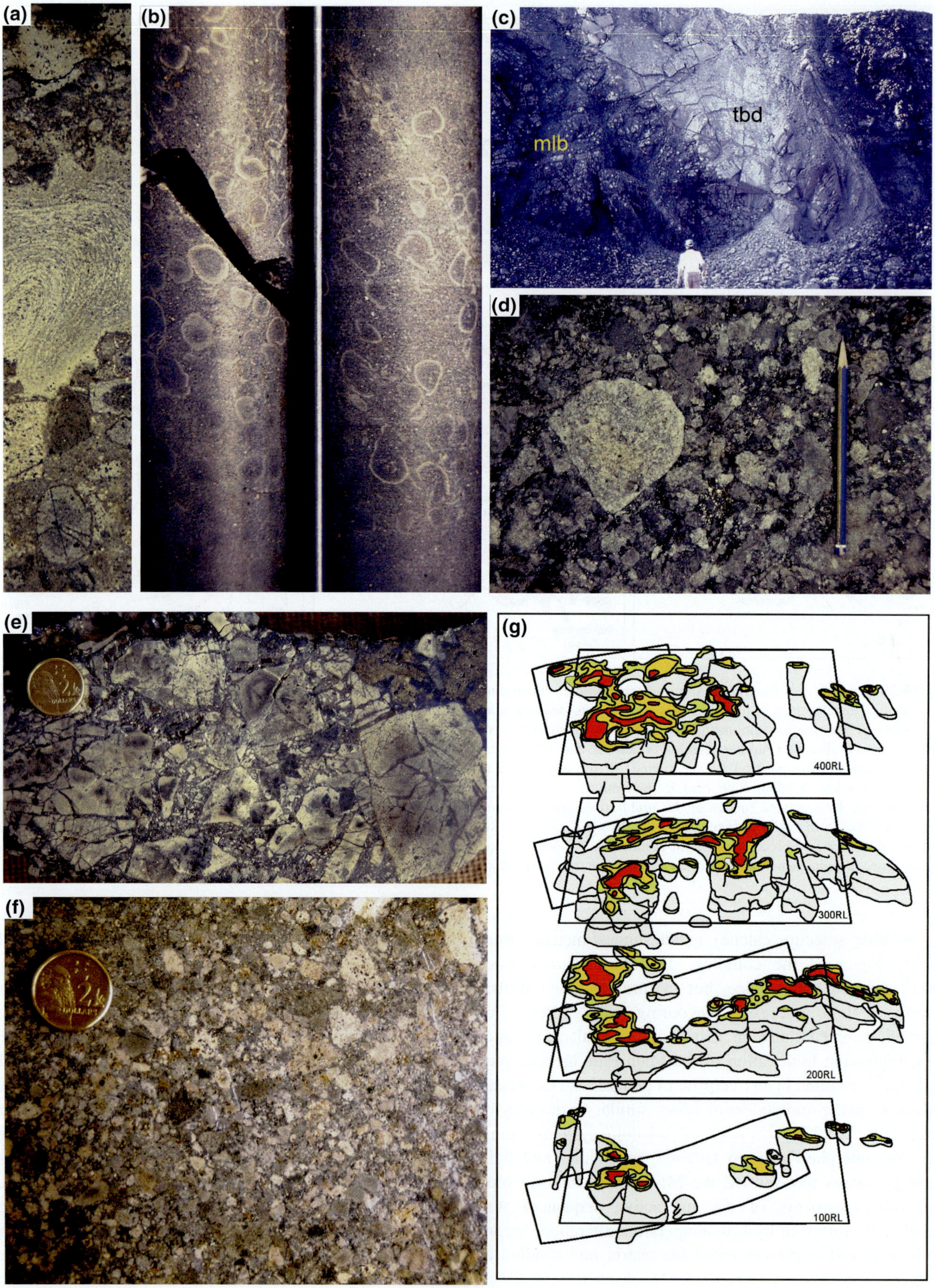
(a)
(b)
(c)
mlb
tbd
(d)
(e)
(f)
(g)
400RL
300RL
200RL
100RL

Fig. 18.26 Features of the Mt Leyshon breccia complex and gold deposit. **a** Coarse-grained sub-facies of Main Pipe breccia cut by late interactive stage, fine-grained (matrix-rich) tuffisite breccia dyke showing "swirl-like" fluidal flow texture and interfingering intrusive contacts. Drill core is ~5 cm wide. **b** Fine-grained sub-facies of Main Pipe breccia containing both intact and broken accretionary lapilli (pellets); diameter of drill core is ~5 cm. In detail accretionary lapilli-rich zones map out as steeply dipping slot- or pipe-like bodies (Fig. 18.24). **c** Pit view (400 m RL) showing fine-grained tuffisite breccia dyke (tbd) intruding Mt Leyshon breccia (mlb); bench height is 10 m. The tuffisite breccia dyke contains up to about 10 g/t Au, although much higher grades ($\geq$100 g/t Au) locally occur in this unit elsewhere. **d** Mineralised (low grade, <1 g/t Au) Mt Leyshon breccia (400 m RL); well mixed and tightly packed breccia (representing zone 4A, Fig. 18.27b). The lightest clasts comprise granite, grey clasts are altered Main Pipe breccia, while darker clasts are dolerite. The most visible sulphides are pyrite. **e** Detail of Mt Leyshon breccia, which consists mostly of fragments of hydrothermally altered Main Pipe breccia. Note the internal fragmented nature of clasts and abundant jigsaw-fits between rebrecciated fragments (representing zone 3, Fig. 18.27b). Little to no mixing with other host rock fragment types has occurred. The fine breccia matrix is replaced by grey quartz–white mica and locally chlorite–pyrite alteration (also partially infilling cavities). The cut surface is bound by an irregular, cross-cutting fault-vein structure infilled with sphalerite, pyrite, and minor chalcopyrite. The bluish minerals are Bi-sulphides and sulphosalts containing gold mineralisation (typically 2 to $\geq$5 g/t Au). **f** Medium-grained tuffisite breccia dyke; note the well-sorted (fines-poor) and clast-supported fabric composed predominantly of cognate rhyodacitic fragments. **g** Perspective view of the Mt Leyshon gold orebody looking due east and showing the internal grade distribution at the 400–100 m RL elevations (3D model after N. Tate and P.J. Womald; From Wormald 1993). Contour intervals are 0.96, 1.44, and >2.4 g/t Au. At lower levels, the orebody consists of linear, en echelon segments trending NE and NW, but at higher levels spreads out to form a wide area of interconnected ore zones. This suggests gold deposition was likely controlled by strike-slip fault segments linked at depth into the regional transcurrent fault system, while in the upper levels by the irregular nature of breccia permeability and porosity. Late reverse fault offsetting the orebody is also shown

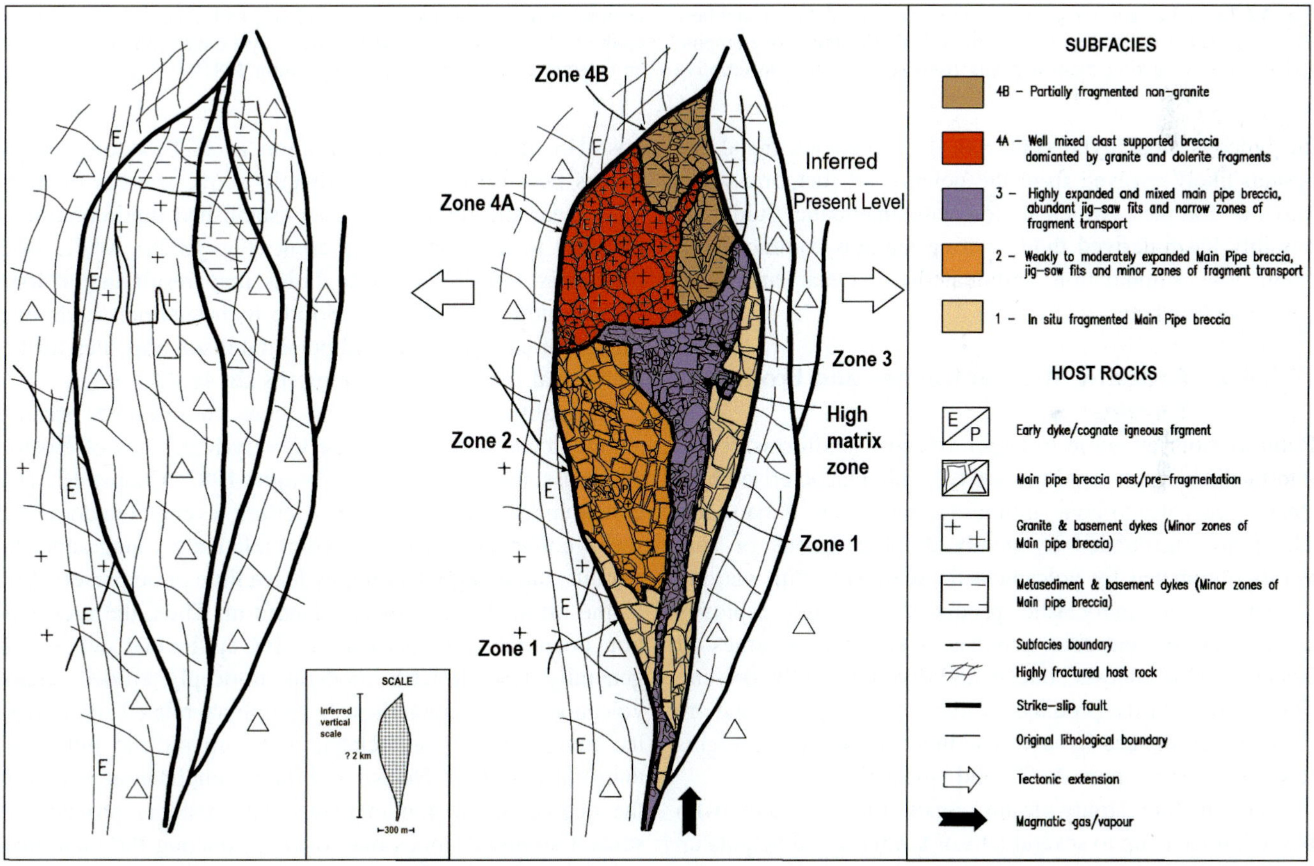

Fig. 18.27 Schematic reconstruction and facies model of the Mt Leyshon breccia pipe. (Left) Pre-pipe host rock configuration, showing the brittle infrastructure and inferred strike-slip fault architecture; note the regions of metasediment and granite within the Main Pipe breccia. (Right) Mt Leyshon breccia pipe; this shows the crude preservation of the host rock configuration following pipe generation, and the effect of host rock type, degree of dilation, and mixing on breccia characteristics. Younger tuffisite breccia dykes omitted. (After Wormald 1993)

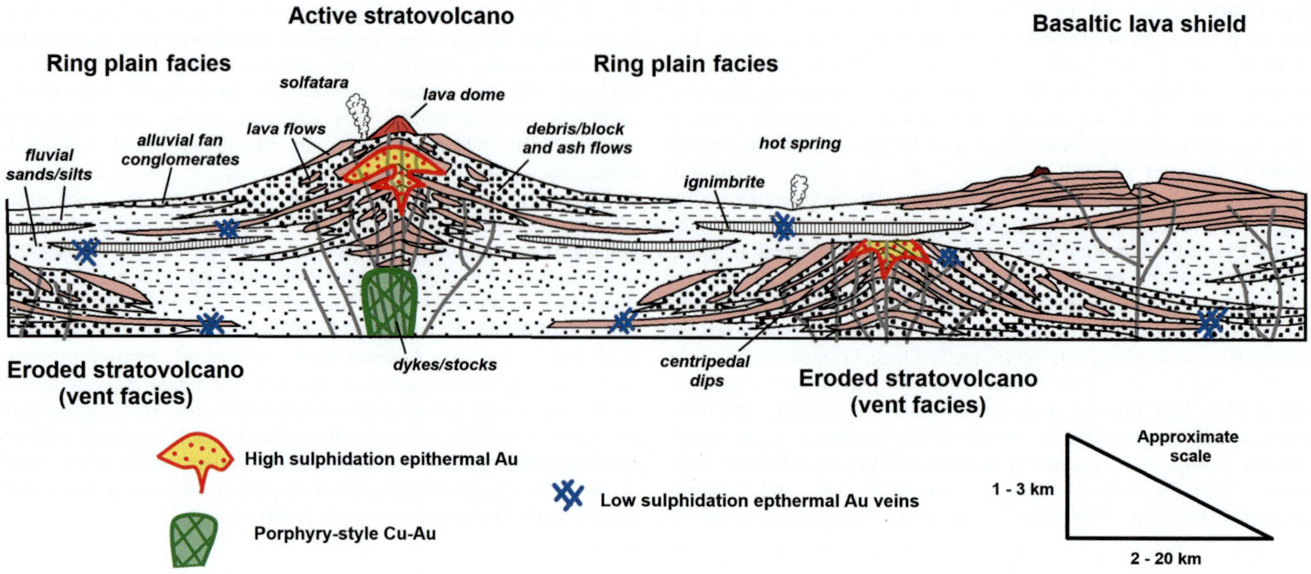

Fig. 18.28 Schematic cross-section showing typical facies architecture of an andesitic stratovolcano terrain. This illustrates overlapping and interfingering ring plain deposits from several separate eruptive centres, nearer vent facies associations, and potential locations of epithermal precious metal and porphyry-style mineralisation. (Based on an original stratigraphic concept by Smith 1987)

the igneous stock/dyke contacts). This precious metal ore system likely evolved from the collapse of meteoric water into the complex, which mixed with magmatic fluid, and possibly basin-derived fluid, sharing features common with deep, low sulphidation carbonate-base metal and gold deposits elsewhere.

18.3.4.5 Andesitic Stratovolcanoes and Eroded Terrains

Returning our attention to larger volcanic landforms, major intermediate eruptive centres are typically characterised by a steep stratovolcano cone around a summit vent (Chap. 16). These are surrounded by gently dipping aprons or *ring plains* dominated by volcaniclastic sediment, with varying proportions of pyroclastic products. Although typically constructed of andesitic magmas, significant volumes of basalt, basaltic andesite, dacite, and rhyolite may be present. These represent the products of numerous eruptions and are generally dominated by lavas and breccias of various origins (e.g. autobreccia, pyroclastic, and epiclastic).

Calk-alkaline stratovolcano cones form edifices with basal diameters up to several tens of km wide and heights up to 4 km (Sect. 16.6.1). These often occur in chains, so the ring plain deposits of one cone may interfinger or overlap with those of another (Fig. 18.28). Feeder dykes or plugs, lava domes, shallow intrusions, and coarse breccias (vent fill talus, proximal pyroclastic fallout) mark former vents. These locations are also the sites of magmatic solfataras and, therefore, probably pervasively hydrothermally altered (Fig. 18.3). In theory, they are also the areas most likely to host HS epithermal mineralisation at shallow-to-moderate depths because they occur directly above sources of magmatic heat and fluids, where an established fluid conduit or pathway to the surface is present (Fig. 18.28). While over time, mixing with meteoric fluids may result in the long lateral outflow of neutral-chloride hydrothermal fluids into ring plain environments and potential distal trap sites for LS epithermal mineralisation (Figs. 18.2b, 18.28).

In reality, the situation is more complicated because stratovolcano cones are prone to very high rates of erosion and mass wastage (Cas and Wright 1987; Chaps. 12, 16), compounded by catastrophic failure events (caldera and sector collapse). Even in active or relatively young andesitic terrains, rapid degradation may leave little or no preservation of the original stratovolcano landform. Thus, the preservation potential of in particularly HS epithermal deposits is generally low. Indeed, moderate-to-deeply eroded stratovolcanoes are characterised by porphyry-related ore deposits, more typically associated with subvolcanic intrusions than volcanic host facies. A further complication is that as the degree of erosion increases the erosional products of several stratovolcanoes may overlap, making the identification of volcanic centres increasingly difficult to decipher. Although some indication may be obtained from the mapped proportion of subvolcanic intrusions (including radial dykes), lava domes, and diatremes/breccia pipes or identification of ring plain facies (quaquaversal dips, paleocurrents, depositional trends), these may be difficult to establish without a considerable amount of regional mapping and analysis, especially in tropical environments. Nevertheless, several HS epithermal gold deposits are interpreted as being hosted by central vent volcanoes, some possibly associated with calderas (Table 18.1).

In the Pacific Rim, these are located in the middle Miocene and younger successions, although more deeply eroded examples of the Late Cretaceous age occur in the Alpine–Tethyan collisional zone of eastern Europe. At Choquelimpie (Chile), the relationship to volcanic landform is unequivocal and gold mineralisation is hosted by a late Miocene stratovolcano comprising dacitic and andesitic volcanic flows and breccias, intruded by lava domes (Gröpper et al. 1991). Erosion of its central cone has exposed a zone of hydrothermal alteration and mineralisation centred upon an andesitic feldspar porphyry, which is probably a dome. The Nansatsu district in southern Kyushu (Japan) is a cluster of Pliocene HS gold deposits associated with andesitic volcanics of similar age, which occur some 25 km west of the present-day active volcanic front. These were important in the early classification of the HS-type and show strong lithological control with gold mineralisation contained within residual and vuggy silica ore bodies that expand or mushroom into permeable pyroclastic units sandwiched between less permeable lava flows (e.g. Hedenquist et al. 1994b). The ore bodies appear to be associated with shallowly eroded central vent volcanoes, possibly emplaced within a larger caldera.

In Southeast Asia, Nena (Frieda River, Papua New Guinea), Nalesbitan, and Tampakan (Philippines) are hosted by variably eroded andesitic stratovolcano complexes. The shallowest is Nalesbitan, where any relation to porphyry is only inferred (Sillitoe et al. 1990), then Nena, which is spatially close to several porphyry copper deposits and a district-wide halo of advanced argillic alteration (some 1–4 km × 12 km long; Espi et al. 2002). The most deeply eroded is the superlarge Tampakan copper–gold deposit, which represents a major, low-grade (0.26 g/t) HS deposit superimposed on an underlying porphyry copper system (Middleton et al. 2004; Fig. 18.29). The deposit is hosted by a sequence of subaerial andesitic lava flows and is associated with and central to a flat-lying, tabular body of pervasive silica, vuggy silica, and silica-clay alteration developed within a district-scale lithocap of advanced argillic and argillic alteration extending ≤ 100 km^2. Mineralisation is mainly developed along an erosional, intra-volcanic unconformity.

The Alpine–Tethyan collisional zone exposes several metallogenic belts which record the development of a number of subduction-related, volcano-plutonic arcs initiated in the late Mesozoic and continuing until the present. In eastern Europe, remnants of a late Cretaceous arc (the Alpuseni-Balkan or "Banatitic" belt) extend from Romania, through Serbia into Bulgaria. These arc remnants are both deeply eroded and structurally dissected by younger collisional thrusting and post-collisional transcurrent faulting. They comprise volcano-sedimentary successions of andesitic to dacitic and trachy-andesitic subvolcanic bodies, lava flows, pyroclastic, and epiclastic rocks, as well as detrital sedimentary rocks derived from the basement. These successions host a number of economically significant porphyry copper deposits, which are spatially associated with clusters of HS epithermal gold deposits, as well as extensive areas of hydrothermal alteration. The principal HS deposits are Bor in Serbia and Chelopech in Bulgaria (Table 18.1). While it has generally been acknowledged that their host rock successions represent those of stratovolcanoes, only recently have more detailed interpretations taking a modern facies approach become available.

At Bor, the main host rocks are coherent and autoclastic andesitic lava facies, which host a deeper porphyry copper system (Borska reka) intimately associated with an overlying HS epithermal system (Jelenković et al. 2016; Fig. 18.11). Copper–gold mineralisation extends vertically over a distance of >1 km. The andesitic lava succession is dated 89.0 Ma, with mineralisation at about 86 Ma indicating they were likely genetically related. At Chelopech, the immediate host rocks, in contrast, appear to form part of a maar-diatreme complex emplaced into a lagoonal environment (Chambefort et al. 2007; Chambefort and Moritz 2014). A subvolcanic andesitic intrusive body and phreatomagmatic and hydrothermal breccias host most copper—gold mineralisation. Massive sulphide bodies are also formed by replacement of sedimentary oolitic, bioclastic, and sandstone layers interbedded with phreatomagmatic beds containing accretionary lapilli and pumice. Although the host succession does not conform to a stratovolcano, the presence of relatively thick hyaloclastite and other volcaniclastic (including pillow) breccias (>1 km thick) may suggest the possibility this could be part of a larger volcanic structure such as the margin of a shallow marine caldera.

Large-scale LS geothermal systems may occur both associated with mature stratovolcanoes and/or in persistently tectonically and volcanically active andesitic terrains in cordilleran settings. In such cases, volcanic host rocks may not be directly contemporaneous with magmatism controlling hydrothermal activity and mineralisation (Fig. 18.28).

The Štiavnica stratovolcano (Slovakia) is the largest volcano in the youngest (Neogene) volcano-plutonic arc in the Carpathian Alpine–Tethys collisional zone (Chap. 17). Detailed volcanic facies reconstruction and radiometric dating have outlined a complex, multistage evolutionary history (Chernyshev et al. 2013; Fig. 18.30). Construction of a large andesitic stratovolcano (3000–4000 m high, with a basal diameter of 35–40 km) occurred between 15.0 and 13.5 Ma. A period of erosion was concluded by caldera collapse (13.5–13.0 Ma). The caldera was infilled initially

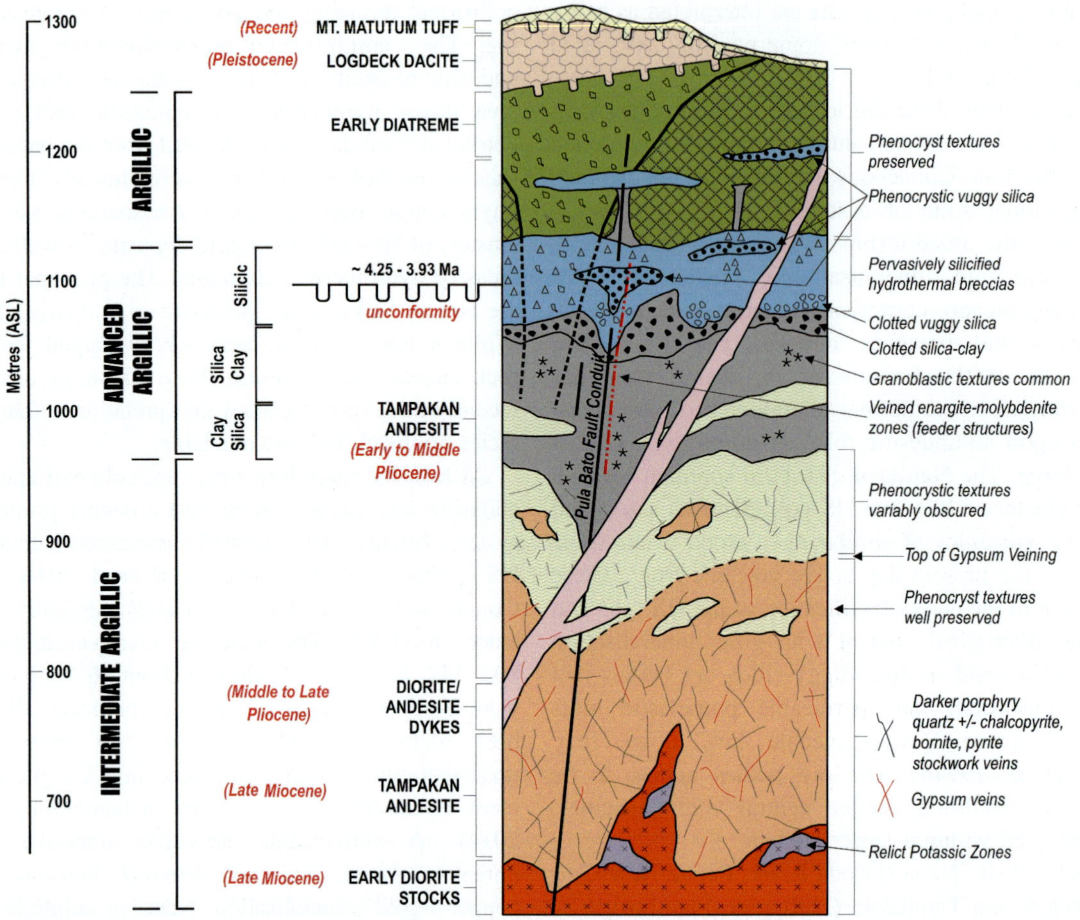

Fig. 18.29 Generalised stratigraphy and hydrothermal alteration and textural changes associated with the Tampakan high sulphidation epithermal copper–gold deposit, southern Mindanao, Philippines. In the ore environment, hydrothermal alteration assemblages have obliterated or overprinted original textures and mineralogy. Zones of silica alteration often display multi-phase brecciation and vuggy porosity resulting from intense acid leaching. Copper sulphide mineralisation typically comprises digenite, bornite, and enargite and to a lesser extent chalcocite and covellite, usually as fractures, veins, and vug-fills or as disseminations at depth. Pyrite is usually present throughout and enargite appears to be preferentially associated with late-stage structures and areas of strong silicification. Porphyry copper-style mineralisation at depth comprises disseminated chalcopyrite, bornite, and pyrite associated with variably developed quartz-stockwork veining. (Modified from Middleton et al. 2004)

with lacustrine and limnic sediments and later effusive eruptions of andesitic lavas and younger explosive products (13.0–12.2 Ma). This was followed by the resurgence of a large horst block in the central part of the caldera, accompanied by rhyolitic volcanic activity (12.2–11.4 Ma), with hydrothermal activity persisting until (10.7 Ma). Epithermal mineralising systems developed at several stages in response to changing topographic, stress field, and hydrologic conditions (Koděra et al. 2005, 2016; Chernyshev et al. 2013; Fig. 18.30). These include (1) a pre-caldera barren HS lithocap, as well as distal LS gold–silver mineralisation, (2) shallow dipping LS gold veins and veinlets associated with a subhorizontal extensional stress field (σ_3 vertical) during initiation of caldera subsidence, and (3) an extensive system of LS precious metal and silver-base metal veins (Hodruša-type) coeval with resurgent uplift of the caldera. The latter were the historically exploited resource (Table 18.1) and occur as multiple quartz veins along northeast–southwest faults cutting the resurgent horst. They were emplaced in a post-caldera stress field with σ_3 oriented NW–SE, and σ_1 variably oriented NE-SW or vertically.

Epithermal mineralisation in Kyushu, the principal gold-producing region in Japan, is predominantly associated with LS hot spring, quartz vein systems (Izawa and Urashima 1989; Watanabe 2005). Volcanism in Kyushu is presently controlled by two linear volcano-tectonic depressions, the Beppu–Shimabara and Kagoshima grabens, which are dominated by Quaternary subaerial andesitic stratovolcanoes and large, in part submarine, calderas. These are built on accreted Cretaceous–Cenozoic terranes of oceanic and

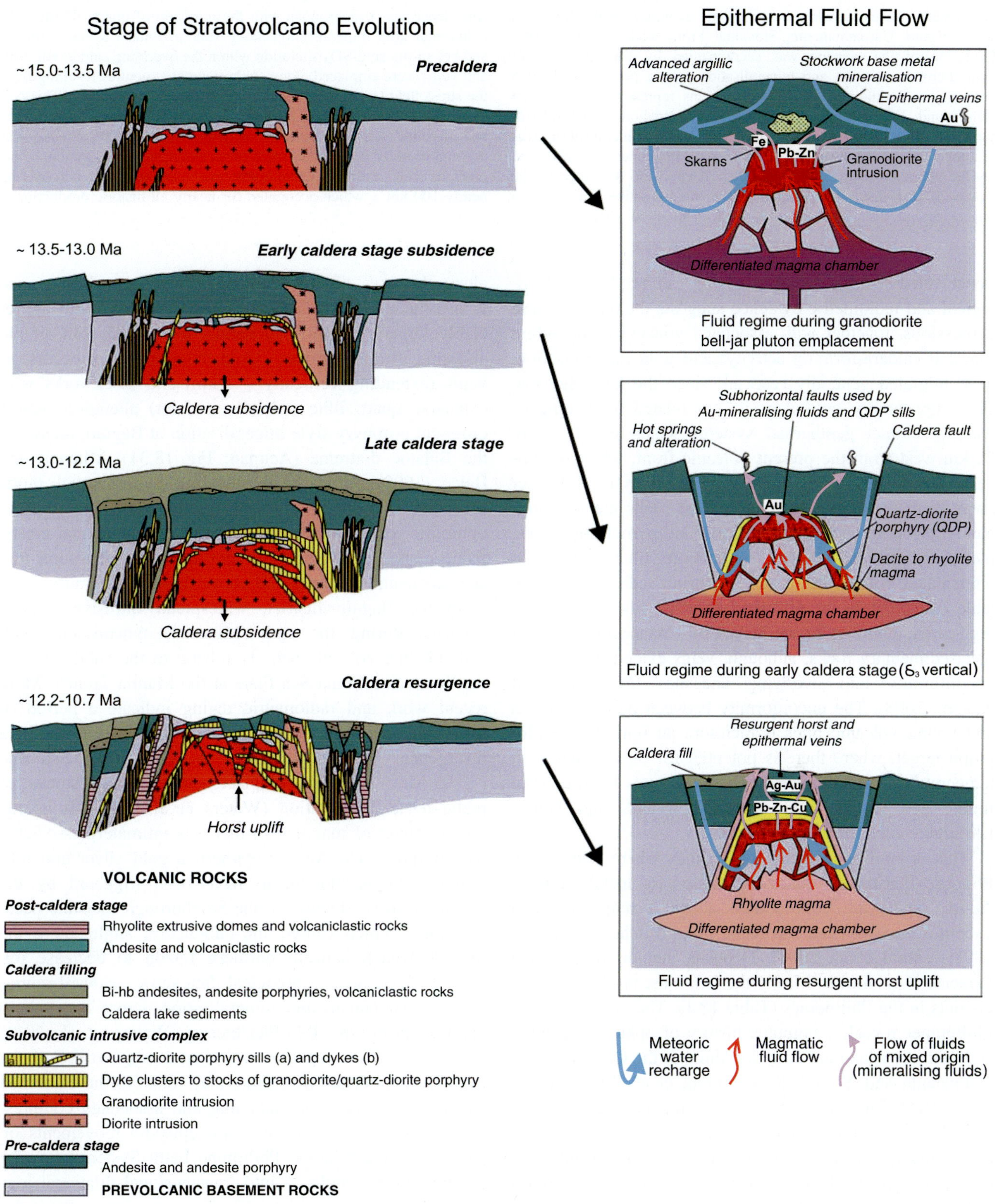

Stage of Stratovolcano Evolution
~15.0-13.5 Ma
Precaldera
~13.5-13.0 Ma
Early caldera stage subsidence
Caldera subsidence
Late caldera stage
~13.0-12.2 Ma
Caldera subsidence
Caldera resurgence
~12.2-10.7 Ma
Horst uplift
VOLCANIC ROCKS
Post-caldera stage
Rhyolite extrusive domes and volcaniclastic rocks
Andesite and volcaniclastic rocks
Caldera filling
Bi-hb andesites, andesite porphyries, volcaniclastic rocks
Caldera lake sediments
Subvolcanic intrusive complex
Quartz-diorite porphyry sills (a) and dykes (b)
Dyke clusters to stocks of granodiorite/quartz-diorite porphyry
Granodiorite intrusion
Diorite intrusion
Pre-caldera stage
Andesite and andesite porphyry
PREVOLCANIC BASEMENT ROCKS
Epithermal Fluid Flow
Advanced argillic alteration
Stockwork base metal mineralisation
Epithermal veins
Au
Fe
Pb-Zn
Skarns
Granodiorite intrusion
Differentiated magma chamber
Fluid regime during granodiorite bell-jar pluton emplacement
Subhorizontal faults used by Au-mineralising fluids and QDP sills
Hot springs and alteration
Caldera fault
Au
Quartz-diorite porphyry (QDP)
Dacite to rhyolite magma
Differentiated magma chamber
Fluid regime during early caldera stage (δ₃ vertical)
Resurgent horst and epithermal veins
Caldera fill
Ag-Au
Pb-Zn-Cu
Rhyolite magma
Differentiated magma chamber
Fluid regime during resurgent horst uplift
Meteoric water recharge
Magmatic fluid flow
Flow of fluids of mixed origin (mineralising fluids)

◀ **Fig. 18.30** Stages showing the schematic evolution of the Štiavnica stratovolcano, Carpathian arc, Slovakia (From Koděra et al. 2005). Evolution of the volcano was accompanied by various types of hydrothermal alteration and mineralisation over a period of about 5 million years, although LS epithermal veins represent the dominant style found. Prior to caldera collapse, early intrusion-related subvolcanic porphyry copper, skarn, and base metal mineralisation was accompanied by the formation of a barren HS lithocap of advanced argillic alteration in overlying andesites. Away from the central cone (some 20 km), distal hot spring LS vein mineralisation belonging to this stage has also been discovered (Koděra et al. 2016). Caldera subsidence was synchronous with the intrusion of quartz diorite sills, epithermal gold mineralisation along subhorizontal LS vein systems, and hot spring acid-SO_4 alteration within the overlying caldera fill. Sills and veins were emplaced along subhorizontal structures resulting from the stress field (σ_3 vertical) which governed caldera collapse. This was followed by the uplift of a resurgent horst in the centre of the caldera and rhyolitic intrusive volcanic activity. Structural resurgence was accompanied by the emplacement of an extensive zoned system of LS precious metal and base metal epithermal veins and veinlets (covering nearly 100 km^2), which accounted for nearly all historic production

trench-filled sediments and overlying fore-arc basins. Geological and radiometric evidence suggests the axes of these depressions, together with associated volcanism (including repeated caldera-forming activity) and gold mineralisation, have migrated generally eastward since the late Miocene. Thus, Quaternary gold deposits are related to waning or recently extinct geothermal systems which occur within 30 km west from the present volcanic front, whereas Pliocene deposits relate to older geothermal systems further away from the active volcanoes. Host rocks are predominantly subaerial andesitic lava flows and pyroclastic facies, including ignimbrites, with some more silicic lavas and intercalated volcaniclastic and lacustrine sediments. A notable exception is the Hishikari deposit, where extremely high-grade quartz veins are hosted by Cretaceous basement metasedimentary rocks, although, veins extend through the unconformity into overlying andesites (Morishita and Nakano 2008). The unconformity between basement rocks and young volcanic piles is therefore an important exploration target, where there is potential for rapid changes in temperature, pH, oxidation conditions of hydrothermal fluids, and mixing with meteoric waters, and hence, favourable sites for gold deposition.

This is confirmed in the Philippines where widespread Pliocene–Pleistocene quartz-carbonate-base metal sulphide Au–Ag veins are concentrated above a major regional unconformity (Mitchell and Balce 1990; Mitchell and Leach 1991; Yumul et al. 2008). Deposits include Baguio and Paracale (Luzon), historically the two most productive gold districts in the Philippines (Table 18.1). The geology of the Philippines reveals a complex history of ophiolite stacking and multiple island arc accretion through the Cenozoic (e.g. Diamalanta and Yumul 2004; Yumul et al. 2008; Suzuki et al. 2016). This has built semi-continuous mountain belts and elevated regions dominated by andesitic volcaniclastic debris and epiclastic successions cut by a series of intrusive rocks, on top of which sit today's presently active stratovolcanoes. Most of the LS epithermal vein systems intersect the unconformity between Cretaceous ophiolitic complex rocks and deformed marine sedimentary and volcanic arc successions. Although the form of the volcanic centres is poorly understood, mineralisation appears concentrated along the axes of late Cenozoic subaerial volcanic arcs, where the geothermal systems were extensive and appreciably larger than those generally associated with simple andesitic stratovolcanoes. Gold mineralisation occurs as veins (extending vertically ≥ 1 km) and stockworks with extensive quartz-illite (adularia is rarer) alteration, which cross-cut porphyry-style mineralisation at Baguio, including the Balatoc diatreme (Acupan; Fig. 18.31). Mitchell and Balce (1990) proposed gold-bearing hydrothermal fluids ascended from depth through fracture-permeable in part ophiolitic rocks. Fluids were focused into anticlinal crests, rising along reactivated basement faults into structural trap sites concentrated along and above the unconformity.

In the Baguio district, widespread intrusive activity occurred during the early Miocene, synchronous with calk-alkaline volcanic activity related to the subduction of the South Sea China Sea Plate at the Manila Trench. More recent work and radiometric dating indicate a period of magmatic quiescence in the late Miocene was followed by a resurgence of magmatism in the Pio-Pleistocene (≤ 3.5 Ma), involving emplacement of small stocks of mafic–felsic composition (Waters et al. 2011). It is this second phase of magmatism that was intimately associated with porphyry Cu–Au and epithermal gold–silver mineralisation. This is thought to have been triggered by the east-directed subduction of the Scarborough aseismic oceanic ridge, which caused the subduction angle along the Manila Trench beneath northern Luzon to decrease (cf. Sect. 17.5.4). Geochronological data show eastward migration of magmatism and mineralisation across the Baguio district during the Plio-Pleistocene (Waters et al. 2011; Fig. 18.31). This, together with increasing fertility of the intrusive complexes with time, may relate to the combined effects of ongoing slab flattening and associated compressional deformation, coupled with episodic movements on splays of the left-lateral Philippine Fault System, allowing magmas to be emplaced into transtensional dilational jogs and strike-slip basins (Chap. 17). Transpression on the major restraining bend in the Philippine Fault to the south of Baguio, coupled with compressional tectonism during shallowing of the subduction zone, caused extreme uplift and exhumation during this period. This led to complex

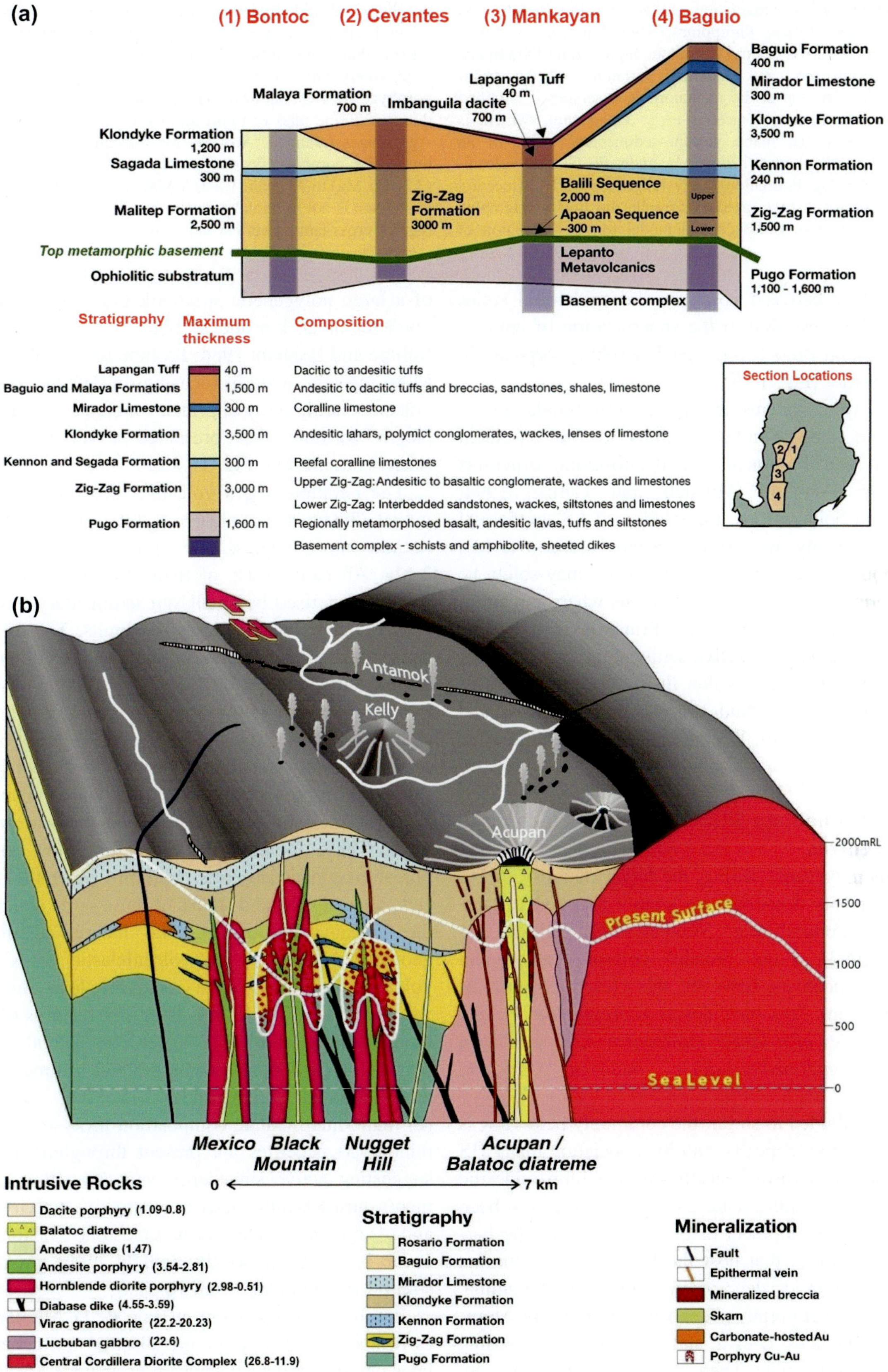
(a)
(1) Bontoc
(2) Cevantes
(3) Mankayan
(4) Baguio
Baguio Formation
400 m
Mirador Limestone
300 m
Klondyke Formation
3,500 m
Kennon Formation
240 m
Zig-Zag Formation
1,500 m
Pugo Formation
1,100 - 1,600 m
Lapangan Tuff
40 m
Malaya Formation
700 m
Imbanguila dacite
700 m
Klondyke Formation
1,200 m
Sagada Limestone
300 m
Malitep Formation
2,500 m
Top metamorphic basement
Ophiolitic substratum
Zig-Zag
Formation
3000 m
Balili Sequence
2,000 m
Apaoan Sequence
~300 m
Upper
Lower
Lepanto
Metavolcanics
Basement complex
Stratigraphy
Maximum thickness
Composition
Lapangan Tuff
40 m
Dacitic to andesitic tuffs
Baguio and Malaya Formations
1,500 m
Andesitic to dacitic tuffs and breccias, sandstones, shales, limestone
Mirador Limestone
300 m
Coralline limestone
Klondyke Formation
3,500 m
Andesitic lahars, polymict conglomerates, wackes, lenses of limestone
Kennon and Segada Formation
300 m
Reefal coralline limestones
Zig-Zag Formation
3,000 m
Upper Zig-Zag: Andesitic to basaltic conglomerate, wackes and limestones
Lower Zig-Zag: Interbedded sandstones, wackes, siltstones and limestones
Pugo Formation
1,600 m
Regionally metamorphosed basalt, andesitic lavas, tuffs and siltstones
Basement complex - schists and amphibolite, sheeted dikes
Section Locations
1
2
3
4
(b)
Antamok
Kelly
Acupan
Present Surface
Sea Level
2000mRL
1500
1000
500
0
Mexico
Black Mountain
Nugget Hill
Acupan / Balatoc diatreme
0
7 km
Intrusive Rocks
Dacite porphyry (1.09-0.8)
Balatoc diatreme
Andesite dike (1.47)
Andesite porphyry (3.54-2.81)
Hornblende diorite porphyry (2.98-0.51)
Diabase dike (4.55-3.59)
Virac granodiorite (22.2-20.23)
Lucbuban gabbro (22.6)
Central Cordillera Diorite Complex (26.8-11.9)
Stratigraphy
Rosario Formation
Baguio Formation
Mirador Limestone
Klondyke Formation
Kennon Formation
Zig-Zag Formation
Pugo Formation
Mineralization
Fault
Epithermal vein
Mineralized breccia
Skarn
Carbonate-hosted Au
Porphyry Cu-Au

◀ **Fig. 18.31** Setting of low suphidation epithermal gold mineralisation in the Baguio district, Luzon, Philippines (After Waters et al. 2011). **a** Stratigraphy of the Luzon Central Cordillera, highlighting lithological composition, correlations, and thickness variations of the major stratigraphic units. The top of the metamorphic basement is signified by the Pugo Formation (Cretaceous–early Paleogene), which is believed to represent the mafic volcano-sedimentary cover of an ophiolite, and is overlain unconformably by andesitic–basaltic volcaniclastics of the Zig-Zag Formation (early Oligocene–early Miocene). The inset map shows districts used to compile the four stratigraphic sections, northern Luzon. **b** Conceptual model for the formation of porphyry and epithermal mineralisation in the Baguio district. Porphyry mineralisation is centred on composite quartz diorite porphyry complexes that have intruded folded Oligocene–Pleistocene volcano-sedimentary units. Low sulphidation epithermal veins cross-cut the porphyry systems, but have been juxtaposed into the porphyry deposits due to extreme rates of uplift and exhumation during the Pleistocene. Age ranges (mainly $^{40}Ar/^{39}Ar$) are indicated for the major intrusive units. The porphyry systems young eastwards from Black Mountain (3.2–2.6 Ma) to Acupan (1.2–0.5 Ma). The age of the Balatoc diatreme at Acupan is not accurately constrained between ≤ 1 Ma and 0.65 Ma (age of cross-cutting epithermal veins)

inter-relationships between porphyry and epithermal styles of mineralisation, resulting in the superposition of epithermal mineralisation onto copper–gold porphyry deposits in the Baguio district (Fig. 18.31).

Major LS vein systems in the western Sunda arc in Indonesia also show similar relations to the basement unconformity and basement domal/antiformal structures (e.g. Gunung Pongkor; Mitchell and Carlile 1994; Table 18.1). In both regions, the relation of LS epithermal mineralisation, if any, to a stratovolcano landform is less easy to demonstrate, and host volcanic facies may solely be providing permeable trap sites at locations where the structure is the overriding control (see Prihatmoko and Idrus 2020). In such tectonically active settings, the contributions of magmatic versus metamorphic fluids and the source of gold in the ore-forming fluid may be more controversial (Mitchell and Balce 1990; Waters et al. 2011; Cooke et al. 2011).

18.3.4.6 Intermediate-Silicic Multivent Volcanic Centres

Regarding their volcanic setting, the larger majority of HS epithermal deposits described are simply ascribed to "*andesitic–dacitic domes*" (e.g. Sillitoe 1999; Sillitoe and Hedenquist 2003). From broader volcanic facies and exploration perspectives, however, this clearly presents us with a conundrum: "*do such domes represent the eroded central vent facies of a larger stratovolcano, or part of a low-standing, intermediate-silicic multivent volcanic centre?*"

As we have alluded to so far, the commonly held view is that porphyry copper deposits, and by association many HS epithermal deposits, form beneath and/or within andesitic stratovolcano successions. One example, which has been important in the development of this concept, was the Farallón Negro volcanics that host the Bajo de la Alumbrera porphyry copper–gold deposit in northwest Argentina. Unlike most Andean porphyry deposits, the Bajo de Alumbrera district includes coeval and co-magmatic volcanic rocks that have not been entirely removed by post-mineralisation erosion. Several studies have interpreted the late Miocene Farallón Negro volcanics to have been part of a large polygenetic stratovolcano, possibly with a basal diameter of 20 km and as high as 4–6 km (Sillitoe 1973; Sillitoe and Bonham 1984; Proffett 2003; Halter et al. 2004). More recently, a study focusing on detailed analysis of the volcanic stratigraphy and facies architecture of the Farallón Negro volcanics has presented a very different picture (Harris et al. 2006; Fig. 18.32).

The Farallón Negro volcanics evolved from effusive to mixed effusive and explosive styles, as magma compositions became more intermediate and silicic over a period of 2 Ma. An early phase of mafic to intermediate volcanism was characterised by small syn-sedimentary intrusions with peperitic contacts and lesser lava units. These are interbedded with thick and extensive successions of coarse-grained, sedimentary breccias deposited from numerous debris and hyper-concentrated flow events (Chap. 12). A later phase of silicic volcanism included both effusive eruptions, forming several aerially restricted nested dome complexes, and explosive eruptions producing more widely dispersed (up to 5 km) pyroclastic units up to 30 m thick in proximal settings. These include block and ash flow and rhyolitic phreatomagmatic deposits. Multiple generations of subvolcanic dykes, sills, and stocks of these same compositions cut the volcanic succession. The nature, distribution, and geometry of volcanic and volcaniclastic facies suggest that volcanism occurred as a relatively low-relief, multivent volcanic field (Fig. 18.32). Four key features of the stratigraphy support this: (1) sheet-like, laterally continuous debris-flow and other coarse-grained sedimentary deposits are dominant, particularly in the lower section; (2) mafic-intermediate composition lavas are volumetrically minor; (3) peperites are present throughout the sequence, suggesting active subsidence and deposition of wet sediments during shallow intrusive/extrusive activity (Chap. 16); and (4) fine-grained lacustrine sandstone–siltstone sequences occur in areas previously thought to be proximal to the summit region of the stratovolcano.

Porphyry emplacement at Bajo de la Alumbrera and the earliest stages of copper mineralisation were coeval with the later phase of silicic volcanism at about 8 Ma. Harris et al. (2006) concluded this developed in an alluvial to ring plain setting distal to larger volcanic edifices, where volcanism,

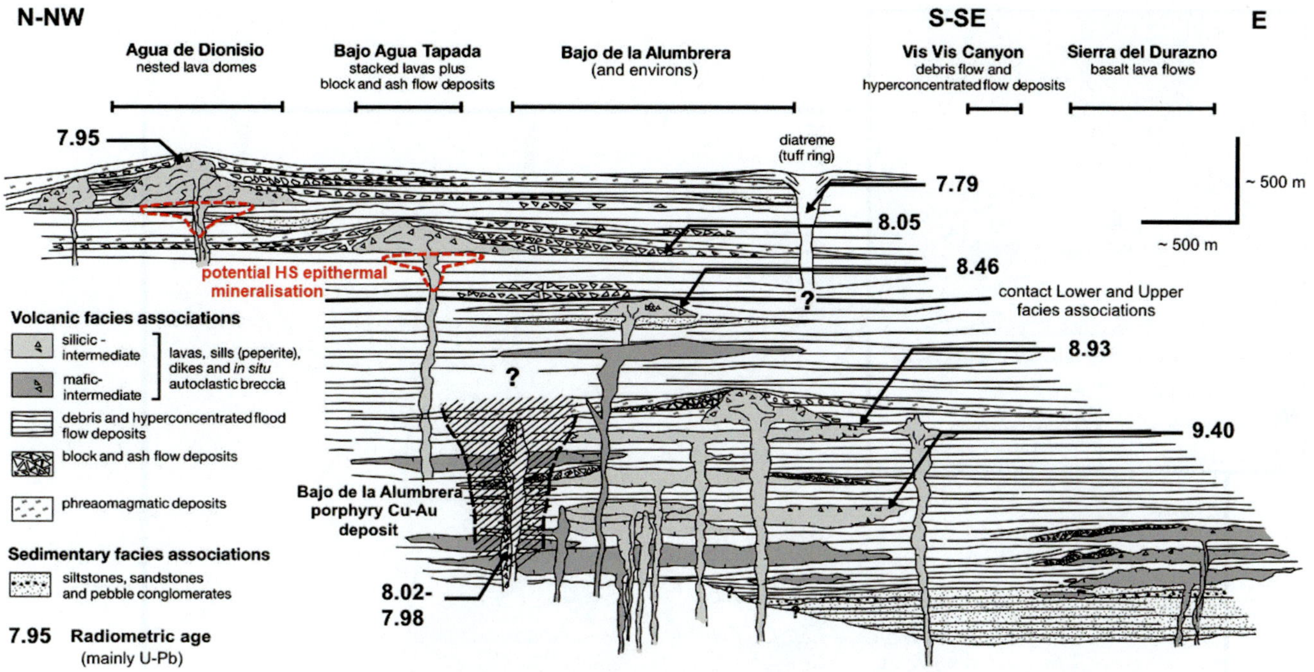

Fig. 18.32 Schematic cross-section showing reconstruction of the facies architecture of the Farallón Negro volcanics in the Bajo de la Alumbrera region, Argentina (From Harris et al. 2006). The paleogeography represented is inferred for the time when the earliest mineralised porphyries were emplaced at Bajo de la Alumbrera at about 8.0 Ma. This shows the development of an evolving topographically subdued, multivent intermediate-silicic volcanic complex, developed across an area of active sedimentation. Eruptions of andesitic to rhyolitic lava flows, nested lava domes, and subvolcanic intrusions followed early mafic effusive eruptions and shallow-level intrusions. The localities of key measured graphic sections important in constraining the facies reconstruction are shown across the top, and the position of Bajo de la Alumbrera porphyry copper–gold deposit. Significant geochronological data of stratigraphically important units are also indicated. A considerable thickness of volcanic and sedimentary cover may have existed (up to 1 km) when the younger Alto de la Blenda stock (not shown) was intruded some 1 Ma later. Active subsidence during the deposition of the Farallón Negro volcanics probably enhanced the preservation of the multivent volcanic complex

porphyry intrusions, and mineralisation were focused along a major structural corridor in an active sedimentary basin, most likely part of a larger foreland basin system concurrent with compressive deformation of the Andes at this time (Chap. 17). Peperites are potentially a key facies indicator separating the successions of low-relief multivent fields (with standing bodies of water and/or water-saturated sediment), from stratovolcanoes.

A similar, relatively topographically subdued, multivent volcanic centre in northern Peru hosts one of the largest known gold deposits. The Yanacocha gold mining district contains past and present resources of >50 Moz (or 1,400 tonnes) of gold in oxidised ore, in over ten discrete HS epithermal deposits. Since the commencement of mining in 1993, this supergiant has been the most productive epithermal gold mine in the world (Plate 18.1). The epithermal deposits also overlie a large, but poorly measured gold and copper resource in sulphidic rocks that includes a copper–gold porphyry (Longo et al. 2010). The Yanacocha mining district is contained within the Yanacocha volcanic field (Fig. 18.33), comprising a cogenetic series of middle–late Miocene andesitic and dacitic lavas, domes, ignimbrites, and minor subvolcanic intrusions, dykes, and breccias covering an area of some 350 km^2, between 3,400 and 4,200 m in elevation. Gold occurs as clusters of HS sulphidation deposits in a northeast-trending corridor of pervasive hydrothermal alteration some 15 km long by 8 km wide (>100 km^2), essentially bisecting the volcanic field. This parallels regionally prominent northeast-striking lineaments which are associated with several significant metal deposits, of which the most southerly and important is Yanacocha (Fig. 18.33).

As a result of its economic importance, a significant amount of detailed geological, geochemical, geochronological, and volcanological data are now available for the Yanacocha volcanic field (Longo et al. 2010; Fig. 18.34). The Yanacocha Volcanics (14.5–8.4 Ma) have a cumulative erupted magma volume of about 90km^3, which evolved from spatially and temporally distinct eruptive centres which are slightly elongated in a northeast–southwest direction for 20 km. Volcanic and subvolcanic intrusive rocks are divided into five mappable units of phenocryst-rich, andesite to

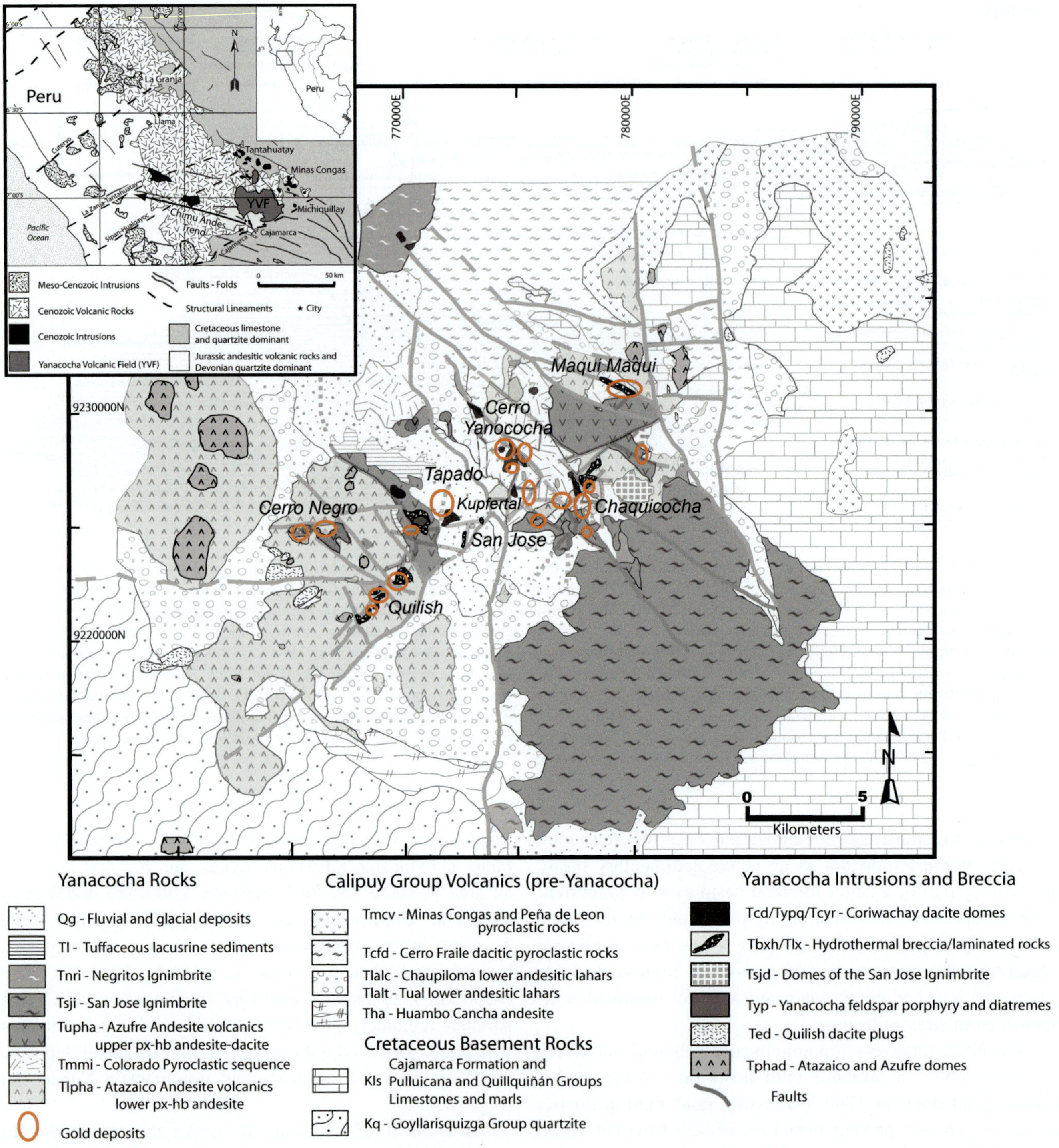

Fig. 18.33 Geological map of the Yanacocha volcanic field and gold mining district, northern Peru. Selected individual gold deposits including those referred to in the text are named. The index map shows the regional location of the Yanacocha volcanic field (YVF). Note the prominent northeast-striking faults, oriented transverse to the Andean Cordillera which define the Cutervo–Cajamarca structural zone, including the four major lineaments named Cajamarca, La Zanja-Tantahuatay, Sipan-Hualgayoc, and Cutervo. The lineaments are associated with several significant metal deposits, of which the most southerly and important is Yanacocha, lying within the Chicama–Yanacocha structural corridor. Approximately 20 km northeast of the Yanacocha district is a northwest-trending belt of eight porphyry Cu–Au deposits that appear to extend 100 km from Michiquillay to La Granja, parallel with the Chimu Andes trend. (Adapted from an original map courtesy and permission of Tony Longo, and Longo et al. 2010)

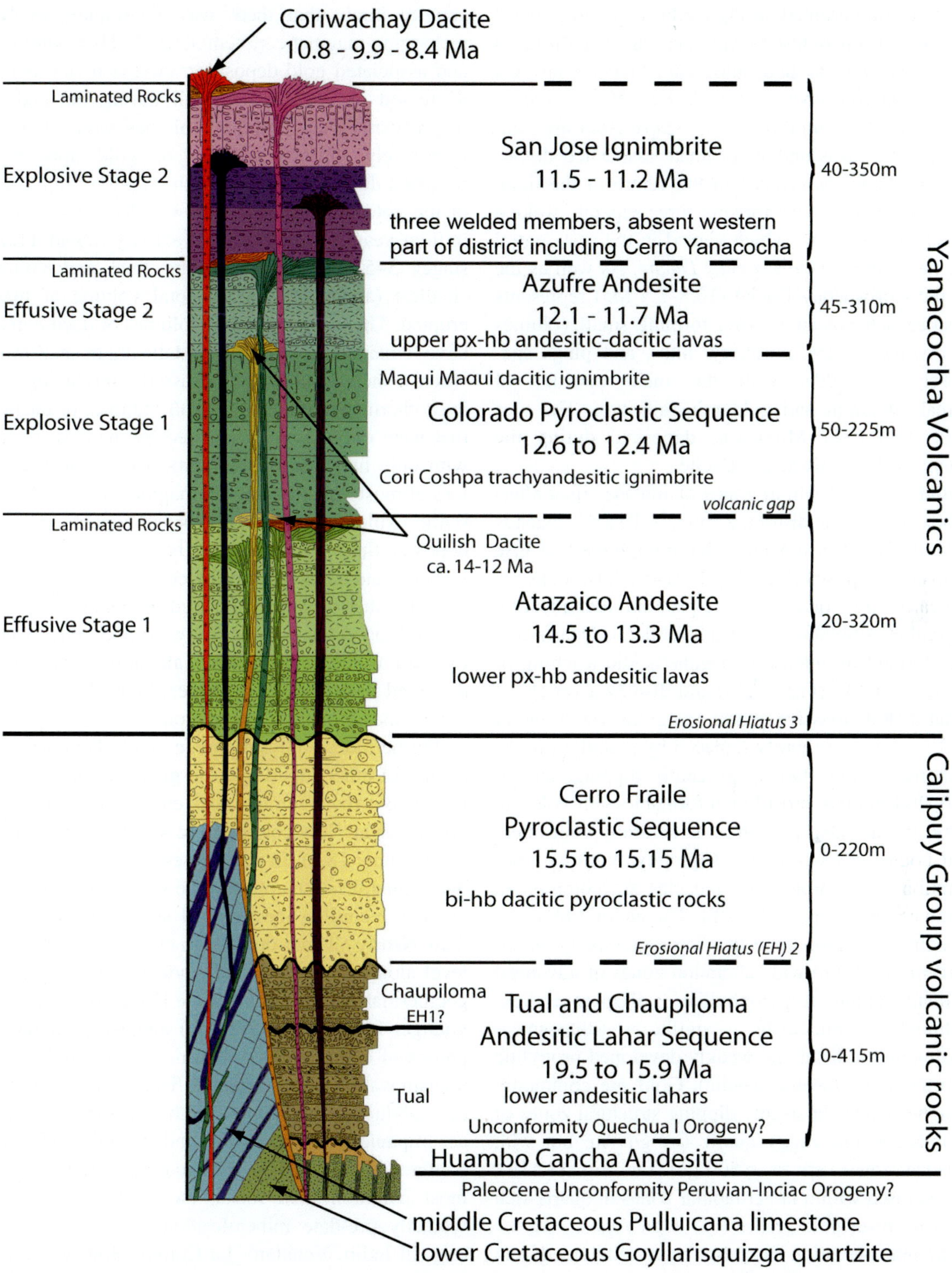

Fig. 18.34 Generalised stratigraphic section for the Yanacocha mining district showing the main volcanic facies of the Miocene Calipuy Group and the overlying ore-bearing Yanacocha Volcanics. Ages are based on detailed $^{40}Ar/^{39}Ar$ determinations, and ranges in thickness are indicated; bi = biotite, hb = hornblende, and px = pyroxene. Laminated rocks consist of a variety of thin-bedded, fine-grained volcano-sedimentary units which may include air fallout and redeposited ashes. These are commonly pervasively hydrothermally altered and may contain thinly laminated chalcedony with local opal and kaolinite, while the upper two horizons are spatially associated with gold mineralisation. Thinly laminated chalcedonic silica has been interpreted as beds of colloidal silica settled from acidic lakes over fumarolic vents in volcanic craters (Longo et al. 2010). These most likely reflect ephemeral lacustrine environments developed within the volcanic (domal) topography. (From Longo et al. 2010)

rhyolite composition erupted during a relatively short period of 6 Ma. These from oldest to youngest are (1) *Atazaico Andesite*—andesitic lavas and autoclastic breccias erupted from small stratovolcanoes; (2) *Colorado Pyroclastics*—andesitic and dacitic ignimbrites; (3) *Azufre Andesite*—andesitic and dacitic lavas and subordinate associated domes and pyroclastic deposits, erupted in the centre of the field; (4) *San Jose Ignimbrite*—comprising three separate welded eruptive units or members, which vented in the east-central part of the field; and (5) *Coriwachay Dacite*. As well as the youngest, the Coriwachay Dacite (10.8–8.4 Ma) represents the most silica-rich volcanic rocks, forming small volumes of dacite–rhyolite domes, porphyry dikes, and plugs spatially closely coincident with the main corridor of hydrothermal alteration and gold mineralisation. The vast majority of gold (>47 Moz) was deposited during the emplacement of the Coriwachay Dacite.

Most gold is hosted within zones of intense silica alteration. In the centre of the district, a zone ($\sim 10\ km^2$) extends southeastwards from Cerro Yanacocha through the San Jose and Chaquicocha deposits (Fig. 18.33), where host rocks are altered to nearly 100% quartz making protolith identification and stratigraphic correlation extremely difficult. Quartz-rich rocks were formed by intense hypogene acidic leaching of original volcanic rocks (Fig. 18.35) and display a variety of textures that include massive quartz, vuggy quartz, granular quartz, and vuggy quartz partly replaced by granular quartz. Quartz-rich rock occurs largely as gently dipping, tabular bodies 100–200 m thick broadly conformable with volcanic stratigraphy, but are also present as structurally controlled, subvertical bodies cross-cutting stratigraphy. Various types and generations of hydrothermal breccias are common in zones dominated by quartz-rich rocks. The quartz-rich rocks define a central alteration zone that is progressively enveloped by 10 m- to 100 m-wide alteration zones of advanced argillic (quartz–alunite $\pm$ pyrophyllite $\pm$ diaspore $\pm$ dickite $\pm$ kaolinite), immediate argillic (illite–smectite $\pm$ pyrite), and a distal and weakly developed propylitic zone (chlorite–illite–smectite–calcite). Gold ore commonly contains higher grades in steeply dipping structural zones of fractured massive and vuggy quartz (>2 g/t Au, but with local bonanza zones up to ≥ 500 g/t; see Fig. 18.11h), which are enclosed within larger tabular zones of quartz-rich alteration with lower gold grades (0.2–1 g/t Au).

Hydrothermal alteration has also been subdivided into five stages based on the $^{40}Ar/^{39}Ar$ dating of hydrothermal alunite. Magmatism and hydrothermal activity progressed through time from southwest to northeast across the district. In a rare opportunity, gold deposition rates have been estimated as a function of the total mass of gold deposited and the magma output rate for the Yanacocha volcanic field (Longo et al. 2010; Fig. 18.36). The compositional diversity of the volcanic rocks and discrete pulses of hydrothermal activity imply that there was a complex, evolving, and recharging magmatic system at depth. Hydrothermal activity and associated gold deposition increased as eruptive output decreased and reached a peak during the end stages of magmatism when eruption volumes were at a minimum. Conversely, little deposition of gold appears to have occurred during periods of high eruption output. Over 85% of the gold ore at Yanacocha was deposited in the last 2.8 million years of hydrothermal activity (about 11.0–8.2 Ma, stages 3–5), during which the final and volumetrically smallest (about 2%) of the total volume of magma was erupted. This suggests larger volumes of dacitic magma may have remained in the crust, crystallising in shallow chambers where their volatiles degassed producing extensive hydrothermal activity. The small volumes of dacite–rhyolite that were erupted are dominated by thick lavas and domes with only minor explosive pyroclastics, consistent with the loss of most volatiles prior to magma eruption. If Yanacocha is the exploration yardstick for supergiant epithermal gold deposits, their formation may be favoured by monogenetic intermediate-silicic multivent centres, with temporally and spatially distinct magmatic–hydrothermal pulses over several million years, periods of low eruption rates accompanied by extensive hydrothermal alteration, the eruption of degassed dacitic–rhyolite domes, high phenocryst contents, and abundant subvolcanic plutons.

Furthermore, physiographic and landform evolution studies of other superlarge and giant HS epithermal deposits throughout the Central Andes endorse the view that stratovolcanoes are less common ore settings (Bissig et al. 2002, 2015; Table 18.1). In most cases, volcanic rocks coincident with mineralisation appear to be volumetrically restricted (or absent), recording the terminal stages of local arc magmatism. Nearly all crop out at high elevation, 3500 m above sea level along the crest of the Andean orogen (i.e. cordilleran geothermal setting). Most of the HS deposits are Miocene or younger in age (ca. 17–5 Ma) and were emplaced during phases of compressional deformation and uplift when magmatism was absent or subdued. This coincides with a zone of flat subduction along the South American continental arc, accompanied by shortening and crustal thickening of the continental crust (Chap. 17). Intermediate volcanics are the most common host rocks, but (unlike Yanacocha) they typically pre-date mineralisation by several million years (e.g. El Indio, Veladaro, La Coipa). However, diatremes or breccia pipes in virtually any rock type, including significantly older plutonic rocks (Pascua-Lama) or quartzites (Lagunas Norte), can host mineralisation. Mineralisation is typically located near the backscarp of pediments or the heads of valleys incising now high elevation, low-relief surfaces. Lowering of the water table and reduction of hydrostatic and lithostatic pressure at these sites of high local relief are believed to have enhanced both boiling and mixing

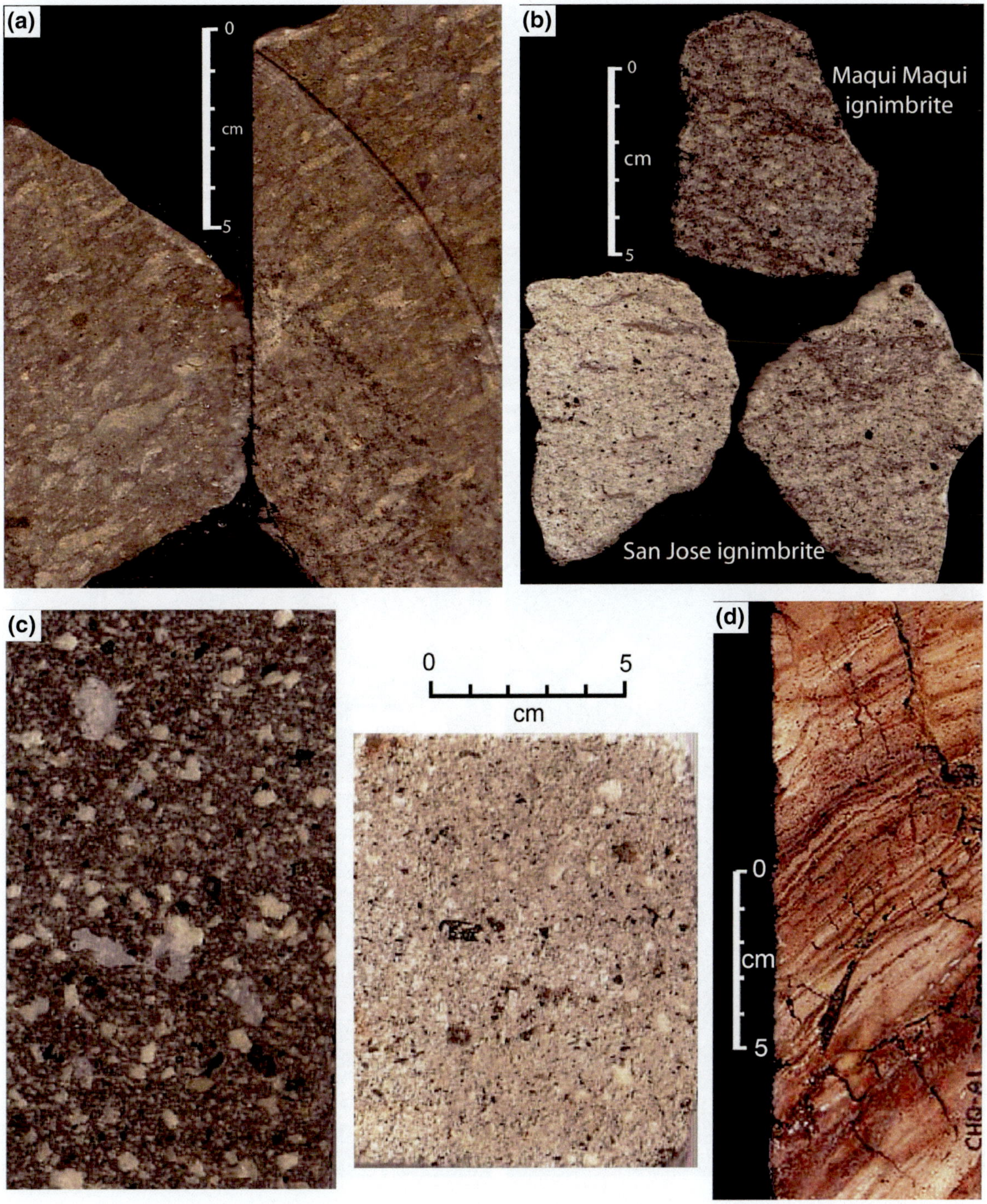

Fig. 18.35 Examples of the immediate host rock succession found within the supergiant Yanacocha high sulphidation gold district (Photos all by Tony Longo; From Longo 2005). **a** Middle San Jose ignimbrite that hosts and overlies gold ore at the San Jose deposit; **b** Comparison of eutaxitic textures in the Maqui Maqui ignimbrite which hosts gold ore at Maqui Maqui, and the San Jose ignimbrite. **c** Late or Coriwachay Dacite, representing the main gold-forming magmatic phase; (left) drill core from ~400 m depth, (right) surface sample. **d** Thinly laminated, chalcedonic silica sediment, Chaquicocha gold deposit. **e** Altered ignimbrite hosting gold ore at Maqui Maqui; the ignimbrite is fines-depleted containing ~50 modal % broken phenocrysts in the matrix, and 25–30 modal % in the pumice (ppl). **f** Non-welded hornblende ± pyroxene crystal-rich dacitic ignimbrite forming the lower or middle San Jose ignimbrite (ppl). **g** Crystal-rich middle San Jose ignimbrite (x-nicols). **h–j** Hypogene advanced argillic alteration: **h** amorphous silica–kaolinite altered lower Yanacocha Volcanics pyroxene–hornblende andesite lava dome showing perfectly preserved phenocyst shapes, Chaquicocha gold deposit (ppl); **i** irregular-shaped vugs resembling leached phenocrysts in fine-grained quartz ground mass lined with coarse, euhedral alunite laths and "stars", above the Kupfertal gold deposit (~100 m depth; x-nicols); **j** Heterolithic amorphous silica-matrix breccia with fragments of alunite and quartz "pebbles" cut by late quartz and alunite veinlets, southern margins of the Tapado gold deposit (x-nicols)

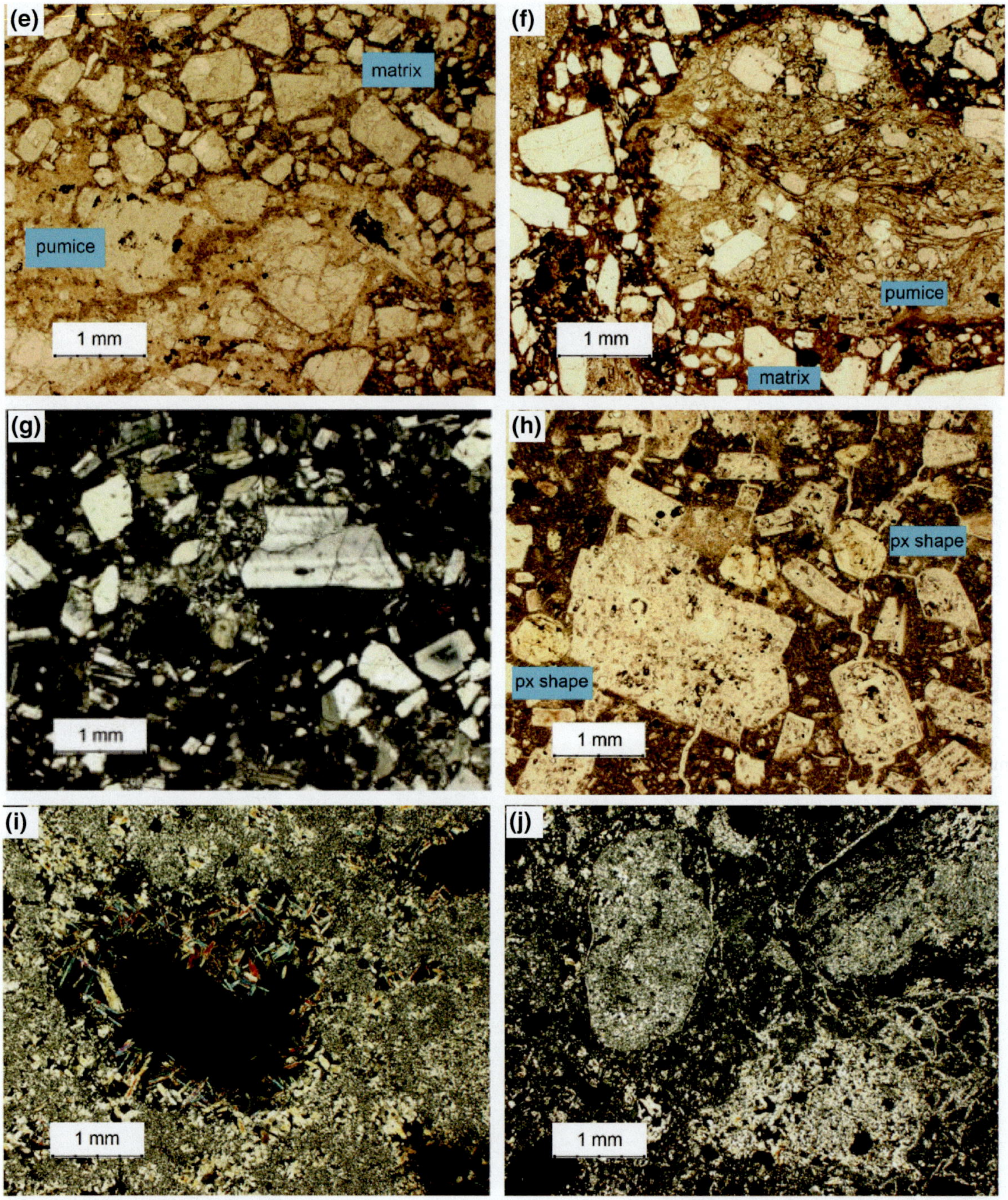

Fig. 18.35 (continued)

of magmatic with meteoric fluids, ultimately enhancing gold deposition. The host rock composition, permeability, and location of the water table control the distribution of alteration zones and ore. The near-surface, acid-leached steam-heated zone (lithocap) can attain a thickness of several hundred metres in dry climates, but is typically poorly developed and less than 20 m thick in humid climates (Bissig et al. 2015).

18.3.4.7 Alkaline Oceanic Volcanoes

Oceanic alkaline (mafic to intermediate) stratovolcanoes or shield volcanoes host several important gold deposits in the Pacific Rim. The Emperor gold–telluride deposit is positioned on the western caldera margin of the 5.2 Ma Tavua volcano on Viti Levu, Fiji (Table 18.1). This is the largest of a series of upper Miocene to Pliocene alkaline (shoshonitic) shield volcanoes along the northeast trending Viti Levu

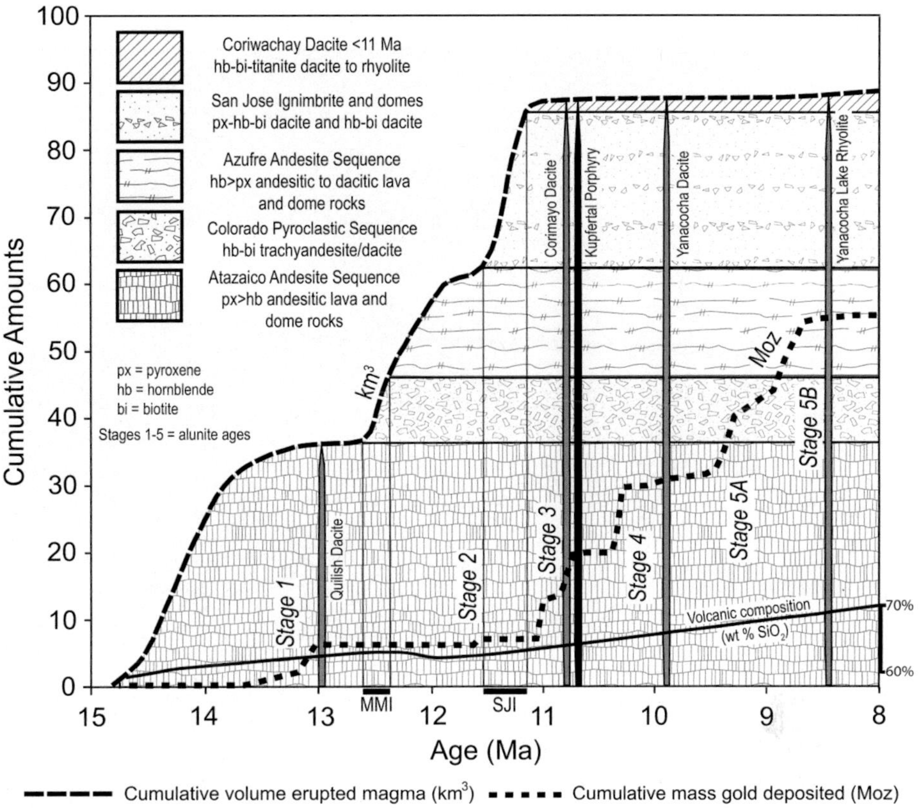

Fig. 18.36 Cumulative volume (km^3) of the Yanacocha Volcanics erupted, their average silica content (wt% SiO_2), hydrothermal alunite stages (1 through 5), and the cumulative amount of gold deposited (Moz) in the Yanacocha mining district as a function of age. SJI = San Jose ignimbrite, MMI = Maqui Maqui ignimbrite. The grey and black vertical bands represent various dacitic to rhyolitic domes and intrusions from the early Quilish Dacite to those associated with the main gold-forming phase of the Coriwachay Dacite. (From Longo et al. 2010)

lineament, which is located where this is intersected by the northwest-trending Nasivi shear zone. The caldera formed at 4.6 Ma by combined faulting and down-sagging (Setterfield et al. 1991). Host rocks comprise a succession of mafic subaerial a'a lavas and minor air-fall tuffs, but mineralisation occurs as vein fills controlled by series of steeply dipping faults and dykes, shallow dipping faults (or "flat-makes"), and complex intersection zones ("shatters"). Mineralisation is associated with higher temperature (porphyry-style) alteration zones in the centre of the caldera along the Nasivi shear zone, and lower temperature, high-level hot spring features further along the caldera margin. Isotopic and fluid inclusion data indicate that gold was introduced by a mixed magmatic, seawater and meteoric fluid at temperatures averaging 200 °C through the Nasivi shear zone and other steep faults (Kwak 1990; Eaton and Setterfield 1993). Gold deposition was caused either by fluid boiling and/or by mixing at dilated structural intersections. The flat-makes are low-angle thrusts whose movements were concurrent with epithermal mineralisation at 3.9 Ma (Begg and Gray 2002). Formation of the flat-makes is consistent with a subhorizontal σ_1–σ_2 plane and vertical extension (σ_3). This can be correlated with rotation and north–south shortening of the Fiji Platform following collision of an east-facing island arc with an oceanic plateau and subduction zone reversal.

The supergiant Ladolam epithermal deposit is second only to Yanacocha in its known gold content (Table 18.1). Positioned within a small caldera or marginal collapse scar open to the sea on the side of Luise volcano, Lihir Island, Papua New Guinea. Gold was discovered in 1982 (Moyle et al. 1990). Lihir is one of four alkaline groups of islands in the Tabar–Feni volcanic chain, northeastern Papua New Guinea (Chap. 17; see Fig. 17.37). Similar to Fiji, these appear to have formed in response to collision of an island arc with an oceanic plateau (Ontong Java) and subduction zone polarity reversal. Some workers have suggested rotation caused extension on faults radial to the bend in the New Britain Trench (Lindley 1988, 2016; Carman 2003). These structures were interpreted as the surface expression of deep-seated tensional gashes radial to the curve in the subducting Solomon Sea slab, which tapped asthenospheric partial melts in the east New Britain region. More recent models suggest a subvertical slab tear formed on the nose of the curve in Solomon slab as it was subducted beneath the New Britain Trench (Holm and Richards 2013; Holm et al. 2016; see Fig. 17.37). The tip of the tear terminates beneath southern New Ireland and is considered to provide a window where asthenosphere could penetrate from the rear of the slab to the front, allowing generation and the rise of alkaline magmas.

Gold mineralisation is predominantly contained within sulphidic breccias (averaging 2.69 g/t Au) hosted by volcanics and monzonitic intrusives (Moyle et al. 1990; Carman 2003). The volcanic rocks have been described as

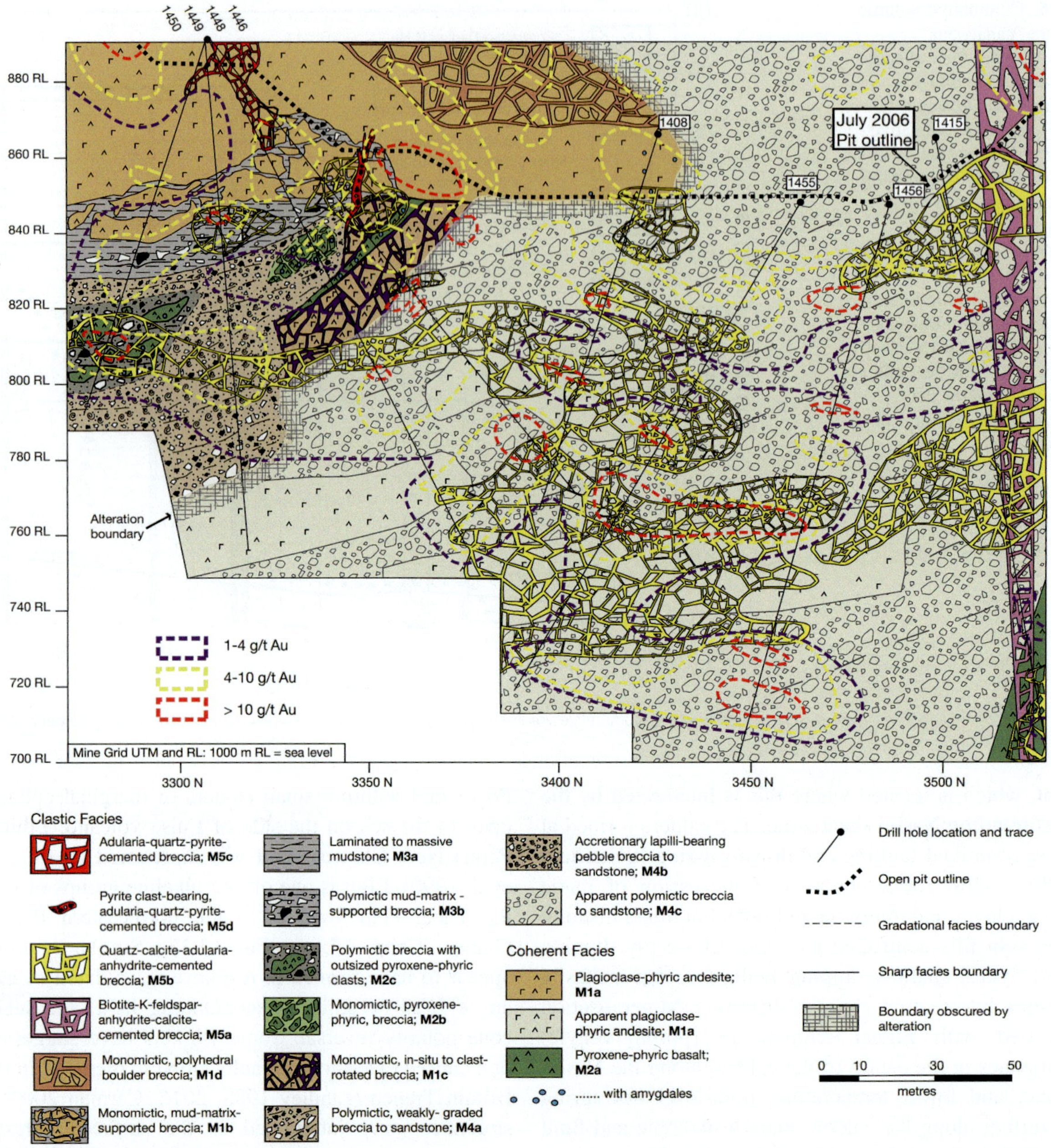

Fig. 18.37 Geological lithofacies cross-section based on detailed open pit mapping and drill hole logging of the Minifie ore zone, Ladolam gold deposit, Lihir Island, Papua New Guinea (From Blackwell et al. 2014). The various different and complex lithofacies identified in this study are illustrated in the legend and were defined using the principles described in Chap. 14. (For comprehensive descriptions, and accompanying images of breccia and other lithofacies, see Blackwell et al. 2014)

mafic-intermediate lavas, lapilli tuffs, and pyroclastic breccias together with subvolcanic breccias, but intense advanced argillic alteration has severely hampered the interpretation of volcanic facies. Epithermal alteration (adularia) is superimposed on earlier porphyry-style potassic alteration and weak mineralisation associated with the intrusion of alkaline stocks about 0.9–0.35 Ma. Caldera or more likely sector collapse occurred between 0.35 and 0.15 Ma. This caused widespread brecciation, possibly due to exposure and explosive depressurisation of the potassic alteration system, and inrushing seawater healed the breccia with anhydrite and calcite. Sealing of the breccia isolated the hydrothermal system from the sea and meteoric waters became the primary source of fluid recharge (see Fig. 18.4a). Downward dissolution of the anhydrite–calcite stockwork by meteoric waters and hydrothermal explosive activity created permeability. Gold was deposited as a result of fluid boiling and/or mixing in the open breccias. Fluctuations in the water table could have led to repeated episodes of Au deposition, which may still be occurring as neutral-chloride waters actively discharging from hot springs are probably similar to those which deposited the bulk of the orebody.

More recently, a facies approach has been taken to analysing the diverse number of host rock breccia types and further resolving the various breccia events at Ladolam (Blackwell et al. 2014). This has entailed detailed lithofacies description (following the methodology described in Chap. 14) and synthesis of lateral and vertical facies variations of the host mineralised breccias based primarily on field-based geological observation of the Minifie ore zone (Figs. 18.37, 18.38). Lithofacies were initially subdivided into over 15 facies/sub-facies types, which could be categorised into three temporally distinctive groups: (1) volcano-sedimentary stratigraphy and lithofacies (i.e. part of the alkaline island stratovolcano, Fig. 18.38a, b), (2) porphyry-style breccia dike and alteration (Fig. 18.38c), and (3) post sector collapse epithermal style brecciation, alteration, and mineralisation (Fig. 18.38d, e). The volcano-sedimentary stratigraphy reflects the transition from subaerial deposition of pyroclastic surge deposits close to vent to a subaqueous, quiet depositional environment into which a cryptodome was emplaced. The stratigraphy is predominantly composed of polymict, matrix-supported breccias, and sandstones interbedded with lavas and shallow intrusions. These are overprinted by three hydrothermally cemented breccia facies: one formed in the porphyry environment (M5a), and two formed in the epithermal environment (M5b and M5c). Porphyry-style alteration assemblages include a ≤ 5 m-wide biotite–K-feldspar–calcite–anhydrite cemented breccia. These are overprinted by shallow- and deeper-level epithermal features. Below the present mining surface (>140 m below sea level), an epithermal vein stockwork is composed of quartz–calcite–adularia–anhydrite cemented breccias and shallow dipping veins with grades of 1 to >60 g/t Au. Above this adularia–quartz–pyrite cemented breccias, veins, and associated alteration host bulk-mineable gold ore (typically >4 g/t Au).

18.3.4.8 Sediment-Hosted

In contrast to the above systems, sediment-hosted disseminated gold (or SHDG) deposits generally do not display any obvious or direct relation to volcanism, although the presence of igneous intrusions in some proximity to deposits is common. These deposits are also known as Carlin-type, since the recognition of large bulk-mineable, sediment-hosted gold orebodies on the Carlin trend in Nevada in the 1960s. They rank as the foremost style of gold producer in the United States with past production and reserves in excess of 100 Moz (or >3000 t) of gold, contained within more than 40 separate deposits (Hofstra and Cline 2000; Teal and Jackson 2002; Fig. 18.39). These form a 60 km long north–northwest alignment of predominantly carbonate-hosted gold deposits, including the super/giant Betze–Post deposit (Table 18.1). The largest operation, and richest part of the trend, is represented by the Goldstrike mining complex which incorporates the large Betze–Post open pit, and the Meikle and Rodeo underground mines; in 2016 Goldstrike produced 1.1 Moz of gold, making it the third largest gold mine in the world.

Gold is, characteristically, exceedingly fine-grained (sub-micron sized) contained within a pyrite or arsenian pyrite; the Betze–Post deposits have an average grade of 6.4 g/t Au and a recovery rate of 65–85%. Mineralisation is Paleogene in age and formed during the mid-Cenozoic extensional collapse of the Cordilleran Orogen (Chap. 17; see Fig. 17.19a); that is, prior to the onset of Basin and Range extension and faulting and Miocene volcanism (described above). While this occurred in the same region, the sedimentary host rocks are considerably older, being Paleozoic (Arehart 1996; Teal and Jackson 2002). Igneous intrusions include larger Jurassic granodiorite and diorite stocks ($\sim$158 Ma, e.g. Goldstrike stock), and Tertiary silicic–intermediate dykes both emplaced along the regional N–NW trend (Fig. 18.39a).

The most significant characteristics of SHDG deposits on the Carlin trend are, firstly, their host rocks are composed of thinly laminated calcareous and/or carbonaceous shales and siltstones. These occur in a complex overthrust terrane, with the majority of deposits occurring in the upper 400–500 m of the footwall sequence, structurally below the Roberts Mountains Thrust (Fig. 18.39a). The host sediments are cut by high-angle normal faults which served as primary fluid conduits, and ore bodies range up to tens of m from these faults. These are essentially pod-like to tabular, stratabound to stratiform zones of silicification (jasperoids) or decalcified zones (Fig. 18.39b). They show multiple episodes of silica veining and brecciation, varying degrees of carbonate dissolution and decalcification (i.e. secondary porosity), the critical factor creating high permeable host rocks and migration of ore fluids. Alteration assemblages consist of quartz + illite $\pm$ carbonate, while intrusive bodies and igneous dykes within or near ore deposits may contain significant gold (as with the Goldstrike stock; Fig. 18.39a). The age of gold mineralisation along the Carlin trend continues to be a topic of debate, but more recent $^{40}Ar/^{39}Ar$ ages from altered, gold-bearing dikes constrain what appears to be the primary period of gold mineralisation to the mid–late Eocene (40–36 Ma; Teal and Jackson 2002; Ressel and Henry 2006).

A picture of Carlin-type mineralisation has emerged showing gold is deposited when deep, upwelling metalliferous brines mix with more dilute cooler meteoric fluids (Berger and Henley 1989; Arehart 1996). There is also some evidence that the gold-bearing fluids interacted with organic

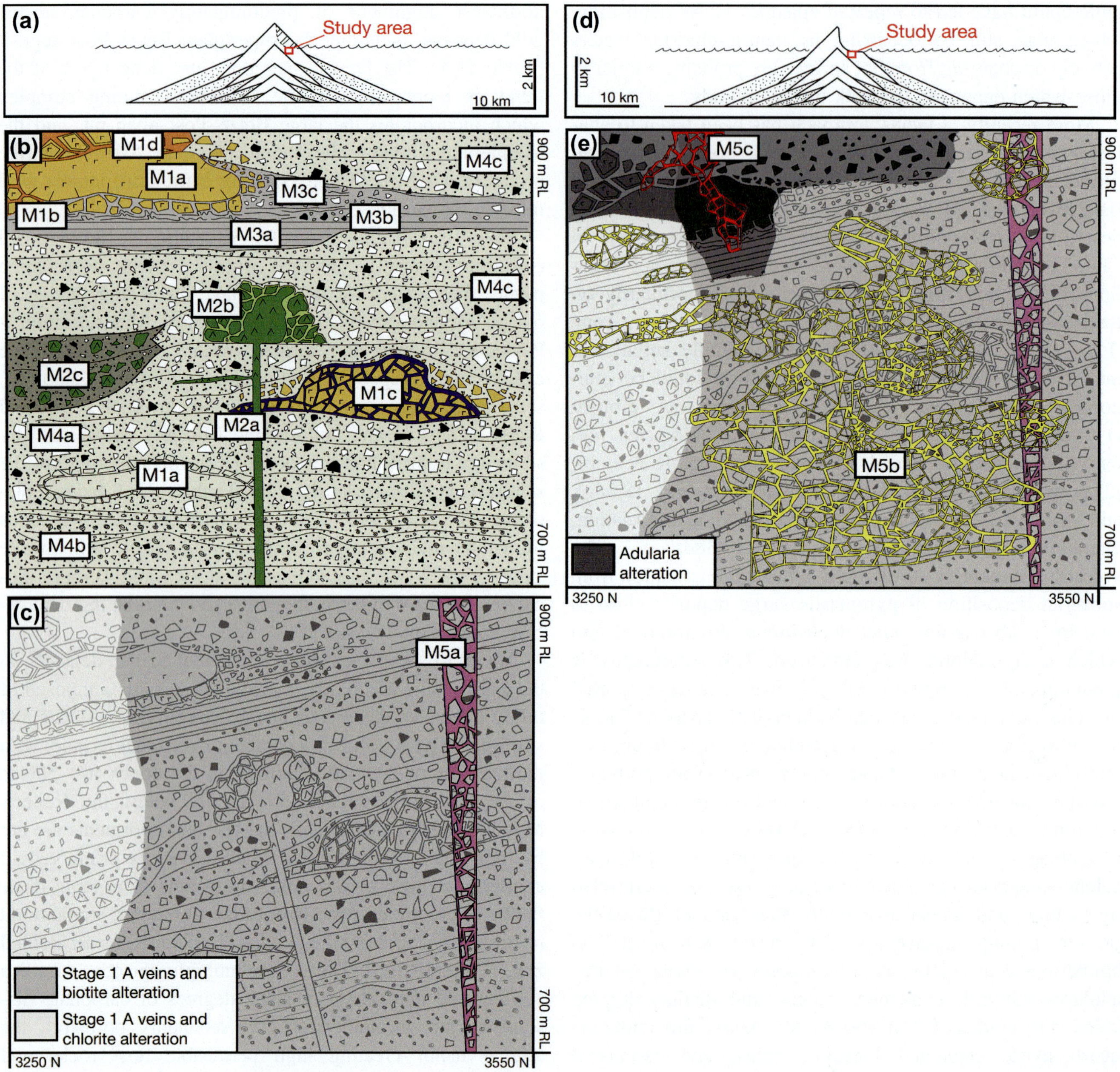

Fig. 18.38 Schematic evolution of the Minifie ore zone, Ladolam gold deposit, Lihir Island, from **a**, **b** volcano-sedimentary stratigraphy, to **c** porphyry-style and associated magmatic fragmentation, to **d** sector collapse, and **e** subsequent epithermal mineralisation and associated hydrothermal fragmentation. See legend Fig. 18.37 and text for further details of lithofacies types. (From Blackwell et al. 2014)

compounds or hydrocarbons, as ore may contain up to 6% organic carbon (Lindblom 1991; Arehart 1996; Hulen and Collister 1999). The close spatial association with young intrusions has been cited as evidence that Carlin-type deposits are the distal products of magmatic–hydrothermal systems (Sillitoe and Bonham 1990). More detailed studies now indicate that the Eocene dykes form distinctive suites and cannot have evolved from a single magma chamber but rather multiple plutons (Ressel and Henry 2006). This requires a plutonic complex about 50 km long, essentially coincident with the northern and central Carlin trend, between 12 and 23 km across, underlying an area of some 1,000 km^2, emplaced at depths between 10 and ≤ 5 km. This complex was emplaced over ~ 4 Ma that coincided with gold mineralisation along the Carlin trend. Although, significantly divergent opinions regarding the origins of the hydrothermal fluids and source of the metals prevail (e.g. Ilchik and Barton 1997; Cline et al. 2005; Kesler et al. 2005;

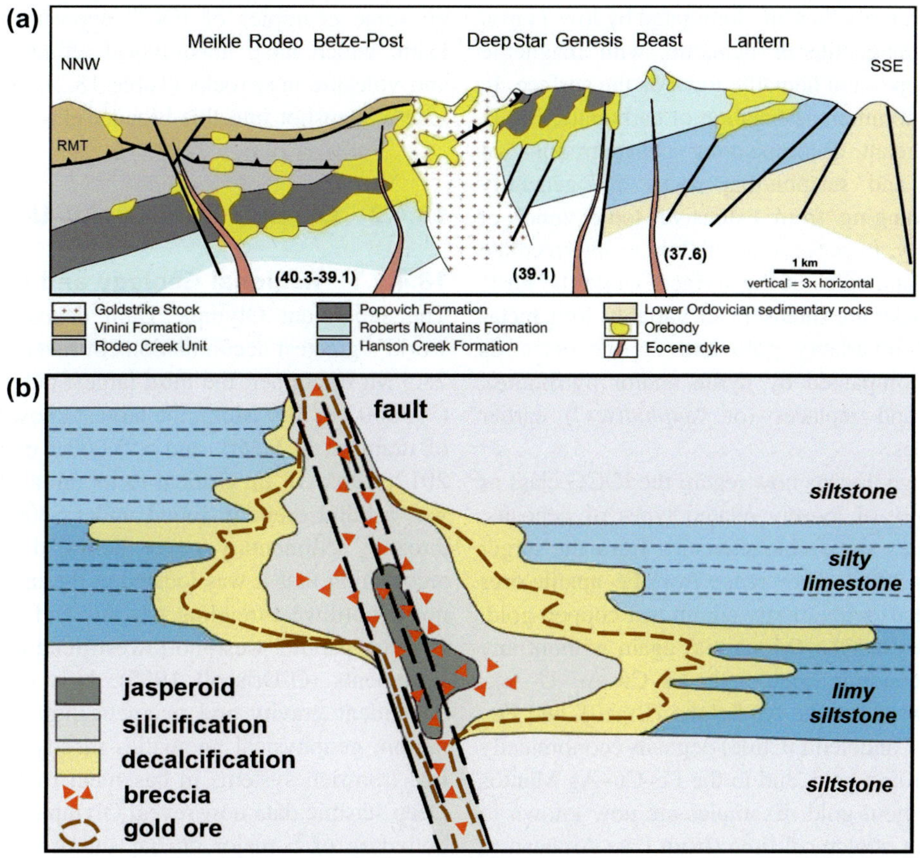

Fig. 18.39 Geological setting of sediment-hosted, disseminated Carlin-type gold deposits located along the Carlin trend, Nevada, USA. **a** Simplified long-section of the northern Carlin trend, showing the position of the Roberts Mountains Thrust (RMT) which effectively forms a major regional seal, with the Paleozoic Vinni Formation (Ordovician) and Rodeo Creek Formation (Devonian) forming cap rocks in the hanging wall of the thrust. In the footwall, the "favourable" calcareous host stratigraphy for gold mineralisation is represented by the Popovich Formation (Devonian) and Roberts Mountain Formation (Siluro-Devonian). Significant amounts of gold are also hosted by the Jurassic Goldstrike stock. The locations of Eocene rhyolitic–andesitic dykes (which may also locally be gold-bearing) are shown schematically with their radiometric ages in Ma; these decrease systematically southwards and are interpreted as the apophyses of multiple, larger plutons successively emplaced at depth over a 4 Ma period (see text). (Modified from Large et al. 2011, After Hofstra and Cline 2002 and others; radiometric ages from Ressel and Henry 2006). **b** Generalised model showing discordant structurally controlled and stratabound gold mineralisation with respect to silicified and decalcified zones in receptive limestone host rocks, as a result of deep hydrothermal fluid migration under the influence of enhanced geothermal gradient. Evidence from deposits of the northern Carlin trend suggests they probably formed at paleodepths no greater than about 1 km. (After Robert et al. 2007)

Large et al. 2011; Muntean et al. 2011), magmatic heat appears abundant in the right place and at the right time, at the very least, to drive the mineralising system.

18.4 Iron Oxide Copper–Gold Deposits

Since the discovery of the supergiant Olympic Dam deposit in South Australia in 1975, iron oxide copper–gold or *IOCG* deposits have become the world's largest producer of uranium, the third highest producer of copper and a significant producer of gold (Zhu 2016). These deposits may also contain high contents rare earth elements (REEs), silver, molybdenum, nickel, cobalt, barium, fluorine, and phosphorus. Consequently, IOCG deposits have been considered highly attractive exploration targets over the past three or more decades.

18.4.1 Diverse Geological Character

Following its discovery, Olympic Dam was initially thought to represent a unique deposit of a new style of mineralisation. However, both academic researchers and exploration geologists soon recognised there were broad similarities between the Olympic Dam and a number of other significant ore bodies around the world. Many of these were of the Proterozoic age, although, they were not confined to this geological age. Such orebodies are dominated by low Ti iron oxides present as magnetite or hematite, with magnetite more abundant at depth and hematite towards the surface. In addition, there is a common association of carbonate, Ba, P, and F with significant accompanying concentrations of REEs. The hosts and surrounding rocks are generally intensely altered, ranging from extensive sodic zones at depth and, regionally, to potassic at shallow to intermediate levels, and sericite and silica at the surface, while the hosts and surrounding rocks are intensely to variably iron metasomatised. Almost invariably gold and copper occur as sulphide (often accompanied by pyrite and/or pyrrhotite), which overprints and replaces (or "*suphidises*") earlier emplaced iron oxides.

Many economic geologists now regard the IOCG class as representing a family of loosely related types of deposits, each of which shares some characteristics from the larger "gene pool" (Porter 2000). They range from Fe–apatite ores as at Kiruna (Sweden) without any significant copper–gold, to Fe–REE–F at Bayan Obo (Mongolia) again without any copper or gold of economic value, to the Fe–Cu–Au–U–Ag–REE of Olympic Dam, Fe–Cu–Au Salobo (Brazil), and Fe–Cu–Au–Zn–Ag La Candeleria (Chile) deposits economically important in copper and gold, and to the Fe–Cu–Ag Mantos Blancos (Chile) without gold. Examples are now known to be found throughout geological time (from Late Archean to Pliocene), have a global distribution, and are located, from a geotectonic perspective, either in intra-cratonic, post-collisional extensional rift settings or along continental margins above subduction zones, although both are not mutually exclusive volcano-tectonic regimes (Chap. 17).

IOCG deposits can form at shallow (Olympic Dam) to relatively deep (Salobo) levels within the first 10 km of crust. Some of these clearly have a close spatial and genetic relation to subvolcanic and volcanic processes and rocks, whereas in others it is not so obvious or not apparent at all. The iron oxides may be an early evolutionary phase of a large and continuing magmatically driven hydrothermal ore-forming system, as possibly at Olympic Dam, or alternatively be little related to mineralisation (e.g. pre-existing ironstones emplaced in or along structures later reactivated as ore-bearing fluid conduits). There is neither consensus about the source of the metals, whether these are magmatic, or have been leached by fluid wall rock reactions with the surrounding country rocks as at Olympic Dam, or evaporites have been important in the formation of the ore fluids.

Many of these aspects are beyond the scope of this chapter. For comprehensive reviews and discussions of the complexities of this family of ore deposits, we direct the reader to Porter (2000, 2002, 2010), Williams et al. (2005), Groves et al. (2010), and Barton (2014). Here we will focus on some examples of IOCG deposits, including Olympic Dam, which have unequivocal affinities with subvolcanic and volcanic host rocks (Table 18.2), but even so will provide a snapshot into this broader class of ore deposit types.

18.4.2 Olympic Dam Cu–Au–U–Ag Deposit

18.4.2.1 Regional Geology and Orebody Setting

The supergiant Olympic Dam deposit represents one of world's greatest accumulations of metals (9.5 Gt), including 78.5 Mt of copper, the third largest published gold resource ($\sim$3000 t or 100 Moz), the largest known natural repository of uranium (2.1 Mt), and >13,000 t of silver (Ehrig et al. 2012; Table 18.2). Critical to its initial discovery, in that it was a blind orebody found under >300 m of late Neoproterozoic sedimentary cover below the Stuart Shelf, was recognition that it was located at the intersection of a major north–northwest-trending tectonic and gravity corridor, and two sub-parallel west–northwest-trending gravity and photo lineaments (O'Driscoll 1985). It is one of a number of coincident gravity and magnetic highs in the Stuart Shelf region, geophysical anomalies mostly linked to hydrothermal iron-rich systems in basement rocks (see Fig. 18.41a). Deep seismic data now reveal Olympic Dam sits close to the boundary of a major crustal suture between Archean and Paleoproterozoic rocks of the Gawler Craton, with the deposit located within the Paleoproterozoic basement (Fig. 18.40). The mid-Proterozoic tectonic evolution of the Gawler Craton was characterised by copious late-post-orogenic (anorogenic) magmatic activity forming a large silicic igneous province (Allen et al. 2008; Betts et al. 2009; Wright 2008; McPhie et al. 2011a), possibly associated with delamination and loss of the lithospheric root (Fig. 18.40; Chap. 17). Extrusion of the bimodal Gawler Range Volcanics (ca. 1592 Ma), dominated by very extensive rhyolitic lavas (Chap. 6), was accompanied by emplacement of co-magmatic A-type granites forming the Hiltaba Suite (1600–1580 Ma; Creaser 1996; Budd et al. 1998; see Fig. 15.34b).

The orebody is hosted entirely within a hydrothermal breccia system (Olympic Dam Breccia Complex) surrounded by the Roxby Downs Granite (Fig. 18.41), which forms part of a Hiltaba Suite granite batholith. This is constrained by—Pb zircon intrusive ages of 1598–1588 Ma. The breccia system is broadly zoned, comprising an irregular barren core of hematite–quartz breccias, surrounded by hematite-rich and granite-rich breccias, passing into a wider, rhomb-shaped zone in plan (some 7 × 5 km, and locally deeper than 2 km) of granite breccias and fractured granite. The hematite-rich breccias are the main host rock to Cu–Au–U–Ag mineralisation.

Table 18.2 Selected features of iron oxide copper–gold (IOCG) deposits referred to in the text

Location	Age mineralisation (Ma)	Major host rock	Structural control	Mineralisation style	Main ore assemblage[a]	Main ore stage temperature (°C)	Resource
Olympic Dam, South Australia	~1590	Granite, minor felsic, and mafic volcanics	WNW regional tectonic and gravity lineament, dextral strike-slip jog	Breccia pipe, disseminated sulphides in multi-phase hematitic breccias and veins	Co–bn–cpy–mag–hem + pitchblende	250–150	9,576 Mt @ 0.82% Cu, 0.31 g/t Au, 0.26 kg/t U_3O_8, 1.39 g/t Ag
Prominent Hill, South Australia	~1585	Argillaceous and dolomitic sediments, basaltic–dacitic volcanics, dykes, and sills	Overthrust E–W rift basin margin fault, with coincident gravity anomaly	Disseminated sulphides in hematitic breccias	Cpy–bn–co–hem	475–110	278 Mt @ 0.98% Cu, 0.75 g/t Au, 2.5 g/t Ag
Candelaria, Chile	115–114	Andesite breccias, conglomerates, lava	NW–NNW sinistral faults, NNE (Candelaria) shear zone	Veins, breccias, and massive magnetite replacement bodies	mag ± hem, cpy, py	500–300	470 Mt @ 0.95% Cu, 0.22 g/t Au, 3.1 g/t Ag
Punta del Cobra, Chile	116–114, ? 112–110	Andesite–dacite lava	NW–NNW sinistral faults	Veins, breccias, and massive magnetite replacement bodies	mag ± hem, cpy, py		>120 Mt @ 1.5% Cu, 0.4 g/t Au, 5 g/t Ag
Cerro Negro, Chile	~100	Andesitic lava, volcaniclastic sediments	NNE dextral faults	Breccia mantos, stockworks, veins	hem–mag–cpy–bn + bitumen		250 Mt @ 0.4% Cu
Manto Verde, Chile	123–117	Andesitic lava, volcaniclastics	NNW sinistral faults	Tabular breccia bodies	hem–mag–py–cpy	500–180	>600 Mt @ 0.5% Cu, 0.11 g/t Au
El Soldado, Chile	103	Rhyodacitic–andesitic lava and volcaniclastics	N-S to NNW dilational zone sinistral strike-slip fault	Disseminations, veins, porosity infills in residual petroleum reservoir	py–cpy–bn–co–hem + bitumen	300	>200 Mt @ 1.4% Cu
Mantos Blancos, Chile	150–140	Rhyolitic–andesite lavas, sills, and dykes	WNW lineament	Disseminations and fracture infills forming tabular mantos	co–bn–cpy–py–hem	355–285	220 Mt @ 1.2% Cu, 12 g/t Ag
Mina Justa, Peru	160–154	Andesite and epiclastic sediments, porphyry sill	NE listric normal fault	Irregular replacement body	mag–cpy–bn–co–py		210 Mt @ 0.9% Cu
Raúl-Condestable, Peru	115	Andesitic lava, volcaniclastics, limestone, porphyry sills, and dykes	NW, NE faults	Veins, mantos, disseminated bodies	cpy–py–prh–mag		>30 Mt @ 1.7% Cu, 0.3 g/t Au, 6 g/t Ag

[a] Mineral abbreviations: bn = bornite; co = chalcocite; cpy = chalcopyrite; hem = hematite; mag = magnetite; prh = pyrrhotite; py = pyrite. This summary is based on published sources included as text references

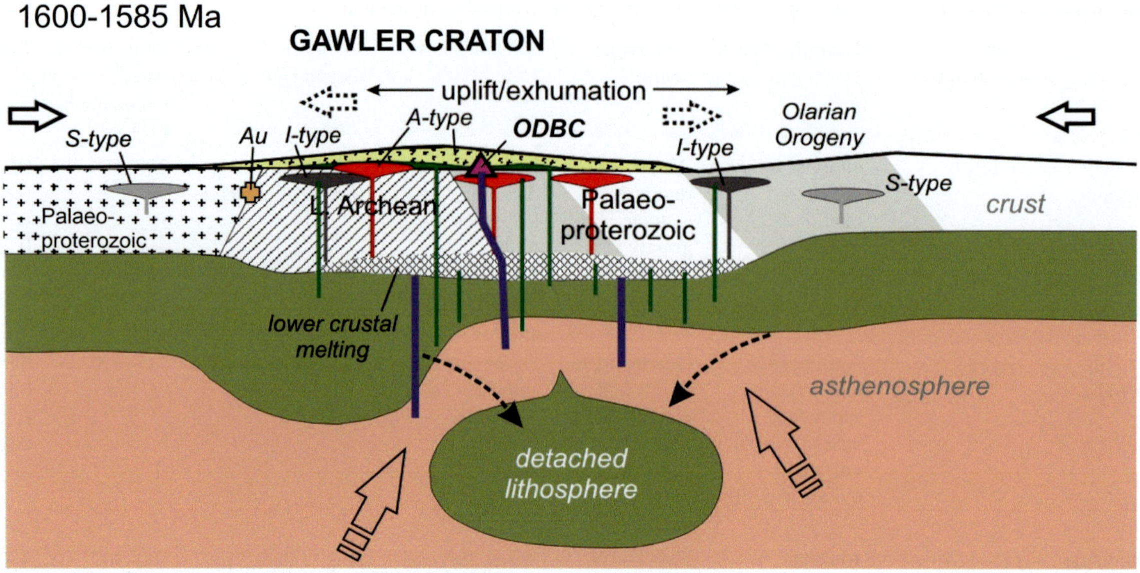

Fig. 18.40 Regional geodynamic setting of the Olympic Dam Breccia Complex (ODBC), and iron oxide copper–gold province, South Australia, based on deep seismic interpretation. Detachment of lithospheric slab or root leads to extensional orogenic collapse and copious late-post-orogenic A-type magmatism, including the eruption of the bimodal Gawler Range Volcanics (~ 1592 Ma) and emplacement of co-magmatic A-type granites forming the Hiltaba Suite (1600–1580 Ma). Other tectonic models suggest extension and magmatic flare-up resulted from slab retreat and roll-back over a mantle plume (Betts and Giles 2006; Betts et al. 2009). Purple dykes schematically represent the possible connection with mafic–ultramafic magma chambers which may have been important sources of copper ± gold (see text). (From Skirrow et al. 2009; *Source* Geoscience Australia, through Creative Commons Attribution-Share Alike 4.0 International licence)

Generalised alteration and mineralisation patterns are summarised in Fig. 18.42. In detail, these are highly variable and complex, reflecting the polycyclic nature of the hydrothermal activity. The characteristic alteration assemblage is sericite–hematite, with less abundant chlorite, silica, siderite, and magnetite. In general, the degree of alteration intensity is directly related to the degree of brecciation (Fig. 18.43). Magnetite cores within hematite grains suggest the earliest phase of iron oxide alteration was magnetite. Magnetite was subsequently overprinted by widespread hematite alteration and is now only preserved at depth or in some peripheral breccias. Copper minerals (principally chalcopyrite, bornite, and chalcocite) occur as disseminated grains, veinlets, and fragments within the breccia zones (Plate 18.1b). Mineralisation primarily occurs in the matrix of the breccias although repeated lithification and rebrecciation resulted in mineralised clasts. Gold and silver are associated with the copper sulphides. The main uranium mineral is pitchblende, generally occurring as fine-grained disseminations within hematite breccias intergrown with sulphides and hematite. Fluorite is locally abundant within mineralised breccias (up to 1–2%) as disseminations veinlets and clasts. Minor barite occurs throughout most of the hematite–granite breccias as disseminations and veinlets, but is locally abundant (2–5%) in the hematite–quartz core (Fig. 18.43d).

18.4.2.2 Facies and Interpretation

The principal components of the Olympic Dam Breccia Complex are derived from brecciation of the Roxby Downs Granite, with lesser contributions from Gawler Range Volcanics, bedded clastic facies (described below), as well as hydrothermal sources. As indicated above, brecciation and alteration intensity increase from the edges of the Olympic Dam Breccia Complex towards the deposit centre. Because of the intense, texturally destructive hydrothermal alteration, subdivision of the breccias, based on identifying primary rocks and features, has proven to be complicated. What is known is that breccia composition and textural characteristics are gradational and most form a continuum, ranging from sericite-altered Roxby Downs Granite to sericite-altered, granite-rich breccias through to hematite-rich breccias in the deposit centre. At the deposit scale, compositional changes are systematic, arranged in order of increasing hematite concentration (Fig. 18.43e). For mine logging and mapping purposes, the granite-derived breccias are subdivided according to their hematite content, and not primary facies and textural characteristics.

The breccias form an irregular arrangement of elongate, steeply dipping, locally interconnecting, variably mineralised zones (Fig. 18.41b, c). These are controlled by a complex array of high-angle brittle faults trending N-S, NW, WNW, and NE which were in place pre-and/or

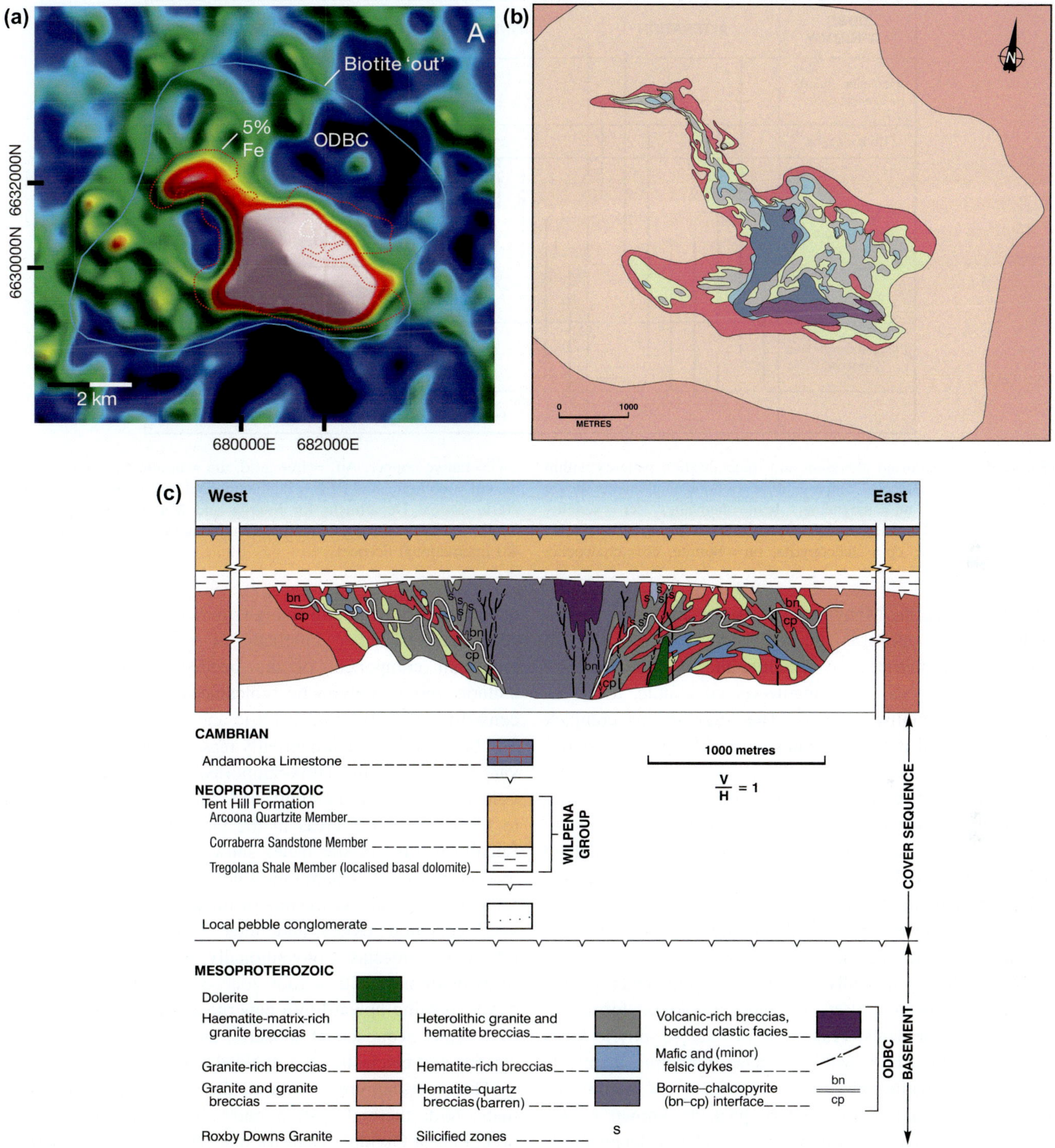

Fig. 18.41 Geological setting of the supergiant Olympic Dam IOCG deposit, South Australia. **a** Residual Falcon gravity image. The blue line (presence of igneous biotite) represents the contact between unaltered Roxby Downs Granite (outside) and the Olympic Dam Breccia Complex (inside), while the inner red line is the 5% weight % Fe contour. These relationships show that Fe metasomatism rapidly increases to the centre of the deposit. (From Ehrig et al. 2012). **b**, **c** Simplified geological plan and cross-section, respectively. These show the general distribution of major breccia types from the host granite at the margins of the complex to progressively more hematite-rich lithologies at the centre. Note the position of felsic volcanic-rich breccias and bedded clastic facies discussed in the text. (From Reynolds 2001; *Source* Department for Energy and Mining, Government of South Australia, through Creative Commons Attribution-Share Alike 4.0 International licence)

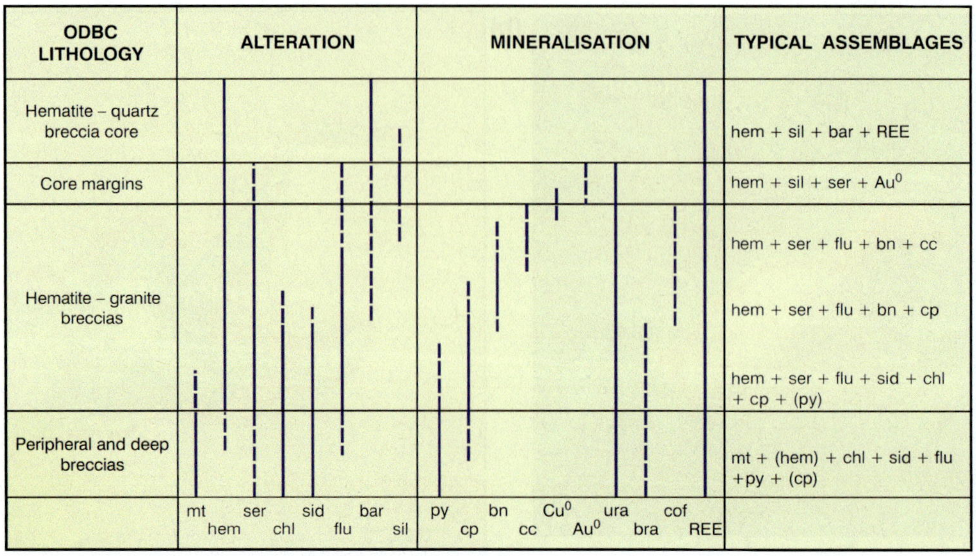

Fig. 18.42 Generalised alteration and mineralisation patterns within the Olympic Dam Breccia Complex. More common components shown in solid lines; mt = magnetite, hem = hematite, ser = sericite, chl = chlorite, sid = siderite, flu = fluorite, bar = barite, sil = silicification, py = pyrite, cp = chalcopyrite, bn = bornite, cc = chalcocite, Cu^0 = native copper, Au^0 = free gold, ura = uraninite, bra = brannerite, cof = coffinite, and REE = lanthanum and cerium. (From Reynolds 2001; *Source* Department for Energy and Mining, Government of South Australia, through Creative Commons Attribution-Share Alike 4.0 International licence)

syn-mineralisation. Many individual hematite-rich breccia bodies grade outwards through granite breccias with hematite-rich matrices, to hematite-altered granite/granite breccia and/or localised hematite-veined granite, to essentially intact sericitised granite. This suggests the complex interfingering of pipe-like bodies with antler geometries (cf. Fig. 18.24). In the less texturally overprinted granite breccias, clasts are typically angular ranging in size from microscopic (finely comminuted granite rock flour matrix) to m, or tens of m. They range from clast-supported to matrix-supported breccias. In matrix poor granite breccia evidence of hydraulic fracturing and jigsaw-fit textures are extensive. Layering and graded bedding interpreted as "gas streaming" fabrics are locally developed.

Minor, but genetically significant components of the breccias are volcanic clasts and other lithologies formed in surficial environments. Fragments of feldspar–phyric lavas have been found both within breccias (including mappable bodies of hematite-rich breccia consisting of porphyritic felsic clasts), and a sedimentary package known as the "bedded clastic facies". Breccias rich in felsic volcanic clasts and intervals of bedded clastic facies (up to 350 m thick) occur mainly near the top of the breccia complex and immediately below the Neoproterozoic unconformity (Fig. 18.41c), although they have also been located at deeper levels. While contact relations with the hematite breccias are often obscured by intense alteration, these have well-defined planar beds and overall finer grain size. They are dominated by well-bedded and laminated red hematite- or green chlorite-altered sandstones and mudstones, which have been described as containing a high pyroclastic ash and lapilli component. Thicker interbeds of conglomerate are an important component of the bedded clastic facies and include minor polymictic volcanic conglomerate-breccia consisting of well-rounded felsic and mafic volcanic clasts. The beds are thick and internally massive, poorly sorted, and vary from clast- to matrix-supported. The felsic and mafic volcanic clasts are texturally and compositionally identical to felsic and mafic lavas in the Gawler Range Volcanics. Locally bedded clastic facies show syn-sedimentary deformation including disharmonic folding near intra-formational faults (Fig. 18.44). Remnants of intensely altered mafic and syn-brecciation dykes and their fragmental equivalents intrude the breccias. Conventionally, these were believed to form diatremes, with a root zone of ultramafic and mafic dykes, passing upward into peperites/tuffisites, and overlain by a crater sequence of subaerial and/or subaqueous pyroclastics and epiclastics (i.e. bedded clastic facies).

The maar-diatreme setting of the upper part of the magmatic–hydrothermal system underpinned geological interpretations at Olympic Dam for over 20 years and influenced interpretations of fluid compositions, metal sources, and the age of the deposit. More recently, however, detailed sedimentological-based analysis has shown the bedded clastic facies extends continuously over a larger area than previously thought and cannot be limited to small maar-like craters (McPhie et al. 2011b; Ehrig et al. 2012). This and other detailed observations imply the bedded clastic facies are remnants of a sedimentary basin that was present at Olympic Dam prior to the intrusion of the breccia complex

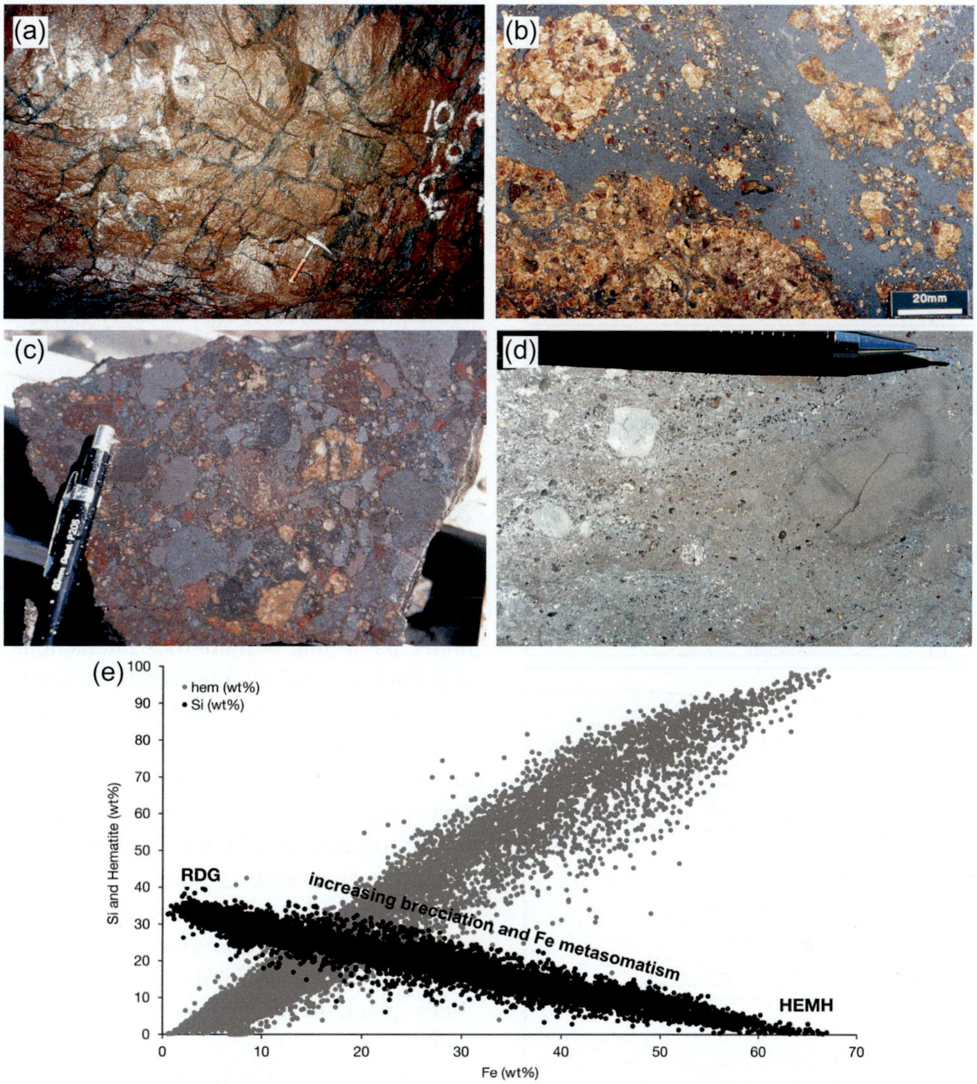

Fig. 18.43 Progressive brecciation and alteration of the Roxby Downs Granite. **a** Weakly brecciated Roxby Downs Granite; the breccia matrix consists mainly of sericite and hematite and occupies steep, dilated joints and cross-cutting fractures. **b** Typical granite-rich breccia with hematite-rich matrix; note the partial disaggregation of the larger granite clasts. **c** Heterolithic hematite breccia containing a variety of hematised granitic clast types as a result of repetitive lithification and rebrecciation. Heterolithic breccias are the most common haematite-rich breccia type and can contain a wide variety and proportion of different haematite and altered granite clasts, as well as other lithologies occurring as minor clast types, including porphyritic volcanics correlated with the Gawler Range Volcanics. **d** Hematite–quartz breccia containing hematite as clasts and matrix; minor white disseminations are barite. (Photos from Reynolds 2001; *Source* Department for Energy and Mining, Government of South Australia, through Creative Commons Attribution-Share Alike 4.0 International licence). **e** Scatter plot (10,000 analyses) showing the relationship between whole-rock iron, silicon, and hematite concentration from (left to right) unaltered Roxby Downs Granite (RDG) through to intensely Fe-metasomatised, heterolithic hematite-rich breccias (HEMH) (From Ehrig et al. 2012)

(Fig. 18.44). The texture, contact relationships, distribution, and non-stratified character of breccias are consistent with subsurface fragmentation involving a combination of tectonic and hydrothermal processes. This implies the magmatic–hydrothermal-driven system operated beneath and partly within an active sedimentary basin was not confined to maar craters and did not vent. Clast shapes previously interpreted as peperite suggest that both brittle failure and dissolution (by aggressive fluorine-rich fluids) of mainly felsic lava from Gawler Range Volcanics flooring the sedimentary basin were important (Fig. 18.44; McPhie et al. 2011a).

18.4.2.3 Ore Formation

Notwithstanding the above, this was clearly a dynamic, high-level system displaying evidence of both magmatically

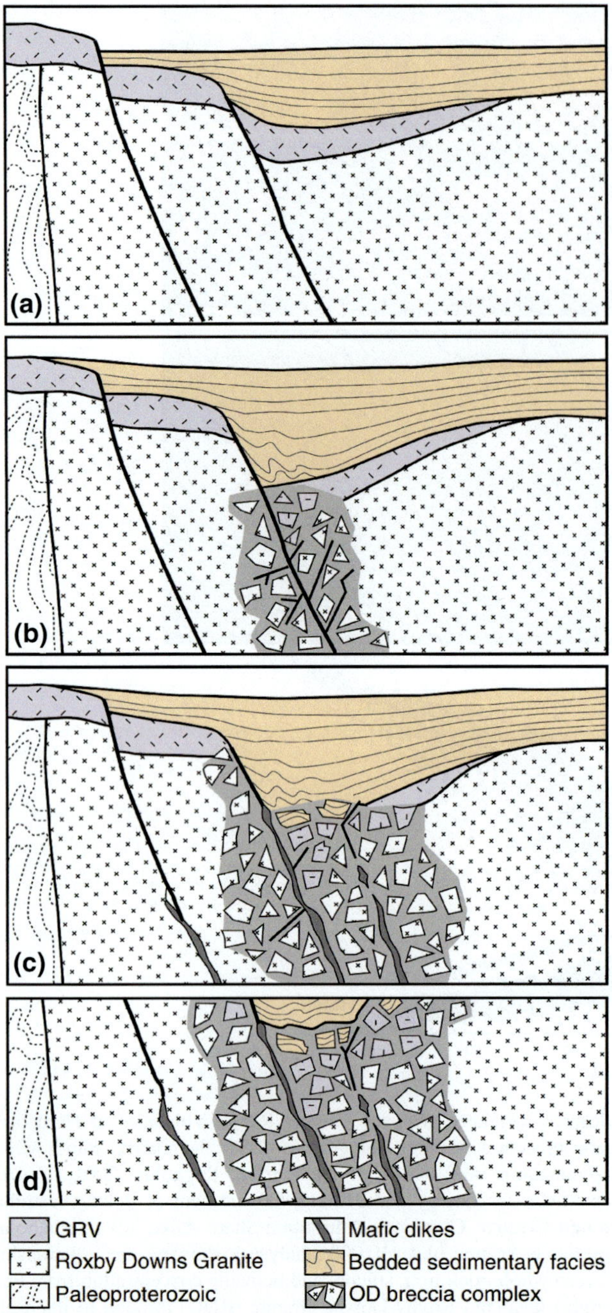

Fig. 18.44 Schematic geological evolution Olympic Dam Breccia Complex (From McPhie et al. 2011b). **a** Bedded clastic facies accumulated in fault-bounded basin above Mesoproterozoic Gawler Range Volcanics felsic lavas and Roxby Downs Granite. **b** Tectonic–hydrothermal-driven subsurface fragmentation propagates upwards from a deeper site in the Roxby Downs Granite. This forms the ODBC dominated by granite-derived breccias at depth, but rich in fragments of porphyritic lava and syn-sedimentary deformed, bedded clastic facies at shallower levels. **c** Mafic dikes intruded granite and breccia. **d** Erosion unroofed breccia complex, removing almost all traces of the presence of sedimentary basin facies. *Note* Rocks previously described as peperites (i.e. hydrovolcanic fragmentation) occur in sections dominated by felsic-volcanic breccias. While they may show angular or ragged, amoeboid, and digitate clast shapes, these are texturally identical to those found in monomict granite breccia, suggesting that both brittle failure and dissolution of pre-existing wall rock were responsible (McPhie et al. 2011a, b)

and seismically driven hydrothermal activity. Faulting, brecciation, alteration, dyke injection, and metal precipitation were interdependent and multistage. Much of the brecciation occurred in a relatively near-surface environment. The complex pattern of structures and related veins recognised within the deposit shows evidence of a long history of episodic fault activity. Earlier mine-based studies interpreted these as a dextral dilational jog or pull-apart suggesting the initial distribution of the breccia complex and subsequent evolution was strongly influenced by strike-slip activity on a regional west–northwest-trending fault zone. Regional analysis shows northwest-trending lineaments and structures, such as those in the vicinity of Olympic Dam, were likely to have accommodated dilation associated with strike-slip motions at the time of mineralisation (Hand et al. 2007).

Over the past 40 years, a number of models for ore formation have been published. Key differences relate to the role of Hiltaba granite versus other sources of metals and fluids, the timing of fluid–rock interaction, and which of the two fluids carried the ore metals. Several data sets indicate ore deposition in the range of 150–350 °C. Reeve (1990) proposed Cu–Au–U deposition resulted from the interaction of ascending hot, relatively reduced, iron and metal-rich brines with cooler, oxidised, meteoric waters, producing a broad zonation from lower chalcopyrite ± pyrite to upper bornite ± chalcocite. Other studies have proposed magnetite, formed by early high-temperature (magmatic) fluids, produced a reduced ironstone precursor, later overprinted by hematite during the influx of low-temperature, oxidised fluids precipitating the ore species (Oreskes and Einaudi 1992).

Haynes et al. (1995) also advocated fluid mixing as the depositional mechanism for both Fe-oxides and ore but favoured the transport of Cu–Au–U in cooler descending meteoric waters (modelled at 150 °C). Gawler Range Volcanics basalts were preferred as the copper source, leached by saline waters derived from the overlying maar or playa lake. A significant problem with this model, however, is that it demands a relatively thick local basalt source rock succession (presumed now eroded). Alternatively, if we consider the presence of an overlying sedimentary basin (Fig. 18.44), this introduces the potential for significant volumes of oxidised, saline fluids circulating and descending through fault networks into the breccia complex. Leaching of uranium from felsic lavas and granitic breccias rich in fluorine (McPhie et al. 2011a, b), and/or of detrital grains in the overlying sedimentary cover (Cherry et al. 2017), could therefore have been the principal source(s) of uranium at Olympic Dam. Copper and gold species were likely introduced by ascending, hotter, reduced, and magmatic-related brines, as more generally proposed. Nd isotopic data from the ore indicates a strong mantle influence, suggesting exsolution of particularly copper from mafic–ultramafic chambers may have been important (Johnson and McCulloch 1995; Skirrow et al. 2007; Fig. 18.40).

18.4.3 Prominent Hill Cu–Au Deposit

As well as at Olympic Dam, widespread IOCG alteration and mineralisation occurs throughout the Gawler Craton, of which several deposits are described as volcanic-hosted. These are generally considered to be related to a common Hiltaba Suite–Gawler Range Volcanics anorogenic magmatic event that affected the central and eastern part of the Gawler Craton (e.g. Skirrow et al. 2002, 2007). One of the most economically significant of these is Prominent Hill discovered in 2001 (2.7 Mt Cu, 7 Moz Au; Table 18.2). The main host rock types are weakly metamorphosed argillaceous sediments and dolomitic limestone, but include basaltic, andesitic, and dacitic volcanic rocks, subvolcanic dykes and sills, fragmental volcanic rocks (variously termed agglomerates, lapilli tuffs, and peperites), and finer-grained volcaniclastic rocks or greywacke (Belperio et al. 2007; Schlegel and Heinrich 2015). Superimposed on these rocks is an extensive progressive hydrothermal iron-oxide alteration assemblage, characterised by very large-scale replacement (several km^3) by iron oxides, sericite, and silica, as well as physical fragmentation of host rocks by hydrothermal fluids creating a variety of breccia units (Fig. 18.45).

As observed at Olympic Dam, brecciation includes a spectrum of weakly brecciated host rock through to clast-supported breccias with jigsaw-fits, to matrix-supported heterolithic breccias. Heterolithic (sedimentary ± volcanic clasts), hematitic breccias form subvertical, tabular, sheet-like bodies which can be confidently correlated between drill holes and are the main hosts to economic grades of copper and gold. Mineralised breccias and host volcaniclastic and sedimentary rocks at Prominent Hill are strongly influenced by major regional crustal structures. Smaller, locally active faults and shear zones control the shape of the distribution of lithological units, breccias, and mineralisation, indicating that they were active as syn-sedimentary basin-controlling features, as well as reactivated during mineralisation. Iron oxide sericite–silica alteration is pervasive within and marginal to the main breccias and is surrounded by a wider zone of less intense alteration. Iron, copper, and gold are strongly correlated with barium, fluorine, uranium, and REEs.

Mineralisation has been interpreted as synchronous with volcanism and sedimentation within a narrow east–west-trending graben developed at ~1585 Ma (Belperio et al. 2007). More recent investigation has shown the hematite breccias and deposition of copper developed after tilting of the host sequence into its present steep orientation, as indicated by geopetal structures within the breccia matrix (Schlegel and Heinrich 2015). Thus, ore mineralisation appears to have occurred late in the deformational history of the region, after extrusion of the Gawler Range Volcanics and later over-thrusting (Fig. 18.46). Hydrothermal alteration and mineralisation at Prominent Hill involved the pervasive and texturally destructive replacement of formerly calcareous, dolomitic, and siliciclastic breccia components. Detailed mineralogical studies indicate similar ore depositional controls to Olympic Dam despite the lithologically different host rocks.

18.4.4 Central Andes Coastal IOCG Province

The youngest, economically significant IOCG province worldwide occurs as a series of semi-continuous belts extending some 2,000 km along the coastal interiors of Peru and northern Chile (Sillitoe 2003; Fig. 18.47). These broadly parallel the Cenozoic Andean porphyry copper belt and taken together form the earth's foremost region endowed in copper. This is also an important iron ore province, where many large, apatite-bearing magnetite ± hematite deposits are associated with extensively alkali-metasomatised rocks. These iron oxide-rich ore systems lie west of and are present at generally lower elevations than the many large Andean porphyry copper deposits. However, the coastal IOCG province also contains porphyry-style deposits, which are commonly gold-rich compared to most of the younger porphyry deposits of the Andes. The province is further characterised by numerous Cu and Cu–Ag deposits in breccia bodies and mantos (Maksaev and Zentilli 2002), which are stratabound or broadly concordant replacement bodies (e.g. Mantos Blancos and El Soldado; Fig. 18.47; Table 18.2). Some of these may be genetically as well as spatially related to the IOCG deposits, while in terms of historic copper production, they are second only in importance to the Tertiary porphyry systems (Sillitoe 2003). IOCG mineralisation styles include veins, hydrothermal breccias, replacement mantos, and calcic skarns, with the larger deposits generally exhibiting several of these in combination (Sillitoe 2003).

The IOCG and apatite-bearing iron ore systems are diachronous, in that, both appear to have accompanied the eastward migration of arc magmatism in the middle–late Jurassic through the early Cretaceous (Fig. 18.47). The arc magmas were predominantly calc-alkaline and emplaced mainly during extension and transtension of the western South American continental margin, possibly induced by subduction zone roll-back and orogenic collapse (Sillitoe 2003; Chap. 17). This produced a series of ensialic marginal or back-arc basins, which broadly evolved over time from

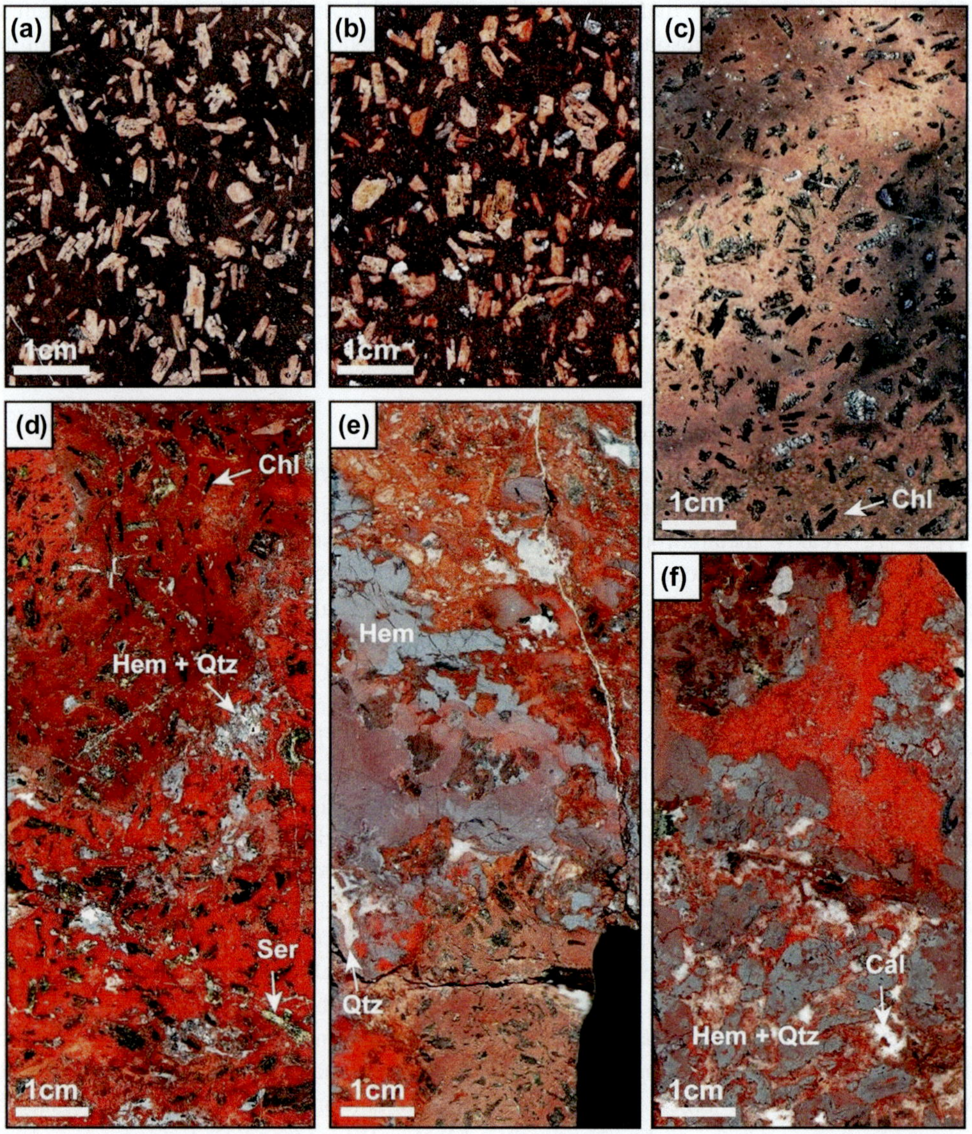

Fig. 18.45 Regional iron oxide (or "*red rock*") alteration very typical of volcanic host rocks in the Gawler Craton IOCG Province, South Australia. Illustrated from the Triton prospect, Prominent Hill Cu–Au deposit. Photographs of rock slabs are taken from a single drill hole over a 10 m interval and show the progressive alteration sequence of plagioclase–phyric andesite from hematite–chlorite–sericite–fluorite–barite–kaolinite to hematite–quartz–fluorapatite–barite assemblages. (From Schlegel and Heinrich 2015). **a** Least altered andesite precursor, 203.7 m. **b** Hematite–chlorite to sericite–altered plagioclase phenocrysts, 204.8 m. **c** Plagioclase crystals are nearly completely altered to chlorite and sericite, 209.2 m. **d** Chlorite–sericite–kaolinite-bearing example of hematite–quartz alteration replacing plagioclase–phyric volcanic rock, 214.2 m. **e** Pseudo-brecciated interval showing hematite- and quartz-replaced rock next to chlorite- and sericite-altered patches, 212.4 m. **f** Nearly complete replacement of andesite by hematite–quartz alteration, 213.2 m. Mineral abbreviations: Cal = calcite, Chl = chlorite, Hem = hematite, Qtz = quartz, and Ser = sericite

deeper marine to alternating shallow marine and continental depositional environments. These generated alternating cycles of deeper marine turbidites, shallow marine carbonates and evaporites, and continental sediments intercalated with considerable volumes of basaltic and felsic volcanic rocks. As the basins evolved, these produced bimodal submarine volcanic successions of predominantly basaltic to dacitic composition intercalated with shales, sandstones, and limestones deposited within the marginal basin, and thicker (up to several km) continental successions dominated by voluminous dacitic ignimbrites, basaltic flood basalts, and andesitic lavas intercalated with terrestrial sedimentary facies. Plutonism was fairly continuous and overlapped with volcanism, although this appears to have reached a peak in mid-Cretaceous (ca. 120–100 Ma), producing a series of granitoid plutons which form part of the Coastal Batholith.

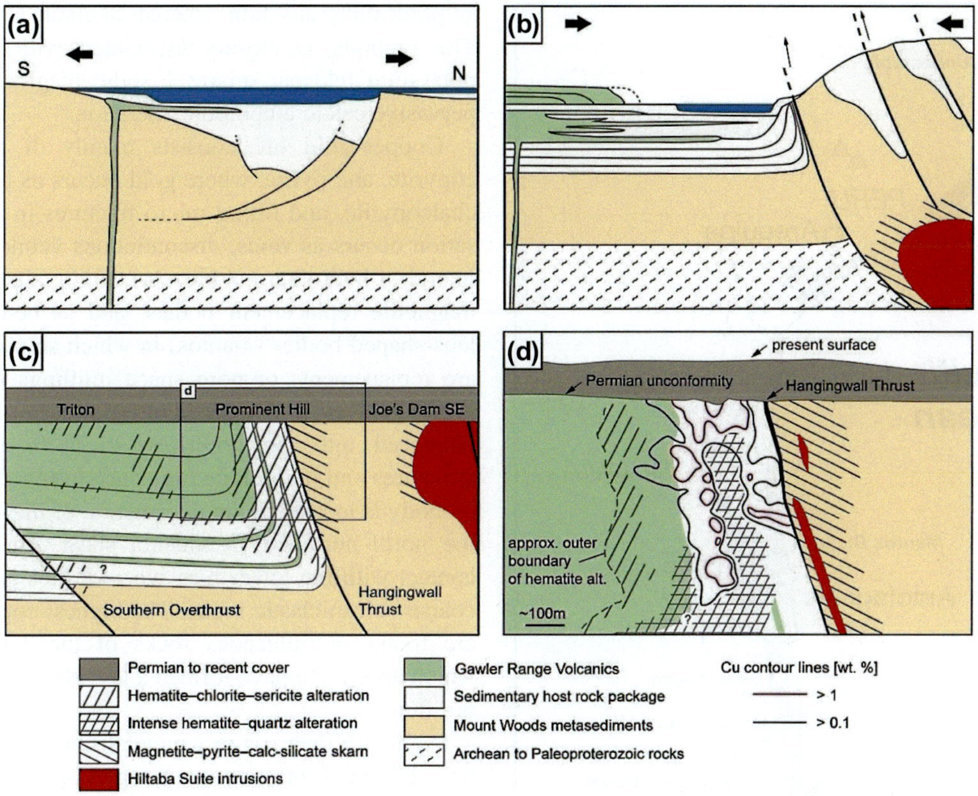

Fig. 18.46 Schematic evolution of the Prominent Hill Cu–Au deposit and peripheral prospects. **a** Sedimentation of host rock succession and eruption of Gawler Range Volcanics lavas. **b** Continued extrusion of mafic-intermediate volcanics, intrusion of co-magmatic Hiltaba Suite granites, and regional compression, tilting, and brecciation of host rocks. **c** Area of Prominent Hill following compression, intrusion of subvolcanic dikes, alteration, and mineralisation, followed by exhumation, erosion, and deposition of post-Permian cover sequence. **d** Simplified cross-section through the deposit area, showing upward-convex alteration profile and Cu precipitation front defining current economic orebodies. (From Schlegel and Heinrich 2015)

18.4.4.1 La Candelaria Cu–Au–Zn–Ag Deposit

La Candelaria deposit (450 Mt Cu, 3.5 Moz Au) was discovered in 1987 and is the largest operating iron oxide copper–gold mine in the province, located in the Punta del Cobra IOCG district, Chile (Fig. 18.47; Table 18.2). This represents an early Cretaceous continental volcanic arc/marine back-arc basin terrain. The volcanic arc and marine carbonate back-arc sequences are intruded by early Cretaceous granitoid plutons. IOCG mineralisation occurs along the eastern margin of the Copiapó Batholith either within the contact metamorphic aureole as at Candelaria, or just outside, and is hosted primarily by volcanic and volcaniclastic rocks (Marschik and Fontboté 2001; Williams et al. 2005; Fig. 18.48a). These form a lower succession of mainly massive andesite rocks with more locally distributed altered dacitic domes and lava flows (Geraldo-Negro member), which effectively form the footwall of the mine sequence (Fig. 18.48b). These are overlain by a thick succession of volcaniclastic breccias, conglomerates, and so-called tuffaceous rocks with lenses of massive andesitic volcanic rock (Algarrobos member), which form the main zone of ore mineralisation. In their upper part, these contain ammonites suggesting deposition at least in part within significant water depths. These pass vertically and laterally into alternating carbonate and volcaniclastic beds, which are generally unmineralised.

Three horizons serve as local markers within the main ore zone. These are the "basal breccia" (up to 25 m), described as a red polymictic volcaniclastic breccia which in places is conglomeratic and contains sandstone lenses, and of sedimentary origin. The Trinidad Siltstone (up to 60 m) is mainly siltstones, sandstones, and cherts which locally contain large clasts of brecciated limestone, and fine-grained tuffaceous beds, which again appear of sedimentary origin. These contain high copper–gold grades, and the top of the Trinidad Siltstone represents the upper limit of economic mineralisation. This is overlain by the "upper lavas" (up to 45 m), which form a discontinuous horizon defined by lenses of altered basaltic andesite lava flows which grade laterally and vertically into volcanic breccias, tuffs and reworked tuffs, siltstones, cherts, and carbonate rocks.

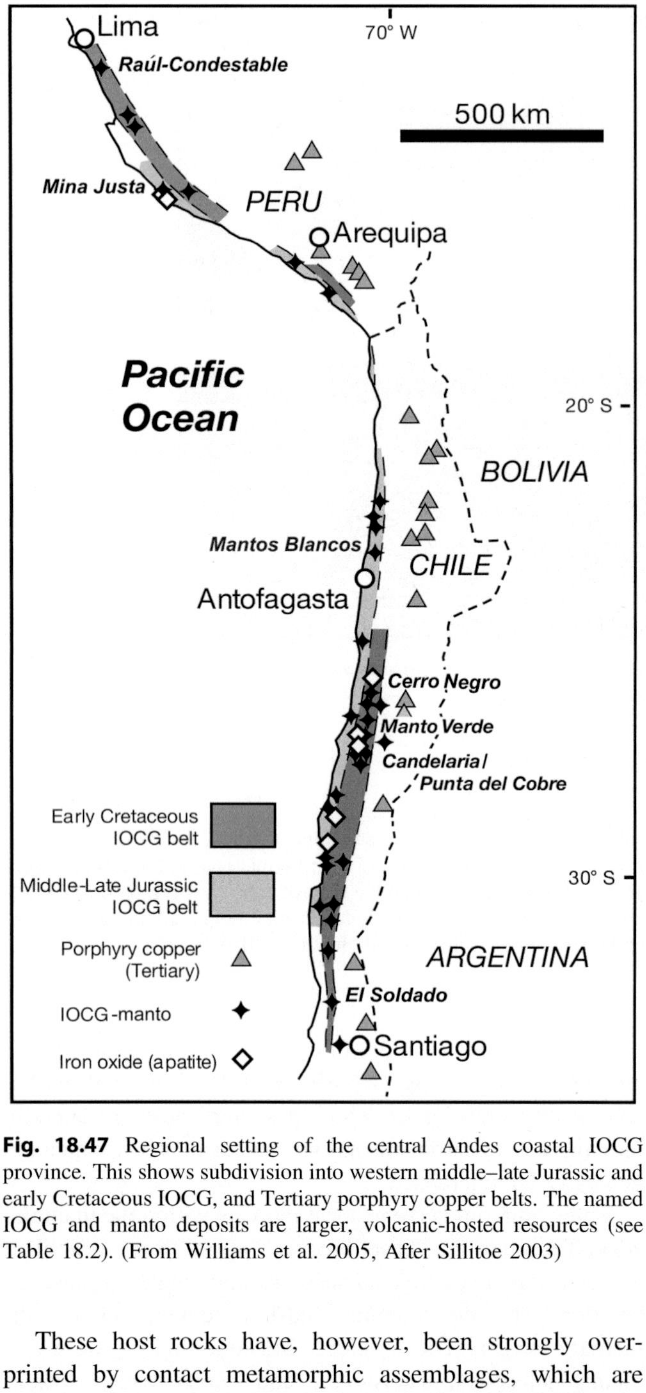

Fig. 18.47 Regional setting of the central Andes coastal IOCG province. This shows subdivision into western middle–late Jurassic and early Cretaceous IOCG, and Tertiary porphyry copper belts. The named IOCG and manto deposits are larger, volcanic-hosted resources (see Table 18.2). (From Williams et al. 2005, After Sillitoe 2003)

These host rocks have, however, been strongly overprinted by contact metamorphic assemblages, which are zoned west to east outboard of the Copiapó batholith, and have been superimposed by pervasive metasomatic hydrothermal alteration, largely obliterating original volcanic textures (Fig. 18.48c). The earliest phase of hydrothermal alteration caused a widespread albitic alteration, commonly overprinted by a pervasive phase of potassic alteration. Hydrothermal alteration spatially associated with copper–gold mineralisation caused intense iron metasomatism in the volcanic and volcaniclastic host rock succession and mainly pre-dates ore formation. Magnetite is ubiquitous and occurs as massive magnetite bodies with or without superimposed sulphide mineralisation, whereas hematite is generally scarce. The orebody is closely associated with intense biotite–potassium feldspar–quartz ± sodic plagioclase and locally pervasive calcic amphibole alteration.

Copper–gold ore consists mainly of magnetite, chalcopyrite, and pyrite, where gold occurs as tiny inclusions in chalcopyrite, and filling micro-fractures in pyrite. Mineralisation occurs as veins, discontinuous veinlets and stringers, matrix of hydrothermal breccias cutting altered host rocks or magnetite replacement bodies, and as bedding-concordant lens-shaped bodies (mantos) in which sulphide ore minerals are replacements or pore space infillings mimicking sedimentary layering. Larger veins (up to 1–1.5 m wide) were emplaced into the northwest to north–northwest brittle structures and high-grade ore zones follow these trends. The orebody is located at the intersection of these structures with the north–northeast Candelaria shear zones at the upper contact with the lower massive andesites and the overlying coarse volcaniclastic rocks. The highest copper–gold grades are found in tuffaceous rocks of the Trinidad Siltstone, which appears to have formed a less permeable cap or seal to ore fluid.

Although data suggest mineralisation was pulsed, it can be subdivided into two main superimposed stages: an early iron oxide, and a later copper sulphide stage. The main iron oxide (magnetite) phase of mineralisation took place at temperatures of about 600–500 °C. While the main copper sulphide mineralisation at temperatures between 500 and 300 °C, with subsequent cooling of the hydrothermal to 180 °C indicated by homogenisation late-stage calcite. Isotopic data indicate a mainly magmatic sulphur source in the ore-forming fluids. Age dating indicates that the hydrothermal activity spanned a period of several million years (116–110 Ma), broadly coeval with the emplacement of the Copiapó batholith and regional uplift, while stratigraphic arguments imply that burial during mineralisation did not exceed 2–3 km. Taken together, the data suggest metal-bearing magmatic fluids were channelled through faults into a succession of more permeable volcaniclastic sediments. There, cooling and possible mixing with external fluids caused ore minerals to precipitate. The nature of these non-magmatic fluids remains yet to be determined.

18.4.4.2 El Soldado Cu–Ag Deposit

El Soldado deposit (2.7 Mt Cu) is the largest known volcanic-hosted manto-type deposit in northern Chile (Fig. 18.47; Table 18.2). The orebody is located within lower Cretaceous basalt to rhyodacite host rocks, which cap a succession of marine carbonaceous shales and siltstones (Fig. 18.49). What is particularly interesting about this orebody is not only its relative size and grade (in terms of total metal content, it is equivalent to a medium-size porphyry copper ore deposit), but also the close association of solid

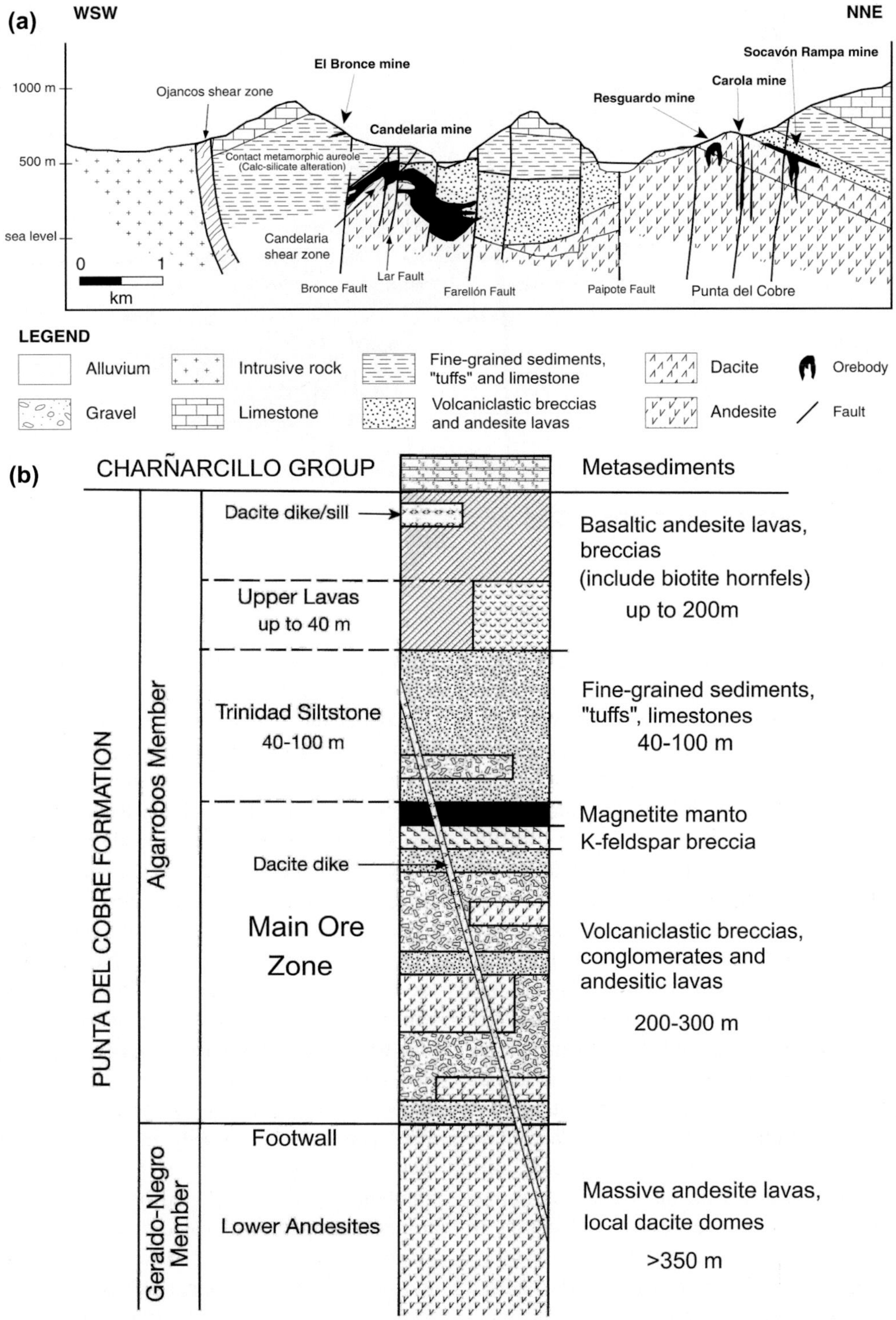

Fig. 18.48 Geological setting of the Candelaria IOCG deposit, Chile. **a** Geological cross-section through Candelaria mine and wider Punta del Cobra mining district (From Williams et al. 2005, after Marschik and Fontboté 2001). **b** Lithostratigraphic column for the Punta del Cobra Formation at the Candelaria mine showing the position of the main ore zone (From Marschik and Fontboté 2001). **c** Cross-section through Candelaria orebody showing the distribution of iron oxide minerals and alteration types (From Williams et al. 2005, after Marschik and Fontboté 2001). Within the ore zone, any primary volcanic textures are destroyed

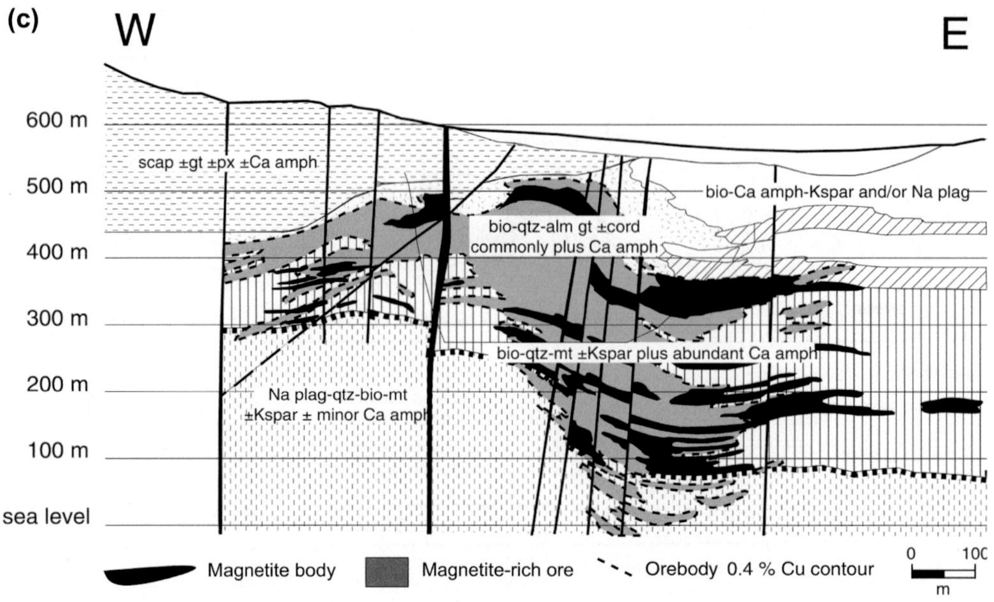

Fig. 18.48 (continued)

bitumen (residual petroleum) with sulphide mineralisation (Wilson and Zentilli 1999; Boric et al. 2002). Solidified petroleum is known to be intimately associated with other manto-style (both sediment- and volcanic-hosted) copper ore deposits in northern Chile (Maksaev and Zentilli 2002), but El Soldado is considerably larger in tonnage than any of these other deposits.

The volcanic host rocks are described as a 500 m thick sequence of interlayered basaltic and rhyodacitic lava flows and domes, closely associated volcanic breccias, and pyroclastic and sedimentary volcaniclastic rocks with marine limestone lenses (Upper Prado Formation; Fig. 18.49). These overlie a thick (1500 m) sequence of organic-rich calcareous shales and siltstones, and thin volcaniclastic sandstones (Lower Prado Formation), which contain marine ammonites and bivalves, as well as plant debris suggesting proximity to coastal environments. While pillow lavas have not been reported from the mine sequence, they have been described locally and, together with intercalated limestone lenses, suggest that volcanic host succession was marine or paralic. The mine sequence is then conformably overlain by a very thick succession (5–7 km thick) of red oxidised subaerial basaltic–andesite lava flows and breccias, intercalated tuffs, sandstones, and lacustrine siltstones with abundant plant remains (Veta Negra Formation). In addition, there are subvertical dykes that form both feeders to rhyodacitic flows in the host rocks, and the overlying basaltic–andesite flows. Most significant as a copper host are the rhyodacite flows which are thicker above the feeder dykes, forming domes with spectacular columnar jointing and locally autobrecciated tops. Away from the feeder dykes, the rhyodacites grade into coarse to fine volcaniclastic rocks.

Copper mineralisation occurs in both basaltic and rhyodacite units as disseminations, veins and infilling amygdales, fractures, and flow top breccias. Mineralisation is strongly structurally controlled and consists of a number of clustered copper orebodies which are mostly vein-like and discordant. Individual orebodies are zoned from an external and deeper zone of pyrite–chalcopyrite, followed inwards by a zone of chalcopyrite–bornite, and an internal and uppermost bornite–chalcocite–hematite zone. Intimately associated with these ore bodies are abundant bitumen and framboidal pyrite which occur as pore filling cements, veinlets, and fracture fills which are replaced by copper sulphides. The bitumen represents biodegraded oil most likely generated from the underlying carbonaceous shales, which migrated into primary and fracture porosity in the overlying volcanic host rock succession (about 130–125 Ma). Low-temperature framboidal pyrite developed by bacterial sulphur reduction around and within the paleo-petroleum reservoir(s). A later hydrothermal stage (ca. 103 Ma) introduced higher temperature (ca. 300 °C) copper-rich fluids into the degraded reservoir, whereby pre-existing pyrite and bitumen were replaced to form zoned chalcopyrite, bornite, and chalcocite orebodies.

From the perspective of ore models and their classification, there has been much debate among researchers as to whether or not the El Soldado copper deposit should be included within the IOCG family. Locally, hematite is significant and associated with the copper mineralising phase,

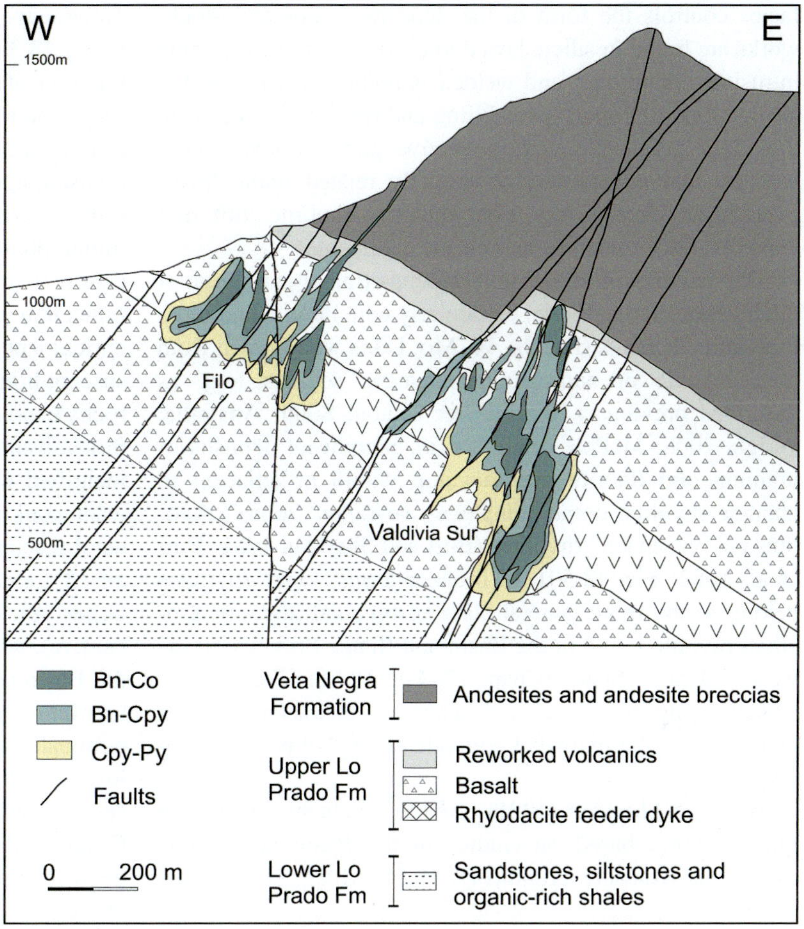

Fig. 18.49 Geological cross-section through the El Soldado volcanic-hosted, manto-type Cu–Ag deposit, Chile. This shows simplified sulphide zonation of main copper orebodies accumulated in degraded petroleum reservoirs. Ore mineralogy: Cpy = chalcopyrite, Py = pyrite, Bn = bornite, and Co = chalcocite. (After Wilson and Zentilli 1999)

which regional evidence suggests was concurrent with the emplacement of the Cretaceous Batholith into the early Cretaceous back-arc basin. However, El Soldado differs from many recognised members of the IOCG family (Olympic Dam and Candelaria) in that there is no direct evidence for a magmatic contribution to the mineralisation process. Indeed, isotopic evidence suggests a more definitive link to basinal or metamorphic-derived fluids (Boric et al. 2002; Wilson and Zentilli 2003), and processes generally associated with sediment ($\pm$ volcanic) hosted copper ore deposits (e.g. Maghfouri et al. 2017; Rabiee et al. 2019; 2022). Nevertheless, a distal magmatic component to the fluids cannot be ruled out.

18.5 Volcanic-Hosted Uranium Deposits

Felsic volcanic rocks are arguably the source of uranium for many styles of uranium mineralisation; however, most volcanic-hosted uranium deposits are relatively small and low grade. This is especially so when compared to the classic styles of unconformity-related and sediment-hosted uranium mineralisation (e.g. Cuney and Kyser 2009). Nonetheless, significantly larger and higher-grade volcanic-hosted resources are known particularly in the Transbaikal uranium province (Khomich et al. 2016). This includes the Saddle Hills Basin (northeastern Mongolia) and the supergiant Streltsovska caldera uranium deposit (southeast Russia). Streltsovska represents the largest known volcanogenic deposit, if not the earth's second-largest uranium accumulation, that is, after Olympic Dam, South Australia (Sect. 18.4.2).

18.5.1 Volcanic Setting, Facies, and General Ore Model

Uranium deposits are hosted in similar volcanic facies and in some cases the same terrestrial successions as epithermal gold systems associated with subaerial silicic volcanism in high-strain extensional tectonic settings, namely silicic domes and lavas, maar-type volcanoes and diatremes, and caldera successions (Nash 2010). Magmas at depth are similarly thought to drive hydrothermal fluid circulation, as well as the volatile component which transports metals in solution. The range of volcanic host facies is comparable and, generally, some combination of structure and host rock facies controls the form of the deposits. Veins and stockworks are found in silicic lava domes, flows and subvolcanic intrusions, diatremes, and welded ignimbrites, whereas disseminations and open space filling and replacement occur in the more permeable and/or reactive parts of lavas, flow breccias and intercalated or spatially related mafic lavas, pyroclastic deposits (e.g. non-welded ignimbrite, tuff ring deposits), and epiclastic and clastic sediments.

The association of uranium minerals (mainly uraninite or pitchblende, and coffinite) particularly with fluorite and other lithophile elements (Be, Sn, W, Mo, Li) is well known from post-orogenic or anorogenic magmatic settings (e.g. as described previously, the Gawler Craton, South Australia; Wright 2008; McPhie et al. 2011a). Significant deposits of uranium and fluorite may develop within or adjacent to alkaline rhyolitic dome complexes, so-called "topaz" rhyolites or "rare metal" rhyolites (Burt and Sheridan 1981; Burt et al. 1982; Fig. 18.50). In broader terms, topaz rhyolites belong to the more general class of bimodal high-silica or alkali rhyolite found in the Basin and Range Province in the western United States (Chap. 17; Fig. 18.51). Their extensional setting as well as a geochemical character suggests that they are the extrusive equivalents of A-type or anorogenic granites.

Burt and Sheridan (1981) provide a valuable model for mineralisation based on studies in the Basin and Range extensional volcanic terrain (Fig. 18.50) which, as described previously, is a major epithermal gold–silver province (Sect. 18.3.4.2). Critical aspects of this model are the combination of a fluorine- and uranium-enriched magma and a high water table or availability of a shallow aquifer at the time of the eruption. This allows for low-temperature mineralisation resulting from near-surface convective circulation. The availability of shallow water also promotes an initial phase of phreatomagmatic activity. Rhyolitic maar crater or tuff ring pyroclastic deposits commonly host significant uranium mineralisation as a result of their initial enrichment in lithophile elements and their permeability and susceptibility to alteration (Fig. 18.50). Other reactive "trap" rocks exposed in the volcanic vent area may host significant mineralisation. These may include reactive carbonate rocks identified with the aquifer, although, probably more significant could be mafic lava and volcaniclastic breccias (e.g. autobreccias and interbedded scoria fall deposits). Mineralised topaz rhyolite dome complexes may form part of larger volcanic structures such as calderas (e.g. McDermitt caldera). Although rhyolitic rocks are a major component of the Basin and Range Province few recognised volcanogenic uranium deposits have resources or reserves that are economically viable and are commonly <1000 t U with very low to low grades (0.02–0.1% U). The better-known prospects include Spor Mountain, Utah, and Aurora on the Oregon-side of the McDermitt caldera, and the Peña Blanca deposit in Chihuahua, northern Mexico (Nash 2010; Plate 18.1c).

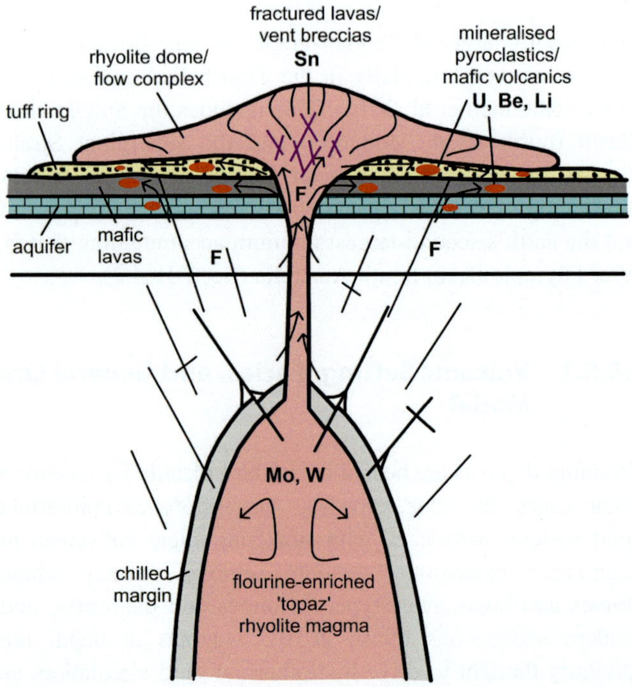

Fig. 18.50 Schematic model for uranium-lithophile element mineralisation associated with fluorine-rich rhyolitic volcanic rocks (Adapted from Burt and Sheridan 1981)

18.5.2 Saddle Hills Basin Uranium Deposits, Mongolia

The Saddle Hills Basin is one of a belt of late Jurassic to early Cretaceous rift basins extending through Mongolia and northern China (Mironov 2005). Dramatic changes in the geodynamic evolution of Mongolia commenced in the Late Jurassic (ca. 160 Ma), when rifting, bimodal (alkaline) magmatism and syn-rift sedimentation were concurrent with the extensional collapse of the mid-Jurassic Mongol–Okhotsk orogenic fold belt created by the continental collision of Mongolia and Siberia (Fig. 18.51). The Mesozoic extensional basins are characterised by typical syn-rift geometries with high-angle normal faults bounding half-grabens (Johnson et al. 2001). These are infilled (to 3 km thick) with grossly upward fining sequences of coarse alluvial clastic sediments intercalated with rhyolitic to basaltic volcanic flows and high-level intrusions, finer-grained fluvial and lacustrine facies, and resedimented pyroclastic deposits. Although early rift volcanism appears to be widespread in the formation of these basins, more copious post-orogenic collapse volcanism in eastern Mongolia and Russia occurred along the line of the Mongol–

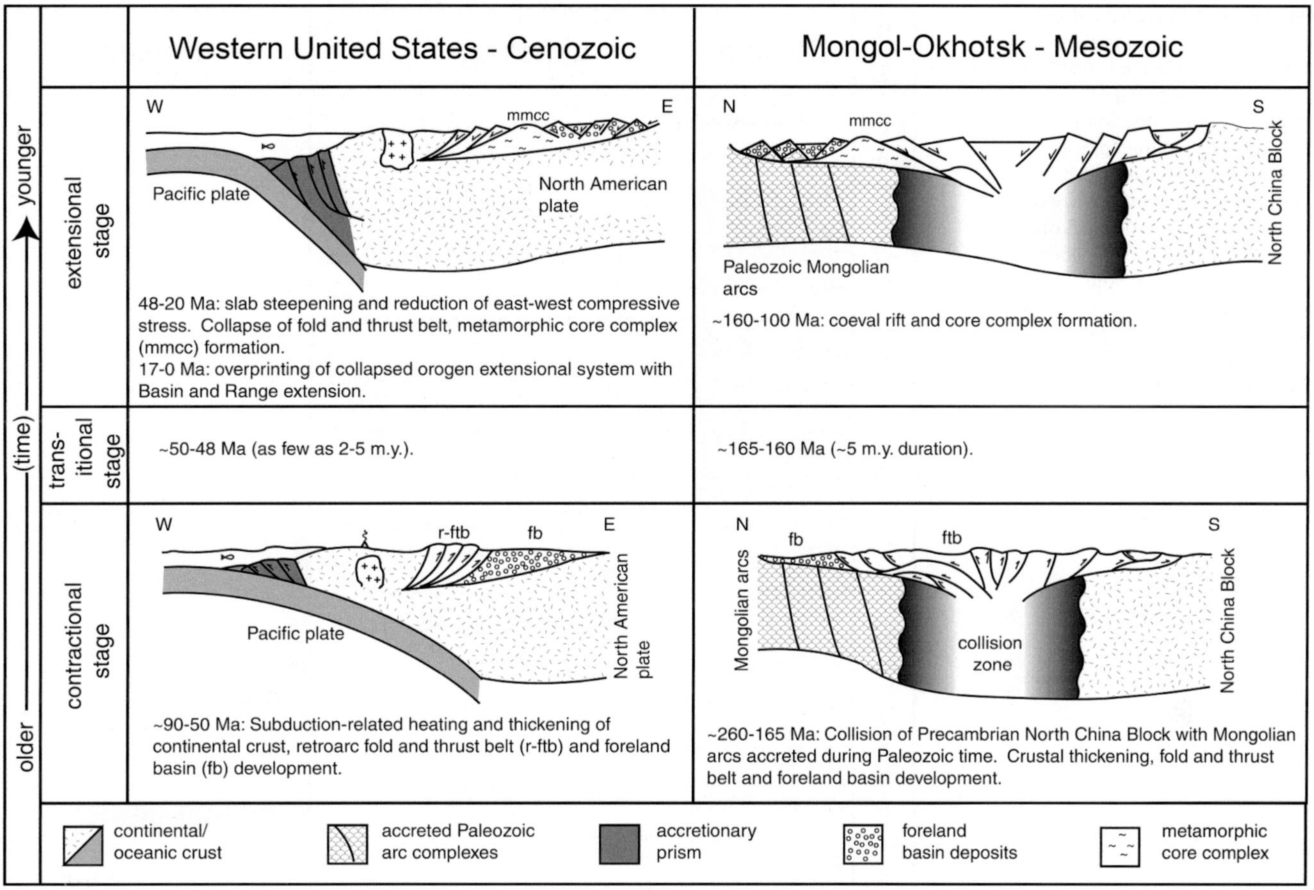

Fig. 18.51 Comparison of extensional orogenic collapse processes in the Cenozoic Basin and Range Province of the western United States, and the Mesozoic of the Mongolia–Transbaikal region (From Johnson et al. 2001). Such tectonic processes appear to have been important in the development of volcanic-hosted uranium mineralisation in both regions. More recent studies have also now inferred the involvement of the western Paleo-Pacific plate in Mesozoic extension, magmatism, and metallogenesis in northeastern Asia (Kusky et al. 2014; Khomich et al. 2016)

Okhotsk Suture Zone (Meng 2003; Seltmann et al. 2005; Khomich et al. 2016).

Identification of early Cretaceous metamorphic core complexes and zones of high-strain extensional detachment faulting draw comparisons with the Basin and Range Province (Fig. 18.51). Both regions show similarities in terms of the close timing between periods of shortening and extension, and the formation of metamorphic core complexes and rift provinces. However, in the western United States, rifting occurred of the long stabilised Precambrian crust of the North American continent that was heated and thickened during Mesozoic arc and/or back-arc compressive tectonism. In contrast, in Mongolia, highly extended regions and basins appear to have formed primarily along suture zones or terrane boundaries on crust previously thickened during Paleozoic orogenic accretion.

Within the Saddle Hills Basin, the volcano-sedimentary succession is subdivided into three formations (Mironov 2005; Fig. 18.52). The lowest reaches a maximum thickness of 800 m comprising alternating sedimentary and mafic to intermediate volcanic rocks, while some quartz feldspar porphyries may be sills. The middle formation (2,000 m thick) comprises almost entirely felsic volcanics including rhyolitic lava flows, ignimbrites, volcaniclastics, and volcanic sediments. The upper formation (200 m thick) consists of lava flows again of a more intermediate to the mafic character. Uranium mineralisation is hosted within the lower–middle parts of the succession (Fig. 18.52). This predominantly occurs along the contact between the lower and middle formations, which separates more reactive, and/or permeable basaltic andesite breccias and sedimentary and rhyolitic tuffaceous units on both sides of the contact, which are capped by massive rhyolites and welded tuffs. Mineralisation is generally stratabound, but steeply dipping and structurally controlled vein and fracture filling mineralisation is also found.

The thick successions of alkaline volcanic rocks are intercalated with coarse alluvial sediments (conglomerates

Fig. 18.52 Schematic volcanic stratigraphy and localisation of uranium mineralisation, Saddle Hills Basin, northeastern Mongolia. Mineralisation is focused along stratigraphical and/or structural contact between more permeable and/or reactive basaltic-andesites, volcaniclastics, and tuffaceous sediments and overlying massive rhyolite and welded tuffs of middle formation. See Fig. 18.54a for the location of Saddle Hills Basin. (Modified from Mironov 2005)

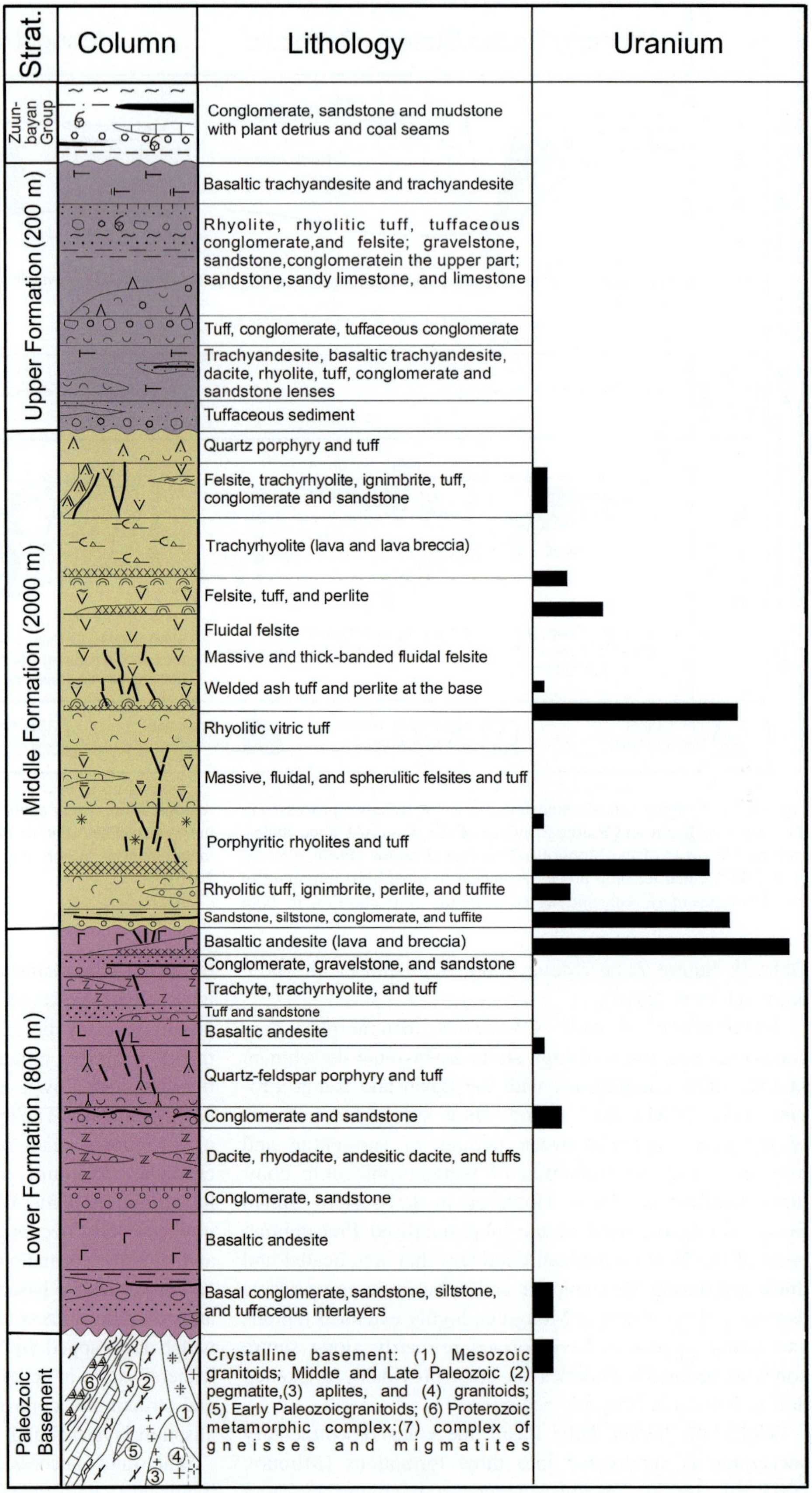

with carbonaceous material). These likely interdigitate with and are overlain by lacustrine mudstones in basin centres, and coal-rich alluvium towards the basin margins (now eroded). Cross-cutting intrusions ranging from dykes to pipes of variable dimensions occur. These are mostly steeply dipping to vertical and trend along the dominant faults. Their compositions are similar to the flows and probably represent the feeder system(s). Rifting and half-graben formation occurred along northeast–southwest-trending normal faults, with segmentation along northwest transfer faults (Wright 2006). Inversion has caused cross-folding on similar axes which may have provided preferred channel ways for uranium mineralisation.

The focus of Russian exploration activities was the Dornod area following its discovery in 1977, and production between 1989 and 1995 totalled a modest 627 t uranium (Harper 2005). Present estimates indicate a resource for the Saddles Hill Basin of some 65,000t uranium (grading 0.1–0.2% U) contained in two main deposits, Dornod and Gurvanbulag. At Dornod, the main host rocks are strongly altered, basaltic andesite (? autoclastic) boulder-breccias. Geochemically, in comparison with little-altered rocks, they show an increase in silica and lower contents of Mg, Ca, and Fe. In contrast to the sedimentary and felsic volcanic rocks, they contain significantly more dispersed titanium oxide (1.5–2.5%), which in strongly altered ore-bearing areas may reach 4–5%. Uranium is contained in U–titanate and pitchblende. Mineralisation at Gurvanbulag comprises low-angle fracture fills, vein and disseminated pitchblende and coffinite, and fracture coating uranophane (Fig. 18.53). The predominant host rocks are felsic volcanic rocks including tuffs or volcanic sediments with lesser mineralisation in volcanic breccias or intermediate to mafic volcanic rocks. The main mineralised horizon appears to be along a shallow structure that dips at 10–30°, paralleling the contact between the lower volcanic formation and massive rhyolites. Russian cross-sections describe this as a crush and intensely hydrothermally altered zone (20–40 m thick), representing a major detachment fault that controlled mineralisation (Mironov 2005).

18.5.3 Streltsovka Uranium Deposit, Russia

Streltsovka caldera has been Russia's principal uranium resource (>90% of production) since processing operations began in 1968 (Ischukova 1997). Uranium mineralisation is widely dispersed in 19 separate intra-caldera deposits, although, past production and reserves (>350,000 t U) are contained in five larger deposits within an area <100 km^2 (Fig. 18.54; Chabiron et al. 2003; Khomich et al. 2016). This makes Streltsovka, if not larger, comparable in size with the largest individual uranium deposits in the Athabasca Basin, Canada. The caldera (ca. 20 km in diameter) is infilled with a succession of late Jurassic–early Cretaceous basalt, andesite, trachy-dacite, rhyolite, and coal-bearing sediments, locally 1400 m thick. Rhyolites form the youngest erupted volcanics and are dated 142 $\pm$ 7 Ma. These are strongly hydrothermally altered and, although mainly preserved in the centre of the caldera, may have formed about one-third of the original infill. The basement is mainly composed of Hercynian granites ($\sim$300 Ma). The location of the caldera and uranium mineralisation is strongly structurally controlled, at the intersection of north–south, and northeast and northwest-trending basement faults (Fig. 18.54a), which form part of a transcurrent fault system traced in seismic data to a depth of 14 km (Khomich et al. 2016).

Discovery (in 1963) was only made following nearly 15 years of exploration when drilling commenced below (>200 m) known fluorite veins which had minor associated uranium and radioactive anomalies. None of the uranium deposits are exposed at the surface; most occur at 400–900 m depth and are associated with intense argillic alteration of wall rocks (Figs. 18.54b, 18.55). Uranium mineralisation has been located within all volcano-sedimentary units that comprise the caldera fill, and in drilling down to 2,400 m into the granite basement. Generally, mineralisation occurs as subvertical veins or stockworks, but sometimes infills porosity and forms replacement bodies along sandstones, conglomerates, and volcaniclastic units. Uranium contents generally average 0.2%, but are up to 0.6% in large stockworks and rise to 1.0% in veins although grades may locally exceed 10%. The major individual deposits are the mined-out Tulukuevskoe open pit and the presently operating Streltsovskoe and Antei underground mines (Fig. 18.54). The Tulukuevskoe mineralisation occurred as a stockwork and was restricted to the volcano-sedimentary pile and had a mined resource of >35,000 t U. The Streltsovskoe-Antei deposit is vein-style mineralisation hosted in the volcanic rocks and basement (Fig. 18.55), with a mined resource of >100,000 t of uranium. Molybdenum, of unknown grade and tonnage, has also been produced.

Detailed age determinations indicate uranium formed closely following the cessation of volcanic activity and was closely associated with pervasive hydrothermal alteration of both the caldera infill pile and basement (Chabiron et al. 2003). Most intensive hydrothermal alteration (stage 2) corresponds with the formation of hydromica (Fig. 18.56). In deep drilling of the Streltsovkoe-Antei deposit, illite, phengite, and/or chlorite occur along the entire length of the drill core (Fig. 18.55). Illite is dominant to 1,800 m and phengite at greater depths. Quartz veins $\pm$ carbonate, pyrite, and hydromica formed during stage 3 without uranium. The main uranium deposition occurred during quartz–molybdenite–pitchblende mineralisation of stage 4. The most common uranium mineral is pitchblende (dated 133 $\pm$ 4 Ma). Fluorite

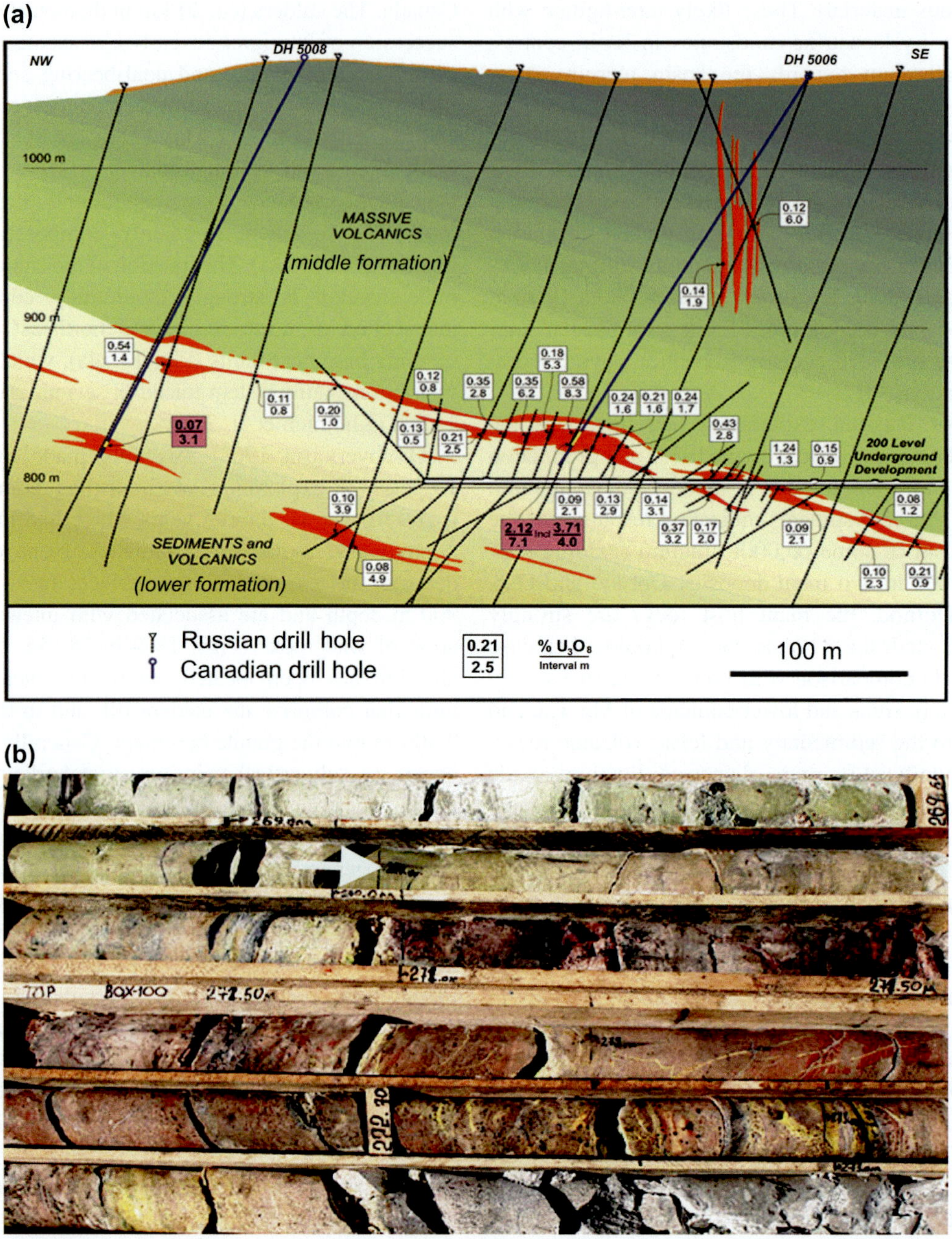

Fig. 18.53 Geological setting of the Gurvanbulag volcanic-hosted uranium deposit, Saddle Hills Basin (*Source* Western Prospector Group Ltd.). **a** Schematic cross-section showing main mineralised horizon along shallow dipping detachment fault at the base of massive rhyolite lavas and welded tuffs. **b** High-grade uranium mineralisation from drill hole 5006 commencing at 270 m (below arrow), which contains 7.1 m @ 2.12% U_3O_8, including 4 m @ 3.71% U_3O_8. Note yellow uranophane coating low-angle fracture surfaces; the drill core is in 1 m intercepts

was deposited during syn-mineralisation stage 4 and with calcite during post-mineralisation stage 5.

Detailed geochemical studies show uranium and fluorine were sourced primarily from the leaching of rhyolitic lavas by convecting acid magmatic fluids above a high-level caldera magma chamber (Chabiron et al. 2003). Mass balance calculations demonstrate the amounts of uranium and fluorine leached from the rhyolites potentially greatly exceed the amounts concentrated in the known ore deposits (26,000 t U and 52 Mt F per km^3). Alteration of uranium-bearing minerals in the basement granites shows that a considerable amount of uranium was also liberated from this source. Any direct contribution of uranium from magmatic volatiles was relatively minor. Illitic alteration was caused by

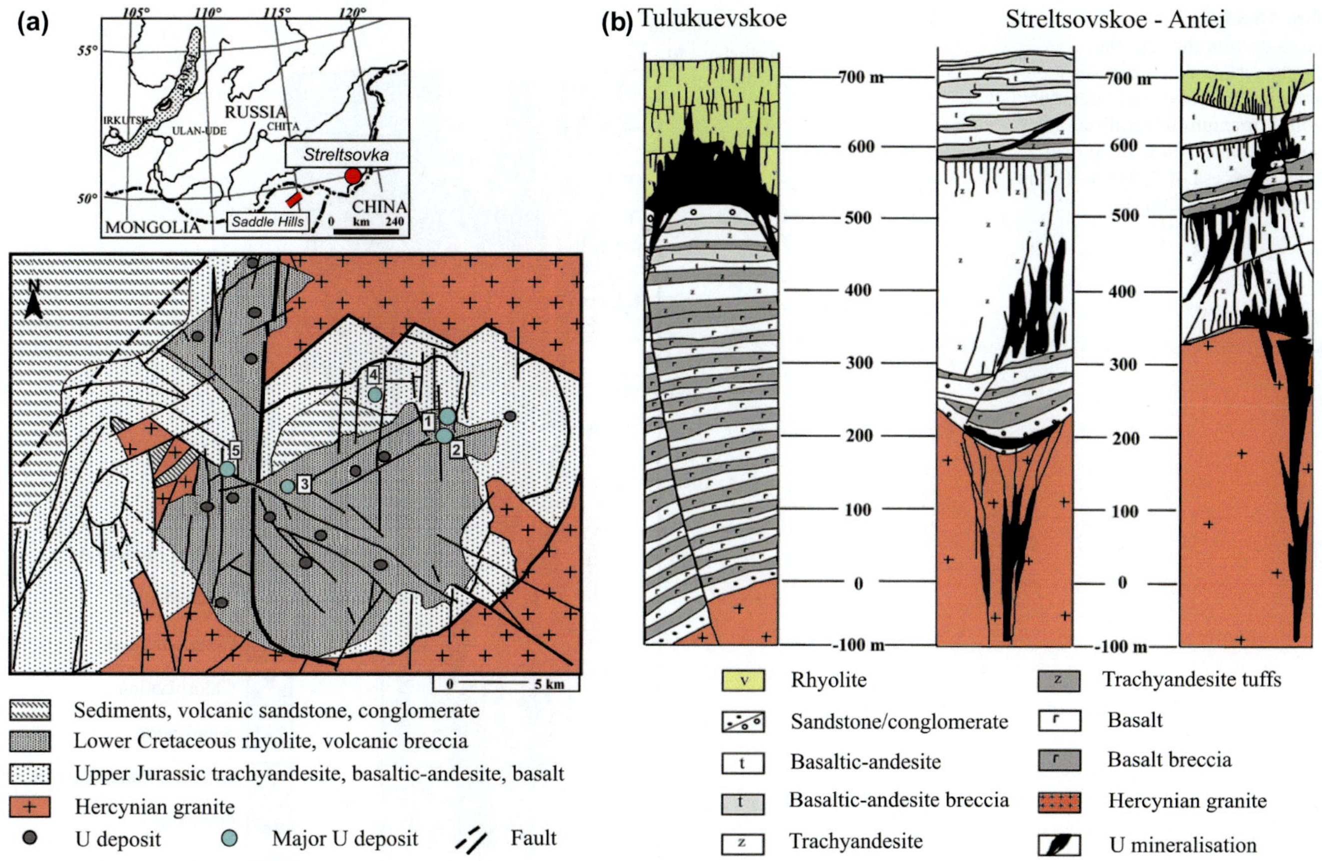

Fig. 18.54 Geological setting of the supergiant Streltsovska caldera uranium deposit, southeast Russia. **a** Location and simplified geological map. Major individual uranium deposits that define past and present resources: 1–2 = Streltsovskoe-Antei, 3 = Tulukuevskoe, 4 = Octiabrskoe, and 5 = Argunskoe. **b** Simplified geological sections of the Tulukuevskoe and Streltsovskoe-Antei deposits displaying strong structural control of uranium mineralisation. (From Chabiron et al. 2003)

high-temperature (~300 °C) acidic and reducing fluids, probably partly of magmatic origin. Because of the reducing characteristics of these fluids, uranium liberated from the rhyolite glass matrix, accessory mineral structures, and other labile uranium sources during pervasive hydrothermal alteration was probably not significantly mobilised. Uranium dissolution and deposition therefore likely occurred later, at lower temperatures, possibly when hydrothermal convection cells established above the magma chamber–hydrothermal system were invaded by oxidised meteoric fluids, which also infiltrated the granite basement.

The setting of uranium mineralisation at the Streltsovka caldera appears to be very different not only in age, but also volcanologically and metallogenically from Olympic Dam. The juxtaposition of two major uranium sources: (1) intra-caldera mildly peralkaline rhyolites, and (2) older U-rich subalkaline granites, is believed to be the major factor contributing factor to the origin of the large uranium resource at Streltsovka. At Streltsovka, the role in mobilisation of uranium by cooler, descending, oxidised meteoric fluids infiltrating the felsic volcanic pile is also emphasised. This may be the common link, if there is any, between these two supergiant uranium accumulations (McPhie et al. 2011a, b; Sect. 18.3.4.1). From a wider geodynamic perspective, it has been speculated that Streltsovka and other large uranium deposits in southeast Russia and east Mongolia occur along the frontal boundaries of a northwest-leading stagnant, Paleo-Pacific oceanic slab observed in seismic tomography data (Khomich et al. 2016). These authors speculate slab dehydration and influx of fluids into the asthenospheric mantle, and their ensuing interaction with the lithosphere caused metasomatic reworking of Th–U–REE in the Archean–Proterozoic basement. These elements were then concentrated in upwelling magmas and fluids from the marginal parts of the slab, leading to the region's exceptional endowment in uranium.

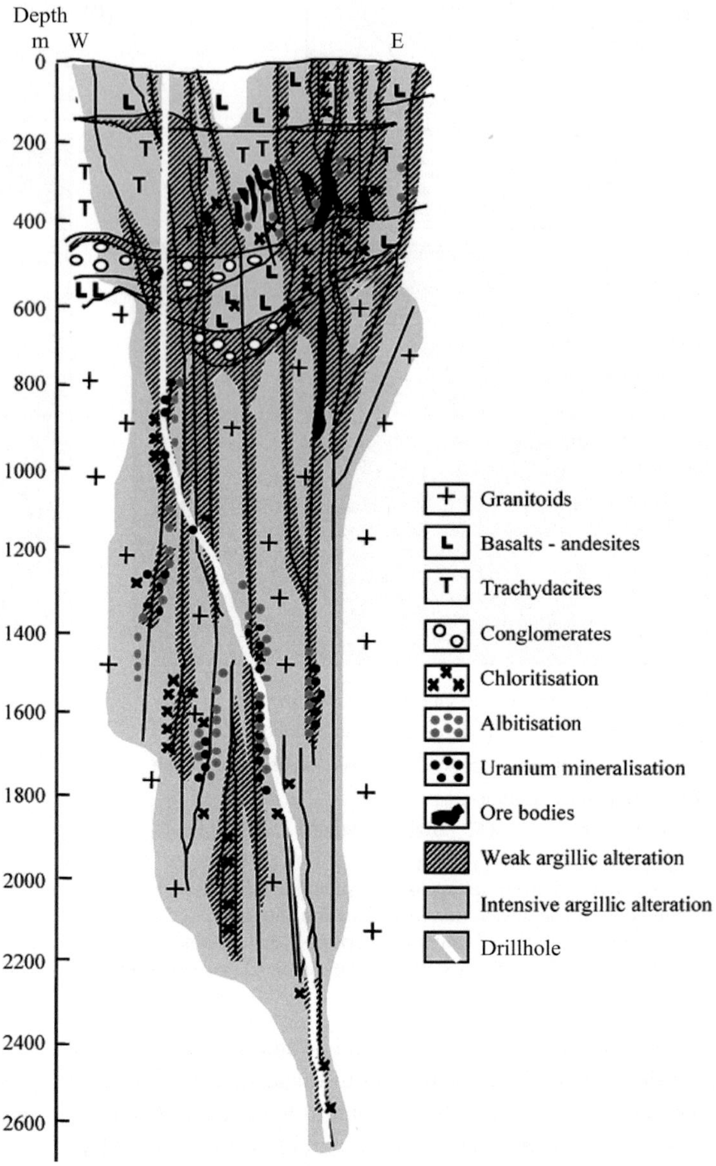

Fig. 18.55 Geological cross-section through the Streltsovskoe-Antei deposit showing hydrothermal alteration zones, uranium mineralisation, and orebodies. Note deep drill hole in excess of 2,600 m. (From Chabiron et al. 2003)

18.5.4 Related Styles of Mineralisation: New Exploration Concepts

Recently, Jowitt et al. (2017) have drawn attention to the exploration potential of high-silica or topaz rhyolites as bulk tonnage sources for REE metals. As described earlier, such rhyolites (Fig. 18.50) are often enriched in other critical and/or economically important metals (e.g. Be, Li, F, Sn, Rb, Th, and U). Although they often erupt as small rhyolite dome complexes, Jowitt et al. (2017) emphasise the fact that rocks of such highly fractionated rhyolitic compositions also form some of the largest known eruptive units contemporaneous with large igneous province (LIP) evolution. These can occur as exceptionally large lava flows (e.g. Gawler Range Volcanics; Allen et al. 2008), as well as large ignimbrite and intra-caldera eruption fill sequences several hundreds of m thick (Jowitt et al. 2017). The latter may form thick eutaxitic to rheomorphic lava-like units and be associated with coeval subvolcanic rhyolite intrusions (e.g. Mesoproterozoic Kathleen Ignimbrite, Musgrave Province of Western Australia; Medlin et al. 2015). That they also frequently crop out at the Earth's surface makes them ideally suited for bulk open pit extraction. Jowitt et al. (2017) propose that large volume, high-silica REE-enriched rhyolites should be considered potential REE analogues of bulk tonnage, low-grade porphyry Cu deposits, warranting further investigation to determine whether they are a viable new source of the REE and possible targets for future mineral exploration.

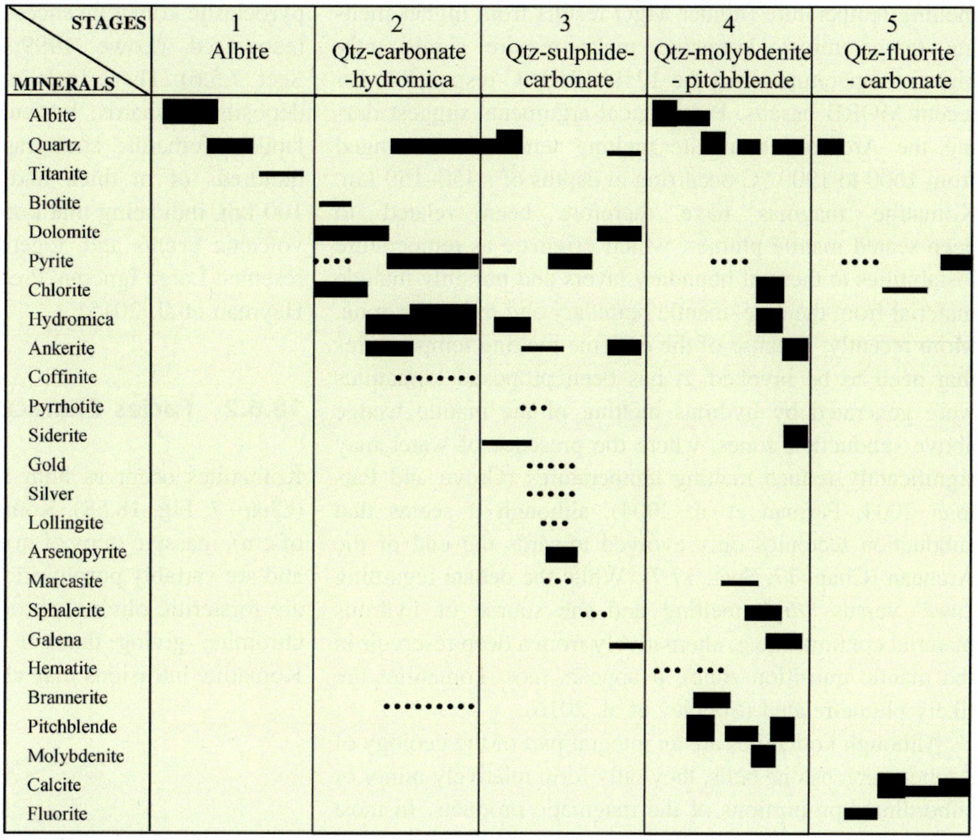

Fig. 18.56 Paragenetic scheme for uranium mineralisation at Streltsovka caldera. Symbol size depends on the mineral abundance and the length of time of its deposition in the mineralogical sequence. (From Chabiron et al. 2003)

18.6 Komatiites and Their Nickel Sulphide Deposits

Komatiites are rare, very low silica (SiO_2 <45 wt%), and very high MgO (>18–45 wt%) ultramafic rocks (Chap. 7). They are characterised by spinifex and olivine cumulate textures. Spinifex textures are defined by arrays of sub-parallel or randomly oriented, platy, bladed, or skeletal crystals of olivine, and sometimes pyroxene set in an originally glassy groundmass, that is now invariably serpentinised and talc-carbonate altered. Although some Proterozoic and Phanerozoic examples are known, komatiites were formed almost exclusively in the Archean, where they have been considered to be derived from high-temperature melts (with eruption temperatures of around 1600 °C) related to mantle plume events in the Earth's early history (Chap. 17). Volcanologically komatiites are fascinating because they represent one of the lowest viscosity magmas known on Earth (Chap. 7). Both komatiite lavas and komatiite intrusions occur, and both may locally host significant orthomagmatic nickel sulphide (Ni–Cu–PGE) mineralisation (McPhie and Cas 2015). Indeed, komatiites represent one of the world's major sources (18%) of nickel production.

18.6.1 Geological Setting

Following the identification of komatiites in the Archean Barberton Mountain Land greenstone belt in South Africa (Viljoen and Viljoen 1969a, b), similar ultramafic lavas were soon identified in other Archean or Paleoproterozoic greenstone belts, particularly in Canada, Australia, and Africa. Most are Neoarchean, but they extend in age from early Archean (Chap. 17, see Sect. 17.7) through into the Phanerozoic, and have been discovered as young as late Cretaceous on Gorgona Island, Colombia (Echeverria 1982; Kerr et al. 1996; see Fig. 7.1a), which is an accreted fragment of the Caribbean–Colombian oceanic plateau (Chaps. 16, 17). However, komatiites have no known modern analogues. The predominance of komatiites in the Archean, their diminishing occurrence in the Proterozoic, and their rarity in the Phanerozoic have been interpreted to reflect secular cooling of the mantle by up to 400 °C (Nesbet et al. 1993; Arndt et al. 2008; Dostal 2008).

Conventional wisdom among petrologists is that komatiite magmas were high-temperature, low-viscosity melts produced by high degrees (50–30%) of anhydrous melting of mantle plumes (Campbell et al. 1989; Arndt et al. 1998, 2008; Dostal 2008; Chap. 7). Because MgO contents of anhydrous magmas are proportionally related to their melting temperature (higher MgO results from higher melting temperatures), komatiite melts require significantly higher temperatures than the 1250–1350 °C responsible for recent MORB basalts. Petrological arguments suggest during the Archean, komatiite melting temperatures ranged from 1600 to 1900 °C occurring at depths of >450–150 km. Komatiite magmas have therefore been related to deep-seated mantle plumes, which originate as temperature instabilities at thermal boundary layers and possibly include material from the core–mantle boundary and transition zone. More recently, because of the extreme melting temperatures that need to be invoked, it has been proposed komatiites were generated by hydrous melting of the mantle wedge above subduction zones, where the presence of water may significantly reduce melting temperatures (Grove and Parman 2004; Parman et al. 2004), although it seems that subduction tectonics only evolved towards the end of the Archean (Chap. 17, Sect. 17.7). While the debate regarding "*wet*" versus "*dry*" melting and the source of hydrous material continues (e.g. alternatively from a deep reservoir in the mantle transition zone), it appears most komatiites are likely plume-related (Sobolev et al. 2016).

Although komatiites are an integral part of the geology of Archean greenstone belts, they only form relatively minor or subordinate proportions of the magmatic products. In most greenstone belts, typically they constitute <10% of the total volume of volcanic rocks, though, in some areas they may form up to 30% of the volcanic succession (Fig. 18.57). Almost all known Archean komatiites are preserved as lavas or sills in deep submarine volcano-sedimentary successions, and as intrusions in the subjacent crust that forms part of Archean cratons (Chap. 7). Komatiite lavas were almost certainly erupted in subaerial and shallow marine environments as well, but they have been eroded. Although some vesicular komatiite lavas exist (Beresford et al. 2000, 2002), pyroclastic komatiite successions have been only very rarely recognised (Lowe 1999; Stiegler et al. 2008, 2011; Sect. 7.5.6). These include primary and variably reworked deposits of shards, vesicular pyroclasts, and accretionary lapilli. Komatiite stratigraphic units can be tens or even hundreds of m thick and traced continuously for up to 100 km, indicating that komatiite eruptions were significant volcanic events and, together with associated basalts, represented Large Igneous Provinces (LIPS; Barnes et al. 2012; Hayman et al. 2015).

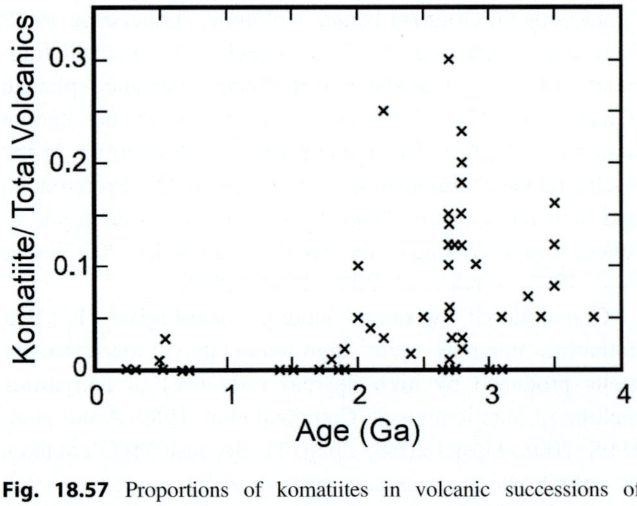

Fig. 18.57 Proportions of komatiites in volcanic successions of greenstone belts as a function of age (From Dostal 2008)

18.6.2 Facies Characteristics and Interpretation

Komatiites occur as both lavas and subvolcanic intrusions (Chap. 7; Fig. 18.58). Komatiite lavas range from thin (tens of cm), massive (tens of m) to thick (>100 m) eruptive units and are variably porphyritic. The most common phenocrysts are forsteritic olivine, chrome-rich pyroxene, anorthite, and chromite, giving them a peridotitic mineral assemblage. Komatiite intrusions may also be porphyritic if they are thin,

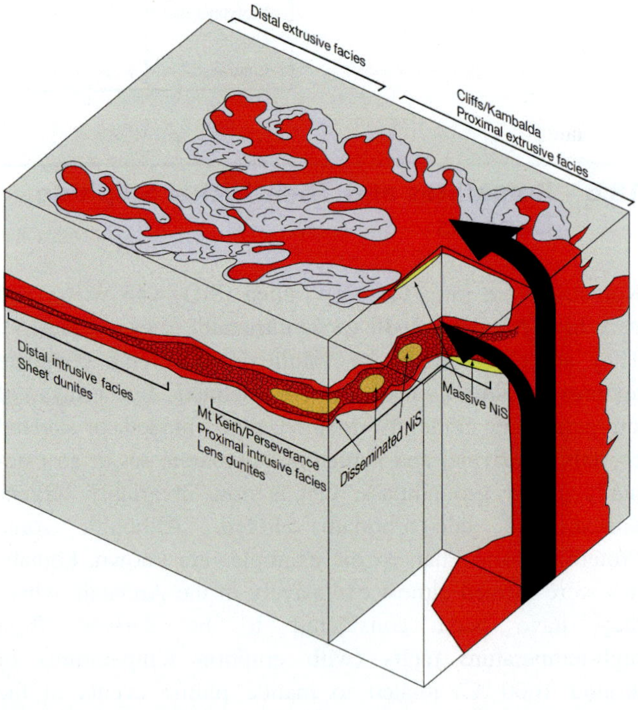

Fig. 18.58 Schematic representation of the spatial relationships between extrusive komatiite lava flow fields and subsurface intrusive komatiites, relative to the magma conduit. Because nickel sulphide is very dense, it is predicted that it will be deposited from the host magmas close to source, whether in lavas, or sill-like intrusions. In sills, the flow simulates a lava tube, and so dense massive sulphide liquid finds the low points on the basal contact to flow into and deposit their massive sulphides. Examples named are from the Norseman-Wiluna greenstone belt in Western Australia which, as described in the text, is the world's premier province for this ore deposit type. (From Fiorentini et al. 2012)

very high-level intrusions. Deeper-seated, thicker (hundreds of m) intrusions of very coarse, holocrystalline, olivine-rich dunites are the intrusive equivalents of komatiite lavas. The rapidly cooled margins of dunitic intrusions are also porphyritic. Being ultramafic, komatiites are prone to variable degrees of hydrothermal and seafloor alteration, metamorphism, and deformation, which all combine to obliterate original textures, contact relations, and mineralogy, in many cases, severely hindering the interpretation of primary facies characteristics (Chap. 13). Common alteration minerals are serpentine, talc-carbonate, and chlorite.

Individual flows may display well-developed layering which typically consists of an upper part characterised by spinifex textures (zone A), and a lower part (zone B) containing a high proportion of equant olivine crystals resembling peridotite to dunite cumulate rocks (for detailed discussion and references, see Sect. 7.3). Both zones A and B are further subdivided into several subzones (see Figs. 7.1, 7.2, 7.5), although individual subzones may not be continuous and occasionally may be absent. Some thin and even some thick komatiite flows are not layered and do not exhibit spinifex zones. Instead, these may be massive textured with olivine phenocrysts commonly concentrated in the centre of such flows. Massive komatiite sheets >10–50 m thick, traceable, and compositionally and texturally homogeneous laterally >10 km have been reported. Massive lavas may pass laterally into flows with well-developed spinifex textures. The variation in the textures and layering of komatiite units has been attributed to variations in flow volume, lava flow velocity, effusion rate, topography, and cooling rate (Dann 2001; Dann and Grove 2007; Cas et al. 1999; Cas and Beresford 2001; Chap. 7), all of which may also be related to distance from an eruptive centre.

In addition, jigsaw-fit to clast-rotated monomict breccia may be present at the tops and bases of lavas (Cas and Beresford 2001; see Sect. 7.5, Fig. 7.6a–c). These breccias are autoclastic hyaloclastite produced by quench fragmentation. Thin syn-depositional komatiite sills that were intruded into unconsolidated sediment may have all of the above zones as well as sediment-matrix komatiite breccia (peperite) along the contacts (Beresford and Cas 2001; Fig. 7.6b). Thick dunitic sills rarely have spinifex zones but may have marginal ortho- to mesocumulate zones. The main dunitic textural zone is holocrystalline and mostly consists of polygonal olivine without groundmass, and is called adcumulate. Adcumulate texture is diagnostic of very thick sills that have cooled slowly and crystallised fully (see Sect. 7.7.3, Fig. 7.15). Komatiites are also closely associated with komatiitic basalts (12–18 wt% MgO), which have similar flow facies and textures as komatiites, except that the dominant phenocrysts are pyroxenes. Their differentiated flows contain upper pyroxene spinifex zones and lower cumulate zones.

The spinifex textural zone has been interpreted as a chilled lava crust, with the various skeletal spinifex zones reflecting the extremely rapid chilling of this crust due to the subaqueous environment and the very high-temperature gradient between the lava surface and the water (see discussion Sect. 7.3). Given that many komatiite lavas are entirely coherent from base to top, an insulating steam layer must have been created at the interface between lava and seawater, preventing quench fragmentation. However, lavas with hyaloclastite at their tops also imply that steam film collapse occurred and resulted in quench fragmentation. The underlying cumulate textural zone is interpreted to represent a lava tube insulated from contact with water by the spinifex-textured lava crust. The lava tube allowed magma to be transferred from the vent to the propagating lava front without losing heat. Slow cooling of stagnant lava in the tube allowed crystallisation, forming ortho- and mesocumulate textural zones. Preserved orthocumulate and mesocumulate textures indicate that crystal settling was not efficient, most likely because of the low-density contrast between olivine crystals and the interstitial melt.

Komatiite magmas were very high-temperature magmas (1400–1650 °C), with ultralow viscosities (0.1–10.0 Pa s) and very high density (2800 kg m^{-3}) (Huppert and Sparks 1985; Chap. 7). These physical properties and their deep marine setting influenced flow behaviour and morphology. Active lavas had very thin flow fronts, perhaps only cm-thick which spread as very thin sheets and lobes (Cas et al. 1999). The slightest seafloor slope would have created a hydraulic head effect from vent to flow front. Thicker lava units resulted from ponding in seafloor depressions or inflation under a chilled crust, as happens for modern basalt lavas. Komatiite lavas could have been turbulent but mostly only near vent, becoming laminar farther away (Cas et al. 1999). Because of their very high density, komatiite magmas were very prone to forming intrusions, both high-level intrusions into unconsolidated sub-seafloor sediments, and deeper crustal intrusions (Beresford and Cas 2001; Rosengren et al. 2005; Houlé et al. 2008; Fig. 18.59).

The nature of the near-surface country rocks (i.e. consolidated lavas and sediment versus unconsolidated sediment and volcaniclastics) will govern whether komatiitic magmas are emplaced as lava flows, invasive flows, or sills. Where near-surface strata are competent, komatiitic magma will erupt predominantly as flows to form extensive, very low-angle submarine lava shields (e.g. Kambalda, Western Australia). When near-surface strata are not competent, komatiites will be emplaced as high-level sills. However, in time this increases the bulk density of the volcano-sedimentary pile, eventually allowing the eruption of lava and construction of volcanic lava shield and flow field complexes with associated subvolcanic intrusive complexes (Houlé et al. 2008; Fig. 18.59), seemingly a common

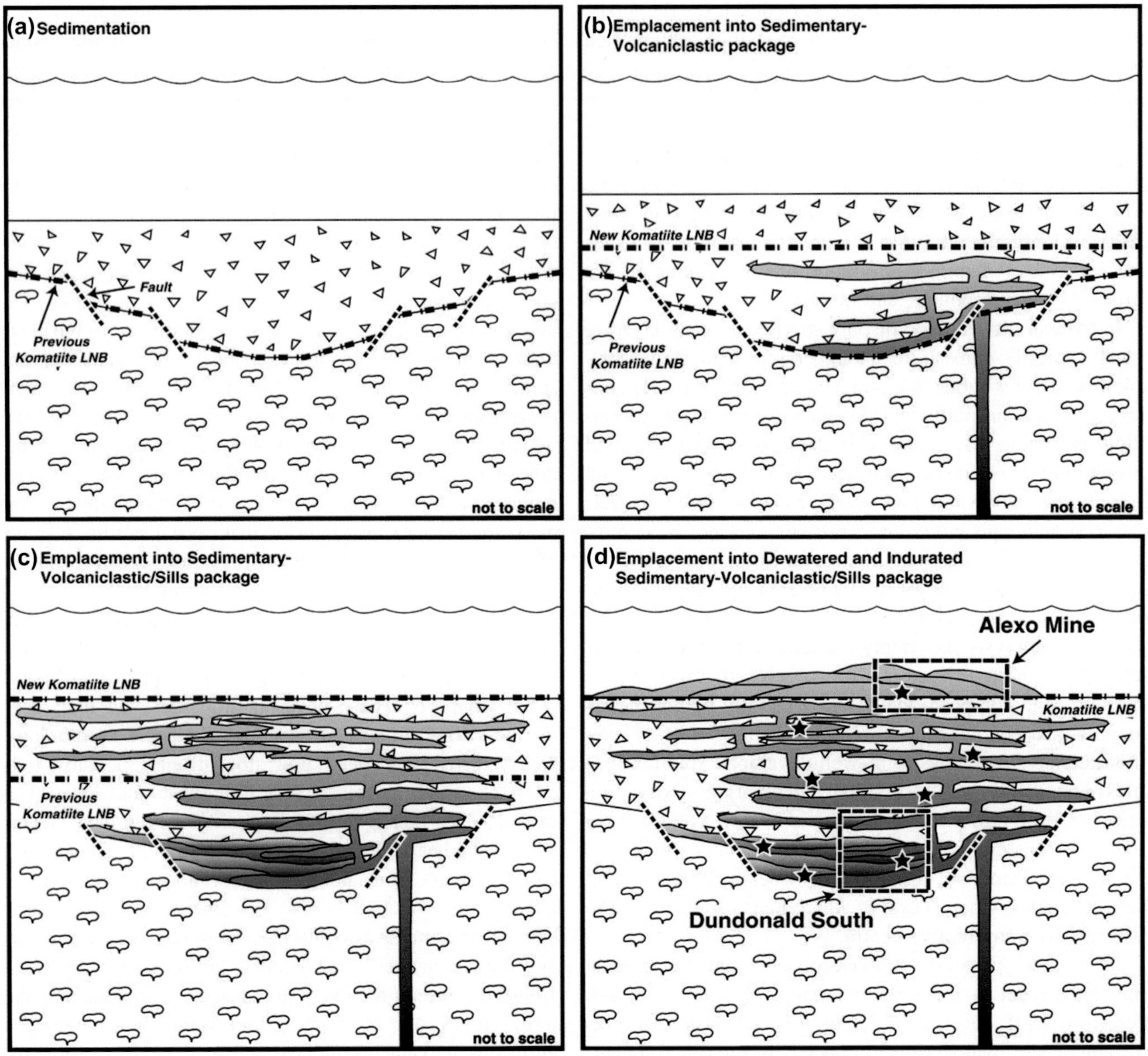

Fig. 18.59 Idealised model showing the evolution **a–d** of komatiitic subvolcanic and/or volcanic architecture within a volcaniclastic- and/or sediment-dominated komatiite volcanic environment, e.g. Alexo–Dundonald and Raglan Ni–Cu–(PGE) deposits, Canada (see Table 18.3). Dashed boxes in (**d**) represent the approximate location of the Alexo mine and Dundonald South deposits within the komatiite volcano. Stars represent the potential location of komatiite-associated nickel sulphide mineralisation in sediment- and/or volcaniclastic-dominated successions. LNB = level of neutral buoyancy. (From Houlé et al. 2008)

evolutionary pattern in some Canadian Archean and Proterozoic examples (Table 18.3).

18.6.3 Origin of Nickel Sulphide Mineralisation

Komatiites may be mineralised or barren. Nickel sulphide deposits occur in both komatiite lavas and subvolcanic intrusions, precipitated directly from the ultramafic magmas. However, it is commonly agreed that komatiite magmas had relatively high mantle-derived nickel contents (>400 ppm), but low sulphur contents. Therefore, to precipitate sulphide minerals, komatiite magmas must have assimilated sulphur and attained sulphide saturation (Lesher 1989). Sulphur isotope data indicate the sulphur is not mantle-derived. For mineralised komatiite intrusions, the sulphur has to be added by the assimilation of sulphidic crustal source rocks through which the magma has intruded. For komatiite lavas, sulphur may also have been derived from subsurface crustal rock assimilation within the conduit, and/or through assimilation

Table 18.3 Selected features of komatiite-hosted nickel sulphide deposits referred to in the text

Deposit	Age (Ga)	Greenstone belt	Komatiite host	Main ore Type	Total resource (Mt Ni)	Grade (%Ni)
Forrestania, Western Australia	2.90	Southern Cross	Flows	1	0.45	2.0
Kambalda, Western Australia	2.70	Norseman	Flows, invasive flows (proximal)	1	1.13	3.0
Black Swan, Western Australia	2.71		Sill (proximal)[a]	2	0.17	0.6
Mount Keith (MKD5), Western Australia	2.70	Agnew-Wiluna	Sill (proximal)[a]	2	2.77	0.6
Perseverance, Western Australia	2.72		Sill (proximal)[a]	2	1.05	0.9
Yakabindie, Western Australia	2.70		Sill (proximal)[a]	2	1.20	0.6
Honeymoon Well, Western Australia	2.70		Sill (proximal)[a]	1	1.02	0.7
Cliffs, Western Australia	2.70		Flow (proximal)	1	0.13	2.3
Alexo-Dundonald, Canada	2.71	Abitibi	Sills and flows (?ponded)	1	0.01	3.6
Shaw Dome-Langmuir, Canada	2.71		Sills, flows	1	0.06	2.2
Hunter's Road, Zimbabwe	2.83	Midlands-Kwekwe	Intrusive dyke, flows (proximal)[a]	2	0.24	0.6
Trojan, Zimbabwe	2.69	Shamva	Flows and sills	2	0.16	0.9
Shangani, Zimbabwe	2.7	Shangani	Lava channel, sill	2	0.06	0.4
Thompson, Canada	1.88	Thompson	Sills (channelised, proximal)	1	3.49	2.3
Raglan, Canada	1.92	Cape Smith	Sills and flows (channelised, proximal)	1	0.67	3.1

[a] Extreme bimodal association with dacitic–rhyodacitic volcanics and subvolcanic intrusive complexes. This summary is based on various published sources not all of which are included in the text references

of sulphidic seafloor substrate (e.g. sulphidic sediments, VMS deposits, or sulphur-rich, felsic volcanic rocks) over which komatiite lavas flowed (Fiorentini et al. 2012; Le Vaillant et al. 2016; Fig. 18.60).

Nickel, copper, and platinoid elements as well as iron are scavenged by the sulphur to form liquid immiscible, mono-sulphide drops which then separate into various component sulphide minerals and settle (termed "batch segregation") in the komatiite melt. The principal sulphide minerals associated with komatiites are pyrrhotite ($Fe_{1-x}S$), pentlandite ($[FeNi]_9S_8$), and chalcopyrite ($CuFeS_2$). There are two styles of komatiite-hosted mineralisation, type 1 (or "*contact ore*") massive nickel sulphide and type 2, disseminated (or "*blebby*") nickel sulphide (Fig. 18.61). In komatiite lavas, type 1 high-grade nickel sulphide is the most important and most commonly occurs as lenses up to several m thick and several tens of m or more in lateral extent, at the base of the lowest lava in a sequence of komatiite lavas, at its contact with the substrate (Figs. 18.61a, b, 18.62). This has led to suggestions that during lava flow an immiscible nickel sulphide liquid separated from the silicate komatiite magma and segregated to the bottom of the lava (Lesher 1989). Disseminated nickel sulphide bodies in komatiite lavas are typically of such low grade and they are very cost-sensitive to prevailing global nickel prices and/or proximity to high-grade massive sulphide ore. Komatiite intrusions can host both type 1 and type 2 nickel sulphides, although, type 2 disseminated nickel sulphide is more common (Fig. 18.61d, e). In addition, it has become apparent that nickel sulphide is prone to be remobilised hydrothermally and tectonically (Stone et al. 2005; Cas et al. 2013; Vaillant et al. 2016), and thus replace and become interleaved with substrate units.

Ore models for komatiite-hosted nickel–copper–PGE have historically emphasised the role of dynamic, high-temperature, potentially turbulent komatiitic lava flows in forming highly channelised flows with basal thermo-mechanical scours, which became important trap sites for segregating sulphides (e.g. Kambalda, Western Australia; Fig. 18.60). However, recent volcanological

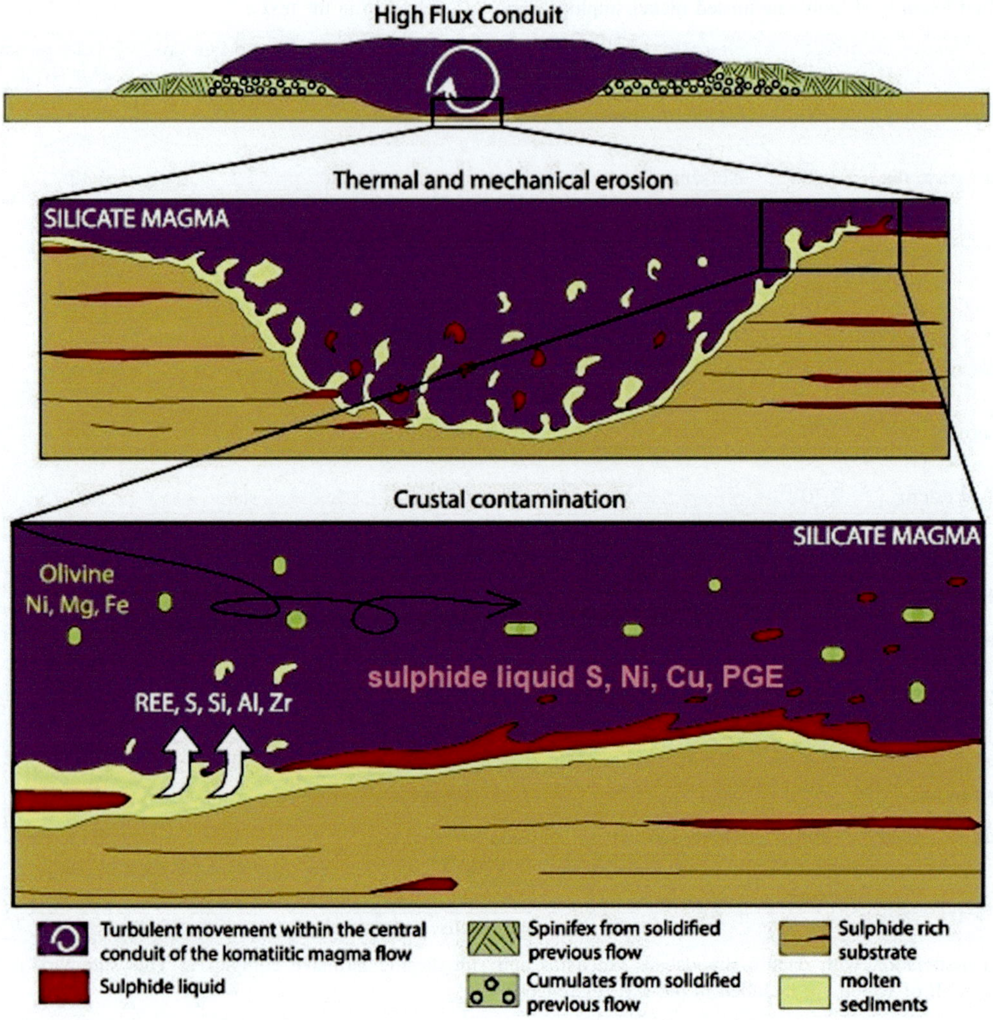

Fig. 18.60 Typical schema for the formation of komatiite-hosted "*contact*" (or type 1) massive nickel sulphide ore deposits, characteristic of high-grade mineralisation at the base of a sequence of komatiite lavas. To precipitate sulphide minerals, komatiite magmas must assimilate sulphur by contamination from sulphidic crustal source rocks through which the magma has intruded or, for komatiite lavas, assimilation within the conduit, and/or through erosion of sulphidic seafloor substrate over which komatiite lavas flowed. While thermal or thermo-mechanical erosion and even melting of the substrate by komatiite lavas is commonly proposed in the economic geology literature, the weight of volcanological evidence now suggests purely, physical erosion processes are more likely to have been responsible; for a critical discussion, see Sect. 7.9.2. (From Le Vaillant et al. 2016)

studies (based on less deformed and modified successions) show the bases of komatiite lavas are perhaps more generally planar and have questioned these tenets (read below, and for a detailed analysis, see Sect. 7.9). These studies have also indicated that komatiitic sills are more common in these successions than previously appreciated and may host significant and world-class ore bodies (Table 18.3). Both proximal komatiite lavas and intrusions associated with sulphide-bearing felsic (lava) successions are now considered to be highly prospective for nickel sulphide ore deposits (Fig. 18.61f). Volcanic facies analysis and an understanding of emplacement processes and proximal-to-distal relationships are ever more a very important exploration tool for komatiite-hosted nickel sulphide ore deposits. In the following section, we will highlight these advances with specific reference to world-class deposits in the world's premier komatiite-hosted nickel belt in Western Australia.

18.6.4 World-Class Ore Bodies

18.6.4.1 Skewed Distribution

Whereas komatiites and komatiitic basalts may be found in most Archean and some Proterozoic greenstone terrains, nickel sulphide mineralisation and especially the occurrence of larger, economic-grade ore bodies have a much more restricted distribution (Table 18.3; Fig. 18.63). World-class (>1 Mt contained Ni metal) Archean deposits are essentially

(a)
C
B
A
(b)
P
P
P
(c)
(d)
(e)
(f)

◀ **Fig. 18.61** Selected features of komatiitic-hosted nickel sulphide mineralisation (see Fig. 18.62 and text for further discussion). **a** Type 1 or "*contact ore*" from the base of a komatiite lava flow, Kambalda mine, Western Australia (Photo by David Mole, CSIRO). Conventionally, this is interpreted as follows: A is the underlying Lunnon Basalt showing "evidence" of melting by the overlying Kambalda Komatiite Formation; B is massive, high-grade nickel sulphide ore pooled at the base of the lowest komatiite lava; and C, the overlying komatiite lava flow. **b** Massive nickel sulphide and interpillow ore hosted within hyaloclastite breccias at top of Lunnon Basalt, Foster mine, Kambalda. **c** Massive nickel sulphide overlying inter-spinifex ore, Lunnon shoot, Kambalda (Source Department of Mines and Petroleum, Western Australia). Inter-spinifex ore occurs at the top of a lava that underlies the base of a mineralised komatiite lava flow in a hanging-wall setting and results from replacement of spinifex (see Fig. 18.62). **d** Type 2 "blebby" or disseminated nickel sulphide in the komatiitic sill, Alexo mine, Canada. **e** Disseminated pentlandite in coarse adcumulate komatiite sill, Mt Keith, Western Australia; the purple streaks are patches of violarite, which is an alteration mineral of pentlandite; section of core, the field of view is 5 cm × 8 cm. **f** Footwall dacitic subvolcanics showing variable replacement by tectonically remobilised massive pentlandite, Black Swan mine, Western Australia; drill core is 5 cm wide. While massive or disseminated nickel sulphide occurs at the contact or in the overlying komatiite (interpreted as a younger sill complex), mineable quantities also occur in the footwall felsic succession (Cas et al. 2013)

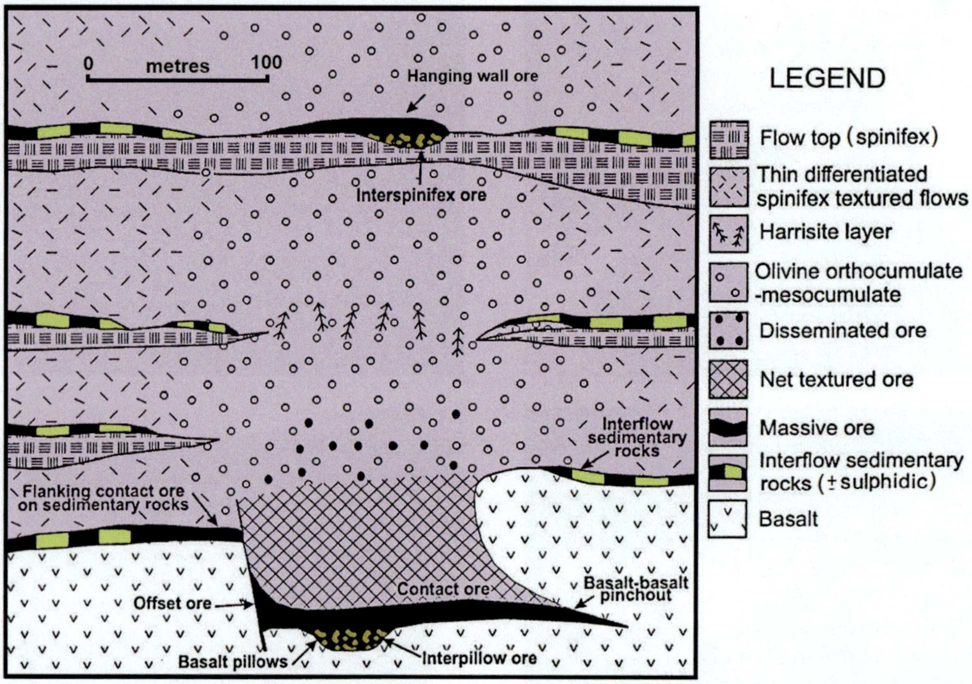

Fig. 18.62 Simplified cross-section illustrating the major textural types of komatiitic lava and associated nickel sulphide orebodies in the Kambalda Ni–Cu Camp, Western Australia, based on the lower Silver Lake Member of the Kambalda Komatiite Formation (see Fig. 18.64a). Note that the embayment or trough only partially confines the main contact orebody and is largely a structural feature. In some cases, faulting has been superimposed upon the margins of original erosional volcanic channels (or even lava tubes) to produce structural troughs. However, not all mineralisation nor all troughs coincide with original erosional channels. In places, mineralisation formed within the basal parts of komatiites with planar basal contacts. Commonly, faulting and folding have offset planar contacts, also producing trough structures (e.g. Stone et al. 2005; Sect. 7.9.2). Inter-spinifex ore occurs in the groundmass between spinifex crystals at the top of a lava that underlies the base of a mineralised komatiite in a hanging-wall setting and has been regarded to be indicative of partial melting (Groves et al. 1986; Barnes et al. 2016). However, inter-spinifex ore also occurs isolated within spinifex zones and below the bases of komatiites that are not mineralised at all and, in places, spinifex and equant crystals are replaced by sulphide. We suggest that remobilisation of ore and replacement during metamorphism offer more likely explanations for some inter-spinifex ore. (Modified from Cas and Beresford 2001, After Cowden and Roberts 1990)

known only from the Norseman-Wiluna greenstone belt (2.7 Ga) in Western Australia. This forms part of the Kalgoorlie Terrane within the Eastern Goldfields Superterrane, eastern Yilgarn craton (Chaps. 7, 17; see Figs. 7.10a, 17.63 a). Somewhat limited-tenor deposits of similar age occur in the Canadian Abitibi greenstone belt of the Superior province (Alexo–Dundonald, Shaw Dome–Langmuir, and the large but very low-grade Dumont deposit) and greenstone belts of the Zimbabwe craton (Hunter's Road, Trojan, Shangani). Larger Archean deposits have also more recently been discovered in the Southern Cross-Forrestania greenstone belt, within the southern part of the Youanmi Terrane (western central Yilgarn craton; Figs. 7.10a, 17.63a), which are somewhat older (ca. 2.9 Ga). Younger Paleoproterozoic

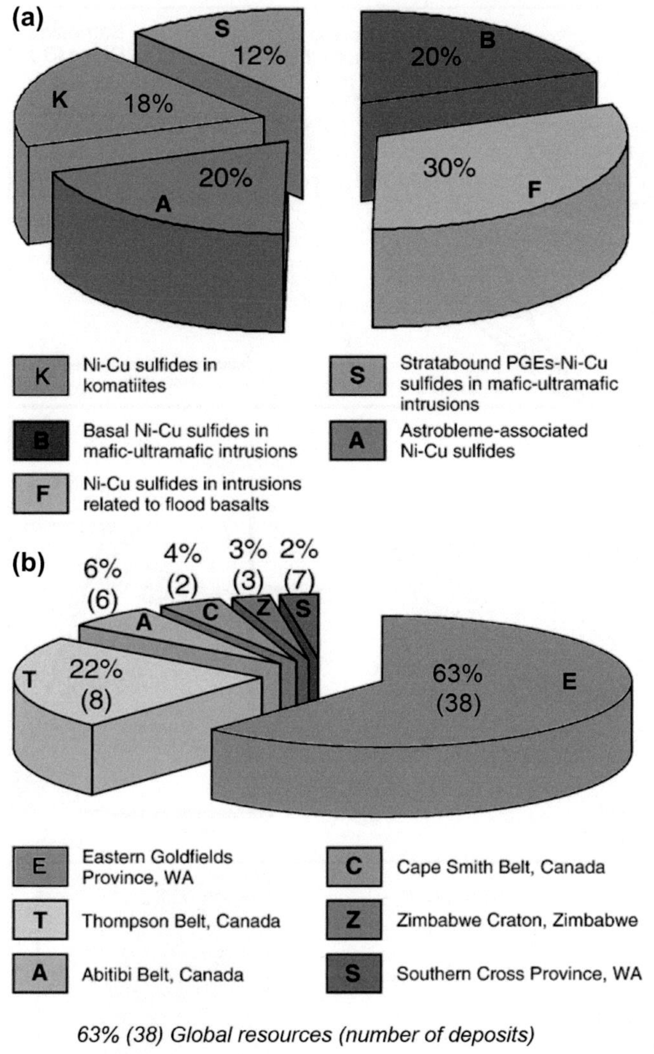

Fig. 18.63 Global resources of nickel metal by **a** type of nickel sulphide ore deposits, and **b** major komatiite provinces/belts (Adapted from Hoatson et al. 2006)

(ca. 1.9 Ga) occurrences include Canada's world-class Thompson nickel belt (northern Manitoba), and the Raglan deposit (northern Quebec), occurring in rifted cratonic settings along the margin of the Archean Superior Province. These occurrences are thought to have been caused by mantle over-turn events associated with mantle plumes or larger superplumes and early cratonisation of the Earth's crust (Chap. 17, Sect. 17.7). Barnes et al. (2012) and Mole et al. (2014) proposed this cratonisation process drove the hottest and most voluminous komatiite eruptions to the edge of established continental blocks, where the eruption of komatiite would likely have been facilitated by plume-related extension and rifting at this interface. Ultramafic magmas rapidly ascended from the mantle to the surface without significant ponding or differentiation in the lithosphere, forming giant, high-flux komatiite flow fields containing large nickel sulphide deposits. This occurred at Forrestania at 2.9 Ga, while between 2.9 and 2.7 Ga, the focus of hot and voluminous komatiite magmatism shifted to the east towards the Kalgoorlie Terrane, following the margin of the growing continent.

In terms of nickel endowment, the Norseman-Wiluna greenstone belt accounts for over 60% (or ~11 Mt Ni) of the global komatiite-associated nickel sulphide resources and is overwhelmingly the world's premier province for this ore deposit type (Fig. 18.63). It hosts at least five world-class ore bodies, namely at Kambalda (1.1 Mt Ni) in the south-central portion (Fig. 7.10b), and Mt Keith (2.8 Mt Ni), Perseverance (1.1 Mt Ni), Yakabindie (1.2 Mt Ni), and Honeymoon Well (1.0 Mt Ni) in the north-northwest (Table 18.3), as well as a large number of other economic and sub-economic deposits. In particular, the northwestern portion, or Agnew–Wiluna greenstone belt (Fig. 7.9), represents the most nickel-rich, komatiitic-hosted region known in the world, containing nearly 80% of the past and present resources of the whole belt (some 8.5 Mt Ni). Within this Western Australian context, nickel deposits have been variously grouped into camps, districts, and domains, although, not always consistently. Camps are an association of genetically related individual deposits, and/or a group that could be expected to be processed through one mill. A district represents a larger-scale association of deposits and camps that are genetically associated, while a domain is an association of camps and districts found within a significantly different tectonostratigraphic setting or sub-terrane. Thus, the Kambalda Domain contains 46 known nickel sulphide deposits located in four camps of which Kambalda (23 deposits) is the most economically significant. The Kambalda Domain, as opposed to the Kambalda camp, accounts for 1.6 Mt of nickel metal.

18.6.4.2 Kambalda Ni–Cu Camp

The discovery of massive nickel–copper sulphides at Kambalda in the Eastern Goldfields in 1966 and the ensuing nickel boom, which persisted up until 1971, heralded the beginning of Australia's nickel industry. Since then it has become the type locality for Archean komatiite-hosted nickel sulphide deposits and a world-class deposit with pre-mined resources of ~35 Mt containing 3% nickel (Table 18.3; Plate 18.1d). The mafic–ultramafic volcano-sedimentary sequence has been intruded by multiple suites of intermediate-felsic intrusions, metamorphosed to upper greenschist-lower amphibolite facies, and complexly faulted and folded during at least four phases of deformation (e.g. Cowden and Roberts 1990; Stone et al. 2005). Nickel sulphide mineralisation occurs as a number of ore "shoots" located around the Kambalda dome, which is a doubly plunging antiformal structure (Fig. 18.64). Despite regional metamorphism igneous textures are locally well preserved and can, with some caution, be confidently interpreted.

(a)

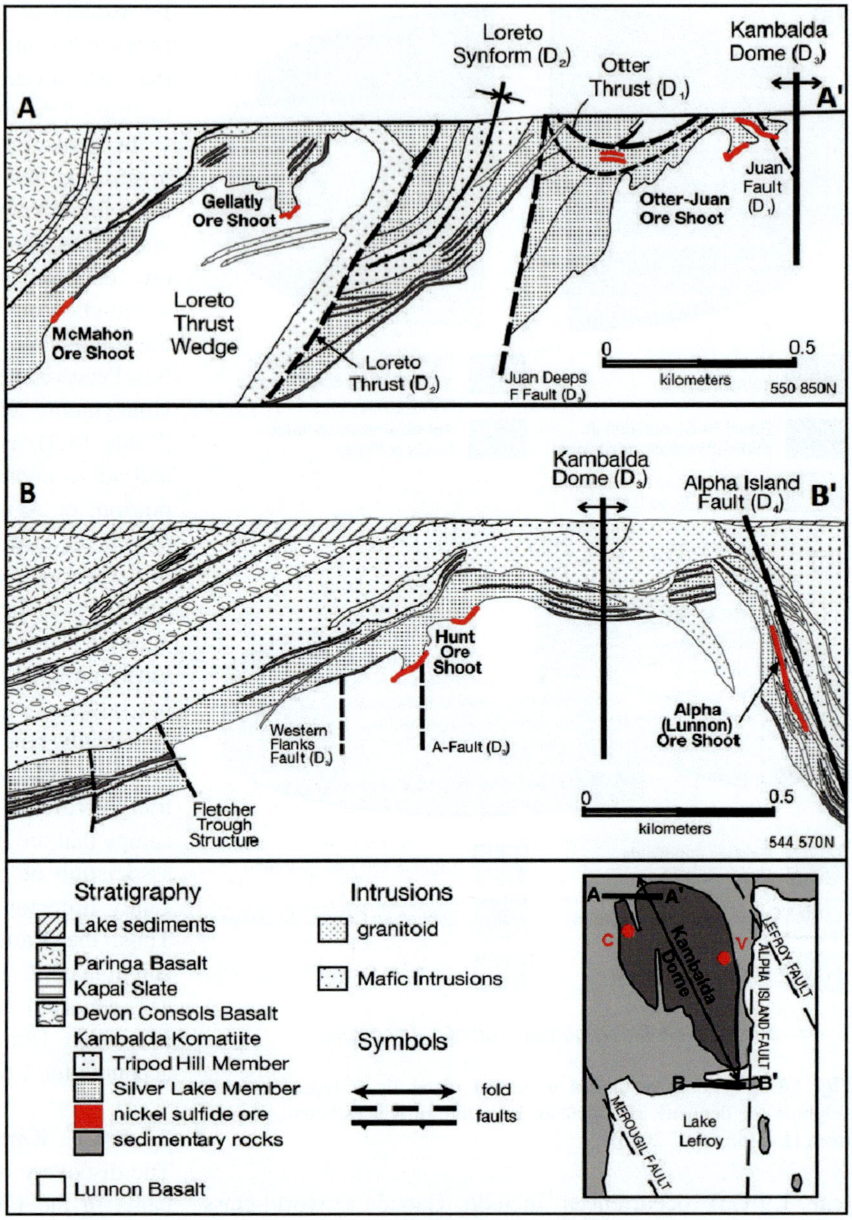

(b)

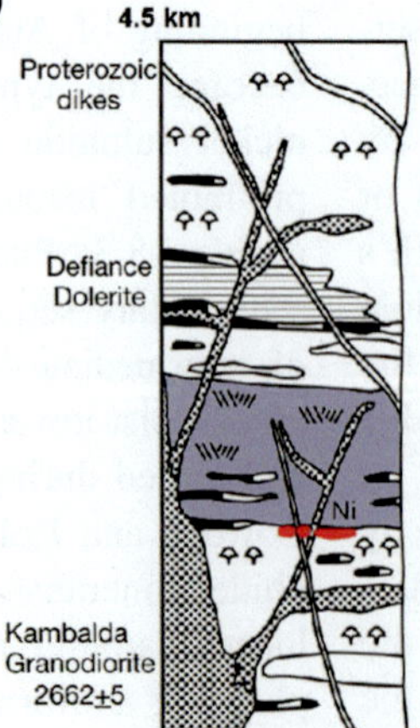

Formation	Age	Lithology
Paringa Basalt	2690±5	High-MgO pillowed and massive metabasalt
Kapai Slate	2692±2	Siliceous metasedimentary unit
Devon Consols Basalt		High-MgO variolitic metabasalt
Kambalda Komatiite	2702±4	Variably serpentinised or talc-carbonate altered metakomatiite, with intercalated siliceous metasedimentary rocks. Nickel mineralisation associated with basal units
Lunnon Basalt		A sequence of thin pillowed and massive tholeiitic metabasaltic flows and intercalated siliceous sedimentary units. Rare doleritic sills

Fig. 18.64 Structural and stratigraphic relations of the Kambalda dome. **a** Cross-sections of the northwest flank across the McMahon, Gellatly, and Otter-Juan ore shoots (west-dipping reverse faults have formed a series of wedges of the Lunnon Basalt footwall) and across the southern part of the dome through the Hunt and East Alpha (Lunnon) ore shoots. The thickness of the nickel sulphide ore shoots and sedimentary units was exaggerated for clarity. Locations of the less deformed Victor and Coronet orebodies discussed in the text are shown in the inset. (Modified from Stone et al. 2005). **b** Graphic representation of stratigraphy showing the favoured location of Kambalda-type or "*contact*" orebodies at the contact between the Kambalda Komatiite Formation and Lunnon Basalt; ages refer to U–Pb SHRIMP dating (From Beresford et al. 2005)

The stratigraphy of the Kambalda dome has been established from exploration, mining development, and detailed research over the past 50 years (Fig. 18.64). At the base of the sequence, a metabasalt unit (Lunnon Basalt) is overlain by a serpentinised and variably talc-carbonate altered komatiite unit (Kambalda Komatiite Formation). Varieties of metasedimentary rocks occur at the contact between the metabasalt and komatiite and/or are intercalated with komatiite. Contact (type 1) nickel sulphide ore is completely or partially confined within embayments or trough-like structures at the base of the thick basal komatiitic lava flow and the upper surface of the metabasalt (e.g. Gresham and Loftus-Hills 1981; Cowden and Roberts 1990; Stone and Archibald 2004; Stone et al. 2005; Figs. 18.61a, b, 18.62). There is generally an absence of sedimentary units in the troughs or ore environments, although there are numerous exceptions, suggesting sediment may have been eroded from the ore environment. Thickened sections (up to 100 m) of MgO-enriched (up to 45% MgO) komatiitic rocks in the immediate hanging wall are thought to be genetically important lava channel facies.

A long-standing issue at Kambalda has been an understanding of the spatial and genetic association of these troughs with ore. Historically they were considered to represent either primary volcanic topography (Gresham and Loftus-Hills 1981) or, alternatively, to have resulted from post-volcanic deformation (Cowden and Archibald 1987; Cowden 1988). Volcanic models of ore genesis have since emphasised the role of thermal erosion of a sulphidic sediment substrate to turbulent lava channels as the key process in ore genesis (Groves et al. 1986; Lesher 1989; Williams et al. 1998; Lesher et al. 2001; Barnes 2006; Staude et al. 2016). Consequently, exploration models emphasise dynamic lava channels or magma pathways (i.e. ore shoots) identified on the basis of (1) deep footwall embayments (trough structures), (2) the presence of thick, high-Mg komatiite channel facies flows in the hanging wall flanked by sheet flow facies (non-ore environment), (3) absence of sedimentary rocks, and (4) systematic variations in nickel tenor (Ni content in 100% sulphides).

Nonetheless, komatiite-substrate contacts have been critically reassessed at Kambalda, both volcanologically and structurally (Stone and Archibald 2004; Stone et al. 2005; Sect. 7.9.2). Detailed volcanic facies analysis undertaken in areas of low strain (in particular, the Victor and Coronet deposits) has rather suggested the majority of lower contacts between komatiite lava and sedimentary substrate are planar and rarely show evidence of erosion, or marginal autoclastis and peperite formation (Cas et al. 1999; Cas and Beresford 2001; Beresford et al. 2002, 2005; Fig. 18.65). Cas et al. (1999) proposed that the absence of sediment in trough locations could also be explained by physical erosion by turbulent komatiite flow fronts, and/or by thermally induced turbulence in the water mass immediately ahead of the advancing flow front due to heat transfer from the advancing 1600 °C, submarine komatiite lava flows. Sometimes the komatiites possess irregular and peperitic upper margins, with apophyses intruded into the surrounding sedimentary rocks that are locally soft-sediment-deformed, indicating a shallow, syn-sedimentary intrusive origin. The restriction of these syn-sedimentary intrusive bodies to domains of the thicker sedimentary substrate, the lack of komatiitic dykes, and the abundance of komatiite lava flows in the Kambalda region suggest that the syn-sedimentary intrusive bodies are of invasive flow origin. That is, they formed due to the burrowing of lava flows into the underlying wet, unconsolidated to semi-consolidated sedimentary substrate (Beresford and Cas 2001; Beresford et al. 2005). Invasive flows can be explained by the much higher density of komatiitic lava (up to 2.75 g cm^{-3}) compared to unconsolidated water-saturated pelitic sediments (1.5–2.0 g cm^{-3}). The higher density and resultant negative buoyancy caused komatiites to load and burrow into the unconsolidated sediments and should be a common feature of similar successions (Chap. 7). Variations in thickness of the Kambalda Komatiite Formation across the camp indicate that the komatiite pile either filled irregular pre-existing seafloor topography (e.g. half-grabens), and/or formed low slope, seafloor lava shield piles.

As indicated above the komatiitic host succession and, therefore, the nickel sulphide ore shoots have been variably affected by polyphase deformation. Mine-based structural analysis and reverse 3D modelling show the major controls on the shoots at Kambalda are structural and metamorphic, and the main orebodies occur in the most deformed rocks (Stone and Archibald 2004; Stone et al. 2005). However, these reflect a continuum based on the degree of deformation and metamorphic overprinting. In the least deformed deposits (including Victor and Coronet), ore shoots show some control by primary volcanic features (i.e. shallow troughs), and these are associated with mainly serpentinisation of the hanging-wall rocks. The majority (some 70%) of ore shoots occur in structurally controlled troughs where host rocks are intensely altered to talc and carbonate. The trough structures are fault-bounded, transgress ore and represent hinge zones of parasitic synclines that formed on the flanks of the Kambalda dome.

Stone et al. (2005) conclude that, following magmatic emplacement, nickel ore shoots surrounding the Kambalda dome were structurally deformed, and hanging-wall rocks altered. Strain localisation within and about mechanically weak sulphide ore and komatiite resulted in sulphide remobilisation and structural modification of original ore-hosting features. While some original trough features may have been partly preserved, their finite geometry reflects

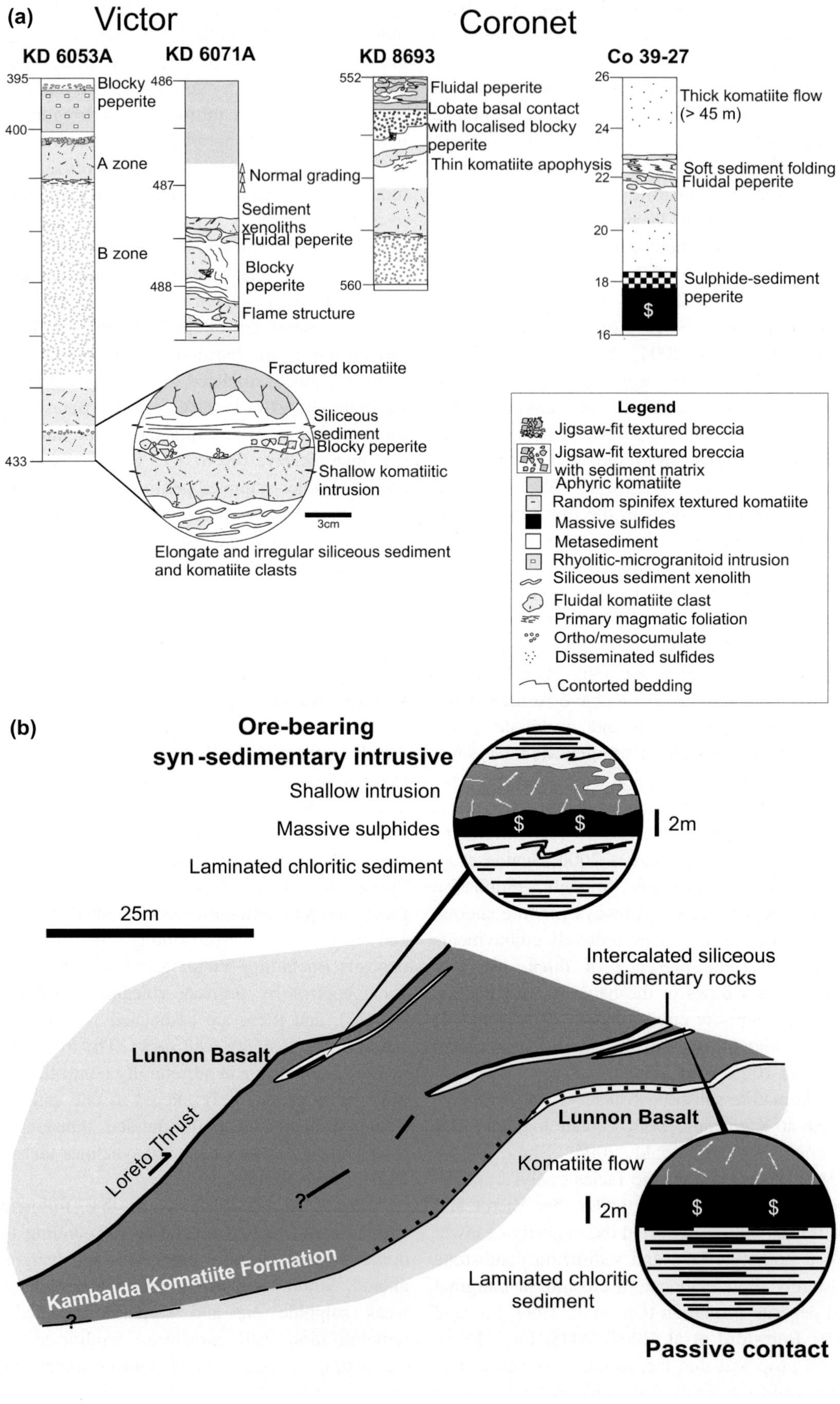
(a)
Victor
Coronet
KD 6053A
KD 6071A
KD 8693
Co 39-27
Blocky peperite
A zone
B zone
Normal grading
Sediment xenoliths
Fluidal peperite
Blocky peperite
Flame structure
Fluidal peperite
Lobate basal contact with localised blocky peperite
Thin komatiite apophysis
Thick komatiite flow (> 45 m)
Soft sediment folding
Fluidal peperite
Sulphide-sediment peperite
Fractured komatiite
Siliceous sediment
Blocky peperite
Shallow komatiitic intrusion
3cm
Elongate and irregular siliceous sediment and komatiite clasts
Legend
Jigsaw-fit textured breccia
Jigsaw-fit textured breccia with sediment matrix
Aphyric komatiite
Random spinifex textured komatiite
Massive sulfides
Metasediment
Rhyolitic-microgranitoid intrusion
Siliceous sediment xenolith
Fluidal komatiite clast
Primary magmatic foliation
Ortho/mesocumulate
Disseminated sulfides
Contorted bedding
(b)
Ore-bearing syn-sedimentary intrusive
Shallow intrusion
Massive sulphides
Laminated chloritic sediment
2m
25m
Intercalated siliceous sedimentary rocks
Lunnon Basalt
Loreto Thrust
Lunnon Basalt
Komatiite flow
2m
Laminated chloritic sediment
Kambalda Komatiite Formation
Passive contact

◀ **Fig. 18.65** Coronet and Victor areas, Kambalda, Western Australia. **a** Selected graphic logs of shallow, syn-sedimentary intrusive bodies in the Coronet and Victor areas, Kambalda; note the different scales. These illustrate irregular margins and localised peperitic contacts with apophyses of komatiite and fluidal komatiite clasts. The host sedimentary unit is commonly massive adjacent to the komatiite units, but domains of contorted or planar lamination are present up to the contacts. Structureless sedimentary rocks associated with peperites suggest fluidisation of the unconsolidated to semi-cohesive, fine-grained sediment at the time of komatiite emplacement. Jigsaw-fit textured domains are restricted to areas with large fractures, which extend into the coherent komatiite. These fractures are inferred to be quench-related, associated with blocky peperite. In the Coronet area, syn-sedimentary intrusive bodies contain 1–2 m of Ni-massive sulphide at their basal margins and are commonly overlain by a thick high-Mg flow with disseminated sulphides. This association suggests that the thick high-Mg flow is genetically related to the underlying intrusive bodies, which formed as invasive or burrowing komatiite flows. **b** Schematic cross-section through the western flank of the Coronet ore shoot (beneath the Loreto Thrust; Fig. 18.64a). This shows the position of syn-sedimentary komatiitic intrusions with Ni-massive sulphide bodies along their basal margins. Ore is also hosted by thin komatiite flows with passive lower contacts with underlying sediments. (Modified from Beresford and Cas 2001)

the regional tectonostratigraphic and metamorphic history that, in addition to volcano-stratigraphic controls, must be considered in exploration. The term "*structural trough*" should be used for fault-bounded depressions in the base of komatiites, and these should be distinguished from erosionally formed volcanic channels, for which clear field evidence must be documented (Cas and Beresford 2001; Sect. 7.9.2).

18.6.4.3 Agnew–Wiluna Greenstone Belt

In the Agnew–Wiluna greenstone belt, stratigraphic relationships are best preserved at Mt Keith, where three horizons of komatiites are present (Fig. 18.66): (1) the Mt Keith ultramafic unit, (2) the Cliffs ultramafic unit, and (3) the Monument ultramafic unit. All three ultramafic units are highly deformed with steep to vertical dips and isoclinal folding. The three ultramafic units are separated by a variably deformed sequence of felsic and mafic rocks, ranging in composition from dacite to Mg-rich tholeiitic basalt. The Mt Keith and Cliffs ultramafic units are mineralised, whereas the Monument ultramafic unit is regarded as barren and unprospective. Outside the Mt Keith area only two ultramafic units occur, a basal cumulate-rich unit (equivalent to the Mt Keith ultramafic unit) and an overlying spinifex-bearing komatiite unit (Fig. 18.66).

The Mt Keith ultramafic unit comprises >500 m-thick spinifex-free, olivine adcumulate-textured dunite (>90% olivine) pods and lenses, which are flanked by laterally extensive ortho- and mesocumulate-textured units (see Fig. 7.15). Disseminated nickel sulphides are concentrated in the lensoidal areas, which include the very large, low-grade type 2 Mt Keith MKD5 deposit (Figs. 18.61e, 18.67), as well as the Yakabindie and Honeymoon Well deposits (Table 18.3). This represents a strike length of >80 km for the single komatiite unit. Controversy however surrounds the precise volcanological setting (Sect. 7.7.3). The dunite lenses have generally been interpreted as integral parts of extrusive lava flows, coinciding with komatiite thickening (Barnes et al. 1998; Barnes 2006). More recently, these have been reinterpreted as parts of very thick intrusive sills (Rosengren et al. 2005, 2007, 2008; Duuring et al. 2012; Fig. 18.67, and see Figs. 7.16, 7.17). This is based on the very coarse and thick adcumulate development, requiring long slow cooling and complete crystallisation, the extraordinary thickness given the magma had the viscosity of olive oil, and the presence of apophyses and peperite along the top margin; nonetheless, arguments to the contrary persist (Gole et al. 2013; Sect. 7.7.3).

Conversely, the Cliffs and the Monument ultramafic units consist of multiple, typical, texturally differentiated, spinifex-bearing komatiite lava flows with basal cumulate zones. The Cliffs ultramafic unit is locally more than 150 m thick and comprises a sequence of differentiated olivine cumulates and spinifex-textured flow units. The basal unit is the thickest and contains basal (type 1) massive nickel sulphide mineralisation, including the moderate Cliffs deposit (Table 18.3). The Monument ultramafic unit comprises a series of spinifex-bearing thin flows (i.e. <5–10 m thick), which are thought to have been deposited in a more distal setting, rather than a more prospective, proximal-to-vent location (Fig. 18.58).

The geology of the Perseverance area is dominated by the Perseverance ultramafic complex, which comprises a central nickel sulphide-poor, 700 m thick dunite lens stratigraphically underlain by a nickel sulphide-bearing sequence dominated by layers of olivine orthocumulate and adcumulate. The stratigraphy is complexly poly-deformed, dips steeply, and is overturned. The ultramafic complex has been interpreted as a thick sequence of komatiite flows that eroded underlying footwall rocks until they stratigraphically overlay a nickel sulphide-bearing komatiite (Barnes et al. 1988). However, these komatiite sequences show discordant relationships, intrude compositionally similar rhyodacitic hanging and footwall country rocks, and are reinterpreted as a subvolcanic sill (Trofimovs et al. 2003; Duuring et al. 2010). Mineralisation at the Perseverance mine is strongly remobilised, and evidence of primary volcanic control is difficult to decipher. Ores in contact with komatiite at or close to the dunite lens margin are controlled by mylonites developed in more ductile rocks along the footwall margin. As a result of progressive folding and shearing of the footwall contact, massive and disseminated primary nickel sulphide orebodies

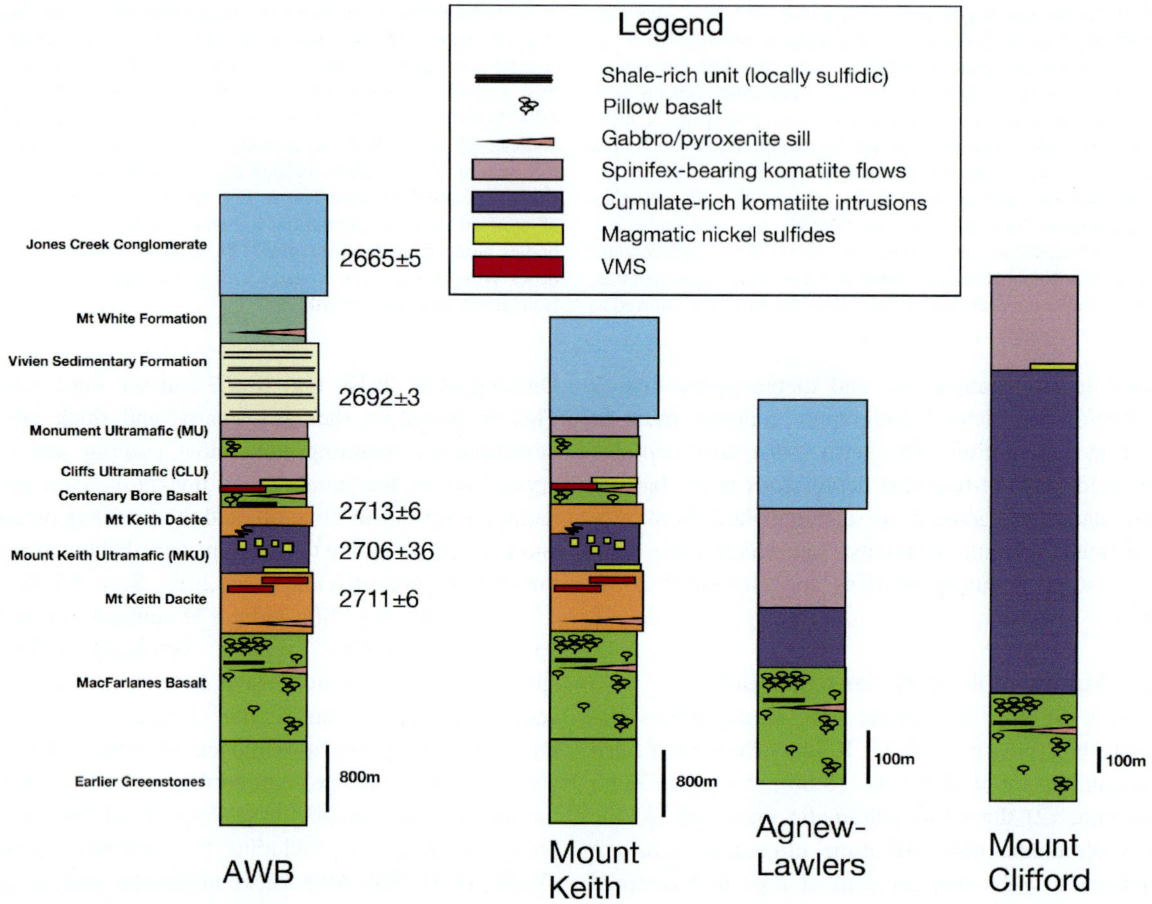

Fig. 18.66 Generalised stratigraphic sections for the Agnew–Wiluna greenstone belt (left) and selected domains as labelled, Western Australia. The Agnew–Wiluna greenstone belt comprises a distinctive, but not unique, trimodal volcanic association of felsic, mafic, and ultramafic rocks. Stratigraphic relationships are best preserved in the Mt Keith Domain which contains three komatiitic horizons: (1) the Mt Keith ultramafic unit (MKU), (2) the Cliffs ultramafic unit (CLU), and (3) the Monument ultramafic unit (MU). These are variably mineralised, the most important being the Mt Keith ultramafic unit which comprises >500 m-thick, spinifex-free, adcumulate-textured pods and lenses. Disseminated nickel sulphides occur interstitial to former olivine crystals and are concentrated in the lensoidal areas, which have generally been interpreted as integral parts of extrusive lava flows, coinciding with komatiite thickening. More recently these have been reinterpreted as parts of very thick intrusive komatiitic sills (see text for discussion, and Fig. 18.58). (From Fiorentini et al. 2012; for location, see Fig. 7.15)

were physically remobilised into multiple generations of fold hinges and shear zones. Massive sulphides were physically remobilised up to 20 m from their primary footwall position into deposit-scale fold hinges to form orebodies.

Lastly, it is important to consider why nickel endowment within the Norseman-Wiluna greenstone belt (or Eastern Goldfields Superterrane) is considerably richer than that of any other known komatiite belt (Fig. 18.63). A notable feature and, indeed, of many komatiite-hosted nickel areas worldwide appears to be a close spatial association of large deposits with felsic volcanic rocks (Rosengren et al. 2008; Fiorentini et al. 2012; Table 18.3), while it has previously been acknowledged the origin and timing of crustal sulphur addition are likely critical to generate world-class deposits. In the case of the Agnew-Wiluna greenstone belt, multiple, separate (at least two and locally three) horizons of komatiite are intercalated with coeval dacitic lavas and volcaniclastics, as well as barren (pyritic) volcanic massive sulphide lenses and sulphidic carbonaceous shales (Fig. 18.66). Multiple sulphur isotope data now suggest extensive assimilation occurred during the passage of komatiite magma through the seafloor substrate and upon emplacement, whereby hot, turbulent magma thermo-mechanically eroded and assimilated volcanic massive sulphides located within or close to their vents, and is thought to be the primary sulphur source for larger, higher tenor nickel deposits (Fiorentini et al. 2012; Chap. 7). In more distal areas, flows and sills interacted rather with organic-rich sulphidic shales resulting in smaller, lower tenor deposits. This also implies sulphide-bearing felsic lava domes utilised the same crustal conduits and mark the same proximal-to-vent areas as prospective komatiitic flows and sills. The lateral persistence

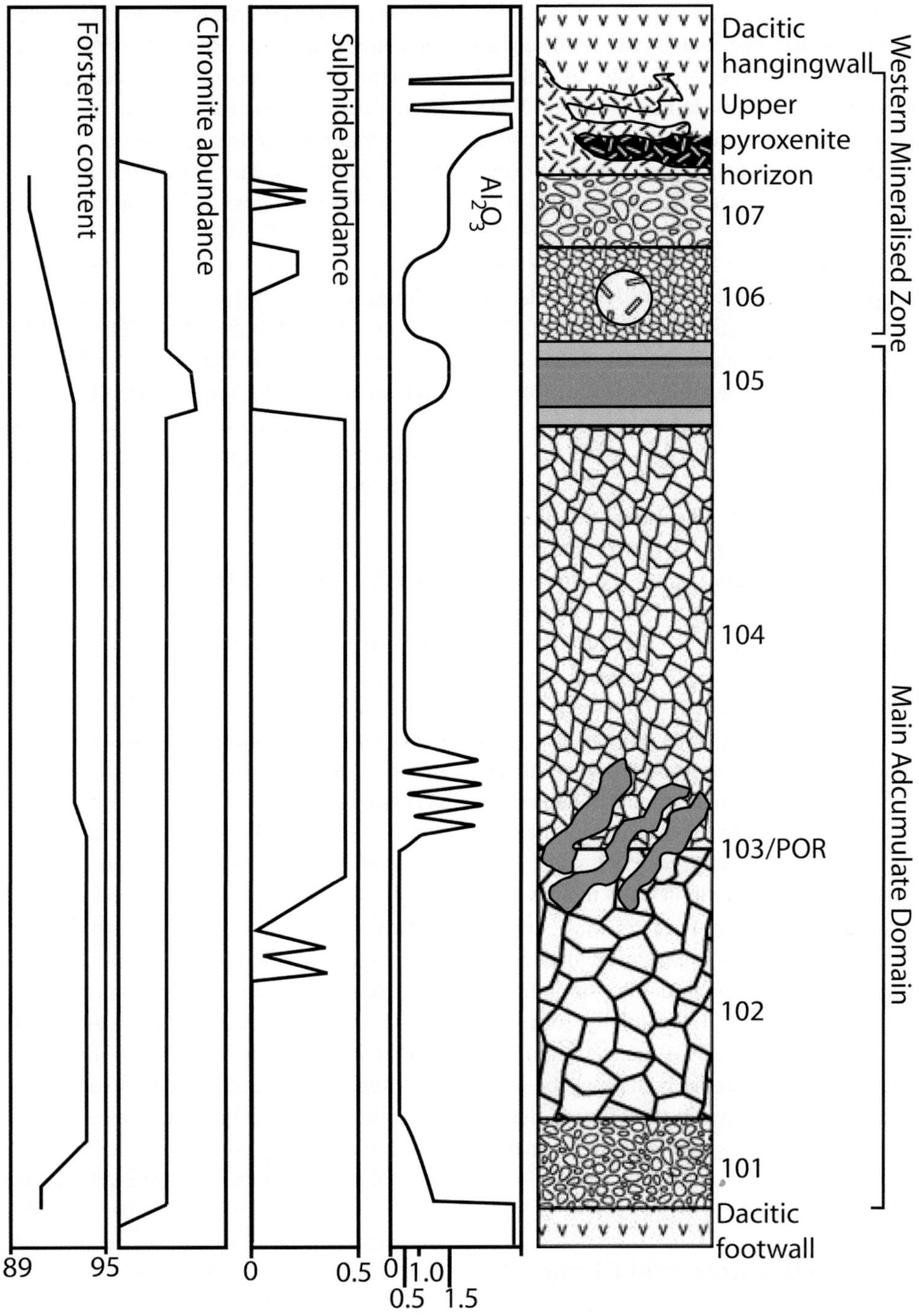

Fig. 18.67 Stratigraphic column for the Mt Keith ultramafic unit, Mt Keith open pit. This shows internal textural subdivisions, schematic traces for Al_2O_3, sulphide and chromite abundance, and forsterite content; the total thickness represented varies between about 300–600 m. The trace of Al_2O_3 reflects the increasing amount of interstitial melt phase, generally associated with a gradual transition from olivine adcumulate to olivine orthocumulate. The main disseminated MK5D orebody is hosted within the major zone of adcumulate (unit 104), which is located approximately in the centre of the komatiite body. Unit 103 (POR) cross-cuts units 102 and 104 and represents late-stage ultramafic pegmatites (visible as peaks on the Al_2O_3 profile). Distinct orthocumulate textures are discernible within 1 m of the hanging-wall contact. The final metre of ultramafic rock between the cumulate zone and the hanging-wall contact consists of very fine-grained to aphyric serpentinite, with no evidence of shearing or other deformation-induced textural modification. The hanging-wall lithology above the Mt Keith ultramafic unit contact is dacitic in composition. Texturally it is an in situ jigsaw-fit breccia with visible (3-mm) phenocrysts of quartz and plagioclase, which grades into a massive variably feldspar porphyritic and amygdaloidal dacite body. This is interpreted as quench fragmented hyaloclastite grading into coherent dacite sill (Chap. 7). (From Rosengren et al. 2007)

of proximal felsic volcanic facies along the axis of the Agnew-Wiluna greenstone belt suggests it may represent an inverted paleo-rift. Within this framework, the spatial and genetic association among felsic domes, related VMS accumulations, and komatiite channels is largely controlled by the same syn-rift architecture. The Agnew–Wiluna komatiites may have been emplaced at exceptionally high eruption rates, giving rise to turbulent erosion and cannibalisation of hydrothermal sulphide-rich wall rocks and proximal channelised flows and sills generally regarded as essential for forming large deposits (see Fig. 18.58). The Kambalda camp, lacking significant coeval felsic volcanics, however, appears a very, notable exception to this model.

18.6.4.4 Extra Terrestrial

As well as the possible eruption of ultramafic magmas during the 1997 activity of Io (Chap. 7), a wide number of data sets (e.g. orbiting satellite, Mars Lander and Rover, and Martian meteorite) indicate that the Martian geological record comprises widespread igneous activity showing striking mineralogical, petrographic, and chemical similarities with terrestrial komatiites. In an interesting study, Baumgartner et al. (2015) have evaluated whether such igneous activity may have led to the formation of nickel sulphide mineralisation. Given the high sulphur inventory of Martian crustal reservoirs, ranging from sulphide-bearing magmatic rocks to sulphate-rich soils, regolith, and sediments, it is likely that metal-rich mantle-derived melts would have assimilated significant amounts of crustal sulphur during ascent and emplacement. It is proposed that channelised lava flows emplaced and incised into sulphur-rich regolith and crustal lithologies could potentially form Ni–Cu–PGE orebodies. However, while PGE concentrations in Martian and terrestrial primitive mantle melts may be comparable, lower Ni and Cu abundances in the Martian mantle and derived melts may be a limiting factor to the formation of solar system-class deposits.

18.7 Kimberlites, Lamproites, and Diamonds

Kimberlites and lamproites are again rare magmas, but economically important as they transport diamonds to the Earth's surface. Diamonds are not magmatic crystals, but xenocrysts derived from ultra-metamorphism deep in the Earth's sub-cratonic mantle, having formed some 1–2 billion years after planetary accretion from primordial carbon, then scavenged and carried by kimberlite magmas. Kimberlite magmas are considered to have had very low viscosities and risen at very high velocities to the Earth's surface. From a wider perspective, some diamonds are natural antiques formed in pre-solar supernovae by carbon vapour deposition and in asteroidal impacts and meteorite craters by shock metamorphism (e.g. Haggerty 1999). Diamond is therefore an extraordinary recorder of astrophysical and geodynamic events that extend from the far reaches of space to Earth's deep interior. Diamonds were first recorded in India 4,000 years ago, but it was not until much later that discoveries were made in Brazil (1720) and South Africa (1866). Major commercial producers today are Africa, Russia, Australia, and Canada. Over 500 metric tonnes of diamonds have been mined worldwide since 1900, and forecasts for 2020 are about 24 t or 120 million carats (1 carat = 0.2 g), significantly lower than a peak of nearly 180 Mct in 2005. Primary diamonds are mostly located in pipe-like volcanic structures, or diatremes, which when eroded may result in placer (alluvial, beach) resources only briefly mentioned here.

18.7.1 Kimberlite and Lamproite Magmas

Kimberlite was the term given to the diamond host rock at the type locality of Kimberley, also known as "the Big Hole", discovered in South Africa in 1871. Many kimberlite occurrences are known worldwide; although some are diamondiferous, the greater majority (80%) are barren, with perhaps as few as 1% ever becoming mines. Kimberlite was considered the only important primary source of diamonds for over a century. This was until the 1979 discovery of the Argyle (AK1) deposit in lamproite host rocks in Western Australia, which has been the world's richest primary diamond mine in history (see Table 18.4). These two magma types represent the only two commercial sources of primary diamonds. Both magma types are products of intraplate continental alkaline volcanism. Indeed, many kimberlites and lamproites occur in geographic and age clusters, much like small intraplate basaltic monogenetic volcanic provinces and fields (Brown and Valentine 2013). When compared to other alkaline rocks, they are relatively small in volume. Reviews on the nature of kimberlite and lamproite magmas and rocks are provided by Mitchell (1986, 1995), Bergman (1987), Mitchell and Bergman (1991), Russell et al. (2012) and Giuliani et al. (2023).

Kimberlite is ultramafic and has extremely low SiO_2 content (<35 wt%), very high MgO (>15 wt%), extremely low Al_2O_3 (<5 wt%), and very low Na_2O (<0.5 wt%) and Na_2O/K_2O (<0.5). Lamproite is an undersaturated, ultra-potassic basaltic magma, with those hosting diamonds generally olivine lamproite. Apart from diamond, the main minerals found in kimberlite are forsteritic olivine, chrome pyrope garnet, chrome diopside, and spinel. Some kimberlites also contain phlogopite. However, kimberlites contain a "mantle cargo" comprising large volumes of xenoliths and xenocrysts, including most of the olivine population and

Table 18.4 Selected features of some of the world's largest producing diamond mines and others referred to in the text

Deposit	Discovery	Age (Ma)	Crustal setting	Structural control	Magma Composition	Volcanic Feature	[a]Reserves/ *Total resource* (Mct)	Grade (ct/t)
Ekati, Lac de Gras, Northwest Territories, Canada	1991	55–51	Central Slave Craton	Margin alkaline dyke corridor (010°) coincides mantle-depth structure	Kimberlite	Cluster of 6 pipes	*170*	4.71–0.4
Diavik, Lac de Gras, Northwest Territories, Canada	1994	56–55	Central Slave Craton	Margin alkaline dyke corridor (010°) coincides mantle-depth structure	Kimberlite	Cluster of 4 pipes	46	4.78–2.7
Orapa, Botswana	1967	93	Zimbabwe Craton	Margin of Paleoproterzoic mobile belt	Kimberlite	2 converging pipes	150	0.75
Catoca, Angola	1965	~90	Congo Craton (Lucapa Graben)	Major structural corridor traced the Mid-Atlantic transform	Kimberlite	Pipe	110	0.6
Kimberley, South Africa	1871	87	Kaapvaal Craton (Ventersdorp Graben)	Intersection major NE-basin-forming fault and NW-transfer fault	Kimberlite	3 coalescing pipes	*≥50*	1.0
Star-Orion South, Fort à la Corne, Saskatchewan, Canada	2000	103–95	Glennie Domain (Trans-Hudson Orogen)	Major high-grade collisional zone with reworked Archean crust	Kimberlite	Volcano-sedimentary complex with discordant bodies (feeder pipes)	*66*	0.14
Jericho, Nunavat, Canada	1995	173	Northern Slave Craton	Adjacent (010°) alkaline dyke and potential mantle depth structure	Kimberlite	Complex pipe with 4 coalescing lobes (vents)	6	0.84
Jwaneng, Botswana	1974	235	Kaapvaal Craton	Deformed Paleo–Mesoproterozoic basin	Kimberlite	3 pipes	150	1.3
Yubileinaya (Jubilee), Yakutia, Siberia, Russia	1975	358	Siberian Craton (Kotuykan shear zone)	Margin Paleoproterozoic Kotuykan suture zone	Kimberlite	Pipe plus 2 satellite bodies	*170*	0.5
Udachnaya, Yakutia, Siberia, Russia	1955	376	Siberian Craton (Kotuykan shear zone)	Margin Paleoproterozoic suture zone	Kimberlite	2 coalescing pipes	*225*	1.3
Mir (Mirny), Yakutia, Siberia, Russia	1955	360	Siberian Craton (Kotuykan shear zone)	Intersection Kotuykan suture zone and Yiluyi aulocogen	Kimberlite	Pipe	100	2
Venetia, South Africa	1980	519	Limpopo Belt (between Zimbabwe and Kaapvaal Cratons)	Major high-grade collisional zone with reworked Archean crust	Kimberlite	Cluster of 14 pipes and dykes	*170*	1.2
Snap Lake, Northwest Territories, Canada	1997	523	Southern Slave Craton	Margin alkaline dyke corridor (015°) coincides mantle-depth structure	Kimberlite	Multi-phase sill/dyke complex	28	1.46
Argyle, Western Australia	1979	1,178	Halls Creek Mobile Zone	Major Paleoproterozic sinistral fault zone	Olivine lamproite	Pipe	*865*	6.8–3.0
Guaniamo, Venezuela	1982	1,732	Guyana Craton	Regional deep-seated faults, megalineaments	Kimberlite	Dykes, sills	*>30* (in alluvials)	

[a] Total resource information (past rough diamond production plus reserves) is generally difficult to obtain for many diamond mines, so where unavailable an indication of "size" is expressed here as present reserves; one million carats = 0.2 tonne (t). The grade for economic diamond deposits ranges from 0.02 to 10 carats per tonne (c/t) and is dependent on many variable factors; unlike metals, there is not a single value for diamonds. This summary is based on multiple sources including current global industry reviews, company reports, and text references

diamonds. Kimberlite volcanism involves the ascent of low viscosity (0.1–1 Pa s) and volatile-rich (CO_2 and H_2O) ultramafic magmas from depths of ≥ 150 km (Sparks et al. 2006; Sparks 2013; Russell et al. 2012; Jones et al. 2019). These travelled rapidly and turbulently through the lithosphere in a matter of hours, reaching the surface at speeds in the range >4–20 m s^{-1}; a necessity to inhibit the transformation of diamond to graphite or the combustion of diamond to a carbon gas species. Although there is much debate about the genesis of kimberlite magmas, it is now thought kimberlite magmas originated as carbonatitic melts which assimilated mantle xenoliths, especially orthopyroxene-bearing xenoliths, thereby transforming them into silicate compositions (Russell et al. 2012). Assimilation releases large amounts of carbon dioxide, giving the magmas extra buoyancy which helps them accelerate to the Earth's surface. Some kimberlite and lamproite magmas were hydrous, as evidenced by the presence of phlogopite, so H_2O also likely contributed to the ascent and eruption of the magmas (Sokol et al. 2013). Very few fresh kimberlites have been analysed, and the actual volatile content at the time of eruption is unknown.

18.7.2 Geological Setting

18.7.2.1 Cratons and Diamond Formation

By the early 1960s, it had been recognised that African diamondiferous kimberlite pipes were restricted to older Archean cratonic areas. An empirical observation that was to direct diamond exploration worldwide for many years. While the discovery of the Argyle deposit in 1979 was a result of routine stream sampling for diamond indicator minerals, nonetheless important was its location at the margin of a craton, within a Proterozoic mobile belt. Economic kimberlites are still considered to be restricted to Archean cratons, but diamondiferous lamproites occur both within and outside cratonic margins. This one, albeit a major discovery, effectively expanded the potential areas for diamond exploration from off the Archean cratons, to include adjacent Proterozoic and even younger terrains. Nonetheless, the proximity at the surface or in the subsurface of Archean or Paleoproterozoic cratonic crust, and hence subcratonic mantle, is an important first order guide to finding kimberlites and lamproites, and especially diamondiferous ones.

The formation of diamond-bearing magmas relates uniquely to the earth's sub-cratonic mantle. Studies of mantle xenoliths and diamond xenocrysts, together with experimental petrology, indicate that such magmas originate from depths of at least 150 km from the base of the cratonic lithosphere, or deeper. Beneath cratons, the lithospheric mantle may extend to depths of 250–300 km. Because of the downward protruding shape and the long-term attachment of this mantle to the continental crust of the craton, this portion of the mantle has been termed "mantle keel" (Fig. 18.68). Critically most of this region lies within the diamond stability field (Haggerty 1994, 1999; Shirey et al. 2013). Archean lithospheric keels are more melt-depleted and deeper than Proterozoic and younger continental mantle keels, and therefore are the most, if not only, fertile diamond source region. As well as within the mantle keel, under the right fO_2, diamonds can form in the convecting mantle and the subducting slab, but with few exceptions, diamonds are erupted only by kimberlite and lamproite magmas. Most diamondiferous magmas were emplaced into the crust that is >2.5 Ga old, some lamproites intrude crust formed from 1.6 to 2.5 Ga, but no primary diamond deposits are recognised intruding crust younger than 1.6 Ga.

Another remarkable attribute of diamonds is their age. Although diamonds cannot be directly dated, the ages of mineral inclusions in diamonds can be determined radiometrically. These data imply that diamonds are about 3.5–1.5 Ga old and geologically ancient (Gurney et al. 2010). In contrast, most of the host magmas which brought diamonds to the surface are considerably younger (<50 million years to about 1.2 Ga). Thus, the age and compositions of mineral inclusions imply that most diamonds were formed in an early event and then stored in the mantle keels of cratons for long periods of geological time. Eclogite and peridotite are regarded as the main parental rocks of diamonds. Loose monocrystalline diamonds seen in kimberlite are considered to have been released from eclogitic or peridotitic rocks by alteration and mechanical disaggregation during entrainment by kimberlitic magma in the lithosphere, or early transport by the kimberlite (Gurney et al. 2010; Shirey et al. 2013). Both diamondiferous and diamond-free eclogites often survive transport by kimberlite, whereas diamondiferous peridotite is exceptionally rare, and nearly all peridotite xenoliths are diamond-free.

18.7.2.2 Superchrons and Superplumes

It has been increasingly recognised that some xenoliths and mineral inclusions in diamonds have origins close to or in the mantle transition zone (Haggerty 1999; Shirey et al. 2013; Fig. 18.68). This requires that magmas transporting such samples of superdeep or sub-lithospheric diamonds must originate either in the transition zone or lower mantle (>660–2,900 km). Several compelling arguments now suggest that diamondiferous magmas may be related to plumes that sample the lower and upper mantle and are emplaced synchronously on a global scale (Haggerty 1994; Torsvik et al. 2010). Plumes may transgress the entire vertical extent of the mantle, possibly as deep as the core–mantle boundary at 2,900 km. In so doing, they have the opportunity to sequester substantial concentrations of incompatible

Fig. 18.68 Schematic relationship between a continental craton, its lithospheric mantle keel and diamond stable regions in the keel, and the convecting mantle. Under the right fO_2, diamonds can form in the convecting mantle, the subducting slab, and the mantle keel. G = graphite, D = diamond, and LAB = lithosphere/asthenosphere boundary. (From Shirey et al. 2013, After Tappert and Tappert 2011)

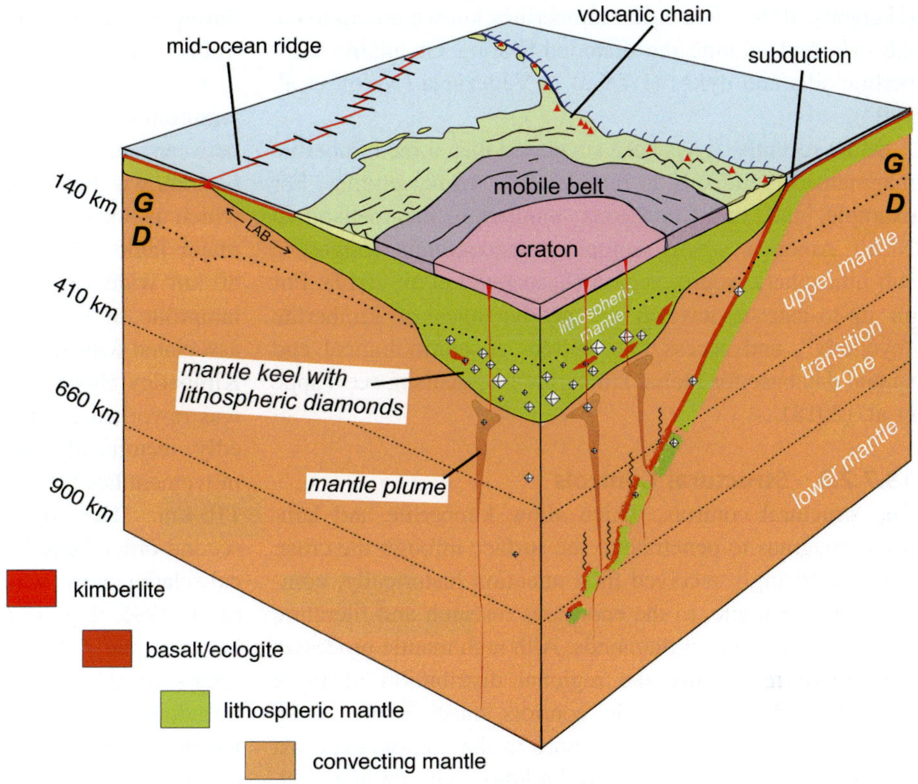

elements, including carbon, along the extensive plume path. Persistent, sub-lithospheric magmatism could therefore potentially add diamonds to the interior of cratons from below. The true amount of diamond crystallising at depth below the lithosphere is not known, but possibly is much larger than what has been erupted in kimberlites and lamproites (Shirey et al. 2013). The oldest diamonds dated are thought to be intimately associated with the first plume event that produced depleted mantle keel, either during plume upwelling and attendant melting, or some associated process of recycling or advective thickening.

Plumes have been correlated with long periods of persistent, unidirectional behaviour of Earth's geomagnetic field (or *superchrons*), and both are considered to be related to the emplacement of kimberlite and lamproite and their closely allied carbonatitic magmas. Several major diamondiferous eruptive episodes occurred between approximately 1 Ga and 50 Ma, and at least five of these globally synchronous episodes correlate with superchron events of Earth's magnetic field (Haggerty 1994, 1999). The theory holds that on disruption of the core–mantle boundary, heat is released to the mantle creating more vigorous convection in the liquid outer core. This in turn causes instabilities and lower intensities of the geomagnetic field, resulting in constant directions rather than high frequencies in reversal rates. The diamondiferous events were often preceded by or are coincident with the eruption of large igneous provinces of voluminous basalt (both continental flood basalt provinces and oceanic lava plateaus) ascribed to extensive melting of the mantle by plumes.

The youngest of these episodes (early Tertiary) is prominent in Canada (Table 18.4). The second in the mid–late Cretaceous (120–80 Ma) is the most widespread and well documented, with diamond-bearing kimberlites found extensively throughout Africa, North America, Brazil, and Siberia. This episode is the largest known diamondiferous global event and was coincident with the break-up of the Gondwanaland supercontinent. Rifting was initiated at about 150–140 Ma followed by the extrusion of continental flood basalts and oceanic lava plateaus and increased mid-ocean ridge volcanism, which also resulted in global oceanic anoxia (e.g. Kerr 2005; Sect. 16.10.2). Other episodes are recorded in the Jurassic (Botswana, Canada, Swaziland, and Tanzania), Permo-Triassic (Siberia), Devonian (Siberia and North America), and Ordovician (Russia, China, Canada, South Africa, and Zimbabwe). There are also prominent populations of kimberlites at 1.2–1.1 Ga in all major diamond provinces, including the Argyle lamproites which have an emplacement age of 1.18 Ga. These are broadly contemporaneous with the final break-up of the Columbia supercontinent at about 1.3–1.2 Ga and similar in age to the Keweenanwan flood basalts and mid-continental rifting in North America. Although the geomagnetic data are ill-defined, this episode is predicted to be a superchron

(Haggerty 1994). The oldest kimberlites known are up to ca. 2.85 Ga and include the diamond-bearing Guaniamo kimberlitic sills and dykes (1.7 Ga) in Venezuela (Nixon et al. 1992).

More recently, it has been suggested that some kimberlite magmatism is unlikely to be related to mantle plumes. For example, the mid-Cretaceous kimberlite magmatism in North America where isotopic considerations suggest a sub-lithospheric source region (Kjarsgaard et al. 2017). For an up-to-date review on the geodynamics of kimberlite "cyclicity", and access to a current geochronological and spatial database for global kimberlite magmatism, see Tappe et al. (2018).

18.7.2.3 Structural Controls

The structural controls, which allow kimberlite and lamproite magmas to penetrate to the surface through the crust, have surprisingly received little attention historically, compared, for example, to the enormous research and literature on the petrogenesis of diamonds. Although mantle processes may serve to localise the regional distribution of these magmas, the main tectonic controls must involve deep fractures, faults, and/or shear zones in the continental crust that potentially travel deep into the lithosphere (White et al. 1995; Jelsma et al. 2009). For many years, diamond geologists (with the exception of Russian) largely ignored the potential exploration significance of such structures, and even the existence of deeply penetrating structures within the continental lithosphere was controversial. This however changed in the late-1980s with the availability of deep seismic reflection data, which effectively showed such structures are commonly imaged.

For some time, it has been proposed that linear zones of alkaline intrusions (carbonatites and kimberlites) in West Africa occurred along the on-land extensions of oceanic transform fracture zones (Williams and Williams 1977). Similar relations have been found elsewhere, such as with Jurassic kimberlites in South Australia, but it seems rather than the transform fracture simply reactivating the old structure, it is the old continental structures that may control the position of the younger transforms (Lister et al. 1986). Thus, carbonatites of the Pan African Damara mobile belt appear spatially related to a mid-Atlantic fracture zone. The Damara mobile belt is a major, long-lived (Proterozoic) zone of crustal weakness that can be traced over 2000 km across Africa, with a counterpart in Brazil. While a relationship between present-day oceanic transforms and old continental faults and mobile belts is apparent for Mesozoic kimberlites and lamproites, not all occur on the continuations of oceanic transforms. In the Miocene Ellendale field (southwest of the Argyle mine), sub-economic lamproites are associated with a Paleozoic–Mesozoic rift basin centred above an older Proterozoic mobile belt. Here reactivated basement faults during basin extension, and especially transfer faults, controlled magma emplacement (White et al. 1995; Fig. 18.69).

It is clearly not possible to relate older kimberlites and lamproites to oceanic transforms. However, a relationship between mobile belts and these magmas exists, the best-known being at Argyle, in northwestern Australia, which is located within the Halls Creek Mobile Zone, a major fault zone over 1000 km long and between 50 and 80 km wide (Fig. 18.69). As well as hosting the Argyle lamproite, a number of kimberlitic pipes and dykes are associated with splays along its northwestern border with the Kimberley Block. It is a fault zone in which the basement was reworked in the Paleoproterozoic and has been repeatedly tectonically active into post-Devonian times. The movement has been dominantly sinistral, cumulatively some 110 km. The internal structure is characterised by second-order faults/shears with consistent Riedel geometrical relationships, which is well displayed at Argyle (White et al. 1995; Fig. 18.69c). Here the fault zone is confined between the Greenvale and Halls Creek Faults. Internal shears fit an overall Riedel pattern with the Argyle pipe intruding an area of transtensional Riedel shears (Carr Boyd, Revolver, and Glenhill Faults), which produced local extension during sinistral strike-slip faulting. The root of the pipe itself sits in a pull-apart zone at the intersection of the Razor Ridge Fault and the extension of the Glenville Fault through Devonian cover sediments and occurs in Mesoproterozoic sediments. Kimberlite bodies along the border with the Kimberley Block are probably younger (ca. 800 Ma), emplaced during a period of younger sinistral movement.

In their review of structural controls of kimberlite and lamproite emplacement, White et al. (1995) concluded two types of regional structures were important: (1) major mobile belts or fracture/fault corridors, representing long established features that have been frequently reactivated, and (2) linear rifts or grabens and aulacogens sited above major mobile zones, which again show frequent reactivation. The connection is that they are deep structures, possibly trans-lithospheric. The rule of thumb is deep structures are also likely to be long, major regional faults in plan view. They are intruded by kimberlites and lamproites when transcurrent or extensional deformation occurs, or a combination of both. This usually follows a period of major oceanic or continental rifting, but not orogenic collapse (Chap. 17). Diamondiferous kimberlites occur when there is an association of the above and Archean crust within an area of the thickened lithosphere. Diamondiferous lamproites occur where the old basement has been tectonically reworked in the early Proterozoic. Mesozoic diamondiferous kimberlites are often associated with continental structures that give rise to oceanic transform fracture zones; these may display contemporary kimberlite magmatism in separate continents after a break-up (Jelsma et al. 2009).

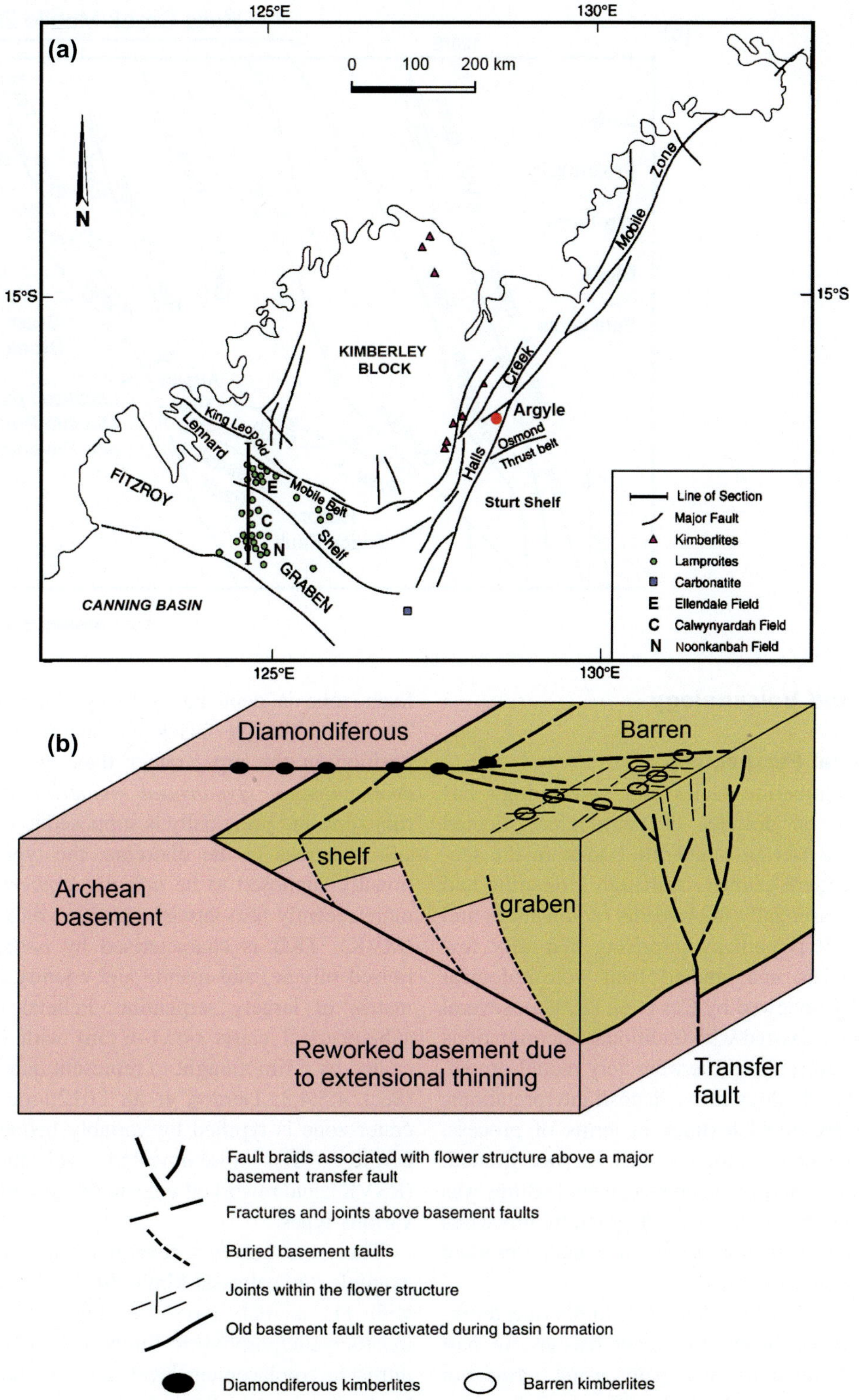

Fig. 18.69 Structural setting of kimberlites and lamproites of the Kimberley region, northwestern Australia (Adapted from White et al. 1995). **a** Regional tectonics and location map. **b** Schematic cross-section illustrating potential relationships between basin faults, basin structures, and kimberlites in a rift basin setting as based on geological relationships observed along the line of the section indicated in (**a**). The cartoon shows that as the depth of sediment increases, basement faults or shears become less distinct and more dispersed, and consequently, the kimberlite (or lamproite) clusters also become more dispersed**.** **c** Internal fault architecture within the Halls Creek Mobile Zone in the environs of the Argyle lamproite pipe. The pipe is closely associated with motions along the small Razor Ridge fault and an extension through sediments of the Glenhill fault; these form part of a wider sinistral Reidel array affecting the Halls Creek Mobile Zone as a whole (White et al. 1995)

Fig. 18.69 (continued)

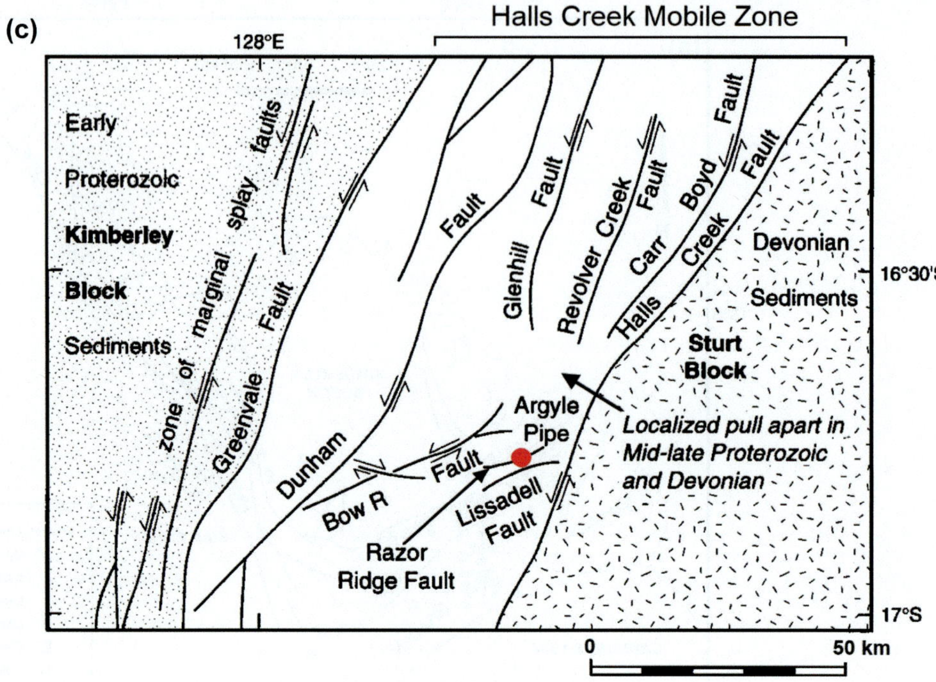

18.7.3 Facies and Volcanology

18.7.3.1 Historical Perspective

Understanding of kimberlite facies and volcanology has been based mainly on decades of petrological-oriented research on southern African kimberlite bodies in the second half of the twentieth century. Although kimberlite and lamproite pipes are essentially the remains of volcanoes and their vent systems, it is perhaps surprising that very few physical volcanologists had studied their volcanological aspects. As critically reviewed by Cas et al. (2008a), several significant issues have existed with traditional interpretations of their volcanic geology: (1) these were very model-driven, (2) highly genetic terminology drove deposit interpretations, (3) understanding preserved textures in terms of process, especially in fragmental kimberlite facies, was limited, (4) effects of alteration on original emplacement textures was underestimated, and (5) some inferred eruption processes and deposits models were not based on actual, observed modern volcanological processes.

As highlighted previously (Chap. 14), kimberlite terminology can be very confusing for many reasons, in part because it evolved down its own path, using terms not commonplace in mainstream physical volcanology (see also Sparks et al. 2006; Cas et al. 2008b, 2009; Sparks 2013). Three kimberlite pipe models have been proposed. The traditional, southern African pipe model (or Class 1) consists of a steep tapering pipe with three morphological parts: a deep, relatively narrow root zone; a middle, steep-sided diatreme zone; and an upper, shallow, flared crater zone (Fig. 18.70). Each zone is depicted as being dominated by distinctive "facies", although facies are incorrectly named by their position in the pipe, rather than their distinctive deposit characteristics. *Hypabyssal kimberlite* (HK) occurs in the root zone and principally is supposed to consist of dykes and sills, whereas in the diatreme, the typical facies is traditionally supposed to be *tuffisitic kimberlite breccia* (TKB), more recently also termed massive volcaniclastic kimberlite (MVK). TKB is characterised by *pelletal lapilli*, serpentinised olivine, and mantle and country rock xenoliths in a matrix of largely serpentine. Pelletal lapilli are discrete sub-spherical clasts (<0.1–6 cm) with a central fragment coated by a rim thought to represent quenched juvenile melt (Sect. 4.5.4.3; Gernon et al. 2012; see Fig. 18.74h). The crater zone is typified by variably bedded pyroclastic kimberlite (PK), resedimented volcaniclastic kimberlite (RSVK), and reworked volcaniclastic kimberlite (RWVK) of various types.

There are however several issues with this model. For example, "hypabyssal kimberlite" has become synonymous with (1) a rock type (i.e. massive, coherent, variably macrocrystic/porphyritic kimberlite), (2) a shallow crustal, intrusive emplacement level, and (3) the origin of tuffisitic kimberlite breccia through fluidisation-induced fragmentation of HK. It thus has considerable genetic significance. However, although coherent kimberlite can be emplaced at hypabyssal levels in the crust, it can also be emplaced at different levels in the pipe (e.g. Koala pipe in the Ekati kimberlite field, Lac de Gras, Canada; Nowicki et al. 2004) and as lavas or sills in the crater rim succession (e.g. Igwisi

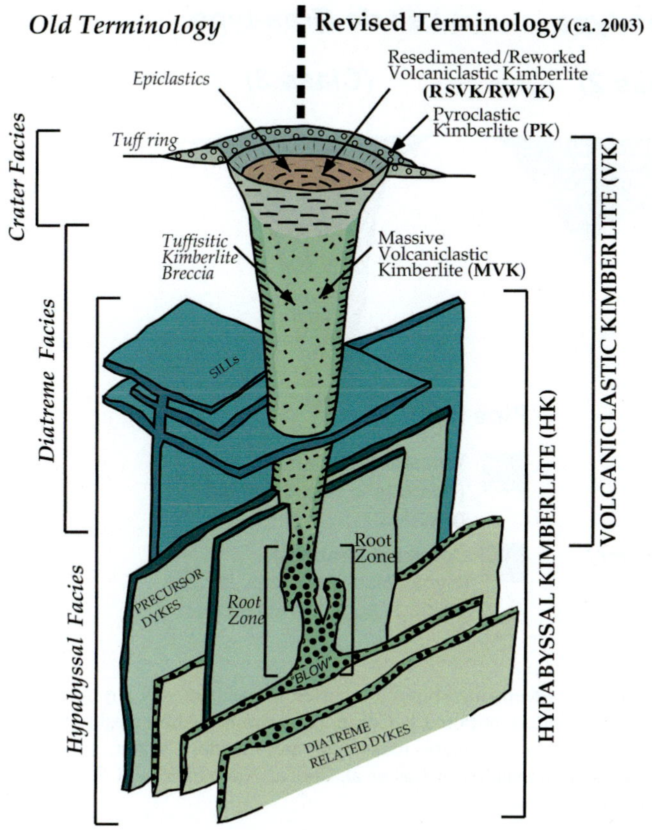

Fig. 18.70 The "classic southern African model" of a kimberlite pipe. Older terminology (e.g. Mitchell 1986) is shown on left, with a revised terminology (ca. 2003) on right. As discussed in the text, while the generalised pipe structure provides a reasonably accurate picture for some pipes, the application of terminology is problematic because it is a mixture of descriptive and genetic terminology and there are so many variations in different pipes. For example, "hypabyssal" is better renamed "coherent" kimberlite as it can occur anywhere from deep in a pipe to erupted lavas in and outside the crater (e.g. Igwisi Hills; Fig. 18.72). Conversely, some resedimented volcaniclastic can occur very deep in a pipe/vent that was empty at the end of the eruption (e.g. Koala pipe, Fig. 18.73a). A more systematic approach (as illustrated in Fig. 18.71) is to identify the deep intrusive feeder/root zone, the diatreme zone, and the crater–crater rim zone, without any implications of what facies will be found in the diatreme and crater and extra-crater zone. These in different cases could contain widely varying proportions of coherent, pyroclastic, resedimented, and/or reworked volcaniclastic kimberlite. (Modified from Kjarsgaard 2007)

Hills, Tanzania; see below). Secondly, tuffisitic kimberlite breccia also has many genetic process connotations associated with it. The term tuffisite is generally not used by modern volcanologists, although, it does have some historic connotations. As defined by Cloos (1941), tuffisite forms in a subvolcanic setting, from gas fracturing of country rock, and streaming of gas through a magma (fluidisation) that causes it to fragment and mix-in country rock debris (see Fig. 18.24, where the term is used in a correct context). In pipes where TKB contains clasts of supracrustal cover sequences, or where tuffisitic kimberlites are directly overlain by crater-derived bedded sequences, the vent was undoubtedly open when the TKB formed. The term TKB has little textural meaning to most volcanologists and is better replaced by a descriptive name, such as massive, poorly sorted lapilli tuff (see Chap. 14). Fragmentation is traditionally, simply ascribed to "fluidisation", but this fragmentation mechanism has not been critically assessed in many cases and is poorly understood in the context of TKB.

The diamond fields of Canada discovered in the early 1990s produced two new kimberlite model types (Fig. 18.71; Field and Scott-Smith 1999): shallow, open, bowl-shaped Prairies-type, based on Fort à la Corne (or Class 2), and narrow, steep-sided Lac de Gras-type (Class 3). Class 2 kimberlites lack coherent and tuffisitic facies and are dominated by *volcaniclastic kimberlite*. Volcaniclastic kimberlite is massive to bedded (Pittari et al. 2008), xenolith poor, and dominated by a mixture of juvenile lapilli and olivine crystals, including fresh olivine, in a matrix of serpentine, carbonate, and magnetite. While some bodies of Class 2 kimberlites are crater infills, others may have been positive landforms (see Igwisi Hills, below), indicating that the previous facies model is an over-simplification. Lac de Gras bodies are mostly filled with volcaniclastic and coherent kimberlite. Country rock xenoliths are apparently less common than in African pipes but cover sequence sedimentary xenoliths can be common. TKB-like rocks, including pelletal lapilli and microlitic clinopyroxene, have been reported in some pipes. However, RSVK, RWVK, and PK may occur in the diatreme zone (Nowicki et al. 2004), indicating that the pipe was unfilled at the end of the eruption, but was subsequently infilled through resedimentation of crater rim material, and by tephra from nearby vents.

To conclude here, historically and more recently with the discoveries in Canada, three end-member models for diamondiferous kimberlite pipes had evolved by the beginning of the new millennium. However, these were little constrained by modern understandings of facies analysis and volcanological processes. Facies or features considered to be typical of a particular morphological pipe zone can also occur in different pipe zones. Care must therefore be used in assigning particular process significance and pipe context significance to particular deposit types, components, and characteristics. The variability of kimberlite pipe morphology and deposit types is controlled by several factors, which we will develop a different understanding from a volcanological perspective in the next section. For a comprehensive discussion and critique of earlier kimberlite models, the reader is directed to Cas et al. (2008a) and the references therein.

18.7.3.2 A Modern Volcanological Reassessment

A major problem in understanding the volcanic eruption style of kimberlite volcanoes is that no eruption has been

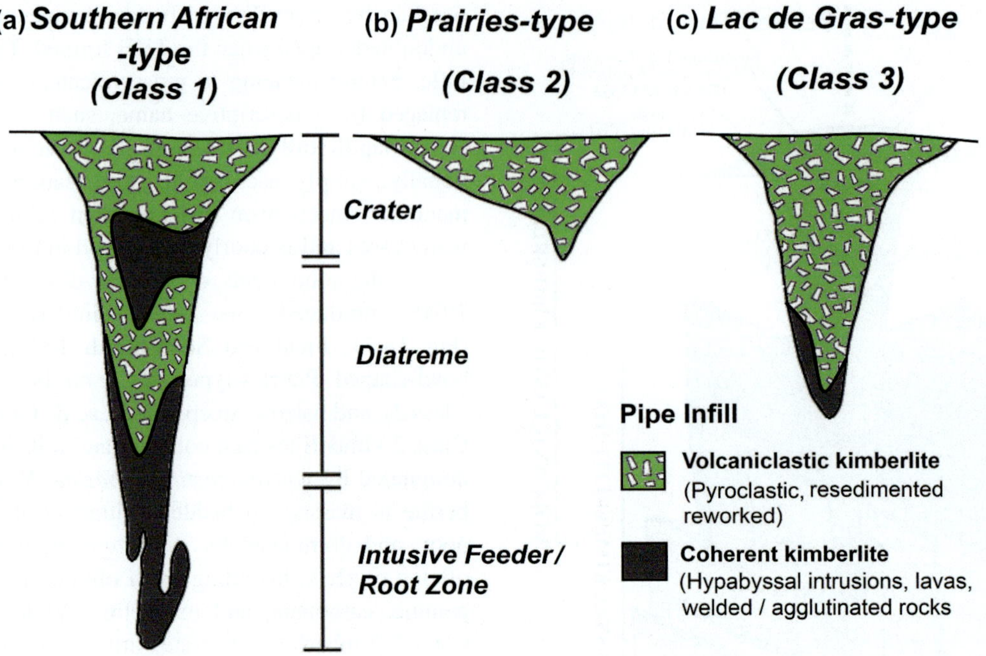

Fig. 18.71 Comparison of the three conventional kimberlite pipe models. **a** Narrow, tapering, steep-sided southern African-type kimberlite model (Class 1). **b** Open bowl-shaped or flaring Canadian Prairies-type (Fort à la Corne) kimberlite body (Class 2). **c** Dual tapering to flaring Canadian Lac de Gras-type kimberlite pipe model (Class 3). Steep-sided pipes are hosted in mechanically strong country rocks, whereas, at Fort à la Corne, pipes are hosted by weakly lithified Cretaceous sedimentary rocks and are therefore wider, open, and flaring. (Adapted from Cas et al. 2008a, After Field and Scott Smith 1998)

witnessed. The Igwisi Hills in Tanzania are not only the youngest (ca. 10 ka) kimberlitic rocks known, but uniquely preserve their surface deposits and volcanic landforms (Brown et al. 2012). The volcanoes are located on the western side of the Tanzanian Craton and were emplaced through a granitic gneiss basement dated 2.5 Ga. The next youngest kimberlites are Paleogene in age at Kundelungu (32 Ma), Democratic Republic of Congo, and the slightly older Ekati kimberlite field, Lac de Gras, northwestern Canada (75–45 Ma; Sarkar et al. 2015). The youngest lamproites are found in the Leucite Hills, Wyoming (3–1 Ma), and the Murcia-Almeria area, southeastern Spain (8.3–6.7 Ma), which both present well-preserved volcanic centres (e.g. Gernon et al. 2015). The youngest diamondiferous lamproites are the early Miocene (24–17 Ma) Ellendale field, Western Australia (Fig. 18.69a). Because of their geological age, therefore, original kimberlite and lamproite volcanic landforms and surface deposits are only rarely preserved.

The Igwisi Hills comprise three partially eroded kimberlite volcanoes (Northeast, Central, and Southwest volcanoes). These consist of small (<40 m high), relatively steep-sided pyroclastic cones with late-stage small volume lavas, reflecting the eruption of residual degassed kimberlite, after an initial phase of explosive volcanism (Fig. 18.72). Craters range from 100 to 200 m in diameter. The cones consist of variably bedded, relatively well-sorted lapillistone, lapilli tuff, and tuff, in which the pyroclasts include non-vesicular to poorly vesicular (<20% vesicles) juvenile clasts, pelletal lapilli, country rock xenoliths, and xenocrystic olivine. Brown et al. (2012) interpreted the pyroclastics as fall deposits, much like a basaltic scoria cone succession, produced by repeated small explosive eruptions. The lavas are variably vesicular and variably olivine xenocryst-rich and represent the final eruption of degassed kimberlite magma, again common in monogenetic basaltic scoria cones. The small size of the cones and craters indicates low explosive intensity and, therefore, low volatile content. All vents in the three Igwisi Hills cones were filled with lava and were not left as open deep craters, as appears to be the case with many older kimberlite pipes. This fact again indicates that the explosive intensity of the Igwisi Hills volcanoes was relatively weak. Thus, from a modern analogue viewpoint, there seems little correspondence with their older counterparts, which appear to have been largely explosive in their behaviour.

Most of our understanding of kimberlite volcanic processes therefore must come from the study of the deposits that fill kimberlite pipes, which are accessible where mined for diamonds or drilled for exploration (Plate 18.1e; Fig. 18.73). The diameter of the pipes ranges from a few tens of m to over 400 m, and they can extend to depths of 2 km. Many appear to have been monogenetic volcanoes, although there are also pipe complexes. These consist of coalesced or semi-overlapping pipes or lobes (sometimes aligned along structures), and pipe-in-pipe structures resulting from a succession of eruption events, though the temporal spacing of these is often unclear. In general, however, each eruption phase appears to have been short-lived, much like many monogenetic volcanoes. Where they are hosted by older, mechanically competent rock, the pipes are steep-sided ($\sim 70°$) and taper with depth (southern African- and Lac de Gras-type pipes, Classes 1 and 3; Fig. 18.71a, b). The volume of country rock xenoliths in the volcaniclastic pipe fill is too small to account for the volume of the country rock that must have been explosively excavated. This deficit of country rock can only be accounted for by explosive ejection from the pipe. However, the Fort à la Corne kimberlite bodies (Class 2) are intercalated with contemporaneous Cretaceous shallow marine and coastal plain sedimentary facies that were poorly lithified when kimberlites erupted explosively through them. Because the country rock had little strength, the resulting craters were shallower, more open, and flaring in geometry (Cas et al. 2008a, b; Fig. 18.71b). Notably, these pipes preserve some of the surface form (buried under thick glacial till) of the pipes, as either shallow wide craters or positive relief tuff cones or rings (e.g. Orion Central kimberlite complex, Pittari et al. 2008; Star and Orion South cluster kimberlites, Zonneveld et al. 2004; Kjarsgaard et al. 2008).

The principal facies in kimberlite pipes are essentially only two-fold (Figs. 18.74, 18.75): (1) *coherent kimberlite* (CK), and (2) *volcaniclastic kimberlite* (VK). However, because of their ultramafic composition kimberlites are very prone to alteration, and even these most basic facies divisions may in some situations be very difficult to distinguish (e.g. Stripp et al. 2006; Hayman et al. 2008, 2009; Hayman and Cas 2011a, b; Pittari et al. 2015); for a comprehensive review of the effects of alteration on kimberlites, see Chap. 13 (Sect. 13.7, and references therein). In many kimberlite pipes, most of the pipe consists of syn-eruptive pyroclastic deposits (volcaniclastic kimberlite) and intrusions (coherent kimberlite). Coherent kimberlite commonly intrudes the deep root zone of the pipe (e.g. Koala, Fig. 18.74c; Porritt and Cas 2011) but may also be found higher into the volcaniclastic pipe fill (see Figs. 18.71, 18.73). Lithologically, it consists of abundant olivine and other xenocrysts of highly variable sizes, mantle xenoliths, and country rock xenoliths, in a groundmass composed of serpentine, carbonate, and clinopyroxene microlites (Fig. 18.75a–d). Xenocrysts vary from sub-angular to round, indicating magmatic resorption in the magma. In some coherent kimberlites, there is a population of evenly sized, small, round olivine crystals that may represent magmatic crystals (Fig. 18.75c, d). Coherent kimberlite usually represents the last phase of degassed kimberlite magma intruding the volcaniclastic pipe-fill succession.

The predominant volcaniclastic kimberlite facies that occurs in most pipes is massive, poorly sorted, matrix-supported volcaniclastic kimberlite breccia (Fig. 18.74), which ranges from m to hundreds of m thick. The principal components are the same as in coherent kimberlite, except that instead of a groundmass there is a fine (<2 mm) matrix of these components variably altered to serpentine, carbonate, and clinopyroxene microlites, and the presence of pelletal lapilli (Figs. 18.74h, 18.75e–h). Units of massive volcaniclastic facies can be subhorizontal to subvertical (Fig. 18.73). This facies is the product of the explosive eruption phase of kimberlite pipe formation. Cas et al. (2008a) noted the similarity of the texture of massive volcaniclastic kimberlite to ignimbrite and suggested that, because of the high density of all pyroclasts in a kimberlite explosive eruption column, such columns would be more prone to gravitational collapse than for any other magma type. If the vents are large, a significant part of the collapsing column would fall back into the vent producing deposits very similar to ignimbrites but containing pelletal lapilli (Fig. 18.74d–g). The texture of massive volcaniclastic kimberlite has also been explained by fluidisation, but this process is not effective at the scale of entire vents and requires separate stages for vent excavation and vent filling (Porritt et al. 2008a, b).

A further origin for some, but not all massive volcaniclastic kimberlite, may be through injection from below driven by suppressed weak explosions in the root zone of the pipe generating tephra jets that try to punch their way upwards through the pipe. These can homogenise some of the pipe fill and can produce subvertical contacts in the pipe as observed in some maar-diatremes (e.g. Lefebvre et al. 2013), but perhaps more commonly observed in breccia pipes (see Fig. 18.24). Although tephra jets can explain some of the contact features formed late during the waning activity of a kimberlite pipe, they do not explain the initial explosive opening of the vent, its excavation and enlargement, and the ejection of a large volume of country rock. During this early waxing stage much of the debris that was ejected upwards out of the vent in a vulcanian-type of the eruption column, almost certainly fell back into the crater and vent through column collapse, accounting for much of the massive volcaniclastic facies (previously called tuffisitic kimberlite) that fills many kimberlite pipes.

In the Fort à la Corne kimberlite field (103–95 Ma; Kjarsgaard et al. 2017), the central massive volcaniclastic facies becomes progressively finer and thinly bedded, laminated, locally normally graded, and cross-laminated on the gently sloping flanks and margins of the pipes (Fig. 18.74k, l). Some isolated thin-graded beds (<40 cm) of volcaniclastic kimberlite interbedded in the basinal sedimentary facies appear to be distal fall deposits or shelf turbidites (e.g. Orion Central kimberlite complex; Pittari et al. 2008, Star

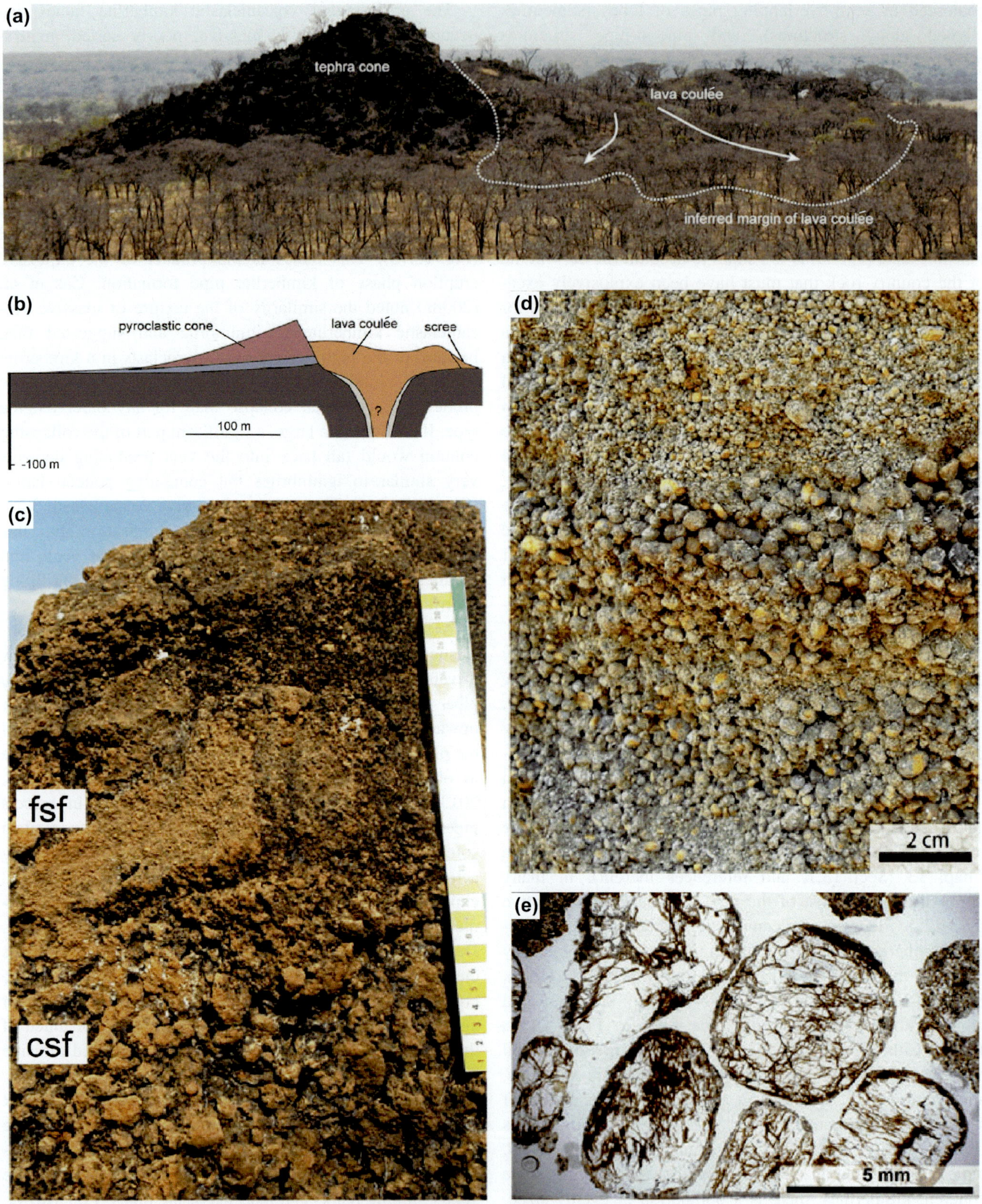

Fig. 18.72 The late Pleistocene–Holocene (ca. 10 ka) Igwisi Hills kimberlite volcanoes, Tanzania. **a** Panorama of the Central volcano showing the partial pyroclastic cone on the west of the volcano and the lava coulée that fills the inferred Central volcano crater. **b** Cross-section through the Central volcano. **c** Coarse-grained scoria fall deposit (csf) overlain by fine-grained scoria fall deposit (fsf). **d** Bedded pelletal-clast-rich lapillistone from the Northeast volcano; the deposit is comprised predominantly of clast-supported pelletal lapilli. **e** Extracted pelletal lapilli from the pelletal-clast-rich lapillistone. (Adapted from Brown et al. 2012)

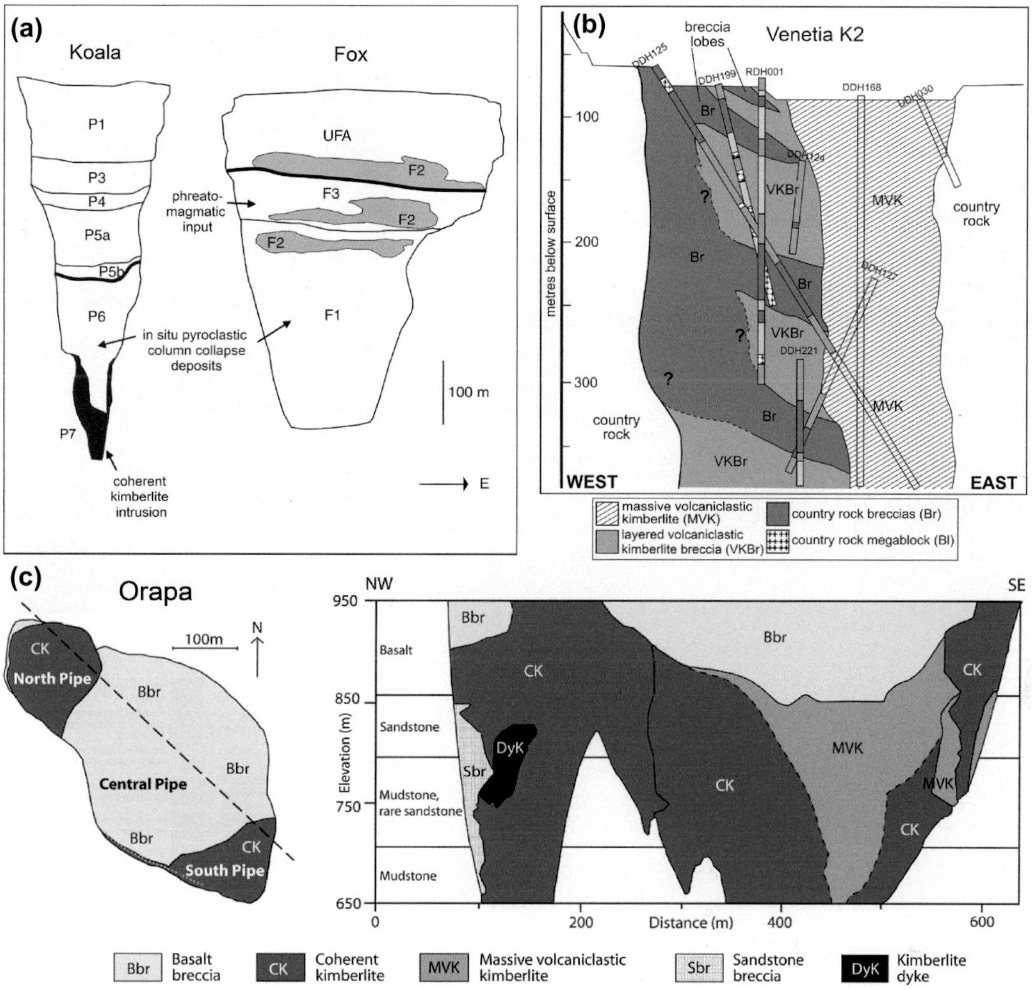

Fig. 18.73 Examples of kimberlite pipes reconstructed from mining operations and drill core intercepts. These highlight variations in internal facies organisation; note several consist of coalesced or semi-overlapping pipes or lobes, sometimes aligned along structures (Table 18.4). **a** Cross-sections showing the Eocene Koala and Fox pipes, Ekati kimberlite field, Lac de Gras, Northwest Territories, Canada (Modified from Porritt et al. 2011). These pipes are characterised by mainly fragmental deposits, which in the Koala pipe form a distinct subhorizontal layered sequence, comprising a high proportion of resedimented or reworked material (above thicker black line). *Koala:* P1, mega-graded bed; P3, multiple normally graded beds; P4, wood-rich silty fine sandstone; P5a, massive poorly sorted, mud-rich mass-flow deposit; P5b, granodiorite boulder breccia; P6, massive poorly sorted, olivine-rich column collapse lapilli tuff; P7, macrocrystic coherent kimberlite. *Fox:* UFA, weakly bedded, sandy mudstone; F3, massive to weakly bedded, poorly sorted accretionary lapilli bearing, column collapse lapilli tuff; F2, granodiorite cobble-boulder breccia; F1, massive, poorly sorted, granodiorite-rich, column collapse lapilli tuff. **b** Cross-section of the Venetia K2 pipe, one of 14 known pipes forming the Cambrian Venetia kimberlite cluster, South Africa (From Brown et al. 2009). The upper half of the pipe (K2 West) comprises a series of gently dipping layered volcaniclastic breccias (VKBr) and country rock breccias (Br); below the 350 m level, the succession also includes coherent kimberlite interpreted as welded pyroclastic kimberlite (see Brown et al. 2008a). Late-stage explosivity excavated a large portion of K2 West and deposited massive volcaniclastic kimberlite to form K2 East. **c** Simplified geological map and cross-section of the Cretaceous Orapa B/K9 kimberlite body, Botswana (From Brown et al. 2008a); note the major units of coherent kimberlite which are reinterpreted as largely composed of welded or agglutinated kimberlitic pyroclasts. **d** The Triassic Jwaneng kimberlite cluster, Botswana. (Left) Cross-section through the Centre pipe showing the inferred positions of the major country rock units and internal facies architecture (From Brown et al. 2008b). The preferential occurrence of Karoo megablocks along the northern side of the Centre pipe may result from the prior excavation of Karoo surface stratigraphy to the south by the South Pipe eruption. The presence of centroclinal-dipping bedding, surface megablocks, and sheared marginal breccias indicates substantial syn-eruptive subsidence of the pipe fill. (Right) 3D geological models of Jwaneng mine (From Field et al. 2008, after Farrow 2004, unpubl. data). The upper model shows the position of kimberlite pipes (green) emplaced within the country rock, where the blue is Transvaal dolomite, yellow (QS) is Transvaal quartzitic shales including the Bevets conglomerate marker, pink is Transvaal laminated shale, red is Transvaal carbonaceous shale, cyan is diabase dykes, and the light brown and brown are Kalahari-aged calcrete and aeolian sands; approximate dimensions of the model are 2.5 km × 1.6 km × 1 km (deep). The lower model shows three pipes that comprise Jwaneng mine; different colours represent various geological zones and facies (for detailed descriptions, see Field et al. 2008), while the grey areas at the bottom are inferred; approximate maximum radii of the three pipes are 360 m, 490 m, and 325 m for the North, Centre, and South pipes, respectively

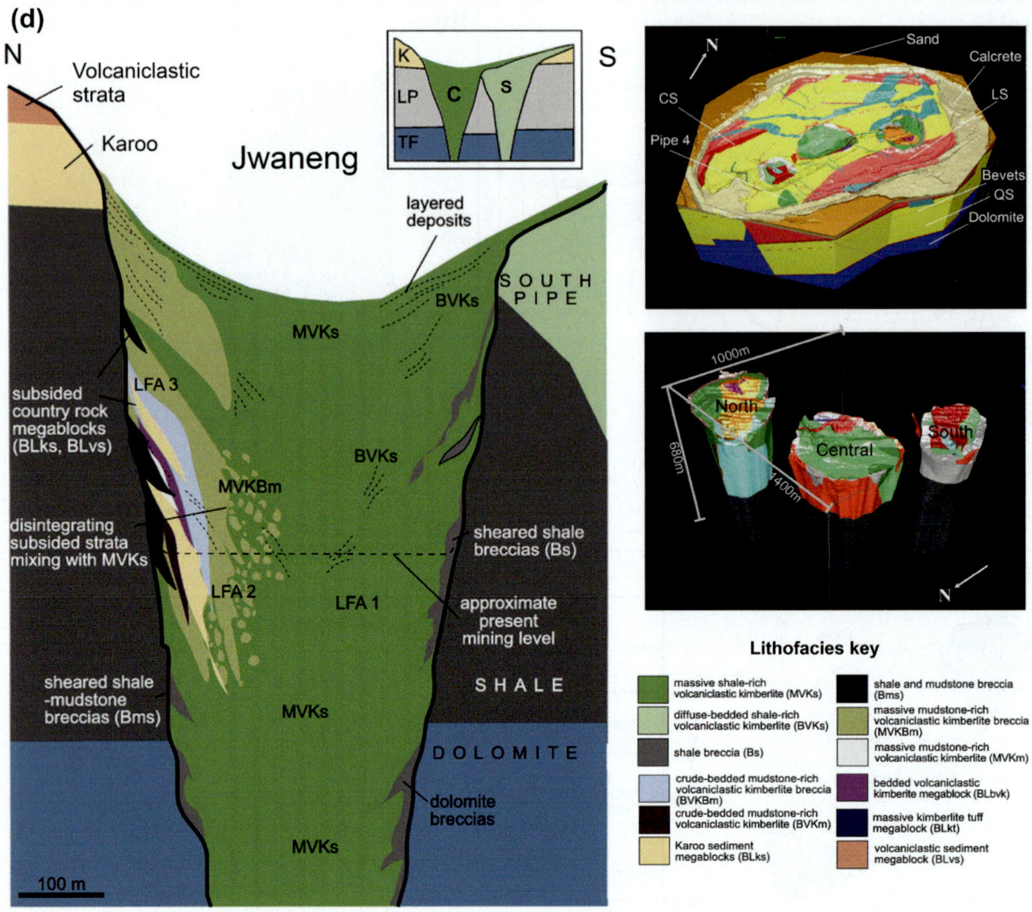

Fig. 18.73 (continued)

and Orion South cluster kimbelites, Zonneveld et al. 2004, Kjarsgaard et al. 2008). Some pipes are dominated by post-eruptive volcaniclastic and sedimentary facies up to hundreds of m thick, indicating that a very deep, open vent existed temporarily (Figs. 18.73a, 18.74h–j). These may be filled with volcaniclastic kimberlite resedimented from the crater rim succession, and terrigenous sediment (including carbonaceous mudstone, peat, and lignite) eroded and transported into the crater by normal surface processes. In the Ekati kimberlite field, tree logs have been found within deep pipes/craters, and in some pipes, possible primary volcaniclastic kimberlite facies are intercalated with the crater-fill sediments (Fig. 18.74h, i). These units were not produced by explosive activity from the pipes in which they occur, because the host successions are undisturbed, but probably supplied by eruptions and transported by pyroclastic flows from nearby kimberlite vents.

Where pipes are mostly filled by resedimented volcaniclastic facies, the pipe-forming explosive eruption appears to have been a short-lived major explosive event, whereas in other cases, explosive activity may have been repeated and of intermediate intensity. In these cases, smaller subsequent eruptions can core smaller, vent-in-vent conduits through part of the infill so establishing some intra-vent complexity in the fill, with subvertical bodies and contacts (Fig. 18.73). In other cases, the explosive activity was relatively weak, as in the Igwisi Hills volcanoes. Such variations in eruption intensity would have been controlled largely by magma ascent rate and volatile contents. Base surge deposits and accretionary lapilli have been recognised and described from a number of kimberlite and lamproite pipes (e.g. Fig. 18.74a; including Argyle, Boxer et al. 1989). This has led some workers to propose phreatomagmatic explosive activity is *both* essential and the major driving mechanism of their formation (e.g. Lorenz 1975; Lorenz et al. 1999; Lorenz and Kurszlaukis 2007; Kurszlaukis and Lorenz 2008). If phreatomagmatic deposits are present then clearly, external water (via an aquifer or surface water) had access to the vent during the phase of the eruption that produced those deposits (e.g. in the Ekati kimberlite field at Fox, Fig. 18.73a, and the Diavik A418 pipe; Porritt et al. 2013). However, our studies (Monash Kimberlite Research Group) suggest there is little evidence that phreatomagmatism creates all or even a majority of pipes, and while some kimberlite eruptions may

Fig. 18.74 Selected facies and textural characteristics of diamondiferous kimberlite pipes. **a** Surge dune bedforms draping large ballistic boulder (outlined in yellow), Cretaceous Orapa pipe, Botswana. **b** Crater wall collapse/rock fall breccia derived from local country rock, Orapa (= Basaltic breccia in Fig. 18.73c). **c** Coarsely porphyritic coherent kimberlite (CK) facies from the Eocene Koala pipe, Ekati kimberlite field, Lac de Gras, Northwest Territories, Canada; note the very large and irregular nature of most of the large olivine xenocrysts. **d** Poorly sorted, massive volcaniclastic breccia (VK) facies which in this case represent an intra-crater pyroclastic flow deposit, Orapa pipe. **e** Poorly sorted, massive VK, Jurassic Jericho pipe, Nunavat, Canada; the lithic clasts are granite and limestone. **f** VK with pelletal lapilli (PL) showing lithic fragment with the rim of quenched juvenile melt, Jurassic Muskox pipe, Nunavat, Canada. **g** VK with granite, limestone, and mantle (eclogite clast outlined in red) xenoliths and chrome-diopside, garnet, and eclogitic xenocrysts set within a serpentinised matrix, Jurassic Muskox pipe, Nunavat. **h** Crystal-rich VK forming the uppermost post-eruptive facies at the Koala pipe. This comprises a ~100 m thick, poorly moderately sorted, mega-graded bed (P1, Fig. 18.73a), which grades almost continuously (right to left) from a coarse olivine-rich lapillistone at the base to fine olivine sand and quartz siltstone at the top. Its origin is enigmatic, and perhaps either generated by the mass collapse of a Koala tephra cone or as favoured here, input from a pyroclastic flow sourced from a nearby volcano and modified by deposition into a crater lake (Porritt et al. 2011). **i** Coarse, olivine-rich, diamondiferous lapillistone at the base of P1 mega-graded bed, Koala pipe. **j** Horizontal alignment of elongated clasts in a coarse VK mega-turbidite, Orion South complex, Cretaceous Fort á la Corne kimberlite field, Saskatchewan, Canada. **k** and **l** Laminated and cross-laminated, reworked fine-grained VK, from kimberlite body 158 and Orion Central complex, Fort á la Corne kimberlite field. All drill core is ~5 cm in diameter

Fig. 18.74 (continued)

have had phreatomagmatic phases, involvement of external water is not a necessary or fundamental driving mechanism (see further discussion below).

Examination of the country rock around kimberlite pipes shows that although minor pipe margin breccias and closely fractured country rock can be found locally (Fig. 18.74b), generally, the country rock is massive and surprisingly coherent and unfractured, or only weakly so (Fig. 18.73). A notable exception is the Venetia K2 (West) pipe, marked on its preserved outer margin by a thick (>300 m) sequence of inward dipping, country rock breccia lobes (showing jigsaw-fits), and rotated megablocks of gneissic country rock (Brown et al. 2009; Fig. 18.73b). This is consistent with large magma overpressures and pressure differences resulting in tensile failure of pipe wall rocks, possibly associated with explosive cratering and widening, although, this appears to be a phenomenon perhaps more commonly observed in breccia pipes (Fig. 18.24). Country rock contacts vary from sharp to gradational, depending on whether the loose, fractured country rock was undisturbed or subsequently removed (Barnett 2008). Exposures of the exhumed vent walls in mining often show smooth, striated, and faceted faces and sharp contacts between the country rock and pipe fill, consistent with a sustained period of the gas jet (plus solid pyroclasts)-driven erosion, abrasion, ablation, and smoothing. Some pipes include megaclasts of country rock (up to tens of m) that collapsed from the vent walls (Barnett 2008, e.g. Jwaneng Centre pipe, Fig. 18.73d).

For the systematic approach and terminological scheme, we propose describing kimberlitic rocks specifically, and volcanic rocks in general, see Chap. 14. This involves a step-by-step progression from descriptive to genetic terminology and can be employed for both coherent and volcaniclastic (fragmental kimberlites). The key steps to this approach are summarised in two key flow charts (Fig. 14.1a, b); for further discussion, see Cas et al. (2008b, 2009).

18.7.3.3 Model of Kimberlitic Explosive Volcanism

As proposed by Cas et al. (2008a), explosive kimberlitic eruptions generally evolve through six stages (Fig. 18.76a–f):

Stage 1—Conduit Formation

A rising column of kimberlite magma and gas penetrates through the upper mantle and crust, preferentially via some trans-lithospheric zone of weakness (i.e. fault zone). The fluid can then either dilate existing fractures or hydraulically fracture new fracture networks by tip crack propagation. Increasing degrees of fracturing occur in the zone of fluid propagation, especially close to the Earth's surface where the lithostatic pressure becomes low, and where a separated gas phase occurs. Much of this hydraulically fractured zone of country rock will be removed and excavated during the succeeding two stages. Decompression of the magma as it rises will lead to saturation of the magma with regards to CO_2 and H_2O which exsolves (Sparks et al. 2006; Sparks 2013).

Stage 2—Vent Opening

When fractures reach the surface, sudden decompression of the magma-gas system to atmospheric pressure generates a very large gas overpressure. This triggers a high-energy ballistic gas explosion, perhaps akin to a vulcanian explosion, excavating and violently ejecting a volume of country rock equivalent to the size of the initial vent. The shape of the pipe will be controlled at this stage by the mechanical competency of the country rock, the volatile content of the magma, the depth of volatile exsolution, and the depth of the fragmentation surface during the eruption. If the latter are deep and the country rock competent, a narrow steep-sided pipe will be formed. If they are shallow and the country rock mechanically weak, then a more flared pipe, with shallow sloping walls, will form. This stage would leave an apron of ballistic country rock debris around the vent.

Stage 3—Vent Clearing, Reaming, and Sculpting

Sudden decompression of the rising magma on the vent opening will lead to the acceleration of volatile exsolution and explosive fragmentation of the magma (Chap. 4). This produces a waxing (or accelerating) gas jet carrying both juvenile and accessory lithic pyroclasts, feeding a buoyant eruption column. Most of the excavated country rock debris is ejected from the vent during this highly explosive phase of activity. Simultaneously, the levels at which vesiculation and explosive fragmentation occur in the magma column will migrate downwards into the magma column as it decompresses, allowing deeper excavation of the vent/diatreme (Sparks et al. 1994). Little infilling of the vent is likely to have occurred during this stage of the eruption, although, pyroclastic fallout deposits would form downwind. The fragmentation processes for the erupting kimberlite are described further in Chap. 4 (Sect. 4.5.4.3).

Stage 4—Pyroclastic Vent Infilling

As the eruption wanes, the gas–particulate eruption column rising above the vent quickly becomes too dense to remain buoyant, because unlike normal explosive eruptions of other magma types, nearly all pyroclasts are dense. As a result, the column rapidly collapses at a very high mass flux rate. As well as forming extra-vent pyroclastic flows (Chap. 10), in this case, a significant part of the column is likely to collapse into the vent, which acts as a trap or micro-basin (Porritt

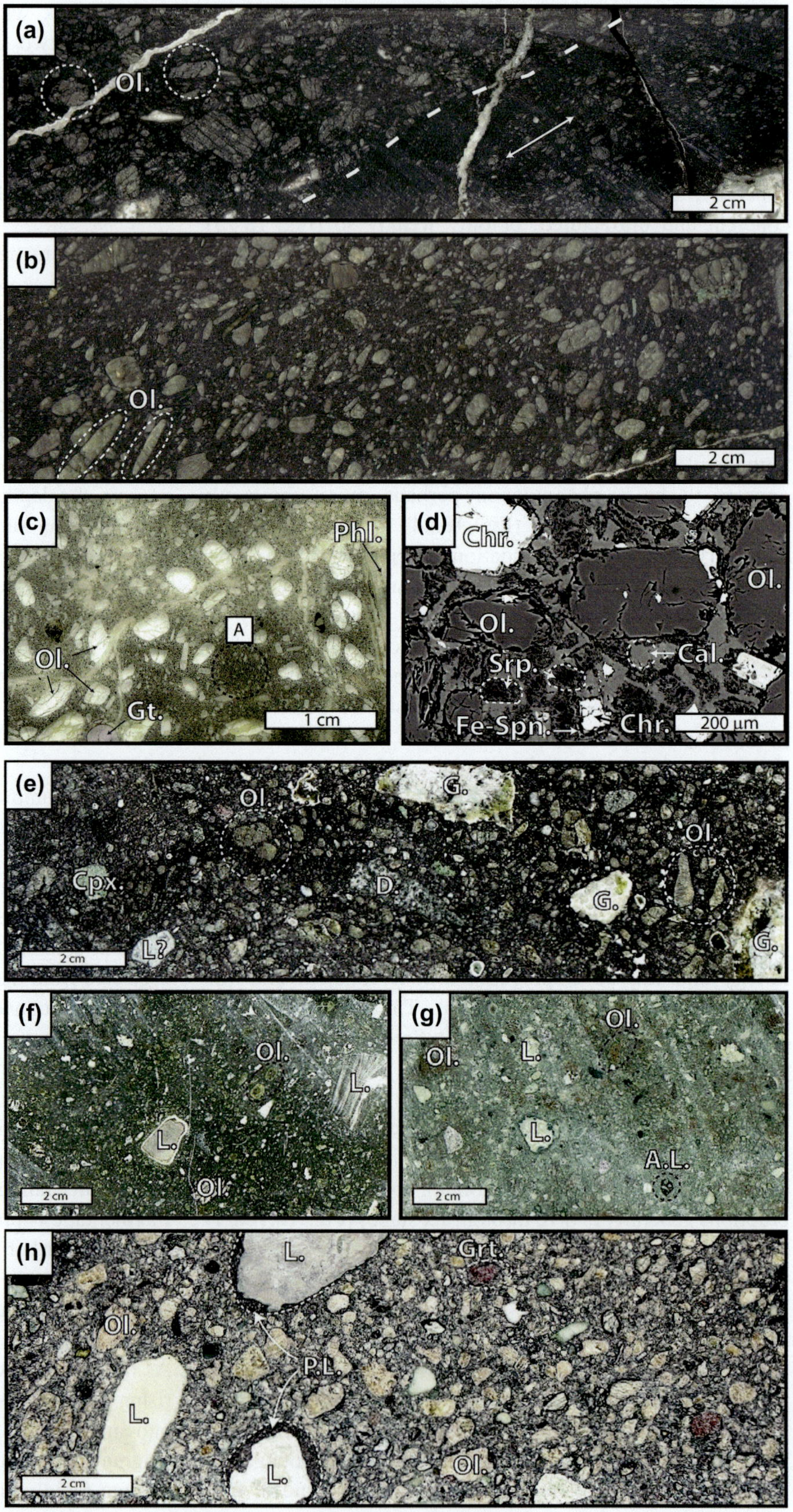

Fig. 18.75 Detailed textural characteristics of the two principal facies found in kimberlite pipes, illustrated from the Jurassic Jericho kimberlite pipe, Nunavat, Canada. **a–d** Coherent kimberlite (CK) facies: **a** polished slab showing contact (dashed line) between coarse- and fine-grained CK (arrow indicates the direction of crystal alignment; **b** polished slab of coarse-grained CK illustrating the abundance of olivine crystals and paucity of country rock xenoliths (note the olivine crystals with high aspect ratios = dashed white circles); **c** thin section image of CK illustrating the even distribution of olivine and better-preserved groundmass (marked "A"); **d** SEM image of CK groundmass illustrating even distribution of olivine crystals and serpentine (Srp) domains, which together are contained in calcite (note the homogeneous chromite (Chr) cores and the thin Fe-rich spinel (Fe-Spn) rims). (From Hayman and Cas 2011b). **e–h** Volcaniclastic kimberlite (VK) facies: **e** poorly sorted olivine-rich VK (note the presence of angular olivine crystals, e.g. white dashed circle, and country rock lithic clasts); **f** poorly sorted country rock-rich VK; **g** poorly sorted south vent VK; **h** olivine-rich VK with densely packed olivine crystals and pelletal lapilli (PL). Ol = olivine, L = limestone, G = granodiorite, Cpx = clinopyroxene, AL = accessory lithic clast, Grt = garnet, D = diabase. (From Hayman and Cas 2011a)

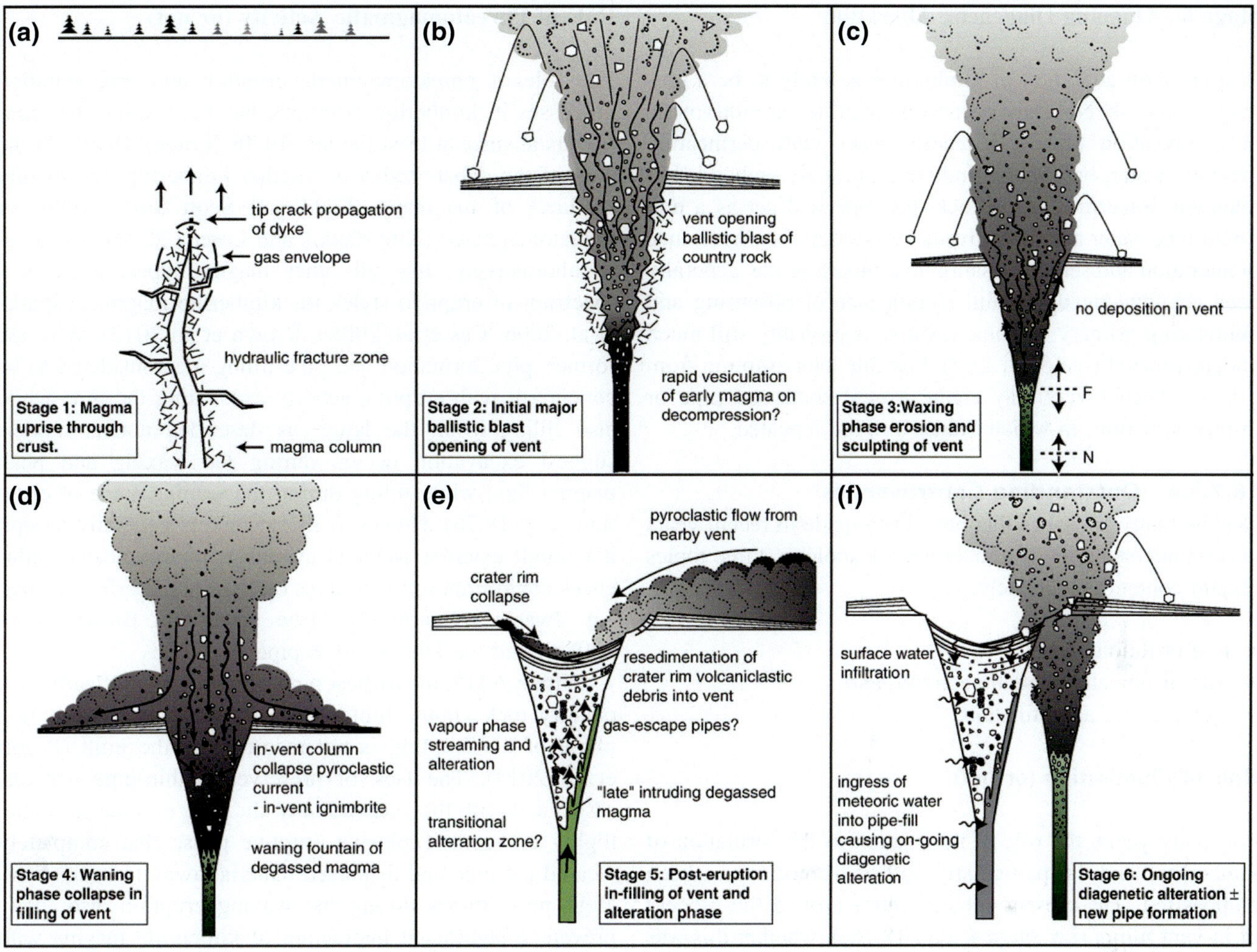

Fig. 18.76 Stages (1–6) during an explosive kimberlite eruption; see text for explanations of each stage. **a** Uprise and propagation of magma through the crust (*Stage 1*). **b** Initial major ballistic blast opening of a vent to the surface (*Stage 2*). **c** Waxing phase of eruption intensity resulting in net erosion and sculpting of the vent walls (*Stage 3*); abbreviations: N = bubble nucleation level, F = magma fragmentation surface. **d** Waning phase of eruption intensity, the collapse of the eruption column, and infilling of the vent (*Stage 4*). **e** Post-eruption processes resulting in further infilling of vent and deposit alteration (*Stage 5*). **f** Ongoing diagenetic alteration of pipe-fill deposit and possible later eruption creating a new coalesced pipe (*Stage 6*), in which Stages 1–6 are repeated. (From Cas et al. 2008a)

et al. 2008a, b). This produces a unique deposit, formed by in-vent ponding of the eruption column collapse debris, which is texturally identical to poorly sorted, pyroclastic flow deposits (i.e. massive lapilli tuff, *aka* TKB). Elutriation of kimberlitic ash (Porritt and Russell 2012) above the collapsing eruption column at this stage may be critical in producing an economic deposit. Deflection of collapsing material off the steep walls in the narrow vent may result in pulsing deposition and the formation of crude stratification.

Once the vent was largely filled with ponded pyroclastic debris, the explosive eruption would likely be stifled, although minor subsequent eruptions could core smaller, vent–in-vent conduits through part of the infill caused by muted tephra jets punching their way vertically through the pipe fill.

Stage 5—Late-Stage Filling of the Pipe

Stratified deposits filling the vent or pipe above the massive lapilli tuff can only result from the collapse of debris from the crater rim, minor explosive events in part of the crater, or pyroclastic deposits being sourced from other nearby vents and transported by pyroclastic fallout, flow, or surge. Coherent kimberlite may represent the final degassed magma intruding the base of the vent fill sequence or a later separate magma batch. This stage also involves syn- to post-depositional alteration of the highly reactive kimberlite minerals, glass and lithic components by magmatic vapour phase alteration, hydrothermal fluids, and meteoric waters, which will quickly pool in the vent under the right climatic conditions.

Stage 6—Ongoing Diagenetic Alteration

Post-eruption alteration of kimberlites is likely to be a protracted process. Not only because kimberlites are ultramafic in composition and will readily react with permeating meteoric water, but where kimberlite pipes are embedded in relatively impermeable bedrock, the pipe will act as a trap for surface water and groundwater. Although the importance of alteration and serpentinisation in kimberlites has generally been acknowledged, the full significance of obscuring and obliterating primary volcanic textures is probably still much underestimated (see Sect. 13.7). Possible later eruption from a new adjacent vent may create a new coalesced pipe or in-pipe structure, in which Stages 1–6 are repeated.

18.7.3.4 Outstanding Controversies

Despite many advances that have been made in recent years in our understanding of kimberlite volcanology, three topics remain contentious, namely:

- role of fluidisation,
- role of phreatomagmatic activity, and
- welding and agglutination.

Role of Fluidisation (or not)

For many years, the role of fluidisation in the formation of kimberlite, and in particularly tuffisite breccia has been emphasised. The present debate centres around the process of in-vent filling (i.e. *stage 4*, Fig. 18.76e): whether this was due to vent widening and overloading of the eruption column with dense pyroclasts, or an alternative model involving waning deposition and large-scale fluidisation of pyroclasts (see Porritt et al. 2008a, b; Brown et al. 2008c; Porritt and Cas 2009, 2011; Gernon et al. 2009; Sparks 2013). Immediately following the eruption, degassing of the pipe could likely cause local elutriation of fine ash through fluidisation of deposits ponded in the vent (e.g. gas escape pipes), although more significant gas streaming and escape of gas from the vent could potentially homogenise parts of the vent infill. However, we have found no evidence that vent-wide fluidisation during the eruption contributes significant deposits to the pipe fill. Another significant issue is that preserved textures have been misunderstood. Many serpentine-altered fragmental deposits in pipes show poorly sorted packstone to wackestone-like textures in thin section (cf. Dunham 1962; Cas et al. 2008a; Chap. 14), indicating an original ash matrix existed prior to serpentinisation, consistent with the mechanism of in-vent column collapse and the poorly sorted ignimbrite-like textures that are preserved.

Role of Phreatomagmatic Activity (or not)

The role of phreatomagmatic eruption and fragmentation processes in kimberlite eruptions has been somewhat controversial since at least the late 1970s (Lorenz 1975). At the core of the debate today is whether kimberlites are mainly products of magmatic (Field and Scott Smith 1999) or phreatomagmatic (Kurszlaukis and Lorenz 2008) processes, or alternatively, like all other magma types, there is a spectrum of eruptive styles for kimberlite magmas (Sparks et al. 2006; Cas et al. 2008a; Brown et al. 2012). With the former, pipe formation and pipe-filling are considered to be concurrent with eruptive activity excavating the pipe while also filling it. In the latter, as described above, models suggest excavation occurs during the waxing and peak magma flux, while filling during the waning stage of eruption (Fig. 18.76). However, most workers generally accept, if enough external water is available, hydrovolcanic influences could impose themselves on rising kimberlite magma, e.g. Jwaneng Centre pipe (see discussion Brown et al. 2008b), and the Diavik A418 pipe.

Within A418, the earliest pipe-filling deposits clearly lack country rock granite lithics, indicating the pipe excavation phase of the eruption is not recorded by the infill (Porritt et al. 2013). The lack of preserved within-pipe deposits associated with the excavation of the pipe requires an initial highly energetic explosive eruptive phase that completely cleared the pipe and dispersed material away from the vent. High mass fluxes during the waxing eruption also likely prevented significant interaction of kimberlite magma with any surface or groundwater, pressuring the vent and preventing in-vent deposition of pyroclasts and influx of groundwater. A decrease in mass flux, during the waning of the eruption, enabled the interaction of the kimberlite magma with the ground and meteoric water, wet sediment, or combination. The decrease in flux marks the onset of pipe-filling by base surge-like deposits with abundant accretionary lapilli. The depth at which hydrovolcanic explosions occur, as well as the competency of the country rock, will be important, leading to many possible permutations and combinations influencing both pipe shape and deposit characteristics. However, the presence of accretionary lapilli is critical to assigning a phreatomagmatic origin; as far as we are aware, large (m-scale) pipe- or slot-like bodies containing accretionary lapilli have never been described from kimberlite pipes, but are known from true phreatomagmatic diatremes and breccia pipes (cf. Figs. 18.22, 18.24 and 18.26b).

Welding and Agglutination

Given the low viscosity of kimberlite magmas and the near-ubiquitous presence of pelletal lapilli in some volcaniclastic kimberlite facies (i.e. traditional TKB), it would not be surprising that welded and/or agglutinated rocks, formed from bursts of late, degassed lava fountaining or jetting, might be present in some pipes (Cas et al. 2008a). Sparks et al. (2006) suggested that kimberlite rock types previously interpreted as hypabyssal intrusions in some kimberlite pipes might be welded pyroclastic rocks. These rocks, as we have discussed, are now more generally termed coherent kimberlite. Much of the present debate hinges on the fact that while in some cases welded and agglutinated rocks may be present (e.g. the Orapa B/K9 and Venetia K2 pipes, Brown et al. 2008a; Fig. 18.73), not all coherent kimberlite has such origins. As we have stressed above, all kimberlites are pervasively altered rocks where even distinguishing the most basic facies types can be difficult. While we agree, in some cases agglutination (welding) may occur, in our experience, especially from the Canadian kimberlite pipes, this is not a common phenomenon, and agglutination and clastogenic textures in coherent kimberlite have to be convincingly assessed on a case-by-case basis (e.g. van Straaten et al. 2011).

18.7.4 Diamond Resources

Diamonds are mined for both their gem and industrial attributes. Selected features of some of the world's largest producing mines or others referred to in the text are given in Table 18.4. The Argyle (AK1) lamproite pipe has been the world's largest producer of rough diamonds in history (Shigley et al. 2001; Fig. 18.77). Output peaked in 1994 (42.8 million carats, or 8.6 t), which was 40% (although only 7% by value) of the diamonds produced worldwide in that year. In 2018, it produced some 18 Mct or 12% of world production, but with limited reserves, the mine closed at the end of 2020 having produced more than 865 Mct since it opened in 1983. The rough diamonds were sorted based on size, shape, colour, and "purity" (clarity). At Argyle production consisted of 5% gem, 25% industrial, and the remainder termed "near-gem" which received substantial processing to extract a normally low-value, polished diamond. Gem quality diamonds were divided into brown (72%), near colourless-to-pale yellow (27%), and rare pink-to-red (<<1%), with some rare blue and green diamonds. A large number of brown diamonds led to a major marketing campaign in the mid–late 1980s promoting "champagne" and "cognac" diamonds. Of more scientific interest, most Argyle diamonds show evidence of having been plastically deformed following formation in the earth's mantle or during their ascent to the surface, which has been thought to cause both brown and pink colours. Other major producing mines (Table 18.4) include Jwaneng and Orapa (Botswana), Catoca (Angola), Venetia (South Africa), Yubileinaya, Udachnaya, and Mir (Russia), and Ekati and Diavik (Canada).

Diamonds range in size up to the famous Cullinan, recovered in South Africa in 1905, at 3106.75 carats (621.35 g). Diamonds greater than 500 carats are rare and sizes decrease exponentially as a function of frequency. Diamonds are typically clear but as at Argyle commonly vary in body colour. Green transparent diamonds exhibit radiation damage, whereas green opaque or green opaque-coated diamonds contain abundant mineral and fluid inclusions. Yellow and greenish-yellow diamonds contain nitrogen substituting for carbon, and semiconducting blue diamonds have trace amounts of boron. Not all kimberlite or lamproite bodies are diamond-bearing. This is either because the mantle through which their magmas passed was not fertile, or diamonds that were incorporated into the magma as xenocrysts were resorbed by the magma during ascent. Many diamonds show signs of resorption, which affects their commercial value. Nearly all commercial production of primary diamonds has been from pipes, with a few exceptions solely mining sills and dykes (Table 18.4).

Porritt et al. (2011) assessed factors that influence diamond grades in the Ekati kimberlite pipes (Plate 18.1e). They found higher diamond grades correspond with facies that contain high concentrations of large olivine crystal fragments and low abundances of country rock xenoliths. Olivine is dense and behaves hydraulically in a similar way to diamonds during transport. Removal of ash by elutriation above a collapsing eruption column and during any fluidisation process affecting the in-vent fill could be critical in creating an economic deposit (Fig. 18.76d). Evidence from the Ekati mine suggests removal of 50% of fine ash at this stage could have initially doubled the grade relative to the erupting magma (e.g. unit P6, Koala; Fig. 18.73a). Porritt et al. (2011) also found that volcaniclastic kimberlite potentially sourced and transported as pyroclastic flows from a nearby pipe (mega-graded bed P1; Fig. 18.74i) may contain as high diamond grades as in-vent, ponded massive volcaniclastic kimberlite facies within the Koala pipe. In contrast, large amounts of country rock xenoliths can have the effect of diluting diamond grade (e.g. Fort à la Corne where kimberlites were emplaced into unconsolidated sediments). Understanding lithofacies and stratigraphic architecture, eruption, transportation, depositional processes, and the provenance of every deposit in a pipe is therefore critical to understanding the total grade distribution within a kimberlite pipe. This could be important in accessing areas for

Fig. 18.77 Selection of rough diamonds produced at the Argyle mine, Western Australia. This illustrates the range of colours produced as well as the variety of crystal shapes, including a relatively small proportion of octahedra and macles (flat, triangular-shaped), and abundant rounded or irregular shapes. These crystals vary from about 0.5–1 carat; as such, they are larger than the crystals typically recovered from the mine, which have a mean size of less than 0.1 carat. (From Shigley et al. 2001)

selective mining, particularly with regard to marginal deposits. More recently, Jones et al. (2019) have shown experimentally how the ascent rate and path may also affect the attrition rate of the mantle xenocryst cargo in kimberlites. Deep-seated melts and sustained rapid ascent favour pervasive attrition due to a high-energy environment and a long residence time; this results in heavily rounded olivine grains and potentially enhanced diamond grades.

Erosion of primary kimberlite and lamproite host rocks can lead to diamond placers deposited in alluvial, beach, and more rarely offshore environments. Sorting of diamonds during sedimentary transport is well known from detrital diamonds, where diamonds generally decrease in size away from the source and are often found concentrated in bedrock traps. Sedimentary transport also affects the quality of the diamonds due to preferential breakage and loss of low-quality diamonds with flaws or inclusions. Significant resources (producing >1 Mct/year) can therefore be found in some diamond placer deposits.

For a comprehensive global overview of diamond resources including placer deposits, see Kjarsgaard et al. (2022).

18.8 Volcanic (and Intrusive)-Hosted Massive Sulphide Deposits

Volcanic-hosted massive sulphide (VMS) deposits represent a distinctive, if not highly variable, submarine magmatic–hydrothermal ore deposit type (Fig. 18.1). They are polymetallic in nature and result from the precipitation of sulphide minerals in deep water, marine environments from seafloor hydrothermal systems. These are often associated with black or white "smoker" chimneys (e.g. Fig. 18.78, and see Fig. 16.87), although, they may also emanate from more diffuse hydrothermal vent systems spread over larger areas of the seafloor. The deep-water environment is a major factor in the ore formation process, due to the hydrostatic pressure

Fig. 18.78 Features of modern seafloor "black smoker chimneys", illustrated by the entirely submarine Brothers volcano in the Kermadec arc, north of New Zealand. These testify to the formation of VMS deposits at depths of ~1,600 m b.s.l. **a** Beehive-type structure at the top of an ~15 cm high chimney that occurs at the top of a ~4 m high chimney complex. The maximum vent temperature from this chimney is 302 °C. The vertical field of view is ~0.5 m. **b** Cross-sectional view through left-hand chimney shown in (**a**) dominated by chalcopyrite. **c** Black smoker vent of a 0.5 m high broken-off chimney emitting dark hydrothermal fluid at 292 °C. **d** Top 1 m of a ~4 m high cluster of inactive chimneys. (From de Ronde et al. 2011)

effect on the boiling behaviour of hydrothermal fluids. This is also a critical control on the volcanic eruption styles and the resulting volcanic rock associations that host VMS deposits, as discussed by Cas (1992), Cas and Giordano (2014), Cas and Simmons (2018), and in Chap. 3.

Historically this type of resource was viewed as being a hydrothermal deposit essentially precipitated "*syn-genetically*" on the seafloor as so-called exhalites, derived from thermal-driven plumes or smokers released above hydrothermal vents. However, over the past three decades, it has become increasingly clear that all VMS deposits also have a sometimes very substantial, subsurface, subvolcanic sulphide deposit component. This is precipitated "*epi-genetically*" in open fracture and in fragmental volcanic rocks, within inter- and intra-clast porosity, as well as resulting from the replacement of existing host rocks around the hydrothermal fluid conduit. In fact, in some cases, there may be no or only a minor volcanic component, with the whole process driven by a high-level, syn-depositional intrusive system that generates a subsurface hydrothermal system through interaction with saline pore water in strata below the seafloor. Such predominantly or wholly sub-seafloor sulphide deposits we call here *intrusive-hosted massive sulphide deposits* (IMS).

As a global metal resource, VMS and IMS deposits account cumulatively for 22% of the world's zinc production, 9.7% of lead, 6% of copper, 8.7% of silver, and 2.2% of gold, as well as minor amounts of other metals (Co, Sn, Se, Cd, In, Bi, Te, Ga). Approximately 1,100 VMS deposits have been identified worldwide (from Archean to Cenozoic) with a total estimated resource of about 10×10^9 t (Mosier et al. 2009, http://pubs.usgs.gov/of/2009/1034/).

In this section, we briefly review the environmental and tectonic settings of known modern submarine hydrothermal systems and associated sulphide deposits, the physical and chemical constraints on ore formation, and the nature of volcanic, intrusive, and sedimentary host rock successions to both modern and ancient VMS deposits. Other studies that have discussed volcanic eruption processes, styles, and deposits associated with VMS deposits include Cas (1992), Gibson et al. (1999), and Morgan and Schulz (2012).

18.8.1 Water Depths of Known Modern Seafloor Sulphide-Forming Hydrothermal Systems

Modern seafloor hydrothermal systems have been known for decades building on the development of submarine technology to explore, survey, and image the deep-seafloor (Chap. 16). Most modern seafloor hydrothermal systems that produce massive sulphide deposits (also referred to as seafloor massive sulphides, or SMS) occur in water depths >1000 m (Hannington et al. 2005, 2011; Monecke et al. 2014; Fig. 18.79). Of the 110 hydrothermal fields listed by Hannington et al. (2005), only 13 occur at water depths <1000 m, and only 8 at <800 m water depth. It also seems that as water depths decrease, the mineralising system tends towards becoming epithermal, such that there are some deposits described as transitional or hybrid VMS-epithermal at depths of a few hundred m (Galley et al. 2007), and some that are epithermal in water depths of tens of m (Hannington et al. 2005).

18.8.2 Tectonic Settings of Seafloor Sulphide-Forming Hydrothermal Systems

The majority of modern seafloor hydrothermal systems are associated with mid-oceanic and back-arc spreading ridge systems, and the summits of submarine subduction-related arc volcanoes (Fig. 18.79; Hannington et al. 2005, 2011; Monecke et al. 2014; de Ronde et al. 2011). To initiate submarine hydrothermal systems first requires large volume magma masses to be intruded at shallow levels in the crust that then act as the heat engines to drive hydrothermal circulation of mixed mostly saline seawater and lesser magmatic fluids in the overlying crust. Secondly, abundant subvertical fault/fracture systems must exist or be propagated to allow upward fluid migration. Almost all VMS deposits appear to have formed in extensional tectonic settings in seafloor spreading ridge systems, volcanic arcs, and back-arc basins. Together with the interpreted settings of

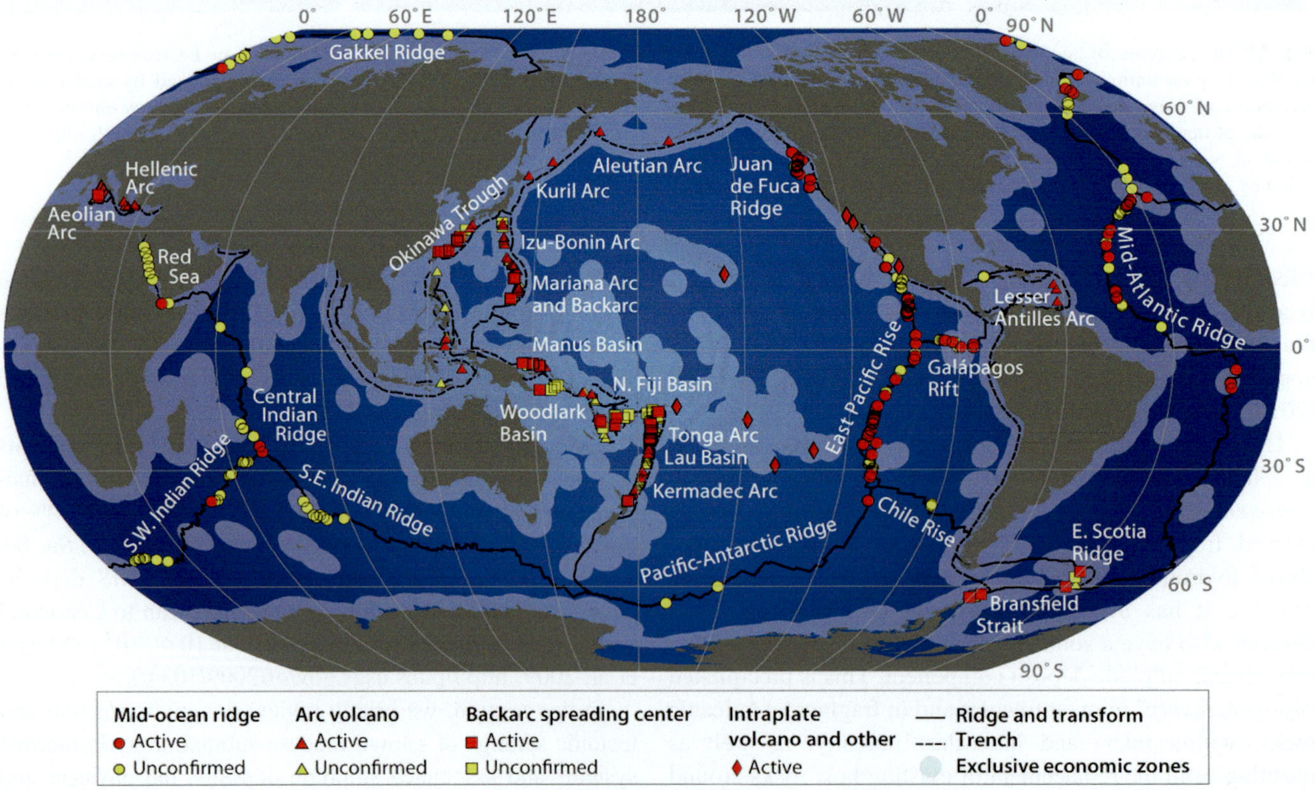

Fig. 18.79 Location of known modern hydrothermal seafloor systems and associated sulphide mineralisation (From Hannington et al. 2011; *Source* InterRidge Vents Database: http://vents-data.interridge.org/). This database contains information on >550 sites of seafloor hydrothermal activity (confirmed and unconfirmed) and inactive deposits. About 300 are sites of high-temperature hydrothermal venting; 165 are confirmed sites of massive sulphide accumulation

some ancient deposits, this has led to the following settings now identified for VMS deposits (Hannington et al. 2005, 2011).

18.8.2.1 Mid-oceanic Spreading Ridges

These mostly lie at water depths >2000 m and are associated with large high-temperature magma sources, which induce the formation of common, high-temperature geothermal or hydrothermal systems. Ridge volcanoes are mostly lava shield, pillow cone, and lava flow field volcanoes (Chap. 17), which typically erupt mid-oceanic ridge tholeiitic basalt (MORB). Seafloor hydrothermal fields associated with fast-spreading ridges host most high-temperature, polymetallic black smoker hydrothermal vents, and they tend to be short-lived and small because they are constantly buried and dismembered by frequent seafloor eruptions and tectonic deformation in the ridge crest zone. By comparison, hydrothermal systems associated with slow-spreading ridge systems tend to be long-lived and larger (e.g. TAG or "Trans-Atlantic Geotraverse" hydrothermal field, Mid-Atlantic Ridge; Hannington et al. 1998, 2005). Where spreading ridges are shallow (<500 m water depth), they lack sulphide-forming seafloor hydrothermal fields and are often marked by Fe–Mn crusts and metalliferous sediments.

It is also interesting to note recent geothermal drilling on the landward extension of the Mid-Atlantic (Reykjanes) Ridge in southwest Iceland has located geothermal fluids similar in composition and temperature to those of black smokers and the fluids precipitate base-metal sulphide deposits similar to black smoker chimneys (Hardardóttir et al. 2010; Friðleifsson et al. 2018; Sect. 18.10.5). The most significant difference between typical black smoker vents and the Reykjanes geothermal system is the pressure difference at the surface, which results in subsurface boiling and deposition of metals at Reykjanes, in contrast to chimney formation by cooling during mixing with seawater on the seafloor.

18.8.2.2 Intra-oceanic Back-Arc Basins

These vary from narrow, newly rifted basins formed by rifting of an intra-oceanic arc block and floored by very young oceanic crust produced by newly formed back-arc spreading ridges (e.g. ends of the Marianas Trough behind the Marianas arc; see Fig. 17.7) to wide back-arc basins with long-established back-arc spreading ridges and oceanic crust (e.g. Central Marianas Trough, North Fiji Basin). Back-arc ridge volcanic rocks vary from arc-like in narrow basins to mid-oceanic basalts in mature basins. Features of hydrothermal systems are similar to those of oceanic spreading ridges.

18.8.2.3 Sediment-Covered Spreading Ridges and Related Rifts

These occur where spreading ridges converge with continental land masses (e.g. Guaymas Basin in the Gulf of California, and Juan de Fuca and Gorda Ridges, offshore the western United States), from which terrigenous sediment is derived and can bury ridges with hundreds of m of sediment (Gieskes et al. 1982; Koski et al. 1985; Goodfellow et al. 1993; Zierenberg et al. 1993; Goodfellow and Zierenberg 1999; Hannington et al. 2005; James et al. 1998). The Guaymas Basin is also characterised by the rapid deposition of biogenic organic-rich sediments derived from highly productive overlying waters. Volcanic eruptions are rare, but syn-depositional intrusions into the sediment pile are common, often leading to cryptodome up-doming of the seafloor (Chap. 8). Although some low-temperature hydrothermal vents occur associated with these, it is thought that most sulphide precipitation occurs below the seafloor in the sediment pile, involving pore filling and replacement processes. The hydrothermal fluids can carry large quantities of petroleum derived from the thermal alteration of organic matter during hydrothermal circulation in the subsurface (Simoneit and Lonsdale 1982; Simoneit et al. 1992; Teske et al. 2018; Simoneit 2018). This is transported in the venting hydrothermal fluid and occurs in pore spaces in cavities in chimneys, mounds, and subsurface replacement bodies (Peter et al. 1990).

18.8.2.4 Submarine Volcanoes in Marine Volcanic Arcs

Marine volcanic arcs can form either on the pre-existing oceanic crust (e.g. Izu–Bonin arc) or submerged continental or older evolved arc crust (e.g. the southern part of the Kermadec arc; Chap. 17). Marine arc volcanoes are typically stratovolcanoes, spaced at ~50 km apart, up to 30 km in basal diameter and are 1–2 km high above the surrounding seafloor (see Fig. 17.34a). Young volcanoes and arcs are typically tholeiitic and mafic, but as they evolve they become more calc-alkaline and intermediate and some erupt dacites and even rhyolites. Many possess summit calderas generally up to 5 km in diameter, although slightly larger examples occur, and many have active polymetallic hydrothermal systems.

Active hydrothermal systems occur along caldera walls, associated with caldera margin fractures, and at the summits of post-caldera-forming, intra-caldera cones, and lava domes. Although some caldera floors occur at depths of ca. 1500 m, we note here that contrary to many previous interpretations, it is unlikely that any explosive caldera-forming explosive eruptions occurred at those depths due to hydrostatic pressure constraints (Chaps. 3, 4). If explosive eruptions were involved in caldera formation (they can also form through lateral flank lava eruptions leading to summit collapse), the eruptions would have occurred from the summit of the pre-collapse volcano at much shallower water depths than the current caldera floors. Therefore, the caldera rim water depths represent the maximum possible explosive water depth, and if a significant part of the summit was removed explosively, the eruption water depths would have been even shallower. The current depth of caldera floors represents the depths to which they collapsed after the eruption that formed them (see Sect. 18.8.7).

The most informative studies on arc volcanoes and their hydrothermal fields have been undertaken in the Kermadec Arc, north of New Zealand. These have variously involved: seafloor surveys and mapping, petrology and geochemistry (e.g. Wright et al. 2006; de Ronde et al. 2007), detection of hydrothermal plumes in the water mass above submarine volcanoes using light scattering and He detection technology (de Ronde et al. 2007), detailed bathymetry surveys and imaging of the seafloor of selected volcanoes (e.g. Wright and Gamble 1999; de Ronde et al. 2005; Embley et al. 2012), and studies of the mineralising systems and processes (e.g. de Ronde et al. 2005, 2011; Leybourne et al. 2012a, b).

18.8.2.5 Rifts Associated with Arc Volcanoes

These rifts occur in pre-existing arc crust, and so magmas vary from high-K calc-alkaline to MORB. They may eventually evolve into back-arc spreading ridge basins, but in early stages, the magma dynamics is strongly influenced by the rifted arc crust, as are the hydrothermal systems, with distinctions between young more primitive intra-oceanic arcs and older more evolved arcs. Hydrothermal deposits tend to be polymetallic sulphides with higher Au, Ag, As, Sb, Pb, and Ba contents than those formed in back-arc basin settings. For example, the massive sulphide hydrothermal vent fields in the eastern Manus Basin, Papua New Guinea (PACMANUS and SuSu Knolls-Suzette; Moss and Scott 2001), have high precious metal contents compared with central Manus Basin hydrothermal fields.

Where oceanic arcs merge with continental crust, such trends also occur (e.g. southern end of the Kermadec arc where it merges into New Zealand in the Bay of Plenty; Fig. 18.79). This category seems to merge into a category called rifting of continental margins proposed by Hannington et al. (2005), because the examples cited also correspond with the rifts associated with arc volcanoes.

18.8.2.6 Intraplate or Off-Axis Volcanoes

These are mostly shield, pillow cone, and lava flow field alkaline basalt (OIB) volcanoes and tend to produce low-temperature hydrothermal systems that are enriched in iron (Hannington et al. 2005).

18.8.2.7 Intra-continental and Back-Arc Intra-continental Submarine Rifts

These are the initial continental rifts from which oceanic spreading ridges will eventually form. The Red Sea rift is a type of example beginning from the initial crustal rifting to the present-day incipient ridge segments. Hydrothermal activity is associated with brine pools, formed by the melting of pre-existing evaporite deposits. Principal precipitates are Fe–Mn oxy-hydroxides hosted by metalliferous sediment (Hannington et al. 2005). When arcs rift from pre-existing continental or evolved island arcs, resulting basins are floored by subsided continental crust, which plays a role in the types of ore deposits that can form; for example, the Miocene Green Tuff Belt, northwestern Japan which hosts the type Kuroko VMS deposits (Ohmoto and Skinner 1983), and the modern Okinawa Trough, southern Japan where similar massive sulphide deposits are actively forming today (Tsuji et al. 2012; Yeats et al. 2017).

18.8.3 Relationship Between Volcanic Setting, Water Depth, and Hydrothermal Deposit Type

Although there are some general relationships that link tectonic setting, water depth, and mineralisation types, there are also exceptions, because sulphide precipitation is influenced by many variables, as discussed further below. Nonetheless, these relationships include the following:

(a) VMS deposits formed at water depths >1000 m contain more Cu than those formed in shallower water, where >2 wt% is rare. This is consistent with precipitation from high-temperature (>300 °C) hydrothermal vents under relatively high hydrostatic pressure due to water depth (Monecke et al. 2014).
(b) Zn and Pb can be abundant at any water depth, but Zn content >5 wt% commonly occurs at water depths of <1,000 m. High Pb and Zn are consistent with precipitation from lower temperature hydrothermal vents (150–250 °C). However, significant Cu could also be found in the deeper stockwork to such deposits, having been precipitated from the fluids that boiled when they were at a higher temperature at depth, and under higher combined hydrostatic and lithostatic pressures.
(c) Some shallow marine massive sulphide deposits are enriched in Au, but other factors also contribute to Au content, including source rocks with which hydrothermal fluids have interacted.
(d) As, Sb, and Hg are enriched in shallower water deposits.
(e) VMS deposits associated with deep-water mid-oceanic spreading ridges and spreading ridges in wide, mature

back-arc basins tend to be Cu and pyrite rich (e.g. TAG mound, Mid-Atlantic Ridge, Hannington et al. 1998). This is because they precipitate at significant water depths (>2000 m) and high hydrostatic pressures, which prevented high-temperature fluids from boiling before reaching the seafloor.

(f) VMS deposits associated with young, narrow, back-arc basin spreading ridges are more polymetallic, reflecting the influence of the arc substrate/crust and setting in sourcing metals.

(g) Similarly, deposits forming at submarine arc volcanoes and in rifted continental and arc crust are mostly polymetallic (e.g. JADE and Iheya North, VAMP area, middle Okinawa Trough; Glasby and Notsu 2003; Yeats et al. 2017; and, Miocene Kuroko deposits, Japan Ohmoto and Skinner 1983).

(h) Recent work has shown shallower water deposits (<several hundred m deep) appear to become more epithermal in character (e.g. Conical Seamount, Lihir, Papua New Guinea, Petersen et al. 2002; Müller et al. 2003; Sect. 18.3.4.7; Aeolian Islands, Tyrrhenian Sea, Petersen et al. 2014; Milos, Aegean Sea, Alfieris et al. 2013; and, Bay of Plenty, New Zealand, Hocking et al. 2010).

Although the majority of known modern seafloor hydrothermal systems occur at mid-oceanic spreading ridge systems (65%), it is the hydrothermal systems associated with back-arc basin ridges (22%) and arc volcanoes (12%) that are the more likely to be analogues for the setting of ancient VMS deposits (Hannington et al. 2005, 2011). This is because most normal oceanic crust is generally subducted, whereas slices of back-arc crust and arc volcanoes may be obducted and amalgamated into continental or older arc crust. VMS deposits formed on developing volcanically active continental crust (e.g. Archean cratons, Paleozoic orogens) or failed rifts in developed arc crust (e.g. Kuroko Green Tuff Belt in northwestern Japan) are those commonly preserved in the rock record.

In all the above scenarios including mid-oceanic ridges (e.g. TAG mound, Hannington et al. 1998; Petersen et al. 2000), substantial massive sulphide deposits also form below the seafloor in the stockwork fracture system and in porous host rocks such as autoclastic volcanic breccias, and through replacement of sub-seafloor host rocks.

18.8.4 Constraints on the Formation of VMS Deposits

The two principal requirements for the formation of a VMS deposit are (1) relatively deep water (>1000 m = high hydrostatic pressure), and (2) a relatively shallow, large sub-seafloor intrusion (<2 km from the seafloor). The latter generates a hydrothermal circulation cell that mixes seawater and magmatic fluids, which then leach metal ions from the volcanic and sedimentary substrate (Fig. 18.80). When these circulating, hot, metal-enriched hydrothermal fluids then encounter cold seawater or boil, thus cooling, they precipitate metals as sulphide species.

The constraints that contribute to determining the location, processes of formation, and mineralogy of sulphide ore deposits include several factors (Hannington et al. 2005; Shanks 2012a; Tornos et al. 2015) summarised below.

18.8.4.1 Magmatic Heat Source, Its Volume, and Depth Below the Seafloor

These factors control how effective a magma body is as a heat engine driving the hydrothermal system. A high-level, shallow intrusion into the seafloor crust induces a strong thermal gradient that causes a buoyant uprise of crustal fluids around the intrusion (mostly intra-stratal seawater) along crustal fractures to the seafloor (Fig. 18.80a, b). Such discharging high-temperature fluids then induce the recharge of seawater back into the crust, creating a convective fluid cell system that consists mostly of recycled seawater with variable additions of magmatic fluids (Fig. 18.80b, c; Franklin et al. 2005). Ohmoto (1996) suggested that typically a 10 km^3 size pluton intruded less than 2 km below the seafloor could generate a convective hydrothermal circulation cell around the pluton with a radius of $\sim$5 km. Based on the location of earthquake hypocentres, Dziak et al. (2008) have constrained the depth of a magma body beneath seafloor hydrothermal systems at the wholly submarine Brothers volcano in the Kermadec Arc, north of New Zealand to be >2.5 km.

Such a heat source also causes pervasive alteration of the country rock around the magma body and the hydrothermal fluid conduit above the pluton, as well as around any massive sulphide ore body that may form. Given both the temperature and pressure regime, the typical alteration halo above and around a VMS ore deposit is similar to a greenschist facies metamorphic aureole (Fig. 18.80a, d). This shows intense (Fe-) chlorite (-quartz) alteration developed closest to the ore body, and outwards a gradation into a (Fe-) chlorite–sericite zone, and then into a (Mg-) chlorite–phengite alteration zone (Goodfellow et al. 2003; Franklin et al. 2005; Galley et al. 2007; Shanks 2012b).

18.8.4.2 Magma Type

Lesher et al. (1986) demonstrated that the geochemistry of Archean felsic volcanics in host rock successions to VMS deposits helped to discriminate those that are fertile and host VMS deposits, from those that don't.

Their FI suite, marked by steep REE patterns with weakly negative to moderately positive Eu anomalies, high Zr/Y, low abundances of high-field-strength elements (e.g. HREE, Y, Zr, Hf), and high abundances of Sr, was considered to have originated from deep crustal sources, were neither modified by crystal fractionation nor generated hydrothermal circulation systems, and are barren of VMS deposits. The FII suites are rhyodacites and rhyolites characterised by gently sloping REE patterns with variable Eu anomalies, moderate Zr/Y, and intermediate abundances of HFS elements and Sr and host some VMS deposits. The FIII suite consists of rhyolites and high-silica rhyolites characterised by relatively flat REE patterns, which are further subdivided into two types. FIIIa felsic metavolcanic rocks exhibit variable negative Eu anomalies, low Zr/Y, and intermediate abundances of HFS elements and Sr. In contrast, FIIIb felsic metavolcanic rocks exhibit pronounced negative Eu anomalies, low Zr/Y, high abundances of HFS elements, and low abundances of Sr. FIII suite felsic volcanics are the most evolved and host numerous significant VMS deposits.

Lentz (1998) and Hart et al. (2004) added a further suite, FIV, to accommodate subtle differences in some Phanerozoic felsic volcanics associated with VMS deposits. The FIV suite are rhyolites and high-silica rhyolites characterised by flat to slightly LREE-depleted REE patterns and low REE and HFSE abundances and are common hosts to VMS deposits.

Mafic volcanics that are associated with VMS deposits reflect the specific tectonic settings in which they formed, e.g. mid-oceanic ridge = MORB basalt; back-arc basin basalt = BABB, etc.; Piercey 2011.

18.8.4.3 Hydrothermal Fluid Chemistry and Sources

Hydrothermal fluid analyses from direct sampling, analysing fluid inclusions in VMS deposits, and through experimental studies have demonstrated that VMS-forming hydrothermal fluids are mostly in a reduced state and are saline fluids (2–6 wt% NaCl), with salinity levels similar to, or higher than seawater ($\sim$3.2 wt% NaCl; Hannington et al. 2005; Shanks 2012b; Bodnar et al. 2014). Most fluids are acidic with low pH values (de Ronde et al. 2011). Hydrothermal fluids emitted from spreading ridge systems (mafic associations) are low sulphidation fluids, whereas those emitted from arc volcanoes, with felsic associations, can be high sulphidation fluids, especially those that may be transitional into epithermal systems (Sillitoe et al. 1996; Dubé et al. 2007; de Ronde et al. 2011; Mercier-Langevin et al. 2011, 2015).

Seawater contains Na^+, Cl^-, $Mg_2{}^+$, $Ca_2{}^+$, and $SO_4{}^{2-}$ ions. When heated to 150 °C anhydrite ($CaSO_4$) is precipitated, often beneath the seafloor. Magmatic fluids are the source of CO_2, He, H_2S, SO_2, CH_4, H_2, HCl, and HF (de Ronde et al. 2011; Shanks 2012a). Stable isotope data indicates that most S in sulphide minerals in VMS deposits comes from degassing of magmas (Tornos et al. 2015). Though some magmatic fluids can contain metals (Hedenquist et al. 1993; Hedenquist and Lowenstern 1994), a major source of metals in VMS deposits appears to be the host volcanic rocks through which hydrothermal fluids flux. Magmatic volatiles have low Cl contents, indicating that most of the mineralising fluids are seawater that has been cycled into and through the crust, convected by the heat of the relatively shallow magma intrusions (Fig. 18.80; Hannington et al. 2005; Shanks 2012a). Magmatic fluids released from the magma body are mixed with the recycling seawater in the crust. Based on oxygen isotopic data, magmatic fluid input represents a small volume of the resulting hydrothermal fluid in most systems (Bodnar et al. 2014), the exception being in high sulphidation systems where the magmatic component may be significant, even dominant (Sillitoe et al. 1996; de Ronde et al. 2011). Using strontium and sulphur isotope data, Roberts et al. (2003) propose that the mineralising fluids responsible for different sulphide deposits at the PACMANUS site in the eastern Manus Basin, Papua New Guinea, reflect varying degrees of interaction between hydrothermal, magmatic, and seawater fluids.

As heated hydrothermal fluid fluxes through the crust around and above the subsurface intrusive magma body, it leaches elements from the volcanic and sedimentary rocks, including metals (Ohmoto 1996; Shanks 2012a). Where these strata are fragmental volcanic rocks (e.g. hydraulically fractured rock, and/or porous and unconsolidated autoclastic volcanic rocks and sediments), the surface area for, and degree of fluid–rock interaction, is optimal. According to Shanks (2012a), hydrothermal fluids with the salinity of seawater can carry metal ions readily at temperatures above 200 °C. Sulphides are precipitated if those fluids are cooled by mixing with cold seawater, or through boiling. As hydrothermal fluids cool from 350 °C, they successively precipitate sulphides, then sulphates, and then silicate minerals at lower temperatures (Shanks 2012a).

18.8.4.4 Interactions with the Host Rocks

The composition of the substrate through which the hydrothermal fluids flux and the degree to which elements are leached from the substrate into the hydrothermal fluids is a significant factor in determining the metal species that are mobilised. Yang and Scott (2002) summarise the relationship between metal abundances in volcanic rocks of different compositions at the submarine PACMANUS site in the eastern Manus Basin, Papua New Guinea. They found that metals in melt inclusions and groundmass glass in basalts and basaltic andesites were dominated by Ni, Cu, Zn, and Fe, whereas in andesite Cu, Zn, and Fe were abundant, Cu and Fe were dominant in dacite, Fe in rhyodacite, and

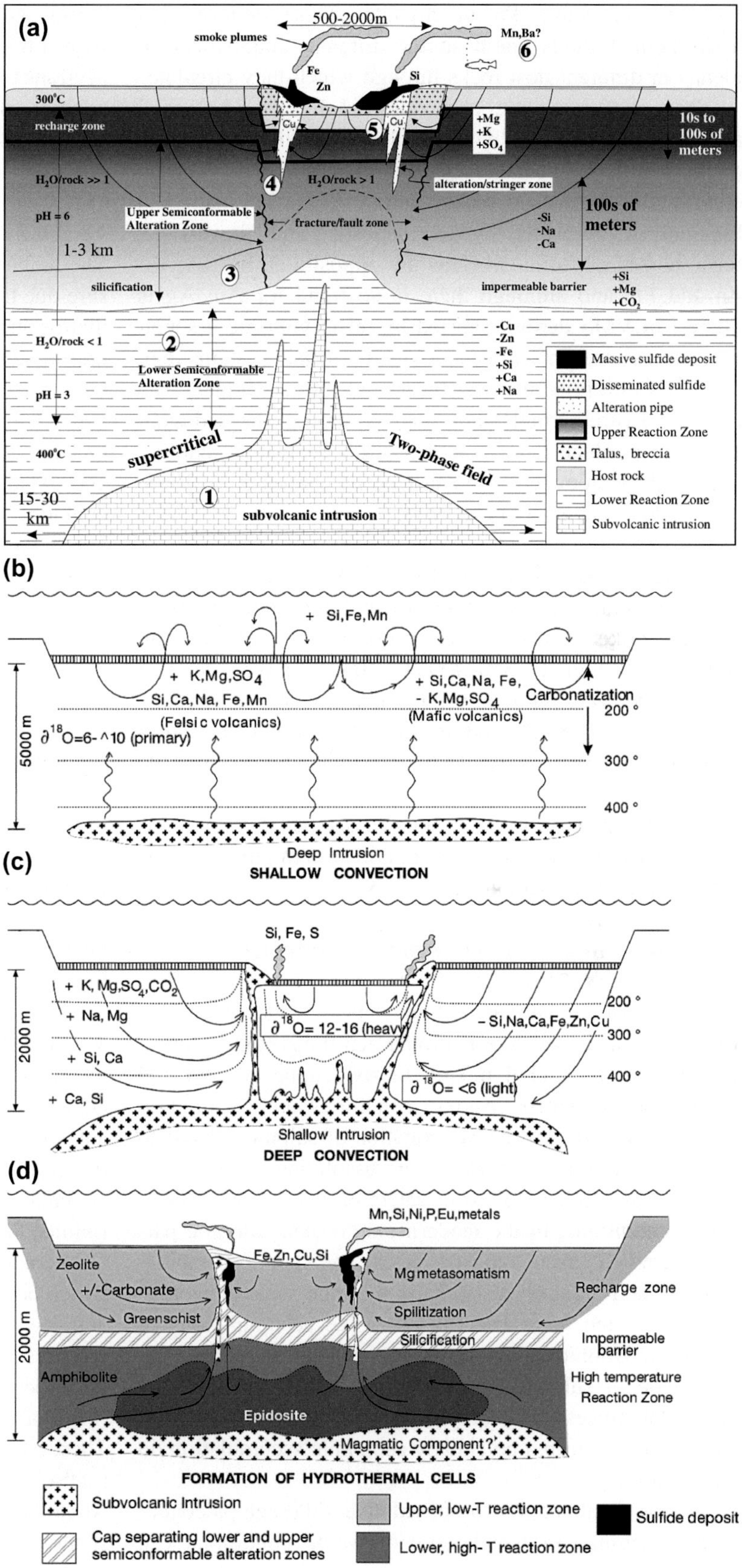

Fig. 18.80 Schematic representations of submarine hydrothermal systems that produce VMS and IMS deposits. **a** Relatively shallow/high-level intruding magma body that acts as the heat engine. This induces a strong thermal gradient causing the buoyant uprise of crustal fluids along crustal fractures, which in turn induces the recharge of seawater back into the crust, creating a convective fluid cell system that consists mostly of recycled seawater with variable additions of magmatic fluids. **b** The uprise of heated crustal fluid and released magmatic fluid above a shallow, high-level intrusion, leading to element exchange and diffuse fluid discharge on the seafloor. **c** Advanced development of convection cells with continued element exchange between circulating fluids and crustal rocks, leading to the formation of hydrothermal vent systems on the seafloor above subvertical fractures. **d** The advanced state of the hydrothermal system that has also induced greenschist facies alteration zones above and around the hydrothermal system. (From Franklin et al. 2005)

Fe, Zn, and some Pb in rhyolite. This suggests that hydrothermal fluids could source different abundances of metals in different host rocks through which they circulate.

These variations are also reflected by average metal abundances associated with VMS deposits hosted by the major host rock association types discussed herein. Franklin et al. (2005) show that Cu abundance is highest in mafic and bimodal mafic-hosted VMS deposits and lowest in felsic-hosted deposits, whereas the opposite is the case for Zn and Pb, and although there is no trend for Au, Ag is highest in VMS deposits hosted by felsic volcanic association successions. In sediment-covered volcanic systems (e.g. Gulf of California, Juan de Fuca, and Gorda Ridges), the effect of hydrothermal fluids fluxing through the sediment pile is to produce more zinc-rich VMS deposits, including many sub-seafloor replacement deposits (James et al. 1998; Goodfellow and Zierenberg 1999).

18.8.4.5 Hydrothermal Fluid Temperature

Measured temperatures of hydrothermal fluids emitting from "black smoker" hydrothermal vents are at >280 °C, and some are as high as 350–400 °C. Black smokers tend to produce deposits with higher Cu content. White smokers eject fluids at <280 °C and are richer in Zn and Pb, and at very low temperatures, Ba, and then Mn and Fe deposits (de Ronde et al. 2011; Tornos et al. 2015). When hydrothermal fluids mix with cold seawater, this causes metal compounds to precipitate because they are soluble at high temperatures, but not at low temperatures.

18.8.4.6 Ambient Environment Chemistry

Tornos et al. (2015) highlight that most preserved VMS deposits form in anoxic bottom waters, made anoxic by venting of anoxic seafloor hydrothermal fluids, and basins in which there is limited mixing of oxic upper waters and deeper anoxic waters. Anoxic conditions are required to preserve sulphides, whereas, oxidation of sulphides leads to their alteration to sulphates and dissolution. Venting of hydrothermal fluids into oxic waters produces poor preservation conditions. In the modern oxic ocean, sulphide particles emitted from modern black smokers are rapidly oxidised and redissolved (97% on the time scale of months to a year), and large laterally extensive accumulations of stratiform sulphides are therefore unlikely to form. The deposition of sulphides from buoyant plumes is a highly inefficient process, and only long-lived hydrothermal systems could result in a potentially economic accumulation on the seafloor.

Tornos et al. (2015) proposed five different paleogeographic settings in which anoxic marine conditions can occur: (1) the anoxic Meso-Archean to Paleoproterozoic; (2) other regionally extensive periods of oceanic anoxia, e.g. during the early Paleozoic and Mesozoic; (3) localised small basins, including submarine rifts and calderas, or large vents, with limited circulation; (4) isolated basins with poor connections to the open sea (e.g. Black Sea); and (5) emission of anoxic, saline hydrothermal fluids on the seafloor that then pond in small sub-basins and may even form seafloor brine pools (e.g. Red Sea).

18.8.4.7 Ambient Pressure

The ambient pressure is a combination of hydrostatic pressure due to water depth for seafloor hydrothermal vents, and hydrostatic plus lithostatic pressure, for fluids in hydrothermal conduits below the seafloor. This controls the boiling behaviour of hot hydrothermal fluids. Boiling is exothermic, and when boiling hydrothermal fluids mix with seawater, this causes a temperature drop, commonly increases the pH and oxidation state of the fluids, and also leads to the precipitation of sulphides (Drummond and Ohmoto 1985; Tornos et al. 2015). Where hydrothermal systems are located in very deep water (>2.5 km), the hydrostatic pressure alone is enough to prevent the boiling of fluids below the seafloor. Hence, hydrothermal fluids reach the seafloor, and through cooling by mixing with cold seawater and/or boiling on being emitted onto the seafloor, they can precipitate massive sulphides on the seafloor, often forming sulphide mounds (e.g. TAG field, Mid-Atlantic Ridge). However, in shallower water (<2000 m), hydrothermal fluids may begin to boil in the sub-seafloor conduits or stockwork of fractures, causing early precipitation of sulphides in the stockwork, especially Cu sulphides, before the fluids reach the seafloor. Those fluids will have cooled by the time they reach the seafloor and may precipitate Zn and Pb sulphides on venting on the seafloor.

Other factors that contribute to the pressure-depth regime at which hydrothermal fluids will boil are the initial fluid temperature, their salinity, and their CO_2 content. Haas (1976) showed that salinity affects the boiling behaviour of aqueous saline fluids, with increasing salinity lowering the pressure at which boiling begins, so that boiling could begin at shallower water depths in the oceans compared with low salinity fluids. CO_2 has the opposite effect, with increasing CO_2 content and partial pressure increasing the pressure at which boiling of hydrothermal fluids will begin (Hedenquist and Henley 1985b). Hydrothermal fluids with high CO_2 contents boil at greater depths and pressures in the crust or the ocean than fluids with low CO_2 contents (Pisutha-Arnond and Ohmoto 1983; Bodnar et al. 2014). These variations will determine if boiling occurs below the seafloor in the stockwork fracture system or at the seafloor when the fluids vent onto the seafloor. Where the seafloor is shallow, more sulphide mineralisation is likely to form in the stockwork below the seafloor or in laterally extensive porous–permeable lithofacies horizons, than as seafloor sulphide mounds and stratiform deposits.

18.8.4.8 Availability of Subvertical Fractures

Subvertical fractures allow hydrothermal fluids to rise and usually coincide with extensional tectonic and volcanic crustal settings (Piercey 2011; Fig. 18.80a, c, d). Regarding an ancient example, that is relatively undeformed, Archean VMS deposits in the Strelley Belt, Pilbara Craton, Western Australia, show discordant high-angle, deep penetrating faults in the footwall, which splay close to the zones of massive sulphide mineralisation (Vearncombe et al. 1998). These faults do not extend far into the overlying sedimentary cover, indicating their syn-volcanic and syn-mineralisation timing. Mineralisation is located at the edge of elevated grabens, at the site of intersecting syn-volcanic extensional faults. As with their subaerial counterparts, narrow, linear submarine rift basins are likely to be controlled by steep normal basin bounding and intra-rift faults and detachments, compartmentalised by oblique transfer faults (Wright 1994; Chap. 17). In the deformed Paleoproterozoic Vargfors Basin in the central Skellefte mining district, Sweden (Table 18.3), detailed mapping and structural analysis have now established a pattern of SE–NW-striking basin-forming normal faults and interlinked NE–SW-striking transfer faults (Bauer et al. 2011). These define distinct fault-bound compartments, each with an individual structural geometry and stratigraphy. The VMS deposits show a close spatial relationship to faults indicating that the ore-forming hydrothermal fluids utilised the syn-extensional faults as fluid conduits. Subsequently, the deposits were transposed into the inverted faults during later crustal shortening (Sect. 18.8.5.4).

18.8.4.9 Availability of Other Permeable Horizons

These are necessary for hydrothermal fluids to percolate and interact with seawater. Fluids flow laterally and up-dip along permeable horizons within the volcano-sedimentary sequence (e.g. Tsuji et al. 2012). Such horizons can include autoclastic deposits in the substrate (e.g. hyaloclastite breccias) and pumice deposits of local or distal origin. Such permeable horizons will be saturated with cold seawater, and as hot hydrothermal fluids pass through them, they mix and cool, and may precipitate significant sub-seafloor pore-filling sulphide deposits, as well as initiate host rock replacement to form sulphide replacement deposits.

18.8.4.10 Time

The duration or longevity of seafloor hydrothermal systems can be highly variable but is also surprisingly short in geological terms. Modern hydrothermal vent fields are ephemeral features on the seafloor and become inactive when seafloor-spreading moves them away from a given body of magma or when, in the absence of recurrent hydrothermal hydraulic fracturing and/or tectonic restructuration, they become "sealed" by precipitation of sulphide and other secondary mineral phases. Some vent fields may remain active for >10,000 years, but individual vents are much shorter-lived, as discussed in Sect. 18.8.6.

18.8.5 VMS Resources and Mineral Deposit Characteristics

18.8.5.1 Metal Resources

Ancient VMS deposits have been mined principally for their base metal contents, especially zinc or copper, although, also in combination with their precious metal credits, in particular gold (Table 18.5), which in recent years have continued to make Au-rich deposits an economically viable exploration target. Most VMS deposits are generally low tonnage ranging between 2 and 10 Mt, low grade (<10% combined metal), and relatively small exploration targets, which can detract from their value as priority targets for many major exploration companies. For example, the average size of the archetypal Kuroko deposits in the Green Tuff Belt, Japan, is only 1 Mt; the largest deposit represented by the Matsumine deposit is 30 Mt, containing about 7% combined metal, while most of the other mined Kuroko deposits are significantly smaller (Table 18.5). However, VMS deposits are often found clustered together in larger districts or camps which increase the probability of further discovery. They may also average between 1 and 4 g/t Au, while Au-rich zones may contain considerably higher grades (e.g. 5–30 g/t at Que River, Tasmania), or they may be intrinsically Au-rich (e.g. Boliden, Sweden, and Eskay Creek, British Columbia).

Nonetheless, as a deposit type, they also comprise some giant (>100 Mt) and supergiant (>100 Mt) deposits (Table 18.5) that are anomalous either because of their size (e.g. Rio Tinto, Windy Craggy, Mt Lyell), their size plus base-metal content (e.g. Kidd Creek, Neves Corvo, Brunswick No. 12), or their size plus precious metal content (e.g. Horne, LaRonde Penna). Rio Tinto and Neves Corvo represent some of the largest accumulations of metal sulphides known on Earth. These contribute to make the Iberian Pyrite Belt in Spain and Portugal the greatest concentration of volcanogenic massive sulphides (>1,500 Mt) on Earth (Sáez et al. 1996; Leistel et al. 1998; Soriano and Martí 1999), albeit largely uneconomic with very low commercial metal grades. Windy Craggy, in northern British Columbia, Canada, is an enormous copper–cobalt–gold massive sulphide deposit hosted within a succession of late Triassic sedimentary and mafic volcanic to intrusive rocks (Peter and Scott 1999). The Horne VMS deposit, located within the Archean Noranda Volcanic Complex of the Abitibi greenstone belt, Quebec, Canada, represents a superlarge gold deposit in its own right and produced some 280 t (9 Moz) of Au between 1927 and 1989 (Gibson et al. 2000). Similarly, the LaRonde Penna VMS deposit, which is located some 40 km east, is the largest Au mine currently operating in Canada making this the world's foremost region, present and past, for Au-rich VMS deposits (Mercier-Langevin et al. 2007). Mining of VMS deposits has been historically significant in Canada accounting for 27% of Cu production, 49% Zn, 20% Pb, 40% Ag, and 3% Au, where exploration interest remains high, and in Australia.

More recently, the possibility of mining fossil massive sulphide deposits on the modern seafloor has promoted increasing interest and assessment of their potential commercial development and the risks. Although more than 300 sites of high-temperature hydrothermal venting have been identified since the discovery of black smokers, significant massive sulphide accumulations have been found at only 165 of these sites (Hannington et al. 2011; Fig. 18.79). Global estimates of the total of ridge, arc, and back-arc spreading centres indicate the amount of massive sulphide in the easily accessible neo-volcanic zones to be of the order of 600 Mt, containing resources of ~30 Mt of copper and zinc. While this is wholly inadequate to satisfy the growing global demand for these metals (Hannington et al. 2011), significant accumulations of metals from hydrothermal vents are found in the oceans, such as the metalliferous sediments in the Atlantis II Deep of the Red Sea (91.7 Mt @ 2.06% Zn and 0.46% Cu; Guney et al. 1988). Recent developments, however, suggest that the first commercial deep-sea mining will likely be from the much smaller, Au-rich, and higher-grade massive sulphide deposits in the neo-volcanic zones of volcanic arcs and their back-arc basins. The most advanced venture to date is the Solwara project (SuSu Knolls and Suzette hydrothermal vent fields) in the eastern Bismarck Sea, Papua New Guinea, where feasibility studies for a commercial-scale mining operation to extract deep-sea metallic resources are well advanced (Plate 18.1f). At Solwara-1, which covers an area of moderately deep ocean up to 2,000 m deep within which there are several fossil and semi-active seafloor mounds, a preliminary resource of 1.44 Mt @ 8.2% Cu and 6.1 g/t Au is estimated (Lipton et al. 2018).

As a footnote, we point out that significantly more metal is believed to be contained in manganese nodules, e.g. the global resource of Cu found in nodules has been estimated at 700 Mt (Hannington et al. 2011).

18.8.5.2 Mineral Deposit Styles

VMS deposits were traditionally depicted as being stratiform and lens-like bodies (Lambert and Sato 1974), consisting of >60% sulphide minerals (mainly pyrite and/or pyrrhotite). This evolved into a model of seafloor sulphide mounds, fed by a cross-cutting hydrothermal stockwork and formed through venting of hydrothermal fluids on the seafloor from black and white smoker sulphide and sulphate chimneys (Fig. 18.81; Lydon 1984; Tornos et al. 2015). However, it has become increasingly clear that VMS deposits can have a range of morphologies and spatial and temporal relationships to their volcanic, intrusive, and sedimentary host rocks, particularly since the realisation that many deposits include a large sub-seafloor pore-filling and replacement component. Indeed, some deposits may be predominantly or entirely sub-seafloor replacement deposits (Fig. 18.81a–c; Lydon 1984; Large 1992; Doyle and Allen 2003; Galley et al. 2007; Tornos et al. 2015). Furthermore, from an exploration point of view, it is only such largely replacement deposits that are likely to provide the tonnage and grade to create economic value.

Tornos et al. (2015) identify four main end-member styles and paleo-environmental settings of VMS deposits (Fig. 18.81c), which in many ways mimic the categories also proposed by Large (1992): (I) mounds and black smoker chimneys in oxic environments, (II) brine pools, (III) mounds and stratiform deposits in regional anoxic environments, and (IV) sub-seafloor replacement deposits.

Sulphide Mounds

Sulphide mounds build through the precipitation of sulphide crystals from the plume of hot hydrothermal fluid rising and ejected into cold seawater directly above the smoker vents, in the subsurface around the fluid conduit beneath the growing mound, and in the smoker chimneys. Chimney or spire heights often range up to about 20 m in height, but have been observed as high as 40–60 m, which are susceptible to gravitational collapse especially in seismically active settings. Lofted, dense sulphide particles in the hydrothermal plume fallout around the hydrothermal vents, while others are dispersed by the buoyant, hot hydrothermal plume and ambient submarine currents. These form fallout layers of exhalite sulphide sediment on the seafloor away from the mound (Fig. 18.81a). Modern seafloor sulphide mounds can attain dimensions up to a few tens of m in height and hundreds of m in diameter and, when active, are frequented by thriving thermo-tolerant benthic communities and microbial mats.

One of the best studied is the large TAG mound on the Mid-Atlantic Ridge, which is 200 m in diameter and stands as a 50 m-high constructional feature on the seafloor (Hannington et al. 1998; Petersen et al. 2000). Many mounds are relatively small sites of hydrothermal venting, as observed in the Okinawa Trough (Ishibashi and Urabe 1995; Glasby and Notsu 2003; Ishibashi et al. 2015; Yeats et al. 2017). In the eastern Manus Basin, the hydrothermal vent fields at PACMANUS and SuSu Knolls-Suzette are associated with larger, topographic features forming ridges and domes of basaltic to rhyodacitic lavas and breccias (Binns and Scott

Table 18.5 Selected features of volcanic (and intrusive)-hosted massive sulphide deposits referred to in the text

Location	Age mineralisation (Ma)	Volcano-tectonic setting	Deposit	Tonnage (Mt)	Zn (%)	Pb (%)	Cu (%)	Ag (g/t)	Au (g/t)	Immediate host rocks	Water depth
Green Tuff Belt, Hokuroko Basin, Honshu, Japan	16–14	Bimodal back-arc intra-continental submarine rift	Fukazawa	3	15.4	3.3	1.1	93	0.6	Resedimented pumice breccias and sandstones, carbonaceous mudstones	500–2800 m
			Hanaoka-Matsumine	30	3.6	1	2.4	57	0.5		
			Hanaoka-Shakanai	7	2.25	0.6	1.35	30	0.4		
Stikine Terrane, northern Cordillera, British Columbia	Mid-Jurassic	Bimodal submarine rift	Eskay Creek	2	5.2	3.2	0.7	2,652	60	Carbonaceous argillite	1500 m (anoxic)
Alexander Terrane, northern Cordillera, British Columbia	~225	Submarine mafic-intermediate rift	Windy Craggy	300	0.25		1.4	3.8	0.2	Pillow basalt, mafic sills, and graphitic to calcareous siltstone and argillite	≥2000 m (anoxic)
Iberian Pyrite Belt, Spain, and Portugal	354–346	Bimodal submarine (trans)extensional basin	Rio Tinto	335	0.34	0.12	0.39			Dacitic–rhyolitic porphyritic sills, peperites, and carbonaceous mudstones	Deep
			Neves Corvo	220	4.1	0.74	3.2			Rhyolitic lava, hyaloclastites, pseudo-fiamme-rich sandstone, and carbonaceous mudstone	Deep
Bathurst Mining Camp, New Brunswick	~465	Restricted bimodal back-arc basin	Brunswick No. 12	128	8.8	3.4	0.35	105		Carbonaceous mudstones (silicified)	Deep (anoxic)
Bathurst Mining Camp, Maine	Early Ordovician	Synvolcanic submarine caldera or graben	Bald Mountain	30	1.1		1.0	14.4	0.5	Crystal-poor pumiceous rhyolitic ignimbrite	1400 m
Mt Read Volcanic Belt, Tasmania	~502	Bimodal submarine extensional basin	Rosebery	32	14.3	4.4	0.6	146	2.3	Resedimented pumiceous sandstone and breccia, carbonaceous mudstones	500–1000 m
			Hercules	3.4	17.3	5.4	0.4	169	2.8		
			Hellyer	16	13	7	0.4	160	2.3	Dacite, autoclastic breccia, resedimented volcanic breccia, and sandstone	
			Que River	3.1	13.5	7.5	0.6	200	3.4		
			Mt Lyell	312			1		0.3	Pyritic sericite–chlorite–silica altered schists derived from felsic–andesitic volcanic precursors	<500 m
Skellefte District, Sweden	1894–1891	Back-arc intra-continental submarine rift	Boliden	8	0.9	0.3	1.4	48.3	15.1	Proximal rhyolitic dome and cryptodome–hyaloclastite–tuff cone deposits	Shallow
Bryah Basin, Capricorn Orogen, Western Australia	~2030	Submarine mafic rift	De Grussa	6.6			5.0		1.7	Carbonaceous mudstones, and basalt sills with peperitic basalt-sediment breccias	Deep
Eastern Goldfields Superterrane, Yilgarn Craton, Western Australia	2690	Bimodal submarine rift	Teutonic Bore	1.7	10.7		3.5	140		Pillow basalt and andesite lava/sills, peperites, carbonaceous mudstone, and turbiditic sandstones	Deep
			Bentley	3	9.8	0.7	3.1	115			
			Jaguar	1.6	11.3	0.6	2.0	139	0.7		

(continued)

Table 18.5 (continued)

Location	Age mineralisation (Ma)	Volcano-tectonic setting	Deposit	Tonnage (Mt)	Zn (%)	Pb (%)	Cu (%)	Ag (g/t)	Au (g/t)	Immediate host rocks	Water depth
Southern Youanmi Terrane, Yilgarn Craton, Western Australia	2746	Bimodal submarine rift	Hollandaire	2.8	1.6			5	0.4	Turbidites and mudstones, coherent rhyodacite sills	Deep
Golden Grove Camp, northern Youanmi Terrane, Yilgarn Craton, Western Australia	2950	Bimodal submarine rift	Gossan Hill	15.9	1.5	0.2	2.6	21	0.6	Felsic pumice-bearing turbidites, resedimented hyaloclastites	Deep
			Scuddles	10.5	11.7	0.8	1.2	89	1.1		
Abitibi Greenstone Belt, Superior Craton, Quebec	2702–2698	Bimodal submarine caldera (Noranda Cauldron Camp)	Norbec	4.5	4.8		2.8	44.3	0.9	Rhyolitic flows, domes, and cryptodomes intercalated with pillowed andesites and basalts	Deep
			Amulet	4.5	4.8		2.8	44.3	0.9		
			Millenbach	3.6	4.3		3.6	56.2	1.0		
			Ansil	1.6	0.9		7.2	26.5	1.6		
			Quémont	14	1.8		1.3	19.5	5.4		
		Bimodal submarine rift (±caldera)	Horne	55			2.2	13	6.1	Rhyolitic flows/hyaloclastites, volcaniclastic mass flows, and argillite	Deep
			La Ronde Penna	60	2.17		0.33	45.0	4.31	Dacite–rhyodacite lavas/breccias, mafic sills, and dykes	
Abitibi Greenstone Belt, Superior Craton, Ontario	2717–2710	Ultra-bimodal submarine rift (with komatiites)	Kidd Creek	150	6.4	0.2	2.9	92	0.05	Volcaniclastic mass-flow units derived footwall rhyolite dome, cryptodome complex	Deep
Sturgeon Lake Greenstone Belt, Superior Craton, Ontario	2735	Bimodal submarine caldera	Mattabi	12.5	8.3	0.85	0.7	88		Bedded mass flow volcaniclastics including pumiceous units, mafic surtseyan tuff cone deposits	<500 m

This summary is based on various published sources not all of which are included in the text references

1993; Moss and Scott 2001; Lipton et al. 2018). At the Solwara 1 prospect (Suzette), the main mound of massive sulphide forms a feature roughly 100 m in diameter ×25 m high; however, this is part of a wider area of mineralisation, possibly formed by coalescing feeder zones, extending some 400 m × 250 m (Lipton et al. 2018).

Ancient examples include the giant Archean Windy Craggy deposit in Canada (Peter and Scott 1999), the much smaller Cambrian Hellyer deposit in Tasmania, Australia (Large et al. 2001; Gemmell and Fulton 2001), and individual sulphide mounds in the Miocene Hokuroko Basin, Japan (Ishihara 1974). Mounds formed in relatively shallow, oxygenated seas are prone to oxidation, dissolution effects, and gravitational collapse as they weaken. Mounds formed in deeper anoxic environments are more likely to survive (Tornos et al. 2015).

Sulphide Deposits Formed in Brine Pools

These are hosted within a dense saline bottom layer ponded on the seafloor in a restricted basin. Dense highly saline hydrothermal fluids venting in such a setting pond at the seafloor and can carry large amounts of metal oxides, which commonly precipitate at the interface between the saline bottom layer and the overlying seawater mass. The present-day Atlantis II Deep in the Red Sea is the largest example of this type of deposit, which is rich in zinc and forms a stratiform deposit (Guney et al. 1998; Tornos et al. 2015). Although some massive sulphide deposits in the Iberian Pyrite Belt (Spain-Portugal) may have formed in ocean-bottom brine pool settings, proving such settings for ancient deposits is difficult (Tornos et al. 2015).

Regionally Anoxic Settings

These are difficult to identify in the modern oceans because the Earth's oceans are not experiencing anoxia at present. VMS deposits resulting from regional oceanic anoxia are thought to be largely restricted to the geological past when global oceanic anoxia events occurred (e.g. Kerr 2005). During these events numerous VMS deposits formed, including during the Archean before 2.4 Ga, in the

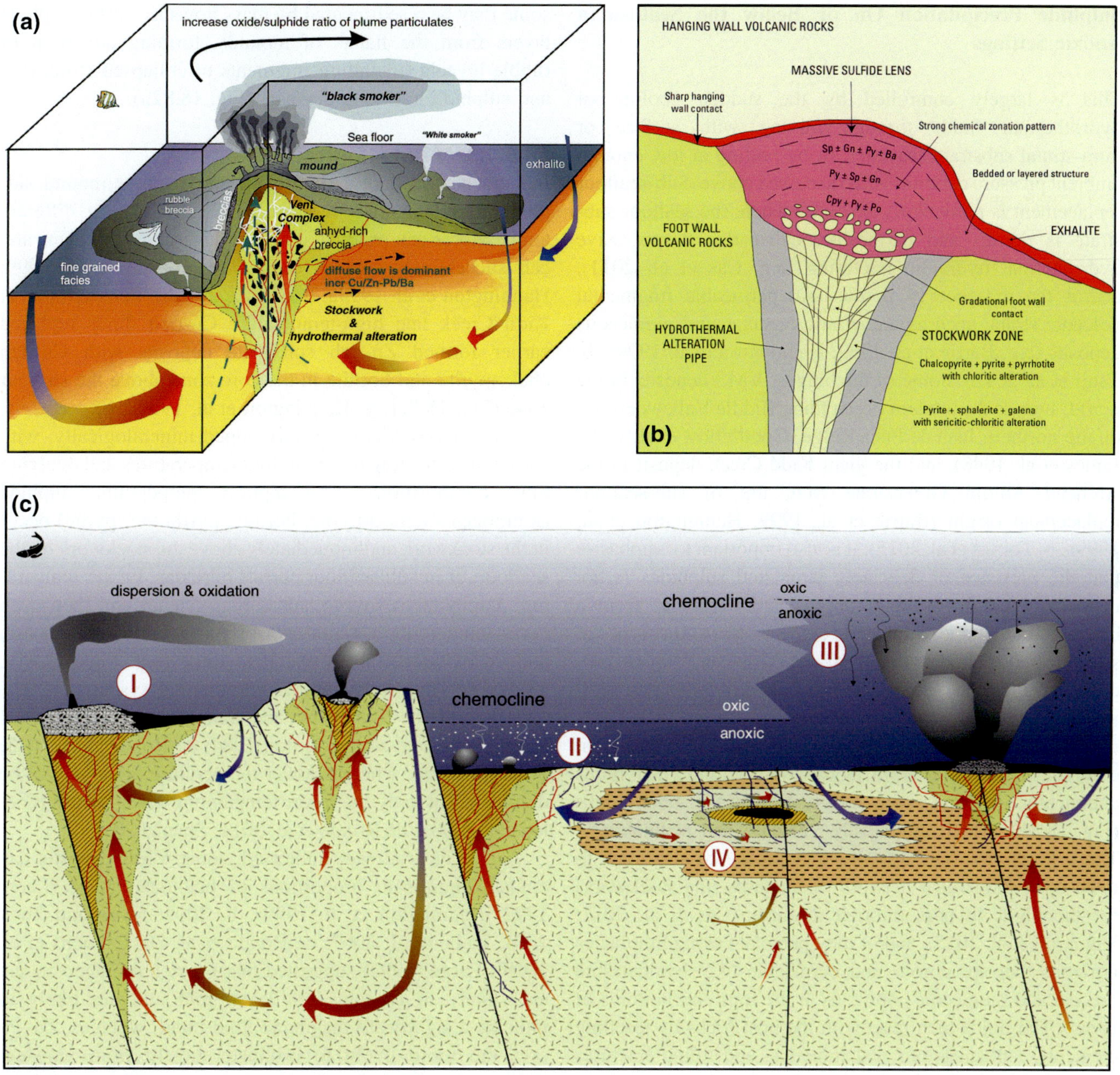

Fig. 18.81 VMS deposit geometry and context in different submarine settings. **a** Schematic depiction of a seafloor mound, showing the subsurface hydrothermal fluid conduit through the stockwork and alteration zone, high-temperature black smoker chimneys building the crest of the mound upwards, peripheral lower-temperature white smoker hydrothermal vents and chimneys on the apron of the mound, and dispersed fine-grained exhalite suspension deposits on the surrounding seafloor (From Tornos et al. 2015). **b** Schematic cross-section through classical seafloor VMS sulphide mound showing internal mineralogical architecture, the relationship between the mound and the subsurface stockwork, and the alteration mineral associations in the host rock (From Koski 2012, After Lydon 1984). **c** Mineralisation styles and chemical environment in the oceans: I. mound style in oxic oceanic settings and water depths; II. brine pools (e.g. Red Sea), which are stratiform deposits in restricted basins with a highly saline seawater bottom layer; III. mounds and stratiform deposits in regional anoxic submarine basin settings; and IV. sub-seafloor replacement deposits (From Tornos et al. 2015)

Paleoproterozoic from 1.9 to 1.8 Ga, during the late Cambrian to Devonian, from the Devonian to the end of the Carboniferous, the late Triassic, and through the Jurassic and Cretaceous (Tornos et al. 2015). Examples include the Ordovician Brunswick No. 12 deposit, in eastern Canada (Lentz and Goodfellow 1996; Goodfellow et al. 2003; Table 18.5). The association with organic-rich shales also introduces an important new dynamic: the presence and possible role of hydrocarbons in the mineralising process (Simoneit and Lonsdale 1982; Sect. 18.9.4.3).

Sulphide Precipitation On or Below the Seafloor in Anoxic Settings

This is largely controlled by the sudden cooling of hydrothermal fluids through mixing with surface or intra-stratal substrate seawater, or by boiling at low enough ambient pressure conditions, as discussed above. Sub-seafloor replacement is particularly favoured where the seafloor substrate is porous and permeable, either due to pervasive hydrothermal hydraulic fracturing (e.g. Cas et al. 2011), and/or the presence of porous and permeable fragmental volcanic strata, such as autoclastic breccias and pumiceous deposits (e.g. Barrie et al. 1999; Hannington et al. 1999a, b; Tsuji et al. 2012). Some of the biggest VMS deposits in the world, such as the modern ocean floor Middle Valley deposit, on the northern Juan de Fuca Ridge (Goodfellow et al. 1993; James et al. 1998), and the giant Kidd Creek deposit in the Archean Abitibi Greenstone Belt, are of sub-seafloor replacement origin (Barrie et al. 1999; Hannington et al. 1999a, b; Tornos et al. 2015). It is also important to emphasise that the presence of delicately laminated sulphides is not necessarily confirmation of an exhalative (fallout from a hydrothermal plume) origin, and evidence from other types of shale-hosted orebodies demonstrates these can form by processes of subsurface replacement of already laminated sediments (e.g. Broadbent et al. 1998).

All VMS deposits will have a sub-seafloor replacement component, resulting from aggressive hydrothermal fluids chemically reacting with original seafloor host rocks (Fig. 18.81), which are mostly volcanic and sedimentary. These fluids first hydraulically fracture the country rock to open up a subvertical network of fractures (the stockwork) for the fluids to pass through upwards (Cas et al. 2011; Vignaroli et al. 2015), and/or they percolate along lithofacies horizons that are porous and permeable (e.g. hyaloclastite breccia and pumiceous, horizons at the tops and bottoms of lavas). The fluids then react with the host rocks by leaching elements, replacing ions and minerals, and filling fractures and pore spaces. Both volcanic and sedimentary host rocks can be replaced. The replacement can be partial, so that original host rock textures are still visible, or at the extreme, the replacement is pervasive, with all original textures being overprinted (Cas et al. 2011; Vignaroli et al. 2015). Mineralisation in the stockwork can range from vein-like, called stringer mineralisation (Fig. 18.82a, b), to patchy where the replacement has been irregular (Fig. 18.82c), to massive where the replacement has been more or less pervasive (Fig. 18.8d–f). In seafloor mounds and stratiform sheet-like deposits, the mineralisation is mostly massive, although some may be resedimented by mass flows as clastic sulphide layers from the flanks of mounds, forming an apron of rubble breccia (including fragments of collapsed chimneys) and sulphidic turbidite deposits (Fig. 18.82h).

18.8.5.3 Mineralogical Variation

In both sub-seafloor replacement and seafloor mound sulphide deposits, different generations of fluids at different temperatures can produce different mineral assemblages and can successively replace each other (Barrie et al. 1999; Hannington et al. 1999a, b; Shanks 2012b). For example, at Kidd Creek late high-temperature Cu-rich fluids replaced earlier formed Zn-rich sphalerite deposits with Cu-rich chalcopyrite and bornite in the core zone above the stringer zone (Fig. 18.82f, g; Hannington et al. 1999a, b).

VMS deposits are typically zoned mineralogically, with each zone reflecting different fluid temperatures and evolving fluid compositions. The highest temperature sulphide assemblages (chalcopyrite, bornite, pyrrhotite, pyrite) occur in the stockwork and immediately above the stockwork on the seafloor, or in sub-seafloor porous horizons. Lower temperature assemblages (pyrite, sphalerite, galena) and then even lower temperature assemblages of pyrite, barite, manganese, and Fe oxide occur peripheral to these (e.g. Figs. 18.81b, 18.82a–f). High-temperature hydrothermal systems (>280 °C) will likely be copper-rich and contain all of these mineral assemblages and zones (Lydon 1984; de Ronde et al. 2011; Shanks 2012a), whereas lower temperature systems will be dominated by only the lower temperature assemblages (Galley et al. 2007; de Ronde et al. 2011; Shanks 2012a).

As we have already indicated, some VMS deposits can contain significant proportions of precious metals (Large 1992; Hannington et al. 1999a, b, 2005; Moss and Scott 2001, Dubé et al. 2007; de Ronde et al. 2011; Fuchs et al. 2019). Large (1992) suggested that Cu-rich ore deposits are also likely to contain significant gold and are likely to indicate a significant magmatic fluid input, whereas, Zn–Pb ores are more likely to originate mostly from seawater leaching of metals from substrate sources. There will be a spectrum of deposit types reflecting varying ratios of magmatic and substrate-leaching metal input and hydrothermal fluid temperatures. Gold-rich VMS deposits tend to be high sulphidation deposits, they seem to form in shallower water, and may be transitional to epithermal mineralisation (Petersen et al. 2002; Müller et al. 2003; Dubé et al. 2007; Mercier-Langevin et al. 2011, 2015; Hollis et al. 2017a). Host volcanic rocks are more generally felsic-intermediate, and as a result, gold-rich systems should be more common in arc volcanoes (Hannington et al. 2005).

Fig. 18.82 Sulphide mineralisation styles. **a** Vein-like stringer pyrite and minor chalcopyrite mineralisation cross-cutting silicified submarine rhyolite, Archean Hollandaire deposit, Cue, Yilgarn Craton, Western Australia. **b** Chalcopyrite and pyrite stringer veins filling hydraulic fractures, Archean Kidd Creek mine VMS deposit, Abitibi Greenstone Belt, Ontario, Canada. **c** Pyrite and minor chalcopyrite mineralisation incompletely replacing black mudstone, Archean Teutonic Bore VMS deposit, YilgarnCraton, Western Australia. **d** Massive sphalerite and chalcopyrite ore resulting from the replacement of black mudstone, with bedding fabric still visible. Archean Brookland Lens, Bentley VMS deposit, Yilgarn Craton, Western Australia. **e** Pyrite and chalcopyrite mineralisation almost completely replacing basalt hyaloclastite peperite breccia, with pseudo-morphed breccia texture still visible. Unreplaced hyaloclastite breccia with black mudstone matrix is visible in the right drill core, with patches of mudstone matrix also visible in the left core. Archean Teutonic Bore VMS deposit. **f** Massive, banded chalcopyrite and sphalerite ore, Archean Kidd Creek VMS deposit and mine, Abitibi Greenstone Belt, Canada. **g** Bornite ore, Archean Kidd Creek mine VMS deposit. **h** Graded bedded sulphide turbidite beds from the Miocene Furutobe mine VMS deposit, Kuroko Belt, Japan; note the flame structure at the base of the second turbidite (Photo by W Sacco; *Source* Frontispiece, Ohmoto and Skinner 1983). All drill core is ~5 cm wide

Fig. 18.82 (continued)

18.8.5.4 Post-depositional Deformation and Metamorphic Effects

Many ancient examples of VMS occur in basins that have suffered major inversion, especially those that have been accreted along continental margins (e.g. Paleozoic Lachlan Fold Belt in Eastern Australia, northern Cordilleran margin of Canada, and Paleoproterozoic Svecokarelian margin, Sweden), or within Archean greenstone belts. In these cases, the extensional faults/fractures controlling magmatism and mineralisation are very likely to have been reactivated. It is therefore not surprising, especially considering the pervasive hydrothermal alteration and presence of ductile sulphides, that the geometry of many orebodies is both complex and structurally controlled. This includes the development of mylonitic fabrics and discordant contacts in ore zones, the remobilisation of sulphide ore, and the formation of alteration "*pseudo-fiamme*" in both hyaloclastite and coherent lava/intrusive rocks (Chap. 13). Many so-called exhalites in ancient settings (e.g. silica-, sericite-, chlorite-, hematite-rich rocks) are better attributed to hydrothermal alteration and deformation.

As early, basin-forming, extensional faults will be reactivated as compressional structures, knowledge of such a basin's original architecture and deformational history are going to be critical in exploration (Wright 1994). For example, in the Vargfors Basin, central Skellefte mining district (Sweden) crustal shortening at ~1.87 Ga caused the inversion of normal and transfer faults and the formation of upright, fault-bound synclines and anticlines. Detailed studies show VMS deposits were transposed along the faults (Bauer et al. 2011, 2014). Analysing the shape and orientation of VMS ore bodies shows how their deformation pattern mimics those of the hosting structures and results from the overprinting deformation. Furthermore, regional structural transitions are imitated in the deformation patterns of the ore bodies. A comparison of the size of VMS deposits with their location shows that the smallest deposits are not related to known high-strain zones, while the largest deposits are associated with regional-scale high-strain zones. The comparison of distribution and size with the pattern of high-strain zones provides an important tool for regional-scale mineral exploration in the Skellefte district, whereas the analysis of ore body shape and orientation can aid near-mine exploration activities.

More generally, in highly inverted basins, ore lenses are now likely to be structurally controlled and located in structural traps in compressional fault zones, such as (1) dilational jogs in shear zones, (2) lateral ramps and tear faults (along reactivated transfer faults), (3) extensional "flats" in thrusts, and (4) extensional parts of folds and domal structures. In highly deformed settings, the ore lens geometry, therefore, may bear little relationship to original stratigraphy, with lenses recumbently overturned, boudinaged, and/or structurally repeated. This also means no autochthonous sequences may actually exist in the mine area, which may open the possibility for other exploration targets.

In regionally metamorphosed volcanic terrains, VMS ores along with their host successions can be variably metamorphosed from lower greenschist through upper greenschist, to amphibolite to granulite (± contact metamorphic) facies. With the increasing metamorphic grade, as well as the accompanying deformation, this overprint can make the identification of both primary volcanic lithologies and the effects of hydrothermal alteration even more ambiguous (Chap. 13). The use of hydrothermal alteration haloes and geochemical analysis as an exploration vector can be difficult in many ancient terrains.

18.8.6 Longevity of Seafloor Hydrothermal Systems and Time Scales for Formation of VMS Deposits

Summarising studies of the TAG (Mid-Atlantic Ridge), Middle Valley (northern Juan de Fuca Ridge), and Atlantis II Deep (Red Sea) VMS massive sulphide deposits,

Hannington et al. (2005) proposed that massive sulphide deposits ($\geq$2–100 Mt in size) at modern spreading ridges are formed on time scales of 10^3 to 10^5 yr. However, hydrothermal activity may be intermittent, leading to even lower actual time periods of hydrothermal activity, with sulphide accumulation rates of 500–1,000 t yr^{-1}. Individual black smoker chimneys are thought to be active for <100 yr (Hannington et al. 2005; de Ronde et al. 2011). These calculations are constrained by known ages of the young seafloor on which the deposits have accumulated.

Based on numerical modelling studies, Schardt and Large (2009) proposed that the time required for the formation of an "average-sized" 1.2 Mt Zn and Cu VMS deposit was ~5,000 yr and for "giant" deposits >5,000 yr, which are both instantaneous geologically. However, these average estimates are also affected by fluid discharge rate and metal concentrations in the venting fluids. Super/giant deposits either need a much longer time to form (>35,000 yr) or at least 100 ppm of metal in solution. Results indicate that super/giant deposits may only form in certain geotectonic environments where the longevity and preservation potential of the hydrothermal system are high.

De Ronde et al. (2011) have proposed that the character of VMS forming hydrothermal systems changes as they mature or age. This was based on studies of an older hydrothermal system located along caldera wall ring faults, compared with younger hydrothermal fields associated with the summits of two young cones within the submarine caldera of the wholly submarine Brothers volcano in the Kermadec intra-oceanic arc, north of New Zealand. Geochemical and morphological studies suggest that the longer-lived caldera wall hydrothermal field, which is producing metallic massive sulphide deposits from high-temperature black smokers, represents a hydrothermal system that originates from magmatic fluids sourced from an older magma body (sill?) at >2.5 km beneath the volcano summit (Fig. 18.83). These magmatic fluids mix with seawater during slow ascent along hydraulic fractures and faults in the older part of the edifice and leach metals from the volcanic substrate (Fig. 18.83). In contrast, the lower temperature hydrothermal fields at the top of the young cones are dominated by fluids that are more directly derived from a younger magma body, then rise quickly through young magmatic conduits, and mix with seawater in the permeable cones. These have produced mostly barite and anhydrite from diffuse zones of discharge at the cone summits. In other words, it can take time for submarine hydrothermal fields to evolve geochemically, thermally, and physically before they can produce economic metallic VMS deposits.

18.8.7 Some Volcanological Considerations of the Nature of VMS Host Rock Successions

Cas (1992) pointed out that at the water depths at which VMS deposits are likely to form, explosive volcanism is constrained by the hydrostatic pressures that are important for VMS formation. This suggests that most VMS host rock successions should contain limited pyroclastic (pumiceous) deposits. However, for various reasons, many VMS host rock successions *do contain* voluminous deposits of pumice. Cas (1992), Cas and Giordano (2014), and Cas and Simmons (2018) summarised the possible volcanic eruption styles and volcanic deposit types that could form in the submarine water depths at which VMS deposits form (generally >1,000 m). They indicate that at about 1,000 m water depth, there could be a significant transition from explosive volcanism in shallow water depths to progressively non-explosive volcanism in deeper water depths (also see Chap. 3). Where VMS systems form in shallower waters, the volcanic system could experience some explosive eruptions. As water depth increases, explosive options become less common, and effusive eruptions become predominant (Cas and Simmons 2018; Chap. 3).

Nonetheless, there are other factors that are important in understanding the preserved host rock successions and in interpreting eruption and depositional processes in deep water. First, pumice is not just formed through explosive eruptions. It can also form through slow, steady vesiculation of erupting seafloor lavas, which if subject to quench fragmentation can produce large volumes of in situ pumice breccias (Chaps. 4, 6). Secondly, such autoclastic pumice, as well as pumice that forms explosively in shallow water or even in subaerial environments can fallout into deep water (Chap. 9) and can be resedimented by mass-flow processes into deep water (Chaps. 11, 12). Thirdly, where VMS deposits occur in submarine calderas, and pumice is a significant component of the host volcanic succession and was erupted during the caldera-forming event, the pumice deposits may have been erupted from vent(s) that were located at considerably shallower water depths prior to caldera collapse (Chaps. 11, 16). In contrast, the VMS deposits could have formed after caldera collapse, from post-caldera collapse hydrothermal systems, and at water depths much greater than the eruption water depths of the host pumiceous deposits, which have subsided during caldera collapse.

Pumice deposits in deep water host rock successions therefore do not necessarily indicate that the host volcanic system was behaving explosively at the time of VMS formation.

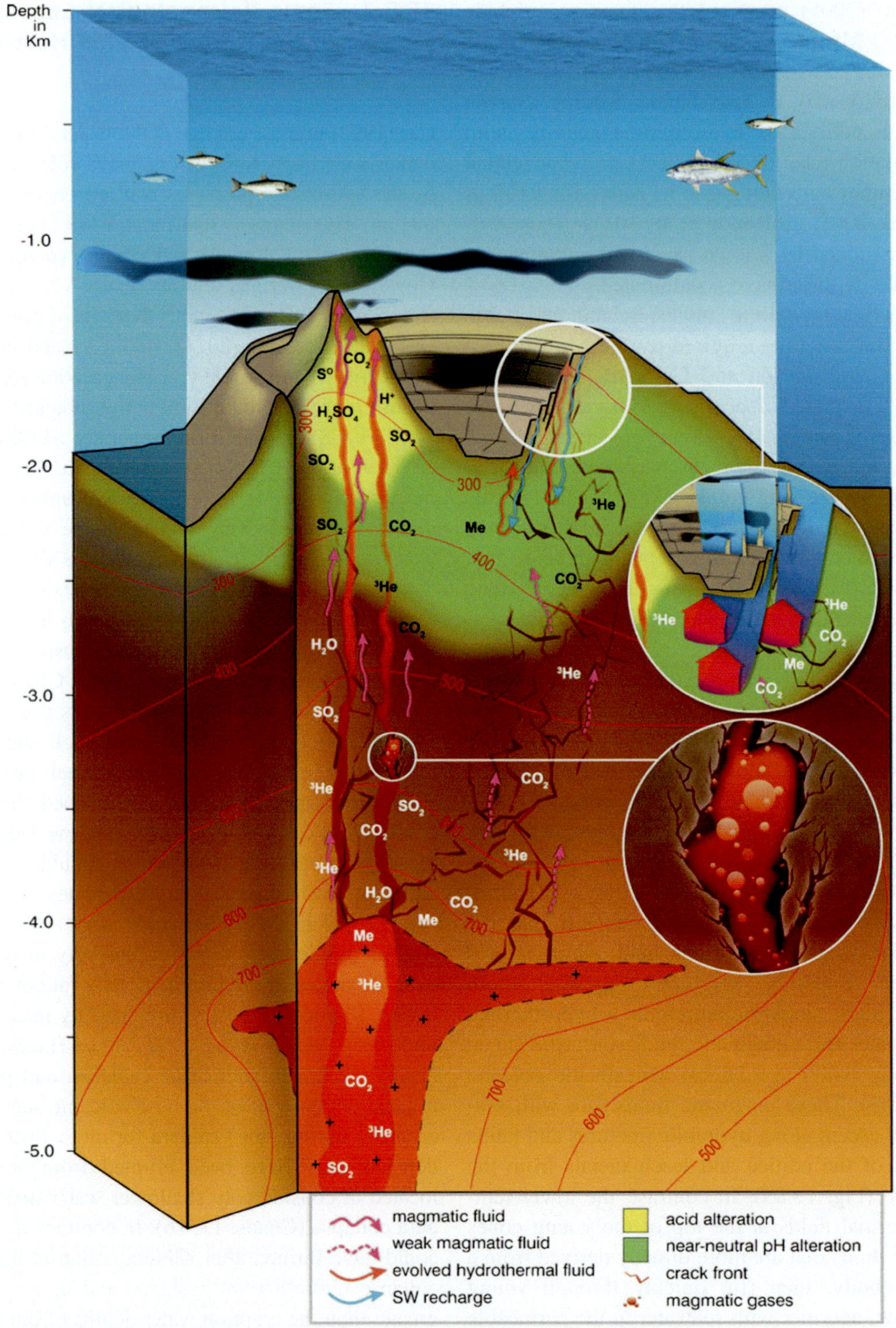

Fig. 18.83 Schematic representation of the context of two submarine hydrothermal fields at the summit of the submarine Brothers volcano, Kermadec arc, north of New Zealand. The older, longer-lived, and high-temperature hydrothermal field has formed above caldera margin faults (right) and has produced massive sulphide mounds from black smokers. The younger hydrothermal field at the top of two young cones (left) is of lower temperature and is represented by diffuse discharge points over the cone summit area. (From de Ronde et al. 2011)

18.8.7.1 The Significance of Pumice: A Case Study of Misinterpretation

An interesting case study, which serves to highlight many of these issues and the consequences of misinterpreting the volcanological context of VMS deposits in general, is the archetypal Kuroko-type VMS deposits found in the Miocene Green Tuff Belt, Japan.

Submarine volcanic rocks within the main Hokuroko mining district, northeastern Honshu (see Fig. 18.102), are dominated by a bimodal association of lavas, hyaloclastite, and autobreccias which form the footwall sequence (21–14 Ma), whereas the younger succession, post-dating the M2 mudstone and ore horizon, contains voluminous pumice deposits (Cas 1992; Yamada and Yoshida 2011; Yamada et al. 2012; see Fig. 18.88). In the benchmark monograph describing many aspects of the geology of the Kuroko deposits Ohmoto and Skinner (1983), and in particular Ohmoto and Takahashi (1983), proposed that the host volcanic centres were highly explosive, deep-water caldera volcanoes. This was based on (1) the presence of "tuff-breccias" in the host volcanic successions and abundant pumice deposits, especially in the hanging-wall successions to some deposits, which were interpreted as in situ explosive deposits; (2) the assessment by some American micro-paleontologists that the ecological affinities of foraminifera in mudstones in the host rock successions indicated water depths of 3.5–4.1 km; (3) that clusters of dacite bodies interpreted as lava domes, defined circular structures that must define the margins of submarine calderas; and (4) that the existing model for explosive subaerial calderas was also valid at depths of 4 km water depth.

There are several problems with these assumptions. Firstly, as already discussed, the presence of pumice in submarine settings does not necessarily indicate an in situ explosive origin at great water depths (Chap. 3; Cas 1992; Cas and Giordano 2014; Cas and Simmons 2018). Secondly, the origin of the "tuff-breccias" was not interrogated in the original research and, as Cas (1992) demonstrated, many are of quench fragmentation origin, i.e. they are non-explosive hyaloclastite breccias, and *not* explosive pyroclastic deposits. Thirdly, although Guber and Merrill (1983) considered the paleoecological affinities of Miocene foraminifera to indicate deeper paleo-water depths, Japanese micro-paleontologists Kitazato (1979) and Hasegawa et al. (1989) interpreted the water depths to be 0.5–2.5 km and 0.7–2.8 km, respectively. The great water depths (3.5–4.1 km) proposed by Guber and Merrill (1983) make it theoretically impossible for the explosive activity to occur at those water depths (Cas and Simmons 2018), questioning the origin of the pumice deposits; whereas, the shallower ends of the water depth spectrum proposed by the Japanese workers would allow some explosive activity. Most of the pumice deposits occur in graded beds that range from tens of cm to tens of m thick, indicating a mass-flow resedimentation origin from shallower settings (Cas 1992). Estimates of the water depth at the time of formation of the Kuroko deposits based on sulphide fluid inclusion paleo-barometry data suggest a water depth of ~1,750 m (Pisutha-Arnond and Ohmoto 1983). Fourthly, there is no data to support the presence of ring fracture faults and extremely thick intra-caldera pyroclastic deposits (cf. subaerial calderas, Chap. 16) in the Kuroko region. The definition of circular features (supposed calderas) defined by lava domes (Ohmoto and Takahashi 1983) is arbitrary and unconvincing. Ours and more recent studies have indicated that volcanism more generally took place in deep-water extensional rift basins, presumably from multivent sources (Cas 1992; Allen et al. 2002; Chap. 16).

Yamada and Yoshida (2011) have also now provided a plausible explanation for why the footwall volcanic sequences are dominated by coherent and autoclastic volcanic rocks and are different from the hanging-wall rock sequences comprising significant pumice deposits. The pre-ore felsic volcanic sequence is a high aluminium, low calcium, and high niobium volcanic succession in a bimodal basalt-dacite association that was erupted in a back-arc rift deep-water setting. The felsic rocks are aphyric to weakly plagioclase-phyric, highly differentiated, and derived from sources at crustal depths of 10–15 km. In contrast, post-ore pumiceous volcanics (<12 Ma) are lower aluminium, high calcium, and low niobium, quartz and plagioclase-phyric volcanic rocks, which become progressively clinopyroxene, hornblende, and biotite phyric with decreasing age. These were derived from shallower magma source regions (~5 km) and coincided with a change in the regional tectonic context of Japan to a subduction-generated volcanic arc, which was accompanied by regional uplift (Yamada and Yoshida 2011). There was also a decline in erupted volumes of volcanic deposits from pre-ore to post-ore times.

Furthermore, Yamada and Yoshida (2011) have correlated mineralisation with the eruption of the most differentiated "white rhyolites" in the deep-water, bimodal back-arc rift setting. The VMS mineralisation thus formed in deep water before the formation of the hanging-wall pumice deposits. These younger pumice deposits are related to much shallower water to emerging or uplifting, subduction-related settings that would have facilitated explosive volcanism, as well as being sourced from perhaps more volatile-rich arc magmas. Such fundamental geotectonic changes offer an intriguing explanation of the apparent contradiction between the deep-water VMS deposit setting and the very pumice-rich hanging-wall succession, which formed under different paleo-environmental, volcanological, and tectonic settings. It is salient to note that these changes occurred within a 2 Ma window of time, which in the even older Paleozoic and Archean geological record is unlikely to be decipherable because of the lower resolution of age dating methods.

18.8.7.2 Role of Submarine Calderas

Whereas the role of submarine calderas in the formation of the classic Kuroko-type deposits in the Hokuroko Basin may have been diminished, they cannot be excluded as potential ore-forming centres (Chap. 16). As described above, many summit calderas of submarine volcanoes of the modern Kermadec arc north of New Zealand have active seafloor hydrothermal vents and are known to host massive sulphides (Hannington et al. 2011; Leybourne et al. 2012a, b). Several ancient VMS districts and deposits have also been attributed to submarine caldera environments, in particular, those found within Archean greenstone belts of the Superior Province, Canada (Table 18.5), although the link between caldera and origins of the ore deposits often remains poorly understood.

The Sturgeon Lake district contains several, exploited Zn–Cu deposits within the Sturgeon Lake greenstone belt, northwestern Ontario, Canada. Mineralisation is hosted within a 9 km thick, dominantly bimodal package of basalt–rhyolite volcanic rocks. The succession was divided into three depositional cycles by Morton et al. (1991) and related to the formation of the 30 km diameter, 2,735 Ma submarine Sturgeon Lake cauldron. Below the largest VMS deposit at Mattabi, subaerial and shallow subaqueous mafic hyalotuffs, lava flows, flow breccias, felsic lava flows, and pyroclastic flows and falls form a 2 km thick succession (Groves et al. 1988). These pyroclastic deposits formed thin- to thick-bedded hyalotuffs that contain highly vesicular and quenched juvenile fragments. Intermediate to felsic pyroclastic flow deposits form the immediate mine footwall strata and include massive basal beds and overlying bedded ash tuffs (? turbidites), and massive pumiceous pyroclastic (? hyaloclastite) units. Massive sulphide mineralisation is interpreted to have formed in a shallow subaqueous environment in <500 m of water.

Mueller and Mortensen (2002) proposed a lava-dominated submarine caldera setting for the Archean Hunter Mine Group in the northern Abitibi greenstone belt, Ontario, with caldera subsidence being attributed to Hawaiian shield volcano-like summit caldera subsidence driven by lateral flank rift eruptions of lavas (Chap. 16). A 5–7 km-wide, N-trending felsic dyke swarm, coupled with abundant syn-volcanic and syn-sedimentary faults, chaotic breccias, and mineralisation suggest a central volcanic depression, which is consistent with a caldera structure. However, it is not clear from the description of Mueller and Mortensen (2002) that caldera ring faults have been defined. The succession is also certainly consistent with an extensional, volcano-tectonic graben-like volcanic centre, in which lava-dominated edifices have grown.

Perhaps the best-documented example is the Noranda cauldron or caldera in the southern Archean Abitibi greenstone belt in Quebec (Fig. 18.84; Gibson and Kerr 1993; Gibson and Galley 2007; Gibson et al. 2007). The caldera contains a 7-to-9-km-thick succession of bimodal mafic–felsic, tholeiitic to calc-alkaline volcanic rocks called the Noranda Volcanic Complex (NVC), which forms part of the more extensive Blake River Group (2704–2695 Ma), believed to have formed in a primitive rifted arc setting in oceanic lithosphere.

The NVC were erupted during five major cycles of volcanic activity. The NVC hosts 30 significant VMS metal occurrences (Noranda camp) including at least 20 deposits that have been mined in the past (e.g. Norbec, Amulet, Millenbach, Ansil, Quémont; Fig. 18.84; Table 18.5). Most deposits are located within the centrally placed Noranda cauldron, which is a 15×20 km area of syn-volcanic subsidence and, with the exception of the gold-rich Quémont deposit, are individually less than 5 Mt in size. These are hosted by mafic or felsic lava flow deposits. They typically consist of concordant lenses of massive sulphides that are cone or mound-shaped and underlain by vertically extensive, discordant stockworks of stringer mineralisation and vertically extensive well-defined alteration pipes. More than 90% of the past producing VMS deposits occur within the subsidence structure, which may indicate it was an important factor in contributing to the high concentration of deposits and hydrothermal activity that occurred coevally with subsidence (Gibson and Galley 2007; Gibson et al. 2007). However, the largest, and most gold-rich deposit at Horne is located outside, although, close to the southern margin of the suspected caldera (Fig. 18.84). This is hosted in a succession of predominantly felsic volcaniclastics and was largely emplaced by sub-seafloor replacement processes (Gibson and Galley 2007; Monecke et al. 2008). The locations of most VMS deposits in the NVC are controlled by syn-volcanic fault subsidence, some of which may have accommodated the subsidence. These faults were later replaced or intruded by swarms of dikes and sills of varying composition.

The Noranda caldera has been compared in size and tectonic origin to the modern Monowai caldera located at the mid-point along the Tonga–Kermadec arc (Gibson and Galley 2007; Leybourne et al. 2012b). The Monowai volcanic centre is composed of a large shield-like volcano (≥ 500 km^2) which includes a large conical stratovolcano and a separate elongate caldera 7 km $\times$ 5 km (~ 35 km^2), as well as other cones and parasitic vents. Mafic volcanic rocks dominate; basalts are the most common rock type recovered from the cone, whereas, basaltic andesites are

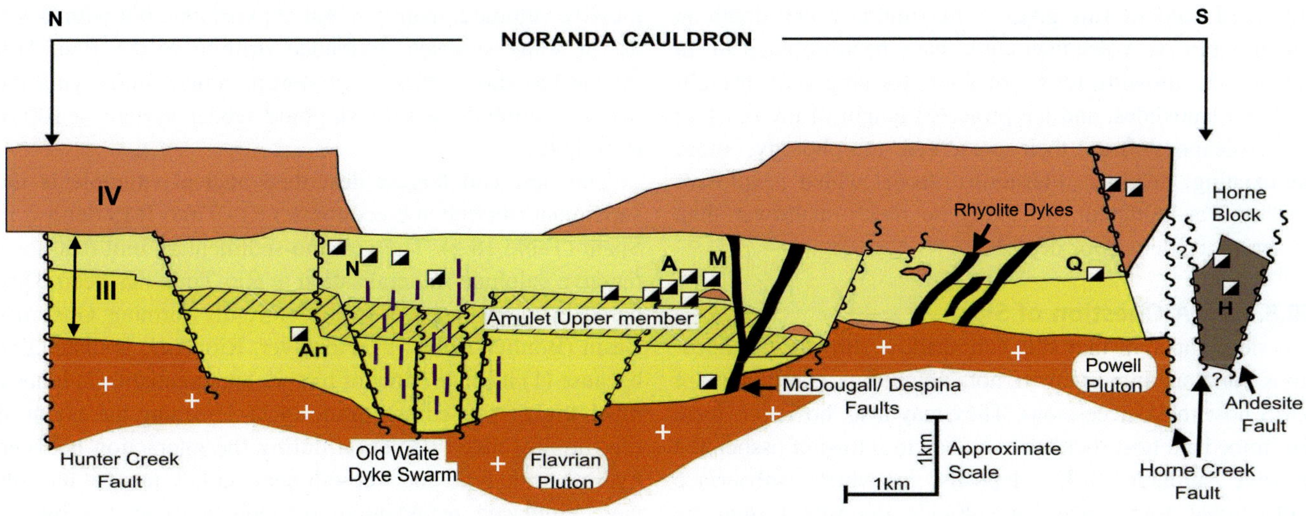

Fig. 18.84 Geological cross-section and location of VMS deposits of the Archean submarine Noranda cauldron, Abitibi greenstone belt, Quebec, Canada. The Noranda Volcanic Complex (NVC) is subdivided into five conformable cycles (cycles I through V) which young to the east; each cycle consists of an andesitic/basaltic lower unit and a bimodal, andesitic/basaltic, and rhyolitic upper unit. The Mine Sequence (or Cycle III) hosts one of the highest concentrations of VIMS deposits in the world, including N = Norbet, A = Amulet, M = Millenbach, An = Ansil, and Q = Quémont (Table 18.5), and is interpreted to have erupted during caldera subsidence ca. 2698 Ma. The superlarge Horne gold-rich VMS deposit (H) is located adjacent to but south of the inferred southern margin of the subsidence structure and is separated from the Mine Sequence by the Horne Creek Fault. The Flavian and Powell plutons that core the NVC are interpreted to be the intrusive equivalents of the magma chamber that fed this volcanic edifice. (Adapted from Gibson et al. 2007)

common within the caldera. The rim of the caldera lies at a depth of about 500–1,000 m, and the caldera floor at about 1,600 m. This appears to have formed within an older caldera structure with a surface area of 84 km^2, which would be the largest, such structure in the Kermadec arc. More recently, it has also been proposed that the Noranda cauldron may be included within two older and larger caldera structures named the Blake River mega-caldera complex (Pearson and Daigneault 2009). The outer caldera structure (called the Misema caldera) is defined by a double ring fault structure 65 km × 45 km across (~3,000 km^2), which would incorporate both the Horne and LaRonde Penna gold-rich VMS deposits. From a paleogeographic perspective, this is 2–3 orders of magnitude larger than most mafic submarine calderas identified on the present-day seafloor.

18.8.7.3 Assessing Paleo-water Depths

Apart from regional scale uplift or subsidence that could raise or lower the seafloor above or below the possible water depth for explosive volcanism, caldera collapse or sector collapse (Chaps. 16, 11) could change the depth at which explosive volcanism occurs, and the depth at which VMS formation on the post-collapse deepened seafloor occurs. Explosive eruption vents that lead to caldera collapse will be located at the existing water depths *before* caldera collapse occurs. If caldera collapse involves hundreds to thousands of m of subsidence, and VMS deposits later form in the floor of the collapsed caldera or contiguous structures from post-collapse hydrothermal systems, there will be a significant difference between in ambient hydrostatic pressure under which explosive volcanism took place and that which hydrothermal venting occurred on the much deeper caldera floor.

During the formation of the Ordovician Bald Mountain caldera, Maine, eastern United States, Kessel and Busby (2003) and Busby et al. (2003) proposed that both explosive, caldera-forming volcanism and VMS mineralisation occurred at water depths of ~1,400 m. This was based on paleo-barometry data of fluid inclusions in the sulphide deposits as a proxy for the hydrostatic pressure and water depth. However, the VMS deposits sit above a pumice-rich "footwall ignimbrite" succession that is up to 1 km thick (Busby et al. 2003). This suggests to us it is conceivable that at least hundreds of m of caldera collapse subsidence may have occurred before the VMS deposits formed, and that the explosive eruption occurred in water depths very much shallower than 1,400 m. Thus, the preserved caldera rims represent the maximum depth at which eruption and caldera collapse started, and allowing for erosion, gravitational scarp collapse, and explosive removal, water depths were probably considerably shallower (Chap. 16).

This problem of assessing water depths of explosive volcanism is just as relevant in modern submarine volcanic settings, where reconstructing the pre-eruption bathymetry and water depths preceding a major pyroclastic sequence is essential to understanding at what water depth the explosive eruption occurred at. The post-eruption bathymetry is irrelevant to this assessment, except that the depth of the preserved caldera rim gives a maximum water depth at which explosive volcanism could have been initiated in. In most cases, allowing for post-caldera-forming wall instabilities and landslides, and the projected height of pre-existing stratovolcanoes from their preserved morphology, those pre-existing seafloor elevations from which explosive activity was initiated will likely be much shallower than preserved caldera floor depths.

18.8.7.4 A Question of Sills?

It is now apparent that sills and shallow subvolcanic intrusions form an important, if not, ubiquitous component of VMS host rock successions. They may also, however, form the immediate host rocks that are key to intrusion-associated massive sulphide (IMS) deposits, in which, subsurface replacement forms massive sulphide deposits hosted by sediments in close proximity to sills and other intrusive bodies. Elsewhere in the stratigraphy, some volcanic rocks may be present, but these are not the immediate or principal hosts.

Type examples of modern IMS deposits are the Guaymas Basin, the Gulf of California, and the Middle Valley on the Juan de Fuca Ridge. As described above, these represent mineralisation associated with sediment-covered spreading ridges and related rifts. In the Guaymas Basin intrusion of sills into wet sediments appears to be the main mechanism of new ocean floor creation (Fig. 18.85). DSDP drilling suggests that the thicknesses of such sills can be greater than 100 m, but more generally smaller thicknesses occur (Gieskes et al. 1982). Two types of hydrothermal activity can be recognised (Fig. 18.85a), either associated with (1) basaltic sill intrusions at relatively shallow depths (<50 m) into highly porous sediments or (2) larger-scale magmatic intrusions at greater depths in the sediments. Linked to the former are contact metamorphism effects (recrystallisation of clays, dissolution of opaline silica, and recrystallisation of carbonates) and chemical changes in interstitial waters, which indicate temperatures of 100–200 °C. In the latter, higher temperature hydrothermal activity leads to substantial thermal (albite–epidote–chlorite–quartz, i.e. greenschist facies) alteration of sediments, which must occur in an open system at temperatures of $\sim$300 °C. Recharge of basin bottom waters and discharge of heated fluids occur through fault zones, with sills acting as permeability "lids" on these hydrothermal systems, thus forcing fluid flow in lateral directions towards faults. Higher temperature, metal-rich fluids emanating from faults and fissures will lead to the build-up of surface mounds and subsurface replacement deposits, while deeper replacement bodies may form below capping sills (Fig. 18.85a). Mineralisation can be extensive (e.g. ODP drill-hole 856H in Middle Valley, at the northern extremity of the Juan de Fuca spreading ridge; James et al. 1998). This intersected some 90 m of sediment-hosted massive sulphide, mainly massive pyrrhotite, but with 2–8% Zn in some samples. Extended drilling at the Bent Hill mound has shown there is an abrupt contact underlying the massive sulphide with a sulphide feeder system $\geq$100 m thick below.

The best and largest described ancient example is the supergiant (though sub-economic) Rio Tinto IMS deposit in Spain (Table 18.5). The volcano-sedimentary unit that hosts massive sulphide mineralisation at Rio Tinto has been compared in many respects to the currently forming Guaymas Basin (Boulter 1993a, b). However, Rio Tinto is distinctive because (1) it is a product of bimodal magmatism, (2) it has a much higher sill/sediment ratio (4:1), lessening the extent of ore-forming metal depletion during the interaction between hydrothermal fluid and wet sediment, and (3) most of the sills were emplaced penecontemporaneously as shown by the ubiquitous development of peperitic (Chap. 4) sill margins. These data together with evidence for local extrusive behaviour imply the rapid formation of the complex at exceptionally high levels of emplacement (Boulter 1993a, b). Crucial to the mineralisation process was the sheet-like form of the sill complex, which, by acting as a cap, focused considerable discharge at zones of structural weakness and faulting (as illustrated in Fig. 18.85a). The widespread occurrence of peperitic facies in the Iberian Pyrite Belts suggests that most silicic and mafic magmas were emplaced as shallow intrusions into wet, unconsolidated sediments at depths of $\leq$500 m, although some minor extrusions also likely occurred (Soriano and Martí 1999). This suggests other massive sulphide deposits in the Iberian Pyrite Belt likely have intrusion-related, hydrothermal origins.

Another supergiant IMS deposit is the cobalt-rich late Triassic Windy Craggy deposit, British Columbia, which is located within a sequence of intercalated basalt flows and sills and turbiditic argillite (Peter and Scott 1999; Table 18.5). Gabbroic sills predominate the lower part of the host rock succession which grades stratigraphically upwards into pillowed to massive tholeiitic basalt lava flows. Mineralisation occurs at the transition from graphitic to calcareous siltstone and argillite to the overlying volcanic rocks and extends along strike for 1.6 km, at least 600 m vertically, and is >200 m wide. On a much smaller scale, is the Paleoproterozoic DeGrussa IMS deposit in Western Australia as based on our interpretation, although, Hawke et al. (2015) favoured a VMS mound-style deposit (see Fig. 18.90). A final example is the interesting discovery of possible IMS massive sulphide mineralisation associated with deep sills in the Tertiary North Atlantic Igneous Province (Corseri et al. 2017). This was located during magnetotelluric survey of the Vøring continental margin, offshore Norway, which identified a highly conductive (? pyrrhotite) body formed within the contact metamorphic aureole of intruded sediments at depths of 8–13 km.

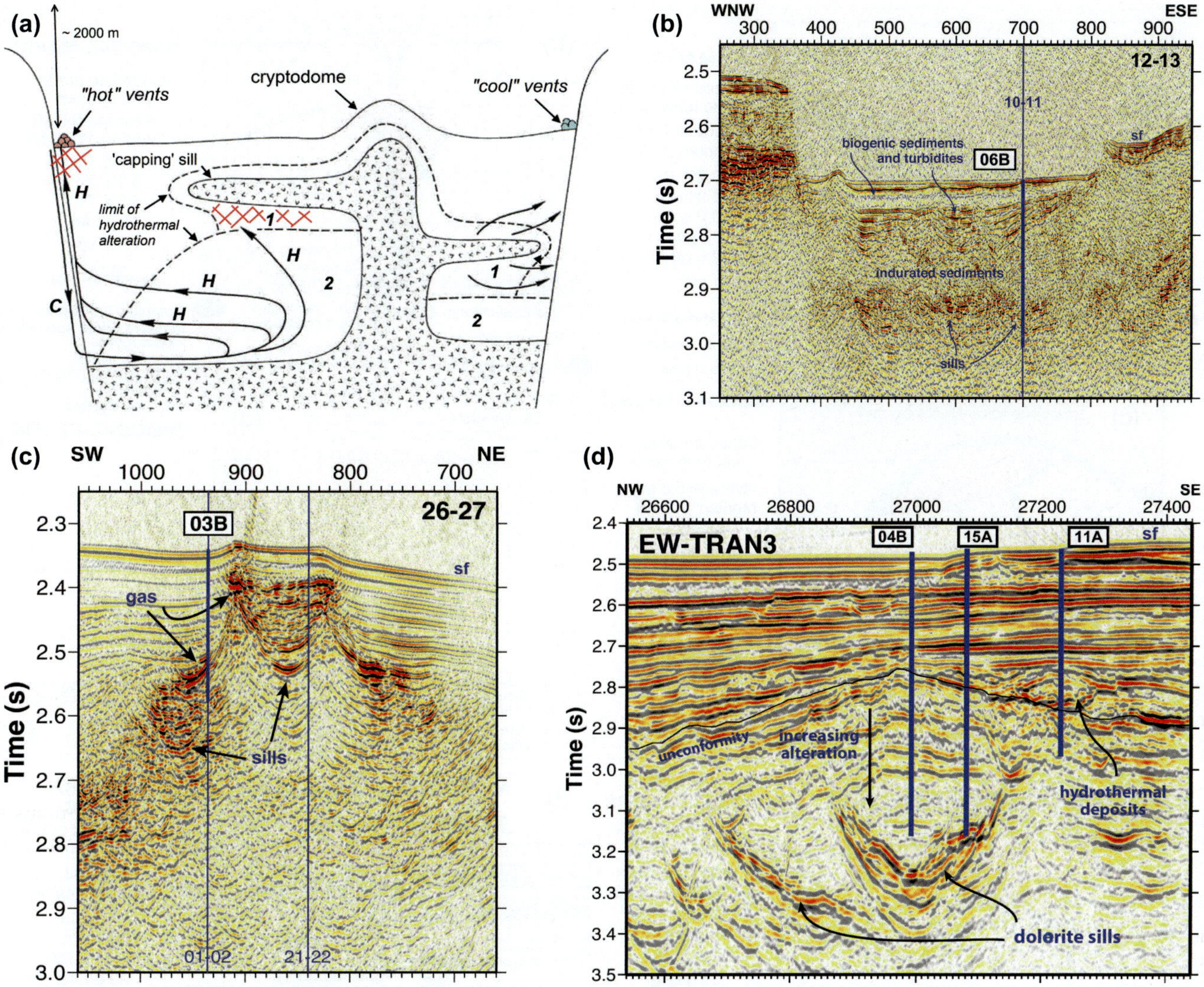

Fig. 18.85 Setting of IMS mineralisation in the Guaymas Basin, Gulf of California. **a** Conceptual model of two active hydrothermal systems. Small-scale circulation (right) is related to shallow-level mafic sills intruded along the rift axis into highly porous sediments, which causes low-temperature contact metamorphism and alteration (1) at temperatures <200 °C. Deep circulation of seawater (left) penetrates larger mafic sills and intrusions beneath the sediment pile to a depth of 500 m. This forms greenschist facies mineral assemblages (2) during extensive interaction at high temperatures (>300 °C) with sediment. Cooled shallow sill acts as a "lid" capping deeper hydrothermal system. Such larger-scale convection is fed by the recharge of seawater through the sediments and fault zones bordering the rift axis; C = cold water advected down through fault zones, and H = hot water advected through sediments and fault zones. The latter forms massive sulphide deposits (red crosshatch), either as seafloor mounds and subsurface replacements localised by fault zones below high-temperature vent sites (<325 °C) or deeper replacement bodies trapped by capping sills. (Modified from Gieskes et al. 1982). **b–d** Examples of shallow seismic profiles (≤ 400 mbsf) visualising some sill, sediment, and hydrothermal relations. Blue vertical lines represent planned drill holes IODP Expedition 385, 2019; sf is present seafloor. **b** Rift axis position, northern Guayama spreading centre. This illustrates the position of indurated and organic carbon-depleted sediments and sills which begin at 170 m b.s.f. at nearby DSDP Site 481. **c** Ring-vent location, on the ridge flanks 28.5 km northwest of the northern Guaymas spreading centre. Here, a still-active shallow sill at about 200 m b.s.f. drives seafloor methane seepage and hydrothermal activity. The topographically elevated ring coincides precisely with active seepage, tubeworm colonies, microbial mats, authigenic mineral precipitates, and conspicuous thermal gradients of 5 °C m^{-1}. **d** Off-axis position 29 km southeast of the northern Guaymas spreading centre. A saucer-shaped sill has intruded predominately terrigenous turbidite sediments near a depositional fan from the Sonoran continental margin. Note increasing alteration is expected to be greatest near the lip of the sill, and "eye structure" may represent seafloor hydrothermal deposits formed from the time the system was active. (From Teske et al. 2018; *Source* IODP through Creative Commons Attribution-Share Alike 4.0 International licence)

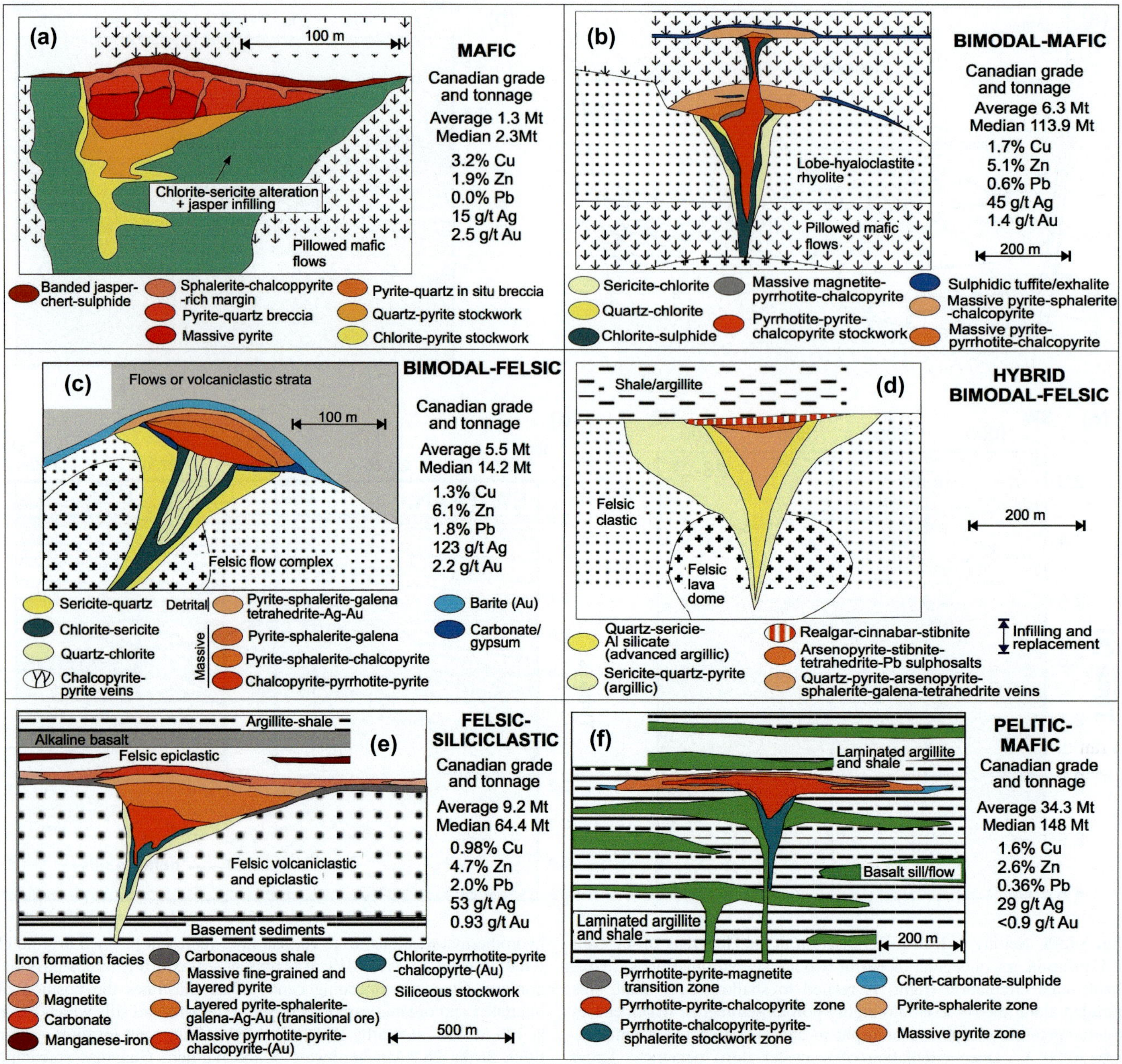

Fig. 18.86 Schematic representation of the VMS host rock lithological-facies associations, ore deposit styles, alteration mineral associations, ore mineral associations, and typical ore grades of Canadian VMS deposits in each type. **a** Mafic association; **b** bimodal mafic association; **c** bimodal felsic association; **d** hybrid bimodal felsic association, a sub-type of (**c**) that may host transitional VMS-epithermal deposits; **e** felsic–siliciclastic association; and **f** pelitic–mafic association. See Fig. 18.87 for a graphic representation of each facies and stratigraphic architecture. (From Galley et al. 2007)

18.8.8 Classification of "Type" Host Rock Associations to VMS Deposits

Traditionally VMS deposits have been classified into a number of deposit "types" based on broad similarities of associated volcanic rocks, compositions, and to a limited extent tectonic settings (e.g. Cyprus-, Besshi- Kuroko-type, etc.). In recent years, a more systematic approach has been used. The foregoing volcanic and tectonic settings and eruption processes that have been described lend themselves to their grouping or classification by common host rock associations (Figs. 18.86, 18.87). This was originally based on host rock composition, with emphasis on the pre-alteration composition of coeval volcanic hosts rocks (Barrie and Hannington 1999), and further elaborated and tied to the tectonic settings described earlier in Sect. 18.8.2

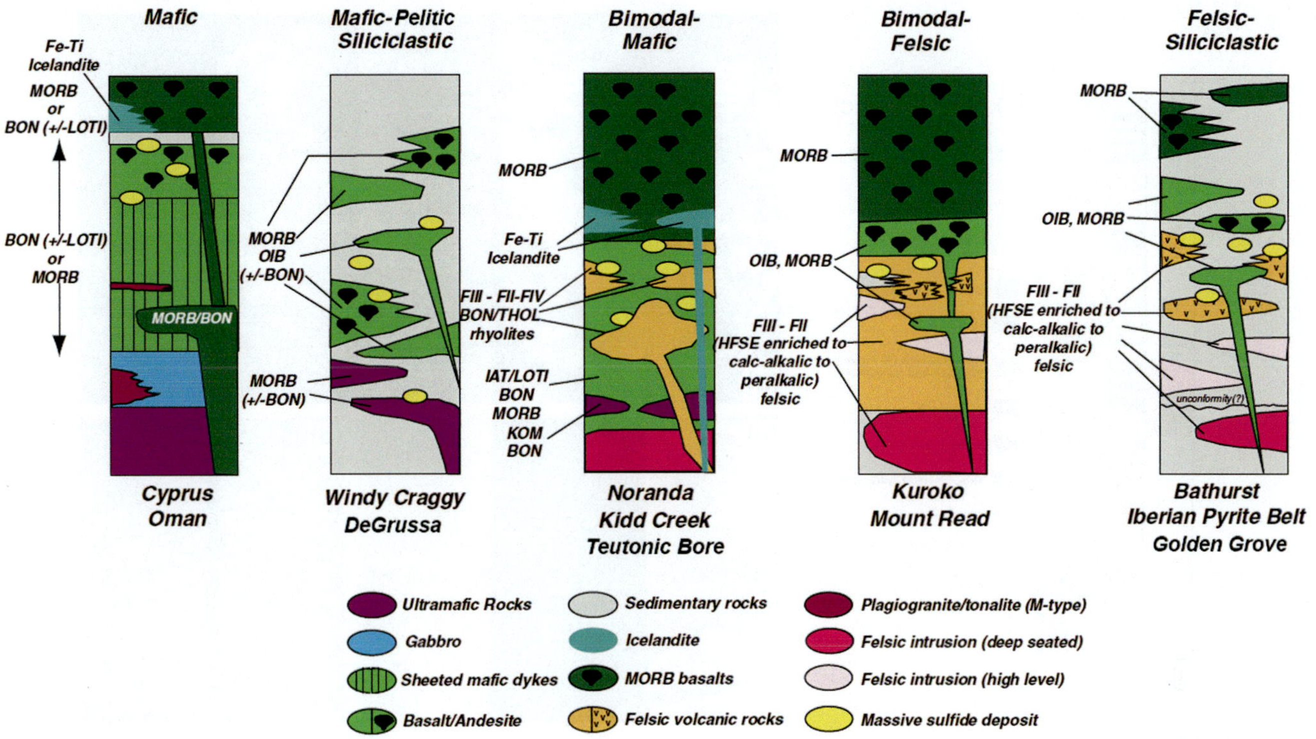

Fig. 18.87 Schematic representation of the facies and stratigraphic architecture found within the main VMS host rock association types (From Piercey 2011). See text for discussion and description of examples

(Franklin et al. 2005; Galley et al. 2007; Gibson et al. 2007; Piercey 2011; Morgan and Schulz 2012; Ross and Mercier-Langevin 2014).

Although some of these links are clear, particularly in modern plate tectonic settings (Fig. 18.79), there are many cases in the ancient geological record where, based on the rock association, a particular tectonic setting is inferred, but the tectonic elements of that setting are not preserved in the rock record, e.g. an apparent oceanic crustal setting where the clear ophiolitic oceanic crustal substrate cannot be proven. Readers are therefore cautioned against making tectonic implications where the geological evidence is ambiguous. In the following summaries of the major rock associations that are known to host VMS deposits, it should be remembered that these represent generalised end members and that any particular host rock succession can preserve its own idiosyncrasies or variations:

(a) *Mafic*: This association is dominated by basalt (both pillowed and massive sheet) lavas, with subordinate autoclastic hyaloclastite and breccias, and mafic intrusions, including sills, dykes and laccoliths, associated peperite breccias, and minor pelagic and hemi-pelagic sediments (Figs. 18.86a, 18.87a). It is typical of mid-oceanic spreading crust (e.g. TAG, Mid-Atlantic Ridge) and back-arc basin spreading ridge oceanic crust (e.g. Marianas Trough, Lau Basin, and Fiji Basin) in the modern oceans, and therefore the typical oceanic crust or ophiolite stratigraphic profile would be expected (Fig. 18.87a; Chap. 17). In the rock record, such an association is more likely to be obducted back-arc basin crust (e.g. Troodos ophiolite, Cyprus, and Semail ophiolite, Oman) incorporated into a developing orogenic belt crust, rather than true oceanic crust.

(b) *Mafic–pelitic (siliciclastic)*: In this association mafic volcanic and intrusive rocks and sedimentary facies, including a large pelitic component, occur in sub-equal proportions (Figs. 18.86f, 18.87b). Mafic sills in the sedimentary intervals are common (<25%). This association is representative of mature back-arc basin settings or rift basins in continental crust, and those that could evolve into narrow spreading ridge basins bounded by continental margins and subject to high levels of terrigenous sedimentation (e.g. Guaymas Basin, Gulf of California; Paleoproterozoic DeGrussa VMS deposit in an intra-continental rift setting, Western Australia).

(c) *Bimodal mafic*: This association is dominated by basaltic rocks but may have <25% felsic volcanics (Figs. 18.86b, 18.87c). Both compositional types are represented by lavas (pillow and massive for basalts), subvolcanic intrusions, associated autoclastic rocks, and volcanic sediments derived from nearby volcanoes. This association is the product of early rifted intra-oceanic volcanic arcs or submerged, intra-continental crust rifts (e.g. Archean Teutonic Bore camp, Western Australia).

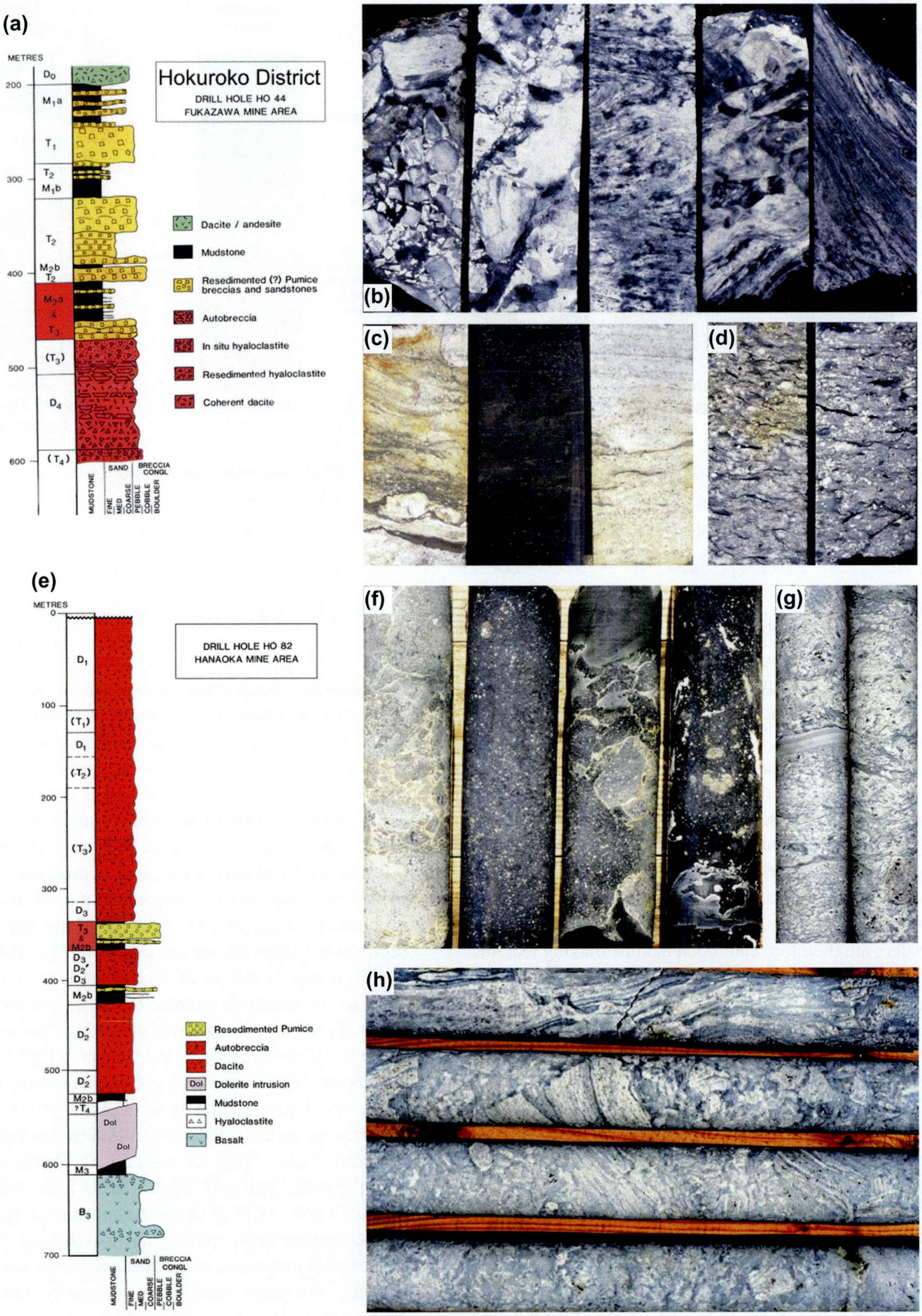
(a)
METRES
Hokuroko District
DRILL HOLE HO 44
FUKAZAWA MINE AREA
Dacite / andesite
Mudstone
Resedimented (?) Pumice breccias and sandstones
Autobreccia
In situ hyaloclastite
Resedimented hyaloclastite
Coherent dacite
MUDSTONE
SAND
FINE
MED
COARSE
BRECCIA CONGL
PEBBLE
COBBLE
BOULDER
(b)
(c)
(d)
(e)
DRILL HOLE HO 82
HANAOKA MINE AREA
Resedimented Pumice
Autobreccia
Dacite
Dolerite intrusion
Mudstone
Hyaloclastite
Basalt
(f)
(g)
(h)

◀ **Fig. 18.88** Some VMS host rock lithofacies from the Miocene Kuroko-type deposits of the Green Tuff Belt, Hokuroko Basin, Honshu, Japan. **a** Graphic log of diamond drill hole HO44 through the Fukazawa mine sequence, noting the dominance of coherent and autoclastic dacite in the footwall and interbedded mass-flow pumice deposits and mudstones in the hanging wall. The ore horizon is at the interface between the T_3 breccias and the M_2 mudstones. **b** Footwall succession of coherent flow banded dacite (cores 1, 3), to plastic accommodation autobreccia (cores 2, 4), and finer blocky hyaloclastite (core 5). **c** Drill core of pumiceous mass-flow deposits (left, right) and organic-rich mudstone (middle) at the ore horizon. **d** Drill core of massive mass-flow pumice breccia deposits in the hanging wall to the Fukazawa mine sequence. The pumice has been altered to clays and the dark "pseudo-fiamme" are diagenetically and altered pumice clasts. They are a not welded texture which is confirmed in thin section. **e** Graphic log of diamond drill hole HO82 through the Hanaoka–Shakanai mine sequence, noting coherent and autoclastic basalt in the footwall and the dominance of coherent and autoclastic dacite at the ore horizon and in the hanging wall. Interbedded mass-flow pumice deposits and mudstones also occur at the ore horizon. The ore horizon is at the interface between the T_3 breccias and the M_2 mudstones. **f** Hatakizoa basalt in the footwall, showing coherent lava facies, blocky hyaloclastite, and silicified mudstone-breccia peperite. **g** Mass-flow T3 dacitic pumice breccia and ash above the ore horizon level. **h** Coherent flow banded, autobrecciated to quench fragmented dacitic hyaloclastite breccia in the hanging wall to the Hanaoka mine. All drill core is $\sim$5 cm wide

(d) *Bimodal felsic*: This association is dominated by felsic volcanic and intrusive rocks relative to mafic rocks and also includes sedimentary facies of mostly volcanic provenance but also of continental crustal derivation (Figs. 18.86c, 18.87d). This association is typical of continental margin arcs and back-arc rift basins, or intra-continental rift settings (e.g. Kuroko VMS deposits).

(e) *Hybrid bimodal felsic*: This is a subtype of the bimodal–felsic association that hosts so-called hybrid VMS-epithermal ore deposits (Galley et al. 2007). It consists of higher proportions of sedimentary and felsic volcaniclastic rocks (Fig. 18.86d) and coincides with shallower water settings, whereas all other associations represent deep-water settings. This could thus include transitional or hybrid VMS-epithermal ore deposits (e.g. Boliden, Sweden, Allen et al. 1996; Mercier-Langevin et al. 2013; LaRonde Penna, Abitibi Belt, Quebec, Mercier-Langevin et al. 2007; and Eskay Creek, northern British Columbia, Sherlock et al. 1999).

(f) *Siliciclastic–felsic (or felsic–siliciclastic)*: In this association, continental crust-derived terrigenous epiclastic facies dominate, felsic volcaniclastic and lesser coherent volcanics and intrusions (<25%) are prominent, and mafic volcanic and intrusive rocks represent a minor component (<10%; Figs. 18.86e, 18.87e). The setting is consistent with a mature epi-continental back-arc basin setting.

18.8.9 Some Case Study Examples of VMS Host Rock Successions

In this section, we will briefly illustrate some typical rock associations mainly hosting ancient, ore-producing VMS deposits that we have experience in undertaking some research on, mainly from Australia.

18.8.9.1 Miocene Kuroko Deposits, Hokuroko District, Honshu, Japan

The well-studied Kuroko VMS deposits, as described above (Sect. 18.8.7), are marked by a footwall submarine volcanic succession dominated by coherent, bimodal lavas and autoclastic deposits (Fig. 18.88a, b, e, f). These are overlain stratigraphically by the ore horizon(s) intercalated with anoxic, organic-rich mudstones and pumiceous mass-flow deposits (Fig. 18.88a, c, e). Yamada and Yoshida (2011) relate these facies to a deep-water submarine rift setting (Hokuroko Basin) in which the VMS deposits formed. The younger, post-ore, hanging-wall sequence is more pumiceous and consists of variably graded, mass-flow deposits of pumice interbedded with variable amounts of organic-rich shales (as at the Fukazawa mine; Fig. 18.88a, d), although may include autobrecciated to quench fragmented hyaloclastite breccia (as at the Hanaoka mine; Fig. 18.88e, h). This succession relates to the transition to a subduction-related arc setting, shallowing of the marine basin(s), and more explosive volcanism in shallower water and emergent environments (Yamada and Yoshida 2011; Yamada et al. 2012; Sect. 18.8.7). The Kuroko host rock succession is classified as a bimodal-felsic association. For further description of the Hokuroko Basin and the role of organic-rich mudstones, see Sect. 18.9.4.3 below).

18.8.9.2 Cambrian Rosebery and Hercules Deposits, Mt Read Volcanic Belt, Tasmania, Australia

The Cambrian Mt Read Volcanic Belt in western Tasmania is a wholly submarine belt of volcanic rocks that hosts numerous VMS deposits, including Rosebery, Hercules, Hellyer, Que River, and Mt Lyell (Large 1992; Large et al. 2001; McPhie and Allen 1992; Table 18.5). Volcanism occurred in a submarine extensional basin called the Dundas Trough. The volcanics are a bimodal association of basalts, dacites and rhyolites, and minor andesites (Corbett 1992, 2001; McPhie and Allen 1992). The felsic volcanics vary from coherent to autoclastic, and in some of the VMS host rock successions, there are also abundant deposits of pumice, transported and deposited by mass-flow processes (Fig. 18.89a–c). In the Rosebery and Hercules mine sequences, there are abundant deposits of highly altered and deformed mass-flow deposited, pumice deposits in both the

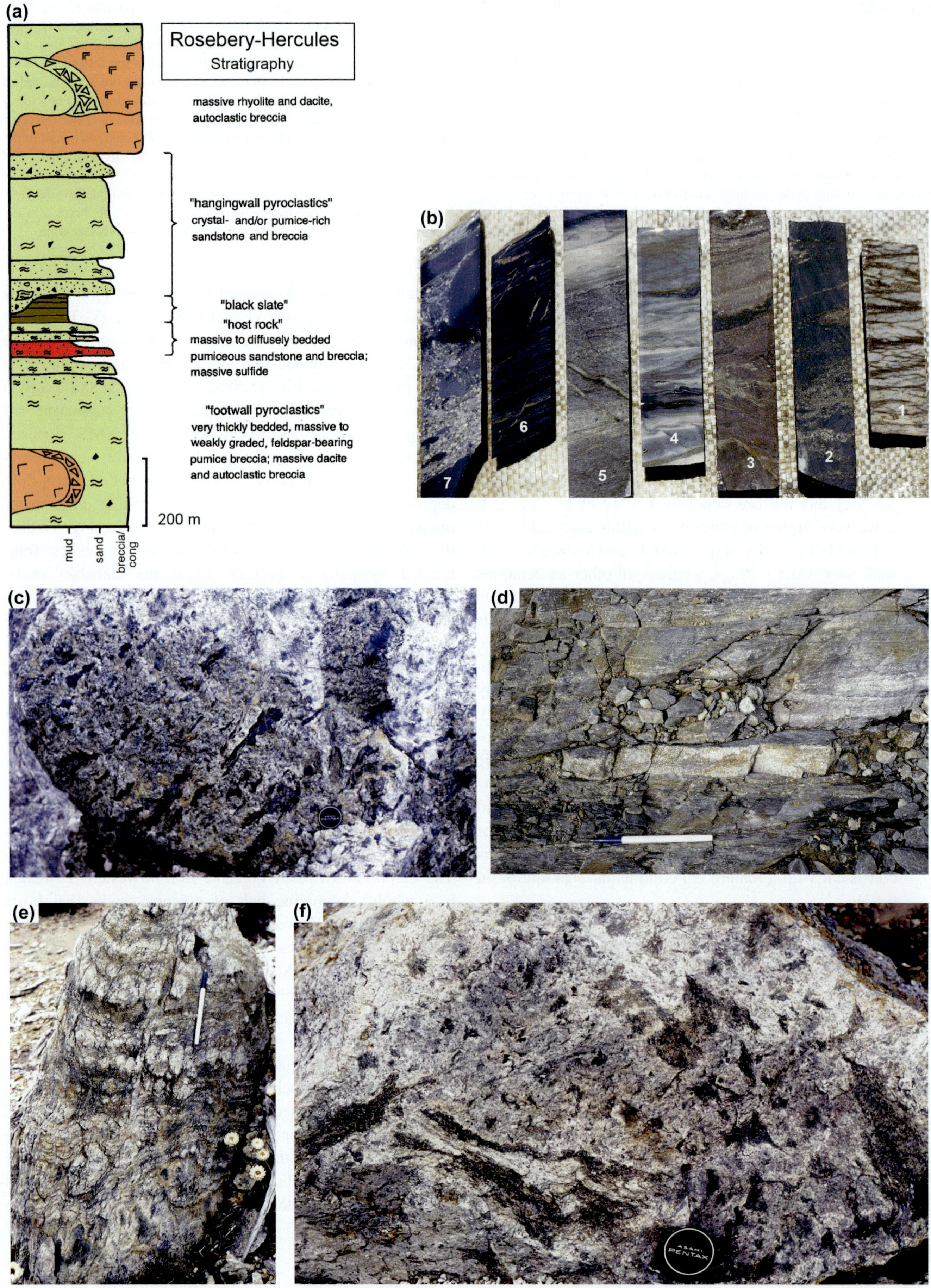
(a)
Rosebery-Hercules
Stratigraphy
massive rhyolite and dacite, autoclastic breccia
"hangingwall pyroclastics"
crystal- and/or pumice-rich sandstone and breccia
"black slate"
"host rock"
massive to diffusely bedded pumiceous sandstone and breccia; massive sulfide
"footwall pyroclastics"
very thickly bedded, massive to weakly graded, feldspar-bearing pumice breccia; massive dacite and autoclastic breccia
200 m
mud
sand
breccia/
cong
(b)
7
6
5
4
3
2
1
(c)
(d)
(e)
(f)

◀ **Fig. 18.89** Characteristics of the Cambrian Rosebery and Hercules VMS mine sequences, western Tasmania, Australia. **a** Generalised lithofacies and stratigraphic architecture noting the dominance of graded bedded pumice deposits throughout the succession, so-called "pyroclastics" (Modified from Large et al. 2001). **b** Samples from diamond drill hole 85R through the footwall and Rosebery mine host sequence. Core 1 = highly cleaved and altered footwall mass-flow pumice breccia; 2, 3 = sulphide ore; 4, 5 = pumiceous sediments; 6 = laminated, pyritic back mudstone; and 7 = polymictic pebbly sandstone. All drill core is 5 cm wide. **c** Massive, altered footwall pumice breccia from the Hercules mine sequence, showing pseudo-eutaxitic or pseudo-fiamme texture, in a rock that microscopically shows no evidence of welding textures. **d** Hercules ore horizon bedded volcaniclastic sediments. **e** Foliated, flow banded hanging-wall dacite. **f** Dacite autoclastic breccia in the hanging wall of the Hercules VMS deposit

footwall and hanging-wall successions, derived from unknown vent locations (Fig. 18.89c). These are intercalated with sedimentary mudstones and turbidite sandstones and breccias at the ore horizon and coherent and autoclastic dacites high in the hanging wall at Hercules (Fig. 18.89d–f). The Mt Read Volcanics vary from being bimodal-mafic at the Que River and Hellyer deposits to bimodal-felsic in affinity at Rosebery Hercules, where mafic volcanics are rare to absent, and Mt Lyell possibly represents a large, hybrid bimodal felsic system (Corbett 2001).

18.8.9.3 Paleoproterozoic DeGrussa Deposit, Capricorn Orogen, Western Australia

The Paleoproterozoic DeGrussa IMS host rock succession is dominated by basalts, dolerites, mudstones, and turbiditic sandstones and conglomerates, which were emplaced in the Bryah Basin, on the northern margin of the Yilgarn craton (Hawke et al. 2015). The mafic units include coherent basalt and autoclastic basalt breccia and are largely intrusive, being intercalated with the sediments (Fig. 18.90a). Contacts with the sediments are commonly irregular (Fig. 18.90b), often with peperitic breccias (Fig. 18.90c, d) indicating syn-depositional intrusive emplacement of mafic units into wet unconsolidated sediments. Some coarser dolerites (Fig. 18.90e) have planar cross-cutting margins and indicate late emplacement into a lithified succession. Although Hawke et al. (2015) indicate that pillow lavas occur in the succession, these have not been personally observed, whereas, spheroidally weathered outcrops have been seen. Nonetheless, the presence of peperite breccias indicates high-level emplacement of the early basalts in an unconsolidated seafloor sediment sequence. There is abundant textural evidence for the sulphide mineralisation being of replacement origin, replacing mudstones, basalts, and peperitic basalt–sediment breccias (Fig. 18.90c, d). The DeGrussa host rock succession is a mafic–pelitic association dominated by sills, and seems to be an example of an IMS VMS deposit.

18.8.9.4 Archean Teutonic Bore Camp, Eastern Goldfields Superterrane, Eastern Yilgarn Craton, Western Australia

The Archean Teutonic Bore VMS camp consists of multiple, small, relatively higher-grade VMS ore deposits, including Teutonic Bore, Bentley, Jaguar deposits, and the Triumph prospect. The host rock succession consists of footwall and hanging-wall rhyolites and a more mafic–intermediate basalt, andesite, black mudstone, and turbiditic sandstone host rock succession, also cross-cut by late dolerite intrusions (Fig. 18.91a; Hallberg and Thompson 1985; Chen et al. 2015). Although well-preserved pillow lavas occur (Fig. 18.91b), there are also many peperitic contacts to basalts and andesite bodies indicating both effusive and syn-depositional intrusive emplacement (Fig. 18.91d). As for the DeGrussa deposit, most of the massive sulphide mineralisation shows evidence for the replacement of black mudstones and basaltic peperite breccias, indicating that the majority of the mineralisation is sub-seafloor replacement (Fig. 18.91e). The footwall rhyolite hosts stringer-style sulphide mineralisation and is highly altered, including the formation of pseudo-eutaxitic textures (Fig. 18.91c) in coherent rhyolite; this was likely emplaced as a sill given the absence of autoclastic breccias at its top. The Teutonic Bore camp can be considered to be a bimodal-mafic succession.

18.8.9.5 Archean Golden Grove Camp, Northern Youanmi Terrane, Yilgarn Craton, Western Australia

The Meso-Archean Golden Grove camp hosts two main VMS deposits, Gossan Hill and Scuddles. The succession is part of a long-lived basin system, but the ore deposits are hosted by an andesitic to rhyodacitic volcaniclastic sediment succession overlain by a package of coherent to autoclastic lavas and intrusions (Fig. 18.92a; Clifford 1992; Sharpe and Gemmell 2002). The volcaniclastic sediments are turbidite facies (Fig. 18.92b), which include well-preserved outlines of altered vesicular pumice (Fig. 18.92c) and even glass shards, indicating derivation by mass flow from shallow water vents and volcanic centres. The hanging-wall succession includes amygdaloidal lavas (Fig. 18.92d) and hyaloclastite breccias which are variably resedimented (Fig. 18.92e). Mineralisation is largely of sub-seafloor replacive style, primarily formed by replacement of the bedded volcaniclastic turbidite succession (Sharpe and Gemmell 2002; Fig. 18.92f). The Golden Grove camp can be considered to be a felsic–siliciclastic succession.

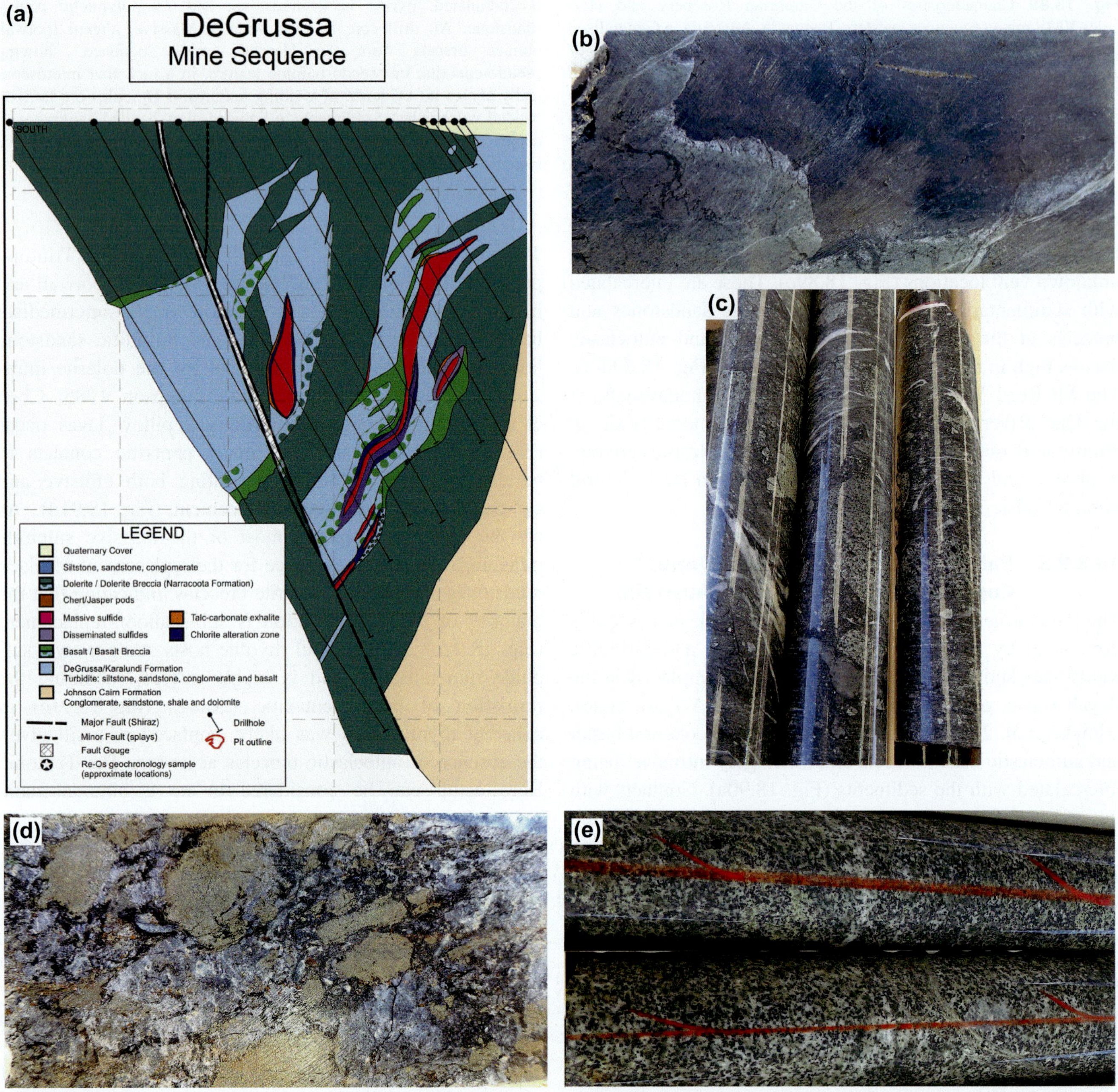

Fig. 18.90 Host rock succession to the Paleoproterozoic DeGrussa VMS deposit, Bryah Basin, Western Australia. **a** North–south cross-section through the mine sequence showing the intercalation of basalts, dolerites, peperitic breccias, and sediments (From Hawke et al. 2015). **b** Irregular margin of intrusive basalt showing incipient quench fragmentation. **c** Basaltic peperite breccia showing grey basalt clasts variably replaced by buff-coloured pyrite, with interstitial black mudstone between clasts. **d** Close-up view of buff-coloured pyrite variably replacing basalt clasts and mudstone matrix in peperite breccia. **e** Coarse, late leucocratic dolerite. All drill core is ~5 cm wide

18.8.10 Summary of Volcanic Eruption Styles in VMS Host Rock Successions

This brief overview indicates that volcanic eruption styles associated with the formation of the volcanic and sedimentary successions hosting VMS deposits can range from effusive in very deep water, to effusive and moderately explosive at water depths of ca. 1,000 m, to effusive and more explosive at shallower water depths (Cas 1992; Cas and Giordano 2014; Cas and Simmons 2018). Effusive eruptions or eruption phases will produce lavas. For basaltic effusive eruption events, the lavas include massive, lobate, and pillow lavas (Chap. 5) and accompanying autoclastic breccias (autobreccia, hyaloclastite, pillow fragment breccia;

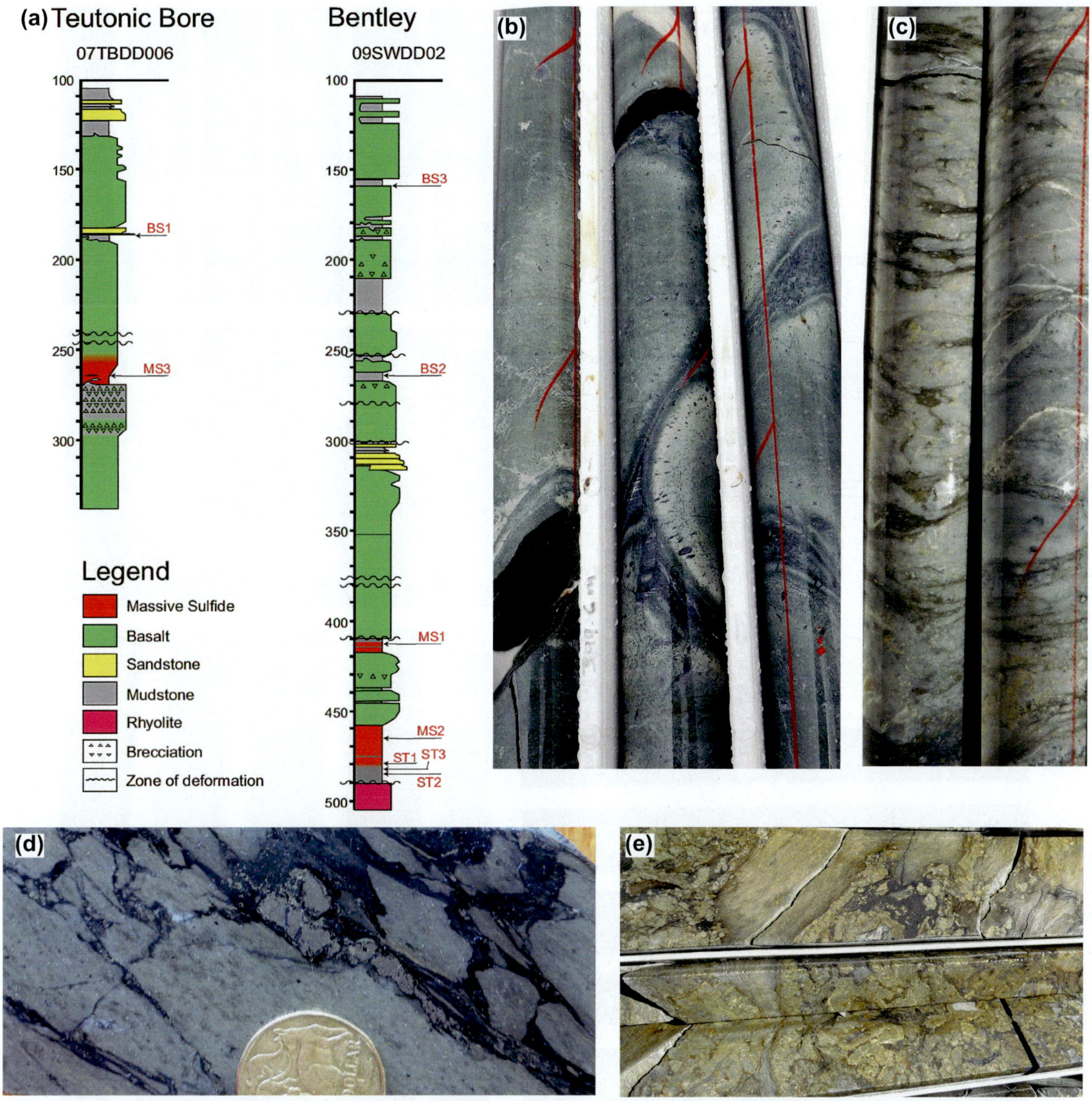

Fig. 18.91 Host rock succession to the Archean Teutonic Bore camp VIMS deposits, Western Australia. **a** Generalised host rock stratigraphy of the Teutonic Bore and Bentley deposits showing the footwall rhyolite, and an association of massive, pillowed, and quench fragmented basalts and andesites, and sediments as the immediate hosts to the mineralisation (From Chen et al. 2015). **b** Spectacular pillow lava preserved in drill core from the Triumph VMS prospect. **c** Highly polyphase altered (silicification, chlorite) coherent footwall rhyolite showing "pseudo-fiamme", Triumph prospect. **d** Blocky in situ basaltic peperite breccia with a black mudstone matrix, Jaguar mine. **e** Sulphide replacing basaltic peperite breccia and the interstitial back mudstone and pseudo-morphing some basalt clasts. All drill core is ~5 cm wide

Chap. 4). For felsic lavas, sheet, lobate, and lava dome forms may form (Chap. 6), which are also accompanied by variable proportions of autoclastic breccias (autobreccia, hyaloclastite; Chap. 4). Submarine fire fountaining and spatter deposits may also be formed by both basaltic eruptions (e.g. Simpson and McPhie 2001; Cas et al. 2003) and deep-water felsic eruptions, where hydrostatic pressure and retention of magmatic volatiles render the erupting magma of low enough viscosity to facilitate this style of activity in mineralised settings (e.g. Mueller and White 1992; Busby et al. 2003; Busby 2005; Chaps. 3, 4, 9).

In addition, syn-depositional and post-lithification intrusions representing all compositions of the magma plumbing system (dykes, sills, dills, cryptodomes, laccoliths; Chap. 8)

Fig. 18.92 Host rock succession to the Golden Grove VMS camp, Murchison Domain, Yilgarn Craton, Western Australia. **a** Stratigraphy, showing the host rock succession to the mineralisation, which is hosted by andesitic–rhyodacitic volcaniclastic sediments, overlain by a hanging-wall succession of dacite, rhyodacite, and andesite lavas and intrusions (Modified from Hollis et al. 2017b, After Clifford 1987). **b** Bedded andesitic–rhyodacitic, resedimented volcaniclastic turbidites of the footwall succession. **c** Photomicrograph of highly vesiculated andesitic–rhyodacitic pumice clasts in the resedimented volcaniclastic footwall sediments (Image by B. Clifford; width of the field of view 2.5 mm, ppl). **d** Porphyritic and amygdaloidal andesite lava of the hanging-wall succession. **e** Blocky hyaloclastite breccia, which appears to have been resedimented given the lack of jigsaw-fit textures. **f** Sulphide diffusely dispersed through and replacing laminated fine volcaniclastic turbidite facies sediments. All drill core is 5 cm in diameter

should occur in most VMS host rock successions and dominate IMS host rock successions. In some, subvolcanic sills and the widespread occurrence of peperitic facies (both bottom and top contacts) form a major proportion of the igneous component. Given the shallow level of intrusion, some emplacement units may show evidence for being both intrusive and extrusive at different locations (Chap. 8; also see Fig. 18.65).

Identifying true, in situ pyroclastic deposits in submarine settings is not easy because, as discussed in Chaps. 11, 12, and 16, pumiceous deposits can easily be resedimented into deep water from shallower eruptions settings, and gaseous pyroclastic flows can transform into water-supported mass flows which can travel tens of km from the source. In situ subaqueous fallout deposits will be better hydraulically sorted than subaerial counterparts (Chap. 9). Near-vent subaqueous pyroclastic flow deposits will be similar to subaerial counterparts if they have been insulated from interaction with the ambient water mass, whereas if they have transformed into water-supported megaturbidity currents they will be better hydraulically sorted developing a mega-graded turbidite bedform (Chap. 11). Nevertheless, many pumice deposits in deep-water settings are quench fragmented hyaloclastic pumice breccias, not pyroclastic. Thinner, normal-graded turbidite beds, as well as pelagic deposits, exhalite layers, and hemi-pelagic mudstones may also occur in the host successions to VMS deposits. Mudstones are likely to be black, organic-rich shales, reflecting the increasingly anoxic conditions with increasing water depths.

18.9 Volcanic-Hosted Reservoirs in Hydrocarbon Exploration

Volcanic and volcaniclastic rocks are commonly considered to have poor petroleum reservoir potential as a result of burial diagenesis, and the alteration of glassy components and unstable grains which leads to an abundance of clay and zeolite minerals that occlude porosity (Chap. 13). Nonetheless, for over fifty years in Japan, commercial oil and gas have been produced from precisely the same basaltic and silicic submarine extrusive facies that host volcanic massive sulphide deposits in the Green Tuff Belt described in the previous section. Although such volcanic reservoirs have been generally of a smaller scale, in recent years, a number of larger oil and gas fields hosted by volcanic rocks have been discovered worldwide.

Thus, as an unconventional type of petroleum reservoir, increasing interest in volcanic facies has been generated in the past fifteen years, representing a new frontier field of exploration (e.g. Petford and McCaffrey 2003; Wu et al. 2006; Farooqui et al. 2009; Jia 2012; Zou 2013; Liu et al. 2012; Seubert 2015; Jackson et al. 2017). For example, Mesozoic volcanic reservoirs have yielded oil and gas productive systems, especially in deeper basins, where they better retain porosity compared to normal sediments, and high flow rates of hydrocarbons can be gained after synthetic fracturing. With rebounding oil prices after the 2014 global crash attention given to volcanic reservoirs is therefore likely to continue, until renewables become the principal energy source worldwide.

18.9.1 Some Basic Principles of Hydrocarbon Exploration

The processes controlling the generation and migration of hydrocarbons (oil and gas) are now generally well understood. During the burial of sedimentary rocks, the onset of oil generation (that is the thermal cracking of kerogen, whereby heavy hydrocarbon molecules are broken into lighter molecules) commences at about 60 °C (Fig. 18.93). Source rocks are overmature or spent of any oil generation capacity at about 130 °C, after which natural gas (primarily methane) may continue to be generated up to 180–200 °C. At normal geothermal gradients of 25–30 °C, the "*oil window*" corresponds to burial depths of about 1.5–4 km, with peak oil generation occurring at about 3 km. At lower geothermal gradients oil will be generated at deeper depths, while at higher gradients oil may be generated at shallower depths. Depending on the geothermal gradient, the gas generation window corresponds generally to burial depths of between 4 and 6 km. Methane may also be generated at much lower temperatures and pressures by biological processes. This is referred to as "*biogenic*", in contrast to thermogenic gas involving the thermal cracking of kerogen.

The area where a volume of source rock reaches appropriate conditions of temperature and pressure to generate oil and/or gas is termed the petroleum "*kitchen*". Compaction (due to sediment loading) expels water and hydrocarbon fluids either vertically upwards or downwards (*primary migration*) from fine-grained, low-permeability source rocks (shales) into permeable carrier beds (sandstones, porous and/or fractured limestones). Hydrocarbon fluids then migrate (*secondary migration*) laterally and up-dip along the carrier beds driven by the water potential gradient (created by sediment compaction) and buoyancy effects. Hydrocarbons will migrate along lateral carrier beds until trapped, unless they intersect low-permeability, cross-strata pathways or conduits such as faults, or stratigraphic "chimneys" (e.g. carbonate reefs or steeply stacked, basin margin clastic fans). On meeting these, they rise until ultimately trapped or continue to form natural oil seeps on reaching the land surface or seafloor. In this manner, it is not uncommon for hydrocarbons to migrate >2 km vertically and over 100 km laterally from the kitchen to the reservoir.

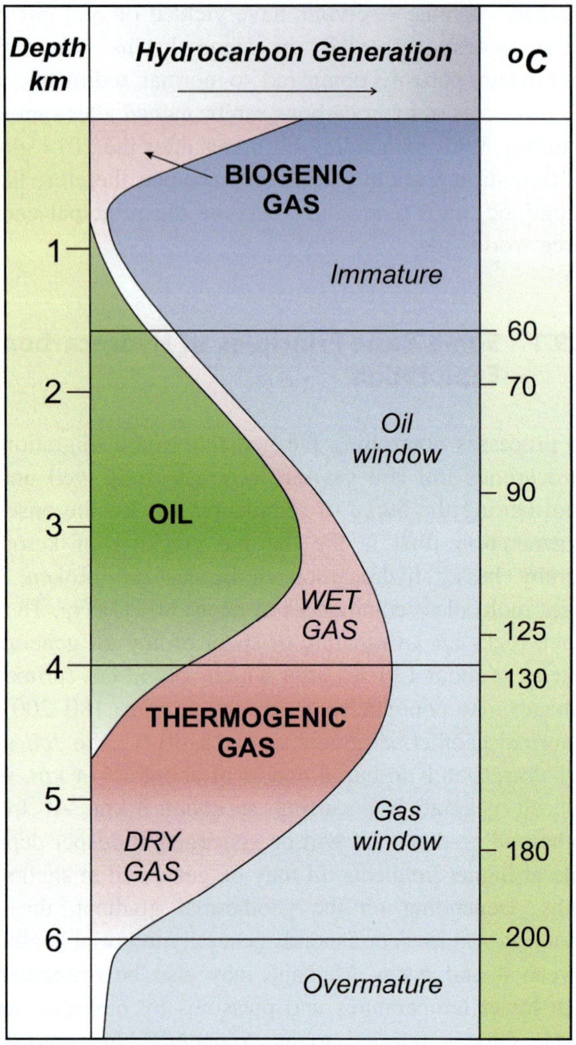

Fig. 18.93 Schematic representation of the depth/pressure and temperature conditions determining the oil and gas generation windows

This whole process of generation–migration–accumulation is referred to as a "*petroleum system*" (Fig. 18.94). The essential stages and elements must occur together in time and space, so that organic matter in a source rock can be converted to a petroleum accumulation. A complete or successfully working petroleum system exists wherever these essential stages and elements follow each other in the continuum. This is in large part controlled by sedimentary basin evolution. Different types of sedimentary basins have very different evolutionary patterns and thermal histories. Extensional (rift) and foreland basins are the two most significant types in terms of petroleum exploration. Important variants of extensional basins are transtensional basins and passive margin basins. The latter form when rifting leads to the actual break-up of the continental plate, with the formation of new oceanic crust, with the term referring to the

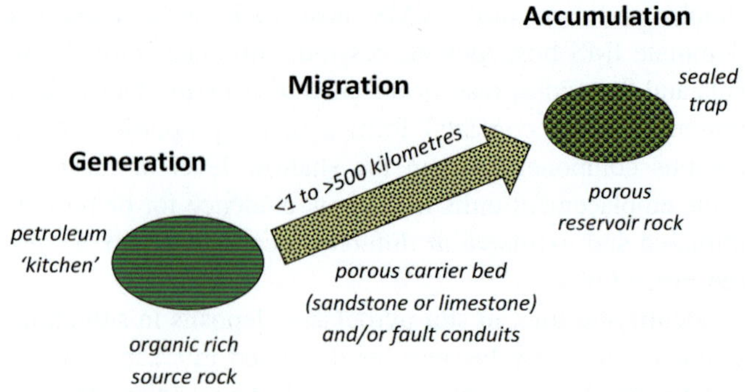

Fig. 18.94 Essential stages and elements in the generation of a petroleum system

thick (up to 20 km) wedge or prism of sediment and break-up volcanics deposited adjacent to the new continental margin (Chap. 17).

18.9.2 Volcanic Petroleum Systems

It is now recognised that volcanism and volcanic processes can potentially affect all aspects of the petroleum system: (1) promoting deposition of organic-rich source rocks, (2) accelerating maturation, (3) aiding fluid migration, and (4) creating traps, reservoirs, and seals.

Volcanic activity may contribute to organic source richness in a number of ways, and on differing scales. On a global scale, greater amounts of submarine basaltic volcanism as a result of increased spreading rates at divergent oceanic plate margins and/or mantle hot spot activity creating submarine oceanic plateaus or oLIP (Chaps. 4, 16), can be closely correlated with worldwide anoxic events (e.g. Kerr 1998, 2005). These in turn correlate with the deposition of some of the planet's most productive petroleum source rocks (e.g. Cenomanian–Turonian organic-rich black shales; Jarvis et al. 2011). As described in Chap. 17, this is illustrated by the formation of the Caribbean–Colombia Oceanic Plateau during precisely the same period of time as the deposition of the prolific La Luna Formation, which is the major source rock in the Southern Caribbean petroleum province (see Figs. 17.38 and 17.39). The input of volcanic ash into marine basins can promote algal blooms with enhanced organic growth, while sudden input of volcaniclastic debris into restricted marine or lacustrine environments during an eruption would immediately kill most living organisms, but also preserve organic matter from oxidation. Subaerial volcanism especially in rift basin settings (e.g. East African Rift System; Chap. 17 and Sect. 18.9.4.5) may create restricted lakes and swamps with organic-rich sediments and where geothermal waters could promote nutrient growth further enhancing the production of organic material.

In addition, altered volcanic glass rich in zeolites can catalyse the decomposition of organic matter and expulsion of mature hydrocarbons at a lower temperature than conventional thermal models predict (Liu et al. 2012; Seubert 2015). Similarly, in restricted submarine rift basins (such as the Green Tuff Belt), the hydrothermal activity could promote the deposition of organic-rich source rocks intimately associated spatially with potential volcanic reservoir facies.

By directly adding heat, subvolcanic intrusive bodies can accelerate hydrocarbon maturation. Larger intrusive bodies such as thick dykes and sills can cause local over-maturation of source rocks, while extrusive bodies generally have negligible impact. However, the circulation of hydrothermal and geothermal fluids associated with intrusive bodies could have more considerable and regional thermal effects, again especially in extensional rift basin settings. While this may lead to over-maturation, in some basins this could promote early (shallower) burial maturation of source rocks and, rather than the millions of years normally required, where the process could take hundreds of thousands of years to generate oil. Convecting geothermal fluids are also capable of dissolving and mobilising hydrocarbons and depositing them in volcanic rocks. Furthermore, if the vapour pressure in a subvolcanic intrusion becomes low enough during cooling, hydrocarbons may be drawn into pore spaces. Igneous intrusions can, themselves, also directly form closed traps within surrounding sediments. Often quoted is the Omaha Dome field (Illinois Basin, United States), where the trapping structure is an impermeable, ultramafic Christmas tree-like laccolith body, intruded into limestones and sandstones (Farooqui et al. 2009).

More generally, once hydrocarbons are generated, they migrate elsewhere to become trapped structurally and/or stratigraphically by conventional basinal processes. Volcanic rocks can both provide the trap and/or reservoir. The best example of the former is the giant Forties oil field (5 billion barrels or Bbbl in place), which is the largest oil field discovered in the UK sector of the North Sea (Wills 1991; Chap. 17). Here a major paleovolcanic high, composed of middle Jurassic basalts, is draped by Paleocene submarine fan reservoir sandstones forming a broad antiform with four-way-dip closure (90 km^2 in area). As with typical sandstone and carbonate reservoirs, successful efforts to discover hydrocarbons in igneous and volcanic rocks will depend on the coincidence of porosity preservation and the timing of hydrocarbon migration and trap formation (Fig. 18.94).

18.9.3 Porosity and Its Controls

As a result of their greater variability in mineralogy and composition, volcanic rock reservoirs are more complicated than those of normal clastic sedimentary rocks. Put simply, this is because of (1) their higher proportion of more reactive and unstable minerals and volcanic glass, (2) more complicated and exaggerated textural characteristics, and (3) the potential for rapid, extensive changes during burial or thermal diagenesis and tectonic evolution. Contrary to conventional thinking, primary porosity and permeability in volcanic rocks can be relatively high (e.g. Lenhardt and Götz 2011). However, there have been very few systematic studies in petroleum-producing basins aimed at establishing the factors controlling porosity and permeability (*poroperms*) of volcanic rocks and the sequence of processes occurring during cooling and post-cooling stages and burial which can substantially modify their original petrophysical characteristics.

18.9.3.1 Types of Porosity in Volcanic Rocks

Volcanic rocks develop primary and secondary porosity and permeability depending both on the lithology and the sequence of syn- and post-eruptive processes involved in their formation. *Primary porosity* is principally controlled by the style and degree of vesiculation and fragmentation of magma or volcanic rock (e.g. magmatic, phreatomagmatic, quench, or epiclastic; Chap. 4) and emplacement mechanism (e.g. pyroclastic flow, fallout or surge, lava flow or epiclastic). As emphasised in Chap. 4, volcanic materials, whether they are erupting magmas, chilled volcanic glass, or already crystallised volcanic rock, can be fragmented by a spectrum of processes, ranging from primary, syn-eruptive volcanic, to post-eruptive processes, producing a range of primary to secondary volcaniclastic deposits and clast types (see Fig. 4.1). Primary fragmentation processes include autoclastic, non-explosive processes associated mainly with emplacement of lavas and high-level, syn-depositional intrusions, to explosive (pyroclastic) processes. Post-eruptive processes include the spectrum of normal surface sedimentary processes including gravitational collapse, mass movement, weathering, erosion, and tractional transport, as well as subsurface hydrothermal and tectonic fragmentation processes (Chaps. 12, 13). Pumice-bearing pyroclastic deposits and especially ignimbrites can develop protracted cooling histories because of their high heat-retention capacity, during which welding and volatile activity (i.e. deuteric alteration) can have a considerable effect on porosity and permeability distributions (Chap. 10). In contrast, subaerial lavas are rapidly quenched and have relatively short stages of deuteric alteration, although, not when emplaced in submarine environments where hydrothermal circulation incorporating seawater may be created (Chaps. 6, 13; Sect. 18.8).

Subsequently, *secondary porosity* develops and generally results either from the chemical leaching of minerals and/or the generation of a fracture system. It has been widely recognised in petroleum exploration that secondary porosity can be the predominant form of effective porosity in the reservoir sandstones of many oil and gas fields (e.g. Surdam et al. 1984, 1989). This has been linked principally to reactions occurring during burial which, as well as changes to organic phases (i.e. maturation), also promote and affect diagenetic changes in mineral phases. During burial increasing pressure (sediment load) and temperature in general promote a compactional reduction in porosity and permeability. Clay minerals show that with increasing burial both illite crystallinity and illite content increase, due to the addition of K^+ ions from the degradation of smectite. However, the reactions in sandstones can be more complex and not only related to pressure and temperature gradient, but to changing fluid regime (both flow rate and composition). Dissolution of unstable framework grains (plagioclase, labile rock fragments, and/or volcanic glass) or of earlier carbonate cements (calcite, dolomite, and siderite) in sandstone carrier beds and potential reservoirs can actually substantially increase porosity and permeability. The decarbonitisation process is very significant because this results from decarboxylation reactions involving the combination of organic phases with formation waters to produce acidic solutions. Evidence suggests that the main phase of hydrocarbon generation follows closely the culmination of decarboxylation reactions, and hydrocarbon migration may follow a front of acid waters. This would facilitate both the dissolution of both aluminosilicate minerals and carbonate cements (Surdam et al. 1989; Seubert 2015).

In epiclastic sandstones (e.g. feldspathic wacke, feldspathic litharenite, and lithic wacke), the reactions will follow the same paths as those affecting typical immature sandstones. However, the final product will be dependent on many variable factors, including (1) initial composition, (2) environment of deposition, (3) nature of interbedded sediment, (4) composition of formation waters, and (5) depth of burial. Most terrestrial volcaniclastics (both primary and epiclastic deposits) have good primary porosity. Generally, this porosity is often significantly reduced during early diagenesis, when unstable volcanic components alter rapidly to pore-filling phases and carbonate cements. Their alteration to zeolite minerals (e.g. heulandite, laumontite, and pumpellyite) plays a key role in the diagenetic sequence of events, from the modification of framework grains to obliteration of primary and/or generation of secondary porosity (Surdam and Boles 1979; Chap. 13, see Sect. 13.7). The formation of zeolites however appears more dependent on pressure, temperature, pH, and ions in the formation water than the composition of detrital framework minerals. Development of extensive secondary porosity has also been reported in non-marine volcanic sandstones (up to 40%) during shallow burial diagenesis (<900 m), as a result of high pore-fluid flow rates in shallow alluvial deposits (Mathisen 1984). It is important to stress that in many instances, burial and diagenesis of volcanic rocks are detrimental to, rather than enhancing, porosity and permeability. As well as dissolution, fracturing can be an important process enhancing porosity especially in competent volcanic rocks. This can vary from cooling jointing in welded tuffs, lavas, and subvolcanic intrusions to hydraulic fracturing and tectonic faulting (Chap. 4).

18.9.3.2 Austral and Neuquén Basins, Argentina

As indicated above, there have been few systematic studies in petroleum basins of porosity development and its controls in volcanic rocks. One exception has been studies of Mesozoic (Jurassic) volcanic reservoirs from the Austral and Neuquén Basins, which are the two main oil- and gas-producing basins in Argentina. Although oil and gas fields have been producing since 1960, it is only in the past 20 years that exploration has focused on volcanic units as potential reservoirs (Sruoga et al. 2004; Sruoga and Rubinstein 2007). The reservoir rocks encompass a wide spectrum of volcanic lithologies, ranging from non-welded ignimbrites to dense volcanic glasses and lavas, all of which show different types of primary and secondary porosity development (Fig. 18.95; Table 18.6). The relative contribution of each type of porosity to overall poroperms is highly variable.

In non-welded to moderately welded deposits, pore space is represented by former gas bubbles, loose shard packing (*inter-shard porosity*), and non-collapsed pumice fragments (*intra-pumice porosity*). As the degree of welding increases, porosity values are reduced proportionally (Fig. 18.95). Typical intra-pumice and inter-shard types in non-welded ignimbrites show high porosities (up to 26%), medium values (14%) in moderately welded ignimbrites, whereas in welded zones porosity decreases to low values (4.8%; Table 18.6). Although some late dissolution processes overprint primary porosity, it is evident that welding is the first-order control. Primary permeabilities are generally low because of pore disconnection, with values of less than 1 md for moderately welded, and 1–16 md for non-welded ignimbrite (Table 18.6). However, the most porous and permeable lithology identified was non-welded ignimbrite with well-developed gas segregation pipes, which imply gas streaming and elutriation of fine ash matrix (Cas and Wright 1987; Chap. 10). Thin section study shows this pipe-related porosity was accompanied by the formation of an efficient channel network providing interconnected permeability. High porosity (26%) and high permeability (42–164 md) are recorded in such non-welded zones with gas pipes (Table 18.6). Gas pipes and the elutriation process may therefore have a significant effect on permeability (Sruoga et al. 2004; Sruoga and Rubinstein 2007).

Taken together, the results of these studies show that the highest poroperms occur in rocks with quench fractured

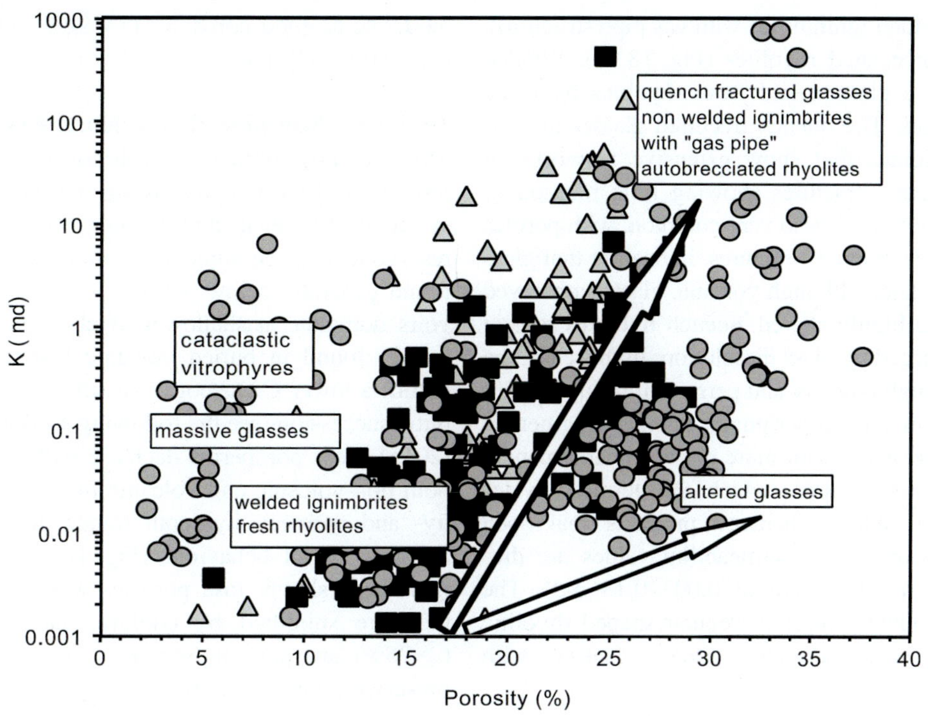

Fig. 18.95 Porosity versus permeability plot showing the main volcanic lithologies studied in the Austral Basin, Argentina (From Sruoga et al. 2004; also see Table 18.6). Cerro Norte gas field = black squares, Campo Bremen gas field = triangles, and Océano oil and gas field = circles. Increasing trends in porosity and permeability are shown by arrows. The highest poroperm values correspond to quench fractured glasses and non-welded ignimbrite with gas segregation pipe structures, followed by autobrecciated rhyolites; lowest values are represented by welded ignimbrites, massive glasses, and fresh rhyolites. Lower trend corresponds to altered (devitrified) glasses and reflects the increasing development of secondary porosity concomitant with the dissolution of primary minerals (± precipitation of authigenic minerals), variously described as spongy-cavernous (depending on size), sieve to mouldic, and/or lacy (Table 18.6; also see Zheng et al. 2018)

Table 18.6 Summary of poroperm characteristics of Jurassic volcanic petroleum reservoirs from the Austral and Neuquén Basins, Argentina (From Sruoga and Rubinstein 2007)

Field	Depth (m)	Porosity (%)	Permeability (md)	Lithology	Types of porosity[a]
Cerro Norte (AB; g)	1698.00–1706.52	13–28	0.001–6.7	Rhyolites	Intracrystalline sieve-mouldic, interclast, fracture
Campo Bremen (AB; g)	1728.00–1787.00	4.8–26	0.001–164	Non-welded–welded ignimbrites, epiclastics	Inter-shard, intra-pumice, vesicular, gas segregation pipes, intracrystalline sieve-mouldic, shattered crystal, intergranular
Océano (AB; o&g)	1175.00–1375.00	9.4–37.6	0.003–762	Vitrophyres, obsidians, hyaloclastites, ignimbrites, epiclastic breccias	Quench fracture, inter-shard, secondary spongy-cavernous, fracture, intergranular
Barranca de los Loros (NB; o&g)	979–997	10.7–18.56	0.85–15.76	Non-welded–welded ignimbrites	Secondary spongy-cavernous, lacy, intra-pumice, intracrystalline sieve-mouldic, secondary sieve-mouldic, fracture
	1055.80–1061.60	9.72–21.27	0.003–205.3	Andesites	Drusy breccia, secondary sieve-mouldic, cavitation voids
	1157–1161.7			Welded ignimbrites	Intracrystalline sieve-mouldic, vesicular, gas segregation pipes, drusy breccia
	1232–1249	12.7–23.05	>1	Rhyolites	Interclast, vesicular, interflow laminar and tension cracks

[a] For detailed description of porosity types and photomicrographs, the reader is referred to Sruoga and Rubinstein (2007). AB = Austral Basin, NB = Neuquén Basin; o&g = oil and gas, g = gas

glasses and non-welded ignimbrites with gas pipe structures, followed by autobrecciated rhyolites (Fig. 18.95). Welded ignimbrites, massive glasses, and fresh rhyolites have the lowest permeabilities. The quench fractured glasses are vitrophyres and obsidians that show extensive networks of polyhedric, curviplanar fractures showing little rotation of fragments. Perlitic texture is also very common, with perlites enclosed by the quench micro-fractures, indicating that glass hydration occurred late. Although volcanic glass is observed to be moderately to highly altered, quench fractures remain open and even enlarged by glass dissolution. In this case, the rocks can achieve high porosity and permeability (up to 38% and 762 md). From core descriptions quench fragmented glasses appear intercalated with more extremely fragmented units with jigsaw-fit textures, which are termed hyaloclastites (Chap. 4). Where alteration products seal open spaces in hyaloclastite units, permeability values are dramatically reduced (in the range of 0.003–0.18 md). The autobrecciated rhyolites comprise irregular shaped rhyolitic fragments in a hydrothermally altered matrix with moderate-to-good porosities (13–28%), but low permeabilities (0.001–6.7 md).

18.9.3.3 Songliao Basin, Northeast China

More recently, poroperm evolution in Cretaceous volcanic reservoirs of the Songliao Basin (northeast China) has been systematically evaluated (Wang and Chen 2015). These reservoirs host significant thermogenic gas accumulations found generally below 3000 m, whereas sedimentary reservoirs dominate at shallower levels. Gas accumulations are mostly found in buried volcanic highs or domal features within a lower Cretaceous syn-rift bimodal basaltic to rhyolitic succession (see the following section). Wang and Chen (2015) show poroperms decrease with increasing depth for both non-volcanic and volcanic lithologies, but their porosity- and permeability-depth trends diverge as a result of differing burial behaviour (Fig. 18.96). Lava and welded ignimbrite slowly lose porosity with burial depth because they are solidified by cooling, and their groundmasses ($\geq 75\%$) are poor in secondary quartz and calcite, thus preserving pore space. In contrast, sandstone, conglomerate,

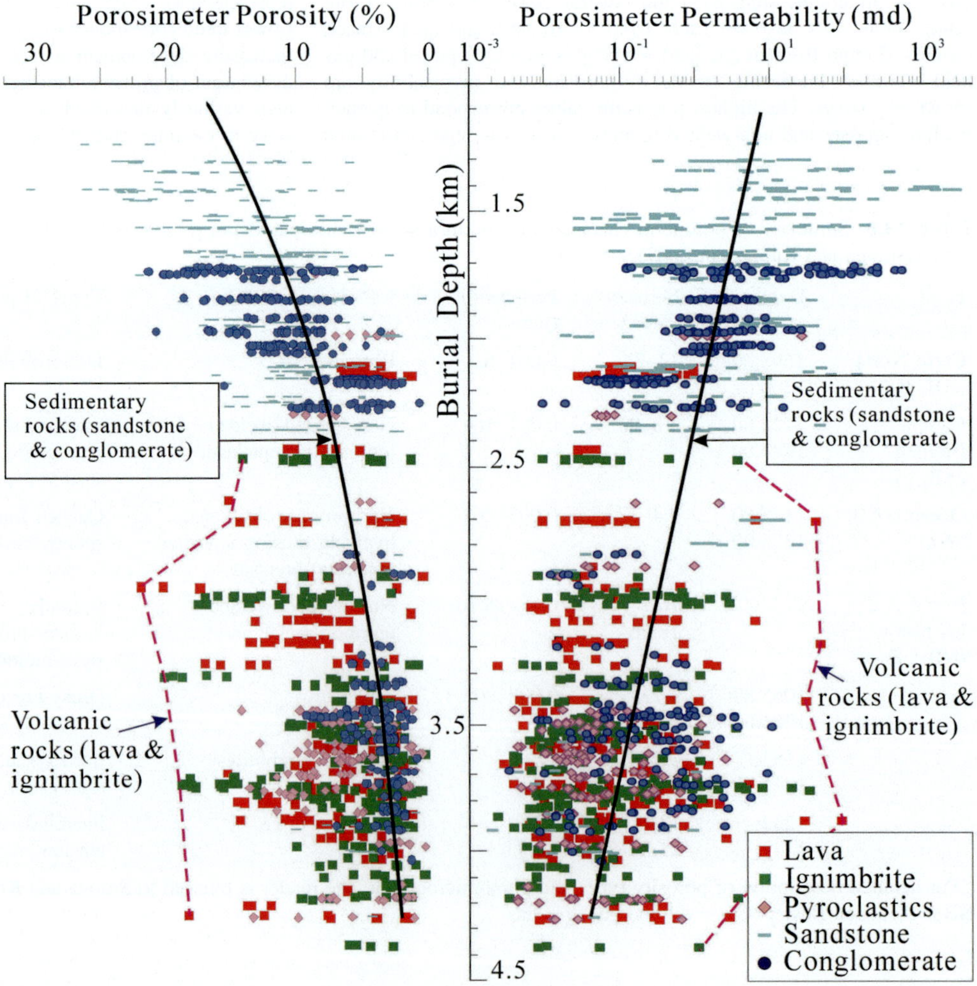

Fig. 18.96 Trends in porosity (%) and permeability (md) versus burial depth for volcanic (lava, welded ignimbrites, and pyroclastics) and sedimentary (sandstone and conglomerate) rocks for the Songliao Basin, northeast China. Porosity and permeability of the sedimentary rocks show a sharp decline between 1000 and 3000 m; below 3000 m, both porosity and permeability decrease slightly and show generally low values. In contrast, the porosity and permeability of the volcanic rocks are rather independent of depth. As a result of these two different trends, volcanic rocks have generally greater porosity and permeability than sandstones and conglomerates below ~3000 m. (From Wang and Chen 2015)

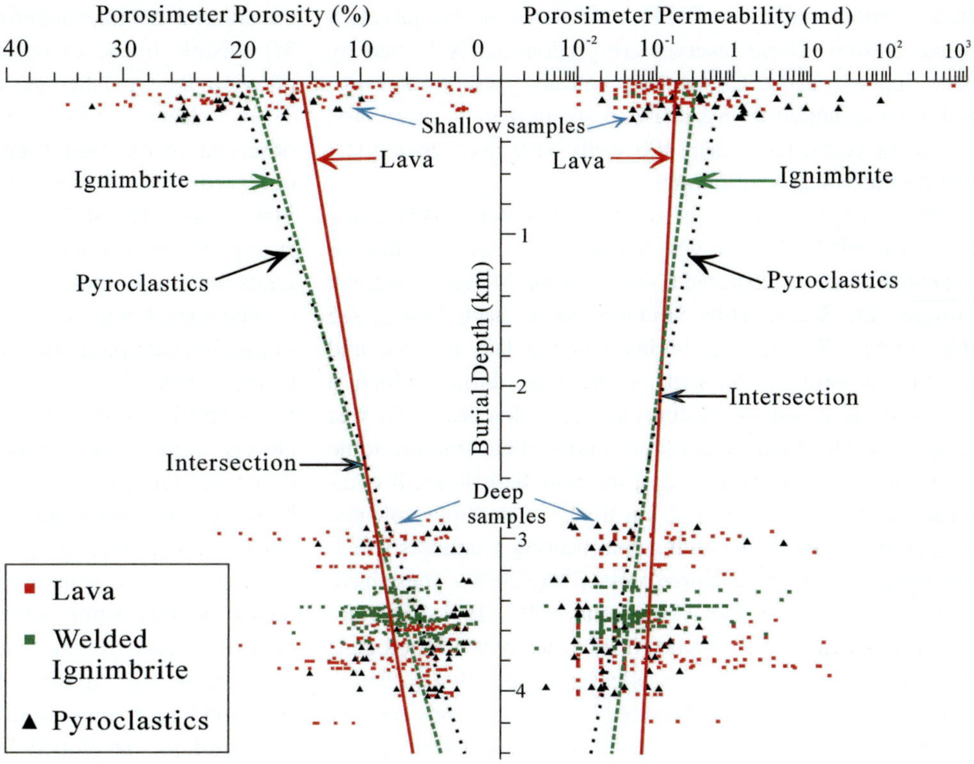

Fig. 18.97 Trends in porosity (%) and permeability (md) *versus* burial depth among lavas, welded ignimbrites, and pyroclastics in the Songliao Basin, northeast China. For burial depths shallower than 2000 m, both porosity and permeability are in the order of lava < welded ignimbrite < pyroclastic rocks, whereas below a depth of approximately 3000 m, they change and reverse order, that is lava > welded ignimbrite > pyroclastic rocks. (From Wang and Chen 2015)

and pyroclastic (tuffaceous) lithologies are more sensitive to overburden pressure, suffering more intense compaction and cementation. As a result, the porosity and permeability of lava and welded ignimbrite exceed that of the other rocks and are the best potential reservoir rocks below burial depths of about 3000 m.

Volcanic facies also respond differently to burial depth in the Songliao Basin. Whereas porosity and permeability both decrease for lavas, welded tuffs, and other pyroclastic deposits, the trends are significantly different. Lavas show the slowest decrease, whereas pyroclastic deposits display the fastest decrease from shallow to a greater depth. The difference in both porosity and permeability between shallow and deep samples increases systematically from lava to ignimbrite to other pyroclastics (Fig. 18.97), indicating an increasing impact of overburden pressure on their reservoir parameters. For burial depths shallower than 2000 m, both porosity and permeability are in order of lava < welded ignimbrite < pyroclastic rocks, whereas below about 3000 m, they change to lava > welded ignimbrite > pyroclastics. Textural evidence indicates that vesicles and other primary porosity textures of the lavas are preserved in deeply buried samples because vesicles are actually interlaminated with flow bands. In contrast, deeply buried pyroclastic deposits are much more densely compacted than shallow samples and have also subsequently undergone cementation of the pyroclasts.

18.9.4 Volcanic-Hosted Oil and Gas Fields

We will now briefly describe some volcanic-hosted oil and gas fields, concluding with a section that identifies some frontier regions and basins, which, in terms of reserve additions, hold potential for hydrocarbon discoveries in volcanic reservoirs.

18.9.4.1 Qingshen Gas Field, Songliao Basin, Northeastern China

The first oil and gas field reservoir within volcanic rocks in eastern China was discovered in the Bohai Bay Basin in the 1970s. This was followed by successes in the Liaohe Basin, and later in the Songliao Basin. However, it was drilling in 2002 of Xushen-1 in the Xujiaweizi rift (Songliao Basin) and the discovery of the Qingshen gas field that was to change exploration perceptions in China (Feng 2008; Wang and Chen 2015). This well encountered a daily gas flow of >1 million m^3 from volcanic and volcaniclastic reservoirs and demonstrated the enormous potential for deep gas (≥ 3 km) exploration in volcanic rocks. Since then, the proven natural gas reserves in the Qingshen field have exceeded 3.5 billion cubic feet, while in the wider Xujiaweizi rift proven gas reserves have been recently estimated at >10 trillion cubic feet (TCF), or in terms of oil equivalent some 2 Bbbl. These reserves are predominately hosted by late Jurassic–early Cretaceous volcanic reservoirs, with subordinate amounts of gas also in conglomeratic reservoirs. In recent years, more than 400 wells have been drilled targeting the volcanic reservoirs.

The Songliao Basin is in one of several Mesozoic–Cenozoic rift basins that developed in northeastern China in response to the extensional collapse of the Mongol–Okhotsk orogen (cf. Saddle Hills uranium basin; Sect. 18.4.2, see Fig. 18.51). Rifting was initiated in the late Jurassic and continued through the early Cretaceous during which a bimodal succession of alkaline basaltic to rhyolitic rocks was emplaced. The Xujiaweizi rift is a half-graben structure some 4–5 km deep near the centre of the Songliao Basin. It comprises a rift succession (1–2 km thick) of multiple volcanic and clastic intervals, including coal-bearing sediments which are the main gas-prone source rocks (Feng 2008). These have total organic carbon (TOC%) contents of 0.1–16%, and other organic geochemical parameters indicate type III kerogens are dominant with good to excellent gas generative potential. These sediments were deposited in progressively deeper water depth, with braided delta and shallow lacustrine facies evolving into deep-water lacustrine facies upwards. The syn-rift volcano-sedimentary successions are overlain by super-rift (or "sag") sedimentary sequences of the early–late Cretaceous age. These also contain two prolific lacustrine oil shale hydrocarbon source rocks and fluvial–deltaic clastic sediments, which form the major reservoir units at Daqing, the largest onshore oil field in China. Thus, the Songliao Basin hosts significant gas accumulations below 3000 m in syn-rift volcanic and volcaniclastic reservoirs, whereas oil-bearing sedimentary reservoirs dominate at shallower levels in the overlying sag succession. Although the charging history was complicated, the main episode of gas generation and accumulation occurred during the mid-Cretaceous.

Most of the main high-yield producing wells drilled in the Qingshen gas field were drilled on paleovolcanic highs which are mapped both geologically and seismically (Fig. 18.98). The play evolved as part of a three-pronged exploration strategy (Wang and Chen 2015): (1) establishment of volcanic facies models based on outcrop observations, (2) interpretation of downhole well logs and core lithologies and volcanic facies, and (3) interpretation of 2D and 3D seismic data. Mean dip angles of 3D seismic data are also frequently used to map buried volcanoes, which can delineate both cone-shaped (? presence of tuff cone), or non-cone-shaped (? simple lava dome) centres. Multi-dome complexes may show the development of crater features and facies (Chen et al. 2014; Fig. 18.99a), but fault systems coupled with truncation surfaces are likely the major paths for gas migrating into the volcanic reservoirs (Huafeng et al. 2015). Proximal (or conduit), medial, and distal facies associations are transitional and interfingering with each other (Fig. 18.98a; a 3D seismic image of buried volcanic rhyolite dome complexes in the Qingshen gas field is also shown in Fig. 16.81). These volcanic map patterns are remarkably similar to those observed in exposed highly extended rift basins such as observed in the Basin and Range Province, western United States (Sect. 18.3.4.2; Chap. 17); in this case, they suggest strong structural control by north–northwest basin-forming faults, and east–northeast transfer faults.

The early Cretaceous volcanic reservoir rocks (Yingchen Formation) are predominantly silicic lavas and welded tuffs. Chinese geologists have classified volcanic lithofacies into 5 facies and 15 sub-facies (Liu et al. 2012; Fig. 18.99b), which closely conform to western models of subaerial rhyolite dome and tuff cone/ring complexes (Chap. 16). This should be no surprise considering their volcano-tectonic setting, but what is more interesting is that they occupy precisely the same position as a number of metal-bearing volcanic host rocks, notably some epithermal gold–silver and uranium deposits (Sects. 18.3.4.2 and 18.4). From proximal to distal settings, the size, density, style, and arrangement of porosity are found to change significantly. The highest porosity is preferentially developed in proximal rhyolitic lava facies near the central part of each dome or dome complex. In these regions, primary porosity is better preserved, e.g. larger and more abundant vesicles and laminar flow voids, more development of breccias of various origins, and the presence of cooling joints is more common. In the medial facies association vesicles and voids are smaller and cooling joints less well developed, while distally many, primary pore spaces such as vesicles or cooling fractures are absent. The mean primary porosity in proximal, medial, and distal facies associations is 7.1, 5.9, and 4.3%, respectively. Proximal facies are also more favourable locations for the development of secondary porosity, including devitrification pores associated with the development of spherulites and lithophysae (Zheng et al. 2018), and particularly along faults and fractures due to post-eruption weathering, leaching, and dissolution, and hydrothermal brecciation. In proximal positions (paleo-topographic highs), wells may contain >100 m of effective gas reservoirs, whereas in distal locations these are generally <30 m thick. In addition, detailed facies analysis has identified several volcanic or eruptive cycles which provide the opportunity for multiple gas pay zones, where gas intervals are mainly distributed in volcanic reservoirs near the top of the cycle (Feng 2008; Chen et al. 2014; Wang et al. 2015).

18.9.4.2 Karamay Oil Field, Junggar Basin, Northwest China

Although volcanic rock and reservoir formations of many Mesozoic and Cenozoic basins in east and northeast China have received much attention in recent years, oil and gas

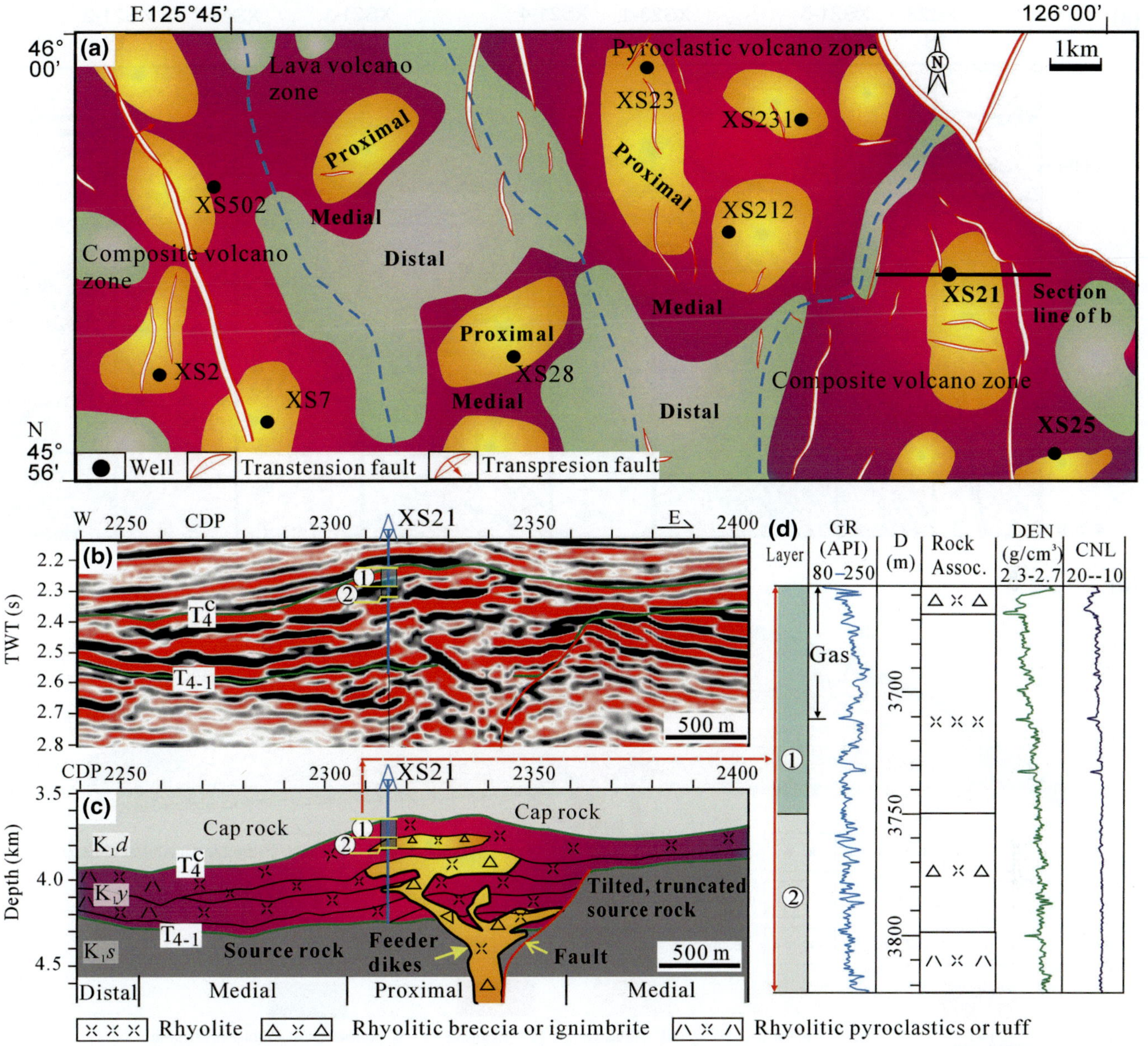

Fig. 18.98 Representative features of volcanic reservoirs of the Qingshan gas field, Songliao Basin, northeast China. **a** Plan view of buried volcanic complexes of the early Cretaceous Yingcheng Formation (K_1y) with facies distribution and fault patterns from 3D seismic integration (also see Fig. 16.81). **b** Seismic profile and well Xushen 21 (XS21) located in (**a**). **c** Interpreted section showing volcanic facies associations and petroleum system. **d** Well logs, lithological section, and gas accumulation. Symbols K_1s, K_1y (syn-rift), and K_1d (super-rift) represent local lower Cretaceous stratigraphy; those prefixed by T are seismic reflectors. (From Wang and Chen 2015)

were first identified in China within Carboniferous volcanic reservoirs of the Junggar Basin, northwest China. The giant Karamay field discovered in 1955 has proven oil reserves of 3 Bbbl and was China's first commercial oil field. The oil field is located on the northwestern structural edge of the basin, within the Kebai fault zone, a regional, deep penetrating, and long-lived structural culmination (Chen et al. 2016a). The main host rocks are described as andesites and tuffs. The reservoirs are strongly structurally controlled and occur in fault-bounded traps, or weathered volcanic rock beneath unconformities, the most important being a regional Carboniferous–Permian erosion surface (Fig. 18.100). The main controlling factors for oil distribution in the Carboniferous volcanic reservoirs are faulting, fracture development, and degree of weathering when subaerially exposed in the Permian. The area in which these faults and fractures developed is the primary area of oil accumulation with high yields (>135 bbl/day).

The present-day Junggar Basin is principally a foreland basin formed during Himalayan compression during the

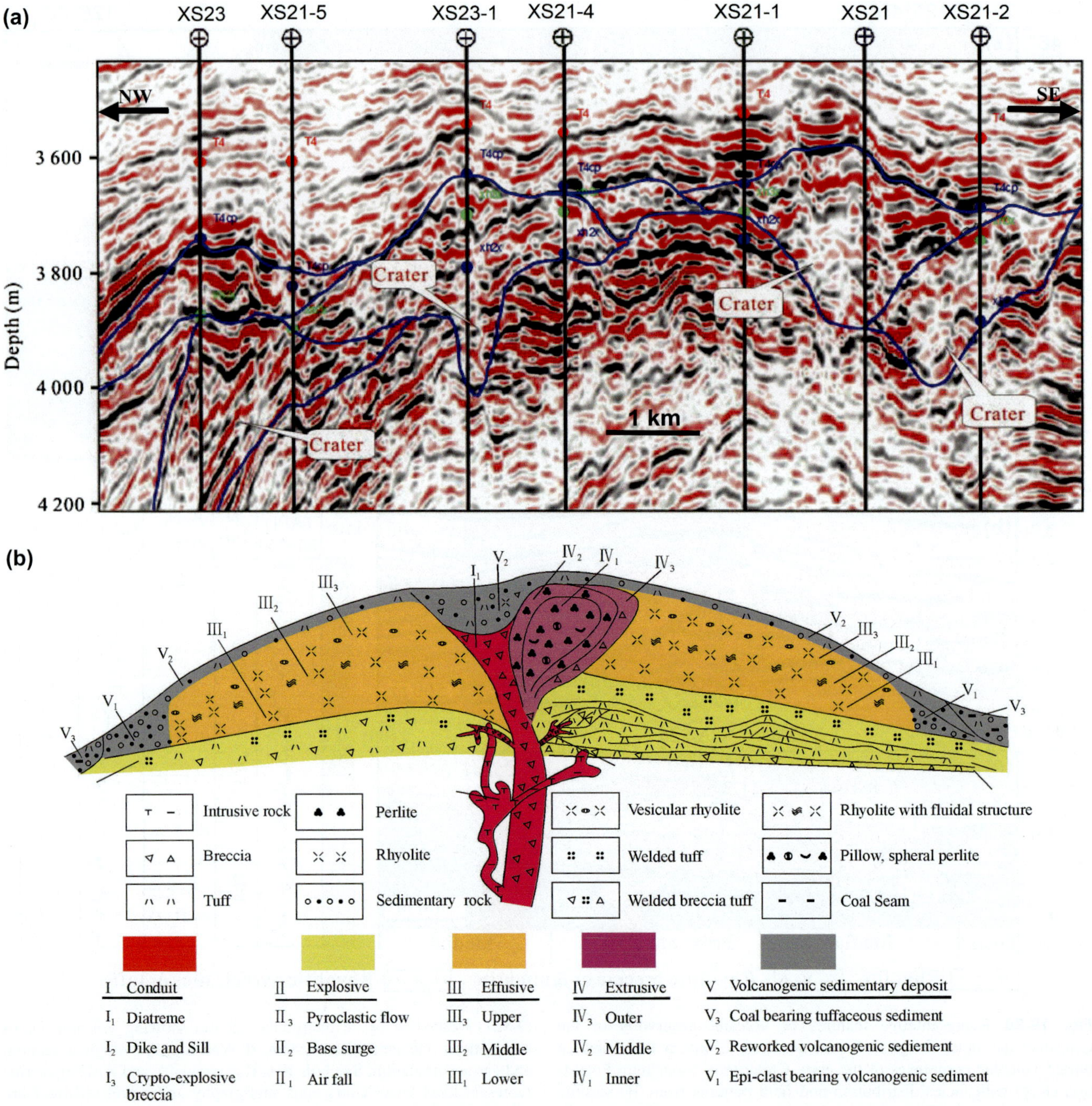

Fig. 18.99 **a** Semi-regional seismic correlation within the Yingcheng Formation, eastern Xushen gas field, Songliao Basin, northeast China. The blue line boundaries are believed to define a number of overlapping silicic volcanic centres and craters; note the position of the gas-producing well Xushen 21 (XS21) described in Fig. 18.98. (From Chen et al. 2014). **b** Classification scheme employed for petroleum exploration of silicic volcanic facies in the Songliao Basin. The scheme distinguishes 5 facies and 15 sub-facies; in detail, this incorporates features such as transportation mechanism and origin, diagenesis style, representative lithology and structure, facies sequence and cycle, and potential reservoir quality. The classification emphasises the relationship between lithofacies and reservoir capability, facilitating recognition of volcanic lithofacies at the various scales of outcrop section, drill core, well cuttings, and thin section. (From Liu et al. 2012; *Source* InTechOpen through Creative Commons Attribution 3.0 Unported licence)

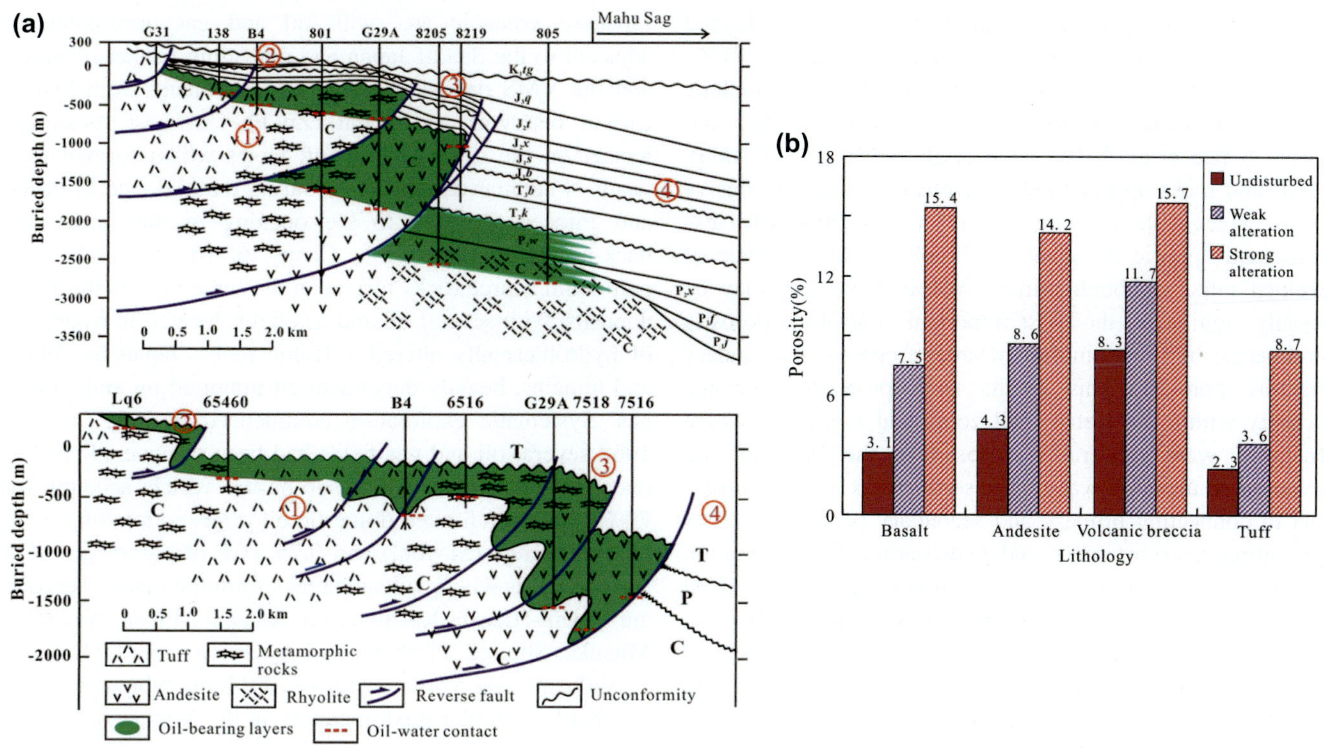

Fig. 18.100 **a** Schematic cross-sections showing Carboniferous fault block reservoirs (upper) and weathered crust reservoirs (lower) of volcanic rock in the Kebai fault zone, western Junggar Basin, northwest China. The circled numbers indicate (1) the principal part of the nappe, (2) thinning out of strata in the hanging wall, (3) fault block at the edge of the nappe, and (4) the footwall of the nappe; C = Carboniferous, P = Permian, T = Triassic, and K = Cretaceous, while suffix represent local stratigraphy. (From Chen et al. 2016a). **b** Distribution of average porosity for volcanic rocks with different lithology and degree of Permian weathering, highlighting enhanced values of the volcanic weathered crust reservoirs. The lower limit of effective porosity is 6%, and the lower limit of permeability is 0.5 md of the proven reserves of Carboniferous volcanic reservoirs. (From Hou et al. 2015)

Cenozoic, but it has had a complex, multi-phase evolutionary history. Rifting during the late Carboniferous–early Permian deposited a thick volcanic-sedimentary succession (generally 500–1500 m thick) within the Kebai fault zone, which appears to have formed the basin edge to a half-graben (Chen et al. 2016a). Paleo-geographic reconstructions suggest proximal ("crater") facies were distributed along and at the confluence of the main faults. The volcanic rocks are believed to have been erupted in an environment which gradually shallowed from deep marine to shallow marine, and then to continental by the early Permian, but most eruptions supposedly happened under water. During the Permian, a period of inversion was associated with the development of a northeast-trending thrust belt. The area then underwent substantial subsidence from the late Permian (Mahu Sag), during which Permian source rocks were buried sufficiently deep to generate oil in two main phases during the early and late Jurassic (ca. 200 and 160 Ma; Chen et al. 2016a). Subsidence was punctuated by a series of plate collisions in the central Asian (Altaids) region which caused several episodes of reactivation along the Kebai fault zone, culminating in Himalayan tectonic movements and uplift through the Cenozoic.

The Kebai fault zone is a mesh of reverse and thrust faults, which divide the Carboniferous rift volcanics into several blocks and faulted steps. These deeply seated, large-throw faults control strata development and directly connect with oil source rocks in the adjacent Mahu Sag to the southeast (Fig. 18.100a). Most of the oil accumulations are structurally controlled with oil being found near large faults in inverted horst blocks and generally buried deeper (fault block reservoirs). In addition, there are sub-unconformity stratigraphic or structural traps that occur where weathered volcanic rocks are distributed in the hanging wall of a large thrust fault. These Carboniferous structural highs are long-lived features experiencing extensive weathering and pervasive leaching under a hot-warm climate during the Permian. The volcanic rocks are preserved at shallow levels, and form a high-quality, unconventional "*weathered crust*" reservoir with a fracture-pore-dominant porosity (Fig. 18.100b; Hou et al. 2015; Chen et al. 2016b).

Available published descriptions of the volcanic reservoir rocks unfortunately generally lack detailed modern facies analysis and, unlike with those in the Songliao Basin, it is more difficult to reconcile these in terms of modern western nomenclature and paleo-environmental models. As indicated above, they are generally described as andesites (~50%) and tuffs (35%), while less common lithologies include volcanic breccias, "sedimentary tuff", conglomerate, and basalt (Chen et al. 2016a; Hou et al. 2015; Fig. 18.100b). Andesitic rocks appear to show typical textures of effusive lavas, whereas the tuffs include rocks with vitroclastic textures and "plastic deformation", implying the presence of welded tuffs. Although primary volcanic porosity may be locally significant, the most important control on porosity and permeability within the volcanic reservoirs is secondary fracture porosity. This results from protracted tectonic activity within the Kebai fault zone, and especially where fractures were subaerially exposed in the Permian and exposed to meteoric waters. These fractures are also essential in connecting primary and secondary pores, as well as providing reservoir space and hydrocarbon fluid migration pathways (Chen et al. 2016a, b; Hou et al. 2015). Dissolved phenocryst pore–fracture combinations occur in rhyolite, andesite, and basalt reservoirs, whereas micro-fractures occur primarily in tuffaceous rocks. The fractures are crucial to improving the permeability of volcanic rocks and contribute primarily to commercial flow in the reservoirs.

Elsewhere in the Junggar Basin, primary volcanic porosity may be a more decisive factor. One post-unconformity (early Permian), oil-producing rhyolitic reservoir rock is described as "vesicular welded tuff", in which porosity is defined by lithophysal cavities and voids within ignimbrite cooling units (Zhu et al. 2012; Chap. 10). Although the burial depth of the reservoirs exceeds 4000 m, lithophysae are well preserved (Fig. 18.101), resulting in an average porosity of 14.6% (30.1% maximum) and average permeability of 0.203 md, producing high oil flow rates following fracturing. Lithophysal cavities and vesicular textures in welded ignimbrites result from volatile activity, or vapourisation of surface or groundwater (e.g. Wright and Coward 1977; McArthur et al. 1998; Chap. 11) and appear to significantly improve their reservoir quality compared to other rhyolitic and pyroclastic lithologies (Fig. 18.101d). In this case, primary porosity has not been subject to collapse or formation damage during burial. Welded ignimbrites with abundant lithophysae deposited in paleo-depressions with shallow lakes or limited groundwater are considered important oil reservoir targets in this part of the basin (Zhu et al. 2012).

18.9.4.3 Niigata and Akita Oil and Gas Fields, Hokuroko Basin, Honshu, Japan

The Green Tuff Belt in northern Honshu is Japan's most important area for both mineral and petroleum resources. Many Kuroko-type VMS deposits and oil and gas fields are located in the Neogene Hokuroko Basin (Fig. 18.102; Kikuchi et al. 1991; Magara 2003; Glasby et al. 2004). These are distributed in two distinct belts along the northeast Japanese volcanic arc, with oil and gas accumulations adjacent to the Sea of Japan and extending further offshore, and the VMS deposits located further inland. Both hydrocarbon reservoir and metal (Zn–Pb–Cu) host rocks are located in the same Green Tuff facies which formed as a result of submarine volcanism during the mid–late Miocene and Pliocene, related to the opening of the Japan Sea back-arc basin (Sect. 18.8; Chaps. 6, 16). Generally, the petroleum source rocks formed at the same time as the VMS deposits, with several oil and gas fields located in reservoirs of hydrothermally altered volcanic rocks. Japan has been, and remains, heavily dependent on imported oil and natural gas. Systematic exploration commenced in 1955, and by 1968 several oil and gas fields had been discovered centred on the Niigata and Akita sub-basins which account for 70% of Japan's limited domestic oil reserves, totalling ~50 million barrels (MMbbl) and 0.75 TCF of natural gas. Oil and gas fields where volcanic rocks form the main reservoirs are Yoshii-Higashi Kashiwazaki, Katakai-Minami Nagaoka, Mitsuke, Shiunji, Yurihara, and Ayukawa (Fig. 18.102a).

Volcanic reservoirs are formed by both submarine dacitic–rhyolitic (found predominantly in the Niigata sub-basin) and basaltic (found in the Akita sub-basin) facies. The general hydrocarbon framework is illustrated in Fig. 18.102b. The Neogene section is generally several km thick, although it may reach more than 8 km in parts of the Niigata sub-basin. The silicic reservoirs are mostly composed of domes, domal clusters, and effusive facies that erupted on or were emplaced at very shallow levels below the seafloor (Komatsu et al. 1984; Sato 1984; Kikuchi et al. 1991; Magara 2003; Watanabe 2006; Yagi et al. 2009). As described, pillow breccia and lava facies form the main producing reservoirs (Fig. 18.103a). These retain higher porosities (10–30%) and permeabilities (1–150 md) and contain primary pore spaces (vesicles and perlitic cracks), secondary pore spaces (vughs, interstitial pores, and leached pores), and fracture porosity (Hoshi et al. 1992; Okubo 2001). The volcanic reservoirs are commonly encountered at depths between 3000 and >4500 m. Hyaloclastite facies, which are also hydrocarbon-bearing, are mostly not sufficiently productive because of their low permeability.

Most hydrocarbons are thought to have been generated from contemporaneous fine-grained, marine, organic-rich shales and mudstones adjacent to and/or intercalated with the volcanic reservoir units (e.g. Sakata et al. 1994). These migrated towards the upper levels of the domes and accumulated there under a cap or seal of fine-grained sediment. During compaction, hydrocarbon fluids could have also migrated downward as these fine-grained cap rocks acted as pressure seals as well as capillary seals over the oil/gas saturated zones below (Magara 2003). However, there is some evidence for the involvement of older and deeper source rocks (?Eocene or basement; Tomaru et al. 2009).

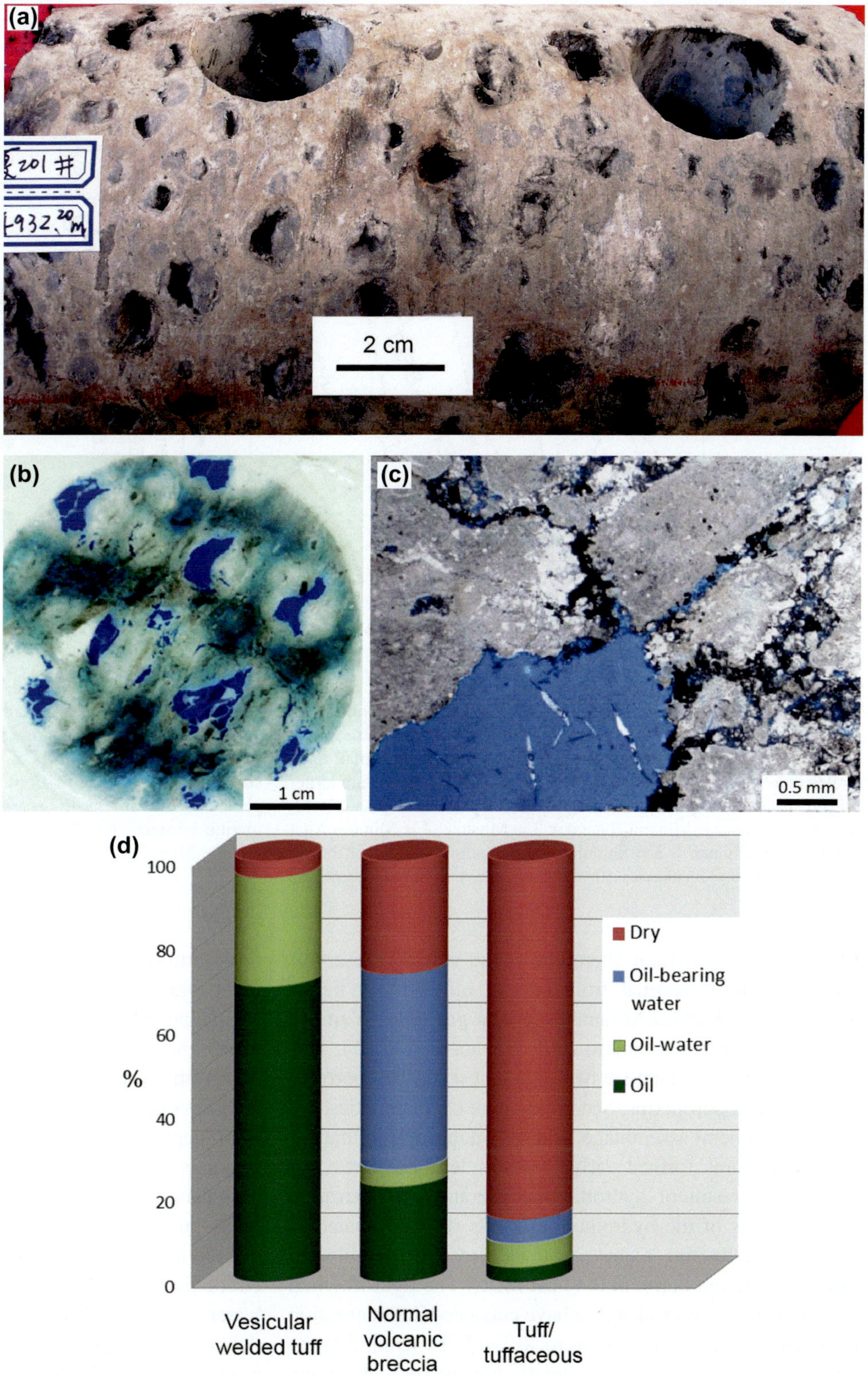

Fig. 18.101 Vesicular cavities and voids (lithophysae) in early Permian rhyolitic welded ignimbrite reservoirs of the Fengcheng Formation, Wu-Xia area, Junggar Basin, northwestern China. **a** Lithophysae in drill core from petroleum well Xia 201 (at 4932 m depth). Lithophysae are usually about 1 cm in diameter and their volume varies from 20 to 70%, reflecting vertical zoning patterns within a cooling unit, which increase at first, then decrease from the bottom to the top. **b** Sliced and impregnated core showing porosity development (blue); from well Xia 202 (4826 m). **c** Photomicrographs indicating lithophysae are interconnected by shrinkage (cooling) micro-fractures and -jointing; from well Xia 201 (4934 m). **d** Probability histogram of oil presence in cores in a series of four wells (including Xia 201 and 202; accumulated perforation thickness is 82 m). Based on the presence of oil layers (70%), this demonstrates vesicular welded tuffs are the best reservoir target; tuff (air fallout) or tuffaceous layers generally only show oil fluorescence or dry layers, while other volcanic breccias are intermediate. (Adapted from Zhu et al. 2012)

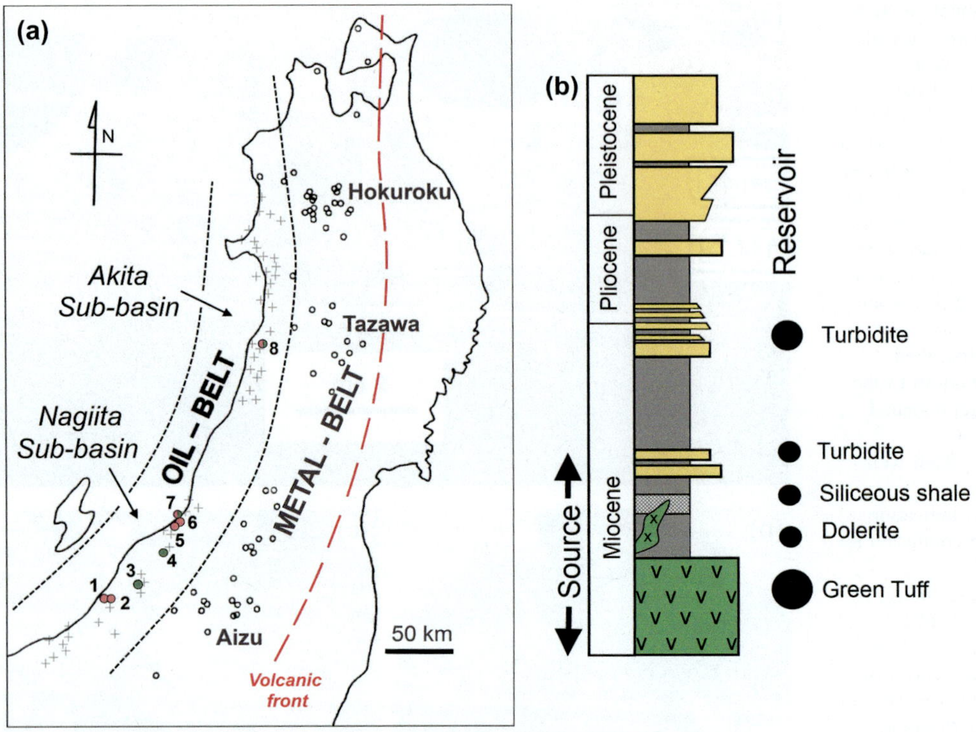

Fig. 18.102 **a** The Neogene Green Tuff Belt, Hokuroko Basin in northern Honshu, Japan, showing the paired distribution of volcanic-hosted Kuroko-type polymetallic sulphide mineralisation and petroleum fields and relation to present-day volcanic front. Small open circles indicate Kuroko-type metal deposits. Coloured circles are larger oil (green) and gas (pink)-producing fields: 1 = Yoshii-Higashi Kashiwazaki; 2 = Katakai-Minami Nagaoka; 3 = Mitsuke; 4 = Minami Aga; 5 = Higashi-Niigata; 6 = Shiunji; 7 = Iwafune-oki (offshore); 8 = Yurihara; 9 = Ayukawa; 10 = Yabase. Cross symbols show the locations of the other oil and gas fields and shows. (Modified from Glasby et al. 2004). **b** Schematic petroleum system. The volcanic reservoirs comprise submarine rhyolitic and basaltic lavas and dolerite intrusives. Relative volumes of petroleum (oil and gas) indicated by the size of black circles. Oils from both volcanic and clastic reservoirs are geochemically similar to each other and derived from normal (type II) marine kerogen (e.g. Sakata et al. 1994). (Adapted from Hoshi and Okubo 2010)

Extensive hydrothermal alteration of the volcanic reservoir rocks has been reported from several oil and gas fields, including the Katakai-Minami Nagoaka gas field (Yamada and Uchida 1997; Kawamoto 2001; Yahata et al. 2007). This resulted in the development of significant secondary porosity and enhanced poroperms. Secondary pore characteristics, the alteration mineral assemblage, and $\delta^{34}S$ data suggest that the interstitial pores formed early, whereas leached pores followed as a result of hydrothermal alteration by seawater. Temperatures of the hydrothermal fluids were estimated to have been 200–300 °C based on the presence of chlorite, illite, calcite, and barite in the host rocks. The hydrothermal activity was also considered to have enhanced the maturation of organic matter in the source rocks, with the result that the oil and gas migrated into the altered volcanic rock reservoirs immediately following the expiration of hydrothermal activity.

In the Akita sub-basin, volcanic rock reservoirs in the Yurihara oil and gas field comprise Miocene submarine basaltic lavas (Hoshi et al. 1992; Inaba 2001). Well correlation shows that they change laterally from proximal sheet flows erupted near a vent, to pillow lavas, hyaloclastites, and redeposited (epiclastic) hyaloclastite facies (Fig. 18.103b). Vesicular pillow lavas and sheet flows provide the most important primary porosity; however, highly altered basalt forms the principal reservoir which is characterised by well-developed secondary leached and fracture porosity. Four depth zones are defined by characteristic authigenic minerals (Hoshi et al. 1992): (1) smectite, (2) thomsonite–corrensite, (3) laumontite–albite, and (4) laumontite–prehnite–pumpellyite–chlorite. Sub-seafloor alteration is considered to have immediately followed basalt eruptions by the circulation of hydrothermal fluids composed mainly of seawater (see Fig. 18.80). High-temperature (150–220 °C) dehydration alteration may have also prevented porosity reduction caused by low-temperature alteration. The main oil production wells are located in the prehnite zone in areas dominated by pillow-flow basalt at some 2200 m depth. In the Ayukawa oil and gas field, sheeted dolerite dykes form shallow (1200–1800 m) oil reservoirs (Hoshi and Okubo 2010). Dolerite sheets intrude into Neogene sediments and form stratigraphic and diagenetic traps. Secondary porosities

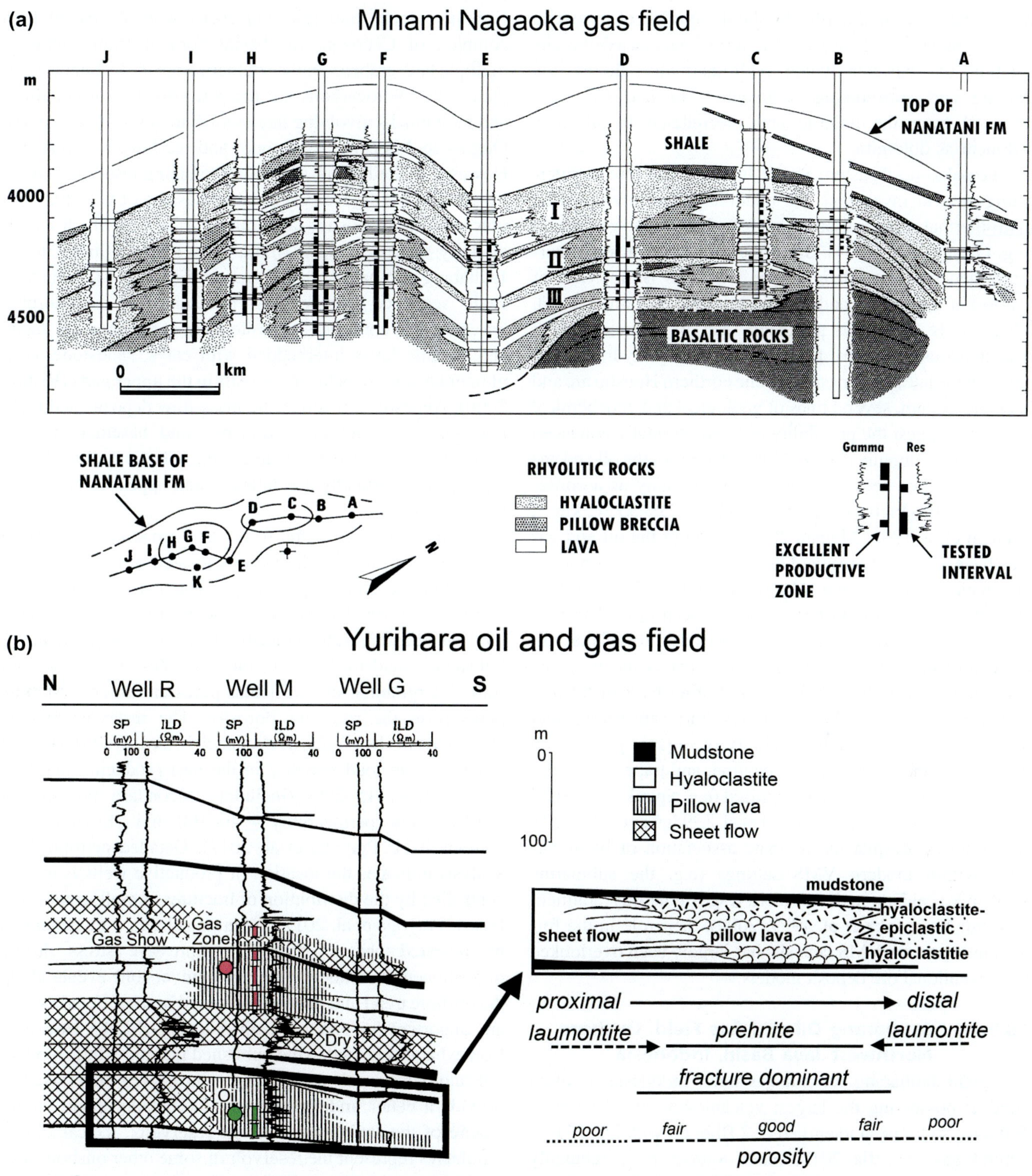

Fig. 18.103 **a** Well log and volcanic facies correlation across the Minami Nagaoka gas field, Nagiita sub-basin, northern Honshu. Rhyolite reservoirs having a total thickness of 1000 m and a gas column of more than 800 m occur at depth of 3900–4600 m below the surface. The rhyolite is subdivided into lava, pillow breccia, and hyaloclastite facies, of which pillow breccia and lava form favourable reservoirs. I = hyaloclastite zone; II = lava + pillow breccia zone (lava > pillow breccia); III = mixed zone (lava < pillow breccia). (Modified from Sato 1984, After Komatsu et al. 1984). **b** Well log and volcanic facies correlation across the shallower Yurihara oil and gas field, Akita sub-basin, northern Honshu. The primary reservoir is basalt lava, with vesicular pillow flows and more proximal-to-vent sheet flows forming the best reservoirs between about 2000–2300 m depth subsurface. Oil-producing wells are located in a pillow flow-dominant area, where highly altered basalt forms the main reservoir that is characterised by well-developed secondary leached and fracture. (Adapted from Inaba 2001)

were created in the dolerites by the dissolution of pyroxene and plagioclase, as heat from the intrusions caused the circulation of seawater which hydrothermally altered the dolerite and surrounding sediments. Mg and Ca were removed from the dolerite and precipitated in siliceous sediment as dolomite.

The paleo-northeastern Sea of Japan is therefore an interesting area to investigate the relations between submarine hydrothermal mineralisation and hydrocarbon generation (e.g. Glasby et al. 2004). Since the discovery of hydrothermally generated petroleum in the Guaymas Basin, Gulf of California (Simoneit and Lonsdale 1982; Simoneit et al. 1992), it is clear that petroleum can be formed almost instantaneously in terrestrial and submarine hydrothermal areas. In the late Miocene to Pliocene northern Honshu arc and back-arc system, several lines of geological and geochemical evidence suggest the possibility of hydrothermally enhanced maturation of organic matter. Although most of the oil and gas appears to have been generated conventionally as a consequence of the high geothermal gradients (38–55 °C km^{-1}; Kikuchi et al. 1991; Magara 2003), some of the oil and gas may have formed as a result of the extensive hydrothermal and magmatic activity (e.g. Yamada and Uchida 1997). Hydrothermal alteration has, however, been critical in promoting the secondary porosity and permeability of the volcanic reservoirs. On the other hand, what is the potential contribution of hydrocarbons in promoting base metal precipitation in similar volcanic settings? In recent years, clear relationships indicating the mobility of organic phases and their role in the transport, precipitation, and trapping of base metal sulphides have been recognised (e.g. Spirakis and Heyl 1993; Giordano 1994; Broadbent et al. 1998; Chi et al. 2017). Nevertheless, despite the intimate association of hydrocarbons within modern VMS settings (e.g. the submarine Wakamiko caldera, in the northern Kagoshima Bay, southern Kyushu; Yamanaka et al. 1999, 2000), their role in the formation of ancient VMS deposits has largely been overlooked in conventional ore deposit models.

18.9.4.4 Jatibarang Oil and Gas Field, Onshore Northwest Java Basin, Indonesia

The giant Jatibarang oil and gas field, Indonesia, is often cited as possessing the largest volcanic-hosted oil reserves that have so far been identified (1.2 Bbbl oil with 2.7 TCF of natural gas, e.g. Jia 2012; other sources have reputedly placed these as high as 4 Bbbl and 3 TCF gas). The field was discovered in 1967 and development began in 1973; it is the largest petroleum accumulation onshore in northwest Java. The northwest Java Basin is a grouping of several offshore and onshore Cenozoic basins and sub-basins and is one of Indonesia's most prolific hydrocarbon provinces (Bishop 2000; Doust and Noble 2008). The geothermal gradient is 40–50 °C km^{-1}.

Source rocks and reservoir rocks were deposited in a complex of Cenozoic rift basins formed from single or multiple half-grabens on the southern edge of the Sunda Shelf plate. The hydrocarbons are trapped primarily in anticlines and fault blocks involving late-syn-rift to post-rift (sag) phase Oligocene–Miocene sandstone and carbonate reservoirs. However, at Jatibarang, the main producing reservoirs are an Eocene–early Oligocene syn-rift succession of volcanics and volcaniclastic deposits (Jatibarang Formation) which is particularly well developed (more than 1200 m thick) in the Jatibarang sub-basin or half-graben (Fig. 18.104). The volcanic sequence overlies the pre-Cenozoic metamorphic basement and comprises andesite lavas at the base and dacite and basaltic lavas interbedded with clays, sandstone, conglomerate, and pyroclastic deposits in the upper parts (Bishop 2000). Andesitic volcaniclastic mass-flow deposits and tuffs, reworked volcaniclastic sediments, and basement-derived sediments have also been described. Clastic facies change rapidly both vertically and laterally and appear to be mostly fluvial in origin.

Depth to the top of the reservoir formation ranges from 2700–4000 m, with hydrocarbons located in both structural and lithologic–stratigraphic traps. The reservoir rocks are variously described as fractured volcanic rock, including lava (andesite–basalt), tuff, volcanic breccias, "agglomerates", tuffaceous sandstones, and conglomerates. Fractures with some intergranular and vesicular pores and inter-crystalline pores form the reservoir porosity. The reservoir rock is characterised by high heterogeneity and well-developed fractures with total porosity in the best-producing intervals ranging from 15–20% (including secondary porosity of 2–12%), and permeability $\geq$40–100 md (Ernando and Fathoni 2011; Prayoga et al. 2017). Detailed petrophysical analysis indicates the location of productive wells is mainly controlled by the distribution of fractures as well as volcanic facies (Prayoga et al. 2017). Fracture subsets were generated by repeated phases of N–S compressive stress during mid-Miocene to Pliocene collision (inversion) events along the onshore trend and have been studied in some detail as they are important for enhanced oil recovery (EOR). However, from what is reflected in the published literature, little modern volcanological analysis appears to have been undertaken to provide a better understanding of the detailed facies architecture of the Jatibarang volcanic reservoirs. Volcaniclastic lithofacies represent the reservoir in some other onshore Java fields, whereas oil and gas have also been tested within the Jatibarang Formation in several non-commercial hydrocarbon accumulations in the offshore area.

The Jatibarang Formation also contains organic-rich lacustrine shales interbedded with the volcanics. Together with thick, oil- and gas-prone coal units of the overlying Oligocene to Miocene Talang Akar Formation, these form the main source rocks in the onshore region (Fig. 18.104).

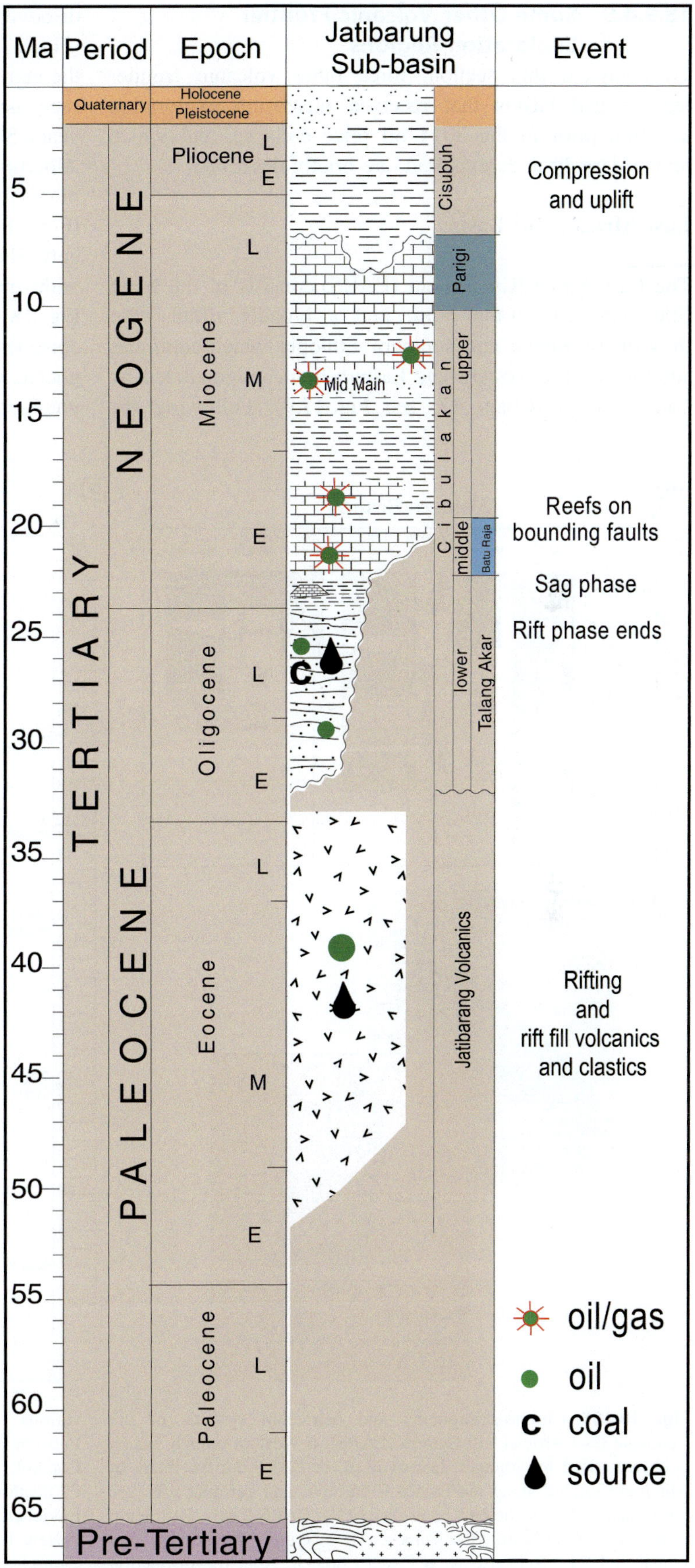

Fig. 18.104 Generalised stratigraphy and petroleum system of the Jatibarang Sub-basin, onshore northwest Java Basin, Indonesia. The Jatibarang Formation volcanic reservoirs are dated 60–34 Ma. (Adapted from Bishop 2000)

18.9.4.5 Some Other Volcanic Frontier Exploration Regions

To complete this section, three other volcanic frontier regions and basins that received increasing exploration attention prior to the 2014 oil price collapse, and which remain largely underexplored, are briefly described.

East African Rift Basins

The East Africa Rift System (EARS) consists of the Main Ethiopian Rift in the north and a broadly rifted zone in northern Kenya and southern Ethiopia which continues southward into Western and Eastern branches separated by Lake Victoria (Chap. 17; see Fig. 17.9). Following the discovery of oil in Lake Albert or Albertine in 2006 (Fig. 17.11a), the mainly amagmatic Western branch and the central broad overlap zone in Kenya and Ethiopia have seen increasing levels of hydrocarbon exploration in recent years. Since 2006, over 4 Bbbl of oil have been proven in the Albertine and South Lokichar (northern Kenya) Rifts, with more than 20 oil fields discovered in these two basins (Fig. 18.105; Davison et al. 2014). Both rift basins are largely filled by fluvial–deltaic and lacustrine clastic deposits, with more limited volcanic units (Macgregor 2015; Fig. 18.105 a, b). These however span two distinct rifting episodes: EARS 1, more evolved rifts initiated in the Oligocene with peak rifting in the early Miocene; and EARS 2, younger, less evolved rifts initiated as depressions with

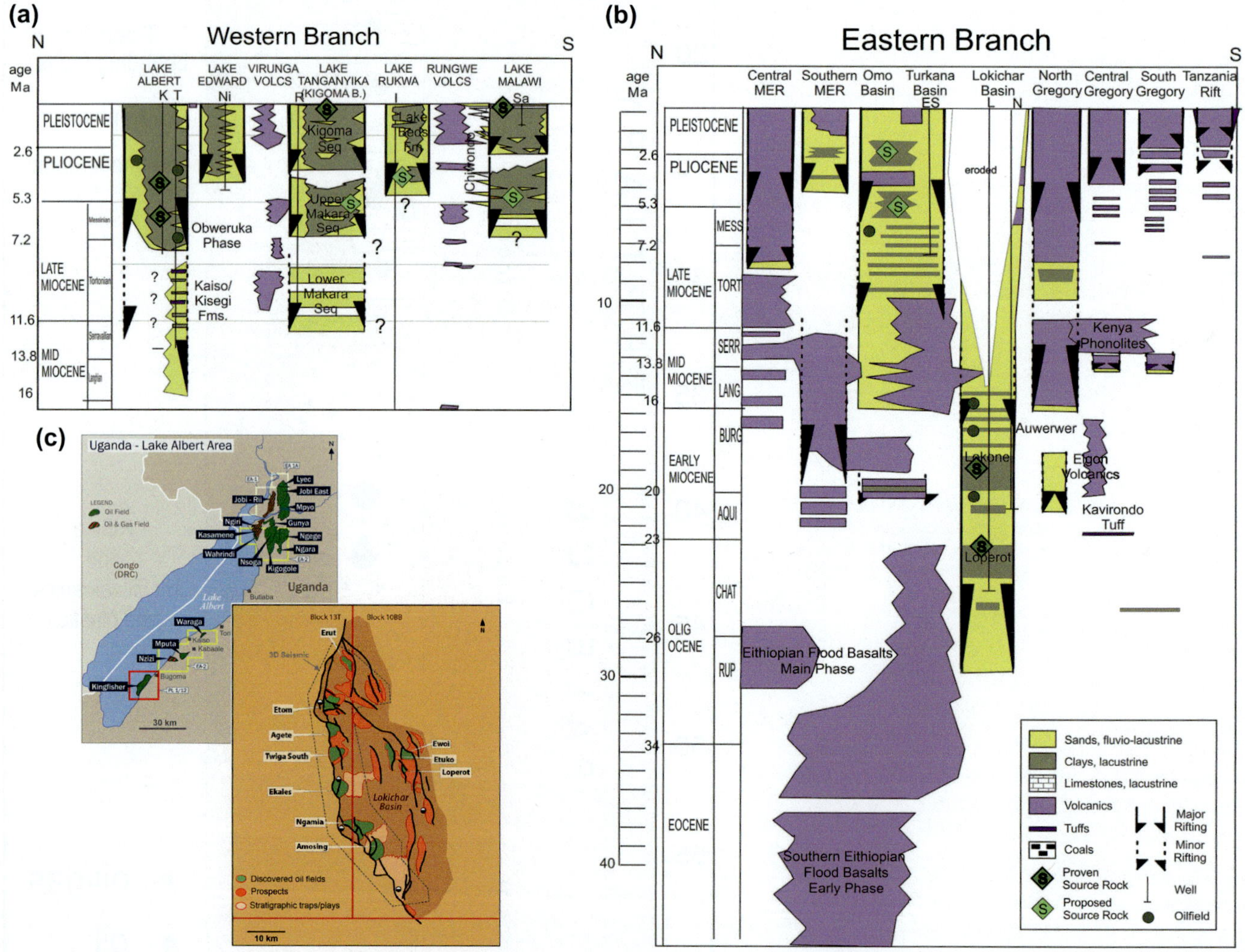

Fig. 18.105 Chrono-stratigraphy and petroleum systems of the Cenozoic East African Rift System (EARS). **a** Western branch, which is characterised by younger, less evolved, or EARS 2 rifts, many of which are located along reactivated Cretaceous and Permian rifts (see text). Lake Albert is the best controlled of the EARS 2 rifts. Petroleum wells: K = Kingfisher, T = Turaco, Ni = Nyuni, R = Ruzizi, I = Ivuna, and Sa = Syracuse A. **b** Eastern branch. Note the distinction between EARS 1 rifts initiated in Oligocene with peak rifting in Early Miocene, and EARS 2 rifts initiated as depressions with minor rifting at various times in Mid-Late Miocene with peak rifting in Pliocene-Present Day. Petroleum wells: ES = Eliye Springs, L = Loperot, and N = Ngamia. (**a**, **b** adapted from Macgregor 2015). **c** Location of recent petroleum discoveries in the Albert and Southern Lokichar rift basins. The first major discovery was in 2006 at Mputa-1, closely followed by Waraga-1 and Kingfisher-1 in the Albert Basin. (*Sources* upper map, Tullow Oil PLC; lower map, Africa Oil Corporation)

minor rifting at various times in the mid–late Miocene, and peak rifting occurring in the Pliocene–Present.

The older Eastern branch of the EARS was and presently is much hotter than the Western branch with large thicknesses of lavas and volcaniclastic deposits infilling the rifts (e.g. Gregory Rift, Chap. 17; Fig. 18.105b). Sandstone sequences are present in all the rifts, and these can reach 1 km in thickness near the main sediment entry points (often controlled transfer faults and transverse structures; Chap. 17). In the volcanic areas of the Eastern branch and the central broad overlap zone, sandstones often have a large proportion of volcanic detritus, which can be detrimental to reservoir quality. However, volcaniclastic reservoirs with 15% porosity can still produce at economic flow rates in an onshore situation.

A key factor in the prospectivity of these basins has been favourable conditions for source rock development, including a rapid rifting phase with hot, humid climatic conditions. In such conditions, lakes will be deep, fresh to slightly alkaline, and with stable stratification suitable for the preservation of organic material. Hot humid conditions also promote dense vegetation growth, and large amounts of organic material will be washed into the basin. Intense chemical weathering enhances the nutrient content of the run-off water to promote organic productivity, with some of present-day lake bottom sediments containing up to 20% total organic carbon (TOC). However, heat flow can be so high that the source rocks can be over-cooked within a few hundred m of burial (e.g. the Main Ethiopian Rift has a geothermal gradient >200 °C km^{-1} in the volcanic segments). Outside the volcanic segments, the geothermal gradient is still high and source rocks can be mature at 1–2 km burial.

Two effective petroleum systems have so far been established within the EARS system (Macgregor 2015; Fig. 18.105). The older *South Lokichar petroleum system* is thought to be charged from an early Miocene lacustrine source rock deposited during the peak of the older EARS 1 rift cycle. This affects only this and a few adjoining basins in the central overlap zone; reservoirs are also mainly of early Miocene age. The *Albertine petroleum system* is representative of the more widespread EARS 2 rift cycle, with a source rock thought to be of late Miocene–early Pliocene age, and oil-bearing reservoirs ranging from late Miocene–late Pliocene in age. Both petroleum systems seem to show evidence of an active present-day charge system.

The EARS will attract future exploration interest because it is a very large, accessible onshore basin system, which has only recently yielded large hydrocarbon discoveries (Macgregor et al. 2018). It remains highly underexplored with only two rift segments partially explored, although, most other basins, particularly in the Western branch have similar geological histories to those with proven hydrocarbons. As now recorded in other rift settings described above, especially in their deeper sections, volcanic rocks can better preserve primary porosity and develop secondary porosity and should not be ignored as a potentially significant reservoir target.

North Atlantic Igneous Province, Northwest Europe

The North Atlantic Igneous Province (NAIP) is one of the larger known Large Igneous Provinces (or LIPs; Chap. 17, see Fig. 17.23). Initial magmatic activity can be divided into two main phases: (1) commencing at $\sim$63 Ma, a pre-break-up phase of continental flood basalt volcanism and associated subvolcanic intrusions and plutonism, with magmas mainly erupted through and onto continental crust; and (2) a main, more voluminous phase of thick lava flows, large subvolcanic intrusions, and plutonic complexes that began at $\sim$56 Ma, which was coeval with the break-up of the North Atlantic and formation of a volcanic passive margin (VPM; Chap. 17). Within this broad extensional setting, there are a number of Cenozoic and older sedimentary basins, which lie landwards of an oceanic crustal domain where water depths rapidly deepen from the continental shelf edge (ca. 350 m) to 4000 m. Although these represent some of the largest basins in northwest Europe, because of the harsh environment on the eastern margin of the Atlantic Ocean and the geological evolution of the region, exploration has been slow in this high-risk, deep-water setting. From a geological perspective, the presence of thick volcanic successions was regarded very negatively. The basin sediments interdigitate with and are covered by an extensive succession of volcanic rocks, with up to 5 km of continental flood basalts, hyaloclastites, and interbedded sedimentary rocks. In particular, lavas and sills within the sedimentary section severely attenuate and disperse the seismic signal, degrading the subsurface imaging at depth which severely hindered petroleum investigations, especially in the earlier phases of exploration (the early to late 1990s).

Commercially viable oil reserves were first discovered in Faroe–Shetland Basin in 1992 (Foinavan field; 1 Bbbl of oil in place), hosted by Paleocene basin-fan sands. However, it was not until 2004 that the Rosebank discovery (500 MMbbl) successfully proved that intra-basalt, siliciclastic sandstones may form attractive hydrocarbon reservoirs in the Faroe–Shetland Basin (Duncan et al. 2009; Poppitt et al. 2018; Plate 18.1g). The discovery well encountered two oil and gas accumulations with a total net pay of $\geq$50 m (Fig. 18.106 and see Fig. 16.39); an elongate anticline forms the structural trap, approximately 20 km long and 5 km wide, with dips of up to 68° on the flanks. The two reservoirs comprise (1) intra-basalt late Paleocene–early Eocene siliciclastic sandstones (Flett Sandstones) forming the principal

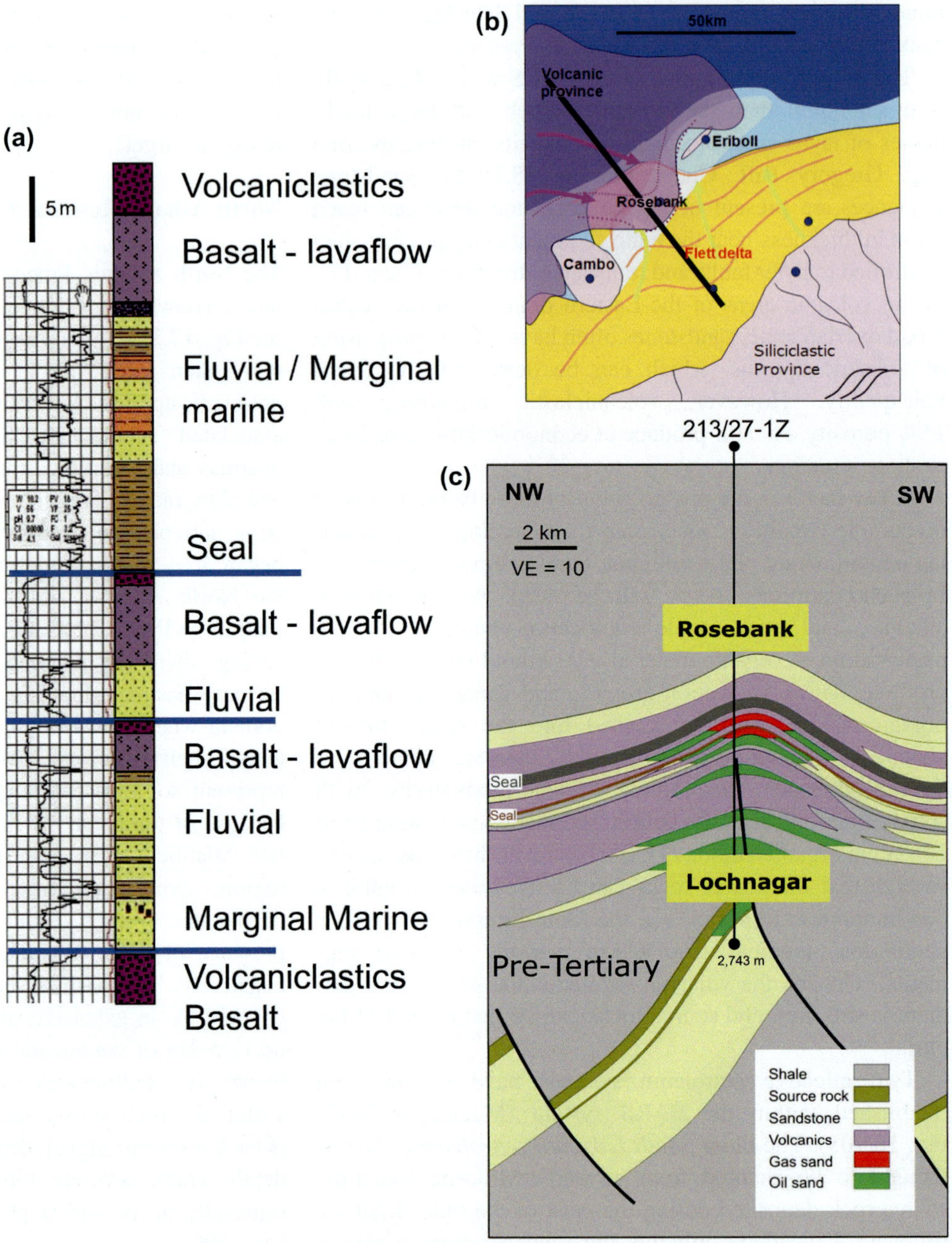

Fig. 18.106 Rosebank oil and gas field, Faroe-Shetland Basin, North Atlantic Igneous Province. **a** Litholog illustrating intra-basaltic sedimentary cycles and main reservoir units. **b** Paleogeographic reconstruction showing the interaction between easterly advancing Paleocene–early Eocene flood basalts and Flett delta system prograding to northwest; blue circles = other petroleum wells including the Cambo oil and gas field reservoired in Paleocene sandstones; thick black line = cross-section shown in Fig. 16.39. **c** Schematic cross-section over the Rosebank structure showing the petroleum system. (Adapted from Duncan et al. 2009)

Rosebank reservoirs, and (2) a deeper, sub-basalt Jurassic sandstone reservoir. Well-log and core data indicate that the intra-basalt reservoir quality is excellent, with porosities of 19–23% and very high permeabilities of 2–5 d (Poppitt et al. 2018).

The siliciclastic sandstones are locally mixed with volcaniclastic sandstones, which are often in direct contact with basalt. The high-quality reservoirs are not usually contaminated by basaltic detritus but, where it has occurred, reservoir quality can be reduced (see Fig. 18.107). At about 56 Ma, the modern Rosebank structure was located at the juxtaposition of an easterly advancing volcanic system of rift-fed lavas and volcaniclastic deposits (several km thick), and continental-derived fluvio-deltaic systems (Flett delta) prograding to the northwest (Fig. 18.106b; Fig. 16.39). A mixed succession of basalt lava flows, hyaloclastite breccias, and volcaniclastic sediments are separated by a series of intra-basaltic fluvial to marginal marine clastic sequences, which form the reservoir sandstones. Since the discovery, improved seismic imaging techniques have improved reservoir definition and enabled the separation of distinct volcanic sub-units and the visualisation of complex stacking patterns within individual volcanic layers (Poppitt et al. 2018). This was supplemented by comparative studies of the Paleogene break-up sequence onshore southeast Greenland, and the Columbia River Flood Basalt Province,

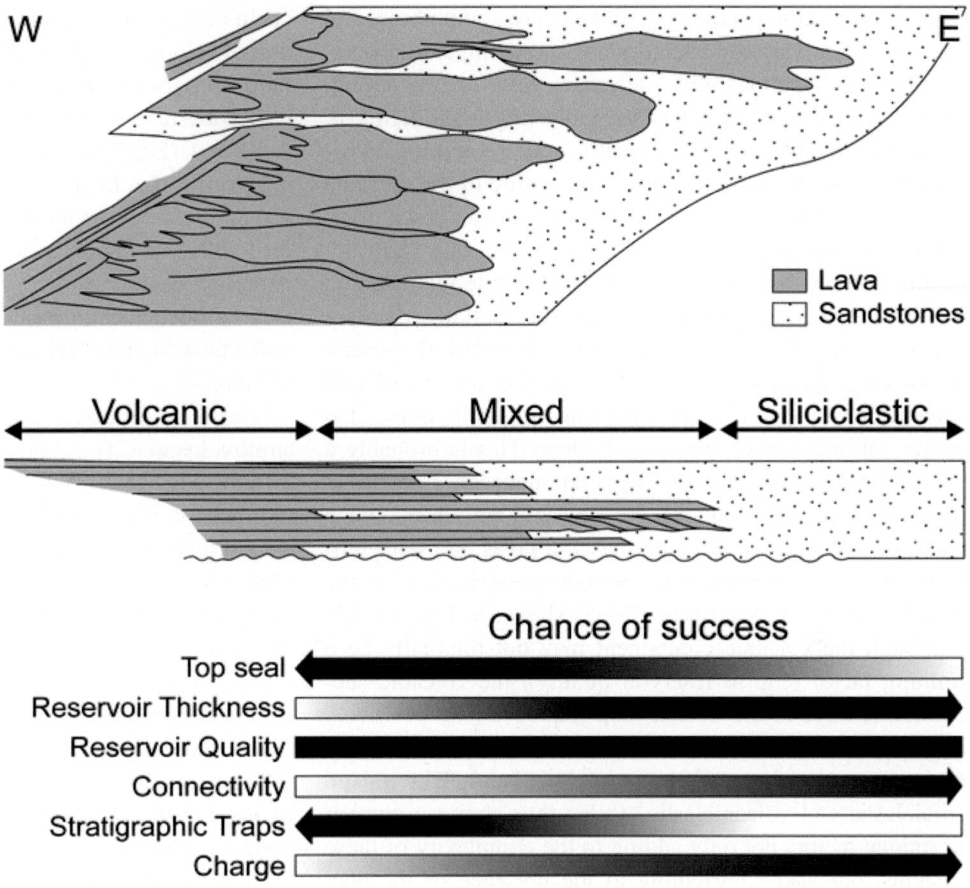

Fig. 18.107 Conceptual model of the intra-basalt petroleum play, Faroe-Shetland Basin, North Atlantic Igneous Province. Key risk factors affected by the volcanic system are illustrated by arrows showing the direction of increasing the chance of success. (From Larsen et al. 2018)

western United States (Chaps. 5, 16, and 17). These are seen as critical to improving reservoir characterisation and optimising future development of the field.

Following the Rosebank discovery, increased exploration activities were promoted along the southeastern, seawards edge of the NAIP. These mainly focused on the area where the thickness of the volcanic succession is ≤ 1000 m, to allow for seismic imaging of intra-basalt and pre-basalt pre-volcanic sequences and facilitate drilling. To understand the complexity of the intra-basaltic play also increased the demand for new exploration methodologies, advanced seismic imaging and processing, and seismic stratigraphic techniques for the interpretation of volcanic successions (e.g. Jerram et al. 2009; Cannon and Ellis 2014; Millett et al. 2015, 2016; Abdelmalak et al. 2016; Planke et al. 2017).

With these advances in our understanding of volcanic passive margins, targeted 3D seismic data over key volcanic areas has provided invaluable insights into how these and rifted margin basins develop through time. For example, the Vøring marginal basin (offshore mid-Norway) has attracted attention for a number of years with several sub-commercial gas discoveries. Recent industry-standard 3D data has been acquired over the Vøring Marginal High to image sub-basalt sedimentary rocks (Planke et al. 2017; Chap. 17). As described in Fig. 17.22, these have provided an unprecedented opportunity for imaging top- and intra-basalt structures, which, in turn, reveal new insights into understanding the seismic volcano-stratigraphic and geomorphological development of rifted margin basins (Sect. 17.2.4). These include the prominent, km-high Vøring Escarpment, which formed when eastward-flowing lava reached the coastline, and the significance of seaward dipping reflectors (SDRs). High-resolution seismic data has also been able to image the 3D nature of subvolcanic intrusions and associated hydrothermal vent complexes (e.g. Planke et al. 2005; Hansen and Cartwright 2006; Rateau et al. 2013; Schofield et al. 2017; Kjoberg et al. 2017; see Figs. 8.9, 8.13).

Despite these technical advances exploration results have so far been mixed, with further development and exploration stalled by the 2014 oil crash. As recently expressed by Levell and Bowman (2018), prospectivity along the North Atlantic margin would be deemed greater if the industry could gain confidence in reservoirs within the volcanic sequences. Analysis of intra-basalt reservoir sandstones from offshore wells and sediments onshore in East Greenland show their mineralogical compositions are highly variable and range from siliciclastic to purely volcaniclastic (e.g. Larsen et al. 2018). Volcaniclastic sandstones are composed entirely of basaltic phenocrysts (plagioclase, amphibole, and pyroxene) and variably altered glass and authigenic phases. Porosity and permeability decrease with an increasing amount of volcaniclastic detritus, indicating mafic-rich volcaniclastic sandstones are generally poor reservoirs. However, this also strongly depends on the nature of the volcanic fragments. Glass-rich (?palagonitised) volcanic rock fragments are typically completely altered and this probably results in more rapid porosity reduction, while feldspar and mafic mineral grains have the potential to result in secondary porosity depending on a sandstone's detailed diagenetic (subsidence/uplift) history. Conversely, the amount of volcaniclastic material, even in sandstone units fully embedded in the volcanic succession, may be low. This is probably a consequence of access to large amounts of siliciclastic material from areas outside the volcanic province, and a small degree of erosion of the basaltic lavas and palagonitised hyaloclastite sequences. A conceptual model for the intra-basalt "Flett" sandstone play is shown in Fig. 18.107. Although the volcanic component may not implicitly be a limiting factor (e.g. to reservoir quality), the volcanic succession relative to the overall rock sequence may positively contribute to, or even be critical to, the formation of a hydrocarbon-bearing system (e.g. seal, trap). The mixed siliciclastic and volcaniclastic system thus shows a number of unique factors not only adding to the complexity of these systems, but also contributing to the presence of the play (Larsen et al. 2018).

Rosebank remains the UK's biggest undeveloped oil and gas field, and has received final government approvals as part of it's North Sea Transition policy (to provide energy security and investment while supporting a net zero emissions target), with production expected to commence between 2026-28.

Northern Taranaki Graben, Taranaki Basin, New Zealand

The Taranaki Basin, located along the central west coast of New Zealand's North Island, is a very special frontier region for volcanic-hosted petroleum exploration. Conventional exploration has been successful along both the on- and offshore flanks of the basin, which contains all of New Zealand's producing fields. The basin is a Late Cretaceous rift containing as much as 9 km of Cretaceous–Cenozoic sediments which are locally overlain by Quaternary volcanic rocks of the onshore Taranaki Peninsula. Horst, tilted fault block, inversion features, and thrust anticlines have been the traditional targets, but the industry has shown increased interest in relatively more difficult plays involving turbiditic, volcaniclastic, and diagenetic reservoirs, as well as within a chain of more than 25 buried Miocene stratovolcanoes in the deeper, Northern Taranaki graben. Long avoided by exploration programs, such volcanic edifices were found to possess good reservoir characteristics and entrap hydrocarbons at Kora-1 (Bergman et al. 1992). Drilled in 1988, the well penetrated the flank of the Kora stratovolcano, where it serendipitously encountered hydrocarbons in volcaniclastic reservoirs, en route to a deeper Eocene sandstone target. Although deemed sub-commercial at the time because of the lack of adequate top seal, Kora-1 demonstrated significant volcaniclastic reservoir properties and a viable petroleum system.

Kora is a submarine, Miocene andesitic polygenetic stratovolcano (20–8 Ma, with a basal diameter of about 10 km; Fig. 18.108) located on the western edge of the Northern graben, which is now buried by 1–2 km of Pliocene–Pleistocene sediments. Lithological (100 m of core, plus sidewall cores) and well data from the original drilling campaign (4 wells) have been comprehensively described by Bergman et al. (1992) and Adams (1998). Miocene volcanic rocks intersected at Kora (up to 800 m) are subduction-related arc andesites and comprise a sequence composed entirely of volcaniclastic deposits. The top of the volcanic section contains benthonic foraminifera indicative of upper bathyal water depths, suggesting the volcano was entirely submarine.

The volcanic deposits were originally described as subaqueous pyroclastic flow and fall deposits, epiclastic debris flows, and channelised volcanic conglomerates (Bergman et al. 1992). Following more detailed facies analysis of the core, these have been described as interbedded tuffs, "hyaloclastite tuffs", volcanic conglomerates, and tuff breccias lacking vesiculated volcanic clasts, and the potential role of phreatomagmatic or hydrovolcanic activity in their formation was emphasised (Adams 1998; see Chap. 4). Within this scenario, it is proposed the sequence was largely generated by the collapse of primary submarine eruption columns, which fed a succession of subaqueous mass and grain flows (e.g. White 2000; Chaps. 9, 11). Although the precise volcanological origin of these facies is still equivocal, in the 50 m interval at the top of the reservoir section (1783–1830 m), from an exploration perspective, all major cored volcanic or epiclastic lithologies contained residual or free oil (Bergman et al. 1992). The so-called "epiclastic tuffs" (? crystal and lithic tuffs of Adams 1998) and channelised volcanic conglomerates form the best reservoir horizons, with porosity ranging from 20 to 26%, and permeability from 170–240 md. However, pore geometry is complex and highly variable, and reservoir distribution is vertically and laterally discontinuous, complicated by diagenesis (Bergman et al. 1992).

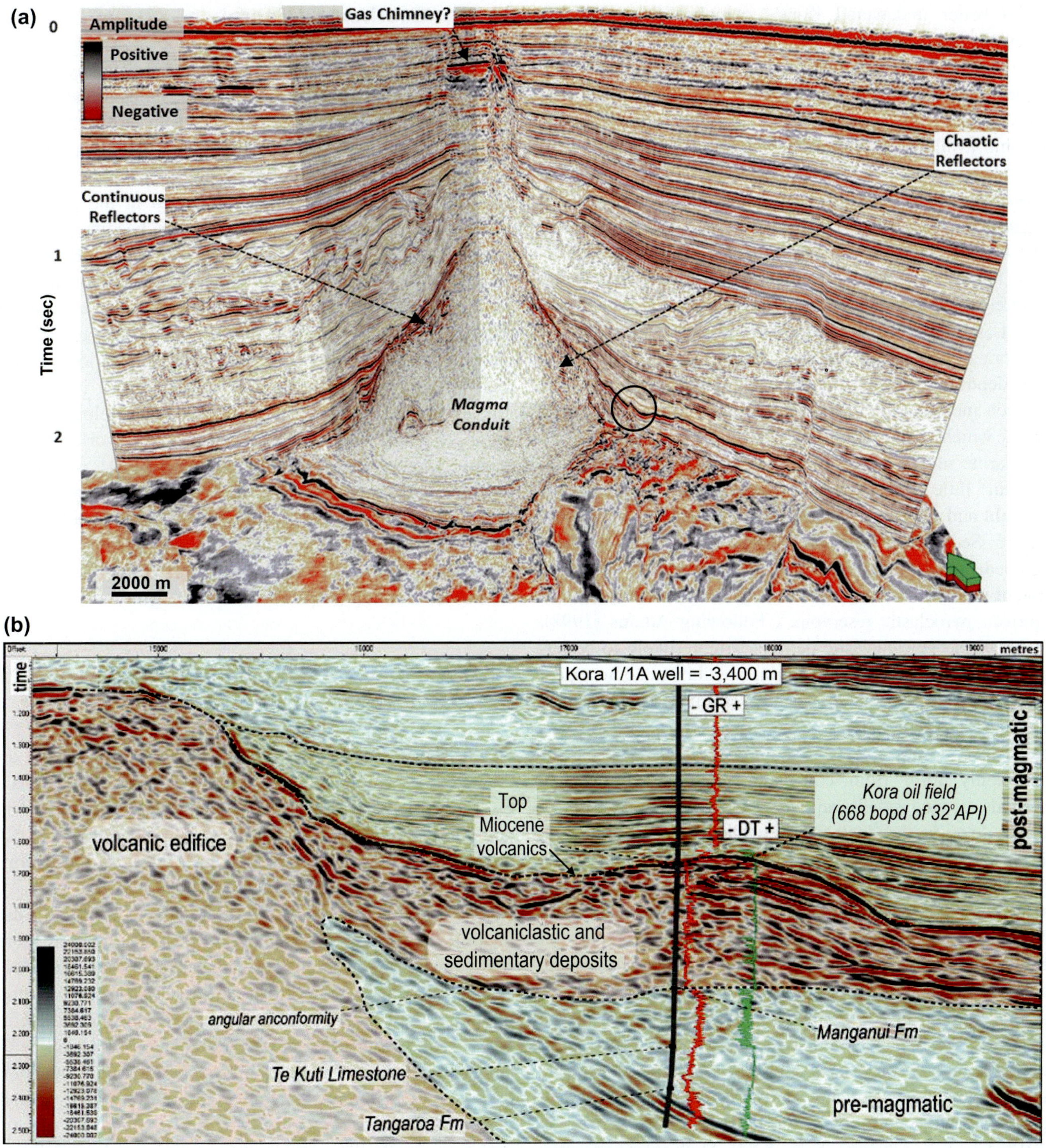

Fig. 18.108 **a** Representative 3D composite seismic data showing the Miocene Kora stratovolcano and overlying Pliocene–Pleistocene sedimentary section, Taranaki Basin, offshore west-central New Zealand. Circle is the approximate position of the Kora oil field. (From Infante-Paez and Marfurt 2017). **b** Profile illustrating seismic characteristics of different stages of volcanism and the location and setting of the sub-commercial Kora oil field. Kora-1 discovered oil near the top of the Miocene, in a sequence of andesitic volcaniclastics (at 1783 m b.s.l.). This succession is described as principally composed of primary pyroclastic flow deposits. However, considering the wholly submarine nature of Kora, these are more likely to be dominated by the products of hydrovolcanic activity and quench fragmented detritus (including hyaloclastite), and redeposited debris (see text and Chap. 4). Kora-1 well: red = gamma ray, green = density. The Miocene volcanic reservoir section was tested by the sidetrack Kora-1A; a long-term production test flowed 668 bbl/day of light crude for 254 h. Petroleum source rocks are located in the subvolcanic early Miocene Manganui Formation and possibly underlying Eocene mudstones at the base of the well. (Modified from Bischoff 2018)

To better understand Kora's complexities and better access its petroleum potential, modern 3D seismic data was acquired in 2006 and has provided a somewhat unique opportunity to image a buried larger volcanic edifice seismically. Recently some of these spectacular profiles have become available in the published literature (Infante-Paez and Marfurt 2017; Bischoff et al. 2017; Fig. 18.108). The two studies focus on the geomorphology of Kora, its internal seismic and stratigraphic architecture, and building predictive models. Infante-Paez and Marfurt (2017) have seismically mapped two distinct facies, pyroclastic and lava flows. Referencing Bergman et al. (1992) and well-completion reports, they indicate the main reservoir rocks, as intersected in Kora-1, are a thick sequence (>1000 m) of "pyroclastic flow deposits", which is not the case. These clearly represent a much more complex sequence of andesitic volcaniclastic facies which, considering the wholly intermediate to deep submarine setting and lack of an obvious caldera, probably contain little truly, pyroclastically fragmented detritus (Wright and Gamble 1999; Wright et al. 2006; Chaps. 4, 9, 11, 16; Sect. 18.8.7). Bischoff et al. (2017) provide a more detailed picture of the stratigraphic architecture and evolution of the Kora volcano, nevertheless, they similarly refer to "proven pyroclastic reservoirs". Following Adams (1998), much of the volcaniclastic detritus is more likely to be largely quench fragmented in origin, possibly to some extent formed within subaqueous eruption columns, or as hyaloclastites and their redeposited debris. Indeed, modern volcaniclastic rocks on submarine arc volcanoes commonly form from collapsing proximal pillow lava mounds, around small eruptive vents, or larger sector collapses (e.g. in the southern Kermadec arc; Wright et al. 1996, 2006; Chap. 16).

The thick, layered extrusive volcanic and volcaniclastic sequences (perhaps $\geq 10{,}000$ km^3) within the Northern Taranaki graben remain a mostly unexplored region of frontier exploration interest. Entombed volcanic edifices along the graben's axis are more deeply buried by Pliocene sediments than at Kora, and sealing may be improved. In addition, subvolcanic intrusions and feeder pipes have uplifted underlying Eocene and early Miocene reservoirs into large domal structures, drag folds, and upturned pinch-outs (Bischoff et al. 2017; Bischoff 2018; Morley 2018; Chap. 8, see Figs. 8.26, 8.27). Such intrusions may have also induced complex patterns of diagenetic alteration in sandstones proximal to igneous centres, creating possible diagenetic traps for hydrocarbons in more porous reservoirs down dip. In the overlying post-magmatic sequence, potential reservoirs are also present as a result of erosion and burial of the Kora volcano, e.g. submarine channels and fans, and debris avalanche aprons enclosing the edifice.

18.10 Geothermal Energy

Previously in Sect. 18.3.1 we have described many aspects of the geology of modern geothermal systems in the context of epithermal mineralisation (e.g. volcanic setting, typology and classification, hydrology, fluid chemistry, hydrothermal alteration assemblages, and evolution). The application of principles learned and models developed in the course of exploration for geothermal resources have become powerful tools for guiding mineral exploration, especially for precious metals in active and ancient volcanic arcs. Nonetheless, geothermal energy resources in themselves are becoming increasingly of economic interest as they are among the renewable energy resources that contribute to low-carbon policies aimed at reducing the anthropogenic impact on global climate change (https://sustainabledevelopment.un.org/topics/energy). Crude assessment of the world's continental geothermal resources suggests that potentially exploitable magmatic systems have the capacity to produce 15×10^{24} J that is equivalent to 2,400,000 Bbbl of oil equivalent, several orders or magnitude greater than the estimated global oil reserves (data from Geothermal Resources Council (https://geothermal.org)!

18.10.1 Types of Geothermal Resources and Scope

Geothermal resources are commonly subdivided, based on their average temperature, as low (<90 °C), medium (<150 °C), and high (>150 °C) enthalpy resources, even though this definition is rather incorrect as the concept of enthalpy is not only associated with the temperature, but with the energy stored in the system and is expressed in Joules [kg m^2 s^{-2}]. *Conventional* geothermal resources are associated with the identification of reservoirs made of volumes of permeable rocks hosting hydrothermal to supercritical fluids and covered by impermeable cap rocks that seal the reservoirs and limit the dissipation of heat towards the atmosphere. *Unconventional* geothermal resources are associated with hot dry rocks where permeability stimulation and injection of fluids are essential to recover the thermal energy, and these systems are commonly known as *Enhanced Geothermal Systems* (EGS), and together with supercritical fluids and magma, they are known as *Super-Hot Geothermal Systems* (SHGS).

High and medium enthalpy geothermal resources are used for electric power generation, whereas low enthalpy resources are best suited for direct uses of heat, as for example urban district heating. Depending on the local geothermal gradient, high and medium enthalpy resources can be found either at great depths or in tectonically and magmatic active regions, where the heat flow can be several times higher than normal (50–60 mW m^{-2}). Current drilling and extraction technologies limit the economic interest for geothermal resources to the first 4–5 km from the surface, and shallower if possible. For these reasons, the best sites for electric power generation from geothermal energy are active or very recent magmatic systems, where the largest part of the current production at the global scale is located. These are therefore located in active geodynamic areas where volcanoes are common, such as active rift zones and at convergent arcs (Chap. 17), which conform to the two end-member types of basinal (silicic) or stratovolcano (andesitic) systems (Fig. 18.2). Commercial producing systems are all dominated by meteoric waters, that is, evolved neutral-chloride fluids found in basinal (rift or silicic caldera) or mature stratovolcano settings which develop over time as meteoric recharged waters are heated and convect (Sect. 18.3.1).

A detailed description of exploration methods for geothermal resources is well beyond the scope of this book and requires the integration of stratigraphic, structural, petrologic, geochemical, and geophysical approaches. Herein, we therefore focus on some key aspects of volcanological studies applicable to geothermal energy exploration, partly because these aspects are often undervalued in industrial projects where engineering and geotechnical approaches can dominate development. Based on our experience in the field, we highlight how volcanological studies can impact on the following:

- identification, geometry, depth, temperature, volume, age, and recharge history of the magmatic source(s) and their impact on the modelling of the heat flow entering at the base of geothermal reservoirs,
- 3D architecture of volcanic and subvolcanic rock bodies and their hydrothermal alteration, including the role of volcano-tectonic structures in the development and location of secondary porosity/permeability, and
- development of conceptual geological–volcanological models of the evolution of geothermal fields over time, and hence assessment of their potential productivity and sustainability.

For further aspects of the geochemistry of hydrothermal fluids and alteration assemblages, surface features, and fluid flow conditions, see Sect. 18.3.1 (and references therein). The application of geophysical techniques is also a fundamental aspect of geothermal exploration, which should be continuously integrated with the field volcanological studies to improve and develop the best model of any geothermal field. For a more comprehensive review of exploration methods, the reader is referred to Wohletz and Heiken (1992), IGA Service GmbH (2014), Cumming (2016), and Di Pippo (2016).

18.10.2 Magmatic Heat Sources

An empirical approach to estimate the first-order potential of magmatic heat sources takes into account the integrated volume and age of magma chambers as the two critical factors that define active versus fossil (barren) geothermal systems (Fig. 18.109). The greatest potential is intuitively related to the active and large volume of magmas emplaced at shallow crustal levels, and as cooling rates are more efficient for small bodies, larger ones can consequently project their thermal anomaly for much longer times (Smith and Shaw 1978). This kind of approach is of course rather simplistic as it does not explicitly include the notion of magma chamber depth, magma composition(s) and thermal state, the stratigraphy and structure of the country rocks, regional geothermal gradient, longevity and recharge of the magma chamber, its possible complex geometrical configuration, and so forth. Nevertheless, Fig. 18.109 conveys a straightforward concept that is important to the definition of the magma heat source to understand the potential of the geothermal resource (Muffler and Cataldi 1978).

To detail many of the above characteristics of the magma heat source, there are several volcanological studies that can be performed, which we briefly analyse below.

18.10.2.1 Crustal Magma Reservoirs

The presence of upper crustal magma reservoirs is crucial for the existence of a high to medium enthalpy exploitable geothermal resource. This is because it is the persistence over time of a shallow heat source that allows the upward migration of isotherms and the local modification of the regional geothermal gradient. The most basic requirement is the presence of volcanic products that are silicic in composition, either formed by crystal fractionation and/or anatexis. In both cases, silicic chemical magma compositions indicate that time was necessary for magma formation by crustal differentiation and/or crustal assimilation (see Chap. 2), hence, for magma to be stored and/or produced at crustal levels.

The next step is to assess the volume of the magma plumbing system and its longevity. This information requires more detailed stratigraphic investigations and a complex petrological approach. This commonly requires analysis of a suite of differentiated volcanic products (e.g. at calc-alkaline arc volcanoes), the mapping of volcanic units, the detailed definition of their different chemical compositions, associated geothermobarometry, and the

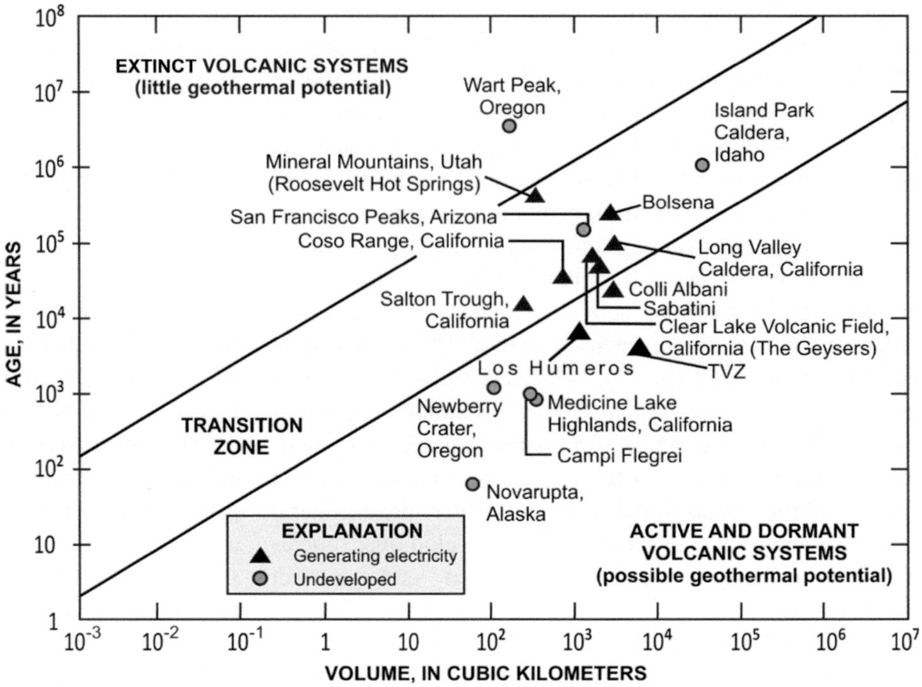

Fig. 18.109 The first-order geothermal potential of magmatic systems has traditionally been approached by comparing the age and volume of magma reservoirs, the larger and the younger being clearly more perspective. Here some of the best-known geothermal fields are plotted based on the average age and volume of the related magmatic system. (After Carrasco-Núñez et al. 2018)

reconstruction of their volume and age. These are essential to compute the amount of parent basic magma intruded necessary to generate the erupted differentiated products, as well as identifying the complex geometry of the plumbing system (Giordano and Caricchi 2022).

Quantitative assessment of the amount of mass fractionated during the process of magma differentiation can be retrieved from compositional analyses, and in particular by studying the partition behaviour of chemical elements during crystallisation. The general case for fractionation is

$$C_r = C_{ri} \times F^{(D-1)} \quad (18.1)$$

where C_r is the concentration of an element in the residual melt (to be estimated for example with microprobe analyses on the differentiated product), C_{ri} is the concentration of that element in the primitive melt (which can be obtained from the geochemical analysis of the most primitive products erupted), D is the partition coefficient between liquid and crystal(s), and F is the fraction crystal/melt in the investigated process of differentiation.

If we consider the chemical elements with a strong affinity for the melt phase (e.g. U, Th, REE), D is <<1, then the ratio C_r/C_{ri} gives the proportion of crystals fractionated. In order to translate this proportion into the volume of intruded and crystallised magma, we need to calculate the Dense Rock Equivalent (DRE) volume of the erupted differentiated products. The DRE volume is equal to the volume of the reconstructed deposits for lavas, while it should be corrected for porosity and lithic content for pyroclastic deposits, as well as other sources of uncertainties (see discussion in Giordano and Cas 2021 and in Chap. 10). There are more accurate ways to assess the fractionated volumes by mass balance calculations, and these approaches can significantly help the understanding of the magma heat sources (e.g. Ferriz and Mahood 1984). However, these calculations must take into account that differentiation does not occur in just one magma chamber, but across a significant depth interval that is the extension of the transcrustal magmatic plumbing system of the volcano. So the coupling of mass balance calculations with geothermobarometry is essential to reconstruct the geometry of the plumbing system and identify the sources of the shallow heat anomaly (e.g. Giordano and Caricchi 2022; Smithies et al. 2023).

18.10.2.2 Intrusion Rate

Once the total volume of the plumbing complex is reasonably estimated, the next question to consider is the rate at which it was intruded, as the rate of intrusion determines the thermal impact on the country's crustal rocks. For example, based on numerical simulations, Annen (2009) proposed that large volume active magma chambers can only be formed at average magma recharge rates exceeding 10^{-2} km^3 yr^{-1}. At lower average magma recharge rates, the primitive batches of magma that progressively reach crustal levels may be too small and time-spaced, so that they cool and never coalesce, inducing a much lower thermal impact on the regional geothermal gradient. In addition, at constant intrusion rates, the thermal and chemical reactivity of the crust changes over time, so that early magma intrusions likely freeze at depth, whereas with time the crust progressively heats up and is modified by intrusive complexes that allow much longer residence of melts. The rate of intrusion is therefore a critical parameter that can be evaluated only by accurately assessing the geochronology of the eruption sequences (e.g. Carrasco-Núñez et al. 2018). In this way, it is possible to combine erupted volumes, geochemistry, and the evaluation of fractionation processes to infer the intrusion versus eruption rates.

Recent developments in single crystal radiometric age determinations now also allow more refined histories of the pulses of magma assembly to be determined (e.g. as has been possible by the combined use of U–Th and/or U–Pb ages of magmatic zircons, that indicate the beginning and evolution of crystallisation history, and $^{40}Ar/Ar^{39}$ ages of feldspars). Although conventional views and data generally suggest large magma reservoirs form over protracted periods (e.g. Bachmann et al. 2007; Lipman 2007; Costa 2008; Deering et al. 2016; Lucci et al. 2018), there are several recent studies that contradict this tenet. These show that the assembly of even large volumes of magma in the shallow crust can be very rapid, of the order of a few millennia, centuries, or even decades (Reid and Coath 2000; Wark et al. 2007; Reid 2008; Allan et al. 2013; Cooper and Kent 2014; Carrasco-Núñez et al. 2018; Pamukçu et al. 2020). These also suggest that the thermal flux may be pulsatory over similar time scales (Di Renzo et al. 2016; Deb et al. 2022).

18.10.2.3 Heat Flux

How the heat source determines the heat flux at geothermal systems strongly depends on the thermal state, geometry, and depth(s) of the magma reservoir(s) (e.g. Di Renzo et al. 2016; Deb et al. 2022). The thermal state can be broadly assigned based on the chemical composition and crystal content of erupted products at temperatures comprised between the liquidus (for aphyric products) and the solidus (at high crystal cargos). More precise determinations should be made, if possible, by the application of crystal–crystal (e.g. Fe–Ti oxides, Andersen et al. 1991) or crystal–glass geothermometers (e.g. clinopyroxene–glass, Putirka et al. 2003; plagioclase–glass, Putirka 2005), also taking into account the H_2O contents.

The average depth of the magma reservoir can be determined by geobarometric studies on the crystal cargo of erupted products, commonly done on crystal solid solutions, such as clinopyroxene and plagioclase (Putirka 2008), or from melt inclusion analyses (Lowenstern 2003). The availability of large mineral geobarometric datasets now also allows the use of accurate machine learning approaches (Higgins et al. 2022; Jorgenson et al. 2022). Recent views of the geometry of plumbing systems, grounded on an increasing number of volcanological and petrologic data, are profoundly modifying the classic conceptual model of magma chambers. Rather than a single, "balloon-shaped", magma chamber, they are perceived as a more complex multi-storey, transcrustal plexus of melt lenses and pockets sitting in larger crystal mush domains (e.g. Cashman and Giordano 2014; Cashman et al. 2017; Giordano and Caricchi 2022; Chaps. 8, 16). It is likely that rather than a single average depth, a range of depths can be retrieved from geobarometric studies, which can provide a way to reconstruct the architecture of the plumbing system, by also coupling geothermometers and geochemistry of erupted products. Once volume(s), depth(s), temperature(s), and age(s) of the magma reservoir(s) have been determined, their lateral extension can be inferred, for example in caldera systems, by identifying volcano-tectonic ring faults, where the shape and main axes of the structural caldera rim can be safely used as a first-order approximation of the top view of the minimum size of the main magma reservoir (Chap. 16). More generally, the inference from the plan view geometry of the plumbing system at stratovolcanoes, although the smaller overall dimensions of the system, is that the horizontal extension of the magma reservoirs rarely exceeds a few km across. However, it must be taken into account that the lithology and structure of the shallow crust may significantly affect the distribution of geothermal fluids above a magmatic heat source and as a consequence the shape of the heat flux, as lateral advection may occur along preferential pathways controlled by primary and secondary permeability (e.g. Giordano et al. 2013, 2014; Tescione et al. 2021).

Case Study: Los Humeros Caldera, Mexico

An excellent case study, where the configuration of the magmatic plumbing system has been reconstructed in detail to define the heat sources of the local geothermal system, is at the Los Humeros volcano, Mexico (Fig. 18.110a). Los Humeros is an active caldera system, whose evolution has been subdivided into three main stages by Carrasco-Núñez et al. (2018; also see Fig. 16.63b). An initial, long period of pre-caldera volcanism (ca. 700–164 ka) shows evidence of an extended building phase leading eventually to the establishment of the large-volume rhyolitic reservoir. This fed the Xaltipan rhyolitic ignimbrite eruption at 164 ka but was likely in existence for only a very restricted period of time (as short as 5 ka in comparison with U/Th ages of zircons; Carrasco-Núñez et al. 2018).

The Caldera stage includes two major caldera collapse events, the first occurred at 164.0 ± 4.2 ka (Carrasco-Núñez et al. 2018; Fig. 16.63b) with the eruption of the 115 km^3 (DRE) Xaltipan ignimbrite and the formation of ca. 18 km across Los Humeros caldera. The second occurred at 69 ± 16 ka with the eruption of the 15 km^3 (DRE), chemically zoned Zaragoza ignimbrite and the formation of the nested and smaller (9 km in diameter) Los Potreros caldera (Carrasco-Núñez et al. 2018).

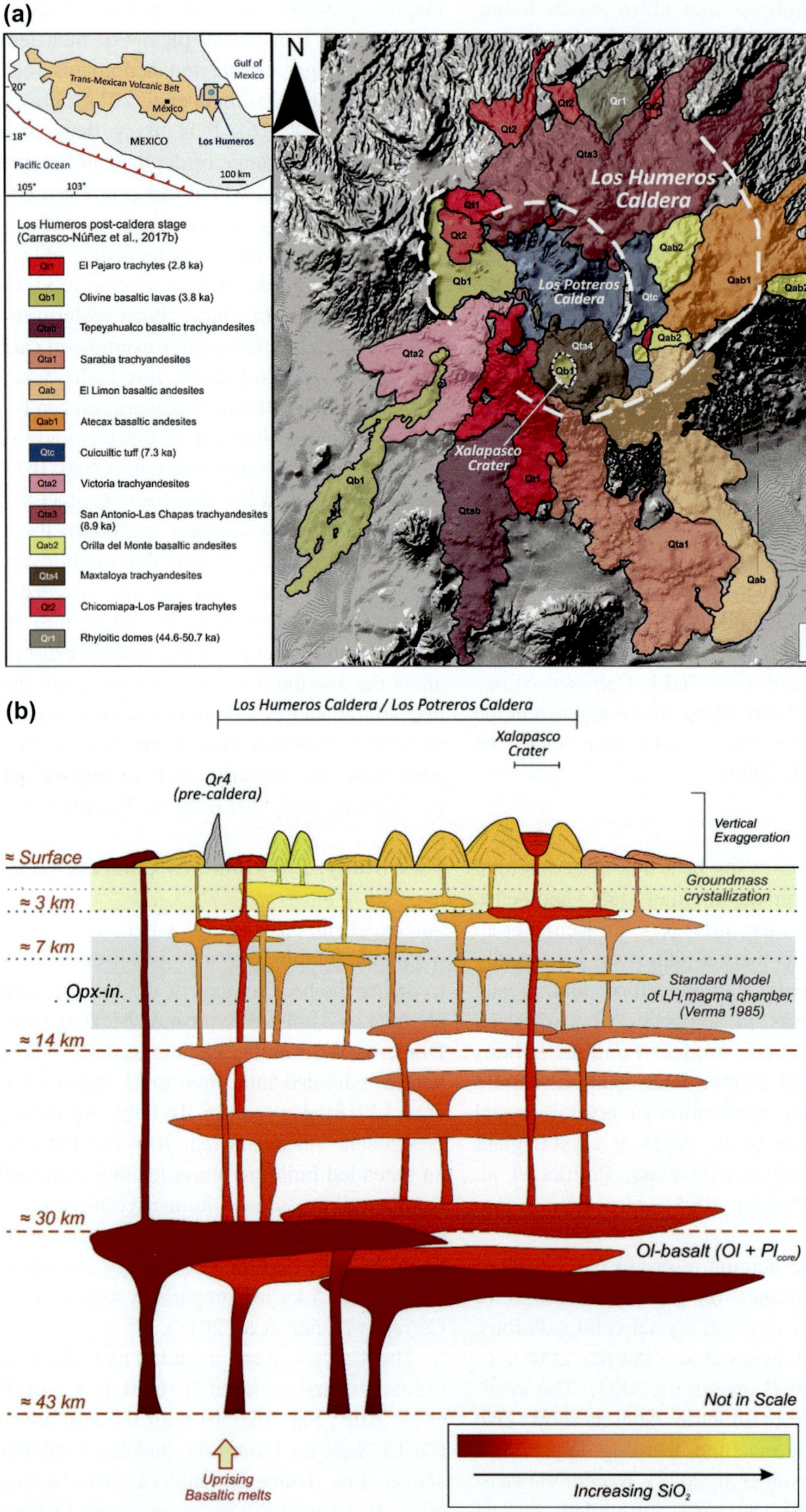

Fig. 18.110 Los Humeros caldera, central Mexico. **a** Simplified geology of post-caldera volcanism. **b** Reconstruction of the configuration of the post-caldera plumbing system (<69 ka) from geothermobarometric data. This indicates that even if a large rhyolitic reservoir was in existence at the time of the caldera formation ca. 164–69 ka, the present heat source is instead dominated by small batches of magma intruded at different depths, each undergoing differentiation independently. The main consequence of such a type of plumbing system is that the heat source cannot be modelled as a single large volume "entity" or "event". (From Lucci et al. 2020)

The Post-caldera stage (<69 ka) has been subdivided into a resurgent phase and a Holocene phase characterised by intra-caldera, bimodal (basaltic to trachyte and rhyolite), and monogenetic volcanism (Fig. 18.110a), which suggests that the configuration of the plumbing system at present is not characterised by the occurrence of a large and homogenised magma reservoir. The integration of geothermic–barometric data (Lucci et al. 2020) and the eruption location, composition, volume, and age of post-caldera volcanic products also support this inference. Therefore, at Los Humeros, the large magma rhyolitic reservoir that was in existence during the Caldera stage modelled at ca. 5 km depth (Ferriz and Mahood 1984; Verma 1985; Verma et al. 2011) is not anymore in existence and does not represent a rheological boundary to the direct uprise of basaltic magmas from deep mantle sources (Fig. 18.110b). By contrast, such basalts have either directly erupted during Holocene times within the caldera area or fed the establishment of shallow, small volume reservoirs (of the order of a few km^3, based on erupted volumes), at various depths (as shallow as <3 km deep). Each of these evolves independently and is able to feed a variety of small-volume (rarely >1 km^3) effusive and explosive eruptions from basaltic andesite to trachytic and rhyolitic in composition (Fig. 18.110b; Lucci et al. 2020).

The implications of such detailed volcanological studies are fundamental for the modelling of the heat source of the Los Humeros geothermal system and the assessment of its medium- to long-term sustainability. For example, the heat source at Los Humeros has conventionally been considered a single, giant magma chamber modelled as the one that fed the caldera-forming Xaltipan ignimbrite (e.g. Ferriz and Mahood 1984; Verma et al. 2011). However such a model failed to accurately evaluate the heat flow inside the caldera leading to overestimation of the geothermal potential. The postcaldera volcanism indicates instead that after the last caldera collapse the magma chamber has not been reestablished, and been substituted by a complex array of smaller and shallower intrusions as heat sources (Lucci et al. 2020; Urbani et al. 2020). These are able to much better explain the complex, laterally highly variable local geothermal gradient (Arellano et al. 2003; Deb et al. 2022), suggesting that the reconstruction of the plumbing system and its evolution over time through phases of magma assembly and magma withdrawal are essential to define the geothermal potential of volcanic systems (e.g. Di Renzo et al. 2016; Giordano and Caricchi 2022).

18.10.2.4 Geophysical Studies

Geophysical studies (e.g. magnetotelluric, seismic, gravity, aeromagnetic) can, on the one hand, greatly benefit from the volcanological–petrological reconstruction of the magma plumbing system, and, on the other side, can provide very important geometrical and physical property constraints to that reconstruction. For these reasons, it is essential that conceptual models for magmatic heat sources are developed by integrating volcanological, petrological, geochronological, and geophysical data, rather than by performing sectorial and non-interrelated studies (see Giordano and Caricchi 2022). The discussion of geophysical methods applied to the reconstruction of magmatic plumbing systems deserves a book on its own and is well beyond the scope of this one, but a comprehensive review of those methods can be found in Magee et al. (2018).

The reconstruction of the present configuration and past evolution of the magmatic heat source can be translated into and tested by the numerical modelling of heat transfer to the country's crustal rocks. Conductive–convective models of heat transfer have classically represented magma sources as single bodies (i.e. magma chambers), instantaneously installed in the crust (Fig. 18.111), possibly chemically and thermally stratified (e.g. Verma et al. 1990), and possibly recharged over time (e.g. Wohletz et al. 1999). More advanced approaches to numerical modeling of the heat flux that take into account the evolution over time of the entire transcrustal magma plumbing system (or at least of its shallow portions) and that include the geophysical-petrological-volcanological reconstruction of the changes in geometry of magma storage regions along with the alternation of periods of magma recharge, transfer and withdrawal are still in their beginnings (Di Renzo et al. 2016; Deb et al. 2022). However, as just discussed for the Los Humeros case study, such more complex approaches not only better reflect our current understanding of how magma plumbing systems work (Chaps 8 and 16), but may result in much more accurate evaluations of the geothermal potential, which are very relevant for correctly plan profitable and sustainable industrial exploitation of the geothermal resources.

Case Study: Campi Flegrei, Italy

One of the few attempts we know in the literature, besides that just described for Los Humeros, of translating the volcano eruption stratigraphy, petrology and geochronology, and geophysical constraints into a thermal model for the heat source of a geothermal system is at Campi Flegrei, southern Italy. Here, Di Renzo et al. (2016; also see Fig. 16.63c, d) segmented the eruption history into four phases, each associated with a changing configuration of the plumbing system:

(a) During the first phase, estimated by U–Th data at 44–39 ka, the main trachyte magma chamber was built at 6 km depth. This was simulated by 5 inputs of 50 km^3 of magma at 950 °C every 1 kyr in a geometry

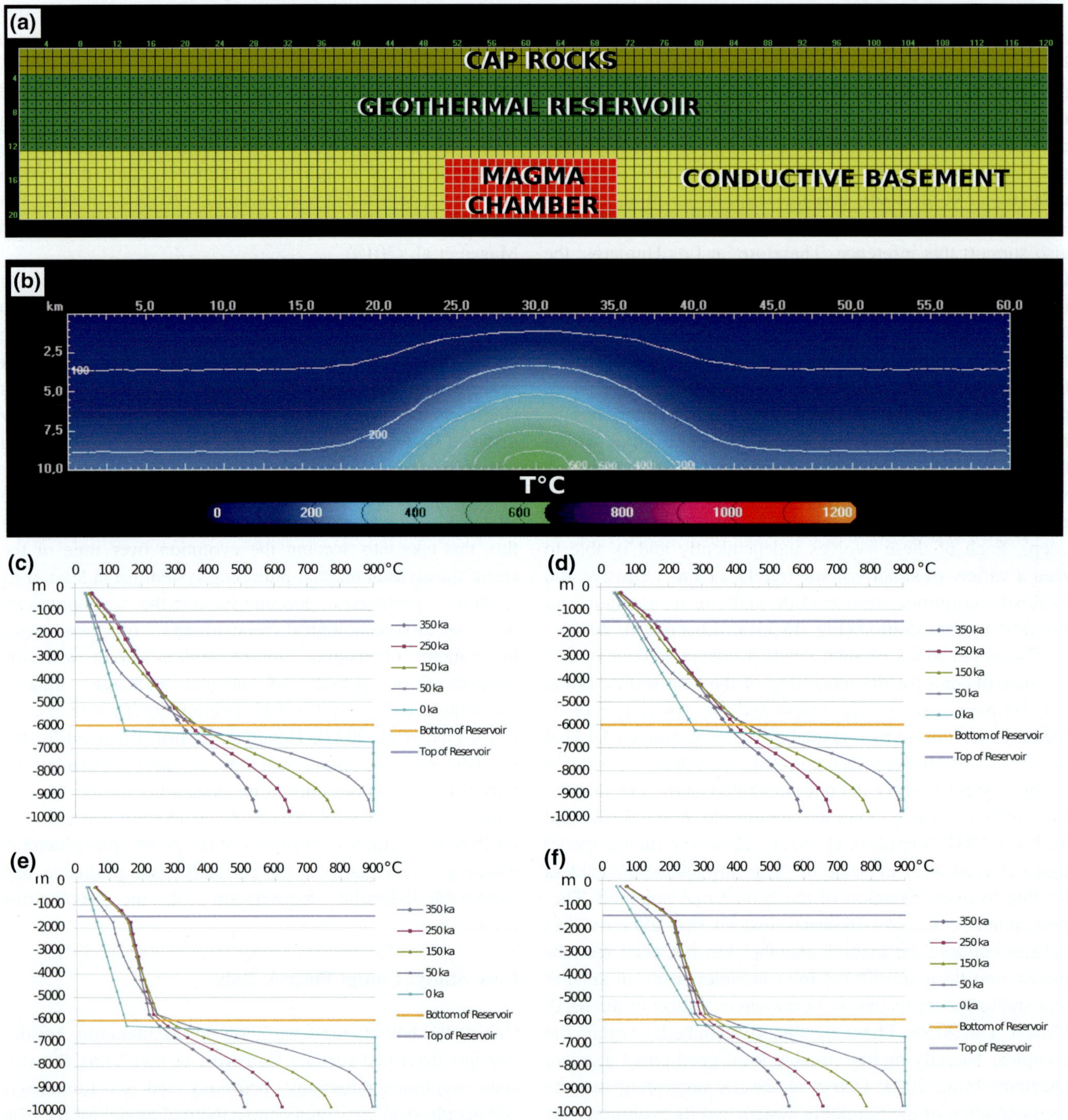

Fig. 18.111 Heat 3D numerical simulation of the caldera-related geothermal systems of Roman Geothermal Province, Latium region, central Italy. **a** Computational domain reconstructed by volcanological, petrological, structural, and geophysical data: yellow = metamorphic basement; green = geothermal reservoir (carbonate rocks); brown = cap rocks (clayey rocks); red = magma chamber of intermediate composition; the numbers indicate the 500 m × 500 m cell position. **b** 2D configuration of isotherms generated by the instantaneous instalment of the magma chamber within a regional geothermal gradient of 20 °C km^{-1} and its conductive thermal relaxation (cooling) for 250 ka, that is the time from the last period of significant output of the Colli Albani volcano (the last eruption, instead, occurred at <23 ka). **c** Time evolution of modelled conductive geothermal gradients above the centre of the magma chamber, in the case of an original regional gradient of 20 °C km^{-1}. **d** Same as previously in the case of an original regional gradient of 40 °C km^{-1}. **e** Time evolution of modelled geothermal gradients above the centre of the magma chamber, in the case of an original regional gradient of 20 °C km^{-1} with convection within the geothermal reservoir. **f** Same as previously in the case of an original regional gradient of 40 °C km^{-1}. (From Baffioni 2017)

determined by the horizontal dimensions of the Campi Flegrei caldera (16 km × 16 km), resulting in a thickness of each input of 0.2 km.

(b) During the second phase, the 39 ka, caldera-forming Campanian ignimbrite eruption, evaluated at 300 km^3 DRE, was simulated as the total emptying of the magma chamber built up previously during the first phase, and its substitution for the following 20 kyr with collapsed roof rocks.

(c) During the third phase, estimated to have begun at 19 ka, build-up of the magma chamber of the Neapolitan Yellow Tuff occurred. This was simulated by 3 inputs of latitic magma at 1000 °C in a smaller domain 7.2 km 7.2 km, constrained by the diameter of the Neapolitan Yellow Tuff caldera, which is nested within the Campanian ignimbrite caldera (see Fig. 16.63c).

(d) The fourth phase started with the 50 km^3 (DRE), caldera-forming eruption of the Neapolitan Yellow Tuff at 15 ka, which partially emptied the magma chamber, leaving an estimated 7.2 × 7.2x0.2 km^3 of resident magma in contact with roof rocks, which collapsed within the space left by the erupted magma volume. Such configuration was then let cool for the next 2 ka, before the renewal of scattered and small volume volcanism within the caldera floor (total integrated volume of 4–5 km^3), simulated as smaller-sized injections of magma constrained at shallower depths of 2–4 km by melt inclusion studies.

The important aspect of the approach proposed by Di Renzo et al. (2016) is that field volcanological studies, integrating stratigraphy, geochronology, structure, petrology, and geochemistry of volcanoes, can provide essential input data to constrain the integral heat flux across a geothermal system (cf. Deb et al. 2022). This is from source to surface, as well as for the modelling of the history and evolution of build-up and cooling phases that characterise the system. Such modelling of the magmatic heat source can be cross-verified by the inverse modelling of bore-hole data in exploited geothermal reservoirs (e.g. O'Sullivan et al. 2001), such as the actual reservoir temperature configuration and its time-evolution constrained by radiometric dating of hydrothermal mineral alteration assemblages.

18.10.3 Geothermal Reservoirs in Volcanic Areas

Conventional geothermal systems are associated with the presence at depth of one or more volumes of rocks that are permeable. In much the same way as for hydrocarbon reservoirs described in the previous section, these are controlled by primary porosity and/or secondary porosity (in particular hydraulic fracturing; Chap. 4), which in turn control the circulation of geothermal fluids (e.g. Muffler and Cataldi 1978). Geothermal reservoirs in volcanic areas can be volcanic or non-volcanic rock hosted.

The identification of permeable domains at depth is a major and fundamental challenge in geothermal exploration. Surface field geology can provide a first-order assessment of the mechanical stratigraphy of potential geothermal reservoirs, which relies on the presence of exhumed reservoir rocks, and the reconstruction of their petrophysical properties at depth (e.g. Giordano et al. 2014). As in oil and gas exploration, porosity and permeability (poroperms) associated with specific rock formations can be reasonably confidently determined from laboratory analyses (e.g. Dobson et al. 2003), including the prediction of how that porosity changes in depth with increasing lithostatic load (e.g. Ramm 1992; Ehrenberg and Nadeau 2005; see Fig. 18.96). However, much more complex is the evaluation of the localisation and intensity of secondary porosity in fractured domains (e.g. Rossetti et al. 2010; Giordano et al. 2014; Filipovich et al. 2022), as well as, specifically in active volcanic areas, how fluid circulation can modify permeability by processes of mineral alteration, dissolution (leaching), and/or cementation (e.g. Weydt et al. 2022).

The archetypal geothermal system is Larderello, Italy, where the geothermal reservoir is hosted by fractured regional basement rocks. At Larderello, as well as for most of the other high enthalpy, magmatic geothermal systems in central Italy, the igneous heat source is located at a few km depth within a conductive metamorphic basement. This heats an overlying regional reservoir system made by fractured Mesozoic–Cenozoic carbonate rocks, thrusted and faulted since Miocene times, and sealed by thick sequences of marine to continental clayey rocks (Giordano and Caricchi 2022; Fig. 18.112). Tertiary orogens dominated by thick sedimentary marine sequences, such as the predominantly carbonate Apennine system in Italy and elsewhere (e.g. Laramide orogen in Mexico), offer excellent potential reservoirs as the stratified sedimentary sequences are folded, thrusted, and variably faulted conferring a secondary permeability to significant thicknesses of rocks, which can reach 3000 m and more if tectonically superposed. Nevertheless, the distribution of fractures, while very widespread if taken at the regional scale, can vary substantially at the local scale, which is the main interest for the localisation of productive drillings. In such settings, petroleum exploration can provide excellent benchmark approaches for the reconstruction and prediction of the reservoir geometry and permeability distribution based on the integration of studies on exhumed reservoir rocks and geophysics, and specifically seismic survey.

A good example describing the application of such an integrated approach is for the Rosario de La Frontera

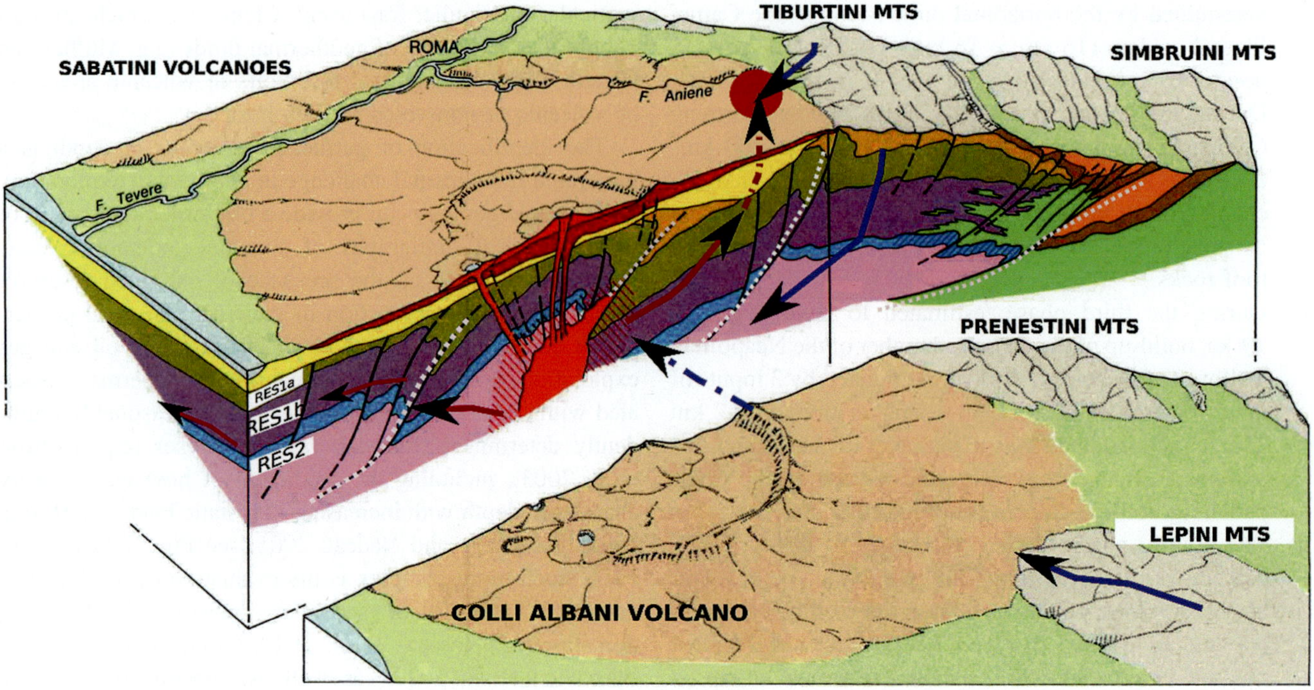

Fig. 18.112 3D block diagram illustrating a conceptual model for the Colli Albani geothermal system, central Italy. This also applies more generally to all Larderello-type systems characterised by a crustal magmatic source set at a few km depths within a low permeability metamorphic basement, covered by fractured reservoir rocks, sealed by impermeable cap rocks, and recharged by meteoric waters. Legend: pink = Paleozoic metamorphic basement, and Triassic evaporites; blue = Lower Jurassic shelf carbonates; purple and green (s) = Jurassic-Tertiary carbonate sequence; brown(s) = Tertiary siliciclastic flysch; yellow = Pliocene–Pleistocene marine-continental post-orogenic sediments; red(s) = Pleistocene volcanic rocks and magma reservoir; RES (1a, 1b, 2) = geothermal reservoir(s); blue arrows = meteoric recharge flow paths; red arrows = deep geothermal fluids advection paths; white dashed lines = fault compartments; red circle = geothermal manifestation. (From Giordano et al. 2014)

low-medium enthalpy geothermal field (northwest Argentina; Invernizzi et al. 2014; Maffucci et al. 2015). The geothermal reservoir in this case is Cretaceous sandstone, which unconformably overlies impervious Paleozoic low-grade metamorphic basement, and capped by Tertiary, low permeability rock sequences. The structure is the northern pericline termination of a large frontal, cylindrical anticline of the Santa Barbara chain (Fig. 18.113), where reservoir rocks are in outcrop and provide the recharge for the deep hydrothermal reservoir. The exhumed structure was investigated to determine the distribution across the anticline of fractures and faults by orientation, kinematics, aperture, and spacing in order to generate a predictive Discrete Fracture Network (DFN) model by geostatistical stochastic analysis (e.g. Chiles 1988; Bonnet et al. 2001; Xu and Dowd 2010). Such DFN models can then be used to predict fracture flow and heat transfer in the reservoir (e.g. Cacas et al. 1990; Kolditz 1995).

However, the application of geostatistical modelling of fracture distribution and fracture flow modelling is much more complex in volcanic environments. Volcano-tectonic deformation associated with magma intrusions, caldera collapses and/or resurgence, gravitational instabilities of edifices, and cycles of pressurisation and hydrofracturing of the geothermal systems can generate very complex patterns of fractures, which can be difficult to model in terms of their permeability behaviour (e.g. Garden et al. 2017). Geothermal fluid circulation along permeable fractured and/or porous pathways is always accompanied by alteration of rocks which can substantially modify the permeability and promote cycles of self-sealing and re-opening of rock domains (Fig. 18.114). This is very likely in volcanic environments where the most paroxysmal expressions of such cycles are phreatic/hydrothermal eruptions (e.g. Browne and Lawless 2001; Rowland and Simmons 2012; Vignaroli et al. 2015; Mayer et al. 2016; see Figs. 18.14 and 18.15).

Besides the identification of the volumetric distribution of interconnected fractures, another critical issue is the identification of the role of faults in controlling focused flow and compartmentalisation of the reservoirs. An excellent case study of localised fluid flow is provided by the Boccheggiano extensional fault in the Larderello area. Here the exhumed part of the fault hosts low sulphidation, adularia/illite-type epithermal, and sulphide ore deposition in localised veins related to precipitation from hydrothermal fluids channelled within the anisotropic permeability

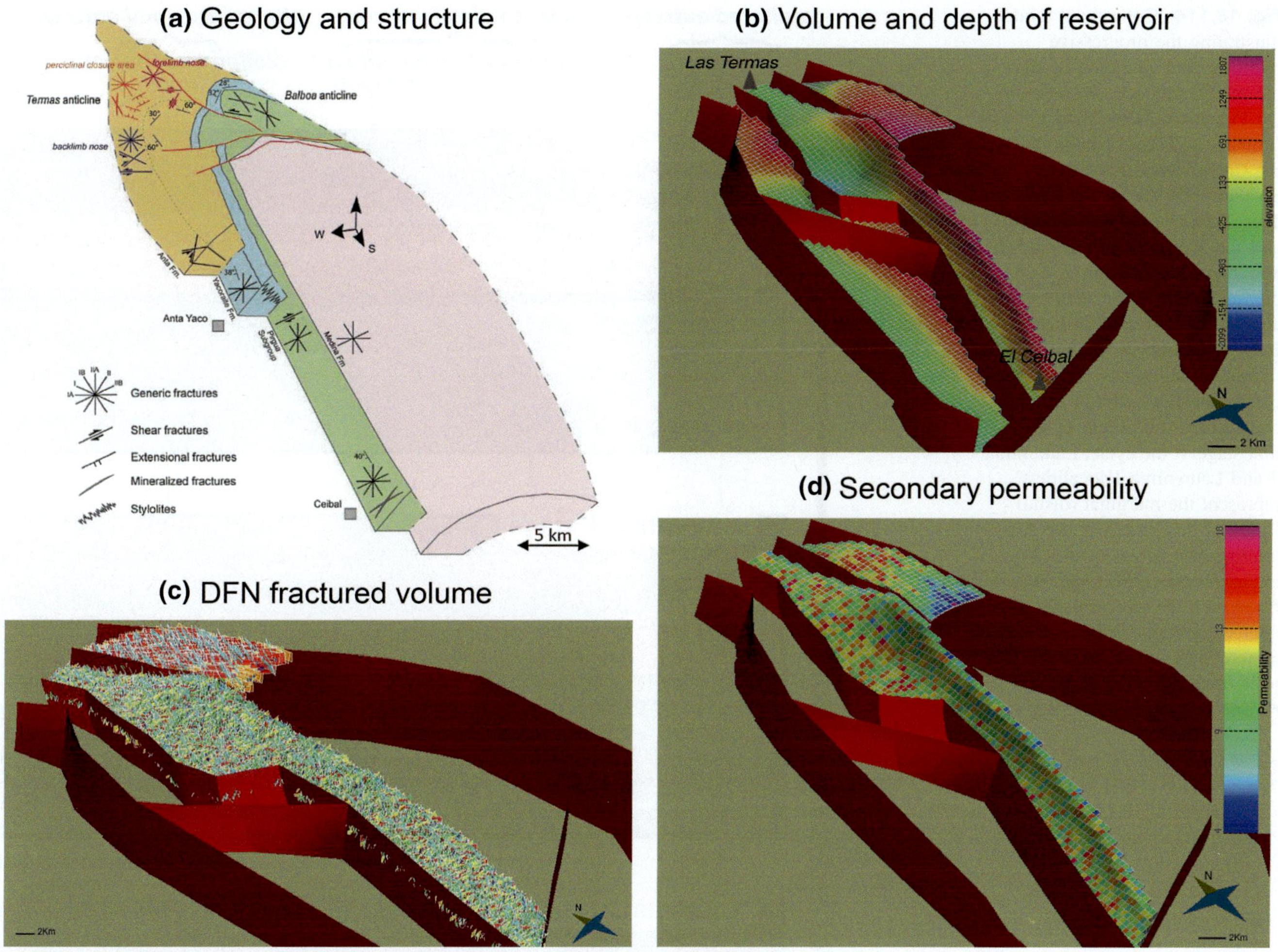

Fig. 18.113 Workflow to model the permeability of a deep geothermal reservoir based on the fracture distribution and geometry of the exhumed part of the structure at the Rosario de La Frontera geothermal system, Argentina. **a** The reservoir is hosted in a Cretaceous clastic formation (Piragua Subgroup in green) resting on a Paleozoic low-grade metamorphic basement (Medina Formation in pink) and capped by a Tertiary low permeability clastic sequence (Anta Formation in brown). The structure is an anticline with the Paleozoic basement exhumed at its core; structural analysis at the outcrop scale allows the definition of structural domains and patterns associated with the different sectors of the anticline. **b–d** Modelling (Move® software) results showing in 3D: **b** the volume of the buried reservoir, **c** the distribution of fractures by the construction of a Discrete Fracture Network, and **d** the associated secondary permeability. (Modified from Maffucci et al. 2015)

structure of the fault (Rossetti et al. 2010). The intensity of the deformation has created a low permeability barrier along the fault core, which focuses fluid flow within the hanging wall where fractures enhance permeability.

Vignaroli et al. (2013) have also investigated the importance of structural compartmentalisation in geothermal reservoirs in the case of the medium enthalpy Torre Alfina geothermal field, Italy. The Torre Alfina geothermal field is located just to the north of the Quaternary Bolsena caldera with an average geothermal gradient of 0.1 °C m^{-1} and was exploited during the 1980s. Fifteen deep bore holes were drilled into the identified fractured carbonate reservoir both in the structural high and lows and down to 4,500 m from the surface (Buonasorte et al. 1988). Contrary to expectations, productivity proved to be highly variable, and only a few boreholes provided economically interesting production of up to 500,000 kJ hr^{-1}. Moreover, the geothermal gradient also was very variable laterally. The initial conceptual model considered the reservoir identically permeable everywhere, which based on the engineering of the reservoir was designed to drill productive holes in the structural, high and to reinject the geothermal fluid after production in the structural low. Vignaroli et al. (2013) instead showed that the distribution of permeability at Torre Alfina is very variable, based on the surface expression of faulting and its relationship with hydrothermal alteration along faults. Structural data showed that fault cores have low permeability, whereas the damage zones are permeable, so

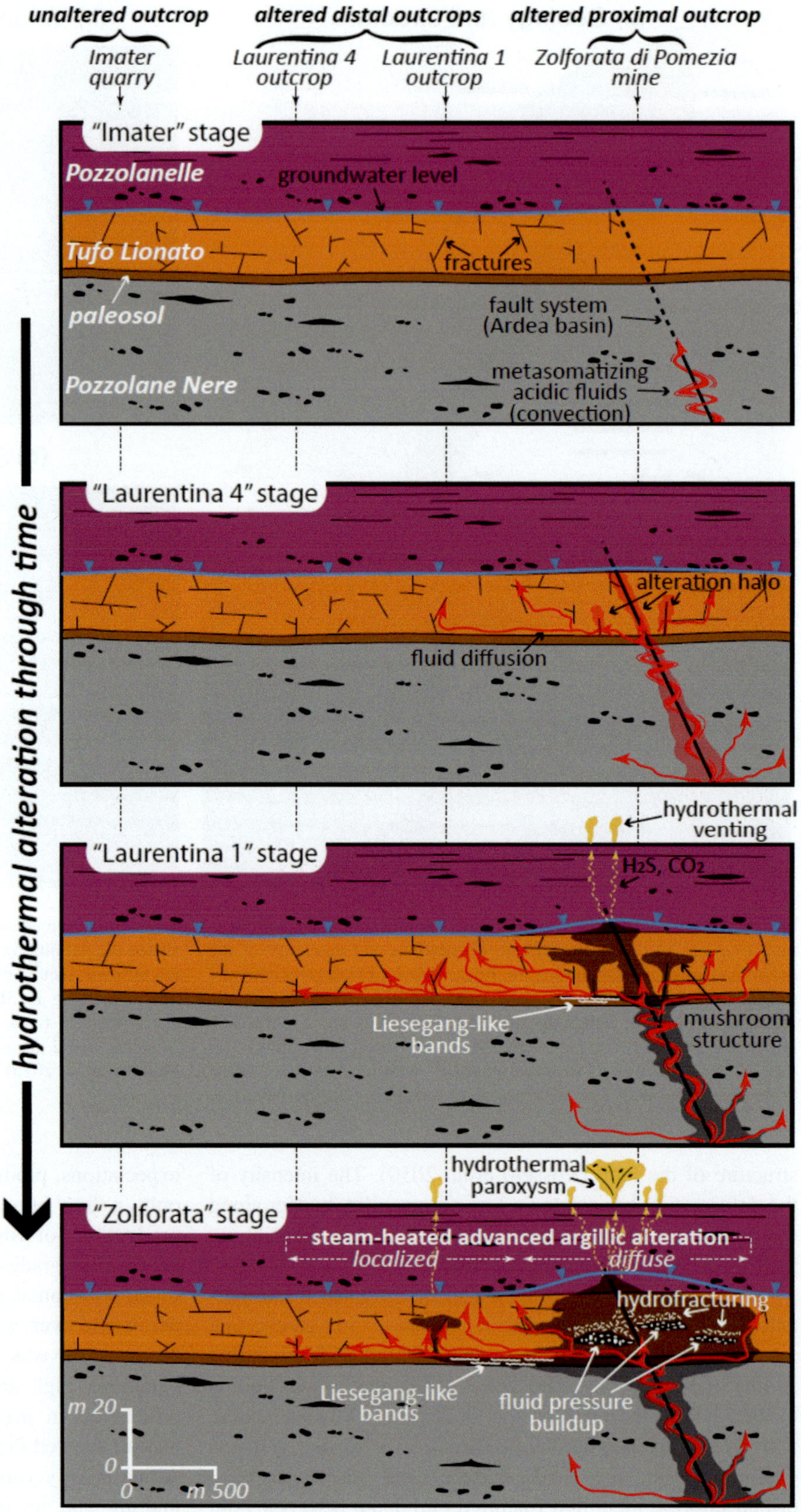

Fig. 18.114 Conceptual model illustrating the progressive hydrothermal alteration of an ignimbrite sequence (comprising the Pozzolane Nere, Tufo Lionato, and Pozzolanelle) within the Colli Albani active geothermal system (Fig. 18.112). This leads to argillic alteration and the formation of a cap rock, allowing for cycles of pressurisation of the shallow aquifer oversaturated in deep CO_2, hydrofracturing, and hydrothermal explosions. Imater quarry locality is the unaltered end member and represents the first stage of the cycle; Laurentina 4 and Laurentina 1 localities represent the marginal (distal) facies of the alteration and the intermediate stages of the alteration cycle; Zolforata di Pomezia mine represents the fully altered proximal facies and the mature stage of the cycle leading to hydrothermal explosions. (From Vignaroli et al. 2015)

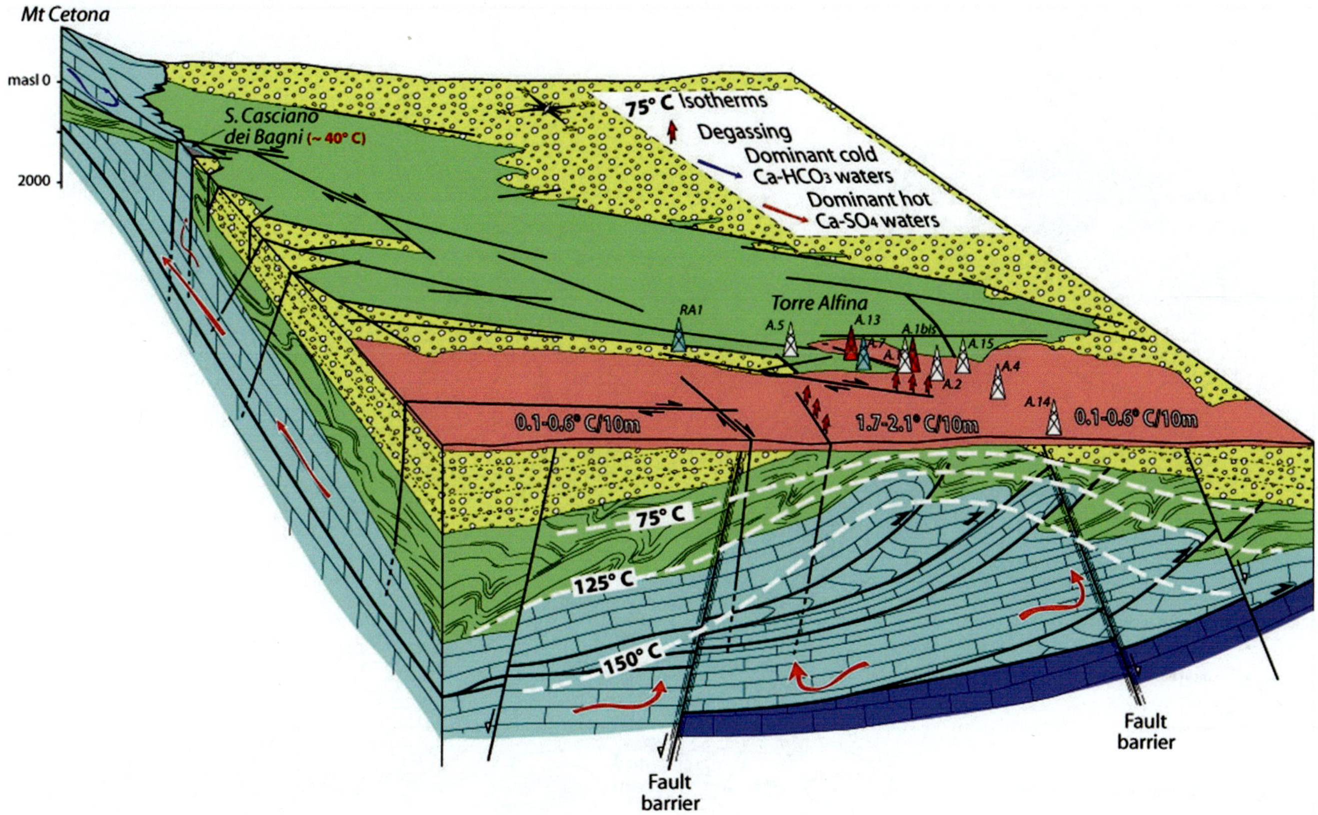

Fig. 18.115 3D block diagram of the Torre Alfina geothermal system, central Italy. Fault systems control the compartmentalisation of the reservoir and the lateral advection of geothermal fluids, focusing the upwelling of hot fluids towards the structural high and along preferential fractured domains (see Fig. 18.116 for regional setting). Legend: deep blue = Mesozoic-Tertiary Umbro-Sabina carbonate sequence (reservoir); pale blue = Mesozoic-Tertiary Tuscan carbonate sequence (reservoir); green = Liguride allochthonous flysch (cap rock); yellow = Quaternary marine to continental post-orogenic sediments; white triangles = wells; red triangles = productive wells. (From Vignaroli et al. 2013)

that the structural high is in fact hydraulically separated by the nearby lows by the master faults (Fig. 18.115). Such types of reconstructions are very important as both natural and reinjected recharge of the reservoir depend on the possible pathways controlled by the anisotropic distribution of fractures and their permeability. In the case of Torre Alfina, permeability controls the lateral advection of geothermal fluids and heat along a narrow corridor from the Bolsena caldera (Tescione et al. 2021; Fig. 18.116).

Caldera faults may play different roles in limiting or allowing leakage of geothermal fluids away from the caldera, depending on their permeability (e.g. Garden et al. 2017). In the example of the Roman Geothermal Province, caldera faults are largely permeable and allow lateral advection of geothermal fluids along carbonate geothermal reservoirs that are cross-cut by calderas (Giordano et al. 2014). Such later advection allows the presence of high to low enthalpy geothermal resources even tens of km away from the caldera area (Fig. 18.116), with fluids that are progressively mixed with meteoric recharge.

Another case study of a structurally controlled geothermal field is the Cerro Tuzgle–Tocomar field in the Argentinian Andes (Altiplano Puna), where the circulation of hydrothermal fluids is controlled by fracturing in otherwise impervious Permian–Ordovician basement rocks (Giordano et al. 2013, 2016; Chiodi et al. 2019; Filipovich et al. 2022). In this geothermal field, the integration of structural field survey and the study of morphotectonic lineaments from remote sensed satellite images and Digital Elevation Model (DEM) data sets (Chap. 15) proved to be successful in identifying both the main trends of fractures, but also their anisotropic distribution in terms of km of lineaments per km^2 (Fig. 18.117). This is essential to predict areas where there is prospectivity for geothermal circulation at depth, which can then be imaged by geophysical surveys (e.g. magnetotellurics; Ahumada et al. 2022).

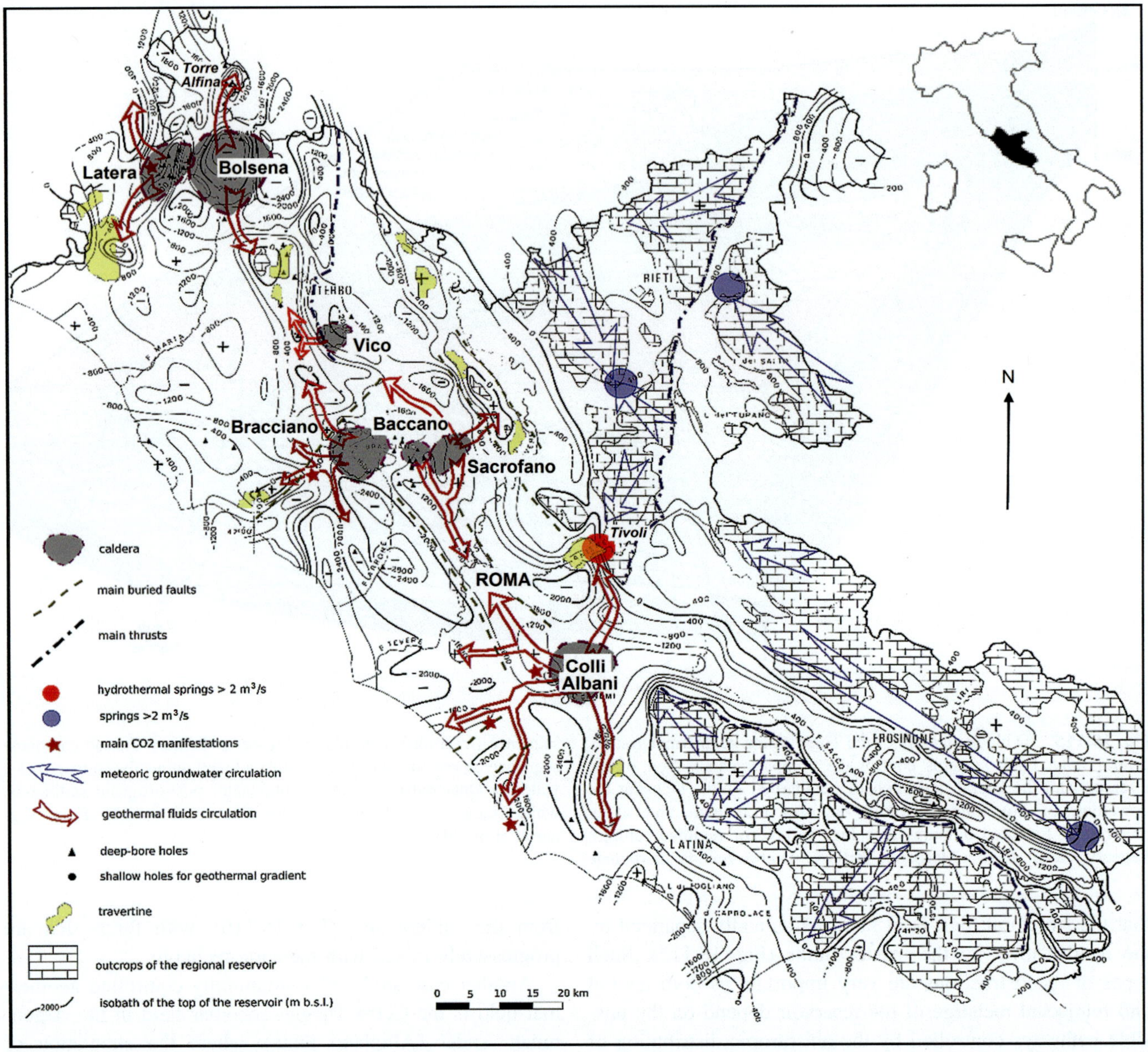

Fig. 18.116 Scheme of the deep circulation in the Roman Geothermal Province, central Italy. The brickwork pattern indicates the outcrops of the main regional carbonate reservoir; these are the main meteoric recharge areas for both the shallow hydrogeological systems (blue arrows for main groundwaters flow paths and blue dots for main springs) and the deep regional geothermal reservoirs (the depth of the top of which is contoured). The grey areas are the main Quaternary calderas of the Roman Volcanic Province (see Fig. 16.56), and the red arrows indicate the main lateral advection paths of geothermal fluids away from the calderas along structurally controlled corridors. Note that most of the surface geothermal manifestations (hydrothermal springs, CO_2 vents, and travertine deposits) are located along these paths. The thermal (or thermogene) travertine deposits or "plateau" are some of the largest known in the world of the Quaternary age; they show evidence of recent-active fault and fissure control, thus opening pathways for the ascent of geothermal fluids towards the surface, e.g. at Tivoli (De Filippis et al. 2013). (After Giordano et al. 2014)

A similar approach was applied to the Colli Albani caldera geothermal field, Italy (Giordano et al. 2014). In this case, the analysis of the distribution of lineament orientation and the frequency of their length distinguished two distinct populations, longer and shorter than 2 km, respectively (Fig. 18.118a–e). The long lineaments showed mostly NW- and NE-trends parallel to the main regional fault patterns, therefore, reflecting the main fracture patterns in the regional geothermal reservoir. By contrast, the short lineaments are mostly random in orientation and localised in space, suggesting much shallower deformation sources. Areas with high densities of short lineaments correspond to sites of

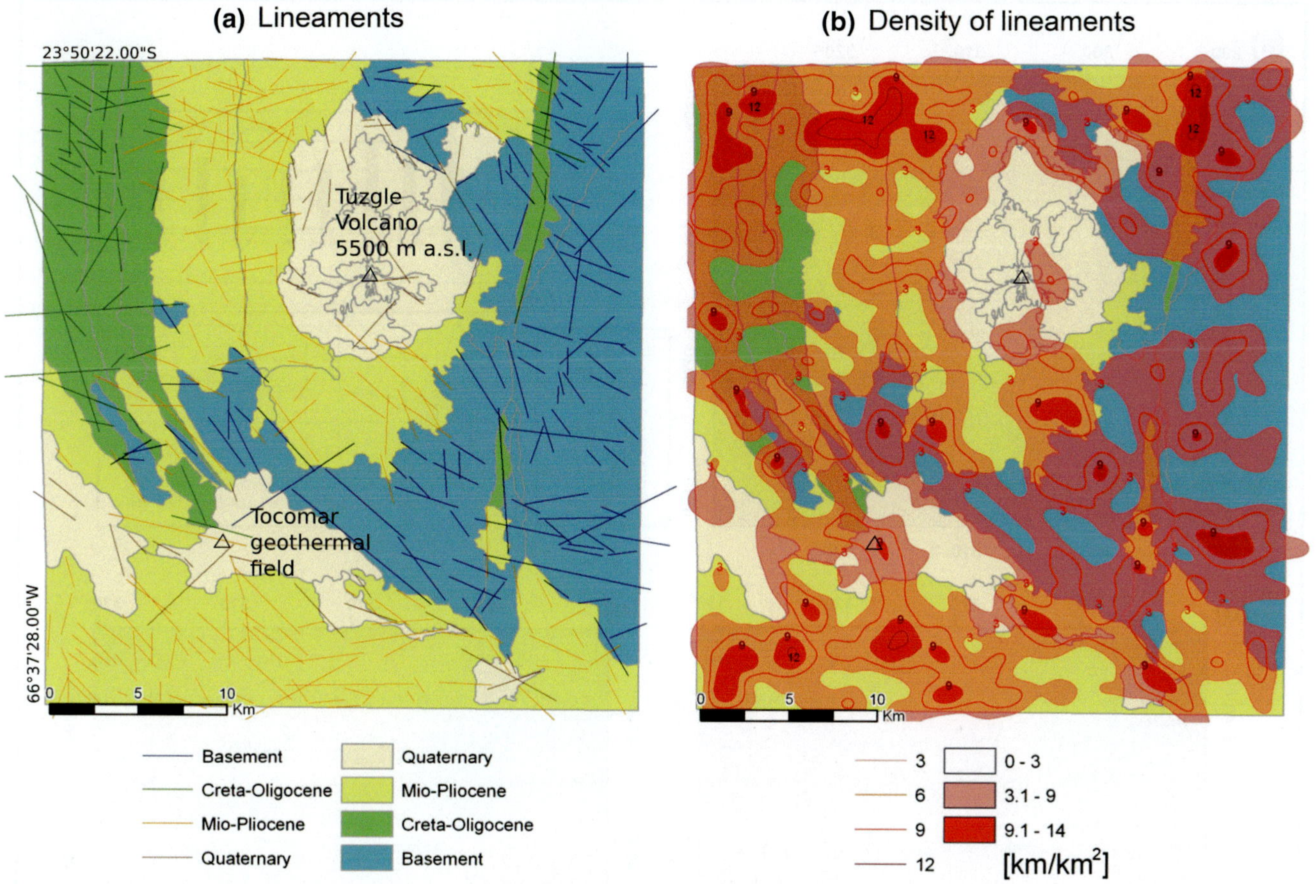

Fig. 18.117 **a** Morpho-structural lineament analysis from satellite images and DEMs for the geothermal area of Tuzgle–Tocomar, Argentinian Puna. **b** As most potential reservoir rocks are permeable for fracturing, the map of the density of km of lineaments per square kilometre allows to identify areas where lineaments overlap and are more prone to develop secondary permeability. (From Giordano et al., 2013)

active and fossil upwelling of geothermal fluids and strong rock alteration (Fig. 18.118f, g). By overlaying morphotectonic lineament densities with maps of carbon dioxide concentration, temperature, and electric conductivity of groundwaters, as well as structural maps and locations of active and fossil hydrothermal alteration, areas that correspond to the most permeable volumes of the geothermal reservoir were identified.

Furthermore, this approach allowed a much more refined estimate of the energetic potential than calculated by the traditional "*volume method*" (e.g. Muffler and Cataldi 1978; Doveri et al. 2010). The volume method implies the identification of a productive volume and calculates the energetic potential with the following equation:

$$\begin{aligned} H_t &= H_r + H_f \\ &= [(1-\varphi)\,\rho_r C_r V\,(T - T_o)] + [(\varphi)\,\rho_f C_f V\,(T - T_o)] \end{aligned} \tag{18.2}$$

where H_t is the total thermal energy in Joules, stored in a volume V of the geothermal reservoir at average temperature T with respect to mean surface temperature T_o, partitioned in H_r that is the energy stored in the rock of density ρ_r and specific heat C_r, and characterised by an effective porosity φ, and in H_f that is the energy stored in the geothermal fluid of density ρ_f and specific heat C_f.

Usually, the effective porosity, which includes the interconnected primary pores and fractures, is <0.2 and the recovery factor which expresses the exploitable heat with respect to the stored heat is at best 0.5 for liquid-dominated systems and 0.15 for vapour-dominated systems. However, the generic identification of a geothermal reservoir volume may be commercially misleading unless qualified for its internal anisotropic distribution of permeability and possible compartmentalisation. All of which can be qualitatively and quantitatively approached by volcanologists by integrating field surveys, remote sensed analysis, and geophysical and geochemical surveys.

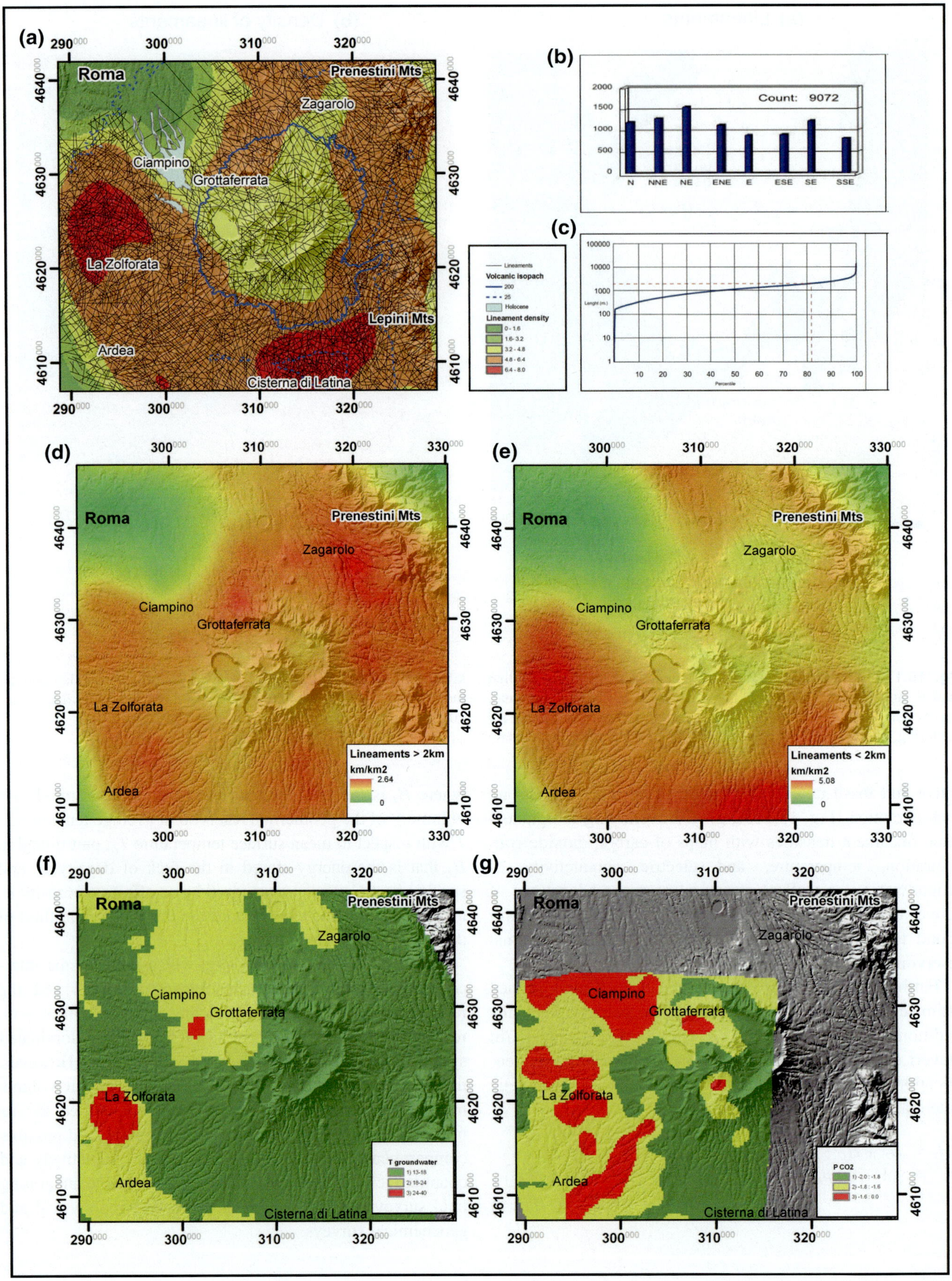
(a)
Roma
Prenestini Mts
Zagarolo
Ciampino
Grottaferrata
La Zolforata
Lepini Mts
Ardea
Cisterna di Latina
Lineaments
Volcanic isopach
200
25
Holocene
Lineament density
0 - 1.6
1.6- 3.2
3.2 - 4.8
4.8 - 6.4
6.4 - 8.0
(b)
Count: 9072
N NNE NE ENE E ESE SE SSE
(c)
Lenght (m.)
Percentile
(d)
Lineaments > 2km
km/km2
2.64
0
(e)
Lineaments < 2km
km/km2
5.08
0
(f)
T groundwater
1) 13-18
2) 18-24
3) 24-40
(g)
P CO2
1) -2.0 : -1.8
2) -1.8 : -1.6
3) -1.6 : 0.0

◀ **Fig. 18.118** Integration of surface indicators of permeability of the geothermal reservoir at Colli Albani volcano, Roman Geothermal Province, central Italy. **a** The density of morpho-structural lineaments derived from the interpretation of aerial stereo photos. The isopach 200 m of volcanic deposits and the outline of Holocene deposits are indicated to take into account the differential response to fracturing as a function of rock rheology (thickness of unconsolidated volcanic deposits) and age, together with the urbanised cover of Rome. **b** Frequency of lineaments by orientation. **c** Percentage of lineaments by length. These show an inflection at 2 km length indicating the presence of two families: lineaments longer than 2 km are parallel to the main regional tectonic trends (mostly NW-, NE-, and N-trending), whereas the short lineaments are mostly random and suggest shallower pressure sources, associated with the upwelling of hydrothermal fluids and show different areal density distributions (**d**, **e**). **f** Temperature and **g** concentration of CO_2 of groundwaters indicate areas of leakage of geothermal fluids and largely coincide with the highest density of short lineaments (**e**). (From: Giordano et al. 2014)

18.10.4 Cap Rocks

The final essential ingredient for a sustainable geothermal resource is the presence of impervious cap rock(s) that both hydrodynamically seal the geothermal reservoir and prevent the fast, convective dispersion of the heat stored. Cap rocks have low permeability, and may be primary in origin, such as marine clays, or formed by the hydrothermal alteration of rocks (e.g. Facca and Tonani 1967). This is especially so for vitric volcanic rocks where glass is easily altered to clay minerals such as kaolinite during advanced argillic alteration by acidic fluids (e.g. Browne, 1970, 1978; Alfaro et al. 2015; Sect. 18.3, and references therein). The effectiveness of cap rocks depends on the lateral and vertical extension of the low permeability package and on the presence of fractures and faults. This can be studied by analysing the temperature-dependent mineralogy of clay minerals, in particular illite–smectite (e.g. Corrado et al. 2014; Maffucci et al. 2016), as well as mapping the effects of diffuse degassing (e.g. Carapezza et al. 2015). In volcanic environments such as calderas and stratovolcanoes, hydrothermal alteration of volcanic rocks largely controls the development, location, and extension of cap rocks (e.g. Izquierdo et al. 2000; see Figs. 18.2, 18.3 and 18.4).

18.10.5 Enhanced Geothermal Systems and Super-Hot Geothermal Systems in Magmatic Areas

Some final remarks are necessary with respect to the frontiers of geothermal exploration. Conventional geothermal resources, as described in the previous paragraphs, rely on the identification of permeable volumes of rocks with enough natural fluids to extract heat. The exploration of such resources is therefore limited to areas where effective permeability and efficient recharge are in co-existence. However, most geothermal energy is actually contained in dry rocks, a fact that has prompted studies on resources known as Enhanced Geothermal Systems (EGS), where the heat stored is extracted by inducing hydraulic fracturing and injecting fluids (e.g. Olasolo et al. 2016; Lu 2018).

In these systems, geological exploration is directed towards stratigraphic and structural analysis aimed at defining the mechanical architecture of the hot rock volumes that are to be stimulated. Public concerns for these kinds of exploitation projects are related to induced seismicity, which is generally low intensity, but in some cases has reached magnitudes that have affected local populations (e.g. Deichmann and Giardini 2009). Such kinds of accidents have now prompted innovations in fracturing methods as well as a shift to the use of CO_2 instead of water as the injected fluid. EGS do not need to be located in volcanic areas, and indeed most of the presently active projects are located in non-volcanic areas (Lu 2018).

A recent development that involves active volcanic regions is under implementation in Iceland aimed at tapping supercritical fluids (Super-Hot Geothermal Systems or SHGS) along the mid-Atlantic Ridge high heat flow rift system (Chap. 17). To drill into a hydrothermal system where both temperature and pressure are greater than the critical point of water not only requires reaching the relevant temperatures, but also hydrostatic pressures found at depths $\geq$4–5 km. The Iceland Deep Drilling Project (IDDP) has included exploratory drilling projects at Krafla caldera, in northern Iceland, and on the landward extension of the Reykjanes Ridge in southwest Iceland (Elders et al. 2014; Friðleifsson et al. 2018; Plate 18.1h). Drilling at Krafla intersected an active rhyolitic magma pocket unexpectedly encountered at just 2.1 km depth, with the potential to release an enormous quantity of geothermal energy (Elders et al. 2014). While at the Reykjanes geothermal field, deep drilling (to $\sim$4,500 m vertical depth) demonstrated that it is possible to drill a supercritical geothermal well, with flow tests planned to determine the potential energy output and commercial viability (Friðleifsson et al. 2018). At present, the development of SHGS is in its infancy, but clearly, this has the largest potential for the production of electric power, although the sites prone to such developments are invariably in active magmatic systems. Here there may also be potential hazard issues related to the accidental triggering of volcanic phenomena, making these kinds of prospects poorly acceptable socially in densely inhabited regions (e.g. Campi Flegrei; De Natale et al. 2016).

References

Abdelmalak MM, Planke S, Faleide JI, Jerram DA, Zastrozhnov D, Eide S, Myklebust R (2016) The development of volcanic sequences at rifted margins: new insights from the structure and morphology of the Vøring Escarpment, mid-Norwegian Margin. J Geophys Res Solid Earth 121(7):5212–5236

Adams JDB (1998) Kora—a study of a Miocene, submarine arc-stratovolcano, north Taranaki Basin, New Zealand. MSc thesis, Massey University, New Zealand, p 142

Ahumada MF, Guevara L, Favetto A, Filipovich R, Chiodi A, Viramonte JG, Giordano G (2022) Electrical resistivity structure in the Tocomar geothermal system obtained from 3-D inversion of audio-magnetotelluric data (Central Puna, NW Argentina). Geothermics 104: 102436

Alfaro C, Ponce P, Monsalve ML, Ortiz I, Franco JV, Ortega A, Torres R, Gómez D (2015) A preliminary conceptual model of Azufral geothermal system, Colombia. In: Proceedings World Geothermal Congress 2015

Alfieris D, Voudouris P, Spry PG (2013) Shallow submarine epithermal Pb–Zn–Cu–Au–Ag–Te mineralization on western Milos Island, Aegean Volcanic Arc, Greece: mineralogical, geological and geochemical constraints. Ore Geol Rev 53:159–180

Allan AS, Morgan DJ, Wilson CJ, Millet MA (2013) From mush to eruption in centuries: assembly of the super-sized Oruanui magma body. Contribs Mineral Petrol 166(1):143–164

Allen SR, McPhie J, Ferris G, Simpson C (2008) Evolution and architecture of a large felsic igneous province in western Laurentia: the 1.6 Ga Gawler Range Volcanics, South Australia. J Volcanol Geotherm Res 172(1–2):132–147

Allen RL, Weihed P, Global Project VMS Research Project Team (2002) Global comparisons of volcanic-associated massive sulphide districts. Geol Soc Lond Spec Publ 204(1):13–37

Allen RL, Weihed P, Svenson SA (1996) Setting of Zn-Cu-Au-Ag massive sulfide deposits in the evolution and facies architecture of a 1.9 Ga marine volcanic arc, Skellefte District, Sweden. Econ Geol 91(6):1022–1053

Ambrosio M, Doveri M, Fagioli MT, Marini L, Principe C, Raco B (2010) Water-rock interaction in the magmatic-hydrothermal system of Nisyros Island (Greece). J Volcanol Geotherm Res 192 (1–2):57–68

Andersen DJ, Bishop FC, Lindsley DH (1991) Internally consistent solution models for Fe-Mg-Mn-Ti oxides: Fe-Mg-Ti oxides and olivine. Am Mineral 76(3–4):427–444

Annen C (2009) From plutons to magma chambers: thermal constraints on the accumulation of eruptible silicic magma in the upper crust. Earth Planet Sci Lett 284(3–4):409–416

Arehart GB (1996) Characteristics and origin of sediment-hosted disseminated gold deposits: a review. Ore Geol Rev 11(6):383–403

Arellano VM, García A, Barragán RM, Izquierdo G, Aragón A, Nieva D (2003) An updated conceptual model of the Los Humeros geothermal reservoir (Mexico). J Volcanol Geotherm Res 124 (1–2):67–88

Arndt N, Ginibre C, Chauvel C, Albarede F, Cheadle M, Herzberg C, Jenner G, Lahaye Y (1998) Were komatiites wet? Geology 26 (8):739–742

Arndt N, Lesher M, Barnes S (2008) Komatiite. Cambridge University Press, p 468

Arndt N, Lesher CM, Czamanske GK (2005) Mantle-derived magmas and magmatic Ni-Cu-(PGE) deposits. Econ Geol 100th Anniversary Volume 34:5–24

Arribas A (1995) Characteristics of high-sulfidation epithermal deposits, and their relation to magmatic fluid. Mineral Assoc Canada Short Course 23:419–454

Bachmann O, Oberli F, Dungan MA, Meier M, Mundil R, Fischer H (2007) $^{40}Ar/^{39}Ar$ and U-Pb dating of the Fish Canyon magmatic system, San Juan Volcanic field, Colorado: evidence for an extended crystallization history. Chem Geol 236(1–2):134–166

Baffioni C (2017) Numerical exploration for medium-low enthalpy fluids in the Roman Geothermal Province. PhD thesis, Università Roma Tre, Italy

Baker EM, Andrew AS (1991) Geologic, fluid inclusion, and stable isotope studies of the gold-bearing breccia pipe at Kidston, Queensland, Australia. Econ Geol 86(4):810–830

Baker EM, Kirwin DJ, Taylor RG (1986) Hydrothermal breccia pipes. James Cook University North Queensland, Townsville, EGRU Contribution 12, 45 pp

Barnes SJ (2006) Komatiites: petrology, volcanology, metamorphism, and geochemistry. In: Barnes SJ (ed) Nickel deposits of the Yilgarn Craton: geology, geochemistry, and geophysics applied to exploration. Soc Econ Geol Spec Publ 13:13–49

Barnes SJ, Beresford SW, Le Vaillant M (2016) Interspinifex Ni sulfide ore from the Coronet shoot, Kambalda: characterization using microbeam X-ray fluorescence mapping and 3-D X-ray computed tomography. Econ Geol 111(6):1509–1517

Barnes SJ, Hill RET, Gole MJ (1988) The Perseverance ultramafic complex, Western Australia: the product of a komatiite lava river. J Petrol 29(2):305–331

Barnes SJ, Van Kranendonk MJ, Sonntag I (2012) Geochemistry and tectonic setting of basalts from the Eastern Goldfields Superterrane. Aust J Earth Sci 59(5):707–735

Barnett WP (2008) The rock mechanics of kimberlite volcanic pipe excavation. J Volcanol Geotherm Res 174(1–3):29–39

Barrie CT, Hannington MD (1999) Classification of volcanic-associated massive sulfide deposits based on host-rock composition. Reviews Econ Geol 8:1–11

Barrie CT, Hannington MD, Bleeker W (1999) The giant Kidd Creek volcanic-associated massive sulfide deposit, Abitibi Subprovince, Canada. Reviews Econ Geol 8:247–269

Barton MD (2014) Iron oxide(-Cu-Au-REE-P-Ag-U-Co) systems. In: Holland H, Turekian K (eds) Treatise of geochemistry, vol 13. Elsevier, pp 515–536

Bauer TE, Skyttä P, Allen RL, Weihed P (2011) Syn-extensional faulting controlling structural inversion–insights from the Palaeoproterozoic Vargfors syncline, Skellefte mining district, Sweden. Precamb Res 191(3–4):166–183

Bauer TE, Skyttä P, Hermansson T, Allen RL, Weihed P (2014) Correlation between distribution and shape of VMS deposits and regional deformation patterns, Skellefte district, northern Sweden. Miner Deposita 49(5):555–573

Baumgartner RJ, Fiorentini ML, Baratoux D, Micklethwaite S, Sener AK, Lorand JP, McCuaig TC (2015) Magmatic controls on the genesis of Ni-Cu±(PGE) sulphide mineralisation on Mars. Ore Geol Rev 65:400–412

Begg G, Gray DR (2002) Arc dynamics and tectonic history of Fiji based on stress and kinematic analysis of dikes and faults of the Tavua Volcano, Viti Levu Island, Fiji. Tectonics 21(4):1–14

Belperio A, Flint R, Freeman H (2007) Prominent Hill: a hematite-dominated, iron oxide copper-gold system. Econ Geol 102(8):1499–1510

Beresford SW, Cas RAF (2001) Komatiitic invasive lava flows, Kambalda, Western Australia. Canad Mineral 39(2):525–535

Beresford SW, Cas RAF, Lambert DD, Stone WE (2000) Vesicles in thick komatiite lava flows, Kambalda, Western Australia. J Geol Soc Lond 157(1):11–14

Beresford SW, Cas R, Lahaye Y, Jane M (2002) Facies architecture of an Archean komatiite-hosted Ni-sulphide ore deposit, Victor, Kambalda, Western Australia: implications for komatiite lava emplacement. J Volcanol Geotherm Res 118(1):57–75

Beresford S, Stone WE, Cas R, Lahaye Y, Jane M (2005) Volcanological controls on the localization of the komatiite-hosted Ni-Cu-(PGE) Coronet deposit, Kambalda, Western Australia. Econ Geol 100(7):1457–1467

Berger BR, Henley RW (1989) Advances in the understanding of epithermal gold-silver deposits. In: Keays RR, Ramsay WRH, Groves, DI (eds) The geology of gold deposits: the perspective in 1988. Econ Geol Monogr 6:405–423

Bergman SC (1987) Lamproites and other potassium-rich igneous rocks: a review of their occurrence, mineralogy and geochemistry. In: Fitton JG, Upton BGJ (eds) Alkaline igneous rocks. Geol Soc Lond Spec Publ 30(1):103–190

Bergman SC, Talbot JP, Thompson PR (1992) The Kora Miocene submarine andesite stratovolcano hydrocarbon reservoir, northern Taranaki Basin, New Zealand. In: Proceedings of the 1991 New Zealand oil exploration conference. Ministry of Commerce, Wellington, New Zealand, pp 178–206

Betts PG, Giles D (2006) The 1800–1100 Ma tectonic evolution of Australia. Precamb Res 144(1–2):92–125

Betts PG, Giles D, Foden J, Schaefer BF, Mark G, Pankhurst MJ, Forbes CJ, Williams HA, Chalmers NC, Hills Q (2009) Mesoproterozoic plume-modified orogenesis in eastern Precambrian Australia. Tectonics 28:TC3006

Binns RA, Scott SD (1993) Actively forming polymetallic sulfide deposits associated with felsic volcanic rocks in the eastern Manus back-arc basin, Papua New Guinea. Econ Geol 88(8):2226–2236

Bischoff AP (2018) Architectural elements of buried volcanic systems and their impact on geoenergy resources. PhD thesis, University of Canterbury, Christchurch, New Zealand, p 226

Bischoff AP, Nicol A, Beggs M (2017) Stratigraphy of architectural elements in a buried volcanic system and implications for hydrocarbon exploration. Interpretation 5(3):141–159

Bishop MG (2000) Petroleum systems of the northwest Java Province, Java and offshore southeast Sumatra, Indonesia. US Geol Surv Open-file Rept 99-50R:34

Bissig T, Clark AH, Rainbow A, Montgomery A (2015) Physiographic and tectonic settings of high-sulfidation epithermal gold-silver deposits of the Andes and their controls on mineralizing processes. Ore Geol Rev 65:327–364

Bissig T, Clark AH, Lee JK, Hodgson CJ (2002) Miocene landscape evolution and geomorphologic controls on epithermal processes in the El Indio-Pascua Au-Ag-Cu belt, Chile and Argentina. Econ Geol 97(5):971–996

Blackwell JL, Cooke DR, McPhie J, Simpson KA (2014) Lithofacies associations and evolution of the volcanic host succession to the Minifie ore zone: Ladolam gold deposit, Lihir Island, Papua New Guinea. Econ Geol 109(4):1137–1160

Bodnar RJ, Lecumberri-Sanchez P, Moncada D, Steele-MacInnis M (2014) 13.5 Fluid inclusions in hydrothermal ore deposits. In: Holland HD, Turekian KK (eds) Treatise on geochemistry, vol. 13. Elsevier, pp 119–142

Bogie I, Lawless JV (1987) Controls on the hydrology of large volcanically hosted geothermal systems: implications for exploration for epithermal mineral deposits. In: Proceedings Pacrim Congress, Gold Coast, Australia. Australasian Institute of Mining and Metallurgy, pp 57–60

Bogie I, Lawless JV, Rychagov S, Belousov V (2005) Magmatic-related hydrothermal systems: classification of the types of geothermal systems and their ore mineralisation. In: Rychagov S (ed) Geothermal and mineral resources of modern volcanism areas. Proceedings of the International Kuril-Kamchatka Field Workshop, pp 51–73. http://web.ru/conf/kuril_kam2005/art3.pdf

Bonham HF (1989) Bulk mineable gold deposits of the western United States. In: Keays RR, Ramsay WRH, Groves DI (eds) The geology of gold deposits: the perspective in 1988. Econ Geol Monogr 6:193–207

Bonnet E, Bour O, Odling NE, Davy P, Main I, Cowie P, Berkowitz B (2001) Scaling of fracture systems in geological media. Rev Geophys 39(3):347–383

Boric R, Holmgren C, Wilson NS, Zentilli N (2002) The geology of the El Soldado manto type Cu (Ag) deposit, central Chile. In: Porter TM (ed) Hydrothermal iron oxide copper-gold and related deposits: a global perspective. PGC Publishing, Adelaide, pp 163–184

Boulter CD (1993a) Comparison of Rio Tinto, Spain, and Guaymas Basin, Gulf of California: an explanation of a supergiant massive sulfide deposit in an ancient sill-sediment complex. Geology 21 (9):801–804

Boulter C (1993b) High-level peperitic sills at Rio Tinto, Spain: implications for stratigraphy and mineralization. Trans Inst Min Metall, Sect b. Appl Earth Sci 102:B30–B38

Boxer G, Lorenz V, Smith CB (1989) The geology and volcanology of the Argyle (AK1) lamproite diatreme, Western Australia. Geol Soc Aust Spec Publ 14:140–152

Broadbent GC, Myers RE, Wright JV (1998) Geology and origin of shale-hosted Zn-Pb-Ag mineralization at the Century deposit, northwest Queensland, Australia. Econ Geol 93(8):1264–1294

Brothers DS, Driscoll NW, Kent GM, Harding AJ, Babcock JM, Baskin RL (2009) Tectonic evolution of the Salton Sea inferred from seismic reflection data. Nat Geosci 8:581

Brown KL (1986) Gold deposition from geothermal discharges in New Zealand. Econ Geol 81(4):979–983

Brown RJ, Buse B, Sparks RSJ, Field M (2008a) On the welding of pyroclasts from very low-viscosity magmas: examples from kimberlite volcanoes. J Geol 116(4):354–374

Brown RJ, Field M, Gernon T, Gilbertson M, Sparks RSJ (2008c) Discussion of 'In-vent column collapse as an alternative model for massive volcaniclastic kimberlite emplacement: an example from the Fox kimberlite, Ekati Diamond Mine, NWT, Canada' by Porritt et al. J Volcanol Geotherm Res 178:847–850

Brown RJ, Gernon T, Stiefenhofer J, Field M (2008b) Geological constraints on the eruption of the Jwaneng Centre kimberlite pipe, Botswana. J Volcanol Geotherm Res 174(1–3):195–208

Brown RJ, Manya S, Buisman I, Fontana G, Field M, Mac Niocaill C, Sparks RSJ, Stuart FM (2012) Eruption of kimberlite magmas: physical volcanology, geomorphology and age of the youngest kimberlitic volcanoes known on earth (the Upper Pleistocene/Holocene Igwisi Hills volcanoes, Tanzania). Bull Volcanol 74 (7):1621–1643

Brown RJ, Tait M, Field M, Sparks RSJ (2009) Geology of a complex kimberlite pipe (K2 pipe, Venetia Mine, South Africa): insights into conduit processes during explosive ultrabasic eruptions. Bull Volcanol 71(1):95–112

Brown RJ, Valentine GA (2013) Physical characteristics of kimberlite and basaltic intraplate volcanism and implications of a biased kimberlite record. Geol Soc Amer Bull 125(7–8):1224–1238

Browne PR (1970) Hydrothermal alteration as an aid in investigating geothermal fields. Geothermics 2:564–570

Browne PR (1978) Hydrothermal alteration in active geothermal fields. Ann Rev Earth Planet Sci 6(1):229–248

Browne PR, Lawless JV (2001) Characteristics of hydrothermal eruptions, with examples from New Zealand and elsewhere. Earth Sci Rev 52(4):299–331

Budd A, Wyborn L, Bastrakova I (1998) Exploration significance of the Hiltaba Suite, South Australia. Aust Geol Surv Res Newslett 29:1–4

Buonasorte G, Cataldi R, Ceccarelli A, Costantini A, d'Offizi S, Lazzarotto A, Ridolfi A, Baldi P, Barelli A, Bertini G, Bertrami R (1988) Ricerca ed esplorazione nell'area geotermica di Torre Alfina (Lazio-Umbria). Boll Soc Geol Ital 107(2):265–337

Burnham CW (1967) Hydrothermal fluids at the magmatic stage. In: Barnes HL (ed) Geochemistry of hydrothermal ore deposits, vol 1. Wiley, pp 34–76

Busby C (2005) Possible distinguishing characteristics of very deepwater explosive and effusive silicic volcanism. Geology 33 (11):845–848

Busby CJ, Kessel LO, Schulz KJ, Foose MP, Slack JF (2003) Volcanic setting of the Ordovician Bald Mountain massive sulfide deposit, northern Maine. Econ Geol Monogr 11:219–244

Burt DM, Sheridan MF (1981) Model for the formation of uranium/lithophile element deposits in fluorine-rich volcanic rocks. In: Goodell PC, Waters AC (eds) Uranium in volcanic and volcaniclastic rocks. Amer Assoc Petrol Geol Studies 13:99–109

Burt DM, Sheridan MF, Bikun JV, Christiansen EH (1982) Topaz rhyolites; distribution, origin, and significance for exploration. Econ Geol 77(8):1818–1836

Cacas MC, Ledoux E, Marsily GD, Tillie B, Barbreau A, Durand E, Feuga B, Peaudecerf P (1990) Modeling fracture flow with a stochastic discrete fracture network: calibration and validation: 1. The flow model. Water Resour Res 26(3):479–489

Campbell IH, Griffiths RW, Hill RI (1989) Melting in an Archean mantle plume: Heads it's basalts, tails it's komatiites. Nature 339 (6227):697

Cannon SJC, Ellis D (eds) (2014) Hydrocarbon exploration to exploitation west of Shetlands. Geol Soc Lond Spec Publ 397:223

Carapezza ML, Ranaldi M, Gattuso A, Pagliuca NM, Tarchini L (2015) The sealing capacity of the cap rock above the Torre Alfina geothermal reservoir (Central Italy) revealed by soil CO_2 flux investigations. J Volcanol Geotherm Res 291:25–34

Carman GD (2003) Geology, mineralization, and hydrothermal evolution of the Ladolam gold deposit, Lihir Island, Papua New Guinea. Soc Econ Geol Spec Publ 10:247–284

Carrasco-Núñez G, Bernal JP, Davila P, Jicha B, Giordano G, Hernández J (2018) Reappraisal of Los Humeros volcanic complex by new U/Th zircon and $^{40}Ar/^{39}Ar$ dating: implications for greater geothermal potential. Geochem Geophys Geosyst 19(1):132–149

Carrasco-Núñez G, Branney MJ (2005) Progressive assembly of a massive layer of ignimbrite with a normal-to-reverse compositional zoning: the Zaragoza ignimbrite of central Mexico. Bull Volcanol 68(1):3

Cas RAF (1992) Submarine volcanism; eruption styles, products, and relevance to understanding the host-rock successions to volcanic-hosted massive sulfide deposits. Econ Geol 87(3):511–541

Cas RAF, Beresford S (2001) Field characteristics and erosional processes associated with komatiitic lavas: implications for flow behaviour. Canad Mineral 39:505–524

Cas RAF, Giordano G (2014) Submarine volcanism: a review of the constraints, processes and products, and relevance to the Cabo de Gata volcanic succession. Ital J Geosci 133(3):362–377

Cas RAF, Giordano G, Balsamo F, Esposito A, Lo Mastro S (2011) Hydrothermal breccia textures and processes: Lisca Bianca islet, Panarea volcano, Aeolian Islands, Italy. Econom Geol 106(3): 437–450

Cas RAF, Hayman PC, Pittari A, Porritt L (2008a) Some major problems with existing models and terminology associated with kimberlite pipes from a volcanological perspective, and some suggestions. J Volcanol Geotherm Res 174(1–3):209–225

Cas RAF, Marks K, Perazzo S, Beresford SW, Trofimovs J, Rosengren N (2013) Were intercalated komatiites and dacites at the Black Swan nickel sulphide mine, Yilgarn Craton, Western Australia, emplaced as extrusive lavas or intrusive bodies? The significance of breccia textures and contact relationships. Precambrian Res 229: 133–149

Cas RAF, Porritt L, Pittari A, Hayman PC (2008b) A new approach to kimberlite facies terminology using a revised general approach to the nomenclature of all volcanic rocks and deposits: descriptive to genetic. J Volcanol Geotherm Res 174(1–3):226–240

Cas RAF, Porritt L, Pittari A, Hayman PC (2009) A practical guide to terminology for kimberlite facies: a systematic progression from descriptive to genetic, including a pocket guide. Lithos 112:183–190

Cas RAF, Self S, Beresford S (1999) The behaviour of the fronts of komatiite lavas in medial to distal settings. Earth Planet Sci Lett 172 (1):127–139

Cas RAF, Simmons JM (2018) Why deep-water eruptions are so different from subaerial eruptions. Front Earth Sci 6:198. https://doi.org/10.3389/feart.2018.00198

Cas RAF, Wright JV (1987) Volcanic successions: modern and ancient. Allen & Unwin, London, p 528

Cas RAF, Yamagishi H, Moore L, Scutter CR (2003) Miocene submarine fire fountain deposits, Ryugazaki Headland, Oshoro Peninsula, Hokkaido, Japan: implications for submarine fountain dynamics and fragmentation processes. In: White JDL, Smellie JL, Clague DA (eds) Explosive subaqueous volcanism. Amer Geophys Union Monogr 140:299–316

Cashman KV, Giordano G (2014) Calderas and magma reservoirs. J Volcanol Geotherm Res 288:28–45

Cashman KV, Sparks RS, Blundy JD (2017) Vertically extensive and unstable magmatic systems: a unified view of igneous processes. Science 355(6331):eaag3055

Chabiron A, Cuney M, Poty B (2003) Possible uranium sources for the largest uranium district associated with volcanism: the Streltsovka caldera (Transbaikalia, Russia). Miner Deposita 38(2):127–140

Chambefort I, Moritz R (2014) Subaqueous environment and volcanic evolution of the Late Cretaceous Chelopech Au-Cu epithermal deposit, Bulgaria. J Volcanol Geotherm Res 289:1–13

Chambefort I, Moritz R, von Quadt A (2007) Petrology, geochemistry and U-Pb geochronology of magmatic rocks from the high-sulfidation epithermal Au-Cu Chelopech deposit, Srednogorie zone, Bulgaria. Miner Deposita 42(7):665–690

Chen H, Hu Y, Jin J, Ran Q, Yan L (2014) Fine stratigraphic division of volcanic reservoir by uniting of well data and seismic data—taking volcanic reservoir of member one of Yingcheng Formation in Xudong area of Songliao Basin for an example. J Earth Sci 25 (2):337–347. https://doi.org/10.1007/s12583-014-0424-8

Chen M, Campbell IH, Xue Y, Tian W, Ireland TR, Holden P, Cas RAF, Hayman PC, Das R (2015) Multiple sulfur isotope analyses support a magmatic model for the volcanogenic massive sulfide deposits of the Teutonic Bore Volcanic Complex, Yilgarn Craton, Western Australia. Econ Geol 110(6):1411–1423

Chen Z, Liu W, Zhang Y, Yan D, Yang D, Zha M, Li L (2016a) Characterization of the paleocrusts of weathered Carboniferous volcanics from the Junggar Basin, western China: significance as gas reservoirs. Mar Petrol Geol 77:216–234

Chen Z, Wang X, Zha M, Zhang Y, Cao Y, Yang D, Wu K, Chen Y, Yuan G (2016b) Characteristics and formation mechanisms of large volcanic rock oil reservoirs: a case study of the Carboniferous rocks in the Kebai fault zone of Junggar Basin, China. Amer Assoc Petrol Geol Bull 100(10):1585–1617

Chernyshev IV, Konečný V, Lexa J, Kovalenker VA, Jeleň S, Lebedev VA, Goltsman YV (2013) K-Ar and Rb-Sr geochronology and evolution of the Štiavnica stratovolcano (central Slovakia). Geol Carpath 64(4):327–360

Cherry AR, McPhie J, Kamenetsky VS, Ehrig K, Keeling JL, Kamenetsky MB, Meffre S, Apukhtina OB (2017) Linking Olympic Dam and the Cariewerloo Basin: was a sedimentary basin involved in formation of the world's largest uranium deposit? Precamb Res 300:168–180

Chi G, Xue C, Sun X, Lai J, Luo P, Song H, Li S, Zeng R (2017) Formation of a giant Zn-Pb deposit from hot brines injecting into a shallow oil-gas reservoir in sandstones, Jinding, southwestern China. Terra Nova 29(5):312–320

Chiles JP (1988) Fractal and geostatistical methods for modeling of a fracture network. Math Geol 20(6):631–654

Chiodi A, Tassi F, Báez W, Filipovich R, Bustos E, Galli MG, Vaselli O et al (2019) Preliminary conceptual model of the Cerro Blanco caldera-hosted geothermal system (Southern Puna, Argentina): Inferences from geochemical investigations. J South Am Earth Sci 94: 102213

Clifford BA (1992) Facies and palaeoenvironment analysis of the Archaean volcanic sedimentary succession hosting the Golden Grove Cu-Zn massive sulphide deposits, Western Australia. PhD thesis, Monash University, Melbourne, Australia

Cline JS, Hofstra AH, Muntean JL, Tosdal RM, Hickey KA (2005) Carlin-type gold deposits in Nevada: critical geologic characteristics and viable models. Econ Geol 100th Anniversary Volume 451–484

Cloos H (1941) Bau und Tatigkeit von Tuffschloten. Untersuchungen an dem Schwäbischen Vulkan. Geol Rundsch 32:705–800

Conrad JE, McKee EH, Rytuba JJ, Nash JT, Utterback WC (1993) Geochronology of the Sleeper Deposit, Humboldt County, Nevada; epithermal gold-silver mineralization following emplacement of a silicic flow-dome complex. Econ Geol 88(2):317–327

Cooke DR, Deyell CL, Waters PJ, Gonzales RI, Zaw K (2011) Evidence for magmatic-hydrothermal fluids and ore-forming processes in epithermal and porphyry deposits of the Baguio district, Philippines. Econ Geol 106(8):1399–1424

Cooke DR, McPhail DC (2001) Epithermal Au-Ag-Te mineralization, Acupan, Baguio district, Philippines: numerical simulations of mineral deposition. Econ Geol 96(1):109–131

Cooke DR, Simmons SF (2000) Characteristics and genesis of epithermal gold deposits. Rev Econ Geol 13:221–244

Cooper KM, Kent AJ (2014) Rapid remobilization of magmatic crystals kept in cold storage. Nature 506(7489):480

Corbett GJ (2002) Epithermal gold for explorationists. Aust Inst Geoscientists J 2002–01:26

Corbett GJ (2013) Pacific Rim epithermal Au-Ag. In: Proceedings World Gold, Brisbane, vol 9/13. Australasian Institute of Mining and Metallurgy, pp 5–13

Corbett KD (1992) Stratigraphic-volcanic setting of massive sulfide deposits in the Cambrian Mount Read Volcanics, Tasmania. Econ Geol 87(3):564–586

Corbett KD (2001) New mapping and interpretations of the Mount Lyell mining district, Tasmania: a large hybrid Cu-Au system with an exhalative Pb-Zn top. Econ Geol 96(5):1089–1122

Corrado S, Aldega L, Celano AS, De Benedetti AA, Giordano G (2014) Cap rock efficiency and fluid circulation of natural hydrothermal systems by means of XRD on clay minerals (Sutri, Northern Latium, Italy). Geothermics 50:180–188

Corseri R, Senger K, Selway K, Abdelmalak MM, Planke S, Jerram DA (2017) Magnetotelluric evidence for massive sulphide mineralization in intruded sediments of the outer Vøring Basin, mid-Norway. Tectonophys 706:196–205

Costa F (2008) Residence times of silicic magmas associated with calderas. In: Gottsmann J, Martí J (eds) Developments in volcanology, vol 10. Elsevier, Amsterdam, pp 1–55

Cowden A (1988) Emplacement of komatiite lava flows and associated nickel sulfides at Kambalda, Western Australia. Econ Geol 83 (2):436–442

Cowden A, Archibald NJ (1987) Massive-sulfide fabrics at Kambalda and their relevance to the inferred stability of monosulfide solid-solution. Canad Mineral 25(1):37–50

Cowden A, Roberts DE (1990) Komatiite hosted nickel sulphide deposits, Kambalda. In: Hughes FE (ed) Geology of the mineral deposits of Australia and Papua New Guinea. Australasian Institute of Mining and Metallurgy Monogr 14(1):567–581

Creaser RA (1996) Petrogenesis of a Mesoproterozoic quartz latite-granitoid suite from the Roxby Downs area, South Australia. Precamb Res 79(3–4):371–394

Cumming W (2016) Resource conceptual models of volcano-hosted geothermal reservoirs for exploration well targeting and resource capacity assessment: construction, pitfalls and challenges. Geotherm Res Council Trans 40:623–637

Cuney M, Kyser TK (eds) (2009) Recent and not-so-recent developments in uranium deposits and implications for exploration. Mineral Assoc Canada and Soc Geol Appl Mineral Deps 39:279

Cunningham CG, McNamee J, Pinto Vásquez J, Ericksen GE (1991) A model of volcanic dome-hosted precious metal deposits in Bolivia. Econ Geol 86(2):415–421

Dann JC (2001) Vesicular komatiites, 3.5-Ga Komati Formation, Barberton Greenstone Belt, South Africa: inflation of submarine lavas and origin of spinifex zones. Bull Volcanol 63(7):462–481

Dann JC, Grove TL (2007) Volcanology of the Barberton Greenstone Belt, South Africa: inflation and evolution of flow fields. Dev Precambr Geol 15(5.4):527–570

Davies AG, Cooke DR, Gemmell JB, van Leeuwen T, Cesare P, Hartshorn G (2008) Hydrothermal breccias and veins at the Kelian gold mine, Kalimantan, Indonesia: genesis of a large epithermal gold deposit. Econ Geol 103(4):717–757

Davison I, Steel I, Taylor M, Beirne EO, Faull T (2014) Exploration in the East African Rift System. GeoExPro 11(5)

De Filippis L, Anzalone E, Billi A, Faccenna C, Poncia PP, Sella P (2013) The origin and growth of a recently-active fissure ridge travertine over a seismic fault, Tivoli, Italy. Geomorphology 195:13–26

De Natale G, Troise C, Mark D, Mormone A, Piochi M, Di Vito MA, Isaia R, Carlino S, Barra D, Somma R (2016) The Campi Flegrei Deep Drilling Project (CFDDP): new insight on caldera structure, evolution and hazard implications for the Naples area (Southern Italy). Geochem Geophy Geosys 17(12):4836–4847

de Ronde CE, Baker ET, Massoth GJ, Lupton JE, Wright IC, Sparks RJ, Bannister SC, Reyners ME, Walker SL, Greene RR, Ishibashi J (2007) Submarine hydrothermal activity along the mid-Kermadec Arc, New Zealand: large-scale effects on venting. Geochem Geophys Geosys 8(7). https://doi.org/10.1029/2006GC001495

de Ronde CE, Fornari DJ, Ferrini VL, Walker SL, Davy BW, LeBlanc C, Tontini FC, Kukulya AL, Littlefield RH (2016) The Pink and White Terraces of Lake Rotomahana: what was their fate after the 1886 Tarawera Rift eruption? J Volcanol Geotherm Res 314:126–141

de Ronde CE, Hannington MD, Stoffers P, Wright IC, Ditchburn RG, Reyes AG, Baker ET, Massoth GJ, Lupton JE, Walker SL, Greene RR (2005) Evolution of a submarine magmatic-hydrothermal system: Brothers volcano, southern Kermadec arc, New Zealand. Econ Geol 100(6):1097–1133

de Ronde CE, Massoth GJ, Butterfield DA, Christenson BW, Ishibashi J, Ditchburn RG, Takai K et al (2011) Submarine hydrothermal activity and gold-rich mineralization at Brothers Volcano, Kermadec Arc, New Zealand. Mineralium Deposita 46: 541–584

Deering CD, Keller B, Schoene B, Bachmann O, Beane R, Ovtcharova M (2016) Zircon record of the plutonic-volcanic connection and protracted rhyolite melt evolution. Geology 44 (4):267–270

Deichmann N, Giardini D (2009) Earthquakes induced by the stimulation of an enhanced geothermal system below Basel (Switzerland). Seismol Res Lett 80(5):784–798

Deyell CL, Leonardson R, Rye RO, Thompson JF, Bissig T, Cooke DR (2005) Alunite in the Pascua-Lama high-sulfidation deposit:

constraints on alteration and ore deposition using stable isotope geochemistry. Econ Geol 100(1):131–148
Di Pippo R (ed) (2016) Geothermal power generation. Elsevier, Amsterdam, p 654
Di Renzo V, Wohletz K, Civetta L, Moretti R, Orsi G, Gasparini P (2016) The thermal regime of the Campi Flegrei magmatic system reconstructed through 3D numerical simulations. J Volcanol Geotherm Res 328:210–221
Dimalanta CB, Yumul GP (2004) Crustal thickening in an active margin setting (Philippines): the whys and the hows. Episodes 27 (4):260–264
Dobson PF, Kneafsey TJ, Hulen J, Simmons A (2003) Porosity, permeability, and fluid flow in the Yellowstone geothermal system, Wyoming. J Volcanol Geotherm Res 123(3–4):313–324
Dostal J (2008) Igneous rock associations 10. Komatiites. Geosci Canada 35(1)
Doust H, Noble RA (2008) Petroleum systems of Indonesia. Mar Petrol Geol 25(2):103–129
Doveri M, Lelli M, Marini L, Raco B (2010) Revision, calibration, and application of the volume method to evaluate the geothermal potential of some recent volcanic areas of Latium, Italy. Geothermics 39(3):260–269
Doyle MG, Allen RL (2003) Subsea-floor replacement in volcanic-hosted massive sulfide deposits. Ore Geol Rev 23(3–4):183–222
Drummond SE, Ohmoto H (1985) Chemical evolution and mineral deposition in boiling hydrothermal systems. Econ Geol 80(1):126–147
Dubé B, Dunning GR, Lauziere K (1998) Geology of the Hope Brook Mine, Newfoundland, Canada; a preserved late Proterozoic high-sulfidation epithermal gold deposit and its implications for exploration. Econ Geol 93(4):405–436
Dubé B, Gosselin PA, Mercier-Langevin P, Hannington M, Galley A (2007) Gold-rich volcanogenic massive sulphide deposits. In: Goodfellow WD (ed) Mineral Deposits of Canada: a synthesis of major deposit-types, district metallogeny, the evolution of geological provinces, and exploration methods. Geol Assoc Canada Mineral Deps Div Spec Publ 5:75–94
Duncan l, Helland-Hansen D, Dennehy C (2009) The Rosebank discovery: a new play type in intra basalt reservoirs of the North Atlantic Volcanic Province. In: Abstracts 6th European production and development conference and exhibition (DEVEX), Aberdeen, Scotland. http://www.devex-conference.org/pdf/Presentations_2009/
Dunham RJ (1962) Classification of carbonate rocks according to depositional textures. In: Ham WE (ed) Classification of carbonate rocks. Amer Assoc Petrol Geol Mem 108–121
Duuring P, Bleeker W, Beresford SW, Fiorentini ML, Rosengren NM (2012) Structural evolution of the Agnew-Wiluna greenstone belt, Eastern Yilgarn Craton and implications for komatiite-hosted Ni sulfide exploration. Aust J Earth Sci 59(5):765–791
Duuring P, Bleeker W, Beresford SW, Hayward N (2010) Towards a volcanic-structural balance: relative importance of volcanism, folding, and remobilisation of nickel sulphides at the Perseverance Ni-Cu-(PGE) deposit, Western Australia. Miner Deposita 45 (3):281–311
Dziak RP, Haxel JH, Matsumoto H, Lau TK, Merle SG, de Ronde CEJ, Embley RW, Mellinger DK (2008) Observations of regional seismicity and local harmonic tremor at Brothers volcano, south Kermadec arc, using an ocean bottom hydrophone array. J Geophys Res 113:B08S04. https://doi.org/10.1029/2007jb005533
Eaton PC, Setterfield TN (1993) The relationship between epithermal and porphyry hydrothermal systems within the Tavua Caldera, Fiji. Econ Geol 88(5):1053–1083
Echeverria LM (1982) Komatiites from Gorgona Island, Colombia. In: Arndt NT, Nisbet EG (eds) Komatiites. Allen & Unwin, London, pp 199–209
Ehrenberg SN, Nadeau PH (2005) Sandstone vs. carbonate petroleum reservoirs: a global perspective on porosity-depth and porosity-permeability relationships. Amer Assoc Petrol Geol Bull 89(4):435–445
Ehrig K, McPhie J, Kamenetsky VS (2012) Geology and mineralogical zonation of the Olympic Dam iron oxide Cu-U-Au-Ag deposit, South Australia. Soc Econ Geol Spec Publ 16:237–267
Elders WA, Friðleifsson GÓ, Albertsson A (2014) Drilling into magma and the implications of the Iceland Deep Drilling Project (IDDP) for high-temperature geothermal systems worldwide. Geothermics 49:111–118
Embley RW, de Ronde CE, Merle SG, Davy B, Tontini FC (2012) Detailed morphology and structure of an active submarine arc caldera: Brothers volcano, Kermadec Arc. Econ Geol 107(8):1557–1570
Ernando Z, Fathoni A (2011) Volcanic reservoir characterization of Jatibarang formation based on an integrated study of petrography, core, FMI, and well log. In: Proceedings of the 35th annual Indonesian Petrol Assoc convention and exhibition, Jakarta. Digital paper IPA11-G-166
Espi JO, Kajiwara Y, Hawkins MA, Bainbridge T (2002) Hydrothermal alteration and Cu-Au mineralization at Nena high sulfidation-type deposit, Frieda River, Papua New Guinea. Resource Geol 52 (4):301–313
Etoh J, Izawa E, Watanabe K, Taguchi S, Sekine R (2002) Bladed quartz and its relationship to gold mineralization in the Hishikari low-sulfidation epithermal gold deposit, Japan. Econ Geol 97 (8):1841–1851
Facca G, Tonani F (1967) The self-sealing geothermal field. Bull Volcanol 30(1):271–273
Farooqui MY, Hou H, Li G, Machin N, Neville T, Pal A, Shrivastva C, Wang Y, Yang F, Yin C, Zhao J (2009) Evaluating volcanic reservoirs. Oilfield Rev 21(1):36–47
Feng ZQ (2008) Volcanic rocks as prolific gas reservoir: a case study from the Qingshen gas field in the Songliao Basin, NE China. Mar Petrol Geol 25(4–5):416–432
Ferriz H, Mahood GA (1984) Eruption rates and compositional trends at Los Humeros volcanic center, Puebla, Mexico. J Geophys Res Solid Earth 89(B10):8511–8524
Field M, Scott Smith BH (1999) Contrasting geology and near-surface emplacement of kimberlite pipes in southern Africa and Canada. In: Proceedings of the 7th international kimberlite conference, Cape Town, South Africa, vol 1, pp 214–237
Field M, Stiefenhofer J, Robey J, Kurszlaukis S (2008) Kimberlite-hosted diamond deposits of southern Africa: a review. Ore Geol Rev 34(1–2):33–75
Fiorentini M, Beresford S, Barley M, Duuring P, Bekker A, Rosengren N, Cas RAF, Hronsky J (2012) District to camp controls on the genesis of komatiite-hosted nickel sulfide deposits, Agnew-Wiluna greenstone belt, Western Australia: insights from the multiple sulfur isotopes. Econ Geol 107(5):781–796
Filipovich R, Chiodi A, Báez W, Ahumada MF, Invernizzi C, Taviani S, Giordano G et al (2022) Structural analysis and fluid geochemistry as tools to assess the potential of the Tocomar geothermal system, Central Puna (Argentina). Geothermics 98: 102297
Fournier RO (1999) Hydrothermal processes related to movement of fluid from plastic into brittle rock in the magmatic-epithermal environment. Econ Geol 94(8):1193–1211
Franklin JM, Gibson HL, Jonasson IR, Galley AG (2005) Volcanogenic massive sulfide deposits. Econ Geol 100th Anniversary Volume 523–560

Friðleifsson GÓ, Elders WA, Zierenberg RA, Fowler AP, Weisenberger TB, Mesfin KG, Sigurðsson Ó, Níelsson S, Einarsson G, Óskarsson F, Guðnason EÁ (2018) The Iceland Deep Drilling Project at Reykjanes: drilling into the root zone of a black smoker analog. J Volcanol Geotherm Res. https://doi.org/10.1016/j.jvolgeores.2018.08.013

Fuchs S, Hannington MD, Petersen S (2019) Divining gold in seafloor polymetallic massive sulfide systems. Miner Deposita 54(6):789–820

Galley AG, Hannington MD, Jonasson IR (2007) Volcanogenic massive sulphide deposits. In: Goodfellow WD (ed) Mineral deposits of Canada: a synthesis of major deposit-types, district metallogeny, the evolution of geological provinces, and exploration methods. Geol Assoc Canada Mineral Deps Div Spec Publ 5:141–161

Garden TO, Gravley DM, Kennedy BM, Deering C, Chambefort I (2017) Controls on hydrothermal fluid flow in caldera-hosted settings: evidence from Lake City caldera, USA. Geosphere 13 (6):1993–2016

Gemmell JB, Fulton R (2001) Geology, genesis, and exploration implications of the footwall and hanging-wall alteration associated with the Hellyer volcanic-hosted massive sulfide deposit, Tasmania, Australia. Econ Geol 96(5):1003–1035

Gernon TM, Brown RJ, Tait MA, Hincks TK (2012) The origin of pelletal lapilli in explosive kimberlite eruptions. Nat Commun 3:832

Gernon TM, Gilbertson MA, Sparks RSJ, Field M (2009) The role of gas-fluidisation in the formation of massive volcaniclastic kimberlite. Lithos 112:439–451

Gernon TM, Spence S, Trueman CN, Taylor RN, Rohling E, Hatter SJ, Harding IC (2015) Emplacement of the Cabezo María lamproite volcano (Miocene, SE Spain). Bull Volcanol 77(6):52

Gibson HL, Allen RL, Riverin G, Lane TE (2007) The VMS model: advances and application to exploration targeting. In: Milkereit B (ed) Proceedings of exploration 07: fifth decennial international conference on mineral exploration, Toronto, Canada, pp 713–730

Gibson HL, Galley AG (2007) Volcanogenic massive sulfide deposits of the Archean, Noranda District, Quebec. In: Goodfellow WD (ed) Mineral deposits of Canada: a synthesis of major deposit-types, district metallogeny, the evolution of geological provinces and exploration methods. Geol Assoc Canada Mineral Deps Div Spec Publ 5:552–553

Gibson HL, Kerr DJ (1993) Giant volcanic-associated massive sulphide deposits; with emphasis on Archean deposits. In: Whiting BH, Hodgson CJ, Mason R (eds) Giant ore deposits. Soc Econ Geol Spec Publ 2:319–348

Gibson HL, Kerr DJ, Cattalani S (2000) The Horne mine: geology, history, influence on genetic models, and a comparison to the Kidd Creek mine. Explor Min Geol 9(2):91–111

Gibson HL, Morton RL, Hudak G (1999) Submarine volcanic processes, deposits, and environments favorable for the location of volcanic-associated massive sulfide deposits. In: Barrie CT, Hannington MD (eds) Volcanic-associated massive sulfide deposits-Processes and examples in modern and ancient settings. Rev Econ Geol 8:13–51

Gibson P (2020) New Zealand Whakaari volcano eruption and occupational hygiene responses. Occup Health S Afr 26(1):29

Gieskes JM, Kastner M, Jolla L (1982) Hydrothermal activity in the Guaymas Basin, Gulf of California: a synthesis. Init Rep Deep Sea Drill Program 64:1159–1167

Giordano G, Ahumada MF, Aldega L, Baez WA, Becchio RA, Bigi S, Caricchi C, Chiodi AL, Corrado S, De Benedetti AA, Favetto AB (2016) Preliminary data on the structure and potential of the Tocomar geothermal field (Puna plateau, Argentina). Energy Procedia 97:202–209

Giordano G, Caricchi L (2022) Determining the state of activity of transcrustal magmatic systems and their volcanoes. Ann Rev Earth Planet Sci. 50: 231–259

Giordano G, Cas RAF (2021) Classification of ignimbrites and related pyroclastic flows. Earth Sci Rev 220:103697. https://doi.org/10.1016/j.earscirev.2021.103697

Giordano G, De Benedetti AA, Bonamico A, Ramazzotti P, Mattei M (2014) Incorporating surface indicators of reservoir permeability into reservoir volume calculations: application to the Colli Albani caldera and the Central Italy Geothermal Province. Earth Sci Rev 128:75–92

Giordano G, Pinton A, Cianfarra P, Baez W, Chiodi A, Viramonte J, Norini G, Groppelli G (2013) Structural control on geothermal circulation in the Cerro Tuzgle-Tocomar geothermal volcanic area (Puna plateau, Argentina). J Volcanol Geotherm Res 249:77–94

Giordano TH (1994) Metal transport in ore fluids by organic ligand complexation. In: Pittman ED, Lewan MD (eds) Organic acids in geological processes. Springer, Berlin, Heidelberg, pp 319–354

Giuliani A, Schmidt MW, Torsvik TH, Fedortchouk F (2023) Genesis and evolution of kimberlites. Nature Rev Earth Envir 4:738–754

Glasby GP, Notsu K (2003) Submarine hydrothermal mineralization in the Okinawa Trough, SW of Japan: an overview. Ore Geol Rev 23 (3–4):299–339

Glasby GP, Yamanaka T, Yamamoto J, Sato H, Notsu K (2004) Kuroko and hydrocarbon deposits from northern Honshu, Japan: a possible common hydrothermal/magmatic origin? Resource Geol 54(4):413–424

Goodfellow WD, Grapes KA, Cameron B, Franklin JM (1993) Hydrothermal alteration associated with massive sulfide deposits, Middle Valley, northern Juan de Fuca Ridge. Canad Mineral 31 (4):1025–1060

Goodfellow WD, McCutcheon SR, Peter JM (2003) Massive sulfide deposits of the Bathurst Mining Camp, New Brunswick and Northern Maine: introduction and summary of findings. Econ Geol Monogr 11:1–16

Goodfellow WD, Zierenberg RA (1999) Genesis of massive sulfide deposits at sediment-covered spreading centers. Rev Econ Geol 8:297–324

Gole MJ, Robertson J, Barnes SJ (2013) Extrusive origin and structural modification of the komatiitic Mount Keith ultramafic unit. Econ Geol 108(7):1731–1752

Gresham JJ, Loftus-Hills GD (1981) The geology of the Kambalda nickel field, Western Australia. Econ Geol 76(6):1373–1416

Gröpper H, Calvo M, Crespo H, Bisso CR, Cuadra WA, Dunkerley PM, Aguirre E (1991) The epithermal gold-silver deposit of Choquelimpie, northern Chile. Econ Geol 86(6):1206–1221

Grove TL, Parman SW (2004) Thermal evolution of the Earth as recorded by komatiites. Earth Planet Sci Lett 219(3–4):173–187

Groves DA, Morton RL, Franklin JM (1988) Physical volcanology of the footwall rocks near the Mattabi massive sulphide deposit, Sturgeon Lake, Ontario. Canad J Earth Sci 25(2):280–291

Groves DI, Bierlein FP, Meinert LD, Hitzman MW (2010) Iron oxide copper-gold (IOCG) deposits through Earth history: implications for origin, lithospheric setting, and distinction from other epigenetic iron oxide deposits. Econ Geol 105(3):641–654

Groves DI, Goldfarb RJ, Robert F, Hart CJ (2003) Gold deposits in metamorphic belts: overview of current understanding, outstanding problems, future research, and exploration significance. Econ Geol 98(1):1–29

Groves DI, Korkiakoski EA, McNaughton NJ, Lesher CM, Cowden A (1986) Thermal erosion by komatiites at Kambalda, Western Australia and the genesis of nickel ores. Nature 319(6049):136–139

Guber AL, Merrill S (1983) Paleobathymetric significance of the foraminifera from the Hokuroko district. In: Ohmoto H, Skinner BJ (eds) The Kuroko and related volcanogenic massive sulfide deposits. Econ Geol Monogr 5:55–70

Guido D, De Barrio R, Schalamuk I (2002) La Marciana Jurassic sinter: Implications for exploration for epithermal precious-metal deposits in Deseado Massif, southern Patagonia, Argentina. Appl Earth Sci 111(2):106–113

Guney ME, Al-Marhoun MA, Nawab ZA (1988) Metalliferous sub-marine sediments of the Atlantis-II-Deep, Red Sea. Canad Inst Min Metall Bull 81:33–39

Gurney JJ, Helmstaedt HH, Richardson SH, Shirey SB (2010) Diamonds through time. Econ Geol 105(3):689–712

Haggerty SE (1994) Superkimberlites: a geodynamic diamond window to the Earth's core. Earth Planet Sci Lett 122(1–2):57–69

Haggerty SE (1999) A diamond trilogy: superplumes, supercontinents, and supernovae. Science 285(5429):851–860

Hallberg A (1994) The Enåsen gold deposit, central Sweden. Miner Deposita 29(2):150–162

Hallberg JA, Thompson JFH (1985) Geologic setting of the Teutonic Bore massive sulfide deposit, Archean Yilgarn block, Western Australia. Econ Geol 80(7): 1953–1964

Halter WE, Bain N, Becker K, Heinrich CA, Landtwing M, VonQuadt A, Clark AH, Sasso AM, Bissig T, Tosdal RM (2004) From andesitic volcanism to the formation of a porphyry Cu-Au mineralizing magma chamber: the Farallón Negro Volcanic Complex, northwestern Argentina. J Volcanol Geotherm Res 136(1): 1–30

Hames W, Unger D, Saunders J, Kamenov G (2009) Early Yellowstonehotspot magmatism and gold metallogeny. J Volcanol GeothermRes 188(1–3):214–222

Hand M, Reid A, Jagodzinski L (2007) Tectonic framework and evolution of the Gawler craton, southern Australia. Econ Geol 102 (8):1377–1395

Hannington M, Jamieson J, Monecke T, Petersen S, Beaulieu S (2011) The abundance of seafloor massive sulfide deposits. Geology 39 (12):1155–1158

Hannington MD, Bleeker W, Kjarsgaard I (1999a) Sulfide mineralogy, geochemistry, and ore genesis of the Kidd Creek deposit: Part I. North, Central, and South orebodies. In: Hannington MD, Barrie CT (eds) Volcanic-associated massive sulfide deposits: processes and examples in modern and ancient settings. Econ Geol Monogr 10:163–224

Hannington MD, Bleeker W, Kjarsgaard I (1999b) Sulfide mineralogy, geochemistry, and ore genesis of the Kidd Creek deposit: Part II. The bornite zone. In: Hannington MD, Barrie CT (eds) Volcanic-associated massive sulfide deposits: processes and examples in modern and ancient settings. Econ Geol Monogr 10:225–266

Hannington MD, de Ronde CD, Petersen S (2005) Sea-floor tectonics and submarine hydrothermal systems. In: Hedenquist JW, Thompson JFH, Goldfarb RJ, Richards JP (eds) Econ Geol 100th Anniversary Volume 111–141

Hannington MD, Galley AG, Herzig PM, Petersen S (1998) Comparison of the TAG mound and stockwork complex with Cyprus-type massive sulfide deposits. Proc ODP Sci Results 158(28):389–415

Hansen DM, Cartwright J (2006) Saucer-shaped sill with lobate morphology revealed by 3D seismic data: implications for resolving a shallow-level sill emplacement mechanism. J Geol Soc Lond 163 (3):509–523

Hardardóttir V, Hannington M, Hedenquist J, Kjarsgaard I, Hoal K (2010) Cu-rich scales in the Reykjanes geothermal system, Iceland. Econ Geol 105(6):1143–1155

Harper G (2005) Saddle Hills area uranium prospects, Dornod Aimag, eastern Mongolia. Report to Western Prospector Group, Vancouver, Canada, p 237

Harris AC, Bryan SE, Holcombe RJ (2006) Volcanic setting of the Bajo de la Alumbrera porphyry Cu-Au deposit, Farallón Negro volcanics, northwest Argentina. Econ Geol 101(1):71–94

Hart TR, Gibson HL, Lesher CM (2004) Trace element geochemistry and petrogenesis of felsic volcanic rocks associated with volcanogenic massive Cu-Zn-Pb sulfide deposits. Econ Geol 99 (5):1003–1013

Hasegawa S, Akimoto K, Kitazato H, Matoba Y (1989) Late Cenozoic paleobathymetric indices based on benthic foraminifers in Japan. Geol Soc Jpn Mem 32:241–253

Hayba B (1985) Geologic, mineralogic and geochemical characteristics of volcanic hosted epithermal precious metal deposits. In: Geology and geochemistry of epithermal systems. Rev Econ Geol 2:129–167

Hawke ML, Meffre S, Stein H, Hilliard P, Large R, Gemmell JB (2015) Geochronology of the DeGrussa volcanic-hosted massive sulphide deposit and associated mineralisation of the Yerrida, Bryah and Padbury Basins, Western Australia. Precamb Res 267:250–284

Hayman PC, Cas RA, Johnson M (2009) Characteristics and alteration origins of matrix minerals in volcaniclastic kimberlite of the Muskox pipe (Nunavut, Canada). In: Proceedings of the 9th international kimberlite conference, Frankfurt, Germany. Lithos 112:473–487

Hayman PC, Thébaud N, Pawley MJ, Barnes SJ, Cas RAF, Amelin Y, Sapkota J, Squire RJ, Campbell IH, Pegg I (2015) Evolution of a ~ 2.7 Ga large igneous province: a volcanological, geochemical and geochronological study of the Agnew Greenstone Belt, and new regional correlations for the Kalgoorlie Terrane (Yilgarn Craton, Western Australia). Precamb Res 270:334–368

Hayman PC, Cas RAF (2011a) Reconstruction of a multi-vent kimberlite eruption from deposit and host rock characteristics: Jericho kimberlite, Nunavut, Canada. J Volcanol Geotherm Res 200 (3–4):201–222

Hayman PC, Cas RAF (2011b) Criteria for interpreting kimberlite as coherent: insights from the Muskox and Jericho kimberlites (Nunavut, Canada). Bull Volcanol 73(8):1005–1027

Hayman PC, Cas RAF, Johnson M (2008) Difficulties in distinguishing coherent from fragmental kimberlite: a case study of the Muskox pipe (Northern Slave Province, Nunavut, Canada). J Volcanol Geotherm Res 174(1–3):139–151

Hayman PC, Cas RAF, Squire RJ, Campbell IA, Chen M, Doutch D (2019) Emplacement origins of coarsely-crystalline mafic rocks hosted in greenstone belts: examples from the 2.7 Ga Yilgarn Craton, Western Australia. Precamb Res 324:236–252

Haynes DW, Cross KC, Bills RT, Reed MH (1995) Olympic Dam ore genesis; a fluid-mixing model. Econ Geol 90(2):281–307

Heald P, Foley NK, Hayba DO (1987) Comparative anatomy of volcanic-hosted epithermal deposits; acid-sulfate and adularia-sericite types. Econ Geol 82(1):1–26

Heap MJ, Gravley DM, Kennedy BM, Gilg HA, Bertolett E, Barker SL (2020) Quantifying the role of hydrothermal alteration in creating geothermal and epithermal mineral resources: the Ohakuri ignimbrite (Taupo Volcanic Zone, New Zealand). J Volcanol Geotherm Res 390:106703

Hedenquist JW, Arribas A (2021) Exploration implications of multiple formation environments of advanced argillic minerals. Econ Geol. https://doi.org/10.5382/econgeo.4880

Hedenquist JW, Arribas A, Gonzalez-Urien E (2000) Exploration for epithermal gold deposits. In: Gold in 2000. Rev Econ Geol 13 (2):45–77

Hedenquist JW, Arribas A, Aoki M (2017) Zonation of sulfate and sulfide minerals and isotopic composition in the Far Southeast porphyry and Lepanto epithermal Cu–Au deposits, Philippines. Resour Geol 67(2):174–196

Hedenquist JW, Arribas A, Reynolds TJ (1998) Evolution of an intrusion-centered hydrothermal system; Far Southeast-Lepanto porphyry and epithermal Cu-Au deposits, Philippines. Econ Geol 93(4):373–404

Hedenquist JW, Harris M, Camus F (eds) (2012) Geology and genesis of major copper deposits and districts of the world: a tribute to Richard H. Sillitoe. Soc Econ Geol Spec Publ 16:619

Hedenquist JW, Henley RW (1985a) Hydrothermal eruptions in the Waiotapu geothermal system, New Zealand; their origin, associated breccias, and relation to precious metal mineralization. Econ Geol 80(6):1640–1668

Hedenquist JW, Henley RW (1985b) The importance of CO_2 on freezing point measurements of fluid inclusions: evidence from active geothermal systems and implications for epithermal ore deposition. Econ Geol 80(5):1379–1406

Hedenquist JW, Lowenstern JB (1994) The role of magmas in the formation of hydrothermal ore deposits. Nature 370(6490):519

Hedenquist JW, Simmons SF, Giggenbach WF, Eldridge CS (1993) White Island, New Zealand, volcanic-hydrothermal system represents the geochemical environment of high-sulfidation Cu and Au ore deposition. Geology 21(8):731–734

Hedenquist JW, Taran YA (2013) Modeling the formation of advanced argillic lithocaps: volcanic vapor condensation above porphyry intrusions. Econ Geol 108(7):1523–1540

Henley RW (1985) The geothermal framework of epithermal deposits. In: Berger BR, Bethke PM (eds) Geology and geochemistry of epithermal systems. Rev Econ Geol 2:1–24

Henley RW, Ellis AJ (1983) Geothermal systems ancient and modern: a geochemical review. Earth Sci Rev 19(1):1–50

Henley RW, Hughes GO (2000) Underground fumaroles: "Excess heat" effects in vein formation. Econ Geol 95(3):453–466

Henry CD, Elson HB, McIntosh WC, Heizler MT, Castor SB (1997) Brief duration of hydrothermal activity at Round Mountain, Nevada, determined from Ar^{40}/Ar^{39} geochronology. Econ Geol 92 (7–8):807–826

Higgins O, Sheldrake T, Caricchi L (2022) Machine learning thermo-barometry and chemometry using amphibole and clinopyroxene: a window into the roots of an arc volcano (Mount Liamuiga, Saint Kitts). Contrib Mineral Petrol 177(1): 10.

Hoatson DM, Jaireth S, Jaques AL (2006) Nickel sulfide deposits in Australia: characteristics, resources, and potential. Ore Geol Rev 29 (3–4):177–241

Hocking MW, Hannington MD, Percival JB, Stoffers P, Schwarz-Schampera U, de Ronde CE (2010) Clay alteration of volcaniclastic material in a submarine geothermal system, Bay of Plenty, New Zealand. J Volcanol Geotherm Res 191(3–4):180–192

Hofstra AH, Cline JS (2000) Characteristics and models for Carlin-type gold deposits. In: Gold in 2000. Rev Econ Geol 13:163–220

Hollis SP, Mole DR, Gillespie P, Barnes SJ, Tessalina S, Cas RA, Hildrew C, Pumphrey A, Goodz MD, Caruso S, Yeats CJ (2017a) 2.7 Ga plume associated VHMS mineralization in the Eastern Goldfields Superterrane, Yilgarn Craton: insights from the low temperature and shallow water, Ag-Zn-(Au) Nimbus deposit. Precamb Res 291:119–142

Hollis SP, Yeats CJ, Wyche S, Barnes SJ, Ivanic TJ (2017b) VMS mineralization in the Yilgarn Craton, Western Australia: a review of known deposits and prospectivity analysis of felsic volcanic rocks. Geol Surv Western Australia Rept 165:68

Hoshi K, Saga H, Minowa H, Inaba M (1992) Alteration and reservoir property of the Green Tuff Rocks in Akita and Niigata oil fields. J Jpn Assoc Petrol Techn 57(1):77–90

Hoshi K, Okubo S (2010) Hydrothermally dissolved dolerite reservoir in the Akita Basin, Japan. In: Amer Assoc Petrol Geol conference and exhibition, Calgary, Canada. Search and Discovery Article 50370

Holm RJ, Rosenbaum G, Richards SW (2016) Post 8 Ma reconstruction of Papua New Guinea and Solomon Islands: Microplate tectonics in a convergent plate boundary setting. Earth Sci Rev 156:66–81

Holm RJ, Richards SW (2013) A re-evaluation of arc-continent collision and along-arc variation in the Bismarck Sea region, Papua New Guinea. Aust J Earth Sci 60(5):605–619

Hou L, Yang C, Yang F, Lou X, Wei Y, Batten-Hender K (2015) Petroleum geology of Carboniferous volcanic weathered crust in northern Xinjiang. China. Canad Petrol Geol Bull 63(2):171–191

Houlé MG, Gibson HL, Lesher CM, Davis PC, Cas RAF, Beresford SW, Arndt NT (2008) Komatiitic sills and multigenerational peperite at Dundonald Beach, Abitibi greenstone belt, Ontario: volcanic architecture and nickel sulfide distribution. Econ Geol 103 (6):1269–1284

Huafeng T, Cryton P, Youfeng G, Yulong H, Weihua B (2015) Types and characteristics of volcanostratigraphic boundaries and their oil-gas reservoir significance. Acta Geol Sin 89(1):163–174

Hulen JB, Collister JW (1999) The oil-bearing, carlin-type gold deposits of Yankee Basin, Alligator Ridge District, Nevada. Econ Geol 94(7):1029–1049

Huppert HE, Sparks RSJ (1985) Komatiites. I: eruption and flow. J Petrol 26(3):694–725

IGA Service GmbH (2014) Best practices guide for geothermal exploration. Bochum, Germany, p 196

Ilchik RP, Barton MD (1997) An amagmatic origin of Carlin-type gold deposits. Econ Geol 92(3):269–288

Inaba M (2001) Basalt reservoir in the Yurihara Oil and Gas Field, northeast Japan. J Jpn Assoc Petrol Techn 66(1):56–67

Infante-Paez L, Marfurt KJ (2017) Seismic expression and geomorphology of igneous bodies: a Taranaki Basin, New Zealand, case study. Interpretation 5(3):121–140. https://doi.org/10.1190/INT-2016-0244.1

Invernizzi C, Pierantoni PP, Chiodi A, Maffucci R, Corrado S, Baez W, Tassi F, Giordano G, Viramonte J (2014) Preliminary assessment of the geothermal potential of Rosario de la Frontera area (Salta, NW Argentina): insight from hydro-geological, hydro-geochemical and structural investigations. J South Amer Earth Sci 54:20–36

Ischukova LP (1997) The Streltsovskoye uranium district. Int Atomic Energy Agency TECDOC 961:237–249

Ishibashi JI, Ikegami F, Tsuji T, Urabe T (2015) Hydrothermal activity in the Okinawa Trough back-arc basin: geological background and hydrothermal mineralization. In: Subseafloor biosphere linked to hydrothermal systems. Springer, Tokyo, pp 337–359

Ishibashi JI, Urabe T (1995) Hydrothermal activity rel ated to arc-backarc magmatism in the Western Pacific. In: Taylor B (ed) Backarc basins: tectonics and magmatism. Plenum Press, New York, pp 451–495

Ishihara S (ed) (1974) Geology of the Kuroko deposits. Society of Mining Geologists Japan, Special Issue 6:437

Izawa E, Urashima Y (1989) Quaternary gold mineralization and its geologic environments in Kyushu, Japan. In: Keays RR, Ramsay WRH, Groves DI (eds) The geology of gold deposits: the perspective in 1988. Econ Geol Monogr 6:233–241

Izquierdo G, Arellano VM, Aragón A, Portugal E, Martínez I (2000) Fluid acidity and hydrothermal alteration at the Los Humeros geothermal reservoir, Puebla, Mexico. In: Proceedings of the World Geothermal Congress, pp 1301–1306

Jackson CA, Holford SP, Magee C, Schofield N, Sun Q (eds) (2017) Subsurface expression of igneous systems and their impacts on petroleum systems. Soc Explor Geophys and Amer Assoc Petrol Geol, Interpretation 5(3)

Jackson RG, Chevillon CV (2007) 3D Geochemical and mineralogical model of the Sleeper low sulphidation gold system, Nevada. Exploration '07, Poster Presentation, Toronto, Canada. http://www.dmec.ca/ex07-dvd/E07/pdfs/18P.pdf

Jacobson HS, Murillo C, Ruiz L, Tapia O, Zapata H, Alarcón H, Delgadillo E, Velasco C (1969) Geology and mineral deposits of the San Cristobal district, Villa Martin province, Potosí, Bolivia. US Geol Surv Bull 1273:22

James RH, Duckworth RC, Palmer MR (1998) Drilling of sediment-hosted massive sulphide deposits at the Middle Valley and Escanaba Trough spreading centres: ODP Leg 169. Geol Soc Lond Spec Publ 148(1):177–199

Jarvis I, Lignum JS, Gröcke DR, Jenkyns HC, Pearce MA (2011) Black shale deposition, atmospheric CO_2 drawdown, and cooling during the Cenomanian-Turonian Oceanic Anoxic Event. Paleocean Paleoclimat 26(3). https://doi.org/10.1029/2010PA002081

Jelenković R, Milovanović D, Koželj D, Banješević M (2016) The mineral resources of the Bor metallogenic zone: a review. Geologia Croatica 69(1):143–155

Jelsma H, Barnett W, Richards S, Lister G (2009) Tectonic setting of kimberlites. Lithos 112:155–165

Jensen EP, Barton MD (2000) Gold deposits related to alkaline magmatism. In: Gold in 2000. Rev Econ Geol 13:279–314

Jerram DA, Single RT, Hobbs RW, Nelson CE (2009) Understanding the offshore flood basalt sequence using onshore volcanic facies analogues: an example from the Faroe-Shetland basin. Geol Mag 146(3):353–367

Jia C (ed) (2012) Characteristics of Chinese petroleum geology. Advanced topics in science and technology in China. Springer, Berlin, Heidelberg, p 591

John DA (2001) Miocene and early Pliocene epithermal gold-silver deposits in the northern Great Basin, western United States: characteristics, distribution, and relationship to magmatism. Econ Geol 96(8):1827–1853

Johnson CL, Webb LE, Graham SA, Hendrix MS, Badarch G (2001) Sedimentary and structural records of late Mesozoic high-strain extension and strain partitioning, East Gobi basin, southern Mongolia. Geol Soc Amer Mem 194:413–434

Johnson JP, McCulloch MT (1995) Sources of mineralising fluids for the Olympic Dam deposit (South Australia): Sm-Nd isotopic constraints. Chem Geol 121(1–4):177–199

Jones TJ, Russell JK, Sasse D (2019) Modification of mantle cargo by turbulent ascent of kimberlite. Front Earth Sci 7:134

Jorgenson C, Higgins O, Petrelli M, Bégué F, Caricchi L (2022) A Machine Learning-Based Approach to Clinopyroxene Thermobarometry: Model Optimization and Distribution for Use in Earth Sciences. J Geophys Res: Solid Earth, 127(4): e2021JB022904

Jowitt SM, Medlin CC, Cas RAF (2017) The rare earth element (REE) mineralisation potential of highly fractionated rhyolites: a potential low-grade, bulk tonnage source of critical metals. Ore Geol Rev 86:548–562

Kawamoto T (2001) Distribution and alteration of the volcanic reservoir in the Minami-Nagaoka gas field. J Jpn Assoc Petrol Techn 66(1):46–55

Keays RR, Lightfoot P (2010) Crustal sulfur is required to form magmatic Ni–Cu sulphide deposits; evidence from chalcophile element signatures of Siberian and Deccan Trap basalts. Mineral Deposita 45:241–257

Kelley KD, Romberger SB, Beaty DW, Pontius JA, Snee LW, Stein HJ, Thompson TB (1998) Geochemical and geochronological constraints on the genesis of Au-Te deposits at Cripple Creek, Colorado. Econ Geol 93(7):981–1012

Kerr AC (1998) Oceanic plateau formation: a cause of mass extinction and black shale deposition around the Cenomanian-Turonian boundary? J Geol Soc Lond 155(4):619–626

Kerr AC (2005) Oceanic LIPs: the kiss of death. Elements 1(5):289–292

Kerr AC, Marriner GF, Arndt NT, Tarney J, Nivia A, Saunders AD, Duncan RA (1996) The petrogenesis of Gorgona komatiites, picrites and basalts: new field, petrographic and geochemical constraints. Lithos 37(2–3):245–260

Kesler SE, Riciputi LC, Ye Z (2005) Evidence for a magmatic origin for Carlin-type gold deposits: isotopic composition of sulfur in the Betze-Post-Screamer deposit, Nevada, USA. Miner Deposita 40 (2):127–136

Kessel LG, Busby CJ (2003) Analysis of VHMS-hosting ignimbrites erupted at bathyal water depths (Ordovican Bald Mountain sequence, northern Maine). In: White JD, Smellie JL, Clague DA (eds) Explosive subaqueous volcanism. Amer Geophys Union Monogr 140:361–379

Khomich VG, Boriskina NG, Santosh M (2016) Geodynamic framework of large unique uranium orebelts in Southeast Russia and East Mongolia. J Asian Earth Sci 119:145–166

Kikuchi Y, Tono S, Funayama M (1991) Petroleum resources in the Japaneses island-arc setting. Episodes 14(3):236–240

Kitazato H (1979) Marine paleobathymetry and paleotopography of the Hokuroku district during the time of the Kuroko deposition, based on foraminiferal assemblages. Min Geol 29(156):207–216

Kjarsgaard BA (2007) Kimberlite diamond deposits. In: Goodfellow WD (ed) Mineral deposits of Canada: a synthesis of major deposit types, district metallogeny, the evolution of geological provinces, and exploration methods. Geol Assoc Canada Mineral Deps Div Spec Publ 5:245–272

Kjarsgaard BA, de Wit M, Heaman LM, Pearson DG, Stiefenhofer J, Janusezcak, Shirey SB (2022) A review of the geology of global diamond mines and deposits. Reviews in Mineral and Geochem 88:1-118

Kjarsgaard BA, Harvey S, Du Plessis P, McClintock M (2008) Comparative volcanology and emplacement styles of the Star and Orion South kimberlites, Fort à la Corne, Canada. In: 9th international kimberlite conference extended abstract no. 9IKC-A-00342

Kjarsgaard BA, Heaman LM, Sarkar C, Pearson DG (2017) The North America mid-Cretaceous kimberlite corridor: wet, edge-driven decompression melting of an OIB-type deep mantle source. Geochem Geophys Geosyst 18(7):2727–2747

Kjoberg S, Schmiedel T, Planke S, Svensen HH, Millett JM, Jerram DA, Galland O, Lecomte I, Schofield N, Haug ØT, Helsem A (2017) 3D structure and formation of hydrothermal vent complexes at the Paleocene-Eocene transition, the Møre Basin, mid-Norwegian margin. Interpretation 5(3):65–81

Koděra P, Bakos F, Kubač A, Lexa J, Sýkora M (2016) Epithermal Au-Ag mineralisation at the locality Prochot-Lazy in the mantle of the Štiavnica Stratovolcano. Acta Geologica Slovaca 8:149–163

Koděra P, Lexa J, Rankin AH, Fallick AE (2005) Epithermal gold veins in a caldera setting: Banská Hodruša, Slovakia. Miner Deposita 39 (8):921–943

Komatsu N, Fujita Y, Sato O (1984) Cenozoic volcanic rocks as potential hydrocarbon reservoirs. In: Proceedings 11th World Petroleum Congress, London, UK. Wiley, Chichester 2, pp 411–422, 137–146

Kolditz O (1995) Modelling flow and heat transfer in fractured rocks: conceptual model of a 3-D deterministic fracture network. Geothermics 24(3):451–470

Koski RA (2012) Hypogene ore characteristics. In: Shanks, Thurston R (eds) Volcanogenic massive sulfide occurrence model. US Geol Surv Scient Invest Rept 2010–5070-C:133–146

Koski RA, Lonsdale PF, Shanks WC, Berndt ME, Howe SS (1985) Mineralogy and geochemistry of a sediment-hosted hydrothermal sulfide deposit from the Southern Trough of Guaymas Basin, Gulf of California. J Geophys Res Solid Earth 90(B8):6695–6707

Krupp RE, Seward TM (1987) The Rotokawa geothermal system, New Zealand; an active epithermal gold-depositing environment. Econ Geol 82(5):1109–1129

Kurszlaukis S, Lorenz V (2008) Formation of "tuffisitic kimberlites" by phreatomagmatic processes. J Volcanol Geotherm Res 174(1–3):68–80

Kusky TM, Windley BF, Wang L, Wang Z, Li X, Zhu P (2014) Flat slab subduction, trench suction, and craton destruction: comparison of the North China, Wyoming, and Brazilian cratons. Tectonophysics 630:208–221

Kwak TA (1990) Geochemical and temperature controls on ore mineralization at the Emperor gold mine, Vatukoula, Fiji. J Geochem Explor 36(1–3):297–337

Lambert IB, Sato T (1974) The Kuroko and associated ore deposits of Japan; a review of their features and metallogenesis. Econ Geol 69 (8):1215–1236

Large RR (1992) Australian volcanic-hosted massive sulfide deposits: features, styles, and genetic models. Econ Geol 87:471–510

Large RR, Bull SW, Maslennikov VV (2011) A carbonaceous sedimentary source-rock model for Carlin-type and orogenic gold deposits. Econ Geol 106(3):331–358

Large RR, McPhie J, Gemmell JB, Herrmann W, Davidson GJ (2001) The spectrum of ore deposit types, volcanic environments, alteration halos, and related exploration vectors in submarine volcanic successions: some examples from Australia. Econ Geol 96 (5):913–938

Larsen M, Bell B, Guarnieri P, Vosgerau H, Weibel R (2018) Exploration challenges along the North Atlantic volcanic margins: the intra-volcanic sandstone play in subsurface and outcrop. In: Bowman M, Levell B (eds) Petroleum geology of NW Europe: 50 years of learning. Geol Soc Lond, Petroleum Geology Conference Series 8 (1):231–245

Le Vaillant M, Fiorentini ML, Barnes SJ (2016) Review of lithogeochemical exploration tools for komatiite-hosted Ni-Cu-(PGE) deposits. J Geochem Explor 168:1–9

Lefebvre NS, White JD, Kjarsgaard BA (2013) Unbedded diatreme deposits reveal maar-diatreme-forming eruptive processes: Standing Rocks West, Hopi Buttes, Navajo Nation, USA. Bull Volcanol 75 (8):1–7

Leistel JM, Marcoux E, Thiéblemont D, Quesada C, Sánchez A, Almodóvar GR, Pascual E, Sáez R (1998) The volcanic-hosted massive sulphide deposits of the Iberian Pyrite Belt Review and preface to the Thematic Issue. Miner Deposita 33(1–2):2–30

Lenhardt N, Götz AE (2011) Volcanic settings and their reservoir potential: an outcrop analog study on the Miocene Tepoztlán Formation, Central Mexico. J Volcanol Geotherm Res 204(1–4):66–75

Lentz DR (1998) Petrogenetic evolution of felsic volcanic sequences associated with Phanerozoic volcanic-hosted massive sulphide systems: the role of extensional geodynamics. Ore Geol Rev 12 (5):289–327

Lentz DR, Goodfellow WD (1996) Intense silicification of footwall sedimentary rocks in the stockwork alteration zone beneath the Brunswick No. 12 massive sulphide deposit, Bathurst, New Brunswick. Canad J Earth Sci 33(2):284–302

Lesher CM (1989) Komatiite-associated nickel sulfide deposits. In: Whitney JA, Naldrett AJ (eds) Ore deposition associated with magmas. Rev Econ Geol 4:45–101

Lesher CM, Burnham OM, Keays RR, Barnes SJ, Hulbert L (2001) Trace-element geochemistry and petrogenesis of barren and ore-associated komatiites. Canad Mineral 39(2):673–696

Lesher CM, Goodwin AM, Campbell IH, Gorton MP (1986) Trace-element geochemistry of ore-associated and barren, felsic metavolcanic rocks in the Superior Province, Canada. Canad J Earth Sci 23(2):222–237

Levell B, Bowman M (2018) The Petroleum geology of NW Europe: 50 years of learning—a platform for present value and future success. An overview. In: Bowman M, Levell B (eds) Petroleum geology of NW Europe: 50 years of learning. Geol Soc Lond, Petroleum Geology Conference Series 8(1):1–8

Leybourne MI, de Ronde CE, Wysoczanski RJ, Walker SL, Timm C, Gibson HL, Layton-Matthews D, Baker ET, Clark MR, Tontini FC, Faure K (2012a) Geology, hydrothermal activity, and sea-floor massive sulfide mineralization at the Rumble II west mafic caldera. Econ Geol 107(8):1649–1668

Leybourne MI, Schwarz-Schampera U, de Ronde CE, Baker ET, Faure K, Walker SL, Butterfield DA, Resing JA, Lupton JE, Hannington MD, Gibson HL (2012b) Submarine magmatic-hydrothermal systems at the Monowai volcanic center, Kermadec Arc. Econ Geol 107(8):1669–1694

Lightfoot PC, Hawkesworth CJ (1997) Flood basalts and magmatic Ni, Cu, and PGE sulphide mineralization: comparative geochemistry of the Noril'sk (Siberian Traps) and West Greenland sequences. Amer Geophys Union Monogr 100:357–380

Lindblom S (1991) Organic matter and gold deposition in disseminated gold deposits in Nevada. In: Pagel M, Leroy JL (eds) Source, transport and deposition of metals. Balkema, Rotterdam, pp 553–556

Lindley D (1988) Early Cainozoic stratigraphy and structure of the Gazelle Peninsula, east New Britain: an example of extensional tectonics in the New Britain arc-trench complex. J Geol Soc Aust 35 (2):231–244

Lindley ID (2016) Plate flexure and volcanism: Late Cenozoic tectonics of the Tabar-Lihir-Tanga-Feni alkalic province, New Ireland Basin, Papua New Guinea. Tectonophysics 677:312–323

Lipman PW (2007) Incremental assembly and prolonged consolidation of Cordilleran magma chambers: evidence from the Southern Rocky Mountain volcanic field. Geosphere 3(1):42–70

Lipton I, Gleeson E, Munroe P (2018) Preliminary economic assessment of the Solwara project, Bismarck Sea PNG. Report to Nautilus Minerals Niugini Ltd., AMC Consultants, Brisbane, Australia, p 274

Lister GS, Etheridge MA, Symonds PA (1986) Detachment faulting and the evolution of passive continental margins. Geology 14 (3):246–250

Liu J, Wang P, Zhang Y, Bian W, Huang Y, Tang H, Chen X (2012) Volcanic rock-hosted natural hydrocarbon resources: a review. In: Nemoth K (ed) Updates in volcanology. IntechOpen. https://doi.org/10.5772/54587

Longo AA, Dilles JH, Grunder AL, Duncan R (2010) Evolution of calc-alkaline volcanism and associated hydrothermal gold deposits at Yanacocha, Peru. Econ Geol 105(7):1191–1241

Lorenz V (1975) Formation of phreatomagmatic maar-diatreme volcanoes and its relevance to kimberlite diatremes. Phys Chem Earth 9:17–27

Lorenz V, Kurszlaukis S (2007) Root zone processes in the phreatomagmatic pipe emplacement model and consequences for the evolution of maar-diatreme volcanoes. J Volcanol Geotherm Res 159(1–3):4–32

Lorenz V, Zimanowski B, Büttner R, Kurszlaukis S (1999) Formation of kimberlite diatremes by explosive interaction of kimberlite magma with groundwater: field and experimental aspects. In: Proceedings of the 7th international kimberlite conference, Cape Town, South Africa, vol 2, pp 522–528

Lowe DR (1999) Shallow-water sedimentation of accretionary lapilli-bearing strata of the Msauli Chert: evidence of explosive hydromagmatic komatiitic volcanism. In: Lowe D, Byerly G (eds) Geologic evolution of the Barberton Greenstone Belt, South Africa. Geol Soc Amer Spec Pap 329:213–232

Lowenstern JB (2003) Melt inclusions come of age: volatiles, volcanoes, and Sorby's legacy. In: Developments in volcanology, vol 5. Elsevier, Amsterdam, pp 1–21

Lu SM (2018) A global review of enhanced geothermal system (EGS). Renew Sustain Energy Rev 81:2902–2921

Lucci F, Carrasco-Núñez G, Rossetti F, Theye T, White JC, Urbani S, Azizi H, Asahara Y, Giordano G (2020) Anatomy of the magmatic plumbing system beneath Los Humeros Caldera (Eastern Trans Mexican Volcanic Belt, Central Mexico): implications for geothermal systems. Solid Earth 11:125–159. https://doi.org/10.5194/se-11-125-2020

Lucci F, Rossetti F, Becchio R, Theye T, Gerdes A, Opitz J, Baez W, Bardelli L, De Astis G, Viramonte J, Giordano G (2018) Magmatic Mn-rich garnets in volcanic settings: age and longevity of the magmatic plumbing system of the Miocene Ramadas volcanism (NW Argentina). Lithos 322:238–249

Lydon JW (1984) Volcanogenic massive sulphide deposits Part I: a descriptive model. In: Roberts RG, Sheenan PA (eds) Ore deposit models. Geosci Canada Reprint Series 3:145–153

Mathisen ME (1984) Diagenesis of Plio-Pleistocene nonmarine sandstones, Cagayan Basin, Philippines: early development of secondary porosity in volcanic sandstones: Part 2. Aspects of porosity modification. In: McDonald DA, Surdam RC (eds) Clastic diagenesis. Amer Assoc Petrol Geol Mem 37:177–215

McArthur AN, Cas RAF, Orton GJ (1998) Distribution and significance of crystalline, perlitic and vesicular textures in the Ordovician Garth Tuff (Wales). Bull Volcanol 60(4):260–285

McCallum ME (1985) Experimental evidence for fluidization processes in breccia pipe formation. Econ Geol 80(6):1523–1543

Macgregor D (2015) History of the development of the East African Rift System: a series of interpreted maps through time. J Afr Earth Sci 101:232–252

Macgregor D, Argent J, Sansom P (2018) Introduction to the thematic set: tectonics and petroleum systems of East Africa. Petrol Geosci 24(1):3–7

McKibben MA, Andes JP, Williams AE (1988) Active ore formation at a brine interface in metamorphosed deltaic lacustrine sediments; the Salton Sea geothermal system, California. Econ Geol 83(3):511–523

McLellan JG, Oliver NH, Hobbs BE, Rowland JV (2010) Modelling fluid convection stability in continental faulted rifts with applications to the Taupo Volcanic Zone, New Zealand. J Volcanol Geothermal Res 190(1–2):109–122

McPhie J, Allen RL (1992) Facies architecture of mineralised submarine volcanic sequences; Cambrian Mount Read Volcanics, western Tasmania. Econ Geol 87(3):587–596

McPhie J, Cas RAF (2015) Volcanic successions associated with ore deposits: facies characteristics and ore host relationships. In: Sigurdsson H, Stix J, Houghton BF, McNutt SR, Rymer H (eds) Encyclopedia of volcanoes. Academic Press, San Diego, pp 865–880

McPhie J, Kamenetsky V, Allen S, Ehrig K, Agangi A, Bath A (2011a) The fluorine link between a supergiant ore deposit and a silicic large igneous province. Geology 39(11):1003–1006

McPhie J, Kamenetsky VS, Chambefort I, Ehrig K, Green N (2011b) Origin of the supergiant Olympic Dam Cu-U-Au-Ag deposit, South Australia: was a sedimentary basin involved? Geology 39 (8):795–798

Maffucci R, Bigi S, Corrado S, Chiodi A, Di Paolo L, Giordano G, Invernizzi C (2015) Quality assessment of reservoirs by means of outcrop data and "discrete fracture network" models: the case history of Rosario de La Frontera (NW Argentina) geothermal system. Tectonophysics 647:112–131

Maffucci R, Corrado S, Aldega L, Bigi S, Chiodi A, Di Paolo L, Giordano G, Invernizzi C (2016) Cap rock efficiency of geothermal systems in fold-and-thrust belts: evidence from paleo-thermal and structural analyses in Rosario de La Frontera geothermal area (NW Argentina). J Volcanol Geotherm Res 328:84–95

Magara K (2003) Volcanic reservoir rocks of northwestern Honshu Island, Japan. In: Petford N, McCaffrey K (eds) Hydrocarbons in crystalline rocks. Geol Soc Lond Spec Publ 214(1):69–81

Maghfouri S, Hosseinzadeh MR, Moayyed M, Movahednia M, Choulet F (2017) Geology, mineralization and sulfur isotopes geochemistry of the Mari Cu (Ag) Manto-type deposit, northern Zanjan, Iran. Ore Geol Rev 81:10–22

Maksaev V, Zentilli M (2002) Chilean strata-bound Cu- (Ag) deposits: an overview

Marschik R, Fontboté L (2001) The Candelaria-Punta del Cobre iron oxide Cu-Au (-Zn-Ag) deposits, Chile. Econ Geol 96(8):1799–1826

Mayer K, Scheu B, Montanaro C, Yilmaz TI, Isaia R, Aßbichler D, Dingwell DB (2016) Hydrothermal alteration of surficial rocks at Solfatara (Campi Flegrei): petrophysical properties and implications for phreatic eruption processes. J Volcanol Geotherm Res 320:128–143

Medlin CC, Jowitt SM, Cas RAF, Smithies RH, Kirkland CL, Maas RA, Raveggi M, Howard HM, Wingate MT (2015) Petrogenesis of the A-type, mesoproterozoic intra-caldera rheomorphic Kathleen ignimbrite and comagmatic Rowland suite intrusions, West Musgrave Province, Central Australia: products of extreme fractional crystallization in a failed rift setting. J Petrol 56(3):493–525

Meng QR (2003) What drove late Mesozoic extension of the northern China-Mongolia tract? Tectonophysics 369(3–4):155–174

Mercier-Langevin P, Dubé B, Hannington MD, Davis DW, Lafrance B, Gosselin G (2007) The LaRonde Penna Au-rich volcanogenic massive sulfide deposit, Abitibi greenstone belt, Quebec: part I. Geology and Geochronology. Econ Geol 102(4):585–609

Mercier-Langevin P, Hannington MD, Dubé B, Bécu V (2011) The gold content of volcanogenic massive sulfide deposits. Miner Deposita 46(5–6):509–539

Mercier-Langevin P, Hannington MD, Dubé B, Piercey SJ, Peter JM, and Pehrsson SJ (2015) Precious metal enrichment processes in volcanogenic massive sulphide deposits—a summary of key features, with an emphasis on TIGI-4 research contributions. In: Peter JM, Mercier-Langevin, P (eds) Targeted Geoscience Initiative 4: contributions to the understanding of volcanogenic massive sulphide deposit genesis and exploration methods development. Geol Surv Canada Open File 7853:117–130

Mercier-Langevin P, McNicoll V, Allen RL, Blight JH, Dubé B (2013) The Boliden gold-rich volcanogenic massive sulfide deposit, Skellefte district, Sweden: new U-Pb age constraints and implications at deposit and district scale. Miner Deposita 48(4):485–504

Middleton C, Buenavista A, Rohrlach B, Gonzalez J, Subang L, Moreno G (2004) A geological review of the Tampakan copper-gold deposit, southern Mindanao, Philippines. In: Proceedings Pacrim Congress, Adelaide, Australia. Australasian Institute of Mining and Metallurgy, pp 173–187

Millett JM, Hole MJ, Jolley DW, Schofield N, Campbell E (2015) Frontier exploration and the North Atlantic Igneous Province: new insights from a 2.6 km offshore volcanic sequence in the Faroe-Shetland Basin. J Geol Soc Lond 173(2):320–336

Millett JM, Wilkins AD, Campbell E, Hole MJ, Taylor RA, Healy D, Jerram DA, Jolley DW, Planke S, Archer SG, Blischke A (2016) The geology of offshore drilling through basalt sequences: understanding operational complications to improve efficiency. Mar Petrol Geol 77:1177–1192

Mironov YB (2005) Uranium of Mongolia. In: Seltmann R, Dolgopolova AV (eds) Centre for Russian and Central EurAsian Mineral Studies (CERCAMS), Natural History Museum, London, p 241

Mitchell AH, Balce GR (1990) Geological features of some epithermal gold systems, Philippines. J Geochem Explor 35(1–3):241–296

Mitchell AH, Carlile JC (1994) Arc-related gold-copper mineralization, basement domes and crustal extension. J SE Asian Earth Sci 10(1–2):39–50

Mitchell AH, Leach TM (1991) Epithermal gold in the Philippines: island arc metallogenesis, geothermal systems and geology. Academic Press, p 457

Mitchell RH (1986) Kimberlites: mineralogy, geochemistry, and petrology. Plenum Press

Mitchell RH (1995) Kimberlites, orangeites, and related rocks. Springer Science & Business Media

Mitchell RH, Bergman SC (1991) Petrology of lamproites. Springer Science & Business Media

Mole DR, Fiorentini ML, Thebaud N, Cassidy KF, McCuaig TC, Kirkland CL, Romano SS, Doublier MP, Belousova EA, Barnes SJ, Miller J (2014) Archean komatiite volcanism controlled by the evolution of early continents. Proc Nat Acad Sci 111(28):10083–10088

Moncada D, Mutchler S, Nieto A, Reynolds TJ, Rimstidt JD, Bodnar RJ (2012) Mineral textures and fluid inclusion petrography of the epithermal Ag–Au deposits at Guanajuato, Mexico: application to exploration. J Geochem Explor 114:20–35

Mondal SK, Griffin WL (eds) (2018) Processes and ore deposits of ultramafic-mafic magmas through space and time. Elsevier, Amsterdam, p 382

Monecke T, Gibson H, Dubé B, Laurin J, Hannington MD, Martin L (2008) Geology and volcanic setting of the Horne deposit, Rouyn-Noranda, Quebec: initial results of a new research project. Geol Surv Canada Curr Res 2008–9:16

Monecke T, Petersen S, Hannington MD (2014) Constraints on water depth of massive sulfide formation: evidence from modern seafloor hydrothermal systems in arc-related settings. Econ Geol 109 (8):2079–2101

Morgan LA, Schulz KJ (2012) Physical volcanology of volcanogenic massive sulfide deposits. In: Shanks WCP, Thurston R (eds) Volcanogenic massive sulfide occurrence model. US Geol Surv Sci Invest Rept 2010-5070-C:61-100

Morley CK (2018) 3-D seismic imaging of the plumbing system of the Kora Volcano, Taranaki Basin, New Zealand: the influence of syn-rift structure on shallow igneous intrusion architecture. Geosphere 14(6):2533–2584

Morishita Y, Nakano T (2008) Role of basement in epithermal deposits: the Kushikino and Hishikari gold deposits, southwestern Japan. Ore Geol Rev 34(4):597–609

Morton RL, Walker JS, Hudak GJ, Franklin JM (1991) The early development of an Archean submarine caldera complex with emphasis on the Mattabi ash-flow tuff and its relationship to the Mattabi massive sulfide deposit. Econ Geol 86(5):1002–10011

Mosier DL, Berger VI, Singer DA (2009) Volcanogenic massive sulfide deposits of the world: database and grade and tonnage models. US Geol Surv Open-file Rept 009-1034:46. http://pubs.usgs.gov/of/2009/1034/

Moss R, Scott SD (2001) Geochemistry and mineralogy of gold-rich hydrothermal precipitates from the eastern Manus Basin, Papua New Guinea. Canad Mineral 39(4):957–978

Moyle AJ, Doyle BJ, Hoogvliet H, Ware AR (1990) Ladolam gold deposit, Lihir island. In: Hughes FE (ed) Geology of the mineral resources of Australia and Papua New Guinea. Australasian Institute of Mining and Metallurgy Monograph 14(2):1793–1805

Mueller AG, Hall GC, Nemchin AA, O'Brien D (2008) Chronology of the Pueblo Viejo epithermal gold-silver deposit, Dominican Republic: formation in an Early Cretaceous intra-oceanic island arc and burial under ophiolite. Miner Deposita 43(8):873–889

Mueller WU, White JDL (1992) Felsic fire-fountaining beneath Archean seas: pyroclastic deposits of the 2730 Ma Hunter Mine Group, Quebec, Canada. J Volcanol Geotherm Res 54(1):117–134

Mueller WU, Mortensen JK (2002) Age constraints and characteristics of subaqueous volcanic construction, the Archean Hunter Mine Group, Abitibi Greenstone Belt. Precamb Res 115(1–4):119–152

Muffler P, Cataldi R (1978) Methods for regional assessment of geothermal resources. Geothermics 7(2–4):53–89

Müller D, Franz L, Petersen S, Herzig PM, Hannington MD (2003) Comparison between magmatic activity and gold mineralization at Conical Seamount and Lihir Island, Papua New Guinea. Mineral Petrol 79(3–4):259–283

Muntean J, Hall G, Kesler S, Mueller A, O'Brian D, Russell N (2007) Pueblo Viejo high-sulfidation epithermal gold-silver deposit, Dominican Republic: a new model of formation beneath barren limestone cover—a discussion. Econ Geol 102(4):755–757

Muntean JL, Cline JS, Simon AC, Longo AA (2011) Magmatic-hydrothermal origin of Nevada's Carlin-type gold deposits. Nat Geosci 4(2):122

Murphy WM, Pickett DA (2005) Episodic release and transport at the repository analog site at Peña Blanca. Presentation 15th annual Goldsmith conference, Session S-44 Keynote, Moscow, Idaho, United States

Naldrett AJ (2013) Magmatic sulfide deposits: geology, geochemistry and exploration. Springer Science & Business Media, p 744

Nash JT (2010) Volcanogenic uranium deposits: geology, geochemical processes, and criteria for resource assessment. US Geol Surv Open-File Rept 1001(2010):110

Nash JT, Utterback WC, Saunders JA (1989) Geology and geochemistry of the Sleeper gold-silver deposit, Humboldt County, Nevada; an interim report. US Geol Surv Open-File Rept 89–476:39

Nelson CE (2000) Volcanic domes and gold mineralization in the Pueblo Viejo district, Dominican Republic. Miner Deposita 35 (6):511–525

Nisbet EG, Cheadle MJ, Arndt NT, Bickle MJ (1993) Constraining the potential temperature of the Archaean mantle: a review of the evidence from komatiites. Lithos 30(3–4):291–307

Nixon PH, Davies GR, Rex DC, Gray A (1992) Venezuela kimberlites. J Volcanol Geotherm Res 50(1–2):101–115

Nowicki T, Crawford B, Dyck D, Carlson J, McElroy R, Oshust P, Helmstaedt H (2004) The geology of kimberlite pipes of the Ekati property, Northwest Territories, Canada. Lithos 76(1–4):1–27

O'Driscoll EST (1985) The application of lineament tectonics in the discovery of the Olympic Dam Cu-Au-U deposit at Roxby Downs, South Australia. Global Tect Metall 3:43–57

Ohmoto H (1996) Formation of volcanogenic massive sulfide deposits: the Kuroko perspective. Ore Geol Rev 10(3–6):135–177

Ohmoto H, Skinner BJ (eds) (1983) The Kuroko and related volcanogenic massive sulfide deposits. Econ Geol Monogr 5:604

Ohmoto H, Takahashi T (1983) Geologic setting of the kuroko deposits, Japan: part III Submarine calderas and kuroko genesis. In: Ohmoto H, Skinner BJ (eds) The Kuroko and related volcanogenic massive sulfide deposits. Econ Geol Monogr 5:39–54

Okubo S (2001) Factors controlling volcanic-rock reservoirs in Japan: generation of pores and their preservation. J Jpn Assoc Petrol Techn 66(1):69–80

Olasolo P, Juárez MC, Morales MP, Liarte IA (2016) Enhanced geothermal systems (EGS): a review. Renew Sustain Energy Rev 56:133–144

Oreskes N, Einaudi MT (1992) Origin of hydrothermal fluids at Olympic Dam; preliminary results from fluid inclusions and stable isotopes. Econ Geol 87(1):64–90

Orth K, Cas RA, Wright JV (1989) Facies analysis and facies associations in the recognition of volcanic centres in silicic terranes: an example from the Early Devonian of Australia. Aust J Earth Sci 36(2):167–188

O'Sullivan MJ, Pruess K, Lippmann MJ (2001) State of the art of geothermal reservoir simulation. Geothermics 30(4):395–429

Pamukçu AS, Wright KA, Gualda GA, Gravley D (2020) Magma residence and eruption at the Taupo Volcanic Center (Taupo Volcanic Zone, New Zealand): insights from rhyolite-MELTS geobarometry, diffusion chronometry, and crystal textures. Contrib Mineral Petrol 175: 1–27

Parman SW, Grove TL, Dann JC, De Wit MJ (2004) A subduction origin for komatiites and cratonic lithospheric mantle. S Afr J Geol 107(1–2):107–118

Paull PL, Hodkinson IP, Morrison GW, Teale GS (1990) Mount Leyshon gold deposit In: Hughes FE (ed) Geology of the mineral deposits of Australia and Papua New Guinea. Australasian Institute of Mining and Metallurgy Monograph 14(2):1471–1481

Pearson V, Daigneault R (2009) An Archean megacaldera complex: the Blake River Group, Abitibi Greenstone Belt. Precamb Res 168(1–2):66–82

Peter JM, Scott SD (1999) Windy Craggy, northwestern British Columbia; the world's largest besshi-type deposit. Rev Econ Geol 8:261–295

Peter JM, Simoneit BR, Kawka OE, Scott SD (1990) Liquid hydrocarbon-bearing inclusions in modern hydrothermal chimmeys and mounds from the southern trough of Guaymas Basin, Gulf of California. Appl Geochem 5(1–2):51–63

Petersen S, Herzig PM, Hannington MD (2000) Third dimension of a presently forming VMS deposit: TAG hydrothermal mound, Mid-Atlantic Ridge, 26 N. Miner Deposita 35(2–3):233–259

Petersen S, Herzig PM, Hannington MD, Jonasson IR, Arribas A (2002) Submarine gold mineralization near Lihir Island, New Ireland fore-arc, Papua New Guinea. Econ Geol 97(8):1795–1813

Petersen S, Monecke T, Westhues A, Hannington MD, Gemmell JB, Sharpe R, Peters M, Strauss H, Lackschewitz K, Augustin N, Gibson H (2014) Drilling shallow-water massive sulfides at the Palinuro volcanic complex, Aeolian island arc, Italy. Econ Geol 109 (8):2129–2158

Petford N, McCaffrey K (eds) (2003) Hydrocarbons in crystalline rocks. Geol Soc Lond Spec Publ 214:236

Phillips GN, Powell R (1993) Link between gold provinces. Econ Geol 88(5):1084–1098

Phillips GN, Powell R (2009) Formation of gold deposits: review and evaluation of the continuum model. Earth-Sci Rev 94(1–4):1–21

Phillipson SE, Romberger SB (2004) Volcanic stratigraphy, structural controls, and mineralization in the san cristobal Ag-Zn-Pb deposit, southern Bolivia. J South Amer Earth Sci 16(8):667–683

Piercey SJ (2011) The setting, style, and role of magmatism in the formation of volcanogenic massive sulfide deposits. Miner Deposita 46(5–6):449–471

Pisutha-Arnond V, Ohmoto H (1983) Thermal history, and chemical and isotopic compositions of the ore-forming fluids responsible to the Kuroko massive sulfide deposits in the Hokuroku district of Japan. In: Ohmoto H, Skinner BJ (eds) The Kuroko and related volcanogenic massive sulfide deposits. Econ Geol Monogr 5:523–558

Pittari A, Cas RAF, Lefebvre N, Kurszlaukis S (2015) Alteration Styles in the Orion Central Volcanic Complex, Fort à la Corne Kimberlite Field, Saskatchewan, and their effects on primary volcaniclastic textures: implications for facies mapping and diamond exploration. Econ Geol 110(1):147–171

Pittari A, Cas RAF, Lefebvre N, Robey J, Kurszlaukis S, Webb K (2008) Eruption processes and facies architecture of the Orion Central kimberlite volcanic complex, Fort à la Corne, Saskatchewan; kimberlite mass flow deposits in a sedimentary basin. J Volcanol Geotherm Res 174(1–3):152–170

Planke S, Millett JM, Maharjan D, Jerram DA, Abdelmalak MM, Groth A, Hoffmann J, Berndt C, Myklebust R (2017) Igneous seismic geomorphology of buried lava fields and coastal escarpments on the Vøring volcanic rifted margin. Interpretation 5 (3):1–17

Planke S, Rasmussen T, Rey SS, Myklebust R (2005) Seismic characteristics and distribution of volcanic intrusions and hydrothermal vent complexes in the Vøring and Møre basins. In: Geol Soc Lond, Petroleum Geology Conference Series 6(1):833–844

Poppitt S, Duncan LJ, Preu B, Fazzari F, Archer J (2018) The influence of volcanic rocks on the characterization of Rosebank Field—new insights from ocean-bottom seismic data and geological analogues integrated through interpretation and modelling. In: Bowman M, Levell B (eds) Petroleum geology of NW Europe: 50 years of learning. Geol Soc Lond, Petroleum Geology Conference Series 8 (1):373–384

Porritt LA, Cas RAF (2009) Reconstruction of a kimberlite eruption, using an integrated volcanological, geochemical and numerical approach: a case study of the Fox Kimberlite, NWT, Canada. J Volcanol Geotherm Res 179(3–4):241–264

Porritt LA, Cas RAF (2011) The influence of complex intra-and extra-vent processes on facies characteristics of the Koala Kimberlite, NWT, Canada: volcanology, sedimentology and intrusive processes. Bull Volcanol 73(6):717–735

Porritt LA, Cas RAF, Ailleres L, Oshust P (2011) The influence of volcanological and sedimentological processes on diamond grade distribution in kimberlites: examples from the EKATI Diamond Mine, NWT, Canada. Bull Volcanol 73(8):1085–1105

Porritt LA, Cas RAF, Crawford BB (2008a) In-vent column collapse as an alternative model for massive volcaniclastic kimberlite emplacement: an example from the Fox kimberlite, Ekati Diamond Mine, NWT, Canada. J Volcanol Geotherm Res 174(1–3):90–102

Porritt LA, Cas RAF, Crawford BB (2008b) Reply to: Discussion by Brown et al. on "In-vent column collapse as an alternative model for massive volcaniclastic kimberlite emplacement: an example from the Fox kimberlite, Ekati Diamond Mine, NWT, Canada". J Volcanol Geotherm Res 178:851–854

Porritt LA, Russell JK (2012) Kimberlite ash: fact or fiction. Phys Chem Earth, Parts A/B/C 45:24–32

Porritt LA, Russell JK, McLean H, Fomradas G, Eichenberg D (2013) A phreatomagmatic kimberlite: the A418 kimberlite pipe, Northwest Territories, Canada. In: Proceedings of 10th international kimberlite conference. Springer, New Delhi, pp 97–107

Porter TM (ed) (2000) Hydrothermal iron oxide copper-gold and related deposits: a global perspective, vol 1. PGC Publishing, Adelaide, Australia, p 349

Porter TM (ed) (2002) Hydrothermal iron oxide copper-gold and related deposits: a global perspective, vol 2. PGC Publishing, Adelaide, Australia, p 377

Porter TM (ed) (2010) Hydrothermal iron oxide copper-gold and related deposits: a global perspective, volumes 3/4: advances in understanding of IOCG deposits. PGC Publishing, Adelaide, Australia, p 600

Prayoga OA, Alvrida DA, Taslim M, Ginanjar AR, Nurhadi DR, Barkah A (2017) Integrated workflow for fracture characterization using post-stack geometrical attributes, borehole image, core and wireline log: case study for fractured volcanic reservoir on the Java area, north west Java Basin, Indonesia. In: SPE/IATMI Asia Pacific Oil & Gas conference and exhibition, Jakarta, Indonesia. Soc Petrol Engineers: SPE-186968-MS

Prihatmoko S, Idrus A (2020) Low-sulfidation epithermal gold deposits in Java, Indonesia: characteristics and linkage to the volcano-tectonic setting. Ore Geol Rev 121:103490

Proffett JM (2003) Geology of the Bajo de la Alumbrera porphyry copper-gold deposit, Argentina. Econ Geol 98(8):1535–1574

Putirka KD (2005) Igneous thermometers and barometers based on plagioclase+ liquid equilibria: tests of some existing models and new calibrations. Amer Mineral 90(2–3):336–346

Putirka KD (2008) Thermometers and barometers for volcanic systems. Rev Min Geochem 69(1):61–120

Putirka KD, Mikaelian H, Ryerson F, Shaw H (2003) New clinopyroxene-liquid thermobarometers for mafic, evolved, and volatile-bearing lava compositions, with applications to lavas from Tibet and the Snake River Plain, Idaho. Amer Mineral 88(10):1542–1554

Rabiee A, Rossetti F, Lucci F, Lustrino M (2022) Cenozoic porphyry and other hydrothermal ore deposits along the South Caucasus-west Iranian tectono-magmatic belt: A critical reappraisal of the controlling factors. Lithos: 106874

Rabiee A, Rossetti F, Tecce F, Asahara Y, Azizi H, Glodny J, ... Selby D (2019) Multiphase magma intrusion, ore-enhancement and hydrothermal carbonatisation in the Siah-Kamar porphyry Mo deposit, Urumieh-Dokhtar magmatic zone, NW Iran. Ore Geol Rev 110: 102930.

Ramm M (1992) Porosity-depth trends in reservoir sandstones: theoretical models related to Jurassic sandstones offshore Norway. Mar Petrol Geol 9(5):553–567

Rateau R, Schofield N, Smith M (2013) The potential role of igneous intrusions on hydrocarbon migration, West of Shetland. Petrol Geosci 1:2012–2035

Reeve JS (1990) Olympic Dam copper-uranium-gold-silver deposit. In: Hughes FE (ed) Geology of the Mineral Resources of Australia and Papua New Guinea. Australasian Institute of Mining and Metallurgy Monograph 14(2):1009–1035

Reid MR (2008) How long does it take to supersize an eruption? Elements 4(1):23–28

Reid MR, Coath CD (2000) In situ U-Pb ages of zircons from the Bishop Tuff: no evidence for long crystal residence times. Geology 28(5):443–446

Ressel MW, Henry CD (2006) Igneous geology of the Carlin trend, Nevada: development of the Eocene plutonic complex and significance for Carlin-type gold deposits. Econ Geol 101(2):347–383

Reynolds LJ (2001) Geology of the Olympic Dam Cu-U-Au-Ag-REE deposit. MESA J 23:4–10

Rice CM, Ashcroft WA, Batten DJ, Boyce AJ, Caulfield JB, Fallick AE, Hole MJ, Jones E, Pearson MJ, Rogers G, Saxton JM (1995) A Devonian auriferous hot spring system, Rhynie, Scotland. J Geol Soc Lond 152(2):229–250

Rice CM, Steele GB, Barfod DN, Boyce AJ, Pringle MS (2005) Duration of magmatic, hydrothermal, and supergene activity at Cerro Rico de Potosi, Bolivia. Econ Geol 100(8):1647–1656

Rice CM, Trewin NH, Anderson LI (2002) Geological setting of the Early Devonian Rhynie cherts, Aberdeenshire, Scotland: an early terrestrial hot spring system. J Geol Soc Lond 159(2):203–214

Richards JP (1995) Alkalic-type epithermal gold deposits—a review. Mineral Assoc Canada Short Course 23:367–400

Richards JP, Kerrich R (1993) The Porgera gold mine, Papua New Guinea; magmatic hydrothermal to epithermal evolution of an alkalic-type precious metal deposit. Econ Geol 88(5):1017–1052

Robert F, Brommecker R, BourneBT, Dobak PJ, McEwan CJ, Rowe RR, and Zhou X (2007) Models and exploration methods for major gold deposit: types. In: Proceedings exploration '07, fifth decennial int conference on mineral exploration, Toronto, pp 691–711

Roberts S, Bach W, Binns RA, Vanko DA, Yeats CJ, Teagle DA, Blacklock K, Blusztajn JS, Boyce AJ, Cooper MJ, Holland N (2003) Contrasting evolution of hydrothermal fluids in the PACMANUS system, Manus Basin: the Sr and S isotope evidence. Geology 31(9):805–808

Rosengren NM, Cas RAF, Beresford SW, Palich BM (2008) Reconstruction of an extensive Archaean dacitic submarine volcanic complex associated with the komatiite-hosted Mt Keith nickel deposit, Agnew-Wiluna greenstone belt, Yilgarn Craton, Western Australia. Precamb Res 161(1):34–52

Rosengren NM, Grguric BA, Beresford SW, Fiorentini ML, Cas RAF (2007) Internal stratigraphic architecture of the komatiitic dunite-hosted MKD5 disseminated nickel sulfide deposit, Mount Keith Domain, Agnew-Wiluna greenstone belt, Western Australia. Miner Deposita 42(8):821–845

Rosengren NM, Beresford SW, Grguric BA, Cas RAF (2005) An intrusive origin for the komatiitic dunite-hosted Mount Keith disseminated nickel sulfide deposit, Western Australia. Econ Geol 100(1):149–156

Ross PS, Núñez GC, Hayman P (2017) Felsic maar-diatreme volcanoes: a review. Bull Volcanol 79(2):20

Rossetti F, Aldega L, Tecce F, Balsamo F, Billi A, Brilli M (2010) Fluid flow within the damage zone of the Boccheggiano extensional fault (Larderello–Travale geothermal field, central Italy): structures, alteration and implications for hydrothermal mineralization in extensional settings. Geol Mag 148(4):558–579

Rowland JV, Sibson RH (2001) Extensional fault kinematics within the Taupo Volcanic Zone, New Zealand: soft-linked segmentation of a continental rift system, New Zealand. J Geol Geophys 44(2):271–283

Rowland JV, Sibson RH (2004) Structural controls on hydrothermal flow in a segmented rift system, Taupo Volcanic Zone, New Zealand. Geofluids 4(4):259–283

Rowland JV, Simmons SF (2012) Hydrologic, magmatic, and tectonic controls on hydrothermal flow, Taupo Volcanic Zone, New Zealand: implications for the formation of epithermal vein deposits. Econ Geol 107(3):427–457

Russell JK, Porritt LA, Lavallée Y, Dingwell DB (2012) Kimberlite ascent by assimilation-fuelled buoyancy. Nature 481(7381):352

Russell N, Polanco J, Kesler SE (1986) Geology of the Monte Negro gold-silver deposit, Pueblo Viejo district, Dominican Republic. In: Gold '86. Geol Assoc Canada Proceedings Volume, Toronto, pp 497–503

Russell N, Kesler SE (1991) Geology of the Pueblo Viejo maar-diatreme complex, Los Ranchos Formation, Dominican Republic. In: Mann P, Draper, Lewis, JF (eds) Geologic and tectonic development of the North America-Caribbean plate boundary in Hispaniola. Geol Soc Amer Spec Pap 262:202–219

Sáez R, Almodóvar GR, Pascual E (1996) Geological constraints on massive sulphide genesis in the Iberian Pyrite Belt. Ore Geol Rev 11(6):429–451

Sakata S, Matsuhisa Y, Igari SI (1994) Organic geochemistry of petroleum from the Green Tuff basin, Japan. Organic Geochem 21 (1):27–34

Sarkar C, Heaman LM, Pearson DG (2015) Duration and periodicity of kimberlite volcanic activity in the Lac de Gras kimberlite field, Canada and some recommendations for kimberlite geochronology. Lithos 218:155–166

Sato O (1984) Rock facies and pore spaces on volcanic-rock reservoirs—especially rhyolite reservoirs in the Minami Nagaoka gas field. J Jpn Assoc Petroleum Techn 49(1):11–19

Schalamuk IB, Zubia M, Genini A, Fernandez RR (1997) Jurassic epithermal Au-Ag deposits of Patagonia, Argentina. Ore Geol Rev 12(3):173–186

Schardt C, Large RR (2009) New insights into the genesis of volcanic-hosted massive sulfide deposits on the seafloor from numerical modeling studies. Ore Geol Rev 35(3–4):333–351

Schlegel TU, Heinrich CA (2015) Lithology and hydrothermal alteration control the distribution of copper grade in the Prominent Hill iron oxide-copper-gold deposit (Gawler craton, South Australia). Econ Geol 110(8):1953–1994

Schofield N, Holford S, Millett J, Brown D, Jolley D, Passey SR, Muirhead D, Grove C, Magee C, Murray J, Hole M (2017)

Regional magma plumbing and emplacement mechanisms of the Faroe-Shetland sill complex: implications for magma transport and petroleum systems within sedimentary basins. Basin Res 29(1): 41–63

Seltmann R, Gerel O, Kirwin DJ (eds) (2005) Geodynamics and metallogeny of Mongolia with a special emphasis on copper and gold deposits. Centre for Russian and Central EurAsian Mineral Studies (CERCAMS), Natural History Museum, London, IAGOD Guidebook Series 11, p 225

Setterfield TN, Eaton PC, Rose WJ, Sparks RS (1991) The Tavua Caldera, Fiji: a complex shoshonitic caldera formed by concurrent faulting and downsagging. J Geol Soc 148(1):115–127

Seubert BW (2015) Volcaniclastic petroleum systems-theory and examples from Indonesia. In: Proceedings of the 39th Annual Indonesian Petrol Assoc Convention and Exhibition, Jakarta. Digital paper IPA-15-G026

Shanks WCP (2012a) Theory of deposit formation. In: Shanks WCP, Thurston R (eds) Volcanogenic massive sulfide occurrence model. US Geol Survey Sci Invest Rept 2010-5070-C:289–303

Shanks WCP (2012b) Hydrothermal alteration. In: Shanks WCP, Thurston R (eds) Volcanogenic massive sulfide occurrence model. US Geol Survey Sci Invest Rept 2010-5070-C:165–180

Sharpe R, Gemmell JB (2002) The Archean Cu-Zn magnetite-rich Gossan Hill volcanic-hosted massive sulfide deposit, Western Australia: genesis of a multistage hydrothermal system. Econ Geol 97(3):517–539

Sherlock RL, Lehrman NJ (1995) Occurrences of dendritic gold at the McLaughlin Mine hot-spring gold deposit. Miner Deposita 30(3–4):323–327

Sherlock RL, Roth T, Spooner ET, Bray CJ (1999) Origin of the Eskay Creek precious metal-rich volcanogenic massive sulfide deposit; fluid inclusion and stable isotope evidence. Econ Geol 94(6):803–824

Sherlock RL, Tosdal RM, Lehrman NJ, Graney JR, Losh S, Jowett EC, Kesler SE (1995) Origin of the McLaughlin Mine sheeted vein complex; metal zoning, fluid inclusion, and isotopic evidence. Econ Geol 90(8):2156–2181

Shigley JE, Chapman J, Ellison RK (2001) Discovery and mining of the Argyle diamond deposit, Australia. Gems Gemol 37(1):26–41

Shimizu T (2014) Reinterpretation of quartz textures in terms of hydrothermal fluid evolution at the Koryu Au-Ag deposit, Japan. Econ Geol 109(7):2051–2065

Shirey SB, Cartigny P, Frost DJ, Keshav S, Nestola F, Nimis P, Pearson DG, Sobolev NV, Walter MJ (2013) Diamonds and the geology of mantle carbon. Rev Mineral Geochem 75(1):355–421

Sibson RH (1987) Earthquake rupturing as a mineralizing agent in hydrothermal systems. Geology 15(8):701–704

Sillitoe RH (1973) The tops and bottoms of porphyry copper deposits. Econ Geol 68(6):799–815

Sillitoe RH (1985) Ore-related breccias in volcanoplutonic arcs. Econ Geol 80(6):1467–1514

Sillitoe RH (1989) Gold deposits in western Pacific island arcs: the magmatic connection. In: Keays RR, Ramsay WRH, Groves DI (eds) The geology of gold deposits: the perspective in 1988. Econ Geol Monogr 6:274–291

Sillitoe RH (1993) Epithermal models: genetic types, geometrical controls and shallow features. In: Kirkham RV et al (eds) Mineral deposit modeling. Geol Assoc Canada Spec Pap 40:403–417

Sillitoe RH (1994) Erosion and collapse of volcanoes: causes of telescoping in intrusion-centered ore deposits. Geology 22(10): 945–948

Sillitoe RH (1995) Exploration of porphyry copper lithocaps In: Proceedings Pacrim Congress, Auckland, New Zealand. Australasian Institute of Mining and Metallurgy, pp 527–532

Sillitoe RH (1999) Styles of high-sulphidation gold, silver and copper mineralisation in porphyry and epithermal environments. In: Proceedings Pacrim Congress, Bali, Indonesia. Australasian Institute of Mining and Metallurgy, pp 29–44

Sillitoe RH (2002) Some metallogenic features of gold and copper deposits related to alkaline rocks and consequences for exploration. Miner Deposita 37(1):4–13

Sillitoe RH (2003) Iron oxide-copper-gold deposits: an Andean view. Miner Deposita 38(7):787–812

Sillitoe RH (2010) Porphyry copper systems. Econ Geol 105(1):3–41

Sillitoe RH (2015) Epithermal paleosurfaces. Miner Deposita 50 (7):767–793

Sillitoe RH, Angeles CA, Comia GM, Antioquia EC, Abeya RB (1990) An acid-sulphate-type lode gold deposit at Nalesbitan, Luzon, Philippines. J Geochem Explor 35(1–3):387–411

Sillitoe RH, Baker EM, Brook WA (1984) Gold deposits and hydrothermal eruption breccias associated with a maar volcano at Wau, Papua New Guinea. Econ Geol 79(4):638–655

Sillitoe RH, Bonham HF (1984) Volcanic landforms and ore deposits. Econ Geol 79(6):1286–1298

Sillitoe RH, Bonham Jr, HF (1990) Sediment-hosted gold deposits: Distal products of magmatic-hydrothermal systems. Geology 18(2): 157–161

Sillitoe RH, Grauberger GL, Elliott JE (1985) A diatreme-hosted gold deposit at Montana Tunnels, Montana. Econ Geol 80(6):1707–1721

Sillitoe RH, Hall DJ, Redwood SD, Waddell AH (2006) Pueblo Viejo high-sulfidation epithermal gold-silver deposit, Dominican Republic: a new model of formation beneath barren limestone cover. Econ Geol 101(7):1427–1435

Sillitoe RH, Hall DJ, Redwood SD, Waddell AH (2007) Pueblo Viejo high-sulfidation epithermal gold-silver deposit, Dominican Republic: a new model of formation beneath barren limestone cover—a reply. Econ Geol 102(4):758–760

Sillitoe RH, Hannington MD, Thompson JF (1996) High sulfidation deposits in the volcanogenic massive sulfide environment. Econ Geol 91(1):204–212

Sillitoe RH, Hedenquist JW (2003) Linkages between volcanotectonic settings, ore-fluid compositions, and epithermal precious metal deposits. Soc Econ Geol Spec Publ 10:315–343

Sillitoe RH, Steele GB, Thompson JF, Lang JR (1998) Advanced argillic lithocaps in the Bolivian tin-silver belt. Miner Deposita 33 (6):539–546

Simmons SF, Brown KL, Browne PR, Rowland JV (2016) Gold and silver resources in Taupo Volcanic Zone geothermal systems. Geothermics 59:205–214

Simmons SF, Christenson BW (1994) Origins of calcite in a boiling geothermal system. Amer J Sci 294(3):361–400

Simmons SF, White NC, John DA (2005) Geological characteristics of epithermal precious and base metal deposits. Econ Geol 100: 485–522

Simoneit BR (2018) Hydrothermal petroleum. In: Wilkes H (ed) Hydrocarbons, oils and lipids: diversity, origin, chemistry and fate. Handbook of hydrocarbon and lipid microbiology, Springer International Publishing, pp 1–35

Simoneit BR, Lonsdale PF (1982) Hydrothermal petroleum in mineralized mounds at the seabed of Guaymas Basin. Nature 295 (5846):198

Simoneit BR, Leif RN, Sturz AA, Sturdivant AE, Gieskes JM (1992) Geochemistry of shallow sediments in Guaymas Basin, Gulf of California: hydrothermal gas and oil migration and effects of mineralogy. Organic Geochem 18(6):765–784

Simpson K, McPhie J (2001) Fluidal-clast breccia generated by submarine fire fountaining, Trooper Creek Formation, Queensland, Australia. J Volcanol Geotherm Res 109(4):339–355

Skirrow RG, Bastrakov E, Davidson GJ, Raymond OL, Heithersay P (2002) The geological framework, distribution and controls of Fe-oxide Cu-Au mineralisation in the Gawler Craton, South Australia. Part II-Alteration and mineralisation. In: Porter TM (ed) Hydrothermal iron oxide copper-gold and related deposits: a global perspective, vol 2. PGC Publishing, Adelaide, pp 33–47

Skirrow RG, Bastrakov EN, Barovich K, Fraser GL, Creaser RA, Fanning CM, Raymond OL, Davidson GJ (2007) Timing of iron oxide Cu-Au-(U) hydrothermal activity and Nd isotope constraints on metal sources in the Gawler craton, South Australia. Econ Geol 102(8):1441–1470

Skirrow RG, Jaireth S, Huston DL, Bastrakov EN, Schofield A, van der Wielen SE, Bar C (2009) Uranium mineral systems: processes, exploration criteria and a new deposit framework. Geosci Aust Rec 2009(20):44

Smith JG (1987) New compilation geologic map of the Cascade Range in Washington. Trans Geotherm Res Coun 11:309–314

Smith RL, Shaw HR (1978) Igneous-related geothermal systems. In: Muffler LJP (ed) Assessment of geothermal resources of the United States. US Geol Surv Circ 790:12–17

Smithies SL, Harmon LJ, Allen SM, Gravley DM, Gualda GA (2023) Following magma: The pathway of silicic magmas from extraction to storage during an ignimbrite flare-up, Taupō Volcanic Zone, New Zealand. Earth Planet Sci Lett 607: 118053

Sobolev AV, Asafov EV, Gurenko AA, Arndt NT, Batanova VG, Portnyagin MV, Garbe-Schönberg D, Krasheninnikov SP (2016) Komatiites reveal a hydrous Archaean deep-mantle reservoir. Nature 531(7596):628

Sokol AG, Kupriyanov IN, Palyanov YN (2013) Partitioning of H_2O between olivine and carbonate-silicate melts at 6.3 GPa and 1400 C: Implications for kimberlite formation. Earth Planet Sci Lett 383: 58–67

Song X, Wang Y, Chen L (2011) Magmatic Ni-Cu-(PGE) deposits in magma plumbing systems: features, formation and exploration. Geosci Front 2(3):375–384

Soriano C, Martí J (1999) Facies analysis of volcano-sedimentary successions hosting massive sulfide deposits in the Iberian pyrite belt, Spain. Econ Geol 94(6):867–882

Sparks RSJ (2013) Kimberlite volcanism. Ann Rev Earth Planet Sci 41:497–528

Sparks RSJ, Baker L, Brown RJ, Field M, Schumacher J, Stripp G, Walters A (2006) Dynamical constraints on kimberlite volcanism. J Volcanol Geotherm Res 155(1–2):18–48

Sparks RSJ, Barclay J, Jaupart CP, Mader HM, Phillips JC (1994) Physical aspects of magma degassing; 1, Experimental and theoretical constraints on vesiculation. Rev Mineral Geochem 30 (1):413–445

Spirakis CS, Heyl AV (1993) Organic matter (bitumen and other forms) as the key to localisation of Mississippi Valley-type ores. In: Parnell J, Kucha H, Landais P (eds) Bitumens in ore deposits. Soc Geol Appl Miner Deps Spec Publ 9:381–398

Spycher NF, Reed MH (1989) Evolution of a Broadlands-type epithermal ore fluid along alternative PT paths; implications for the transport and deposition of base, precious, and volatile metals. Econ Geol 84(2):328–359

Sruoga P, Rubinstein N (2007) Processes controlling porosity and permeability in volcanic reservoirs from the Austral and Neuquén basins, Argentina. Amer Assoc Petrol Geol Bull 91(1):115–129

Sruoga P, Rubinstein N, Hinterwimmer G (2004) Porosity and permeability in volcanic rocks: a case study on the Serie Tobífera, South Patagonia, Argentina. J Volcanol Geotherm Res 132(1):31–43

Staude S, Barnes SJ, Le Vaillant M (2016) Evidence of lateral thermomechanical erosion of basalt by Fe-Ni-Cu sulfide melt at Kambalda, Western Australia. Geology 44(12):1047–1050

Stiegler MT, Lowe DR, Byerly GR (2008) Abundant pyroclastic komatiitic volcanism in the 3.5–3.2 Ga Barberton greenstone belt, South Africa. Geology 36(10):779–782

Stiegler MT, Lowe DR, Byerly GR (2011) Fragmentation and dispersal of komatiitic pyroclasts in the 3.5–3.2 Ga Onverwacht Group, Barberton greenstone belt, South Africa. Geol Soc Amer Bull 123 (5–6):1112–1126

Stone WE, Archibald NJ (2004) Structural controls on nickel sulphide ore shoots in Archaean komatiite, Kambalda, WA: the volcanic trough controversy revisited. J Struct Geol 26(6):1173–1194

Stone WE, Beresford SW, Archibald NJ (2005) Structural setting and shape analysis of nickel sulfide shoots at the Kambalda Dome, Western Australia: implications for deformation and remobilization. Econ Geol 100(7):1441–1455

Stripp GR, Field M, Schumacher JC, Sparks RSJ, Cressey G (2006) Post-emplacement serpentinization and related hydrothermal metamorphism in a kimberlite from Venetia, South Africa. J Metamorph Geol 24(6):515–534

Surdam RC, Boles JR (1979) Diagenesis of volcanic sandstones. Soc Econ Palaeont Mineral Spec Publ 26:227–242

Surdam RC, Boese SW, Crossey LJ (1984) The chemistry of secondary porosity: Part 2. Aspects of porosity modification. In: McDonald DA, Surdam RC (eds) Clastic diagenesis. Amer Assoc Petrol Geol Mem 37:127–149

Surdam RC, Crossey LJ, Hagen ES, Heasler HP (1989) Organic-inorganic interactions and sandstone diagenesis. Amer Assoc Petrol Geol Bull 73(1):1–23

Suzuki S, Peña RE, Tam III TA, Yumul Jr GP, Dimalanta CB, Usui M, Ishida K (2016) Development of the Philippine Mobile Belt in northern Luzon from Eocene to Pliocene. J Asian Earth Sci 142: 32–44

Tappe S, Smart K, Torsvik T, Massuyeau M, de Wit M (2018) Geodynamics of kimberlites on a cooling Earth: clues to plate tectonic evolution and deep volatile cycles. Earth Planet Sci Lett 484:1–14

Tappert R, Tappert MC (2011) Diamonds in nature: a guide to rough diamonds. Springer Science & Business Media

Teal L, Jackson M (2002) Geologic overview of the Carlin trend gold deposits. Gold deposits of the Carlin trend. Nevada Bur Mines Geol Bull 111:9–19

Tescione I, Todesco M, Giordano G (2021) Geothermal fluid circulation in a caldera setting: The Torre Alfina medium enthalpy system (Italy). Geothermics 89: 101947

Teske A, Lizarralde D, Höfig TW (2018) Expedition 385 scientific prospectus: Guaymas Basin tectonics and biosphere. Internat Ocean Discovery Program. https://doi.org/10.14379/iodp.sp.385.2018

Thompson TB, Trippel AD, Dwelley PC (1985) Mineralized veins and breccias of the Cripple Creek district, Colorado. Econ Geology 80 (6):1669–1688

Tingley JV, Berger BR (1985) Lode gold deposits of Round Mountain, Nevada. Nevada Bur Mines Geol Bull 100:62

Tomaru H, Lu Z, Fehn U, Muramatsu Y (2009) Origin of hydrocarbons in the Green Tuff region of Japan: 129I results from oil field brines and hot springs in the Akita and Niigata Basins. Chem Geol 264(1–4):221–231

Tornos F, Peter JM, Allen R, Conde C (2015) Controls on the siting and style of volcanogenic massive sulphide deposits. Ore Geol Rev 68:142–163

Torsvik TH, Burke K, Steinberger B, Webb SJ, Ashwal LD (2010) Diamonds sampled by plumes from the core-mantle boundary. Nature 466(7304):352

Trofimovs J, Tait MA, Cas RAF, McArthur A, Beresford SW (2003) Can the role of thermal erosion in strongly deformed komatiite–Ni–Cu–(PGE) deposits be determined? Perseverance, Agnew-Wiluna Belt, Western Australia. Aust J Earth Sci 50(2):199–214

Tsuji T, Takai K, Oiwane H, Nakamura Y, Masaki Y, Kumagai H, Kinoshita M, Yamamoto F, Okano T, Kuramoto SI (2012) Hydrothermal fluid flow system around the Iheya North Knoll in the mid-Okinawa trough based on seismic reflection data. J Volcanol Geotherm Res 213:41–50

Urbani S, Giordano G, Lucci F, Rossetti F, Acocella V, Carrasco-Nunez G (2020) Estimating the depth and exolution of intrusions at resurgent domes: Los Humeros (Mexico). Solid Earth 11(2); 527-545.

van Straaten BI, Kopylova MG, Russell JK, Smith BS (2011) A rare occurrence of a crater-filling clastogenic extrusive coherent kimberlite, Victor Northwest (Ontario, Canada). Bull Volcanol 73 (8):1047–1062

Vearncombe S, Vearncombe JR, Barley ME (1998) Fault and stratigraphic controls on volcanogenic massive sulphide deposits in the Strelley Belt, Pilbara Craton, Western Australia. Precamb Res 88(1–4):67–82

Verma MP, Verma SP, Sanvicente H (1990) Temperature field simulation with stratification model of magma chamber under Los Humeros caldera, Puebla, Mexico. Geothermics 19(2):187–197

Verma SP (1985) On the magma chamber characteristics as inferred from surface geology and geochemistry: examples from Mexican geothermal areas. Phys Earth Planet Inter 41(2–3):207–214

Verma SP, Gómez-Arias E, Andaverde J (2011) Thermal sensitivity analysis of emplacement of the magma chamber in Los Humeros caldera, Puebla, Mexico. Int Geol Rev 53(8):905–925

Vignaroli G, Aldega L, Balsamo F, Billi A, De Benedetti AA, De Filippis L, Giordano G, Rossetti F (2015) A way to hydrothermal paroxysm, Colli Albani volcano, Italy. Geol Soc Amer Bull 127(5–6):672–687

Vignaroli G, Pinton A, De Benedetti AA, Giordano G, Rossetti F, Soligo M, Berardi G (2013) Structural compartmentalisation of a geothermal system, the Torre Alfina field (central Italy). Tectonophys 608: 482–498

Viljoen MJ, Viljoen RP (1969a) Evidence for the existence of a mobile extrusive peridotitic magma from the Komati Formation of the Onverwacht Group. Geol Soc S Afr Spec Publ 2:87–112

Viljoen RP, Viljoen MJ (1969b) The geological and geochemical significance of the upper formations of the Onverwacht Group. Geol Soc S Afr Spec Publ 2:113–152

Wang P, Chen C, Zhang Y, Gao Y, Qu X, Yi J (2015) Characteristics of volcanic reservoirs and distribution rules of effective reservoirs in the Changling fault depression, Songliao Basin. Nat Gas Ind B 2 (5):440–448

Wang P, Chen S (2015) Cretaceous volcanic reservoirs and their exploration in the Songliao Basin, northeast China. Amer Assoc Petrol Geol Bull 99(3):499–523

Wark DA, Hildreth W, Spear FS, Cherniak DJ, Watson EB (2007) Pre-eruption recharge of the Bishop magma system. Geology 35 (3):235–238

Watanabe Y (2005) Late Cenozoic evolution of epithermal gold metallogenic provinces in Kyushu, Japan. Miner Deposita 40 (3):307–323

Watanabe Y (2006) Distribution and characteristic features of volcanic rocks in and around the Minami Nagaoka gas field from the view point of reservoir geology. J Japan Assoc Petrol Technol 71(1):76–84

Waters PJ, Cooke DR, Gonzales RI, Phillips D (2011) Porphyry and epithermal deposits and $^{40}Ar/^{39}Ar$ geochronology of the Baguio district, Philippines. Econ Geol 106(8):1335–1363

Weydt LM, Lucci F, Lacinska A, Scheuvens D, Carrasco-Núñez G, Giordano G, Sass I et al (2022) The impact of hydrothermal alteration on the physiochemical characteristics of reservoir rocks: the case of the Los Humeros geothermal field (Mexico). Geother Energy 10(1):20

White JDL (2000) Subaqueous eruption-fed density currents and their deposits. Precambrian Res 101(2-4):87–109.

White SH, De Boorder H, Smith CB (1995) Structural controls of kimberlite and lamproite emplacement. J Geochem Explor 53(1–3):245–264

White SM, Crisp JA, Spera FJ (2006) Long-term volumetric eruption rates and magma budgets. Geochem Geophys Geosys 7(3)

Williams DA, Kerr RC, Lesher CM (1998) Emplacement and erosion by Archean komatiite lava flows at Kambalda: revisited. J Geophys Res Solid Earth 103(B11):27533–27549

Williams HR, Williams RA (1977) Kimberlites and plate-tectonics in West Africa. Nature 270(5637):507

Williams PJ, Barton MD, Johnson DA, Fontboté L, De Haller A, Mark G, Oliver NH, Marschik R (2005) Iron oxide copper-gold deposits: geology, space-time distribution, and possible modes of origin. Econ Geol 100th Anniversary Volume 371–405

Wills J (1991) The Forties Field, block 21/10, 22/6a, UK North Sea. Geol Soc Lond Mem 14(1):301–308

Wilson CJ, Houghton BF, McWilliams MO, Lanphere MA, Weaver SD, Briggs RM (1995) Volcanic and structural evolution of Taupo Volcanic Zone, New Zealand: a review. J Volcanol Geotherm Res 68(1–3):1–28

Wilson CJ, Gravley DM, Leonard GS, Rowland JV (2009) Volcanism in the central Taupo Volcanic Zone, New Zealand: Tempo, styles and controls. In: Thordarson T et al (eds) Studies in volcanology: the legacy of George Walker. Spec Publ IAVCEI and Geol Soc Lond 2:225–247

Wilson CJ, Rowland JV (2016) The volcanic, magmatic and tectonic setting of the Taupo Volcanic Zone, New Zealand, reviewed from a geothermal perspective. Geothermics 59:168–187

Wilson NS, Zentilli M (1999) The role of organic matter in the genesis of the El Soldado volcanic-hosted manto-type Cu deposit, Chile. Econ Geol 94(7):1115–1135

Wilson NS, Zentilli M, Reynolds PH, Boric R (2003) Age of mineralization by basinal fluids at the El Soldado manto-type copper deposit, Chile: $^{40}Ar/^{39}Ar$ geochronology of K-feldspar. Chem Geol 197(1–4):161–176

Witt WK, Cassidy KF, Lu YJ, Hagemann SG (2020) The tectonic setting and evolution of the 2.7 Ga Kalgoorlie–Kurnalpi Rift, a world-class Archean gold province. Miner Deposita 55(4):601–631

Wohletz K, Civetta L, Orsi G (1999) Thermal evolution of the Phlegraean magmatic system. J Volcanol Geotherm Res 91(2–4):381–414

Wohletz K, Heiken G (1992) Volcanology and geothermal energy. University California Press, Berkeley, p 432

Wormald PJ (1993) Geology of the Mt. Leyshon gold deposit, Australia: a study of breccia pipe formation, facies and brecciation mechanisms. PhD thesis, James Cook University North Queensland, Townsville, Australia, p 201. https://researchonline.jcu.edu.au/47307/

Wormald PJ (2017) The Mt. Leyshon gold deposit. In: Phillips GN (ed) Australian ore deposits. Australasian Institute of Mining and Metallurgy Monograph 32:697–704

Wormald PJ, Orr TO, Hodkinson IP (1991) The Mount Leyshon gold mine (NE Queensland), an intrusive breccia and igneous complex. In: Proceedings World Gold, Cairns, Australia. Australasian Institute of Mining and Metallurgy 2/91:223–232

Wright IC (1996) Volcaniclastic processes on modern submarine arc stratovolcanoes: sidescan and photographic evidence from the Rumble IV and V volcanoes, southern Kermadec Arc (SW Pacific). Mar Geol 136(1–2):21–39

Wright IC, Gamble JA (1999) Southern Kermadec submarine caldera arc volcanoes (SW Pacific): caldera formation by effusive and pyroclastic eruption. Mar Geol 161(2–4):207–227

Wright IC, Worthington TJ, Gamble JA (2006) New multibeam mapping and geochemistry of the 30–35 S sector, and overview, of southern Kermadec arc volcanism. J Volcanol Geotherm Res 149 (3):263–296
Wright JV (1990) Basins and styles of basin, epithermal and other high crustal level mineralisation. James Cook University of North Queensland, EGRU contribution, vol 35, 158 pp
Wright JV, (1994) Guides to metals exploration in basins. Short Course presented to Normandy Exploration Ltd. (Darwin, Australia; Trabzon, Turkey), 360 pp (unpubl)
Wright JV (1996) Mineralisation associated with volcanic terrains: the Mesozoic-Cenozoic Alpine collisional zone. In: Normandy mining group geological conference, 18–25 November, Darwin, Northern Territory, Australia. Rept no: CR21049, 12 pp (unpubl)
Wright JV (2006) An exploration assessment of the uranium prospectivity of Mongolia. Report to Quasar Resources, Adelaide, Australia, 60 pp (unpubl)
Wright JV (2008) Uranium prospectivity of the Gawler Craton: final report. Report to Quasar Resources, Adelaide, Australia, 77 pp (unpubl)
Wright JV, Coward MP (1977) Rootless vents in welded ash-flow tuffs from northern Snowdonia, North Wales, indicating deposition in a shallow water environment. Geol Mag 114(2):133–140
Wu C, Gu L, Zhang Z, Ren Z, Chen Z, Li W (2006) Formation mechanisms of hydrocarbon reservoirs associated with volcanic and subvolcanic intrusive rocks: examples in Mesozoic-Cenozoic basins of eastern China. Amer Assoc Petrol Geol Bull 90(1):137–147
Xu C, Dowd P (2010) A new computer code for discrete fracture network modelling. Comput Geosci 36(3):292–301
Yagi M, Ohguch T, Akiba F, Yoshida T, Tiba T (2009) The Fukuyama volcanic rocks: submarine composite volcano in the Late Miocene to Early Pliocene Akita-Yamagata back-arc basin, northeast Honshu, Japan. Sed Geol 220(3–4):243–255
Yahata M, Inaba M, Okamura S (2007) The role of hydrothermal alteration in the formation of gas reservoir in the Katagai gas field, Niigata, Japan. J Jpn Assoc Petrol Techn 72(3):290–303
Yamada R, Yoshida T (2011) Relationships between Kuroko volcanogenic massive sulfide (VMS) deposits, felsic volcanism, and island arc development in the northeast Honshu arc, Japan. Miner Deposita 46(5–6):431–448
Yamada R, Yoshida T, Kimura JI (2012) Chemical and Isotopic characteristics of the Kuroko-forming volcanism. Resour Geol 62 (4):369–383
Yamada Y, Uchida T (1997) Characteristics of hydrothermal alteration and secondary porosities in volcanic rock reservoirs, the Katakai gas field. J Jpn Assoc Petrol Techn 62:311–320
Yamanaka T, Ishibashi J, Hashimoto J (2000) Organic geochemistry of hydrothermal petroleum generated in the submarine Wakamiko caldera, southern Kyushu, Japan. Organic Geochem 31(11):1117–1132
Yamanaka T, Mizota C, Murae T, Hashimoto J (1999) A currently forming petroleum associated with hydrothermal mineralization in a submarine caldera, Kagoshima Bay, Japan. Geochem J 33(6):355–367
Yang K, Scott SD (2002) Magmatic degassing of volatiles and ore metals into a hydrothermal system on the modern sea floor of the eastern Manus back-arc basin, western Pacific. Econ Geol 97 (5):1079–1100
Yeats CJ, Hollis SP, Halfpenny A, Corona JC, LaFlamme C, Southam G, Fiorentini M, Herrington RJ, Spratt J (2017) Actively forming Kuroko-type volcanic-hosted massive sulfide (VHMS) mineralization at Iheya North, Okinawa Trough, Japan. Ore Geol Rev 84:20–41
Yılmaz H (2002) Ovaçik gold deposit: an example of quartz-adularia-type gold mineralization in Turkey. Econ Geol 97 (8):1829–1839
Yilmaz H (2003) Exploration at the Kuscayiri Au (Cu) prospect and its implications for porphyry-related mineralization in western Turkey. J Geochem Explor 77(2–3):133–150
Yilmaz H, Oyman T, Arehart GB, Colakoglu AR, Billor Z (2007) Low-sulfidation type Au-Ag mineralization at Bergama, Izmir, Turkey. Ore Geol Rev 32(1–2):81–124
Yumul GP, Dimalanta CB, Tam TA, Ramos EG (2008) Baguio Mineral District: an oceanic arc witness to the geological evolution of northern Luzon, Philippines. Island Arc 17(4):432–442
Zheng H, Sun X, Wang J, Zhu D, Zhang X (2018) Devitrification pores and their contribution to volcanic reservoirs: a case study in the Hailar Basin, NE China. Mar Petrol Geol 98:718–732
Zhu S, Zhu X, Liu J, Yao Y, Xian B, Niu H, Zhao C (2012) Genesis and hydrocarbon significance of vesicular welded tuffs: a case study from the Fengcheng Formation, Wu-Xia area, Junggar Basin, NW China. Petrol Explor Develop 39(2):173–183
Zhu Z (2016) Gold in iron oxide copper-gold deposits. Ore Geol Rev 72:37–42
Zierenberg RA, Koski RA, Morton JL, Bouse RM (1993) Genesis of massive sulfide deposits on a sediment-covered spreading center, Escanaba Trough, southern Gorda Ridge. Econ Geol 88(8):2069–2098
Zonneveld JP, Kjarsgaard BA, Harvey SE, Heaman LM, McNeil DH, Marcia KY (2004) Sedimentologic and stratigraphic constraints on emplacement of the Star Kimberlite, east–central Saskatchewan. Lithos 76(1–4):115–138
Zou C (ed) (2013) Volcanic reservoirs in petroleum exploration. Petroleum Industry Press, Elsevier Inc., p 204

Index

R. Cas et al., *Volcanology*, Springer Textbooks in Earth Sciences, Geography and Environment,
https://doi.org/10.1007/978-3-319-66613-6

E

H

J

K

M

N

O

T

U

GPSR Compliance

The European Union's (EU) General Product Safety Regulation (GPSR) is a set of rules that requires consumer products to be safe and our obligations to ensure this.

If you have any concerns about our products, you can contact us on ProductSafety@springernature.com

In case Publisher is established outside the EU, the EU authorized representative is:

Springer Nature Customer Service Center GmbH
Europaplatz 3
69115 Heidelberg, Germany

Batch number: 10733514

Printed by Printforce, the Netherlands

Springer Textbooks in Earth Sciences, Geography and Environment

The Springer Textbooks series publishes a broad portfolio of textbooks on Earth Sciences, Geography and Environmental Science. Springer textbooks provide comprehensive introductions as well as in-depth knowledge for advanced studies. A clear, reader-friendly layout and features such as end-of-chapter summaries, work examples, exercises, and glossaries help the reader to access the subject. Springer textbooks are essential for students, researchers and applied scientists.

Ray Cas • Guido Giordano •
John V. Wright

Volcanology

Processes, Deposits, Geology and Resources

Volume 1

Springer

Ray Cas
School of Earth, Atmosphere and Environment
Monash University
Clayton, Victoria, Australia

Ray Cas
CODES, School of Natural Sciences
University of Tasmania
Sandy Bay, Tasmania, Australia

Guido Giordano
Dipartimento di Scienze, Sezione Geologia
Università Roma Tre
Rome, Italy

John V. Wright
Consultant
Spanish Water, Curaçao

ISSN 2510-1307 ISSN 2510-1315 (electronic)
Springer Textbooks in Earth Sciences, Geography and Environment
ISBN 978-3-319-66612-9 ISBN 978-3-319-66613-6 (eBook)
https://doi.org/10.1007/978-3-319-66613-6

Cover Part of the fissure vent magma fountaining eruption at Bárðarbunga during the Holuhraun eruption, Iceland, 2014. Photo by Gro Birkefeldt Møller Pedersen

This Springer imprint is published by the registered company Springer Nature Switzerland AG
The registered company address is: Gewerbestrasse 11, 6330 Cham, Switzerland

Paper in this product is recyclable.

(a)
(b)
(c)
(d)
(e)
(f)
(g)
(h)

◀ **Frontispiece** Images illustrating the diversity of volcanic activity occurring this twenty-first century—so far! **a** Lava lake at the summit crater of the Nyiragongo volcano ($\sim$3000 m a.s.l). The ultramafic composition results in eruptions with unusually fluid lava flows which, on September 2002 and May 2021, caused loss of life and destruction in the city of Goma, Democratic Republic of the Congo. (Photo by Roger Scoon, taken in March 2020). **b** Magma or fire fountaining on 4 October 2021, during the September to December eruption of Cumbre Vieja shield volcano on La Palma, Canary Islands, Spain. This built a new basanitic scoria cone, while lava flows reached the sea forming a lava delta. **c** Explosive eruption column and downwind plume from the Calbuco stratovolcano, which erupted on 22 April 2015 for the first time in four decades. The eruption lasted 90 minutes and sent a plume of volcanic ash >15 km into the atmosphere. View from Puerto Montt city, southern Chile. (Photo by Carolina Barría Kemp; *Source* Wikimedia, through Creative Commons Attribution-Share Alike 2.0 Generic licence). **d** The former village of San Miguel Los Lotes, buried by a pyroclastic flow on 3 June, 2018, erupted from the Fuego stratovolcano. It is now designated as an official national cemetery by the Government of Guatemala. (Photo by Capt. Bombero; *Source* Wikimedia, through Creative Commons Attribution-Share Alike 4.0 International licence). **e** Intra-subglacial volcanic eruption of Eyjafjallajökull volcano on 17 April, 2010, southern Iceland. The basaltic ash plume rose to a height of over 10 km causing widespread disruption of European air traffic. (Photo by Boaworm; *Source* Wikimedia, through Creative Commons Attribution 3.0 Unported licence). **f** Satellite view of the dramatically expanding, ash plume above the explosive submarine basaltic andesite eruption of Hunga Tonga–Hunga Ha'apai volcano on 15 January 2022, Tonga. The top of the main umbrella region rose to a height of at least 40 km and shows a series of concentric, outward-moving, "gravity waves" of gas, water vapour, and ash. The lower pale grey cloud is of uncertain origin but could be a radially spreading cloud of fine ash and ice particles that nucleated on fine ash particles at an altitude of $\sim$15 km (Andrew Prata, pers. comm.) and could also be related to a buoyant ash cloud generated at sea level by near vent pyroclastic flows flowing over and into the sea. (*Source* NOAA GOES-West satellite). **g** Small, weak explosion near the summit of the submarine West Mata volcano ($\sim$1200 m water depth) ejects ash and rock, with magma glowing red. This volcanic activity was the first observed to erupt a lava of boninite composition, northeastern Lau Basin, northern end of the Tonga volcanic arc. (*Source* NSF/NOAA; ROV image taken May 2009). **h** Pristine, glassy, coherent basaltic pillow lava erupted on the seafloor at the Marianas back-arc spreading centre, western Pacific Ocean. This occurred between 2013–2015 at $\sim$4000 m water depth. (*Source* NOAA; ROV image taken April 2016)

"We can't understand what we are looking at until we understand what we are looking for."

(Monash University Volcanics Shortcourse, 1982)

Preface

Since the publication of *Volcanic Successions* in 1987 (Cas and Wright, Allen and Unwin) there have been many advances in the field of volcanology. *Volcanic Successions* was written as a senior undergraduate, postgraduate, and industry professional-level book. Our aim here with this new book is to provide a major advance on that book at a higher level, by increasing the scope; providing much more detail on key concepts, processes, and deposit characteristics; and providing a comprehensive bibliography of each of the major topics addressed. We hope that this new book is as well received and useful as the first. In the following Introduction, we provide an overview of what we cover, the approach, and the structure of the book.

Melbourne, Australia
January 2024

Ray Cas
Guido Giordano
John V. Wright

Acknowledgements

First, we want to thank our families for their support and understanding about the time we have had to spend on the "book project", sometimes at the expense of time with them. We also thank our Institutions and Schools at Monash University, University of Tasmania, and Università Roma Tre for the opportunities, facilities, and financial support to undertake the decades of research that have led to the preparation of this book. We also thank the many corporate entities and government agencies both in Australia and from overseas and the individual course participants who have supported Monash University's annually run Volcanics Shortcourse, which since its inception in May 1982 provided the essential financial backing to complete this book. In addition, although there are too many to name specifically, we thank the many work colleagues and present and past research colleagues all over the world, in both academia and industry, for their collaborations and their encouragement to write this book. We greatly appreciate the efforts of many research students and postdoctoral research fellows we have worked with for their significant contributions to research in volcanology; some have become ongoing research collaborators. Ray Cas specifically thanks former students and postdocs Jack Simmons, Nathan Bowman, and Jozua van Otterloo for their technical assistance in drafting figures, compiling reference lists, and some proof reading of draft chapters at Monash University. Tom Orr (Orr & Associates) provided drafting facilities for several figures in Chaps. 17 and 18. In addition, we have used many figures drafted by Draga Gelt for the predecessor of this book, *Volcanic Successions* (1987). We also acknowledge the individuals who have kindly provided figures and images that we have incorporated into this book. Finally, we thank Springer for publishing our book, but which unfortunately involved multiple delays during Springer's production process, ultimately causing a significant delay in its publication.

Ray Cas
Guido Giordano
John V. Wright

Introduction: Aims, Scope, Approach, and Structure of the Book

Background

Volcanic eruptions have been one of the most fundamental influences on the Earth's evolution throughout its 4.54 Ga history. Volcanic eruptions formed the Earth's crust, were the source for the atmosphere and oceans and are arguably the basis for life on Earth. Huge eruptions have almost certainly caused climate change and influenced mass extinction events. The dynamic and varied nature of volcanic eruptions in subaerial, subaqueous, and subglacial environments and the processes by which eruption products are dispersed across the Earth's surface, the spectrum of volcanic deposit types, and their potential enormous volumes and dispersal distances make volcanic hazards among the most dangerous and threatening natural hazards on Earth. Throughout Earth's history, volcanism has produced a huge spectrum of natural resources that our society depends on, including base metals (Cu, Pb, Zn, Ni), precious metals (Au, Ag), platinum group elements (PGE), diamonds, and in modern times, it is the source for geothermal energy.

Understanding eruption styles and dispersal processes and interpreting the origins of volcanic deposits are not simple because volcanic systems and processes can be highly variable in terms of physical and chemical processes, as well as their spatial and time scales. Although we learn much about volcanic processes by witnessing modern eruptions and studying their deposits, we learn at least as much from the prehistoric volcanic rock and deposit record. Studies of the rock record also provide the database for reconstructing Earth's history. In addition, in terms of volcanic hazard evaluation, understanding the stratigraphic record of eruptions and deposits of any particular active volcano provides the best insights into the possible future eruption behaviour. However, the uncertainty of the depositional setting and its effects on eruption processes and deposit types, the huge spectrum of possible primary products and secondary deposits, and the effects of hydrothermal alteration, metamorphism, and deformation make the reconstruction of ancient volcanic successions and processes extremely challenging. The book by Cas and Wright (*Volcanic Successions* 1987) was the first to develop a facies analysis approach to the documentation and interpretation of modern and ancient volcanic successions. In this new book, we progress that approach considerably further, incorporating new approaches and developments in a total revision of the 1987 edition of *Volcanic Successions*.

Aims of the Book

The aim of this book is to help develop a better understanding of the spectrum of eruption styles in different environments and tectonic settings. In addition, we will provide a comprehensive account of the enormous range of deposit types (facies) preserved in modern and ancient volcanic terrains, their characteristics, the processes involved in their eruption and deposition, the facies and stratigraphic architecture, their depositional and tectonic setting, and their resource potential. We will also touch on the natural hazards associated with the spectrum of processes operating in volcanic terrains, although that topic is worthy of a book in its own right, and so our consideration is not comprehensive. A major aim of the book is to provide students and geoscientists with the approach and basis for making their own interpretations and reconstructions of modern, historical and ancient eruption sequences. Therefore, the book will also briefly address methods and approaches that may be useful in documenting and analysing both modern and ancient successions. The book is not an encyclopaedia, and although we present a balanced overview and review of processes and interpretations, we also present our own personal insights and assessments as research scientists with considerable experience in the range of topics discussed.

Structure, Scope, and Approach of the Book

The first-order organisation of the book is into six major parts that contain groupings of chapters related to the themes of each part: Part A: *Laying the Foundations*; Part B: *Effusive Eruptions, Lavas and Sub-Volcanic Intrusions*; Part C: *Explosive Eruptions, Processes, and Pyroclastic Deposits*; Part D: *Secondary Processes in Volcanic Settings*; Part E: *Terminology and Approaches to Understanding the Geology of Volcanoes*; and Part F: *Bringing Everything Together: The Big Picture*. This progression leads us from specific processes and deposit types to assessing the larger-scale architecture of volcanoes, their global setting, and their resources.

In Part A, many important introductory and background concepts are introduced. In Chap. 1, "Introductory Concepts", the diversity of volcanoes, eruption styles, the relative time relationships between volcanic and other processes in volcanic terrains, the facies concept, essential facies characteristics and some terminology that are important in the successful documentation and interpretation of facies are introduced. In Chap. 2, "Properties of Magmas", the variable physical and chemical properties of magmas are reviewed as the basis for understanding the nature of volcanic eruptions and the behaviour of lava flows and intrusions. In Chap. 3, "Influence of Environment on Magma Properties, Eruption Processes, and Deposits", we consider the influence that different ambient environments (atmosphere, oceans/lakes, ice, lithosphere) can have on magma properties, eruption styles, and eruption products. Finally, in Chap. 4, "Fragmentation Processes in Magmas and Volcanic Rocks", we review the diverse ways (autoclastic, explosive, hydraulic fracturing, epiclastic) by which magmas and volcanic rocks can be fragmented and the characteristics of the clast types and aggregates that characterise those different origins.

In Part B, *Effusive Eruptions, Lavas and Sub-Volcanic Intrusions*, we summarise the processes that produce largely coherent volcanic rocks, lavas, and their subvolcanic counterparts, high-level intrusions. These include Chap. 5, "Basaltic and Other Low Viscosity Lavas"; Chap. 6, "High Viscosity Silicic Lavas (Rhyolite, Dacite and Andesite"; and Chap. 7, "Komatiite Lavas and Intrusions". In each case, we consider the behaviour of these lava flow types in subaerial and subaqueous environments and the facies characteristics and facies architecture of lavas in these different settings. In Chap. 7, we also briefly address other very low-viscosity lavas, including carbonatites, and sulphur lavas. In Chap. 8, "Sub-volcanic Intrusions", we briefly consider the principles controlling the behaviour of magmas rising to the Earth's surface, the factors that determine if they will erupt or cease to rise and be emplaced as intrusions. We also assess the criteria for distinguishing intrusions, particularly sills from lavas in the geological record, which can be difficult in ancient deformed terrains but is important for understanding the chronology of events and likely resource types and settings.

Part C deals with *Explosive Eruptions, Processes, and Pyroclastic Deposits*. In Chap. 9, "Explosive Eruption Styles, Columns, and Pyroclastic Fallout Deposits", we review the characteristics and behaviour of different explosive eruption styles and columns, the dispersal processes that produce pyroclastic fallout deposits, and the deposit characteristics. We also tackle the problem of classifying or subdividing pyroclastic fallout-forming eruption styles and deposit types, which used to be simple, but is increasingly becoming more difficult, as the range of variables that affect eruption styles and deposit characteristics is better understood. As a result, we propose that existing classification schemes (D-F, VEI, Magnitude, Intensity) be recalibrated, with essential revisions proposed. In Chap. 10, "Subaerial Pyroclastic Density Currents (Pyroclastic Flows and Surges)", we assess the spectrum of pyroclastic density current types, including distinctive pyroclastic flow to pyroclastic surge types, and directed blasts. We discuss in detail the flow and sedimentation processes and the huge variety of resulting deposits and facies. We also propose a new classification scheme for ignimbrites. In Chap. 11, "Subaqueous Flows of Pyroclastic Debris and Their Deposits", we consider what happens when pyroclastic density currents enter or are erupted under water. Do they retain integrity as hot gas-supported and fluidised pyroclastic density currents, or do they ingest water and transform into water-supported transport systems? Are they still "pyroclastic", and how are such syn-eruptive deposits distinguishable from post-eruptive, resedimented deposits of pyroclastic debris and from subaerial counterparts?

This provides a segue into Part D, *Secondary Processes in Volcanic Settings*. Chapter 12, "Surface Sedimentary (Epiclastic) Processes and Deposits in Volcanic Environments", begins with an overview of the importance of surface sedimentary processes in transporting and depositing volcanic debris in both subaerial and subaqueous volcanic settings through the effects of gravity and the involvement of ice, water, or air. We consider the duration of periods of repose and their surface sedimentary processes relative to the duration of eruption episodes during the history of polygenetic volcanoes. Surface processes also pose major natural hazards in most volcanic terrains, often more so than eruption-generated primary phenomena. In Chap. 13, "Syn- to Post-emplacement Alteration Processes", we briefly address the ways in which hydrothermal, diagenetic, metamorphic alteration, and deformational processes can affect original rock textures, sometimes producing misleading pseudo-textures.

In Part E, *Terminology and Approaches*, we begin with the vexed question of "Terminology for Volcanic Deposits and Rocks" in Chap. 14, and we propose a systematic, step-by-step approach, beginning with descriptive terminology and culminating in the application of genetic terminology. In Chap. 15, "Documenting the Geology of Volcanoes, Volcanic Terrains, and Constructing Facies Models", we consider the importance of constructing facies models for all volcanoes and ancient volcanic successions, the approaches and data sets that are useful, and some of the difficulties.

In Part F, *Bringing Everything Together: The Big Picture*, we integrate concepts from preceding chapters to address some "big-picture" issues. In Chap. 16, "The Geology of Volcanoes and Their Facies Models", the facies and stratigraphic architecture and geology of different types of volcanoes and volcanic settings are considered, based on both modern and ancient examples, as well as introducing their economic significance and summarising their potential hazards. In Chap. 17, "Volcanism and Tectonic Setting", the modern plate tectonic framework is used to assess where volcanism occurs and the distinctive characteristics of volcanoes and their deposits in those settings. We then explore the tectonic regime for volcanism from the Earth's beginnings at 4.54 Ga during the Hadean and Archean and summarise how different the Earth was then (no plate tectonics until the latest Archean!) and how unique some volcanic and plutonic rocks were. Finally, in Chap. 18, "Volcanic-Hosted Resources", the spectrum of important natural resources that have formed on Earth in volcanic terrains are considered, in particular epithermal deposits, volcanic-hosted iron oxide copper–gold and uranium deposits, magmatic nickel (PGE) sulphide deposits, VMS base metal deposits, diamonds, and volcanic reservoirs hosting both hydrocarbon and geothermal energy.

We hope that you find the book to be a valuable reference volume and enjoy it as much as we have enjoyed producing it over many years.

Contents

Part A
Laying the Foundations

Aims, Scope, and Structure
Part A introduces the important concepts that are the foundations to understanding volcanic eruption processes and dynamics, deposit characteristics, facies, terminology and documentation. We advance specific concepts in detail in Parts B to E, and further elaborate on them in Part F which describes the larger-scale aspects of volcanic systems and terrains. Central to our approach is to treat the diversity of deposits in volcanic terrains as facies, and use facies analysis and interpretation as a means of constructing facies models for different volcanic settings and their resources. In Chap. 1 the importance of volcanism through Earth history, the diversity of volcano types and the complex process spectrum that operates in volcanic settings are discussed. Magma properties are reviewed next in Chap. 2 in order to understand how magmas behave as they rise to the Earth's surface, why magmas erupt coherently as lavas or explosively as pyroclastic deposits, and how they behave subsequently. We also provide a brief introduction to lava flows and discuss the different controls on lava flow morphology, prior to their detailed discussion in Part B. Building on these themes, in Chap. 3, we consider the properties of the different environments through which magmas pass, namely the Earth's lithosphere, atmosphere, hydrosphere, and glaciosphere to assess the influence these different environments have on eruption processes. In Chap. 4, we consider the many processes by which magma and volcanic rock can be fragmented. Irrespective of their origin, all of these of these fragmental products can, in the first instance, be termed volcaniclastic deposits.

1 Introductory Concepts

Significance of Volcanism in Earth History, the Process Spectrum in Volcanic Terrains, and Describing Deposits

Ray Cas, John V. Wright, and Guido Giordano

R. Cas et al., *Volcanology*, Springer Textbooks in Earth Sciences, Geography and Environment,
https://doi.org/10.1007/978-3-319-66613-6_1

Plate 1.1 **a** Deformed pillow lava in the oldest known volcanic rocks on Earth, 3.7–3.8 Ga Isua Greenstone Belt, Greenland, highlighting the importance of volcanism during early Earth history. The pillows have been deformed and stretched, and metamorphosed to amphibolite facies metamorphic grade, but the younging direction can still be determined from the geometry of the pillows as being towards the top left. **b** Spectacular 300 m high magma or "fire" fountain producing a lava flow, Mauna Ulu, Hawai'i (*Source* United States Geological Survey; Photo taken 30–12-1969)

Abstract

In this chapter, the importance of volcanism throughout Earth history, the diversity of volcano types and the complex process spectrum that operates in volcanic settings are discussed. Volcanoes occur in a variety of forms and sizes, their life spans vary enormously, and the processes, deposits, and stratigraphic successions produced by different types of volcanoes can be extremely variable. Since there is such a diversity of rock and deposit types in the complex environments of volcanic terrains, the interpretation of their possible origins (lavas, intrusions, associated autoclastic, pyroclastic, epiclastic, and chemical deposits) can be difficult. This is especially so for ancient rock sequences, which may have been lithified, hydrothermally altered, metamorphosed, deformed, and tectonically dismembered. As well as some essential genetic terms, we also introduce the facies approach to documenting deposit characteristics. Understanding the spatial and age relationships between facies and associations of facies is crucial in deciphering the stratigraphic architecture, and chronology of events. By doing this, models of the deposits produced by particular events can be developed, and eruptive histories can be reconstructed in both modern and ancient successions. We also urge caution in assessing the tectonic setting of Earth's early volcanism during the Hadean and Archean (45% of the Earth history) because plate tectonics did not operate through most of this huge period of time (see Chap. 17 for discussion of this intriguing topic).

1.1 Introduction

Volcanism has occurred continuously throughout geologic time (Plate 1.1 a, b) and has been one of the most fundamental and important processes to affect the evolution of the Earth system during its history. There are many different types and sizes of volcanoes (see Frontispiece; Fig. 1.1), they have enormously varying life spans, and the processes, deposits, and stratigraphic successions can vary greatly. Since there is a huge range of rock and deposit types in volcanic terrains, understanding their possible origins (lavas, intrusions, associated autoclastic, pyroclastic, epiclastic, and chemical deposits) is challenging. This is particularly the case for ancient terrains, which may have been lithified, hydrothermally altered, metamorphosed, deformed, and tectonically dismembered.

In both modern and ancient successions, interpretation is dependent on careful documentation and description of deposits, for which the approach to facies description and analysis used in sedimentology is very valuable. Understanding the spatial and age relationships between facies and associations of facies is crucial to understanding whether or not spatially related facies are genetically related, and in deciphering the facies and stratigraphic architecture, and chronology of events. By doing this, models of the deposits produced by particular events can be developed, eruptive histories can be reconstructed, clearer pictures of the emplacement processes and environments emerge, and hazards and resource potential can be evaluated.

In this chapter, we provide an introduction to the key concepts that underpin our understanding of volcanic processes and deposits. We introduce the importance of volcanism in the Earth history, the diversity and types of volcanoes, the bigger picture tectonic context of volcanoes through Earth history (it hasn't always been a plate tectonic Earth!), the effect of environment on eruption processes and deposits, the process spectrum in volcanic environments. We also briefly introduce some essential terminology. Finally, we introduce the facies approach advocated by Cas and Wright (1987) for documenting the characteristics of all volcanic deposits as the basis for constructing the facies and stratigraphic architecture of both modern and ancient volcanic deposits and rock successions, and interpreting their origins.

1.2 The Importance of Volcanism During Earth History and Today

Volcanism associated with mantle plumes, hot spots, spreading ridges, and subduction zones has been the fundamental source of the Earth's crust. Although the release of volatiles during impacts of planetesimals during Earth accretion contributed to the formation of the Earth's earliest atmosphere, out-gassing associated with volcanism throughout geological time has been responsible for producing most of the Earth's atmosphere, the oceans, and indirectly through these, life on Earth (Kasting 1993). Photosynthetic organisms then further modified the composition of the atmosphere. Volcanism has significantly affected global climate (e.g. Rampino and Self 1992; McCormick et al. 1995; Ernst and Youbi 2017), evidence of which is locked up in rock successions of the geological record. Major climate-affecting eruptions have had dramatic effects on the biosphere, and almost certainly have caused major extinctions. The 74 ka eruption of Toba caldera in Indonesia is thought to have caused the Earth's climate to regress back into a glacial stage, decimating the human population almost to extinction. The 1783–84 basalt eruption of Laki volcano in Iceland had catastrophic effects on agricultural productivity in Europe and Asia, causing widespread famine and loss of life. Major extinctions coinciding with major volcanic events, such as flood basalt events and explosive

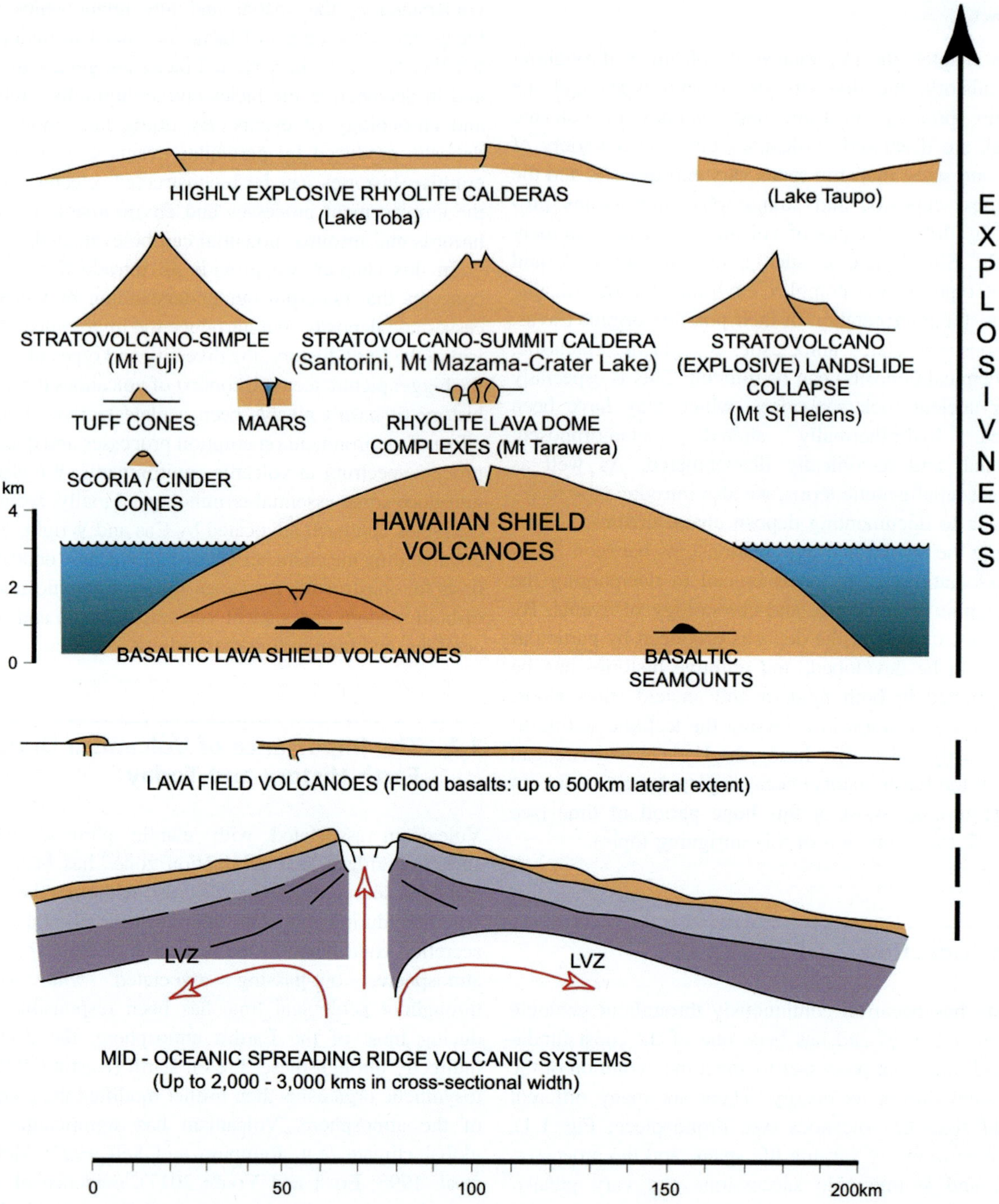

Fig. 1.1 Schematic representation of the principal types of volcano, with an indication of their size and relative explosiveness. Not indicated are the huge areas and distances beyond and between volcanoes that are also covered by volcanic products

super-eruptions, are well documented (e.g. Rampino and Self 1992; Robock 2004; Self 2006; Ernst and Youbi 2017).

Volcanic rocks constitute important stratigraphic components in tracing the history of the Earth. The oldest supra-crustal rocks on Earth, the 3.7–3.8 Ga Isua Greenstone Belt, of southwestern Greenland, are dominated by variably metamorphosed, submarine, mafic volcanic rocks, including pillow lavas (Plate 1.1a; Appel et al. 1998). These supra-crustal rocks contain evidence for the existence of the earliest known life forms (e.g. Rosing 1999). Some think "life" originated in sub-terranean volcanic, hydrothermal systems (e.g. Davies 1998) or around seafloor hydrothermal springs (e.g. Nisbet 1995; Dodd et al. 2017). Volcanism is also thought to have contributed to, and in some cases caused major faunal extinction events (e.g. Rampino and Stothers 1988; Ernst and Youbi 2017). Volcanism has shaped many of the landform features of the Earth's surface since its beginning, but for many modern communities, volcanism represents an everyday hazard to life and property.

Socio-economically, volcanoes are important because of the hazard impacts on regions, nearby communities, and the environment. Different volcano types pose different types of hazards because of a tendency for specific styles of eruption and surface process events to be associated with each volcano type. However, volcanoes cannot be stereotyped. To do so in hazard assessment is to invite unforeseen disaster. Volcanoes do have histories and understanding the past behaviour of a volcano is an essential tool in predicting and anticipating possible future events and hazards. This requires careful, thorough fieldwork, involving detailed mapping, section logging, establishing a field stratigraphy, a chronology and frequency of events, sampling, and follow-up laboratory work, such as petrography, geochemistry, and geochronology. There are no shortcuts in this work. Surprisingly, of the 1,551 volcanoes that are classified as having been active in the Holocene, there is adequate geological information that can be used to assess the history and frequency of volcanic activity in Volcanic risk and hazard assessment for only 328 ($\sim$20%; Loughlin et al. 2015). That means that adequate hazard assessment cannot be undertaken for the remaining 80% of the world's active volcanoes until basic geological mapping and associated studies have been completed.

In addition, volcanoes are also important socio-economically because of the varied resource potential of different types of volcanoes (Chap. 18). The resources of volcanoes range from geothermal energy, to precious stones (e.g. kimberlite–lamproite hosted diamonds) to a variety of metallic ore deposits (epithermal precious metals—Au, Ag; porphyry copper, volcanic hosted base metals—Cu, Pb, Zn, Au, Ag), to magmatic komatiite hosted nickel sulphide and layered intrusion platinum group element deposits, to oil and gas reservoirs (Chap. 18), to even simple industrial aggregate.

Volcanism can occur in any surface environment, including subaerial, submarine, lacustrine, and glacial. At least 70% of the Earth's present surface is oceanic crust formed by the long-term volcanism from mid-oceanic spreading ridges, as well as from intraplate mantle hot-spot and plume sources. In addition, there are vast areas of continental crust that have been affected by volcanism. In the environments where volcanism occurs, the equilibrium of normal sedimentary surface processes is interrupted and overwhelmed by the sometimes, catastrophic effects of episodic, short-lived, voluminous volcanic events. Volcanic terrains are therefore the most complex surface environments on Earth, and therefore contain a greater variety of deposit types, or facies (see below) than any other surface environment on Earth. Volcanism can also occur in virtually every sedimentary basin type and tectonic setting.

Detailed study of the physical processes and deposits of volcanic terrains was for a long-time subordinate to studies of geochemistry, mineralogy, and petrogenesis. However, the growing need to understand better the natural hazards associated with volcanic terrains as civilisation has encroached on them, the natural resource potential, and environmental and climatic impacts of volcanism have driven the need to develop a better understanding of the physical processes operating in volcanic regions. This occurred in parallel with major advances in sedimentology, and in the technology used to observe and monitor volcanic terrains and processes. Studies of modern and ancient, subaerial, subaqueous, and subglacial volcanic settings have contributed to the huge growth in knowledge.

1.3 Volcanoes and Their Diversity

There are many different types of volcanoes with a variety of forms and sizes (Fig. 1.1; Frontispiece; Chap. 16). Volcanoes are influenced by a complex spectrum of processes ranging from eruption phenomena (Chaps. 5–11) to sedimentary (Chap. 12) and tectonic (Chap. 17) events, as well as significant subsurface intrusive (Chap. 8) and hydrothermal events (Chap. 18). These together produce an enormous range of possible products. Many small volcanoes are monogenetic and simple because they experience only one relatively short, eruption phase and style, and are generally sourced by one magma batch (e.g. scoria or cinder cones, some lava domes, small lava shield volcanoes). Other monogenetic volcanoes are more complex with multiple vents, and may even be poly-magmatic (Brenna et al. 2010; van Otterloo et al. 2014; Báez et al. 2017). Polygenetic and complex volcanoes, experience multiple eruption phases, spread over an extended period of time (e.g. stratovolcanoes, large lava-shield volcanoes, explosive caldera volcanoes). They may experience varying eruption styles and are sourced by multiple magma batches. Volcanoes can have variable life spans, from as little as hours to days to months (scoria cones, maars) to tens of thousands of years (e.g. stratovolcanoes, large shield volcanoes), to hundreds of thousands to millions of years (e.g. explosive felsic calderas, regional lava field volcanoes). The biggest volcanoes on Earth are the mid-oceanic spreading ridges and flood basalt lava field volcanoes.

The smallest volcanoes include subaerial basaltic spatter cones and scoria cones, with dimensions from tens of m in diameter and height, to a km in diameter and hundreds of metres high (Chap. 16). Subaerial phreatomagmatic maars with wide craters, usually lakes, are up to 4 km in diameter and have surrounding maar rim or rampart sequences up to tens of metres high. Small subaerial, subaqueous, and subglacial basaltic shield volcanoes are up to several km in diameter and tens to several hundred m high (Chap. 16). Subaerial and submarine felsic, dome lava complexes, are up to several km in diameter and one km in height (Chap. 16). At the larger end of the spectrum there are longer lived, subaerial, and marine stratovolcanoes (Chap. 16), the compositions of which are commonly variable, ranging from basaltic to "intermediate" andesitic/trachytic/phonolitic to dacitic. They may be 60 km in diameter and up to 3 km high, and experience multiple eruptions separated by considerable repose periods, and variable eruption styles from explosive (Chaps. 9–11) to effusive (Chaps. 5, 6). Highly explosive rhyolitic, dacitic, and sometimes even basaltic explosive caldera volcanoes can be tens of km in diameter (e.g. Toba Caldera, Sumatra, Indonesia, at almost 100 km long), and a km high (Chap. 16). The biggest volcanic edifice on Earth is the marine basaltic shield volcano complex edifice of Mauna Loa, Mauna Kea and Kilauea in the Hawaiian Islands. Its basal diameter at a depth of about 4.5 km on the Pacific seafloor is 180 km, and from its base, to its summit at just over 4 km above sea level, the edifice is some 9 km high, making it almost as high as Mt Everest. However, in addition, the Mauna Loa edifice has an isostatically subsided root of basaltic lavas that is $\sim$8 km thick, making the total thickness of the edifice $\sim$17 km (Chap. 16). Its dimension is thus comparable to the Olympus Mons edifice on Mars, which has dimensions of about 250 km diameter and 25 km high and is thought to be the largest single "point source" volcano in the solar system. The volcanoes with the greatest diameter or lateral dimension are extensive lava field volcanoes, especially flood basalt lava field volcanoes and oceanic plateaus, both also called large igneous provinces or LIPS, which can have lava fields extending over 500 km from vent. However, Earth's mid-oceanic spreading ridges, with their continuous fissure vent system many thousands of kilometres long and several thousand kilometres wide (Chap. 16), are the biggest volcanic systems on Earth, and in the solar system.

The form of volcanoes is a direct reflection of the eruption processes, magma volumes, and magma properties. Some are cone-like (e.g. small scoria cones, large stratovolcanoes), reflecting a predominance of eruption processes that limit the dispersal of products to largely near-vent areas, so building up a cone around the vent. For pyroclastic processes, this suggests relatively low-intensity explosive eruptions. Where lava eruptions are involved in producing the cone-like morphology, it indicates repeated, relatively low magma discharge rate and small volume lava eruptions. On the other hand, the formation of large lava shield volcanoes and extensive lava field volcanoes result largely from sustained lava-forming eruptions, lasting tens of years, and involving huge volumes of fluidal, low-viscosity magma. Very explosive volcanoes such as maars and explosive caldera volcanoes disperse their products beyond the immediate limit of the vent and so develop only a very low profile, dominated by a relatively wide crater, and a fringing, gently sloping, pyroclastic shield (calderas) or rampart (maars).

Where volcanoes are closely spaced, or overlap, forming nested complexes, or where volcanic fields are regionally extensive, volcanic activity may last in that volcanic field for hundreds of thousands or even millions of years.

1.4 Volcanic Provinces: Their Tectonic and Paleoenvironmental Context Through Geological Time

Volcanoes never occur in isolation; they form parts of widespread volcanic fields or even larger regionally extensive volcanic provinces that have life spans of millions to tens of millions of years. Volcanic provinces are marked by multiple, co-genetic, pene-contemporaneous volcanoes. The products of individual volcanoes may be dispersed widely throughout the volcanic province, and beyond, and be interstratified in the stratigraphic successions of such provinces and associated sedimentary basins, and beyond these. Volcanic provinces may be confined within a single sedimentary basin or may overlap the margins of one or more basins. Volcanic provinces are always long-lived, with active life spans of millions to tens of millions of years. Phanerozoic volcanic provinces are frequently related to distinct lithospheric plate tectonic settings, including divergent, mid-oceanic spreading ridges, back-arc basin settings, and continental rift zones, or convergent subduction-related island arc or continental margin volcanic arc settings. However, they can also occur anywhere within plates, in intraplate oceanic or continental settings, sourced by mantle hotspots and plumes (Fig. 1.2a).

However, plate tectonics may have occurred on Earth for only the latter half of Earth history, from <2.5 Ga (e.g. Cawood et al. 2018). Prior to the transition to plate tectonics

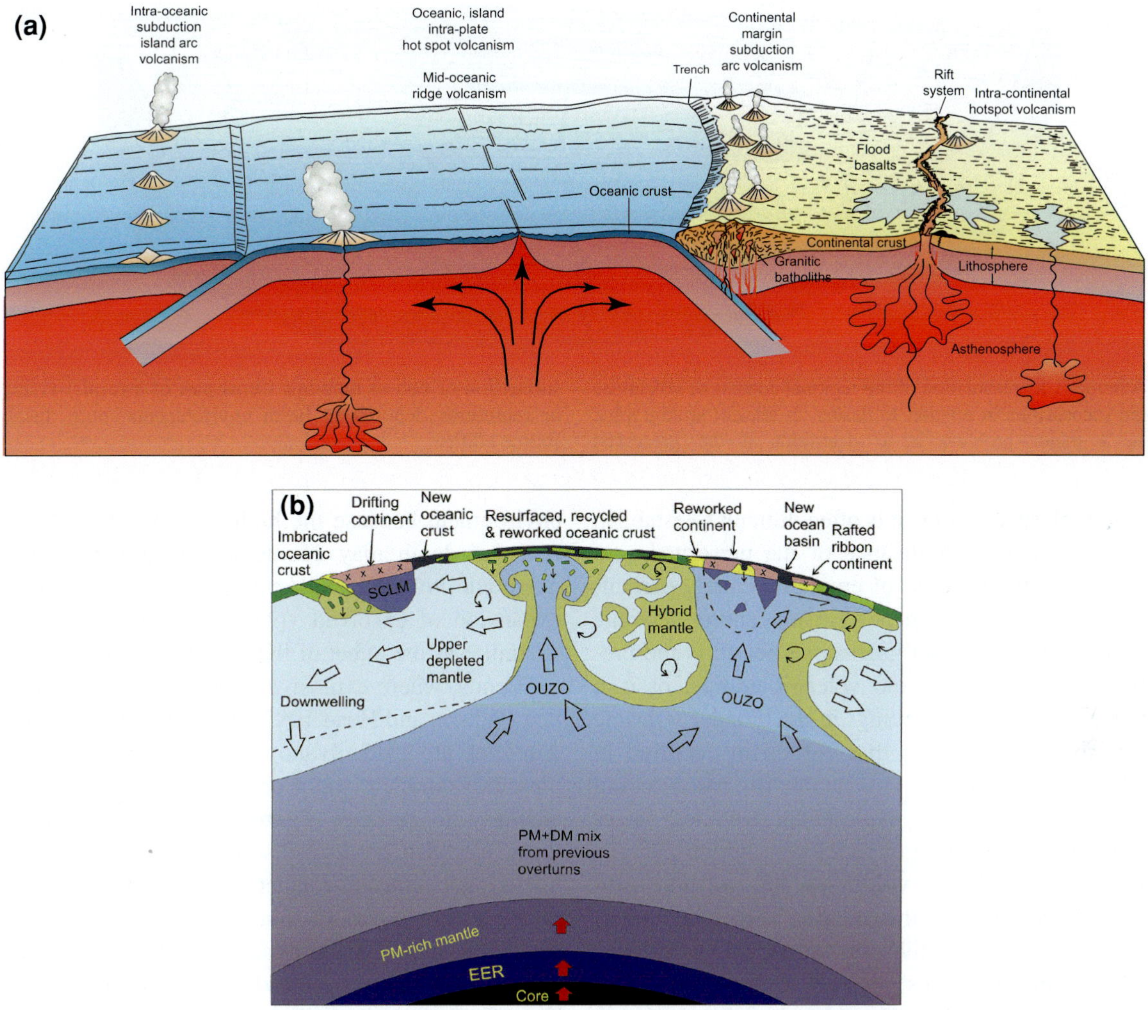

Fig. 1.2 Tectonic setting of volcanism through Earth history. **a** Plate tectonic setting of volcanism on Earth since the late Archean, depicting volcanism at the divergent, globe-encircling, mid-oceanic, spreading ridges, and continental rift zones, at convergent, oceanic island arc, and continental margin arc systems, as well as intra-plate, hotspot, and plume related volcanism in oceanic and continental settings. **b** Late Archean plume dominated magma-tectonic regime with a single lithospheric shell or "stagnant lid" without plate tectonic recycling, as envisaged by Bédard (2018). OUZO = Overturning upwelling mantle zones; SCLM = Sub-continental lithospheric mantle; PM = Primitive mantle; DM = Depleted upper mantle; EER = Early enriched mantle

at the end of the Archean, the nature and tectonic context of volcanic provinces during the Hadean and Archean was very different (Chap. 17, Sect. 17.7). The Earth's dynamics were plume-dominated, and vertical mantle convection was dominant. Instead of lithospheric plates, a continuous crust or "lid" or lithospheric girdle seems to have existed, broken by convection-driven rifting and vertical mantle plume and granitoid pluton diapirism (Fig. 1.2b; e.g. Bleeker 2002; Nebel et al. 2018; Bédard 2018). In this pre-plate tectonic regime, magmatism was not necessarily organised in linear belts, as it is under the current spreading ridge and arc magmatism of the plate tectonic regime. The absence of identifiable accretionary prisms, high-pressure blueschist metamorphic belts, and clear ophiolite slabs in Archean rock successions, which are the only viable physical geological evidence of plate tectonic processes, supports this assessment of the onset of plate tectonics only in the latter half of Earth history. However, in spite of this, there is a very significant record of volcanism and plutonism during the Archean.

However, during the Archean, there was much less subaerial continental or island land area than today, as well as less continental crust (Taylor and McLennan 1995; Hawkesworth et al. 2019). Submarine volcanic and sedimentary successions dominate the Archean rock record. Therefore, during the Hadean and Archean (=45% of the Earth history), the principal physical constraints on eruption processes were those that exist in modern submarine environments (e.g. Cas and Simmons 2018; see Chap. 3 for discussion), and subaerial volcanism was much less common.

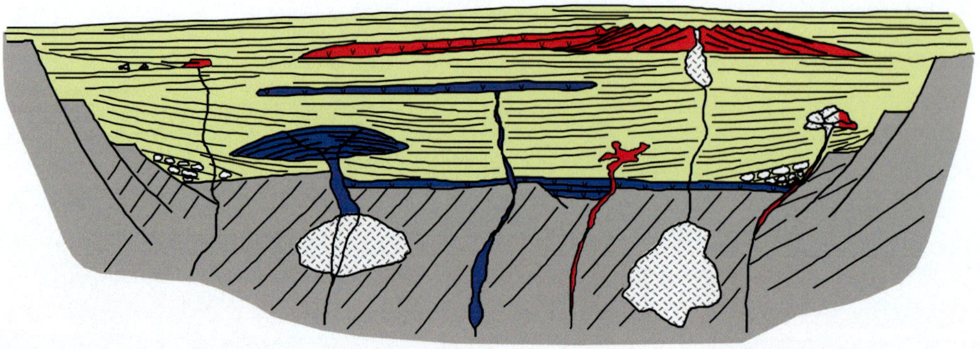

Fig. 1.3 Schematic representation of the regional context of volcanoes and volcanic successions in relation to basins. In ancient stratigraphic successions, a major problem is to determine if the preserved succession or belt of outcrops forms part of a former volcanic edifice or represents deposits in basin environments more distal from the source edifice

In ancient rock successions, it is often difficult to establish the spatial context and significance of the preserved succession relative to the location of eruption centres, particularly if there has been significant erosion and/or tectonic dismemberment (e.g. Archean greenstone belts). Is the preserved succession part of an old volcanic edifice, or is it representative of more distal settings in the broader volcanic province or sedimentary basin that volcanism occurred in (Fig. 1.3)? This distinction is crucial in paleo-setting reconstructions and resource exploration. Detailed facies analysis and an understanding of facies distribution patterns related to facies models for volcanoes and volcanic provinces become important in this regard. Therefore, understanding the scale perspective of individual volcanoes, volcanic fields, and provinces, as well as the spectrum of facies types in modern volcanoes, volcanic fields, and provinces is particularly important for people working in ancient volcanic terrains and rock successions, trying to reconstruct the preserved rock record.

1.5 Effect of Environment on Eruption Styles, Processes, and Volcano Type

Volcanoes can of course occur in both subaerial and subaqueous environments or settings. Since ~70% of the Earth's surface is occupied by the oceans there are many submarine volcanoes that we know little or nothing about. Environment not only influences the types of surface processes that affect a volcanic terrain (e.g. weathering, erosion, and deposition) but also has an effect on volcanic eruption style, the products and even the type of volcano that develops (Chap. 3; Cas and Simmons 2018). Eruption style is obviously influenced by magma properties, especially volatile content and viscosity. However, a magma that might erupt explosively subaerially because it has a relatively high volatile content, may erupt non-explosively in a deep marine environment because the high ambient hydrostatic pressure at water depth may counteract the pressure of exsolving magmatic volatiles, and suppress exsolution and explosive expansion of exsolved volatiles (Chaps. 3, 4). This effect is similar to the effect of the very high atmospheric pressure on Venus, where explosive eruptions are thought to be limited (Wilson and Head 1983; Head et al. 1992; Frankel 1996; Airey et al. 2015; Ivanov et al. 2015). At low ambient pressures, ambient water and hot magma can interact spectacularly to produce hydrovolcanic explosions. Ambient water can also produce distinctive volcanic products when hot magma comes in contact with a substantial body of water through non-explosive quench fragmentation (Chap. 4), producing hyaloclastite deposits, which are unique to aqueous environments and are discussed further in Chaps. 4–8. The effect of environment on volcanism is discussed more thoroughly in Chap. 3.

1.5.1 The Inaccessibility of Ocean Floor Volcanism, and the Technological Developments That Now Allow It

We know much more about subaerial volcanism than we do about submarine and subglacial volcanism, because subaerial eruptions can be witnessed directly and with remote technologies, recorded digitally, and deposits can be examined and sampled during and immediately after eruptions. Only two submarine eruptions have been directly observed and imaged underwater (Chadwick et al. 2008; Resing et al. 2011) because the ocean floor is not readily accessible to direct observation. However, understanding the nature of submarine and, similarly, subglacial volcanic processes (there are many similarities) is important for understanding the evolution of the Earth, evolution of the atmosphere and oceans, climate change, major extinctions, many volcanic hosted mineral deposits, and by analogy, will also shed considerable light on understanding many planetary volcanic processes.

However, the development of technology to explore the seafloor and undertake real-time monitoring of some submarine volcanoes and volcanism has allowed huge advances in our understanding of processes and settings. In particular, manned deep-sea submersible vehicles, such as *Alvin*, can reach almost any water depth now and allow direct observations of small areas of the seafloor, continuous filming of seafloor traverses, and high-precision sampling of rocks and even drilling of limited-length drill cores. In addition, programmed, autonomous underwater vehicles (e.g. *Sentry*), mounted with high-resolution side-scan sonar and bathymetry instrumentation can now map huge areas of the seafloor at increasingly higher levels of resolution, leading to the production of extremely high-resolution digital elevation maps (DEM) of seafloor topography and of individual volcanic edifices (e.g. Carey et al. 2018; Ikegami et al. 2018; Clague et al. 2019; e.g. Fig. 16.76b). DEM imagery is in places so detailed as to be able to map lava flow fields, lobes, and domes and some of their surface features (Clague et al. 2011, 2019; Ikegami et al. 2018), and even transported megaclasts. Remotely operated vehicles (ROVs, e.g. *Jason*, *Medea*) also provide opportunities for photographing, videoing, and sampling of the seafloor and specific sites on volcanoes. Towed chemical detectors can detect chemistry anomalies in the ocean that can only be explained by release from active, seafloor, volcanic hydrothermal springs (e.g. de Ronde et al. 2001). At the same time, ongoing deep-sea drilling of the oceanic crust has continued to provide huge amounts of information about the seafloor through first the JOIDES and subsequently the IODP drilling programs.

Often, submarine eruptions go largely undetected unless satellite imagery happens to capture an eruption, such as the spectacular eruption of Hunga Tonga–Hunga Ha'apai volcano, Tonga, southwest Pacific Ocean, on 15 January 2022, or there is a visible manifestation on the sea surface (e.g. floating rafts of pumice, a buoyant rising steam (and ash) plume; e.g. Jutzeler et al. 2014, submarine Havre volcano eruption, Kermadec Arc, in 2012; Carey et al. 2018).

Technological developments have now also allowed at least one submarine volcano, the Axial Seamount volcano on the Juan de Fuca mid-oceanic spreading ridge, northwest Pacific Ocean, to be instrumented and monitored in real time. An array of instruments (e.g. seismometers, tilt and pressure meters, video cameras, seawater and hydrothermal fluid samplers, fluid mass spectrometer analyser, etc.) have been installed on the seafloor in and around the summit caldera, as part of a seafloor observatory that is connected by a 480 kms long cable to an observatory on land in the city of Portland to the east (Manalang and Delaney 2016; https://oceanobservatories.org/array/cabled-axial-seamount/). This cabled seafloor observatory has detected three eruptions that have produced basalt lava flow fields. Subsequent ultra-detailed, 1 m spacing, bathymetric surveys, have allowed the lava flow fields to be mapped in detail (Caress et al. 2012; Chadwick et al. 2013; Clague et al. 2013). This is in our opinion, one of the major achievements in modern volcanology.

There are both benefits and limitations in trying to better understand subaqueous volcanism by studying the modern seafloor. The benefits include:

- the morphology of young seafloor volcanic edifices is generally well preserved, with only limited effects of erosion, such as mass wasting,
- volcanic deposits are generally preserved without being reworked by waves, currents, and tides, except in shallow water,
- the physical setting is well-constrained (e.g. water depth, topography, tectonic setting), and
- rarely have there been opportunities to directly observe and record active seafloor eruptions (e.g. Chadwick et al. 2008; Resing et al. 2011).

However, there are also many limitations in doing research on subaqueous volcanism on the modern seafloor, including:

- limited access, requiring availability to oceanographic research ships,
- expensive and time-consuming,
- extremely limited visibility of processes and deposits,
- limited technology crucial to "visualising" volcanic forms, deposits, and sampling,
- difficulty of accessing stratigraphy, and 3D and 4D stratigraphic perspectives,
- uncertainty of edifice evolution through time (e.g. changing water depths, topography),
- uncertainty about eruption styles, source vent locations, depth of the vent at the time of the eruption, and the degree to which depth has been modified post-eruptively by edifice sector collapse or caldera collapse; we cannot just assume that the eruption occurred from the post-eruptive preserved summit of an edifice, and
- uncertainty of the provenance of many deposits (e.g. pumice–indigenous or exotic, vent location, transport, and depositional processes). Deposits are not necessarily in situ due to mass-flow resedimentation and oceanic currents.

1.5.2 Benefits and Limitations of Studying Uplifted, Ancient Subaqueous Volcanic Successions

The traditional way of learning more about submarine volcanism has been to study uplifted ancient subaqueous successions. This remains an important approach because of the generally easier and cheaper access. However, there is a complication in studying ancient volcanic successions, in that first we have to reconstruct the nature of that environment before we can make sense of the volcanic processes, because environment has a major influence on volcanic eruption styles and deposits (Chap. 3). We need to establish if the environment was subaerial or subaqueous, and if the latter, marine or non-marine, shallow, intermediate depth or deep water. This requires a sedimentological approach, involving detailed facies analysis, as discussed in Sect. 1.8 below, and an understanding of sedimentary transport and depositional processes as discussed in Chap. 12.

The benefits are:

- old rock successions may preserve stratigraphy and 4D perspective,
- contacts between units and depositional textures may be well and widely preserved,
- usually accessible,
- relatively cheap to do fieldwork (cf. marine research cruises), and
- sampling is easy.

However, there are also limitations in working on ancient volcanic successions, including:

- outcrop can be poor or discontinuous,
- can be time-consuming if significant mapping is required,
- possible effects of uplift, deformation, metamorphism, erosion, and hydrothermal alteration,
- physical setting is not well constrained (e.g. water depth, topography), and
- precise paleo-water depths are extremely difficult to ascertain.

1.6 Process Spectrum in Volcanic Settings

Volcanic eruptions occur in many different eruption styles, and even single eruptions can vary in their style quite markedly. For simplicity, volcanic eruptions can be subdivided into two essential styles: *lava forming or effusive,* and *explosive* (Fig. 1.4). Lava-forming or effusive eruptions in turn consist of two types, those in which continuous mildly explosive *magma or "fire" fountaining* of fluidal magma, usually mafic to peralkaline in composition, feeds a lava flow (e.g. Plate 1.1b), and those in which *passive magma effusion* from the vent, without a fountain, directly supplies a lava. Both fluidal and viscous magma can experience effusive eruptions.

Explosive eruptions also vary in style from those forming low fountains of pyroclasts and gas, to those forming very high, buoyant columns of pyroclasts and gas, to those punctuated by ballistic blasts (Chap. 9). The solid fragments or pyroclasts ejected from a volcanic vent during explosive eruptions can then be dispersed by a variety of processes (Fig. 1.4): *ballistic blasts* from the vent, *fallout* from the eruption column or a higher downwind, eruption plume (Chap. 9), and by a variety of *pyroclastic density currents* (Chap. 10). The latter are gravity-controlled flows of gas, ash, pumice, scoria, and lithic debris, capable of flowing large distances from vent and can originate in several ways (Chaps. 10, 11). The dispersal of pyroclastic debris, therefore, involves a complex spectrum of processes and deposits, which are discussed comprehensively in Chaps. 9–11.

In addition to primary eruption phenomena, there are many other processes, which affect volcanic terrains during and between eruptions (Fig. 1.4). In most volcanic terrains, the duration of these other processes is significantly longer than the actual eruptive events, because most eruptions are short-lived, and with the exception of monogenetic volcanoes, may represent less than 1% of the history of volcanoes (Chap. 12). These other processes include those resulting directly from eruptive activity, such as gravitational collapse of the margins of lava domes, stratovolcanoes, and lava shields, producing landslides and rock avalanches (Chap. 12).

In addition, long-term sedimentary surface processes such as weathering, erosion, mass wasting and flow, tractional reworking, wind-blown and aqueous suspended transport (Chap. 12) are huge influences on the evolution of volcanic edifices. Also, chemical hydrothermal alteration and ore formation processes (Chaps. 13, 18), and magma-tectonically influenced regional and local tumescence (uplift) and detumescence (subsidence) of the land surface (Chap. 17), are all ongoing processes both during and between eruptions. Both primary eruption processes and other surface and subsurface processes contribute to making volcanic environments the most complex and dynamic surface environments on Earth.

1.7 Introduction to Terminology

Some widely used, general terms have already been introduced above and require brief explanation here before many individual terms are introduced in individual chapters, and a more comprehensive treatment of the topic of terminology

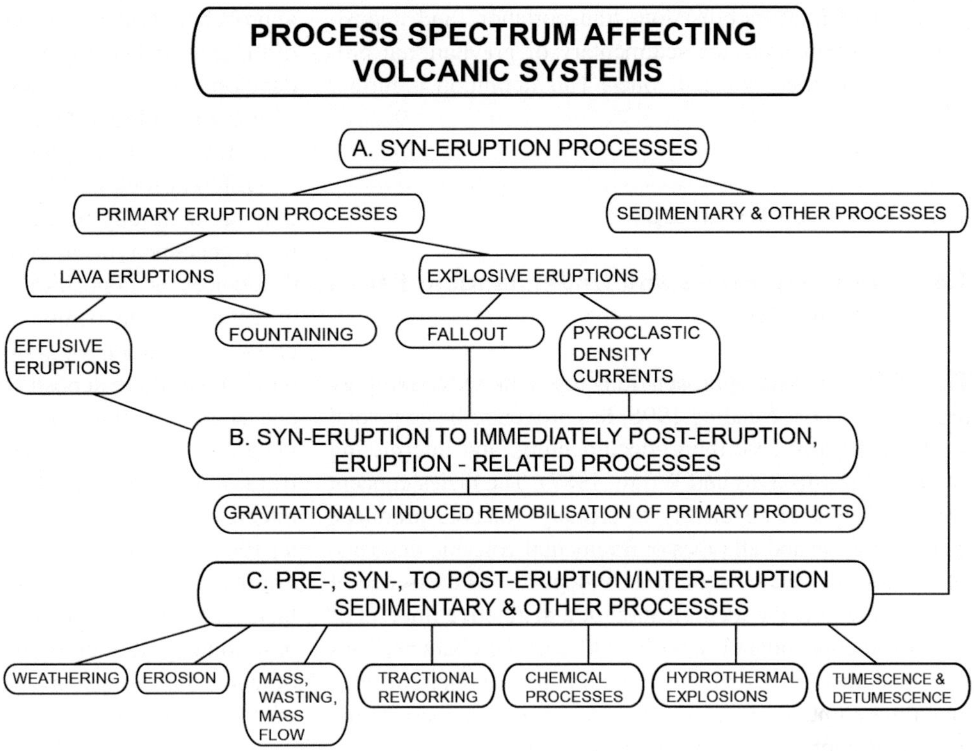

Fig. 1.4 Schematic representation of the spectrum of eruption and dispersal processes of volcanic materials in volcanic terrains. Processes are subdivided into **a** syn-eruption processes including both primary eruption processes (lava-forming, explosive) as well as sedimentary and other processes, **b** those that are syn-eruption to immediately post-eruption, eruption-related processes, and **c** those that are pre-, syn-, and post-eruption sedimentary and other processes

or nomenclature is presented in Chap. 14. Some of these terms are highly genetic and should not be used unless the characteristics of deposits have been carefully analysed against the characteristics of deposits with known origins. The terms are defined here for information only at this stage, and readers should work through Chaps. 2–12, before applying these terms to deposits; otherwise, application of incorrect genetic terminology may lead to mistaken ideas of deposit origins.

The term "deposit" can be used in a general sense to encompass all products of volcanism and sedimentation in volcanic terrains. It can thus be used for an unconsolidated aggregate of debris or lithified rock, whether it be crystalline, or a lithified clastic rock. In volcanic terrains, rocks are not necessarily old and can form instantaneously during eruption through cooling and crystallisation of lava or welding of pyroclastic deposits. Conversely, unconsolidated deposits may remain unconsolidated for millions of years after formation.

The term "*volcaniclastic*" is a non-genetic term for any fragmental aggregate of volcanic debris, regardless of origin (Fisher 1961). There are three types of supra-crustal volcaniclastic deposits, based on their mode of origin: pyroclastic, autoclastic and epiclastic (surface process) deposits and rocks.

"*Pyroclastic*" refers only to those deposits that are/were:

(a) fragmented by explosive volcanic activity, and were
(b) transported and finally deposited by processes resulting uniquely from explosive eruptive activity (pyroclastic fallout, pyroclastic density currents, which include pyroclastic flows and pyroclastic surges; see Chaps. 9–11).

The term "*tephra*" is all encompassing and used to describe unconsolidated *pyroclastic deposits* of whatever origin.

"*Autoclastic*" refers to deposits produced either by the flow fragmentation or *autobrecciation* of the viscous solidifying margins of flowing magma, usually lava but sometimes intrusions, which produces a breccia of lava rubble called *autobreccia*, or by the sudden cooling induced *quench fragmentation* of magma when it comes in contact with cold water, producing an aggregate of glassy debris called *hyaloclastite* (see Chap. 4 for details of processes and deposits characteristics).

"*Epiclastic*" deposits are traditionally defined as those fragmented or deposited, or both, by normal surface sedimentary processes involving weathering and/or erosion of pre-existing older rock. "Epiclastic" is used a little more liberally here to describe a deposit or rock in which the debris is derived from an older or contemporaneous source by normal surface sedimentary processes (weathering, physical erosion, and gravitational collapse) and which is finally deposited by normal surface sedimentary processes. Our usage of this term, therefore, allows the term "epiclastic" to be applied to deposits derived from any source, irrespective of age or type (old crustal rocks of diverse types, or contemporary young lava rocks, young or even old unconsolidated pyroclastic deposits), as long as the processes of final mobilisation, transportation, and deposition were normal surface sedimentary or gravitational processes, not primary volcanic ones. This definition is further discussed in Chap. 12.

1.8 The Facies Concept

1.8.1 Defining Facies and Understanding Facies Variations

The facies analysis approach long used in sedimentology (e.g. Selley 2004; Reading 1996) has proven to be as pivotal in systematically describing, documenting, and interpreting volcanic deposits (Cas and Wright 1987). The facies concept can be applied as effectively to primary volcanic products, including lavas and all types of fragmental volcanic deposits, as it can to associated sedimentary deposits. The essence of facies analysis is the identification of a distinctive interval of rock or unconsolidated deposit with specific characteristics that lend themselves to the interpretation of their origins, including eruption processes, transportation and depositional or emplacement processes, and environments of deposition. Success is dependent on an awareness of the possible diversity and complexity of basic physical, chemical, and sedimentological processes and deposit types. It is difficult to understand what you are looking at until you understand the possible spectrum of processes and deposits and their characteristics. The aim of this book therefore is to provide that understanding.

A facies therefore is a distinctive deposit or body of rock that has a unique definable character that distinguishes it from other deposits, or intervals of rock (Cas and Wright 1987). The definable character may be compositional or textural, or the depositional structures or fossils present. A facies is the product of a unique set of conditions and processes in the depositional environment, whether they be volcanic or not. These processes and conditions may be physical, chemical, or biological in origin, and may include such factors as the topography and bathymetry; the mechanisms and rates of material release, transportation, and deposition; the climate and weather; the nature/composition of the source materials (both chemically and physically); the prevailing chemical conditions; and the floral–faunal influences.

The understanding of the origins of facies (i.e. specific deposit types) involve two approaches: a descriptive lithological approach and a genetic approach. The genetic approach involves a direct judgement being made of the origins of a deposit. However, since the origins of individual rocks or deposits, or groups of them, are often not clear at first inspection, especially in (but not limited to) old rock successions, it is important to develop a careful descriptive approach first. Such an approach involves first the description of the essential characteristics of a deposit or rock in order to develop a data set that can then be used to analyse the range of possible origins, finally leading to a considered, genetic interpretation. This approach is outlined further in Chap. 14.

Different rock or deposit types are distinguished because they are texturally or mineralogically different in outcrop or hand specimen or in thin section, or because they have differences in depositional structures (e.g. the presence or absence of depositional structures such as layering, cross-stratification, or grain size grading). Alternatively, perhaps two or three deposit types that are regularly interbedded and contain distinctive internal depositional structures may have an overall unique appearance that distinguishes them from adjacent deposits. Distinguishing unique facies or facies intervals therefore allows one to focus on essential characteristics that contribute to an understanding of the process origins and conditions under which each facies formed. The facies approach is also a convenient way of identifying, describing, and interpreting distinctive facies or associations of facies that recur many times in a stratigraphic succession, without having to describe in detail each bed.

A facies can be defined at any scale. At a regional level, stratigraphic units such as groups, formations, or members are effectively facies because they have an overall lithological character that distinguishes them from other groups, formations, or members. At a more local scale or outcrop scale, facies may be defined by an individual bed or an interval of several beds which are basically uniform, or by both. The level of detail to which a stratigraphic succession is subdivided into facies will be controlled largely by the aims of the study, the information available, the time available, and the level of understanding that is sought.

Variations in facies and facies characteristics can occur vertically and laterally at all scales. Vertically through a stratigraphic succession, facies may change as a result of changing depositional processes, conditions, or environments over time. Similarly, within a single bed or deposit, the features of the deposit can change as a result of changing flow conditions or energy regimes (e.g. declining current velocities associated with the waning stages of a subaqueous turbidity current or a river flood will cause a change in the sedimentary structures of their deposits from bottom to top; Chap. 12). This is also the case for pyroclastic flows and their deposits (Chap. 10). Facies characteristics can also change laterally within a succession of beds or within a single bed, again because of changing transportation and depositional conditions laterally over a distance.

Even though associated facies may be different, they may still be genetically related as parts of the same depositional or eruptive event. For example, to name but a few possible causes, a single pyroclastic flow deposit or ignimbrite may contain several facies resulting from changing flow processes, proximal to distal changes in flow behaviour, multiple flow lobes or depositional pulses, differential interaction between the host pyroclastic flow and the underlying topography, changing nature of material fed into the flow system at the vent during the course of an eruption (Chap. 10). The May 18, 1980 eruption of Mt St Helens produced at least the following deposits in the eight hour eruption period: blast-flow deposits, pumice flow deposits, debris avalanche deposits, debris flow, and hyper-concentrated flood-flow deposits, air fall deposits, pumice flow deposits, and the beginnings of a crater lava dome (Fig. 1.5a; Lipman and Mullineaux 1981; Tilling et al. 1990). In addition, in the next 5 months, further pyroclastic

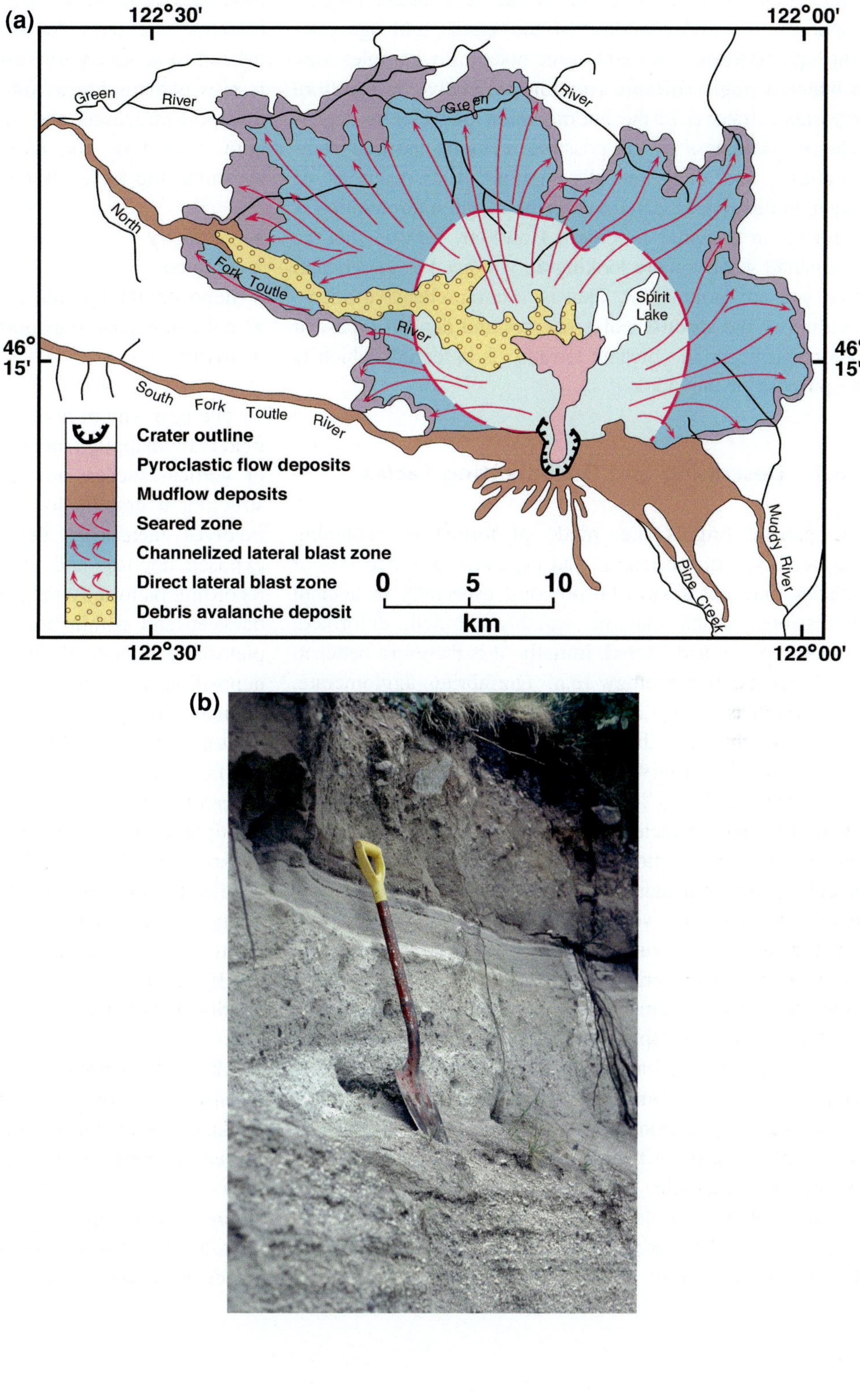

Fig. 1.5 Numerous deposit types or facies can be produced during one eruption. **a** Simple map of the facies/deposit types produced during the 18 May 1980 eruption of Mt St Helens, Cascade Arc, USA. Not shown is the dispersal area for fallout deposits. (Adapted from Tilling et al. 1990; red arrows indicate flow path of the initial blast flow, After Kieffer 1981). **b** Some of the different deposit types produced during the 700 yr BP Kaharoa eruption of the Mt Tarawera Dome Complex, including pyroclastic flow, fall, and surge deposits. Rifle Butts Road, North Island, New Zealand

fall and flow deposits, dacite lava domes, dome-collapse block, and ash-flow deposits and reworked fluvial deposits formed. Similarly, the 700BP Kaharoa eruption of Mt Tarawera, New Zealand, produced numerous deposits (Fig. 1.5b; Nairn et al. 2001). Historically, the different deposits in each of these cases may have been interpreted as the products of discrete, separate eruptions or events where preserved in the rock record, rather than all being linked to a single eruption event, its various phases, and surface process responses. An understanding of the spatial and age relationships between facies and of the potentially complex way in which a single volcanic eruption can evolve, is therefore important. Success in the interpretation of facies is dependent on awareness of the possible relationships and complications, and of genetically significant associations of facies. In this regard, careful description and interpretation is therefore an essential first step. However, beyond that, to understand what you are looking at, you need to understand what you are looking for. That is, a comprehensive understanding of the spectrum of processes and deposit types, or facies that occur in volcanic terrains, is essential, which is the purpose of this book.

1.8.2 Describing and Documenting Facies

The genetic origins (i.e. mode of formation, including eruption style, transportation and depositional processes) of a facies may not always be obvious, especially in ancient successions, which may be variably lithified, deformed, metamorphosed and altered. Initially, it is therefore better to avoid genetic terminology (e.g. ignimbrite, agglomerate, tuff), which is highly interpretive, until the deposits have been thoroughly described and the origins have been clearly thought out. Descriptive terms, such as "matrix-supported dacite breccia facies", are preferable initially (also see Chap. 14). Such a facies may be an ignimbrite (or part of one), a block and ash-flow deposit, a hydrothermal explosion breccia, a debris avalanche deposit, a debris or mud flow deposit, to name only some of the possibilities. To evaluate which possibility is most likely requires careful examination of the facies characteristics and the relationships between superposed facies. This should involve description of a combination of outcrop (or drill core), hand specimen, and microscopic observations, as well as consideration of the lateral extent, geometry, and facies relationships. On-the-spot application of genetic names at a single outcrop may, more often than not, lead to erroneous interpretations. For example, just because a facies consists of angular volcanic clasts, doesn't mean it's an explosion breccia or located near vent. Assuming it's near vent (when it's not) could cause major errors in understanding an eruption or a volcano, or lead to (expensive) errors in developing an exploration strategy for mineral deposits.

The systematic and logical approach to facies description and analysis proposed by Selley (2004) is adopted here. Selley nominated five essential facies descriptors: geometry of a deposit, its lithology, sedimentary structures, paleocurrents, and fossils. In volcanic terrains, distinctive depositional structures can be produced by other than sedimentary processes (e.g. lavas, pyroclastic flows, and blasts), and therefore the term "depositional structures" will be used instead of sedimentary structures. The term "paleocurrents" is also perhaps not as informative as "paleoflow or paleotransport directions and patterns", which will be preferred here. Therefore, the five principal facies descriptors for systematising facies descriptions are:

- geometry,
- lithology,
- depositional structures,
- paleoflow/paleotransport directions and patterns, and
- fossils.

The best way of documenting facies characteristics in the field is by mapping, and detailed graphic or pictorial logging of vertical successions in outcrop or in drill core. Constructing a graphic log or measured stratigraphic column involves measuring the thickness of a section through exposed deposits and all the component beds, as well as recording pictorially the grain sizes and grain size variations between and within beds. Other features that are recorded pictorially include the principal component types, sorting, depositional structures, unit/bed contact relationships, and geometry of beds or units. A graphic log is therefore an accurate visual or pictorial representation of the features of the facies and facies succession (e.g. Fig. 1.6). More significantly, lateral variations can be captured by measuring multiple graphic logs or sections, correlating them, and constructing 2D or 3D fence diagrams and computer-generated 3D facies models of the facies architecture. In young volcanic settings where the original volcanic topography is preserved, facies distribution patterns can be draped on 3D digital elevation models (DEMs), to document the relationships between facies and topography.

1.8.2.1 Geometry

Geometry describes the 3D form or shape (including the thickness) of a facies and its component strata. The preserved geometry of a facies is controlled by:

- pre-depositional relief on the depositional surface,
- volume of material deposited and the way the topography accommodates that volume,

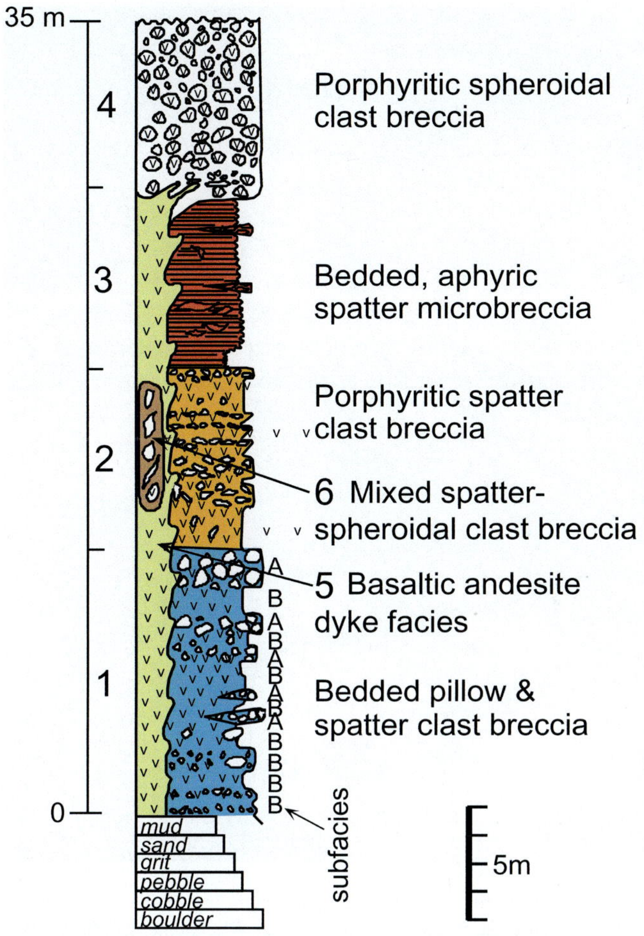

Fig. 1.6 Example of a pictorial graphic log of a Miocene volcanic succession, Ryugazaki, Hokkaido, Japan. Different facies are depicted and named descriptively. The thickness of units is depicted vertically, and the grain size of different units and grain size variations between units is depicted by the changing profile units from left to right. Depositional structures can be depicted as well as the nature of contacts between units. (Adapted from Cas et al. 2003)

- physical properties of the transporting and depositional agent,
- post-depositional erosion, and
- subsequent deformation.

Pre-depositional relief

The relief of the depositional surface is controlled by the ongoing interplay between erosion and deposition on the Earth's surface. Erosion will predominate where slopes are high and the relief is significantly above the base level towards which erosion is working (e.g. sea level, lake level, ocean floor, and alluvial plain). Erosion will produce valleys, rivers, creeks, gullies, channels, and canyons, separated by ridges, plains, and plateaux. Most depositional units deposited in such a terrain will be confined within topographic lows and have valley-filling geometries, but some, such as air-fall deposits, may drape over irregular topography and large volume events can produce deposits that overspill valley margins onto local topographic highs (e.g. see examples of both in Fig. 1.7). Where the topography is flat or has been filled in by earlier depositional events, subsequent deposits will in most instances have more tabular geometries, more uniform thickness and be more laterally continuous (Fig. 1.7). Relief may also be affected by contemporaneous tectonic activity and the emplacement of units with very positive primary topographic relief, usually due to high bulk viscosity and internal strength (e.g. viscous lava domes, debris flows, and rock avalanches).

Where deposits are successively channelled through and confined almost entirely within a steep-sided valley or canyon, peculiar spatio-temporal and stratigraphic relationships can develop, that can complicate the understanding of the stratigraphy, especially where exposure of contacts is poor (e.g. in the rock record; Figs. 1.8, 1.9). For example, the eruption of Lascar volcano in northern Chile in 1993, produced small-volume pyroclastic flow deposits, some of which were deposited in the bottom of a deep canyon eroded into the 26.5 ka Soncor ignimbrite, which itself is ponded in an older canyon eroded into the 4.1 Ma Atana ignimbrite (Fig. 1.9). Younger ignimbrites therefore sit topographically lower than the tops of older ones and are entirely confined within older deposits. In old, poorly exposed sequences, the contacts and age relationships may not be exposed, and because one deposit sits lower in the apparent sequence than another, the natural inclination is to assign an older age and stratigraphic position to the unit sitting lower, when the opposite may be true.

Volume deposited and accommodation by topography

If the volume of material deposited is small compared with the topographic relief and the volume accommodation capacity of the topography into which it is deposited, then this volume will be entirely contained by the topographic depression (Fig. 1.10a). If the volume of the depositional unit is large compared with the volume capacity of topographic depressions, then the deposit will overspill onto the adjacent topographic highs and produce major variations in thickness (e.g. lavas, debris, or pyroclastic flow deposits which infill a valley and spill over onto the confining ridge interfluves as a thinner veneer; Fig. 1.10b). In some cases, it will be impossible to distinguish between these situations because of post-depositional erosion (Fig. 1.10c).

Fig. 1.7 Complex succession of pyroclastic fall and flow deposits from the Quaternary Okataina Volcanic Centre, New Zealand. Complex geometries and stratigraphic relationships result from the erosion of the oldest (9,000 yr BP) Pukerimu pyroclastics (lower right) and infilling of the paleovalley by two "younger" pyroclastic flow deposits of the Mamaku ignimbrite (7,500 yr BP), and mantling of the paleotopography by pyroclastic fall deposits of the Whakatane (5,000 yr BP), Rotokawau (4,000 yr BP; dark basaltic fall layer near the top of succession) and Kaharoa (700 yr BP) eruptive intervals. (From Cas and Wright 1987)

Physical properties of the transporting and depositional agent

The principal consideration here is the rheological properties of the transporting agent (Chap. 2). For example, low-viscosity depositional agents will, topography permitting, spread their load as a broad, thin sheet (e.g. turbidity currents, pyroclastic flows, pyroclastic surges, and basaltic lavas). High-viscosity materials or those with high strength (e.g. rhyolitic lava domes) will produce a mound-like depositional unit with very significant positive relief, which markedly changes the topography on the depositional surface.

Post-depositional erosion

The majority of volcanic terrains have relatively high relief and slopes, which are subject to severe degradation by the agents of erosion: gravity, water, ice, and wind (Chap. 12). As a consequence, many facies units emplaced during active volcanism, whether they be lavas, pyroclastic flow or fall deposits, or redeposited volcaniclastic sediment intervals, can have their original depositional geometry dramatically modified by post-depositional erosion. Some can even have the record of their emplacement completely removed. In particular, unconsolidated deposits such as non-welded pyroclastic fall and flow deposits, and sedimentary units are prone to erosion. Therefore, the stratigraphic record, especially in subaerial successions, is often incomplete, missing the record of numerous eruption events, and preferentially preserving erosionally more resistant units, such as lavas and welded pyroclastic fall and flow deposits (Chap. 15).

Deformation

Volcanoes and volcanic provinces commonly occur in tectonically active regions, and may be subject to orogenic and volcano-tectonic deformation, the former being related to the regional far stress-field (e.g. orogenic compressional, extensional, and transcurrent tectonic regimes; see Chap. 17), while the latter is related to the local stress-field associated

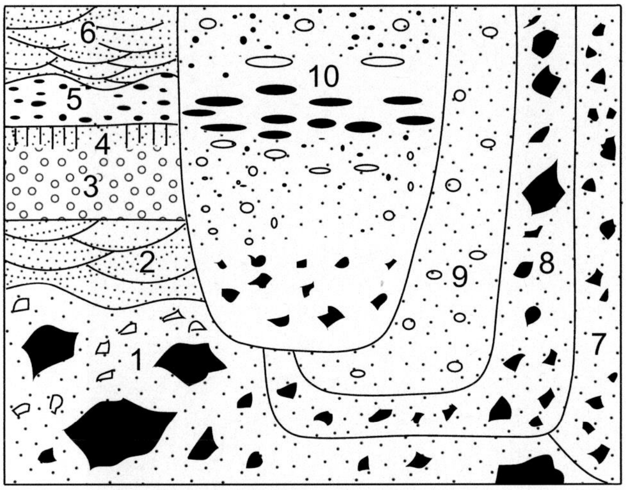

Fig. 1.8 Schematic representation of the possible effects of repeated infilling and excavation of old canyons on preserved depositional geometry, facies, and stratigraphic relationships, which are highly irregular and abrupt, and make correlation and depositional sequence very difficult to determine (From Cas and Wright 1987)

with magma pressure (e.g. during intrusive or caldera collapse events). Deformation may have a marked effect on the preserved geometry of a facies. The effects may range from simple block-faulting to gentle folding to tight folding with high strain, and post-emplacment faulting leading to displacement and dismemberment. However, caution is required when considering the origins of dipping and apparently folded strata, because volcanic strata may be emplaced with initially high angles of repose or dips (<35°; Fig. 1.11a). They may also irregularly drape syn-eruptive topography, giving the impression of being folded and deformed (e.g. Fig. 1.11b). Deformation may also significantly modify original deposit textures, as discussed in Chap. 13.

1.8.2.2 Lithology

The lithology of a deposit or facies has three principal characteristics:

- physical components,
- composition, and
- texture.

In non-volcanic, sedimentary successions and facies, all three properties can be very important in elucidating sediment sources and genesis, and in reconstructing the nature of the depositional environment (Tucker 2001; Boggs 2012). The presence of certain physical components (e.g. shelly fragments, oolites, and distinctive lithic fragments) can be diagnostic of depositional conditions and settings. The presence of distinctive compositional grain types (e.g. glauconite, phosphorite) or sediment types (e.g. radiolarian cherts, evaporites) may be significant enough to establish the depositional setting and conditions. Textural features may also reveal the depositional setting or origin of the facies (e.g. very well-rounded beach or aeolian sands, mudstone-supported debris-flow deposits).

Similarly, primary volcanic facies can have distinctive lithological components and characteristics (shards, accretionary lapilli, eutaxitic texture, and jigsaw-fit breccia clast textures) that contribute to the understanding of their genesis and depositional setting. In ancient successions, these features may be more difficult to identify and quantify because of the effects of lithification, and perhaps metamorphic and structural overprinting. With increasing metamorphic grade the original volcanic components, minerals and textures become progressively overprinted, and more difficult to identify.

Physical Components of Volcanic Facies

Volcanic successions consist of varying proportions of lava flows, sub-volcanic intrusions, and fragmental or clastic rocks. The principal physical components of young lavas and high crystal-level intrusions are large crystals or megacrysts, in a finer grained groundmass of microscopic crystals, called microlites or microcrystals, and uncrystallised solidified lava (volcanic glass) between the phenocrysts, and vesicles (gas and/or liquid bubbles). The megacrysts are either primary magmatic crystals called phenocrysts, or xenocrysts, which are crystal fragments the magma has incorporated from country rock that it is passing through. Xenoliths are foreign rock fragments incorporated from country rock (e.g. wall rock to the magma chamber or conduit). All of these components can be present in varying combinations and proportions. In older volcanic rocks, the glassy groundmass is often altered or devitrified (Chap. 13).

The volcaniclastic facies in volcanic successions can have diverse origins: pyroclastic, quench fragmentation, autobrecciation, and normal surface sedimentary or epiclastic processes, as defined above in Sect. 1.7 and in Chap. 4. They generally consist of aggregates of igneous fragments, foreign lithic clasts, and crystals of varying grain size. The larger clasts constitute the *framework grains or clasts*, and if there is a distinctly finer grain size population as well, dispersed between the framework grains, this is called the

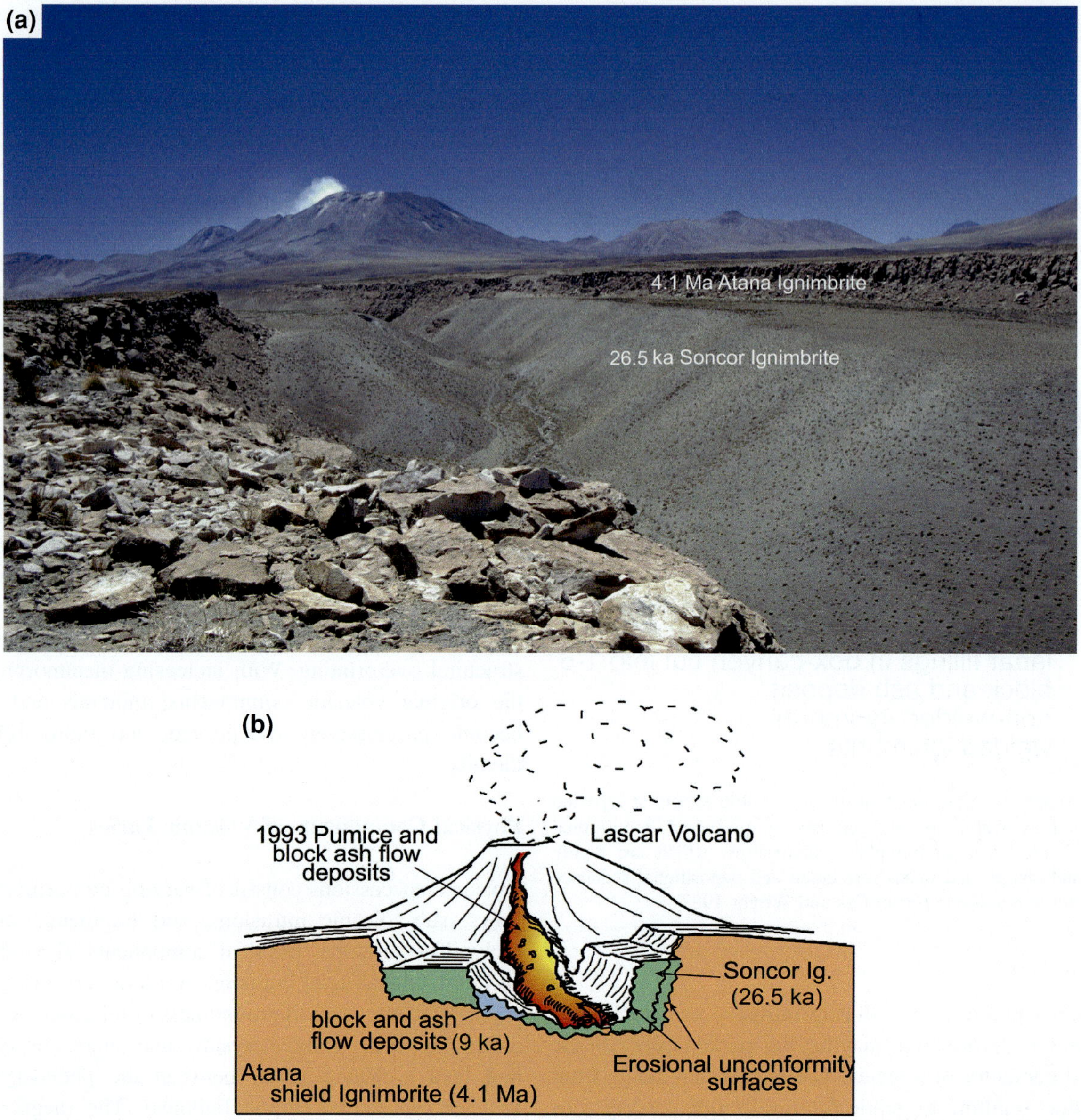

Fig. 1.9 Ignimbrite deposits and relationship to topography. **a** Photo of Lascar volcano, Chile, on the horizon, perched on an ignimbrite plateau consisting dominantly of the 4.1 Ma Atana ignimbrite erupted from a nearby caldera. The plateau has been incised by a deep canyon draining from Lascar, which was almost infilled by the 26.5 ka Soncor ignimbrite, the top of which forms a terrace near the top of the canyon. The Soncor ignimbrite was also incised and other pyroclastic flow deposits from Lascar eruptions were then deposited into this canyon-in-canyon topography. The latest was from the 1993 eruption of Lascar volcano the deposits of which are very small in volume and sit in the bottom of the canyon. **b** Sketch of these stratigraphic-topography relationships

(clastic) *matrix*. If the deposit is lithified, this could be due to the effects of compaction and the electrostatic attraction between fine clay or glassy ash particles, thermal welding, or to the post-depositional precipitation of a chemical cement or alteration minerals in pore spaces within the deposit (Chap. 13). The igneous clasts may be *juvenile fragments* (i.e. a fragment of the erupting magma) or older volcanic or plutonic clasts. The juvenile clasts may vary in vesicularity from dense lava fragments to vesiculated pumice and scoria, and finer ash including cuspate glass shards. The juvenile fragments may be glassy or variably crystallised, and of varying grain size, ranging from large blocks many metres in diameter to micrometre-sized grains.

Breakage of fragments, irrespective of the type, may have occurred during primary volcanic eruption (explosive, autobrecciation, and quench fragmentation), during primary volcanic transport (lava flow, pyroclastic dispersal processes), or post-eruptively by surface sedimentary or gravitational processes. At any time after initial formation, primary volcanic deposits may be eroded, and the clastic components

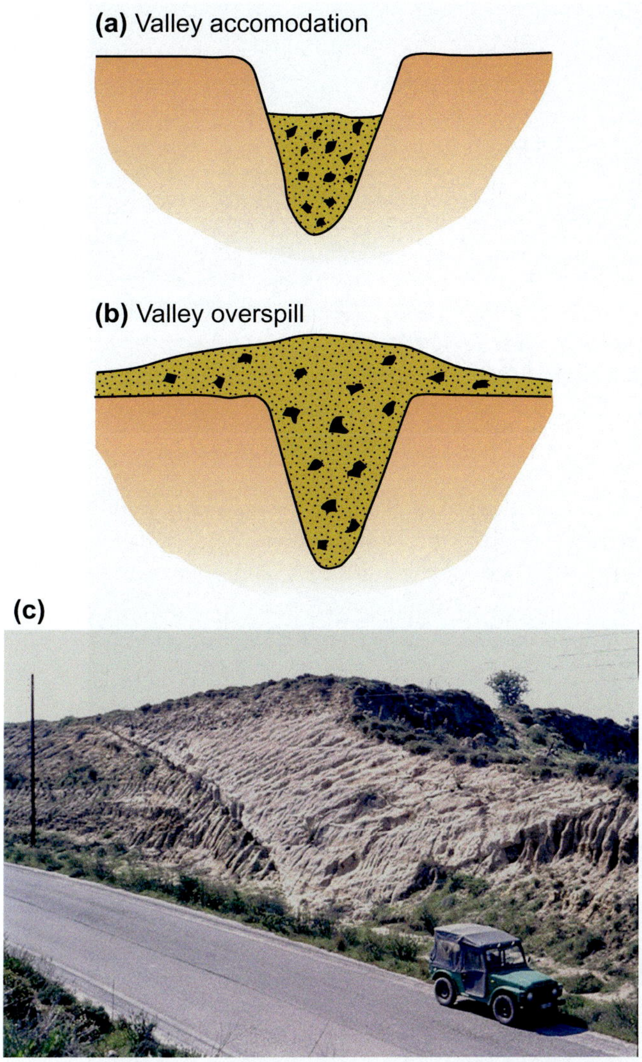

Fig. 1.10 Possible relationships between the volume of topographic lows such as valleys and canyons, and the volume of material emplaced within the topographic low during a depositional or eruptive event. **a** A small-volume debris flow or pyroclastic flow deposit is entirely accommodated or "valley-ponded" within a larger deep valley, producing a thick, laterally discontinuous, valley-confined deposit. **b** The volume of the debris-flow deposit is too large to be accommodated by the valley. The deposit has an extremely variable thickness and a wider extent due to overspill onto the valley margins. **c** Valley ponded 161 ka Kos Plateau ignimbrite, Mastichari Road, Kos, Greece. (**a** and **b** From Cas and Wright 1987)

redeposited by normal surface processes. This can occur minutes after initial deposition or millions of years afterwards. Primary volcanic deposits may remain unconsolidated for millions of years after initial formation or at the other end of the spectrum, they may be indurated hours after emplacement. During sedimentary erosion and transport, juvenile magmatic clasts and crystal fragments, xenoliths, and xenocrysts may be incorporated into volcanic sediments. They may be mixed with varying proportions of older igneous lithic and basement-derived lithic and crystal debris, which may be variably altered, deformed, and metamorphosed. In addition to all of the above supracrustal fragmentation processes, fragmental volcanic rocks may form due to subsurface hydraulic fragmentation in hydrothermal environments (Chap. 4), as well as by common cataclastic deformational processes during tectonic and volcano-tectonic events.

Composition

Composition of a facies refers to the geochemical, mineralogical, and petrological character of a volcanic deposit, irrespective of whether it is a lava, pyroclastic, or redeposited volcaniclastic deposit. The final composition may be the end-result of a complex history of processes causing chemical and physical changes. These processes include pre-eruptive magmatic processes, and syn- and post-eruptive processes. The pre-eruptive processes are both chemical and dynamic, such as chemical differentiation and fractionation, volatile exsolution, and magma mingling. Syn-eruptive processes involve physical sorting and separation of glassy ash from crystals and lithic clasts in an explosive eruption column and during transport in pyroclastic density currents; Chaps. 9, 10). Post-eruptive processes involve physical sorting during normal surface sedimentary processes. In older volcanic successions, hydrothermal activity (Chaps. 13, 17) and, weathering, diagenesis, and metamorphism (Chap. 13), may further alter the chemistry of volcanic rocks. In Chap. 2, we briefly consider the effects of magma composition and physical properties on eruptive behaviour, and also approaches to classifying magmas and volcanic rocks according to chemical and mineralogical composition, which is also further discussed in Chap. 14.

Texture

The term "texture" encompasses the physical characteristics of the components of a facies, deposit or rock, and also the bulk properties of an aggregate of debris or of a coherent rock. The textural properties of components or of an aggregate are a reflection of inherited characteristics from the source, of the mode of fragmentation, and of characteristics developed during or after transportation and deposition. Any one of these influences may produce a distinctive textural character. Aspects of texture that will be considered here in terms of their process significance or environmental significance are:

Fig. 1.11 Apparently deformed volcanic strata. **a** Stratified, resedimented hyaloclastite deposits dipping at the angle of repose, forming the steeply dipping submarine volcaniclastic apron of a marine volcano. Miocene, Kabuto Cape, Hokkaido, Japan. Yellow notebook on large clast for scale. **b** An apparently folded lava-like layer in a scoria cone sequence, Auckland Volcanic Field, New Zealand. The lava-like layer is an agglutinated-spatter fallout layer that mantled and was deposited on the irregular morphology of a growing scoria cone during a strombolian eruption

- grain size,
- rounding,
- sorting,
- shape, and
- fabric.

For more detailed discussion of these, readers are referred to the sedimentology texts by Boggs (2012), Leeder (2011), and Tucker (2001).

Grain size is one of the first textures or characteristics noticed in an outcrop of a deposit, whether it be the size of fragments in a fragmental aggregate, or the sizes of crystals and lithic inclusions in a coherent igneous rock. The preserved grain size of a fragmental deposit is a reflection of a combination of factors, including the type and efficiency of fragmentation at source and/or during transport (Chap. 4), the competency of the transporting and depositing medium to carry clasts of different grain sizes, and the degree of physical abrasion of transported debris during transportation and deposition. These factors all apply for pyroclastic and epiclastic aggregates, but for autoclastic deposits only the physical state of the magma at the time of fragmentation, and the mechanics of fragmentation are important (Chap. 4). For pyroclastic or epiclastic aggregates, grain size is therefore not necessarily a reflection of proximity to source or eruption point. For example, huge boulders, m in diameter, can be transported tens of km from source or eruption point by pyroclastic flows, debris flows, rock avalanches, or glaciers. None of these transporting agents necessarily produce any significant signs of abrasion or rounding, although they may. Equally significant, very fine grain sizes may be preserved at vent in air-fall and pyroclastic flow deposits if the explosive fragmentation during eruption has been very efficient (e.g. during some hydro-volcanic eruptions; Chap. 4).

For lavas, the size of the phenocrysts reflects physico-chemical conditions in the magma reservoir prior to eruption, whereas groundmass microlites and small crystals reflect the conditions in the conduit during the ascent of the magma, and in the magma after emplacement. Factors that influence crystal size include cooling rate and time, melt composition and structure, nucleation kinetics of each mineral type, and sorting processes such as crystal settling.

Determining the grain size characteristics for an unconsolidated deposit is most commonly done by dry or wet sieving, settling tube experiments, and for very fine grain sizes, coulter-counter and laser diffraction techniques. For lithified aggregates imaging of polished rock slabs and/or thin sections (Sorrentino et al. 2011; van Otterloo et al. 2013), image analysis, using programs or apps such as Image J or FOAMS (e.g. Shea et al. 2010), and stereology techniques (e.g. Jutzeler et al. 2012) are used most commonly. From this information and its graphical representation, statistical grain size parameters, such as median and mean grain size, sorting and hydraulic equivalence can be calculated. Such data can then be used to assess aspects of fragmentation processes, eruption styles, and transport or dispersal dynamics, including distinguishing different types of pyroclastic deposits (e.g. Walker 1971, 1973, 1984a, b; Chaps. 9 and 10).

Grain size analysis is more difficult and time-consuming for lithified deposits than for modern unconsolidated deposits because it is not possible to disaggregate the rock into all of its original grains, preserving their shape and size. Image analysis techniques can be applied, as long as grain boundaries are visible, but even then a size limit that is resolvable mesoscopically or microscopically is reached, and the sizes beyond the minimum resolvable size have to be lumped together in a "finer than" bin size population, or better still extrapolated into finer bin sizes using fining trends in the resolvable grain size population. The problem becomes greater as diagenesis, metamorphism, and hydrothermal alteration mask the finer grain size components. For such deposits, only semi-quantitative and in some cases only qualitative estimates of grain size and grain size parameters are practical. One then usually has to rely on the field outcrop facies characteristics in order to determine the genesis and the paleovolcanological and paleogeographic significance.

An important texture is *sorting,* which is a reflection of the degree to which the transporting agent has been capable of separating grains of different hydraulic properties and depositing together grains that are hydraulically equivalent. This is called hydraulic sorting. The hydraulic behaviour of a particle is the way the particle responds when acted on, or entrained within, a moving fluid, whether the fluid be water, wind, mud, or volcanic gas, and/or the way that it settles (relative to other particles) through a static fluid. Particles that are free-falling vertically through a fluid column that is long enough, accelerate to a maximum fall velocity, called their *terminal fall velocity*. Particles are in hydraulic equivalence if they have the same terminal fall velocity, even though they may vary in terms of shape, size, and density. A deposit may therefore be hydraulically well-sorted even though it appears to be poorly sorted according to size if the densities or shapes of the fragments are variable. In normal epiclastic or terrigenous or siliciclastic sediments most grains, being mineral or rock fragments, have approximately equal densities, and therefore the better the hydraulic sorting, the more uniform the grain size becomes. The uniformity of grain size is therefore a good indicator or proxy of the degree of hydraulic sorting in siliciclastic sediments.

However, once sediment populations of differing shape, density, and weight are mixed (e.g. a beach sediment consisting of rounded quartz grains and blade-like shell

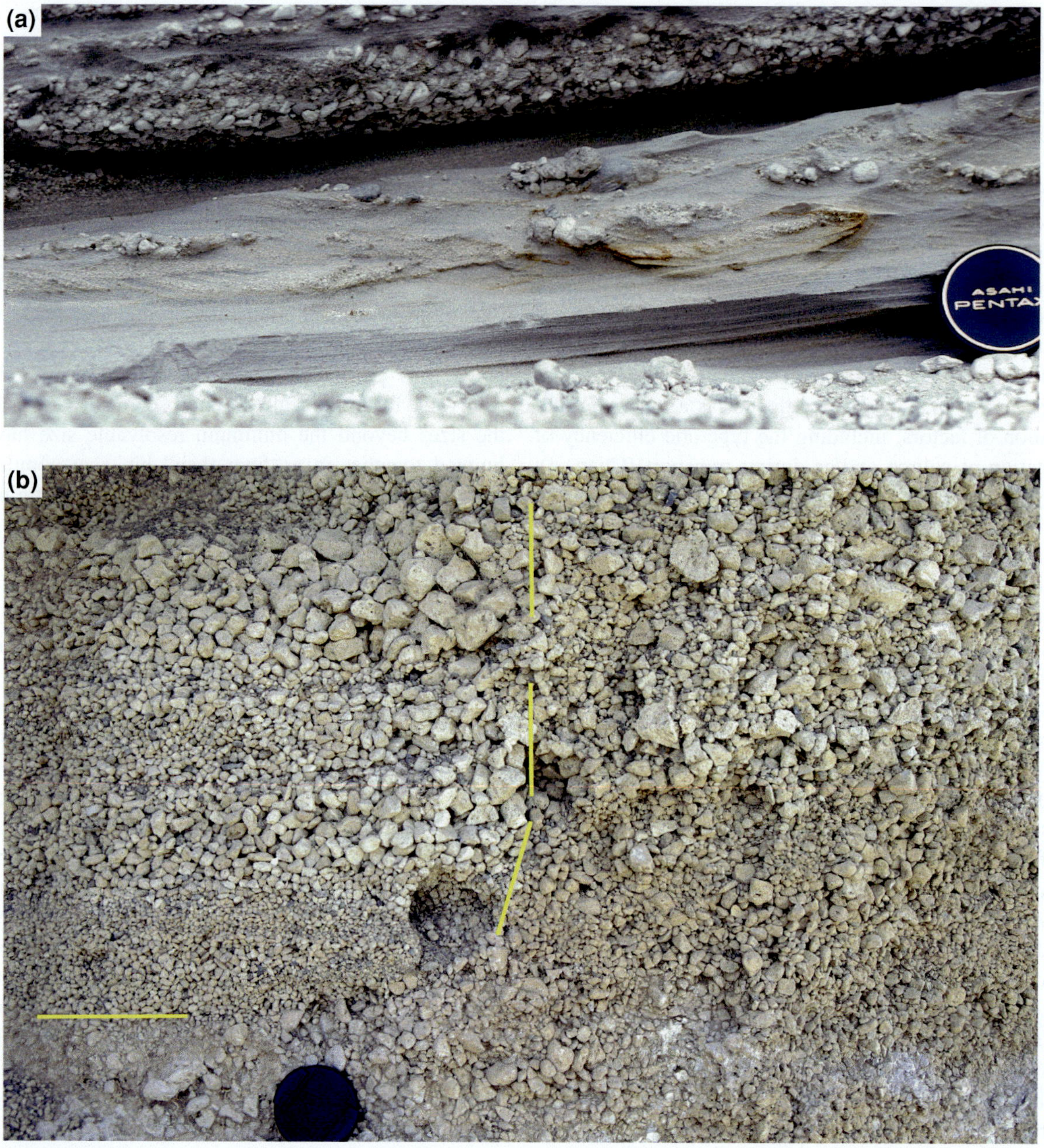

Fig. 1.12 Comparison between hydraulic and size sorting. **a** Large low-density pumice clasts in pockets in a finer grained pumice–crystal–lithic sand made up of denser grains, in a hydraulically well-sorted, tractionally reworked volcanic sediment. Terrace, Lake Taupo, New Zealand. **b** Comparison of grain size sorting between fluvially reworked, stratified, rounded pumice conglomerate in a gully fill (left; only partially outlined in yellow to show how cryptic contacts can be occasionally) eroded into and compared with coarser, more poorly sorted angular pumice pyroclastic fallout deposits (right). Tenerife, Canary Islands, Spain

fragments) well-developed size sorting becomes impossible, even though the populations may be hydraulically well sorted. Many volcaniclastic deposits consist of components with major variations in the shape and density (e.g. crystals, shards, low-density vesiculated pumice, and dense lithic fragments). As a result, volcaniclastic aggregates, whether they are pyroclastic or reworked and/or redeposited volcaniclastic sediments, are likely to be poorly sorted according to size, but may be well sorted hydraulically (e.g. Fig. 1.12a). It is well known that sorting in pyroclastic deposits is poorer than in many epiclastic sediments of the same overall grain size class (e.g. Fig. 1.12b). This is not only a function of the hydraulic behaviour of particles with different densities but also because of the short duration of many pyroclastic transportation processes. In pyroclastic air-fall deposits the diameter of pumice clasts is usually significantly greater than the diameter of associated, equant lithic fragments because the density of the lithic clasts is greater than pumice. Large pumice clasts will therefore fallout at the same terminal fall velocity as smaller, denser, lithic clasts because they have the same hydraulic equivalence or settling velocity. Fluid drag, caused by cross-sectional surface area, and fluid type also play a role. Water-settled fallout deposits will have different sorting characteristics compared with subaerial atmospheric fallout deposits (Cashman and Fiske 1991). However, when lithic clasts in a fallout deposit are elongated and blade- or rod-like in shape, their long dimension may be as great if not greater than that of the low-density pumice clasts.

Hydraulic sorting, however, can only occur if the particles are sufficiently dispersed in a fluid (i.e. the concentration is low enough) to be able to act freely, unhindered by nearby particles. In depositional systems where particle concentrations are high, and there is mutual particle interference during transport and settling (a state called hindered settling; e.g. Druitt 1995), hydraulic sorting may be significantly reduced, if it occurs at all. Debris flow and pyroclastic flow deposits are poorly sorted because of this. If the transport system is a cohesive system (particles are adhering to each other through ionic electrostatic attraction between particles and/or fluid, or thermal, plastic state adhesion; e.g. vitric ash-rich pyroclastic flows, clay-rich debris flows) sorting will be poor. If the transport mechanics lacks cohesive forces (all particles are freely dispersed in a low-viscosity fluid) then higher levels of hydraulic sorting are likely.

The *shape of fragments* is an assessment of the three-dimensional form of a grain, whether it be a mineral grain, lithic clast, or vitric fragment (e.g. pumice, glass shard, and blocky hyaloclastite fragment). For non-volcanic epiclastic sediments, the shape is largely inherited from the grain shape or fabric in the source rock, including original crystal shape, cleavage (mineral and tectonic), and layering, whether it be sedimentary, igneous, or metamorphic. For pyroclastic and autoclastic aggregates, the mode of fragmentation may also impart distinctive shape–morphology properties. As discussed above, during hydraulic sorting, and shape can play a significant role in determining the overall nature of the grain size population.

The *rounding of fragments* is the degree to which sharp corners and edges have been abraded during transportation or deposition, or both. Generally, rounding is best developed in sediments that have been subjected to constant and long-term energy levels and abrasion during tractional reworking. Tractional reworking causes both rounding and smoothing of the surfaces of clasts (e.g. beach sands and gravels; Fig. 1.13a). However, rounding can also be caused by pyroclastic processes (Chaps. 5–8), but generally, such processes are less effective in producing high levels of smoothing of clast surfaces, corners, and edges compared with sedimentary processes (Fig. 1.13b, c). (*Note* most sedimentological texts refer to "roundness" rather than "rounding". "Rounding" is preferred here because it implies process effects, whereas "roundness" has more of a shape connotation.)

Fabric is a consideration of the relationships between, and arrangement or packing of grains in an aggregate. Fabric is clearly a reflection of the transporting mode and depositional conditions (mass flow or tractional, with or without cohesion, etc.). It is more fully discussed in the relevant chapters considering the facies characteristics of particular facies groups and transportational and depositional processes, and in the sections of the book on terminology and classification (Chap. 14).

1.8.2.3 Depositional Structures

Depositional structures are probably the most important analytical tools in facies analysis, especially in terms of understanding transportation and depositional processes and conditions (Allen 1982, 1997; Cas and Wright 1987; Collinson and Thompson 1989; Leeder 2011; Nichols 2009). They are produced before deposition (e.g. erosional features), during deposition (e.g. current-generated structures, such as scours, cross-bedding, and imbrication) and after deposition (e.g. soft-sediment deformation, bioturbation, gas and water escape structures), and can be referred to as being pre-, syn-, and post-depositional in timing, respectively. They are extremely important because they, together with textural aspects, most immediately reflect the depositional conditions, and the modes of transportation and deposition. If the structures are produced by fluid flow, then they are especially important because they reflect the fluid dynamics of the host environment and its transportation and depositional agents.

As discussed in Chap. 12, fragments of mineral, rock, or fossil can be transported in particulate fashion (i.e. one by one) or by mass movement (i.e. bulk aggregates of particles moved instantaneously as one). Particulate movement of granular sediment (coarser than clay) by a fluid through rolling, sliding or saltation often produces an assemblage of tractional depositional structures (ripples, dunes, cross-stratification, etc.). Mass-movement processes frequently deposit a massive, structureless to diffusely layered aggregate. However, the trailing tails and margins of mass flows may also produce tractional sedimentary structures during waning stages of mass flow (e.g. the Bouma B and C divisions of deep-water turbidites, Chap. 12; ash-cloud surge deposits of pyroclastic flow deposits, Chap. 10). Tractional sedimentary structures are therefore not exclusively associated with processes involving just particulate sediment transport, or with environments that are "shallow-water" in aspect. Each case has to be evaluated on its merits. In addition, many linear and asymmetrical sedimentary

Fig. 1.13 Some rounded clast deposits in volcanic settings. **a** Rounded pumice clast conglomerate with pumice clasts with smooth surfaces, in a tractionally-reworked Quaternary lake shoreline succession, Kutchan, Hokkaido, Japan. **b** Fines-depleted, pumice clast "conglomerate" lens, on the lee-side of a ridge, Taupo ignimbrite, New Zealand. Note that although the clasts have developed a rounded form due to mutual clast collisions and abrasion in turbulent eddies formed on the lee-side of a ridge in the pyroclastic flow, the surfaces are not smooth, compared with the sedimentary tractionally reworked pumice clasts in (**a**). **c** Crudely sub-rounded to sub-angular andesitic lithic clasts in a layer in the dipping, cone sequence of Pukeonake cone, Tongariro National Park, New Zealand. The sub-rounding of the clasts is inferred to have occurred in the vent during a period when vent discharge energy was insufficient to eject dense clasts, which were tumbled, and "milled" in the vent until explosive energy was sufficient to eject the milled, sub-rounded debris from the vent

structures can be used to determine the original current paleoflow direction. Since all volcanic terrains are influenced by normal sedimentary surface processes, the analysis of volcanic sediment intervals will be important in reconstructing the hydrodynamics and paleoenvironment(s) of the volcanic terrain, especially in ancient successions.

However, depositional structures are not just produced by sedimentary processes. Primary volcanic processes also produce distinctive depositional or emplacement structures. Pyroclastic processes also involve particulate and mass-movement transportation of clastic aggregates (Chaps. 9, 10). The grain types and shapes, their temperature, and their physical state (solid or plastic or viscous fluid) are different, as are frequently the transporting media, their volumes, the scale and intensity of the processes, and their fluid-dynamic properties, compared with those of sedimentary surface process regimes. Such differences should produce distinctive differences in the types of structures and textures produced, and these will be highlighted in Chaps. 9 and 10. Most importantly, because of such differences, care should be taken in trying to translate the physical principles of sediment transport to volcanic transport processes. In some cases, process regimes may be similar, in others not. In particular, it is over-simplistic and incorrect to equate the processes, depositional structures, and flow regime concepts in aqueous transport systems with those in gas-driven pyroclastic transport systems, because we know surprisingly little about gas–particulate transport systems. Examples of depositional structures in primary volcanic facies that require fluid-dynamic interpretation include the variations in depositional structures in pyroclastic surge deposits (massive, dune-bedded, or planar stratification), the massive to stratified nature of ignimbrites, and the multiple facies of some ignimbrites (Chap. 10). Depositional structures also occur in lavas, and these can at times be used to infer the flow state (Chaps. 2, 5–8) and can also provide useful information about the paleoenvironment (Chaps. 5–8) and even the paleoflow directions (see below).

1.8.2.4 Depositional Structures and Paleoflow/Paleotransport Directions and Patterns

The directions of current flow or sediment transport directions can be measured where asymmetrical depositional structures such as ripples, dunes, angle of repose cross-stratification and imbrication, and sole structures, such as flute casts, can be used to determine local directions of movement of sediment, or paleocurrent flow. Over a larger area, numerous readings can be used to project back towards the direction where the sediment source area may have been, to reconstruct the paleogeography, and to trace paleogeographic changes as they have influenced current flow and sediment transport pathways. Furthermore, distinctive regional paleocurrent patterns develop in certain sedimentary environments (e.g. radial patterns for alluvial and submarine fans and deltas, bimodal for nearshore marine settings, etc.; Selley 2004).

Flow direction indicators in primary volcanic facies can also be used to determine original flow directions. However, the structures will be different, and because they are rarely used as tools in basin analysis, some are briefly highlighted here. The use of such transport or flow directional indicators not only helps to interpret local dispersal pathways but may also help to identify the direction of vent or of the source volcanic edifice.

Paleoflow Direction Indicators in Lavas

For *lavas,* the following examples are useful:

(a) Waters (1960) recognised that vertical pipe vesicles at the bases of basaltic lavas were often deflected in a downstream direction (Fig. 1.14a).
(b) Spherical vesicles may become stretched to produce a flow-parallel, stretched vesicle lineation (Fig 1.14b).

Fig. 1.14 Depositional structures in lavas that can be used to determine paleoflow direction. **a** Vertical pipe vesicles developed at the bases of Miocene low-viscosity basalt lavas in the Columbia River Plateau flood basalt province, with tops dragged over in a down-current direction within the interior of the flow that was travelling faster than the base. The grey "a" zone is an open pipe-like feature, whereas the patterned "b" zone represents a trail of vesicles in groundmass basalt. (Adapted from Waters 1960). **b** Stretched vesicles in a basalt lava, stretched in the local direction of flow, so defining a flow lineation, but not an absolute direction of flow. Pliocene lava flow, Bacchus Marsh, Newer Volcanics Province, Victoria, southeast Australia. **c** Crudely aligned feldspar phenocrysts in trachyandesite lava, Heard Island, Australia. **d** Eddies in flow banding on the down-current side of a xenolith inclusion in a rhyolite lava, Scrugham Peak, Nevada, USA (Adapted from Cummings 1964). **e** Schematic model of subaerial basalt lava flowing into a body of water. Inclined foreset beds of hyaloclastite debris and lava lobes, forming "flow-foot breccias", develop at the shoreline where lava enters water, producing a prograding lava and hyaloclastite delta, with the direction of dip of the foreset beds indicating the direction of propagation of the lava. (From Cas and Wright 1987, After Jones and Nelson 1970)

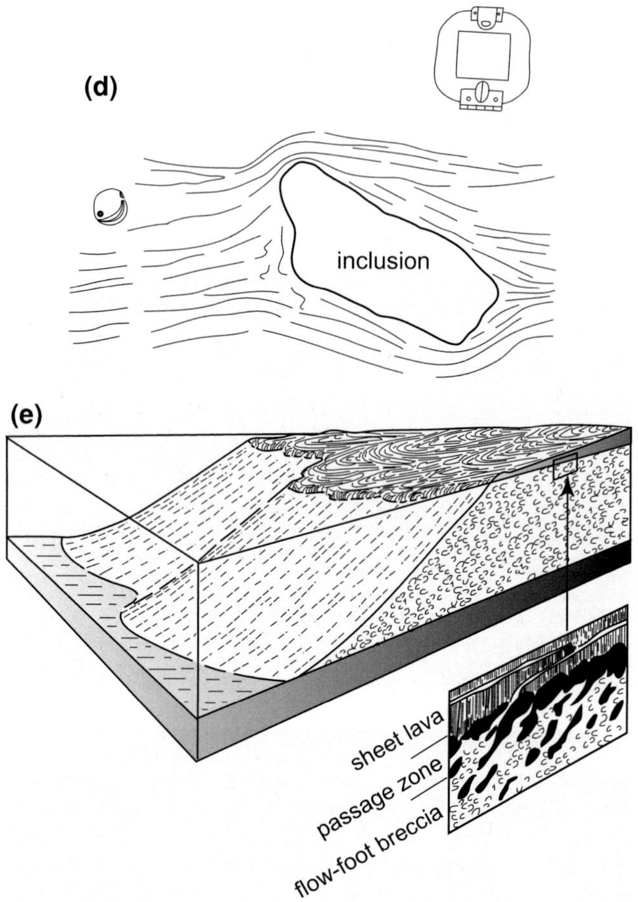

Fig. 1.14 (continued)

(c) During laminar flow in lavas (Chap. 2), elongate phenocrysts may become aligned producing a lineation parallel to flow direction (Fig. 1.14c).

(d) Cummings (1964) described eddies in flow banding that develop on the downstream side of inclusions and obstacles (e.g. tree trunks) in rhyolitic lavas (Fig. 1.14d).

(e) Waters (1960) and Jones and Nelson (1970) recognised that inclined foreset beds which form in prograding lava deltas formed where lava enters a body of water causing quench fragmentation, and foreset-bedded aprons of hyaloclastite to prograde (e.g. Fig. 1.14e). These can be used to assess local lava flow directions for basaltic lavas.

(f) As successive pillows bud in an advancing submarine basaltic pillow lava they do so asymmetrically, producing a convex front, and an angular re-entrant at the point of budding (Chap. 5).

(g) Ropey flow folds on the surfaces of mafic pahoehoe lavas develop convex arc shapes in a down-current direction (Chap. 5). These may only reflect very local, micro-topographically controlled flow directions, and much flow fold orientation data must be collected to develop a larger scale of overall flow directions of the lava flow field relative to source vent.

(h) Most rhyolitic to dacitic lavas propagate or spread laterally and downslope away from the vent by ramping over earlier-erupted parts. Ramp surfaces and often flow banding will dip back towards the vent, and ramp wedges will often thicken away from vent, and will thin asymptotically towards vent (Chap. 6).

(i) Sometimes large-scale lava flow folds or pressure ridges may develop a consistent asymmetry or vergence away from the vent, i.e. with axial planes dipping back to the vent and fold axes perpendicular to local flow directions. However, these criteria are less reliable than others and must be mapped out over a significant distance (see Chap. 6) because, unlike tectonic folding which occurs under the influence of an orientated, confined stress regime, flow folding occurs in a largely unconfined, plastic regime.

(j) Lateral facies changes in lavas may indicate vent direction. For example, in Archean basaltic-andesitic submarine lava flows, Dimroth et al. (1978) noted lateral changes from massive lavas near the inferred eruption point, to pillow lavas and/or hyaloclastites at the distal margins (Chap. 5).

Paleodispersal Direction Indicators in Pyroclastic Fallout Deposits

For *young pyroclastic fallout deposits,* there are some standard methods for locating the vent area, and by inference, the dispersal directions. The first two are rarely useful in older successions because of the effects of erosion, deformation, and tectonic dismemberment leave only limited exposure of deposits. Methods include:

(a) Map the thickness variations of the fall deposits regionally and produce a contour map (isopach map) in which the thickness maximum usually coincides with the vent area (Fig. 1.15a), although not always if displaced by strong wind effects, or fallout induced by heavy rainfall through an eruption plume (Chap. 9).

(b) Determine the distribution of the largest five lithic and pumice clasts in a single deposit at different localities and produce a contour map of average maximum grain size (isopleth contour map) in which the largest clast isopleth contour will usually coincide with the vent area (Chap. 9).

(c) Assess the direction of the vent from the direction and angle of incidence reflected by asymmetrical ballistic impact sag structures (Fig. 1.15b).

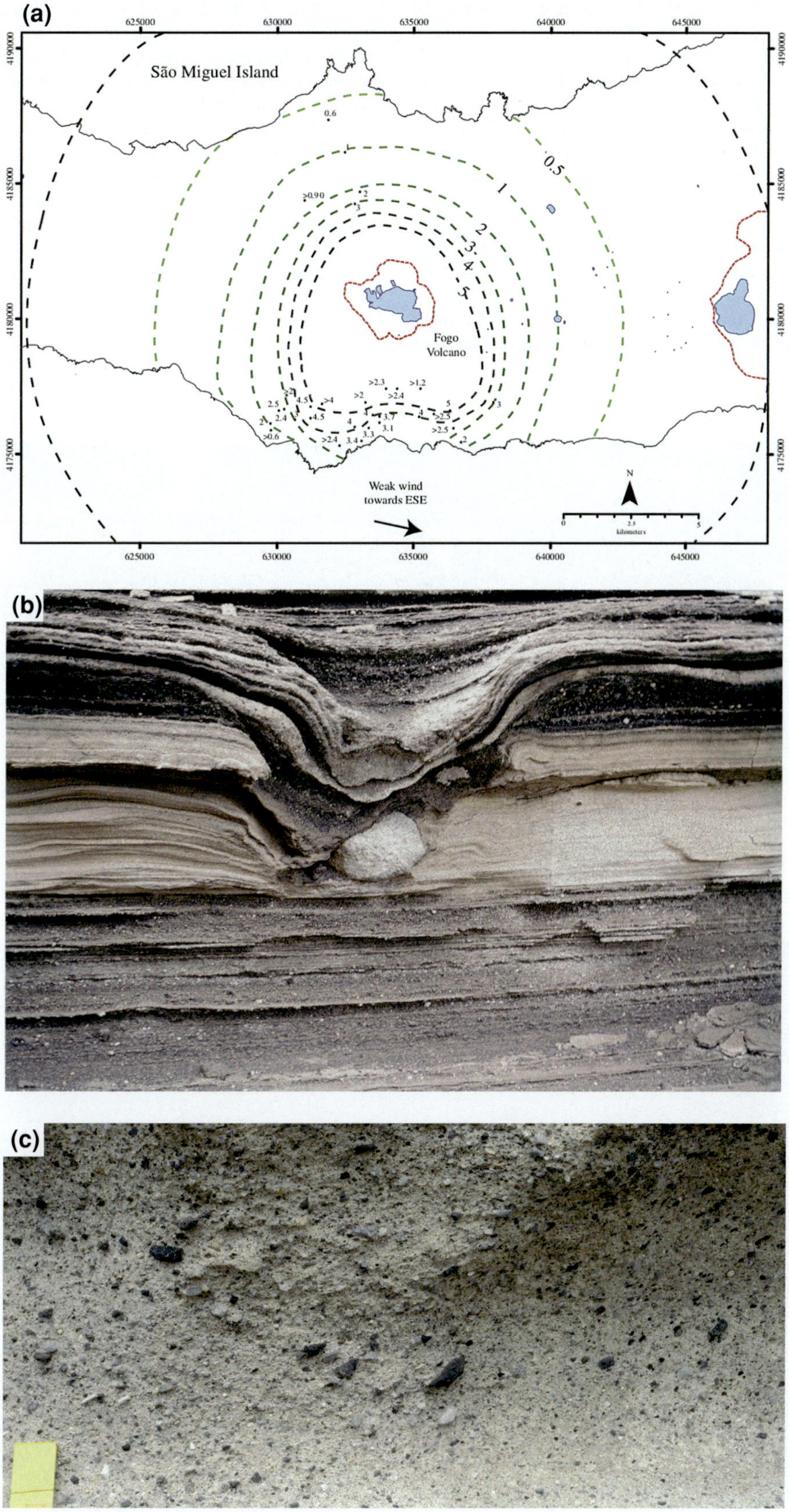
(a)
625000
630000
635000
640000
645000
4190000
4185000
4180000
4175000
São Miguel Island
0.5
1
2
3
4
5
Fogo
Volcano
Weak wind
towards ESE
N
0
2.5
5
kilometers
(b)
(c)

Fig. 1.15 Some depositional structures in pyroclastic deposits that can be used for determining dispersal or paleoflow directions. **a** Isopach map of the near vent deposits of the lower part of the 4.6 ka Fogo A pumice fallout deposit, São Miguel, Azores, Portugal, showing decreasing thickness of the deposit radially away from vent, depicted by decreasing thickness contours away from vent). The nearly radially symmetric pattern indicates that very weak winds were blowing during this phase of the eruption. The irregularity along the southern margin is due to erosion by subsequent intra-plinian pyroclastic flows that flowed to the south. (From Pensa et al. 2015; data for the outer margin of the deposit are from Walker and Croasdale 1971). **b** Asymmetrical ballistic impact sag structure, Pleistocene Tower Hill maar, Newer Volcanics Province, Victoria, southeast Australia. **c** Imbrication of lithic clasts in phonolitic ignimbrite, Abrigo Ignimbrite, La Mareta, Tenerife, Canary Islands, Spain. Flow from right to left. **d** Asymmetrical load casts of lithic clasts at the base of a lithic clast-rich ignimbrite, Abrigo ignimbrite. The development of the load casts indicates that the underlying ignimbrite deposit was still gas-rich and unconsolidated at the time the upper, lithic clast-rich ignimbrite was deposited. The asymmetry of the load casts is due to syn-depositional flow drag. Flow from right to left. **e** Northeast tree blow-down zone produced by the 18 May 1980 lateral blast flow (Chap. 10) from Mt St Helens. The thickest tree trunks are almost 1 m in diameter. **f** Orientation of tree logs in the 232 AD Taupo ignimbrite, New Zealand. The lineations of log alignments project back towards Horomatangi Reef in Lake Taupo, the inferred vent location for the Taupo eruption. (After Frogatt et al. 1981). **g** Base-surge dune bedforms and cross-bedding, indicating flow from right to left. Pleistocene Purrumbete maar, Newer Volcanics Province, Victoria

Fig. 1.15 (continued)

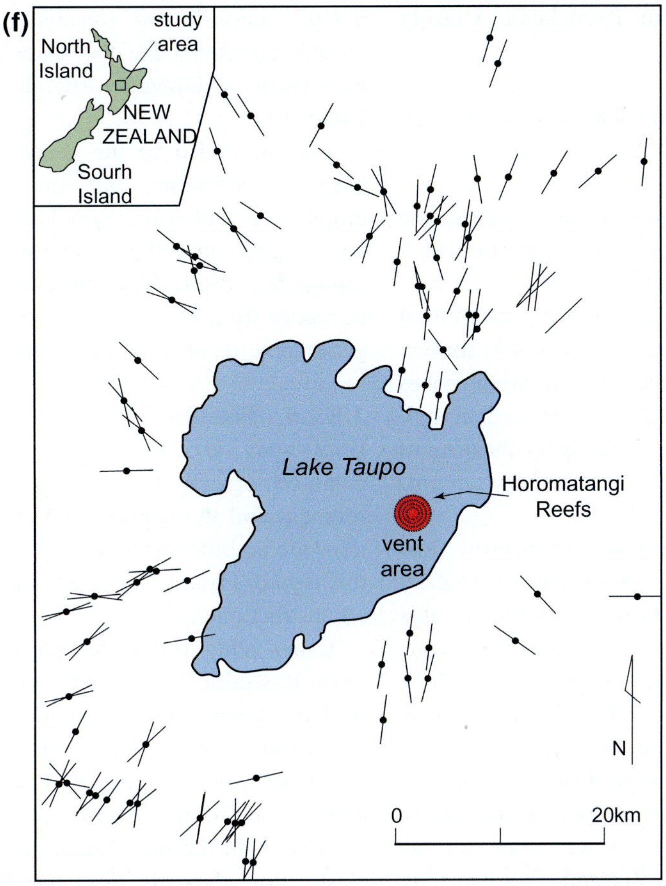

Fig. 1.15 (continued)

Paleoflow Direction Indicators in Pyroclastic Density Current Deposits

For *pyroclastic flows*, transport directions have been variously determined by:

(a) Pumice and lithic clast alignment and imbrication (Elston and Smith 1970; Kamata and Mimura 1983; Fig. 1.15c).
(b) Asymmetrical load structures sheared in the direction of flow, and groove and flute structures sometimes form at the bases of deposits in lithic concentration zones (Pittari and Cas 2004; Fig. 1.15d); asymmetrical substrate deformational structures produced by shearing of the substrate by the base of pyroclastic density currents (LaBerge et al. 2006).
(c) Alignment of logs in relatively young pyroclastic flow deposits erupted in forested, temperate to tropical environments. The most spectacular example of this was the alignment of tree trunks in the blowdown zone produced by the blast flow from the eruption of Mt St Helens on May 18, 1980 (Fig. 1.15e). This has been successfully used to determine detailed flow paths (Waitt 1981; see Fig. 1.5a). For pre-historic examples, the orientation of logs has to be mapped in order to be able to infer the vent location, as was done for the Taupo AD 186 ignimbrite, New Zealand (Froggatt et al. 1981; Fig. 1.15f).
(d) Contouring the regional distribution of average maximum lithic clast sizes can also be used to locate vent area (Kuno et al. 1964; Walker and Croasdale 1971; Walker 1981; Pensa et al. 2015).
(e) Determining the anisotropic magnetic susceptibility fabric (AMS) of magnetic minerals in the ash matrix to determine flow direction of pyroclastic flows (e.g. Fisher et al. 1993; Baer et al. 1997; Porreca et al. 2003; LaBerge et al. 2009; Willcock et al. 2014; Chap. 10).

For *base surges* (Chap. 7), and the low particle concentration ash-cloud surges of pyroclastic flows, depositional structures such as dunes, low-angle cross-stratification (e.g. Fig. 1.15g), and chute-and-pool structures (Fisher and Waters 1970; Schmincke et al. 1973; Chap. 10) are useful if present.

Paleoflow (Current) Direction Indicators in Volcanic and Terrigenous Sediments

Sediment transport directions for volcanic and terrigenous sediment successions are largely (paleo-)topographically controlled, including the location and orientation of river valleys, submarine channels, and canyons, regional slopes). Typical depositional structures that indicate paleoflow/current directions are the same as those found in normal terrigenous sedimentary successions and are discussed in Chap. 12.

This may also be the general case for lavas and pyroclastic successions, but the flow mechanisms during emplacement of some pyroclastic flows and surges may be so energetic that they may largely ignore and surmount topographic highs. Nevertheless, flow directions will mainly be radial from the vent, and they may be useful in paleogeographic reconstruction of the volcanic centre.

1.8.2.5 Fossils

Both body and trace fossils can be extremely useful paleo-ecologically for helping to reconstruct the paleoenvironment and the depositional conditions, and volcanic terrains are no different to normal sedimentary environments in this regard. Fossils can also be used for paleovolcanological reconstructions.

Using body fossils for paleoenvironmental reconstructions, it is critical to establish whether the fossils are in situ or have been transported and redeposited, especially in marine successions, where offshore and downslope redeposition are common (Fig. 1.16a). Marine fossils, however, may also be found in subaerial pyroclastic deposits, as was the case with the occurrence of corals in the on-land pyroclastic flow deposits from the 1883 eruption at Krakatau (Simkin and Fiske 1983). These corals were explosively ejected from their marine habitat and then incorporated into and deposited onshore by subaerial pyroclastic flows. They illustrate the problems that can lead to erroneous interpretations of older volcanic successions. Redeposited fossil remains may only be useful in indicating the nearby environmental conditions, and only if the ecological affinities of the organisms are known (e.g. Bull and Cas 1989). Beyond this, the robustness of faunal remains, associations of organisms, and especially the facies characteristics of associated sediments will be useful in deciphering the nature of the paleoenvironment.

Trace fossils are the burrows, grazing tracks, and resting imprints of organisms and may be less directly useable for paleoenvironmental reconstructions than body fossils (Fig. 1.16b). Nonetheless, significant research has been conducted on the paleoenvironmental affinities of trace fossils to make them useful adjuncts in paleoenvironmental reconstructions (e.g. Bull and Cas 1989). As a general rule, vertically orientated burrows, especially were preserved in coarse clastics, are indicative of shallow, high-energy environments, but not necessarily marine settings, as is commonly assumed. By contrast, sub-horizontally inclined burrows and grazing tracks are more common in offshore, sub-wave base settings. Exceptions to general rules always

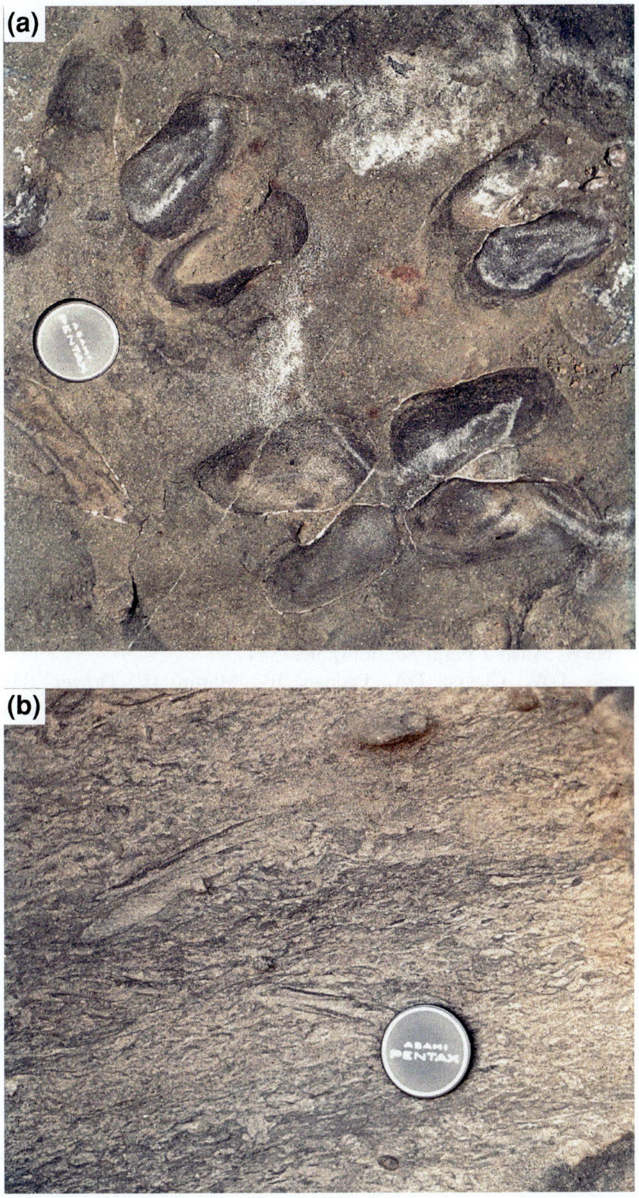

Fig. 1.16 Fossils in volcanic successions. **a** Still-articulated bivalves in offshore storm turbidite volcanic sandstone beds in the Permian Gerringong Volcanics, indicate that the organisms were not transported far from living position, and can be used both for biostratigraphic age dating and for paleo-ecological, paleoenvironmental and paleo-volcanological reconstructions. Kiama, New South Wales, Australia. **b** Sub-horizontal burrows and grazing trails suggest an offshore, sub-fair-weather wave base deposition setting for the host bioturbated volcanic sandstone and mudstone sequence. Permian Wesley Park Sandstone, Kiama, New South Wales

occur, however. For example, vertical tube trace fossils, such as the common *Skolithos* sp., are known to be preserved in fluvial successions as well as their more commonly known occurrence in shallow marine successions. Other exceptions to such general rules include the production of vertical tube escape burrows in offshore settings by organisms suddenly buried by, for example, a shelf storm turbidite bed. Fossils can also be used to reconstruct the volcanological scenario and occur in primary eruptive products as well as in associated sedimentary deposits. The anomalous occurrence of corals in the subaerial pyroclastic flow deposits of Krakatau cited above is one example. Although somewhat morbid, the occurrence of human fossils at Pompeii, buried in the pyroclastic flow deposits of the Vesuvius eruption of 79 AD, often in sleeping positions, help to reconstruct the state of the pyroclastic flows at the time they overwhelmed Pompeii. Similarly, the remarkably good preservation of the buildings of Pompeii (archaeological fossils) testify to the relatively low power of the pyroclastic flows.

The remains of wood and timber can be used to reconstruct flow directions and paleoenvironment, they can be used to date pyroclastic and associated sedimentary deposits less than 50,000 years old, using radiocarbon dating methods, and they can be used for assessing the relative and absolute temperatures of the pyroclastic deposits they occur in (Giordano et al. 2018).

1.9 Facies Analysis, Interpretation, and the Importance of Associations of Facies

Having carefully documented and described individual facies, it is also necessary to consider the association of facies and the relationships between them to evaluate to what degree they are temporally and genetically related. Different facies may be the products of one event (e.g. the diverse facies of ignimbrite-forming eruptions, Chap. 10). Assessment of the spatial and age relationships is therefore important before general models of the deposits and the sequence of events can be formulated. On a larger scale, particular types of volcanic centres and settings may consistently produce similar associations of facies which can then be used to formulate general facies and stratigraphic models for those settings (Chap. 15).

1.10 The Importance of Geological Mapping and Establishing Stratigraphy

Establishing the geological context of any facies or facies association provides the framework for understanding the origin of the facies, the paleoenvironmental setting, the timing of events, and any trends in volcanism or the paleoenvironmental setting. This can only be achieved by undertaking detailed geological mapping (Chaps. 15, 16). With the advent of technology, mapping of volcanoes and volcanic terrains is now most effectively done using a combination of remote sensing geophysical techniques and datasets followed by targeted on-ground mapping. Remote sensing (satellite or aerial imagery, e.g. Aster, LiDar), and digitial geophysical data sets (e.g. aeromagnetic, gravity, radiometric) facilitate the generation of digital elevation models (DEMs), identification of distinctive depositional and rock units, and first-order maps showing the distribution of major units. However, these MUST be followed by careful, on-ground, geological field work to ground-truth vertical and lateral geological relationships and establish sequences of depositional or rock units. This also involves determining in the field, from contact relationships,whether or not depositional units or facies are intrusive or extrusive in origin.

Only after such mapping can a relative stratigraphy (a relative order of emplacement of rock/deposit units) be established. Stratigraphic units consist of one, or an association of, facies that are distinctive from other stratigraphic units and can be mapped as distinctive map units (Chaps. 15, 16). Once a stratigraphy has been established for a particular area or volcano, sampling for geochemistry and geochronology can be undertaken, because it is only then that the rock unit relationships and context of each rock unit can be understood. Undertaking geochemistry or geochronology studies without such a stratigraphic framework will minimise understanding of results and can lead to erroneous interpretations of geochemical trends and the meaning of radiometric ages. Both can be affected by post-emplacement chemical changes (e.g. excess argon effects, metamorphism and alteration, contamination by xenocrystic and xenolithic material), making results and chronology of events misleading. Establishing a chrono-stratigraphy (stratigraphy constrained by radiometric ages of some or all rock units) cannot be achieved by just grabbing samples in the field, without understanding the nature and origins of the sampled units and their contact relationships (e.g. *are coherent basalt units lavas or high-level sill intrusions and what is the evidence?*). “Grab sampling” without understanding the geological framework and stratigraphy leads to poor and often erroneous scientific outcomes.

References

Airey MW, Mather TA, Pyle DM, Glaze LS, Ghail RC, Wilson CF (2015) Explosive volcanic activity on Venus: the roles of volatile contribution, degassing, and external environment. Planet Space Sci 113:33–48

Allen JRL (1982) Sedimentary structures, their character and physical basis. Elsevier Science, p 663

Allen PA (1997) Earth surface processes. Blackwell, p 404

Appel PWU, Fedo CM, Moorbath S, Myers JS (1998) Recognizable primary volcanic and sedimentary features in a low-strain domain of the highly deformed, oldest known (~3.7–3.8 Gyr) Greenstone Belt, Isua, West Greenland. Terra Nova 10(2):57–62

Baer EM, Fisher RV, Fuller M, Valentine GA (1997) Turbulent transport and deposition of the Ito pyroclastic flow: determinations using anisotropy of magnetic susceptibility. J Geophys Res 102 (B10):22565–22586

Báez W, Nuñez GC, Giordano G, Viramonte JG, Chiodi A (2017) Polycyclic scoria cones of the Antofagasta de la Sierra basin, Southern Puna plateau, Argentina. Geol Soc Lond Spec Publ 446 (1):311–336

Bédard JH (2018) Stagnant lids and mantle overturns: implications for Archaean tectonics, magmagenesis, crustal growth, mantle evolution, and the start of plate tectonics. Geosci Front 9(1):19–49

Bleeker W (2002) Archaean tectonics: a review, with illustrations from the Slave craton. Geol Soc Lond Special Publ 199(1):151–181

Boggs S (2012) Principles of sedimentology and stratigraphy (Fifth Edition). Pearson Prentice Hall, New Jersey, p 600

Brenna M, Cronin SJ, Smith IEM, Sohn YK, Németh K (2010) Mechanisms driving polymagmatic activity at a monogenetic volcano, Udo, Jeju Island, South Korea. Contrib Miner Petrol 160 (6):931–950

Bull SW, Cas RAF (1989) Volcanic influences in a storm- and tide-dominated shallow marine depositional system: The late permian broughton formation, southern Sydney Basin, Kiama, NSW. Austr J Earth Sci 36(4):569–584

Caress DW, Clague DA, Paduan JB, Martin JF, Dreyer BM, Chadwick WW Jr, Denny A, Kelley DS (2012) Repeat bathymetric surveys at 1-metre resolution of lava flows erupted at Axial Seamount in April 2011. Nat Geosci 5(7):483

Carey R, Soule SA, Manga M, White JD, McPhie J, Wysoczanski R, Jutzeler M, Tani K, Yoerger D, Fornari D, Caratori-Tontini F (2018) The largest deep-ocean silicic volcanic eruption of the past century. Sci Adv 4(1):e1701121

Cas RAF, Simmons J (2018) Why deep-water eruptions are so different from subaerial eruptions. Front Earth Sci 6:198

Cas RAF, Wright JV (1987) Volcanic successions: modern and ancient. Allen & Unwin, London, p 528

Cas RAF, Yamagishi H, Moore, L, Scutter C (2003) Miocene Submarine Fire Fountain Deposits, Ryugazaki Headland, Oshoro Peninsula, Hokkaido, Japan: Implications for Submarine Fountain Dynamics and Fragmentation Processes. In: White JDL, Smellie JL, Clague DA (eds) Submarine Explosive Volcanism. Amer Geophys Union Monograph 140:299–316

Cashman KV, Fiske RS (1991) Fallout of pyroclastic debris from submarine volcanic eruptions. Science 253(5017):275–280

Cawood PA, Hawkesworth CJ, Pisarevsky SA, Dhuime B, Capitanio FA, Nebel O (2018) Geological archive of the onset of plate tectonics. Philos Trans Roy Soc Math Phys Eng Sci 376 (2132):20170405

Chadwick WW, Cashman KV, Embley RW, Matsumoto H, Dziak RP, De Ronde CEJ, Lau TK, Deardorff ND, Merle SG (2008) Direct video and hydrophone observations of submarine explosive eruptions at NW Rota-1 volcano, Mariana arc. J Geophys Res Solid Earth 113(B8)

Chadwick WW Jr, Clague DA, Embley RW, Perfit MR, Butterfield DA, Caress DW, Paduan JB, Martin JF, Sasnett P, Merle SG, Bobbitt AM (2013) The 1998 eruption of Axial Seamount: new insights on submarine lava flow emplacement from high-resolution mapping. Geochem Geophys Geosyst 14(10):3939–3968

Clague DA, Dreyer BM, Paduan JB, Martin JF, Chadwick WW, Caress DW, Portner RA, Guilderson TP, McGann ML, Thomas H, Butterfield DA (2013) Geologic history of the summit of Axial Seamount, Juan de Fuca Ridge. Geochem Geophys Geosyst 14 (10):4403–4443

Clague DA, Paduan JB, Caress DW, Moyer CL, Glaser BT, Yoerger DR (2019) Structure of Loihi Seamount, Hawai'I and lava flow morphology from high resolution mapping. Front Earth Sci 7:58. https://doi.org/10.3389/feart.2019.00058

Clague DA, Paduan JB, Caress DW, Thomas H, Chadwick WW, Merle SG (2011) Volcanic morphology of West Mata Volcano, NE Lau Basin, based on high-resolution bathymetry and depth changes. Geochem Geophys Geosyst 12(11)

Collinson JD, Thompson DB (1989) Sedimentary structures (second edition). Chapman & Hall, London, p 208

Cummings D (1964) Eddies as indicators of local flow direction in rhyolite. US Geol Surv Prof Pap 475D:70–72

Davies P (1998) The fifth miracle. The search for the origin of life. The Penguin Press, p 320

de Ronde CEJ, Baker ET, Massoth GJ, Lupton JE, Wright IC, Feely RA, Greene RR (2001) Intra-oceanic subduction-related hydrothermal venting, Kermadec volcanic arc, New Zealand. Earth Planet Sci Lett 193(3):359–369

Dimroth E, Cousineau P, Leduc M, Sanschagrin Y (1978) Structure and organization of Archean subaqueous basalt flows, Rouyn-Noranda area, Quebec, Canada. Can J Earth Sci 15(6):902–918

Dodd MS, Papineau D, Grenne T, Slack JF, Rittner M, Pirajno F, O'Neil J, Little CT (2017) Evidence for early life in Earth's oldest hydrothermal vent precipitates. Nature 543(7643):60

Druitt TH (1995) Settling behaviour of concentrated dispersions and some volcanological applications. J Volcanol Geoth Res 65(1–2):27–39

Elston WE, Smith EI (1970) Determination of flow direction of rhyolitic ash-flow tuffs from fluidal textures. Geol Soc Am Bull 81 (11):3393–3406

Ernst RE, Youbi N (2017) How Large Igneous Provinces affect global climate, sometimes cause mass extinctions, and represent natural markers in the geological record. Palaeogeogr Palaeoclimatol Palaeoecol 478:30–52

Fisher RV (1961) Proposed classification of volcaniclastic sediments and rocks. Geol Soc Am Bull 72(9):1409–1414

Fisher RV, Orsi G, Ort M, Heiken G (1993) Mobility of a large-volume pyroclastic flow—emplacement of the Campanian ignimbrite, Italy. J Volcanol Geoth Res 56(3):205–220

Fisher RV, Waters AC (1970) Base surge bed forms in maar volcanoes. Am J Sci 268(2):157–180

Frankel C (1996) Volcanoes of the solar system. Cambridge University Press, p 232

Froggatt PC, Wilson CJN, Walker GPL (1981) Orientation of logs in the Taupo Ignimbrite as an indicator of flow direction and vent position. Geology 9(3):109–111

Giordano G, Zanella E, Trolese M, Baffioni C, Vona A, Caricchi C, De Benedetti AA, Corrado S, Romano C, Sulpizio R, Geshi N (2018) Thermal interactions of the AD79 Vesuvius pyroclastic density currents and their deposits at Villa dei Papiri (Herculaneum archaeological site, Italy). Earth Planet Sci Lett 490:180–192

Hawkesworth C, Cawood PA, Dhuime B (2019) Rates of generation and growth of the continental crust. Geosci Front 10(1):165–173

Head JW, Crumpler LS, Aubele JC, Guest JE, Saunders RS (1992) Venus volcanism: Classification of volcanic features and structures, associations, and global distribution from Magellan data. J Geophys Res Planets 97(E8):13153–13197

Ikegami F, McPhie J, Carey R, Mundana R, Soule SA, Jutzeler M (2018) The eruption of submarine rhyolite lavas and domes in the deep ocean–Havre 2012, Kermadec Arc. Front Earth Sci 6:147. https://doi.org/10.3389/feart.2018.00147

Ivanov MA, Crumpler LS, Aubele JC, Head JW (2015) Volcanism on venus. In: The encyclopedia of volcanoes. Academic Press, pp 729–746

Jones JG, Nelson PHH (1970) The flow of basalt lava from air into water—its structural expression and stratigraphic significance. Geol Mag 107(01):13–19

Jutzeler M, Proussevitch AA, Allen SR (2012) Grain-size distribution of volcaniclastic rocks 1: a new technique based on functional stereology. J Volcanol Geoth Res 239–240:1–11

Jutzeler M, Marsh R, Carey RJ, White JD, Talling PJ, Karlstrom L (2014) On the fate of pumice rafts formed during the 2012 Havre submarine eruption. Nat Commun 5:3660

Kamata H, Mimura K (1983) Flow directions inferred from imbrication in the Handa pyroclastic flow deposit in Japan. Bull Volcanol 46 (3):277–282

Kasting JF (1993) Earth's early atmosphere. Science 259(5097):920–926

Kieffer SW (1981). Fluid dynamics of the May 18 blast at Mount St. Helens. In: Lipman PW, Mullineaux DR (eds) The 1980 eruptions of Mount St. Helens, Washington. US Geological Survey, Washington, pp 379–400

Kuno H, Ishikawa T, Katsui Y, Yagi K, Yamasaki M, Taneda S (1964) Sorting of pumice and lithic fragments as a key to eruptive and emplacement mechanism. Jpn J Geol Geogr 35:223–238

LaBerge RD, Giordano G, Cas RAF, Ailleres L (2006) Syn-depositional substrate deformation produced by the shear force of a pyroclastic density current: an example from the Pleistocene ignimbrite at Monte Cimino, northern Lazio, Italy. J Volcanol Geoth Res 158(3–4):307–320

LaBerge RD, Porreca M, Mattei M, Giordano G, Cas RAF (2009) Meandering flow of a pyroclastic density current documented by the anisotropy of magnetic susceptibility (AMS) in the quartz latite ignimbrite of the Pleistocene Monte Cimino volcanic centre (central Italy). Tectonophysics 466(1–2):64–78

Leeder MR (2011) Sedimentology and sedimentary basins: from turbulence to tectonics (second edition). Blackwell Science Ltd., Oxford, p 784

Lipman PW, Mullineaux DR (1981) The 1980 eruptions of Mount St. Helens, Washington. US Geological Survey, Washington, p 844

Loughlin SC, Sparks RSJ, Brown SK, Jenkins SF, Vye-Brown C (2015) Global volcanic hazards and risks. In: Cambridge University Press, 410 p

Manalang D, Delaney JR (2016) Axial seamount-restless, wired and occupied: a conceptual overview of resident AUV operations and technologies. In: OCEANS 2016 MTS/IEEE Monterey, pp 1–7

McCormick MP, Thomason LW, Trepte CR (1995) Atmospheric effects of the Mt Pinatubo eruption. Nature 373(6513):399–404

Nairn IA, Self S, Cole JW, Leonard GS, Scutter C (2001) Distribution, stratigraphy, and history of proximal deposits from the c. AD 1305 Kaharoa eruptive episode at Tarawera Volcano, New Zealand. New Zeal J Geol Geophys 44(3):467–484

Nebel O, Capitanio FA, Moyen JF, Weinberg RF, Clos F, Nebel-Jacobsen YJ, Cawood PA (2018) When crust comes of age: on the chemical evolution of Archaean, felsic continental crust by crustal drip tectonics. Philos Trans Roy Soc A: Math Phys Eng Sci 376(2132):20180103

Nichols G (2009) Sedimentology and stratigraphy. Wiley-Blackwell

Nisbet EG (1995) Archaean ecology: a review of evidence for the early development of bacterial biomes, and speculations on the development of a global-scale biosphere. Geol Soc Lond Special Publ 95 (1):27–51

Pensa A, Cas RAF, Giordano G, Porreca M, Wallenstein N (2015) Transition from steady to unsteady Plinian eruption column: the VEI 5, 4.6 ka Fogo A Plinian eruption, São Miguel, Azores. J Volcanol Geoth Res 305:1–8

Pittari A, Cas RAF (2004) Sole Marks at the base of the late Pleistocene Abrigo Ignimbrite, Tenerife: implications for transport and

depositional processes at the base of pyroclastic flows. Bull Volcanol 66(4):356–363
Porreca M, Mattei M, Giordano G, De Rita D, Funiciello R (2003) Magnetic fabric and implications for pyroclastic flow and lahar emplacement, Albano maar, Italy. J Geophys Res Solid Earth 108 (B5):2264–2277
Rampino MR, Self S (1992) Volcanic winter and accelerated glaciation following the Toba super-eruption. Nature 359(6390):50–52
Rampino MR, Stothers RB (1988) Flood basalt volcanism during the Past 250 million years. Science 241(4866):663–663
Reading HG (1996) Sedimentary environments: processes, facies and stratigraphy. Blackwell Science Ltd., Oxford, p 704
Resing JA, Rubin KH, Embley RW, Lupton JE, Baker ET, Dziak RP, Baumberger T, Lilley MD, Huber JA, Shank TM (2011) Active submarine eruption of boninite in the northeastern Lau Basin. Nat Geosci 4(11):799–806
Robock A 2004. Climatic impact of volcanic emissions. In: Sparks RSJ, Hawkesworth CJ (eds) State of the planet, vol 150. American Geophysical Union, Geophysical Monograph, pp 125–134
Rosing MT (1999) 13C-depleted carbon microparticles in >3700-Ma sea-floor sedimentary rocks from West Greenland. Science 283 (5402):674–674
Schmincke H-U, Fisher RV, Waters AC (1973) Antidune and chute and pool structures in the base surge deposits of the Laacher See area, Germany. Sedimentology 20(4):553–574
Self S (2006) The effects and consequences of very large explosive volcanic eruptions. Philos Trans R Soc A 364:2073–2097
Selley RC (2004) Ancient sedimentary environments: and their sub-surface diagnosis (fourth edition). Routledge, London, p 320
Shea T, Houghton BF, Gurioli L, Cashman KV, Hammer JE, Hobden BJ (2010) Textural studies of vesicles in volcanic rocks: an integrated methodology. J Volcanol Geoth Res 190(3–4):271–289
Simkin T, Fiske RS (1983) Krakatau 1883, the volcanic eruption and its effects. Smithsonian Institute Press, Washington DC, p 464
Sorrentino L, Cas RAF, Stilwell JD (2011) Evolution and facies architecture of Paleogene Surtseyan volcanoes on Chatham Islands, New Zealand, Southwest Pacific Ocean. J Volcanol Geoth Res 202 (1–2):1–21
Taylor SR, McLennan SM (1995) The geochemical evolution of the continental crust. Rev Geophys 33(2):241–265
Tilling RI, Topinka LJ, Swanson DA (1990) Eruptions of Mount St. Helens: past, present, and future, 2nd edn. US Geological Survey, Reston, VA, p 56
Tucker ME (2001) Sedimentary Petrology: an Introduction to the origin of sedimentary rocks. Blackwell Science Ltd., Oxford, p 262
van Otterloo J, Cas RAF, Sheard MJ (2013) Eruption processes and deposit characteristics at the monogenetic Mt. Gambier Volcanic Complex, SE Australia: implications for alternating magmatic and phreatomagmatic activity. Bull Volcanol 75(8):737–758
van Otterloo J, Raveggi M, Cas RAF, Maas R (2014) Polymagmatic activity at the monogenetic Mt Gambier volcanic complex in the Newer Volcanics Province, SE Australia: new insights into the occurrence of intraplate volcanic activity in Australia. J Petrol 55 (7):1317–1351
Waitt RB (1981) Devastating pyroclastic density flow and attendant air fall of May 18—stratigraphy and sedimentology of deposits. In: Lipman PW, Mullineaux DR (eds) The 1980 eruptions of Mount St. Helens, Washington. US Geological Survey, Washington, pp 439–458
Walker GPL (1971). Grain-size characteristics of pyroclastic deposits. J Geol 696–714
Walker GPL (1973) Explosive volcanic eruptions—a new classification scheme. Int J Earth Sci 431–446
Walker GPL (1981) New Zealand case histories of pyroclastic studies. In: Self S, Sparks RSJ (eds) Tephra studies. Springer, pp 317–330
Walker GPL (1984a) Characteristics of dune-bedded pyroclastic surge bedsets. J Volcanol Geoth Res 20:281–296
Walker GPL, Croasdale R (1971) Two Plinian-type eruptions in the Azores. J Geol Soc 127(1):17–55
Walker RG (1984b) Facies models. In: Geological association of Canada. Geoscience Canada Report
Waters AC (1960) Determining direction of flow in basalts. Am J Sci 258:350–366
Willcock M, Mattei M, Hasalova P, Giordano G, Cas RAF, Morrelli C (2014). Flow behaviour in the intra-caldera setting: an AMS study of the large (>1290 km^3) Permian Ora ignimbrite. In: Ort M, Porreca M, J. G (eds) The use of paleomagnetism and rock magnetism to understand volcanic processes. Geol Soc London Spec Publ 396. https://doi.org/10.1144/SP396.3
Wilson L, Head JW (1983) A comparison of volcanic eruption processes on Earth, Moon, Mars, Io and Venus. Nature 302 (5910):663–669

2 Properties of Magmas

Relevance to Eruption Processes and Lava Flow

Guido Giordano, Ray Cas, and John V. Wright

R. Cas et al., *Volcanology*, Springer Textbooks in Earth Sciences, Geography and Environment,
https://doi.org/10.1007/978-3-319-66613-6_2

Plate 2.1 Varying behaviour of erupting basaltic magma during the 1969–1974 eruption of Mauna Ulu shield volcano, Hawai'i, showing lava over-spilling from the crater of Mauna Ulu (top right), into a secondary collapse "trench" and ponding there as a lava lake (bottom left). Lava is flowing in a laminar state over the edge of the lava fall but is in a turbulent state in the plunge pool at the bottom of the falls. (*Source* United States Geological Survey; Photo taken 7-2-1972)

Abstract

In this chapter the properties of magma are reviewed in order to understand how magmas behave as they rise to the Earth's surface, why magmas erupt coherently as lavas or explosively as pyroclastic deposits, and how they behave subsequently. The process of magma production and transport to the Earth's surface is an important mechanism through which the internal heat of the Earth is dissipated, ultimately to space. The dissipation of thermal energy occurs through the transport of heat, mass, and momentum to the surface. The efficiency of this heat and mass transfer depends on the characteristics of magmas, the volumes transferred, and the history of transport and cooling through the lithosphere. This leads to the wide spectrum of eruption styles observed on Earth. Igneous rocks are the products of the cooling of magma, and this can occur at any stage during the ascent of the magma. Usually, only a small proportion of the magma generated at the source will actually reach the surface. In this chapter, we briefly introduce the compositional variability of magmas and their classification, and then consider specific properties that are relevant to their rheological and eruption behaviour. Of particular importance in understanding the rheology of magmas is their viscosity, which is controlled by many variables, including pressure, temperature, chemical composition (especially volatile and silica contents), crystal content, and bubble content which can affect their fluid flow states. Finally, we provide a brief introduction to lava flows and the factors that control their flow or rheological behaviour and morphological forms, prior to their detailed discussion in Part B.

2.1 Introduction

Magmas may erupt as coherent lavas and then flow coherently or fragment during flow (Plate 2.1), or they may erupt explosively to form a range of pyroclastic deposits. At the time of eruption, the volcanic products may range in character from pure magmatic liquid to "solids". Magmas also intrude the lithosphere through which they pass, and large volumes of rising magma will stall and solidify in the crust as sub-volcanic intrusions in all volcanic settings.

The formation of magma and transport to the Earth's surface allows the internal heat of the Earth to be dissipated, ultimately to space. The thermal energy of magma is dissipated through the transfer of heat, mass, and momentum to the surface (Fig. 2.1; e.g. Dobran 1992; Valentine 1992; Spera 2000). The efficiency of this heat and mass transfer depends on the properties of magmas, the volumes transferred, and the history of transport and cooling through the lithosphere. As a result of these variables a wide range of eruption styles is observed on Earth. These vary from effusive, lava flow forming, to extremely explosive. Buoyant explosive eruption columns of gas, ash, and rock debris then transfer heat and mass to the Earth's atmosphere. Eruptions can vary in volume from a few tens of cubic m to thousands of cubic km, and in duration from a few seconds to tens of years during a single event.

The timescale associated with magmatic processes may vary from milliseconds (e.g. the nucleation of a bubble within a rising magma), to seconds (e.g. an explosive blast at Stromboli volcano), to millions of years (e.g. a thermal anomaly associated with a mantle plume; the lifespan of an explosive caldera; the lifespan of a mid-oceanic spreading ridge). The wide range of time scales of processes is matched by the length scale of volcanic processes.

Magma is the carrier of thermal energy due to its high heat capacity (Dingwell 1998; Spera 2000). The rate of heat transfer is mostly related to the thermal conductivity of magma, which relates to its chemical and physical state, and changes therein (e.g. Gottsmann et al. 2004; Potuzak et al. 2008; Whittington et al. 2009). The chemical-physical changes that magma experiences from source to surface depend on its initial composition, the history of ascent, including the timescale of the cooling process, and the effects of progressively decreasing lithostatic pressures, especially on volatile solubility (Fig. 2.2). These changes in turn control the magma volume that rises to different levels in the lithosphere. Other variables controlled by chemical-physical changes include the density and the viscosity, the degree of differentiation and assimilation during ascent, the storage at progressively shallower levels (if any), the vesiculation and fragmentation (if any), the magma rise or ascent rate, the eruption style (effusive *versus* explosive), and finally, the cooling or quenching of erupted materials to surface temperatures.

Igneous rocks are the products of the cooling of magma, and this can occur at any stage during the ascent of the

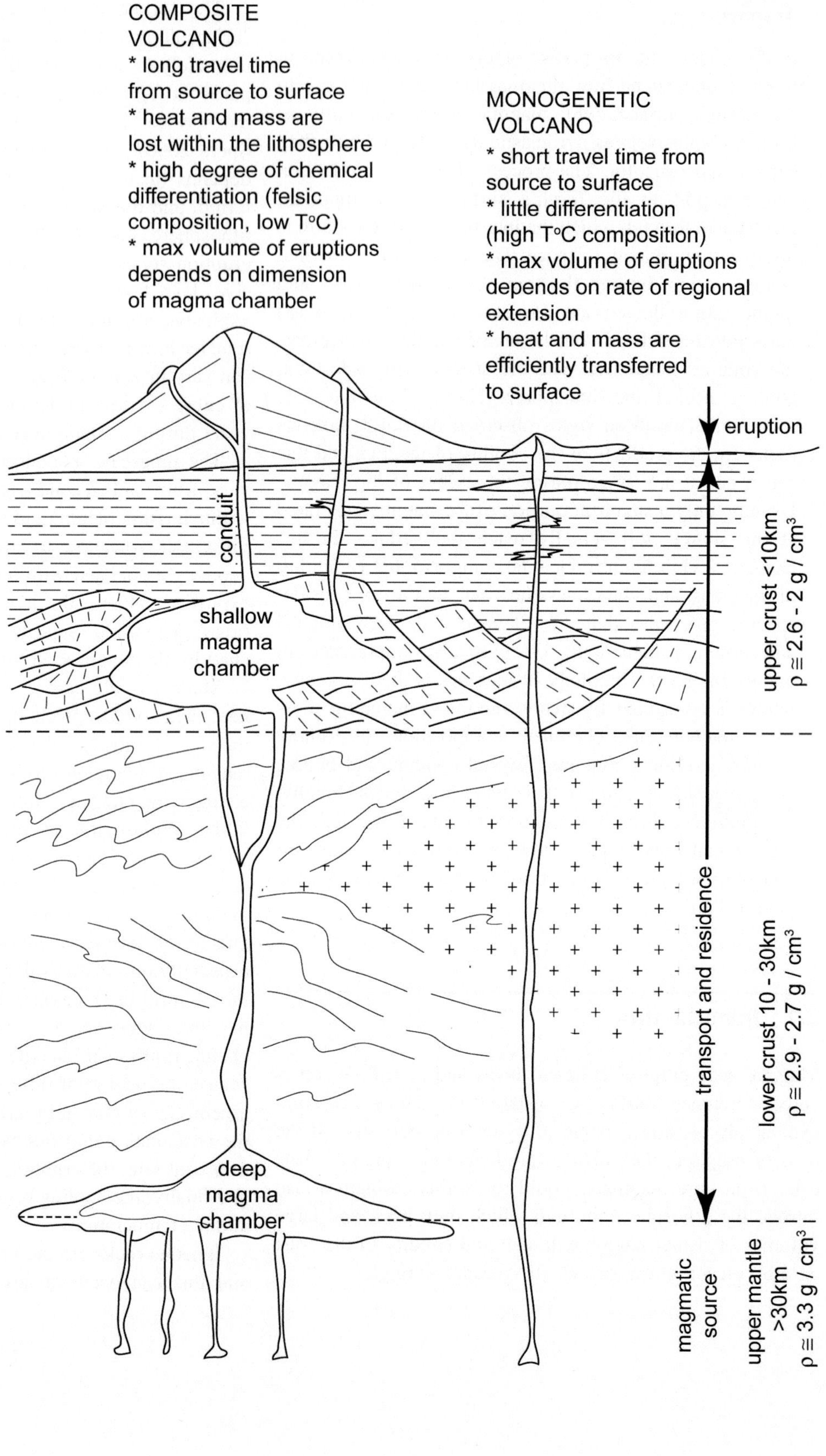

Fig. 2.1 The history and rate of magma ascent from source region to surface through the lithosphere may be very different in different geotectonic settings and these modulate the transfer of heat, mass, and momentum. The chemistry of magma, the type and size of eruptions and of the resulting volcanoes are some of the most significant outcomes of magma ascent processes

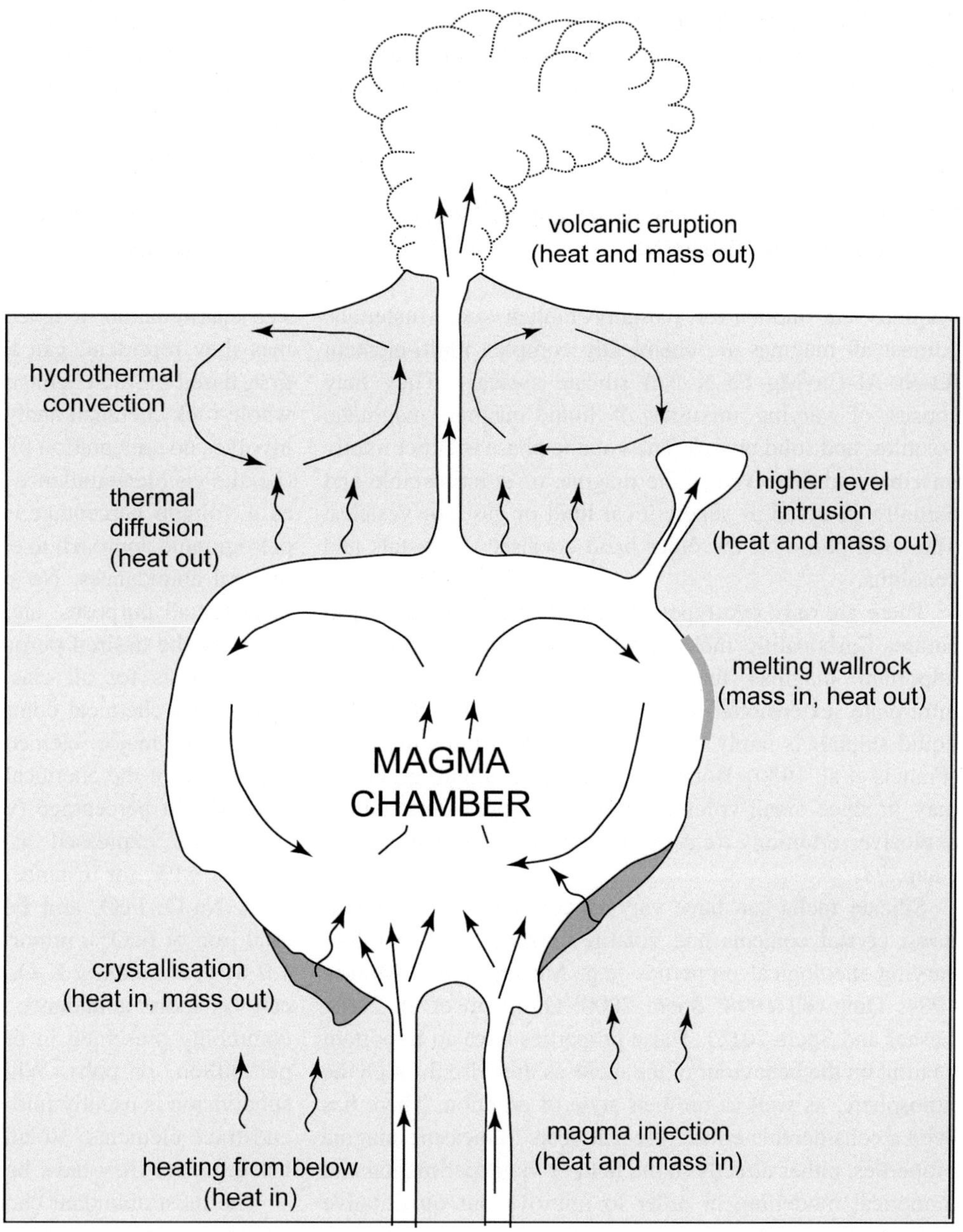

Fig. 2.2 Schematic section of the upper crust region, illustrating how the rates of magma transport *versus* magma storage produce a series of physical and chemical processes that may occur before eruption. These include crystallisation and differentiation of magma, assimilation of wall rocks, intrusions and formation of geothermal and hydrothermal systems. These processes, relayed from the mantle to the lower crust to the upper crust and to the Earth's surface, contribute to heat transfer to the crust and ultimately to space. (From Valentine 1992)

magma. Usually, only a small proportion of the magma generated at the source will actually reach the surface; the bulk is emplaced as intrusions at various levels in the lithosphere. The ratio of erupted to intruded magma volumes is generally assumed to be about 1:9 (Smith 1960; Fisher and Schmincke 1984), but this will vary from case to case (White et al. 2006).

In this chapter we briefly introduce the compositional variability of magmas and their classification, and then consider specific properties that are relevant to their rheological and eruption behaviour. *Rheology* is the study of the deformational behaviour of materials. When magmas flow, they deform. The factors that affect the rheological behaviour of magmas include their composition, temperature (particularly important are the liquidus and the glass transition temperature ranges), density, viscosity, yield strength and the mechanical or tensile strength.

The study of the properties of magmas is one of the most rapidly expanding at the moment. The development of experimental modelling as well as of the numerical modelling is one of the new frontiers in volcanology. Excellent summaries of the properties of magmas that affect their rheology are given by Dingwell (1998), Spera (2000), Cashman (2004); Mysen and Richet (2005), Gonnermann and Manga (2007), Giordano et al. (2008), Mader et al. (2013), and Lesher and Spera (2015).

2.2 Magmas—An Introduction to Their Chemical Diversity

Magmas are molten or partially molten rock materials. Almost all magmas are chemically complex multi-element (O–Si–Al–Ca–Mg–Fe–K–Na) silicate systems. They may consist of varying mixtures of liquid magma, magmatic volatiles, and solid phases. The volatile phase is either totally miscible and dissolved in the magma, or is immiscible and partially exsolved as supercritical fluid or gas into vesicles. The solid phase on the other hand consists of crystals and xenoliths.

There are only two types of non-silicate melts known in nature. Carbonatitic melts (>50% carbonate) are rare but important magmas that are generally associated with intra-plate, extensional settings (see Chap. 16). Secondly, liquid sulphur is rarely found emitting from fumarole vents (Francis et al. 1980). Both carbonatite and sulphur eruptions may produce small-volume lavas, and tephra during mild explosive eruptions (e.g. Delmelle and Bernard 2000; Chap. 7).

Silicate melts can have varying compositions, temperatures, crystal contents and volatile contents, and therefore varying rheological properties (e.g. McBirney and Murase 1984; Dingwell 1998; Spera 2000; Giordano et al. 2008; Lesher and Spera 2015). These properties have an important bearing on the behaviour of the melts as they rise through the lithosphere, as well as on their style of eruption. There has been a considerable effort in recent years to measure magma properties, either directly in the field or via experimental and numerical modelling in order to improve our quantitative understanding of magmatic processes.

Many factors control the compositional and mineralogical characteristics of magma. These include the nature of the subsurface source rocks where melting occurs, the earlier history of that source region in terms of previous thermal, metamorphic and melting events, and the degree of partial melting of the source rocks. They also include the degree of crystallisation in the magma, the extent of segregation or fractionation of crystals from the magma, the amount of contamination by the wall rocks of magma chambers and conduits, and the degree of magma mixing before eruption. Discussion of all of these is beyond the scope of this book, but as a result of these factors, volcanic rocks can have a diversity of chemical and mineralogical compositions, and physical characteristics.

2.2.1 Classification of Magmas and Igneous Rocks

The classification of igneous rocks, and hence of the magmas they represent, can be approached in two ways. The first, based on the chemistry of the magma or rock, requires whole-rock chemical analyses to be undertaken. The second involves documentation of the observable modal mineralogy (i.e. the visible abundance of the minerals present, expressed as a volume percentage, vol%), which requires a field or petrographic approach to estimate, or accurately point-count, mineral abundances. No single scheme can be regarded as ideal for all purposes, and the approach used will be governed by the desired purpose of making the classification.

The basis for all classification schemes based on the whole-rock chemical composition is the relative abundance of various major elements. Conventionally, the relative abundance of the chemical elements is expressed as oxides, on a weight percentage (wt%) basis. Igneous rock compositions are expressed as major elements (i.e. elements with >1 wt%, for instance, SiO_2, Al_2O_3, MgO, FeO, Fe_2O_3, CaO, Na_2O; FeO, and Fe_2O_3 are commonly presented as total iron or FeO*), minor elements (i.e. elements with 0.1–1.0 wt%, commonly K_2O, TiO_2, MnO, P_2O_5, although K_2O can be more abundant), and trace elements (<0.1 wt%, commonly presented in parts per million, or ppm, or parts per billion, or ppb). With improved analytical precision, subdivision is usually into major (including minor elements) and trace elements. Volatile species abundances are rarely listed, unless they have been specifically analysed.

The most abundant chemical component in most silicate magmas and igneous rocks is SiO_2, called silica (not silicon), which can range from <30 to >80 wt% in abundance in different magma types. Volcanic rocks can be initially divided into:

(a) High-silica types (>63 wt% SiO_2), which are called silicic or felsic. These commonly contain visible or modal quartz and feldspars ± biotite ± amphiboles (±pyroxenes), unless they are aphyric or aphanitic. The older term "acidic" is used less commonly because it is strictly incorrect.

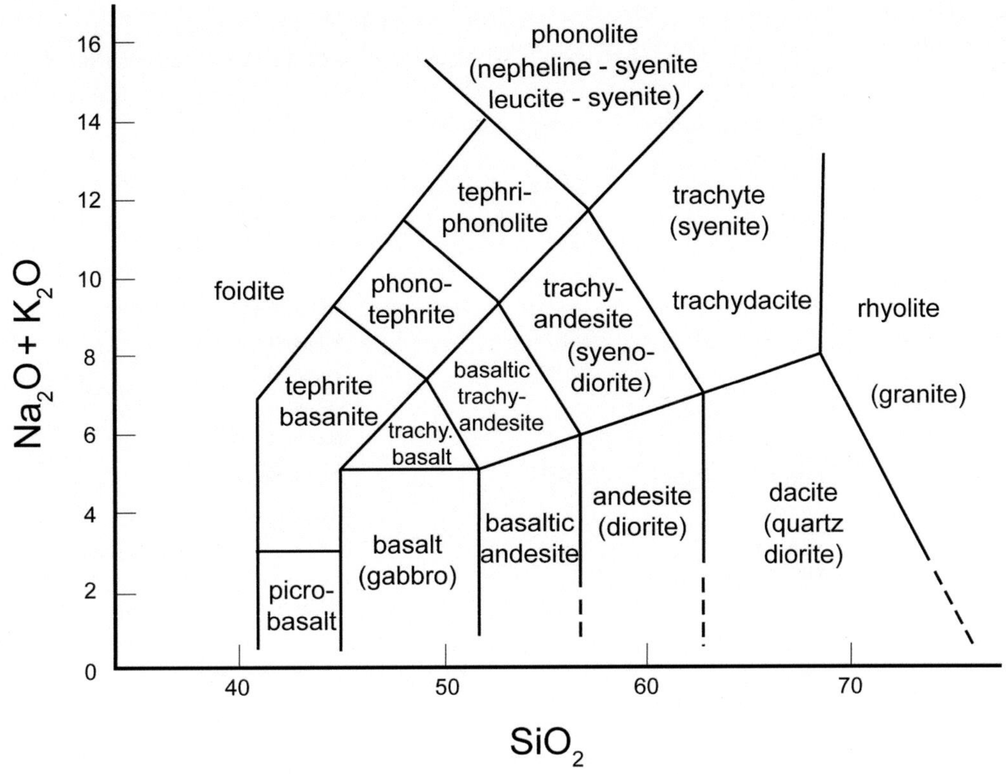

Fig. 2.3 Total Alkali versus Silica (TAS) classification diagram for volcanic and intrusive rocks (in brackets) (After Le Bas et al. 1986)

(b) Intermediate types (52–63 wt% SiO_2), which contain modal felspars, minor quartz if any, ± biotite ± amphiboles ± pyroxenes).
(c) Low-silica types (>45–<52 wt% SiO_2), which are called mafic and lack free quartz crystals, but commonly contain modal plagioclase felspars + pyroxenes + olivine ± amphiboles ± biotite ± feldspathoid minerals. The older term "basic" is decreasingly being used.
(d) Very low silica types or ultramafic rocks (<45 wt% SiO_2), which are dominated by olivine crystals ± pyroxenes. Again, the term "ultrabasic" is going out of favour.

Several rock types occur within each of these silica classes. They can be distinguished on the basis of their variable alkali contents, which generally show positive correlation with silica content. A common, simple classification scheme is the Total Alkali versus Silica (TAS) diagram of le Bas et al. (1986) and Le Maitre et al. (2002; Fig. 2.3). The TAS classification scheme is useful for relatively unaltered, fresh rocks, but is not recommended where regional metamorphism or hydrothermal alteration clearly have had a significant effect on chemical composition, because silicon and alkalis can be mobile under these conditions. In these cases, classification schemes utilising immobile elements (including Al, Ti, P, Zr, Y, and others) and element ratios are recommended (also see Rollinson 1993).

Another scheme based largely on chemistry involves the classification of igneous rocks into peralkaline, alkaline, and sub-alkaline, with further subdivision of the sub-alkaline field into calc-alkaline and tholeiitic (Irvine and Barragar 1971). Furthermore, depending on the variable content in Na_2O versus K_2O, alkaline magmatic rocks can be divided into sodic, potassic, and high-K types (Middlemost 1975). Discriminating different basaltic rocks can be done on the basis of normative mineral calculations (Cross et al. 1903) from the whole-rock chemical composition and then plotting on the basalt tetrahedron of Yoder and Tilley (1962).

Igneous rocks, especially silica-rich to intermediate rocks, can also be subdivided chemically using the relative abundances of molecular Al_2O_3 to total alkalis and calcium (Na_2O + K_2O + CaO), i.e. the degree of aluminium saturation (Cox et al. 1979). The common volcanic rocks are categorised in Table 2.1 using this approach. *Peraluminous*

Table 2.1 A simple chemical classification for the common volcanic rock types using the ratio of Al_2O_3 to total alkalis

	SiO_2 (wt%)	Al_2O_3 saturation classes			
		Peraluminous*	Metaluminous°	Subaluminous§	Peralkaline#
High Silica (Acidic)	>68	Rhyolite			Pantellerite comendite
	63–68	Rhyodacite			
		Dacite			
			Latite	Trachyte	
Intermediate	57–63		Andesite	Phonolite	
	52–57		Mugearite		
			Tholeiitic basalt		
			Hawaiite		
Low Silica (Basic)	45–52		Alkali basalt		
			Basanite		
Very Low Silica (Ultrabasic)	<45			Nephelinite	
				Leucitite	

* Molecular Al_20_3 > ($Na_20 + K_20$ + CaO)
° Molecular Al_20_3 < ($Na_20 + K_20$ + CaO), and Al_20_3 > ($Na_20 + K_20$)
§ Molecular Al_20_3 = ($Na_20 + K_20$)
# Molecular Al_20_3 < ($Na_20 + K_20$)
Note Basaltic rocks cover a wide compositional range and can be further subdivided. For more comprehensive chemical classification schemes see Yoder and Tilley (1962), Green and Ringwood (1967) and Irvine and Baragar (1971)

rocks [Al_2O_3 > ($Na_2O + K_2O$ + CaO)] are always rich in silica and mineralogically are characterised by muscovite and Al-minerals. *Aluminium-saturated rocks* [Al_2O_3 = ($Na_2O + K_2O$ + CaO)] are mainly silica-rich but some can also be intermediate or silica-poor, and mineralogically are characterised by feldspars ± quartz ± biotite ± orthopyroxenes ± cummingtonite ± olivine, i.e. mafic minerals with no Ca because Ca is completely used in plagioclase by combining with Al. *Metaluminous rocks* [Al_2O_3 < ($Na_2O + K_2O$ + CaO), but Al_2O_3 > ($Na_2O + K_2O$)] are characterised by Ca-bearing minerals such as clinopyroxenes and/or hornblende and plagioclase; they are mainly intermediate and silica-poor rocks. *Sub-aluminous rocks* [Al_2O_3 = ($Na_2O + K_2O$)] are characterised by alkali-feldspars, but no plagioclase, and may contain clinopyroxenes ± hornblende; they are intermediate or silica-rich rocks. *Peralkaline rocks* [Al_2O_3 < (Na_2O + K_2O)] contain Na-pyroxenes and Na-amphiboles, and are silica-rich and intermediate rocks.

Variation in silica and alkali contents is reflected in the mineralogy, particularly the quartz content, and the types and abundances of feldspars, feldspathoids, and mafic minerals. Silica-rich and intermediate rocks such as dacites and andesites are dominated by plagioclase feldspars, whereas rhyodacites and latites have sub-equal proportions of potassium feldspars and plagioclase, and rhyolite and trachyte are dominated by potassium feldspars. Pantellerites and comendites are alkali-rich (peralkaline) sodic rhyolites which, compared with more aluminous rhyolites, develop Na-rich feldspars with Na-amphibole or pyroxene.

Rocks with >57% SiO_2 (i.e. silica-rich and more silicic intermediate rocks) are usually silica-oversaturated, and silica-rich rocks in particular will contain free, visible quartz crystals, i.e. modal quartz. In magmas that are undersaturated in SiO_2, and have abundant alkalis, feldspathoid minerals (e.g. analcite, nepheline, leucite; also abbreviated to foids) may occur at the expense of some feldspar. Examples include phonolites, which have potassium-rich or alkali feldspar and minor feldspathoids, basanites with plagioclase, alkali feldspar and feldspathoids like nepheline, and nephelinites and leucitites, which are feldspar-free ultramafic rocks, with abundant nepheline and leucite respectively. Within the range of basalts, tholeiites may be slightly over-saturated, while alkali basalts tend to be slightly under-saturated.

When magma cools and differentiates or fractionates, different major, minor, and trace elements progressively partition into the succession of minerals that form during the

Table 2.2 The most common minerals in igneous rocks and their chemical composition. Not listed are many minor minerals, including Al-minerals, melilite, calcite, Fe–Ti–Cr-oxides, apatite, zirconium, monazite, tourmaline, garnets. Compositions and properties are available in Deer et al. (1992)

Mineral group	Mineral and chemical composition
Quartz	Quartz: SiO_2
Alkali-feldspars	Sanidine, Orthoclase, Microcline: $KAlSi_3O_8$
Plagioclase	Albite: $NaAlSi_3O_8$ to Anorthite $CaAl_2Si_2O_8$ series
Feldspathoids	Nepheline: $NaAlSiO_4$, Leucite: $KAlSi_2O_6$, Kalsilite $KAlSiO_4$
Olivine	Forsterite Mg_2SiO_4 to Fayalite Fe_2SiO_4 series
Orthopyroxenes	Enstatite $Mg_2Si_2O_6$ to Ferrosilite $Fe_2Si_2O_6$ series
Clinopyroxenes	Diopside $CaMgSi_2O_6$ to Hedenbergite $CaFeSi_2O_6$ series
Amphiboles	Hornblende $Ca_2(Mg,Fe)_4Al(Si_7Al)O_{22}(OH,F)_2$
Micas	Biotite $K(Mg,Fe)_3(OH,F)_2(AlSi_3O_{10})$; Muscovite $KAl_2(AlSi_3O_{10})(F,OH)_2$

Table 2.3 CIPW Normative mineral assemblages hypothetically expected for different rocks in relation to the whole rock chemical composition. The normative mineralogy is useful for representing the degree of silica saturation. Normative minerals are often but not always present as crystals in real rocks depending on the cooling history of the magma

Oversaturated/high silica types
Quartz + orthopyroxene + albite + k-feldspar → *quartz disappears* →
Saturated/intermediate types
Orthopyroxene + albite + k-feldspar
Orthopyroxene + olivine + albite + k- feldspar → *orthopyroxene disappears* →
Undersaturated/low silica types
Olivine + albite + k-feldspar
Olivine + albite + k-feldspar + nepheline
Olivine + k-feldspar + nepheline
Olivine + k-feldspar + nepheline + leucite
Olivine + nepheline + leucite
Olivine + nepheline + leucite + kalsilite
Olivine + nepheline + kalsilite

crystallisation sequence, so depleting the residual magma of those elements. Understanding the composition of the minerals in an igneous rock, therefore helps to understand both the chemical dynamics of the magma as it crystallises, and also, as a consequence, the changing physical properties of the crystallising magma. Table 2.2 summarises the chemical compositions of the common igneous rock forming minerals.

In addition to noting the actual visible mineral assemblage present in an igneous rock, the theoretical assemblage of minerals that would form if a magma of known composition were to follow pre-determined theoretical rules for crystallising, can be calculated (see Rollinson 1993), producing a CIPW normative mineral composition (Kelsey 1965). Table 2.3 summarises the predicted normative mineral assemblages for magmas of different silica saturation levels. In practice, crystallisation processes in magmas and the minerals that form are controlled by many variables, including temperature, cooling rate, pressure, rate of pressure change, volatile types, and contents. Igneous rocks therefore may or may not contain the calculated normative mineral assemblages. The actual visible minerals in an igneous rock and their abundance is referred to as the modal mineral composition.

For studies based on field observations of volcanic rocks, an entirely chemical approach to classification is not practical. It also ignores useful mineralogical and textural information. A chemical scheme is also of limited value where rocks have undergone alteration due to hydrothermal or fumarolic activity, weathering or, in many ancient terrains, regional metamorphism. In many cases, a more tangible means of classifying rocks based on mineralogy is therefore required. Even in chemically altered rocks, the primary mineralogy and textures can still be identified in many cases.

Mineral types and abundances not only reflect magma chemistry, but their textural relations give additional information about the eruptive and cooling history of volcanic rocks. Useful atlases of igneous rocks and their textures include MacKenzie et al. (1982), and detailed interpretations of textures can be found in Cox et al. (1979), Winter (2010), Philpotts and Ague (2009), McPhie et al. (1993a, b) and in this book.

Rock names have traditionally been given according to mineral content, and many classification schemes have evolved. The most common and standard scheme based on mineralogy is that presented by Streckeisen (1979) and Le Maitre et al. (2002), and endorsed by the International Union of Geological Sciences (IUGS), shown in Fig. 2.4a. Here, volcanic rock types are allocated fields on the QAPF diagram and are classified according to the relative modal abundances of their felsic minerals, ignoring groundmass abundance. This presents problems in distinguishing the members of the mafic–ultramafic rock spectrum. These are thus distinguished by the abundance of mafic minerals. An equivalent QAPF classification for corresponding plutonic rocks is shown in Fig. 2.4b (Streckeisen 1976; Le Maitre et al. 2002).

For many volcanic rocks modal mineral contents cannot be accurately determined because of the microcrystalline or glassy texture of the groundmass. Where phenocrysts are the only recognisable minerals, and if rock names are based on these alone, then the prefix 'porphyritic' should be added, or the suffix 'phyric' can be added after listing the phenocryst mineral types. Thus, a rock containing plagioclase and quartz phenocrysts in an undetermined groundmass would be a 'porphyritic dacite', or a 'plagioclase-quartz phyric dacite'. Where plagioclase is the sole felsic phenocryst, the rock is a 'porphyritic andesite' or 'porphyritic basalt', even though a complete modal or chemical analysis could show it to be a dacite. On the other hand, rocks with abundant plagioclase together with clinopyroxene phenocrysts may either be andesite or tholeiitic basalt, depending on the total proportion of mafic minerals present. Some very glassy rocks (80–100% glass) are given special names such as obsidian, for glass of rhyolitic or dacitic composition. Rocks without visible phenocrysts are called 'aphyric', whereas those in which all crystals are too small to identify are called 'aphanitic'.

In ancient volcanic terrains metamorphosed or metasomatised volcanic rocks in which feldspars are sericitised or albitised and mafic minerals replaced by chlorite, epidote, serpentine or talc, etc., may be given the prefix 'meta-' whenever original textures and mineralogy can be determined. Some special names, which have been used in the past for rocks in ancient terrains, include spilite (albite-chlorite rock) for a meta-basalt, and keratophyre and quartz keratophyre (albite, quartz and minor chlorite, epidote and iron oxides) for meta-andesite or dacite, respectively. However, the rock names in Figs. 2.3 and 2.4 should be retained where possible, regardless of alteration state or geological age.

2.2.2 Magmatic Associations

Magmas form as a result of partial melting of a mantle or crustal source. Few magmas, however, reach the Earth's surface immediately after formation without some degree of modification to the original starting melt composition. Most rise slowly enough so that they interact with surrounding country rock and become variably contaminated compositionally by some degree of assimilation of the country rock, or by temporarily stopping, they undergo some degree of differentiation and/or crystal fractionation. Magmas that originate from the mantle, and rise rapidly without becoming contaminated or without differentiating, are called primitive magmas and share a close chemistry with the source rocks. Mantle rocks have a relatively narrow range of compositions, depending on the tectonic settings and on the history of the mantle at that particular location, and produce silica-poor (mafic) and very silica-poor (ultramafic) melts, which differ in their initial silica, alkali, and volatile content. According to these initial differences, it is the nature of magma ascent that defines how many chemical and physical transformations will occur prior to eruption. The slower the ascent, the more extensive will be the chemical and physical processes associated with the cooling of the magma and interaction with the country rocks and fluids. The chemical composition of magma therefore can be regarded as the first-order indicator of the efficiency of mass and heat transport from source to surface. Magma with a silica-poor composition and a low viscosity can rise quickly to the Earth's surface, and can transfer mass and heat efficiently to the surface, whereas a magma with silica-rich composition and high viscosity is likely to have lost a large part of its heat and mass during intrusion on its way through the crust because it rises slowly (Fig. 2.1).

Igneous rocks can show large chemical differences, but can still be grouped into various compositionally related 'associations', 'series', or 'suites' based on petrological and chemical characteristics that relate to a common parental magma. Examples of magmatic associations include tholeiitic, alkaline, and calc-alkaline associations. These may show preferred distributions related to specific volcano-tectonic settings. For many years petrologists and geochemists have attempted to relate the petrogenesis of modern and ancient volcanic rocks to their tectonic setting using, in particular, their trace element and isotope chemistry. For a detailed consideration of the petrological characteristics of volcanic rocks, see igneous petrology texts such as Rollinson (1993), Wilson (1989), Best (2009), Philpotts and Ague (2009), and Winter (2010).

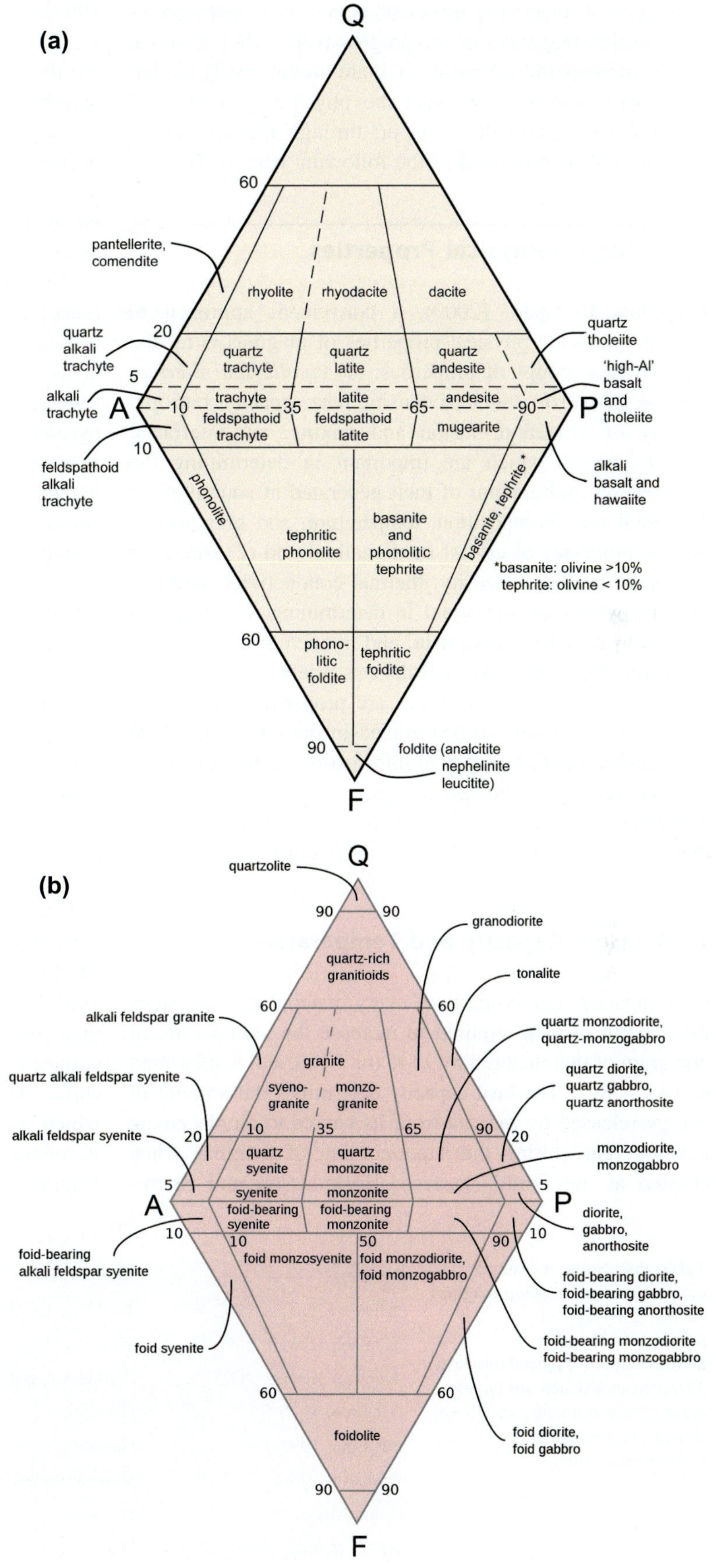

Fig. 2.4 Classification, names and modal mineral (volume percentage) compositions of volcanic and plutonic rocks recommended by the IUGS Sub-commission on the Systematics of Igneous Rocks. (Modified from Streckeisen 1979). **a** Classification for volcanic rocks (From Cas and Wright 1987). Most mafic volcanic rocks plot along the P and F sides of the QAPF diagram. **b** Equivalent classification for plutonic rocks (*Source* Wikimedia Commons through Creative Commons CC0 licence). Minerals at the corners of the QAPF diagram are Q = silica minerals (usually quartz); A = alkali feldspar (including albite); P = plagiociase; F = feldspathoids (e.g. analcite, nepheline, leucite; also called foid). Rock names are determined by ignoring mafic mineral contents and recalculating quartz, plagioclase, alkali feldspar and feldspathoid contents to 100%. The sum of mafic mineral modes (M) is used to distinguish basalt (M = 35–90) from andesite (M = 0–35), and basanite (>10% olivine) from tephrite (<10% olivine). Mafic minerals include olivine, pyroxenes, amphiboles, and micas

In general, magmatic associations consist of petrogenetically related magmas that are progressively richer in silica, alkalis, incompatible elements, and are progressively cooler. These variations greatly affect the physical properties of magmas that control the transport through the lithosphere and that will be discussed in the following paragraphs.

2.3 Magma Physical Properties

According to Spera (2000), a convenient approach to understanding the physical properties of magmas is to distinguish two groups of properties: (1) *equilibrium thermodynamic properties*, such as density, heat capacity, entropy, entropy of formation, fusion and mixing, and interfacial surface energy, which are important in determining the composition and amount of melt generated at source given the initial rock composition, temperature, and pressure, as well as processes of crystal fractionation; and (2) *transport properties* such as viscosity, thermal conductivity, and diffusivity, which are important in determining the ability of magma to transport mass, heat, and momentum.

Of the first group, we will discuss briefly heat capacity, temperature, and density, which are properties that can be measured in the field or determined in the laboratory from rocks and modelling, and give information on the history of magma ascent. In the second group, we will discuss in particular factors influencing the viscosity of magma and its rheology, and ultimately the style of an eruption.

2.3.1 Heat Capacity and Temperature

Heat capacity is a property of every material that measures the amount of heat required to increase the temperature of one gram of that material by 1° K (or 1 °C), and is measured in J kg^{-1} K^{-1}. The heat capacity determines the amount of energy released by magma from its source to emplacement location (e.g. during the super-cooling of magmas when erupted at the Earth's surface; Höskuldsson and Sparks 1997). Average values for heat capacity vary between 1.2×10^3 J kg^{-1} K^{-1} for basalt and 1.04×10^3 J kg^{-1} K^{-1} for rhyolites. However, the heat capacity is not a constant and depends on temperature. During volcanic processes, the heat capacity of magmas therefore varies depending on the heating and cooling rates, defining the temperature at which magma crosses the glass transition temperature (e.g. Gottsmann et al. 2004). Measurement of the heat capacity requires laboratory calorimetric studies.

Temperature is an easier property to measure and is related to the energy of a system. Magma temperatures may be determined in a variety of ways. Direct measurements on lavas may be made, either using a thermocouple probe inserted into a lava flow or lake, or by means of an optical pyrometer, which is especially useful for measuring the temperature of lava fountains and open channel lavas. During the last few years, the use of remote sensing thermal cameras, including satellite, airborne and hand-based, has greatly improved the ability to measure the temperature of lavas and explosive eruption columns (e.g. Rothery and Pieri 1993; McGuire et al. 1995; Calvari et al. 2005; Harris et al. 2005; Bombrun et al. 2018).

A large amount of direct temperature measurement information is available for Hawaiian lavas, and some of this is summarised by MacDonald (1972) and Hon et al. (1994). Hawaiian tholeiitic basalts approach the surface at temperatures between about 1050 and 1200 °C (Table 2.4). For silicic magmas, there are fewer data available because effusive eruptions of such compositions have not been frequently observed. An optically determined temperature of 725 °C for the 1940 dacite dome of Santa Maria (MacDonald 1972) and a thermocouple measurement of the Mt. St. Helens 1980 dacite dome of 850 °C (Friedman et al. 1981) indicate substantially lower eruption temperatures than for basaltic lavas (Table 2.4). Thermal imagery of the surface of the 2008–9 Chaitén rhyolite lava domes indicate a surface temperature of up to 400 °C (Bernstein et al. 2013), which is clearly well below the eruption temperature. A summary of the typical eruption temperatures of the major magma types is given in Table 2.5.

Table 2.4 Some measured temperatures of erupting magmas measured by optical pyrometer (*) and thermocouple (°) placed into lava. Those from Kilauea are largely optical measurements, and those from Etna are largely thermocouple ones

Volcano	Composition	Temperature (°C)	Source
Kilauea, Hawaii, 1952–63*°	Tholeiitic basalt	1050–1190	MacDonald (1972)
Kilauea, Hawaii, 1990	Tholeiitic basalt	1070–1142	Hon et al. (1994)
Kilauea, Hawaii, 2005	Tholeiitic basalt	1155	www.hvo.wr.usgs.gov
Mt Etna, 1970–75*°	Hawaiite	1050–1125	Archambault and Tanguy (1976)
Mt Etna, 1991–93°	Hawaiite	1060–1080	Barberi et al. (1993)
Paricutin, 1944*	Basaltic andesite	943–1057	Bullard (1947)
Santa Maria, 1940*	Dacite	725	MacDonald (1972)
Mt St Helens, 1980°	Dacite	850	Friedman et al. (1981)

Table 2.5 Summary of estimates of typical eruption temperatures for the major volcanic rock types

Rock type	Temperature (°C)
Rhyolite	700–900
Dacite	800–1100
Andesite	950–1200
Basalt	1000–1200
Komatiite	1200–1650

Laboratory experiments may also be used to estimate magma temperatures. Almost all volcanic rocks contain crystals in varying amounts. Therefore, by experimentally determining the liquidus temperature of a magma (the temperature above which no crystals exist, or the temperature below which crystallisation begins in a slowly cooling magma) and the solidus temperature (the temperature below which there is no liquid phase, or the temperature at which melting begins) the maximum and minimum temperature limits for the existence of silicate magma of varying compositions can be determined. However, magma can solidify to rock at temperatures between the solidus and liquidus (the solidification temperature) if cooled rapidly, producing a rock with suspended phenocrysts in a groundmass of microlites and chilled glass (Vona and Romano 2013). Another uncertainty is introduced by H_2O content, which can strongly affect the liquidus and solidus temperatures. For rocks with only a few suspended phenocrysts, the liquidus temperature may be a good approximation of the eruption temperature, whereas for rocks with no suspended crystals it represents the minimum eruption temperature.

Magma crystallisation temperatures can be estimated by mineral geothermometry, which makes use of temperature-dependent compositional relationships between co-existing minerals that are considered to be in mutual equilibrium. Frequently used mineral pairs considered to be good geothermometers and geobarometers include magnetite-ilmenite (e.g. Andersen and Lindsley 1988; Ghiorso and Sack 1991), olivine-ilmenite (Frost et al. 1988) and two pyroxenes (e.g. Davis and Boyd 1966; Wood and Banno 1973; Brey and Köhler 1990; Nimis and Grütter 2010; Liang et al. 2013). A very useful numerical approach to assess the equilibria between these phases is found in QUILF (Andersen et al. 1993; Xirouchakis et al. 2001). More recently developed geothermometry and geobarometry techniques include those based on the glass (or liquid; Helz and Thornber 1987; Montierth et al. 1995; Yang et al. 1999), or equilibria of plagioclase-liquid (e.g. Kudo and Weill 1970; Putirka 2008), two feldspars (e.g. Barth 1951; Stormer 1975; Putirka 2008), olivine-glass (e.g. Häkli and Wright 1967; Beattie 1993; Sisson and Grove 1993; Putirka et al. 2007), clinopyroxene (Nimis 1995; Nimis and Taylor 2000), clinopyroxene-liquid (Putirka et al. 1996, 2003), orthopyroxene-liquid (Wood 1974; Beattie 1993; Putirka 2008), olivine-augite (Häkli and Wright 1967; Loucks 1996), and zircon-rutile (Watson et al. 2006). In addition, single mineral and element in single mineral studies, such as Ti and Zr in rutile (e.g. Zack et al. 2004; Watson et al. 2006; Hayden and Watson 2007; Tomkins et al. 2007; see also Meinhold 2010), Al in olivine and spinel (Wan et al. 2008), Ti in zircon (e.g. Watson et al. 2006; Ferry and Watson 2007; Reid et al. 2011), and Ti in quartz (Wark and Watson 2006; Hayden and Watson 2007). An excellent overview of the different geothermometers and geobarometers that are applicable to volcanic systems is given by Putirka (2008) and to granitic systems by Anderson et al. (2008). Recently, machine learning approaches have also been developed to fully exploit the increasing number of available datasets (Petrelli et al. 2020).

However, laboratory estimates do not necessarily reflect emplacement temperatures. For example, lava flows may have been erupted at higher temperatures than those indicated by single mineral or mineral pair data, which may have crystallised after eruption, or conversely, minerals that formed in the magma in the subsurface source region, magma chamber or conduit may give higher temperatures than the magma eruption temperature. In addition, many magmas when erupted are supercooled through the glass transition temperature, T_g, which is about two thirds of the magma liquidus temperature, to volcanic glass. The nature of volcanic glass and how it forms are discussed in Sect. 4.4.2 in Chap. 4. Also, the final emplacement temperature of hot pyroclastic flows may be governed more by the extent of mixing with air and country rock than by initial magma temperature.

Fe–Ti oxide geothermometers applied to the pumice erupted at Mt Pinatubo (Philippines) in 1991 have indicated a temperature of 780 °C for the dacitic magma in the magma chamber localised between 7 and 11 km at depth (Rutherford and Devine 1996). Pyroxene geothermometry has given a temperature of ca. 860 °C, at 5 km depth for the andesite magma that erupted between 1995 and 1999 at Soufrière Hills volcano, Monserrat (Murphy et al. 2000). Pressure–temperature experiments on crushed pumice from the 2008–2009 Chaitén rhyolite eruption in Chile, to simulate the conditions of crystallisation of plagioclase and orthopyroxene indicate magma temperatures of about 800 °C at depths of 5 km, which is supported by Fe-Ti oxide data (Castro and Dingwell 2009). However, high inferred magma rise rates of 0.5 m s^{-1}, suggest eruption temperature was probably not much lower.

2.3.2 Density

Magma moves up through the lithosphere and ultimately erupts because of its buoyancy, controlled by the density contrast between the magma and the country rocks. Magma density is defined as

$$\rho = \sum X_i M_i / \sum X_i V_i \tag{2.1}$$

where X is the mole fraction of the ith component of the magma, M its molecular weight and V its partial molar volume. More generally magma density can also be defined as $\rho = m/V$, where m is the bulk mass and V is the bulk volume, both of which can vary as the temperature and pressure change during magma ascent and eruption. The magma density depends on the chemical composition, the temperature of the magma, and the pressure conditions in the host rock (Fig. 2.5); it can be determined using the Equation of State (see Dingwell 1998; Spera 2000), by approximating equilibrium conditions. Given that the process of magma ascent implies changes in confining pressure, and, depending on the history of the ascent, changes in chemical composition and temperature, as well as the changing degree of vesiculation, the density of magma changes from source to surface (See Chap. 3, Sect. 3.2.1, Fig. 3.2a, b). This will thus affect the efficiency of the transfer of momentum. Densities of silicate melts can also be determined experimentally from dilatometric measures (Dingwell 1998, and references therein).

Magma densities are markedly different for different magma compositions. At volcanic conditions ($P_{atm} = 10^{-4}$ GPa; T_{volc} = 1650–700 °C) densities range from ~2900 kg m^{-3} (komatiite) to ~2350 kg m^{-3} (rhyolite; Fig. 2.5; Spera 2000; Lesher and Spera 2015). Iacovino and Till (2019) have developed a program, Density X, which allows calculation of density of both dry and hydrous magma, using variables of pressure, temperature and 10 major element oxide abundances (wt%), including H_2O. The effect of density on the fluid dynamical behaviour of magmas can be an important petrological variable affecting the chemical and mineralogical characteristics of magma before eruption. For example, the density of magmas during crystal fractionation controls the settling behaviour of crystals and thus, in turn, the chemical zonation of magma chambers.

Mixing of dense magmas with less dense magmas has been modelled to be a major cause of magma chamber dynamics (see also McBirney 1980; Huppert and Sparks 1984; Folch and Martí 1998; Perugini et al. 2002; Troll et al. 2004; Morgavi et al. 2013). Examples include mixing ultrabasic magmas with less dense, more fractionated basaltic magmas in mid-ocean ridge magma chambers (e.g. Huppert and Sparks 1980, 1981), and of dense, basic magma with less dense, more silicic magma within high-level chambers typical of stratovolcanoes (e.g. Sparks et al. 1977; Huppert et al. 1982; Turner et al. 1983; Blake and Ivey 1986a, b; Pallister et al. 1992; Wolf and Eichelberger 1997; Piochi et al. 2009; Morgavi et al. 2016).

The variation in density of a volume of magma during decompression may be regarded as the effect of the volume stress drop applied to the magma body (Fig. 2.5b). The decrease in density is, therefore, one way of measuring the volume strain associated with decompression (Dingwell 1998).

However, the largest change in magma density occurs when volatiles start exsolving from the magma (e.g. Sparks 1978; Fowler and Spera 2010). Transition of magma from the liquid phase, or liquid + crystal phases, to the liquid (+crystals) + gas phases causes a dramatic reduction of the bulk density, as the volume fraction of bubbles increases during decompression. Nucleation of bubbles is the major factor controlling the increase in buoyancy that controls the rise rate of magma and the style of explosive eruptions (e.g. Cashman et al. 2000), and will be discussed in Chap. 4. Advances in density determinations for silicic melts can be found in Clark et al. (1987), Lange (1997), and Ochs and Lange (1999).

2.3.3 Viscosity and Yield Strength

In lay terms, viscosity is a measure of the fluidity of a substance. Scientifically, it is a measure of the internal resistance to flow in a substance when shear stress is applied. In pure fluids, this resistance to flow is caused by molecular or ionic cohesion. In multi-component and multi-phase natural melts or magmas, viscosity is controlled by composition, volatile content, and temperature, and it is complicated by the presence of solids (crystals) and gas bubbles (McBirney 1980; Dingwell 1998; Manga et al. 1998; Cashman et al. 2000; Saar et al. 2001). Furthermore, the processes of magma ascent and pressure release, crystallisation, cooling, and degassing cause the viscosity of all magmas to change during their history.

Whereas the volume increase affecting magma as it rises and decompresses causes its density to decrease, the shear stresses applied during magma flowage in conduits and on the Earth's surface result in a large spectrum of boundary processes that affect the magma ascent dynamics, the rate of vesiculation, and processes that may eventually lead to

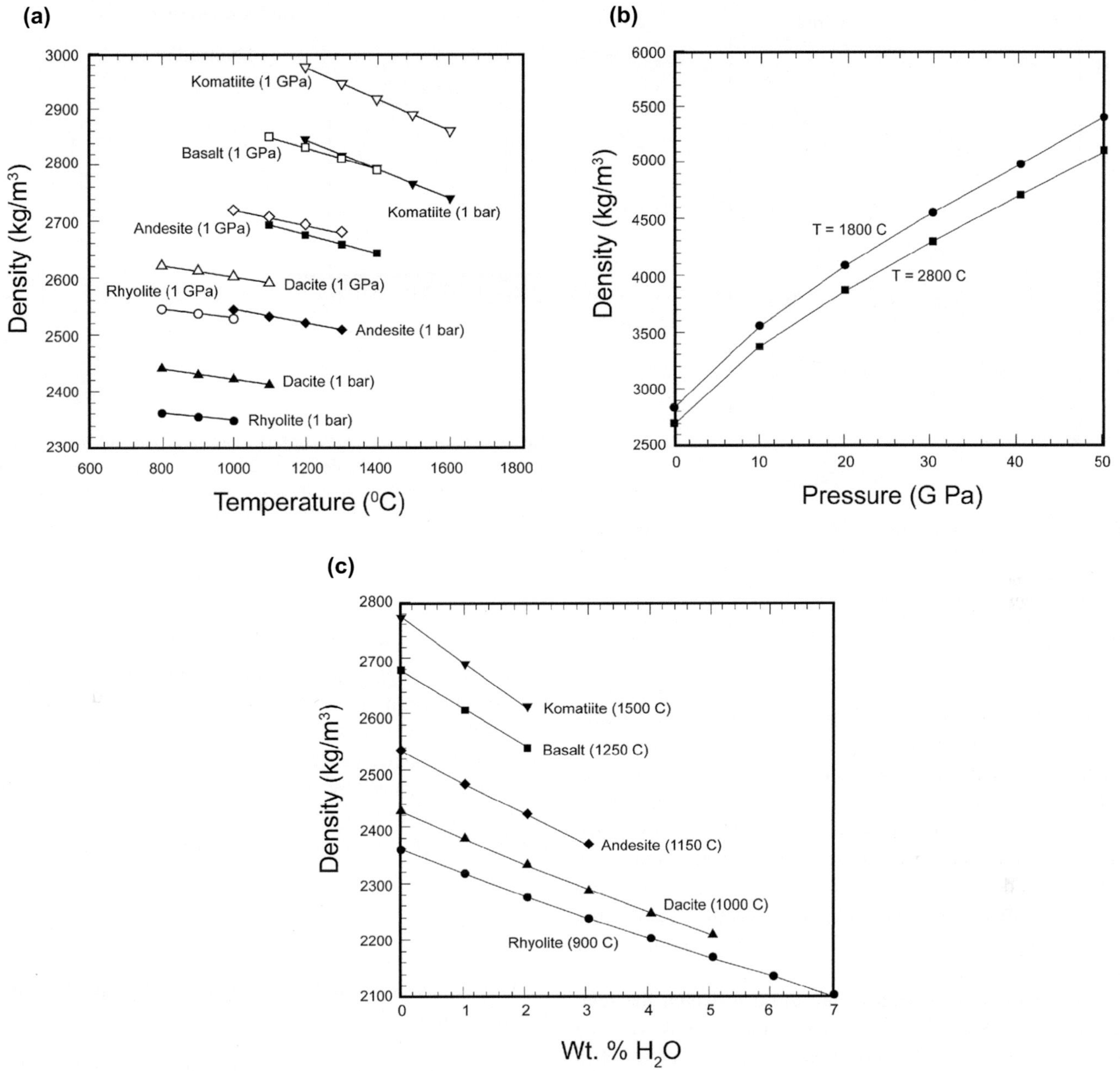

Fig. 2.5 Densities of common magmas or different compositions with varying **a** temperature and pressure (1 bar = 0.1 MPa—filled symbols; 1 GPa—open symbols), **b** pressure for model basalt at different temperatures, and **c** water content at given temperatures (Adapted from Spera 2000)

fragmentation (Chap. 4; see also Dingwell 1998; Papale et al. 1998; Cashman et al. 2000; Gonnerman and Manga 2007), and the mobility and form of erupted lavas (Sect. 2.6; Griffiths 2000). The viscosity of magma is the property that modulates momentum transfer (Spera 2000). These applications will be discussed further in subsequent chapters. The relevance of viscosity to fluid flow states is discussed in Sect. 2.5.

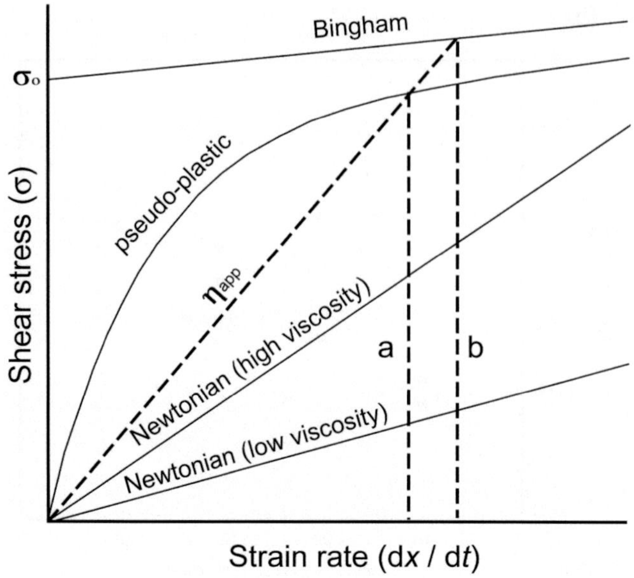

Fig. 2.6 Shear stress–strain rate relationships for Bingham, pseudo-plastic, and two Newtonian substances. Apparent viscosity (η_{app}) is the viscosity of the pseudo-plastic substance at strain rate a, and of the Bingham substance at strain rate b; σ_o is the yield strength of the Bingham substance. Stress and strain rate are explained in text. (From Cas and Wright 1987, After Wolff and Wright 1981a, b)

To define viscosity quantitatively, we first need to consider the rheology of fluids, i.e. how they flow. Some fluids, such as air and water, will flow (i.e. deform) when an infinitesimally low shear stress, such as gravitational force, is applied; these are called Newtonian fluids. For Newtonian fluids, the relationship between the shear stress and strain rate (rate of deformation) is linear (Fig. 2.6). Only some very fluidal lavas with low concentrations of crystals behave in a Newtonian fashion, such as high-temperature carbonatites or Archean komatiites.

Non-Newtonian substances are those in which the relationship between shear stress and strain rate is variable (Fig. 2.6; called *pseudo-plastic* substances), or in which a *yield strength* must be exceeded, after which the relationship between shear stress and strain rate may be linear or non-linear (Fig. 2.6). Substances for which an initial yield strength must be exceeded, and after which the relationship between shear stress and strain rate is linear, are called *Bingham* substances (Fig. 2.6).

The viscosity of a substance can be quantitatively defined as the ratio of the shear stress to the rate of strain. For a foam cube, for example (Fig. 2.7), if shear stress, σ, is applied, the viscosity of the foam, η, is given by

$$\eta = \frac{\sigma}{(d\phi/dt)} \tag{2.2}$$

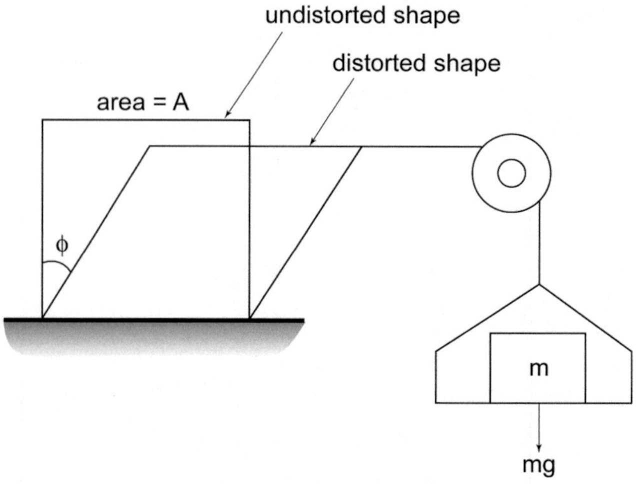

Fig. 2.7 Distortion of a cube of foam rubber by an applied stress, *mg*. Strain is a measure of the degree of deformation, measured here as the angle ϕ. The rate of strain is the rate at which the foam cube deforms, and can be represented by $d\phi/dt$. (From Cas and Wright 1987, After Allen 1970)

in which $d\phi/dt$ is the rate at which the angle between the sides of the cube change as it is deformed. The unit of viscosity is the Pascal second or Pa s (= 10 dyn s cm^{-2} = 10 poise).

For fluids, it is not practical to measure $d\phi/dt$. A more practical parameter to measure is the vertical velocity gradient, du/dy, induced by applying shear stress, σ_σ (= F/A), to the upper plate of two plates of known equal area (A), which confine a fluid (Fig. 2.8). The viscosity of the fluid can be written as

$$\eta = \frac{\sigma_\sigma}{(du/dy)} \tag{2.3}$$

This equation is valid for pure Newtonian fluids. Equation 2.2 can also be written as

$$\sigma_{\sigma tot} = \sigma_o + \eta\,(du/dy)^n \tag{2.4}$$

where $\sigma_{\sigma\ tot}$ is the total shear stress applied and σ_o is the stress required to initiate flow. For pure Newtonian substances $\sigma_o = 0$ and $n = 1$ (i.e. Eq. 2.2); for pseudo-plastic materials $\sigma_o = 0$ and $n < 1$; and for Bingham substances σ_o (= yield strength) has a finite value and $n = 1$. Shear stress and the yield strength of the material are usually expressed in Newtons per square metre (1 N m^{-2} = 10^{-1} dyn cm^{-2}); strain rate is expressed in reciprocal seconds (s^{-1}), since strain itself is dimensionless.

There have been few attempts to estimate the viscosities of magmas in the field. MacDonald (1972) summarised a

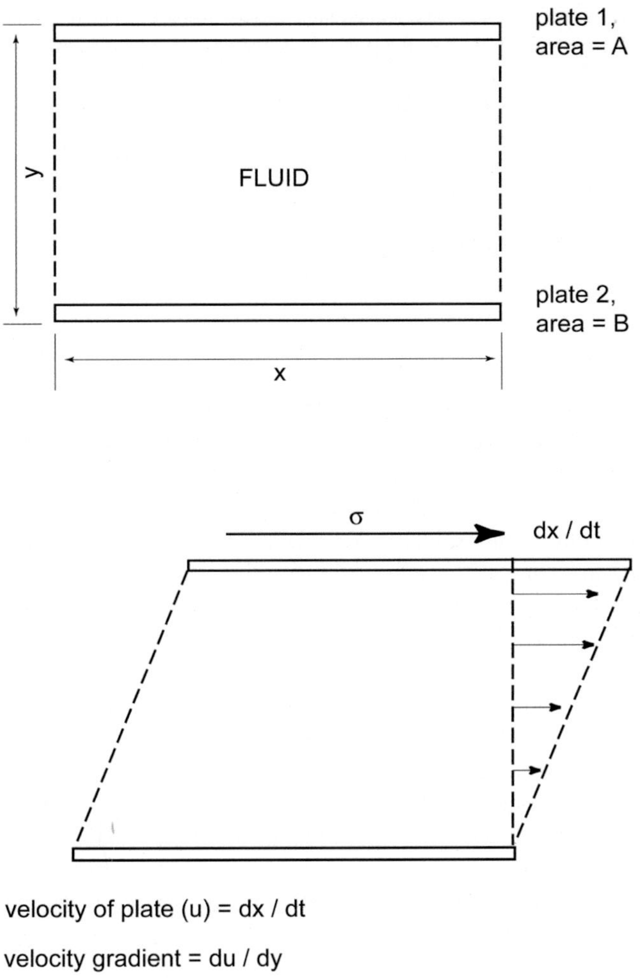

Fig. 2.8 Diagrammatic representation of parameters used to define viscosity in a liquid. Application of shear stress, σ, to the upper plate confining a liquid induces a velocity gradient du/dy. (From Cas and Wright 1987)

number of measurements from lava flows. However, nearly all of these assume Newtonian rheology. The viscosities of lavas assumed to have Newtonian rheology have been determined using *Jeffrey's equation* (e.g. Kilburn 1996; Keszthelyi and Self 1998; Cas et al. 1999):

$$\eta = \frac{g\,\rho \sin\alpha\, d^2}{n\,v} \tag{2.5}$$

where the viscosity, η, can be empirically calculated by knowing the density, ρ, of the lava, the slope angle of the terrain, α, the velocity of flow, v, and the thickness of flow, d. g is the acceleration due to gravity and n is a constant; for broad flows $n = 3$, and for narrow flows $n = 4$.

Field and laboratory measurements have indicated that at sub-liquidus temperatures and at either very low or very high strain rates, lavas and common igneous melts generally have non-Newtonian rheologies (Robson 1967; Shaw et al. 1968; Shaw 1969; Murase and McBirney 1973; Pinkerton and Sparks 1978; McBirney and Noyes 1979; McBirney and Murase 1984; Lejeune and Richet 1995; Dingwell 1998; Lleewellin and Manga 2005; Giordano et al. 2008). This behaviour is due to the presence of dispersed crystals and gas bubbles, ionic and molecular forces, and the visco-elastic nature of magmas. At above-liquidus (supra-liquidus) temperatures Newtonian rheology is applicable.

By assuming Newtonian behaviour, lower apparent viscosities (Fig. 2.6) are calculated by Jeffrey's equation for faster-moving flows. For non-Newtonian lava, which is say pseudo-plastic, its apparent viscosity decreases with increasing strain rate (Fig. 2.6). Hence, a fast-moving lava flow, for example due to flow over a steep slope, will appear to be less viscous or more fluid than when moving more slowly.

The most accurate published field determinations of lava viscosity are given in Table 2.6. Shaw et al. (1968) used a rotating shear vane to measure the viscosity in the tholeiite of the 1968 Makaopuhi lava lake, Hawaii. Although Pinkerton and Sparks (1978) used a variety of methods to measure the rheological properties of small lava flows that erupted on Mt Etna in 1975, the results in Table 2.6 are only from a specially developed penetrometer. Results from both Makaopuhi lava lake on Hawaii and Etna indicated that the lavas behaved in a pseudo-plastic manner, but could be approximated closely to a Bingham model with a definite yield strength (Table 2.6). Bingham or plastic viscosities of the lavas have been determined (Eq. 2.4), and these are also shown in Table 2.6. There are no field measurements of viscosities for more felsic or silicic lavas, which are considered to be Bingham in nature.

In a significant advance in understanding the rheology of lava flows, Castruccio et al. (2014) modified the approach of Hulme (1974) who modelled lavas as Bingham substances, by treating them as viscoplastic Herschel–Bulkley materials. Herschel–Bulkley rheology takes into account downflow changes in lava properties, such as viscosity, which is reflected in changing yield strength, which can be significant in some compositional lava types. Changing viscosity due to shear-thinning behaviour during flow is also considered.

For shear-thinning or shear-thickening non-Newtonian rheologies (Herschel–Bulkley fluids), Castruccio et al. (2014) proposed a further modification of the Jeffreys equation:

$$v = \left(\frac{d^2 \rho\, g\, sin\alpha}{3K}\right)\left[\left(\frac{3n}{d^3 n+1}\right)\left(\frac{\rho\, g\, sin\alpha}{K}\right)^{\left(\frac{1-n}{n}\right)}\right] \times \left[d^{\left(\frac{2n+1}{n}\right)}\left(1-\frac{n}{2n+1}\right)\right] \tag{2.6}$$

Table 2.6 Results of field measurements of physical properties of basaltic lavas. Data from the Makaopuhi lava lake are from Shaw et al. (1968), from Etna 1975 are from Pinkerton and Sparks (1978), and from Etna 1991–93 are from Barberi et al. (1993). Note that viscosity* in this last case was calculated with the method of Shaw (1972)

	Makaopuhi lava lake, Hawaii, 1968	Etna, 1975	Etna, 1991–93
Composition	Tholeiitic basalt	Hawaiite	Hawaiite
Temperature (°C)	1130–1135	1086	1060–1080
Phenocryst content (vol%)	25–35	45	30–33
Yield strength (N m^{-2})	70–120	370	
Bingham viscosity (Pa s)	6.5–7.5 × 10^2	9.4 × 10^3	4.5–6.2 × 10^2*

where K and n are the rheological parameters that describe the non-linear dependency of stress and strain in Herschel–Bulkley fluids and K is the consistency (viscosity) and n is a power law exponent. Herschel–Bulkley rheology involves a viscoplastic regime that includes changes in viscosity during flow that Bingham rheology does not (Castruccio et al. 2014).

Shear-thinning rheologies are those in which, at increasing stress, the strain increases at progressively higher rates. For example, the onset of a shear-thinning behaviour can be due to the progressive coalescence of large bubbles during flow which become a weakness zone and can then accommodate much larger strain for a given stress applied (such as below the upper crust of pahoehoe lava; see Sect. 5.4.2); or similarly due to the cushioning effect of spherical crystals (e.g. leucite; Campagnola et al. 2016). By contrast, the onset of shear-thickening rheologies in lavas can be due to the clustering of elongated crystals that require an increase in the rate of stress applied to allow further deformation.

Additionally, empirical methods have been developed for calculating the viscosities of silicate magmas from their chemical composition and temperature (Bottinga and Weill 1972; Shaw 1972; Hess and Dingwell 1996; Giordano et al. 2008).

2.3.4 Strength

Volcanic products, like all substances, can deform in a ductile or brittle manner when subjected to stress (also see Chap. 4). Given the great diversity of physical states of erupting volcanic materials (liquid magmas to solids), the deformation takes many forms, including highly plastic as reflected by lava flow, to essentially brittle, as reflected by explosive disruption or autobrecciation of a rhyolite lava dome (Chaps. 4, 6). Given the wide range of physical states, volcanic materials therefore have very diverse mechanical strengths. The mechanical strength of rock decreases rapidly as the degree of partial melting increases (van der Molen and Paterson 1979; Shaw 1980), the converse also being true. Therefore, as magma crystallises, its strength increases as the proportion of crystals increases and its temperature decreases (McBirney and Murase 1984; Rosenberg et al. 2007). Consideration of mechanical strength is most significant in the context of explosive fragmentation, because the greater the mechanical strength of the erupting material, the greater the tendency to resist explosive disruption. In this regard, fragmentation can be affected by both tensile stresses and shear stresses.

The former occurs for example during explosive growth of bubbles when the gas pressure exceeds the tensile strength and the surface tension of the magma. The latter can occur, for example, when vesiculated magma is crushed by velocity-induced shear stress that exceeds the shear strength of the erupting mass during high magma discharge rates. The mechanics of fragmentation will be discussed further in Chap. 4. The unit of strength is the same as the unit for stress, Pascals (Pa, or dyn cm^{-2}).

Experimental determinations of the tensile strength of rhyolite glass fibres have shown that the presence of H_2O bubbles may lower the tensile strength by two orders of magnitude (10^6 Pa; Romano et al. 1996) with respect to dry conditions (10^8 Pa; Webb and Dingwell 1990). Furthermore, silicic magmas have very low tensile strength under rapid unloading and decompression (Spieler et al. 2003), a property that has important implications during lava dome collapse events.

2.4 Factors Controlling Viscosity in Magmas

Not all of the factors that control viscosity in magmas are well understood, nor is the way in which they all interact. The principal factors that contribute to viscosity include.

- chemical composition,
- volatile content, especially dissolved water content,
- temperature,
- pressure,
- bubble content, and
- crystal content.

2.4.1 Chemical Composition

The chemical composition of magma affects its viscosity in a complex fashion (Mysen and Richet 2005; Giordano et al. 2008). The elements in magma can be divided into network formers and non-network formers. Si^{4+} and to a lesser extent Al^{3+} and Fe^{3+} are the principal cation network formers. The silica content is important in contributing to the viscosity of magmas, because Si–O bonds are the strongest cation–anion bonds in a magma, mineral, or rock. Even well above its liquidus temperature, magma has a well-defined structure (Burnham 1979; Hess 1980; Mysen et al. 1982; Mysen and Richet 2005), and its strength and shear resistance can be attributed to intermolecular bonds, and particularly Si–O bonds. Al-O bonds are also important, since they are also much stronger than other cation-oxygen bonds, though not as strong as Si–O bonds. O^{2-} is both a network and non-network former. In the former role, its principal function is to form cation-oxygen tetrahedra with Si^{4+}, Al^{3+}, and Fe^{3+}. Mysen et al. (1982) proposed that silicon-oxygen tetrahedra represent the basic building block to a range of network structural units, these being SiO_4^{4-} (monomers), $Si_2O_7^{6-}$ (dimers), $Si_2O_6^{4-}$ (chains), $Si_2O_5^{2-}$ (sheets), and three-dimensional units. Overall, this sequence of units corresponds to increasing degrees of polymerisation and viscosity in magma.

The type of network structural unit present depends on the ratio of non-bridging oxygens to silicon (NBO: Si) and the types of non-tetrahedrally co-ordinated cations present, which are called network modifiers (Mysen et al. 1982; Mysen and Richet 2005), such as monovalent alkalis and divalent alkaline earth elements. If the NBO:Si ratio decreases, for example by decreasing the dissolved water content, magmas will be more polymerised. Similarly, the higher the field strength of the network modifying cations, the more polymerised the melt will be at a given NBO:Si ratio (Mysen et al. 1982).

Peralkaline magmas with relatively high Na^+ and K^+ ion concentrations will be of relatively low viscosity because of the effect of Na^+ and K^+ in lowering the degree of melt polymerisation. Similarly, low silica magmas (higher overall NBO:Si and more network modifiers) will have lower viscosities than high silica ones at the same temperature and volatile contents. Some minor elements can have opposing effects. For example, TiO_2 reduces silica activity and the degree of polymerisation, whereas P_2O_5 increases silica activity and the degree of polymerisation (Ryerson and Hess 1980; Mysen et al. 1982; Mysen and Richet 2005).

Shaw (1972) proposed that the relationships between magmatic composition and viscosity can be calculated by the so-called Arrhenius mixture rule, that illustrates how the viscosity of a mixture (i.e. a magma) made of n components (i.e. SiO_2, Al_2O_3, NaO, etc.…), is equal to the sum of the characteristic viscosity contribution of each ith component of mole fraction X_i. However, more recent research indicates that over a large range of conditions melts instead show non-Arrhenian behaviour (Hess and Dingwell 1996; Richet et al. 1996; Giordano and Dingwell 2003; Romano et al. 2003; Giordano et al. 2008; Di Genova et al. 2013). Giordano et al. (2008) have presented a model for calculating the viscosity of magmatic liquids, which predicts the non-Arrhenian viscosity of silicate melts as a function of temperature and melt composition.

Magma composition, therefore, remains the major factor controlling viscosity. Calculated values for dry melts at eruption conditions vary from up to 10^{14} Pa s for rhyolitic compositions to 1 Pa s or less for ultramafic komatiite and kimberlite compositions (Dingwell 1998; Spera 2000; Giordano et al. 2008). Very low melt viscosities related to magma composition can have major and fatal hazard effects. An example is the ultra-low viscosities of lavas erupted from Nyiragongo volcano in the Democratic Republic of the Congo. Several eruptions from Nyiragongo in 1977, 2002 and 2021 produced high velocity (10 m s^{-1}), very low viscosity lavas that flowed into the city of Goma, leading to significant loss of life. The Nyiragongo lavas are mafic–ultramafic ($\sim$30 wt% SiO_2) foidite, with very high alkali compositions (Na_2O 5.47 wt%; K_2O 5.14 wt%) had viscosities of $\sim$33 Pa s at liquidus temperature ($\sim$1220 °C). Maintenance of low viscosity is enhanced by low degrees of post-eruptive crystallisation (<5%; Morrison et al. 2020).

2.4.2 Volatile Content

Dissolved magmatic water content has a marked effect on the viscosity of magmas (Fig. 2.9; McBirney and Murase 1970; Williams and McBirney 1979; Richet et al. 1996; Dingwell 1998; Papale et al. 1998; Spera 2000; Cashman 2004; Mysen and Richet 2005). At fixed temperatures, the viscosity of a particular magma becomes lower with increasing dissolved water content, especially for more silicic magmas (Shaw 1963; Spera 2000).

The solubility of water in magmas is controlled by temperature, pressure, the presence of other volatiles, especially CO_2 (Williams and McBirney 1979; Wallace and Anderson 2000; Wallace et al. 2015) and the presence of exchangeable cations (those not in tetrahedral co-ordination sites; Burnham 1979; Mysen et al. 1982). The solubility of water in magma increases with decreasing temperature and increasing pressure, and decreases with increasing abundance of other volatiles (Figs. 2.9 and 2.10).

Burnham (1979) indicated that the effect of water on a silicate melt is to depolymerise the melt by breaking

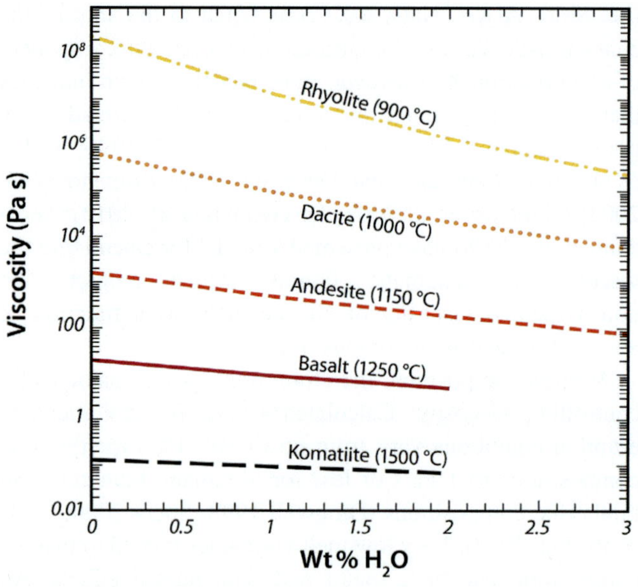

Fig. 2.9 Relationship between viscosity and water content for magmas of different composition, showing that for any specific magma composition, viscosity changes significantly as the proportion of dissolved magmatic H_2O changes. In addition, at a specific water content, the viscosity of different magma types is very different, indicating that magma composition is a major factor in determining magma viscosity. (From Lesher and Spera 2015)

Si–O–Si bridges. An H^+ ion exchanges with any cation that is not in a tetrahedral co-ordination site (e.g. Na^+), so hydrolysing one of the tetrahedral co-ordinating oxygens to OH^-. Small concentrations of water may lower the viscosity of magma by several orders of magnitude (Fig. 2.9) because water is 8/9 oxygen by mass (Spera 2000).

Stolper (1982) showed that dissolved water exists in both hydroxyl (OH^-) and molecular (H_2O) forms in silicate glass, and by inference in silicate melts. At low total water contents (<3 wt %), the rate at which hydroxyl ion concentration increases and melt viscosity decreases are both high. However, at higher water contents, as water is added, the rate of increase of hydroxyl ion concentration falls, as does the rate of decrease of melt viscosity, but the concentration of molecular water increases significantly (Stolper 1982). The effect of dissolved water on magma viscosity is therefore due more to the concentration of network-breaking hydroxyl ions than to the total dissolved water content.

Experiments show that water, on a weight percentage basis, is more soluble in silicic melts than in mafic ones (Burnham 1979; Carroll and Holloway 1994; Wallace and Anderson 2000; Wallace et al. 2015; Mysen and Richet 2005). However, on an equi-molal basis the solubility of water is the same in all magma types. Furthermore, because water solubility increases with decreasing temperatures (Williams and McBirney 1979), most silicic magmas, because of their lower eruption temperatures, could contain more water than mafic ones do, if adequate water is available in the subsurface source area. Other factors probably also influence the actual water content of magmas, such as, H_2O is geochemically incompatible with most silicate systems, and is more concentrated in more evolved magmas.

Pressure is an important control on the solubility of water in magma (Fig. 2.10a; Williams and McBirney 1979; Carroll and Holloway 1994; Wallace and Anderson 2000; Mysen and Richet 2005; Wallace et al. 2015). As magma rises in the subsurface and the confining pressure decreases, water solubility decreases, water will begin to exsolve from the magma, and crystallisation of anhydrous minerals occurs (Cashman 1992, 2004 and references therein). The pressure dependence of water solubility is particularly important because the influence of water exsolution on viscosity with decompression becomes increasingly higher at low pressures and low abundances of H_2O (1.5–0.1 wt%). This means that small pressure changes, ΔP, at shallow depth may induce large rheological changes even in low silica magmas (Cashman 2004).

The property that controls the rate of volatile mobility in a melt, and ultimately exsolution into bubbles, is the volatile diffusivity (see Dingwell 1998 and references therein). The effect of exsolution is to increase the viscosity and the strength of the magma, which may then be sufficient to stop the magma from disrupting explosively, as discussed in Chap. 5. The volatile content of magma can be directly determined by the study of melt inclusions in phenocrysts using Fourier transform infra-red spectroscopy (FTIR), Raman spectroscopy, and microprobe analysis (Lowenstern 1995; Roggensack et al. 1997; Wallace et al. 1999; Métrich et al. 2001; Schmitt 2001; Signorelli et al. 2001; Cashman 2004; Kent 2008; Métrich and Wallace 2008; Mercier et al. 2010; Frezzotti et al. 2012).

However, notwithstanding all of the foregoing, it is significant that the effect of dissolved water in lowering viscosity is greater for silica-rich magmas than for silica-poor ones (Fig. 2.9) simply because there are more Si–O bonds to break. An erupting silicic magma with a low dissolved water content (e.g. one that has degassed) will be more viscous than a silica-poor one with an equivalent weight percentage water content at the same temperature.

The exact effects of other volatiles on magma viscosity are less well known, being dependent on their solubilities and abundances. Chlorine and fluorine have a marked effect on magma rheology (Giordano et al. 2004). Peralkaline rocks have high Cl and F contents, which are thought to considerably reduce viscosities and yield strengths of magmas of these compositions (Schmincke 1974; Wolff and Wright 1981a, b). The supra-liquidus viscosities of pantellerites are typically two orders of magnitude below those of calc-alkaline rhyolites (Wolff and Wright 1981a; Dingwell 1998; Di Genova et al. 2013).

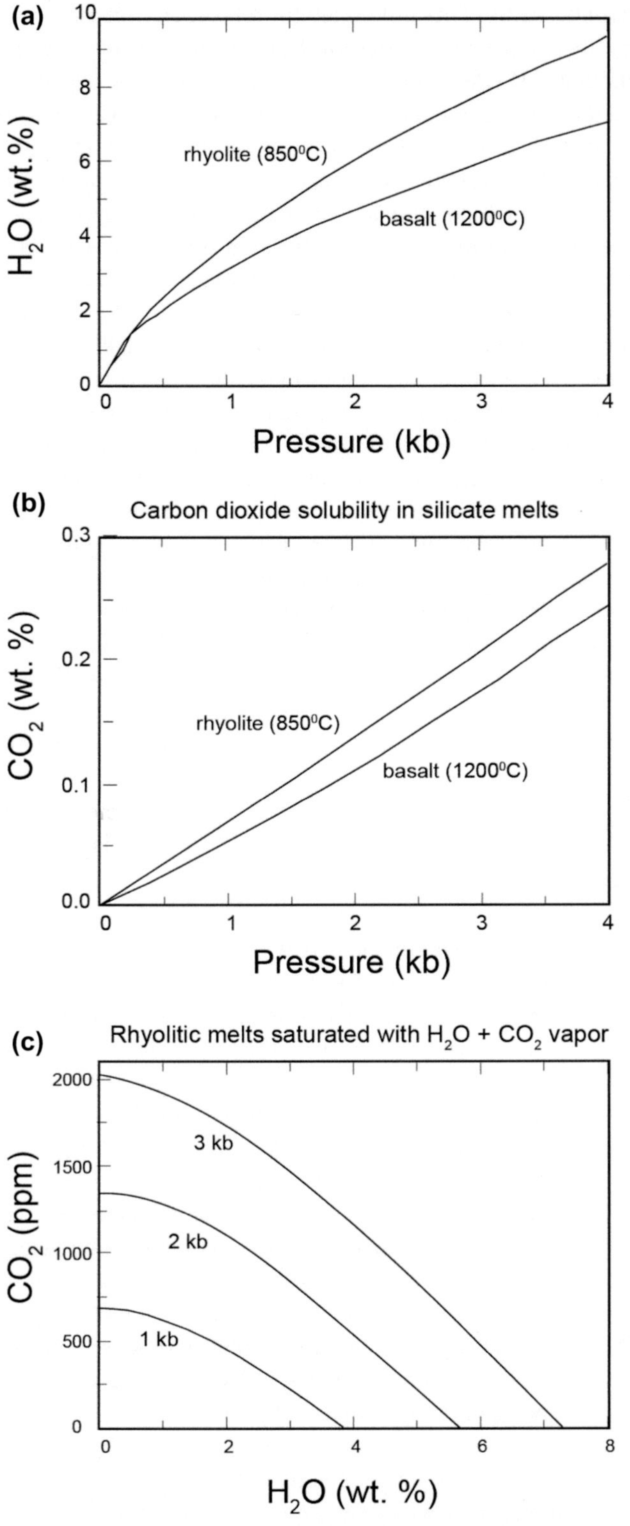

Fig. 2.10 Solubility conditions of H_2O and CO_2 in rhyolite and basalt magmas. **a** H_2O solubility in rhyolite (850 °C) and basalt (1200 °C) as a function of pressure. **b** CO_2 solubility in rhyolite (850 °C) and basalt (1200 °C) as a function of pressure. **c** H_2O and CO_2 solubility in volatile-saturated basaltic melt at 1200 °C at varying pressures. (Adapted from Wallace et al. 2015)

Carbon dioxide has a low solubility at low pressures, but its solubility increases markedly in the presence of H_2O (Fig. 2.10b, c; Mysen 1977; Burnham 1979). Several authors have suggested the involvement of CO_2 in highly explosive mafic magmas, such as in kimberlites (e.g. Mitchell 1986) and potassic foidites (e.g. Freda et al. 1997; Cross et al. 2014), although the exact role played by CO_2 on the overall viscosity of mafic magmas is still to be fully understood. For example, CO_2 has been considered to cause increased polymerisation, and therefore viscosity, in melts by forming CO^{3-} complexes (Eggler and Rosenhauer 1978; Mysen et al. 1982). By contrast, some studies have shown that CO_2 in K-rich melts has qualitatively similar effects to water, that is, to lower the viscosity through the formation of stabilised carbonate ions with free potassium (Bourgue and Richet 2001).

2.4.3 Temperature and Glass Transition

Viscosity is very dependent on the temperature of the magma (Fig. 2.11a). Both field and experimental data show that the viscosity of all magmas increases significantly on cooling (Shaw 1972; Williams and McBirney 1979; Hess and Dingwell 1996; Spera 2000), partly due to crystallisation. The modelling of magmatic silicate liquids as Arrhenian substances (i.e. mixtures for which properties result from the cumulative properties of individual components; Shaw 1972) has recently been questioned for the relationship of dependence of viscosity on temperature, based on a large number of experimental data (Dingwell 1998). Non-Arrhenian models for predicting the variation of viscosity with temperature are approached experimentally for conditions relevant to volcanic eruptions for a number of natural magma types (Giordano et al. 2008; Kolzenburg et al. 2016).

The temperature dependence of viscosity is very important for the rheological behaviour of magma approaching temperatures at which magma becomes glass, known as the glass transition. The glass transition involves a narrow range of temperatures (about 60 °C), when a melt that is cooling too rapidly to pass through the "solidus" curve, passes from the state of a liquid to that of a glass by super-cooling. Glass is essentially a super-cooled liquid that "freezes" molecular complexes in the glass before they have time to aggregate into well-ordered crystal forms. The viscosity of magma increases enormously approaching the glass transition.

Dingwell (1998) considered the physical phenomena involved in the transition from magma to glass, and the following summary is derived from that paper. Magmas are visco-elastic materials. They respond to an applied stress with an initially immediate to short-term elastic response that results from the rearrangement of the melt structure, and can

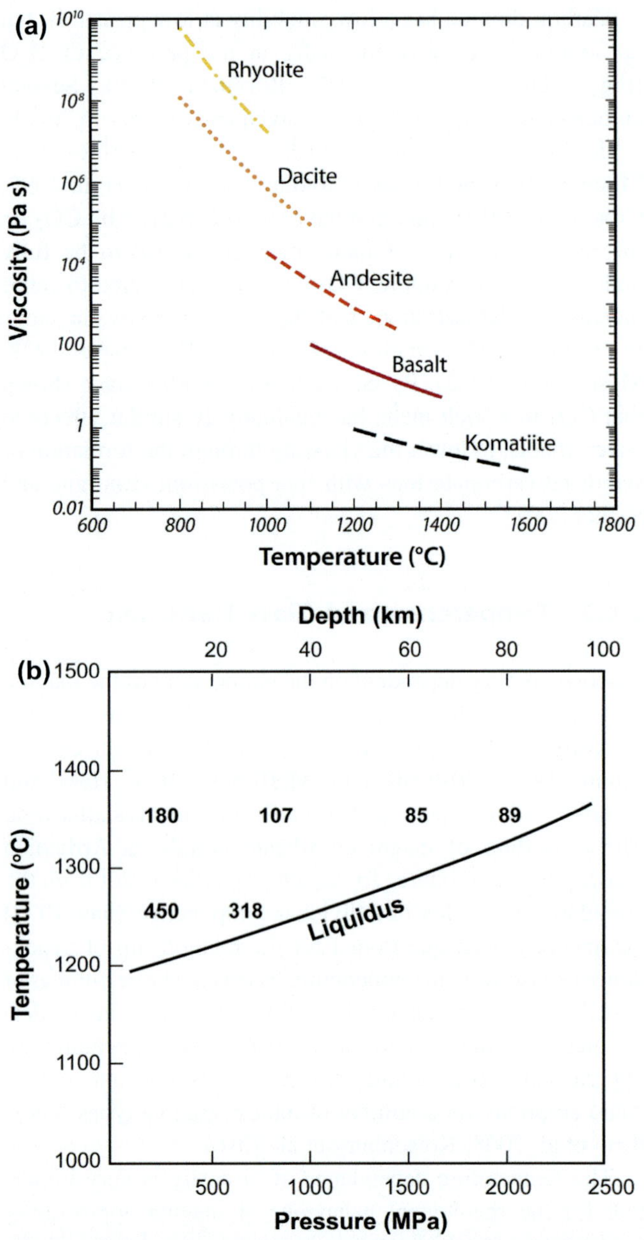

Fig. 2.11 Temperature and pressure controls on viscosity of major magma types. **a** Relationship between viscosity and temperature at eruption conditions for common magma types (from Lesher and Spera 2015). **b** Viscosity of an andesite melt with varying temperature and pressure. Numbers indicate the viscosity of the melt in Pa s; the 180 and 450 Pa s viscosities were calculated at 1 atm pressure using the method of Bottinga and Weill (1972). (From Cas and Wright 1987, After Kushiro et al. 1976)

be recovered when the stress applied is removed. This elastic strain is generally very small compared to the strain attained due to the viscous component, which by contrast cannot be recovered. Because magmas are visco-elastic materials, the actual type of deformational behaviour depends on the time scale of the relative elastic and viscous responses to an applied stress. This time scale (or relaxation time t) is given by the ratio between the viscosity η (Pa s) and the elastic modulus M (Pa). Both η and M are temperature dependent. At increasing temperature, the viscosity decreases and the relaxation time becomes increasingly small and the deformational behaviour visco-elastic. The reverse happens with decreasing temperature and increasing viscosity approaching the glass transition, i.e. the relaxation time increases. Stress rate thus becomes more important in determining elastic-brittle versus visco-elastic deformational behaviour as the temperature drops close to the glass transition. If the stress is applied rapidly, for example during acceleration in a conduit during magma ascent, then visco-elastic deformation is expected for short relaxation times associated with low viscosities, whereas elastic-brittle deformation is expected for long relaxation times associated with greater viscosities.

The heat capacity and expansivity are also significantly higher in the liquid state than in the glass state and this simple fact is enormously important for the processes that arise when magma crosses the glass transition (e.g. Gottsmann and Dingwell 2002; Gottsmann et al. 2004; Potuzak et al. 2008), influencing quenching, fragmentation and crystallisation processes that will be described in the following chaps.

2.4.4 Pressure

In a series of experiments carried out in a piston–cylinder apparatus at supra-liquidus temperatures, Kushiro (1976, 1978, 1980, 1986) and Kushiro et al. (1976) showed that in natural and synthetic melts, the viscosity of magma decreases with increasing pressure, especially at high pressures (Fig. 2.11b). Significant specific findings of these experiments were.

(a) As pressure increases at constant temperature, the rate at which the viscosity becomes lower in basaltic magma is less than the rate in andesitic magma (the viscosity of basaltic magma is initially significantly lower anyway).
(b) The viscosity of an andesitic melt with 4 wt% H_2O is lower by a factor of 20 than in its anhydrous equivalent at the same temperature and pressure.
(c) The viscosity of a $NaAlSi_2O_6$ (jadeite) melt decreased by an order of magnitude more than a melt of $Na_2Si_3O_7$ did with increasing pressure, which Kushiro relates to a change in the co-ordination number of Al from four to six, implying that the melt structure is an important influence on viscosity. (Note: although the viscosity became lower, the density increased with an increase in pressure.).

2.4.5 Bubble Content

The effect of bubbles or vesicles on the bulk viscosity of the magma can be variable, depending on the degree of vesiculation, the size and distribution of bubbles, and the viscosity of the magma interstitial to the bubbles (e.g. Sparks 1978). As described above, dissolved water contributes significantly to lowering magma viscosity (Figs. 2.9 and 2.12a). When the water exsolves, the magma viscosity begins to increase. However, the exsolved phases, are very low viscosity fluids, which may affect the overall bulk viscosity (Fig. 2.12b). In low viscosity magmas such as basalt, exsolution of volatiles may have relatively little effect on the bulk viscosity, because the low viscosity is largely due to the effects of temperature and composition. However, the presence of abundant fluid bubbles may decrease the already low viscosity even further. In more silicic magmas however the viscosity of the magma is initially high, and may be significantly affected by exsolution. Vesiculated rhyolite, for example, will have a very high bulk viscosity, irrespective of the degree of vesiculation, unless it is peralkaline in character. By contrast, the mechanical strength of the rhyolite may be low due to the high vesicle content and overall physical heterogeneity, especially if bubble walls are thin. Hence, in spite of the overall high bulk viscosity, such vesiculated rhyolite, may be very susceptible to mechanical, explosive fragmentation, as discussed in Chap. 4.

Advances in modelling bubble-bearing magmas (Manga et al. 1998; Lejeune et al. 1999; Manga and Loewenberg 2001; Rust and Manga 2002; Pal 2003; Llewellin and Manga 2005; Vona et al. 2016) have shown that a convenient way to describe the rheological behaviour is the capillary number (*Ca*):

$$Ca = \frac{r S \eta}{\sigma_t} \quad (2.7)$$

where r is the radius of the bubble, S is the shear rate applied, η is the viscosity, and σ_t is the surface tension that restores the shape of the bubble. This equation illustrates that a bubble-bearing magma will reduce its effective viscosity for high-shear rates, when it overcomes the surface tension. At low shear rates, by contrast, bubbles may not deform, increasing the effective viscosity (Rust and Manga 2002; Pal 2003). The effect of the reduced effective viscosity may enhance boundary processes such as shear-induced fragmentation (Papale 2001; Chap. 4).

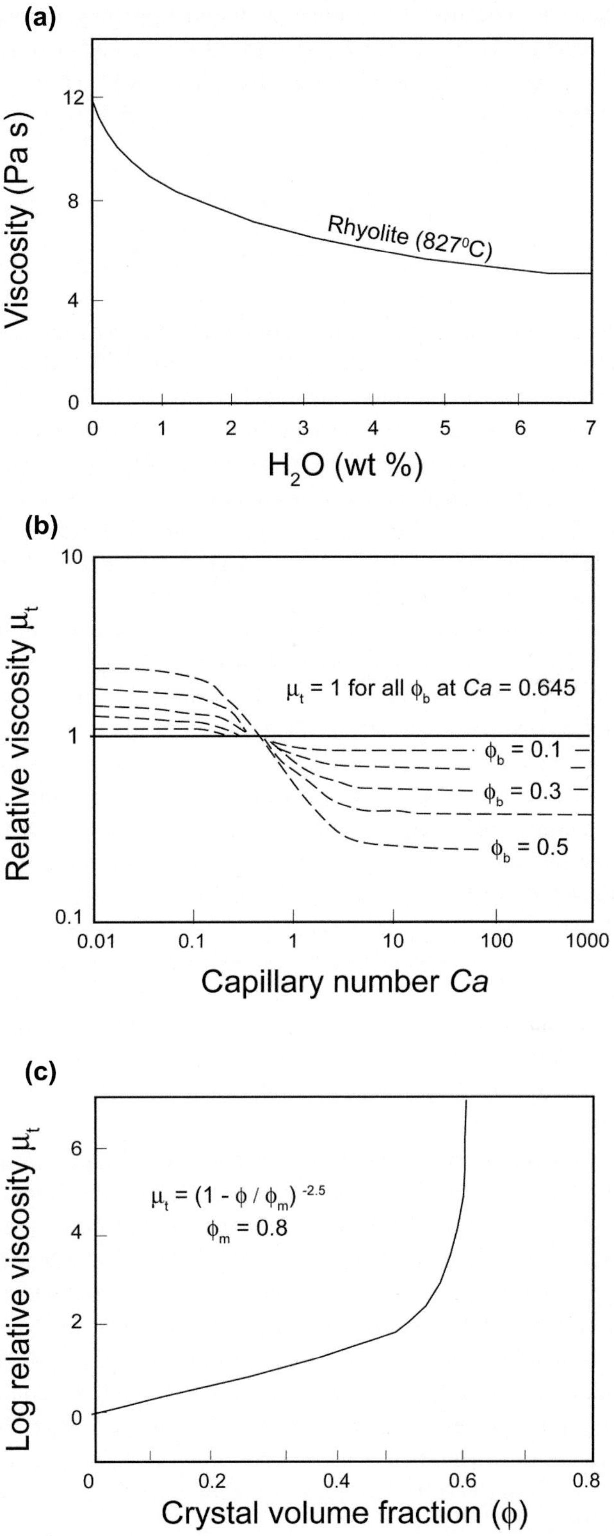

Fig. 2.12 Viscosity variations in bubble- and crystal-bearing magmas. **a** Melt viscosity as a function of dissolved H_2O (After Dingwell 1998). **b** Variation in relative viscosity (μ_t) with capillary number *Ca* for different bubble volume fractions (Φ_b) (After Rust and Manga 2002 and Pal 2003). **c** Variation in log μ_t with increasing crystal volume fraction (φ) according to the simple Roscoe–Einstein relation for $\Phi_m = 0.6$ (e.g. Marsh 1981) (After Cashman 2004)

2.4.6 Crystal Content

The effect of crystals suspended in a magma is to increase the effective or bulk viscosity of the magma (Fig. 2.12c;

Chap. 4), because the volume of liquid that may deform under stress is effectively less (Dingwell 1998). The effective bulk viscosity can be estimated from the Stokes–Einstein relation and the Einstein–Roscoe equation (McBirney and Murase 1984):

$$\eta = \eta_0(1 - R\Phi)^{-2.5} \tag{2.8}$$

where η is the effective viscosity of a liquid with a volume fraction of Φ suspended solids, η_0 is the viscosity of the liquid alone and R is a constant for which the best estimated value for lavas is 1.67 (Marsh 1981; McBirney and Murase 1984). However, results calculated using Eq. 2.7 are frequently at variance with values of lava viscosities measured in the field; this is due to the larger sizes, different shapes, and higher concentrations of crystals than those for which the equation was designed (McBirney and Murase 1984; Lejeune and Richet 1995; Vona et al. 2011, 2017; Campagnola et al. 2016; Kolzenburg et al. 2016).

Interactions between crystals become important above $\Phi = 25$ vol% and the rheology of magma is expected to be non-Newtonian (Ryerson et al. 1988; Lejeune and Richet 1995; Caricchi et al. 2007; Costa et al. 2009). The effective viscosity of a magma approximates that of the solid at maximum packing crystal-volume fraction Φ, that is commonly assumed at 0.6 (Fig. 2.12c; Marsh 1981), although this is for spherical shapes and may vary with elongated crystals (Cimarelli et al. 2011; Vona et al. 2011; Mader et al. 2013). In natural magmas, furthermore, the actual rheological behaviour is complicated because crystals range in size from microns to cm, have very variable aspect ratios and may be present together with bubbles (Bagdassarov et al. 1994; Dingwell 1998; Caricchi et al. 2007; Lavallée et al. 2007; Mueller et al. 2010, 2011; Vona et al. 2011, 2017).

In experimental studies on samples of foidite basalt from eruptions of Nyiragongo volcano, Democratic Republic of the Congo, Morrison et al. (2020) found that undercooling of magma of up to 75 °C below the liquidus temperature, little crystallisation occurred (<5%) and had no discernible effect on the viscosity of the magma, and thus lava mobility. Once undercooling was greater and crystallisation reached >5%, viscosity increased enough to reduce lava flow mobility. Recent studies also suggest that the formation of nanolites (nanoscale crystals) may have a significant impact on the rheology of magmas (Di Genova et al. 2020).

2.5 Fluid Flow State

The physical properties of magmas, and of aggregates of pyroclastic and epiclastic debris, will control to a large extent the nature of resultant flows of lava and debris away from the source. The viscosity, or bulk viscosity (for heterogeneous aggregates), is especially important in controlling not only flow mobility and form, but also the fluid flow state. There are essentially two fluid flow states: laminar flow and turbulent flow.

In *laminar flow*, fluid streamlines are smooth and parallel, and no mixing of streamlines occurs. The fluid is therefore free of eddies and vortices. In *turbulent flow*, streamlines are highly irregular and are dominated by eddying, so high degrees of mixing of the fluid occur. In 1883, the English physicist Reynolds first defined the criteria distinguishing the two flow states in experiments carried out by passing Newtonian fluids through pipes (Fig. 2.13).

The defining parameter for Newtonian fluids, the Reynolds Number (*Re*), is given by

$$Re = \frac{U D \rho}{\eta} \tag{2.9}$$

where U is the average velocity, D the pipe diameter, ρ the fluid density and η the fluid viscosity; Re is dimensionless. For open channel flow (e.g. rivers, lavas), D is replaced by R, where R is the hydraulic radius = A (cross-sectional area)/P (wetted channel perimeter), so that

$$Re = \frac{U R \rho}{\eta} \tag{2.10}$$

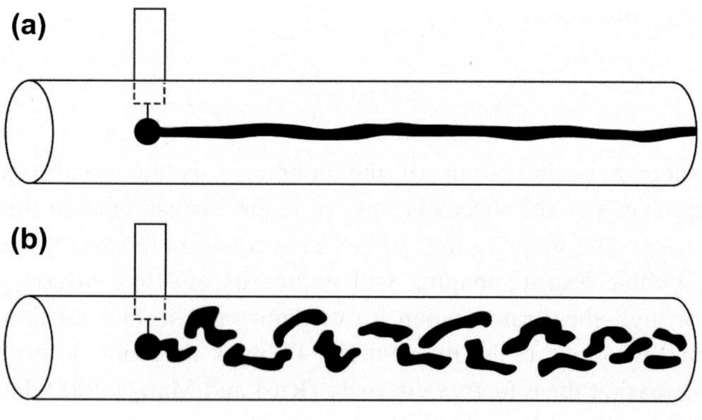

Fig. 2.13 Reynolds' experiment, in which a dye streak is injected into a steady flow of water through a glass tube to define **a** laminar flow and **b** at increased flow velocity above a critical value, turbulent flow develops in which eddies disperse the dye and eventually colour the whole flow (From Cas and Wright 1987)

For flows in both pipes and channels, laminar flow at $Re < 500$, turbulent flow at $Re > 2000$, and the transition between laminar and turbulent flow lies between Re values of 500 and 2000.

For non-Newtonian Bingham substances, such as debris flows and lavas, the Reynolds Number criterion for turbulence is inadequate because of the high strength and viscosity of the substance (Hampton 1972; Hiscott and Middleton 1979). For such substances the criterion for turbulence is

$$Re \geq 1000\,B \tag{2.11}$$

where B is the Bingham Number, which is given by

$$B = \frac{\tau_\chi D}{\eta U} \tag{2.12}$$

where τ_χ is the strength of the substance and the other parameters are as defined in Eq. 2.8. For Bingham fluids, which most lavas are, the strength is the yield strength, which is the minimum shear stress required to induce flow. Equations 2.8 and 2.11 may be combined to produce a criterion for turbulence in Bingham substances, known as the Hampton Number (Hiscott and Middleton 1979):

$$\frac{\rho U^2}{\tau_\chi} \geq 1000 \tag{2.13}$$

The critical point about the Reynolds Number is that it is inversely proportional to the viscosity, and directly proportional to the velocity and the thickness of the flow. Because of this, lavas may move by laminar flow or turbulently, but generally, because of their relatively high viscosities, most lavas flow in laminar fashion. Only very low viscosity, highly fluidal magmas experience turbulent flow, and then usually only where relatively high terrain slopes cause acceleration to relatively high velocities (Cas et al. 1999; Chap. 7 on Archean komatiite lavas). High viscosity lavas move at low velocities, even on steep slopes, because of their high internal yield strength. Flow banding, which is characteristic of rhyolites (Fig. 2.14a), is a reflection of laminar flow. It is usually best preserved in very viscous, high-silica flows, less commonly in lower viscosity lavas such as andesite, and rarely in basalt. Folding of flow bands, or flow folds (Fig. 2.14b) does not reflect turbulent flow, but simply plastic deformation of flow bands during laminar flow. In less viscous flows, laminar flow may be reflected by alignment of platy or elongate phenocrysts (Fig. 2.15).

Fig. 2.14 Flow banding and flow folds in rhyolite lavas. **a** Sub-vertical flow-banding in spherulitically devitrified obsidian dome rhyolite. Holocene Haparangi Rhyolite, Okataina Complex, North Island, New Zealand. **b** Flow folds in variably devitrified, early Pleistocene obsidian rhyolite lava, Palmarola Island, Italy

Fig. 2.15 Flow-aligned feldspar phenocrysts in Ordovician trachy-andesite, Goonumbla Volcanics, New South Wales, Australia. These aligned phenocrysts reflect laminar flow paths and state

Johnson (1970) suggested that lavas (Chaps. 5–7) may have rheological similarities to cohesive debris flows (Chap. 12) and should flow similarly. Many debris flows and lavas approximate Bingham substances, and both frequently develop interior channels flanked by non-moving 'dead zone' margins, which are frequently levée-like in form and so act to contain the more active part of the flow (Fig. 2.16a, b).

The formation of levées and steep margins in lavas (Chaps. 5, 6), debris flows (Chap. 12) and even in pyroclastic flow deposits (Chap. 10; Fig. 2.16c) is attributed to their cohesive character and the existence or development of a yield strength during flow (Johnson 1970; Hulme 1974; Vallance and Iverson 2015). Observations of natural debris flows, laboratory experiments, and theoretical modelling have shown that the flow character of many debris flows cannot be described simply as laminar or turbulent (Vallance and Iverson 2015). As a consequence of a finite yield strength, the bulk of the flow moves as an essentially internally inert 'plug' upon a zone of laminar flow, which is a zone of very high shear between the coherent plug and the channel walls (Johnson 1970; Figs. 2.17, 2.18).

However, flows can be turbulent given high velocities and low bulk viscosities. Where flows move within valleys incised into bedrock, the bedrock constitutes the channel walls, but where flow occurs on an unconfined flat surface, the lateral edges of the flow solidify and become the de facto channel walls. More will be said about lava flow behaviour, types and resultant characteristics in Chaps. 5–7, about pyroclastic flows and their deposits in Chap. 10, and about epiclastic mass flows, including debris flows, in Chap. 12.

2.6 Introduction to Lava Flows and Their Rheological Behaviour

Lava flows are highly variable in composition, physical properties, size, shape, and in their surface and internal textures. In this introduction to lava flows we describe these variations and discuss the different controls on lava morphology, such as effusion rate, physical properties (particularly viscosity and yield strength), and environmental factors. Once formed, by either passive effusion or more spectacular magma fountaining, lavas may remain coherent, or be subjected to autobrecciation and in subaqueous settings, also to quench fragmentation (Fig. 2.19). Post-emplacement, they may become hydrothermally altered and fractured, forming secondary breccia textures. Many of the differences between lava flows can be treated by dividing them into low viscosity (low silica; Chap. 5) and high viscosity (high silica; Chap. 6) types. The characteristics of the flows are thereafter controlled by the nature of the environment into which they are extruded, or the nature of the environment into which they then flow, or both. In addition, we consider ultra-low viscosity lavas in Chap. 7.

2.6.1 Prerequisites for Lava Forming Eruptions

The principal requirement for magma to be erupted as coherent lava in subaerial settings is that the magma at the time of eruption contains an insufficient abundance of exsolved volatiles to drive explosive fragmentation of the magma. That can be due to (1) magma having an inherently low volatile content from the beginning, (2) magma degassing from a subsurface reservoir from fumaroles and hydrothermal springs before eruption, or (3) most of the volatile components having been used to drive an initial explosive eruption phase, leaving residual degassed magma that is then erupted effusively and non-explosively late in the eruption. Low viscosity magmas are more prone to sub-surface degassing, because gas bubbles can form easily and rise rapidly through the magma, so the resulting eruptions are more often effusive compared to those associated with high viscosity magmas. In subaqueous environments, these factors are also valid, especially in shallow water. However, in deep water settings the hydrostatic pressure may suppress the exsolution of volatiles from the magma (Chap. 4). Therefore, magma with a high volatile content that would erupt explosively in a subaerial setting is more likely to erupt effusively in deep-water.

The exsolution and degassing of volatiles from magma (Sect. 2.4.2) will increase its viscosity and yield strength,

Fig. 2.16 Levees in lava flows and pyroclastic flows. **a** Active channel and lateral levees in 2007 a'a lava flow, Mt. Etna, Sicily. **b** Steep-sided levee and channel structure of the phonolitic a'a lava flows of the 1492 AD eruption at Pico de Teide, Tenerife, Canary Islands, Spain, which Columbus may have witnessed. **c** Levees at the front and sides of a 50 cm thick pumice and ash flow deposit, Mt St Helens, USA, defined by elevated ridges of concentrations of large pumice clasts and a lower central channel zone

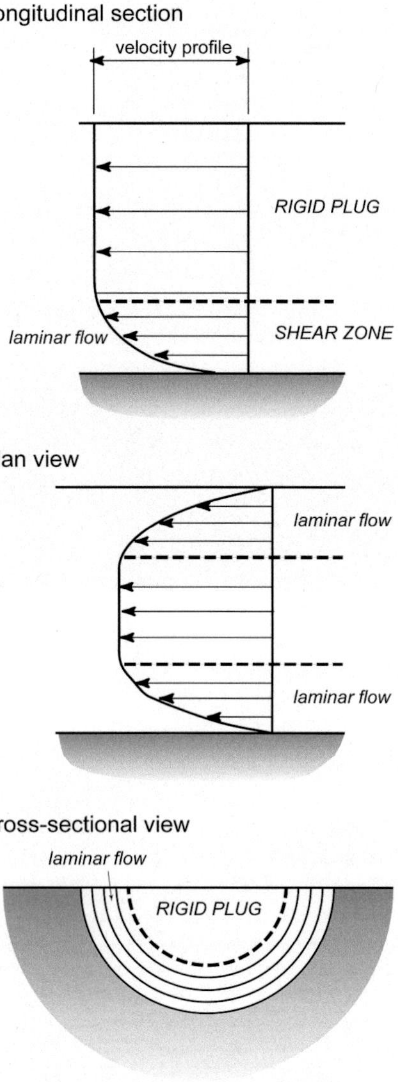

Fig. 2.17 Velocity profiles for flow of a Bingham substance in a subaerial setting. Length of the arrows is proportional to fluid velocity. **a** Longitudinal, vertical profile in a plane parallel to flow direction, **b** in plan view, and **c** cross-section perpendicular to flow direction. (From Cas and Wright 1987, After Johnson 1970)

which will affect the mobility, flow distance, and thickness of any lava flow. However, compared with silica-rich lavas, most mafic lavas are fluidal and mobile, irrespective of their water content. This is because mafic magmas are less polymerised than silicic magmas (Sect. 2.4.1). In general, the volume and length of lava flows decrease as the SiO_2 content and viscosity increase. Thus, large basalt flows are much more voluminous and longer than large rhyolite and dacite lava flows.

2.6.2 Aspect Ratio of Lavas

A parameter frequently used in the description of lava geometry is aspect ratio. This can be defined as the ratio of the average thickness or height (T) to the horizontal extent where horizontal extent is taken as the diameter of a circle (D) with a surface area equal to that of the lava (Fig. 2.20). Aspect ratio is therefore defined as T/D. Walker (1973) distinguished low viscosity, low aspect ratio lavas (mainly basalts) from high viscosity, high aspect ratio lavas (trachytes, andesites, dacites, and rhyolites), as summarised in Fig. 2.21. Aspect ratio can also be quantified using physical properties of the magma and the effusion rate (see Eq. 2.14 below).

2.6.3 Factors Affecting the Form of Lavas

Walker (1973) proposed that the form or shape of lava flows (area, thickness, length, or flow distance from vent) can be related to three main controls: effusion rate, physical properties of the magma, and slope of the terrain that the lava is flowing over. However, in reality, the factors controlling lava shapes are many, and complexly inter-related. For instance, effusion or lava discharge rate is itself dependent on a large number of factors such as vent shape and dimensions, magma viscosity, yield strength, and magma pressure gradient within the volcano.

2.6.3.1 Effusion Rate

Walker (1973) considered effusion rate to be the most important controlling factor on lava form. He showed that the distance travelled by lava flows was proportional to the effusion rate. This is due to the effect of cooling, which increases viscosity and limits mobility. Lavas erupted at higher rates stay hotter for longer and travel further before cooling, whereas lavas erupted at low rates, cool quickly and this limits their flow mobility (Pinkerton and Wilson 1994).

The effusion rate of lavas is not always straightforward to measure in the field (Harris et al. 2007). The effusion rate observed in lava flows of mafic compositions ranges from <0.5 to <5–10 $m^3 s^{-1}$ for average Hawai'ian pahoehoe lavas and >5–10 $m^3 s^{-1}$ for a'a lavas (Walker 1971; Rowland and Walker 1990). For the 1783–84 Laki eruption in Iceland, the estimated maximum effusion rate was ~9,000 $m^3 s^{-1}$ (Thordarson and Self 1993; Table 5.1). The magma discharge rate for the spectacular 2013–2014 Bardabunga–Holuhraun eruption in Iceland fluctuated between 50 and 350 $m^3 s^{-1}$ (Pedersen et al. 2017; Table 5.1).

Fig. 2.18 Cross-sectional view of a basaltic volcaniclastic debris flow of late Eocene-early Oligocene age, Waiareka–Deborah volcanics, Bridge Point, South Island, New Zealand. The lower stratified succession into which the debris flow has eroded consists of primary and redeposited pyroclastic deposits. The cross-sectional view is similar to that in Fig. 2.17c, with a semicircular channel form, a lower laminar flow layered zone, and an inner massive plug core in which large clasts have been concentrated

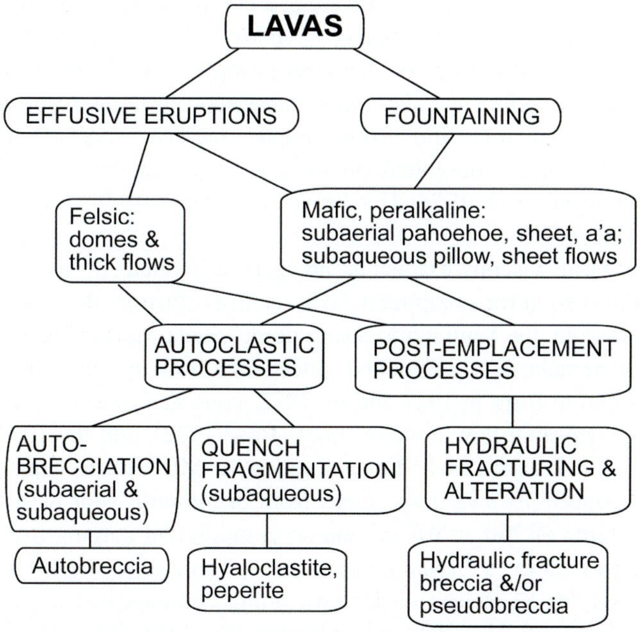

Fig. 2.19 Lava flow process spectrum depicting eruption styles, lava flow types, fragmentation processes, and post-emplacement alteration processes

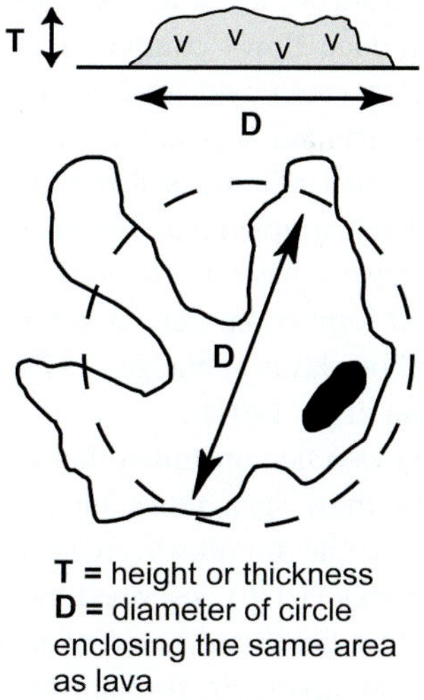

Fig. 2.20 Diagram to illustrate the concept of aspect ratio: average maximum thickness of the lava, T, to the diameter of a circle with the same area as the lava, D

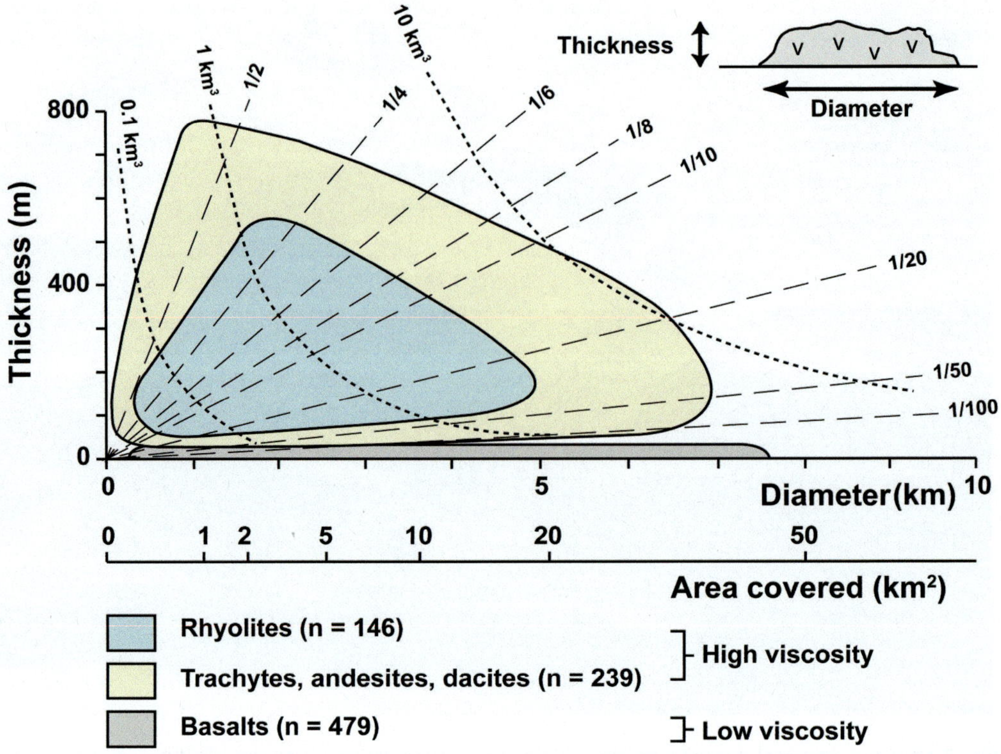

Fig. 2.21 Aspect ratios of the main compositional lava types, highlighting the low aspect ratio of low-viscosity mafic lavas compared with the high aspect ratio of higher viscosity intermediate and high silica lavas (Modified from Cas and Wright 1987, After Walker 1973)

Walker (1971) suggested that basaltic lavas extruded at high effusion rates form far-reaching flows composed of a single flow unit, termed *simple lavas,* which are internally massive, and often sheet-like, whereas those extruded at low effusion rates produce flows composed of multiple, small flow lobes forming *compound lavas* (Fig. 2.22). The Laki 1783–84 flow in Iceland was extruded at a high effusion rate, and is composed of only a few flow units, so it can almost be considered simple lava. The historic lava flows of Mt Etna and Hawai'i have been erupted at much lower effusion rates, and form compound lava flows or flow-fields, including pahoehoe lavas (Wadge 1978; Pinkerton and Sparks 1976; Hon et al. 1994).

Wadge (1978) also demonstrated that the maximum distance attained by these lava flows increases linearly with increasing effusion rate for flows greater than 1 km long. Pinkerton and Sparks (1976) described the formation of the 1975 Etna lava flow-field (composed of many thousands of flow units/lobes) as involving formation of new break-out flow lobes commonly formed at the fronts of mature flow units which had otherwise ceased to flow. Subaerial pahoehoe and subaqueous pillow lava are examples of complex (or compound) lavas. Griffiths and Fink (1992) found experimentally that subaqueous basaltic sheet (simple) lavas form under high effusion rates, whereas pillow-like lavas form under lower effusion rates.

Large single flow unit flood basalt lavas found in the geological record are believed to have been erupted at high, sustained effusion rates ($\sim$4,000 m^3 s^{-1}; Self et al. 2015; Table 5.1) from very long (>100 km long) fissure vent systems. Nonetheless, some flood lavas that appear to be massive sheets, may also have basal pahoehoe lobes and surface lobes or rises, indicating a complex character, giving rise to a suitable name, *complex lavas*, for lavas that show hybrid features between simple and compound. Such transitions are also found subaqueously in the form of lavas that transition from simple, massive sheet lavas near vent to compound pillow lava facies at the margins (Dimroth et al. 1978; Hargreaves and Ayres 1979; Ballard et al. 1979).

More viscous extrusions are generally small in volume and also form compound lavas at low effusion rates. For example, the Santiaguito dacite dome, Santa Maria volcano, Guatemala, is a compound silicic lava flow system which began to form in 1922 and by 1972, consisted of at least 14 recognisable flow units or lobes (Rose 1972), and the 1991 to 1995 dacite lava dome eruption at Mt. Unzen, Japan, produced 13 lava dome lobes from the summit vent with a volume of 200 $\times$ 10^6 m^3 and an eruption rate ranging from 0.1 to 7.5 m^3 s^{-1} (Nakada et al. 1999; Table 6.1). The 1980–1984 Mt St Helens dacite lava dome complex had a final volume of 74 $\times$ 10^6 m^3, whereas the 2004–2008 Mt St Helens dome system had a final volume of 92 $\times$ 10^6 m^3, with an effusion rate that waned from 5.9 to 0.7 m^3 s^{-1} (Calder et al. 2015). The 2008–2009 Chaiten rhyolite dome complex in Chile had a volume of 800 $\times$ 10^6 m^3, at an average (very high) eruption rate of 45 m^3 s^{-1}, due to the low viscosity of the magma (Calder et al. 2015; Table 6.1).

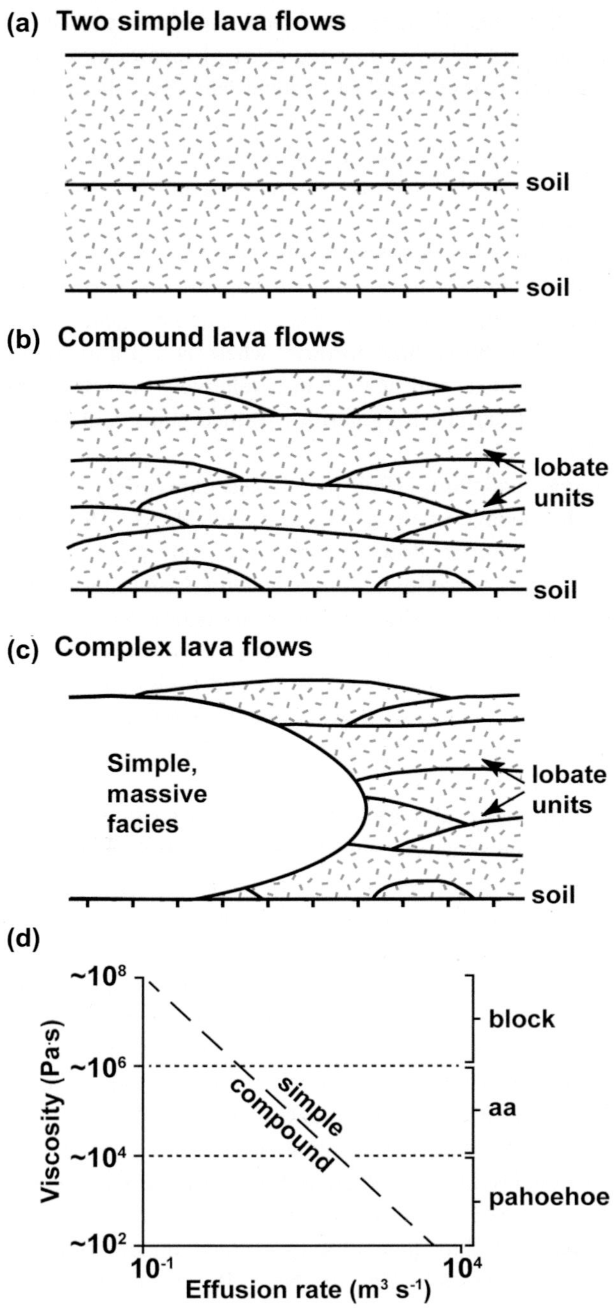

Fig. 2.22 Schematic representation of **a** simple, massive lavas, **b** multi-lobate compound lava, **c** complex lavas showing transition between simple and lobate forms **d** the control of viscosity versus effusion rate (Modified from Cas and Wright 1987, After Walker 1971)

Fink and Griffiths (1990, 1992) and Griffiths and Fink (1992) showed experimentally using polyethylene glycol wax as an analogue material for lavas that the surface form and features of experimental lava flows changed in response to the rate at which a surface crust formed, which was related to the extrusion rate and cooling rate.

2.6.3.2 Physical Properties

The physical properties of magma and its rheology (Sect. 2.3) play a major part in determining the flow behaviour and characteristics of lavas, including aspect ratio. Low viscosity lavas can travel far and fast if the effusion rate is high enough, whereas high viscosity lavas move more slowly and less far. As a result, low viscosity mafic lavas (Chap. 5) can flow very large distances (e.g. flood basalts can flow hundreds of km from vent if the eruption volumes are large and the discharge rate is constant), and be relatively thin, whereas viscous felsic lavas (Chap. 6) struggle to flow large distances but can be hundreds of m thick, commonly forming high aspect lava domes. Under sustained eruption, they can form lava flows or coulees (de Silva et al. 1994).

However, physical properties of magmas for a given composition may vary significantly along flow as the lava cools, degases, and crystallises (Cashman et al. 1994; 1999; Cashman and Mangan 2014). In addition, the margins and sometimes the entire lava may be supercooled to form volcanic glass, the character of which and the processes by which it forms are discussed in Sect. 4.4.2 in Chap. 4. In general, these variations all promote stiffening and transition from ductile to brittle behaviour, slowing and eventually leading to the cessation of movement of the flow front. A major phenomenon that helps to preserve the initial physical properties of the lava along the flow path is the formation of a visco-elastic crust, so insulating a lava tube underneath it from significant heat loss, and prolonging the flow life (Hon et al. 1994; Keszthelyi 1995; Calvari and Pinkerton 1998; Ball et al. 2008). All far reaching flows have moved under such conditions.

Hulme (1974) modelled lavas as Bingham substances (Fig. 2.6) and indicated that the principal factor governing their shape was their non-Newtonian rheology. His theoretical analysis and experiments with kaolin suspensions, which are close to Bingham substances in rheology, showed that aspect ratio was mainly dependent on yield stress. For a Bingham substance to flow downhill, it must form a layer deep enough for the shear stress at the base to exceed the yield strength. Close to the lateral margins the depth is not great enough for downhill flow to occur, and dead zones of stationary fluid form levees along margins. The depth and width of a flow, and the width of each dead zone, are related to five independent initial parameters: effusion rate (F), the slope (α), and three properties of the fluid—Bingham viscosity (η), yield strength (σ_o), and specific weight ($g\rho$, where g is the acceleration due to gravity and ρ is the density of the fluid). The critical depth (d_c) which must be exceeded for any flow to occur is given by

$$d_c = \frac{\sigma_o}{g\,\rho\,sin\alpha} \tag{2.14}$$

For lavas with higher yield strengths, d_c and the thickness of the lava flow are greater.

The aspect ratio of a lava flow can be predicted (Hulme 1974) from

$$\text{Aspect ratio } (AR) = \frac{\sigma_o}{(F\,\eta)^{0.25}(g\,\rho)^{0.75}} \quad (2.15)$$

Aspect ratio therefore depends mainly on yield strength. Equation 2.14 predicts that lavas with low yield strengths, such as basalts, give rise to flows with lower aspect ratios, and more-silicic lavas with higher yield strengths will form higher aspect ratio flows, which is in general agreement with field observations (Walker 1973, Fig. 2.21).

From Eq. 2.14, aspect ratio seems to be insensitive to changes in effusion rate, but in reality, this is more complicated because of the effect of temperature variations and slope. A change in effusion rate will lead to a change in the temperature within flowing lava at any given point, and this will have a marked effect on the yield strength, because an increase in temperature will therefore reduce the yield strength, at any particular point in the lava.

Fink and Griffiths (1990, 1998) and Griffiths (2000) summarised numerous years of analogue experimental and numerical modelling on the behaviour of Bingham-like lava flows, and developed a dimensionless parameter, ψ, that influences the morphology of lavas and takes into account physical properties of the magma. ψ is defined as the ratio of the time that it takes for a crust to solidify, t_s, to the time that it takes for material to advect laterally, t_A:

$$\psi = \frac{t_s}{t_A} \quad (2.16)$$

where t_A can be defined as

$$t_A = \left(\frac{\eta}{\rho\,g}\right)^{3/4} Q^{1/4} \quad (2.17)$$

where Q is the eruption rate ($cm^3\ s^{-1}$). This can be rearranged for Bingham fluids in a new dimensionless parameter ψ_B

$$\psi_B = \left(\frac{g\,\rho}{\sigma_o}\right)^3 Q\,t_s \quad (2.18)$$

On relatively flat slopes (<10°) if ψ is low (<3, indicating relatively rapid surface cooling and crust formation, multi-lobate flows (cf. pahoehoe, pillow) flows formed, where $3 < \psi < 13$ (= lower cooling rates) platey, "rift"-like crusts formed, where $13 < \psi < 25$, the crust developed surface folds, and at the lowest crust cooling rates, $\psi > 25$, levée-like marginal ridges formed; these transitions coincide with increasing eruption rates (Griffiths 2000).

Griffiths and Fink (1993) found experimentally that the strength (and thickness) of the crust of lava domes significantly affects their ability to spread laterally, whereas lavas with weak crusts are prone to breakouts of lobes (e.g. pahoehoe, pillow lava). These different behaviours correspond to high magma viscosity and low viscosity, respectively.

Even with viscous lava analogues (cf. rhyolite, dacite lavas) differences in ψ reflect differences in surface morphology. Where $\psi = 0.09$, a spiny surface formed, where $\psi = 0.79$ a lobate flow formed, where ψ 1.3 a segmented surface formed, and at $\psi = 30$ a relatively smooth surface with small surface irregularities formed (Fink and Griffiths 1998).

Gregg and Fink (1995) conducted experiments to simulate the behaviour of subaqueous lava flows. They found generally similar flow features with pillows forming at $\psi < 3$. They found that pillows become larger and flatter as eruption rate and slope increase transforming flows into lineated sheet and slabby rift flows ($3 < \psi < 13$). At higher effusion rates, flow surfaces developed folds ($13 < \psi < 25$) and at $\psi > 25$, leveed flows formed, and these conditions may also account for "jumbled sheet flows" with irregular broken slabs of crust.

As discussed above (Sect. 2.3.3), Castruccio et al. (2014) modelled lavas as viscoplastic Herschel–Bulkley materials, which allows downflow changes in lava properties to be considered, such as viscosity. Changing viscosity causes yield strength to change, and can be significant in some compositional lava types. Bulk lava flow viscosity changes can be caused by shear-thinning behaviour during flow.

Also reported above, experimental studies on samples of foidite basalt from eruptions of Nyiragongo volcano, Democratic Republic of the Congo, by Morrison et al. (2020) found that undercooling of magma of up to 75 °C below the liquidus temperature, little crystallisation occurred (<5%) and had no discernible effect on the viscosity of the magma, and thus lava mobility. Once undercooling was greater and crystallisation reached >5%, viscosity increased enough to reduce lava flow mobility.

2.6.3.3 Slope of Terrain

Flow width varies inversely with ground slope (Hulme 1974). Gregg and Fink (2000) specifically studied the effects of slope on flow morphology in analogue experiments with polyethylene glycol. They found that as substrate slope increased flows became narrower, thinner, and longer. They found that the same sequence of flow surface morphology occurred as summarised above (pillow-lobate, slabby-rifted, folded, levéed) in response to increasing eruption rate and/or increasing slope, both of which have the effect of reducing the cooling rate, and so slowing crust formation.

References

Allen JRL (1970) Physical processes of sedimentation—an introduction. Allen and Unwin, London

Andersen DJ, Lindsley DH (1988) Internally consistent solution models for Fe–Mg–Mn–Ti oxides: Fe–Ti oxides. Amer Miner 73:714–726

Andersen DJ, Lindsley DH, Davidson PM (1993) QUILF: a pascal program to assess equilibria among Fe–Mg–Mn–Ti oxides, pyroxenes, olivine, and quartz. Comput Geosci 19(9):1333–1350

Anderson JL, Barth AP, Wooden JL, Mazdab F (2008) Thermometers and thermobarometers in granitic systems. In: Putirka KD, Tepley FJI (eds) Minerals, inclusions and volcanic processes. Mineral Soc Amer, Virginia, pp 121–142

Archambault C, Tanguy JC (1976) Comparative temperature measurements on Mount Etna lavas: problems and techniques. J Volcanol Geother Res 1(2): 113–125

Bagdassarov NS, Dingwell DB, Webb SL (1994) Viscoelasticity of crystal- and bubble-bearing rhyolite melts. Phys Earth Planet In83 (2):83–99

Ball M, Pinkerton H, Harris AJL (2008) Surface cooling, advection and the development of different surface textures on active lavas on Kilauea. Hawai'i. J Volcanol Geotherm Res 173(1–2):148–156

Ballard RD, Holcomb RT, Andel TH (1979) The Galapagos Rift at 86 W: 3. Sheet flows, collapse pits, and lava lakes of the rift valley. J Geophys Res: Solid Earth 84(B10):5407–5422

Barberi F, Carapezza ML, Valenza M, Villari L (1993) The control of lava flow during the 1991–1992 eruption of Mt. Etna. J Volcanol Geother Res 56(1–2): 1–34

Barth TFW (1951) The feldspar geologic thermometers. Neu J Miner 82:143–154

Beattie P (1993) Olivine-melt and orthopyroxene-melt equilibria. Contrib Miner Petrol 115(1):103–111

Bernstein M, Pavez A, Varley N, Whelley P, Calder ES (2013) Rhyolite lava dome growth styles at Chaitén volcano, Chile (2008–2009): interpretation of thermal imagery. Andean Geol 40 (2):295–309

Best MG (2009) Igneous and metamorphic petrology. Blackwell Science Ltd., Oxford, p 756

Blake S, Ivey GN (1986a) Density and viscosity gradients in zoned magma chambers, and their influence withdrawal dynamics. J Volcanol Geotherm Res 30(3–4):201–230

Blake S, Ivey GN (1986b) Magma-mixing and the dynamics of withdrawal from stratified reservoirs. J Volcanol Geotherm Res 27 (1–2):153–178

Bombrun M, Jessop D, Harris A, Barra V (2018) An algorithm for the detection and characterisation of volcanic plumes using thermal camera imagery. J Volcanol Geotherm Res 352:26–37

Bottinga Y, Weill DF (1972) The viscosity of magmatic silicate liquids; a model calculation. Amer J Sci 272(5):438–475

Bourgue E, Richet P (2001) The effects of dissolved CO2 on the density and viscosity of silicate melts: a preliminary study. Earth Planet Sci Lett 193(1–2):57–68

Brey GP, Köhler T (1990) Geothermo-barometry in four-phase Lherzolites II. New thermobarometers, and practical assessment of existing thermobarometers. J Petrol 31(6):1353–1378

Bullard FM (1947) Studies on Paricutin volcano, Michoacan, Mexico. Geol Soc Am Bull 58(5): 433–450

Burnham CW (1979) The importance of volatile constituents. In: Yoder HS (ed) The evolution of the igneous rocks. Princeton University Press, New Jersey, pp 439–482

Calder ES, Lavallee Y, Kendrick JE, Bernstien M (2015) Lava dome eruptions. In: Sigurdsson H, Stix J, Houghton BF, McNutt SR, Rymer H (eds) Encyclopedia of volcanoes. Academic Press, San Diego, pp 343–362

Calvari S, Spampinato L, Lodato L, Harris AJL, Patrick MR, Dehn J, Burton MR, Andronico D (2005) Chronology and complex volcanic processes during the 2002–2003 flank eruption at Stromboli volcano (Italy) reconstructed from direct observations and surveys with a handheld thermal camera. J Geophys Res: Solid Earth 110. B2(B2):23

Campagnola S, Vona A, Romano C, Giordano G (2016) Crystallization kinetics and rheology of leucite-bearing tephriphonolite magmas from the Colli Albani volcano (Italy). Chem Geol 424:12–29

Caricchi L, Burlini L, Ulmer P, Gerya T, Vassalli M, Papale P (2007) Non-Newtonian rheology of crystal-bearing magmas and implications for magma ascent dynamics. Earth Planet Sci Lett 264(3–4):402–419

Carroll MR, Holloway JR (1994) Volatiles in magmas. In: Reviews in mineralogy. Mineralogical Society of America, Washington, D.C.

Cas RAF, Self S, Beresford S (1999) The behaviour of the fronts of komatiite lavas in medial to distal settings. Earth Planet Sci Lett 172 (1–2):127–139

Cas RAF, Wright JV (1987) Volcanic successions: modern and ancient. Allen and Unwin, London, p 528

Cashman KV (1992) Groundmass crystallization of Mount St. Helens dacite, 1980–1986: a tool for interpreting shallow magmatic processes. Contrib Miner Petrol 109(4):431–449

Cashman KV (2004) Volatile controls on magma ascent and eruption. In: Sparks RSJ, Hawkesworth C (eds) The state of the planet: frontiers and challenges in geophysics. American Geophysical Union, Washington, D.C., pp 109–124

Cashman KV, Mangan MT (2014) A Century of Studying Effusive Eruption in Hawaii. In: Poland MP, Takahashi TJ, Landowski CM (eds) Characteristics of hawaiian volcanoes. Geol Surv Prof Pap 1801:357–396

Cashman KV, Mangan MT, Newman S (1994) Surface degassing and modifications to vesicle size distributions in active basalt flows. J Volcanol Geotherm Res 61(1–2):45–68

Cashman KV, Sturevant B, Papale P, Navon O (2000) Magmatic fragmentation. In: Sigurdsson H, Stix J, Houghton BF, McNutt SR, Rymer H (eds) Encyclopedia of volcanoes. Academic Press, San Diego, pp 421–430

Cashman KV, Thornber C, Kauahikaua JP (1999) Cooling and crystallization of lava in open channels, and the transition of Pāhoehoe Lava to'A'ā. Bull Volcanol 61(5):306–323

Castro J, Dingwell DB (2009) Rapid ascent of rhyolitic magma at Chaitén volcano, Chile. Nature 461:780–783

Castruccio A, Rust AC, Sparks RSJ (2014) Assessing lava flow evolution from post-eruption field data using Herschel-Bulkley rheology. J Volcanol Geotherm Res 275:71–84

Cimarelli C, Costa A, Mueller S, Mader HM (2011) Rheology of magmas with bimodal crystal size and shape distributions: insights from analog experiments. Geochem Geophys 12(7)

Clark S, Spera FJ, Yuen DA (1987) Steady state double-diffusive convection in magma chambers heated from below. In: Mysen BO (ed) Magmatic processes: physicochemical principles: a, vol in. honor. Yoder Jr. Geochemical Society, Washington, D.C., pp 289–305

Costa A, Caricchi L, Bagdassarov N (2009) A model for the rheology of particle-bearing suspensions and partially molten rocks. Geochem Geophys 10(3)

Cox KG, Bell JD, Pankhurst RJ (1979) The interpretation of igneous rocks. Allen and Unwin, London, p 396

Cross JK, Tomlinson EL, Giordano G, Smith VC, De Benedetti AA, Roberge J, Manning CJ, Wulf S, Menzies MA (2014) High level triggers for explosive mafic volcanism: Albano Maar, Italy. Lithos 190:137–153

Cross W, Iddings JP, Pirsson LV, Washington HS (1903) Quantitative classification of igneous rocks. University of Chicago Press, Chicago, p 315

Davis BTC, Boyd FR (1966) The join Mg2Si2O6-CaMgSi2O6 at 30 kilobars pressure and its application to pyroxenes from kimberlites. J Geophys Res 71(14):3567–3576

Deer WA, Howie RA, Zussman J (1992) An introduction to the rock-forming minerals. Longman, Hong Kong, p 696

Delmelle P, Bernard A (2000) Volcanic Lakes. In: Sigurdsson H, Stix J, Houghton BF, McNutt SR, Rymer H (eds) Encyclopedia of volcanoes. Academic Press, San Diego, pp 877–896

Di Genova D, Brooker RA, Mader HM, Drewitt JW, Longo A, Deubener J, Neuville DR, Fanara S, Shebanova O, Anzellini S, Arzilli F (2020) In situ observation of nanolite growth in volcanic melt: A driving force for explosive eruptions. Sci adv 6(39): p. eabb0413

Di Genova D, Romano C, Hess K-U, Vona A, Poe BT, Giordano D, Dingwell DB, Behrens H (2013) The rheology of peralkaline rhyolites from Pantelleria Island. J Volcanol Geotherm Res 249:201–216

Dimroth E, Cousineau P, Leduc M, Sanschagrin Y (1978) Structure and organization of Archean subaqueous basalt flows, Rouyn-Noranda area, Quebec, Canada. Canad J Earth Sci 15(6):902–918

de Silva SL, Self S, Francis PW, Drake RE, Carlos RR (1994) Effusive silicic volcanism in the Central Andes: the Chao dacite and other young lavas of the Altiplano-Puna volcanic complex. J Geophys Res: Solid Earth 99(B9):17805–17825

Dingwell DB (1998) Recent experimental progress in the physical description of silicic magma relevant to explosive volcanism. In: Gilbert JS, Sparks RSJ (eds) The physics of volcanic eruptions. Geological Society, London, pp 9–26

Dobran F (1992) Nonequilibrium flow in volcanic conduits and application to the eruptions of Mt. St. Helens on May 18, 1980, and Vesuvius in AD 79. J Volcanol Geotherm Res 49(3–4):285–311

Eggler DH, Rosenhauer M (1978) Carbon dioxide in silicate melts; II, solubilities of CO_2 and H_2O in $CaMgSi_2O_6$ (diopside) liquids and vapors at pressures to 40 kb. Am J Sci 278(1):64–94

Ferry JM, Watson EB (2007) New thermodynamic models and revised calibrations for the Ti-in-zircon and Zr-in-rutile thermometers. Contrib Miner Petrol 154(4):429–437

Fink JH, Griffiths RW (1990) Radial spreading of viscous-gravity currents with solidifying crust. J Fluid Mech 221:485–509

Fink JH, Griffiths RW (1998) Morphology, eruption rates, and rheology of lava domes: insights from laboratory models. J Geophys Res Solid Earth 103(B1):527–545

Fisher RV, Schmincke H-U (1984) Pyroclastic rocks. Springer-Verlag, Berlin, p 472

Folch A, Martı (1998) The generation of overpressure in felsic magma chambers by replenishment. Earth Planet Sci Lett 163(1–4):301–314

Fowler SJ, Spera FJ (2010) A metamodel for crustal magmatism: phase equilibria of giant ignimbrites. J Petrol 51(9):1783–1830

Francis PW, Thorpe RS, Brown GC, Glasscock J (1980) Pyroclastic sulphur eruption at Poas volcano, Costa Rica. Nature 283 (5749):754–756

Freda C, Gaeta M, Palladino DM, Trigila R (1997) The Villa Senni Eruption (Alban Hills, central Italy): the role of H_2O and CO_2 on the magma chamber evolution and on the eruptive scenario. J Volcanol Geotherm Res 78(1–2):103–120

Frezzotti ML, Tecce F, Casagli A (2012) Raman spectroscopy for fluid inclusion analysis. J Geochem Explor 112:1–20

Friedman JD, Olhoeft GR, Johnson GR, Frank D (1981) Heat content and thermal energy of the June dacitic dome in relation to total energy yield May-October 1980. In: Lipman PW, Mullineaux DR (eds) The 1980 eruptions of Mount St. Helens, Washington. US Geol Surv Prof Pap 1250:557–567

Frost BR, Lindsley DH, Andersen DJ (1988) Fe-Ti oxide silicate equilibria: assemblages with fayalitic olivine. Amer Miner 73:727–740

Ghiorso MS, Sack O (1991) Fe-Ti oxide geothermometry: thermodynamic formulation and the estimation of intensive variables in silicic magmas. Contrib Miner Petrol 108(4):485–510

Giordano D, Dingwell DB (2003) Non-Arrhenian multicomponent melt viscosity: a model. Earth Planet Sci Lett 208(3–4):337–349

Giordano D, Romano C, Dingwell DB, Poe B, Behrens H (2004) The combined effects of water and fluorine on the viscosity of silicic magmas. Geochim Cosmochim Acta 68(24):5159–5168

Giordano D, Russell JK, Dingwell DB (2008) Viscosity of magmatic liquids: a model. Earth Planet Sci Lett 271(1–4):123–134

Gonnermann HM, Manga M (2007) The fluid mechanics inside a volcano. Annu Rev Fluid Mech 39:321–356

Gottsmann J, Dingwell DB (2002) The thermal history of a spatter-fed lava flow: the 8-ka pantellerite flow of Mayor Island, New Zealand. Bull Volcanol 64(6):410–422

Gottsmann J, Harris AJL, Dingwell DB (2004) Thermal history of Hawaiian pāhoehoe lava crusts at the glass transition: implications for flow rheology and emplacement. Earth Planet Sci Lett 228(3–4):343–353

Green DH, Ringwood AE (1967) The genesis of basaltic magmas. Contrib Mineral Petrol 15(2): 103–190

Gregg TKP, Fink JH (1995) Quantification of submarine lava-flow morphology through analogue experiments. Geology 23(1):73–76

Gregg TKP, Fink JH (2000) A laboratory investigation into the effects of slope on lava flow morphology. J Volcanol Geotherm Res 96 (3):145–159

Griffiths RW (2000) The dynamics of lava flows. Annu Rev Fluid Mech 32(1):477–518

Griffiths RW, Fink JH (1992) Solidification and morphology of submarine lavas: a dependence on extrusion rate. J Geophys Res Solid Earth 97(B13):19729–19737

Griffiths RW, Fink JH (1993) Effects of surface cooling on the spreading of lava flows and domes. J Fluid Mech 252:667–702

Häkli TA, Wright TL (1967) The fractionation of nickel between olivine and augite as a geothermometer. Geochim Cosmochim Acta 31(5):877–884

Hampton MA (1972) The role of subaqueous debris flow in generating turbidity currents. J Sediment Res 42(4):775–793

Hargreaves R, Ayres LD (1979) Morphology of Archean metabasalt flows, Utik Lake. Manitoba. Canad J Earth Sci 16(7):1452–1466

Harris AJ, Dehn J, Calvari S (2007) Lava effusion rate definition and measurement: a review. Bull Volcanol 70:1–22

Harris A, Dehn J, Patrick MR, Calvari S, Ripepe M, Lodato L (2005) Lava effusion rates from hand-held thermal infrared imagery: an example from the June 2003 effusive activity at Stromboli. Bull Volcanol 68(2):107–117

Hayden LA, Watson EB (2007) Rutile saturation in hydrous siliceous melts and its bearing on Ti-thermometry of quartz and zircon. Earth Planet Sci Lett 258(3–4):561–568

Helz RT, Thornber CR (1987) Geothermometry of Kilauea Iki lava lake. Hawaii. Bull Volcanol 49(5):651–668

Hess KU, Dingwell DB (1996) Viscosities of hydrous leucogranitic melts: a non-Arrhenian model. Am Miner 81:1297–1300

Hess PC (1980) Polymerization model for silicate melts. In: Hargraves RB (ed) Physics of magmatic processes. Princeton University Press, Princeton, New Jersey, pp 3–48

Hiscott RN, Middleton GV (1979) Depositional mechanics of thick-bedded sandstones at the base of a submarine slope, Tourelle Formation (Lower Ordovician), Quebec, Canada. In: Doyle LJ,

Pilkey OH (eds) Geology of continental slopes. Society of Economic Paleontologists Mineralogist, pp 307–326

Hon K, Kauahikaua J, Denlinger R, Mackay K (1994) Emplacement and inflation of pahoehoe sheet flows: Observations and measurements of active lava flows on Kilauea Volcano. Hawaii. Geol Soc Amer Bull 106(3):351–370

Höskuldsson A, Sparks RSJ (1997) Thermodynamics and fluid dynamics of effusive subglacial eruptions. Bull Volcanol 59 (3):219–230

Hulme G (1974) The interpretation of lava flow morphology. Geophys J Int 39(2):361–383

Huppert HE, Sparks RSJ (1980) Restrictions on the compositions of mid-ocean ridge basalts: a fluid dynamical investigation. Nature 286 (5768):46–48

Huppert HE, Sparks RSJ (1981) The fluid dynamics of a basaltic magma chamber replenished by influx of hot, dense ultrabasic magma. Contrib Miner Petrol 75(3):279–289

Huppert HE, Sparks RSJ (1984) Double-diffusive convection due to crystallization in magmas. Ann Rev Earth Planet Sci 12(1):11–37

Huppert HE, Sparks RSJ, Turner JS (1982) Effects of volatiles on mixing in calc-alkaline magma systems. Nature 297(5867):554–557

Iacovino K, Till CB (2019) DensityX: A program for calculating the densities of magmatic liquids up to 1,627 C and 30 kbar. Volcanica 2(1):1–10

Irvine TN, Barragar WRA (1971) A guide to the chemical classification of common volcanic rocks. Can J Earth Sci 8:523–548

McPhie J, Doyle M, Allen R (1993a) Volcanic Textures—a guide to the interpretation of textures in volcanic rocks. CODES - Univ Tasmania, Hobart, p 198

McPhie J, Doyle M, Allen R (1993b) Volcanic textures—a guide to the interpretation of textures in volcanic rocks. CODES – Univ Tasmania, Hobart, p 198

Keszthelyi L, Self S (1998) Some physical requirements for the emplacement of long basaltic lava flows. J. Geophys. Res: Solid Earth 103(B11):27447–27464

Johnson AM (1970) Physical processes in geology: a method for interpretation of natural phenomena; intrusions in igneous rocks, fractures, and folds, flow of debris and ice. Freeman, Cooper, p 577

Kelsey CH (1965) Calculation of the CIPW norm. Miner Mag 34:276–282

Kent AJR (2008) Melt inclusions in basaltic and related volcanic rocks. In: Putirka KD, Tepley FJI (eds) Minerals, inclusions and volcanic processes. Mineralogical Society America, Chantilly, Virginia, pp 273–331

Kilburn CRJ (1996) Patterns and Predictability in the Emplacement of Subaerial Lava Flows and Flow Fields. Monitoring and Mitigation of Volcano Hazards. Springer, Berlin Heidelberg, pp 491–537

Kolzenburg S, Giordano D, Cimarelli C, Dingwell DB (2016) In situ thermal characterization of cooling/crystallizing lavas during rheology measurements and implications for lava flow emplacement. Geochim Cosmochim Acta 195:244–258

Kudo AM, Weill DF (1970) An igneous plagioclase thermometer. Contrib Miner Petrol 25(1):52–65

Kushiro I (1976) Changes in viscosity and structure of melt of $NaAlSi_2O_6$ composition at high pressures. J Geophys Res 81 (35):6347–6350

Kushiro I (1978) Density and viscosity of hydrous calc-alkalic andesite magma at high pressure. Carnegie Institution of Washington Yearbook 77:675–677

Kushiro I (1980) Viscosity, density and structure of silicate melts at high pressures, and their petrological applications. In: Hargraves RB (ed) Physics of magmatic processes. Princeton Univ Press, Princeton, NJ, pp 93–120

Kushiro I (1986) Viscosity of partial melts in the upper mantle. J Geophys Res: Solid Earth 91(B9):9343–9350

Lange RA (1997) A revised model for the density and thermal expansivity of K_2O-Na_2O-CaO-MgO-Al_2O_3-SiO_2 liquids from 700 to 1900 K: extension to crustal magmatic temperatures. Contrib Miner Petrol 130(1):1–11

Lavallée Y, Hess K, Cordonnier B, Dingwell DB (2007) Non-Newtonian rheological law for highly crystalline dome lavas. Geology 35(9):843–846

Le Bas MJ, Le Maitre RW, Streckeisen AL, Zanettin B (1986) A chemical classification of volcanic rocks based the total alkali-silica diagram. J Petrol 27:745–750

Le Maitre RW, Streckeisen AL, Zanettin B, Le Bas MJ, Bonin B, Bateman P, Bellieni G, Dudek A, Efremova S, Keller J, Lameyre J, Sabine PA, Schmid R, Sørensen H, Wooley AR (2002) Igneous rocks: a classification and glossary of terms. Cambridge University Press, Cambridge, UK

Lejeune AM, Bottinga Y, Trull TW, Richet P (1999) Rheology of bubble-bearing magmas. Earth Planet Sci Lett 166(1–2):71–84

Lejeune A-M, Richet P (1995) Rheology of crystal-bearing silicate melts: an experimental study at high viscosities. J Geophys Res: Solid Earth 100(B3):4215–4229

Lesher C, Spera FJ (2015) Thermodynamic and transport properties of silicate melts and magma. In: Sigurdsson H, Houghton B, McNutt SR, Rymer H, Stix J (eds) The Encyclopedia of Volcanoes Academic Press Elsevier, pp 113–141

Liang Y, Sun C, Yao L (2013) A REE-in-two-pyroxene thermometer for mafic and ultramafic rocks. Geochim Cosmochim Acta 102:246–260

Llewellin EW, Manga M (2005) Bubble suspension rheology and implications for conduit flow. J Volcanol Geotherm Res 143(1–3):205–217

Loucks RR (1996) A precise olivine-augite Mg-Fe-exchange geothermometer. Contrib Miner Petrol 125(2–3):140–150

Lowenstern JB (1995) Applications of silicate melt inclusions to the study of magmatic volatiles. In: Thompson JFH (ed) Magmas, Fluids and Ore Deposits. Mining Association Canada, Québec, pp 71–99

MacDonald GA (1972) Volcanoes. Prentice-Hall, Englewood Cliffs, NJ, p 510

MacKenzie WS, Donaldson CH, Guilford C (1982) Atlas of igneous rocks and their textures. John Wiley & Sons, New York

Mader HM, Llewellin EW, Mueller SP (2013) The rheology of two-phase magmas: a review and analysis. J Volcanol Geotherm Res 257:135–158

Manga M, Castro J, Cashman KV, Loewenberg M (1998) Rheology of bubble-bearing magmas. J Volcanol Geotherm Res 87(1–4):15–28

Manga M, Loewenberg M (2001) Viscosity of magmas containing highly deformable bubbles. J Volcanol Geotherm Res 105(1–2):19–24

Marsh BD (1981) On the crystallinity, probability of occurrence, and rheology of lava and magma. Contrib Miner Petrol 78(1):85–98

McBirney AR (1980) Mixing and unmixing of magmas. J Volcanol Geotherm Res 7(3–4):357–371

McBirney AR, Murase T (1970) Factors governing the formation of pyroclastic rocks. Bull Volcanol 34(2):372–384

McBirney AR, Murase T (1984) Rheological properties of Magmas. Ann Rev Earth Planet Sci 12(1):337–357

McBirney AR, Noyes RM (1979) Crystallization and layering of the Skaergaard intrusion. J Petrol 20(3):487–554

McGuire WJ, Murray JB, Kilburn CJ (1995) Monitoring active volcanoes: strategies. University College London Press, London, Procedures and Techniques, p 352

Meinhold G (2010) Rutile and its applications in earth sciences. Earth Sci Rev 102(1–2):1–28

Mercier M, Di Muro A, Métrich N, Giordano D, Belhadj O, Mandeville CW (2010) Spectroscopic analysis (FTIR, Raman) of

water in mafic and intermediate glasses and glass inclusions. Geochim Cosmochim Acta 74(19):5641–5656
Métrich N, Bertagnini A, Landi P, Rosi M (2001) Crystallization driven by decompression and water loss at Stromboli volcano (Aeolian Islands, Italy). J Petrol 42(8):1471–1490
Métrich N, Wallace PJ (2008) Volatile abundances in basaltic magmas and their degassing paths tracked by melt inclusions. In: Putirka KD, Tepley FJI (eds) Minerals, inclusions and volcanic processes. Mineralogical Society of America, Chantilly, Virginia, pp 363–402
Middlemost EAK (1975) The basalt clan. Earth Sci Rev 11:337–364
Mitchell RH (1986) Kimberlites: mineralogy, geochemistry, and petrology. Plenum Press, New York
Montierth C, Johnston AD, Cashman KV (1995) An empirical glass-composition-based geothermometer for Mauna Loa lavas. In: Rhodes JM, Lockwood JP (eds) Mauna Loa revealed: structure, composition, history and hazards. American Geophysical Union, San Francisco, pp 207–217
Morgavi D, Arzilli F, Pritchard C, Perugini D, Mancini L, Larson PB, Dingwell DB (2016) The Grizzly Lake complex (Yellowstone Volcano, USA): Mixing between basalt and rhyolite unraveled by microanalysis and X-ray microtomography. Lithos 260:457–474
Morgavi D, Perugini D, De Campos CP, Ertl- W, Lavallée Y, Morgan L, Dingwell DB (2013) Interactions between rhyolitic and basaltic melts unraveled by chaotic mixing experiments. Chem Geol 346:199–212
Morrison AA, Whittington A, Smets B, Kervyn M, Sehlke A (2020) The rheology of crystallizing basaltic lavas from Nyiragongo and Nyamuragira volcanoes, DRC. Volcanica 3(1):1–28
Mueller S, Llewellin EW, Mader HM (2010) The rheology of suspensions of solid particles. Proceed Roy Soc A: Math Phys Eng Sci 466(2116):1201–1228
Mueller S, Llewellin EW, Mader HM (2011) The effect of particle shape on suspension viscosity and implications for magmatic flows. Geophys Res Lett 38(13):L13316
Murase T, McBirney AR (1973) Properties of some common igneous rocks and their melts at high temperatures. Geol Soc Amer Bull 84 (11):3563–3592
Murphy MD, Sparks RSJ, Barclay J, Carroll MR, Brewer TS (2000) Remobilization of andesite magma by intrusion of mafic magma at the Soufriere Hills volcano, Montserrat. West Indies. J Petrol 41 (1):21–42
Mysen BO (1977) The solubility of H2O and CO2 under predicted magma genesis conditions and some petrological and geophysical implications. Rev Geophys 15(3):351–361
Mysen BO, Richet P (2005) Silicate glasses and mets: properties and structure. Elsevier, Amsterdam, p 544
Mysen BO, Virgo D, Seifert FA (1982) The structure of silicate melts: implications for chemical and physical properties of natural magma. Rev Geophys 20(3):353–383
Nakada S, Shimizu H, Ohta K (1999) Overview of the 1990–1995 eruption at Unzen volcano. J Volcanol Geotherm Res 89(1–4):1–22
Nimis P (1995) A clinopyroxene geobarometer for basaltic systems based on crystal-structure modeling. Contrib Miner Petrol 121 (2):115–125
Nimis P, Grütter H (2010) Internally consistent geothermometers for garnet peridotites and pyroxenites. Contrib Miner Petrol 159 (3):411–427
Nimis P, Taylor WR (2000) Single clinopyroxene thermobarometry for garnet peridotites. Part I. Calibration and testing of a Cr-in-Cpx barometer and an enstatite-in-Cpx thermometer. Contrib Miner Petrol 139(5):541–554
Ochs FA, Lange RA (1999) The density of hydrous magmatic liquids. Science 283(5406):1314–1317
Pal R (2003) Rheological behavior of bubble-bearing magmas. Earth Planet Sci Lett 207(1–4):165–179
Pallister JS, Hoblitt RP, Reyes AG (1992) A basalt trigger for the 1991 eruptions of Pinatubo volcano? Nature 356(6368):426–428
Papale P (2001) Dynamics of magma flow in volcanic conduits with variable fragmentation efficiency and nonequilibrium pumice degassing. J Geophys Res: Solid Earth 106(B6):11043–11065
Papale P, Neri A, Macedonio G (1998) The role of magma composition and water content in explosive eruptions: 1. Conduit ascent dynamics. J volcanol Geotherm Res 87(1–4):75–93
Pedersen GBM, Höskuldsson A, Dürig T, Thordarson T, Jonsdottir I, Riishuus MS, Óskarsson BV, Dumont S, Magnússon E, Gudmundsson MT (2017) Lava field evolution and emplacement dynamics of the 2014–2015 basaltic fissure eruption at Holuhraun, Iceland. J Volcanol Geotherm Res 340:155–169
Perugini D, Poli G, Gatta GD (2002) Analysis and simulation of magma mixing processes in 3D. Lithos 65(3–4):313–330
Petrelli M, Caricchi L, Perugini D (2020) Machine learning thermo-barometry: application to clinopyroxenebearing magmas. J Geophys Res Solid Earth 125(9):e2020JB020130
Philpotts AR, Ague JJ (2009) Principles of igneous and metamorphic petrology. Cambridge University Press, Cambridge
Pinkerton H, Sparks RSJ (1976) The 1975 sub-terminal lavas, Mount Etna: a case history of the formation of a compound lava field. J Volcanol Geotherm Res 1(2):167–182
Pinkerton H, Sparks RSJ (1978) Field measurements of the rheology of lava. Nature 276(5686):383–385
Pinkerton H, Wilson L (1994) Factors controlling the lengths of channel-fed lava flows. Bull Volcanol 56(2):108–120
Piochi M, De Astis G, Petrelli M, Ventura G, Sulpizio R, Zanetti A (2009) Constraining the recent plumbing system of Vulcano (Aeolian Arc, Italy) by textural, petrological, and fractal analysis: the 1739 A.D. Pietre Cotte lava flow. Geochem Geophys Geosyst 10(1):Q01009. https://doi.org/10.1029/2008GC002176
Potuzak M, Nichols ARL, Dingwell DB, Clague DA (2008) Hyperquenched volcanic glass from Loihi Seamount. Hawaii. Earth Planet Sci Lett 270(1–2):54–62
Putirka KD (2008) Thermometers and barometers for volcanic systems. In: Putirka KD, Tepley FJI (eds) Minerals, inclusions and volcanic processes. Miner Soc Am, Chantilly, Virginia, pp 61–120
Putirka KD, Johnson M, Kinzler R, Longhi J, Walker D (1996) Thermobarometry of mafic igneous rocks based on clinopyroxene-liquid equilibria, 0–30 kbar. Contrib Miner Petrol 123(1):92–108
Putirka KD, Mikaelian H, Ryerson F, Shaw H (2003) New clinopyroxene-liquid thermobarometers for mafic, evolved, and volatile-bearing lava compositions, with applications to lavas from Tibet and the Snake River Plain, Idaho. Am Miner 88 (10):1542–1554
Putirka KD, Perfit M, Ryerson FJ, Jackson MG (2007) Ambient and excess mantle temperatures, olivine thermometry, and active versus passive upwelling. Chem Geol 241(3–4):177–206
Reid MR, Vazquez JA, Schmitt AK (2011) Zircon-scale insights into the history of a Supervolcano, Bishop Tuff, Long Valley, California, with implications for the Ti-in-zircon geothermometer. Contrib Miner Petrol 161(2):293–311
Richet P, Lejeune A-M, Holtz F, Roux J (1996) Water and the viscosity of andesite melts. Chem Geol 128(1–4):185–197
Robson GR (1967) Thickness of Etnean Lavas. Nature 216 (5112):251–252
Roggensack K, Hervig RL, McKnight SB, Williams SN (1997) Explosive basaltic volcanism from Cerro Negro volcano: influence of volatiles on eruptive style. Science 277(5332):1639–1642
Rollinson H (1993) Using geochemical data: evaluation, presentation, interpretation. Pearson Education Ltd., Harlow, p 352

Romano C, Giordano D, Papale P, Mincione V, Dingwell DB, Rosi M (2003) The dry and hydrous viscosities of alkaline melts from Vesuvius and Phlegrean Fields. Chem Geol 202(1–2):23–38

Romano C, Mungall JE, Sharp T, Dingwell DB (1996) Tensile strengths of hydrous vesicular glasses: an experimental study. Am Miner 81(9):1148–1154

Rose WI (1972) Santiaguito volcanic dome, Guatemala. Geol Soc Am Bull 83(5):1413–1434

Rosenberg CL, Medvedev S, Handy MR (2007) Effects of melting on faulting and continental deformatiom. Dahlem Workshop Rep 95:357–401

Rothery DA, Pieri DC (1993) Remote sensing of active lava. In: Kilburn CRJ, Luongo G (eds) Active lavas: monitoring and modelling. Univ College London Press, London, pp 203–233

Rowland SK, Walker GPL (1990) Pahoehoe and aa in Hawaii: volumetric flow rate controls the lava structure. Bull Volcanol 52 (8):615–628

Rust AC, Manga M (2002) Effects of bubble deformation on the viscosity of dilute suspensions. J Non-Newton Fluid Mech 104 (1):53–63

Rutherford MJ, Devine JD (1996) Preeruption pressure-temperature conditions and volatiles in the 1991 dacitic magma of Mount Pinatubo. In: Newhall CG, Punongbayan RS (eds) Fire and mud: eruptions and lahars of Mount Pinatubo, Philippines. Univ Washington Press, Seattle, pp 751–766

Ryerson FJ, Hess PC (1980) The role of P2O5 in silicate melts. Geochim Cosmochim Acta 44(4):611–624

Ryerson FJ, Weed HC, Piwinskii AJ (1988) Rheology of subliquidus magmas: 1. Picritic compositions. J Geophys Res Solid Earth 93 (B4):3421–3436

Saar MO, Manga M, Cashman KV, Fremouw S (2001) Numerical models of the onset of yield strength in crystal–melt suspensions. Earth Planet Sci Lett 187(3–4):367–379

Schmincke HU (1974) Volcanological aspects of peralkaline silicic welded ash-flow tuffs. Bull Volcanol 38(2):594–636

Schmitt AK (2001) Gas-saturated crystallization and degassing in large-volume, crystal-rich dacitic magmas from the Altiplano-Puna, northern Chile. J Geophys Res: Solid Earth 106(B12):30561–30578

Self S, Coffin MF, Rampino MR, Wolff JA (2015) Large igneous provinces and flood basalt volcanism. In: Sigurdsson H, Stix J, Houghton B, McNutt S, Rymer H (eds) Encyclopedia of volcanoes. Academic Press, San Diego, pp 441–458

Shaw HR (1963) Obsidian-H_2O viscosities at 1000 and 2000 bars in the temperature range 700° to 900°C. J Geophys Res 68 (23):6337–6343

Shaw HR (1969) Rheology of basalt in the melting range. J Petrol 10 (3):510–535

Shaw HR (1972) Viscosities of magmatic silicate liquids; an empirical method of prediction. Am J Sci 272(9):870–893

Shaw HR (1980) The fracture mechanisms of magma transport from the mantle to the surface. In: Hargraves RB (ed) Physics of magmatic processes. Princeton University Press, Princeton, NJ, pp 201–264

Shaw HR, Wright TL, Peck DL, Okamura R (1968) The viscosity of basaltic magma; an analysis of field measurements in Makaopuhi lava lake, Hawaii. Am J Sci 266(4):225–264

Signorelli S, Vaggelli G, Romano C, Carroll MR (2001) Volatile element zonation in Campanian Ignimbrite magmas (Phlegrean Fields, Italy): evidence from the study of glass inclusions and matrix glasses. Contrib Miner Petrol 140(5):543–553

Sisson TW, Grove TL (1993) Temperatures and H_2O contents of low-MgO high-alumina basalts. Contrib Miner Petrol 113 (2):167–184

Smith RL (1960) Ash Flows. Geol Soc Am Bull 71(6):795–841

Sparks RSJ (1978) The dynamics of bubble formation and growth in magmas: a review and analysis. J Volcanol Geotherm Res 3:1–37

Sparks RSJ, Sigurdsson H, Wilson L (1977) Magma mixing: a mechanism for triggering acid explosive eruptions. Nature 267:315–318

Spera FJ (2000) Physical properties of magma. In: Sigurdsson H, Stix J, Houghton BF, McNutt SR, Rymer H (eds) Encyclopedia of volcanoes. Academic Press, San Diego, pp 171–190

Spieler O, Alidibirov M, Dingwell DB (2003) Grain-size characteristics of experimental pyroclasts of 1980 Mount St. Helens cryptodome dacite: effects of pressure drop and temperature. Bull Volcanol 65 (2–3):90–104

Stolper E (1982) Water in silicate glasses: an infrared spectroscopic study. Contrib Miner Petrol 81(1):1–17

Stormer JC (1975) Practical two-feldspar geothermometer. Am Miner 60:667–674

Streckeisen AL (1976) To each plutonic rock its proper name. Earth Sci Rev 12:1–33

Streckeisen A (1979) Classification and nomenclature of volcanic rocks, lamprophyres, carbonatites, and melilitic rocks: recommendations and suggestions of the IUGS subcommission on the systematics of igneous rocks. Geology 7(7):331–335

Thordarson T, Self S (1993) The Laki (Skaftár Fires) and Grímsvötn eruptions in 1783–1785. Bull Volcanol 55(4):233–263

Tomkins HS, Powell R, Ellis DJ (2007) The pressure dependence of the zirconium-in-rutile thermometer. J Met Geol 25(6):703–713

Troll VR, Donaldson CH, Emeleus CH (2004) Pre-eruptive magma mixing in ash-flow deposits of the Tertiary Rum Igneous Centre, Scotland. Contrib Miner Petrol 147(6):722–739

Turner JS, Huppert HE, Sparks RSJ (1983) An experimental investigation of volatile exsolution in evolving magma chambers. J Volcanol Geotherm Res 16(3–4):263–277

Valentine GA (1992) Magma chamber dynamics. In: Nierenberg WA (ed) Encyclopedia of earth system science. Academic Press, San Diego, pp 1–17

Vallance JW, Iverson RM (2015) Lahars and their deposits. In: Sigurdsson H, Houghton B, McNutt SR, Rymer H, Stix J (eds) The encyclopedia of volcanoes. Academic Press Elsevier, pp. 649–664

van der Molen I, Paterson MS (1979) Experimental deformation of partially melted granite. Contrib Miner Petrol 70(3):299–318

Vona A, Di Piazza A, Nicotra E, Romano C, Viccaro M, Giordano G (2017) The complex rheology of megacryst-rich magmas: the case of the mugearitic "cicirara" lavas of Mt. Etna volcano. Chem Geol 458:48–67

Vona A, Romano C (2013) The effects of undercooling and deformation rates on the crystallization kinetics of Stromboli and Etna basalts. Contrib Miner Petrol 166(2):491–509

Vona A, Romano C, Dingwell DB, Giordano D (2011) The rheology of crystal-bearing basaltic magmas from Stromboli and Etna. Geochim Cosmochim Acta 75(11):3214–3236

Vona A, Ryan AG, Russell JK, Romano C (2016) Models for viscosity and shear localization in bubble-rich magmas. Earth Planet Sci Lett 449:26–38

Wadge G (1978) Effusion rate and the shape of aa lava flow-fields on Mount Etna. Geology 6(8):503–506

Wallace PJ, Anderson AT (2000) Volatiles in magmas. In: Sigurdsson H, Stix J, Houghton BF, McNutt SR, Rymer H (eds) Encyclopedia of volcanoes. Academic Press, San Diego, pp 149–170

Wallace PJ, Anderson AT, Davis AM (1999) Gradients in H2O, CO2, and exsolved gas in a large-volume silicic magma system: Interpreting the record preserved in melt inclusions from the Bishop Tuff. J Geophys Res: Solid Earth 104(B9):20097–20122

Wallace PJ, Plank T, Edmonds M, Hauri EH (2015) Volatiles in magmas. In: Sigurdsson H, Houghton B, McNutt SR, Rymer H, Stix J (eds) The encyclopedia of volcanoes. Academic Press Elsevier, pp 163–184

Walker GPL (1971) Compound and simple lava flows and flood basalts. Bull Volcanol 35(3):579–590
Walker GPL (1973) Lengths of Lava flows. Phil Trans Roy Soc Lond Series a, Math Phys Sci 274(1238):107–118
Wan Z, Coogan LA, Canil D (2008) Experimental calibration of aluminum partitioning between olivine and spinel as a geothermometer. Amer Miner 93(7):1142–1147
Wark DA, Watson EB (2006) TitaniQ: a titanium-in-quartz geothermometer. Contrib Miner Petrol 152(6):743–754
Watson EB, Wark DA, Thomas JB (2006) Crystallization thermometers for zircon and rutile. Contrib Miner Petrol 151(4):413–433
Webb SL, Dingwell DB (1990) Non-Newtonian rheology of igneous melts at high stresses and strain rates: experimental results for rhyolite, andesite, basalt, and nephelinite. J Geophys Res 95 (B10):15695–15701
White SM, Crisp JA, Spera FJ (2006) Long-term volumetric eruption rates and magma budgets. Geochem Geophys Geosys 7(3)
Whittington AG, Hofmeister AM, Nabelek PI (2009) Temperature-dependent thermal diffusivity of the Earth's crust and implications for magmatism. Nature 458(7236):319
Williams H, McBirney AR (1979) Volcanology. Freeman, Cooper, San Francisco
Wilson M (1989) Igneous petrogenesis: a global tectonic approach. Chapman & Hall, London, p 466
Winter JD (2010) Principles of igneous and metamorphic petrology. Prentice Hall
Wolf KJ, Eichelberger JC (1997) Syneruptive mixing, degassing, and crystallization at Redoubt volcano, eruption of December 1989 to May 1990. J Volcanol Geotherm Res 75(1–2):19–37
Wolff JA, Wright JV (1981a) Formation of the Green Tuff, Pantelleria. Bull Volcanol 44(4):681–690
Wolff JA, Wright JV (1981b) Rheomorphism of welded tuffs. J Volcanol Geotherm Res 10(1):13–34
Wood BJ (1974) The solubility of alumina in orthopyroxene coexisting with garnet. Contrib Miner Petrol 46(1):1–15
Wood BJ, Banno S (1973) Garnet-orthopyroxene and orthopyroxene-clinopyroxene relationships in simple and complex systems. Contrib Miner Petrol 42(2):109–124
Xirouchakis D, Lindsley DH, Frost BR (2001) Assemblages with titanite ($CaTiOSiO_4$), Ca–Mg–Fe olivine and pyroxenes, Fe–Mg–Ti oxides, and quartz: Part II. application. Am Miner 86(3):254–264
Yang H, Frey FA, Clague DA, Garcia M, O. (1999) Mineral chemistry of submarine lavas from Hilo Ridge, Hawaii: implications for magmatic processes within Hawaiian rift zones. Contrib Miner Petrol 135(4):355–372
Yoder HS, Tilley CE (1962) Origin of basalt magmas: an experimental study of natural and synthetic rock systems. J Petrol 3(3):342–532
Zack T, Moraes R, Kronz A (2004) Temperature dependence of Zr in rutile: empirical calibration of a rutile thermometer. Contrib Miner Petrol 148(4):471–488

Influence of Environment on Magma Properties, Eruption Processes, and Deposits

3

How Different Ambient Environments (Atmosphere, Oceans/Lakes, Ice, and Lithosphere) Affect Volcanic Eruptions and Deposits

Ray Cas, Guido Giordano, and John V. Wright

R. Cas et al., *Volcanology*, Springer Textbooks in Earth Sciences, Geography and Environment,
https://doi.org/10.1007/978-3-319-66613-6_3

Plate 3.1 Various environments in which volcanism occurs. **a** Subaerial magma or fire fountaining forming lava flows from a series of fissure vents at 3,000–3600 m above sea level, Mauna Loa, Hawai'i (Photo by M. Patrick, taken 29-11-2022; *Source* United States Geological Survey). **b** Subaerial vulcanian explosive eruption column and plume which rose to 12 km high above Soufrière Hills volcano, Montserrat, Lesser Antilles (Photo by B. Voight, taken 6-8-1997; *Source* National Science Foundation). **c** Shallow marine phreatomagmatic explosive eruption, Surtsey volcano, Iceland (Photo by Howell Williams, taken 30-11-1963; *Source* NOAA). **d** Deep submarine effusive pillow lava erupted on the Galapagos Rift spreading centre at ~2,400 m water depth (*Source* NOAA; ROV image taken July 2010). **e** Suppressed deep-water explosive blast from deep submarine vent, West Mata volcano, northeast Lau Basin at ~1,200 m water depth (*Source* NSF/NOAA; ROV image taken May 2009)

Abstract

In this chapter, we consider the properties of the different environments through which magmas pass, namely the Earth's lithosphere, atmosphere, hydrosphere, and glaciosphere, to assess the influence that these different environments have on eruption processes. The physical properties and conditions of the different environments and the ambient fluids through which magma travels are fundamentally important in influencing the properties of magma, the behaviour of volatiles, the styles of eruption, the dispersal of products, and deposit characteristics. Buoyancy is one of the most important factors driving the rise and the rise rate of magmas to the surface, and most magmas will rise as long as they are less dense than the ambient rocks they are rising through, because they have buoyancy. However, dense basalt magmas rise to the Earth's surface all the time, even though they are denser than the crust, and even rise through unconsolidated to weakly consolidated sedimentary basin fill successions with much lower densities. This indicates that other factors also drive magma upwards. These include the pressure difference between the mantle or crustal magma source area, where the pressure regime is lithostatic, and the Earth's surface (atmospheric, hydrostatic, and glaciostatic pressure regimes), upward momentum once upward migration begins, and volatile exsolution and bubble growth as magmas rise, which can significantly reduce the bulk density of magma, making them more buoyant than at the source. Once erupted, the thermal properties of ambient air, water, or ice will interact with the hot magma to determine the final eruption style. In subaqueous and subglacial environments, water/glacial depth and hydrostatic/glaciostatic pressure play fundamental roles in determining eruption and fragmentation processes. These environmental and crustal variables determine if magma reaches the Earth's surface or stalls in the subsurface, and if it erupts, whether the eruption is effusive and lava-forming, or explosive, and whether or not erupting lavas will be subject to quench fragmentation, an autoclastic process that is unique to subaqueous and subglacial environments. Another critical factor in understanding the behaviour of erupting magma and eruption styles and deposit types in subaqueous settings is the state of exsolved volatiles (supercritical, critical, subcritical, and gaseous fluids), as well as that of the external ambient water that magma comes in contact with, which we discuss in detail in this chapter. The environment has a clear effect on the eruption processes and deposits.

3.1 Introduction

Generally, when we consider the factors that determine eruption processes and deposit characteristics, we focus on the physical and chemical properties of erupting magmas. However, the nature of and the different physical properties and conditions of environments through which magma passes (lithosphere) and erupts into (subaqueous, subglacial, and subaerial) are fundamentally important in influencing the properties of magma, the behaviour of volatiles, the styles of eruption, and deposit characteristics (e.g. Plate 3.1).

In this chapter, we will briefly summarise the principal properties of the different lithospheric, aquatic, atmospheric, and glacial environments through which magma passes or into which it erupts and their effects on magma properties, adding to what was presented in Chap. 2. We will assess how these are important in understanding the differences in eruption styles and deposit characteristics in different environments, focusing on the Earth. However, this overview also provides insights into the influences of environmental conditions on eruption processes on the other planets for those working on planetary volcanology.

3.2 Lithospheric Conditions, Properties, and Effects on Magma Behaviour

3.2.1 Density of Earth's Crust, Mantle, Magmas, and Magma Buoyancy

The properties of the Earth's mantle and crust greatly affect processes in a magma reservoir or conduit. First, the density of the lithosphere at different depths (Fig. 3.1; Table 3.1) influences the relative buoyancy of magmas, which is one of the most important factors driving the rise and the ascent rate of magmas towards the Earth's surface. Mantle density ranges from about 5,600 to 3,400 kg m^{-3}, lithified continental crust has an average density of ~2,500–2,200 kg m^{-3}, varying from ~3,100 to <2,200 kg m^{-3}, and oceanic crust has a density of ~2,900 kg m^{-3}. Unconsolidated, usually water-saturated basin sediment successions have average densities of ~1,700 kg m^{-3}, depending on the degree of compaction (Tenzer and Gladkikh 2014; Fig. 3.1; Table 3.1).

In comparison, non-vesiculated, anhydrous magma densities range from ~2,900 to 2,700 kg m^{-3} for komatiites, ~2,850–2,700 kg m^{-3} for basalts, ~2,650–2,450 kg m^{-3} for andesites, and ~2,350–2,200 kg m^{-3} for silicic magmas (Spera 2000; Lesher and Spera 2015; Table 3.1; Chap. 2, Fig. 2.5a). Hydrous magmas of all compositions are ~10%

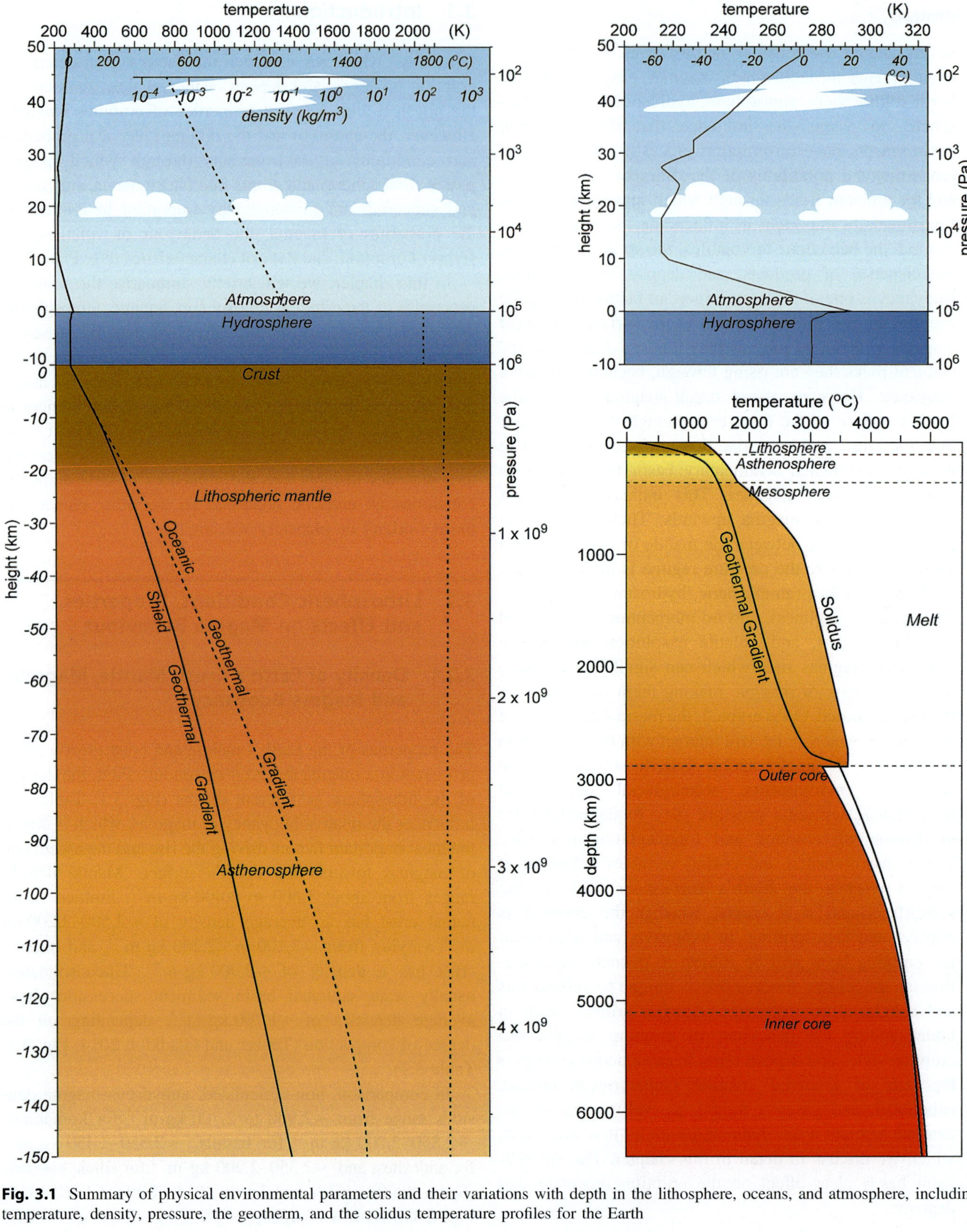

Fig. 3.1 Summary of physical environmental parameters and their variations with depth in the lithosphere, oceans, and atmosphere, including temperature, density, pressure, the geotherm, and the solidus temperature profiles for the Earth

Table 3.1 Densities of and pressure gradients in the atmosphere, surface bodies of water, unconsolidated sediments, continental or arc crust, oceanic basaltic crust, the mantle, and magmas

Physical entity	Density (kg m^{-3})	Pressure gradients (MPa km^{-1})
Air (atmosphere)	1.23–<0.001	−0.009
Water (~ice)	1,000	9.8
Seawater	1030	10.1
Continental/Arc crust	2,500–2,200	24.5
Unconsolidated sediments	>1,700	~16.5
Oceanic crust	2,900	28.5
Mantle (upper)	5600–3400	37.5
Felsic magma (non-vesiculated)	2,200	21.6
Andesite magma (non-vesiculated)	2500	24.5
Mafic magma (non-vesiculated)	2,700	26.5
Komatiite magma (non-vesiculated)	2,900	28.5

less dense at the same pressures (Spera 2000; Chap. 2, Fig. 2.5c), making them more buoyant than anhydrous magmas. As magma rises and decompresses, dissolved volatiles (principally H_2O or CO_2; Chap. 4) become saturated in the magma, exsolving to form low-density fluid-filled immiscible bubbles or vesicles that decrease the bulk density of magma because the magma expands to accommodate them. The density of vesiculated magma varies not only according to the abundance of vesicles, but also depending on the state of the fluid in the bubbles, be it supercritical fluid, dense vapour, or low-density vapour such as steam (Table 3.2a–c; Fig. 3.2a, b). In any case, the density of the magma decreases, making it more buoyant compared to non-vesiculated magma, and the increase in magma volume resulting in volatile bubble formation either leads to expansion of the magma upwards in the conduit, or increased total magma fluid pressure in the magma reservoir, both of which contribute to driving the magma upwards.

Magmas will try to rise if they are less dense than the ambient rocks they are enclosed in, because they are buoyant. They propagate dykes upwards as they rise, as long as the magma fluid pressure exceeds the minimum lithostatic principal stress component and the tensile strength of the country rock (Chap. 8). They will generally cease to rise when they arrive at a level in the crust, called the level of neutral buoyancy, where the host “rocks” have a similar density or lower density than the magma. The magmas are then neutrally (equal density) or negatively buoyant (excess density), respectively, and will be prone to form sill-like intrusions (Chap. 8). However, dense komatiite magmas reached the Earth’s surface in the Archean, and dense basalt magmas frequently rise to the Earth’s surface, even though they are denser than the crust. Basaltic magmas even rise through unconsolidated to weakly consolidated sedimentary basin fill successions with much lower densities (Table 3.2; Fig. 3.2), indicating that other factors also drive magma upwards. These include the pressure difference between the mantle or crustal magma source area (lithostatic), and Earth’s surface (atmospheric, hydrostatic, glaciostatic pressures); continued magma accumulation in magma reservoirs supplied from deeper sources, leading to increasing fluid pressure in the reservoir; upward momentum once upward migration begin; and, volatile exsolution and bubble growth as magmas rise, which can significantly reduce the bulk density of magma (Table 3.2; Fig. 3.2; Chap. 2), making them more buoyant and increasing magma fluid pressure in the conduit.

3.2.2 Confining Lithostatic Pressure, Pressure Gradients, and Effects on Magma Behaviour

Lithostatic pressure increases with depth below the Earth’s surface and varies according to the host country rock type and rock density (Fig. 3.3; Table 3.1). As a result, the lithostatic pressure gradient also varies, and is highest in the upper mantle (~34.3 MPa km^{-1}), lower in mafic oceanic crust (~28.4 MPa km^{-1}), lower again in continental or arc crust (~24.5 MPa km^{-1}), and lowest in water-saturated, unconsolidated sediments (~16.7 MPa km^{-1}; Fig. 3.3; Table 3.1). For comparison, hydrostatic pressure in a body of water increases at a rate of 9.8 MPa km^{-1} of water depth, and magmastatic pressures increase in a column of mafic magma at 26.5 MPa km^{-1}, at 24.5 MPa km^{-1} in a column of andesite, and at 21.6 MPa km^{-1} in a column of rhyolite magma (Fig. 3.3; Table 3.1).

Increasing confining lithostatic pressure with increasing depth below the Earth’s surface plays several very important roles in influencing magma properties and behaviour. These include:

Table 3.2 Densities of magmas with varying degrees of exsolved volatile bubbles[a,b]

(a) Bulk density of felsic magmas (rhyolite, dacite) with H_2O as the volatile phase												
Volatile bubble abundance (Vol. %)	Density of felsic magma ($\rho \approx 2200$ kg m^{-3}) with X % dense hot H_2O ($\rho \approx 800$ kg m^{-3}) bubbles	Density of felsic magma ($\rho \approx 2200$ kg m^{-3}) with X % dense H_2O steam ($\rho \approx$ 100 kg m^{-3}) bubbles	Density of felsic magma ($\rho \approx 2200$ kg m^{-3}) with X % H_2O steam ($\rho \approx 1$ kg m^{-3}) bubbles	Density difference ($\Delta\rho$) between variably vesiculated felsic magma and unconsolidated sediments $\rho = 1700$ kg m^{-3}			Density difference ($\Delta\rho$) between variably vesiculated felsic magma and continental crust $\rho = 2500$ kg m^{-3}			Density difference ($\Delta\rho$) between variably vesiculated felsic magma and oceanic crust $\rho = 2900$ kg m^{-3}		
				H_2O density (kg m^{-3})			H_2O density (kg m^{-3})			H_2O density (kg m^{-3})		
				800	100	1	800	100	1	800	100	1
0	2200	2200	2200	+500	+500	+500	−300	−300	−300	−700	−700	−700
10	2060	1990	1980	+360	+290	+280	−440	−510	−520	−840	−910	−920
20	1920	1780	1760	+220	+80	+60	−580	−720	−740	−980	−1120	−1140
30	1780	1570	1540	+80	−130	−160	−720	−930	−960	−1120	−1330	−1360
40	1640	1360	1320	−60	−340	−380	−860	−1140	−1180	−1260	−1540	−1580
50	1500	1150	1100	−200	−550	−600	−1000	−1350	−1400	−1400	−1750	−1800
60	1360	940	880	−340	−760	−820	−1140	−1560	−1620	−1540	−1960	−2020
70	1220	730	660	−580	−970	−1040	−1280	−1770	−1840	−1680	−2170	−2240
80	1080	520	440	−620	−1080	−1260	−1420	−1980	−2060	−1820	−2380	−2460
(b) Bulk density for mafic magmas with H_2O as a volatile phase (e.g. arc basalts)												
Volatile bubble abundance (Vol. %)	Density of mafic magma ($\rho \approx 2700$ kg m^{-3}) with X % dense hot H_2O ($\rho \approx$ 800 kg m^{-3}) bubbles	Density of mafic magma ($\rho \approx 2700$ kg M^{-3}) with X % dense H_2O steam ($\rho \approx$ 100 kg m^{-3}) bubbles	Density of mafic magma ($\rho \approx$ 2700 kg m^{-3}) with X % H_2O steam ($\rho \approx 1$ kg m^{-3}) bubbles	Density difference ($\Delta\rho$) between variably vesiculated mafic magma and unconsolidated sediments $\rho = 1700$ kg m^{-3}			Density difference ($\Delta\rho$) between variably vesiculated mafic magma and continental crust $\rho = 2500$ kg m^{-3}			Density difference ($\Delta\rho$) between variably vesiculated mafic magma and oceanic crust $\rho = 2900$ kg m^{-3}		
				H_2O density (kg m^{-3})			H_2O density (kg m^{-3})			H_2O density (kg m^{-3})		
				800	100	1	800	100	1	800	100	1
0	2700	2700	2700	+1000	+1000	+1000	+200	+200	+200	−200	−200	−200
10	2510	2440	2430	+810	+740	+730	+10	−60	−70	−390	−460	−470
20	2320	2180	2160	+620	+480	+460	−180	−320	−340	−580	−720	−740
30	2130	1920	1890	+430	+220	+190	−370	−580	−610	−770	−980	−1010
40	1940	1660	1620	+240	−40	−80	−560	−840	−880	−960	−1240	−1280
50	1750	1400	1350	+50	−300	−350	−750	−1100	−1150	−1150	−1500	−1550
60	1560	1140	1080	−140	−560	−620	−940	−1360	−1420	−1340	−1760	−1820
70	1370	880	810	−330	−820	−890	−1130	−1620	−1690	−1530	−2020	−2090
80	1180	620	540	−520	−1080	−1160	−1320	−1880	−1960	−1720	−2280	−2360
(c) Bulk density for mafic magmas with CO_2 as the volatile phase												
Volatile bubble abundance (Vol. %)	Density of mafic magma ($\rho \approx 2700$ kg m^{-3}) with X % supercritical CO_2 ($\rho \approx 800$ kg m^{-3}) bubbles	Density of mafic magma ($\rho \approx 2700$ kg m^{-3}) with X % dense CO_2 ($\rho \approx 450$ kg m^{-3}) bubbles	Density of mafic magma ($\rho \approx$ 2700 kg m^{-3}) with X % CO_2 gas ($\rho \approx 200$ kg m^{-3}) bubbles	Density difference ($\Delta\rho$) between variably vesiculated mafic magma and unconsolidated sediments $\rho = 1700$ kg m^{-3}			Density difference ($\Delta\rho$) between variably vesiculated mafic magma and continental crust $\rho = 2500$ kg m^{-3}			Density difference ($\Delta\rho$) between variably vesiculated mafic magma and oceanic crust $\rho = 2900$ kg m^{-3}		
				CO_2 density (kg m^{-3})			CO_2 density (kg m^{-3})			CO_2 density (kg m^{-3})		
				800	450	200	800	450	200	800	450	200
0	2700	2700	2700	+1000	+1000	+1000	+200	+200	+200	−200	−200	−200
10	2510	2475	2450	+810	+775	+750	+10	−25	−50	−390	−425	−450

(continued)

Table 3.2 (continued)

(c) Bulk density for mafic magmas with CO_2 as the volatile phase

Volatile bubble abundance (Vol. %)	Density of mafic magma ($\rho \approx 2700$ kg m^{-3}) with X % supercritical CO_2 ($\rho \approx 800$ kg m^{-3}) bubbles	Density of mafic magma ($\rho \approx 2700$ kg m^{-3}) with X % dense CO_2 ($\rho \approx 450$ kg m^{-3}) bubbles	Density of mafic magma ($\rho \approx 2700$ kg m^{-3}) with X % CO_2 gas ($\rho \approx 200$ kg m^{-3}) bubbles	Density difference ($\Delta\rho$) between variably vesiculated mafic magma and unconsolidated sediments $\rho = 1700$ kg m^{-3}			Density difference ($\Delta\rho$) between variably vesiculated mafic magma and continental crust $\rho = 2500$ kg m^{-3}			Density difference ($\Delta\rho$) between variably vesiculated mafic magma and oceanic crust $\rho = 2900$ kg m^{-3}		
				CO_2 density (kg m^{-3})			CO_2 density (kg m^{-3})			CO_2 density (kg m^{-3})		
				800	450	200	800	450	200	800	450	200
20	2320	2250	2200	+620	+550	+500	−180	−250	−300	−580	−650	−700
30	2130	2025	1950	+430	+325	+250	−370	−475	−550	−770	−875	−950
40	1940	1800	1700	+240	+100	0	−560	−700	−800	−960	−1100	−1200
50	1750	1575	1450	+50	−125	−250	−750	−925	−1050	−1150	−1325	−1450
60	1560	1350	1200	−140	−350	−500	−940	−1150	−1300	−1340	−1550	−1700
70	1370	1125	950	−330	−575	−750	−1130	−1375	−1550	−1530	−1775	−1950
80	1180	900	700	−520	−800	−1000	−1320	−1600	−1800	−1720	−2000	−2200

[a]This shows values for varying abundances of exsolved volatile bubbles, and volatiles in varying states and densities, including very dense hot H_2O, dense vapour, and expanded low-density vapour, such as steam. The density of supercritical H_2O is ~ 322 kg m^{-3} (Pioro and Mokry 2011). Also shown are the density differences between variably vesiculated magma of differing compositions and unconsolidated basinal sediments, continental or arc crust, and oceanic crust

[b]The density difference data reflect the buoyancy state between magmas and their host, negative values indicating the magma is buoyant, and positive values indicating that the magma is not buoyant. Interestingly, magmas that are not density contrast buoyant still reach the Earth's surface, indicating that other factors than just buoyancy determine if magmas reach the surface (see text)

- changing the density of magma,
- changing the viscosity of magma,
- constraining the depth of exsolution of volatiles,
- controlling the fluid state of exsolved volatiles and explosive behaviour, and
- creating a pressure head in a column of magma.

3.2.2.1 Magma Density Decrease with Lithostatic Pressure Decrease

Confining lithostatic pressure increases with increasing depth below the Earth's surface (Figs. 3.1, 3.3), which causes magma density to increase with increasing depth and pressure (Fig. 2.5a, b; Chap. 2; Kushiro 1980; Stolper and Walker 1980). Conversely, magma density decreases during ascent due to decreasing lithostatic pressure, and magma becomes more buoyant, driving it upwards. The slight volume increase with decreasing density also increases fluid pressure in the confined magma, also helping to drive it upwards.

3.2.2.2 Magma Viscosity Decrease with Lithostatic Pressure Increase

Kushiro et al. (1976) and Kushiro (1980) showed that the viscosity of magma, even in an anhydrous state, decreases with increasing (confining) pressure (Chap. 2, Fig. 2.11b). In addition, high confining pressure prevents or limits the exsolution of magmatic H_2O, which remains dissolved in the magma and in turn has the effect of lowering the magma viscosity. This means that the viscosity of a specific magma is lowest when it is in a reservoir or conduit in the mantle or crust and slowly increases as the magma ascends towards the Earth's surface. In addition, viscosity may also increase slowly as volatiles exsolve from the magma and when volatile exsolution-driven crystallisation (e.g. of microlites in the groundmass magma) occurs (Chaps. 2, 4). Viscosity increases reduce magma mobility during ascent.

3.2.2.3 Lithostatic Pressure as a Control on the Depth of Volatile Exsolution in the Crust, and Magma Viscosity

As magma rises, it is subject to decreasing lithostatic pressures and decompresses, so the likelihood of volatile exsolution occurring increases (Chap. 4). The depth at which exsolution occurs is dependent on volatile concentration, lithostatic pressure (and therefore depth in the crust), and magma properties. Volatiles can exsolve at any depth in the crust or mantle if volatile content is high enough, relative to solubility behaviour in the magma and the pressure constraints on volatile solubility at depth (Chap. 4). The greater the initial magma volatile content, the deeper is the crust that

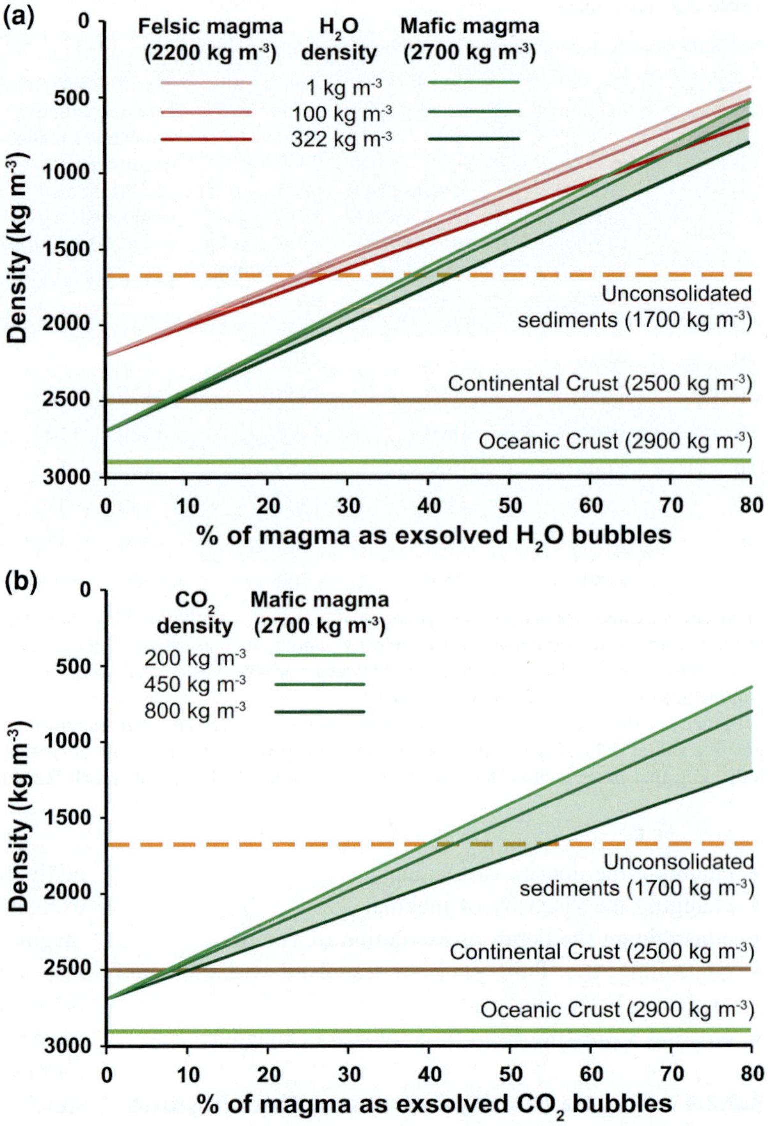

Fig. 3.2 Densities of dense to vesiculated magma relative to crustal densities. **a** Densities of felsic and mafic magmas in varying degrees of vesiculation with H_2O as the exsolved volatile phase (supercritical H_2O, dense steam, and low-density steam), relative to the densities of oceanic crust, continental (or arc) crust, and water-saturated unconsolidated, basin sediments. **b** As for (**a**) but for mafic magma with exsolved CO_2. Regions on the plot where magma densities lie below the lines depicting the density of different hosts indicate that the magmas are not positively buoyant, whereas the fields above the lines depicting particular host density lines indicate that the magmas are buoyant

exsolution is likely to occur. If the confining pressure is higher than the saturation pressure of the volatile phases, volatiles remain dissolved in the magma. Since many volatiles, especially H_2O and halogen gases, make magma less viscous when dissolved in magma (Chap. 2), the magma is therefore more fluidal or less viscous prior to volatile exsolution and becomes more viscous during volatile exsolution. In addition, confining pressure controls the state of the exsolved volatiles, as discussed next, which controls whether or not explosive eruption can occur.

3.2.2.4 Lithostatic Pressure as a Control on the State of Exsolved Volatiles and the Depth of Explosive Potential: The Critical Point of Volatiles

Most volatile species can exist in multiple states or phases, depending on temperature and pressure. For example, water and carbon dioxide occur in solid (ice), liquid, gaseous (steam), critical, and supercritical states (Fig. 3.4a–c). The state that drives explosive eruptions is the (subcritical) gaseous state, which, when at high magmatic temperature and pressure, has a very high expansion coefficient and capability to expand instantaneously and explosively, particularly if the ambient pressure is much lower than the volatile pressure. However, when volatiles first exsolve to form fluid bubbles in magma conduits and during magma ascent to the Earth's surface during decompression, the volatiles are not in a gaseous state, but are mostly significantly denser critical or supercritical fluids (Table 3.3; Fig. 3.4a–c). This is because of the high lithostatic and magmastatic pressures they are subject to beneath the Earth's surface, or in very deep water or under km of ice. A critical fluid forms at the critical point, which is a combination of a critical pressure and critical temperature at which the volatile transforms from a gaseous

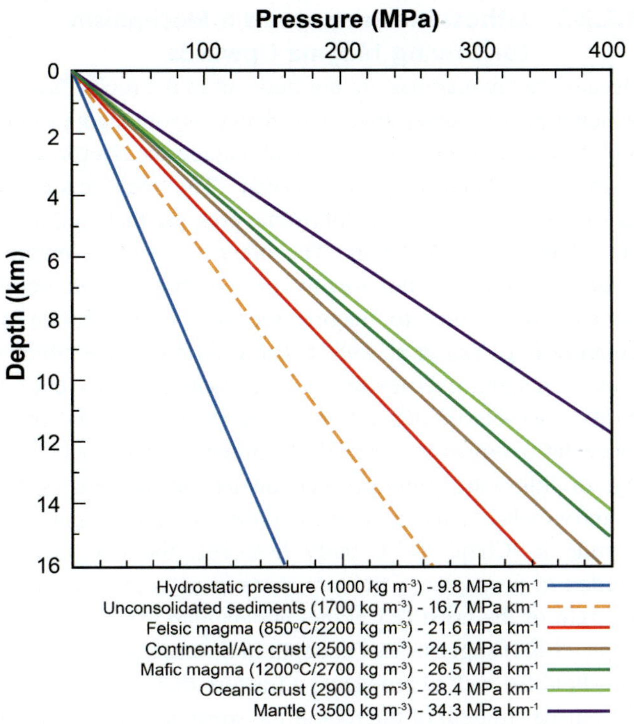

Fig. 3.3 Comparison of lithostatic pressure gradients in the mantle, oceanic crust, continental or arc crust, unconsolidated, water-saturated sediments, with hydrostatic pressure in a body of water, and magmastatic pressures in columns of felsic and mafic magmas (From Cas and Simmons 2018)

state into a critical fluid, or vice versa. Critical fluids have properties of both liquid and gaseous phases (Chap. 4) and have low specific volumes, and very low expansion capability (or expansivity coefficient), virtually precluding explosive expansion behaviour. Supercritical fluids are at pressures and temperatures above their critical point (Fig. 3.4a–c) and are also highly compressed and significantly denser than steam (Table 3.3).

As the magma continues to rise in the conduit, decompressing supercritical fluids in vesicles can boil to gases (subcritical fluids) when the confining pressure becomes less than the critical pressure for a volatile species (Fig. 3.4a–c). As a result of this phase change, the gas can expand, leading to higher gas pressures, and enhancing the potential for explosive eruption behaviour, but not guaranteeing it. The critical pressure of H_2O is ~22 MPa (Fig. 3.4a, c), and for CO_2, 7.4 MPa (Fig. 3.4b). If a magma conduit is closed to the atmosphere, the transition from supercritical H_2O fluid to H_2O gas in volatile bubbles already exsolved from the rising magma occurs at a depth of about 900 m (which is the depth in the crust where the lithostatic pressure is equal to the critical pressure of H_2O). In a conduit open to the atmosphere, the depth in a dense magma column at which this transition occurs is ~833 m in a column of mafic magma ($\rho = 2700$ kg m^{-3}), at which depth the magmastatic pressure equals the critical pressure for H_2O, ~900 m for an

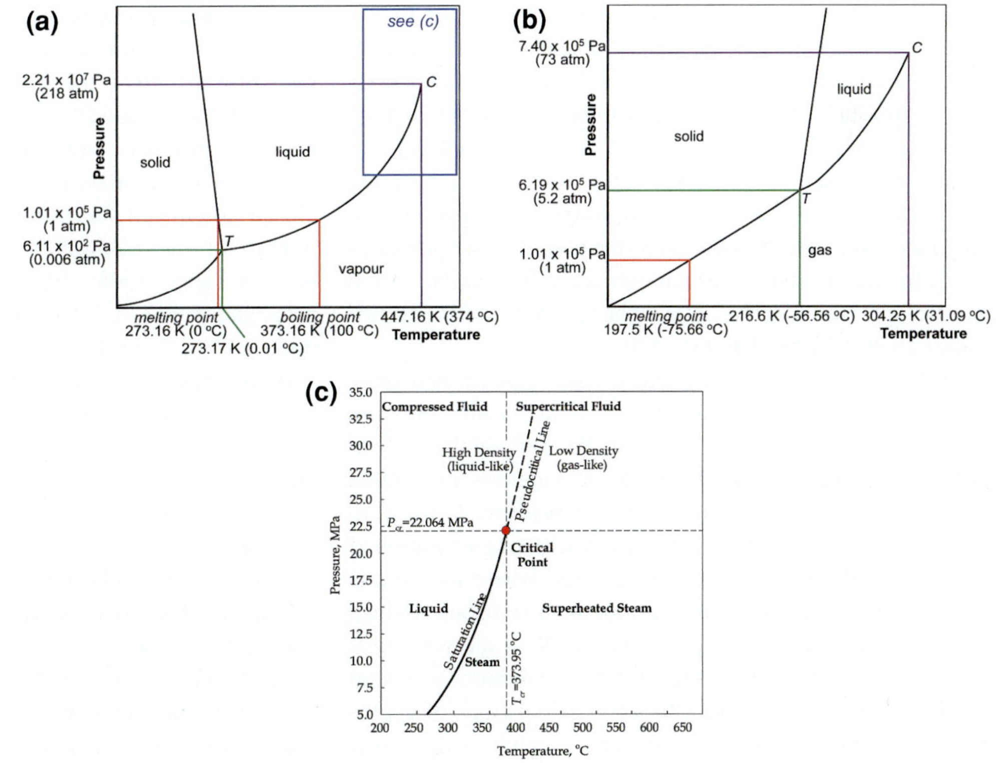

Fig. 3.4 Phase diagrams for **a** H_2O and **b** CO_2, showing the fields for gas, liquid, and solids, and the critical pressures. T is the triple point between the phases, C is the critical point which for water is at a pressure of 22.1 MPa, and temperature of 374 °C, and for carbon dioxide at a pressure of 7.4 MPa and temperature of 31.1 °C. (From Cas and Giordano 2014).
c Expanded area within the blue box in (**a**) shows the phase relations between liquid water, subcritical gaseous water, and supercritical water at high pressures and temperatures leading up to and beyond the critical point of water (From Pioro and Mokry 2011, courtesy of I. Pioro)

andesite magma (ρ = ~2500 kg m^{-3}), and at ~1020 m for rhyolite (ρ = 2200 kg m^{-3}). As the levels of vesiculation increase, the magmastatic pressure decreases because the bulk density decreases (Table 3.2), and so the "boiling" depths in a column of magma are greater; i.e. the exsolution and boiling depths in the vesiculated magma become deeper. The perhaps surprising similarity of depths of volatile boiling from supercritical fluid to vapour or steam in the crust (~900 m) and in an open column of magma (~800–1000 m) is because the density and lithostatic pressure of the crust is similar to the density and magmastatic pressure in an open column of non-vesiculated silicate magma (Table 3.1). However, rapid explosive expansion of gas filled bubbles will only occur when the gas overpressure far exceeds the sum of the magmastatic pressure and the tensile strength of magma in the bubble walls, causing instantaneous bubble volume (specific volume) expansion (see Fig. 3.9). This occurs at pressures of ~10 MPa (see Fig. 3.9), considerably less than the CP and between the CP equivalent depth and the fragmentation surface in the magma column.

If the principal volatile phase is CO_2, the transition from supercritical fluid to gas will only occur in a closed conduit in the crust at depths of ~302 m, and in an open column of mafic magma, at a depth of ~280 m. Gas-driven explosive fragmentation will begin in conduits only at pressures and equivalent depths significantly *below* the critical pressure equivalent because the gaseous phase has to be a low-density compressible phase capable of rapid explosive expansion (Chap. 4). Once exsolution and boiling of volatiles to gas have occurred, and the magma is vesiculated, the magma density above the level of exsolution becomes less, as does the magmastatic pressure in a column of magma and its density, making it more buoyant. The exsolution level, the boiling level, and the fragmentation level in the magma column (Chap. 4) will therefore migrate deeper in the magma column and the conduit as the magmastatic pressures decrease due to volatile exsolution and vesiculation, if the external pressure regime (atmospheric, hydrostatic, glaciostatic) allows (Table 3.1; Fig. 3.3).

Although hydrostatic and glaciostatic pressure and pressure gradients are not as high as lithostatic pressure (Table 3.1; Fig. 3.3), they can also be sufficiently high in subaqueous and subglacial settings to suppress or subdue volatile exsolution, boiling (from supercritical fluid to vapour or steam), and potential explosive fragmentation if the water and ice depth are deep enough where the vent is located (see Sect. 3.5.4). In addition, in subaqueous settings the hydrostatic pressure due to water depth at the vent must be included together with the lithostatic pressure as a total combined confining pressure factor in calculations of depths of volatile exsolution in the magma conduit beneath the seafloor.

3.2.2.5 Lithostatic Pressure as a Mechanism for Driving Magma Upwards

Because many magmas are not buoyant in the rocks through which they pass, other driving mechanisms for magma ascent and propagation must exist. These include the overpressure inside a confined magma reservoir. The overpressure is defined in reference to the lithostatic pressure that confines a closed magmatic reservoir. The pressure inside a magma reservoir builds up dominantly in response to volumetric increase due either to magma recharge or exsolution of volatiles (e.g. Tait et al. 1989). The response of the country rock to magma overpressure is deformation, partly elastic, partly visco-elastic depending on the rheology of the crust. Once this overpressure exceeds the minimum principal stress, σ_3, of lithostatic pressure and the tensile strength of the country rocks, then a fracture can propagate, and magma will intrude it (Chap. 8). Usually however, the overpressure needed to propagate a fracture is taken as twice the tensile strength of the crust (e.g. Pinel and Jaupart 2003). Cañón--Tapia (2008) suggests that at depth in the crust or mantle, the threshold for rock to propagate a fracture may be higher than the tensile strength of the rock if the stress field is not purely lithostatic, but affected by external tectonic stresses, although assessing the critical stress threshold is very difficult.

The process of country rock fracturing and magma intrusion withdraws the magma from the reservoir lowering its overpressure until magmatic pressure becomes equal to the lithostatic pressure according to $P_m = P_L$. In principle, fractures may propagate in any direction from the magma (pressure) reservoir source. However, fractures that propagate upward drive the intruding magma towards regions where the lithostatic pressure confining the propagating dyke becomes progressively lower and lower, i.e. towards the Earth's surface. Pascal's Principle determines that in a closed column of fluid, the highest pressure applied at the bottom of the column is transmitted to the whole fluid (Cañón-Tapia 2008; Fig. 3.5a). Therefore, during dyke propagation, the fluid overpressure inside the magma conduit, derived from the lithostatic pressure at the base of the vertical conduit, is more efficient in propagating fractures upwards at the top and tip of the propagating dyke, where the lithostatic pressure is lower, compared to any other direction.

The pressure gradient in the crust or lithosphere from the bottom (highest lithostatic pressure) of a magma column to its top (lowest lithostatic pressure), therefore, facilitates migration of magma upwards, in addition to any positive buoyancy it may have due to density contrasts between magma and host rock, and in some cases, in spite of negative buoyancy contrasts (e.g. basalt rising through lower density crust). Since lithostatic pressure is lower high in the crust than deeper, it should also be easier for dykes to open

Table 3.3 Properties of the crust, water in its different states, ice, air, and factors that are significant in determining differences in eruption conditions, eruption styles, and eruption products between subaerial and subaqueous environments[a]

Physical property	Crust	Ice	Fresh water	Seawater	Steam (at °C, MPa)	Critical Water (critical point P = 22 MPa; T = 374 °C)	Supercritical water at >P and T than CP	Air/atmosphere	Difference factor (seawater/air)
Heat capacity (kJ kg^{-1} K^{-1})	~2	~2	4.187	~4.0	1.96	~90 (or higher)	50–3	1.005	4x
Thermal conductivity (W m^{-1} K^{-1})	~2.0–2.5	~2.0–2.6	0.58	0.61 at 25 °C	0.02 (100, 0.1)	>0.4 (to infinite?)	0.5–0.12	0.025 at 15 °C	25x
Density (kg m^{-3})	~2500 (cont. or arc crust) ~2900 (oceanic crust)	930	1000	1025	0.6 (100, 0.1) 15 (234, 0.3) 30 (286, 7) 50 (318, 11) 100 (342, 15)	322	100–200	1.23–<0.001	1000x
Viscosity (Pa s)				~0.001	0.000013	~30	25–35	0.000018	55x
Pressure gradient (MPa km^{-1})	24.5 (cont. or arc crust) 28.5 (oceanic crust)	9.3	9.8	~10				−0.009	1100x
Young's/Bulk modulus (MPa)	20,000–70,000	9,700–11,200	2,300	2,340	~0.1			0.142	16,500x

[a]Data from multiple online sources including Engineering Toolbox (www.EngineeringToolbox.com), Pioro (2020), Pioro and Mokry (2011), Sharqawy et al. (2010)

fractures high in the crust than deeper because σ_3 is lower shallower in the crust. Some analogue models show that magma parcels in dykes may be squeezed upwards by the tendency of the bottom of the fracture to close compared to the tip, because of the differential horizontal component of the lithostatic load (e.g. Valentine and Gregg 2008).

If a dyke-conduit breaks to the surface and becomes a vent for an effusive eruption (i.e. for the sake of this discussion, we disregard the most common cases where the upward migration of magma is mostly due to the presence of an exsolved volatile phase that promotes lowering of bulk density and higher buoyancy), we reach open conduit conditions (Fig. 3.5b, c). In this condition, we will have two main phases: (1) a transient eruptive phase, during which the magma overpressure is progressively exhausted controlling the intensity and duration of the lava effusion; in this condition, the magmastatic pressure exerted by the magma column on the base of the reservoir ($\rho_m gh$, where ρ_m is the magma density, g is the acceleration of gravity, and h is the conduit length) is less than the magma pressure inside the reservoir, which has an overpressure component due to elastic/visco-elastic deformation of the crust in response to magma withdrawal, tectonic stresses, or pressure due to volatile exsolution; (2) a static phase when the eruption stops and the pressure of the magma column $\rho_m gh$ fully equilibrates the magmatic pressure inside reservoir level, in turn, equal to the lithostatic pressure at that level, so that $\rho_m gh = \rho_c gh$, where ρ_c is the crustal density. These latter conditions are typical of lava lakes (e.g. Nyiragongo, Democratic Republic of the Congo; Erta Ale, Ethiopia; Mt Erebus, Antarctica), or stationary open conduit volcanoes (e.g. Stromboli, Italy; Villarrica, Chile).

Given that the lithostatic pressure gradient is higher in oceanic crust than in continental or arc crust (Fig. 3.3), the absolute values of confining lithostatic pressure in oceanic crust should be greater than in continental crust at comparable depths. A higher absolute magmatic pressure may result in faster rise rates and longer eruptions of one-phase liquid magma (i.e. again disregarding the role of volatile exsolution which in fact is probably orders of magnitude more important, even for effusive eruptions of basaltic magmas; Chaps. 4, 5). Another consequence may be the

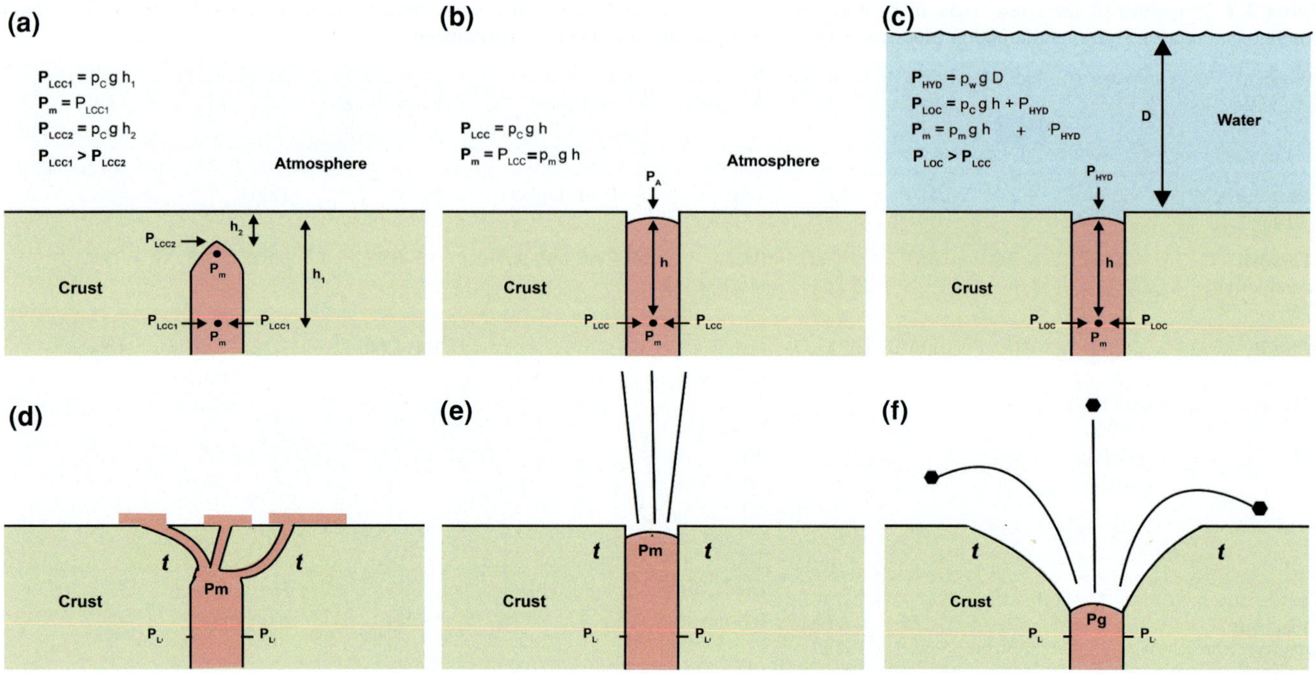

Fig. 3.5 Pressure factors contributing to the pressure regime of a rising column of magma. **a** A column rising through the crust to the Earth's surface in a subaerial environment, with the fluid pressure being equal to the highest lithostatic pressure in continental crust (P_{LCC1}). **b** As for (**a**) in an open vent, with pressure at depth in the magma column being the magmastatic pressure at depth, P_m (=~lithostatic pressure, P_{LCC}). **c** Magma erupting subaqueously through oceanic crust, so that the magma fluid pressure at depth is the sum of the magmastatic (P_m) (=~lithostatic pressure, P_{LOC}) and hydrostatic (P_{HYD}) pressure components. Note that because the lithostatic pressure gradient in oceanic crust is higher than in continental or arc crust, $P_{LOC} > P_{LCC}$, the effect of lithostatic pressure in driving magma upwards should be higher in oceanic crust than in continental crust. **d** Irregular feeder dyke system to lavas where magma pressure Pm >> the country rock tensile strength, *t*, allowing magmas to irregularly find the easiest paths to the surface. **e** Magma pressure, Pm, is sufficient to dilate fractures to open a conduit to the surface where the country rock strength, *t*, is high, but a little lower than Pm. **f** Flaring-shaped vent in weakly lithified country rock of sediments, with a significantly higher magma or gas pressure, Pm or Pg, that can easily explosively excavate sideways due to high gas overpressure relative to low country rock strength, *t*

maximum height of effusive volcanoes, where the building of the edifice progressively extends the length of the magmatic conduit until it reaches magmastatic pressures that largely equilibrate the magmatic pressures inside the reservoir, also equal to the lithostatic pressure at the reservoir level, resulting in equilibrium conditions.

3.2.3 Mechanical Strength of Country Rock

The tensile strength of most crustal rocks varies from 2 to 20 MPa (Perras and Diederichs 2014), with unconsolidated sediments having very low mechanical strength (<2 MPa), weakly lithified sedimentary and tuffaceous rocks having low strengths of <5 MPa, whereas dense igneous and metamorphic rocks have higher strengths. This is significant in several ways for understanding the behaviour of magma in the crust and during explosive eruptions. Firstly, for magma to be able to rise through the crust to erupt on the Earth's surface, it must dilate existing fractures or propagate new ones, creating a magma plumbing system of dykes and sills (Chap. 8). To do this, the magma fluid pressure, *Pm*, must exceed the minimum principal stress component of the lithostatic pressure *and* the tensile strength of the country rock, τ. In a vertically rising column of magma, the minimum principal stress component of lithostatic pressure is lowest at shallower levels in the crust (i.e. at the top of the magma column compared with at the base), and so it is easier for magma to propagate fractures upwards, in the direction of the pressure gradient. Since many magmas reach the Earth's surface, this indicates that magmas can fracture host rocks and dilate existing fractures if their fluid pressure exceeds the above criteria.

Secondly, the strength of crustal host rocks, the homogeneity or heterogeneity of any layering in the crust, and the regional stress field will determine the geometry of intrusions, including the magma plumbing system to volcanoes (Chap. 8). If the country rock is lithified and Pm > *t*, the intrusions and magma conduits are likely to be bound by planar surfaces (lower parts of Fig. 3.5d–f), whereas if the host material is unconsolidated sediment, with little mechanical strength, (*Pm* >>> τ), the intrusions are likely to have very irregular margins and geometries (upper part of Fig. 3.5d; Chap. 8), and there may even be evidence of dynamic interaction and mixing between magma and sediment (e.g. formation of peperite; Chap. 4). Finally, the geometry of an explosive vent or diatreme will be influenced by the mechanical strength of the host crust. In strongly lithified crustal rocks, if magma pressure, Pm, or for an explosive eruption, magma gas pressure, Pg, are slightly higher than the tensile strength of the country rock ($Pg > \tau$), vents will be narrow and steep-sided (Fig. 3.5e), with most of the explosive energy being directed upwards, out of the vent. In weakly consolidated hosts, the vent will be wider and have shallowly dipping sides because the explosive energy can easily excavate sideways as well as upwards (i.e. $Pg >>> \tau$), leading to an open flaring vent (Fig. 3.5f; Cas et al. 2008).

3.2.4 Deformability/Compressibility of Rock—Why Volcanic Vents Are Excavated Upwards and Explosive Energy Goes Up

Solid rock is not readily deformable and compressible when affected by instantaneous pressure such as that generated by an explosive eruption. Young's Modulus (E, ratio of stress to strain below the yield point) for different rocks varies from about 10 to 70 GPa, or 10,000 to 70,000 MPa (e.g. Karagianni et al. 2010), and for a range of volcanic rock types from 2,000 to 40,000 MPa (Heap et al. 2020; the variation is due to varying levels of porosity), which is orders of magnitude higher than for deformable fluids such as water and gas (Table 3.3). Rock is elastic up to its yield point and has a high level of resistance to permanent deformation. It will fail in a brittle fashion when subject to instantaneous pressure (e.g. explosion generated pressure) that exceeds the yield strength/tensile strength. In addition, the confining lithostatic pressure increases downward making rocks progressively more resistant to brittle failure.

In comparison, the bulk modulus of oceanic seawater is ~2,340 MPa, at least an order of magnitude less than most rocks, and that of air is 0.142 MPa, five orders of magnitude less than most rocks (Table 3.3). As discussed further in Sects. 3.3.5 and 3.5.5, explosive magma fragmentation in volcanic conduits generates enormous pressures associated with volatile expansion. Most of the explosive energy is directed upwards, in the direction of the lithostatic pressure gradient, and particularly once a vent has been opened, the explosive energy is directed out of the vent and upwards into the ambient fluid (air, water) medium. Both seawater and air are much more deformable because their bulk moduli are orders of magnitude lower than that of the lithosphere, and in addition, the ambient pressures of the atmosphere, oceans, and lithosphere are significantly different. As a result, the intensity of subaerial explosions is significantly greater than subaqueous ones, which are more intense than those that may occur in a subterranean setting.

When the pressure in a gas–pyroclast mixture in the vent exceeds the ambient pressure, the mixture can also expand laterally, exerting pressure on and eroding the country rocks around the vent. Country rock around the vent can only absorb a limited amount of instantaneous strain resulting from explosions, but this causes it to undergo brittle fracturing, liberating loose accessory lithic clasts into the gas-pyroclastic stream in the conduit (Chap. 4) as well as producing an envelope of remnant in situ fractured country rock (Barnett 2008; Barnett et al. 2011). The typical conical shape of explosive vents thus represents the envelope of the progressive equilibration between gas–pyroclast mixture pressure and the mechanical strength of the country rocks. Rocks can also undergo brittle failure when pressure in the conduit and in the vent drops during waning explosive phases, generating gravitational rock wall collapses, producing further accessory lithics and even clogging up the vent.

3.2.5 Temperature Gradient in the Earth's Crust and Mantle

Temperature increases with increasing depth beneath the Earth's surface (Fig. 3.1). This is important in two ways. Firstly, the mechanical properties of subsurface rocks change from brittle to ductile with increasing depth, causing changes to the physical regime through which the magma migrates during ascent. In the brittle part of the crust, magma must propagate or dilate brittle fractures to allow it to migrate, producing planar dykes and sills. In the deeper ductile crust and mantle, and also around shallow, long-lived, and large crustal magma chambers, the propagation of brittle deformation is not favoured and requires much higher strain rates (e.g. Jellinek and DePaolo 2003). Alternatively, migration paths are likely to be less well-defined and geometrically much more irregular, perhaps even channelised. Secondly, with increasing depth and temperature (Fig. 3.1), crustal and mantle rocks are closer to their solidus temperature. In these conditions several triggers, such as decompression caused by tectonic movements, increasing temperature due to rise of a deeper, hotter magma or metasomatic fluid, or addition of water, could lower the solidus, initiating partial melting at a particular depth.

3.2.6 Fluids in the Crust

Country rock crossed by intrusions of dykes and conduits may, and usually are, the host to different types of fluids, including groundwater, hydrothermal systems, and hydrocarbon reservoirs. These may flow at different rates according to the primary and secondary permeability of the host rocks, i.e. the geothermal/hydrothermal reservoirs, aquifers, and hydrocarbon reservoirs. The degree of interaction of the rising magma with these external fluids may significantly affect the eruption style. The physical principles of magma–water interaction known as Molten Fuel–Coolant Interaction (MFCI) which apply to surface environments, also apply to subterranean settings and are discussed later in this chapter and Chap. 4. At one end of the spectrum, thermal and mechanical interaction between magma and water can cause autoclastic quench fragmentation, whereas when heat exchange becomes very efficient under relatively low confining ambient pressures, explosivity may significantly increase in intensity giving place to an eruption mechanism known as phreatomagmatism. Both are discussed further below. In Chap. 18, we discuss the relationship between magmatism and hydrocarbon generation and reservoirs in volcanic settings.

3.3 Atmospheric Effects

3.3.1 Density and Density Gradient of Air, and Effects on Explosive Eruption Columns

Air has a density of 1.23 kg m^{-3} at ground level, and this rapidly decreases with altitude (Fig. 3.1), being 0.74 kg m^{-3} at 5 km, 0.41 kg m^{-3} at 10 km, 0.09 kg m^{-3} at 20 km, 0.02 kg m^{-3} at 30 km, 0.004 kg m^{-3} at 40 km, and 0.001 kg m^{-3} at 50 km (Fig. 3.1; Table 3.1). By contrast, typical hot volcanic gas has a density of 0.001–0.005 kg m^{-3} at magmatic temperatures, making it very buoyant by itself in the atmosphere. However, during an explosive eruption, volcanic gas will contain variable proportions of suspended ash, which increases the bulk density of the explosive gas-ash eruption column. Counteracting this, cold air that is mixed into the margins of an explosive eruption column will be heated to the volcanic gas temperature, so becoming less dense and thereby increasing the buoyancy of the eruption column. The density of the hot gas and the bulk density of the explosive eruption column relative to that of the atmosphere at different altitudes will influence the buoyancy of gas plumes and explosive eruption columns and downwind plumes (Plate 3.1a, b). This determines the altitude to which the column will rise buoyantly above a vent, before reaching a level of neutral buoyancy, at which it will spread laterally in the atmosphere (Chap. 9). Given that the altitude reached will influence dispersal areas and patterns, the relative densities of the atmosphere, gas plumes, and explosive eruption columns are fundamental to determining to what degree eruption columns are affected by winds, and the dispersal patterns of gas and fallout tephra (Chap. 9). Gas and ash that rise to the level of the stratosphere (≥ 11 km), and into the upper-atmosphere jet stream can be dispersed globally and can even circumnavigate the globe several times. If the volumes of aerosol gases and suspended ash are large enough, they can affect global weather patterns and even long-term climate. If an explosive eruption column becomes significantly denser than the surrounding atmosphere, parts may collapse gravitationally and generate pyroclastic density currents (Chap. 10). In contrast, the atmospheric density on Mars is about one-third that on Earth, making the explosive eruption column less buoyant there, whereas the atmospheric density of Venus is ninety times denser than Earth's, making explosive eruption columns much more buoyant.

3.3.2 Pressure and Pressure Gradient in the Atmosphere

The atmospheric pressure at the Earth's surface is 1 atm or $\sim 1.01 \times 10^5$ Pa (0.101 MPa) and also decreases with altitude (Fig. 3.1) at a rate of an order of magnitude every 15 km, or about 0.009 MPa km^{-1} (Fig. 3.1; Tables 3.1 and 3.3). During explosive eruptions, this atmospheric pressure is so low that it has no suppressing effects on whether or not an explosive eruption occurs, because gas pressures in vesiculated magmas can reach many MPa, and even tens of MPa for very explosive eruptions (Chap. 4). However, because volcanic gases are hot and pressurised, they expand under atmospheric conditions (Plate 3.1a, b). The more the hot gas expands, the less dense and more buoyant it becomes; this is another factor contributing to the buoyancy of explosive eruption columns.

3.3.3 Wind and Effects on Dispersal of Gas and Volcanic Fallout Deposits

Wind strength strongly influences dispersal processes and patterns of fallout deposits from buoyant explosive eruption columns (Chap. 9). Weak winds during an explosive eruption produce a more symmetrical radial dispersal pattern of fallout deposits (Chap. 9). Strong unidirectional winds produce asymmetrical fans of dispersal of fallout deposits, and atmospheric wind direction stratification may cause deposits to be dispersed in different directions from different altitudes in an explosive eruption column (Chap. 9). Wind patterns change significantly across the troposphere and stratosphere, and therefore plinian explosive eruption columns which involve stratospheric injections may significantly extend their dispersal due to the strength of wind jet streams

(Chap. 9). Winds may even influence the flow behaviour of weak, low-concentration pyroclastic density currents, causing them to be deflected in a downwind direction (Chap. 10).

3.3.4 Temperature, Thermal Conductivity, and Heat Capacity of Air: Effects on Explosive Eruption Columns

The temperature of the atmosphere, which generally ranges from –40 to +40 °C, is obviously significantly less than that of freshly erupted volcanic gas, pyroclasts, and magma, at several hundreds of °C magmatic temperatures. The specific heat capacity of air varies according to temperature, but at atmospheric pressure at the Earth's surface, a value of 1.006 kJ kg^{-1} K^{-1} is widely used. This is less than a quarter the heat capacity of water (see Sect. 3.5.1). Air has a low thermal conductivity of 0.025 W m^{-1} K^{-1} at 10 °C, and although this increases to 0.052 W m^{-1} K^{-1} at 400 °C, it is still more than 20 times less than that of water (Table 3.3), and so little heat transfer occurs by conduction in the atmosphere. Heat can be transferred from magma or hot pyroclasts to the atmosphere through radiation, and especially through the convective mixing of hot gas and hot suspended particles with cold air. This is important in determining the dynamics of explosive eruption columns and pyroclastic flows. Cold atmospheric air mixes into hot, gas-rich, explosive eruption columns, becomes heated, and expands, increasing the turbulence levels, decreasing the bulk density of the eruption column, and causing it to become more buoyant (Chap. 9). Air mixed into the margins of pyroclastic density currents, especially longer lived pyroclastic flows, is also instantaneously heated, expands, and increases the turbulence levels of the flow margins, which become more dilute generating ash-cloud surges or buoyant ash-cloud plumes (Chap. 10).

3.3.5 Deformability/Compressibility of Air—Why Subaerial Explosions Are so Intense

Air is very compressible and so absorbs explosive strain during explosive eruptions, allowing the explosive energy and shock waves to be distributed widely away from the vent. The elastic property of a fluid or solid, called the bulk modulus (K; unit is Pa), is a measure of its resistance to changes to its volume. It is defined as the ratio between change in the confining pressure and the dilation related to the change in pressure. The reciprocal of the bulk modulus is the compressibility (Birch 1942a, b). The bulk modulus of pure water is 2.15×10^9 Pa, for seawater it is 2.34×10^9 Pa, for solid rock it is $0.2–2 \times 10^{11}$ Pa, whereas for air the adiabatic bulk modulus is only 1.42×10^5 Pa (Table 3.3), meaning that air is orders of magnitude more compressible and deformable than water and rocks. This is why subaerial eruptions are much more explosive than subaqueous and subterranean explosions, and why in an open vent the explosive energy is directed upwards into the atmosphere.

3.4 Vegetation

Vegetation has no effect on an eruption, but interacts with the dispersal processes of eruptions in a number of ways. A thick forest will contribute to the roughness of the substrate affecting the mobility of lavas, at least at their initial propagation stage, especially pahoehoe lavas (Chap. 5). When lavas invade vegetated areas and forests, they ignite them, and this may lead to weak explosions and preservation of tree moulds. Pyroclastic flows and surges (Chap. 10) are more directly affected by vegetation. Firstly, heavily vegetated areas generally represent surfaces with considerable bed roughness relative to flow on a smooth surface, which may affect the dynamics of particulate pyroclastic density currents, generating significant turbulence and pressure drops. Furthermore, vegetation that is inundated by and incorporated into pyroclastic density currents, ignites and is volatilised (Chap. 10; Pensa et al. 2015, 2019); this contributes extra gas to the flow system, increasing turbulence and providing extra fluidisation to the flow system that may also increase its mobility.

3.5 Subaqueous Environments: Effects on Volcanic Eruption Styles and Deposit Characteristics

The effects of aqueous environments on eruption styles and products are profound and very complex, especially with increasing water depths (Plate 3.1c, d, e). We cannot simply take volcanological concepts from well-understood subaerial volcanic settings, apply them to subaqueous or subglacial settings, and expect them to be the same. The laws of physics apply in both settings, but because the physical properties of the ambient environments and fluids (air vs. water, ice) are so different, the outcomes are different. Therefore, subaqueously, *every* process must be analysed in terms of the effects of the very different physical properties of the subaqueous environment.

The significant differences in eruption processes subaqueously compared with subaerial eruptions, because of their unique physical properties, have previously been summarised by McBirney (1963), Cas and Wright (1987), Cas (1992), White et al. (2003), Cas and Giordano (2014), and Cas and Simmons (2018). In Table 3.3, we compare some of the key physical properties of water and air that affect eruption processes and compare them via a Difference Factor to emphasise how significant those differences are in contributing to differences in eruption processes between subaerial and subaqueous eruptions. Key properties include the thermal properties of water in its various phases (solid [ice], liquid, gas [steam], critical fluid, supercritical fluid), its density, the ambient pressure gradient, and the bulk modulus as an ambient medium, compared with air, all of which contribute (in addition to magma properties) to determining if eruptions will be explosive or effusive, whether or not fragmentation of erupting magma will occur or not, and if it will be largely explosive or autoclastic (Chap. 4).

The aqueous environments could be surface bodies of water such as the sea, lakes and rivers, and snow or ice, or subsurface groundwater aquifers and hydrothermal systems. In any of these environments, the nature of the interaction between magma and water could be passive, involving film boiling at the interface between the magma and the external water that produces an insulating layer of steam (Chaps. 4–6), called the *Leidenfrost effect* (Mills 1984), or dynamic, involving explosive interaction (Plate 3.1; Chaps. 4, 9–11). We will consider the physical properties of ambient bodies of water and consider their impact on eruption processes and deposit characteristics, as largely evaluated by Cas and Giordano (2014) and Cas and Simmons (2018).

3.5.1 Thermal Properties of Water and Steam

The thermal properties of water, steam, and (super)critical water (Fig. 3.4a, c) play a huge part in understanding the nature of the interaction between magma erupting or lava flowing into an aqueous environment, and the ambient water mass. The principal properties of water in this regard are its heat capacity, thermal conductivity, thermal diffusivity, and heat transfer rate or coefficient.

Heat capacity is the amount of heat a substance can store, or the amount of heat required to raise the temperature of a substance by 1 °C or 1 °K. It can be formalised as the *specific heat* of a substance, which is the heat capacity of a given mass (g, kg) of the substance. The SI units of heat capacity are kJ kg^{-1} K^{-1}.

Thermal conductivity reflects the ability of a substance to conduct heat and is defined as

$$k = \frac{Q\,\Delta L}{A\,\Delta T} \qquad (3.1)$$

where k is the thermal conductivity (W m^{-1} K^{-1}; W = watts), Q is the heat transferred through the material (J s^{-1} or W; J = joules), ΔL is the distance (m) that the heat is transferred, A is the surface area perpendicular to the direction of heat transfer (m^2), and ΔT is the temperature difference over the distance L (°K or °C).

Thermal diffusivity, α, is the rate at which heat will flow through a substance (m^2 s^{-1}), which can be expressed as

$$\alpha = \frac{k}{\rho\,C_p} \qquad (3.2)$$

where ρ is the density of the substance (kg m^{-3}), which is water in this case, and C_p is its specific heat capacity (kJ kg^{-1} K^{-1}). The density of a substance is controlled by both its temperature and the pressure it is subjected to, and this can be highly variable for water in its liquid and gaseous (steam) forms. When water is heated, it convects and heat is then transferred by a combination of conduction and bulk motion, or advection.

Fresh water is much more (20x) thermally conductive than air, with a thermal conductivity of 0.58 W m^{-1} K^{-1} at 25 °C, compared with 0.025 W m^{-1} K^{-1} for air (Table 3.3). The thermal conductivity of seawater is 0.61 W m^{-1} K^{-1} at 25 °C, and 0.68 W m^{-1} K^{-1} at 120 °C (Sharqawy et al. 2010). Fresh water also has a high specific heat capacity of 4.187 kJ kg^{-1} K^{-1} (slightly higher than that of seawater ~4 kJ kg^{-1} K^{-1}) and almost four times that of air (1.158 kJ kg^{-1} K^{-1}). Water therefore acts as a heat sink when magma comes in contact with it, causing water to boil at the interface, and a film or layer of steam to form (i.e. Leidenfrost effect; Mills 1984). If this film of steam remains stable for several seconds, the magma at the interface can cool, initially forming a visco-elastic crust, or chill quickly through the glass transition temperature, forming a glass crust. Steam has a much lower thermal conductivity of 0.016 W m^{-1} K^{-1} at 125 °C and a specific heat capacity of 1.996 kJ kg^{-1} K^{-1} than water (Table 3.3), although cited values can vary, depending on the state of the steam.

However, the thermal conductivity of water increases up to 130 °C and then declines as temperature rises further, but it does not change in response to increasing pressure. The thermal conductivity of steam does increase with increasing temperature (e.g. from 24.5 W m^{-1} K^{-1} at 100 °C to 120 W m^{-1} K^{-1} at 900 °C) (Engineering Toolbox, https://www.engineeringtoolbox.com/water-liquid-gas-thermal-conductivity-temperature-pressure-d_2012.html), which is still well below that of water. However, subjecting steam to increasing pressure as temperature increases does cause thermal conductivity to increase to the levels of water because the steam becomes denser with increassing pressure.

The thermal diffusivity of water increases as temperature increases to 100 °C but then also decreases, irrespective of increasing pressure (Engineering Toolbox, https://www.engineeringtoolbox.com/water-steam-thermal-diffusivity-d_2058.html). For steam, the thermal diffusivity increases with increasing temperature, and also with increasing pressure, but as pressure increases, the rate of increase declines (Engineering Toolbox).

The parameters and values cited above, however, relate to static water, whereas in reality when there is a temperature gradient, heat is transferred not just by conduction, but by convection and advection as well. The *heat transfer rate* combines all manners of heat transfer and is measured in W (atts) or Joules per second (J s^{-1}), and formalised for comparison between different substances by a concept called the total heat transfer coefficient (W m^{-2} K^{-1}), which is a measure of the heat transferred by all means (conduction, convection and radiation where relevant) and is more encompassing than simple heat transfer parameters such as thermal conductivity and diffusivity. The heat transfer coefficient, H (W m^{-2} K^{-1}), is defined as

$$H = \frac{Q}{A\,\Delta T} \tag{3.3}$$

where Q is the heat flux (W m^{-2}), A is the surface area (m^2), and ΔT (K) is the temperature difference between the surface of the heat source (lava in our case) and the ambient fluid (liquid water in our case).

The heat transfer coefficient (H) for water in its various states can be highly variable depending on conditions. The H of boiling water is usually two orders higher than for convecting water, and significantly higher than for steam, and it can be highly variable for supercritical water, but is usually substantially less than for boiling water (I. Pioro, pers. comm).

3.5.2 Quench Fragmentation and the Role of Steam and (Super)Critical Fluid Films

Because the heat capacity and thermal conductivity of water are high (Table 3.3), when magma comes in contact with water, water acts as a heat sink. This leads to a rapid transfer and loss of heat from magma and magma fragments to water, in turn causing the magma at the interface to chill and transform to glass (Höskuldsson and Sparks 1997; van Otterloo et al. 2015). The dynamics of heat transfer from magma to water will determine whether the magma (and glass) at the interface will remain coherent, or undergo in situ quench fragmentation-forming hyaloclastite breccia (a non-explosive process; Chap. 4), or if the water becomes superheated and expands explosively, producing phreatic and phreatomagmatic explosions, as discussed in Chap. 4.

The heating of the water by magma at hundreds °C and at hydrostatic pressures and water depths less than the critical point (<22 MPa, ~2,200 m in fresh water; <30 MPa, ~3,000 m in seawater) often causes a vapour film of steam to form instantaneously at the interface through a process called film boiling (Mills 1984; van Otterloo et al. 2015; Chap. 4). The formation of the vapour film lowers the rate of heat transfer from magma to the main water mass beyond the vapour film, because conductive heat transfer, the most efficient heat transfer mechanism, is instantaneously reduced when the film of steam forms (Table 3.3). Heat is also transferred by radiative transfer and by the development of convection in the water mass adjacent to the magma–water interface (van Otterloo et al. 2015; Chap. 4).

3.5.2.1 Quench Fragmentation Mechanisms

The thermal conductivity, heat capacity, and heat transfer coefficient of the steam that constitutes the vapour film are low. If the vapour film remains stable, heat loss from the lava is slow, the lava surface is not subject to high levels of thermal stress and thermal shock. As a result, initially a visco-elastic crust forms which then chills through the glass transition temperature range to form a coherent glass crust, producing a coherent lava crust, and producing coherent lava. If the steam vapour film collapses, due to fluid instabilities (e.g. Kelvin–Helmholtz, Chap. 4), leading to direct contact between the planar magma or hot glass surface and liquid water, the lava surface is subject to thermal shock as the water rapidly absorbs heat from the lava because of its high heat capacity. Thermal shock creates thermal tensile stress in the glass, volumetric contraction, and if the thermal stress or shock is high enough, cracking of the glass. Such cooling contraction fracturing of volcanic glass represents an in situ autoclastic fragmentation process called *quench fragmentation*, which produces an aggregate of glassy fragments termed hyaloclastite breccia (Chap. 4).

Glass has a high coefficient of expansion, or conversely contraction, which is a measure of the change of volume over a temperature range, and is thus a measure of the strain experienced over a temperature range, which can be expressed as a linear relationship:

$$\alpha_L = \frac{\Delta l}{l_0\,\Delta T} \tag{3.4}$$

where Δl is the change in length, l_0 is the original length, and ΔT is the temperature range (Kingery 1955; van Otterloo et al. 2015). The coefficient of thermal expansion of rhyolite glass is 6×10^{-6} K^{-1} (Bagdassarov and Dingwell 1992) and for basalt glass it varies between ~7.6 and 8.9×10^{-6} K^{-1} (Liu et al. 2020). In contrast, it is 1.76×10^{-5} K^{-1} for variably crystallised rhyolite rock, 1.97×10^{-5} K^{-1} for andesite, and 2.08×10^{-5} K^{-1} for basalt (van Otterloo et al. 2015). Basaltic glass has a high coefficient of thermal expansion (or contraction) and therefore experiences more strain, making it more prone to fail and quench fracture.

For glass to undergo quench fragmentation, the contractional or tensile stress in the glass as it supercools must exceed the tensile strength of the glass. For this to occur, the glass must experience a critical temperature difference, ΔT_c, with the ambient fluid. This is enhanced where the ambient fluid is water, which has a high heat transfer coefficient, compared with steam. Experimental results show that the minimum ΔT_c is ~160 °C or °K, at which a few, widely spaced, long and curviplanar fractures form in glass, whereas as ΔT_c increases, quench fractures become shorter, more planar, closely spaced, and interconnected (Bahr et al. 1986; van Otterloo et al. 2015). That is, the glass is more intensely quench fragmented at higher ΔT_c. Although most quench fragmentation occurs as magma is quenched or super-cooled through the glass transition temperature range, Porreca et al. (2014) demonstrated that it can continue below the glass transition temperature.

In addition to surface cooling contraction fracturing once water penetrates the initial contractional fractures, it can cause further cooling contraction fractures to propagate perpendicular to the first-order fractures, and in addition, that "entrapped" water experiences heating, vaporisation, expansion, and pressure build-up. That overpressured water can then further (thermo)hydraulically fracture the crust. Quench fragmentation thus involves *both* cooling contraction fracturing *and* (thermo) hydraulic fracturing.

At water depths up to several hundred metres, where the hydrostatic pressure is only up to several MPa, in addition to causing quench fragmentation, the water that comes into contact with the hot lava through vapour film collapse may be flashed to steam explosively in a process called thermal detonation, readily triggering significant phreatomagmatic explosive activity (Sheridan and Wohletz 1983; Wohletz 1986, 2003; Zimanowski et al. 1991; Zimanowski and Büttner 2003; van Otterloo et al. 2015). However, at greater water depths and higher hydrostatic pressures from 800 to 900 m (8–9 MPa) and deeper, phreatic explosions become progressively suppressed (Zimanowski and Büttner 2003; Cas 1992; Cas and Simmons 2018). This also applies to volcanic explosions that may be driven by exsolving volcanic gases in the erupting magma in deep water.

3.5.2.2 The State of the Fluid Film (Steam, Critical Water, Supercritical Water) and Water Depth Constraints for Quench Fragmentation to Occur

Another key factor in determining the water depths at which quench fracturing will occur is the state of the fluid that forms during film boiling and its thermal properties. In subaqueous environments, boiling of ambient water and the resulting vapour film of steam will exist from the shoreline to water depths where the hydrostatic pressure approaches or is equivalent to the critical pressure of water, although its density progressively increases with increasing hydrostatic pressure as it is progressively compressed (McBirney 1963; Kokelaar 1986; Cas 1992; Cas and Giordano 2014; Cas and Simmons 2018). However, in submarine environments, although it is saline seawater that boils at the interface between the erupting magma surface and the ocean water mass, the steam bubbles and film that form are essentially fresh water, and therefore have the physical properties of fresh water, including the critical point (i.e. 22 MPa, ~2.2 km, rather than 30 MPa for seawater), critical fluid behaviour. At water depths approaching the critical pressure point equivalent (22 MPa), the steam layer transforms into a *critical fluid or critical water*, and beyond that into *supercritical fluid* or *supercritical water* (Fig. 3.4c). Therefore, the focus of the following discussion will be based on the properties of the various phases of fresh water.

As discussed in Sect. 3.2.2.4, *critical and supercritical water* have hybrid properties of both liquid and gaseous (steam) forms of water. Although boiling *sensu stricto* of water cannot occur beyond the CP water depths because steam cannot be produced, pseudo-boiling involving the formation of supercritical fluid bubbles can occur. Such supercritical pseudo-boiling fluid is denser and less compressible than subcritical steam. Initially, nucleate (individual spaced bubbles) boiling off the magma surface, then film pseudo-boiling occurs, forming a continuous supercritical film.

The density of critical water varies according to temperature and pressure, but at the critical point (22 MPa, 374 °C) a value ~322 kg m^{-3} is commonly cited (Reid et al. 1987; Pioro and Mokry 2011; Table 3.3). The density of supercritical fluid (i.e. at P and T above the CP) decreases with increasing T but is higher than for critical water and it increases as the supercritical pressure increases at the same T (Fig. 3.6a; Pioro 2020; Pioro and Mokry 2011). The density of supercritical fluid is many times higher than that for subcritical steam (Table 3.3). Therefore, the deeper an eruption occurs beyond the CP water depth, the greater the density of the supercritical film that forms between the lava and water mass.

At the critical point (P = 22 MPa, T = 374 °C), the specific heat of critical water is at a peak (Fig. 3.6b; Table 3.3; Pioro 2020; Pioro and Mokry 2011), about 40x greater than that of subcritical steam and 20x greater than subcritical water. At temperatures higher than the critical point, the specific heat of critical water declines but remains higher than or about the same as for subcritical water (Fig. 3.6b; Table 3.3). At supercritical temperatures and pressures, supercritical water has higher specific heat than critical water, subcritical steam, and liquid water, but approaches that of subcritical water at high temperatures (Fig. 3.6b; Pioro and Mokry 2011). The specific heat of supercritical water is higher as pressure increases, and at each supercritical pressure experiences a specific heat peak at what is called a pseudo-supercritical point, beyond which

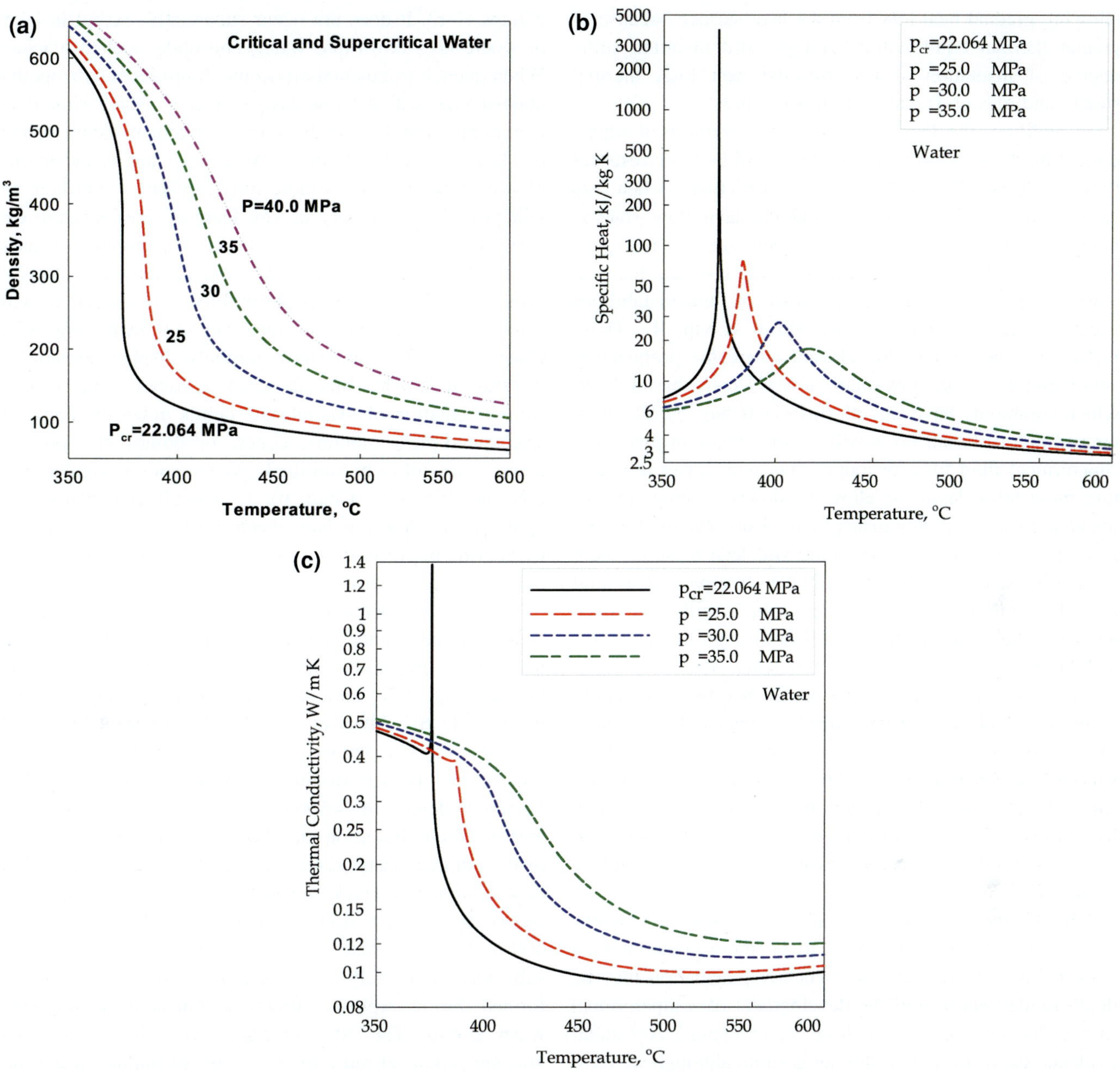

Fig. 3.6 Some properties of critical and supercritical water (see text). The curves labelled 22 MPa represent critical water, and those depicting higher pressures represent supercritical fluid. **a** Density as a function of changing pressure and temperature. **b** Heat capacity as a function of changing pressure and temperature. **c** Thermal conductivity as a function of changing pressure and temperature. (From Pioro and Mokry 2011, courtesy of I. Pioro)

the specific heat decreases (Fig. 3.6b; Pioro and Mokry 2011). These specific heat values are also many times higher than for subcritical steam and water.

The thermal conductivity of critical water is about 30x greater than that of steam and almost that of water (Table 3.3; Fig. 3.6c). Then at higher pressures and temperatures above the critical point, in the regime of *supercritical fluid or water* the thermal conductivity of supercritical water decreases significantly (Fig. 3.6c) but is still substantially higher than for steam at lower pressures and water depths (>10x). Again, at increasing pressures above the critical pressure (i.e. increasing water depth), thermal conductivity of supercritical water increases.

The importance of the preceding assessment is that at water depths beyond the critical point equivalent, the thermal properties of supercritical water are significantly higher than for steam (Table 3.3). This will thus produce more gradational temperature, thermal conductivity, heat capacity, and diffusivity gradients between the lava surface and the liquid water mass across the (super)critical fluid layer, allowing gradual heat loss from the lava surface because of greater thermal equilibration, and yet still maintain some degree of insulation against extreme heat loss, thermal shock, and quench fragmentation occurring.

In addition, the denser, more highly pressurised supercritical fluid film at great water depths will be more resistant to film collapse. This will be the case particularly when lava surfaces are relatively smooth and the lava flow state is laminar, which is mostly the case, reducing the shear gradient at the interface between the moving lava and the host water mass, minimising the likelihood of fluid instabilities such as Kelvin–Helmholtz instabilities (Chap. 4). These factors will be particularly relevant where the propagating submarine lavas are low-relief, low-viscosity basalt lavas which commonly have smooth surfaces (Chap. 5). The high temperature of basalt lavas will also maintain an insulating supercritical fluid film for longer than with lower temperature more felsic lavas, so allowing slower cooling rates of the lava surface and minimising thermal stress in the forming lava crust. Fluid instabilities could still lead to some localised collapse of the high-pressure, denser supercritical fluid film, bringing the ambient water, which will have been heated by heat transfer from the lava through the supercritical layer, into contact with hot magma. This will allow some quench fragmentation to occur, but less commonly than at shallower water depths where low-density, low-pressure compressible steam constitutes the Leidenfrost vapour film. Since supercritical fluid can expand, pressurised supercritical water that penetrates quench fractures can hydraulically fracture and separate quench clasts from each other, without necessarily involving violent explosive dispersal.

This all indicates that lavas erupted in seawater depths greater than 2200 m are likely to be more coherent than quench fragmented, and that quench fragmentation may be depth limited because of the development of a supercritical film at the lava surface at those water depths. Anecdotal evidence seems to support this suggestion, although we have not investigated this exhaustively. Images from papers documenting deep-water basaltic lavas on the seafloor erupted at depths greater than 2200 m do not show abundant hyaloclastite breccia (e.g. Clague et al. 2002, 2017; Chadwick et al. 2016, 2018), and D. Clague and W. Chadwick (pers. comm.) have indicated that they have rarely seen abundant hyaloclastite breccias in lavas erupted in such deep water. Much more research needs to be done to verify this suggestion.

3.5.2.3 Hyaloclastite Breccia: The Product of Quench Fragmentation

Quench fragmentation is not explosive, but is an in situ, contractional cooling and thermo-hydraulic fragmentation or granulation process that is unique to subaqueous settings (Chaps. 4–6). It does not occur subaerially, except in lakes or where lava flows into water, and under melted glaciers. When quench fragmentation occurs, it produces a distinctive deposit type called hyaloclastite breccia, which should be commonly found in shallow to intermediate depth waters (<2 km), as discussed above. We discuss the characteristics of hyaloclastite deposits more fully in Chap. 4, but here we will provide a summary of key features. Hyaloclastite can vary from ash-size exfoliations from the surfaces of lavas to in situ, pervasive block to ash-size hyaloclastite breccias in lavas (Fig. 3.7a–c), or peperite breccias associated with syn-depositional intrusions into wet sediments (Fig. 3.7d; Chaps. 4–6). The deposits commonly display domains of in situ jigsaw-fit breccia clast textures (Fig. 3.7a–c). Resedimentation of hyaloclastite debris down the flanks of steep-sided lava mounds, cones, or domes can produce massive to diffusely stratified, poorly sorted angular clast breccias (Fig. 3.7e; Chap. 6). It is likely that many such hyaloclastite deposits have been mistakenly identified as pyroclastic in origin in the past, simply because they consist of large angular clasts.

Quench fragmentation of highly vesiculated lava can even produce pumice clast deposits or pumice hyaloclastite in subaqueous settings, including *in situ* jigsaw-fit pumice breccia (Fig. 3.7c), not involving explosive fragmentation (Chaps. 4, 6). Allen et al. (2010) interpreted the high vesicular, rhyolite pumice clast breccia of Sumisu submarine volcano on the Izu–Bonin Arc, which erupted from lava domes at depths of ~600–1300 m to be of quench fragmented origin, based on morphological and textural features of the pumice clasts (slabby, polyhedral shape, with polygonal contraction crack networks orthogonal to clast surfaces). In contrast, domes erupted at depths from 90 to 400 m have associated aprons of blocky, subangular to sub-rounded coarsely vesicular pumice, interpreted to have formed during explosive fragmentation at those shallower water depths. The 2012 eruption from Havre submarine volcano produced more than 1.2 km^3 of pumice which has been interpreted as quench fragmented from an effusive dome eruption (Carey et al. 2018; Manga et al. 2018). The resulting giant pumice raft was dispersed in the southern Pacific Ocean and floated for years (Jutzeler et al. 2014; Carey et al. 2018; Manga et al. 2018).

Exfoliated, fine hyaloclastite ash-size debris can be lofted by thermal plumes in the water mass and be dispersed as ash fallout (cf. Barreyre et al. 2011; Chap. 9). Again, this is not a pyroclastic dispersal and fallout process, and perhaps some ash layers in marine successions have been mistaken as distal pyroclastic fallout deposits when they could be fallout deposits of fine hyaloclastite from thermal aqueous plumes.

Clasts ejected from subaqueous vents during subaqueous fountaining of magma or explosive activity may also be subject to cooling contraction quench fragmentation in the

Fig. 3.7 Features of hyaloclastite breccia in field outcrops. **a** Quench fragmented submarine Miocene dacite lava dome with in situ hyaloclastite breccia, Kariba, Hokkaido, Japan. Note the transition from the coherent to in situ, jigsaw-fit brecciated, flow-banded core to progressively more intensively fragmented, finer breccia outwards. **b** Coherent to in situ fractured, Eocene–Oligocene, basaltic pillow lava with black, chilled, polygonally quench fractured glassy margins, and an internal fracture network, separated by bioclastic limestone (pale colour). Boatmans Harbour, Oamaru, New Zealand. **c** In situ to clast-rotated pumice hyaloclastite breccia, Cala Feola, Ponza, Italy. **d** Rhyolite clast (pale) peperite breccia with black mudstone matrix. Bunga Beds outlier, Late Devonian Boyd Volcanic Complex, southern New South Wales, Australia. **e** Clast-rotated, probably resedimented, basaltic hyaloclastite breccia resulting from pervasive quench fragmentation of fluidal Eocene–Oligocene pillow lavas, Oamaru, New Zealand. **f** Fractured and broken, poorly vesicular, submarine fire fountain spatter clast, in a matrix of blocky, coarse ash size, glassy fragments, fragmented by quench fragmentation in the submarine fire fountain. Miocene, Ryugazaki, Oshoro Peninsula, Hokkaido, Japan

eruption column. Quenching of spatter material from subaqueous magma "fire" fountains produces fractured and broken spatter fragments and blocky coarse ash-size glassy fragments that are also deposited by fallout from the water column around the magma "fire" fountain (Fig. 3.6f; Chaps. 4, 9; Cas et al. 2003).

3.5.3 Hydrostatic Pressure and Effects on Volatile Exsolution and Magmatic Explosive Eruption Processes

As water depth increases in a body of water, hydrostatic pressure increases at a rate of 1 bar or 0.1 MPa per 10 m of water depth (or $\sim$10 MPa km^{-1}; Figs. 3.1, 3.3). This means that magmas erupting in deep water are subject to much higher ambient pressures at vent than magmas erupting from subaerial vents. To illustrate that this hydrostatic pressure effect is not insignificant, at a water depth of 100 m erupting magmas would decompress to a hydrostatic pressure of 1.1 MPa (10x atmospheric pressure, AP), at 500 m water depth to 5.1 MPa (50x AP), at 1,000 m water depth to 10.1 MPa (100x AP), at 2,000 m to 20.1 MPa (200x AP), at 3,000 m to 30.1 MPa (3000x AP), at 4,000 m to 40.1 MPa (400x AP), and at 5,000 m to 50.1 MPa (500x AP; Fig. 3.8).

Hydrostatic pressure in a body of water therefore has significant effects on the physical properties of magmas erupting from subaqueous vents, the eruption processes from deep-water vents, and deposit characteristics, including:

- contributing to the pressure regime driving magma to the Earth's surface,
- effect on timing and depth of volatile exsolution in deep-water settings,
- determining the state of exsolved magmatic volatiles,
- constraining water depth of explosive eruptions and their intensity,
- inducing implosions of condensing volatile bubbles and cavities in lavas,
- effects on magma viscosity, lava fluidity, and mobility, and
- the origins of pumice in deep water.

3.5.3.1 Extra Pressure on Columns of Magma Erupting in Deep Water

The (lithostatic) confining pressure in oceanic crust is greater than subaerially in continental/evolved arc crust (Fig. 3.5) because oceanic crust is denser than continental crust. Confining pressure is also greater at vents on the seafloor than at subaerial vents because the hydrostatic pressure adds to the lithostatic pressure, increasing the total confining pressure acting on magma in the crust (Fig. 3.5c). The total magma fluid pressure at the base of the magma column in oceanic crust (the magmastatic pressure) is the sum of the lithostatic pressure and hydrostatic pressure plus or minus

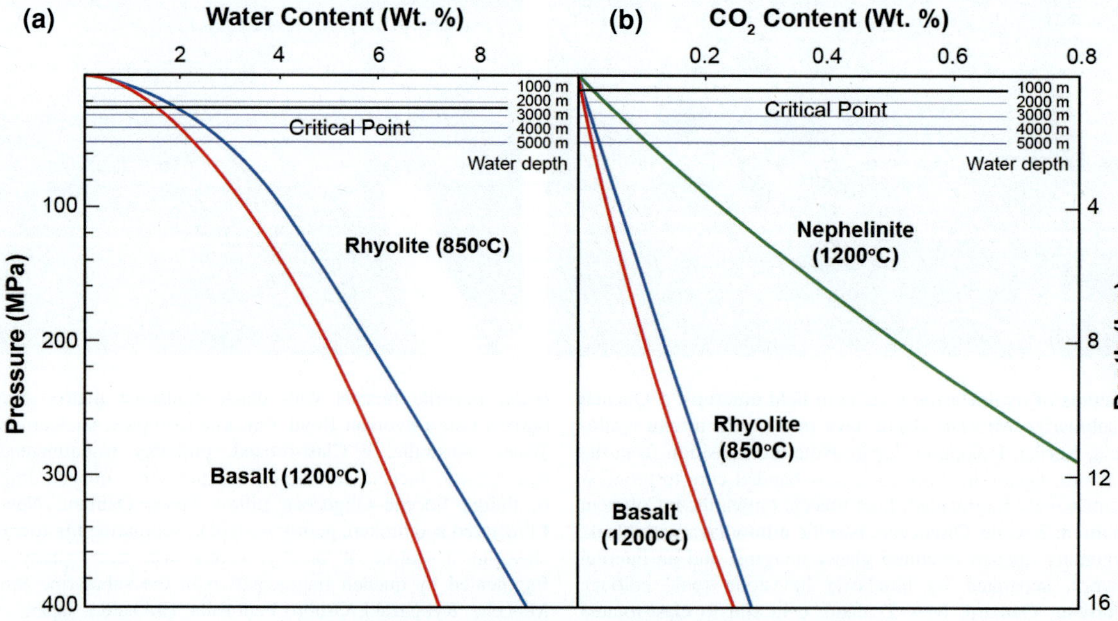

Fig. 3.8 Volatile saturation curves for **a** H_2O and **b** CO_2 as a function of confining pressure and volatile content, together with depth in the crust (right side), and superimposed pressure lines representing increasing water depths in deep-water environments. Volatile exsolution stops in subaqueous environments when magmas erupt at varying water depths and are chilled to solids. The amount of volatile exsolution is therefore less in (deep) subaqueously erupted magmas than for subaerially erupted magmas (pressure = 0.01 MPa and 0 depth). (From Cas and Simmons 2018; Adapted from Wallace et al. 2015)

any excess pressure provided by tectonic forces or volatile exsolution (Fig. 3.5c).

The magma is also less viscous, or more fluidal, and so the effect of more fluidal magma and higher confining pressures driving magma up to the seafloor is that magma rise rates should be higher, magma discharge rates should be higher, and fountaining of magma on the seafloor should be more common relative to comparable magmas erupting from subaerial vents. An analogue is the gushing of fountains of overpressured oil during oil well breakouts that are driven by lithostatic pressure.

3.5.3.2 Hydrostatic Pressure Effect on Timing, Degree, and Depth of Volatile Exsolution in Deep-Water Settings

In subaerial settings, the exsolution of volatiles in magma is controlled by the confining pressure and the decompression of the magma as it rises to the Earth's surface in the conduit through the crust. The lithostatic pressure prevents exsolution until the volatile vapour pressure exceeds the lithostatic pressure, at which point volatiles become (super)saturated, exsolve, modulated by magma viscosity and volatile diffusion rates, form fluid bubbles and have the potential to cause explosive fragmentation of the magma (Chap. 4).

In subaerial settings, once a vent is open to the atmosphere, the magma in the conduit decompresses to atmospheric pressure, and the magma in the conduit will experience only the magmastatic pressure (i.e. the pressure exerted by the weight of the magma column). However, under water, the confining pressure affecting magma rising up through the crust consists of both lithostatic *plus* hydrostatic pressure components. The increasing hydrostatic pressure with increasing water depth in a body of water should also have an effect on the timing and depth in the crust where the exsolution of volatiles occurs (Fig. 3.8; Wright et al. 2003; Cas and Giordano 2014; Cas and Simmons 2018). For a given conduit length, comparing a subaqueous vent with an equivalent case in subaerial settings, the deeper the body of water, (1) the less likely that significant exsolution of volatiles will occur; (2) exsolution will occur closer to the subaqueous basin floor than it would to the Earth's surface in a subaerial environment; and (3) less exsolution of volatiles will occur than for equivalent magmas erupting from subaerial vents, because the exsolution process is short-circuited when the magma erupts subaqueously at the depth of the seafloor and chills to a solid (Fig. 3.6).

This has two effects: the magma remains less viscous because more H_2O is retained in solution (see below), and secondly, it is less prone to erupt explosively relative to subaerial counterparts.

3.5.3.3 Hydrostatic Pressure Effect on Retention of Volatiles and Magma Viscosity

Increasing pressure lowers magma viscosity (Kushiro 1976; Kushiro et al. 1976; Chap. 2). When magma is erupted at considerable water depth, the hydrostatic pressure at the eruption depth contributes to making magma viscosity lower, because volatile exsolution is prevented or limited and pressure naturally reduces magma viscosity. Magma erupting in deep water or under a thick cover of ice (>500 m) will have a lower viscosity than when it erupts from a subaerial vent. Because magma is more fluidal in deep-subaqueous or subglacial environments, it is therefore more prone to erupt effusively or through magma "fire" fountaining; even subaqueously erupting rhyolite or dacite magmas can experience magma fountaining eruption styles in deep water (e.g. Mueller and White 1992). This applies to all magma compositions, including fountaining of silicic magmas, producing silicic spatter deposits (e.g. Mueller and White 1992). Deep submarine lavas should be more fluidal than subaerial counterparts and be able to flow further than subaerial counterparts (Chaps. 5, 6), even though their mobility is affected by the way water interacts with the lava to produce hyaloclastite (see Sect. 3.5.2.1 and Chap. 4).

3.5.3.4 Hydrostatic Pressure and Water Depth Constraints for Subaqueous Magmatic Explosive Eruptions: The State of Exsolved Magmatic Volatiles

Hydrostatic pressure will not only limit the amount of magmatic volatiles that exsolve from erupting magmas in deep water, but if volatile exsolution occurs, it will also influence the state of the exsolved volatiles (gas, critical fluid, supercritical fluid; Table 3.1), and reduce the growth rates of volatile bubbles. As a result, instantaneous explosive growth of vesicles in erupting magma is progressively less likely as water depth of vents increases (Chap. 4). Most exsolved magmatic volatiles are fresh water (not saline) and carbon dioxide, and so the following discussion focuses on the physical behaviour of these volatiles. In the following Sect. 3.5.4, we will discuss factors affecting the explosive superheating of ambient water by magma erupting into it during phreatic and phreatomagmatic explosive eruptions (also see Chap. 4, Sect. 4.6.9.1).

At confining pressures and water depths that are equivalent to or greater than their critical point, volatiles are almost incompressible supercritical fluids, incapable of producing rapid enough volumetric expansion rates, so limiting the potential for explosivity. The critical point of a fluid is the temperature and pressure beyond which there is no distinction between the liquid and gaseous phases; it then forms a supercritical fluid that has the properties of both a liquid and gas. Pure water has a critical pressure, C_p, of about 22 MPa (220 bars), and a critical temperature, C_t, of 374 °C (Fig. 3.4a), whereas for seawater, C_p is about 300 bars or 30 MPa, and C_t is about 400 °C (Wohletz 2003). The density of supercritical water is about 322 kg m^{-3} (Pioro and Mokry 2011), whereas the density of low-density steam is about 1 kg m^{-3}. By comparison, CO_2, has a critical pressure, C_p, of 7.38 MPa or 73.8 bars, and its critical temperature, C_t, is 31.1 °C (Fig. 3.4b).

Volatiles can exsolve from magma at any depth and pressure in the crust if the volatile content and volatile vapour pressures exceed the confining pressures. However, at confining pressures greater than the critical pressures, the exsolved phases will be bubbles of supercritical fluid. In a column of magma rising in the crust without an open vent, the depth/lithostatic pressure in the crust equivalent to the critical pressure of water is about 800 m. In an open vent, with a column of felsic magma, the magmastatic pressure/depth in the magma column equivalent to the critical pressure of water is about 1,000 m, and for a mafic magma, about 750 m.

The specific volume of water (i.e. cm^3 gm^{-1}) at magmatic temperatures varies enormously as a function of pressure (Fig. 3.9; McBirney 1963; Cas and Giordano 2014; Cas and Simmons 2018). At low pressures, the specific volume changes greatly and the state of the water at very high temperatures is steam. This has the potential to drive explosive eruptions, but only if the growth rate of the bubbles is instantaneous, which is also pressure constrained. Specific volume changes in response to changes in pressure are very small at high pressures approaching the critical pressure, where the exsolved water is highly compressed supercritical fluid (Fig. 3.9; Pioro and Mokry 2011). At high pressures therefore, exsolved, superheated water cannot expand quickly enough to be capable of driving explosive eruptions. For carbon dioxide, the limiting pressures and depths are much lower.

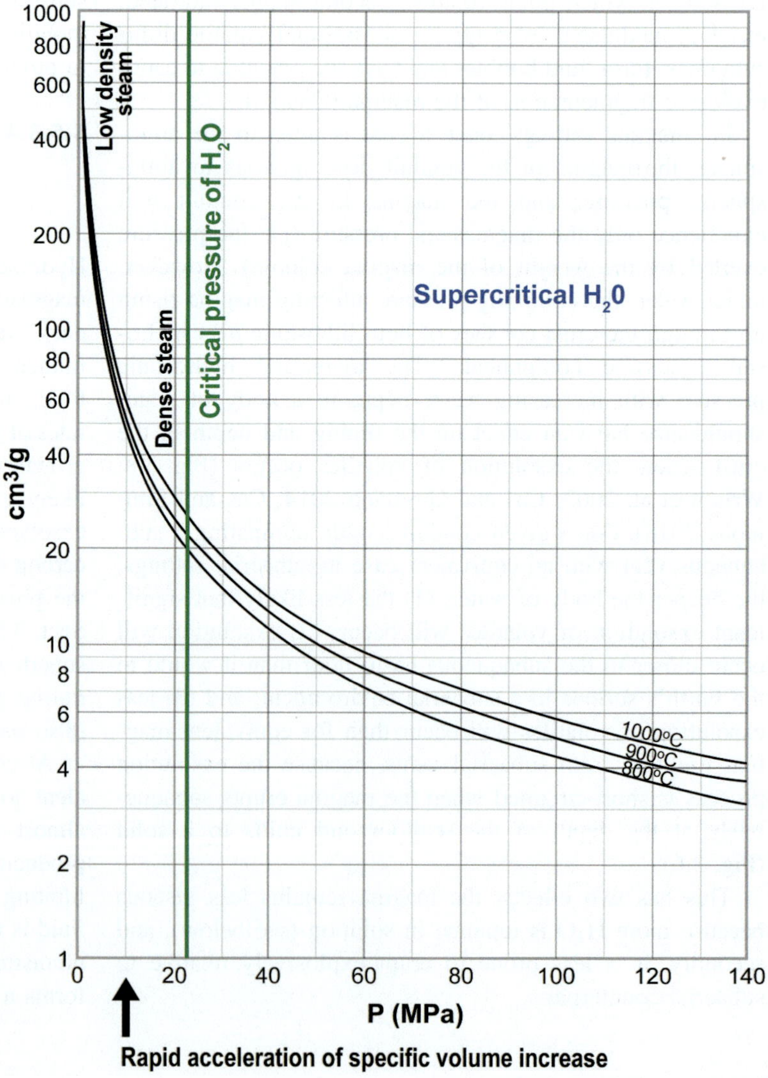

Fig. 3.9 Specific volume of water at magmatic temperatures as a function of pressure (From Cas and Giordano 2014, After McBirney 1963; note the logarithmic scales). Also highlighted is the critical point pressure of H_2O, and the approximate fields of supercritical H_2O, dense steam, and low-density steam, the last of which facilitates explosive eruptions driven by expanding H_2O

In a body of water, the water depth corresponding to the critical pressure of pure water is therefore about 2,200 m, whereas in seawater it is about 3,000 m (Sourirajan and Kennedy 1962; McBirney 1963). Water, whether as a magmatic (fresh water) volatile phase or as an external, ambient, superheated fluid, is therefore unlikely to be able to expand explosively at pressures at and above its critical point. This places theoretical constraints on the water depths at which explosive subaqueous eruptions can occur (Chap. 4). Figure 3.9 indicates that specific volume changes accelerate significantly at pressures less than 10 MPa, where the volatile phase changes from a very dense to lower density steam, and continue to accelerate greatly at lower pressures, especially below 5 MPa. It thus seems that highly explosive eruptions driven by exsolved magmatic H_2O or superheated external water can occur at water depths <1,000 m and especially <500 m, but only if the volatile pressure is high enough, and in deeper water, they become less intense. At depths approaching the critical pressure depths, highly suppressed "explosions" tantamount to volcanic burps can occur, driven by increasingly dense volatiles incapable of the massive expansion and expansion rates that drive very explosive subaerial eruptions (Fig. 3.9).

A caveat for explosive eruptions is that the bubble growth rate in the magma has to be rapid, i.e. almost instantaneous, and bubbles must remain non-interconnected (to prevent permeable gas escape) to be explosive. Explosions can be considered to be a sudden release of thermal energy that ejects and violently disperses clasts from the explosion focus. Other forms of fracturing that do not eject and disperse clasts are in situ (e.g. contractional quench fragmentation and displacive hydraulic fracturing) and not explosive but can displace clasts in situ or lead to gravitational resedimentation and clast rotation during transport.

However, it is also possible to generate very highly vesiculated lavas with pumice carapaces (>70% vesicles) *without* explosive fragmentation occurring in both subaerial (Fig. 3.10a; Fink and Manley 1987; Fink et al. 1992) and subaqueous (Fig. 3.10b; Scutter et al. 1998; Binns 2003; Mann et al. 2004; Allen et al. 2010; Manga et al. 2018) environments. This occurs at slow exsolution and bubble growth rates when gas overpressures are low. The pumice carapaces can be coherent (Fig. 3.10a, b), or given the subaqueous setting, such coherent pumice carapaces can be subject to cooling contraction quench fragmentation, producing breccias of slabby to polygonal pumice clasts (Fig. 3.10c; Mann et al. 2004; Allen et al. 2010). Highly vesicular deep-sea basalt lavas (>56% vesicles) have also been documented (Dixon et al. 1997). Such occurrences, which are by no means rare, indicate that volatile contents and overpressures at ambient (both atmospheric, subaqueous) eruption pressures were below the critical level required to produce highly energetic, explosive expansion of gas bubbles and explosive fragmentation of the erupting magma.

Vesicles grow if the gas overpressure, ΔP (difference between vesicle gas P and ambient confining P), exceeds the surface tension or yield strength of the magma in the bubble walls, allowing that bubble wall magma to begin to flow and stretch. Since the yield strength of magma is only a few Pa to kPa, vesicle growth is relatively easy once exsolution has occurred and the overpressure is high enough.

Vesicles explode or burst if $\Delta P >$ tensile strength of the magma. Mueller et al. (2008) proposed a value of 1.54 MPa for magma tensile strength, whereas for most silicic magmas, the tensile strength is ~ 1 to several MPa, but can be as high as 100 MPa for some magmas (Dingwell 1997). We have adopted a working average of $\sim$ 5–6 MPa. Subaerially, many erupting magmas that have high enough volatile content can reach this threshold overpressure for explosive vesicle growth and bursting because they decompress to low atmospheric P, bubbles grow quickly, and generate high enough bubble wall strain rate for vesicles to burst explosively. However, some don't if volatile overpressure is less than the tensile strength of the magma, which may happen at initial low volatile content, or high magma tensile strength, or through the development of bubble coalescence and permeability that allows gas escape from the bubbles, reducing overpressure. In such cases, bubbles grow slowly at low strain rates, resulting in the eruption of lavas with highly vesicular to pumiceous carapaces (Fig. 3.10a).

Subaqueously, as water depth increases the degree of decompression of magma erupting in deep water decreases, and so exsolution rates, gas overpressure, and bubble growth rates also decrease (McBirney 1963; Cas and Wright 1987; Cas 1992; White et al. 2003; Wright et al. 2003; Cas and Giordano 2014; Cas and Simmons 2018). As a result, many magmas that would erupt explosively if erupting from subaerial vents may erupt effusively from subaqueous vents, especially as water depth increases. Such erupting magmas may form coherent, highly vesiculated lavas, many with pumiceous carapaces (e.g. Fig. 3.10b) even at water depths >1300 m (Allen et al. 2010), because ΔP is insufficient to drive explosive eruptions (Cas and Wright 1987; Cas 1992; Wright et al. 2003; Cas and Giordano 2014; Cas and Simmons 2018). Such pumice carapaces are prone to experience quench fragmentation, producing voluminous pumice hyaloclastite clast breccias autoclastically, not pyroclastically (e.g. Fig. 3.10c–e; Cas 1992; Scutter et al. 1998; Mann et al. 2004; Allen et al. 2010; van Otterloo et al. 2015).

Subaqueously, conditions for $\Delta P > 6$ MPa become increasingly difficult to achieve with increasing water depth, because the ambient pressure becomes so high. At water depths of 500 m and less the hydrostatic pressure is

Fig. 3.10 Coherent and autoclastic pumice carapaces on felsic lavas. **a** Subaerial rhyolitic Rocche Rosse lava, Lipari Island, Italy, showing an in situ, flow-banded, highly vesicular, coherent pumice domain. **b** Highly vesicular, coherent rhyolitic pumice surface of a submarine lava dome at water depth of 1210 m, from Sumisu submarine volcano, Izu-Bonin arc, Japan. Note rough columnar jointing on curviplanar surface on right side. Field of view: 5 m wide. **c** Polygonal giant pumice blocks, Sumisu volcano, water depth of 960 m. The large block is 3 m in diameter and the left-hand planar surface shows small-scale polygonal, cooling jointing. **d** 5–30 cm diameter pumice blocks, some with planar surfaces and polygonal form, Sumisu volcano, water depth of 430 m. **d** Large pumice clasts (>2 m) with rough surfaces from a shallow dome at 360 m water depth. (**b–e** From Cas and Simmons 2018, After Allen et al. 2010)

$\sim$5 MPa and less, and gas overpressures of this magnitude can be readily achieved if volatile contents are high enough. However, at 1,000 m water depth, the hydrostatic pressure is $\sim$10 MPa, so to achieve a gas overpressure of 6 MPa, the gas pressure in vesicles must be >16 MPa. At 1,500 m the vesicle gas pressure must be >21 MPa, equivalent to about the gas pressures during the 3.5 BP Minoan and 232 AD Taupo explosive eruptions (Thomas et al. 1994), and at 2,000 m, gas pressure must be >26 MPa, but this is then about the pressure where exsolved volatiles are supercritical fluids, incapable of producing rapid volumetric expansion rates. But in most of these cases, *non-explosive vesiculation* could occur, even at depths of about 1 km or more, producing coherent pumice/highly vesicular lava (Scutter et al.

1998; Binns 2003; Allen et al. 2010; Carey et al. 2018; Manga et al. 2018; Chap. 6), or even shallower depths (Mann et al. 2004).

For those undertaking planetary research, the atmospheric pressure on the surface of Venus today is about 10 MPa, equivalent to pressure conditions at water depths of 1,000 m on Earth (Head et al. 1992; Airey et al. 2015; Ivanov et al. 2015), suggesting that many/most explosive eruptions are suppressed on that planet relative to subaerial explosive eruptions on Earth.

3.5.4 Hydrostatic Pressure and Water Depth Constraints for Subaqueous Phreatic and Phreatomagmatic Explosive Eruptions: The State of Superheated Ambient Water

Phreatic and phreatomagmatic explosive eruptions can result from the instantaneous explosive superheating of external ambient water by magma erupting into a surface body of water from a subaqueous vent (Plate 3.1c), or where lava flows into such a body of water from a subaerial vent, or where magma intrudes water-saturated, unconsolidated sediments, including subsurface aquifers, or already superheated hydrothermal systems (Chap. 4, Sect. 4.6.9.1). Phreatic explosions are largely just steam explosions, whereas in phreatomagmatic explosions juvenile magmatic clasts and exsolved magmatic volatiles may also play a role in driving the explosions. Hydrothermal explosions are different from phreatic explosions in that the high entropy state of the hydrothermal system is already primed for explosive activity (Chap. 4).

Phreatic/phreatomagmagmatic explosive activity has traditionally been explained using the theory of Molten Fuel–Coolant Interaction (MFCI), considered to be limited to perhaps 700–800 m water depths by high hydrostatic pressure (e.g. Zimanowski and Büttner 2003). However, recently Dürig et al. (2020) have proposed that fuel–coolant explosions can also occur in very deep water of several km in a process they call Induced Fuel–Coolant Interaction (IFCI). We will briefly review these two mechanisms and their constraints.

3.5.4.1 Molten Fuel–Coolant Interaction (MFCI) Explosive Eruptions

During MFCI, the "fuel" or the source of the thermal energy is magma and the coolant is ambient water (e.g. Sheridan and Wohletz 1983; Wohletz 1983, 1986; Wohletz et al. 2013; Zimanowski and Büttner 2003; Zimanowski et al. 1991, 2015). During MFCI, collapse of an unstable steam vapour film produced by film boiling at the interface between the magma and the liquid water mass (see Sect. 3.5 above, and Chap. 4, Sect. 4.6.9.1) causes direct contact between the hot magma surface and the cold liquid water mass, not only causing quench fragmentation, but also in many cases, the instantaneous explosive superheating of the water in a process called thermal detonation or thermo-hydraulic explosions. The water flashes to steam and expands explosively, as long as ambient pressures are low enough. Such activity can become cyclical as a new, unstable steam vapour film forms immediately after an MFCI explosion at the interface of the disrupted magma surface, and collapses repeatedly (Chap. 4, Sect. 4.6.9.1). The explosive activity is considered to be a surface explosivity phenomenon (Dürig et al. 2020).

Zimanowski and Büttner (2003) proposed that "subaqueous volcanic thermo-hydraulic explosions should become increasingly improbable at water depths exceeding 100 m, and practically improbable at water depths in excess of 1 km" because of the suppressing effect of increasing hydrostatic pressure on surface steam explosions. This is because with increasing water depth and hydrostatic pressure the overpressure between superheated steam and the ambient hydrostatic pressure decreases, so reducing the likelihood of instantaneous explosive growth of steam bubbles and steam films. In almost all cases, subaqueous explosive eruptions will be variably subdued due to hydrostatic pressure constraints (e.g. Plate 3.1).

3.5.4.2 Induced Fuel–Coolant Interaction (IFCI) Explosive Eruptions

However, a widely dispersed rhyolite ash deposit resulting from a phase of the deep water ($\sim$1 km) 2012 Havre volcano submarine rhyolite eruption, Kermadec Arc (Carey et al. 2018; Murch et al. 2019, 2020), suggests that some other efficient fragmentation process can occur at water depths at and potentially greater than maximum depth at which MFCI can operate (Dürig et al. 2020). Dürig et al. (2020) proposed a new deep-water explosive process called Induced Fuel–Coolant Interaction (IFCI). The Havre ash is dominated by curviplanar fine ash particles, and because of the similarity in morphology to ash particles produced experimentally through explosive magma–water interaction, the Havre ash deposit has also been interpreted to have formed through explosive submarine activity from a vent at 1 km water depth (Murch et al. 2019; Dürig et al. 2020).

To better understand this explosive mechanism at 1 km water depth, Dürig et al. (2020) conducted laboratory explosion experiments using molten Havre ash as a thermal or fuel source. In the experiments, there were two essential setups. In the first "dry" experiments, injection of pressurised gas at the base of a layer of magma in a crucible caused the magma to arch up, fracture, and disintegrate explosively. In the second, the setup was the same, but a layer of water was superimposed on the magma layer just immediately before injection of pressurised gas at the base of the magma layer ("wet" or IFCI experiments). All experiments were conducted at atmospheric pressure, NOT the high-pressure equivalent of deep-water conditions. In both cases, fragmented magmatic debris was ejected from the crucible, with a force of ~2.9 kN for the dry experiments and ~5.8 kN for the wet experiments (Dürig et al. 2020). The wet IFCI experiments produced 2.25x more ash-sized fragments, and much more fine ash (<125 mμ), than the dry experiments, indicating increased fragmentation efficiency.

In both the dry and wet (IFCI) experiments, fragmentation events rely on an injection of a pressurised inflating gas slug under the magma charge to fracture it. Inflation/expansion created tensile stress, which is equivalent to an internal (e. g. expanding vesicle) magmatic gas driver (cf. magmatic and even some phreatomagmatic explosions). However, tensile stress fracturing also replicates first-order contractual quench fracturing to produce cracks during subaqueous eruptions, although this is not mentioned by Dürig et al. (2020). In the wet IFCI experiments, the superposed water layer definitely has the capacity to cause actual quench cracking and fragmentation, which then allows water to permeate into those cracks to trigger thermo-hydraulic explosions.

According to Dürig et al. (2020), IFCI does not require collapse of a pre-existing steam layer to initiate surface thermo-hydraulic detonation as in MFCI explosions. It "occurs under less restrictive initial and boundary conditions", involving instantaneous explosive vaporisation of water that has permeated into first-order cracks, which they called a distinctive Induced Fuel–Coolant Interaction (IFCI) explosive mechanism. However, this seems to be a contradiction because it is instantaneously expanding steam that drives explosions. Therefore, a momentary micro-vapour film must form in the cracks, which Dürig et al. (2020) also acknowledge may happen. And in the real submarine world how is vapour film boiling at the interface between magma and water prevented from happening?

Dürig et al. (2020) propose that ash particles produced during IFCI experimental explosions developed very fine curviplanar shapes that can be used to fingerprint the IFCI mode of formation. Curviplanar ash particles also occur in much of the 2012 Havre fine ash, which therefore they also attribute to IFCI explosive fragmentation at 1 km water depth, which is deeper than the 700–800 m water depth limit for MFCI explosions. We agree with most of the above, and the suggestion of IFCI crack-hosted explosions as distinct from MFCI surface explosions provides a significant advance in understanding phreatomagmatic explosive eruptions in water up to 1 km depth, although the role of quench fracturing of volcanic ejecta has been largely ignored. We also wonder if in MFCI some degree of quench cracking also allows penetration of water into such cracks, such that IFCI represents an extension of MFCI, but involves a micro-process at a more detailed scale that has not previously been recognised. The concept is a valuable advance in understanding. However, other factors come into play.

In addition, Ikegami et al. (2018) state that the rhyolite lava domes which represent the vent locations of the 2012 Havre volcano eruption range in vesicularity from <5% to >50% vesicles and much of the vesicularity is micro-vesicular. Mitchell et al. (2019) indicate that the giant pumices, thought to have detached from the erupting lava domes, are up to 80% vesicular. However, Dürig et al. (2020) do not discuss the likelihood that the original vesicularity was a significant shape-controlling factor of ash particles. The vesicles in the domes and giant pumice are unrelated to explosive expansion, but due to non-explosive slow degassing and vesiculation during eruption (Manga et al. 2018). Quench fragmentation of such highly vesicular pumice will produce curviplanar shard-like fragments, including abundant ash-size debris (van Otterloo et al. 2015).

In discussing the results of the IFCI experimental runs, Dürig et al. (2020) do not seem to take into account the likelihood that some or even much of the deposit could be the result of quench cooling fracturing by water penetrating the first-order gas pressure–inflation generated cracks. As summarised by van Otterloo et al. (2015), using the materials engineering, ceramics, and glass literature (Bahr et al. 1986) on the fracturing behaviour of glass, during quench fragmentation "curviplanar cracks form if $\Delta T < 217$ K and planar ones if $\Delta T > 217$ K". Curviplanar shards are therefore not just the product of IFCI but can also form through quench fragmentation, especially in a standing body of water such as a lake or the ocean.

Based on their experiments performed at ambient pressure, not high pressures, Dürig et al. (2020) extrapolate the importance of IFCI and explosivity to any depth in submarine environments, including very deep water. However, based on the discussion of varying fluid properties with increasing water depth as detailed in previous paragraphs, while we agree that at shallow depths IFCI may be very important, its significance drastically reduces at greater depths. This is because at water depths >2200 m, steam generated by boiling seawater will immediately compress to a supercritical fluid, not subcritical steam. This implies that a layer of critical and supercritical H_2O fluid forms at the magma–water interface, which has a lower heat transfer coefficient than liquid water. In addition, at and beyond the critical point, supercritical fluid has a limited thermal expansion coefficient (Sect. 3.5.1), significantly lower than boiling water. Therefore, although any fluid can expand many times its original volume as it is heated and under pressure, this does not necessarily lead to explosions. *Explosions only occur if the rate of expansion is instantaneous*, and coefficient of expansion actually does not reflect rate of expansion. In addition, counteracting the pressure in superheated critical and supercritical fluid is the hydrostatic pressure of the water mass at great water depth that produces high supercritical fluid pressure. So, it seems that a significant overpressure situation that is capable of producing an instantaneous explosive expansion rate, as occurs in shallower water, does not occur in very deep water.

The boiling of seawater actually produces an essentially salt-free, freshwater steam, which has a much lower critical point (22 MPa) than seawater (30 MPa). Therefore, at water depths >2200 m steam generated by boiling seawater will also immediately compress to a supercritical fluid, not subcritical steam, again essentially eliminating the likelihood of very deep-water IFCI.

What may happen is that water that penetrates first-order cracks is heated, expands, builds up pressure, and causes the enclosing glass to hydraulically fracture. At depth, such hydraulic fracturing is not explosive but is an in situ fracturing process than can dilate and displace fragments, but not explosively eject clasts. Some clasts may be entrained in escaping overpressured interstitial fluid that rise to the seafloor. In deep water, with higher hydrostatic pressure, the interstitial fluid pressure just hydraulically fractures glass in situ, some quench fragmentation is likely to occur, and buoyant plumes of heated water rising above the erupting magma can carry ash-size debris upwards into the water column where it is dispersed by currents. This can be all non-explosive.

A test of whether or not IFCI eruptions occur commonly in deep water beyond the critical point water depth equivalent is whether or not the seafloor is mantled with fine ash that can be related specifically to a known deep-water eruption. This criterion is important because as discussed in Chap. 9, there are several possible origins for ash deposits on the deep seafloor, not just in situ deep-water eruptions. Given the abundance of deep-water basalt eruptions and the high temperature of such magmas, IFCI ash deposits should be common if IFCI occurs commonly in deep water. To our knowledge this is not the case. In Chap. 5, we speculate that some fine ash-like deposits have resulted from quenching-induced spalling of ash-size glassy particles from the surfaces of propagating submarine basalt lavas, which were dispersed by hot water thermal plumes rising off the surfaces of hot lavas.

So, in summary, we feel that IFCI explosive eruptions can occur at water depths where relatively low-density, expandible steam is generated (<1,000 m, perhaps a little deeper in some cases), but not at water depths approaching and deeper than the critical point of fresh water (>1,500 m).

3.5.5 Implosions in Deep Water

As water depths and hydrostatic pressure increase, the likelihood of implosions or cavitation of cavities increases in erupted volcanic deposits (e.g. drained lava tubes, gas/supercritical fluid cavities in lavas) and even in erupting gas/fluid bubbles. Implosions are the opposite of explosions, in that material is concentrated by collapse compared with being energetically dispersed far from the source during explosions.

In the simple case of cavities in lavas, if the hydrostatic pressure exceeds the pressure in the cavity (e.g. a sealed cavity in which gas or supercritical fluid has condensed, or lava has drained), and the tensile strength of the surface lava crust, the cavity will collapse or cavitate energetically and implosively. This is most likely to occur during or soon after the eruption (e.g. cooling of gases in erupting pillow lava, or perhaps lava draining from lava and pillow tubes; e.g. Moore 1975). Usually, the lava crust of subaqueous lavas will be weakened by in situ, contractional quench fracturing, which can produce visible and invisible networks of cracks, resulting in loss of strength and easy cavitation collapse of the crusts of drained lava tubes and pillows. Once water can permeate into the cavities, the hydrostatic pressure in the cavity will balance that outside, and no implosive collapse will occur, although the gravitational collapse of a weakened crust could occur.

A more complex example of implosive cavitation is the case of gas bubbles in a liquid (Rayleigh 1917; Fujikawa and Akamatsu 1980; Brennen 1995). Again, a requirement is a negative pressure differential between the pressure in the gas bubbles and the external fluid, that is, a gas under-pressure. Because of surface tension, bubbles form as spheres, and this is the most difficult or strongest shape to collapse (Brennen 1995). Compression of gas-filled bubbles normally leads to significant pressure and temperature increase (Brennen 1995). However, condensation can alleviate both (Fujikawa and Akamatsu 1980), and distortion of bubbles by asymmetric pressure gradients enhances bubble collapse (Brennen 1995). Although this has not been specifically addressed to our knowledge in the volcanic realm, we propose that vesiculating magmas erupting on the seafloor, and gas bubbles of magmatic CO_2 and H_2O and superheated bubbles of ambient water being released from deep subaqueous vents, will first decompress to the ambient hydrostatic pressure. This initially involves an expansion of the volatile phase, which is a dense vapour or even a supercritical fluid. Such fluids can expand several-fold at such pressures, but growth rates are relatively slow because of the ambient hydrostatic pressure. The bubbles will be subject to rapid cooling, leading to condensation of gas in the bubbles, a rapid decrease in internal bubble pressure, leading to underpressured bubbles that collapse and potentially may even experience energetic implosions under high hydrostatic pressures. We suggest that some deep-sea events described as explosions could in fact be largely implosions (e.g. some gas release events from the 2009 West Mata submarine eruption of boninite lava, northeastern Lau Basin, at ~1,200 m water depth; Plate 3.1e; Resing et al. 2011).

There are several confusing aspects of implosions. They also trigger an immediate shock or pressure wave and acoustic signals, and although the implosion motion is inwards, there can then be a rebound against solid surfaces leading to limited ejection of solids and disintegrating bubbles away from the implosion source, mimicking explosions, which they are not. Acoustic signals emitted during deep-water basaltic lava-forming eruptions (e.g. 2015 eruption at Axial Seamount volcano, East Pacific Rise at water depths >1,500 m; Chadwick et al. 2016) may have been mistaken as explosions but may have been sonic signals of implosions. Implosive cavitation of drained lava tubes may have triggered first a shock wave into the tube, followed by a reflected shock wave that was directed outwards, and which scattered debris away from collapse cavities a few m. Gas bubbles actually disintegrate through fission (i.e. break into multiple small bubbles) when they implode (Brennen 2002).

3.5.6 Compressibility of Water/Young's and Bulk Modulus of Water, Air, and Rock, and Effect on Explosive Intensity

Water in all its fluid states has an extremely low tensile strength, it is only slightly compressible in the liquid state, and is a Newtonian substance which deforms readily when stress is applied. Both the kinematic and dynamic viscosities of seawater decrease by a factor of 4 as it is heated to 60 °C or more, and by much more as water converts to steam (Sharqawy et al. 2010).

The intensity of an explosive eruption, or the extent to which it affects its surroundings, can be measured by the bulk modulus of the surrounding material, which is its resistance to compressibility, or the rate of change of pressure as volume decreases as pressure is increased.

The Bulk Modulus can be calculated as

$$K = -\frac{dP}{(dV/V_0)} = -\frac{(P_1 - P_0)}{((V_1 - V_0)/V_0)} \quad (3.5)$$

where K = bulk modulus of elasticity (Pa, N m^{-2}), dP = differential change in pressure on the object (Pa, N m^{-2}), dV = differential change in volume of the object (m^3), V_0 = initial volume of the object (m^3), P_0 = initial pressure (Pa, N m^{-2}), P_1 = final pressure (Pa, N m^{-2}), and V_1 = final volume (m^3) (www.EngineeringToolbox.com).

Alternatively, Bulk Modulus can be expressed as

$$K = \frac{d\rho}{(d\rho/\rho_0)} = \frac{(\rho_1 - \rho_0)}{((\rho_1 - \rho_0)/\rho_0)} \quad (3.6)$$

where $d\rho$ = differential change in density of the object (kg m^{-3}), ρ_0 = initial density of the object (kg m^{-3}), and ρ_1 = final density of the object (kg m^{-3}).

Given that the bulk modulus of water is 2,150 MPa (for seawater, it is 2,340 MPa), the adiabatic bulk modulus of air is 0.142 MPa, and Young's modulus of rock is about 2,000–70,000 MPa (Table 3.3), the explosive intensity from subaerial vents is much higher than from submarine vents, because water is four orders of magnitude less compressible and deformable than air. Water has a very significant effect in attenuating explosive energy (Resnyansky and Delaney 2006; Chap. 4).

With increasing water depth, water becomes less compressible, less deformable, and will suppress explosive pressure or intensity more than at shallow water depths because it is denser at greater water depths.

3.5.7 Subaqueous Eruption Columns, Thermal Plumes, and Their Density: Effects of Heat Capacity and Thermal Conductivity of Water

Subaqueous eruption columns and aqueous thermal plumes can form above explosive seafloor vents and above erupting lavas as overlying water is heated. Few explosive seafloor eruptions have been imaged (e.g. 2009 West Mata volcano eruption, Lau Basin, Chadwick et al. 2008; Resing et al. 2011; 2004–2005 eruption, Rota 1 volcano, Mariana volcanic arc, Embley et al. 2006; Deardorff et al. 2011), and these events were small in scale and did not produce eruption columns that breached the sea surface to form substantial subaerial eruption columns. It is only from shallow water vents (<200 m) that explosive eruption columns can breach the sea surface and form substantial subaerial eruption columns (Rowell et al. 2022; Cahalan and Dufek 2021). At greater depths (<500 m), breaching of the sea surface to form subaerial eruption columns can only occur at very high mass eruption rates (Cahalan and Dufek 2021), whereas eruptions at greater depths will not produce a subaerial eruption column or only a plume of steam that rises above the water surface (Newland et al. 2022; e.g. 2012 submarine eruption of Havre volcano, Kermadec Arc, Carey et al. 2018). For this reason, relatively deep-water subaqueous eruption columns will never reach the heights that subaerial eruption columns do, because they are water depth limited, and the two cannot be equated in terms of their dynamics (Chap. 9; Newland et al. 2022).

Subaqueous explosive eruption columns emerge from vents at the seafloor as gas jets driven by explosive momentum, and as momentum wanes, the jets transition into turbulent plumes that mix in external ambient water which is heated, forming a buoyant rising plume of water and steam (Deardorff et al. 2011; Cahalan and Dufek 2021; Newland et al. 2022; Rowell et al. 2022). The buoyancy of the rising plume is driven by its lower bulk density relative to the cold ambient water mass. The dynamics of subaqueous explosive eruption columns and plumes are discussed in detail in Chap. 9 (Sect. 9.9.1). However, the following relationships summarise the role of the density of the columns/plumes relative to the ambient water mass in determining the behaviour of the columns/plumes. The velocity of rising plumes is given by

$$v = \frac{\sqrt{(8r\,g\,\Delta\rho)}}{3\rho_{sw}\,C_D} \tag{3.7}$$

where v is the plume rise velocity, r is the plume radius, g is the acceleration due to gravity, $\Delta\rho$ is the density contrast between the seawater mass and the plume, ρ_{sw} is the density of seawater, and C_D is the drag coefficient (Deardorff et al. 2011).

The bulk density of the rising column, ρ_c, which determines its buoyancy and upward velocity after the initial explosive thrust dissipates can be expressed by

$$\rho_c = \frac{(Q_m + Q_w)}{[(Q_m/\rho_m) + (Q_w/\rho_w)]} \tag{3.8}$$

where Q_m is the mass flux of fragmented magma, Q_w mass flux of water, ρ_m is the density of the magmatic clasts and ρ_w is the water density (Newland et al. 2022).

Bulk column density can also be expressed by

$$\rho_c = \left[\frac{\phi_p}{\rho_p} + \frac{(1-\phi_p)}{\rho_w}\right]^{-1} \tag{3.9}$$

where ϕ_p is the mass fraction of pyroclasts, with density of ρ_p.

Heating of water at the interface with subaqueous lavas, above vents and from subaqueous vapour-rich explosive eruption columns, causes subaqueous thermal plumes and eruption columns to be less dense than water and to rise buoyantly (Barreyre et al. 2011; Murch et al. 2020). Seawater heated to several hundred °C becomes ~10% less dense than cold seawater, changing from ~1,027 to ~900 kg m^{-3} (Safarov et al. 2009; Sharqawy et al. 2010). Resulting buoyant thermal plumes of seawater will rise upwards. However, as they rise, they may be rapidly cooled by the enclosing water mass, because of their high heat capacity and thermal conductivity and due to the mixing in of cold seawater. The rising column or plume and the water mass are miscible, and the heated water, vapours, and gases condense from the margins inwards, and lose their buoyancy. That is, gas- or vapour-rich columns undergo what we call condensation erosion (Chap. 9; Head and Wilson 2003). Columns quickly lose buoyancy momentum, and although hot water may rise to the surface of the water body, columns rarely break through the water surface as gas-rich systems and rise into the atmosphere, unless they originate from shallow water vents (Cahalan and Dufek 2021; Rowell et al. 2022; Newland et al. 2022).

3.5.8 Density of Water and Effects on Eruption Columns, Pumice Clast Dispersal, and Depositional Processes

In the atmosphere, no pyroclasts or autoclasts are buoyant, and they are only transported upwards because of the shear stress applied by hot, turbulent, vertically convecting gas in an eruption column and laterally in a downwind plume. They eventually fallout when their terminal fall or settling velocity is greater than the gas velocity entraining them (Chap. 9). However, in water, every clast with density <1,000 kg m^{-3} (e.g. highly vesiculated pumice) is buoyant in the water mass. Although there may initially be entrainment in rising hot plumes of water or gas, even without these, low-density pumice clasts originating from a subaqueous vent or spalled from the margins of highly vesiculated lava as autoclasts (Chap. 4) will rise buoyantly, and spontaneously because of their density buoyancy until they ingest enough water (Whitham and Sparks 1986) to render them buoyantly neutral or negatively buoyant (e.g. 2012 submarine eruption of Havre volcano, Kermadec Arc; Carey et al. 2018; Fauria et al. 2018; Manga et al. 2018). This does not happen subaerially. Even clasts with density >1,000 kg m^{-3} have a reduced effective weight in water and are more easily transported in water than in air; they thus settle much more slowly in water than in air (Chap. 9).

The settling velocity of a particle in fluids of varying density and viscosity under laminar fluid flow conditions is best represented by the Stokes equation, which is discussed in detail in Chap. 9. Clasts have much lower settling velocities when falling in water than in air because the density difference between clasts and water is much less than between clasts and air, and so they take much longer to settle in water than in air (Chap. 9). Secondly, given that settling velocity is inversely proportional to fluid viscosity, and the viscosity of air (1.8 × 10^{-5} Pa s) is 55 times less than that of water (9 × 10^{-4} Pa s; Table 3.3), the viscous drag on a falling particle in water is greater than in air, so slowing it down (see Sect. 3.5.3). As a result, sorting and grading of subaqueous deposits are much better in subaqueous deposits than in subaerial equivalents. The buoyancy of clasts in water and their slower settling velocities also affects the trajectories and transport distances of ballistics in water.

Some pumice clasts may absorb water quickly when hot and become water-logged and sink, if they have an interconnected network of vesicles (Whitham and Sparks 1986), whereas others don't become water-logged quickly because either they have poor permeability or gas pockets get trapped in interconnected pore spaces (Fauria et al. 2017). Pumice clasts may float for a long time before some infiltration of water occurs and they become water-logged, and then sink very slowly according to their semi-water-logged settling velocities. However, some pumice clasts in water never settle, because they have enclosed trapped vesicles that do not allow water to permeate. However, settling velocities of clasts in water are much slower than in air, so again the two cannot be equated.

Because of the buoyancy of pumice in bodies of water, fallout behaviour is not comparable to fallout from subaerial explosive eruption columns. Large volumes of pumice are rafted away from the eruption site by waves, wind, and currents, often in irregular patterns (e.g. Jutzeler et al. 2014; Carey et al. 2018), and much deposition occurs days, weeks, months, and years (or never) after an eruption. Dispersal is unrelated to the eruption column generated by the eruption, but controlled by waves, wind, and (oceanic) currents.

Furthermore, unlike subaerial fallout deposits, there may be no systematic grain size decrease and improvement in sorting with distance away from the vent; very large pumice clasts could be rafted to great distances and will not fall just around the vent. In large masses of floating pumice, called rafts, the thickness of the pumice mat could be 1 m or more, and during movement, abrasion occurs, producing significant rounding of the pumice clasts in the raft, and producing a lot of fine ash-size fragments of vesicular pumice shreds and shard shapes, which settle to the seafloor, forming a thin fine vitric "ash" layer that is totally unrelated to explosive fragmentation at the vent (e.g. 2012 Havre volcano eruption pumice raft, Jutzeler et al. 2014; Manga et al. 2018).

3.5.9 Viscosity of Water, and Viscous Drag Effects on Clasts and Plumes

The trajectory of ballistic blocks and bombs is affected by a drag force as the ballistics pass through a fluid (Alatorre-Ibargüengoitia and Delgado-Granados 2006). Water is more viscous than air, and so will create higher levels of viscous drag force on fragments passing through it. This will apply to ballistic clasts as they leave a vent, and as they travel downwards, as well as to pumice clasts rising buoyantly and then subsequently settling passively. Even rising eruption columns and thermal water and gas plumes will experience significant viscous drag as they rise through a body of water. Ballistics ejected from subaqueous explosive vents will travel shorter distances from the vent and reach lower heights than ballistics from subaerial vents due to viscous drag effects. The viscous drag force acting on a particle moving through a fluid is defined and discussed in Chap. 9.

3.5.10 The 15 January 2022 Explosive Eruption of Hunga Tonga–Hunga Ha'apai Volcano

The spectacular explosive eruption of Hunga Tonga–Hunga Ha'apai volcano on 15 January 2022, has raised many questions about the origins and processes driving this marine explosive event. Briefly, at the time of writing, we know the following. The volcano is an almost totally submerged or submarine arc stratovolcano in the Tonga volcanic arc, located north of New Zealand. It was built onto the arc oceanic crust at a depth of ~ 2 km by largely submarine eruptions, building to sea level (Fig. 3.11a). Until 2014, the summit of the volcano was marked by two emergent islands, Hunga Tonga and Hunga Ha'apai, from which the oldest dated volcanic deposits are $\sim 1{,}040$–1,180 AD (Cronin et al. 2017; Brenna et al. 2022). These islands are located on the rim of a submarine caldera that was produced during a prehistoric eruption (Fig. 3.11a).

Known historical eruptions occurred in 1912, 1937, 1988, 2009 (Vaughan and Webley 2010), 2014–15 (Fig. 3.11b; Cronin et al. 2017), and then beginning again on 20 December 2021, continuing until the highly explosive 15

Fig. 3.11 Features of Hunga Tonga–Hunga Ha'apai volcano, Tonga, and its recent eruptions. **a** Simple map and bathymetry showing the main features of the summit area of the volcano, including the caldera, the rim, the emergent islands Hunga Tonga and Hunga Ha'apai, and the locations of the 2009 and 2014–15 eruptions and tuff cones that formed (*Source* Cronin 2022). **b** Image showing the surtseyan style of the 2014–15 eruption, originating from a vent in the newly formed tuff cone. See part (**a**) for location. (Photo taken 14–1-2015; *Source* Global Volcanism Program, Smithsonian Institute, courtesy of the New Zealand High Commission, Tonga). **c** Surtseyan explosive eruption from a vent between Hunga Tonga and Hunga Ha'apai islands, the day before the major explosive eruption of 15 January 2022 (*Source* Tonga Geological Services). **d** Satellite view of the 15 January 2022 eruption plume at 5.00 p.m., 50 min after the start of the highly explosive eruption. Note that there are two visible parts of the plume, a lower grey, radially spreading cloud at ~ 15 km altitude and an upper, still rising and radially spreading cloud with lobate structure driven by gravity waves, and at an altitude of at least 25 km (A. Prata, pers. comm.). The lower grey cloud is probably dominated by fine ice particles formed through sublimation of water moisture in the eruption column, much of it derived from the initial at least phreatomagmatic blast, and carrying some very fine ash (A. Prata, pers. comm.), whereas the upper cloud could be more driven by magmatic gases (our interpretation). (*Source* Himawari-8 weather satellite imagery) ▶

(a)
175°24'W
175°23'W
175°22'W
Approximate Caldera Margin
Hunga Tonga
2015 Cone
Hunga Ha'apai
2009 Cone
Caldera centre
20°32'S
20°33'S
20°34'S
20°35'S
Water Depth
<50 m bsl
<100 m bsl
<150 m bsl
<200 m bsl
1 km
(b)
(c)
(d)
Himawari-8 AHI FLDK true_color 15-Jan-2022 / 2022015 @ 05:00:00 UTC UW-SSEC CIMSS

January 2022 eruption (Fig. 3.11c, d; Global Volcanism Program, Smithsonian Institution, https://volcano.si.edu/volcano.cfm?vn=243040).

The historical eruptions up to 15 January were low-intensity phreatomagmatic, surtseyan explosive eruptions, erupting from vents at or just below sea level causing a pulsing eruption with jets of rock debris and ash being ejected from the shallow water vent in rapid succession (e.g. Fig. 3.11b). Small columns of ash and steam rose several km into the atmosphere above the vent (Vaughan and Webley 2010; Cronin et al. 2017). The 2009 and 2014–15 eruptions built small tuff cones above sea level around the vents (e.g. Fig. 3.11c), and these were variably eroded by ocean waves and rain in succeeding years. The 2014–2015 cone merged with the two old islands (Fig. 3.11a, c; Cronin et al. 2017; Brenna et al. 2022). From 20 December 2021, the same style of eruptions occurred again and intermittently into 2022 including Friday, 14 January (Fig. 3.11d).

Then at 4.08 p.m. (UTC) on Saturday, 15 January 2022, there was a sudden very intense explosive eruption that sent a column or plume of ash and gas at least 40 km into the stratosphere based on early assessments (Fig. 3.11e, f), and perhaps as high as 57 km, into the mesosphere according to NASA (https://earthobservatory.nasa.gov/images/149474/tonga-volcano-plume-reached-the-mesosphere). The most intense explosive phase of the eruption lasted for 2–3 h, and then waned for a further 8 h.

The eruption destroyed much of the previously merged island, it dispersed pumice and ash widely over the Tongan islands, it generated atmospheric shock waves that circumnavigated the globe at least 3 times, and it produced tsunamis that spread in all directions across the Pacific Ocean, causing minor to moderate damage in many countries, and some deaths as far away as the Peruvian coast. According to K. Mackay from New Zealand's National Institute for Water and Atmospheric Research (NIWA) (https://www.abc.net.au/news/2022-11-21/sea-floor-map-reveals-huge-cavity-left-behind-by-tonga-volcano/101674708) and Seabrook et al. (2023), the eruption ejected $\sim$ 10 km^3 of material, and deepened the pre-existing caldera at a water depth of $\sim$150 m to 850 m deep. S. Cronin (https://theconversation.com/a-year-on-we-know-why-the-tongan-eruption-was-so-violent-its-a-wake-up-call-to-watch-other-submarine-volcanoes-175734) reports that the caldera is now almost 4 km in diameter.

It is clear that something very different occurred at Hunga Tonga–Hunga Ha'apai volcano on 15 January 2022, compared with the low-intensity surtseyan activity in the previous decade. This sudden much more explosive event suggests that a significant volume of new magma with perhaps a very high dissolved magmatic gas content rose quickly, which S. Cronin (see above) suggests interacted with two other magmas at depth, mingling with them and triggering rise of the magma to the seafloor. Preliminary geochemical analyses of pyroclasts confirm that the average magma composition was basaltic andesite with $\sim$57% SiO_2, compared with the andesitic composition of the surtseyan phase of volcanism (S. Barker et al., Victoria University Wellington, online seminar Geological Society of Australia, 24-2-2022). S. Cronin has proposed that there may have been 3 different magmas involved and that mingling trigggered magma uprise (https://theconversation.com/a-year-on-we-know-why-the-tongan-eruption-was-so-violent-its-a-wake-up-call-to-watch-other-submarine-volcanoes-175734). Although some phreatomagmatic explosive energy must have contributed to the eruption, given that the vent was probably submerged, it is not clear that the energy released by the explosive eruption was all of phreatomagmatic origin. The vent was probably quite shallow (several tens of m?), so hydrostatic pressure would not have exerted any influence in suppressing explosive intensity, nor vesiculation of rising and erupting magma. Juvenile pyroclasts apparently have a variable vesicularity, with some very vesicular pumice ($\sim$70%+?; S. Barker et al., as reported above) indicating that magma volatile content was high.

At present, the eruption style can be classified as phreatoplinian, based on using the eruption column height of $\sim$40 km as a proxy, although if the eruption column was 57 km high, it would be a phreato-ultraplinian eruption. The erupted volume of $\sim$ 10 km^3 (or 10^{10} m^3) is also consistent with the eruption being classified as a VEI 5 to 6 phreato-ultraplinian eruption using the Volcanic Explosivity Index of Newhall and Self (1982). Unfortunately, other parameters and data that are usually used to classify eruption style (erupted volume, area of dispersal, grain size, and degree of fragmentation; Chap. 9) are not yet available, given that much of deposit will have been deposited in the ocean. This includes a significant volume of deposits by syn-eruptive submarine pyroclastic or mass flow density currents (see Chap. 11) that flowed over the seafloor up to 80 to 100 km from the volcano and caused submarine telecommunication cables to be broken (K. Mackay, NIWA, see above; Seabrook et al. 2023).

There is an ongoing discussion on whether the 15 January 2022 column height was 40 km (and similar in height to the 1991 Pinatubo eruption column height), or alternatively up to a maximum height of 57 km, but how is that defined? We caution over the column height estimate, because it is not clear if "eruption column height" should be defined on the basis of the height to which ash is projected upwards and dispersed laterally from that height from the spreading umbrella region, and/or the height to which gas carrying only minor very fine ash rises and overshoots at the top of the plume. S. Cronin (see above) indicates that the peak height of the column was 57 km, but that the column then settled at a height of 40 km as it spread radially. In terms of the dispersal of deposits, it is only the ash component that is important, but gas may be more buoyant and rise higher than solid ash particles. Another factor is the resolution of satellite technology, and how that has changed over the decades. Does the satellite technology that has detected the 15 January 2022 Hunga Tonga–Hunga Ha'apai eruption column behaviour and height have the same resolution as that during the Pinatubo eruption or is it better? Therefore, are comparisons between assessments of column height, properties, and behaviour made in 1991 and 2022 valid? Are we comparing apples with apples?

Finally, there is much still to learn about this eruption and the causes of its effects (also see Chap. 9). Much of this understanding will depend on:

(a) Collecting and undertaken textural (vesicle properties, abundance) and geochemical analysis (pre-eruption volatile content from glass inclusions).
(b) New detailed deep-sea bathymetry to evaluate whether the size and shape of the caldera have changed as a result of the eruption and to assess if the eruption was caldera-forming.
(c) Employing new deep-sea bathymetry to assess if the generation of the tsunamis was due to seafloor displacement associated with caldera collapse and/or due to a submarine landslide on the submarine slopes of the volcanic edifice. If neither of these alternatives proves to be correct, does this indicate that the tsunamis were generated by explosive displacement of the water mass around the vent, and/or by shock waves propagating through the water mass, as they did through the atmosphere, which could also have generated meteo-tsunamis?
(d) Evaluating how much of the explosive eruption was driven by phreatomagmatic interaction between the newly injected magma and ocean water, versus how much by magmatic volatiles, and if and how the driving parameters changed during the eruption.

3.6 Glacial Ice

As the solid state of water, ice also influences volcanism by way of exerting a glaciostatic pressure on rising and erupting magma, depending on the thickness of the glacial ice. Petrovic (2003) reviewed the mechanical properties of ice and snow, and Pringle et al. (2007) summarised the thermal conductivity of glacial and sea ice. Lucas et al. (2022) undertook thermo-mechanical finite element modelling to assess the effects of varying thicknesses of ice caps above magma bodies on the magma bodies and the potential for eruptions. They modelled varying ice cap thickness (1–3 km), magma chamber depths, geometry, magma flux, and seasonal variations in ice cover thickness. They found that repose periods between eruptions increased with ice cap thickness. Lucas et al. (2022) did not model the effects of glaciostatic pressure on volatile exsolution, but it would suppress volatile exsolution and thus make magma less buoyant than magma which has vesiculated. This may contribute to longer repose periods. In addition, the glaciostatic (and hydrostatic pressures where base of glacial melting occurs and meltwater ponds form—see below) can exert constraints on whether or not volatile exsolution and magmatic and phreatomagmatic explosive volcanism occur, as discussed above for water (also see Chap. 4). Since the density of ice and water are similar, the pressure constraints are similar in subaqueous and subglacial settings.

Ice also affects volcanism in slightly different ways to liquid water. Of course, it melts very quickly when in contact with magma, and in glacial environments this can lead to major flood events (Chap. 12). However, field evidence suggests that in subglacial settings, even when some degree of melting of the ice occurs, forming subglacial meltwater ponds, thick bodies of ice can act as buttresses to the lateral flow of subglacial lavas and lateral growth of subglacial volcanoes. This leads to the formation of steep-sided piles and volcanic edifices of lava, called tuyas (e.g. Jakobsson and Gudmundsson 2008; Smellie 2000; Russell et al. 2014; Edwards et al. 2015). Basal sequences are typical subaqueous volcanic deposits, including pillow and sheet lavas, hyaloclastite breccias (e.g. Chap. 5), and transforming upwards into subaqueous to emergent (i.e. above water level) phreatomagmatic deposits, sometimes capped by subaerial volcanic successions (Jones 1969; Smellie 2006; Russell et al. 2014). A coherent ice cap, while it lasts, also acts as a "lid" on the uprise of magma fountains, eruption columns, and thermal aqueous plumes, but only until melting from the base of the glacial ice upwards breaks through to the top of the ice mass, after which the eruption conditions become subaqueous.

3.7 Summary: Why Deep Water and Subglacial Volcanic Processes Are so Different from Subaerial Volcanism

The foregoing raises many issues about understanding subaqueous eruptions, and particularly makes it clear that when eruptions occur in deep water or under thick ice, the thermal and other physical properties of the water mass and the higher hydro/glaciostatic pressure regime create different physical constraints on subaqueous and subglacial eruptions, deposit types, and dispersal processes compared with subaerial eruptions. Some of these constraints can also apply on other planets with higher atmospheric pressure regimes at present (e.g. Venus), or in the past. In summary, some of the key findings from this review are:

(a) Magma physical properties, especially viscosity, are affected by eruption in deep water under high ambient (hydrostatic) pressure, which not only inherently reduces magma viscosity but also restricts volatile exsolution, rendering magma less viscous than a subaerial counterpart because more volatiles remain dissolved in the magma. Magmas and lavas may thus be more fluidal.
(b) Deep-water submarine magma “fire” fountains are likely to be more fluidal, and probably also driven by exsolved supercritical fluid-induced buoyancy of magma (i.e. a fluidal gushing magma fountain analogous to an overpressured, gushing oil well breakout and fountain, that is not explosive).
(c) Subaqueously erupted lavas may remain coherent if stable steam film boiling insulates the surface of the lava from sudden heat loss, allowing a coherent visco-elastic skin and then a coherent glass crust to form.
(d) Alternatively, if steam vapour film collapse occurs, direct contact between the hot lava surface and liquid water causes instantaneous thermal shock, cooling contraction fracturing, and granulation, called quench fragmentation. This produces a glassy breccia (hyaloclastite) that is unique to subaqueous and subglacial environments. Quench fragmentation cannot occur in subaerial environments except where there are surface bodies of water (lakes, rivers) or glaciers.
(e) It has previously been assumed that quench fragmentation can occur at all water depths. However, thermal shock requires the collapse of a highly compressible steam layer, but as hydrostatic pressure increases with increasing water depth, the steam vapour film compresses into a high-pressure critical and supercritical fluid. Critical and supercritical fluids have higher heat transfer properties than steam, allowing a steadier, higher rate of heat loss and thermal gradient to form from the lava to the water mass that reduces or largely eliminates thermal shock. In addition, the high pressure of the critical/supercritical fluid also mitigates film collapse and thermal shock. It thus seems that quench fragmentation is likely to be depth limited to largely <2000 m. Anecdotal observations of the deep ocean floor seem to suggest that hyaloclastite breccia is rare.
(f) Formation of critical and supercritical fluid also affects the depths at which explosive eruptions occur.
(g) The exsolution of magmatic volatiles initially manifests as bubbles of supercritical fluid. Supercritical fluid bubbles under high ambient pressures can expand but cannot expand at explosive rates. Once magma has decompressed to water depths where supercritical fluid transforms to steam, the bubbles can expand at a higher rate, but still not necessarily explosively. It is only if magmatic volatile contents are high enough and hydrostatic pressures are low enough to permit explosive magmatic eruptions. Hydrostatic pressure generally constrains magmatic explosive eruptions to water depths of <1000 m, and occasionally greater. The intensity of explosive eruptions decreases with increasing water depths.
(h) Vesiculation therefore does not necessarily lead to explosive eruptions because vesicles/bubbles are initially filled with supercritical fluid when exsolution occurs. The depth and pressure at eruption will determine the rate of growth and evolution of those initial vesicles. Rapid magma decompression, which can trigger explosive growth of exsolving magmatic volatile vesicles in many subaerial explosive eruptions and shallow marine settings due to low ambient pressures, becomes less effective with increasing water depth.
(i) Subaqueous lavas can be highly vesicular and even form coherent pumice carapaces and interiors, which are prone to quench fragmentation, forming aggregates of pumice of autoclastic hyaloclastite pumice breccia deposits, rather than being of explosive origins (Chaps. 4, 9).
(j) Quench fragmentation of pumiceous submarine lavas can produce voluminous pumice clast deposits that did not form by explosive eruptions.
(k) Pumice deposits in the deep ocean can form by quench fragmentation, dispersal and settling from shallower water explosive eruptions, resedimentation from subaerial and shallower water settings, and raft transport by oceanic currents from far away volcanic eruptions, including submarine ones.
(l) Similarly, the superheating of external water during explosive water–magma phreatomagmatic eruptions is constrained by the state of the fluid film at the interface between erupting magma and the water mass, which is controlled by hydrostatic pressure.
(m) Molten Fuel–Coolant Interaction (MFCI) explosions depend on the existence of a compressible, collapsible steam vapour film. As discussed for quench fragmentation, steam film collapse instantaneously brings cold liquid water in contact with the magma surface. This causes quench fracturing, but if the ambient hydrostatic pressure is low enough (<7–8 MPa), it can also trigger cyclical thermal detonation explosive activity as the liquid water is superheated and flashed to steam explosively. At greater pressures and depths, this won’t occur because the steam compresses and at great depths transforms into critical–supercritical fluid. It also becomes more pressurised, which counteracts the ambient hydrostatic pressure inhibiting film collapse.

(n) A newly identified variant of MFCI, called Induced Fuel–Coolant Interaction (IFCI) occurs when water penetrates cracks in the surface of the lava, formed by quench contraction fracturing, and that intra-crack water is also flashed to steam explosively. Although it has been proposed that this can also occur at water depths beyond the critical point of water, we have discussed why this is unlikely to impossible because at those depths boiling of the intra-crack water produces critical and supercritical water, and the *expansion rate* of critical–supercritical fluid is constrained by the extremely high ambient hydrostatic pressure. However, the pressurised supercritical fluid could hydraulically fracture the enclosing glass. Again, there is little anecdotal evidence that IFCI occurs at great depths in the form of widespread deep-water ashes. Deep-water ashes can also be formed in other ways, as discussed in Chap. 9.

(o) Many authors have noted that there are fewer pyroclastic deposits with increasing depth in the modern oceans (e.g. Gregg and Fink 1995; Grosfils et al. 2000; Wright et al. 2003, 2006), indicating an effect of water depth and hydrostatic pressure in limiting explosive eruptions. Although the seafloor is dominated by basalt lavas, even voluminous dacite lavas are now being found at 3–4 km water depth in the northern Lau Basin where the crust is oceanic (Embley and Rubin 2018), and felsic eruption products are common in marine arc volcanoes (e.g. Fiske et al. 1998, 2001; Wright et al. 2003, 2006; Carey et al. 2018).

(p) Vesiculation can occur at any water depth if magmatic volatile content is high enough, but the state of the exsolved fluids (steam, supercritical fluid) varies according to confining hydrostatic pressure, and this will affect the vesiculation rate and whether or not explosive eruptions occur or pumiceous or non-vesiculated coherent lavas erupt.

(q) Submarine eruption columns behave differently to subaerial ones (Chap. 9). They range from heated plumes of water above non-explosively erupting lavas to steam-rich plumes above explosive vents. They are inherently buoyant in the oceans because they are less dense.

(r) Submarine eruption columns may however undergo mixing with ambient water as they rise making them denser, and cooling and condensation of steam and supercritical fluid bubbles (condensation erosion) can also render them less buoyant.

(s) Submarine eruption columns can entrain all clasts that are less dense than the bulk density of the column, and even some that are slightly denser due to upward momentum and if the rise velocity of the columns exceeds the settling velocity of the clasts.

(t) Unlike subaerial explosive eruptions, during and after submarine explosive eruptions and also during quench fragmentation of the pumice carapaces of effusive lavas, the properties of pumice clasts entrained into buoyant thermal eruption plumes of hot water or steam-rich eruption columns or simply rising buoyantly through the water mass can change. Two processes can affect such pumice clasts: quench fragmentation, so reducing their size, and infiltration of seawater into the vesicle networks as the gas cools and condenses. Pumice clasts may thus become water-logged, increasing their bulk density and affecting their buoyancy. They sink or continue to rise and float, or flat and then sink (Chap. 9).

(u) Buoyant pumice clasts are mostly dispersed by oceanic currents which can vary at different water depths and can be very irregular in their paths as they are affected by oceanic gyre and regional wind patterns and weather systems (Chap. 9).

References

Airey MW, Mather TA, Pyle DM, Glaze LS, Ghail RC, Wilson CF (2015) Explosive volcanic activity on Venus: the roles of volatile contribution, degassing, and external environment. Planet Space Sci 113–114:33–48

Alatorre-Ibargüengoitia MA, Delgado-Granados H (2006) Experimental determination of drag coefficient for volcanic materials: calibration and application of a model to Popocatépetl volcano (Mexico) ballistic projectiles. Geophys Res Lett 33(11)

Allen SR, Fiske RS, Tamura Y (2010) Effects of water depth on pumice formation in submarine domes at Sumisu, Izu-Bonin arc, western Pacific. Geology 38(5):391–394

Bagdassarov NS, Dingwell DB (1992) A rheological investigation of vesicular rhyolite. J Volcanol Geotherm Res 50(3):307–322

Bahr HA, Fischer G, Weiss HJ (1986) Thermal-shock crack patterns explained by single and multiple crack propagation. J Mater Sci 21:2716–2720

Barnett WP (2008) The rock mechanics of kimberlite volcanic pipe excavation. J Volcanol Geotherm Res 174(1–3):29–39

Barnett WP, Kurszlaukis S, Tait M, Dirks P (2011) Kimberlite wall-rock fragmentation processes: Venetia K08 pipe development. Bull Volcanol 73(8):941–958

Barreyre T, Soule SA, Sohn RA (2011) Dispersal of volcaniclasts during deep-sea eruptions: settling velocities and entrainment in buoyant seawater plumes. J Volcanol Geotherm Res 205(3):84–93

Binns RA (2003) Deep marine pumice from the Woodlark and Manus basins, Papua New Guinea. In: White JDL, Smellie JL, Clague DA (eds) Explosive subaqueous volcanism. Amer Geophys Unon Monogr 140:329–343

Birch F (1942a) Density at high pressure; compressibility. Geol Soc Amer Spec Pub 36:39–62

Birch F (1942b) Elasticity (except compressibility). Geol Soc Amer Spec Pub 36:63–86

Brenna M, Cronin S, Smith I, Pontesilli A, Tost M, Barker S, Tongaonevai S (2022) Post-caldera volcanism reveals shallow priming of an intra-ocean arc andesitic caldera: Hunga volcano, Tonga, SW Pacific. Lithos 412–413:106614

Brennen CE (1995) Cavitation and bubble dynamics. Oxford University Press, Oxford, p 304
Brennen CE (2002) Fission of collapsing cavitation bubbles. J Fluid Mech 472:153–166
Cahalan RC, Dufek J (2021) Explosive submarine eruptions: the role of condensable gas jets in underwater eruptions. J Geophys Res Solid Earth 126(2):e2020JB020969
Cañón-Tapia E (2008) How deep can be a dyke? J Volcanol Geotherm Res 171(3–4):215–228
Carey R, Soule SA, Manga M, White JDL, McPhie J, Wysoczanski R, Jutzeler M, Tani K, Yoerger D, Fornari D (2018) The largest deep-ocean silicic volcanic eruption of the past century. Sci Adv 4 (1):e1701121
Cas RAF (1992) Submarine volcanism; eruption styles, products, and relevance to understanding the host-rock successions to volcanic-hosted massive sulfide deposits. Econ Geol 87(3):511–541
Cas RAF, Giordano G (2014) Submarine volcanism: a review of the constraints, processes and products, and relevance to the Cabo de Gata volcanic succession. Ital J Geosci 133(3):362–377
Cas RAF, Hayman P, Pittari A, Porritt L (2008) Some major problems with existing models and terminology associated with kimberlite pipes from a volcanological perspective, and some suggestions. J Volcanol Geotherm Res 174(1–3):209–225
Cas RAF, Simmons J (2018) Why deep-water eruptions are so different from subaerial eruptions. Front Earth Sci 6:198
Cas RAF, Wright JV (1987) Volcanic successions: modern and ancient. Allen and Unwin, London, 528 pp
Cas RAF, Yamagishi H, Moore L, Scutter CR (2003) Miocene submarine fire fountain deposits, Ryugazaki Headland, Oshoro Peninsula, Hokkaido, Japan: implications for submarine fountain dynamics and fragmentation processes. In: White JDL, Smellie JL, Clague DA (eds) Explosive subaqueous volcanism. Amer Geophys Union Monogr 140:299–316
Chadwick WW, Cashman KV, Embley RW, Matsumoto H, Dziak RP, De Ronde CEJ, Lau TK, Deardorff ND, Merle SG (2008) Direct video and hydrophone observations of submarine explosive eruptions at NW Rota-1 volcano, Mariana arc. J Geophys Res Solid Earth Solid Earth 113(B8)
Chadwick WW Jr, Merle SG, Baker ET, Walker SL, Resing JA, Butterfield DA, Anderson MO, Baumberger T, Bobbitt AM (2018) A recent volcanic eruption discovered on the central Mariana back-arc spreading center. Front Earth Sci 6:172
Chadwick WW, Paduan JB, Clague DA, Dreyer BM, Merle SG, Bobbitt AM, Caress DW, Philip BT, Kelley DS, Nooner SL (2016) Voluminous eruption from a zoned magma body after an increase in supply rate at Axial Seamount. Geophys Res Lett 43(23):12063–012070
Clague DA, Paduan JB, Caress DW, Chadwick WW Jr, Le Saout M, Dreyer BM, Portner RA (2017) High-resolution AUV mapping and targeted ROV observations of three historical lava flows at Axial Seamount. Oceanography 30(4):82–99
Clague DA, Uto K, Satake K, Davis AS (2002) Eruption style and flow emplacement in the submarine North Arch Volcanic Field, Hawaii. In: Hawaiian volcanoes: deep underwater perspectives. Amer Geophys Union Monog 128:65–84
Cronin S (2022) Why the volcanic eruption in Tonga was so violent, and what to expect next. The Conversation, January 16, 2022. https://mail.google.com/mail/u/0/#search/Shane+Cronin/FMfcgzGmtXCRxPtdgDXpmkstLfQGmjNv
Cronin SJ, Brenna M, Smith IE, Barker S, Tost M, Ford M (2017) New volcanic island unveils explosive past. Eos 98:18–23
Deardorff ND, Cashman KV, Chadwick WW (2011) Observations of eruptive plume dynamics and pyroclastic deposits from submarine explosive eruptions at NW Rota-1, Mariana Arc. J Volcanol Geotherm Res 202(1):47–59
Dingwell DB (1997) The brittle–ductile transition in high-level granitic magmas: material constraints. J Petrol 38(12):1635–1644
Dixon JE, Clague DA, Wallace P, Poreda R (1997) Volatiles in alkalic basalts form the North Arch Volcanic Field, Hawaii: extensive degassing of deep submarine-erupted alkalic series lavas. J Petrol 38 (7):911–939
Dürig T, White JD, Murch AP, Zimanowski B, Büttner R, Mele D, Dellino P, Carey RJ, Schmidt LS, Spitznagel N (2020) Deep-sea eruptions boosted by induced fuel–coolant explosions. Nature Geosci 13(7):498–503
Edwards BR, Gudmundsson MT, Russell JK (2015) Glaciovolcanism. In: Sigurdsson H, Houghton B, McNutt SR, Rymer H, Stix J (eds) The encyclopedia of volcanoes. Academic Press Elsevier, Cambridge, pp 377–394
Embley RW, Chadwick WW, Baker ET, Butterfield DA, Resing JA, De Ronde CEJ, Tunnicliffe V, Lupton JE, Juniper SK, Rubin KH (2006) Long-term eruptive activity at a submarine arc volcano. Nature 441(7092):494–497
Embley RW, Rubin KH (2018) Extensive young silicic volcanism produces large deep submarine lava flows in the NE Lau Basin. Bull Volcanol 80(4):36
Fauria KE, Manga M, Wei Z (2017) Trapped bubbles keep pumice afloat and gas diffusion makes pumice sink. Earth Planet Sci Lett 460:50–59
Fink JH, Anderson SW, Manley CR (1992) Textural constraints on effusive silicic volcanism: Beyond the permeable foam model. J Geophys Res: Solid Earth 97(B6):9073–9083
Fink JH, Manley CR (1987) Origin of pumiceous and glassy textures in rhyolite flows and domes. Geol Soc Amer Spec Pub 212:77–88
Fiske RS, Cashman KV, Shibata A, Watanabe K (1998) Tephra dispersal from Myojisho, Japan, during its shallow submarine eruption of 1952–1953. Bull Volcanol 59:262–275
Fiske RS, Naka J, Iizasa K, Yuasa M, Klaus A (2001) Submarine silicic caldera at the front of the Izu-Bonin arc, Japan: voluminous seafloor eruptions of rhyolite pumice. Geol Soc Amer Bull 113:813–824
Fujikawa S, Akamatsu T (1980) Effects of the non-equilibrium condensation of vapour on the pressure wave produced by the collapse of a bubble in a liquid. J Fluid Mech 97(03):481–512
Gregg TKP, Fink JH (1995) Quantification of submarine lava-flow morphology through analog experiments. Geology 23(1):73–76
Grosfils EB, Aubele J, Crumpler L, Gregg TKP, Sakimoto S (2000) Volcanism on Earth's seafloor and Venus. In: Environmental effects on volcanic eruptions. Springer, New York City, pp 113–142
Head JW, Crumpler LS, Aubele JC, Guest JE, Saunders RS (1992) Venus volcanism: classification of volcanic features and structures, associations, and global distribution from Magellan data. J Geophys Res Solid Earth Planets 97(E8):13153–13197
Head JWI, Wilson L (2003) Deep submarine pyroclastic eruptions: theory and predicted landforms and deposits. J Volcanol Geotherm Res 121(3–4):155–193
Heap MJ, Villeneuve M, Albino F, Farquharson JI, Brothelande E, Amelung F, Got JL, Baud P (2020) Towards more realistic values of elastic moduli for volcano modelling. J Volcanol Geotherm Res 390:106684
Höskuldsson A, Sparks RSJ (1997) Thermodynamics and fluid dynamics of effusive subglacial eruptions. Bull Volcanol 59 (3):219–230
Ikegami F, McPhie J, Carey R, Mundana R, Soule A, Jutzeler M (2018) The eruption of submarine rhyolite lavas and domes in the deep ocean—Havre 2012, Kermadec arc. Front Earth Sci 6:147
Ivanov MA, Crumpler LS, Aubele JC, Head JW (2015) Volcanism on venus. In: Sigurdsson H, Houghton B, McNutt SR, Rymer H, Stix J (eds) The encyclopedia of volcanoes. Academic Press Elsevier, Cambridge, pp 729–746

Jakobsson SP, Gudmundsson MT (2008) Subglacial and intraglacial volcanic formations in Iceland. Jökull 58:179–196

Jellinek AM, DePaolo DJ (2003) A model for the origin of large silicic magma chambers: precursors of caldera-forming eruptions. Bull Volcanol 65(5):363–381

Jones JG (1969) Intraglacial volcanoes of the Laugarvatn region, south-west Iceland—I. Quarterly J Geol Soc 124(1–4):197–211

Jutzeler M, Marsh R, Carey RJ, White JDL, Talling PJ, Karlstrom L (2014) On the fate of pumice rafts formed during the 2012 Havre submarine eruption. Nat Commun 5:1–10

Karagianni A, Karoutzos G, Ktena S, Vagenas N, Vlachopoulos I, Sabatakakis N, Koukis G (2010) Elastic properties of rocks. Bull Geol Soc Greece 43(3):1165–1168

Kingery WD (1955) Factors affecting thermal stress resistance of ceramic materials. J Amer Ceram Soc 38:3–15

Kokelaar P (1986) Magma-water interactions in subaqueous and emergent basaltic. Bull Volcanol 48(5):275–289

Kushiro I (1976) Changes in viscosity and structure of melt of $NaAlSi_2O_6$ composition at high pressures. J Geophys Res Solid Earth 81(35):6347–6350

Kushiro I (1980) Viscosity, density and structure of silicate melts at high pressures, and their petrological applications. In: Hargraves RB (ed) Physics of magmatic processes. Princeton University Press, Princeton, pp 93–120

Kushiro I, Yoder HS, Mysen BO (1976) Viscosities of basalt and andesite melts at high pressures. J Geophys Res Solid Earth 81 (35):6351–6356

Lesher C, Spera FJ (2015) Thermodynamic and transport properties of silicate melts and magma. In: Sigurdsson H, Houghton B, McNutt SR, Rymer H, Stix J (eds) The encyclopedia of volcanoes. Academic Press Elsevier, Cambridge, pp 113–141

Liu J, Chang Z, Wang L, Xu J, Kuang R, Wu Z (2020) Exploration of basalt glasses as high-temperature sensible heat storage materials. Amer Chem Soc Omega 5(30):19236–19246

Lucas LC, Albright JA, Gregg PM, Zhan Y (2022) The impact of ice caps on the mechanical stability of magmatic systems: implications for forecasting on human timescales. Front Earth Sci 10:868569

Manga M, Fauria KE, Lin C, Mitchell SJ, Jones M, Conway CE, Degruyter W, Hosseini B, Carey R, Cahalan R (2018) The pumice raft-forming 2012 Havre submarine eruption was effusive. Earth Planet Sci Lett 489:49–58

Mann CP, Stix J, Vallance JW, Richer M (2004) Subaqueous intracaldera volcanism, Ilopango Caldera, El Salvador, Central America. Geol Soc Amer Spec Pap 375:159–174

McBirney AR (1963) Factors governing the nature of submarine volcanism. Bull Volcanol 26(1):455–469

Mills AA (1984) Pillow lavas and the Leidenfrost effect. J Geol Soc Lond 141(1):183–186

Mitchell SJ, Houghton BF, Carey RJ, Manga M, Fauria KE, Jones MR, Soule SA, Conway CE, Wei Z, Giachetti T (2019) Submarine giant pumice: a window into the shallow conduit dynamics of a recent silicic eruption. Bull Volcanol 81:1–21

Moore JG (1975) Mechanism of formation of pillow lava: pillow lava, produced as fluid lava cools underwater, is the most abundant volcanic rock on earth, but only recently have divers observed it forming. Amer Sci 63(3):269–277

Mueller S, Scheu B, Spieler O, Dingwell DB (2008) Permeability control on magma fragmentation. Geology 36(5):399–402

Mueller W, White JDL (1992) Felsic fire-fountaining beneath Archean seas: pyroclastic deposits of the 2730 Ma Hunter Mine Group, Quebec, Canada. J Volcanol Geotherm Res 54(1):117–134

Murch AP, White JD, Barreyre T, Carey RJ, Mundana RE, Ikegami F (2020) Volcaniclastic dispersal during submarine lava effusion: the 2012 eruption of Havre Volcano, Kermadec arc, New Zealand. Front Earth Sci 8:237. https://doi.org/10.3389/feart.2020.00237

Murch AP, White JDL, Carey RJ (2019) Characteristics and deposit stratigraphy of submarine-erupted silicic ash, Havre volcano, Kermadec arc, New Zealand. Front Earth Sci 7:1. https://doi.org/10.3389/feart.2019.00001

Newhall CG, Self S (1982) The volcanic explosivity index (VEI) an estimate of explosive magnitude for historical volcanism. J Geophys Res 87(C2):1231–1238

Newland EL, Mingotti N, Woods AW (2022) Dynamics of deep-submarine volcanic eruptions. Sci Rept 12(1):3276

Perras MA, Diederichs MS (2014) A review of the tensile strength of rock: concepts and testing. Geotech Geol Eng 32(2):525–546

Petrovic JJ (2003) Review mechanical properties of ice and snow. J Mater Sci 38(1):1–6

Pensa A, Capra L, Giordano G (2019) Ash clouds temperature estimation. Implication on dilute and concentrated PDCs coupling and topography confinement. Sci Repts 9(1):1–4

Pensa A, Giordano G, Cas R, Porreca M (2015) Thermal state and implications for eruptive styles of the intra-Plinian and climactic ignimbrites of the 4.6 ka Fogo A eruption sequence, São Miguel, Azores. Bull Volcanol 77(11):99

Pinel V, Jaupart C (2003) Magma chamber behavior beneath a volcanic edifice. J Geophys Res Solid Earth Solid Earth 108(B2)

Pioro IL (2020) Supercritical-fluids thermophysical properties and heat transfer in power-engineering applications. In: Pioro I (ed) Advanced supercritical fluids technologies, Chapter 1, 42 pp, IntechOpen

Pioro I, Mokry S (2011) Chapter 22 Thermophysical properties at critical and supercritical pressures. In: Belmiloudi A (ed) Heat transfer-theoretical analysis, experimental investigations and industrial systems. InTech, pp 573–592

Porreca M, Cifelli F, Soriano C, Giordano G, Romano C, Conticelli S, Mattei M (2014) Hyaloclastite fragmentation below the glass transition: an example from El Barronal submarine volcanic complex (Spain). Geology 42(1):87–90

Pringle DJ, Eicken H, Trodahl HJ, Backstrom LGE (2007) Thermal conductivity of landfast Antarctic and Arctic sea ice. J Geophys Res Solid Earth Oceans 112(C4)

Rayleigh L (1917) VIII. On the pressure developed in a liquid during the collapse of a spherical cavity. Lond Edinb Dubl Phil Mag 34 (200):94–98

Reid RC, Sherwood TK, Prasnitz JM, Poling BE (1987) The properties of gases and liquids, 4th edn. McGraw-Hill, 741 pp

Resing JA, Rubin KH, Embley RW, Lupton JE, Baker ET, Dziak RP, Baumberger T, Lilley MD, Huber JA, Shank TM, Butterfield DA et al (2011) Active submarine eruption of boninite in the northeastern Lau Basin. Nature Geosci 4(11):799–806

Resnyansky AD, Delaney TG (2006) Experimental study of blast mitigation in a water mist. Aust Govt Defence Sci Technol Org Tech Rept DSTO-TR-1944, 41 p

Rowell CR, Jellinek AM, Hajimirza S, Aubry TJ (2022) External surface water influence on explosive eruption dynamics, with implications for stratospheric sulfur delivery and volcano-climate feedback. Front Earth Sci 10:788294

Russell JK, Edwards BR, Porritt L, Ryane C (2014) Tuyas: a descriptive genetic classification. Quaternary Sci Revs 87:70–81

Safarov J, Millero F, Feistel R, Heintz A, Hassel E (2009) Thermodynamic properties of standard seawater: extensions to high temperatures and pressures. Ocean Sci 5(3):235–246

Scutter C, Cas RAF, Moore CL, De Rita D (1998) The facies architecture and origin of a submarine rhyolite lava flow-dome complex, Ponza, Italy. J Geophys Res Solid Earth 103:27551–527566

Seabrook S, Mackay K, Watson S, Clare M, Hunt J, Yeo I, Lane E, Clark M, Wysoczanski R, Rowden A, Hoffmann L, Armstrong E, Williams M (2023) Pyroclastic density currents explain far-reaching

and diverse seafloor impacts of the 2022 Hunga Tonga Hunga Ha'apai eruption. Research Square Springer Nature, 20p (in press)
Sharqawy MH, Lienhard JH, Zubair SM (2010) Thermophysical properties of seawater: a review of existing correlations and data. Desalin Water Treat 16(1–3):354–380
Sheridan MF, Wohletz KH (1983) Hydrovolcanism: basic considerations and review. J Volcanol Geotherm Res 17(1–4):1–29
Smellie J (2006) The relative importance of supraglacial versus subglacial meltwater escape in basaltic subglacial tuya eruptions: an important unresolved conundrum. Earth Sci Rev 74(3–4):241–268
Smellie JL (2000) Subglacial eruptions. In: Sigurdsson H, Houghton B, McNutt SR, Rymer H, Stix J (eds) Encyclopedia of volcanoes. Elsevier, New York, pp 403–419
Sourirajan S, Kennedy GC (1962) The system H_2O-NaCl at elevated temperatures and pressures. Am J Sci 260(2):115–141
Stolper E, Walker D (1980) Melt density and the average composition of basalt. Contrib Mineral Petrol 74(1):7–12
Tait S, Jaupart C, Vergniolle S (1989) Pressure, gas content and eruption periodicity of a shallow, crystallising magma chamber. Earth Planet Sci Lett 92(1):107–123
Tenzer R, Gladkikh V (2014) Assessment of density variations of marine sediments with ocean and sediment depths. Sci World J 2014
Thomas N, Jaupart C, Vergniolle S (1994) On the vesicularity of pumice. J Geophys Res Solid Earth 99(15):615–633
Valentine GA, Gregg TK (2008) Continental basaltic volcanoes—processes and problems. J Volcanol Geotherm Res 177(4):857–873
van Otterloo J, Cas RAF, Scutter CR (2015) The fracture behaviour of volcanic glass and relevance to quench fragmentation during formation of hyaloclastite and phreatomagmatism. Earth Sci Rev 151:79–116
Vaughan RG, Webley PW (2010) Satellite observations of a surtseyan eruption: Hunga Ha'apai, Tonga. J Volcanol Geotherm Res 198(1–2):177–186
Wallace PJ, Plank T, Edmonds M, Hauri EH (2015) Volatiles in magmas. In: Sigurdsson H, Houghton B, McNutt SR, Rymer H, Stix J (eds) The encyclopedia of volcanoes. Academic Press. Elsevier, New York City, pp 163–184
White JDL, Smellie JL, Clague DA (2003) Introduction: a deductive outline and topical overview of subaqueous explosive volcanism. In: White JDL, Smellie JL, Clague DA (eds) Explosive subaqueous volcanism. Amer Geophys Union Monogr 140:1–23
Whitham AG, Sparks RSJ (1986) Pumice. Bull Volcanol 48 (4):209–223
Wohletz KH (1986) Explosive magma-water interactions: thermodynamics, explosion mechanisms, and field studies. Bull Volcanol 48 (5):245–264
Wohletz KH (2003) Water/magma interaction: physical considerations for the deep submarine environment. In: White JDL, Smellie JL, Clague DA (eds) Explosive subaqueous volcanism. Amer Geophys Union Monogr 140:25–49
Wohletz K, Zimanowski B, Büttner R (2013) Magma-water interactions. In: Fagents SA, Gregg TKP, Lopes RMC (eds) Modeling volcanic processes. Cambridge University Press, pp 230–256
Wright IC, Gamble JA, Shane PA (2003) Submarine silicic volcanism of the Healy caldera, southern Kermadec arc (SW Pacific): I–volcanology and eruption mechanisms. Bull Volcanol 65:15–29
Wright IC, Worthington TJ, Gamble JA (2006) New multibeam mapping and geochemistry of the 30–35 S sector, and overview, of southern Kermadec arc volcanism. J Volcanol Geotherm Res 149 (3):263–296
Zimanowski B, Büttner R (2003) Phreatomagmatic explosions in subaqueous volcanism. In: White JDL, Smellie JL, Clague DA (eds) Explosive subaqueous volcanism. Amer Geophys Union Monogr 140:51–60
Zimanowski B, Büttner R, Dellino P, White DL, Wohletz KH (2015) Magma-water interaction and phreatomagmatic fragmentation. In: Sigurdsson H, Houghton B, McNutt SR, Rymer H, Stix J (eds) The encyclopedia of volcanoes. Elsevier, Amsterdam, pp 473–484
Zimanowski B, Fröhlich G, Lorenz V (1991) Quantitative experiments on phreatomagmatic explosions. J Volcanol Geotherm Res 48(3–4):341–358

4 Fragmentation Processes in Magmas and Volcanic Rocks

Autoclastic, Explosive, Hydraulic Fracturing—Characterising Clast and Aggregate Properties

Ray Cas, Guido Giordano, and John V. Wright

R. Cas et al., *Volcanology*, Springer Textbooks in Earth Sciences, Geography and Environment,
https://doi.org/10.1007/978-3-319-66613-6_4

Plate 4.1 Some fragmentation processes and deposits. **a** Autobreccia at the surface of the 1925 dacite lava erupted from Nea Kameni, Santorini, Greece. **b** Submarine, dacite lava dome hyaloclastite breccia and transition to subvertical coherent feeder dyke, Miocene, Kariba, Hokkaido, Japan. **c** Multiple vent magmatic explosive eruption from the summit of Mt Etna, Italy, in 2003. **d** Ballistic tephra jet phreatomagmatic explosive blast from the shallow marine basaltic Surtsey volcano, Iceland, November 1963 (Photo by Sigurdur Thorarinsson, courtesy of Sven Sigurdsson). **e** In situ hydrothermal hydraulic fracture breccia from the basaltic andesite Lisca Bianca islet, Aeolian Islands, Italy

Abstract

In this chapter, we consider the many processes by which magma and volcanic rock can be fragmented. This comprehensive review describes the processes, dynamics, and characteristics of the diverse fragmental products, especially clast textures. Irrespective of their origin, all of these products can in the first instance be called volcaniclastic deposits. Volcaniclastic deposits can be produced by a spectrum of processes, ranging from primary, syn-eruptive autoclastic and explosive processes, syn- to post-eruptive surface sedimentary processes, subsurface hydrothermal processes, and even tectonic processes. Autoclastic processes involve non-explosive, syn-eruptive fragmentation of magma, usually associated with the flow of lavas in subaerial and subaqueous settings. The products include monomictic lava autobreccia, and in aqueous settings, quench fragmented hyaloclastite breccia. There are many different situations in which explosive eruptions and fragmentation of erupting magma and country rock can occur. Explosive eruptions can be driven by the exsolution of magmatic volatiles from magma, and the explosive growth of the resulting vesicles (magmatic explosive eruptions), or the explosive superheating of external water (phreatic explosive), or a combination of the two (phreatomagmatic explosive eruptions), or by overpressured hydrothermal fluids (hydrothermal explosive eruptions). Pyroclastic deposits and their pyroclasts can be enormously variable in grain size, clast and aggregate texture, and componentry. Secondary fragmentation can be caused by surface sedimentary processes. Volcanic terrains are often very dynamic surface sedimentary regimes, and therefore syn- to post-eruptive sedimentary processes produce a wide range of volcaniclastic deposit types, from mass wasting to mass-flow resedimented to tractionally reworked. In addition, post-depositional hydrothermal or hydraulic brecciation or fracturing of volcanic rocks, driven by sub-surface, overpressured hydrothermal or metamorphic fluids, and tectonic fracturing of volcanic rocks commonly produce volcanic breccias in volcanic settings. Although all the processes are distinctive, the characteristics of the resulting deposits can often be very similar, and not always easily distinguishable, especially in older volcanic successions. It is therefore important to establish criteria to distinguish the deposits and the processes from each other.

4.1 Introduction—Understanding the Spectrum of Syn-eruptive to Post-eruptive Fragmentation Processes in Volcanic Settings

Volcanic materials, whether they are erupting magmas, chilled volcanic glass, or already crystallised volcanic rock, can be fragmented by a spectrum of processes, ranging from primary, syn-eruptive volcanic, to post-eruptive secondary processes (Plate 4.1; Fig. 4.1). These produce a range of

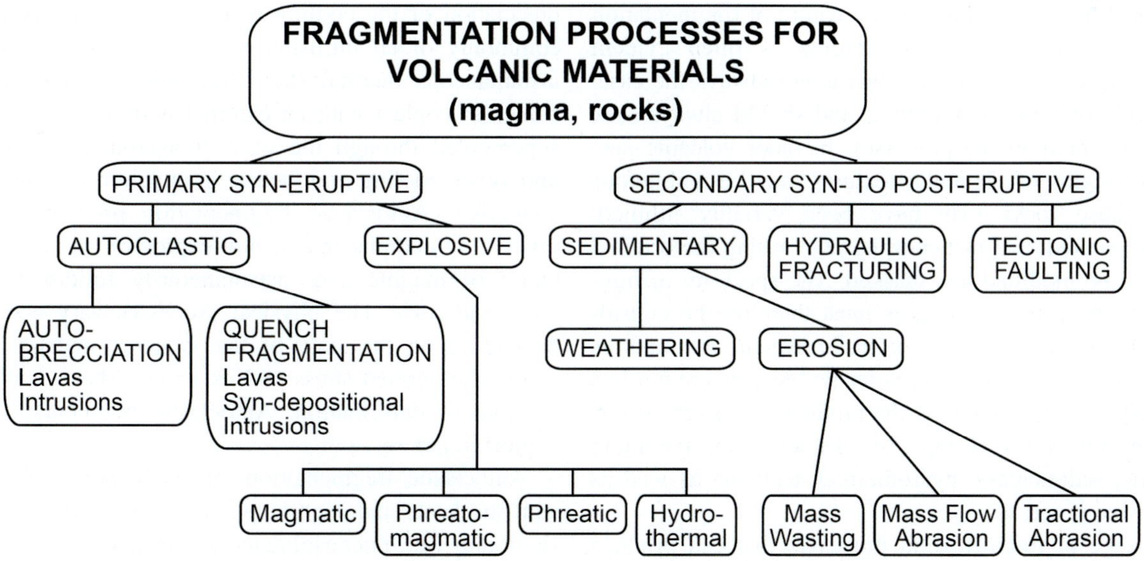

Fig. 4.1 Process spectrum summarising the fragmentation processes affecting erupting and intrusive magmas, and volcanic rocks, ranging from syn- to post-eruptive, and including autoclastic (autobrecciation, quench fragmentation), explosive (magmatic, phreatomagmatic, phreatic, hydrothermal), sedimentary or surface process (weathering, erosion, including mass wasting, mass-flow abrasion, tractional abrasion), hydrothermal hydraulic fracturing, and tectonic fault brecciation

primary-to-secondary volcaniclastic deposits and clast types (Fig. 4.1). All fragmental volcanic deposits and rocks in the first instance are volcaniclastic, which is a non-genetic, descriptive term that encompasses all aggregates of fragmental volcanic debris, irrespective of their origins. Although we could review each of the primary and secondary fragmentation processes in separate chapters, we have decided to review all fragmentation processes and their clast and mesoscopic aggregate characteristics together in one chapter for the purpose of comparison of processes and deposit characteristics, because it is often unclear what the origins of a volcaniclastic deposit were, and because the characteristics of different types of clasts and aggregates may share some similarities.

In this chapter, we summarise the different processes by which volcanic materials can be fragmented, we discuss the principles of each process, and we list some of the characteristic features of the clasts and aggregates produced. These characteristics are further investigated in the succeeding chapters that address the origins of the major primary volcanic and secondary sedimentary facies and deposit types that are found in modern and ancient volcanic successions and settings.

Primary fragmentation processes include syn-eruptive autoclastic, non-explosive processes associated mainly with emplacement of lavas (Chaps. 5, 6, and 7) and high-level, subvolcanic intrusions (Chap. 8), to explosive processes (Chaps. 9, 10, and 11). Post-eruptive fragmentation processes include the spectrum of normal surface sedimentary processes including gravitational collapse, mass movement, weathering, erosion and tractional transport (Chap. 12), as well as subsurface hydrothermal and tectonic fragmentation processes (Chap. 13). Although the nature of fragmentation processes in modern volcanic settings is often evident because they can be witnessed, this is not always the case. We should not make assumptions and should always seek evidence to interpret the processes. In older volcanic systems, the origins of particular textures may be ambiguous because those rocks may have been variably lithified, deformed, altered, and metamorphosed, causing the original features to be modified and masked. The spectrum of fragmentation processes and origins must therefore be considered before a conclusion can be made. Again, assumptions cannot be made. For example, volcanic breccias are not just pyroclastic in origin; they can be produced by over 20 different specific ways (Chap. 14, Table 14.1), including autoclastic, sedimentary, hydrothermal, tectonic, as well as by explosive processes.

The range of fragmentation processes that can produce volcaniclastic deposits includes

PRIMARY, SYN-ERUPTIVE

- Autoclastic:
  - autobrecciation, and
  - quench fragmentation.

- Explosive:
  - magmatic explosive,
  - phreatic explosive,
  - phreatomagmatic explosive:
    - "dry" phreatomagmatic, and
    - "wet" phreatomagmatic.
  - hydrothermal explosive,
  - impact shattering.

SECONDARY, SYN- TO POST-ERUPTIVE

- Sedimentary surface process/epiclastic:
  - weathering, and
  - physical erosion.
- Hydraulic fracturing (subsurface):
  - magmatic intrusive,
  - hydrothermal, and
  - metamorphic.
- Tectonic:
  - faulting/shearing related.

4.2 Autoclastic Fragmentation Processes

Autoclastic fragmentation involves either flow-induced brecciation of the cooling margins of moving lava or less commonly of an intrusion, called *autobrecciation*, or the instantaneous thermal shock fracturing as magma, which comes in contact with an external body of water or ice, is supercooled through the glass transition temperature, and undergoes cooling contraction granulation or *quench fragmentation*. Autoclastic fragmentation processes therefore involve syn-emplacement, non-explosive, in situ fragmentation of magma and instantaneously solidified volcanic glass and rock. The physical processes vary enormously, depending on the state of the volcanic material (plastic, solid), the applied stress and its nature (shear, tensile), and the rate of deformation. Autoclastic fragmentation is not explosive *per se*.

Autoclastic fragmentation occurs in subaerial and subaqueous lavas and high-level intrusions, particularly intrusions into wet, unconsolidated sediments. In erupting lavas and intrusions, which have both a temperature and viscosity gradient from the cooled, sometimes solidified outer margins to the still hot and plastic interior, there will be a spectrum of deformational behaviour, from possibly brittle at the margins to plastic or ductile in the interior, depending on the strain rate, viscosity, and stress type (shear, tensile).

The properties of magma are summarised in Chap. 2, and readers should review that chapter. A key concept is that magmas behave differently under stress at different viscosities (Chap. 2). Such concepts also apply in the glass and ceramics chemical engineering literature (e.g. Bouška et al. 1993; Shelby 1997; Shackelford 1990), from which volcanologists can learn a great deal (van Otterloo et al. 2015). The origin of stress in a melt can be varied and complex. It includes the shear stress produced by differential flow rates between the cooled, more viscous margins or crust of a lava or intrusion and the still plastic, flowing interior. The crust may deform or break by extension as the shear stress produced by flow tries to drag it along or by compression as a flow encounters an obstacle or break in slope. In hot magma that is supercooled through contact with water or ice, contractional or tensile stress develops as the magma rapidly passes through the glass transition temperature range as it chills or cools and solidifies.

A material may deform elastically or plastically. In elastic deformation, atomic bonds stretch but recover, whereas in plastic deformation, which is permanent, bonds break and then new bonds form (Shackelford 1990). At low viscosity, melts behave plastically, and can even behave like Newtonian fluids, deforming (i.e. flowing) at a constant rate (strain rate) in response to an applied stress. More viscous melts deform plastically and behave in a pseudo-plastic manner with changing strain rate as stress is increased, leading to changing viscosity. High-viscosity melts behave elastically under stress, and therefore are prone to fracture in a brittle manner. Some melts that are more viscous than pseudo-plastic melts, but less so than high-viscosity melts, behave in an intermediate fashion, involving both elastic and viscous behaviour, and may be called elasto-viscous or visco-elastic materials. Such materials resist deformation with increasing stress (elastic regime) up to an initial yield strength, at which point they begin to deform or flow (plastic regime). Materials which have a yield strength that must be exceeded before they flow, are called Bingham materials.

Magmas of different compositions all change viscosities as they cool and can behave in all of these ways as their viscosity changes. However, once a melt is supercooled to a glass, its behaviour is essentially elastic, which means it can absorb a certain level of stress without visible deformation, up to its yield point, at which it fails through brittle fracturing (Chap. 2). However, the breaking of magma is not only stress dependent, but also strain rate dependent, and even a viscous, plastic substance can break in a brittle manner if the strain rate (rate at which it is deforming or flowing) is high enough.

4.3 Autobrecciation

4.3.1 Processes

As a newly erupted lava flows away from its eruption point, its outer margin is cooled by the ambient medium (air, water, and/or ice), it becomes more viscous as it cools, initially forming a visco-plastic skin or crust, and then as it continues to cool, the crust goes through the transition from melt to glass in its glass transition temperature range, T_g (Chap. 2), forming a solid crust. The crust, however, has both temperature and viscosity gradients from the outside to the inside. Continued flow of the liquid lava under the crust, for example, in a lava tube (Chap. 5), imposes a shear stress on the crust, which may cause it to deform plastically if it is still visco-plastic, or to break into slabs or blocks through ductile stretching in the visco-elastic regime, or even break by brittle fracturing, if the product of the strain rate and viscosity exceeds the tensile strength of the plastic or solidified elastic crust (e.g. basaltic a'a lavas, Fig. 4.2a, b; also see Chap. 5). This process of brecciation of a lava due to its own flow is called *autobrecciation* and the aggregate of clasts is called *autobreccia* (e.g. Plate 4.1a).

In low-viscosity lavas, such as basaltic pahoehoe (Chap. 5), Archean komatiite, and carbonatite lavas (Chap. 7), the viscosity of the lava under the solidified crust is so low that the crust and internal, still molten lava are not strongly frictionally coupled. The lava under the crust in the lava tube therefore does not exert a significant shear stress on the crust, which may remain coherent and may be subject to ropey fold formation, and is not subjected to autobrecciation (e.g. smooth or ropey folded pahoehoe basalt lava; Chap. 5). As lava viscosity increases through cooling, and the coupling between interior lava and crust increase, such ropey textured crust may break visco-elastically, producing irregular slabby blocks (Fig. 4.2c). With more viscous magma, such as basaltic a'a (Fig. 4.2a, b) and lavas of more evolved magmas (Chaps. 5 and 6), there is a stronger thermo-frictional coupling between solid crust and liquid interior, leading to shear-induced autobrecciation of the crust in the visco-elastic regime to brittle regime (Fig. 4.2d–i). The contact between the autobreccia and the interior of the lava is commonly gradational (Fig. 4.2e).

The textural characteristics of the autobreccia clasts will reflect the rheological state of the crust of the lava at the time of autobrecciation. Where the crust is visco-elastic or plastic,

(a)
(b)
(c)
(d)
(e)
(f)
(g)
57736
cm
(h)
(i)

◀ **Fig. 4.2** Types and features of autobreccias and their clasts. **a** Basaltic a'a lava breccia forming in Hawai'i, showing the spiney, irregular form of typical a'a autobreccia crust in the making, and the red-hot coherent interior (*Source* United States Geological Survey; Photo taken 29-6-2016). **b** Coarse, fines poor clast-supported texture and the highly irregular spinose nature of autobreccia clasts in a profile through an a'a lava on Mt Etna, Sicily, Italy. Note the coherent core or the lava, representing the lava feeder tube, the gradational transition from coherent core to autobreccia and the irregular nature of this transitional boundary, including apophyses from the core into the autobreccia. **c** Slabby pahoehoe autobreccia developed in the channel of a 1972 pahoehoe lava flow in Hawai'i where the lava flow is flowing over a scarp. The coherent part of the lava is visible upper left. The flow is flowing over an a'a lava. **d** 1925 dacite lava autobreccia, Nea Kameni Island, Santorini, Greece, showing the coarse block nature of the autobreccia formed in silicic lavas, with autobreccia blocks mostly bound by curviplanar brittle fracture surfaces. The grey interior of the blocks is relatively dense dacite, the buff-coloured margins are highly vesicular pumiceous domains. The large block, top left is ~0.75 m in diameter. **e** Gradation from inner coherent lava to outer autobrecciated dacite dome lava, to clast-rotated autobreccia, summit dome, Mt Unzen, Japan (Photo by D. Uehara). **f** Subaerial clast-rotated, flow banded, rhyolite lava autobreccia, Late Devonian Boyd Volcanic Complex, Eden, southeastern Australia. **g** Flow-banded and autobrecciated submarine Miocene dacite lava, showing plastic clast accommodation; drill hole HO82, Hanaoka mine sequence, Hokuroko Basin, Green Tuff Belt, Japan. **h** Coarse block rhyolite lava autobreccia in a flow banded, submarine lava, Miocene Shinzan Rhyolite, Kamoasa Bay, Oga Peninsula, Green Tuff Belt, Japan. The finer breccia domains between the large autobreccia blocks are probably quench fragmented hyaloclastite breccia. **i** Crumble breccia blocks, resulting from post-eruptive gravitational collapse of original lava spines, summit dome complex, Mt Unzen (Photo by D. Uehara)

the breakage of the crust is by viscous shearing and fracturing, producing plastically deformed clasts with irregular margins (Fig. 4.2a–c), whereas when the crust is solid, fracturing of the crust will be by brittle fracturing, producing blocks with smooth, curviplanar fracture surfaces (Fig. 4.2 d–i). In some thin lavas, almost the whole lava may be autobrecciated (e.g. Fig. 4.2b), whereas in thick lavas, usually only the outer carapace of the lava consists of breccia, which grades internally into coherent or non-brecciated lava (Fig. 4.2e). Autobreccia may also occur at the base of the lava, through basal frictional coupling with the substrate, or roll-over of surface autobreccia as lava flow front rolls over as it advances (cf. tracks of a bulldozer; Chaps. 5 and 6).

4.3.2 Characteristics of Autobreccia Clasts and Deposits

Autobreccia consists of aggregates of angular-to-irregular-shaped volcanic clasts of the same composition and texture as the source lava, that is, they are monomictic breccias (Fig. 4.2). They are relatively coarse, fines absent, boulder to pebble-sized aggregates, with little or no sand size and finer matrix (Fig. 4.2). Autobreccia zones may vary from tens of cm to many m in thickness. No quantitative grain size studies, as undertaken for pyroclastic and sedimentary deposits, have been carried out on autobreccias. In both subaerial and subaqueous environments, the autobreccias of low-viscosity lavas consist of a significant population of irregular, plastically distorted clasts, including irregular, spiney, and hackly clasts, as in a'a lava autobreccia (Fig. 4.2a, b; Chap. 5) or ropey textured slabs in slabby pahoehoe autobreccia (Fig. 4.2c). For more viscous lavas, the autobreccia clasts are blocky in form, bounded by curviplanar, brittle fracture faces (Fig. 4.2d–i). In some lavas where the surface of the lava crust is solidified but becomes plastic inwards, autobreccia blocks may show curviplanar brittle fractures at the top but irregular, plastically stretched surfaces at the base.

Textures in the autobreccia zone of a lava flow will grade from the coherent core of a lava, with little or no fracturing, outwards towards the top, sides and bottom of the flow into a zone of jigsaw-fit relationships between adjacent clasts (Fig. 4.2a, b, e), indicating that there has been no relative movement between them, and then into an outer zone of clast-rotated textures (Fig. 4.2c–i), resulting from rotation and movement of clasts relative to each other and the core, as a lava continues to flow. Clasts will generally show no signs of rounding or abrasion and will therefore generally be angular to irregular in morphology (Fig. 4.2). The blocks may be free to rotate or tumble (Fig. 4.2d–f, i), or they may be sintered together, plastically moulding into each other's shape (Fig. 4.2g), or be enclosed in non-fragmented magma. Autobrecciation should operate subaqueously (e.g. Fig. 4.2g, h) as well as subaerially, but it may be accompanied by contemporaneous quench fragmentation.

Thus, in subaqueous successions, the recognition of autobreccia is complicated by the likely co-existence of abundant quench fragmented debris (hyaloclastite; Fig. 4.2h; see next section). De Rosen Spence et al. (1980) suggested that in the Archean submarine rhyolite flows at Rouyn-Noranda, Canada, an outer coarse blocky breccia represents the autobreccia, whereas the finer interior micro-breccia represents quench fragmented hyaloclastite. Initial autobrecciation around the margins of the flow would have opened up fractures into which water could have permeated, and then triggered quench fragmentation internally, as discussed below. In other cases, big blocks, perhaps representing initial autobrecciation of the lava crust, are separated by finer breccia domains (Fig. 4.2h), which appear to represent subsequent quench fragmentation forming hyaloclastite. Even in narrow diamond drill core, autobreccia can be recognised (Fig. 4.2g).

Although the surfaces of many viscous felsic lavas and domes may consist of very large blocks, they are not always formed as autobreccia. The surfaces of viscous lava flows and domes may consist of irregular, steep, and high spines (Chap. 6). For years after the cessation of flow, those spines may collapse gravitationally forming what Macdonald (1972) called "crumble breccia". For example, this process has occurred continuously since the cessation of the Mt Unzen lava dome eruption in 1995, such that now, relatively few parts of the original spinous lava domes' surface are preserved. The summit is now marked largely by crumble breccia (Fig. 4.2i). On the flanks of steep sides of lava domes, post-eruptive gravitational collapse through ongoing rock falls can build up an apron of talus. Distinguishing clast-rotated autobreccia from such crumble breccias and talus breccias may be difficult because the clast morphology may be the same in all three, and they are all monomictic breccias.

Finally, if a lava has undergone volatile exsolution and vesiculation, vesicles can often rise or be concentrated in the upper parts of crust of such lavas (Chaps. 5 and 6). The level of vesiculation may produce a coherent pumice carapace on the lava, and if that is subject to autobrecciation, a pumice breccia can be produced autoclastically, not explosively. This can occur in both subaerial and subaqueous settings (Chaps. 3 and 6), and in the latter is complicated because the pumice carapace may also be subject to quench fragmentation. Pumice breccias are therefore not just produced by explosive eruptions.

4.4 Quench Fragmentation

When hot magma comes in contact with ambient water or ice, it may be subjected to supercooling, instantaneously transforming the magma through the glass transition into a glass. During this transformation, the glass is affected by cooling-induced contractional or tensile stress, which may cause contractional thermal shock fractures to propagate inwards from the cooling surface (Plate 4.1b; Chap. 3; van Otterloo et al. 2015, and references therein). The magma fragments to an in situ aggregate of glassy debris. This fragmentation process is called *quench fragmentation*, or *cooling contraction granulation/fragmentation*, or *thermal shock fracturing*. The deposit of glassy debris is called "*hyaloclastite*" and the individual clasts, *hyaloclasts* (Rittmann 1962, Pichler 1965, Kokelaar 1986, Cas and Wright 1987; Cas 1992; McPhie et al. 1993; Scutter 1999; Batiza and White 2000; Cas and Giordano 2014; van Otterloo et al. 2015; Cas and Simmons 2018). Quench fragmentation *per se* is not explosive, although in shallow water, some explosive activity may also occur simultaneously, as ambient water is superheated and flashed to steam explosively under the low hydrostatic pressure conditions in shallow water.

Quench fragmentation can occur where subaerial lavas enter water (Fig. 4.3; Fuller 1931; Waters 1960; Jones 1969; Moore et al. 1973; Umino et al. 2006; Watton et al. 2013), where magma is erupted under water (Plate 4.1b; Pichler 1965; de Rosen-Spence et al. 1980; Yamagishi 1987, 1991; Cas et al. 2001; Batiza and White 2000; Goto and McPhie

Fig. 4.3 Quench fragmentation of lavas entering the sea. **a** Entry of the Banana basalt lava into the sea on 25 July 2004, Hawai'i, showing steam rising where quench fracturing of the lava surface is occurring. Note the in situ, blocky texture of the flow margin where it has been quench fragmented in the lower right hand outcrops. **b** Steaming, hot blocks of glassy basalt, spalled off the advancing front of the Banana lava flow as it enters the sea (as in **a**). (From van Otterloo et al. 2015; *Source* United States Geological Survey)

2004; Goto and Tsuchiya 2004; Soriano et al. 2013, 2016), or sub-glacially (Jones 1969; Furnes et al. 1980; Edwards et al. 2002, 2015; Smellie 2000, 2008; Tuffen et al. 2001; Tuffen 2007), forming subaqueous and subglacial lavas and hyaloclastite, where lavas flow over water-saturated sediments (Waters 1960; Schmincke 1967a), and where magma intrudes water-saturated sediments or country rock (Pichler 1965; Kokelaar 1982; Brooks et al. 1982; Brooks 1995; Hanson and Schweikert 1982; Hanson and Wilson 1993; Busby-Spera and White 1987; Cas and Wright 1987; Cas et al. 1990). However, quench fragmentation does not necessarily affect all lavas or syn-depositional intrusions or all parts of the margins of these, if effective film boiling and steam envelope insulation (Leidenfrost effect; see below) of those margins have occurred (Mills 1984). The margins of subaqueous lavas could be variably coherent or quench fragmented.

In this section, the processes by which magma is supercooled to glass by contact with external water, the processes by which it fractures, and the characteristics of the resulting hyaloclasts and hyaloclastite deposits are reviewed.

4.4.1 Film Boiling at the Interface Between Magma and External Water

4.4.1.1 Film Boiling at Subcritical Point Pressure and Water Depths

When hot magma, at temperatures ranging from 700 to 1,650 °C, comes into contact with cold water or water-saturated sediment, the magma at the interface rapidly cools and loses heat, and conversely, the water at the interface is instantaneously superheated to a vapour (Fig. 4.3). When water is heated by an immiscible heat source, such as magma, several heat transfer phenomena can occur (Whitaker 1983). At low heat flux (measured in Watts m^{-2}; W m^{-2}), below that required to initiate boiling, heat is transferred from the heat source to the water by simple conduction and is then convected away from the interface. Where heat flux is high enough to cause steam bubbles to form, the steam bubbles nucleate as individual bubbles at the surface of magma at the interface with the external water mass and are convected away or float away from it. This is called the nucleate or heterogeneous boiling stage. At higher heat flux, very dense nucleate boiling occurs, but there are still domains of liquid water in contact with the heat source. This stage is called the partial nucleate boiling and unstable film boiling stage. Finally, at very high heat flux, known as the critical superheat, ΔT_C, a continuous vapour layer, separates the heat source from the water (Fig. 4.4a), and this stage is called the stable film boiling or homogeneous boiling stage. Heat transfer across the film is by radiation, and from the outer margin of the film to the water by conduction.

Maintenance of a stable vapour film requires a minimum heat flux, q''_L from the magma, which is dependent on the ambient pressure, P_a, of the coolant water (Kovalev 1966; van Otterloo et al. 2015):

$$q''_L = 2.77\sqrt{P_a} \tag{4.1}$$

The minimum temperature for maintaining q''_L is called the Leidenfrost temperature, T_L. If the temperature of the magma exceeds T_L, and the degree of sub-cooling (the difference between the boiling temperature and actual temperature of the ambient water), T_{SC}, is low to moderate, the vapour film will be unstable and susceptible to collapse (Freud et al. 2009). Although T_L depends on the geometry of the magma body, its size, and the magma type, it can be approximated (Dhir and Purohit 1978; Freud et al. 2009) using

$$T_L = 201 + 8T_{SC} \tag{4.2}$$

Stable film boiling occurs when ΔT_C (= temperature of the heat source − saturation temperature of water at a given pressure) $\geq$ 40 °C. In the case where magma comes in contact with water, the homogeneous boiling stage will be established instantaneously because ΔT_C will always be greater than 40 °C no matter what the pressure. The water at the interface therefore flashes to steam and a vapour film will form.

Film boiling involves a rapid phase transformation of water to steam at pressures and equivalent water depths that are less than the critical pressure of water, which are 22 MPa and 2,200 m in fresh water, and 30 MPa and 3,000 m in seawater, respectively (Chap. 3; Mills 1984; Wohletz 1986; Zimanowski and Büttner 2003; Head and Wilson 2003; Cas and Giordano 2014; Cas and Simmons 2018). The vapour film of steam may form a stable boundary layer only millimetres to centimetres thick (Fig. 4.4a), lasting for seconds or longer, or it may collapse, leading to direct surface contact between the magma surface and liquid water (Fig. 4.4b), or in shallow water it may instantaneously expand explosively, as discussed further in Sect. 4.6 on phreatomagmatic explosive activity.

A stable boundary layer of steam between the magma surface and the body of water (Fig. 4.4a) insulates the magma surface and allows it to cool or chill at a slow enough rate initially to form a coherent visco-plastic crust that may deform plastically (e.g. ropey textures), and finally cool to a glassy crust. The insulating effect of the vapour film has been called the *Leidenfrost effect* (Mills 1984). Cooling of the magma occurs by radiative heat loss and convection across the vapour film, which is slow. The crust then also

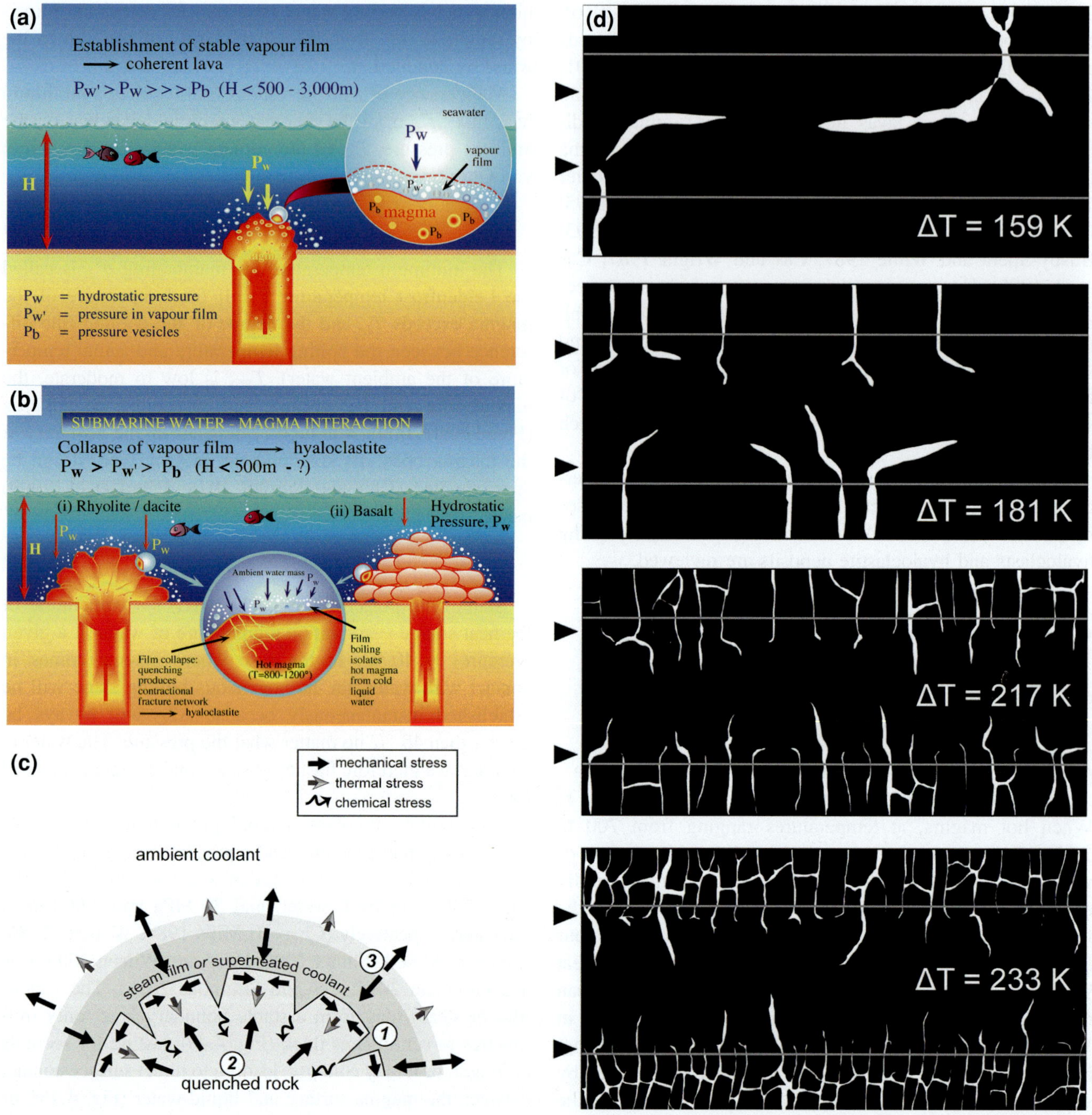

Fig. 4.4 Processes and stages associated film boiling and quench fragmentation. **a** Schematic representation of stable or homogenous film boiling at the interface between the surface of a subaqueously erupting lava and the ambient water body, showing the formation of a stable vapour film at the interface that insulates the lava from supercooling and thermal stresses, leading to formation of a coherent lava crust. **b** Schematic representation of vapour film collapse, leading to direct contact between magma and liquid water, causing rapid chilling or supercooling of the magma surface to glass, thermal stressing, and cooling contraction fractures to propagate in the chilled glass, producing jigsaw-fit, in situ hyaloclastite breccia. **c** Representation of the stresses at the interface between hot magma or glass and ambient water as quench chilling and fracturing occurs. These include thermal stress perpendicular to the cooling surface, tensile stress parallel to the cooling surface and perpendicular to forming cracks, and chemical stress along propagating cracks due to the presence of water at the crack tip. (From van Otterloo et al. 2015). **d** Thermal shock patterns in sintered glass as a function of temperature differential between hot glass and the external ambient fluid. Note the wide fracture spacing and curviplanar form of fractures at low ΔT, whereas at high ΔT, fractures are closely spaced and mostly planar. Arrowheads represent the supposed depth of unstable cracking. (From van Otterloo et al. 2015, After Bahr et al. 1986)

insulates the magma under the crust from heat loss and direct contact with water. Stable film boiling and slow cooling of the magma produces coherent subaqueous lava flows such as coherent basaltic pillow and massive sheet lavas (Mills 1984), and even coherent felsic lava flows and domes (Binns 2003; Carey et al. 2018; Ikegami et al. 2018) on the seafloor, without explosive interaction.

However, the steam vapour film, being a dynamic fluid, may experience natural instabilities such as Taylor, Landau, and Kelvin–Helmholtz instabilities (e.g. Wohletz 1986), leading to collapse of the vapour film and causing direct contact between liquid water and magma (Fig. 4.4b). Vapour film instabilities develop at the outer margin of the vapour film (Wohletz 1986). The outer margin undergoes a natural oscillation towards and away from the magma surface on a timescale of microseconds. *Taylor instability* results from the large accelerations of the film–water interface during the oscillation described above, which may cause the magma surface to distort into nodes if the magma viscosity is low enough. Large droplets of magma may thus detach into the film and could mix into the water. *Landau instability* arises from evaporation at the film–water interface, which causes vorticity and irregularities in the interface. This may lead to large droplets of water penetrating and even detaching into the film. *Kelvin–Helmholtz instability* involves acceleration of the film–water interface parallel to the magma surface, caused, for example, by turbulent eddies or currents in the water mass, or high magma flow velocities. As a result, the film-water interface undulates or develops waves, film collapse may occur in the troughs (Fig. 4.4b), and this may trigger collapse of the vapour film over large surface areas. These instabilities are further discussed in the section on phreatomagmatic explosive fragmentation (Sect. 4.6.9).

Steam vapour film collapse that leads to direct contact between the magma surface and liquid water dramatically increases the cooling rate, called supercooling, of the magma at the interface, because heat transfer occurs directly through efficient conduction and convection from magma to liquid water. As discussed in Chap. 3, water has a higher heat capacity and thermal conductivity than a steam vapour film (see Table 3.3), and so heat transfer (and loss) is much more efficient and faster than by radiative heat loss through, and convective heat loss in, a steam vapour film. Instantaneous heat loss and cooling of the lava surface causes rapid cooling, transition to glass, which is subject to thermal shock, represented by contractional stress, high strain rates leading to brittle fracturing called quench fragmentation. The process is discussed further in Sect. 4.4.3.

As hydrostatic pressure increases, the steam vapour film becomes more compressed and denser, especially at depths and pressure approaching the critical point and should be less prone to collapse, particularly where lava surfaces are smooth. As mentioned in Chap. 3, an interesting twist in submarine environments is that when saline seawater boils, the resulting steam is largely fresh water. That steam then becomes subject to the physical constraints of freshwater steam, including the critical point, rather than of seawater.

4.4.1.2 Film Boiling at the Critical Point Pressure and Water Depths and Beyond

At and beyond the critical point (22 MPa) water depth of water (2,200 m), the resultant film fluid is critical to supercritical fluid, not steam, irrespective of whether the ambient body of water is fresh water or seawater. Critical and supercritical fluids have the properties of both gaseous and liquid states.

Critical and supercritical H_2O fluids have higher thermal conductivity, heat capacity, and thermal transfer rates than steam, but less so than water, and are less compressible than steam (see Table 3.3, Chap. 3). As a result, they are less prone to collapse, the thermal gradient from the lava surface to the water mass across the critical–supercritical fluid layer is gradational, and a higher gradual rate of heat loss from the lava surface is likely to occur, limiting the degree of thermal stress that may develop (Chap. 3). This is particularly the case where lava flow regime is laminar and the lava flow surface is relatively smooth, facilitating a stable film that is less likely to collapse. This suggests that quench fragmentation is likely to occur at water depths beyond the freshwater critical point depth (>2,200 m) than at subcritical water depths (<2,200 m). This seems to be borne out anecdotally by observations that the deep seafloor is not covered by widespread and large volumes hyaloclastite (D. Clague, W. Chadwick, pers. comm; e.g. Clague et al. 2002, 2017; Chadwick et al. 2016, 2018), in spite of the main eruption products being high-temperature basalt lavas with an extremely high-temperature differential between the erupting basalt lavas and seawater, which theoretically represent conditions that should be conducive to quench fragmentation (e.g. Bahr et al. 1986).

That is, it appears that widespread quench fragmentation is limited to water depths where compressible steam vapour film exists at the interface between an erupting lava and the liquid water mass (<2,200 m), so facilitating film collapse, thermal shocking of the lava surface, and cooling contraction quench fracturing (Chap. 3). In deeper water, at and beyond the critical point pressure point, film collapse is less likely because of the high-pressure state of the supercritical fluid film, and thermal shock will be less because the magma and liquid water have thermally equilibrated more because of the more gradational temperature gradient across the supercritical fluid film. Nonetheless, some quench fragmentation could occur if hot magma is suddenly ejected into cold seawater, without time for a vapour film to form. However, much more research needs to be done to verify this suggestion.

4.4.2 Formation of Glass Through Supercooling, and Its Properties

Glass is a non-crystalline material that forms by effectively freezing or solidifying a supercooled silicate liquid through the glass transition temperature range into a vitreous state (Zarzycki 1991; van Otterloo et al. 2015). It contains dispersed molecular and ionic networks but lacks structural order. Rapid heat loss, chilling, or supercooling of silicate magma through contact with external water leads to instantaneous phase transformation from hot ($\sim$1,200 °C for basalt; $\sim$900 °C for rhyolite) liquid silicate magma to solid silicate glass at the interface, at high glass transition temperatures (T_g).

Glass is a non-crystalline type of ceramic, which undergoes transition to glass from a molten state, both with the same chemical composition. Ceramics are made by devitrifying glass, involving crystallisation of glass into a microcrystalline state (Shackelford 1990). Much experimental and theoretical research has been undertaken on the conditions required to stop glass and ceramics from fracturing in the field of materials science, and volcanologists can use this research to understand the behaviour of volcanic glasses that are subject to thermal stress during interaction between magma and ambient water (van Otterloo et al. 2015). The synthetic glass most similar to volcanic glass is a soda-lime-silica glass, consisting of 15% Na_2O, 10% CaO, and 70% SiO_2. Although glass has no crystalline structure, it does contain nanometre-scale SiO_4^{4-} tetrahedra, with the rest consisting of disordered, non-bonded cations and anions (Shelby 1997). The tendency for a liquid to cool to a non-crystalline glass rather than a crystalline substance depends on its composition, the atomic bonding energy, the liquidus temperature, the cooling rate, and the crystallisation kinetics of minerals that might form from the liquid (Bouška et al. 1993). High viscosity and rapid increase in viscosity during cooling, as well as a low liquidus temperature, favour formation of glass because ions cannot migrate quickly enough in a viscous melt at the time scale of cooling to form ordered aggregates of ions that form crystals (Kingery et al. 1976; Shelby 1997). Glass has a higher enthalpy or chemical energy than a crystallised material, making it metastable, because it has many unbalanced electrostatic charges.

Rapid or supercooling causes the melt to become a supercooled liquid just below the liquidus temperature, in which state it is metastable and viscous (Bouška et al. 1993; Shackelford 1990; Shelby 1997; Scutter 1999). Further rapid cooling will cause the melt to enter the glass transition temperature range, leading to a significant rise in the viscosity. In the glass state, the viscosity is $>10^{12}$ Pas (Bottinga et al. 1995), ions become immobile, and the structure of the liquid is frozen into the glass (Zarzycki 1991; Bouška et al. 1993; Shelby 1997). For any substance, a specific glass transition temperature, T_g, is an approximation only, because T_g varies according to the cooling rate for a specific magma. Faster cooling rates produce higher T_g than slower cooling rates. T_g therefore actually represents a temperature range over which a magma can transition to glass. T_g varies from one magma to another, depending on composition. T_g is commonly estimated to be two-thirds of the liquidus temperature, T_{LQ}. Magmas with low liquidus temperatures are more likely to form glass than high-temperature melts, although very high cooling rates can form glass in high-temperature melts (Kingery et al. 1976), such as basaltic glass.

The minimum cooling rate required to form glass and prevent crystallisation is called the critical cooling rate, which varies according to melt composition. For most silicate melts, the critical cooling rate is between 10^{-2} and 10^{-3} K s^{-1} (Giordano et al. 2005).

The heat capacity of a melt (the amount of heat or energy input required to raise the temperature of the melt by 1 °C) and specific heat (energy required to raise the temperature of 1 g by 1 °C) are temperature dependent, increasing as the temperature increases, but dependent on the material and its volume. This means that higher temperature melts store more heat or energy, and conversely, they release this heat when cooled. Heating a glass increases the vibrational energy of atoms and causes their spacing to increase, leading to a volume increase, which is assessed using its thermal expansion coefficient (Shelby 1997). Conversely, when it cools, it contracts proportionally. Mastin and Ghiorso (2000) have written freeware to calculate the specific heat of different magma compositions.

The thermal expansion coefficient reflects the rate of change in volume in response to changing temperature (Shelby 1997). Volume changes are significant at and approaching the glass transition temperature. Cooling leads to volume decrease, which induces tensile stresses, strain and contractional-tensile fracturing. Because volume and volume changes are difficult to measure, a linear thermal expansion coefficient is used:

$$\alpha = \frac{\Delta l}{l_0 \Delta T} \quad (4.3)$$

where Δl is the change in length in the sample, ΔT is the specified temperature interval, and l_0 is the original length of the sample in a given direction (van Otterloo et al. 2015).

The ability of a material to transfer heat is defined as the thermal conductivity, k, and occurs when a temperature gradient exists:

$$k = \frac{(\delta Q/\delta t)}{[A(\delta T/\delta x)]} \quad (4.4)$$

where $\delta Q/\delta t$ is the rate of heat transfer, A is the area of the body, and $\delta T/\delta x$ is the temperature gradient (cf. Chap. 3, Eqn. 3.1). The units are Watts (J s^{-1}). The thermal conductivity of glass reflects the rate at which it can gain or lose heat. Thermal conductivity varies according to temperature, with thermal conductivity decreasing with increasing temperature, and vice versa. Höskuldsson and Sparks (1997) report k-values for basalt of 2.9 W m^{-1} K^{-1}, and for rhyolite/obsidian, 6.9 W m^{-1} K^{-1}.

4.4.3 Quench Fragmentation or Thermal Shock Fracturing of Glass by Ambient Water or Ice

Quench fragmentation of magma and volcanic glass through supercooling is caused by thermal shock, as the magma is more or less instantaneously supercooled from a silicate liquid at magmatic temperatures to a solid glass during contact with water or ice at ambient temperatures. Although volcanologists are interested in the way supercooled volcanic glass fractures, very few volcanological laboratory experiments have been undertaken to understand the physics of this and the nature of fracture propagation (e.g. Büttner et al. 1999; Dürig et al. 2012; van Otterloo et al. 2015). However, materials scientists have long studied the physical properties of, and the manner in which industrial glass and ceramics fracture, and more specifically, the converse, what is required to prevent fracturing and failure of industrial glass and ceramics. van Otterloo et al. (2015) have thoroughly reviewed the physics of the fracturing of glass and ceramics from the materials science literature, from which the following is a summary.

Extreme cooling rates cause rapid volume contraction at the magma surface and produces very high thermal stress in the magma and resulting glass. The thermal stress is effectively contractional-tensile stress (Fig. 4.4c). If the tensile stress exceeds the cohesive strength of the glass, and the rate of elastic recovery of the glass, the glass fractures in a brittle manner (Zarzycki 1991). The resultant very high contractional or tensile strain rates in the magma/glass lead to the formation of many closely spaced brittle contractional cooling cracks to accommodate the volume contraction. The spacing of the contractional cracks will be controlled by the cooling rate, magma temperature and rheology, and the temperature difference between magma/glass and ambient water (Fig. 4.4d). Because glass is an elastic material, it fractures in a brittle manner when it fails. The deformation and nature of fracture formation in glass is controlled by the stress imposed on the glass and the strain within the glass. Although glass may experience degrees of plastic deformation or creep at high temperatures, at low temperatures and conditions of tensile stress, the behaviour is almost perfectly elastic (Broek 1989). When it fails by fracture or rupture the elastic strain cannot be recovered. The resistance of a substance to elastic strain is called the modulus of elasticity or Young's Modulus, E, which is the ratio of the stress (σ) to strain (ε), with the unit being Pa:

$$E = \frac{\sigma}{\varepsilon} \quad (4.5)$$

Glass is relatively weak under tensile strength, with experimentally measured values of E/100 to E/1000 (Kingery et al. 1976), which Griffith (1921) attributed to the presence of flaws in the glass (see below).

Thermal stress occurs in a material when a temperature gradient forms in it, whereas thermal shock occurs when a sudden change in temperature (ΔT) is imposed on a material (Kingery 1955). The emplacement of magma into water causes super-cooling, or quenching, and the onset of thermal shock, potentially leading to contractional, thermal shock fracturing. Kingery et al. (1976) found that during quenching in water a minimum temperature drop or difference (ΔT) of 100 °C was required to induce thermal shock fracturing. Even if water is heated to 100 °C through contact with magma, there is still a huge temperature difference between the magma and the heated water sufficient to cause quench fragmentation. The resistance of a substance to thermal shock is called the thermal shock or stress resistance, which is influenced by the size and shape of the body, the modulus of elasticity, the strength, which is influenced by the internal constitution of the substance, such as whether or not some crystallisation has occurred, Poisson's ratio, and the thermal conductivity (Kingery 1955).

Glasses have inherent flaws, associated with heterogeneous ion distributions and bond strengths in disordered glass, called Griffiths flaws or cracks, which are often elliptical in form (Griffith 1921). These intensify stresses when they form and readily lead to tensile failure of glass, that is, brittle failure and crack formation (Shackelford 1990). Extension of inter-atomic bonds at the tips of forming cracks leads to tensile strain and development of tensile stress (Knott 1973). Griffith (1921) proposed that cracks propagate by breaking of simple inter-atomic bonds, but Hillig (1983) suggested this would allow easy healing when stress is eased, proposing instead that for cracks to propagate and remain open, chains of atoms must be broken, leading to lattice trapped cracks. It is also known that cracks do not propagate as a single crack tip, but as a crack front, with multiple diffuse micro-cracks propagating simultaneously (Shelby 1997).

Krausz and Krausz (1988) proposed that at the time of fracture the stress level in a material is inversely proportional to the crack length, and that the stress required to propagate a large crack is significantly less than for a small crack. That is, once cracks form, it becomes easier for them to propagate as they lengthen.

Cracks also grow by coalescence of nanometre-scale cavities ahead of the crack, associated with stress intensification ahead of the advancing crack (Wilkinson and Vitek 1981; Evans and Blumenthal 1983). At failure, the stress intensification factor reaches a critical level at the crack tip, defined by

$$K_{lc} = \sigma_f \sqrt{\pi c} \tag{4.6}$$

where σ_f is the stress at failure and c is the length of a surface crack or half the length of an internal crack. The units of K_{lc} are MPa $m^{0.5}$. The stress intensification around a propagating crack increases with crack length, leading to acceleration of the crack-propagation rate (Broek 1986, 1989). K_{lc} for most glasses and ceramics is less than 5 MPa $m^{0.5}$ (Shackelford 1990). Crack propagation velocities of one-third the speed of sound ($\sim$85 m s^{-1}) maintain a steady stress and strain field around the leading crack tip, but when the propagation velocity is at and greater than half the speed of sound (>170 m s^{-1}), the maximum tensile stress migrates to a plane at 60° to the original plane and causes crack bifurcation (Broek 1986).

Cracks are initially oriented perpendicular to the axis of maximum tensile stress (Fig. 4.4c, d). They may change direction and become curviplanar if there is rotation of orientation of the maximum tensile stress, which may occur if a crack approaches another crack and if multiple cooling fronts propagate from different directions (Kepple and Wasylyk 1994; Lawn and Marshall 1994; Scutter 1999; van Otterloo et al. 2015). A crack propagating from another must initially propagate at right angles to the original crack. Since stress cannot be transmitted across an existing free surface or crack, cracks intersecting existing cracks will terminate (Kepple and Wasylyk 1994).

The spacing between cracks decreases with increasing temperature difference between the body and the water (ΔT; Bahr et al. 1993; Fig. 4.4d), leading to finer grainsize in the fractured glass. Crack length is also influenced by ΔT, which influences the maximum tensile strength. Bahr et al. (1986) showed that low ΔT cracks tend to be curviplanar, whereas higher ΔT leads to straighter fractures, and fewer curviplanar ones, with the change occurring at ΔT = 217 K (Fig. 4.4d).

In addition to physical-stress-crack-induced propagation, cracks can also propagate through the reaction of glass with external water *during the quench fragmentation process*, i.e. syn-eruptively, while the glass is still hot. Fresh glass is very reactive to water, which plays the role of breaking atomic bonds in the glass, and at temperatures of 150 °C or greater, the reaction is too rapid to measure (Shackelford 1990; Scutter 1999; van Otterloo et al. 2015). The reaction between glass and water causes a chemical bond weakening and breaking process due to chemical stress (Krausz and Krausz 1988). The energy required to break Si–O bonds is 20 times less in the presence of water than in a dry state (Michalske and Bunker 1987). H^+ ions substitute for alkali cations in the glass, liberating OH- ions. When pH > 9, the rate of reaction between water and silicate networks becomes very rapid. The process becomes self-sustaining because the propagation of a crack exposes new glass to water (Krausz and Krausz 1988). Even in the absence of thermal stress, cracks can propagate through the effects of chemical reaction between water and glass in a process called static fatigue or dissociative chemisorption (Michalske and Bunker 1987). Porreca et al. (2014) also demonstrated that quench fragmentation can keep occurring at relatively low temperatures that are hundreds of °C below T_g, for as long as the volumetric contraction rate is larger than the relaxation time of the glass. This type of fragmentation below T_g can explain the huge thickness of hyaloclastites associated with subaqueous felsic lavas (see Chap. 6), as a continuous process of water percolation through the lava thickness occurs along developing contraction cracks.

Where water is available in a crack, the crack accelerates, but where it is dry, crack-propagation rate decelerates (Michalske et al. 1978; Michalske 1994). Both are common during quench fragmentation of magma by water, because water gains access to crack tips irregularly. The acceleration–deceleration of cracks causes curves and kinks to form in the cracks, and the wetting–drying of cracks will lead to cyclical acceleration and deceleration of cracks during quench fragmentation.

The presence of concentric spheroidal or ovoid shell-like cracks in glass, called perlitic cracks, has previously been thought to result from the effects of water, during hydration of glass. The timing of formation of perlite cracks is unclear. It commonly occurs in dense glass, including coherent obsidian and even densely welded vitrophyric ignimbrites (McPhie et al. 1993). In the latter case, the formation of perlite cracks appears to have nothing to do with subaqueous quench fragmentation. However, von Aulock et al. (2013) demonstrated through calorimetric FTIR analysis of subaerial obsidian that perlite cracks form at high temperatures, not low diagenetic temperatures. They proposed that the perlite fractures formed at the time that spherulite growth occurred from temperatures above T_g (i.e. >600 °C) down to 400 °C. From about 450 °C, cracks radial to the spherulites and perlite cracks were formed, and were hydrated at these temperatures. They concluded that water was released from the obsidian during anhydrous crystallisation of spherulites and contributed to crack propagation. They also consider that perlite cracks can form during subaqueous quench fragmentation, that is, they result from thermal shock in an isotropic medium and are not related to late, cold-state hydration of volcanic glass.

When thermal shock first occurs, a large number or a high density of cracks will form at the surface of the glass and will be randomly distributed (Bahr et al. 1996). The originally formed cracks will arrest close to the surface, and only a few major cracks will propagate deep into the body (Bahr et al. 1993). However, these major cracks may bifurcate to form new cracks, and the access of water to these deep cracks will allow water to permeate into the interior, potentially triggering further thermal shock and cracking inside thick magma bodies, creating a second-order network of fractures at right angles to the primary fractures. These may then nucleate a third-order network of fractures, etc. In time crack length and crack spacing become equal (Bahr et al. 1993), so defining the grainsize of the fractured glassy aggregate. The network of brittle fractures may propagate inwards from the surface into the body of the magma as the magma-glass phase transition front propagates inward, especially if water penetrates into the network of fractures. This causes the cooling front to accelerate into the interior of the magma.

4.4.4 Effects of Crystals and Bubbles in Glass on Quench Fragmentation

Glass with suspended crystals, lithic fragments, and vesicles experiences complex fracture development during thermal shock, because it is influenced by the varying behaviour of the different components, including the glass, crystals, and vesicles. The bulk strength of a poly-component glass or magma will be stronger and has a higher resistance to thermal fracturing than an aphyric magma or glass (cf. crystalline ceramics). Biswas and Bhaduri (1983) found that the resistance to thermal stress of glass increased in proportion to the abundance of crystals, whereas Lange (1971) considered that the size of inclusions influenced the increase in resistance to thermal stress. Ohji et al. (1998) found that the more crystalline a ceramic or glass is, the more stress resistant it is, and only shorter cracks will be formed.

The interface between crystals and glass groundmass are regions of high energy and may act to nucleate fractures (Krausz and Krausz 1988). Crystals are stronger than glass, and so, many fractures should propagate along the interface between crystals and enclosing glass (Erdogan 1974). Crystals should also be expected to fracture along planes of weakness, such as crystal cleavage planes. Although this does happen, crystals also show irregular fractures not controlled by cleavage, suggesting shock fractures propagated so quickly that they could not conform into cleavage planes.

The behaviour of porous media or vesicular glass to thermal stress is less well understood. However, Biswas and Bhaduri (1983) proposed that increasing pore (vesicle) size at a given porosity (vesicularity) led to a decrease in the thermal stress resistance, or an increase in the potential for thermal fracture propagation during thermal shock. When a crack first propagates, it will be diverted to and accelerate towards a pore if it is large enough and not too far from the crack because of the high stress intensity around the pore (Green et al 1977; Broek 1986). It will then try to bifurcate to try to by-pass the pore. If the propagation rate of the crack is low or the pore is large, the crack is likely to arrest, especially if it intersects the pore. If the rate of propagation is high, the crack may propagate through the pore and incorporate it into the crack (Broek 1986). The strength of vesicular magma is significantly less than non-vesicular magma, particularly in the presence of water (Mungall et al. 1996; Romano et al. 1996; Vona et al. 2013, 2016), making it more prone to intense quench fragmentation. As a result, pumice clast deposits can form from the subaqueous autoclastic quench fragmentation of vesicular magma, without involving any explosive activity at all (Cas 1992; Allen et al. 2010; Cas and Giordano 2014; van Otterloo et al. 2015; Cas and Simmons 2018; Manga et al. 2018). Intense quench fragmentation of pumice will produce curviplanar ash-sized hyaloclastite fragments, many mimicking cuspate pyroclastic glass shards. Such ash size fragments can easily be entrained into buoyantly rising plumes of hot water rising above the vent or a propagating lava flow, from which they can be dispersed by oceanic currents (e.g. Barreyre et al. 2011).

4.4.5 Quench Fragmentation of Pyroclasts During Phreatomagmatic Explosive Activity and Submarine Fire Fountaining

As discussed further in Sect. 4.6.9, when magma and external water interact explosively during phreatomagmatic explosive activity, the magma is not only fragmented by explosive shock waves as the water is instantaneously and explosively superheated by the heat of the magma, but in addition, the cooling effects of the external water on the magma cause cooling contraction fracturing of the magma and glass fragments (Kokelaar 1986; Cas and Wright 1987; Zimanowski et al. 1997; Büttner et al. 1999; Dürig et al. 2012; Liu et al. 2015; van Otterloo et al. 2015; Colombier et al. 2019). During the thermal detonation stage of phreatomagmatic molten fuel–coolant interaction, as discussed in Sect. 4.6, the explosive expansion of superheated water transmits compressive shock waves through the magma-glass causing first-order major damage cracks. Perpendicular to these, shorter, second-order cracks, called "normal cracks", are formed as a result of cooling contraction when cold water permeated along major cracks and chilled the still hot glass. These second-order fractures continue to propagate, forming a fragmentation front, and accounting for a lot of very fine ash in phreatomagmatic pyroclastic deposits (Zimanowski et al. 2015 and references therein).

Colombier et al. (2019) also propose that many ash-rimmed or "enclosed" phreatomagmatic particles have resulted from in situ quench fragmentation of the particles, leading to a rind of in situ quench fragmented fine ash, rather than from ash accreting to an ash particle in the explosive eruption column. Both processes almost certainly occur, but proof of in situ quench fragmentation of the rims of phreatomagmatic particles exists on the form of in situ jigsaw-fit textures in at least the edges of such particles (Colombier et al. 2019).

As discussed in Sect. 4.6.8, in submarine fire fountains large proportions of spatter clasts ejected into the water mass will be subject to quench fragmentation, forming a bimodal deposit of fractured remnant spatter clasts and ash to lapilli size blocky glass fragments which have resulted from quench fragmentation and not explosive fragmentation (Cas et al. 2003).

4.4.6 Summary of Applications of Glass Fracture Theory to Formation of Hyaloclastite

There are many important points from the materials science literature on glass and ceramics, as reviewed above (and discussed in much greater detail and more quantitatively by van Otterloo et al. 2015), that can be applied to understanding quench fragmentation of volcanic glass and the characteristics of hyaloclastite deposits. This summary is taken directly from Soriano et al. (2016), based on an assessment in van Otterloo et al. (2015):

(a) Cooling contraction cracks propagate perpendicular to the cooling surfaces, which initially are the margins of the magma body at its interface with water or water-saturated sediments.

(b) The distance between cooling cracks depends on the cooling rate, the temperature differential and gradient between magma and water, and the level of thermal stress. This determines the first-order grain size of the resultant hyaloclastite breccia and explains why grain size is smaller at the margins of a magma body, compared with the interior, where cooling rates and temperature gradients are lower.

(c) Once a cooling crack forms, it also becomes a cooling surface, and other cracks will form perpendicular to it, then others perpendicular to the second-order fractures, etc.

(d) The spacing between propagating cracks is usually equal to the length of the crack.

(e) When a crack approaches another crack, it will try to change direction and approach at right angles.

(f) When a crack intersects another crack, it will usually stop because it cannot propagate across the space.

(g) Crack formation results in release of both thermal and mechanical energies, and cracks may stop if the tensile stress falls below the strength of the glass, which increases as glass cools.

(h) Crack shape, and thus fragment shape, depends on cooling rate, the crack-propagation velocities, and the orientation of the cooling surfaces. At high crack velocities, which occur when the temperature differential is high, cracks tend to be straight, and form under what is called the critical crack regime. Crack velocities can be >600 ms^{-1}. At crack velocities greater than 0.25 times the speed of sound in glass, cracks may bifurcate, and form wedge to splinter-shaped fragments. Low-temperature differentials lead to low crack velocities (subcritical regime; <cms s^{-1}), and cracks may become curviplanar.

(i) Water not only plays a physical role in facilitating crack propagation by causing cooling contraction tensile stresses at the crack tip, but also plays a chemical role in breaking chemical bonds in the glass in a propagating crack tip.

(j) Heterogeneous glass, with dispersed crystals, lithic fragments, and vesicles, experiences complex fracture formation because of the variable properties of the various components of the glass.

(k) Glass with abundant dispersed crystals is stronger than aphyric glass, is more resistant to quench fragmentation, and should it fragment, it is likely to form coarse hyaloclastite breccia.

(l) The presence of abundant vesicles in glass weakens it and makes it more prone to quench fragmentation, also producing curviplanar hyaloclastite fragments.

(m) Pumice hyaloclastite formed through non-explosive quench fragmentation of the pumiceous carapace of subaqueous lavas should be quite common in submarine settings. If the deposit is still in situ, it may still preserve jigsaw-fit textures. However, if it is resedimented, it should be bedded and preserve characteristics of a variety of subaqueous mass-flow processes (e.g. grain-flow, turbidity currents, debris flow, debris avalanches, rafted suspension transport).

In addition to conventional, in situ quench fragmentation of subaqueous lavas and syn-depositional intrusions, fine ash size fragments can spall off the chilled glass surface of propagating lavas due to cooling-induced thermal stressing. These can then become lofted in hot thermal plumes of seawater rising above the lava and then be dispersed as fallout ash size deposits on the surrounding sea floor (cf. Barreyre et al. 2011). Deposits of ash in deep water, such as around the Axial volcano on the East Pacific Rise at depths >1,500 m, are typically interpreted as pyroclastic, suggesting deep-water explosive activity (e.g. Chadwick et al. 2016). This is unlikely given the water depth and hydrostatic pressure (see below), but they are more likely fallout deposits of lofted fine ash size hyaloclastite particles.

As discussed in Sect. 4.6.8, submarine fire fountains can occur on the deep seafloor driven by the buoyancy of hot, low-viscosity magma (cf. gushing fountains of oil from breakouts at oil wells, or fountain mud volcanoes), rather than explosively expanding volatiles. Spatter clasts ejected into the water column will be subject to quench fragmentation, producing glassy ash size particles that can be dispersed as fallout deposits of hyaloclastic ash by thermal plumes rising above the fire fountain (Cas et al. 2003).

4.4.7 Hyaloclasts, Hyaloclastite Deposits, and Their Characteristics

The fragments produced by quench fragmentation are called *hyaloclasts*, and aggregates are called *hyaloclastite breccia,* although finer ash size debris can also be generated. Hyaloclastite consists of aggregates of monomictic glassy to devitrified fragments of varying sizes, frequently splintery to blocky and polygonal in shape, and frequently with sharp edges and corners, and curviplanar surfaces (Fig. 4.5a–k). This reflects the cooling contractional, brittle fracturing origin of the host, supercooled, glassy lava or intrusion due to contact with ambient water. In thin section they consist of angular fragments of glass, often having perlite cracks. In older rocks, original glass has usually been altered, devitrified, or metamorphosed, leaving only the clast shapes and aggregate textures as indicators of the quench fragmentation origins. Sideromelane basaltic glass initially becomes hydrated and alters to yellow-orange palagonite, but with diagenesis and metamorphism alters to zeolites, clays, and chlorite, whereas silicic glass alters initially to montmorillonite and then to sericite and even chlorite during metamorphism and hydrothermal alteration (Chap. 13; Gifkins et al. 2005). Archean komatiite hyaloclastite is typically altered to chlorite (Cas and Beresford 2001).

In outcrop, hyaloclastite deposits can vary from clast-supported (Fig. 4.5a–g, i–k) to matrix-supported breccias (Fig. 4.5h), but even clast-supported breccias can have significant proportions of finer matrix between the larger framework clasts. Hyaloclastite deposits show gradations from massive coherent rock, with flow banding (Fig. 4.5b, c, i) or without (Fig. 4.5a, d–g), in the interior of a lava or intrusion, outwards into in situ fractured, jigsaw-fit hyaloclastite breccia (see Figs. 3.6a and 4.5c, g), which may be clast-supported to matrix-supported, with varying proportions of finer granular glassy matrix grains, into an upper or outer domain of clast-rotated breccia and matrix (Fig. 4.5c–e, h–j; also see Fig. 6.16 for examples of this transition; Furnes et al. 1980; Yamagishi 1987; Hanson and Wilson 1993; McPhie et al. 1993; Scutter et al. 1998; Cas and Beresford 2001; Watton et al. 2013; Soriano et al. 2012, 2013). Large clasts may show radial, closely spaced mini-columnar jointing (Fig. 4.5h), indicating that such large clasts acted as individual cooling units once formed. Clast-rotated textures indicate varying degrees of rotation of clasts at the margins of a moving lava, or downslope clast movement and resedimentation on aprons at the margins of a high relief lava (De Rita et al. 2001; Soriano et al. 2012). The fine matrix in some hyaloclastite on close inspection often also shows in situ jigsaw-fit texture, reflecting intense in situ quench fragmentation, even down to silt-size grains (Scutter et al. 1998; cf. autobreccia, which lacks a fine matrix).

Full-scale resedimentation of hyaloclast aggregates should produce well-defined beds, which may be massive to normally or inverse graded, or reworked, and may make up the aprons of debris around high aspect ratio subaqueous lava flows and domes, or where lava has flowed into water, forming a prograding delta of bedded breccia and intercalated lava lobes (e.g. Fig. 1.14e, Jones and Nelson 1970; Porębski and Gradziński 1990; Watton et al. 2013; Fig. 6.16 e, f). Resedimented hyaloclastite consists of a bedded sequence of monomictic aggregates of angular splintery to blocky volcanic clasts with a clast-rotated texture. Clasts will not have jigsaw-fit relationships to each other because they have been transported away from the site where in situ quench fragmentation occurred. However, some clasts may have internal networks of intact contractional quench fractures, indicating they have not been fully broken into individual clast fragments during transportation. Low-relief lavas will not generate resedimentation of hyaloclasts. The overall architecture of subaqueous lava flows with significant hyaloclastite is further addressed in Chaps. 5 and 6.

Hyaloclasts and hyaloclastite can be variably vesicular, including deposits that are truly pumice hyaloclastite (Fig. 4.5i, j; Dimroth and Lichtblau 1979; Furnes et al. 1980; Cas and Wright 1987; Kurokawa 1991; Cas 1992; McPhie et al. 1993; Scutter et al. 1998; Binns 2003; Allen et al. 2010; Soriano et al. 2013; Carey et al. 2018; Manga et al. 2018). This simply means that subaqueously erupting lava was vesiculated at the time it was subject to quench fragmentation

(a)
cm
(b)
(c)
(d)
(e)
(f)
(g)
(h)
(i)

◀ **Fig. 4.5** Features of hyaloclastite deposits. **a** In situ, jigsaw-fit (lower right), to slightly clast-rotated hyaloclastite breccia, subaqueous rhyolite cryptodome, Late Devonian Bunga Beds, Boyd Volcanic Complex, southern New South Wales, Australia. Clasts are blocky and polygonal to splintery in shape. **b** Jigsaw-fit (left half), to clast-rotated, clast-supported, rhyolite hyaloclastite breccia, signified by the parallelism (left side) and dislocation (right side) of flow banding in adjacent clasts. Subaqueous rhyolite lava or shallow syn-depositional sill. Location as for (**a**). **c** Transition from the coherent, flow banded core of a Miocene submarine dacite lava downwards into in situ, clast-supported jigsaw-fit breccia into matrix-supported hyaloclastite breccia. All hyaloclasts are blocky clasts with curviplanar margins. Tsukhotan, Oshoro Peninsula, Hokkaido, Japan. **d** Jigsaw-fit to clast-rotated basaltic hyaloclastite breccia, showing splintery to blocky clasts. Some clasts preserve in situ jigsaw-fit texture (bottom centre). Mt Fraser, Archean Whim Creek Belt, Pilbara Craton, Western Australia. **e** In situ basaltic pillow margin and inter-pillow hylaoclastite breccia, Archean Teutonic Bore mine, Western Australia. Drill core is ~5 cm in diameter. **f** Basaltic hyaloclastite in drill core, showing blocky-to-splinter-shaped fragments with local jigsaw-fit domains, and wide clast size range. Archean Paringa Basalt, Kambalda, Western Australia. Drill core is ~5 cm in diameter. **g** Massive in situ quench fragmented dacite lava lobe with lobate apophyses, enclosed in jigsaw-fit to clast-rotated dacite hyaloclastite breccia, Miocene Green Tuff Belt, Shoiwa, southwest Hokkaido, Japan. **h** Radially jointed dacite hyaloclast block in poorly sorted, matrix-rich, Miocene dacite hyaloclastite breccia, Mt Olympia, Sapporo, southwest Hokkaido, Japan. **i** In situ jigsaw-fit to clast-rotated flow banded and 2 Ma rhyolite tube pumice hyaloclastite breccia, submarine lava flow, Cala Feola, Ponza Island, Tyrrhenian Sea, Italy. The vesicles are stretched tube vesicles oriented parallel to the flow banding. **j** Same body as in (**i**) showing clast-rotated tube pumice breccia, disorientated tube vesicle fabric visible in different clasts. **k** Coherent, flow banded (lower half), and quench fragmented hyaloclastite breccia base (top right), of the Proterozoic, densely welded, rheomorphic, rhyolitic Kathleen ignimbrite, in the West Musgraves Province, Central Australia

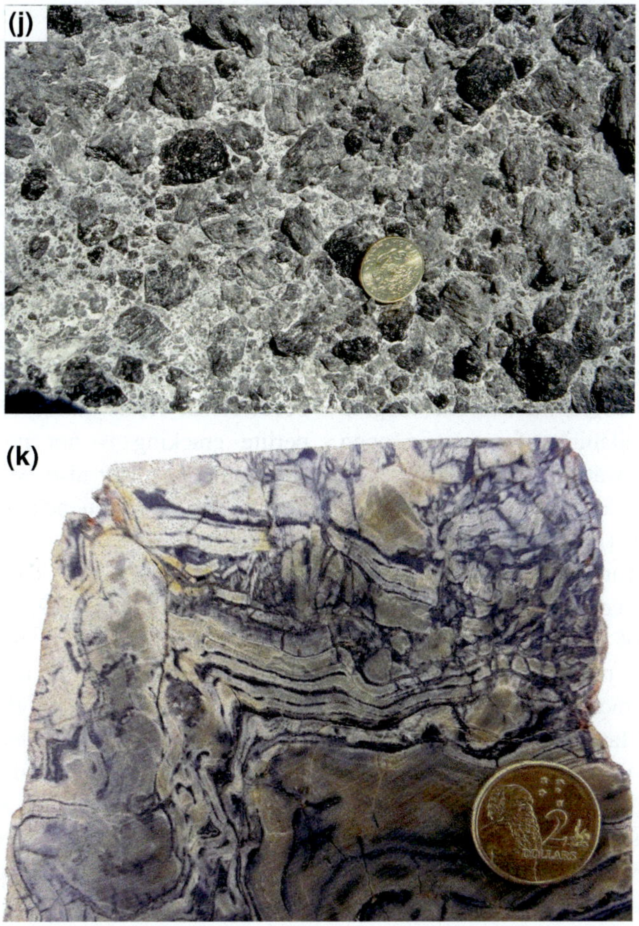

Fig. 4.5 (continued)

and that film boiling effectively insulated the vesiculating magma from quench fragmentation by ambient water until it had vesiculated. Given that subaerial lavas can have an outer carapace of very vesiculated coherent pumice resulting from slow, non-explosive vesiculation even to levels of 70% (Fig. 4.5i, j; Chap. 6), this can also occur in subaqueous settings, as discussed in Chap. 3. The condition for this to occur is that the gas overpressure in the vesicles should exceed the ambient pressure and the yield strength of the magma in the growing vesicle walls only enough to cause the vesicle walls to stretch plastically as the vesicles slowly grow and remain coherent, but not so high that the bubbles grow so rapidly that they burst explosively. Pumice deposits viewed, cored, or dredged from the sea floor or found in ancient submarine volcanic and sedimentary successions and therefore cannot just be assumed to be the product of explosive volcanism. They could also be in situ or resedimented pumice hyaloclastite (e.g. Scutter et al. 1998; Mann et al. 2004; Allen et al. 2010; Cas and Giordano 2014; Cas and Simmons 2018; Manga et al. 2018).

Pumice hyaloclasts can also vary from regular blocky to polygonal to irregular in shape (Fig. 4.5i, j). Existing fabrics, such as a vesicle stretching lineation or foliation, may affect quench fracture orientations and geometry and determine clast shapes, which could then be elongate, even platy (Manga et al. 2018). Pumice hyaloclastite also shows gradations from coherent non-vesicular or vesicular rock to in situ, jigsaw-fit pumice breccia, to variably clast-rotated pumice breccia (Fig. 4.5i, j; Scutter et al. 1998; Scutter 1999; Binns 2003).

The grain size in hyaloclastite deposits can range from boulder down to silt size (cf. fine ash size), the latter testifying to very intense fragmentation caused by cooling contraction stresses in glass. Coarse aggregates presumably represent "slower" cooling rates, and finer aggregates, instantaneous cooling rates. However, the degree of groundmass crystallisation prior to quench fragmentation is also important in building thermal resistance to fracturing, as discussed above. Other factors could include the magma composition, its rheology, the availability of water (e.g. open body of water or availability of water in a water-saturated sediment), and the temperature differential between the magma and the external water (Bahr et al. 1986). The availability of water is important in intrusive situations and in the interior of a magma, once quench fractures have formed, with the permeability of the network of fractures controlling the rate of delivery of water to the inwards propagating cooling front, with crack propagation controlled by both thermal shock and static fatigue or chemisorption as discussed above. Water that has penetrated quench cracks is quickly heated, expands, and exerts a hydraulic pressure that could further fracture the glass and/or dilate an aggregate of fragments, so displacing them from each other. In shallow water at low-ambient hydrostatic pressures such heated intra-crack water could thermally detonate producing phreatic explosions (Dürig et al. 2020).

Few studies have documented the detailed textural characteristics of hyaloclasts. However, Scutter (1999) undertook detailed textural studies of rhyolitic hyaloclastites of the uplifted Pliocene–lower Pleistocene submarine rhyolite lava flow succession now exposed on the island of Ponza, Italy, and from the Miocene Green Tuff Belt of Japan, in Hokkaido. Scutter (1999) found that overall these hyaloclastites were finer grained at the margins of lavas, closest to the interface with the water, than in the interior. She found that pumiceous hyaloclastite on Ponza (Fig. 4.5i, j) was finer than non-vesicular hyaloclastite, and was finest where the lava was most vesiculated, reflecting the susceptibility of vesiculated magma to fragment more finely when subjected to quench fragmentation. Cuspate shard-like textures were visible in thin sections. Hyaloclastite does not show distinctive grain size characteristics compared with other volcaniclastic aggregates, such as pyroclastic deposits, with significant overlap between these. Clast shapes include equant and polygonal, tetrahedral/triangular, spheroidal, blade-like, discoid, splintery, and irregular (Scutter 1999; Fig. 4.5). All clasts are angular except for the spheroidal ones, and where the matrix content of finely granulated hyaloclastite debris increased. Large clasts in such deposits commonly have more sub-round shapes, due to resistant core blocks being enclosed by a network of closely spaced, encroaching quench fractures that produced the in situ matrix and took "short-cuts" across angular sharp corners of large first-order hyaloclast blocks. Equant polygonal clasts are dominant where the lava is isotropic in its fabric. Elongate clasts occur more commonly, but not exclusively, where the lava is anisotropic in its fabric, such as flow banding defined by microlite banding, or where a vesicle stretching lineation or foliation occurs. Scutter (1999) found that spheroidal, perlite crack networks and jigsaw-fit clasts occurred in fresh non-hydrated obsidian, suggesting that perlite cracking is not necessarily just related to glass hydration, but may also be formed by thermal shock fracturing, that is, by quench fragmentation (von Aulock et al. 2013).

In Miocene Green Tuff Belt submarine volcanics in Hokkaido, Scutter (1999) also found that the grain size of hyaloclastites in aphyric to weakly crystallised submarine lavas was finer than in lavas with a variably crystallised groundmass, reflecting both the cooler nature of the crystallised lavas, but also reflecting the increased resistance to thermal stress of a crystallised material compared with a non-crystallised glass, as noted by Biswas and Bhaduri (1983). She also found that clast surfaces in crystallised hyaloclastites are more irregular than in glassy hyaloclastites because of this, whereas in the latter they are smooth and planar to curviplanar. Irregular clast surfaces were also noted by de Rosen Spence et al. (1980), Hanson (1991), and Hanson and Wilson (1993).

Pichler (1965) pointed out that for submarine viscous, silicic lavas the processes of quench fragmentation and autobrecciation (see above) may be closely interrelated and contemporaneous. Initial autobrecciation may open fractures in the lava crust that allow water to penetrate into the interior of a subaqueous lava, triggering the onset of quench fragmentation (e.g. Fig. 4.2h).

In addition, when pyroclastic flows enter the sea and produce densely welded ignimbrites, even some that are rheomorphic, and which flow like lavas (Chap. 10), their welding bases have the same rheological properties as lavas, and can also be subjected to quench fragmentation, producing typical blocky, hyaloclastite breccias (Fig. 4.5k).

Cas (1992), Cas et al. (2003), van Otterloo et al. (2015), Porreca et al. (2016), and Soriano et al. (2016) summarised the characteristics of hyaloclasts and hyaloclastite deposits and the conditions under which they form. The following is taken from Soriano et al. (2016):

(a) Equant, blocky clast aggregates are likely to have formed under steady, slow, subcritical crack-propagation conditions. Equant clasts with curviplanar margins are likely to have formed at low temperatures and low-temperature differentials.
(b) Elongate, splinter, and wedge-shaped clasts are likely to have formed under high crack-propagation velocity, in critical crack-growth regimes, at high-temperature differentials.
(c) Perlite cracking may be a product of thermal cooling, stressing, and quenching, under slow-cooling, subcritical conditions, at temperatures $< T_g$.
(d) Fine hyaloclastite breccia is the product of a fast cooling rate and high-temperature differential, which is why hyaloclastite grainsize may vary from fine at the margins of a body to coarser in the interior, and even grade into a coherent interior.
(e) Jigsaw-fit textures are indicative of in situ, non-explosive, cooling contraction quench fragmentation.
(f) Clast-rotated textures indicate post-crack formation movement as a result of continued lava or intrusion movement, or downslope resedimentation.
(g) The grainsize of quench fragmented crystalline glass hyaloclastite is likely to be coarser than for aphyric glass hyaloclastite.
(h) Pumice hyaloclastite should not be unusual in subaqueous settings, and not all pumice is pyroclastic in origin.
(i) Ash size glassy fragments can also form from spalling of fine glass fragments from the chilled glass surface of propagating lavas. These particles can then become lofted by hot thermal seawater plumes that rise buoyantly above the lava, dispersing and depositing fallout ash on the surrounding seafloor. This hyaloclastic fallout ash material is not pyroclastic in origin.
(j) Quench fragmentation of spatter clasts in submarine fire fountains produces fractured spatter clasts, and ash size blocky glass fragments that can also be dispersed by thermal plumes producing "fallout" deposits of hyaloclasts, not pyroclasts.

Therefore, identifying the specific processes or mechanisms of fragmentation in subaqueous settings can be very difficult because of the complexity and multiple processes that may occur. Full understanding of deposit characteristics at all scales, establishing facies relationships and context, is essential to minimise ambiguities.

4.4.8 Limu o Pele—Pyroclastic or Hyaloclastic?

For some time, small sand size to large platy, curviplanar glass clasts, called limu o Pele, that look like the walls of very large gas bubbles, but can also be contorted, have been found associated with subaerial to deep-water submarine basalt lavas (e.g. Clague et al. 2000; Maicher et al. 2000; Maicher and White 2001; Schipper and White 2010; Schipper et al. 2013). Limu o Pele were thought to be explosive in origin based on subaerial observations in Hawai'i that they form where lava flows over and entraps pockets of water that expand and burst with very weak explosive intensity. Discovery of Limu o Pele in the deep ocean led to a similar interpretation of these, in spite of the very high hydrostatic pressures in deep waters. However, Maicher and White (2001), Schipper and White (2010), and Schipper et al. (2013) have demonstrated that they result from the entrapment of pockets of water at the base of lavas flowing over the seafloor, which then expand, forming large vapour-filled balloons with magma walls. The magma walls are then quench fragmented, to produce sand size to mega-shard-like clasts of glass. Although limu o Pele in subaerial settings may form by very weak bursting of the gas-filled magma balloons, in deep subaqueous settings they are a type of hyaloclast, and aggregates are hyaloclastite, called sheet hyaloclastite by Maicher and White (2001), not pyroclastic deposits.

4.4.9 Peperite—A Special Type of Hyaloclastite

When magma intrudes water-saturated, unconsolidated sediment, or a lava flows over wet, unconsolidated sediment, quench fragmentation, or thermal shock fracturing of the margin of the magma may also occur, forming hyaloclastite. The same principles of heat transfer, or loss, between magma and water, in this case pore water in the sediment, and of glass fracturing apply as for normal hyaloclastite, as outlined in previous sections. However, the pore water in the sediment will be heated, expand and begin to convect, and can lead to dynamic fluidisation of the sediment and mixing of the sediment and hyaloclasts, forming a chaotic mixture of the two called *peperite* (Kokelaar 1982; Fisher and Schmincke 1984; Cas and Wright 1987). However, sediment–water mixtures contain less water, they have different physical or rheological properties, and may be less efficient in absorbing heat from the magma compared with pure water. The presence of some coherent, non-fragmented contacts between lava, intrusions, and sediment indicate that the sediment was water poor, and/or that film boiling of water in the pore spaces in the sediment may have insulated some parts of the magma margin producing a coherent contact.

Where pore water is abundant, the pore water in the sediment at and near the direct interface with the magma is heated, will expand, convect, and even boil, perhaps sometimes explosively. In addition, the pore pressure in the water will increase as it is heated and expands. The convecting water may fluidise the sediment, and if it loses strength, the sediment–water mix is liquefied and can mix with hyaloclasts forming a chaotic mixture of hyaloclasts with sediment and finely granulated hyaloclasts as a matrix (Fig. 4.6a; Kokelaar 1982). Fluidised sediment–water mixes may flow intra-stratally near the contact or may be more dynamically expelled away from it as a pressurised jet or slurry of water and sediment. In such cases, the fluidised sediment can inject into hyaloclastite fractures, forming well-defined sediment dyke-like intrusions (Fig. 4.6b). Any previous sedimentary structures in the sediment, such as bedding, will be destroyed by such dynamic mixing. Chaotic mixtures of hyaloclasts with a matrix of sediment that pass gradationally into jigsaw-fit hyaloclastite breccia are called *peperite* (e.g. Fig. 4.6a–i). Peperite is a special type of hyaloclastite resulting from dynamic interaction between a syn-depositional intrusion and contemporaneous, water-saturated, unconsolidated sediment, or the base of a lava flowing over a substrate of water-saturated, unconsolidated sediment. The history of the term "peperite" was summarised by Cas and Wright (1987) and White et al. (2000), and the processes and various deposit characteristics were reviewed by Cas and Wright (1987), Cas (1992), McPhie (1993), Skilling et al. (2002), and van Otterloo et al. (2014).

If the overpressured water is injected into the initial quench fractures, this could trigger an inward propagating quench fragmentation front within a syn-depositional intrusion or lava. Jigsaw-fit arrangement of hyaloclasts, separated by sediment stringers or dykes, are a common reflection of this (Fig. 4.6b–d). Such features also demonstrate that the hydraulic fluid pressure is able to displace fragments from each other without involving explosive processes, just hydraulic wedging.

Peperite occurring around the margins and within a syn-depositional intrusion, or at the base of a lava, is indicative of syn-depositional emplacement of the lava or intrusion, i.e. when the sediment was still water-saturated and unconsolidated. Peperite may occur within cm and up to a few m from the margin of the source intrusion or base of a lava.

As for normal hyaloclastite, in peperite there should be a gradation, from a coherent domain of the intrusion or lava, to an in situ, jigsaw-fit monomictic breccia of hyaloclastite (Fig. 4.6a). Beyond that, parts of the jigsaw-fit breccia will have stringers of injected sediment between the jigsaw-fit clasts (Fig. 4.6a–d), and these textural domains will grade simply or complexly into a chaotic clast-rotated monomictic hyaloclast–sediment mix (Fig. 4.6a, e, f). Without this context of gradational relationships from a coherent source to a chaotic volcanic clast-sediment breccia, it is difficult to unequivocally distinguish chaotically mixed peperite from a volcanic debris flow deposit; both consist of volcanic clasts dispersed in a sedimentary matrix. In addition, in peperite the clast population will be monomictic, whereas in debris flows, the clast population can be variably polymictic. Occurrences of sedimentary clasts enclosed in coherent igneous bodies as xenoliths are also not peperites, but are intrusive inclusions.

The characteristics of peperite hyaloclasts vary from angular, blocky, polygonal, and splintery (Fig. 4.6a–e) to irregular and fluidal (Fig. 4.6e, f), according to the rheological properties of the magma and the constitution of the host sediment. With viscous magmas the hyaloclasts in peperite tend to be "blocky" or polygonal, reflecting instantaneous brittle fracturing of the magma as it is supercooled and quench fragmented. However, Busby-Spera and White (1987) noted that where basalt intruded coarse volcaniclastic deposits, blocky hyaloclasts (i.e. polygonal, planar to curviplanar bounded clasts) had formed (e.g. Fig. 4.6a), whereas where the same intrusion invaded fine-grained mud sediment, the hyaloclasts were fluidal in form (e.g. Fig. 4.6f). In the case of the coarse sediment, the magma had access to abundant pore water, leading to very rapid normal cooling quench fragmentation and brittle fracturing, whereas with the fine mud sediments, limited availability of water and the rheology of the mud may have caused slower cooling of the magma and dynamic mixing of fluidal clots of the low-viscosity magma and the muddy sediment before solidification of the fluidal magma clasts. Such deposits appear not to be the products of true quench fragmentation (i.e. thermal shock of glass), but more likely of dynamic mixing of two fluids because the fluidal form indicates that fragmentation of the magma occurred above the glass transition.

(a)
(b)
(c)
57392
cm
(d)
(e)
(f)
(i)
57730.
(g)
(h)

◀ **Fig. 4.6** Textural features of peperite **a** Transition from the coherent to fractured core of a basalt sill (pale colour, left) intruding unconsolidated, wet, coarse sandstone and mudstone (dark), to in situ jigsaw-fit basalt breccia with sediment injections to clast-rotated and mixed basalt clast–sediment peperite breccia with sediment matrix between angular blocky-to-splinter-shaped hyaloclasts (right). Late Devonian Bunga Beds, Boyd Volcanic Complex, southern New South Wales, Australia. **b** Same location as in (**a**), with sediment dykes (dark grey) injected between basalt hyaloclasts (light grey) within a syn-depositional, intrusive basalt. Note the jigsaw-fit nature of the hyaloclasts, separated by the sediment stringers. **c** Slab of peperite from base of a shallow marine, trachy-andesite lava, Permian Gerringong Volcanics, Kiama, New South Wales. Note blocky-to-splintery-shaped hyaloclasts, jigsaw-fit arrangement of hyaloclasts, even when separated by red mudstone. **d** Peperite at the base of an Archean pillow basalt along contact with black mudstone as seen in diamond drill core ($\sim$5 cm wide), Teutonic Bore, Western Australia. **e** In situ to clast-rotated, rhyolite clast-mudstone peperite. Both jigsaw-fit and clast-rotated textural domains can be seen. Location as for (**a**). **f** Fluidal peperite, with irregular hyaloclasts of basalt mixed with black mudstone, Archean base metal Bentley Mine host succession, Yilgarn Craton, Western Australia. Drill core is $\sim$5 cm in diameter. **g** In situ peperite at the base of the Ordovician Garth Tuff (ignimbrite), Capel Curig, North Wales, showing injection of underlying substrate sandstone into the quench fractures. The ignimbrite was emplaced into shallow shelf paleoenvironment. **h** Peperite breccia from the base of the Ordovician Garth Tuff, showing blocky clasts, jigsaw-fit and clast-rotated texture, and a matrix of mudstone. **i** "Motoyama explosion breccia" from the host rock succession to the Motoyama massive sulphide deposit, Miocene Green Tuff Belt, northwestern Honshu, Japan. The texture is identical to that of the peperite at the base of the Garth Tuff (previous image), showing angular, blocky clasts, some with internal fractures, in a dark (sedimentary) matrix. *Could this be a peperite at the margin of a dome or cryptodome?*

Peperites can also be pumiceous if the magma has vesiculated at the time of interaction with water-saturated unconsolidated sediments (e.g. Hunns and McPhie 1999; Gifkins et al. 2002). The hyaloclasts are variably vesiculated and in some cases more irregular in shape than with dense clast blocky peperite.

Peperite can also form at the base of very densely welded, and rheomorphic ignimbrites that have flowed into water, and been quench fragmented by the wet, water-saturated, unconsolidated sediment substrate (Fig. 4.6g, h). In addition, quench fragmentation and peperite formation can at times be a more viable explanation for some volcanic breccias which have been misinterpreted as "explosion breccias" because of poor understanding of volcanic processes. For example, the Motoyama "explosion breccia" from the Kosaka massive sulphide mine in the Miocene Green Tuff Belt, Japan (Fig. 4.6i), has long been interpreted as a phreatic explosion breccia because it is dominated by blocky, angular poorly vesicular dacite debris (Horikoshi 1969). It is more likely that the dacite breccias are in situ to resedimented hyaloclastite, and those with mudstone matrix (Fig. 4.6i) are peperite associated with eruption and emplacement of dacite lava domes.

The presence of peperite is very important in understanding the timing relationships and emplacement processes in ancient volcanic and sedimentary successions. Peperite can only form at the base of a lava that flows over water-saturated unconsolidated sediments, but not at its top and margins, because only the base is in contact with sediment when the lava was emplaced. However, peperite should generally occur around all margins of syn-depositional intrusions of magma into water-saturated, unconsolidated sediments, but won't occur at the margins of intrusions into already lithified host rocks.

4.5 Explosive Fragmentation

4.5.1 Introduction

Explosive volcanic eruptions are caused by numerous factors and occur in varied settings or circumstances, each of which leaves its imprint on the pyroclasts and deposits that are produced. Ultimately, the driving force for all volcanic explosive eruptions is the enormous thermal energy of magmas. However, only a tiny fraction of the total thermal energy of a magma reservoir is expended in driving an explosive eruption at the Earth's surface, for several reasons. First, only a very small fraction of the initial volume of magma that is formed in the subsurface is ever erupted. This volume of "eruptible magma" could be as low as 10% or less of the available magma in the subsurface. Evidence of this lies in the huge volumes of batholiths and regionally extensive sills preserved in the rock record, which represent the non-eruptible magma fraction. Secondly, only a fraction of the heat in an erupting magma is used in driving exsolution and heating of exsolving gases. All magmatic pyroclasts are still at liquidus or sub-liquidus temperature when erupted. Thirdly, only a fraction of the volatile content in a magma exsolves, and is available to drive and explosive eruption. Even with phreatomagmatic explosive eruptions, only a fraction of the thermal energy of a rising magma goes into superheating external water that it comes in contact with explosively.

All fragments that are ejected explosively from a vent are called *pyroclasts*, irrespective of whether they are juvenile magmatic pyroclasts, juvenile magmatic or xenocrystic crystal fragments, or country rock fragments. In this section, we first discuss the processes associated with the different mechanisms that can cause explosive fragmentation of erupting magma and country rock, namely:

- magmatic explosions,
- phreatomagmatic explosions,
- phreatic explosions, and
- hydrothermal explosions.

Understanding the specific explosive mechanism is important as an indicator of the the possible explosive eruption style (Chap. 9), and the volcanic hazards associated with distinctive explosive fragmentation mechanisms and eruption styles.

4.5.2 Magmatic Explosive Fragmentation and Magmatic Volatiles

Magmatic explosive eruptions are driven by the volatiles that are dissolved in the magma and exsolve from it (that is, they separate from the magma as an immiscible fluid phase) as the magma ascends towards the Earth's surface (Plate 4.1c). The solubility of volatiles in a magma is controlled by confining pressure and the volatile content (e.g. Figs. 2.10 and 3.8). As magma rises through the crust and decompresses, the solubility of volatile phases decreases because volatile solubility is pressure dependent. Therefore, as the magma rises and decompresses it may become saturated to supersaturated with regard to particular volatile species. When the volatile saturation pressure exceeds the confining pressure, fluid bubbles or vesicles of exsolving volatiles form in the magma (Sparks 1978; Gardner et al. 2000). The exsolved fluid is initially a supercritical fluid (Chap. 3), but as further decompression occurs as the magma rises in the conduit, the fluid boils to a highly pressurised gas once the confining pressure is less than the critical point (pressure) of the volatile phase (Chap. 3). Even then, the formation of vesicles does not necessarily lead to explosive eruptions (Fink 1983; Fink and Manley 1987; Eichelberger et al. 1986). Explosive fragmentation only occurs if the gas (over) pressure is considerably higher than the confining pressure and can overcome various magma properties, such as viscosity, surface tension, and tensile strength. A very rapid growth rate of the vesicles is required to cause explosive disintegration of the magma, whereas if growth rates are slow, due to low volatile contents, gas pressure that exceeds the yield strength of the magma in the bubble walls but not its tensile strength, and/or low ascent velocities and decompression rates, a highly vesicular, coherent lava may be erupted without explosive activity. Dissolved magmatic volatiles therefore not only influence the physical properties of magmas, such as viscosity and melting temperature, but they also determine whether volcanic eruptions will be explosive or non-explosive (Chap. 2).

Dissolved magmatic water is responsible for explosive disruption of most magmas, especially in intermediate to silicic magmas (Sparks et al. 1994) and some arc and back-arc basin basalts (Cashman and Mangan 1994). CO_2 is also an important volatile in some tholeiitic and intraplate alkaline basalts (Carroll and Holloway 1994; Cashman and Mangan 1994; Wallace et al. 2015), and during explosive eruptions of carbonatites and kimberlitic magmas. CO_2 is less soluble in silicate melts than H_2O and exsolves before water (Macpherson 1984; Cashman and Mangan 1994; Vogel and Aines 1996; Lowenstern 2001; Wallace and Anderson 2000; Wallace et al. 2015). Other important gases are SO_2 and H_2S, which may affect the solubility and degassing behaviour of H_2O in some magmas (e.g. Carroll and Rutherford 1985, 1987; Wallace et al. 2015). The solubility of each volatile species in silicate melts depends on the partial pressure of other dissolved species (Fig. 2.10c, Chap. 2), as well as on magma composition, and is higher in rhyolitic melts than in basaltic melts (Wallace et al. 2015; Fig. 2.10a, b, Chap. 2).

During magma ascent, the solubility of volatiles decreases as the magma rises and decompresses to shallower crustal levels with lower confining pressures (Figs. 2.10 and 3.8). At a certain depth, that depends on and varies with the concentration of volatiles, the solubility will decrease sufficiently, first of CO_2, which is less soluluble in silicate melts, and then of H_2O (Fig. 3.8). They can then begin to exsolve when they become saturated in the magma. The depth at which this occurs depends on the magma type, the actual volatile content, and the vapour pressure of the dissolved water and carbon dioxide relative to the confining pressure (e.g. Burnham 1972; Lowenstern 2001; Wallace and Anderson 2000; Wallace et al. 2015). The higher the magmatic volatile content, the higher the vapour pressure exerted, and the greater the depth at which volatile saturation and subsequent exsolution will begin (Fig. 3.8).

The concentrations of volatiles in magmas involved in explosive eruptions can vary enormously depending on magma type, volatile solubilities, and pre-eruptive magma history (Wallace and Anderson 2000; Wallace et al. 2015). Determining the volatile content of magma immediately prior to eruption is not easy and is complicated by the very significant degassing and volatile loss in the conduit and during eruption. Analysing gases emitted from active vents provides valuable data on the principal volatiles being degassed from the magma and their relative proportions but does not indicate their pre-eruption concentrations. Analyses of the volatile content of unaltered volcanic groundmass glass in erupted products will, in general, only provide an indication of the degassed post-eruptive magma volatile content. However, analyses of unaltered glass (or "melt" as they are sometimes euphemistically called) inclusions trapped within crystals may provide accurate data on the pre-eruption magma volatile contents in the magma chamber and/or conduit prior to the onset of crystallisation, but even this is not necessarily so (Vogel and Aines 1996; Wallace et al. 2015). The techniques used for analysing the volatile contents of volcanic glasses include Fourier Transform Infrared Spectroscopy (FTIR), Secondary Ion Mass Spectrometry (SIMS), Raman Spectroscopy, and microprobe analyses (e.g. Ihinger et al. 1994).

Johnson et al. (1994) and Wallace et al. (2015) summarised the estimated pre-eruption volatile content of various magma types. Oceanic tholeiite (N-MORB) typically has H_2O content of less than 0.4 wt% and E-MORB <1.5 wt%. Oceanic seamount (e.g. Hawaii) basalts typically have H_2O content of less than 1 wt% for tholeiitic basalts and <1.5 wt% for alkalic basalts. For Kilauea basalts, the original CO_2 content may have been about 0.5 wt%. Dixon et al. (1997) report H_2O contents from 1.3 to 2 wt% and CO_2 values of 2 wt% to an unusually high, 5.4 wt% in submarine alkali basalts from the North Arch lava field north of Oahu, Hawaii, whereas Loihi submarine basalts have <1 wt% H_2O and <0.63 wt% CO_2 (Dixon and Clague 2001; Garcia et al. 2006). Back-arc basin basalts commonly have H_2O contents of 1–3 wt%, whereas arc basalts can have H_2O contents <6 wt% (Wallace et al. 2015).

Silicic arc magmas generally have H_2O contents <6 wt% and CO_2 contents <400 ppm, but generally silicic magmas involved in explosive eruptions have H_2O contents higher than 3 wt% (Thomas et al. 1994) and <6 wt% H_2O (Sparks et al. 1994; Wallace et al. 2015). Pre-eruption water content for intermediate magmas is intermediate between basalts and silicic end members. The systematics of CO_2 in magmas, including the effects of its co-existence with H_2O (Fig. 2.10c; Wallace and Anderson 2000; Wallace et al. 2015), are still not fully understood because of limitations in measuring its concentration (Blank and Brooker 1994; Holloway and Blank 1994; Johnson et al. 1994); however, it is less soluble than water in silicate melts and will therefore have largely degassed from magmas before eruption from subaerial vents, but may be retained in magmas erupting in deep water (Wallace et al. 2015). H_2O will also be in higher abundance in magmas erupted in deep water compared with subaerial counterparts, because of the confining hydrostatic pressure at considerable water depths, which constrains exsolution (Wallace et al. 2015).

An important point to keep in mind in modelling explosive pressures is that only a fraction of initial, pre-eruptive volatile content exsolves and contributes to vesiculation and explosive fragmentation, typically about 60 to 70% in subaerial settings, because volatiles will only diffuse into vesicles as long the volatile vapour pressure in the magma exceeds that in the vesicles. The volatile contents cited above therefore cannot be used to calculate vesicle gas pressures at the time of explosive fragmentation. Studies need to be undertaken of the melt inclusion pre-eruptive volatile content, and residual groundmass glass volatile content in the post-eruptive deposits (e.g. pumice clasts) to determine how much of the original volatiles have been released and used during the explosive eruption. Even then, however, the groundmass glass may have been altered and hydrated post-depositionally, which may be detectable using FTIR water speciation analyses (e.g. McIntosh et al. 2017), and the magma may have degassed non-explosively in the conduit during ascent.

Explosive magmatic vesiculation can occur in the following situations:

- within open vents,
- from erupting lava flows and domes,
- from sealed vents and magma chambers, and
- triggered by subsurface magma mixing.

4.5.3 Volatile Exsolution and Nucleation of Volatile Bubbles in an Open Vent/Conduit

For magmatic volatiles to drive explosive eruptions, magmas must first become saturated with volatiles, which then exsolve from the magma forming immiscible bubbles in a process called vesiculation. There are two main processes by which magmas can become saturated in volatiles driving them to exsolve volatile bubbles: (1) *decompressional exsolution–vesiculation* during ascent of a magma from its source towards the Earth's surface, and (2) *crystallisation-induced exsolution–vesiculation,* as a consequence of crystallisation of a magma. Decompressional exsolution–vesiculation has previously also been called *first boiling,* whereas crystallisation-induced vesiculation has been called *second boiling* (Turner and Verhoogen 1960; Burnham 1972). In addition, the readiness with which volatile bubbles or vesicles form will depend on whether or not the magma is aphyric or has crystals suspended in it. Where vesicles form in an aphyric magma, the process is called *homogenous nucleation*, whereas if bubbles form in a magma with suspended crystals the process is called *heterogeneous nucleation.* The immediate effect of exsolution of volatile H_2O from a magma is to increase the viscosity of the magma, because as water exsolves from the melt it no longer contributes to lowering the viscosity of the magma (Chap. 2). Exsolution of volatile bubbles also has the effect of lowering the bulk density of the magma plus vesicles, making it more buoyant in the crust than dense, unvesiculated magma and causing it accelerate as it rises.

4.5.3.1 Decompressional Exsolution and Vesiculation

In a magma conduit that has not yet opened a vent, the magma is subject to confining lithostatic pressure, whereas in an already open conduit and vent magma volatile exsolution is no longer constrained by lithostatic confining pressure, but is subject to the pressure due to the mass of magma in the conduit above it (Fig. 4.7a; Blake 1981a), called *magmastatic pressure*. In effect, however, magmastatic pressure in non-vesiculated magma is not much less than lithostatic pressure (Chap. 3) but does decrease as volatile exsolution and vesiculation occur above the exsolution level in a magma conduit make the vesiculated magma less dense. The magmastatic pressure gradient in a column of non-vesiculated silicic magma is ~21.6 MPa km^{-1}, and for a column of mafic magma, it is ~26.5 MPa km^{-1}, compared with a lithostatic pressure gradient of ~24.5 MPa km^{-1} for continental or arc crust, and ~28.4 MPa km^{-1} for oceanic crust (Fig. 3.3, Chap. 3).

As a particular batch of magma rises in an open conduit, it is subject to lower and lower magmastatic pressure. Because pressure is a major factor in determining whether or not volatiles remain dissolved or exsolve from magma, the magma will reach a level in the conduit where it is saturated with a particular volatile species (e.g. H_2O), called the saturation level (labelled *S* in Fig. 4.7a). The volatile vapour saturation pressure of the volatiles in the magma begins to exceed the confining magmastatic pressure because the magma does not equilibrate with the decreasing external pressures rapidly enough due to retarding factors such as viscous strength (Sparks 1978). The pressure and depth at which saturation occurs will vary for different volatiles, and also depend on the volatile content and magma composition and rheology (see Fig. 2.10). At low volatile contents, saturation will only occur at very shallow crustal levels, in some cases even at ground level, whereas at high-volatile contents (e.g. 4–6 wt%), saturation pressures could be as high as 100–200 MPa, corresponding to depths of 4–8 km (Fig. 3.8).

However, even at the point of saturation, most volatiles will not exsolve immediately because of resisting forces or properties of the magma, such as viscosity. The magma will continue to rise until it becomes supersaturated, at which point bubbles of a supercritical fluid, only microns in size, spontaneously nucleate and exsolve from the magma, which is called the nucleation level (*N* in Fig. 4.7a, inset i). The

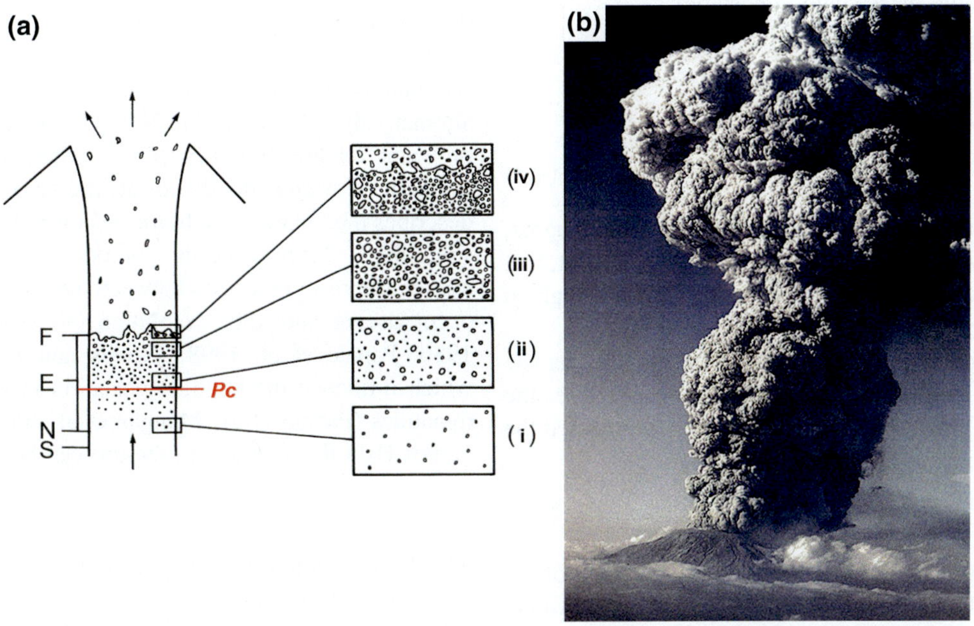

Fig. 4.7 Stages of explosive fragmentation in an open vent. **a** Schematic profile through an open volcanic conduit, vent, and magma column showing the transition from a deep, rising unvesiculated magma through the level of explosive fragmentation. S is the level of gas saturation, N is the level at which bubble nucleation begins, E is the interval of bubble growth and exsolution, and F is the magma fragmentation level or surface. Degrees of progressive exsolution and bubble growth stages are represented schematically in insets (i)–(iv): (i) at early stages nucleation, N, at and above the volatile saturation level S, occurs and bubbles grow uninhibited; (ii) bubble nucleation and growth continues in the exsolution depth range E; the larger bubbles begin to interfere with one another's growth; (iii) the magma froth is saturated with bubbles and growth rates are retarded and eventually cease; (iv) the fragmentation surface, F, forms in the froth and propagates down by the bursting of bubbles. *Pc* is a schematic representation of the level of the critical pressure of the volatile phase in the magma column. Below *Pc*, the exsolved phase is supercritical fluid, above *Pc* it is expandible gas. (From Cas and Wright 1987, After Sparks 1978). **b** Magmatic plinian eruption column from the 18 May 1980 eruption of Mt St Helens, USA, showing the convective column of gas and pyroclasts rising buoyantly above the vent (*Source* United States Geological Survey)

process is called decompressional exsolution–vesiculation or first boiling. This represents the stage where the volatiles become immiscible in the magma. They will continue to nucleate, exsolve, and grow above N, in a depth zone called the exsolution depth range, E (Fig. 4.7a, ii). During passage through the E depth range, more volatiles will diffuse from the magma into already formed bubbles (also called vesicles) as well as forming new ones. As the magma and bubbles rise, the volatile supercritical fluid in the bubbles boils to a gas when the confining magmastatic pressure falls below the critical point pressure, P_c, for H_2O, at 22 MPa, which is at a lithostatic equivalent pressure depth in the crust of ~900 m, for a closed conduit in continental or arc crust or ~750 m in oceanic crust, or at <~1020 m in an open column of rhyolite magma, ~900 m for andesite, and ~833 m for a mafic magma in a conduit open to the atmosphere (Chap. 3). From this point on, as the vesiculated magma continues to rise and bubbles expand, the volume of the gas bubbles will increase exponentially (McBirney 1973; Fig. 4.7a, iii), and the pressure in the bubbles will also increase dramatically.

Volatile bubbles grow in magma if the exsolved gas bubble pressure, P_g, exceeds the yield strength, y_s, of the magma (which for most magmas is several (kilo)pascals; kPa), and the ambient pressure, P_A (magmastatic, atmospheric, hydrostatic, glaciostatic, lithostatic):

$$P_g > y_s + P_A \qquad (4.7)$$

where the yield strength is given by

$$y_s = \sigma - \eta\,\gamma^n \qquad (4.8)$$

where σ is the total shear stress applied to a flowing magma, η is the viscosity, γ is the strain rate, and n is a constant, that is, <1 for pseudo-plastic behaviour and = 1 for Bingham behaviour (Allen 1970).

If the gas pressure significantly exceeds the tensile strength of the magma and the magmastatic pressure, the bubbles burst explosively, at a level in the conduit called the fragmentation surface, F (Fig. 4.7a, iv), that is:

$$P_g \ggg \tau_m + P_m \qquad (4.9)$$

where P_g is the gas pressure, τ_m is the tensile strength of the magma (typically <5–6 MPa), and P_m is the magmastatic pressure. Papale (1999) has expressed this criterion for fragmentation in terms of viscosity, η and total strain rate, γ (which includes both the strain related to bubble expansion and elongational strain related to velocity gradients in the conduit):

$$\eta\,\gamma > X \qquad (4.10)$$

where X is a pressure factor analogous to tensile strength.

The depth of the fragmentation surface is dependent on the volatile content and magma properties, and could vary from being at depths of hundreds of metres in the conduit, to almost at ground level. At the fragmentation level, the magma changes from a bubbly flow below F to an overpressurised particulate gas flow above F (Fig. 4.7a, iv) with suspended pyroclasts. The latter constitutes the in-conduit part of the gas-thrust zone at the base of the buoyant eruption column that typically rises above the vent in magmatic explosive eruptions (Fig. 4.7b).

The rapid decompression experienced by the magma during ascent results in the transition from a melt ± crystals system to a melt ± crystals + bubble system. Decompression and exsolution of volatiles, which is exothermic, also trigger crystallisation of groundmass microlites, and even (further) growth of phenocrysts (Blundy and Cashman 2005; Cashman 2004; Cashman and Scheu 2015).

For rhyolitic magmas, Wilson et al. (1980) proposed that the depth of exsolution of magmatic water can be determined using the following relationship:

$$\rho_{cr}\, g\, h_e = (n/s)^2 - P_s \qquad (4.11)$$

where ρ_{cr} is the crustal density (kg m^{-3}), g is the acceleration due to gravity (9.8 m s^{-2}), h_e the depth of exsolution (km), n the total weight fraction of exsolved volatile water, P_s the pressure at the Earth's surface (0.1 MPa), and s is a constant (= 0.0013, if pressure is expressed in bars, or alternatively = 4.1×10^{-6} m N$^{-1/2}$) found from the solubility of water relationship $n_d = sP^{0.5}$, where n_d is the weight fraction of water dissolved at pressure P. This model describes the volatile exsolution as controlled by equilibrium solubility. More recent works account also for non-equilibrium processes between melt and volatiles, which occur during and after exsolution (Vergniolle and Jaupart 1986; Papale et al. 1998; Woods and Koyaguchi 1994), which infiuence the bubble nucleation kinetics and growth dynamics (Papale 2001; Mangan et al. 2004a, b).

For H_2O the saturation pressure can be determined by

$$C_S = k\,P^n \qquad (4.12)$$

where C_S is concentration at saturation level (= magmatic water content expressed as the mass fraction, not wt%), k the solubility constant (= 4.11×10^{-6} for rhyolitic magmas, 6.8×10^{-8} for basaltic magmas), n = 0.5 for rhyolitic magmas and 0.7 for basaltic magmas, and P the pressure (Tait et al. 1989). This equation can therefore be used to calculate the saturation vapour pressure if the original magmatic water content is known. Theoretically, exsolution and vesiculation will occur in a rising magma if the combined vapour pressure, V_p, of dissolved volatiles begins to exceed the pressure in the magma column, P_m, plus the

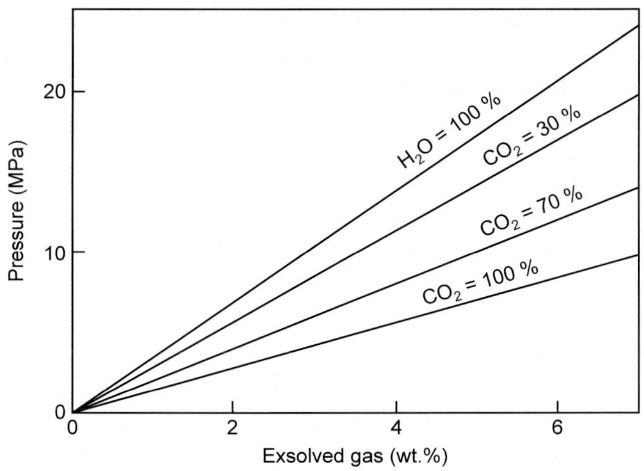

Fig. 4.8 Gas pressure exerted by HO_2/CO_2mixtures at the point of explosive disruption of vesiculated rhyolite magma as a function of the total weight percentage of exsolved gases (From Cas and Wright 1987, After Wilson et al. 1980)

ambient pressure, P_A, which in a vent open to the atmosphere, is only 0.1 MPa (1 bar). However, even at saturation pressures (i.e. when $V_p = P_m + P_A$) volatiles will not exsolve because viscosity, which controls diffusion rates of volatiles from magma to bubbles, retards nucleation and growth (Wilson et al. 1980; Thomas et al. 1994; Sparks et al. 1994; Hurwitz and Navon 1994). Toramaru (1995, 2006) proposed that the principal controls on bubble nucleation are the volatile content, or saturation pressure, magma viscosity, and magma decompression rate.

In deep-water submarine basalts at least, CO_2 will exsolve and form vesicles before water does, which reflects its lower solubility in silicate melts (Macpherson 1984; Head and Wilson 2003). However, its concentration is lower than water in most magmas and the gas pressure it exerts is less than that for an equivalent wt% magmatic water (Wilson 1980; Fig. 4.8). Varying mixtures of CO_2 and H_2O generate lower combined gas pressures than pure water (Fig. 4.8).

4.5.3.2 Crystallisation-Induced Exsolution and Vesiculation

During the ascent of magma in the conduit, or if it temporarily stalls and ponds in a magma reservoir somewhere in the plumbing system of a volcano (e.g. Carrigan 2000; Blundy and Cashman 2001), it may lose heat to the country rock and begin to crystallise. Degassing releases heat and could thus initiate crystallisation. Crystals could form phenocrysts or groundmass microlites (Blundy and Cashman 2005). Cooling and crystallisation have the effect of concentrating the dissolved volatile components in the remaining liquid magma between the crystals, and this will result in higher volatile vapour pressures in the groundmass magma than prior to crystallisation. The latent heat released through crystallisation will also have a feedback effect, helping to maintain high temperatures and keep vapour pressures high. If the vapour pressure exceeds the magmastatic or lithostatic pressure, then the magma can become saturated and supersaturated, and exsolution or "boiling" of volatiles may occur. This *crystallisation-induced exsolution–vesiculation* has been called *second boiling* (Turner and Verhoogen 1960; Burnham 1972). Such second boiling could then drive major explosive eruptions if the magma system is sufficiently overpressured relative to lithostatic pressure and country rock tensile strength constraints (Stock et al. 2016).

Vesiculation in a magma could be produced by both first and second boiling. Burnham (1972) calculated that an andesitic magma with no more than 2.8 wt% water will become saturated with water at about 2 km below the Earth's surface ($\sim$50 MPa). If the magma rises further, cools, or cools and crystallises, then the water will separate as a high temperature supercritical fluid or steam (or both). The exsolved phase causes an enormous increase in pressure within the magma chamber and the potential increase in volume of the system during this crystallisation and boiling can be up to 53% (Burnham 1972) under low confining pressures. The magma pressure build-up can trigger crack propagation in the country rocks, seismic activity, and possibly the initiation of magma rise towards the surface, eventually perhaps resulting in eruptions (e.g. Mt St Helens, USA, 1980; Mt Unzen, Japan, 1990).

Moitra et al. (2018), Arzilli et al. (2019a), Valdivia et al. (2022), and Marshall et al. (2022) have proposed that for basaltic magmas that would normally erupt effusively, highly explosive plinian eruptions can be triggered by rapid ascent of basaltic magma at <1,100 °C (Arzilli et al. 2019a) that leads to rapid undercooling, and rapid groundmass microlite growth (30–40% volume or greater). Sudden microlite growth leads to a sudden magma viscosity increase (10^5 Pa s is the minimum required; Arzilli et al. 2019a), leading to rapid gas bubble overpressure increase, and explosive fragmentation by brittle fragmentation as the magma deforms under visco-elastic behaviour and the extensional strain rate in the bubble walls reaches 1 s^{-1} or higher (Moitra et al. 2018). Major explosive eruptions of mafic magmas thus do not necessarily require unusually high magmatic water content, although that helps, and it also increases the rate of magma undercooling during magma ascent as volatile exsolution occurs.

The unexpected sub-plinian explosive eruption of Calbulco volcano in Chile in April 2015 has been attributed to closed system second boiling in a basaltic andesite magma at a depth of $\sim$8.5 km by Arzilli et al. (2019b). They proposed that late-stage groundmass crystallisation of 25-to-250-mm-sized microlites, representing $\sim$20% of the total crystal content of the erupted scoria, in addition to $\sim$25% pre-existing phenocrysts in magma saturated with

<6.5 wt% water led to rapid increase in magma viscosity and overpressurisation of the system, perhaps as little as 3 h before the eruption.

However, exsolution of volatiles from a magma does not necessarily lead to explosive fragmentation. Explosive fragmentation only occurs if the gas bubbles can grow instantaneously under very high gas overpressures at strain rates that overcome the tensile strength of magma.

4.5.3.3 Homogeneous and Heterogeneous Bubble Nucleation

Experimental work by Hurwitz and Navon (1994) on the nucleation of H_2O bubbles in natural obsidians at temperatures of 780–850 °C, high pressures, and decompression rates ranging from about 2 to 10 MPa s^{-1} indicate that super-saturation pressures of at least tens of MPa may be typical to produce the high nucleation densities (number of volatile bubbles per unit volume) required to account for bubble number densities in pumice. Hurwitz and Navon (1994) found that for aphyric magmas (magmas without crystals), little or no nucleation occurred below super-saturation pressures of 10 MPa, moderate nucleation occurred between super-saturation pressures of 10–70 MPa, and very high levels of nucleation above 70 MPa. However, for magmas with crystal microlites, very high levels of nucleation were recorded at super-saturation pressures of greater than 5 MPa. Nucleation rates of fluid bubbles were found to be as high as 10^{11} m^{-3} s^{-1} for aphyric magma and 10^{12} m^{-3} s^{-1} for magmas with microlites (Hurwitz and Navon 1994). For the Mt St Helens May 18, 1980, plinian eruption (see Chap. 9 for definition of "plinian"), nucleation rates may have reached 2×10^6 m^{-3} s^{-1} and for the 3,600 years BP Minoan eruption of Santorini, Greece, 10^{11} m^{-3} s^{-1} based on determinations of vesicle abundances in pumices from these deposits (Klug and Cashman 1994).

It is thus easier for vesicles to nucleate in a magma when there are crystals to nucleate on to, than in an aphyric magma (Fig. 4.9a). The nucleation of bubbles in magmas with suspended crystals is called *heterogeneous nucleation*,

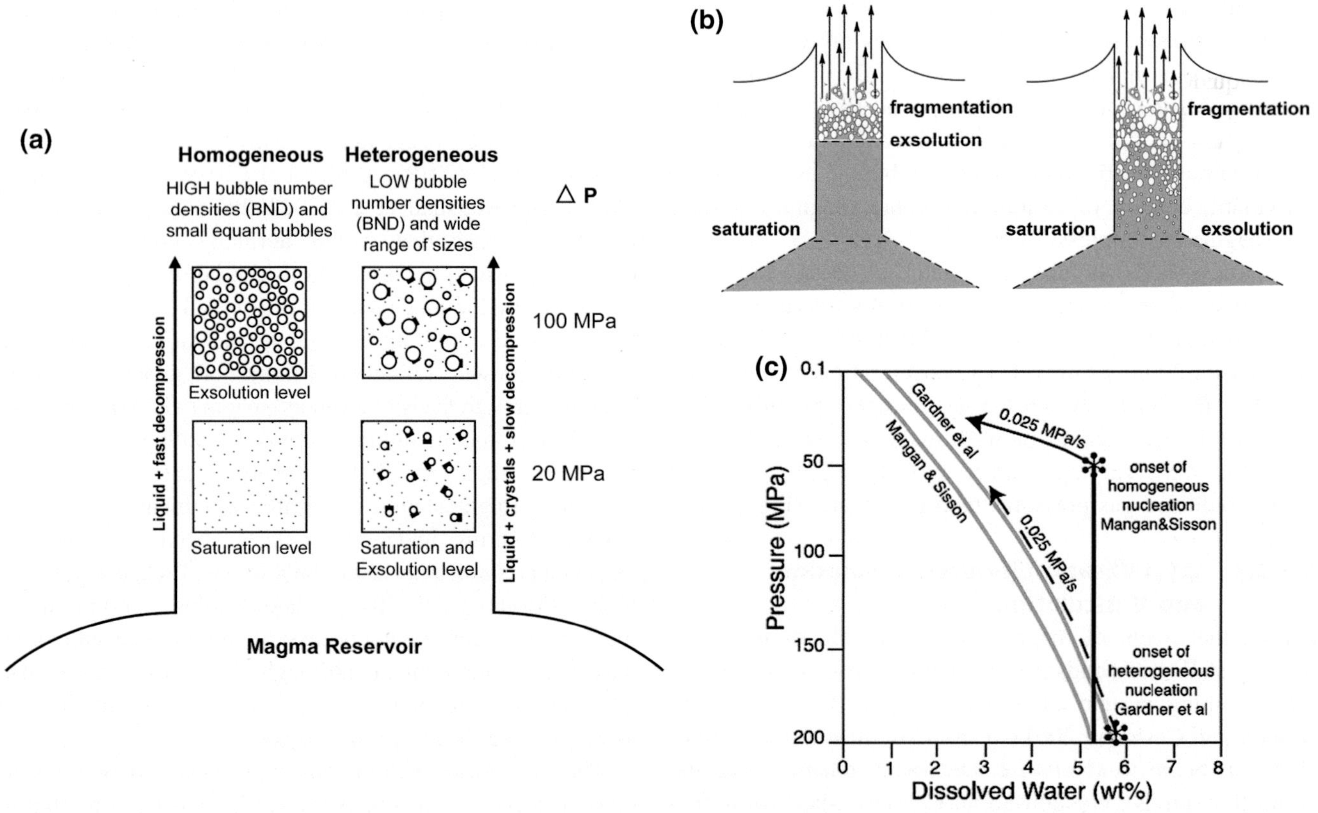

Fig. 4.9 Schematic representation of the difference between homogeneous and heterogeneous bubble nucleation. **a** On the left, with homogeneous nucleation, bubble nucleation occurs late and higher in the conduit than with heterogeneous nucleation (right) which facilitates early nucleation due to a combination of low viscosity of the melt, slow ascent rates and the presence of nucleation sites such as crystal surfaces. **b** As in (**a**) but taking a conduit-scale view of the differences in timing and depth between homogeneous nucleation (left) and heterogeneous nucleation (right) (From Cashman et al. 2000). **c** Exsolution–vesiculation–degassing paths for homogeneous (solid line) and heterogeneous (dashed line) during decompression at rates of 0.025 MPa s^{-1}, showing the exsolution level for each, and indicating the high degree of volatile supersaturation overpressure reached before nucleation in the case of disequilibrium homogeneous nucleation; grey lines show the experimental equilibrium conditions from Gardner et al. 1999, and Mangan and Sisson (2000). (From Mangan et al. 2004a)

whereas in aphyric magmas it is called *homogenous nucleation* (Cashman et al. 2000). Given that nucleation of vesicles is retarded in aphyric magmas until very high-volatile supersaturation pressures have developed, once nucleation has occurred, very rapid, and in many cases, very explosive growth rates can occur as a result of homogeneous nucleation.

Nucleation of bubbles therefore depends largely on the rate of decompression (e.g. Mangan et al. 2004a, b; Toramaru 2006), the presence of nucleation sites (e.g. Gardner et al. 1999), and the kinetics of volatile diffusion from the melt into the bubble (e.g. Toramaru 1995; Prousevitch et al. 1993; Proussevitch et al. 1993; Proussevitch and Sahagian 1998). At slow ascent rates, phenocrysts or groundmass crystals (or other "imperfections" in the melt) may effectively act as nucleation sites and focus volatile exsolution. Therefore during ascent, the total number of bubbles tends to remain constant, while their size increases (Gardner et al. 1999), as long as the actual groundmass crystals are not rare. This type of bubble nucleation represents near-equilibrium volatile exsolution and the presence of nucleation sites favours bubble nucleation close to the theoretical exsolution depth (Fig. 4.9a; see Sparks 1978; Mangan and Sisson 2000).

By contrast, at high ascent rates and/or with magmas near their liquidus temperature (i.e. with few to no crystals), nucleation of bubbles requires high levels of volatile super-saturation and ΔP of ~ 100 MPa (Mangan and Sisson 2000; Cashman 2004). A very large number of bubbles nucleate almost instantaneously at a much shallower depth than that required by equilibrium exsolution (Fig. 4.9b), and are uniformly distributed within the melt. This type of bubble nucleation represents non-equilibrium volatile exsolution (Fig. 4.9c; Mangan and Sisson 2000; Mangan et al. 2004a, b; Mourtade-Bonnefoi and Laporte 2002). Mangan et al. (2004a, b) found that in experimental runs using a high silica rhyolite saturated with H_2O, heterogeneous, equilibrium bubble nucleation occurred at volatile overpressures <5 to 20 MPa, producing 10^5–10^7 bubbles cm^{-3}, which had a typical Gaussian size distribution. In contrast, homogeneous, disequilibibrium bubble nucleation occurred at volatile overpressures >120–150 MPa, producing bubble number densities of 10^6–10^9 bubbles cm^{-3}, with exponential bubble size distributions. Residual water contents in the magma at vent were 0.5–1.0 wt% higher than in the heterogeneous experiments. Ascent velocities, conduit pressures, exit pressures, and mass flux were greater in homogeneous than heterogeneous experiments. In real cases, it is likely that a spectrum of processes intermediate between the two end-member types occurs (Mangan and Sisson 2000; Cashman et al. 2000; Cashman and Scheu 2015).

4.5.3.4 The State of Exsolved Volatiles (Supercritical, Gas) and Impact on Magmatic Explosive Fragmentation

As briefly mentioned in Sect. 4.5.1, and in Chap. 3, when magmatic volatiles exsolve from magma they are likely to exsolve above their critical point pressure as supercritical fluid bubbles. This occurs when their saturation pressure in the magma exceeds the ambient pressure (magmastatic, lithostatic, hydrostatic, glaciostatic) and the ambient pressure exceeds the critical pressure of the volatile phase. That is, the exsolved volatile when it first exsolves is *not* gas, a fact that seems to be little recognised in the volcanological literature.

Supercritical fluid has properties that are hybrid between liquid and gaseous states, but they are denser than gas and much less compressible or conversely less expandible. It is only when the rising magma and suspended exsolved supercritical fluid bubbles decompress to pressures below the critical pressure for that volatile in the magma column that the supercritical fluid boils to gas (see Fig. 4.7a, line *Pc*). Supercritical fluid cannot expand quickly enough to be explosive until it decompresses below its critical pressure, and usually well below its critical pressure where it is a gas which is more expandible and prone to expand explosively, although even then, not necessarily. For explosive magmatic fragmentation, the exsolved magmatic voilatile has to be a gas with a high gas overpressure relative to the ambient pressure that drives exponential gas bubble growth rates. McBirney (1963) demonstrated that the specific volume of water at magmatic temperatures (i.e. gaseous or steam state) begins to increase exponentially from confining pressures of ~ 10 MPa and less (Fig. 4.10a; Chap. 3).

As discussed in Chap. 3, the critical pressure of H_2O is ~ 22 MPa (Fig. 3.4a, c), and for CO_2, 7.4 MPa (Fig. 3.4b). If a magma conduit is closed to the atmosphere, the transition from supercritical H_2O fluid to H_2O gas in volatile bubbles already exsolved from the rising magma occurs at a depth of about 900 m (which is the depth in the crust where the lithostatic pressure is equal to the critical pressure of H_2O). In a conduit open to the atmosphere, the depth in a dense magma column at which this transition occurs is ~ 833 m in a column of mafic magma (ρ = 2,700 kg m^{-3}), at which depth the magmastatic pressure equals the critical pressure for H_2O, ~ 900 m for an andesite magma (ρ = $\sim 2{,}500$ kg m^{-3}), and at $\sim 1{,}020$ m for rhyolite (ρ = 2,200 kg m^{-3}). As the levels of vesiculation increase, the magmastatic pressure decreases because the bulk density decreases (Table 3.2), and so the "boiling" depths in a column of magma are greater, i.e. the exsolution and boiling depths in the vesiculated magma become deeper.

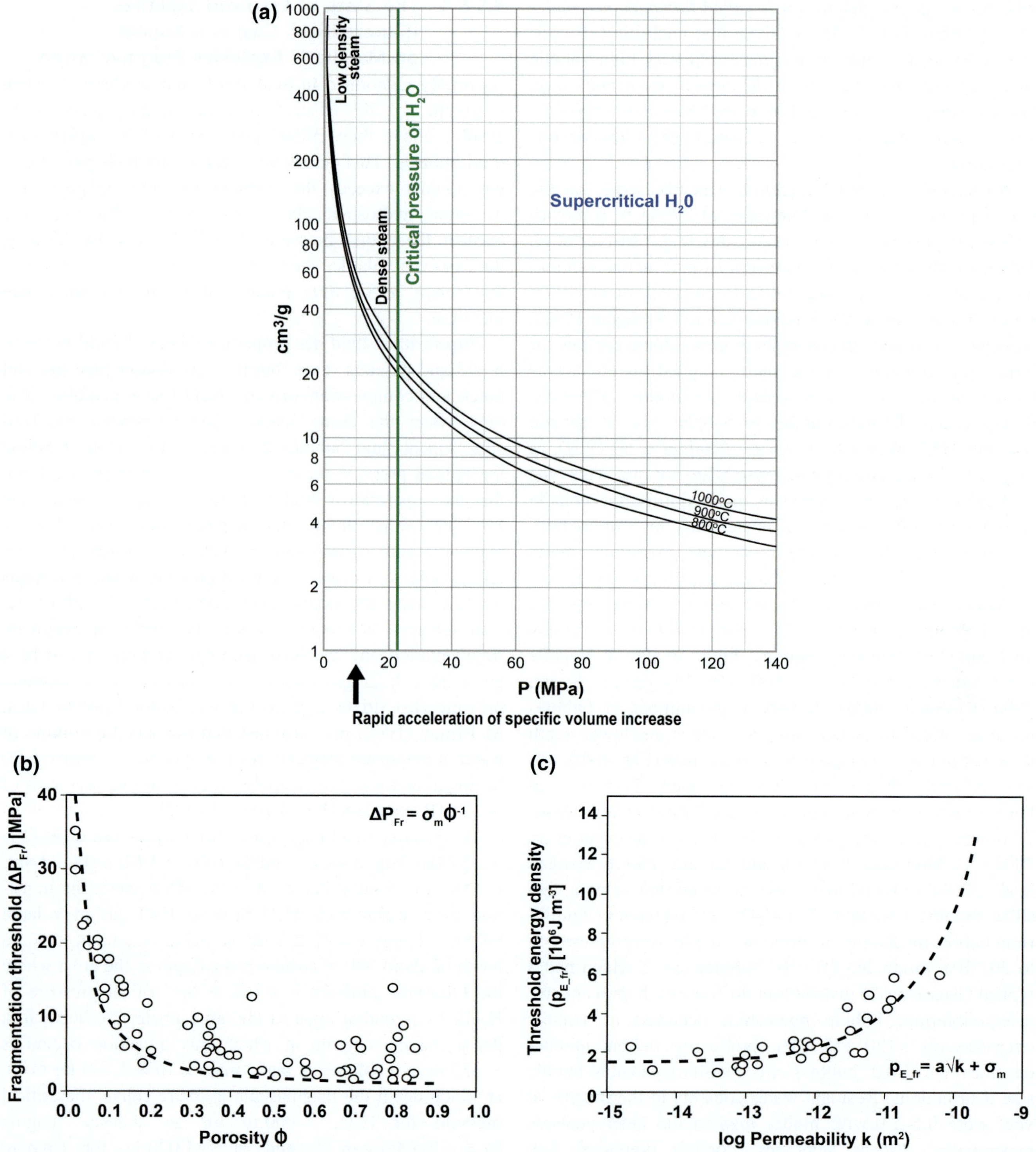

Fig. 4.10 Criteria for magmatic explosive fragmentation. **a** Changes in specific volume of water (steam) at magmatic temperatures as a function of decreasing ambient pressure. Expansion rates only become explosive where the curve gradient becomes subvertical. Also highlighted is the critical point pressure of H_2O, and the approximate fields of supercritical H_2O, dense steam, and low-density steam, the last of which facilitates explosive eruptions driven by expanding H_2O. (From Cas and Giordano 2014, After McBirney 1963). **b** Experimentally derived gas overpressure threshold, ΔP_{fr}, for magmatic explosive fragmentation versus porosity as a volume fraction. The points above the criterion dashed line reflect samples with increasing permeability. **c** Threshold energy density levels for magmatic explosive fragmentation plotted against permeability coefficient, demonstrating that required energy levels to induce fragmentation increase significantly as permeability increases. (From Mueller et al. 2008)

In the following sections on exsolved bubble or vesicle growth and magmatic explosive fragmentation, the focus is on the behaviour of subcritical gases, unless otherwise stated.

4.5.4 Diffusional and Decompressional Growth of Gas Bubbles

During nucleation and the subsequent growth of bubbles, the volatile phase(s) will diffuse from the magma into the volatile bubbles (Prousevitch et al. 1993) as long as the vapour pressure of the volatile phase in the magma exceeds the pressure of the volatile phase in the bubbles. The rate at which a particular volatile diffuses through magma, represented by the diffusion coefficient, is dependent on the viscosity, and can be expressed as follows (Sparks 1978):

$$-log_{10}D = log_{10}\eta + 5.82 - 4100/T \qquad (4.13)$$

where D is the diffusion coefficient for that magma, η is viscosity, and T is the temperature in degrees Kelvin. This equation was derived from the experimental data of Scholze and Mulfinger (1959) using artificial silicate melts at 1 atm pressure (101 kPa) and at temperatures between 1,000 and 1,400 °C. Sparks (1978) comments that the relationship should be used with care because of the particular compositions, viscosities, and temperatures used. Nevertheless, it indicates that diffusion is dependent on viscosity, and that diffusion rates in mafic magmas should be higher than in silicic melts. Sparks (1978) calculated that for basaltic magmas at 1,100–1,300 °C, $\log_{10} D = 10^{-5}$–10^{-6} $cm^2\ s^{-1}$, whereas with rhyolitic melts at 750–850 °C, $\log_{10} D = 10^{-7}$–10^{-8} $cm^2\ s^{-1}$ for H_2O.

More recent approaches to calculating diffusivity (Zhang and Behrens 2000; Behrens et al. 2004; Toramaru 2006) take into account other magma properties, including temperature, silica content, water content, which all influence viscosity (Chap. 2), as well as pressure. The relationship for diffusivity of H_2O, $D_{(ZB)}$, proposed by Zhang and Behrens (2000) as represented in Toramaru (2006), is

$$D_{(ZB)} = X \cdot \exp(m) \cdot \left\{1 + \exp\left[56 + m + X \cdot \left(-34.1 + \frac{44620}{T} + \frac{57.3\left(P \times 10^{-6}\right)}{T}\right) - \sqrt{X} \cdot \left(0.091 + \frac{4.77 \times 10^6}{T^2}\right)\right]\right\} \qquad (4.14)$$

where X (mole fraction of water) = $0.0167C_w$, $P = 2 \times 10^{10} \times X^2$, T is the melt temperature in °K, C_w is the initial water content (wt%), and

$$m = -20.79 - \frac{5030}{T} - \frac{1.4\left(P \times 10^6\right)}{T}$$

Behrens et al. (2004) proposed that diffusivity of H_2O, $D_{(BZX)}$, could be defined as

$$D_{(BZX)} = 10^{(-0.757-0.0868C_S) + \left(\frac{-14785+131.7C_S}{T}\right) + \left(\frac{(3.079-0.049C_S)(P\times 10^{-6})}{T}\right)} \qquad (4.15)$$

where C_s is the SiO_2 content (wt%).

Toramaru (2006) considered the two variants could be combined to give a more definitive assessment of diffusivity, as follows:

$$D_T = \frac{D_{(BZX)} \times D_{(ZB)f}}{D_{(ZB)1wt\%}} \qquad (4.16)$$

where $D_{(BZX)}$ uses the initial water content in wt% and $D_{(ZB)1wt\%}$ uses 1 wt% water content. Readers are referred to all three papers for the derivation of Eqs. 4.11, 4.12, 4.13, but particularly Toramaru (2006) for the final reconciliation of the relationships in Eq. 4.16.

Vesicles will grow through a combination of diffusive growth and decompressional growth. The growth of a gas bubble is favoured by volatile abundance, but limited by melt viscosity. For a given magma composition, bubble growth is controlled by (a) the diffusion of volatiles dissolved in the magma into the bubble; (b) the rate at which the confining pressure falls as the bubble or the magma, or both, rise; and (c) the variation in properties of the magma during degassing, that is, viscosity, surface tension, density, and the solubility and diffusion coefficient of the gas in the magma. Growth rates due to diffusion are controlled by the composition, solubility, concentration, and the degree of supersaturation of the volatiles. Vesicle size is related to diffusivity by

$$R = 2\beta\,(D\,t)^{0.5} \qquad (4.17)$$

where R is the radius, β is the growth rate constant, D is the diffusion coefficient, and t is time. The values for D cited for basaltic and rhyolitic magmas suggest that bubbles in basalt should grow to a radius 10–50 times larger than in rhyolite in a fixed time (Sparks 1978). This is supported by direct measurement of vesicle sizes in basaltic scoria compared with rhyolitic pumice (e.g. Cashman and Mangan 1994). During the initial stages of bubble growth, the diffusion of gas towards the bubble from surrounding regions may be fast, and therefore bubble growth is only limited by the melt viscosity (*viscosity-limited*). As the process of bubble growth progresses, volatiles must diffuse towards the bubble from farther regions in the melt, and therefore the growth rate progressively slows at decreasing volatile contents and diffusion rates, as well as at increasing bubble number densities (number of bubbles per volume), since nearby vesicles "drain" volatiles from the surrounding magma (*diffusion-limited*). Bubble growth during the diffusion-limited stage is essentially due to decompression.

Decompressional growth of bubbles is controlled by the rise velocity of the magma, the rate at which the magma is disrupted and removed at the free surface in the vent, and by the rise of the bubbles within the magma body (Sparks 1978). In practice, both diffusion and decompression may operate together. In an open vent and conduit, the confining pressure on gas bubbles is due to the weight of the magma in the conduit above, that is, the magmastatic pressure, rather than the lithostatic pressure. Expansion of vesicles leading to explosive disruption is not steady and progressive during uprise in the magma conduit, especially for magmas with viscosities greater than 10^6 Pa s, but rapid and accelerating, occurring during the latter stages of the ascent of the magma in the conduit (Sparks et al. 1994).

Three factors contribute to this. First, the pressure dependence of water vapour volume expansion is not linear but controlled by a power law that leads to accelerating growth rates with decreasing pressures (e.g. McBirney 1963; Sparks 1978; Prousevitch et al. 1993; Proussevitch et al. 1993; Sparks et al. 1994; Proussevitch and Sahagian 1998). Secondly, the viscous resistance in the magma limits the growth rate of bubbles in a rising decompressing magma (Thomas et al. 1994). Volatile exsolution increases the viscosity of the melt by up to four orders of magnitude, and will lead to a reduction in the diffusion rate of volatiles from magma to bubbles. Thomas et al. (1994) calculated that viscous resistance affects nucleation and bubble growth above magma viscosities of 10^6 Pa s, it becomes significant in retarding growth rates above viscosities of 10^7 Pa s, and prevents bubble expansion above a viscosity of 10^9 Pa s. Initial bubble growth rates are slow and accelerating, approximately following a parabolic law. Final growth rates are decelerating due to the reduction in vesicle gas pressure as the bubble expands, producing an overall sigmoidal growth pattern (Prousevitch et al. 1993).

Bubbles cannot continue to grow infinitely, because they do not grow in isolation from each other. Bennett (1974) suggested that bubbles growing near to each other cease growing because high-viscosity magma liquid has to be forced through intricate channels between bubbles, which are trying to expand. Sparks (1978) predicted that magma viscosities greater than 10^5 Pa s limit the bubble growth and produce large overpressures. This implies that viscous rhyolitic magmas will generally form vesicles that are smaller than in low-viscosity basaltic magmas, and will develop greater overpressures in the bubbles, so that explosive eruption styles are more explosive. In low-viscosity basalt magma, vesicularity can exceed 80% in reticulite, in which to accommodate each other, interfering bubbles develop a polygonal shape.

Furthermore, the rapid exsolution of volatiles causes a rapid increase in the viscous resistance to growth of bubbles, and increases the tensile strength of the magma (McBirney 1973; Sparks 1978; Wilson et al. 1980). This effect can also be largely enhanced by the syn-eruptive nucleation of microlites, which form as a result of degassing, and which can increase the viscosity by orders of magnitude, along with the presence of phenocrysts (Costa and Dungan 2005; Caricchi et al. 2007).

Thirdly, the growth rate of vesicles is influenced most by the rate at which magmas decompress as they rise to the vent and not by post-exsolution diffusion (Thomas et al. 1994). Thomas et al. (1994) calculated that based on the time available for vesiculation (seconds to minutes), diffusion rates of water still dissolved in the magma into vesicles would be too low and would add only 10% of the gas mass in the bubbles between the initiation of vesiculation in the magma and solidification of pyroclasts by air quenching in the eruption column.

There is now general agreement that the principal trigger for explosive vesiculation is sudden, even instantaneous decompression of magma supersaturated with H_2O resulting from magma ascent rates of metres per second which are too rapid for the vapour pressure to equilibrate (Wilson et al. 1980; Thomas et al. 1994; Sparks et al. 1994; Cashman and Mangan 1994; Toramaru 1995, 2006; Cashman and Scheu 2015). This drives instantaneous nucleation and explosive growth of vesicles at a high level in the magma conduit, perhaps only tens of metres below the fragmentation level where the magmastatic pressure is low (Sparks et al. 1994). Calculations for the pre-vesiculation magma ascent rate for the May 18, 1980 Mt St Helens plinian eruption, (see Chap. 9 for definition of "plinian") based on the duration of the eruption, known erupted volume, and depth of the magma chamber at 6–7 km, suggest magma rise rates between 2 m s^{-1} (Carey and Sigurdsson 1985) and 15 m s^{-1} (Sparks et al. 1994), and volume uprise rates of 10^4 m^3 s^{-1} (Sparks et al. 1994). The dacitic magma with 4.6 wt% H_2O would have become saturated at a lithostatic pressure of 125 MPa (Carey and Sigurdsson 1985) (or 110 MPa; Sparks et al. 1994), equivalent to a depth of about 4,560 m, with the fragmentation level being at a depth of 580 m below the vent (Carey and Sigurdsson 1985). Klug and Cashman (1994) calculated the time for vesiculation from exsolution to fragmentation to be 16 min, whereas Sparks et al. (1994) suggest 5–10 min and probably shorter. Vent gas and ash exit velocities were about 330 m s^{-1} (Carey and Sigurdsson 1985). For the major plinian Bishop Tuff, Taupo, Hatepe, and Minoan eruptions, Thomas et al. (1994) calculated that the total decompressions ranged from at least 18 MPa (Minoan) to perhaps as much as 27 MPa (Bishop Tuff), and that the time available from the point of vesiculation to reaching the vent was perhaps only seconds, giving decompression rates in the order of 12 MPa s^{-1}.

It is unclear how much of the original magma volatile content, recorded in melt inclusions in pre-eruptive phenocrysts, contributes to the explosive energy and gas overpressure during explosive eruptions. Although the calculated difference between melt inclusion data and residual groundmass glass volatile content from pyroclasts may give an approximation, it is unknown how much the magma may have outgassed to the country rock prior to final explosive fragmentation. There are surprisingly few studies that record both melt inclusion volatile content data and groundmass glass residual volatile content. This requires certainty that the "phenocrysts" are truly juvenile magmatic phenocrysts grown in the erupting magma batch, rather than xenocrysts or antecrysts from a previous magma batch, and that they have not leaked volatiles since formation. Secondly, groundmass glass in pyroclasts is prone to rehydrate from atmospheric or meteoric water sources post-depositionally (e.g. Giachetti and Gonnermann 2013), raising uncertainties about the true values of residual magmatic water content in groundmass glass. The few data that are available suggests for silicic pumices that the proportion of original magmatic H_2O content that is lost through pre-eruptive degassing and/or used to drive explosive eruptions is perhaps about two-thirds to three-quarters (e.g. Wright et al. 2003; Blundy and Cashman 2005). Calculations of explosive pressures and dynamics should therefore not be based on pre-eruptive magma volatile content.

Rapid decompression as a control of instantaneous explosive vesiculation also appears to be supported by bubble or vesicle size/volume distribution patterns. It has long been noticed that vesicle size distribution patterns are usually bimodal (e.g. Whitham and Sparks 1986). Although early interpretations were that the less numerous large vesicle population represented nucleation in the magma chamber and the more numerous smaller population, nucleation in the conduit, Klug and Cashman (1994) and Cashman and Mangan (1994) point out that many of the large vesicles are aggregated, irregularly shaped coalesced smaller bubbles. It appears that pumice clasts continue to expand between ejection from the vent and air quenching in the subaerial eruption column (Thomas et al. 1994). Klug and Cashman (1994), following the interpretation of Orsi et al. (1992), therefore propose that many of the large coalesced vesicles are the result of post-eruption expansion and coalescence of smaller bubbles. The smaller and by far more numerous and relatively uniformly sized vesicles would therefore represent essentially one short-lived vesiculation event that is a response to rapid decompression high in the magma conduit, with rapid vesicle growth resulting from late equilibration between excessive vapour phase pressures and the low pressures above the fragmentation level in the conduit.

The importance of rapid decompression has also been highlighted by the experimental simulations of Mader et al. (1994a, b) and Sugioka and Bursik (1995). Mader et al. (1994a, b) summarised the results of two experiments. In the first, CO_2 saturated water was instantaneously decompressed by 7 MPa leading to bubble densities of 10^9 m^{-3}, velocities in the vesiculating liquid of 14 m s^{-1} and accelerations of 200 *g* (where *g* is the acceleration due to gravity). Bubble or vesicle number density remained constant from the onset of nucleation. In the second experiment, a 6-molar saturated solution of K_2CO_3 was injected into 12 molar concentrated HCl leading to a violent chemical reaction that produced voluminous CO_2. Nucleation of CO_2 bubbles was not instantaneous because the injection was not instantaneous, and because chemical reaction was dependent on the mixing time. Nonetheless, CO_2 supersaturation pressures of 9 MPa (equivalent to decompression pressures) were generated, and velocities of 30 m s^{-1} and accelerations of 150 g were recorded. The fragmentation of the vesiculating fluids in both experiments resulted from expansion of bubbles and the rapid acceleration of vesiculating fluids.

In the experiments of Sugioka and Bursik (1995), highly volatile liquids containing glass spheres were pressurised above the liquid vapour pressure in a pressure vessel and were then instantaneously decompressed. At *relatively* low decompression rates (ratio of amount of decompression, ΔP, to volatile vapour pressure, $P_V < 0.45$; note this value is relevant to the experiment context only, not necessarily magmatic systems), nucleation, bubble growth, explosive disintegration of the liquid-solids mixture, and explosive ejection of the solids were virtually instantaneous. At $\Delta P/P_V > 0.45$ nucleation was also instantaneous throughout the fluid at lower rates of decompression, but only near the top of the fluid–particle mixture at very high rates of decompression. Bubble growth and explosive fragmentation did not occur until a disruption front propagated downwards from the top at velocities of 2–4 m s^{-1}. Compression ahead of the downwards propagating front prevented bubble expansion and explosive fragmentation ahead of the front. However, as the front passed, rarefaction and instantaneous explosive expansion occurred in its wake, leading to sequential ejection of solids at progressively deeper levels in the test mixture.

The pressure, P_b, in growing bubbles in vesiculating magma in an open subaerial vent is at least equal to the sum of the forces retarding bubble growth (Sparks 1978; Sparks et al. 1994), and is given by the Raleigh-type equation:

$$P_b = \rho_m g h + \frac{2\sigma_t}{r} + \rho_m\left(R'R'' + \frac{3}{2R'^2}\right) + \frac{4\eta R'}{r} + P_A \tag{4.18}$$

where $\rho_m gh$ = magma pressure due to the depth of variably vesiculated magma in the magma conduit/chamber (magmastatic pressure), $2\sigma_t/r$ = tension in the magma bubble walls, $\rho_m(R'R'' + 3/2R'^2)$ = inertia of the magma in the bubble walls, $4\eta R'/r =$ viscous resistance of the magma in the bubble walls, P_A = ambient pressure, ρ_m is magma density, g is acceleration due to gravity, h is depth below magma surface, σ_t is the surface tension of the magma, r is radius of the vesicle(s), R' is the bubble growth rate, R'' is the acceleration rate, and η is the dynamic viscosity (after Sparks 1978; Sparks et al. 1994).

Equation 4.18 can be simplified by ignoring the inertial force and surface tension effects which, relative to viscous resistance, are minor (Thomas et al. 1994; Sparks et al. 1994). The viscous resistance to vesicle growth may be up to 1 MPa (Prousevitch et al. 1993). For subaerial vents P_A may also be ignored given it is only 0.1 MPa (1 bar). Thomas et al. (1994) represent P_b as follows:

$$P_b = P_e + \left(\frac{4\eta}{3} \cdot \frac{1}{\varepsilon} \cdot \frac{d\varepsilon}{dt}\right) \tag{4.19}$$

where P_e is the external pressure (= magmastatic pressure plus ambient pressure), η is the magma viscosity, ε is the void (vesicle) fraction, and $d\varepsilon/dt$ is the rate of increase of void fraction.

If the gas pressure is greater than the forces resisting growth, an *excess or overpressure* exists, which may be great enough to cause bubble growth. If the viscous resistance in silicic magmas can reach 1 MPa (Prousevitch et al. 1993), then this could be taken as the minimum overpressure required to cause vesicle growth. Gas pressures will clearly have to be significantly above this to cause growth at any significant depth in a magma column. If there is a very large excess or overpressure, then the vesicles will grow at explosive rates, perhaps no more than tens of m below the fragmentation or disruption surface as defined by Sparks (1978).

Sparks et al. (1994) propose that the region of rapid explosive degassing in the conduit is only tens of m thick, and that it is a zone of extremely high-pressure gradients. It will normally occur in the conduit tens of m to several hundred m below the vent depending on magma composition and volatile content (e.g. Mt St Helens, 580 m; Carey and Sigurdsson 1985). Below this zone the subjacent magma is rising in a supersaturated state under laminar flow conditions at velocities of m s^{-1}. At the top of the zone, pressure release to virtually atmospheric pressure occurs through instantaneous, explosive bubble growth. The fragmented particulate–gas mixture accelerates up to the vent in a highly turbulent state at accelerations of perhaps 8 m s^{-2} and vent exit velocities of 200–500 m s^{-1} (Wilson 1980; Dobran 1992; Sparks et al. 1994).

The main physical effect of the nucleation and growth of gas bubbles within a magma is the rapid increase in volume of the magma and the consequent rapid decrease in density, which will favour the upward acceleration through the conduit (e.g. Sparks 1978; Papale et al. 1998). The acceleration experienced by the rising magma will depend on the rate of bubble growth and nucleation, which in turn depends on the diffusivity of gas at the magma–bubble interface, the viscosity of the melt, and on whether bubbles will remain coupled with the magma (*closed degassing*), or will escape from the system (*open degassing*). Gas bubbles will have in fact the tendency to rise much more rapidly than magma, and will escape if the rheology of the magma phase allows that. That will depend on the bulk chemistry and viscosity, the volatile content and species, the crystal content, and particularly by the rate of microlite formation during ascent, which may substantially change the viscosity.

As a broad general rule, low-viscosity magmas (or slow ascent rates) will tend to allow gas escape, whereas viscous magmas (or very rapid ascent rates) will not. However, the extent of gas escape also depends on the degree of gas bubble coalescence, which may form a continuous and therefore permeable network of voids through the magma which will favour the upward and sideward streaming of gas away from the magma (Klug and Cashman 1996; Rust et al. 2003; Cashman and Scheu 2015). Similarly, for viscous magmas, both coalescence and non-explosive autoclastic fragmentation along the conduit walls may promote magma outgassing (Gonnermann and Manga 2003; Okumura et al. 2008; Shields et al. 2014).

Cashman (2004) and Cashman and Scheu (2015) suggest that if magma rises slowly and conditions are open to degassing (permeable magma foam) then the eruption behaviour will be effusive or weakly explosive (strombolian *sensu stricto*), whereas for closed-degassing conditions (poorly permeable magma foam), eruptions will tend towards explosive, i.e. Hawaiian in basaltic and plinian in rhyolitic magmas. Furthermore, volatile expansion becomes increasingly important at decreasing depths, therefore bubble expansion associated with homogeneous nucleation processes occurring at shallow depths and at high-volatile supersaturations may lead to magma fragmentation and explosive eruptions (Cashman et al. 2000).

Upward acceleration within a conduit generates an upward convex velocity profile, highest in the centre, lowest against the conduit walls, which relates to frictional effects at the conduit wall. The shape of the profile depends on the bulk rheology of the rising magma, which may vary substantially during ascent and degassing, ranging from the typical parabolic profile for Newtonian rheologies, to piston-like flows for non-Newtonian ones (see Chap. 2). Although progressive degassing and increasing crystal content (both phenocrysts and groundmass microlites) increase the magma bulk viscosity (Fig. 2.12c), bubbles will have a variable influence on the bulk viscosity of the rising magma depending not just on their volume, but on their size and on the competing melt viscosity and shear rate (Rust and Manga 2002; Pal 2003). The capillary number (cf. Chap. 2, Eq. 2.7) expresses these relationships as

$$Ca = \frac{r S \eta}{\sigma_t} \tag{4.20}$$

where r is the bubble radius, S is the shear rate, η is the melt shear viscosity, and σ_t is the surface tension. Small capillary numbers <1 indicate the presence of small bubbles and low shear rates; in these cases, bubbles remain spherical and act as rigid bodies in the magma, increasing the bulk viscosity (Fig. 2.12). By contrast, capillary numbers >1 indicate that shear rates are high enough to deform bubbles, then the net effect of their presence is to reduce the bulk viscosity (Fig. 2.12c), such are the cases for large bubbles and low surface tension. Changes across the conduit of shear rates and bubble size distribution induce the departure from the usually assumed homogeneity of the rising foam (e.g. Wilson et al. 1980; Dobran 1992; Papale et al. 1998). Under certain conditions, the foam may develop a yield strength, and strain may be partitioned inside weak zones, where shear-thinning behaviour may favour strain-induced brittle fragmentation (Marti et al. 1999; Cashman 2004; Wright and Weinberg 2009), if the rate of strain rate exceeds that of the relaxation time at glass transition temperatures (Dingwell and Webb 1989).

Spieler et al. (2004a, b) and Mueller et al. (2008) investigated the relationship between gas overpressure, ΔP, magma porosity (vesicularity, ϕ), permeability, k, and magma tensile strength σ_m, and determined that the fragmentation threshold pressure difference, ΔP_{fr}, required for fragmentation is

$$\Delta P_{fr} = \frac{a\sqrt{k} + \sigma_m}{\phi} \tag{4.21}$$

where a is a constant (8.21×10^5 MPa m^{-1}) and σ_m has a value of 1.54 MPa (Mueller et al. 2008). At low permeability the threshold overpressure for fragmentation is determined by the magma tensile strength and porosity. The relationship between threshold gas overpressure and porosity, and threshold energy and permeability are shown in Fig. 4.10a, b.

An important consequence of the growth of bubbles and crystals during magma decompression is the significant change in rheological properties that affect both the bulk behaviour and the localisation of strain (e.g. Caricchi et al. 2007; Pistone et al. 2012; Costa et al. 2009; Vona et al. 2016; Ryan et al. 2019), in turn modulating fragmentation processes.

4.5.5 Magmatic Explosive Fragmentation Processes

During fragmentation the potential and thermal energy of magma transform into mechanical energy resulting in explosive gas bubble expansion and the disruption of the vesiculated magma in a particulate mixture of gas with liquid fragments which cool into solids (Cashman et al. 2000). Fragmentation fundamentally changes the dynamics of ascent, because at the fragmentation level the *continuum* changes from being the magma with dispersed gas bubbles, to gas with dispersed magma fragments.

Magmatic explosive fragmentation occurs fundamentally associated with two types of fragmentation regimes (Cashman et al. 2000; Cashman 2004; Cashman and Scheu 2015): (1) *dynamic fragmentation* and (2) *static fragmentation*. Dynamic fragmentation occurs as magma rises in a conduit, decompresses, and experiences syn-eruptive, closed system exsolution degassing. The magma accelerates up the conduit because of vesiculation and bubble growth which drives bulk volume expansion and increases buoyancy upwards. Explosive disintegration of the foam occurs when either or both (i) the internal gas pressure inside the bubbles causes them to expand and burst forming ash shards, and (ii) the strain in the foam exceeds the strength of the magma, forming pumice clasts (Fig. 4.11a). Sustained explosive eruptions, such as plinian eruptions, experience this style of fragmentation. During static fragmentation episodes of sudden decompression of the volcanic edifice occur, such as a dome collapse event during which the sudden removal of the upper or outer part of a dome immediately reduces the pressure on the inner part of the dome. This leads to open system degassing and propagation of a decompression wave downward into already vesiculated magma (Fig. 4.11b), leading to sudden explosive fragmentation, such as during vulcanian eruptions.

4.5.5.1 Fragmentation by Rapid Acceleration

As discussed previously, depending on magma rheology, bubbles can be more or less coupled with the magma during acceleration in the conduit (Fig. 4.11a). Rhyolitic magmas produce bubbly flow (Fig. 4.11c), i.e. a foam made largely of pervasive and size-limited vesicles growing in and unable to escape from a deforming but progressively stiffening magma (bubbly flow; Fig. 4.11c), before fragmentation by bubble bursting and rapid acceleration and shearing, at which stage the flow in the conduit becomes a disperse flow,

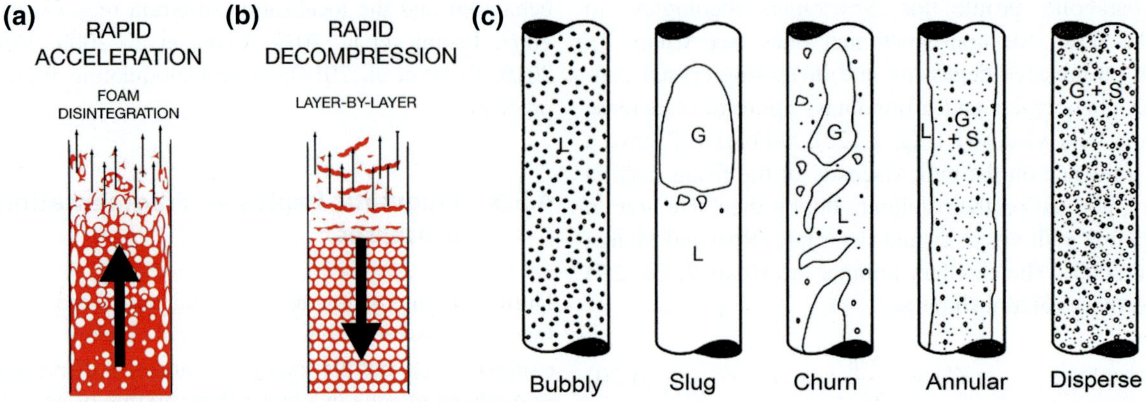

Fig. 4.11 **a**, **b** End-member models of magma fragmentation. **a** Rapid upward acceleration of magma operates from bottom-up and promotes fast decompression, bubble growth and high strain rates that drive melt to cross its glass transition and to fail either through inertial processes or in a brittle manner, representing dynamic fragmentation. **b** Instantaneous decompression propagates from top-down and induces brittle fragmentation of vesicular magma that is unable to relax as the decompression wave induces the sudden expansion of gas inside the bubbles, representing static fragmentation. (Modified from Cashman and Scheu 2015). **c** Distinction between bubbly, slug, churn, annular, and disperse or mist flow in vesiculated magma rising in a conduit. L = magma liquid, G = gas, S = solids suspended in gas. (From Brennen 2005)

or particulate flow of gas with suspended pyroclasts (Fig. 4.11c). In contrast, low-viscosity basaltic magmas allow the growth and coalescence of bubbles (bubbly flow) and their behaviour becomes increasingly decoupled, exhibiting slug flow with slugs of gas in the magma to churn flow, to annular flow of magma around the conduit wall with a core of gas-driven disperse flow to ultimately fully disperse or particulate flow regime in the upper part of the conduit (Fig. 4.11c). These transitions depend on the relative momentum of the liquid (magma) versus the gas phase (e.g. Jaupart 2000; Brennen 2005; Mudde 2005). After pervasive explosive fragmentation of the magma in the conduit, the pyroclastic jet consists of a jet of gas with dispersed magma fragments, called disperse of mist flow (Fig. 4.11c; Brennen 2005).

The magma disrupts or fragments explosively at its free surface in the vent, because it is there that a high-pressure gradient exists between the vesiculating magma and the atmosphere. The sequence is shown schematically in Fig. 4.7a. Gas pressures at the fragmentation surface are controlled by the proportion of gas exsolved and their type. According to Sparks (1978) and Wilson et al. (1980), magma fragments when the void fraction in the magma is approximately 0.77 (based on examination of modern pumice fragments), a value which has been later assumed by many more complex models, such as those by Dobran (1992), Papale et al. (1998), Mastin (2002). However, Thomas and Sparks (1992) propose that 60% is a more realistic vesiculation level at the point of explosive failure, and that some post-fragmentation exsolution and bubble growth can occur in pyroclasts leading to the higher levels of vesiculation preserved in pyroclasts. Cashman and Scheu (2015) suggest that fragmentation could occur at lower vesicularity levels in impermeable magmas, which leads to enhanced bubble gas pressures. This is confirmed by some studies (e.g. Valdivia et al. 2022, average vesicularity of mafic pyroclasts ~54 vol%; Marshall et al. 2022, mafic pyroclasts ~52 vol%). As the vesiculating and fragmenting magma accelerates upwards through the conduit to speeds ranging from subsonic to supersonic, the shear stress produced at the high velocities and, most importantly, the rate at which this stress and the strain rate increase may be sufficient to overcome the short-term tensile strength of the magma in the bubble walls, leading to disruption (Bennett 1974; Wilson et al. 1980; Dingwell 1996; Mader et al. 1996; Papale 1999; Valdivia et al. 2022). The sequence shown in Fig. 4.7a implies that the fragmentation, surface, F, should migrate downwards at a velocity of several tens of m s^{-1} (Spieler et al. 2004a, b). However, counteracting this is the rise velocity of the magma, resulting from volume increases associated with continued exsolution and expansion of volatiles at depth in a conduit of limited volume, and by continued supply of magma from deeper levels (Fig. 4.11a).

4.5.5.2 Fragmentation by Rapid Decompression

It is presently recognised that not only the void fraction developed during ascent determines the condition for fragmentation but that a complex interplay exists between fast decompression (e.g. Alidibirov and Dingwell 1996), strain rate (e.g. Dingwell 1996; Mader et al. 1996; Papale 1999), vesicle size distribution and connectivity (e.g. Klug and Cashman 1996; Marti et al. 1999; Rust and Manga 2002; Spieler et al. 2004a, b; Mueller et al. 2008; Wright and Weinberg 2009), crystal size distribution of both phenocysts and syn-eruptive microlites (e.g. Mangan 1990; Klug and Cashman 1994; Cashman and Blundy 2000; Costa and Dungan 2005; Caricchi et al. 2007) in determining the rheology of magma (e.g. Dingwell and Webb 1999; Vona et al. 2016), and transient conditions for fragmentation.

The fragmentation by the rapid decompression end-member defines the role of the rapid development of pressure gradients that propagate inside a magma body, by the sequential inward disruption and removal of layers, which is able to progressively decompress the interior of the magma body (Fig. 4.11b). Spieler et al. (2004a, b) demonstrated experimentally that the differential pressure threshold required for fragmentation, ΔP_{fr} depends on the overpressure in the bubbles, which can originate, in nature, for example, by the flank failure of a dome, and on porosity which lowers the threshold. This means that vesiculation is a pre-requisite for efficient fragmentation in rapid decompressional events.

Connected porosity, by contrast, favours degassing at faster velocities (300 m s^{-1}), and the downward propagation of decompression waves (tens of m s^{-1}). Efficient degassing may induce the transition to an effusive phase of an eruption or its cessation.

4.5.5.3 Mechanisms of Magma Fragmentation During Explosive Eruptions of Normal Silicate Magmas

As magma vesiculates and bubbles expand, the walls of the bubbles expand and the bubble walls stretch in a ductile manner at low strain rates if the gas overpressure, ΔP, exceeds the yield strength of magma. The yield strength of magma is usually only a few pascals and is less than the tensile strength of the magma, which is usually <5–6 MPa. The walls of the bubbles can then fail through plastic stretching and thinning at low strain rates to the point where the bubble walls run out of matter (Fig. 4.12a). This happens not only in low-viscosity magmas, such as basalts, which have very short melt relaxation times, but can also happen in higher viscosity magmas at low strain rates (Gonnermann 2015).

At higher gas overpressures, the bubble growth rate and the strain rate may be high enough to push the magma through the glass transition, so that the tensile strength of the magma in the bubble walls is exceeded and the bubble walls break by brittle failure (Gonnermann 2015; Fig. 4.12b). Viscous magmas have longer melt relaxation times, and are particularly prone to brittle failure at high gas overpressures and strain rates. However, even low-viscosity mafic magmas can experience brittle fragmentation during highly explosive eruptions. Moitra et al. (2018), Arzilli et al. (2019a), Valdivia et al. (2022), and Marshall et al. (2022) proposed that if mafic magma ascends quickly through the conduit, it experiences rapid undercooling, which initiates rapid groundmass microlite growth, and an order of magnitude increase in viscosity (to 10^5 Pa s minimum; Arzilli et al. 2019a). That leads to a gas overpressure in bubbles because of the increased tensile strength of the magma, which begins to behave as a visco-elastic material. Mueller et al. (2008) have calculated the gas overpressure, ΔP, required to cause explosive fragmentation as a function of the porosity (vesicularity) and permeability of the magma. They propose that where vesicularity is 80 vol% ΔP is only a few MPa, similar to the tensile strength of ductile magma. However, as vesicularity decreases ΔP to 40 vol%, ΔP rises to 10 MPa or more, consistent with an increasing magma viscosity related to groundmass microlite crystallisation.

Glass shards with sharp very thin pointy ends have formed through ductile bubble wall thinning (Fig. 4.12a), as

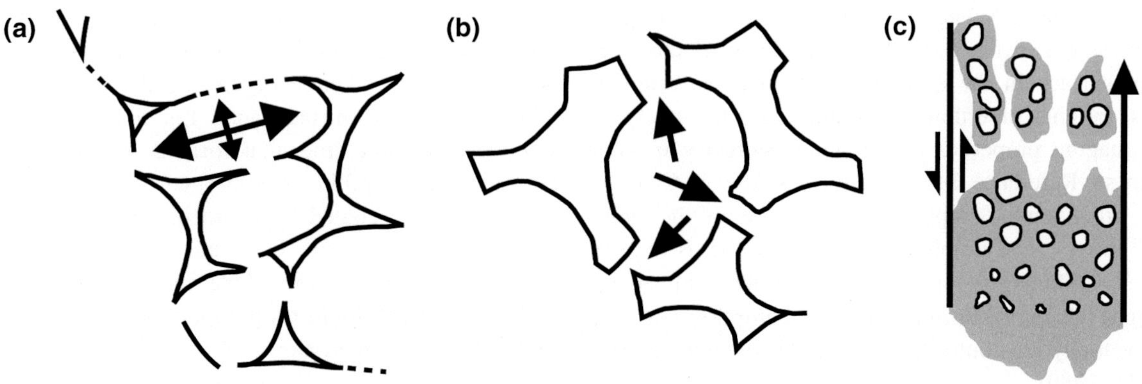

Fig. 4.12 Fragmentation mechanisms for vesiculating magma. **a** Ductile stretching and bubble wall thinning as the bubbles expand at relatively low strain rates, and with gas overpressures above the yield strength of magma, and below the tensile strength of the magma. **b** Brittle bubble wall bursting at high strain rates due to gas overpressures being above magma tensile strength, which is usually below 5–6 MPa. **c** Shearing of magma at the contact with the conduit walls, leading to formation of large vesiculating magma fragments that solidify into pumice or scoria pyroclasts

have Pele's hairs and tears, whereas shards with blocky ends (Fig. 4.12b) have failed through brittle fracturing. The manner in which bubble walls fail is thus strain rate dependent as well as being dependent on magma properties, especially viscosity.

In addition, large pumice and scoria clasts could form in the above ways in areas of permeable foam related to behaviour of the walls of large gas vughs in the magma during ascent, or due to shearing effects within a vertically accelerating frothy magma, especially near the conduit walls (Gonnermann and Manga 2003; Gonnermann 2015; Fig. 4.12c). Shear coupling between the conduit walls and the magma establishes a shear gradient in the magma that causes it to shear apart, liberating clots of vesiculated magma in a gas stream that produces large pumice and scoria clasts (Klug and Cashman 1996; Papale 1999; Palladino et al. 2008; Gonnermann 2015). Such large-scale fracturing increases the permeability of the degassing, decompressing magma, freeing hot expanding gas that then drives the solid pyroclasts upward and out of the vent. Stretched or tube vesicles are evidence of the shearing of vesiculated magma in the conduit and development of domains of considerable permeability. As a result, the final total grain size distribution during explosive eruptions is reflective also of the permeability of the foam both before and at the time of fragmentation (e.g. Rust and Cashman 2011).

Many other aspects of explosive magmatic eruptions could be discussed, given space. Many of these have been modelled by increasingly complex thermo-dynamic numerical simulators, which include the solution of the momentum energy and mass transport for multi-phase magmatic systems, in near-real computational domains made of magma chamber-conduit-vent-atmosphere in separate or continuous environments (Wilson et al. 1980; Wilson and Head 1981; Vergniolle and Jaupart 1986; Dobran 1992; Woods and Koyaguchi 1994; Papale et al. 1998; Legros and Kelfoun 2000; Mastin 2002; Sahagian et al. 2005). These models allow assessment of the controls on the depth of the fragmentation surface in the conduit, eruption rates, eruption velocities, heights of eruption columns, and the interplay between magma properties and conduit geometry in controlling these. Reference is made to these studies in Chap. 9. By way of example, eruption rates can be in the order of $\times\ 10^6\ m^3\ s^{-1}$, and eruption velocities can be as high as $600\ m\,s^{-1}$, producing eruption columns up to 50 km maximum height (Chap. 9). Eruption intensity, eruption rate, and eruption velocity are controlled by volatile contents (Wilson et al. 1978, 1980).

4.5.5.4 Mechanisms of Magma Fragmentation During Explosive Eruptions of Ultra-Low-Viscosity Magmas (Some Basalts, Kimberlites, Carbonatites)

The deposits of many ultra-low-viscosity magmas contain fluidal pyroclasts that range in diameter from a few mm to a few cm (Fig. 4.13a). These include pele's tears, achneliths and pele's hairs in Hawai'ian magma or fire fountain deposits (e.g. Macdonald 1972; Jones et al. 2019; see Fig. 4.20a), *pelletal lapilli* in kimberlite and lamproite deposits (Fig. 4.13; Mitchell 1997; Chap. 18), and spherical clasts in carbonatite deposits (Hay 1989). The viscosities of magmas considered here are as low as 1 Pa s (kimberlite, carbonatite) to 100 Pa s (alkaline and under-saturated basalts, lamproites), compared with 10^5 to 10^9 Pa s for calc-alkaline magmas such as andesites to rhyolites (Chap. 2). For low viscosity melts (mafic, ultramafic, carbonatitic), Jones et al. (2019) proposed that fragmentation may also occur due to extension of the melt, which they call fluid induced break-up. At short time scales (rapid ascent) basaltic melts and even spatter like pyroclasts can undergo viscous instabilities during extensional deformation or stretching as they emerge from vents in magma or fire fountains and jets, whereas very low viscosity melts such as carbonatites, kimberlites and some fire fountain basaltic magmas fail through capilliary failure (Jones et al. 2019). These processes can produce spatter, acneliths, pelletal lapilli and pele's hairs.

The spheres form as a result of the fragmentation of a fountaining low-viscosity magma, probably by shear and tensile stretching of the magma in the fountain (Chap. 9), producing low-viscosity droplets. The spherical form is interpreted to be the result of centrifugal spinning and surface tension effects in the low-viscosity magma, as also suggested by Scott-Smith (1996), Cas et al. (2008), Gernon et al. (2012) and Jones et al. (2022a) for kimberlite pelletal lapilli (Fig. 4.13b). Surface tension is due to the inter-molecular forces in a liquid, which in a free liquid droplet has the effect of organising the liquid into a form with the smallest surface area. For a liquid with a short viscous relaxation time, that form is a sphere. Most kimberlitic pelletal lapilli consist of a core grain (olivine crystal, lithic fragment) coated by a rim or layer of coherent kimberlite, so that the rim layer has been subject to surface tension forces, giving the whole clast a spherical form.

In a spherical liquid droplet, the surface tension is given by the Laplace equation:

$$\wp = \frac{R(P_i - P_o)}{2} \tag{4.22}$$

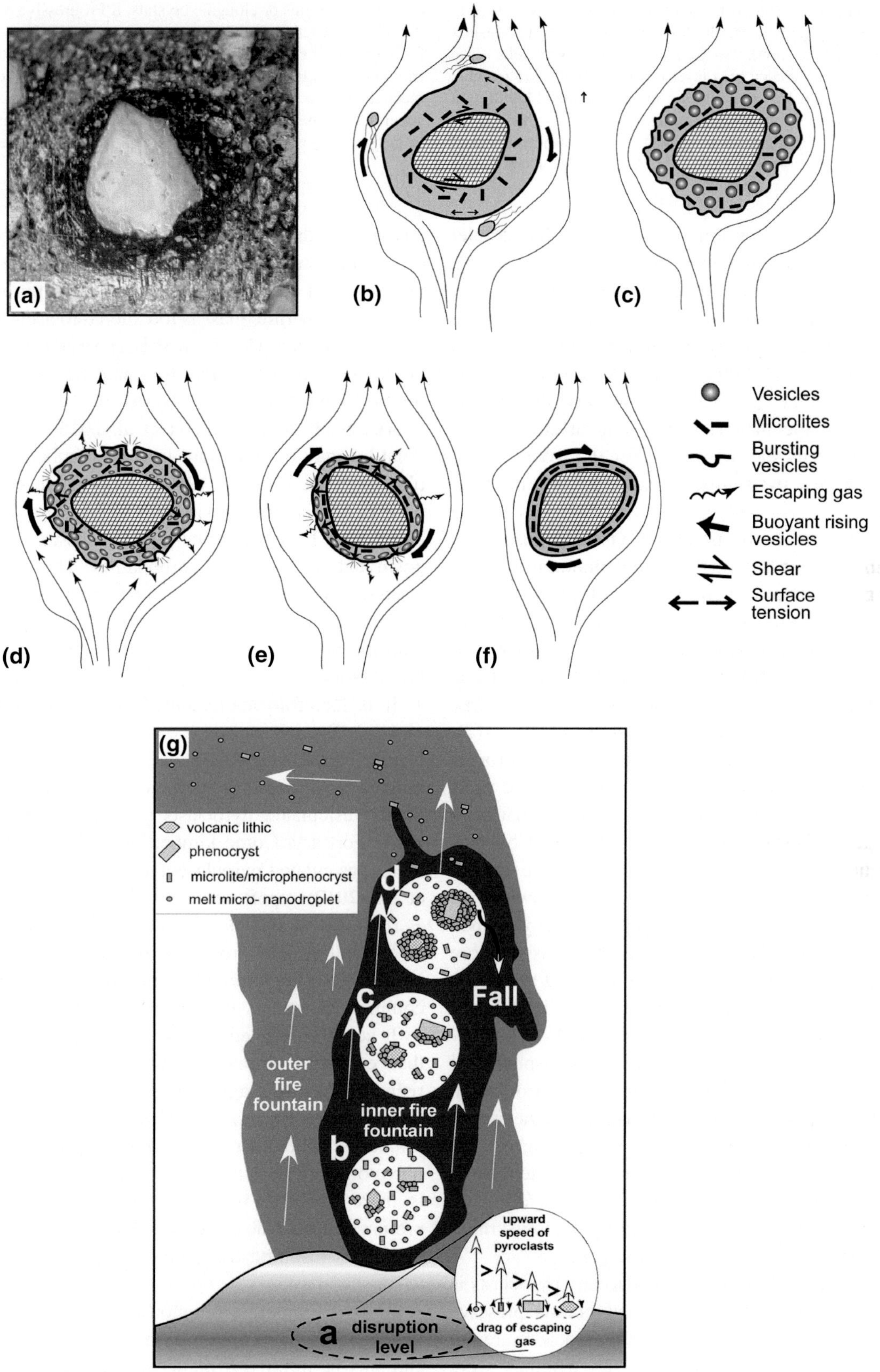
(a)
(b)
(c)
(d)
(e)
(f)
Vesicles
Microlites
Bursting vesicles
Escaping gas
Buoyant rising vesicles
Shear
Surface tension
(g)
volcanic lithic
phenocryst
microlite/microphenocryst
melt micro- nanodroplet
d
c
b
Fall
outer fire fountain
inner fire fountain
upward speed of pyroclasts
drag of escaping gas
a disruption level

◀ **Fig. 4.13** Formation of spherical pyroclasts in ultra-low-viscosity pyroclasts. **a** Spherical pelletal lapillus in kimberlite, Jurassic Muskox kimberlite pipe, Nunavut, northwest Canada, showing a nucleus of an olivine crystal fragment, coated by a layer or rim of kimberlite magma. The core is ~5 cm wide. **b** Formation of a spherical pelletal lapilli with a coherent magma rim in a gas–magma fragment fountain through the combined effects of centrifugal spinning, and surface tension. **c** Vesiculation in a low-viscosity pelletal lapilli rim. Also shown are randomly oriented microlites or elongate crystals. **d** Progressive boiling off of gas bubbles. **e** Deflation of the magma rim as a result of loss of gas bubbles. **f** Alignment of elongate crystals as a result of deflation/flattening of the rim, and perhaps shear drag between the rim and the core. **g** A second mechanism for forming pelletal lapilli by agglutination of hot glass spherules, formed as in (**a**) to large nucleus fragments, and to each other in gas–magma clast fountains (From Carracedo-Sánchez et al. 2015)

Where $\wp$ is the surface tension (N m^{-1}), R is the droplet radius (m), P_i is the pressure inside the droplet (Pa), and P_o is the pressure on the surface of the droplet (Pa).

The density of the liquid plays a role since the pressure inside a droplet is affected by density, and viscosity also contributes (Kaully et al. 2007). The pyroclasts of viscous magmas do not assume spherical forms because the viscous relaxation time of the magma is too long at atmospheric pressure, and exceeds the time taken for the pyroclasts to quench to solid pyroclasts, resulting in irregular morphologies.

If a newly formed pyroclast of low-viscosity magma is spinning in the conduit flow and eruption column of gas and pyroclasts, smaller droplets may be flung off from the margins of the larger pyroclasts like satellites, so producing small fine ash-sized spherical droplets in a gaseous fluid (Fig. 4.13b; Cas et al. 2008; Jones et al. 2022a called this process magma stripping) that is akin to an aerosol spray (Scott-Smith 1996; Cas et al. 2008; Gernon et al. 2012), leaving behind a larger pyroclast that surface tension forces are pulling into a spherical form. Unfortunately, the morphology of small juvenile fine ash particles in the matrix of fragmental kimberlites is almost always obscured by pervasive alteration, especially by serpentine, but we predict that spherical forms should be common (e.g. Brown et al. 2008; Chap. 18), compared with cuspate to blocky shaped glass shards that form during explosive eruptions or more viscous magmas. However, Carracedo Sánchez et al. (2015) have imaged the micro-texture of spherical pelletal lapilli from under-saturated basaltic deposits (see below), recognising that the rims consist of aggregated microspheres of glass, supporting the above scenario. Junqueira-Brod et al. (1999) ascribed the spherical form of juvenile pyroclasts in kamafugitic deposits in Brazil mostly to spinning effects, and although this is a contributing factor, surface tension is likely to be the principal force. We will focus the following discussion on kimberlite pyroclasts, and by analogy, rheologically similar magmas and pyroclasts.

Another intriguing issue about low-viscosity pyroclasts is that relatively few juvenile kimberlite pyroclasts preserve vesicles, and highly vesiculated kimberlite pyroclasts are not recorded. Does this mean that kimberlite magmas were volatile-poor and their fragmentation was not driven by explosive vesiculation of magmatic volatiles, or have most gas bubbles that formed been able to escape from pyroclasts of magma once formed because of the ultra-low magma viscosity and before they quench to solid pyroclasts?

Most kimberlitic glass is too altered to accurately measure volatile contents. However, it is generally agreed that kimberlite magmas are volatile-rich with carbon dioxide content up to 20 wt% and water up to 10 wt% (Mitchell 1986; Sparks et al. 2006; Willcox et al. 2015; Chap. 18). Other volatiles may be important in some kimberlites. For example, Kamenetsky et al. (2007) report high chlorine (up to 6 wt%) and carbon dioxide (up to 14 wt%) contents in the Udachnaya East kimberlite pipe in Russia, which has compositional similarities to carbonatites. The presence of phlogopite and carbonate in many kimberlites, and pervasive serpentine in all kimberlites (Mitchell 1986) is consistent with the presence of considerable magmatic water and carbon dioxide.

It is therefore anomalous that few pyroclasts preserve vesicles. Even if carbon dioxide exsolved early in the rise history of the magma because it is less soluble in silicate magma than water, pyroclasts should still be vesiculated. Some vesiculated pyroclasts are reported from the Cretaceous Fort à la Corne kimberlite field in Canada, but even there, vesiculated pyroclasts are far from ubiquitous. Brown et al. (2012) report some vesiculated pyroclasts from the ~10 ka Igwisi Hills kimberlite cones in Tanzania, with 10–20% small vesicles from 0.02 to 3 mm in diameter. Pelletal lapilli with a fine-grained coating of coherent kimberlite are common (Brown et al. 2012; Chap. 18).

It could be argued that the absence of vesicles suggests that explosive phreatomagmatic interaction between the rising kimberlite magma and external water short-circuited the vesiculation process. However, for the majority of kimberlite pipes there is no other supporting evidence for phreatomagmatic activity. In fact, phreatomagmatic activity is more likely to produce pyroclasts with some vesicles preserved due to the early quenching of magmaclasts to solid pyroclasts resulting from the cooling effects of the external water.

Since a kimberlite eruption has not been witnessed, our understanding of eruption style and fragmentation processes can only be inferred from deposit characteristics. We do not really know in what state of vesiculation the magma is in at the time of fragmentation, when pyroclasts are produced.

The high crystal content of many kimberlites represents ideal conditions for heterogeneous gas bubble nucleation to occur, especially in a magma with high-volatile contents. Given that kimberlite magma is likely to have had a very low viscosity in the order of 1 Pa s (Sparks et al. 2006; Chap. 18), vesiculation, once initiated, is likely to have occurred quickly and bubble growth is also likely to have been rapid, given that viscous resistance would be low. In addition, gas bubbles will be very buoyant in very low-viscosity, dense magma.

The rate at which gas bubbles can rise buoyantly through a liquid can be calculated using the Stokes Settling Velocity Law, commonly used to determine the settling velocity of dense sediment particles in low-density and low-viscosity water. In this case, we will use the Stokes Law to calculate the rate at which low-density gas bubbles rise (i.e. have negative settling velocity) in a dense, but low-viscosity magma. The Stokes Law can be represented by

$$-V_t = \frac{g\,d^2\left(\rho_g - \rho_m\right)}{18\eta} \qquad (4.23)$$

where $-V_t$ is the gas bubble rise velocity in kimberlite magma (m s^{-1}), g is the acceleration due to gravity (9.8 m s^{-2}), d is the diameter of a gas bubble (m), ρ_g is the density of the gas bubble (100 kg m^{-3}), ρ_m is the density of kimberlite magma (2900 kg m^{-3}; Sparks et al. 2006), and η is the viscosity of kimberlite magma (1 Pa s; Sparks et al. 2006).

We have calculated the gas bubble rise velocity relative to magma rise velocity (Fig. 4.14a). The rise velocity will vary according to bubble size. For example, a 1 cm bubble will rise at 15.24 cm s^{-1}, a 1 mm bubble will rise at 0.15 cm s^{-1}, and a 0.1 mm bubble will rise at 0.0015 cm s^{-1}. At relatively low magma rise velocity, gas bubbles can escape from a rising magma column. However, at high magma rise velocity, gas bubbles remain coupled with the rising magma, indicating that in some cases juvenile pyroclasts could therefore be vesicular at the point of explosive fragmentation. And of course, magmas may rise in stages undergoing several velocity changes related to stalling and deep degassing.

The time that it takes for gas bubbles of different sizes to escape from the magma rinds of pyroclasts (Fig. 4.13c, d), assuming their maximum escape distance is from the inner margin of the rim of the pyroclast can be calculated (Fig. 4.14b). Not surprisingly, in such low-viscosity magma, it only takes a fraction of a second for large bubbles (e.g. 1 cm, 1 mm diameter bubbles) to escape from most pyroclasts. Small bubbles (e.g. 0.1 mm) by contrast could take many seconds to escape from the rim of a pyroclast (Fig. 4.14b) which suggests there should be many (micro) vesicular juvenile pyroclasts. These escape times, however, assume that once formed a gas bubble will remain the same size, which in an ultra-low-viscosity magma, like kimberlite magma, is unlikely. Gas bubbles will expand rapidly once formed, and, in addition, will also coalesce and increase in size through this, so causing accelerated rise/escape velocity rates. The calculated times represented in Fig. 4.14b are therefore minimum escape times. It is therefore not surprising with these short escape times that very few kimberlite juvenile pyroclasts are vesicular. Through the process of buoyant gas bubble escape and bubble collapse in low-viscosity kimberlite magma in the rinds of pelletal lapilli, the magma rind must deflate to form a vesicle-poor or vehicle-free coherent rind around the core grain (Fig. 4.13e, f). The only other factor that could hinder gas bubble escape is the surface tension of the juvenile magma droplets.

A concentric fabric of concentrically aligned elongate crystals and microlites of olivine is not uncommon in the rim of pelletal lapilli. In the proposed escape of gas bubbles from the magma rind outlined above, this must lead to deflation of that magma rind from a thick, vesicular, expanded layer, to a thinner deflated non-vesicular layer (Fig. 4.13d–f). During that process, the orientation of the axis of any elongate crystals in the rind will become more parallel to the margin of the core grain and the outer rim of the pelletal lapilli as the magma rind deflates (Fig. 4.13e, f), unless the crystal axis is exactly orthogonal to the core grain margin. It is also possible that if the pyroclasts are spinning, then a differential rotational torque will develop between the solid core grain and the magma rind. This may impose a simple shear on the magma rind, causing it to shear or flow relative to the solid core grain, thus causing elongate crystals to align along shear or flow lines.

Gernon et al. (2012) have proposed that some pelletal lapilli form by aggregation of largely glassy microspherules and crystal fragments in a gas-magma fountain but did not provide evidence. However, Carracedo Sánchez et al. (2015) imaged such micro-spherule components in the rim of pelletal lapilli from alkaline basaltic deposits from the Cabezo Segura volcano, Calatrava Volcanic Field, south-central Spain (Fig. 4.13g). They propose agglutination of such microspheres to larger core grains and to each other in the gas-magma fountain (Fig. 4.13g). Their images and model are convincing for the pelletal lapilli described, but we see no reason why both models could not apply in specific cases, and even within the same eruption system, with pelletal lapilli with coherent magma rims forming low in a fountain and pelletal lapilli with clastic rims forming in the more gas-rich part of a fountain. Magma rheology will play a large role in determining if one or other or both mechanisms of formation apply, and there is clearly a need for further study of these types of processes and deposits.

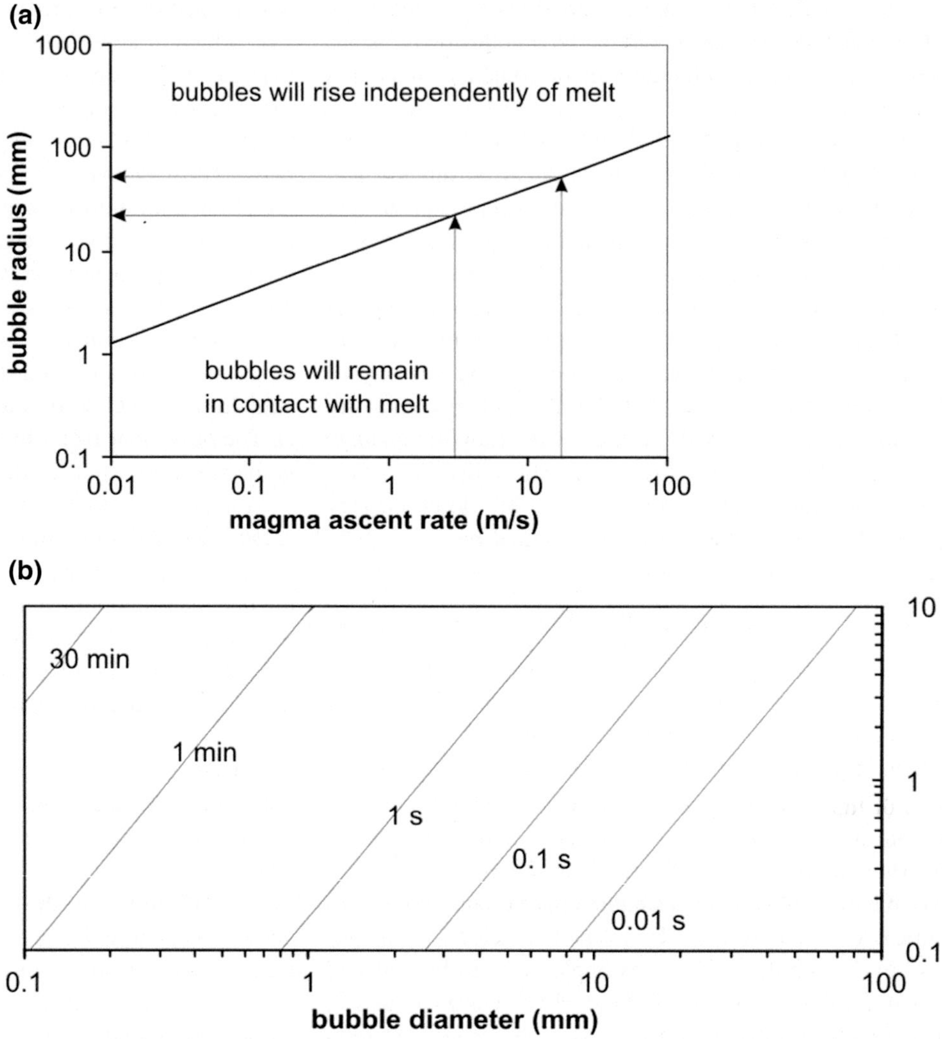

Fig. 4.14 Escape time of gas bubbles from magma. **a** In a rising column of kimberlite magma. **b** Of gas bubbles of different sizes from the magma rim of kimberlite pelletal lapilli

For a discussion of kimberlite and lamproite magmatism in relation to their diamondiferous pipes and volcanic host rock facies, the reader is referred to Chap. 18 (Sect. 18.7), Cas et al. (2008), and Sparks (2013).

4.5.6 Magmatic Explosions from Lava Domes and Flows

Explosive eruptions can also occur from lavas or very high-level intrusive cryptodomes because internally they continue to undergo volatile exsolution after erupting effusively or intrusion (e.g. Santiaguito volcano, Guatemala; Mt St Helens volcano, USA; Soufrière Hills volcano, Montserrat; Merapi volcano, Indonesia; and Colima volcano, Mexico). The effusive eruption of magma at atmospheric pressure indicates that the volatile content of the magma was too low in the conduit to trigger explosive vesiculation of the magma in the conduit at the time of lava eruption. As long as volatile vapour pressure remains less than the tensile strength of the magma explosions will be suppressed, although slow steady vesiculation may occur, even producing highly vesicular pumice domains in silicic lavas (Chap. 6) and reticulite in basaltic lavas (Chap. 5).

However, explosions can occur if the gas pressure in vesicles in the lava increases to the point where it exceeds the tensile strength of the magma. There is actually relatively little data on the tensile strength of vesiculating magma. Carey and Sigurdsson (1985) suggest a tensile strength of a few MPa for the 18 May 1980, Mt St Helens plinian dacitic magma, which had a viscosity of about 10^6 Pa s. Sparks et al. (1994) suggest a tensile strength of perhaps up to 10 MPa for lava with viscosity of 10^9 Pa s, the limiting quench viscosity of Thomas et al. (1994) for rhyolitic magma. Most researchers now agree that a tensile strength of <5 to 6 MPa is an appropriate value for the tensile strength of most magmas. By contrast Webb and Dingwell (1990), Griffiths and Fink (1993), and Mungall et al. (1996) suggest a range from about 1 to 2×10^8 Pa (= 100–200 MPa) for solidified silicic dome lava at the glass transition temperature. This greatly exceeds, by an order of magnitude, the gas pressures at fragmentation calculated by Thomas et al. (1994) for some magmas of the most explosive plinian eruptions known.

Mungall et al. (1996) point out that there is an obvious contradiction in such domes still being able to undergo explosive fragmentation, but they do (e.g. Santiaguito dome, Guatemala, Rose 1973; Mt St Helens 1980, 2004 domes; Soufrière Hills lava dome, Woods et al. 2002). Mungall et al. (1996) propose that as vesiculated glassy lava cools the gas pressure in the vesicles drops rapidly. Continued but slowing diffusion of water from the glass in the immediate vesicle walls into the vesicles in response to gas pressure drop leads to dehydration of the glass, and shrinkage. The resulting tangential stresses in the bubble walls due to shrinkage may exceed the weakened tensile strength of the magma in the bubble walls, so forming many micro-fractures (Romano et al. 1996). Such cooling-induced micro-fracture weakening of the vesiculated glassy lava may lower the tensile strength, especially where the shapes of vesicles are irregular (e.g. stretched vesicles, coalesced vesicles, vesicles whose shapes are affected by crystals in the magma). Gas overpressure may then be enough to cause explosive disintegration. The explosive fragmentation in this instance is clearly by brittle failure of the glass. Mungall et al. (1996) suggest that rhyolitic lavas with between 0.4 and 1.85 wt% H_2O may undergo this type of cooling micro-fracturing, which will make them susceptible to explosive fragmentation at gas pressures as low as 1.5 MPa.

This fragmentation process is distinctive from, but could occur in concert with, the process proposed by Fink and Manley (1989) and Fink et al. (1992) for the onset of explosive fragmentation in rhyolitic lavas distal to the vent. They point out that the relatively slow cooling of the interior of very thick rhyolites, perhaps hundreds of m thick, may take many tens of years, and may allow crystallisation. This concentrates the volatile fraction in the residual groundmass, and may cause vesiculation (second boiling), leading to vesicles as much as 1 mm in diameter. Simultaneously, the surface of the lava may be vesiculating, forming a fragile pumiceous carapace with greater than 50% vesicles and many m thick (e.g. the 1980 Mt St Helens dome; Fink et al. 1992). Upward migration of exsolving volatiles may result in gas-rich zones beneath the fragile pumiceous carapace, with gas pressures greater than the tensile strength of the vesiculated carapace, so leading to local explosive outbursts that may even generate extra-vent, distal pyroclastic flows (Fink et al. 1992).

4.5.7 Magmatic Explosions from Sealed Subaerial Vents

There are several different ways in which explosive eruptions can be triggered from sealed conduits. When a magma chamber or conduit is sealed, build-up of pressure in the magma under the seal or crustal cap may exceed the minimum principal stress and tensile strength of the roof of the magma reservoir or the sealed cap rock (e.g. solidified lava crust or vent collapse debris) causing it to fail and fracture. This could cause sudden decompression, triggering an explosive event, which could be a short-lived vulcanian blast (e.g. Stromboli volcano, Giordano and De Astis 2021), or which could evolve into a sustained plinian-style event (e.g. Soufrière Hills volcano, Montserrat 1995–2010; Chap. 9).

Magma pressure build-up could occur at shallow levels due to steady, slow exsolution, or the continued uprise or recharge of magma into a high level of the conduit, magma chamber, or cryptodome from a deeper reservoir (e.g. 1980 Mt St Helens cryptodome; 2019 vulcanian blasts at Stromboli), or the onset of crystallisation, leading to second boiling as outlined in Sect. 4.5.2.2. As an example, for an andesite with $\sim$2.8 wt% H_2O, Burnham (1972) calculated that it will become saturated with water at about 2 km below the Earth's surface. As crystallisation and cooling proceed, the internal pressure of the magma reservoir should theoretically increase, until at the point of full crystallisation (700 °C) the internal pressure will have reached 130 MPa. This would be far in excess of the confining pressure of 50 MPa and the typical tensile strength of the country rock (15 MPa). Therefore, before this stage, fracture of the roof of the chamber will have occurred, allowing magma to rise to the surface, probably leading to explosive disruption of the magma and the volcanic edifice. Burnham (1983) suggested that the 18 May 1980 eruption of Mt St Helens may have been triggered by crystallisation-induced vesiculation.

Decompression of a magma can also occur through the gravitational collapse of an overlying volcanic edifice. It is now widely accepted that the 18 May 1980 Mt St Helens eruption was initiated by the collapse of a large sector of the edifice following a magnitude 6 earthquake, and the consequent instantaneous decompression of the main magma chamber at a depth of 7 km (Chap. 6). This allowed instantaneous vesiculation of the dacitic magma and triggered the initial blast, so leading to the sustained day-long pyroclastic eruption (e.g. Carey and Sigurdsson 1985; Chap. 9).

4.5.8 Sub-surface Magma Mingling, Mixing, and Gas Sparging in Triggering Magmatic Explosions

Because most magmas are not simple closed systems, their chemistry and eruptive history are controlled by the mixing of separately formed or separately evolved magma batches (Sparks et al. 1977; O'Hara and Matthews 1981; Huppert et al. 1982).

Evidence that explosive eruptions may be triggered by magma mingling or mixing occurs worldwide in the occurrence of mixtures of mafic and felsic pumice populations and compositionally *banded pumice* fragments in largely silicic pumice fall and ignimbrite deposits (e.g. Sparks et al. 1977; Eichelberger 1980; Druitt and Bacon 1989; Edgar et al. 2002, 2017; Carrasco-Núñez et al. 2012; Morgavi et al. 2016). Banded pumice is pumice that contains a streaky, fluidal layering of compositionally different layers, produced by the mingling of felsic and more mafic magmas. The mechanics of magma mingling and mixing are discussed by Huppert et al. (1982) and Morgavi et al. (2013).

Magma mingling and mixing can trigger explosive eruptions when two magmas at different temperatures and with different volatile contents come into contact in a subsurface magma reservoir. This could involve a higher temperature mafic magma being injected into a cooler silicic magma chamber, or even a high-temperature silicic magma being injected into a cooler more crystallised silicic magma body from a deeper reservoir as a recharge magma batch (e.g. Lindsay et al. 2001; de Silva et al. 2008; Wright et al. 2011). All these examples relate to major explosive eruptions of crystal-rich magmas by injection of higher temperature, crystal poor magmas from major caldera volcanoes in the Andean volcanic arc of South America.

Where a rhyolitic magma with a moderate volatile content is injected by a hotter (perhaps mafic) magma from below, the superheating of the lowest part of the rhyolitic magma will induce convective uprise of this magma. Also, such superheating of the rhyolite will by itself (to a certain extent) increase the volatile vapour pressure in the magma, even without decompression, and may cause further exsolution of the remaining volatiles. As magma rises, it decompresses, volatiles exsolve, and the gas pressure and the total fluid pressure in the chamber may rise sufficiently to cause explosive eruption. The rising injecting mafic magma may contain relatively high-volatile contents, which exsolve during uprise and are transferred by convection, diffusion, and mixing into a low-volatile rhyolitic magma, leading again to fluid pressure build-up and explosive eruption.

Bachmann and Bergantz (2003) have also suggested that even if magma mixing does not occur, the transfer of volatiles upward from a magma intruding into the base of an existing magma chamber, in a process called *gas sparging*, may cause the existing magma to become saturated in volatiles, leading to vesiculation, and perhaps explosive eruption. They considered that the giant Fish Canyon Tuff ignimbrite eruption from the La Garita caldera in the San Juan Mountains Volcanic Province, USA, was triggered by the migration of H_2O and CO_2 from a subjacent mafic magma source upward into the crystal-rich rhyolitic Fish Canyon magma chamber.

4.6 Subaqueous/Subglacial Explosive Eruptions and the Constraints

In Chap. 3, many of the constraints on explosive eruptions in subaqueous and subglacial settings compared with subaerial settings were outlined and will also be discussed in this section. Although there have been many qualitative studies of subaqueous explosive eruption processes and their deposits, there have been relatively few attempts to quantify the processes and the physical constraints (e.g. McBirney 1963; Kokelaar 1986; Cas and Wright 1987; Cas 1992; Wohletz 1986, 2003; Wohletz et al. 2013; Head and Wilson 2003; White et al. 2003; Zimanowski and Büttner 2003; Downey and Lentz 2006; Cas and Giordano 2014; Cas and Simmons 2018). The approaches used either assess the processes from the point of view of magmatic gas exsolution and explosive behaviour under conditions of increased ambient pressure due to hydrostatic pressure, or from the phreatomagmatic explosive magma–water interaction point of view. It is highly likely that any subaqueous explosive eruption will involve some, if not major, phreatomagmatic explosive activity, which is discussed in Sect. 4.6.9.

To understand the role of magmatic volatiles in subaqueous/subglacial settings, the constraints on magmatic volatile exsolution as drivers for subaqueous/subglacial explosive eruptions are discussed. Although previous studies have provided enormous advances in understanding processes, there are still major deficiencies in our understanding because the variety of ways in which the water mass affects the processes have not been adequately addressed previously. In particular, apart from the paper by McBirney (1963) and the reviews of White et al. (2003), Cas and Giordano (2014), and Cas and Simmons (2018), no studies have adequately considered the properties and effects of the ambient water mass on subaqueous explosive eruptions, especially in deep water (e.g. Chap. 3). Most studies have simply adopted the principles of volatile exsolution and explosive expansion for open subaerial conduits and vents, and applied them conceptually subaqueously, even using similar eruption style names. However, in comparing the differences in the properties of the two ambient mediums (air, water), it becomes obvious that they are physically very different (Chap. 3, Table 3.3). The atmosphere has a low atmospheric pressure and pressure gradient, low density, low thermal conductivity and heat capacity, very high compressibility with very low bulk modulus, and very low energy attenuating properties. In subaqueous settings, every property just listed for air has an antithetic value for water (Table 3.3). Therefore, the physics of every subaqueous eruption, especially in deep water, will be very different to subaerial ones, or at least very attenuated. First, however, we should assess the effects of high ambient hydrostatic pressure and pressure gradient on submarine explosions.

4.6.1 Effects of High Confining Pressure on Magmas Erupting on the Seafloor in Deep Water

As discussed in Chap. 3, when magma reaches the Earth's surface on the deep-sea floor, it is in a more highly pressurised state than a magma erupting subaerially because the magma is decompressing to the hydrostatic pressure exerted by the ambient water mass at the water depth of eruption. One consequence of this is that the magma will have a lower viscosity and will erupt with a higher exit velocity subaqueously compared with subaerial settings. Magmas are therefore more prone to erupt in a fountaining style than magmas in a subaerial setting. These effects progressively decrease in shallower waters. In addition, the hydrostatic pressure and water depth affect the timing of volatile exsolution and the state of the exsolved volatiles (gas, supercritical fluid; Chap. 3).

4.6.2 Effects of High Ambient Hydrostatic/Glaciostatic Pressure on Exsolution and Vesiculation of Magmas in Subaqueous/Subglacial Environments

As discussed for subaerially erupting magmas, there is no constraint on the depth in the crust at which volatile exsolution can occur, this being dependent on the magma volatile content, the vapour pressure it exerts and the counterbalancing confining pressure, which in subaqueous and subglacial settings include a component of hydrostatic and glaciostatic pressure, in addition to lithostatic pressure in the conduit. Magmas can vesiculate to high degrees at considerable crustal depths, but because the vesicles are filled with supercritical fluids, they won't have the potential to drive explosive eruptions unless they are erupted at water depths considerably shallower than critical pressure equivalent water depths.

In subaqueous environments where a subaqueous vent is open to the water mass, the ambient hydrostatic pressure ($=P_w$) increases at a rate of ~ 1 MPa per 100 m of water depth. This is significantly higher than the atmospheric pressure regime for a subaerial vent, and leads to dramatically higher ambient pressures with increasing water depths (e.g. at 500 m water depth, P_w at the seafloor is 5 MPa (50× atmospheric), at 1,000 m water depth, 10 MPa (100× atmospheric), at 2,000 m, 20 MPa (200× atmospheric), at 3,000 m, 30 MPa (300× atmospheric), at 4,000 m, 40 MPa (400× atmospheric, and at 5,000 m, 50 MPa (500× atmospheric; Fig. 4.15). That is, magmas decompress to much higher ambient pressures when they erupt on the seafloor in deep water, relative to subaerial settings.

Since all aspects of magma volatile exsolution, bubble expansion and growth rates, and explosive fragmentation are controlled by ambient pressure (lithostatic, magmastatic, and, in this case, hydrostatic and glaciostatic as well), the high ambient pressure in deep subaqueous environments will affect all of these processes. The required (super-) saturation pressure required to trigger volatile exsolution will be the same as for subaerially erupting magmas, but the amount of decompression will be reduced by the value of the hydrostatic pressure at the eruption water depth. Magmas will therefore rise higher in a magma conduit in the submarine crust before reaching a (super-)saturated state and undergoing exsolution than they would in a conduit in subaerially exposed crust (Fig. 4.15).

Equation 4.11, which assesses the depth at which exsolution occurs in a rhyolitic magma in the crust, can be reconfigured to

$$\rho_{cr}\, g\, h_e + \rho_w\, g\, h_w = (n/s)^2 \tag{4.24}$$

where ρ_{cr} is the crustal density (kg m^{-3}), ρ_w is the density of water (kg m^{-3}), g is the acceleration due to gravity, h_e is the depth at which exsolution occurs in the crust (km), h_w is the water depth (km), n is the total weight fraction of exsolved volatile water, and s is a constant (= 0.0013, if pressure is expressed in bars).

Because the absolute decompression and therefore the decompression rates are less, the nucleation rate must also be lower than for a subaerial counterpart. Because magmas will reach the seafloor earlier in their exsolution history than they would at subaerial vents, because they decompress to higher ambient pressures, and because the erupting magmas are cooled more quickly than subaerially due to the high thermal conductivity and heat capacity of water, the process for full exsolution is short-circuited and less complete than in subaerial settings. That is, in most cases, less of the volatile content will exsolve in a deep subaqueous setting compared with a subaerial setting (Fig. 4.15). This has implications for

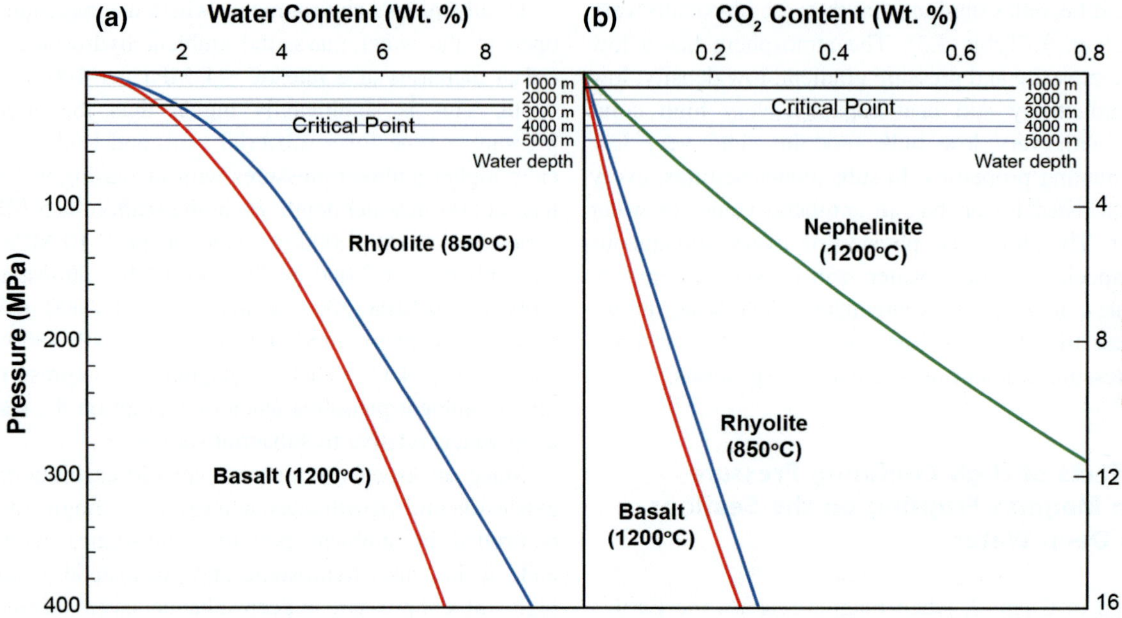

Fig. 4.15 Volatile saturation curves for **a** H_2O and **b** CO_2 as a function of confining pressure and volatile content, together with depth in the crust (right side) and superimposed horizontal isobar or pressure lines representing increasing water depths in deep-water environments. Volatile exsolution stops in subaqueous environments when magmas erupt at varying water depths and are chilled to solids. The amount of volatile exsolution is therefore less in (deep) subaqueously erupted magmas (pressure from 0.1 to 50 MPa and depth from 1 to 5,000 m) compared with subaerially erupted magmas (pressure = 0.1 MPa and 0 depth). (From Cas and Simmons 2018; Adapted from Wallace et al. 2015)

the intensity of deep subaqueous/subglacial explosive eruptions compared with subaerial counterparts.

4.6.3 Effects of Hydrostatic Pressure on the State of Exsolved Magmatic Volatiles in Deep Water

Exsolved magmatic water is essentially fresh, not highly saline. McBirney (1963) pointed out that water becomes largely incompressible at its critical point pressure and temperature, and if it exsolves as a magmatic volatile at pressures at or above the critical point, it will exsolve as supercritical fluid and be incapable of expanding rapidly and explosively, i.e. instantaneously. It can still expand, but much less than steam which drives explosive eruptions at subcritical pressures and water depths. The critical pressure of fresh water is 22 MPa (Fig. 3.4a; = 2,200 m water depth), and therefore no explosive activity should be possible at such water depths if the exsolved volatile phase is H_2O (McBirney 1963; Kokelaar 1986; Cas and Wright 1987; Cas 1992; Cas and Giordano 2014). For CO_2, the critical pressure is 7.4 MPa (Fig. 3.4b), equivalent to about 740 m water depth. In addition to ambient hydrostatic pressure, volatiles in a magma conduit that has an open vent in the subaqueous environment is also subject to the magmastatic pressure of the magma column, and in a closed conduit to the lithostatic pressure, which all contribute to constraining the depth at which magmatic volatiles exsolve from a rising magma in subaqueous environments (Chap. 3).

4.6.4 Effects of Hydrostatic Pressure on Rate of Expansion of Exsolved Volatiles and Explosive Eruptions

Once vesiculation has commenced, the hydrostatic pressure will also suppress the growth rate of vesicles relative to subaerially erupting counterparts, simply as a function of the reduced pressure differential (reduced gas overpressure) between gas pressure and the ambient hydrostatic pressure. That is, the amount of decompression and the volatile (gas) overpressure will be less by the amount of the hydrostatic pressure at the eruption water depth compared with subaerial eruptions. McBirney (1963) showed that the rate of specific volume change of water at magmatic temperatures (800 °C and higher) increased exponentially with decreasing pressure (i.e. at shallower water depths; Fig. 3.9). Supercritical and subcritical fluids (e.g. dense steam) can expand but at very much lower rates than gases.

Conversely exsolved bubble growth rates will therefore decrease with increasing hydrostatic pressures and water depth and will become progressively less explosive. Any explosions will be weak at 1,500 m and will increase in intensity rapidly from pressures of ~10 MPa (= 1,000 m water depth) and less (Chap. 3, Fig. 3.9). However, even at relatively shallow water depths (<1 km), some magmas will not erupt explosively, even though they may be highly vesicular (50–80% vesicles), due to low volatile contents, low gas overpressures, and vesicle growth rates constrained by the hydrostatic pressure (e.g. Cas 1992; Allen et al. 2010; Cas and Giordano 2014; Cas and Simmons 2018; Carey et al. 2018; Manga et al. 2018; Chap. 3).

The explosive intensity will thus decrease rapidly with increasing water depth, and if the water depths and hydrostatic pressures are great enough, all explosive activity will be suppressed in a subaqueously erupting magma that subaerially would erupt extremely violently (Cas and Wright 1987; Cas 1992; Cas and Giordano 2014; Cas and Simmons 2018).

If explosive fragmentation occurs, vesiculated pyroclasts should be slightly less vesiculated or have smaller vesicles than subaerial equivalents because they will have decompressed less and be quenched through the glass transition temperature by the ambient water mass more quickly than in air, because of the higher thermal conductivity and heat capacity of water relative to air.

Head and Wilson (2003) generally agree that increasing hydrostatic pressure should limit deep subaqueous explosive eruptions; however, they suggest that there may be some situations where deep-water magmatic explosive eruptions may occur, citing the submarine eruptions of Loihi at depths of 1.2 km (Clague et al. 2003) as an example. However, a potential problem arises from their assumption (and others also) that the volatiles are gas when they exsolve in deeper water or in the crust under deep water. At the pressures existing at water depths of >2,200 m, exsolving volatiles will be supercritical fluids, or even in a liquid state, not gas. A further problem exists in their analysis in assuming that to achieve fragmentation underwater, a vesicle volume of 75% is required. As discussed above, although magma may be fragmented by dynamic vesicle bubble expansion and shearing, very importantly subaqueously, magmas may vesiculate to high levels non-explosively, and then be fragmented by quench fragmentation (Cas 1992; Manga et al. 2018).

Quench fragmentation could produce large volumes of variably vesiculated hyaloclastite debris, even voluminous pumice debris in magmas in which vesiculation has been steady, but not explosive, without involving explosive bubble expansion and bursting (e.g. Binns 2003; Rotella et al. 2013; Manga et al. 2018). Just because vesiculation levels reach 75% doesn't mean that explosive eruptions will occur (Cas and Simmons 2018), as demonstrated by pumice carapaces in many subaerial (e.g. Fink and Manley 1987) and subaqueous lavas (e.g. Binns 2003; Allen et al. 2010; Ikegami et al. 2018; Manga et al. 2018; Fig. 3.10). And, fragmentation can also occur at vesicularities of significantly less than 75% through shearing and quench fragmentation. Subaqueous magma or fire fountains could be generated by exsolution of volatiles that increases magma column buoyancy, and magma uprise velocities, but once in the water column, the fire fountain could be fragmented largely by shearing and quench fragmentation effects as the fountain expands and flares into the water column (Cas et al. 2003).

However, Schipper and White (2010) and Schipper et al. (2013) have demonstrated unequivocally that previous theories about limu o Pele clasts (see Fig. 5.53) being formed by magmatic explosive activity are incorrect, and that they form non-explosively by entrapment of seawater into the margins (bases) of submarine lava flows (or magma erupting from vents?), expansion of water trapped in the magma bounded vughs due to heating of the water by the lava, and quench fragmentation and collapse of the vugh magma walls. We suggest that the collapse of the magma walls could also have been triggered by implosions, as the trapped water fluid (supercritical water, dense steam?) cooled, condensed, and lost pressure relative to the ambient hydrostatic pressure, which leads to cavitation of such cavities (see Sect. 4.6.5), and perhaps explains some of the folded limu o Pele clasts.

Head and Wilson (2003) also propose that at deep-water levels, CO_2 is probably more important than H_2O in the explosive disruption of basaltic magmas. CO_2 is less soluble than H_2O and usually will exsolve early, mostly in the conduit and can decouple from the magma, unless hydrostatic pressure is high. However, since the critical pressure of CO_2 is only 7.4 MPa, equivalent to 740 m water depth, even if CO_2 remains coupled with the magma, it will remain a critical fluid, incapable of expanding explosively at water depths greater than 740 m, and, in most cases, significantly shallower than this.

Dixon et al. (1997) document an H_2O content of 1.9 wt% and unusually high CO_2 content of 5.4 wt% in glass from the rinds of alkali pillow basalts at water depths of 3,900–4,300 m on the seafloor north of Hawai'i. Although vesicularity levels up to 53% were documented, no evidence of explosive volcanism was recorded, although glassy spherules imply seafloor fire fountaining occurred nearby. Such volatile contents would drive moderate explosive activity from subaerial eruptions, but subaqueously they did not, indicating that explosive activity was suppressed by high ambient hydrostatic pressure of ~40 MPa.

Head and Wilson's (2003) modelling suggest that magma or fire fountaining of such a magma is possible to depths of ~2 km, which we don't doubt given that subaqueous fire fountaining involves a jet of variably vesiculated, low-viscosity magma in which much of the fragmentation could be due to shearing of the magma fountain as it spreads or flares upon emergence from the vent, and especially due to quench fragmentation of fire fountain ejecta lofted into the water column (Cas et al. 2003; Sect. 4.6.8). We see no reason why magma fountaining can't occur at even greater water depths. This process is capable of producing abundant ash size fragments that can be dispersed around the vent by thermal plumes in the water mass and produce fallout deposits of quench fragmented ash size material (Chaps. 3 and 9).

By contrast, Downey and Lentz (2006), using the modelling program CONFLOW, suggested that explosive eruptions driven by exsolved magmatic volatiles can occur at very deep-water depths, based on an analysis of Burnham (1983) for the eruption of an explosive caldera at water depths of 3.5 km in the Hokuroko District of Japan, the type setting for Kuroko volcanic-hosted massive sulphide (VMS) deposits. There are several problems with their analysis and assumptions. First, there is no volcanological evidence of deep-water explosive calderas in the Hokuroko District (Cas 1992). Secondly, Japanese micropaleontologists have proposed significantly shallower water depths of 0.5–2.8 km (Kitazato 1979; Hasegawa et al. 1989). The so-called tuff-breccias in the footwall are mostly autoclastic, not pyroclastic, and the pumiceous deposits of pyroclastic debris in the hanging wall (Cas 1992) relate to explosive activity in the basin during the transition from extensional back-arc basin setting to shallowing, compressional arc setting (Yamada and Yoshida 2004). The fluid inclusion data from massive sulphides suggest water depths of formation of about 1,700 m for the Kuroko VMS mineral deposits. Downey and Lentz (2006) also make the mistake of assuming that exsolved volatiles will generate an explosion. Supercritical fluids can expand but not rapidly enough to generate explosions. Explosive expansion of volatiles requires instantaneous, orders of magnitude expansion. They also erroneously assume that when magma reaches a vesicularity of 68 to 83% explosive fragmentation will occur. Again, it is not the absolute level of vesicularity, but the rate of vesicle growth that determines if explosive fragmentation will occur.

There are many examples where high levels of vesiculation in subaqueously erupting magmas have not produced explosive fragmentation (e.g. Binns 2003; Rotella et al. 2013; Ikegami et al. 2018; Manga et al. 2018). So just because a body of magma has the thermal potential to drive a major explosive eruption, it can't do so unless it can translate that thermal energy into mechanical energy that drives explosive eruptions by exsolution of sufficient volatiles. But even then, those volatiles are subject to the laws of physics in terms of their phase behaviour.

The 2012 eruption of Havre volcano in the Kermadec Arc is the largest historic silicic submarine eruption recorded, which produced a >1 km^3 giant raft of buoyant pumice (Jutzeler et al. 2014; Carey et al. 2018; Manga et al. 2018), quench fragmented once the unbroken magma foam reached the seafloor at $\sim$900 m water depth. Manga et al. (2018) demonstrate that the hindered exsolution of volatiles due to the hydrostatic pressure maintained a low magma viscosity (5×10^6 Pa s, for this rhyolitic magma), hence allowing its effusive extrusion, followed by quench fragmentation. These kinds of pumice hyaloclasts reach above 80% of vesicularity at low strain rates but are unable to trigger pyroclastic fragmentation, whereas they form through quench fragmentation at the time of foam extrusion (van Otterloo et al. 2015). However, Murch et al. (2019a) propose that a phase of the 2012 Havre submarine eruption was explosive, producing a largely ash and fine pumice lapilli deposit. This is entirely possible at water depths of 900 m, and not inconsistent with other phases of the eruption being effusive, as also occurs during many subaerial eruptions.

According to Schipper et al. (2010a, b), because of the effects of high hydrostatic pressure in inhibiting deep subaqueous eruptions, the optimum conditions for deep subaqueous explosive eruptions occur when there is coupled magma-volatile bubble rise (i.e. closed system exsolution, represented by pyroclasts with isolated spherical vesicles and minor non-ruptured coalesced vesicles) from source to the seafloor. Under these conditions, there is minimal gas loss and the lack of permeability enhances volatile pressure build-up in the vesicles, as was the case in Lo'ihi seamount explosive eruptions at depths of $\sim$1,000 m (Schipper et al. 2010a, b).

The view that explosive eruptive activity is suppressed by hydrostatic pressure with increasing water depth is supported empirically by numerous studies by IC Wright on the submarine volcanoes of the Kermadec Arc in the southwest Pacific (e.g. Wright 1996; Wright and Gamble 1999; Wright et al. 2002, 2003, 2006). Wright found that for various volcanoes, and allowing for variations in magma compositions, between water depths of 450 to 1,500 m there was a transition from explosive to non-explosive volcanism, represented by a transition from pyroclastic pumice deposits in shallower water, to increasingly lava sequences in deep water. Similarly, Moore et al. (1982) noted that lavas dominate eruption products on the deep-sea floor.

4.6.5 Effects of Bulk Modulus and Shock Wave Attenuation in Water

In addition to overcoming the effects of hydrostatic pressure, bubbles in the magma that are trying to expand explosively have to overcome the incompressibility of the water body within which they are trying to explode. The bulk modulus of water, a measure of its resistance to being compressed, is approximately 2.15×10^9 MPa (Chap. 3, Table 3.3), higher for seawater at 2.34×10^9 Pa (= 2,340 MPa), and this increases with increasing pressure. The bulk modulus of water therefore has a very significant effect on suppressing subaqueous explosions, which has not been taken into account in other studies. Subaerial explosions are not affected by this factor since air at atmospheric pressure is highly compressible (bulk modulus ~0.142 MPa; Cas and Simmons 2018). Steam however has a much lower bulk modulus than water, so if a column of steam rises from a subaqueous vent, or if a subaqueous vent is occupied by a cupola of steam due to the vaporisation of seawater within a vent, as proposed by Kokelaar (1986), the constraining effects of bulk modulus are reduced and suppressed explosions may be possible, although then still constrained by the high bulk modulus of the enclosing water mass. Such a column or cupola of steam would become more compressed and denser with increasing water depth and hydrostatic pressure. Although buoyant steam bubbles should expand as they rise in a water mass and decompress, they also lose heat quickly to the surrounding water mass, which has high heat capacity. Such bubbles would be subject to condensation (erosion) very quickly in the cold ambient water mass, causing shrinkage of the bubbles. Such effects will be minimal in shallow water, but significant in deep water. In addition, water attenuates explosive or shockwave energy by 80% compared with an explosion in air, and it attenuates exponentially with distance from source (e.g. Resnyansky and Delaney 2006).

In summary, hydrostatic pressure, P_w bulk modulus, and shockwave attenuation properties of water will reduce the explosive intensity of an erupting subaqueous magma driven by exsolving magmatic volatiles, relative to a subaerial counterpart with the same properties, certainly suppressing the intensity of any magmatic-volatile-driven explosions, and perhaps totally excluding explosive fragmentation where water depths are great enough. As alluded to by Wohletz (1986) and Kokelaar (1986), if a decompression can be generated above a vent by a shock wave, then perhaps some element of explosive vesiculation of magma in a vent could occur. Possible triggers for shock waves are seismic and explosive, but the latter becomes the chicken or the egg. So, there is still much uncertainty about this topic, and plenty of room for debate! In Sect. 4.6.9, phreatomagmatic explosive eruptions are considered.

4.6.6 Implosions

As discussed in Chap. 3, Sect. 3.5.4.5, with increasing water depths and hydrostatic pressure, implosions, or cavitation of cavities in drained lava tubes, and gas cavities in lavas, and even in erupting gas bubbles, is likely to occur. Implosions are the opposite of explosions, in that debris collapses inwards and is concentrated by collapse rather than being energetically dispersed away from source, as in the case of explosions.

Implosions occur if the hydrostatic pressure exceeds the pressure in a cavity and the tensile strength of the surface lava crust. This is most likely to occur during or soon after eruption (e.g. cooling of gases in erupting pillow lava, Moore 1975, or perhaps lava draining from pillow tubes). Usually the lava crust of subaqueous lavas will be weakened by in situ, contractional quench fragmentation, which can produce visible and invisible networks of cracks, resulting in easy cavitation collapse of the weakened crusts of drained lava tubes and pillows. However, if water percolates into the cavities, the hydrostatic pressure in the cavity will balance that outside, and no collapse will occur.

Gas bubbles can also experience implosive cavitation (Rayleigh 1917; Fujikawa and Akamatsu 1980; Brennen 1995) if a negative pressure differential develops between the pressure in the gas bubbles and an external fluid. Compression of gas-filled bubbles normally leads to significant pressure and temperature increase (Brennen 1995). However, condensation can alleviate both (Fujikawa and Akamatsu 1980), and distortion of bubbles by asymmetric pressure gradients enhances bubble collapse (Brennen 1995). Although this has not been specifically addressed in the volcanic realm to our knowledge, a proposed scenario for such deep-water events is as follows. As the erupting vesiculated magma or gas bubbles of magmatic CO_2 and H_2O and superheated bubbles of meteoric water being released from deep subaqueous vents arrive at the seafloor, they decompress to the ambient hydrostatic pressure. This initially involves an expansion of the volatile phase, which is a dense vapour or even a supercritical fluid. Such fluids can expand several fold at such pressures, but growth rates are relatively slow. As they are cooled at the interface with the water body, the volatile phases rapidly cool and condense or just contract, leading to underpressured bubbles, which collapse through implosion, at the high hydrostatic pressure.

Implosions trigger an immediate shock or pressure wave and acoustic signals, and although the implosion motion is inwards, there can then be a rebound against solid surfaces leading to limited ejection of solids and disintegrating bubbles away from the implosion source, mimicking explosions, which they are not. We suggest that some deep-sea events described as explosions (e.g. sonic signals of implosions, scattered debris away from collapse cavities) could in fact be largely implosions (e.g. some gas release events from the 2009 West Mata submarine eruption, Lau Basin at 1,200 m water depth; Resing et al. 2011). Gas bubbles actually disintegrate through fission when they implode (Brennen 2002).

4.6.7 The Nature of Observed Deep-Water "Explosive" Eruptions at N.W. Rota 1, (Marianas Arc), and West Mata (Lau Basin) Volcanoes

Only two "explosive" eruptions have been witnessed on the modern deep-sea floor. The N.W. Rota 1 volcano eruptions in the Marianas arc were observed from 2004–2006 (Embley et al. 2006; Chadwick et al. 2008; Deardorff et al. 2011) at a water depth of ~540 m. The eruptions varied from effusive to pulsing explosive. They occurred from small vents up to 15 m in diameter and produced low-intensity eruption columns dominated by steam and basaltic ejecta. Although described as strombolian like in character (Chadwick et al. 2008; Deardorff et al. 2011), which is a subaerial, magmatic, discrete fountain spray style of eruption (Chap. 9), the pulsating nature of the imaged eruptions was more reminiscent of phreatomagmatic explosive activity. It is extremely difficult to envisage external sea or crustal meteoric water not being involved in a deep subaqueous explosive eruption.

The 2009 eruption of West Mata volcano in the Lau Basin, southwest Pacific, occurred from small vents (a few metres in diameter) at a depth of 1,200 m (Resing et al. 2011). The eruption fluctuated between effusive, producing pillow lava, and pulsing "explosive". Some explosive activity originated from vents, others from lava flows (Resing et al. 2011). Although some very small intensity explosive events occurred, some phenomena could have been implosions of drained lava tubes or condensed gas-filled vughs, or, of slowly expanding gas slugs emanating from vents, which were then cooled and condensed on contact with seawater, before imploding.

4.6.8 Is Deep-Water Magma ("Fire") Fountaining Necessarily Explosive?

Eruptions of magma through a fountaining style are the result of the rapid uprise rate of low-viscosity magma with a high total fluid pressure as it rises through the crust. This may be caused by lithostatic load pressure differential between the magma source at depth in the crust and vent, and closed system exsolution of magmatic volatiles as a low-viscosity magma rises through the crust, with resulting volume increase, transforming it into lower density magma with increased buoyancy, which causes it to accelerate upwards. Deep-water submarine fire fountain deposits have been found on the modern seafloor at water depths where magmatic gas explosions could not be the driver of fire fountaining (e.g. Smith and Batiza 1989; Gill et al. 1990; Davis and Clague 2003) and in the geological rock record (e.g. Simpson and McPhie 2001; Cas et al. 2003; Fujibayashi et al. 2003). They are commonly marked by variably fractured and broken, fluidal spatter bombs with chilled rims, dispersed in a finer matrix of blocky, angular glass fragments, ranging from sand to pebble size (e.g. Fig. 3.7f).

Smith and Batiza (1989) recognised the importance of deep-water magma or "fire" fountaining in documenting the characteristics of spatter deposits on Pacific Ocean seamounts at water depths of 2,000 m and greater. Gill et al. (1990) suggested that variably vesiculated basaltic agglutinate deposits discovered in water depths of about 2,000 m in the Sumisu Basin of the Izu Bonin Arc were the products of deep-marine explosive fire fountain eruptions. However, the vesicularities of about 35% they recorded are inconsistent with vesiculation and explosive fragmentation driven by exsolving magmatic volatiles, irrespective of whether or not they are water and/or carbon dioxide. At 2,000 m, growth rates and specific volume changes of H_2O in vesicles will be too low for explosive expansion to occur (e.g. McBirney 1963; Chap. 3) especially given the low water contents of basalts and the low gas pressures that would be generated. Even though CO_2 will exsolve before water because of its lower solubility, its abundance in most basalts is low, its critical point is less than that of water, and given the lower gas pressures exerted relative to magmatic water (Wilson 1980), its explosive potential is less than that of magmatic water. Exsolution and vesiculation can occur at any water depth (e.g. Moore 1965) and is simply dependent on the vapour pressure of the volatile phase in the magma being greater than the ambient hydrostatic pressure. Vesiculation does not signify explosive fragmentation, as apparently inferred by Gill et al. (1990). Many vesiculated pillow basalts can develop levels of vesiculation greater than 50% at water depths of 4 km or more, without erupting explosively (Dixon et al. 1997).

In subaerial hawai'ian-style magma fountains, the fragmentation mechanisms include both explosive expansion of gas-filled vesicles and shearing and tensile extension as the fountain expands as it decelerates, flares, and spreads out of the vent. With subaqueous (fire) fountaining explosive vesicle bursting will become less significant with increasing water depth because exsolving volatiles at great water depths are a supercritical fluid or even a liquid, and are constrained from expanding explosively by ambient conditions, including hydrostatic pressure and bulk modulus of the water mass. However, shearing and also quench fragmentation will also cause significant fragmentation of variably vesiculated magma and as it fountains and flares out of the vent into the enclosing water mass (Cas et al. 2003; Chap. 9, Sect. 9.9.4.1, Fig. 9.75). Magma spatter clasts ejected above the vent in the fountain into the overlying water will also be subject to quench fragmentation, producing abundant ash size glass fragments that are not pyroclastic in origin (Cas et al. 2003).

4.6.9 Phreatic, Phreatomagmatic and Hydrothermal Explosive Eruptions

4.6.9.1 Introduction

The interaction between hot magma and external, ambient water produces what can be collectively called *hydrovolcanic activity*, which encompasses a spectrum of processes from coherent lava and intrusion formation to autoclastic fragmentation, to phreatomagmatic, phreatic, and hydrothermal explosive activity. These can produce a range of deposit types, including a variety of volcaniclastic facies. The external water could be a substantial surface body of water or ice (e.g. ocean, lake, crater lake, beneath glacier) or a subsurface groundwater aquifer, or water that is already in a volcanic conduit, including hydrothermal systems. The interaction could be passive, producing coherent subaqueous or subglacial lavas (Chaps. 5, 6 and 7) or coherent intrusions into water-saturated unconsolidated sediments (Chap. 8) through effective, stable film boiling of water at the interface between magma and water (Sect. 4.4.1). Alternatively, the interaction could lead to quench fragmentation, without explosive activity (Sect. 4.4.3), or dynamic explosive interaction, involving explosive phreatic to phreatomagmatic activity, or hydrothermal explosions, as discussed here. In all of these, film boiling plays a fundamental role. *Phreatic* (from the Greek word *phreos*, for well, and *phreati*, for groundwater) *explosions* are steam explosions, resulting from the instantaneous explosive superheating of cold external water when suddenly heated by intruding or erupting magma, with little or no solids ejected, or just country rock debris (Montanaro et al. 2022). *Phreatomagmatic explosions* derive their explosive energy from a combination of explosively expanding magmatic volatiles and external superheated steam (Plate 4.1d) and involve ejection of both magmatic pyroclasts and accessory country rock debris, and steam. *Hydrothermal explosions* result from the build-up of pressure and development of a seal in an already high temperature, high pressure hydrothermal system (Montanaro et al. 2022). In addition, we add the common scenario of the addition of heat, usually from intruding magma, into a hydrothermal system that is already overpressurised and has a high heat content (enthalpy) driving the hydrothermal water system to boil explosively. Phreatic explosions are different from hydrothermal explosions in that the former involve cold external water, whereas the latter involves hot water with high enthalpy that is already thermally primed to explode with extra thermal or pressure input. Hydrothermal explosions therefore require very different hazard assessments and involve risks which are always present for the life of the hydrothermal system, whereas phreatic explosions can be impulsive short-lived events. Hydrothermal explosions will be discussed further in Sect. 4.6.10.6.

Hydrovolcanic explosive eruptions can be as devastating as magmatic ones, and can have major socio-economic impacts, as shown by the 2010 eruption of Eyjafjallajökull volcano in Iceland (Dellino et al. 2012; Gudmundsson et al. 2012), which caused air traffic in Europe to be grounded for almost a week (Chap. 16).

4.6.9.2 Fuel–Coolant Magma–Water Explosive Interaction (MFCI)

The essential process in explosive water–magma interaction is the instantaneous superheating of ambient water (lakes, oceans, rivers, glaciers, groundwater, permafrost) to steam and its explosive expansion, when the two come in contact, called *molten fuel–coolant interaction* (MFCI; Chap. 3). As discussed in Sect. 4.4.1 and Chap. 3, when magma comes in contact with external water, water at the interface is superheated and a thin vapour film of superheated steam forms through a process called film boiling. Although water has a high heat capacity (Chap. 3), when a steam vapour film forms heat transfer from the magma surface to the liquid water, mass across the steam film is significantly reduced because steam has a much lower thermal conductivity and heat capacity than liquid water. If the vapour film persists or remains stable for seconds or longer, slow cooling of the magma surface due to the insulating effects of the steam film allows coherent lava to form. In water depths down to several hundred metres, if the stable vapour film between the two becomes unstable and suddenly collapses due to boundary instabilities (see below), leading to direct contact between the magma surface and liquid water, this causes thermo-hydraulic detonation of the instantaneously superheated water and explosive expansion of the steam (Board et al. 1975; Buchanan and Dullforce 1973; Wohletz 1986, 2003; Cas and Wright 1987; Büttner and Zimanowski 1998; White and Houghton 2000; Zimanowski and Büttner 2003; Wohletz and McQueen 1984; Wohletz et al. 2013; Zimanowski et al. 2015).

During explosive interaction between external water and magma, the heat energy of the magma is transformed into the mechanical energy driving the explosive expansion of the superheated steam. In the nuclear energy industry, the nuclear energy source is called the fuel, and the surrounding cooling agent, usually water, is called the coolant. The role of the coolant is to reduce the heat generated by nuclear fission in the core of a reactor and to facilitate the transfer of heat to electrical power generators. This requires a very balanced process, because if the coolant water is explosively superheated and damages the reactor infrastructure, major leakage of nuclear contaminant can occur, as happened at the

Chernobyl power plant in the Ukraine in 1986, and the Fukashima power plant in Japan in 2011 as a consequence of the magnitude 9 Tohoku earthquake and tsunami.

In the case of the interaction between magma and external water, the fuel is the molten magma, and the external ambient water is the coolant. During unstable, explosive magma–water interaction, the molten fuel–coolant interaction (MFCI) can become cyclical (Board et al. 1975; Wohletz and McQueen 1984; Wohletz 1986; Morrissey et al. 2000; Zimanowski et al. 2015). As water at the interface is instantaneously superheated and expands explosively, this uses the available water at the interface, so that explosive interaction wanes temporarily for several seconds until water recharges to the interface. This is then again superheated, and generates the next explosion, a process that can be repeated cyclically over an extended period (Wohletz 1986; Morrissey et al. 2000). This cyclical, pulsing MFCI explosive style is typical of sustained phreatomagmatic explosive eruptions and is represented visually by closely spaced pulses of "cock's tail" like ballistic blasts emanating from the vent (Plate 4.1d; see Chap. 9).

Instabilities in the vapour film can be caused by movement of the vapour film relative to the magma surface, called Kelvin–Helmholtz, Raleigh–Taylor, and Landau instabilities (Wohletz 1986; Morrissey et al. 2000; Fig. 4.16a). Undulations in the melt surface can develop, some forming plumes of magma. These can detach in the vapour film as magma particles. Simultaneously, the vapour–water interface begins to oscillate, and lobes of water accelerate towards the magma surface, initiating collapse of the insulating vapour film and direct contact between liquid water and magma. Instantaneous superheating and explosive expansion (detonation) of steam at the magma–water interface is called surface contact explosivity. If the water accelerates with sufficient force towards the magma surface, penetrates it, becomes enclosed in the magma, and is then explosively superheated, this is called bulk interaction explosivity (Kokelaar 1986). Büttner and Zimanowski (1998) found that the explosive energy release is highest with bulk interaction explosivity because all the entrapped water is superheated explosively, whereas with surface contact explosivity much of the energy and heat is dissipated by the water mass.

There is common agreement that volcanic MFCI can be subdivided into four stages (Wohletz 1986; Morrissey et al. 2000; Zimanowski et al. 2015; Fig. 4.16b):

(a) *Stage 1 or pre-mix stage*: During this stage stable film boiling occurs between coherent magma and water, producing a vapour film that keeps the two separated (Fig. 4.16b, Stage 1).
(b) *Stage 2 or trigger stage*: Due to thermally induced instability (see Fig. 4.14a), or an external trigger (e.g. fracturing of host rock, seismic event), the vapour film collapses and becomes laterally discontinuous (Fig. 4.16b, Stage 2).
(c) *Stage 3 or the direct contact phase*: Direct contact between the melt and cold liquid water causes heat flux to increase by two orders of magnitude (Zimanowski et al. 2015), so that the water tries to expand, increasing pressure on the magma/glass. The melt goes through the glass transition, and the mechanical stress results in propagation of long leading cracks. These leading cracks are followed by a front of smaller fractures that result from quench fragmentation due to the thermal shock caused by the immediate melt–water contact. Leading cracks propagate at supersonic speeds (van Otterloo et al. 2015; Zimanowski et al. 2015; Fig. 4.16b, Stage 3). This represents the onset of thermo-hydraulic detonation.
(d) *Stage 4 or the expansion stage*: Superheating of the water converts it to steam which expands explosively, dispersing the glass fragments. Explosive shock waves may further fragment the magma/glass fragments, which may also be subjected to further quench fragmentation (Fig. 4.16b, Stage 4).

The intensity of the explosive activity is controlled by the water–magma mass ratio and the amount of superheating of the water (Fig. 4.17; Colgate and Sigurgeirsson 1973; Peckover et al. 1973; Sheridan and Wohletz 1981, 1983; Wohletz 1983, 1986, 2003; Wohletz and Sheridan 1983; Wohletz et al. 2013; Kokelaar 1986; Zimanowski et al.

Fig. 4.16 **a** Schematic representation of the development of instability between liquid water, vapour film, and magma in a subaqueous setting. The bottom sequence shows the transition from stable vapour film (left) to undulating magma surface (centre and enlarged at top) to detachment of magma lobes or plumes to form fragments with a surface-tension-influenced spherical morphology. (From Wohletz 1986). **b** Stages in molten fuel–coolant interaction (MFCI) n between magma and external water. *Stage 1 or Premix Stage*: Stable film boiling between magma and water. *Stage 2 or Trigger Stage*: Due to heat-transfer-induced instability (see **a**), or an external trigger the vapour film collapses. *Stage 3 or the Direct Contact Phase*: Direct contact between the melt and cold liquid water cause the water to expand instantaneously, causing shockwave-induced mechanical stress to propagate through the melt. The melt goes through the glass ▶ transition, and the mechanical stress results in long leading cracks; these leading cracks are followed by a front of smaller fractures that result from quench fragmentation due to the thermal shock caused by the immediate melt–water contact. *Stage 4 or the Expansion Stage*: Superheating of the water converts it to steam which expands explosively, dispersing the glass fragments. Explosive shock waves may further fragment the magma/glass fragments, which may be subjected to further quench fragmentation. These complex and multiple fragmentation processes cause a higher degree of fragmentation of phreatomagmatic pyroclastic deposits than magmatic explosive events, producing a much higher proportion of ash size fragments. (From van Otterloo et al. 2015)

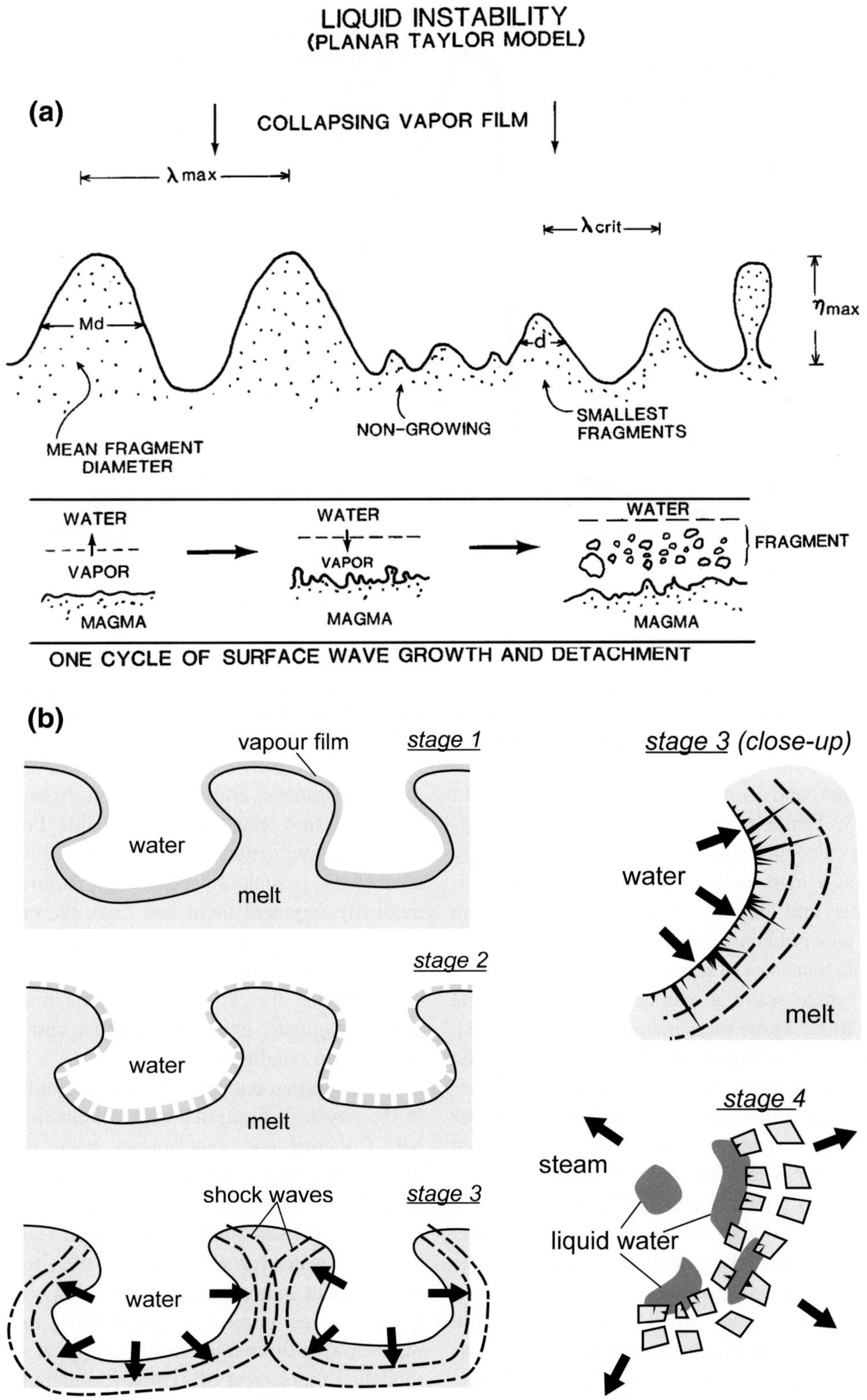
LIQUID INSTABILITY
(PLANAR TAYLOR MODEL)
(a)
COLLAPSING VAPOR FILM
λ max
λcrit
Md
d
ηmax
MEAN FRAGMENT DIAMETER
NON-GROWING
SMALLEST FRAGMENTS
WATER
VAPOR
MAGMA
WATER
VAPOR
MAGMA
WATER
FRAGMENT
MAGMA
ONE CYCLE OF SURFACE WAVE GROWTH AND DETACHMENT
(b)
vapour film
stage 1
water
melt
stage 2
water
melt
shock waves
stage 3
water
stage 3 (close-up)
water
melt
stage 4
steam
liquid water

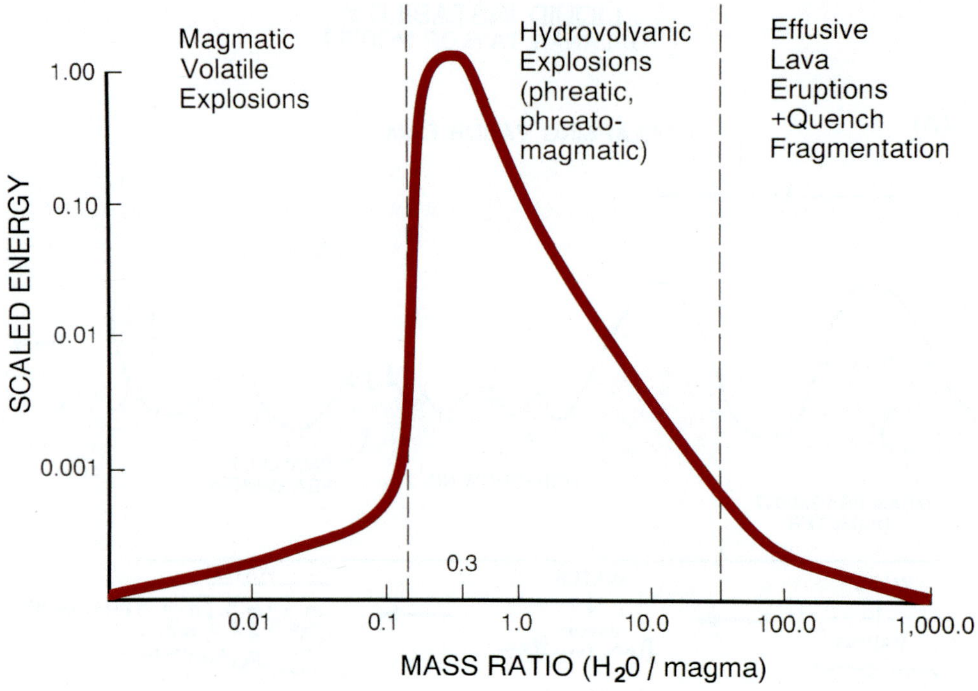

Fig. 4.17 Efficiency of transfer of thermal energy to mechanical energy resulting from the interaction of hot magma with cold water as a function of the mass ratio of magma to water, based on thermite-water experiments. The curve also shows approximate median grain size of melt fragments. The maximum on the curve corresponds to maximum efficiency of energy transfer and therefore explosiveness (water–magma mass ratio ~0.3). For ratios less than 0.1, fragmentation of magma by external water will be minimal, and either magmatic explosive activity will dominate, or subaerial lavas will form. (Adapted from Sheridan and Wohletz 1981, and Wohletz 1983)

1997, 2015; Büttner and Zimanowski 1998; Zimanowski and Büttner 2003; White et al. 2003). Colgate and Sigurgeirsson (1973) calculated that optimum explosive energy release occurred at a mass ratio of 0.25. Based on experiments using water and thermite, Sheridan and Wohletz (1981) proposed an optimum mass ratio of 0.3 (Fig. 4.17). Volumetrically this translates to about 12% water by volume to 88% magma where water is entrapped within magma (Morrissey et al. 2000). However, Zimanowski et al. (1991), and White (1996), proposed that efficient phreatomagmatic explosive activity could occur at lower water–magma mass ratios, perhaps even an order of magnitude lower. Schipper et al. (2011) evaluated experimentally the efficiency of heat transfer through water with varying proportions of suspended sediment when a melt was poured into such water, finding that heat transfer was inefficient when the water contained >20% suspended sediment, irrespective of the grain size. However, where vents are already filled with a slurry of pyroclasts and water, called an impure coolant, solid particles can help to nucleate steam bubbles, much as crystals in magma help to nucleate vesicles of magmatic volatiles during heterogeneous nucleation (Sect. 4.5.2.3). Hajimirza et al. (2022) calculated the water : magma mass ratio that would be required to finely fragment already magmatically vesiculated hot mafic pyroclasts mixed with ejected glacial meltwater in the 17 to 18 km high phreatomagmatic eruption column from the 10th century Eldgjá fissure eruption in Iceland. For this secondary explosive fragmentation event, they calculated the surface area of the pyroclasts, the energy required to phreatomagmatically fragment them and drive the eruption column so high, proposing that the water: magma mass ratio was between 1 and 2. It thus seems that there is no magic number that accounts for the water: magma mass ratio for every phreatomagmatic explosive eruption and that each eruption has its own conditions.

An important concept that has not yet been fully explored in the physics of magma–water interaction is that while the MFCI theory and experiments have been developed to describe the static contact between fuel and coolant, explosive phreatomagmatic activity occurs in nature at variable magma flow rates through either static (lake, sea) or flowing bodies of water (groundwater, ice-melt water). This implies that efficient magma–water interaction should be described more than just in terms of mass ratios, but also in terms of mass-flow or discharge rates. Aravena et al. (2018a, b) have modelled this aspect of explosive magma–water interaction involving aquifer water and rising magma interaction. They found that effective surface contact interaction could only occur where aquifers are highly permeable and magma discharge rates (MDR) are low. For trachytic magmas MDR is 1×10^6 kg s^{-1}, and for dacitic magma, 3×10^6 kg s^{-1}.

At high MDR and with high-viscosity magma such as rhyolite, it seems that effective surface water explosive magma–water interaction is required. This has implications for the fragmentation setting in erupting magma for high MDR eruptions (e.g. phreatoplinian explosive eruptions). Aravena et al. (2018a, b) propose that high MDR phreatomagmatic eruptions must have occurred through the involvement of surface bodies of water (lake, ocean) into which magma was erupting. They also propose that pyroclast vesicularity reflects vesiculation levels in the magma at the time of fragmentation. Hajimirza et al. (2022) have proposed that the fine grained nature of the phreatomagmatic ash deposits from the 10th century Eldgjá fissure eruption in Iceland resulted from the mixing of already vesiculated mafic pyroclasts that emerged from the subglacial vent, with glacial meltwater that had been ejected into the 17 to 18 km high eruption column, in the eruption column. The phreatomagmatic fragmentation was thus a secondary fragmentation process involving both quench fragmentation and explosive superheating of the ejected meltwater in the eruption column (Hajimirza et al. 2022).

In addition to critical mass ratios, Zimanowski et al. (2015) proposed that a critical surface contact area between magma and water needs to be maintained, citing experimental results of 1 m^2 per 1 m^3 of mixed magma and water. This can be accomplished by physical mingling of low-viscosity magma and water, or by permeation of water along fractures that develop in rapidly cooling magma/glass near the boundary between the two. Hajimirza et al. (2022) calculated the surface area of erupted tephra from the 10th century Eldgjá eruption to calculate the water: magma ratio that would be required to generate the energy driving the secondary phreatomagmatic explosive fragmentation and the height of the eruption column.

Peckover et al. (1973) and Buchanan and Dullforce (1973) calculated maximum water depths for hydrovolcanic explosions of ~700 m where heterogeneous nucleation of steam bubbles is involved, and 130 m where homogeneous nucleation of steam bubbles occurs based on limiting confining hydrostatic pressures of 6.75 and 1.3 MPa, respectively. Zimanowski and Büttner (2003) agree with this assessment on the basis that at depths of several hundred metres, any vapour film that forms is so compressed and dense that sufficient acceleration and impact of the collapsing water surface will not occur to generate thermal detonations. They propose that "subaqueous volcanic thermo-hydraulic explosions should become increasingly improbable at water depths exceeding 100 m, and practically improbable at water depths in excess of 1 km" (Zimanowski and Büttner 2003). Wohletz et al. (2013) suggested that "although the (magma-water) interactions extend to pressures exceeding critical, it is not yet clear if explosive expansion can initiate at these high pressures (equivalent to ~2.2 km ocean depth)". This is based on the fact that even at or beyond the critical pressure supercritical fluid can expand. However, the key issue with volcanic explosions is not just that the volatile phase can expand, but that the rate of expansion has to be instantaneous to be explosive. This requires a high gas or steam overpressure relative to the ambient hydrostatic pressure, and a fluid that is a gas that expands orders of magnitude at the time of explosive expansion. It is because of these constraints, that we agree with the assessment of Zimanowski and Büttner (2003) that steam-generated explosions are almost impossible at depths greater than 1 km, and even at shallower water depths of several hundred metres they would still be highly attenuated.

At water depths approaching those where the hydrostatic pressure approaches or is equivalent to the critical pressure of water, which is 22 MPa and 2,200 m for fresh water and 30 MPa and 3,000 m for seawater and at greater water depths, it is inconceivable that superheated ambient water can expand rapidly enough to be explosive. The density of steam progressively increases with increasing hydrostatic pressure as it is progressively compressed (McBirney 1963; Kokelaar 1986; Cas 1992; Cas and Giordano 2014; Cas and Simmons 2018). In submarine environments, boiling of saline seawater steam bubbles and steam vapour film boiling produce freshwater steam bubbles, which therefore have the physical properties of fresh water, including the critical point (i.e. 22 MPa, ~2,200 m, rather than 30 MPa as for seawater). At water depths approaching the critical pressure point equivalent (22 MPa), the steam layer transforms into a *critical fluid* or *critical water* at the critical point (CP), and beyond that into *supercritical fluid* or *supercritical water* (Fig. 3.4c, Chap. 3).

As discussed in Sect. 3.2.2.4, Chap. 3, critical and supercritical water have hybrid properties of both liquid and gaseous (steam) forms of water. Although boiling *sensu stricto* of water cannot occur beyond the CP water depths because steam cannot be produced, pseudo-boiling involving formation of supercritical fluid bubbles can occur. Such supercritical pseudo-boiling fluid is denser and less compressible than subcritical steam. Initially nucleate (individual spaced bubbles) boiling off the magma surface, then film pseudo-boiling occurs, forming a continuous supercritical film.

The density of critical water varies according to temperature and pressure, but at the critical point (22 MPa, 374 °C) it is ~322 kg m^{-3} (Pioro and Mokry 2011; Table 3.3). This is 300× higher than that of low-density steam, making it much less compressible than steam.

The denser more highly pressurised supercritical fluid film at great water depths will be more resistant to film collapse. This will be the case particularly when propagating subaqueous lava surfaces are relatively smooth and the lava flow state is laminar, reducing the shear gradient at the interface between the moving lava and the host water mass, minimising the likelihood of fluid instabilities such as Kelvin–Helmholtz instabilities. Fluid instabilities could still lead to some localised collapse of the high-pressure, denser supercritical fluid film, bringing the ambient water, which will have been heated by heat transfer from the lava through the supercritical layer, into contact with hot magma, facilitating some quench fragmentation, but less commonly than at shallower water depths where low-density, low-pressure compressible steam constitutes the Leidenfrost vapour film. Although supercritical fluid can expand, it seems inconceivable that it could expand at instantaneous explosive rates against an ambient hydrostatic pressure regime that exceeds the critical pressure of water and depths of 2,200 m. In practice, this constraint will be effective even for water depths greater than 1,000 m where subcritical steam becomes more compressed and denser.

4.6.9.3 Induced Fuel–Coolant Interaction (IFCI)

It is therefore interesting that Dürig et al. (2020) have proposed that very deep-water explosive eruptions can occur through a newly proposed explosive mechanism they call Induced Fuel–Coolant Interaction (IFCI), based on experiments using molten rhyolitic ash erupted at water depths of 900 m during the 2012 eruption of the wholly submarine Havre volcano. This proposal has also been discussed in Chap. 3, from which much of the following is taken.

In the experiments, there were two essential set-ups. In the first "dry" experiments, injection of pressurised gas at the base of a layer of the magma in a crucible caused the magma to arch up, fracture, and disintegrate explosively. In the second, the set-up was the same, but a layer of water was superimposed on the magma layer just immediately before injection of pressurised gas at the base of the magma layer ("wet" or IFCI experiments). Importantly, all experiments were conducted at atmospheric pressure, NOT the high-pressure equivalent of deep-water conditions. In both cases, fragmented magmatic debris was ejected from the crucible, with a force of ~ 2.9 kN for the dry experiments and ~ 5.8 kN for the wet experiments (Dürig et al. 2020). The wet IFCI experiments produced 2.25 more ash-sized fragments, and much more fine ash (<125 mm), than the dry experiments, indicating increased fragmentation efficiency.

In both the dry and wet (IFCI) experiments, fragmentation events rely on an injection of a pressurised inflating gas slug under the magma charge to fracture it. Inflation/expansion created tensile stress, which is equivalent to an internal (e.g. expanding vesicle) magmatic gas driver (cf. magmatic and even some phreatomagmatic explosions). However, tensile stress fracturing also replicates first-order contractual quench fracturing to produce cracks during subaqueous eruptions, although this is not mentioned by Dürig et al. (2020). In the wet IFCI experiments, the superposed water layer definitely has the capacity to cause actual quench cracking and fragmentation, which then allows water to permeate into those cracks to trigger thermo-hydraulic explosions.

According to Dürig et al. (2020), IFCI does not require collapse of a pre-existing steam layer to initiate surface thermo-hydraulic detonation as in MFCI explosions. It "occurs under less restrictive initial and boundary conditions", involving instantaneous explosive vaporisation of water that has permeated into first-order cracks. However, this seems to be a contradiction because it is instantaneously expanding steam that drives explosions. Therefore, a momentary micro-vapour film must form in the cracks, which Dürig et al. (2020) acknowledge may happen.

Dürig et al. (2020) propose that ash particles produced during IFCI explosions developed very fine curviplanar shapes that can be used to fingerprint the IFCI mode of formation. Curviplanar ash particles also occur in much of the 2012 Havre fine ash, which therefore they also attribute to IFCI explosive fragmentation at 1 km water depth, which is deeper than the 700–800 m water depth limit for MFCI explosions. However, curviplanar fragments can also be produced during quench fracturing of glass and of pumiceous lava (van Otterloo et al. 2015). So, we largely agree that IFCI crack-hosted explosions as an extension of MFCI surface explosions provides a significant advance on understanding phreatomagmatic explosive eruptions to water depths of ~ 1 km depth, but the role of quench fracturing of volcanic ejecta has been largely ignored. During MFCI some degree of quench cracking (e.g. Colombier et al. 2019) allows penetration of water into such cracks, such that IFCI represents an extension of MFCI, but involving a micro-process at a more detailed scale that has not previously been recognised.

However, based on the discussion of varying fluid properties with increasing water depth as detailed in previous paragraphs, we feel that the significance of IFCI is drastically reduced at greater depths. This is because at water depths >2200 m steam generated by boiling of seawater will immediately compress to a supercritical fresh water fluid, not subcritical steam. This implies that a layer of critical and supercritical H_2O fluid forms at the magma–water interface, which has a lower heat transfer coefficient than liquid water. In addition, at and beyond the critical point, supercritical fluid has a limited thermal expansion coefficient (Sect. 3.5.1, Chap. 3), significantly lower than boiling water. Therefore, although any fluid can expand many times its original volume as it is heated and under pressure, this does not necessarily lead to explosions. *Explosions only occur if the rate of expansion* is instantaneous, and coefficient of expansion actually does not reflect rate of expansion. In addition, counteracting the pressure in superheated critical and supercritical fluid is the hydrostatic pressure of the water mass at great water depth that produces the high supercritical fluid pressure. So, it seems that a significant overpressure situation that is capable of producing an instantaneous explosive expansion rate in a film of supercritical fluid, as it occurs in shallower water, but does not occur in very deep water.

What may happen is that when water, which penetrates first-order cracks, is heated it expands, builds up pressure, and causes the enclosing glass to hydraulically fracture. At depth such hydraulic fracturing is not explosive but is an in situ fracturing process than can dilate and displace fragments, but not necessarily explosively eject clasts. In deep water, with higher hydrostatic pressure, the interstitial fluid pressure just hydraulically fractures glass in situ, some quench fragmentation is likely to occur and buoyant plumes of heated water rising above the erupting magma can carry ash size debris upwards into the water column where it is dispersed by currents. This can be all non-explosive.

A test of whether or not IFCI eruptions occur commonly in deep water beyond the critical point water depth equivalent is whether or not the seafloor is mantled with fine ash that can be related specifically to a known deep-water eruption. This criterion is important because as discussed in Chap. 9 there are several possible origins for ash deposits on the deep seafloor, not just in situ deep-water eruptions, nor just deep-water explosive eruptions. Given the abundance of deep-water basalt eruptions and the high temperature of such magmas, IFCI ash deposits should be common if IFCI occurs commonly in deep water. To our knowledge this is not the case. In Chap. 5, we speculate that some fine-ash-like deposits have resulted from quenching-induced spalling of ash size glassy particles from the surfaces of propagating submarine basalt lavas, which were dispersed by hot water thermal plumes rising off the surfaces of hot lavas.

So, in summary, we conclude that IFCI explosive eruptions can occur to water depths where relatively low-density, expandible steam is generated (<1,000 m, perhaps a little deeper in some cases), but not at water depths approaching and deeper than the critical point of fresh water (>1,500 m).

4.6.9.4 Fragmentation of Magma During Phreatomagmatic Explosive Activity

If the water–magma mass ratio is lower than the optimum ratio, whatever that is (e.g. $<\sim 0.2$ in Fig. 4.17), then the external water mass contributes very little to the explosive fragmentation of the magma, leading to what is commonly called "dry phreatomagmatic". In such a case, the explosive activity involves relatively little steam, low levels of fragmentation, producing coarse pyroclastic debris. Such eruption phases can transition into brief phases driven only by magmatic volatiles (magmatic explosions) before fluxing back to phreatomagmatic activity.

At the optimum water–magma ratio (~ 0.3?; or better, at optimum mass-flow rates ratio; Aravena et al. 2018a, b), phreatomagmatic eruptions experience efficient, very high degrees of magma fragmentation (Fig. 4.17; e.g. Self and Sparks 1978), producing a lot of wet, very fine ash (e.g. the 2010 eruption of Eyjafjallajökull volcano in Iceland; Gudmundsson et al. 2012; Dellino et al. 2012). Such eruptions are commonly called "wet phreatomagmatic". At higher water–magma ratios, explosive intensity declines rapidly, and quench fragmentation becomes more prominent (Fig. 4.17).

Typically during phreatomagmatic explosive activity, a range of fragmentation processes occur, from viscous detachment of fluidal particles from the magma surface (Fig. 4.16a) to the effects of shockwaves that propagate through the magma when explosions occur (mechanical fragmentation), to vesicle bubble expansion, to the effects of thermal stress and cooling contraction due to contact between magma, magma fragments, and water (quench fragmentation, also called thermal granulation; Sect. 4.4; Kokelaar 1986; Cas and Wright 1987; Zimanowski et al. 1997; Büttner et al. 1999; Dürig et al. 2012; Liu et al. 2015; van Otterloo et al. 2015; Colombier et al. 2019). In addition, as proposed above during IFCI, water that permeates quench cracks can be heated and expand creating an hydraulic fluid overpressure that can hydraulically fracture glass, without necessarily being explosive.

These complex and multiple fragmentation processes cause a higher degree of fragmentation during phreatomagmatic explosive eruptions than during magmatic explosive events, producing a much higher proportion of ash size fragments. Van Otterloo et al. (2015) and Zimanowski et al. (2015) point out that if magma is already vesiculated at the time of magma–water interaction, it requires less energy to fragment the magma through MFCI than if it was non-vesiculated, and the resulting deposits also contain more very fine ash. Abundant fine ash in a deposit cannot simply be used as an indicator of very intense phreatomagmatic explosions, because magma properties, including level of vesiculation, also play a role in determining final grainsize.

For magmas that are already vesiculated at the time of interaction with external water (e.g. Surtsey 1963–1964 tephra, Thorarinsson et al. 1964; Houghton and Carey 2019), significant explosive energy may also come from magmatic volatile explosive release, as well as superheating of the external water. Fragmentation may therefore involve several factors including gas-driven vesicle bursting as well as cooling-quenching fragmentation caused by the vaporised external water. Although Houghton and Carey (2019) propose these processes be considered as primary and secondary, respectively, this may give the impression that these processes occur sequentially, whereas they may be occurring concurrently. Most importantly they are both syn-eruptive fragmentation processes.

4.6.10 Situations in Which Phreatic–Phreatomagmatic–Hydrothermal Explosive Eruptions Occur

Hydrovolcanic explosions can occur in several situations (Fig. 4.18), including:

- from subaqueous vents in the ocean or within lakes,
- subglacial volcanic eruptions,
- subsurface explosive interaction between magma rising through the crust and ground water or permafrost,
- lava flowing into a body of water or over water-saturated sediment,
- pyroclastic flows moving into a body of water or over water-saturated sediment, and
- hydrothermal explosions from over-pressurised hydrothermal systems.

Fig. 4.18 Different settings for phreatic-phreatomagmatic explosive eruptions. **a** Phreatomagmatic explosive eruption in 1 July 2021 from the Main Crater Lake, Volcano Island, within Lake Taal caldera, Philippines. This shows dark, solids-rich "cocks-tail" tephra jets. Note also the just visible, outward moving shock wave to the right of the vent captured in the image. (*Source* PHIVOLCS https://www.phivolcs.dost.gov.ph/). **b** Phreatomagmatic eruption through the thick ice sheet at the summit of Eyjafjallajökull volcano Iceland, 16 April 2010. The ice is covered in brown, fine wet ash. (Photo by M. Fulle; *Source* Stromboli OnLine https://www.swisseduc.ch/stromboli/ with permission R. Carniel). **c** Phreatomagmatic eruption of Ukinrek maars in Alaska in 1977, resulting from magma rising through high-level permafrost in the subsurface (*Source* United States Geological Survey). **d** Phreatic explosions at the point of entry of basalt lavas into the sea, southern coast of Hawai'i in 1988 (Photo by J.D. Griggs; *Source* United States Geological Survey). **e** Phreatic explosion craters blasted through pyroclastic flow deposits that have ponded in the Toutle River Valley, at Mt St Helens, USA, 18 May 1980 (*Source* United States Geological Survey). **f** Hydrothermal explosion crater, surrounded by hydrothermally altered ground and substrate. Nisyros caldera, southeastern Aegean Sea, Greece

4.6.10.1 Phreatomagmatic Explosions from Subaqueous Vents

Explosive activity may result from the explosive boiling of ambient water in an open vent, such as a tuff cone in the ocean or a lake with a breached crater rim and a water-filled crater, or of interstitial water within volcaniclastics around such a vent, superheated by rising magma. If the interaction between hot magma, quenched debris, and water is explosive, prolonged, cyclical, pulsing, and self-sustaining, fuel–coolant interaction occurs as discussed above (Sects. 4.6.9.2 and 4.6.9.3). The typical eruption style is called *surtseyan explosive eruptions* (Chap. 9). There have been many historic examples of explosive hydrovolcanic or phreatomagmatic eruptions from subaqueous vents in the ocean (e.g. the 1952 eruption of Myojinsho volcano, Izu-Bonin Arc, Japan, Fiske et al. 1998, see Fig. 9.80; the 1957–1958 eruption at Capelinhos, Azores, Cole et al. 2001, see Fig. 10.100a–d; the 1963–1964 eruption at Surtsey, Iceland (Fig. 4.18a; Thorarinsson et al. 1964, Plates 3.1b, 4.1d; the 2009, 2014–2015, 20 December 2021–15 January 2022 eruptions of Hunga Tonga-Hunga Ha'apai, Tonga, see Figs. 3.11b, c, 9.80, and 10.100f) and from lakes (e.g. the 1966 and 2020–2022 eruptions at Taal volcano, Philippines, Moore 1967, Fig. 4.18a; the 2007 eruption at Ruapehu, New Zealand; Kilgour et al. 2010).

As discussed above, the explosive expansion of superheated external water is constrained by the increasing hydrostatic pressure with increasing water depth. The pressure gradient in a body of pure water is 1 MPa (10 bars) per 100 m of water depth (Fig. 3.3). McBirney (1963) demonstrated that the specific volume changes of water at magmatic temperatures of 800–1,100 °C (Fig. 3.9; only likely at the immediate interface with magma in the film boiling layer, and in vesicles) are very dependent on pressure (Peckover et al. 1973). Substantial growth rates of steam bubbles or specific volume increases only occur at pressures less than 10 MPa (100 bars; equivalent to water depths less than 1 km), the rate of change accelerating significantly at lower pressures to what could be considered explosive rates (Fig. 3.9; Peckover et al. 1973).

If water is superheated isobarically (at the same pressure = water depth) by sudden interface with an erupting magma, it would only undergo rapid (explosive) volume expansion above temperatures of 800 °C if the pressure is less than 10–15 MPa (Fig. 3.9).

In addition, unlike magmatic volatiles enclosed in hot silicate magma, steam bubbles in seawater are subject to rapid cooling and condensation. Steam bubbles rising buoyantly through the water column above the vent would decompress slowly as they rise. However, their explosive potential is limited by slow decompression and condensation erosion, which would lead to decreasing gas pressure during uprise, and collapse/dissipation of the bubbles.

Clearly, magmas with significant magmatic volatile contents may experience explosive thrust from explosive exsolution of volatiles (Sect. 4.5) as well as dynamic interaction with superheated seawater as discussed here, so leading to phreatomagmatic explosive activity. The relative intensity of the phreatic and magmatic explosive components will vary depending on water depth and magmatic volatile content. Seawater (or lake water) may act as a catalyst in triggering explosions in already vesiculated magma which by itself had insufficient gas pressure to trigger magmatic explosions against the viscous resistance of the magma and the hydrostatic pressure of the water mass. Quench fragmentation of the vesiculated magma (Sect. 4.4) would overcome the viscous force and perhaps allow incipient explosions to be initiated. Secondly, such quenching fracturing would allow access of external water into the hot interior of the magma, causing the water to superheat, perhaps explosively, as discussed above during induced fuel–coolant interaction (IFCI; Sect. 4.6.9.3).

The 15 January 2022 spectacular explosive eruption of Hunga Tonga–Hunga Ha'apai volcano in Tonga, southwest Pacific (see Chaps. 3 and 9 for description and discussion), suggests initially that the eruption was probably a phreatomagmatic one, driven by explosive interaction between newly erupting magma and the seawater that was in the shallow submarine vent. However, the intensity of that explosive eruption, which was much greater than the typical explosive style of the 2009, 2014–2015, and 20 December 2021–14 January 2022 surtseyan eruptions, suggests that a high magmatic volatile content in a new batch of magma probably also played a part. This highlights the difficulty of simplistically classifying explosive eruptions as phreatomagmatic; there is a spectrum of explosive energy inputs from phreatomagmatic *sensu stricto* to increasing degrees of magmatic volatile input.

4.6.10.2 Subglacial to Emergent Phreatomagmatic Explosive Eruptions

Where magmas erupt under glaciers, regional ice sheets, or ice covered craters, the initial effect is to melt the base of the glacial ice, creating a subglacial meltwater pond or vault into which subaqueous lavas are erupted, producing pillow lavas and associated hyaloclastite (e.g. Jones 1966, 1969; Skilling 1994; Smellie 2000; Russell et al. 2013; Edwards et al. 2015; Chap. 16, Sect. 16.9). Under thick ice the glaciostatic pressure of the ice may be high enough to suppress explosive eruptions initially (Edwards et al. 2015), as discussed above for subaqueous settings, and involving similar pressure gradients. If the ice melts to the surface, or fractures in the ice propagate to the surface lowering confining pressure, hydrovolcanic explosive eruptions may begin in a meltwater pond that is open to the atmosphere. The numerous eruptions of Grimsvötn volcano (Gudmundsson et al. 1997; Smellie and Hole 1997; Jude-Eton et al. 2012) and the 2010 eruption Eyjafjallajökull volcano (Fig. 4.18b; Gudmundsson et al. 2012; Edwards et al. 2015; Chap. 16) and the 10th century Eldgjá eruption, (Hajimirza et al. 2022), all in Iceland, are excellent examples.

These eruptions are marked by the production of large volumes of fine ash that can be widely dispersed with potentially severe social and economic consequences, as in the case of the Eyjafjallajökull eruption in Iceland in 2010 (Gudmundsson et al. 2012; Chap. 16). In addition, sudden glacial meltwater floods called jökulhlaups may break-out from the base or margins of the glacier or ice sheet (Chap. 12). The explosive activity may build a surtseyan tuff cone that emerges above water level (Chap. 16), through a passage zone representing transition from subaqueous to subaerial. Once this has happened subaerial lava can be erupted, which may then lead to a lava-hyaloclastite delta forming around the cone margins. Such ash-forming subglacial phreatomagmatic events have been common in Iceland and include the quite regular eruptions of Grimsvötn volcano (e.g. 2004, 2011; e.g. Jude-Eton et al. 2012) and the 2010 eruption of Eyjafjallajökull volcano (e.g. Gudmundsson et al. 2012; Dellino et al. 2012; Chap. 16). Because of the unique englacial setting and influences, we have proposed in Chap. 9 that such explosive eruptions be called *glacio-surtseyan explosive eruptions.*

4.6.10.3 Explosive Interaction Between Rising Magma and Aquifers or Permafrost

In this situation, magma rising through the crust encounters a groundwater aquifer or frozen groundwater table or permafrost, leading to explosive superheating of the groundwater (e.g. Ukinrek maar eruptions, Alaska, 1977, Self et al. 1980; Kienle et al. 1980; Fig. 4.18c). For further examples and descriptions, also see White and Houghton 2000; Giordano et al. 2002; Pardo et al. 2009; Ort and Carrasco-Núñez 2009; White and Ross 2011; Geshi et al. 2011; Jordan et al. 2013; van Otterloo et al. 2014; Zimanowski et al. 2015).

For purely phreatic explosions, the gas pressure produced by superheating of the groundwater and the resultant expansion of the ground water as it changes phase from water to steam must equal or exceed the minimum principal stress in the country rock and the tensile strength of the country rock before failure of the overburden and explosive eruption can occur. Again, however, involvement of magmatic volatiles, leading to a phreatomagmatic eruption, could occur. This could be triggered by initial opening of a vent through phreatic explosions, which then decompresses the magma so that it vesiculates explosively, or the magma may already be well vesiculated at the time it encounters a shallow aquifer. The succeeding eruption continues as a combination of both steam explosions from the groundwater aquifer and magmatic explosions, but may fluctuate from "wet" to "dry" phreatomagmatic, to dry magmatic explosive activity as a function of variations in the water permeability or recharge rate in the aquifer, and/or the magma supply rate from the conduit (Cas et al. 2017).

This style of phreatomagmatic explosive eruption has some similar features to open-water surtseyan eruptions (pulsing, cyclical, ballistic blasts) but is constrained by a limited subsurface aquifer water source. Such eruptions typically form subaerial maar and tuff ring volcanoes, and sometimes tuff cones (Chap. 16), and we call them *ukinrekian explosive eruptions* (Chap. 9). Once a phreatic or phreatomagmatic explosive event has excavated a crater or deeper diatreme, water can flow into the diatreme from the aquifer, and explosive activity can then be activated within the diatreme and its fill, depending on the depth in the diatreme that water and magma interact. That depth can be determined using the accessory lithic clasts in the phreatomagmatic deposits (e.g. De Bennedetti et al. 2010; Danese and Mattei 2010; Jordan et al. 2013; van Otterloo et al. 2013; Chap. 16).

In order to achieve an efficient interaction of magma with groundwater, fulfilling the right conditions as described in Sect. 4.6.9.2, magma rising through the volcanic conduit needs to pre-mix with water at rates that are optimum at optimum rise speed, involving the full section of aquifer in the conduit. Such conditions are dependent on the transmissivity of the aquifer. As transmissivity depends on the permeability and thickness of the aquifer (measured in $m^2\ s^{-1}$), efficient phreatomagmatic interaction will occur preferentially at slow ascent rates (depending on aquifer permeability; White and Houghton 2000) and narrow conduit widths, i.e. most likely for, although not restricted to, basaltic eruptions. From the aquifer perspective, pressurised, permeable, and thick aquifers are most likely to induce efficient magma–water interaction relative to thin and poorly permeable aquifers. Excellent examples of the former, associated with extensive phreatomagmatism, are the deep karstic hydrothermal aquifers in the Roman Magmatic Province, Italy (Giordano et al. 2014).

Phreatomagmatic explosive activity associated with groundwater is responsible for formation of spectacular continental maars, tuff rings and tuff cones, which are documented in detail in Chap. 16 (Sect. 16.3; e.g. Cas et al. 2017). These include the 1977 Ukinrek maar forming eruptions in Alaska (Kienle et al. 1980; Ort et al. 2018; Fig. 4.18c) and the Lago Albano maar, Colli Albani volcano, Roman Magmatic Province (Anzidei et al. 2008; De Benedetti et al. 2010). Maar volcanoes and their deposits are well preserved in larger regions of intraplate alkaline volcanism such as in the Eiffel area of Germany (Lorenz, 1973, 1986), and the basaltic Newer Volcanics Province of western Victoria and southeastern South Australia (Joyce 1975; Cas et al. 1993, 2011, 2017; Lesti et al. 2008; Boyce 2013; Jordan et al. 2013; van Otterloo et al. 2013; Chaps. 15–17), and the Roman Magmatic Province (Giordano et al. 2002, 2010; De Benedetti et al. 2008; Sottili et al. 2012).

4.6.10.4 Explosions Resulting from Lava Flowing into Water or Over Water-Saturated Sediment

The flow of hot lava into a body of water or over water-saturated sediments could produce local explosive activity as trapped water (bulk steam explosivity; Kokelaar 1986) is superheated and expands rapidly and explosively. During the 1973 eruption on Heimaey in Iceland, the entry of basalt lavas into the sea in the harbour produced local hydrovolcanic explosions (Colgate and Sigurgeirsson 1973), as did the numerous basalt flows entering the sea from the 1983 to 2018 eruption along Kilauea volcano's East Rift Zone, Hawai'i (Fig. 4.18d; Mattox and Mangan 1997; Chap. 5). Explosive activity resulting from the flow of lavas into water has, in some cases, produced secondary cones or craters called littoral cones and pseudo-craters (Thorarinsson 1953; Fisher 1968; Hamilton et al. 2010; Chaps. 5 and 16). These are also called rootless vents and cones because they are not sourced directly from and connected to a subjacent magma conduit. Similar explosions may occur also when lava flows onto snow, such as occurred on 16 March 2016 at Mt Etna, when a sudden explosion of snow-melt water trapped beneath the advancing lava flow injured ten tourists.

4.6.10.5 Explosions Resulting from Pyroclastic Flows Moving into Water or Over Water-Saturated Sediment

When pyroclastic flows flow into water bodies (sea, lakes, rivers; Chap. 11; e.g. pyroclastic flows flowing into the sea at Montserrat 1995–2010, Edmonds and Herd 2005) or over water-saturated substrate (e.g. pyroclastic flows in the Toutle River Valley, 18 May 1980, Mt St Helens, Rowley et al. 1981; Fig. 4.18e), spectacular phreatic explosions can occur.

During explosive eruptions from the Soufrière Hills volcano on Montserrat, pyroclastic flows frequently reached the coast and entered the sea (Edmonds and Herd 2005; Dufek et al. 2007; Chap. 11). The dynamic interaction ranged from discrete phreatic explosions to littoral blasts leading to the generation of base surges, to vigorous boiling of water. In some cases, the surge ash cloud at least was seen to flow over the sea surface for up to hundreds of metres offshore, as well as onshore, while the body of the flows entered the sea and transformed into subaqueous volcaniclastic turbidity currents (e.g. Edmonds and Herd 2005; Trofimovs et al. 2008). This was also observed during the 2019 explosive activity of Stromboli, Italy (Giordano and De Astis 2021; Chap. 11). Such littoral explosions could also generate rootless vents, and their formation has previously been discussed for Paleozoic welded ignimbrites in northern Snowdonia, North Wales (Wright and Coward 1977; Kokelaar and Koniger 2000; Chap. 11). The 18 May 1980 pyroclastic flow deposits of Mt St Helens contain numerous phreatic explosion craters resulting from the passage of the pyroclastic flows over water-saturated debris and ground within the devastated watershed of Spirit Lake and the Toutle River (Rowley et al. 1981). The superheating of the river water and ground water trapped under the instantaneously deposited very hot pyroclastic deposits, which are tens of m thick in the valleys, produced numerous phreatic explosions and craters (Fig. 4.18e), although no major eruption columns are known to have developed.

4.6.10.6 Hydrothermal Explosions

Hydrovolcanic explosive eruptions involving hydrothermal fluids occurred during the famous 1886 eruption of Mt Tarawera in New Zealand and destroyed the famous Pink and White sinter terraces in the Lake Rotomahana Area (de Ronde et al. 2016; Pittari et al. 2016). Although it appears that the interaction of basaltic magma with water-saturated lake sediments and lake waters may have been the source of some of the explosive activity, as evidenced by the mixture of basaltic fragments, lake sediment, and hydrothermally altered clasts in the eruption deposits (Nairn 1979; de Ronde et al. 2016; Pittari et al. 2016), the Rotomahana Area had a long history of high-level hydrothermal activity before the 1886 eruption. This had produced the extensive sinter deposits (formerly an eighth wonder of the world), hot springs, boiling mud pools, and numerous hydrothermal explosion craters (Chap. 18).

For the Rotomahana eruption, Nairn (1979) proposed that if the hydrothermal fluids at depth were already near the boiling point when hot basaltic magma was injected from below, then the extra thermal input from the basaltic magma that caused the 1886 explosive eruption would have been sufficient to boil the hydrothermal fluids. Flashing of these fluids to steam and induced steam pressures greater than the lithostatic load would have produced explosive activity. Lowering of overburden pressure by initial explosions would produce instantaneous boiling in the highly pressured, high-temperature, subsurface hydrothermal reservoir system, so producing significant, wholly hydrothermal explosions. Because of the already high temperature of subsurface hydrothermal fluids, the efficiency of energy transfer from thermal to mechanical energy would be high.

Although some hydrothermal explosions are fuelled by rising magma, others are not, such as those associated with the Kawerau geothermal field in New Zealand (Nairn and Wiradiradja 1980; Browne and Lawless 2001; Rowland and

Simmons 2012; Chap. 18). These explosions were apparently due to a reduction of fluid pressure in the hydrothermal system, leading to the flashing to steam of the superheated hydrothermal water. Products include a massive to diffusely stratified breccia of hydrothermally altered country rock fragments (e.g. ignimbrite, rhyolite) in a silty hydrothermal clay matrix.

Hydrothermal explosions are common in geothermal areas (Muffler et al. 1971; Nairn 1979; Marini et al. 1993; Morgan et al. 2009; Ambrosio et al. 2010; Vignaroli et al. 2015; Montanaro et al. 2016, 2022). The island volcano Nisyros in Greece has a large summit caldera, half of the floor of which is pervasively hydrothermally altered. A series of hydrothermal explosions has produced a cluster of spectacular craters (Fig. 4.18f; Marini et al. 1993; Ambrosio et al. 2010), the rims of which consist of massive to diffusely stratified breccias (see Fig. 4.31g). These, like the New Zealand examples, consist of highly hydrothermally altered blocks of country rock supported by a pervasive clay-rich matrix of hydrothermal clays and gypsum, and finely comminuted country rock fragments.

Whakaari/White Island volcano, an offshore marine stratovolcano north of the Taupo Volcanic Zone in New Zealand, has one of the most active hydrothermal systems in a volcanic crater, anywhere in the world, and episodically experiences both violent magmatic and hydrothermal explosions (e.g. 2016 and 2019; Kilgour et al. 2019; also see Chap. 18). The devastating explosive eruption of Whakaari/White Island volcano in New Zealand on 9 December 2019, which killed 22 people and severely injured 25 others, was also a hydrothermal explosive eruption, originating from the very active hydrothermal system under the crater floor of the island stratovolcano. Most of the historical eruptions of Whakaari/White Island volcano, which occur on average every 2 to 3 years, involved explosive input from the hydrothermal system (Mayer et al. 2015). These explosive eruptions have been mistakenly called phreatic explosions by some New Zealand authors, who seem not to understand the different enthalpy levels, and the different and very acute hazard and risk issues associated with hydrothermally active volcanoes, compared with "normal" magmatic stratovolcanoes, and one-off phreatic explosive events. Hydrothermal systems are already primed at a high temperature (almost magmatic) and high pressure, and require very little extra thermal input or a pressure trigger for explosive activity to be activated. In their qualitative classification scheme for phreatic, phreatomagmatic and hydrothermal explosions, Montanaro et al. (2022) classified the 9 December 2019 Whakaari explosive eruption as mostly phreatic, which is very strange given the constantly primed state of the hydrothermal-geothermal system on the island. The hydrothermal system can erupt unexpectedly with virtually no detectable precursor signals once in this state. Why then were uneducated tourists allowed to go to this very remote, marine stratovolcano and go right into the enclosed crater right up to the vent? It was a death trap and to date, 3 years after the eruption, no answers have been forthcoming from New Zealand authorities, organisations or scientists as to how this was allowed to happen.

Although hydrothermal explosions eject a lot of variably altered country rock debris as ballistic blasts, such eruptions can also generate hot density currents such as base surges, and if the volume is sufficiently high, pyroclastic flows. These can be deadly, as demonstrated by the 9 December 2019 eruption of Whakaari/White Island volcano and documented for the 2016 eruption (Kilgour et al. 2019). The deposits are dominated by hydrothermally altered rock debris and ash.

Hydrothermal explosion craters and their associated hot spring phenomena are potential sites of epithermal, especially precious metal (gold, silver), mineralisation, as well as geothermal energy, and their importance as such are described in detail in Chap. 18 (Sect. 18.3).

4.7 Fragmentation of Country Rock During Explosive Fragmentation

Most research on explosive fragmentation has focussed on the way magma fragments during explosive eruptions (Sects. 4.5 and 4.6). However, many pyroclastic deposits contain abundant accessory lithic clasts sourced from the country rock that makes up the walls of the vent. In a series of seminal papers, Barnett has provided insights into the way country rock can be fragmented in vent settings (Barnett 2004, 2008; Barnett et al. 2011). Barnett used exposures in the walls of kimberlite pipes, exposed through open pit mining. During mining of kimberlite pipes for diamonds, mine excavation work cuts down through former explosive vents and conduits, and so provide unique insights into the way those vents have formed and the nature of the interaction between the vent/pipe and the country rock that hosts the vent and is being excavated by explosive activity (Chap. 18).

Barnett found that the factors that contribute to country rock fragmentation include: (1) the strength of the country rock, (2) the nature of pre-existing structures in that country rock, (3) the degree to which early explosions fracture or pre-condition the country rock, and (4) the pressure regime in the conduit during an eruption. If country rock is strong, with few weaknesses before an eruption, any early explosive activity will propagate shock waves that cause coarse, in situ brittle, pre-conditioning fracturing, producing only a small component of fine grain sizes. Under-pressure in the vent relative to the confining pressure in the country rock will enhance block caving of large blocks, spalling of already fractured rock debris and gravity-induced slumping into the crater and vent, often resulting in shear-induced secondary fracturing and size comminution, and even a degree of abrasion-induced rounding (e.g. Campbell et al. 2013). An overall finer grainsize and an increase in the proportion of fine grainsizes is characteristic. This lithic clast debris is then available for explosive ejection as ballistic blocks and buoyantly entrained finer lithic debris during the next explosive phase, which then further fractures and breaks down the lithic debris already in the vent. Pre-conditioned and already fractured country rock (e.g. cleaved, jointed, faulted) produces more initial finer debris which can be further fractured also be scoured from the conduit walls by the vertically ejecting gas-pyroclast stream.

The study of the types and abundance of the accessory lithic component throughout a pyroclastic sequence is an essential tool to reconstruct the evolution of the vent and eruption (e.g. De Benedetti et al. 2010; Valentine et al. 2014) and both fluctuations in the level of fragmentation within the conduit and of exit pressure (see Chap. 9). Recent work has also pointed out how the geometry of the conduit which is usually very far from the ideal cylinder shape usually envisaged may also contribute to modulate the lithic excavation from the conduit walls (Massaro et al. 2018).

4.8 Introduction to Pyroclasts—The Products of Explosive Eruptions

All fragments that result from an explosive eruption and are explosively ejected from a vent are called *pyroclasts*. Pyroclasts include three different types of components: (1) juvenile clasts of the erupting magma, called "juvenile or magmatic pyroclasts"; (2) crystal fragments, which may be juvenile magmatic crystals or xenocrysts; and (3) country rock lithic fragments, or xenoliths. Their characteristics can tell us much about the specific processes that caused the explosive eruption and the style of that event (Benet et al. 2024). Once ejected from a vent, pyroclasts are transported and deposited by different processes, including ballistic ejection and transport, and convective transport in the buoyant explosive eruption column rising above the vent, from which pyroclastic fallout occurs (Chap. 9), or by pyroclastic flow and pyroclastic surge (Chap. 10). The latter two represent end members of a spectrum of particulate gravity or density current flows of gas and pyroclasts, called pyroclastic density currents. We discuss pyroclastic transport processes and the meso- or outcrop-scale characteristics of the resultant deposits in Chaps. 9 to 11. The principal aim here is to introduce the components of pyroclastic deposits, pyroclasts, the different types, and their characteristics.

Pyroclasts can have a wide range of sizes, irrespective of their origin. Fragments greater than 64 mm in diameter are called bombs if they are fluidal and irregular in shape, or blocks if they are "blocky" and bound by more regular curviplanar surfaces; those between 64 and 2 mm in diameter are called lapilli; and those less than 2 mm in diameter are called ash. These terms should be strictly confined to pyroclasts that are finally transported and deposited by pyroclastic transport processes. Pyroclasts that are subsequently reworked or resedimented by normal surface processes should be named by their sedimentary size class terms or qualified by terms such as "resedimented" or "reworked". The terminology for pyroclastic deposits is further treated in Chap. 14. "*Tephra*" is a collective term for all unconsolidated pyroclastic deposits, including the deposits of pyroclastic flows, surges, and falls (Self and Sparks 1981).

First, we will consider ballistic clasts, which are ejected from a vent through enormous explosive momentum and are unaffected by the turbulent convective motion of buoyant gas in the eruption column above the vent because they are too heavy. Therefore, they are large and/or dense and usually in the bomb and block size range. We will then consider smaller pyroclasts of a range of compositions, which are subject to convective entrainment into the explosive eruption column above the vent.

4.8.1 Ballistic Bombs and Blocks

Because ballistic bombs and blocks are ejected on ballistic, usually parabolic trajectories, they are most commonly found associated with fallout deposits at distances <4 km from the vent (Chap. 9). Although not affected by the rising, buoyant convective eruption column, if that column becomes overloaded with solids and becomes too dense to remain buoyant, it collapses *en masse*, and as it does so many ballistic bombs and blocks may become entrained into the collapsing column and become part of a dynamic pyroclastic flow, which is capable of transporting them many km to tens of km from the vent.

Ballistic bombs represent a juvenile or magmatic fluidal pyroclast that is fluidally shaped and contorted as it twists and turns through the air. They can form from any magma composition if the viscosity is low enough. *Ballistic blocks* are derived from an already solidified source rock in the conduit or from its country rock walls, and when explosively detached from the source tend to be angular and bound by (curvi)planar surfaces. They can consist of cognate lithics (chilled, solidified clasts of the erupting magma, e.g. chilled conduit lining, lava crust in a lava lake, lava dome), or an accessory country rock lithic clast of any rock type that has been explosively excavated and ejected from the vent derived from the country rock that bounds the vent.

Bombs are fluid dynamically shaped during ballistic flight out of the vent and vary in shape from twisted cow-pat spatter bombs (Fig. 4.19a) to fusiform spindle shapes

(a)
(b)
(c)
(d)
(e)
(f)
(g)

◀ **Fig. 4.19** Ballistic bombs and blocks. **a** Contorted basaltic spatter bomb, Pico Viejo volcano, Tenerife, Canary Islands, Spain. **b** Fusiform or "spindle" bomb, Antofagasta scoria cone, southern Puna Plateau, Argentina. **c** Fusiform bombs, Mt Fuji, Japan. **d** Spheroidal bomb with breadcrust texture on the surface, and **e** twisted, elephant-trunk-like ballistic bomb with breadcrust texture. Basaltic Mt Noorat volcano, Newer Volcanics Province, Victoria, southeastern Australia. **f** Vesiculated interior of spatter bombs, basaltic Red Rock volcanic complex, Newer Volcanics Province, Victoria. **g** Dunite mantle xenoliths inside basaltic bomb, Mt Noorat volcano. **h** Vesiculated and bread-crusted rhyolitic pumice bomb, Mt Tarawera lava dome complex, New Zealand. **i** Lava-coated, lithic block, ejected 3 July 2019 eruption, Stromboli, Italy. **j** Ballistic lithic block field, July–September 2000 eruption of Miyakejima volcano, Miyakejima, Izu Islands, Japan. Lithic blocks are up to 2 m in diameter. **k** Ballistic older basalt lava block embedded in maar rim deposits, Red Rock volcanic complex. Note the impact sag in underlying tephra layers

Fig. 4.19 (continued)

(Fig. 4.19b, c), sometimes with two symmetrical tapering fins or wings, and more spheroidal forms (Fig. 4.19d). Others may be more irregular and contorted, ribbon-like, soft-state deformed flaps, and even long and contorted elephant trunk-like (Fig. 4.19e). Jones et al. (2018) distinguished between primary near vent spatter, from primary near vent, but late spatter (erupted late in a fountaining phase), and secondary, spatter, erupted from a secondary fracture vent in a lava away from a vent in Hawai'i. They found that they could be distinguished by vesicularity levels and patterns as well as vesicle/bubble number density values. The surfaces of bombs may preserve a network of rectilinear cracks, called breadcrust cracks. These result from the chilling of the surface of the bomb to glass during flight through the air which deforms and cracks during flight, while the interior remains fluidal. The fluidal interior may continue to exsolve volatiles and form vesicles, which causes the interior to expand, which in turn causes the surface crust to stretch through tensile stresses, forming the cracks (Fig. 4.19e, h). The interior of such bombs can be highly vesiculated, whereas the glassy chilled rim preserves fewer small vesicles, indicating that interior vesiculation was ongoing after the magma clast was formed and ejected from the vent (Fig. 4.19f). Other surface crack networks can result from impact of partially solidified bombs onto a hard surface or other bombs and blocks. In such cases, the fracture network shows no signs of expansion (Fig. 4.19d).

Bombs may consist of a single magma blob that has been fluidally contorted or consist of two or more coatings of magma. In addition, some bombs consist of agglutinated clasts, down to microns in size, so constituting composite bombs (Carracedo Sánchez et al. 2009). Juvenile bombs and cognate juvenile blocks may be cored by or contain xenoliths of mantle derivation (dunite lherzolite, harzburgite, wehrlite, pyroxenite, eclogite, anorthite; Fig. 4.19g), or from crustal sources (including deep-crustal metamorphic, mid crustal plutonic to upper crustal sedimentary and volcanic). Mantle xenoliths may show rounding due to magmatic or chemical resorption during their ascent (Fig. 4.19g), as well as the effects of mechanical abrasion (Sasse et al. 2020). Magmatic resorption reflects instability due to the changing temperature pressure regime that the magma is subjected to as it rises, causing some minerals and xenoliths to start dissolving back into the magma, so accounting for the rounding of crystal corners and edges, and rounded xenoliths. The presence of mantle xenoliths indicates that the host magma rose quickly though the crust, so preventing complete resorption or sedimentation in transit.

Ballistic clasts travel on approximately parabolic trajectories and could travel up to 4 kms from vent (Chap. 9). Around vent, coherent spatter bombs of magma may be so hot and fluidal when deposited, that they fuse together into lava, which flows into or away from the vent (Chap. 5). Other such lumps may retain identity and accumulate as aggregates of poorly defined spatter fragments called *agglutinate*. Magma "fire" fountaining can produce both spatter-fed lavas and agglutinate deposits around a vent, which could grow into spatter cones and even shield edifices.

Although we usually associate fluidal bomb forms with low-viscosity magmas such as basalts and phonolites, more silica rich magmas also form bombs that can be highly vesiculated, including rhyolites. In pumice fallout deposits, it is not unusual to find "outsize" pumice clasts that are unlikely to have been transported by convective transport in an eruption column. Some pumice bombs can also show breadcrust textures (Fig. 4.19h). Juvenile ballistic pyroclasts formed through vulcanian explosions generally have low-to-moderate levels of vesicularity (Morrissey and Mastin 2000; Wright et al. 2007; although cf. Fig. 4.21c), and bread-crusting of the surface of large bombs is common (Wright et al. 2007).

Country rock accessory ballistic blocks ejected from a vent can be of any rock type (sedimentary, metamorphic, plutonic), as well as penecontemporaneous volcanic rocks that the rising magma encounters in the conduit to the Earth's surface (Fig. 4.19i–k; also see Fig. 1.15b). Accessory lithics are almost exclusively liberated explosively, and so they are a good source of information about the subsurface geology of a volcano. If the depth of particular rock units below the Earth's surface is known from regional studies, they are also a good indicator of the depth of the fragmentation surface during an eruption, particularly early in an eruption, and/or the rock types exposed in the conduit above the fragmentation surface. Once an eruption has been underway for some time, there is increasing likelihood of recycling of clasts in the conduit, and gravitational collapse of debris from the crater rim or high in the conduit, so producing a more diverse assemblage of rock types in both extra- and intra-vent deposits.

Diagnostic criteria to recognise ballistic clasts are their unusually large size relative to smaller clast populations (lapilli, ash) around them, and impact sags in the substrate underneath (e.g. Fig. 4.19i–k and 1.15b).

4.8.2 Juvenile Magmatic (Non-ballistic) Pyroclasts

Juvenile pyroclasts or fragments represent samples of the erupting magma. They may therefore be partially crystallised, or uncrystallised, glassy and aphyric, depending on the pre-fragmentation history of the magma. On eruption they rapidly chill through the glass transition temperature to partially crystallised or aphyric glassy fragments. Because different magma compositions have different densities, viscosities and fluidity, pyroclasts from different magmas will develop varying morphologies during explosive ejection from the vent. Other factors controlling the morphology of particles include the degree of vesiculation of the magma, whether the style of fragmentation is magmatic or phreatomagmatic, and ductile or brittle, and the mode of transport (Chaps. 9 and 10). Benet et al. (2024) have produced an open file database illustrating characteristic features of > 6,300 pyroclasts of a variety of types (juvenile, free crystal, lithic, altered) and by quantifying 33 features of shape, texture and colour, they have assigned (statistically significant) eruption styles.

4.8.2.1 Mafic and Other Low-Viscosity Juvenile Pyroclasts

General Description

Mafic magmatic pyroclasts are dark in colour (black, dark grey, brown, and reddish to orange if oxidised and/or hydrated) and can have very varied morphologies because of the low-viscosity, very fluidal nature of basaltic, phonolitic, and other low-viscosity magmas (Macdonald 1972). They may be aphyric or porphyritic and have microlites in the groundmass. Principal magmatic crystals in mafic pyroclasts include olivine, pyroxene, and calcic plagioclase, and as the composition becomes less mafic, hornblende, biotite, and more sodic plagioclase become more prevalent. Fluidally shaped pyroclasts include large fluidal spatter and spindle bombs, and smaller smooth coated glassy achneliths, pele's tears, and pele's hairs, whereas vesiculated pyroclasts are called scoria (or cinders).

Achneliths, Pele's Tears, and Pele's Hairs

The morphology of lapilli and ash-sized particles derived from very fluid basaltic eruptions producing a large proportion of lava spray (e.g. magma "fire" fountaining style eruptions) often have smooth glassy surfaces moulded by surface tension in magma droplets (Fig. 4.20a, b). Such fragments include irregular smooth surfaced glassy *achneliths* (Walker and Croasdale 1972; Fig. 4.20a), and spherical to drop-like shapes called *Pele's tears* (Fig. 4.20b) and, in an extreme form, the long threads of glass or clusters of threads called *Pele's hair* (Macdonald 1972; Fig. 4.20b, c). Extreme vesiculation of low-viscosity magma under conditions of low gas overpressures can also produce *reticulite*, during which naturally spherical gas bubbles, as they continue to expand, are squeezed more and more together to the point that the shape that best accommodates the bubble volume is polyhedral (Fig. 4.20d). The vesicularity of reticulite varies from 70 to 90%. Jones et al. (2022b) have documented a new pyroclast type associated with the transition to or from low intensity fire fountaining, which they called inflated pyroclasts. They are spherical or ovoid, range from 0.2 to 5 cm in diameter, have a smooth outer shell enclosing several large vesicles, with smaller vesicles in the walls between the large vesicles and occur in association with large scoria deposits near the summit of small scoria and spatter cones in British Columbia, Canada. They envisage they were erupted as small magma droplets with low viscosity, at a temperature above the glass transition temperature, with a short trajectory before landing, followed by expansion of vesicles inside the droplets, in situ in the deposit (Jones et al. 2022b). They could represent an expansion of clasts which otherwise would be considered to be achneliths, which are often also quite vesiculated internally.

Scoria and Cuspate Glass Shards

Magmatic explosive eruptions of basalt and basaltic andesite, as well as other mafic compositions such as tephrites to phonolites, are dominated by lapilli-sized vesiculated fragments, called *scoria* (or cinders in the USA) (Fig. 4.21). Scoria is a general term for all dark-coloured vesiculated fragments of mafic to intermediate volcanic glass. Scoria forms in the conduit or vent as magmatic volatiles exsolve from rising mafic magmas forming gas bubbles or vesicles. Large clots of vesiculated magma form at the fragmentation

Fig. 4.20 Variety of ultra-low-viscosity basaltic pyroclasts whose morphology is caused by surface tension effects. **a** Smooth-surfaced achneliths with a glass coating. These clasts can be highly vesiculated internally. 1959 Kilauea Iki eruption, Hawai'i. **b** Pele's spherules, tears, and hairs, Mt Rouse volcano, Newer Volcanics Province, Victoria, southeastern Australia. Field of view is ~3 cm wide. (Photo courtesy of J. Boyce). **c** Pele's hairs, Halema'uma'u crater area, Kilauea, Hawai'i (*Source* United States Geological Survey; Photo taken 9-11-2016). **d** Extremely vesiculated basaltic reticulite, Hawai'i

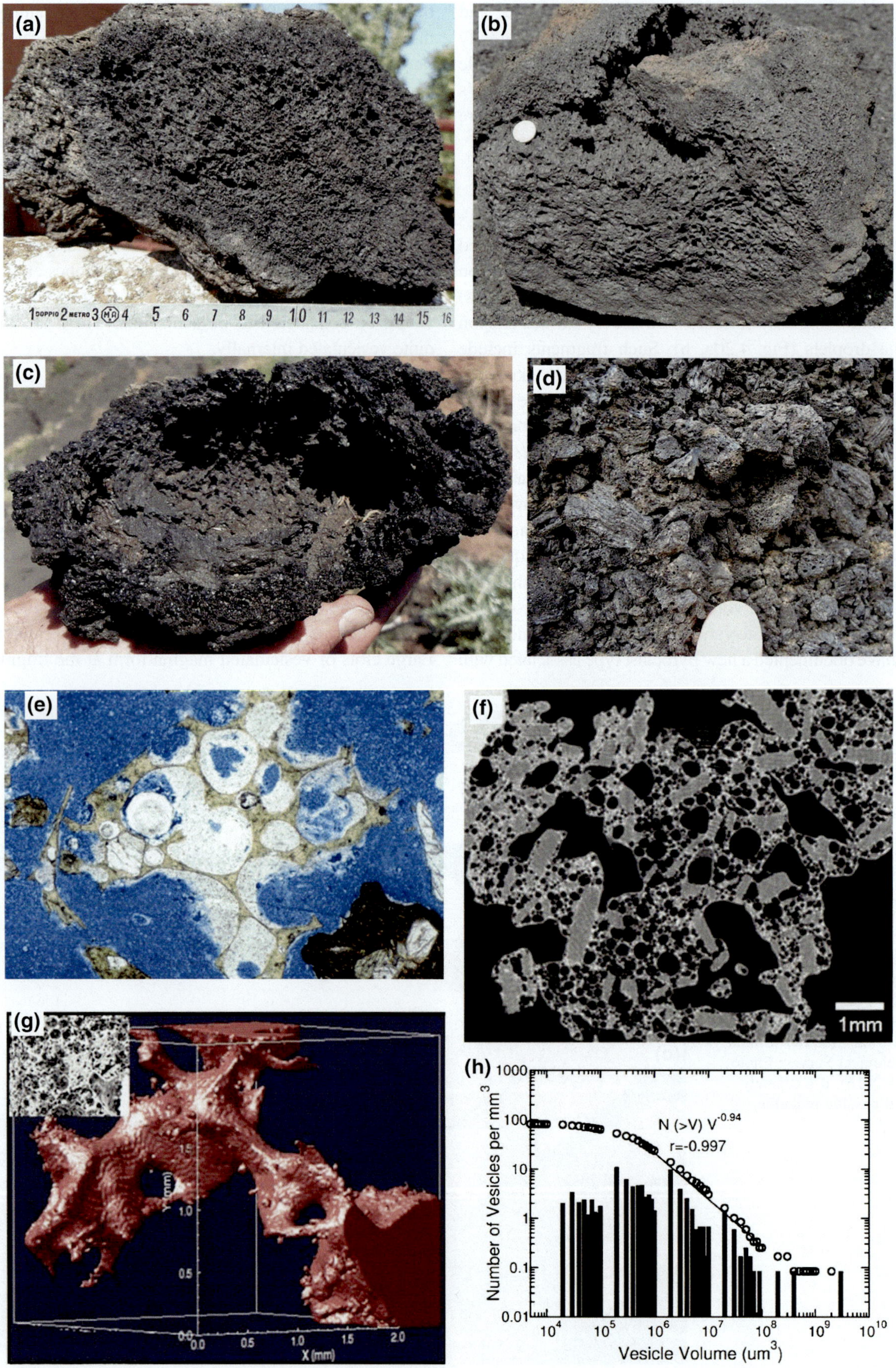
(a)
1 DOPPIO 2 METRO 3 4 5 6 7 8 9 10 11 12 13 14 15 16
(b)
(c)
(d)
(e)
(f)
1mm
(g)
Y (mm)
0.5
1.0
0.0
0.5
1.0
1.5
2.0
X (mm)
(h)
N (>V) V^-0.94
r=-0.997
Number of Vesicles per mm^3
1000
100
10
1
0.1
0.01
10^4
10^5
10^6
10^7
10^8
10^9
10^10
Vesicle Volume (um^3)

◀ **Fig. 4.21** Features of scoria pyroclasts. **a** Scoriaceous bomb ejected during strombolian-style activity, 3 July 2019, Stromboli, Italy. **b** Highly vesiculated scoria bomb, Etna, Sicily, Italy. Note large internal gas cavity. **c** Highly vesiculated scoriaceous bomb with cavernous interior, ejected during 2019 paroxysmal (vulcanian-style; Chap. 9) explosions at Stromboli. **d** Highly vesiculated scoria lapilli fallout deposit, Tower Hill volcanic complex, Newer Volcanics Province, Victoria, southeastern Australia. Note several clasts with stretched tube vesicles giving the clasts a stringy texture. **e** Photomicrograph of highly vesiculated scoria, Red Rock volcanic complex, Newer Volcanics Province, Victoria. Most vesicles are more or less spherical although mutual interference is beginning to distort them. In the centre, two vesicles have coalesced, producing a multi-lobate coalesced vesicle. **f–h** Vesicular characteristics of scoria erupted during typical activity at Stromboli in 2005 (From Polacci et al. 2008). **f** Tomographic image of vesicular scoria clast. Black features are vesicles, dark grey laths are plagioclase crystals, and light grey crystals are pyroxenes. Note population of small, spherical-to-sub-spherical vesicles compared with population of larger, coalesced vesicles with more complex irregular shapes and multi-lobate margins. **g** 3D view of large vesicle. Most large vesicles have irregular, tortuous, channel-like shapes generated by extensive coalescence of smaller vesicles and adjustment around the porphyritic crystal framework. **h** Cumulative (circles) and non-cumulative (bars) showing vesicle volume distribution. Vesicle volume distributions of scoria are comparable among different samples, span many orders of magnitude in size, and are characterised by a power law with an exponent near 1

surface through shearing or explosive bursting of large gas cavities in the magma, forming isolated vesiculated clasts in a gas stream. The gas bubbles may burst as they continue to expand and the bubble walls thin through viscous shear thinning, forming small, vesiculated shreds of scoria and curviplanar to *glass shards* of ash to fine ash sizes (e.g. Vinkler et al. 2012).

When the groundmass glass of scoria is black or dark brown, due to small crystals of iron oxide, usually magnetite, it is called tachylite glass. Pale brown mafic glass (called sideromelane) scoria fragments also occur. These lack magnetite and are considered to have quenched to glass at higher temperatures than tachylite. The term mafic pumice can also be used for scoria, but is uncommon. Scoria can show a wide range in vesicularity, and hence density. The larger fragments may have a ropey or stringy surface texture, but commonly scoria has very irregular margins characterised by truncated spherical vesicles (Fig. 4.21a–c). Scoria usually has 60–80% vesicles by volume (Fig. 4.21a–c), although even more highly vesiculated scoria can occur, which is known as reticulite (Fig. 4.20d).

However, it has become apparent that the scoria produced by highly explosive plinian eruptions may have vesicularities below 60 vol%, even ranging down to 45–50 vol% (Valdivia et al. 2022; Marshall et al. 2022). Such juvenile pyroclasts appear to have been fragmented in a brittle manner, resulting from rapid magma ascent, significant magma undercooling, groundmass microlite growth, sudden increased magma viscosity, and gas bubble overpressure, leading to brittle failure at high strain rates (Moitra et al. 2018, Arzilli et al. 2019a; Valdivia et al. 2022; Marshall et al. 2022). Such pyroclasts may be more blocky in shape, especially glass shards.

When vesicles first form, they are always spherical due to surface tension forces in the magma around the gas bubble. They may become elongated, ovoid, or tube like (Fig. 4.21d) due to shearing of the magma in the conduit as it rises towards the Earth's surface or distorted as they grow due to mutual interference. At this stage they may coalesce due to rupturing or shear thinning of the very thin bubble walls between large bubbles (Fig. 4.21e, f). Multi-lobate vesicles suggest coalescence of adjacent vesicles and suggest a longer phase of exsolution and bubble growth compared with a population of small, uniform, spherical bubbles (Fig. 4.21f–h; Polacci et al. 2008). Coalescence enhances open system degassing as permeability increases (Fig. 4.21h). Also, the usually extensive syn-eruptive crystallisation that occurs during decompression in the conduit and cooling of the pyroclasts may significantly affect the shape of bubbles and their coalescence (e.g. Shea et al. 2010a, b; Cimarelli et al. 2010; Vinkler et al. 2012; Voltolini et al. 2011).

4.8.2.2 Felsic Juvenile Pyroclasts—Pumice and Cuspate Glass Shards

Felsic magmatic pyroclasts (Fig. 4.22) originate from more silicic and viscous magmas and therefore tend to be light in colour (white, pale yellow, pale grey), often highly vesicular, and are called *pumice*. They do not form highly fluidal pyroclasts, except where the magma is peralkaline in composition, in which case pyroclasts can have similar morphologies to mafic pyroclasts, as described above. Most felsic pyroclasts tend to be roughly equant (Fig. 4.22a), but the margins may be irregular due to the morphology of truncated vesicles, or planar where they have fractured in brittle fashion in the eruption column or on impact with the ground. Platy magmatic pyroclasts may also form when the magma has developed a pre-existing planar fabric such as flow banding or tube vesicle textures before fragmentation. Pyroclasts range from blocks to lapilli and ash size fragments.

The products of highly explosive silicic to intermediate explosive eruptions are dominated by highly vesiculated, low-density juvenile *pumice* (Fig. 4.22a–h) and a large proportion of more finely fragmented vitric ash, usually in the form of cuspate glass shards and pumice shreds (Fig. 4.22i–m). Pumice clasts typically contain between 60 and 80% vesicles. The vesicles vary in shape from simple spherical (Fig. 4.22a, b), to coalesced and multi-lobate (Fig. 4.22c) to highly stretched tube vesicles (Fig. 4.22d–g). Simple spherical vesicles indicate relatively late exsolution, vesiculation in the conduit, and little degassing (closed

(a)
(b)
(c)
(d)
(e)
(f)
(g)
(h)
1 mm
1 mm

◀ **Fig. 4.22** Felsic pumice clasts and shards. **a** Equant, blocky, to subrounded highly vesiculated trachytic pumice clasts, some with irregular margins due to broken and truncated vesicles at the margin, others with planar margins suggesting brittle fracturing either in the eruption column or on impact on the ground. Some pumice clasts have complex coalesced vesicles (near finger), indicating a prolonged exsolution history, others (bottom) have simple spherical vesicles, suggesting late vesiculation in the conduit. Others have tube vesicles. Sub-rounding would have occurred in the conduit and eruption column. Ribeira Cha Formation, São Miguel, Azores, Portugal. **b** Photomicrograph of highly vesiculated pumice, with simple (sub-)spherical vesicles. Blue is dyed epoxy glue. Hatepe plinian pumice fallout deposit, Taupo 232 AD eruption, New Zealand. **c** SEM back-scatter image of highly vesiculated pumice from the 2.08 Ma Cerro Galan ignimbrite, northwestern Argentina. Many vesicles have coalesced and are distorted due to mutual interference, and some shearing of the vesiculating magma in the conduit. (Image by C. Folkes). **d** Spectacular tube pumice deposit, showing long tubes in longitudinal faces and small equidimensional vesicles in cross-section. Blocky form of pumice clasts suggests in-conduit brittle fracturing. Late Miocene (~6.6 Ma) Corte Blanco Tuff, Ramadas volcano, northwest Argentina. **e** SEM image of cross-section of tube vesicles in tube pumice, Corte Blanco Tuff (Image by M. Tait). **f** SEM image of longitudinal view of tube vesicles in tube pumice, Corte Blanco Tuff. Small spherical vesicles formed after the shearing of magma in the conduit that produced the tube vesicles. (Image by M. Tait). **g** 3-billion-year-old tube pumice clasts in the Cistern Formation, Red Hill, Pilbara Region, Western Australia. **h** Compositionally and/or texturally banded trachytic pumice, 4.6 ka Fogo A plinian fallout deposit, São Miguel, Azores, Portugal. **i** Photomicrograph of bubble wall, Y and cuspate rhyolitic glass shards, Early Devonian Devil's Pulpit Member, New South Wales, Australia. **j** SEM image of Y-shaped shard from the rhyolitic 184 ka Lower Pumice Formation, Santorini (Image by J. Simmons). **k** SEM image of pumice shred with truncated walls of many vesicles. Lower Pumice Formation, Santorini (Image by J. Simmons). **l** Plastically deformed, welded bubble wall, Y pumice shred glass shards, in the 360 ka Rangitaiki ignimbrite, Taupo Volcanic Zone, New Zealand. **m** Angular, blocky, and platey ash shards from the matrix of 1991–1995 block and ash flows, Mt Unzen, Japan (Image by D. Uehara)

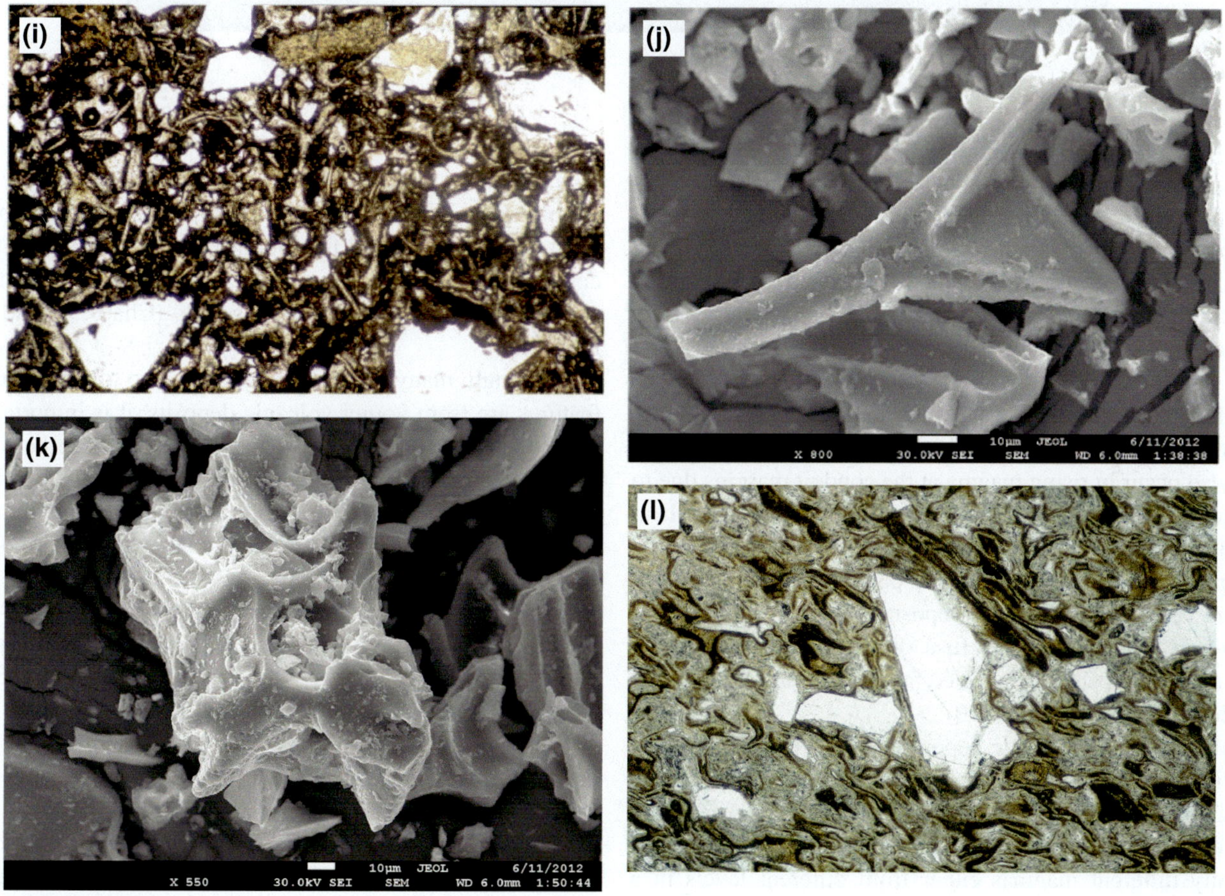

Fig. 4.22 (continued)

Fig. 4.22 (continued)

system), whereas complex, coalesced vesicles indicate early and prolonged exsolution and vesiculation, which is often corroborated by co-existing late, small spherical vesicles in the same pumice clasts (e.g. Fig. 4.22a). Stretched and tube vesicles indicate late shearing of originally spherical vesicles in the conduit, indicating very high shear in the magma and very high velocities through the conduit and exit velocities. Pumice with stretched or tube vesicles is common in highly explosive plinian pyroclastic deposits. Stretching of vesicles can also occur in rising magma at the conduit margins due to frictional coupling with the country rock, leading to a shear gradient from the conduit margin into the centre of the rising magma column, especially where conduits are narrow (e.g. fissure vents). Pumice clasts deposited at or above the glass transition temperature may fuse or weld together in some pyroclastic deposits (e.g. welded fallout deposits, Chap. 9, or welded ignimbrites, Chap. 10).

Also, many pyroclastic deposits contain banded pumice, consisting of fluidal layering of compositionally and/or texturally alternating light (more silicic?) and darker (more mafic?) magma (Fig. 4.22h). Mingling of two compositionally different magmas either from different levels in a compositionally zoned magma chamber or through injection of one magma from one source into the reservoir of another before eruption can produce banded or mixed pumice (Blake 1981b). The banding indicates that mingling occurred under laminar flow conditions in the source reservoir.

Very finely fragmented vitric ash is marked by spherical bubble wall, Y and cuspate shapes, representing shards of burst vesicle walls, and fine pumice shreds (Fig. 4.22i–l). If deposited in a hot enough and plastic state, shards may deform plastically and weld or fuse together (Chaps. 9 and 10; Fig. 4.22l).

Although many of the images used to illustrate these features are from young volcanic deposits, it is important to stress that pumice clasts and glass shards can also be found preserved in ancient volcanic successions (Fig. 4.22g, i), in spite of the effects of time, lithification, deformation, and alteration, especially in low strain domains. Figure 4.22g shows 3-billion-year-old tube pumice deposit from the Archean Pilbara Craton, Western Australia! Figure 4.20i shows well-preserved Devonian bubble wall and cuspate glass shards, even though the glass is now replaced by low-grade metamorphic minerals such as sericite.

Although pumice clast vesicularity and density can vary considerably, pumice clasts commonly have densities less than or slightly greater than 1.0 g cm^{-3} reflecting vesicularity levels of 60–80%. As a result, they are easily transported upwards convectively in the explosive eruption column and may float if deposited or erupted in water. Whitham and Sparks (1986) showed that hot pumice can become water-logged quickly if it falls into water, because the hot gases in the vesicles quickly condense, create a vacuum, or at least a pressure gradient that sucks water into the vesicles if there is a permeability network of vesicles (e.g. Fauria et al.

2017). When water-logged this increases the bulk density of pumice to the point where it sinks. However, other pumice clasts either lack permeability or have trapped air or gas bubbles in vesicle networks that block later infiltration by water and make them prone to float for a long time.

Primary pyroclastic pumice fragments may be angular or roughly rounded, the rounding being due to abrasion in the vent, eruption column, or during pyroclastic flow transport (see Chap. 7). However, pumice (and even shards) can withstand sedimentary reworking, mass-flow resedimentation, or long-distance flotation transport, which can cause significant further abrasion and smooth-rounded surfaces (Chaps. 3 and 12).

Although strictly not explosive in origin in many cases, the ash particles of block and ash flows, resulting from felsic lava dome collapse (Chap. 10), are also juvenile "pyroclasts". The 1991–1995 eruption of Mt Unzen in Japan was marked by about 10,000 gravitational collapse events from the growing margins of lava domes. The hot, brittle collapsing dome blocks shattered on impact with the substrate, producing blocks to ash-sized fragments in hot flows of debris called block and ash flows (e.g. Uehara et al. 2015; Chap. 10). The ash fragments are angular, blocky, and platey, resulting from brittle fracturing of large tumbling dome lava blocks, but are largely non-vesicular (Fig. 4.22m), because the source lava domes were generally poorly vesiculated.

4.8.2.3 Ultramafic, Low-Viscosity Juvenile Pyroclasts (Kimberlite, Lamproite, Kamafugite)

Understanding the origin of ultra-low-viscosity pyroclasts has been problematic, largely because most deposits are ultramafic in composition and prone to rapid alteration. This is particularly the case for kimberlites, which usually show pervasive alteration, overprinting, and replacement by secondary serpentine, carbonate, diopside, and chlorite, making it difficult to ascertain original clast morphologies and internal textures (Chap. 18). Pyroclasts include abundant country rock and mantle-derived accessory lithic clasts; abundant crystal fragments such as olivine, garnet, pyroxene, spinel, as well as juvenile magmatic pyroclasts, and occasionally accretionary lapilli. In this section, we will focus on juvenile clasts (e.g. Junqueira-Brod et al. 1999, 2005).

Although there are some suggestions that small juvenile spatter-like fragments occur (e.g. Fig. 4.23a), these could be an artefact of alteration of clasts and matrix around the margins of apparent clasts. Some of these clasts are vesicular (Fig. 4.23b), but most kimberlite pyroclasts are non-vesicular. Nonetheless, there is no doubt that kimberlite pipe forming eruptions were explosive (e.g. Sparks et al. 2006; Cas et al. 2008; Sect. 4.5.3; Chap. 18). Some authors have suggested that due to similarity of kimberlite pipe morphology to phreatomagmatic maar diatreme morphology that most kimberlite pipes and deposits are phreatomagmatic in origin, even in the absence of many of the characteristics of phreatomagmatic explosive deposits (e.g. abundant accretionary lapilli, abundant base surge deposits). However, it seems more likely that the explosions were driven by magmatic gases, especially CO_2 and perhaps some H_2O (Chap. 18). The absence of abundant vesicles in most juvenile pyroclasts is thus explained by the boiling off of gas bubbles from the extremely low-viscosity (~ 1 Pa s ?) magma and magma clasts during and after fragmentation, as discussed above in Sect. 4.5.5.4.

Most ultramafic pyroclastic deposits contain spherical pyroclasts called *pelletal lapilli* (Fig. 4.23c–e; also see Fig. 4.13a, and Sect. 4.5.5.4 for discussion of the mechanism of formation). They vary in diameter from mm to several cm, and usually they have a visible core or nucleus grain, often an olivine or pyroxene crystal or a lithic fragment (Fig. 4.23c–e). These core grains have a fine-grained rim that may be millimetres to a centimetre or more in thickness. The rims may appear to be massive or diffusely concentrically layered, and in some, microlitic crystals are aligned concentrically, parallel to the rim margin (Fig. 4.23d, e). In thin section, the nature of many of the fine components of the rim are not discernible due to the fine grainsize and the effects of alteration. Although it is commonly assumed that they consist of coherent solidified magma, which is probably true in many cases, in other cases, the rims could be fragmental consisting of micron-sized ash particles. As described previously (Sect. 4.5.5.4), Carracedo Sánchez et al. (2015) have documented microspheres of glass and crystal fragments in high-resolution SEM images in the rims of Pliocene ultrabasic pelletal lapilli in the deposits of the Cabezo Segura volcano, Spain (Fig. 4.23f, g). They propose an agglutination origin for the pelletal rims and dismiss an accretionary lapilli origin due to the absence of any evidence for a phreatomagmatic eruption origin, although some kimberlite pipes may contain true accretionary lapilli (e.g. Porritt and Cas 2009).

Although many kimberlite and lamproite pyroclasts are pelletal lapilli, others are more irregular to angular (Fig. 4.23h). In addition, many pyroclasts consist of olivine crystal fragments (mostly mantle xenocrysts) and country rock lithic fragments.

4.8.3 Phreatomagmatic Pyroclasts

Phreatomagmatic pyroclastic deposits are dominated by juvenile magmatic pyroclasts, and accessory lithic fragments. Many also contain accretionary lapilli, which are almost unique to explosive eruptions involving external water–magma interaction.

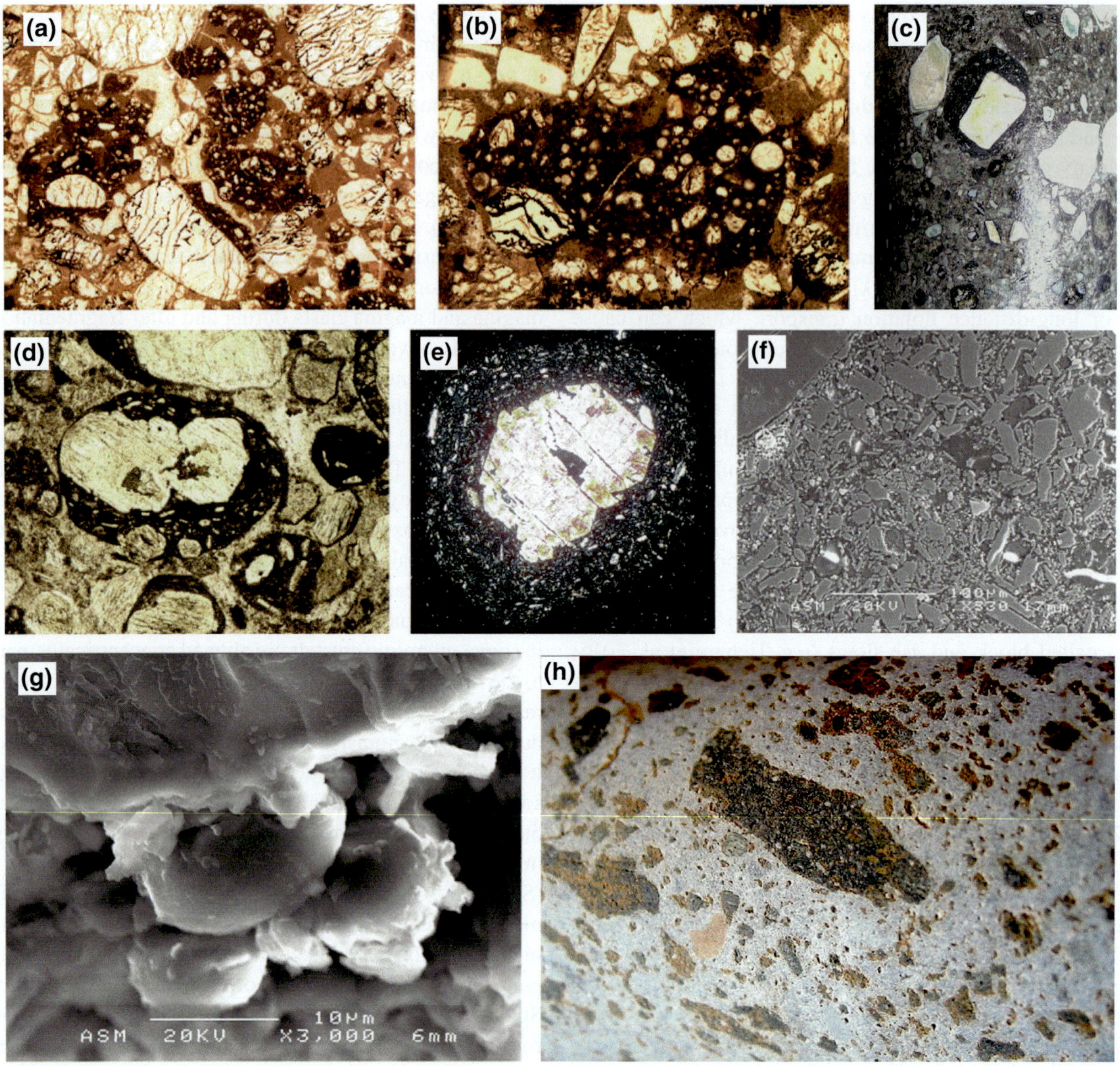

Fig. 4.23 Ultramafic (kimberlite, lamproite) pyroclasts. **a** Possible spatter-like clast (dark brown, right side) in the Cretaceous Orion Central kimberlite complex, Fort à la Corne kimberlite field, Saskatchewan, Canada. The pale brown area is pervasive serpentine which may be replacing both original matrix and clasts. Field of view is 13 mm across. (Image by A. Pittari). **b** Weakly vesicular kimberlite pyroclast, with infilling amygdaloidal spherical crystalline carbonate. Orion Central kimberlite complex, Fort à la Corne. **c** Pelletal lapilli with a core or nucleus grain of limestone mantled by fine-grained kimberlite, in (?) Jurassic Anuri kimberlite pipe, Nunavut, Canada. The pelletal lapilli occur in massive fragmental kimberlite. Drill core is ~5 cm in diameter. **d** Photomicrograph of pelletal lapilli in the Cretaceous Koffiefontein kimberlite pipe, South Africa. Note the concentric alignment of microlite crystals in the rim of the large lapillus, around the core olivine grain. Width of field of view is ~5 mm. **e** Photomicrograph of pelletal lapilli from the Pliocene Cabezo Segura ultramafic volcano, south central Spain, showing a core of olivine crystal and a rim with concentrically aligned microlites, which Carracedo-Sánchez et al. (2015) interpret as consisting or very fine, agglutinated ash. Width of field of view is 1.25 mm. **f** Back-scatter SEM thin section image a core olivine crystal (top left) and the rim of a pelletal lapillus dominated by clinopyroxene and titanomagnetite microcrystal fragments. Field of view is 350 microns across. (From Carracedo-Sánchez et al. 2015). **g** SEM image of microspheres of glass adhering to the margins of a core olivine crystal (From Carracedo-Sánchez et al. 2015). **h** Irregular juvenile pyroclasts in the lamproitic, 1.2 Ga Argyle diamond mine, Western Australia. Drill core is ~5 cm in diameter

4.8.3.1 Juvenile Phreatomagmatic Pyroclasts

Phreatomagmatically fragmented juvenile pyroclasts are frequently less fluidal and less vesicular than the pyroclasts of magmatic explosions. Ash-sized particles are commonly more blocky, and less cuspate (Heiken 1972, 1974; Self and Sparks 1978; Self et al. 1980; Wohletz 1983; Büttner et al. 1998; Morrissey et al. 2000; Zimanowski et al. 2015; Fig. 4.24a, b), unless the magma had vesiculated before phreatomagmatic explosive interaction with external water occurred, as happened during the 1963–1964 eruption of Surtsey, Iceland (Thorarinsson et al. 1964) and eruptions at Askja caldera, Iceland and Taupo caldera, New Zealand (Houghton and Carey 2019). In such cases (Fig. 4.24c, d), distinction between magmatic and phreatomagmatic deposits may not be simple (see below), because the deposits are hybrid, deriving explosive energy from both ends of the spectrum. There is a spectrum from purely magmatic to "dry" phreatomagmatic to "wet" phreatomagmatic deposits and eruption styles, and the characteristics of the juvenile pyroclasts reflect the different circumstances of eruption and fragmentation.

Phreatomagmatic basaltic pyroclasts are often fawn coloured to pale brown sideromelane glass (Fig. 4.24a), because basaltic magma is quenched to glass at high temperature before significant groundmass crystallisation has occurred, especially of magnetite which would render the glass a black colour. Sideromelane glass in "wet" phreatomagmatic deposits and in subaqueous settings alters by hydration and oxidation to orange-yellow, yellowish-brown to reddish brown palagonite, a heterogeneous mix of isotropic gel, various types of clays and zeolites, which often forms an initial rim around the margin of the sideromelane glass shards (Fig. 4.24a) but may pervasively alter such pyroclasts. Palagonite also forms readily through normal weathering (Hay and Iijima 1968; MacDonald 1972; Chap. 16). By contrast, silicic glass either devitrifies into a fine equigranular mosaic of fine quartz and feldspar crystals or alters to clays, such as montmorillonite or kaolinite.

Houghton and Wilson (1989) noted that phreatomagmatic pyroclasts generally have lower levels, and a greater range of vesicularities than magmatic pyroclasts, although Aravena et al. (2019) and Houghton and Carey (2019) point out that similar ranges of vesicularity can also be found in dry magmatic pumice deposits. As a result, vesicle walls are thicker, and therefore ash-sized particles will often be poorly vesicular, and generally blocky and angular in shape (Fig. 4.24a, b). However, some vesiculated and cuspate bubble wall forms can also occur (Fig. 4.24c, d), as was the case at Surtsey mentioned above, and many other volcanoes

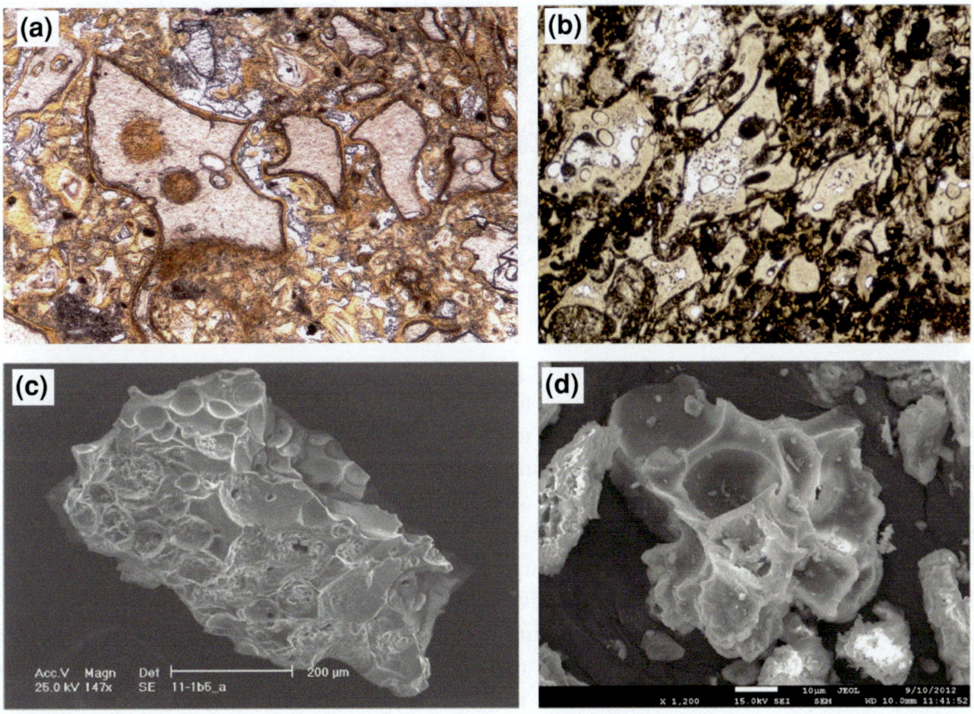

Fig. 4.24 Phreatomagmatic basaltic glass shards. **a** Photomicrograph of blocky, poorly vesiculated, basaltic sideromelane glass shards with orange palagonite alteration around the rims, Cape Bridgewater volcano, Newer Volcanics Province, Victoria, southeastern Australia. **b** Blocky, basaltic, slightly vesiculated glass shards, basaltic tuff cone, Late Devonian Boyd Volcanic Complex, Eden, southern New South Wales, Australia. **c** SEM image of micro-vesiculated ash particle, Asososca maar, Nicaragua (From Pardo et al. 2009). **d** SEM image of micro-vesiculated ash particle, Purrumbete maar, Newer Volcanics Province, Victoria (Image by S. Jordan; From Jordan et al. 2014)

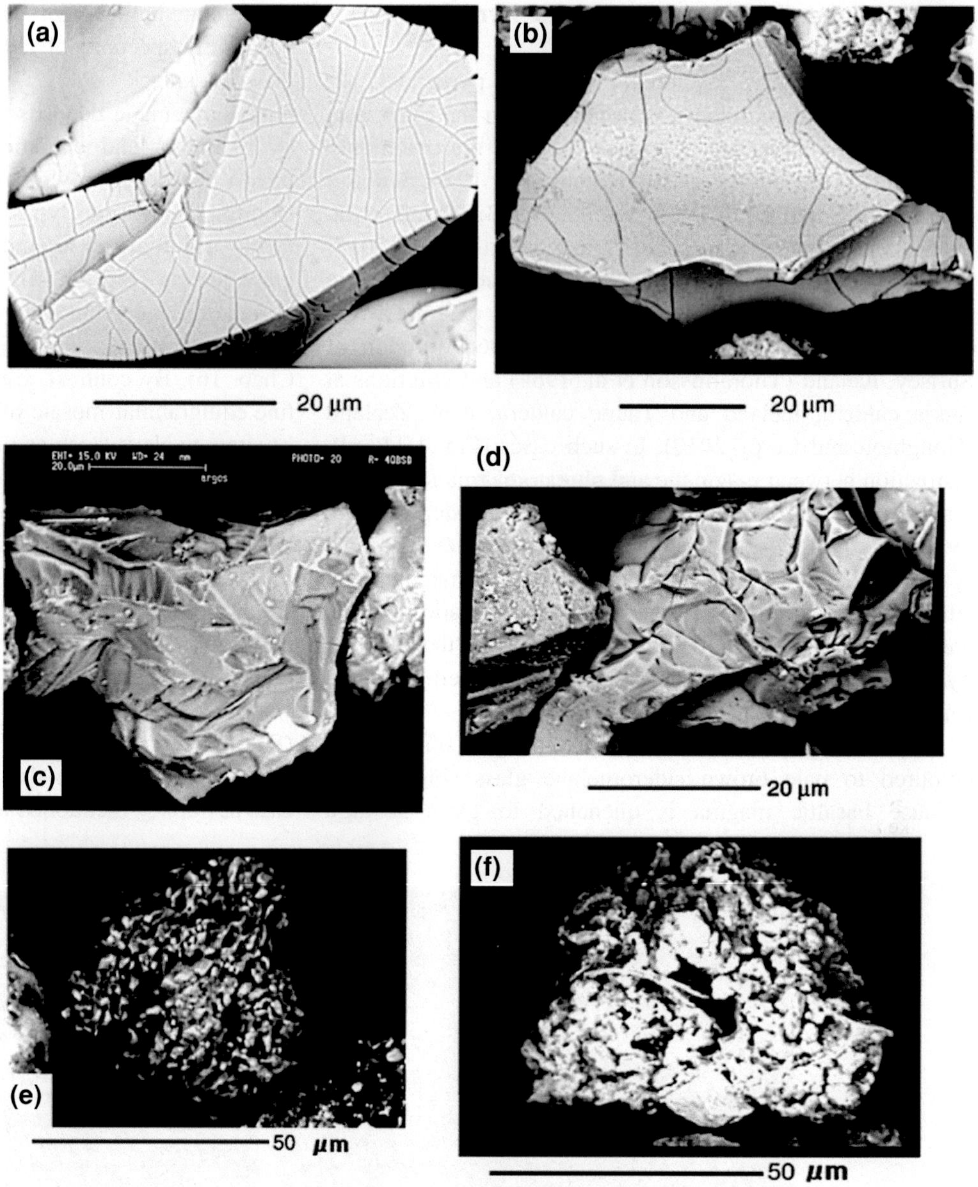

Fig. 4.25 SEM images of the features of "active" phreatomagmatic ash particles (Images all by P. Dellino; From Zimanowski et al. 2015). **a** Blocky fragment with curviplanar surfaces, and prominent cooling contraction quench cracks, from MFCI experiment. **b** As for (**a**) but a natural ash pyroclast. **c** Blocky basaltic ash particle with irregular step or chiselled features. **d** As for (**c**) but rhyolitic in composition. **e** Ash particle with moss-like or broccoli-like surface texture, from MFCI experiment. **f** As for (**e**) but a natural ash particle

dominated by phreatomagmatic explosive behaviour (e.g. Thorarinsson et al. 1964; Pardo et al. 2009; Jordan et al. 2014). SEM imaging by Heiken (1972, 1974), Büttner et al. (1999), Morrissey et al. (2000), Pardo et al. (2009), Dellino et al. (2012), Dürig et al. (2012), and Zimanowski et al. (2015) documented systematically this variation in morphology and texture compared with the more uniform cuspate bubble wall shards formed during magmatic explosive eruptions. . These studies and more recent ones, including experimental studies, have identified distinctive groups of juvenile ash particle morphology, which can be related to mechanisms of fragmentation (Wohletz 1983; Büttner et al. 1999; Morrissey et al. 2000; Dellino et al. 2012; Dürig et al. 2012; Zimanowski et al. 2015). In particular, "*active*" ash particles, which result from direct contact between magma and external water during explosive thermo-hydraulic explosive MFCI and IFCI interaction, have been distinguished from "*passive*" particles (those that are fragmented after thermo-hydraulic explosions) using both shape and surface textural features of ash particles.

For active ash particles, the principal ash clast morpho types (Zimanowski et al. 2015; Fig. 4.25) are:

(a) *Type A1: Blocky shards* with curviplanar surfaces and low vesicularity, showing quench fracture networks (Fig. 4.25a, b). These blocky shards may form through the propagation of stress (shock) waves through the magma near its surface and when it is, or is almost a solid, and through cooling contraction granulation during phreatomagmatism (e.g. van Otterloo et al.

2015; Colombier et al. 2019). These are known from both mafic and felsic deposits.

(b) *Type A2*: Irregular, blocky, angular ash particles with irregular, polyhedral step-like features (Fig. 4.25c, d), resulting from extreme brittle fracturing. These are known from both mafic and felsic deposits.

(c) *Type A3*: Moss-like or broccoli-like particles, consisting of irregular aggregates of fused very small particles (Fig. 4.25e, f). They are known only from basaltic ashes.

Passive ash particles (Fig. 4.26) include:

(d) *Type P1*: Spherical, pele's tears-like particles (Fig. 4.26a, b), known only from basaltic ashes.

(e) *Type P2*: Fluidal, elongate, pele's hair-like fragments (Fig. 4.26c), known only from basaltic ashes.

(f) *Type P3*: Blocky fragments without quench crack networks (Fig. 4.26d), known mostly from basaltic ashes, although also reported in deep water ash resulting from a deep water rhyolitic eruption (Murch et al. 2019b).

(g) *Type P4*: Platey fragments (Fig. 4.26e), known from both mafic and felsic deposits.

(h) *Type P5:* Curviplanar fragments (Fig. 4.26f), known from both mafic and felsic deposits.

Types P1-2 form in free air by ejection of melt not involved in active magma–water interaction. Types P3-5 fragments are the product of cooling contraction or thermal granulation of magma by water, i.e. quench fragmentation,

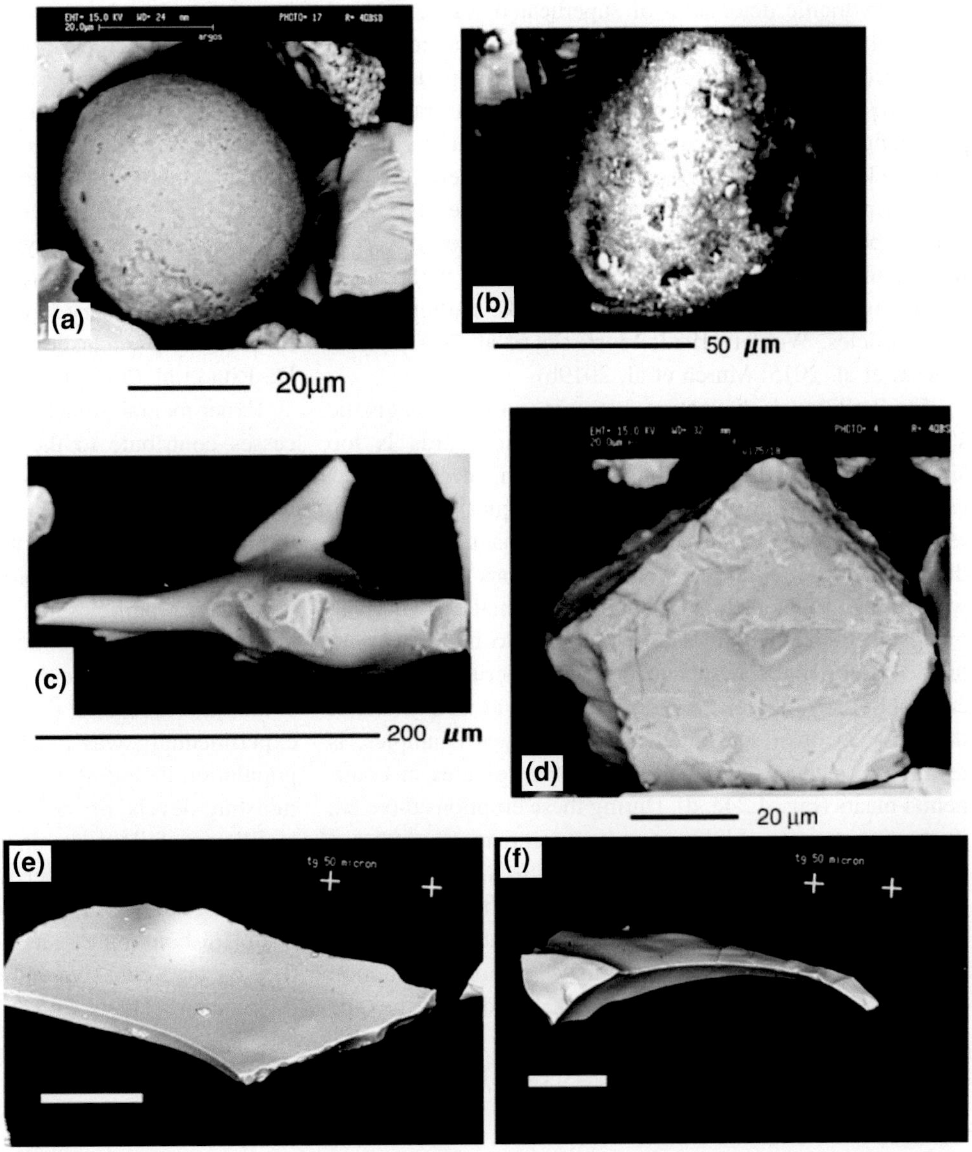

Fig. 4.26 SEM images of the features of "passive" phreatomagmatic ash particles (Images all by P. Dellino; From Zimanowski et al. 2015). **a** Rounded Pele's tear-like ash particle. **b** Subrounded Pele's tear-like ash particle. **c** Pele's hair-like fragment. **d** Blocky ash particle. **e** Platey ash particle. **f** Cuspate ash particle

although P3 could also form through explosive shockwave propagation.

In ancient successions, the three-dimensional form of shards may not be visible, but there may be good preservation of phreatomagmatic shard forms in thin section (e.g. variably vesiculated blocky to cuspate forms preserved in a Devonian basaltic tuff cone in the Late Devonian, Boyd Volcanic Complex, southeastern Australia; Fig. 4.24b).

The shape of the juvenile shards produced during phreatomagmatic explosions depends on the viscosity, surface tension, yield strength and vesicularity of the melt, and the rate of heat energy transfer between magma and water, which depends on the nature of the interface and phase nature of the water (liquid, steam) at the interface, which changes cyclically during MFCI and IFCI. Fragmentation may be of a brittle or ductile nature (Wohletz 1983; Zimanowski et al. 2015). Brittle fracturing occurs during the propagation of explosive shock waves during thermo-hydraulic detonation of superheated water, producing highly irregular blocky fragments, as well as moss-like or broccoli-like aggregates of very fine ash particles. In addition, quenching causes cooling contraction granulation producing blocky fragments with internal cooling contraction cracks and the disaggregated finer components, as well as curviplanar and platey fragments (passive particles). Ductile behaviour results from the surface tension effects in the magma in the absence of direct contact with water, producing drop-like and elongated hair-like particles (passive particles; Wohletz 1983; Morrissey et al. 2000; Zimanowski et al. 2015; Murch et al. 2019b).

The traditional distinction between cuspate magmatic shards and more blocky phreatomagmatic shards is too simplistic. In practice, a complete spectrum of shard types can form during any phreatomagmatic explosive eruption, reflecting varying degrees of direct water–magma interaction during an MFCI cycle, as well as transitions from optimal water–magma interaction to purely magmatic gas-driven explosive fragmentation. The moderately to highly vesiculated shards from the subaqueous to subaerial marine tuff cone at Surtsey, in Iceland (Thorarinsson et al. 1964), one of the classical open sea phreatomagmatic type examples, is testimony to this, as is diverse pyroclast textures in continental maars (Fig. 4.24c, d). During these eruptions there are fluctuating degrees of interaction between rising magma and surface water or groundwater aquifers (e.g. van Otterloo et al. 2013; Jordan et al. 2013, 2014; Murch et al. 2019b), leading to fluctuations between "wet" phreatomagmatic, to "dry" phreatomagmatic, to magmatic explosive eruption styles, and even effusive eruption phases during the one eruption.

4.8.3.2 Accretionary Lapilli and Pellets

A common pyroclast type of phreatic, phreatomagmatic, and hydrothermal explosive eruptions consists of spherical aggregates of ash called *accretionary lapilli* and *ash pellets* (Fig. 4.27a–c; Moore and Peck 1962; Gilbert and Lane 1994; Schumacher and Schmincke 1991, 1995; Thordarson 2004; Brown et al. 2010; Mueller et al. 2016). Accretionary lapilli usually have a massive core of ash and one or more outer concentric layers (Fig. 4.27a, c, f, g), whereas ash pellets are massive and unlayered (Fig. 4.27f). They can both range in diameter from a fraction of a mm to 20 or 30 mm and occasionally up to 50 mm (Fig. 4.27a–g). Some may have a large nucleus or core clast, such as a pumice or lithic clast, coated in massive ash or concentric laminae of fine ash (sometimes termed cored lapilli; Fig. 4.27d, 10.105b). Although the ash particles that accrete are very fine, under high-resolution petrographic and SEM microscopy, individual ash particles, including glass shards, can be distinguished, even in old pyroclastic deposits (e.g. Fig. 4.27e). Accretionary lapilli and ash pellets form most commonly where water in the forms of steam, liquid, or even ice are present in a turbulent cloud of ash. This could be in an explosive eruption column above a vent, or in the ash cloud of pyroclastic density currents, such as surges or pyroclastic flows. Brown et al. (2010) proposed that with those associated with pyroclastic density currents, massive ash pellets formed in ash clouds above pyroclastic flows, whereas accretionary lapilli (i.e. concentric laminated rims) form in the pyroclastic flows. (Also see Fisher et al. 1980; De Rita et al. 2002; Giordano et al. 2008).

Experimental studies have demonstrated that many processes contribute to the progressive aggregation of accretionary lapilli and ash pellets. These include ash particles with adhered moisture droplets adhering to each other though surface tension and capillary forces; aggregation through electrostatic forces between ash particles, and between ash particles and water droplets or ice particles (e.g. high in an eruption column); and even chemical adhesion due to reactivity of acidic volatiles in the ash cloud. Mueller et al. (2016) found that formation of accretionary lapilli experimentally was enhanced by a polydisperse ash-sized population including ash <40 and up to 300 μm, and water moisture levels of <45% relative humidity. Accretionary lapilli can fallout as unconsolidated ash aggregates that disintegrate upon impact with the ground, or they may be cohesive, but ductile, or through either baking or freezing or chemical bonding effects of NaCl adhered to ash particles, they become cohesive and can be deposited as solid pellets. At moisture levels >45% relative humidity, non-cohesive ash aggregates form (Mueller et al. 2016) that can

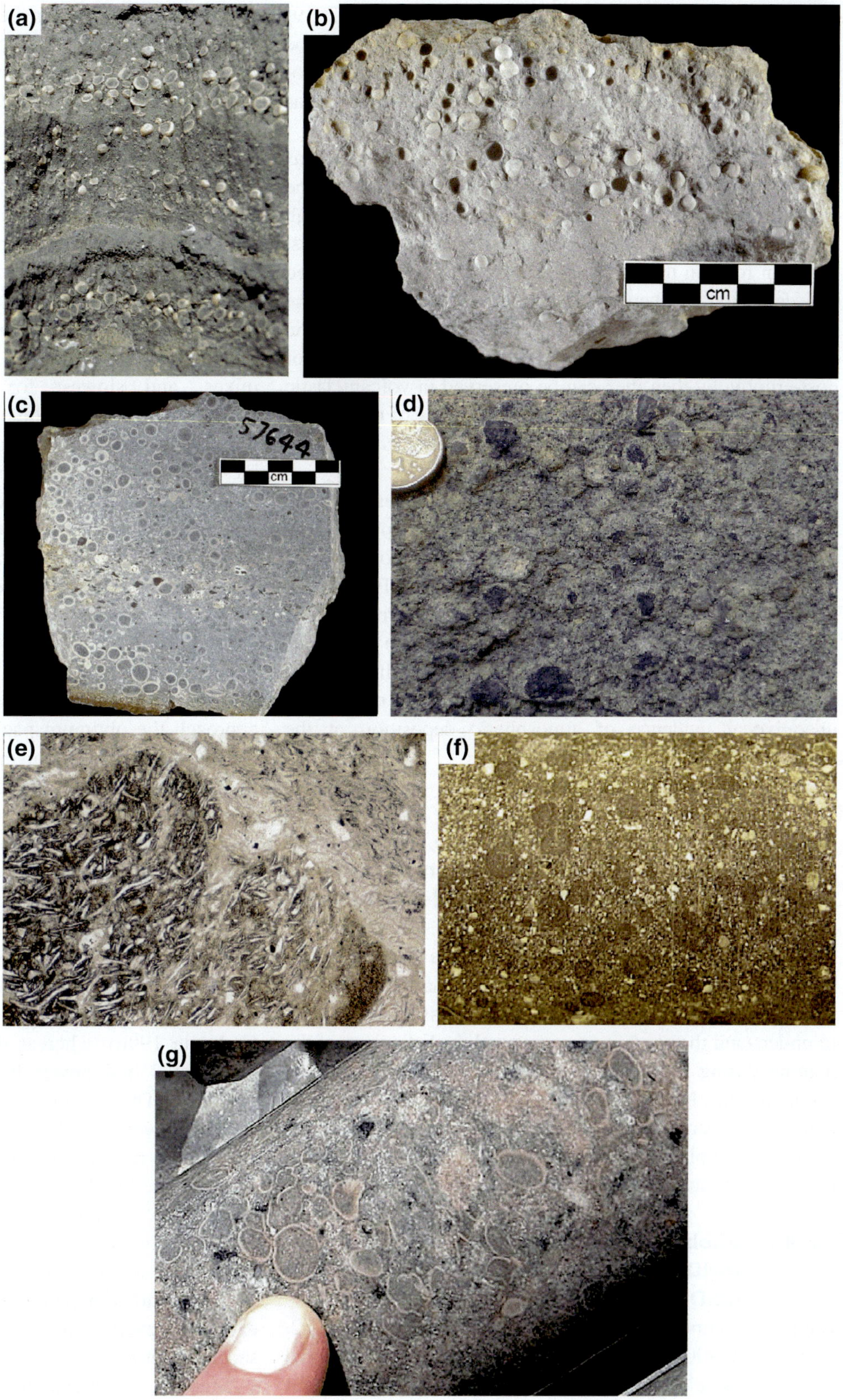

Fig. 4.27 Accretionary lapilli. **a** Basaltic accretionary lapilli in base surge deposits from the 1965 phreatomagmatic eruption of Taal volcano, Philippines. **b** Rhyolitic accretionary lapilli in base surge deposits in a tuff ring sequence, Early Devonian Snowy River Volcanics, Victoria, Australia. **c** Slabbed section through **b** showing the fine ash rims around massive ash cores in the rhyolitic accretionary lapilli. **d** Basaltic-cored lapilli from the Cape Bridgewater tuff cone complex, Newer Volcanics Province, Victoria. The black cores of the lapilli are variably vesiculated basalt scoria clasts. **e** Photomicrograph of cross-section through part of fragmented accretionary lapilli, showing abundant bubble wall glass shards. The original lapilli rims are the dark, fine-grained layer to the bottom right. The deposit is a rhyolitic phreatomagmatic ignimbrite. Late Devonian Boyd Volcanic Complex, Eden, southern New South Wales, Australia. Field of view ~3 mm wide. **f** Accretionary lapilli and massive ash pellets in the Eocene Fox kimberlite pipe, Ekati kimberlite field, Lac de Gras, Northwest Territories, Canada. Drill core is ~5 cm diameter. **g** Accretionary lapilli in 2.68 Ga volcaniclastic deposits, Kanowna Belle gold mine host rock succession, Kalgoorlie, Western Australia. Drill core is ~5 cm diameter

disintegrate on impact with a surface. Morgavi et al. (2019) propose that the formation of accretionary lapilli may be enhanced in ash clouds developing by interaction of pyroclastic flows with seawater, as salt water droplets may act as an aggregation and cementing agent during the accretion.

Accretionary lapilli and ash pellets can be deposited as fallout clasts with higher terminal fall velocities than the component ash particles, they can be transferred through eruption column collapse into surges and pyroclastic flows, they can form in the ash clouds of these and be transported and deposited by them, including long distances from the source vent (Brown et al. 2012). If such primary pyroclastic deposits are eroded and reworked, either syn-eruptively or post-eruptively, then they can be preserved in volcaniclastic sediments if they were already hard at the time of reworking. They are therefore not indicative of proximity to vent or of a particular deposit type.

Accretionary lapilli and ash pellets can form during hydrovolcanic explosive activity involving any magma composition, from basaltic (Fig. 4.27a, d) to rhyolitic (Fig. 4.27b, c, e) to ultramafic (e.g. kimberlite; Fig. 4.27f). They can be found in Recent (Fig. 4.27a) to Precambrian deposits (Fig. 4.27g). Similar structures are also known to occur in breccia pipes and complexes (McCallum 1985; Chap. 18, Fig. 18.26b), so they are not necessarily unequivocal evidence of atmospheric deposition at the Earth's surface.

4.8.4 Bubble Number, Bubble Size, and Crystal Size and Number Distribution Data (BND, BSD, CSD, CND) from Juvenile Pyroclasts

There has been much study of the vesicle populations in scoria and more silicic pumice, as well as the fine microlites that are often found in the glassy groundmass of such clasts, to understand the conduit processes and the behaviour of the magma during its ascent to the surface. It is beyond the scope of this book to review these topics in detail, but readers are referred to the key references listed here to follow up on the application of this relatively new approach to better understanding explosive eruption dynamics.

4.8.4.1 Bubble Number Density (BND) and Bubble Size Distributions (BSD)

Bubble exsolution has been discussed above in Sect. 4.5.2. Toramaru (1990, 1995, 2006) first recognised the potential power of vesicle size and vesicle number density data for interpreting eruption parameters, and others have embraced the approach. Here we will focus on the factors that influence the bubble size distribution (BSD) and bubble or vesicle number density (BND; VND) which can be used to interpret eruption parameters such as decompression rate, eruption column height (Toramaru 1990, 1995, 2006), and bubble nucleation rates, which relate to eruption styles (Mangan and Cashman 1996). Shea et al. (2010a, b) have provided an excellent summary of the methods for documenting BSD data and how that data can be interpreted (also see Sect. 9.6.2.5).

Because of surface tension effects, exsolving volatile bubbles in magmas are spherical, initially small (microns), and filled with supercritical fluid, not gas. Initially, therefore, the growth of exsolved bubbles depends on the diffusion rate of specific volatiles from the magma into the newly formed bubbles, which is highest for H_2O, lower for H_2O–CO_2 mixes, and slowest for CO_2 (Lautze et al. 2011). Diffusion-controlled bubble growth is slow at the supercritical fluid stage, but once the supercritical fluids in the bubbles boil to gas as a result of decompression of magma (e.g. rising in a conduit) as it rises to depths less than the critical pressure depth equivalent for a particular volatile species, the growth of bubbles accelerates in response to the rate of decompression. Bubble growth then becomes more controlled by decompression than diffusion, but this also depends on the magma rise rate and thus the decompression rate. The growth of bubbles through these processes is called (Ostwaldt) bubble "ripening". As the levels of vesiculation approach the optimum packing level for spherical bubbles ($\sim$66% or higher in polydisperse bubble populations), further bubble growth leads to bubble wall ductile stretching, thinning, ductile rupture, and bubble coalescence, or, bubble wall brittle fracturing if growth rates and strain rates are high. Once exsolution of volatiles begins it continues until explosive fragmentation occurs and pyroclasts are chilled through the glass transition.

Steady, continuous volatile exsolution, or equilibrium exsolution produces a continuous unimodal BSD as in low-intensity explosive events (e.g. reticulite, Mangan and Cashman 1996), whereas during disequilibrium exsolution exponential and power law bubble size distributions are common. These coincide with more explosive deposits (BSD; Blower et al. 2002, e.g. scoria, Mangan and Cashman 1996), and reflect continuous and usually multiple exsolution events. Small bubble populations with a simple unimodal size distribution are consistent with a single, late, rapid exsolution event (Klug et al. 2002; Alfano et al. 2012). Polymodal populations, particularly with large coalesced vesicles, and a population of small vesicles suggest two exsolution events, one deep in the conduit and early, and the second high and late.

Shea et al. (2010a, b) summarised the ways in which different BSD profiles could be interpreted, including kinematics of nucleation, bubble nucleation and growth rates, bubble ripening, and coalescence. Bubble number density (BND) is controlled by magma properties, volatile concentration and overpressure, exsolution style (homogenous, heterogeneous), timing of exsolution during magma ascent, magma ascent rate, decompression rate, and bubble size. A sample with small bubbles is likely to have a higher BND than a sample with large bubbles. Toramaru (2006) proposed that BND could be used as a measure of decompression rate, as explained in Sect. 9.6.2.5, and using the following equation:

$$dP/dt = a_0\, D\, \sigma^2\, P_w^{-0.33}\, T^{-0.5} N_v^{0.67} \quad (4.25)$$

where a_0 is a constant (= 1 × 10^{15}), D is the diffusivity of water in a silicate magma ($m^2\ s^{-1}$; see Eqs. 4.11–4.13), σ is surface tension ($N\ m^{-1}$), P_w is water saturation pressure (MPa), T is the magma temperature (in °K), and N_V is the number density of vesicles in the size range 1–10 µm, which are considered to be the last formed vesicle population resulting from rise in the conduit and decompression. Given that only a fraction of the volatile content of the magma contributes to the final small vesicle population, Simmons et al. (2017a) corrected the initial water content using the vesicle content of >10 µm vesicles, and applying the equilibrium models of Gardner et al. (1999) and Shea et al. (2011).

Studies of bubble number densities indicate a correlation between BND and explosive eruption intensity and style, including magma decompression rate and magma mass discharge rate (Table 4.1; Alfano et al. 2012). Lautze and Houghton (2007) evaluated BND for weakly explosive basaltic andesite events on Stromboli in 2002 at ~10^6 cm^{-3}, whereas Toramaru (2006) calculated BND for sub-plinian rhyolitic fallout deposits erupted from Towada volcano, Japan, of ~10^7 cm^{-3}, and 10^8 cm^{-3} for the 1980 Mt St Helens deposits. Simmons et al. (2017a) calculated BND ~10^9 for the 184 ka Lower Pumice 1 rhyodacitic plinian fallout, and succeeding ignimbrites on Santorini, increasing to 10^{10} cm^{-3} for the climactic co-ignimbrite lag breccia deposits (Chap. 10). Vinkler et al. (2012) calculated BND of 10^5 cm^{-3} for the tephri-phonolitic scoria clasts in the basal plinian fallout of the ~355 ka Villa Senni eruption sequence from Colli Albani volcano, Italy, and 10^8 cm^{-3} for the first major ignimbrite, and then waning values for higher eruption units. However, Alfano et al. (2012) point out that the relationship between BND and eruption intensity and style is not simple, but seems to vary with magma composition, which is to be expected given the varying rheological behaviour of magmas of different compositions.

BND data has also been used to calculate decompression rates, and infer magma mass eruption rates and eruption column heights (e.g. Toramaru 2006; Alfano et al. 2012; Simmons et al. 2017a; Table 4.1). Simmons et al. (2017a) used a combination of pumice clast density, BSD and BND data (Fig. 4.28) to trace the evolution of the 184 ka Lower Pumice 1 plinian eruption sequence on Santorini demonstrating that decompression rates and mass eruption rates progressively increased to very high levels at the culminating stage of caldera collapse (Fig. 4.29). Using a similar approach, similar results were obtained for the 4.1 ka Agnano Monte Spina plinian eruption at Campi Flegrei by Romano et al. (2020), where the pumice associated with the opening plinian phases of each of the three subsequent pulses of the eruption (units A1, B1, D1) shows a progressive increase in BND from 2 × 10^8 cm^{-3} to 3.6 × 10^8 cm^{-3}, corresponding to calculated decompression rates ranging from 0.13 to 0.21 MPa s^{-1}, which match with the reconstructed increase in plume height from 5 to 27 km. Furthermore, the BND drastically drops to values around 10^7 cm^{-3} in pumice of the associated collapsing pyroclastic flow phases (A2, B2, D2), as a result of a decrease in ascent rates.

Although BND data is now widely being used, there are some limitations that need to be investigated to make the method an even more robust measure of magma dynamics in the conduit. First, the method assumes essentially equilibrium degassing, which is clearly often not the case. Secondly, different scales of bubble size analysis have been used based on accessible instrumentation and resolution levels, placing limitations on the degree to which the results of different studies can be compared. Thirdly, more work needs to be done to compare the results from viscous silicic magmas, with slow diffusion rates, and low-viscosity (mafic) magmas. Toramaru (2006) and Shea et al. (2010a, b) recommend assessing BND only for vesicles <0.01 mm in diameter to ensure that only the last vesicles that nucleated in the conduit are considered in assessing decompression rates, so providing insights into the final ascent behaviour of the magma during eruption only.

4.8.4.2 Crystal Microlite Size Distributions and Crystal Microlite Number Density

Studies of crystallisation kinetics, including crystal size distributions and number densities, have long been used to better understand the petrological evolution of coherent igneous bodies (e.g. Cashman 1988; Marsh 1988, 1998; Mangan 1990; Cashman and Mangan 1994; Cashman and Marsh 1988; Higgins 1996). Early studies focussed on crystallisation kinetics in magma chambers, dykes and lava lakes and lavas, and on magma cooling and ascent rates in the conduit during lava eruptions. Subsequently, interest has focussed on the microlite populations in pyroclasts (e.g. plagioclase, Fe–Ti oxides, pyroxene), their nucleation, growth, and cooling rates to better understand magma ascent rates and effects on processes during explosive eruptions.

The fundamental controls on crystal nucleation and growth, including microlites, are decompression, exsolution of H_2O, and cooling rates (Cashman and Blundy 2000). Crystals can form under homogeneous crystallisation regimes in which magmas are at liquidus temperatures, with

Table. 4.1 Compilation of bubble number density (BND), decompression rate, inferred magma mass discharge rate, and eruption column height for varying intensity and style explosive eruptions

	Eruption		Composition, SiO_2 wt%	Eruption style	H_t, km	BND ($N^m{}_V$), cm^{-3}	MDR, $kg\ s^{-1}$	Decompression rate ($\Delta P/\Delta t$), $MPa\ s^{-1}$
1	Askja 1875 (Iceland)	Unit B	55–70	Sub-Plinian	8	9.0×10^8	2.6×10^6	
		Unit C		Plinian	23	2.0×10^9	6.8×10^7	
		Unit D		Plinian	26	1.4×10^9	3.5×10^7	
2	Chaitén May 6, 2008 (Chile)		74.2	Sub-Plinian	19	1.3×10^8	4.2×10^7	7.9 ± 2.8
3	Cotopaxi (Ecuador)	Layer 1	56.7	Plinian	33	1.1×10^8	1.6×10^7	
		Layer 2	59.1		28	4.8×10^8	8.1×10^7	
		Layer 3	57.9		29	2.7×10^8	9.3×10^7	
4	Etna 122 BC (Italy)	Unit C	49	Plinian	26	1.0×10^8	8.5×10^7	
		Unit E	49		26	8.7×10^7	8.5×10^7	
5	Fontana lapilli 60 ka (Nicaragua)	Microlite-rich	53	Plinian	32	8.7×10^7	1.4×10^8	
		Microlite-poor				5.8×10^7		
6	Izu Oshima (Japan)	1986 B	55	Sub-Plinian	16	2.1×10^7	2.1×10^7	4.9
7	Kilauea Iki (Hawaii)	1959	50	Hawaiian		1.8×10^7	1.0×10^5	
8	9.4 ka Kilian volcano (Chaine des Puys, France)	U1 base	62–66	PDC phase		1.6–3.8×10^9		5.2–9.2
		U1 middle		PDC phase		2.7×10^9		7.3
		U2 base		PDC phase		3.4×10^9		8.5
		K1 base		Plume/fallout		5.4×10^9		11.5
		K1 top		Plume/fallout		1.1–4.4×10^9		4.0–10.0
9	184 ka Lower Pumice 1 (Santorini, Greece)	LP1-A1	68.5	Plinian		1.2–1.7×10^9		7–9
		LP1-B4	67.1	PDC phase	N/A	1.3–2.1×10^9		7–10
		LP1-C1	68.1	PDC phase/caldera collapse	N/A	4.5×10^9–1.1×10^{10}		15–28
10	172 ka Lower Pumice 2 (Santorini, Greece)	LP2-A2-1	69	Plinian		1.7–5.1×10^9		7–18
		LP2-A2-2				3.0–6.2×10^9		13–20
		LP2-A3			~36	6.2×10^9–1.0×10^{10}	~10^8	19–29
11	7.7 ka Mt Mazama (Crater Lake, USA)	CP	70	Plinian		4.9×10^9		
		PF		PDC phase/caldera collapse	N/A	1.0×10^{10}	~ 10^9	
12	Mt St Helens 1980 (USA)		66.0	Plinian	19	8.2×10^9	1.9×10^7	1.4×10^2
13	Novarupta 1912 (Alaska, USA)	Ep. II	73–78	Plinian	22–25	9.6×10^8	2.0×10^8	
		Ep. III			17–23	2.0×10^9	4.0×10^7	
14	Stromboli 2002 (Italy)	May		Strombolian Weak, ash-rich eruptions		9.9×10^2		
		Sept./Oct		Spatter-rich eruptions		1.9×10^2–1.7×10^3		
		December		Power explosions		7.4×10^2–3.4×10^3		
15	Tarawera 1886 (New Zealand)		51	Plinian	>28	4.0×10^6	1.8×10^8	

(continued)

Table. 4.1 (continued)

	Eruption		Composition, SiO_2 wt%	Eruption style	H_t, km	BND ($N^m{}_V$), cm^{-3}	MDR, kg s^{-1}	Decompression rate ($\Delta P/\Delta t$), MPa s^{-1}
16	Taupo 232 AD (New Zealand)	Unit 2	74	Plinian		1.0×10^9	1.0×10^7	
		Unit 3		Phreatoplinian		4.0×10^9	1.0×10^6	
		Unit 4		Phreatoplinian		3.0×10^8	1.0×10^6	
		Unit 5		Plinian	<55	3.0×10^9	1.0×10^8	
		Unit 6		PDC phase		5.0×10^9	1.0×10^9	
17	915 AD—1 ka Towada (Japan)	A	71.4	Plinian	30	6.0×10^8	2.6×10^8	64
		B	74.2	Sub-Plinian	19	1.2×10^8	4.2×10^7	18
		C	66.7	Plinian	32	3.7×10^8	3.4×10^8	70
		D	67.8	Sub-Plinian	16	1.9×10^8	2.1×10^7	41
		E	65.7	Plinian	25	2.2×10^8	1.3×10^8	56
		F	61.1		23	2.0×10^8	9.0×10^7	91
		G	66.1	Sub-Plinian	12	9.0×10^6	6.7×10^6	6.3
18	Vesuvius 79 AD (Italy)	EU1 fall	54.1	Sub-Plinian	15	3.3×10^9	1.0×10^7	0.4
		EU2 fall	55.4	Plinian	26	1.5×10^{10}	8.0×10^7	1.1
		EU3 max	54.7		29	6.4×10^9	1.0×10^8	6.2
		EU4 fall	54.1		23	3.3×10^9	4.0×10^9	0.4
19	Vesuvius 512 AD (Italy)	U5	59	Sub-Plinian	7–15	1.4–6.3×10^8	5.0×10^6	
		U7			6–9	1.2–8.9×10^8	1.0×10^6	
20	Villarica 2004 (Chile)	Scoria	54.1	Hawaiian	≥ 0.08	$0.41.6 \times 10^6$	1.0×10^4	
		Golden Pumice				0.1–0.7×10^6		
21	Villa Senni 355 ka (Colli Albani, Italy)	VSN0 (SC1)		Plinian		10^5–10^6		
		VSN0 (SC2)		Plinian		10^6		
		VSN1 (SC1)		PDC		10^7–10^8		
		VSN1 (SC2)		PDC		10^6–10^8		
		VSN1 (BSC2)		PDC/Caldera collapse		10^5		
		VSN2 (BSP)		PDC/Caldera collapse		10^5–10^7		
		VSN2		PDC		10^5–10^6		
		VSN3		Fallout		10^5		
22	Agnano Monte Spina 4.6 ka (Campi Flegrei, Italy)	Unit A1	61.92	Sub-Plinian	5	2×10^8	2.07×10^5	0.13
		Unit B1	62.18	Plinian	23	3.13×10^8	8.89×10^7	0.16
		Unit D1	60.07	Plinian	27	3.58×10^8	1.68×10^8	0.21

Sources of data 1. Carey et al. (2009), Carey.(2010), [Askja 1875]; 2. Alfano et al. (2012) [Chaitén 2008]; 3. Barberi et al. (1995); Costantini et al. (2010); Biass and Bonadonna (2011); Pistolesi et al. (2011) [Cotopaxi]; 4. Coltelli et al. (1998); Houghton et al. (2004); Sable et al. (2006, 2009) [Etna 122 BC]; 5. Costantini et al. (2010) [Fontana Lapilli 60 ka]; 6. Toramaru (1990); Blower et al. (2003); Toramaru (2006); [Izu Oshima 1986]; 7. Helz (1987); Parfitt (1998); Wallace and Anderson (1998); Parfitt and Wilson (1999); Stovall et al. (2011); [Kilauea Iki 1959]; 8. Colombier et al (2017) [Kilian 9.4 ka]; 9. Simmons et al. (2016, 2017a) [Lower Pumice 1 184 ka]; 10. Simmons et al. (2017b) [Lower Pumice 2 172 ka]; 11. Klug et al. (2002); [Mt. Mazama 7.7 ka]; 12. Klug and Cashman (1994); Toramaru (2006) [Mt. St. Helens]; 13. Fierstein and Hildreth (1992); Adams et al. (2006a, b) [Novarupta 1912]; 14. Metrich et al. (2001); Lautze and Houghton (2005); Patrick (2005); Lautze and Houghton (2007, 2008) [Stromboli 2002]; 15. Walker (1984); Sable et al. (2009); [Tarawera 1886]; 16. Wilson and Walker (1985); Wilson (1993); Sutton et al. (1995, 2000); Houghton et al. (2010) [Taupo 1.8 ka]; 17. Toramaru (1990, 2006); Blower et al. (2003) [Towada]; 18. Carey and Sigurdson (1987), Gurioli et al. (2005); Shea et al. (2010a, b), (2011) [Vesuvius 79AD]; 19. Cioni et al. (2011) [Vesuvius 512AD]; 20. Witter et al. (2004); Gurioli et al. (2008) [Villarica 2004]; 21. Vinkler et al. (2012) [Villa Senni 335 ka]; 22. Romano et al. (2020) [Agnano Monte Spina 4.6 ka]

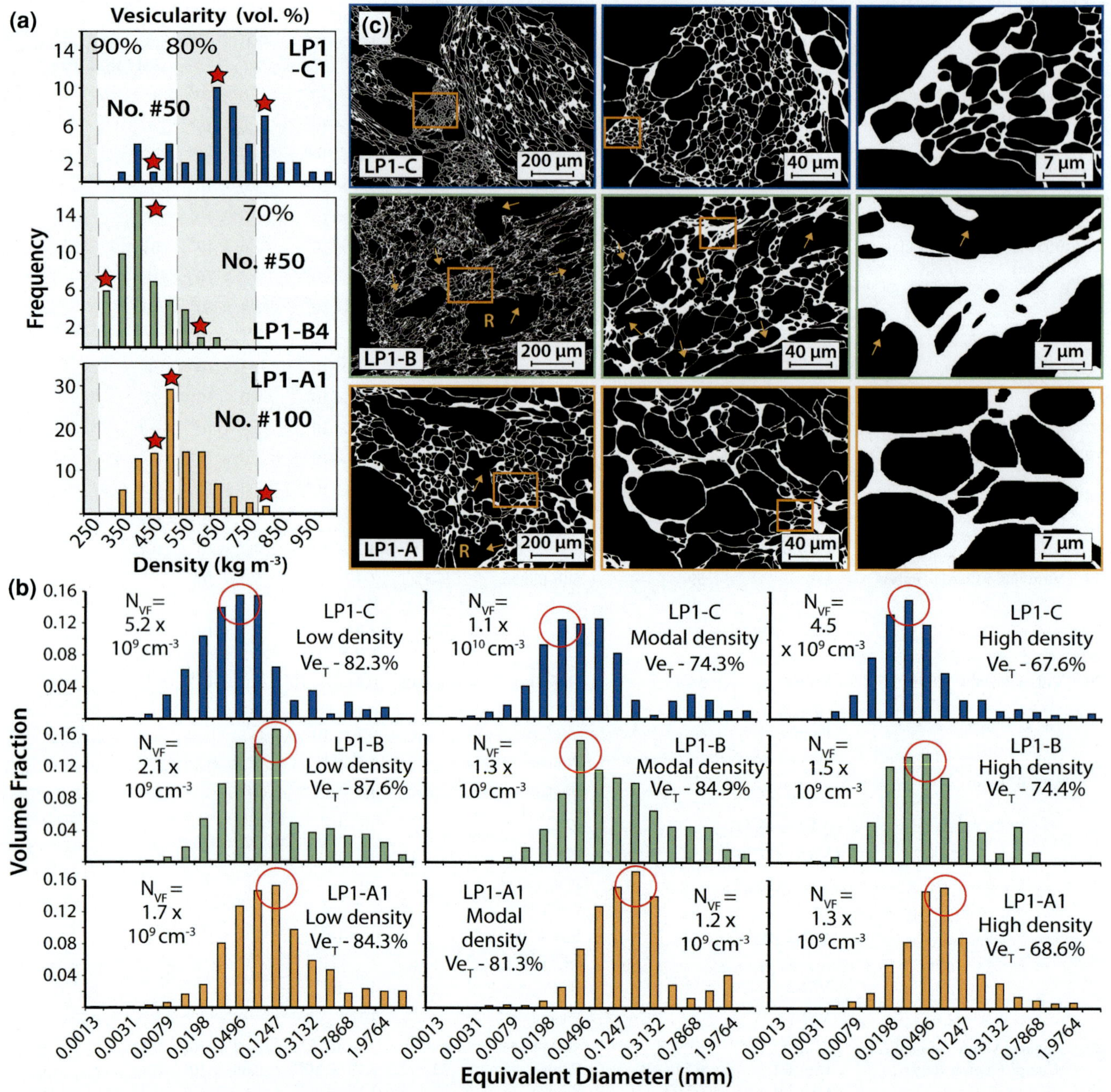

Fig. 4.28 Textural characteristics, density, BSD, and BND data for the main units from the 184 ka Lower Pumice 1 rhyodacitic plinian fallout (LP1-A1), ignimbrites (LP1-B), and co-ignimbrite lag breccias (LP1-C), Santorini, Greece. (a) Histograms illustrating the contrasting density distributions for rhyodacitic pumices from the fallout deposit (-A; orange), ignimbrite (-B; green) and lithic lag breccia (-C; blue). Bin size 50 kg m^{-3}. Stars denote designated density classes (low, modal and high densities) from which clasts were selected for 2D image analysis (Data from Simmons et al. 2016). (b) Vesicle volume distributions for low, modal and high density pyroclasts from each eruptive unit. (c) Characteristic vesicle textures for pyroclasts from the fallout deposit, ignimbrite and lithic lag breccia. From the left to the right, each image denotes low (×100), medium (×500) and high (×3000) magnifications, respectively. Juvenile pyroclasts from the lithic lag breccia are distinguished from the fall deposit and ignimbrite by having larger vesicle number densities and a significant fraction of small, spherical vesicles. No clast preserves evidence for magma outgassing and bubble collapse. R ruptured vesicle/bubble walls. (From Simmons et al. 2017a)

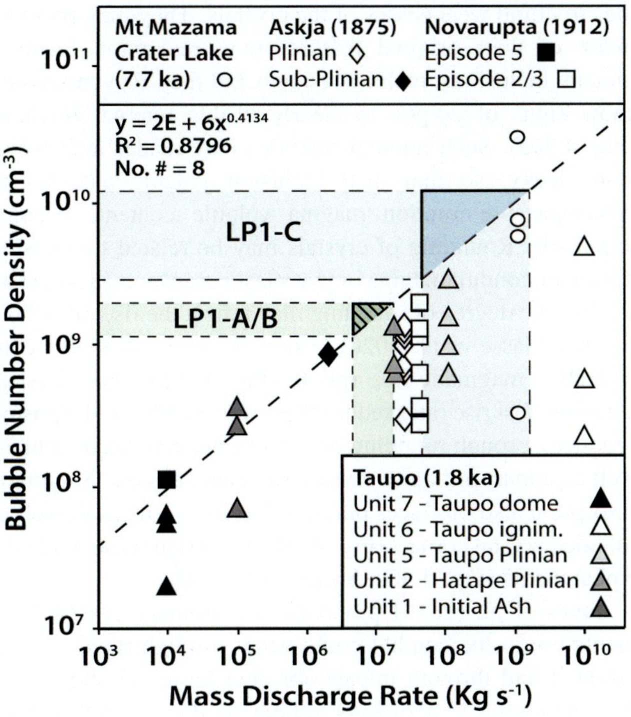

Fig. 4.29 Inferred relationship between Vesicle/Bubble Number Density (BND) and mass discharge rate for the Lower Pumice 1 eruption units, Santorini, compared with other plinian- and ignimbrite-forming eruption sequences. For detailed commentary of the Taupo 232 AD eruption, see Chaps. 9 and 10 (in particular, Sect. 10.9.1.1). Also see Fig. 4.22b for photomicrograph showing vesicularity of Hatepe plinian pumice fallout deposit. (From Simmons et al. 2017a)

no or few dispersed silicate network nuclei as crystal building blocks, or at a heterogeneous, sub-liquidus temperature and crystallisation regime, in which the magma has abundant silicate network nuclei. Unless magma ascent rates are very high for supra-liquidus magmas, most crystallisation occurs in the heterogeneous crystallisation regime, facilitating crystal growth in the conduit. Exsolution of volatiles and degassing of magma are exothermic and induce cooling in the magma. Loss of H_2O increases the stability of anhydrous mineral phases, such as plagioclase, enhancing crystallisation rates and levels if the magmas are supersaturated chemically to precipitate those minerals (Cashman and Blundy 2000). The size and morphology (whole or skeletal) of the microlites reflect the cooling rate (steady vs. undercooling) and growth rate of crystals. According to Cashman and Blundy (2000), the average crystal size, L, and the crystal number density per unit volume, N_{cv}, are proportional to the average nucleation rate, J, and growth rate, G, as follows:

$$L \approx (J/G)^{-0.25} \quad (4.26)$$

$$N_{CV} \approx (J/G)^{0.75} \approx 1/L^3 \quad (4.27)$$

Magmas with high ascent velocities grow few microlites, and have relatively low crystal density numbers, reflecting low nucleation and growth rates. Conversely, high crystal number densities reflect high rates of microlite nucleation and growth, which occur at slow ascent velocities, or perhaps more likely, initial very high, rise rates interrupted by a period of stalling during ascent, enhancing high degrees of under/supercooling and crystallisation (Cashman and Blundy 2000). Cashman and Blundy (2000) indicate that there is a direct relationship between CND and crystal size, indicating that if high levels of crystal nucleation occur, growth rates are also high. High levels of volatile exsolution and microlite growth have the effect of increasing magma viscosity, which will in turn slow ascent velocities, but will also lead to an increase in bubble/vesicle gas pressure. Cashman (1992) indicated that the plagioclase growth rates are 5–10 times slower in dacite magma such as Mt St Helens dacite, compared with Hawaiian basalts, reflecting a control of viscosity, which is also implied by Alfano et al. (2012). Cashman and Blundy (2000) also propose that phenocrysts may grow quickly in the conduit, and are not necessarily just the product of deep, magma reservoir crystallisation.

Importantly though, more than the total CND, it is the aspect ratio of crystals that plays the major role in controlling viscosity. For a given crystallinity, low aspect ratio (i.e. elongated such as plagioclase) crystals result in higher viscosities compared with high aspect ratio (e.g. equant, such as quartz) crystals (e.g. Vona et al., 2011, 2017). Campagnola et al. (2016) even suggested that spherical crystals such as leucite may locally decrease the viscosity due to shear localisation.

Toramaru et al. (2008) proposed that microlite formation is less controlled by cooling than decompression rate and the effect of exsolution of H_2O, which raises the liquidus. They proposed that microlite number density could be used as a measure of the rate of exsolution of H_2O (CND or "MND water exsolution rate meter") to calculate decompression rate and magma ascent velocity. They argue that since decompression triggers H_2O exsolution, which in turn affects the liquidus temperature, this will have a greater effect than cooling on microlite formation. However, a degree of cooling will also induce crystallisation. Toramaru (2006) relate CSD or MND, N_v (m^{-3}), to the rate of exsolution of water as

$$N_v^{0.66} = \frac{dC_w/dt}{a^{0.66}b} \quad (4.28)$$

where C_w is the H_2O content (wt%), a is a kinematic factor, and b a thermo-dynamic factor. They relate CSD or MND to decompression rate as

$$\frac{dP}{dt} = \frac{c}{b}\left[\frac{N_v}{a}\right]^{0.66} \qquad (4.29)$$

where c is defined as $11.2 \times 10^6 \times C_w$.

On the assumption of equilibrium exsolution conditions and steady flow rates, Toramaru (2006) found marked differences in the behaviour of lava dome-forming magmas at Mt Unzen, Japan, and explosive eruption generating magmas at Izu-Oshima. For Mt Unzen, he calculated water exsolution rates ranging from 6.1×10^{-6} to 2.8×10^{-5} (wt% s^{-1}), plagioclase MND from 10^{14} to 10^{15} m^{-3}, decompression rate of 240–1100 Pa s^{-1}, and magma ascent velocity in the range of 0.014–0.068 m s^{-1}. For Izu-Oshima they calculated water exsolution rates ranging from 1.1×10^{-3} to 1.1×10^{-1} wt% s^{-1}, pyroxene MND of 10^{15} to 10^{17} m^{-3}, decompression rate of 1.2×10^4 to 1.3×10^6 Pa s^{-1}, and magma ascent velocity 1.3 to 133 m s^{-1}.

4.8.5 Crystal Fragment Pyroclasts—Magmatic and Xenocrystic, How Do They Form?

Free crystals and angular fragments of crystals are released during the explosive disruption and breakage of porphyritic magmas and juvenile fragments. They form a discrete juvenile component from the vesiculated or non-vesiculated magmatic fragments in the pyroclastic deposit, which could be a fallout, surge, or pyroclastic flow deposit, or, in some cases, deposited inside a conduit. Selvedges of glassy groundmass may be attached in some cases, but the rheological contrast between rigid crystals and the enclosing magma or groundmass acts as a strain partition and often seems to promote fracturing along the interface between the two.

Crystals can be present in both magmatic and phreatomagmatic pyroclastic deposits. Some non-magmatic or older magmatic crystals may also be derived by fragmentation of country rock and accessory lithics (see below), and these are called xenocrysts. To understand the nature of the erupting magma, the geochemistry of pyroclastic deposits, and particularly before undertaking geochronological studies, it is important to determine which crystal fragments are juvenile magmatic and which are xenocrystic, their proportions, and, if necessary, separate them by hand picking, sometimes preceded by limited crushing of samples. Free crystal fragments may be whole and euhedral, subhedral and incompletely crystallised crystals, or broken and angular, anhedral crystal fragments of once whole euhedral phenocrysts (Fig. 4.30a–f). The morphology of the crystals reflects the pre-eruption crystallisation of a magma, the amount and rate of decompression it experiences during its rise in a conduit, intensity of the explosive fragmentation, and the structural weaknesses of the crystals. They may preserve many of their original features (e.g. resorption features, mineral inclusions, melt inclusions), but they also commonly show signs of cryptic to clearly visible internal fractures (Fig. 4.30a). Such internal fractures may render melt inclusions leaky, so that melt inclusion volatile analyses to determine pre-eruption magma volatile contents become unreliable. Rounding of crystals may be related to abrasion during in-conduit magma or pyroclastic transport (Sasse et al. 2020). The degree of rounding may reflect the rise rate of the magma (Sasse et al. 2020). Slow rise rates allow for considerable magmatic resorption (Fig. 4.30a) and physical abrasion, fast rise rates reduce these effects. Physical abrasion produces a rough rounding of corners and crystals overall, as well as impact pitting of crystal fragment surfaces. Magmatic resorption also produces pitting of crystal surfaces as well as rounded crystal corners (Fig. 4.30a, b), and in some minerals, embayments (Fig. 4.30a; Sasse et al. 2020).

Jones et al. (2016) proposed that primary crystal fragmentation in the conduit could occur through rapid decompression and through impact abrasion based on shock tube decompression experiments similar to those conducted by Alidibirov and Dingwell (1996), Spieler et al. (2004a, b), and Scheu et al. (2008). Jones et al. (2016) and van Zalinge et al. (2018) demonstrated that significant fracturing and fragmentation of phenocrysts (Fig. 4.30a) in magmas and pyroclasts could occur if sudden decompression of tens of MPa affected rising magma in the conduit. Tait (1992), Best and Christiansen (1997) and van Zalinge et al. (2018) attributed most of the fracturing to sudden expansion of volatiles within melt inclusions trapped in the (cryptic fractures in the) phenocrysts during rapid decompression as magma rises in a conduit, which is thus a form of micro-hydraulic fracturing, not explosive. Van Zalinge et al. (2018) reported that such fracturing led to a 12% expansion of the crystal and its aggregate of crystal fragments as these dilated, accommodated by fractures in the crystals. Jones et al. (2016) and Mueller et al. (2015) have pointed out that impacts and abrasion between clasts in the conduit also cause primary fragmentation of crystals.

In addition, van Zalinge et al. (2018) propose that further in situ fracturing of crystals in welded ignimbrites can occur due to compactional stresses during welding and also shear stress associated with lateral rolling and movement during welding.

In some deposits, such as kimberlites, olivine crystal fragments are usually the most abundant pyroclast type. These, however, are overwhelmingly xenocrystic, derived from the mantle and brought to the Earth's surface by kimberlite melts, before mostly erupting explosively. In transit, those olivines are usually subject to extensive magmatic resorption and physical abrasion (Sasse et al. 2020),

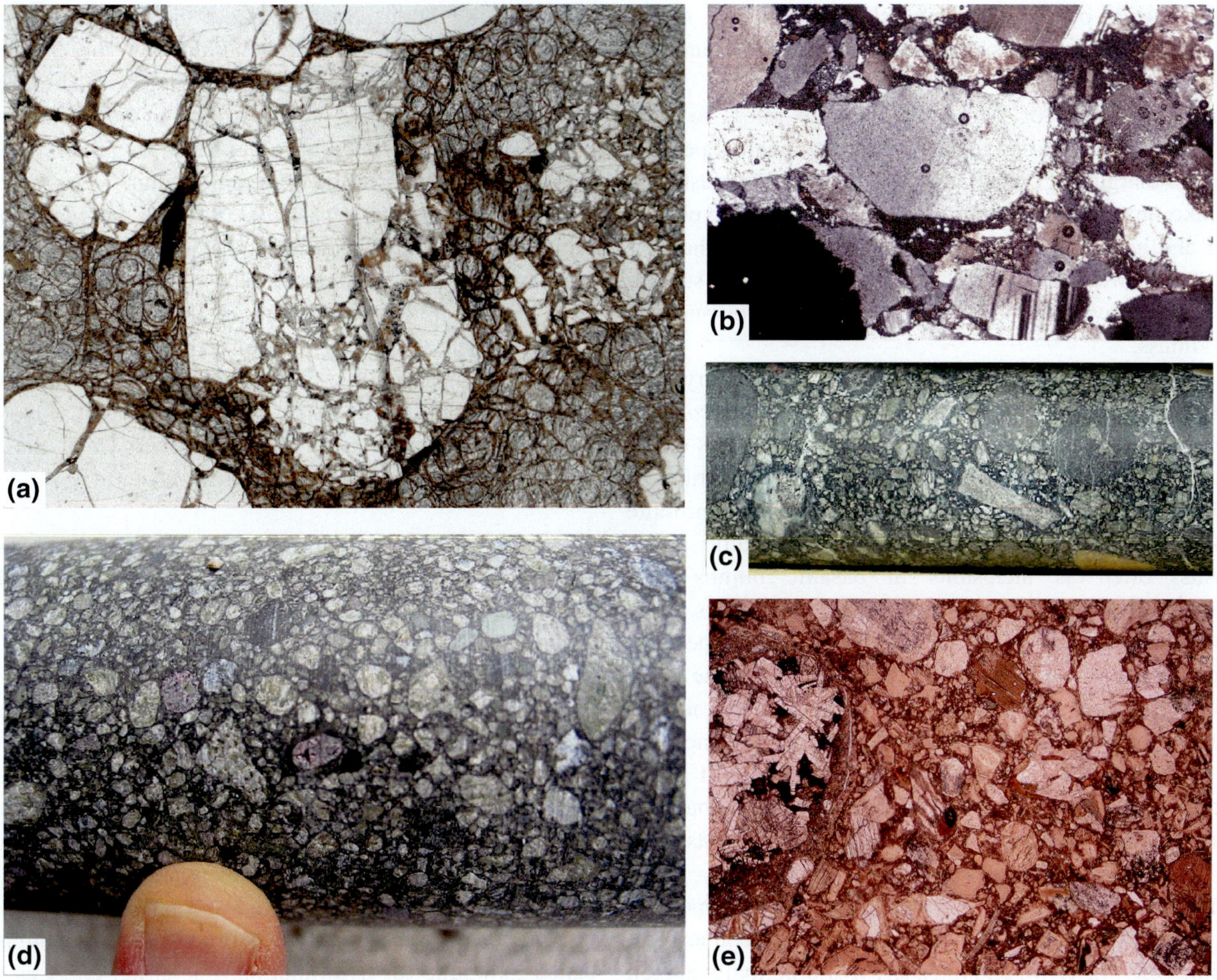

Fig. 4.30 Features of crystal fragments in pyroclastic deposits. **a** Euhedral, subhedral, and fragmented phenocryst fragments of volcanic quartz (rounded, embayed) and sanidine (centre, fractured) in densely welded ignimbrite with perlite fractures. Permian Ora welded ignimbrite, Bolzano, northern Italy. Field of view 5 mm across; ppl. (Image by M. Willcock). **b** Fractured bipyramidal volcanic quartz and variably twinned plagioclase crystal fragments at the base of a syn-eruptive, deep-water megaturbidite, Lower Devonian Merrions Tuff, New South Wales, Australia. Field of view ~5 mm across; x-nicols. **c** Drill core showing abundant olivine crystal fragments (medium greenish grey, rounded to angular with some internal fractures), together with highly angular basement clasts and rounded juvenile clasts in fragmental kimberlite from the Cretaceous Orion South kimberlite complex, Fort à la Corne kimberlite field, Saskatchewan, Canada. Round, spherical clasts are pelletal lapilli, the large irregular clasts are cognate lithic clasts. Diamond Drill hole FALC 141-14, 182.5 m. Diameter of core ~5 cm. **d** Olivine xenocryst (pale grey-green)-rich fragmental kimberlite, from a mega-graded bed in the Eocene Koala kimberlite pipe, Ekati kimberlite field, Lac de Gras, Northwest Territories, Canada. In addition, there are pale granite crustal xenoliths and pink garnet xenocrysts. Diameter of core ~5 cm. **e** Photomicrograph of mostly angular xenocrystic olivine (white to altered brown, resorption rounded to angular)-rich fragmental kimberlite, Eocene Fox kimberlite pipe, Ekati kimberlite field, Lac de Gras, Northwest Territories, Canada. Also present are a large mantle xenolith-cored pelletal lapilli (left), and brown cleaved phlogopite mica

leading to typically rounded olivine grains which may then be broken during explosive eruption (Fig. 4.30c–e). Even diamonds, which are sub-cratonic, mantle-derived xenocrysts, can be subject to magmatic resorption if the rise rate of kimberlite or lamproite magmas is slow. Some kimberlite magmas may even have had all of their diamond crystal cargo dissolved away through such a process under very slow magma rise rates. Sometimes it doesn't pay to be too slow!

4.8.6 Lithic Fragment Pyroclasts

The term "lithic clasts" or "lithics" generally describes the dense rock fragments in a pyroclastic deposit. Lithics may be subdivided into non-vesiculated juvenile magmatic fragments (*cognate lithics*, or also called *autoliths* in the kimberlite literature), country rock lithic fragments that have been explosively ejected during explosive eruption (*accessory lithics*) and clasts picked up locally from the ground surface outside of a vent by pyroclastic flows and surges (*accidental lithics*). Accessory and accidental lithics are also called *xenoliths*.

"Cognate" lithic or "autolith" therefore refers to fragments that have solidified from the erupting magma, perhaps as chilled margins of conduit magma, or solidified crusts of a lava lake, that are then broken up by an energetic explosive phase and are ejected from the vent (Fig. 4.31a). They are thus usually fresh and unaltered in young volcanic successions and may have a glassy or variably crystallised groundmass.

Accessory lithics can be derived from any country rock that the rising magma encounters in the conduit *en route* to the Earth's surface, including mantle or crustal rocks or the metamorphosed equivalents of these (e.g. De Benedetti et al. 2010). These can include upper mantle-derived peridotite xenoliths, which are common in alkaline basaltic scoria deposits, often as inclusions in basaltic bombs (e.g. Fig. 4.19f), but they may also occur as free pyroclasts. They are common not only in monogenetic scoria and some maar successions (e.g. van Otterloo et al. 2013; Valentine et al. 2014), but also in kimberlite pipes, where they are accompanied by constituent mineral xenocrysts (e.g. olivine, garnet, pyroxene; Sasse et al. 2020). Metamorphosed upper mantle clasts such as eclogite, accompanied by abundant olivine, garnet, and other mantle-derived xenocrysts (e.g. Fig. 4.31b), also occur in some kimberlite deposits.

Crust-derived accessory xenoliths of sedimentary (Fig. 4.31c, d), volcanic (see Fig. 4.19h), metamorphic, plutonic (Fig. 4.31e, f), and intensely hydrothermally altered lithics (Fig. 4.31g) are common accessory lithic clasts in all types of pyroclastic deposits where the upper crustal geology contains these lithological units. Accessory lithic clasts are usually angular (Fig. 4.31c, d). However, in some circumstances, large lithics, which cannot be ejected from a vent initially because of their mass, may become rough rounded due to in-vent abrasion or "milling" (Campbell et al. 2013) during eruption. Some may eventually be ejected onto the extra-vent pyroclastic apron (Fig. 4.31e), but some may be trapped and deposited in the vent (Fig. 4.31h).

Accessory lithic clasts or xenoliths provide valuable information about the subsurface geology and stratigraphy around a vent, and depth of the fragmentation surface if subsurface geology is known (De Benedetti et al. 2010; Valentine et al. 2014). If systematic changes in the accessory lithic clast population can be identified vertically through a pyroclastic succession, it can reflect the changing depth of the fragmentation surface, especially early in an eruption, during waxing explosive eruptions (e.g. Simmons et al. 2016, 2017b) or eruption phases (e.g. Massaro et al. 2018). Where lateral changes in accessory lithic clast populations around an explosive vent can be demonstrated, it can be used to trace the location and migration of vents (e.g. Pittari et al. 2008). Using lithic clasts to assess the depth of the fragmentation surface becomes more difficult the longer the eruption lasts, and during the waning stages of an explosive eruption, due to mixing and recycling of lithic clasts from multiple levels in conduit, and collapse of lithics from the crater rim and high in the eruption conduit.

Accidental lithics consist of any rock fragment that is lying on the Earth's surface outside of the vent and is incorporated into and deposited by pyroclastic density currents during flow and deposition. If they were originally volcanic then they may be difficult to distinguish from accessory volcanic lithic clasts. Sometimes it is possible to trace accidental lithic clasts back to an exposed source outcropping upslope from the deposit in which the accidental lithics are now preserved.

4.9 Hydraulic Fracturing of Volcanic Rocks—How to Distinguish These Breccias from Primary Volcanic Breccias

In volcanic settings, hydrothermal fluid systems have enormous potential for causing hydraulic fracturing and brecciation (e.g. Plate. 4.1e), because the magma releases volatiles to the country rock and heat from the magma system can induce convective circulation of meteoric waters in the country rock, which mix with magmatic volatiles. These fluids can become overpressured and have the potential to fracture the country rock, in a process that is commonly called hydraulic (or hydrothermal fracturing, e.g. Cas et al. 2011, Rowland and Simmons 2012, Chaps. 15, 18). We prefer to use the term hydraulic fracturing, rather than hydrothermal fracturing to avoid confusion with hydrothermal explosion breccias. In this section, we focus on hydraulic fracturing, alteration, and their products, because of their common occurrence in most volcanic settings.

The fracturing of rock in the subsurface occurs under a range of stress regimes, from those that are tectonic in origin, to those associated with the intrusion of magma to those associated with hydrothermal and other over- and under-pressured fluid regimes. In volcanic terrains, all of these regimes may prevail. Subsurface fragmentation processes may pre-date, be synchronous with, or post-date volcanic activity (Chap. 15). Several authors have reviewed the spectrum of processes and stress regimes operating in subsurface settings, both volcanic and non-volcanic, and summarised the textural and compositional characteristics of resultant breccias.

Sibson (1977, 1986) distinguished between deep-crustal breccias, with random fabrics, and upper crustal breccias, which are frequently foliated. Sillitoe (1985; also see other papers in the same volume) summarised the characteristics of a range of breccia types associated with mineralised

(a)
(b)
(c)
(d)
(e)
(f)

◀ **Fig. 4.31** Lithic clast types in pyroclastic deposits **a** Cognate, slab-like basalt lithic clast ejected ballistically but embedded at the base of a "wet" phreatomagmatic base surge deposit in a phreatomagmatic maar rim succession of interstratified wet and dry phreatomagmatic and magmatic scoria deposits, Red Rock volcanic complex, Newer Volcanic Province, Victoria, southeastern Australia. The basalt slab is interpreted as a lava lake crust fragment blasted out at the start of a new phreatomagmatic explosive cycle. **b** Eclogite accessory lithic clast with garnet and greenish pyroxene (outlined in red) within massive lapilli-tuff in the Jurassic Muskox kimberlite pipe, Nunavut, Canada. Other lithic clasts include white limestone, mottled granodiorite, in association with bright green chrome diopside and grey olivine xenocryts. Diameter of core ~5 cm. **c** Tertiary limestone bomb with impact sag structure in the maar rim succession of the 35 ka Tower Hill maar volcano, Newer Volcanics Province, Victoria. **d** Fossiliferous accessory limestone clast in massive lapilli-tuff in the Muskox kimberlite pipe. Also visible are white limestone clasts, bright green chrome diopside, pink garnet, and grey olivine xenocrysts. Diameter of core ~5 cm. **e** Granite accessory lithic clasts in the Burrumbeet maar rim succession, Newer Volcanics Province, Victoria. The rough rounded form could be due to vent milling. **f** Granite and quartzite (grey) metasedimentary accessory lithic clasts in massive lapilli-tuff in the Eocene Fox kimberlite pipe, Ekati kimberlite field, Lac de Gras, Northwest Territories. Diameter of core ~5 cm. **g** Hydrothermally altered lithic blocks from an 1873 hydrothermal explosion breccia forming the rim of a hydrothermal explosion pit (see Fig. 4.18f) at an active geothermal field in the caldera floor of Nisyros volcano, Greece. **h** Huge rounded vent-milled accessory quartzite boulder found within the explosive 1.2 Ga Argyle lamproite diamond pipe during mining, Kimberley Craton, Western Australia

Fig. 4.31 (continued)

volcanic and plutonic rock settings, including both supracrustal volcaniclastic breccias and subsurface breccias. Laznicka (1988) provided a comprehensive review of all types of breccias and proposed both genetic and non-genetic classifications.

Hydraulic fracturing in volcanic settings involves both physical and chemical processes and fracturing of volcanic, intrusive, and other rock types by overpressured and chemically reactive hydrothermal fluids in the volcanic edifices and the crust below the Earth's surface (e.g. Cas et al. 2011; Vignaroli et al. 2015). Jébrak (1997) reviewed the processes associated with the formation of hydrothermal breccias in hydrothermal vein-type ore deposit settings, and these were summarised by Cas et al. (2011) in a study of hydraulic breccia textures on the volcanic island of Lisca Bianca, in the Aeolian Arc, southern Italy. These processes will apply in any setting where hydrothermal fluids are active. The physical processes can be called "*fluid-assisted brecciation*", and the chemical processes are called chemical "*corrosive wear*" or simply hydrothermal alteration (see Sect. 4.10).

Physical brecciation occurs when the pressure applied by hydrothermal fluids exceeds the brittle resistance or strength of the materials, whereas chemical brecciation involves selective or pervasive replacement or dissolution of original minerals, and precipitation of new hydrothermal minerals.

4.9.1 Physical Hydraulic Brecciation

If a fluid develops a pressure, P_f, which exceeds the tensile strength of the host rock, τ, and the minimum principal stress component of the lithostatic pressure regime, σ_3:

$$P_f > \tau + \sigma_3 \tag{4.30}$$

then the country rock can fracture, opening a network of tensile fractures (Secor 1969; Phillips 1972; Shaw 1980; Cas and Wright 1987). The rock is fractured in situ in brittle, tensile fashion by the overpressured fluids, which then significantly lowers the effective pressure. The resultant fractures need not involve shear motion, and so the breccia fragments and the associated wall rock need not be pervasively foliated or sheared nor moved relative to each other.

"*Fluid-assisted brecciation*" involves two steps, hydraulic fracturing and critical fracturing. Where rock systems are saturated with fluid, the normal lithostatic pressures are effectively lowered by the fluid pressure, P_f, producing effective stress components (Sibson 1990; Clark and James 2003):

$$\sigma_1' = (\sigma_1 - P_f) > \sigma_2' = (\sigma_2 - P_f) > \sigma_3' = (\sigma_3 - P_f) \tag{4.31}$$

where σ_1', σ_2', and σ_3' are fluid pressure-modified principal stress components. According to Sibson (1990) at any depth, z, where crust density is ρ, the fluid pressure, P_f, is related to the vertical stress, σ_v, by a pore fluid factor, λ_v, using the relationship

$$\lambda_v = \frac{P_f}{\sigma_v} = \frac{P_f}{\rho g z} \tag{4.32}$$

The effective vertical stress then becomes

$$\sigma_v' = (\sigma_v - P_f) = \rho g z(1 - \lambda_v) \tag{4.33}$$

Critical fracturing occurs when the balance between fluid pressure and regional stress is destroyed by hydraulic fracturing, which causes a sudden pressure reduction, especially in the fracture filling fluid. As a result, spontaneous spalling (implosion) of rock debris from the country rock (Jébrak 1997; Cas et al. 2011), much like "rock bursts" into open drives in deep mines, may occur. However, before critical crack failure, subcritical crack growth can occur, which allows cracks to propagate below the strength limit of fresh, unaltered rock by the combined effects of chemical and physical processes.

Hydraulic fracturing and critical fracturing are closely related, and both produce jigsaw-fit breccias with little clast rotation, although critical fracture breccias may produce some clast rotation if there is space (Cas et al. 2011). In some hydraulic breccia zones, clasts may have been moved relative to each other, significantly rotated, and even transported through the system, leading in some cases to a mixing of clast types, although usually the clast population is monomictic.

4.9.2 Chemical Processes

"*Hydrothermal alteration* is a general term embracing the mineralogical, textural and chemical response of rocks to a changing thermal and chemical environment in the presence of hot water, steam or gas" (Henley and Ellis 1983; Chaps. 13 and 18). The upper limits of this process are ill-defined and are gradational into low-grade burial metamorphism. Hydrothermal alteration involves ion-exchange reactions, mineral phase transformations, mineral dissolution, and new mineral growth. Although the original rock composition has some effect on the secondary alteration mineralogy, the main influences are the permeability of the rock pile, temperature, the composition of the fluids moving through the rock pile, and the degree of chemical disequilibrium between these fluids and the host rock. Hydrothermal alteration has also been called "*corrosive wear*" (Jébrak 1997) to describe the chemical reaction between hydrothermal fluids and host rock, causing extensive alteration and dissolution in the zones of reaction. Hydrothermal alteration and its effects on rock texture is extremely important in most volcanic systems.

All hydraulic fracture breccias are chemically and mineralogically altered. This is reflected in the hydrothermal alteration of the host rock, especially in and adjacent to the hydraulic fracture zones, as well as the infilling of the hydraulic fracture network with secondary minerals. Chemical alteration weakens host rock and host rock fragments and makes them prone to physical fracturing.

Chemical processes involve selective or pervasive, element replacement, dissolution, and new (hydrothermal) mineral formation, both through in situ replacement of host rock and infilling of hydraulic fractures by precipitation of vein-forming minerals (Jébrak 1997; Cas et al. 2011; Vignaroli et al. 2015). Intense alteration can begin to mask original textures, as well as hydraulic fracture clast margins, and may produce misleading pseudo-volcanic textures that give the appearance of an alternative origin to the actual hydraulic fracture origin, especially autoclastic breccias, which also commonly produce in situ jigsaw-fit breccias (see above; Cas and Wright 1987; McPhie et al. 1993; Chap. 13).

4.9.3 Formation of Hydraulic Fracture Zones

Hydraulically fractured and hydrothermal alteration zones can be relatively narrow focussed zones, or more widely distributed zones where subsurface hydrothermal systems are large. Jébrak (1997), Cas et al. (2011) and Vignaroli

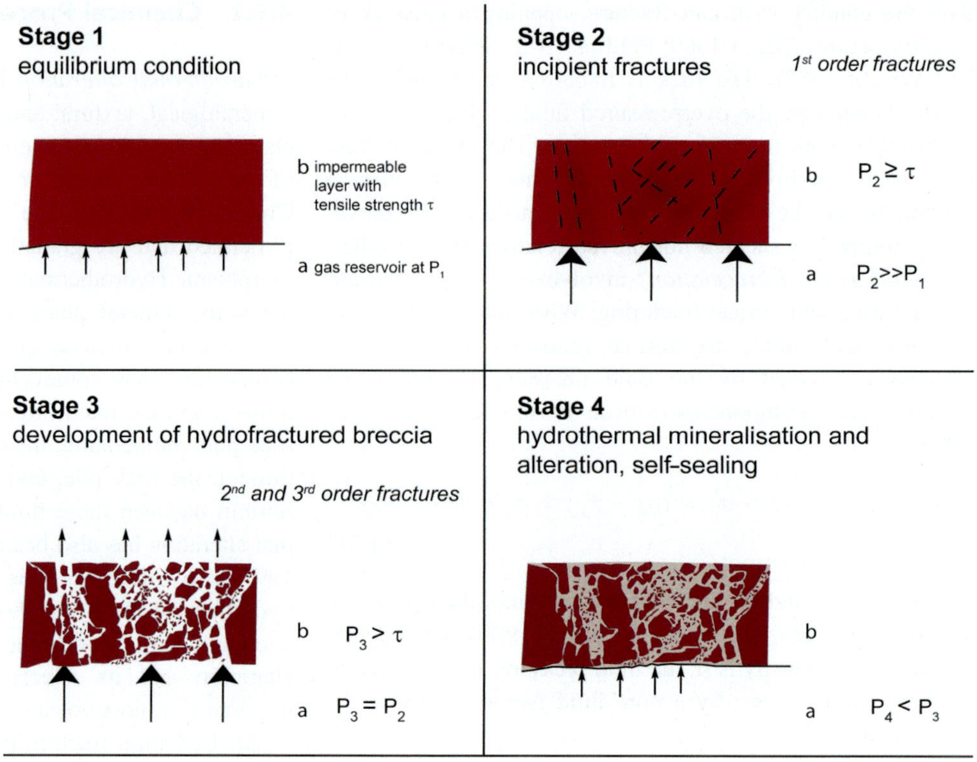

Fig. 4.32 Stages in the formation of hydraulic fracture breccias and associated hydrothermal alteration. (1) Coherent rock and pressure equilibrium stage. (2) Hydraulic fracturing stage. (3) Fluid percolating and hydro-fractured stage. (4) Hydrothermal vein filling and self-sealing stage. P represents fluid pressure regimes and τ the tensile strength of the rock. (From Cas et al. 2011)

et al. (2015) propose that hydraulic/hydrothermal breccia evolution can be complex, but may involve several stages, progressing from (1) a coherent rock stage, during which fluid pressure builds up, but the system is in pressure equilibrium; to (2) hydraulic fracturing rock stage involving brittle failure of the rock due to high fluid pressure which exceeds the tensile strength of the rock; to (3) a fluid percolating stage leading to a breccia with permeable fluid pathways; and to (4) a hydrothermal mineralisation (or precipitation) and alteration stage, leading to self-sealing of the fracture network (Fig. 4.32; Plate 4.1e Cas et al. 2011). During stage 2 the rock becomes a non-continuous solid, called a mechanical discontinuity threshold, and during stage 3, it becomes a hydraulic continuity threshold, where the fluid forms a continuously connected phase throughout a fracture system. In a hydrothermal system, the four stages can repeat in a cyclical way, through crack and seal processes, allowing for a very complex brecciation history to occur, with multiple generations of fractures. This cyclical behaviour appears to be caused by infilling of fractures through precipitation of hydrothermal minerals, which then causes fluid pressure to build up again. This may then initiate a new cycle of hydraulic fracturing, similar to crack-seal vein formation in deformational zones (Ramsay 1980). As a result, hydraulic breccia systems may contain multiple generations of superposed and cross-cutting veins, and multiple generations of brecciated clasts.

4.9.4 Hydraulic Fracture Breccia Characteristics

In isolation and out of context, an exposure of hydraulic fracture breccias could be confused with some deposits resulting from supracrustal volcanic fragmentation processes (e.g. quench fragmented hyaloclastite and autobreccia deposits), as well as various types of fault breccias, especially dilational zone implosion breccias. Distinguishing hydraulic fracture breccias from these other origins can be a significant problem for those working in old, eroded, and variably deformed and altered volcanic terrains.

Where observable in outcrop, hydraulic fracture breccia zones should be distinguishable by their cross-cutting character and confinement to a breccia zone that could be tens of cm to m, to tens of m wide, sometimes even more, especially if it has a multi-phase origin. In addition, hydrothermal alteration will be pervasive. Jigsaw-fit textures are prevalent, and clast-rotated textures are generally absent or minor in abundance (Cas et al. 2011; Fig. 4.33a–d), although where there is a very long-lived and large hydrothermal system, wider fluid pathways may develop, allowing clast rotation and even clast transport (Fig. 4.33a). Where there is transition into hydrothermal explosive activity (e.g. some upper crustal epithermal clast systems), rotation and transport become more abundant in a breccia pipe, including mixing of clasts from different stratigraphic units or levels (e.g. Rowland and Simmons 2012; Blackwell et al. 2014; Chap. 18).

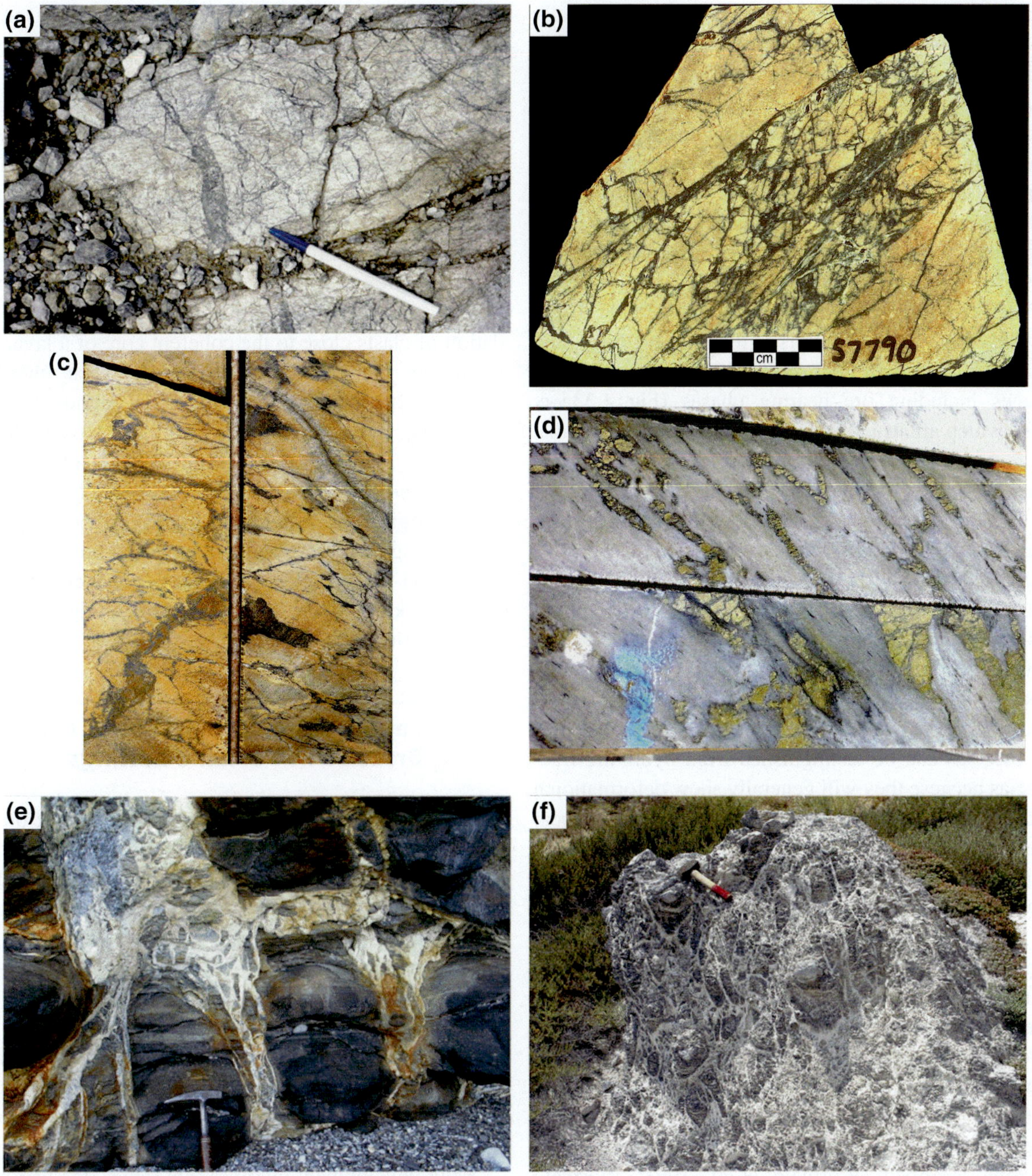

Fig. 4.33 Examples of hydraulic fracture breccias. **a** Hydraulically fractured Cambrian dacite, hanging wall of the Hercules VMS base metal mine, Cambrian Mt Read Volcanics Belt, western Tasmania, Australia. Note the pervasive fracture network, with all fractures filled with dark chlorite, whereas the dacite is sericitically altered. The large central, vertical chlorite-filled fracture has been wide enough for clast transport through the fracture by hydrothermal fluids. **b** Close-up, polished slab of **a**, showing monomict jigsaw-fit clast textures, the angular blocky hydraulic fracture clasts and dark chlorite filling every fracture. **c** Hydraulic fracture network in the footwall host rock succession to the 2,692 Ma Teutonic Bore massive sulphide mine succession, Western Australia. The fractures are filled with chlorite and pyrite, which in places are beginning to invade and replace the host basalt, producing diffuse margins to the hydraulic fracture clasts. Width of core ~5 cm. **d** Hydraulic fracture breccia with jigsaw-fit texture, with fractures infilled with pyrite veins. Archean Nimbus epithermal-VMS deposit, Kalgoorlie, Western Australia. Width of core ~5 cm. **e** Coarse, hydraulic fracture, in situ monomictic, andesite clast breccia, with a subvertical clast orientation and fracture-vein fabric. The vertical fabric is parallel to nearby, well-defined columnar joints, whereas a horizontal diffuse flow banding fabric is also visible. Splintery fragments have spalled off sharp corners, to begin forming rounded margins to the large core blocks. Quaternary, Lisca Bianca Islet, Aeolian Arc, southern Italy. (From Cas et al. 2011). **f** From the same location as **e**, monomictic, in situ boulder–cobble–pebble andesite breccia, with a subvertical clast elongation fabric, indicating vertical hydrothermal flow and fracturing patterns. The vein network is very complex with several cross-cutting generations, mostly filled with natro-alunite. Note some rounded forms due to spalling of splintery fragments off sharp corners

Hydraulic and critical fracture breccia clasts are usually of relatively uniform size but vary from boulder to fine pebble size (Fig. 4.33a–f). Fragments are usually angular, to blocky and splintery in shape (Fig. 4.33a–f), and their shapes may reflect pre-existing preferred fabrics (e.g. layering, foliation), or be random with an isotropic fabric. According to Sillitoe (1985), pressure fluctuations may cause "hypogene exfoliation", or spalling of slivers from core fragments, producing round core clast forms, although this usually also involves chemical effects as well (Fig. 4.33f; Chap. 13; Jébrak 1997; Cas et al. 2011). Breccias may have high dilation factors in dilational settings, and fractures are infilled with new secondary alteration mineral assemblages including ore bearing phases (Fig. 4.33a–e; Chap. 18), which are usually precipitated from hydrothermal fluids.

Because hydraulic fracture breccia zones are usually the sites of significant hydrothermal alteration and mineralisation (e.g. Phillips 1972; Vignaroli et al. 2015; Chap. 18), misleading or pseudo-textures can be produced (Chap. 13; Cas and Wright 1987; McPhie et al. 1993). In addition, where significant weakening of the rock has resulted from these alteration processes, such zones may preferentially absorb strain and become highly deformed during subsequent tectonic movements.

Fault breccias can be distinguished from hydraulic fracture breccias because they will generally show deformational preferred strain fabrics, especially in attrition breccia shear zones and zones of distributed crushing. Fault dilational jog zone implosion breccias may be very difficult to distinguish from hydraulic fracture breccias (*sensu stricto*) in isolation because they show little clast deformational fabric. However, the larger scale contextual relationship to a fault zone should indicate the fault origin.

However, we stress these processes are not mutually exclusive and probably the best representation is given by subvolcanic breccia pipes, for which the Permo-Carboniferous Mt Leyshon breccia pipe and gold deposit in northeast Queensland, Australia is one of the most comprehensively described (Wormald et al. 1991; Wormald 1993, 2017; Chap. 18). Detailed facies analysis, charting the progressive fragmentation and transport (expansion) of brecciated lithologies and their geometrical relations, has demonstrated fault-derived brittle fracturing coupled to magma decompression and gas discharge at depth drove fragmentation. Strike-slip fault motions accompanied all stages of breccia complex evolution, and likely controlled pre-breccia basement fragmentation, magma recharge, vapour discharge into dilated host rock, hydraulic fracturing, and brecciation, which were synchronous or stepwise interactive processes (Chap. 18). The facies characteristics of the breccia units suggest processes of gas/solids mobilisation and fluidisation controlled their formation. Domains of jigsaw-fit relationships between clasts are ubiquitous on all scales.

Hyaloclastite (Sect. 4.4.6) will show widespread jigsaw-fit to clast-rotated textures and should show a gradation into the coherent volcanic or high-level intrusion core of the host emplacement unit. Hyaloclastite will occur within a supracrustal, stratified succession, either as part of the margins of a stratiform lava or the peperitic margins of syn-depositional sills or dykes. Autobreccia (Sect. 4.3.2) will show similar features but may show plastic accommodation of clasts close to the coherent core of a lava. There is no reason why hyaloclastite or autobreccia should show evidence of hydrothermal alteration, although near-vent facies (e.g. felsic lava domes) could well have hydrothermal alteration superimposed on hyaloclastite. Abundant clast-rotated textures are not consistent with hydraulic fracturing origins, and could indicate hydrothermal explosive origins (e.g. Davies et al. 2008). Superimposition of processes and textural effects creates the greatest challenge for those working in the rock record in terms of identifying facies origins.

4.10 Fault Zone Brecciation and Fault Breccias in Volcanic Settings

Sibson (1977, 1986, 1990, 1996) reviewed fault brecciation processes and proposed a classification for fault breccias based on textural features of clasts, their internal deformation features, clast size distribution, and clast and matrix compositions. These concepts were not developed specifically for volcanic terrains, but apply to any terrain with fault zones. In old volcanic terrains that have undergone tectonic deformation, fault zones through the volcanic stratigraphy and cataclasite fault breccia domains consisting of faulted volcanic rocks and fault breccias of volcanic clasts will be common and must be distinguished from primary and other secondary volcanic breccias.

Fault development involves two stages, the first being the formation of an *en echelon* set of fractures, the second the joining of these by other fractures to form a large fault zone. The spacing of fractures follows a power law, resulting in a highly variable clast size range. Sibson (1986) proposed three different mechanisms of brecciation in fault zones related to the setting and morphology along the fault zone. Attrition brecciation occurs along planar segments of faults, where interaction between hanging wall and footwall blocks is relatively uniform and involves shear. The physical processes involve wear abrasion which is the physical attrition of fragments by frictional forces as they move relative to each other in fault zones (Jébrak 1997). Sibson (1986) identified a spectrum of brittle to ductile processes involved in attrition brecciation in such settings, including brittle shear, asperity tip plasticity and melting, sidewall cracking and plucking, and inter- and intra-grain cracking. Fragments include extensive rolled clasts, clasts with high degrees of internal strain, a preferred fabric which is well developed in places, the clast aggregate is poorly to moderately sorted, the clast population may be varyingly polymictic, and the matrix composition is the same as the clasts (Sibson 1986).

Distributed crush brecciation occurs over broad areas where movement along faults is impeded where footwall horsts occur. This produces significant stress build-up in the "lock" zone, resulting in widespread micro-cracking and faulting in both footwall and hanging wall (Sibson 1986). The clast zone is marked by widespread micro-fracturing, moderate to high degrees of clast strain (deformation), poorly sorted aggregate texture, monomictic clast assemblage, and a matrix with a similar composition to the clasts, but also including minor interstitial hydrothermal minerals (Sibson 1986).

According to Sibson (1986), implosion breccias occur where there is a dilational jog in the fault. Jébrak (1997) calls these volume expansion breccias. The trans-tensional stress regime in these dilational jogs causes the extensional opening of cracks, which causes a sudden decrease in fluid pressure in the cracks. This produces a significant pressure gradient between country rock and fluid-filled cracks, which triggers implosive spalling of country rock debris. In these zones, jigsaw textures are common and clasts show low degrees of deformation. Domains of *en echelon* clast slivers or blades may be common, the clast size distribution is often uniform, the clast population is monomictic, and the matrix may include abundant, often exotic hydrothermal mineral assemblages and epithermal-to-mesothermal-style mineralisation (Sibson 1986, 1987).

Blenkinsop (1991) suggested that the main size reduction mechanisms in fault zones are alteration, which is most active in dilatant zones; constrained comminution, which is common in shear zones; and selective fracture of larger particles, where there is increased strain and confining pressure. There may be an evolutionary progression of these mechanisms during cataclasis. Fragments may develop round forms and grainsize may decrease through frictional breaking or attrition (Blenkinsop 1991). Deformation may transgress the brittle–ductile transition, in which dissolution and recrystallisation may add further textural complexity, including the development of widespread veining. There is little dilation, and clast size range can be variable.

Where hydrothermal activity is associated with fault zones, the processes may be complicated by the influence of both fault-related and hydrothermal-fluid-related processes.

Faults in solid rock volcanic successions can be sharp planes, bounding displaced fault blocks, to brittle-ductile breccia zones in which country rock has been subject to intense strain (Fig. 4.34a), to zones with brecciation of country rock and high strain deformation of fault breccia clasts, cross-cut and variably replaced by quartz and carbonate veins resulting from fluid flow along the fault zone, sometimes accompanied by precious metals and sulphide minerals, and variably overprinted by hydrothermal and/or metamorphic alteration minerals (Fig. 4.34b). In weakly lithified volcaniclastic successions (e.g. young unlithified pyroclastic, autoclastic, or sedimentary deposits), fault zones may be diffuse, with fault parallel dismemberment of host rock, and anastomosing seams of fine country rock ("rock powder") and secondary alteration minerals (Fig. 4.34c).

4.11 Epiclastic or Surface Process Fragmentation

In most volcanic terrains, huge volumes of volcaniclastic debris are also produced and transported by surface sedimentary processes, which are not of primary volcanic origin, i.e. they are not vent-related processes. Such processes are collectively called epiclastic (epi = upon, referring to processes at the Earth's surface) or surface sedimentary processes, and include gravitational collapse or mass wastage; chemical and physical weathering; and transportation processes involving wind, water, or ice; or just simply gravity as the driving agent for transport (Cas and Wright 1987). More details of epiclastic surface processes and products in volcanic terrains will be discussed in Chap. 12.

The processes range from remobilisation of already fragmental primary volcaniclastic deposits (autoclastic, pyroclastic), and the secondary physical abrasion and breaking of those primary clasts, to physical and chemical erosion of coherent and lithified volcaniclastic rocks, so generating epiclastic *sensu stricto* volcanic sediments.

The significance of sedimentary surface processes should never be underestimated in volcanic terrains. It is clear from some of the major eruptions of recent times (Mt St Helens 1980, USA; Mt Pinatubo 1991, Philippines; Mt Unzen 1991–1995, Japan; Soufrière Hills 1995–2015, Montserrat) that sedimentary surface processes can be significant agents of resedimentation and reworking immediately from the time of initial deposition of volcanic materials by primary volcanic processes (Chap. 12). One of the complications in this case is that the clasts may still preserve their primary juvenile volcanic morphology inherited/preserved from the primary volcanic mode of fragmentation (explosive, autobrecciation, quench fragmentation), whereas the aggregate character could reflect the effects of resedimentation and/or reworking (note the distinction) by mass-flow or tractional, reworking processes (e.g. Mt Unzen, Uehara et al. 2015; Chap. 12).

Fig. 4.34 Fault breccias in volcanic rock successions. **a** Highly strained basalt breccia clasts, some of which have been folded. The clasts were originally sub-equant, variably rounded, resedimented mass-flow breccia before being deformed in a high-strain deformational zone in a fault zone. Archean Agnew gold camp, Yilgarn Craton, Western Australia. Width of core ~5 cm. **b** High-strain deformation of fault breccia clasts and abundant quartz veining in a fault zone, St Ives gold camp, Yilgarn Craton, Western Australia. Width of core ~5 cm. **c** Diffuse fault zone in 2 Ma weakly lithified hyaloclastite breccia, Ponza. Tyrrhenian Sea, Italy. Note the anastomosing seams of finely fragmented grey hyaloclastite and alteration zone

Mass flow processes by themselves are generally too short-lived to produce identifiable abrasive rounding of clasts and often even sorting. Only tractional reworking (e.g. in the bed of a river or in the shoreline zone) will cause enough abrasive rounding of clasts and appreciable sorting to begin imparting the earmarks of surface sedimentary processes. Such reworking will also often begin to cause mixing of materials from different source rocks (provenance). So texturally and compositionally mass-flow resedimented deposits may not look all that different from primary source deposits. Reworked deposits should begin to show signs of rounding and sorting of debris, as well as often an increasingly polymictic compositional character. But remember, we are dealing with volcanic terrains, so nothing is ever clear cut and dried, as unfortunately too many people at times think! What we hope is that the depositional facies characteristics (Chaps. 1 and 12) will also help to distinguish primary volcaniclastic deposits from epiclastic ones.

Epiclastic processes are capable of producing fragments of all sizes (house-size boulders to clay-size particles) from older lithified volcanic source rocks, and of transporting them, as well as fragments derived from primary volcaniclastic deposits (the size of debris will be inherited from that source), well away from source. Therefore, a volcanic composition by itself does not mean that the deposit is pyroclastic or near-vent, no matter what the size of the fragments or the angularity of the fragments (Chaps. 1, 12, and 14).

References

Adams NK, Houghton BF, Fagents SA, Hildreth W (2006a) The transition from explosive to effusive eruptive regime: the example of the 1912 Novarupta eruption, Alaska. Geol Soc Am Bull 118(5–6):620–634

Adams NK, Houghton BF, Hildreth W (2006b) Abrupt transitions during sustained explosive eruptions: examples from the 1912 eruption of Novarupta, Alaska. Bull Volcanol 69(2):189–206

Alfano F, Bonadonna C, Gurioli L (2012) Insights into eruption dynamics from textural analysis: the case of the May, 2008, Chaitén eruption. Bull Volcanol 74(9):2095–2108

Alidibirov M, Dingwell DB (1996) Magma fragmentation by rapid decompression. Nature 380(6570):146–148

Allen JRL (1970) Physical processes of sedimentation - an introduction. Allen and Unwin, London. 248p.

Allen SR, Fiske RS, Tamura Y (2010) Effects of water depth on pumice formation in submarine domes at Sumisu, Izu-Bonin arc, western Pacific. Geology 38(5):391–394

Ambrosio M, Doveri M, Fagioli MT, Marini L, Principe C, Raco B (2010) Water-rock interaction in the magmatic-hydrothermal system of Nisyros Island (Greece). J Volcanol Geotherm Res 192(1–2):57–68

Anzidei M, Carapezza ML, Esposito A, Giordano G, Lelli M, Tarchini L (2008) The Albano Maar Lake high resolution bathymetry and dissolved CO2 budget (Colli Albani volcano, Italy): constrains to hazard evaluation. J Volcanol Geotherm Res 171(3–4):258–268

Aravena A, Vitturi MD, Cioni R, Neri A (2018a) Physical constraints for effective magma-water interaction along volcanic conduits during silicic explosive eruptions. Geology 46(10):867–870

Aravena A, Vitturi MD, Cioni R, Neri A (2018b) Reply: physical constraints for effective magma-water interaction along volcanic conduits during silicic explosive eruptions. Geology 47(5):462

Arzilli F, La Spina G, Burton MR, Polacci M, Le Gall N, Hartley ME, Di Genova D, Cai B, Vo NT, Bamber EC, Nonni S (2019a) Magma fragmentation in highly explosive basaltic eruptions induced by rapid crystallization. Nat Geosci 12(12):1023–1028

Arzilli F, Morgavi D, Petrelli M, Polacci M, Burton M, Di Genova D, Spina L, La Spina G, Hartley ME, Romero JE (2019b) The unexpected explosive sub-Plinian eruption of Calbuco volcano (22–23 April 2015; southern Chile): triggering mechanism implications. J Volcanol Geotherm Res 378:35–50

Bachmann O, Bergantz GW (2003) Rejuvenation of the Fish Canyon magma body: a window into the evolution of large-volume silicic magma systems. Geology 31(9):789–792

Bahr H-A, Fischer G, Weiss H-J (1986) Thermal-shock crack patterns explained by single and multiple crack propagation. J Mater Sci 21 (8):2716–2720

Bahr HA, Bahr U, Balke H, Maschke H, Petzold A, Weiß HJ (1993) Multiple crack propagation under thermal load. In: Schneider G, Petzow G (eds) Thermal shock and thermal fatigue behavior of advanced ceramics. Springer, Netherlands, pp 143–153

Bahr H-A, Bahr U, Gerbatsch A, Pflugbeil I, Vojta A, Weiss H-J (1996) Fracture mechanical analysis of morphological transitions in thermal shock cracking. In: Bradt RC, Hasselman DPH, Munz D, Sakai M, Shevchenko VY (eds) Fracture mechanics of ceramics. R-curve behaviour, toughness determination, and thermal shock, vol 11. Plenum Press, New York, pp 507–522

Barberi F, Coltelli M, Frullani A, Rosi M, Almeida E (1995) Chronology and dispersal characteristics of recently (last 5000 years) erupted tephra of Cotopaxi (Ecuador): implications for long-term eruptive forecasting. J Volcanol Geotherm Res 69 (3):217–239

Barnett WP (2004) Subsidence breccias in kimberlite pipes—an application of fractal analysis. Lithos 76(1):299–316

Barnett WP (2008) The rock mechanics of kimberlite volcanic pipe excavation. J Volcanol Geotherm Res 174(1–3):29–39

Barnett WP, Kurszlaukis S, Tait M, Dirks P (2011) Kimberlite wall-rock fragmentation processes: Venetia K08 pipe development. Bull Volcanol 73(8):941–958

Barreyre T, Soule SA, Sohn RA (2011) Dispersal of volcaniclasts during deep-sea eruptions: settling velocities and entrainment in buoyant seawater plumes. J Volcanol Geotherm Res 205(3):84–93

Batiza R, White JDL (2000) Submarine lavas and hyaloclastite. In: Sigurdsson H, Stix J, Houghton BF, McNutt SR, Rymer H (eds) Encyclopedia of volcanoes. Academic Press, San Diego, pp 361–382

Behrens H, Zhang Y, Xu Z (2004) H_2O diffusion in dacitic and andesitic melts. Geochim Cosmochim Acta 68(24):5139–5150

Benet D, Costa F, Widiwijayanti C, Pallister J, Pedreros G, Allard P, Humaida H, Aoki Y, Maeno F (2024) VolcAshDB: a Volcanic Ash DataBase of classified particle images and features. Bull Volcanol 86(1):1–30

Bennett FD (1974) On volcanic ash formation. Am J Sci 274 (6):648–661

Best MG, Christiansen EH (1997) Origin of broken phenocrysts in ash-flow tuffs. Geol Soc Am Bull 109(1):63–73

Biass S, Bonadonna C (2011) A quantitative uncertainty assessment of eruptive parameters derived from tephra deposits: the example of two large eruptions of Cotopaxi volcano, Ecuador. Bull Volcanol 73 (1):73–90

Binns RA (2003) Deep marine pumice from the Woodlark and Manus Basins, Papua New Guinea. In: Explosive subaqueous volcanism, vol 140. American Geophysical Union Monograph, pp 329–334

Biswas DR, Bhaduri SB (1983) Thermal shock resistance in glass-nickel composites. In: Bradt RC, Evans AG, Hasselman DPH, Lange FF (eds) Fracture mechanics of ceramics. Measurements, transformations, and high-temperature fracture, vol 6. Plenum Press, New York, pp 475–485

Blackwell JL, Cooke DR, McPhie J, Simpson KA (2014) Lithofacies associations and evolution of the volcanic host succession to the Minifie ore zone: Ladolam gold deposit, Lihir Island, Papua New Guinea. Econ Geol 109(4):1137–1160

Blake S (1981a) Eruptions from zoned magma chambers. J Geol Soc 138(3):281–287

Blake S (1981b) Volcanism and the dynamics of open magma chambers. Nature 289(5800):783

Blank JG, Brooker RA (1994) Experimental studies of carbon dioxide in silicate melts; solubility, speciation, and stable carbon isotope behavior. Rev Mineral Geochem 30(1):157–186

Blenkinsop TG (1991) Cataclasis and processes of particle size reduction. Pure Appl Geophys 136(1):59–86

Blower JD, Keating JP, Mader HM, Phillips JC (2002) The evolution of bubble size distributions in volcanic eruptions. J Volcanol Geotherm Res 120(1-2):1–23

Blower JD, Keating JP, Mader HM, Phillips JC (2003) The evolution of bubble size distributions in volcanic eruptions. J Volcanol Geotherm Res 120(1):1–23

Blundy J, Cashman KV (2001) Ascent-driven crystallisation of dacite magmas at Mount St Helens, 1980–1986. Contrib Mineral Petrol 140(6):631–650

Blundy J, Cashman KV (2005) Rapid decompression-driven crystallization recorded by melt inclusions from Mount St. Helens volcano. Geology 33(10):793–796

Board SJ, Hall RW, Hall RS (1975) Detonation of fuel coolant explosions. Nature 254:319–321

Bottinga Y, Richet P, Sipp A (1995) Viscosity regimes of homogeneous silicate melts. Am Mineral 80(3–4):305–318

Bouška V, Borovec Z, Cimbálníková A, Kraus I, Lajčáková A, Pačesová M (1993) Natural glasses, 2nd edn. Ellis Horwood Ltd., West Sussex, U.K., p 354p

Boyce J (2013) The Newer Volcanics Province of southeastern Australia: a new classification scheme and distribution map for eruption centres. Aust J Earth Sci 60(4):449–462

Brennen CE (1995) Cavitation and bubble dynamics. Oxford University Press, Oxford, p 304

Brennen CE (2002) Fission of collapsing cavitation bubbles. J Fluid Mech 472:153–166

Brennen CE (2005) Fundamentals of multiphase flow. Cambridge University Press, 345p

Broek D (1986) Elementary engineering fracture mechanics. Kluwer Academic Publishers, Dordrecht, The Netherlands, p 516p

Broek D (1989) The practical use of fracture mechanics. Kluwer Academic Publishers, Dordrecht, The Netherlands, p 522p

Brooks ER (1995) Paleozoic fluidization, folding, and peperite formation, northern Sierra Nevada, California. Can J Earth Sci 32 (3):314–324

Brooks ER, Wood MM, Garbutt PL (1982) Origin and metamorphism of peperite and associated rocks in the Devonian Elwell Formation, northern Sierra Nevada, California. Geol Soc Am Bull 93 (12):1208–1231

Brown RJ, Buse B, Sparks RSJ, Field M (2008) On the welding of pyroclasts from very low-viscosity magmas: examples from kimberlite volcanoes. J Geol 116(4):354–374

Brown RJ, Branney MJ, Maher C, Dávila-Harris P (2010) Origin of accretionary lapilli within ground-hugging density currents: evidence from pyroclastic couplets on Tenerife. Geol Soc Am Bull 122 (1–2):305–320

Brown RJ, Bonadonna C, Durant AJ (2012) A review of volcanic ash aggregation. Phys Chem Earth Parts A/B/C 45:65–78

Browne PR, Lawless JV (2001) Characteristics of hydrothermal eruptions, with examples from New Zealand and elsewhere. Earth Sci Rev 52(4):299–331

Buchanan DJ, Dullforce TA (1973) Mechanism for vapour explosions. Nature 245:32–34

Burnham CW (1972) The energy of explosive eruptions. Earth Mineral Sci (Pennsylvania State University) 41:69–70

Burnham CW (1983) Deep submarine pyroclastic eruptions. Econ Geol Monogr 5:142–148

Busby-Spera CJ, White JDL (1987) Variation in peperite textures associated with differing host-sediment properties. Bull Volcanol 49 (6):765–776

Büttner R, Zimanowski B (1998) Physics of thermohydraulic explosions. Phys Rev E 57(5):5726

Büttner R, Dellino P, Zimanowski B (1999) Identifying magma–water interaction from the surface features of ash particles. Nature 401 (6754):688–690

Campagnola S, Vona A, Romano C, Giordano G (2016) Crystallization kinetics and rheology of leucite-bearing tephriphonolite magmas from the Colli Albani volcano (Italy). Chem Geol 424:12–29

Campbell ME, Russell JK, Porritt LA (2013) Thermomechanical milling of accessory lithics in volcanic conduits. Earth Planet Sci Lett 377–378:276–286

Carey S, Sigurdsson H (1987) Temporal variations in column height and magma discharge rate during the 79 AD eruption of Vesuvius. Geol Soc Am Bull 99(2):303–314

Carey RJ, Houghton BF, Thordarson T (2009) Abrupt shifts between wet and dry phases of the 1875 eruption of Askja Volcano: microscopic evidence for macroscopic dynamics. J Volcanol Geotherm Res 184(3):256–270

Carey RJ, Houghton BF, Thordarson T (2010) Tephra dispersal and eruption dynamics of wet and dry phases of the 1875 eruption of Askja Volcano, Iceland. Bull Volcanol 72(3):259–278

Carey R, Soule SA, Manga M, White JDL, McPhie J, Wysoczanski R, Jutzeler M, Tani K, Yoerger D, Fornari D (2018) The largest deep-ocean silicic volcanic eruption of the past century. Sci Adv 4 (1):e1701121

Carey S, Sigurdsson H (1985) The May 18, 1980 eruption of Mount St. Helens: 2. Modeling of dynamics of the Plinian phase. J Geophys Res Solid Earth 90(B4):2948–2958

Caricchi L, Burlini L, Ulmer P, Gerya T, Vassalli M, Papale P (2007) Non-Newtonian rheology of crystal-bearing magmas and implications for magma ascent dynamics. Earth Planet Sci Letts 264(3–4):402–419

Carrasco-Núñez G, McCurry M, Branney MJ, Norry M, Willcox C (2012) Complex magma mixing, mingling, and withdrawal associated with an intra-Plinian ignimbrite eruption at a large silicic caldera volcano: Los Humeros of central Mexico. Geol Soc Am Bull 124(11–12):1793–1809

Carrecedo Sánchez MC, Sarrionandia F, Arostegui J, Larrondo E, Gil Ibarguchi JIG (2009) Development of spheroidal composite bombs by welding of juvenile spinning and isotropic droplets inside a mafic eruption column. J Volcanol Geotherm Res 186(3):265–279

Carrecedo Sánchez MC, Sarrionandia F, Arostegui J, Gil Ibarguchi JIG (2015) Silicate glass micro and nanospherules generated in explosive eruptions of ultrabasic magmas: Implications for the origin of pelletal lapilli. J Volcanol Geotherm Res 293:13–24

Carrigan CR (2000) Plumbing systems. In: Sigurdsson H, Stix J, Houghton BF, McNutt SR, Rymer H (eds) Encyclopedia of volcanoes. Academic Press, San Diego, pp 219–235

Carroll MR, Holloway JR (eds) (1994) Volatiles in magmas, vol 30. Reviews in Mineralogical Society of America, 517p

Carroll MR, Rutherford MJ (1985) Sulfide and sulfate saturation in hydrous silicate melts. J Geophys Res Solid Earth 90(S02):C601–C612

Carroll MR, Rutherford MJ (1987) The stability of igneous anhydrite: experimental results and implications for sulfur behavior in the 1982 El Chichon trachyandesite and other evolved magmas. J Petrol 28(5):781–801

Cas R, Porritt L, Hayman P, Pittari A (2008) Pyroclast formation processes during explosive kimberlite eruptions. In 9th International Kimberlite Conference: Extended Abstracts (Vol. 9), Frankfurt, Germany.

Cas RAF (1992) Submarine volcanism; eruption styles, products, and relevance to understanding the host-rock successions to volcanic-hosted massive sulfide deposits. Econ Geol 87(3):511–541

Cas RAF, Beresford SW (2001) Field characteristics and erosional processes associated with komatiitic lavas: implications for flow behavior. Can Mineral 39(2):505–524

Cas RAF, Giordano G (2014) Submarine volcanism: a review of the constraints, processes and products, and relevance to the Cabo de Gata volcanic succession. Ital J Geosci 133(3):362–377

Cas RAF, Simmons J (2018) Why deep-water eruptions are so different from subaerial eruptions. Front Earth Sci 6:198

Cas RAF, Wright JV (1987) Volcanic successions: modern and ancient. Allen and Unwin, London, p 528p

Cas RAF, Allen RL, Bull SW, Clifford BA, Wright JV (1990) Subaqueous, rhyolitic dome-top tuff cones: a model based on the Devonian Bunga Beds, southeastern Australia and a modern analogue. Bull Volcanol 52(3):159–174

Cas RAF, Edgar C, Allen RL, Bull S, Clifford BA, Giordano G, Wright JV (2001) Influence of magmatism and tectonics on sedimentation in an extensional lake basin: the Upper Devonian Bunga Beds, Boyd Volcanic Complex, south-eastern Australia. In: White JDL, Riggs NR (eds) Volcaniclastic sedimentation in lacustrine settings. Blackwell Science Ltd., Oxford, pp 83–108

Cas RAF, Hayman P, Pittari A, Porritt L (2008) Some major problems with existing models and terminology associated with kimberlite pipes from a volcanological perspective, and some suggestions. J Volcanol Geotherm Res 174(1–3):209–225

Cas RAF, Giordano G, Balsamo F, Esposito A, Mastro SL (2011) Hydrothermal Breccia textures and processes: Lisca Bianca Islet, Panarea Volcano, Aeolian Islands, Italy. Econ Geol 106(3):437–450

Cas RAF, Sato H, Simpson CJ (1993) Newer volcanics province-processes and products of phreatomagmatic activity. IAVCEI Scientific Assembly Field Guide. Aust Geol Surv Org, Canberra, Bull 64

Cas RAF, Yamagishi H, Moore L, Scutter CR (2003) Miocene submarine fire fountain deposits, Ryugazaki Headland, Oshoro Peninsula, Hokkaido, Japan: implications for submarine fountain dynamics and fragmentation processes. In: White JDL, Smellie JL, Clague DA (eds) Explosive subaqueous volcanism, vol 140. American Geophysical Union Monograph, pp 299–316

Cas RAF, van Otterloo J, Blaikie TN, van den Hove J (2017) The dynamics of a very large intra-plate continental basaltic volcanic province, the Newer Volcanics Province, SE Australia, and implications for other provinces, vol 446. Geological Society of London Special Publications, SP446. 448

Cashman KV (2004) Volatile controls on magma ascent and eruption. In: Sparks RSJ, Hawkesworth C (eds) The state of the planet: frontiers and challenges in geophysics. American Geophysical Union, Washington, D.C., pp 109–124

Cashman KV, Blundy J (2000) Degassing and crystallization of ascending andesite and dacite. Philos Trans R Soc Lond A Math Phys Eng Sci 358(1770):1487–1513

Cashman KV, Mangan MT (1994) Physical aspects of magmatic degassing; II, Constraints on vesiculation processes from textural studies of eruptive products. Rev Mineral Geochem 30(1):447–478

Cashman KV, Marsh BD (1988) Crystal size distribution (CSD) in rocks and the kinetics and dynamics of crystallization II: Makaopuhi lava lake. Contrib Mineral Petrol 99(3):292–305

Cashman KV, Scheu B (2015) Magmatic fragmentation. In: Sigurdsson H, Houghton B, McNutt SR, Rymer H, Stix J (eds) The encyclopedia of volcanoes. Academic Press Elsevier, Cambridge, pp 459–471

Cashman KV, Sturevant B, Papale P, Navon O (2000) Magmatic fragmentation. In: Sigurdsson H, Stix J, Houghton BF, McNutt SR, Rymer H (eds) Encyclopedia of volcanoes. Academic Press, San Diego, pp 421–430

Cashman KV (1988) Crystallization of Mount St. Helens 1980–1986 dacite: a quantitative textural approach. Bull Volcanol 50 (3):194–209

Cashman KV (1992) Groundmass crystallization of Mount St. Helens dacite, 1980–1986: a tool for interpreting shallow magmatic processes. Contrib Mineral Petrol 109(4):431–449

Chadwick WW Jr, Merle SG, Baker ET, Walker SL, Resing JA, Butterfield DA, Anderson MO, Baumberger T, Bobbitt AM (2018) A recent volcanic eruption discovered on the central Mariana back-arc spreading center. Front Earth Sci 6:172

Chadwick WW, Cashman KV, Embley RW, Matsumoto H, Dziak RP, De Ronde CEJ, Lau TK, Deardorff ND, Merle SG (2008) Direct video and hydrophone observations of submarine explosive eruptions at NW Rota-1 volcano, Mariana arc. J Geophys Res Solid Earth 113(B8)

Chadwick WW, Paduan JB, Clague DA, Dreyer BM, Merle SG, Bobbitt AM, Caress DW, Philip BT, Kelley DS, Nooner SL (2016) Voluminous eruption from a zoned magma body after an increase in supply rate at Axial Seamount. Geophys Res Lett 43(23):12,063–012,070

Cimarelli C, Di Traglia F, Taddeucci J (2010) Basaltic scoria textures from a zoned conduit as precursors to violent Strombolian activity. Geology 38(5):439–442

Cioni R, Bertagnini A, Andronico D, Cole PD, Mundula F (2011) The 512 AD eruption of Vesuvius: complex dynamics of a small scale subplinian event. Bull Volcanol 73(7):789–810

Clague DA, Paduan JB, Caress DW, Chadwick WW Jr, Le Saout M, Dreyer BM, Portner RA (2017) High-resolution AUV mapping and targeted ROV observations of three historical lava flows at Axial Seamount. Oceanography 30(4):82–99

Clague DA, Uto K, Satake K, Davis AS (2002) Eruption style and flow emplacement in the submarine North Arch Volcanic Field, Hawaii. In: Hawaiian volcanoes: deep underwater perspectives, vol 128. American Geophysical Union Monograph, pp 65–84

Clague DA, Batiza R, Head JW, Davis AS (2003) Pyroclastic and hydroclastic deposits on Loihi Seamount, Hawaii. In: White JDL, Smellie JL, Clague DA (eds) Explosive subaqueous volcanism, vol 140. American Geophysical Union Monograph, pp 73–95

Clark C, James P (2003) Hydrothermal brecciation due to fluid pressure fluctuations: examples from the Olary Domain, South Australia. Tectonophysics 366(3):187–206

Cole PD, Guest JE, Duncan AM, Pacheco J-M (2001) Capelinhos 1957–1958, Faial, Azores: deposits formed by an emergent surtseyan eruption. Bull Volcanol 63(2–3):204–220

Colgate SA, Sigurgeirsson T (1973) Dynamic mixing of water and lava. Nature 244:552–554

Colombier M, Scheu B, Kueppers U, Cronin SJ, Mueller SB, Hess KU, Wadsworth FB, Tost M, Dobson KJ, Ruthensteiner B, Dingwell DB (2019) In situ granulation by thermal stress during subaqueous volcanic eruptions. Geology 47(2):179–182

Colombier M, Gurioli L, Druitt TH, Shea T, Boivin P, Miallier D, Cluzel N (2017) Textural evolution of magma during the 9.4-ka trachytic explosive eruption at Kilian Volcano, Chaîne des Puys, France. Bull Volcanol 79(2):17

Coltelli M, Del Carlo P, Vezzoli L (1998) Discovery of a Plinian basaltic eruption of Roman age at Etna volcano, Italy. Geology 26 (12):1095–1098

Costa F, Dungan M (2005) Short time scales of magmatic assimilation from diffusion modeling of multiple elements in olivine. Geology 33(10):837–840

Costa A, Caricchi L, Bagdassarov N (2009) A model for the rheology of particle-bearing suspensions and partially molten rocks. Geochem Geophys Geosyst 10(3)

Costantini L, Houghton BF, Bonadonna C (2010) Constraints on eruption dynamics of basaltic explosive activity derived from chemical and microtextural study: the example of the Fontana Lapilli Plinian eruption, Nicaragua. J Volcanol Geotherm Res 189 (3–4):207–224

Danese E, Mattei M (2010) The sedimentary substrate of the Colli Albani volcano. The Colli Albani Volcano, vol 3. Geological Society of London Special Publications-IAVCEI, pp 141–151

Davies AGS, Cooke DR, Gemmell JB, Simpson KA (2008) Diatreme breccias at the Kelian gold mine, Kalimantan, Indonesia: precursors to epithermal gold mineralization. Econ Geol 103(4):689–716

Davis AS, Clague DA (2003) Hyaloclastite from Miocene seamounts offshore central California: compositions, eruption styles, and depositional processes. In: White JDL, Smellie JL, Clague DA (eds) Explosive subaqueous volcanism, vol 140. American Geophysical Union Monograph, pp 129–142

De Benedetti AA, Funiciello R, Giordano G, Diano G, Caprilli E, Paterne M (2008) Volcanology, history and myths of the Lake Albano maar (Colli Albani volcano, Italy). J Volcanol Geotherm Res 176(3):387–406

De Benedetti AA, Caprilli E, Rossetti F, Giordano G (2010) Metamorphic, metasomatic and intrusive xenoliths of the Colli Albani volcano and their significance for the reconstruction of the volcano plumbing system. The Colli Albani Volcano, vol 3.

Geological Society of London Special Publications-IAVCEI, pp 153–176

De Rita D, Giordano G, Cecili A (2001) A model for submarine rhyolite dome growth: Ponza Island (central Italy). J Volcanol Geotherm Res 107(4):221–239

De Rita D, Giordano G, Esposito A, Fabbri M, Rodani S (2002) Large volume phreatomagmatic ignimbrites from the Colli Albani volcano (Middle Pleistocene, Italy). J Volcanol Geotherm Res 118 (1–2):77–98

de Ronde CE, Fornari DJ, Ferrini VL, Walker SL, Davy BW, LeBlanc C, Tontini FC, Kukulya AL, Littlefield RH (2016) The Pink and White Terraces of Lake Rotomahana: what was their fate after the 1886 Tarawera Rift eruption? J Volcanol Geotherm Res 314:126–141

de Rosen-Spence AF, Provost G, Dimroth E, Gochnauer K, Owen V (1980) Archean subaqueous felsic flows, Rouyn-Noranda, Quebec, Canada, and their quarternary equivalents. Precamb Res 12(1–4):43–77

de Silva S, Salas G, Schubring S (2008) Triggering explosive eruptions—the case for silicic magma recharge at Huaynaputina, southern Peru. Geology 36(5):387–390

Deardorff ND, Cashman KV, Chadwick WW (2011) Observations of eruptive plume dynamics and pyroclastic deposits from submarine explosive eruptions at NW Rota-1, Mariana Arc. J Volcanol Geotherm Res 202(1):47–59

Dellino P, Gudmundsson MT, Larsen G, Mele D, Stevenson JA, Thordarson T, Zimanowski B (2012) Ash from the Eyjafjallajökull eruption (Iceland): fragmentation processes and aerodynamic behavior. J Geophys Res Solid Earth 117(B9)

Dhir VK, Purohit GP (1978) Subcooled film-boiling heat transfer from spheres. Nucl Engin Des 47(1):49–66

Dimroth E, Lichtblau AP (1979) Metamorphic evolution of Archean hyaloclastites, Noranda area, Quebec, Canada. Part I: comparison of Archean and Cenozoic sea-floor metamorphism. Can J Earth Sci 16 (7):1315–1340

Dingwell DB (1996) Volcanic dilemma-flow or blow? Science 273 (5278):1054–1055

Dingwell DB, Webb SL (1989) Structural relaxation in silicate melts and non-Newtonian melt rheology in geologic processes. Phys Chem Minerals 16(5):508–516

Dingwell DB, Webb SL (1990) Relaxation in silicate melts. Eur J Mineral 4:427–449

Dixon JE, Clague DA (2001) Volatiles in basaltic glasses from Loihi Seamount, Hawaii: evidence for a relatively dry plume component. J Petrol 42(3):627–654

Dixon JE, Clague DA, Wallace P, Poreda R (1997) Volatiles in alkalic basalts form the North Arch Volcanic Field, Hawaii: extensive degassing of deep submarine-erupted alkalic series lavas. J Petrol 38 (7):911–939

Dobran F (1992) Nonequilibrium flow in volcanic conduits and application to the eruptions of Mt. St. Helens on May 18, 1980, and Vesuvius in AD 79. J Volcanol Geotherm Res 49(3–4):285–311

Downey WS, Lentz DR (2006) Igneous associations 6. Modelling of deep submarine pyroclastic volcanism: a review and new results. In: Geoscience. Canada, pp 5–24

Druitt TH, Bacon CR (1989) Petrology of the zoned calcalkaline magma chamber of Mount Mazama, Crater Lake, Oregon. Contrib Mineral Petrol 101(2):245–259

Dufek J, Manga M, Staedter M (2007). Littoral blasts: pumice-water heat transfer and the conditions for steam explosions when pyroclastic flows enter the ocean. J Geophys Res Solid Earth 112 (B11)

Dürig T, Mele D, Dellino P, Zimanowski B (2012) Comparative analyses of glass fragments from brittle fracture experiments and volcanic ash particles. Bull Volcanol 74(3):691–704

Dürig T, White JD, Murch AP, Zimanowski B, Büttner R, Mele D, Dellino P, Carey RJ, Schmidt LS, Spitznagel N (2020) Deep-sea eruptions boosted by induced fuel–coolant explosions. Nat Geosci 13(7):498–503

Edgar CJ, Cas RAF, Olin P, Wolff JA, Marti J, Simmons JM (2017) Causes of complexity in one of the largest known Plinian fallout eruption sequences: the 312 ka Fasnia member of the Diego Hernandez Formation, Tenerife, Spain. J Volcanol Geotherm Res 345:21–45

Edgar CJ, Wolff JA, Nichols HJ, Cas RAF, Martí J (2002) A complex quaternary ignimbrite-forming phonolitic eruption: the Poris Member of the Diego Hernández Formation (Tenerife, Canary Islands). J Volcanol Geotherm Res 118(1–2):99–130

Edmonds M, Herd RA (2005) Inland-directed base surge generated by the explosive interaction of pyroclastic flows and seawater at Soufriere Hills volcano, Montserrat. Geology 33(4):245–248

Edwards B, Russell J, Anderson R (2002) Subglacial, phonolitic volcanism at Hoodoo Mountain volcano, northern Canadian Cordillera. Bull Volcanol 64(3–4):254–272

Edwards BR, Gudmundsson MT, Russell JK (2015) Glaciovolcanism. In: Sigurdsson H, Houghton B, McNutt SR, Rymer H, Stix J (eds) The Encyclopedia of Volcanoes. Academic Press Elsevier, Cambridge, pp 377–394

Eichelberger JC (1980) Vesiculation of mafic magma during replenishment of silicic magma reservoirs. Nature 288:446–450

Eichelberger JC, Carrigan CR, Westrich HR, Price RH (1986) Non-explosive silicic volcanism. Nature 323(6089):598–602

Embley RW, Chadwick WW, Baker ET, Butterfield DA, Resing JA, De Ronde CEJ, Tunnicliffe V, Lupton JE, Juniper SK, Rubin KH (2006) Long-term eruptive activity at a submarine arc volcano. Nature 441(7092):494–497

Erdogan F (1974) The interaction between inclusions and cracks. In: Bradt RC, Hasselman DPH, Lange FF (eds) Fracture mechanics of ceramics. Volume 1: Concepts, flaws and fractography. Plenum Press, New York, pp 245–267

Evans AG, Blumenthal W (1983) High-temperature failure in ceramics. In: Bradt RC, Evans AG, Hasselman DPH, Lange FF (eds) Fracture mechanics of ceramics. Volume 6: Measurements, transformations, and high-temperature fracture. Lawrence Berkeley Lab., CA (USA), Plenum Press, New York City, pp 423–448

Fauria KE, Manga M, Wei Z (2017) Trapped bubbles keep pumice afloat and gas diffusion makes pumice sink. Earth Planet Sci Lett 460:50–59

Fierstein J, Hildreth W (1992) The plinian eruptions of 1912 at Novarupta, Katmai National Park, Alaska. Bull Volcanol 54 (8):646–684

Fink JH (1983) Structure and emplacement of a rhyolitic obsidian flow: Little Glass Mountain, Medicine Lake Highland, northern California. Geol Soc Am Bull 94(3):362–380

Fink JH, Anderson SW, Manley CR (1992) Textural constraints on effusive silicic volcanism: beyond the permeable foam model. J Geophys Res Solid Earth 97(B6):9073–9083

Fink JH, Manley CR (1987) Origin of pumiceous and glassy textures in rhyolite flows and domes, vol 212. Geological Society of American Special Publications, pp 77–88

Fink JH, Manley CR (1989) Explosive volcanic activity generated from within advancing silicic lava flows. In: IAVCEI proceedings in volcanology, pp 169–179

Fisher RV (1968) Puu Hou littoral cones, Hawaii. Geol Rundsch 57 (3):837–864

Fisher RV, Schmincke H-U (1984) Pyroclastic rocks. Springer, Berlin, p 472p

Fisher RV, Smith AL, Roobol MJ (1980) Destruction of St. Pierre, Martinique, by ash-cloud surges, May 8 and 20, 1902. Geology 8 (10):472–476

Freud R, Harari R, Sher E (2009) Collapsing criteria for vapor film around solid spheres as a fundamental stage leading to vapor explosion. Nuclear engineer design 239(4):722–727
Fujibayashi N, Sakai U (2003) Vesiculation and eruption processes of submarine effusive and explosive rocks from the Middle Miocene Ogi Basalt, Sado Island, Japan. In: White JDL, Smellie JL, Clague DA (eds) Explosive subaqueous volcanism, vol 140. American Geophysical Union Monograph, pp 259–272
Fujikawa S, Akamatsu T (1980) Effects of the non-equilibrium condensation of vapour on the pressure wave produced by the collapse of a bubble in a liquid. J Fluid Mech 97(03):481–512
Fuller RE (1931) The aqueous chilling of basaltic lava on the Columbia River Plateau. Am J Sci 124:281–300
Furnes H, Fridleifsson IB, Atkins FB (1980) Subglacial volcanics—on the formation of acid hyaloclastites. J Volcanol Geotherm Res 8 (1):95–110
Garcia MO, Caplan-Auerbach J, Eric H, Kurz MD, Becker N (2006) Geology, geochemistry and earthquake history of Lōihi Seamount, Hawaii's Youngest Volcano. Chemie Der Erde-Geochemistry 66 (2):81–108
Gardner JE, Hilton M, Carroll MR (1999) Experimental constraints on degassing of magma: isothermal bubble growth during continuous decompression from high pressure. Earth Planet Sci Lett 168 (1):201–218
Gardner JE, Hilton M, Carroll MR (2000) Bubble growth in highly viscous silicate melts during continuous decompression from high pressure. Geochim Cosmochim Acta 64(8):1473–1483
Gernon TM, Brown RJ, Tait MA, Hincks TK (2012) The origin of pelletal lapilli in explosive kimberlite eruptions. Nat Commun 3:832
Geshi N, Németh K, Oikawa T (2011) Growth of phreatomagmatic explosion craters: a model inferred from Suoana crater in Miyakejima Volcano, Japan. J Volcanol Geotherm Res 201(1):30–38
Giachetti T, Gonnermann HM (2013) Water in volcanic pyroclast: rehydration or incomplete degassing? Earth Planet Sci Lett 369:317–332
Gifkins CC, McPhie J, Allen RL (2002) Pumiceous rhyolitic peperite in ancient submarine volcanic successions. J Volcanol Geotherm Res 114(1–2):181–203
Gifkins CC, Herrmann W, Large RR (2005) Altered volcanic rocks: a guide to description and interpretation. Centre for Ore Deposit Research, University of Tasmania, 275p
Gilbert JS, Lane SJ (1994) The origin of accretionary lapilli. Bull Volcanol 56(5):398–411
Gill J, Torssander P, Lapierre H, Taylor R, Kaiho K, Koyama M, Kusakabe M, Aitchison J, Cisowski S, Dadey K, Fujioka K, Klaus A, Lovell M, Marsaglia K, Pezard P, Taylor B, Tazaki K (1990) Explosive deep water basalt in the Sumisu backarc rift. Science 248(4960):1214
Giordano G, De Astis G (2021) The summer 2019 basaltic Vulcanian eruptions (paroxysms) of Stromboli. Bull Volcanol 83(1):1–27
Giordano G, De Rita D, Cas R, Rodani S (2002) Valley pond and ignimbrite veneer deposits in the small-volume phreatomagmatic 'Peperino Albano'basic ignimbrite, Lago Albano maar, Colli Albani volcano, Italy: influence of topography. J Volcanol Geotherm Res 118(1–2):131–144
Giordano D, Nichols ARL, Dingwell DB (2005) Glass transition temperatures of natural hydrous melts: a relationship with shear viscosity and implications for the welding process. J Volcanol Geotherm Res 142(1–2):105–118
Giordano G, Porreca M, Musacchio P, Mattei M (2008) The Holocene Secche di Lazzaro phreatomagmatic succession (Stromboli, Italy): evidence of pyroclastic density current origin deduced by facies analysis and AMS flow directions. Bull Volcanol 70(10):1221–1236
Giordano G, De Benedetti AA, Bonamico A, Ramazzotti P, Mattei M (2014) Incorporating surface indicators of reservoir permeability into reservoir volume calculations: application to the Colli Albani caldera and the Central Italy Geothermal Province. Earth Sci Rev 128:75–92
Giordano G and the CARG Team (2010) Stratigraphy, volcano tectonics and evolution of the Colli Albani volcanic field, vol 3. Geological Society of London Special Publications-IAVCEI, pp 43–97
Gonnermann HM (2015) Magma fragmentation. Ann Rev Earth Planet Sci 43:431–458
Gonnermann HM, Manga M (2003) Explosive volcanism may not be an inevitable consequence of magma fragmentation. Nature 426 (6965):432–435
Goto Y, McPhie J (2004) Morphology and propagation styles of Miocene submarine basanite lavas at Stanley, northwestern Tasmania, Australia. J Volcanol Geotherm Res 130(3):307–328
Goto Y, Tsuchiya N (2004) Morphology and growth style of a Miocene submarine dacite lava dome at Atsumi, northeast Japan. J Volcanol Geotherm Res 134(4):255–275
Green DJ, Nicholson PS, Embury JD (1977) Crack shape studies in brittle porous materials. J Mater Sci 12(5):987–989
Griffith AA (1921) The phenomena of rupture and flow in solids. Philos Trans R Soc Lond Ser A 163–198
Griffiths RW, Fink JH (1993) Effects of surface cooling on the spreading of lava flows and domes. J Fluid Mech 252:667–702
Gudmundsson MT, Sigmundsson F, Björnsson H (1997) Ice–volcano interaction of the 1996 Gjálp subglacial eruption, Vatnajökull, Iceland. Nature 389(6654):954–957
Gudmundsson MT, Thordarson T, Höskuldsson Á, Larsen G, Björnsson H, Prata FJ, Oddsson B, Magnússon E, Högnadóttir T, Petersen GN (2012) Ash generation and distribution from the April-May 2010 eruption of Eyjafjallajökull, Iceland. Sci Rep 2:572
Gurioli L, Houghton BF, Cashman KV, Cioni R (2005) Complex changes in eruption dynamics during the 79 AD eruption of Vesuvius. Bull Volcanol 67(2):144–159
Gurioli L, Harris AJL, Houghton BF, Polacci M, Ripepe M (2008) Textural and geophysical characterization of explosive basaltic activity at Villarrica volcano. J Geophys Res Solid Earth 113(B8)
Hajimirza S, Jones TJ, Moreland WM, Gonnermann HM, Thordarson T (2022) Quantifying the water-to-melt mass ratio and its impact on eruption plumes during explosive hydromagmatic eruptions. Geochem, Geophys, Geosyst 23(5):e2021GC010160.
Hamilton CW, Thordarson T, Fagents SA (2010) Explosive lava–water interactions I: architecture and emplacement chronology of volcanic rootless cone groups in the 1783–1784 Laki lava flow, Iceland. Bull Volcanol 72(4):449–467
Hanson RE (1991) Quenching and hydroclastic disruption of andesitic to rhyolitic intrusions in a submarine island-arc sequence, northern Sierra Nevada, California. Geol Soc Am Bull 103(6):804–816
Hanson RE, Schweickert RA (1982) Chilling and brecciation of a Devonian rhyolite sill intruded into wet sediments, Northern Sierra Nevada, California. J Geol 90(6):717–724
Hanson RE, Wilson TJ (1993) Large-scale rhyolite peperites (Jurassic, southern Chile). J Volcanol Geotherm Res 54(3–4):247–264
Hasegawa S, Akimoto K, Kitazato H, Matoba Y (1989) Late Cenozoic paleobathymetric indices based on benthic foraminifers in Japan. Mem Geol Soc Japan 32:241–253
Hay RL (1989) Holocene carbonatite-nephelinite tephra deposits of Oldoinyo Lengai, Tanzania. J Volcanol Geotherm Res 37(1):77–91
Hay RL, Iijima A (1968) Nature and origin of palagonite tuffs of the Honolulu Group on Oahu, Hawaii, vol 116. Geological Society of America Memoirs, pp 331–376

Head JW, Wilson L (2003) Deep submarine pyroclastic eruptions: theory and predicted landforms and deposits. J Volcanol Geotherm Res 121(3):155–193

Heiken G (1972) Morphology and petrography of volcanic ashes. Geol Soc Am Bull 83(7):1961–1988

Heiken G (1974) An atlas of volcanic ash. Smithson Contrib Earth Sci 12

Henley RW, Ellis AJ (1983) Geothermal systems ancient and modern: a geochemical review. Earth Sci Rev 19(1):1–50

Higgins MD (1996) Magma dynamics beneath Kameni volcano, Thera, Greece, as revealed by crystal size and shape measurements. J Volcanol Geotherm Res 70(1):37–48

Hillig WB (1983) An inquiry into the irreversibility of brittle fracture. In: Bradt RC, Evans AG, Hasselman DPH, Lange FF (eds) Fracture Mechanics of ceramics. Measurements, transformations, and high-temperature fracture, vol 6. Plenum Press, New York City, pp 157–164

Holloway JR, Blank JG (1994) Application of experimental results to COH species in natural melts. In: Carroll MR, Holloway JR (eds) Volatiles in magmas, vol 30. Reviews in Mineralogical Society of America, p 187

Horikoshi E (1969) Volcanic activity related to the formation of the Kuroko-type deposits in the Kosaka district, Japan. Mineralium Deposita 4:321–345

Höskuldsson A, Sparks RSJ (1997) Thermodynamics and fluid dynamics of effusive subglacial eruptions. Bull Volcanol 59 (3):219–230

Houghton BF, Carey RJ (2019) Comment physical constraints for effective magma-water interaction along volcanic conduits during silicic explosive eruptions. Geology 47(5):461

Houghton BF, Wilson CJN (1989) A vesicularity index for pyroclastic deposits. Bull Volcanol 51(6):451–462

Houghton BF, Wilson CJN, Del Carlo P, Coltelli M, Sable JE, Carey R (2004) The influence of conduit processes on changes in style of basaltic Plinian eruptions: Tarawera 1886 and Etna 122 BC. J Volcanol Geotherm Res 137(1):1–14

Houghton BF, Carey RJ, Cashman KV, Wilson CJN, Hobden BJ, Hammer JE (2010) Diverse patterns of ascent, degassing, and eruption of rhyolite magma during the 1.8 ka Taupo eruption, New Zealand: evidence from clast vesicularity. J Volcanol Geotherm Res 195(1):31–47

Hunns SR, McPhie J (1999) Pumiceous peperite in a submarine volcanic succession at Mount Chalmers, Queensland, Australia. J Volcanol Geotherm Res 88(4):239–254

Huppert HE, Sparks RSJ, Turner JS (1982) Effects of volatiles on mixing in calc-alkaline magma systems. Nature 297(5867):554–557

Hurwitz S, Navon O (1994) Bubble nucleation in rhyolitic melts: experiments at high pressure, temperature, and water content. Earth Planet Sci Lett 122(3–4):267–280

Ihinger PD, Hervig RL, McMillan PF (1994) Analytical methods for volatiles in glasses. In: Carroll MR, Holloway JR (eds) Volatiles in magmas, vol 30. Reviews in Mineralogical Society of America, pp 67–121

Ikegami F, McPhie J, Carey R, Mundana R, Soule A, Jutzeler M (2018) The eruption of submarine rhyolite lavas and domes in the deep ocean-Havre 2012, Kermadec Arc. Front Earth Sci 6:147

Jaupart C (2000) Magma ascent at shallow levels. In: Sigurdsson H, Stix J, Houghton BF, McNutt SR, Rymer H (eds) Encyclopedia of volcanoes. Academic Press, San Diego, pp 237–245

Jébrak M (1997) Hydrothermal breccias in vein-type ore deposits: a review of mechanisms, morphology and size distribution. Ore Geol Rev 12(3):111–134

Johnson MC, Anderson AT, Rutherford MJ (1994) Pre-eruptive volatile contents of magmas. In: Carroll MR, Holloway JR (eds) Volatiles in magmas, vol 30. Rev Mineral. Mineralogical Society of America, pp 281–330

Jones JG (1966) Intraglacial volcanoes of south-west Iceland and their significance in the interpretation of the form of the marine basaltic volcanoes. Nature 212:586–588

Jones JG (1969) Intraglacial volcanoes of the Laugarvatn region, south-west Iceland—I. Q J Geol Soc 124(1–4):197–211

Jones JG, Nelson PHH (1970) The flow of basalt lava from air into water—its structural expression and stratigraphic significance. Geol Mag 107(01):13–19

Jones TJ, Houghton BF, Llewellin EW, Parcheta CE, Hoeltgen L (2018) Spatter matters–distinguishing primary (eruptive) and secondary (non-eruptive) spatter deposits. Scientific Repts 8(1):9179.

Jones TJ, Le Moigne Y, Russell JK, Williams-Jones G, Giordano D, Dingwell DB (2022b) Inflated pyroclasts in proximal fallout deposits reveal abrupt transitions in eruption behaviour. Nature Comms 13(1):2832.

Jones TJ, McNamara K, Eychenne J, Rust AC, Cashman KV, Scheu B, Edwards R (2016) Primary and secondary fragmentation of crystal-bearing intermediate magma. J Volcanol Geotherm Res 327:70–83

Jones TJ, Reynolds CD, Boothroyd SC (2019) Fluid dynamic induced break-up during volcanic eruptions. Nature Comms 10(1):3828.

Jones TJ, Russell JK, Brown RJ, Hollendonner L (2022a) Melt stripping and agglutination of pyroclasts during the explosive eruption of low viscosity magmas. Nature Comms 13(1):992.

Jordan SC, Cas RAF, Hayman PC (2013) The origin of a large (> 3 km) maar volcano by coalescence of multiple shallow craters: lake Purrumbete maar, southeastern Australia. J Volcanol Geotherm Res 254:5–22

Jordan SC, Dürig T, Cas RAF, Zimanowski B (2014) Processes controlling the shape of ash particles: results of statistical IPA. J Volcanol Geotherm Res 288:19–27

Joyce EB (1975) Quaternary volcanism and tectonics in southeastern Australia. In: Suggate RP, Cresswell MM (eds) Quaternary studies. Bull R Soc NZ 13:169–176

Jude-Eton TC, Thordarson T, Gudmundsson MT, Oddsson B (2012) Dynamics, stratigraphy and proximal dispersal of supraglacial tephra during the ice-confined 2004 eruption at Grímsvötn Volcano, Iceland. Bull Volcanol 74(5):1057–1082

Junqueira-Brod TC, Brod JA, Thompson RN, Gibson SA (1999) Spinning droplets—a conspicuous lapillisize structure in Kamafugitic diatremes of Southern Goiàs Brazil. Rev Brasil Geosci 29(3):437–440

Junqueira-Brod TC, Gaspar JC, Brod JA, Kafino CV (2005) Kamafugitic diatremes: their textures and field relationships with examples from the Goiás alkaline province, Brazil. J South Am Earth Sci 18(3–4):337–353

Jutzeler M, Marsh R, Carey RJ, White JDL, Talling PJ, Karlstrom L (2014) On the fate of pumice rafts formed during the 2012 Havre submarine eruption. Nat Commun 5:1–10

Kamenetsky VS, Kamenetsky MB, Sharygin VV, Golovin AV (2007) Carbonate-chloride enrichment in fresh kimberlites of the Udachnaya-East pipe, Siberia: a clue to physical properties of kimberlite magmas? Geophys Res Lett 34(9):L09316. https://doi.org/10.1029/2007GL029389

Kaully T, Siegmann A, Shacham D, Marmur A (2007) The effect of viscosity on surface tension measurements by the drop weight method. J Appl Polym Sci 106(3):1842–1846

Kepple JB, Wasylyk JS (1994) Fracture of glass containers. In: Bradt RC, Tressler RE (eds) Fractography of glass. Springer, New York, pp 207–252

Kienle J, Kyle PR, Self S, Motyka RJ, Lorenz V (1980) Ukinrek Maars, Alaska, I. April 1977 eruption sequence, petrology and tectonic setting. J Volcanol Geotherm Res 7(1):11–37

Kilgour G, Manville V, Della Pasqua F, Graettinger AH, Hodgson KA, Jolly GE (2010) The 25 September 2007 eruption of Mount Ruapehu, New Zealand: directed ballistics, surtseyan jets, and ice-slurry lahars. J Volcanol Geotherm Res 191(1):1–14

Kilgour G, Gates S, Kennedy B, Farquhar A, McSporran A, Asher C (2019) Phreatic eruption dynamics derived from deposit analysis: a case study from a small, phreatic eruption from Whakāri/White Island, New Zealand. Earth Planets Space 71(1):36

Kingery WD (1955) Factors affecting thermal stress resistance of ceramic materials. J Am Ceram Soc 38(1):3–15

Kingery WD, Bowen HK, Uhlmann DR (1976) Introduction to ceramics. Wiley, New York, p 1032

Kitazato H (1979) Marine paleobathymetry and paleotopography of the Hokuroku district during the time of the Kuroko deposition, based on foraminiferal assemblages. Mining Geol 29(156):207–216

Klug C, Cashman KV (1994) Vesiculation of May 18, 1980, Mount St. Helens magma. Geology 22(5):468–472

Klug C, Cashman KV (1996) Permeability development in vesiculating magmas: implications for fragmentation. Bull Volcanol 58(2–3):87–100

Klug C, Cashman KV, Bacon CR (2002) Structure and physical characteristics of pumice from the climactic eruption of Mount Mazama (Crater Lake), Oregon. Bull Volcanol 64(7):486–501

Knott JF (1973) Fundamentals of fracture mechanics. Butterworths, London, U.K.

Kokelaar BP (1982) Fluidization of wet sediments during the emplacement and cooling of various igneous bodies. J Geol Soc Lond 139(1):21–33

Kokelaar BP (1986) Magma-water interactions in subaqueous and emergent basaltic volcanism. Bull Volcanol 48:275–289

Kokelaar BP, Koniger S (2000) Marine emplacement of welded ignimbrite: the Ordovician Pitts Head Tuff, North Wales. J Geol Soc Lond 157(3):517-536

Kovalev SA (1966) An investigation of minimum heat fluxes in pool boiling of water. Int J Heat Mass Transf 9(11):1219–1226

Krausz AS, Krausz K (1988) Mechanical behaviour of materials. Volume 1: Fracture kinetics of crack growth. Kluwer Academic Publishers, Dordrecht, Netherlands, 181p

Kurokawa A (1991) Formation of fesic pumiceous hyaloclastites a case study from Tadami district, Fukushima Prefecture, Japan. J Mineral Petrol Econ Geol 86(10):439–458

Lange FF (1971) Fracture energy and strength behavior of a sodium borosilicate glass-Al2O3 composite system. J Am Ceram Soc 54 (12):614–620

Lautze NC, Houghton BF (2005) Physical mingling of magma and complex eruption dynamics in the shallow conduit at Stromboli volcano, Italy. Geology 33(5):425–428

Lautze NC, Houghton BF (2007) Linking variable explosion style and magma textures during 2002 at Stromboli volcano, Italy. Bull Volcanol 69(4):445–460

Lautze NC, Houghton BF (2008) Single explosions at Stromboli in 2002: use of clast microtextures to map physical diversity across a fragmentation zone. J Volcanol Geotherm Res 170(3):262–268

Lautze NC, Sisson TW, Mangan MT, Grove TL (2011) Segregating gas from melt: an experimental study of the Ostwald ripening of vapor bubbles in magmas. Contrib Mineral Petrol 161(2):331–347

Lawn BR, Marshall DB (1994) Indentation fractography. In: Bradt RC, Tressler RE (eds) Fractography of glass. Springer, New York, pp 1–35

Laznicka P (1988) Breccias and coarse fragmentites: petrology, environments, associations, ores. Elsevier, 832p

Legros F, Kelfoun K (2000) On the ability of pyroclastic flows to scale topographic obstacles. J Volcanol Geotherm Res 98(1–4):235–241

Lesti C, Giordano G, Salvini F, Cas R (2008) Volcano tectonic setting of the intraplate, Pliocene-Holocene, Newer Volcanic Province (southeast Australia): role of crustal fracture zones. J Geophys Res Solid Earth 113(B7)

Lindsay JM, Schmitt AK, Trumbull RB, de Silva SL, Siebel W, Emmermann R (2001) Magmatic evolution of the La Pacana caldera system, Central Andes, Chile: compositional variation of two cogenetic, large-volume felsic ignimbrites. J Petrol 42(3):459–486

Liu EJ, Cashman KV, Rust AC, Gislason SR (2015) The role of bubbles in generating fine ash during hydromagmatic eruptions. Geology 43(3):239–242

Lorenz V (1973) On the formation of maars. Bull Volcanol 37 (2):183–204

Lorenz V (1986) On the growth of maars and diatremes and its relevance to the formation of tuff rings. Bull Volcanol 48 (5):265–274

Lowenstern JB (2001) Carbon dioxide in magmas and implications for hydrothermal systems. Mineralium Depos 36(6):490–502

MacDonald GA (1972) Volcanoes. Prentice-Hall, Englewood Cliffs, NJ, p 510p

Macpherson GJ (1984) A model for predicting the volumes of vesicles in submarine basalts. J Geol 92(1):73–82

Mäder UK, Percival JA, Berman RG (1994a) Thermobarometry of garnet–clinopyroxene–hornblende granulites from the Kapuskasing structural zone. Can J Earth Sci 31(7):1134–1145

Mader HM, Zhang Y, Phillips JC, Sparks RSJ, Sturtevant B, Stolper E (1994b) Experimental simulations of explosive degassing of magma. Nature 372:85–88

Mader HM, Phillips JC, Sparks RSJ, Sturtevant B (1996) Dynamics of explosive degassing of magma: observations of fragmenting two-phase flows. J Geophys Res Solid Earth 101(B3):5547–5560

Maicher D, White JD (2001) The formation of deep-sea Limu o Pele. Bull Volcanol 63:482–496

Maicher D, White JD, Batiza R (2000) Sheet hyaloclastite: density-current deposits of quench and bubble-burst fragments from thin, glassy sheet lava flows, Seamount Six, Eastern Pacific Ocean. Marine Geol 171(1-4):75–94

Manga M, Fauria KE, Lin C, Mitchell SJ, Jones M, Conway CE, Degruyter W, Hosseini B, Carey R, Cahalan R (2018) The pumice raft-forming 2012 Havre submarine eruption was effusive. Earth Planet Sci Lett 489:49–58

Mangan MT (1990) Crystal size distribution systematics and the determination of magma storage times: the 1959 eruption of Kilauea volcano, Hawaii. J Volcanol Geotherm Res 44(3–4):295–302

Mangan MT, Cashman KV (1996) The structure of basaltic scoria and reticulite and inferences for vesiculation, foam formation, and fragmentation in lava fountains. J Volcanol Geotherm Res 73 (1):1–18

Mangan M, Sisson T (2000) Delayed, disequilibrium degassing in rhyolite magma: decompression experiments and implications for explosive volcanism. Earth Planet Sci Lett 183(3):441–455

Mangan MT, Mastin LG, Sisson TW (2004a) Gas evolution in eruptive conduits: combining insights from high temperature and pressure decompression experiments with steady-state flow modeling. J Volcanol Geotherm Res 129(1):23–36

Mangan MT, Sisson TW, Hankins WB (2004b) Decompression experiments identify kinetic controls on explosive silicic eruptions. Geophys Res Lett 31(8)

Mann CP, Stix J, Vallance JW, Richer M (2004) Subaqueous intracaldera volcanism, Ilopango Caldera, El Salvador, Central America. Geological Society of America Special Paper, vol 375, pp 159–174

Marini L, Principe C, Chiodini G, Cioni R, Fytikas M, Marinelli G (1993) Hydrothermal eruptions of Nisyros (Dodecanese, Greece). Past events and present hazard. J Volcanol Geotherm Res 56(1–2): 71–94

Marsh BD (1988) Crystal size distribution (CSD) in rocks and the kinetics and dynamics of crystallization. Contrib Mineral Petrol 99 (3):277–291

Marsh BD (1998) On the interpretation of crystal size distributions in magmatic systems. J Petrol 39(4):553–599

Marshall AA, Brand BD, Martínez V, Bowers JM, Walker M, Wanless VD, Andrews BJ, Manga M, Valdivia P, Giordano G (2022) The mafic Curacautín ignimbrite of Llaima volcano, Chile. J Volcanol Geotherm Res 421:107418

Marti J, Soriano C, Dingwell DB (1999) Tube pumices as strain markers of the ductile–brittle transition during magma fragmentation. Nature 402(6762):650–653

Massaro S, Costa A, Sulpizio R (2018) Evolution of the magma feeding system during a Plinian eruption: the case of Pomici di Avellino eruption of Somma-Vesuvius, Italy. Earth Planet Sci Letts 482:545–555

Mastin LG (2002) Insights into volcanic conduit flow from an open-source numerical model. Geochem Geophys Geosyst 3(7): 1–18

Mastin LG, Ghiorso MS (2000) A numerical program for steady-state flow of magma-gas mixtures through vertical eruptive conduits. Department of the Interior Washington Document, pp 1–53

Mattox TN, Mangan MT (1997) Littoral hydrovolcanic explosions: a case study of lava–seawater interaction at Kilauea Volcano. J Volcanol Geotherm Res 75(1):1–17

Mayer K, Scheu B, Gilg HA, Heap MJ, Kennedy BM, Lavallée Y, Letham-Brake M, Dingwell DB (2015) Experimental constraints on phreatic eruption processes at Whakaari (White Island volcano). J Volcanol Geotherm Res 302:150–162

McBirney AR (1963) Factors governing the nature of submarine volcanism. Bull Volcanol 26(1):455–469

McBirney AR (1973) Factors governing the intensity of explosive andesitic eruptions. Bull Volcanol 37(3):443–453

McCallum ME (1985) Experimental evidence for fluidization processes in breccia pipe formation. Econ Geol 80(6):1523–1543

McIntosh IM, Nichols AR, Tani K, Llewellin EW (2017) Accounting for the species-dependence of the 3500 cm− 1 H2Ot infrared molar absorptivity coefficient: Implications for hydrated volcanic glasses. Am Miner 102(8):1677–1689

McPhie J (1993) The Tennant Creek porphyry revisited: a synsedimentary sill with peperite margins, Early Proterozoic, Northern Territory. Aust J Earth Sci 40(6):545–558

McPhie J, Doyle M, Allen R (1993) Volcanic textures—a guide to the interpretation of textures in volcanic rocks. CODES—University of Tasmania, Hobart, p 198p

Métrich N, Bertagnini A, Landi P, Rosi M (2001) Crystallization driven by decompression and water loss at Stromboli volcano (Aeolian Islands, Italy). J Petrol 42(8):1471–1490

Michalske TA (1994) Fractography of stress corrosion cracking in glass. In: Bradt RC, Tressler RE (eds) Fractography of glass. Plenum Press, New York City, pp 111–142

Michalske TA, Bunker BC (1987) The fracturing of glass. Sci Am 257 (6):122–129

Michalske TA, Varner JR, Frechette VD (1978) Growth of cracks partly filled with water. In: Bradt RC, Hasselman DPH, Lange FF (eds) Fracture mechanics of ceramics. Crack growth and microstructure, vol 4. Plenum Press, New York City, pp 639–649

Mills AA (1984) Pillow lavas and the Leidenfrost effect. J Geol Soc Lond 141(1):183–186

Mitchell RH (1986) Kimberlites: mineralogy, geochemistry, and petrology. Plenum Press, 442p

Mitchell RH (1997) Kimberlites, orangeites, lamproites, melilitites, and minettes: a petrographic atlas. Almaz Press, Ontario

Moitra P, Gonnermann HM, Houghton BF, Tiwary CS (2018) Fragmentation and Plinian eruption of crystallizing basaltic magma. Earth Planet Sci Letts 15(500):97–104

Montanaro C, Mick E, Salas-Navarro J, Caudron C, Cronin SJ, de Moor JM, Scheu B, Stix J, Strehlow K (2022) Phreatic and Hydrothermal Eruptions: From Overlooked to Looking Over. Bull Volcanol 84(6):64.

Montanaro C, Scheu B, Gudmundsson MT, Vogfjörd K, Reynolds HI, Dürig T, Strehlow K, Rott S, Reuschlé T, Dingwell DB (2016) Multidisciplinary constraints of hydrothermal explosions based on the 2013 Gengissig lake events, Kverkfjöll volcano, Iceland. Earth Planet Sci Letts 434:308–319

Moore JG (1965) Petrology of deep sea basalt near Hawaii. Am J Sci 263(1):40–52

Moore JG (1967) Base surge in recent volcanic eruptions. Bull Volcanol 30(1):337–363

Moore JG (1975) Mechanism of formation of pillow lava: pillow lava, produced as fluid lava cools underwater, is the most abundant volcanic rock on earth, but only recently have divers observed it forming. Am Sci 63(3):269–277

Moore JG, Peck DL (1962) Accretionary Lapilli in volcanic rocks of the western continental United States. J Geol 70(2):182–193

Moore JG, Phillips R, Grigg RW, Peterson DW, Swanson DA (1973) Flow of lava into the sea, 1969–1971, Kilauea Volcano, Hawaii. Geol Soc Am Bull 84(2):537–546

Moore JG, Clague DA, Normark WR (1982) Diverse basalt types from Loihi seamount, Hawaii. Geology 10(2):88–92

Morgan LA, Shanks WC, Pierce KL (2009) Hydrothermal processes above the Yellowstone magma chamber: large hydrothermal systems and large hydrothermal explosions, vol 459. Geological Society of America, Special Paper, pp 1–95

Morgavi D, Perugini D, De Campos CP, Ertl-Ingrisch W, Lavallée Y, Morgan L, Dingwell DB (2013) Interactions between rhyolitic and basaltic melts unraveled by chaotic mixing experiments. Chem Geol 346:99–212

Morgavi D, Arzilli F, Pritchard C, Perugini D, Mancini L, Larson P, Dingwell DB (2016) The Grizzly Lake complex (Yellowstone Volcano, USA): mixing between basalt and rhyolite unraveled by microanalysis and X-ray microtomography. Lithos 260:457–474

Morgavi D, Valentini L, Porreca M, Zucchini A, Di Michele A, Ielpo M, Costa A, Rossi S, Perugini D (2019) Volcanic ash aggregation enhanced by seawater interaction: the case of the Secche di Lazzaro phreatomagmatic deposit (Stromboli). Ann Rev Geophys 61(6 sup):673

Morrissey MM, Zimanowski B, Wohletz K, Buettner R (2000) Phreatomagmatic fragmentation. In: Sigurdsson H, Stix J, Houghton BF, McNutt SR (eds) Encyclopedia of volcanoes. Academic Press, San Diego, pp 431–462

Morrissey MM, Mastin LG (2000) Vulcanian eruptions. In: Sigurdsson H, Stix J, Houghton BF, McNutt SR, Rymer H (eds) Encyclopedia of volcanoes. Academic Press, San Diego, pp 463–475

Mourtada-Bonnefoi CC, Laporte D (2002) Homogeneous bubble nucleation in rhyolitic magmas: an experimental study of the effect of H2O and CO2. J Geophys Res Solid Earth 107(B4)

Mudde RF (2005) Gravity-driven bubbly flows. Ann Rev. Fluid Mech 37:393–423

Mueller S, Scheu B, Spieler O, Dingwell DB (2008) Permeability control on magma fragmentation. Geology 36(5):399–402

Mueller SB, Lane SJ, Kueppers U (2015) Lab-scale ash production by abrasion and collision experiments of porous volcanic samples. J Volcanol Geotherm Res 302:163–172

Mueller SB, Kueppers U, Ayris PM, Jacob M, Dingwell DB (2016) Experimental volcanic ash aggregation: internal structuring of accretionary lapilli and the role of liquid bonding. Earth Planet Sci Lett 433:232–240

Muffler LP, White DE, Truesdell AH (1971) Hydrothermal explosion craters in Yellowstone National Park. Geol Soc Am Bull 82 (3):723–740

Mungall JE, Bagdassarov NS, Romano C, Dingwell DB (1996) Numerical modelling of stress generation and microfracturing of vesicle walls in glassy rocks. J Volcanol Geotherm Res 73(1):33–46

Murch AP, White JDL, Carey RJ (2019a) Characteristics and deposit stratigraphy of submarine-erupted Silicic Ash, Havre Volcano, Kermadec Arc, New Zealand. Front Earth Sci 7:1. https://doi.org/10.3389/feart.2019.00001

Murch AP, White JD, Carey RJ (2019b) Unusual fluidal behavior of a silicic magma during fragmentation in a deep subaqueous eruption, Havre volcano, southwestern Pacific Ocean. Geology 47(5):487–490

Nairn IA (1979) Rotomahana—Waimangu eruption, 1886: base surge and basalt magma. NZ J Geol Geophys 22(3):363–378

Nairn IA, Wiradiradja S (1980) Late Quaternary hydrothermal explosion breccias at Kawerau geothermal field, New Zealand. Bull Volcanol 43(1):1–13

O'Hara MJ, Mathews RE (1981) Geochemical evolution in an advancing, periodically replenished, periodically tapped, continuously fractionated magma chamber. J Geol Soc 138(3):237–277

Ohji T, Jeong YK, Choa YH, Niihara K (1998) Strengthening and toughening mechanisms of ceramic nanocomposites. J Am Ceram Soc 81(6):1453–1460

Okumura, S, Nakamura M, Tsuchiyama A, Nakano T, Uesugi K (2008) Evolution of bubble microstructure in sheared rhyolite: formation of a channel-like bubble network. J Geophys Res Solid Earth 113(B7)

Orsi G, Gallo G, Heiken G, Wohletz K, Yu E, Bonani G (1992) A comprehensive study of pumice formation and dispersal: the Cretaio Tephra of Ischia (Italy). J Volcanol Geotherm Res 53(1–4):329–354

Ort MH, Carrasco-Núñez G (2009) Lateral vent migration during phreatomagmatic and magmatic eruptions at Tecuitlapa Maar, east-central Mexico. J Volcanol Geotherm Res 181(1–2):67–77

Ort MH, Lefebvre NS, Neal CA, McConnell VS, Wohletz KH (2018) Linking the Ukinrek 1977 maar-eruption observations to the tephra deposits: new insights into maar depositional processes. J Volcanol Geotherm Res 360:36–60

van Otterloo J, Cas RAF, Sheard MJ (2013) Eruption processes and deposit characteristics at the monogenetic Mt. Gambier Volcanic Complex, SE Australia: implications for alternating magmatic and phreatomagmatic activity. Bull Volcanol 75(8):737–758

Pal R (2003) Rheological behavior of bubble-bearing magmas. Earth Planet Sci Lett 207(1–4):165–179

Palladino DM, Simei S, Kyriakopoulos K (2008) On magma fragmentation by conduit shear stress: evidence from the Kos Plateau Tuff, Aegean Volcanic Arc. J Volcanol Geotherm Res 178 (4):807–817

Papale P (1999) Strain-induced magma fragmentation in explosive eruptions. Nature 397(6718):425–428

Papale P (2001) Dynamics of magma flow in volcanic conduits with variable fragmentation efficiency and nonequilibrium pumice degassing. J Geophys Res Solid Earth 106(B6):11043–11065

Papale P, Neri A, Macedonio G (1998) The role of magma composition and water content in explosive eruptions: 1. Conduit ascent dynamics. J Volcanol Geotherm Res 87(1–4):75–93

Pardo N, Macias JL, Giordano G, Cianfarra P, Avellán DR, Bellatreccia F (2009) The ~1245 yr BP Asososca maar eruption: the youngest event along the Nejapa-Miraflores volcanic fault, Western Managua, Nicaragua. J Volcanol Geotherm Res 184(3):292–312

Parfitt EA (1998) A study of clast size distribution, ash deposition and fragmentation in a Hawaiian-style volcanic eruption. J Volcanol Geotherm Res 84(3):197–208

Parfitt EA, Wilson L (1999) A Plinian treatment of fallout from Hawaiian lava fountains. J Volcanol Geotherm Res 88(1):67–75

Patrick M (2005) Strombolian eruption dynamics from thermal (FLIR) video imagery. University of Hawaii, Hawaii

Peckover RS, Buchanan DJ, Ashby D (1973) Fuel–coolant interactions in submarine vulcanism. Nature 245(5424):307–308

Phillips WJ (1972) Hydraulic fracturing and mineralization. J Geol Soc 128(4):337–359

Pichler H (1965) Acid hyaloclastites. Bull Volcanol 28(1):293–310

Pioro I, Mokry S (2011) Chapter 22 Thermophysical properties at critical and supercritical pressures. In: Belmiloudi A (ed) Heat Transfer-Theoretical Analysis, Experimental Investigations and Industrial Systems. InTech, pp 573–592

Pistolesi M, Rosi M, Cion R, Cashman KV, Rossotti A, Aguilera E (2011) Physical volcanology of the post-twelfth-century activity at Cotopaxi volcano, Ecuador: behavior of an andesitic central volcano. Bull Geol Soc Am 123(5):1193–1215

Pistone M, Caricchi L, Ulmer P, Burlini L, Ardia P, Reusser E, Arbaret L (2012) Deformation experiments of bubble-and crystal-bearing magmas: rheological and microstructural analysis. J Geophys Res Solid Earth 117(B5)

Pittari A, Cas RAF, Wolff JA, Nichols HJ, Larson PB, Martí J (2008) The use of lithic clast distributions in pyroclastic deposits to understand pre-and syn-caldera collapse processes: a case study of the Abrigo Ignimbrite, Tenerife, Canary Islands. Dev Volcanol 10:97–142

Pittari A, Briggs RM, Bowyer DA (2016) Subsurface geology, ancient hydrothermal systems and crater excavation processes beneath Lake Rotomahana: evidence from lithic clasts of the 1886 AD Rotomahana Pyroclastics. J Volcanol Geotherm Res 314:110–125

Polacci M, Baker DR, Bai L, Mancini L (2008) Large vesicles record pathways of degassing at basaltic volcanoes. Bull Volcanol 70 (9):1023–1029

Porębski SJ, Gradziński R (1990) Lava-fed gilbert-type delta in the polonez cove formation (lower oligocene), King George Island, West Antarctica. In: Colella A, Prior DB (eds) Coarse-grained deltas, vol 10. International Association of Sedimentologists Special Publication, Blackwell, Oxford, pp 333–351

Porreca M, Cifelli F, Soriano C, Giordano G, Romano C, Conticelli S, Mattei M (2014) Hyaloclastite fragmentation below the glass transition: an example from El Barronal submarine volcanic complex (Spain). Geology 42(1):87–90

Porritt LA, Cas RAF (2009) Reconstruction of a kimberlite eruption, using an integrated volcanological, geochemical and numerical approach: a case study of the Fox Kimberlite, NWT, Canada. J Volcanol Geotherm Res 179(3-4):241–264

Prousevitch AA, Sahagian DL, Anderson AT (1993) Dynamics of diffusive bubble growth in magmas: Isothermal case. J Geophys Res Solid Earth 98(B12):22283–22307

Proussevitch AA, Sahagian DL (1998) Dynamics and energetics of bubble growth in magmas: analytical formulation and numerical modeling. J Geophys Res Solid Earth 103(B8):18223–18251

Proussevitch AA, Sahagian DL, Kutolin VA (1993) Stability of foams in silicate melts. J Volcanol Geotherm Res 59(1–2):161–178

Ramsay JG (1980) The crack-seal mechanism of rock deformation: Nature 284:135–139

Rayleigh L (1917) VIII. On the pressure developed in a liquid during the collapse of a spherical cavity. Lond Edinb Dubl Philos Mag 34 (200):94–98

Resing JA, Rubin KH, Embley RW, Lupton JE, Baker ET, Dziak RP, Baumberger T, Lilley MD, Huber JA, Shank TM (2011) Active submarine eruption of boninite in the northeastern Lau Basin. Nat Geosci 4(11):799–806

Resnyansky AD, Delaney TG (2006) Experimental study of blast mitigation in a water mist. Aust Govt Defence Sci Technol Org Tech Rept DSTO-TR-1944, 41p

Rittmann A (1962) Volcanoes and their activity. Interscience Publications, Wiley, New York, p 305p

Romano C, Mungall JE, Sharp T, Dingwell DB (1996) Tensile strengths of hydrous vesicular glasses: an experimental study. Am Mineral 81(9):1148–1154

Romano, C, Vona A, Campagnola S, Giordano G, Arienzo I, Isaia R (2020) Modelling and physico-chemical constraints to the 4.5 ka Agnano-Monte Spina Plinian eruption (Campi Flegrei, Italy). Chem Geol 532:119301

Rotella MD, Wilson CJN, Barker S, Wright IC (2013) Highly vesicular pumice generated by buoyant detachment of magma in subaqueous volcanism. Nat Geosci 6(2):129–132

Rowland JV, Simmons SF (2012) Hydrologic, magmatic, and tectonic controls on hydrothermal flow, Taupo Volcanic Zone, New Zealand: implications for the formation of epithermal vein deposits. Econ Geol 107(3):427–457

Rowley PD, Kuntz MA, McLeod NS (1981) Pyroclastic flow deposit. In: Lipman PW, Mullineaux DR (eds) The 1980 eruption of Mount St. Helens, Washington. US Geological Survey Professional Paper, vol 1541, pp 489–512

Russell JK, Edwards BR, Porritt LA (2013) Pyroclastic passage zones in glaciovolcanic sequences. Nature Commun 4(1):1788

Rust AC, Manga M (2002) Effects of bubble deformation on the viscosity of dilute suspensions. J Non Newt Fluid Mech 104(1):53–63

Rust AC, Manga M, Cashman KV (2003) Determining flow type, shear rate and shear stress in magmas from bubble shapes and orientations. J Volcanol Geotherm Res 122(1–2):111–132

Rust AC, Cashman KV (2011) Permeability controls on expansion and size distributions of pyroclasts. J Geophys Res Solid Earth 116 (B11)

Ryan A, Russell JK, Heap MJ, Kolzenburg S, Vona A, Kushnir ARL (2019) Strain-dependent rheology of silicate melt foams: importance for outgassing of silicic lavas. J Geophys Res Solid Earth 124 (8):8167–8186

Sable JE, Houghton BF, Wilson CJN, Carey RJ (2006) Complex proximal sedimentation from Plinian plumes: the example of Tarawera 1886. Bull Volcanol 69(1):89–103

Sable JE, Houghton BF, Wilson CJN, Carey RJ (2009) Eruption mechanisms during the climax of the Tarawera 1886 basaltic Plinian eruption inferred from microtextural characteristics of the deposits. Studies in volcanology: the legacy of George Walker Special Publications, vol 2. IAVCEI, pp 129–154

Sahagian DL (2005) Volcanic eruption mechanisms: insights from intercomparison of models of conduit processes. J Volcanol Geotherm Res 143(1):1–15

Sasse D, Jones TJ, Russell JK (2020) Transport, survival and modification of xenoliths and xenocrysts from source to surface. Earth Planet Sci Letts 548:116499.

Scheu B, Kueppers U, Mueller S, Spieler O, Dingwell DB (2008) Experimental volcanology on eruptive products of Unzen volcano. J Volcanol Geotherm Res 175(1–2):110–119

Schipper I, White JDL (2010) No depth limit to hydrovolcanic limu o Pele: analysis of limu from Lōihi Seamount, Hawaii. Bull Volcanol 72(2):149–164

Schipper CI, White JDL, Houghton BF, Shimizu N, Stewart RB (2010a) Explosive submarine eruptions driven by volatile-coupled degassing at Lōihi Seamount, Hawaii. Earth Planet Sci Lett 295 (3):497–510

Schipper CI, White JDL, Houghton BF, Shimizu N, Stewart RB (2010b) "Poseidic" explosive eruptions at Loihi Seamount, Hawaii. Geology 38(4):291–294

Schipper CI, White JDL, Zimanowski B, Büttner R, Sonder I, Schmid A (2011) Experimental interaction of magma and "dirty" coolants. Earth Planet Sci Lett 303(3):323–336

Schipper CI, Sonder I, Schmid A, White JDL, Dürig T, Zimanowski B, Büttner R (2013) Vapour dynamics during magma–water interaction experiments: hydromagmatic origins of submarine volcaniclastic particles (limu o Pele). Geophys J Int 192(3):1109–1115

Schmincke H-U (1967) Fused tuff and pépérites in south-central Washington. Geol Soc Am Bull 78(3):319–330

Scholze HV, Mulfinger HO (1959) Der Einbau des Wassers in Glasern, V. Die Diffusion des Wassers in Glasen bei hohen Temperaturen. Glastech 3:381–386

Schumacher R, Schmincke H-U (1991) Internal structure and occurrence of accretionary lapilli—a case study at Laacher See Volcano. Bull Volcanol 53(8):612–634

Schumacher R, Schmincke H-U (1995) Models for the origin of accretionary lapilli. Bull Volcanol 56(8):626–639

Scott-Smith BH (1996) Chapter 10. Kimberlites. In: Mitchell RH (ed) Undersaturated alkaline rocks: mineralogy, petrogenesis and economic potential. Mineralogical Association of Canada Short Course Series, pp 217–243

Scutter CR, Cas RAF, Moore CL, de Rita D (1998) Facies architecture and origin of a submarine rhyolitic lava flow-dome complex, Ponza, Italy. J Geophys Res Solid Earth 103(B11):27551–27566

Scutter CR (1999) Felsic hyaloclastites of Ponza, Italy and Hokkaido, Japan: eruption, fragmentation and textures. Unpublished PhD thesis, Monash University, Australia, 550p

Secor DT (1969) Mechanics of natural extension fracturing at depth in the earth's crust. Research in tectonics. Geological Survey of Canada Special Paper, pp 68–52

Self S, Sparks RSJ (1981) Tephra studies. D. Reidel, Dordrecht, 481p

Self S, Kienle J, Huot J-P (1980) Ukinrek Maars, Alaska, II. Deposits and formation of the 1977 craters. J Volcanol Geotherm Res 7 (1):39–65

Self S, Sparks RSJ (1978) Characteristics of widespread pyroclastic deposits formed by the interaction of silicic magma and water. Bull Volcanol 41(3):196–212

Shackelford JF (1990) Introduction to materials science for engineers. MacMillan Publishing Co., New York, 601p + 6 appendicies

Shaw HR (1980) The fracture mechanisms of magma transport from the mantle to the surface. In: Hargraves RB (ed) Physics of magmatic processes. Princeton University Press, Princeton, pp 201–264

Shea T, Gurioli L, Larsen JF, Houghton BF, Hammer JE, Cashman KV (2010a) Linking experimental and natural vesicle textures in Vesuvius 79AD white pumice. J Volcanol Geotherm Res 192(1):69–84

Shea T, Houghton BF, Gurioli L, Cashman KV, Hammer JE, Hobden BJ (2010b) Textural studies of vesicles in volcanic rocks: an integrated methodology. J Volcanol Geotherm Res 190(3–4):271–289

Shea T, Gurioli L, Houghton BF, Cioni R, Cashman KV (2011) Column collapse and generation of pyroclastic density currents during the AD 79 eruption of Vesuvius: the role of pyroclast density. Geology 39(7):695–698

Shelby JE (1997) Introduction to glass science and technology. Royal Society of Chemistry, Cambridge, U.K., p 244p

Sheridan MF, Wohletz KH (1981) Hydrovolcanic explosions: the systematics of water-pyroclast equilibration. Science 212 (4501):1387–1389

Sheridan MF, Wohletz KH (1983) Hydrovolcanism: basic considerations and review. J Volcanol Geotherm Res 17(1–4):1–29

Shields JK, Mader HM, Pistone M, Caricchi L, Floess D, Putlitz B (2014) Strain-induced outgassing of three-phase magmas during simple shear. J Geophys Res Solid Earth 119(9):6936–6957

Sibson RH (1977) Fault rocks and fault mechanisms. J Geol Soc Lond 133(3):191–213

Sibson RH (1986) Earthquakes and rock deformation in crustal fault zones. Ann Rev Earth Planet Sci 14(1):149–175

Sibson RH (1987) Earthquake rupturing as a mineralizing agent in hydrothermal systems. Geology 15(8):701–704

Sibson RH (1996) Structural permeability of fluid-driven fault-fracture meshes. J Struct Geol 18(8):1031–1042

Sibson RH (1990) Faulting and fluid flow. In: Nesbitt BE (ed) Fluids in tectonically active regimes of the continental crust. Mineralogical Association of Canada Short Course Handbook, pp 93–132

Sillitoe RH (1985) Ore-related breccias in volcanoplutonic arcs. Econ Geol 80(6):1467–1514

Simmons JM, Cas RAF, Druitt TH, Carey RJ (2017b) The initiation and development of a caldera-forming Plinian eruption (172 ka Lower Pumice 2 eruption, Santorini, Greece). J Volcanol Geotherm Res 341:332–350

Simmons JM, Cas RAF, Druitt TH, Folkes CB (2016) Complex variations during a caldera-forming Plinian eruption, including precursor deposits, thick pumice fallout, co-ignimbrite breccias and climactic lag breccias: the 184ka Lower Pumice 1 eruption sequence, Santorini, Greece. J Volcanol Geotherm Res 324:200–219

Simmons JM, Carey RJ, Cas RAF, Druitt TH (2017a) High magma decompression rates at the peak of a violent caldera-forming eruption (Lower Pumice 1 eruption, Santorini, Greece). Bull Volcanol 79(6):42

Simpson K, McPhie J (2001) Fluidal-clast breccia generated by submarine fire fountaining, Trooper Creek Formation, Queensland, Australia. J Volcanol Geotherm Res 109(4):339–355

Skilling IP (1994) Evolution of an englacial volcano: Brown Bluff, Antarctica. Bull Volcanol 56(6):573–591

Skilling IP, White JDL, McPhie J (2002) Peperite: a review of magma–sediment mingling. J Volcanol Geotherm Res 114(1–2):1–17

Smellie JL (2000) Subglacial eruptions. In: Sigurdsson H, Stix J, Houghton BF, McNutt SR, Rymer H (eds) Encyclopedia of volcanoes. Academic Press, San Diego, pp 403–419

Smellie JL (2008) Basaltic subglacial sheet-like sequences: evidence for two types with different implications for the inferred thickness of associated ice. Earth Sci Rev 88(1):60–88

Smellie JL, Hole MJ (1997) Products and processes in Pliocene-Recent, subaqueous to emergent volcanism in the Antarctic Peninsula: examples of englacial Surtseyan volcano construction. Bull Volcanol 58(8):628–646

Smith TL, Batiza R (1989) New field and laboratory evidence for the origin of hyaloclastite flows on seamount summits. Bull Volcanol 51(2):96–114

Soriano C, Riggs N, Giordano G, Porreca M, Conticelli S (2012) Cyclic growth and mass wasting of submarine Los Frailes lava flow and dome complex in Cabo de Gata, SE Spain. J Volcanol Geotherm Res 231:72–86

Soriano C, Giordano G, Cas R, Riggs N, Porreca M (2013) Facies architecture, emplacement mechanisms and eruption style of the submarine andesite El Barronal complex, Cabo de Gata, SE Spain. J Volcanol Geotherm Res 264:210–222

Soriano C, Cas RAF, Riggs NR, Giordano G (2016) Chapter 12 Submarine volcanism of the Cabo de Gata Magmatic Arc in the Betic-Rif Orogen, SE Spain: processes and products. In: Nemeth K (ed) Updates in volcanology—from volcano modelling to volcano geology. InTech, Rijeka, pp 325–354

Sottili G, Palladino DM, Gaeta M, Masotta M (2012) Origins and energetics of maar volcanoes: examples from the ultrapotassic Sabatini Volcanic District (Roman Province, Central Italy). Bull Volcanol 74(1):163–186

Sparks RSJ (1978) The dynamics of bubble formation and growth in magmas: a review and analysis. J Volcanol Geotherm Res 3:1–37

Sparks RSJ (2013) Kimberlite volcanism. Ann Rev Earth Planet Sci 41:497–528

Sparks RSJ, Sigurdsson H, Wilson L (1977) Magma mixing: a mechanism for triggering acid explosive eruptions. Nature 267:315–318

Sparks RSJ, Baker L, Brown RJ, Field M, Schumacher J, Stripp G, Walters A (2006) Dynamical constraints on kimberlite volcanism. J Volcanol Geotherm Res 155(1–2):18–48

Sparks RSJ, Barclay J, Jaupart C, Mader HM, Phillips JC (1994) Physical aspects of magmatic degassing I. Experimental and theoretical constraints on vesiculation. In: Carroll MR, Holloway JR (eds) Volatiles in magmas, vol 30. Reviews in Mineralogical Society of America, pp 413–443

Spieler O, Dingwell DB, Alidibirov M (2004a) Magma fragmentation speed: an experimental determination. J Volcanol Geotherm Res 129(1):109–123

Spieler O, Kennedy B, Kueppers U, Dingwell DB, Scheu B, Taddeucci J (2004b) The fragmentation threshold of pyroclastic rocks. Earth Planet Sci Lett 226(1):139–148

Stock MJ, Humphreys MC, Smith VC, Isaia R, Pyle DM (2016) Late-stage volatile saturation as a potential trigger for explosive volcanic eruptions. Nat Geosci 9(3):249

Stovall WK, Houghton BF, Gonnermann HM, Fagents SA, Swanson DA (2011) Eruption dynamics of Hawaiian-style fountains: the case study of episode 1 of the Kīlauea Iki 1959 eruption. Bull Volcanol 73(5):511–529

Sugioka I, Bursik M (1995) Explosive fragmentation of erupting magma. Nature 373(6516):689

Sutton AN, Blake S, Wilson CJN (1995) An outline geochemistry of rhyolite eruptives from Taupo volcanic centre, New Zealand. J Volcanol Geotherm Res 68(1–3):153–175

Sutton AN, Blake S, Wilson CJN, Charlier BLA (2000) Late Quaternary evolution of a hyperactive rhyolite magmatic system: Taupo volcanic centre, New Zealand. J Geol Soc 157(3):537–552

Symonds RB, Rose WI, Bluth GJS, Gerlach TM (1994) Volcanic-gas studies; methods, results, and applications. In: Carroll MR, Holloway JR (eds) Volatiles in magmas, vol 30. Reviews in Mineralogical Society of America, pp 1–66

Tait S (1992) Selective preservation of melt inclusions in igneous phenocrysts. Am Mineral 77(1–2):146–155

Tait S, Jaupart C, Vergniolle S (1989) Pressure, gas content and eruption periodicity of a shallow, crystallising magma chamber. Earth Planet Sci Letts 92(1):107–123

Thomas N, Jaupart C, Vergniolle S (1994) On the vesicularity of pumice. J Geophys Res 99(15):633–615,633

Thomas RME, Sparks RSJ (1992) Cooling of tephra during fallout from eruption columns. Bull Volcanol 54(7):542–553

Thorarinsson S (1953) The crater groups in Iceland. Bull Volcanol 14 (1):3–44

Thorarinsson S, Einarsson T, Sigvaldason G, Elisson G (1964) The submarine eruption off the Vestmann Islands 1963–64. Bull Volcanol 27(1):435–445

Thordarson T (2004) Accretionary-lapilli-bearing pyroclastic rocks at ODP Leg 192 Site 1184: a record of subaerial phreatomagmatic eruptions on the Ontong Java Plateau. Geological Society of London Special Publications, pp 275–306

Toramaru A (1990) Measurement of bubble size distributions in vesiculated rocks with implications for quantitative estimation of eruption processes. J Volcanol Geotherm Res 43(1):71–90

Toramaru A (1995) Numerical study of nucleation and growth of bubbles in viscous magmas. J Geophys Res Solid Earth 100 (B2):1913–1931

Toramaru A (2006) BND (bubble number density) decompression rate meter for explosive volcanic eruptions. J Volcanol Geotherm Res 154(3–4):303–316

Toramaru A, Noguchi S, Oyoshihara S, Tsune A (2008) MND (microlite number density) water exsolution rate meter. J Volcanol Geotherm Res 175(1):156–167

Trofimovs J, Sparks RSJ, Talling PJ (2008) Anatomy of a submarine pyroclastic flow and associated turbidity current: July 2003 dome collapse, Soufrière Hills volcano, Montserrat, West Indies. Sedimentology 55(3):617–634

Tuffen H (2007) Models of ice melting and edifice growth at the onset of subglacial basaltic eruptions. J Geophys Res Solid Earth 112: B03202. https://doi.org/10.01029/02006JB004523

Tuffen H, Gilbert J, McGarvie D (2001) Products of an effusive subglacial rhyolite eruption: Bláhnúkur, Torfajökull, Iceland. Bull Volcanol 63(2–3):179–190

Turner FJ, Verhoogen J (1960) Igneous and metamorphic rocks. McGraw-Hill Book Company, New York, NY, p 694p

Uehara D, Cas RAF, Folkes CB, Takarada S, Oda H, Porreca M (2015) Using thermal remanent magnetisation (TRM) to distinguish block and ash flow and debris flow deposits, and to estimate their emplacement temperature: 1991–1995 lava dome eruption at Mt. Unzen Volcano, Japan. J Volcanol Geotherm Res 303:92–111

Umino S, Nonaka M, Kauahikaua J (2006) Emplacement of subaerial pahoehoe lava sheet flows into water: 1990 Kūpaianaha flow of Kilauea volcano at Kaimū Bay, Hawaii. Bull Volcanol 69(2): 125–139

Valdivia P, Marshall AA, Brand BD, Manga M, Huber C (2022) Mafic explosive volcanism at Llaima Volcano: 3D x-ray microtomography reconstruction of pyroclasts to constrain shallow conduit processes. Bull Volcanol 84(1):1–8

Valentine GA, Graettinger AH, Sonder I (2014) Explosion depths for phreatomagmatic eruptions. Geophys Res Letts 41(9):3045–3051

van Otterloo J, Raveggi M, Cas RAF, Maas R (2014) Polymagmatic activity at the monogenetic Mt Gambier volcanic complex in the Newer Volcanics Province, SE Australia: new insights into the occurrence of intraplate volcanic activity in Australia. J Petrol 55 (7):1317–1351

van Otterloo J, Cas RAF, Scutter CR (2015) The fracture behaviour of volcanic glass and relevance to quench fragmentation during formation of hyaloclastite and phreatomagmatism. Earth Sci Rev 151:79–116

van Zalinge ME, Cashman KV, Sparks RS (2018) Causes of fragmented crystals in ignimbrites: a case study of the Cardones ignimbrite, Northern Chile. Bull Volcanol 80(3):22

Vergniolle S, Jaupart C (1986) Separated two-phase flow and basaltic eruptions. J Geophys Res Solid Earth 91(B12):12842–12860

Vignaroli G, Aldega L, Balsamo F, Billi A, De Benedetti AA, De Filippis L, Giordano G, Rossetti F (2015) A way to hydrothermal paroxysm, Colli Albani volcano, Italy. Geol Soc Am Bull 127 (5–6):672–687

Vinkler AP, Cashman KV, Giordano G, Groppelli G (2012) Evolution of the mafic Villa Senni caldera-forming eruption at Colli Albani volcano, Italy, indicated by textural analysis of juvenile fragments. J Volcanol Geotherm Res 235:37–54

Vogel TA, Aines R (1996) Melt inclusions from chemically zoned ash flow sheets from the Southwest Nevada Volcanic Field. J Geophys Res Solid Earth 101(B3):5591–5610

Voltolini M, Zandomeneghi D, Mancini L, Polacci M (2011) Texture analysis of volcanic rock samples: quantitative study of crystals and vesicles shape preferred orientation from X-ray microtomography data. J Volcanol Geotherm Res 202(1–2):83–95

Von Aulock FW, Nichols ARL, Kennedy BM, Oze C (2013) Timescales of texture development in a cooling lava dome. Geochim Cosmochim Acta 114:72–80

Vona A, Di Piazza A, Nicotra E, Romano C, Viccaro M, Giordano G (2017) The complex rheology of megacryst-rich magmas: The case of the mugearitic "cicirara" lavas of Mt. Etna volcano. Chem Geol 458:48–67

Vona A, Romano C, Giordano D, Russell JK (2013) The multiphase rheology of magmas from Monte Nuovo (Campi Flegrei, Italy). Chem Geol 346:213–227

Vona A, Ryan AG, Russell JK, Romano C (2016) Models for viscosity and shear localization in bubble-rich magmas. Earth Planet Sci Letts 449:26–38

Vona A, Romano C, Dingwell DB, Giordano D (2011) The rheology of crystal-bearing basaltic magmas from Stromboli and Etna. Geochim Cosmochim Acta 75(11):3214–3236

Walker GPL (1972) Crystal concentration in ignimbrites. Contrib Mineral Petrol 36(2):135–146

Walker GPL (1984) Characteristics of dune-bedded pyroclastic surge bedsets. J Volcanol Geotherm Res 20:281–296

Walker GPL, Croasdale R (1971) Two Plinian-type eruptions in the Azores. J Geol Soc 127(1):17–55

Wallace PJ, Anderson AT Jr (1998) Effects of eruption and lava drainback on the H2O contents of basaltic magmas at Kilauea Volcano. Bull Volcanol 59(5):327–344

Wallace PJ, Anderson AT (2000) Volatiles in magmas. In: Sigurdsson H, Stix J, Houghton BF, McNutt SR, Rymer H (eds) Encyclopedia of volcanoes. Academic Press, San Diego, pp 149–170

Wallace PJ, Plank T, Edmonds M, Hauri EH (2015) Volatiles in magmas. In: Sigurdsson H, Houghton B, McNutt SR, Rymer H, Stix J (eds) The encyclopedia of volcanoes. Academic Press Elsevier, Cambridge, pp 163–184

Waters AC (1960) Determining direction of flow in basalts. Am J Sci 258:350–366

Watton TJ, Jerram DA, Thordarson T, Davies RJ (2013) Three-dimensional lithofacies variations in hyaloclastite deposits. J Volcanol Geotherm Res 250:19–33

Whitaker S (1983) Fundamental principles of heat transfer. R. E. Krieger Publishing Company, Malabar, Florida, p 572

White JDL (1996) Impure coolants and interaction dynamics of phreatomagmatic eruptions. J Volcanol Geotherm Res 74(3–4): 155–170

White JDL, Ross P-S (2011) Maar-diatreme volcanoes: a review. J Volcanol Geotherm Res 201(1):1–29

White JDL, McPhie J, Skilling I (2000) Peperite: a useful genetic term. Bull Volcanol 62(1):65–66

White JDL, Houghton B (2000) Surtseyan and related phreatomagmatic eruptions. Encyclopedia of volcanoes. Academic, San Diego, pp 495–513

White JDL, Smellie JL, Clague DA (2003) Introduction: a deductive outline and topical overview of subaqueous explosive volcanism. In: White JDL, Smellie JL, Clague DA (eds) Explosive subaqueous volcanism, vol 140. American Geophysical Union Monograph, pp 1–23

Whitham AG, Sparks RSJ (1986) Pumice. Bull Volcanol 48(4): 209–223

Wilkinson DS, Vitek V (1981) The stress analysis and diffusional growth of microvoids ahead of a crack. In: Gittus J (ed) Cavities and cracks in creep and fatigue. Applied Science Publishers, London, U.K., pp 243–258

Willcox A, Buisman I, Sparks RSJ, Brown RJ, Manya S, Schumacher JC, Tuffen H (2015) Petrology, geochemistry and low-temperature alteration of lavas and pyroclastic rocks of the kimberlitic Igwisi Hills volcanoes, Tanzania. Chem Geol 405:82–101

Wilson L (1980) Relationships between pressure, volatile content and ejecta velocity in three types of volcanic explosion. J Volcanol Geotherm Res 8(2–4):297–313

Wilson CJN (1993) Stratigraphy, chronology, styles and dynamics of late Quaternary eruptions from Taupo volcano, New Zealand. Philos Trans R Soc Lond A Math Phys Eng Sci 343(1668):205–306

Wilson L, Head JW (1981) Ascent and eruption of basaltic magma on the Earth and Moon. J Geophys Res Solid Earth 86 (B4):2971–3001

Wilson L, Sparks RSJ, Huang TC, Watkins ND (1978) The control of volcanic column heights by eruption energetics and dynamics. J Geophys Res Solid Earth 83(B4):1829–1836

Wilson CJN, Walker GPL (1985) The Taupo eruption, New Zealand I. General aspects. Philos Trans R Soc Lond A Math Phys Eng Sci 314(1529):199–228

Wilson L, Sparks RSJ, Walker GPL (1980) Explosive volcanic eruptions—IV. The control of magma properties and conduit geometry on eruption column behaviour. Geophys J Int 63 (1):117–148

Witter JB, Kress VC, Delmelle P, Stix J (2004) Volatile degassing, petrology, and magma dynamics of the Villarrica Lava Lake, Southern Chile. J Volcanol Geotherm Res 134(4):303–337

Wohletz KH (1983) Mechanisms of hydrovolcanic pyroclast formation: grain-size, scanning electron microscopy, and experimental studies. J Volcanol Geotherm Res 17(1):31–63

Wohletz KH (1986) Explosive magma-water interactions: thermodynamics, explosion mechanisms, and field studies. Bull Volcanol 48:245–264

Wohletz KH, McQueen RG (1984) Experimental studies of hydromagmatic volcanism. Studies in geophysics-explosive volcanism: inception, evolution, and hazards. National Academy Press, Washington D.C., pp 158–169

Wohletz KH, Sheridan MF (1983) Hydrovolcanic explosions; II, evolution of basaltic tuff rings and tuff cones. Am J Sci 283 (5):385–413

Wohletz K, Zimanowski B, Büttner R (2013) Magma-water interactions. In: Fagents SA, Gregg TKP, Lopes RMC (eds) Modeling volcanic processes. Cambridge University Press, pp 230–256

Wohletz KH (2003) Water/magma interaction: physical considerations for the deep submarine environment. In: White JDL, Smellie JL, Clague DA (eds) Explosive subaqueous volcanism, vol 140. American Geophysical Union Monograph, pp 25–49

Woods AW, Koyaguchi T (1994) Transition between explosive and effusive eruptions of silicic magmas. Nature 370:25

Woods AW, Sparks RSJ, Ritchie LJ, Batey J, Gladstone C, Bursik MI (2002) The explosive decompression of a pressurized volcanic dome: the 26 December 1997 collapse and explosion of Soufrière Hills Volcano, Montserrat. Geol Soc Lond Mem 21(1):457–465

Wormald PJ, Orr TO, Hodkinson IP (1991) The Mount Leyshon gold mine (NE Queensland), an intrusive breccia and igneous complex. In: Proceedings World Gold, Cairns, Australia, vol 2/91. Australian Institute of Mining and Metallurgy, pp 223–232

Wormald PJ (1993) Geology of the Mt. Leyshon gold deposit, Australia: a study of breccia pipe formation, facies and brecciation mechanisms. Unpubl PhD thesis, James Cook University North Queensland, Townsville, Australia, 201p. https://researchonline.jcu.edu.au/47307/

Wormald PJ (2017) The Mt. Leyshon gold deposit. In: Phillips GN (ed) Australian ore deposits, vol 32. Australian Institute of Mining and Metallurgy Monograph, pp 697–704

Wright IC (1996) Volcaniclastic processes on modern submarine arc stratovolcanoes: sidescan and photographic evidence from the Rumble IV and V volcanoes, southern Kermadec Arc (SW Pacific). Mar Geol 136(1–2):21–39

Wright JV, Coward MP (1977) Rootless vents in welded ash-flow tuffs from northern Snowdonia, North Wales, indicating deposition in a shallow water environment. Geol Mag 114(02):133–140

Wright IC, Gamble JA (1999) Southern Kermadec submarine caldera arc volcanoes (SW Pacific): caldera formation by effusive and pyroclastic eruption. Mar Geol 161(2):207–227

Wright HMN, Weinberg RF (2009) Strain localization in vesicular magma: implications for rheology and fragmentation. Geology 37 (11):1023–1026

Wright IC, Stoffers P, Hannington M, De Ronde CEJ, Herzig P, Smith IEM, Browne PRL (2002) Towed-camera investigations of shallow–intermediate water-depth submarine stratovolcanoes of the southern Kermadec arc, New Zealand. Mar Geol 185(3):207–218

Wright IC, Gamble JA, Shane PA (2003) Submarine silicic volcanism of the Healy caldera, southern Kermadec arc (SW Pacific): I—volcanology and eruption mechanisms. Bull Volcanol 65(1):15–29

Wright IC, Worthington TJ, Gamble JA (2006) New multibeam mapping and geochemistry of the 30–35 S sector, and overview, of southern Kermadec arc volcanism. J Volcanol Geotherm Res 149 (3):263–296

Wright HMN, Cashman KV, Rosi M, Cioni R (2007) Breadcrust bombs as indicators of Vulcanian eruption dynamics at Guagua Pichincha volcano, Ecuador. Bull Volcanol 69(3):281–300

Wright HMN, Folkes CB, Cas RAF, Cashman KV (2011) Heterogeneous pumice populations in the 2.08-Ma Cerro Galán Ignimbrite: implications for magma recharge and ascent preceding a large-volume silicic eruption. Bull Volcanol 73(10):1513–1533

Yamada R, Yoshida T (2004) Volcanic sequences related to Kuroko mineralization in the Hokuroku district, Northeast Japan. Resour Geol 54(4):399–412

Yamagishi H (1991) Morphological and sedimentological characteristics of the Neogene submarine coherent lavas and hyaloclastites in Southwest Hokkaido, Japan. Sediment Geol 74(1–4):5–23

Yamagishi H (1987) Studies on the Neogene subaqueous lavas and hyaloclastites in southwest Hokkaido. Report of the geological survey of Hokkaido, vol 59, pp 55–117

Zarzycki J (1991) Glasses and the vitreous state. Cambridge University Press, Cambridge

Zhang Y, Behrens H (2000) H2O diffusion in rhyolitic melts and glasses. Chem Geol 169(1):243–262

Zimanowski B, Fröhlich G, Lorenz V (1991) Quantitative experiments on phreatomagmatic explosions. J Volcanol Geotherm Res 48 (3):341–358

Zimanowski B, Büttner R, Lorenz V, Häfele HG (1997) Fragmentation of basaltic melt in the course of explosive volcanism. J Geophys Res Solid Earth 102(B1):803–814

Zimanowski B, Büttner R, Dellino P, White DL, Wohletz KH (2015) Magma-water interaction and phreatomagmatic fragmentation. In: Sigurdsson H, Houghton B, McNutt SR, Rymer H, Stix J (eds) The encyclopedia of volcanoes. Elsevier, Amsterdam, pp 473–484

Zimanowski B, Büttner R (2003) Phreatomagmatic explosions in subaqueous volcanism. In: White JDL, Smellie JL, Clague DA (eds) Explosive subaqueous volcanism, vol 140. American Geophysical Union Monograph, pp 51–60

Part B
Effusive Eruptions, Lavas and Sub-Volcanic Intrusions

Aims, Scope, and Structure

Part B summarises the principal processes by which lavas are erupted and emplaced in both subaerial and subaqueous environments, as well as the principal characteristics of lavas of different compositions. They can be erupted from subaerial or subaqueous vents, or flow from subaerial into subaqueous environments, and once formed lavas may remain coherent, or be subjected to autobrecciation and in subaqueous settings, also to quench fragmentation. As we have previously described in Chap. 2, lava flows are highly variable in composition, physical properties, size, shape, and in their surface and internal textures. Many of the differences between lava flows can be treated by initially dividing them into low viscosity (low silica) and high viscosity (high silica) types. Firstly in Chap. 5, we summarise the eruption processes, flow behaviour and characteristics of low viscosity, basaltic lavas, which come from primitive magmas formed in the mantle, and whose eruptions are the most frequent and voluminous on Earth, especially in submarine environments. In contrast, high viscosity felsic and silicic lavas are considered next in Chap. 6. Crustal melting, magma differentiation and country rock assimilation, usually in crustal magma reservoirs, generate felsic magma bodies with lower temperature, increasing silica content and higher viscosity. In Chap. 7 we describe ultra-low viscosity lavas which are represented principally by Archean ultramafic komatiites (and their sub-volcanic intrusions) which form integral components of greenstone belts, but also share some physical properties with other very low viscosity lavas such as carbonatites and sulphur lavas. Before reaching the Earth's surface, magma passes through the lithosphere as a viscous melt through a complex sub-volcanic magma plumbing system of dykes, sills, laccoliths, cryptodomes and other intrusions which are considered in Chap. 8.

5 Basaltic and Other Low-Viscosity Lavas

Eruption, Flow Behaviour, and Features of Subaerial and Subaqueous Lavas

Ray Cas, John V. Wright, and Guido Giordano

R. Cas et al., *Volcanology*, Springer Textbooks in Earth Sciences, Geography and Environment,
https://doi.org/10.1007/978-3-319-66613-6_5

Plate 5.1 Features of subaerial and subaqueous basalt lavas. **a** Typical, ropy-textured, pahoehoe lava, Pu'u O'o eruption, East Rift Zone, Kilauea volcano, Hawai'i (*Source* United States Geological Survey; Photo taken 23-11-2016). **b** Typical autobrecciating a'a lava flow, Pu'u O'o eruption, showing the spinose autobreccia crust and the red incandescent core of the lava (*Source* United States Geological Survey; Photo taken 29-1-2008). **c** Stacked, extensive, sheet basalt lava flows of the late Tertiary, Taos Plateau volcanic field, Rio Grande Rift, exposed in the Rio Grande River Canyon, New Mexico, USA. The canyon is about 240 m deep. **d** Pillow lava on the modern seafloor, offshore Hawai'i (*Source* NOAA). **e** Spectacular radial columnar jointing in a solidified Miocene lobate to pillow basalt lava, Muriwai, New Zealand. **f** Diamond drill core intersection through extraordinarily well-preserved 2,692 Ma Archean pillow basalt lava, Teutonic Bore Mine sequence, Yilgarn Craton, Western Australia

Abstract

This chapter summarises the eruption processes, flow behaviour, and characteristics of low-viscosity basalt lavas. Basalts are primitive magmas formed in the mantle, and their eruptions are the most frequent and voluminous on Earth, especially in submarine environments. They also form the largest volcanic features on Earth, such as the mid-ocean spreading ridges, flood basalt provinces, oceanic plateaus, and very large shield volcanoes. Here we first discuss the different factors affecting subaerial basalt lava behaviour, such as eruption temperature and effusion rate, which are complexly interrelated to a number of physical factors, principally magma viscosity and yield strength. In turn these control basaltic eruption styles, flow characteristics, and lava morphology. The two main types of subaerial basaltic lava are called pahoehoe and a'a, which are now best viewed as end-member types with a range of transitions in between. Tholeiitic flood basalt lavas are the largest and farthest flowing lava flows known, and consist of extensive sheet pahoehoe lavas with maximum flow distances of $\geq 1{,}000$ km. The characteristics of the flows are thereafter controlled by the nature of the environment into which they are extruded, or the environment into which they then flow, or both. Modern high-precision sonar bathymetric surveys now allow large areas of the seafloor to be mapped to the detail of defining individual lava flow fields, lava flows, and lobes. There are, however, major differences in the effects of an ambient aqueous medium, especially in deep-water environments, on subaqueous basaltic eruption processes, and the deposit types produced.

5.1 Introduction: Occurrence and General Characteristics of Basalt Lavas

Eruptions of basaltic magmas are the most frequent and voluminous on Earth (Plate 5.1), especially in submarine environments along the mid-ocean ridges, where they are largely unobserved. Indeed, basaltic volcanism also represents a major feature of the other planets and satellites in the Solar System, such as the Moon.

Basalts are primitive magmas and form in the mantle. Their eruption is therefore mostly associated with divergent plate margins and other extensional geodynamic settings where the crust is stretched and thinned, and volcanism can be voluminous, such as along the Eastern Africa Rift Zone and flood basalt provinces (White and McKenzie 1989; Geoffroy 2005; Chorowicz 2005; Chap. 17). Intraplate basaltic volcanism can occur in other extensional and trans-tensional settings experiencing lower degrees of extension, where volumes and average mass eruption rates are low, usually generating fields of monogenetic volcanoes, such as the Newer Volcanics Province in southeastern Australia (Hare and Cas 2005; Lesti et al. 2008; Cas et al. 2017; Chap. 17). Basalts are more rarely erupted in convergent settings, where the presence of a thickened crust has generally been inferred to usually prevent the direct withdrawal of primitive or contaminated magmas from the mantle (e.g. Kay et al. 1994). Interestingly, the advent of global seismic tomography has suggested that some intraplate basaltic volcanism (including flood basalts) may be related to the upwelling of hot asthenospheric materials associated with the deep subduction and stagnancy of the down-going plate where it penetrates along the mantle seismic transition zone at depths between 410 and 660 km, for example, the Pacific slab under East Asia (e.g. Zhao 2012; Wang et al. 2015; Chap. 17), and the Pacific Plate under Eastern Australia (Mather et al. 2020; Chap. 17).

Basalts form the largest volcanic features on Earth, such as the mid-ocean spreading ridges, flood basalt provinces, oceanic plateaus, and very large shield volcanoes, such as the 9-km-high or 17-km-thick volcanic pile forming the Mauna Loa–Kilauea shield complex on Hawai'i (Chap. 16, Sect. 16.5.1). With a basal diameter of nearly 200 km on the seafloor at a depth of 5 km, this volcanic edifice is of the same order of magnitude as Olympus Mons on Mars and it is still active.

Basaltic effusive eruptions in Iceland represent the longest and largest in historic times. These occurred at Eldgjá in 939–40 AD and Laki in 1783–84 AD, the first 20 km^3 in volume and 70 km long, and the latter 14.7 km^3 and 65 km long (Thordarson and Self 1993; Oppenheimer et al. 2018; Fig. 5.1). The amount of SO_2 injected into the atmosphere during both eruptions caused severe climate effects and famine in the northern hemisphere (Sigurdsson 1982; Thordarson and Self 1993; Oppenheimer et al. 2018). On Hawai'i, individual eruptions may last tens of years, such as the more than 30-year-long eruption from Pu'u O'o crater and elsewhere along the East Rift Zone of Kilauea volcano. This started in 1983 (Heliker et al. 2003; Orr et al. 2012) and ended in 2018 as a result of the flank eruption that drained the Kilauea system and affected the inhabited lower slope of the volcano (Neal et al. 2019).

In spite of these already impressive numbers, by far the largest and longest flows are known from the rock record in the Jurassic–Cretaceous flood basalts, where maximum flow distances reach 1,000 km (e.g. the Rajahmundry flow in the Deccan Traps flood basalt province, India, Self et al. 2008; Fig. 5.1a) and volumes of individual eruptions may exceed 1,000 km^3 (e.g. the Roza Member of the Columbia River Plateau flood basalt province, western USA; Thordarson and Self 1998; Fig. 5.1b). Self et al. (2008) suggest that some

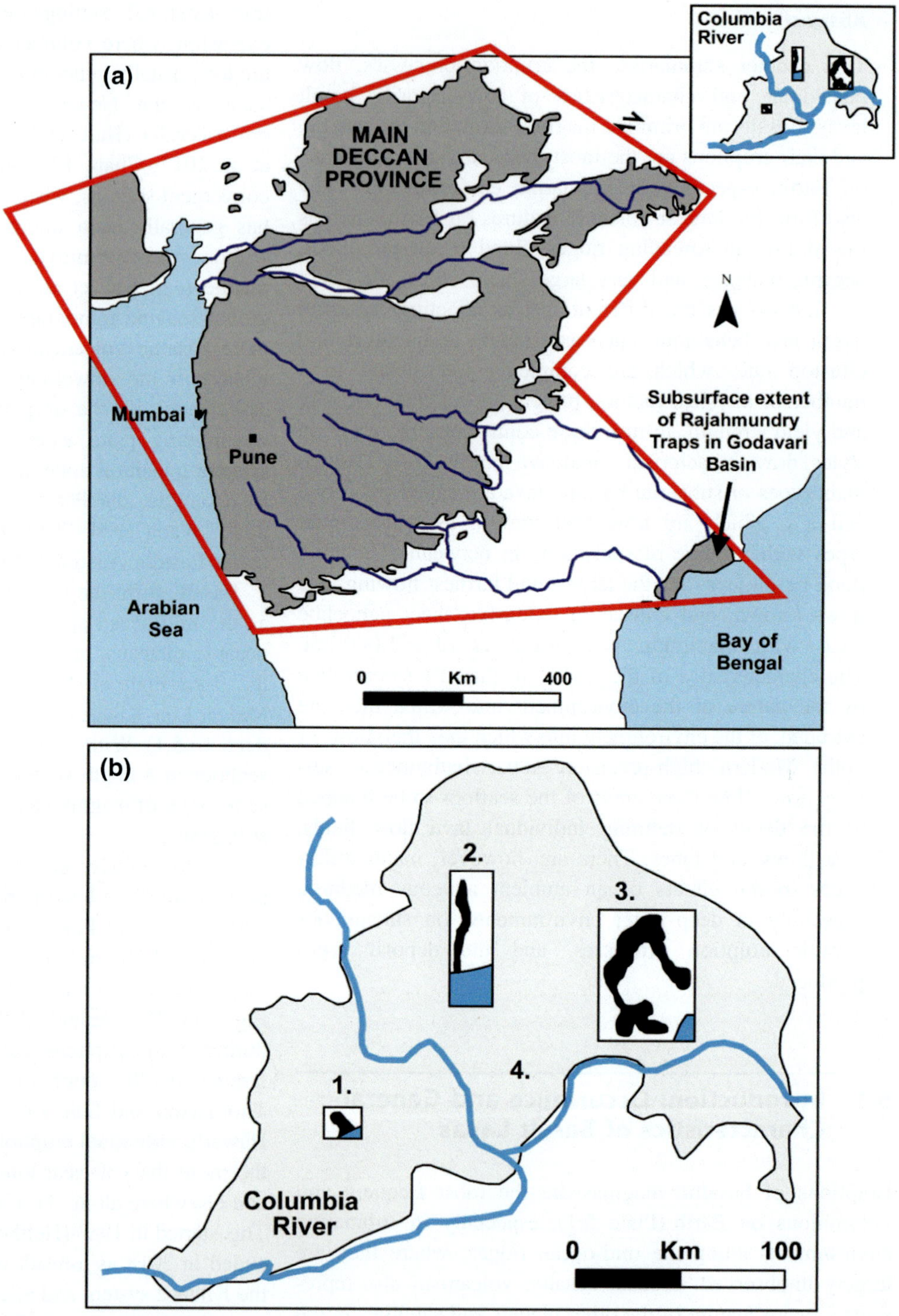

Fig. 5.1 A schematic comparison of dimensions of basaltic lava flows, from the largest flood basalts to small Hawai'ian flows. **a** Deccan flood basalt province, showing the extent of two flood lavas, and in the inset at the same scale, the Roza Member flood basalt lava, as well as smaller lava flows, is shown enlarged in (**b**) (Adapted from Self et al. 2008; Self et al. 1997). **b** Enlargement of the inset in (**a**), showing 4 = Roza Member flood basalt lava flow, Columbia River flood basalt province, compared with 1 = Kupaianaha 1986–1992 flow, Kilauea, Hawai'i, 2 = 1859 Mauna Loa flow, Hawai'i, and 3 = 1783–84 AD Laki lava flow field, Iceland (After Self et al. 1997)

flows in the Deccan Traps may have volumes as high as 5,000 km^3. The longest known, individual Quaternary basalt lava flow is the 160 km long Undara flow in northern Australia (Stephenson et al. 1998).

Basalt lavas have dominated Earth history, from before the formation of continental crust, at a time when ultramafic komatiite compositions with even lower silica contents than basalts, and mainly restricted to the Archean (see Chaps. 7, 17), were also being erupted.

Most of our knowledge about processes associated with basaltic lava eruptions, transport, and hazards comes from Hawai'i, Etna, and Iceland, where eruptions are frequent and many historical and instrumental observations are available (excellent reviews can be found in Cashman and Mangan

2014; Harris and Rowland 2015; for an older review of basaltic-volcano systems, see Walker 1993). The main physical processes associated with the flow behaviour of basaltic lava flows and their characteristics relate to their low viscosity. Similar viscosities, and therefore also lava characteristics, can also be typical of other magma compositions, usually alkaline in character (see Chap. 2), such as basanites, tephrites, foidites, and phonolites, that are rarer but common at some volcanoes such as Vesuvius (Ventura and Vilardo 2008; Campagnola et al. 2016). In spite of the relatively low viscosity ($\sim$100–1000 Pa s), most basalt lavas flow in a laminar state, except at the bases of high magma fountains and at the base of cascades over steep topography (Plate 2.1).

In terms of hazards and direct impact, basaltic lava flows are usually destructive against built infrastructure and buildings, but rarely fatal to humans, as towns are generally far enough from common source vents to enable the local population to escape. However, some unfortunate exceptions have occurred where high temperature, unusually low-viscosity lava was erupted near towns along the low flanks of volcanoes, as was the case of the 2002 Nyiragongo eruption in the Democratic Republic of the Congo, which caused numerous deaths (Tedesco et al. 2007; Chap. 16). In addition, as the global population increases, human settlements are being established closer and closer to volcano vents. Some attempts to protect properties by building earth-dams and excavating channels to change the path of lava flows, as happened during the 1991–93 eruption at Etna (Barberi et al. 1993), and spraying the surface and margins of advancing lavas to stop them (e.g. Heimaey, Iceland 1973), have had some apparent success.

There are two distinctive types of subaerial basalt lavas, *pahoehoe* and *a'a*, which are the Hawai'ian names that relate to their typical surface features. In Hawai'ian "*hoe*" means to paddle, which produces swirls on the water surface (cf. ropy texture), while *"a'a"* means to burn, which probably relates to the red-hot interior and perhaps flame-like nature of the irregular tops of a'a lavas. Pahoehoe lavas are characterised by smooth, billowy to ropy surface textures and lobate "toe" structures (Plate 5.1a) whereas a'a lavas have exceedingly rough, spinose, and fragmented surfaces (Plate 5.1b); there are also transitional types that will be discussed later in this chapter. Small volume pahoehoe and a'a lavas commonly erupt from low effusion rate volcanoes such as in Hawai'i, whereas very large volume, high effusion rate, massive, pahoehoe *sheet flood basalt lavas* are characteristic of continental Large Igneous Provinces (or cLIPS; Plate 5.1c). In subaqueous environments, *pillow lavas* (Plate 5.1d, e, f) are equivalent to subaerial pahoehoe-type lavas, a'a lavas are rare and massive sheet lavas are also found, including voluminous lavas of oceanic plateaus or submarine Large Igneous Provinces (oLIPs).

In the following discussion, we will first consider the processes and controls on the formation of subaerial basalt lavas, and then describe their characteristics before considering subaqueous basalt lavas. The processes and characteristics are also applicable to some compositionally very different but rheologically similar lavas, such as peralkaline lavas.

5.2 Subaerial Basalt Lavas: Factors Affecting Their Behaviour and Features

The eruption duration, transport, and final shape of basalt lava flows can be related to several main controls, which are principally the initial temperature and effusion rate, the type of environment in which the lava flows, cooling rate, degassing and crystallisation, transient physical properties, and slope. However, there are many controls which are complexly interrelated. For example, effusion or lava discharge rate depends on a number of factors, including vent shape and dimensions, magma viscosity, yield strength, and magma pressure gradient in the conduit. Similarly, along the length of a flow the cooling rates will be affected by changing environmental conditions (e.g. air, water, or ice), topographic confinement, and the variable development of an insulating crust. In the following sections, we will briefly review the main controls on basalt lava flow processes.

5.2.1 Initial Magma Temperature and Eruption Rates

Basalts are primitive magmas and encompass chemical compositions that range from tholeiite to alkali-basalt (see Chap. 2), which can be related to varying geodynamic settings, mantle source conditions (composition, degrees of metasomatism, history of prior melting, pressure and temperature), and degree of source partial melting. These in turn lead to distinctive compositional differentiation series (e.g. calc-alkaline to alkaline series). Such compositional variations are reflected in the varying physical properties, such as eruption temperature, viscosity and effusion rates.

The temperature of active lavas at source vents has been measured mostly with the use of in situ thermocouples, or remotely with the use of thermal cameras. On Hawai'i, measured temperatures range between 1100 and 1170 °C (Lipman and Banks 1987; Cashman and Mangan 2014). At Etna, temperatures are slightly lower, estimated to be around 1070–1090 °C (Calvari et al. 1994; Harris and Rowland 2015 and references therein). Basaltic shoshonites at Stromboli are estimated to erupt at 1000 °C (Harris et al. 2005). These temperature differences usually also reflect a different crystallinity of the lavas, with lower crystallinity (0–30%) for lavas at high temperatures, and higher crystallinity (up to 45–50%) for lavas at the lower temperatures. Eruption temperatures can be estimated in the rock record by the use of mineral geothermometers (e.g. Blundy and

Table 5.1 Effusion rates for basalt-lava-forming eruptions

Eruption	Effusion rate ($m^3\ s^{-1}$)	References
Laki 1783–84 and flood basalt lavas	1×10^3–5×10^3	Thordarson and Self (1993), Self et al. (2015)
Etna 1865–1975 (n = 17)	15–45	Wadge (1978)
Etna 1980–1999 (n = 18)	0.4–220	Harris et al. (2000)
Etna 2012	0.1–0.24	Slatcher et al. (2015)
Krafla 1975–1984 (n = 9) (f)	<20–120	Harris et al. (2000)
Mauna Loa 1843–1984 (n = 33)	35–140[1]	USGS (2021)
Mauna Loa 1984	950–1500[2]	Lipman and Banks (1987)
Askja 1961 (f)	600 (2000)[3]	Thorarinsson (1968)
Parícutin 1943–1952[4]	1.5–14.2	Larrea et al. (2017)
Bárdarbunga (Holuhraun) 2014–2015 (f)	20–600	Bonny et al. (2018)
Pu'u O'o (Kilauea ERZ) 1983–1986 (lf)	10–390	Heliker and Mattox (2003)
Kupaianaha (Kilauea ERZ) 1991	1–1.9	Mangan et al. (1995)
Puna (Kilauea lower ERZ) 2018 (f)	100–1000	Dietterich et al. (2018)

(f) fissure eruption; (lf) lava fountaining dominated this first 3-year period of eruption; [1]average effusion rates for different sectors of the volcano over the past 200 yr; [2]maximum effusion rates over a 4-day period from the main vent outlet; [3]average of first 8.5 h and in brackets maximum for this period; [4]eruption evolved from basaltic andesite to andesite during the eruption

Cashman 2008; Putirka 2008) and glass geothermometers (e.g. Cashman et al. 1999; Soule et al. 2004; Riker et al. 2009).

Determinations of effusion rates for mafic lava-forming eruptions are based on a variety of techniques, ranging from ground-based techniques such as manual surveying, photography, thermal imaging terrestrial laser scanning (TLS), to satellite-based techniques such as Landsat Thematic Mapper, radar, thermal radiometer data such as MODVOLC, TADR, MODIS spectral, and spatial imagery (e.g. Harris et al. 1998, 2005; Slatcher et al. 2015; Wright 2016; Bonny et al. 2018). The at-vent effusion rates observed for normal, relatively small volume pahoehoe and a'a lava flows (<1 km^3) generally range from <0.5 to 50 $m^3\ s^{-1}$ (e.g. Wadge 1981; Calvari and Pinkerton 1998; Harris and Rowland 1998; Harris et al. 2000), although this can reach 100–1,000 $m^3\ s^{-1}$ during initial eruption stages, especially when fissure eruptions are involved (Table 5.1, e.g. Wadge 1981; Lipman and Banks 1987; Harris et al. 2000; Bonny et al. 2018). For large volume historic eruptions, such as the 1783 Laki eruption in Iceland, magma effusion rates of 1×10^3 to 4×10^3 $m^3\ s^{-1}$ could be calculated using historical records of the duration of phases of the eruption and calculated eruption volumes (lavas and pyroclastics; Thordarson and Self 1993). For flood basalt eruptions, similar high effusion rates to those of the Laki eruption, perhaps $\sim 5 \times 10^3$ $m^3\ s^{-1}$ based on regionally extensive fissure vent systems, have been assumed, with flood basalt eruptions potentially lasting tens of years (Table 5.1; Self et al. 2015). Effusion rates can also be calculated knowing the approximate volume of a lava flow and using an empirical relationship between (minimum) flow duration and lava crust thickness proposed by Hon et al. (1994) for pahoehoe sheet lava lobes in Hawai'i (see Eq. 5.6).

Such large variability in effusion rates depends on magma viscosity, driving overpressure (including the presence of volatiles), and the geometry of the source vents, which may vary from single point source to fissure vents. Walker (1973) considered effusion rate to be the most important controlling factor on lava processes. He showed that the distance travelled by lava flows was proportional to the effusion rate, with lavas erupted at higher rates travelling further before cooling, which would increase their viscosity and inhibit movement. Harris and Rowland (2009) agree with this assessment, with the proviso that composition, degree of insulation of the lava in the feeder tube and morphology remain constant.

Large, single flow unit flood basalt lavas found in the geological record are now thought to be erupted at very high effusion rates of 4 or 5×10^3 $m^3\ s^{-1}$ (Self et al. 2015). By comparison, historic, more silicic, higher viscosity lavas (see Chap. 6) show average effusion rates that are generally a few orders of magnitude lower than those for mafic lavas.

5.2.2 Transient Lava Rheology: Cooling, Crystallisation, and Viscosity

Basaltic magmas at vent have a low viscosity, usually varying between < 10^2 and 10^3 Pa s depending on composition, eruption temperature, volatile, and crystal contents. Despite having a low viscosity, basalts are non-Newtonian fluids and their behaviour has been modelled as that of a Bingham fluid, characterised by a low yield strength (Chap. 2).

As soon as basaltic lavas are erupted subaerially (but see Chap. 3 and Sect 5.8 for conditions that apply to subaqueous environments) both radiative and convective cooling start from the top and sides inward, whereas conductive cooling occurs from the base. Degassing at atmospheric pressure will induce gas bubble formation, migration, and escape, as well as crystallisation within the flowing lava (Fig. 5.2a–c). All these processes will promote the development of rapidly changing rheological profiles, especially in viscosity, both in cross-section and longitudinally (Fig. 5.2d). These changes will eventually result in the development of channels, tubes, lateral and frontal levees, as well as the typical appearance of the basaltic lava surfaces such as "clinkery" a'a or coherent pahoehoe crust that we will describe in the next sections. These changes will also ultimately control the areal extent of lavas and their 3D form.

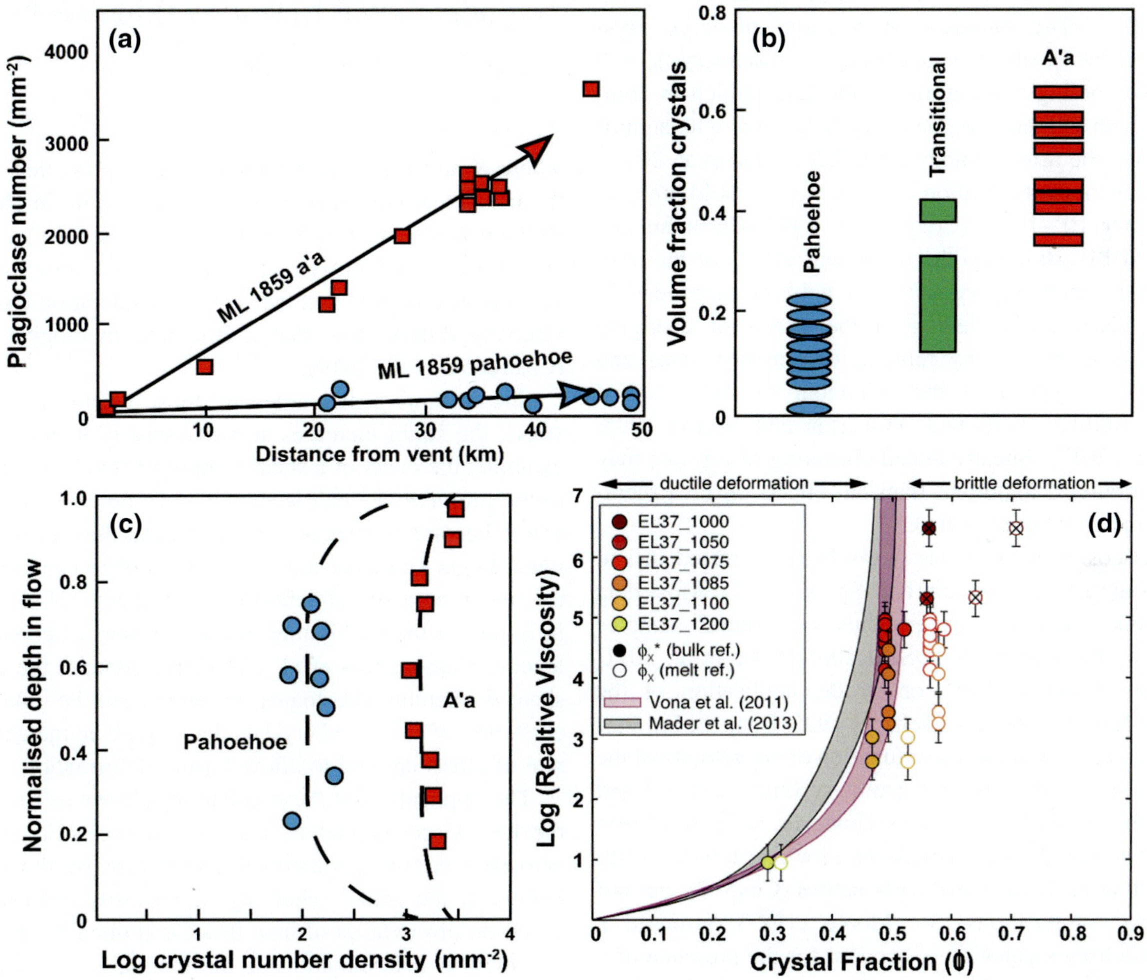

Fig. 5.2 Some crystal abundance parameters in pahoehoe and a's lavas. **a** Plagioclase microlite number density versus distance from vent in pahoehoe and a'a lavas from Mauna Loa volcano, Hawai'i, showing the much higher microlite content and higher rate of increase of microlite numbers in a'a lavas than pahoehoe lavas with increasing distance from vent (After Riker et al. 2009; Cashman and Mangan 2014). **b** Groundmass crystal volume fraction in different types of basalt lavas from the eruption of Pu'u O'o, Kilauea volcano's East Rift Zone, Hawaii, showing significantly higher crystal content in a'a lavas than pahoehoe lavas (After Cashman and Mangan 2014). **c** Crystal number density in the vertical profile of pahoehoe and a'a lavas in Hawai'i showing higher CND in a'a lavas than pahoehoe lavas, an absence of microlites in the glassy crusts of pahoehoe lavas (After Katz and Cashman 2003; Cashman and Mangan 2014). **d** Effect of increasing crystal content on lava viscosity. The colour bands showed projected trends from different studies. Crystal fraction is expressed as melt-referenced (i.e. ϕ_X = empty symbols) and bulk-referenced (i.e. ϕ_X = closed symbols). (From Vona et al. 2017)

Crystallisation strongly affects the rheology of lava flows, especially when degrees of crystallisation reach close to the maximum packing fraction, i.e. close to 40–50%, depending on crystal shapes (e.g. Vona et al. 2011; Mader et al. 2013). Measured down-flow variations of temperature and crystallinity of lavas on Hawai'i (Fig. 5.2a–d; Lipman and Banks 1987; Cashman et al. 1999) result in exponential increase in viscosity up to 10^6 Pa s (Fig. 5.2d) and yield strengths of 10^3–10^4 Pa (Harris and Rowland 2015, and references therein). Given the low viscosity of basalt lavas, nucleation and growth of syn-emplacement microlites is significant and plays a role in increasing the viscosity when the total crystal content rises above a threshold of 30–35% by volume (Fig. 5.2b, d), at which crystals start to interact (e.g. Costa et al. 2009; Cimarelli et al. 2011; Vona et al. 2011, 2017). The presence of a cargo of large crystals (Fig. 5.2a–d), of either phenocrysts or xenocrysts, will reduce the cooling window inside the lava, which in combination with syn-emplacement crystallisation will significantly affect the flow rheology (Fig. 5.2d; Vona et al. 2017). Time scales for crystallisation to reach the threshold limit are in the order 10^1–10^2 h (Crisp et al. 1994; Cashman and Mangan 2014). Based on the non-linear effects of the multiple phases, including crystals and bubbles suspended in lava, on viscosity, modelling of the effects of changing viscosity is best done by relating deformation rates and applied stress typical of the behaviour of the so-called Herschel–Bulkley fluids (e.g. Pinkerton and Norton 1995; Vona et al. 2017). Shear-induced clustering of crystals may also contribute to important transient effects on lava rheology (Campagnola et al. 2016).

The viscosity of lavas has rarely been measured in situ (Shaw et al. 1968; Pinkerton and Sparks 1978). Most data come from laboratory experiments on natural samples (Pinkerton and Norton 1995; Sato 2005; Vona et al. 2011; Vona and Romano 2013), or by the application of the empirical Jeffrey's equation (Jeffreys 1925; Chap. 2, Eq. 2.5) that expresses the viscosity as a function of the velocity of the lava for a given density and geometry (cross-section) and slope of the channel it flows into. However, this formulation is for Newtonian fluids, which is valid for very low-viscosity lavas such as komatiites and carbonatites (Chap. 7), but not other lava compositions. So, Moore (1987) proposed a modified Jeffrey's equation to account for the presence of a yield strength in basaltic lavas. For very wide Bingham rheology (see Chap. 2) lava flows, the relationship is

$$v = \left(\frac{d^2 \rho\, g \sin\theta}{3\eta}\right)\left[1 - 1.5\left(\frac{\tau_0}{\tau}\right) + 0.5\left(\frac{\tau_0}{\tau}\right)^3\right] \tag{5.1}$$

For narrow, channelised Bingham rheology lava flows:

$$v = \left(\frac{d^2 \rho\, g \sin\theta}{8\eta}\right)\left[1 - 1.33\left(\frac{\tau_0}{\tau}\right) + 0.33\left(\frac{\tau_0}{\tau}\right)^4\right] \tag{5.2}$$

where v is the measured flow velocity, d is the lava thickness, ρ is the lava density, g is the acceleration of gravity, θ is the slope or inclination of the surface on which flow is occurring, η is the lava viscosity, τ is the basal shear stress, and τ_0 is the yield strength (Moore 1987).

For shear-thinning or shear-thickening non-Newtonian rheologies (Herschel–Bulkley fluids), Castruccio et al. (2014) proposed a further modification of Jeffrey's equation (Chap. 2, Eqn. 2.6):

$$v = \left(d^2 \rho\, g \sin\theta / 3K\right)\left[(3n/d^3 n + 1)(\rho\, g \sin\theta / K)^{(1-n/n)}\right] \times \left[d^{(2n+1/n)}(1 - (n/2n+1)\right] \tag{5.3}$$

where K and n are the rheological parameters that describe the non-linear dependency of stress and strain in Herschel–Bulkley fluids and K is the consistency (viscosity) and n is a power law exponent. Herschel–Bulkley rheology involves a visco-plastic regime that takes into consideration changes in viscosity during flow that a Bingham rheology does not (Castruccio et al. 2014).

Shear-thinning rheologies are those where, at increasing stress, the strain increases at progressively larger rates. For example, the onset of a shear-thinning behaviour can be due to the progressive coalescence of large bubbles during flow which become a weakness zone and can then accommodate much larger strain for a given stress applied (such as below the upper crust of pahoehoe lava; see Sect. 5.4.2), or similarly due to the cushioning effect of spherical crystals (e.g. leucite; Campagnola et al. 2016). By contrast the onset of shear-thickening rheologies in lavas can be due to the clustering of elongated crystals that require an increase in the rate of stress applied to allow further deformation.

The application of these equations allows us to calculate the lava viscosity and its changes by measuring the flow velocity and the geometrical parameters of the channel. However, the actual reliability of viscosity determinations based on observation of lava flows is limited by the lack of accuracy in determination of the thickness (d) of the lava flow and the assumption of a constant temperature across the channel, which is not the case, as will be discussed in the following sections. For these reasons, determinations of viscosities that use Jeffreys' or Moore's equations can be considered only apparent.

5.2.3 Crust Formation and Deformation, Open Channels, Tubes, and Lava Inflation

5.2.3.1 Cooling

Cooling patterns of flowing lavas are quite complex and depend on the competition between the cooling agents, which are the radiative and the convective heat losses, and the factors that contribute to the conservation of the heat along flow, which are the mass-flow rate and the release of the latent heat of crystallisation as the lava degasses and crystallises.

A generalised equation that relates all those factors to determine the down-flow variation of temperature *dT/dx* is given by Harris and Rowland (2015):

$$\frac{dT}{dx} = \frac{(Q_{rad} + Q_{conv})}{E_r\, \rho\, (c_p + L\, df/dT)} \tag{5.4}$$

where Q_{rad} is the radiative heat loss, Q_{conv} is the convective heat loss, E_r is the volumetric flow rate, ρ is the lava density, c_p is the lava specific heat, L is the latent heat of crystallisation, and *df/dT* is the rate of crystallisation as temperature decreases.

However, measured cooling rates are far from simple and appear to increase gradually downslope and become very high near the flow front (Lipman and Banks 1987; Cashman et al. 1999). More complex patterns are shown in cases of formation of an insulating crust, where cooling rates within lava tubes can be as low as 0.35 °C km^{-1} (e.g. Clague et al. 1999).

The rate of cooling in a lava also determines the styles of degassing and crystal nucleation and growth, which in turn feed back into the dynamic rheology (Cashman and Mangan 2014). In particular, fast cooling in internally stirring a'a lavas tends to favour nucleation over crystal growth, whereas the reverse is true for slow-cooling systems like those associated with the formation of tubes (see next paragraph; Katz and Cashman 2003).

5.2.3.2 Crust Formation

The formation of a *surface crust* is an important process associated with the very fast heat loss by radiation from the lava surface, which begins almost immediately during flow away from the vent. Based on field data, Hon et al. (1994) identified distinct layers within active lavas determined by rheological variations associated with the inward progression of cooling (Fig. 5.3). From the top, the first layer to develop is the surface crust defined as bounded at its base by the isotherm 800 °C, below which the behaviour is brittle for strain rates appropriate to flowing lavas. The lava surface may cool to 800 °C just a few seconds after issuing from the vent and Hon et al. (1994) proposed an empirical equation to describe the increase in thickness of the crust with time, by tracking the inward propagation of the 800 °C isotherm with time *t*:

$$D_{800^\circ C} = 0.0473\sqrt{t - 0.0233} \tag{5.5}$$

Progressing inward, an upper visco-elastic layer is sandwiched between the 800 °C isotherm above and the 1070 °C isotherm below, wherein the lava is partially molten. The visco-elastic behaviour of this layer determines the velocity gradient between the rigid crust and the fully molten and flowing lava core. Towards the base the lava core transitions again into a lower visco-elastic layer, in turn, grading into a basal crust, which has lost heat to the substrate over which the lava is flowing.

The crust and the visco-elastic layers thicken or grow from the inside of the lava flow at different rates as the lava flow cools downslope. As the hotter and low-viscosity inner core will tend to flow much faster than the upper and lower layers, both velocity gradients and differential shear stresses will develop across the layers, inducing variable strain rates, but also in response to down-flow changes in volumetric flow rates, slope angle, and flow velocity.

Hon et al. (1994) proposed a relationship between the time, t_c, that it takes for the solid crust on an inflating

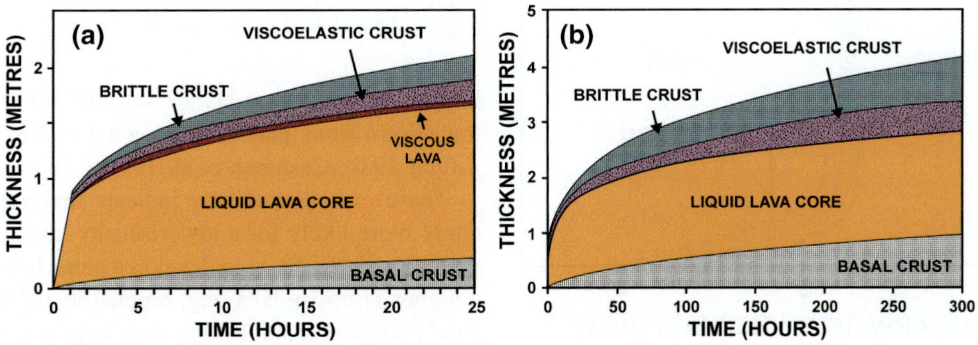

Fig. 5.3 Section through the April 1990 Kupaianaha lava flow, Kilauea volcano's East Rift Zone, Hawai'i, showing the growth of the internal layers: **a** after 25 h, and **b** over 300 h. The boundary between brittle and visco-elastic upper crust is the position of the 800 °C isotherm, and the base of the visco-elastic upper crust is the 1070 °C isotherm. (After Hon et al. 1994)

pahoehoe sheet lava flow to form, which is effectively the active flow duration, and its thickness, *Hc*, based on observations in Hawai'i in 1988:

$$t_c = 164.8\,H_c \quad (5.6)$$

Vye-Brown et al. (2013) then used that empirical relationship to estimate the minimum flow duration of the thickest sheet flow lobes in Miocene Columbia River Plateau flood lavas, suggesting durations of 10–20 years.

In Sect. 5.4.2, we discuss the internal textural zones and layers that form in pahoehoe lava crusts.

5.2.3.3 Crust Deformation: Controls on Pahoehoe Versus A'a Lava Formation

The fate of the upper crust thus depends on the competition between the changing viscosity of the lava due to the developing cooling profile and the shear rate (Fig. 5.4; Soule and Cashman 2005; Cashman and Mangan 2014). In very low-viscosity basalts (e.g. pahoehoe lavas), the visco-elastic layer accommodates significant strain, acting as a detachment between the rigid upper crust and the molten core. The crust is not broken even where the lava interior is still moving quickly and can form the typically smooth and unbroken pahoehoe surface (Plate 5.1a; Fig. 5.5a, c). Similarly, the crust can be preserved at low strain rates, for example, along flat topography where the lava is moving slowly. By contrast, where the strain rates on the developing crust are high (Fig. 5.3), such as along steep slopes, or at increased volumetric flux (e.g. Rowland and Walker 1990), the crust is torn apart and breaks plastically and in brittle fashion to form the typical clinkery surface of slabby pahoehoe and a'a lavas (Plate 5.1b; Fig. 5.5b, c). This also happens as lava viscosity increases (Fig. 5.4), which can

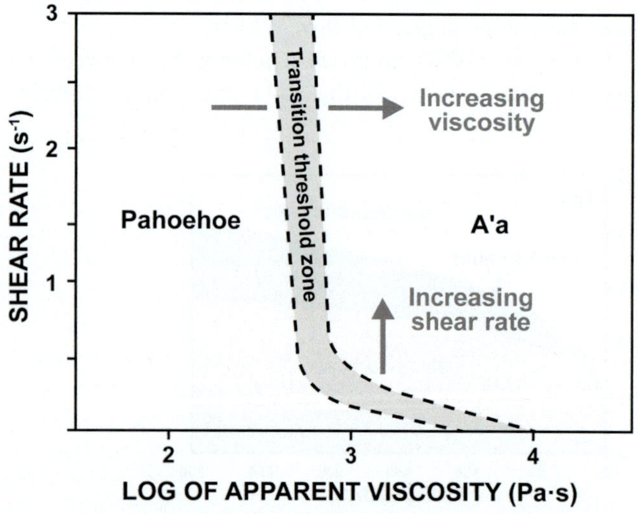

Fig. 5.4 Plot of apparent viscosity versus shear rates where fields for formation of pahoehoe and a'a lava morphologies are defined based on empirical data (After Cashman and Mangan 2014)

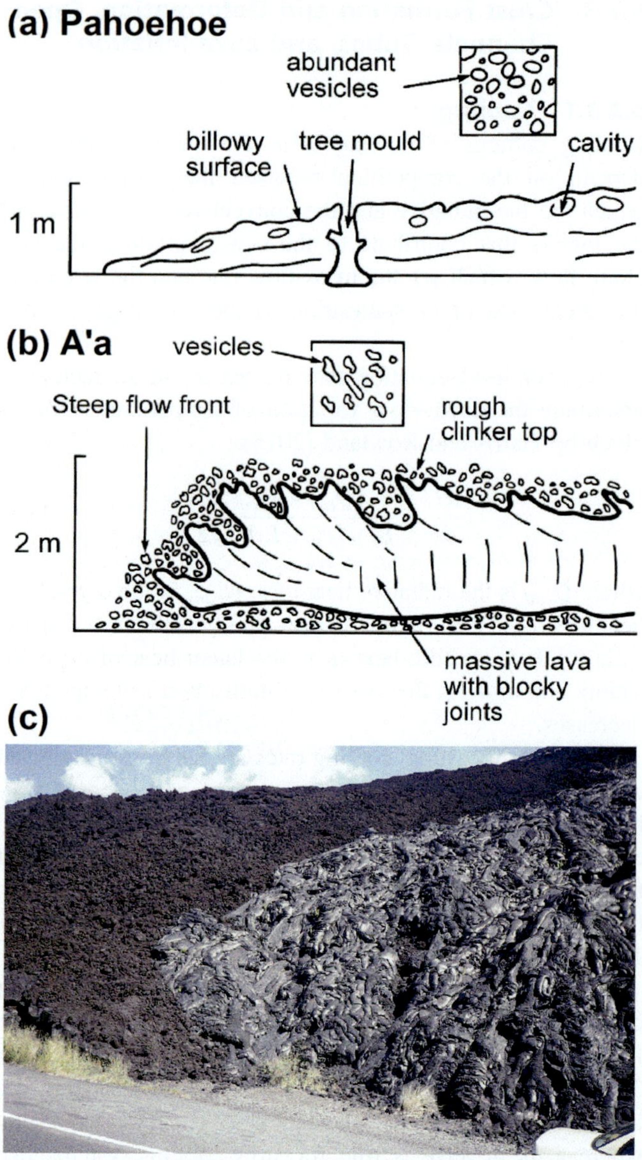

Fig. 5.5 Differences between pahoehoe and a'a lava. **a** Schematic representation of the lower aspect ratio nature of pahoehoe lava with the coherent, smooth crust, and **b** higher aspect ratio a'a lava with a hackly autobrecciated surface and margin (From Cas and Wright 1987). **c** Typical brecciated a'a lava (left) and smooth pahoehoe toe lava (right). Mauna Ulu, Kilauea volcano, Hawai'i

happen during flow as lavas cool, and can be exemplified by transitions from pahoehoe to a'a lava (e.g. Peterson and Tilling 1980; Cashman et al. 1999).

These examples clearly indicate that although it is certainly more likely for a low-viscosity lava (e.g. erupted at a viscosity of 10^2 Pa-s) to develop a pahoehoe surface, and for a higher viscosity lava (e.g. erupted at 10^3 Pa-s) to develop an a'a surface, a single lava flow may develop both pahoehoe and a'a surfaces based on the interplay between volumetric flux, down-flow cooling history and slope, all of which contribute to changing its apparent viscosity and shear rates (Fig. 5.4). Between the typical very smooth pahoehoe and the fully developed clinkery a'a lava surfaces, transitional conditions are quite commonly experienced, producing lavas with hybrid surface features (e.g. *slabby pahoehoe*, *rubbly pahoehoe*; Peterson and Tilling 1980; Duraiswami et al. 2003; Harris et al. 2017; Pedersen et al. 2017; Fig. 5.4).

5.2.3.4 Open Channels and Lava Levees

As soon as lavas flow away from the vent area, they begin to cool. Usually their margins cool first, become more viscous, slow down, and solidify, while the centre of the flow remains molten and continues to flow. The flow thus develops a clear *lava channel* bounded by stationary, solidified marginal *levees* (Figs. 5.6a–c and 5.7 a–d). During the early stages of the lava flow, close to vent, all lavas focus into open channels, especially when guided by topography (Fig. 5.6a). Close to vent, the levees bounding channels are frequently built up by the superposition of thin coherent overflow lobes (Fig. 5.6b). Away from vent, as lavas flow onto flat, relatively unconfined topography, they can spread, and multiple braided channels can form, bounded by the stationary parts of the lava, which act as levees (Fig. 5.6c).

However, levees can form in multiple ways. In a study of the way in which levees formed at Mt Etna, Sparks et al. (1976) found four principal types of levees (Fig. 5.7a–d). *Initial levees* (Figs. 5.6a, 5.7a and 5.8a) are formed because of the increasing yield strength of the lava at the flow margins as the margins cool. This causes them to solidify and form stationary walls or levees to the flowing lava in the channel, so confining the lava, as indicated by the studies of Hulme (1974). These form in both pahoehoe and a'a flows. *Accretionary levees* (Fig. 5.7b) were observed near boccas and consist of piles of clinker accreted to smooth, open lava channels (Fig. 5.8b). The clinkers weld themselves together to form a steep, solid levee. If, and where down-flow, lavas develop an a'a surface, i.e. where the lava viscosity and strain rates are both high enough to continuously break the cooling surface, clasts of various shapes and sizes, called "clinker", form on its top and are carried or tumble down-flow. These are pushed to the side and front of the

Fig. 5.6 Lava channels and levees. **a** Near-vent, open channel lava flow during the 1973 Heimaey eruption, Iceland, showing solidified levees with small overspill lobes, spilling onto the near levee and so building it up (Photo by George Carman). **b** Small overspill lava lobes (pale grey) at the margins an open lava channel during the 1983 lava eruption at Mt Etna, Sicily, Italy (Photo by Pierre). **c** Braided, lobate Puna a'a lava channel system, Kilauea volcano, Hawai'i, bounded by stationary parts of the lava that forms levees. The flow was erupted from newly opened fissure vents along the lower East Rift Zone, southeast of Pu'u O'o volcano and destroyed much of the Leilani Estate. (*Source* United States Geological Survey; Photo taken 19-5-2018)

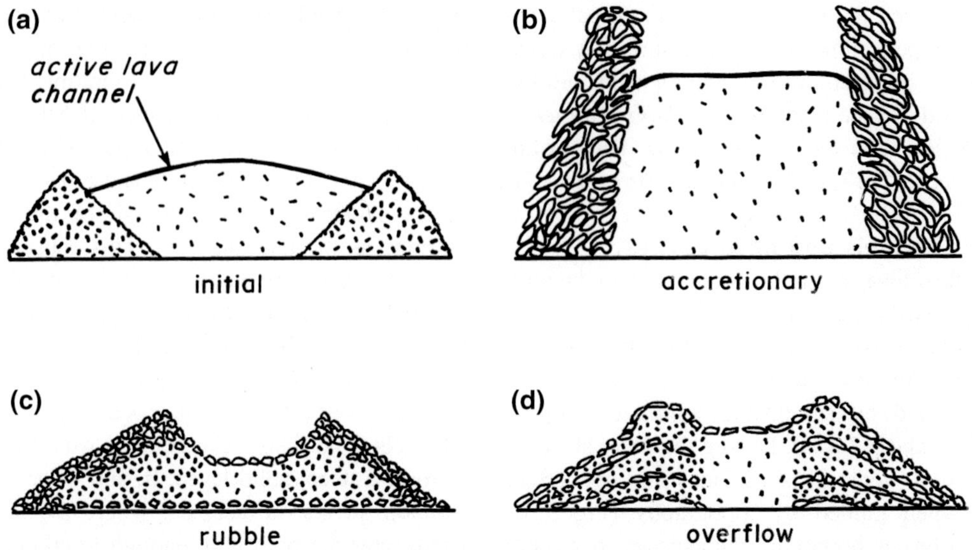

Fig. 5.7 Cross-sections through four different types of lava levee observed on Mt Etna, Sicily, Italy. Heavier stipple is massive lava; sparsely stippled areas represent flowing lava. **a** Initial levees consisting of coherent, solidified lava. **b** Accretionary levees consisting of accreted spatter and clinker that welds to the side of the channel. **c** Rubble levees, built up by a'a rubble from the lava surface tumbling down the levee slopes. **d** Overflow levees, resulting from overspill of thin lava lobes over the top and down the sides of the levees. (From Cas and Wright 1987, After Sparks et al. 1976)

flow, where they accumulate. This process of breaking of the crust is called autobrecciation (see Chap. 4). The lateral zones made up of the piled up lava clinker eventually rise as lateral autobreccia levees or *rubble levees* (Figs. 5.7c and 5.8c), and contribute to the containment of the lava channel, also allowing it to rise and thicken during the flow duration. In flows where a'a lava has developed fully, the flow front advances and the sides also expand by avalanching of a'a debris (Fig. 5.8c) or by overspills of a'a lava lobes onto and down the outer slopes of the levees (Fig. 5.8d), forming rubble levees. These rubble levees are at angles of repose of 30–35°. The fourth type, *overflow levees* (Fig. 5.7d), is formed when lava repeatedly floods over existing levees, producing a multi-lobate pile that builds up and bounds the open lava channel (Fig. 5.8d). Most levees on Mt Etna are hybrids of two or more of the four types.

5.2.3.5 Lava Tubes—The Magma Plumbing System Inside Lavas

Where the lava crust builds over the lava channel from the levees and remains unbroken the lava may form an enclosed *lava tube* under the surface crust (Fig. 5.9). A tube is an alternative lava distributary system compared to open channels. Once a visco-elastic crust becomes thick enough it freezes or solidifies at the edge of the channel and, step by step, a solid, mechanically strong roof spreads over the channel from the side. The moving molten lava in the tube shears under the crust, detaches mechanically from it, and continues to flow downslope under the solid crust towards the flow front with minimal loss of heat to the atmosphere. Peterson and Swanson (1974) observed lava tubes forming during the 1970–71 activity of Mauna Ulu, Hawai'i. In addition to the crust forming by cooling of a lava surface from the levees inwards and downstream, it has also been observed forming by gradual roofing-over of a lava stream from its levees inwards by the accretion of lava spatter along the edges.

The roof of the tube, or parts of it may collapse at any time during an eruption if lava supply rate wanes and the lava level in the tube drops leaving a partially evacuated tube, or after the eruption has stopped when lava has drained down the tube to the flow front, again causing the lava level in the tube to drop. The collapse openings in the roof of the tube are called *skylights*. Skylights formed during the eruption provide a window to look into the lava tube, providing a view of the flowing red-hot lava (Fig. 5.9a–c). Skylights provide an opportunity to sample the lava flowing through the tube and even photograph the lava flow stream in the active lava tube, and processes that occur in it, such as lava stalactites dripping from the tube roof (Fig. 5.9c). Even when skylights don't exist to confirm the presence of lava tubes in an active lava flow, thermal imagery can be used to highlight where the tube crust is thinnest and hottest, so identifying the internal lava pathways within the lava.

Lava tubes range in size from less than 1 m in diameter to large tubes >30 m wide and 15 m high and can form large distributary networks which can allow lava to flow for distances of km, tens of km, and even hundreds of km (flood basalts) beneath the nearly stationary lava surface crust. With thick pahoehoe lava flows, particularly when confined

Fig. 5.8 Lava levees. **a** Initial levees consisting of the solidified margins of the coherent pahoehoe Harman Valley lava flow, Newer Volcanics Province, Wallacedale, Victoria, southeastern Australia. The blocks are columnar joint blocks. **b** Lateral levees made of the agglutination of clinker to the sides of the open channel during the 2004 a'a lava flow at Mt Etna, Sicily, Italy, some 500 m downslope of the main bocca. The flow is about 2 m thick. **c** Rubble levees from the 1380 AD phonolitic lava flows of the El Teide stratovolcano, Tenerife, Canary Islands. Note the multiple lava flow lobes, each with well-defined central channel and steep, angle of repose a'a rubble levees. **d** Channelised a'a lava with a'a rubble levees, being built up by overspill a'a lobes from the main channel, "61 g" lava flow, Pu'u O'o volcano, Hawai'i. Note an older pahoehoe lava in the foreground. (*Source* United States Geological Survey; Photo taken 29-6-2016)

by topography, once the magma supply at vent stops, the magma still in the tube continues to drain downslope through the tube to the lava front. If the surface crust is thick and strong enough, the draining lava produces a cave as the level of the lava in the tube drops. In old lava flows such lava caves have been traced tens to >100 km (e.g. 160-km-long Undara lava flow, north Queensland, Australia, Atkinson et al. 1975). Alternatively, where the mass flux increases, either by an increase of the mass eruption rate or through changes in the local geometry of the tube, the roof may be forced to break which then accommodates lava breakouts from the lava surface crust.

Lava tubes are important because the roof crust inhibits radiative heat losses from the molten lava in the interior of a flow, which therefore maintains a low viscosity, thus enabling the flow to travel long distances. The heat loss can be as low as 0.001 °C s^{-1} (Harris and Rowland 2009). Keszthelyi and Self (1998) calculated that tube-fed lavas longer than 100 km could be achieved if heat loss was <0.5 °C km^{-1}, at eruption rates of at least several tens of m^3 s^{-1}, and at flow velocities as low as 0.2–1.4 m s^{-1}. Tube-fed pahoehoe flows can achieve lengths much greater than a'a flows of equivalent effusion rate. The formation of tubes is therefore a major issue in lava hazard studies, as it can substantially increase the distance that a lava flow can reach. During the 1991–93 eruption at Mt Etna, Sicily, for example, the lava formed a tube and the distal portion of the lava field, some 8 km away from vent, and kept being fed at temperatures close to the eruptive temperature ($\sim$1070 °C), threatening the local village of Zafferana Etnea (Calvari and Pinkerton 1998). Successful attempts to decrease the magma supply rate to the flow front involved the clogging of the tube with blocks of concrete and eventually the diversion of the flow into an artificial open channel by blasting the lateral wall of the tube (Barberi et al. 1993). During the 1983 to 2018 eruption of Pu'u O'o on Kilauea volcano, Hawai'i, lava tubes allowed lavas to flow 12 km to the coastline, in many cases, threatening, destroying, and burying settlements (see Fig. 5.18b).

Fig. 5.9 Lava tubes. **a** Lava tube formed by cooling of the surface of lava in an open channel, forming a crust that forms a roof over the underlying lava in the tube. Part of the roof has collapsed to open a "skylight" providing a view into the red-hot lava in the tube. The active lava stream in the tube is orange. Mauna Ulu 1969–1970 lava flow, Kilauea volcano, Hawai'i. (*Source* United States Geological Survey; Photo by J. Judd, taken 21-10-1970). **b** Skylight through the pahoehoe crust on the feeding tube of lava flow from the Kupaianaha vent, east of Pu'u O'o crater, Kilauea volcano, Hawai'i. Note in the distance the white, steam plume shows the active flow entering in the sea from the end of the tube system. (*Source* United States Geological Survey; Photo taken 19-6-1989). **c** Sideview taken through a skylight in the tube of Pu'u O'o pahoehoe lava flow. This shows the active lava stream in the bottom of the tube (yellow), mottled red, and grey congealed lava against the walls of the tube, and active lava drips forming stalactites. (*Source* Poland et al. 2014; Photo by M.R. Patrick, taken 19-2-2008). **d** The solid core of basalt enclosed in a'a lava breccia represents the solidified lava tube of the a'a lava. Mt Etna, Sicily, Italy

Lava tubes form easily in pahoehoe flows, although they can also occur in transitional and a'a lavas, such as at Etna, represented by solid, coherent cores enclosed in a'a lava breccia (e.g. Fig. 5.9d). The clinker that accumulates at the front of an a'a lava is overridden by the flowing lava front which rolls over like the caterpillar tracks of a bulldozer, and is then found in the rock record as a basal autobreccia beneath the solidified core of the lava tube, which propagates over the basal a'a breccia and is in turn overlain by the upper or surface a'a breccia crust.

5.2.3.6 Lava Flow Inflation

The development of a solid and insulating crust in pahoehoe lavas allows the process of *flow inflation* or thickening to occur over time (Hon et al. 1994; Self et al. 1998; Hoblitt et al. 2012). This process typically occurs where the volumetric flow rate exceeds the advance rate of the flow front, such as happens at breaks in slope of a volcano and causes the surface crust to rise as more magma is injected into the lava tube under the crust. In addition, during flow inflation, the presence of a rigid confining crust causes the pressure in the magma in the tube to increase, at times leading to brittle rupture of the crust if magma fluid pressure exceeds the strength of the crust. Fractures or ruptures occur either from the lava front (Fig. 5.10a), surface, or laterally at the margins, feeding *breakouts* of lava (Fig. 5.10a–d) which cause the flow front to branch into several toes and larger lobes.

Although inflation and breakouts are best documented for pahoehoe lavas, they can also occur in a'a lavas. Where the front of the a'a lava slows down due to cooling or slope reduction, the accumulation of the frontal autobreccia contributes to resistance to the ongoing flow and leads to considerable thickening of the lava flow (Fig. 5.10e). As a result of the build-up of pressure behind the stationary, "dammed-up" lava front, breakouts of incandescent red lava lobes may occur through the front, the marginal levees or from the surface of the lava behind the front.

Fig. 5.10 Lava breakouts. **a**–**d** Examples from the Pu'u O'o eruption, Kilauea volcano's East Rift Zone, Hawai'i. **a** Breakout at front of lava flow, with smaller breakouts issuing through an already formed crust-forming propagating toes. The brittle crust broke in response to the inflation and build-up of magma pressure in the molten lava core. Thanksgiving Eve Breakout tube system. (Photo by Andrew Cheesman, taken 15-6-2009). **b** Breakout lobe from the advancing Pahoa pahoehoe lava flow front above Pahoa Village (*Source* United States Geological Survey; Photo taken 28-10-2014). **c** Breakout of lava spills down the side of inflating Banana flow and slows as slope decreases, causing tangle of folded crust to form. Lava emerges from under crust near top of image. Width of new lava stream ~1 m (*Source* United States Geological Survey; Photo taken 19–5-2004). **d** Further instance showing breakout from Banana flow (*Source* United States Geological Survey; Photo taken 16-5-2004). **e** Slow advancing front of the thickened 1983 a'a lava flow at Mt Etna, Sicily, Italy, showing red-hot incandescent lava in the process of breaking out as a'a rubble tumbles and slides down the flow front (Photo by Pierre)

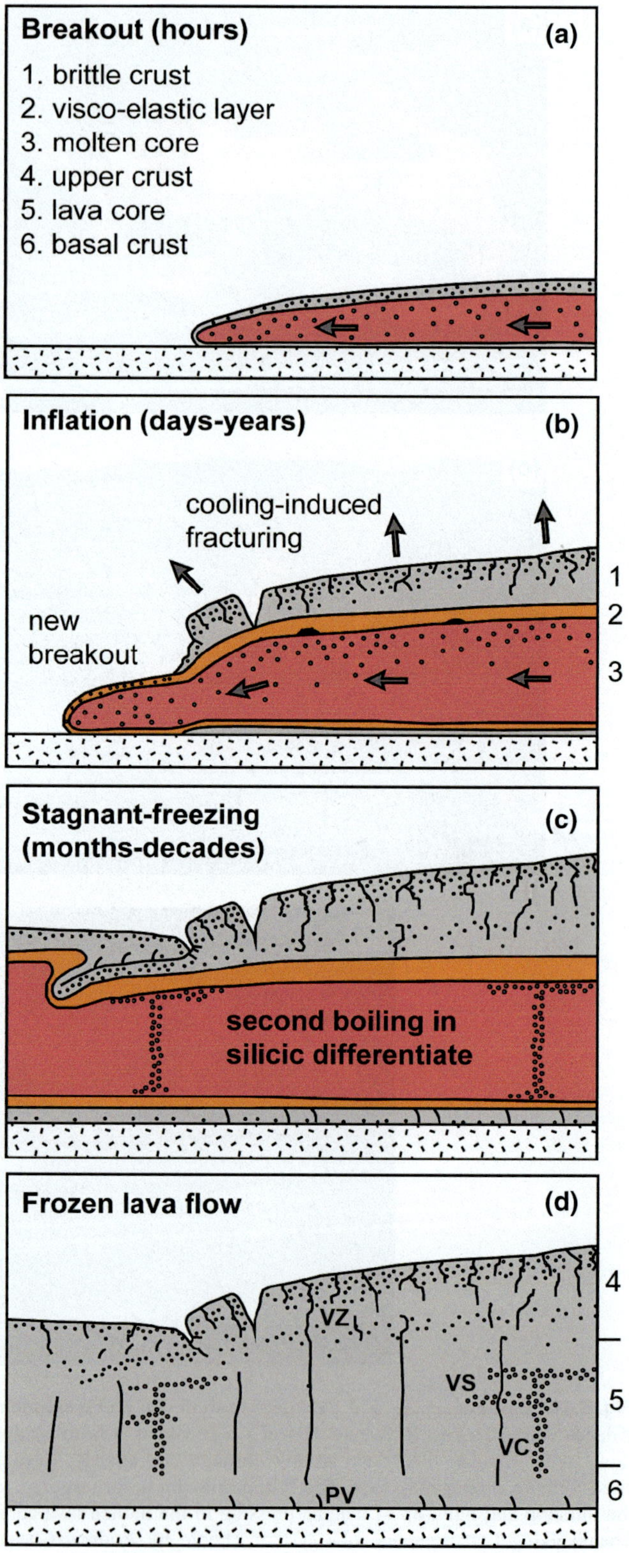

Fig. 5.11 Schematic evolution of the inflation process of pahoehoe lava (After Self et al. 1996). **a** Initial propagation of a frontal toe, just a few dm thick. **b** Development of a thick crust and thickening of the flow by inflation (or endogenous growth), which raises the crust as magma continues be fed into the underlying tube. During active inflation, the rise of bubbles from liquid magma form semi-continuous vesicular zones along the rheological boundary (visco-elastic layer) at the base of the crust. **c** Maximum extent of the tube system, and stagnation of the lava core that crystallises and degasses (second boiling). This allows the diapiric rise of bubbles through the groundmass crystal mush, forming cylindrical vertical segregation pipes and horizontal sheets, and is accompanied by the precipitation of interstitial silicic glass or mesostasis. **d** Cooled lava which shows a distinctive three-part cross-section of lower or basal crust, core, and upper crust. VZ = horizontal vesicular zones; VC = vertical segregation cylinders or pipes; VS = horizontal vesicle sheets; PV = pipe vesicles. For Hawai'ian pahoehoe lavas, the vertical scale varies from 1 to 5 m, while for flood basalt flows such as forming the Columbia River Plateau 5–50 m

The stages associated with the inflation of pahoehoe lavas has been well documented (Fig. 5.11; Hon et al. 1994; Thordarson and Self 1998; Self et al. 1996, 1998). Pahoehoe lava toes are usually only some decimetres in thickness when they form (Fig. 5.11a). However, they quickly develop a visco-elastic crust, while the interior is still molten, allowing lava to be transferred from the rear to the propagating flow front (Fig. 5.11b). At this stage, inflation and uplift of the upper crust can occur as a result of the resistance of the slow moving and cooling flow front. Thickening of the inflating lava can reach several to tens of m (Fig. 5.11c, d). The inflation process leads to the three-part subdivision of the pahoehoe flow lobes into the lower or basal crust, the liquid core, and surface crust. Observations on Hawai'i suggest 20–30-cm-thick pahoehoe toes can repeatedly inflate and coalescence into 3–5-m-thick pahoehoe sheets with lateral dimensions of hundreds of m over a period of days to weeks.

If the frontal lobes coalesce they may form a rather large tube system wherein inflation can occur over a significant width, up to several m or tens of m. These are called *lava rises* (Walker 2009), resulting from the smooth pahoehoe crust being wholesale uplifted or "jacked up" as lava is injected underneath. The rises are bounded laterally by fractures and up-tilted slabs of lava crust (Fig. 5.12a). Other surface evidence for inflation are areas of localised doming of the surface crust caused by magma pressure in the tube which exploits weaknesses in the crust forcing it upwards forming usually radially fractured blister-like surface bulges or longitudinal, up-tilted slabs of crust on the surface of the lava, known as *tumuli* (Fig. 5.12b, c). The radial fractures

Fig. 5.12 Lava rises and tumuli. **a** Lava rise of the top of an inflated pahoehoe lobe. The rise is bounded by up-arched slabs of lava broken along extensional fractures that allowed the inflation to occur. Mývatn, northern Iceland. **b** Lava tumulus in a tumuli field, Harman Valley lava flow, Newer Volcanics Province, Wallacedale, Victoria, southeastern Australia. Note the radial orientation of columnar joints, and collapse of part of the crest of the main tumulus. (From Cas et al. 2017). **c** Single tumulus with upheaved layered lava crust that has been subject to extensional fracturing during upheaval. Approximate diameter ~10 m. Hafnir, Reykjanes Peninsula, Iceland. (Photo by Hansueli Krapf; *Source* Wikimedia through Creative Commons Attribution 3.0 Unported licence)

form as the solidified crust is subject to axisymmetric extension as the tumulus crust is pushed up from beneath by the magma, which may sometimes extrude lobes along the fractures.

It is now thought that the large thickness of individual lava units typical of huge flood basalt systems is the result of lava inflation (Self et al. 1997, 2015; Thordarson and Self 1998; Vye-Brown et al. 2013; Sect. 5.6). For this reason, it is not possible to use equations such as Eqs. 5.1–5.4 to retrieve lava paleoflow velocities, as the final preserved thickness of a lava unit may bear no relationship with that of the moving flow front. Where flow inflation is likely to have occurred, Jeffreys' equation can only be used by applying an estimated thickness of the advancing flow front at the time of flow, based largely on observation of modern flowing lavas (e.g. flow front thicknesses of modern active flowing pahoehoe and a'a lava flows).

5.3 Subaerial Basaltic Lavas

5.3.1 Basalt Lava Forming Eruption Styles and Vents

Basaltic lavas are erupted from either *fissure vents* or *central vents* (also called point source) (Fig. 5.13a, b). Fissure vents are laterally continuous fractures in the crust through which magma rises (Fig. 5.13c) forming dykes to the Earth's surface and can be called line sources. Recent examples include the 2014 eruption from Bárdarbunga volcano in Iceland, and the system of *en echelon* fissure vents that opened up spectacularly in May 2018 in the Leilani Estate along the lower East Rift Zone, Kilauea volcano, Hawai'i. Basaltic lavas erupt in two main styles:

(a) *Magma or "fire" fountaining* producing aggregates of spatter around the vent that can then agglutinate or fuse together, reconstituting into lava that then flows away from the vent area (Fig. 5.13a, b).
(b) *Passive effusion* of coherent lava from small vents or "mouths", termed *boccas* in Italian (Fig. 5.13d, e), or from the overspill or breaching of a lava lake ponded in a crater (Fig. 5.13f), or from secondary breakout boccas from the surfaces or margins of lavas away from vent.

Magma or "fire" fountains associated with typical Hawai'ian-style eruptions range from tens to hundreds of m high (Fig. 5.13a–c) and produce gas and ash plumes that may rise several to >10 km into the atmosphere (Mangan et al. 2014). The Bárdarbunga or Holuhraun magma fountain eruption in Iceland in 2014 produced a fountain up to 100 m high (Fig. 5.13) and a gas and ash plume up to 3 km high (Schmidt et al. 2015; Witt et al. 2018). In contrast, the Miocene Roza member fissure eruption in the Columbia River Plateau flood basalt province, western United States (Chap. 16, see Fig. 16.37) is estimated to have produced a fountain >1,500 m high, giving rise to a gas and ash plume ~13–15 km high (Self et al. 2014).

In a detailed study of the geometry of the 1969 fissure vent system of Mauna Ulu along Kilauea's East Rift Zone,

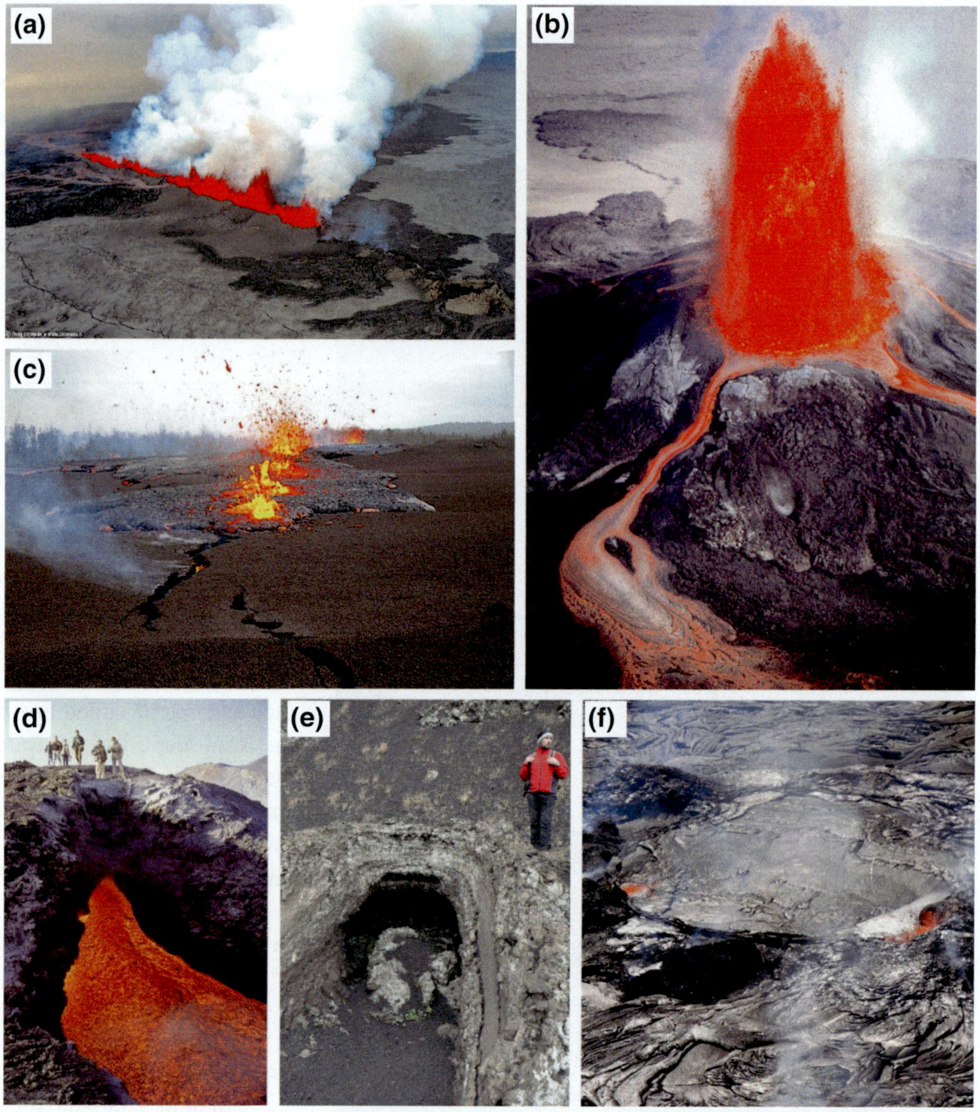

Fig. 5.13 Types of basaltic effusive vent and eruption styles. **a** Fissure fire fountaining eruption at Bárdarbunga volcano in 2014, Eastern Volcanic Zone, southeastern Iceland (Photo by Thordis Högnadóttir, University of Iceland). **b** Central vent, high fountaining eruption feeding lava flows at Pu'u O'o volcano, Hawai'i (*Source* United States Geological Survey; photo by J.D. Griggs, taken 28-7-1984). **c** Propagating fissure vent, marked by cracks in the ground with magma rising in the fracture ahead of the advancing line of fire fountaining vents that are feeding a lava, Pu'u O'o volcano, Hawai'i (*Source* United States Geological Survey; Photo taken 5-3-2011). **d** Passive effusion of lava from a bocca on the slopes of Mt Etna, Sicily, Italy during 1983 eruption (Photo by Pierre). **e** Inactive bocca, 2004 eruption of Mt Etna. **f** Passive effusion of lava from a summit crater lake at Pu'u O'o volcano, Hawai'i (*Source* United States Geological Survey)

Hawai'i, Parcheta et al. (2015) demonstrated that the 800-m-long western part of the fissure system that is still accessible consists of at least five *en echelon* segments that were linked in the subsurface during the eruption. They note that this geometry is common for basaltic fissure vent systems on Hawai'i (e.g. Fig. 5.13c) and elsewhere (e.g. Re et al. 2015), and attribute it to a combination of tensile and shear cracking of the uppermost crust. Each segment is represented by multiple closely spaced vents at the surface. A similar set of *en echelon* fissure vents opened up sequentially towards the southeast in May 2018 in the Leilani Estate, on the lower East Rift Zone, feeding spectacular fire fountains and lava flows (e.g. Fig. 5.13c). Fissure vent systems that feed large flood basalts may be >100 km in length (Fig. 5.14). Central vents are isolated vents, more or less cylindrical in geometry, that are typical of larger basaltic and andesitic volcanoes, scoria cones, and other types of smaller basaltic volcanoes. However, even these smaller vents are commonly aligned along fissure conduits or dykes, and even on the large volcanoes fissures may control flank eruptions. Many eruptions of basaltic lava may begin along a large length of a fissure, but activity quickly localises to a few point sources or "nodes" (see Fig. 5.16b; Wilson and Head 1981; Delaney and Pollard 1982; Höskuldsson et al. 2015; Witt et al. 2018). This seems to be true even for large flood basalt lavas.

Many eruptions of basalt lava flows begin with a phase of magma fire fountaining (usually gas rich), succeeded by the passive effusion of coherent gas-poor magma. There may also be periods when lava is issuing as coherent flows and from fountains at the same time, either from the same vent or separately along a fissure vent system. Flows formed from agglutination of lava spatter are associated with spatter cones and spatter ramparts (Chaps. 9 and 16). Lavas in which the

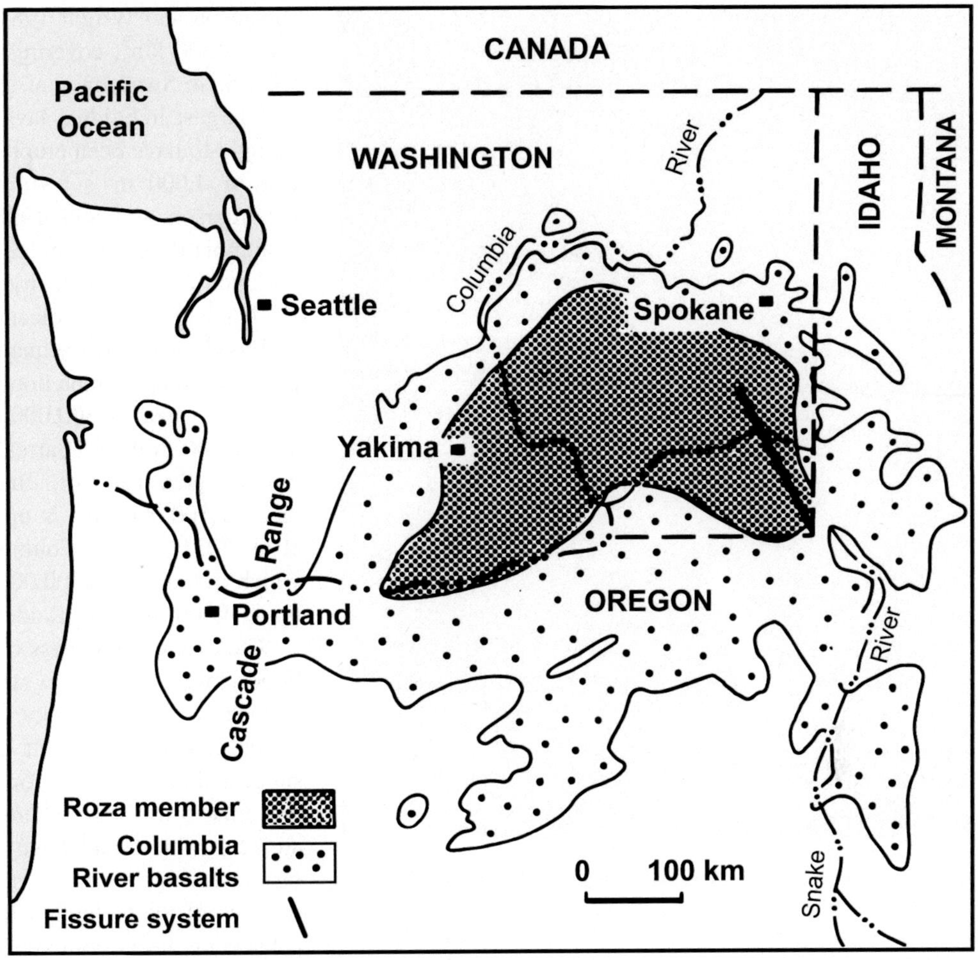

Fig. 5.14 The Columbia River flood basalt province with the distribution of the Roza Member and its fissure vent system (From Cas and Wright 1987, After Swanson et al. 1975)

outlines of spatter fragments are still visible can be called *clastogenic lavas* (Fig. 5.15). Such spatter fragments will be flattened, stretched, and deformed as in some welded pyroclastic deposits, and in many ways they form by an analogous mechanism to welded air-fall lapilli-tuffs through agglutination of spatter falling out of a fire fountain (Chap. 9), but they lack an ash matrix. Detailed topographic analysis at Bárdarbunga fissure shows the spatter cones remained similar in shape but increased in size as the eruption progressed (Witt et al. 2018). In addition, changes in morphology may have affected lava ponding in the crater, which in turn strongly influenced the lava fountain heights.

The loss of volatiles during magma fountaining and early stages of lava flow has an important effect on the rheology of the lava (Cashman and Mangan 2014). Degassing of the lava leads to considerable undercooling, rapid growth of quench crystallites (e.g. Fig. 5.2a), a rapid increase in the viscosity of the lava (Fig. 5.2d), and the development of a high yield strength. Thus, gas-charged magmas giving rise to intense lava fountaining could generate higher viscosity basaltic flows with higher yield strengths due to efficient degassing. Conversely, magmas with lower initial gas contents could form more fluidal lavas from less-vigorous magma fountains or effusive over-spilling of lava lakes because volatiles are retained. The lavas erupted in 1961 at Askja in Iceland changed from higher viscosity and higher yield strength a'a to pahoehoe flows later in the eruption as the intensity of fire fountaining waned (Pinkerton and Sparks 1978).

High-intensity magma or "fire"-fountaining (see Chap. 9) associated with large volume effusive eruptions may also contribute to large volume injection of volcanic aerosol gases into the atmosphere with potentially severe effects on climate (Stothers et al. 1986; Oman et al. 2006; Self et al. 2015).

5.3.2 Dimensions: Flow Distances and Volumes

Individual basalt lava flows can range from hundreds of m^3 to hundreds of km^3 or more in volume. The largest known lava flows are *flood basalt lavas*. These can occur in both continental settings and in submarine environments from oceanic plateaus and mid-ocean ridges (e.g. Iceland). Rarer, large volume, Archean komatiite lava flows (Chap. 7) and

Fig. 5.15 Ghost outline of fused vesiculated spatter clasts in core of a clastogenic basalt lava, ~5 ka Mt Schank volcano, Newer Volcanics Province, South Australia. The vesiculated domains represent the cores of spatter clasts which have become flattened and stretched during flow

extensive phonolite and trachyte flows that have been described from the East African Rift can also be considered as flood lavas. The extensive phonolites and trachytes in East Africa (King 1970) are, however, very different compositionally (alkaline and peralkaline affinities) suggesting that these lavas may have been erupted with viscosities close to those of basalts.

Flood lavas are erupted from large fissure vents (Fig. 5.14). The largest historic basalt lava-forming eruptions have occurred in southeastern Iceland at Eldgjá in 939–40 AD and Laki in 1783–84 AD (Fig. 5.16; Thorarinsson 1969; Thordarson and Self 1993; Oppenheimer et al. 2018). The Eldgjá lava had a volume of ~20 km^3 and was 70 km long, and the Laki lava was 14.7 km^3 and 65 km long. However, much larger flood basalts over 600 km long are known in the geological record, and perhaps ~1,000 km long (Self et al. 2008). In the mid-Miocene Columbia River Plateau of the western USA, flows of the Roza MemberC travelled <600 km and have volumes of >700 km^3 (Swanson and Wright 1981). The Roza Member has a volume >1300 km^3, covering a minimum area of 40,000 km^2 (Fig. 5.14; Swanson et al. 1975; Thordarson and Self 1998). The largest individual lava flows of the Roza Member are thought to have been erupted over ~10 years at an eruption rate of 4,000 $m^3\ s^{-1}$ from the 150-km-long fissure vent (Thordarson and Self 1998). This and other lavas in the Columbia River Plateau (e.g. Swanson and Wright 1981) are volumetrically three to five times larger than many of the world's largest stratovolcanoes (Chap. 16).

Flood basalt lava sequences can be extremely thick. For example, in the Paraná flood basalt province in Brazil, which has an area >1,300,000 km^2, the sequence is up to ~1,700 m thick (Barreto et al. 2014). The Deccan flood basalt province of India, which originally covered >1,500,000 km^2 is up to 3,000 m thick or more (Self et al. 2021). In the Columbia River flood basalt province, which has an area >200,000 km^2, the lava sequence may be up to 4,000 m thick (Reidel et al. 2013).

The average thickness of single lava flows in continental flood basalt provinces, which are overwhelmingly sheet pahoehoe or hummocky or compound pahoehoe lavas, varies considerably (Self et al. 2021). Self et al. (2021) suggest that compound or hummocky pahoehoe lavas are less than 5 m thick, whereas simple sheet pahoehoe lavas are >5 m thick, and have a median thickness of 8–12 m, rarely reaching up to 80 m (Self et al. 2021). In measured sections of the Columbia River Plateau flood basalts, sheet pahaoehoe lavas commonly range between 15 and 35 m thick (Waters 1961; Reidel et al. 2013; Vye-Brown et al. 2013; Self et al. 2021). Comparable lava flow thicknesses are found in the flood basalts of the Deccan Traps in western and central India (Choubey 1973; Subbarao and Sukheswala 1981; Bondre et al. 2004; Self et al. 2021), with modes of ~20 m being common. In the flood basalts of eastern Iceland, Walker (1963) found that the average flow thickness was 17 m (e.g. Fig. 5.17). All these thick basaltic lavas are thought to have attained their thickness by inflation (e.g. Thordarson and Self 1998; Self et al. 2021).

Other basaltic lavas can be erupted from central volcanoes, point-source vents, or small restricted fissures. These lava flows are much smaller in volume and cover much smaller areas. For example, since 1535, Mt Etna in Italy has erupted <4 km^3 of lava; the largest single flow was 0.6 km^3 (Branca et al. 2015). On Mauna Loa, historic flank eruptions of basalt along the southwest and northeast rifts range up to 0.5 km^3 in volume (Lipman 1980; Rowland and Walker 1990). According to Self et al. (2021), plains-type basalt lava flows are considerably thinner than continental flood basalt lavas.

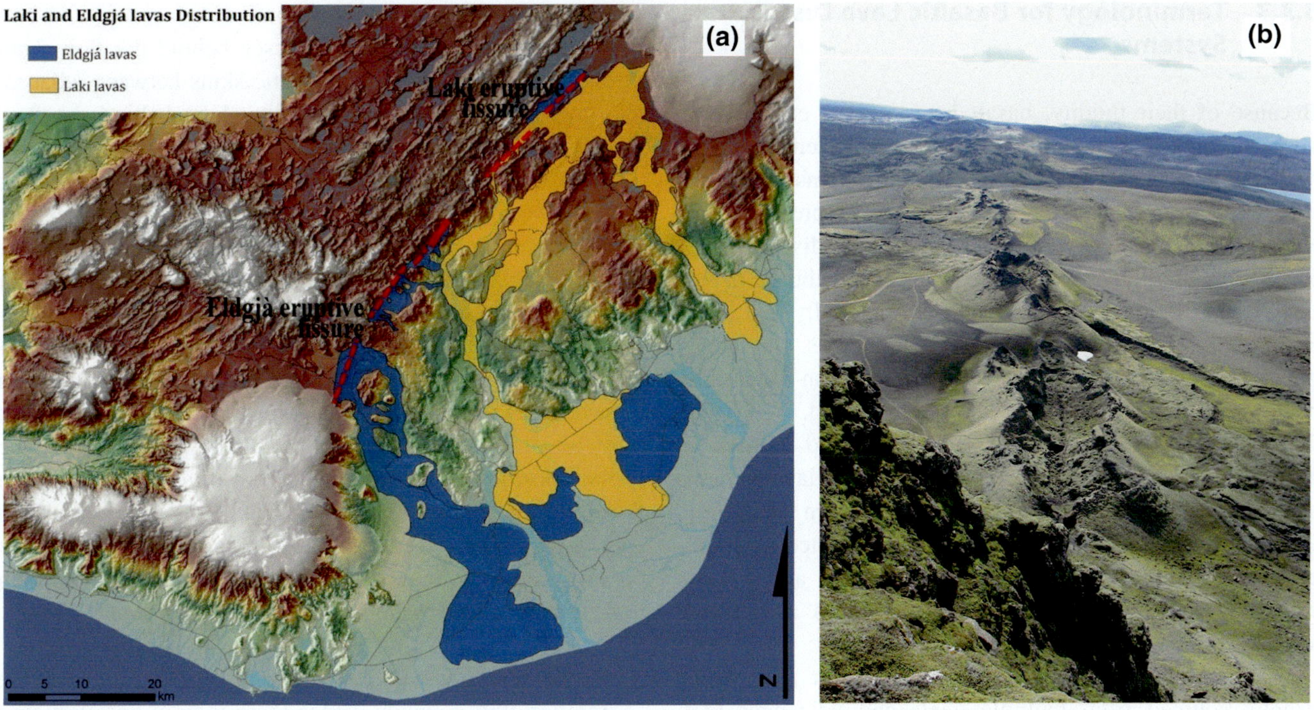

Fig. 5.16 Fissure vent systems in Iceland. **a** Map of the fissure systems (red dashed line) and extent of lavas from the largest historical basaltic eruptions in Iceland, from Eldgjà (930–38 AD) and Laki (1783–84 AD). Eastern Volcanic Zone, southeastern Iceland. (Map courtesy of A. Pinton). **b** View towards the SW of the Laki fissure system showing the alignment of spatter cones defining part of the fissure system

Fig. 5.17 Spectacular view of the Dettifoss Canyon, Northern Volcanic Zone, northeastern Iceland, showing several superimposed large volume Quaternary basaltic lava flows, each up to several tens of m thick

5.3.3 Terminology for Basaltic Lava Distributary Systems

Because of their fluidity, basalt lava systems can develop very complex architecture and architectural elements at all scales. At the largest scale, basalt lava systems form *lava provinces* (e.g. Figs. 5.1a and 5.14). Lava provinces are long-lived systems, developed through many discrete eruptions from multiple vents and volcanoes distributed over a large area regionally, leading to a regionally extensive sequence of lava flows, of variable compositions, and locally preserving pyroclastic deposits. The life span is usually millions of years. Flood basalt provinces (e.g. Self et al. 2015; Chap. 16), the basaltic-lava-dominated system of Hawai'i (e.g. Poland et al. 2014), and plains lava systems such as the Snake River Plain in the western USA (e.g. Greeley 1982) or the Newer Volcanics Province of south-eastern Australia (Cas et al. 2017; Chaps. 16 and 17) are examples of basalt-dominated lava provinces at different scales.

Within a province there are multiple *lava flow fields*, usually sourced from multiple vents and volcanoes (e.g. Fig. 5.18a). Lava flow fields are constituted of multiple individual *lava flows* (e.g. Fig. 5.18a, b), which could originate from one eruption, or from multiple eruptions or eruption phases, as in the case of the 35-year-long Pu'u O'o eruption of Kilauea volcano on Hawai'i which began in 1983 and ended in 2018. Single lava flow forming eruptions could last from days to years to tens of years, such as flood basalt forming eruptions. For example, Vye-Brown et al. (2013) estimate minimum emplacement timescales of 8–19 years for three lava flow fields from the Columbia River Plateau based on the thickness of the lava crusts, using an empirical relationship proposed by Hon et al. (1994) for pahoehoe sheet lava lobes in Hawai'i.

At the next scale down, each lava flow could consist of multiple *lava lobes* (e.g. Fig. 5.18b, c; Fig. 5.6c) each representing a lava breakout event and/or a change in effusion rate and conditions during an ongoing eruption. These breakout lava lobes can cause a sudden increase and diversion of the propagation of the lava. Individual lava lobes can extend for hundreds of m to km from the breakout point. At the smallest scale, the front of pahoehoe lavas propagate by the breakout of small *pahoehoe flow toes* which inflate and may bud to form new breakout flow toes (Plate 5.1a; Figs. 5.5c, 5.10a–d, 5.18d). Pahoehoe flow toes are up to tens of cm in diameter and of similar length to perhaps several m. They immediately form a visco-elastic skin that wrinkles as the toe propagates, forming the classical pahoehoe ropy texture. The duration of pahoehoe flow toe formation is seconds to minutes. These are the architectural elements that are so characteristic of pahoehoe lava flows. They may be preserved at the front, margins, top, and bottom of flows but they often coalesce behind the flow front through assimilation of visco-elastic skins between adjacent toes to form a more sheet-like flow lobe behind the flow front and with a continuous crust.

Walker (1971) suggested that the basic unit for lavas is the *cooling unit*, so that in the field a cooling unit can be discriminated from the unit below and the one on top by evidence for differential cooling, such as a crust, a breccia, or columnar joints of varying dimensions (see following sections). However, as stated, individual effusive events, especially long-lasting events, may produce the superimposition of several flows and lava lobes over time. For this reason, it may be difficult in the field to distinguish between the superimposition of cooling units from different eruptions to flow/cooling units associated with the one eruption (Fig. 5.19). Where it is possible to map out and reconstruct cooling unit(s) over the entire areal extent of the eruption unit, it should be possible to distinguish between:

- simple lava flows, that is, where the cooling unit and the eruption unit coincide, hence there is only one flow unit for that eruption, and
- lava flow fields that enclose all products of a single eruption and consist of several different partly overlapping flows.

However, simple lava flows are rare and only result from very small volume eruptions, where the maximum distance is reached more or less at the time that the eruption ends. Compound, complex, and lava flow fields are much more common. However, especially for high eruption rate events, such as the Laki 1783–84 AD eruption, or with flood basalts, inflation and breakout flow lobes result in individual and very thick lava (cooling) units that can be traced laterally for tens of kms (Figs. 5.16a and 5.17; Vye-Brown et al. 2013).

5.3.4 Subaerial Basalt Lava Types

There is a range of basalt lava flow types. The history of nomenclature and typical characteristics has been recently reviewed by Harris et al. (2017) and Pedersen et al. (2017). *Pahoehoe* (Plate 5.1a) and *a'a* (Plate 5.1b) lavas are now best viewed as end-member types with all transitions between them, which will be briefly discussed in Sect. 5.4.7. Nonetheless, as discussed in Sect. 5.2.3.3, whether pahoehoe or a'a lava forms depends on the relationship between magma viscosity, controlled by eruption temperature and crystallinity, and shear rate (Fig. 5.4), which is controlled by flow velocity and terrain slope. Although some lavas consist entirely of pahoehoe or a'a texture, others develop

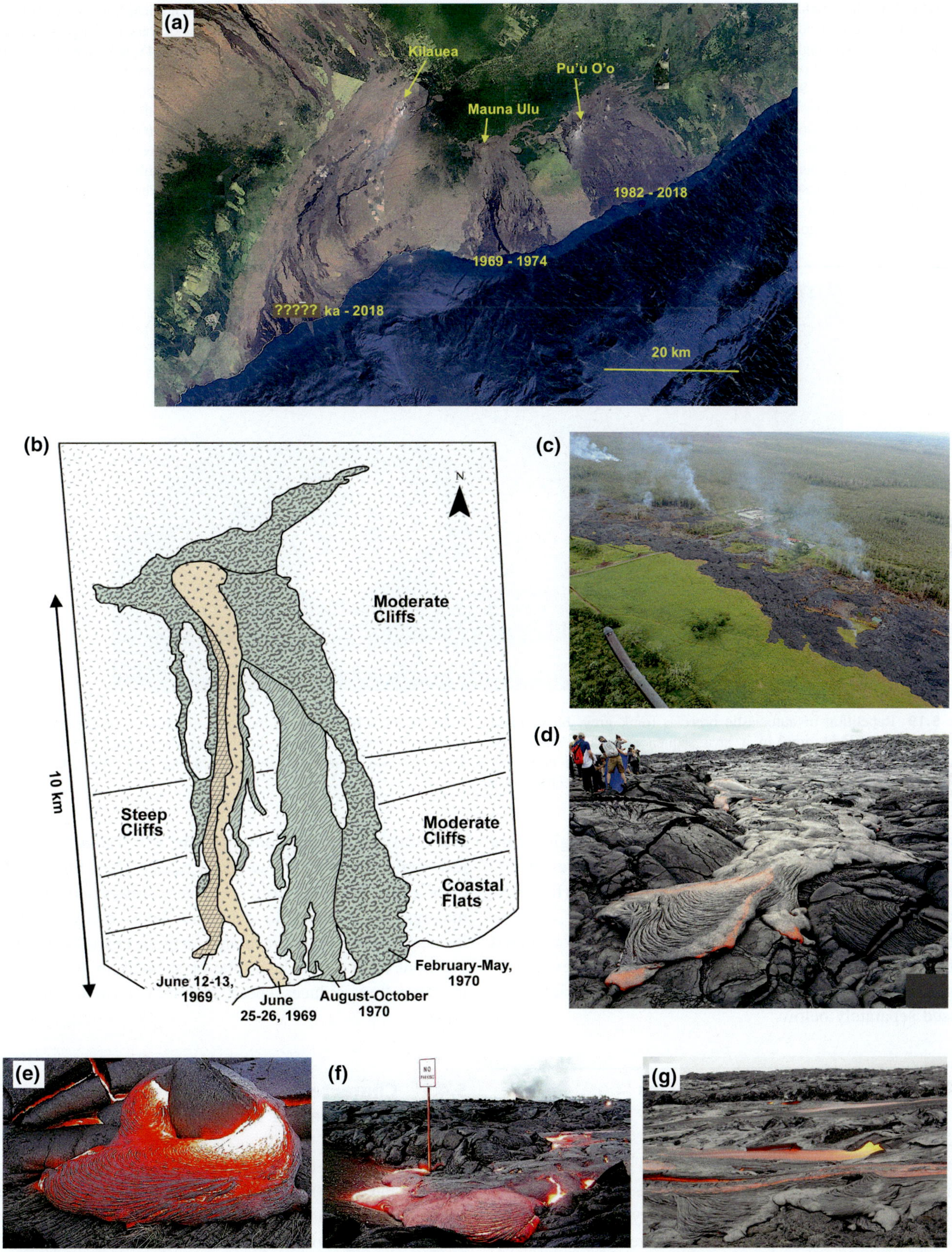
(a)
Kilauea
Pu'u O'o
Mauna Ulu
1982 - 2018
1969 - 1974
????? ka - 2018
20 km
(b)
N
Moderate Cliffs
10 km
Steep Cliffs
Moderate Cliffs
Coastal Flats
June 12-13, 1969
June 25-26, 1969
August-October 1970
February-May, 1970
(c)
(d)
(e)
(f)
(g)

◀ **Fig. 5.18** Lava flow fields, Hawai'i. **a** Google Earth image of the three lava flow fields produced in historical times along Kilauea volcano's East Rift Zone. Although age ranges of activity are known for the Mauna Ulu and Pu'u O'o flow fields, the oldest activity in the Kilauea flow fields is uncertain. Some individual lava flows can be deciphered in each lava flow field by differences in colour. **b** Lava flow field of the 1969–1970 eruption period of the monogenetic Mauna Ulu lava shield, showing several discrete lava flows, including both pahoehoe (green) and a'a (tan) erupted during the eruption (Modified from Keszthelyi et al. 1998, After Swanson et al. 1979). **c–g** Lava flow fields of the 1983–2018 eruption of Pu'u O'o volcano, the longest and most voluminous known period of activity on the East Rift Zone in more than 500 years (*Source* United States Geological Survey). **c** Part of the Pahoa lava flow showing multiple lobes of varying lengths branching from the main flow (Photo taken 15-11-2014). **d** Pahoehoe lava flow lobe showing multiple pahoehoe toe lobes breaking out from the front (Photo taken 12-8-2016). **e** Pahoehoe lava flow toe breakout (Photo taken 23-6-2002). **f** Lobate pahoehoe lava with Pu'u O'o crater visible in the background, Mother's Day flow, Chain of Craters Road (Photo taken 1-4-2003). **g** Multiple pahoehoe flow lobes ~250 m below breakout point on the Quarry flow lava tube above Pulama pali, Kupaianaha fissure vent eruption (Photo by T.R. Orr, taken 22-12-2010)

Fig. 5.19 Repetition of autoclastic breccias (dark grey, brown) and coherent zones (light grey) in a lava sequence without significant evidence of time breaks (e.g. paleosoils and/or reworked deposits), which makes it difficult in the field to distinguish the emplacement of a series of lobes belonging to the same eruption, from the emplacement of different lava units from closely spaced eruptions. Stromboli, Italy. There are also some obliquely cross cutting light grey dykes

transitions from one to the other, such that a single entire lava can consist of both pahoehoe and a'a textures at different places along the flow path. As such the surface textural features of these two end-member types, and the transitional forms to be discussed below, can be considered to be facies of basalt lavas. Flood basalt lavas are extremely large volume pahoehoe lava flows that will be briefly discussed separately below.

5.4 Pahoehoe Lava Flow Facies Characteristics

Pahoehoe lavas are mostly coherent and multi-lobate, and therefore compound, especially when effusion rates are low. At higher effusion rates, they are internally massive, simple, and sheet like. Both effusion rate types are characterised by coherent ropy-textured surfaces. They result from the combination of high eruptive temperatures (i.e. low viscosities), low effusion rate eruptions (<10 m^3 s^{-1}; Table 5.1), and/or low angle slopes (i.e. low strain rates). They form in their own right at vent from relatively low-viscosity basalt magma. Only rarely do pahoehoe lavas evolve from a'a lavas, and only where insulated magma emerges through a breakout from an a'a lava tube or at break in slopes, where the lava flow reduces speed and may form a continuous crust.

5.4.1 Characteristic Surface Features of Pahoehoe Lavas

The characteristic facies features of pahoehoe lavas, particularly at the flow fronts, are cylindrical to pillow-like lobate pahoehoe toes tens of cm in diameter and up to a few m long, with ropy flow folds on the surface crust (Plate 5.1a, Figs. 5.18d–g and 5.20a). The flow toes (Fig. 5.18d–g) represent the toe-by-toe manner by which slow moving, low

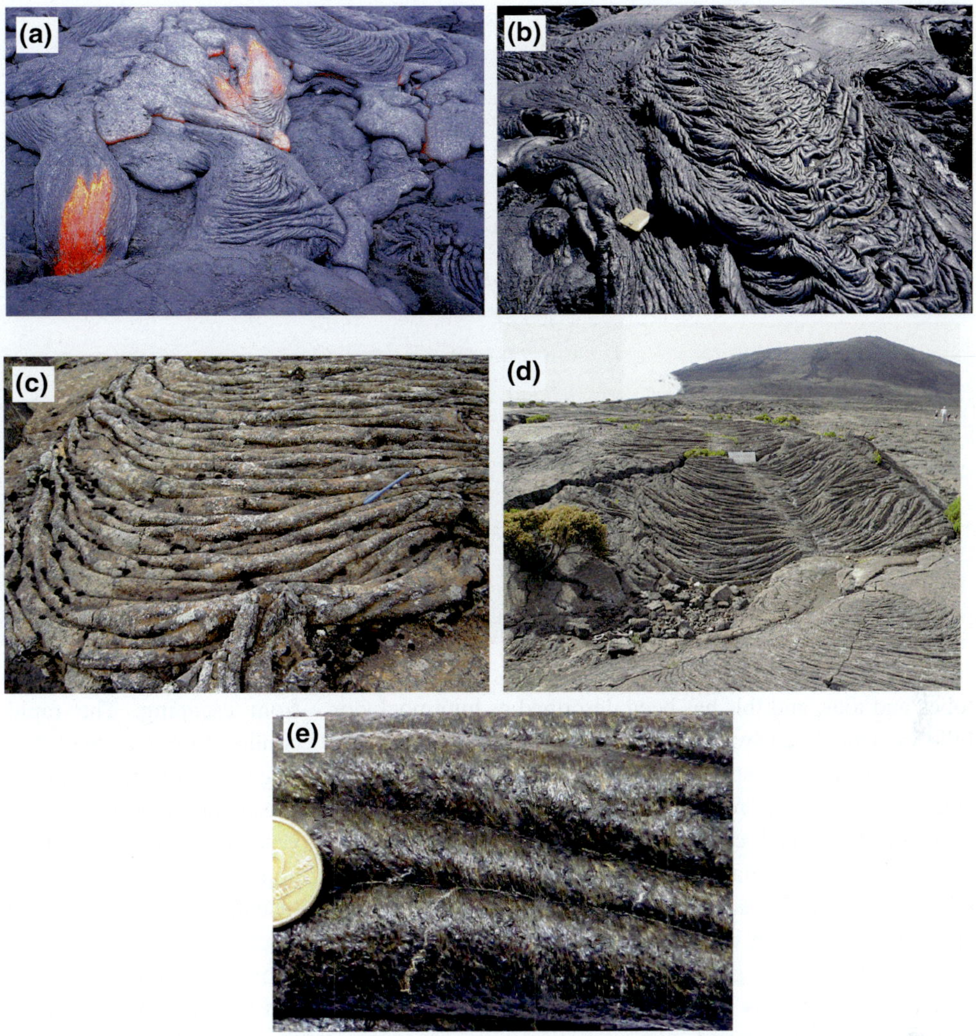

Fig. 5.20 Pahoehoe ropy flow textures. **a** Pahoehoe lava lobate flow toes, some with ropy surfaces and some with stretching lineation of the surface skin, 1989 lava flow, Pu'u O'o volcano, Hawai'i. Examples of ropy surfaces of pahoehoe lavas at: **b** Mauna Ulu, 1969–1970 lava, Hawaii; **c** Iceland; and **d** Reunion Island. **e** Close-up view of glassy ropy surface texture and stretching lineation of 1989 pahoehoe lava, Pu'u O'o volcano (also see Fig. 5.18d–g)

magma flux pahoehoe lavas propagate downslope, by breakouts of small flow toes from the surface visco-elastic crust. As each toe breaks out from the lava margins, it immediately forms a ductile, but coherent crust, which gains strength as it cools. The magma that is being fed into the toe from the lava feeder tube behind the flow front exerts pressure on the visco-elastic skin or crust of the toe lobe, eventually causing a rupture of the skin, and another lava toe to breakout, so extending the flow margins step by step.

The surface of the toes and the lava in general is typically *ropy* flow folded as the propagating toe slows down, and the visco-elastic crust wrinkles up (Plate 5.1a, Figs. 5.18d–g and 5.20a–e). At high strain rates induced by higher magma flux or steep slopes, the forming crust can be deformed in the ductile regime as it begins to cool and become more viscous, but before becoming fully brittle (Rowland and Walker 1987). These conditions form the typical ropy type of pahoehoe (Fig. 5.20). The ropes consist of a regular train of corrugations a few cm in height and wavelength, their long axes being perpendicular to or convex to the local flow direction (Plate 5.1a, Figs. 5.18d–g and 5.20a–e). Fink and Fletcher (1978) conducted a structural analysis of ropy surfaces and interpreted them as folds which develop at the surface of a fluid whose viscosity decreases with depth and which is subjected to compression as the front of the toe lobe slows down or comes to a stop. The braided appearance and more complex structures found in many pahoehoe flows can be explained by the superimposition of two or more episodes of folding. The surface of pahoehoe lobes may also preserve a *stretching lineation,* resulting from stretching of gas bubbles that are at the surface as the pahoehoe toe is propagating (Fig. 5.20a, e).

The crust can be fully smooth where the lava has propagated and inflated on flat topography under almost stationary conditions. At such localities, *lava rises* are common, and are the result of flow inflation. They represent very low relief bulges of the surface crust, usually m across, which fracture under tension during inflation, and are bounded by polygonal fractures (Figs. 5.12a and 5.21). More irregular surface morphology results from the superimposition of smooth

Fig. 5.21 Pahoehoe lava rise cut by polygonal fractures. Mt Etna, Sicily, Italy

lobes and toes, and this has been described as hummocky or billowy pahoehoe (Swanson 1973).

The surface crust of all pahoehoe lavas is glassy (Fig. 5.22a) as a result of the very high rates of radiation-dominated cooling. This glassy surface is very fragile, spalling off glassy splinters that usually litter the top of active lava flows (Harris and Rowland 2015). This spontaneous spalling occurs in response to the build-up of stress in the surface glassy crust during the rapid contractional cooling, and is an example of rare, subaerial air driven quench fragmentation.

These pahoehoe surface features generally have a low preservation potential in the geological record, but may be preserved, especially if quickly covered over by another lava flow or flow lobe. If found, convex trains of pahoehoe ropes can be used as paleoflow direction indicators, although caution is required in determining flow direction based on only one or two occurrences, since some pahoehoe ropes may be a response to local disturbances of the flow surface (Macdonald 1972).

5.4.2 Vesicle Abundance, Types, Zonation, Crust Textural Layers, and Pahoehoe Lava Sub-Types

When lava is flowing in an open channel, exsolving gases can rise buoyantly through the lava and escape into the atmosphere. However, the rapid formation of the crust (see Sect. 5.2.3.2) has an important consequence on the style of degassing of pahoehoe lavas, as it acts as a permeability barrier and a physical trap that stops buoyant gas bubbles from escaping. The rapid downward propagation of the cooling front (see Sect. 5.2.3.2 and Eq. 5.5) during the early stages of flow traps and freezes forming gas bubbles in the glassy surface crust, so that pahoehoe crusts may be highly vesicular (Figs. 5.11a–d and Fig. 5.22a, b). Bubbles in the upper part of the chilled crust are as a consequence small (usually <0.5 cm), rounded, and evenly distributed, and may reach up to >40% by volume. Such vesicular facies is known as *spongy* or *S-type pahoehoe* (Fig. 5.21a; Wilmoth and Walker 1993; Harris et al. 2017). This characteristic also occurs at the base, where the lower crust is also usually vesicular. Some S-type pahoehoe lava is so gas rich that the interior of pahoehoe toes contains large caverns, giving the lava a brittle, treacherous character because the crust is prone to collapse when walking on it (Fig. 5.22c, d). This type of pahoehoe lava is called *shelly pahoehoe lava*.

In addition to the ropy flow folds, S-type pahoehoe lava surfaces often display a stretching lineation that is parallel to the direction of propagation of the toe lobes (Fig. 5.20a, e). It results from the stretching of gas bubbles on the visco-elastic crust or skin as the toe lobe propagates. On ropy-textured surfaces the lineation is usually perpendicular to the axes of the ropes in the direction of flow.

Other pahoehoe lavas are less vesicular, reflected both in the crust, which is largely non-vesicular, and the interior, which may have few, large coalesced vesicles and pipe vesicles, especially near the base of the lava or lava toes (Fig. 5.22e). These have been called *P-type pahoehoe* (Walker 1987; Wilmoth and Walker 1993).

In sheet pahoehoe flows (including flood basalt lobes), bubbles rise and accumulate at the rheological boundary below the crust (see Fig. 5.11) and if large enough they can

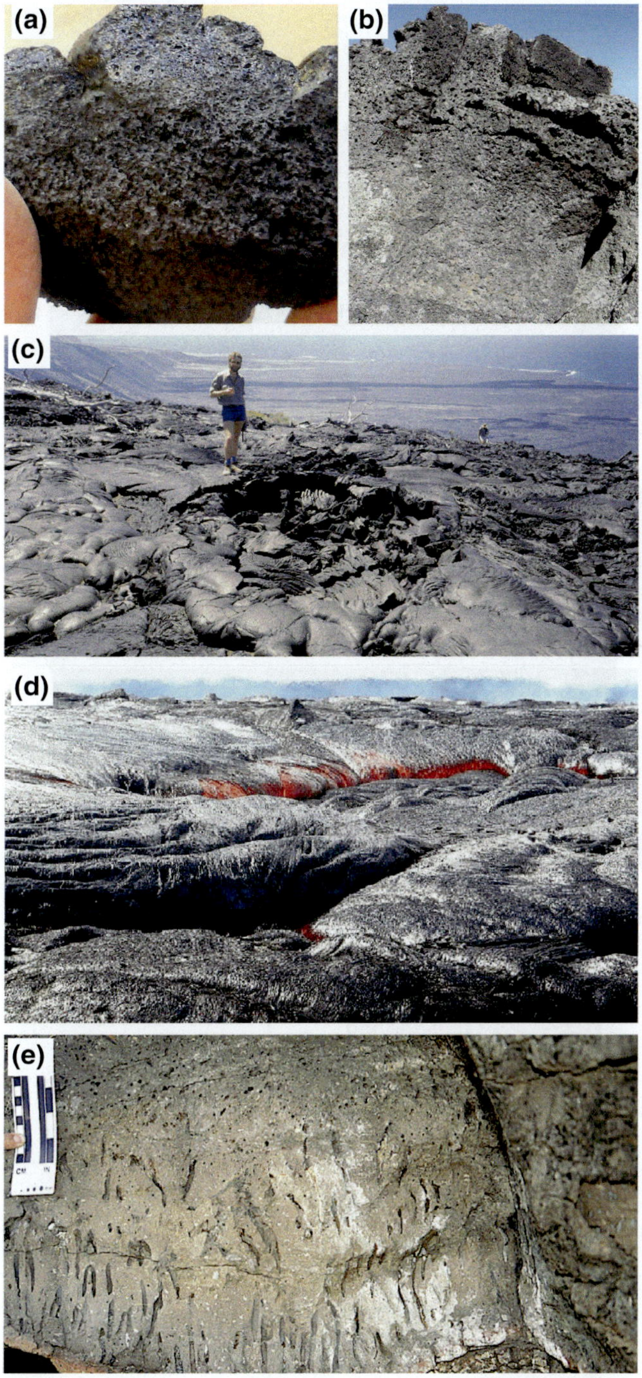

Fig. 5.22 Vesicle and gas cavity features in pahoehoe lavas. **a** Highly vesicular crust of S-type pahoehoe lava, Pu'u O'o volcano, Hawai'i. **b** Highly vesicular crust of a pahoehoe lava, Harman Valley flow, Newer Volcanics Province, Wallacedale, Victoria, southeastern Australia. **c** Cavernous, shelly pahoehoe lava, Mauna Ulu, Hawai'i. **d** Active shelly pahoehoe lava flow, Pu'u O'o volcano. The billow-like features are 0.3–1 m high and hollow. (*Source* United States Geological Survey; Photo taken 1-3-2002). **e** Pipe vesicles in P-type pahoehoe lava toe, Deccan Traps flood basalt province, Ganure, western India. Height of the image ~40 cm. (From Bondre et al. 2004)

flatten and deform (i.e. at high capillary number; see Chap. 4) under their own buoyancy or as they are sheared by the flowing lava (Fig. 5.23a). The inward progression of the cooling front keeps trapping the bubbles, and as there are injections of new hot magma under the crust, each such injection undergoes gas exsolution, adding a new layer of vesicular magma to the interior of the lava causing the crust to thicken by incremental addition at its base. As a result, a series of alternating, somewhat gradational, vesicle-rich (>10 vol.%) and vesicle-poor layers form in the upper lava crust, which may be up to 1–2 m thick for typical simple pahoehoe sheet lava, although commonly as thick as 10–20 m in some flood basalt flows (Fig. 5.23b–e; Self et al. 1998, 2021; Thordarson and Self 1998; Vye-Brown et al. 2013). Such vesicle-rich layers are termed *horizontal vesicular zones*. Large (>1 cm), deformed vesicles can also be observed resulting from coalescence of smaller vesicles. Larger bell- or dome-shaped voids with flat floors or *megavesicles* several to tens of cm wide may also occur, which are valuable younging direction indicators in ancient deformed mafic successions. A similar layered vesicular division is more rarely seen at the base above the vesicular basal crust, as the cooling front advances at a reduced rate at the base (usually assumed at 70% of the upper front), allowing bubbles to grow and rise or escape from the lower cooling front.

Degassing of the thermally insulated core of the lava occurs at much slower cooling rates. During the cooling and crystallisation of the interiors of thick basalt lavas, crystallisation-induced boiling (second boiling) can occur, producing vesicle populations that segregate and collect in the groundmass magma before it solidifies and rises diapirically through the groundmass magma crystal mush (see Fig. 5.11c; Self et al. 1996; Thordarson and Self 1998). These form broadly cylindrical, *vertical segregation vesicle pipes* (Figs. 5.23c–e and 5.24), which also get trapped and may spread horizontally within the lava or under the crust, when they are called *horizontal vesicle sheets* (Fig. 5.24a), and are composed of vesicular segregated material. Such pipes can be simple (Figs. 5.23c and 5.24a, c), or complex and anastomosing (Fig. 5.24b, d), but are usually defined by sharp contacts with surrounding crystallised lava.

Thus, two distinctive types of vesicle populations are found in sheet pahoehoe lava and flood basalt flows: (1) gas bubbles interpreted as having formed during eruption and flow of molten magma, which are sometimes referred to as "primary" and (2) "secondary" bubbles formed during crystallisation of the inner lava core. The former are associated with active lava flow inflation, while the latter are thought to form after the flow has stagnated and are post-emplacement, although, some workers suggest that

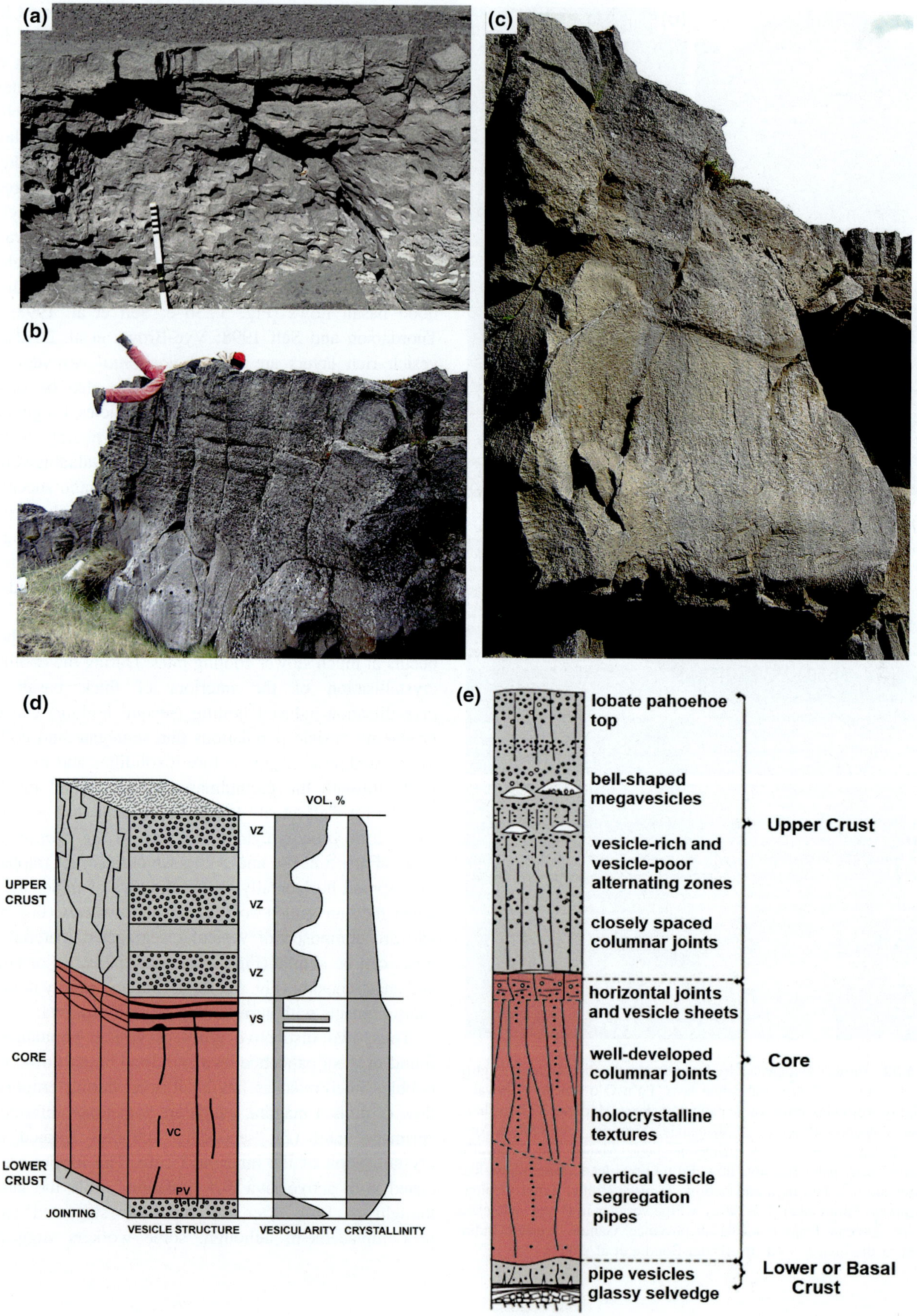
(a)
(b)
(c)
(d)
VOL. %
VZ
VZ
VZ
VS
UPPER CRUST
CORE
LOWER CRUST
VC
PV
JOINTING
VESICLE STRUCTURE
VESICULARITY
CRYSTALLINITY
(e)
lobate pahoehoe top
bell-shaped megavesicles
vesicle-rich and vesicle-poor alternating zones
closely spaced columnar joints
Upper Crust
horizontal joints and vesicle sheets
well-developed columnar joints
holocrystalline textures
vertical vesicle segregation pipes
Core
pipe vesicles
glassy selvedge
Lower or Basal Crust

◀ **Fig. 5.23** Vesicle distribution and zonation in inflated pahoehoe lavas. **a** Vesicle-rich zone below the crust of a pahoehoe lava flow. Note the elongation of the large bubbles (which were bell-shaped with a flat floor) deformed by the flow underneath the crust. Mt Etna, Sicily, Italy. **b** Section through the interior of a pahoehoe lobe showing a layered vesicle-rich division just below the topmost 10–30 cm of crust transitioning downward into the coherent core. Iceland. **c** Profile through the interior of a massive, inflated pahoehoe lava lobe from base to top, showing a layered vesicle-rich zone that is ~1 m thick below the surface crust, and narrow, vertical vesicle segregation pipes in the coherent interior. Note the base is in contact with an underlying columnar jointed crust, which is part of a different cooling unit, and similarly for the top contact Iceland. **d** Conceptual cross-section through an inflated pahoehoe lobe. This is divided into three parts (lower crust, core and upper crust) based on vesicle structure, joint patterns, and crystal texture, which closely relate to the mechanism of inflation (as discussed in text and Fig. 5.11). VZ = horizontal vesicle zones; VC = vertical vesicle segregation pipes; VS = horizontal vesicle sheets; PV = pipe vesicles. (Adapted from Self et al. 1998). **e** Idealised lithological log. This highlights multiple vesicle concentration zones and "bell-shaped" megavesicles which characterise the upper crust above the massive, columnar jointed lava core or interior. The lava core may also contain vertical vesicle segregation pipes which sometimes feed horizontal vesicle sheets at the boundary with the upper crust (see Fig. 5.24a). The thin lower or basal crust is characterised by vesicles, pipe vesicles, and a thin chilled, glassy selvedge. Total thickness varies from ≥1–2 m for thin pahoehoe lava to ≥50 m for flows of flood basalt proportions. (Adapted from Vye-Brown et al. 2013)

Fig. 5.24 Vesicle segregation pipes in lavas. **a** Vertical vesicle segregation pipes within the core of a pahoehoe unit, feeding a sub-horizontal vesicle sheet, Iceland. **b** Anastomosing vertical vesicle segregation pipes. **c** Cross-section through a vertical vesicle segregation pipe. **d** Irregular vesicle segregation pipes, orientation uncertain. (**b-d**) are all from quarried slabs of basalt lava, Newer Volcanics Province, Victoria, southeastern Australia

their formation is a continuous process initiated during flow (Thordarson and Self 1998). Nonetheless the base of horizontal vesicle sheets is generally thought to mark the boundary between the upper crust and lava core at the time when input of fresh magma ceased, and the liquid core became stagnant (Figs. 5.11c, d and 5.23d, e). The thickness of that crust can be used to calculate the minimum flow duration (Eq. 5.6; Hon et al. 1994; Vye-Brown et al. 2013).

5.4.3 Crystals and Xenoliths

Depending on the history of the magma, lavas may contain large macrocrysts (phenocrysts and xenocrysts) and small groundmass crystals. Phenocrysts grow in the source magma reservoir and may be enlarged during magma ascent in the conduit. The groundmass crystal fraction grows rapidly in the conduit during ascent and at the surface inside the lava under the influence of rapid subaerial cooling (Fig. 5.2a). At the microscopic scale, the cooling history is reflected by the nucleation and growth of microlites. The very topmost and basal crusts of lavas are usually glassy and aphyric, or aphanitic, and the most vesicular parts of the lava (see above); microlites may not be observed in the topmost few cm (Fig. 5.2c). As cooling rates decrease exponentially inward (see Eq. 5.5), microlites form in the thickening crust both at the base and top. Under these conditions, nucleation of microlites is favoured over growth and the microscopic texture is typically characterised by a high number of very small crystals (Fig. 5.2a, c; Katz and Cashman 2003). However, within the core of the lava, the much slower cooling rates allow crystal growth to dominate over

Fig. 5.25 Crystal and mantle xenolith content of some lavas. **a** Xenocryst-rich pahoehoe lava crust, Iceland. The crystal cargo does not significantly affect the rheology of lava below 30% crystal content, but it shortens the time available for in-flow cooling-crystallisation that would reach maximum packing crystal fraction (>50%), which would render the flow immobile. **b** Concentration of mantle-derived peridotite xenoliths in the base of a basalt lava, Hawai'i (Photo by T. Ivanic)

nucleation (Cashman and Marsh 1988; Fig. 5.23d). In thick basalt lavas, the groundmass may even develop relatively coarse dolerite (diabase) textural zones due to the slower cooling and longer crystallisation times, although in some rubbly flows the groundmass texture is finer grained, explained by rapid magma ascent and cooling rates (Barreto et al. 2014). Core textures are typically holocrystalline dominated by plagioclase and pyroxene, with sub-ophitic and dyktitaxitic (irregular intersertal vesicles) being common (Barreto et al. 2014).

In addition, such textural changes are mirrored by compositional changes of the interstitial glass or mesostasis, which progressively increases in SiO_2 content downwards to the lava crust-core boundary. In the Roza Member, Columbia River Plateau, the mesostasis composition ranges from basaltic andesite to dacite (Thordarson and Self 1996). Within the core, $\leq 5\%$ modal glass mesostasis of rhyolite composition occurs, which has been termed silicic residuum or differentiate (see Fig. 5.11c).

A significant "cargo" of large phenocrysts and/or xenocrysts originating from the magma reservoir and subsurface host rocks (Fig. 5.25a) as well as significant concentrations of (mantle) xenoliths (Fig. 5.25b) can affect the cooling regime and rheology of the lavas, as discussed in Sect. 5.2.2.

5.4.4 Columnar Jointing

Cooling of the lava surfaces is accompanied by contraction of the surface crust, which causes a network of cooling contractional fractures to form. Surface energy considerations of cooling isotropic media determine that the most efficient fracture configuration to conserve energy is polygonal, usually with five or six sides. Cooling then progresses from the cooling surfaces (principally the top and bottom of the flow) inwards. The tensional stresses set up during contraction of the surface produce regular joint sets perpendicular to the cooling surfaces (top and bottom of the lava) and isotherms inside the lava that progressively propagate inwards into the lava interior as it progressively cools, forming polygonal columns (Fig. 5.26a, b; Spry 1962; Budkewitsch and Robin 1994). Spry (1962) provided a detailed analysis of the physics and geomechanics of cooling columnar joint formation and propagation, and proposed that by analysing the fracture systems, several orders of fractures could be identified. The first-order fractures are widely spaced and form early during cooling with irregular orientations, so defining large initial cooling cells on the lava surface. As the surface cools further, smaller cooling cells or prisms form, bounded by contractional fractures that propagate perpendicular to the first-order fractures. Third-order factures can form perpendicular to these, and so on. Wright et al. (2011) found a similar hierarchy of cooling columnar joints in the Cerro Galan ignimbrite, Argentina.

Well-defined intersecting joint sets produce regular polygonal columns (Fig. 5.26), which are usually vertical to subvertical in orientation, producing colonnade columnar joint sets. The joint faces (and columns) propagate inwards from the cooling surfaces as the "cooling front" advances inwards. The variation in cooling rates across a pahoehoe unit results in varying width and therefore density of cooling joints (number m^{-2}). The upper and lower crusts of a lava are usually characterised by small cm-to-dm-large columns (Fig. 5.26c). These coalesce inwards producing dm-to-m-wide columns progressing inwards where the thick lava cores show only one population of vertical columns (Figs. 5.17, 5.23c and 5.26c). This relatively simple type of joint pattern may reflect the three-part vertical division of

Fig. 5.26 Columnar jointing in lavas. **a** Plan view, cross-sectional view through top of polygonal columnar jointed basalt. Eocene Flinders Volcanic Province, Flinders, Mornington Peninsula, Victoria, southeastern Australia (Photo by J. Simmons). **b** Regular, vertical colonnade columnar jointed basalt lava, with horizontal segmentation ("chisel marks"), presenting the step-by-step inward propagation of the columns from the top and bottom of the lava flow. Organ pipes National Park, Newer Volcanics Province, Victoria, southeastern Australia. Columns are ~30 cm across. **c** Columns at the base of a lava, where the cooling rate is high, are small and upwards into the lava interior where the cooling rate is lower, they amalgamate to form larger columns. Eocene Flinders Volcanic Province, Flinders, Mornington Peninsula. (Photo by J. Simmons). **d** Plumose surface texture on sub-horizontal cooling segments on a vertical columnar joint plane. ~6000 yr BP Thjorsa lava, Iceland. **e** Lower, vertical colonnade columnar joint sets and upper, highly irregular entablature columnar joint sets, Iceland. **f** Lower, vertical colonnade columnar joint sets and upper, highly irregular entablature columnar joint sets in 2 Ma Campaspe River lava flow, Newer Volcanics Province, Victoria. Person circled for scale

other internal structures, particularly vesicularity and crystallinity as described above, which directly result from pahoehoe flow inflation during emplacement (Fig. 5.11). More complicated joint patterns as described below probably result from the cooling effects of water following lava stagnation.

The progressive downward propagation of the joints may be reflected by complementary sub-horizontal "platy" or transverse joints within columns (especially at the contact between upper crust and core; Fig. 5.23d, e), or by a segmentation pattern ("chisel marks") or plumose appearance on the surface of the vertical joints, reflecting successive propagation stages (Fig. 5.26b–d).

Columnar jointing can exhibit a two- or three-tiered arrangement (Scrope 1872; Spry 1962; Macdonald 1942; Long and Wood 1986; Budkewitsch and Robin 1994). The bottom consists of thick, usually well-formed, vertical column sets normal to the base of the flow, called *colonnade* (Fig. 5.26b, c, e, f). This occurs in the lower one-half to two-thirds of flow thickness and corresponds in general to the position of the lava core. Above this colonnade, a layer of thinner, less regular, often chaotic columns, which are essentially normal to the flow top, but highly irregular in structure, is found, called *entablature* (Fig. 5.26e, f). Fracture patterns are in places cuboid to pseudo-pillow in form (Forbes et al. 2014). There may be an upper colonnade above this. Entablature horizons are thought to be associated with fractures and preferential cooling paths, caused by development of irregular, in places localised isotherm sets in the centre of lavas (Spry 1962; De Graff et al. 1989). Several workers have noted that the occurrence of entablatures often coincides with where lavas were emplaced into valleys and result from more complex cooling histories involving ingress of meteoric water. This could also be initiated where surface water gains access into the interior of a lava, e.g. during a rainstorm in a valley confined lava, or perhaps more likely by repeated flooding events caused by obstructed drainage (Long and Wood 1986; Forbes et al. 2014). Similar complex cooling-joint structures are observed in areas where the lava is cross-cut by rootless cones developed through lava–water interaction where lava crosses marshes and ponds (cf. Hamilton et al. 2010; Chap. 16).

Radiating "rosette"-like column clusters could also represent original lava tubes which remained active even while the top and bottom crusts were cooling, and cooled later after magma supply from the vent through the tube ceased.

5.4.5 Preserved Lava Tubes and Caves

The enclosed, insulated tubes in lavas that transfer erupting magma from the vent to the propagating lava flow front under the lava crust (Fig. 5.9a–c) can drain in proximal-to-medial areas. This occurs when the eruption stops, magma supply into the lava tube stops and the slope and the low magma viscosity allows the magma still in the tube to continue flowing downslope for a certain time, so draining a large part of the tube. As the magma level in the tube drops it creates *lava caves* roofed by the lava crust (Fig. 5.27a–c, e, h). Lava tube caves can be ephemeral as parts of their roof eventually collapse to produce large open "skylights" (Fig. 5.27c; also called collapse pits), open canals, and depressions on the surface of older flows. For this reason, preserved and collapsed lava cave tubes are known at active volcanoes, less commonly in the Quaternary rock record, and never in the older geological record. Some of the best examples that have been described are from the Quaternary basaltic lava provinces in Australia (e.g. Fig. 5.27a, c; Ollier and Brown 1965; Ollier 1969; Atkinson et al. 1975). Atkinson et al. (1975) report a system of lava tubes with numerous collapsed caves in the Undara lava flow in north Queensland which may have extended for more than 160 km from the vent (Fig. 5.27a). Other well-known examples include the Kipuka Kanohina Cave System (Fig. 5.27e, h) and Nahuku (Thurston) lava cave in Hawai'i, the Lava Beds National Monument in California, and the Manjanggul Cave System on Jeju Island, South Korea (Fig. 5.27b). Lava tube caves can sometimes be re-occupied and flooded and infilled by younger lavas that cascade from the surface into open skylights (Fig. 5.27i).

Where preserved, the interiors of drained lava tubes show a variety of spectacular features related to the changing volumetric flow rate over time. These include: (1) horizontal striations along the lateral walls, related to magma flowage (Fig. 5.27d) and "tide marks", reflecting changing lava levels in the tube (Fig. 5.27b); (2) lava driblet structures hanging down from the roofs of caves, called lava stalactites (Figs. 5.9c and 5.27e); (3) lava driblet marks down the walls of a lava tube, frozen at the time of lava drainage from the tube (Fig. 5.27f); (4) vertical spires of spatter-like globs of lava that fall from the roof lava stalactites, and build up the spire, called a lava stalagmite, on the floor of a draining lava tube (Fig. 5.27g); (5) rare lava balls that rolled along in the active lava stream and are then left stranded in the tube when the lava drains out of it (Fig. 5.27h); and (6) carving of channels below the original tube base, due to thermal erosion of the substrate at the time of prolonged, high flow rates (Fig. 5.27i).

5.4.6 Surface Features Related to Changes in Pressure Inside Tubes

As tubes are rigid structures confined by the brittle lavas, several features are observed at the top of pahoehoe and transitional lavas that reflect increase in pressure within the tubes. These are hornitos, pressure ridges, tumuli, lava rises, and tree moulds.

Hornitos are small, rootless spatter cones or spires up to several metres high that form above windows or fractures in the roof of a lava tube and result from spattering due to gas release or surging in flow rate in the tube (Fig. 5.28a, b). Hornitos can form in series aligned along the tube direction. Larger *rootless cones* (termed pseudo-craters and littoral cones; Chap. 16) are mainly present where lavas either flow over surface water or wet sediments (Fig. 5.28c), or flow into water (Fig. 5.28d) and generate secondary explosions that build small variably bedded spatter cones (Fig. 5.28e, f).

Fig. 5.27 Features of lava tubes and caves. **a** Drained lava tube with a flat floor. Barkers Cave, Undara lava flow, North Queensland, Australia, which can be traced discontinuously for 160 km. (From Atkinson and Atkinson 1995; Photo courtesy of the authors). **b** Tide marks and lava benches left as the lava in the tube drained at the end of the eruption, Manjanggul cave system, Jeju Island, South Korea (Photo by John Brush). **c** Lava tube cave exposed by collapse of part of the roof of the drained lava tube, Harman Valley lava flow, Newer Volcanics Province, Byaduk, Victoria, southeastern Australia. **d** Horizontal striations along the lateral walls of a collapsed tube. Flow from left to right. Mt Etna, Sicily, Italy. **e** Lava stalactites hanging from the roof of an evacuated lava tube, *"Chocolate Passage"*, Kipuka Kanohina cave system, Mauna Loa Hawai'i (Photo by John Brush). **f** Lava dribble marks down the side of a lava tube, The Shaft, Budj Bim (Mt Eccles) volcano, Newer Volcanics Province, Victoria (Photo by John Brush). **g** Lava stalagmite, Theatre Cave, Byaduk Lava Caves, Victoria (Photo by John Brush). **h** Lava ball jammed in drained lava tube, *"Escher's Ball's"*, Cordwinder Cave, Kipuka Kanohina cave system, Mauna Loa, Hawaii (Photo by John Brush). **i** Lava tube entrenched in a succession of thin lava flows, with its base well below that of the source lava, which is the thick dark unit at the top. Could this entrenched tube have resulted from thermal erosion of the substrate through prolonged focussed flow? The layering is a laminar flow, shear layering in the narrow tube. Kaulau volcano, Oahu, Hawai'i. (Photo by Steve Self)

Fig. 5.27 (continued)

Pressure ridges are elongated uplifts of the lava crust, usually oriented perpendicular to flow direction in central portions of lavas (Fig. 5.29a), but subparallel to the flow direction at flow margins. They are thought to be due to upward pressure from still liquid lava flowing beneath the solidifying surface. They are like large open flow folds of the lava crust (Fig. 5.29b, c). *Tumuli* are small mounds or dome-like blisters or crust heave-ups up to 20 m or more in diameter that form on the crust of a lava flow (Figs. 5.29d, e; 5.12b, c), again caused by pressure from lava in the tube under the crust, and/or pressure associated with volatilisation of groundwater.

It is not uncommon to find tree-trunk casts resulting from trees that burnt in place when encased by a lava flow, leaving cylindrical holes inside the lava (Fig. 5.30a). Elsewhere, cylindrical structures protruding above the surface of a lava and with a central hole with carbonised wood inside represent lava tree moulds that also record higher levels of the flow surface before the end of the eruption and after which lava in the tube drained and the lava surface deflated (Fig. 5.30a, b).

5.4.7 Transitional Lava Types Between Pahoehoe and A'a Lava

It is common for pahoehoe lava to transition to a'a lava away from the vent as the lava cools and/or the groundmass begins to crystallise, causing the magma viscosity and shear rate to increase, and so facilitating transition from pahoehoe to a'a lava (Fig. 5.4; Cashman et al. 1999; Cashman and Mangan 2014). Intermediate forms also occur. *Slabby pahoehoe lavas* and *rubbly pahoehoe lavas* have autobrecciated surfaces, as do a'a lavas, but they are less spinose than a'a lava. Slabby pahoehoe lava surfaces are dominated by ripped apart, relatively thin slabs of crust with ropy surface textures (Fig. 5.31a; Harris et al. 2017; Pedersen et al. 2017). They may form when pahoehoe lava speeds up over a steep slope where pahoehoe lava transitions to a'a lava as a result of increasing viscosity and/or shear rate (Fig. 5.31b) or alternatively where they are forced against topographic barriers, causing the pahoehoe to pile up and the ropey crust to break-up. During the early days of the 2007 Stromboli eruption, a pahoehoe lava field formed at 600 m a.s.l.

Fig. 5.28 Hornitos and spatter cones associated with some lavas. **a** Hornito spatter cone above abandoned lava tube, showing agglutinated spatter clasts that built the hornito stack, Pu'u O'o volcano, Hawai'i (*Source* United States Geological Survey; Photo taken 2-12-2015). **b** Hornito spatter cone on the surface of pahoehoe lava, showing spatter and entrail pahoehoe that built the hornito. Piton de la Fournaise volcano, Reunion Island, France. **c** Line of rootless spatter and scoria cones on a lava surface where it flowed over water-saturated ground, Iceland. **d** Rootless cones formed where lava from Mauna Loa entered the sea in 1868, Pu'u Ki, southernmost coast of Hawai'i. **e** Welded spatter beds, Pu'u Ki cones. **f** Bedding surface showing agglutinated lava spatter, Pu'u Ki cones

on a relatively flat surface laterally transitioning to slabby crust where the flow was forced against a topographic barrier (Giordano and Porreca 2009).

Rubbly pahoehoe lavas are marked by autobreccia clasts that are thicker (i.e. derived from a thick crust) more equant in form and have mostly curviplanar to rough round surfaces reflecting brittle fracturing of a solid lava crust. Rubbly pahoehoe forms from late-stage inflation of a lava flow that has already developed a thick crust. However, Barreto et al. (2014) proposed that thick basaltic lavas at the top of the Paraná flood basalt province in Brazil are rubbly sheet flows with a thick upper autobrecciated zone up to 8 m thick. They documented the facies relationships within such flows, and clast morphologies. Flows are about 50 m thick, have a smooth vesiculated base, which is marked by spherical-to-stretched vesicles, but lack pipe vesicles. This passes upwards into a massive aphanitic core which becomes increasingly more vesicular upwards, then an upper vesiculated section and a brecciated rubbly top. Clasts are 5 to 30 cm in diameter, subrounded to subangular and variably massive to vesiculated. Barreto et al. (2014) interpreted these flows as high effusion rate, thick inflating sheet flows in which the crust was autobrecciated as inflation and flow continued. Unlike rubbly pahoehoe lavas, the clasts do not appear to show ropy pahoehoe textures, and hence the flows are just called rubbly flows (Barreto et al. 2014).

Pahoehoe and a'a lava surfaces commonly form in the same lava flow in response to changing viscosity and shear rate down-flow (Cashman et al. 1999). As previously stated, pahoehoe lava may change downslope to a'a due to cooling, increasing viscosity and where the lava increases its velocity (shear rate) due to steepening of the slope. The opposite transition is rarely observed but can result from the breakout of the lava from lava tubes under an a'a lava crust, hence allowing the lava in the tube to preserve high temperature and low viscosity. This case is frequently observed at Etna, such

Fig. 5.29 Pressure ridges and tumuli. **a** Pressure ridges on the surface of a 2 ka phonolitic lava flow, Montaña Blanca, Tenerife, Canary Islands, Spain. **b** Pressure ridge on basaltic lava surface, Mývatn, northern Iceland. **c** Pressure ridges, pahoehoe basalt lava flow, Piton de la Fournaise volcano (summit on the horizon), Reunion Island, France. Note the continuous crack along the crest of the ridge to the left. **d** Tumuli field on the surface of 1969–70 pahoehoe lava flow, Mauna Ulu, Hawai'i. **e** Tumulus on the surface of a lava in Iceland. Width of the field of view is 8 m

Fig. 5.30 Tree moulds in lavas. **a** Tree trunk mould in pahoehoe lava crust, Mt Etna, Sicily, Italy. **b** Cylindrical tree mould protruding above the surface of the 1969–70 pahoehoe lava flow, Mauna Ulu, Hawai'i

as during the 1991–93 eruption, when lava tubes formed under a thick a'a crust (see Sect. 5.5.2), and secondary boccas in the lava crust formed local, pahoehoe textured breakout lava lobes, which as they flowed and cooled could transform back into a'a lava. In a recent important study, Belousov and Belousova (2018) have documented a consistent transition from a'a lava close to the vent to pahoehoe lava away from the vent during the 2012–2013 eruption of Tolbachik volcano in Kamchatka, Russia. During the eruption, the viscosity of both the pahoehoe and a'a' lava remained the same, but as the lava spread, the magma supply rate to the spreading distal margins decreased. This resulted in a lower flow velocity of the flow front (as discussed below for the transition from massive sheet lava to pillow lava for submarine lavas; see

Fig. 5.31 Pahoehoe lava transitions. **a** Transformation of pahoehoe lava (top right) into slabby pahoehoe lava and a's lava at a steepening of slope, Mauna Ulu shield volcano, Hawai'i. **b** Slabby pahoehoe to a'a lava transition, Piton de la Fournaise volcano, Reunion Island, France

Fig. 5.51a), a lower internal shear rate, and hence a transition from a'a to pahoehoe lava (Belousov and Belousova 2018).

These distinctive lithological-textural features in pahoehoe and a'a lava interiors allow a lava-type origin to be assigned to basalt lavas in the rock record, even where surfaces are poorly preserved and exposed.

5.4.8 Other Aspects

The surface features of pahoehoe lavas are all susceptible to weathering and erosion and are rarely preserved in the rock record unless they have been covered by younger lavas that have protected the surface features from destruction. However, the distinctive patterns of vesiculation, crystallisation, and jointing can be used to characterise the interior of pahoehoe lavas and cooling units to help discriminate pahoehoe from a'a lavas. A distinctive feature of pahoehoe lava units is the absence of basal and top autobreccias so that successive units directly overlie the smooth crust of the previous lava, lava lobe, or flow toe.

Pahoehoe lavas are most likely to be preserved in the rock record if they are thick, inflated flows, m or even tens of m in thickness. In the field, we recognise simple units, characterised by tabular pahoehoe units with flat tops and bases (Fig. 5.17) that may persist for many km and correspond to sheet flows and large tube systems. Although the ropy surface textures may be removed by weathering and erosion, coherent basalt lava units without autobreccia at the top or bottom will be indicative of a pahoehoe lava origin. If preserved, lateral levees and frontal zones, as well as the bases of some flows could preserve superimposed smaller lobes and toes that are ovoid in shape. These are compound units and typically on-lap and fill the micro-topography made up by the previously emplaced lobes (e.g. Fig. 5.18c). This characteristic produces the typical pointy re-entrants at the base of the lobe or toe unit, much as with pillow lavas (see below), allowing definition of the facing of successions in the ancient rock record.

Pahoehoe toe lobe lava can be distinguished from pillow lavas (Macdonald 1953). Pahoehoe lava toes tend to have a concentric internal layering (e.g. vesicles, flow layering) and frequently have internal gas and drainage cavities. Pillow lava lobes commonly have a very well-developed glassy crust, and internal radial jointing and associated hyaloclastite debris or breccias interstitial to pillow lobes and at the margins, all due to the chilling and quenching effects of a body of cold water. In addition, pillow lavas will have associated subaqueous sediments ([hemi]pelagic in deep water) above and below and commonly in the interstices between pillows where sediment has filtered downwards.

5.5 A'a Lava Flow Facies Characteristics

A'a lavas may be erupted from vent as a'a lavas or they can form during flow as pahoehoe lavas cool, crystallise (microlites), become more viscous, and transform into a'a lavas downstream from the vent. A'a lavas can consist of two principal facies: (1) coherent core and (2) marginal spiny zone to autobreccia (Plate 5.1b), although the coherent core may not always be visible. Effusion rates for a'a lavas are commonly >10 $m^3\ s^{-1}$ (e.g. Wadge 1978; Table 5.1).

5.5.1 A'a Lava Sub-Type Facies

A'a lavas are characterised by highly irregular, spiny (Plate 5.1b, Fig. 5.32a-c) to rubbly surfaces (Figs. 5.5 and 5.32d–f) that form during flow by the ductile stretching and breakage of the viscous cooling surface crust at the brittle-ductile transition. Some a'a lavas preserve in situ spiny protrusions

Fig. 5.32 Features of a'a lava surfaces. **a** Active basaltic a'a lava flow front ~4 m thick with spiny to rubbly top, Banana flow, Pu'u O'o volcano, Hawai'i (*Source* United States Geological Survey; Photo taken 2-5-2004). **b** Spiny basaltic andesite a'a' lava, Tongariro volcano, New Zealand. **c** Spiny phonolitic a'a lava, Tenerife, Canary Islands, Spain. **d** Rubbly a'a lava from Piton de la Fournaise volcano, Reunion Island, France. **e** Irregular a'a clinker clasts in the autobreccia at the front of the 19 May 2016 lava at Mt Etna, Sicily, Italy. **f** As for **e**, rounded a'a lava clasts shaped at the ductile—brittle transition

from the top and margins, which we call *spiny a'a lava* (Plate 5.1b, Fig. 5.32a–c), and Kilburn (1990) called cauliflower lava. Other a'a lavas are dominated by loose to rooted, irregularly shaped rubble, and have been called *rubbly a'a lava* (Fig. 5.32d–f; Kilburn and Guest 1993; Harris et al. 2017).

The spiny protrusions grade into and are rooted to the massive core of the lava. Since these spines are still connected to the massive core, they are not strictly autobreccia, although they superficially look like it, and may form as clots of spinose magma in an a'a lava aggregate while still plastic (Cashman et al. 1999). They represent an extreme contortion and upward protrusion of the volatile-rich crust of the lava as a result of flow of the viscous lava (Macdonald 1953). As such lavas continue to flow, these spines break (i.e. autobrecciate), and add hackly autobreccia rubble to the lava surface and margins. Downstream, many a'a lavas become dominated by loose a'a lava autobreccia rubble. However, even the spiny a'a lava has some associated autobreccia rubble. The irregular spines of spiny a'a lava can range from 1 cm to 2 m high.

The morphology of the autobreccia rubble, clinker, or blocks varies from equant to irregular, angular to variably deformed, spinose to scoriaceous, and they vary in size from cm to m (Fig. 5.32d, e) leading to a spectrum of a'a lava subtypes that are summarised by Harris et al. (2017) and Pedersen et al. (2017). Although monomictic, a'a lava autobreccias may include clasts with varying degrees of vesicularity that may be difficult to distinguish from, or may be confused with, truly pyroclastic scorias that may litter the top of lavas where lava effusion is accompanied at source by strombolian or microplinian activity (cf. Chap. 9). In low-viscosity a'a lava, some blocks may become rough rounded by rolling and tumbling in the flow to form sub-spheroidal clasts (Fig. 5.32f).

5.5.2 A'a Lava Tubes and Channels

Even though inflation also occurs in a'a lavas, they are generally already thick to start with (from 2 to 3 m, up to about 20 m) and advance in open channels (Fig. 5.8b, d), bounded by well-developed lateral levees (Fig. 5.33a, b). The levees are accretionary features and accommodate the changes in lava flux, which causes the surface of the lava in the channel to rise and break-out, causing overflows onto the levees (e.g. Fig. 5.8d). When the eruption stops the lava

Fig. 5.33 A'a lava channels and tubes. **a** Channel between elevated levees, a'a lava from the 1983 eruption, Mt Etna, Sicily, Italy. **b** As for (**a**), from pre-historic Keamoku a'a flow, Mauna Loa, Hawai'i. **c** A'a lava tube and channel, Mt Etna

continues to drain downslope, so that the lava level in the channel drops below the tops of the levees (Fig. 5.33a, b).

Although development of interior lava tubes is most commonly associated with pahoehoe lavas, lava tubes can also form in a'a lavas (Fig. 5.9d) and are also instrumental in allowing long-distance flow of a'a lavas relative to those that flow in open channel style, as described on Mt Etna by Calvari and Pinkerton (1998). The efficiency of these tubes is lower than for pahoehoe lavas because the clastic nature of the crust usually limits its mechanical stability, increases mechanical coupling between crust and tube lava, and makes detachment from the underlying flow in the tube difficult. This is often reflected by the visible movement of the a'a crust on the interior lava flow. Nevertheless, at Etna, tubes may form below a'a surfaces by the inward progression of the cooling front below the top breccia most likely where the flow slows at breaks in slopes and shear rates decrease (Fig. 5.33c). Lavas at Etna often show transitional characters especially close to vent and within the gently sloping Valle del Bove depression where flows can form tubes, and typical, complex, stacked toothpaste-like structures (Rowland and Walker 1987).

5.5.3 A'a Lava Facies Architecture

A'a lavas advance slowly and form very steep, thick, and jagged flow fronts (Plate 5.1b, Figs. 5.10e and 5.32a) that creep forward and steepen until a section becomes unstable and breaks off, tumbling down the front and sides of the lava. Collapse is repeated as the flow interior slowly advances in caterpillar-track fashion over a layer of autobrecciated debris that has collapsed from the flow front, resulting in an irregular, coherent flow core facies (the feeder tube) enclosed above, below, and laterally by a'a clinker breccia facies, represented schematically in a facies model in Fig. 5.5b, and in cross-sectional outcrops in Figs. 5.9 d and 5.34a, b. The contacts between the coherent interior and the upper and lateral breccias may be complex, highly irregular, and gradational, as a result of the rough morphology determined by pressure ridges that form in response to the balance between incoming lava flux and flow advancement (Fig. 5.34a, b). It is therefore quite common to see upwards and laterally directed injections of coherent domains that may evolve into spines, ridges, or breakouts (Fig. 5.34a). Columnar jointing in a'a lavas is usually less well developed

Fig. 5.34 Sections through a'a lavas. **a** Thin a'a lava showing the top and bottom clinker breccia enclosing the interior coherent lava tube massive basalt, and passage into another overlying a'a lava. The highly irregular upper part represents potential spines from which clinker blocks break. Oia, Santorini, Greece. **b** Massive core of an a'a lava in irregular contact with over- and underlying a'a breccia. Mt Ruapehu, New Zealand. **c** Succession of a'a lavas with thick coherent, columnar jointed lava tube core zones separated by a'a clinker breccia horizons, representing tops and bottoms of superposed lava flows. Note the lateral thickness variations of coherent domains, the poorly developed and rather irregular cooling joints. Quarry, Mt Etna, Sicily, Italy

and more irregular compared with pahoehoe lava and is confined to the coherent core zone (Fig. 5.34c).

5.5.4 A'a Lava Vesicle Character and Zonation, and Crystallinity

Internally, a'a lava is characterised by a coherent, usually poorly vesicular (<20%) core zone facies, with irregular elongate vesicles that are stretched in response to internal flow. The lower part of the lava tube core zone may be less vesiculated than the upper part (Fig. 5.35a–c). This poorly vesicular character is developed in response to the continuous internal stirring that promotes an efficient syn-eruptive and flow degassing. Furthermore, the usual open channel transport and rubbly crust facilitates faster degassing and cooling, and the internal nucleation of microlites, with a higher number density compared with pahoehoe lavas (Fig. 5.2a–c; Cashman and Marsh 1988; Cashman et al. 1999; Katz and Cashman 2003). Cashman et al. (1999) documented an increased level of microlite crystallisation in lavas from Pu'u O'o volcano in Hawai'i as they transformed from pahoehoe lava to a'a lava downstream.

5.6 Continental Flood Basalt Lavas

Flood basalt systems consist mostly of extensive simple sheet pahoehoe lavas with very low aspect ratios and compound or hummocky pahoehoe lavas (Figs. 5.1, 5.14, 5.17, 5.36). A'a lavas are very rare (Self et al. 2021), but have been identified, and rubbly sheet lavas may also occur (Barreto et al. 2014). Many features of flood basalt lavas will be those of pahoehoe lava flows as already described in Sect. 5.4, except of course their dimensions and eruption rates, which are much greater. The summary here will thus be brief. Flood basalt lavas are known in the rock record to form the major component of continental Large Igneous Provinces (cLIPs), including those at volcanic passive margins (VPMs; also see Chaps. 16 and 17). These originated before and during some of the major extensional continental rifting events that occurred on Earth, such as the Jurassic–Cretaceous break-up of Gondwana (e.g. White and McKenzie 1989; Courtillot et al. 1999; Mahoney and Coffin 1997; Geoffroy 2005; Bryan et al. 2010; Bryan and Ferrari 2013; Ernst 2014a; Self et al. 2015). The most recent reviews of current knowledge about cLIPs can be found in Self et al. (2015), Bryan et al. (2010), Bryan and Ferrari (2013), and Ernst (2014a). These focus mostly on "younger"

Fig. 5.35 Vesicle patterns in a'a lavas. **a** Section through an a'a lava showing a very thick interior coherent core zone with stretched large vesicles near the top, overlain and underlain by a'a clinker breccia. Reykjanes Peninsula, Iceland. **b** Profile through an a'a lava core zone showing increasing vesicle content upwards, sheared stretched vesicles near the base and large irregular vesicles upwards. Mt Napier volcano, Newer Volcanics Province, Victoria, southeastern Australia. **c** Vesicle zonation in a'a lava seen in diamond drill core at Mt Napier volcano, showing large irregular vesicles high in the flow. Upper two core intervals show top of the lava, while lower two bottom of the lava; drill core ~5 cm wide

Fig. 5.36 Exposures of sequences inflated pahoehoe flood basalt lavas. **a** Stacked, sheet-like Miocene flood basalt lavas from the Picture Gorge Basalt (17–16 Ma), Columbia River flood basalt province, unconformably overlying pale-coloured, terrestrial volcaniclastic sediments, ash fallout deposits, and ignimbrite of the 39–18 Ma John Day Formation, Oregon, USA. **b** Stacked, sheet flood basalt lavas of the Columbia River flood basalt province, exposed in the Grande Ronde River Valley, Washington (Photo by Steve Self). **c** Stacked, sheet flood basalt lavas of the Columbia River flood basalt province, exposed in the Three Devils grade Moses Coulee Cattle Feed Lot, Washington. The topmost lava is the 15.8 Ma Roza Member (see Fig. 5.14 for aerial extent). (Photo by William Borg; *Source* Wikimedia Commons). **d** Sheet-like lavas of the Deccan Traps flood basalt province, Satara, western India (From Bondre et al. 2004)

flood basalt systems. Older examples in the rock record that are variably preserved include the Mesoproterozoic Keweenawan Province of North America, which still conserves valuable characteristics of lava flows and their varying dimensions (e.g. Green 1989). Continental LIPs cover hundreds of thousands of km^2 (Figs. 5.1, 5.4 and 5.14), and consist of many stacked individual sheet-like flood basalt lavas (Fig. 5.36a–d), and in some cases anomalously large silicic lavas as well. For a more detailed discussion of vent systems within flood basalt plateaus, see Chap. 16, Sect. 16.6.

Although generalised descriptions of flood basalt lava sequences give the impression that the lava sequences are relatively uniform and consist of either high effusion rate massive sheet or lower effusion rate lobate pahoehoe lavas, it is not until detailed logging of sections and their facies are undertaken that some trends may begin to emerge. For example, in the southern part of the Lower Cretaceous Paraná flood basalt province in Brazil, Baretto et al. (2014) noted that flows at the base of the 370-m-thick succession are largely lobate compound lavas, which pass upwards into an interval of largely simple sheet pahoehoe lavas, which are overlain by an interval of largely rubbly sheet lavas (Fig. 5.37a). They interpret this vertical stratigraphic progression as reflecting an increasing magma effusion rate through the 1 Myr history of the Paraná flood basalt province. This vertical transition also coincides with an increase in lava flow thicknesses, with compound lavas being 2–3 m

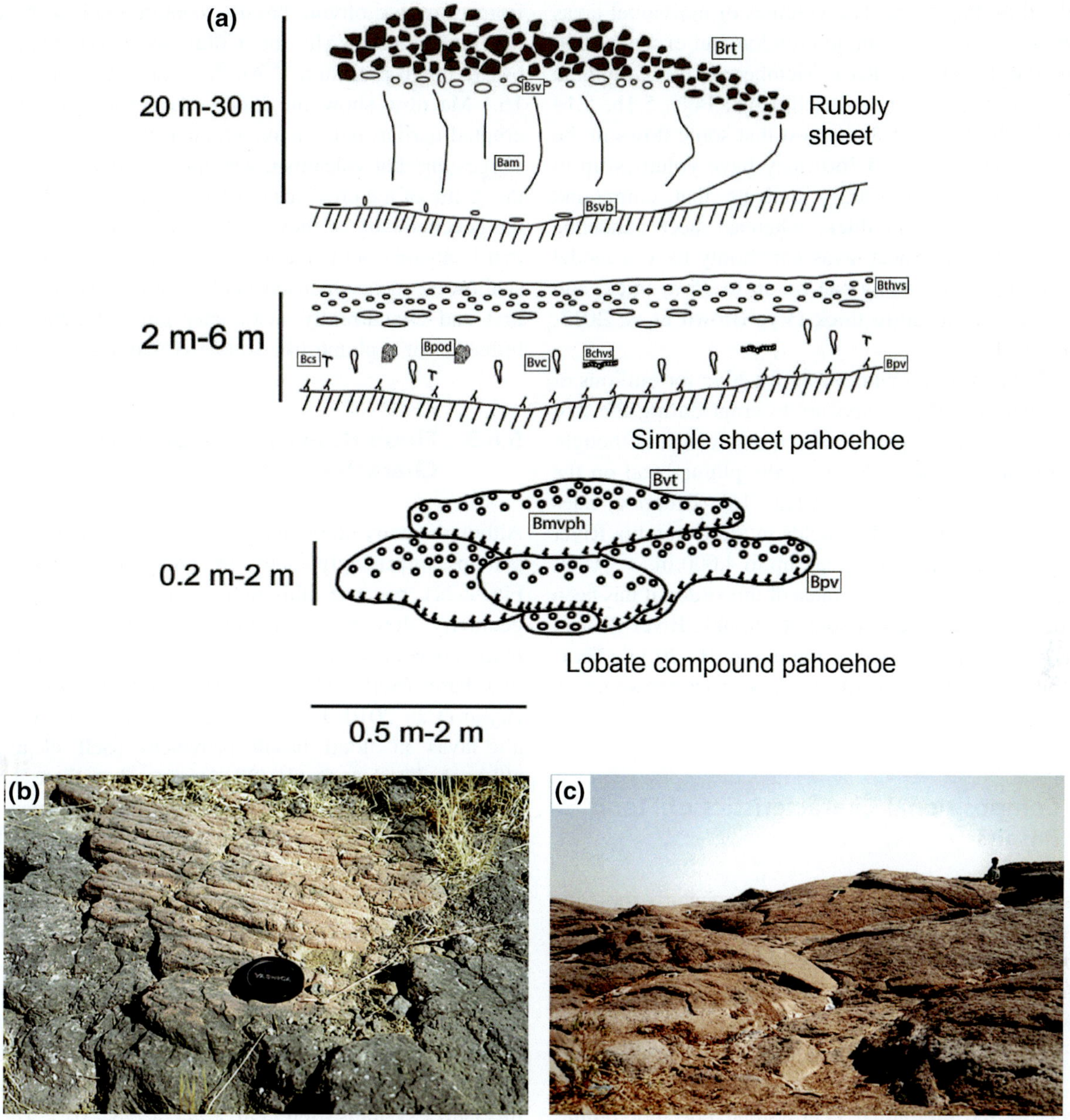

Fig. 5.37 Types of flood basalt lava types and their facies features. **a** Subdivision of flood basalt types and their facies features found in the southern part of the Lower Cretaceous Paraná flood basalt province, Brazil (Adapted from Barreto et al. 2014). Stratigraphically (representing ~1 Ma) upwards from: bottom, lobate compound pahoehoe lava; middle, simple sheet pahoehoe lava; and top, rubbly sheet lava). Acronyms: Bpv = basal pipe vesicle zone; Bmvph = micro-vesiculated massive phaneritic basalt; Bvt = vesiculated top crust; Bvc = vesicle cylinders; Bcs = C-S vesicle cylinders; Bpod = vesicle pods; Bchvs = core horizontal vesicle sheets; Bthvs = top vesicle sheet; Bsvb = slightly vesicular base with spherical-to-stretched vesicles; Bam = aphanitic and holocrystalline massive basalt; Bsv = aphanitic basalt, sparse vesicles; Brt = flow top breccia. **b** Pahoehoe ropy texture preserved on the surface of a flood basalt lava, Kondhwa Hill, western India. **c** Lava lobes or rises in pahoehoe lava, Daund, western India. Both (**b** and **c**) are from the Late Cretaceous Deccan Traps flood basalt province, extending over western and central India (see Fig. 5.1a). (From Bondre et al. 2004)

thick, simple sheet pahoehoe lavas being 1–6 m thick, and the uppermost rubbly sheet flows being 50 m thick. It also corresponds with a change from primitive olivine basalts in the basal compound lavas to more evolved basalts and even basaltic andesites in the higher sheet pahoehoe and sheet rubbly flows (Barreto et al. 2014).

5.6.1 Flood Basalt Lava Dimensions, Geometry, Eruption Rates, and Durations

Flood basalt lavas are the largest and farthest flowing lava flows known, with maximum flow distances of 1,000 km or more (e.g. the Rajahmundry flow in the Deccan Traps, India,

Self et al. 2008; Fig. 5.1a). The volumes of individual lavas or lava flow fields from a single eruption can exceed 1,000 km^3 (e.g. the 1,300 km^3 Roza Member of the Columbia River Plateau, Thordarson and Self 1998; Figs. 5.1b, 5.14 and 5.36c). Self et al. (2008) suggest that some flows in the Deccan Traps (Figs. 5.1a, 5.36d) may have volumes up to 5,000 km^3. Self et al. (2021) indicate that compound pahoehoe lavas are <5 m thick, whereas sheet pahoehoe lavas are >5 m thick. Sheet lavas commonly have a modal thickness of 15–20 m, many can be up to 35 m thick, and they are rarely up to 80 m thick (Vye-Brown et al. 2013; Self et al. 2021).

While flood basalt provinces can be active for millions of years, the majority of lava appears to erupt during the first million years, or "main phase" of activity, which is thought to represent impingement of the mantle plume head on the lithosphere (e.g. Ernst 2014a; Chap 16). Eruption rates estimated for single event flow fields in the Columbia River flood basalt system in the USA are up to 4,000 $m^3\ s^{-1}$ (Self et al. 2015). Based on eruption rates of this order, it has been estimated that the duration of some Columbia River Plateau eruptions would have lasted a minimum 10–20 yr (Thordarson and Self 1998; Self et al. 2014; Vye-Brown et al. 2013).

5.6.2 Compositional Characteristics of Flood Basalts

Compositionally, flood basalt lavas are dominantly tholeiitic basalts (e.g. Swanson and Wright 1981; Peate 1997; Self et al. 2015), although they can also be alkali basalts (e.g. in the Ethiopian Yemen Plateau, Mohr 1983; Deccan Traps of India, Krishnamurthy and Udas 1981). In the Columbia River Flood Plateau, basaltic andesites are also common (Self et al. 2015), resulting from crustal contamination and the effects of the migrating Yellowstone hot spot plume. Within specific flood basalt stratigraphies, distinguishable chemical variations can be identified (Thordarson and Self 1998; Jerram and Widdowson 2005; Barreto et al. 2014). Self et al. (2008) found that compositionally distinctive lava sets could be correlated over distances of $\sim 1{,}000$ km in the Deccan Traps, India, using trace element and Sr isotopic data. The Deccan lavas are generally high Ti tholeiitic basalts, but distinctive "chemotypes" could be correlated from the main Deccan Traps lava succession in northwestern India to the Rajahmundry Traps in east-central India using stable trace elements Ba, Y, Zr, and Nb, and Sr isotopic characteristics, constrained by the magnetic polarity chron timescale. Some continental flood basalt systems also contain silicic flood lavas (Bryan et al. 2010; Chap. 6).

Barreto et al. (2014) found that the southern Paraná flood basalt province were low-Ti basalts, but change vertically from primitive olivine basalts to more evolved basalts and basaltic andesites. In the Columbia River Plateau flood basalts high-precision $^{40}Ar/^{39}Ar$ ages between 17.2 and 16.1 Ma now show the Picture Gorge Basalt (Fig. 5.36a) erupted earlier and for longer than other main-phase units, suggesting that volcanism was initiated over a broad area in the centre of the province (Cahoon et al. 2020). The ages and the geochemistry of these early lavas reflect a mantle source that is distinct both spatially and temporally, indicating that shallowest metasomatised back-arc-like mantle was tapped first and concurrently, with later units showing increased influence of a plume-like source (Cahoon et al. 2020).

5.6.3 Flood Basalt Lava Flow Facies Characteristics

Although many flood basalt lavas are simple sheet lavas, perhaps surprisingly many are compound lavas (Chap. 2, Fig. 2.22) because although overall they have a sheet geometry, they consist of multiple flow lobes and in many places, pahoehoe flow toes are also preserved, particularly at their bases (Self et al. 1997, 2014, 2021; Bondre et al. 2004; Bryan et al. 2010; Figs. 5.11, 5.23). There are virtually no a'a lavas in flood basalt provinces (Self et al. 2021), although Barreto et al. (2014) note that there are some documented in the Paraná flood basalt province, Brazil. Flood basalt lavas generally consist of sheet, lobe, and pahoehoe toe facies. Sometimes ropy surface features (Fig. 5.37b), lobate rises (Fig. 5.37c), pahoehoe toes, and pipe vesicles (Fig. 5.22e) can be preserved (Self et al. 1997; Bondre et al. 2004). Flood basalt lavas commonly show a vertical zonation of vesicle concentration and types (e.g. Fig. 5.23d, e; Self et al. 1998; Vye-Brown et al. 2013). Many of the larger flows of this type reach in excess of 30 m in thickness which can be related to inflation within large tube/sheet systems (Self et al. 1997, 1998, 2015; Thordarson and Self 1998; Bondre et al. 2004; Vye-Brown et al. 2013; Barreto et al. 2014; Self et al. 2021).

In their facies analysis studies of the flood basalt lava sequence in the southern part of the Lower Cretaceous Paraná flood basalt province, Brazil, Barreto et al. (2014; Fig. 5.37a) found that there are changes vertically and through time in the lava types and facies of the succession (Fig. 5.37a). The basal part of the stratigraphy consists of lobate or compound pahoehoe lavas in which both S- and P-type lobes occur. S-type lobes are vesicular particularly at the top, with spherical vesicles and lack pipe vesicles (Fig. 5.37a) whereas P-type lobes are marked by pipe vesicles at their bases (e.g. Fig. 5.22e) and spherical vesicles near the tops (Fig. 5.37a). Both can have smooth or ropy-textured tops (Fig. 5.37b). Internally, lobes are massive and phaneritic. An interval of simple sheet pahoehoe lavas overlies the compound pahoehoe sequence. Lavas are also marked by basal pipe vesicles (e.g. Fig. 5.22e) and a vesicle concentration zone in the crust, but in addition they have abundant subvertical vesicle segregation cylinders and pods (e.g. Fig. 5.24a–d), and horizontal vesicle zones. The massive core can be holocrystalline. The upper rubbly flows, which are up to 50 m thick, consist of a thin basal vesicular zone (<20 cm thick) with spherical and stretched vesicles, but no pipe vesicles, passing up into a massive, jointed aphanitic core (~ 30 m thick), then into an upper vesicular zone with spherical-to-stretched vesicles (Fig. 5.37a). The top is marked by a flow top breccia that can be up to 20 m thick and consists of subangular-to-subrounded clasts of variably massive and vesicular fragments, interpreted as autobreccia. No pahoehoe ropy textures have been noted in rubbly autobreccia clasts (Barreto et al. 2014). Although Barreto et al. (2014) did not record any slabby pahoehoe lava breccias, Vye-Brown et al. (2013) did in the Columbia Flood Basalt lavas.

Lava tubes, lava channels, and other flow features are generally difficult to observe in flood basalts because of the large dimension of the feeding systems and the widespread sheets and lobes. Large, circular down-sag structures have been described, which may result from magma withdrawal. As in smaller scale pahoehoe units, vertical colonnade columnar jointing is ubiquitous (Figs. 5.17 and 5.36c), especially in the lower one-half to two-thirds of lava flows (corresponding to the lava core; Sect. 5.4.4). In the Columbia River Plateau, these are commonly overlain by narrower and more irregular columns (upper crustal zone; Fig. 5.23d, e). However, locally more complicated upper entablatures may be developed which are probably related to influx of significant amounts of water during major rainfall or flood events (Long and Wood 1986).

The typical geometry of flood basalt provinces is that of the piling of hundreds of m of multiple, laterally extensive, sub-horizonal lava units (e.g. Fig. 5.36a–d), usually associated with highly productive long fissure systems (e.g. Fig. 5.14; Swanson et al. 1975; Vye-Brown et al. 2013; Self et al. 2015). However, at a more detailed scale of observation, the internal stratigraphy may be more complex and involves also chemical variations (Jerram and Widdowson 2005; Self et al. 2008; Barreto et al. 2014). Jerram and Widdowson (2005) demonstrated that on close examination, many sheets wedge out and overlap with other lava units. Such wedging lavas may represent the margins of individual sheets or sheet-like lobes in larger flows.

Because most flood basalt lavas are pahoehoe lavas they are overwhelmingly coherent, but quench fragmentation hyaloclastite and peperite breccias have been recorded where subaerial lavas flowed into water (e.g. Schmincke 1967; Jerram et al. 2009; Planke et al. 2000, 2017; Vye-Brown et al. 2013). The relationship between coherent lava and hyaloclastite in the succession is very important for understanding the evolution of volcanic passive margins where seismic data has now provided valuable insights in the past two decades into how these develop (Chap. 17, Sect. 17.2.4; Jerram et al. 2009; Planke et al. 2000, 2017).

Paleoflow direction in flood basalts can be determined if pipe vesicles or *spiracles* are present. These small, curved pipe-like structures are found at the base of flows, with the tops dragged in a down-flow direction (see Fig. 1.14a). They form when steam from heated ground water rises into the base of the lava-forming vertical pipe vesicles, the tops of which are then stretched in the direction of flow as the interior lava continues to move (Waters 1960; Macdonald 1967; Walker 1987). However, in consideration of the very mild paleotopography and the huge extent of the lava units, paleoflow determinations must be used with care as they may simply reflect a local direction. For a non-traditional assessment and comprehensive review suggesting pipe vesicles form by the sinking of dense immiscible melt droplets, see Sheth (2020).

For more detailed assessments of proximal-to-distal facies characteristics and relations found in flood basalt lavas, we recommend Thordarson and Self (1998) and Vye-Brown et al. (2013).

5.6.4 Environmental and Climatic Effects of Flood Basalt Eruptions

It is beyond the scope of this book to address this major topic, which has been the source of much discussion for decades. The most recent reviews by Self et al. (2014) and Ernst and Youbi (2017) provide thorough summaries of our state of understanding, the issues, and the uncertainties. Briefly, however, there is a strong correlation between the timing of at least four flood basalt events in the last 300 Ma with mass extinction events (Self et al. 2014; Ernst and Youbi 2017). The Roza flood basalt eruption is calculated to have released 12,000 Mt of the climate-affecting aerosol SO_2 gas over the ~ 15 yr long estimated duration of the eruption, a rate of ~ 100 Mt per year (Self et al. 2014). Every 1 km^3 of flood basalt magma erupted contains 6 ± 1.7 Mt of SO_2, which is much greater than for an equivalent volume of silicic magma erupted during a major explosive eruption (Self et al. 2014). So, if major explosive super-eruptions caused climate effects (e.g. Toba, Yellowstone), flood basalt eruptions are even more likely to have done so (Ernst and Youbi 2017).

5.7 Subaerial Basaltic Lavas Flowing into Water and Over Wet Sediments: Lava Deltas

When basaltic lavas: flow from land into a lake (Tucker and Scott 2009), or from an emergent volcanic island or landmass into the sea (e.g. Hawai'i, Reunion, La Palma; Jones and Nelson 1970; Moore et al. 1973; Porębski and Gradziński 1990; Mattox and Mangan 1997; Skilling 2002; Watton et al. 2013; Soule et al. 2021), or from an emergent volcanic island that has grown in englacial meltwater ponds formed during the eruption of intra-glacial volcanoes (Jones 1970; Edwards et al. 2015; Chap. 16, Sect. 16.10), lava deltas often build out from the entry point at a shoreline (Fig. 5.38a–e). In general, such deltas consist of a prograding shoreline fan marked by an emergent subaerial lava flow field (Fig. 5.38a, b) that is building over a lower subaqueous part consisting of bedded, prograding, mass-flow resedimented, palagonitised hyaloclastite breccias, gravitational collapse (pillow) breccias, and pillow lava lobes characterised by steeply dipping foreset beds (up to 40°; Fig. 5.38c–e), which have been termed flow-foot breccias (Figs. 1.14e, 5.38e; Jones and Nelson 1970; Skilling 2002; Watton et al. 2013). Resedimentation processes include slides, slumps, grain-flows, debris flows, and turbidity currents (Chap. 12), which can operate syn-eruptively to post-eruptively (Skilling 2002; Watton et al. 2013; Soule et al. 2021). In addition, in some situations, the interaction between lava flowing into shallow water can trigger violent phreatomagmatic explosive activity (Mattox and Mangan 1997; Watton et al. 2013), producing lapilli-tuffs that mantle or are interbedded with the other lithofacies.

The lower lava delta foreset bedded succession is overlain by nearly flat-lying massive lavas (Fig. 5.38a, b, e), which originate from a vent that is located above the water level and which prograde over the underlying bedded succession in time. A "passage zone" between these, represented by dipping lobes of basalt, some pillow-like, protruding downwards from the base of the coherent lava and intercalated within the top of the dipping, foreset bedded breccia pile, marks the approximate water level at that time (Figs. 1.14e and 5.38e; Jones 1970; Jones and Nelson 1970; Skilling 2002; Watton et al. 2013). The subaerial lava may transform into a subaqueous pillow lava if it flows beyond the delta edge into and under water (Moore et al. 1973). The June 1969 flow from Mauna Ulu, Kilauea volcano, entered the sea as a narrow flow of a'a lava which travelled underwater for several hundred m without a lava delta being constructed (Moore et al. 1973). In contrast, the lava flows erupted in March–May 1971 from Mauna Ulu, which were lower viscosity pahoehoe flows, constructed a lava delta (Moore et al. 1973), including pahoehoe lava on the subaerial segment and a combination of pillow lava lobes and hyaloclastite debris in the submarine part of the delta which had a steep slope.

Soule et al. (2021) also compared the offshore lava deltas formed from the entry of largely low-viscosity pahoehoe lavas during eruptions from Pu'u O'o volcano over the last decade from 2006, and the more viscous a'a forming eruption from Kilauea in 2018 (Fig. 5.38d). They used pre- and post-eruption bathymetry to assess volumes and seaward reach of the deltas as well as ROV observations of deposit characteristics. The 2018 a'a lava eruption produced four entry point deltas along a 10 km stretch of coastline. The 2018 a'a lava deltas are up to 3 km wide, extend up to 1.2 km offshore to a depth of 700 m, are < 250 m thick and have slopes up to 35°. The surface area of deltas is 108×10^5 m^2, the volume of delta material is 760×10^6 m^3 (representing >60% of the total erupted volume), and the eruption rate was $>3000 \times 10^6$ m^3 yr^{-1}. The deposits viewed by seafloor imagery are dominated by coarse breccias and coherent lava. In contrast, the Pu'u O'o pahoehoe lava deltas which have an area of 360×10^5 m^2, a volume of 309×10^6 m^3, a thickness <20 m, and extend up to 5 km seaward, were formed under an eruption rate of 43×10^6 m^3 yr^{-1}, two orders of magnitude less than the 2018 a'a eruption, and have slopes 15–20°. The deposits are dominated by fine sand size debris.

Soule et al. (2021) attribute the low degrees of fragmentation (coarse breccias) for the 2018 a'a lava entries being due to the combination of high magma flux and high viscosity of the a'a lavas as they entered the sea, with most fragmentation being caused by gravitational collapse and mass−flow resedimentation due to slopes exceeding angle of repose at the advancing lava delta front (Fig. 5.38d). Conversely, they suggest that the fine−grained nature of the Pu'u O'o pahoehoe deltas being due to high degrees of fragmentation of the low−viscosity pahoehoe lava entering the sea at low magma flux, allowing efficient quench fragmentation and phreatic explosive interaction at the shoreline. Soule et al. (2021) suggest that the lower slope and greater seaward extent of the pahoehoe deltas are due to greater ease of gravitational collapse of the finer deposits on the upper delta slope, and then greater transport distance due to high intergranular pore pressure in the finer grained debris and turbidity current mass flows (Fig. 5.38d).

Jones and Nelson (1970) showed how relative movements of water level and a volcanic pile or terrain (e.g. through gravitational subsidence of part of the lava delta, tidal sea level changes) over a period could be deciphered

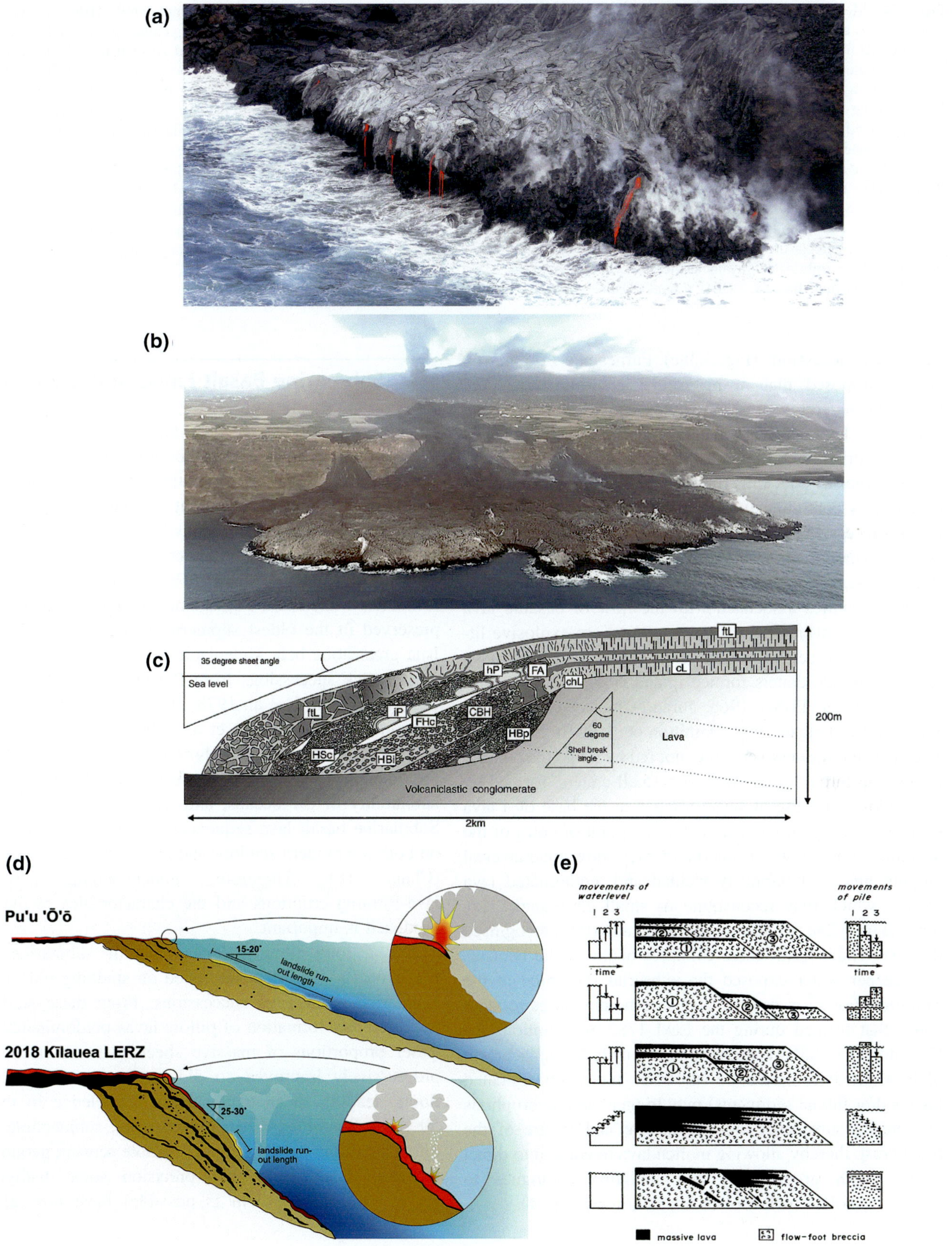
(a)
(b)
(c)
35 degree sheet angle
Sea level
ftL
hP
FA
chL
cL
iP
FHc
CBH
HBp
HSc
HBi
60 degree
Shell break angle
Lava
200m
Volcaniclastic conglomerate
2km
(d)
Pu'u 'Ō'ō
15-20˚
landslide run-out length
2018 Kīlauea LERZ
25-30˚
landslide run-out length
(e)
movements of waterlevel
movements of pile
1 2 3
time
massive lava
flow-foot breccia

◀ **Fig. 5.38** Morphology and architecture of lava deltas prograding into water bodies. **a** Pahoehoe lava flow delta built at the coastline by a lava flow erupted from Pu'u O'o, Kilauea volcano's East Rift Zone, Hawai'i (*Source* United States Geological Survey; Photo taken 2–05-2013). **b** Lobate lava delta prograding into the Atlantic Ocean, fed by several lavas flowing from Cumbre Vieja volcano (looming in background), which erupted from 19 September to 13 December 2021, La Palma, Canary Islands, Spain (*Source* IGME-CSIC; video frame taken 6-10-2021). **c** Schematic longitudinal section of the facies architecture through a late Pliocene–Pleistocene lava-fed delta deposited in a marine embayment, Stóri-Nupur, Iceland (From Watton et al. 2013). Acronyms: ftL = flow top hyaloclastite and gravitational collapse breccia; FHc, HSc = hyaloclastite breccia, channelised; iP = isolated pillow breccia; HBi = imbricated hyaloclastite breccia; hP = pillow lava and hyaloclastite breccia; CBH = hyaloclastite breccia, mostly clast-supported; FA = tephra and quench lapilli; HBp = planar bedded hyaloclastite breccia; chL = curved columnar jointed and hackly fractured lava; cL = columnar jointed sheet pahoehoe lava. **d** Differences in the geometry and facies associations of lava deltas fed by low magma flux pahoehoe lavas entering the sea from Pu'u O'o volcano from 2006 onwards, and from the entry of higher magma flux a'a lavas entering the sea during the 2018 Kilauea fissure eruption along the lower East Rift Zone, Hawai'i. Note the finer grained nature, lower slope, and greater seaward extent of the Pu'u O'o delta compared with the Kilauea LERZ delta, as discussed in the text. (From Soule et al. 2021). **e** Effects of changing relative water level on the facies architecture and relationships between topset coherent lava and foreset flow-foot breccia facies (From Cas and Wright 1987, After Jones and Nelson 1970)

from such successions (Fig. 5.38e). Furnes and Fridleifsson (1974) proposed that in macrotidal environments (tidal ranges of several metres), the rising tide would overstep advancing lava flows, producing complex relationships between different lithologies. Hyaloclastite and resedimented breccias inter-fingering and alternating with massive lava bodies continuously build up to high tide level. The whole succession is then covered by massive subaerial lava as progradation and build-up of the delta continues (Figs. 1.14e, 5.38a–e).

Other features associated with the flow of basaltic lava into water or over water-saturated ground are explosive littoral cones formed at the shoreline which may be a lava delta, or pseudo-craters formed inland (Fig. 5.28c, d; Thorarinsson 1953, Fisher 1968; Jurado-Chichay et al. 1996; Mattox and Mangan 1997; Hamilton et al. 2010; Sect. 6.4). The vents are rootless vents (i.e. not fed by a magma conduit coming up through the crust) with small craters, formed by the explosive release of steam trapped at the base of a lava flow or flooding into a lava tube at the point of entry of the lava into a body of water. Internally, they show quaquaversal dipping layers of variously welded and agglutinated lava spatter, which may reconstitute as small clastogenic lava flows (Fig. 5.28e, f). Mattox and Mangan (1997) recognised that they could form where collapses of lava delta benches into deeper water exposed the subaerial lava tube at the shoreline. For an in-depth analysis of the numerous rootless cones that formed during the Laki 1783–84 eruption, see Hamilton et al. (2010). These authors conclude emplacement of lava above compressible substrates (i.e. unconsolidated lake and/or fluvial sediments) may trigger rootless eruptions by causing subsidence-induced flexure and failure of the basal crust, thereby allowing molten lava to come into direct contact with groundwater and initiating analogues to explosive molten fuel–coolant interactions (Chap. 4).

5.8 Submarine Basalt Lavas and Eruptions

Submarine basalt lavas cover almost 70% of the Earth's surface, making up layer 2 of the Earth's oceanic crust. They also constitute much of the stratigraphy of the thousands of submarine seamounts, emergent oceanic island volcanoes, and huge oceanic Large Igneous Province plateaus (oLIPs). They therefore represent an extremely significant component of the geology of our planet. They have contributed to the formation of the Earth's crust since the Earth formed and are preserved in the oldest supracrustal rocks, the 3.7–3.8 Ga Isua greenstone belt, Greenland (see Plate 1.1a) and in old crustal rock successions from Precambrian to Phanerozoic time (e.g. Dimroth et al. 1978; Hargreaves and Ayres 1979; Corcoran 2000; Hayman et al. 2015; Squire et al. 1998). Volatile exhalations from submarine basaltic eruptions have the potential to significantly affect water chemistry and the habitability of the oceans, and hence it's organic content. Submarine basalt lava sequences can host mineral deposits on both the modern seafloor and in ancient rock successions (Chap. 18). Therefore, understanding submarine lava-forming eruptions and the characteristics of the lavas produced is important.

For a long time understanding of submarine eruption processes and deposits relied on studying old, uplifted submarine basalt lava successions. From these studies, it appeared that formation of pillow lavas predominated, with lesser proportions of massive sheet lavas. As technology progressively improved so did exploration of the ocean floors, leading to an explosion of knowledge in the last 50 years. Sampling, direct photography, submersible expeditions, and the development of remote sensing geophysical technology, especially high-precision sonar bathymetric surveys (resolution <1 m is possible) have now allowed

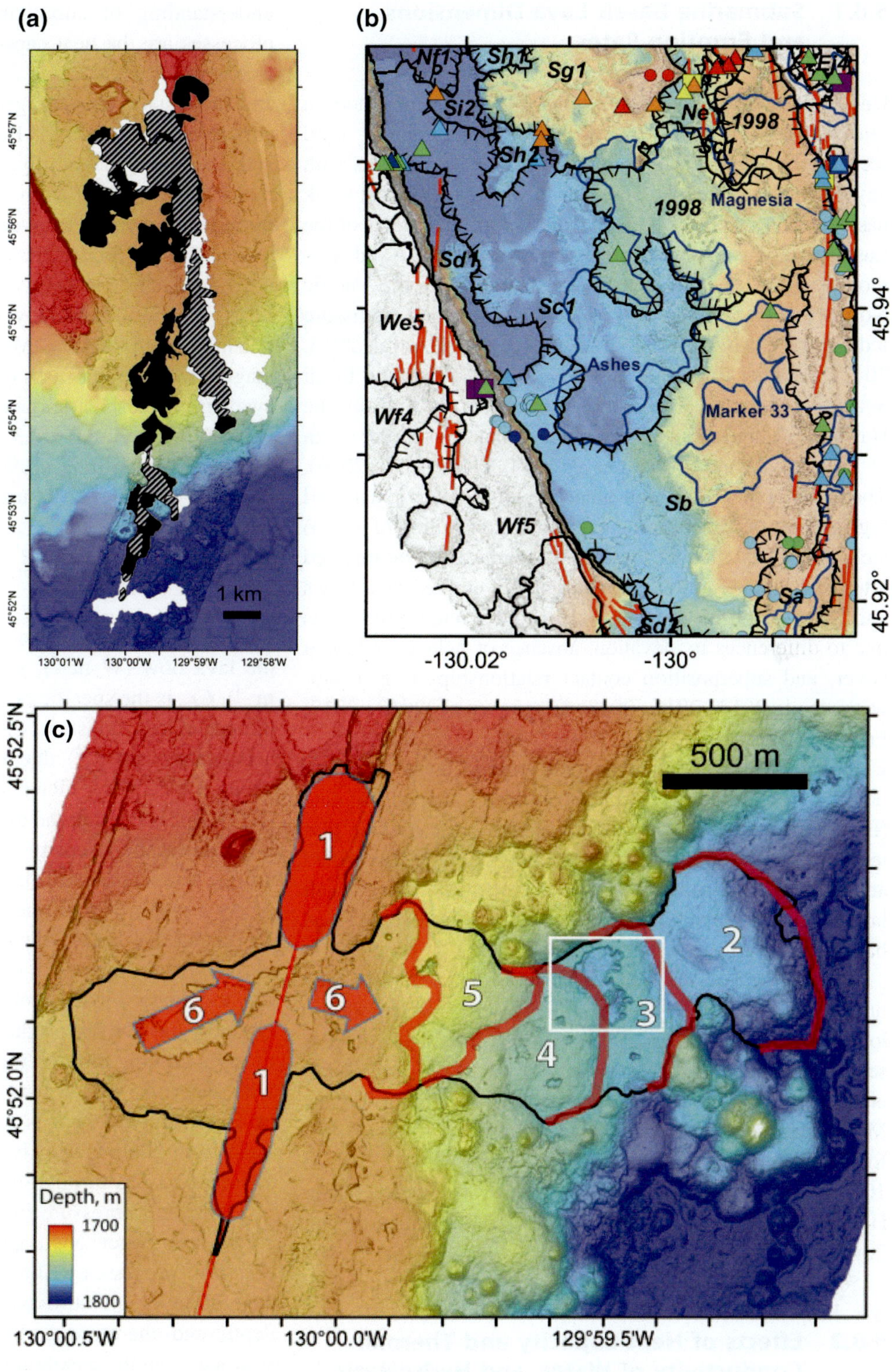

Fig. 5.39 Maps of recent submarine lava flow fields from the Axial Seamount region, Juan de Fuca Ridge, northeast Pacific Ocean. **a** Showing in white the extent of the 1998 lava flow field, and in black extent of the 2011 lava flow field; the hachured pattern depicts where the 2011 flows are superimposed over the 1998 lava flows. Colour ramp is between 1,500–1,700 depth m b.s.l. (From Chadwick et al. 2013). **b** Area of the western margin of the Axial Seamount caldera, showing mapped margins of individual lava flows (From Clague et al. 2013). **c** Individual lava at the southern end of the 1998 lava field, showing the relative chronology of individual lobes, from 1 the earliest (fissure eruption that produced a ridge of pillow and lobate lavas), to 6 the youngest. In the eastern part of the lava, lobes are back-stepping back towards the vent area, presumably in response to waning magma discharge rate. Colour ramp is between 1,500 and 1,570 m depth b.s.l. (From Chadwick et al. 2013)

large areas of the seafloor to be mapped to the detail of defining individual lava flow fields, lava flows, and lobes (e.g. Fig. 5.39a–c; Clague et al. 2002, 2013, 2017, 2019; Fundis et al. 2010; Chadwick et al. 2013, 2016).

There is still much to learn, but it seems that the hierarchy and terminology for lava provinces, fields, flows, lobes, and flow toes, discussed above for subaerial basalt lava systems, can also be applied to subaqueous lavas. First, however, we should consider the known dimensions of submarine lavas and briefly the impact of the subaqueous physical environment on lava-forming eruptions and the behaviour of subaqueous/subglacial basalt lavas, as discussed in Chap. 3.

5.8.1 Submarine Basalt Lava Dimensions and Eruption Rates

Understanding of the dimensions of individual modern seafloor lava flows is very limited, constrained, until recently, by the inability to map the full extent of lavas with any certainty, especially large, far-flowing lavas. Most work has been done on lavas erupted from specific segments of the East Pacific Rise spreading ridge, including the Axial Seamount region on the Juan de Fuca Ridge off the Pacific Northwest coast of North America, most of which are fissure fed lavas (Chadwick et al. 2013, 2016; Clague et al. 2013, 2017). Another important area of study has been the Loihi Seamount which is the youngest active volcano in the Hawai'ian-Emperor seamount chain, located offshore, south of Kilauea volcano (Fornari et al. 1988; Clague et al. 2019). The development of high-precision side scan sonar and high-resolution bathymetric mapping (to resolution of <1 m) technology has allowed these ridge segments to be mapped after each new eruption, allowing newly erupted lava flows and fields to be distinguished from older underlying lavas due to differences in elevation, absence of pelagic sediment cover, and superposition contact relationships (e.g. Chadwick et al. 2013, 2016, 2018; Clague et al. 2013, 2017; Fundis et al. 2010).

However, only relatively small volume lavas and flow fields of small areal extent have been mapped to date (Fig. 5.39a–c). Given that there are also huge oceanic plateaus (that unfortunately have been only sporadically drilled and cored), it is assumed that there are also submarine basalt lavas of the dimensions and flow distances of subaerial flood basalts as described earlier in this chapter.

The longest lava flows that have been confidently mapped on the modern seafloor from the East Pacific are tens of km long (Gregg and Fornari 1998). More commonly with present constraints, flows have been mapped to several km either side and along the axis of the spreading ridges. Magma discharge rates calculated for submarine basalt lava forming eruptions are generally in the order of 10^2 to 10^3 m^3 s^{-1}, comparable to eruption rates from recent Hawai'ian eruptions (Chadwick et al. 2013; Clague et al. 2017; Soule et al. 2005).

5.8.2 Effects of Heat Capacity and Thermal Conductivity of Water, and Hydrostatic Pressure on Submarine Basaltic Eruptions

As discussed in Chap. 3, there are major differences in the effects of an ambient aqueous medium, especially deep-water environments, relative to air, on subaqueous eruption processes and the deposit types produced (Cas and Simmons 2018). The principal factors that affect our understanding of submarine forming lava eruptions and processes are the heat capacity and thermal conductivity of water, which are 4× and 25× higher, respectively, than air, and hydrostatic pressure gradient (10,000× greater than the atmosphere; Table 3.3). Because of its high heat capacity and thermal conductivity, water acts as a heat sink and rapidly absorbs heat from the surface of a subaqueous lava, through conduction, convection, and even radiation, even if vapour film boiling occurs at the interface between water and magma (Chap. 3). As a result, the margins of subaqueous lavas will cool extremely quickly, pass through the glass transition, and solidify. Most of the heat loss/transfer from magma to the water mass occurs through convection of heat away from the lava surface by the water mass (Gregg and Fornari 1998; Gregg 2013; Cas and Simmons 2018).

Fink and Griffiths (1990) proposed that the convective heat flux from the lava surface can be represented by.

$$F_{CONV} = \rho_{sw} C_{sw} \gamma \left[\frac{g' \alpha_{sw} K_{sw}}{v_{sw}} \right]^{1.33} (T_c - T_a)^{1.33} \quad (5.7)$$

where F_{CONV} is the convective heat flux from the surface of the lava flow (W m^{-2}), ρ_{sw} is the density of seawater (kg m^{-3}), C_{sw} is the specific heat of seawater (J kg^{-1} K^{-1}), γ is a dimensionless constant (0.1), g' is reduced gravitational acceleration (m s^{-2}) due to the immersion of magma in water, α_{sw} is the coefficient of thermal expansion of seawater (K^{-1}), K_{sw} is the thermal diffusivity of seawater (m^2 s^{-1}), v_{sw} is the kinematic viscosity of seawater (m^2 s^{-1}), and T_a and T_c are the temperatures of the seawater and lava crust (°K).

As a result of rapid heat loss, Griffiths and Fink (1992) calculated that basaltic lava at 1,000 °C would form a solid glassy crust within 0.7 s on contact with seawater. According to Gregg (2013), submarine mid-oceanic ridge lavas typically have a rind of glass that is ~2 cm thick as a result of rapid cooling through the glass transition through contact with water. An effect of such a considerable solid outer crust on a subaqueous lava is to insulate the magma under the crust from cooling quickly, which enhances the mobility of submarine lavas through enclosed tube-fed processes.

Hydrostatic pressure in subaqueous environments, especially deep-water ones, can also have a major effect on eruption processes (Chaps. 3 and 4). Because volatile exsolution is pressure constrained, the greater the water depth and therefore hydrostatic pressure, the more constrained volatile exsolution becomes relative to subaerial environments. If volatile exsolution is suppressed, dissolved magmatic H_2O that is retained in the magma makes the magma less viscous than a subaerial counterpart. High hydrostatic pressure also ensures that the viscosity of magmas erupted on the deep seafloor is less viscous than subaerial counterparts, since magma viscosity decreases as confining pressure increases (Chap. 3). Again, this will enhance the potential for lava mobility. For these reasons also, subaqueous a'a lavas do not seem to form, i.e. basaltic magma in deep water is not viscous enough (Gregg 2013). Exceptions are where subaerial a'a lavas have flowed into the sea, such as during the 1969 phase of the Mauna Ulu eruption, when a subaerial a'a lava that had entered the sea was observed to have flowed at least several hundred m offshore beyond the 70 m water depth level (Moore et al. 1973).

Another important effect of hydrostatic pressure is to inhibit vesicle growth with increasing water depth (Chaps. 3 and 4). As a result, both vesicle size and abundance decrease with depth as first demonstrated by Moore (1965) from dredge samples collected off the coast of Hawai'i, and Moore and Schilling (1973) from the Reykjanes Ridge, Iceland. While other effects such as basalt composition and volatile content are important in controlling vesicle development, this method can provide a useful paleoenvironmental indicator of relative seafloor depth at the time of eruption (e.g. Corcoran 2000; Furnes et al. 2011; Schnur and Gilbert 2012). As a rule of thumb, tholeiitic lavas erupted in water depths of greater than 1,000 m have less than 5–10% vesicles with mean diameters of <1–2 mm, whereas vesicularity approaches zero at an eruption depth of about 4,000 m. However, we stress some caution is required, as vesiculation can occur at any depth if the magmatic water content is sufficiently high (Cas and Simmons 2018). In magmas erupted at water depths (or under thick ice) and pressures where the rate of decompression, and the level of gas overpressure are too low to drive explosive fragmentation, highly vesicular lavas (even with rhyolitic compositions; Chap. 6) can form because gas bubbles will grow more slowly than under subaerial atmospheric conditions. Slow, non-explosive vesiculation of magma at high pressures can therefore produce highly vesicular lavas, e.g. deep submarine-erupted alkalic basalt lavas (56% vesicles) described from the North Arch Volcanic Field, north of Hawai'i in water depths of 3900–4480 m (Dixon et al. 1997).

As summarised in Chaps. 3 and 4 and discussed in detail by van Otterloo et al. (2015), when hot magma comes in contact with water, it causes the water at the interface to boil producing a vapour film, in a process called film boiling (Mills 1984). If the vapour film remains stable for a few seconds the heat loss from the surface of the magma across the vapour film by radiative and convective heat loss is slow enough to form a coherent visco-elastic crust that may then ensure that the lava remains coherent. If, however, the vapour film collapses and there is direct contact between the water and still liquid magma, the magma at the interface will instantaneously chill through the glass transition because of the very high rates of conductive and convective heat loss to the water mass. The chilled glass will be highly stressed because it contracts and as a result it shatters into an aggregate of angular glassy debris or hyaloclastiteby a process called quench fragmentation (Sect. 5.12; Chap. 4).

It is important to emphasise that quench fragmentation *is a mostly non-explosive process*, especially in deep water, although in shallow water where hydrostatic pressure is low, phreatic explosions may accompany quench fragmentation. In particular, cooling contraction quench processes can produce in situ cooling contraction fracture networks as well as pervasive quench fragmentation (Chaps. 3 and 4; Cas 1992; Cas and Giordano 2014; van Otterloo et al. 2015) resulting in large volumes of autoclastic hyaloclastite breccia and finer, granulated hyaloclastite sandstone. For more details on the physics of formation of hyaloclastite and clast characteristics, see Chap. 4 and van Otterloo et al. (2015).

Since exploration of the seafloor began utilising seafloor photography, it is striking how much of the seafloor is covered with coherent basalt lavas, indicating that vapour film insulation has been very effective in protecting propagating lavas from quench fragmentation. This is enhanced by slow lava propagation or flow velocities, which ensures a stable boundary steam layer at the interface between the lava surface and the water mass. It is also likely that as the hydrostatic pressure increases with water depth that the vapour film changes from steam in shallower waters (<1,500 m) and becomes progressively dense steam, then progressively denser supercritical fluid from 2,000 to 3,000 m water depth and deeper. Such a supercritical fluid has a higher thermal conductivity than steam and so would reduce the thermal gradient across the magma–water interface, and the thermal shock potential that is fundamental to both phreatic explosive activity and quench fragmentation (Chaps. 3 and 4). It is likely that at water depths of >3,000 m less quench fragmentation and hyaloclastite forms than in shallower water depths.

5.8.3 Effects of Slope and Magma Discharge Rate on Submarine Lava Morphology

It has long been suggested that magma effusion or discharge rate influences lava flow morphological type, with massive sheet lavas forming at high effusion rates, and pillow lavas corresponding at low effusion rates (Ballard et al. 1979; Bonatti and Harrison 1988; Perfit and Chadwick 1998). Perfit and Chadwick (1998) also suggested that high spreading rate mid-oceanic ridges promoted high magma effusion rates, and so sheet lavas would predominate, whereas at low spreading rate ridges, low magma effusion rates would prevail, so producing more pillow lavas.

In experimental assessments of the effects of slope on lava morphology, Gregg and Fink (2000) found that as substrate slope increases, experimental analogue flows became narrower, thinner, and longer and that a succession of morphological flow types occurred from pillow-lobate, slabby, folded, and channelled-leveed in response to increasing slope and/or eruption rate, both of which have the effect of reducing the cooling rate, and so slowing crust formation. In a study of the Puna Ridge, Hawai'i, Gregg and Smith (2003) found that sheet flows only form on slopes <15°, lobate and pillow basalts on slopes <25°, and on slopes >25° flows disintegrated into rubble flows. They proposed that experimental results may be a little misleading because there are tensile strength differences between the analogue materials used and basalt magma. Fundis et al. (2010) were more equivocal about the effect of slope on lava morphology based on mapping the 2005–2006 lava flow field erupted on the modern seafloor on a segment of the East Pacific Rise between the Clipperton (~ 10°N) and Siqueiros (~ 8°30′N) transform faults, concluding that lava effusion rates were a more important control.

5.8.4 Lava Flow Inflation, Tube Feeding, and Channel Propagation

Like subaerial counterparts, subaqueous basalt lavas also propagate and advance through the formation of an insulating crust and, under the crust, supply of magma from the vent to the advancing flow front through a tube feeding mechanism (Fornari 1986). Interconnected lava pillows and lobes (Fig. 5.40a, b; also see Figs. 5.43 and 5.44b) represent an interconnected feeder tube network. Collapse pits or cavities on the surfaces of some lavas on the modern seafloor indicate that lava drainage has created vacant tubes under some lava crusts, and high hydrostatic pressure then causes cavitation or implosions of the crust into the empty tubes beneath (Chap. 3; Cas and Simmons 2018). Within some of these collapse pits, relicts of the roof crust are still supported by lava pillars (Fig. 5.40c), the vertical sides of which are marked by horizontal layers reflecting the "tide" marks of the rising and falling lava in the original tubes (Chadwick 2003). Such collapse pits may also contain debris collapsed from the tube roof crust (Fig. 5.40f). Evidence of inflation and deflation was also recorded when a pressure gauge was engulfed by a lava flow erupting on the sea floor at Axial Seamount (Chadwick 2003). Applegate and Embley (1992) also documented tumuli on the surface of lavas at Axial Seamount as evidence of lava flow inflation in submarine flows.

In addition, several studies have recognised that submarine basalt lavas can also propagate by the development of morphologically defined channels, which are roofed over with a crust that has subsided relative to more elevated channel margins (e.g. Fornari 1986; Soule et al. 2005; Chadwick et al. 2016; Clague et al. 2017, 2019).

Fig. 5.40 1998 Axial Seamount eruption submarine basalt lava types and surface textures (From Chadwick et al. 2013). **a** Pillow lava on older lineated sheet flow. **b** Lobate morphology of lava where it is >1–2 m thick. **c** Relict of drained and collapsed lava tube roof supported by pillars under lobate morphology sheet flow. The horizontal layering on the pillars reflects lowering "tide" marks as the lava level dropped. **d** Lineated sheet flow. **e** Jumbled sheet flow, with autobrecciated crust that features broken pieces of ropy-textured crust. **f** Collapsed crust breccia

5.8.5 Subaqueous Basalt Lava Types and Facies

Pillow lavas (Plate 5.1d, f; Fig. 5.40a, also see Fig. 5.44) were regarded as the archetypal submarine basaltic lava type for a long time, based on studies of exposed, uplifted ancient subaqueous basalt lava successions. However, as soon as visual observations of the modern seafloor using cameras and submersible vessels and deep-sea drilling became possible, major advances in understanding deep ocean floor volcanic processes were made (Cas 1992; Rubin et al. 2012; Gregg 2013; White et al. 2015). It was soon realised that *massive sheet lavas* were also common (Fig. 5.40c–f; Ballard et al. 1979; van Andel and Ballard 1979), in many places they were lobate or hummocky, and there was a range in surface textures, including smooth, rippled/ropy, lineated, jumbled or chaotic (Fig. 5.40b–e), and hackly (Ballard et al. 1979; van Andel and Ballard 1979; Fox et al. 1988; and Embley et al. 1990; Chadwick et al. 1999).

Although there are some similarities between subaerial and some aqueous basalt lava flows, there are also significant differences. These differences result from very different physical properties of air and water as ambient fluids into which magmas are erupted, as discussed in Chap. 3, and above. A'a lavas are very rare in subaqueous settings, which Gregg (2013) attributes to the lower viscosity of erupted deep submarine basaltic magmas resulting from suppression of volatile exsolution due to hydrostatic pressure. Where they are known they occur in shallow water they entered the water from subaerial vents. Most subaqueous lavas are therefore essentially pahoehoe lavas. However, due to the varying nature of interaction of the erupting lavas with the water mass, the lavas can be coherent, and/or quench fragmented. Variations in effusion rate at the vent, changing magma supply rates at the distal margins of spreading lavas, changes in the slope of the subaqueous environment, and changing viscosity can result in distinctive lava flow types. These are now commonly subdivided into pillow (Fig. 5.40a), lobate (Fig. 5.40b) and sheet (Fig. 5.40b–f) flows (Ballard et al. 1979; Fox et al. 1988; Gregg and Fink 1995; Fundis et al. 2010; Chadwick et al. 2013; White et al. 2015; Clague et al. 2017, 2019), but on steep slopes, rubble flows also form (Lonsdale and Batiza 1980; Gregg and Smith 2003). In addition, distinctive surface textural features have been observed on modern seafloor basalt lavas, including smooth, rippled (or ropy; Plate 5.1d; Fig. 5.40a), lineated (Fig. 5.40c), jumbled or chaotic (Fig. 5.40e), and hackly (Fox et al. 1988; Griffiths and Fink 1992; Gregg and Fink 1995; Gregg and Fink 2000; Fundis et al. 2010; Chadwick et al. 1999, 2013), which have all been duplicated experimentally under varying conditions (Gregg 2013).

5.9 Sheet Lavas and Surface Textures

Sheet lavas have been observed on the sea floor (Fig. 5.40c; e.g. Ballard et al. 1979; Clague et al. 2002, 2013; Chadwick et al. 2013), intersected in numerous DSDP/ODP drill holes, and have been documented in ancient submarine successions, commonly intimately associated with pillow lavas (e.g. Figs. 5.41 and 5.42; Dimroth et al. 1978; Hargreaves and Ayres 1979; Schmincke and Sunkel 1987; Squire et al. 1998; Corcoran 2000). Bonatti and Harrison (1988) proposed that below the maximum depth for explosive volatile exsolution, the eruption style of basalt is controlled by

Fig. 5.41 Irregular base of an Archean massive submarine sheet lava. The breccia at the base is hyaloclastite, ~2,710 Ma Lunnon Basalt, 1839 level, Otter-Juan Mine, Kambalda, Western Australia. (From Squire et al. 1998)

viscosity (dependent on eruption temperature, composition, volatile content, crystallinity; Chap. 3) and magma discharge rate. They proposed that the formation of submarine lava "lakes" (ponded lavas) and sheet lavas is enhanced where the eruption temperature and magma discharge rate are high, and the viscosity and crystal-to-liquid ratio of the magma are low.

Ballard et al. (1979) also proposed that submarine sheet lavas are erupted at high magma discharge rates producing an internally massive facies (i.e. not lobate or pillowed), whereas pillow lava is erupted at lower discharge rates and is tube fed (see also, e.g. Dimroth et al. 1978; Schmincke and Sunkel 1987).

Massive sheet lavas can form extensive and thick submarine lava fields (Fig. 5.42; Hargreaves and Ayres 1979; Cas 1992; Squire et al. 1998; Clague et al., 2002, 2013; Gregg and Fornari 1998), with the most extensive lavas so far confidently traced on the modern seafloor extending >100 km (Gregg and Fornari 1998). They also contribute to the growth of volcanic edifices (e.g. Holcomb et al. 1988). Such flows may cover hundreds to thousands of km^2 of the seafloor, they may be multi-lobed and multi-tiered, have smooth or brecciated surfaces, and may have pillowed margins. They propagate through interior lava tubes and channels analogous to subaerial lavas (Fornari 1986; Cas 1992). Although sheet lavas and/or sheet lava sequences may be tens of m thick (Fig. 5.42), individual flows are commonly less than 10 m thick (Clague et al. 2013).

Sheet lavas may have smooth surfaces, but may also have ropy surfaces, and lineated surfaces, defined by closely spaced ridges or ropy folds on lava crusts that are 1 to 2 cm high and have a wavelength of 1–3 cm (Fig. 5.40a, d, e; Chadwick et al. 1999, 2013; Gregg 2013). Lineated surfaces (Fig. 5.40d) are often found where a flow is channelised and the lineation is parallel to the margins of the flow. In addition, in places they have autobrecciated or "jumbled" or "hackly" surfaces, defined by clasts of broken crust which are irregularly oriented and piled up (Fig. 5.40e), resembling subaerial slabby pahoehoe lava. They may also be subject to quench fragmentation at their margins, including the base (Fig. 5.41).

According to Chadwick et al. (2013), at a larger observational scale, sheet lavas on the surface appear to consist of large flat lobes that have coalesced to form a sheet with an integrated *sheet magma tube system* beneath the crust. Lobes are not visible internally because of the coalescence of the lobes, indicating that the thin original lobe margin crusts have fused and assimilated. After eruption has ceased, magma in the *sheet tube system* continues to drain to the flow front causing evacuation of part of the tube as the lava level drops. The crust may then collapse (Fig. 5.40c), producing collapse pits, and many such collapses are likely to represent implosions of the crust under the effect of very high hydrostatic pressure, filling the tube floor with collapse debris (Fig. 5.40f).

5.10 Lobate Flows

Lobate flows consist of visible large lobes 1–3 m in diameter (Fig. 5.40b). These lobes may coalesce and fuse together to form sheets (Gregg 2013; Chadwick et al. 2013), but otherwise the lobes are preserved. Chadwick et al. (2013) use the term "*inflated lobate flows*" for the near-vent or proximal morphologies that result from high effusion rate eruptions, which distally pass into "*inflated pillow flows*", "inflated" refers to the presence of sheet-like flows as described above that had a molten interior beneath a solidified crust during emplacement analogous to inflated flows on land (Hon et al. 1994; Sect. 5.2.3.6). Inflated flows are initially thin and subsequently increase in thickness after they stop advancing laterally but while molten lava is still being added to the fluid interior of the flow from the eruptive vents. After this inflation phase, the still molten interior of such flows may drain-out downslope. Where drain-out of magma occurs from tubes in submarine inflated flows, it leads to collapse of the (now unsupported) upper solidified crust (Fig. 5.40c). Francheteau et al. (1979) recognised collapse breccias near the axis of the East Pacific spreading ridge.

5.11 Pillow Lava Flows

Pillow lavas can form in their own right (i.e. the whole lava flow may consist of pillow lava), or as now commonly documented, they occur as the marginal distal facies of sheet and/or lobate flows or facies domains that transition into pillow lava at the distal margins (Dimroth et al. 1978; Hargreaves and Ayres 1979; Fundis et al. 2010; see Fig. 5.50). A single lava flow could thus consist of all three facies as a transition from proximal-to-distal margins. These transitions reflect changing magma eruption rates at vent and/or changing magma supply rates to the advancing flow front as a lava spreads, increases surface area, and magma supply rate to the flow front declines.

5.11.1 Pillow Lava Propagation

Observations of the formation of pillow lava in shallow water from offshore Hawai'i (e.g. Moore et al. 1973; Moore 1975), and in deep water at 1200 m water depth at West Mata volcano, Lau Basin (e.g. Resing et al. 2011), indicate that when a new pillow forms a lobe of incandescent magma propagates from a precursor lobe (Fig. 5.43a).

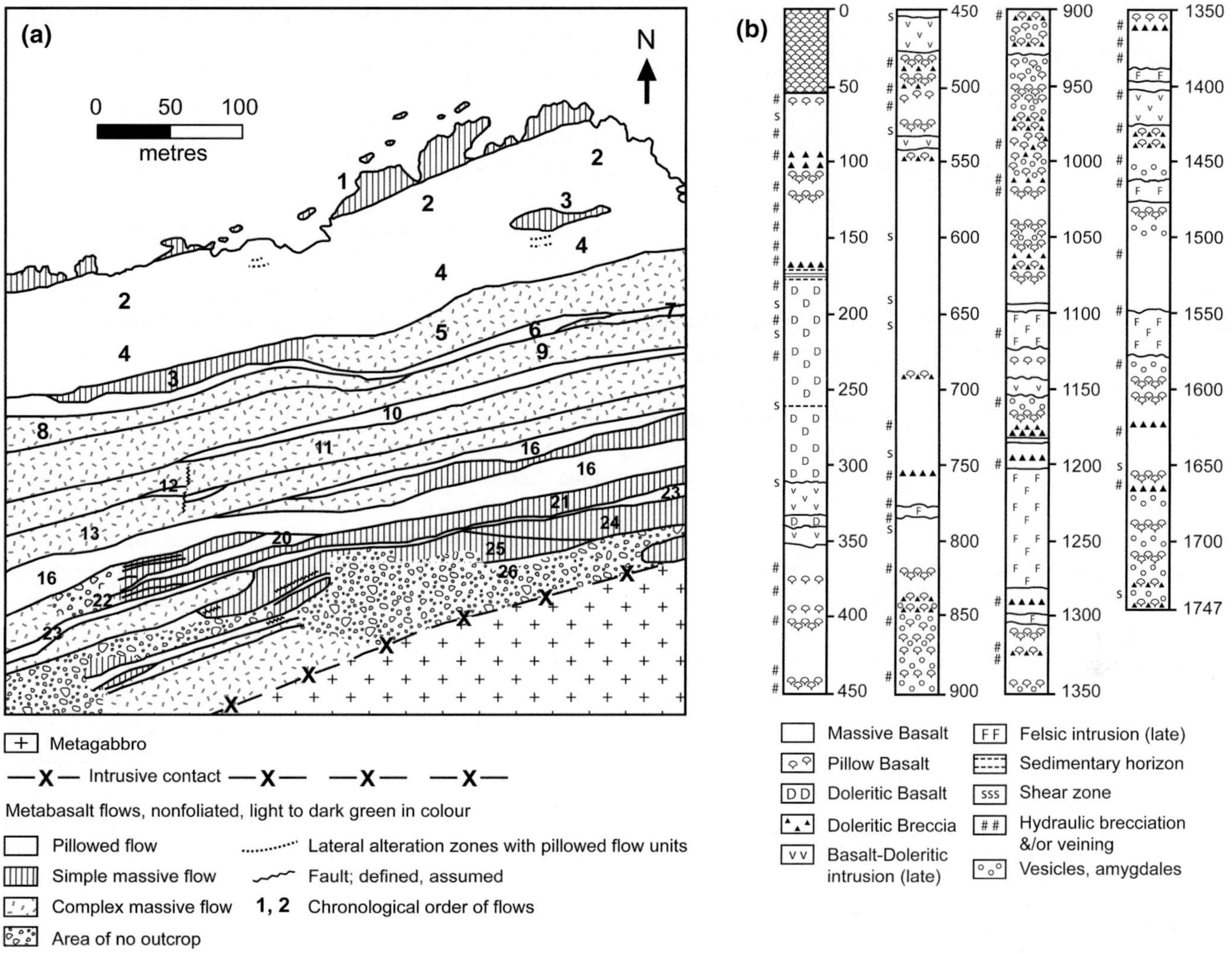

Fig. 5.42 Map and graphic log of Archean massive submarine basalt sheet lava sequences. **a** Association of massive sheet and pillow lavas, Manitoba, Canada (From Hargreaves and Ayres 1979). **b** Log of succession of massive and pillow lavas, and associated hyaloclastite breccias, diamond drill hole KD1029, >1,700 m thick, ~2,710 Ma Lunnon Basalt, Kambalda, Western Australia (After Squire et al. 1998)

A visco-elastic crust forms instantaneously as the pillow propagates and expands as magma is injected under the crust through the pillow tube. The pillow may become stationary as the skin or crust chills and solidifies to glass and constrains further expansion, until the magma pressure inside the pillow exceeds the tensile strength of the crust, causing it rupture and allowing a new pillow bud to breakout from the pillow, so repeating the pillow-forming process (Fig. 5.43b). The new pillow buds or lobes originate from cracks in the pillow skin called longitudinal and transverse spreading cracks (Fig. 5.44b; Yamagishi 1985). As a result of this lobe- or tube-like mode of propagation, pillow lavas thus become a network of interconnected elongate tubes or lobes, which in longitudinal view are like stacks of connected sausages (Figs. 5.43c, 5.44b and 5.45a; Yamagishi 1985), and in cross-section look like a stack of balloons or inverted tear drops, with a convex up inflated top and a lower angular re-entrant where the bottom of each pillow moulds into the shape of the space between the convex tops of two underlying pillows (Figs. 5.44a and 5.45b, d, e). Pillow lavas are therefore tube-fed, pahoehoe-like lavas which allow magma to be transferred from the vent through the internal interconnected pillow tube system to the advancing pillow lava front.

Observations from shallow seas offshore from Hawai'i also showed that pillows could easily become detached from the feeder tube system by gravitational collapse and tumble down steep slopes (steep lava sides or subaqueous slope) as clasts of detached whole and/or broken pillows (Moore et al. 1973). They also imploded as magma drained downslope from inside a pillow tube. Associated quench fragmentation also contributed to the formation of associated pillow fragment debris.

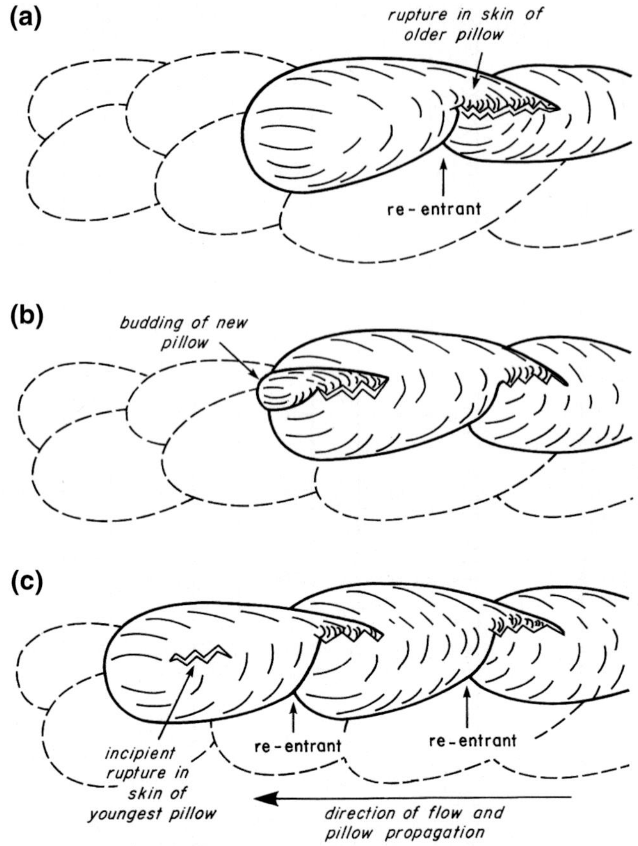

Fig. 5.43 Schematic representation of lobe-by-lobe propagation of pillow lava. **a** The surface of a newly propagated pillow cools, forming a visco-elastic skin, as more magma is injected under the skin into the pillow, causing it to expand. **b** When the skin solidifies it can no long stretch, so it ruptures and a new small pillow bud bursts out, expands, and develops a new visco-elastic skin. **c** The process is repeated, forming an interconnected pillow tube system for transferring magma from the vent to the advancing lava flow front. (From Cas and Wright 1987)

5.11.2 Morphology of Pillows

The morphology and features of pillow lavas have been well documented from both modern and ancient successions (e.g. Jones 1968; Moore et al. 1973; Ballard and Moore 1977; Dimroth et al. 1978, 1985; Wells et al. 1979; Staudigel and Schmincke 1984; Yamagishi 1985, 1989; Busby-Spera 1987; Yamagishi 1985, 1991; Yamagishi et al. 1989; Kawachi and Pringle 1988; Walker 1992; Bear and Cas 2007; Schnur and Gilbert 2012). Well-preserved modern (Plate 5.1d, Fig. 5.40a) and ancient pillow lavas (Fig. 5.45) show the three-dimensional and interconnected nature of pillow lava lobes (Figs. 5.44b and 5.45a) as observed by Jones (1968) in field studies in Iceland and Wales, as well as by other authors in the Canadian Archean (Hargreaves and Ayres 1979); the Eocene–Oligocene at Oamaru, New Zealand (Kawachi and Pringle 1988); and the Miocene at Muriwai, New Zealand (Bear and Cas 2007).

The diameter of pillows can vary from flow to flow, or within single flows, and appear to be controlled by composition and emplacement temperature (which both influence viscosity), magma eruption rate, and slope (Furnes and Fridleifsson 1979; Dimroth et al. 1978; Staudigel and Schmincke 1984; Walker 1992; Schnur and Gilbert 2012). They usually vary from 0.5 to 1 m in diameter (Fig. 5.45b–g), and each interconnected, budded lobe is 1–3 m long, forming a sausage-like chain of tube-like pillow lobes (Fig. 5.44b). Pillows can branch and change direction (Fig. 5.45a, c), and they are usually stacked on top of each other (Figs. 5.44a and 5.45b, d–g), representing a prolonged period of lobe by lobe, and/or simultaneous lobe propagation and emplacement. A relatively small pillow size indicates either a relatively high effusion rate and low pillow viscosity, or a steep depositional slope. A steeper slope leads not only to smaller pillows, but to higher aspect ratios (i.e. rounded rather than flattened). Distinctive tube-like morphology, with pillows elongated parallel to flow direction, is also evidence of a steep slope. A steeper slope also results in the increased collection of interstitial quench fragmented glass fragments (hyaloclastite), spalled from the glassy crusts of the interfacing pillows.

In cross-section, individual pillows have bulbous, nearly circular shapes with an upper convex form (Figs. 5.44a and 5.45b–g) and a lower margin that is often a V-shaped, angular re-entrant, which reflects accommodation of the base of the pillow into the space between two underlying pillows (Figs. 5.44a and 5.45b, d). In old, exposed pillow lava successions, these distinctive asymmetrical shapes of pillows can be used to determine the younging directions of the stratigraphic succession hosting them (e.g. Fig. 5.45d, e), including structurally overturned successions (Fig. 5.45e). The median cross-sectional aspect ratio (V/H, where H is the horizontal dimension parallel to dip and V is the vertical dimension, orthogonal to H) of most basaltic pillows is ~0.5–1 (Fig. 5.45b, d; Walker 1992; Schnur and Gilbert 2012), indicating round (Fig. 5.45b) to slightly flattened (Fig. 5.45c) pillows elongated in the flow direction. This also reflects an apparent, moderately high viscosity, although it is largely controlled by the strength of the chilled glassy crust or selvedge. This glassy crust is 1–3 cm thick and is sometimes duplicated (Kawachi and Pringle 1988), representing multiple phases of propagation and magma injection through the pillow tubes. The interior is not wholly glassy, but has a microcrystalline groundmass resulting from slower cooling of the pillow interior.

5.11.3 Pillow Lava Surface and Internal Features

Yamagishi (1985) and Yamagishi et al. (1989) illustrated a range of features on the surfaces of pillows (e.g. ropy wrinkles, corrugations, extensional and polygonal, cooling, contractional cracks) (Fig. 5.44a, b). Whereas the surface of the glassy crust has a network of narrow, to cryptic, polygonal, cooling contraction cracks, the interior is marked by (slightly) more widely spaced, radial, small-scale, cooling joint (Figs. 5.44 and 5.45b, f) reflecting the slower cooling rate of the interior of pillows. The interior of pillows can be vesicular, with vesicles varying from spherical (Fig. 5.45d) to pipe shaped (Plate 5.1f). Vesicles are larger near the margins, and pipe vesicles are located near the margins. This reflects lower pressures and higher levels of degassing within the marginal zone of pillows (Plate 5.1f). In addition, some pillows may contain sub-horizontal gas cavities with flat floors and convex upper margins (Figs. 5.44g and 5.45a). These may be up to 15 cm long and several cm high, and represent gas vughs, and/or partial drainage cavities resulting from outflow of magma from a pillow tube. A special case of such features has been termed *"pyjama-stripe pillows"*, which consist of interlayered, altered (green or red) basalt cavity "shelves" and secondary pale-coloured, hydrothermal minerals (calcite, quartz, zeolite) within the cavity forming a striped pattern (Palinkaš et al. 2008; Kiss et al. 2008; see Fig. 5.50h). These are thought to form within water-saturated sediments and represent the alternating

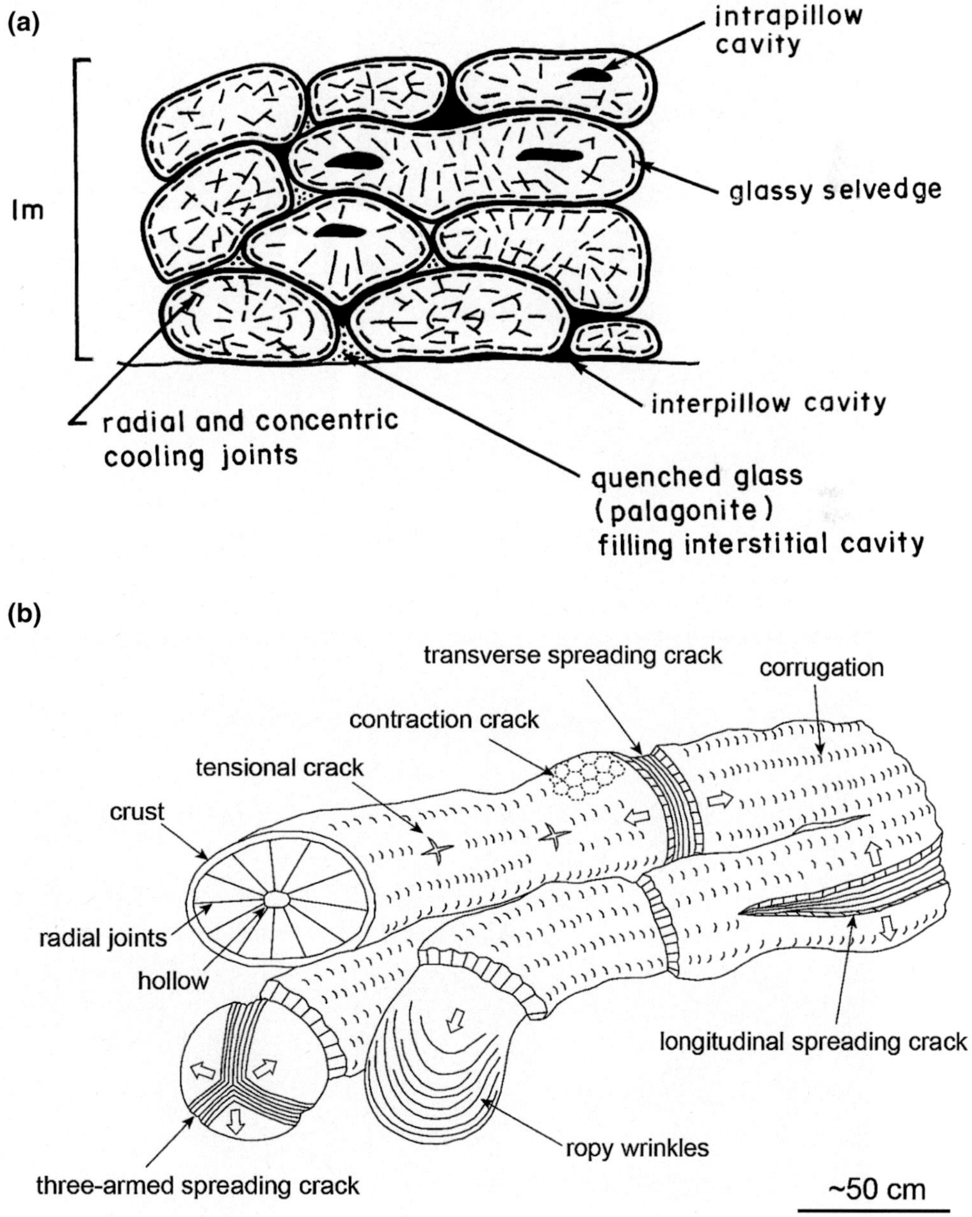

Fig. 5.44 Schematic representation of the features of pillow lavas. **a** Schematic cross-section of pillow lava, depicting the mutual accommodation geometry with upper convex form of pillows and their angular re-entrant base, chilled glassy crust, internal radial, and, in some cases, concentric cooling joints or fractures, and internal drainage cavities (From Cas and Wright 1987). **b** Schematic representation of the morphological and surface textural features of pillow lavas (From Goto and McPhie 2012, After Yamagishi 1985)

(a)
(b)
(c)
(d)
(e)
(f)
(g)

◀ **Fig. 5.45** Features of pillow lavas in outcrop. **a** Longitudinal perspective of uplifted Miocene pillow lava at Maori Bay, Muriwai, New Zealand, showing the interconnected nature of the train of sausage-like pillow lobes (middle-to-lower part of outcrop). Outcrop is ~3 m high. **b** Cross-sectional perspective through an Eocene–Oligocene pillow lava, Boatman's Harbor, Oamaru, New Zealand. Pillows are about 0.5–0.75 m in diameter. They are defined by a black glassy crust, with a tortoise shell, closely spaced, polygonal, expansion, tensional crack network on the surface, more widely spaced, crudely radiating cooling fractures internally, and upper convex surface, and for many pillows a lower "v"-shaped re-entrant shape where the base of the pillow mimics the surface relief on underlying convex surfaces of pillows. The convex upper surface and lower angular re-entrant can be used in ancient pillow lava successions to determine the younging and structural facing direction (see Sects. 5.11.1 and 5.11.2). **c** Plan view of wave platform exposure of pillow lava underlying the Mt Etna edifice, at Castello, Sicily, Italy, showing a branching and internally quench fractured pillow lava. **d** Siluro-Devonian pillow lava in the southern Hill End Trough, Tuena, central New South Wales, Australia, showing shape accommodation, and a diffuse, concentric layering defined by amygdales. **e** Slightly overturned ~2,690 Ma Archean pillow basalt lava, Paringa Basalt, Mt Hunt, Boulder–Kambalda Road, Western Australia. The upper convex surface shape (to the lower left) and the basal re-entrant shape (towards the top right) indicate the stratigraphic younging and structural younging are to the lower left of the field of view. **f** Internal, spaced cooling fracturing in Miocene pillows at Maori Bay, Muriwai. **g** Elongate cavities with flat floors inside a pillow, reflecting magma drainage, Eocene–Oligocene pillow lava, Oamaru

draining of lava from the burrowed pillow lobe and invasion of sea water into the emptied cavity, followed by rapid precipitation deposition of hydrothermal minerals from heated seawater.

At Muriwai, North Island, New Zealand, cross-sectional coastal exposure of an uplifted Miocene submarine pillow lava pile preserves a complex association of normal pillow lava, enclosing 20 m diameter, radially jointed master feeder tubes (Plate 5.1e), together with several sheet lavas (Fig. 5.46a; Bear and Cas 2007). The basal pillows are larger than those higher in the pile (Fig. 5.46a), reflecting an initial higher effusion rate. Strikingly, the base of the basal pillows is flat (Fig. 5.46b), and perfectly conformable with the bedding in the underlying sedimentary succession. There is no evidence of scouring, or dynamic peperite formation, reflecting the very slow propagation velocity of the pillow lava system, and/or less likely, the compacted, largely dewatered state of the sediments at the time of eruption. The Muriwai pillow lava(s) also taper rapidly from thick to thin at the margins (Fig. 5.46c).

Ancient pillow lavas preserve many of the features of modern pillows, but deformation can impose deformational strain, which highly distorts the shape of pillows (Fig. 5.47a, b, c). In addition, alteration transforms the glassy crust into secondary phyllosilicate minerals (e.g. chlorite) and can mask many features. In some ancient lavas, hydraulic fracturing and chlorite alteration along hydraulic fractures can sometimes mimic pillow rims. They can usually be distinguished as "pseudo-pillows" because they are irregular in form (Fig. 5.47d) rather than smooth, continuous, and spheroidal (Fig. 5.47e).

Many older pillows contain spherical features up to 2 cm in diameter in the outer originally glassy rinds, defining a concentric zone of these features. These features often coalesce (Fig. 5.47f, g) and often form trains. They are devitrification spherulites (Furnes 1973; Dimroth et al. 1978; Fowler et al. 1987; see Chaps. 6 and 13), which are often called *varioles* or *variolites* in mafic lavas. They consist mostly of fibrous, radially distributed plagioclase crystals, and sometimes pyroxene, but in older rocks these are altered to secondary sericite and carbonate, and the surrounding glass could be altered to chlorite under greenschist metamorphic conditions. They typically occur within non- to very sparsely vesicular pillows and for these reasons are thought to be indicators of deep-water (>2000 m) eruption under high hydrostatic pressure (Furnes 1973; Sandstå et al. 2011). Large varioles appear to be particularly common in Archean basalt and komatiite lavas. However, they are not restricted to the Precambrian and have been reported from Paleozoic pillow lavas, and from many deep-sea dredge sites, as well as the Ontong Java and Caribbean-Colombian Oceanic Plateau (see Fig. 15.55d).

A clear similarity exists between subaqueous pillow lava and subaerial pahoehoe lava (Jones 1968). Nevertheless, the two can be distinguished because pillow lobes usually have a higher aspect ratio, thick glassy crust, closely spaced radial cooling jointing, and associated (hemi-)pelagic sediment, which filters into spaces between pillows, none of which occur with pahoehoe toe lavas. In addition, the spaces between pillows may contain interstitial quench fragmented glass crust fragments (hyaloclastite).

In addition, as discussed in Sect. 5.13, pillow lavas can vary significantly in facies characteristics both upwards and laterally. Dimroth et al. (1978, 1985) found that Archean submarine basalt flows at Noranda, Quebec, show lateral transitions from thick massive basalt up to 100 m thick in proximal areas into pillow lava via a transition zone of megapillow tubes or lobes, and then at the distal end, into pillow breccia many m thick (see Fig. 5.51). Busby-Spera (1987) noted that pillow size decreased in a distal direction, as well as upwards. Pillow size reduction within single flows is attributed to declining magma discharge rate during an eruption (Dimroth et al. 1978; Staudigel and Schmincke 1984; Busby-Spera 1987), or due to a declining effective magma supply rate at the distal margins of flows as their surface area increases, even though the absolute magma

Fig. 5.46 Features of the Miocene pillow lava at Maori Bay, Muriwai, New Zealand, interbedded with bedded submarine mass-flow conglomerates and turbidites (also see Fig. 5.45a, f). **a** 20-m-thick succession of pillow lava, with largest pillows at the base, with well-defined radial cooling jointing, and two thin colonnade columnar jointed sheet lavas near the top. To the left is the margin of a large feeder tube. The flat **b** Flat base of basal pillows sitting conformably of bedded sediments underneath. Pillows show higher levels of vesiculation around the margins, including some pipe vesicles. **c** Margin of a pillow lava lobe, near the top of the cliff, tapering to one to two pillows thick

discharge rate at the vent stays constant (Dimroth et al. 1978). Lateral facies changes may thus be significant in pillow lavas, involving in some cases changes from massive to pillowed lava, and at the distal margins, into wholly pillow fragment breccia and hyaloclastite (Sect. 5.13). Vertical facies changes in the size of pillows, vesicularity, and proportion of interbedded massive/sheet, flows and hyaloclastite and pillow fragment breccia may also record the growth and shoaling of a seamount (Corcoran 2000; Schnur and Gilbert 2012).

5.11.4 Pillow Lava Mounds and Ridges

Observations from high-resolution bathymetry of recently active basaltic lava fields in the deep ocean (both mid-oceanic ridges, and flank eruptions of seamounts) have revealed the common occurrence of small volcanic morphological forms comprising pillow mounds and ridges. These are thought to form during eruptions with very low effusion rates, as pillow lava piles up directly over the vent(s). In doing so, they can build cone-like structures generally having basal diameters between 50 and 300 m and heights of 20 and 200 m, or ridges extending several km long and many tens of m thick (Yeo et al. 2012; Chadwick et al. 2013; Caress et al. 2012; Clague et al. 2017, 2019). Clague et al. (2017) modified the terminology to include "*hummocky flows*", which are mostly pillow lavas that form mounds or coalesced mounds or ridges on or close to eruptive fissures. In contrast to the "pillow mounds" proposed and illustrated by Chadwick et al. (2013) that lack molten cores, hummocky flows have molten cores as is evident from off-fissure growth of lava flows fed from on-fissure mounds, collapse pits in their summits, or by complex levee-enclosed lava ponds and tumuli on their surfaces.

Morphological and textural features suggest as well as pillow lava, pillow mounds and ridges also comprise pillow fragment breccia talus deposited at the base of steep-sided hummocky pillow mounds and ridges; these form as pillows cascade down the steep slopes and over near-vertical scarps. Such mounds formed during eruptions in the northeastern Pacific Ocean on the North Gorda Ridge in 1996 (Paduan et al. 2014), and in 1998, 2011, and 2015 at Axial the Seamount on the Juan de Fuca Ridge (Fig. 5.48a; Caress et al. 2012; Chadwick et al. 2013; Clague et al. 2017). Two examples of larger pillow cones (one up to ~300 m tall) have also been identified from near the base of the south rift on the flanks of the Loihi Seamount (Fig. 5.48b; Clague et al. 2019); these occur at 4,800 m below sea level, some 25 km away from the summit plateau and caldera of the seamount which are at ~1000 m depth).

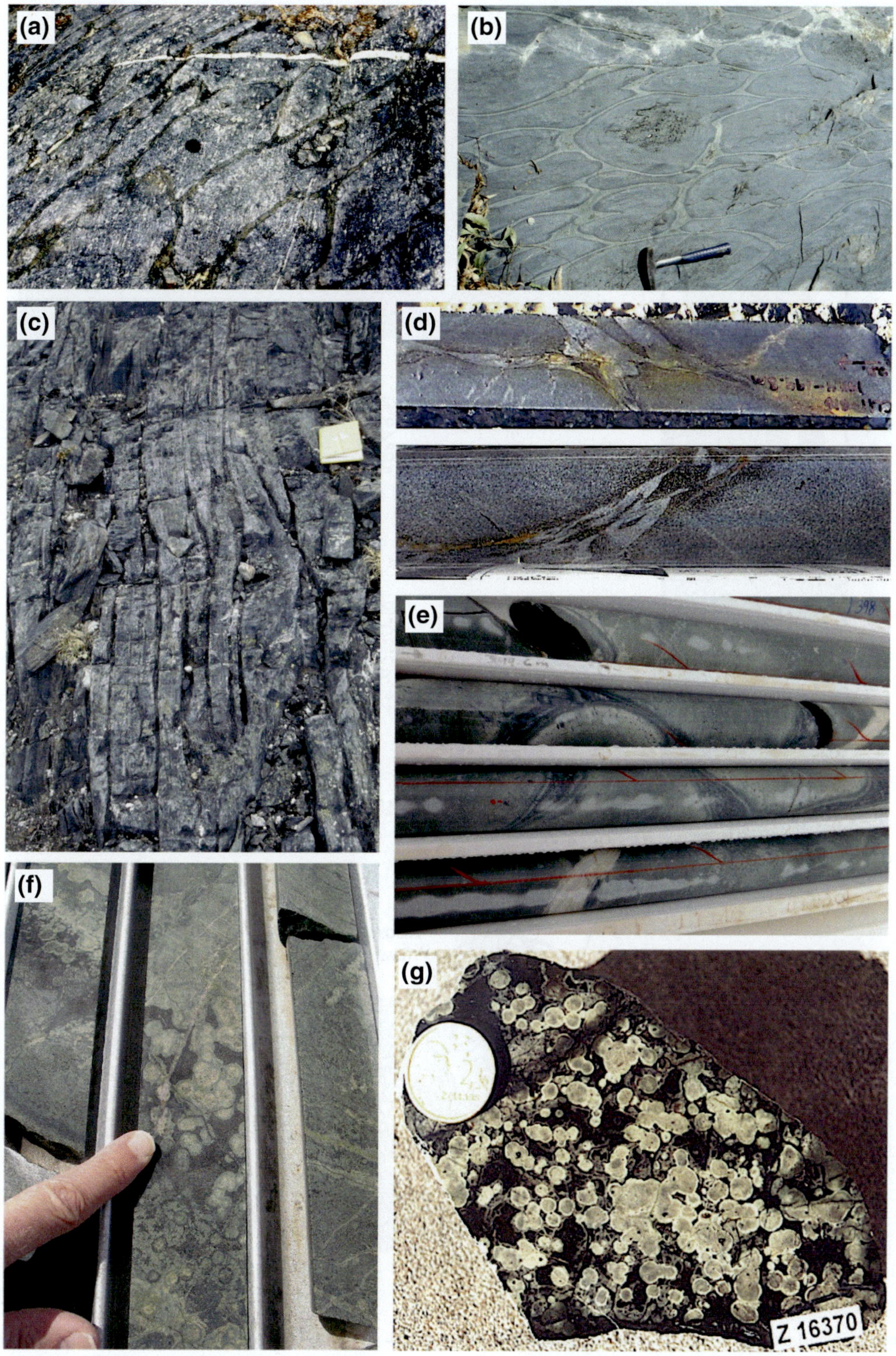

Fig. 5.47 Deformationally strained old pillow lavas. **a** Strong foliation cross-cutting Archean pillows, with associated stretching of the pillows. Pillow facing is to the top right. Lake Miranda, Yilgarn Craton, Western Australia. **b** Stretched Siluro-Devonian pillows, southern Hill End Trough, Copperhania Creek, New South Wales, Australia. Facing is to the bottom left. **c** Extremely stretched pillows, 3.7–3.8 Ga Isua Greenstone Belt, Greenland. **d** Hydraulic fractures in Archean basalt with chlorite alteration along the fractures, giving a superficial impression of pillows margins. However, the fractures are irregular, not smoothly curved. Cave Rocks Mine, Kambalda, Western Australia. Core width ~5 cm. **e** Drill core through Archean pillow lava, showing smooth curved, pillow outlines, chloritically altered originally glassy rims, and marginal vesicles. Core width ~5 cm. Jaguar Mine, Western Australia. **f** and **g** Spherical varioles in chloritically altered ~2,690 Ma Archean Paringa Basalt in diamond drill core, most of which are coalesced. Kambalda, Western Australia

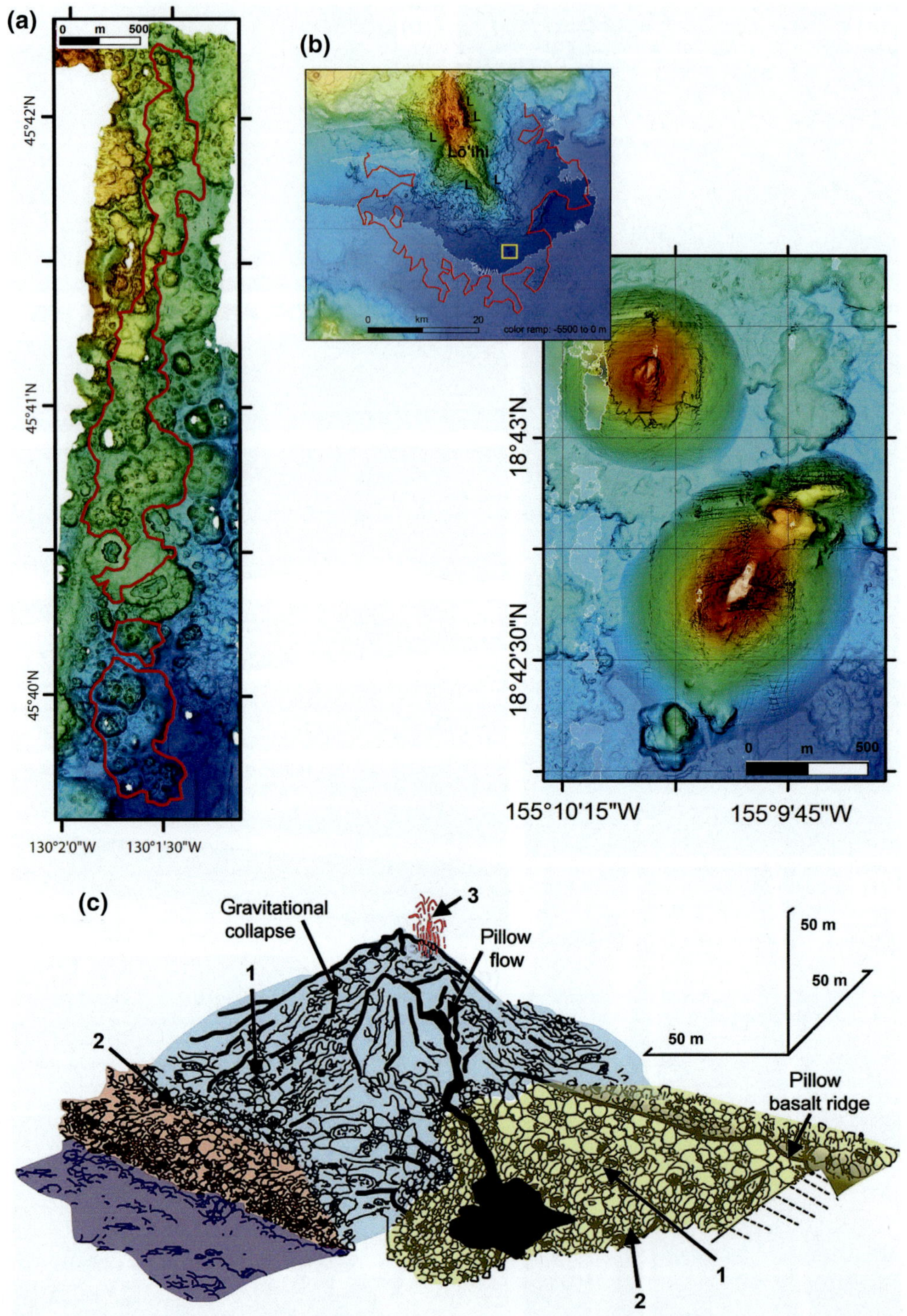

Fig. 5.48 Pillow mounds and ridges. **a** Axial Seamount, Juan de Fuca Ridge. Lower south rift pillow mounds erupted in 2011 showing AUV bathymetry at 1 m resolution. Colour ramp is between 2,350 and 1,800 m depth b.s.l. Flow outline is show red; the eruptive fissure is assumed to underlie the new pillow mounds and ridges. (From Clague et al. 2017, through Creative Commons Attribution 4.0 International licence). **b** Loihi Seamount, Hawai'i Detail of two large pillow mounds (cones) erupted on the extension of the south rift; resolution is 2 m. Colour ramp is between 5,070 and 4,770 m depth b.s.l. The southern one is $\sim$300 m high and 160 $\times$ 260 m across at the base. Inset shows regional map of Loihi Seamount and location of large pillow mounds (yellow box). Red line is inferred limit of young, far-travelled lavas from Loihi; L = locations of flank seaslides. (Adapted from Clague et al. 2019). **c** Reconstruction of Macquarie Island (Southern Ocean) pillow mound facies associations in a mid-oceanic spreading-ridge setting (From White et al. 2015, After Dickinson et al. 2009). Model shows generation of basaltic pillow fragment breccias and associated rocks on the on either the footslopes of the ocean floor (purple), or in a basin close to a pillow mound (pale blue) with a lateral pillow ridge (yellow). Processes include: (1) in situ and/or talus accumulation of pillow fragment breccia; (2) grain flow deposits derived from unconsolidated pillow fragment breccia and pillow basalt above; and (3) minor submarine lava fountaining deposits. Gravity collapse from destabilisation of the erupting vent may generate thick hyaloclastite and pillow fragment breccia aprons (orange) along pillow cone slopes. Advancing pillow lava tubes (labelled pillow flow) may also bulldoze accumulated pillow fragment breccia debris down slope

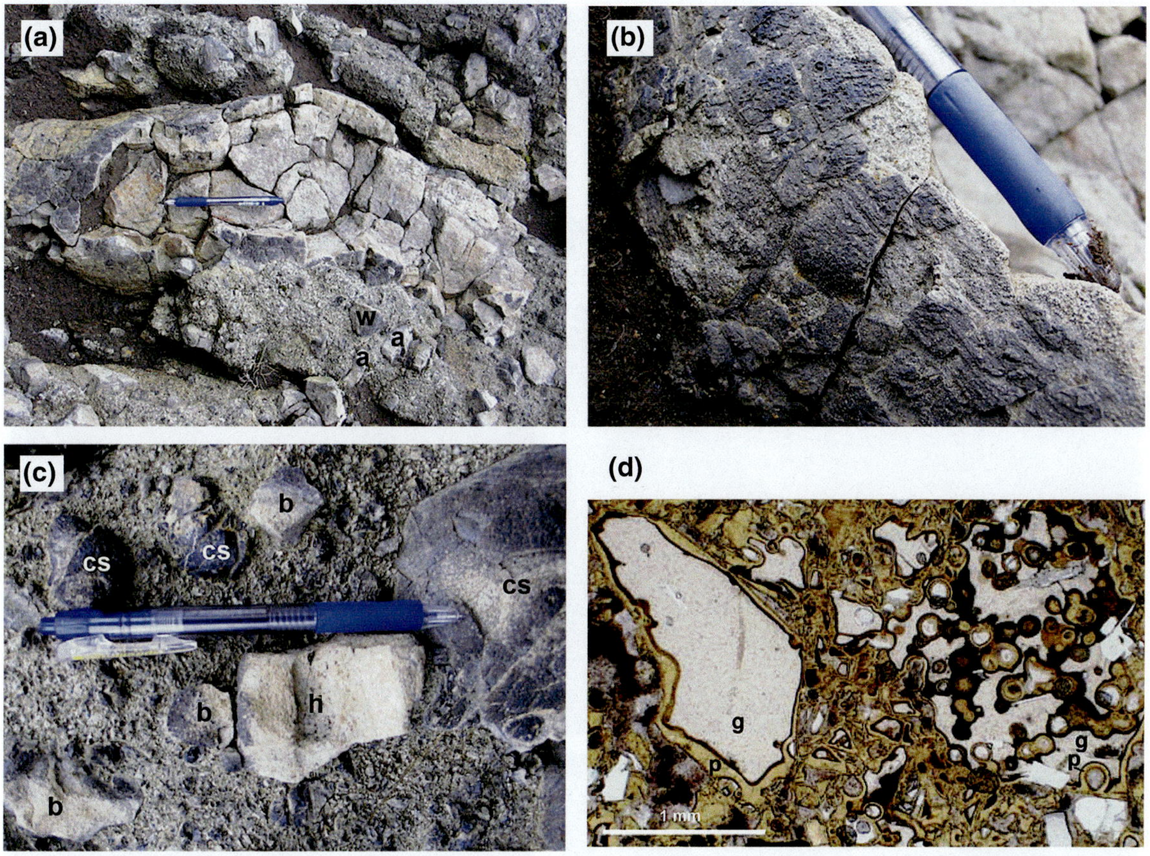

Fig. 5.49 Features of tectonically exhumed, deep oceanic (abyssal), late Miocene Macquarie Island (Southern Ocean) pillow fragment breccias (also see Fig. 5.48c). **a** Pillow fragment breccia with cross-section through a pillow tube. Note crystalline basalt clasts embedded in glassy matrix with wedge (w) and arcuate (a) shapes. Pillow tube shows fracturing patterns of chilled margin and core. Elongate amygdale-filled vesicles are observed in the chilled margin parallel to the pillow tube elongation. **b** Striations on outer glassy surface of pillow fragment. **c** Clasts and glassy matrix of hyaloclastite breccia. One clast exhibits a cylindrical cavity or hollow (h), while others are blocky (b) or exhibit curved surfaces (cs). **d** Thin section (ppl) of granular hyaloclastite matrix showing basaltic shards with low vesicularity (left) and high vesicularity (right); pale yellowish-grey fresh glass (g) is devitrified to dark yellow palagonite (p), and semi-opaque Fe-Ti oxides along shard and vesicle margins. (**a-c** From Dickinson et al. 2009; **d** From Portner et al. 2010)

Rare exhumed exposures of younger pillow mounds and pillow fragment breccias (12–6 Ma) are also described from Macquarie Island, a tectonically uplifted mid-oceanic spreading ridge, in the Southern Ocean (Dickinson et al. 2009; Portner et al. 2010). Pillow fragment breccia can be clast to matrix supported. It is composed of variably fragmented pillow fragments with a prominence of in situ jigsaw-fit textures and includes semi-intact aphanitic and porphyritic ovoid-shaped pillows with rare glassy striated selvedges (Fig. 5.49a, b). The pillow fragment clasts are pebble to cobble sized and have wedge-like or arcuate "pie-slice" morphologies indicative of radial and concentric cooling fracture patterns (Fig. 5.49c). The matrix is mainly composed of glassy grains resulting from the cooling contraction granulation of quenched pillow lava rinds (Fig. 5.49d). Pillow fragment breccia grades into a hyaloclastite breccia which is more matrix supported and composed of rotated arcuate, wedge, and blocky forms, and less common ovoid bodies. Dickinson et al. (2009) concluded the pillow fragment breccias were deposited in a pillow mound setting (Fig. 5.48c). Gravity collapse from destabilisation of the erupting vent initiated the generation of thick hyaloclastite and pillow fragment breccia aprons along pillow cone slopes. Destabilisation may possibly have been triggered by seismic activity, dyke intrusion, or over steepening of slopes during a voluminous eruption episode.

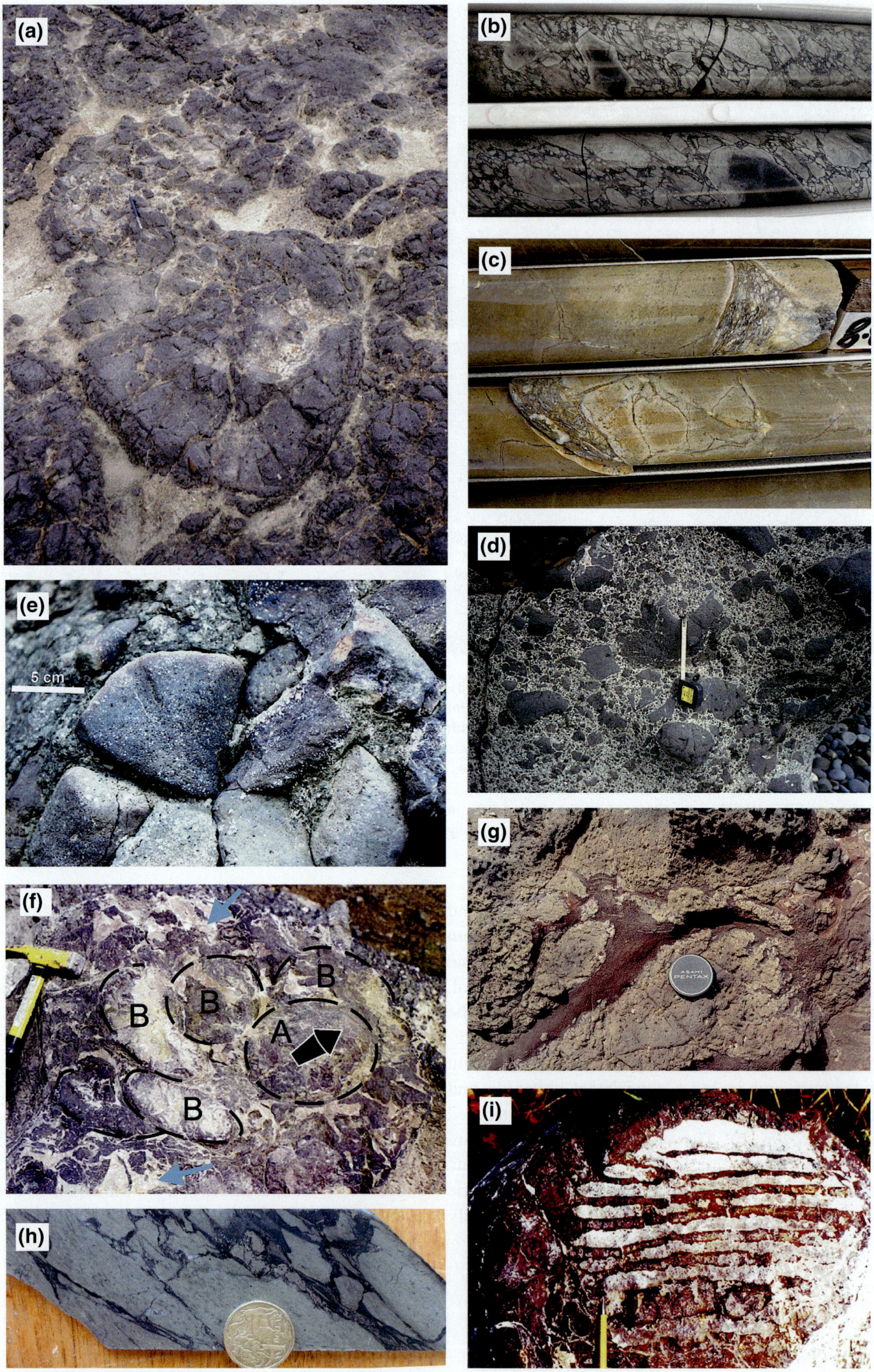
(a)
(b)
(c)
(d)
(e)
5 cm
(g)
(f)
B
B
B
A
B
(h)
(i)

◀ **Fig. 5.50** Hyaloclastite associated with submarine basalt and invasive lava. **a** Pervasively, internally quench fragmented Miocene pillow lava, Sakaehawa, Hokkaido, Japan. The matrix between the pillows is finely granulated hyaloclastite spalled from the glassy rims of the pillows. **b** Coarse blocky Archean basaltic hyaloclastite in diamond drill core showing jigsaw-fit to clast-rotated texture, ~2,690 Ma Paringa Basalt, Kambalda, Western Australia. Drill core is ~5 cm wide. **c** Archean pillow lava with hyaloclastite breccia in pillow interstices and at the margins of pillows. Breccia in interstices results from the quench spalling of pillow glassy rims. Teutonic Bore Mine, Western Australia. Drill core is ~5 cm wide. **d** Clast-rotated basaltic hyaloclastite breccia associated with Eocene–Oligocene pillow lava, Boatman's Harbour, Oamaru, New Zealand. The matrix is very finely granulated glassy debris. **e** Pillow fragment breccia showing characteristic *"pie-slice"*-shaped fragment formed by breakage along the concentric and radial cracks of pillows, Pliocene seamount series, La Palma, Canary Islands (From Staudigel and Koppers 2015). **f** Peperite formation associated with burrowing pillow-like lava buds (B) into lime mud (blue arrows) from centrally placed feeder tube (A), mid-late Triassic submarine basalt lava-flow complex, Hruškovec, northwest Croatia (From Palinkaš et al. 2008). **g** Invasive pillow-like lava intruded into unconsolidated sediments, Permian Gerringong Volcanics, Kiama, New South Wales. Note the irregular shape, highly irregular margins, and detached hyaloclastic pieces (From Cas and Wright 1987). **h** Peperite in margin of Archean submarine basalt, with jigsaw-fit hyaloclastite fragments separated by black mudstone. Teutonic Bore Mine, Western Australia. Diamond drill core is ~5 cm wide. **i** Rounded "pyjama-stripe style pillow" derived by burrowing of lava within unconsolidated, water-saturated lime mud at Hruškovec, northwest Croatia. This shows parallel bands ("shelves") of red, oxidised lava and calcite which infills an almost spherical lava tube. The striped texture is thought to form by the repeated draining of lava from the feeder tube, followed by invasion of seawater into the emptied cavity, and deposition of hydrothermal calcite rapidly precipitated from the heated sea water (also cf. Fig. 5.45g). (From Palinkaš et al. 2008)

Flowing pillow lava tubes and subsurface lava lobes may also have assisted by pushing or "bulldozing" accumulated pillow fragment breccia debris down slope.

5.12 Hyaloclastite and Peperite Associated with Subaqueous and Invasive Basalt Lava

Although some subaqueous lavas will be essentially coherent (as a result of effective film boiling), others consist of both coherent and hyaloclastite domains (Chap. 4; van Otterloo et al. 2015). Indeed, most subaqueous lavas will be associated with some degree of hyaloclastite development, ranging from coarse, angular, in situ (Fig. 5.50a–c) to clast-rotated breccia that is boulder to pebble size, but also includes finer glassy sand- or ash-sized debris (Fig. 5.50a–d). The latter could give the appearance of being phreatomagmatic pyroclastic in origin, but consists of very fine-grained, quench-granulated vitric fragments, spalled off the surface of cooling lava crusts (see Fig. 5.49d). These may form a granular matrix to the coarse hyaloclastite clasts, while the finest spalled debris could be dispersed by the warm water thermal plumes that will rise above propagating subaqueous lavas, to fall back infilling surface irregularities in the lavas or older lavas. Hyaloclastite may occur at the surface, or at the base of lavas, and may propagate well into the interior of the lavas. With pillow lavas, hyaloclastite aggregates consisting of slabby to curviplanar blade-like fragments may occur interstitial to pillows, having originated by spalling of debris from the crusts bounding interstices between the pillows (Fig. 5.50a, c).

Quench fragmentation can be minor and local in significance, or pervasive through thick lava. Schmincke et al. (1978) documented a 90 m interval of basaltic hyaloclastite sandwiched between pillow lava flows and pillow fragment breccia horizons in DSDP Hole 396B near the Canary Islands, in 13 Ma old oceanic crust. The 90 m section consists almost totally of sand size, blocky grains of basaltic glass and minor larger pillow fragments, which have been locally resedimented, but were interpreted by Schmincke et al. (1979) to have formed contemporaneously with the enclosing pillow lavas by quench fragmentation. Thicker successions of hyaloclastite and pillow fragment breccia may be associated with the formation of pillow mounds and ridges (Sect. 5.11.4), and seamounts. Where hyaloclastite forms near the summits of larger seamounts, it may be syn-eruptively resedimented as debris or rubble flows, sometimes called *hyaloclastite stone streams* (Lonsdale and Batiza 1980; Smith and Batiza 1989; Gregg and Smith 2003) down the sides of the edifice (Fig. 5.48c). Lonsdale and Batiza (1980) traced debris flow deposits back into broken pillows and pillow lavas at the seamount summit. Although they suggested that they were formed in part by hydrovolcanic explosions, in deep water it is more likely that they are formed by quench fragmentation and simple gravitational collapse. Furnes and Fridleifsson (1979) distinguished between gravitational collapse pillow breccias and quench fragmented hyaloclastite breccia in a subglacial environment whereas on steep slopes of a seamount both are likely to occur in combination and it may be difficult to distinguish them.

Although hyaloclastite can form around all margins of a propagating seafloor or lacustrine lava flow (as discussed in Chap. 4), another important facies type associated with submarine lavas, is *peperite*. Peperite is hyaloclastite debris that has been dynamically mixed with substrate sediment due to heating, fluidisation, and convective mixing of pore water in the sediments (Cas and Wright 1987; Skilling et al. 2002). Peperite can only form at the base of a "true" lava flow because only the base interfaces with sediment

(Palinkaš et al. 2008; Tucker and Scott 2009). However, invasive lavas or lobes that intrude water-saturated, unconsolidated sediments can experience quench fragmentation and formation of peperite around all margins, because all margins interface with unconsolidated wet sediment.

Byerly and Swanson (1978) recognised that some subaqueous basalt lavas may burrow down into underlying unconsolidated sediments over which they are flowing, especially when these are water saturated, forming what they called "*invasive lavas*". This occurs due to the gravitational instability of dense lava flowing over less dense, water-saturated sediments. The invasive flows (or syn-depositional intrusions; see Chap. 7) can be many m thick. Their margins can be lobate, and lobes may be superficially pillow like in form, but in detail can have very irregular geometry and margins (Fig. 5.50f, g; Bull and Cas 1989; Palinkaš et al. 2008; Tucker and Scott 2009). Irregular apophyses, including very narrow ones, can branch off the intrusive pillow-like lobes into the sediments. Invasive lava flows and their lobes are enclosed in sediment, and in places can be quench fragmented, also forming peperite through dynamic fluidisation-driven mixing of hyaloclasts into the host sediment or injection of water-fluidised sediment into fractures in hyaloclastite (Fig. 5.50h; Chap. 4). Peperite could occur around all margins of invasive lavas and lobes, whereas it can only form at the base of lava flows. Pyjama-stripe-style pillows (Sect. 5.11.3) may also be paleoenvironmentally distinctive features of invasive basaltic lavas (Fig. 5.50i; Palinkaš et al. 2008; Kiss et al. 2008).

5.13 Lateral Transitions in Complex Lavas: Massive Sheet Lava Facies to Lobate and Pillow Lava to Hyaloclastite Facies

Ballard et al. (1979) recognised that submarine basalt lavas on the modern seafloor can transform laterally from sheet flows into lobate and pillow lavas. Transitions from one to the other in a single lava flow either laterally (Fig. 5.51a–c) or vertically (Fig. 5.51b, c) have also been recognised in the geological rock record for a long time (Hargreaves and Ayres 1979; Cas 1992; Squire et al. 1998; Dimroth et al. 1978, 1985; Carracedo Sánchez et al. 2012). Transitions from massive sheet facies to pillow lava facies within a single lava suggest either declining magma discharge rates from the vent during a single submarine eruption, or more likely, declining rates of magma supply to the advancing flow front as the area of a lava progressively increases while the magma discharge rate at the vent remains high and constant. As the lava surface area increases, the effective magma supply rate to the distal margins of the flow declines, and so lava form may change from sheet/massive to pillowed (Fig. 5.51a–c). Sometimes, in vertical profiles, the bases of lavas may be pillowed but upwards the lava becomes massive and sheet like (Carracedo Sánchez et al. 2012), indicating initial low magma discharge that then increases (Fig. 5.51d).

5.14 Submarine Basaltic Magma Fountaining and Limu O Pele

Submersible studies of modern ocean floor, including seamounts and their summit calderas, have shown the existence of fluidal basalt clasts, including "cow dung bombs", and glass shards at depths of up to 4,000 m and greater (Fig. 5.52a, b; Smith and Batiza 1989; Clague et al. 2000, 2003a, b; Davis and Clague 2006; Schipper et al. 2011, Kelly et al. 2014). The fluidly deformed bombs and shards often occur in quench fragmented deposits which are inferred to have been produced by basaltic magma fountaining or in shallow water depths, strombolian-like (magmatic CO_2 driven) eruptions (cf. Cas and Simmons 2018; and see discussion Chap. 4, Sect. 4.6.8). They are commonly marked by variably fractured and broken, fluidal spatter bombs with chilled rims, dispersed in a finer matrix of blocky, angular variably vesicular glass fragments, ranging from sand to pebble size (Fig. 5.52c; Chap. 9). These tend to produce small-diameter cones probably of monogenetic origin and spatter ramparts that are indicative of brief, small volume eruptions. An interesting feature of the Macquarie Island (Southern Ocean) pillow fragment breccias (Sect. 5.11.4) is that they also contain rare sub-spherical (globule) glass pebbles (Fig. 5.52d; Dickinson et al. 2009). Some of these exhibit radial fractures, lapilli-sized glass droplets, bowl-shaped glassy grains, and sub-spherical crystalline basalt clasts surrounded by concentric glassy shells. These features are characteristic of submarine fire fountain activity in the vicinity of a pillow mound vent area on the deeper ocean floor (Fig. 5.48c).

Another unusual occurrence of an agglutinated scoriaceous breccia that also must have erupted on the ocean floor (at some 2,000 m water depth) has been described by Gill et al. (1990, 1992) from ODP Leg 126 (Site 791) in a back-arc rift (Simisu Basin), behind the Izu-Bonin Island Arc, Japan. Eruption of tholeiitic basalt (1 Ma) is thought to have produced frothy agglutinate (or basaltic *"mousse"*) which welded either on the seafloor or in a submarine eruption column resulting in 135-m-thick deposit. A similarly exotic deep-water Archean tholeiitic breccia (2.71 Ga),

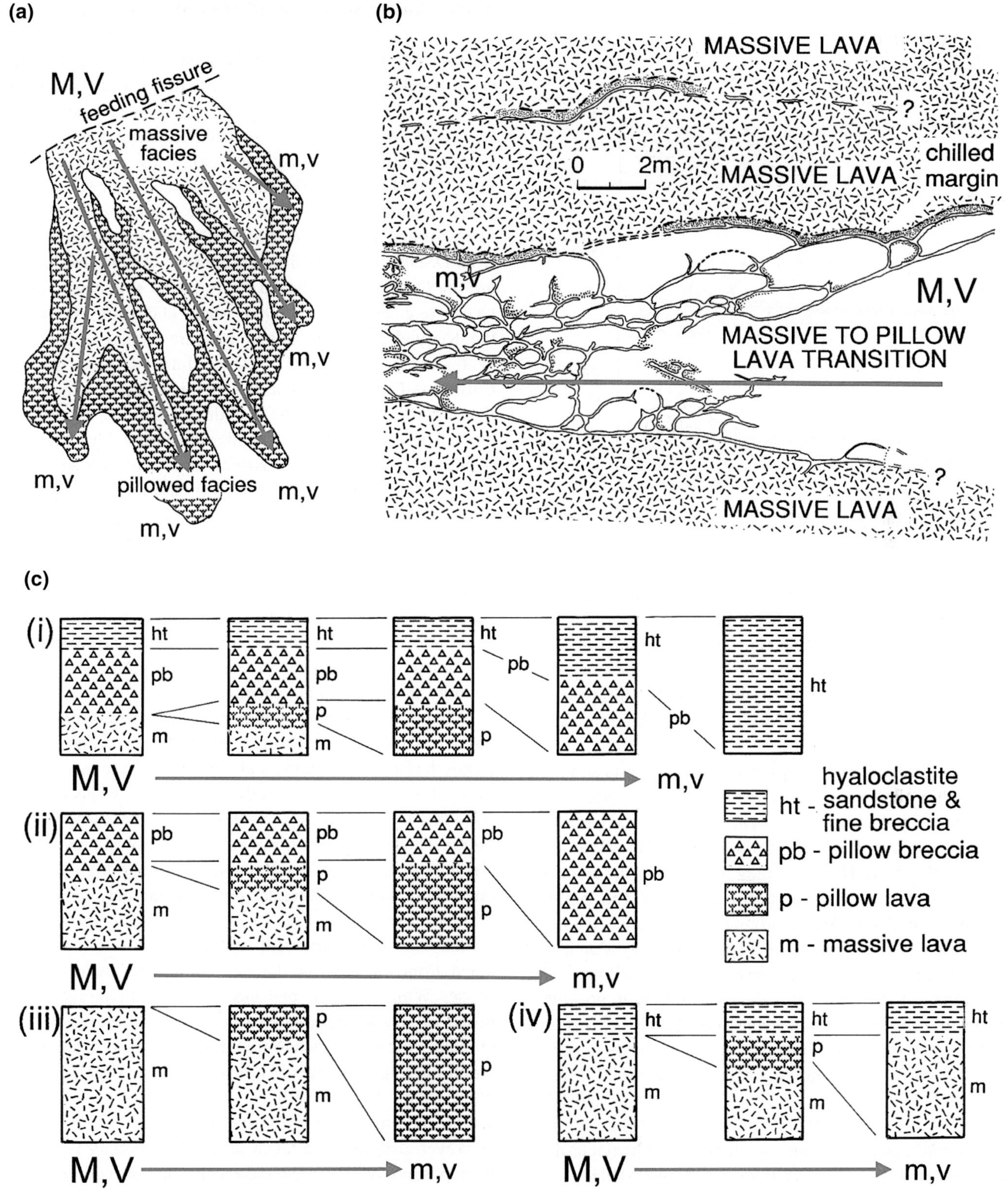

Fig. 5.51 Lateral and vertical changes in submarine Archean basalt lavas. In (**a–c**), M is the magma mass flux at the vent; m is the magma supply rate around the advancing flow front, where m < M; V is the magma lava flow rate at the vent; and v is the flow front velocity, where v < V. **a** Lateral changes from massive sheet lava at vent to pillow lava at the distal margins (Modified from Cas 1992, After Dimroth et al. 1978). **b** Lateral transition from massive to lobate to pillow lava (Modified from Cas 1992, After Hargreaves and Ayres 1979). **c** Representation of range of lateral facies changes, with (i) to (iv) representing possible lateral facies variations in four different submarine lavas (Modified from Cas 1992, After Dimroth et al. 1978). **d** Vertical variations (From Cas and Wright 1987, After Hargreaves and Ayres 1979)

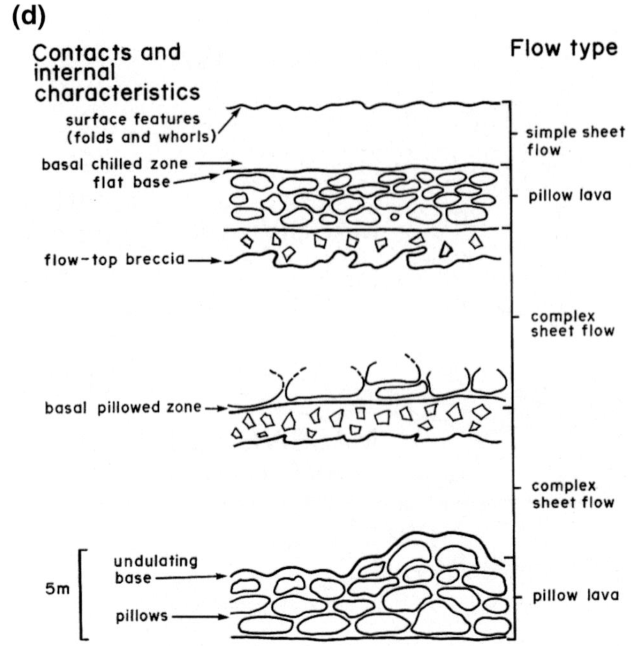

Fig. 5.51 (continued)

consisting of agglutinated basalt clasts and bombs, has been described from the Potter Mine sequence, Abitibi Greenstone Belt, Ontario, Canada (Fig. 5.52e; Houlé et al. 2010; Chap. 9). Other ancient examples of submarine magma or "fire" fountaining have been described by Simpson and McPhie (2001), Cas et al. (2003), Fujibayashi et al. (2014), and Hosoi and Amano (2017).

Platy to curviplanar to variably contorted clasts and flakes of basaltic glass named *limu o Pele* (Chap. 4, Sect. 4.4.8) have been known to occur associated with littoral explosions when basalt lava flows enter water during Hawai'ian eruptions (Fig. 5.53a). However, they may also occur associated with some deep submarine sheet basalt lavas (Fig. 5.53b–d; Clague et al. 2009). While originally considered to form in submarine fire fountains or explosive (CO_2 dominated) eruptions, they are now recognised (as their shallower water counterparts) to result from the entrapment of pockets of water into the base of advancing sheet basalt lavas, formation of water-filled lava balloons, and their steady expansion and quench fragmentation (e.g. Clague et al. 2000; Maicher et al. 2000; Maicher and White 2001; Schipper and White 2010; Schipper et al. 2013). Even strands of Pele's hair have been found associated with such limu o Pele hyaloclastite in deep water (Fig. 5.53e; Schipper and White 2010).

5.15 Submarine Flood Basalt Lavas

As on continental land masses, there are extremely large flood basalt provinces found on the deep ocean floor (e.g. Kerr 2014; Ernst 2014b; Fig. 5.54). These are commonly known as oceanic plateaus or oceanic Large Igneous Provinces (oLIPs), covering areas >1 × 10^6 km^2 and rising $\geq$ 2,000 m above the surrounding sea floor (Chap. 16). Most oceanic plateaus found in the present ocean basins were formed in the Cretaceous. Two of the better known from DSDP and ODP drilling programs, and also the largest, are the Ontong Java Plateau (OJP) in the southwest Pacific Ocean (e.g. Fitton et al. 2004; Mann and Taira 2004), and the Kerguelen Plateau in the southern Indian Ocean (e.g. Wallace et al. 2002). The third largest is the Caribbean-Colombian Plateau which is somewhat unique in that large sections have been obducted and are exposed on land. These show typical submarine basalt lava features, including pillow and massive lava and feeder lobes (Fig. 5.55a–c), varioles (Fig. 5.55d), and hyaloclastites and pillow fragment breccias (Fig. 5.55e–f). Spectacular pillow fragment breccias are located on the island of Aruba (Fig. 5.56) and are very similar in their facies characteristics to those described from the uplifted Macquarie Island submarine ridge (Sect. 5.11.4). Nonetheless, little is known about the nature, extent, or surface area, and volume of individual lavas erupted on oceanic plateaus or the morphology of their source vent regions.

What we do know is that oceanic plateaus appear to consist of giant piles (10 to $\geq$ 30 km thick), dominated by pillow basalt, sheet flows, and sills. Seismic reflection data from the upper part of OJP shows many lava flows, including hyaloclastites, several tens of m thick (Inoue et al. 2008). Thicker massive flows extending several hundred km are thought to have been generated during high effusion rates, while low effusion rates produced thinner pillow flows of lesser extent (see Fig. 16.91). Inoue et al. (2008) concluded that significant vent and/or fissure eruptions emanated from the crest of the OJP during the main eruptive phase. While oceanic plateaus are overwhelmingly composed of submarine flood basalts ± sills, some subaerially erupted volcanics, including unusually thick accretionary lapilli-bearing sequences (Thordarson 2004, Wright et al. in prep; Chap. 16), as well as rhyolite lavas and subaerial sediments have been recognised (Wallace et al. 2002). However, care must be taken in that accretionary lapilli can be deposited, or reworked and redeposited into deeper water

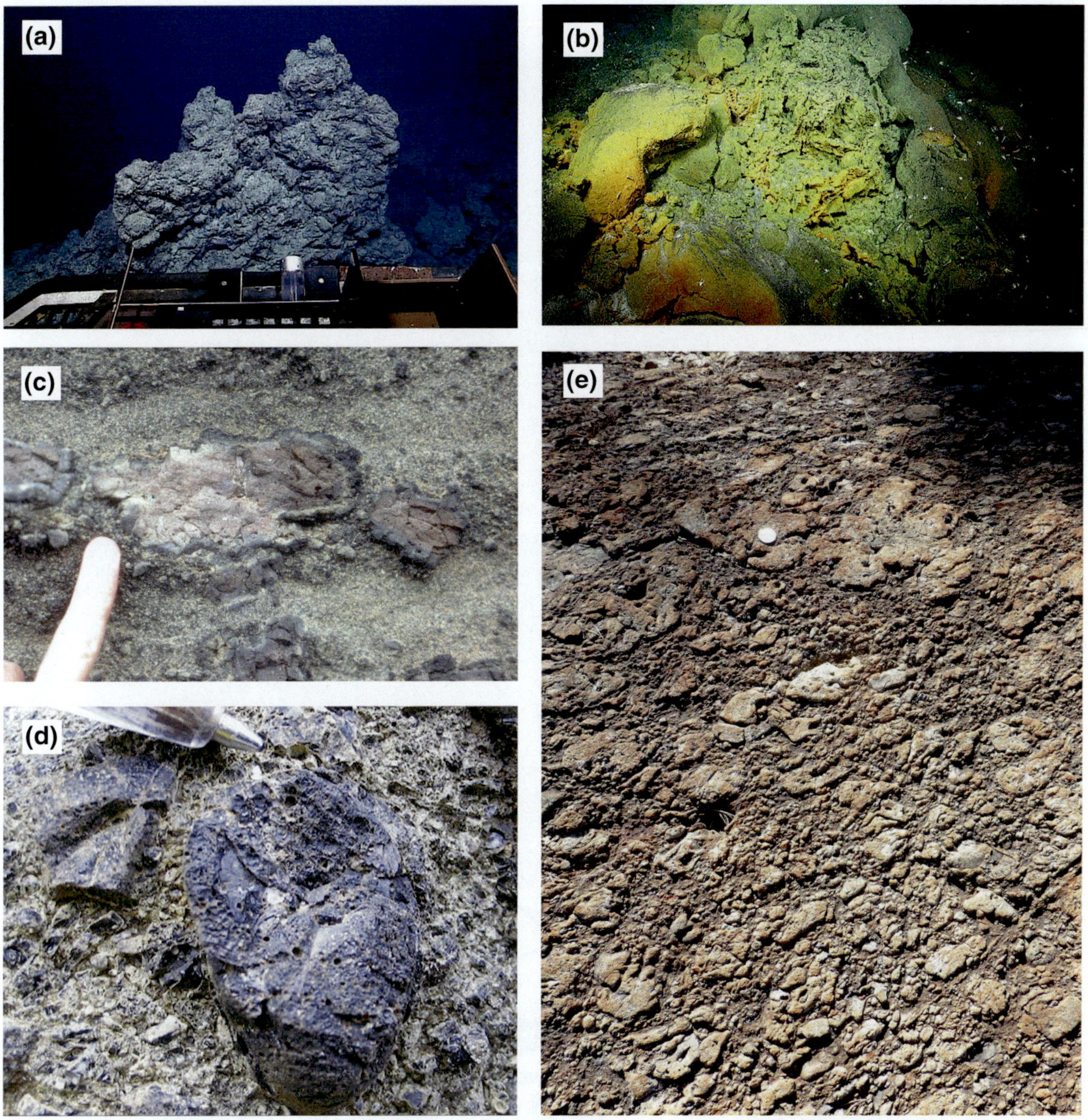

Fig. 5.52 Submarine basaltic magma or "fire" fountain deposits. **a** Spatter-like deposits at the summit of vent sites of the historic submarine eruption of Foerstner volcano, northwest of the island of Pantelleria, Italy, October 1891. The eruption took place at a water depth of 250 m. Width of the outcrop is about 2.2 m. (From Kelly et al. 2014). **b** Near-vent spatter deposits with yellow hydrothermal sediment from vent site of the 2012–2016 eruption of West Mata volcano, northeast Lau Basin, southwestern Pacific. Water depth is 1,300 m. (From Chadwick et al. 2019). **c** Fluidal bomb with chilled (tachylitic) rind within finer matrix of blocky, angular variably vesicular glass fragments, Miocene, Ryugazaki Headland, Oshoro, Peninsula, Hokkaido, Japan. **d** Subspherical glassy, globular "pebble" (3 × 5 cm) with 3 mm pitted rind in hyaloclastite matrix, late Miocene Macquarie Island (Southern Ocean) pillow fragment breccias. See Fig. 5.49e for photomicrograph of granular hyaloclastite matrix. (From Dickinson et al. 2009). **e** Archean agglutinated lava spatter deposit containing globular, lapilli-size agglutinate and fluidal bombs, Potter Mine sequence, Abitibi Greenstone Belt, Ontario, Canada

a long way from vent (Chap. 14) and hyaloclastites, in particular, pillow fragment breccias are not misinterpreted as fluvial or shallow marine epiclastic sediments, or their redeposited equivalents (e.g. Fig. 5.56), as this can ultimately lead to wrong assumptions regarding the exhumation and subsidence histories of oceanic plateaus (cf. Kerr 2014).

Despite its size, the OJP basalts are remarkably homogeneous in composition with the most abundant type being a uniform low-K tholeiite (Fitton et al. 2004). The Kerguelen volcanics are more diverse and although the basalts are similarly largely tholeiitic, there are also more evolved trachytic to rhyolitic lavas (Wallace et al. 2002). The

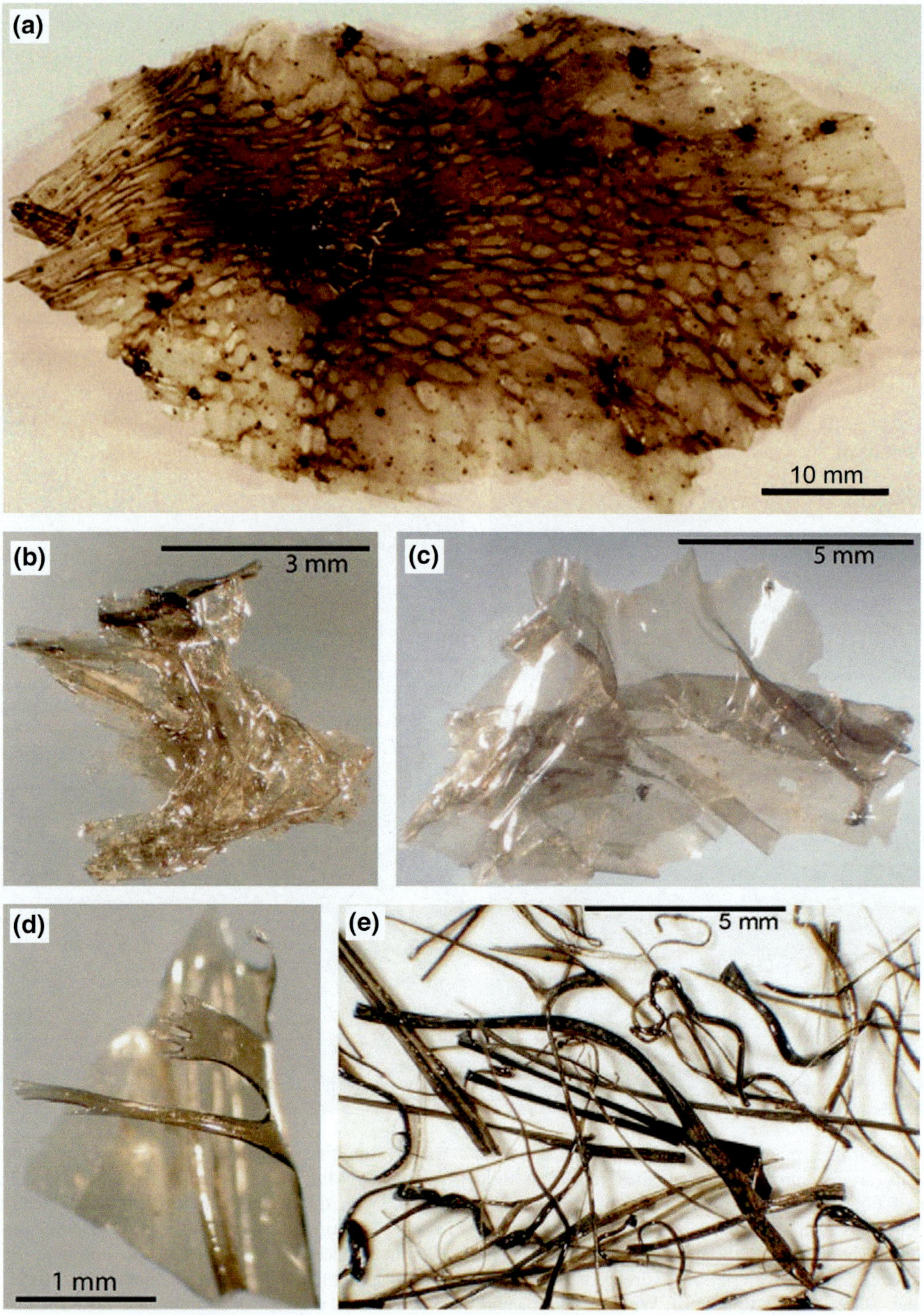

Fig. 5.53 Features of limu o Pele. **a** Near-colourless volcanic glass, commonly resembling seaweed, formed as lava erupted from Kilauea volcano, Hawai'i, entered the Pacific Ocean (*Source* US Geological Survey; image by JD Griggs). **b–e** Examples from submarine eruptions found along ocean spreading ridges. **b**, **c** Intensely folded limu o Pele fragments only 5 μm thick; **d** thin curved limu o Pele fragment which was stretched beyond the tensile strength of the melt and pulled apart; **e** Pele's hair from the summit of Axial Seamount on the Juan de Fuca Ridge, showing many strands are curved and some are "tack-welded" to themselves. (**b–e** from Clague et al. 2009)

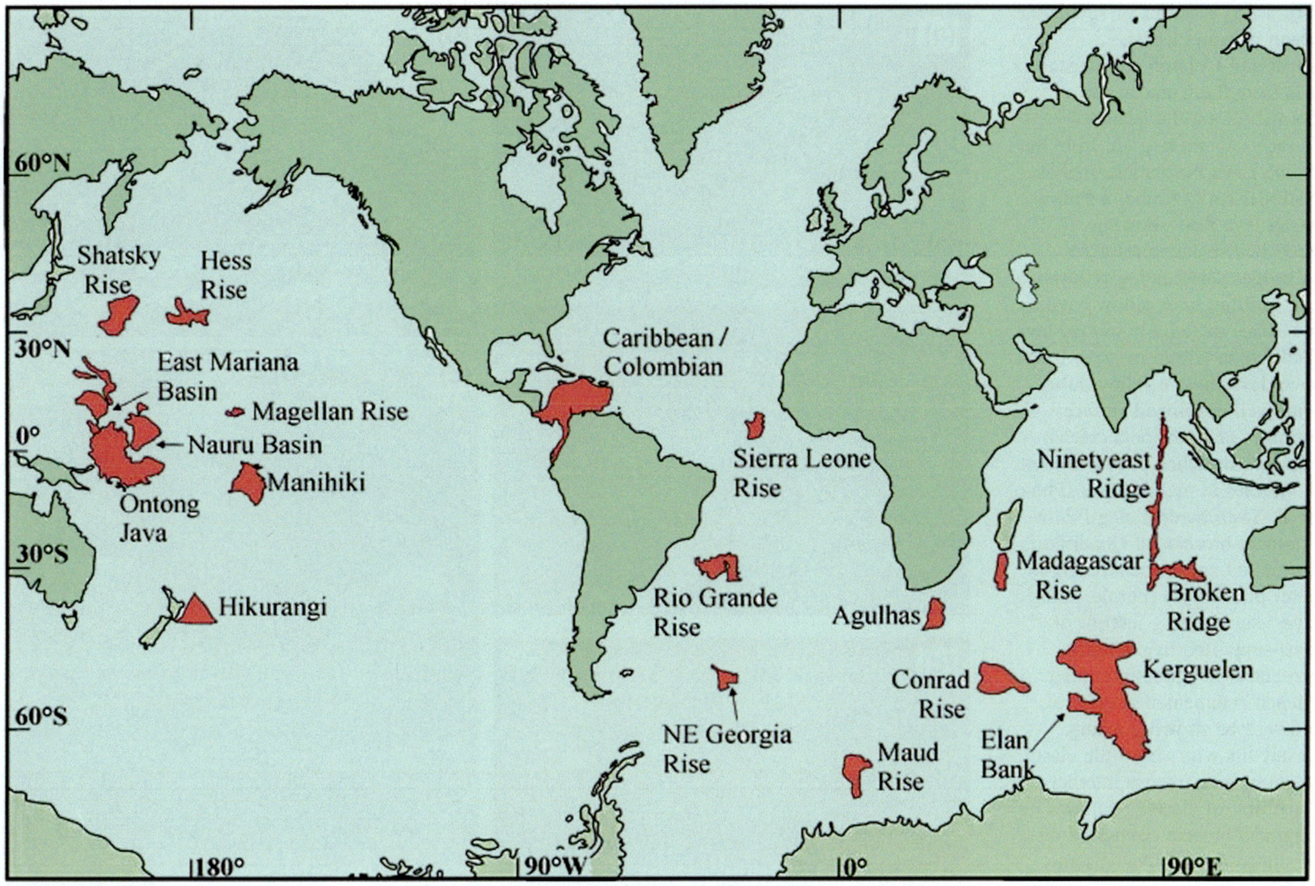

Fig. 5.54 Distribution of major oceanic plateaus formed within the last 150 Ma (From Kerr 2014)

Caribbean-Colombian Plateau includes rare komatiites and more abundant picrites representing more primitive mantle plume magmas (Kerr et al. 1996, 1998).

Submarine flood basalt eruptions and the construction of oceanic plateaus correlate with major global ocean anoxia and mass extinction events (e.g. Kerr 2005). There is also increasing consensus that construction of oceanic plateaus may have been more common over geological time than previously thought. Some Archean and Proterozoic greenstone belts may represent ancient oceanic plateaus that have been transferred by subduction or collision processes onto continental crust (e.g. Desrochers et al. 1993; Puchtel et al. 1998; Mann and Taira 2004; Chap. 16), while the number of proposed oceanic plateaus in the Phanerozoic record is growing (e.g. the Triassic Wrangellia Oceanic Plateau in the Pacific Northwest of North America; Greene et al. 2010; Chap. 16). Lastly, here, it is important to note that not all submarine flood basalts appear related to the proximal construction of an oceanic plateau. Others are associated with volcanic passive margins formed during the break-up and separation of continents, or simply termed ocean basin flood basalts (Mann and Taira 2004; Coffin et al. 2006; Ernst 2014b). Many passive margins worldwide are now considered "volcanic", with the best studied examples in the North and South Atlantic (e.g. Eldholm et al. 2000; Geoffroy et al. 2007; Chap. 16). Formation of volcanic passive margins produces a rapid succession of flood basalts over large areas, but unlike oceanic plateaus, a large proportion is erupted subaerially. Ocean basin flood basalts are less well defined but are extensive submarine lava flows accumulating at abyssal depths in the ocean basins that do not form topographic plateaus (e.g. Nauru Basin, western Pacific Ocean; Mochizuki et al. 2005). In current usage, they are usually grouped with oceanic plateaus since they are thought to be genetically linked (Kerr 2014; Ernst 2014b).

Fig. 5.55 Features of submarine ocean plateau basalts, Caribbean-Colombian Plateau; **a** is from the Curaçao Lava Formation, on the island of Curaçao, while **b–g** are from the Aruba Lava Formation, Arikok National Park, Aruba. **a** Pillow basalt lava flow showing quenched (palagonite) glass selvedges surrounding pillows and infilling inter-pillow cavities. **b** Surface of radially jointed lava feeder lobe within massive basalt sheet lava flow. **c** Pillow tube with well-developed surface wrinkles and tensional cracks; note two detached pillows on left. **d** Varioles in massive lava (Photo by F. Sastrohardjo). **e–g** Pillow fragment breccias at Quadirikiri; **e** ruptured and brecciated glassy outer pillow skin (purple, striated layer) surrounding interior of clast-supported hyaloclastite breccia; **f** polygonally jointed, quench fragmented interior of pillow lobe showing mainly jigsaw-fits with very little clast rotation; note intense internal fracturing of clasts. **g** Pillow fragment breccia composed of in situ to rotated hyaloclastite. The basaltic clasts show a range of textural and colour types but are essentially monomictic. The matrix consists of fine-grained, quench fragmented blocky- to splinter-shaped basalt, which has subsequently been infilled by sub-seafloor (pale-coloured) cement now mainly composed of a cryptocrystalline aggregate quartz, plagioclase, chlorite, and epidote

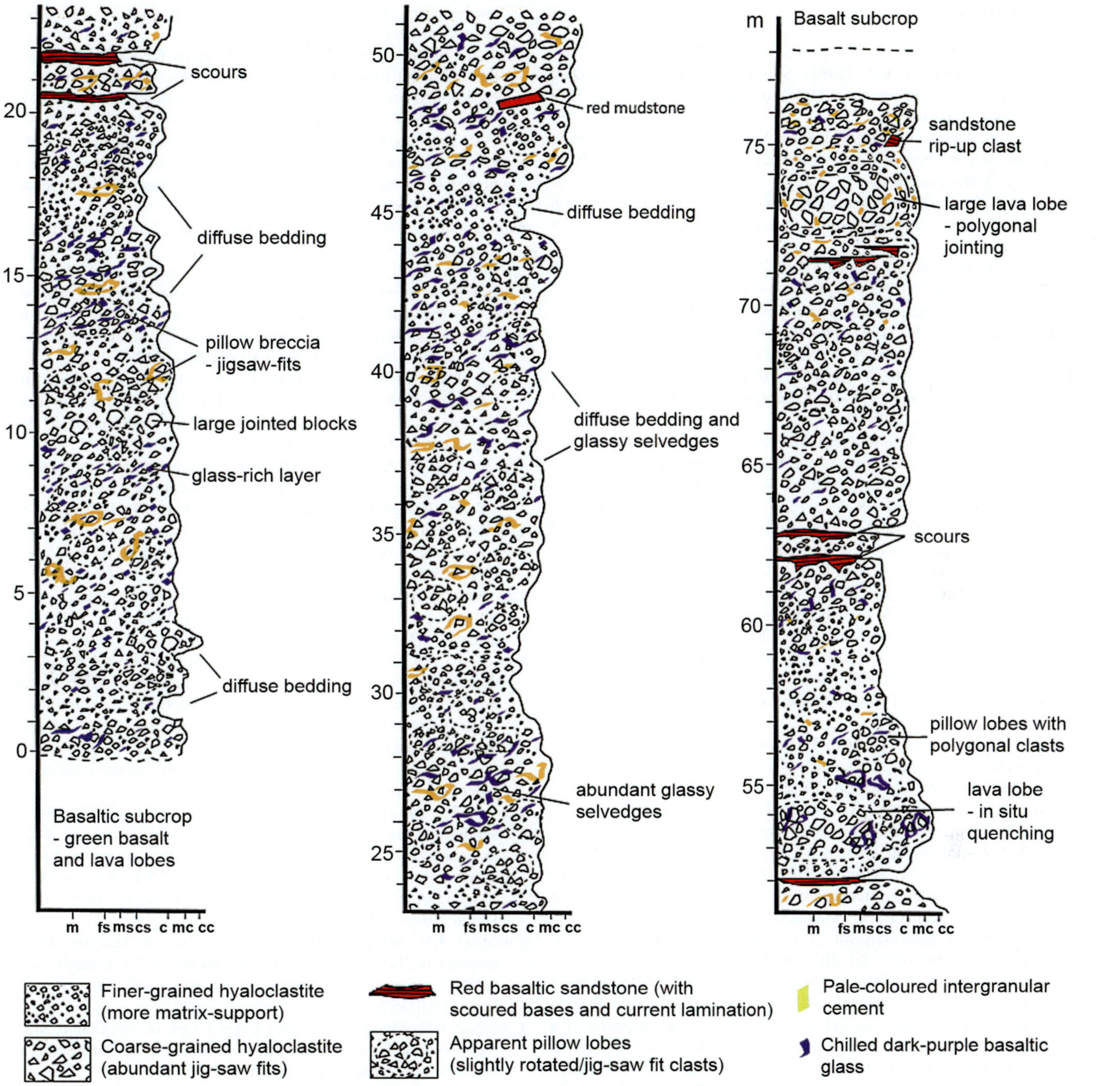

Fig. 5.56 Graphic log of pillow fragment breccias, Caribbean-Colombian ocean plateau succession, Quadirikiri, Arikok National Park, Aruba. These are composed of in situ hyaloclastites with jigsaw-fits, and variably rotated and matrix-supported hyaloclastite which sometimes show diffuse bedding or layering. Blocky-to-curviplanar-shaped pebble to boulder-size clasts (including "pie-slice" shapes) are common throughout and, while these may vary both texturally and in colour, they are all composed of sparsely vesicular basalt. Fragments of purple-dark volcanic glass are ubiquitous through the succession. Larger lava lobes are also found; the largest (approximately 6 × 2 m occurs at 73–75 m) comprises a polygonal-jointed, but essentially coherent basaltic core surrounded by zones of variously fragmented or polygonal basalt grading into normal hyaloclastite breccia. These are interpreted as major lava feeder tubes, and confirm the lava flow origin of the breccias. The only sedimentary structures occur within minor, thin (5–20 cm thick) reddish-coloured, current-laminated sandstones, which have scoured bases and appear to drape larger basaltic fragments. Microscopically, they consist mostly of palagonitised and altered basaltic glass shards and lithic fragments, and are interpreted as fine-grained hyaloclastite, formed by the cooling contraction granulation of pillow lava selvedges. Their origin is uncertain, but they may represent enhanced bottom currents produced by thermal convection and/or gravitational collapse events during eruption. This sequence of pillow fragment breccias on Aruba has previously been described as fluvial and/or shallow marine conglomerates, supposedly signifying a substantial period of volcano emergence and erosion. (Digitised by F. Sastrohardjo)

References

Appelgate B, Embley RW (1992) Submarine tumuli and inflated tube-fed lava flows on Axial Volcano, Juan De Fuca Ridge. Bull Volcanol 54(6):447–458

Atkinson A, Atkinson V (1995) Undara Volcano and its lava tubes: a Geological Wonder of Australia in Undara Volcanic National Park, North Queensland, p 85

Atkinson A, Griffin TJ, Stephenson PJ (1975) A major lava tube system from Undara Volcano, North Queensland. Bull Volcanol 39 (2):266–293

Ballard RD, Holcomb RT, Andel TH (1979) The Galapagos Rift at 86 W: 3. Sheet flows, collapse pits, and lava lakes of the rift valley. J Geophys Res Solid Earth 84(B10):5407–5422

Ballard RD, Moore JG (1977) Photographic atlas of the Mid-Atlantic Ridge rift valley. Springer, Berlin, p 114

Barberi F, Carapezza ML, Valenza M, Villari L (1993) The control of lava flow during the 1991–1992 eruption of Mt. Etna. J Volcanol Geotherm Res 56(1–2):1–34

Barreto CJ, de Lima EF, Scherer CM, Rossetti LD (2014) Lithofacies analysis of basic lava flows of the Paraná igneous province in the south hinge of Torres Syncline, Southern Brazil. J Volcanol Geotherm Res 285:81–99

Bear AN, Cas RAF (2007) The complex facies architecture and emplacement sequence of a Miocene submarine mega-pillow lava flow system, Muriwai, North Island, New Zealand. J Volcanol Geotherm Res 160(1):1–22

Belousov A, Belousova M (2018) Dynamics and viscosity of 'a'a and pahoehoe lava flows of the 2012–2013 eruption of Tolbachik volcano, Kamchatka (Russia). Bull Volcanol 80(1):6

Blundy J, Cashman K (2008) Petrologic reconstruction of magmatic system variables and processes. Rev Mineral Geochem 69(1):179–239

Bonatti E, Harrison CGA (1988) Eruption styles of basalt in oceanic spreading ridges and seamounts: effect of magma temperature and viscosity. J Geophys Res Solid Earth 93(B4):2967–2980

Bondre NR, Duraiswami RA, Dole G (2004) Morphology and emplacement of flows from the Deccan Volcanic Province, India. Bull Volcanol 66(1):29–45

Bonny E, Thordarson T, Wright R, Höskuldsson A, Jónsdóttir I (2018) The volume of lava erupted during the 2014 to 2015 eruption at Holuhraun, Iceland: a comparison between satellite and ground-based measurements. J Geophys Res Solid Earth 123. https://doi.org/10.1029/2017JB015008

Branca S, Azzaro R, De Beni E, Chester D, Duncan A (2015) Impacts of the 1669 eruption and the 1693 earthquakes on the Etna Region (Eastern Sicily, Italy): an example of recovery and response of a small area to extreme events. J Volcanol Geotherm Res 303:25–40

Bryan SE, Ferrari L (2013) Large igneous provinces and silicic large igneous provinces: progress in our understanding over the last 25 years. Geol Soc Am Bull 125(7–8):1053–1078

Bryan SE, Peate IU, Peate DW, Self S, DIA J, Mawby MR, Marsh JS, Miller JA (2010) The largest volcanic eruptions on Earth. Earth Sci Rev 102(3–4):207–229

Budkewitsch P, Robin PY (1994) Modelling the evolution of columnar joints. J Volcanol Geotherm Res 59(3):219–239

Bull SW, Cas RAF (1989) Volcanic influences in a storm- and tide-dominated shallow marine depositional system: the late permian broughton formation, southern Sydney Basin, Kiama, NSW. Aust J Earth Sci 36(4):569–584

Busby-Spera CJ (1987) Lithofacies of deep marine basalts emplaced on a Jurassic backarc apron, Baja California (Mexico). J Geol 95 (5):671–686

Byerly G, Swanson DA (1978) Invasive Columbia River basalt flows along the northwestern margin of the Columbia Plateau, north-central Washington. In: Geological Society of America abstracts with programs, p 98

Cahoon EB, Streck MJ, Koppers AA, Miggins DP (2020) Reshuffling the Columbia River Basalt chronology—Picture Gorge Basalt, the earliest-and longest-erupting formation. Geology 48(4):348–352

Calvari S, Coltelli M, Neri M, Pompilio M, Scribano V (1994) The 1991–1993 Etna eruption: chronology and lava. Acta Vulcanol 4:1–14

Calvari S, Pinkerton H (1998) Formation of lava tubes and extensive flow field during the 1991–1993 eruption of Mount Etna. J Geophys Res Solid Earth 103(B11):27291–27301

Campagnola S, Vona A, Romano C, Giordano G (2016) Crystallization kinetics and rheology of leucite-bearing tephriphonolite magmas from the Colli Albani volcano (Italy). Chem Geol 424:12–29

Carracedo Sánchez M, Sarrionandia F, Juteau T, Gil Ibarguchi JI (2012) Structure and organization of submarine basaltic flows: sheet flow transformation into pillow lavas in shallow submarine environments. Int J Earth Sci 101(8):2201–2214

Caress DW, Clague DA, Paduan JB, Martin JF, Dreyer BM, Chadwick WW Jr, Denny A, Kelley DS (2012) Repeat bathymetric surveys at 1-metre resolution of lava flows erupted at Axial Seamount in April 2011. Nature Geosci 5(7):483

Cas RAF (1992) Submarine volcanism; eruption styles, products, and relevance to understanding the host-rock successions to volcanic-hosted massive sulfide deposits. Econ Geol 87(3):511–541

Cas RAF, Giordano G (2014) Submarine volcanism: a review of the constraints, processes and products, and relevance to the Cabo de Gata volcanic succession. Italian J Geosci 133(3):362–377

Cas RAF, Simmons JM (2018) Why deep-water eruptions are so different from subaerial eruptions. Frontiers Earth Sci 6:198. https://doi.org/10.3389/feart.2018.00198

Cas RAF, van Otterloo J, Blaikie TN, van den Hove J (2017) The dynamics of a very large intra-plate continental basaltic volcanic province, the Newer Volcanics Province, SE Australia, and implications for other provinces. Geol Soc Lond Special Publications 446(SP446):448

Cas RAF, Wright JV (1987) Volcanic successions: modern and ancient. Allen and Unwin, London, p 528

Cas RAF, Yamagishi H, Moore L, Scutter C (2003) Miocene submarine fire fountain deposits, Ryugazaki Headland, Oshoro Peninsula, Hokkaido, Japan: implications for submarine fountain dynamics and fragmentation processes. In: White JDL, Smellie JL Clague DA (eds), *Explosive Subaqueous Volcanism.* Amer Geophys Union Geophys Mon 140:299–316

Cashman KV, Mangan MT (2014) a century of studying effusive eruption in Hawaii. In: Poland MP, Takahashi TJ, Landowski CM (eds) Characteristics of Hawaiian Volcanoes. US Geol Surv Prof Paper 1801:357–396

Cashman KV, Marsh BD (1988) Crystal size distribution (CSD) in rocks and the kinetics and dynamics of crystallization II: Makaopuhi lava lake. Contrib Mineral Petrol 99(3):292–305

Cashman KV, Thornber C, Kauahikaua JP (1999) Cooling and crystallization of lava in open channels, and the transition of pāhoehoe lava to a'ā. Bull Volcanol 61(5):306–323

Castruccio A, Rust AC, Sparks RSJ (2014) Assessing lava flow evolution from post-eruption field data using Herschel-Bulkley rheology. J Volcanol Geotherm Res 275:71–84

Chadwick Jr WW, Rubin KH, Merle SG, Bobbitt AM, Kwasnitschka T, Embley RW (2019) Recent eruptions between 2012 and 2018 discovered at west mata submarine volcano (NE Lau Basin, SW Pacific) and characterized by new ship, AUV, and ROV data. Front Marine Sci 2019:495

Chadwick WW (2003) Quantitative constraints on the growth of submarine lava pillars from a monitoring instrument that was caught in a lava flow. J Geophys Res Solid Earth 108(B11)

Chadwick WW, Clague DA, Embley RW, Perfit MR, Butterfield DA, Caress DW, Paduan JB, Martin JF, Sasnett P, Merle SG (2013) The

1998 eruption of axial seamount: new insights on submarine lava flow emplacement from high-resolution mapping. Geochem Geophys 14(10):3939–3968
Chadwick WW Jr, Merle SG, Baker ET, Walker SL, Resing JA, Butterfield DA, Anderson MO, Baumberger T, Bobbitt AM (2018) A recent volcanic eruption discovered on the central Mariana back-arc spreading center. Frontiers Earth Sci 23(6):172
Chadwick WW, Paduan JB, Clague DA, Dreyer BM, Merle SG, Bobbitt AM, Caress DW, Philip BT, Kelley DS, Nooner SL (2016) Voluminous eruption from a zoned magma body after an increase in supply rate at Axial Seamount. Geophys Res Letts 43(23):12–063
Chadwick WW, Gregg TKP, Embley RW (1999) Submarine lineated sheet flows: a unique lava morphology formed on subsiding lava ponds. Bull Volcanol 61(3):194–206
Chorowicz J (2005) The east African rift system. J Afr Earth Sci 43(1–3):379–410
Choubey VD (1973) Long-distance correlation of Deccan basalt flows, central India. Geol Soc Am Bull 84(8):2785–2790
Cimarelli C, Costa A, Mueller S, Mader HM (2011) Rheology of magmas with bimodal crystal size and shape distributions: insights from analog experiments. Geochem Geophys 12(7)
Clague DA, Batiza R, Head JW, Davis AS (2003a) Pyroclastic and hydroclastic deposits on Loihi Seamount, Hawaii. In: White JDL, Smellie JL Clague DA (eds) Explosive subaqueous volcanism. Amer Geophys Union Geophys Mon 140:73–95
Clague DA, Davis AS, Dixon JE (2003b) Submarine strombolian eruptions on the Gorda mid-ocean ridge. In: White JDL, Smellie JL Clague DA (eds) Explosive subaqueous volcanism. Amer Geophys Union Geophys Mon 140:111–128
Clague DA, Davis AS, Bischoff JL, Dixon JE, Geyer R (2000) Lava bubble-wall fragments formed by submarine hydrovolcanic explosions on Lō'ihi Seamount and Kīlauea Volcano. Bull Volcanol 61 (7):437–449
Clague DA, Dreyer BM, Paduan JB, Martin JF, Chadwick WW, Caress DW, Portner RA, Guilderson TP, McGann ML, Thomas H (2013) Geologic history of the summit of Axial Seamount, Juan De Fuca Ridge. Geochem Geophys 14(10):4403–4443
Clague DA, Hagstrum JT, Champion DE, Beeson MH (1999) Kīlauea summit overflows: their ages and distribution in the Puna District, Hawai'i. Bull Volcanol 61(6):363–381
Clague DA, Paduan JB, Davis AS (2009) Widespread strombolian eruptions of mid-ocean ridge basalt. J Volcanol Geotherm Res 180 (2–4):171–188
Clague DA, Paduan JB, Caress DW, Chadwick WW Jr, Le Saout M, Dreyer BM, Portner RA (2017) High-resolution AUV mapping and targeted ROV observations of three historical lava flows at Axial Seamount. Oceanography 30(4):82–99
Clague DA, Paduan JB, Caress DW, Moyer CL, Glazer BT, Yoerger DR (2019) Structure of Loihi Seamount, Hawai'i and lava flow morphology from high-resolution mapping. Frontiers Earth Sci 7:58. https://doi.org/10.3389/feart.2019.00058
Clague DA, Uto K, Satake K, Davis AS (2002) Eruption style and flow emplacement in the submarine North Arch Volcanic Field, Hawaii. In: Hawaiian volcanoes: deep underwater perspectives. Amer Geophys Union Monog 128:65–84
Coffin MF, Duncan RA, Eldholm OL, Fitton JG, Frey FA, Larsen HC, Mahoney JJ, Saunders AD, Schlich RO, Wallace PJ (2006) Large igneous provinces and scientific ocean drilling: status quo and a look ahead. Oceanography 19(4):150–160
Corcoran PL (2000) Recognizing distinct portions of seamounts using volcanic facies analysis: examples from the Archean Slave Province, NWT, Canada. Precambr Res 101(2–4):237–261
Costa A, Caricchi L, Bagdassarov N (2009) A model for the rheology of particle-bearing suspensions and partially molten rocks. Geochem Geophys 10(3)
Courtillot V, Jaupart C, Manighetti I, Tapponnier P, Besse J (1999) On causal links between flood basalts and continental breakup. Earth Planet Sci Lett 166(3):177–195
Crisp J, Cashman KV, Bonini JA, Hougen SB, Pieri DC (1994) Crystallization history of the 1984 Mauna Loa lava flow. J Geophys Res Solid Earth 99(B4):7177–7198
Davis AS, Clague DA (2006) Volcaniclastic deposits from the North Arch volcanic field, Hawaii: explosive fragmentation of alkalic lava at abyssal depths. Bull Volcanol 68(3):294–307
DeGraff JM, Long PE, Aydin A (1989) Use of joint-growth directions and rock textures to infer thermal regimes during solidification of basaltic lava flows. J Volcanol Geotherm Res 38(3–4):309–324
Delaney PT, Pollard DD (1982) Solidification of basaltic magma during flow in a dike. Am J Sci 282(6):856–885
Desrochers J-P, Hubert C, Ludden JN, Pilote P (1993) Accretion of Archean oceanic plateau fragments in the Abitibi, greenstone belt, Canada. Geology 21(5):451–454
Dietterich HR, Patrick MR, Diefenbach AK, Parcheta C, Lev E, Foks NL (2018) Lava flow hazard modeling and the assessment of effusion rates and topographic change with UAS and lidar during the 2018 Kilauea lower East Rift Zone eruption. Amer Geophys Union Fall Meeting Abstracts 2018 Dec.
Dickinson JA, Harb N, Portner RA, Daczko NR (2009) Glassy fragmental rocks of Macquarie Island (Southern Ocean): mechanism of formation and deposition. Sed Geol 216(3–4):91–103
Dimroth E, Cousineau P, Leduc M, Sanschagrin Y (1978) Structure and organization of Archean subaqueous basalt flows, Rouyn-Noranda area, Quebec, Canada. Canad J Earth Sci 15(6):902–918
Dimroth E, Imreh L, Cousineau P, Leduc M, Sanschagrin Y (1985) Paleogeographic analysis of mafic submarine flows and its use in the exploration for massive sulphide deposits. Geol Assoc Canada Spec Pap 28:203–222
Dixon JE, Clague DA, Wallace P, Poreda R (1997) Volatiles in alkalic basalts form the North Arch Volcanic Field, Hawaii: extensive degassing of deep submarine-erupted alkalic series lavas. J Petrology 38(7):911–939
Duraiswami RA, Dole G, Bondre N (2003) Slabby pahoehoe from the western Deccan Volcanic Province: evidence for incipient pahoehoe–aa transitions. J Volcanol Geotherm Res 121(3):195–217
Edwards BR, Gudmundsson MT, Russell JK (2015) Glaciovolcanism. In: Sigurdsson H, Houghton B, McNutt SR, Rymer H, Stix J (eds) The encyclopedia of volcanoes. Academic Press Elsevier, Cambridge, pp 377–394
Eldholm O, Gladczenko TP, Skogseid J, Planke S (2000) Atlantic volcanic margins: a comparative study. Geol Soc London Spec Publ 167(1):411–428
Embley RW, Murphy KM, Fox CG (1990) High-resolution studies of the summit of Axial Volcano. J Geophys Res Solid Earth 95 (B8):12785–12812
Ernst RE (2014a) Continental flood basalts and volcanic rifted margins. In: Ernst RE (ed) Large Igneous Provinces. Cambridge University Press, pp 56–89
Ernst RE (2014b) Oceanic LIPs: oceanic plateaus and ocean-basin flood basalts and their remnants through time. In: Ernst RE (ed) Large Igneous Provinces. Cambridge University Press, pp 90–110
Ernst RE, Youbi N (2017) How Large Igneous Provinces affect global climate, sometimes cause mass extinctions, and represent natural markers in the geological record. Palaeogeogr Palaeoclimatol Palaeoecol 478:30–52
Fink JH, Fletcher RC (1978) Ropy pahoehoe: surface folding of a viscous fluid. J Volcanol Geotherm Res 4(1):151–170
Fink JH, Griffiths RW (1990) Radial spreading of viscous-gravity currents with solidifying crust. J Fluid Mech 221:485–509
Fisher RV (1968) Puu Hou littoral cones, Hawaii. Geol Rundsch 57 (3):837–864

Fitton JG, Mahoney JJ, Wallace PJ, Saunders AD (2004) Origin and evolution of the Ontong Java Plateau: introduction. Geol Soc Lond Spec Publ 229(1):1–8

Forbes AE, Blake S, Tuffen H (2014) Entablature: fracture types and mechanisms. Bull Volcanol 76:820–833

Fornari DJ (1986) Submarine lava tubes and channels. Bull Volcanol 48(5):291–298

Fornari DJ, Garcia MO, Tyce RC, Gallo DG (1988) Morphology and structure of Loihi seamount based on seabeam sonar mapping. J Geophys Res Solid Earth 93(B12):15227–15238

Fowler AD, Jensen LS, Peloquin SA (1987) Varioles in Archean basalts; products of spherulitic crystallization. Canad Mineral 25 (2):275–289

Fox CG, Murphy KM, Embley RW (1988) Automated display and statistical analysis of interpreted deep-sea bottom photographs. Marine Geol 78(3–4):199–216

Francheteau J, Juteau T, Rangan C (1979) Basaltic pillars in collapsed lava-pools on the deep ocean-floor. Nature 281(5728):209–211

Fujibayashi N, Asakura K, Hattori T, Allen S (2014) Pillow lava and spasmodic submarine fire fountaining in the middle Miocene marginal basin, Sado Island, Japan. Island Arc 23(4):344–364

Fundis AT, Soule SA, Fornari DJ, Perfit MR (2010) Paving the seafloor: volcanic emplacement processes during the 2005–2006 eruptions at the fast spreading East Pacific Rise, 9 50′ N. Geochem Geophys 11(8)

Furnes H (1973) Variolitic structure in Ordovician pillow lava and its possible significance as an environmental indicator. Geology 1 (1):27–30

Furnes H, de Wit MJ, Robins B, Sandstå NR (2011) Volcanic evolution of the upper Onverwacht Suite, Barberton greenstone belt, South Africa. Precambr Res 186(1–4):28–50

Furnes H, Fridleifsson IB (1974) Tidal effects on the formation of pillow lava/hyaloclastite deltas. Geology 2(8):381–384

Furnes H, Fridleifsson IB (1979) Pillow block breccia—occurrences and mode of formation. Neues Jahrb Geol Paläontol 3:147–154

Geoffroy L (2005) Volcanic passive margins. CR Geosci 337 (16):1395–1408

Geoffroy L, Aubourg C, Callot J, Barrat J (2007) Mechanisms of crustal growth in large igneous provinces: the north Atlantic province as a case study. In: Foulger GR, Jurdy DM (eds) Plates, plumes, and planetary processes. Geol Soc Amer Spec Pap 430:747–774

Gill J, Torssander P, Lapierre H, Taylor R, Kaiho K, Koyama M, Kusakabe M, Aitchison J, Cisowski S, Dadey K, Fujioka K (1990) Explosive deep water basalt in the Sumisu backarc rift. Science 248 (4960):1214–1217

Gill J, Seales C, Thompson P, Hochsteadter, Dunlap C (1992) Petrology and geochemistry of Pliocene-Pleistocene volcanic rocks from the Izu Arc, Leg126. In: Taylor B, Fujioka K et al (eds) Proc ODP Sci Results 126:383–404

Giordano G, Porreca M (2009) Field observations on the initial lava flow and the fracture system developed during the early days of the Stromboli 2007 eruption. J Volcanol Geotherm Res 182(3–4):145–154

Goto Y, McPhie J (2012) Morphology and formation of spreading cracks on pillow lavas at Cape Grim, northwestern Tasmania, Australia. Bull Volcanol 74(7):1611–1619

Greeley R (1982) The Snake River Plain, Idaho: representative of a new category of volcanism. J Geophys Res: Solid Earth 87 (B4):2705–2712

Green JC (1989) Physical volcanology of mid-Proterozoic plateau lavas: The Keweenawan North Shore volcanic group, Minnesota. Geol Soc Am Bull 101(4):486–500

Greene AR, Scoates JS, Weis D, Katvala EC, Israel S, Nixon GT (2010) The architecture of oceanic plateaus revealed by the volcanic stratigraphy of the accreted Wrangellia oceanic plateau. Geosphere 6(1):47–73

Gregg TKP (2013) Deep-sea eruptions. Modeling volcanic processes: the physics and mathematics of volcanism. Cambridge University Press, New York, pp 258–274

Gregg TKP, Fink JH (1995) Quantification of submarine lava-flow morphology through analog experiments. Geology 23(1):73–76

Gregg TKP, Fink JH (2000) A laboratory investigation into the effects of slope on lava flow morphology. J Volcanol Geotherm Res 96 (3):145–159

Gregg TKP, Fornari DJ (1998) Long submarine lava flows: observations and results from numerical modeling. J Geophys Res Solid Earth 103(B11):27517–27531

Gregg TKP, Smith DK (2003) Volcanic investigations of the Puna Ridge, Hawai′ i: relations of lava flow morphologies and underlying slopes. J Volcanol Geotherm Res 126(1):63–77

Griffiths RW, Fink JH (1992) Solidification and morphology of submarine lavas: a dependence on extrusion rate. J Geophys Res Solid Earth 97(B13):19729–19737

Hamilton CW, Thordarson T, Fagents SA (2010) Explosive lava–water interactions I: architecture and emplacement chronology of volcanic rootless cone groups in the 1783–1784 Laki lava flow, Iceland. Bull Volcanol 72(4):449–467

Hare AG, Cas RAF (2005) Volcanology and evolution of the Werribee Plains intraplate, basaltic lava flow-field, Newer Volcanics Province, southeast Australia. Aust J Earth Sci 52(1):59–78

Hargreaves R, Ayres LD (1979) Morphology of Archean metabasalt flows, Utik Lake, Manitoba. Canad J Earth Sci 16(7):1452–1466

Harris A, Dehn J, Patrick MR, Calvari S, Ripepe M, Lodato L (2005) Lava effusion rates from hand-held thermal infrared imagery: an example from the June 2003 effusive activity at Stromboli. Bull Volcanol 68(2):107–117

Harris AJ, Flynn LP, Keszthelyi L, Mouginis-Mark PJ, Rowland SK, Resing JA (1998) Calculation of lava effusion rates from Landsat TM data. Bull Volcano 60(1):52–71

Harris AJ, Murray JB, Aries SE, Davies MA, Flynn LP, Wooster MJ, Wright R, Rothery DA (2000) Effusion rate trends at Etna and Krafla and their implications for eruptive mechanisms. J Volcanol Geotherm Res 102(3–4):237–269

Harris AJL, Rowland SK (2009) Effusion rate controls on lava flow length and the role of heat loss: a review. Studies in volcanology: The legacy of George Walker. 2:33

Harris AJL, Rowland SK (2015) Lava flows and rheology. In: Sigurdsson H, Stix J, Houghton BF, McNutt SR, Rymer H (eds) Encyclopedia of volcanoes. Academic, London, pp 321–342

Harris AJL, Rowland SK, Villeneuve N, Thordarson T (2017) Pāhoehoe, ʻaʻā, and block lava: an illustrated history of the nomenclature. Bull Volcanol 79(1):7

Hayman PC, Thébaud N, Pawley MJ, Barnes SJ, Cas RAF, Amelin Y, Sapkota J, Squire RJ, Campbell IH, Pegg I (2015) Evolution of a ~ 2.7 Ga large igneous province: a volcanological, geochemical and geochronological study of the Agnew Greenstone Belt, and new regional correlations for the Kalgoorlie Terrane (Yilgarn Craton, Western Australia). Precamb Res 270:334–368

Heliker C, Mattox TN (2003) The First Two Decades of the Pu ʻu ʻO'ö-Küpaianaha. US Geol Surv Prof Pap 1676(1676):1

Heliker CC, Swanson DA, Takahashi TJ eds (2003) The Puʻu ʻŌʻō-Kūpaianaha eruption of Kīlauea Volcano, Hawaiʻi: the first 20 years. US Geol Surv Prof Pap 1676:206p

Hoblitt RP, Orr TR, Heliker C, Denlinger RP, Hon K, Cervelli PF (2012) Inflation rates, rifts, and bands in a pāhoehoe sheet flow. Geosphere 8(1):179–195

Holcomb RT, Moore JG, Lipman PW, Belderson RH (1988) Voluminous submarine lava flows from Hawaiian volcanoes. Geology 16 (5):400–404

Hon K, Kauahikaua J, Denlinger R, Mackay K (1994) Emplacement and inflation of pahoehoe sheet flows: observations and measurements of active lava flows on Kilauea Volcano, Hawaii. Geol Soc Am Bull 106(3):351–370

Höskuldsson A, Jónsdóttir I, Riishus MS, Pedersen G, Gudmundsson MT, Thordarson T, Drouin V, Futurevolc IES (2015) Magma discharge and lava flow field growth in the Nornahraun/Bardarbunga eruption Iceland. In: EGU General Assembly conference abstracts, p 12755

Hosoi J, Amano K (2017) Submarine basaltic fountain eruptions in a back-arc basin during the opening of the Japan Sea. J Asian Earth Sci 149:103–109

Houlé G, Lesher CM, Préfontaine S, Ayer JA, Berger BR, Taranovic V, Davis PC, Atkinson B (2010) Stratigraphy and physical volcanology of komatiites and associated Ni-Cu-(PGE) mineralization in the western Abitibi Greenstone Belt, Timmins area, Ontario: a field trip for the 11th International Platinum Symposium. Ontario Geol Surv Open File Report 6255:99p

Hulme G (1974) The interpretation of lava flow morphology. Geophys J Int 39(2):361–383

Inoue H, Coffin MF, Nakamura Y, Mochizuki K, Kroenke LW (2008) Intrabasement reflections of the Ontong Java Plateau: implications for plateau construction. Geochem Geophys 9(4)

Jeffreys H (1925) The flow of water in an inclined channel of rectangular section. Lond Edinb Dub Phil Mag J Sci 49(293):793–807

Jerram DA, Widdowson M (2005) The anatomy of Continental flood basalt provinces: geological constraints on the processes and products of flood volcanism. Lithos 79(3):385–405

Jerram DA, Single RT, Hobbs RW, Nelson CE (2009) Understanding the offshore flood basalt sequence using onshore volcanic facies analogues: an example from the Faroe-Shetland basin. Geol Mag 146(3):353–367

Jones JG (1968) Pillow lava and pahoehoe. J Geol 76(4):485–488

Jones JG (1970) Intraglacial volcanoes of the Laugarvatn Region, Southwest Iceland, II. J Geol 78(2):127–140

Jones JG, Nelson PHH (1970) The flow of basalt lava from air into water—its structural expression and stratigraphic significance. Geol Mag 107(01):13–19

Jurado-Chichay Z, Rowland SK, Walker GPL (1996) The formation of circular littoral cones from tube-fed pāhoehoe: Mauna Loa, Hawai'i. Bull Volcanol 57:471–482

Katz MG, Cashman KV (2003) Hawaiian lava flows in the third dimension: identification and interpretation of pahoehoe and′ a′ a distribution in the KP-1 and SOH-4 cores. Geochem Geophys 4(2)

Kawachi Y, Pringle IJ (1988) Multiple-rind structure in pillow lava as an indicator of shallow water. Bull Volcanol 50(3):161–168

Kay SM, Coira B, Viramonte J (1994) Young mafic back arc volcanic rocks as indicators of continental lithospheric delamination beneath the Argentine Puna plateau, central Andes. J Geophys Res Solid Earth 99(B12):24323–24339

Kelly JT, Carey S, Pistolesi M, Rosi M, Croff-Bell KL, Roman C, Marani M (2014) Exploration of the 1891 Foerstner submarine vent site (Pantelleria, Italy): insights into the formation of basaltic balloons. Bull Volcanol 76(7):844

Kerr AC (2005) Oceanic LIPs: the kiss of death. Elements 1(5):289–292

Kerr AC (2014) Oceanic plateaus. In: Turekian KK, Holland HD (eds) Treatise on geochemistry, 4: the crust. Elsevier, pp 631–667

Kerr AC, Tarney J, Marriner GF, Klaver GT, Saunders AD, Thirlwall MF (1996) The geochemistry and petrogenesis of the late-Cretaceous picrites and basalts of Curaçao, Netherlands Antilles: a remnant of an oceanic plateau. Contrib Mineral Petrol 124(1):29–43

Kerr AC, Tarney J, Nivia A, Marriner GF, Saunders AD (1998) The internal structure of oceanic plateaus: inferences from obducted Cretaceous terranes in western Colombia and the Caribbean. Tectonophysics 292(3–4):173–188

Keszthelyi L, Self S (1998) Some physical requirements for the emplacement of long basaltic lava flows. J Geophys Res Solid Earth 103(B11):27447–27464

Kilburn C (1990) Surfaces of aa flow-fields on Mount Etna, Sicily: morphology, rheology, crystallization and scaling phenomena. Lava flows and domes. Springer, New York, pp 129–156

Kilburn CRJ, Guest JE (1993) aa lavas of Mount Etna, Sicily. Active lavas: monitoring and modelling. UCL Press, London, pp 73–106

King BC (1970) Vulcanicity and rift tectonics in East Africa. In: Clifford TW, Gass IG (eds) African magmatism and tectonics. Oliver and Boyd, Edinburgh, pp 263–283

Kiss G, Molnár F, Palinkaš LA (2008) Volcanic facies and hydrothermal processes in Triassic pillow basalts from the Darnó Unit, NE Hungary. Geologia Croatica 61(2–3):385–394

Krishnamurthy P, Udas GR (1981) Regional geochemical characters of the Deccan trap lavas and their genetic implications. In: Deccan volcanism. Mem Geol Soc India, pp 394–418

Larrea P, Salinas S, Widom E, Siebe C, Abbitt RJ (2017) Compositional and volumetric development of a monogenetic lava flow field: the historical case of Paricutin (Michoacan, Mexico). J Volcanol Geotherm Res 348:36–48

Lesti C, Giordano G, Salvini F, Cas R (2008) Volcano tectonic setting of the intraplate, Pliocene-Holocene, Newer Volcanic Province (southeast Australia): role of crustal fracture zones. J Geophys Res Solid Earth 113(B7)

Lipman PW (1980) Rates of volcanic activity along the southwest rift zone of Mauna Loa volcano, Hawaii. Bull Volcanol 43(4):703–725

Lipman PW, Banks NG (1987) A'a flow dynamics, Mauna Loa 1984. US Geol Surv Prof Pap 1350:1527–1567

Long PE, Wood BJ (1986) Structures, textures, and cooling histories of Columbia River basalt flows. Geol Soc Amer Bull 97(9):1144–1155

Lonsdale PF, Batiza R (1980) Hyaloclastite and lava flows on young seamounts examined with a submersible. Geol Soc Am Bull 91 (9):545–554

Macdonald GA (1953) Pahoehoe, aa, and block lava. Am J Sci 251 (3):169–191

Macdonald GA (1967) Forms and structures of extrusive basalt rocks. In: Hess HH, Poldervaart A (eds) The Poldervaart Treatise on rocks of basaltic composition. Interscience, New York, pp 1–61

MacDonald GA (1972) Volcanoes. Prentice-Hall, Englewood Cliffs, NJ, p 510

Mader HM, Llewellin EW, Mueller SP (2013) The rheology of two-phase magmas: a review and analysis. J Volcanol Geotherm Res 257:135–158

Mahoney JJ, Coffin MF (1997) Large igneous provinces: continental, oceanic, and planetary flood volcanism. Amer Geophys Union Monograph 100:450p

Maicher D, White JD (2001) The formation of deep-sea Limu o Pele. Bull Volcanol 63(7):482–496

Maicher D, White JDL, Batiza R (2000) Sheet hyaloclastite: density-current deposits of quench and bubble-burst fragments from thin, glassy sheet lava flows, Seamount Six, Eastern Pacific Ocean. Marine Geol 171(1):75–94

Mangan MT, Heliker CC, Mattox TN, Kauahikaua JP, Helz RT (1995) Episode 49 of the Pu'u'O'o-Kupaianaha eruption of Kilauea volcano-breakdown of a steady-state eruptive era. Bull Volcanol 57(2):127–135

Mangan MT, Cashman KV, Swanson DA (2014) The dynamics of Hawaiian-style eruptions: a century of study: Chapter 8. In:

Poland MP, Takahashi TJ, Landowski CM (eds) Characteristics of Hawaiian volcanoes. US Geol Surv Prof Pap 1801:323–354

Mann P, Taira A (2004) Global tectonic significance of the Solomon Islands and Ontong Java Plateau convergent zone. Tectonophys 389 (3–4):137–190

Mather BR, Müller RD, Seton M, Ruttor S, Nebel O, Mortimer N (2020) Intraplate volcanism triggered by bursts in slab flux. Sci Adv 6(51):eabd0953

Mattox TN, Mangan MT (1997) Littoral hydrovolcanic explosions: a case study of lava–seawater interaction at Kilauea Volcano. J Volcanol Geotherm Res 75(1):1–17

Mills AA (1984) Pillow lavas and the Leidenfrost effect. J Geol Soc Lond 141(1):183–186

Mohr PA (1983) Ethiopian flood basalt province. Nature 303:577–584

Moore HJ (1987) Preliminary estimates of the rheological properties of 1984 Mauna Loa lava. US Geol Surv Prof Pap 1350:1569–1588

Moore JG (1965) Petrology of deep sea basalt near Hawaii. Am J Sci 263(1):40–52

Moore JG (1975) Mechanism of formation of pillow lava: pillow lava, produced as fluid lava cools underwater, is the most abundant volcanic rock on earth, but only recently have divers observed it forming. Am Sci 63(3):269–277

Moore JG, Phillips R, Grigg RW, Peterson DW, Swanson DA (1973) Flow of lava into the sea, 1969–1971, Kilauea Volcano, Hawaii. Geol Soc Am Bull 84(2):537–546

Mochizuki K, Coffin MF, Eldholm O, Taira A (2005) Massive Early Cretaceous volcanic activity in the Nauru Basin related to emplacement of the Ontong Java Plateau. Geochem Geophys Geosyst 6(10)

Neal CA, Brantley SR, Antolik L, Babb JL, Burgess M, Calles K, Dotray P et al (2019) The 2018 rift eruption and summit collapse of Kīlauea Volcano. Science 363(6425):367–374

Ollier CD (1969) Volcanoes. Australian National University Press, Canberra

Ollier CD, Brown MC (1965) Lava Caves of Victoria. Bull Volcanol 28(1):215

Oman L, Robock A, Stenchikov GL, Thordarson T, Koch D, Shindell DT, Gao C (2006) Modeling the distribution of the volcanic aerosol cloud from the 1783–1784 Laki eruption. J Geophys Res Atmospheres 111(D12)

Oppenheimer C, Orchard A, Stoffel M, Newfield TP, Guillet S, Corona C, Sigl M, Di Cosmo N, Büntgen U (2018) The Eldgjá eruption: timing, long-range impacts and influence on the Christianisation of Iceland. Clim Change 147(3–4):369–381

Orr T, Heliker C, Patrick M (2012) The ongoing Pu'u O'o eruption of Kilauea Volcano, Hawai'i—30 years of eruptive activity. US Geol Surv Fact Sheet 3127

Paduan JB, Clague DA, Caress DW (2014) Constructional talus: built during eruption, not by later tectonism. In: Proc American Geophysical Union fall meeting, San Francisco

Palinkaš LA, Bermanec V, Šoštarić SB, Kolar-Jurkovšek T, Palinkaš SS, Molnar F, Kniewald G (2008) Volcanic facies analysis of a subaqueous basalt lava-flow complex at Hruškovec, NW Croatia—evidence of advanced rifting in the Tethyan domain. J Volcanol Geotherm Res 178(4):644–656

Parcheta C, Fagents SA, Swanson DA, Houghton BF, Ericksen T (2015) Hawaiian fissure fountains: quantifying vent and shallow conduit geometry, Episode 1 of the 1969–1974 Mauna Ulu eruption. In: Carey R, Cayol V, Poland M, Weis D (eds) Hawaiian volcanoes: from source to surface. Amer Geophys Union Mon 208:369–391

Peate DW (1997) The Paraná-Etendeka Province. In: Mahoney JJ, Coffin MF (eds) Large igneous provinces: continental, oceanic, and planetary flood volcanism. Amer Geophys Union Mon 100:217–245

Pedersen GBM, Höskuldsson A, Dürig T, Thordarson T, Jonsdottir I, Riishuus MS, Óskarsson BV, Dumont S, Magnússon E, Gudmundsson MT (2017) Lava field evolution and emplacement dynamics of the 2014–2015 basaltic fissure eruption at Holuhraun, Iceland. J Volcanol Geotherm Res 340:155–169

Perfit MR, Chadwick WW (1998) Magmatism at mid-ocean ridges: constraints from volcanological and geochemical investigations. In: Faulting and magmatism at mid-ocean ridges. AGU Mono, pp 59–115

Peterson DW, Swanson DA (1974) Observed formation of lava tubes. Stud Speleology 2(6):209–222

Peterson DW, Tilling RI (1980) Transition of basaltic lava from pahoehoe to aa, Kilauea Volcano, Hawaii: field observations and key factors. J Volcanol Geotherm Res 7(3–4):271–293

Pinkerton H, Norton G (1995) Rheological properties of basaltic lavas at sub-liquidus temperatures: laboratory and field measurements on lavas from Mount Etna. J Volcanol Geotherm Res 68(4):307–323

Pinkerton H, Sparks RSJ (1978) Field measurements of the rheology of lava. Nature 276(5686):383–385

Planke S, Symonds PA, Alvestad E, Skogseid J (2000) Seismic volcanostratigraphy of large-volume basaltic extrusive complexes on rifted margins. J Geophys Res Solid Earth 105(B8):19335–19351

Planke S, Millett JM, Maharjan D, Jerram DA, Abdelmalak MM, Groth A, Hoffmann J, Berndt C, Myklebust R (2017) Igneous seismic geomorphology of buried lava fields and coastal escarpments on the Vøring volcanic rifted margin. Interpretation 5(3):1–17

Poland MP, Takahashi TJ, Landowski CM (2014) Characteristics of Hawaiian volcanoes. In: US Geol Surv Prof Pap 1801

Porębski SJ, Gradziński R (1990) Lava-Fed Gilbert-Type Delta in the Polonez Cove Formation (Lower Oligocene), King George Island, West Antarctica. In: Colella A, Prior DB (eds) Coarse-grained deltas. Int Assn of Sediment Spec Pub 10, Blackwell, Oxford, pp 333–351

Portner RA, Daczko NR, Dickinson JA (2010) Vitriclastic lithofacies from Macquarie Island (Southern Ocean): compositional influence on abyssal eruption explosivity in a dying Miocene spreading ridge. Bull Volcanol 72(2):165–183

Puchtel IS, Hofmann AW, Mezger K, Jochum KP, Shchipansky AA, Samsonov AV (1998) Oceanic plateau model for continental crustal growth in the Archaean: a case study from the Kostomuksha greenstone belt, NW Baltic Shield. Earth Planet Sci Letts 155 (1):57–74

Putirka KD (2008) Thermometers and barometers for volcanic systems. In: Putirka KD, Tepley FJI (eds) Minerals, inclusions and volcanic processes. Mineralogical Society of America, Chantilly, Virginia, pp 61–120

Re G, White JDL, Ort MH (2015) Dikes, sills, and stress-regime evolution during emplacement of the Jagged Rocks complex, Hopi Buttes Volcanic field, Navajo Nation, USA. J Volcanol Geother Res 295: 65–79

Reidel SP, Camp VE, Tolan TL, Martin BS, Ross ME, Wolff JA, Wells RE (2013) The Columbia River flood basalt province: stratigraphy, areal extent, volume, and physical volcanology. The Columbia River Flood Basalt. Geol Soc Amer Spec Paper 497:1–43

Resing JA, Rubin KH, Embley RW, Lupton JE, Baker ET, Dziak RP, Baumberger T, Lilley MD, Huber JA, Shank TM (2011) Active submarine eruption of boninite in the northeastern Lau Basin. Nat Geosci 4(11):799–806

Riker JM, Cashman KV, Kauahikaua JP, Montierth CM (2009) The length of channelized lava flows: insight from the 1859 eruption of

Mauna Loa Volcano, Hawai'i. J Volcanol Geotherm Res 183 (3):139–156
Rowland SK, Walker GPL (1987) Toothpaste lava: characteristics and origin of a lava structural type transitional between pahoehoe and aa. Bull Volcanol 49(4):631–641
Rowland SK, Walker GPL (1990) Pahoehoe and aa in Hawaii: volumetric flow rate controls the lava structure. Bull Volcanol 52 (8):615–628
Rubin KH, Soule SA, Chadwick WW Jr, Fornari DJ, Clague DA, Embley RW, Baker ET, Perfit MR, Caress DW, Dziak RP (2012) Volcanic eruptions in the deep sea. Oceanography 25(1):142–157
Sandstå NR, Robins B, Furnes H, De Wit M (2011) The origin of large varioles in flow-banded pillow lava from the Hooggenoeg Complex, Barberton Greenstone Belt, South Africa. Contrib Mineral Petrol 162(2):365–377
Sato H (2005) Viscosity measurement of subliquidus magmas: 1707 basalt of Fuji volcano. J Mineral Petrol Sci 100(4):133–142
Schipper CI, Sonder I, Schmid A, White JDL, Dürig T, Zimanowski B, Büttner R (2013) Vapour dynamics during magma–water interaction experiments: hydromagmatic origins of submarine volcaniclastic particles (limu o Pele). Geophys J Int 192(3):1109–1115
Schipper CI, White JDL (2010) No depth limit to hydrovolcanic limu o Pele: analysis of limu from Lōihi Seamount, Hawaii. Bull Volcanol 72(2):149–164
Schipper CI, White JD, Houghton BF (2011) Textural, geochemical, and volatile evidence for a Strombolian-like eruption sequence at Loihi Seamount, Hawaii. J Volcanol Geotherm Res 207 (1–2):16–32
Schmidt A, Leadbetter S, Theys N, Carboni E, Witham CS, Stevenson JA, Birch CE, Thordarson T, Turnock S, Barsotti S (2015) Satellite detection, long-range transport, and air quality impacts of volcanic sulfur dioxide from the 2014–2015 flood lava eruption at Bárðarbunga (Iceland). J Geophys Res Atmospheres 120(18):9739–9757
Schmincke H-U (1967) Fused Tuff and Pépérites in South-Central Washington. Geol Soc Am Bull 78(3):319–330
Schmincke H-U, Robinson PT, Ohnmacht W, Flower MFJ (1979) Basaltic hyaloclastites from Hole 396B, DSDP Leg 46. Proceedings of the ocean drilling program: initial report. National Science Foundation, p 341
Schmincke H-U, Sunkel G (1987) Carboniferous submarine volcanism at Herbornseelbach (Lahn-Dill area, Germany). Geol Rundsch 76 (3):709–734
Schnur SR, Gilbert LA (2012) Detailed volcanostratigraphy of an accreted seamount: implications for intraplate seamount formation. Geochem Geophys Geosyst 13(12)
Scrope GP (1872) Volcanoes. London
Self S, Coffin MF, Rampino MR, Wolff JA (2015) Large igneous provinces and flood basalt volcanism. In: Sigurdsson H, Stix J, Houghton B, McNutt S, Rymer H (eds) Encyclopedia of volcanoes. Academic Press, San Diego, pp 441–458
Self S, Jay AE, Widdowson M, Keszthelyi LP (2008) Correlation of the Deccan and Rajahmundry Trap lavas: are these the longest and largest lava flows on Earth? J Volcanol Geotherm Res 172(1):3–19
Self S, Keszthelyi L, Thordarson T (1998) The importance of pāhoehoe. Ann Rev Earth Planet Sci 26(1):81–110
Self S, Mittal T, Jay AE (2021) Thickness characteristics of pāhoehoe lavas in the Deccan Province, Western Ghats, India, and in continental flood basalt provinces elsewhere. Front Earth Sci 8:720. https://doi.org/10.3389/feart.2020.630604
Self S, Schmidt A, Mather TA (2014) Emplacement characteristics, time scales, and volcanic gas release rates of continental flood basalt eruptions on Earth. Geol Soc Am Spec Pap 505:SPE505-516
Self S, Thordarson T, Keszthelyi L (1997) Emplacement of continental flood basalt lava flows. In: Mahoney JJ, Coffin MF (eds) Large Igneous Provinces: continental, oceanic, and planetary flood volcanism. Amer Geophys Union Mon 100:381–410
Self S, Thordarson T, Keszthelyi L, Walker GP, Hon K, Murphy MT, Long P, Finnemore S (1996) A new model for the emplacement of Columbia River basalts as large, inflated pahoehoe lava flow fields. Geophys Res Letts 23(19):2689–2692
Shaw HR, Wright TL, Peck DL, Okamura R (1968) The viscosity of basaltic magma; an analysis of field measurements in Makaopuhi lava lake, Hawaii. Am J Sci 266(4):225–264
Sheth H (2020) "Pipe vesicles" in basalt; trails left by dense immiscible melt droplets sinking through a viscous basal thermal boundary layer. Earth-Sci Rev 201:103031
Sigurdsson H (1982) Volcanic pollution and climate: the 1783 Laki eruption. Eos Trans AGU 63(32):601–602
Simpson K, McPhie J (2001) Fluidal-clast breccia generated by submarine fire fountaining, Trooper Creek Formation, Queensland, Australia. J Volcanol Geotherm Res 109(4):339–355
Skilling IP (2002) Basaltic pahoehoe lava-fed deltas: large-scale characteristics, clast generation, emplacement processes and environmental discrimination. Geol Soc Lond Spec Pub 202(1):91–113
Skilling IP, White JD, McPhie J (2002) Peperite: a review of magma–sediment mingling. J Volcanol Geotherm Res 114(1–2):1–7
Slatcher N, James MR, Calvari S, Ganci G, Browning (2015) Quantifying effusion rates at active volcanoes through integrated time-lapse laser scanning and photography. Remote Sens 7 (11):14967–14987
Smith TL, Batiza R (1989) New field and laboratory evidence for the origin of hyaloclastite flows on seamount summits. Bull Volcanol 51(2):96–114
Soule SA, Cashman KV (2005) Shear rate dependence of the pāhoehoe-to-'a 'ā transition: analog experiments. Geology 33 (5):361–364
Soule SA, Cashman KV, Kauahikaua JP (2004) Examining flow emplacement through the surface morphology of three rapidly emplaced, solidified lava flows, Kīlauea Volcano, Hawai'i. Bull Volcanol 66(1):1–14
Soule SA, Fornari DJ, Perfit MR, Tivey MA, Ridley WI, Schouten H (2005) Channelized lava flows at the East Pacific Rise crest 9–10 N: the importance of off-axis lava transport in developing the architecture of young oceanic crust. Geochem Geophys 6(8)
Soule SA, Zoeller M, Parcheta C (2021) Submarine lava deltas of the 2018 eruption of Kīlauea volcano. Bull Volcanol 83(4):1–16
Sparks RSJ, Pinkerton H, Hulme G (1976) Classification and formation of lava levees on Mount Etna. Sicily. Geology 4(5):269–271
Spry A (1962) The origin of columnar jointing, particularly in basalt flows. J Geol Soc of Australia 8(2):191–216
Squire RJ, Cas RAF, Clout JMF, Behets R (1998) Volcanology of the Archaean Lunnon Basalt and its relevance to nickel sulfide-bearing trough structures at Kambalda, Western Australia. Aust J Earth Sci 45(5):695–715
Staudigel H, Koppers AA (2015) Seamounts and island building. In: Sigurdsson H, Stix J, Houghton BF, McNutt SR, Rymer H (eds) Encyclopedia of Volcanoes. Academic Press, San Diego, pp 405–421
Staudigel H, Schmincke HU (1984) The pliocene seamount series of La Palma/Canary Islands. J Geophys Res: Solid Earth 89(B13):11195–11215
Stephenson PJ, Burch-Johnston AT, Stanton D, Whitehead PW (1998) Three long lava flows in north Queensland. J Geophys Res: Solid Earth 103(B11):27359–27370
Stothers RB, Wolff JA, Self S, Rampino MR (1986) Basaltic fissure eruptions, plume heights, and atmospheric aerosols. Geophys Res Lett 13(8):725–728
Subbarao KV, Sukheswala RN (1981) Deccan volcanism and related basalt provinces in other parts of the world. Geol Soc India Mem 3

Swanson DA (1973) Pahoehoe flows from the 1969–1971 Mauna Ulu eruption, Kilauea volcano, Hawaii. Geol Soc Am Bull 84(2):615–626

Swanson DA, Wright TL (1981) The regional approach to studying the Columbia River Basalt Group. In: Subbarao KV, Sukheswala RN (eds) Deccan volcanism and related basalt provinces in other parts of the world. Geol Soc India Mem 3:58–80

Swanson DA, Wright TL, Helz RT (1975) Linear vent systems and estimated rates of magma production and eruption for the Yakima Basalt on the Columbia Plateau. Am J Sci 275(8):877–905

Swanson DA, Duffield WA, Jackson DB, Peterson DW (1979). Chronological narrative of the 1969-71 Mauna Ulu eruption of Kilauea volcano, Hawaii. US Geological Survey Professional Paper 1056, 55 pp

Tedesco D, Vaselli O, Papale P, Carn SA, Voltaggio M, Sawyer GM, Durieux J, Kasereka M, Tassi F (2007) January 2002 volcano-tectonic eruption of Nyiragongo volcano, Democratic Republic of Congo. J Geophys Res Solid Earth 112(B9)

Thorarinsson S (1953) The crater groups in Iceland. Bull Volcanol 14 (1):3–44

Thorarinsson S (1968) On the rate of lava-and tephra production and the upward migration of magma in four Icelandic eruptions. Geol Rundsch 57(3):705–718

Thorarinsson S (1969) The Lakagigar eruption of 1783. Bull Volcanol 33(3):910–929

Thordarson T (2004) Accretionary-lapilli-bearing pyroclastic rocks at ODP Leg 192 Site 1184: a record of subaerial phreatomagmatic eruptions on the Ontong Java Plateau. In: Geol Soc Spec Pub, pp 275–306

Thordarson T, Self S (1993) The Laki (Skaftár Fires) and Grímsvötn eruptions in 1783–1785. Bull Volcanol 55(4):233–263

Thordarson T, Self S (1996) Sulfur, chlorine and fluorine degassing and atmospheric loading by the Roza eruption, Columbia River Basalt Group, Washington, USA. J Volcanol Geotherm Res 74(1–2):49–73

Thordarson T, Self S (1998) The Roza Member, Columbia River Basalt Group: a gigantic pahoehoe lava flow field formed by endogenous processes? J Geophys Res Solid Earth 103(B11):27411–27445

Tucker DS, Scott KM (2009) Structures and facies associated with the flow of subaerial basaltic lava into a deep freshwater lake: the Sulphur Creek lava flow, North Cascades, Washington. J Volcanol Geotherm Res 185(4):311–322

van Andel TH, Ballard RD (1979) The Galapagos Rift at 86 W: 2. Volcanism, structure, and evolution of the rift valley. J Geophys Res Solid Earth 84(B10):5390–5406

van Otterloo J, Cas RAF, Scutter CR (2015) The fracture behaviour of volcanic glass and relevance to quench fragmentation during formation of hyaloclastite and phreatomagmatism. Earth Sci Rev 151:79–116

Ventura G, Vilardo G (2008) Emplacement mechanism of gravity flows inferred from high resolution Lidar data: the 1944 Somma-Vesuvius lava flow (Italy). Geomorphology 95(3):223–235

Vona A, Di Piazza A, Nicotra E, Romano C, Viccaro M, Giordano G (2017) The complex rheology of megacryst-rich magmas: The case of the mugearitic "cicirara" lavas of Mt. Etna Volcano. Chem Geol 458:48–67

Vona A, Romano C (2013) The effects of undercooling and deformation rates on the crystallization kinetics of Stromboli and Etna basalts. Contrib Mineral Petrol 166(2):491–509

Vona A, Romano C, Dingwell DB, Giordano D (2011) The rheology of crystal-bearing basaltic magmas from Stromboli and Etna. Geochim Cosmochim Acta 75(11):3214–3236

Vye-Brown C, Self S, Barry TL (2013) Architecture and emplacement of flood basalt flow fields: case studies from the Columbia River Basalt Group, NW USA. Bull Volcanol 75(3):1–21

Wadge G (1978) Effusion rate and the shape of a'a lava flow-fields on Mount Etna. Geology 6(8):503–506

Wadge G (1981) The variation of magma discharge during basaltic eruptions. J Volcanol Geotherm Res 11(2–4):139–168

Walker GPL (1963) The Breiddalur central volcano, eastern Iceland. Quarterly J Geol Soc 119(1–4):29–63

Walker GPL (1971) Compound and simple lava flows and flood basalts. Bull Volcanol 35(3):579–590

Walker GPL (1973) Mount Etna and the 1971 eruption-lengths of lava flows. Phil Trans Royal Soc Lond Series A, Math Phys Sci 274 (1238):107–118

Walker GPL (1987) Pipe vesicles in Hawaiian basaltic lavas: Their origin and potential as paleoslope indicators. Geology 15(1):84–87

Walker GPL (1992) Morphometric study of pillow-size spectrum among pillow lavas. Bull Volcanol 54(6):459–474

Walker GPL (1993) Basaltic-volcano systems. Geol Soc Lond Spec Pub 76(1):3–8

Walker GPL (2009) The endogenous growth of pahoehoe lava lobes and morphology of lava-rise edges. In: Studies in volcanology: the legacy of George Walker. Special Publications of IAVCEI, Geological Society, London, vol 2, pp 17–32

Walker GPL, Huntingdon AT, Sanders AT, Dinsdale JL (1973) Lengths of lava flows [and Discussion]. Philos Trans Royal Soc A Math Phys Sci 274(1238):107–118

Wallace PJ, Frey FA, Weis D, Coffin MF (2002) Origin and evolution of the Kerguelen Plateau, Broken Ridge and Kerguelen archipelago. J Petrol 43:1105–1413

Wang XC, Wilde SA, Li QL, Yang YN (2015) Continental flood basalts derived from the hydrous mantle transition zone. Nature Commun 6(1):1–9

Waters AC (1960) Determining direction of flow in basalts. Am J Sci 258:350–366

Waters AC (1961) Stratigraphic and lithologic variations in the Columbia River basalt. Am J Sci 259(8):583–611

Watton TJ, Jerram DA, Thordarson T, Davies RJ (2013) Three-dimensional lithofacies variations in hyaloclastite deposits. J Volcanol Geotherm Res 250:19–33

Wells G, Bryan WB, Pearce TH (1979) Comparative morphology of ancient and modern pillow lavas. J Geol 87(4):427–440

White JDL, McPhie J, Soule SA (2015) Submarine lavas and hyaloclastite. In: Sigurdsson H, Stix J, Houghton BF, McNutt SR, Rymer H (eds) Encyclopedia of volcanoes. Academic Press, San Diego, pp 363–376

White R, McKenzie D (1989) Magmatism at rift zones: the generation of volcanic continental margins and flood basalts. J Geophys Res Solid Earth 94(B6):7685–7729

Wilmoth RA, Walker GPL (1993) P-type and S-type pahoehoe: a study of vesicle distribution patterns in Hawaiian lava flows. J Volcanol Geotherm Res 55(1):129–142

Wilson L, Head JW (1981) Ascent and eruption of basaltic magma on the Earth and Moon. J Geophys Res Solid Earth 86(B4):2971–3001

Witt T, Walter TR, Müller D, Guðmundsson MT, Schöpa A (2018) The relationship between lava fountaining and vent morphology for the 2014–2015 Holuhraun eruption, Iceland, analyzed by video monitoring and topographic mapping. Front Earth Sci 6:235

Wright HMN, Lesti C, Cas RAF, Porreca M, Viramonte JG, Folkes CB, Giordano G (2011) Columnar jointing in vapor-phase-altered, non-welded Cerro Galán Ignimbrite, Paycuqui, Argentina. Bull Volcanol 73(10):1567–1582

Wright R (2016) MODVOLC: 14 years of autonomous observations of effusive volcanism from space. In: Harris AJL, De Groeve T, Garel F, Carn SA (eds) Detecting, modelling and responding to effusive eruptions. Geol Soc Lond Spec Publ 426(1):23–53. https://doi.org/10.1144/SP426.12

Wright JV, Sastrohardjo F, Curet EA in prep. Volcanic facies analysis of the Aruba Lava Formation, Leeward Antilles: Eruption and depositional processes on the distal submarine flank of an emergent oceanic plateau volcano

Yamagishi H (1985) Growth of pillow lobes—evidence from pillow lavas of Hokkaido, Japan, and North Island, New Zealand. Geology 13(7):499–502

Yamagishi H (1991) Morphological and sedimentological characteristics of the Neogene submarine coherent lavas and hyaloclastites in Southwest Hokkaido, Japan. Sediment Geol 74(1–4):5–23

Yamagishi H, Sakamoto I, Ishii J (1989) Internal structures of pillow lobes of Cenozoic and Mesozoic pillow lavas in and around Hokkaido. Proc Hokkaido Tokai Univ Sci Eng 2:107–118

Yeo I, Searle RC, Achenbach KL, Le Bas TP, Murton BJ (2012) Eruptive hummocks: building blocks of the upper ocean crust. Geology 40(1):91–94

Zhao D (2012) Tomography and dynamics of western-Pacific subduction zones. Environ Earth Planets Monograph 1(1):1–70

High-Viscosity Silicic Lavas (Rhyolite, Dacite, and Andesite)

6

Eruption, Flow Behaviour, and Features of Subaerial and Subaqueous Lavas

Guido Giordano, Ray Cas, and John V. Wright

R. Cas et al., *Volcanology*, Springer Textbooks in Earth Sciences, Geography and Environment,
https://doi.org/10.1007/978-3-319-66613-6_6

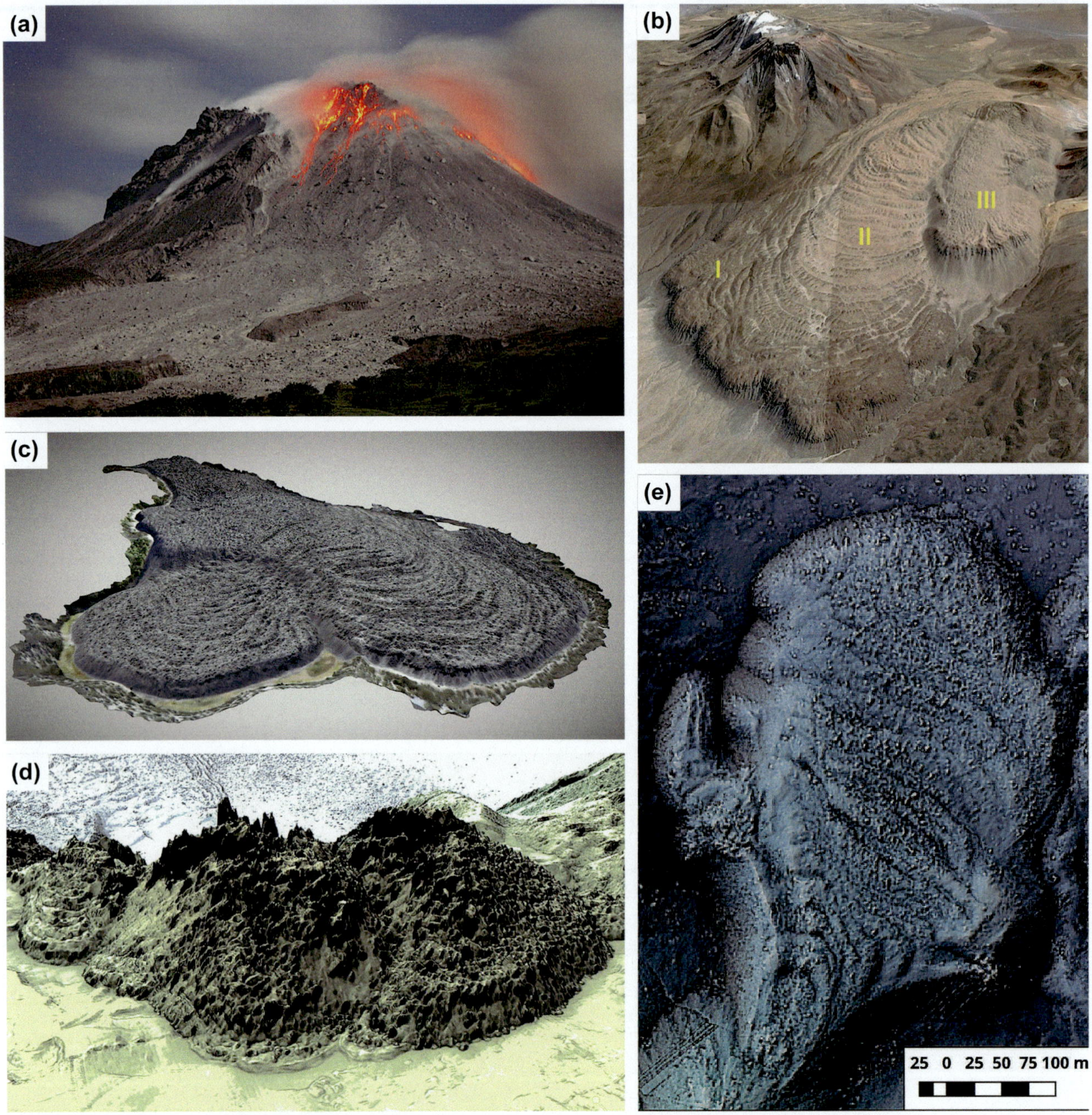

Plate 6.1 Morphology of subaerial and submarine felsic lavas. **a** Enhanced night-time view of glowing Soufrière Hills volcano summit lava dome, 28 January 2010, Montserrat, Lesser Antilles. Note the high aspect ratio morphology, and the bouldery volcaniclastic slopes and apron around the dome, consisting of lava dome debris transported by rock falls and block and ash flows. (Photo by Richard Roscoe; *Source* Photovolcanica). **b** Oblique Google Earth image of the <100 ka Chao dacite lava, northern Chile. This shows at least three (I–III) high aspect ratio lava flow units (main lobes or *coulées*) which individually are up to 400 m thick, and cumulatively are up to ~1,000 m thick, with steep sides and talus aprons, surface pressure ridges (or *ogives*), and multi-lobate flow front and sides. The largest flow (l) is 14 km long (de Silva et al. 1994; also see Fig. 6.8c). To the NW is the stratovolcano Cerro Paniri. (*Source* Google Earth). **c** Drone-derived 3D digital terrain model of the ~690 AD Big Obsidian Flow, Newberry caldera, Oregon, USA, showing two main flow lobes and well-defined convex ogives. View is looking south to the southern caldera margin where the well-defined vent was located; maximum length of lava flow is 2.9 km. (*Source* created by Graham Andrews, West Virginia University Volcanological and Petrology Lab; through Creative Commons Attribution 4.0 International licence). **d** High-resolution digital elevation model (DEM) of submarine rhyolite lava domes G, H, and I (left to right) erupted in July 2012 from a ring fracture fissure vent of the summit caldera of the Havre submarine stratovolcano at a depth of ~900 m, based on very high-resolution AUV bathymetry surveys. Note the high aspect ratio of the lava domes, their spiney irregular surfaces. **e** Plan view DEM of submarine rhyolite lava flow A, also erupted from the rim fissure of Havre volcano in July 2012 at a water depth of ~900 m. This lava flowed down the inner caldera slope to the floor of the caldera at a water depth of ~1400 m where it spread to form the multi-lobate lava in the image. Also note the pressure ridges on the lava surface. (**d** and **e** from Ikegami et al. 2018, through Creative Commons Attribution 4.0 International licence)

Abstract

The eruption processes, flow behaviour, and characteristics of high-viscosity felsic and silicic lavas are now considered. Magma differentiation and country rock assimilation, usually in crustal magma reservoirs, generate felsic magma bodies with progressively lower temperature, increasing silica content and higher viscosity. These in turn affect the mobility, flow distance of any erupted lava flow which generally decreases as the SiO_2 content and viscosity increase. Such lavas are characterised by their high aspect ratio, forming very thick lava domes, mesa lavas, or short lava flows (coulées), depending on their viscosity and the topography onto which they flow, or intrude in the case of cryptodomes. Subaerial domes and lavas are commonly texturally zoned from stony or crystallised interiors to obsidian, coherent pumice, a carapace of autobrecciated obsidian and pumice clasts outwards, and an apron of block and ash flow deposits and resedimented talus laterally. The most common settings for eruption of evolved, high-viscosity lavas are stratovolcanoes and calderas at convergent volcanic arcs, rift and hot spot or plume settings. Here they are often erupted subaerially from point-source vents or along caldera ring fractures. Such domes and lavas may generate explosive eruptions, producing a range of deposit types from rings of phreatic lithic breccias to pumice-tuff cones surrounding the lava and lateral aprons of block and ash flow and pumice and ash flow deposits. Felsic domes are also commonly erupted in subaqueous settings in lakes, including caldera lakes, subglacially, and in the oceans. Modern seafloor felsic domes and lavas are found at the summits of submarine arc volcanoes, including those with summit calderas, as well as on the seafloor of back-arc basins. They can be distinguished from subaerial counterparts by abundant hyaloclastite breccia and may have significant coherent pumice carapaces and pumice hyaloclastite breccias. We highlight high-resolution bathymetry studies of the rhyolite domes and short lavas erupted during the 2012 eruption of the wholly submarine Havre volcano, in the Kermadec Arc. In deep subaqueous and subglacial settings, explosive exsolution of volatiles is suppressed by hydrostatic and glaciostatic pressures, leading to eruption of coherent domes and lavas, and their autoclastic (autobreccia, quench fragmented hyaloclastite) carapaces. For both subaerial and subaqueous domes, lavas, and cryptodomes, we summarise the facies characteristics and provide facies models.

6.1 Introduction: General Characteristics, Occurrences, and Dimensions of Felsic Lavas

Felsic intermediate to silicic magmas can be generated in a variety of geodynamic settings, mainly at convergent magmatic arcs, but also along active rifts, such as in Iceland and Ethiopia, and associated with intraplate hot spot or plume settings (e.g. Yellowstone-Snake River Plain). Typical compositions of the resulting high-viscosity lavas and domes include the andesite-dacite-rhyolite series, and also latites and trachytes, which we collectively refer to in this discussion as felsic lavas (Plate 6.1).

Most commonly, historic eruptions of felsic lava domes and lava flows have been of intermediate composition (e.g. andesitic Soufrière Hills lava domes, Montserrat, Watts et al. 2002; Wadge et al. 2014; Plate 6.1a; Merapi volcano summit lava domes, Java, Indonesia, Voight et al. 2000). In the last century, only three known rhyolite lava-forming eruptions have been witnessed, in the Tuluman Islands in 1954–1957 in Papua New Guinea (Reynolds and Best 1976; Reynolds et al. 1980; Johnson and Smith 1974), at Chaitén volcano in Chile in 2008–2009 (Castro and Dingwell 2009; Bernstein et al. 2013; Pallister et al. 2013), and during the 2011–2012 eruption of Puyehue Cordón Caulle volcano in Chile (Castro et al. 2013; Schipper et al. 2013; Tuffen et al. 2013). In historic times, rhyolite lavas were also erupted at Newberry caldera, Oregon, USA in ~690 AD (Plate 6.1c), from Monte Pilato pumice cone on Lipari, Aeolian Islands, Italy between 776 and 1220 AD (Forni et al. 2013), and the Pietre Cotte obsidian lava on Vulcano volcano, Aeolian Islands, Italy, which is dated at 1739 AD (Arrighi et al. 2006; De Astis et al. 2013). Historic dacite lava eruptions are more common, including Mt St Helens, USA, 1980–86, 2004–06 (Swanson et al. 1987; Schilling et al. 2008; Sherrod et al. 2008; Vallance et al. 2008), Mt Unzen, Japan from 1991 to 1995 (Nakada et al. 1999), on and around the Kameni Islands' lavas, Santorini, from Roman times to 1950 (Pyle and Elliott 2006; Nomikou et al. 2014), and Santiaguito dome complex, Santá Maria volcano, Guatemala (Harris et al. 2003).

In addition, felsic lavas and domes can also be erupted subglacially, such as in Iceland (Furnes et al. 1980; Tuffen et al. 2001) and the Rocky Mountains in Canada (Edwards et al. 2002), from the summits of submarine arc volcanoes, such as the Izu Bonin Arc (Fiske et al. 1998; Allen et al. 2010) and the Kermadec Arc (Wright et al. 2006; Carey et al. 2018; Rotella et al. 2013; Ikegami et al. 2018; Manga et al. 2018), and on the floors of deep ocean back-arc basins, such as the Woodlark Basin (Binns 2003) and the northeast Lau Basin (Embley and Rubin 2018). Perhaps the largest historical rhyolitic lava-forming eruption on Earth occurred in 2012, during a wholly submarine eruption from the summit of the Havre volcano in the Kermadec Arc, at water depths of 900 to 1300 m (Carey et al. 2018; Ikegami et al. 2018; Manga et al. 2018). Although the 2012 Havre volcano eruption was not directly witnessed, the pumice raft it produced, which was widely dispersed around the southwest Pacific after the eruption, and subsequent submarine survey and sampling expeditions have shed significant light on this recent submarine rhyolite eruption.

The processes of magma differentiation and country rock assimilation in crustal magma reservoirs generate felsic magma bodies with progressively lower temperature, increasing silica content, and higher viscosity. As high-viscosity magmas move slowly through the crust they stop and start. High-viscosity, evolved magmas generally have high-volatile contents and when they are extruded subaerially as lavas they often also generate explosive eruptions. However, such magmas can also degas at depth and be erupted effusively as lavas, without significant explosive activity (e.g. 1991–1995 Mt Unzen lava dome eruptions; Nakada et al. 1999). In addition, in deep subaqueous and subglacial settings, explosive exsolution of volatiles is suppressed by hydrostatic and glaciostatic pressures (Chap. 3), leading to eruption of coherent domes and lavas, and their autoclastic carapaces.

The most notable characteristic of high-viscosity lavas is the high aspect ratio that reflects their generally low mobility (Plate 6.1, Fig. 2.21). High-viscosity lavas form very thick lava domes, and thick lava flows (or coulées; Fink 1987; Fink and Anderson 2000; Plate 6.1, Fig. 6.1a–f), depending on their viscosity and the topography onto which they flow. Domes basically just extrude from the vent, bury it, and don't flow away from the vent. Lava flows or coulées can flow moderate-to-long distances. The term coulées is not commonly used anymore, and most volcanologists talk about lava domes and lava flows, referring to coulées as short lava flows. The viscosity of high aspect ratio lavas is generally $>10^5$ Pa s and can exceed 10^{14} Pa s, such as for the degassed andesite of the 1995–1999 eruption at Soufrière Hills, Montserrat (Sparks et al. 2000; Calder et al. 2015).

In this chapter, we will review the eruption processes, flow behaviour, and characteristics of subaerial and subaqueous (and subglacial) felsic (silicic) domes and lava flows. Other recent reviews include Calder et al. (2015) for subaerial systems, White et al. (2015) for subaqueous systems, and Edwards et al. (2015) for subglacial systems. McPhie et al. (1993) also provide excellent images of the range of textures we discuss and illustrate in this chapter. High-viscosity lavas and closely related high-level intrusions are important host rocks to and spatially, if not genetically, associated with a number of ore deposit types, in particular, epithermal precious metal, volcanic-hosted uranium, and, in submarine environments, volcanogenic massive sulphide (VMS) deposits. They may also be drivers of geothermal fluids and even act as petroleum reservoirs. These geological and exploration aspects are described and developed in detail in Chap. 18.

6.1.1 Settings

The most common settings for evolved, high-viscosity lavas are stratovolcanoes and calderas at convergent volcanic arcs. They may be erupted from point-source vents, such as the 1912 Novarupta lava dome, Alaska (Adams et al. 2006; Hildreth and Fierstein 2012; see Fig. 6.6c) or along caldera ring fractures, often forming arcuate distributions. Excellent examples of this latter type are at Valles caldera, New Mexico (Smith et al. 1970; Heiken et al. 1986; Wolff and Gardner 1995; see Figs. 10.22d; 16.66) and La Primavera caldera, Mexico (Fig. 6.2a; Mahood 1980; Clough 1981; Clough et al. 1982). At Mono Craters, California, an arcuate line of rhyolite lavas is thought to represent activity over part of an actively developing ring fracture system around the foundering roof of a large crustal magma body in a pre-caldera stage of evolution (Hermance 1983; Rundle and Eichelberger 1983). At Los Humeros caldera (Mexico), rhyolite and trachyte domes occur close in space and time to basaltic to trachy-andesite lavas, along ring fractures that surround and cross the resurgent caldera floor (Carrasco-Nuñez et al. 2018; Lucci et al. 2020). Intra-caldera domes may also occur as solitary bodies, as is the case of the active Chaitén Obsidian Dome, Chile (Carn et al. 2009), or align along major regional tectonic trends that may extend beyond caldera margins, such as for the Roccamonfina intra-caldera latite domes, Italy (De Rita and Giordano 1996), and the Nea Kameni dacite lava and dome complex, Santorini, Greece (Fig. 6.2b, c; Pyle and Elliot 2006; Nomikou et al. 2014).

Lava domes and thick lava flows are commonly extruded at the summit of stratovolcanoes, usually after major explosive phases, such as at Mt St Helens in the USA. After the initial 18 May 1980 plinian eruption, dacite domes formed after each of several explosive eruptions and then during two main periods, in 1981–1986 and in 2004–2006 (Fig. 6.1e; Swanson et al. 1987; Sherrod et al. 2008; Vallance et al. 2008). Similarly, the large crater formed at the top of Santá Maria volcano (Guatemala) after the 1902 Plinian eruption is being filled by nested domes of the Santiaguito dacite dome complex (Fig. 6.1f; Harris et al. 2003).

The volatile content and crumbly nature of many high-viscosity lavas, together with the very highly erosional subaerial setting, ensures a very low preservation potential

Fig. 6.1 High aspect ratio morphology of high-viscosity lava domes and lava flows. **a** Trachytic Puy de Dome (right); other hills are scoria cones. Auvergne Region, France. **b** Multiple domes of the Tarawera rhyolite dome complex, New Zealand. **c** Thick autobrecciated andesitic block lava flow, Tuzgle volcano, Andes, Argentina. **d** Basaltic andesite to dacitic Panarea Dome volcano, Aeolian Islands, Italy. **e** Dacite lava dome in the amphitheatre crater of Mt St Helens, USA, during the 1980 eruption (*Source* United States Geological Survey). **f** Four dacitic to andesitic lava domes constituting the Santiaguito cluster at the summit of Santá Maria stratovolcano, Guatemala: El Caliente (left), La Mitad, El Monje, El Brujo (Photo by R. W. Sanderson)

for lava domes at the summits of subaerial stratovolcanoes in the geological rock record. They are frequently partially or wholly destroyed during major explosive phases, often occurring after long repose times, but also during the dome-building phase, such as at andesitic calc-alkaline stratovolcanoes like Merapi (Java, Indonesia; Hammer et al. 2000; Voight et al. 2000) and Soufrière Hills (Montserrat, Caribbean; http://www.mvo.ms/about-volcanoes/Soufrière-hills-volcano/chronology-of-current-eruption), where several pulses of dome growth were interrupted by major dome collapse and explosive phases. Secondly, because of their elevation they are prone to erosion after eruptive activity of a stratovolcano ceases. By contrast, farther reaching thick lava flows emplaced a few or more km away from the summit crater, along the volcano slopes, have a much higher preservation potential, and actually form the bulk of

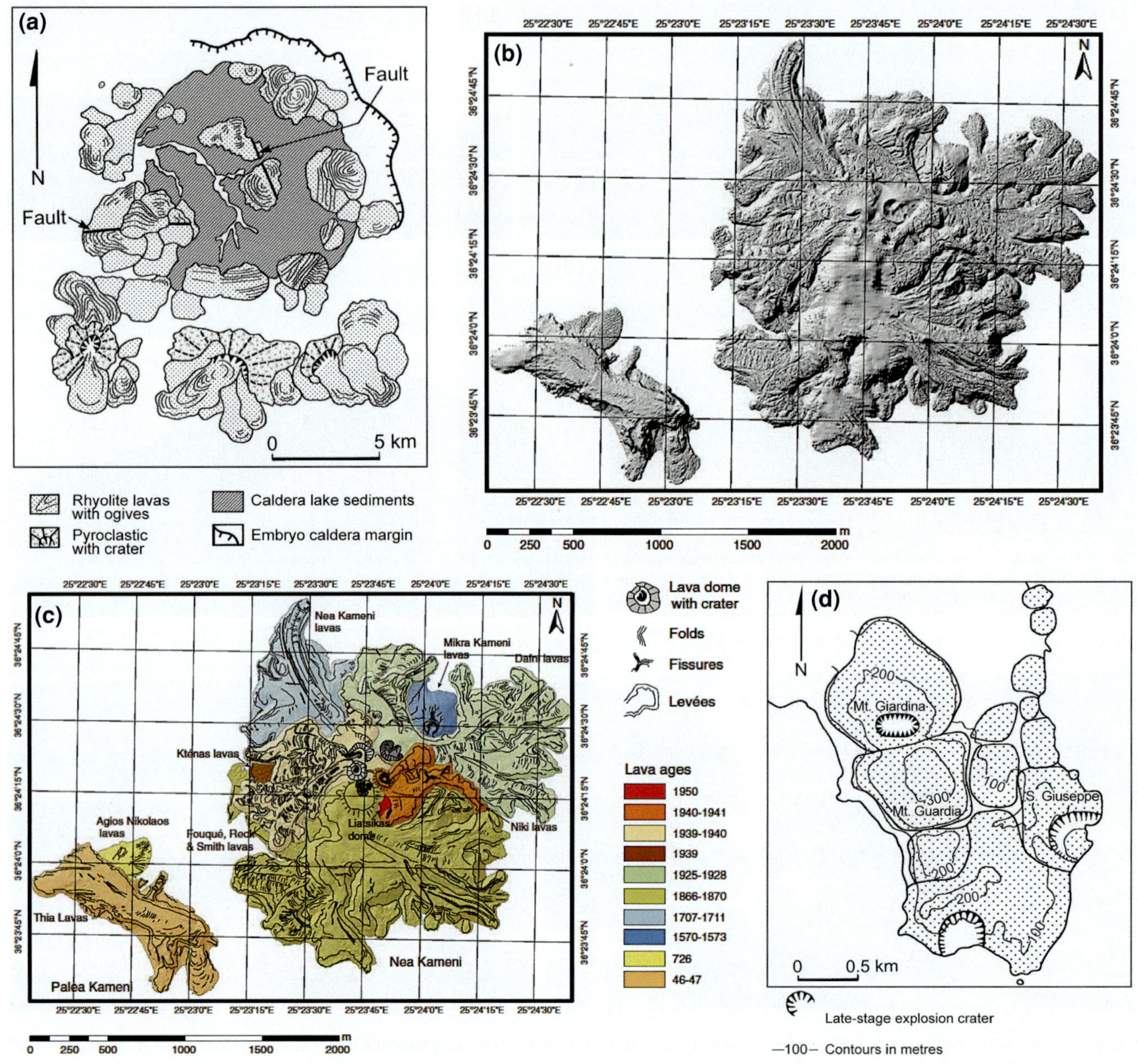

Fig. 6.2 Features of rhyolite and dacite domes and lavas. **a** Map of the distribution and surface features of the rhyolite lava flows of La Primavera volcano, Mexico. Some of the linear features are faults, while curved features are *ogives* or pressure ridges. (From Cas and Wright 1987, After Clough 1981). **b** Digital elevation model of the historical dacite lavas and domes erupted from the Kameni Islands volcano in the Santorini caldera, Greece (From Pyle and Elliot 2006). **c** Map of the historical dacite lavas and domes erupted from the Kameni Islands volcano in the Santorini caldera, Greece (From Pyle and Elliot 2006). **d** Map of the rhyolite domes of southern Lipari. Aeolian Islands, Italy (From Cas and Wright 1987, After Richardson 1978). **e** Map of the cluster of lava domes erupted from the Tarawera rhyolite dome complex (T, top left inset) during several eruption periods along an NNE trending fault-fissure within the Okataina Volcanic Complex (OVC). This includes the most recent ~1315 AD Kaharoa eruption, which produced four domes (stippled pattern) and associated pyroclastic deposits, labelled Kaharoa Pyroclastics. Taupo Volcanic Zone, North Island, New Zealand. (From Nairn et al. 2004). **f** Oblique aerial view of the Tarawera dome complex showing the steep-sided Wahanga (left), Ruawahia (middle), and Tarawera (right) rhyolite lava domes erupted during the 1315 AD Kaharoa eruption from Tarawera dome complex. The arrow points to the basaltic rift fissure vent formed during the 1886 basaltic plinian eruption (Chap. 9). (From Nairn et al. 2001). **g** The tops of massive jointed 1315 AD Kaharoa rhyolite domes in (**f**) draped by diffusely bedded pyroclastic deposits (white), and in turn draped by bedded 1886 basaltic pyroclastic deposits (dark) exposed in the walls of the 1886 basaltic fissure vent which cut through the 1315 AD rhyolite domes, Tarawera dome complex

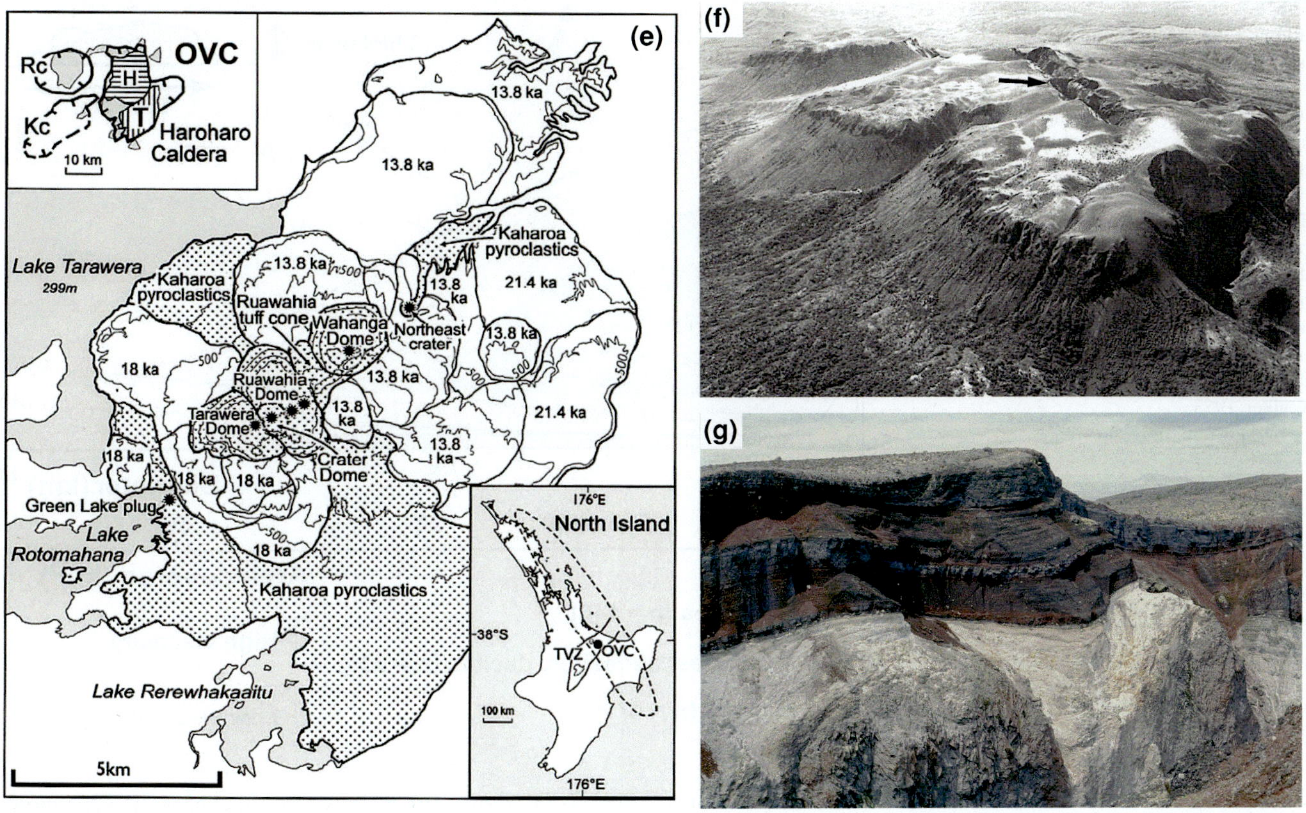

Fig. 6.2 (continued)

stratigraphic record of many andesitic-dacitic stratovolcanoes (Fig. 6.1c).

On other volcanoes, viscous lavas are not necessarily obviously associated with a summit vent system, such as the spectacular concentration of rhyolite domes in the southern part of Lipari in the Aeolian Islands, Italy (Fig. 6.2d; Tranne et al. 2002; Forni et al. 2013). The emplacement of high-viscosity lavas may occur more rarely outside edifices from "leaky" basement fracture conduits, both as solitary domes and as dome complexes (clustered domes), in association with shallow intrusions. Examples of dome clusters are the Cimini rhyodacitic dome complex, central Italy (Cimarelli and De Rita 2006); the Guayaques rhyolite-dacite dome complex in the Andes, in Chile-Bolivia (de Silva and Francis 1991); and the Tarawera rhyolite domes, New Zealand, part of the Okataina Volcanic Complex, New Zealand (Cole 1970; Nairn et al. 2001, 2004; Figs. 6.1b, 6.2e–g).

In addition, viscous lavas may also be emplaced as cryptodomes. A cryptodome is a dome-like uplift of the ground surface in a volcanic area, seemingly caused by a very shallow, near-surface intrusion. Some of the best documented occurrences of cryptodomes are those described from Usu volcano, Japan (Minakami et al. 1951; Goto and Tomiya 2019). Cryptodomes will be discussed further below (Sect. 6.3.5; also see Chaps. 8, 15). Cryptodomes and their intrusion depth can be identified by the shape and degree of the deformation of the ground surface (usually upbulging) that they induce (e.g. Goto and Tomiya 2019; Urbani et al. 2020).

In subaqueous settings, felsic domes and lavas are also commonly erupted in lakes (e.g. Cas et al. 1990), subglacially (e.g. Furnes et al. 1980; Tuffen et al. 2001; Edwards et al. 2002, 2015) and in the oceans, where they are found at the summits of submarine arc volcanoes (e.g. Fiske et al. 1998; Allen et al. 2010; Soriano et al., 2012, 2013), including those with summit calderas (Wright et al. 2006; Carey et al. 2018; Rotella et al. 2013, 2015; Ikegami et al. 2018; Manga et al. 2018), as well as on the seafloor of back-arc basins (e.g. Binns 2003; Embley and Rubin 2018;). The rock record contains many examples of ancient subaqueous domes and lavas, although the geological and tectonic context is not always so clear (e.g. Cas 1978, 1992; Yamagishi and Dimroth 1985; Cas et al. 1990, 2001; Yamagishi 1991; Pensa et al. 2019).

6.1.2 Flow Distance, Volume, and Size

Viscous lava domes and flows extrude and form very thick bodies of magma above the vent. They generally move

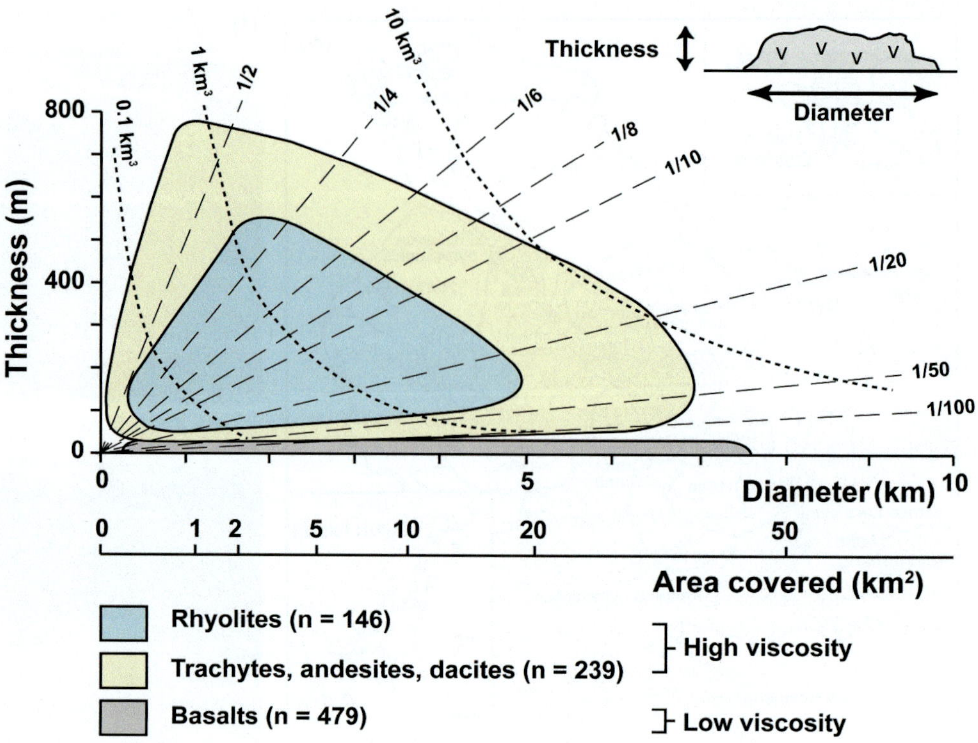

Fig. 6.3 Dimensions of lavas of different compositions and their aspect ratios. The vertical y-axis shows the mean thickness (*T*), whereas the two scales along the x-axis give the area covered by the extrusion (*A*) and the diameter of a circle having this area (*D*). The broken lines give the aspect ratio *T/D*. The dotted lines give the volumes of circular disc-like bodies of the dimensions shown as a rough guide to the volumes of the lava extrusions. (Modified from Cas and Wright 1987, After Walker 1973)

slowly and not very far, developing the high aspect ratio (ratio of the average thickness or height of the lava, T, to the diameter, D, of a circle that has the same area as the lava flow, i.e. aspect ratio is T/D; see Fig. 2.20, and Sect. 2.6.2 for definition) characteristic of felsic domes and lavas (Plate 6.1; Figs. 6.1, 6.3), although some large volume felsic lava flows can flow several to tens of km if eruption is sustained. Flow behaviour depends on total volume extruded, the effusion rate, rheology of the magma (Bingham), topography, and cooling rate.

Typical flow distances for viscous lavas can be as short as a few hundred m and up to a few km, and their volume usually ranges from fractions of km^3 to rarely 10's and even 1,000 km^3. Heights of individual viscous lava flows are in the order of 10's to 100's m. Rhyolite and dacite lavas have a wide range in thickness from <50 m to >500 m (Fig. 6.3). However, the average thickness is probably about 100 m. The 1980–1986 dacite lava dome at Mt St Helens was approximately 300 m high and 1000 m in diameter (Tilling et al. 1990). The Chao dacite flow from Chile (Plate 6.1b; also Fig. 6.8c), which travelled 14 km is over 400 m thick, and has a volume of 24 km^3 (Guest and Sanchez 1969; de Silva et al. 1994). This lava flow is considered to be unusually extensive and voluminous, but much farther flowing (tens of kms) and larger volume flow (up to 1,000 km^3) are also known, especially from Large Igneous Provinces (e.g. Cas 1978; Bonnichsen and Kauffman 1987; Henry et al. 1988, 1990; Allen et al. 2003, 2008; McPhie et al. 2008; see Sect. 6.8). Flow dimensions and distances in subaqueous environments are comparable to those in subaerial settings.

6.2 Subaerial Felsic (Silicic) Domes and Lavas—Processes and Constraints

Here we describe the main processes associated with the extrusion of subaerial viscous lavas, although many of those processes apply also to eruptions in subaqueous settings, which will be described in Sect. 6.10.

6.2.1 Processes

6.2.1.1 Prerequisites for Effusion of Viscous Subaerial Magmas

The principal requirement for the effusive eruption of viscous magma as coherent lava is that its volatile content (Chap. 2) is low before eruption. For most felsic magmas, this involves loss of volatiles from the source magma prior to effusive eruption, which can be achieved by the following processes:

(a) Direct escape (outgassing or open degassing) of exsolving volatiles, either gradually from the conduit and through country rocks from the vent, hydrothermal springs and fumaroles or, more rapidly, by episodic hydrothermal (or steam) explosions. Open degassing conditions (Chap. 4) depend on magma permeability at a given decompression rate (i.e. ascent rate along the conduit) and for viscous magmas, an efficient outgassing at shallow depths is associated with very slow rise rates (in the order of $\leq 10^{-4}$ m s^{-1}; e.g. Cashman et al. 2008) and the development of a connected network of both fractures along the conduit walls and interconnected bubbles/fractures within the magma (Gonnerman and Manga 2003; Rust and Cashman 2004, Okumura et al. 2010). Magma fracturing (and degassing) during ascent has been suggested as a trigger for volcanic earthquakes (Goto 1999; Tuffen and Dingwell 2005; Neuberg et al. 2006).

(b) A previous or contemporaneous phase of explosive activity during which most of the volatiles of the magma and magma reservoir have been released. This suggests that lava eruption could be a terminal event of many explosive eruptions and, indeed, this often seems to be the case. Examples of this include: the 1980 Mt St Helens eruption (Fig. 6.1e); the ~690 AD effusion of the Big Obsidian Flow, Newberry caldera, Oregon (Gardner et al. 1998); the 1220 AD Rocche Rosse obsidian lava, Lipari, Italy, after the explosive formation of the Monte Pilato pumice cone (Fig. 6.6e; Dellino and La Volpe 1995); effusion of the Chao dacite lavas, Chile, after a pumice cone building explosive phase (de Silva et al. 1994; see Fig. 6.8c); eruption of the Tarawera lava dome complex, preceded and interspersed with explosive activity, New Zealand (Cole 1970; Nairn et al. 2001, 2004; see Fig. 6.21); and lastly, the rhyolite dome at Novarupta, Alaska plugged the vent of the 1912 Valley of Ten Thousand Smokes ignimbrite eruption that drained magma some 10 km westwards away from below the Mt Katmai summit area, eventually leading to its caldera collapse (Fig. 6.6c; Chap. 10).

The exsolution and degassing of volatiles from magma will increase its viscosity and yield strength (Chaps. 2 and 4), possibly associated with microlite crystallisation, which will affect the mobility, flow distance, and thickness of any lava flow. If continuing volatile exsolution, in turn induced by progressive cooling and crystallisation, is not compensated by efficient, open degassing, magma will become volatile supersaturated through second boiling (Chap. 2). For this reason, domes may behave explosively during and after eruption, and usually the volume of volcaniclastics associated with the eruption of viscous magmas far exceeds the volume of coherent lavas (Wadge et al. 2009).

6.2.1.2 Effusion Rates, Vents, and Shallow Plumbing System

Walker (1973) considered effusion rate to be the most important factor controlling lava form. He showed that the distance travelled by lava flows was proportional to the effusion rate. It was thought that this was due to the competing effect of cooling. Lavas erupted at lower effusion rates would travel less because cooling rates would rapidly increase their viscosity and limit movement. By contrast higher effusion rates would increase the thermal input to the lava flow, allowing for a slower effect of cooling on lava rheology and mobility. A survey of historic silicic, high-viscosity lavas shows that effusion rates (Table 6.1) are usually between 0.02 and 2–3 m^3 s^{-1} (e.g. Watts et al. 2002; Dirksen et al. 2006), generally much lower than those for mafic lavas (Chap. 5). Higher effusion rates up to 54 m^3 s^{-1} have also been documented (Table 6.1).

Cyclicity of dome-forming phases is common and involves pulsing or variable effusion rates at various time scales, which may range from short-term pulsatory activity which last hours to days, to longer term pulsatory activity over weeks to years, also involving repose periods (Swanson et al. 1987; Barmin et al. 2002; Melnik and Sparks 2005). Long-term pulsations might be expected to form compound lavas at low extrusion rates. For example, the Santiaguito dacite dome complex at the summit of Santá Maria volcano, Guatemala, is a compound silicic lava complex, which began to form in 1922 and continues to grow to date (Fig. 6.1f; Harris et al. 2003). De Silva et al. (1994) estimated that the Chao dacite lava flow, Chile, was likely emplaced over a period of between 100 and 150 yr. It comprises at least three flow units, or main lobes, forming a compound coulèe (Plate 6.1b, Fig. 6.8c), displaying little evidence of significant breaks in activity, although there may have been variations in the effusion rate. This suggests a steady supply of magma would be required to keep the flow moving.

Most rhyolite lavas seem to reach the surface through a cylindrical conduit even though many may be aligned along fissure-fracture systems. However, some dome lavas in intra-caldera settings are elongate, and surface ridges are parallel to the caldera ring-fault (e.g. at La Primavera caldera), suggesting that rhyolite lavas may also be extruded along fissure vents (Fig. 6.2a; Clough 1981). The Circle Creek rhyolite in Nevada is thought to have erupted through a large fissure system, marked by multiple aligned vents and has an area of 130 km^2 (Coates 1968). These trends are thought to represent fissure segments, which closed at different points along the fissure, forming a series of aligned sub-cylindrical vents as lava was extruded, in a similar manner to that observed along basaltic fissure eruptions.

Table 6.1 Eruption rates for felsic-lava-forming eruptions

Eruption and composition	Effusion rate ($m^3 s^{-1}$)	References
Havre (submarine) 2012, rhyolite	0.14–14.4	Ikegami et al. (2018)
Puyehue Cordón-Caulle, Chile, 2011–2012, rhyolite	20–70	Bertin et al. (2015)
Merapi, Indonesia, 2010, andesite	25 (peak before explosive eruption)	Surono et al. (2012)
Chaitén, Chile, 2008–2009, rhyolite	45	Pallister et al. (2013)
Colima, Mexico, 2007, andesite	0.02	Lavallée et al. (2011)
Mt St Helens, USA, 2004–2008, dacite	45–60	Calder et al. (2015)
Soufrière Hills, Montserrat, 1995–2008, andesite	0.5–20	Ryan et al. (2015)
Shiveluch, Russia, 2001–2004, andesite	0.2–7	Dirksen et al. (2006)
Colima, Mexico, 1995–1996, andesite	0.05	Lavallée et al. (2011)
Shiveluch, Russia, 1993–1995, andesite	3–14	Dirksen et al. (2006)
Unzen, Japan 1991–1995, dacite	0.1–7.5	Nakada et al. (1999)
Mt St Helens, USA, 1980–1984, dacite	0.4–40.3	Anderson and Fink (1990)
Usu, Japan, 1977–1983, dacite	0.6	Newhall and Melson (1983)
Augustine, USA, 1976, andesite	11.6	Newhall and Melson (1983)
Lamington, Papua-New Guinea, 1951–1956, andesite	5.8	Newhall and Melson (1983)
Bezymianny, 1955–1983, andesite	1.8	Newhall and Melson (1983)
Usu, Japan, 1943–1945, dacite	1.2	Newhall and Melson (1983)
Usu, Japan, 1910, dacite	3.5	Newhall and Melson (1983)
Santiaguito, Chile, 1922–1993	0.6–2.2	Anderson et al. (1995)
Merapi, Indonesia 1902–1992 averaged, andesite	0.04	Siswowidjoyo et al. (1995)
Chao, Chile, < 100 ka, dacite	<25	de Silva et al. (1994)
Nea Kameni, Santorini, (some submarine) 1707–1711, 1866–1870, 1925–1928, and 1939–1941, dacite	2–3	Nomikou et al. (2014)

It was suggested that the driving force for this extrusion was the weight of fissured crust that down-sagged into a magma chamber, thus forming a sag-basin as opposed to a caldera.

The Tarawera Volcanic Complex of New Zealand (Figs. 6.1b, 6.2e–g; Cole 1970; Nairn et al. 2001, 2004; Shane et al. 2007; Ashwell et al. 2018) consists of a cluster of rhyolite domes and associated pyroclastic deposits erupted in episodes over a 22 kyr period, the last being the 1315 AD Kaharoa eruption (see Fig. 6.21). The dome complex defines a NNE trend (Fig. 6.2e), which reflects a crustal fissure-fracture zone within the Okataina Volcanic Centre. This fissure vent system was activated again in 1886 when a purely basaltic explosive fissure eruption occurred along a 17 km length of the NNE trending fissure, cutting through the rhyolite dome complex (Fig. 6.2e, f; Nairn 1979; Walker et al. 1984; Houghton et al. 2004; Carey et al. 2018), and providing cross-sectional exposure of some of the domes and associated pyroclastic deposits in the walls of the fissure (Fig. 6.2g).

Results of the drilling the conduit in the Unzen Scientific Drilling Project (Nakada et al. 2005) have shown that the shallow plumbing system at Mt Unzen volcano Dome system is made of a series of dyke intrusions over a zone about 500 m wide, with individual dykes/conduits being 7–40 m wide. Combined volcanological observation and modelling suggest that the conduit of the 2001–2004 dome eruption at Shiveluch volcano was some 53–71 m wide (Dirksen et al. 2006).

6.2.1.3 Lava Properties, Rheology, and Flow State

Hulme (1974) modelled lavas as Bingham substances (Chap. 2) and indicated that the principal factor governing their shape or form was their non-Newtonian rheology. Given their very high-viscosity (10^5–10^9 Pa s) felsic domes and lavas invariably flow in a laminar flow state. Lavallée et al. (2007) and Cordonnier et al. (2009) demonstrated experimentally that dome lavas have a non-Newtonian rheology, especially if they are crystal rich, and also that the apparent viscosity of dome lavas, especially crystal-rich ones, decreases as it experiences internal fracturing and cracking (autobrecciation at a range of scales from macro to micro) during flow in response to flow shear and increasing strain rate (Calder et al. 2015). Theoretical analysis and experiments with kaolin suspensions, which are close to Bingham visco-plastic substances in rheology, showed that aspect ratio was mainly dependent on yield strength (σ_0) (Blake 1990; Fink and Griffiths 1998; Balmforth et al. 2000).

For a (near-)Bingham substance to flow downhill, it must form a layer deep enough for the shear stress at the base to exceed the yield strength. Close to the lateral margins the depth is not great enough for downhill flow to occur, and dead zones of stationary magma form natural lava levees along the margins, especially as those margins cool. Within the channel confined by the levees, the active flow thickness is greatest, and flow occurs as yield strength is exceeded. The depth and width of a flow, and the width of each dead zone, are related to five independent initial parameters: effusion rate (F), the slope (α), and three properties of the fluid—Bingham viscosity (η), yield strength (σ_0), and specific weight ($g\,\rho$, where g is the acceleration due to gravity and ρ is the density of the fluid). The critical depth (d_c) or thickness that must be exceeded for movement of the flow to occur is given by

$$d_c = \frac{\sigma_0}{g\,\rho\,sin\alpha} \quad (6.1)$$

For lavas with higher yield strengths, the thickness of the lava flow is greater, and d_c is therefore higher.

The aspect ratio of a lava flow can be predicted using effusion rate, slope, and magma properties (Hulme 1974) from

$$\frac{T}{D} = \frac{\sigma_0}{(F\,\eta)^{0.25}(\rho\,g)^{0.75}} \quad (6.2)$$

where T is the flow depth or thickness, D is the width of the flow, and other parameters are defined above.

Blake (1990) considered the shape characteristics of experimental lava domes and demonstrated a relationship between dome height or thickness (H), yield strength (σ_0), radius (R), and density (ρ):

$$H = 1.76\left(\frac{\sigma_0\,R}{\rho\,g}\right)^{0.5} \quad (6.3)$$

where R is the radius of the dome. Aspect ratio therefore depends mainly on yield strength. Equation 6.2 predicts that lavas with low viscosity and yield strengths give rise to flows with lower aspect ratios, and more viscous lavas with higher yield strengths will form higher aspect ratio flows, which is in general agreement with field observations (Fig. 6.3; Walker 1973). Furthermore, the preceding equations imply that the aspect ratio of a lava dome characterised by a given yield strength tends to be maintained during its growth.

From Eqs. 6.2 and 6.3, aspect ratio would seem to be insensitive to changes in effusion rate, but in reality, this is more complicated because of the effect of temperature variations, which control viscosity. A change in effusion rate will lead to a change in the temperature within flowing lava at any given point, and this will have a marked effect on the yield strength. An increased effusion rate would maintain a higher lava temperature, and therefore reduce the yield strength, at any particular point in the lava. Eruption temperatures of andesite to rhyolite lavas range approximately between 1,000 °C (+) and 750 °C, respectively (Chap. 2).

The viscosity of high aspect ratio lavas may range over several orders of magnitude, generally from $>10^5$ Pa s to $<10^{14}$ Pa s. Temperature-dependent viscosity and yield strength may increase substantially during the eruption and the emplacement of a viscous lava due to the cooling rate and the eventual solidification of a crust (Griffiths and Fink 1992, 1993, 1997). Numerical models for bulk cooling of obsidian flows suggest an exponential temperature decay with time, by some hundreds of degrees over tens of years (Manley 1992). Cooling rates estimated from relaxation temperature of glass for the Rocche Rosse rhyolite frontal flow ramps at Lipari range between 0.2 and 0.03 °K min^{-1} (Gottsmann and Dingwell 2001), which can equate to several tens to hundreds of degrees Kelvin per day.

Cooling rates control the glass transition temperature (Chap. 2) and consequently the development of a solid crust. In turn, the formation of a solid crust has a substantial influence on the possibility and style of further lateral spreading of the growing dome and on the formation of an autoclastic carapace and a talus apron of resedimented autoclastic debris. A crust will help to insulate the still hot and plastic interior from heat loss, maintain a low viscosity, and enhance flow mobility, whereas the talus apron also contributes to the bulk yield strength of the dome and hence hinders its capacity to spread laterally (Blake 1990).

In summarising a major body of analogue experimental research on the behaviour of viscous lava flows, Fink and Griffiths (1990, 1998) and Griffiths (2000) proposed that the rate of cooling and rate of formation of the surface crust of lava flows strongly affected the shape, surface morphology, and flow behaviour. They developed a dimensionless parameter, ψ, that takes account of the physical properties of the magma, where ψ is the ratio of the time that it takes for a crust to solidify, t_s, to the time that it takes for material to advect laterally, t_A:

$$\psi = \frac{t_s}{t_A} \tag{6.4}$$

where t_A is defined as

$$t_A = \left(\frac{\eta}{\rho\, g}\right)^{3/4} Q^{1/4} \tag{6.5}$$

in which Q is the eruption rate in cm^3 s^{-1}. This relationship can be rearranged for Bingham fluids in a new dimensionless parameter ψ_B, so that

$$\psi_B = \left(\frac{g\,\rho}{\sigma_o}\right)^3 Q\, t_s \tag{6.6}$$

On flat slopes of <10°, in experiments using viscous lava analogues (cf. rhyolite, dacite lavas) when $\psi = 0.09$, indicating a rapid cooling rate and rapid crust formation, a spiny surface formed on the analogue lava (Fig. 6.4a), when $\psi = 0.79$ a lobate flow formed (Fig. 6.4b), when $\psi = 1.3$ a segmented surface formed (Fig. 6.4c), and at $\psi = 30$ (equivalent to a slow-cooling rate) a relatively smooth surface with small surface irregularities formed (Fig. 6.4d; Fink and Griffiths 1998).

From both experimental simulations and data from actual growing lava domes (Mt St Helens 1981–1986; Soufrière 1979; Pinatubo 1991; Unzen 1991–1995), Fink and Griffiths (1998) and Griffiths (2000) demonstrated a relationship between lava dome height, H (metres), and eruption and dome growth duration, t (days):

$$H = t^{0.25} \tag{6.7}$$

This relationship is under-pinned by the development of a crust with yield strength with increasing time that limits the lateral flow of the domes, enhancing vertical growth above the vent. In each of the listed lava domes, the yield strength of the crusts was assessed as $\sim 10^8$ Pa.

Rheological properties of dome lavas also vary substantially over time in response to degassing-induced crystallisation and related stiffening (Chaps. 2, 4; Sparks et al 2000; Cashman et al. 2008; Cashman 2004; Massol and Jaupart 2009). Furthermore, viscous lavas may show shear-thinning rheologies, i.e. a decreasing viscosity at increasing shear rates. Therefore, changing eruption rates may also induce changes in the time scale of deformation and in dome behaviour, and also triggering of explosive phases (e.g. Lavallée et al. 2011). Rheological properties are also influenced by the non-explosive development of vesicles and fractures in the erupting lava. Vesicularity may be domainal, it can reach >70% (Fink et al. 1992; Ramsay and Fink 1999) and is another important factor in determining the deformation style and eventually fragmentation processes during explosive phases (Mueller et al. 2005). The rheology of a vesicular lava depends on the bubble nucleation rate, number density, size distribution, growth/resorption rate (dependent on volatile diffusion rate), and shear rate (see Sect. 2.4.5, Chap. 2). In general, small bubbles do not deform at low shear rates, and play the role of rigid bodies, increasing the bulk viscosity, whereas larger bubbles may deform more easily, especially at increasing shear rates, causing a decrease in the bulk viscosity of the lava (Rust and Manga 2002; Rust et al. 2003). However, recent studies emphasise that the most important effect of bubbles in magma is not just on the bulk rheology but in developing localised shear bands where deformation is largely accommodated (Vona et al. 2016; Ryan et al. 2019).

Finally, because of the very high viscosity of felsic lavas, the flow state is almost always laminar, even on steep slopes. Only very peralkaline, intermediate lavas, such as phonolites, may have low enough viscosities, and high enough flow velocities on steep slopes to locally experience turbulent flow, as do some basaltic lavas, but at relatively low Reynolds numbers.

6.2.1.4 Effects of Pre-eruptive Topography

One of the factors which determines the shape of viscous lava flows is pre-existing, pre-eruptive topography. The presence of a confining crater built by earlier explosive activity (e.g. at Mt St Helens and Novarupta) may confine smaller domes, but larger domes may exceed the critical crater volume, overspill it, and flow away from the crater area laterally (e.g. at Merapi, Mt Unzen). Many domes, on the other hand, do not appear to be associated with an earlier constructional crater (preceding pyroclastic material being more widely dispersed from the vent), although it is possible that such a feature could have been completely buried beneath the succeeding dome.

Viscous lavas erupted on sloping topography will form thick lava flows. Flow width varies inversely with ground slope (Hulme 1974). A viscous analogue lava can change from being almost axisymmetric on very low slope (<10°; Fig. 6.4e) to clearly elongated on steeper slopes (Fig. 6.4f; Osmond and Griffith 1998; Griffith 2000). Lava thickness will decrease with increasing slope as the lava volume is spread over a larger surface area as it flows downslope relative to flows on flat terrain. However, Walker (1973) considered the effect of slope on lava length to be less important compared with other factors, such as magma effusion rate.

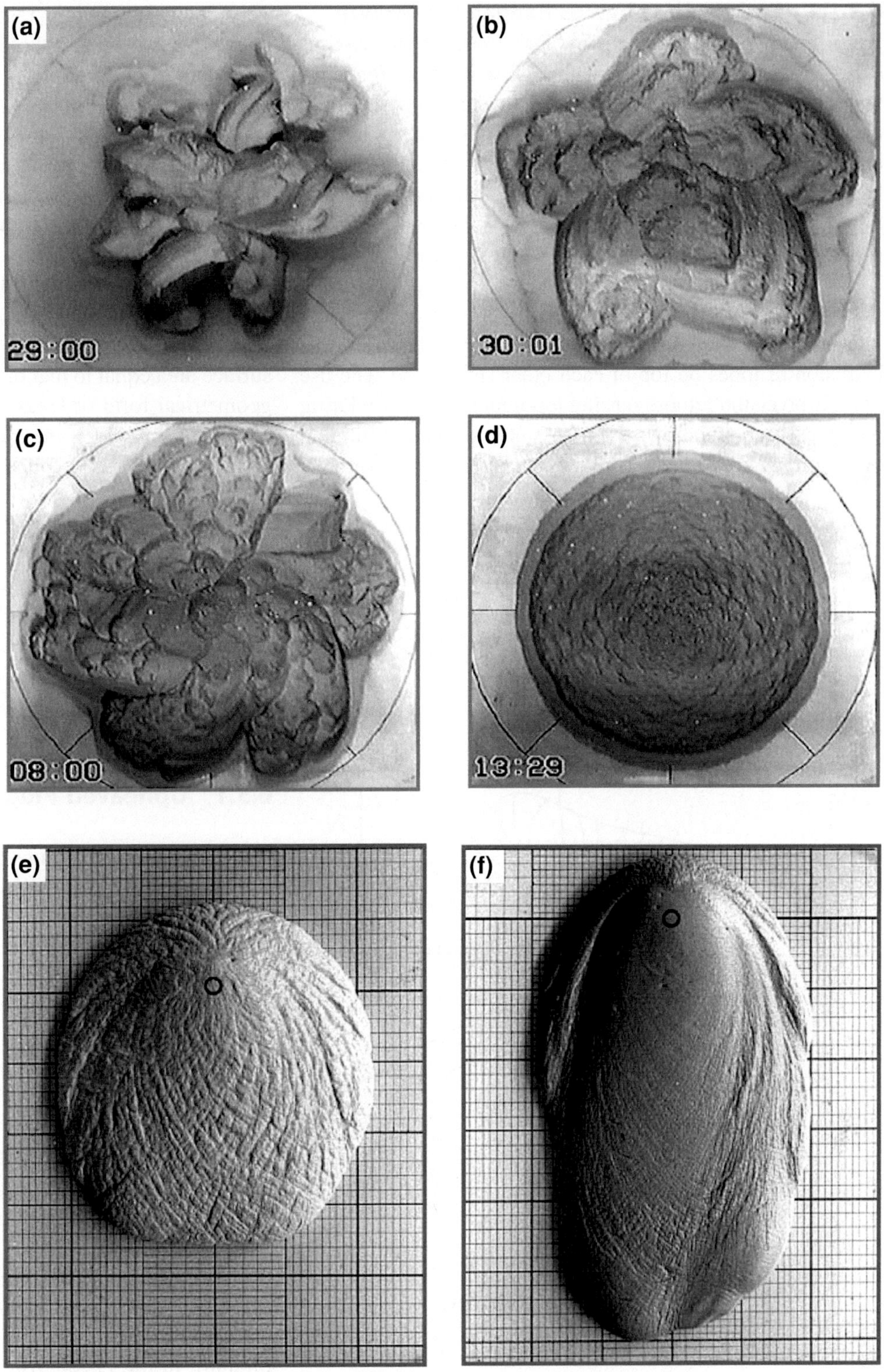

Fig. 6.4 Experimental simulations of the surface features and morphology of viscous lava domes, using a slurry of kaolin in polyethylene glycol. **a** Spiny (or Pelean) lava dome, with $\psi = 0.09$ (i.e. fast cooling rate). **b** Multi-lobate flow, with $\psi = 0.79$. **c** Poly-, micro-lobate flow, with $\psi = 1.3$. **d** Relatively smooth surfaced, or wrinkled, axisymmetric flow at $\psi = 30$ (i.e. slow-cooling rate, thin crust). **e** Kaolin-water slurry flow with a yield strength of 92 Pa released from an emission point beneath the black circle on a slope of 12°. **f** Kaolin-polyethylene glycol slurry with a yield strength of 84 Pa, released on a slope of 18°. (**a–d** from Fink and Griffiths 1998; and **e**, **f** from Osmond and Griffiths 1998)

6.2.1.5 Endogenous Versus Exogenous Growth

Early ideas suggested that rhyolite domes were viscous extrusions from a central vent and grew in an *endogenous* way, that is, formed by inflation and thickening by addition of magma into the interior of the lava body under a surface crust or carapace (Fig. 6.5a, c). Ideally, such bodies should have a concentric structure (cf. Goto and McPhie 1998), and this was produced experimentally in Williams (1932) by squeezing a viscous substance through a narrow aperture. However, in most cases, domes grow through a series of pulses that generate both endogenous growth of the dome interior as a whole (Fig. 6.5a, c) and by the *exogenous* extrusion of lobes on top of each other (Fig. 6.5b). The use of high-precision remote sensing topographic surveys during ongoing eruptions allows the assessment of the partitioning of magma into endogenous and exogenous pulses, which is very relevant as they regulate both the style of magma degassing and the morphology of the dome (Fink et al. 1990; Bernstein et al. 2013).

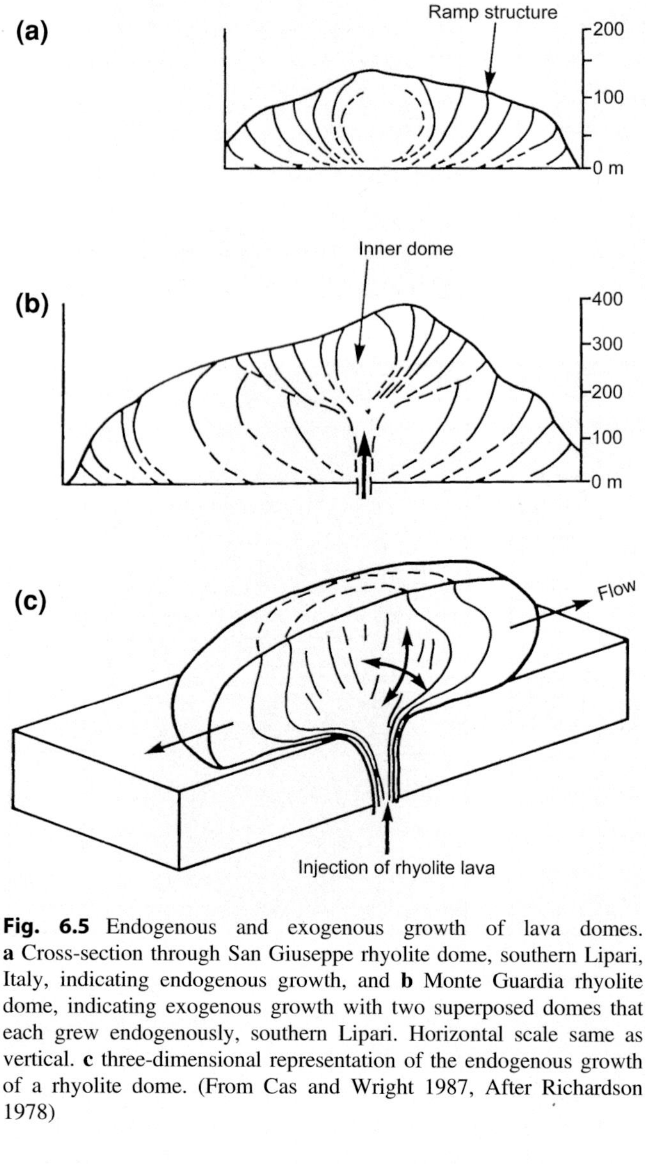

Fig. 6.5 Endogenous and exogenous growth of lava domes. **a** Cross-section through San Giuseppe rhyolite dome, southern Lipari, Italy, indicating endogenous growth, and **b** Monte Guardia rhyolite dome, indicating exogenous growth with two superposed domes that each grew endogenously, southern Lipari. Horizontal scale same as vertical. **c** three-dimensional representation of the endogenous growth of a rhyolite dome. (From Cas and Wright 1987, After Richardson 1978)

6.3 Types of Felsic Extrusions Based on Aspect Ratio, Geometry, and Morphology

We have already introduced the concept of aspect ratio (ratio of the average thickness (T) to the horizontal extent (D), i.e. T/D, where D is taken as the diameter of a circle with a surface area equal to that of the lava; Fig. 6.3) to describe the geometrical form of lavas (Chap. 2). Thickness and surface area are measured from contoured topographic maps of the lava. Walker (1973) found that viscous lavas are typically high aspect ratio lavas (trachytes, andesites, dacites, and rhyolites; Fig. 6.3). Classification of viscous lava flows is based on their geometry, including aspect ratio and morphology. Types include those that are approximately circular in plan with a small surface area and high aspect ratio, usually named upheaved plugs and Pelean domes, and then in order of decreasing aspect ratio, low domes, and longer, more areally extensive mesa lavas, and lava flows (Fig. 6.6).

6.3.1 Upheaved Plugs

Upheaved plugs or "pitons" are very steep and cylindrical to conical in shape, very tall with respect to their basal diameter, usually several tens to hundreds of m in height *versus* tens of m of basal diameter. They form directly above the conduit, usually during the early phases of an eruption or of an eruptive pulse and have a very low preservation potential as they tend to either spread out with time to form lower aspect ratio domes or crumble or collapse gravitationally. These lavas must have been extremely viscous and have had very high yield strengths; some even have striated and gouged margins, showing that they were nearly solid when extruded (e.g. the 1902 dacitic spine of Mt Pelée, Martinique; Fig. 6.6a). Other very good examples are Gros and Petit Piton, erupted within the Qualibou Depression on St Lucia, Caribbean (Chap. 16). These upheaved plugs generally have very high crystal contents (up to 60% phenocrysts), accounting in large part for their high viscosity and yield strength. In the Lesser Antilles volcanic arc, their extrusion may follow sector or flank collapse events due to edifice unloading, which allows the rise of such viscous magma (Boudon et al. 2013). Their extrusion is presumably driven by pressure from beneath driven by rising deeper seated buoyant magma.

(a)
(b)
(c)
(d)

◀ **Fig. 6.6** Felsic extrusion types based on morphology. **a** The 1902 dacitic spine of Mt Pelée, Martinique, Lesser Antilles, representing an *upheaved plug*. Note the striations on the spine formed as the lava extruded. (Photo by A. Lacroix; *Source* Tempest Anderson Photographic Archive, York Museum Trust, UK; through Creative Commons Attribution-ShareAlike 3.0 Unported licence). **b** The Mt Unzen lava dome, Japan, April 1992, a typical multi-lobate *Pelean dome* (*Source* Geological Society of Japan). **c** *Low lava dome*, from 1912 Novarupta eruption, Alaska, USA (Photo by M. Fritz; *Source* Katmai National Park and Preserve). **d** Corona *mesa lava*, near Lascar volcano, northern Chile. Note the central spine which represents the last intruded–extruded magma batch, or upheaved spine. **e** Oblique Google Earth image of the 1220 AD Rocche Rosse tongue-like, obsidian rhyolite lava flow, Lipari, Italy, which originated from the breached crater of the Monte Pilato pumice cone (white cliffs) and extends to the coast. Several flow lobes are visible together with marginal levees, and channel pressure ridges. (*Source* Google Earth)

Fig. 6.6 (continued)

6.3.2 Pelean Domes

Pelean domes are usually multi-lobate, formed over time by a series of exogenous and endogenous pulses that juxtapose highly viscous lava lobes around a central vent to form an irregular stellate outline with spires (Fig. 6.6b; also see Fig. 6.4a). Their dimensions may reach several hundreds of metres in both height and basal diameter, and they usually have very rough upper surfaces.

6.3.3 Simple Domes and Mesa Lavas

Simple domes are inflated, generally axisymmetric, with tops that may have some spines and low ridges. They range from high (aspect ratio), steep-sided domes (Figs. 6.1e, f, 6.2f) to low domes (Fig. 6.6c) and spreading mesa lavas (Fig. 6.6d). Individual domes usually range from tens to hundreds of m high, up to 1 km in diameter, and usually <0.1 km^3 in volume. Low aspect ratio domes and mesa lavas have wider bases compared with their heights and show relatively regular circular base areas, with gentle slopes and smooth upper surfaces (Fig. 6.6c, d). They originate from individual or continuous pulses of relatively low-viscosity lavas and relatively high eruption rates, with a dominant endogenous growth.

6.3.4 Lava Flows (or Coulées)

Lava flows (originally called coulées de laves in French) form when lava flows away from the vent (Figs. 6.6e, 6.8a, c, f). Although not essential, flow is usually asymmetric and concentrated to one side of the vent because of pre-existing topography, producing a lava that is elongated in plan view (Figs. 6.6e, 6.8c; also see Plate 6.1b, c).

Lava flows that form where topography is relatively steep or viscosity is relatively low (Figs. 6.1c, 6.6e), and/or from a prolonged and sustained eruption, create enough of a hydraulic head to drive lateral spreading of the lava, producing elongate lobate lavas (e.g. 1220 AD Rocche Rosse, Fig. 6.6e, and 1739 AD Pietre Cotte obsidian lava flows, Lipari and Vulcano Islands, respectively, Italy) and have aspect ratios lower than domes. Continued flow is also enhanced where a surface crust or autobreccia carapace insulates the hotter interior from heat loss and maintains the viscosity at a low enough threshold to sustain flow through an internal feeder tube (Manley 1992). Some of the thinnest rhyolite lavas known are aphyric and have been ascribed to unusually low viscosities. For example, the early rhyolites of Long Valley caldera (Bailey et al. 1976) contain some flow units only 50 m thick, which have flowed up to 6 km. The aphyric condition suggests an extremely high near-liquidus magma temperature and relatively low viscosity at the time of eruption as the cause of the increased fluidity and flow mobility. Other thin rhyolite lavas recorded are the mixed silicic-mafic lavas of Iceland (Gibson and Walker 1963), having an average thickness of 60 m. These lavas may have had reduced yield strengths and viscosity due to superheating on contact with mafic magma.

6.3.5 Cryptodomes (Subaerial)

A cryptodome is a dome-like uplift or bulge of the ground surface or subaqueous basin floor, caused by a very shallow, near-surface intrusion. Usu volcano, Japan has an historic record of major ground surface changes. During 1910, 1943–1945 (Fig. 6.7a–c), and 1977–8, areas of up to 1 km diameter were uplifted 150–200 m by the intrusion of magma at a shallow depth. The 1943–1945 formation of the Showa-Shinzan cryptodome was well documented by the local postmaster (Fig. 6.7b; Minakami et al. 1951). The precise thickness of sediment overlying the 1943–1945 Showa-Shinzan cryptodome is uncertain, although in the final stages of its growth the hypersthene dacite lava could be seen glowing through large cracks in a thin mantling layer of soil and substrate 3–10 m thick (Fig. 6.7c).

At La Primavera volcano, in Mexico, porphyritic rhyolite lavas that occur on or within an inner caldera ring fracture show many features which suggest these lavas were emplaced as cryptodomes into the wet caldera lake sediments (Fig. 6.2a; Clough et al. 1982). Clough et al. (1982) showed that porphyritic lavas and cryptodomes associated with caldera lake sediments are smaller and have higher aspect ratios than aphyric lavas extruded on dry land (generally to the south, Fig. 6.2a), which was attributed to their special environment of emplacement. Some porphyritic lavas extruded on dry land outside the region occupied by the caldera lake are, however, the thinnest on the volcano. The Devonian Boyd Volcanic Complex in southeastern Australia shows excellent examples of such intrusive rhyolite bodies (Chap. 8; Cas et al. 1990), emplaced syn-depositionally into lacustrine sediments, involving formation of significant domains of peperites around the margins (Chaps. 4, 8; Cas et al. 2001; Giordano and Cas 2001).

The significant surface deformation induced by the emplacement of a shallow cryptodome inside a stratovolcano can be a major trigger for volcano edifice instability and sector collapse. This was the case of the 18 May 1980, Mt St Helens lateral blast, induced by the intrusion of a dacitic cryptodome under the northern flank of the volcano. Tilling et al. (1990) reported a 2-month period of seismicity and phreatic activity associated with the rising magma, preceding the 18 May eruption. The actual shallow intrusion of the cryptodome magma started around 27 March 1980, when the north flank began to swell and crack in an area that became known as "the bulge" (see Chap. 8). By 12 May 1980 some parts of the bulge were uplifted by more than 100 m, but also significant was the horizontal component of the total displacement, which caused the summit to extend and form an axial graben. The total volume of the intrusion was estimated at 0.1 km^3 and on 18 May the north flank failed, causing one of the best ever documented landslide-debris avalanche events, which in turn led to instantaneous decompression of the cryptodome, triggering explosive release from the cryptodome, that produced the famous lateral blast pyroclastic flow (see Chap. 10, Sect. 10.3.4) and a sequence of subsequent plinian explosive eruption phases.

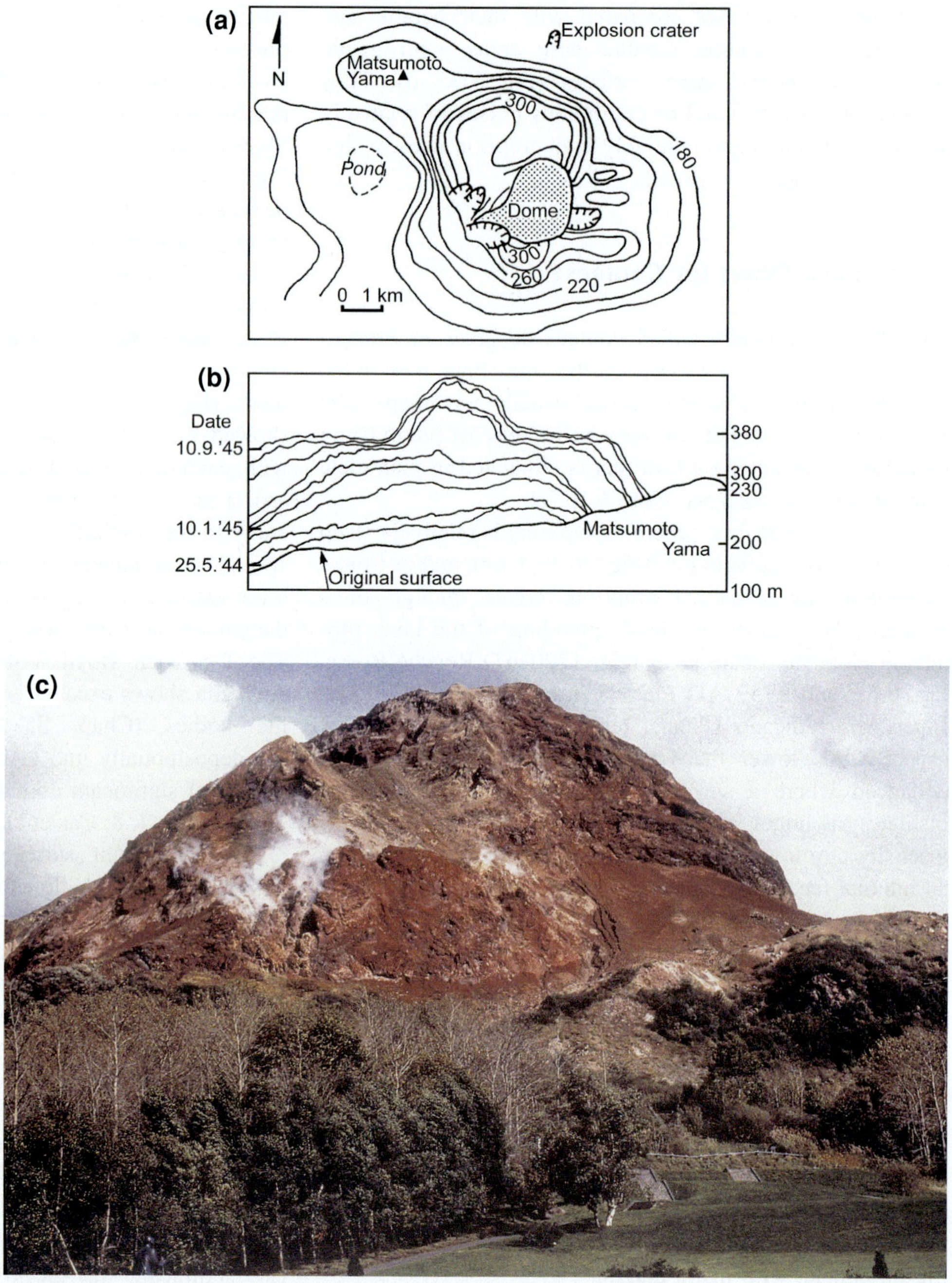

Fig. 6.7 Features of the 1944–1945 Showa-Shinzan cryptodome of Usu volcano, Hokkaido, Japan. **a** Topographic contour map, and **b** profiles showing growth of the cryptodome during 1944 and 1945 as recorded by local postmaster Minakami by drawing the changing profile relative to a mesh of horizontal strings as a reference (From Cas and Wright 1987, After Minakami et al. 1951). **c** The 1944–45 Showa-Shinzan cryptodome that is still releasing gases today. A diffusely stratified sequence perched on the top right of the dome represents uplifted sediments and soil. Part of the interior of the dome is exposed where volcanic gases are being emitted

6.4 Surface Features of Subaerial Felsic Domes and Lavas

The upper surfaces of young viscous felsic domes and lavas are typically highly irregular, blocky, and rough (Figs. 6.6, 6.8), and are marked by autobreccia (Figs. 6.6b, c, 6.8a) and irregular, protruding spines (Fig. 6.8b), ogives (Plate 6.1b, c; Fig. 6.8c), small domes (Fig. 6.6d), and whalebacks that form during the active eruption period.

6.4.1 Autobreccia (Monomictic Coarse Breccia)

The surfaces of viscous felsic domes and lavas are characteristically marked by very coarse monomictic breccia that are commonly interpreted as auto- or flow breccias. The cooling, highly viscous (usually $\sim 10^{10}$–10^{12} Pa s depending on strain rate) crust breaks into boulder-size blocks through a combination of cooling contraction fracturing and dynamic stressing-induced fracturing as the dome or lava continues to expand and/or flow (Figs. 6.8a, e; 4.2d–g;

(a)
(b)
(c)
I
PC
D
III
II
I
PF
(d)
(e)

◀ **Fig. 6.8** Surface features of viscous lavas. **a** Blocky autobrecciated surface of the 1925 dacite lava flow, Nea Kameni, Santorini (also see (**e**); and Plate 4.1a, Fig. 4.1b). **b** 15 m high spine preserved on the surface of the 1991–1995 Mt Unzen lava dome, Japan, which is surrounded by autobreccia and collapse crumble breccia (Photo by D. Uehara). **c** Landsat 8 Earth Explorer image of the upper surface of the <100 ka Chao dacite lava flow field, Chile, which is marked by 30 m high, curved ogives or pressure ridges, with flow direction from north to south. Also note the steep southern margin of the main-phase flow units (I and II) which has a relief of 400 m. The longest flow is over 14 km long. PF = early pumice flow deposits; PC = truncated pumice cone which was in part coeval with flow units I and II and marks the vent site; D = collapsed dome of flow unit III. (Image by R. Simmon; *Source* NASA Earth Observatory and United States Geological Survey). **d** Crevasses on the surface of the 1220 AD Rocche Rosse obsidian lava flow, Lipari, Italy. **e** The margin of the 1928 Nea Kameni dacite lava, Santorini, Greece, with multiple coherent lobes protruding through the autobrecciated margin. **f** Plan view photomap highlighting breakout lobes (in green) around the margin of the rhyolite obsidian lava erupted in 2011–2012 from Cordón Caulle volcano, Puyehue-Cordón Caulle basaltic-rhyolitic volcanic complex, Chile. Also shown are the traces of ogives in black. (From Tuffen et al. 2013). **g** Oblique photo of marginal breakout lobes (1-3) on the northeast edge (area located in **(f)** by white arrow and box) of the 2011–2012 rhyolite obsidian lava, Cordón Caulle volcano, Chile. Scale bar (white) is 30m high. (From Tuffen et al. 2013). **h** Talus apron of Soufrière Hills dome in 2002, showing hot rock avalanches, Montserrat (*Source* United States Geological Survey)

Fig. 6.8 (continued)

Chap. 4, Sect. 4.3). Autobrecciation occurs in response to both compression and extension during flow at strain rates that induce fracturing in regimes close to the brittle–ductile transition or in a brittle state (Chap. 4). Given that the surface crust will consist of both a pumice carapace and an underlying obsidian domain, the autobreccia blocks will consist of pumice blocks, obsidian blocks, and blocks with both facies. Clast size varies from metres to centimetres in diameter, and finer sizes are very rare. Blocks will be bounded by curviplanar brittle and irregular ductile surfaces (Chap. 4). Although much of the blocky surface of lavas originates as autobreccia, spires that develop particularly on the surfaces of Pelean domes will also crumble gravitationally to form aggregates of crumble breccia (Figs. 6.8b, 4.2i). This occurs both during the eruption, and continues for years after the eruption has ceased, because being quite glassy, the spines develop fractures and become fragile. The "crumble breccia" (Fig. 4.1 h) may be difficult to distinguish from autobreccia, except that autobreccia should preserve domains of in situ jigsaw-fit breccia (e.g. Fig. 4.2d, e). Autobrecciation can also occur in the outer boundary zone of high-level cryptodomes (e.g. Goto and Tomiya 2019).

Observation of the changing surface of the Mt Unzen dacite domes, erupted between 1991 and 1995, shows that since the eruption stopped numerous steep spires, whalebacks and small surface domes and ridges have crumbled. The current top of the lava domes is now largely a very coarse block breccia, consisting of both original autobreccia (Fig. 4.2e), post-eruptive crumble breccia (Fig. 4.2i), and relict spires (Figs. 6.8b, 4.2e). Although this process can be recognised, because there is a photographic record of the Mt Unzen dome surfaces from the time of the eruption to the present day, distinguishing crumble breccias in the rock record may be very difficult.

6.4.2 Ogives or Pressure Ridges

Viscous flows often have curved, concentric surface ridges called *ogives* or *pressure ridges* (Fig. 6.8c, Plate 6.1b, c), which have spacings or wavelengths of tens of m, and can be up to 30 m in height. The axes of ogives are more or less perpendicular to flow direction and are curved and convex in shape in a down-flow direction (Plate 6.1b, c; Fig. 6.8c, f), which can be used to ascertain the general flow direction if preserved. Ogives on rhyolite lavas have been variously interpreted as ramp structures (MacDonald 1972), open anticlines within the flow foliation caused by longitudinal compression of the lava stream due to external resistance in their advance (Loney 1968), or else folds on the surface of the lavas with a similar origin to pahoehoe ropes (Fink 1980a). The wavelength and amplitude of the ridges/folds are dependent on the temperature gradient, the rheological contrast between the surface crust and interior including viscosity of the crust ($\sim 10^{10}$–10^{12} Pa s) and interior ($\sim 10^{8}$ Pa s) (Gregg et al. 1998) and yield strength, and the ratio of the compressive stress (due to flow) to the gravitational stress (due to the height and weight of lava). Fink's (1980a) analysis allows calculation of the minimum viscosity of many flows, for which such data are available. In many of these cases, ridges form in response to compressional forces parallel to flow, developed as the flow front slows down and cools at faster rates than the rear. Once the lava surface is cold, it will behave in a brittle manner, and fracturing and ramping could then be important and superimposed on earlier folding features. As a result, ogives can form through both folding of the crust as it compresses in response to "backing-up" behind the slowing and stiffening flow front as it cools, and ramping from within the lower viscosity interior in a flow deformational event that is progressive and changing during flow (Fink 1980a; Gregg et al. 1998; Castro and Cashman 1999; Castro et al. 2002; Cioni and Funedda 2005; Latutrie et al. 2017; Bullock et al. 2018).

For a novel study that challenges the common interpretation that ogives originate as folds on the surface of silicic lavas, the reader is referred to Andrews et al. (2021). Through analysis of new field observations, viscometry experiments, and stress modelling, these authors deduce that ogives are brittle-tensile features that record extensional brittle failure and stretching of the upper surface as the lava advances and spreads. As the upper parts of lava cool, vesiculate and outgas, the resulting vitric and pumiceous lithofacies have very low tensile strengths leading to pervasive vesiculation-driven tensile failure. Ogives therefore reflect the strength and thickness of the brittle upper surface stretching over a ductile interior, and analysis of fold geometry (wavelength, amplitude, etc.) will result in miscalculations of lava flow rheology according to Andrews et al. (2021).

6.4.3 Surface Domes and Pumice Diapirs

Other types of ridges can be found that are "squeeze-up" extrusions through cracks during stretching of the flow surface, which locally may form small-scale spires of coherent lava "emerging" from the autoclastic carapace (Fig. 6.8b). In addition, small and low domes (Fig. 6.6d) and associated pumice diapirs may also occur, which Fink (1980b, 1983) described in detail in the Big and Little Glass Mountain obsidian flows, northern California. These diapirs rose from the basal pumiceous layer of the flows, due to the gravitational instability caused by the density inversion inherent in the flow. Dome spacings of between 50 and 70 m were measured, and elongation and surface folding of the diapirs indicates they emerged before the flows stopped moving and before surface ogives had formed.

6.4.4 Creases, Crevasses, and Striations

Other surface features include both longitudinal and parallel to flow narrow troughs called *creases* and *crevasses* (Fig. 6.8d) that may form by thermal contraction during cooling of the surface and as extensional structures where the outer and cooled crust of the lava is forced to spread. These troughs may be m to several tens of m wide and long (Fink and Anderson 2000). Usually, where the interior of the coherent domains is exposed, such as along spines or crevasses, the lava flow banding and lobe surfaces may show *striations* indicating shear direction (see Fig. 6.15a).

6.4.5 Frontal and Marginal Lobes, Levees, Autobreccia, and Talus Aprons

Lava domes and flows have steep flow fronts and margins (Figs. 6.1e, 6.8a, c), which consist of accumulations of bouldery autobreccia (Fig. 6.8a, e, g; also see Plate 4.1a, Fig. 4.2d, e) and scree or talus (Fig. 6.8h) made of blocky and pumiceous lava rubble. Expansion of the flow front as the lava moves requires the outward bulldozing of the upper and marginal blocky rubble, which is eventually overridden by the moving core of the lava as the lava moves like a conveyor belt, so forming a basal autobreccia layer. Lava flows also develop levees (Plate 6.1b; Figs. 6.6e, 6.8c) which become stationary or slow moving and confine the active lava channel through which faster flowing lava moves to the flow front. The margins of large lava flows are also marked by multiple lava lobes (e.g. Chao dacite lava, Chile, Guest and Sanchez 1969; de Silva et al. 1994; Plate 6.1b; Fig. 6.8c; 2011–2012Puyehue Cordón Caulle rhyolite lava, Chile; Tuffen et al. 2013; Fig. 6.8e–g;), which like basalt lava flows allow the felsic lava to spread (Tuffen et al. 2013). Magma is injected into the collar of rubble in the flow front and along levees, protruding from them, sometimes as largely coherent lobes (Fig. 6.8e), and sometimes as autobrecciated lobes (Fig. 6.8f, g). Tuffen et al. (2013) highlighted the similarity between numerous marginal breakout lobes from the margins of basalt lavas and felsic lavas, proposing that both could be considered to be compound lavas as a result. There are however differences between the spatial and time scales of propagation between the two systems.

Lava domes are laterally contiguous with variably developed marginal *talus aprons* of resedimented autoclastic debris (Plate 6.1a; Figs. 6.1e, f, 6.8h), formed by the continuous collapse of the growing, steep lava dome margins, and gravitational remobilisation of the autobreccia (Chap. 4) carapace as the dome grows and propagates laterally away from the vent. The talus aprons (Plate 6.1a; Figs. 6.1e, f; 6.8h) may also include pyroclastic block and ash flow deposits resulting from gravitational and explosive mobilisation of hot dome margin debris (Chap. 10). Where domes grow close to each other along fissures they may eventually partially merge forming dome clusters, where the talus apron may be composite (Fig. 6.1f).

6.5 Lithofacies and Textural Zones of Felsic Domes and Lavas

Internally, subaerial (and subaqueous) felsic domes and lavas have distinctive internal lithofacies, which show lithofacies organisation (Sect. 6.6). One of the most distinctive, autobreccia, has already been discussed above as a surface feature (Sect. 6.4.1).

6.5.1 "Stony Rhyolite"

A significant lithological component of the interior of most thick viscous lavas, especially domes, is massive to flow foliated or flow banded lava with a *microcrystalline groundmass texture,* also called "*stony*" or lithoidal lava facies (Fig. 6.9a, b; *note*: in the following cited examples, we use images from both subaerial and subaqueous lavas to illustrate the textures, which can be common to both settings). Stony texture forms in the interior of a thick dome or thick lava by slow, late crystallisation of the melt or groundmass to a finely crystalline rock as the interior slowly cools during and after eruption and emplacement. Even if some glassy groundmass remains after cooling-induced crystallisation, this will invariably undergo secondary crystallisation during diagenesis (Chap. 13). The groundmass texture may consist of microgranular quartz and feldspar for silicic lavas (Fig. 6.9b) or microlitic plagioclase for intermediate lavas. In addition, late spherulitic crystallisation may contribute in the outer margin during cooling (Chap. 13).

Large phenocrysts, representing slow, crystal growth in the source magma chamber before the magma rises up through the conduit and before eruption, may be dispersed through this microcrystalline groundmass (Fig. 6.9b–d), together with some xenocrysts. Phenocrysts may be euhedral, subhedral, or angular and broken (Fig. 6.9b–d), with angular broken crystals most likely breaking in the conduit due to magma rise-related decompressional shock fracturing along pre-existing crystal lattice weaknesses (e.g. vacuole trails) and due to expansion of volatiles in melt inclusions trapped in phenocrysts (Allen and McPhie 2003; van Zalinge et al. 2018), representing a micro-scale form of hydraulic fracturing. In lavas of intermediate composition, elongate phenocrysts such as plagioclase phenocrysts may become aligned parallel to each other during laminar flow (Fig. 6.9d).

Fig. 6.9 Lithologies of viscous lavas. **a** "Stony", flow banded porphyritic rhyolite lava, Siluro-Devonian Rocklands Rhyolite, Victoria, Australia. **b** Microcrystallised stony groundmass texture of a submarine dacite lava, Early Devonian Merrions Tuff, New South Wales, Australia. Note the rounded, embayed volcanic quartz and sericite altered speckled plagioclase phenocrysts in the microcrystalline groundmass. Field of view ~6 mm across; ppl. **c** Subaqueous porphyritic rhyolite with large "dark" to clear volcanic quartz and milky coloured feldspar phenocrysts that vary from euhedral, to subhedral to angular, in a fine-grained sericite altered groundmass. Late Devonian Bunga Beds, Boyd Volcanic Complex, Aragunnu Bay, southern New South Wales, Australia. **d** Flow aligned plagioclase phenocrysts in an Ordovician andesite, Goonumbla Volcanics, Parkes, New South Wales, Australia. **e** Obsidian glass with cryptic flow banding, ~690 AD Big Obsidian lava flow, Newberry caldera, Oregon, USA. **f** Flow-banded obsidian, Taupo Volcanic Zone, North Island, New Zealand. **g** Microscopic view (x-nicols) of isotropic glass, with suspended euhedral phenocrysts, lath-shaped groundmass microlites and curved, concentrically shaped cryptic perlite cracks, submarine Miocene Shinzan Rhyolite, Oga Peninsula, Honshu, Japan. Field of view is ~ 5 mm across. **h** Inter-banded obsidian (black), and finely vesiculated pumice layers (fawn colour). The white spots in the obsidian layers are spherulites. 1220 AD Rocche Rosse rhyolite lava flow, Lipari, Italy

Although the interior of thick lava domes and flows are almost invariably made of stony facies, with many young rhyolite lavas, little of this rock is seen because erosion will not have had time to cut through and expose the interiors. In addition, thin obsidian lava flows may not develop a stony interior because they cool too quickly; they thus consist wholly of obsidian in the interior with perhaps a pumice carapace on top (e.g. 1220 AD Rocche Rosse and 1739 AD Pietre Cotte obsidian lava flows, Lipari and Vulcano Islands, respectively, Italy; see below). Conversely, in old lava flows only crystallised stony rhyolite may be preserved because the surface textural domains of autobreccia, coherent pumice and obsidian glass (see below) will have been eroded, and any remaining glass will have devitrified and undergone secondary groundmass crystallisation, making it indistinguishable from stony rhyolite unless devitrification textures such as spherulites are preserved.

The "stony" crystal-rich lithofacies associated with the slow cooling of the interior of viscous lava flows of course is not restricted to rhyolitic compositions. Ischia Island, for example, hosts a series of Holocene, post-caldera, alkali-trachytic lava domes and thick flows which are a "stony" lithofacies and can reach 90–100% crystallisation by combining phenocrysts and microlites (Selva et al. 2019). Coherent stony facies can be massive, or diffusely or well flow banded, reflecting varying degrees of internal laminar shear flow during emplacement.

6.5.2 Obsidian

Black, vitreous *obsidian* glass generally forms from high silica rhyolite, dacite, and some alkaline intermediate lavas. It mostly occurs in the outer parts of thick domes and lava flows, but it may also constitute protruding spines, spires, and upheaved plugs, and may constitute almost the entire thickness of some thin lavas. Obsidian is dense, dark homogenous glass (Fig. 6.9e–g), generally aphyric or with low phenocryst content, low levels of vesicularity, with conchoidal and perlitic fractures (Fig. 6.9g; see below), and it is transparent when sliced thinly. Because it splinters into extremely sharp slivers along conchoidal fractures when broken by hammer or other implement, it was a highly valued cutting tool among early hominids.

Obsidian forms when high silica magma close to the liquidus temperature is cooled very rapidly through the glass transition temperature range, so freezing the magma before it has time to undergo steady crystallisation (Gottsmann and Dingwell 2001). Obsidian glass is therefore a supercooled or frozen magma in which ions and molecules are dispersed in an unordered manner through the glass rather than being organised into crystals. Any phenocrysts that occur in the glass will have formed in the pre-eruption magma reservoir, under conditions of moderate lithostatic pressure and slow sub-liquidus cooling. The groundmass of the obsidian commonly contains some microlites (Fig. 6.9g), formed in the conduit and during eruption during rapid cooling of the lava. Degassing the magma may also contribute to microlite growth.

Obsidian may be massive or flow banded (Fig. 6.9e, f). Flow banding may be cryptic and difficult to recognise (Fig. 6.9e, f). Obsidian is often inter-banded with or has lenses of highly vesiculated coherent pumice layers (Fig. 6.9h). This layering, or flow foliation, is frequently folded. Obsidian therefore usually forms a chilled glassy carapace around the core of stony rhyolite and other silicic viscous lavas and has cooled too quickly for primary groundmass crystallisation to occur. The obsidian carapace is commonly several metres to tens of m thick over the top and around the flow front, with a thinner layer along the base. Some of the thinner "obsidian flows", and "glass flows", may be obsidian throughout their interiors.

There has been a long-standing debate about whether or not obsidian can be erupted in its own right as a dense magma, or if it results from collapse of a permeable foam (i.e. highly vesiculated) magma in the conduit and during eruption. Eichelberger et al. (1986) proposed the model of foam collapse based on drilling of the rhyolitic Obsidian Dome in the Inyo Domes chain in California. The drill cores showed that at the distal margin where the dome is 30 to 50 m thick, the 10-m-thick pumice carapace was underlain by dense obsidian, which is itself 10 to 20 m thick, passing into a more crystallised zone, which is then underlain by more obsidian. Around the vent, where the dome is 100 m thick, there is no obsidian, just vesicular lava, and in the underlying feeder dyke, the rock is mostly crystallised except for 10-cm-wide chilled obsidian margins. Water content in the lava dome obsidian is 0.1 wt %, whereas fragments of glass derived from the feeder at depths of 290 and 370 m contain $\sim$1.2 wt% water, and phase relations indicate an original water content of >3 wt% (Eichelberger et al. 1986). Eichelberger et al. (1986) propose that the magma vesiculated high in the conduit, but simultaneous degassing of the vesiculating magma in the upper conduit allowed escape of volatiles, eventually resulting in collapse of vesicle walls under conditions of under-pressure relative to magmastatic pressure of the rising magma. Other authors (e.g. Martel and Iacono-Marziano 2015; Shields et al. 2016) also support this model.

However, this model does not explain why the Obsidian Dome as well as most other rhyolitic domes still have well-preserved pumice carapaces, which Fink et al. (1992) addressed in their counter-proposal. They proposed that rhyolitic magma has largely degassed prior to eruption, either through earlier explosive events and/or during ascent in the conduit, and that once erupted into the low-ambient pressure atmosphere the residual volatiles can exsolve to form pumice carapaces and some internal pumice zones. It seems to us that both models could apply in different circumstances and sometimes together or sequentially in others. Shields et al. (2016) propose that the collapse of the highly vesiculated foam occurs through degassing along shears that form during shear flow in the magma in the conduit, and after forming dense magma, it could revesiculate through exsolution of the residual low volatile content during further conduit flow or once erupted as a lava flow, so forming pumice layers found in most obsidian lavas. Foaming experiments with rhyolites demonstrate that H_2O content as low as 0.11 wt% dissolved in the melt at ambient pressure can exsolve resulting in vesicularities up to 70% (Ryan et al. 2015)!

6.5.3 Spherulites and Devitrification of Obsidian

Obsidian is however metastable at the Earth's surface and begins to crystallise crude, fibrous-like crystals as it cools and during later diagenesis. The process has generally been called *devitrification*, implying conversion of solid glass to a crystallised aggregate. However, the process of fibrous crystallite crystallisation actually begins during cooling of the magma and through the glass transition into obsidian as it forms during eruption and emplacement. Transformation of glass into a crystallised rock then continues after the lava has solidified into cold obsidian during diagenetic regime crystallisation induced by ion diffusion and migration in the glass by gases and meteoric fluids. Because most of the crystal growth occurs in a very viscous magma and hot solid, not a fluidal magma, the crystals are poorly formed and irregular, not euhedral. At high temperatures, they range from needle or fibre like, to clusters of such crystals that form "bow-tie"-like aggregates to spherical radiating clusters, called spherulites, and eventually at low diagenetic temperatures to blocky, equant micro-crystals (Lofgren 1971).

Spherulites are spherical aggregates (Fig. 6.10a) of radiating needle-like or fibrous plagioclase or alkali feldspar crystals (Fig. 6.10b), with or without quartz in the form of cristobalite and tridymite and are commonly found in the glassy carapace of obsidian lavas (Lofgren 1971). Spherulite growth is associated with advection and diffusion of water through magma/glass ahead of the advancing spherulite growth front of anhydrous crystals (Castro et al. 2008; Watkins et al. 2009). Spherulites begin to form as obsidian undercools and grow at time scales of hours to weeks at high temperatures at about or above the glass transition (Castro et al. 2008). Fenn (1977) demonstrated that at low undercooling of $\Delta T < 40$ °C, small tabular feldspar crystals form, at moderate undercooling, ΔT of 75–145 °C, coarse crystallised spherulites formed, whereas at high undercooling, ΔT of 245–395 °C, finely crystallised, dense spherulites formed. Watkins et al. (2009) and Clay et al. (2013) propose that growth rates are magma/glass temperature dependent, with spherulite growth occurring both when the magma is above the glass transition, as well as below it in the glass state down to temperatures of about 400 °C. Watkins et al. (2009) calculated that a 2.6 mm diameter spherulite would take about 300 days to grow at 800 °C, about 10 years at 600 °C and about 300 years at 400 °C. Clay et al. (2013) suggested average growth durations of ~ 5 days, with further modifications for ~ 400 years in the solid state. Spherulites often grow along specific flow layers, forming trails of anastomosed or isolated spherulites. Higher water contents in some layers could promote growth rates of spherulite fibres locally, but Bullock et al. (2017) highlight that flow-induced shear and later brittle deformation can also trigger spherulite nucleation and crystallisation.

Spherulites can also be superimposed on flow banding, and in some cases the flow foliation can be traced through, and is not deflected by the spherulites, showing that crystallisation of at least some spherulites took place after flowage of the lava had (nearly) ceased (Watkins et al. 2009). In other cases, flow fabrics such as microlite alignments can be deflected by spherulite growth, indicating growth during flow (Bullock et al. 2017). Spherulites commonly have diameters of 0.1–2 cm (Lofgren 1971) but can be much larger and occasionally grow into *megaspherulites* up to nearly 3 m (Smith et al. 2001; Fig. 6.10f). Some spherulitic growths are called *lithophysae*, which are radiating spherulitic aggregates of fibrous crystals, which have formed around an expanding vesicle or gas vugh in the very viscous magma while it is still (just) above the glass transition temperature (Fig. 6.10a, f, g). Because the spherulitic rim of the gas vugh is solid, the rim splits as the gas vugh expands in the very viscous magma, often producing irregular star-shaped vughs or cavities (Fig. 6.10b, d, f, g). These vughs have thus formed in a melt while it was still capable of deforming (McArthur et al. 1998), and they can be subsequently infilled by secondary minerals such as quartz in some cases. Broken lithophysae testify to post-formation laminar shear during ongoing flowage of the lava. Further discussion of devitrification and spherulite growth can be found in Chap. 13, Sect. 13.2.

6.5.4 Perlitic Cracks and Perlite

Domains of *perlite*, consisting of spheroidal-to-ellipsoidal kernels of variably hydrated glass, defined by enclosing networks of sub-concentric, conchoidal cracks (Fig. 6.11a–c; also see Fig. 6.9g) in obsidian lavas, are formed during

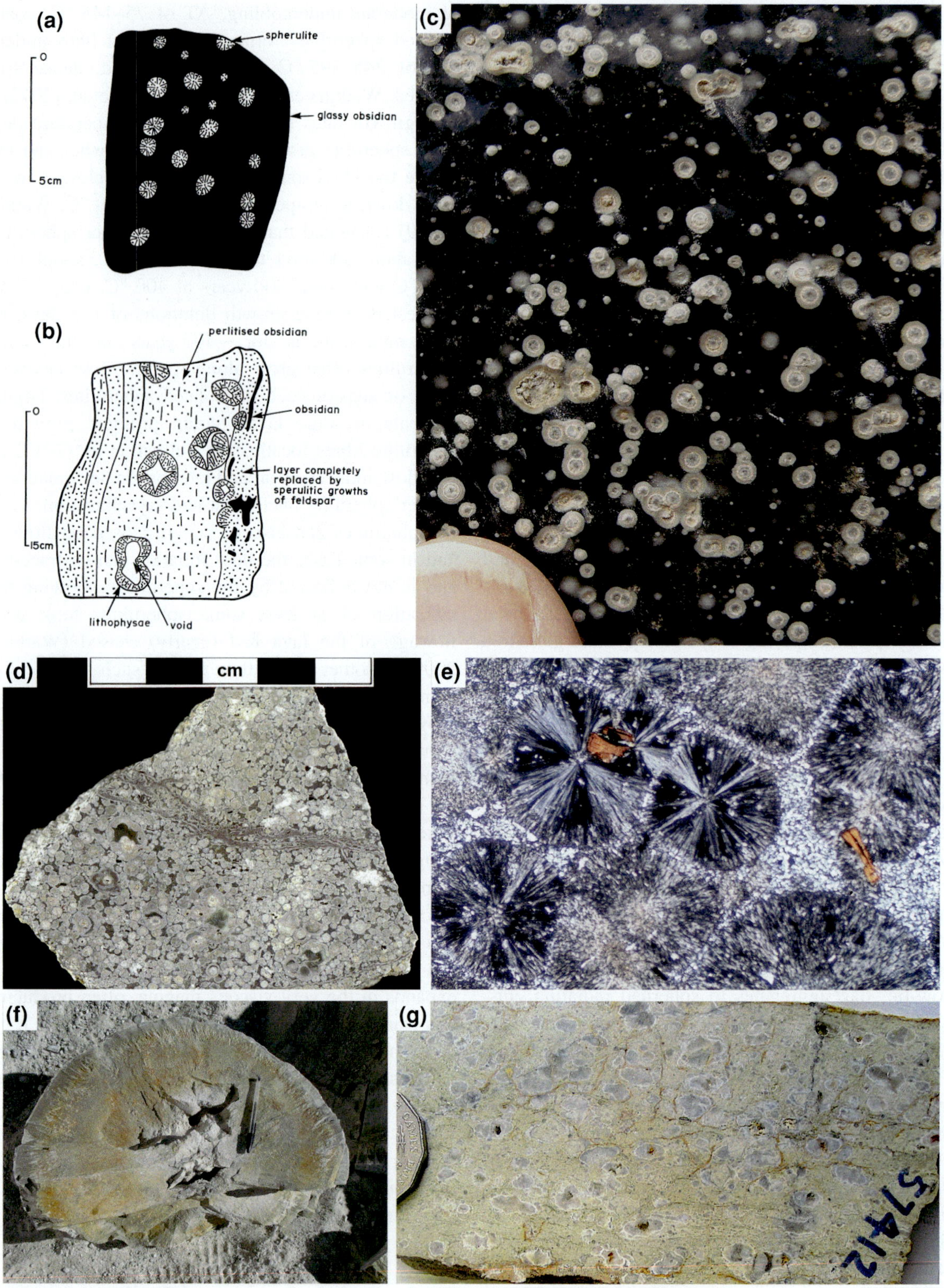

Fig. 6.10 Features of spherulites in felsic lavas. **a** Sketch showing radially textured spherulites in obsidian. **b** Lithophysae, which are large spherulites with central gas cavities. (**a** and **b** from Cas and Wright 1987). **c** Small spherulites dispersed in obsidian glass, showing a concentric layering, some anastomosed into clusters, 1220 AD Rocche Rosse rhyolite lava flow, Lipari, Italy. **d** Slab of almost entirely spherulitically crystallised, submarine Miocene Shinzan Rhyolite, Oga Peninsula, Honshu, Japan. The dark patches represent original obsidian glass that has been altered to chlorite, as well as several irregular star-like lithophysae infillings. The irregular bands represent alteration veins. **e** Photomicrograph of (**d**) showing the spectacular radial fibrous crystallite texture within spherulites dispersed in a groundmass with later developed interstitial microgranular crystalline texture. Field of view 10 mm across; x-nicols. **f** Megaspherulite, with lithophysae cavity, Quiron lava dome, Quevar volcano, Andes, Chile. **g** Lithophysae in hydrothermally altered subaqueous rhyolite, Late Devonian Bunga Beds, Boyd Volcanic Complex, southern New South Wales, Australia

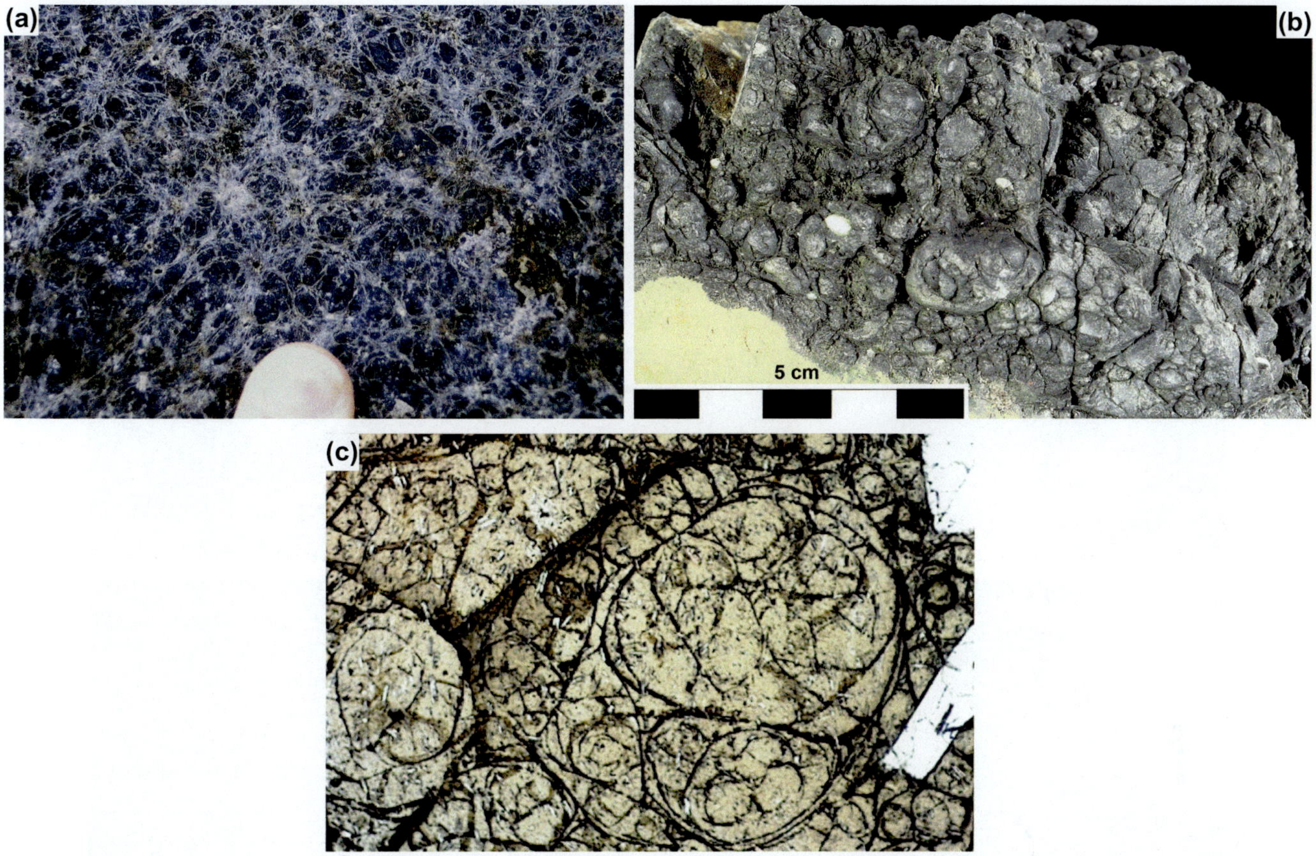

Fig. 6.11 Perlite and perlitic cracks. **a** Perlite showing typical ovoid kernels and concentric perlite fractures, 1739 AD Pietre Cotte obsidian lava flow, Vulcano, Italy. **b** Handspecimen of perlite texture in Miocene submarine obsidian, Kawashira, Japan, showing multiple spheroidal perlite kernels bounded by one or more concentric perlite cracks. **c** Photomicrograph of perlite fractures in hydrated Miocene submarine obsidian lava, Kawashira, Japan, showing multiple orders of concentric ovoid perlite cracks. Field of view is ~10 mm across

cooling contraction below the glass transition temperature range. The cracks, called *perlitic cracks*, are created by thermal stresses set up during cooling of the lava at temperatures of ~400 °C (von Aulock et al. 2013). Water then diffuses along and from these cracks into the bordering glass, and a hydration front advances into the interior of the residual kernel of non-hydrated obsidian (von Aulock et al. 2013). The hydration front is visible in thin section because the refractive index of the hydrated glass is lower than that of the non-hydrated glass.

Obsidian readily adsorbs water from the atmosphere and meteoric sources (Seligman et al. 2016), forming a hydrated layer which thickens with time as the water diffuses into the glass. From measurements of the thickness of the hydration rind on artefacts collected from archaeological sites and experimental studies (Friedman and Long 1976), it is known that the square of the rind thickness is approximately proportional to time and varies from less than 0.5 μm^2 per 1000 yr to as much as 30 μm^2 per 1000 yr. This variation is partly due to differences in composition of the obsidian (trachyte obsidian hydrates much more rapidly than rhyolite), and partly to differences in climate. In older rhyolites, the obsidian is divided into small rounded kernels of structureless obsidian about 1 cm across, surrounded by concentric cracks in light grey friable perlite, although perlite kernels can be many cm in maximum dimension. The formation of perlite cracks, as well as cracks associated with growth of spherulites in lava domes at temperatures of 450–400 °C, together with contemporaneous hydration of surrounding glass, weakens lava domes and makes them prone to collapse gravitationally (von Aulock et al. 2013), generating block and ash flows (Chap. 10).

6.5.5 Coherent Pumice Domains and Carapaces, and Pumice Autobreccia in Felsic Lavas

As discussed above, viscous lavas and, especially, rhyolites can also develop substantial, variably vesicular, *coherent pumice* carapaces and domains in their outer margins (Fig. 6.12a). Fink and Manley (1987) recognised two horizons of coherent pumice in a typical felsic lava or dome. The surface of the lavas is often marked by finely vesicular *pumiceous lava* facies in which volatile exsolution has

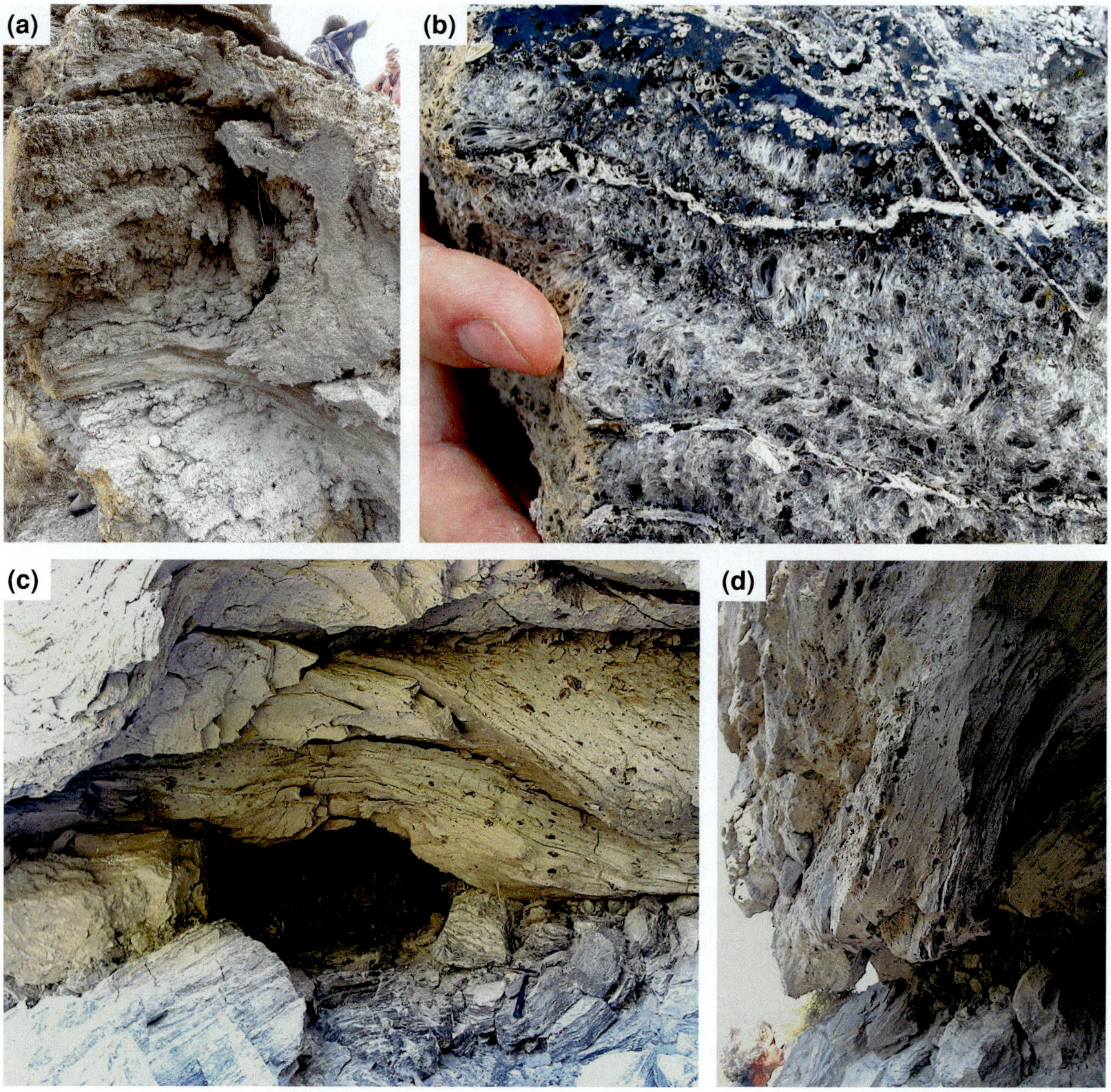

Fig. 6.12 Coherent pumice carapaces and domains in rhyolitic lavas. **a** Coherent, diffusely flow banded, pumice facies domain forming a surface pumice carapace on the 1220 AD Rocche Rosse obsidian lava flow, Lipari, Italy. **b** Close-up view of the very coarsely vesicular nature of some of the pumice carapace of the Rocche Rosse lava, showing also an interface with obsidian which contains patches or domains of vesiculated pumice. **c** The 8-m-thick frontal part of the 1739 AD Pietre Cotte obsidian lava flow, Vulcano, showing the diffusely flow banded pumice carapace (fawn colour) on top of the flow banded obsidian interior of the lava. Note the geological hammer for scale. **d** Close-up view of the actual front of the Pietre Cotte lava showing where the pumice carapace has rolled over the front of the lava causing the flow layering to become subvertical. This outcrop is located at the base of the slope of the source Fossa Cone down which the lava flowed and from which it was erupted from a flank fissure

occurred after eruption due to the low magmastatic pressure at the surface of the lava. However, we have seen coarsely vesicular coherent pumice carapaces on the surfaces of both the 1220 AD Rocche Rosse lava, Lipari, (Fig. 6.12a, b) and the 1739 AD Pietre Cotte obsidian lava flow, Vulcano, Italy (Fig. 6.12c, d), both of which are thin obsidian lavas (<30 m thick). Vesicles grow slowly because of the low volatile content of the magma, and their growth is quickly arrested by the chilling, and solidification of the lava surface, so preserving intact, coherent pumice, without any explosive fragmentation involved. The second zone of vesiculation identified by Fink and Manley (1987) occurs beneath the obsidian horizon or interlayered with it (e.g. Fig. 6.9h) above the stony rhyolite zone if the lava is thick enough to develop a stony interior. Vesicles are large, and vesicularity in these domains can be >70% producing densities <1 g cm^{-3} which are the vesicularity and density of pyroclastic pumice. However, these domains are coherent and demonstrate that even viscous magma can vesiculate to very high levels non-explosively due to a relatively low gas overpressure that is high enough to cause slow steady vesicle growth but not cause explosive fragmentation.

This coarsely vesicular pumice results from volatile exsolution as the core of the lava crystallises, with the volatiles and gas bubbles permeating and rising upwards to the base of the solidifying surface crust. Sometimes the coarsely pumiceous domains can rise diapirically through the crust, where, under low atmospheric pressure, they can even source unexpected explosions that form small craters in the surface of the lava flow. Coarsely pumiceous layers or lenses may occur interlayered with obsidian and spherulitic layers. Bubble dimensions may vary from mm to cm in size, or more, defining *micro-* or *finely vesicular* to *coarsely vesicular* pumice domains (Fink 1983; Fink and Manley 1987). The shape of deformed bubbles has been used to infer the type of shear (pure versus simple shear) and the flow type (Rust et al. 2003). Because of the low confining pressure, pumice domains are more likely to develop (or migrate) towards the surface of a lava. Shields et al. (2016) proposed that the obsidian bands in the 1220 AD Rocche Rosse lava flow on Lipari Island, Italy, formed by development of shear planes during flow in the conduit in pumiceous rhyolite, allowing degassing of the magma and leading to bubble wall collapse and welding to eventually form dense magma. The degassed magma could then revesiculate again during rise in the upper conduit and during lava flow. They suggested that such cycles of shear flow, degassing, bubble collapse, and welding could occur several times, eventually forming some obsidian. Uncollapsed domains are preserved as pumiceous lava.

Consequently, the surface *autobreccia* carapace of many rhyolite lavas also contains blocks of pumice or pumiceous lava (e.g. Fig. 4.2d). Within these pumiceous breccias, and even within the vesicles, crumbled smaller pieces of pumice and ash-sized bubble wall shards may be lodged, giving the impression that this horizon could be pyroclastic, whereas it is autoclastic. In old flows, the pumice domains and autobreccia are unlikely to be preserved because of the effects of weathering and erosion of the elevated tops of lava domes after eruption ceases. Pumiceous breccias formed during flow can also be found at their bases, as the flow front advances and rolls over like a conveyor belt. However, co-eruptive pumiceous pyroclastic deposits, which represent periods of explosive growth of exsolving gases, can also occur in the deposits of lava dome volcanoes. Because of this fine balance between non-explosive and explosive volatile exsolution in lava domes during and even ongoing after eruption seems to have stopped erupting and recently erupted viscous domes and lavas remain dangerous and unpredictable for years after the eruption, until they have cooled and degassed sufficiently.

6.5.6 Mafic Inclusions

Many rhyolite lavas contain a small, but significant, proportion of *mafic inclusions*, such as, the rhyolites in the Aeolian Islands (Fig. 6.13a), at La Primavera volcano and in some from the Late Devonian, Boyd volcanic Complex, southeastern Australia (Fig. 6.13b). These mafic inclusions exhibit varying degrees of original fluidity, including embayed and crenulate margins, as well as phenocrysts of the rhyolite in the mafic inclusion (Fig. 6.13b), indicating that both the inclusion and the host felsic magma were still liquid at the time of magma mingling. In other examples, the mafic inclusions show angular shapes, showing that they had solidified while the lower melting temperature felsic magma was still capable of flowing. The inclusions within La Primavera lavas are very similar to the basaltic-andesite lavas which have been erupted around the volcano (Mahood 1980; Clough 1981). Although these mafic inclusions, with their fluidal characteristics, form only a volumetrically small proportion of their host lavas, their presence indicates that these lavas resulted from the co-existence and mingling of rhyolitic and more mafic magmas in a vertically complex plumbing system (Chap. 2, e.g. Piochi et al. 2009). This bimodal association is characteristic of rhyolite volcanoes and suggests that mantle-derived mantle magmas are the heat engine driving crustal melting to form many felsic magmas (Chap. 16). Size distribution of mafic inclusions within felsic lavas has been used as a proxy for processes of mingling within the magma chamber (e.g. Perugini et al. 2007; Morgavi et al. 2016).

6.6 Facies Models for Subaerial Felsic Lava Domes, Lava Flows, and Cryptodomes

Viscous felsic domes, lavas, and cryptodomes may be variably coherent and brecciated, massive or flow banded, aphyric or variably porphyritic, glassy or variably crystallised, and vesicular to non-vesicular. A single lava or dome that is thick may show all of these characteristics, whereas a thin lava may only show the upper facies zones. Wadge et al. (2009) proposed that the proportion of coherent lava facies produced during a lava dome eruption could be as little 10 to 50%, with the rest of the erupted material being autobreccia, clastic talus, and block and ash flow deposits, based on study of three andesite lava domes erupted from the Soufrière Hills volcano, Montserrat.

It has long been recognised that thick subaerial silicic domes and lavas commonly show a lithological-textural facies zonation (e.g. Fink and Manley 1987; Fig. 6.14) including:

Fig. 6.13 Mafic inclusions in rhyolitic lavas. **a** Mafic inclusions in pumiceous, flow banded 1739 AD Pietre Cotte obsidian lava flow, Vulcano, Italy. **b** Fluidal basalt inclusion in coarsely porphyritic rhyolite, Late Devonian Bunga Beds, Boyd Volcanic Complex, Aragunnu Bay, southern New South Wales, Australia. The irregular margins, and sparse inclusions of large phenocryst crystals from the host rhyolite, indicate that both the basalt and rhyolite were still fluidal at the time of mingling

(1) a core zone inside the lava of massive coherent "stony rhyolite" represented by a crystallised microcrystalline groundmass, passing upwards and outwards, into
(2) a coherent zone of dense obsidian, with some cross-cutting pumiceous diapirs, passing transitionally, into
(3) a coherent zone of interlayered pumice and dense obsidian layers and into more continuous pumice carapace, and then into
(4) an outer zone of obsidian and pumice clast autobreccia, flanked by
(5) variably developed talus aprons of rock-fall, and in some cases block and ash flow pyroclastic deposits (Chap. 10).

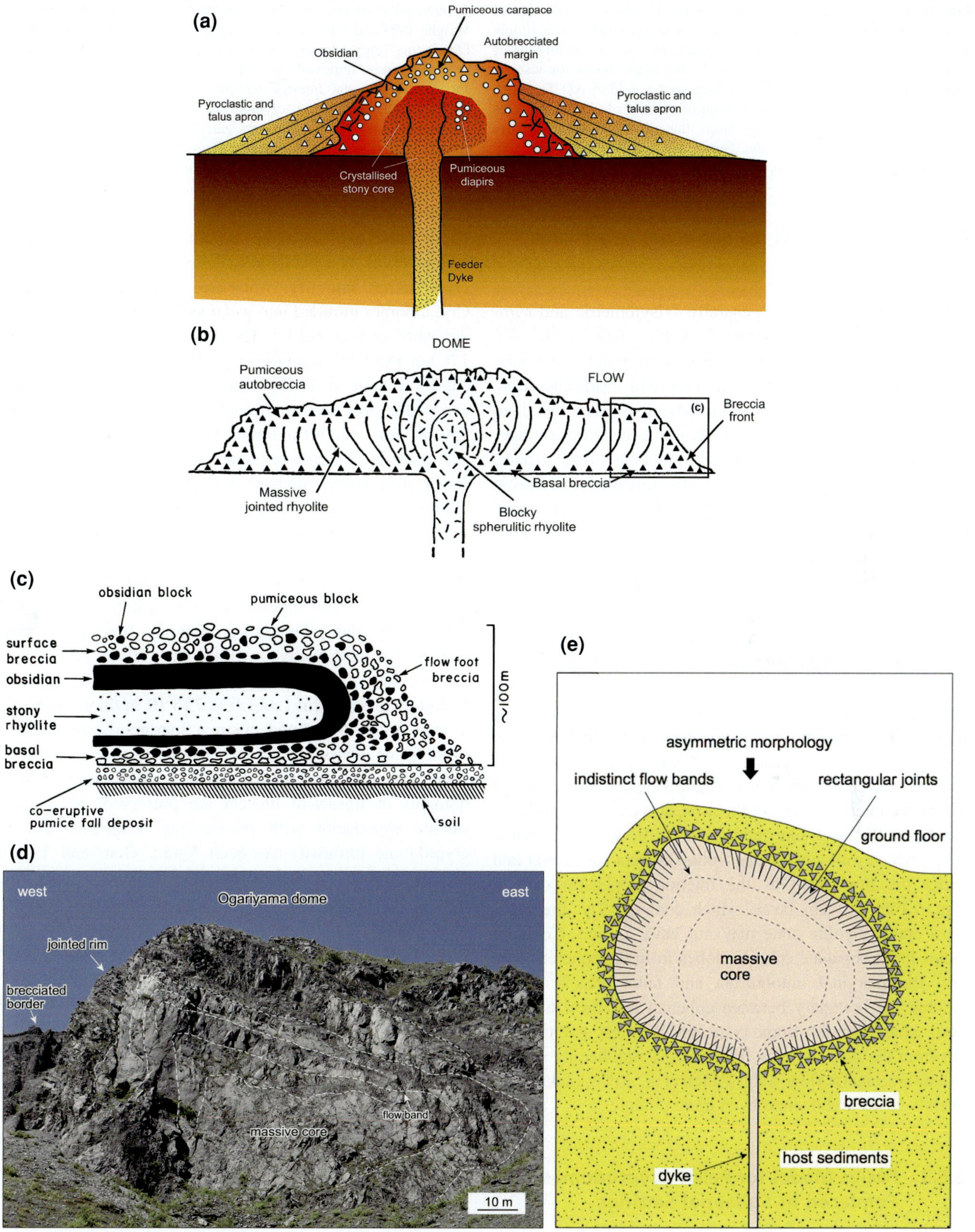
(a)
Pumiceous carapace
Autobrecciated margin
Obsidian
Pyroclastic and talus apron
Pyroclastic and talus apron
Crystallised stony core
Pumiceous diapirs
Feeder Dyke
(b)
DOME
Pumiceous autobreccia
FLOW
(c)
Breccia front
Massive jointed rhyolite
Basal breccia
Blocky spherulitic rhyolite
(c)
obsidian block
pumiceous block
surface breccia
obsidian
flow foot breccia
stony rhyolite
~100m
basal breccia
co-eruptive pumice fall deposit
soil
(d)
west
Ogariyama dome
east
jointed rim
brecciated border
flow band
massive core
10 m
(e)
asymmetric morphology
indistinct flow bands
rectangular joints
ground floor
massive core
breccia
host sediments
dyke

◀ **Fig. 6.14** Schematic facies models of the lithological and textural architecture of felsic domes and lavas. **a** A facies model for a simple viscous lava dome, depicting an asymmetrical zonation of lithologies and textures from the conduit outwards. **b** Facies model for viscous rhyolitic lava flow, based on the geology of the ~1315 AD Tarawera lava dome complex, New Zealand (Modified from Cole 1970). **c** Facies zonation in a lava flow away from the vent, showing a more symmetrical zonation from the centre of the flow upwards and downwards as the flow propagates laterally by endogenous growth through injection into the core from the feeder conduit (From Cas and Wright 1987). **d** Cross-sectional view of the uplifted and exhumed Ogariyama cryptodome which was intruded into overlying dry sediments (now removed by erosion after exhumation), Usu volcano, Japan, showing the internal textural zonation. **e** Emplacement and facies model for subaerial cryptodomes emplaced into dry, water-free sediments based on the exhumed Ogariyama cryptodome, Usu volcano, Japan. (**d** and **e** from Goto and Tomiya 2019, through Creative Commons Attribution 4.0 International licence)

6.6.1 Facies Model for Subaerial Felsic Lava Domes

Although lava domes are generally axisymmetric and symmetrical in cross-section (Plate 6.1a; Figs. 6.6c, 6.14a), the internal textural zonation may be very irregular and gradational from the stony core at the base and inside the dome merging with the top of the underlying conduit and then passing outwards into a semi-circular concentric arrangement of the obsidian, then pumiceous and autobreccia zones, flanked by the talus aprons of resedimented autoclastic debris radially in three dimensions (Fig. 6.14a). In some large domes, coherent lobes could intrude into the autobreccia and talus apron from the core, producing an even more irregular internal facies zone relationship.

6.6.2 Facies Model for Subaerial Felsic Lava Flows (Coulées)

In lava flows (Plate 6.1b; Figs. 6.6e, 6.8c, 6.14b, c), there may be a more symmetrical zonation from the inner core both upwards to the surface and downwards to the ground surface, reflecting a more symmetrical temperature gradient within the interior "tube" of the flow, as well as the effects of the caterpillar-track-like motion of the flow front in some cases, which rolls surface and frontal breccia zones over and under the flow front as the interior propagates over it (Fig. 6.14c). In thin lavas, which cool quickly, the inner most stony rhyolite zone may not have formed. Coherent facies may propagate coherent lobes from the coherent core into the marginal autobreccia and talus breccia facies, making the boundary between coherent and breccia facies even more irregular. Some coherent lobes may even break-out from the flow margins (Fig. 6.8e, f) and then be subject to autobrecciation (Fig. 6.8g; Tuffen et al. 2013), making the overall facies architecture even more complex.

6.6.3 Facies Model for Felsic Cryptodomes Emplaced in Subaerial Settings

Goto and Tomiya (2019) provide a description of the internal textural zonation of a high-level rhyodacite cryptodome (Ogariyama) exposed at the summit of Usu volcano in Japan, which serves as a valuable facies model (Fig. 6.14d, e) for cryptodomes into dry, water-free near-surface sediments. Cryptodomes intruded into wet unconsolidated sediments are described in Sect. 6.13.3. The dome was emplaced in either 1769 or 1822 AD as an unexposed cryptodome, but was then cut by a fault in 1977–78, uplifted and exhumed, which exposed a cross-section through the dome (Fig. 6.14d). It is 150 m across and 80 m high. The core zone is 130 m across and consists of massive coherent non-vesicular rhyodacite (porphyritic with a microcrystalline groundmass) that at the margins of the core is vaguely concentrically flow banded, suggesting endogenous inflation (Goto and Tomiya 2019). The core zone has rectilinear jointing spaced at 50 to 200 cm (Fig. 6.14d). The core is enclosed in an 8-to-12-m-thick rim zone of porphyritic rhyodacite with a pilotaxitic microlitic and slightly glassy textured groundmass, marked by radial rectangular joint patterns with joints 30 to 80 cm apart (Fig. 6.14d). This rim zone passes into the outermost boundary breccia zone, which is 7 to 10 m thick, consisting of angular clasts of rhyodacite in a finer poorly sorted matrix of the same debris, similar in petrological texture to the rim zone. Although clast-rotated texture is most common some domains of jigsaw-fit breccia are preserved. No peperite (mixed rhyodacite with pre-existing sediments that the cryptodome intruded) has been found. Goto and Tomiya (2019) propose a model of intrusion and endogenous inflation of the cryptodome into a dry volcaniclastic sediment host sequence that caused the ground surface to dome (Fig. 6.14e). The cryptodome cooled slowly leading to the outer radial jointing. The boundary breccia was produced by cooling contraction and dynamic stress at the margins of the intruding viscous cryptodome, producing essentially an intrusive autobreccia (Fig. 6.14 d, e).

6.7 Structural Architecture and Dynamic Emplacement Models of Subaerial Lava Domes, Lava Flows, and Cryptodomes

In addition to their facies, viscous felsic lavas deform and develop internal structures that help to elucidate the flow mechanisms and behaviour of these lavas. We will review the important internal structural features, their internal organisation in the lavas, and, together with facies organisation, consider flow and emplacement models.

6.7.1 Internal Deformation and Flow State Structures

As viscous magmas move in the conduit and in lava domes and flows, they deform and this produces a range of deformational features including ductile folds and brittle faults, fractures, and breccias. There have been relatively few studies of the deformational structures in viscous lava flows, but those studies have helped to develop a better understanding of the flow behaviour of viscous lava flows (e.g. Christiansen and Lipman 1966; Fink 1980a; Castro and Cashman 1999; Cioni and Funedda 2005; Latutrie et al. 2017; Bullock et al. 2018).

Viscous lavas flow in a laminar flow state as a result of their high viscosity and low velocity (Chap. 2). The magma flow develops a velocity gradient, especially where it is in contact with a solid surface such as the walls of the magma conduit and the ground surface. This induces a viscous shear regime leading to the development of a layered mass in which different layers or flow bands flow at slightly different velocities relative to each other. Sometimes the boundaries between flow bands may act as permeable surfaces along which gas can segregate. As the flow moves it deforms and may develop ductile folds, which can be harmonic or disharmonic and of varying orientations, some of which can be related to the flow behaviour and location in the flow, and some not. Such folding is continuous and progressive, occurring throughout the duration of the flow, and occurs as the rheology of the flow changes in different places in the flow, through time as parts of it cool and become more viscous, in response to changing magma supply rates at the vent and at its distal margins, and as it responds to changing topography or slope in its flow path. As a result, several "generations" of folding can affect different parts of the lava flow at different times.

The deformational history is also complicated by the transition from plastic or ductile to brittle rheology vertically and laterally within the flow and temporally during its flow duration. In parts where it becomes solid a lava will behave in a brittle manner and fracture and undergo faulting. It will also experience changes from extension to compressional regimes in different locations in the flow and through time, particularly in response of cooling and solidification of the flow front, and to changing slope as it moves over changing topography.

Lava flows, in particular, will be marked by both ductile fold features of varying scales, geometries, and degrees of overprinting, as well as brittle faults. The deformation of the lava flow is thus progressive, and both spatially and temporally non-uniform (Cioni and Funedda 2005; Bullock et al. 2018).

6.7.2 Ductile Features—Flow Banding and Flow Folds

As discussed above, viscous lava flows vary from having a massive internal texture to having a layering defined by varying lithologies and textures, to thin mm to cm thick *flow banding*, and even aligned tabular and platy phenocrysts. Flow banding (Figs. 6.9a, e, f, h, 6.15), and *crystal alignment* (e.g. Fig. 6.9d) are generally attributed to the development of shear gradients, partings, and boundaries, however slight, between adjacent flow layers that develop in response to laminar flow in a viscous medium. Flow banding commonly occurs at the base of lavas, at the margins or fronts and at the tops of lavas but may also be pervasive throughout some lavas. Many lavas however also have massive zones without internal flow banding or crystal alignment, indicating that those zones moved as massive plugs without internal differential shear and velocity gradients.

Flow band boundaries may form discontinuous partings that often allow volatiles to diffuse and concentrate there, forming vesicle trails or elongate vughs of cavities. In ancient rocks, where secondary minerals may have been precipitated in these partings, producing a lensoidal fabric, this can sometimes be confused with eutaxitic and rheomorphic textures developed in densely welded pyroclastic deposits (Chap. 10). Flow band layering in lavas is relatively continuous (e.g. Fig. 6.15), even when plastically deformed and folded due to flow, whereas in densely welded pyroclastic deposits a compactional eutaxitic and rheomorphic layering is often discontinuous (Henry and Wolff 1992). In addition, crystals in true lavas should have mostly euhedral regular shapes (Fig. 6.9b–d), and not be as broken and fragmented as in the pyroclastic rocks (Chap. 4). Flow bands or layers in rhyolite obsidian lavas are defined by subtle partings at the flow band margins, often marked by trails of very small vesicles of gas partings. Flow bands consist of layers of varying groundmass crystallinity and textural style, (obsidian and stony rhyolite), spherulite content, and vesicularity. The latter may form lensoidal domains.

In some cases, different lithological layers are thought to have been batches of physically heterogeneous lava with variations in microlite crystal content, vesicularity, temperature, and perhaps water content, although Castro et al. (2005) discount this. Such heterogeneity might be inherited from magma mingling in the conduit or in the magma chamber (Seaman et al. 1995; Perugini et al. 2004; Castro et al. 2005). Alternatively, repeated processes of magma

Fig. 6.15 Ductile deformation features of viscous lavas. All images with the exception of (**e**) are from the rhyolitic 1220 AD Rocche Rosse obsidian lava flow, Lipari, Aeolian Islands, Italy. **a** Flow lineation striations on exposed flow folded layer planes in pumiceous rhyolite lava facies. **b** Similar fold in pumiceous rhyolite. Note the stretching of vesicles parallel to the axial plane of the fold. **c** Parallel fold in pumiceous rhyolite. **d** Decimetric scale, open, upright parallel fold. **e** Similar, tight, recumbent folds developed in the interior of the 55 ka Basiluzzo rhyolite dome, Aeolian Islands. Height of rock face ~1 m. **f** Parallel, open (right) to isoclinal (centre) upright folds at the top of the Rocche Rosse obsidian flow. **g** Ptygmatic folding

brecciation and subsequent annealing during conduit and surface flow might be responsible for varying water contents (Gonnermann and Manga 2003; Tuffen et al. 2003), as well as processes of local temperature increases in layers, due to shear stresses in the moving lava (Nelson 1981). The prominent foliations are then generated during stretching, shearing, and attenuation during flow.

Layering may be continuous but may also be characterised by detachments along discrete layer planes, especially where a strong rheological contrast develops such as at the sharp transition from obsidian to pumice domains. The surfaces of detached layers are striated displaying the sense of shear (Fig. 6.15a). Flow banding may be folded at any scale because of continuous deformation during flow advance and spreading (Smith and Houston 1994; Cioni and Funedda 2005).

In general, *flow folds* that develop within the interior of a lava flow are influenced by both layer parallel compression and gravity, whereas those at the top and at the front are largely due to simple buckling. Folding occurs in a plastic regime, and so folds often have irregular and highly variable geometry, wavelength, amplitude, and facing (Cioni and Funedda 2005), which distinguishes them from tectonic folds in older rock successions, which tend to have more regular geometrical characteristics. Flow folding does not indicate a turbulent flow state, but plastic deformation of laminar flow state flow bands. If turbulent flow occurred, the flow would be homogenised and all flow banding would be destroyed or just not form.

Large flow folds that affect the crust of a lava may coincide with ogives, and their form and plan view curvature (convex in a down-flow direction) can be used to indicate flow direction (Guest and Sanchez 1969; de Silva et al. 1994; Fink 1980a, b, 1983; Bullock et al. 2018; cf. Andrews et al. 2021).

Flow folds can be similar (Fig. 6.15b) or parallel (Fig. 6.15c; Ramsay 1967) and form at any scale from microscopic to decimetric (Fig. 6.15), or more, indicating that rheology and strain rate vary within flows as well as at the layer scale. Larger scale folds in the surface crust may coincide with and define ogives, which may also be defined by faults. Axial planes may be upright or recumbent and limbs may be open to isoclinal (Fig. 6.15d–f). A very important parameter that controls the type of folding, especially at the meso-scale, is the rheological contrast between different layers (Fig. 6.9h), such as obsidian and variably vesicular pumice layers. Folds can be harmonic, disharmonic, or single layer ptygmatic (Fig. 6.15g) at varying wavelengths depending on thickness and shear viscosity ratios of competent and less competent folded layers (Castro and Cashman 1999).

Similar folds are generally found in the interior of lava flows, where folds are also tight to isoclinal (Fig. 6.15e). By contrast, at the top and distal edges of lava flows, parallel folds are more common and occur with varying degrees of opening and geometrical complexity (Fig. 6.15f), indicating that the complex thermal profile plays an important role in changing the rheology at the local scale. In parallel folds, the layer thickness does not vary and this implies slip along the foliation planes that then show striations (Fig. 6.15a), and the development of spaces at the hinges that experience decompression.

Quite often in multilayered folds the most competent layers buckle without changing thickness, whereas less competent layers flow from the limbs, where shear induces compression, towards the hinges, where there is space to accommodate. The rather ubiquitous presence of coarse vesicles at the hinges of this type of multilayered fold (Fig. 6.15b, c) indicates that extension in the fold hinges causes local decompression leading to volatile exsolution and vesiculation in these sites (e.g. Fink et al. 1992). Alternatively, the reduced or absent vesicularity along the limbs with respect to hinges may be seen as a result of foam collapse and bubble resorption (Eichelberger et al. 1986).

Flow directions, flow type, and strain of viscous lavas can be determined by structural analysis of the orientation of foliations, surface ogives, folds, vesicles, tabular and platy crystals, and stretched-out volcanic glass (Chistiansen and Lipman 1966; Fink 1980a; Smith and Houston 1994; Smith 1996, 2002; Ventura et al. 1996; Castro and Cashman 1999; Ventura 2001; Castro et al. 2002; Rust and Manga 2002; Rust et al. 2003; Cioni and Funedda 2005; Latutrie et al. 2017; Bullock et al. 2018; Andrews et al. 2021). The analysis of deformation of bubbles (Manga et al. 2018; Rust et al. 2003) and rotation of crystals (Ventura et al. 1996; Castro et al. 2002) allows interpretation of the kinematics of the flow. Simple shear produces elongation of vesicles and imbrication of crystals at an angle with respect to flow direction and implies a flow characterised by a velocity profile, more likely to occur towards the base of the flow or in the conduit prior to eruption. Pure shear produces flattening or stretching of both bubbles and crystals perpendicular to maximum stress during inflation and spreading of the flow (Fig. 6.15b, c). Buisson and Merle (2004) suggested that in a flow, pure shear proportionally increases upward with respect to simple shear.

Flow banding in a felsic lava flow cannot be used consistently as a reliable indicator of bedding orientation in ancient volcanic successions. However, the vergence orientation of flow folds and flow banding orientation *may* be useable as an indicator of general flow directions. However, given that deformation occurs in a plastic regime, without a significant confining stress field, deformation structures can also assume random orientations. They must be used with care in any study trying to establish flow directions. Flow banding generally ramps and becomes steeper near the advancing flow front and at the top of the flow. Packages of flow banded lava may be bounded by listric-like detachment faults called ramp faults, which may also coincide with ogives. Flow banding and detachments in such ramp packages therefore dip backwards towards the vent, and although fold orientations may be highly variable, fold vergence and flow lineations are oriented in the direction of flow, and fold axial planes dip back towards the vent (see Fig. 6.17a, b; Christiansen and Lipman 1966; Fink 1983; Latutrie et al. 2017; Bullock et al. 2018). Cioni and Funedda (2005) found that:

- the wavelength of F1 folds decreased away for the vent,
- F1 folds became progressively overturned away from the vent,
- F1 folds were refolded into upright or vertical F2 folds,

- the inter-limb angle of F2 folds decreased away from the vent, and
- the primary S_0 flow band foliation is transformed into a vertical S_2 shear foliation along the margins of the flows.

6.7.3 Brittle Features, Including Autobrecciation

Late-stage movements of cooled lava are dominated by the effects of flow when much, if not most, of the lava, especially its outer crust, is undergoing ductile–brittle transition. Even though strain rates are very low, the liquid undercools through the glass transition temperature to solid glass (Dingwell 1996) and undergoes brittle fracturing (Fig. 6.16). By contrast, fracturing which occurs in hotter and less viscous lava at higher strain rates, such as during conduit flow, may be annealed afterwards (e.g. Gonnermann and Manga 2003).

Brittle fracturing therefore occurs mostly in the crust at the front, margins, and in the upper parts of viscous lava flows, by a combination of cooling contraction and dynamic stressing of the crust as the dome or lava continues to expand and flow. Autobrecciation is represented most prominently by the formation of in situ and clast-rotated, coarse block autobreccia (Sect. 6.4.1, Fig. 6.8 a, e; Chap. 4, Sect. 4.3, Fig. 4.2d–g), and it forms commonly on the surface of subaerial lavas and domes, but also in the boundary zone of cryptodomes (e.g. Goto and Tomiya 2019). Smith (1996) also described faulting and fracturing associated with the development of the basal breccia of the Minyon Falls Rhyolite, Australia, suggesting that the upward directed cooling from the cold ground would induce brittle faulting in the highly sheared base of an advancing lava flow. Cioni and Funedda (2005) suggested that thrust faulting associated with the formation of ramps and many surface folds, especially large-scale ogives, was more likely to be found near the flow front resulting from cooling and stiffening of the distal margins of the flow.

Ductile–brittle features include the development of faults and fractures associated with folds and ramps directed away from the vent (Fig. 6.16a, b). Faults can develop striations, slickensides, and tension gashes, which can be used to reconstruct the local flow kinematics (Fig. 6.16c). Faults can be compressional, strike-slip, or extensional. Extensional fractures may also reflect strain partitioning between layers of different rheological properties (e.g. Fig. 6.16d), and some larger extensional fractures may allow the squeezing out of liquid lava from the interior of the flow. Other types of late-stage brittle features include thermal contraction columnar joints, which propagate perpendicular to the cooling surfaces (i.e. the margins) of the flow. The

Fig. 6.16 Brittle deformation of viscous lavas from the rhyolitic 1220 AD Rocche Rosse obsidian lava flow, Lipari, Aeolian Islands, Italy. **a** Thrust fault plane at the base of a recumbent fold, **b** extensional fault and breccia, **c** striation and tear fractures over a slip-plane developed within pumiceous facies. Note the ductile stretched vesicle textures between the brittle-fractured pumice clasts. **d** Strain partitioning during simple shear: pale layers are dense and show tear fractures, whereas dark layers are pumiceous and vesicles are stretched

observation of relatively small columns merging to form large columns inwards indicates progressive inward propagation of the solidification front at slower rates than at the margins because the rate of cooling of the interior is lower than at the surface. A segmentation pattern on the surfaces of columns perpendicular to the column axes indicates that cooling and contraction occurred step by step over an extended period of time.

6.7.4 Structural Architecture

The simple conceptual facies association models presented above in Sects. 6.6.1 and 6.6.2 for domes and lavas, respectively, may not adequately describe the complex associations of many domes and thick lava units that are made of many internal juxtaposed endogenous and exogenous lobes, or coalesced multiple domes, as a result of continuous *versus* discontinuous eruption and eruption rates, exogenous *versus* endogenous growth, vent migration, varying rheology, etc. It is therefore essential, especially in ancient successions, to map in detail the spatial distribution of lithofacies, their contact relationships, which will help to identify boundaries between juxtaposed lavas and lobes and their substrate, and all structural elements, including flow banding/foliation relationships; fold vergence directions; fold axes orientations; ogive patterns and orientations; and lineation, fracture, and fault patterns and orientations (e.g. Fig. 6.17a, b; Christiansen and Lipman 1966; Fink 1983; Bullock et al. 2018; Andrews et al. 2021).

Mapping of the internal structural architecture of felsic lava flows (Fig. 6.17) provides insight into their flow dynamic evolution. Such studies show that there are lateral changes in the structural architecture from the proximal (above the) vent facies laterally to the medial and distal parts of the lava. Above and adjacent to the vent area, flow banding is subvertical (Figs. 6.5, 6.18a, f, g), but as a lava (dome) begins to spread and flow away from the vent the flow banding leans over and becomes less steeply dipping towards the vent, verging in the down-flow direction (Fig. 6.5a, b), and even becoming sub-horizontal (Fig. 6.18b, c, f, g). In such medial distances from the vent, isoclinal inclined folds with axial planes dipping towards the vent commonly become recumbent folds verging in a down-flow direction (Fig. 6.18b, c, f, g). On sloping terrain, gravity may impose an extensional stress regime on the flow away from the vent; Fig. 6.18g), while at the cooling flow front stiffening of the flow causes it to slow, as less viscous lava behind begins to ramp up onto the slowing front, establishing a compressional stress field, and downslope verging folds, thrust faults, and ramps (Fig. 6.18d, f, g). Upright parallel and chevron folds with steep axial planes (Fig. 6.18d, f) may penetrate the entire thickness of the flow and be represented at the surface by large amplitude (30 m) ogives or pressure ridges.

The proximal facies association above the vent is dominated by vertical-to-subvertical flow banding, as shown on the island of Basiluzzo, in the Aeolian Islands, Italy (Fig. 6.18a), associated with fracturing, brecciation, and hydrothermal alteration. The (sub-)vertical flow banding sits above the conduit and, depending on viscosity and eruption rate, may laterally spread and become progressively folded to feed dome lobes and thick lava flows (Fig. 6.18b). The interiors of lava flows are dominated by tight to isoclinal sheet folding with axial planes with and overall vergence (i.e. statistically at the scale of the lava flow) towards the flow direction (Fig. 6.18b). Vertical flow banding is common instead in the upper parts of lavas and at their fronts. The predominant flow foliation orientation on the surface of La Primavera rhyolite lavas is subvertical (Clough 1981). The same is true of the rhyolite lavas of Lipari (Hall 1978; Bullock et al. 2018) and Basiluzzo (Fig. 6.18a–f). These subvertical structures correspond in cross-section to ramp structures, steep at the top of the flow, curving downwards from the up-flow direction to be asymptotic against the base back towards the vent (Fig. 6.18c, d). As lava is repeatedly injected into the growing dome, the ramp structures move outwards and it is subject to dominantly pure compressional shear, both radially and tangentially (Fig. 6.5). With time, each fraction of lava becomes progressively attenuated, due to stretching and shearing, especially in the inner part of the dome where simple shear dominates. Striations or slickensides found on ramp structures indicate movement directions on the foliation surfaces, which must have cooled to be solid enough to allow such structures to form.

Vertical flow banding may also correspond to tight upright folds, more likely to develop at the flow front (Fig. 6.18d). Therefore, the occurrence of (paleo)vertical flow banding in the rock record should not necessarily be interpreted as a dyke or as a vent facies, because ramps and tight anticlines may also generate large domains of vertical flow banding, away from the vent near the flow front. Therefore, field mapping of the total lava flow is required to establish context within a flow (e.g. Fink 1983; Cioni and Funedda 2005; Bullock et al. 2018; Andrews et al. 2021).

Contacts between juxtaposed flow units and lobes may be associated with the development of autobreccia and pumiceous zones, resulting from discontinuous feeding, fracturing, and brecciation in the vent facies association, and also by sharp changes in flow banding attitude.

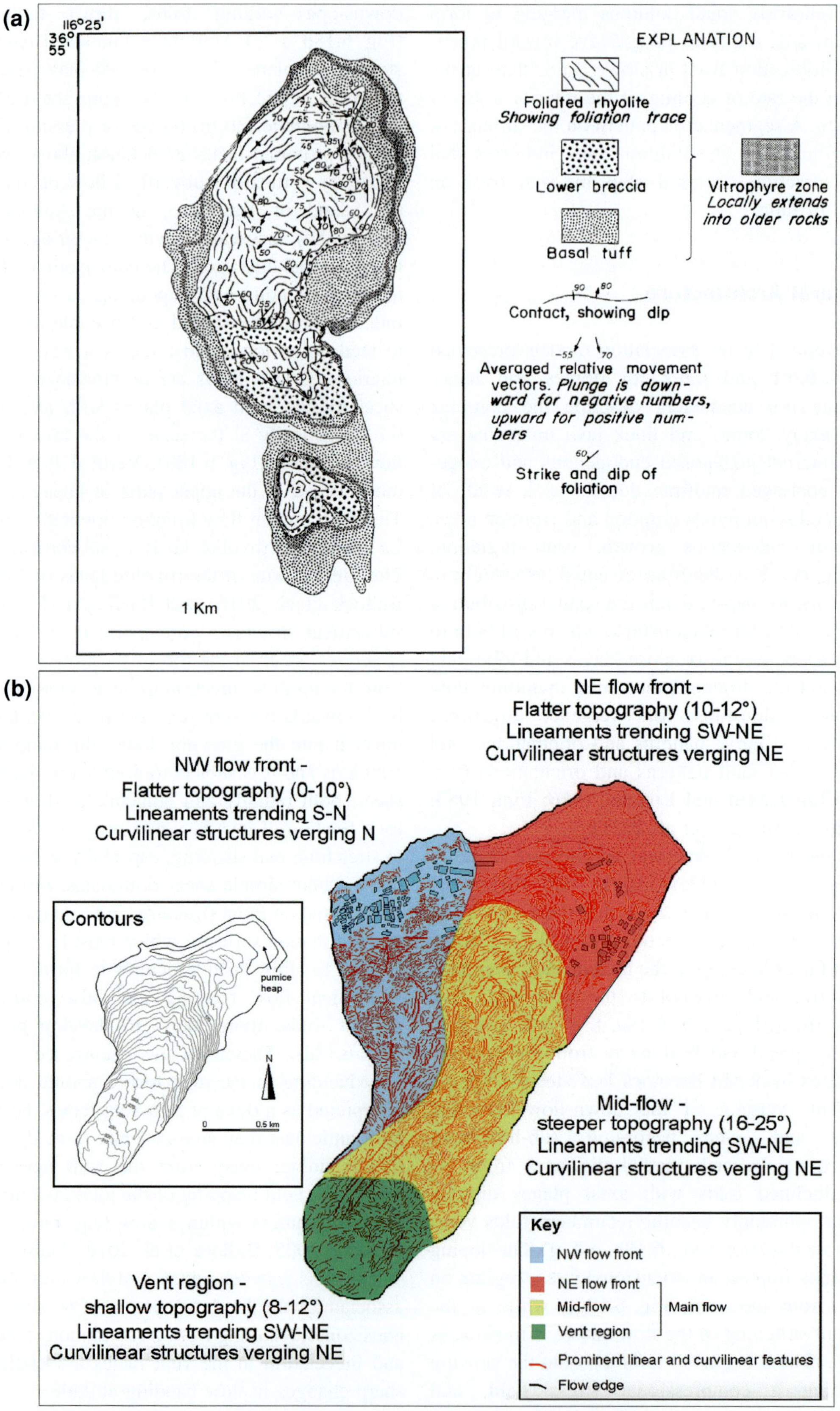

Fig. 6.17 Maps of some felsic lavas. **a** Tertiary Comb Peak rhyolite lava and subjacent co-genetic pyroclastic sequence, showing distribution of the main facies, dips, and trend of main flow foliation and folds, vector data for directions of movement (After Christiansen and Lipman 1966). **b** Plan view map of the two lava lobes (NW-blue; main–green, yellow, red) of the ~1220 AD Rocche Rosse obsidian lava flow, Lipari, Italy, highlighting the trends of curvilinear and linear features on the surface (From Bullock et al. 2018)

(a)
(b)
(c)
(d)
(e)
(f)
W
E
LOBE PILE-UP REGION
vertical pure shear dominant
LATERAL LOBE EXTRUSION AND SPREADING REGION
simple shear dominant
FRONTAL REGION
horizontal pure shear dominant
VENT REGION
ridges(2D)/ogives(3D)
upright chevron to open to ptygmatic folds
autobreccia
columnar joints
isoclinal tight and recumbent folds
1
2
3
5
6
7
9
10
11
8
7
6
4
5
3
3
100 m

◀ **Fig. 6.18** Structural architecture of subaerial felsic lava flows. All images with the exception of (**g**) are from the ~55 ka Basiluzzo rhyolite dome and flow, Aeolian Islands, Italy. **a** Proximal vent facies, characterised by vertical and outward flaring flow banding. **b** Columnar joints cross-cutting horizontal layering in different superposed lobes that form a single cooling unit. Note the recumbent fold at the top of the cliff and down-flow ramping of layering. **c** Intermediate outcrops are dominated by horizontal layering represented largely by highly strained, recumbent isoclinal folds, as seen in the middle of the outcrop in the left-hand side of the photo. **d** Distal outcrops are dominated by upright parallel to chevron folds with verging down-flow and with axial planes dipping steeply back towards the vent. **e** Cross-sectional view of the lava dome and flow cliffs of Basiluzzo Island. **f** Cross-sectional line drawing of the main internal structural architecture and geometries and relationships of discrete emplacement lobes of the Basiluzzo lava dome and flow. Each lobe is bound by listric thrust faults that are steepest near the flow front and become shallow and horizontal back towards the vent. **g** Schematic, interpretative cross-section along the length of the 1220 AD Rocche Rosse obsidian lava flow, Lipari, showing internal structural and facies architecture (From Cas and Wright 1987, After Hall 1978)

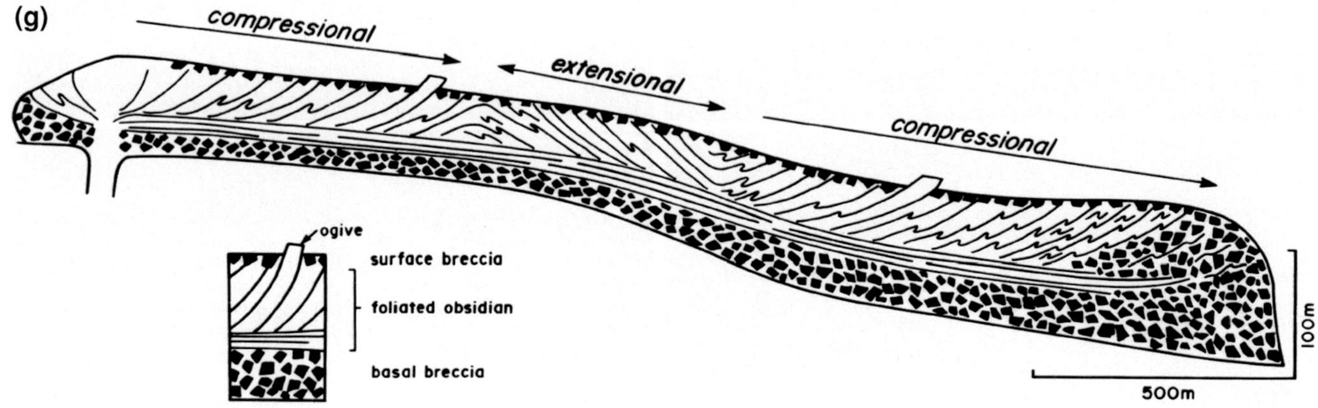

Fig. 6.18 (continued)

6.7.5 A Dynamic Flow Emplacement Model for Subaerial Felsic Lava Domes and Lava Flows

Although low-viscosity basaltic lava flows propagate by flowing with a roll-over front, like the tracks of a bulldozer, and thicken largely by endogenous flow inflation (Chap. 5), more viscous lavas flow in more complex ways. Lower viscosity and thin end members, such as the small volume rhyolitic Pietre Cotte rhyolitic lava on Vulcano, also appear to roll-over at the flow front, as evidenced by the vertically dipping coherent pumice flow front (Fig. 6.12d). This behaviour may have been driven by the steep 25° slope of the Fossa Cone down which the flow moved, which would have increased its flow velocity and internal strain rate, lowered its dynamic viscosity, and led to mostly internal stretching or extension, with little internal compression, except at the flow front at the break in slope, where it rolled over, compared with flows on low slopes.

However, the flow behaviour of more viscous and thicker flows on low slopes has been compared with the motion of glaciers (Christiansen and Lipman 1966; Latutrie et al. 2017) and is commonly accepted to be non-Newtonian in behaviour, ranging from pseudo-plastic to Bingham like (Chap. 2; Hulme 1974; Lavallée et al. 2007; Cordonnier et al. 2009; Calder et al. 2015). This involves a significant component of sliding or plug flow over a substrate, including a basal breccia layer consisting of surface autobreccia and frontal talus apron, which have been rolled over down the flow front to be buried under the advancing flow body, and/or are overridden by breakouts of lava from the flow interior. It seems from the behaviour of the 2011–2012 Puyehue Cordón Caulle volcano obsidian lava in Chile that some flows propagate and spread by breakouts of mega-lobes tens of m or more in thickness from the flow interior, and an interior lava tube (Tuffen et al. 2013). Internally, gliding is facilitated by the development of internal listric thrusts, which dip back towards the vent, and downslope verging folds that transport the lava in a downslope direction away from the vent (Fig. 6.18f, g). Increasing internal fracturing and cracking at increasing strain rates lower the apparent (bulk) viscosity of the flow, contributing to increased mobility or flow distance (Lavallée et al. 2007; Calder et al. 2015).

On steep slopes flows can develop slower moving or stationary marginal levees, facilitating faster moving channel flow between levees. On low slopes or breaks in slope onto low slopes, the flows slow and begin to experience significant compression, especially at the flow front which is cooling, becoming more viscous and slowing down, so that the following flow banks up behind the slowing flow front. Compression therefore causes the lava to deform into tight folds, which verge downslope and down-flow near the flow front, with axes transverse to the flow direction, as well as by stacked, listric thrusts, and ramps which dip back towards the vent or upslope, but which transport lava in a downslope or down-flow direction, as it occurs with glaciers (Fig. 6.18f, g; Latrutrie et al. 2017). Viscous felsic lavas may therefore thicken through internal tube-fed inflation, but also through compressional deformation, and they may advance through roll-over of the flow front, breakouts of thick lobes from the interior tube, as well as sliding over the substrate facilitated by folding and thrusting in a down-flow direction. The flow mechanism may change along the flow path in response to changing topography, magma effusion rate, viscosity, eruption temperature, and changing flow rheology as it cools and its viscosity and yield strength increase.

6.8 Large Volume, Far-Flowing Subaerial Felsic Lavas

Large volume and far-flowing felsic (silicic) lava flows are relatively rare. To form, they require large volumes of degassed or low-volatile-bearing viscous magma that is then erupted in a sustained eruption that continues to feed magma to the spreading flow front under an insulating crust (Manley 1996). The best young example is the late Pleistocene (<100 ka) Chao dacite lava system of Chile (Guest and Sanchez 1969; de Silva et al. 1994; Plate 6.1b, Fig. 6.8c), consisting of 3 main flow units (or main lobes), with a maximum flow distance of 14 kms from the vent. The eruption commenced with an early phase of pumiceous pyroclastic flow activity forming a fan beyond the southern flow front, while the vent is located within a pyroclastic pumice cone which was in part coeval with the main effusive activity (Fig. 6.8c). The pre-existing ground surface slope near the vent was 15 to 20°. The flows are up to 400 m thick, and the flow surfaces are marked by 30 m high ogives, and blocky autobreccia. De Silva et al. (1994) estimated that the magma viscosity was $\sim 10^{10}$ to 10^{12} Pa s, the erupted volume is ~ 25 km^3, the eruption rate may have peaked at ~ 25 m^3 s^{-1}, using Mt St Helens lava dome eruption rates (Swanson and Holcomb 1990), and the eruption duration was at least 20 years, if it was continuous, but perhaps more realistically 100 to 150 yr with much lower, later effusion rates or some pauses in activity.

Even bigger subaerial rhyolite and dacite lavas are preserved in the rock record. Bonnichsen and Kauffmann (1987) have summarised extensive studies of a number of very large volume, very extensive rhyolite lavas in the western Snake River Plain of Idaho, USA. The Snake River Plain forms part of the Columbia River continental Large Igneous Province or cLIP (Chap. 5; also see Table 17.1). Volcanic rocks of the province are compositionally bimodal (basalt–rhyolite), and the rhyolites produce a broadly time-transgressive record since about 16 Ma (late Miocene) along the trend of the Yellowstone "hot spot" (Chap. 17, see Fig. 17.19b). They range in volume from 10 to 200 km^3, in thickness from <25 to 200 m (although generally between 100 and 150 m), and some extend for at least 40 km along strike (Fig. 6.19). They are also interpreted to have been erupted from fissure vents, in some cases, the sources of initial pyroclastic deposits. The lavas are spectacularly columnar jointed, flow-banded, and flow folded, and they have variably devitrified or crystallised interiors, obsidian tops and bottoms, and both top and bottom autobreccias (Fig. 6.19). Although there are ignimbrites in the province, Bonnichsen and Kauffman (1987) present convincing arguments for the documented bodies having been erupted as lavas; also see Branney et al. (2007), Bonnichsen et al. (2008), and Ellis et al. (2013) for further discussion.

Henry et al. (1988, 1990) also described the magnitude and features of Eocene ($\sim$36–38 Ma) rhyolite lavas in the Trans Pecos region of Texas, USA, located in a former continental margin volcanic arc setting (Parker et al. 2017). These lavas have preserved strike lengths of up to 70 km,

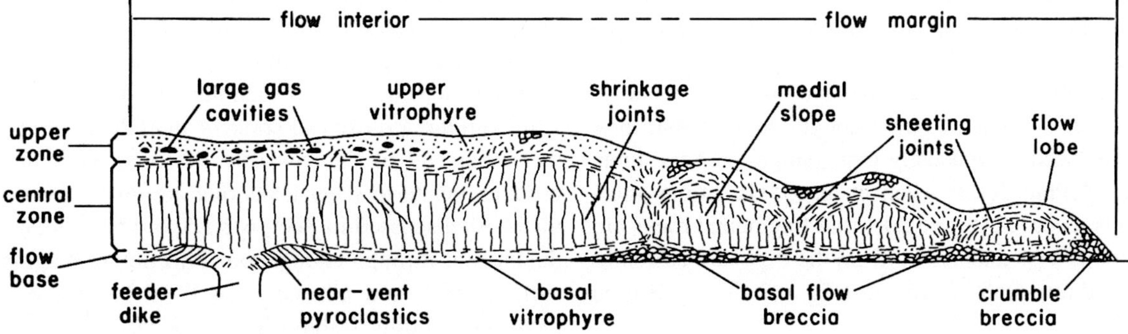

Fig. 6.19 Schematic diagram illustrating the spatial relationships among the zones and physical features (i.e. larger scale facies characteristics) typical of the extensive rhyolite lava flows found in the Snake River Plain Volcanic Province, southwestern Idaho, USA. The longitudinal distance represents 10–40 km, while the vertical height 100–200 m). (From Bonnichsen and Kauffmann 1987)

and range in thickness from 20 to 120 m. They have aspect ratios of 1:500 to 1:700, they have coherent flow banded cores enclosed in obsidian and autobreccia domains at the tops and bottoms, stretched vesicles and show no signs of having originated as rheomorphic ignimbrites (see Henry and Wolff 1992 for distinguishing criteria, as well as Chap. 10), although ignimbrites also occur in the Trans Pecos region. They were erupted from fissure vents, not from calderas, even though there are calderas in the region, and flow distances may have been about 35 kms.

Perhaps the most remarkable large volume, subaerial felsic lava flows are those the Mesoproterozoic (~ 1592 Ma) Gawler Ranges Volcanics (GRV) of South Australia, a felsic or silicic Large Igneous Province or SLIP (Table 17.1; Allen et al. 2003, 2008; Pankhurst et al. 2011). The GRV are divided into lower and upper successions. The lower GRV outcrop in several widely separated (rift-controlled) successions perhaps locally amounting to >1000 m thick, and which may be gentle to steeply dipping. They range in composition from rhyolite to basalt, and comprise small–moderate volume basaltic lavas, rhyolitic dome/tuff cone complex, pyroclastic deposits, and high-level intrusions (cryptodomes) intercalated within and/or intruded into clastic-dominated sediments. These are overlain by three extensive felsic (rhyolitic or dacitic) sheet-like units (each in excess of 1,000 km^3) which largely form the upper GRV. The three main lava units forming the upper GRV (Eucarro Rhyolite, and Pondanna and Moonaree members of the Yardea Dacite; Fig. 6.20a; also see Fig. 15.34b) are very thick (200–300 m), sheet-like, have lobate margins, are variably columnar jointed, mostly massive with some domains of steeply dipping flow banding, and relatively crystal rich (Fig. 6.20b; Allen et al. 2008; Pankhurst et al. 2011). They have a preserved outcrop extent of >12,000 km^2 across the central and southern Gawler Ranges, dipping very gently northwards. Generally, they are unmetamorphosed, but are oxidised and locally pervasively hydrothermally altered giving their characteristic reddish-orange colour (Fig. 6.20c–f). On the basis of their widespread nature and a few samples with apparent vitriclastic textures, historically, they were generally interpreted as densely welded ignimbrites. However, detailed studies of the textural and lithofacies characteristics (e.g. massive to flow banding, elongate crystal alignment) now strongly suggest that they are lavas emplaced as unusually large, single eruptive units (Allen and McPhie 2002; Allen et al. 2003). Some autobrecciation is recorded (Fig. 6.20e), but fragmental rocks are not voluminous as is understood from the exposure currently preserved, and rare mafic fragments (Fig. 6.20f) may also be present.

The upper Gawler Range Volcanics are only one component of the A-type felsic large igneous province which also comprises an extensive, coeval Hiltaba Granite Suite which hosts the supergiant Olympic Dam Cu-Au-U-Ag orebody (Chap. 18, see Sect. 18.4). As well as larger batholithic bodies intruded into basement, the Hiltaba Suite granites form discrete circular plutons (apparent in magnetic data) and sill-like bodies intruded at all stratigraphic levels into the GRV. Field relations confirm that the GRV and Hiltaba Suite were generally coeval, and entrained clasts and megablocks within the lavas include Hiltaba-type granite, in probable near-vent locations (Garner and McPhie 1999; Allen and McPhie 2002). Concentrations of granitic clasts coincide with changes in bulk composition and may record the rupture of wall rock between nearby but separate magmas. Extensive outflow of lava is favoured by eruption from multiple vents aligned along a fissure(s). Although no fissures have been located, Allen and McPhie (2002) suggest the Eucarro Rhyolite was erupted along an east–west trending fissure, suggesting that it could conceivably be no more than a few tens of km from source vents. More recently, anisotropy of magnetic susceptibility (AMS) data has been obtained which suggest flow from a southwest source (McPhie et al. 2008), or from a central cluster of magnetic features to the northeast interpreted as subvolcanic structures, including pipes/feeder complexes and frozen magma chambers (Pankhurst et al. 2011).

The lavas have been interpreted to have resulted from mantle plume activity, producing very high eruption temperatures of >950 °C, causing low magma viscosities (Pankhurst et al. 2011). In addition, the lavas are characterised by elevated primary halogen concentrations (~ 1600 ppm fluorine, ~ 400 ppm chlorine), which would have depolymerised the magma such that temperature-composition-volatile non-Arrhenian melt viscosity modelling suggests they had very low viscosities of <3,100 Pa s, comparable to those of tholeiitic basalts, enhancing their mobility relative to normal calc-alkaline rhyolites and dacites (Pankhurst et al. 2011). Three individual lavas within the upper GRV have volumes of 1,000 to 3,000 km^3 and each cover several thousands of km^2 (Allen et al. 2008), ensuring that both low-viscosity, and long-term sustained eruption of very large volumes of magma produced lavas covering thousands of km^2. They are truly “flood” lavas, but silicic in composition.

The flow behaviour of such huge felsic lavas is not clear. Although the viscosity of some of these lavas seems to be at the upper end of the viscosity of some basalt lavas, they produce lobate to sheet lavas that are up to 300 m or more thick. As discussed in Sect. 6.2.1.5, did they inflate to these thicknesses through endogenous injection of magma under a surface crust, and advance like flood basalts, or did they move in a sliding fashion like more viscous, small volume felsic lavas?

Concordant, laterally extensive, apparently coherent felsic units in the geological rock record, must be considered

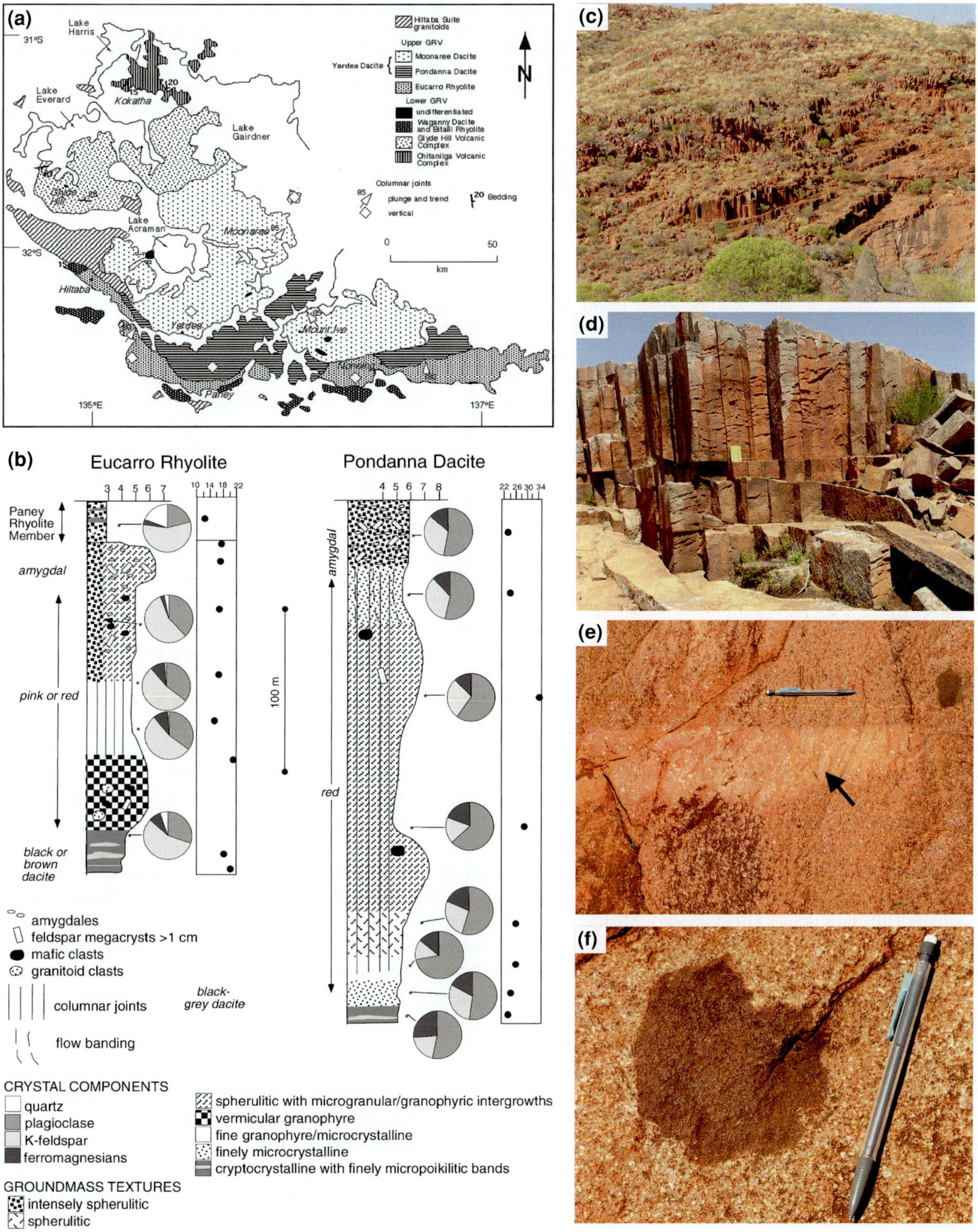
(a)
Lake Harris
31°S
Lake Everard
Kokatha
Lake Gairdner
Glyde Hill
Lake Acraman
Moonaree
32°S
Hiltaba
Yardea
Mount Ive
Paney
135°E
137°E
Hiltaba Suite granitoids
Upper GRV
Yardea Dacite
Moonaree Dacite
Pondanna Dacite
Eucarro Rhyolite
Lower GRV
undifferentiated
Waganny Dacite and Bitalli Rhyolite
Glyde Hill Volcanic Complex
Chitanilga Volcanic Complex
Columnar joints
plunge and trend
vertical
Bedding
0
50
km
N
(b)
Eucarro Rhyolite
Paney Rhyolite Member
amygdal
pink or red
black or brown dacite
Pondanna Dacite
amygdal
red
100 m
black-grey dacite
amygdales
feldspar megacrysts >1 cm
mafic clasts
granitoid clasts
columnar joints
flow banding
CRYSTAL COMPONENTS
quartz
plagioclase
K-feldspar
ferromagnesians
GROUNDMASS TEXTURES
intensely spherulitic
spherulitic
spherulitic with microgranular/granophyric intergrowths
vermicular granophyre
fine granophyre/microcrystalline
finely microcrystalline
cryptocrystalline with finely micropoikilitic bands
(c)
(d)
(e)
(f)

◀ **Fig. 6.20** Features of extensive felsic "flood" lavas of the Mesoproterozoic upper Gawler Ranges Volcanics, Gawler Ranges National Park, South Australia. **a** Map showing the extent of the lavas of the Gawler Range Volcanics Large Igneous Province. Extensive rhyolitic and dacitic lava units dominate the southern and central map area, especially the three youngest and most extensive: the Eucarro Rhyolite (stipple), the Pondanna Dacite (horizontal lines), and Moonaree Dacite (spaced dots) forming the two members of the Yardea Dacite. (Modified from Allen et al. 2008). Note the scale difference between these huge lavas and more normal felsic lavas (cf. Figs 6.2a–e, 6.3, 6.17, 6.18f, g). **b** Stratigraphic sections through the Eucarro Rhyolite and Pondanna Dacite Member of the Yardea Dacite. Each unit preserves a single cooling profile, with a black, formerly glassy base, a granophyric/microcrystalline-spherulitic and columnar jointed interior, and an amygdaloidal outer margin. Contacts among the various groundmass textures are gradational, except where mingled (solid contact); some parts have two varieties of groundmass textures. (Modified from Allen et al. 2003). **c** View to Eucarro Rhyolite at Yadinga Well (exposure ~60 m high). **d** Columnar jointing showing sub-horizontal segmentation due to upward stacking of successively younger crack segments, Pondanna Dacite at Kolay Mirica Falls. **e** Slab of autobrecciated fine-grained (chilled?) felsic lava (arrow) and small mafic inclusion entrained in groundmass with granophyric/microcrystalline texture, Pondanna Dacite. **f** Detail of mafic clast showing irregular margins, Pondanna Dacite

carefully in terms of their origins. They could be large volume lavas, as described here, they could be the remains of very densely, perhaps even rheomorphic ignimbrites (Chap. 10) or they could be regionally extensive sills (Chap. 8). The characteristics of all of these options must be looked for as a test for origin, especially top and bottom features and contact relationships. Thin, coherent felsic units up to a few metres thick are invariably intrusions because felsic lavas are always thick because of their high viscosity, unless they are peralkaline.

6.9 Pyroclastic Deposits Associated with Felsic Lavas

Many viscous lavas are associated with pyroclastic deposits, resulting from preceding, intervening, and post-dating phases of explosive activity. The style of explosive activity can vary from mainly phreatic eruptions, producing rings of lithic breccias surrounding the lava (e.g. Panum Crater, one of the Mono Craters), to magmatic eruptions which can vary from short-lived vulcanian explosions to explosive (micro- to sub-)plinian-style pumice-tuff cone and ignimbrite-forming eruptions (Chaps. 9, 10). Such styles of explosive activity have all been observed during emplacement of the Soufrière Hills lava domes (1995–recent), Montserrat. Even while lavas and domes are growing, explosive eruptions may emanate from the domes, and evidence for this would be unusually large amounts of obsidian ejecta among contemporaneous pyroclastic deposits. Also, visible craters may be present on some domes, and this can be seen on some of the domes of southern Lipari (Fig. 6.2d; Forni et al. 2013). Formation of cratered domes is attributed to the late-stage build-up of gas pressure within the viscous magma inside the lava dome due to ongoing, post-eruptive volatile exsolution. Built-up gas pressure is released in the form of an explosion from a parasitic vent, not necessarily associated with fresh magma effusion.

During both eruption phases of Mt St Helens from 1980 to 1986, and 2004 to 2008, effusion of lava domes was interspersed with explosive phases, some of which destroyed preceding domes, and were followed by new dome growth. Such a cyclicity regularly occurs from the summit dome complex of Merapi stratovolcano, in Java, Indonesia, and was also documented from the eruption products of the ~1315 AD Kaharoa dome and pyroclastic event at Tarawera dome complex, New Zealand (Fig. 6.21; Cole 1970; also see Fig. 6.2e–g). In addition, according to Nairn et al. (2001, 2004) after initial vent opening (mainly plinian) explosive phases along the fissure vent system, followed by lava dome effusion, there were also explosive phases from vents that formed on the erupted domes, as well as block- and ash-flow-forming events. They model the rhyolitic magma chamber as a sill-like body, ~8 km long, 1 km wide, 1.4 km thick, and located at ~6–7 km depth in the upper crust.

During explosive phases preceding lava dome eruptions, pumice cones or tuff rings can be built-up around the vent. For example, the spectacular 776 AD Monte Pilato pumice cone, in northeastern Lipari, Italy, which formed prior to the eruption of the 1220 AD Rocche Rosse obsidian lava flow (Fig. 6.6e; Dellino and La Volpe 1995; Forni et al. 2013); the Tertiary Comb Peak Rhyolite and subjacent pumice cone (Christiansen and Lipman 1966; Fig. 6.17a); and the Chao dacite lava and contemporaneous pumice cone, de Silva et al. 1994; Plate 6.1b, Fig. 6.8c). In other cases, any evidence of the vent area for earlier explosive phases may be covered by lava (e.g. ~690 AD Big Obsidian Flow, Newberry caldera; Plate 6.1c; Gardner et al. 1998). Some lavas may be hot enough to cause fusion and welding of pyroclastic deposits they come in contact with around their vent area (e.g. Christiansen and Lipman 1966).

6.10 Subaqueous and Subglacial Felsic Domes and Lavas—Processes and Constraints

Felsic domes and lavas are also erupted subaqueously in lakes, in shallow and deep marine settings, and subglacially in meltwater ponds beneath glaciers and ice caps. As for subaerial felsic lavas, the eruption forms include the spectrum (Fig. 6.22) from above-vent, high aspect ratio lava

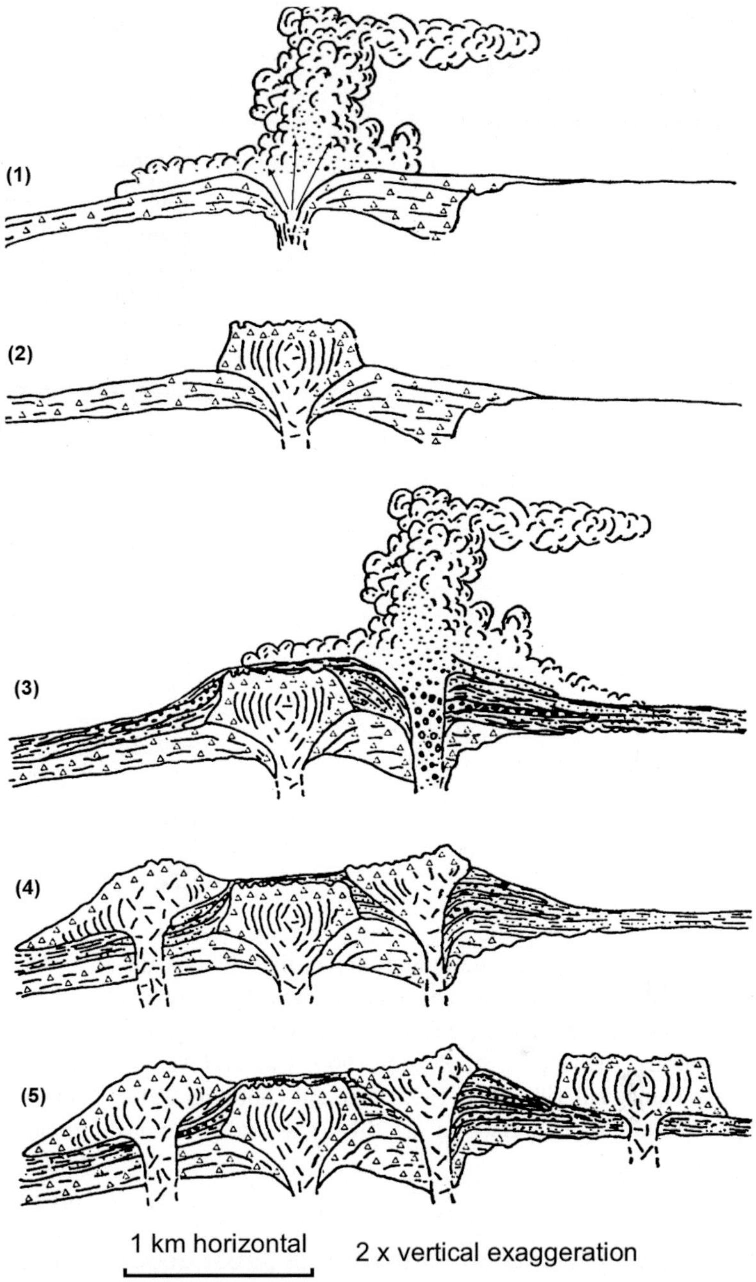

Fig. 6.21 Schematic reconstruction of the eruption sequence at the Tarawera dome complex, New Zealand, during the ~1315 AD Kaharoa rhyolite lava dome and pyroclastic eruption. (1) Explosive excavation of Crater dome vent. (2) Extrusion of Crater dome. (3) Explosive eruption of the Kaharoa Pyroclastics. (4) Extrusion of Ruawahai (centre of profile) and Tarawera (left) domes. (5) Extrusion of Wahanga dome (right). The triangles in the subhorizontally bedded pyroclastic deposits represent ballistic blocks and bombs. In the high aspect domes, triangles represent autobreccia, dashed domains represent feeder dykes and dome cores, and subvertical curved form lines depict endogenous dome growth and lateral dome expansion through displacement. (Modified from Cole 1970)

domes (Pichler 1965; Allen 1992; Cas 1992; Maeno and Taniguchi 2006; Stewart and McPhie 2006; White et al. 2015; Carey et al. 2018; Ikegami et al. 2018), cryptodomes, and sills (Cas et al. 1990; Allen 1992; Goto and McPhie 1998; White et al. 2015), to moderate distance lava flows (de Rosen-Spence et al. 1980; Kano et al. 1991; Soriano et al. 2012, 2013; Embley and Rubin 2018; Ikegami et al. 2018) and rare, far-flowing submarine lavas (Cas 1978).

In subaqueous settings, it is likely that many cryptodomes and high-level intrusions will form (Chap. 8; Cas et al. 1990, 2001; Goto and McPhie 1998; Stewart and McPhie 2003; Fig. 6.22), because magmas rising to the seafloor will frequently pass through water-saturated sediments and volcaniclastic horizons that have a bulk density that is less than that of the rising magma (Chap. 8). The magma has then passed through its level of neutral buoyancy, and because the water-saturated sediments have little strength, the magma can readily intrude sideways rather than continuing to rise to the seafloor (Chap. 8). Domes erupted in relatively shallow water may become explosive as a result of magmatic volatile exsolution under low hydrostatic pressures at shallow water depths, as well as dynamic, explosive water–magma phreatomagmatic activity occurring (e.g. Cas et al. 1990; Fiske et al. 1998; Chap. 4).

However, although some processes and characteristics of subaqueous and subglacial felsic lavas are similar to those

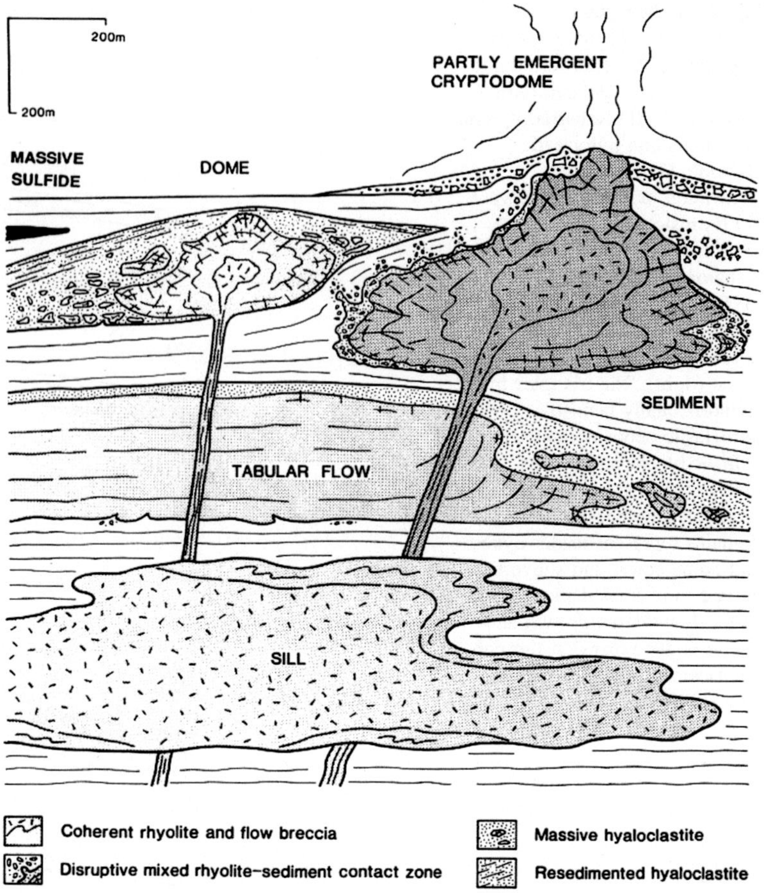

Fig. 6.22 Schematic representation of the form of subaqueous viscous felsic domes, cryptodomes, lavas, and syn-depositional intrusions (From Allen 1992)

erupted subaerially, and described above, the obvious difference in subaqueous and subglacial settings is the cooling and quench fragmentation effects of the ambient aqueous environment (Chaps. 3, 4; Pichler 1965; Cas and Wright 1987; Yamagishi 1991; Cas 1992; Cas and Giordano 2014; White et al. 2015; Soriano et al. 2016; Cas and Simmons 2018).

6.10.1 Constraints on Formation of Subaqueous Domes and Lavas

The principal constraints on the formation of subaqueous and subglacial felsic lavas and domes are that the erupting magma volatile content is low, and/or that water/ice depths are high enough so that the hydrostatic/glaciostatic pressure suppresses explosive expansion of exsolved magmatic volatiles, and also suppresses explosive superheating of external water that the magma is in contact with (McBirney 1963; Cas and Wright 1987; Cas 1992; Cas and Giordano 2014; Cas and Simmons 2018).

As discussed in Chaps. 3 and 4, hydrostatic pressure increases with water depth at a rate of ~ 1 MPa per 100 m of water/ice depth, and as the principal ambient pressure component on the sea or lake floor or subglacially, it can first suppress exsolution of volatiles from magma or limit the amount of volatiles that exsolve. Even if volatiles do exsolve, hydrostatic/glaciostatic pressure will reduce the rate at which volatile bubbles can expand. At water depths equivalent to or near the critical pressure of water (2,100 m, 22.1 MPa) exsolved magmatic water cannot expand rapidly enough to be explosive, and this can even be rare at water depths of 1,000 m (Chaps. 3, 4; McBirney 1963; Cas and Wright 1987; Cas 1992; Cas and Giordano 2014; Cas and Simmons 2018), unless volatile contents are very high. Similarly, superheated external water is also subject to that hydrostatic pressure, and explosive eruptions driven by superheated external water (phreatic) are almost impossible at water depths greater than 1,000 m, generally only in water depths of several hundred metres (Zimanowski and Büttner 2003; Cas and Giordano 2014; Cas and Simmons 2018). As a result, it is more likely that eruptions in deep water/ice will be effusive.

6.10.2 Forming a Coherent Crust Subaqueously

Even in shallower water effusive eruptions forming coherent lava can occur, if magma volatile content is low, and if a stable layer or film of steam forms at the interface between erupting magma and an external body of water through a process called film boiling (Mills 1984; Cas and Wright 1987; Chap. 4, Sect. 4.4.1, Fig. 4.4a). Heat transfer from the lava directly to water can be very rapid because water has a high heat capacity and thermal conductivity (Chap. 3; White et al. 2003; Cas and Giordano 2014; Cas and Simmons 2018. However, the rate of heat transfer can be reduced through formation of a stable, insulating film of steam that reduces heat flux from the lava to the water, called the Leidenfrost effect (Mills 1984; Cas and Wright 1987; Chap. 4, Sect. 4.4.1). At the interface between the lava surface and the water mass, a thin vapour film of steam can efficiently insulate the magma from direct contact with liquid water, initially allowing a visco-elastic crust to form at the surface of the lava. If the visco-elastic layer cools slowly enough under this insulating vapour film to absorb the cooling-induced contractional strain as it cools and contracts, then it forms a coherent, solid crust. This then insulates the interior of the lava from heat loss, allowing it to propagate as a coherent lava. If, however, the vapour film collapses, direct contact between liquid water and magma can cause the magma to cool so rapidly to glass that it experiences thermal shock. Contractional stresses in the glass cause it to shatter in situ through quench fragmentation forming hyaloclastite breccia, as discussed next in Sect. 6.10.3.

At water depths of 2 km in fresh water and 3 km in seawater, the hydrostatic pressure approaches the critical pressure of water. At that pressure, the vapour or steam film discussed above is so compressed it is a supercritical fluid. The steam that forms from the boiling/vapourisation of seawater is largely saline free, and so behaves like highly compressed supercritical fresh water. Its density, heat capacity, and thermal conductivity are intermediate between water and steam and the thermal gradient in the supercritical boundary layer is less than across the steam film at lower pressures and depths. Under these circumstances, the potential for quench fragmentation may be significantly reduced because thermal shock is less likely to occur. Hence, in very deep water, felsic lavas with coherent tops may be common (e.g. Embley and Rubin 2018).

6.10.3 Quench Fragmentation Forming Hyaloclastite Breccia

Once formed, the solid crust of a subaqueous lava may remain coherent or be subjected to autobrecciation if the interior of the lava continues to flow, imposing shear and extensional stresses on the solid crust, and causing it to undergo brittle fracturing into a breccia of blocks.

However, if the vapour film between the lava surface and the water mass collapses due to natural instabilities before a solid coherent crust can form, direct contact between hot magma and cold liquid water causes instantaneous supercooling and phase transformation of the magma surface to solid glass. The contractional strain cannot be accommodated by plastic or ductile stretching, and so, brittle contraction cracks propagate in the outer glassy crust, transforming the coherent magma into an in situ jigsaw-fit breccia of fine to coarse glassy debris called *hyaloclastite breccia*, through a process called quench fragmentation (Chap. 4, Sect. 4.4.1, Fig. 4.4b; Pichler 1965; Cas and Wright 1987; Cas 1992; Cas and Giordano 2014; Van Otterloo et al. 2015; White et al. 2015; Cas and Simmons 2018). If autobrecciation of the surface crust has already occurred, the autobreccia clast margins may also provide pathways for external water to permeate into the lava, which may then trigger an advancing front of quench fragmentation into the interior of the lava (van Otterloo et al. 2015). However, quench fragmentation may occur to varying degrees in subaqueous lavas (De Rosen-Spence et al. 1980). The processes of quench fragmentation and the characteristics of hyaloclastite have been described in detail in Chap. 4 (Sect. 4.4) and won't be repeated here. Although almost all subaerial felsic lavas are marked by a carapace of coarse autobreccia, subaqueous lavas will usually have a more substantial carapace of poorly sorted breccia, consisting of both autobreccia, but mostly abundant, finer grained hyaloclastite breccia and hyaloclastite sand(stone) (e.g. Figs. 4.2h, 4.5; also see Figs. 6.31, 6.32). This represents a significant difference between subaerial and subaqueous felsic lavas. De Rosen-Spence et al. (1980), Yamagishi (1991), and Yamagishi and Dimroth (1985) demonstrated that submarine lavas may consist of up to 60 to 70% hyaloclastite (and autobreccia), and only 30 to 40% coherent domains (see Fig. 6.34b), although others may be mostly coherent. Hyaloclastite breccia successions can form piles tens to hundreds of metres thick across several lava lobes.

Based on thermoremanent magnetisation (TRM) temperature determinations, Porreca et al. (2014) showed that hyaloclastite may form after lava flows stopped flowing and were cooling internally, even at temperatures well below the glass transition (down to 210–390 °C), suggesting that hyaloclastite formation in thick lavas may occur over most of the cooling history, as a result of the progressive access of sea water towards the lava interior by development of a complex network of contraction fractures. This explains why the thickness of felsic hyaloclastite may be 10–100 + m at individual lava complexes (Figs. 6.28).

Subglacial felsic lava eruptions are subject to some unique confining constraints due to the strength of the ice into which magma is erupted subglacially (at least in the initial phases of the eruption, before a significant volume of ice is melted into water), which affects the shape of subglacial volcanic edifices (Smellie 2013, 2018; Edwards et al. 2015). However, in general, there is similarity of processes (hydro- and glaciostatic pressure, cooling-quenching effects of the ice and water) between subaqueous and subglacial settings because of the meltwater envelope and pond that forms when hot magma melts the base of the glacier that it is erupted into (Furnes et al. 1980; de Rosen-Spence et al. 1980; Tuffen et al. 2001; Edwards et al. 2002, 2015; Smellie 2006, 2013, 2018; Chap. 16).

6.10.4 Effusion Rates

There is very little data available specifically on effusion rates of subaqueous felsic lavas, but available rates are comparable to those from subaerial settings (Table 6.1). Based on detailed bathymetric imaging of dacite lavas erupted in historical times (1866–1870, 1925–1928, and 1939–1941) at water depths of ~300 m on the submarine flanks of Nea Kameni dacite lava volcano in the caldera of Santorini volcano, Nomikou et al. (2014) calculated volumetric eruption rates of ~2 to 3 $m^3\ s^{-1}$ (Table 6.1).

Calculations from the 2012 submarine eruption of 11 rhyolite domes and 5 lavas along a fissure vent system around and below the rim of the wholly submarine Havre volcano caldera in the Kermadec Arc, at water depths of ~900 m to 1300 m (Carey et al. 2018; Manga et al. 2018; Ikegami et al. 2018), indicate minimum effusion rates between 0.14 and 14.4 $m^3\ s^{-1}$ (Ikegami et al. 2018; Table 6.1), with lava effusion for the whole field thought to have lasted about 90 days (Carey et al. 2018).

6.10.5 Factors Affecting the Rheology of Subaqueous Felsic Domes and Lavas

In general, the properties of subaqueously erupting felsic magmas are similar to counterparts erupting subaerially. This is because their compositions and eruption temperatures are generally similar. However, two factors could affect their viscosity and therefore the flow behaviour of erupted subaqueous domes and lavas. First, a significant factor affecting the viscosity of all magmas is the dissolved magmatic water content (Chap. 2). The greater the (retained) dissolved water content, the lower the magma viscosity, whereas the more magmatic water that exsolves, the higher the viscosity becomes. Ambient pressure determines how much magmatic water exsolves and remains dissolved in a magma as it ascends and is erupted (Chaps. 2–4). Therefore, the greater the water depth, the higher the ambient hydrostatic pressure at a subaqueous vent. As a result, less magmatic water will have exsolved from the magma at the time of eruption, and the magma viscosity will be less than for a compositional counterpart that erupts subaerially under low-ambient atmospheric pressure. Felsic lavas erupted in deep water should therefore be expected to be more fluidal, be able to flow further and develop lower aspect ratios than subaerial counterparts if erupted in deep water (at least hundreds of m) due to the effects of high hydrostatic pressure.

However, counteracting the effects of high hydrostatic pressure in producing lower than expected viscosity is the cooling effect of the water mass, which will lead to the formation of a solid crust more quickly than in subaerial settings, and should cause the overall cooling of the lava more quickly than in subaerial settings (Ikegami et al. 2018). Since temperature is a major factor in determining the viscosity of magma, this cooling effect, initially, in forming an outer crust and in causing more rapid cooling of the whole lava should cause viscosity and yield strength (Nomikou et al. 2014) of subaqueously erupted lavas to increase more rapidly than for subaerial counterparts. This may also affect the aspect ratio of subaqueously erupted lavas.

6.10.6 Effects of Topography on Lava Rheology, Morphology, and Aspect Ratio

The 2012 submarine rhyolite eruption of the wholly submarine Havre volcano, Kermadec Arc (Carey et al. 2018; Ikegami et al. 2018; Manga et al. 2018), allows us to address this topic. Nine rhyolite lava domes and one lava flow were erupted at water depths ~900 to 1000 m on the rim of the summit caldera, and four rhyolite lavas and one dome were erupted from a fissure at depths of 1200–1300 m on the inner caldera wall on a slope of 40°. This illustrates that as in subaerial settings, topography, especially slope, can have a major effect on the form of erupted lavas (Fig. 6.23a–c; Plate 6.1c, d; Ikegami et al. 2018). Ten rhyolite domes have volumes ranging from 0.0003 to 0.1115 km^3 and five lavas have volumes ranging from 0.0023 to 0.0334 km^3. All but one of the eruptions on the caldera rim produced domes, whereas where magma was erupted onto or reached the slope, they formed lavas that flowed down the slope. Given that the volumes of domes and lavas are comparable, slope was the significant factor in determining if the erupted form was a dome or a lava.

The nine rhyolite domes erupted on the caldera rim have typical lava dome morphologies—they are axisymmetric, steep sided, and have surface spines and high aspect ratios (Plate 6.1d; Ikegami et al. 2018). Ikegami et al. (2018), however, found that one lava dome (G) had a lower relief than the others, and instead of spines it has surface ridges, suggesting lower viscosity and higher effusion rate were

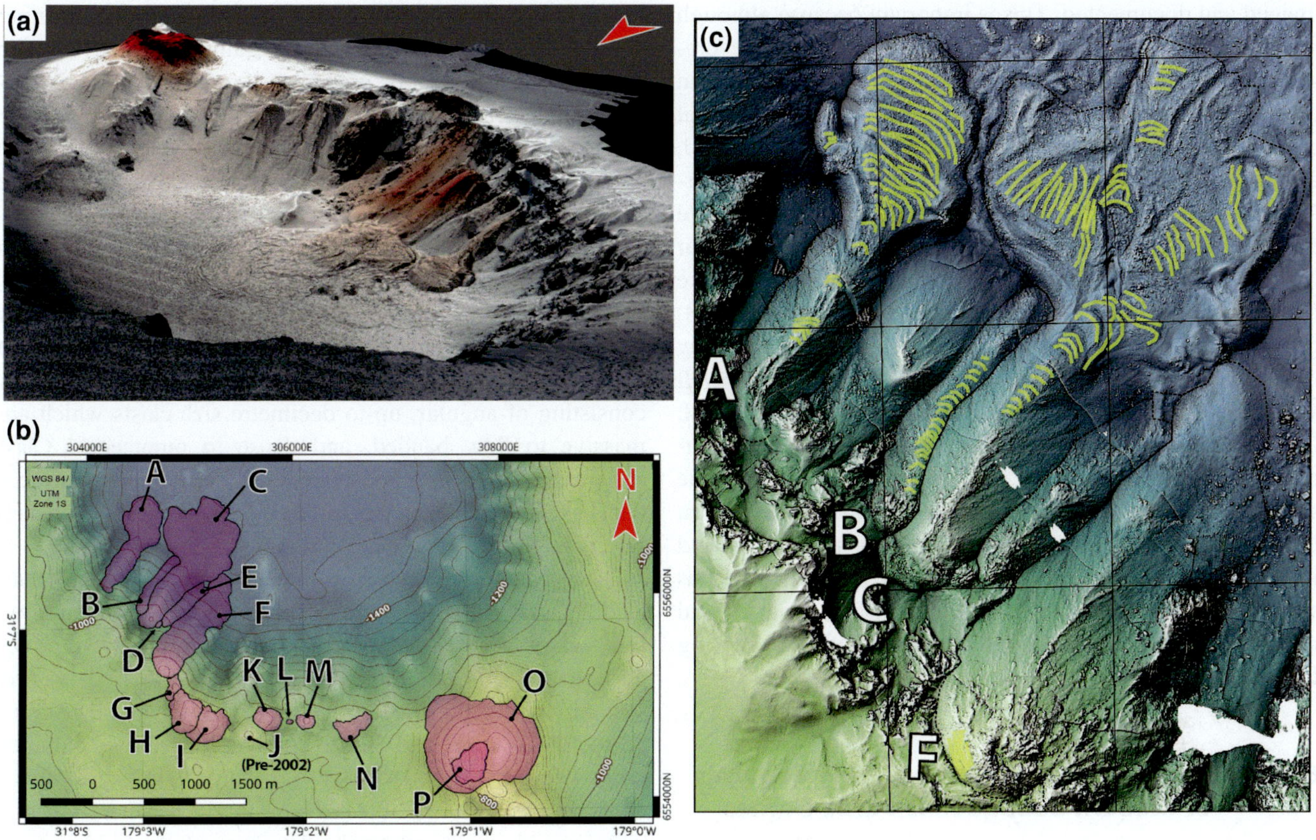

Fig. 6.23 Features of the submarine rhyolite domes and lavas erupted in 2012 from the wholly submarine Havre volcano, Kermadec Arc, southwest Pacific Ocean, part 1. **a** High-resolution digital elevation model (DEM) constructed from very detailed bathymetry data of the southern rim and floor of the submarine Havre volcano summit caldera after the 2012 rhyolite dome and lava flow eruptions. Note high aspect axisymmetric lava domes on the rim and lobate lava flows down the caldera slope and floor. **b** Map and bathymetric contour map of the area shown in (**a**) highlighting the nearly equidimensional lava domes versus the lobate lava flows, especially lavas A and C, which have spread as fans once they flowed from the steep 40° caldera wall slopes onto the flat caldera floor. **c** Contrasting morphology of the lobate lavas, where they are located on the steep caldera wall slope (narrow, steep-sided, levees, and central channel) compared with where lavas A and C fan out on the flat caldera floor. Transverse ridges or ogives are marked in yellow. (From Ikegami et al. 2018, through Creative Commons Attribution 4.0 International licence)

responsible, as simulated experimentally by Fink and Griffiths (1998; see Fig. 6.4d). The domes are generally about 200 m in diameter and <80 m high, although the composite OP domes are ~700 m in diameter and up to 125 m high. The lavas have a maximum thickness ranging from 30 to 100 m and are from 540 to 1350 m long (Ikegami et al. 2018). The lava flows are narrow and steep sided where they flowed down the caldera wall slope of the Havre volcano, (Fig. 6.23c). Two lavas spread in a fan-like form after the break in slope onto the flat caldera floor at a depth of 1500 m (Fig. 6.23c; Plate 6.1e), and lava A has two breakout lobes (Plate 6.1e).

Relatively little accurate data is available on the aspect ratio of subaqueously erupted lavas because of limitations of access to the deep-water settings in modern oceans, as well as the incomplete preservation of ancient subaqueous lavas, due to erosion and deformation of the rock record. However, Nomikou et al. (2014) calculated the aspect ratio T/D of the submarine dacite lava erupted in 1570–1573 AD at Nea Kameni Santorini, Greece to be 0.101. Using data from Ikegami et al. (2018) we have calculated the aspect ratio of domes and lavas from the 2012 wholly submarine eruption of Havre volcano, Kermadec Arc, to range between 0.0625 and 0.263. Both results coincide with data from subaerial felsic lavas (Fig. 6.3), suggesting that the counterbalancing effects of retained magmatic water in deep water (lowering viscosity) and more rapid cooling (increasing viscosity) on magma rheology could more or less cancel each other out.

6.11 Modern Submarine Felsic Domes and Lavas

Unlike modern subaerial felsic domes and lavas, the surface features of modern subaqueous domes and lavas are not readily visible and have only rarely and only recently been imaged and documented. This is important because although many ancient subaqueous felsic domes and lavas are preserved in the rock record, their surfaces are not preserved in three dimensions, and often in two-dimensional cross-section, the features are only intermittently exposed, and in many cases have been eroded or subjected to deformation and shear strain. The recent studies of the rhyolite domes and short lavas erupted during the 2012 eruption of the wholly submarine Havre volcano, in the Kermadec Arc (Carey et al. 2018; Ikegami et al. 2018; Manga et al. 2018) provide the most information based on extremely high-resolution bathymetry ($\sim$5 m contour spacing), as well as photography and sampling. The paper by Ikegami et al. (2018) is a seminal study that has provided in-depth documentation and interpretation of these lavas. In addition, the study of recently discovered dacite lava flows, up to 10 km long on the floor of the northeast Lau Basin (Embley and Rubin 2018), although not as detailed as the Havre lavas study, provides great insight into the surface features and morphology of very deep-water erupted, and far-flowing felsic lavas.

6.11.1 Morphology, Dimensions, and Features of Modern Subaqueous Felsic Domes

The Havre volcano lava domes were erupted and emplaced at water depths of $\sim$900 to 1,000 m, and are generally steep-sided, high aspect ratio domes (Plate 6.1c; Figs. 6.23, 6.24a), which range in height from 50 to 125 m, and in diameter from <50 to 720 m. Their volumes range from 3×10^5 to 11×10^7 m^3. They have irregular tops with steep spines up to 40 m height (Plate 6.1d; Fig. 6.24a; Ikegami et al. 2018). These domes are like subaerial Pelean domes (Ikegami et al. 2018). Only dome G has a lower aspect ratio and low ridges rather than spines (Ikegami et al. 2018). The tops of the domes are all marked by *surface breccia* and *talus aprons* (Fig. 6.24b), but some also expose domains of *coherent rhyolite*, including *coherent pumice carapaces* and *spines*, and slightly denser rhyolite. The coherent pumice has >50% *vesicles* (cf. Fig. 6.24d; Ikegami et al. 2018). Dome N is partially intrusive and partially extruded, forming a dome–cryptodome complex (Ikegami et al. 2018). The dome rocks are weakly *porphyritic* (<5 modal %; plagioclase, pyroxene, Fe-Ti oxides, minor quartz), but have abundant groundmass microlites of mostly plagioclase (>60 modal %), except for the pumiceous carapaces which are microlite poor (<5 modal %) (Ikegami et al. 2018). Geochemically the eruptive products have SiO_2 contents of $\sim$70–72 wt% (Carey et al. 2018).

According to Ikegami et al. (2018), the *surface breccias* on both the 2012 domes (Fig. 6.24b) and lavas (Fig. 6.24d) from the Havre volcano are monomictic rhyolite breccias consisting of angular, up to decimetre size clasts which are massive to flow banded, and dense to pumiceous. Clast shapes are polyhedral and bounded by curviplanar surfaces and have been interpreted to have formed through quench fragmentation and autobrecciation.

The lava flows erupted on the submarine Havre volcano caldera wall slope at depths of 1200–1300 m are narrow, steep sided, and have *levees* and a *central channel*, suggesting a channel or more likely internal tube-fed magma system (Fig. 6.23c; Ikegami et al. 2018). Lavas range from 33 to 100 m thick, and 540 to 1350 m long. Their volumes range from 2.3×10^6 to 33.4×10^6 m^3. *Levees* are rubbly and slope at 30 to 40°. At the base in slope, where lavas A and C reached the relatively flat floor of the caldera at a depth of $\sim$1500 m, they spread as fans and developed prominent *surface ridges* or *ogives* (Fig. 6.23c; Plate 6.1e; Ikegami et al. 2018), which are 1 to 2 m high and have wavelengths of 10 to 30 m (Ikegami et al. 2018). Less prominent ogives are also present on the surfaces of lavas A, B, and C where they are narrow and on the caldera wall slope (Fig. 6.23c). The lavas are also marked by exposed domains of *coherent rhyolite*, including a *pumice carapace* (Fig. 6.24c), but mostly by surface *autobrecciated* and *quench fragmented breccia tops*, and some have marginal *talus aprons*, as well (Fig. 6.24d). Surface samples collected from lava A include both *highly vesiculated* pumice carapace (Fig. 6.24e) and denser, *glassy rhyolite*, showing diffuse *flow banding* (Fig. 6.24f). The mineralogy of the lavas is the same as the domes (see above).

Fig. 6.24 Features of submarine rhyolite domes and lavas erupted in 2012 from the wholly submarine Havre volcano, Kermadec Island Arc, southwest Pacific Ocean, part 2. **a** Spire of coherent pumiceous rhyolite on the surface of lava dome OP (From Carey et al. 2018). **b** Resedimented talus apron of dome OP, consisting of pumice and dense lava debris (From Carey et al. 2018). **c** Coarsely autobrecciated coherent pumiceous carapace of lava flow A (From Ikegami et al. 2018). **d** Marginal talus apron of pumiceous and dense lava debris, lava A. **e** Highly vesicular pumice carapace of lava A. **f** Denser glassy, diffusely flow banded rhyolite from the surface of lava A. (Images **d**–**f** courtesy of F. Ikegami). **g** Stacked, giant 1 m pumice hyaloclast blocks on the caldera floor at 1500 m water depth, resulting from fallout from suspension after limited buoyancy and dispersal (From Carey et al. 2018). **h** 2 m diameter giant pumice block with curviplanar bounding surfaces and a breadcrust surface texture, both resulting from quench fragmentation of an erupting highly vesiculated rhyolite magma at water depths of $\sim$900 m (From Carey et al. 2018). **i** Cartoon representing the quench fragmentation origins of giant pumice blocks derived from the 2012 eruption of a highly vesiculated rhyolite dome. This highlights their buoyancy, flotational dispersal after detachment from the dome margins, and sedimentation from suspension of some around the vent, whereas others floated to the sea surface, where they were rafted away by currents and wind around the southwest Pacific Ocean. (Modified from Manga et al. 2018) ▶

(a)
(b)
(c)
(d)
(e)
(f)
2 cm
2 cm
(g)
(h)
(i)
wind
gas plume
- mostly steam
floating
raft pumice
cooling joint
propagation
buoyantly
rising
pumice
clasts
sinking pumice
clasts
current
erupting
vesicular
magma
<1 m
giant
pumice
>1 m
~15 m/s
~40 m

Binns (2003) also documented the occurrence of outcrops of *highly vesiculated* (66%) pumiceous peralkaline rhyolite lava poking through a sediment fill at depths of 800 to 950 m in the Woodlark Basin, Papua New Guinea. The surface is coherent but in places has a network of fractures, some polygonal in form and representing *columnar jointing*. Some *ropey surfaces* also occur, as well as *spines* and *domes* up to 20 m high (Binns 2003). Ropy textures reflect relatively low viscosity.

Embley and Rubin (2018) described several "young" dacite lava flow fields mapped on the floor of the back arc, (northeast) Lau Basin at water depths of $\sim$2,500 to 2,800 m around West Mata volcano, Niuatahi Ridge and Motutahi volcano, using multibeam sonar, dredge sampling, and seafloor deep tow photography (Fig. 6.25a, b). These lavas are not covered by sediment and so the surface features could be imaged and sampled. The lavas are almost *aphyric*, with <1% plagioclase microlites (Fig. 6.25c), <1% pyroxene microphenocrysts, and they have SiO_2 content of $\sim$65 to 67 wt%. Notably the lavas are *sparsely vesicular* (<1 to 3 modal%), with deeper water ones being less vesicular than shallower water ones. Lavas have *thick glassy (chilled) rinds* (15 to 50 cm thick; Fig. 6.25c), some with ropy surface structures (Fig. 6.25d) and are more crystalline internally.

There are several lava flow fields associated with the flanks of Niuatahi seamount (Fig. 6.25a) that seem to have been erupted in association with caldera formation. Their eruption probably caused caldera subsidence through magma withdrawal from under the seamount. These lavas are multi-lobate, they radiate out from the seamount for up to 6 km, they have areas from 13 to 55 km^2, they are >50 m thick, and some have <10 m high *surface domes* (tumuli?) and *ridges* (Embley and Rubin 2018). Their volumes are orders of magnitude greater than those of the Havre volcano, but the exact dimensions and volumes of individual lavas cannot be determined.

The lava flow fields north of Niuatahi seamount have areas from 11 to 144 km^2, and maximum plan view dimensions up to almost 20 km long (Fig. 6.25a, b). Some of these flow fields seem to consist of stacked lavas or flow lobes, they are *multi-lobate*, and several flows show rounded *pillows* with diameters of several m (Fig. 6.25e), pillow expansion or *spreading cracks* (Fig. 6.25d), flatter lobes and brecciated ropey textures and flow breccias (Embley and Rubin 2018). Coherent *ropey flow surfaces* are also preserved (Fig. 6.25d), suggesting low-viscosity flows. Some *autobrecciated* or *quench fragmented spines* also occur (Fig. 6.25g). *Domes* and *ridges* and some *columnar jointing* were also visible. *Pits* of varying dimensions suggesting lava drain-out and tube formation inside the lavas also occur. Some lava lobes are 40 to 60 m thick, and lava field volumes seem to be from 2.4 to 7.2 km^3 (Embley and Rubin 2018). One lava lake may also have been found.

Based on the aphyric nature, low vesicularity, megapillow forms, and ropy surface textures, Embley and Rubin (2018) propose that the deep-water dacites were erupted as low-viscosity lavas with viscosities about the same as andesite lavas ($\sim 10^4$ Pa s). The eruption temperature would have been near-liquidus temperature and high hydrostatic pressure at water depths of >2,500 m would prevent most of the volatile content from exsolving, so maintaining low magma viscosity. The lavas behaved in some ways like viscous basaltic submarine lavas (Chap. 5), but the preserved features indicate a more viscous state (e.g. mega-pillows, thicknesses of $\sim$50 m). Similar mega-pillowed dacite lavas have also been reported in the rock record, with similar explanations. The large erupted volumes for the northeast Lau Basin dacite lavas ensured long-distance (>10 km, potentially 15 to 20 km) flow. Similar flow morphologies, features, and explanation for the formation of pillows in dacitic lavas in the rock record have been proposed for submarine dacite mega-pillowed lavas and associated hyaloclastite breccia in the Archean Abitibi Greenstone Belt in Canada (Dinel et al. 2008), and submarine Ordovician rhyodacite lavas, and hyaloclastite breccia in North Wales, United Kingdom (Bevins and Roach 1979).

6.11.2 Coherent and Autoclastic Pumice, and Resulting Pumice Rafts

There is growing evidence that many deep-sea felsic and subglacial felsic lavas can have highly vesiculated, coherent to autoclastically fragmented pumice carapaces, without any evidence of associated explosive activity (Furnes et al. 1980; Cas and Wright 1987; Cas 1992; Scutter et al. 1998; Binns 2003; Allen et al. 2010; Carey et al. 2018; Ikegami et al. 2018; Manga et al. 2018). Such pumice can only form through slow, non-explosive exsolution and vesiculation of volatiles under conditions of high hydrostatic pressures due to water and ice depths. Such conditions suppress instantaneous explosive expansion of vesicles but allow slow expansion under conditions of low differential pressure between ambient hydrostatic pressure and gas pressure in lava vesicles (Chaps. 4, 9).

The 800- to 950-m-deep coherent pumice lava surface documented by Binns (2003) in the Woodlark Basin has a vesicularity of 66%. Vesicles range from 20 to 500 μm in size in thin section, but in hand samples some are up to 3 cm across and show signs of coalescence (Binns 2003). Vesicles range from equant to stretched as a result of shear associated with flow.

As described above, the domes and lavas erupted from the submarine Havre volcano have coherent pumice and autoclastic (autobreccia, quench fragmented hyaloclastite) pumice breccia surfaces that have not resulted from

Fig. 6.25 Features of extensive deep marine dacite lava flows from the northeast Lau Basin, southwest Pacific Ocean. **a** Map showing extent of the dacite lava flow fields (blue outlined in red = high back-scatter). N = Niuatahi seamount with its large silicic submarine caldera and central cone Motutahi (Mo); NELSC = Northeast Lau spreading centre; NR = Niuatahi Ridge; MB = Mata Basin; WM = West Mata volcano. The volcanic knoll "BoCo", and the chain of volcanoes to the east of the Mata Basin (including West Mata) have erupted boninite magmas. For details of other symbols (mainly dredge samples and camera tows), see Embley and Rubin (2018). **b** Oblique 3D bathymetric images; white lines show outlines of dacite lava flow fields. Top = view southwest towards Niuatahi seamount with its caldera and inner cones; Bottom = LL-B flow field again looking southwest. **c** Dredge sample of glassy, aphyric, non-vesicular dacite flow top from lava flow field LL-D. Black scale bar is 25 cm long. **d** Dredge sample of ropy surface folds from large lava field LL-B. **e–g** Grey-scale images from camera tows: **e** Dacite pillow lobes up to 2 m in diameter from lava flow field LL-C; **f** Triple point expansion or spreading cracks on the surface at the end of a pillow tube, 3 to 4 m in diameter, lava flow field LL-C; **g** Autobrecciated spinose surface protrusion (pumiceous?), ~5 m across, lava flow field LL-B. (From Embley and Rubin 2018)

explosive fragmentation (Carey et al. 2018; Ikegami et al. 2018; Manga et al. 2018). The pumice has >50 modal% vesicles and has formed through slow, non-explosive exsolution and vesiculation of volatiles under high hydrostatic pressures of ~ 10 MPa, even though the magmatic H_2O content is high (5.8 wt%; Manga et al. 2018).

In addition, Havre lavas A, B, C, and E are capped by a novel deposit of "giant pumice" blocks, but not the domes, indicating that the lavas were erupted first, before the giant pumice-forming event, whereas the domes were erupted after it (Ikegami et al. 2018). The giant pumice deposit also occurs across and outside the caldera floor (Fig. 6.24g; Carey et al. 2018). The giant pumice blocks range from >1 m in dimension to at least 9 m in diameter and have curviplanar surfaces commonly marked by polygonal networks of breadcrust textures and cooling contraction cracks (Fig. 6.24h; Carey et al. 2018). They are not pyroclastic in origin but have been interpreted to have formed by quench fragmentation of an erupting highly vesicular dome, with big, fractured, highly vesicular (gas filled) clasts becoming buoyant, and rising into the water column (Fig. 6.24i; Manga et al. 2018). Some became water-logged and sank to the seafloor inside and outside the caldera, whereas others, and smaller pumice clasts rose to the sea surface and were then dispersed as a floating raft of pumice (Fig. 6.24i) that delivered rafted pumice fallout clasts (*not* pyroclastic fallout) around the southwest Pacific seafloor (Jutzeler et al. 2014), including some reaching the Australian east coast beaches.

Pumice clasts analysed from the 2012 Havre pumice raft, derived from the quench fragmentation of erupting pumiceous rhyolite, have 68 to 89% vesicles, which are spherical to ovoid, a bimodal vesicle size population, vesicles <10 μm are lacking, and vesicles have thin glassy walls which lack microlites (Rotella et al. 2015). These characteristics are similar to other deep-water erupted pumice clasts from Healy, Raoul, and Macauley volcanoes, and they all have low bubble number densities (Rotella et al. 2013, 2015). This contrasts with subaerial pyroclastic pumices, which often have larger irregular coalesced vesicles, have many micro-vesicles and many microlites, and high bubble number densities, suggesting that in deep water, subaqueous hydrostatic pressure and cooling chilling effects limit the exsolution and vesiculation process, in this case also suppressing explosive activity (Rotella et al. 2015).

Allen et al. (2010) also documented both a coherent pumice carapace and pumice hyaloclastite blocks on the summit and slopes of submarine lava domes associated with the Sumisu volcano, Izu-Bonin Arc, as well as some pumice interpreted to have resulted from explosive activity (Fig. 3.9 b–e). Rotella et al. (2013) reported abundant submarine pumice from Macauley volcano in the Kermadec Arc, and also proposed a non-explosive origin for this pumice, proposing that pumice clasts detached from erupting, vesiculated magma/lava domes by ductile stretching and detachment, rather than quench fragmentation.

6.11.3 Internal Structure of Modern Subaqueous Domes and Lavas

Although modern seafloor felsic domes and lavas do not provide views of their internal structure, there is an uplifted historical example that does, in part. The eruption of a rhyolite-dacite lava dome erupted in 1934–37 along the margin of the largely submarine Kikai caldera, Kyushu, Japan, commenced in a submarine setting at a water depth of 300 m, the dome then emerged above sea level and continued to grow both subaerially and subaqueously (Maeno and Taniguchi 2006). The uplifted emergent part preserves ogives, ramp structures, flow folds, a microcrystalline core, a pumice carapace, autobreccia, hyaloclastite, and a fault zone. According to Maeno and Taniguchi (2006), the flow folds, which are expressed as ridges and troughs on the flow surface, coincide with ramp structures and faults. Maeno and Taniguchi (2006) interpret the fault zones at the margins of the lava dome to have been caused by lateral growth and propagation of the lava, onset of ramping and attenuation, much as for subaerial felsic lavas summarised above in Sects. 6.7.1 and 6.7.2 (cf. Fig. 6.18f, g).

In contrast, on the $\sim$4–2.5 Ma uplifted submarine rhyolite complex on the island of Ponza in the Tyrrhenian Sea, Italy, De Rita et al. (2001) described sets of conjugate normal faults located in quench fragmented lavas and domes. The succession consists of older hyaloclastite, representing quench fragmented lavas and domes, intruded and cross-cut by younger obsidian feeder dykes and laccoliths to a younger eroded dome complex (Fig. 6.26; Scutter et al. 1998; De Rita et al. 2001). The sets of conjugate normal faults are located in the older quench fragmented lavas and domes, dipping back towards the younger dykes and laccoliths (Fig. 6.27a, b). At the outcrop scale, these are shown to be anastomosed shear zones (Fig. 6.27c), which allowed fluid escape during faulting and compaction, and were interpreted to have originated from collapse of the summits of domes as they inflated and expanded (Fig. 6.27d).

6.12 Insights About Subaqueous Domes and Lavas from the Rock Record: Lithofacies and Associations

Although we have learned much from the recent studies undertaken on modern seafloor felsic domes and lavas summarised above, including understanding of overall geometry, volumes, surface features, effusion rates, and

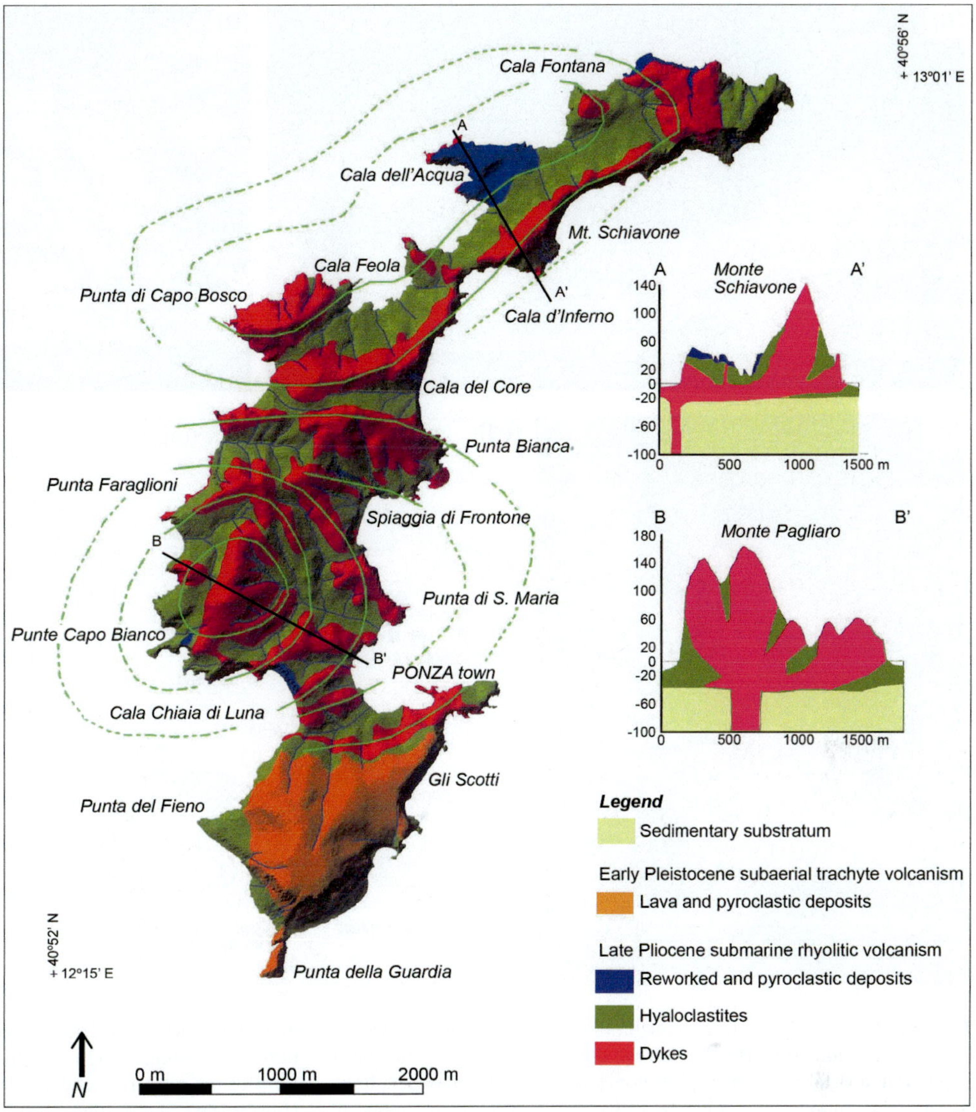

Fig. 6.26 Map and interpretative cross-sections for the Pliocene–Pleistocene Volcanic Island of Ponza, Tyrrhenian Sea, Italy. The rhyolite "dykes" (in red/pink) are late and clearly cross-cut a slightly older, but co-magmatic and penecontemporaneous rhyolite lava flow system (green). The "dykes" range in geometry from subvertical tabular dykes, to flaring cryptodomes and sub-horizontal sills (or "dills"; Chap. 8), and were probably the feeders for exogenously erupted domes, the tops of which have been eroded following uplift of the submarine complex above sea level. Most of the dykes were probably rootless, connected to each other by a branching magma plumbing system. The trachytic volcanics (orange) are subaerial and post-date uplift. (From De Rita et al. 2001)

effects of water depths and water properties in determining textural features and constraints, modern subaqueous domes and lavas cannot provide insights into their internal facies, their relationships, and the overall facies architecture. This requires careful study of the cross-sectional exposures of ancient submarine domes and lavas which have been uplifted and variably eroded to produce cross-sectional exposures. Thankfully there have been many such studies, which have documented the internal facies, their relationships, and the overall facies architecture. This understanding can now be combined with what has been recently learned from modern submarine domes and lavas to produce generalised facies models, as follows.

Although the range of facies and associations of facies are similar, the facies architecture varies spatially between the different subaqueous felsic lava dome types and lava flows from one example to the next (e.g. Fig. 6.22). We will consider the facies and their characteristics together, according to their spatial organisation in domes and lavas.

In addition, given the propensity for obsidian to devitrify in time through diagenetic modification, ancient examples may not preserve the original evidence of the original proportions of stony to obsidian textures. A texture such as spherulites can be used as a proxy for the existence of obsidian originally if it is preserved.

6.12.1 Coherent Facies and Its Relationship to Enclosing Autoclastic Facies

As for subaerial felsic domes and lavas, subaqueous lava domes and lavas are fed by a feeder dyke or cylindrical conduit (Fig. 6.28a–g). The magma in the conduit is largely *coherent*, although it could be subject to both ductile and

Fig. 6.27 Syn-emplacement structural features of the Pliocene–Pleistocene submarine rhyolite lavas, domes, and feeder intrusions exposed in coastal cliffs after uplift, Ponza Island, Tyrrhenian Sea, Italy. **a** Two sets of normal conjugate faults (red arrows show sense of motion) diverging from the centre of the field of view, occurring in alternating grey and white jigsaw-fit, quench fragmented, flow banded rhyolite lava. Each fault set dips towards a major younger dyke (dark brown) at each side of the field of view. **b** Close-up view of the red box area in (**a**) showing a set of normal faults dipping towards a major cross-cutting dyke to the left of the picture. One fault truncates an irregular young brown, intrusive lobe into the older pale grey-white flow banded and quench fragmented lava. De Rita et al. (2001) interpret these faults to result from inflation of the sequence around the major intrusion, leading to gravitational normal fault collapse. The irregular form of the intrusive lobe indicates that the grey hyaloclastite was not consolidated at the time "late" dykes and intrusions were emplaced. **c** Close-up view of one fault, showing cataclastic textures in a wide shear zone. **d** Interpretation of the structural evolution of a dome with a substantial hyaloclastite carapace intruded by dykes, which are often "flower-like" rootless dykes. These dykes established an extensional stress field at the summit of the dome as it inflated, leading to gravitational collapse normal faults. Formation of such fault sets would be enhanced by late magma withdrawal and dyke deflation. (From De Rita et al. 2001)

brittle fracturing (shear fracturing-autobrecciation, limited quench fragmentation), especially along the walls of the conduit. The magma in the conduit and that which erupts is viscous and can also undergo supercooling to form obsidian (Fig. 6.28c, e–g, i, j) and/or undergo crystallisation to form a stony textural domain. The feeder conduit feeds the coherent core of the dome or lava above the basin floor. The outer margin of the erupting dome undergoes autobrecciation and quench fragmentation to varying degrees, producing an outer envelope of autoclastic breccia (Fig. 6.28a–c, h–j). Magma from the conduit and core can then propagate upwards into the quench fragmented hyaloclastite pile of the growing lava dome or flow producing an irregular, wavy contact (Fig. 6.28b), and highly irregular apophyses or lobes may intrude into the hyaloclastite and then themselves be subject to varying degrees of quench fragmentation (Fig. 6.28d–j). This irregular form is due to the lack of strength of the host unconsolidated hyaloclastite pile, and because the feeder intrusion itself is being quench fragmented by the water in the saturated hyaloclastite pile it is intruding.

(a)
(b)
(c)
(d)
(e)
(f)

◄ **Fig. 6.28** Characteristics of coherent feeder dykes, core zones, and lobes in felsic subaqueous lava domes and flows. **a** Profile through uplifted Miocene submarine dacite lava dome with late feeder dyke (dark grey, centre right) cutting across altered hyaloclastite breccia (pale grey) and spreading upwards into a coherent exogenous dome. Note the horizontal columnar joints at the margins of the dyke. Tsukotan, Hokkaido, Japan. **b** Gradation from coherent massive to flow banded feeder dyke of a Miocene submarine dacite lava dome to in situ, jigsaw-fit autobreccia and hyaloclastite breccia. Note the breccia becomes finer away from the coherent core, reflecting more intense quench fragmentation. Kariba, southwestern Honshu, Japan. **c** Top of feeder obsidian rhyolite dyke (dark grey-black) passing upwards into in situ jigsaw-fit hyaloclastite breccia with large (>1 m diameter) obsidian clasts (dark grey) within a finer jigsaw-fit hyaloclastite breccia (white). Cala de Fontana, Ponza Island, Tyrrhenian Sea, Italy. **d** Irregular Miocene andesite feeder dyke with apophyses along the margin, intruding a pile of andesite hyaloclastite breccia sourced from this and other feeder dykes/lobes. Kabuto Cape, Hokkaido, Japan. **e** Sub-vertical obsidian rhyolite dykes and a very small cryptodome (dark) cross-cutting sub-horizontally flow banded, in situ quench fragmented rhyolite lava (pale grey). **f** Irregular late obsidian rhyolite intrusion with apophyses, cross-cutting the older quench fragmented pale grey lava sequence consisting of hyaloclastite breccia that still preserves sub-horizontal flow banding. This body is either the remains of a cryptodome in the older lava hyaloclastite or the base of an exogenous dome erupted onto the older lava. Ponza Island, Italy. **g** Upward flaring obsidian rhyolite intrusion with apophyses (dark) cross-cutting older quench fragmented lava, which consists of pale jigsaw-fit hyaloclastite breccia in which continuous sub-horizontal flow banding is visible. The intrusions illustrated in (**e**)–(**g**) all show a yellow hydrothermal alteration halo extending from the margin of the younger dykes into the adjacent older quench fragmented lava. **h** Dacite lava lobe enclosed in more pervasively fragmented hyaloclastite breccia. Miocene, Gasutoh-jima, southwest Hokkaido, Japan. **i** Map of the rock platform at Kawashira, Hokkaido, Japan, showing the distribution of "necked-off" lobes of obsidian rhyolite surrounded by hyaloclastite breccia in a submarine rhyolite lava dome (From Yamagishi and Dimroth 1985). **j** Overview of part of the rock platform in the map area shown in (**i**), showing the black isolated "necked-off" obsidian lobes surrounded by white-to-grey hyaloclastite breccia

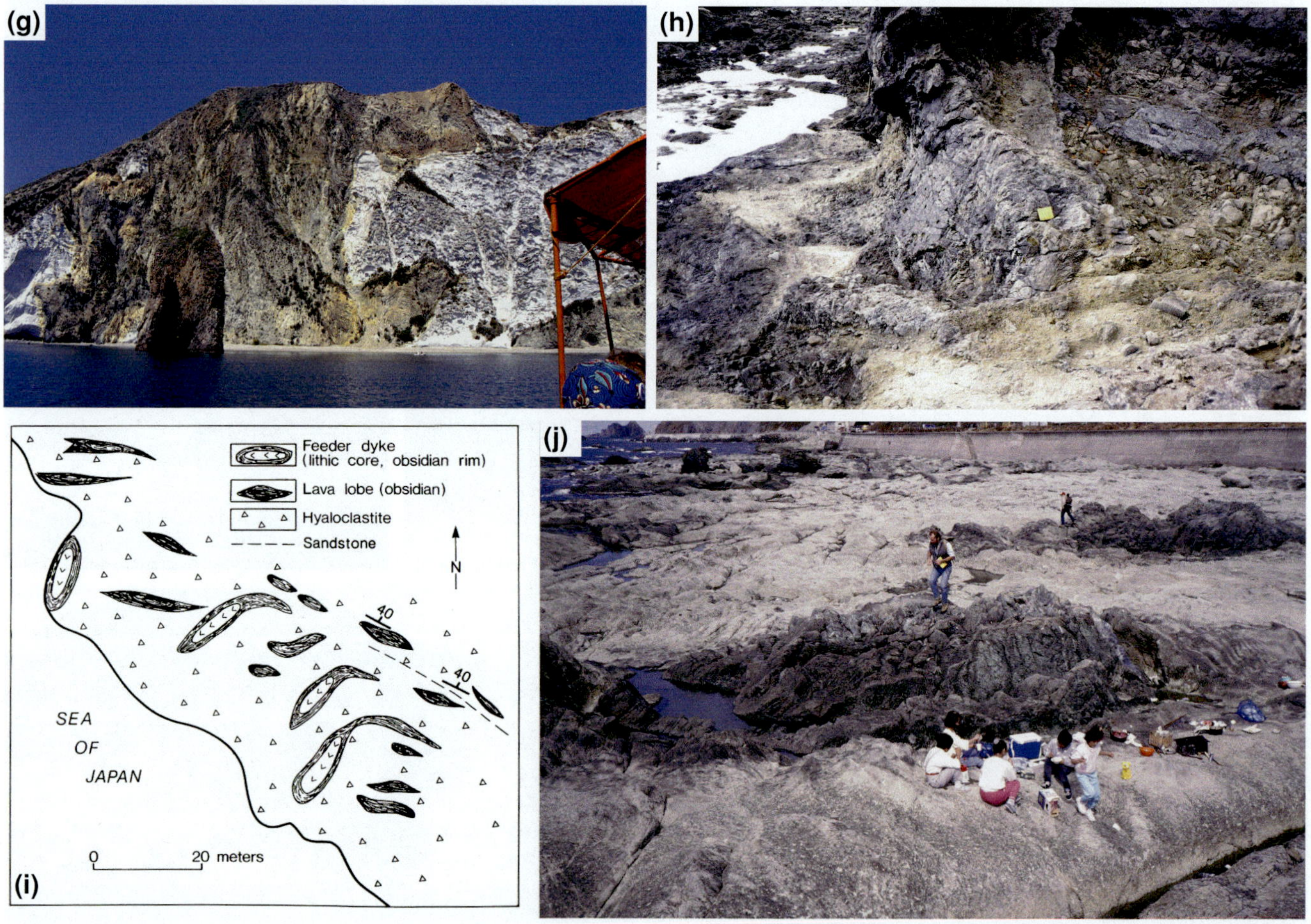

Fig. 6.28 (continued)

The geometry of the coherent facies domain of a subaqueous dome or lava may thus be (highly) irregular, producing a potentially complex architecture of coherent lobes and enclosing breccia, often with gradational contacts (Fig. 6.28b, h; Pichler 1965; Kano et al. 1991; Furnes et al. 1980; de Rosen-Spence et al. 1980; Yamagishi and Dimroth 1985; Yamagishi 1991; Cas 1992). Coherent lobes may be necked-off by quench fragmentation, leaving isolated lobes surrounded by hyaloclastite (Fig. 6.28i, j; Yamagishi and Dimroth 1985). Such transitions can also be documented in diamond drill core drilled through ancient domes during exploration for mineral deposits, such as VMS deposits (e.g. Fig. 6.31d, from the Kuroko deposits in Japan, in the Hokuroko District).

A growing hyaloclastite mound or carapace is a mechanically weak and water-saturated environment that allows significant extension and mechanical readjustment as coherent magma is injected into the unconsolidated hyaloclastite pile (De Rita et al. 2001). This is ideally demonstrated in the Ponza Island (Italy) lava and dome complex (Figs. 6.26, 6.27, 6.28e–g) and results in dykes up to several tens of m in thickness, characterised by irregular margins both in cross-section and in plan view, and irregular lateral apophyses and protrusions into enclosing hyaloclastite (Fig. 6.28e–g). On Ponza there are frequent intrusive cross-cutting contacts of "young", late, feeder dykes and conduits cross-cutting hyaloclastite formed during emplacement of early or "older" lavas and domes. In plan view, the dykes may form outward fanning concentric rings around the conduit region with geometries similar to cone-sheets (Figs. 6.26, 6.27d) and these were named "rootless dykes" by Pichler (1965) as they do not all propagate downward into the basement and do not reflect the orientation of the regional stress field in the Earth's crust at the time (De Rita et al. 2001). Their exposed geometries range from subvertical dykes to irregular cryptodomes (Fig. 6.28e), to upward flaring laccolith-like bodies (Fig. 6.28f, g), to inclined sub-horizontal sills or "dills" as we like to call them (Chap. 8). Due to the low confinement strength of the pre-existing hyaloclastite, dykes experience very complex magma paths, including repeated periods of magma injection and draining, as indicated by studies of their anisotropy of magnetic susceptibility (Aubourg et al. 2002; Porreca et al. 2015).

If erupting lava flows away from the vent area, the subvertical flow banding emerging from the conduit begins to ramp over and becomes more horizontal. As the lava flows, that flow banding may remain planar and horizontal (Fig. 6.29a) or become folded (Fig. 6.29b). Some lavas may remain almost wholly coherent (Fig. 6.29a), whereas others experience varying degrees of quench fragmentation, so that the top, bottom, and margins will consist of in situ jigsaw-fit to clast-rotated hyaloclastite brecccia, which enclose a coherent core (Fig. 6.28b), or even several coherent core lobes (Fig. 6.29c, e.g. Soriano et al. 2013). The coherent core is essentially the feeder tube for magma from the vent to the propagating flow front. During cooling the coherent core could develop columnar jointing that could be oriented perpendicular to the flow margins (Fig. 6.29a, c, d), or forms rosette radial column clusters (Fig. 6.29e). Rhyolitic and dacitic subaqueous lavas can form obsidian coherent domains.

The rock that solidifies in the conduit, the core zone of thick domes and lavas may be largely crystallised and analogous to stony lithofacies in subaerial domes and lavas, with a largely microcrystalline groundmass. It also grades into obsidian, and outwards into coherent pumice. Coherent facies can be massive, diffusely flow banded, or well flow banded, which reflects variations in the degree of internal laminar shear flow and the shear gradients.

If domains of coherent pumice form (Sect. 6.12.4), they could develop flow banding, and vesicles will be stretched tube vesicles (Fig. 6.29f). Both obsidian and pumice domains could also be subject to autobrecciation and quench fragmentation, forming pumice and obsidian autobreccia and hyaloclastite breccia (Fig. 6.29f). De Rosen Spence et al. (1980) have pointed out that the proportion of coherent-to-autoclastic facies can vary greatly from one lava flow to another (e.g. see Fig. 6.34b).

6.12.2 Autobreccia

The autobrecciaof subaqueous felsic domes and lavas will also be monomictic and coarse, like subaerial counterparts, dominated by angular to slightly abraded angular blocks, and angular-lapilli-sized debris (Fig. 6.30a; Chap. 4). In situ autobreccia will show jigsaw-fit clast textures (Fig. 6.30a, b), gradational into coherent lava in the core of the lava, and perhaps into clast-rotated texture at the lava surface, formed as the flowing lava causes clasts in the autobrecciated crust to rotate relative to each other. However, autobrecciation may be overprinted by and masked by cooling contraction quench fragmentation, which tends to produce a network of closely spaced fractures and hyaloclastite breccia that is generally finer grained than autobreccia (Figs. 6.30a, b). Both autoclastic facies may thus be present in subaqueous domes and lavas. Although autobreccia will be confined to the outer crustal zone of the dome or lava, hyaloclastite formation could penetrate deep into domes and lavas (e.g. Fig. 6.28b, c, i, j; de Rosen-Spence et al. 1980; Furnes et al. 1980; Yamagishi and Dimroth 1985; Cas and Wright 1987; Cas 1992; Yamagishi 1991).

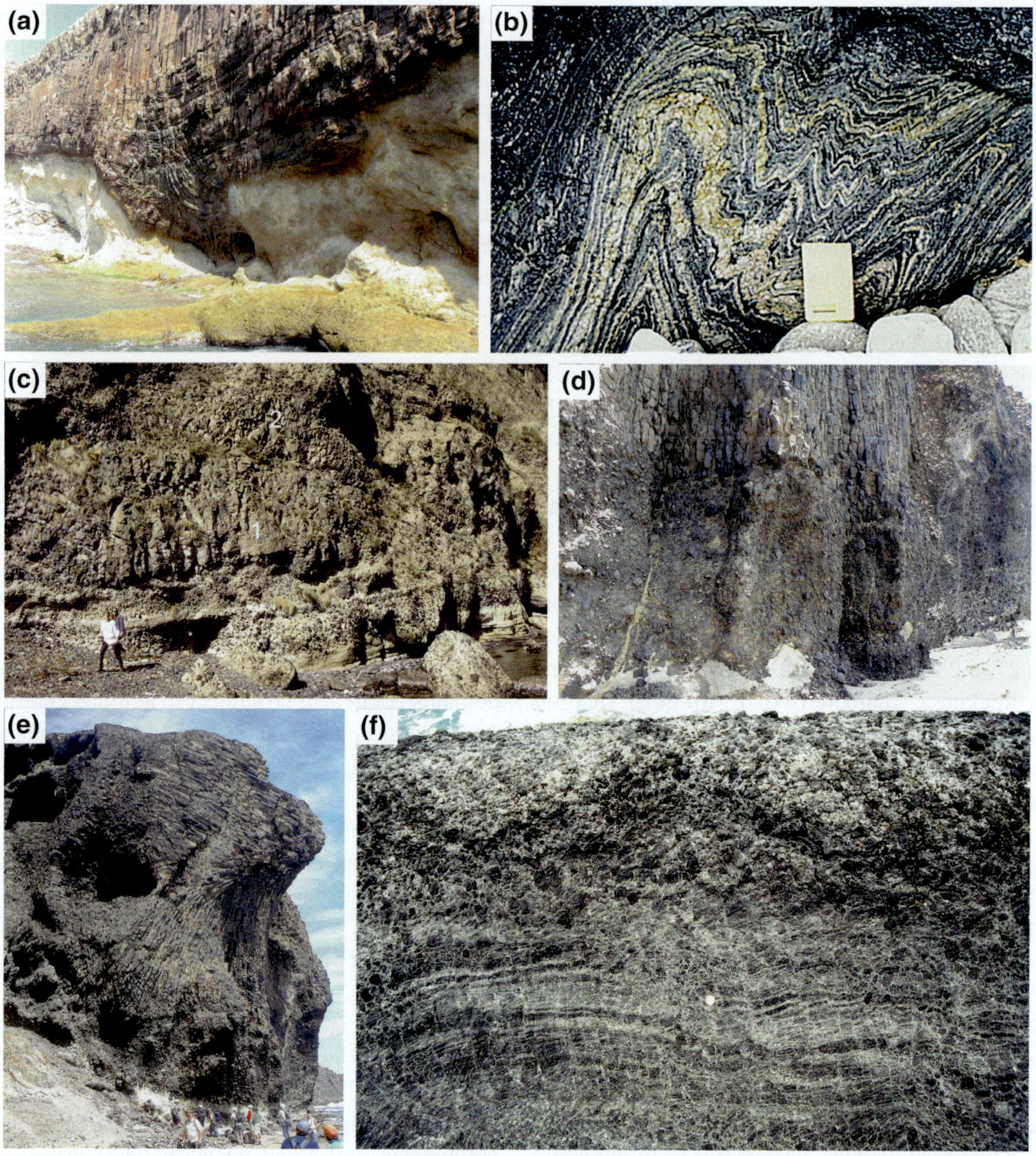

Fig. 6.29 Features of subaqueous felsic coherent lavas and coherent core domains. **a** Largely coherent, submarine, flow banded andesite lava of the ~12 Ma El Barronal Formation (dark), overlying altered ~13 Ma dacite hyaloclastite. The base of the lava has foundered into the hyaloclastite or is filling in a depression. Note the planar flow banding and the nearly vertical columnar jointing except in the depression where it curves to be perpendicular to the contact. Andesite hyaloclastite occurs locally along the base. Punta Colorada, Cabo de Gata, Spain. **b** Flow-banded and flow-folded obsidian from a <2 Ma rhyolite lava, Palmarola Island, Pontine Islands, Tyrrhenian Sea, Italy. **c** Cross-sectional exposure through complex Miocene dacite submarine lava showing two coherent to in situ quench fragmented lobes enclosed in more pervasively fragmented hyaloclastite breccia. The base of the lava overlies a fine-grained diffusely bedded volcaniclastic succession just above sea level. Tsukotan Peninsula, southwest Hokkaido, Japan. **d** Exposed base of the ~12 Ma El Barronal andesite lava, showing white stratified tuffaceous sediments overlain by the peperitic and hyaloclastite breccia base of the lava, with a moderately sharp but irregular contact into the columnar jointed coherent core zone, although the fracturing in it is probably also due to quench fragmentation. Playa del Barronal, Cabo de Gata. **e** Profile through the El Barronal andesite lava in (**d**) showing a spectacular radiating rosette-like columnar jointing pattern in the coherent core of the lava. **f** Coherent sub-horizontally flow banded obsidian tube pumice domain (bottom) to quench fragmented in situ pumice hyaloclastite breccia (top) in subaqueous ~4 Ma rhyolite lava. Cala Feola, Ponza Island, Tyrrhenian Sea, Italy

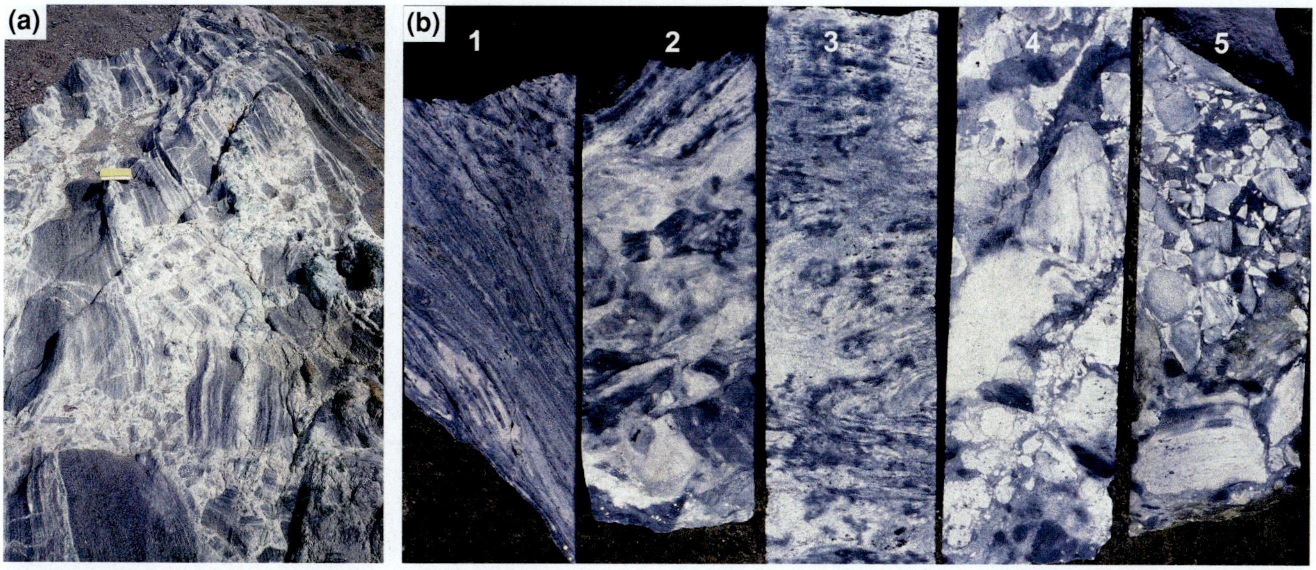

Fig. 6.30 Features of subaqueous (dacite) autobreccia. **a** Coarse block rhyolite lava autobreccia in a flow banded submarine lava, Miocene Shinzan Rhyolite, Kamoasa Bay, Oga Peninsula, Green Tuff Belt, Honshu, Japan. The finer breccia domains between the large autobreccia blocks are probably quench fragmented hyaloclastite breccia. **b** Apparent transitions from coherent flow banded and flow folded dacite (cores 1, 3), to plastic accommodation autobreccia (cores 2, 4), and finer blocky hyaloclastite (core 5) in Miocene submarine dacite lava. Width of core ~5 cm. D4 dacite, Diamond drill hole HO44, 500–600 m, Fukazawa mine footwall sequence, Hokuroko District, Green Tuff Belt, Honshu, Japan

6.12.3 Hyaloclastite Breccia

Hyaloclastite breccia (Chap. 4, Sect. 4.4.7) will also be monomictic, show jigsaw-fit to slightly clast-rotated clast textures near the core zone of the lava (Figs. 6.28b, 6.29d, f, 6.31a–e, 6.32a, b), with gradations to coherent lava in the core and variably clast-rotated textures at the lava margins (Chap. 4, Sect. 4.4.7). Hyaloclastite will have a greater grainsize range than autobreccia, from boulders to sand-sized debris. Clasts could be blocky, splintery, and blade like in shape, with planar to curviplanar surfaces, reflecting instantaneous brittle fracturing in a largely isotropic medium (Chap. 4, Sect. 4.4.7). Hyaloclasts may also preserve internal flow banding, resulting from viscous laminar flow in the conduit and during eruption before quench fragmentation occurred (Fig. 6.31b, c). Hyaloclastite domains could be tens or more metres thick in thick domes and lavas and represent a minor or the major facies in subaqueous felsic lavas (de Rosen-Spence et al. 1980), but these domains could show heterogeneities in clast size, proportions of large clasts to small clasts, and could preserve coherent to in situ quench fragmented lobes (Figs. 6.28h, 6.31e).

6.12.4 Coherent Pumice Carapaces and Pumice Hyaloclastite

In addition, if *coherent pumice* domains form through slow, steady volatile exsolution and vesicle growth, as preserved in both modern (Binns 2003; Ikegami et al. 2018; Manga et al. 2018) and ancient subaqueous felsic domes and lavas (Furnes et al. 1980; Cas and Wright 1987; Cas 1992; Kano et al. 1991; Scutter et al. 1998; Kano 2003; Cas and Giordano 2014; Cas and Simmons 2018), then that pumice can then be subjected to quench fragmentation producing *pumice hyaloclastite* (e.g. Figs. 6.29f, 6.31c; 4.5i, j; Cas and Wright 1987; Cas 1992; Scutter et al. 1998; Binns 2003; Kano 2003; Maeno and Taniguchi 2006; Cas and Giordano et al. 2014; Cas and Simmons 2018; Ikegami et al. 2018; Manga et al. 2018). The vesicles could be stretched by flow to form stretched tube pumice that help to define flow layering domains (Figs. 6.29f, 6.31c). Pumice hyaloclastite breccia will be distinguishable from pyroclastic pumice by preservation of in situ jigsaw-fit clast domains grading into coherent pumice domains and clast-rotated pumice breccia domains (Figs. 6.29f, 6.31c), and pumice clasts with curviplanar faces and polygonal fracture patterns on the surface.

(a)
(b)
(c)
(d)
(e)
(f)

◀ **Fig. 6.31** Features of hyaloclastite facies of subaqueous domes and lavas. **a** Base of lowest coherent lobe in Miocene submarine dacite lava shown in Fig. 6.29c, showing irregular contact between the coherent, flow banded part of the lobe and its quench fragmented base and the enclosing compositionally identical hyaloclastite breccia. Note the basal apophysis, which is coarsely in situ brecciated, and the necked-off apophysis remnant (lower left) enclosed in fine breccia, which has a jigsaw-fit texture. Tsukotan Peninsula, southwest Hokkaido, Japan. **b** Flow-banded in situ to slightly clast-rotated obsidian autobreccia (large clasts) and hyaloclastite breccia (finer domains). Pliocene submarine rhyolite lava complex, Ponza Island, Tyrrhenian Sea, Italy. Height of outcrop ~1.5 m. **c** Transition from in situ jigsaw-fit obsidian pumice hyaloclastite breccia (left side) to variably clast-rotated pumice hyaloclastite breccia (right side) showing increased abundance of finely fragmented hyaloclastite matrix. Cala Feola, Ponza Island, Tyrrhenian Sea, Italy. **d** Change from coherent core of a submarine Miocene dacite lava dome to blocky hyaloclastite breccia, Hokuroko Basin, northern Honshu, Japan. Drill core is ~5 cm in diameter. **e** Relict in situ quench fragmented andesite lobe in more pervasively quench fragmented hyaloclastite breccia of the same composition in a submarine Miocene andesite lava. El Barronal Formation, Playa del Barronal, Cabo de Gata, southeast Spain. **f** Diffusely stratified sequence of reversely graded debris flow deposits of resedimented andesite hyaloclastite forming the submarine apron of a lava dome complex, Kabuto Cape, southwest Hokkaido, Japan. **g** Massive to diffusely stratified, resedimented, dacite boulder Miocene hyaloclastite mass-flow breccia, part of a lava dome debris apron tens of m thick, Setakamuiwa, Japan

(g)

Fig. 6.31 (continued)

6.12.5 Resedimented Aprons of Autoclastic Debris

Resedimented aprons of autoclastic debris will form the margins of subaqueous felsic domes and lavas. In proximal settings, these volcaniclastic aprons will be crudely stratified, poorly sorted, monomictic breccias, and sandstones, with some outsize blocks (Fig. 6.31f, g). Most clasts will have formed through quench fragmentation and were then resedimented downslope. Transportation processes include grain flow slides, debris flow, and, in some cases, major collapses of lava domes could produce debris avalanche deposits (e.g. Soriano et al. 2016; see Chap. 12 for discussion of mass-flow processes). Individual beds will be up to metres in thickness, massive or normal or reverse graded (Fig. 6.31f, g), often depicted by size changes in the coarsest size fraction ("coarse-tail grading"). Clasts will be angular to subangular. Rounded clasts indicate that the lava dome emerged above water level, and that shoreline tractional reworking occurred before resedimentation into deeper water. Distal deposits of such an apron will become gradationally finer grained laterally, thinner bedded, and begin to assume Bouma sequence turbidite characteristics.

6.12.6 Bases of Lava Domes and Lava Flows, and Peperite

Peperite can only form at the bases of lava flows at the time of emplacement (Chap. 4), and not around the other margins because water-saturated sediment, with which hyaloclastite at the base of the lava is in contact, is dynamically mixed through convective mixing to form peperite at the base only, finding peperite at the base of a unit but not at its top is a clear indication that the unit was emplaced as a lava flow underwater while substrate sediment was still water-saturated and unconsolidated. If peperite occurs both at the base and top, as well as the margins, the unit must be an intrusion into wet sediments. The top contact is the key to distinguishing them. However, peperite can be misidentified at the top contact if the top of a lava has been eroded and fragments of the lava are mixed into sediments being deposited on the eroded lava top by currents. Such fragments may show signs of abrasion and (sub-)rounding. As discussed in Chap. 4 (Sect. 4.4.9), peperite at the base of subaqueous lavas consists of angular-to-fluidal fragments of lava mixed into pre-existing sediment by convective heating of sediment pore water as heat from the lava is transferred to the pore water at the base of the flowing lava and as the lava base is being subjected to cooling contraction and quench fragmentation. Fluidised sediment may be injected and mixed into cracks between hyaloclasts in the base of the lava producing a breccia with angular lava clasts with a matrix of sediment and finer hyaloclasts (Fig. 6.32). The peperite breccia can preserve in situ jigsaw-fit clast relationships close to and gradational with the coherent core of the lava (Fig. 6.32a), to clast-rotated textures away at the original base of the lava, at the interface with the substrate sediments (Figs. 6.32b, c, 6.29d).

6.13 Facies Models for Subaqueous Domes, Lavas, and Cryptodomes

As for subaerial lava domes, lava flows, and cryptodomes, there is a fairly consistent organisation of the facies described that allow facies models for subaqueous felsic domes, lavas, and cryptodomes to be proposed (Pichler 1965; Cas and Wright 1987; Cas 1992; McPhie et al. 1993; Goto and Tomiya 2019).

6.13.1 Facies Models of Subaqueous Felsic Lava Domes

Uplifted subaqueous felsic lava domes and lava flows have some of the same textural features as subaerial lavas, as discussed above in the following cited sections, including flow banding (Figs. 6.28b, 6.29a, b, f) and flow folding (Fig. 6.29b; Sect. 6.7.1), autobreccias (Fig. 6.30a, b, Sect. 6.4.1), spherulites (Fig. 6.10d, e) and lithophysae (Fig. 6.10g; Sect. 6.5.3), and perlitic fracturing (Fig. 6.11b, c; Sect. 6.5.4). The major difference is the likely abundance of jigsaw-fit to clast-rotated hyaloclastite and in some cases peperite breccias in subaqueous and subglacial domes, lavas, and cryptodomes (see Chap. 4 for further details of the textures and origins), which don't occur in subaerial lavas unless they have flowed into an aqueous environment.

Pichler (1965) was the first to propose a simple, conceptual facies model for submarine rhyolitic lava domes based on the spectacular coastal exposures of uplifted late Pliocene submarine lavas and domes on the island of Ponza, in the Tyrrhenian Sea, Italy. Coastal cliffs expose an amazing cross-section through this ~4–2.5 Ma lava and dome complex. Pichler's model depicts (a) a simple, coherent feeder dyke, merging into (b) the coherent core to the erupting dome on the seafloor (Fig. 6.33a, e.g. Fig. 6.28a–d), the upper part of which may be pumiceous. The propagating dyke and coherent core are enclosed in and grade variably into a carapace of (c) in situ autobreccia and clast-rotated hyaloclastite breccias (Figs. 6.24b, c, 6.28a–d, 6.31a–e, 6.33a), which are in turn enclosed in (d) a mantle and apron of variably mass-flow resedimented hyaloclastite breccia (Figs. 6.24d, 6.31f, g, 6.33a; Pichler 1965; Stewart and McPhie 2006). The coherent domains, hyaloclastite domains, and resedimented domains can contain substantial

Fig. 6.32 Features of the peperitic base of subaqueous felsic lavas. **a** Slab of peperite from base of a shallow marine, trachy-andesite lava, Permian Gerringong Volcanics, Kiama, New South Wales, Australia. Note blocky-to-splintery-shaped hyaloclasts, jigsaw-fit arrangement of hyaloclasts, even when separated by red mudstone, which has been fluidised and injected and mixed into the basal hyaloclastite of the lava from the underlying shallow marine mudstone. **b** Irregular hyaloclastite with basal peperite overlying white bedded tuffaceous sediments and merging upwards into columnar jointed coherent core of Miocene submarine andesite lava. Note the white sediment has been mixed up or injected upwards several m into hyaloclastite. El Barronal Formation, Playa del Barronal, Cabo de Gata, southeast Spain. **c** Close-up of the basal contact in (**b**) showing passive down-sagging or load casting of several large andesite hyaloclasts into the bedded sediments, and the peperitic mixing of the white tuffaceous sediments into the base of the quench fragmented andesite hyaloclastite breccia where it forms the matrix. Hyaloclasts all have internal quench fracture networks

proportions and lobes of coherent-to-autoclastic pumice, and the coherent spines may protrude up to and through the surface of the dome (Plate 6.1d; Figs. 6.24a, 6.33a; Ikegami et al. 2018).

However, domes may also be complexes of multiple superposed domes and conduits migrating along fissures, leading to a multiple dome and internal cryptodome complex (Fig. 6.33b; Scutter et al. 1998; De Rita et al. 2001; Stewart and McPhie 2006). In addition, internally, domes may have a multi-lobate architecture, with coherent lobes intruding the water-saturated autoclastic and resedimented aprons, and themselves being subject to quench fragmentation and even necking off, producing internal stranded coherent pods or lobes (Figs. 6.28i, j, 6.33c; Yamagishi and Dimroth 1985). The facies in these variants of subaqueous dome models are generally similar but are organised internally in different ways.

Although Pichler (1965) proposed the simple submarine dome facies model based on the geology of Ponza, the island, in fact, preserves a much more complex uplifted

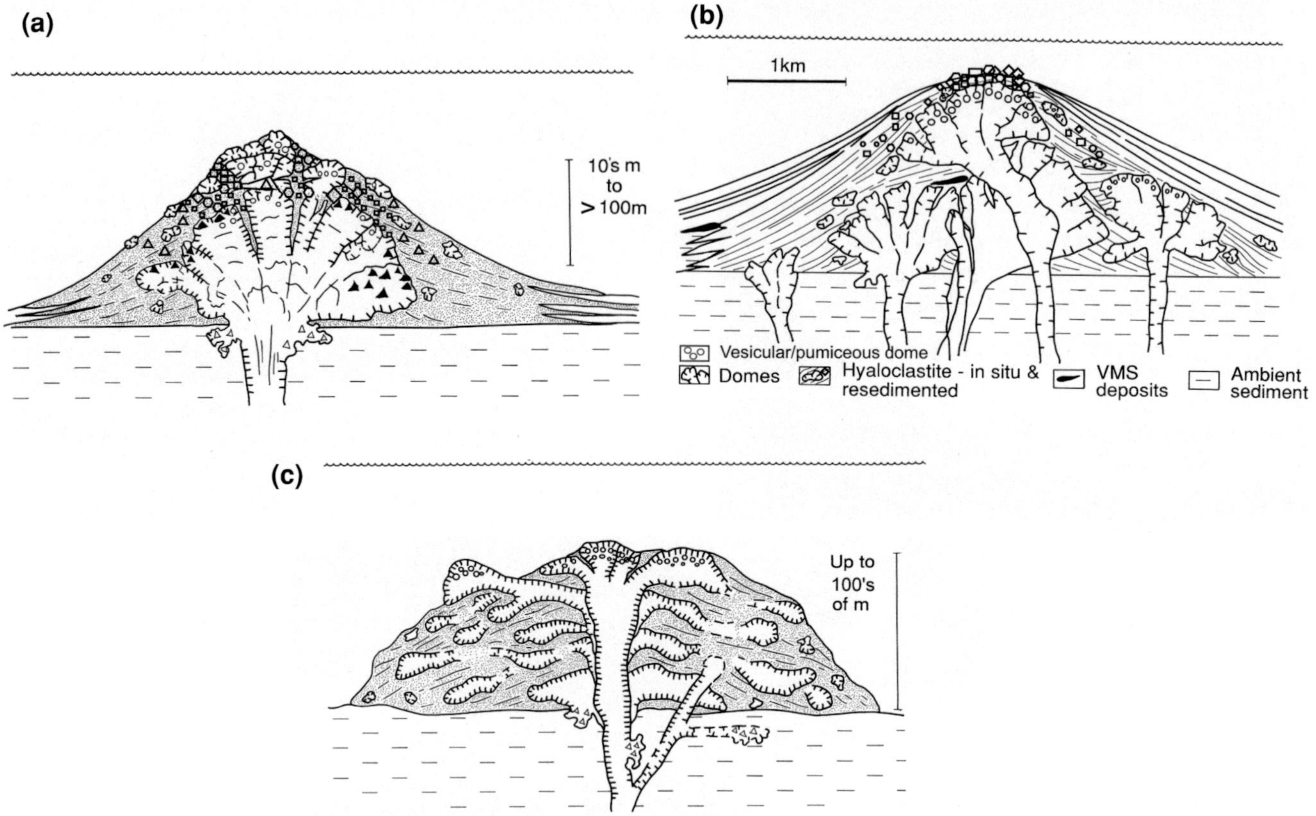

Fig. 6.33 Schematic facies models for subaqueous felsic lava domes, showing variations (From Cas 1992). **a** Facies model of a simple subaqueous lava dome, showing the feeder dyke feeding the coherent to autobrecciated and quench fragmented hyaloclastite core of the dome, enclosed in a carapace and apron of variably resedimented, bedded hyaloclastite and megablocks (After Pichler 1965). **b** Multiple dome and cryptodome complex, enclosed in hyaloclastite and resedimented hyaloclastite. **c** Schematic facies model of a complex, multi-lobate submarine lava dome, showing a complex distribution of endogenous lava lobes enclosed within hyaloclastite breccia, which contains coherent "necked-off" intrusive lobes, isolated by quench fragmentation. For all models refer to legend in (**b**)

submarine lava dome, flow, and feeder dyke complex (Figs. 6.26, 6.28e–g; Pichler 1965; Scutter et al. 1998; De Rita et al. 2001). The Ponza system consists of an early coherent to quench fragmented lava flow and lava dome complex, cross-cut by late, co-magmatic, largely coherent feeder dykes and exogenously superposed late domes, the tops of which have been eroded (Scutter et al. 1998; De Rita et al. 2001). Pichler's model is appropriate for single isolated domes and has been supported by more recent studies elsewhere (Stewart and McPhie 2006; Maeno and Taniguchi 2006; Chap. 16).

6.13.2 Facies Models of Subaqueous Felsic Lavas (Coulées)

Submarine felsic lava flows of moderate flow distance will generally have a similar facies zonation as outlined for domes, but this could be skewed asymmetrically in a down flow direction. They will be essentially block lavas (Fig. 6.34a) with coherent cores (Fig. 6.29a–f) or lobes, sometimes, multiple lobes, enclosed in autoclastic breccias (Figs. 6.29c, 6.31a, e, 6.34a, b), ranging from in situ jigsaw fit to clast rotated, and resedimented in the frontal and marginal apron (Fig. 6.34a; cf. Fig. 6.31f, g; de Rosen-Spence et al. 1980; Kokelaar et al. 1985; Kano et al. 1991; Scutter et al. 1998; Soriano et al. 2013). Although the coherent core is usually enclosed in a cover of autoclastic breccia, consisting of coherent and pumiceous clasts, some coherent dense and pumiceous domains and spines may be exposed at the surface and margins of the lava (Ikegami et al. 2018). The proportion of autoclastic facies to coherent core facies will vary from flow to flow (de Rosen-Spence et al. 1980). In some lavas, autoclastic facies could represent the most significant lithofacies, particularly around the margins, where the entire lava thickness in some places could have been quench fragmented and consist completely of hyaloclastite (Figs. 6.28b, 6.34b; de Rosen-Spence et al. 1980). Alternatively, other lavas may be dominated by coherent facies (Cas 1978; de Rosen-Spence et al. 1980; Embley and

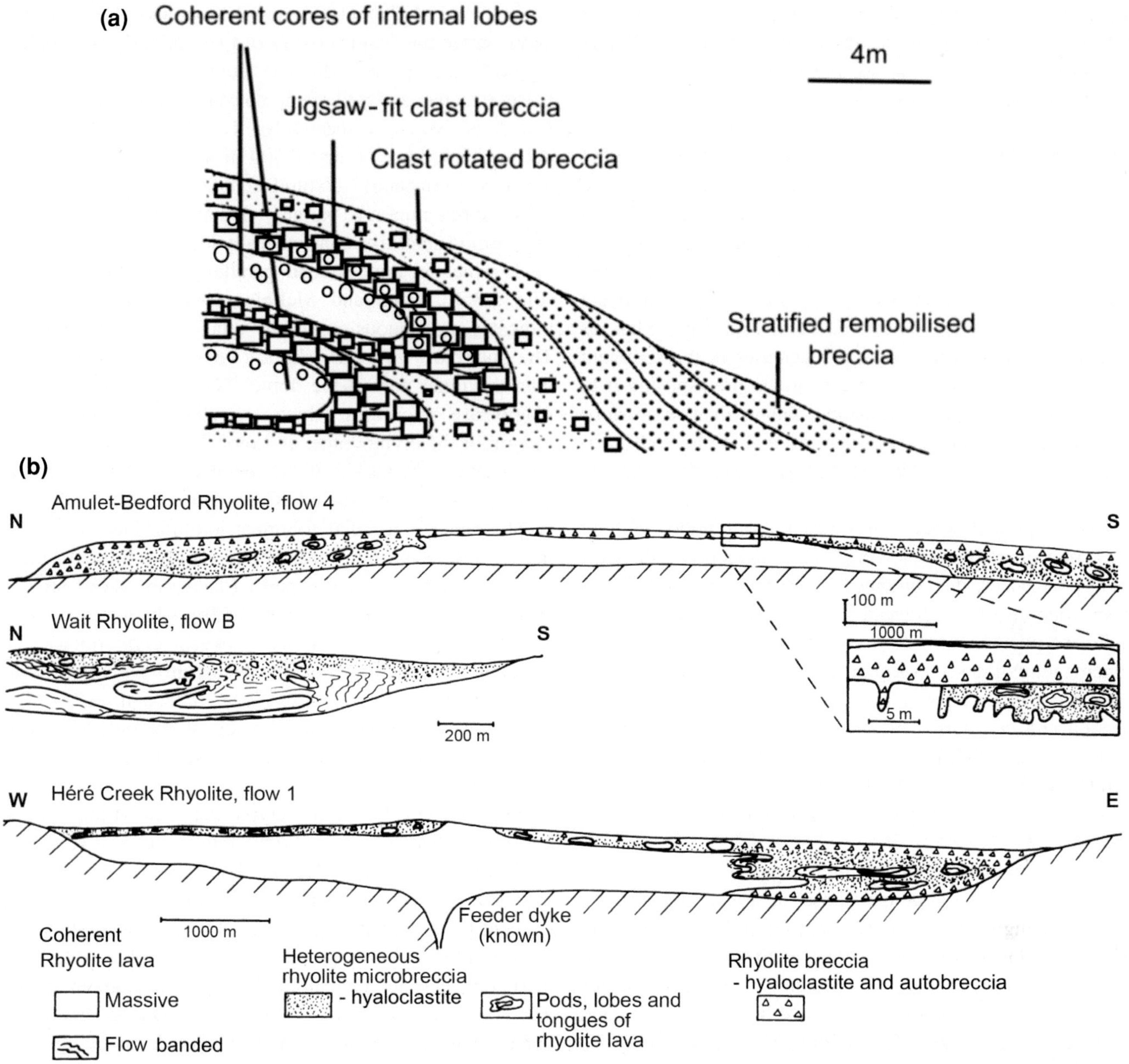

Fig. 6.34 Schematic facies models for subaqueous felsic lavas. **a** Based on Miocene submarine rhyolite lava flow, Shimane Peninsula, Japan, showing the spectrum of facies from coherent core lobes to in situ jigsaw-fit autobreccia and hyaloclastite breccia, to clast-rotated hyaloclastite, to an apron of resedimented autoclastic breccia (Modified from Kano et al. 1991). **b** Cross-sectional representation of the considerable variations in the proportions of coherent and autoclastic rhyolite facies in three Archean submarine rhyolite lava flows in the Rouyn-Noranda area, Abitibi Greenstone Belt, Quebec, Canada (Modified from de Rosen-Spence et al. 1980)

Rubin 2018). Autoclastic facies include autobreccia (e.g. Figs. 4.2, 6.24c, 6.30), hyaloclastite (Figs. 4.5, 6.31), perhaps pumice breccia (Figs. 6.29f, 6.31c), basal peperite (Fig. 6.33a–c), where lavas have flowed over unconsolidated, water-saturated sediments, and marginal aprons of resedimented autoclastic debris (Figs. 6.34a, cf. 6.31f, g). Lithofacies characteristics will be similar to those described for simple domes.

6.13.3 Emplacement, Facies, and Facies Models for Cryptodomes in Subaqueous Settings

Cryptodomes, as we have already described from subaerial settings (Sect. 6.3.5), are high-level intrusions that generally don't quite erupt on the Earth's surface, or only partially. This may occur on the floors of oceans or lakes where a considerable unconsolidated, water-saturated sediment pile provides an ideal host for the rising magma to stop rising if it is moving slowly, given that the magma density will likely exceed the density of the sediment succession (Fig. 6.22, e.g. La Primavera caldera, Mexico; Clough et al. 1982; Devonian Boyd Volcanics, southeastern Australia, Cas et al. 1990, 2001). If the rising domal magma masses reach within several tens of metres from the basin floor, they can cause up-arching of the sediment sequence above them producing a morphological mound on the basin floor (Fig. 6.35a–c), hence the term cryptodome. Cryptodomes and other high-level intrusions should be common where there are thick water-saturated unconsolidated sediment piles as magma density-driven buoyancy decreases in the sediment pile after passing through dense bedrock, predisposing it to form syn-depositional intrusions rather than extrusions (Chap. 8). The details of syn-depositional, high-level intrusions, including cryptodomes and sills, are discussed in Chap. 8.

For cryptodomes to nearly reach basin floors without triggering phreatomagmatic explosive interaction with sediment pore water, the combined hydrostatic pressure due to basin water depth and lithostatic pressure due to depth in the sediment pile must be high enough to suppress explosive superheating of the sediment pore water, and/or the sediment pore water is low, perhaps due to a degree of burial compaction and dewatering prior to magma intrusion. In addition, magma volatile content must be low and/or the combined hydrostatic and lithostatic pressures must be high enough to suppress explosive expansion of any exsolved volatiles.

Cryptodomes have similar high aspect ratios to erupted domes (Fig. 6.35; although those at La Primavera are somewhat higher than for subaerial domes), and generally have a similar internal architecture, with a central coherent core. Cryptodomes may be equant or irregular shapes depending on the rheological states of both the magma and host sediments (Figs. 6.35, 6.36a). The margin may be irregular, with apophyses or lobes of the cryptodome magma intruding the enclosing sediment and fluidised sediment being injected into fractures or clefts in the margin of the cryptodome (Fig. 6.36b–d). A zone of hyaloclastite, and then around most margins at the interface with the wet sediments, domains of peperite occur (Figs. 6.35, 6.36b–d; Hanson and Wilson 1993; Goto and McPhie 1998; Stewart and McPhie 2003). The widths of these zones will depend on the abundance of pore water available in the sediment, which will be determined by the initial porosity, grainsize, degree of compaction, and dewatering of the sediments prior to intrusion of magma. Sometimes the uppermost part of such cryptodomes may breach the sediment surface, causing extrusion of the top of the dome onto the sea or lake floor (Figs. 6.22, 6.35a; also see Fig. 6.38).

Goto and McPhie (1998) and Stewart and McPhie (2003) documented the characteristics of two uplifted and exhumed cryptodomes in the Miocene submarine successions of northern Japan and a Pliocene volcanic system on the Greek Island of Milos, respectively. The Miocene Momo-iwa dacite cryptodome is up to 300 m across and 190 m high and has a concentric textural zonation (Goto and McPhie 1998). It has a massive variably crystallised core zone and an outer onion shell-like concentrically layered zone up to 40 m thick which is flow banded parallel to the dome margins (Fig. 6.35b). Goto and McPhie (1998) suggest that this indicates endogenous growth or inflation of the cryptodome. The outermost layers are slightly vesicular and more glassy than the inner part of the dome. The groundmass contains plagioclase microlites which are aligned parallel to the margin of the cryptodome, and probably became aligned by laminar flow stretching of the outer layers of the intruding magma body as it inflated (Goto and McPhie 1998). The cryptodome has a radial columnar jointing, indicating cooling from the margin inwards. There is only a limited zone of hyaloclastite breccia and peperite (dacite–mudstone breccia) at the margins, <1 m thick (Fig. 6.35b), which they consider to be a significant contrast between cryptodomes and extrusive domes, the latter of which can have very considerable hyaloclastite margins (Figs. 6.28ba–d, 6.33). This may, however, depend on the state of the enclosing sediments, their porosity and water content at the time of intrusion.

The Pliocene Kalogeros cryptodome on Milos is up to 1300 m in diameter, up to 120 m thick and has a volume of 0.12 km^3 (Stewart and McPhie 2003). The dacite dome has intruded pumiceous sediments and mudstone and is also concentrically zoned (Fig. 6.35c). The core is slightly porphyritic (15%); has a groundmass of microlites of plagioclase, opaques, and interstitial glass; and is columnar jointed. The core is enclosed in a concentrically banded zone that is enclosed by a 30-to-40-m-thick banded zone of light and dark layers. This zone is also porphyritic, the grey layers are slightly vesicular ($\sim$5%), whereas the dark layers are barely vesicular, and slightly less porphyritic. Both have a groundmass of plagioclase, opaques, and more groundmass glass than the core. An outer zone, up to 4 m thick consists of jigsaw-fit to clast-rotated banded dacite clast breccia that becomes more matrix rich outwards, forming peperite, and is gradational into the host pumiceous sediment sequence. This dome is also interpreted to have grown by endogenous inflation.

The major facies that distinguish cryptodomes from eruptive domes is the presence of an apron of resedimented hyaloclastite breccia around subaqueous eruptive domes. In addition, the proportion of hyaloclastite associated with subaqueous eruptive domes will generally be higher than

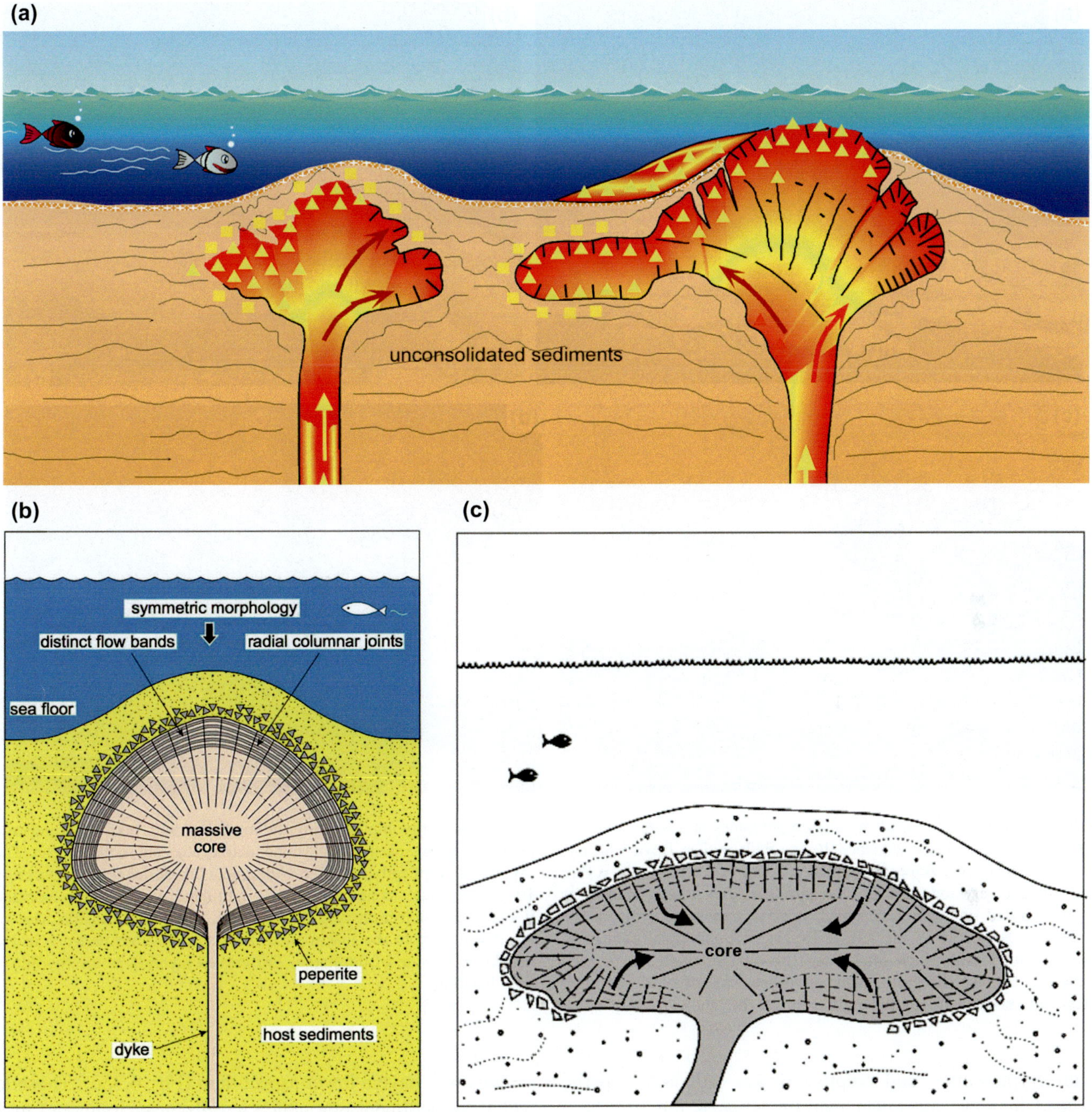

Fig. 6.35 Schematic facies models for syn-depositional felsic cryptodomes intruding wet, unconsolidated basin sediments. **a** Moderate viscosity cryptodomes intruding water-saturated unconsolidated sediments will be irregular, may spawn lateral intrusive sills or lobes into the sediment and be subjected to considerable marginal quench fragmentation forming hyaloclastite (yellow triangles) and peperite formation, with hyaloclasts mixed into marginal sediment (yellow boxes). Some may emerge above the basin flow and extrude lava lobes or a small dome. **b** High-viscosity cryptodomes will have a high aspect ratio and have some marginal hyaloclastite and peperite, based on the Miocene Momo-iwa dome Japan (After Goto and Tomiya 2019). **c** Intermediate viscosity magmas intruding water-poor sediments will produce a more regular, moderate aspect ratio cryptodome with only minor marginal quench fragmentation and peperite formation, based on the Pliocene Kalogeros cryptodome on Milos, Greece (From Stewart and McPhie 2003)

Fig. 6.36 Features of small rhyolite cryptodomes in a Late Devonian subaqueous sedimentary sequence, Bunga Beds outlier, Boyd Volcanic Complex, Aragunnu Bay, southern New South Wales Australia. **a** Profile of two cryptodomes (1, 2) intruded into black mudstones (ms) exposed in two-dimensional coastal exposure. Dip in the sediments is steeply to the left. **b** Highly irregular basal contact of cryptodome 1 with black mudstone. The top, although not as well exposed, is also irregular with apophyses of rhyolite intruding sediment. **c** Close-up of the right side of (**b**), showing very small apophyses of rhyolite intruding the mudstone with pockets of peperite (angular rhyolite clasts in mudstone matrix) adjacent to the contact. The rhyolite is cryptically fractured with a network of quench fractures in a 10–15 cm wide zone along the right-hand side of the contact. **d** Flame-like injection of fluidised black mudstone into the base of the rhyolite. Note the jigsaw-fit hyaloclastite fracture network in the rhyolite each side of the flame. The curved layering in the rhyolite is flow banding

with cryptodomes, and eruptive domes will lack peperite around the side and upper margins whereas this will be a distinctive feature of cryptodomes that intrude water-saturated sediments. However, some cryptodomes may emerge partially above the basin floor and may thus be both eruptive in their upper parts and intrusive beneath the basin floor, giving rise to the characteristics of both effusive domes and intrusive cryptodomes.

Several variants for facies models can be proposed for syn-depositional felsic cryptodomes. Viscous cryptodomes will have a higher aspect ratio and be more regular in shape (Fig. 6.35a, b). Those that intrude host sediments with limited water will have limited development of marginal hyaloclastite and peperite breccias. Those that are not very viscous will have a lower aspect ratio (Fig. 6.35c). Those in which the sediment pile which is fully water saturated may be irregular in form and have a relatively low aspect ratio (Fig. 6.35a) experience significant quench fragmentation around the margins as well as peperite formation (Fig. 6.35a). They may spawn lateral sills or lobes into the sediment pile, and some may partially extrude, producing a limited apron of resedimented hyaloclastite (Fig. 6.35a). Figure 6.22 also illustrates these variations.

Goto and Tomiya (2019) propose that there are differences between cryptodomes intruded into dry, water-free sediments in subaerial settings (e.g. Usu volcano, Japan), and cryptodomes into water-saturated sediments such as lakes and the oceans. In subaqueous cryptodomes, well-defined radial columnar jointing occurs from the core zones to the margins, hyaloclastite and peperite occur at the margins and contact zone with enclosing water-saturated sediments (Fig. 6.35b, c), and the outer margins at least are glassy (Goto and McPhie 1998; Goto and Tomiya 2019). These all reflect the enhanced and more rapid cooling regime in water-saturated sediments and are all lacking in cryptodomes intruding dry sediments (Goto and Tomiya 2019).

6.14 Far-Flowing Subaqueous Felsic Lava Flows—the Lower Devonian Merrions Formation

We discussed subaqueous felsic lava flows above, including some on the modern ocean floor in the northeast Lau Basin, which may have flowed up to 10 km from vent (Embley and Rubin 2018). However, almost no subaqueous counterparts of the huge continental felsic or silicic "flood lavas" discussed above in Sect. 6.8, with flow distances of tens of km or more, have been documented.

However, Cas (1978) documented two such regionally extensive, far-flowing submarine dacite lavas in the Lower Devonian ($\sim$410 Ma) Merrions (Tuff) Formation in the Siluro-Devonian Hill End Trough Basin of southeastern Australia (Figs. 6.37a, 11.15a), which preserves a 5-km-thick deep-water basin succession dominated by terrigenous sandstone turbidites and hemi-pelagic mudstones. At the time, southeastern Australia was experiencing regionally extensive silicic volcanism and plutonism in a continental crustal setting undergoing regional extension, leading to numerous ensialic deep-water basins (Cas 1983). Two lavas, B1 and H, and a more restricted one, B2, are interstratified with co-genetic, very voluminous, syn-eruptive volcaniclastic megaturbidites that constitute the Merrions (Tuff) Formation (Fig. 11.15b; Cas 1979) and represent a huge >1,500 km^3 magmatic event(s). The lavas are perhaps surprisingly largely coherent, as are the extensive dacite lavas reported by Embley and Rubin (2018) at water depths of >2,500 m in the northeast Lau Basin.

The contacts between the Merrions lavas with underlying and overlying strata are seldom exposed, but one basal contact shows centimetre size clasts of the dacite "tear-dropping" off the base of the lava into the underlying mudstone (Fig. 6.37b), indicating that magma viscosity was not very high. The Merrions lavas are internally massive, variably columnar jointed (Fig. 6.36c), and are porphyritic (quartz, plagioclase, minor K-feldspar), with modal phenocryst contents of $\sim$25–40% (Fig. 6.9b). Relatively localised autobreccia and hyaloclastite breccia facies occur at top and bottom. Breccias at the top of the lava have been largely eroded, stripped off by succeeding volcaniclastic megaturbidity currents (Chap. 11), producing matrix-supported breccias of angular clasts of the underlying lava in the base of the overlying very thick volcaniclastic megaturbidite beds (Fig. 6.37d). In a couple of localities lava H may have burrowed into basinal muds, as suggested by irregular contacts between the top of the lava and overlying mudstones.

The two largest lavas are elongate and lobate on a regional scale, with their axes trending northwest (Fig. 6.37a). Their preserved lengths are $\sim$80 km long, and they are at least 20 kms wide, and thus covered an area of at least 1,600 km^2. The axis of the younger lava H is offset to the northeast of the older lava B1. The thickest lava, B1, is up to 390 m thick and the other, lava H, is up to 150 m thick (Fig. 6.37a). The original full extent of the lavas to the southeast is unknown, and with average thicknesses of $\sim$200 m (B1) and 50 m (H), their volumes were B1 > 300 (minimum) to 400 km^3 and H $\sim$ >80 (minimum) to 100 km^3. Given the significant average maximum thicknesses of >300 m (up to 380 m, Fig. 6.37a) and 80 m (maximum 150 m), the aspect ratios of lavas H and B1 are very low, at 0.0017 and 0.0067, respectively, because of their very large areal extent. Because the location of the vent(s) could not be determined, flow distances could not be determined, but were at least tens of kms or more (Cas 1978). Examining isopach contour patterns, lava B1 may have originated from a basin seafloor fissure vent to the west of the current outcrop area, and lava H from a vent to the southeast (Fig. 6.37a), although this requires caution because pre-existing basin floor topography can also cause ponding of lavas in depressions and thickness maxima. The B1 lava, which is generally 200 to 300 m thick along its northwest trending axis (Fig. 6.37a), had sufficient relief above the basin floor to control the distribution and thickness of succeeding the lava H and volcaniclastic megaturbidite units (P, D, E) in the Merrions Formation (Fig. 6.37e). In Fig. 6.37 e, which is an east–west stratigraphic cross-section of the Merrions Formation, lava B1 caused the depocentres (thickest parts) of the succeeding units P, and D volcaniclastic megaturbidites to be displaced northeast of the northwest trending B1 lava (Cas 1977). Lava H then flowed parallel to but northeast of lava B1, presumably guided by a depression that still existed alongside the northeastern flank of lava B1, even after deposition of the unit D megaturbidite deposit, which would have also been at least partly topography controlled. The volcaniclastic unit E also has an antithetic thickness relationship to that of lava unit H.

The volumes of the Devonian submarine Merrions lavas B1 and H are comparable to the subaerial Miocene Snake River Plain rhyolite lavas (10 to 200 km^3; Bonnichsen and Kauffman 1987) and almost of the same order as the huge volume Proterozoic subaerial lavas of the Gawler Range Volcanic Province in South Australia, which as discussed in Sect. 6.8, each $\geq$1,000 km^3 in volume (Allen et al. 2003, 2008), as discussed in Sect. 6.8. As for these subaerial

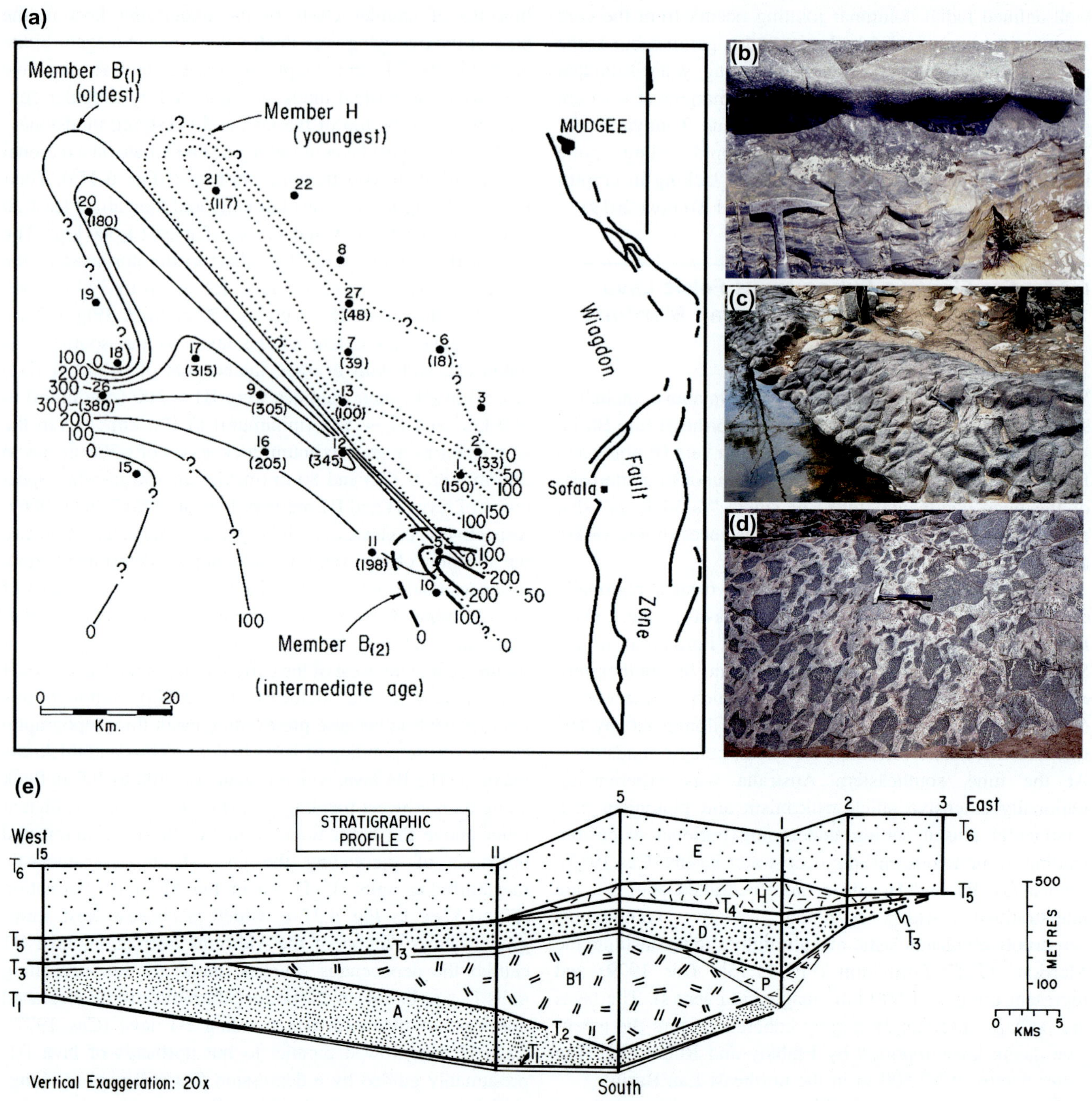

Fig. 6.37 Features of far-flowing subaqueous dacite lavas of the Lower Devonian Merrions Formation, central-western New South Wales, Australia. **a** Isopach contours of two far-flowing and one local dacite submarine lavas. The areal distribution recreates the palinspastically unfolded state of the regional rock succession. Note the northwest trend of the two lobate lavas. (From Cas 1978). **b** The irregular, slightly quench fragmented and peperitic base of submarine dacite lava B1 at its contact with underlying mudstones at Winburndale Rivulet. Note several fragments of dacite appear to have detached from the base of the lava and sagged into the underlying mudstone, which appears to have been homogenised up to 15 cm below the lava base. **c** Columnar jointing in Merrions lava B1 at Pyramul Creek. **d** Matrix-supported breccia of angular hyaloclasts of the underlying lava H in the base of the overlying very thick crystal-rich tuffaceous volcaniclastic megaturbidite bed at Turon River. **e** Stratigraphic west to east correlation profile across the southeastern part of the outcrop area of the Merrions Formation. Random dashes and double dashes are the lavas. Stipple and arrow patterns are volcaniclastic megaturbidite deposits, and blank is mudstone. T_1, T_2, etc. are time planes. Note the eastwards displacement of the depocentres for units D (megaturbidite) and H (lava) in an apparent depression off the eastern flank of lava B1 and the antithetic thinning of units over the crest of the thick unit B lava. (From Cas 1977)

counterparts, large, virtually batholith size magma reservoirs were required to produce the voluminous Merrions submarine lavas. The eruptions would have been sustained, probably for several years, given their far-flowing (tens of kms or more) nature and internally massive nature (no determinable stacked lobes or multiple lava sheets). Eruption would have occurred from submarine fissure vents, so that eruption rates could sustain the long-distance flows. There is no indication that eruptions could have occurred from subaerial vents and then flows entered the sea, such as abundant resedimented hyaloclastite.

Cas (1978) proposed that eruption in deep-water eruption would have suppressed exsolution of volatiles, which would have maintained a lower flow viscosity compared with subaerial counterparts, so facilitating long-distance flow. Nonetheless, the lavas inflated to nearly 400 m thick and 150 m thick, respectively. Interestingly, there is relatively little hyaloclastite preserved, a feature shared with the very deep-water (>2,500 m) dacites of the modern northeast Lau Basin (Embley and Rubin 2018). Little hyaloclastite implies very effective film boiling at the surface of the lava, so insulating the lava crust and interior from undergoing quench fragmentation. If the crust did not shatter very often, a coherent crust at the lava surface would further insulate the lava interior and maintain low lava viscosity.

So, a question arises—why do these very deep-water felsic lavas undergo minimal quench fragmentation? Either the flow propagation velocity was so slow that the vapour film remained stable at the lava surface, and/or the vapour film produced by film boiling is so compressed at water depths (~3 km) and hydrostatic pressure equivalent to the critical pressure point of seawater (~30 MPa, i.e. the vapour film is a supercritical fluid) that there is no sharp temperature jump from lava surface to the liquid water body by a continuous but steep, temperature gradient (see Chap. 3 for more detailed discussion). This may mean that the lava surface cannot be commonly subject to thermal shock that would produce widespread quench fragmentation, as it seems to occur commonly in shallower water depths.

6.15 Explosive Dome-Top Tuff and Pumice Cones

As with some subaerial lava domes, some subaqueous domes, which grow to shallow water depths, can become explosive due to magmatic volatile content and dynamic phreatomagmatic explosive magma–water interaction (Chap. 4, Sect. 4.6.9; Reynolds and Best 1976; Reynolds et al. 1980; Cas et al. 1990; Fiske et al. 1998; Allen and McPhie 2000; Stewart and McPhie 2006). This can lead to development of an accumulation of pyroclastic deposits forming a stratified pumice or tuff cone on the surface of the rising dome (e.g. see Figs. 9.78, 9.79). Cas et al. (1990) described the stages in the evolution of two dome-top tuff and pumice cones preserved in the Late Devonian Boyd Volcanic Complex on the south coast of New South Wales, Australia. These included initial upward migration of a viscous rhyolitic cryptodome through water-saturated, unconsolidated basinal muds and turbidite sandstones. As the domes rose and expanded, the outer margin autobrecciated allowing water-saturated sediment to squeeze into inter-clast cracks. The rising dome caused the sediment surface to arch upwards to the point of destabilising it and causing it to slump off the flanks of the rising cryptodome, producing a chaotic sediment rhyolite clast debris flow breccia. Unroofing of the dome by this event led to decompression of the emerging dome, triggering magmatic volatile exsolution, vesiculation, and both weak magmatic and phreatomagmatic explosive activities that led to the deposition of several tens of m of diffusely bedded pumice and tuff deposits that formed low relief cones around the vent (Fig. 6.38; Cas et al. 1990). Allen and McPhie (2000) have also documented the growth of a shallow water pumice cone on the Greek Island of Yali, discussed in Chap. 9 (Sect. 9.9.4.4; Figs. 9.78, 9.79).

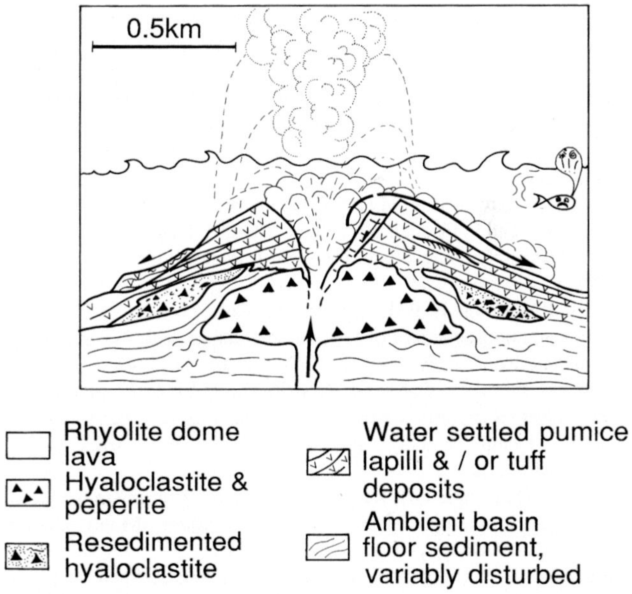

Fig. 6.38 Schematic model for the formation of shallow water dome-top tuff and pumice cones, based on preserved deposits in the Late Devonian Bunga Beds outlier, Boyd Volcanic Complex, Aragunnu Bay, southern New South Wales, Australia (Modified from Cas et al. 1990). The emergence of a cryptodome dislodged sediment and hyaloclastite debris down the rising flanks of the dome, exposing the top of the emerging dome to explosive magma–water interaction that produced hydraulically well sorted, bedded crystal-rich water settled tuff, vitric ash, and pumice deposits around the vent that built a dome-top tuff cone

References

Adams NK, Houghton BF, Fagents SA, Hildreth W (2006) The transition from explosive to effusive eruptive regime: the example of the 1912 Novarupta eruption, Alaska. Geol Soc Am Bull 118(5–6):620–634

Allen RL (1992) Reconstruction of the tectonic, volcanic, and sedimentary setting of strongly deformed Zn-Cu massive sulfide deposits at Benambra, Victoria. Econ Geol 87(3):825–854

Allen SR, McPhie J (2000) Water-settling and resedimentation of submarine rhyolitic pumice at Yali, eastern Aegean, Greece. J Volcanol Geotherm Res 95(1–4):285–307

Allen SR, McPhie J (2002) The Eucarro Rhyolite, Gawler Range Volcanics, South Australia: a > 675 km^3, compositionally zoned lava of Mesoproterozoic age. Geol Soc Am Bull 114(12):1592–1609

Allen SR, McPhie J (2003) Phenocryst fragments in rhyolitic lavas and lava domes. J Volcanol Geotherm Res 126(3–4):263–283

Allen SR, Simpson CJ, McPhie J, Daly SJ (2003) Stratigraphy, distribution and geochemistry of widespread felsic volcanic units in the Mesoproterozoic Gawler Range Volcanics, South Australia. Aust J Earth Sci 50(1):97–112

Allen SR, Fiske RS, Tamura Y (2010) Effects of water depth on pumice formation in submarine domes at Sumisu, Izu-Bonin arc, western Pacific. Geology 38(5):391–394

Allen SR, McPhie J, Ferris G, Simpson C (2008) Evolution and architecture of a large felsic igneous province in western Laurentia: the 1.6 Ga Gawler Range Volcanics, South Australia. J Volcanol Geotherm Res 172(1–2):132–147

Anderson SW, Fink JH, Rose WI (1995) Mount St. Helens and Santiaguito lava domes: the effect of short-term eruption rate on surface texture and degassing processes. J Volcanol Geotherm Res 69(1–2):105–116

Anderson SW, Fink JH (1990) The development and distribution of surface textures at the Mount St. Helens dome. In: Fink JH (ed) Lava flows and domes. Springer, Berlin, Heidelberg, pp 25–46

Andrews GD, Kenderes SM, Whittington AG, Isom SL, Brown SR, Pettus HD, Cole BG, Gokey KJ (2021) The fold illusion: the origins and implications of ogives on silicic lavas. Earth Planet Sci Lett 553:116643

Arrighi S, Tanguy J, Rosi M (2006) Eruptions of the last 2200 years at Vulcano and Vulcanello (Aeolian Islands, Italy) dated by high accuracy archeomagnetism. Phys Earth Planet Inter 159(3–4):225–233

Ashwell PA, Kennedy BM, Edwards M, Cole JW (2018) Characteristics and consequences of lava dome collapse at Ruawahia, Taupo Volcanic Zone, New Zealand. Bull Volcanol 80(5):43

Aubourg C, Giordano G, Mattei M, Speranza F (2002). Magma flow in sub-aqueous rhyolitic dikes inferred from magnetic fabric analysis (Ponza Island, W. Italy). Phys Chem Earth, Parts A/B/C 27(25–31):1263–1272

Bailey RA, Dalrymple GB, Lanphere MA (1976) Volcanism, structure, and geochronology of Long Valley Caldera, Mono County, California. J Geophys Res 81(5):725–744

Balmforth NJ, Burbidge AS, Craster RV, Salzig J, Shen A (2000) Visco-plastic models of isothermal lava domes. J Fluid Mech 403:37–65

Barmin A, Melnik O, Sparks RSJ (2002) Periodic behavior in lava dome eruptions. Earth Planet Sci Lett 199(1–2):173–184

Bernstein M, Pavez A, Varley N, Whelley P, Calder ES (2013) Rhyolite lava dome growth styles at Chaitén Volcano, Chile (2008–2009): Interpretation of thermal imagery. Andean Geol 40(2):295–309

Bertin D, Lara LE, Basualto D, Amigo Á, Cardona C, Franco L, Gil F, Lazo J (2015) High effusion rates of the Cordón Caulle 2011–2012 eruption (Southern Andes) and their relation with the quasi-harmonic tremor. Geophys Res Lett 42(17):7054–7063

Bevins RE, Roach RA (1979) Pillow lava and isolated-pillow breccia of rhyodacitic composition from the Fishguard Volcanic Group, Lower Ordovician, SW Wales, United Kingdom. J Geol 87(2):193–201

Binns RA (2003) Deep marine pumice from the Woodlark and Manus Basins, Papua New Guinea. Explosive subaqueous volcanism. American Geophysical Union, Washington, D.C., pp 329–343

Blake S (1990) Viscoplastic models of lava domes. In: Fink JH (ed) Lava flows and domes: emplacement mechanisms and hazard implications. IAVCEI proceedings in volcanology 2. Springer, Berlin, pp 88–128

Bonnichsen B, Kauffman DF (1987) Physical features of rhyolite lava flows in the Snake River Plain volcanic province, southwestern Idaho. Geol Soc Am Spec Pap 212:119–145

Bonnichsen B, Leeman WP, Honjo N, McIntosh WC, Godchaux MM (2008) Miocene silicic volcanism in southwestern Idaho: geochronology, geochemistry, and evolution of the central Snake River Plain. Bull Volcanol 70(3):315–342

Boudon G, Villemant B, Le Friant A, Paterne M, Cortijo E (2013) Role of large flank-collapse events on magma evolution of volcanoes. Insights from the Lesser Antilles Arc. J Volcanol Geotherm Res 263:224–237

Branney MJ, Bonnichsen B, Andrews GD, Ellis B, Barry TL, McCurry M (2007) 'Snake River (SR)-type' volcanism at the Yellowstone hotspot track: distinctive products from unusual, high-temperature silicic super-eruptions. Bull Volcanol 70(3):293–314

Buisson C, Merle O (2004) Numerical simulation of strain within lava domes. J Struct Geol 26(5):847–853

Bullock LA, Gertisser R, O'Driscoll B (2017) Spherulite formation in obsidian lavas in the Aeolian Islands, Italy. Periodico di Mineralogia 86(1)

Bullock LA, Gertisser R, O'Driscoll B (2018) Emplacement of the Rocche Rosse rhyolite lava flow (Lipari, Aeolian Islands). Bull Volcanol 80(5):48

Calder ES, Lavallee Y, Kendrick JE, Bernstien M (2015) Lava dome eruptions. In: Sigurdsson H, Stix J, Houghton BF, McNutt SR, Rymer H (eds) Encyclopedia of volcanoes. Academic Press, San Diego, pp 343–362

Carey R, Soule SA, Manga M, White JDL, McPhie J, Wysoczanski R, Jutzeler M, Tani K, Yoerger D, Fornari D (2018) The largest deep-ocean silicic volcanic eruption of the past century. Sci Adv 4 (1):e1701121

Carn SA, Pallister JS, Lara LE, Ewert JW, Watt S, Prata AJ, Thomas RJ, Villarosa G (2009) The unexpected awakening of Chaitén volcano, Chile. Trans Am Geophys Union 90(24):205–206

Carrasco-Núñez G, Bernal JP, Dávila P, Jicha B, Giordano G, Hernández J (2018) Reappraisal of Los Humeros volcanic complex by new U/Th zircon and 40Ar/39Ar dating: implications for greater geothermal potential. Geochem Geophys Geosyst 19(1):132–149

Cas RAF (1977) Basin characteristics of the Early Devonian part of the Hill End Trough, New South Wales, based on a stratigraphic analysis of the Merrions Tuff. J Geol Soc Aust 24(7–8):381–401

Cas RAF (1978) Silicic lavas in Paleozoic flyschlike deposits in New South Wales, Australia: Behavior of deep subaqueous silicic flows. Geol Soc Am Bull 89(12):1708–1714

Cas RAF (1979) Mass-flow arenites from a Paleozoic interarc basin, New South Wales, Australia: mode and environment of emplacement. J Sediment Petrol 49(1):29–44

Cas R (1983) A review of the palaeogeographic and tectonic development of the Palaeozoic Lachlan Fold Belt of southeastern Australia. Geol Soc Aust Spec Publ 10:104p

Cas RAF (1992) Submarine volcanism; eruption styles, products, and relevance to understanding the host-rock successions to volcanic-hosted massive sulfide deposits. Econ Geol 87(3):511–541

Cas RAF, Giordano G (2014) Submarine volcanism: a review of the constraints, processes and products, and relevance to the Cabo de Gata volcanic succession. Ital J Geosci 133(3):362–377

Cas RAF, Simmons J (2018) Why deep-water eruptions are so different from subaerial eruptions. Front Earth Sci 6:198

Cas RAF, Wright JV (1987) Volcanic successions: modern and ancient. Allen & Unwin, London, p 528

Cas RAF, Allen RL, Bull SW, Clifford BA, Wright JV (1990) Subaqueous, rhyolitic dome-top tuff cones: a model based on the Devonian Bunga Beds, southeastern Australia and a modern analogue. Bull Volcanol 52(3):159–174

Cas RAF, Edgar C, Allen RL, Bull S, Clifford BA, Giordano G, Wright JV (2001) Influence of magmatism and tectonics on sedimentation in an extensional lake basin: the Upper Devonian Bunga Beds, Boyd Volcanic Complex, south-eastern Australia. In: White JDL, Riggs NR (eds) Volcaniclastic sedimentation in lacustrine settings. Blackwell Science Ltd., Oxford, pp 83–108

Cashman KV (2004) Volatile controls on magma ascent and eruption. In: Sparks RSJ, Hawkesworth C (eds) The state of the planet: frontiers and challenges in geophysics. American Geophysical Union, Washington, D.C., pp 109–124

Cashman KV, Thornber CR, Pallister JS (2008) From dome to dust: Shallow crystallization and fragmentation of conduit magma during the 2004–2006 dome extrusion of Mount St. Helens, Washington (No. 1750–19). US Geological Survey, pp 387–413

Castro J, Cashman KV (1999) Constraints on rheology of obsidian lavas based on mesoscopic folds. J Struct Geol 21(7):807–819

Castro J, Dingwell DB (2009) Rapid ascent of rhyolitic magma at Chaitén volcano, Chile. Nature 461:780–783

Castro JM, Manga M, Cashman KV (2002) The emplacement of Obsidian Dome, California: insights from measurements of crystal orientation. Earth Planet Sci Lett 199:211–226

Castro JM, Dingwell DB, Nichols ARL, Gardner JE (2005) New insights on the origin of flow bands in obsidian. Geol Soc Am Spec Pap 396:55–65

Castro JM, Beck P, Tuffen H, Nichols ARL, Dingwell DB, Martin MC (2008) Timescales of spherulite crystallization in obsidian inferred from water concentration profiles. Am Mineral 93(11–12):1816–1822

Castro JM, Schipper CI, Mueller S, Militzer A, Amigo A, Parejas CS, Jacob D (2013) Storage and eruption of near-liquidus rhyolite magma at Cordón Caulle, Chile. Bull Volcanol 75:702–719

Christiansen RL, Lipman PW (1966) Emplacement and thermal history of a rhyolite lava flow near Fortymile Canyon, southern Nevada. Geol Soc Am Bull 77(7):671–684

Cimarelli C, De Rita D (2006) Structural evolution of the Pleistocene Cimini trachytic volcanic complex (Central Italy). Bull Volcanol 68 (6):538–548

Cioni R, Funedda A (2005) Structural geology of crystal-rich, silicic lava flows: a case study from San Pietro Island (Sardinia, Italy). Geol Soc Am Spec Pap 396:1–14

Clay PL, O'Driscoll B, Gertisser R, Busemann H, Sherlock SC, Kelley SP (2013) Textural characterization, major and volatile element quantification and Ar–Ar systematics of spherulites in the Rocche Rosse obsidian flow, Lipari, Aeolian Islands: a temperature continuum growth model. Contrib Mineral Petrol 165(2):373–395

Clough BJ, Wright JV, Walker GPL (1982) Morphology and dimensions of the young comendite lavas of La Primavera volcano, Mexico. Geol Mag 119(5):477–485

Clough BJ (1981) The geology of La Primavera volcano, Mexico. Imperial College, University of London PhD:452

Coats RR (1968) The Circle Creek Rhyolite, a volcanic complex in northern Elko County, Nevada. Geol Soc Am Mem 116:69–106

Cole JW (1970) Structure and eruptive history of the Tarawera Volcanic Complex. New Zeal J Geol Geophys 13(4):879–902

Cordonnier B, Hess KU, Lavallee Y, Dingwell DB (2009) Rheological properties of dome lavas: case study of Unzen volcano. Earth Planet Sci Lett 279(3–4):263–272

De Astis G, Lucchi F, Dellino P, La Volpe L, Tranne CA, Frezzotti ML, Peccerillo A (2013) Geology, volcanic history and petrology of Vulcano (central Aeolian archipelago). Geol Soc Lond Memoirs 37 (1):281–349

De Rita D, Giordano G (1996) Volcanological and structural evolution of Roccamonfina volcano (Italy): origin of the summit caldera. Geol Soc Lond Spec Pub 110(1):209–224

De Rita D, Giordano G, Cecili A (2001) A model for submarine rhyolite dome growth: Ponza Island (central Italy). J Volcanol Geotherm Res 107(4):221–239

de Silva SL, Francis PW (1991) Volcanoes of the central Andes. Springer, New York, p 216

de Rosen-Spence AF, Provost G, Dimroth E, Gochnauer K, Owen V (1980) Archean subaqueous felsic flows, Rouyn-Noranda, Quebec, Canada, and their Quarternary equivalents. Precambrian Res 12 (1):43–77

de Silva SL, Self S, Francis PW, Drake RE, Carlos RR (1994) Effusive silicic volcanism in the Central Andes: the Chao dacite and other young lavas of the Altiplano-Puna Volcanic Complex. J Geophys Res: Solid Earth 99(B9):17805–17825

Dellino P, Volpe LL (1995) Fragmentation versus transportation mechanisms in the pyroclastic sequence of Monte Pilato-Rocche Rosse (Lipari, Italy). J Volcanol Geotherm Res 64(3):211–231

Dinel E, Saumur BM, Fowler AD (2008) Spherulitic aphyric pillow-lobe metatholeiitic dacite lava of the Timmins area, Ontario, Canada: a new archean facies formed from superheated melts. Econ Geol 103(6):1365–1378

Dingwell DB (1996) Volcanic dilemma-flow or blow? Science 273 (5278):1054–1055

Dirksen O, Humphreys MCS, Pletchov P, Melnik O, Demyanchuk Y, Sparks RSJ, Mahony S (2006) The 2001–2004 dome-forming eruption of Shiveluch volcano, Kamchatka: observation, petrological investigation and numerical modelling. J Volcanol Geotherm Res 155(3–4):201–226

Edwards B, Russell J, Anderson R (2002) Subglacial, phonolitic volcanism at Hoodoo Mountain volcano, northern Canadian Cordillera. Bull Volcanol 64(3–4):254–272

Edwards BR, Gudmundsson T, Russell K (2015) Glaciovolcanism. In: Sigurdsson H, Stix J, Houghton BF, McNutt SR, Rymer H (eds) Encyclopedia of volcanoes. Academic Press, San Diego, pp 377–394

Eichelberger JC, Carrigan CR, Westrich HR, Price RH (1986) Non-explosive silicic volcanism. Nature 323(6089):598–602

Ellis BS, Wolff JA, Boroughs S, Mark DF, Starkel WA, Bonnichsen B (2013) Rhyolitic volcanism of the central Snake River Plain: a review. Bull Volcanol 75(8):745

Embley RW, Rubin KH (2018) Extensive young silicic volcanism produces large deep submarine lava flows in the NE Lau Basin. Bull Volcanol 80(4):36

Fenn PM (1977) The nucleation and growth of alkali feldspars from hydrous melts. Canad Mineral 15(2):135–161

Fink J (1980a) Surface folding and viscosity of rhyolite flows. Geology 8(5):250–254

Fink JH (1980b) Gravity instability in the Holocene Big and Little Glass Mountain rhyolitic obsidian flows, northern California. Tectonophysics 66(1–3):147–166

Fink JH (1983) Structure and emplacement of a rhyolitic obsidian flow: Little Glass Mountain, Medicine Lake Highland, northern California. Geol Soc Am Bull 94(3):362–380

Fink JH, Anderson SW (2000) Lava domes and coulees. In: Sigurdsson H, Stix J, Houghton BF, McNutt SR, Rymer H (eds) Encyclopedia of volcanoes. Academic Press, San Diego, pp 307–319

Fink JH, Griffiths RW (1998) Morphology, eruption rates, and rheology of lava domes: insights from laboratory models. J Geophys Res: Solid Earth 103(B1):527–545

Fink JH, Manley CR (1987) Origin of pumiceous and glassy textures in rhyolite flows and domes. Geol Soc Am Spec Pap 212:77–88

Fink JH, Malin MC, Anderson SW (1990) Intrusive and extrusive growth of the Mount St Helens lava dome. Nature 348(6300):435–437

Fink JH, Anderson SW, Manley CR (1992) Textural constraints on effusive silicic volcanism: beyond the permeable foam model. J Geophys Res: Solid Earth 97(B6):9073–9083

Fink JH (ed) (1987) The emplacement of silicic domes and lava flows. Geol Soc Am Spec Pap 212:146

Fiske RS, Cashman KV, Shibata A, Watanabe K (1998) Tephra dispersal from Myojinsho, Japan, during its shallow submarine eruption of 1952–1953. Bull Volcanol 59(4):262–275

Forni F, Lucchi F, Peccerillo A, Tranne CA, Rossi PL, Frezzotti ML (2013) Stratigraphy and geological evolution of the Lipari volcanic complex (central Aeolian archipelago). In: Lucchi et al. (eds) The Aeolian Islands Volcanoes. Geol Soc Lond Mem 37(1):213–279

Friedman I, Long W (1976) Hydration rate of obsidian. Science 191 (4225):347–352

Furnes H, Fridleifsson IB, Atkins FB (1980) Subglacial volcanics—on the formation of acid hyaloclastites. J Volcanol Geotherm Res 8 (1):95–110

Gardner JE, Carey S, Sigurdsson H (1998) Plinian eruptions at Glacier Peak and Newberry volcanoes, United States: implications for volcanic hazards in the Cascade Range. Geol Soc Am Bull 110 (2):173–187

Garner A, McPhie J (1999) Partially melted lithic megablocks in the Yardea Dacite, Gawler Range Volcanics, Australia: implications for eruption and emplacement mechanisms. Bull Volcanol 61(6):396–410

Gibson IL, Walker GPL (1963) Some composite rhyolite/basalt lavas and related composite dykes in eastern Iceland. Proc Geol Assoc 74 (3):301-IN303

Giordano G, Cas RAF (2001) Structure of the upper Devonian Boyd volcanic complex, south coast New South Wales: implications for the Devonian-Carboniferous evolution of the Lachlan Fold Belt. Aust J Earth Sci 48(1):49–61

Gonnermann HM, Manga M (2003) Explosive volcanism may not be an inevitable consequence of magma fragmentation. Nature 426 (6965):432–435

Goto A (1999) A new model for volcanic earthquake at Unzen Volcano: melt rupture model. Geophys Res Lett 26(16):2541–2544

Goto Y, McPhie J (1998) Endogenous growth of a Miocene submarine dacite cryptodome, Rebun Island, Hokkaido, Japan. J Volcanol Geotherm Res 84(3–4):273–286

Goto Y, Tomiya A (2019) Internal structures and growth style of a quaternary subaerial rhyodacite Cryptodome at Ogariyama, Usu Volcano, Hokkaido, Japan. Front Earth Sci 7:66

Gottsmann J, Dingwell DB (2001) The cooling of frontal flow ramps: a calorimetric study on the Rocche Rosse rhyolite flow, Lipari, Aeolian Islands, Italy. Terra Nova 13(3):157–164

Gregg TKP, Fink JH, Griffiths RW (1998) Formation of multiple fold generations on lava flow surfaces: influence of strain rate, cooling rate, and lava composition. J Volcanol Geotherm Res 80:281–292

Griffiths RW (2000) The dynamics of lava flows. Annu Rev Fluid Mech 32(1):477–518

Griffiths RW, Fink JH (1992) Solidification and morphology of submarine lavas: a dependence on extrusion rate. J Geophys Res: Solid Earth 97(B13):19729–19737

Griffiths RW, Fink JH (1993) Effects of surface cooling on the spreading of lava flows and domes. J Fluid Mech 252:667–702

Griffiths RW, Fink JH (1997) Solidifying Bingham extrusions: a model for the growth of silicic lava domes. J Fluid Mech 347:13–36

Guest JE, Sanchez R (1969) A large dacitic lava flow in northern Chile. Bull Volcanol 33(3):778–790

Hall SH (1978) The stratigraphy of Northern Lipari and the stratigraphy of the Rocche Rosse rhyolite flow and its implications. Unpublish B Sc Thesis, Univesity of Leeds UK

Hammer JE, Cashman KV, Voight B (2000) Magmatic processes revealed by textural and compositional trends in Merapi dome lavas. J Volcanol Geotherm Res 100(1–4):165–192

Hanson RE, Wilson TJ (1993) Large-scale rhyolite peperites (Jurassic, southern Chile). J Volc Geotherm Res 54(3–4):247–264

Harris AJ, Rose WI, Flynn LP (2003) Temporal trends in lava dome extrusion at Santiaguito 1922–2000. Bull Volcanol 65(2–3):77–89

Heiken G, Goff F, Stix J, Tamanyu S, Shafiqullah M, Garcia S, Hagan R (1986) Intracaldera volcanic activity, Toledo Caldera and Embayment, Jemez Mountains, New Mexico. J Geophys Res: Solid Earth 91(B2):1799–1815

Henry CD, Wolff JA (1992) Distinguishing strongly rheomorphic tuffs from extensive silicic lavas. Bull Volcanol 54(3):171–186

Henry CD, Price JG, Rubin JN, Parker DF, Wolff JA, Self S, Franklin R, Barker DS (1988) Widespread, lavalike silicic volcanic rocks of Trans-Pecos Texas. Geology 16(6):509–512

Henry CD, Price JG, Rubin JN, Laubach SE (1990) Case study of an extensive silicic lava: the Bracks Rhyolite, Trans-Pecos Texas. J Volcanol Geotherm Res 43(1–4):113–132

Hermance JF (1983) The Long Valley/Mono basin volcanic complex in eastern California: status of present knowledge and future research needs. Rev Geophys 21(7):1545–1565

Hildreth W, Fierstein J (2012) The Novarupta-Katmai eruption of 1912 — largest eruption of the twentieth century; centennial perspectives. USGS Prof Paper 1791

Houghton BF, Wilson CJ, Del Carlo P, Coltelli M, Sable JE, Carey R (2004) The influence of conduit processes on changes in style of basaltic Plinian eruptions: Tarawera 1886 and Etna 122 BC. J Volcanol Geotherm Res 137(1–3):1–14

Hulme G (1974) The interpretation of lava flow morphology. Geophys J Int 39(2):361–383

Ikegami F, McPhie J, Carey R, Mundana R, Soule SA, Jutzeler M (2018) The eruption of submarine rhyolite lavas and domes in the deep ocean–Havre 2012, Kermadec Arc. Front Earth Sci 6:147. https://doi.org/10.3389/feart.2018.00147

Johnson RW, Smith IE (1974) Volcanoes and rocks of St Andrew Strait, Papua New Guinea. J Geol Soc Aust 21(3):333–351

Jutzeler M, Marsh R, Carey RJ, White JDL, Talling PJ, Karlstrom L (2014) On the fate of pumice rafts formed during the 2012 Havre submarine eruption. Nat Commun 5

Kano K, Takeuchi K, Yamamoto T, Hoshizumi H (1991) Subaqueous rhyolite block lavas in the Miocene Ushikiri Formation, Shimane Peninsula, SW Japan. J Volcanol Geotherm Res 46(3–4):241–253

Kano K (2003) Subaqueous pumice eruptions and their products: a review. In: White JD, Smellie JL, Clague DA (eds) Explosive subaqueous volcanism. Am Geophys Union Monog 140:213–229

Kokelaar BP, Bevins RE, Roach RA (1985) Submarine silicic volcanism and associated sedimentary and tectonic processes, Ramsey Island, SW Wales. J Geol Soc Lond 142(4):591–613

Latutrie B, Harris A, Médard E, Gurioli L (2017) Eruption and emplacement dynamics of a thick trachytic lava flow of the Sancy volcano (France). Bull Volcanol 79(1):4

Lavallée Y, Hess KU, Cordonnier B, Bruce Dingwell D (2007) Non-Newtonian rheological law for highly crystalline dome lavas. Geology 35(9):843–846

Lavallée Y, Varley NR, Alatorre-Ibargüengoitia MA, Hess KU, Kueppers U, Mueller S, Richard D, Scheu B, Spieler O, Dingwell DB (2011) Magmatic architecture of dome-building eruptions at Volcán de Colima, Mexico. Bull Volcanol 74(1):249–260

Lofgren G (1971) Spherulitic textures in glassy and crystalline rocks. J Geophys Res 76(23):5635–5648

Loney RA (1968) Flow structure and composition of the Southern Coulee, Mono Craters, California—a pumiceous rhyolite flow. Geol Soc Am Mem 116:415–440

Lucci F, Carrasco-Núñez G, Rossetti F, Theye T, White JC, Urbani S, Giordano G (2020) Anatomy of the magmatic plumbing system of Los Humeros Caldera (Mexico): implications for geothermal systems. Solid Earth 11(1):125–159

MacDonald GA (1972) Volcanoes. Prentice-Hall, Englewood Cliffs, NJ, p 510

Maeno F, Taniguchi H (2006) Silicic lava dome growth in the 1934–1935 Showa Iwo-jima eruption, Kikai caldera, south of Kyushu, Japan. Bull Volcanol 68(7–8):673–688

Mahood GA (1980) Geological evolution of a pleistocene rhyolitic center—Sierra La Primavera, Jalisco, Mexico. J Volcanol Geotherm Res 8(2–4):199–230

Manga M, Fauria KE, Lin C, Mitchell SJ, Jones M, Conway CE, Degruyter W, Hosseini B, Carey R, Cahalan R (2018) The pumice raft-forming 2012 Havre submarine eruption was effusive. Earth Planet Sci Lett 489:49–58

Manley CR (1992) Extended cooling and viscous flow of large, hot rhyolite lavas: implications of numerical modeling results. J Volcanol Geotherm Res 53(1–4):27–46

Manley CR (1996) Physical volcanology of a voluminous rhyolite lava flow: the Badlands lava, Owyhee Plateau, southwestern Idaho. J Volcanol Geotherm Res 71(2–4):129–153

Martel C, Iacono-Marziano G (2015) Timescales of bubble coalescence, outgassing, and foam collapse in decompressed rhyolitic melts. Earth Planet Sci Lett 412:173–185

Massol H, Jaupart C (2009) Dynamics of magma flow near the vent: Implications for dome eruptions. Earth Planet Sci Lett 279(3-4): 185-196.

McArthur NA, Cas RAF, Orton JG (1998) Distribution and significance of crystalline, perlitic and vesicular textures in the Ordovician Garth Tuff (Wales). Bull Volcanol 60(4):260–285

McBirney AR (1963) Factors governing the nature of submarine volcanism. Bull Volcanol 26(1):455–469

McPhie J, DellaPasqua F, Allen SR, Lackie MA (2008) Extreme effusive eruptions: palaeoflow data on an extensive felsic lava in the Mesoproterozoic Gawler range volcanics. J Volcanol Geotherm Res 172(1–2):148–161

McPhie J, Doyle MG, Allen RL (1993) Volcanic textures: CODES. University of Tasmania, 198p

Melnik O, Sparks RSJ (2005) Controls on conduit magma flow dynamics during lava dome building eruptions. J Geophys Res: Solid Earth 110(B2):B02209

Mills AA (1984) Pillow lavas and the Leidenfrost effect. J Geol Soc 141(1):183–186

Minakami T, Ishikawa T, Yagi K (1951) The 1944 Eruption of Volcano Usu in Hokkaido, Japan. Bull Volcanol 11(1):45–157

Morgavi D, Arzilli F, Pritchard C, Perugini D, Mancini L, Larson P, Dingwell DB (2016) The Grizzly Lake complex (Yellowstone Volcano, USA): mixing between basalt and rhyolite unraveled by microanalysis and X-ray microtomography. Lithos 260:457–474

Mueller S, Melnik O, Spieler O, Scheu B, Dingwell DB (2005) Permeability and degassing of dome lavas undergoing rapid decompression: an experimental determination. Bull Volcanol 67 (6):526–538

Nairn IA (1979) Rotomahana—Waimangu eruption, 1886: base surge and basalt magma. NZ J Geol Geophys 22(3):363–378

Nairn IA, Shane PR, Cole JW, Leonard GJ, Self S, Pearson N (2004) Rhyolite magma processes of the ~AD 1315 Kaharoa eruption episode, Tarawera volcano, New Zealand. J Volcanol Geotherm Res 131(3–4):265–294

Nairn IA, Self S, Cole JW, Leonard GS, Scutter C (2001) Distribution, stratigraphy, and history of proximal deposits from the c. AD 1305 Kaharoa eruptive episode at Tarawera Volcano, New Zealand. NZ J 44(3):467–484

Nakada S, Shimizu H, Ohta K (1999) Overview of the 1990–1995 eruption at Unzen Volcano. J Volcanol Geotherm Res 89(1–4):1–22

Nakada S, Uto K, Sakuma S, Eichelberger JC, Shimizu H (2005) Scientific results of conduit drilling in the Unzen scientific drilling project (USDP). Sci Drilling Rep 1:18–22. https://doi.org/10.5194/sd-5191-5118-2005

Nelson SA (1981) The possible role of thermal feedback in the eruption of siliceous magmas. J Volcanol Geotherm Res 11(2–4):127–137

Neuberg JW, Tuffen H, Collier L, Green D, Powell T, Dingwell D (2006) The trigger mechanism of low-frequency earthquakes on Montserrat. J Volcanol Geother Res 153(1–2):37–50

Newhall CG, Melson WG (1983) Explosive activity associated with the growth of volcanic domes. J Volcanol Geotherm Res 17(1–4):111–131

Nomikou P, Parks MM, Papanikolaou D, Pyle DM, Mather TA, Carey S, Watts AB, Paulatto M, Kalnins ML, Livanos I, Bejelou K (2014) The emergence and growth of a submarine volcano: the Kameni islands, Santorini (Greece). GeoResJ 1:8–18

Okumura S, Nakamura M, Nakano T, Uesugi K, Tsuchiyama A (2010) Shear deformation experiments on vesicular rhyolite: Implications for brittle fracturing, degassing, and compaction of magmas in volcanic conduits. J Geophys Res: Solid Earth 115(B6):B06201

Osmond DI, Griffiths RW (1998) Silicic lava domes on slopes. In: MC Thomson, K Hourigan (eds) Proceedings of the 13th Australasian fluid Mech. Conf., Monash Univ., Melbourne, pp 827–830

Pallister JS, Diefenbach AK, Burton WC, Muñoz J, Griswold JP, Lara LE, Lowenstern JB, Valenzuela CE (2013) The Chaitén rhyolite lava dome: eruption sequence, lava dome volumes, rapid effusion rates and source of the rhyolite magma. Andean Geol 40 (2):277–294

Pankhurst M, Schaefer B, Betts P, Phillips N, Hand MA (2011) Mesoproterozoic continental flood rhyolite province, the Gawler Ranges, Australia: the end member example of the Large Igneous Province clan. Solid Earth 2(1):25–33

Parker DF, White JC, Ren M, Barnes M (2017) Basement control of alkalic flood rhyolite magmatism of the Davis Mountains volcanic field, Trans-Pecos Texas, USA. Lithos 292:234–249

Pensa A, Pinton A, Vita L, Bonamico A, De Benedetti AA, Giordano G (2019) Atlas of Italian submarine volcanic structures. Mem Descr Carta Geol D'it 104:77–183

Perugini D, Ventura G, Petrelli M, Poli G (2004) Kinematic significance of morphological structures generated by mixing of magmas: a case study from Salina Island (southern Italy). Earth Planet Sci Lett 222(3–4):1051–1066

Perugini D, Valentini L, Poli G (2007) Insights into magma chamber processes from the analysis of size distribution of enclaves in lava flows: a case study from Vulcano Island (Southern Italy). J Volcanol Geotherm Res 166(3–4):193–203

Pichler H (1965) Acid hyaloclastites. Bull Volcanol 28(1):293–310

Piochi M, De Astis G, Petrelli M, Ventura G, Sulpizio R, Zanetti A (2009) Constraining the recent plumbing system of Vulcano

(Aeolian Arc, Italy) by textural, petrological, and fractal analysis: the 1739 A.D. Pietre Cotte lava flow. Geochem Geophys 10(1): Q01009

Porreca M, Cifelli F, Soriano C, Giordano G, Romano C, Conticelli S, Mattei M (2014) Hyaloclastite fragmentation below the glass transition: an example from El Barronal submarine volcanic complex (Spain). Geology 42(1):87–90

Porreca M, Cifelli F, Soriano C, Giordano G, Mattei M (2015) Magma flow within dykes in submarine hyaloclastite environments: an AMS study of the Miocene Cabo de Gata volcanic units. Geol Soc Lond Spec Publ 396(1):133–157

Pyle DM, Elliott JR (2006) Quantitative morphology, recent evolution, and future activity of the Kameni Islands volcano, Santorini, Greece. Geosphere 2(5):253–268

Ramsay JG (1967) Folding and fracturing of rocks. McGraw-Hill, New York

Ramsey MS, Fink JH (1999) Estimating silicic lava vesicularity with thermal remote sensing: a new technique for volcanic mapping and monitoring. Bull Volcanol 61(1–2):32–39

Reynolds MA, Best JG, Johnson RW (1980) Summary of the 1953–57 eruption of Tuluman volcano: rhyolitic volcanic activity in the northern Bismarck Sea. Geol Surv Papua New Guinea Mem 7:44p

Reynolds MA, Best JG (1976) Summary of the 1953–57 eruption of Tuluman volcano, Papua New Guinea. In: Johnson RW (ed) Volcanism in Australasia, pp 287–296

Richardson S (1978) The geology of southern Lipari, with particular reference to the rhyolite tholoids in the extreme south of the island. Unpubl BSc Thesis, Univ Leeds

Rotella MD, Wilson CJ, Barker SJ, Wright IC (2013) Highly vesicular pumice generated by buoyant detachment of magma in subaqueous volcanism. Nat Geosci 6(2):129–132

Rotella MD, Wilson CJ, Barker SJ, Schipper CI, Wright IC, Wysoczanski RJ (2015) Dynamics of deep submarine silicic explosive eruptions in the Kermadec arc, as reflected in pumice vesicularity textures. J Volcanol Geotherm Res 301:314–332

Rundle JB, Eichelberger JC (1983) Continental scientific drilling at Long Valley-Mono craters. Trans Am Geophys Union 64(2):12–14

Rust AC, Cashman KV (2004) Permeability of vesicular silicic magma: inertial and hysteresis effects. Earth Planet Sci Lett 228(1–2): 93–107

Rust AC, Manga M (2002) Bubble shapes and orientations in low re simple shear flow. J Colloid Interface Sci 249(2):476–480

Rust AC, Manga M, Cashman KV (2003) Determining flow type, shear rate and shear stress in magmas from bubble shapes and orientations. J Volcanol Geotherm Res 122(1–2):111–132

Ryan AG, Russell JK, Nichols AR, Hess KU, Porritt LA (2015) Experiments and models on H_2O retrograde solubility in volcanic systems. Am Mineral 100(4):774–786

Ryan AG, Russell JK, Heap MJ, Kolzenburg S, Vona A, Kushnir ARL (2019) Strain-dependent rheology of silicate melt foams: importance for outgassing of silicic lavas. J Geophys Res: Solid Earth 124 (8):8167–8186

Schilling SP, Thompson RA, Messerich JA, Iwatsubo Y (2008) Use of digital aerophotogrammetry to determine rates of lava dome growth, Mount St. Helens, Washington, 2004–2005. In: Sherrod DR, Scott WE, Stauffer PH (eds) A volcano rekindled: the renewed eruption of Mount St. Helens, 2004–2006. US Geol Surv Prof Pap 1750, pp 145–167

Schipper CI, Castro JM, Tuffen H, James MR, How P (2013) Shallow vent architecture during hybrid explosive–effusive activity at Cordón Caulle (Chile, 2011–12): evidence from direct observations and pyroclast textures. J Volcanol Geotherm Res 262:25–37

Scutter C, Cas RAF, Moore CL, De Rita D (1998) The facies architecture and origin of a submarine rhyolite lava flow-dome complex, Ponza, Italy. J Geophys Res 103:27,551–527,566

Seaman SJ, Scherer EE, Standish JJ (1995) Multistage magma mingling and the origin of flow banding in the Aliso Seligman AN, Bindeman IN, Watkins JM, Ross AM (2016) Water in volcanic glass: from volcanic degassing to secondary hydration. Geochim Cosmochim Acta 191:216–238

Seligman AN, Bindeman IN, Watkins JM, Ross AM (2016) Water in volcanic glass: from volcanic degassing to secondary hydration. Geochim Cosmochim Acta 191:216–238

Selva J, Acocella V, Bisson M, Caliro S, Costa A, Della Seta M, De Martino P, de Vita S, Federico C, Giordano G, Martino S, Cardaci C (2019) Multiple natural hazards at volcanic islands: a review for the Ischia volcano (Italy). J Appl Volcanol 8(1):5

Shane P, Martin SB, Smith VC, Beggs KF, Darragh MB, Cole JW, Nairn IA (2007) Multiple rhyolite magmas and basalt injection in the 17.7 ka Rerewhakaaitu eruption episode from Tarawera volcanic complex, New Zealand. J Volcanol Geotherm Res 164(1–2):1–26

Sherrod DR, Scott WE, Stauffer PH (2008) A volcano rekindled: the renewed eruption of Mount St. Helens, 2004–2006. US Geol Surv Prof Pap 1750

Shields JK, Mader HM, Caricchi L, Tuffen H, Mueller S, Pistone M, Baumgartner L (2016) Unravelling textural heterogeneity in obsidian: shear-induced outgassing in the Rocche Rosse flow. J Volcanol Geotherm Res 310:137–158

Siswowidjoyo S, Suryo I, Yokoyama I (1995) Magma eruption rates of Merapi volcano, Central Java, Indonesia during one century (1890–1992). Bull Volcanol 57(2):111–116

Smellie JL (2006) The relative importance of supraglacial versus subglacial meltwater escape in basaltic subglacial Tuya eruptions: an important unresolved conundrum. Earth Sci Rev 74:241–268

Smellie JL (2013) Quaternary vulcanism: subglacial landforms. In: Elias SA (ed) The encyclopedia of quaternary science, vol 1. Elsevier, Amsterdam, pp 780–802

Smellie JL (2018) Glaciovolcanism—a 21st century proxy for palaeo-ice. In: Menzies J, van der Meer JJM (eds) Past glacial environments (sediments, forms and techniques). Elsevier, Amsterdam, pp 335–375

Smith JV (1996) Ductile-brittle transition structures in the basal shear zone of a rhyolite lava flow, eastern Australia. J Volcanol Geotherm Res 72(3–4):217–223

Smith JV (2002) Structural analysis of flow-related textures in lavas. Earth Sci Rev 57(3–4):279–297

Smith JV, Houston EC (1994) Folds produced by gravity spreading of a banded rhyolite lava flow. J Volcanol Geotherm Res 63(1–2):89–94

Smith RK, Tremallo RL, Lofgren GE (2001) Growth of megaspherulites in a rhyolitic vitrophyre. Am Mineral 86(5–6):589–600

Smith RL, Bailey RA, Ross CS (1970) Geologic map of the Jemez Mountains, New Mexico. 1:125,000. US Geol Surv Misc Investigations Map I-571

Soriano C, Riggs N, Giordano G, Porreca M, Conticelli S (2012) Cyclic growth and mass wasting of submarine Los Frailes lava flow and dome complex in Cabo de Gata, SE Spain. J Volcanol Geoth Res 231:72–86

Soriano C, Giordano G, Cas RAF, Riggs N, Porreca M (2013) Facies architecture, emplacement mechanisms and eruption style of the submarine andesite El Barronal complex, Cabo de Gata, SE Spain. J Volcanol Geotherm Res 264:210–222

Soriano C, Cas RA, Riggs NR, Giordano G (2016) Chapter 12 Submarine volcanism of the Cabo de Gata magmatic arc in the Betic-Rif Orogen, SE Spain: processes and products. In: Updates in volcanology-from volcano modelling to volcano geology. Intech Open

Sparks RSJ, Murphy MD, Lejeune AM, Watts RB, Barclay J, Young SR (2000) Control on the emplacement of the andesite lava dome of the Soufriere Hills volcano, Montserrat by Degassing-induced crystallization. Terra Nova 12(1):14–20

Stewart AL, McPhie J (2003) Internal structure and emplacement of an Upper Pliocene dacite cryptodome, Milos Island, Greece. J Volcanol Geotherm Res 124(1–2):129–148

Stewart AL, McPhie J (2006) Facies architecture and Late Pliocene—Pleistocene evolution of a felsic volcanic island, Milos, Greece. Bull Volcanol 68(7–8):703–726

Surono JP, Pallister J, Boichu M, Buongiorno MF, Budisantoso A, Costa F, Andreastuti S, Prata F, Schneider D, Clarisse L, Humaida H (2012) The 2010 explosive eruption of Java's Merapi volcano—a '100-year'event. J Volcanol Geotherm Res 241:121–135

Swanson DA, Holcomb RT (1990) Regularities in growth of the mount St. Helens Dacite Dome, 1980–1986. In: Fink J (ed) Lava flows and domes. Springer, Berlin Heidelberg, pp 3–24

Swanson DA, Dzurisin D, Holcomb RT, Iwatsubo EY, Chadwick WW, Casadevall TJ, Ewert JW, Heliker CC (1987) Growth of the lava dome at Mount St. Helens, Washington, (USA), 1981–1983. Geol Soc Am Spec Pap 212:1–16

Tilling RI, Topinka LJ, Swanson DA (1990) Eruptions of mount St. Helens: Past, present, and future, 2nd edn. US Geological Survey, Reston, VA, p 56

Tranne CA, Lucchi F, Calanchi N, Lanzafame G, Rossi PL (2002) Geological map of the island of Lipari (Aeolian Islands). Università di Bologna e INGV, L.A, C. Firenze

Tuffen H, Dingwell D (2005) Fault textures in volcanic conduits: evidence for seismic trigger mechanisms during silicic eruptions. Bull Volcanol 67(4):370–387

Tuffen H, Gilbert J, McGarvie D (2001) Products of an effusive subglacial rhyolite eruption: Bláhnúkur, Torfajökull, Iceland. Bull Volcanol 63(2–3):179–190

Tuffen H, Dingwell DB, Pinkerton H (2003) Repeated fracture and healing of silicic magma generate flow banding and earthquakes? Geology 31(12):1089–1092

Tuffen H, James M, Castro JM, Schipper I (2013) Exceptional mobility of an advancing rhyolitic obsidian flow at Cordón Caulle volcano in Chile. Nat Commun 4:2709

Urbani S, Giordano G, Lucci F, Rossetti F, Acocella V, Carrasco-Nunez G (2020) Estimating the depth and exolution of intrusions at resurgent domes: Los Humeros (Mexico). Solid Earth 11(2); 527–545. https://doi.org/10.5194/se-2019-100

Vallance JW, Schneider DJ, Schilling SP (2008) Growth of the 2004–2006 lava-dome complex at mount St. Helens, Washington. In: Sherrod DR, Scott WE, Stauffer PH (eds) A volcano rekindled: the renewed eruption of Mount St. Helens, 2004–2006. U.S. Geoligical Survey Professional Paper 1750

Van Otterloo J, Cas RAF, Scutter CR (2015) The fracture behaviour of volcanic glass and relevance to quench fragmentation during formation of hyaloclastite and phreatomagmatism. Earth Sci Rev 151:79–116

van Zalinge ME, Cashman KV, Sparks RS (2018) Causes of fragmented crystals in ignimbrites: a case study of the Cardones ignimbrite, Northern Chile. Bull Volcanol 80(3):22

Ventura G (2001) The strain path and emplacement mechanism of lava flows: an example from Salina (southern Tyrrhenian Sea, Italy). Earth Planet Sci Lett 188(1–2):229–240

Ventura G, De Rosa R, Colletta E, Mazzuoli R (1996) Deformation patterns in a high-viscosity lava flow inferred from the crystal preferred orientation and imbrication structures: an example from Salina (Aeolian Islands, southern Tyrrhenian Sea, Italy). Bull Volcanol 57(7):555–562

Voight B, Constantine EK, Siswowidjoyo S, Torley R (2000) Historical eruptions of Merapi Volcano, Central Java, Indonesia, 1768–1998. J Volcanol Geotherm Res 100(1–4):69–138

Von Aulock FW, Nichols ARL, Kennedy BM, Oze C (2013) Timescales of texture development in a cooling lava dome. Geochim Cosmochim Acta 114:72–80

Vona A, Ryan AG, Russell JK, Romano C (2016) Models for viscosity and shear localization in bubble-rich magmas. Earth Planet Sci Lett 449:26–38

Wadge G, Ryan G, Calder ES (2009) Clastic and core lava components of a silicic lava dome. Geology 37(6):551–554

Wadge G, Robertson REA, Voight B (eds) (2014) The eruption of Soufrie`re Hills Volcano, Montserrat from 2000 to 2010. Geol Soc Lond Mem 39

Walker GPL (1973) Lengths of lava flows. Phil Trans Roy Soc Lond Math, Phys Eng Sci 274(1238):107–118

Walker GPL, Self S, Wilson L (1984) Tarawera 1886, New Zealand—a basaltic plinian fissure eruption. J Volcanol Geotherm Res 21 (1–2):61–78

Watkins J, Manga M, Huber C, Martin M (2009) Diffusion-controlled spherulite growth in obsidian inferred from H_2O concentration profiles. Contrib Mineral Petrol 157(2):163–172

Watts RB, Herd RA, Sparks RSJ, Young SR (2002) Growth patterns and emplacement of the andesitic lava dome at Soufrière Hills Volcano, Montserrat. Geol Soc Lond Mem 21(1):115–152

White JDL, McPhie J, Soule SA (2015) Chapter 19 submarine lavas and hyaloclastite. In: Sigurdsson H, Stix J, Houghton BF, McNutt SR, Rymer H (eds) Encyclopedia of volcanoes. Academic Press, San Diego, pp 363–376

White JDL, Smellie JL, Clague DA (2003) Introduction: a deductive outline and topical overview of subaqueous explosive volcanism. In: White JDL, Smellie JL, Clague DA (eds) Explosive subaqueous volcanism. American Geophysical Union Monograph, pp 1–23

Williams H (1932) The history and characters of volcanic domes. Univ Calif Pub Bull Dept Geol Sci 21(5):51–146

Wolff JA, Gardner JN (1995) Is the Valles caldera entering a new cycle of activity? Geology 23(5):411–414

Wright IC, Worthington TJ, Gamble JA (2006) New multibeam mapping and geochemistry of the 30–35 S sector, and overview, of southern Kermadec arc volcanism. J Volcanol Geotherm Res 149 (3–4):263–296

Yamagishi H (1991) Morphological and sedimentological characteristics of the Neogene submarine coherent lavas and hyaloclastites in Southwest Hokkaido, Japan. Sediment Geol 74(1–4): 5–23

Yamagishi H, Dimroth E (1985) A comparison of Miocene and Archean rhyolite hyaloclastites: evidence for a hot and fluid rhyolite lava. J Volcanol Geotherm Res 23(3–4):337–355

Zimanowski B, Büttner R (2003) Phreatomagmatic explosions in subaqueous volcanism. In*:* White JDL, Smellie JL, Clague DA (eds) Explosive subaqueous volcanism, vol 140. American Geophysical Union Monograph, pp 51–60

7 Komatiite Lavas and Intrusions

Understanding Their Geology

Ray Cas, John V. Wright, and Guido Giordano

R. Cas et al., *Volcanology*, Springer Textbooks in Earth Sciences, Geography and Environment,
https://doi.org/10.1007/978-3-319-66613-6_7

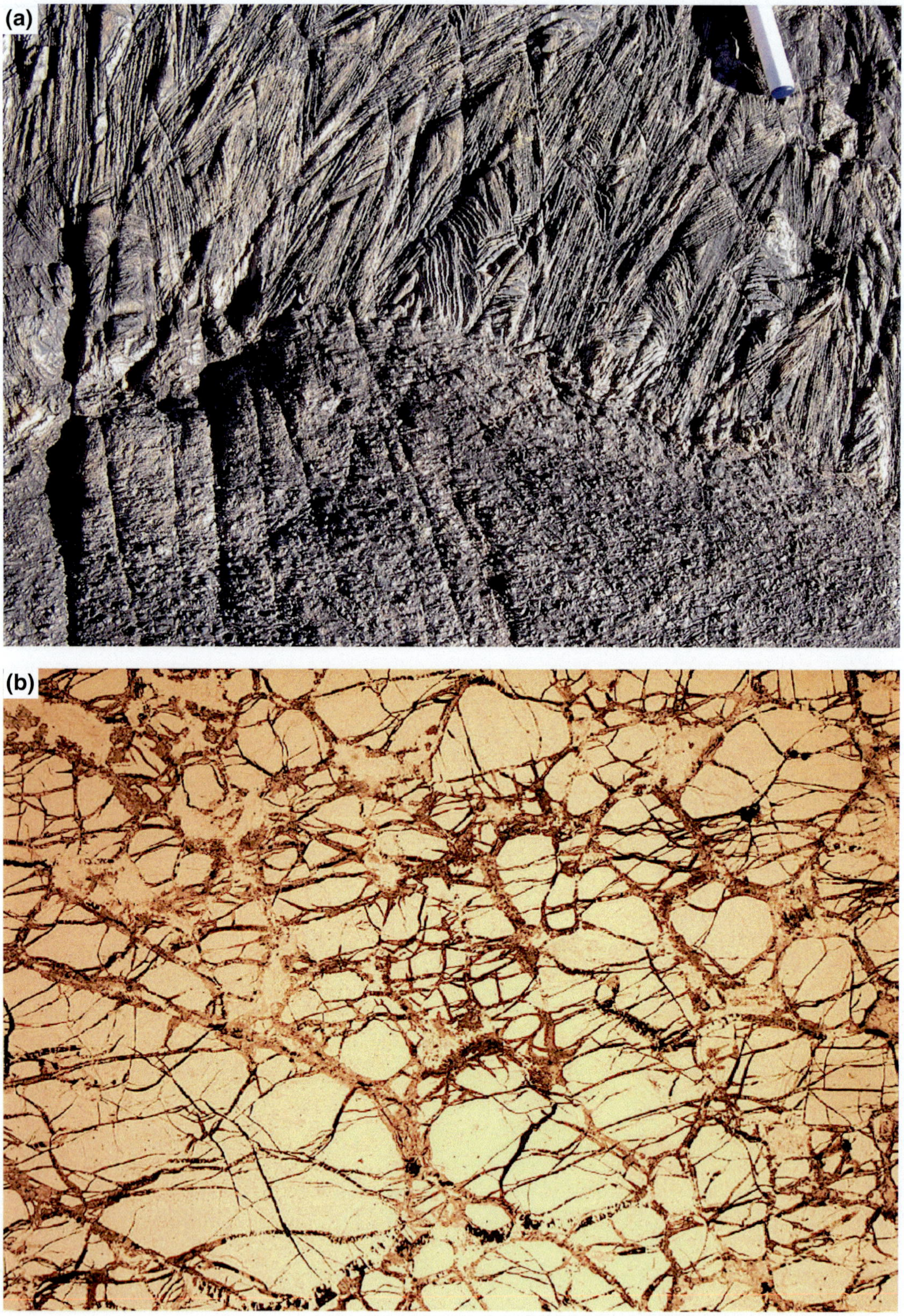

Plate 7.1 **a** Cross-sectional outcrop view through the contact between spectacular etched blade or plate spinifex (zone A3, top) and olivine orthocumulate of zone B2 (the small protruding knobs are olivine crystals, bottom) at the type locality for thin spinifex-cumulate komatiite lava flows, Pyke Hill, Abitibi greenstone belt, Ontario, Canada. **b** Photomicrograph of komatiite intrusion showing adcumulate texture, with no interstitial groundmass between olivine crystals, Black Swan mine, eastern Yilgarn Craton, Western Australia

Abstract

In this Chapter, we consider ultra-low-viscosity lavas. These are represented principally by Archean ultramafic komatiites which form integral components of Archean greenstone belts. Because of their ultramafic, extremely low SiO_2 (<45 wt %) and very high MgO contents (>18 wt %), komatiites are renowned for their olivine-rich mineralogy and unique spinifex and cumulate crystalline textures. We briefly review the characteristics of komatiite magmas, their emplacement origins, and their distinctive textures, based on understanding their geology. They were erupted at very high temperatures (up to 1,650 °C) and were also very low-viscosity magmas ($\sim$1 Pa s, approximately that of olive oil), which has led previous workers to conclude they had the potential to cause substrate melting and to flow in a turbulent flow state. The komatiite rock suite includes both lavas and subvolcanic intrusions, but differentiating them has sometimes been controversial, and we therefore address the main criteria to distinguish them. Many of the facies characteristics of komatiites are based on the morphology and variations in the morphology of olivine and pyroxene crystals through the profile of komatiite units, and we therefore consider the crystallisation dynamics that produce such variations, which reflect flow and cooling processes in komatiite lavas and intrusions. Both komatiite lavas and intrusions are very important as major sources of nickel sulphide ore deposits. Although they represent a unique silicate magma type, komatiites share some physical properties with other very low-viscosity lavas such as carbonatites and sulphur lavas.

7.1 Introduction

Komatiite lavas are ultramafic lavas that were erupted almost exclusively and episodically during the Archean, indicating a petrogenesis tied to the complex, early history of the Earth (Plate 7.1a; Fig. 7.1; Arndt et al. 2008; Herzberg et al. 2010; Condie and O'Neill 2010). Their eruption has never been witnessed and they have only rarely been erupted since the Archean and Paleoproterozoic (e.g. Permo-Triassic from Vietnam, Hanski et al. 2004; Cretaceous from Gorgona Island, Colombia, Echeverria 1982, Kerr et al. 1996). Therefore, understanding their volcanological behaviour depends on fully understanding the geology of preserved komatiite rock successions, which are mostly found within deformed and metamorphosed Archean greenstone belts.

Most komatiite magmas are thought to have originated from deep in the mantle and to be associated with deeply sourced, mantle plume systems (Campbell et al. 1989; Barnes et al. 2012; Mole et al. 2014), although some argue for shallower derivation (e.g. Wyman 2020). Although komatiites no longer erupt today, they are extremely important to society in being a major source of nickel on Earth. They are therefore a major exploration target for nickel sulphide (Ni-Cu-PGE) mineralisation in the mining industry (Chap. 18, Sect. 18.6). Almost all preserved

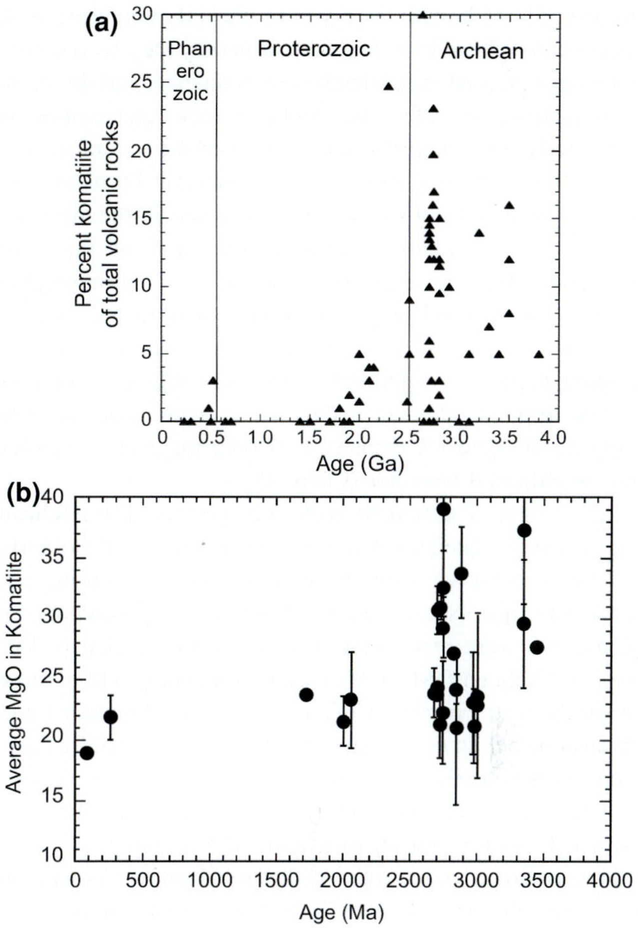

Fig. 7.1 Timing and MgO content of komatiites through Earth history. **a** Frequency of komatiites in greenstone belts as a function of age. Komatiites are thought to track the temperature of the mantle, and more specifically that part of the mantle from which mantle plumes rise (Arndt et al. 2008). An important feature of the graph is that it shows an abrupt rather than a gradual change in komatiite frequency at the end of the Archean; a decrease that does not follow the calculated gradual cooling curve of the mantle as predicted by the decay of radiogenic heat sources. (Modified from Condie and O'Neil 2010, After Dostal 2007, and de Wit and Aswal 1997). **b** Average MgO content versus age for komatiites. The average MgO content of komatiites is one way to track the average eruptive temperature of the magma and of the mantle sources. Each point is the average MgO content for komatiites from a given greenstone belt. The results show a clear decrease in the average MgO content of komatiites at the end of the Archean. As with komatiite frequency, the decrease in MgO at the end of the Archean reflects a decrease in mantle temperatures at this time. (From Condie and O'Neill 2010)

occurrences of komatiites occur in ancient submarine rock successions. There is no logical reason why they would only have been erupted in submarine environments, and therefore their sparsity in subaerial Archean rock successions is attributed to the relative paucity of subaerial environments in the Archean because large areas of emergent buoyant continental crust did not evolve until very late in the Archean (Chap. 17). In addition, uplift and erosion of much of whatever Archean subaerial settings and stratigraphic records that existed may have contributed to the lack of subaerial komatiites. They are very rarely known to have erupted explosively, indicating lack of sufficient volatiles and/or eruptions at water depths where hydrostatic pressure suppressed explosive expansion of both magmatic volatiles and superheated seawater (Chap. 3).

The origin of ultramafic rock units preserved in Archean supracrustal sequences was uncertain for most of the early twentieth century, with most researchers favouring an intrusive origin (see review by Arndt et al. 2008, chap. 1). Komatiites were first interpreted as volcanic rocks by Viljoen and Viljoen (1969a, b), based on mapping and naming of the Komati Formation of the Barberton Mountain Land greenstone belt in northeastern South Africa. Komatiite lavas were subsequently recognised in most Archean cratons around the world, particularly in Canada, Australia, and Africa. However, not all rock units of komatiitic composition were extrusive in origin, as discussed below. As with all magmatic systems, komatiitic volcanic rocks must have subvolcanic intrusive counterparts, especially given that the density of ultramafic magmas is denser than almost all crustal rocks and all other magmas. Although komatiites are an integral part of the geology of Archean greenstone belts, they only form relatively minor or subordinate proportions of the magmatic products. In most greenstone belts, typically they constitute < 15% of the total volume of volcanic rocks although, in some areas, may form up to 30% of the volcanic succession (Fig. 7.1a).

Because of their ultramafic, extremely low SiO_2 (<45 wt %) and very high MgO contents (>18 wt%), komatiites developed olivine-rich mineralogy and unique *spinifex* and *cumulate* crystalline textures (Figs. 7.2a, b, 7.3), which have intrigued geologists for a long time. They were erupted at very high temperatures (up to 1,650 °C), with the potential to cause some substrate melting, although this process was not as widespread as originally proposed. Komatiite magmas were also very low-viscosity magmas ($\sim$0.1–10 Pa s; about that of olive oil) because of their low SiO_2 content and very high eruption temperatures, and therefore could have flowed in a turbulent flow state (Huppert et al. 1984), especially near the vent. However, they are unlikely to have maintained turbulent flow behaviour away from the vent. As flows spread, the flow state would have decayed rapidly to a

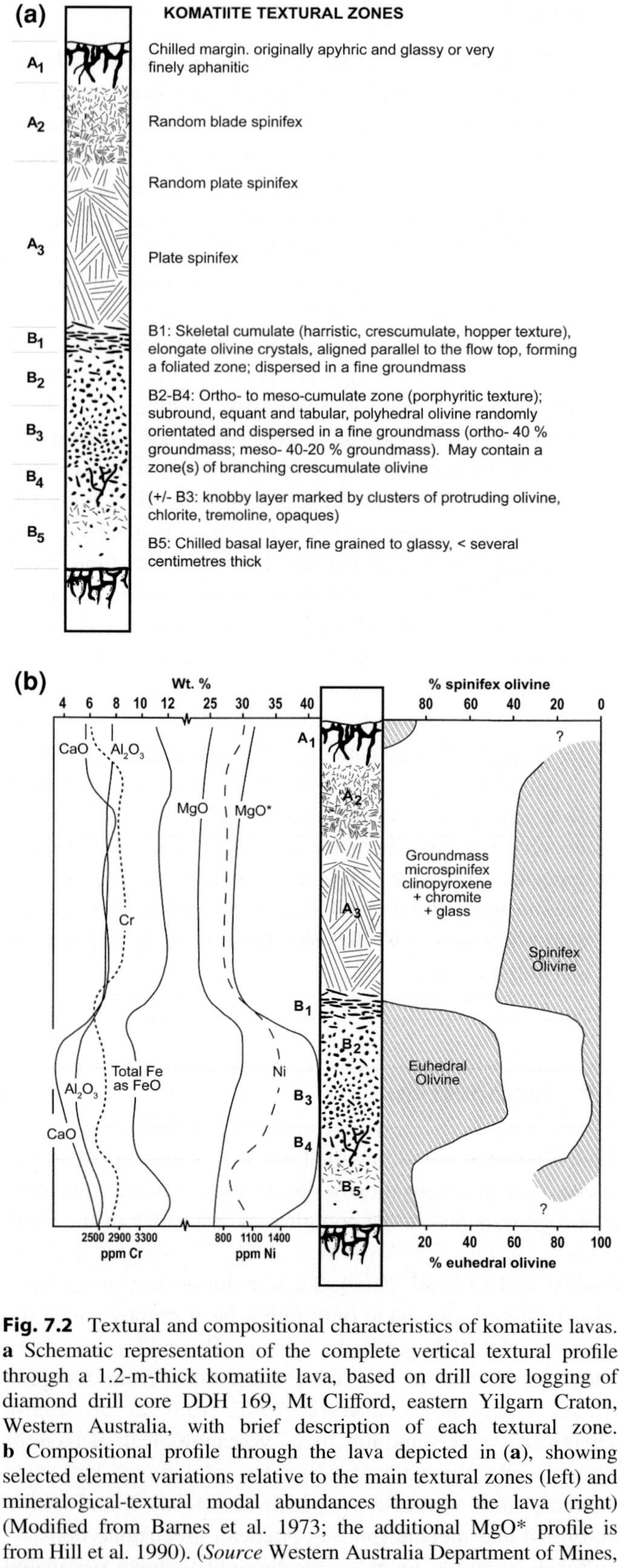

Fig. 7.2 Textural and compositional characteristics of komatiite lavas. **a** Schematic representation of the complete vertical textural profile through a 1.2-m-thick komatiite lava, based on drill core logging of diamond drill core DDH 169, Mt Clifford, eastern Yilgarn Craton, Western Australia, with brief description of each textural zone. **b** Compositional profile through the lava depicted in (**a**), showing selected element variations relative to the main textural zones (left) and mineralogical-textural modal abundances through the lava (right) (Modified from Barnes et al. 1973; the additional MgO* profile is from Hill et al. 1990). (*Source* Western Australia Department of Mines, Industry Regulation and Safety, through Creative Commons Attribution-Share Alike 4.0 International licence)

Table 7.1 Representative geochemical analyses (weight %) of komatiites; komatiitic basalt; and, for comparison, tholeiitic basalt from the Archean Karelian Craton, Finland (From Auvray et al. 1982)

	1	*2*	*3*	*4*	*5*	*6*	*7*	*8*
SiO_2	42.73	42.40	44.44	46.79	47.34	49.13	50.61	50.58
Al_2O_3	2.14	2.49	3.54	7.87	7.75	10.99	14.82	13.06
FeO*	10.56	10.44	12.91	10.54	10.76	11.53	11.07	14.74
MnO	0.16	0.10	0.24	0.17	0.18	0.23	0.22	0.26
MgO	43.53	42.57	34.32	26.04	25.80	14.98	8.31	6.58
CaO	0.61	1.72	4.27	7.73	7.60	10.99	11.70	9.34
Na_2O	–	0.10	–	0.32	0.09	1.23	2.14	2.84
K_2O	–	0.02	0.06	0.03	0.04	0.18	0.25	0.51
TiO_2	0.11	0.13	0.22	0.40	0.37	0.65	0.78	1.91
P_2O_5	0.04	0.02	0.05	0.08	0.08	0.09	0.11	0.18
n=	19	10	10	17	3	16	48	24

n, Number of analyses.
(1) Serpentinites (cumulates) with MgO > 40%
(2) "Soapstones"
(3) Komatiites with serpentine (cumulates) with 30% < MgO < 40%
(4) Komatiites (18% < MgO < 30%)
(5) Komatiites with spinifex texture
(6) Komatiitic basalt (12% < MgO < 18%)
(7) Komatiitic basalt/basalt (MgO < 12%)
(8) Tholeiites

laminar flow state (Cas et al. 1999). Although they represent a unique silicate magma type, komatiites share some physical properties with other very low-viscosity lavas such as carbonatites and sulphur lavas.

In this chapter, we will briefly review the characteristics of komatiite magmas, their emplacement origins, and the distinctive textures produced in the resulting lavas and intrusions (Plate 7.1a, b), based on understanding their geology. A comprehensive review of komatiites was undertaken by Arndt et al. (2008), while shorter but useful reviews have been presented by Dostal (2008) and McPhie and Cas (2015). In Chap. 18, Sect. 18.6, the origin and nature of the magmatic nickel sulphide mineralisation hosted by komatiites is reviewed. Because the main facies characteristics of komatiites are based on the morphology and variations in the morphology of olivine and pyroxene crystals through the profile of komatiite units, we will consider the crystallisation dynamics that produce such variations, which reflect flow and cooling processes in komatiite lavas and intrusions better than in any other compositional magma type. The komatiite rock suite includes both lavas and subvolcanic intrusions, and there has been a substantial amount of debate about how to distinguish them, which we will address.

In Chap. 17, Sect. 17.7, we discuss the importance of komatiites and plume dynamics in understanding the tectonic regime during the Hadean and Archean, which was very different to the plate tectonic regime that has dominated Earth tectonics during the Proterozoic and Phanerozoic.

7.2 Chemical Composition and Physical Properties of Komatiite Magmas

Komatiite lavas have unique ultramafic chemical compositions (Table 7.1). Although komatiites have some superficial compositional similarities to kimberlites, which were erupted on Earth after the Archean, they have some significant differences, such as low incompatible element concentrations and low-to-moderate Al_2O_3 content (which are both high in kimberlites; Arndt 2003). Both magma types are thought to be sourced from deep in the mantle, and both may contain rare diamonds, although commercial quantities have only ever been discovered in kimberlites (Capdevila et al. 1999; Chap. 18).

Conventional wisdom among petrologists is that komatiites are high-temperature, low-viscosity melts produced by high degrees (50–30%) of anhydrous melting of mantle plumes (e.g. Campbell et al. 1989; Arndt 2003; Arndt et al. 1998, 2008; Dostal 2008). Because MgO contents of anhydrous magmas are proportionally related to their melting temperature (higher MgO results from higher melting temperatures), komatiite melts require significantly higher melting temperatures than 1250–1350 °C responsible for recent MORB basalts. Petrological arguments (higher MgO contents; Fig. 7.2b) suggest during the Archean, komatiite melting temperatures ranged from 1600 to 1900 °C occurring at depths of > 450–150 km. Komatiite magmas have therefore been related to deep-seated mantle plumes, which originate as temperature instabilities that formed at the core–mantle boundary (Chap. 17, Sect. 17.7.2.1). More recently, because of the extreme melting temperatures that need to be invoked, it has also been proposed komatiites could have been generated by hydrous melting of the mantle wedge above subduction zones, where the presence of water may significantly reduce melting temperatures (Grove and Parman 2004, Parman et al. 2004). However, this assumes that plate subduction tectonics existed during the Archean, which is reviewed in Chap. 17. While debate regarding "wet" versus "dry" melting and the source of hydrous material continues (e.g. alternatively from a deep reservoir in the mantle transition zone), it appears many, if not most komatiites are likely plume related (Sobolev et al. 2016). There is in fact little or no evidence for a Phanerozoic-style plate tectonic and subduction tectonic regime during the Archean (e.g. Moyen and Laurent 2018; Chap. 17).

The compositions and textures of komatiite rocks can vary significantly from lava to lava (Table 7.1), and vertically through the profile of individual komatiite lavas (Figs. 7.2, 7.3). Cumulate rocks have high olivine crystal content, especially adcumulate rocks, compared with spinifex-textured zones, and as a result their SiO_2 content is low, and MgO content is very high (Fig. 7.2b). In addition, there are rocks called komatiitic basalts, which have higher SiO_2 ($\sim$49–53 wt%) and lower MgO ($\sim$10–18 wt%) contents than komatiites, whereas tholeiitic basalts have MgO contents < 10 wt% (Table 7.1). The compositional differences between komatiites and higher SiO_2 komatiitic basalts are reflected in their mineralogy. In komatiites, clinopyroxene usually only occurs as fine-grained skeletal-spinifex groundmass mineral crystals, whereas in komatiitic basalts it can form abundant well-formed phenocrysts, as well as coarse spinifex crystals, occurring as a major spinifex phase. Tholeiitic basalts don't form spinifex crystal forms, indicating that eruption temperature and chemical composition play a significant role in controlling crystal morphology (Donaldson 1976, 1982a).

Since the liquidus temperature of komatiite is extremely high at $\sim$ 1520 °C (Arndt 1976), and because it has a very low SiO_2 content, the viscosity of komatiite magmas has been estimated to be between 0.1 and 10 Pa s (Huppert and Sparks 1985), 10 to 100 times lower than that of basalt. Eruption temperatures have been estimated to range from 1300 °C to as high as 1650 °C based on preserved mineralogy and the preservation of some aphyric komatiites. These physical properties affected the eruption and flow behaviour of komatiite lavas as discussed further below.

7.3 Mineralogy, Crystal Morphology, and Crystal Textural Profile of Komatiite Lavas

Because of their ultramafic compositions, most komatiites are altered and metamorphosed, resulting from emplacement in subaqueous settings, which facilitate hydration, and subsequent low-to-high-grade regional metamorphism and hydrothermal alteration. They are commonly altered to serpentine, chlorite, metamorphic amphiboles, talc, and carbonate, which commonly overprint or pseudomorph original mineralogy and textures. Nonetheless, there are enough weakly to moderately altered komatiites preserved that illustrate what the original mineralogy and geochemistry of komatiites was. The principal minerals in komatiite are olivine, clinopyroxene, and chromite. The morphological variations of olivine in komatiite lavas led to the recognition that komatiite lavas preserve an asymmetrical textural zonation, consisting of an upper *spinifex* textural zone, called zone A, and a lower *cumulate* textural zone, B (Fig. 7.2). The spinifex A zone is marked by long skeletal plate or

Fig. 7.3 Textural characteristics and zones of komatiite lavas and intrusions. All drill core is $\sim$ 5 cm diameter. **a** Aphyric, quench fragmented zone at top of a komatiite lava (Zone A1), Kambalda, eastern Yilgarn Craton, Western Australia. **b** Random spinifex texture in drill core (Zone A2), Alexo, Abitibi greenstone belt, Ontario, Canada. The dark interval at the left-hand end of the core is also random spinifex but has been altered by chlorite, making the texture cryptic. **c** Random plate spinifex, (Zone A2), etched by weathering, Pyke Hill, Abitibi greenstone belt, Ontario, Canada. **d** Random, dendritic spinifex, (Zone A2), Dundonald, Abitibi greenstone belt, Ontario, Canada. **e** Random spinifex, (Zone A2), Kambalda, eastern Yilgarn Craton, Western Australia. **f** Random spinifex in thin section, (Zone A2), Dundonald, Abitibi greenstone belt, Ontario, Canada. Field of view $\sim$ 3 mm across; ppl. Note the pyroxene microspinifex in the groundmass. **g** Plate spinifex, (Zone A3), Pyke Hill, Abitibi greenstone belt, Ontario, Canada. **h** Plate or blade spinifex, (Zone A3), with a horizon of skeletal hopper olivine in the middle, Dundonald, Abitibi greenstone belt, Ontario, Canada. **i** Plate or blade spinifex, (Zone A3), Nimbus prospect, east of Kalgoorlie, eastern Yilgarn Craton, Western Australia. **j** Plate or blade spinifex, (Zone A3), Cliffs komatiite, Mt Keith, eastern Yilgarn Craton, Western Australia. **k** Skeletal hopper texture (Zone B1), Black Swan mine, eastern Yilgarn Craton, Western Australia. **l** Skeletal cumulate texture, (Zone B1), Dundonald, Abitibi greenstone belt, Ontario, Canada. **m** Orthocumulate, porphyritic texture, (Zone B2–B4), with phenocrysts suspended in groundmass, Dundonald, Abitibi greenstone belt, Ontario, Canada. **n** Mesocumulate texture, (Zone B2–B4) with coarse crystals touching and interstitial groundmass replaced by carbonate, Alexo, Abitibi greenstone belt, Ontario, Canada. **o** Adcumulate texture, with totally interfacing coarse olivine crystals and no groundmass, Mt Keith, eastern Yilgarn Craton, Western Australia. **p** Photomicrograph of adcumulate texture, with no interstitial groundmass between olivine crystals, Black Swan mine, eastern Yilgarn Craton, Western Australia. Field of view $\sim$ 10 mm across; ppl. **q** Dendritic, crescumulate texture in B zone in peridotite, Dundonald, Abitibi greenstone belt, Ontario, Canada. **r** Dendritic, crescumulate texture in gabbro in Centre Hill Complex, Abitibi greenstone belt, Ontario, Canada. **s** Dense dendritic, crescumulate texture in B zone, Dundonald, Abitibi greenstone belt, Ontario, Canada ▶

(a)
(b)
(c)
(d)
(e)
(f)
(g)

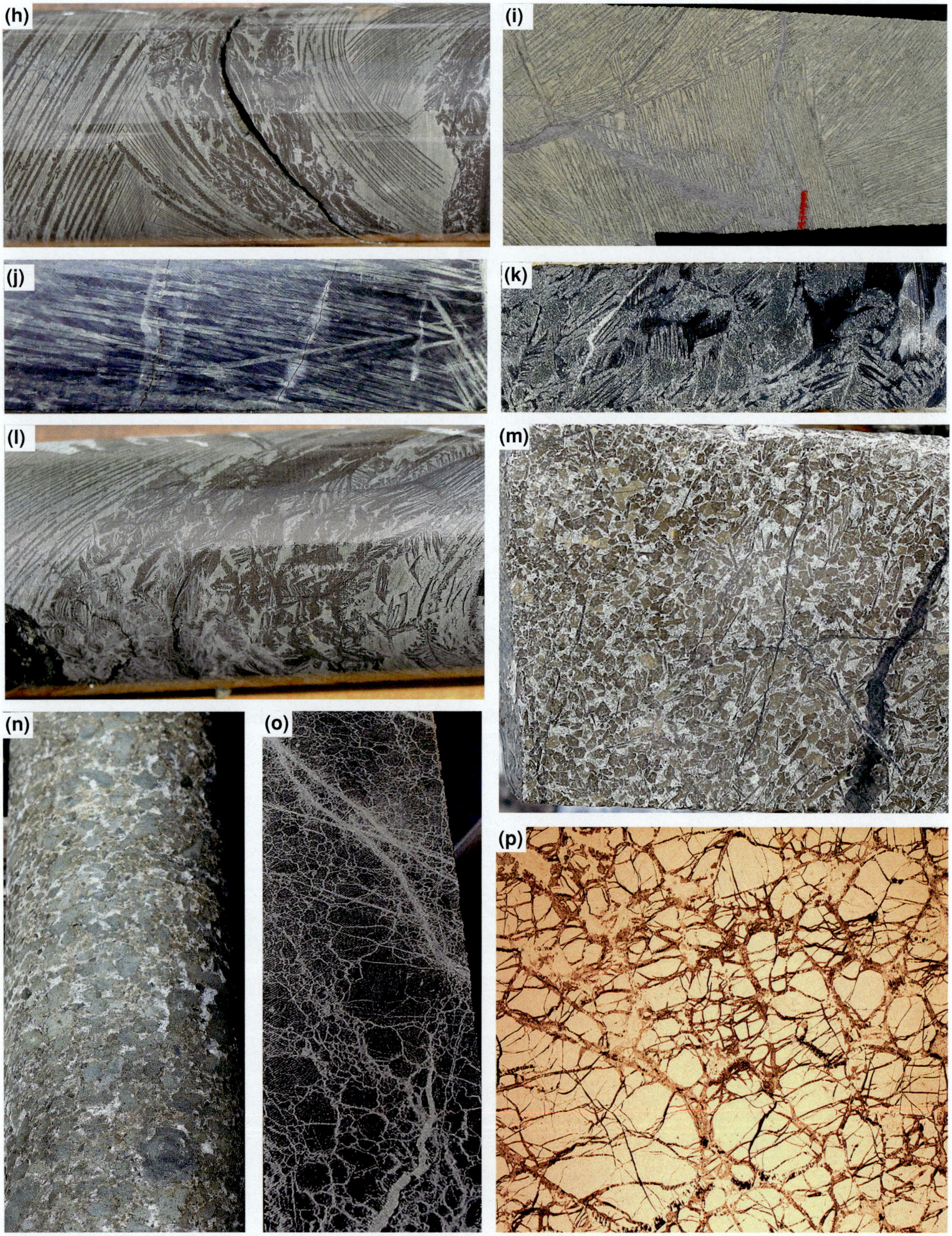

Fig. 7.3 (continued)

Fig. 7.3 (continued)

blade-shaped crystals of olivine (and pyroxene in some cases) and has been subdivided into number of subzones (1–3; Fig. 7.2), whereas the lower cumulate textural zone is marked by equant polyhedral olivine crystals and has also been subdivided into a number of textural zones, 5 in this case (B1–5). However, not all of these textural zones are necessarily present in each komatiite lava, which is controlled by compositional variations, eruption temperature, flow thickness, and internal cooling rate.

7.3.1 Mineralogy and Crystallisation Dynamics

In fresh komatiite lavas, the principal primary magmatic mineral phase is forsterite olivine, which occurs in a number of morphological forms through the profile of the lavas (Figs. 7.2–7.5). The crystal morphology is largely controlled by the initial magma composition, especially MgO content, the eruption and emplacement temperature, the temperature gradient in the komatiite after emplacement, the cooling rate, and the degree of undercooling (also called supercooling; Donaldson 1976, 1982a; Shore and Fowler 1999; Faure et al. 2006). Super- or undercooling is measured as the temperature difference (ΔT) between the liquidus of a magma and the temperature to which it can be cooled below the liquidus temperature without crystallisation of specific mineral phases occurring. Cooling rate is fundamental to degree of supercooling.

Donaldson (1976, 1982a) found experimentally that in peridotitic (komatiitic) and basaltic magmas, the morphology of olivine and pyroxene crystals depends on the cooling rate and the degree of under- or supercooling, ΔT, which is the temperature difference between the liquidus and the crystallisation temperature. He found that each crystal shape has a specific temperature range of stability, and that with increasing ΔT, or more rapid cooling, there was a systematic

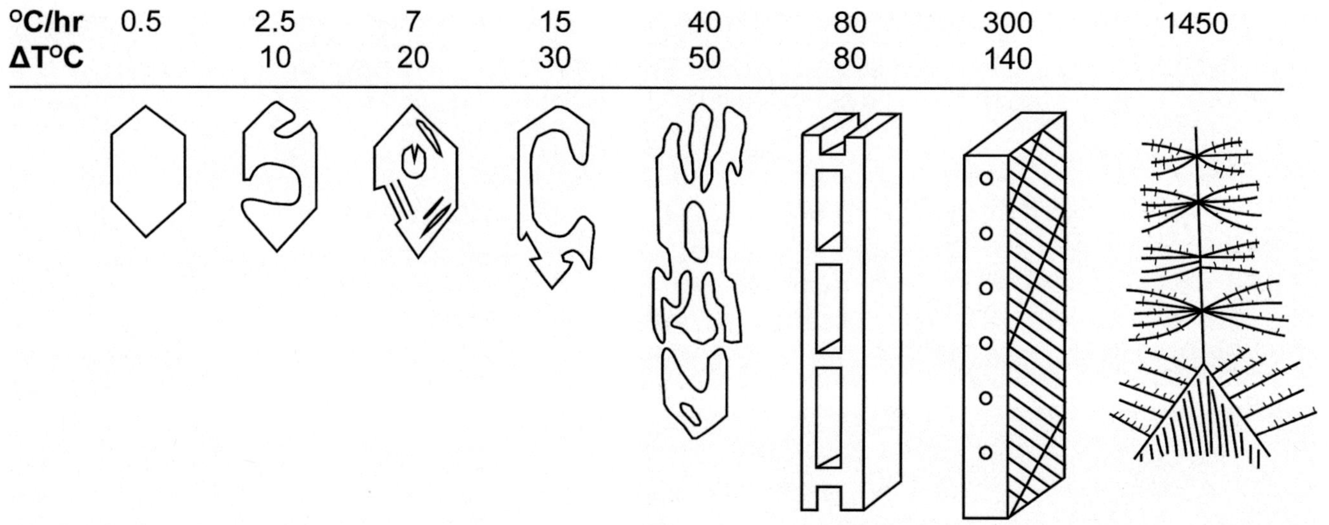

Fig. 7.4 Morphology of olivine crystals from experimental studies, showing cooling rate (°C/hr) and degree of under-/supercooling (ΔT°C) (From Donaldson 1976)

change or order in the morphology of crystals, as follows: Polyhedral or granular, sub-equal and skeletal hopper, elongate hopper, skeletal branching, randomly oriented chain, parallel growth chain, plate or feather (Fig. 7.4). Donaldson (1976, 1982a) proposed that skeletal olivine crystal growth was less controlled by cooling rate than by rapid growth rate caused by high MgO melt content and high degrees of undercooling, which leads to saturation of the melt with regard to olivine. Nonetheless, skeletal crystals (spinifex, hopper, branching, or dendritic) form under conditions of relatively high cooling rates and high degrees of undercooling, which occur near the surface of a lava, compared with whole, euhedral, polyhedral crystals, which grow steadily at low cooling rates and little or no undercooling, within the insulated interior of a lava (Fig. 7.4; Donaldson 1976, 1982a).

Marsh (1988) calculated that granular or polyhedral olivine (like many other common magmatic crystals) grow at rates of 10^{-9} mm s^{-1} under equilibrium conditions, which means a 1 mm diameter, whole, equant olivine crystal will take ~ 30 yr to grow, a 2 mm crystal will take ~ 60 yr, and a 3 mm crystal ~ 90 yr. By contrast, O'Driscoll et al. (2007) found that skeletal olivines grow at rates of 10^{-7} to 10^{-5} mm s^{-1}, giving growth times of hours to days at the most.

Shore and Fowler (1999) proposed that spinifex grows from the cooled, lava surface crust into underlying hot komatiite liquid, across a very high thermal gradient, effected in part by the interface of subaqueous komatiite lavas with ambient water. This provides a thermally constrained crystallisation regime by which heat is transmitted through the growing olivine crystal by conductive and radiative heat flux. The part of the crystal in the cooled crust acts as a heat sink.

Faure et al. (2006) and Welsch et al. (2012) found experimentally that random spinifex forms when the cooling rate (>47 °C hr^{-1}) and the thermal gradient are high (> 20 °C cm^{-1}; Fig. 7.5a), as may be expected in the uppermost crust of komatiite lavas (A2 subzone), especially in submarine environments. However, below this upper chilled crustal layer, where the cooling rate is low (1 to 5 °C hr^{-1}), and low-temperature gradients of between 7 and 35 °C cm^{-1} occur, platy olivine (and needle-like pyroxene) spinifex grew (Fig. 7.5a), with long axes commonly perpendicular to the lava and crust surface (A3 subzone). Olivine spinifex crystals grew when the degree of under-/supercooling, ΔT, was between -37 and -56 °C, whereas for clinopyroxene spinifex, ΔT was $\geq$ -46 °C (Faure et al. 2006; Fig. 7.5b). This is well below the equilibrium liquidus temperature at which, under slow-cooling rates, equant, euhedral, polyhedral olivine crystals grow (Donaldson 1976, 1982a). Welsch et al. (2012) found that skeletal hopper like crystals grow at ΔT between -20 and -60 °C, and whole, polyhedral olivine crystals grew at ΔT of < -20 °C, and low-temperature gradients. The liquidus temperature of olivine in magmas is dependent on the magma composition, especially MgO. In their experiments, Faure et al. (2006) found the olivine liquidus to be ~ 1320 °C, and that of clinopyroxene ~ 1280 °C, at atmospheric pressure, whereas Welsch et al. (2012) cite temperatures of 1230 °C for olivine and 1140 °C for clinopyroxene. Olivine will thus begin to crystallise before clinopyroxene during undercooling, which is further enhanced at high MgO contents. Calculations of crystal growth rates under the cooling rate conditions of the experiments suggest that olivine crystals from 0.5 to 5 mm long grow in time periods ranging from 8 min to 56 days (Welsch et al. 2012).

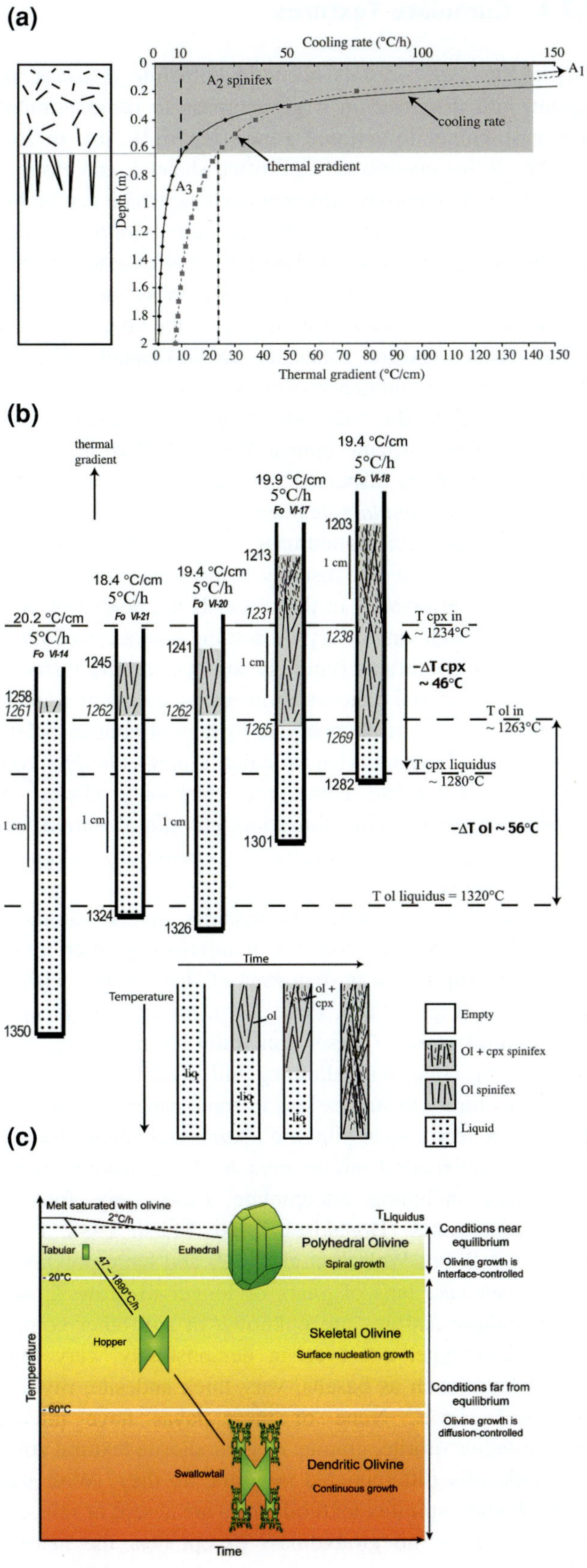

◀ **Fig. 7.5** Temperature conditions for crystallisation of different forms of olivine. **a** Relationship between temperature gradient, cooling rate, and spinifex textural type (From Faure et al. 2006). **b** Diagrammatic representation of the different temperatures at the bottom and top of different experimental charges, the cooling rates, temperature gradients, the temperatures at which spinifex crystallisation occurred, the liquidus temperatures of olivine and clinopyroxene, and the degree of under-/supercooling (From Faure et al. 2006). **c** The relationship between degree of undercooling (temperature axis at left), cooling rate, and olivine crystal morphology (From Welsch et al. 2012)

Welsch et al. (2012) also investigated the crystallisation behaviour and conditions for formation of both skeletal and polyhedral olivine phenocrysts (Fig. 7.5c). They also found that olivine crystal morphologies range from small, randomly oriented, needle-like (but actually plate or blade shaped in 3D) skeletal, spinifex crystals only millimetres to centimetres long that are dispersed in a fine-grained, originally glassy groundmass in zone A2 of a komatiite lava (Fig. 7.2a), to clusters or "books" of parallel olivine plate or blade spinifex crystals that are centimetres to metres in length (zone A3, Fig. 7.2a), to branching, skeletal dendritic tree-like crystals that are tens of centimetres to metres high in a fine groundmass, to elongate "*hopper*" crystals, to equant polyhedral crystals, some of which can also have skeletal forms (zone B1, Fig. 7.2a). By "*skeletal*", we mean that the crystals are incomplete, often with saw-tooth-like edges and internal cavities ("hopper texture"; Fig. 7.3b, l), originally filled with magma and then after solidification by glass. Lower in the komatiite lava profile, equant polyhedral olivine phenocrysts occur in the B2 to B4 zones (Figs. 7.2, 7.3m, p), which are porphyritic textural zones with at least 40% groundmass (Fig. 7.3) and reflect slower, equilibrium crystallisation and growth rates, under conditions of slow-cooling rates without undercooling (Donaldson 1976, 1982a).

Clinopyroxene is another significant mineral in komatiites, but usually occurs as a groundmass phase between larger olivine spinifex and cumulate crystals (Fig. 7.2b), often with a skeletal microspinifex form. The association of olivine with pyroxene defines komatiites as being peridotitic, although they rarely form coarse peridotite in the classical (mantle) sense. In komatiitic basalts, however, because of their lower MgO, and higher CaO and Al_2O_3 contents (Table 7.1), large skeletal clinopyroxene forms spinifex and even euhedral phenocrysts. Orthopyroxene may also occur but is less common (Arndt et al. 2008).

Chromite occurs as very small (usually < 100 microns) skeletal grains in most komatiites and is most abundant in the spinifex and upper parts of cumulate zones (Arndt et al. 2008). Plagioclase is rare in komatiites, except in differentiated gabbroic zones in differentiated sill intrusions, and it becomes a significant groundmass mineral in komatiitic basalts. For further information about the mineralogy of komatiites and komatiitic basalts, readers are referred to Arndt et al. (2008, Chap. 4).

7.3.2 Spinifex Textures

Spinifex crystals lack the normal sub-equant, orthorhombic crystal form or morphology of well-formed igneous olivine, and are considered to be skeletal crystals, marked by an elongate needle to plate-like form, internal cavities, and comb-like margins (Figs. 7.3b, j; 7.4, 7.5). Skeletal spinifex crystals indicate rapid cooling under conditions of high-temperature gradient, high degrees of undercooling, and supersaturation with regard to MgO and olivine (Donaldson 1976), which is a function of the high eruption temperature of komatiite magmas and the high temperature differential between the margins of the komatiite magma and the ambient water medium. However, Donaldson (1976, 1982a) pointed out that spinifex crystal forms are not the result of quench crystallisation, which would form feather-shaped crystals. Nonetheless, they are indicative of rapid, incomplete, or disequilibrium crystallisation from a melt without suspended crystals that could act as nuclei, suggesting homogeneous crystal nucleation and growth conditions.

Spinifex–textured olivine crystals are essentially plate like. Spinifex crystal morphology and textures change downwards through the A zone from small mm to cm size, single, *randomly oriented platy spinifex crystals* (Figs. 7.2, 3b, f), to larger (cm to m) *plate or blade spinifex* crystals which occur in clusters or books of parallel plates, which are initially disoriented, but become increasingly oriented perpendicular to the top of the lava flow unit downwards into the A zone, often with a slight fan-like pattern, leading to the use of the term "*sheaf spinifex*" (Figs. 7.2a, 7.3g, j). In detail, spinifex crystals have an incomplete saw-tooth-like skeletal texture (Fig. 7.3b, e, f). Pyroxene *microspinifex*, which is common in the groundmass of komatiite (Fig. 7.3f), and pyroxene *macrospinifex*, which is the dominant spinifex mineral in komatiitic basalts, is needle like or acicular, rather than platy.

In metamorphosed terrains, care must be taken not to mistake needle to fibrous metamorphic amphiboles (tremolite-actinolite) for random spinifex texture. Close examination with a hand lens or microscope shows that these crystals lack skeletal texture, and these minerals can be readily identified microscopically. Even in other rock types, such needle-like amphiboles have been mistaken for spinifex, and the rocks have been mistakenly interpreted as komatiite.

7.3.3 Cumulate Textures

Olivine cumulate crystals occur in komatiite B zones, are equant, and dispersed in a groundmass in ortho-cumulate and meso-cumulate textured zones (Fig. 7.3k, m). In zone B1, the olivine crystals are still often skeletal, and elongate in shape, they often have internal cavities, which can be box or chevron shaped. Such crystals are called *hopper crystals* and are concentrated just below the spinifex zone in zone B1. Crystals are dispersed in the groundmass and elongated crystals are aligned parallel to the top or the lava or the base of the spinifex textural zone, i.e. sub-horizontally (Fig. 7.2 a), suggesting the influence of flow shear alignment. In the underlying B2 to B4 zones, the dominant crystal morphology is complete, equant, euhedral to subhedral, polyhedral olivine phenocrysts dispersed in a groundmass, in what is called an *orthocumulate* textural zone (Figs. 7.2, 7.3m). There is at least 40% groundmass, but it could be as high as 90%, and there are also usually some skeletal hopper- or chevron-shaped crystals in this zone. If the degree of crystallisation is high, and/or crystal settling has occurred under laminar flow conditions (unlikely to occur during turbulent flow), then a crystal concentration zone may occur near the base of a lava, with phenocrysts in mutual contact (<60% groundmass), forming what is called a *mesocumulate* zone (Fig. 7.3n). Cumulate zones are commonly altered by varying degrees of serpentinisation, chlorite alteration, and talc–carbonate alteration (magnesite is common) and cut by carbonate veins.

In intrusive komatiites containing dunitic domains, almost the whole rock consists of interfacing, interlocking very coarse (up to cm in diameter) olivine crystals with no groundmass (Fig. 7.3o, p; Plate 7.1b), forming an *adcumulate* textural zone. These can only form under very slow-cooling conditions allowing full crystallisation of the magma. Contrary to suggestions in some papers, adcumulate texture is a very coarse plutonic texture and cannot form in lavas. The polyhedral olivine crystals that constitute cumulate zones, including adcumulate zones only form at slow-cooling rates (Donaldson 1976; Marsh 1988; Faure et al. 2003, 2006; Welsch et al. 2012) and large polyhedral crystals can take tens of years or longer to grow. Coarse holocrystalline textures are unknown in any other compositional lava type, not even in demonstrably, very thick ponded lavas such as basalts, very thick andesite, rhyolite, and dacite lavas. None of these lavas have internal holocrystalline diorite, granite, or granodiorite textural zones or significantly differentiated interiors, do they? With such dunitic bodies, spinifex textural zones are absent or rare, and there is little or no groundmass except near the margins where meso- and orthocumulate may occur. Adcumulate textural zones are usually altered through serpentinisation.

In some komatiites and differentiated intrusions, very intricate branching, skeletal *dendritic* networks of olivine crystals can grow in the cumulate zone (Fig. 7.3q, s). These can reach many m in length, and their growth conditions are not clear. We use the term *crescumulate* to apply only to such textures of branching crystals located in cumulate zones.

In the above discussion, we have avoided using the term "harrisite" or "harrisitic". These are historic terms referring to the occurrence of skeletal crystals and a layering of skeletal and polyhedral crystal horizons identified in the intrusions of the island of Rhum, Scotland (Donaldson 1982b). These terms are best used specifically for those rocks and that context, rather than being applied, with dubious and uncertain meanings in komatiites.

7.4 Using the Complete Komatiite Lava Textural Zonation as a Stratigraphic Younging Indicator in Ancient Successions

The complete, vertical, asymmetrical textural zonation identified through komatiite lavas (Figs. 7.2a, b; 7.6; Pyke et al. 1973; Barnes et al. 1973; Arndt et al. 1977, 2008) is preserved only to varying degrees in the rock record because of variable degrees of alteration, metamorphic overprinting, and deformation. However, the textural architecture of komatiite lavas can be well preserved in some surface outcrops and diamond drill core in places, such as Pyke Hill and Alexo in the Abitibi greenstone belt, Ontario, Canada (Fig. 7.6a; Pyke et al. 1973; Arndt et al. 1977, 2008), in

Fig. 7.6 Spinifex and cumulate textural associations and profile through Archean komatiite lavas (see Fig. 7.2a, b) and intrusions. Drill core is ~ 5 cm diameter. **a** Near-complete textural zonation profile through a subvertically dipping thin Pyke Hill komatiite lava, Abitibi greenstone belt, Ontario, Canada. Textural zones are labelled. An A4 very coarse plate spinifex zone is identified to distinguish it from a finer A3 blade spinifex zone. Younging (top) of lava is to the right. The top is marked by zone A1, the bottom by zone B3. **b** Drill core intersection through a 10-m-thick komatiite lava, McCloy ore shoot, Kambalda, Western Australia. Diamond drill hole KD 8783. Stratigraphic younging is from bottom right to top left of the photo. (From Cas and Beresford 2001). **c** Profile through the random spinifex top of a thin komatiitic basalt, Dundonald, Ontario, Abitibi greenstone belt, Canada. Note how the random spinifex becomes coarser downwards, indicating the top of the intrusion is upwards

greenstone belts of the Yilgarn Craton, Western Australia (Fig. 7.6b; Barnes et al. 1973, 1988; Lesher 1989; Hill et al. 1995; Beresford et al. 2002; Cas et al. 2013), and the Barberton greenstone belt, South Africa (e.g. Dann et al. 2001). It is easiest to assess the textural architecture of komatiite lavas where they are relatively thin and well exposed (Fig. 7.6a). The hypothetically complete textural architecture, as depicted in Fig. 7.2, is not often preserved and frequently there are short-circuited variants of that complete sequence.

Where the textural profile is complete or largely complete, it can be used to determine the younging of the stratigraphic sequences that komatiites are part of, much like grain size grading in sedimentary turbidite beds. In the case of komatiites, it is the transition from cumulate textural zone (s) in the basal part of the lava, upwards into coarse parallel blade spinifex, and then up into finer random spinifex near the top that provides a clear indication of upwards younging (Figs. 7.2, 7.6a, b). As discussed further below, the spinifex-textured top is interpreted to represent the crust of the submarine lava, whereas the cumulate zone was the crusted over lava tube through which magma was fed from the vent to the advancing flow front. The preservation of an intact, coherent, or jigsaw-fit flow top breccia has important implications for the flow state and behaviour of the lava flows, which is also discussed below.

However, spinifex texture and zonation are not unique to komatiitic lavas. They can also form at the margins of subvolcanic komatiitic intrusions (e.g. Fig. 7.6c), and can form at both top and bottom margins, so that extreme care in understanding the emplacement origins of komatiitic units must be observed before making interpretations.

7.5 Other Textural Features of Komatiite Lavas

Several other textural features that may be associated with komatiitic lavas, with which the reader should be aware of, are now briefly described and illustrated.

7.5.1 Flow Top Breccias

Komatiite lavas often have a "flow top breccia" consisting of an aggregate of variably jigsaw-fit fragments that have a blocky to curviplanar morphology (Figs. 7.3a, 7.7a, d). Fragments range from tens of cm-size blocks to several cm-to-mm-size fragments. These characteristics are entirely consistent with a quench fragmented, hyaloclastite origin, produced by flow of a lava in a subaqueous environment, as discussed in Chaps. 3 to 6. Key to their interpretation is that such flow top breccias show no evidence of interaction with overlying pre-existing sediments, in the form of peperites (Chap. 4), and therefore represent a true seafloor lava flow.

7.5.2 Peperite: Distinguishing Komatiite Lavas from Syn-Depositional Intrusions and "Invasive Lavas"

Peperite can form at the base of subaqueous lavas, but cannot form at the top, as discussed in Chaps. 4 and 5. However, where peperites occur at the top contact (Fig. 7.7e, f, e.g. Houlé et al. 2008), they are indicative of syn-depositional intrusions, because peperite can only form at the top contact of a komatiite if it is intruding water-saturated, unconsolidated sediments (Chap. 4), which will be discussed in more detail in Chap. 8. Beresford and Cas (2001) and Beresford et al. (2002, 2005) pointed out, however, that when komatiite lavas (density 2,750 kg m^{-3}) flow over underlying, unconsolidated, water-saturated sediments (density 1,200 to 1,500 kg m^{-3}), intrusive or invasive apophyses or sills may burrow down into the sediments from the parent lava, which they called "*invasive lavas*". Some care is required in interpreting vertical and lateral facies variations and relationships based on careful mapping and correlations, especially in drill core. Where low-viscosity magmas such as komatiite intrude muds, the resultant peperite clasts may be irregular and fluidal (Busby-Spera and White 1987).

7.5.3 Vesicles or Amygdales and Segregation Amygdales, Pipes, and Pods

Beresford et al. (2000, 2002) and Dann (2001) drew attention to the relatively common occurrence of vesicles, or more commonly amygdales filled with secondary minerals, such as carbonate, quartz and chlorite, in porphyritic peridotitic komatiite lavas (Fig. 7.8a). Vesicles and amygdales had not previously been recognised in komatiite lavas, and they have now been recognised in shallow intrusions as well (Fig. 7.8b). In their study of the Kambalda (Western Australia) komatiite succession, Beresford et al. (2000) found that vesicles (mostly amygdales) occurred in the spinifex-textured A horizon (<1% volume), beneath the spinifex A and cumulate B contact, and as multiple horizons within thick cumulate zones and lavas (5 to 30% volume). Vesicles/amygdales range up to 2 cm in diameter, they have spherical to ovoid shapes (Fig. 7.8a, b) and depending on the alteration and metamorphic processes are infilled with secondary minerals, including carbonate, chlorite, amphibole, serpentine, and talc. In thick flows, > 50 m thick, at Kambalda, up to four vesicle/amygdale horizons may occur, and the horizons commonly show an upward increase in

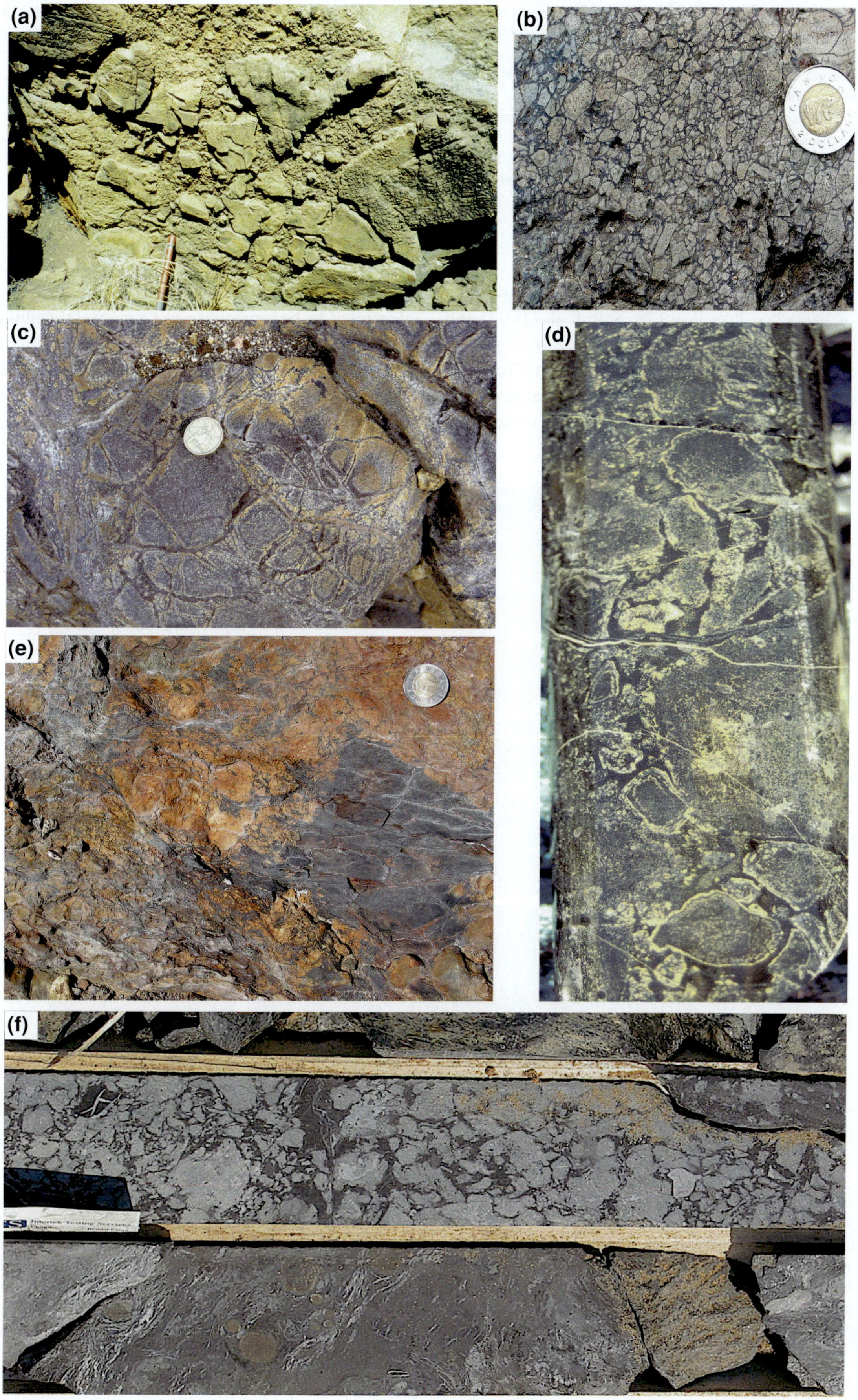
(a)
(b)
(c)
(d)
(e)
(f)

◀ **Fig. 7.7** Archean komatiite hyaloclastite and peperite breccias. **a** Jigsaw-fit to clast-rotated hyaloclastite breccia with angular to irregular hyaloclasts, Barberton Greenstone belt, South Africa (From Cas and Beresford 2001). **b** Komatiite hyaloclastite breccia with distinct jigsaw-fit relationships between blocky hyaloclasts, Abitibi greenstone belt, Guibord Township, Ontario, Canada. The dark seams between fragments are chlorite alteration seams. **c** Jigsaw-fit textured hyaloclastite komatiite breccia, Alunite Locality, Kalgoorlie Terrane, Yilgarn Craton, Western Australia. **d** Jigsaw-fit to clast-rotated komatiite hyaloclastite, Victor East mine, Kambalda, Western Australia. Diamond drill hole KD6068. Dark matrix is secondary chlorite. **e** Peperite breccia at the top of an Archean syn-depositional komatiitic basalt sill intruding and mingling with black mudstone, Dundonald, Abitibi greenstone belt, Ontario, Canada. Hyaloclasts are irregular in shape. **f** Peperite breccia at the top of an Archean syn-depositional komatiitic basalt sill intruding black mudstone which forms the matrix. Hyaloclasts are irregular in shape. Dundonald, Abitibi greenstone belt, Ontario, Canada. Drill core is ~ 5 cm wide

amygdale size. Beresford et al. (2000) interpreted these multiple horizons to reflect several generations of magma injection into the interior of inflating lava flows, as had been suggested for pahoehoe lava flows by Self et al. (1998) and Thordarson and Self (1998) for massive sheet flood pahoehoe basalt flows in the Columbia River Basalt Province, USA (Chap. 5, Sect. 5.4.2, Fig. 7.23d). They interpreted the vesicles/amygdales in the spinifex zone to reflect the presence of syn-eruptive, exsolved magmatic volatiles, indicating the magma was saturated with volatiles. By contrast the vesicular zones within the cumulate B zone were interpreted to be a response to later crystallisation of olivine and pyroxene in the lava, leading to volatile concentration and exsolution in the residual groundmass magma (see Chap. 4, Sect. 4.5.2.2 crystallisation-induced exsolution/vesiculation, or second boiling).

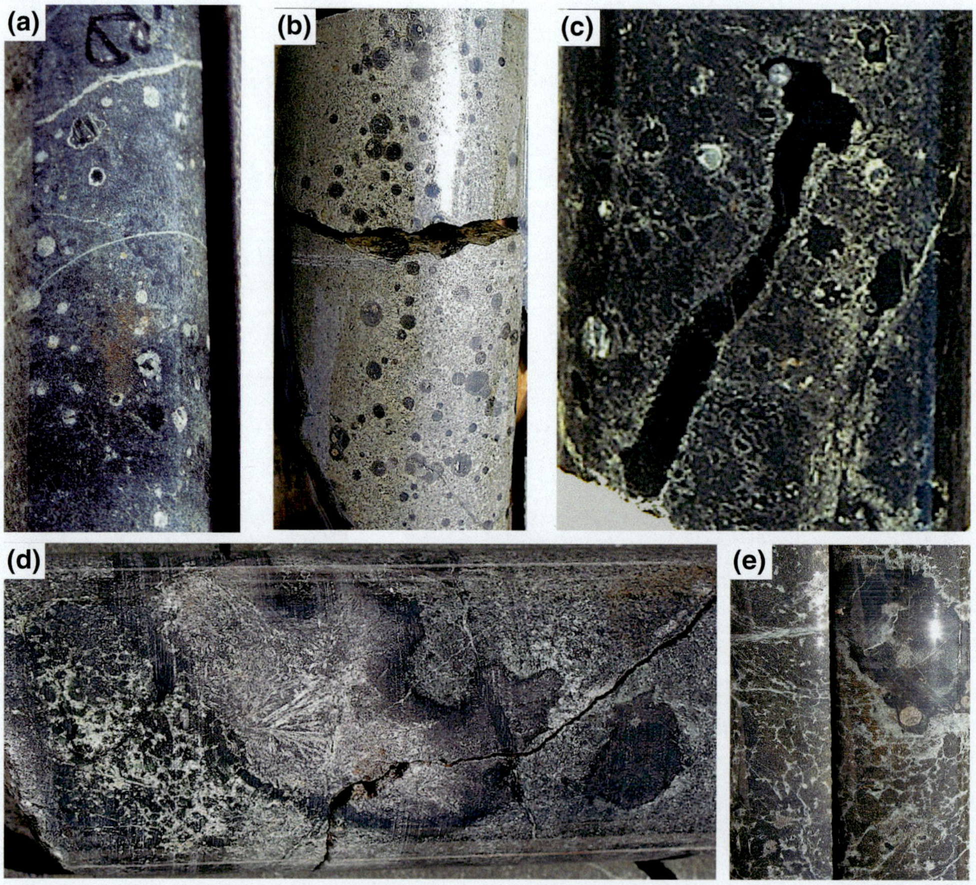

Fig. 7.8 Vesicles, amygdales, and segregation pods in Archean komatiites. All drill core is ~ 5 cm diameter. **a** Amygdales in komatiite from the Victor East mine, Kambalda, Western Australia. Amygdales are filled with carbonate and chlorite. Diamond drill hole KD 6068A. **b** Spherical amygdales filled with chlorite, in komatiitic basalt syn-depositional intrusion, Dundonald, Ontario, Canada. Diamond drill hole HUF. **c** Pipe-like magma segregation structure altered to chlorite, Victor East mine, Kambalda, Western Australia. Diamond drill hole KD 6068A. **d** Irregular magma segregation pod, altered to chlorite, with internal fanning spinifex texture in an intrusive komatiite, Black Swan mine, Western Australia. Diamond drill hole BSD 119 at 126.5 m. The infill is largely altered to chlorite. Relict ortho- to mesocumulate texture is visible in the core to the left of the segregation pod. **e** Irregular chlorite-altered magma segregation pod with internal pyrrhotite spheres, in a coarse mesocumulate groundmass, Black Swan mine, Western Australia. Diamond drill hole BSD 71 at 222 m

Dann (2001) documented komatiite lava flows up to 55 m thick, with upper vesicular/amygdaloidal and spinifex-textured crusts 2 to 15 m thick, in the komatiite lavas of the Komati Formation in the Barberton greenstone belt. Amygdales are up to 4 cm in diameter, but commonly < 0.5 cm, they are filled with serpentine, and reach 35% abundance. Amygdales low in the vesicular crust zone are elongated in a subvertical direction, whereas amygdales higher up are more equant. Amygdales often occur in distinct horizons or layers that are interlayered with spinifex-textured layers. Dann (2001) interpreted these to represent entrapment of multiple generations of vesicles resulting from injections of new magma beneath a surface crust, bringing new volatile-bearing batches of magma into the interior part of the flow.

In addition, segregation vesicles/amygdales, irregular pods, and pipes, filled with chilled komatiite magma that cooled to glass, sometimes with internal spinifex texture, occur in some komatiites (Fig. 7.8c, d; Beresford et al. 2000; Dann 2001). Some are also infilled with spherical aggregates of sulphide (Fig. 7.8e; Beresford et al. 2000; Dann 2001). Some vesicles/amygdales/segregation pods are coalesced, and some may form vertical trains and pipe-like and pod-like structures with bulbous tops (Fig. 7.8c), which are similar to pipe vesicles in basalt lavas (Chap. 5). The infills of these segregation amygdales, pipes, and pods have been interpreted to represent magma that has migrated into vesicles and vughs after the vesicle walls developed strength and were broken by magma pressure in the surrounding groundmass. Some may have expanded into irregular shapes, and the segregation filling magma then chilled to form intra-komatiite spinifex pods. In some cases, these pods themselves contain vesicles/amygdales, and, in some cases, spherical sulphide blebs (Fig. 7.8e). Cas et al. (2013) also documented segregation pipes and pods in some intrusive cumulate-textured komatiite in the Black Swan mine host rock succession north of Kalgoorlie, Western Australia (Fig. 7.8d, e).

The sulphide, mostly pentlandite and pyrrhotite, may represent from 0 to 100% of the amygdale fill, and also constitutes what is known as "*blebby ore*" dispersed or disseminated through the komatiite. Blebby ore occurs in the groundmass of the host komatiite above the location of basal massive nickel sulphide ore horizons and represents late-stage precipitation of sulphide in vesicles from hydrothermal-magmatic fluids in the komatiite (Beresford et al. 2000; Chap. 18).

7.5.4 Tumuli

Excellent exposure and mapping by Dann (2001) in the Barberton greenstone belt in South Africa demonstrated that ancient submarine komatiite lavas can also develop tumuli (Chap. 5, Sect. 5.4.6), resulting from lava flow inflation beneath a vesicular and spinifex-textured lava crust, which ruptures in the process. The lava flows experience inflation and thickening, resulting in the vesicle-spinifex layering, but also in foundered blocks of crust in cumulate-textured zones, which he relates to collapse of the crusts of inflating lava tumuli.

7.5.5 Sedimentary Rip-Up Clasts

Sedimentary rip-up clasts have been documented at the base of komatiite lavas at Kambalda (Beresford and Cas 2001; Beresford et al. 2002; Fig. 7.9a), indicating a physical erosional interaction between komatiite lavas and sedimentary substrate; this will be discussed further below.

7.5.6 Columnar Jointing

Columnar cooling jointing occurs in komatiite and basaltic komatiite lavas (Fig. 7.9b) as it does in other compositional lava types, such as basalt lavas, reflecting progressive inward propagation of cooling fronts from the tops and bottoms of flows after emplacement (Chap. 5, Sect. 5.4.4).

7.5.7 Pyroclastic Komatiite

Pyroclastic komatiite successions have only very rarely been identified. These include primary and variably reworked deposits of shards, vesicular pyroclasts, and accretionary lapilli. The best examples are described from the Barberton greenstone belt, where numerous and extensive (>20 km along strike), thin cherty units, mainly composed of silicified ultramafic tuffs and carbonaceous matter, are interbedded with komatiite lava flows throughout the Onverwacht Group (Lowe 1999; Stiegler et al. 2008, 2011). Most layers are dominated by fine, blocky, poorly to non-vesicular grains, although a few deposits contain substantial bubble wall fragments, and contain armoured and accretionary lapilli. Individual tuff beds are well sorted, massive to normally graded, and range from 1 cm to > 15 m thick. Cross-lamination and flat lamination are widely preserved in tuff beds and at the tops of graded, but otherwise structureless layers, possibly indicating that the volcanic particles were often resedimented by turbidity current mass flows and/or reworked by moving water (Lowe 1999). Stiegler et al. (2008, 2011) suggest the regionally uniform bed thicknesses, great lateral extent, and low lithic contents indicate these were not restricted phreatomagmatic explosions creating maar-like landforms, but large eruptions that generated substantial subaerial plumes during vent shoaling

Fig. 7.9 Some other features of komatiites. **a** Irregular mudstone intra-clasts in komatiite lava, Kambalda, Western Australia. Diamond drill hole KD 6068. Drill core is ~ 5 cm diameter. **b** Columnar jointed komatiitic basalt, Centre Hill complex, near Alexo, Abitibi greenstone belt, Ontario, Canada

in water that was probably < 10 m deep (?), and that dispersed ash over wide areas.

7.6 Komatiite Lavas: Provinces, Correlation, and Typology

7.6.1 Provinces

Komatiites occur in regionally extensive provinces and were erupted seemingly at specific times in the Earth's history (Fig. 7.1a). For example, at 2.7 Ga and 2.9 Ga in the Yilgarn Craton, Western Australia (Fig. 7.10a), 2.7 Ga in the Abitibi greenstone belt, Canada, and in southern Africa, at 3.5–3.2 Ga in the Barberton greenstone belt, and 2.8–2.7 Ga in the Zimbabwe Craton, major episodes of regionally extensive komatiitic magmatism (both volcanic and intrusive) are preserved. There has been some speculation that the komatiites in such provinces could have been erupted from one vent system (cf. Roza Member flood basalt in the Columbia River Basalt Province, see Chap. 5), with dunitic lithofacies representing near-vent settings and spinifex-orthocumulate lithofacies representing distal settings hundreds of km away from the vent area (e.g. eastern Yilgarn Craton; Hill et al. 1995). However, it is now more commonly accepted that locations with (massive) nickel sulphide mineralisation are likely to be proximal to eruption vents (e.g. Beresford et al. 2002; Fiorentini et al. 2012; see below and Chap. 18, Sect. 18.6), indicating that there were multiple eruption sites and lava-subvolcanic intrusive fields in the Eastern Yilgarn Craton (Fig. 7.10b, d).

In the Eastern Goldfields Superterrane (EGST), eastern Yilgarn Craton, this suggests that all the 2.7 Ga komatiites contained within the Norseman–Wiluna greenstone belt of the Kalgoorlie Terrane (Hill et al. 1995), which extend for over 600 km in a north–south direction (Fig. 7.10a, b), represent a single komatiite province. This therefore must consist of multiple lava flow fields and associated subvolcanic intrusions, with eruption points distributed throughout the province, probably involving multiple basement fracture fissure vents. In addition to the komatiites, the plumes also produce and erupt voluminous coeval tholeiitic basalts and high-Mg basalts, including komatiitic basalts (e.g. Fig. 7.10c, d; Campbell et al. 1989; Barnes et al. 2012). The association of komatiites with extensive komatiitic basalts and tholeiitic basalts in the EGST led Barnes et al. (2012) and Hayman et al. (2015) to propose that the province was an Archean Large Igneous Province (oLIP; Chap. 17). To the west of the EGST lies the Southern Cross Domain of the Youanmi Terrane, in the western Yilgarn Craton, that contains another regionally distributed province of komatiites with the slightly older age of 2.9 Ga age (Fig. 7.10 a; Mole et al. 2014; Wyman 2020). Both komatiite provinces have been interpreted to be the product of major mantle plume events (Campbell et al. 1989; Barnes et al. 2012; Mole et al. 2014; Wyman 2020).

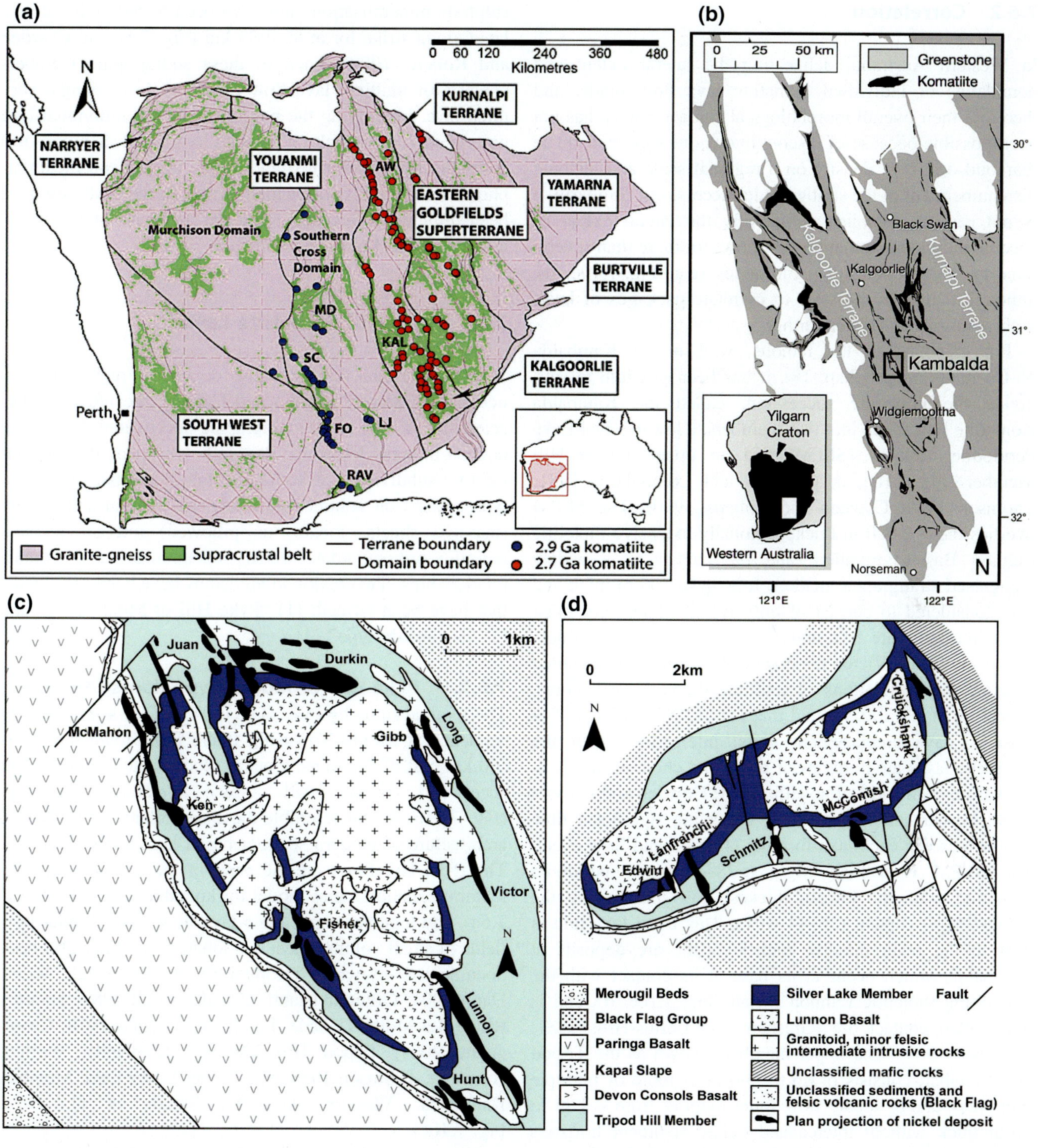

Fig. 7.10 Extent of komatiite provinces and lavas. **a** The 2.7 and 2.9 Ga komatiite provinces in the Yilgarn Craton, Western Australia as understood from present radiometric dating. These show locations of the iconic Norseman–Wiluna greenstone belt (2.7 Ga), host to the world-class Kambalda Ni-Cu mine camp, within the Kalgoorlie Terrane of the Eastern Goldfields Superterrane, eastern Yilgarn Craton, and older greenstone belts (2.9 Ga) which form part of the southern Youanmi Terrane, western Yilgarn craton. The Norseman–Wiluna greenstone belt is often divided into a southern segment (KAL = Kalgoorlie–Kambalda–Norseman), and a northwestern segment (AW = Agnew–Wiluna). The Youanmi Terrane includes the Southern Cross-Forrestania greenstone belt (SC-FO) which hosts economic komatiite-hosted Ni-Cu mineralisation at Forrestania. (From Mole et al. 2014). **b** Extent of komatiites (black) in the southern half of the Kalgoorlie Terrane, where they occur associated with basalts in greenstone belt successions between Kalgoorlie and Norseman (grey), separated by regional batholiths in white (From Barnes et al. 2013). **c** Map of the Kambalda Dome, and **d** the Tramways area 35 km to the south southeast, showing the mapped occurrence of the basal nickel sulphide ore deposits hosted by the komatiitic Silver Lake Peridotite Member (dark blue) and the overlying Tripod Hill Member (pale blue). These two komatiite stratigraphic units are sandwiched between the underlying tholeiitic Lunnon Basalt and overlying high-Mg Devon Consol and Paringa basalts. (From Cowden and Roberts 1990; see Chap. 18)

7.6.2 Correlation

In Western Australia, delineating the lateral extent and boundaries of individual komatiite lava flow fields, and therefore their overall morphological characteristics, has not been possible because of discontinuous outcrop, inadequate diamond drill-core density on a regional scale and tectonic dismemberment of the stratigraphic successions. Similarly, it is not easy to establish definitively the lateral extent of specific individual komatiite lava flows for more than several hundred metres or a few kilometres (e.g. Fig. 7.10c, d), although it may be possible to correlate packages of lavas with a uniform character further.

For example, in the famous, well-studied Kambalda Ni-Cu mine camp (Chap. 18), it has been possible to subdivide the komatiite succession, called the Kambalda Komatiite Formation into two members, a lower Silver Lake Peridodite Member (SLPM) and an upper Tripod Hill Member (Fig. 7.10c, d; Gresham and Loftus-Hills 1981; Thomson 1989; Cowden and Roberts 1990). The SLPM overlies the > 2,000 m thick, regionally extensive tholeiitic Lunnon Basalt Formation, and the contact is a widely recognised stratigraphic marker. It is up to 200 m thick and was considered to consist of only two or three very thick komatiite peridotite lava units, each tens to > 100 m thick. However, in the Victor mine sequence, after very detailed logging and correlations of multiple diamond drill cores, Beresford et al. (2002) found that the SLPM consisted of at least 24 lava emplacement units. Interbedded with the komatiite lavas are horizons of pelite and chert, each < 2 m thick (Gresham and Loftus-Hills 1981; Cowden and Roberts 1990; Lesher and Arndt 1995), indicating significant time breaks between each lava-forming eruption and flow event. The SLPM is a high MgO (>28–44 wt%) peridotite komatiite, with the lowest, thickest flow unit having the highest MgO content (36–44 wt%). This basal lava unit is host to the main massive nickel–copper ore deposits at Kambalda, which occur at its base at its contact with the underlying tholeiitic Lunnon Basalt (Fig. 7.10c, d), and is thus called contact ore (Gresham and Loftus-Hills 1981; Thomson 1989). The upper member, known as the Tripod Hill Member, consists of up to 500 m or more of multiple thin sheet and/or lobate flow units ranging from < 1 m to 20 m thick without intercalated pelitic sediment horizons (Gresham and Loftus-Hills 1981; Thomson 1989; Cowden and Roberts; Lesher and Arndt 1995), indicating a nearly continuous phase of eruption. The Tripod Hill Member is a lower MgO (16–28 wt%) succession of komatiite units and lacks any significant mineralisation.

Thus, at Kambalda, two komatiite members have been distinguished geochemically, in terms of lava unit thicknesses and the presence or absence of nickel–copper sulphide mineralisation, and have been traced and/or correlated along strike for at least 35 km (Fig. 7.10c, d; Cowden and Roberts 1990). Each of these stratigraphic members consists of multiple lava flow units, not just a single very thick lava. Identifying the boundaries of the individual lava units involves painstaking detailed logging of diamond drill cores through a deformed and highly altered and metamorphosed rock succession, which is why in the past they have been difficult to recognise, and the thickness of individual lava flows has been overestimated.

7.6.3 Types of Komatiite Lavas

Within individual mining camps or outcrop regions of particularly Western Australia and Canada, the internal characteristics of komatiite stratigraphic successions have been defined in many locations. Research over many decades has led to a subdivision of komatiites into a number of specific types, based on volume and thickness, internal architecture (massive sheets, lobate, or pillowed) internal textures (spinifex-orthocumulate versus adcumulate) and contact relationships. Two major subdivisions have been recognised that have been termed: (1) "Pyke Hill or Munro type" and (2) "Kambalda type".

7.6.3.1 Pyke Hill- or Munro-Type Komatiite Lavas

The seminal study by Pyke et al. (1973) of the komatiite lava sequence at Munro Township (there is no town there), ~ 60 km east of Timmins, Ontario, documented multiple thin sheet and lobate, small volume komatiite lava flows (Fig. 7.11a, b) that resemble the sheet, lobate, and pillowed architecture of modern seafloor basalt lava flows (Chap. 5). The Pyke Hill flows are the basis of the complete internal textural zonation facies model for komatiite lavas (Figs. 7.2, 7.6a), and together constitute a lava flow field. Similar flow fields of small volume sheet and lobate flows have also been documented elsewhere (Komati Formation, South Africa, Dann 2001; Gilmour Island, Canada, Arndt 1982; Spinifex Ridge, La Motte, Canada, Lajoie and Gélinas 1978, Champagne 2004; Tripod Hill Member, Kambalda, Western Australia, Thomson 1989; Fig. 7.10c; Silver Lake Member, Kambalda, Western Australia, Staude et al. 2020, Fig. 7.10c).

Exposures demonstrate that such lobate systems (Fig. 7.11b, d) are analogous to pillow lavas and that they result from the propagation of lobes of low-viscosity magma, chilling of the outer crust, and inflation of the lobe (Chap. 5), although there are important differences compared with basaltic pillow lavas. By analogy with subaqueous basalt pillow and lobate lavas, such flows indicate low effusion and lava propagation rates. In contrast, massive

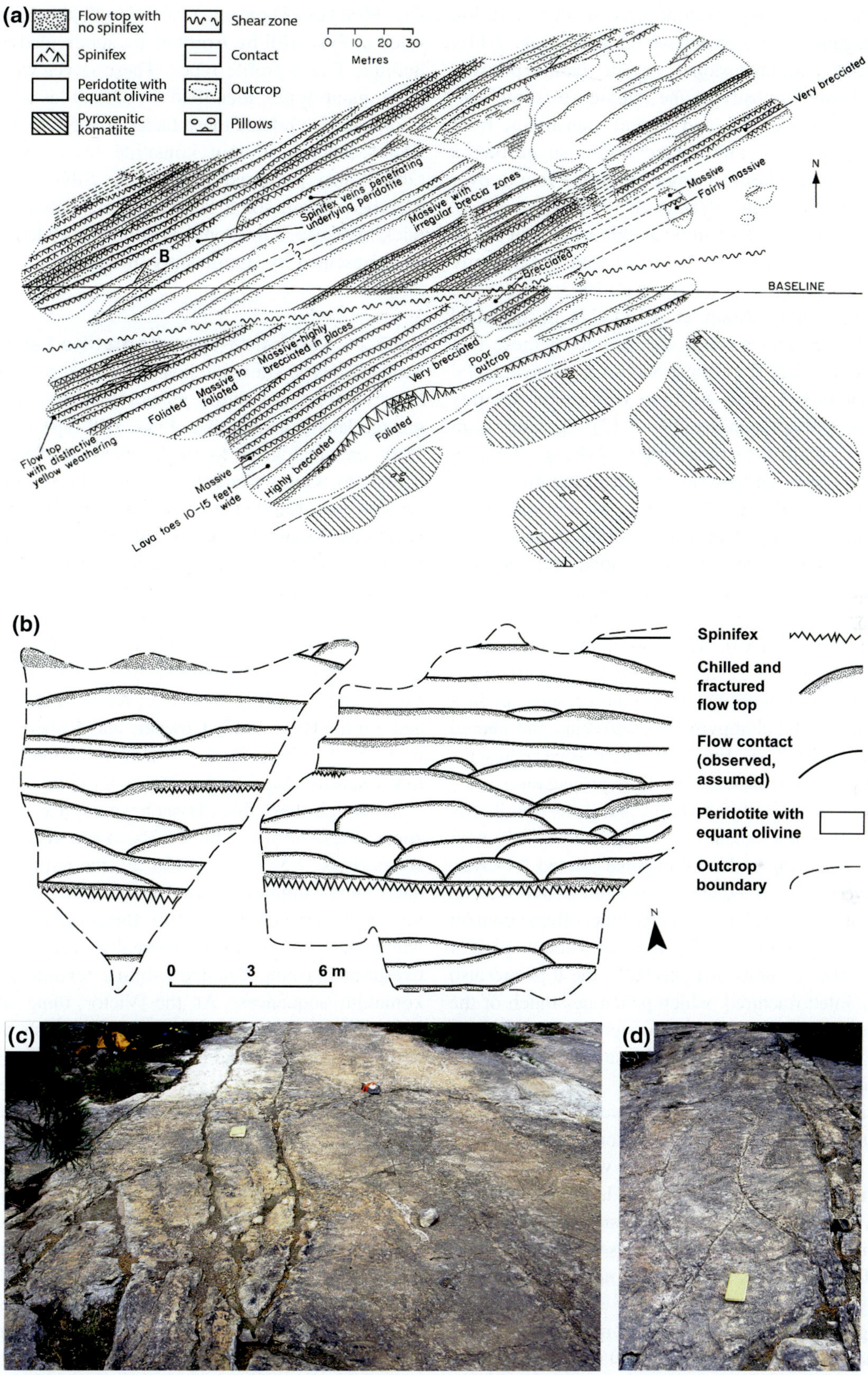

Fig. 7.11 Features of thin Pyke Hill-type komatiite and komatiitic basalt lavas. **a** Map showing the geometry of thin Pyke Hill-type komatiite lavas and flow field, the type locality for thin stacked komatiite lavas, Pyke Hill, Ontario, Canada. Note sheet and lobate flow unit geometries. Succession dips subvertically and youngs to northwest as indicated by spinifex-textured flow tops. (From Arndt et al. 2008, After Pyke et al. 1973). **b** Detailed map of area B in (**a**), showing lobate to pillow-like form of some units (Adapted from Pyke et al. 1973). **c** and **d** Stacked, inflated, lobate, and sheet komatiite flow units, Spinifex Ridge, La Motte, Abitibi greenstone belt, Quebec, Canada. Note that lobes wedge to accommodate each other. Note in (**d**) the inflated convex up top of a lava lobe, which has a flat base and a thin lateral breakout lobe. Younging is to the left

sheet komatiite lavas are the result of higher magma effusion and lava propagation rate. Lobes have flat bases (Fig. 7.11c, d) and are not equidimensional as are basalt pillows, reflecting the lower viscosity of the magma. In some successions, there are vertical and lateral transitions from massive, sheet lavas to multi-lobate flows, suggesting a change in effusion rate during a single eruption event or over the term of a lava flow field (Pyke et al. 1973; Arndt 1982).

Individual lobes or sheets can be as thin as tens of cm to several m thick (Figs. 7.6a, b; 7.11b, d), and occasionally 10 m or more thick (Fig. 7.11a; Pyke et al. 1973; Arndt et al. 1977; Thomson 1989). Arndt et al. (1977) documented variations in the internal textural architecture of these thin flow units at Pyke Hill, from those with a prominent, relatively thick spinifex A zone, to those with minor spinifex texture to those without spinifex (Fig. 7.12). The units with well-developed spinifex textures (Figs. 7.3c, g, 7.6a, b, 7.12a, Plate 7.1) have an aphyric, in situ jigsaw-fit, polygonal fractured top (A1, Figs. 7.2, 7.12a, c), and may have amygdales. This is consistent with cooling quench fragmentation and/or crust extension due to lobe inflation. The spinifex includes random (A2) through to random parallel book spinifex (A2) to large, well-developed, subvertical, parallel sheaf spinifex (A3). The underlying thin hopper crystal B1 textural zone is < 30 cm thick passing downwards into a B2 zone with equant olivine crystals set in a groundmass of skeletal clinopyroxene, chromite, and altered glass (Arndt et al. 1977; see Fig. 7.2). The B zone represents the lava tube under the crust during flow propagation.

In units with minor spinifex (Fig. 7.12b), the spinifex is just random with some equant olivines (phenocrysts?) near the base (A2), occurring under a polygonal-jointed A1 layer, and the B1 layer is absent. Equant olivines may have skeletal overgrowths, and near the base of the B zone, olivine content may reach 60% (Arndt et al. 1977). In units without spinifex (Fig. 7.12c; > 50% of units at Pyke Hill), the tops are also polygonally jointed/fractured, which penetrates much of the unit. Olivines are equant and have overgrowths of skeletal olivine but occur in lower abundance than the other two-zone unit types (Arndt et al. 1977).

Along strike, some units that have well-developed spinifex tops lose those spinifex textural zones, transforming into one of the other types. The units with minor or no spinifex commonly constitute the small lava lobes or toes (cf. pillows), whereas lobate units with significant spinifex are thick (<6 m; Arndt et al. 1977). These features suggest that the smaller units/lobes with little or no spinifex chilled more quickly forming crusts with few crystals, whereas those with significant spinifex still cooled quickly, but not as quickly as the other types, allowing skeletal-spinifex olivine to form under supersaturated, undercooled, homogenous crystallisation conditions. At Kambalda, based on analysis of narrow diamond drill-core intersections through the Tripod Hill Member, Thomson (1989), with the hindsight of the study at Pyke Hill by Arndt et al., subdivided thin units into multiple facies profile types. These cover variations of the three main types identified by Arndt et al. (1977), and Thomson also related these facies changes to down-flow and lateral changes in thin komatiite lavas. Beresford et al. (2002) also proposed that in the Silver Lake Peridotite Member, where some komatiites are tens of m or greater thick in places in the "ore environment", laterally they could be represented by significantly thinner units, as little as 2 m thick.

7.6.3.2 Kambalda-Type Komatiite Lavas

Kambalda-type komatiite lavas are thick, massive, sheet-like lavas, with larger volume, greater thickness, and presumably greater lateral extent than Pyke Hill-type komatiite lavas. They are analogous to thick, massive sheet basalt lavas (Chap. 5). Good examples include the classical komatiite sequence at Kambalda, especially the lower Silver Lake Peridotite Member (SLPM) of the Kambalda Komatiite Formation (Fig. 7.13a, b; Gresham and Loftus-Hills 1981; Cowden and Roberts 1990; Beresford et al. 2002, 2005; Staude et al. 2020), and the komatiitic basalts at Alexo, near Pyke Hill (Fig. 7.13c; Arndt et al. 1977). As indicated above, they are commonly tens of m thick, but some reach in excess of 150 m (Fig. 7.13 a, c; Arndt et al. 1977; Gresham and Loftus-Hills 1981; Cowden and Roberts 1990; Beresford et al. 2002, 2005), which is problematic for very low-viscosity magmas flowing as lavas on a flat seafloor.

Although these thick lavas have surface spinifex textural zones (Figs. 7.3b, e; 7.6b, 7.13b, c) and very thick cumulate zones (Figs. 7.3n, 7.6b, 7.13b, c), the cumulate zones may have more complex internal textural architecture than previously considered (Fig. 7.13b; Beresford et al. 2002, 2005), which may be due to the normally very intense alteration overprint masking detailed original textural zones in most komatiite sequences. At the Victor mine at Kambalda, komatiites in the ore zone are relatively fresh, and Beresford et al. (2002) found that some komatiite units (<30 m thick) have simple spinifex A zone–cumulate B zone architecture (their type I lava unit; Fig. 7.13b, diamond drill hole log KD6024). However, komatiite units that are > 30 m thick exhibited significant textural complexity. In their type II A flow units, the spinifex crust is simple from aphyric to random spinifex textured, but the B cumulate zone shows a layering of cumulate types including intra-cumulate, crescumulate, and skeletal intervals (Fig. 7.13b, KD6042). In type II B units, the A zone consists of a complex layering of random, random parallel plate, chevron, and parallel plate sheaf spinifex, suggesting multiple, incremental phases of spinifex crust formation and thickening. In the B zone, there are very thick intervals of skeletal olivine (Fig. 7.13b, KD 6068). Their type III units are very thick (<150 m) and show

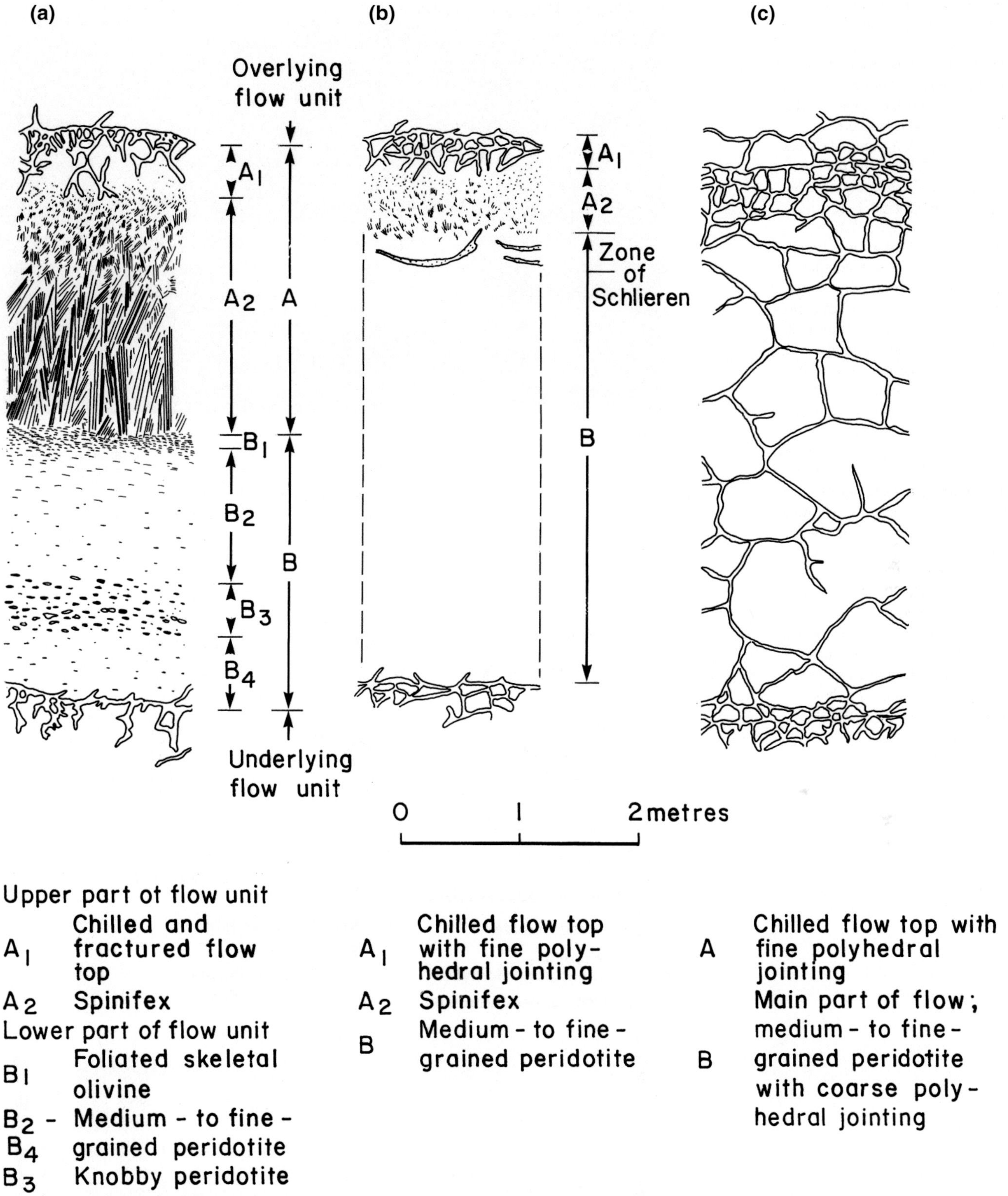

Fig. 7.12 Internal textural variations in thin lava lobes and sheets at the Pyke Hill, Ontario, Canada-type locality for thin spinifex-textured lavas (Modified from Cas and Wright 1987, After Arndt et al. 1977). **a** Lobe with well-developed spinifex texture. **b** Lobe with limited spinifex texture. **c** Lobe without spinifex texture, but abundant polygonal jointing. See text for discussion

Fig. 7.13 Features of thick Kambalda-type komatiite and komatiitic basalt lavas. **a** Overview of the poor outcrop of the Silver Lake Peridotite Member, from the slopes of Red Hill, Kambalda, Western Australia, looking towards Lake Lefroy salt lake in the distance. The poorly outcropping orange outcrops, numbered 1–3, represent silicified interflow sediments and long, time breaks between successive komatiite lavas (grey ground, not outcropping). The two komatiite lavas are each about 20 m thick, whereas sediment interval 1 sits above the basal komatiite, which is about 40 m thick at this locality. The sequence youngs away from the viewer. **b** Graphic logs through the Type I (Diamond drill hole KD 6024), IIA (KD 6042), and IIB (KD 6068) komatiite lava flows, Victor mine, Kambalda, Western Australia (Adapted from Beresford et al. 2002). **c** Graphic log through the very thick (> 120 m) "Fred's flow", a komatiitic basalt lava at Alexo, Ontario, Canada, showing the complex internal textural, chemical, and mineralogical zonation, including a gabbroic zone (From Arndt et al. 2008)

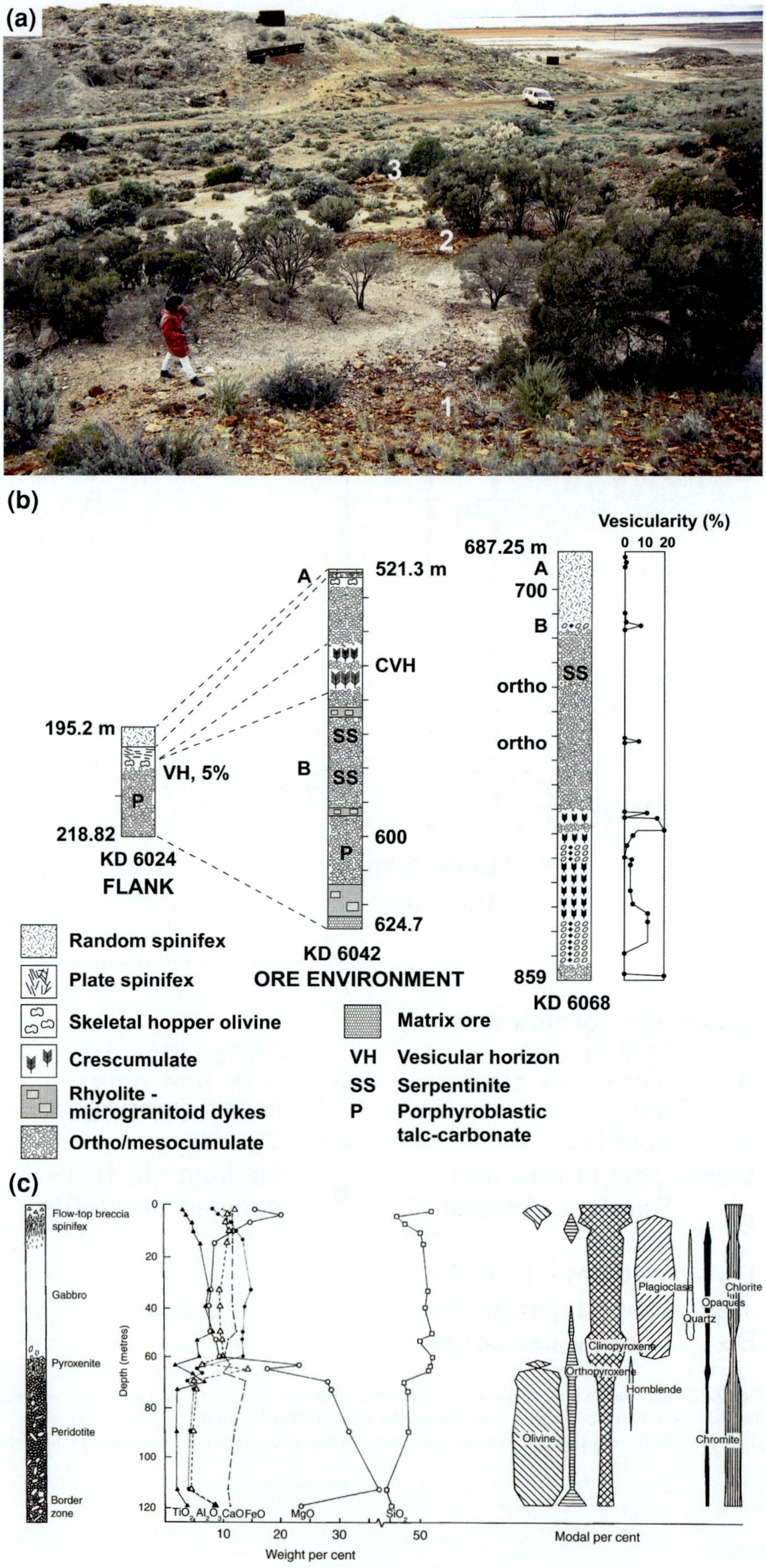

highly differentiated zones in their interior (see further discussion below).

In addition to these complexities, Beresford et al. (2002) found that there are multiple vesicle concentration horizons within the cumulate zone (up to four vesicle horizons with up to 30% vesicles) of individual thick lavas, often associated with crescumulate and skeletal olivine, as well as a low concentration (<1%) of vesicles in the aphyric and spinifex crust (Figs. 7.8a, 7.13 b, KD 6068). Beresford et al. (2002) interpret the intra-cumulate vesicle layers to have resulted from post-emplacement crystallisation-induced exsolution and vesiculation. They could also result from the injection of successive batches of magma into the lava flow interior under its stable (spinifex) crust, resulting in incremental inflation and thickening of the lava flow (cf. multiple vesicle concentration horizons in sheet pahoehoe lavas; Self et al. 1998). Beresford et al. (2002) also found that the basal SLPM in places consists of up to 24 komatiite units and, that contrary to previous suggestions, thin units, which had previously not been recognised, are spatially and stratigraphically associated with very thick units. The thin units were interpreted as thin, "breakout" lobes from the main thick lava flow units.

At Alexo, Dundonald Township (there is also no town there), ~ 45 km northeast of Timmins, Abitibi greenstone belt, Ontario, Arndt et al. (1977, 2008) documented very thick differentiated komatiitic basalt lava flow units, including "Fred's flow" (Fig. 7.13c), which is similar to type III komatiite flows proposed by Beresford et al. (2002). Such thicknesses are much thicker than most flood basalt sheet lavas, suggesting that they are either ponded lavas, perhaps ponded in graben or half-graben tectonic paleotopography, or very shallow sill-like intrusions into a pre-existing lava pile. As documented below, because of the low viscosity of komatiite magma, it is hydro-dynamically impossible for komatiite lavas to be more than several m thick, or if inflated under a chilled crust, more than a few tens of m thick if flowing on an open flat surface (Cas et al. 1999).

7.7 Komatiite Intrusions

As indicated earlier, when dense komatiite lava flowed over low-density, water-saturated, unconsolidated sediments, parts of the lavas may have foundered downwards and burrowed into the sediments below (Beresford and Cas 2001). However, komatiite magma would have been denser than almost all crustal rocks through which they were rising, and therefore, due to their negative buoyancy, would have been prone to forming subvolcanic intrusive complexes, more so than any other magma type. This would particularly have been the case when passing through unconsolidated, water-saturated, sub-seafloor sedimentary successions and fragmental volcaniclastic successions. These would be less dense, as well as mechanically weak lithofacies horizons, which would facilitate intrusion of dense komatiite magma along these sediment and fragmental volcanic horizons. Thus, komatiitic sill intrusions should be omnipresent in every komatiite province.

We can subdivide subvolcanic komatiite intrusions into two types: (1) syn-depositional small volume intrusions or "Dundonald type" and (2) large volume dunitic or "Mt Keith-type" intrusions. Although these will be described in the next chapter on intrusions (Chap. 8), they are also briefly considered in this chapter to ensure that the criteria for distinguishing komatiite lavas from komatiite intrusions are clear. Komatiite feeder dykes are rarely preserved because after peak magma discharge they attenuate because of very low viscosity, and magma pressure dropping below lithostatic pressure. At best, they are preserved as very thin dykes (Rosengren et al. 2005; see Fig. 7.17b) or as discontinuous very thin lenses (Staude and Markl 2019).

7.7.1 Small Volume Intrusions (Dundonald Type)

This komatiite intrusion "type" is named after the syn-depositional komatiitic basalt intrusive complex preserved and exposed at old workings at Dundonald "Beach" (there is no beach there), which is located near Alexo, Dundonald Township (there is no town there either), Ontario (Davis 1999; Arndt et al. 2004; Houlé et al. 2008). The exposed bedrock succession and limited drill core consist of an association of komatiites, komatiitic basalts, intercalated black shales, and rhyodacites (Fig. 7.14).

The komatiites and komatiitic basalts are concordant, tabular to lensing in geometry (Figs. 7.14, 7.15a). They are interleaved with the black shales, and generally consist of units 30 cm to 8 m thick that have irregular margins and geometry (Fig. 7.15a, d). Given the stratiform nature of the units, it is tempting to consider them to be lavas. The bases vary from coherent, sharp, and planar, to sharp and irregular with apophyses into underlying black shales (Fig. 7.15e, f), and the top contacts are also similarly irregular and marked by apophyses into overlying shales (Fig. 7.15e, f). In addition, both tops and bottoms of units both preserve quench fragmented jigsaw-fit to clast-rotated hyaloclastite and peperite (Figs. 7.7e, f; 7.15g). Such characteristics are only consistent with an origin of syn-depositional intrusions into water-saturated, unconsolidated sediments. In particular, the apophyses at the top contact intruding into overlying sediments (Fig. 7.15e, f), and peperitic mixing of hyaloclasts into overlying sediments (Figs. 7.7e, f; 7.15g), as well as mudstone inclusions in the upper parts of some units (Fig. 7.15h), are indicative of syn-depositional intrusion.

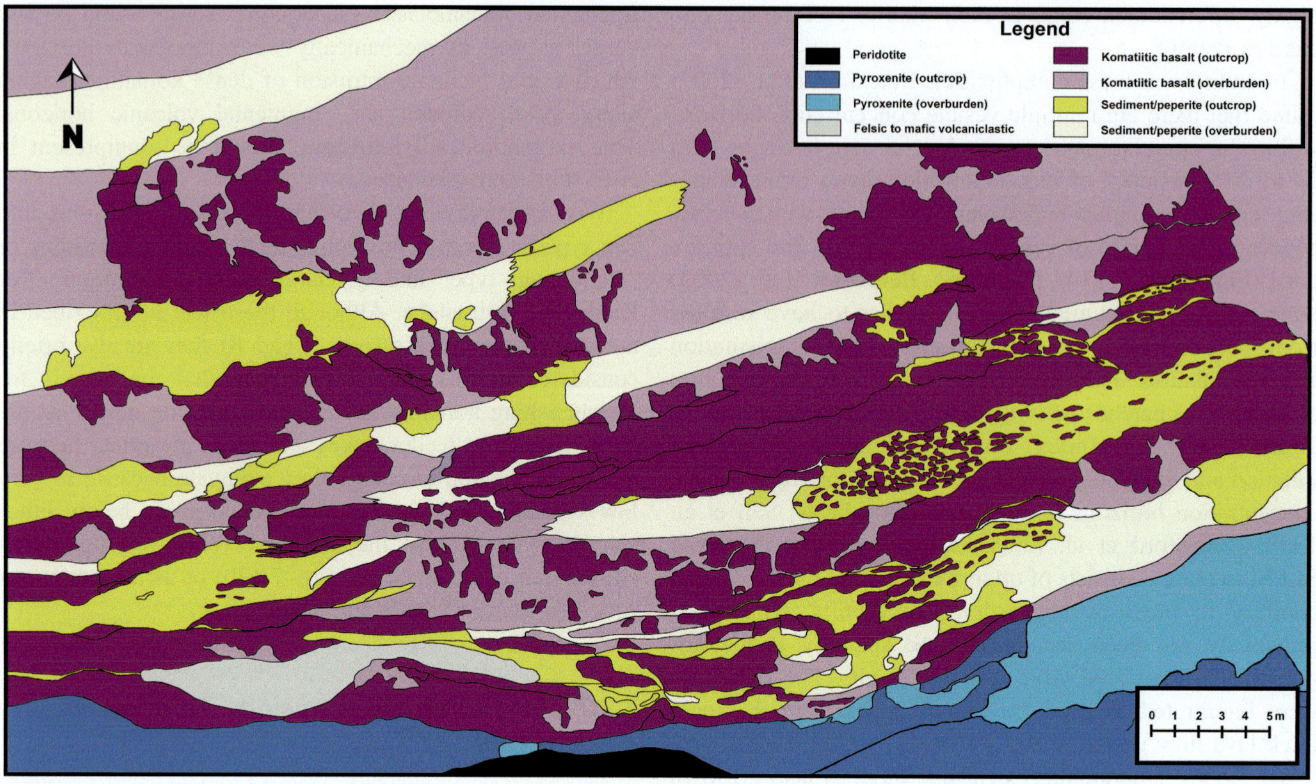

Fig. 7.14 Map of the central part of the Dundonald "Beach" exposure, Dundonald township, Abitibi greenstone belt, Ontario, Canada. This shows the concordant interleaving of variably quench fragmented and peperitic komatiitic basalt and komatiite units with black mudstone. Bedding is subvertical, and the succession youngs southwards. (Unpublished map by R. Cas, S. Beresford, J. Trofimovs, and N. Rosengren)

7.7.2 Large Volume Komatiite Dunite Intrusions (Mt Keith Type)

It has been proposed that komatiite units *hundreds* of m thick, with thick dunitic, adcumulate textural zones hundreds of m thick in northern part of the Norseman–Wiluna (or Agnew–Wiluna) greenstone belt, Kalgoorlie Terrane, Yilgarn Craton, Western Australia (Fig. 7.10a, b), were emplaced as extremely thick, topography ponded komatiite lavas (e.g. Hill et al. 1995, 2004; Barnes et al. 1988; Gole et al. 2013). Strangely, these authors have not identified any subvolcanic intrusive dunitic or other textured komatiite bodies anywhere in the belt, which is anomalous given the high-density komatiite magma relative to crustal rocks should ensure that subvolcanic komatiite intrusions should be as abundant if not more so than other compositional, lower magma density subvolcanic intrusions such as dolerite sills and granites, which abound in the Kalgoorlie Terrane, and elsewhere. It is highly unlikely that all, if any, dunitic bodies were emplaced as lavas given the long, slow cooling required to produce coarse, equigranular polyhedral olivine (Donaldson 1976; March 1988; Faure et al. 2003; Welsch et al. 2012) and highly differentiated and plutonic textured zones in the dunitic bodies (e.g. Rosengren et al. 2007; Fiorentini et al. 2007a, 2012). As a result, many of the bodies previously identified as thick adcumulate-textured lavas have been reinterpreted as subvolcanic intrusions (Trofimovs et al. 2003; Rosengren et al. 2005, 2007; Cas et al. 2013; Fiorentini et al. 2007a, b, 2012). This type of komatiite is best known from the Agnew–Wiluna greenstone belt in the northern part of the Kalgoorlie Terrane (Fig. 7.16a, d), from where the following summary is derived.

Adcumulate-textured komatiite consists almost wholly of very coarse, interlocking olivine crystals up to several centimetres in diameter or long, without any significant groundmass (<5%; Fig. 7.3o, p), making it petrologically a dunite. In other words, it has a plutonic or hypabyssal texture. The very large crystals and the fully crystallised state of dunite indicate slow, equilibrium, cooling, and crystallisation rates (Fig. 7.5c; Donaldson 1976, 1982a; March 1988; Faure et al. 2003; Welsch et al. 2012). Such textural zones are generally lensoidal in geometry and vary significantly in thickness laterally from tens to hundreds of m, in many cases disappearing along strike (Fig. 7.16b, d; Hill et al. 1995; Fiorentini et al. 2007a, b). They are enclosed in ortho- to

(a)
A
A
P
P
P
(b)
(c)
(d)
(e)
(f)
(g)
(h)

Fig. 7.15 Characteristics of syn-depositional intrusive komatiitic basalt units, Dundonald "Beach", Ontario, Canada. **a** Concordant intercalation between tabular to irregular komatiitic basalt units (light) and black shale (dark). **b** Lozenge-shaped, lobate termination of an intrusive unit. **c** Irregular sinuous geometry of two syn-depositional sills intruding black shale. **d** Highly irregular margin of intrusive komatiitic basalt lobe (light orange) at its contact with black mudstone. Note the micro-lobate form of the contact, the radial arrangement of lobes at the propagating head of the lobe (top left) with clefts of mudstone between micro-lobes, and fragments of komatiitic basalt detached from the lobe margin and suspended in the shale, resulting from peperitic mixing of spalled hyaloclasts. **e** Irregular intrusive lobe, with apophyses branching off the top and bottom margins and intruding black shale which is a peperitic black shale–basalt clast mixture. **f** Detailed view of upper margin of the intrusive komatiitic basalt unit in (e) showing an apophysis intruding the overlying mudstone. **g** Irregular, peperitic top of an intrusive komatiitic basalt unit intruding black mudstone. **h** Shale clast inclusions in the top of an intrusive komatiitic basalt unit intruding black mudstone

meso- to orthocumulate-textured zones, into which they can grade vertically and laterally (Fig. 7.16b, d; Hill et al. 1995; Fiorentini et al. 2007a, b). Spinifex can occur at the margins of some bodies (Cas et al. 2013; Gole et al. 2013).

Bodies with adcumulate texture can be hundreds of m thick (Hill et al. 1995; Barnes et al. 1988, 1995; Trofimovs et al. 2003; Rosengren et al. 2005, 2007). Their lower contact may cross-cut substrate stratigraphy with relief on the contacts of at least 150 m (e.g. Perseverance adcumulate komatiite), which has been interpreted as being due to extreme thermal erosion by channelised komatiite lavas (Barnes et al. 1988; Huppert and Sparks 1985). However, in

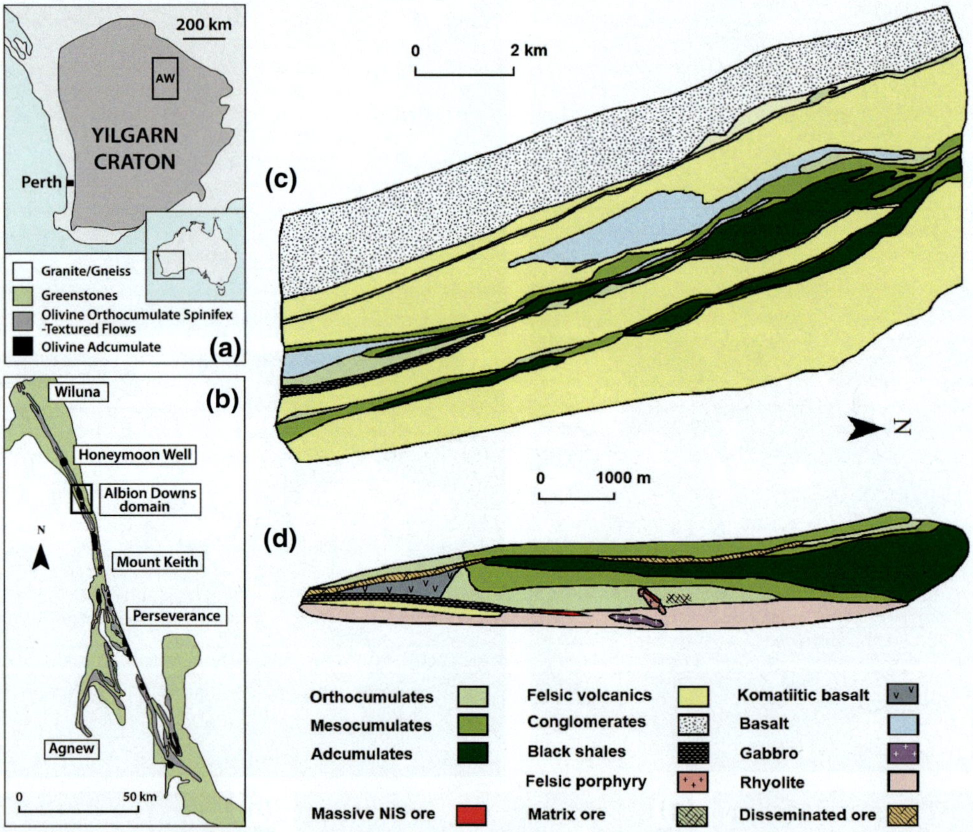

Fig. 7.16 Map-scale lateral variations in Archean intrusive dunitic komatiites, Agnew–Wiluna greenstone belt (AW), northern Kalgoorlie Terrane, Yilgarn Craton, Western Australia. **a** Location of dunite adcumulate-textured komatiites. **b** Regional map showing discontinuous nature of thick adcumulate-textured zones within belt of komatiites. **c** Map showing the along strike lensoidal geometry and lateral changes of the adcumulate bodies within the Albion Downs domain, north of Mt Keith. Three main komatiite horizons can be identified all of which face west have subvertical dips, and can be traced laterally from south of Mt Keith northwards to Honeymoon Well. The komatiites are sandwiched between dacitic-to-rhyolitic volcanic and volcaniclastic rocks which display similar textural and geochemical features to the felsic host rock of the Mt Keith dunite adcumulate (see text). The eastern and central komatiite horizons are generally thick (up to ~ 1500 m) and their internal architecture is made up of adcumulate-textured pods and lenses that are flanked by laterally extensive meso- and orthocumulate-textured units. In addition, no spinifex-textured horizons and no interflow sedimentary rocks have been recognised. The western komatiitic horizon is thin (<100 m), laterally and vertically more homogeneous, and is made up of localised cumulate lenses, flanking low-Mg komatiites and abundant thin spinifex-textured flow units. **d** Detail of along strike variations in the textural architecture of the central komatiite horizon at Albion Downs, controlled by drilling between the Jordan and Boo Boo Well NiS prospects. (Adapted from Fiorentini et al. 2007a, After Fiorentini 2005)

this instance, the contacts are highly tectonised due to strain partitioning between rheologically different adcumulate and host rock lithologies (e.g. felsic intrusions and sedimentary strata; Trofimovs et al. 2003). This makes it very difficult to assess the emplacement origins although, given the very coarse crystalline textures (plutonic), a subsurface intrusive and slow cooling and crystallisation origin is favoured (Trofimovs et al. 2003).

In some adcumulate bodies, such as the Mt Keith dunite, both top and bottom contacts are preserved, and these clearly show apophyses or dykes of komatiite intruding upwards into overlying dacites (Fig. 7.17a) and downwards into the footwall dacites (Fig. 7.17b; Rosengren et al. 2005). The hanging-wall and footwall dacites are geochemically and texturally identical, and the komatiite apophyses contain inclusions of dacite, whereas the dacites do not contain inclusions of komatiite (Rosengren et al. 2005). Although the succession is part of a regionally deformed succession, these contact relationships are not the result of the tectonic mingling in a fault zone as proposed by Gole et al. (2013) because there is no evidence of high-strain tectonic melange mingling in the contacts (Fig. 7.17a, b), even though some outcrops could be part of displaced fault slices. Close examination of Fig. 7.17a reveals that dacite clasts in the komatiite dyke intruding the dacite show no signs of tectonic strain. They are equant to irregular, not highly strained, stretched, and therefore not fault slivers.

The Mt Keith komatiite is also highly differentiated internally, which allowed Rosengren et al. (2007) and Fiorentini et al. (2007a, 2012) to define multiple internal textural and petrological zones (Fig. 7.18a, b). These include from base to top: Basal olivine orthocumulate (subunit 101); very coarse olivine adcumulate with nickel sulphide mineralisation (102); porphyritic komatiite with skeletal olivine crystals and patchy concentrations of plagioclase (103—this facies is irregularly distributed across the 102–104 boundary as pegmatoidal lenses); medium adcumulate with nickel sulphide blebs along olivine crystal boundaries (104); olivine and pyroxene meso- to orthocumulate with oikiocrystic pyroxene rimming the olivine crystals (105); olivine adcumulate with some interstitial sulphide and gabbroic segregations in the middle (106); and topmost, ortho- to mesocumulate pyroxenite with gabbro in more fractionated areas, including the uppermost contact in places, and with highly variable olivine sizes (107). In addition, Gole et al. (2013) found spinifex texture within 2 m of the top contact with the hanging-wall dacite in the uppermost pyroxenite unit in places.

Rosengren et al. (2007) subdivided the Mt Keith komatiite into a lower ultramafic part and an upper fractionated part that includes pyroxenites and gabbros (Fig. 7.18b). They found similar patterns in other differentiated dunitic intrusive sills, including the Dumont sill in Canada and the Shangani sill in Zimbabwe, suggesting a

Fig. 7.17 Intrusive contact relationships of the Archean Mt Keith dunite, northern Kalgoorlie Terrane, Yilgarn Craton, Western Australia. **a** Part of the upper contact of the Mt Keith dunitic komatiite with hanging-wall dacite, showing a komatiite apophysis dyke (dark) containing inclusions of dacite (light), which are clearly not sheared out, high-strain fault slivers. **b** Part of the lower contact of the Mt Keith dunitic komatiite with the footwall dacite, showing irregular apophyses-dykes of komatiite intruding the dacite (light). Although there are some faults offsetting some of the dykes, it is clear from their irregularity that the dykes are not fault slices. (From Rosengren et al. 2005)

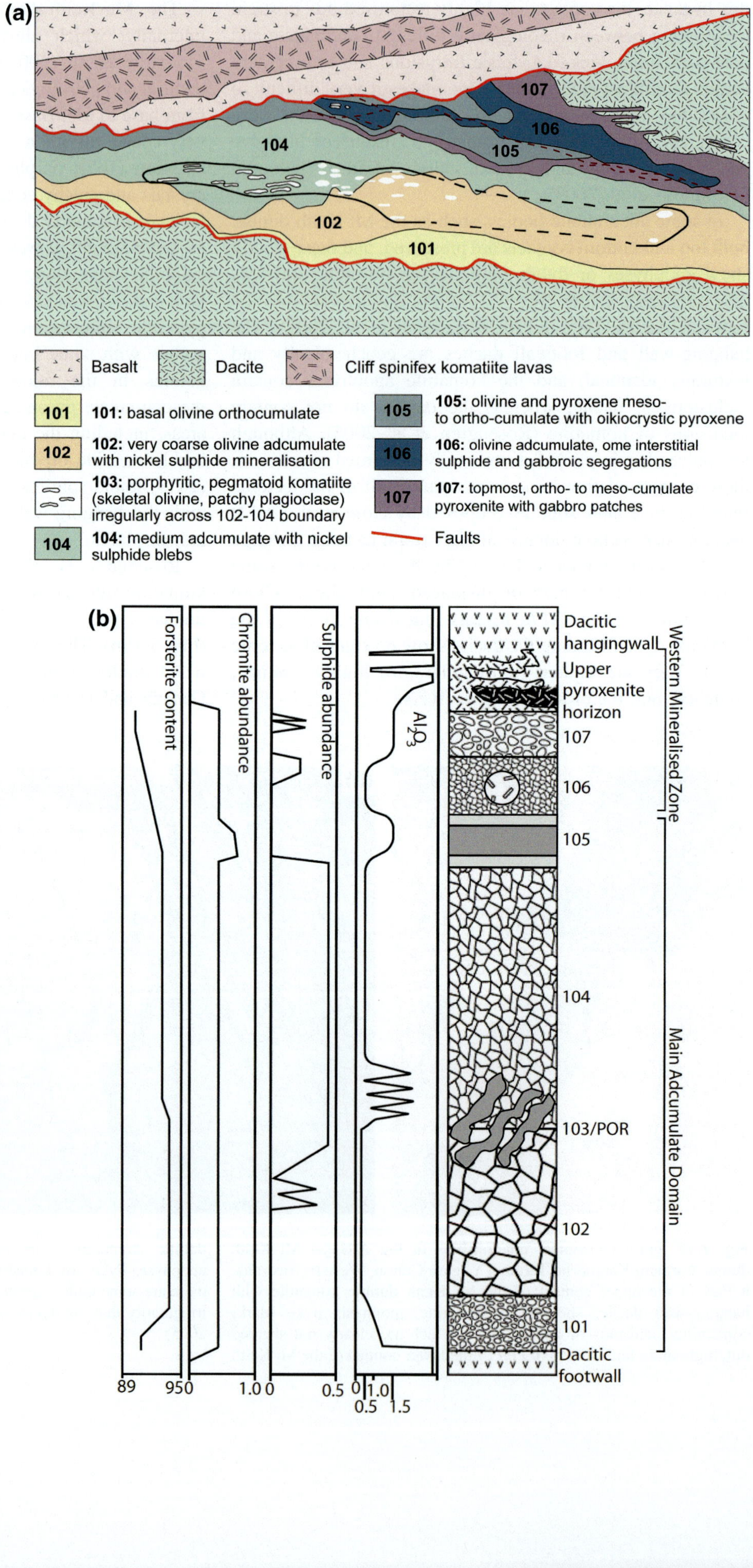

Fig. 7.18 Petrological features of the Mt Keith dunite. **a** Map showing highly irregular along strike and vertical petrological variations. The red lines define faults and major shear zones. Subunits 101–107 define multiple internal textural and petrological zones as described in the text. **b** Schematic stratigraphic column for the entire Mt Keith dunite, indicating the internal vertical petrological divisions, and graphic traces for Al_2O_3, sulphide abundance, chromite abundance, and forsterite content. (From Rosengren et al. 2007)

subvolcanic intrusive origin for the textural zonation. These complex internal textural and mineralogical patterns seem to be most readily explained by complex, multiple magma injection and crystallisation and differentiation events in a very thick differentiated sill rather than a lava flow, however thick (apparently up to 900 m thick!). Rosengren et al. (2007) also divided dunitic sills into lens and sheet dunites. Nickel sulphide mineralisation is a common feature of the lower ultramafic section in both lens and sheet dunites. The lens-shaped Mt Keith and Perseverance dunites in Western Australia host economic accumulations of NiS, whereas the sheet-like bodies at Dumont, Canada (economically marginal) and Shangani, Zimbabwe, contain weakly disseminated mineralisation (Chap. 18, Sect. 18.6).

The alternative view, provided by Gole et al. (2013), proposed that the Mt Keith komatiite was a ponded, very thick lava which implies that it had thermally eroded up to 450 m of the substrate dacite, which interestingly is texturally and chemically identical to the hanging-wall dacite overlying the Mt Keith dunite (Rosengren et al. 2008). Gole et al. (2013) further argued that the Mt Keith komatiite body could not be an intrusion because

(a) The top and bottom komatiite apophyses intruding dacite above and below are fault slivers. However, as Rosengren et al. (2005) have shown (Fig. 7.17a, b), the contacts are too irregular and show only minor deformational fabric to be fault margins.
(b) Dacite inclusions in komatiite apophyses proposed by Rosengren et al. (2005) are fault slivers. However, as we point out above, these inclusions show no evidence of being high strain, sheared out, tectonically mingled, fault slivers, which is clearly visible in Fig. 7.17a.
(c) The presence of spinifex texture near the top contact, the lack of evidence of melting of dacitic country rock in the overlying dacite, the preservation of fine groundmass textures in the overlying dacite near the contact, as well as thermal budget modelling all indicate that, given its thickness of over 400 m, the body must have been a very thick ponded lava. However, the problem with this reasoning is that it assumes that the whole body was emplaced almost instantaneously as a single, > 400-m-thick body of magma at a temperature of 1,500 °C with its total heat budget. This is impossible, and the body must have inflated over time. The rate at which this happens would control the rate of heat flux to substrate, limiting thermal effects.

Our preferred interpretation is that large intrusions will usually be emplaced slowly, initially as a thin wedging sill that propagates by tip crack propagation, and then progressively inflates. If the intrusion rate is slow, the cooling rate is high, and textures in the intrusion margins will reflect rapid cooling rates. It is also not necessary for there to be just one intrusive pulse, but perhaps several, which thus lowers the thermal budget of the intrusive pulses at any time. Nonetheless, the main part of the Mt Keith intrusion must have cooled slowly enough (no undercooling) to form adcumulate texture, and differentiate, unless the basalt/pyroxenite and gabbro textural zones belong to a coeval mafic intrusion, rather than necessarily originating from differentiation and fractionation of a large volume ultramafic magma intrusion.

A further problem for the scenario proposed by Gole et al. (2013) is that Hill et al. (1995) argued that the Mt Keith komatiite thermally eroded (i.e. melted) a channel down into the substrate that is up to 450 m deep. Melting of crustal rocks represents ultra-metamorphism, which should lead to significant metamorphism of that substrate contact, for which there is no evidence. Original volcanic textures are still well preserved in the underlying dacitic host rocks (Rosengren et al. (2005, 2008). Gole et al. (2013) ignored the nature of the contact with the underlying footwall dacite at Mt Keith, which is petrographically and geochemically the same as the hanging-wall dacite (Rosengren et al. 2008). Neither they nor Hill et al. (1995) have demonstrated anywhere in the Yilgarn Craton the existence of anatectic textures and high levels of thermal contact metamorphism at and near the contacts between dunitic bodies and substrate rocks that have been supposedly thermally eroded to a depth of hundreds of m by komatiite "lavas".

In more complex Mt Keith-type intrusive systems, such as at the Black Swan mine, northeast of Kalgoorlie (Fig. 7.10b), a more intricate, interdigitating relationship between komatiite and dacite units is found (Cas et al. 2013). Although this succession had also been interpreted to be a sequence of alternating komatiite lavas and dacite lavas, together with dacitic breccias, taken to be autoclastic (Hill et al. 2004), there are also many problems with that explanation (Cas et al. 2013). First, many of the dacite intervals that are intercalated with komatiite units are very thin, from cm to m in thickness, indicating that they cannot be dacite lavas, which are always tens to hundreds of m thick because of high eruption viscosities (Chap. 6). Secondly, many of the dacite units are coherent at the margins, and some of those are brecciated internally. All felsic lavas have autoclastic margins because of their high-viscosity and flow-induced autobrecciation, or due to significant quench fragmentation in subaqueous settings (Chap. 6). Although Hill et al. (2004) interpreted the dacitic breccias as autoclastic-lava-associated breccias, evidence was not provided. Cas et al. (2013) described the breccias in detail and interpreted them as hydraulic fracture breccias, with textures modified by intense hydrothermal alteration. Cas et al. (2013) also found that the komatiites intruded a pre-existing massive dacite unit, based on inclusions of dacite in komatiite, whereas no komatiite inclusions occur in dacite. On this basis, the komatiite intrusions have split or intrusively dismembered the dacite into multiple units, some only cm thick, producing a succession of alternating komatiite and dacite units, which would be very unusual and difficult to explain as a sequence of alternately erupting ultramafic and dacite lavas from the one volcanic system.

The Black Swan komatiite units also range from cm to tens of m thick. Textures vary from aphyric to random and parallel plate spinifex at the margins, to ortho-, meso-, and adcumulate in the centres, adcumulate occurring where the komatiite is over 100 m thick (Fig. 7.3 p). Skeletal hopper crystal textures also occur (Fig. 7.3k), as well as segregation pods (Fig. 7.8d, e). In places, multiple spinifex-cumulate units are superimposed, giving the impression of stacked, thin (<20 m) lava-like units. In the absence of any other evidence that these units were erupted on the seafloor as lavas, Cas et al. (2013) reinterpreted them as a succession of thin komatiite sills. To achieve this, intrusion rates must have been low, and (under)cooling rates high to allow spinifex texture to form at the margins of individual intrusions. The implication is that the komatiite intrusive event at Black Swan was protracted and a multi-stage event, not just a single event. As a result, the overall architecture of the Black Swan mine sequence is a complex one with interleaving of dacite and komatiite units with very irregular geometries, which is not what would be expected of a low-viscosity komatiite lava flow field.

7.7.3 Spinifex Texture in Some Komatiitic Intrusions: Criteria for Distinguishing Them from Lavas

As we have already described, spinifex texture is best developed in komatiite lavas, most of which are preserved in submarine successions. The ambient water mass of submarine settings, with its high heat capacity and thermal conductivity (Chap. 3), acted as a heat sink that produced a high thermal gradient, cooling rate, and degree of undercooling that allowed spinifex crystallisation to occur in the top of subaqueous komatiite lavas (Fig. 7.6a, b). Nonetheless, both the thin komatiite intrusions at Dundonald and Black Swan mine preserve spinifex textures at their margins, as well ortho- to mesocumulate textures indicating that these textures are not unique to just komatiitic lavas.

Spinifex crystallisation in relatively shallow and small volume intrusions m to tens of m thick, especially into water-saturated sediments (Fig. 7.6c), indicate that the cooling conditions that promote crystallisation of spinifex also exist in the subsurface. Arndt et al. (2004) suggested that because of slower cooling rates in intrusions compared with subaqueous lavas exposed to a water mass, crystals in intrusions will be larger, less elongate, and less skeletal than at the same depth beneath the surfaces of lavas. At Pyke Hill, Arndt et al. (1977) have recognised that some thin lava lobes or pillows are aphyric and lack internal spinifex texture, indicating chilling rates too high to allow spinifex crystallisation, and in turn indicating that the presence or absence of spinifex cannot be used definitively to distinguish lavas from intrusions. The other problem in trying to use this criterion is that post-emplacement metamorphism and alteration can overprint and mask original textures and produce needle-like metamorphic amphibole that have in places been mistaken for spinifex crystals.

The crucial criteria then for distinguishing syn-depositional komatiitic intrusions from lavas are the contact relationships, and especially the upper contact. Syn-depositional intrusions can be cross-cutting or roughly concordant, they will be irregular in geometry along their extent, and both lower, but particularly upper contacts will show evidence of syn-depositional interaction with wet, unconsolidated sediments (irregular apophyses intruding upwards, peperitic mixing into overlying sediments). The presence or absence of spinifex texture is not diagnostic, nor is cumulate texture. Syn-depositional komatiitic intrusions are further discussed in Chap. 8.

In contrast, in sustained intrusions into solid rock which inflate to form very thick intrusions hundreds of m thick, it is unlikely that the thermal gradient and cooling rate will be high, and that undercooling will occur. In the latter situation, even if initial chilling and spinifex growth occurs, it is likely that as the country rock thermally equilibrates with the intrusion, an equilibrium thermal gradient will develop which causes a contact metamorphic aureole to form. Any initial chilled margin that forms when the intrusion first propagates as a thin wedge is likely to be assimilated back into the intrusion. As with every phenomenon in volcanology, there are no rules that determine that every komatiitic lava or intrusion will be texturally and architecturally identical. Every case has to be assessed on its merits, and because of variations in intrusion rates or magma flux, magma volumes, host rock characteristics, (under)cooling rates, each example could be different. Assessing all the contact relationships remains the key criterion to assess. In particular, based on current geological relationships, the cross-cutting lower contacts of dunitic bodies are more likely to be intrusive rather than related to thermal erosion, or melting and removal of hundreds of metres of substrate, unless sound evidence can be found for anatectic and ultra-metamorphic textures at basal contacts and their absence at upper contacts.

7.8 Komatiite Proximal–Distal Relationships

Hill et al. (1995) proposed that adcumulate komatiites (e.g. Mt Keith) of the Agnew–Wiluna greenstone belt in the northern part of the Kalgoorlie Terrane represent proximal near-vent facies of ginormous channelised lava flows while the spinifex-cumulate-textured Kambalda-type komatiites in the southern part of the Kalgoorlie Terrane (Fig. 7.10a, b) represent the distal facies of the same regionally extensive lava flow field (analogous to far-flowing flood basalts). However, this model is predicated on the assumption that all the komatiites represent lavas that they were all erupted from the same vent system, requiring flow distances of at least 500 km for the "distal" Kambalda komatiites. On the other hand, as we have described above, there is significant evidence that the adcumulate-textured Mt Keith-type komatiites were intrusive (Trofimovs et al. 2003; Rosengren et al. 2005, 2007; Cas et al. 2013). At Mt Keith, there is also evidence of multiple komatiite stratigraphic units, including the thin, spinifex-orthocumulate unit lava successions of the Cliffs and the Monument komatiites, which are separated from the adcumulate Mt Keith komatiite by a major fault and basalt units (Rosengren et al. 2005). This suggests that the adcumulate unit is a subvolcanic intrusive complex, whereas the Cliffs and Monument komatiites represent co-magmatic extrusive counterparts, as proposed by Fiorentini et al. (2012; Fig. 7.19).

Nonetheless, it is important to note that not all komatiite intrusions will develop adcumulate textures, this depending on the magma volume intruded, the thickness of intrusions, the rate of magma intrusion, and the cooling rate. Many thin intrusions could thus be spinifex-cumulate-textured bodies, and even the margins of very thick intrusive bodies could preserve some spinifex texture if the intrusion rate was slow, and the cooling rate was high, especially in multi-phase or stage intrusions. The model proposed by Fiorentini et al. (2012; Fig. 7.19) thus allows multiple eruption points or centres along the length of the Kalgoorlie Terrane and elsewhere in the broader Eastern Goldfields Superterrane. We suggest that the location of massive nickel sulphides is a better proxy for proximity to magma conduit (for intrusions) and vent for lavas, because nickel suphide is very dense and is likely to be deposited from the host magmas close to source, whether in lavas, or in sill-like intrusions, as proposed by Fiorentini et al. (2012; Fig. 7.19), and later studies (e.g. Perring 2015).

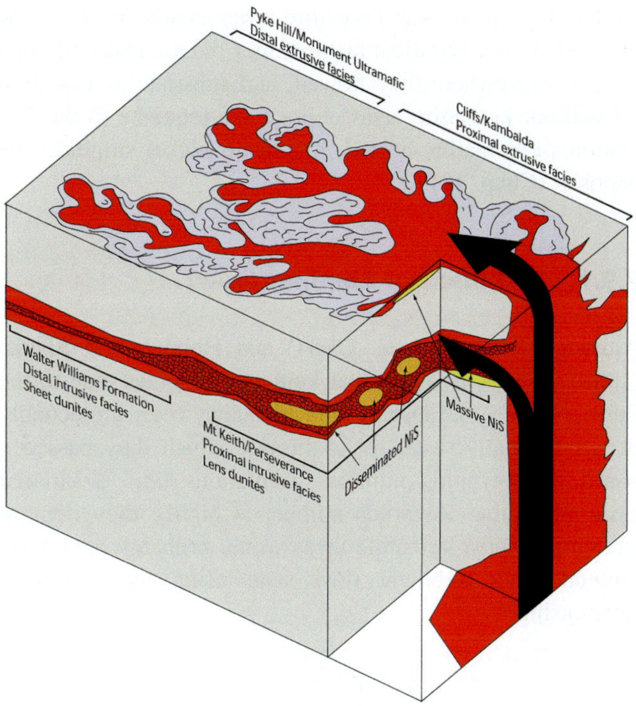

Fig. 7.19 Schematic representation of the spatial relationships between extrusive komatiite lava flow fields and subsurface intrusive komatiites relative to the magma conduit and depicting the locations of nickel sulphide mineralisation. Note the location of the massive sulphide mineralisation in both lavas and sills proximal to the conduit (e.g. Kambalda in the southern Kalgoorlie Terrane, or Cliffs, Mt Keith and Perseverance in the Agnew–Wiluna greenstone belt), rather than distal facies (e.g. Pyke Hill in the Abitibi greenstone belt, the Monument ultramafic unit, Agnew–Wiluna greenstone belt, or the Walter Williams Formation, southern Kalgoorlie Terrane). (From Fiorentini et al. 2012)

7.9 Komatiite Lava Flow Behaviour

Based on numerical modelling, Huppert et al. (1984) and Huppert and Sparks (1985) proposed that because komatiite lavas consisted of very low viscosity (0.1 to 10 Pa s), very high temperature (<1650 °C) magmas, they were erupted and flowed as turbulent lava flows. Such flows were capable of melting and assimilating the substrate over which they flowed, which was instrumental in the formation of the magmatic nickel sulphide ore deposits that some komatiites host (Chap. 18). Although these ideas may be partly correct in some situations, Cas et al. (1999) and Cas and Beresford (2001) demonstrated that geological evidence and further modelling indicate that komatiite lavas mostly did *not* flow everywhere in a turbulent fashion, they did *not* cause thermal erosion everywhere they flowed, and substrate assimilation by seafloor komatiite lavas is *not* a pre-requisite to the formation of komatiite-hosted, magmatic nickel sulphide ore deposits (Chap. 18).

7.9.1 Turbulent or Laminar Flow?

Although Huppert et al. (1984) and Huppert and Sparks (1985) incorrectly assert that komatiite magmas rose to the Earth's surface because of their buoyancy (they are actually denser than all crustal rocks through which they passed!), based on Reynolds number calculations (and a circular assumption that Reynolds number is high!), they demonstrated that most komatiite lavas would probably have been erupted in a turbulent flow state using the following relationship:

$$Q = \left(\frac{g\,\Delta\rho}{k\,\rho_m}\right)^{0.5} d^{1.5} \tag{7.1}$$

where Q is the two-dimensional magma discharge rate, g is the gravitational acceleration constant (9.8 m s^{-2}), $\Delta\rho$ is the density contrast between komatiite magma and host crustal rocks ($\sim$-300 kg m^{-3}), k is a friction coefficient between the magma and the conduit walls ($\sim$0.03), ρ_m is the komatiite magma density (2,800 kg m^{-3}), d is the conduit width (m), and

$$Re = \frac{Q}{\upsilon} \tag{7.2}$$

where Re is the Reynolds number and υ is the kinematic viscosity. The calculated Reynolds numbers were in excess of 2,000, indicating a turbulent state at the time of eruption. However, this does not mean that the flow state remained turbulent throughout the flow once it propagated away from the vent on the seafloor.

Cas et al. (1999) and Cas and Beresford (2001) reviewed these concepts, and assessed the impact of variations in flow viscosity, substrate slope, flow velocity, magma density relative to ambient medium density (subaerial, subaqueous) on flow front thickness of komatiite lavas, and as a result on the Reynolds numbers in the lava flows, using the empirical Jeffreys' equation for Newtonian fluids:

$$v = \frac{d^2 \rho\, g \sin\alpha}{3\eta} \tag{7.3}$$

$$Re = \frac{\rho\, d\, v}{\eta} \tag{7.4}$$

where v is the flow velocity (values used in calculations were 0.1, 0.5, 1.0, 3.0, 5.0, and 10 m s-1), d is the lava (flow front) thickness (to be calculated), ρ is the lava density (2,800 kg m^{-3} in air *BUT* 1,800 kg m^{-3} in water), g is the acceleration of gravity, α is the slope or inclination of the substrate (values used 0.01, 0.1, 1, 2, 3, 5, and 10°), η is the lava viscosity (values used: 0.1, 0.5, 1, and 10 Pa s), and 3 is a geometrical factor for wide flows (it is 4 for narrow flows).

The results (Fig. 7.20) show that near the vent, almost all flows would have been turbulent, but in medial and distal settings, changes in these variables significantly affect flow behaviour (Cas et al. 1999). Very low-viscosity lavas (<0.5 Pa s; $\sim$ [supra]liquidus temperature) would have flow fronts only 10 cm thick at low flow velocities of 0.1 m s^{-1}, and 0.8 m thick at 10 m s^{-1} (36 km hr^{-1}), and would flow in a laminar state on slopes < 1° and at velocities < 1 m s^{-1}, but would flow in a turbulent state on higher slopes and velocities > 2 m s^{-1}. Intermediate viscosity lavas (0.5 to < 1 Pa s) showed similar results. Flows on slopes < 1° flowed in a laminar state at velocities < 1 m s^{-1}, but even on slopes of < 10° and velocities of < 10 m s^{-1}, flow was laminar. For flows with viscosities of < 10 Pa s, flow fronts were still 1 m thick, and they flowed in a laminar flow state on slopes < 10° and at flow velocities of < 10 m s^{-1} (Fig. 7.20; Cas et al. 1999).

In the Archean, the seafloor was dominated by low-viscosity basalt and komatiite lavas, which would have smoothed seafloor slopes to almost 0°, except locally where fault scarps or high aspect ratio felsic lavas occurred. As a result, most basin floors would have had negligible slope, indicating that as lavas spread, and the flow front velocities waned, flow conditions for most komatiite lavas would have quickly transformed to laminar in medial to distal settings. This is because komatiite magma has a very low viscosity, a high density, and is essentially a Newtonian fluid, ensuring that the magma would spread radially from the eruption point as a thin sheet in a laminar flow state, driven by the hydraulic head at the vent.

This assessment is consistent with the geological evidence suggesting that most komatiites were not emplaced as turbulent flows (Cas et al. 1999; Cas and Beresford 2001), including:

(a) The bases of most komatiite lavas in the field are planar, non-erosive bases, indicating that neither physical nor thermal erosion occurred along most of the bases of most flows.
(b) The substrate under many komatiite lavas consists of fine mudstone that is undisturbed (i.e. no loadcast features), and/or intact hyaloclastite tops of underlying komatiite lavas, indicating a very passive emplacement on the substrate.

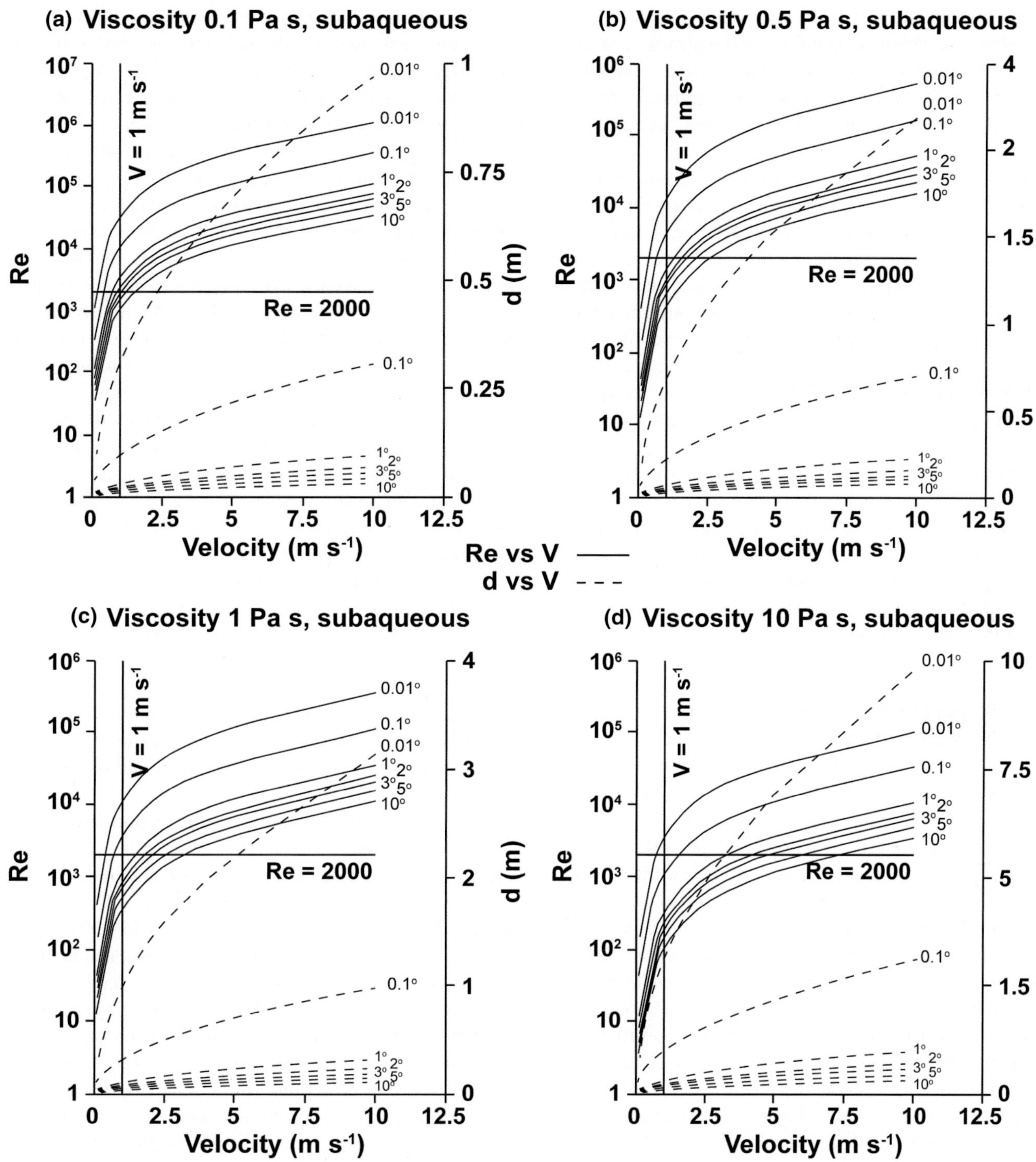

Fig. 7.20 Results of calculations for flow front thickness of active subaqueous komatiite lavas, and Reynolds number, as a function of different flow velocities, substrate slopes, and viscosities. The results show that flow fronts of komatiite lavas of varying viscosities and on slopes up to 10° would have been thin and flowed in a laminar flow state except at the vent. Thick preserved flows must have inflated, been ponded in topography, or are in fact intrusions (see Cas et al. 1999). (From Cas et al. 1999)

(c) All subaqueous komatiite lavas have intact coherent (aphyric to spinifex textured) and/or jigsaw-fit hyaloclastite crusts. If flow was turbulent, crusts would have been constantly autobrecciated during formation, for which there is no geological evidence (i.e. no autobreccias are preserved).
(d) The crystal settling velocity of dense 2 mm diameter olivine crystals (ρ= 3,300 kg m^{-3}) in komatiite magma (ρ= 2,800 kg m^{-3}) using Stokes settling velocity law is ~ 0.11 m s^{-1}, indicating that the maximum flow velocity of a komatiite lava flow would have had to have been only < 0.11 m s^{-1} to allow any crystal settling. This then generates a Reynolds number of 280 for a magma of viscosity 1 Pa s, indicating again, laminar flow conditions for most komatiite lavas, except for ultra-low-viscosity (0.1 Pa s) supra-liquidus lavas.

A simple analogue demonstration of the flow behaviour of komatiite lavas is to take a bucket of olive oil, which has approximately the same viscosity as komatiite magma, and tip it rapidly onto a flat surface from a height. Initially, the oil flows turbulently at the point of impact, but it quickly spreads radially as a thin, smooth sheet of low-viscosity oil flowing in a laminar state.

7.9.2 Physical and Thermal Erosion by Komatiite Lavas?

Komatiite magmas are thought to have been nickel bearing but sulphur poor at source (Huppert et al. 1984; Huppert and Sparks 1985). Therefore, to become sulphur saturated and then precipitate nickel sulphide ore within the flowing komatiite, they need to assimilate an external source of sulphur somewhere between source and final emplacement point on the seafloor. Huppert et al. (1984) and Huppert and Sparks (1985) proposed that komatiite lavas were capable of thermally eroding (i.e. melting) their substrate and assimilating this material, especially sulphidic seafloor mudstones, so helping drive the lavas to sulphur saturation. Their model was also based on the assumption that the lavas flowed in a turbulent state, which would distribute the total heat throughout the lava, so enhancing substrate heating and melting. They also cited the existence of basal trough-like structures at Kambalda, documented by Gresham and Loftus-Hills (1981), which host much the massive nickel sulphide deposits or ore shoots at Kambalda, as evidence of thermal erosion at the base of the seafloor komatiite lavas. These "troughs" had been interpreted by Gresham and Loftus Hills (1981) as seafloor fissures or fault rifts. Many of these troughs lack ambient hemi-pelagic sediments that are preserved laterally along the flanks of the troughs, and they attributed this to removal and assimilation by thermal erosion. Groves et al. (1986) then adopted the thermal erosion model for the trough-like structures at Kambalda (Fig. 7.21), but not the importance of assimilating sulphidic seafloor sediments to account for the nickel sulphide mineralisation.

As demonstrated by Cas et al. (1999), not all komatiites would have flowed in a turbulent fashion. Contrary to the assertion of Huppert et al. (1984), turbulent flow would actually accelerate the rate of heat loss to the ambient ocean mass, which has a high heat capacity, because flow would be open channel, not in enclosed lava tubes with an insulating crust (Chap. 6). Ironically, lava tubes with a stable, insulating surface spinifex-textured chilled crust are most likely to form under laminar flow conditions, and would better limit heat loss than open channel turbulent flow. Furthermore, very few komatiites have basal troughs or channels that are demonstrably of thermal erosional origin (e.g. anatexis textures at contacts), and isotopic geochemical studies have shown that most mineralised komatiites were probably driven to sulphur saturation by assimilating crustal sulphidic source rocks in the conduit through which the magmas were rising to the seafloor (Lesher et al. 1981; Groves et al. 1986; Fiorentini et al. 2012).

We will now consider three key aspects of this topic in more detail:

- origin(s) of nickel-sulphide-bearing troughs,
- numerical modelling of thermal erosion capacity, and
- explaining the "missing sediment" in some ore settings.

7.9.2.1 Origin(s) of Nickel-Sulphide-Bearing Troughs

Detailed studies of trough-like structures at Kambalda have shown that most are faulted structures (Fig. 7.22; Gresham and Loftus-Hills 1981; Stone and Archibald 2004; Stone et al. 2005). The geological setting of the Kambalda nickel sulphide deposits is dominated by a regional structural dome, the Kambalda Dome which is intruded by a granitic batholith and the whole succession has been affected up to four major deformations, D1 to D4. The host rocks include a footwall basalt unit, the Lunnon Basalt, overlain by the variably talc-carbonate and serpentinised Kambalda Komatiite Group. The massive nickel sulphide deposits sit mostly at the contact between the basalt and the komatiite.

Stone et al. (2005) propose that the footwall basalt beneath the komatiite deformed in brittle fashion; the sulphides in ductile fashion; and the komatiite, variably from brittle where serpentinised to ductile where talc-carbonate altered. As a result, there has been major strain partitioning between the three rock types. The basalt and its surface have been folded and frequently brittle faulted and thrust into the komatiite, which has been thickened at ore sites. The

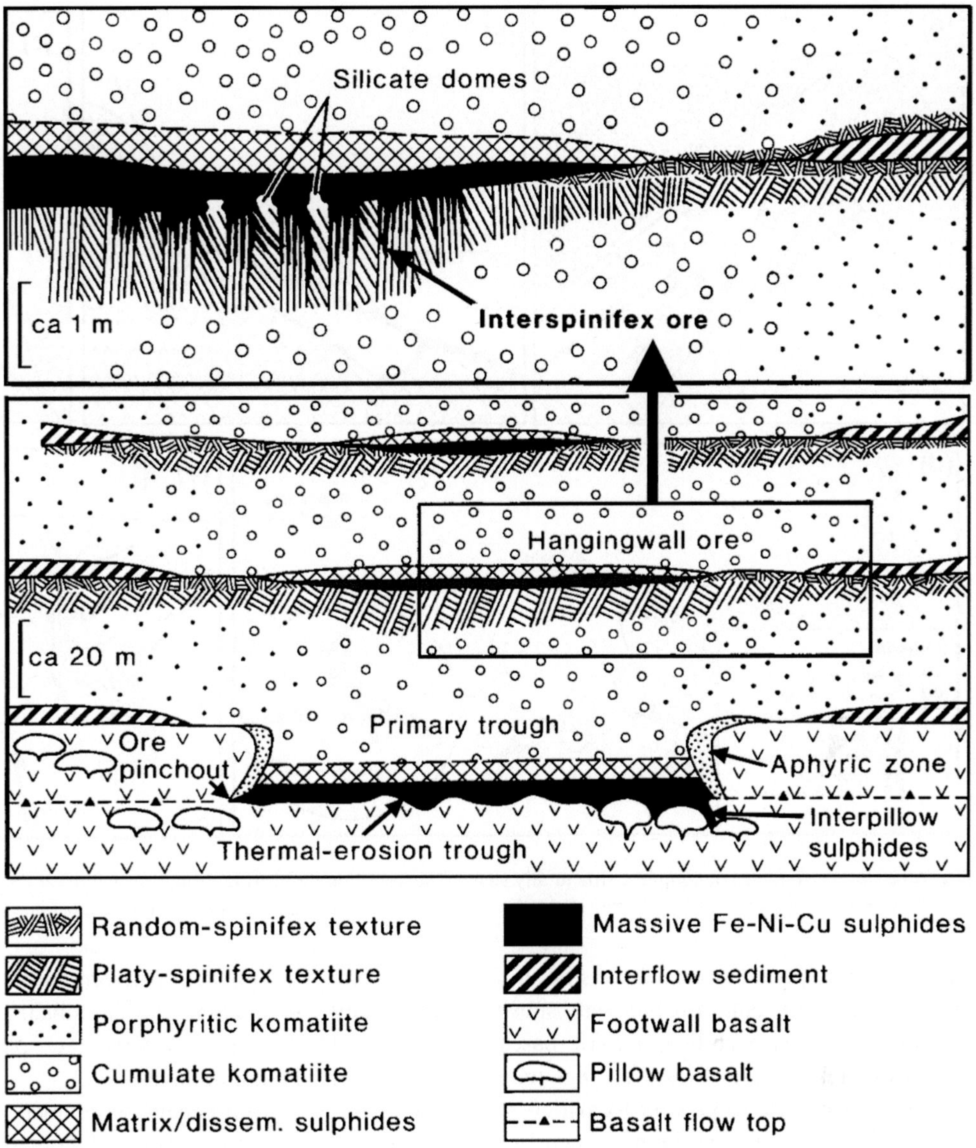

Fig. 7.21 Conceptual thermal erosion and nickel sulphide mineralisation model for trough-like structures at the base of the first komatiite lavas emplaced on the thick Lunnon Basalt at Kambalda. Note that sediment is missing between the basalt and the komatiite in the trough, but is present on the flanks of the trough. (From Groves et al. 1986)

geometry of the ore shoots has been greatly modified and thickened by ductile deformation but also stretched along structures and contacts. The trough structures are thus essentially structural fault bounded and modified features, not primary volcanic features. Only the Durkin ore shoot preserves part of what may have been a largely volcanic morphology, with an open, broad, long depression at least 1,000 m long, several hundred m wide, and up to 50 m deep. It is unclear if that reflects pre-komatiite emplacement volcanic topography on the surface of the Lunnon Basalt (e.g. two separated basalt lava flow lobes separated by a depression between them, rather than a depression created by the flow behaviour of the initial komatiite lavas). However, even the Durkin ore setting has been deformed, by open fold warping and especially by faulting, making even this assessment speculative.

The difficulty in assessing the origin of the trough structures at Kambalda is a sage reminder that great caution is required in reconstructing imaginative volcanological models in ancient volcanic settings without fully appraising the total geological framework and history, including pre- and post-emplacement deformational effects. There is actually no geological evidence that the Kambalda troughs originated by thermal erosion.

7.9.2.2 Numerical Modelling of Thermal Erosion Capacity

Huppert et al. (1984) and Huppert and Sparks (1985) calculated that turbulently flowing submarine komatiite lavas should be able to melt substrate and produce substantial thermal erosion channels in the substrate. This implies that the base of every komatiite lava should show signs of

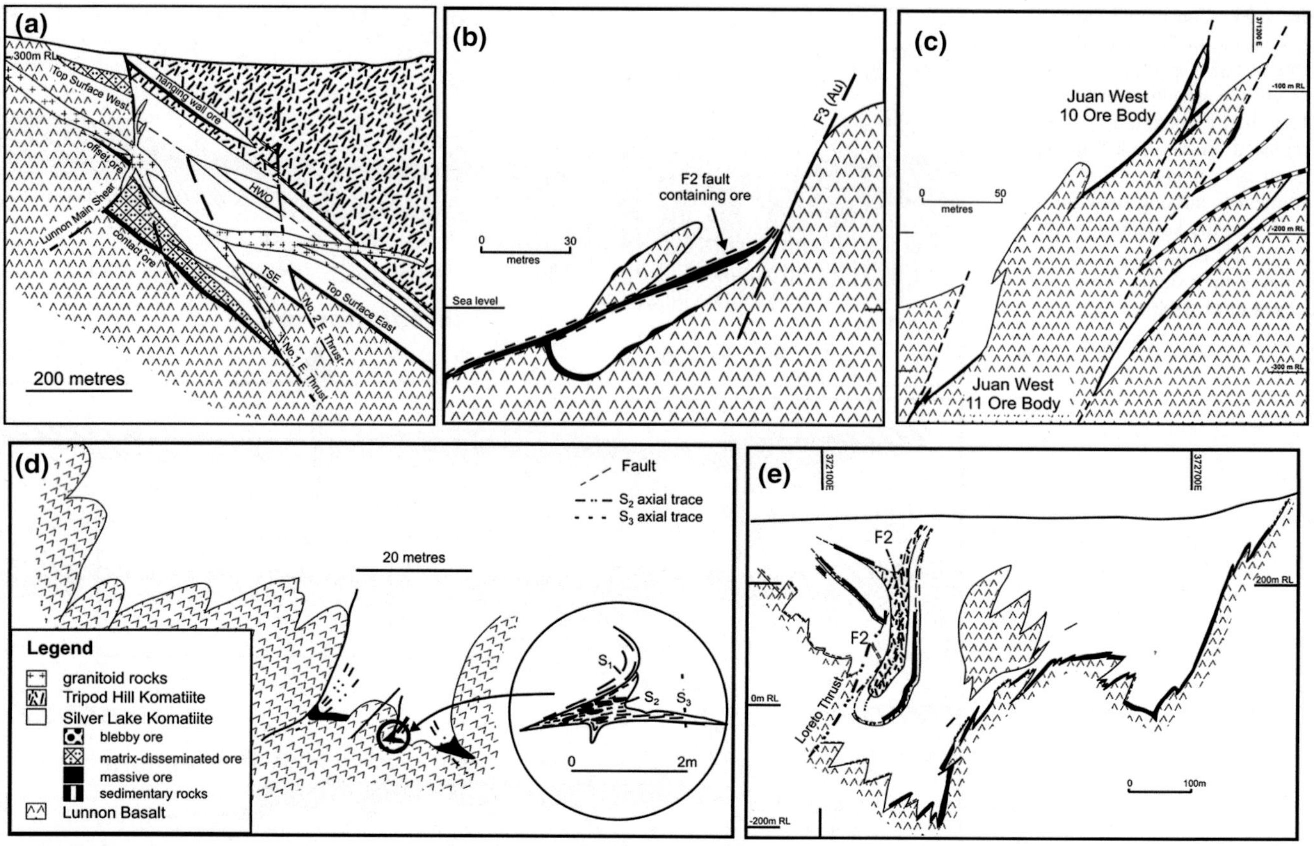

Fig. 7.22 Some examples of complex, structurally shaped, nickel sulphide-bearing, trough structures at the base of the main mineralised komatiite lava flow, Kambalda mine camp, Western Australia. These are not created by primary thermal erosion. (From Stone and Archibald 2004)

substrate anatexis and erosional incision. In fact, this is rare, most contacts are conformable with substrate, and many komatiite lavas have chilled, aphyric bases indicating that the bases of komatiite lavas were not transferring enough heat to the substrate to prevent formation of a chilled base, let alone substrate melting. Perhaps there are some instances where substrate melting could have occurred, but definitive geological evidence has not been presented anywhere.

Williams et al. (1998) extended the approach of Huppert and Sparks (1985) to include a detailed assessment of the thermal properties of different types of substrate, including unconsolidated sediments, consolidated sediments, and basalt, assuming a komatiite lava thickness of 10 m. They also recognised that erosion would be a combination of thermal and mechanical, and proposed the term thermo-mechanical. They proposed that for unconsolidated sediments, most of the erosion would have been by physical fluidisation of the sediment producing mechanical erosion down-cutting rates of 10 to 23 m per day. Whereas, if the sediments were consolidated, down-cutting rates by thermo-mechanical erosion would have been 0.4 to 1.2 m per day, and for a basalt substrate 0.2 to 0.6 m per day. They cite the trough depths at Kambalda as evidence of the validity of the concept of thermal erosion, largely ignoring the impact of deformation of the geometry of these troughs (e.g. Stone and Archibald 2004), and the observational fact that it is a very rare phenomenon along the bases of komatiite lavas.

In a general assessment of the evidence for thermal erosion by basalt and komatiite lavas, Greeley et al. (1998) found that there was some evidence for substrate thermal $\pm$ mechanical erosion by basalt lavas into underlying basalt lavas in Hawai'i, but they questioned if some of the cases often cited as evidence of significant down-cutting wasn't the result of downward intersection into pre-existing drained lava tubes/caves, so reducing the actual rates of thermal erosion and down-cutting. They found that there was no evidence that the largest and longest lived basalt lava events, i.e. flood basalts, thermally eroded their substrate. The best evidence for thermal erosion having occurred was in the Cave Basalt exposure at Mt St Helens, USA, where a basalt lava had caused anatexis and flow of a substrate dacite over which the basalt had flowed; dacite has a much lower solidus temperature than basalt. In evaluating the evidence for komatiites being able to thermally erode substrate in the Yilgarn Craton, Greeley et al. (1998) were circumspect in assessing the evidence, highlighting the difficulties of dealing with the poorly outcropping, deformed, and metamorphosed rock successions there. For the less deformed and metamorphosed Archean Abitibi greenstone belt in Canada, where basal contacts are better exposed and preserved, no evidence of thermal erosion was found.

Furthermore, Fagents and Greeley (2001) analysed the potential for basalt lavas to melt their substrate and highlighted the importance of the interaction between the lava flow dynamics, the contrast in thermal properties of the magma and substrate, and the nature of the thermal gradient at the base of the lava in determining whether or not thermal erosion will occur. Sustained, stable flow conditions are essential, whereas destabilising factors such as breakouts from tubes, pulsing in flow, blockages, and pauses in supply mitigate against thermal erosion. These destabilising factors are common phenomena, and hence thermal erosion is a rare phenomenon.

7.9.2.3 Explaining the Missing Sediment in Some Ore Settings

Nonetheless, nickel sulphide ore did form discrete, perhaps elongate ore bodies or lenses at the bases of komatiites originally (e.g. Fig. 7.10c). Under many of the ore bodies at Kambalda, hemi-pelagic sediment between the top of the Lunnon Basalt and the base of the komatiite lava is absent whereas laterally away from the ore bodies it is present, suggesting it has been removed. Cas and Beresford (2001) suggested that rather than focussing just on thermal erosion as the cause of sediment removal, the flow front of the komatiite could have caused physical erosion, through several mechanisms (Fig. 7.23). If the flow front was turbulent, the dense komatiite lava should have easily scoured unconsolidated, water-saturated, seafloor mudstones (Fig. 7.23a). A turbulent flow front would also generate a turbulent bow wave or zone in the water mass immediately ahead of the lava flow front, leading to lofting of fine sediment ahead of the flow front (Fig. 7.23b). The heat of the komatiite should also cause film boiling in the water mass at the flow front, with convection in that vapour film and in the water mass ahead of that vapour film, again leading to physical erosion of fine sediment particles ahead of the advancing flow front (Fig. 7.23c).

7.9.3 A Model for Eruption and Flow Dynamics of Komatiite Lavas: Tubes and Flow Inflation

Komatiite magmas were probably the densest magmas (2,800 kg m^{-3}) to have been erupted in significant volumes on Earth. As a result, they were not buoyant magmas in

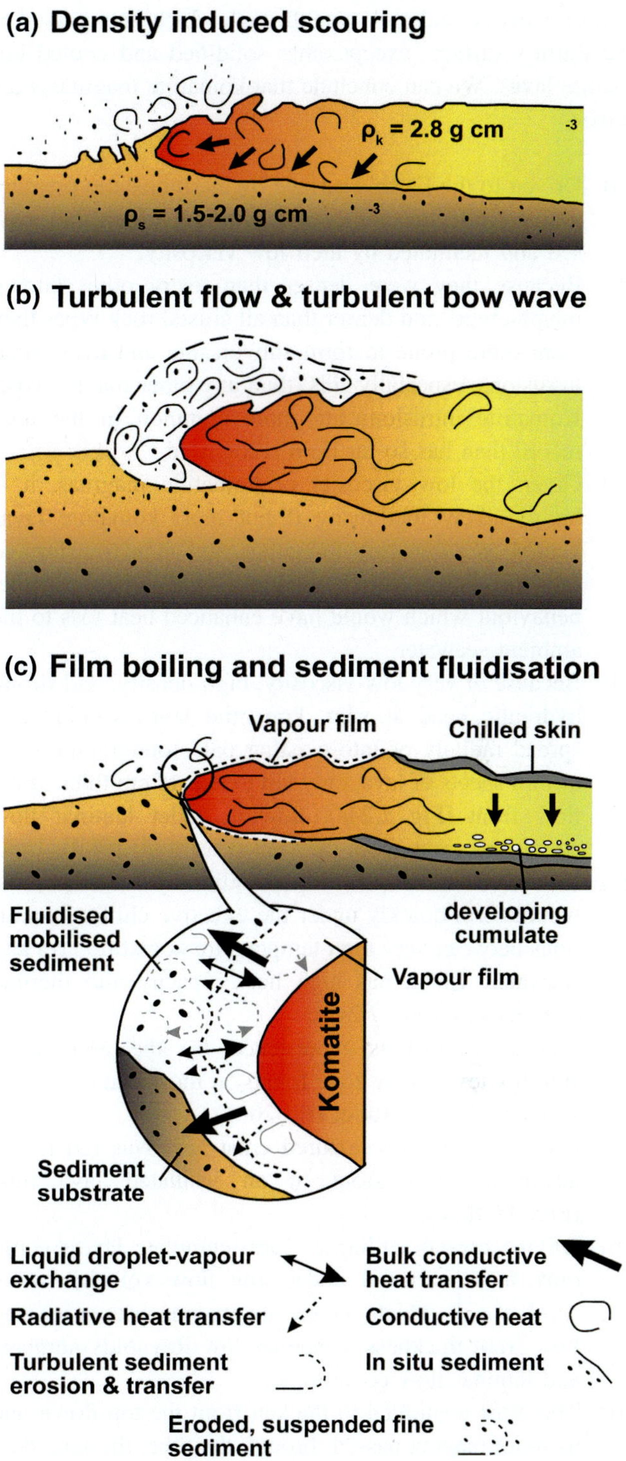

Fig. 7.23 Possible physical erosion processes at the fronts of komatiite lavas (From Cas and Beresford 2001). **a** Density-induced scouring by turbulent lava flow over water-saturated unconsolidated sediments. **b** Turbulent "bow wave" ahead of turbulently flowing komatiite lava flow front. **c** Heat-induced film boiling and thermal convection-induced turbulence in the vapour film around the komatiite lava flow front

almost every crustal rock type through which they passed to the Earth's surface, except other solidified and cooled komatiite lavas. We can conclude that komatiite magmas were thus:

(a) Driven to the Earth's surface by the lithostatic pressure gradient affecting the conduit through which they passed and facilitated by their low viscosity.
(b) Because they were denser than every other known magma type, and denser than all crustal rock types they were more prone to form subvolcanic and even blind intrusions, especially sills, than any other magma type. Komatiite intrusions are more common in the rock record than has so far been recognised.
(c) Given the low viscosity of komatiite magmas, it is highly likely that many if not most komatiite lavas vented as submarine fountains, and then transformed into near-vent turbulent flows, with open channel behaviour which would have enhanced heat loss to the ambient seawater.
(d) Because of very low-viscosity, high-density, and strong hydraulic head at vent, komatiite lavas would have spread radially or into existing paleotopographic lows as thin sheets of lava, only cm to tens of cm thick at the flow front (Fig. 7.24a), flowing under laminar flow conditions away from the vent (Cas et al. 1999).
(e) Under stable laminar flow conditions, a stable crust would form quickly under the effective chilling conditions between very high-temperature komatiite lava and seawater, which has high heat capacity and thermal conductivity (Fig. 7.24b).
(f) The lava crust is represented by the aphyric to spinifex-textured A zone found on most komatiite lava surfaces (Fig. 7.24b, c; Hill 2001).
(g) Under the spinifex-textured crust lava was fed to the advancing flow front in an insulated lava tube (Fig. 7.24b, e).
(h) Olivine crystal settling to form cumulate layers could only have occurred under low flow velocities of < $0.11\ m\ s^{-1}$, which, given the magma properties and flow front thickness, generates low Reynolds numbers and laminar flow conditions.
(i) The crust continued to thicken from the top down and as more magma passed through the tube, the lava flow thickened through inflation (Hill 2001), as basalt lavas are known to do (Fig. 7.24b, f; Cas et al. 1999; Hill 2001). Modelling indicates it is almost impossible that the initial flow front thickness of komatiite lava flows exceeded 10 m (Cas et al. 1999), and therefore all komatiite lavas greater than 10 m, but virtually all komatiite lavas have achieved their preserved flow thickness through tube-fed inflation under an insulating lava crust.
(j) Although thermal erosion of substrate by komatiite lavas was a novel concept, so far no physical geological evidence (e.g. anatectic textures and ultra-metamorphism of substrate) has been presented to support it. Planar conformable bases of komatiite lavas and further numerical modelling indicate that it rarely occurred. It is theoretically possible, but it was not a common process. Where there is evidence of removal of sediment under komatiite lavas, physical erosion processes are more likely to have been responsible (Williams et al. 1998; Cas and Beresford 2001).

As a final point, a significant problem with understanding komatiite flow dynamics is how lavas that were two to three orders of magnitude less viscous than basalt magmas can produce some lavas that are tens to over a hundred m thick, whereas basalt lavas are rarely over 40 m thick, including flood basalt lavas. There are three possible explanations that come to mind. The anomalously thick komatiite lavas are either: (1) ponded in paleotopography and have inflated or thickened in a buttressed depression (e.g. a graben); or (2) actually consist of two of more coalesced lavas, which were emplaced successively while still hot enough for the crust of the lower lavas to be assimilated into the base of higher lavas; or (3) actually sub-seafloor sills that have intruded intra-stratally because of their high density. Walker (1992, 1993) pointed out that large volumes of basaltic shield volcanoes consist of intra-volcanic pile-intruded sheet sill complexes, and as indicated above, by virtue of being the densest magmas to have ever been erupted, komatiites should have more subsurface (sill) intrusions as part of their magma plumbing system, including within their volcanic piles than any other volcanic succession.

7.10 Komatiite-Hosted Nickel Sulphide Mineralisation

It is beyond the scope of this chapter to discuss the nickel sulphide mineralisation associated with komatiites in detail, which is done in Chap. 18, Sect. 18.6. Suffice to say, the Yilgarn Craton and Norseman–Wiluna greenstone belt is overwhelmingly the world's premier province for komatiite-hosted ore deposit types and hosts at least five world-class ore bodies. The most important mineralisation style associated with komatiite lavas is type 1 massive sulphide which is located at the base of some komatiite lavas (e.g. Kambalda), leading to suggestions that during lava flow an immiscible nickel sulphide liquid separated from the

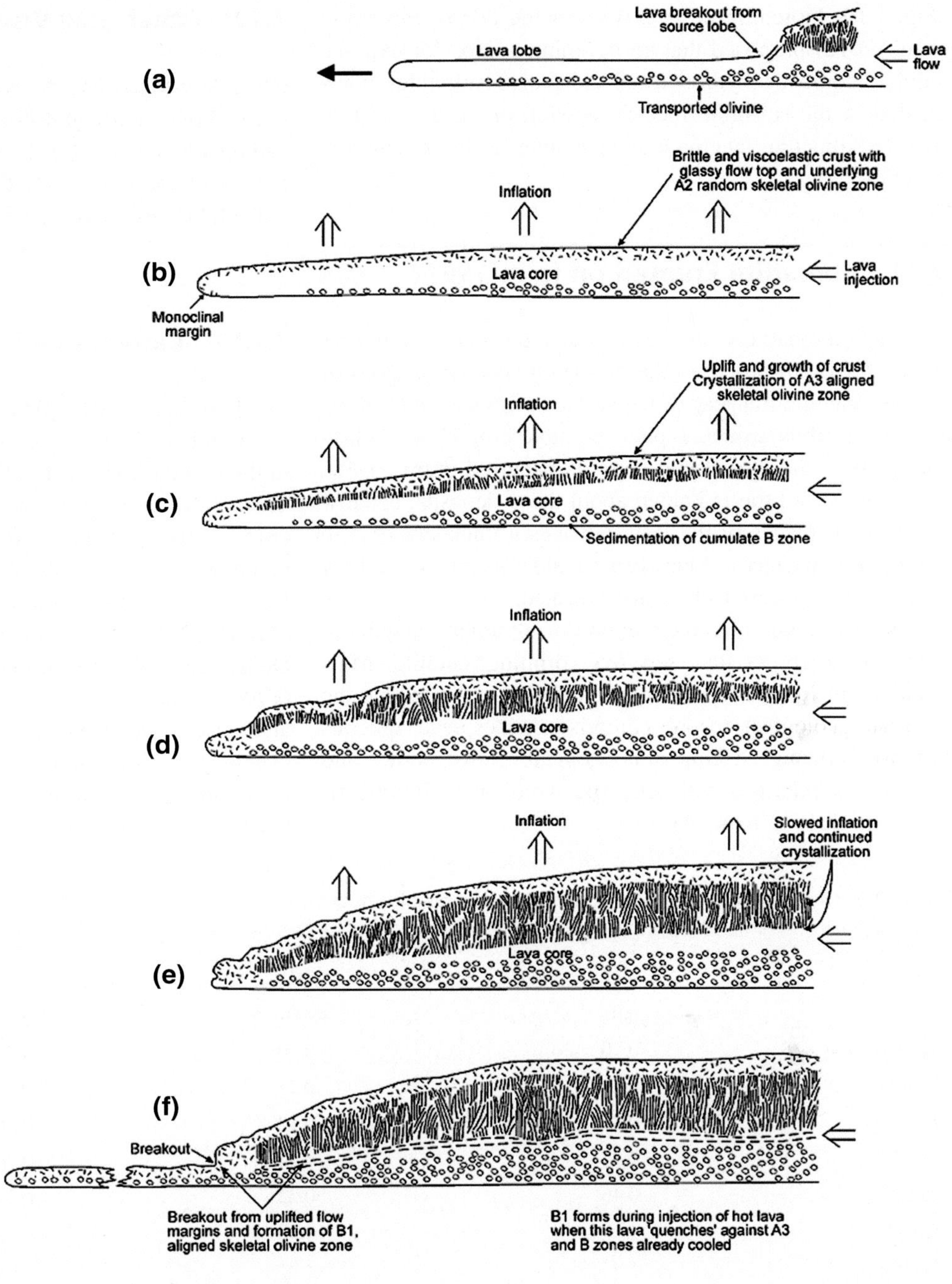

Fig. 7.24 Lava flow inflation model for komatiite lavas, showing initial propagation, crust formation, crust thickening, and flow inflation through lava tube-fed magma supply (From Hill 2001)

silicate komatiite magma and segregated to the bottom of the lava (Lesher 1989). Komatiite lavas may also contain disseminated or "blebby" nickel sulphide dispersed in the cumulate zone, but concentrations are very low and usually uneconomic. Thick dunitic intrusions may also host massive nickel sulphide along basal contacts (e.g. Fig. 7.19; Perseverance and Black Swan). In sills the flow simulates a lava tube, and so dense massive sulphide liquid finds the low points on the basal contact to flow into and deposit their massive sulphides. However, sills are more typically known for type 2 disseminated mineralisation (Lesher 1989), which occurs as small blebs interstitial and sometimes within large olivine crystals in adcumulate textural zones (e.g. Mt Keith).

The acquisition of sulphur, required to promote nickel sulphide precipitation, is now believed to have occurred in the conduit prior to eruption of the komatiite magmas as seafloor lavas (Lesher and Groves 1986; Keays and Lightfoot 2010; Fiorentini et al. 2012; Robertson et al. 2015). Nickel sulphide mineralisation is likely to be found proximal to the magma conduits due to rapid physical and chemical segregation of very dense NiS droplets from the komatiite lava as it spreads and slows down away from the vent

(Fig. 7.19; Fiorentini et al. 2012; Perring 2015). Fiorentini et al. (2012) proposed that an optimum source for external sulphur could be hydrothermal exhalative sulphides intercalated in the substrate rock succession or on the seafloor that the komatiite magma may encounter during ascent and eruption.

7.11 Komatiite Erupted on Io in 1997?

Although komatiites are best known on Earth, Williams et al. (2001a, b) proposed that based on spectral analysis of Galileo satellite imagery of the surface of the satellite moon Io, which orbits around Jupiter, possible eruptions of ultramafic lavas and pyroclastic deposits occurred during 1997. Although very little is known about these deposits, shadow analysis of images of the lava indicates a thickness of 8 to 10 m, flow distances of between 50 and 150 km, and a flow surface that appears to be autobrecciated.

As well as this possible eruption of ultramafic magma in 1997, a variety of data sets (e.g. orbiting satellite, Mars Lander and Rover, and Martian meteorite) indicate that the Martian geological record comprises widespread igneous activity showing striking mineralogical, petrographic, and chemical similarities with terrestrial komatiites (Baumgartner et al. 2015; Chap. 18).

7.12 Other Low-Viscosity Lava Types

Other low viscosity, but volumetrically minor (at least on Earth) flows that could behave like komatiite lavas include carbonatite and sulphur flows. We will only briefly mention these because they are relatively minor phenomena on Earth, but sulphur volcanism could also be a significant process on Io.

7.12.1 Carbonatite Lavas

The best known example of carbonatite volcanism on Earth is Ol Doinyo Lengai volcano in Tanzania, which is located in the Gregory Rift of the East Africa Rift Zone (Chap. 17). This is the only active volcano on Earth that episodically erupts carbonatite magmas, at other times erupting silicate magmas as part of its stratovolcanic history. Krafft and Keller (1989) and Dawson et al. (1990) summarised the key characteristics of carbonatite magmas and lavas from Ol Doinyo Lengai. The carbonatites analysed by Dawson et al. (1990) are alkalic carbonatites, almost totally devoid of SiO_2 and Al_2O_3, and very rich in CaO, Na_2O, K_2O, and CO_2 (Table 7.2). Although some lavas are crystal and vesicle poor, others contain abundant exotic alkali carbonate phenocryst minerals such as nyerereite ($[Na_{0.82}$, $K_{0.18}]_2$Ca

Table 7.2 Geochemistry (in weight %) of carbonatite samples from Ol Doinyo Lengai, Gregory Rift, Tanzania (From Dawson et al. 1990)

	1	2	3	4
SiO2	0.14	0.18	0.10	25.85
TiO2	0.01	<0.01	0.01	0.72
Al2O3	0.10	0.10	0.03	1.98
Fe2O3	0	0.21	0	5.10
FeO	0.10	0.12	0.26	3.20
MnO	0.24	0.45	0.26	0.33
MgO	0.46	0.22	0.27	0.78
CaO	13.04	14.63	14.58	16.24
BaO	1.92	1.05	1.02	0.89
SrO	1.79	1.63	1.47	1.00
Na2O	32.58	32.72	32.73	19.50
K2O	8.64	7.27	6.86	4.30
H2O+	0.23	0.22	0	5.12
P2O5	0.63	1.02	1.26	1.42
SO3	3.07	3.18	2.18	1.24
CO2	29.20	34.30	35.49	10.50
Cl	4.40	1.80	1.26	1.61
F	5.25	1.35	2.56	2.07
Total	101.80	100.45	100.34	101.80

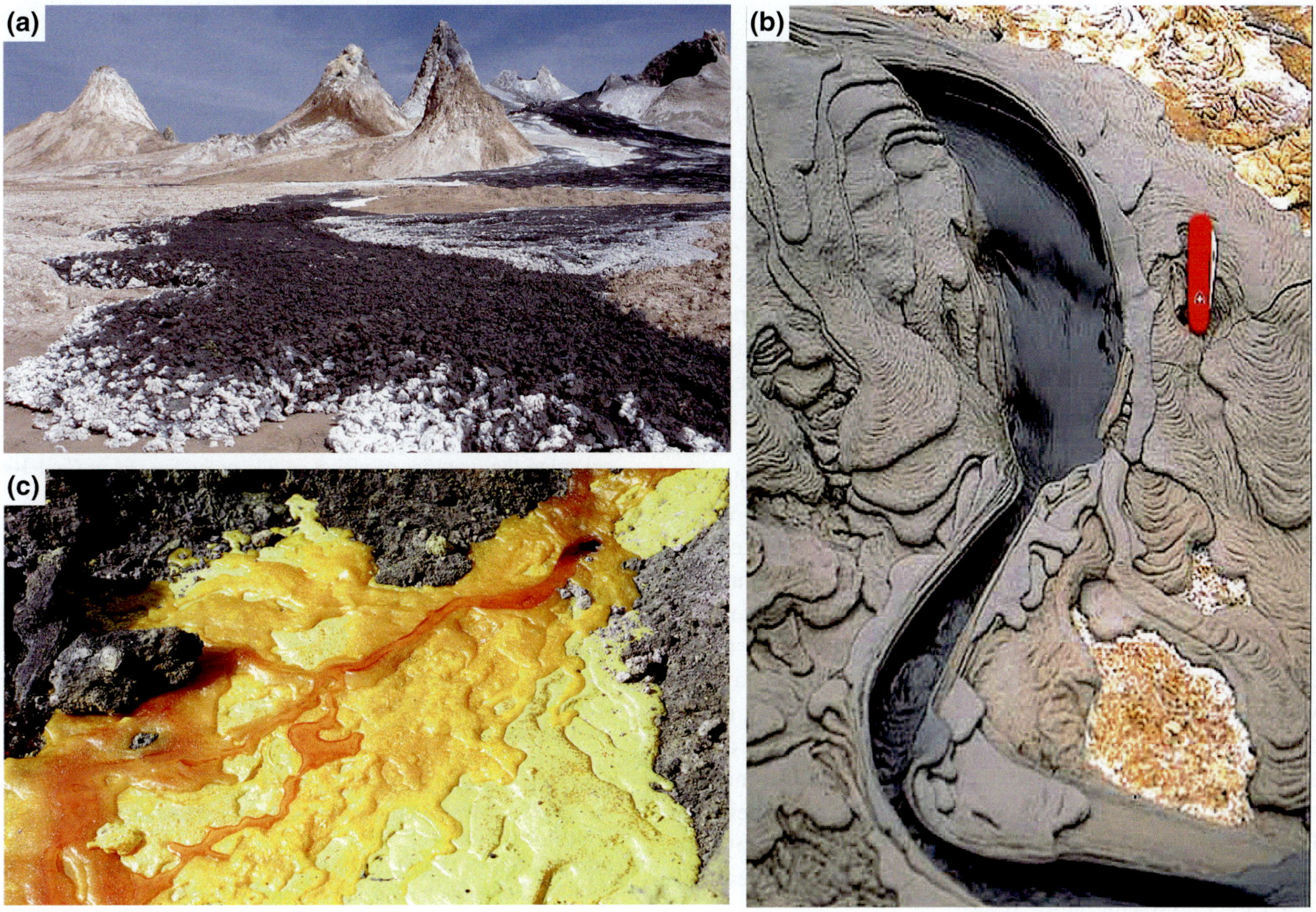

Fig. 7.25 Features of ultra-low-viscosity carbonatite and sulphur lava flows. **a** Overview of the summit of Ol Doinyo Lengai volcano, Tanzania, showing cones up to 10 m high and an a'a-textured carbonatite lava in the foreground (*Source* National Science Foundation). **b** Very thin, small volume carbonatite lava flow, with central open channel with dark carbonatite liquid flowing in a laminar flow state and flanking solidified channel overspill pahoehoe textured flow toes, Oldoinyo Lengai volcano, Tanzania (Photo by S. Beresford). **c** Small sulphur lava flow field, showing small feeder tube, central open channel and multiple flow lobes, Kawah Ijen volcano, east Java, Indonesia. Field of view ~ 1 m wide (*Source* National Science Foundation)

$[CO_3]_2$) and gregorite ($[Na_{0.78}, K_{0.05}]_2Ca_{0.17}[CO_3]$), and are highly vesicular.

Carbonatite lavas are very small volume flows that flow only metres to tens of metres from small source vents and cones that are up to ten metres high (Fig. 7.25a), some of which have summit carbonatitic lava lakes only m in diameter. Some lavas have features of pahoehoe ropy surfaces (Fig. 7.25b), others are a'a like (Fig. 7.25a). Measured eruption temperatures range from 491 to 593 °C (Krafft and Keller 1989; Dawson et al. 1990). Lava flow morphology can be related to magma discharge rate, with a'a lavas erupted at 0.2 to 0.3 $m^3\ s^{-1}$ and pahoehoe lavas at lower rates of 0.04 $m^3\ s^{-1}$. Such rates are comparable to effusion rates during some basalt eruptions, even though eruption volumes are orders of magnitude smaller in the case of carbonatites (Dawson et al. 1990). Measured and calculated (using Jeffrey's equation) viscosities range from 1 to 5 Pa s for gas-free pahoehoe carbonatite lavas and 70 to 120 Pa s for highly vesicular lava at low strain rates, but 0.3 to 3 Pa s at high strain rates (= higher flow velocities). The lower values are comparable to those of komatiite lavas. Lava flow lobes, not surprisingly, are usually very thin (Fig. 7.25), only centimetres to 20 cm thick. Low-viscosity flows may propagate through open channel flow, bounded by overspill pahoehoe flow toes (Fig. 7.25b). Despite their low viscosity, the flow condition away from the immediate vent area is laminar (Fig. 7.25b). Thermal erosion in channels downwards into underlying carbonatite lavas was observed to occur at rates of 2 mm min^{-1} at effusion rates of $10^{-3}\ m^3\ s^{-1}$ under laminar flow conditions (Dawson et al. 1990).

7.12.2 Sulphur Flows

Sulphur flows have been documented from hydrothermal vents on Earth, including Hawai'i (e.g. Greeley et al. 1984), and in the Andes (e.g. Naranjo 1985). Perhaps the most active volcano emitting liquid sulphur flows on Earth at present is Kawah Ijen volcano in East Java, Indonesia (Fig. 7.25c), which has an extraordinarily acidic (pH $\sim$ 0.5 to 1) summit crater lake, with a marginal fumarole field where sulphur is being mined manually. Analysis of spectral imagery from satellite fly-bys of Io also strongly indicates that sulphur volcanism has occurred there on a larger scale than on Earth (Williams et al. 1998).

Naranjo (1985) described two sulphur flows from fumaroles near a summit crater of Lastarria stratovolcano, northern Chile. The flows are 350 and 220 m long, up to 20 cm thick and 1 to 2.5 m wide. They are marked by pahoehoe-like ropy flow folds on the surfaces. Eruption temperatures are estimated to have been < 160 °C (Naranjo 1985). Greeley et al. (1984) describe a sulphur flow originating from fumaroles on a talus slope at Mauna Loa volcano, Hawai'i. It is 27 m long, 14 m wide, and up to 45 cm thick. It preserves a small drained lava tube 30 cm in diameter. Surface features include small anastomosing driblet-like lobes, sheet lobes, and small leveed channels (cf. Fig. 7.25c). The end of the lava flow has a "knobby" or lumpy texture. Estimated flow temperature was between 120 and 150 °C (Greeley et al. 1984).

Although detailed studies of sulphur flows on Io have not been possible, it has long been speculated that sulphur volcanism is a common and volumetrically significant phenomenon on Io based on Voyager satellite imagery (Sagan 1979). The 1999 Galileo satellite spectral data provided a better basis for this interpretation (Williams et al. 2001a, b), but still little detailed data is available. We only note that a sulphur lava flow field up to 370 km long and 190 km wide (!) associated with Emakong Patera volcano has been proposed. This lava flow field shows what appears to be lava channels or tubes. Clearly, if correct, the scale of sulphur lavas on Io is orders of magnitude greater than observed on Earth.

References

Arndt NT (1982) Proterozoic spinifex-textured basalts of Gilmour Island, Hudson Bay. Current research, Part A. Geol Surv Canad Pap:137–142

Arndt NT (1986) Differentiation of Komatiite Flows. J Petrol 27 (2):279–301

Arndt NT (2003) Komatiites, kimberlites, and boninites. J Geophys Res: Solid Earth 108(B6)

Arndt N, Ginibre C, Chauvel C, Albarede F, Cheadle M, Herzberg C, Jenner G, Lahaye Y (1998) Were komatiites wet? Geology 26 (8):739–742

Arndt NT, Lesher CM, Barnes SJ (2008) Komatiite. Cambridge University Press, Cambridge, p 458

Arndt NT, Lesher CM, Houlé MG, Lewin E, Lacaze Y (2004) Intrusion and crystallization of a spinifex-textured komatiite sill in Dundonald Township. Ontario. J Petrol 45(12):2555–2571

Arndt NT, Naldrett AJ, Pyke DR (1977) Komatiitic and iron-rich tholeiitic lavas of Munro Township, northeast Ontario. J Petrol 18 (2):319–369

Auvray B, Blais S, Jahn BM, Piquet D (1982) Komatiites and the komatiitic series of the Finnish greenstone belts. In: Arndt NT, Nisbet EG (eds) Komatiites. George Allen & Unwin, London, pp 131–146

Barnes RG, Lewis JD, Gee RD (1973) Archaean ultramafic lavas from Mount Clifford. In: Geol Surv West Aust Annu Repo pp 59–70

Barnes SJ, Heggie GJ, Fiorentini ML (2013) Spatial variation in platinum group element concentrations in ore-bearing komatiite at the Long-Victor deposit, Kambalda Dome, Western Australia: enlarging the footprint of nickel sulfide orebodies. Econ Geol 108 (5):913–933

Barnes SJ, Hill RET, Gole MJ (1988) The Perseverance ultramafic complex, Western Australia: the product of a komatiite lava river. J Petrol 29(2):305–331

Barnes SJ, Lesher C, Keays R (1995) Geochemistry of mineralised and barren komatiites from the Perseverance nickel deposit. Western Australia. Lithos 34(1–3):209–234

Barnes SJ, Van Kranendonk MJ, Sonntag I (2012) Geochemistry and tectonic setting of basalts from the Eastern Goldfields Superterrane. Aust J Earth Sci 59(5):707–735

Baumgartner RJ, Fiorentini ML, Baratoux D, Micklethwaite S, Sener AK, Lorand JP, McCuaig TC (2015) Magmatic controls on the genesis of Ni-Cu±(PGE) sulphide mineralisation on Mars. Ore Geol Rev 65:400–412

Beresford S, Cas R, Lahaye Y, Jane M (2002) Facies architecture of an Archean komatiite-hosted Ni-sulphide ore deposit, Victor, Kambalda, Western Australia: Implications for komatiite lava emplacement. J Volcanol Geotherm Res 118(1):57–75

Beresford S, Stone WE, Cas R, Lahaye Y, Jane M (2005) Volcanological controls on the localization of the komatiite-hosted Ni-Cu-(PGE) Coronet deposit, Kambalda. Western Australia. Econ Geol 100(7):1457–1467

Beresford SW, Cas RAF (2001) Komatiitic invasive lava flows, Kambalda, western Australia. Canad Mineral 39(2):525–535

Beresford SW, Cas RAF, Lambert DD, Stone WE (2000) Vesicles in thick komatiite lava flows, Kambalda. Western Australia. J Geol Soc 157(1):11–14

Busby-Spera CJ, White JDL (1987) Variation in peperite textures associated with differing host-sediment properties. Bull Volcanol 49 (6):765–776

Capdevila R, Arndt N, Letendre J, Sauvage JF (1999) Diamonds in volcaniclastic komatiite from French Guiana. Nature 399 (6735):456–545

Cas RAF, Beresford SW (2001) Field characteristics and erosional processes associated with komatiitic lavas: Implications for flow behavior. Canad Mineral 39(2):505–524

Cas RAF, Marks K, Perazzo S, Beresford SW, Trofimovs J, Rosengren N (2013) Were intercalated komatiites and dacites at the Black Swan nickel sulphide mine, Yilgarn Craton, Western Australia, emplaced as extrusive lavas or intrusive bodies? The significance of breccia textures and contact relationships. Precambrian Res 229:133–149

Cas RAF, Self S, Beresford S (1999) The behaviour of the fronts of komatiite lavas in medial to distal settings. Earth Planet Sci Lett 172 (1–2):127–139

Cas RAF, Wright JV (1987) Volcanic successions: modern and ancient. Allen and Unwin, London, p 528

Campbell IH, Griffiths RW, Hill RI (1989) Melting in an Archaean mantle plume: heads it's basalts, tails it's komatiites. Nature 339 (6227):697–699

Champagne C (2004) Volcanologie physique et géochimie des komatiites de Spinifex Ridge, formation de La Motte-Vassan, Abitibi. Unpublished Masters Thesis, Université du Québec à Chicoutimi, Canada

Condie KC, O'Neill C (2010) The Archean-Proterozoic boundary: 500 My of tectonic transition in Earth history. Amer J Sci 310(9): 775–790

Cowden A, Roberts DE (1990) Komatiite hosted nickel sulphide deposits, Kambalda. In: Geology of the mineral deposits of Australia and Papua New Guinea. Australasian Institute of Mining and Metallurgy pp 567–581

Dann JC (2001) Vesicular komatiites, 3.5-Ga Komati Formation, Barberton Greenstone Belt, South Africa: inflation of submarine lavas and origin of spinifex zones. Bull Volcanol 63(7):462–481

Davis PC (1999) Physical volcanology of komatiites flows at Dundonald Beach and Alexo Mine, Dundonald and Clergue Townships, Kidd-Munro Assemblage, Ontario portion and the Abitibi Greenstone Belt. Day 2: Guidebook Series, v. 1. Mineral Exploration Research Centre, Laurentian University, pp 27–44

Dawson JB, Pinkerton H, Norton GE, Pyle DM (1990) Physicochemical properties of alkali carbonatite lavas: Data from the 1988 eruption of Oldoinyo Lengai. Tanzania. Geology 18(3):260–263

de Wit MJ, Ashwal LD (1997) Preface: Convergence towards divergent models of greenstone belts. In: de Wit MJ, Ashwal LD (eds) Greenstone Belts. Oxford University Press, Oxford, pp 9–17

Donaldson CH (1976) An experimental investigation of olivine morphology. Contrib Mineral Petrol 57(2):187–213

Donaldson CH (1982a) Spinifex-textured komatiites: a review of textures, compositions and layering. In: Arndt NT, Nisbet EG (eds) Komatiites. Allen & Unwin London, pp 213–244

Donaldson CH (1982b) Origin of some of the Rhum harrisite by segregation of intercumulus liquid. Mineral Mag 45(337):201–209

Dostal J (2008) Igneous rock associations 10. Komatiites. Geoscience Canada 35(1):21–31

Echeverria LM (1982) Komatiites from Gorgona Island, Colombia. In: Arndt NT, Nisbet EG (eds) Komatiites. Allen & Unwin London, pp 199–209

Fagents SA, Greeley R (2001) Factors influencing lava-substrate heat transfer and implications for thermomechanical erosion. Bull Volcanol 62(8):519–532

Faure F, Arndt N, Libourel G (2006) Formation of Spinifex Texture in Komatiites: an Experimental Study. J Petrol 47(8):1591–1610

Faure F, Trolliard G, Nicollet C, Montel JM (2003) A developmental model of olivine morphology as a function of the cooling rate and the degree of undercooling. Contribs Mineral Petrol 145(2):251–263

Fiorentini M, Beresford S, Barley M, Duuring P, Bekker A, Rosengren N, Cas R, Hronsky J (2012) District to camp controls on the genesis of komatiite-hosted nickel sulfide deposits, Agnew-Wiluna greenstone belt, Western Australia: Insights from the multiple sulfur isotopes. Econ Geol 107(5):781–796

Fiorentini ML (2005) Platinum-Group Element geochemistry of komatiites, Agnew-Wiluna Belt (WA): implications for NiS ore genesis and the composition of the Archean mantle. PhD Thesis, University of Western Australia, Perth, Australia

Fiorentini ML, Beresford SW, Barley ME (2007a) Controls on the genesis and emplacement of komatiite-hosted Ni–Cu–PGE-sulphides at Albion Downs (Agnew-Wiluna Belt, Western Australia): a case study on the development of PGE lithogeochemical vectors to Ni–Cu–PGE-sulphide deposits. Appl Earth Sci 116 (4):152–166

Fiorentini ML, Rosengren N, Beresford SW, Grguric B, Barley ME (2007b) Controls on the emplacement and genesis of the MKD5 and Sarah's Find Ni–Cu–PGE deposits, Mount Keith, Agnew-Wiluna Greenstone Belt. Western Australia. Miner Deposita 42(8):847–877

Gole MJ, Robertson J, Barnes SJ (2013) Extrusive origin and structural modification of the komatiitic Mount Keith ultramafic unit. Econ Geol 108(7):1731–1752

Greeley R, Fagents SA, Harris RS, Kadel SD, Williams DA, Guest JE (1998) Erosion by flowing lava: Field evidence. J Geophys Res: Solid Earth 103(B11):27325–27345

Greeley R, Theilig E, Christensen P (1984) The Mauna Loa sulfur flow as an analog to secondary sulfur flows (?) on Io. Icarus 60(1): 189–199

Gresham JJ, Loftus-Hills GD (1981) The geology of the Kambalda nickel field. Western Australia. Econ Geol 76(6):1373–1416

Groves DI, Korkiakoski EA, McNaughton NJ, Lesher CM, Cowden A (1986) Thermal erosion by komatiites at Kambalda, Western Australia and the genesis of nickel ores. Nature 319(6049):136–139

Hanski E, Walker RJ, Huhma H, Polyakov GV, Balykin PA, Hoa TT, Phuong NT (2004) Origin of the Permian-Triassic komatiites, northwestern Vietnam. Contribs Mineral Petrol 147(4):453–469

Hayman PC, Thébaud N, Pawley MJ, Barnes SJ, Cas RAF, Amelin Y, Sapkota J, Squire RJ, Campbell IH, Pegg I (2015) Evolution of a ~ 2.7 Ga large igneous province: a volcanological, geochemical and geochronological study of the Agnew Greenstone Belt, and new regional correlations for the Kalgoorlie Terrane (Yilgarn Craton, Western Australia). Precambrian Res 270:334–368

Herzberg C, Condie K, Korenaga J (2010) Thermal history of the Earth and its petrological expression. Earth Planet Sci Letts 292(1–2):79–88

Hill RET (2001) Komatiite volcanology, volcanological setting and primary geochemical properties of komatiite-associated nickel deposits. Geochem Explor Env Anal 1(4):365–381

Hill RET, Barnes SJ, Dowling SE, Thordarson T (2004) Komatiites and nickel sulphide orebodies of the Black Swan area, Yilgarn Craton, Western Australia. 1. Petrology and volcanology of host rocks. Miner Deposita 39(7):684–706

Hill RET, Barnes SJ, Gole MJ, Dowling SE (1995) The volcanology of komatiites as deduced from field relationships in the Norseman-Wiluna greenstone belt. Western Australia. Lithos 34 (1):159–188

Hill RET, Gole MJ, Barnes SJ (1990) Physical volcanology of komatiites: A field guide to the komatiites between Kalgoorlie and Wiluna, Eastern Goldfields Province, Yilgarn Block. Geological Society of Australia, WA Division, Western Australia

Houlé MG, Gibson HL, Lesher CM, Davis PC, Cas RAF, Beresford SW, Arndt NT (2008) Komatiitic sills and multigenerational peperite at Dundonald Beach, Abitibi greenstone belt, Ontario: volcanic architecture and nickel sulfide distribution. Econ Geol 103 (6):1269–1284

Huppert HE, Sparks RSJ (1985) Komatiites I: Eruption and flow. J Petrol 26(3):694–725

Huppert HE, Sparks RSJ, Turner JS, Arndt NT (1984) Emplacement and cooling of komatiite lavas. Nature 309(5963):19–22

Keays RR, Lightfoot P (2010) Crustal sulfur is required to form magmatic Ni–Cu sulphide deposits; evidence from chalcophile element signatures of Siberian and Deccan Trap basalts. Mineral Deposita 45:241–257

Kerr AC, Marriner GF, Arndt NT, Tarney J, Nivia A, Saunders AD, Duncan RA (1996) The petrogenesis of Gorgona komatiites, picrites and basalts: new field, petrographic and geochemical constraints. Lithos 37(2):245–260

Krafft M, Keller J (1989) Temperature measurements in carbonatite lava lakes and flows from Oldoinyo Lengai. Tanzania. Science 245 (4914):168–171

Lajoie J, Gélinas L (1978) Emplacement of Archean peridotitic komatiites in La Motte township. Quebec. Canad J Earth Sci 15 (4):672–677

Lesher CM (1989) Komatiite-associated nickel sulfide deposits. Rev Econ Geol 4:45–101
Lesher CM, Arndt NT (1995) REE and Nd isotope geochemistry, petrogenesis and volcanic evolution of contaminated komatiites at Kambalda, Western Australia. Lithos 34(1–3):127-
Lesher CM, Groves DI (1986) Controls on the formation of komatiite-associated nickel copper sulphide deposits. In: Friedrich GH (ed) Geology and Metallogeny of Copper Deposits. Springer, pp 63–90
Lesher CM, Lee RF, Groves DI, Bickle MJ, Donaldson MJ (1981) Geochemistry of komatiites from Kambalda, Western Australia; I, Chalcophile element depletion, a consequence of sulfide liquid separation from komatiitic magmas. Econ Geol 76(6):1714–1728
Lowe DR. Shallow-water sedimentation of accretionary lapilli-bearing strata of the Msauli Chert: Evidence of explosive hydromagmatic komatiitic volcanism (1999). In: Lowe D, Byerly G (eds) *Geologic evolution of the Barberton Greenstone Belt, South Africa*: Geol Soc Amer Spec Pap 329: 213–232
Marsh BD (1988) Crystal size distribution (CSD) in rocks and the kinetics and dynamics of crystallization. Contrib Mineral Petrol 99 (3):277–291
McPhie J, Cas RAF (2015) Volcanic successions associated with ore deposits: facies characteristics and ore-host relationships. In: Sigurdsson H, Stix J, Houghton BF, McNutt SR, Rymer H (eds) Encyclopedia of Volcanoes. Academic Press, San Diego, pp 865–880
Mole DR, Fiorentini ML, Thebaud N, Cassidy KF, McCuaig TC, Kirkland CL, Romano SS, Doublier MP, Belousova EA, Barnes SJ (2014) Archean komatiite volcanism controlled by the evolution of early continents. Proc Natal Acad Sci 111(28):10083–10088
Moyen JF, Laurent O (2018) Archaean tectonic systems: a view from igneous rocks. Lithos 302:99–125
Naranjo JA (1985) Sulphur flows at Lastarria volcano in the North Chilean Andes. Nature 313(6005):778–780
O'Driscoll B, Donaldson CH, Troll VR, Jerram DA, Emeleus CH (2006) An origin for harrisitic and granular olivine in the Rum Layered Suite, NW Scotland: a crystal size distribution study. J Petrol 48(2):253–270
Perring CS (2015) Volcanological and structural controls on mineralization at the Mount Keith and Cliffs komatiite-associated nickel sulfide deposits, Agnew-Wiluna belt, Western Australia—implications for ore genesis and targeting. Econ Geol 110(7):1669–1695
Pyke DR, Naldrett AJ, Eckstrand OR (1973) Archean ultramafic flows in Munro township. Ontario. Geol Soc Am Bull 84(3):955–978
Robertson JC, Ripley EM, Barnes SJ, Li C (2015) Sulfur liberation from country rocks and incorporation in mafic magmas. Econ Geol 110:1111–1123
Rosengren N, Grguric B, Beresford SW, Fiorentini M, Cas R (2007) Internal stratigraphic architecture of the komatiitic dunite-hosted MKD5 disseminated nickel sulfide deposit, Mount Keith Domain, Agnew-Wiluna greenstone belt. Western Australia. Miner Deposita 42(8):821–845
Rosengren NM, Beresford SW, Grguric BA, Cas RAF (2005) An intrusive origin for the komatiitic dunite-hosted Mount Keith disseminated nickel sulfide deposit. Western Australia. Econ Geol 100(1):149–156
Rosengren NM, Cas RAF, Beresford SW, Palich BM (2008) Reconstruction of an extensive Archaean dacitic submarine volcanic complex associated with the komatiite-hosted Mt Keith nickel deposit, Agnew-Wiluna greenstone belt, Yilgarn Craton. Western Australia. Precambrian Res 161(1):34–52
Sagan C (1979) Sulphur flows on Io. Nature 280(5725):750–753
Self S, Keszthelyi L, Thordarson T (1998) The importance of pāhoehoe. Annu Rev Earth Planet Sci 26(1):81–110
Shore M, Fowler AD (1999) The origin of spinifex texture in komatiites. Nature 397(6721):691
Sobolev AV, Asafov EV, Gurenko AA, Arndt NT, Batanova VG, Portnyagin MV, Garbe-Schönberg D, Krasheninnikov SP (2016) Komatiites reveal a hydrous Archaean deep-mantle reservoir. Nature 531(7596):628
Staude S, Jones TJ, Markl G (2020) The textures, formation and dynamics of rare high-MgO komatiite pillow lavas. Precambrian Research 343:105729
Staude S, Markl G (2019) Remnant lenses of komatiitic dykes in Kambalda (Western Australia): occurrences, textural variations, emplacement model, and implications for other komatiite provinces. Lithos 342:206–222
Stiegler MT, Lowe DR, Byerly GR (2008). Abundant pyroclastic komatiitic volcanism in the 3.5–3.2 Ga Barberton greenstone belt, South Africa. Geology 36(10):779–782
Stiegler MT, Lowe DR, Byerly GR (2011). Fragmentation and dispersal of komatiitic pyroclasts in the 3.5–3.2 Ga Onverwacht Group, Barberton greenstone belt. South Afr Bull 123(5–6):1112–26
Stone W, Archibald N (2004) Structural controls on nickel sulphide ore shoots in Archaean komatiite, Kambalda, WA: The volcanic trough controversy revisited. J Struct Geol 26(6–7):1173–1194
Stone WE, Beresford SW, Archibald NJ (2005) Structural setting and shape analysis of nickel sulfide shoots at the Kambalda dome, Western Australia: Implications for deformation and remobilization. Econ Geol 100(7):1441–1455
Thomson B (1989) Petrology and stratigraphy of some texturally well preserved thin komatiites from Kambalda. Western Australia. Geol Mag 126(3):249–261
Thordarson T, Self S (1998) The Roza Member, Columbia River Basalt Group: A gigantic pahoehoe lava flow field formed by endogenous processes? J Geophys Res: Solid Earth 103(B11):27411–27445
Trofimovs J, Tait MA, Cas RAF, McArthur A, Beresford SW (2003) Can the role of thermal erosion in strongly deformed komatiite–Ni–Cu–(PGE) deposits be determined? Perseverance, Agnew-Wiluna Belt. Western Australia. Aust J Earth Sci 50(2):199–214
Viljoen MJ, Viljoen RP (1969a) Evidence for the existence of a mobile extrusive peridotitic magma from the Komati Formation of the Onverwacht Group. Geol Soc South Afr Spec Pub 2:87–112
Viljoen RP, Viljoen MJ (1969b) The geological and geochemical significance of the upper formations of the Onverwacht Group. Geol Soc South Afr Spec Pub 2:113–152
Walker GPL (1992) "Coherent intrusion complexes" in large basaltic volcanoes—a new structural model. J Volcanol Geotherm Res 50 (1–2):41–54
Walker GPL (1993) Basaltic-volcano systems. Geol Soc Lond Spec Publ 76(1):3–8
Welsch B, Faure F, Famin V, Baronnet A, Bachèlery P (2012) Dendritic Crystallization: A Single Process for all the Textures of Olivine in Basalts? J Petrol
Williams DA, Davies AGS, Keszthelyi LP, Greeley R (2001a) The summer 1997 eruption at Pillan Patera on Io: Implications for ultrabasic lava flow emplacement. J Geophys Res: Planets 106 (E12):33105–33119
Williams DA, Greeley R, Lopes R, Davies AGS (2001b) Evaluation of sulfur flow emplacement on Io from Galileo data and numerical modeling. J Geophys Res: Planets 106(E12):33161–33174
Williams DA, Kerr RC, Lesher CM (1998) Emplacement and erosion by Archean komatiite lava flows at Kambalda: Revisited. J Geophys Res: Solid Earth 103(B11):27533–27549
Wyman DA (2020) Komatiites from mantle transition zone plumes. Front Earth Sci 8:383

Subvolcanic Intrusions

8

Emplacement, Distinguishing Intrusions from Lavas, and Use as Paleostress Indicators

Guido Giordano, Ray Cas, and John V. Wright

R. Cas et al., *Volcanology*, Springer Textbooks in Earth Sciences, Geography and Environment,
https://doi.org/10.1007/978-3-319-66613-6_8

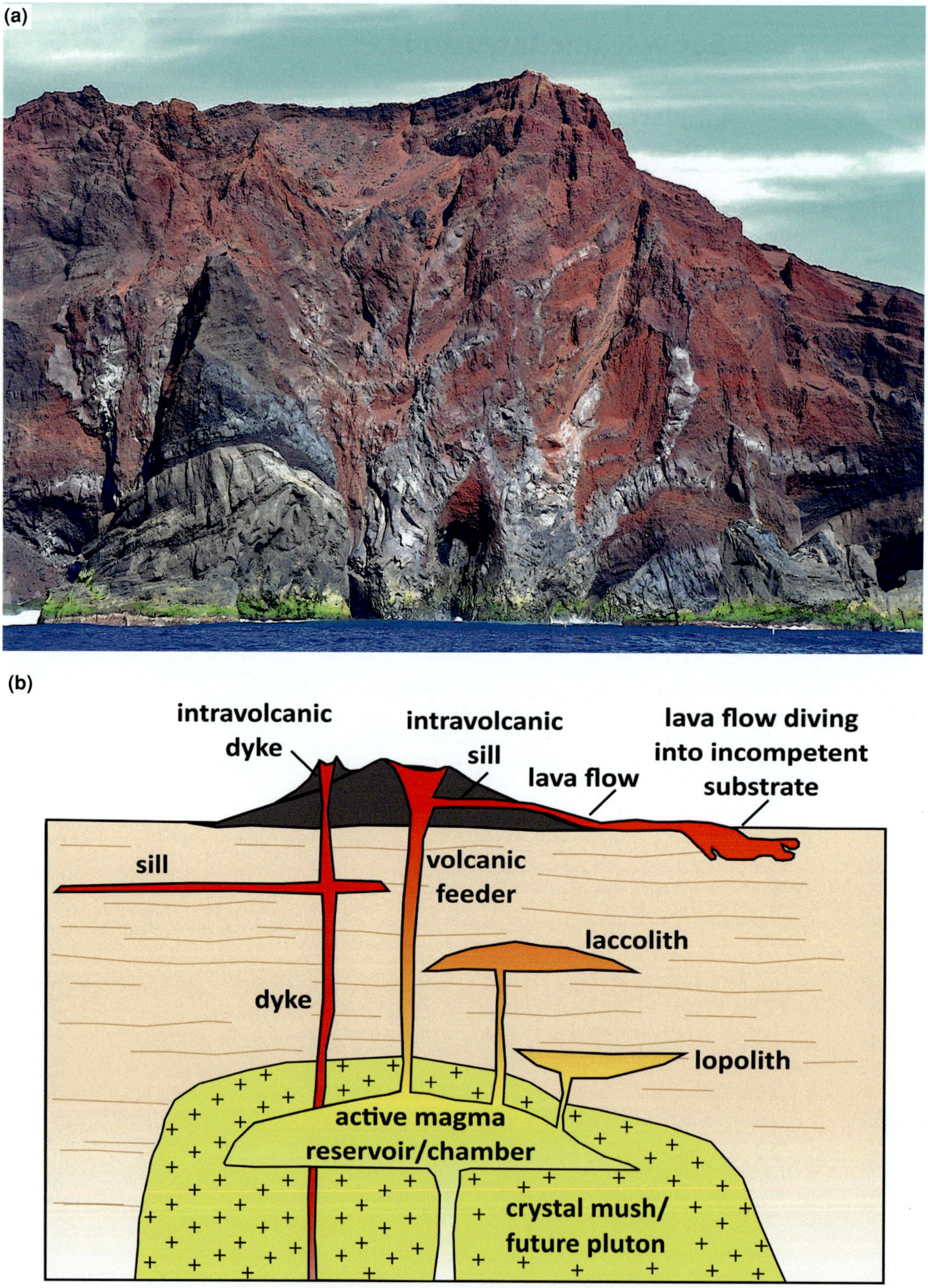

Plate 8.1 The subvolcanic intrusive plumbing system: the conduit for volcanism. **a** Spectacular cross-section through the complex branching magma plumbing system of interconnected dykes, sills and inclined sheets or "dills" (centre) in the late scoria cone (red-black) that grew on top of the early shallow marine surtseyan tuff cone (bedded grey, lower left) during the 1957–1958 Capelhinos eruption, Faial, Azores, Portugal. The irregular geometry resulted from magma propagating through the weakly lithified scoria. Height of sea cliff is 130 m. (Photo by Lisa Buckley). **b** Schematic representation of the different types of subvolcanic intrusions that can form the subsurface magma plumbing system in volcanic terrains (Modified from Breitkreuz and Rocchi 2018)

Abstract

Volcanic plumbing systems are made by complex networks of dykes, sills, laccoliths, cryptodomes, and other intrusions, which may interconnect to form more or less temporary magma reservoirs through which magma is stored and migrates before reaching the Earth's surface. We first summarise the physical factors that determine whether or not magmas rise and erupt, or form subvolcanic intrusions. The competing forces of buoyancy and viscosity determine the ability of magma to move upwards, as well as the structural fabric of country rocks and their mechanical competence. Different types of intrusions are then described. In recent years, the study of petroleum industry 3D seismic reflection data sets from offshore sedimentary basins has revealed that mafic sills are often emplaced as a series of variably interconnected saucer-shaped bodies, fed by inclined sheets at their margins, and composed of magma lobes, similar to lava lobes. This has greatly advanced knowledge of mafic sheet-like complexes. We then consider the architecture and processes within magma plumbing systems in different volcanic systems. Complex magma plumbing systems have been mapped and imaged in stratovolcanoes (e.g. Kora stratovolcano, offshore New Zealand). There are differences in the architecture and processes that occur in thick regionally extensive mafic sills associated with large igneous provinces, and more evolved, and more localised magma reservoirs and chambers associated with caldera volcanoes. Magma plumbing systems not only contribute to the growth of volcanic edifices, but they can also contribute to their destruction. In both young and ancient volcanic terrains, it can be difficult to distinguish lavas from intrusions, especially sills, an ambiguity that can have a significant impact on paleoenvironmental and tectonic reconstructions, and we therefore provide guidelines to differentiate them. The most important criterion that distinguishes intrusions from lavas is the nature of the lower and upper contact relationships with the country rock or host sediments. Finally, we highlight the study of shallow intrusions for paleostress analysis in older and modern volcanic terrains.

8.1 Introduction

Before reaching the Earth's surface, magma passes through the lithosphere as a viscous melt forming a complex subvolcanic magma plumbing system of dykes, sills, laccoliths, cryptodomes, and other intrusions in volcanic edifices (Plate 8.1a), in the crust (Plate 8.1b), and sometimes originating directly from the mantle (Fig. 8.1a). The competing forces of buoyancy that drives the magma upward, and viscosity that retards the movement, will determine the ability of magma to move. Furthermore, whether or not magma erupts at the Earth's surface or stops at a depth also depends on the structural fabric of country rocks, i.e. pre-existing stratigraphic or structural discontinuities (Gudmundsson 2003, 2005), and the mechanical competence, or strength of the country rock. These are controlled by the rock types and their deformational, tectonic, and even compactional history. For example, where magma rises through the crust and then through unconsolidated sediments to the seafloor, the state of lithification and strength of the substrate influences the geometry of the conduits that propagate through them. In solid rock, the magma will dilate existing planar structures (joints, faults) or propagate new ones with planar or tabular geometry, whereas when magma propagates through uncompacted pyroclastic deposits (Plate 8.1a), and especially water-saturated sediments with little strength, resulting magma conduits can be very irregular (Fig. 8.1b). This complex interplay is also very dependent on the volume of magma involved, its cooling history, and the regional and local stress regime acting on country rocks around the magma conduit.

The geology of shallow subvolcanic intrusions is a varied topic that has had considerable attention in recent years. This is largely due to the increasing availability of high-resolution marine seismic reflection data from petroleum exploration of offshore sedimentary basins, which have imaged the 3D geometry of intrusive bodies in considerably more detail than has been previously available. Where a visible more or less radial symmetric volcanic edifice forms, the magma plumbing system is more immediately located under and within the edifice (Plate 8.1a, b), whereas at depth it may be connected to more widespread magma distributary pathways and reservoirs (Plate 8.1b; Fig. 8.1a). In small central vent edifices, such as monogenetic volcanoes, the immediate magma plumbing system may be simple (e.g. Plate 16.1b) or complex (e.g. Plate 8.1a), whereas under larger polygenetic volcanoes, it may be complex (Plate 8.1b). In some regionally extensive volcanic systems, the magma plumbing system can also be far-reaching on the scale of hundreds to perhaps thousands of km (e.g. flood basalt provinces, mid-oceanic spreading ridges). In this complex scenario, a volcano, or an eruption, is just the final product, and in general, the subsurface magma plumbing system of a volcano is volumetrically far larger than the volume of erupted products. Thus, the study of shallow intrusions is important to fully understand the evolution of a volcano or volcanic

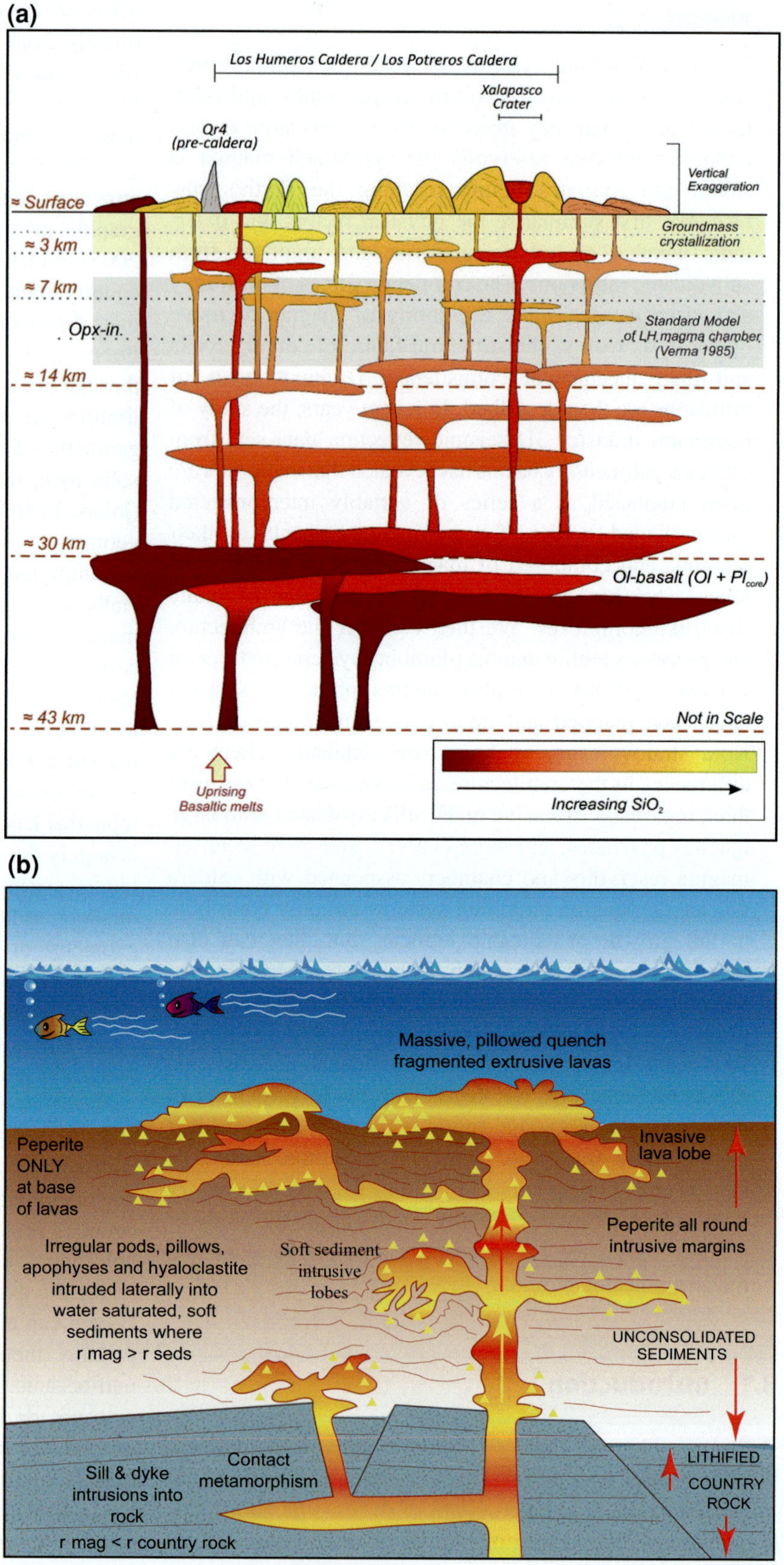

Fig. 8.1 Schematic representation of two examples of magma plumbing system. **a** Reconstruction of the configuration of the post-caldera plumbing system (<69 ka) at Los Humeros caldera, Mexico. Geobarometric data indicate that, even if a large "standard" rhyolitic magma chamber was in existence at the time of the caldera formation ~164–69 ka (e.g. Verma 1985; see discussion Sect. 8.4.3), the present system is instead dominated by small batches of magma intruded at different depths, each undergoing differentiation as independent reservoirs. (From Lucci et al. 2020). **b** Sedimentary basin setting showing magma first passing through lithified crust, and then intruding upwards into water-saturated unconsolidated sediments before erupting on the seafloor. Where magma propagates through lithified crust, it opens planar fractures and produces planar or tabular dykes and sills. Where the magma propagates through unconsolidated sediments, its fluid pressure and density exceed that of the sediment sequence, which has low strength, so allowing the intruding magma to form intrusions with irregular geometry

field, and even the evolution of some, in particular extensional, sedimentary basins associated with continental rifting and break-up (Chap. 17).

However, in ancient volcanic terrains, where much of the volcanic edifice and volcanic context has been removed or modified by erosion, deformation, and metamorphism, and where outcrop is often poor, it can be difficult to distinguish lavas from intrusions, especially sills, and involves time-consuming mapping and geological analysis (see Chap. 7). In addition, in very shallow crustal levels where magma intrudes water-saturated sedimentary successions, dynamic interaction between magma and the sediments can produce voluminous peperitic breccias that can be difficult to distinguish from some breccias associated with subaqueous lavas (Fig. 8.1b). Furthermore, the textures of subvolcanic intrusive rocks vary according to their cooling history, and in some cases, are identical texturally to erupted lavas. The closer intrusions are emplaced to the Earth's surface, the more similar the petrological texture will be to that of extrusive lavas because the cooling regimes of both are relatively similar, making texture largely a non-diagnostic discriminant between lavas and high-level intrusions. The most important criterion to use to distinguish intrusions from lavas is the contact relationships with the country rock, including whether or not the igneous unit is concordant or discordant with stratification in the host rocks, and the details of both the upper and lower contacts, especially the top contacts (Fig. 8.1b).

In addition, the study of shallow intrusions, their orientation, and geometry in the country rock are also very important as they allow intrusions to be used as excellent paleostress indicators (Johnson 1968). In modern volcanically active environments, intrusions are responsible for significant ground movements, which can be geodetically monitored as a tool in eruption and hazard prediction and assessment. Intrusions of magma in the subsurface act as hydraulic jacks, cause an uplift of the crust, and so contribute significantly to the surface relief and elevation of volcanically active regions as well as of individual volcanic edifices. Some intrusions can also be hosts to distinctive mineral deposits (e.g. magmatic nickel sulphide and platinum group elements, epithermal and porphyry-related deposits, iron oxide copper–gold, and base metals; Chap. 18).

In this chapter, we will first review the physical processes that control the emplacement of subvolcanic intrusions. We then describe the different types of intrusions, their settings, characteristics, volcanological significance, and complex plumbing systems. We then assess criteria to be used in the field to discriminate shallow subvolcanic intrusions from lavas, and finally illustrate how subvolcanic intrusions can be used for paleostress analysis in volcanic terrains.

Recent reviews on subvolcanic intrusions, magma plumbing systems, and processes include Charlier et al. (2015), Breitkreuz and Rocchi (2018), Burchardt (2018), and Srivastava et al. (2019).

8.2 Physical Constraints on the Formation of Intrusions

The uprise of magma from mantle and crustal sources to form intrusive complexes, often in a subvolcanic setting leading to volcanic eruptions, is subject to many constraints. Many of these constraints ensure that only a small fraction of the magma that forms actually erupts. As a result, under all volcanic systems, there are huge volumes of unerupted subvolcanic intrusions that are/were part of the magma plumbing system beneath volcanoes. In this section, we will assess the factors that constrain and enhance the emplacement of magma into subvolcanic and volcanic settings.

8.2.1 General Constraints on the Uprise of Magma Through the Crust

There are several physical requirements for magma to rise to the Earth's surface. Firstly, in general, magma will rise through the crust if its density, ρ_m, is less than that of the surrounding country rock, ρ_{CR}, which means that it has positive buoyancy that drives it upwards (Walker 1989, 1992). Magma is less dense than the source rocks from which it is derived, and so it rises. It will continue to rise buoyantly as long as the country rock surrounding the rising magma is denser than the magma. When magma reaches its *neutral buoyancy level*, which is the depth in the mantle or crust where its density is equal to that of the country rock, it should cease to rise through buoyancy alone, and try to intrude laterally into the country rock along weaknesses such as bedding planes, forming sills or cryptodomes. Bulk magma density can become lower if volatiles exsolve during ascent as the magma decompresses, forming vesicles. However, when both H_2O and CO_2 exsolve, they are dense supercritical fluids, not gas, so the initial density decrease due to volatile exsolution is not large (Chaps. 3, 4). It is only at depths in the crust of <800 m that those fluids begin to boil to gas, so this density decrease mechanism only kicks in significantly late during the uprise of the magma. If volatile content is low, this effect can be insignificant. Exsolved H_2O

and CO_2 may also escape from the magma into the country rock during ascent, especially with low magma viscosity and slow ascent rates.

However, some magmas, which are denser than the country rock through which they rise, do continue to rise to the Earth's surface. Basaltic magmas with a density of ~2,750 kg m^{-3} are denser than many continental crustal rocks and sedimentary successions, and yet still erupt as lavas, often forming very voluminous volcanic successions (Chap. 5). Komatiites with a magma density of 2,900 kg m^{-3} were denser than every crustal rock (except other cold komatiite rocks) through which they passed, and still erupted widely during times in the Archean (Chap. 7). Buoyancy alone does not explain this, so a second factor that drives magma upwards is the lithostatic pressure differential or pressure gradient between depth in the crust and the near surface, applied to a long column of low-viscosity magma (Valentine and Gregg 2008). This way magmas that have risen beyond their neutral buoyancy level, become negatively buoyant, and should theoretically cease to rise through the crust and seek to intrude the country rock, but because of upward momentum they can erupt in some cases.

Thirdly, the magma must have enough fluid pressure, inherited from the lithostatic pressure, as well as volatile exsolution and magma recharge, to dilate existing fracture pathways, such as crustal faults, or to fracture the country rock and open new fractures, so that the magma can migrate upwards. The magma fluid pressure must exceed the tensile strength of the country rock, τ, and the minimum principal stress component of the lithostatic confining pressure, σ_3, to do this.

When fractures are opened or propagated by rising magma in the solid crust, they are planar, brittle fractures that have a propagating tip only microns in width. Resulting magma pathways are planar fracture systems forming tabular dykes and sills (Fig. 8.1b). These can dilate or inflate as long as magma fluid pressure exceeds lithostatic pressure. When a magma migration event wanes, and magma fluid pressure falls below the lithostatic pressure, those magma conduits can attenuate or become narrower. They may even close up almost completely without significant amounts of magma being trapped in the attenuated conduits, which now have discontinuous lensoidal forms (Valentine and Gregg 2008; Staude and Markl 2019).

When magma rises through unconsolidated sediment sequences, the magma fluid pressure may exceed the strength of the sediments, allowing the magma to squeeze into the sediment in a relatively unconstrained manner, forming irregular, syn-depositional intrusions and contact relationships (Fig. 8.1b). This is particularly the case when intrusions occur into wet, unconsolidated sediments, which could lead to quench fragmentation of the intrusion margins forming hyaloclastite breccias, and dynamic mixing of hyaloclasts and the host sediments forming peperite breccias (Chap. 4). The details of the physics and mechanics of magma migration through the crust are further discussed in more detail in Sect. 8.2.4.

8.2.2 Conditions and Criteria for Rock Fracturing that Are Precursors to Magma Intrusion: Stress Tensor, Rock Strength and Pore Pressure

Here, we analyse first the stress and the strain conditions acting on rocks (Ramsay and Huber 1987). At each point in the crust of a volcanic region, stress acting on rocks is the complex sum of three stress components (Acocella and Neri 2009):

- gravity and topographic load (vertical),
- regional tectonic stress field, and
- magma pressure within reservoirs or conduits (isotropic).

The stress is therefore anisotropic and can be represented by a tensor with three major orthogonal stress components or vectors: σ_1, which is the maximum principal stress component and is oriented in the direction of maximum compression, σ_2, the intermediate principal stress component, and σ_3, the minimum principal stress component (Fig. 8.2).

The rock *strength* is the value of stress required for a fracture to form, and has the same units as stress, that is, Pascals (Pa = N m^{-2}). Rock strength depends on the lithology, i.e. the degree of grain interlocking, such as in crystalline and clastic rocks, and the degree of cementation that makes the rock coherent or incoherent. At a larger scale, it also depends on the geological properties of a rock succession, such as the presence, nature, density and angle(s) of bedding, foliations, and fracture planes. The strength of a rock will be different under compression, under pure extension, and under shear, so each material has a *compressive strength*, a *shear strength,* and a *tensile strength*. In situ tensile strength measurements of solid rocks at shallow structural levels (Table 8.1) are generally in the range of 0.5 to 9 MPa, and most frequently between 1 and 6 MPa (Gudmundson 2012), and for unconsolidated and water-saturated rocks they can be as low as a few kPa (Nahlawi et al. 2004).

Generally, the tensile strength is lower than the shear strength, which is in turn lower than compressive strength, which can be up to 10 times larger than the tensile strength. This makes it difficult for intrusions to propagate under compressional stress regimes. Furthermore, the strength of a

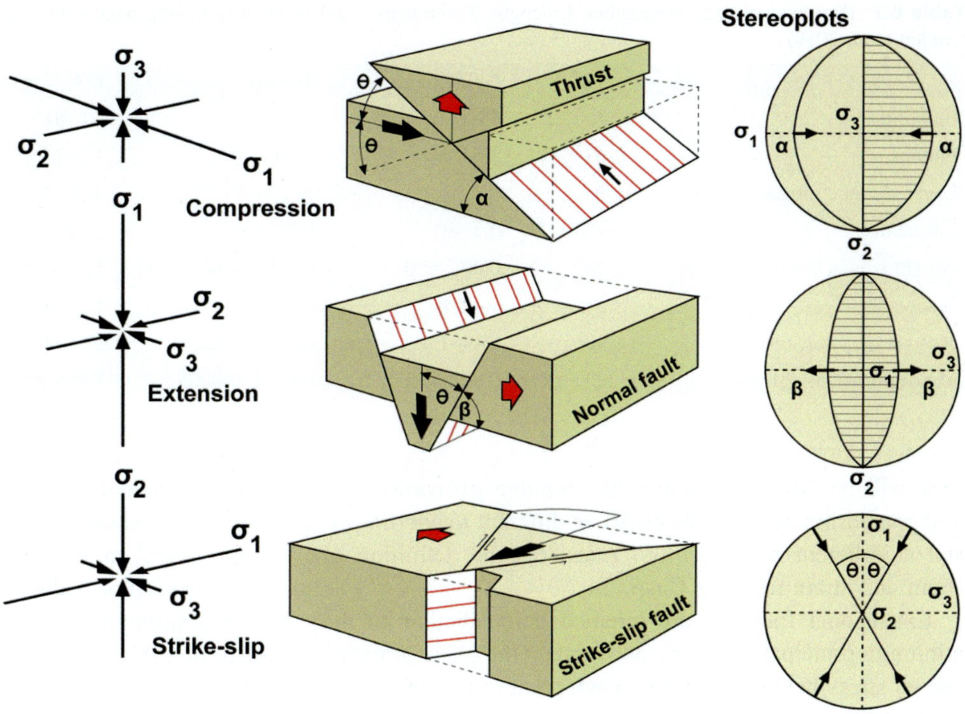

Fig. 8.2 Three principal stress components of the lithostatic stress field, their orientations in 3D space in different tectonic stress fields, the resultant orientations of crustal faults and movement senses of fault blocks, as well as stereo-plots of the stress vectors (Modified from Ramsay and Huber 1987)

Table 8.1 Rock strength properties of common rock types

Rock	Unconfined compressive strength (MPa)	Tensile strength (MPa)	Elastic modulus (GPa)
Igneous			
Granite	100–300	7–25	30–70
Dolerite	100–350	7–30	30–100
Gabbro	150–250	7–30	40–100
Rhyolite	80–160	5–10	10–50
Andesite	100–300	5–15	10–70
Basalt	100–350	10–30	40–80
Sedimentary			
Conglomerate	30–230	3–10	10–90
Sandstone	20–170	4–25	15–50
Shale	5–100	2–10	5–30
Mudstone	10–100	5–30	5–70
Dolomite	20–120	6–15	30–70
Limestone	30–250	6–25	20–70
Metamorphic			
Gneiss	100–250	7–20	30–80
Schist	70–150	4–10	5–60
Phyllite	5–150	6–20	10–85
Slate	50–180	7–20	20–90
Marble	50–200	7–20	30–70
Quartzite	150–300	5–20	50–90

Table 8.2 Dimensional data (N=number, L=length, T=thickness) and power law scaling parameters (a, b) of tabular intrusions (Modified from Cruden et al. 2018)

	N	*L* (km)	*T* (km)	*L/T*	*a*	*b*
Batholiths	4	95–600	3–15	20–90	–	–
Plutons	61	3.5–86	1–10	1.5–17	0.81 ± 0.12	1.08 ± 1.38
Layered mafic intrusions	8	8–550	3–13	2.3–68.7	–	–
Laccoliths	131	0.3 -80	0.03–3	2–50	0.92 ± 0.11	0.12 ± 0.02
All Horizontal sheet intrusions	68	0.001–450	0.0001–0.6	1.9–2400	0.57 ± 0.02	0.4 ± 0.22
Mafic sills	31	1–450	0.01–0.6	10–2400	0.85 ± 0.1	0.01 ± 0.02
Meso-scale intrusions	13	0.14–1.6	0.0005–0.025	37–290	0.49 ± 0.13	0.47 ± 0.33
Minor intrusions (Skye)	17	0.001–0.004	0.0001–0.0005	1.9–9.2	0.91 ± 0.25	0.27 ± 0.04

rock will be different at different *confining pressures* (e.g. with increasing depth in the crust), at different *temperatures*, and at different *strain (stress) rates* (for the definition of strain and strain rate, see Chap. 2).

Extensional fractures are oriented perpendicular to the minimum principal stress, σ_3, and contain the plane of the σ_1 and σ_2 stress field components. Fault planes (shear fractures or thrusts) contain the intermediate stress σ_2 and are at a fixed angle, θ, to σ_1, which is different for each material, but always less than 45° (Fig. 8.2).

The process of fracturing of the country rock to open conduits for magma can be approached with the Mohr–Coulomb criterion (Fig. 8.3; Ramsay and Huber 1987). Each total stress ($\sigma_{1,2,3} = \sigma_1 + \sigma_2 + \sigma_3$) applied to a volume of rock can be divided into a normal component σ_n, and a shear (tangential) component, τ_f, with respect to a plane however oriented through that volume. The Mohr–Coulomb criterion is empirical and is based on the measure of the total $\sigma_{1,2,3}$ applied and the angle θ at which a fracture eventually forms with respect to σ_1, that is the plane of maximum τ for that particular orientation of the total stress applied (σ_1, σ_2, σ_3). Mohr circles show the experimental relationships between the stress applied, σ, and the shear stress τ (Fig. 8.3).

The σ axis plots values of the maximum stress applied, σ_1, and the minimum stress applied, σ_3. As an example, we can start from a lithostatic case where σ_0 is the initial stress at a certain depth in the crust. From this initial state, indicated by a single point along the σ axis in Fig. 8.3, uniform increases in pressure, with increasing depth in the Earth, will only cause the shift of σ_0 to higher values (right in the diagram). A *deviatoric stress* is applied when either the absolute value of stress is increased in one direction, such as when applying a compressional tectonic (or magmatic induced) stress σ_1 (and σ_0 becomes σ_3), or when the confining pressure decreases in one particular direction, as when applying an extensional tectonic force (and σ_0 becomes σ_1), or by removing the flank of a volcano by a sliding collapse. Once a deviatoric stress is applied to a rock, its state is no longer shown by one point on the σ axis but by 2 points corresponding to the absolute values of σ_1 and σ_3.

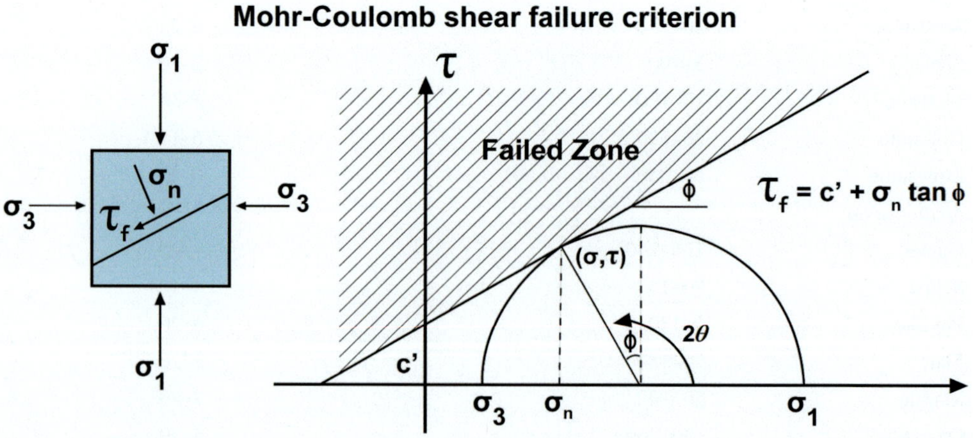

Fig. 8.3 Mohr–Coulomb criterion for fracturing of rocks; see text for explanation (Modified from Ramsay and Huber 1987)

The σ_1–σ_3 distance is the diameter of the Mohr circle, and is defined, at failure, as the *rock strength*. It indicates the limit to which one value can be increased (in the case of compression) or diminished (in the case of extension) with respect to the other before reaching the fracture condition (see Table 8.1). Experimentally, it is seen that at compression, that is, increasing σ, and therefore moving towards the right on the diagram, larger deviatoric stress is required to reach failure than decreasing σ, as in the case of tension. The curve that envelopes all Mohr circles slopes therefore towards the left of the diagram. The rock strength increases by increasing σ_3, that is by increasing the confining pressure. This means that at increasing depths, the deviatoric stress to initiate brittle failure must be higher, but becomes lower and lower as we move closer to the surface.

Pure extensional fractures form where confining pressure is negative, and the angle θ between the maximum stress applied and the fracture is 0°. Extensional fractures are therefore common near the Earth's surface and are typical of ground deformation on active volcanoes. Shear fractures are more common at deeper levels, where the confining pressure is higher. For this reason, the initiation of fractures at depth and of dyke intrusion is well approximated by the classical Coulomb–Mohr criterion (Dvorak et al. 1986; Troise 2001):

$$\tau_f = c' + \sigma_n \tan \phi \qquad (8.1)$$

where, graphically, τ_f is the intercept of the Mohr envelope line at failure on the τ axis, that is the *cohesion* of the rock (or its *internal shear strength, c'*) and *tan* ϕ is the slope angle of the line that is the *coefficient of internal friction* (the diagram also illustrates the relationship between the *angle of internal friction*, ϕ, and θ, the angle between the maximum stress applied and the fracture plane, that is, $\theta = 45° \pm \phi/2$).

Practically all rocks, either lithified or granular, are made of solids + voids, including fractures. Voids may be, and in general are, at least partly filled with fluids, water, gas, oil, and in volcanic settings, magma as well. In equilibrium conditions, the stress (lithostatic + tectonic) applied to a rock is entirely sustained by the solid part along grain contacts, and generally the mechanical strength of the rock is larger than the deviatoric stress applied.

Liquids at equilibrium are subject to hydrostatic conditions, so they experience the pressure of the overlying liquid body, that is

$$P_X = \rho_w \, g \, z \qquad (8.2)$$

where ρ_w is the density of the liquid (kg m^{-3}), g is the acceleration due to gravity (9.8 m s^{-2}), and z is the height (m) of the liquid column. Since liquid is generally less dense than rocks, the *hydrostatic pressure* (gradient) is less than the lithostatic pressure (gradient; Fig. 8.4).

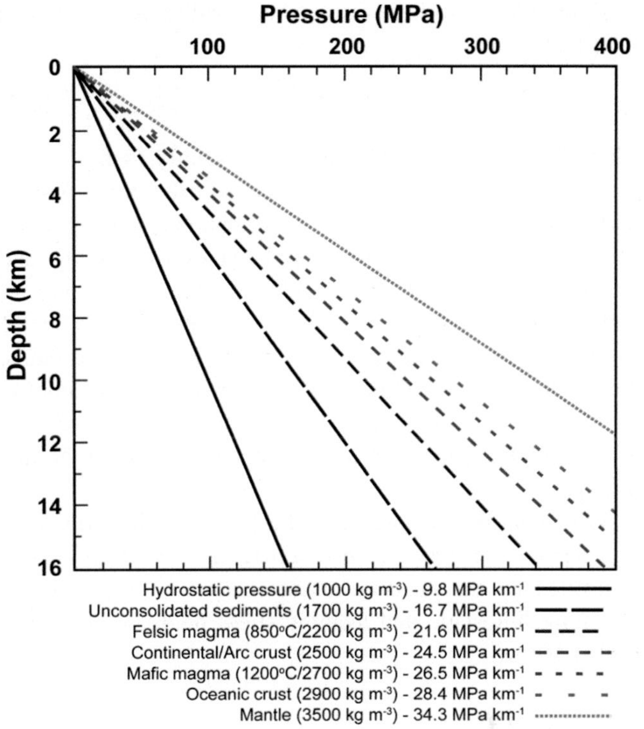

Fig. 8.4 Comparison of hydrostatic pressure gradient with lithostatic (different crust types) and magmastatic (felsic, mafic magmas) pressure gradients (From Cas and Simmons 2018)

This is because either the rock is undersaturated, so voids are partly filled with air, partly with water (or oil, etc.), or because at increasing pressure (lithostatic and tectonic) the rock has enough permeability to allow fluids to escape at the same rate of deformation (elastic or compaction for granular unconsolidated rocks). In this case, volume changes of the rock are allowed and the shear strength of the rock deriving from its increased internal friction due to compaction can be entirely used to counterbalance the applied shear stress on the plane of potential failure.

However, if this is not the case, i.e. if the permeability of the rock does not allow fluids to escape at the same rate as deformation aimed at reducing the voids (e.g. with burial or compressional tectonic forces), then the interstitial (or inter-fracture) liquid will experience an additional pressure, or overpressure. But liquids are incompressible, so despite the overpressure, they will prevent volume changes of the rock and therefore reduce the ability of the solid part to mobilise the shear strength given by friction and cohesion along the plane of potential failure. This condition is therefore more favourable to fracturing.

In this case, the effective stress is reduced so that

$$\sigma_{eff} = \sigma_{tot} - P_X \qquad (8.3)$$

Hydraulic-fracturing occurs when conditions are satisfied such that

$$\sigma_{tot} - P_X > T \tag{8.4}$$

where T is the tensile strength of the country rock.

Pore pressure increases with depth in the crust, that is, with sediment burial, tectonic deformation, rapid agitation (in case of earthquakes or rapid cycles of loading and unloading as during volcanic explosive eruptions), hydrothermal pressuring that forms when the water is heated and builds up pressure, and by addition of magmatic fluids. An understanding of the role of pore pressure is essential to any process of magma intrusion, magma–water interaction, and hydraulic fracturing in geothermal and hydrothermal settings.

Where intruding magma is the principal fluid involved (e.g. deep in the crust), the rise of magma through the crust is controlled by the pressure in the magma column and the confining lithostatic and tectonic stress fields, as discussed further below.

8.2.3 Factors Influencing Uprise of Magma Through the Lithosphere: Buoyancy, Magmatic Pressure, Lithostatic Pressure, and Viscosity Effects

The principal driving force for magma uprise through the lithosphere is *buoyancy*. This is determined by the density contrast, $\Delta\rho$, between the magma and the country rock. The density contrast arises because at melting point, liquid occupies a larger volume than the source solid rock and is therefore less dense than it. This is because in a molten state the constituent ions and molecules that made up minerals in the rock are in a less ordered state, so they occupy more volume. Once all voids in the source rock are magma-saturated, any further increase in volume is counteracted by the lithostatic pressure at that level. In the absence of a physical larger space to be occupied, if lithostatic pressure is enough to maintain the pre-melting volume, the liquid will store a certain amount of strain, in the form of internal deformation of its structure, and its internal pressure will rise. At this stage magma is overpressurised.

Therefore, because of lithostatic confining pressure, volume expansion is constrained, and is expressed by increasing pressure in the magma in the source region. Eventually, when magma overpressure, ΔP, exceeds the tensile strength of the surrounding rocks and propagates hydraulic fractures, or dilates existing fractures, liquid will find a way to expand its volume by expanding up into those fractures, towards regions of lower confining pressure, which in the Earth is generally upwards. The strain stored in the internally deformed magma will be released and will be able to further fracture and deform the surrounding rocks. The larger the $\Delta\rho$ at the source, the larger the magma overpressure is relative to confining pressure, and the larger the force exerted on the country rock. Overpressures, ΔP, of magmas are generally in the order of 10–100 MPa. Such overpressures are important in helping to drive magma upwards to the Earth's surface. Valentine and Gregg (2008) have depicted two scenarios as to how lithostatic pressure helps drive magma upwards (Fig. 8.5). In the first, a rising magma conduit or dyke is connected with its deeper magma reservoir, which is subject to high confining lithostatic pressure (Fig. 8.5a). At the tip of the upwards propagating conduit, volatiles will begin to exsolve, reducing the density of the magma there, increasing its buoyancy and pressure, and allowing it to propagate and extend the fracture up, while it is rising. The combination of increased buoyancy and pressure at depth helps to drive the magma upwards. In scenario 2 (Fig. 8.5b), a rising finite batch of low-viscosity basaltic magma has detached from its source, and as buoyancy at the tip of the conduit/dyke induces the magma to continue to rise as explained in scenario 1, the tail of magma column is squeezed almost closed as magma progressively passes through it at different levels.

A pressure contrast can alternatively be achieved if external forces such as tectonics or subsidence tend to reduce the volume available to magma at any depth. Similar to the description of the fluid pore pressure in the previous paragraph, magma, being incompressible, will accommodate only minor changes in volume unless it finds a way to escape, that is, by fracturing the host rocks. The process is aimed at restoring conditions so that rocks sustain the lithostatic and tectonic load and the magma sustains its own weight, which is also called *magmastatic pressure* or head (Johnson 1992; cf. Fig. 8.4).

Open conduits, such as observed at Stromboli volcano (Italy), and lava lakes, such as Erta Ale (Ethiopia), Pu'u O'o (Hawaii), Mt Erebus (Antarctica), and Villarrica (Chile), are excellent examples of the level of the magmastatic head of a volcanic system. The changes in magma level with time record changes of volume either due to causes internal to the magmatic system (e.g. input of new magma from below or crystal fractionation and cooling) or external causes, such as tectonic squeezing. A general equation for defining the magmatic overpressure, ΔP, in open conduit conditions (Jaupart 2000) is:

$$\Delta P = (\rho_r - \rho_m)\, g\, H + \Delta P_c - \rho_m\, g\, h \tag{8.5}$$

where $(\rho_r{-}\rho_m)gH$ is the buoyancy force that derives from the density contrast between rocks (ρ_r) and magma (ρ_m) at source depth H, ΔP_c is any overpressure due to volume changes at the source (magma input from below, tectonic squeezing, etc.), and $\rho_m gh$ indicates the magmastatic

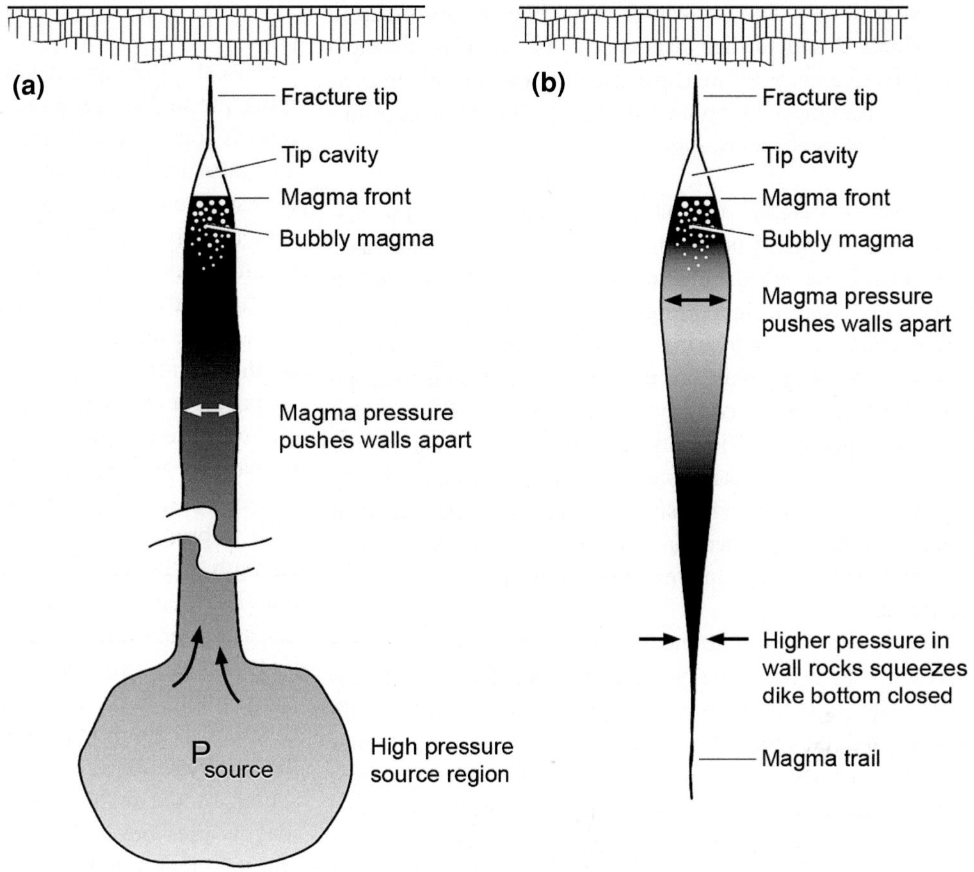

Fig. 8.5 Factors influencing the uprise of basaltic magma in a magma conduit, including volatile exsolution, decreased density and increased buoyancy at the tip of the propagating magma column or dyke, and lithostatic pressure effects on the source magma reservoir at depth in (**a**), and on the tail of a rising magma batch in (**b**), in helping to squeeze the magma up (From Valentine and Gregg 2008)

pressure of a lava flow or dome of height h formed at the vent above the conduit during an effusive eruption.

In actual fact, most magmas, especially mafic magmas, have much higher densities than those of rocks or sediments in the upper crust (Fig. 8.4; also see Fig. 2.4, Chap. 2), especially at the uppermost levels. Above the Moho, the density of the crust decreases rapidly towards the surface. Density values in the upper mantle (below ca 30 km in a "normal crust") are about 3,300 kg m^{-3}, in the lower crust (ca 30–10 km) generally lower to 2,900–2,700 kg m^{-3}, whereas the upper crust (<10 km) values are generally 2,600–2,000 kg m^{-3}. At the Earth's surface, active subsidence (e.g. along passive margins of continents, or active rifts) may account for the presence of low-density, unconsolidated sediments (<1,500–1,800 kg m^{-3}) for up to a few km thick (Fig. 8.4).

Therefore, magmas rarely rise through the entire lithosphere, from source to surface, without encountering a level of neutral buoyancy, where $\rho_m = \rho_r$. This may lead to the formation of a magma chamber at that level, and, with perhaps some differentiation and assimilation occurring that change the chemistry towards more silica-rich and lower density compositions. The base of the crust is the first major density discontinuity that magmas encounter during ascent from the mantle, and they often reside there producing crustal underplating (Cox 1993; Corti et al. 2003). Nevertheless, mafic magmas do reach the surface and this may be explained in two other ways:

(a) Momentum due to rapid ascent related to rapid decompression and strong density contrast with the source rocks. Many mantle xenolith-bearing mafic and ultramafic lavas (generally alkali-basalts and kimberlites) are thought to have reached the surface at a speed larger than the settling velocity of the mantle xenoliths (higher in density) in the basaltic melt (Rutherford and Gardner 2000; Spera 1984). Calculated ascent velocities exceed 2 m s^{-1}, and may reach 30 m s^{-1} for kimberlites. Such high velocities can be attained for low viscosities (see below), and if the volume of magma mobilised is large enough to sustain the momentum (mass x velocity).
(b) Volatile release during decompression, formation of bubbles leading to reduction of bulk density, and an increase in buoyancy. This will affect the top of the rising magma column in particular (Fig. 8.5), leading to increased pressure and the ability to hydraulically fracture the crust above the rising column (e.g. Poppe

et al. 2020). The nucleation of bubbles, as discussed in Chap. 2 and further discussed in Chap. 5, is the most effective way to increase the volume of magma, as for example, 1% volume of gas at 100 MPa becomes 90% at atmospheric pressure.

However, magma rise may be retarded by magma viscosity. As discussed in Chap. 2, magma viscosity increases with increasing silica content, decreasing volatile content (especially water), increasing crystal content, decreasing temperature, and decreasing confining pressure. The viscosity of magma increases as magma rises to the surface and loses heat, decompresses, crystallises, and volatiles exsolve from the magma. Viscosity controls the extent of the boundary effects (shear) that magma experiences during ascent. The larger the viscosity, the larger the internal shear forces, i.e. the magma will have a reduced tendency to move. So, while buoyancy promotes momentum, viscosity reduces momentum.

Melts with Newtonian rheologies (Chap. 2) will flow in a conduit with a parabolic velocity profile. This is true for most magmas at or above liquidus temperatures and/or with low crystal contents. At lower temperatures and higher crystal contents (30–40% volume), rheologies become non-Newtonian and magma may stop (Petford 2003).

A generalised equation for magma flow rates, Q, in a conduit shows that Q is proportional to the radius, R (m), of the conduit and to the magmatic overpressure, ΔP, but is inversely proportional to the viscosity, η (Pa s), and to the length H (m) of the conduit (or dyke) (Jaupart 2000):

$$Q = \frac{\rho_m \pi R^4 \Delta P}{8\eta H} \tag{8.6}$$

These relationships imply that viscous magmas need larger conduits to reach shallow subvolcanic environments or the Earth's surface than fluidal magmas.

8.2.4 Physical Criteria for Propagation of Magma Through the Crust

The intrusion of magma in the crust can only occur if fractures can form and propagate through solid rocks or the displacement of unconsolidated sediments occurs. Hydraulic fracturing and intrusions into solid rocks occur when total magmatic pressure, P_m (i.e. the sum of the lithostatic pressure, P_l, and the excess magmatic pressure, P_e, or overpressure, in the magma reservoir), overcomes the tensile strength of host rocks, T, and the minimum principal stress, σ_3, of the lithostatic pressure at that location (Gudmundsson et al. 1999; Gudmundsson 2012):

$$P_m > T + \sigma_3 \tag{8.7}$$

Pascal's Principle determines that in a closed column of fluid, the highest pressure applied at the bottom of the column is transmitted to the whole fluid (Cañón-Tapia 2008). The pressure inside a magma reservoir builds up dominantly in response to volumetric increase due either to recharge or vesiculation (e.g. Tait et al. 1989). The response of the country rock to magma overpressure is a deformation, partly elastic, partly visco-elastic depending on the rheology of the crust. Once this overpressure overcomes the tensile strength of the country rocks (usually the overpressure needed to propagate a fracture is taken as twice the tensile strength of the crust; e.g. Pinel and Jaupart 2003), then a fracture propagates and magma will intrude it. Cañón-Tapia (2008) suggests that at depth in the crust or mantle, the threshold for rock to propagate a fracture may be higher than the tensile strength of the rock, if the stress field is not purely lithostatic, but is affected by external tectonic stresses, although assessing the critical stress threshold is very difficult.

Dykes can only propagate through country rock if they can continue to propagate a crack or fracture ahead of them, called crack tip propagation. Kavanagh (2018) proposes that the criterion for crack tip propagation can also be represented by the relationship of the overpressure between the magma pressure and the confining lithostatic pressure, ΔP, and the crack toughness, K_c:

$$K_c = \Delta P \sqrt{l} \tag{8.8}$$

where l is the dyke length. Crack toughness, K_c, is a measure of the host rock's ability to resist further fracture when it already contains a crack, but more specifically, it is the energy required to cause a material that contains a crack to fracture further.

At any depth in the crust where a dyke is propagating, the magmatic (over)pressure, P_o, required to allow the dyke to propagate can be expressed as

$$P_o = P_e + (\rho_r - \rho_m)\, g\, h + \sigma_d \tag{8.9}$$

where P_e is the magma pressure in excess of lithostatic pressure in the dyke, ρ_r is the density of the country rock at that depth, ρ_m is the magma density at that depth, g is acceleration due to gravity, h is the depth in the crust at that point, and σ_d is the differential stress between σ_1 and σ_3 (Gudmundsson 2012).

Where sills are the magma reservoir feeding eruptions via a dyke, the pressure regime in the sill, and the manner in which it dissipates during an eruption will depend on whether the whole sill or magma reservoir is a single mass of molten magma or whether it is compartmentalised into multiple compartments of eruptible magma, and the degree to which they are interconnected (Gudmundsson et al. 2012). If the reservoir consists of a single body of magma, the excess pressure, P_e, will dissipate exponentially as the eruption proceeds. If the magma reservoir is compartmentalised, exponential excess pressure decay will apply only to each compartment as it is drained. If multiple compartments provide magma to the eruption, then the excess pressure decay occurs more slowly and is more irregular overall (Gudmundsson 2012).

The value and the orientation of σ_3 changes depending on the presence of a regional tectonic stress σ_r, the stress-induced at the shallow level by mass anisotropy related to topography, σ_t, and by rock anisotropy and weaknesses such as the presence of strata planes and pre-existing fault or fracture planes, σ_a (e.g. Gudmundsson 2002; Gudmundsson and Brenner 2001). Planar intrusions such as dykes can be treated as tensile fractures for the reconstruction of the paleostress. The dyke plane, in general, is (almost) perpendicular to σ_3, that is, the direction of maximum extension.

For this reason, measuring a statistically significant number of strikes and dips of dykes, and the statistical analysis of the orientation data is a powerful tool for reconstructing the orientation of the paleostress field at the time of dyke emplacement (Sect. 8.5). However, if pre-existing fractures, faults, or stratification planes are present, magma may prefer to intrude along those, because of the reduced tensile strength of rocks along those planes, and those orientations may deviate from the theoretical direction associated with the extension at that time. Furthermore, magma may flow vertically or horizontally in a dyke, i.e. the hydraulic fracture can be propagating upward or laterally. This implies that although the orientation of σ_3 can always be determined with some confidence, the orientation of σ_1 and σ_2 generally cannot.

Because magma is a liquid, P_m in a magma reservoir is isotropic. This promotes the formation of radial dykes, at any structural level either branching from magma chambers (Chadwick and Dieterich 1995; Suppe 1985; Gudmundsson et al. 1999; Gudmundsson 2012) or from sub-vertical volcanic conduits, provided that anisotropies induced by external factors, such as σ_r, σ_t, and σ_a, are not important. This is generally true close to the magmatic source, but how “close” will depend on the magnitude of the magmatic pressure. External factors, such as the orientation of the topographic stress, and the regional tectonic stress field, may instead prevail locally to disturb the ideal radial dyke arrangement and may become dominant in determining the orientation of dykes away from the source (e.g. Knopff 1936; Nakamura 1977; McGuire and Pullen 1989; Tibaldi 2015).

In addition to the brittle failure boundary conditions, the tensile strength of a rock during magma ascent can be reduced by:

- anisotropy of the rock, related to the presence of pre-existing discontinuities or newly forming fractures ahead of the rising magma,
- thermal weakening due to heating from the rising magma, and/or
- chemical weakening due to the circulation of aggressive peri-volcanic fluids, such as hydrothermal fluids.

As stated previously, the tensile strength of most rocks is on average 6 MPa, whereas for unconsolidated sediments, it is a few kPa. The effect of this in intrusions is that solid rock fractures in a brittle fashion and produces planar fractures along which the magma migrates, producing tabular dykes and sills. The host rock succession generally experiences elastic deformation, except where it may already be pervasively deformed in which case it can behave inelastically (Kavanagh 2018). Where country rock behaves elastically, the magma overpressure can cause the fracture in country rock to dilate during peak magma flux, and then contract to varying degrees as magma flux wanes, even to the point of closing completely (e.g. Valentine and Gregg 2008), with perhaps only small discontinuous magma selvedges or dykelets preserved. Where magma intrudes unconsolidated sediments, it cannot fracture them in a brittle state, but it pushes irregularly through the sediments, causing inelastic deformation, whereby deformational strain is accommodated by causing grain reorganisation, or if the sediments are water-saturated, they may become fluidised by the heat of the magma, resulting in intrusions with irregular margins and geometry.

Petford et al. (1994) and Cruden and Weinberg (2018), referring to the passage of felsic magma through a dyke feeding a pluton, suggest that the velocity of magma flow in a dyke, U_{av}, can be expressed as

$$U_{av} = \frac{\Delta\rho\, g\, w^2}{12\eta_m} \tag{8.10}$$

where $\Delta\rho$ is the density difference between the magma and country rock, g is the acceleration due to gravity, w is the dyke width, and η_m is the magma viscosity at depth (e.g. 10^3–10^7 Pa s for felsic magmas; Cruden and Weinberg 2018). Cruden and Weinberg (2018) calculated that a felsic magma with a viscosity of 10^5 Pa s flowing through a dyke 1 m wide would have a velocity of 2.45×10^{-3} m s^{-1} equating to 78 days to ascend 15 km, whereas in a 10 m wide dyke the velocity would be 0.245 m s^{-1}, resulting in a rise rate of 0.7 days to rise 15 km. They also propose that the volumetric flow rate in a feeder dyke, Q_A, equals the volumetric filling rate of a pluton or magma reservoir, Q_E, determined by

$$Q_A = Q_E = \frac{\Delta\rho}{12\,\eta_m}\left[g\,w^3\,\beta\right] \tag{8.11}$$

where β is the horizontal length of the dyke (Petford et al. 1994; Cruden and Weinberg 2018). Cruden and Weinberg (2018) calculated that the flow rate for a felsic magma with a viscosity of 10^4–10^8 Pa s flowing through a dyke 1 to 10 m wide ranged from $<10^{-2}$ to $>10^3$ m^3 s^{-1}.

8.2.5 Structures, Fabrics, and Textures Resulting from Subvolcanic Intrusions

Westerman et al. (2018a) have provided a comprehensive overview of the range of structures and textures that can result from the emplacement of intrusions from those that reflect the architecture and geometry of the intrusion to those that form within or at the margins of the intrusion as a result of flow within the intrusion, both in the liquid and solid states, and those that result from the thermal regime and cooling of the intrusion. We cannot list and discuss in detail all features, but readers are referred to Westerman et al. (2018a), and the volume in which that paper occurs, Breitkreuz and Rocche (2018). Here we provide a summary, but further on in this chapter, we discuss some of these features in context.

8.2.5.1 Intrusion-Related Structures

Westerman et al. (2018a) provide a long list of structures that result from intrusion and emplacement of subvolcanic intrusions. These include:

- steps, where an intrusion is segmented into parts that are offset along pre-existing fractures that the intrusion utilises as a conduit to "step up" to another level in the host rock or by-passes them (e.g. Fig. 8.21a–d),
- bridges, which are sections of country rock separating segments of an intrusion (e.g. Fig. 8.11d, e),
- broken bridges, which are ruptured former bridges that an intrusion has cut through,
- multiple sheets, in which an intrusion is composite, consisting of multiple discrete intrusive pulses (e.g. laccoliths, some sills, and dykes),
- internal contacts, representing multiple intrusive pulses within an intrusion that are in contact, with either sharp or cryptic contacts (e.g. Fig. 8.23g),
- fingers, tongues, and lobes, representing multi-lobate propagation at the margins of an intrusion (e.g. Figs. 8.19; cf. pillows, lobes, and mega-lobes in subaqueous lavas, e.g. Chap. 6),
- apophyses (our addition) resulting from relatively narrow magma injections into country rock fractures and bedding planes, and into unconsolidated sediments, and branching from the main larger intrusion (e.g. Figs. 8.13, 8.18),
- marginal faults, caused by lifting of country rock during inflation of sub-horizontal intrusions,
- plume-like *en echelon* fracture sets in country rock along the margins of an intrusion that originate from and show the direction of propagation of intrusion, and
- drag folds in country rock adjacent to intrusion margins that may indicate intrusion propagation and flow direction.

8.2.5.2 Magma Flow-Generated Fabrics

Fabric features generated by flow of magma in intrusive conduits (Westerman et al. 2018a) include:

- lineations and grooves on the intrusion surface, indicating flow direction,
- ropes, waves usually with axes perpendicular to flow direction, and lobelets on the surface of intrusion (e.g. Fig. 8.13g),
- alignment of elongate crystals and stretched vesicles parallel to flow direction,
- trails of xenoliths showing planes of flow,
- marginal autobrecccia (our addition) of magmatic clasts enclosed in coherent intrusive rock occurring locally along intrusion margin, resulting from the shearing of near-solid marginal parts of the intrusion and detachment of clasts by the still moving part of the intrusion; relatively rare,
- intrusive breccia of ripped-up country rock clasts in the intrusion, usually along the margins,
- crystal abundance increasing from intrusion margins to the interior, produced by shear flow-induced segregation of large crystals out of the zone of flow shear at intrusion margins (common in porphyritic dykes), along with the effects of the developing cooling profile (our addition),
- flow banding (our addition) resulting from laminar shear flow in intrusion, particularly along the margins where there is a coupling of the cooling intrusion margins and the country rock,
- flow folds resulting from ductile deformation of all planar fabrics listed above, and
- anisotropy of magnetic susceptibility (AMS) fabric, reflecting the alignment of tiny magnetic minerals along a flow fabric; it may be useful to reconstruct flow directions in intrusion.

8.2.5.3 Brittle Fabrics

As an intrusion cools and crystallises, it can with increasing difficulty continue to deform in an almost solid state, producing (Westerman et al. 2018a):

- cataclastic bands at the margins of intrusions or within intrusions between static and moving parts of an intrusion.

8.2.5.4 Thermally Induced Features

Many fabric and textural features in an intrusion result from its thermal state, which can change dramatically from the time of initial propagation to final cooling and crystallisation. These include (Westerman et al. 2018a):

- chilled margins or rims, often glassy in thinner intrusions, resulting from rapid heat loss to the country rock; they may be absent in large volume, high heat budget intrusions which have thermally equilibrated with country rock, perhaps assimilating initial chilled margins formed during initial stages of intrusion when the body was thin and the country rock cold,
- columnar jointing perpendicular to intrusion margins (e.g. Fig. 8.9h); it may increase in spacing inwards or be absent in very large volume intrusions; result of progressive inward cooling of an intrusion with step-by-step inward propagation of a polygonal network of cooling-induced contractional fractures; it may be difficult to distinguish in very big intrusions because thermal equilibration with country rock occurred,
- crystal settling, producing a layering of crystal sizes,
- chemical differentiation (our addition) in large, thick, deep-seated intrusions, leading to compositionally zoned intrusions,
- vesicles (our addition) in relatively shallow intrusions, resulting from upper crustal volatile exsolution; in dykes, they tend to increase in abundance inwards away from chilled margins; in sills, vesicles/amygdales increase upwards, and some very shallow, syn-depositional sills, they may even be pumiceous at the top (Orth and McPhie 2003); mariolitic cavities may occur in deeper seated and larger intrusions, resulting from late, crystallisation-induced volatile exsolution,
- hyaloclastite and peperite breccias (Chaps. 3, 4; e.g. Figs. 8.13g, 8.18a–h) formed at the margins of intrusions into water-saturated, unconsolidated sediments; fluidised peperite breccia may be injected into cooling fractures in the margins of intrusions as intrusive clastic dykes,
- devitrification and cooling textures, and zonation in felsic sills (e.g. Orth and McPhie 2003), involving groundmass granophyric or equigranular textures, increasing proportions of spherulitic and lithophysae crystallisation inwards; mafic sills may show varying degrees of alteration, from palagonite to chlorite, and
- hydrothermal breccia pipes and domains, produced by hydrothermal fluid fracturing and brecciation of the intrusion mass and sometimes country rock (Chaps. 4, 13, 18).

8.2.6 Dimensional Characteristics of Intrusions

Cruden et al. (2018) propose that more or less tabular intrusions can be distinguished (with some overlap) on a continuum plot of log L, the diameter of a circle of equivalent area (often elliptical) of horizontal intrusions, against log T, the mean vertical thickness (Fig. 8.6a; Table 8.1). L/T is thus the *inverse* of the aspect ratio of lavas we defined in Chap. 2 and apply to mafic and felsic lavas in Chaps. 5 and 6, and to ignimbrites in Chap. 10. For dykes, L is the horizontal trace of a dyke whereas T is the horizontal thickness or width of the dyke. T and L are related by a power law (Cruden et al. 2018):

$$T = b\,L^{a} \tag{8.12}$$

where b is a constant and a is power law exponent (Table 8.2). Cruden et al. (2018) also propose that T can be related to the intrusion volume, V, by another power law:

$$T = d\,V^{c} \tag{8.13}$$

where d is a constant and c is the power law exponent.

On the log T versus log L plot (Fig. 8.6a), the exponent a distinguishes different groups of intrusions, with minor and mesoscale sheets, sills, and dykes falling along a line for which a is ~ 0.5, laccoliths along a line with $a \sim 1.5$, for plutons $a \sim 0.8$, whereas thick layered mafic intrusions and batholiths have $a \sim 0$ (Cruden et al. 2018). There is a bifurcation between minor and mesoscale sheets, sills, and dykes ($a \sim 0.5$) and the other types of intrusions where $L \sim$ 500–1000 m and $T \sim$ 10–30 m (Fig. 8.6a).

On the log L versus log V plot (Fig. 8.6b), there is a similar bifurcation at $V \sim 10^7$–10^8 m^3 (0.01–0.1 km^3), distinguishing sills with an exponent $c \sim 0.3$ and laccoliths with $c \sim 0.4$. Laccoliths, layered mafic intrusions, and plutons have a similar $c \sim 0.38$. The plot of L/T versus L plot (Fig. 8.6c) distinguishes intrusions that undergo lateral spreading (high L/T; $a < 1$) over thickening (lower L/T; $a > 1$).

Mafic dykes are characterised by $a \sim 0.5$ and bounded by lines with b in the range 0.03–0.25 (Fig. 8.6a). Mafic dykes are similar to mafic sills, minor and mesoscale sheets, but have a greater L/T ratio for a given length (Fig. 8.6c; Cruden et al. 2018). Mafic sills, minor and mesoscale sheets have $a \sim 0.5$, and are bounded by lines with b between 0.25

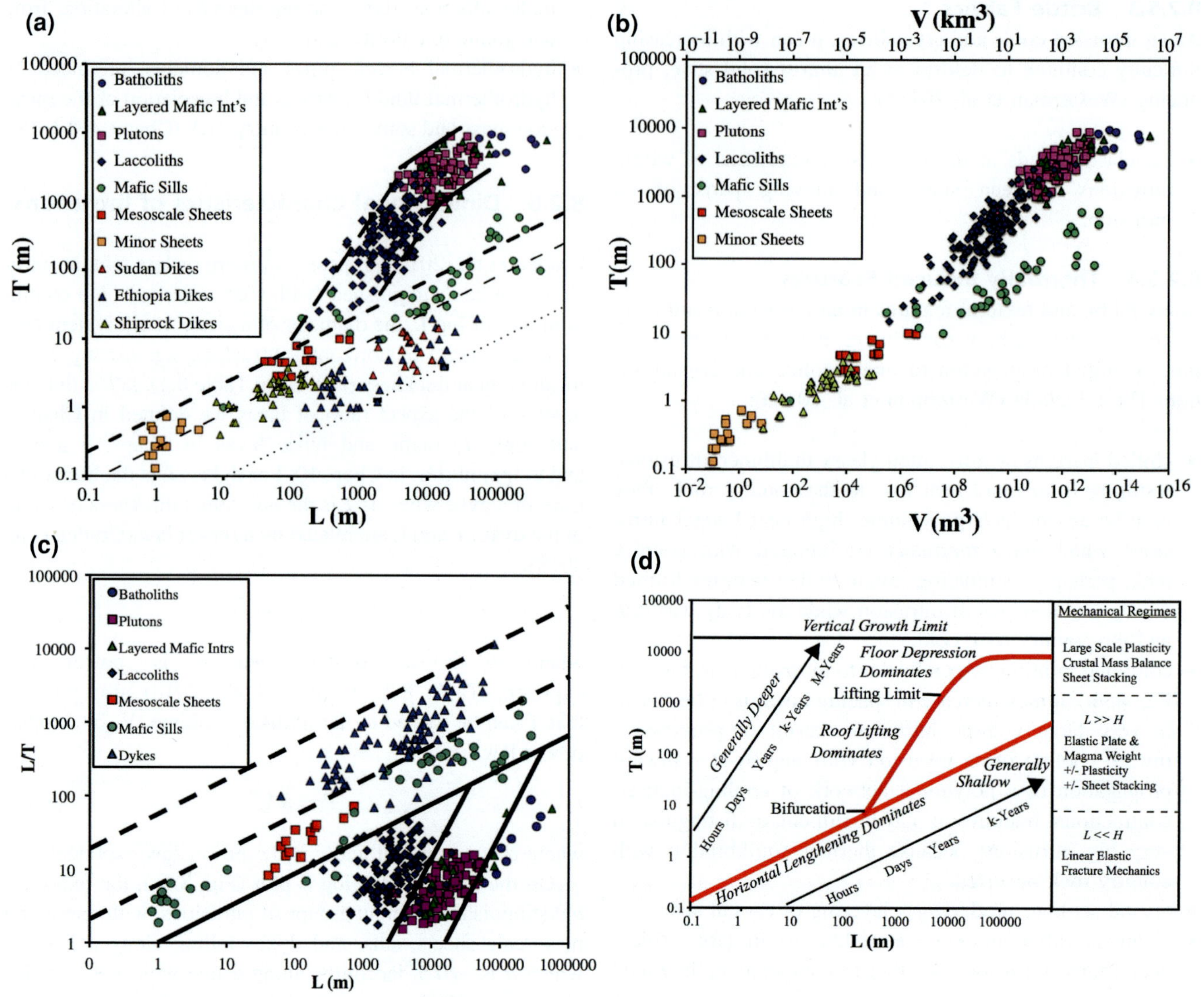

Fig. 8.6 Distinguishing different subvolcanic intrusion types based on dimensional aspects of tabular intrusions. **a** Relationship between mean thickness T and a length dimension L (see text) on a log–log plot. **b** Relationship between T and intrusion volume V on a log–log plot. **c** Relationship between L/T versus L on a log–log plot. **d** Interpretative significance of T versus L log plot (**a**) in terms of depth of intrusion in the crust, crustal rheological regime, and how that influences thickening of intrusions in the shallow crust compared with lengthening in the deeper crust. (From Cruden et al. 2018)

and 0.7 (Fig. 8.6a). The L/T ratio ranges from 2 for $L \sim 1$ m to 2000 for $L \sim 20{,}000$ m (Fig. 8.6c). For laccoliths $a > 1$, indicating they tend to thicken rather than lengthen, and as a result, L/T ratios range from 30 to 3 between $L \sim 500$ and ~ 8000 m (Fig. 8.6c).

Transitioning to plutons and small layered mafic intrusions, $a = 0.81$, and b is bounded by curves with b between 0.37 and 3.13 (Fig. 8.6a). These intrusions show a propensity to lengthen relative to thickening, and small plutons have L/T ratios of ~ 5, whereas for larger ones, $L/T \sim 10$. The transition from laccoliths to small plutons coincides with a change in a from >1 to <1 (Cruden et al. 2018). Batholiths and large layered intrusions are dimensionally the largest intrusive bodies, with thicknesses ranging from 2000 to 9700 m, and lengths from 57 to 550 km. They reflect lengthening growth with $a < 1$, and a maximum thickness of ~ 10 km (Fig. 8.6a).

Cruden et al. (2018) interpret these dimensional characteristics not in terms of magma rheological properties such as viscosity, temperature, and density, but on the depth of emplacement in the crust, and fracture toughness and rock stiffness of host crustal rocks (Fig. 8.6d). This is because intrusions are constrained by the physical properties and rheology of the crust they pass through and are emplaced into. In contrast, lavas are erupted into low-density and easily displaced air and water, and therefore it is the physical properties of the erupting magma and eruption dynamics that determine the dimensional characteristics of lavas (Chaps. 5, 6). Cruden et al. (2018) found that increasing the depth of emplacement in the crust has the effect of increasing intrusion length more than thickness, leading to increased *L/T*, that is, lower aspect ratio. However, where the crustal architecture is multilayered, thickness may increase to a greater degree than in an isotropic crustal rock mass, because the deformational strain can be accommodated by interstratal shear or sliding, reducing resistance to intrusion inflation. *L/T* is therefore lower for intrusions into layered crust than into isotropic crust. Increasing rock fracture toughness leads to thicker intrusions because propagating fractures laterally (leading to lengthening) becomes more difficult. Rock stiffness, reflected by Young's Modulus, appears to be inversely proportional to intrusion thickness, with increasing stiffness leading to an increase in intrusion length relative to thickness (i.e. thinner intrusions relative to low-stiffness crust; Cruden et al. 2018).

8.3 Types of Subvolcanic Intrusions: Settings, Characteristics, and Volcanological Significance

Intrusions are commonly classified according to their geometry and contact relationships with the country "rocks", which could be lithified rocks, unconsolidated sediments, volcanic deposits, or other intrusive rocks. The present review refers specifically to shallow (upper crustal level), subvolcanic intrusions, for which an old term, *hypabyssal*, is also sometimes used, although we prefer the more indicative term subvolcanic. The term hypabyssal has also been used to indicate that shallow intrusive rocks have specific "hypabyssal" textures, but these textures can be identical to effusive rocks, making the term ambiguous, which is why we also prefer not to use it. There are many types of subvolcanic intrusions which can occur in the entire spectrum of volcano-tectonic settings, including dykes, sills, laccoliths, lopoliths, layered intrusions, cryptodomes, plugs, and stocks, (e.g. Plate 8.1b, Fig. 8.7a, b; Breitkreuz and Rocchi 2018; Galland et al. 2018a; Jerram and Bryan 2018; Rocchi and Breitkreuz 2018; Chap. 17). In addition, we suggest the novel term "dill" for inclined sheets. We do not consider deeper crustal intrusions such as batholiths, which consist of multiple plutons, which are individual intrusions of significant size and volume.

Unlike lavas, where low-viscosity basalts (10–10^3 Pa s) are thinner (Chap. 5) than high-viscosity felsic lavas ($\sim 10^9$ Pa s; Chap. 6), subsurface intrusions they can be thin or thick irrespective of composition, because the constraints are different. Intrusions are constrained by country rock strength, temperature and lithostatic pressure effects, and magma properties such as viscosity (e.g. 10^3–10^7 Pa s for felsic magmas at crustal depths; Cruden and Weinberg 2018), whereas lavas are constrained by their viscous strength and their gravitational potential. Therefore, both mafic and felsic intrusions can be thin or thick. First, we describe the main subvolcanic intrusion types, and then in Sect. 8.4 consider the processes that occur in some, as well as their role in edifice building and collapse.

8.3.1 Dykes

The term "*dyke*" is a synonym for "wall" and originates from Scotland, where the morphology of the countryside is dominated by the prominent eroded remnants of Tertiary magmatic dykes, which are more resistant to erosion than the host rocks (Suppe 1985). Farmers historically used the dykes as fences to subdivide properties, so saving work and time. A *dyke* is a tabular, sub-vertical, sheet-like, cross-cutting intrusion of magma, with width usually much smaller than length and vertical dimension (W << L, H). The extent, orientation, and thickness of dykes may vary enormously in different volcanological settings (Figs. 8.8, 8.9) as discussed below.

8.3.1.1 Locations of Dyke Systems

Dykes represent the main part of any magma plumbing system that facilitates the rise of magma from the source to the Earth's surface. They may therefore constitute the main magma conduit, but they may also branch out from it. In central volcanic edifices, dykes radiate from the main magma feeder conduit, forming radial and concentric dyke complexes (Figs. 8.7a, 8.8a–c). Excellent examples of radial dykes emanating from magma conduits include those associated with stratovolcanoes at shallow levels (Fig. 8.8a; Nakamura 1977; Wadge 1976; McGuire and Pullen 1989; Acocella and Neri 2009; Tibaldi 2015), and those radiating from older exhumed stocks, such as at Spanish Peaks, Colorado (Fig. 8.8b; Knopff 1936; Johnson 1968; Suppe 1985).

Radial dyke sets result from the influence of a central magma conduit, whereas concentric dykes occur mostly around the rim regions of both explosive calderas and the non-explosive calderas of shield volcanoes, which Acocella and Neri (2009) attribute to the presence of a high-level magma reservoir and the stress field that it generates above the reservoir (Fig. 8.8a). In older exhumed magma plumbing systems, radial dyke systems radiate from a central stock or magma chamber and have been traced to extend up to tens of km from the source (Fig. 8.8b; Knopff 1936; Tibaldi 2015), and concentric dyke systems form inward dipping *cone sheet* systems around a central stock or magma chamber (Fig. 8.8c; e.g. Schirnick et al. 1999; Tibaldi et al.

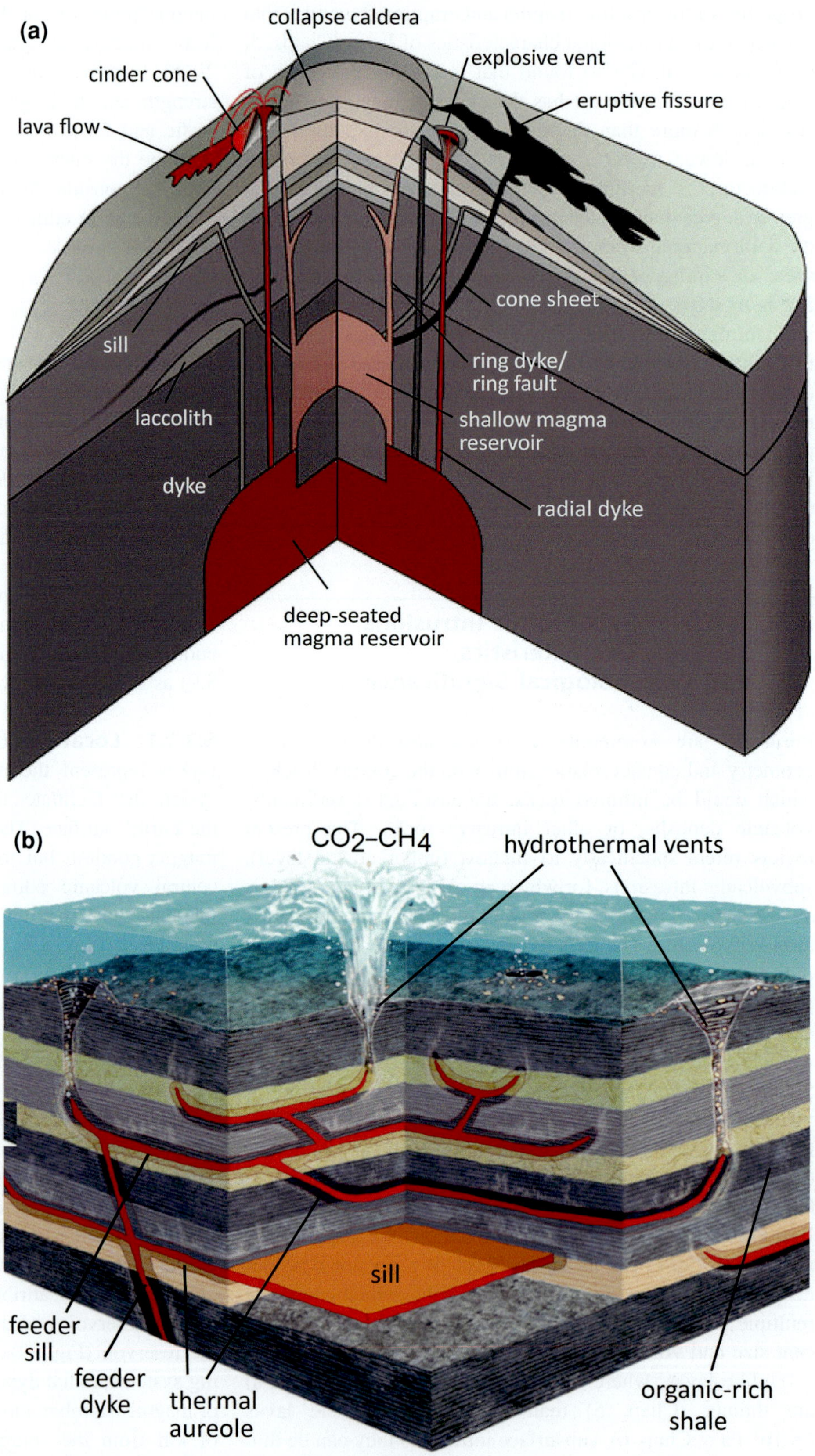

Fig. 8.7 Schematic representation of subvolcanic intrusions in two contrasting volcano-tectonic settings. **a** *Stratovolcano*: these include dykes, cone sheets, sills, laccoliths, caldera-related structures, and intrusions, and magma–fault interactions forming the internal plumbing system of the edifice (From Galland et al. 2018a). **b** *Sedimentary basin*: these include both a dyke feeding sills, and sills feeding sills, and this also illustrates the implications of sill emplacement within organic-rich sediments, with the formation of hydrothermal vents and potential release of greenhouse gases into the atmosphere (see text) (Modified from Galland et al. 2018b)

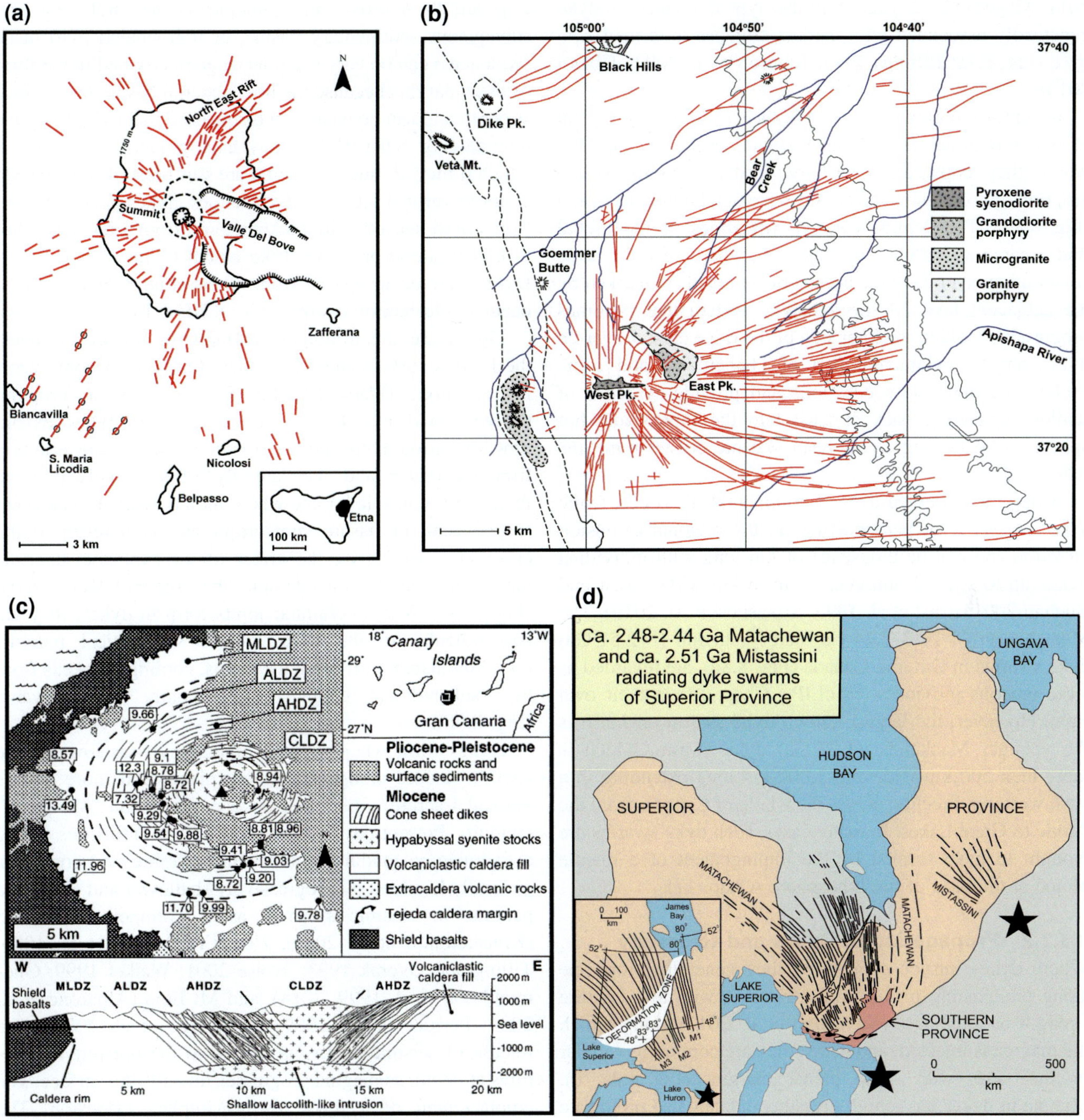

Fig. 8.8 Dyke types related to setting. **a** Radial distribution of fissure vents and dykes at Mt Etna, Sicily, Italy. Fissures high on the edifice are controlled by the gravitational stress field of the edifice, whereas those lower on the edifice are also affected by the regional stress field and structures in the basement. (Adapted from McGuire and Pullen 1989). **b** Radial system of ~25 to 21 Ma dykes focused on several stocks at Spanish Peaks, Colorado, USA, and representing several phases of magmatism. The asymmetry reflects the influence of a regional stress field. (Adapted from Knopf 1936). **c** Map and cross-section showing the cone sheet complex developed around a central stock that intruded the Miocene Tejeda Caldera, Gran Canaria, Canary Islands, Spain. Triangle marks central axis of radial symmetry; CLDZ = central low dyke-density zone, AHDZ = annular high dike-density zone, ALDZ = annular low dike-density zone, MLDZ = marginal low dike-density zone; boxed numbers are radiometric ages of intrusive and extrusive rocks (Ma). (From Schirnick et al. 1999). **d** Giant radiating ~2.48–2.44 Ga Matachewan and ~2.51 Ga Mistassini dyke swarms of the Superior Province associated with mantle plume and subsequent rift activity. The stars indicate the foci of the swarms and postulated plume head centres. Black polygons south of the Matachewan swarm are associated with volcanic rocks and layered intrusions. The Kapuskasing Structural Zone (KSZ) is a zone of deformation that cuts across the three main sub-swarms of the Matachewan swarm. The inset shows these sub-swarms (labelled M1, M2, and M3) to be more continuous after the effects of deformation are removed. (Map courtesy of Ken Buchan; Adapted from Ernst and Buchan 2001, and Buchan et al. 2007)

2011; Magee et al. 2012a). A further type of concentric dyke genetically associated with caldera collapse is termed *ring dyke* (Fig. 8.7a; Billings 1943; Johnson et al. 2002; Tibaldi 2015).

In contrast, regionally extensive *sheet dyke sets* are typical of major regional extensional tectonic settings (Chap. 17) where they constitute long-lived feeder conduits to the eruption of voluminous lavas and can extend hundreds of km (Fig. 8.8d). The central volcanic rift zones of the global mid-oceanic spreading ridge systems where seafloor lavas constituting layer 2a of the ocean crust erupted are marked by the subjacent lava 2b sheeted dyke system that becomes progressively older on either side of the ridge crest. Layer 2b dykes are a fundamental part of Phanerozoic and some Proterozoic ophiolite successions and provide evidence of seafloor spreading back at least into the Proterozoic but appear to be lacking in Archean oceanic crustal remnants (Chap. 17).

Along the exhumed axes of continental rift zones, mafic *dyke swarms* form perpendicular to the regional extension direction and can be thousands of km long with individual dykes up to several hundreds of m to km wide (Ernst and Buchan 2001; Ernst et al. 1995; Srivastava et al. 2019). The Matachewan ($\sim$2.48–2.44 Ga) and Mistassini ($\sim$2.51 Ga) dyke swarms in Ontario, Canada (Fig. 8.8d), are regarded as large igneous provinces (or cLIPs; Chap. 17) in their own right. However, the largest known dyke swarm on Earth is the younger Mackenzie (1.27 Ga) dyke swarm, which is more than 500 km wide and 3,000 km long extending in a northwesterly direction across the whole of Canada, from the Arctic to Great Lakes. In many cases, such dyke swarms are thought to have formed by the impingement of a mantle plume at the base of the lithosphere (Chap. 17).

8.3.1.2 Propagation of Dykes and Geometry

Where dykes intrude solid rock, they generally propagate along pre-existing planar brittle fractures with low fracture strength or toughness (Kavanagh et al. 2017) along which the magma is injected, and hence they are commonly tabular or sheet-like, with sharp planar margins (Fig. 8.9a, b). Magma in dykes has a high pressure and can also propagate new planar, brittle fractures, the orientation of which can be controlled by the prevailing lithostatic stress field orientation. Although propagation of tabular dykes with uniform thickness can be considered to result from propagation in an isotropic elastic regime, where country rock has pre-existing deformational fabrics and heterogeneities, the country rock can behave inelastically, resulting in dykes that can change thickness significantly along their extent (Kavanagh and Sparks 2011; Kavanagh 2018).

Similarly, very high-level dykes that propagate through semi-lithified strata (e.g. those making up a volcanic edifice) become increasingly less regular (Plate 8.1a; Fig. 8.9c, d, f, g, j), and those cross-cutting unconsolidated strata (e.g. into subaqueous sedimentary basins, or non-welded pyroclastic sequences) can be very irregular in geometry and in the form of their margins because the host material has no mechanical strength and can be pushed and squeezed inelastically by the magma (Fig. 8.9j). Dykes may have branching apophyses, which if they are sub-horizontal are sills (described next). On exposed vertical sections of volcanoes, dykes generally vary from cm to tens of m in thickness, and from a few tens of m to km in lateral or along-strike extent (Annen et al. 2001). Dykes may occur spaced from others (Fig. 8.9a–c, f, g), or occur in clusters or swarms (Fig. 8.7d, e, h).

Dykes are the principal conduits for magma to reach the Earth's surface and where they do, dykes become vents for eruptions (Plate 8.1; Fig. 8.9f, g). As an eruption proceeds and a volcanic edifice is built, dykes extend upwards through the edifice to feed directly the eruption from the crater, and also laterally, often radially into the flanks of a volcanic edifice where they may feed flank point source vent, or fissure vent eruptions. As a result, if the eruption is explosive, the crater fill of explosive deposits may transition downwards into the coherent feeder dyke (Fig. 8.9g). Where columnar joints form in dykes, they are perpendicular to the strike of the sub-vertical margins (and flow banding, if present); that is, cooling proceeds from the (sub-)vertical contact with country rock inwards (Fig. 8.9h).

Dykes form when a batch of magma rises towards the Earth's surface along a relatively planar and more or less vertical fracture, which can be regarded as an "elongated" conduit. Present-day evidence of dyke intrusions being active magma conduits includes linear fissure vent eruptions, generally heralded by progressively shallower and elongated patterns of seismic swarms, as is common in Iceland (Konstantinou et al. 2000), Hawai'i (Dvorak et al. 1986; Tilling and Dvorak 1993; Troise 2001; Walker 1999; Gudmundsson et al. 1999, 2014), and Mt Etna (Alparone et al. 2004; Ferrucci et al. 1989; Lanzafame et al. 2003). The release of seismic energy is related to the opening of fractures at depth, as magma progresses towards the surface. The emplacement of a dyke always implies extension. The duration of seismic swarms is generally days or months, which reflects ongoing magma uprise and intrusion (e.g. Gudmundsson et al. 2014). Repeated intrusion of batches of the same magma into the same fissure alongside and/or within each other results in the formation of *compound* or *multiple dykes* (Daly 1905; Fig. 8.10a; e.g. Gudmundsson et al. 1999; Poppe et al. 2020). A compound dyke composed of two or more compositionally different batches of magma is termed a *composite dyke* (Daly 1905; Judd 1893; Gibson and Walker 1963; Fig. 8.10b).

Poppe et al. (2020) described well-exposed Permian tephrite–basanite dykes in the Oslo Rift, Norway, which

(a)
(b)
(c)
(d)
(e)
(f)
(g)
(h)

◀ **Fig. 8.9** Geometry of dykes. **a** Basalt dyke cutting through Devonian granite with sharp planar contacts, Moruya, New South Wales, Australia (Photo by Ian Nicholls). **b** Eroded remnant of a narrow dyke cross-cutting a bedded scoria sequence, La Palma, Canary Islands, Spain; the sharp planar margins indicate that the scoria sequence was lithified at the time of dyke intrusion (Photo by Sam Thiele). **c** Wider dyke cross-cutting a lithified scoria sequence, Madeira Island, Portugal. **d** Variations in dyke width in swarm of dykes cutting a scoria cone sequence, Chatham Island, New Zealand. **e** Swarm of dykes of varying width, Valle de Bove, Mt Etna, Sicily, Italy. **f** Dyke cutting bedded scoria cone sequence to the crater, Madeira Island, Portugal. **g** Feeder dyke to spatter deposit (top eroded unit) that is part of a tuff and spatter cone, Miocene, Ryugazaki, Hokkaido, Japan. There is a transition from coherent basalt to spatter-filled conduit at about the level of the person (circled). **h** Andesite dyke (left centre), with columnar jointing perpendicular to the dyke margins (= cooling surfaces), in the submarine Cabo de Gata Volcanic Complex, southeast Spain. This late dyke is in contact with another dyke to the right that also has sub-horizontal columnar joints at its left margin, but grades into a hyaloclastite breccia to the right, which indicates it was a syn-depositional intrusion into water-saturated hyaloclastite. **i** Flow banding at the edge of 15 m wide vertical rhyolite dyke, Late Devonian Bunga Beds, Boyd Volcanic Complex, southern New South Wales, Australia. **j** Eocene basaltic dyke complex with very irregular shapes and margins cross-cutting which was at the time an unconsolidated scoria cone sequence, Chatham Island, New Zealand

Fig. 8.9 (continued)

intrude bedded lithified sandstones, siltstones, and carbonates, and show the terminations of thin dykes that range from 0.5 to 30 cm thick. Dykes varied from planar and continuous to branching with apophyses to discontinuous and offset across bedding planes, which was *not* due to post-emplacement lateral fault emplacement (Fig. 8.11a) but due to the deflection of the dyke by the bedding plane into another tectonic fracture nearby on the other side of the bedding plane. Dyke terminations include pointy crack tips filled with magma and secondary calcite (Fig. 8.11b–d), indicating original propagating crack tips, as well as "stubby-nosed terminations" surrounded by fractured country rock, and often offset and separated from another dyke segment by a country rock bridge (Fig. 8.11e). Dykes were intruded along pre-existing deformational structural fractures (Fig. 8.11a), some of which were deformed by the advancing dykes, wrapping around the dyke front. Poppe et al. (2020) considered the dykes with tapered crack tips to have been emplaced under opening mode deformation, involving warping the country rock into monoformal folds around the advancing crack tip and widening dykes (Fig. 8.12a, b), whereas the stub-nosed dykes were emplaced under shear mode deformation, causing shear along fractures ahead of the advancing dyke front (Fig. 8.12a, b).

Where dykes have crack tips (Figs. 8.11b, c, 8.12a, b), the very narrow tapered crack tips are not filled with magma, but with secondary calcite or a mixture of hydraulically fractured country rock and calcite seams (Fig. 8.11b, c). A magma meniscus exists several cm from the crack tip (Fig. 8.11b, c, 8.12a, b), suggesting that magma viscosity and surface tension prevented the magma from filling the crack void that it had opened ahead of it.

Fig. 8.10 Compound or multiple basaltic dyke in northern Iceland. **a** The dyke is composed of at least three separate dykes and is some 50 m thick; vertical arrows indicate the dyke's outer margins; person for scale circled. **b** Composite dyke at Streitishvarf in eastern Iceland. The outer parts of the dyke are of basalt, the central part of rhyolite; the total thickness of the dyke is 26 m, the rhyolite part being 13 m, the western basalt part 7.5 m, and the eastern part 5 m. The dyke can be traced for about 14 km. (From Gudmundsson et al. 1999)

Many dykes terminate along one fracture and then continue along another just cm away, creating country rock *bridges* between the two segments (Fig. 8.11d, e), some of which may fracture or be broken bridges. It seems that the magma was not able to continue propagating along the initial fracture, and simply found another nearby fracture to propagate along and jumped across, leaving a solid or fractured country rock bridge. This requires the propagation of a fracture sideways to propagate along in 3D, before finding a suitable fracture to propagate upwards again (Poppe et al. 2020).

Finally, dykes dilate to their maximum width at peak magma discharge when the magma fluid pressure exceeds σ_3 of the lithostatic pressure. However, during waning discharge, dykes may attenuate, especially if magma viscosity is low (basalts, komatiites, kimberlites) and magma pressure declines below σ_3. Such dykes can become very narrow, close up completely, or be represented by a trail of disconnected, lensoidal, "orphaned" dykelets (Valentine and Gregg 2008; Staude and Markl 2019).

8.3.1.3 Textures in Dykes

All magma types can intrude the lithosphere to form dykes, irrespective of the magma composition and viscosity (i.e. from komatiites and kimberlites to basalts to rhyolites). The magma solidified in a dyke may show textures varying from that of porphyritic lava to holocrystalline depending on the cooling history and depth of emplacement. Narrow dykes emplaced and cooled quickly at upper crustal levels are likely to be texturally identical to a lava, with phenocrysts (if present in the magma) suspended in a fine-grained glassy to aphanitic groundmass. Such dykes may also be vesicular resulting from volatile exsolution during magma ascent through the crust. If the magma is porphyritic and viscous, phenocrysts carried from the deeper source magma reservoir may occur right to the dyke margin but in some low-viscosity magmas, the largest phenocrysts may be worked into the centre of the dyke due to laminar shear flow effects at the dyke margin. The centre of dykes usually has coarser groundmass textures due to slower cooling rates allowing more advanced crystallisation than at the margins, particularly in dykes that are tens of cm to many m thick.

Dykes usually have fine chilled margins at the contact with the country rock (Figs. 8.11, 8.12b), which may be glassy or consist of an extremely finely crystallised groundmass, which becomes progressively coarser within the dyke away from the margin. In the Oslo Rift dykes, Poppe et al. (2020) noted that the chilled margins show similarities in Ti and Ni abundances but differences in other elements compared with the coarser crystalline cores of the dykes. In addition, there are element variations in the country rock at contacts with the chilled margins, suggesting element diffusion and exchange have occurred between the dyke margins and its core, as well as with the country rock through contact thermal and/or hydrothermal effects. In older dykes in the rock record, chilled margins and textures may be difficult to recognise because of alteration and metamorphism of the original glassy margins, and due to deformational strain along dyke margins.

Dykes may show a vertical flow banding (e.g. Fig. 8.9c, f, i), especially along the margins due to frictional coupling between the dyke margins and the country rock. Flow banding may be faint in low-viscosity magma dykes, perhaps just represented by planar trails of variably stretched vesicles, or alignment of elongate phenocrysts or groundmass microlites. Flow banding becomes more prominent in more viscous magma dykes, again being most prominent along dyke margins, although in narrow silicic dykes, the entire dyke width may be flow banded. In thick dykes, the

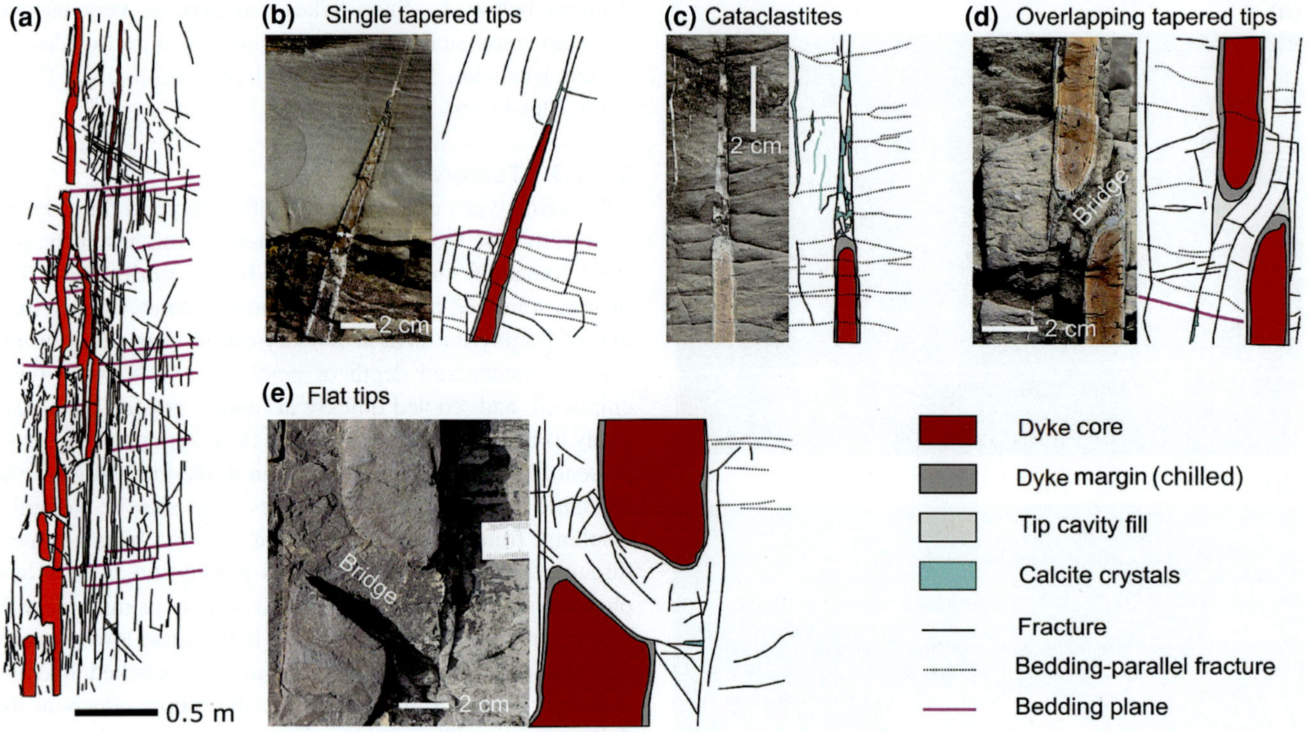

Fig. 8.11 Characteristics of thin compound dykes, from studies of Permian tephrite–basanite dykes in the Oslo Rift, Norway. **a** Field sketch of parallel and side-stepping dykes in red. Black lines represent structural foliation planes and purple lines represent sedimentary bedding planes. **b** Very thin tapered dyke tip with calcite filling the very tapered cavity at the tip. **c** Very thin tapered dyke tip with calcite filling cavities between fractured slivers of country rock. **d** Offset segments of dykes with tapered tips. **e** Offset segments of dykes with flat or stubby noses, separated by a bridge of country rock. (From Poppe et al. 2020)

Fig. 8.12 Schematic representation of propagation of tapered tip and stubby nose dykes. **a** Overview, and **b** details of propagating tapered dyke tip in opening mode deformation, showing monoformal warping or displacement of country rock (left) and fracturing of country rock around the propagating front of a stubby nose dyke in shear mode deformation. (Adapted from Poppe et al. 2020)

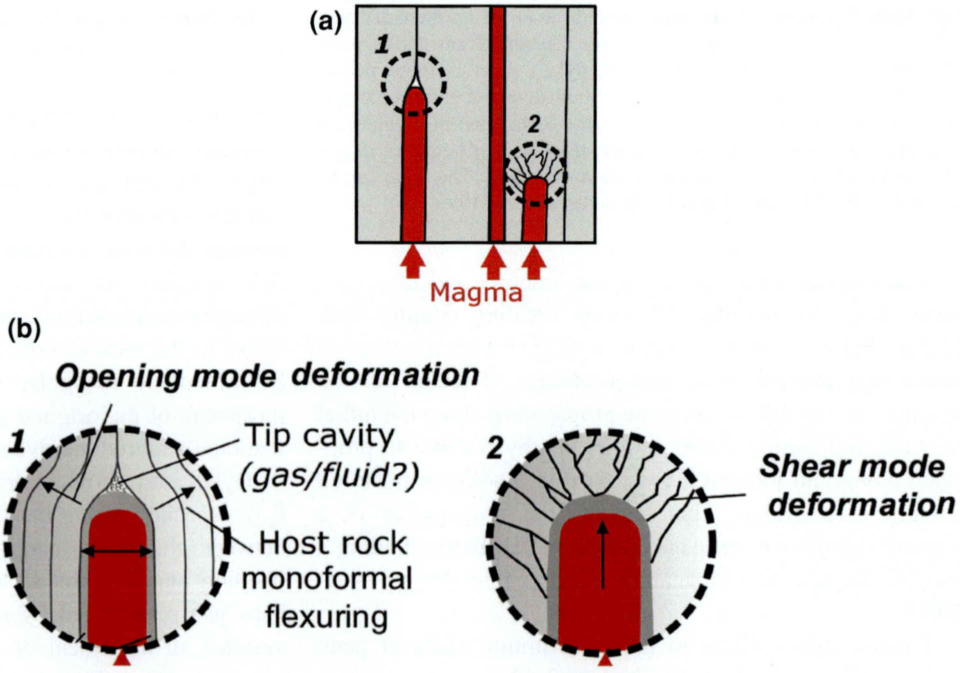

central part is commonly massive, and suggests that as in lavas, there is a central channel zone where the magma moves through the dyke at a uniform velocity, whereas along the margins, where shear flow coupling with country rock is greatest, a flow velocity gradient develops which then leads to differential flow between flow bands. In cross-section, dykes usually have a symmetrical textural and structural profile. In addition, in some viscous magma dykes autoclasts of the dyke may occur along the margin, representing localised autobrecciation, and many dykes will contain varying proportions of xenoliths, which could include deep mantle-derived xenoliths in mafic to ultramafic dykes, or crustal xenoliths resulting from the propagation of dykes through the crust. Some dykes can develop cooling columnar jointing that is perpendicular to the dyke margins (Fig. 8.9h).

Considering the vertical extent of a dyke, texture may be expected to vary with decreasing depth from the base (close to the source) upward (towards the surface), although very little data is available on this topic. Coarse to fine-grained holocrystalline textures are associated with deep-seated pegmatitic and aplitic dykes, granitic compositions and related to the migration of highly differentiated fluids enriched in incompatible elements, which may be related to certain types of ore deposition (e.g. tin–tungsten, lithium, rare earth minerals, and gemstones). As dykes are (almost) vertical in origin, the usual definition includes the concept that dykes generally show cross-cutting contact relationships with the host rocks. However, this may (albeit rarely) not be the case if a dyke intrudes a stratification plane of a deformed, upturned to vertical rock succession. Further complications may arise if the intruded succession is subsequently folded or tilted back to horizontal, in which case the contact and geometrical relationships may match those of a sill (see below). The definition of “dyke” thus should include both the geometry (W << L, H) and the contact relationships, as well as the reconstruction of the vertical magma flow paths with respect to the source.

8.3.1.4 Determining Flow Patterns and Directions in Dykes

Studies have been undertaken to assess flow direction patterns in magma in dykes measuring textural fabrics, such as aligned vesicles and phenocrysts, and other directional features (Coward 1980; Shelley 1985; Rickwood 1990; Harp and Valentine 2018). Paleomagnetism, specifically anisotropic magnetic susceptibility (AMS) analysis, has also been used to detect flow vectors resulting from alignment and even imbrication of very fine magnetic particles in the magma (e.g. Knight and Walker 1988; Herrero-Bervera et al. 2001; Callot et al. 2001; Geoffroy et al. 2002, 2007; Poland et al. 2004; Aubourg et al. 2002; Magee et al. 2012a). These AMS studies include those on basaltic shield volcano settings in Hawai'i (Knight and Walker 1988), continental rift-related settings in the North Atlantic Igneous Province (Herrero-Bervera et al. 2001; Callot et al. 2001; Geoffroy et al. 2007; Magee et al. 2012a), silicic stratovolcano settings in the USA (Poland et al. 2004), and submarine lava dome complexes in Italy (Aubourg et al. 2002) and Spain (Porreca et al. 2015). All show that where a magnetic imbrication fabric can be detected allowing absolute flow directions to be defined (usually along dyke margins), in most dykes, flow directions are more or less (sub)horizontal, indicating lateral propagation of dykes from the source magma body or conduit. Close to the main source, some flow directions may be sub-vertical and directed away from the source, and in some cases, a downward flow direction can be detected, consistent with waning magma supply and magma drawdown back into the source.

Harp and Valentine (2018) measured the orientations of aligned crystals in exposed radial dykes around the 33 Ma Summer Coon volcano, Colorado, USA, where they found two principal flow patterns. In dyke samples close to the conduit, a steep inward plunging crystal lineation was common, which they interpreted to represent upward and outward directed dyke propagation and magma flow, derived from the main deep magma source, whereas, in dykes in medial to distal settings, the crystal lineation often plunged downwards away from the centre of the volcano, which they interpreted as resulting from dykes that were propagating radially from the conduit high in the edifice into the flanks of the volcano. While imaging sub-vertical dykes has previously been extremely difficult, even in high-resolution seismic data, novel developments are now making it possible to map and quantify the emplacement kinematics of dyke swarms in 3D in some sedimentary basins; e.g. the newly discovered Exmouth dyke swarm on the North West Australia Shelf (Magee and Jackson 2020).

8.3.1.5 Clastic Dykes in Volcanic Settings

Although our focus of interest here is in subvolcanic igneous dykes, in basin settings, including volcanically active ones, *clastic dykes* can also occur (see Fig. 8.37a, c) which are terrigenous in componentry and texture. When sediment pore fluid pressure in a particular sediment horizon sandwiched between impermeable layers increases due to load compaction, seismicity, or due to heating when intruded by an igneous body, and it exceeds the tensile strength of the host succession and σ_3, it can propagate fractures upwards, accompanied by injection of the fluidised sediments (Giordano and Cas 2001; Ross and White 2005). Clastic dykes, pipes, and other fluidisation phenomena are now recognised to be common in volcanically active sedimentary basins spatially and causatively associated with the intrusion of saucer-shaped sills (e.g. Svensen et al. 2006; Blažić and Moreau 2017; see below), and other intrusions.

8.3.2 Sills

Sills are also tabular intrusions of magma, but the magma intrudes laterally, horizontally along horizontal planar conduits, and largely concordantly into a layered rock succession, which is bedding in a sedimentary succession (Plate 8.1b; Figs. 8.7a, b, 8.13a–i), or more irregular autoclastic horizons at contacts between lavas, or into pyroclastic and/or epiclastic deposits in volcanic successions (Figs. 8.1b, 8.18a–h). Sills must be fed by one or more (sub) vertical feeder dykes or conduits, which may interconnect multiple sills at different levels of the stratigraphy, forming a complex magma plumbing system (Figs. 8.7b, 8.13e). Sills differ from dykes in that their width and length are much larger than the vertical dimension (W, L >> V), giving them a sheet-like geometry. The dimensions of sills vary from cm thick and m long to hundreds of m thick, and several km to hundreds of km long. Both mafic and felsic sills can be of just local extent (<1 km to a few km), but whereas mafic sill complexes can extend hundreds of km, felsic sills usually seem to extend hundreds of m to a few km, and only rarely tens of km (Cruden et al. 2018).

In some cases, interconnected mafic sills and dykes can form parts of large igneous provinces (cLIPs), erupting flood basalt lavas (e.g. Jurassic Ferrar Large Igneous Province, Antarctica, Fig. 8.13a–e; Chap. 5). Such sill systems can extend >> 1,000 km laterally (Muirhead et al. 2012; Svensen et al. 2012; Magee et al. 2016; Table 8.3). This does not indicate the magma was necessarily transported such a distance because many extensive sills may be fed by regionally extensive dykes that trend parallel to the sill system (Muirhead et al. 2012). Large sill provinces are commonly associated with continental rifting and break-up, and the formation of sedimentary basins and generation of large, mafic melt volumes along volcanic passive margins (Chap. 17).

The sheet-like geometry of sills generally requires low-viscosity magmas that can propagate laterally along weaknesses in the crust (e.g. Fig. 8.14; Kavanagh et al. 2017). Sills are therefore commonly associated with mafic or ultramafic intrusions (e.g. especially laterally extensive sills as found within the early Jurassic (183 Ma) Ferrar Dolerite sill complex, Antarctica (Fig. 8.13a–e) and coeval Karoo Basin, South Africa (see Fig. 8.15b, c), which were emplaced during the initial break-up of the Gondwana supercontinent (Table 8.3). Reconstructions suggest that the Ferrar sill complex extends for ≥ 4000 km from the proposed site of a plume head source, rivalling that of the 3000 km long, Mackenzie dyke swarm in Canada (Ernst and Buchan 2001).

Felsic sills can be common in proximal volcanic settings, but they are limited in lateral extent (e.g. Orth and McPhie 2003). However, even silicic magmas can form extensive sills (e.g. Early Proterozoic Tennant Creek rhyodacitic porphyry, Northern Territory, Australia, McPhie 1993, >60 km along strike) if magma viscosity is low enough (e.g. 10^3–10^7 Pa s for felsic magmas at depth; Cruden and Weinberg 2018) and host rocks are penetrable (e.g. weakly lithified and/or water-saturated sedimentary successions). Sills which result from the successive intrusion of more than one kind of magma are termed composite sills (Daly 1905; Meade et al. 2009).

8.3.2.1 Sill Propagation

The most established mechanism for the intrusion of sills is that they propagate as narrow tips along tensile linear cracks and then inflate behind the tip of the crack; this requires the hydraulic uplift of the overlying rocks indicating that the magma fluid pressure exceeds the vertical lithostatic pressure component (Pollard 1973; Galland et al. 2018b; Fig. 8.14a, b). In some settings where sill complexes are individually or cumulatively hundreds of m thick, this has caused considerable vertical uplift or "jack-up" of the ground surface; for example, the Ferrar Dolerite sill complex which is cumulatively 1,200 m thick caused this much uplift of the host Beacon Supergroup sandstone succession (Fig. 8.13; Muirhead et al. 2012). Similarly, in the Jameson Land Basin, East Greenland, an equal ratio of volumetric expansion and uplift of the overburden has occurred, while the host rock succession remains remarkably undeformed (Eide et al. 2017). However, not all sill emplacement is necessarily accompanied by one-for-one uplift or updoming, which is dependent on the lithology and competence of the intruded host. Counteracting intrusive uplift, there could also be contemporaneous subsidence of the crust due to deflation of the deep magma reservoir as magma migrates upward. This in turn reflects the effects of both the depth of emplacement, degree of lithification (i.e. cementation), and amount of accommodating intra-formational deformation, sometimes termed *"forced folding"*, in the overlying succession (Table 8.3; Hansen and Cartwright 2006a; Magee et al. 2012b, 2017; Jackson et al. 2013; Omosanya et al. 2017; Eide et al. 2017).

Galland et al. (2018b) have summarised the various scenarios that allow magma sills to propagate laterally dependent on host rock succession properties, ranging from coherent rigid elastic host rock dilation along a fracture (Fig. 8.14a, b) to brittle to ductile host rock fracturing at the propagating dyke front (Fig. 8.14c), to fold deformation (Fig. 8.14d), to fluidisaton of cohesionless sedimentary hosts (Fig. 8.14e). Sills can be important components of the upper crust, especially where the lithostatic load is low (e.g. extensional settings), and in sedimentary basin successions where rising magmas exceed their level of neutral buoyancy within the sedimentary basin fill, and so seek to inject laterally, interstratally, rather than continuing upwards.

Neutral buoyancy may, however, not provide the complete explanation, and as pointed out by Thompson and Schofield (2008), for a sill to climb and then feed another shallower sill, it must imply that they are emplaced at a level

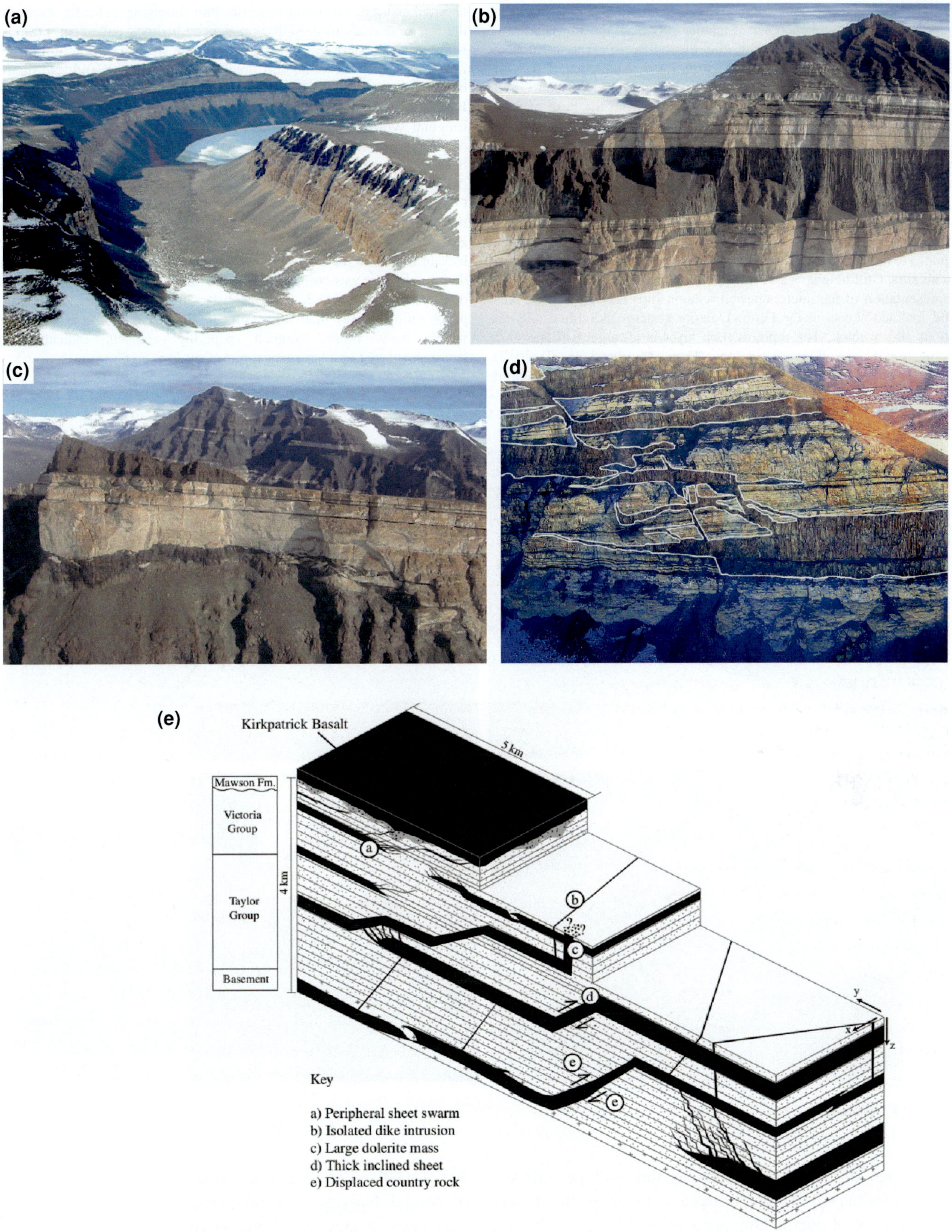
(a)
(b)
(c)
(d)
(e)
Kirkpatrick Basalt
5 km
Mawson Fm.
Victoria Group
Taylor Group
Basement
4 km
a
b
c
d
e
y
x
z
Key
a) Peripheral sheet swarm
b) Isolated dike intrusion
c) Large dolerite mass
d) Thick inclined sheet
e) Displaced country rock

◀ **Fig. 8.13** Geometry and contact features of sills. **a–c** Spectacular images taken from a helicopter of tabular sills from the Jurassic Ferrar Dolerite sill complex intruding the Devonian to Triassic Beacon Supergroup sandstones, northern Victoria Land, Antarctica. **a** Regionally extensive tabular dolerite sill (dark) intruded into Beacon Group sandstone succession. **b** Apophyses branching from the base of a sill (dark) into the underlying sandstones. Cliff height ~400 m. (Photo by Jean Bédard). **c** Irregular upper contact of a sill (dark) with apophyses branching from the top and spalling slabs of light-coloured sandstone which remain perched at the top of the sill. Cliff height ~400 m. (Photo by Jean Bédard). **d** Interconnected sills, dykes, and inclined sheets (darker brown) forming a complex magma plumbing network. Note the wedging geometry and tip of several sills, near vertical apophyses from the tops of sills, and the well-defined columnar jointing in all sills. Cliff height ~200 m. (Photo by James White). **e** Schematic representation of the interconnected relationships between sills, dykes, and inclined sheets in the Ferrar Dolerite system. Individual sills are about 300 m thick. The topmost dark layer is a co-genetic late-stage flood basalt lava flow succession. (From Muirhead et al. 2012). **f** Lobate apophysis from a ~2.7 Ga komatiite sill which has a spinifex textured margin, intruding upwards into overlying coherent dacite in the foreground, Gordon Sidhar breakaway bedrock pavement, north of Kalgoorlie, Western Australia. **g** Ropy flow structure at the top of the Late Carboniferous Whin Sill complex, near Holy Island, Northumberland coast, United Kingdom (*Source* British Geological Survey; From Stone et al. 2010). **h** Concordant intercalation between tabular to irregular syn-depositional komatiitic basalt sills (light) and black shale (dark), with apophyses marked by "A" intruding black shale above and below. The shales contain peperitic black shale–basalt clast mixtures (P). Younging (rotated "Y") is to the right, Dundonald Township, Ontario, Canada. **i** Sharp planar basal contact between a > 80 m thick rhyolite sill and black basinal mudstone, Late Devonian Bunga Beds, Boyd Volcanic Complex, southern New South Wales, Australia. The lack of evidence for soft sediment interaction (e.g. peperite), indicates that the rhyolite was emplaced after the mudstone had compacted, been dewatered and lithified. This post-depositional sill is more highly porphyritic (>30%) than other sills in the Boyd Volcanic Complex which have more irregular peperitic contacts indicating a syn-depositional emplacement timing (see Fig. 8.18b, c)

Fig. 8.13 (continued)

below that of neutral buoyancy as this will provide the impetus for the sill to climb to shallower stratigraphic levels. The presence of ductile horizons within sedimentary basins, especially overpressured shales and coaliferous beds, may permit sill formation below the level of neutral buoyancy (Fig. 8.14f; Thompson and Schofield 2008; Schofield et al. 2012a; Galland et al. 2018b). Emplacement below the level of neutral buoyancy would permit sill inflation, and the associated fracturing of the country rock would permit magma to exploit these pathways in order to ascend to shallower levels and feed shallower intrusions. The initial contact of the magma with shale may also result in the

Table 8.3 Mafic sill complexes and host sedimentary basins referred to in text (Adapted from Magee et al. 2016)

Location sill complex	Main data source	Age	Host rock lithology	Area ($km^2 \times 10^4$)	Volume ($km^3 \times 10^4$)	Volcano-tectonic setting	Magmatic event
Bass Basin, offshore southeastern Australia	Seismic	Miocene	Sandstone, shale, and coal	<2		Post-rifting hot spot	East Australia plume
Ceduna Sub-basin, offshore South Australia	Seismic	Eocene	Sandstone, shale, and coal	<2		Post-rifting hot spot	Australia–Antarctica conjugate margin
Jameson Land Basin, East Greenland	Field	Eocene	Sandstone and shale	1.8	1.8–2.7	Rifted passive margin	North Atlantic Igneous Province
Møre and Vøring Basins, offshore mid-Norway	Seismic	Paleocene–Eocene	Shale, basement	8	0.9–2.8		
Faroe-Shetland Basin, northern North Sea, UK	Seismic	Paleocene–Eocene	Shale	2.25			
North Rockall Basin, northern North Sea, UK	Seismic	Paleocene–Eocene	Shale				
North Carnarvon Basin, North West Shelf, Australia	Seismic	Late Jurassic–Early Cretaceous	Shale	7.5	0.5–1.5	Rifted passive margin	Break-up of India from Australia
Ferrar Sill complex, Antarctica	Field	Early Jurassic	Sandstone, shale, and coal	56	<12.5	Rifted passive margin	Break-up of Gondwanaland
Karoo Basin, South Africa	Field and mining operations	Early Jurassic	Sandstone, shale, and coal	300	>34		
Coal Measure basins, northern UK	Field and mining operations	Late Carboniferous	Sandstone, shale, and coal	[1]4	0.3–0.5	Continental rift	Extensional collapse of Variscan Orogen

[1]Includes estimate of offshore area and volume

volatilisation of the pore fluid, which dramatically increases the pore fluid pressure causing fluidisation, so reducing the effective stress and thus aiding further intrusion into the sediment and dynamic mixing (Fig. 8.14e). In addition, if a sill encounters a pre-existing fault/fracture, then this may be exploited as it is the mechanically easier option (Fig. 8.14f; Magee et al. 2013). Furthermore, once such pathways develop, continued sill inflation and thickening may be halted or at least retarded as the magma has alternative pathways to move along.

Sills may be subject to multiple phases or pulses of intrusion (Gudmundsson 2012) and may preserve internal contacts between different intrusive phases, including internal chilled margins (e.g. Fig. 8.23g). This may lead to sills with multiple magma compartments (Gudmundsson 2012).

8.3.2.2 Saucer-Shaped Sills

Although sills are conventionally depicted as tabular sheets (i.e. strata-concordant), and many are (as represented in Fig. 8.13), research in both young sedimentary basins such as in the northern North Sea and North Atlantic Igneous Province, and older basins such as the Permian–Jurassic Karoo Basin, South Africa, and the Carboniferous coal-bearing basins of northern England and Scotland has shown that many sills may be broadly radial, or bilaterally symmetrical in form and concave-up, or "*saucer-shaped*", and variably interconnected or "*transgressive*" (Fig. 8.15a–c). In large sill complexes, the transgressive parts of sills result in the interconnection of underlying and overlying sills and are known to transgress over several km of strata (e.g. the Vøring Basin, offshore mid-Norway; Figs. 8.15a, 8.16a). Hence, magma does not necessarily have to move along vertical feeder pipes or dikes, but can *"step up"* by moving along a series of horizontal and inclined pathways (or stairs) in interconnected sill-feeding-sill networks that are regionally extensive (Francis 1982; Francis and Walker 1987; Marsh, 2004; Cartwright and Hansen 2006; Muirhead et al. 2012; Magee et al. 2016; see Fig. 8.7b). The inclined pathways or feeders have been termed *inclined sheets*. Saucer-shaped sills have been identified by a combination of high-resolution, 3D seismic reflection imaging in young basins (Hald and Tegner 2000; Planke et al. 2005; Cartwright and Hansen 2006; Hansen 2006; Hansen and Cartwright 2006a, b; Magee et al. 2016), and mapping, drill core

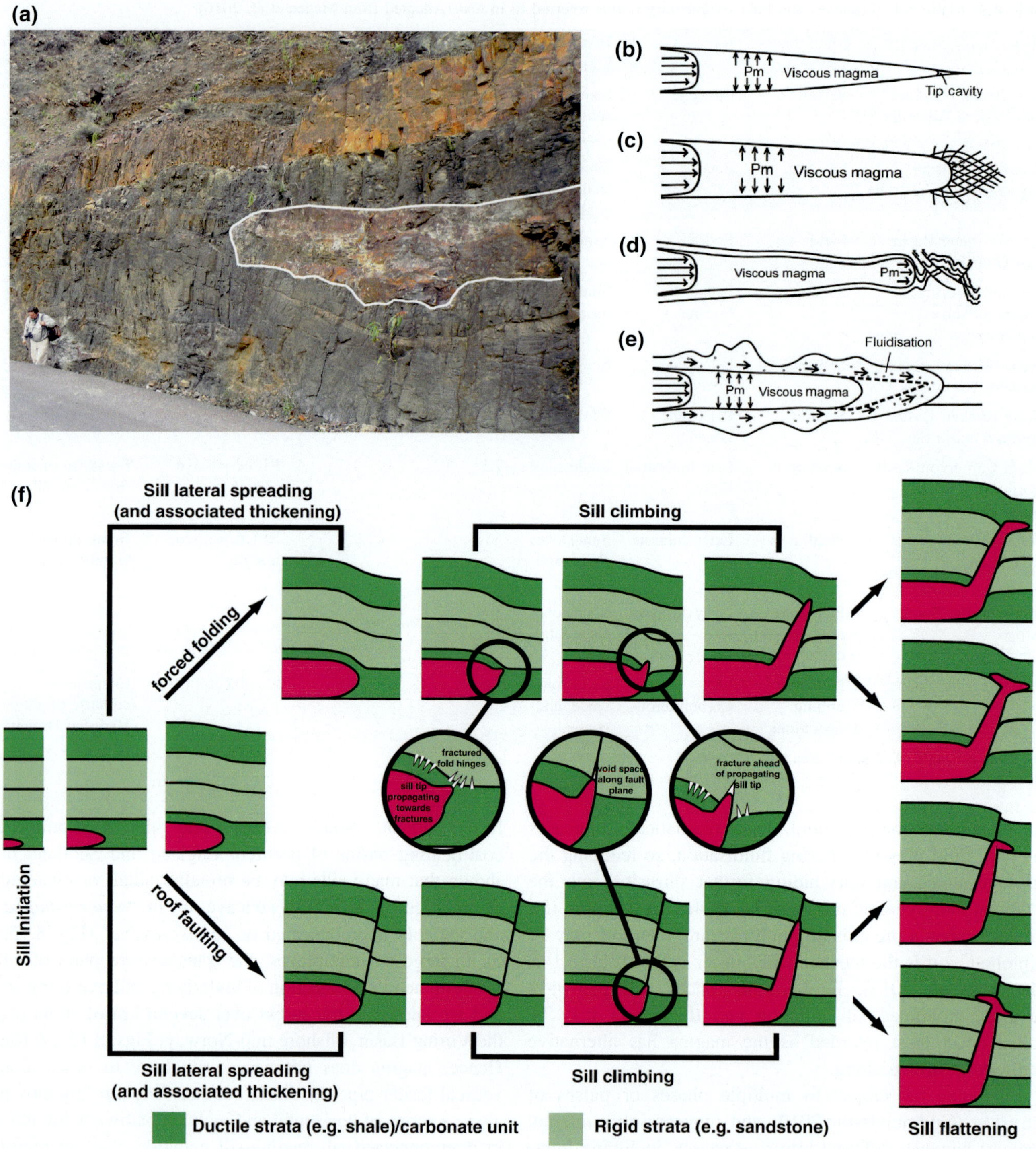
(a)
(b)
Pm
Viscous magma
Tip cavity
(c)
Pm
Viscous magma
(d)
Viscous magma
Pm
(e)
Fluidisation
Pm
Viscous magma
(f)
Sill lateral spreading
(and associated thickening)
Sill climbing
forced folding
fractured fold hinges
sill tip propagating towards fractures
void space along fault plane
fracture ahead of propagating sill tip
Sill Initiation
roof faulting
Sill lateral spreading
(and associated thickening)
Sill climbing
Ductile strata (e.g. shale)/carbonate unit
Rigid strata (e.g. sandstone)
Sill flattening

◀ **Fig. 8.14** Propagation and development of sills. **a** Thin sill (reddish) with a blunt tip, intruding sequence of basalt lavas, Reunion Island, France. Sill is emplaced along tensile parting within a single basalt flow. **b–e** Some existing sill propagation mechanisms (From Galland et al. 2018b). **b** Cross-section showing propagation of sill by tensile failure of the host rock as in (**a**) as a dyke propagates a narrow tip crack in country rock. Pm = magma fluid pressure which must be greater than the vertical lithostatic pressure. Note the sharp tip of the propagating sill and the presence of a tip cavity between the magma front and the sill tip (e.g. Pollard 1973). This cavity is expected to be filled with volatiles either exsolved from the magma or from the host rock. **c** Brittle to ductile fracturing of country rock at the propagating front of a sill. **d** Folding of ductile strata at the propagating front of a sill. **e** Thermal fluidisation model where sill propagation is triggered by rapid boiling of pore fluids as magma intrudes a layered sedimentary succession (e.g. Thomson and Schofield 2008; Schofield et al. 2012a). **f** Schematic model for sill development within a sedimentary basin. A sill initiates when magma encounters a ductile horizon such as shale. Lateral spreading and associated thickening occur. Contact of magma with shales results in lowering the effective stress within the horizon and with potential fluidisation permit spreading. As the intrusion expands and thickens, forced folding and/or roof faulting of the overburden occurs. Sill climbing is initiated at discrete localities along the sill periphery when magma makes contact with either open fractures associated with forced fold hinges or a fault. This forms an inclined sheet of magma. On reaching a shallower horizon suitable for intrusion, the sill will flatten. This may result in the development of a flat outer rim, or symmetrical intrusion about the steep feeder, and ultimately a new radially symmetrical saucer may develop (see Sect. 8.3.2.2). (From Thomson and Schofield 2008)

logging, and correlation in older basin systems (Fig. 8.15b, c; Francis 1982; Francis and Walker 1987; Chevallier and Woodford 1999; Polteau et al. 2008a, b; Svenson et al. 2012).

Although tabular sills appear to be emplaced at deeper, more lithified levels, saucer-shaped sills are generally found within sedimentary successions that are broadly contemporaneous with magmatism, and which are not strongly cemented (Table 8.4). Here, igneous rocks have greater densities and seismic velocities compared to the surrounding sedimentary strata resulting in a high acoustic impedance contrast, which are therefore generally very well imaged as high-amplitude reflections. Seismic evidence from the Norwegian Møre and Vøring Basins suggests the shallowest saucer-shaped sills are narrower, steeper-sided with a rough seismic character, whereas deeper saucer–sills are wider but again possess a rough seismic character (Planke et al. 2005). In contrast, the deepest sills (>3 km) are layer-parallel sills with a smoother seismic character and a large degree of continuity that also intrude pre-Cretaceous sedimentary and crystalline basements up to depths of 8–13 km. Similar observations are also found in the field in the Karoo Basin, where saucer-shaped sills are common at higher stratigraphic levels, while the deepest part of the basin fill is characterised by dominantly tabular or layer-parallel intrusions (Chevallier and Woodford 1999; Fig. 8.15b). Although some saucer-shaped sills are concordant within syn-formal strata sets, most saucer-shaped sills cross-cut bedding at their margins along their upward intruding inclined sheets (Fig. 8.15d–g), confirming their saucer shapes are clearly not the result of post emplacement deformation. In some cases saucer shapes could be related to syn-emplacement subsidence, and in older successions it is possible, if sills are thin, that saucer shapes may result from later, post-emplacement tectonic folding.

It is now recognised saucer-shaped sills are a fundamental morphological form and are most common in shallow-to-intermediate depths below the paleosurface in undeformed sedimentary basins (Table 8.4). Emplacement occurs generally shallower than ca. 3 km depending on the degree of lithification of the sedimentary host rocks. Conditions that promote the formation of saucer-shaped sills (Planke et al. 2005; Thomson and Schofield 2008; Polteau et al. 2008b; see Fig. 8.14f) include:

- intrusion of low-viscosity magma into undeformed sedimentary successions at crustal depths where the host rock has some elastic strength,
- a sill initiates when magma reaches neutral buoyancy, or before when it encounters a ductile horizon such as shale, and lateral spreading and thickening occur,
- strata above the sill are updomed as the intruding melt inflates causing forced folding or roof faulting,
- low tensile strength of host sedimentary succession facilitates hydraulic fracturing near the sill tip as the sill propagates (possibly aided by magma–pore water interaction and fluidisation at shallow levels) and intrudes horizontally,
- as the deformation and uplift of the overburden increases, the stress field at the sill tip changes from symmetrical (isotropic) to asymmetrical (anisotropic), causing the sill to branch upwards, and
- continued intrusion of melt forms the marginally inclined sheets that define the saucer–sill margins and overall saucer-shape morphology.

The saucer-shaped sill then comprises three components: the inner sill, the inclined sheet margins, which may interconnect stacked saucers, and if the sill margin then levels out and propagates horizontally again, an outer sill (Chevallier and Woodford 1999). Injection of a new batch of magma forms a new saucer-shaped sill and the cycle repeats itself adding to the construction of a larger sill complex (until the magma supply stops). Lastly, it is important to note sill geometries are also strongly influenced by structures and heterogeneities such as fault blocks, layering, and deformed strata.

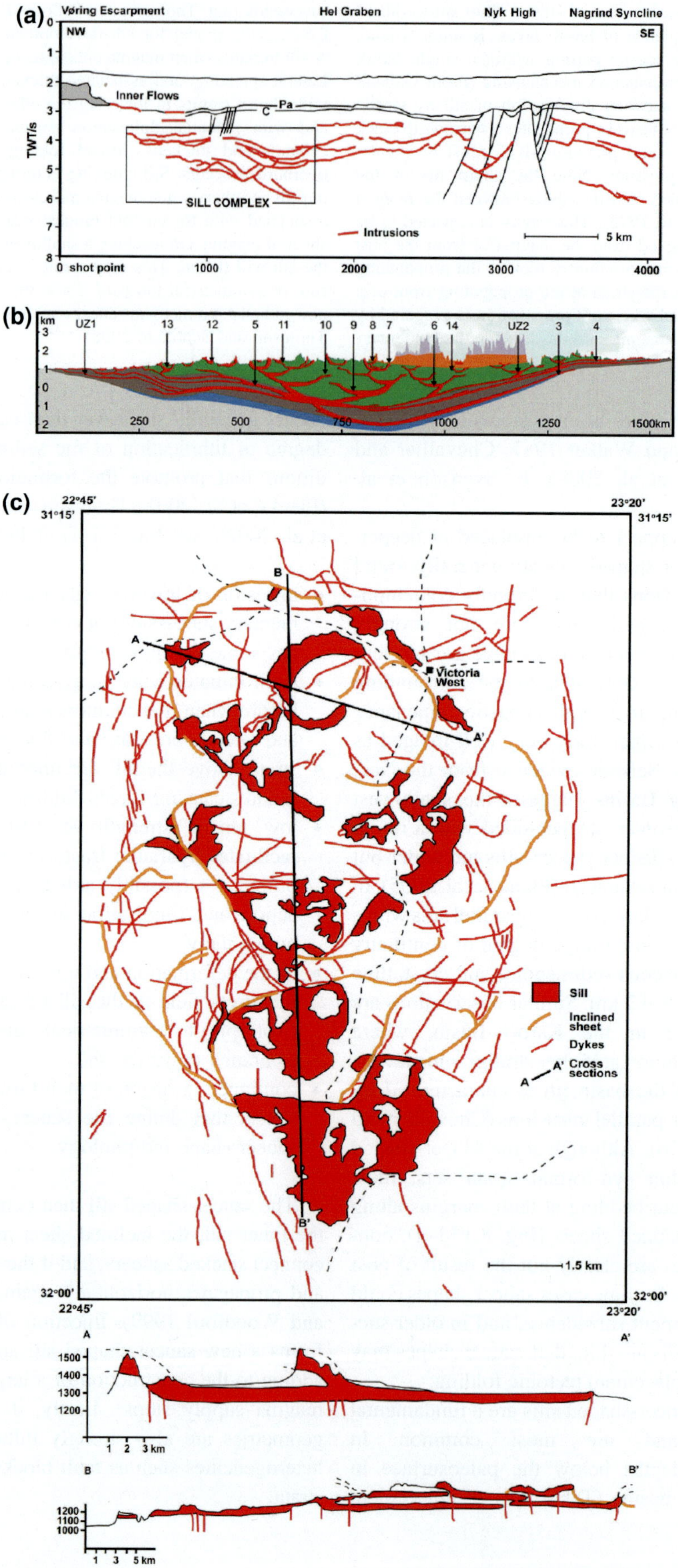
(a)
Vøring Escarpment
Hel Graben
Nyk High
Nagrind Syncline
NW
SE
TWT/s
Inner Flows
Pa
C
SILL COMPLEX
Intrusions
5 km
0 shot point
1000
2000
3000
4000
(b)
km
UZ1
13
12
5
11
10
9
8
7
6
14
UZ2
3
4
250
500
750
1000
1250
1500km
(c)
22°45'
31°15'
23°20'
31°15'
B
A
Victoria West
A'
Sill
Inclined sheet
Dykes
Cross sections
B'
1.5 km
32°00'
22°45'
32°00'
23°20'
A
A'
1500
1400
1300
1200
1 2 3 km
B
B'
1200
1100
1000
1 3 5 km

◀ **Fig. 8.15** Saucer-shaped sill systems in sedimentary basin successions. **a** Regionally extensive Tertiary sill complex, Vøring Basin, offshore mid-Norway. Line drawing of seismic reflection cross-section; C = top Cretaceous, Pa = top Paleocene. From Berndt et al. 2000). **b** Interpreted geometry of regionally extensive Jurassic sill complex (red), intruding late Carboniferous to early Jurassic sedimentary strata, Karoo Basin, South Africa. Numbered arrows identify sample locations that have provided high-resolution U–Pb zircon (and baddeleyite) ages of sills and dykes, some 1100 km across the basin. These give remarkably coherent ages ranging from 183.0 ± 0.5 to 182.3 ± 0.6 Ma, indicating that emplacement took place within an interval of about 0.5 Ma, which was coeval with eruption of the Ferrar Dolerite sill complex in Antarctica and initial break-up of the Gondwana supercontinent. (From Svensen et al. 2012). **c** Details of the architecture of the Victoria West saucer–sill complex in the Karoo Basin (From Chevallier and Woodford 1999). **d** Solsikke compound or multiple sill, Møre Basin, offshore mid-Norway, in interpreted seismic reflection cross-section. This shows a set of disconnected to semi-connected, saucer-shaped sills or sill segments. Although they appear to occur more or less at the same stratigraphic level, the sill complex cross-cuts bedding, as seen in their relationships to bedding reflection horizons B and C. Vertical scale ~700 m. (From Hansen and Cartwright 2006b). **e** Two seismic sections (lines X and Y) highlighting details of the Solsikke sill complex. These show cross-cutting bedding relationships and "capture" of faults by the sills, along which they intrude for some height. Vertical scale ~500 m. (From Hansen and Cartwright 2006b). **f** Two seismic sections (lines A and B) from the Judd Sub-basin, Faroe-Shetland Basin, northern North Sea. The lower part of both profiles consists of a number of generally concordant gently dipping sills. These sills are separated from an upper group, which are also broadly concordant and gently dipping, by a number of more steeply dipping, concave-upwards transgressive sills. The transgressive middle group of sills appears to originate from the concordant lower sills; note how the sills climb to the SE with "Sill B" making contact with the base of "Sill A" in order to feed the shallower intrusion. (From Thompson and Schofield 2008). **g** Saucer-shaped sill observed in seismic data from the Exmouth Sub-basin, North West Shelf, Australia (From Magee et al. 2016)

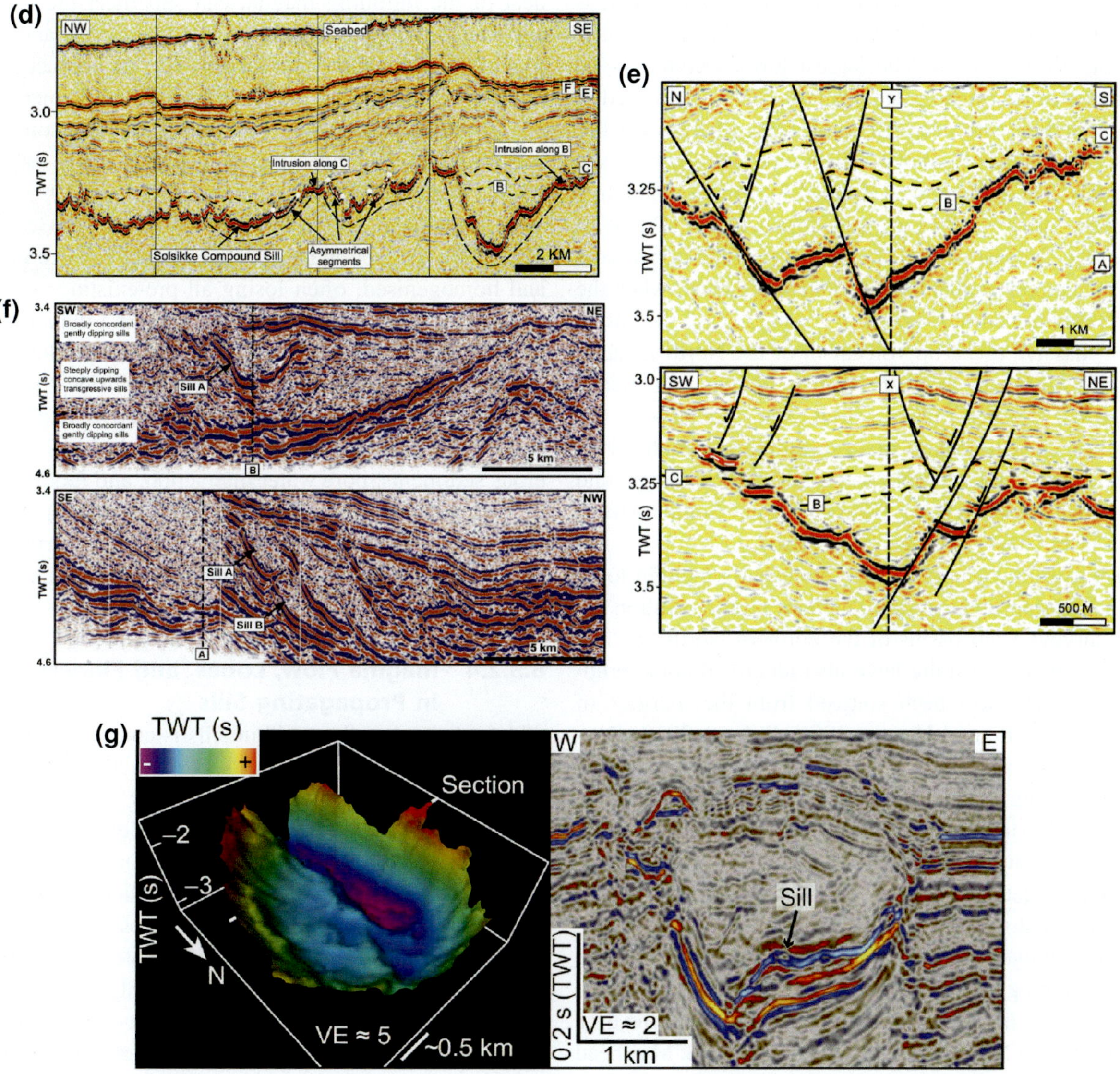

Fig. 8.15 (continued)

Detailed 3D seismic reflection surveys particularly in basins of the northeastern North Atlantic margin have shown that hydrothermal vents and volcano vent complexes are common features associated with the upturned margins or tips of saucer-shaped sills (e.g. Planke et al. 2005; Grove 2013; Galland et al. 2018b; Omosanya et al. 2018; Fig. 8.16a–c). These are preferentially emplaced along favourable stratigraphic levels such as organic-rich shales, where heat provided by the magma triggers rapid degassing of the sedimentary host rock formations, fluidisation, and the explosive release of CO_2 and methane into the ocean and atmosphere (Svensen et al. 2004, 2007; also see Fig. 8.7b). In the Møre and Vøring Basins, such vent complexes are regionally extensive in the subsurface, where over 700 have been identified (Fig. 8.16a; Planke et al. 2005). These are mainly located in the central and outer parts of both basins. In general, small and intermediate vent complexes (<2–5 km wide) dominate the outer parts of the basins, whereas large vent complexes (>5 km wide) are found near the inner part of the Vøring Basin. The largest formed craters ≥ 10 km diameter occur on the paleo-seafloor; these are located in regions underlain with deep sill complexes, whereas smaller features are located above shallower sill complexes (Fig. 8.16a). These have been divided into three distinctive geometrical forms: crater-, eye-, or dome-shaped (Fig. 8.16 b, c; see also Omosanya et al. 2018). Most of the upper parts of the vents are located at the Top Paleocene level, when the intrusive event occurred just prior to, or during, the initial phase of massive break-up volcanism of the North Atlantic at 55.0–55.8 Ma. In ancient basinal settings, analogous hydrothermal vents have been be identified by inward-dipping sedimentary strata surrounding a central vent complex comprising multiple sandstone dykes, pipes, and hydrothermal breccias (e.g. Karoo Basin, South Africa; Jamtveit et al. 2004; Svensen et al. 2006).

Where as true volcanic edifices are usually thought to be fed by a subjacent dyke system, modern high-resolution 3D seismic surveys from basins of the northern North Sea and offshore southern Australia have also identified some eruptive centres that have been sourced from the margins of saucer-shaped sills (e.g. Hansen 2006; Jackson 2012; Reynolds et al. 2018; Fig. 8.17a–d), which in some cases have propagated along basin faults (Fig. 8.17d). This serves to confirm legacy field mapping in the Carboniferous Coal Measure basins in the northern UK, which showed a close correlation of vent structures and "necks" with the margins of sills and basin structures (e.g. Midland Valley sill complex; Francis and Hopgood 1970). The significance of sill and dyke magma plumbing networks as feeders to subaerial monogenetic volcanic fields has also been proposed for some time (e.g. Németh and Martin 2007; Keating et al. 2008; Muirhead et al. 2016; Bishoff et al. 2019a, b). It is therefore quite probable that extremely complex lithological and facies relations could result from the interplay between hydrothermal venting and the development of sub-vent sill/volcanic complexes which are now being revealed in extensional sedimentary basins in various volcano-tectonic settings (e.g. Blažić and Moreau 2017; Medialdea et al. 2017).

8.3.2.3 Sill–Sediment Interaction

Sills intruded into water-saturated and unconsolidated sediments usually have irregular geometries and contacts (Fig. 8.18a, b, and see Fig. 8.1b), because such sediments lack mechanical strength and can be deformed during an intrusion. The magma pressure is able to push magma irregularly into the sediments, forming irregular margins and apophyses at both top and bottom contacts, including apophyses intruding upwards into overlying sediments (Figs. 8.13g, 8.18a, c). In addition, if the sediments are water-saturated, magma can be quench fragmented, and the heat of the intruding magma can superheat the pore water and cause dynamic magma–water interaction and fluidisation of the sedimentary succession near the contact forming a mixture of sediment and hyaloclastite fragments called *peperite* (Fig. 8.18a, c–h; Kokelaar 1982; Hanson and Schweickert 1982; Francis and Walker 1987; Cas and Wright 1987; Hanson 1991; Hanson and Wilson 1993; McPhie 1993; Rawlings 1993; Orth and McPhie 2003; Chap. 4; Fig. 8.1b). The sediment near the contact can become mixed and homogenised, often losing all pre-existing sedimentary bedding and structures (Fig. 8.18a, e), and dynamic mixing of sediment, and hyaloclast fragments can produce peperite breccias around the margins and even inside the intrusion (Fig. 8.18a–h; Chap. 4). As indicated above, exploitation of ductile horizons such as overpressurised shales and coaliferous sediments, pore water interaction, and fluidisation may be an important mechanism in sill propagation (e.g. Figs. 8.14e, 8.21c), especially at shallower crustal levels ($<\sim 2$ km; Walker and Francis 1987; Schofield et al. 2012a).

8.3.2.4 Magma Flow, Lobes, and Fingers in Propagating Sills

Although magma flow within sills is commonly depicted as sheet-like, in some mafic sills at least, there appear to be preferred pathways or lobes and tubes (e.g. Polteau et al. 2008a; Schofield et al. 2012a; Magee et al. 2012b, 2016; Roni et al. 2014; Galland et al. 2019), and some mafic sills may propagate as a series of sub-parallel magma lobes or fingers through the sedimentary succession. Examples of saucer-shaped sills with complex, lobate architecture have been dramatically revealed through highly detailed analysis, interpretation, and enhancement of high-resolution 3D seismic survey data sets in young sedimentary basins, from which flow paths can be reconstructed (Fig. 8.19a–c; Thomson and Hutton 2004; Hansen and Cartwright 2006b;

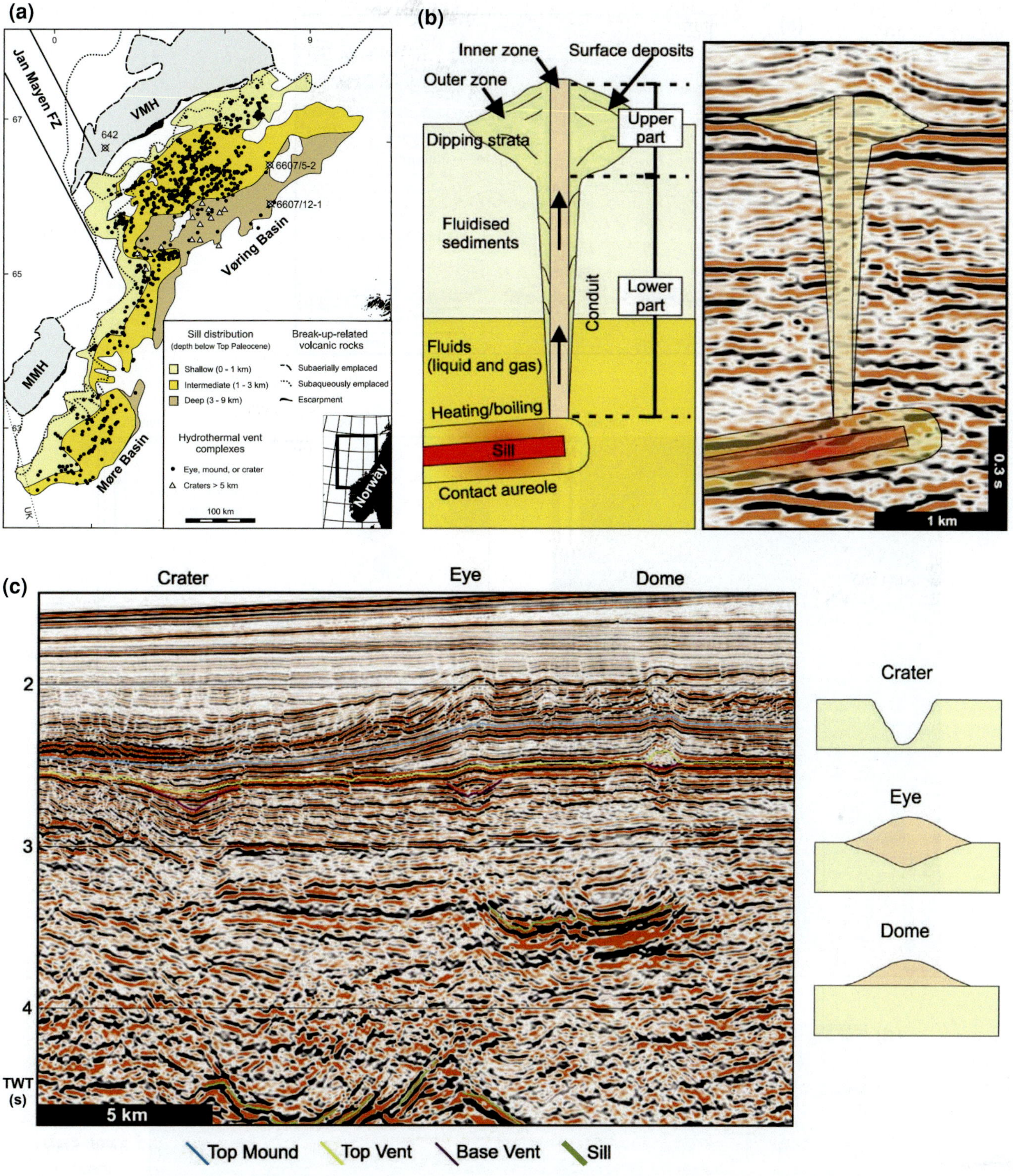

Fig. 8.16 Distribution of sill intrusions and location of subsurface hydrothermal vent complexes in the Vøring and Møre Basins, offshore mid-Norway, northeastern North Atlantic margin. **a** Map showing the extent of the sill complex which is at least 80,000 km^2, with an estimated total volume of 0.9–2.8 $\times$ 10^4 km^3. The sheet intrusions are saucer-shaped. Over 700 hydrothermal vent complexes have been identified, although it is estimated that up to some 3,000 vent complexes may be present in both basins. MMH = Møre Marginal High, VMH = Vøring Marginal High. **b** Interpretive and seismic cross-section of a dome-shaped hydrothermal vent complex. The complex consists of an upper and a lower part. The upper part can be crater-, eye-, or dome-shaped and is connected to the termination of a sill by a cylindrical or pipe-like conduit with disturbed seismic data in the lower part. **c** Seismic profile from the SE Vøring Basin showing the relationships between sill complexes, hydrothermal vent complexes, and mounds; Top Vent = Top Paleocene (55.0–55.8 Ma)**.** Mounds and associated seismic seep anomalies are present in the sediments overlying some vent complexes, suggesting these may have been re-used as pathways for vertical fluid migration (including hydrocarbons) long after their formation during the intrusive event at the onset of North Atlantic break-up. (Modified from Planke et al. 2005)

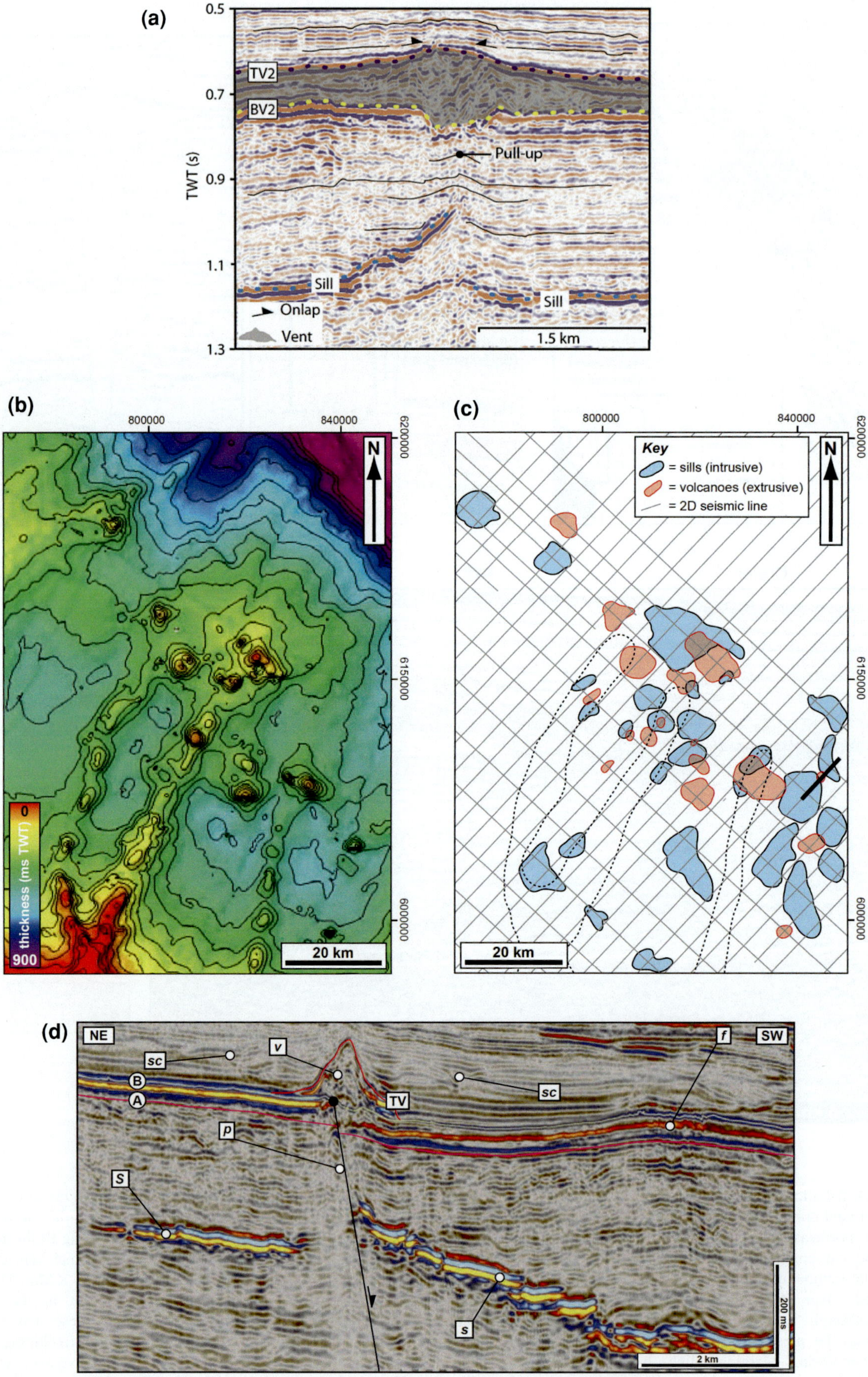
(a)
0.5
0.7
0.9
1.1
1.3
TWT (s)
TV2
BV2
Pull-up
Sill
Sill
Onlap
Vent
1.5 km
(b)
800000
840000
6200000
6150000
6000000
N
0
thickness (ms TWT)
900
20 km
(c)
800000
840000
6200000
6150000
6000000
Key
= sills (intrusive)
= volcanoes (extrusive)
= 2D seismic line
N
20 km
(d)
NE
SW
sc
v
f
B
A
TV
sc
p
S
s
200 ms
2 km

◀ **Fig. 8.17** Volcanic edifices associated with and fed by the upturned margins of sills. **a** Miocene volcano and diatreme (seismic pull-up zone) interpreted to have formed above the edge of a saucer-shaped sill, offshore Bass Basin, southeast Australia. TV2 = top volcanic, BV2 = base volcanic. (From Reynolds et al. 2018). **b** Isochron map showing the interpreted location of subsurface mid-Eocene submarine volcanoes imaged in 3D seismic surveys, Ceduna Sub-basin, Great Australian Bight, Australia. **c** Interpretation of the spatial relationship between volcanoes and subjacent sills imaged in (**b**); dashed lines mark the position of present-day seabed canyons, and thick line the seismic profile shown in (**d**). **d** Example of volcano located above the upturned edge of a sill which terminates against a normal fault; the vent overlies the upper tip of this fault implying the fault may also have been a magma conduit or pipe. ATV = top vent isochron, A = top Cretaceous, B = top mid Eocene; s = sill, v = vent, f = forced fold, p = pipe, sc = scour. (From Jackson 2012)

Magee et al. 2016). Such architecture is reminiscent of lobate subaerial pahoehoe lavas (Chap. 5, Sect. 5.4), as highlighted by Hansen and Cartwright (2006b), or lobate submarine basaltic lavas observed on the modern ocean floor (Chap. 5, Sects. 5.10, 5.11). Multiple lobe geometry suggests the sills grow by the incremental emplacement of discrete magma injections. As magma lobes propagate ahead of the main sill body, they expand laterally as they move downstream and come to overlap vertically along their lateral margins. Thus, a network of lobes exploiting a variety of weak horizontal strata which, with sustained magma input, inflate and progressively coalesce allowing the main sill body to move downstream as a continuous sheet intrusion. The orientation of the axes of individual lobes indicates the local propagation direction of each lobe or finger, which for a spreading sill may thus define a more or less radial or fanning pattern.

In addition to the recognition of lobate sill margins in young poorly to unconsolidated sedimentary basin successions, Galland et al. (2019) found similar structures in Miocene sills that had intruded mechanically weak, Late Cretaceous carbonaceous shales in the Mendoza Basin, Argentina. They found that individual adjacent lobes may inject the host sedimentary succession at slightly different stratal levels, creating a stepped architecture. Lobes may inflate upwards or downwards, have deformed sediments caught and squeezed between adjacent expanding lobes, and may develop surface structures such as slickensides, which indicate local lateral expansion, inflation, and propagation directions that may be different from the orientation of lobe axes, which record the overall lobe propagation directions.

Mapping the long axes of lobes as flow direction indicators, including lobe–lobe contacts, therefore, provides insights into magma flow patterns within individual intrusions and allows magma input zones to be inferred (e.g. Fig. 8.19c). Other seismic-to-outcrop-scale magma flow indicators include *intrusive steps, bridge structures,* and *magma fingers*; in seismic data, these manifest themselves as minor vertical offsets and corresponding amplitude variations (Fig. 8.19c; Schofield et al. 2012a, 2018; Magee et al. 2016; Westerman et al. 2018a, b).

Step features in sills result from the development of slightly offset or *en echelon* segmented intrusion tips (Fig. 8.20a; Schofield et al. 2012a; Magee et al. 2016). Thus, segments are parallel to the intrusion direction, which in cross-section results in a stepped appearance that increases in the amount of offset in the direction of magma flow. The increasing frequency of steps away from the magma source suggests they develop as magma supply wanes and the sill must exploit preferential horizons to continue to propagate. In seismic data, such step-like features are predominantly observed on the outer rim of saucer-shaped intrusions, creating their rough seismic character (Sect. 8.3.2.2). Bridge and "broken bridge" structures develop when two or more overlapping lobes of magma contemporaneously propagate, on vertically offset horizons, creating a "bridge" of sedimentary host rock between them (Fig. 8.20b; also see Fig. 18.13d; Hutton 2009, Schofield et al. 2012b). With magma propagation, vertical inflation of each of the magma lobes will eventually cause the bridge to undergo brittle failure, forming a "broken bridge" represented by large xenoclasts or rafts of sediment contained within the igneous rock. The sill is characterised by separate distinct magma lobes in the areas most distal to the magma source, but proximal to the source the lobes coalesce into a single but stepped sheet (Fig. 8.20 b). In contrast to intrusive steps and bridge structures which develop via brittle fracturing, magma fingers form through the non-brittle deformation of the host sediment (Table 8.4). Magma fingers propagate at the margins of sills or inclined sheets where they intrude ductile host rocks (Fig. 8.20c; Schofield et al. 2012a, b; Magee et al. 2016). They have a finger-like morphology and elliptical cross-sections, and appear to develop as magma is intruded into poorly consolidated or unlithified sediment, which is mobilised by fluidisation processes associated with magma–pore water interaction. If magma supply continues, the fingers may eventually coalesce to form a continuous sheet.

Apophyses in a traditional sense have been regarded simply as discordant offshoots from a sill (or other intrusive body). On a small (cm to m) scale, they may form a network of branching "magma veins, veinlets" or dykelets of igneous rock along the lower and/or upper margins of thicker intrusions. On a larger scale, they may change direction because of bending and thinning from a horizontal concordant sheet at the top to a discordant inclined form that bends again to pass into a lower horizontal concordant sheet, or vice versa (e.g. Fig. 8.13). However, from the previous discussion, it is now quite apparent that larger scale apophyses (a.k.a. *splays*, *leaves*) likely form an integral part of the flow and emplacement mechanism of magma lobes

(a)
(b)
(c)
(d)
(e)
(f)

◀ **Fig. 8.18** Examples of syn-depositional sills into wet, unconsolidated sediments, showing irregular geometry and contact relationships. **a** Eocene–Oligocene basalt sill into bioclastic calcarenite limestone, Airey's Inlet, Otway Basin, Victoria, Australia. The top of the sill is highly irregular, with lobate, irregular apophyses intruding upwards into the limestone, with local preservation of peperitic hyaloclastite breccia at the top contact and also in seams within the interior of the sill. The latter indicate that water-saturated bioclastic sediment was injected along fractures into the hot interior of the sill, triggering quench fragmentation in its interior. **b–d** are from the Late Devonian Bunga Beds, Boyd Volcanic Complex, southern New South Wales, Australia. **b** Rhyolite sill with highly irregular margins intruded into deeper-water, water-saturated black mudstone. The entire thickness of the sill has been quench fragmented, hence the hyaloclastite breccia texture. The enclosing mudstone also contains pockets of rhyolite-mudstone peperite breccia. **c** Lobate apophysis from the top of a syn-depositional rhyolite sill intrusion (dark) intruding up into contorted bedded mudstone with peperite breccia (rhyolite clasts in mudstone matrix) in the lower half of the photo. **d** Peperite of faintly flow banded rhyolite blocks with a matrix of black mudstone. Some clusters of blocks are in almost jigsaw-fit relationship, while others are separated and slightly rotated. **e** Peperite breccia from Mesoproterozoic Gold Creek Volcanics, southern McArthur Basin, northwest Queensland, Australia. The peperites comprise dark brown, altered vesicular basalt clasts in a pale fine-grained sandstone matrix, and occur at the margins of coherent dolerite sills which were historically mapped as lavas (cf. Rawlings 1993). **f** Coherent sill (top) to splintery quench fragmented, peperitic vesicular basaltic breccia in drill core through the Gold Creek Volcanics. **g** Irregular top of 2.68 Ga Golden Mile Dolerite sill intruded into black basinal mudstones and grey siltstones. The top consists of several apophyses and peperitic domains. Diamond drill hole SE18W, at 1,874–1,884 m depth. **h** Irregular base and peperitic domain at the base of the Golden Mile Dolerite sill. Diamond drill hole SE18, at 1,460 m depth. All drill core is ~5 cm wide

Fig. 8.18 (continued)

into the host rock. Although apophyses could have several origins, in some cases they may represent embryonic magma lobes that did not inflate or coalesce (e.g. due to a local cutoff or withdrawal of magma supply). For example, as magma lobes attempt to seek out a mechanically favourable horizon, they may be propagating and splitting perhaps over several separate layers. Once an advantageous layer is encountered, this may focus magma and become the dominant flow pathway and inflate, so starving much of the subsequent magma supply from other lobes. The final geometry therefore becomes one thick sill surrounded by several shorter, thinner apophyses (Fig. 8.13). In other cases, apophyses may represent magma fingers.

8.3.2.5 Determining Magma Flow Patterns in Sills

Magma flow patterns in sills have also been analysed using anisotropic magnetic susceptibility (AMS) methods to detect flow vectors resulting from alignment and even imbrication of very fine magnetic particles in the magma (e.g. Dragoni et al. 1997; Ferré et al. 2002; Polteau et al. 2008a; Magee et al. 2012b; Roni et al. 2014; Westerman et al. 2018b). This

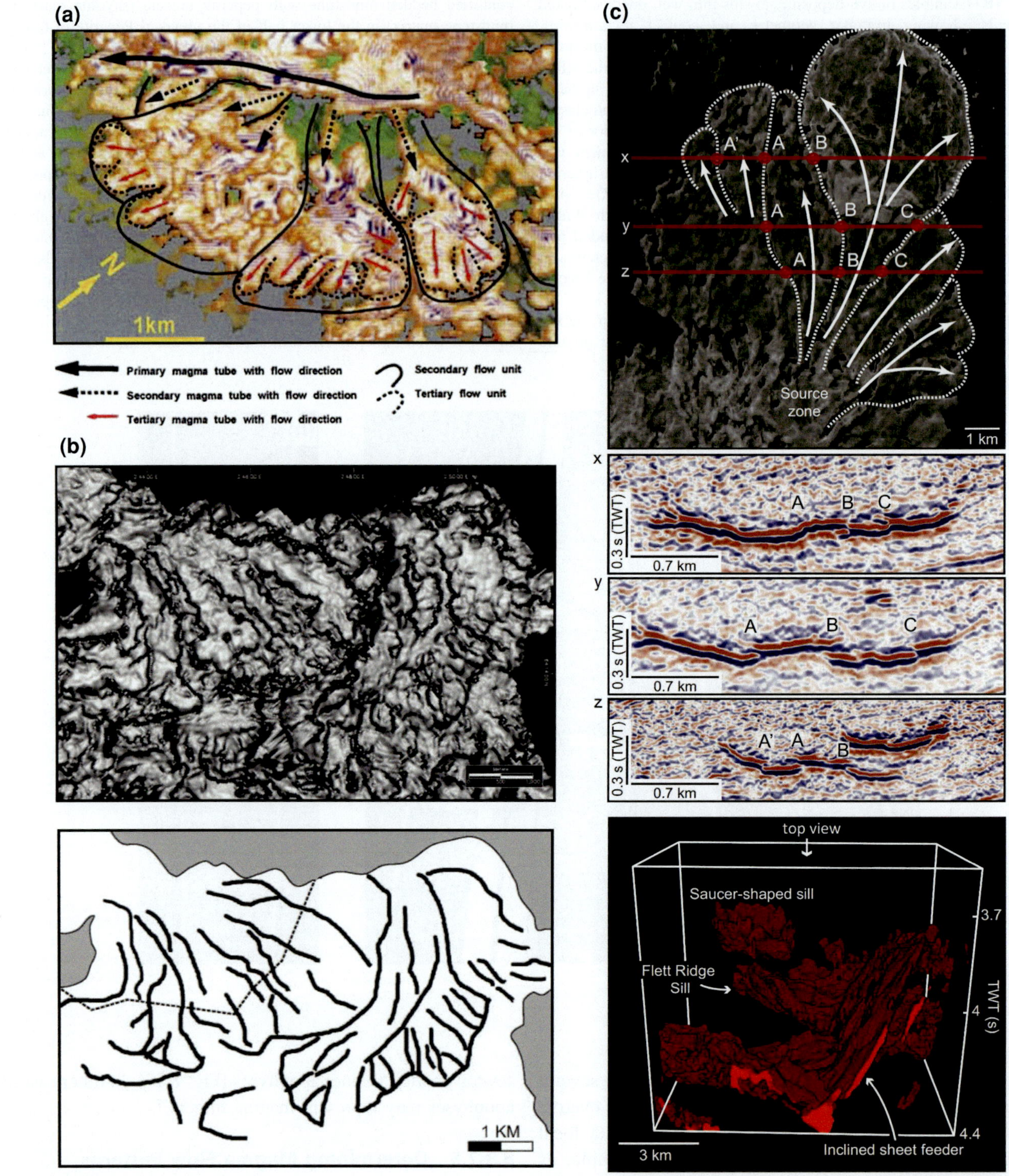

Fig. 8.19 Enhanced 3D seismic renditions of early Tertiary sills in northern North Sea basins of the North Atlantic Igneous Province. These show clear lobate architecture, and interpreted flow or lobe propagation directions. **a** Part of an early Tertiary lobate sill, North Rockall Basin, northern North Sea, UK (From Thomson and Hutton 2004). **b** Part of the Solsikke compound sill complex, Møre Basin, showing enhanced image (top), and interpreted flow lobe patterns (below) (From Hansen and Cartwright 2006b). **c** Plan view of lobate nature of Flett Ridge Sill and corresponding seismic sections documenting intrusive step and broken bridge growth along lobe–lobe contacts, Flett Sub-basin, Faroe-Shetland Basin, northern North Sea, UK. A, A’, B, and C correspond to intrusive steps or bridge structures observed in profiles (x), (y), and (z). Bottom is 3D interpretation of the Flett Ridge Sill showing the magma lobes imaged in the top (plan) view trace back to, and were likely fed from, a fault-hosted inclined sheet. (From Magee et al. 2016, After Schofield et al. 2012b)

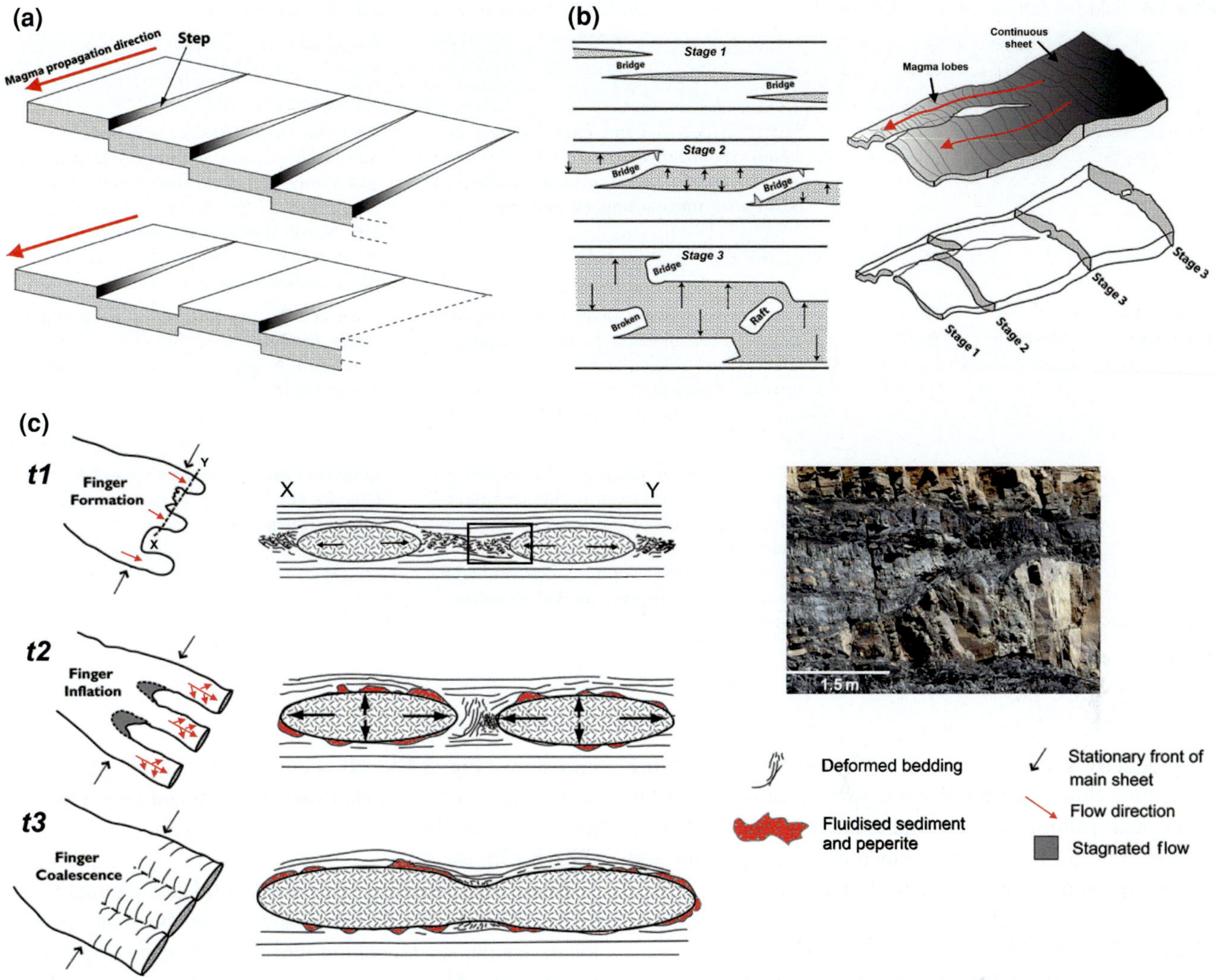

Fig. 8.20 Other seismic- to outcrop-scale magma flow indicators in sills. **a** Schematic diagram showing the development of en echelon steps within a sill, highlighting the increase in offset between sill segments in a downflow direction. Top, steps all trend in one direction. Bottom, steps trend in different directions, as may be expected if a sill is seeking out preferential horizons for intrusion. **b** Left, development of a broken bridge and its relationship to an intruding body. *Stage 1*, magma propagates as a series of offset but overlapping bodies (magma flow out of page towards reader). *Stage 2*, as the magmatic bodies begin to inflate, the bridge of strata sandwiched between two sill segments starts to bend and deform. In zones where flexure of the bridge is at its maximum, tensile fractures form. *Stage 3*, if inflation continues, one of the cross-fractures will eventually connect causing failure in the bridge, leading to a broken bridge. If failure occurs at both ends of the bridge, a raft of host rock strata can become fully detached. Right, 3D reconstruction showing the various stages of broken bridge development during sill emplacement. **c** Evolution of magma fingers; oblique view (left) and cross-section (right). At *t1*, instigation of ductile deformation and fluidisation of the surrounding country rock; normal hydrofracture gives way to a non-brittle relationship between the magma and host rock, leading to the formation of an instability and magma fingers. Photo inset shows field outcrop of magma finger surrounded by convoluted and fluidised coaliferous sediments, including peperitic fragments of dolerite, mid-Cenozoic basin, Colorado, USA. If magma flow rate remains high enough to prevent choking of the propagating fingers, at *t2* they will continue to inflate both vertically and laterally (arrows). If magma supply continues, the fingers will eventually coalesce to form a continuous sheet (*t3*). Propagation of magma can stall at any time, and will reach the final stage (i.e. a coalesced sheet) only if magma run-through rate is high enough to prevent solidification of fingers. Even after the fingers have coalesced, they can remain open and act as conduits feeding magma to other sills. (Modified from Schofield et al. 2012a, based in part after Rickwood 1990, Pollard et al. 1975, Hutton 2009. Photo inset from Magee et al. 2016)

Table 8.4 Selected features of mafic sill complexes observed in sedimentary basins relative to depth of emplacement

Depth below paleosurface	Host rock cementation	Host rock competence	Sill-related features	Examples[1]
Shallow ($\leq$ 1.5 km)	Non-Early	Ductile	Saucer-shaped sill complexes; inclined sheets feeding monogenetic cones and hydrothermal vents; accommodation or forced folding of host rocks; magma fingers; peperites	Bass and Otway Basins, and Ceduna Sub-basin, offshore southern Australia; Møre and Vøring Basins, offshore mid-Norway; North Rockall Basin, northern North Sea, UK; Karoo Basin, South Africa; Coal Measure basins, northern UK; McArthur Basin, northeastern Australia
Intermediate (1.5-3 km)	Early –Moderate	Ductile–Brittle	Transgressive saucer-shaped sill complexes, magma lobes and flow patterns; magma fingers; step, bridge, and broken bridge structures; peperites; hydrothermal vents; normal fault influence; forced folding of host rocks	Møre and Vøring Basins; Faroe-Shetland Basin, northern North Sea, UK; Exmouth Sub-basin, North West Australian Shelf; Karoo Basin
Deep ($\geq$ 3 km)	Extensive	Brittle	Tabular or planar sheet-like sill geometries; step, bridge, and broken bridge structures; vertical 1:1 uplift of host rock overburden; limited host rock deformation; contact metamorphic and melting effects; intrusion pre-basinal sedimentary and/or crystalline basement	Jameson Land Basin, East Greenland; Ferrar Dolerite sill complex, Antarctica; Møre and Vøring Basins

[1]Referred to or illustrated in text

has produced some fascinating reconstructions involving propagating flow and magma withdrawal or drain-back flow phases and patterns in the interconnected sill and dyke systems (Fig. 8.21a–d), and branching and upwards and downwards motions in plan view (Fig. 8.21e).

8.3.2.6 Thin to Thick, Large Mafic Sills: The Origins of Internal Compositional and Textural Architecture

Mafic sills vary from thin, only cm thick to tens to hundreds of m thick, with the thickest being layered mafic intrusions (Sect. 8.3.4). In spite of the low viscosity of mafic magma, mafic sills can inflate to hundreds of m thick because of the mechanical constraint of the country rock. Sills that are several metres or more thick develop relatively coarse groundmass textures because they cool and crystalise slowly and are commonly known as dolerite (diabase) sills. However, very thick mafic sills (>80 m) can even develop gabbroic and more felsic differentiated interiors. Sills intruded into lithified country rock have planar concordant to slightly discordant contacts (e.g. Figs. 8.13c–e, 8.14a) unless they branch of step up or down (Fig. 8.13d, e). Country rock inclusions ranging from cm size to house size in large intrusions (Fig. 8.13c, d) occur along margins. Apophyses branching up or down also vary in width and length and also tend to be rectilinear in geometry in lithified country rocks (Fig. 8.13b, e).

Mafic sills intruding wet, unconsolidated sediments tend to be irregular in thickness and along the margins, with lobate apophyses branching into the sediments (Figs. 8.13g, 8.18a, f, g). The margins of such syn-depositional sills are variably quench fragmented, preserving transitions from coherent to in situ jigsaw-fit angular clast hyaloclastite breccias to hyaloclasts mixed into sediment matrix, forming peperite breccias (Fig. 8.18d–g; Chap. 4). Fluidised injections of such peperite breccias can form clastic dykes injected into the margins of the sill. An excellent Archean komatiitic basalt example of mafic syn-depositional sills from Dundonald Township, Ontario, Canada, has been described in Chap. 7, Sect. 7.7.1 and by Houlé et al. (2008). There, multiple, thin (0.5– <10 m) thick sills are interleaved with black, deep-water mudstones (Fig. 8.13h). The margins of the sills are all irregular, with apophyses intruding both upwards from the sill tops and downwards from their bases, into the enclosing mudstones. The sills vary from being coherent to variably quench fragmented hyaloclastite breccia along strike, and commonly the margins are peperite breccias. The peperite textures range from in situ jigsaw-fit with slivers of mudstone between clasts to thorough mixtures of blocky to fluidal komatiitic basalt clasts in a mudstone matrix. Clearly, the muds were very water-saturated and unconsolidated at the time of intrusion by dense magma that split into multiple finger-like or lobate sheets and easily propagated laterally.

Textures of rocks forming mafic sills vary from those similar to porphyritic lava to coarser groundmass and holocrystalline textures, depending on the history of cooling. Thin sills may have chilled margins and internally coarser textured groundmass, as in thin dykes. Large-volume, very thick sills may undergo very slow cooling, differentiation, and crystal fractionation, leading to intrusions that are texturally and compositionally layered (Marsh 2015). Sills may also experience a continuous injection of magma or pulse-like or multiple phases of magma injection, resulting in complex cooling and crystallisation histories and patterns. Marsh (2000, 2015) has summarised the cooling and crystallisation processes associated with thick sills, which he treats as simple large magma reservoirs (Sect. 8.2.5), based largely on studies of thick dolerite sills, such as the Jurassic Ferrar sill and large igneous complex in Antarctica (Fig. 8.13a–e). This summary is derived mainly from Marsh (2000, 2015). As a sill cools it undergoes the following processes (Fig. 8.22a, b):

(a) A solidification front gradually advances from the margins inwards, forming a rigid crust. In a sill, this occurs from both the top and bottom simultaneously.
(b) At the contact with the country rock, the forming crust cools quickly due to rapid heat loss (quenching) to the country rock. Rapid cooling of the magma at the contact rapidly increases the viscosity of the magma, lowering element diffusion rates to the point that crystals can no longer grow, and preventing the settling of crystals already in the magma, which become trapped at the contact. Rocks in the contact zones will be fine-grained (Fig. 8.22a).
(c) The large crystals (phenocrysts, xenocrysts) and their abundance in the margins of the rigid crust reflect the crystal population that was already in the magma at the time of its injection into the sill (Fig. 8.22b). The groundmass crystal-size population reflects crystallisation conditions in the sill as it cools (Fig. 8.22a).
(d) As the cooling and solidification front advances inwards from the contact into the sill interior, the temperature gradient and cooling rate decrease because the inner margin of the rigid crust is more thermally equilibrated with the liquid magma temperature than at the time of initial intrusion.
(e) Groundmass crystallisation can occur to an advanced state to the point that groundmass crystals are coarser, touch, and interact, called the critical crystallinity level (abundance $\sim$50–55%), hindering crystal movement and settling. Marsh has called this a mush zone.
(f) Inwards of the mush zone at a level where crystallisation of $\sim$25% suspended crystals has caused a viscosity increase that is ten times that of the magmatic core of the intrusion, settling of crystals becomes difficult. This level is called the capture front, which also migrates into the interior of the sill as it cools. There are two capture fronts in each sill.
(g) The core of the intrusion is called the suspension zone. The temperature could vary from liquidus to sub-liquidus, and so the magma viscosity is low. If phenocrysts have been carried in, or groundmass crystals form they are dispersed to a level of 25% abundance, and depending on convection conditions and velocities, large dense crystals could settle downwards into the lower capture zone.
(h) In the lower capture zone, settled crystals cease to settle, but ongoing groundmass crystallisation will transform the capture zone into a mush zone as the lower solidification front migrates upwards from the base of the sill, and eventually as the magma cools to the solidus temperature, the entire rock is crystallised. With very thick sills, the core can be coarsely crystallised gabbro.
(i) Sills with suspended large phenocrysts and low magma viscosities can undergo crystal settling and crystal fractionation in the sill adding crystals to the lower mush zone at an accumulation level. Cumulate zones (pyroxenite, peridotite in mafic sills) can thus form in the lower half of sills, which lead to a compositional differentiation and zonation. This is very difficult to occur in phenocryst-poor or free magmas.
(j) As a sill crystallises, the magma interstitial to the crystals becomes more evolved, silica-rich, and less dense. Any instability that affects the sill (e.g. injection of new hot magma, foundering of part of the dense(r), cooled solidification front from the roof of the sill into the more liquid-rich part of the sill, or simply downward compaction of crystals within the sill) may cause the evolved interstitial melt to be squeezed out of the capture and suspension zone, up towards the top of the still hot, plastic sill, forming more siliceous segregation veins (Marsh 2002; Zavala et al. 2011; Figs. 8.22c, 8.23f). These segregated melt veins can have the composition of andesite or even dacite. They can form very coarse, leucocratic crystal zones or domains or veins, called granophyre zones, that can crystallise feldspar, quartz, magnetite, and zircon.

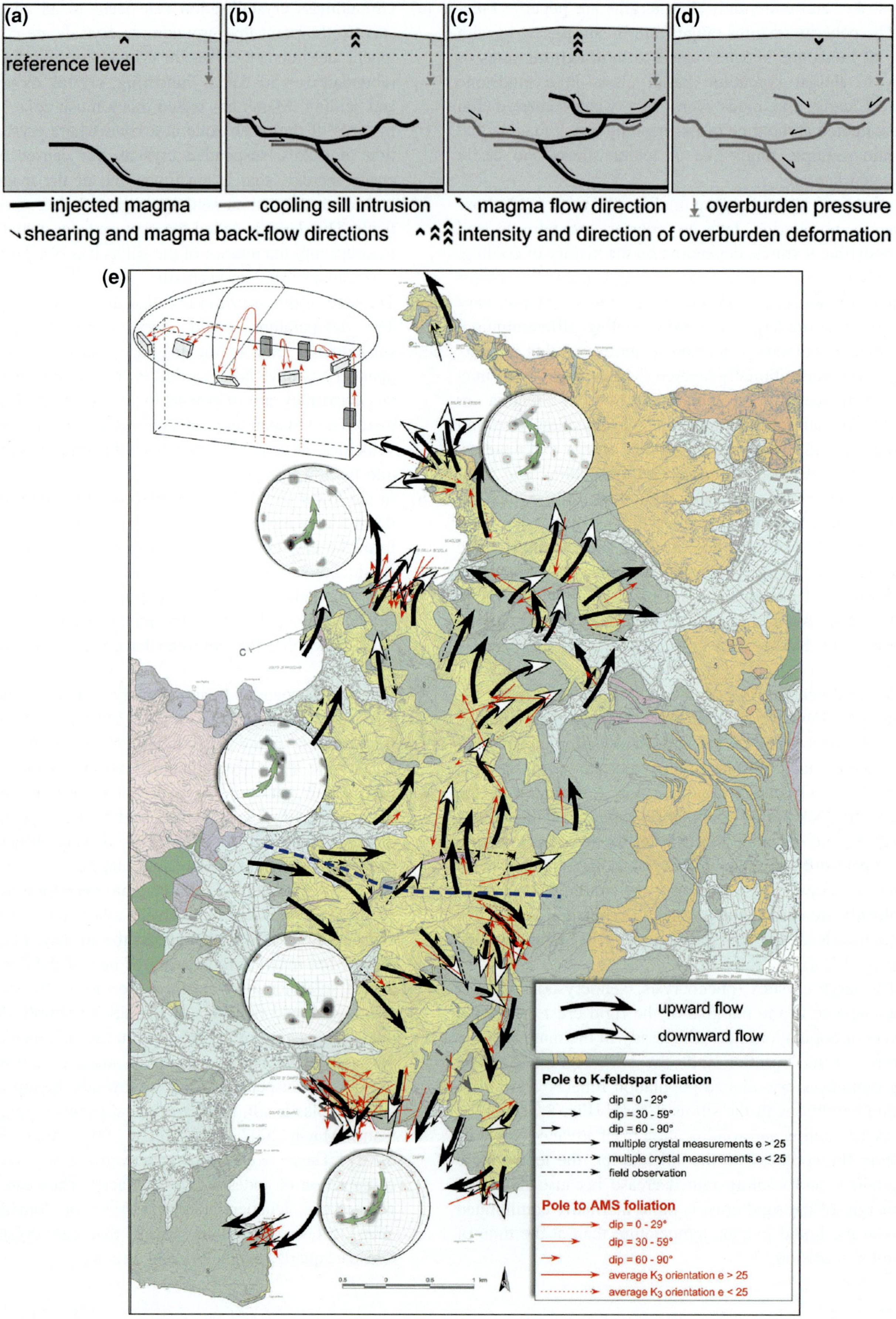
(a)
(b)
(c)
(d)
reference level
injected magma
cooling sill intrusion
magma flow direction
overburden pressure
shearing and magma back-flow directions
intensity and direction of overburden deformation
(e)
upward flow
downward flow
Pole to K-feldspar foliation
dip = 0 - 29°
dip = 30 - 59°
dip = 60 - 90°
multiple crystal measurements e > 25
multiple crystal measurements e < 25
field observation
Pole to AMS foliation
dip = 0 - 29°
dip = 30 - 59°
dip = 60 - 90°
average K3 orientation e > 25
average K3 orientation e < 25
0.5 0 0.5 1 km

◀ **Fig. 8.21** Representation of propagation and flow patterns in sills based on paleomagnetic directional AMS analysis. **a–d** Interconnected (saucer) sills of the Jurassic Golden Valley Sill system, Karoo Basin, South Africa. Schematic basinal cross-section showing phases of lateral and upward propagation and magma flow (**a–c**) can be followed by drain-back reverse flow, especially when the magma supply wanes (**d**). The reference level allows comparison of the overburden doming between successive phases. (From Polteau et al. 2008a). **e** Map showing reconstruction of internal flow directions in plan view, including upwards or downwards, of the 7.4 Ma San Martino laccolith, Elba Island, Italy (From Westerman et al. 2018b)

In their simplest form, thick dolerite sills therefore vary petrologically from basalt at the rapidly chilled margins, to coarser dolerite inwards to very coarse gabbro (Fig. 8.23a–c). The coarse core reflects the post-emplacement cooling and slow crystallisation of the residual magma in the centre, which at the time of emplacement and magma flow through may have been largely crystal-free. Mafic to ultramafic olivine and pyroxene cumulate layers may be present near the base or in the lower half (Fig. 8.23d), and in the upper parts, siliceous granophyre and leucocratic domains and veins may occur (Fig. 8.23e, f), which have elevated SiO_2 and Zr, and reduced MgO contents. In addition, in some differentiated sills, the gabbro may even crystallise quartz (quartz gabbro). However, this zonation can be further complicated by ongoing magma injection and inflation of the sill, changes in the crystal content of injecting magma, internal convection, the thickness of the sill and how that affects temperature gradients and cooling rates, and also by pulsatory or multi-phase episodic magma injections, which can undergo separate cooling and crystallisation histories, depending on their thickness and the spacing between injection events. With multiple injections of magma batches, internal contacts can occur between older slightly cooled and crystallised magma and new hot magma which is chilled by contact with the older cooler magma or rock (Fig. 8.23g). In such cases, intra-clasts of the "older" dolerite can be incorporated into the "younger" basalt through magmatic erosion as the new magma batch intrudes into the older, cooler part of the sill (Fig. 8.23h). Clearly, any combination of these could lead to very complex sill architecture and petrological variations.

In addition, the understanding that sills can propagate by multiple lobe advance (Sect. 8.3.2.5) increases the likelihood that the eruptible magma in many sills may be separated into compartments of magma (e.g. Gudmundsson 2012). Depending on the degree to which these are interconnected will determine how uniform or heterogeneous internal crystallisation processes and distribution of textural domains may be.

Thick ultramafic sills also experience internal differentiation (Chaps. 7, 18), including crystal settling effects because of the low viscosity of the magma, but felsic sills are not prone to extensive internal differentiation because crystal settling is retarded by the high magma viscosity.

8.3.2.7 Felsic Sills: Architecture and Textures

Felsic sills can also vary from very thin to very thick. Ironically, even though felsic magmas are significantly more viscous than mafic magmas and form very thick high aspect ratio lavas and domes (Chap. 6), in subsurface magma conduits under the effects of lithostatic pressure, felsic (and all) magma is less viscous than when erupted, and confining pressure also restrains exsolution of volatiles, making magma less viscous than after volatile exsolution (e.g. 10^3–10^7 Pa s [Cruden and Weinberg 2018] cf. 10^8–10^9 for felsic lavas; Chaps. 2, 6). Felsic magma can squeeze through narrow conduits, producing in places narrow sills and dykes. Laterally, felsic sills generally do not extend as far as mafic counterparts, although some sills that extend tens of km have been documented (e.g. ~60 km long, ~1,800 Ma Tennant Creek porphyry/rhyolite, Northern Territory, Australia, McPhie 1993). Contrary to original interpretations that the ~2,450 Ma Woongarra Rhyolite, Hammersley Basin, Western Australia, represents a regionally extensive sill extending ~500 km along strike (Trendall 1995), reassessment of this unit through detailed facies analysis suggests that it is in fact dominated by multiple rhyolite, rhyodacite, and dacite submarine lavas and domes with associated autoclastic facies, syn-eruptive pumiceous volcaniclastic deposits, and resedimented volcaniclastic sediments intercalated with banded iron formation and terrigenous turbidites, representing a silicic large igneous province (Doyle et al. 2001). This is a sobering reminder that detailed informed mapping and facies analysis can often lead to important new perspectives that overturn interpretations that have established an incorrect perception of significance, in this case that silicic sills can extend hundreds of km regionally. It now seems they can't.

As with mafic sills, felsic sills can also show a range of contact relationships, depending on whether or not the country "rock" was lithified or unconsolidated, water-saturated sediment. With lithified country rock, felsic sill margins are concordant or largely so with bedding (Fig. 8.13f), and flow banding is also largely sub-horizontal. Chilled margins may occur in relatively thin sills (glassy, aphyric to aphanitic groundmass) but may not be visible in thick sills which have thermally equilibrated with the country rock. Chilled textures are only visible in the groundmass, and large phenocrysts may occur in the chilled

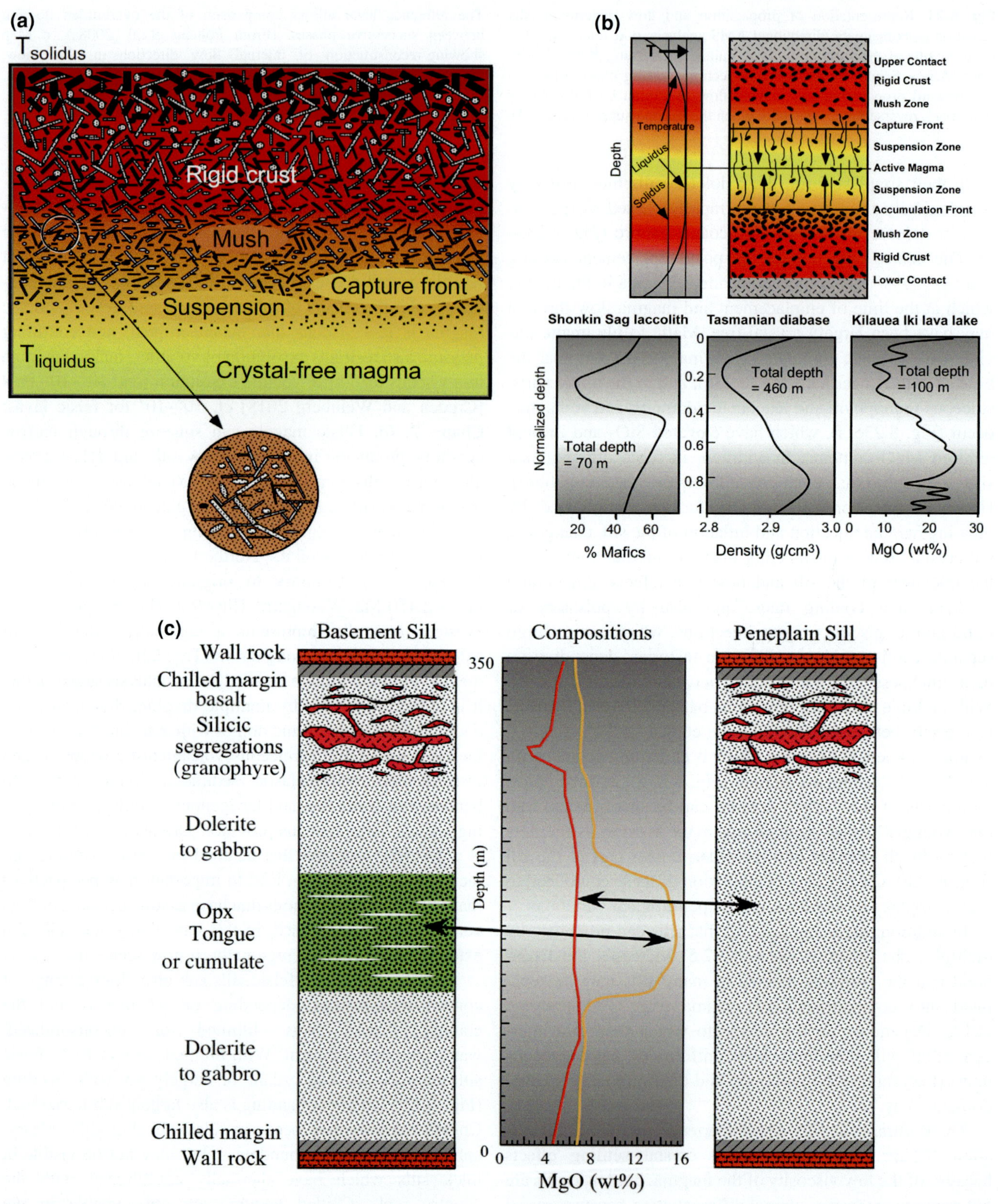

Fig. 8.22 Solidification, crystallisation, and differentiation processes in cooling sills. **a** Cooling and crystallisation zones from the top contact down through the solidification front into the core of the intrusion. **b** Internal zones within a sill from top to bottom, reflecting cooling and crystallisation processes. Also shown are compositional profiles through dolerite sills and the 1959 Kilauea Iki ponded basalt lava, Hawai'i. **c** Profiles through two separate Jurassic Ferrar Dolerite sills, Antarctica, showing the location of siliceous segregations high in the sills, and showing the petrological zonation from chilled margin basalt to dolerite and (in some cases gabbro) in the centre. The Basement Sill (left) also shows an orthopyroxene-rich zone in the middle which Marsh (2015) attributes to the late injection of pyroxene-rich magma. In some sills, such a layer (pyroxenite, peridotite) in the lower half of the sill could represent a cumulate zone resulting from the settling of dense crystals through the melt. (Adapted from Marsh 2015)

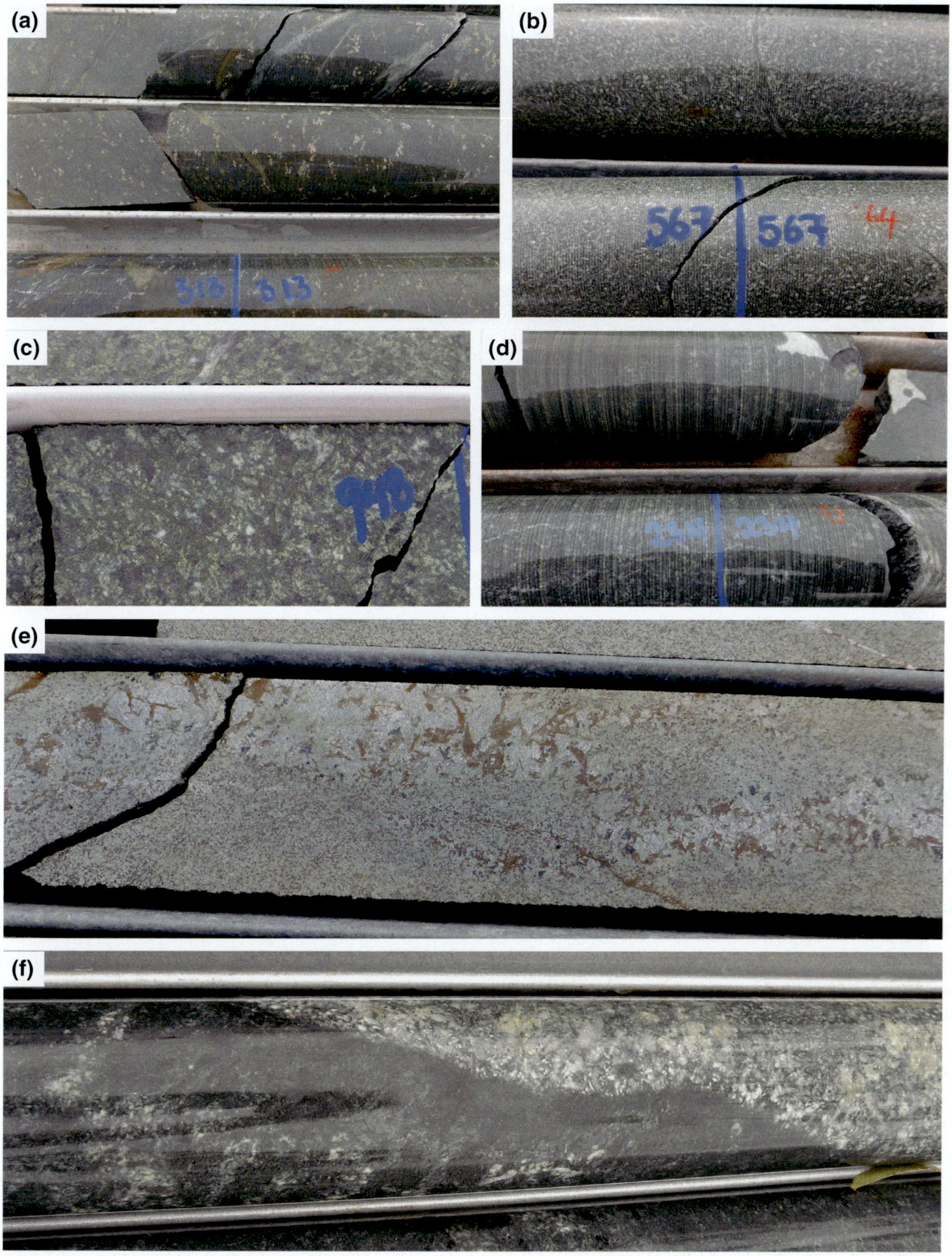

Fig. 8.23 Textural zones within mafic sills. **a–e**, **g**, **h** 2.68 Ga Golden Mile Dolerite, Kalgoorlie, Western Australia. Diamond drill hole JUGD011; depths indicate the depth in the diamond drill hole, not the true depth in the sill. **a** Altered basalt with leucoxene blebs at 313 m; **b** typical dolerite texture at 567 m; **c** altered gabbro at 948 m. **d** Pyroxenite that occurs near the base of the sill at 234 m; **e** coarse granophyre veins with feldspar, magnetite, minor quartz, and pyroxene that occur high in the sill at 757 m. **f** Leucocratic veins in chloritically altered dolerite (Condensor Dolerite), St Ives gold camp, Kambalda, Western Australia. **g** Internal contact between a younger basalt injection into a slightly cooler "older" dolerite in the Golden Mile Dolerite at a core depth of ~530 m. **h** Rip-up intra-clasts of "older dolerite" in "younger" injected basalt in the Golden Mile Dolerite, at a core depth of ~529 m. All drill core is ~5 cm wide

Fig. 8.23 (continued)

rim up to the edge of the intrusion. In older sills, groundmass chilling textures may be overprinted by devitrification textures (Chap. 13), or metamorphic and alteration mineral assemblages and textures. With sills intruding unconsolidated sediments, contacts can be highly irregular with lobes and irregular apophyses intruding the sediments (Fig. 8.18b, c). Similarly, contacts could be marked by hyaloclastite to peperite transitions (Fig. 8.18b, c). Several examples reflect the differences between sills intruded into lithified and unlithified hosts.

Syn-Depositional and Post-Lithification Sills of the Late Devonian Boyd Volcanic Complex, New South Wales, Australia

The Boyd Volcanic Complex (BVC) represents a bimodal rhyolite-basalt volcanic system generated by regional extension within a continental rift setting (see Sect. 8.6.1; Cas et al. 1990, 2001; Giordano and Cas 2001). Although there are widespread subaerial rhyolitic (ignimbrites, lavas) and basaltic (mostly lavas) volcanic sequences, in a contemporaneous subaqueous basin setting most of the rhyolitic bodies are sills and cryptodomes, with only minor locations of basin floor volcanic successions. The sills consist of two groups, weakly porphyritic (<15% small quartz and feldspar) and highly porphyritic (>30% coarse quartz and feldspar) suggesting the latter group represents a slightly more evolved, crystallised magma batch. This petrological difference is reflected in the difference in the timing of the emplacement of these different petrological sill types and their internal features and contact relationships.

The weakly porphyritic sill types are syn-depositional, have irregular margins with apophyses into overlying mudstones, quench fragmented hyaloclastite interiors and margins grading into peperite breccia margins at both the top and bottom (Fig. 8.18b, c). Flow banding in these sills can be highly variable in orientation, with flow folds of amplitude and wavelength from <1 m to >50 m, and columnar jointing is only weakly developed. Such sills range from ~ 1 m to tens of m thick. The irregular, interactive contacts (hyaloclastite, peperite, irregular apophyses) and the variable orientation of the flow banding indicate the weakly porphyritic sills were emplaced early, when the host sand and mud sediments were water-saturated and lacked confining strength and the intruding magma had a relatively low viscosity.

In contrast, late, post-lithification highly porphyritic rhyolite sills (as well as laccoliths or cryptodomes) have more planar margins, and completely lack marginal hyaloclastite and peperite (Fig. 8.13f). The best-preserved late sill at Bunga Head, which was at least 80 m thick, has a sharp planar base that is concordant with bedding in the underlying sedimentary succession. Flow banding is commonly developed in this sill and is almost always sub-horizontal, parallel to the sill base. Columnar jointing is very well developed everywhere, roughly perpendicular to the sill base. The basal 1 m of the sill has well-developed spherulites and siliceous nodules. Along the preserved southern margin of the sill, swarms of fluidal basalt inclusions or xenoliths occur, indicating the co-existence of bimodal rhyolite and basalt magmas during the magmatic and depositional history of this continental rift basin (Cas et al. 2001). The lack of interactive margins (i.e. no hyaloclastite, peperite, and fluidal apophyses) with the enclosing sedimentary host (sandstones, mudstone) indicates that those host sediments had compacted, dewatered, and lithified by the time the coarsely porphyritic rhyolite sills were intruded. The post-lithification sill is thus texturally much simpler than the syn-depositional sills.

The Early Proterozoic Tennant Creek Porphyry: a Laterally Extensive Syn-Depositional Sill

McPhie (1993) interpreted the emplacement of the rhyolitic Tennant Creek Porphyry as being a syn-depositional sill. It extends >80 km along strike, is in places >100 m thick, and in places splits into multiple sill horizons separated by the host turbidite sandstones and shales. The porphyry is largely massive, coarsely porphyritic, with large euhedral phenocrysts of quartz, K-feldspar, and plagioclase, dispersed in a fine-grained microgranular mosaic groundmass of fine quartz and feldspar with dispersed micro-spherulites and relict perlite cracks in places, testifying to an original glassy character (McPhie 1993). The contacts with the host sedimentary succession, including the top contact are highly irregular, lobate in form, and include clasts of sediment in the margins. In addition, in places the margins consist of jigsaw-fit textured breccias of porphyritic clasts, with transitions to peperite breccias of porphyry clasts dispersed in sandstone and shale matrix. These characteristics are typical of syn-depositional intrusions. This example indicates that even highly porphyritic and relatively high-viscosity felsic magma can propagate laterally for tens of km through unconsolidated basinal sediments.

Small Volume, Syn-Depositional Paleoproterozoic Onedin Rhyolitic Sill, Western Australia

Felsic sills are more commonly less extensive than the Tennant Creek Porphyry. A more local example, also from northwestern Western Australia, has been very well documented by Orth and McPhie (2003). The aphyric rhyolite sill is 52 m thick, extends <4 km along strike, and was intruded into water-saturated unconsolidated sediments of Paleoproterozoic age ($\sim$1,840 Ma). The sill was subdivided into 5 textural zones by Orth and McPhie (2003): (1) The 1–2 m thick basal contact zone of perlitic rhyolite has up to 5% amygdales and contains white mudstone pods and tongues up to 0.5 m $\times$ 0.2 m in a dimension similar to the white mudstone beneath the sill. This passes upwards into (2) the $\sim$10 m thick, amygdaloidal lower zone, which also contains white mudstone intra-clasts and is very spherulitic (<3 mm diameter) in the lower 2 m, grading into fewer small spherulites (<1 mm) dispersed in a microgranular quartz–feldspar mosaic groundmass upwards. This zone grades into (3) the $\sim$20 m thick middle zone, in which amygdales (<2 cm diameter) increase in abundance from $\sim$5% to 20% upwards. (4) The upper zone is $\sim$18 m thick and marked by a diffuse layering 10–50 cm thick defined by variations in abundances of amygdales, spherulites, lithophysae, and perlite, with some layers containing up to 40% amygdales. (5) The upper contact zone is also $\sim$2 m thick, and is represented by a breccia of aphyric, perlitic rhyolite clasts dispersed in a matrix of mudstone and fine sandstone. Away from the contact, the sediments are laminated. The rhyolite clasts vary from blocky to fluidal in shape. Orth and McPhie (2003) interpreted the upper contact zone as being peperite resulting from the intrusion of aphyric rhyolite into wet sediments causing quench fragmentation and peperitic mixing between rhyolite hyaloclasts and fluidised sediment. They considered the depth of emplacement of the sill to have been several hundred metres in the sediment pile to allow for vesiculation, particularly in the upper zone. The variation in devitrification textures internally was explained by variations in cooling rates in the different zones. Interestingly, the lower contact zone has not been interpreted as peperite, although tongues and inclusions of sediment occur in the rhyolite. This together with the relatively narrow upper peperitic contact zone my indicate that the host sediment at the time of intrusion was partially dewatered due to diagenetic compaction but was not lithified.

Small volume, Cambrian Syn-Depositional Pumiceous Rhyolite Sill and Peperite, Mt Read Volcanics, Tasmania, Australia

In the Cambrian Mt Read Volcanics, Tasmania, Australia, Gifkins et al. (2002) describe the occurrence of pumiceous rhyolite peperite forming the margins of the syn-depositional rhyolite sill that intruded stratified resedimented pumiceous sediment. The peperite is gradational with the coherent pumiceous rhyolite sill and consists of tube pumice clasts that are bound by curviplanar fracture surfaces. The pumiceous peperite preserves some jigsaw-fit textural domains and has a matrix of pre-existing resedimented pumiceous sediment. This complicated scenario was deciphered by the differences between the weakly porphyritic (3%) tube pumice of intrusive rhyolite and pumiceous peperite, and the more porphyritic (10–20%) pumice in the resedimented pumice sediments.

Other examples of syn-depositional felsic sills include the submarine Ordovician Highway Reward volcanic centre, Queensland, Australia (Doyle and McPhie 2000), and the Devonian sills of the Sierra Nevada described by Hanson and Schweikert (1982) and Hanson (1991).

8.3.2.8 Guidelines for Distinguishing Sills from Lavas

Since both lavas and sills are concordant with the associated stratigraphy, there may be difficulties in distinguishing them in the geological rock record. This may be particularly the case in older deformed successions but may also be true in young volcanic edifices (e.g. in mafic shield volcanoes, stratovolcanoes, and basin settings). Distinguishing them is important because misidentification may lead to incorrect stratigraphy and event chronology determinations. Understanding the relative order of units and events is crucial in terms of developing strategies for geochronological, geochemical, and paleomagnetic sampling. This then helps in the identification of petrogenetic trends and the temporal scales of such trends. It cannot be assumed that every radiometric or exposure age determination is correct, because deviations from normal systematic decay processes can occur (e.g. excess argon in some minerals, abundant xenocrysts, leakage of decay products, or resetting of the decay event by a younger external event such as metamorphism), and so understanding field relationships through detailed field studies that allow for a relative assessment of age relationships between units is always essential.

The following are important guidelines for discriminating sills from lavas in the field:

(a) If units are felsic and thin, they must be intrusions because the viscosity and yield strength of felsic magmas determine that all felsic lavas, except some peralkaline lavas, are tens of m or more, thick.

(b) Thick felsic units may be lavas or intrusions:
   - if they are lavas, they *must* have marginal autobreccia if subaerial, and/or hyaloclastite if subaqueous, because of the high magma viscosity and tendency to autoclastically fragment, but
   - if they are totally coherent, then they *must* be intrusions.

(c) Thin to moderately thick mafic units may be lavas or intrusions, because both can inflate to thicknesses of several tens of m.

(d) Very thick (>50m), high aspect ratio mafic units are most likely intrusions or locally, topography ponded lava because low-viscosity lavas cannot inflate to such thicknesses without physical constraints, and normally spread as thin sheets due to their hydraulic head when unconfined on relatively flat topography.

(e) Sills intruded into solid or cemented rock will generally have concordant planar or broadly curviplanar margins but will still usually have some discordances at both their base and top (e.g. tabular sills, Fig. 8.13c, d), whereas sills intruded into unconsolidated sediments may show major discordances and cross-cutting relationships (e.g. saucer–sills, Figs. 8.15, 8.16, 8.17).

(f) Sills intruded into unconsolidated sediments will have irregular top and bottom margins, closely associated with evidence of fluidised sediment (e.g. clastic dykes and pipes), and peperitic textures (Fig. 8.18a–h).

(g) Sills will generally have some apophyses (in some cases magma lobes and/or magma fingers) branching from *both* the top and bottom of the main intrusion and intruding the country rock or host sediments above and below (Figs. 8.13, 8.18).

(h) Whereas sills will usually have some apophyses intruding from the top contact into the overlying solid rock or sediments (e.g. Figs 8.13c–f, 8.18a–c respectively), lavas cannot because overlying strata post-date lava emplacement.

(i) Although sills will usually also have apophyses intruding substrate rocks or sediments (e.g. Fig. 8.13b), lavas will usually have bottom apophyses only if the substrate consists of unconsolidated sediments.

(j) At map scale, only sills will show transgressive contacts and local "step ups" in stratigraphy and may incorporate larger xenoclasts or rafts (i.e. broken bridge structures) of the sedimentary stratigraphy or country rock (Fig. 8.13d).

(k) Sills as well as lavas can show the development of flow lobes (Fig. 8.19), and ropey-like flow crenulations on their surfaces (Fig. 8.13g).

(l) However, only syn-depositional sills can develop peperite along their top contact (Fig. 8.18) because this requires sediments to already be present at the top contact at the time of intrusion.

(m) Both lavas and sills may develop chilled margins; note, however, chilled margins can contain coarse phenocrysts, even though the groundmass may show evidence of chilling (very finely crystallised or glassy).

(n) Although sills are typically depicted as producing contact metamorphic effects in the host rock at the contacts, this is not always the case, or may not be obviously visible.

(o) Large, thick sills cause contact metamorphism if the host rocks are fully lithified, but this may not be significant with small or thin sills because of their low heat budget and short emplacement duration does not allow for thermal equilibration of the country rock with the sill.

(p) Even very thick intrusions may show little or no evidence of contact metamorphism if the intrusion history was protracted and especially if it was a multi-phase pulsing or incremental event.

(q) Syn-depositional intrusions into sediments are unlikely to show any evidence of contact metamorphism because unconsolidated sediments are poor conductors of heat and if they contain pore water, the water is a very good heat sink.
(r) On the other hand, water–rock interaction and low-temperature alteration and even veining of the enclosing sediments and/or sills may be promoted by hydrothermal circulation.
(s) Thick sills are more likely to crystallise more coarsely and differentiate significantly (Fig. 8.23) because the cooling regime in the Earth's crust is slower compared to lavas, especially those emplaced under water (water has a higher heat capacity and high thermal conductivity than silicate rocks).
(t) Even relatively thin syn-depositional sills may show an asymmetrical cooling profile and more advanced internal groundmass crystallisation (e.g. granophyric or equigranular and spherulitic crystallisation) near the base compared with lavas (Orth and McPhie 2003).
(u) Although some shallow, syn-depositional felsic sills may develop a high-level vesicle concentration zone, as lavas do, they may be distinguished from lavas by having flow banding that is parallel to the top and bottom of the sill, whereas felsic lavas have a sub-vertical ramping banding near the top (Chap. 6), and syn-depositional sills are likely to have peperitic tops (Orth and McPhie 2003).
(v) Modern lavas are not known to develop very coarse internal (plutonic) textural or differentiated zones. Therefore bodies with coarse holocrystalline and differentiated textural zones were emplaced as sills.

To elaborate further on this final point, there are no known basalt lavas with coarse gabbroic textures or differentiation to granophyre that have been documented. Even the 1959 basalt lava from the summit of Kilauea which flowed into and ponded to some 120 m thickness within Kilauea Iki crater, Hawaii, is largely a porphyritic, fine-grained basalt. While drill core shows some zones of higher olivine crystal concentrations resulting from partial crystal settling, cross-cut in places by dolerite textured segregation veins, and minor very thin pegmatitic rhyolitic to andesitic segregation veins in the upper parts, no coarse gabbroic textures and only minor differentiation has been observed (Helz 1987). Even large-volume basaltic flood basalt lavas, which may be 60 m thick, have interiors that have fine to medium crystals, i.e. textures no coarser than dolerite (Self et al. 1997). Similarly, there are no known felsic lavas, even lava domes hundreds of m thick, that have crystallised granitic textural zones in their interior. The lesson is simple: lavas, especially subaqueous lavas, *cool too quickly* to crystallise coarse textural zones and to differentiate. By contrast, crustal intrusions into lithified host rocks, especially large intrusions, cool slowly, and experience slow crystallisation that facilitates differentiation, crystal settling, and formation of coarse (plutonic) crystal textural zones (e.g. Fig. 8.23c–f).

Where this issue has become particularly contentious is with regard to thick dunite adcumulate bodies found within Archean komatiite successions. As discussed in Chap. 7, differentiated, regionally extensive, layered komatiite bodies hundreds of m thick, such as the nickel sulphide-rich Mt Keith sill in Western Australia, have been interpreted by some authors as thick, ponded, ultramafic komatiite seafloor lavas. However, these show a complex compositional and textural zonation, including very coarse, holocrystalline adcumulate olivine textural zones, and differentiated gabbro in some textural zones. We interpret these as thick subvolcanic sills (Rosengren et al. 2007; Fiorentini et al. 2007, 2012).

8.3.3 When Dykes Aren't Dykes and Sills Aren't Sills, Are They Dills?

Dykes are defined as sub-vertical cross-cutting tabular intrusions, and sills as sub-horizontal, concordant intrusions. However, many tabular intrusions are inclined at angles of ± 45° (Plate 8.1a; Fig. 8.24). Some authors have called these inclined sheets, particularly in the context of saucer-shaped sills (Sect. 8.3.2.2), but could we simply apply the hybrid term "*dill*"? We will leave it to readers to decide if this is a useful term that could be applied, especially in the field, for bodies that they can't decide if they are dykes or sills.

8.3.4 Laccoliths and Lopoliths

If a lateral intrusion into layered country rock inflates and has a higher aspect ratio than a sill, that is, vertical dimension or thickness is sub-equal to width and length (W, L > H), then the intrusion is called a *laccolith*. A laccolith generally has a flat base and a broad upward convex top and is more or less concordant with the enclosing rocks (Plate 8.1b; Gilbert 1877; Corry 1988). The diameters of laccoliths are generally from km to tens of km. The geometry of laccoliths is generally suggestive of higher viscosity magmas compared with sills. The increased aspect ratio of a laccolith allows for slower cooling rates at the core of the intrusion, where textures are holocrystalline.

Recent studies suggest that many laccoliths grow by the emplacement of successive magma sheets and pulses

(Schmiedel et al. 2014; Breitkreuz et al. 2018; Horsman et al. 2018; Morgan 2018). Morgan (2018) proposed that many laccoliths grow by the injection and lateral spreading of multiple stacked sill-like sheets until such time as the internal magma pressure exceeds the lithostatic load and causes the intrusive system to start inflating. This is driven by a Pascal Principle inherited pressure from deeper in the crust where the magma migrates from, which is transmitted through the feeder dyke to the growing laccolith. Such inflation causes up-arching of the overlying strata and may trigger gravitational sliding off the flanks of the rising topographic dome at the surface, so reducing the lithostatic confining pressure and enhancing the inflation process. Depending on crustal lithologies and densities, laccoliths may form at depths <3–4 km.

Well-studied examples include the Eocene–Oligocene laccolith complexes in the Henry Mountains of Utah where Gilbert (1877) first defined laccoliths as an intrusive morpho-type. Recent research by Horsman et al. (2018) summarised and refined the understanding of the laccolith complexes there. They range from simple essentially single laccoliths to more complex systems with the main laccolith body surrounded by smaller satellite intrusive bodies (Fig. 8.25a–d). Their volumes range from 18–100 km^3, their diameters range from 9–20 km, and their maximum vertical thickness varies from 1200–2500 m. The maximum dips of adjacent up-arched sedimentary strata range from 25–90°. Limited geochronology indicates an emplacement duration of ~1 Myr for individual laccoliths (Horseman et al. 2018). The cross-sections (Fig. 8.25b–d) are instructive in demonstrating complex internal architecture and up-arching of the overlying stratigraphy, which in a number of places is accompanied by marginal faults that have accommodated the deformational strain along the margins. The laccoliths are thought to have been intruded at depths of 2–4 km below the Earth's surface.

On the island of Elba in the Tyrrhenian Sea, Italy, excellent coastal exposure provides sections through three laccolith complexes emplaced over a period of ~1.5 Myr from ~8.5 Ma (Westerman et al. 2018b). These also show a multiple sheet architecture. The Permo-Carboniferous rhyolitic Halle Volcanic Complex, Saale Basin, Germany, also preserves laccoliths although not well exposed (Schmiedel et al. 2014; Breitkreuz et al. 2018). There, adequate well and drill core intersections have allowed 3D modelling of laccoliths to be undertaken.

If the base and top of a sheeted intrusion such as a sill or a laccolith are instead convex downward, suggestive of down-sagging, then the name used is *lopolith* (Plate 8.1b; Corry 1988). Lopoliths are generally thought to form deeper in the crust than laccoliths, where confining lithostatic pressure is greater, and lopolith inflation is accommodated by down-sagging of the substrate which complements the removal of magma volume from the deeper source magma reservoir. Both laccoliths and lopoliths may be sources of volcanism, or not.

8.3.5 Cryptodomes

If the aspect ratio of an intrusion is higher still, resulting in an almost equidimensional shape (W = L = H), and the intrusion comes very close to the Earth's surface, deforming it and up-arching the ground surface or depositional surface in subaqueous settings, this type of intrusion is called a *cryptodome* (Chap. 6). In order to produce the space for such an intrusion, contact relationships are cross-cutting with the enclosing rocks (Fig. 8.26a, b) to up-arching of overlying strata. Depending on the nature of the enclosing rocks, contact relationships vary from particulate displacement and soft sediment deformation in enclosing unconsolidated deposits to irregular apophyses intruding into unconsolidated deposits (Fig. 8.26c) and peperites at the margins, to fault bounded in coherent to semi-coherent rocks.

Cryptodomes that are emplaced very close to the Earth's surface in subaerial settings may cause it to bulge upwards forming a morphological (crypto)dome (e.g. Minakami et al. 1951; Goto and Tomiya 2019), or on the slopes of a volcanic edifice such as a stratovolcano, to bulge outwards (e.g. Fig. 8.26d, e). The significant bulging of rocks overlying a cryptodome on the steep slopes of a volcanic edifice may produce dramatic gravitational responses, such as those that occurred with the eruption triggering landslide and debris avalanche that began the paroxysmal phase of the 18 May 1980 eruption at Mt St Helens, USA (Chap. 12; Lipman and Mullineaux 1981; Tilling 1984; see also Chap. 15). The reduced lithostatic pressure at shallow levels prevents cryptodomes from growing indefinitely because they will breach the surface and begin to erupt through and onto it, forming a lava dome (Chap. 6). The definition of a cryptodome thus also implicitly includes a dimensional limit that is in the order of ~1 km. The high aspect shape of cryptodomes is generally indicative of intrusions of viscous magmas. For a more detailed discussion of cryptodomes, see Chap. 6 (Sects. 6.3.5, 6.6.3, and 6.13.3) and Chap. 15.

Cryptodomes can form when magma rises through water-saturated unconsolidated sediments as rising magma reaches or exceeds its level of neutral buoyancy. The margin may be irregular (Fig. 8.26a–c), with apophyses or lobes of the cryptodome magma intruding the enclosing sediment and fluidised sediment being injected into fractures or clefts in the margin of the cryptodome (Fig. 8.26c). In cryptodomes in subaqueous settings a zone of hyaloclastite, and then around most margins at the interface with the wet sediments, domains of peperite occur (Chap. 6, Sect. 6.13.3; Kokelaar et al. 1985; Hanson and Wilson 1993; Goto and McPhie 1998; Doyle and McPhie 2000; Stewart and McPhie 2003). The widths of these zones will depend on the abundance of pore water available in the sediment, which will be determined by the initial porosity, grain size, degree of compaction, and dewatering of the sediments prior to the intrusion of magma. Sometimes, the uppermost part of such cryptodomes may breach the sediment surface, causing extrusion of the top of the dome onto the sea or lake floor. This may result in the hyaloclastite and peperitic margins of the cryptodome being up-arched and exposed on the emerging dome, triggering mass-flow resedimentation of these intrusive facies as aprons of debris onto the basin floor (e.g. Kokelaar et al. 1985; Cas et al. 1990; Hanson and Wilson 1993; Doyle and McPhie 2000). The emerging dome then becomes an effusive dome, and so characteristics of both subaqueous cryptodomes and effusive subaqueous domes may develop.

8.3.6 Volcanic Plugs or Necks

A volcanic *plug* or *neck* is the remnant of a cylindrical vertical intrusion or the solidified fill of a volcanic conduit feeding a vent, around which the strata of the host volcanic edifice, classically a basaltic scoria cone as in the Massif Central of France, or the remnants of a viscous extruding plug (e.g. Mt Pelée 1902), have been eroded away leaving a resistant, protruding, usually coherent body above the landscape (Fig. 8.27; Chap. 16). Such plugs may vary in diameter from the m scale to the hundreds of m.

8.3.7 Stocks

Stocks are more irregular in geometry than dykes, sills, and plugs, and generally represent the deeper parts of a volcanic conduit (several km deep) linking a large source magma reservoir to the volcanic conduit (Daly 1905). Strictly, it has a plan view area of <100 km^2, which makes it smaller than say a batholith. It has a sub-vertical orientation and may widen with depth as it approaches a bigger reservoir. It may have numerous dykes and sills branching from it. Stocks are typically exposed through the erosional degradation of a central volcanic edifice such as a stratovolcano (e.g. several stocks exposed at the degraded Spanish Peaks volcanic edifice, USA; Fig. 8.8b).

8.4 Architecture of Some Complex, Large, Upper Crustal Subvolcanic Plumbing Systems and Some of the Processes

Knowledge about the architecture of the magma plumbing of volcanic systems is still very limited. There is now a general agreement on envisaging volcanic plumbing systems as complex and evolving transcrustal networks of storage and transport zones at various depths in the crust (Cashman and Giordano 2014; Cashman et al. 2017; Giordano and Caricchi 2022). Depending on magma production at source and the longevity of magmatism, plumbing systems may evolve from relatively simple, such as in intraplate monogenetic volcanoes, where individual dykes rapidly transfer primary basaltic magma directly from the mantle to surface, to highly complex multistorey systems where primary magmas rarely make it to the surface and differentiated magmas form, migrate and eventually erupt (Chap. 16, Fig. 16.1; Giordano and Caricchi 2022). The shallow portion of such complex plumbing systems (down to 5-10 km from surface) may be the site of repeated but ephemeral fast transfer and accumulation of even large volumes of eruptible magma (as short as decades, e.g. Druitt et al. 2012; Pamukçu et al. 2020), which can be almost fully evacuated during eruptions, especially caldera forming eruptions (e.g. Lucci et al. 2020; Fig. 8.1a). Exhumed plutonic systems represent therefore only the final long-term integral picture of everything that had not erupted during the course of magmatism, but may tell very little of the short-term cycles of build-up and destruction of the shallow subvolcanic plumbing systems that are in existence when a volcano is active (Giordano and Caricchi 2022). Our current knowledge is based on inference from outcrops of exhumed ancient intrusion complexes and drill cores from diamond and ore exploration (e.g. Figs. 8.9, 8.10, 8.11, 8.12, 8.13, 8.18, 8.23, 8.24, 8.26, and 8.28), limited mapping of ancient erosionally degraded volcanoes (e.g. Fig. 8.8), and more recently the application of geophysics (e.g. Figs. 8.15, 8.16, 8.17, 8.19, 8.29, 8.30), laboratory experiments, and numerical modelling of processes (e.g. Kavanagh 2018).

For small, short-lived monogenetic volcanic edifices (e.g. scoria and tephra/tuff cones, maars, and small shield volcanoes; Kereszturi and Nemeth 2012), the magma plumbing system is relatively simple with often a single sub-vertical conduit that may have some minor branches (dykes, small sills) in the uppermost crust or in the edifice (e.g. Plate 8.1 a and Fig. 16.1). Such volcanoes usually have a single or at the most two magma sources. At a larger scale, that of an intraplate continental mafic volcanic province, the regional-scale magma plumbing system is more complex,

Fig. 8.24 Swarm of inclined sheets, or dykes or dills (?), Jurassic Ferrar Dolerite sill complex, Antarctica; height of cliff ~200 m (Photo by James White)

consisting of the combined magma plumbing for each monogenetic volcano, as well as regionally ponded reservoirs, some as deep as underplated sill and pluton complexes at the base of the lithosphere and the crust. At a larger scale again, we have briefly described the regionally extensive network of interconnected sills and dykes that represent the magma plumbing system for flood basalt and large igneous provinces (Muirhead et al. 2014; Sects. 8.3.2 and 8.3.2.6), based on outstanding exposures of the Jurassic Ferrar dolerite and large igneous province in Antarctica. Layered igneous complexes, discussed further below, can also represent a component of such plumbing systems.

However, for long-lived large polygenetic volcanoes, the magma plumbing system may be much more complex, often with several magma sources in separate magma reservoirs (Chap. 16; Fig. 16.1). Subvolcanic intrusive complexes are extremely common within the stratigraphy of larger volcanic edifices and contribute significantly to the upbuilding of the edifice from within, through the formation of a complex internal plumbing system of dykes, sills, laccoliths, and cryptodomes (Plate 8.1b; Figs. 8.7a, b, 8.29, and 8.30). These intrusion types represent a network of vertical, inclined, and horizontal conduits that channel magma towards the surface, and a series of chambers or reservoirs where magma can be stored, reflecting a process of endogenous growth or addition from within the edifice (Walker 1992; Matthieu and van Wyk de Vries 2009; Tibaldi 2015; Morley 2018; Breitkreuz and Rocchi 2018). Intrusion of sills, laccoliths, and cryptodomes causes significant geodetically detectable ground movements on young active volcanoes and volcanic fields (e.g. Biggs et al. 2010; Freymueller et al. 2015; Magee et al. 2017). From a stratigraphic point of view, it can be difficult to distinguish sills from lavas in older volcanic successions (see Sect. 8.5), which is important from the point of view of understanding the stratigraphic evolution, sequence of events, and for determining sampling strategies for geochronology and geochemical research programs.

In the rest of this section, we will focus on the subsurface magma plumbing systems of some large complex magmatic systems including shield volcanoes, stratovolcanoes, and the magma reservoirs and chambers associated with explosive caldera volcanoes. We will also briefly introduce layered mafic intrusions. All involve complex, variably interconnected sill–dill, dyke, laccolith, and deeper stock and pluton/magma reservoir networks in the crust (e.g. Plate 8.1b; Figs. 8.7, 8.15, and 8.16). A whole book could (and should) be written on the architecture and processes of the subvolcanic plumbing system of different types of volcanoes, but due to space constraints, the following is a brief overview only.

8.4.1 Magma Plumbing Systems of Large Shield Volcanoes

Walker (1992) highlighted the complex nature of the magma plumbing system of large mafic shield volcanic edifices (also see Chap. 16). These are marked by summit vents and sometimes deflation calderas induced by effusive magma withdrawal that occur from satellite flank, often fissure vent systems at elevations lower than the magma reservoir.

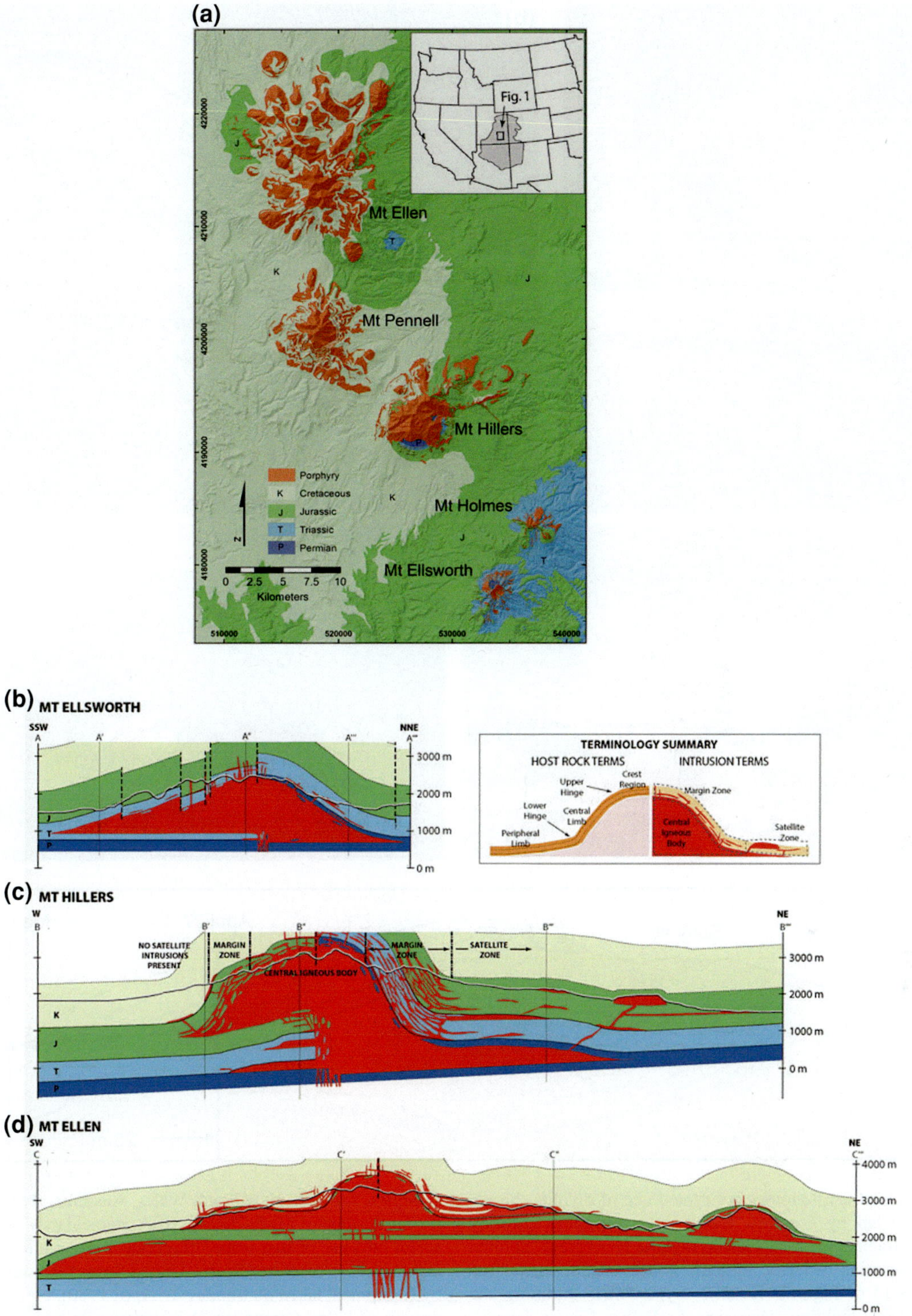

Fig. 8.25 Features of the Oligocene Henry Mountains laccolith province, southern Utah, USA. **a** Map showing the areal extent and distribution of the main laccolith complexes, which were intruded into the generally flat-lying Mesozoic stratigraphy of the Colorado Plateau (CP; see location inset). **b** Cross-section through the relatively simple Mt Ellsworth laccolith showing some vertical dyke apophyses off the main body. **c** Cross-section through the more complex Mt Hilllers laccolith complex, showing a network of multiple sills (sheets) above and in the marginal zone to the central body, as well as more distant sills and small laccoliths in the satellite zone. **d** Cross-section through the complex Mt Ellen laccolith complex, showing its multiple sheet and laccolith architecture. Note that all laccoliths have caused up-arching of the overlying strata, and all laccoliths were interpreted by Horsman et al. (2018) to have formed through intrusion of multiple magma pulses. (From Horsman et al. 2018)

Fig. 8.26 Features of cryptodomes. **a** Cryptodome of rhyolite (brown mushroom shape) fed by a narrow feeder dyke, intruding diffusely stratified quench fragmented hyaloclastite rhyolite lava (grey), uplifted 2 Ma Ponza submarine volcanic system, Ponza Island, Italy. The cryptodome is ~15 m high. **b** Basaltic cryptodome, intruding the bedded, basaltic cone-building succession, La Palma, Canary Islands, Spain. The cryptodome is ~30 m high. (Photo by Sam Thiele). **c** Highly irregular margin to a Late Devonian rhyolite cryptodome intruding deep-water, basinal black mudstones, with pockets of peperite breccia mixed into the mudstone. Bunga Beds outlier, Boyd Volcanic Complex, southern New South Wales, Australia. **d** Detectable bulge on the north slope of Mt St Helens, USA, immediately prior to the 18 May sector collapse of that slope, triggering a devastating debris avalanche and explosive eruption (*Source* United States Geological Survey; Photo by P. Lipman, taken on Sunday 27 April 1980). **e** Field sketch showing growth of bulge (cryptodome) as magma pushed up within the peak. Angle and slope-distance measurements to the bulge indicated it was growing at a rate of up to 1.5 m per day, and by 17 May part of the volcano's north side had been pushed upwards and outwards over 135 m. (*Source* United States Geological Survey)

Fig. 8.27 The volcanic plug of Le Puy en Velay, central France, representing the eroded remains of a ~17 Ma volcano (Photo by D. Gifford; *Source* Wikimedia through Creative Commons Attribution-ShareAlike 3.0 Unported licence)

Hawai'i is of course the archetypical (marine) large mafic shield volcano (last flank eruption at the time of writing occurred in 2018), but other examples include Reunion volcanic edifice in the Indian Ocean and the Canary Island shield volcanoes in the Atlantic Ocean. The complex system of summit and flank parasitic vents, which include point source, radial fissure, and "rift zone" vent systems, indicates that the subsurface plumbing system consists of a complex network of dykes, sills, and magma staging bodies such as large inflated sills and laccoliths (Fig. 8.28a–c; Walker 1988; 1992; Denliger and Morgan 2014; Poland et al. 2014; Chap. 16, Sect. 16.5).

8.4.2 Magma Plumbing Systems of Stratovolcanoes: Case Study of the Kora Stratovolcano, Taranaki Basin, New Zealand

The magma plumbing system of stratovolcanoes can be very complex. In plan view, it looks simple with a radial array of dykes spreading from a central conduit, as mapped on both modern (e.g. Mt Etna, Fig. 8.8a) and relicts of ancient stratovolcanoes (Fig. 8.8b). Stratovolcanoes have a main, central summit vent fed by a cylindrical conduit, which at depth is often connected to a major lopolith, or dyke. In

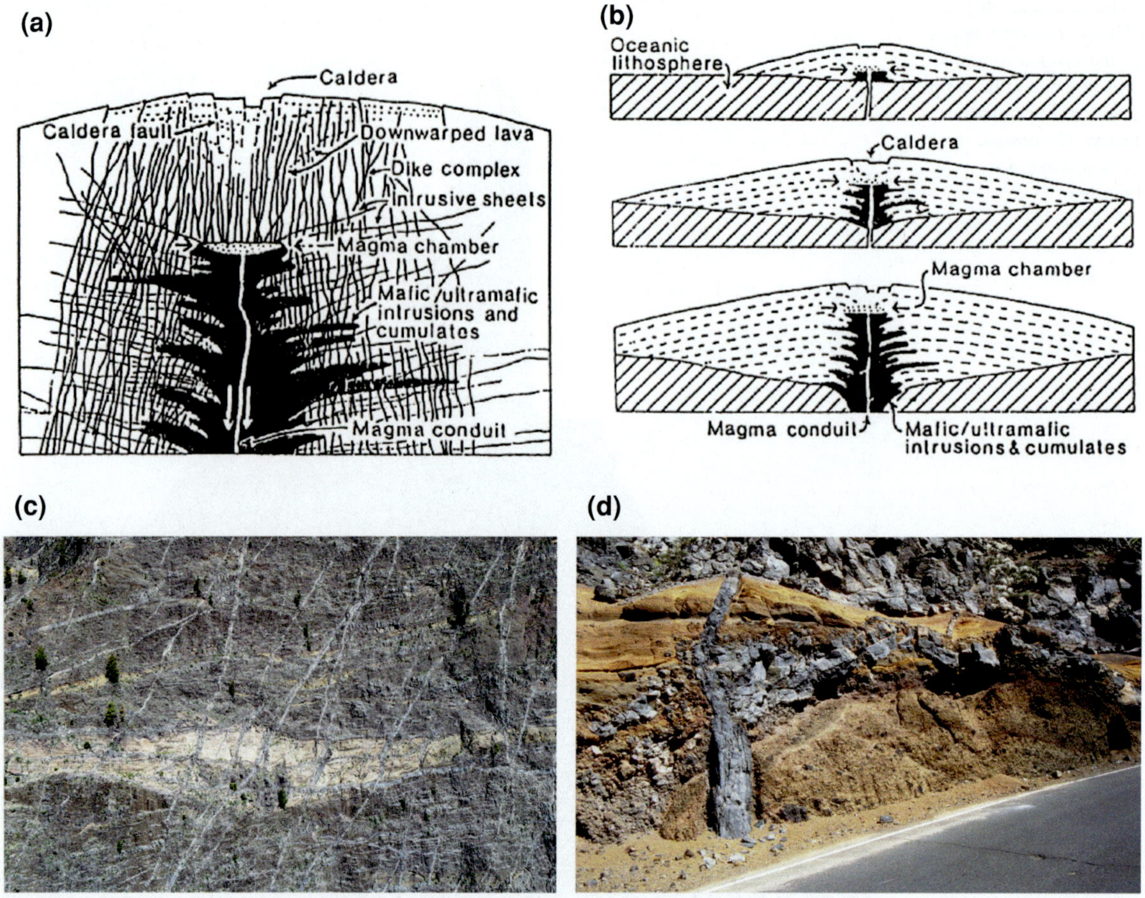

Fig. 8.28 Sill and dyke complexes in basaltic shield volcano edifices. **a**, **b** Schematic representation of the magma plumbing system in a marine basaltic shield volcano, showing a complex network of sills and dykes emanating from a major central conduit (From Walker 1989). **c** Network of sills and criss-crossing dykes contributing to the volcanic edifice of La Palma, Canary Islands, Spain. Height of cliff ~200 m. (Photo by Sam Thiele). **d** Feeder dyke through a basaltic scoria cone, sourcing a sill branching irregularly to the right above the road, and a lava flow at the top of the field of view, which is 20 m high, Dorsal Ridge, Tenerife, Canary Islands, Spain

addition, flank and parasitic vents are also common, and these can take the form of small parasitic cones, fissure vents, or lava domes, all resulting from dykes, or sills that radiate from the central conduit. High-resolution 3D seismic studies of buried volcanic edifices in marine sedimentary basins are revealing more detail, including a "Christmas tree"-like magma plumbing system.

Kora is a submarine, Miocene (20–8 Ma) andesitic polygenetic stratovolcano with a basal diameter of about 10 km, buried by 1–2 km of Pliocene–Pleistocene sediments in the Taranaki Basin, offshore New Zealand. It is unusual in that it has been the location of hydrocarbon exploration and is well-imaged by seismic reflection data (Bischoff et al. 2017; Infante-Paez and Marfurt 2017; Morley 2018; Chap. 18). The Taranaki Basin is a Late Cretaceous rift containing as much as 9 km of Cretaceous–Cenozoic sediments, controlled by half-graben formation, large-displacement normal faults, and thick sequences of syn-tectonic sedimentation. 3D imaging has provided a rare opportunity to examine in detail how the plumbing system of a larger volcanic edifice interacted with the syn-rift structuring and affected the shallow igneous intrusion architecture (Morley 2018). However, the seismic coverage is not perfect and, passing towards the core, data is degraded by the high density of igneous intrusions, energy adsorption, and the presence of steeply dipping beds (Fig. 8.29a). Consequently, in about a 5 km diameter region beneath the Kora volcano there is little coherent seismic reflection data. Hence, information on intrusion geometry must be projected from the well-imaged parts of the seismic data into that area in order to develop a model for the volcano plumbing system. The following summary is taken from Morley (2018).

The shallow (<8 km depth) architecture of the subvolcanic intrusive complex is dominated by stacked saucer-shaped and transgressive sills, and laccolith complexes distributed in an oval shape around the volcano. The thickest sill complexes occur in the deepest (fault-controlled) syn-rift depocentres (Fig. 8.29b). There is also local elongation of sill complexes in a north–south direction following structural trends. Numerous sills locally follow segments of large normal faults, or terminate at the faults. Secondary volcanic (flank) cones have stacked sill complexes beneath them that are focused on faults. Laterally extensive sill complexes tend to lie close to the base of the syn-rift/top of pre-rift contact for distances up to 15 km away from the Kora volcano (Fig. 8.29b). A general pattern is that the lowest level sills are the most laterally extensive, and the shallower sills lie closer to the volcano. Shallow sills form a complex of concentric sills that dip away from the volcano centre (or "ring sills"). "Forced" folding is associated with some but not all sills around Kora; some of the folds are domal, while others elongate along the main structural grain (N-S or NE-SW). In addition to the short wavelength forced folds around the outer regions of the volcano, there is significant uplift and folding directly under the volcano. 3D seismic data shows that extensive stacks of sills are present immediately adjacent to the volcano. The zone over which the sills occur thickens passing towards the volcano (Fig. 8.29b). It seems clear that the sills preferentially intruded the thick, predominantly fine-grained, probably organic-rich syn-rift section, and that a similar, but more intense intrusion pattern continues beneath the Kora volcano as schematically illustrated in Fig. 8.29b.

Sills are interpreted to have been fed by intrusions following faults that intersect a broad upper crustal magma chamber with a calculated area of ~ 350 km^2 (Fig. 8.30). Mapping of the syn-rift section adjacent to the volcano and projection of the extensional fault-controlled section beneath the volcano indicates there is a thick syn-rift section beneath the volcano. Hence, evidence from both the syn-rift morphology, and the thickness trends of the sill-affected intervals, permit the proposition that stacks of sills/laccoliths connected by short dikes, transgressive sills, and probably pipes dominate the intrusions style beneath the volcano as shown in Fig. 8.29b. It is suggested that the syn-rift section stratigraphy favoured the development of sills, and inclined transgressive intrusions, over the development of a simple pipe with dikes type of feeder system (Fig. 8.30). Strong preference for sill/laccolith stacking in the thick syn-rift section appears to have had four effects: (1) in the upper 6 km, there probably was not a simple single pipe feeding the volcano, instead the volcano is inferred to have been fed by a network of sills linked by short dikes and pipes; (2) the extensive intrusions resulted in folding/doming of the Paleogene section in several places, but most significantly under the central part of the volcano; (3) much of the magma was intruded within the syn-rift section and the underlying magma chamber (~ 125 km^3), rather than extruded (~ 40 km^3); (4) edifice construction and broad doming related to subsurface intrusions may have caused partial subaerial emergence of the crest of the volcano.

8.4.3 Mafic and Ultramafic Layered Intrusions

Thicker still than mafic sill complexes that represent part of the conduit system to flood basalt provinces (Sects. 8.3.2 and 8.3.2.6) are mafic to ultramafic layered intrusions, which also form regionally extensive sub-horizontal sills, sheet-like bodies, or lopoliths. Such intrusions can be many km^2 in area, covering from around 100 km^2 to over 50,000 km^2, and can be several hundred m to several km thick. Most layered intrusions are Archean to Proterozoic in age; for example, the Bushveld complex (~ 2.0 Ga), which is the largest known layered intrusion and crops out around the upturned margins of the Transvaal Basin, South Africa; the Great Dyke, Zimbabwe (~ 2.6 Ga), which is not a dyke but a lopolith and Y-shaped in cross-section; Voisey Bay (~ 1.3 Ga), Labrador, Canada; and, in the USA, the Stillwater Complex (~ 2.7 Ga) and the Duluth Complex (~ 1.1 Ga), which is a composite gabbroic sill up to 16 km thick! However, layered intrusions may be of any stratigraphic age, such as the Panzhihua layered intrusion (Permian) in southwest China, the Norilsk Complex (end-Permian), Russia, and the Tertiary Skaergaard intrusion, east Greenland. Many of these intrusions are spatially and genetically associated with continental flood basalts and represent major magmatic events (e.g. at Norils'k within the Siberian Traps, Panzhihua in central part of the Emeishan large igneous province, and Duluth Complex in the Keweenawan mid-continental rift), generally inferred to have mantle plume origins (Chap. 17). Although most layered intrusions are ultramafic to mafic in composition, the Ilimaussaq Complex (~ 1.2 Ga) of Greenland is an alkalic intrusion. For a current review of the geology of layered intrusions, see Charlier et al. (2015).

Layered intrusive complexes exhibit evidence of fractional crystallisation and crystal segregation from a melt, during which orthomagmatic mineral ore deposits may form (Chap. 18). These can be very important economically for the deposition of metals such as nickel, copper, and platinum group elements, and chromium and iron as well as other elements such as palladium, rhodium, ruthenium, iridium, osmium, titanium, vanadium, and gold (Naldrett 1997, 1999, 2004; Lightfoot and Hawkesworth 1997; Arndt et al. 2005; Lightfoot and Keays 2005; Song et al. 2011; Keays et al. 2012; Lesher and Lightfoot 2012; Chap. 18). At Norils'k, several giant deposits are hosted within sill-like feeder

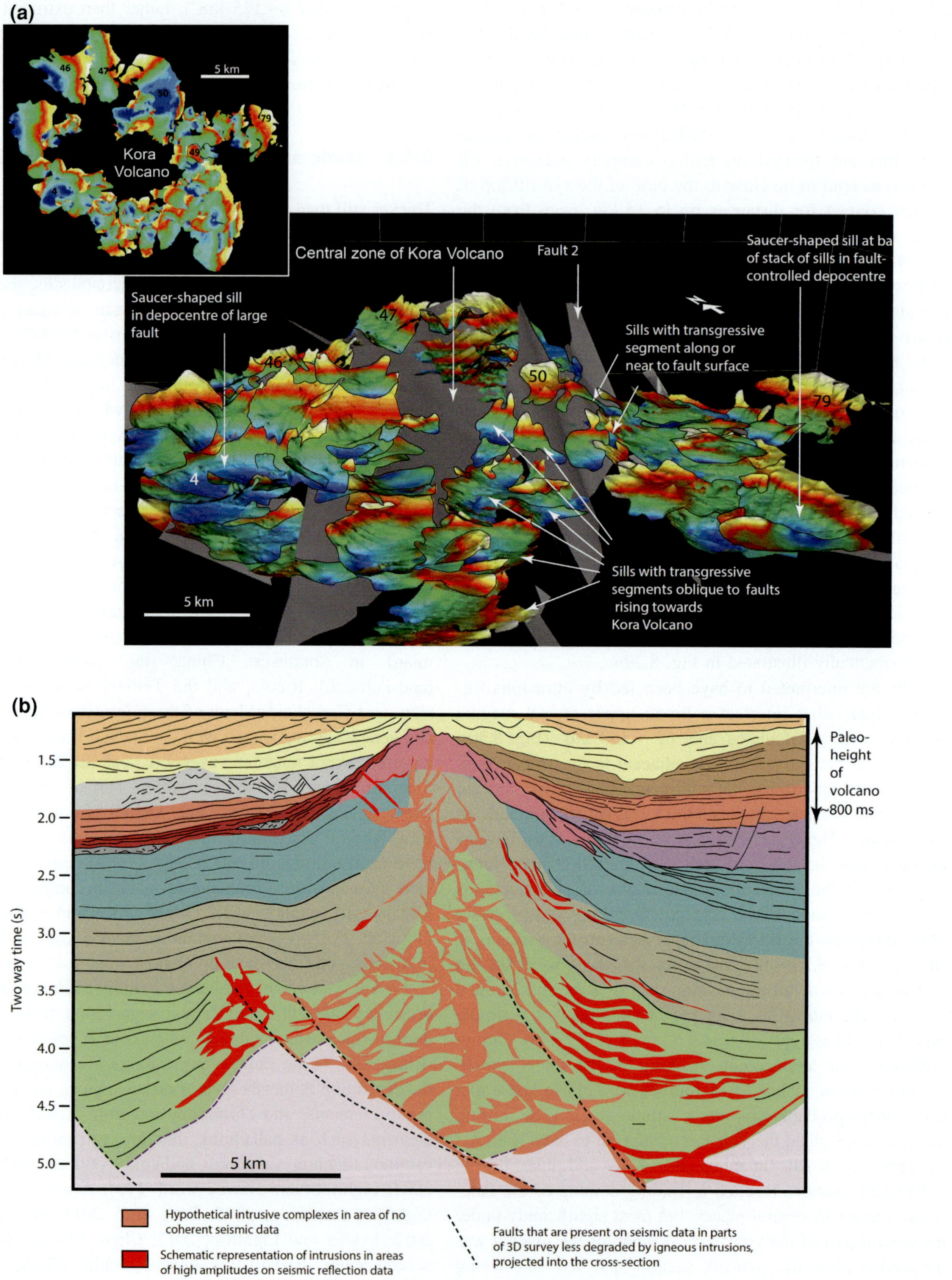
(a)
46
47
5 km
50
79
49
Kora Volcano
Central zone of Kora Volcano
Fault 2
Saucer-shaped sill at ba
of stack of sills in fault-
controlled depocentre
Saucer-shaped sill
in depocentre of large
fault
Sills with transgressive
segment along or
near to fault surface
4
5 km
Sills with transgressive
segments oblique to faults
rising towards
Kora Volcano
(b)
1.5
2.0
2.5
3.0
3.5
4.0
4.5
5.0
Two way time (s)
Paleo-
height
of
volcano
~800 ms
5 km
Hypothetical intrusive complexes in area of no
coherent seismic data
Schematic representation of intrusions in areas
of high amplitudes on seismic reflection data
Faults that are present on seismic data in parts
of 3D survey less degraded by igneous intrusions,
projected into the cross-section

◀ **Fig. 8.29** Sills mapped in 3D seismic data in the area of Kora stratovolcano, Taranaki Basin, offshore New Zealand. **a** Oblique perspective view to the northeast. The sills occur at different levels in the basin, hence the colours are only relative values where blue is the deepest part of the sill and yellow is the highest part. There are 108 mapped sills, hence in many places the sills are stacked and it is not possible to see all sills. Inset shows map view; selectively numbered sills (4, 46, 47, 50, 79) provide location reference. Note, towards the volcano imaging of the sills decreases due to degradation of the seismic reflection data (see text), and in about a 5 km diameter central zone there is little coherent seismic data. **b** Interpretation of WNW-ESE seismic line. The poorly imaged central zone under the Kora volcano is interpreted as one dominated by sills, which are linked by transgressive sills, short dykes, and possibly a pipe system that is offset by sills. This is based on projections of intrusion geometry from the well-imaged parts of the seismic data. (Adapted from Morley 2018)

systems to the overlying flood basalt succession (Naldrett et al. 1995; Fig. 8.31a). Feeder dykes are also important at Voisey Bay in delivering magma to the main intrusive "chamber" complex and location of Ni–Cu sulphide mineralisation (Saumur and Cruden 2016). The overall dyke system has an upward-curving geometry such that their dip shallows upward, which is where the complex mineralisation is located (Fig. 8.31b). Some komatiite successions also contain thick, differentiated ultramafic (dunite) sills which host nickel sulphide mineralisation (Chap. 7, and see discussion in Sect. 8.5). Larger mafic and ultramafic intrusions form both sulphide (Ni–Cu–PGE–Au) and oxide (Cr–Fe–Ti–V) orebodies, while komatiites solely host sulphidic orebodies.

8.4.4 Architecture and Processes in Upper Crustal, Explosive Caldera Magma Reservoirs

One of the most discussed topics over the past decade in volcanology has been the reconsideration of the concept of magma chambers. The classic view of a magma chamber, that has satisfied petrologists for over a hundred years (e.g. Bowen 1915), is that it comprises a large pool of liquid (eruptible) magma which homogeneously and permanently occupies a volume within the crust, hosted within the country rocks (Fig. 8.32a). This concept of a magma chamber or "*standard*" model (so defined in Gualda and Ghiorso 2013) is largely based on the following evidence: (1) the most common eruption of crystal-poor magmas (<30% phenocrysts); (2) petrological evidence for crystal fractionation, envisaged as crystals forming and settling through a liquid magma; (3) petrologic and textural evidence for mixing and mingling of recharge (more primitive) magmas and resident (more evolved) magmas; (4) the plutonic equivalent, where batholiths have been seen as long-lived crystallised magma chambers; (5) emplacement of zoned ignimbrites interpreted as the reverse magma withdrawal from a single magma chamber zone by composition, density, and temperature (see Chap. 10); (6) caldera collapses (Chaps. 10, 16), which occur as a result of rapid magma withdrawals, which are envisaged as emptying the magma chamber cavity.

However, over the years, other lines of evidence have progressively accumulated regarding the eruption of caldera-forming ignimbrites (Chaps. 10, 16), which have developed and shaped the so-called *magmatic mush* model (e.g. Hildreth 2004; Allan et al. 2013; Bergantz et al. 2015; Kennedy et al. 2018; Fig. 8.32b), as well as information from mafic volcanoes and intrusive complexes (Marsh 2004, 2015; Fig. 8.32c). The main lines of evidence for the magmatic mush model are summarised as (1) lack of geophysical evidence (mostly from seismic studies) for the existence of regions of the crust largely melted, even below the largest and hottest volcanic centres, such as the Yellowstone super-volcano caldera (e.g. Chu et al. 2010; Chap. 16), or even below active mid-oceanic spreading ridges; (2) textural and petrographic evidence for the concurrent presence of different magma types within the same erupted product indicating the syn-eruptive activation of different magma batches (e.g. Cashman and Giordano 2014; Kennedy et al. 2018); (3) petrographic and textural evidence for eruptible magma being stored within crystal mushes (Bachman and Bergantz 2004); (4) geochronological and mineralogical evidence for rapid to very rapid (less than thousands, down to hundreds of years or even less) extraction, upward migration and assembly of large volumes of eruptible magmas in shallow reservoirs prior to eruption (e.g. Gualda et al. 2012; Druitt et al. 2012; Pamukçu et al. 2020), strongly contrasting with the long-lived geochronology of intrusive bodies (e.g. Curry et al. 2021; Glazner et al. 2004); (5) polybaric signatures of magmas erupted during the same event (e.g. Roberge et al. 2013).

It must be noted also that recent studies of intrusive complexes have progressively shown how large and long-lived plutons are assembled incrementally in discrete pulses (e.g. Curry et al. 2021; Glazner et al. 2004), which may coalesce to form voluminous magma chambers only at average magma fluxes above 10^{-2} km^3 yr^{-1} (Annen 2009, 2011). At the same time, in order to assemble large volumes of eruptible magma, the plumbing system must become thermally and chemically mature, which modeling shows requires in the order of 5-10x10^6 yr (Giordano and Caricchi 2022).

At present, there is still no general consensus on the frequency or importance of either model, although it is

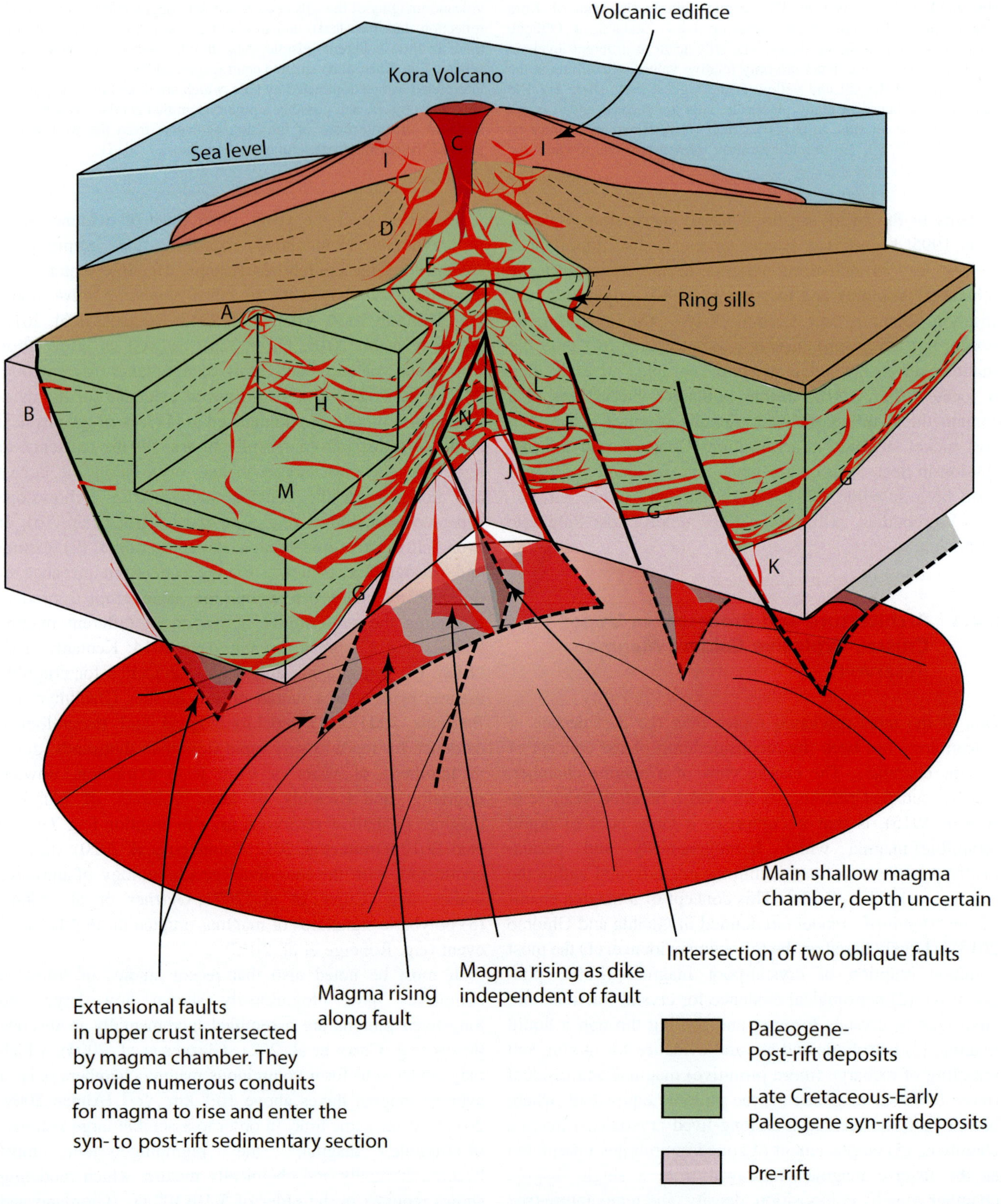

Fig. 8.30 Schematic 3D representation illustrating key features of the Kora volcano plumbing system, and influence of the syn-rift faults on intrusion geometry and distribution. A = intrusions causing forced folding; B = intrusions along boundary fault curving into sills in the basin; C = main volcanic pipe; D = ring sills intruding forced folds around core of the volcano; E = complex of sills at depth forming feeder system to the volcano; F = sills within syn-rift depocentre; G = sills tending to build up extensively along the syn-rift to pre-rift unconformity; H = sills stacked in particular localities, location H lies between two such stacks where the sills tend to curve upward; I = high-level sill and feeder dyke complexes that lead to parasitic volcanic cones; J = intrusions along fault planes feeding magma from the magma chamber to sill complexes within the syn-rift section; K = feeder intrusions along fault planes to location of G sills; L = sills curving upward following fault propagation fold-related dips in the upper part of syn-rift section; M = saucer-shaped sill; N = laccolith-type intrusions in core of volcano feeder system. (From Morley 2018)

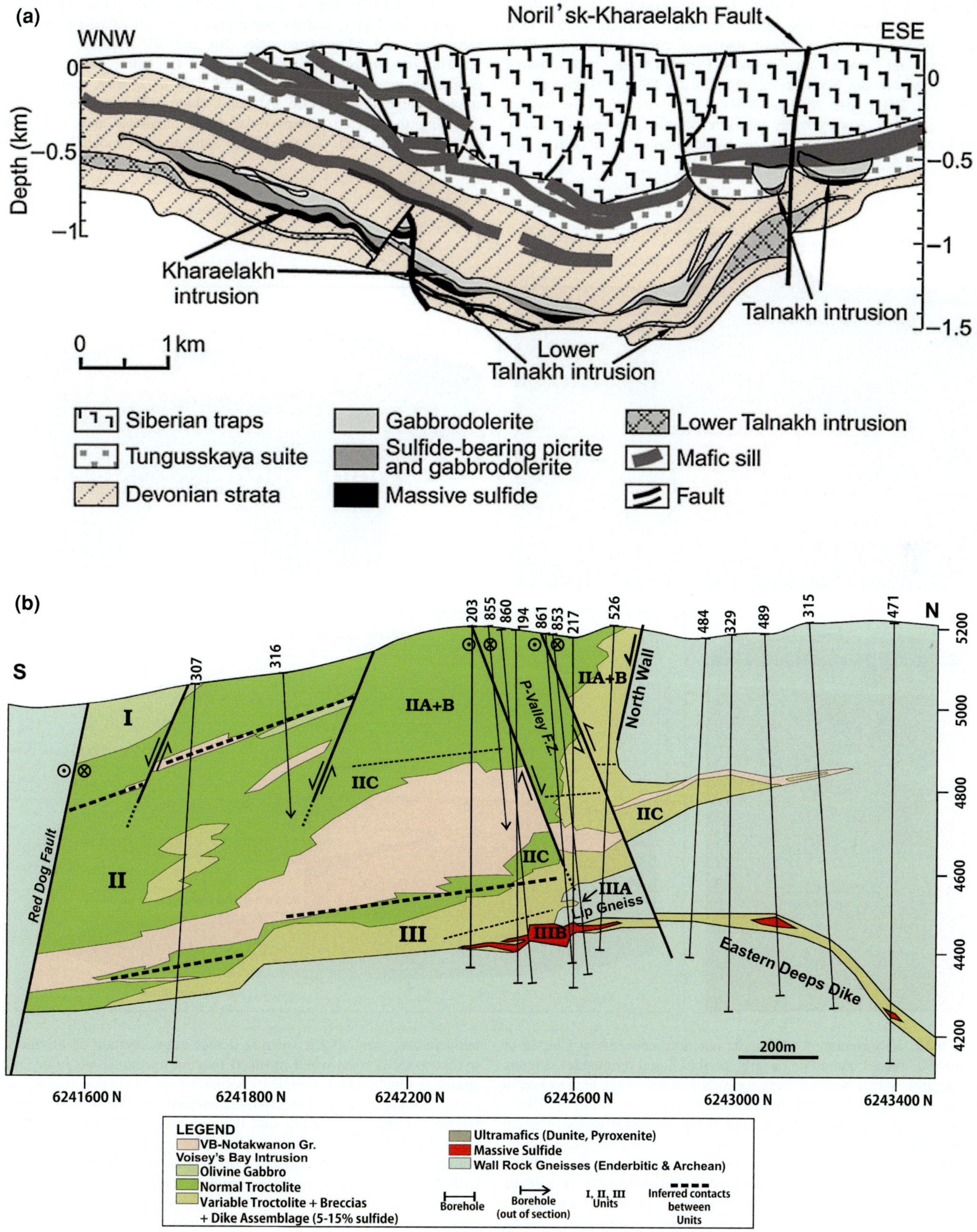

(a)
WNW
ESE
Noril'sk-Kharaelakh Fault
Depth (km)
0
−0.5
−1
−1.5
Kharaelakh intrusion
Talnakh intrusion
Lower Talnakh intrusion
0 1 km
Siberian traps
Tungusskaya suite
Devonian strata
Gabbrodolerite
Sulfide-bearing picrite and gabbrodolerite
Massive sulfide
Lower Talnakh intrusion
Mafic sill
Fault
(b)
S
N
Red Dog Fault
I
II
IIA+B
IIC
IIIA
IIIB
III
P-Valley F.Z.
North Wall
Lip Gneiss
Eastern Deeps Dike
307
316
203
855
860
194
861
853
217
526
484
329
489
315
471
5200
5000
4800
4600
4400
4200
200m
6241600 N
6241800 N
6242200 N
6242600 N
6243000 N
6243400 N
LEGEND
VB-Notakwanon Gr.
Voisey's Bay Intrusion
Olivine Gabbro
Normal Troctolite
Variable Troctolite + Breccias + Dike Assemblage (5-15% sulfide)
Ultramafics (Dunite, Pyroxenite)
Massive Sulfide
Wall Rock Gneisses (Enderbitic & Archean)
Borehole
Borehole (out of section)
I, II, III Units
Inferred contacts between Units

◀ **Fig. 8.31** Layered mafic intrusions hosting worldclass Ni–Cu–(PGE) deposits. **a** Schematic cross-section showing distribution and field relationships of the three largest intrusions containing Ni–Cu–PGE sulphides, and contemporaneous latest Permian "trap" succession of flood basalts in the Noril'sk-Talnakh district, Siberia, Russia (From Song et al. 2011, After Naldrett 2004). **b** Cross-section of the Eastern Deeps Chamber, Voisey's Bay intrusive complex, Labrador, Canada. The Eastern Deeps Chamber can be divided into six units (I, IIA, IIB, IIIC, IIIA, and IIIB) that are distinguished based on geometrical styles, dominant lithologies, and the occurrence of bodies of wall rock partly separating units. Mineralised units IIIA and IIIB from the lowermost level are irregularly shaped and consist of variable troctolite, breccias containing cm-sized xenoliths of host rock paragneiss, and strongly sulphide-mineralised rocks. At least three different feeder dykes including the Eastern Deeps dyke deliver magmas associated with mineralised unit IIIB. (Adapted from Saumur and Cruden 2016)

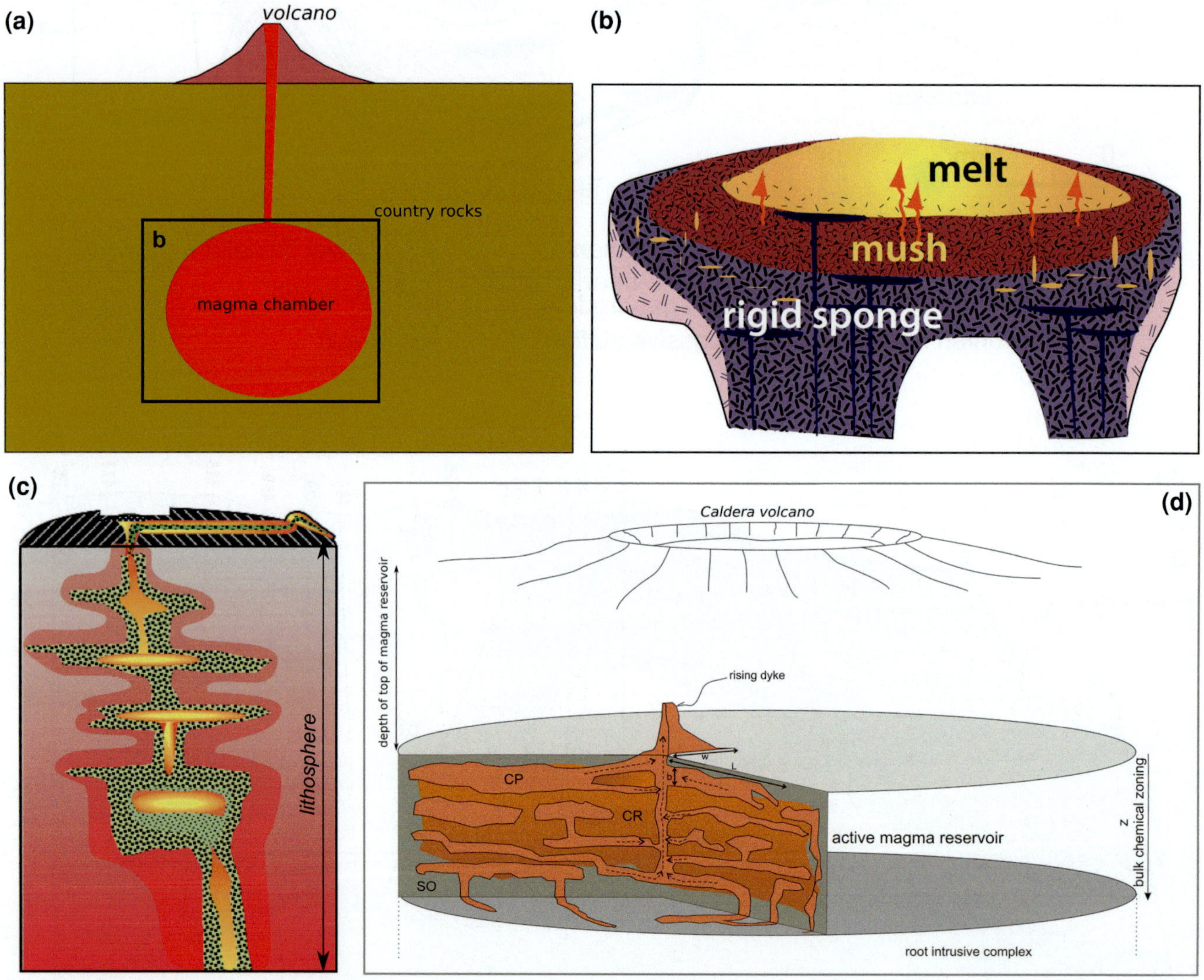

Fig. 8.32 Representation of magma chamber concepts. **a** Classic or "standard" model comprising a large pool of liquid (eruptable) magma. **b** Magmatic mush model occupying approximately same area as box in (**a**) for large silicic magma chambers as proposed by Hildreth (2004). Eruptions are fed from a single melt body contained within a much larger and more crystalline system comprising a crystal mush (>50% crystals) and surrounding rigid sponge (~65% crystals). (From Cashman and Giordano 2014, Modified from Hildreth 2004). **c** Mafic magmatic mush column comprising a series of interconnected magma chambers as envisioned for hawai'ian-type shield volcanos and mafic sill complexes (From Marsh 2004). Colour hotness portrays magma temperature; small black squares depict large crystals of olivine or pyroxene. Sills form when repeated magma injections are sufficiently spaced in time to allow complete cooling between injection events; melt lenses dominate when new inputs are added to systems that are still partially molten, or within sills that are sufficiently thick to allow internal redistribution of melt. **d** Model for syn-eruptive assembly and withdrawal of eruptible crystal-poor magma (CP) from a plexus of melt-dominated lenses (sills) and dykes (each characterised by an aperture b, a width w, and a length L), dispersed with a rigid crystal mush (CR), and embedded in a solidified intrusive complex (SO) (Modified from Cashman and Giordano 2014)

certainly possible that both may represent end members which may co-exist at different time and length scales. For example, it is likely that magma chambers occurring at frequently erupting volcanoes, such as at hot spots or in mafic to intermediate volcanic arc settings (Chaps. 16, 17), may be relatively small and look like the standard model at the time of eruption. Alternatively, larger systems in arc- and extensional settings may more likely relate to mush models with large crystallised domains and localised melt-dominated regions (see Fig. 8.1a). At the same time, each localised melt-dominated region within a mush zone, called compartments by Gudmundsson (2012), in itself is a classic magma chamber, so there is no real need to choose between one model or another, but to understand case by case what fits the evidence. Current thinking suggests the magma systems that occur below large silicic caldera volcanoes involve crystallisation of mafic to intermediate forerunners in the lower crust, from which evolved magmas are extracted and migrate towards the upper crust yielding an intermediate to silicic crystal mush (e.g. Allan et al. 2013). In particular, Cashman and Giordano (2014; Chap. 16) stressed that melt-dominated regions may be dispersed within the crystal mush and can be tapped simultaneously during an eruption (also see Gudmundsson 2012; Gualda and Ghiorso 2013; Kennedy et al. 2018). An essential point to note here is that magma reservoirs (or chambers) should not be seen as permanently present in the upper crust, but undergoing stages of fast replenishement and withdrawal, as compared to the long time-scales of magma production in the deep hot roots of the magmatic plumbing system (e.g. Pamukçu et al. 2020; Giordano and Caricchi 2022). We propose the use of *magma reservoir* to describe such complex systems (Fig. 8.32d).

The main volcanological implications of treating magma chambers as magma reservoirs can be summarised as follows:

(a) Magmas can accumulate and differentiate over time within rheologically rigid large crystal mush-dominated regions, forming a plexus of more or less interconnected melt-dominated regions.
(b) The actual eruptible magma volume, composition(s), and crystallinity depend on the immediately pre-eruptive and/or syn-eruptive extent of the interconnection among the melt-dominated regions.
(c) As eruptible magmas are scattered throughout the rigid mush, their vertical and lateral extent can be much larger than that enclosed in the related Dense Rock Equivalent volume erupted.
(d) During caldera-forming events, the strength of the rigid mush framework can modulate the timing of the caldera collapse (see Chaps. 10, 16).
(e) Occasionally, crystal mush can be erupted wholesale, forming crystal-rich ignimbrites ($\geq 50\%$ crystals; e.g. Willcock et al. 2015; Chap. 10).
(f) Volcanic conduits can extend inside the reservoir, tapping various lenses of eruptible melt of not necessarily similar composition at once, instead of classically starting from its top.
(g) With regard to pressure drivers of the eruption, initially associated with pre-eruptive magma pressurisation, it is possible to envisage a progression of decompressional events during the course of an eruption associated with magma withdrawal, which are able to activate and interconnect with deeper and/or more (laterally) distant melt-dominated regions.
(h) The mechanics of later, post-eruptive relaxation and/or re-pressurisation of the magma reservoir are controlled by the rheological behaviour of the mush (e.g. Galetto et al. 2017).

8.4.5 The Role of High Crustal Level Magma Reservoirs and Dykes in Caldera Collapse

Many volcanic edifice types, including caldera volcanoes, stratovolcanoes, and shield volcanoes (see Chap. 16), are characterised by the presence of large collapsed areas, called calderas, usually larger than 2–3 km, and up to 90 km across, which make them clearly distinguishable from craters, which are smaller. According to the classical mechanism of caldera "collapse", collapse of the crustal roof block above a high-level magma reservoir occurs because of the depressurisation of the reservoir related to rapid magma withdrawal that causes progressive underpressure (Fig. 8.33; Druitt and Sparks 1984; Martì et al. 2009; Geshi et al. 2014). "Collapse" of the roof block involves its subsidence into the underlying magma reservoir as magma is evacuated.

Rapid evacuation of magma not compensated at the reservoir level by recharge is possible only during the most voluminous explosive eruptions (usually involving minimum volumes of at least several and up to thousands of km^3 (e.g. 79 AD eruption at Vesuvius, Italy, the 1815 eruption at Tambora, Indonesia, and the 1883 eruption at Krakatoa, Indonesia; Chaps. 10 and 16), or by effusive drainage of shallow basaltic magma chambers at shield volcanoes (Chap. 16) by the propagation of lateral dykes and opening of vents along the edifice flank (e.g. Galapagos, Hawaii; see Sect. 8.3.1).

Collapse of the roof block is facilitated by sub-vertical faults that form inside the areal extent of the underlying magma reservoir and bound the subsiding block. These faults are commonly called "ring faults", which is

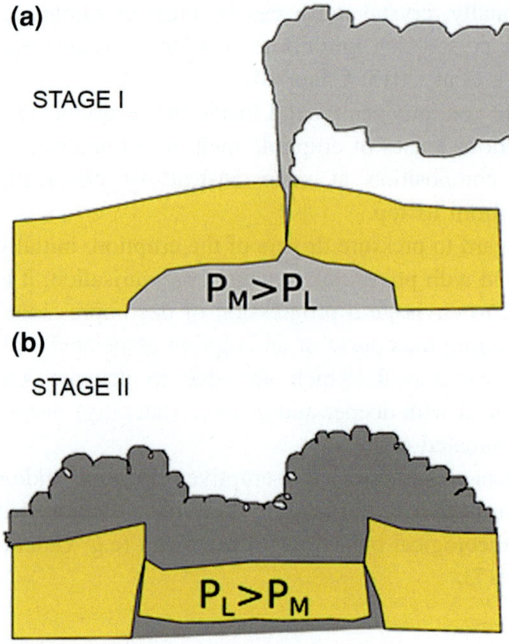

Fig. 8.33 Stages of caldera collapse. Beginning with **a** Stage I, magma overpressure, P_M, greater than lithostatic pressure, P_L, causing tumescence and initial stage of magma chamber roof block failure, often associated with opening of a point source vent. **b** Stage II, underpressure, and collapse of the chamber roof block into the magma chamber as magma is evacuated from it along dykes hosted by an outward dipping ring fracture fault and conduit system. (Modified from Druitt and Sparks 1984)

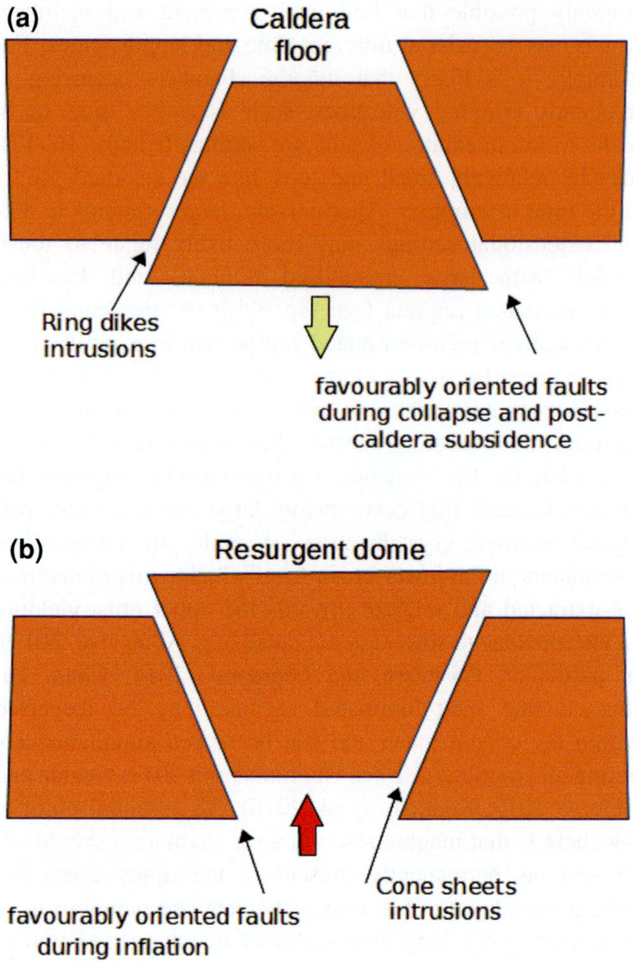

Fig. 8.34 The geometry of collapse faults controls the caldera subsidence and intrusive history. **a** In the case of caldera to post-caldera subsidence, outward-directed faults can further dilate and accommodate dyke intrusions that form *ring dyke* complexes. **b** In the case of resurgence, outward-directed faults lock, while new or reactivated inward-directed faults accommodate uplift. Intrusions along such faults form *cone sheet* complexes

appropriate for circular calderas, but they can be complex in geometry depending on the pre-existing structural fabric of the roof rocks, and often there are multiple faults involved.

Large explosive calderas, with areas >100 km^2, are thought to originate from an early overpressure in the magma reservoir causing tumescence or doming of the crustal roof block overlying the magma reservoir (Fig. 8.33a, Stage I). As magma is evacuated through eruption the roof block initially downsags and inward dipping ring faults form. For caldera collapse to occur however, new faults must form that are outward-directed to allow subsidence of the caldera block (Figs. 8.33b, 8.34a; Acocella et al. 2000; Roche et al. 2000). Such faults, while geometrically inverse or reverse, are kinematically normal because they form in response to a main stress σ_1 which is vertically oriented. But they should not be confused with tectonic reverse faults in which σ_1 is horizontally oriented (Fig. 8.34a, b). During tumescence, the initial faults can become magma conduits, and through the initial stages of eruption, the magma system transforms from overpressured to underpressured, as magma is evacuated from the magma reservoir (Fig. 8.33b, Stage II). At this stage, outward dipping ring and other faults form to mechanically accommodate subsidence, and these also become magma conduits as the roof block subsides and the outward dipping ring faults dilate. Dykes along such caldera-related faults are defined as *cone sheets* when intruding the inward dipping faults and *ring-dikes* when intruding outward dipping faults (Anderson 1937a, b). They can thus not only facilitate the eruption of huge volumes of magma during times of overpressures and underpressures, but also facilitate caldera collapse during underpressure (Fig. 8.33b, Stage II).

Cashman and Giordano (2014) suggested that the timing of caldera collapse and geometry of the vent system reflect not only the crustal roof aspect ratios but dominantly the style of magma extraction from the reservoir. Their hypothesis suggests that eruptible magma does not necessarily only accumulate at the top of the reservoir of crystal-rich magma as envisaged in the standard model, but

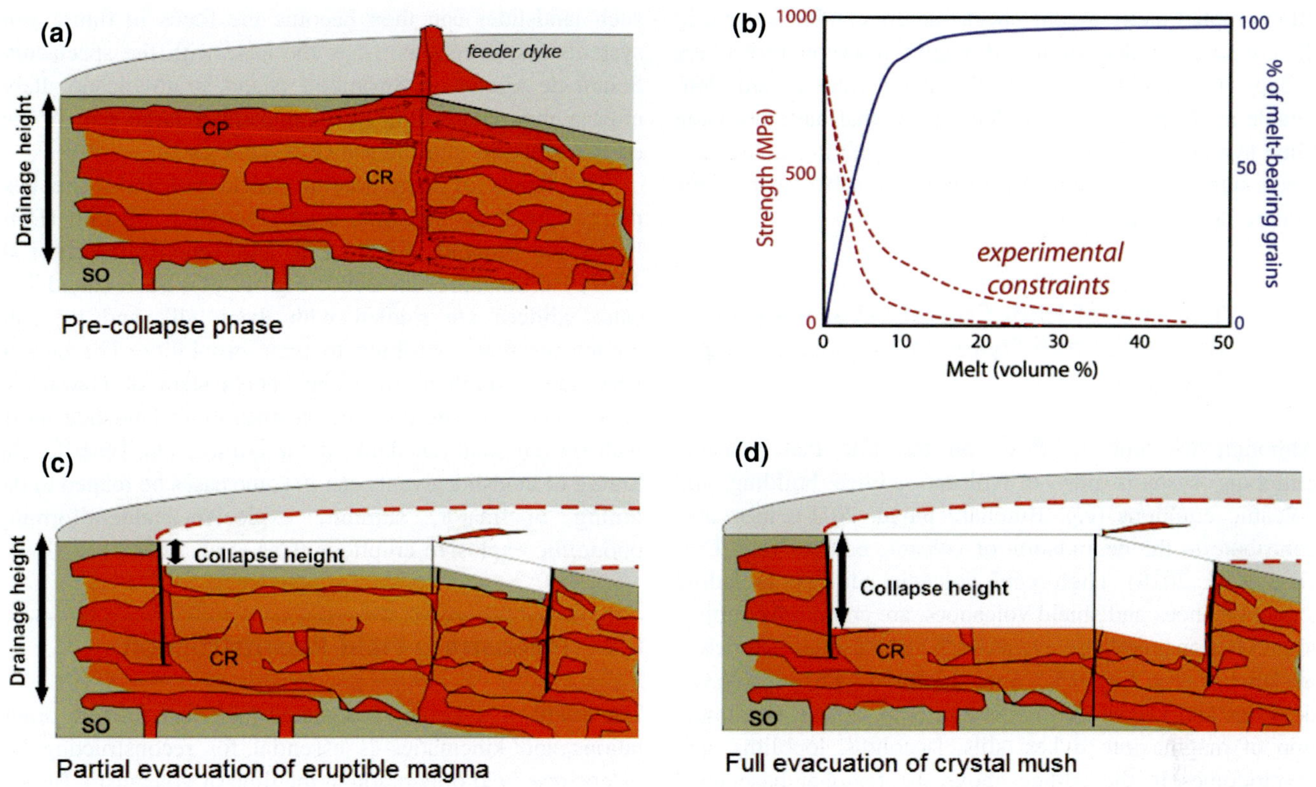

Fig. 8.35 An alternative model for caldera collapse modulated by the strength of the magma reservoir (Modified from Cashman and Giordano 2014). **a** Conceptual model for magma reservoirs made of variably interconnected and interconnectable pods, lenses, dykes, and sills of eruptible (usually) crystal-poor magma (CP) set in a framework of progressively crystal-richer mush (CR) and intrusive rocks (SO). **b** Variation of strength as a function of percentage of melt (dashed lines) versus crystal framework (solid line) illustrates that the driving pressures needed to mobilise magma are progressively larger at higher crystal contents (After Rosenberg et al. 2007). **c** For withdrawals that leave a substantial part of the crystal-rich framework, the strength of the framework contributes to modulating the timing of collapse along with the strength and aspect ratio of the roof; in this case, which is different from classic models, the collapse height is different from the drainage height, i.e. the depth range from which the magma is mobilised. **d** For withdrawals that involve the wholesale mobilisation of the magma reservoir, including its crystal-rich framework, the model coincides with classic models where collapse height and drainage heights are the same; this would be the case for eruption of crystal-rich ignimbrites and possibly justifies very early caldera collapses

rather in a plexus of dykes, sills, lenses, and pods within a much larger and thermally stable crystal mush (Fig. 8.35a, c, d), collectively called a magma reservoir. The conduit therefore may extend deeper into the magma reservoir and tap (or not) different melt batches. This explains common complex chemical zoning patterns, and the usual very large depth ranges from which magma is drained (drainage height) recognised from melt inclusions (e.g. Hildreth and Wilson 2007). It is noticeable that while drainage heights (Fig. 8.35a) known for large volume ignimbrite eruptions are usually in the order of several km, the associated caldera collapse heights (Fig. 8.35c, d) are much smaller, generally < 1 km (Geyer and Martì 2008).

Within this new conceptual model, it is the strength of the reservoir crystalline framework or mush that also dictates the timing and style of collapse. Thus, early mobilisation of only volatile-rich and crystal-poor interconnected melt networks results in single-point source eruptions, which are common at the start of caldera-forming eruptions (Fig. 8.33a). If and when the reservoir strength is overcome by the load of the overlying roof block rocks (Fig. 8.35b), collapse occurs (also see Fig. 16.62c). Alternatively, early mobilisation of the entire crystal mush, such as during the eruption of very large volume, crystal-rich ignimbrites (e.g. Hildreth 1981; Maughan et al. 2002; Willcock et al. 2015), could result in a caldera collapse beginning at the start of the eruption and continuing throughout (i.e. piston collapse; Gregg et al. 2012).

After caldera-forming eruptions cease, and the magma conduits and vent systems have been sealed by solidified dykes and sills, further injection or recharge of magma into shallow crustal magma chambers can cause resurgent uplift or doming of the caldera floor, forming resurgent domes (Fig. 8.34b; Smith and Bailey 1968; Kennedy et al. 2012, 2018; Hildreth et al. 2017), which may be accompanied by small volume, usually effusive lava dome-forming eruptions, often erupted along a cone sheet fracture conduit system in the collapsed caldera floor and magma reservoir roof block (Chap. 16). In addition, given that many calderas are identifiable as deep topographic depressions that have not been filled syn-eruptively by intra-caldera ignimbrite, it suggests that a significant amount of post-eruption subsidence of the caldera floor occurs.

8.4.6 The Role of High Crustal Level Magma Reservoirs and Dykes in Volcano Edifice Collapse

Although we tend to focus on the role that volcanic plumbing systems play in volcano edifice building and volcanic eruptions (e.g. Burchardt et al. 2018), they also contribute to the degradation of volcanic edifices (e.g. Delcam et al. 2018). High-relief volcanic edifices, including stratovolcanoes and shield volcanoes, are commonly subject to gravitational collapse events such as landslides and debris avalanches (e.g. Chap. 12, Sect. 12.4.4, 12.4.5) (van Wyck de Vries and Davies 2015; Delcam et al. 2018). The injection of magma into dykes, sills, laccoliths, lopoliths, and cryptodomes in the edifice above the regional base level exerts an internal magmatic pressure and causes endogenous inflation, which augment the gravitational potential and stresses of the volcanic edifice, and destabilises it. Ground surface uplift is commonly a precursor to eruptions, caused by the effects of inflation as magma rises close to the earth's surface, and/or the effects of re-energising hydrothermal systems by injection of magma into them.

The surfaces of sub-vertical dykes and steep-sided cryptodomes can then also act as structural weaknesses along which detachment structures can be initiated, leading to landslide collapses (Delcam et al. 2018). Triggers for such detachments could include magma injection leading to pressure build-up, and over-steepening of the flanks of the edifice, and earthquakes, either of regional or subvolcanic origin. The spectacular collapse of Mt St Helens stratovolcano, USA, on 18 May 1980, was triggered by a magnitude 5 earthquake associated with the growth of a dacite cryptodome under the north flank of the volcano, which over-steepened, and as a result, collapsed gravitationally in a dramatic landslide and debris avalanche event (Voight et al. 1983; Figs. 12.8d, 12.10a–d), which removed a large part of the summit and north flank of the volcano, leaving a spectacular amphitheatre-like scar (Fig. 12.8e). The sudden removal of the landslide mass above the cryptodome caused immediate decompression of the magma, in turn triggering a plinian-scale explosive eruption (Chap. 9). The headwall of such landslides can then become the focus of future vent systems and eruptions, as is the case with the spectacular Sciara de Fuoco on Stromboli island stratovolcano, Italy, after a major landslide also removed a large part of the edifice that slid into the sea (Fig. 12.8 g).

The flanks of large marine shield volcanoes are also marked by huge landslide scars, including Hawai'i (Denlinger and Morgan, Fig. 12.11a) and Tenerife (Martí et al. 1997; Hürlimann et al. 2000; Fig. 12.8f). Such shield volcano edifices are riddled with dyke, sill, and laccolith complexes that contribute to their instability. The current very active southern rift zone vent system of Hawai'i is almost certainly the trace of an impending landslide headwall on the southern flank of the edifice. On Tenerife, the timing of landslide events can in some cases be related to the timing of major, summit, explosive caldera-forming phonolitic explosive eruptions (Cas et al. 2022).

8.5 Paleostress Field Reconstruction

The study of dyke and associated fault and fracture orientations and kinematics is essential for reconstructing the paleostress field orientation at the time of volcanism. Below, we describe two case studies, firstly of reconstruction of paleostress field, an ancient volcanic succession in southeastern Australia and, secondly, the stress field operating during an actual eruption at Etna volcano in Sicily. Italy.

8.5.1 Late Devonian Boyd Volcanic Complex, Southeastern Australia

The Late Devonian Boyd Volcanic Complex (BVC), NSW, provides a case study for reconstructing the paleostress field associated with volcanism in ancient terrains (Giordano and Cas 2001). The BVC comprises both subaerial and subaqueous volcanic and sedimentary facies, deposited within localised, rapidly subsiding basins. It is preserved over a distance of 100 km at the southern end of a Late Devonian belt of largely continental red beds and bimodal volcanic succession representing a north–south trending continental rift extensional basin (Cas 1983; Giordano and Cas 2001). The bimodal volcanics comprise rhyolitic lava flows and domes, welded and unwelded ignimbrites, basaltic spatter, and scoria cones in subaerial environments, while coherent domes, cryptodomes, and hyaloclastite, as well as dykes and sills, characterise subaqueous environments (Cas et al. 1990, 2001). The BVC underwent later deformational events, involving east–west compressional during the Early

Carboniferous (Cas 1983), and east–west extension during the Jurassic–Cretaceous opening of the Tasman sea (Giordano and Cas 2001).

Understanding paleostress at the time of volcanism is very important in order to define the geodynamic environment. The BVC is, in fact, one of several scattered, relatively small, and short-lived volcanic provinces that developed across the subduction-related, Cambrian–Carboniferous, Lachlan Fold Belt (LFB) between major compressional events (Cas 1983; Gray 1997). The important question is then whether the occurrence of magmatism indicates major rifting events or other types of thermal events. The key features of the BVC that were used to reconstruct the regional paleostress patterns during volcanism were the orientations of rhyolite, basalt, and clastic dykes, as well as faults and fractures (Giordano and Cas 2001; e.g. Fig. 8.36).

The principal observations, restored to syn-volcanic emplacement orientations, are:

- rhyolitic and basaltic dykes (Fig. 8.37a, b) are preferentially oriented along 320° to 340° trending fractures, that is, indicating NE-SW extension (Fig. 8.36),
- 330° trending dykes are always associated with 010° and 290° trending fracture sets (e.g. Fig. 8.36),
- some ~010° trending basaltic dykes and clastic dykes associated with magmatic intrusions show left-lateral syn-emplacement movement along such fractures (Fig. 8.37c),
- syn-depositional and syn-volcanic faults are ~330° trending normal faults. 040° trending normal faults are also present,

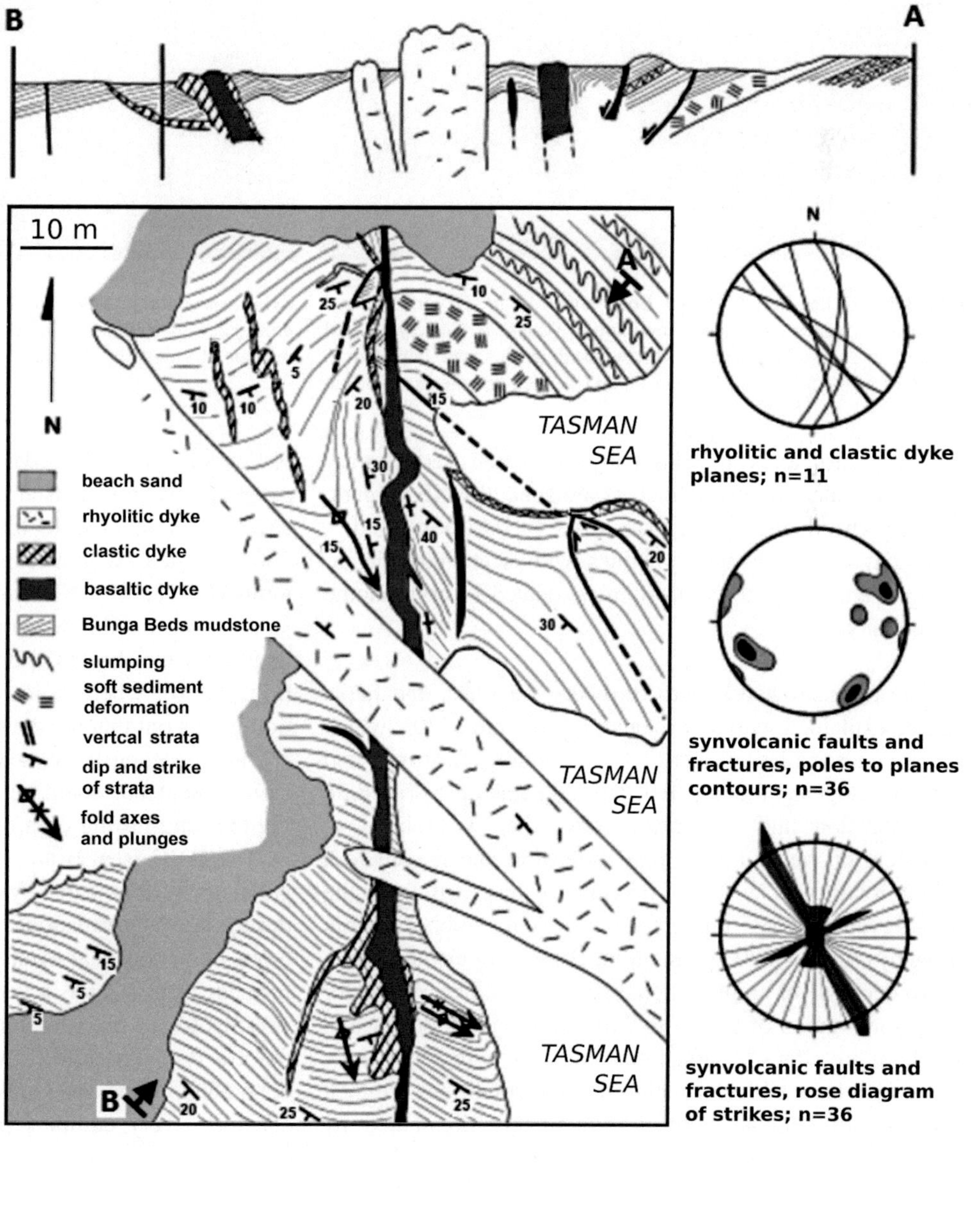

Fig. 8.36 Detailed geological and structural map, and cross-section of southern Bunga Beach, north of Bunga Head, Bunga Beds outlier, Late Devonian Boyd Volcanic Complex, southern New South Wales, Australia. The 330°-trending rhyolitic dyke and its 300°-trending apophysis cross-cut 350 to 010°-trending basaltic and clastic dykes, some of which show kinematic evidence for left-lateral trans-tension during emplacement. Syn-volcanic faults and fractures, as well as intra-formational slumping and peperite are developed. Carboniferous Kanimblan compression resulted in mild deformation except next to the dykes, where tight, small-scale anticlines, some with vertical limbs, are developed. Also shown are lower hemisphere plots of planes of dykes and fractures along Bunga Beach, north of Bunga Head. (From Giordano and Cas 2001)

Fig. 8.37 Dykes in the Bunga Beds outlier, Late Devonian Boyd Volcanic Complex, southern New South Wales. **a** Closely spaced rhyolite, basalt, and clastic dykes exposed in the area depicted in Fig. 8.28 cross-cutting folded and faulted bedded sediments. **b** Basalt dyke cross-cutting folded and faulted bedded sediments. **c** Clastic dyke cross-cutting folded and faulted bedded sediments, and showing left-lateral (sinistral) shear during emplacement

- slump structures in plastically deformed sedimentary rocks related to dyke intrusions are generally NE vergent, i.e. perpendicular to the main strike of intrusions, and
- brachy-anticlines and synclines with scattered orientations are associated with doming related to cryptodome intrusions.

The reconstructed pattern of dykes, fractures, and faults can be explained with a horizontal σ_1, oriented 330°, which would cause dilation along 330° trending fractures and sinistral movement along 010° trending fractures (Fig. 8.38 a). Extension and volcanism could then be related to trans-tension rather than a pure extension and related to a minor rotation of a still horizontal σ_1 inherited from the mid-Devonian E-W compressional event (Tabberabberan orogeny), to a ~330° orientation, contemporaneous with a sharp decrease of compressional stress. This then allows for the re-orientation of the σ_3 to 240° trending and horizontal. Furthermore, trans-tensional tectonics on a regional scale could also favour block rotation if some conditions are met. The eastern Lachlan Fold Belt is largely dominated by N-trending and NE-trending lineaments, and is also characterised by mid-crustal detachments (Gray 1997). This structure is favourable for block rotation. In particular, a 330° oriented σ_1 induces sinistral strike-slip on N-trending faults (Fig. 8.38b). NE-oriented blocks rotate anticlockwise, so allowing for the development of rapidly subsiding, scattered release basins ponded along NE-trending faults (Fig. 8.38b), where volcanism can occur. This reconstruction points out that the development of basins and volcanism are not necessarily related to big changes in the tectonic environment but could be interpreted as episodes punctuating the long-lasting compressional history of the LFB.

The BVC is also an excellent example of how the 3D geometry of a volcanic system influences strain partitioning during later deformational events. The following Carboniferous compressional event (Kanimblan orogeny; Cas 1983) formed thrusts and reverse faults trending ~160°, and verging to the E. Rhyolite, basalt, and sedimentary rocks deformed in very different ways during the Kanimblan event, depending on their original geometry (e.g. dyke vs. sill), and causing a very complex pattern of deformation where the

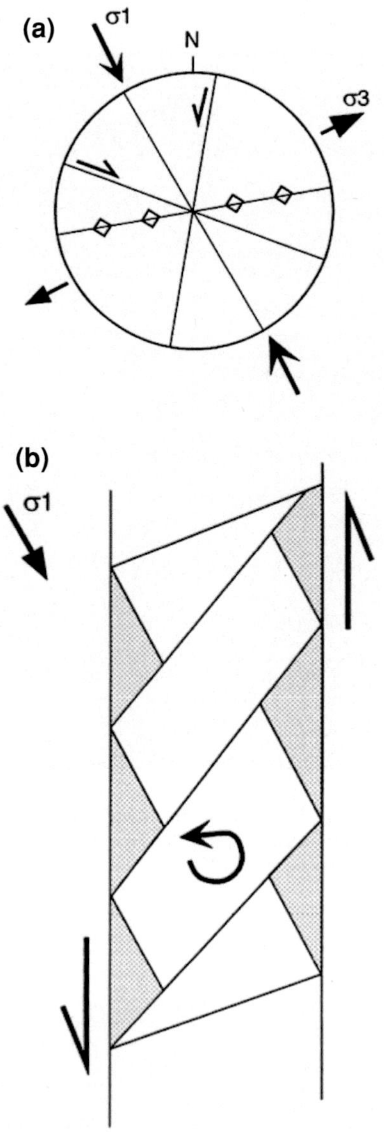

Fig. 8.38 Regional paleostress field reconstruction for the south coast of New South Wales, during volcanism of the Late Devonian Boyd Volcanic Complex. **a** Lower hemisphere diagram of the syn-volcanic deformation recognised in the Boyd Volcanic Complex and inferred orientations of σ_1 and σ_3. **b** Block rotation diagram with formation of release basins (shaded areas). (From Giordano and Cas 2001)

three lithologies are close to each other. Rhyolite domes and lava flows behaved as rigid plugs and deformed mainly by gentle flexing, with low-angle dips, and the development of mainly thrusts and quartz veins, allowing for mild tilting of large-scale blocks. Rhyolitic dykes were weakly deformed or undeformed during the Kanimblan deformation, whereas deformation was concentrated along dyke margins producing local tight folding, steep to vertical dips, and even overturning of the enclosing sediments. By contrast, basalt dykes or sills, that are much thinner than the rhyolitic intrusions, deformed more easily and can be found folded.

8.5.2 Dyke Emplacement at Mt Etna in 2001–02: Interplay of Lithostatic, Tectonic, and Magmatic Pressure

Ancient successions, like the Boyd Volcanic Complex described in the previous section, generally show exhumed structural levels where the geometry of intrusions and associated kinematics can be safely related to the orientation of the regional stress tensor and, if proximal, to the stress field of the magmatic source. In modern volcanic settings, the stress field associated with topographic and structural anomalies is also important, so that dyke orientations may vary significantly. The following example from the 2001 eruption at Mt Etna is a good example showing the interplay of lithostatic, tectonic, and magmatic pressures on dyke orientation.

The July–August 2001 eruption at Mt Etna was one of the most spectacular of the last decades, with the contemporaneous activity of several effusive and mildly to strongly explosive vents along a 6 km long, north–south trending fracture system (Fig. 8.39a, b; Behncke and Neri 2003; Lanzafame et al. 2003). Structural, seismic, and ground deformation data have been interpreted as reflecting the emplacement of a 2–3 m wide dyke along a north–south trending regional structure, from 3,000 m below sea level up to the surface (Billi et al. 2003; Lanzafame et al. 2003). Two narrow fracture zones radiating from the summit towards Torre del Filosofo and Valle del Leone (Fig. 8.39b) are thought to represent the lateral propagation of radial dykes from the Southeastern crater (CSE; Lanzafame et al. 2003), while the Valle del Leone is a linear valley that has an approximate WNW-ESE orientation, heading along a section of the Etna summit caldera ring fault (Fig. 8.39b).

On 20 July, two converging sets of extensional fractures surfaced along the CSE cone slope at an elevation of 2,800 m a.s.l., and propagated downslope, towards the Valle del Leone. The first and most prominent set of fractures was oriented 030° (Fig. 8.40a). It defines a narrow subsiding strip, 1.5–3 m in width and 1–2 m in depth. The extension direction, obtained by joining back the edges of extensional fractures, is oriented 130°, i.e. perpendicular to the strike of fractures. The second set of fractures trend 060° (Fig. 8.40a) and are arranged as a right-stepping, *en echelon* set of three segments, each some tens of m long. These fractures define a similar but narrower subsiding strip with respect to the 030°-trending fractures. The 030° direction is perfectly radial to the CSE and intersects at a low angle with the strike of the topography (Fig. 8.40a). The two fracture sets meet at 2,680 m a.s.l., forming one single 6–8 m wide and 4 m deep graben, the direction of which trends 055°, and from which a short-lived, spattering lava fountain fed a small lava flow (Fig. 8.40a, b; vent 1 in Fig. 8.41).

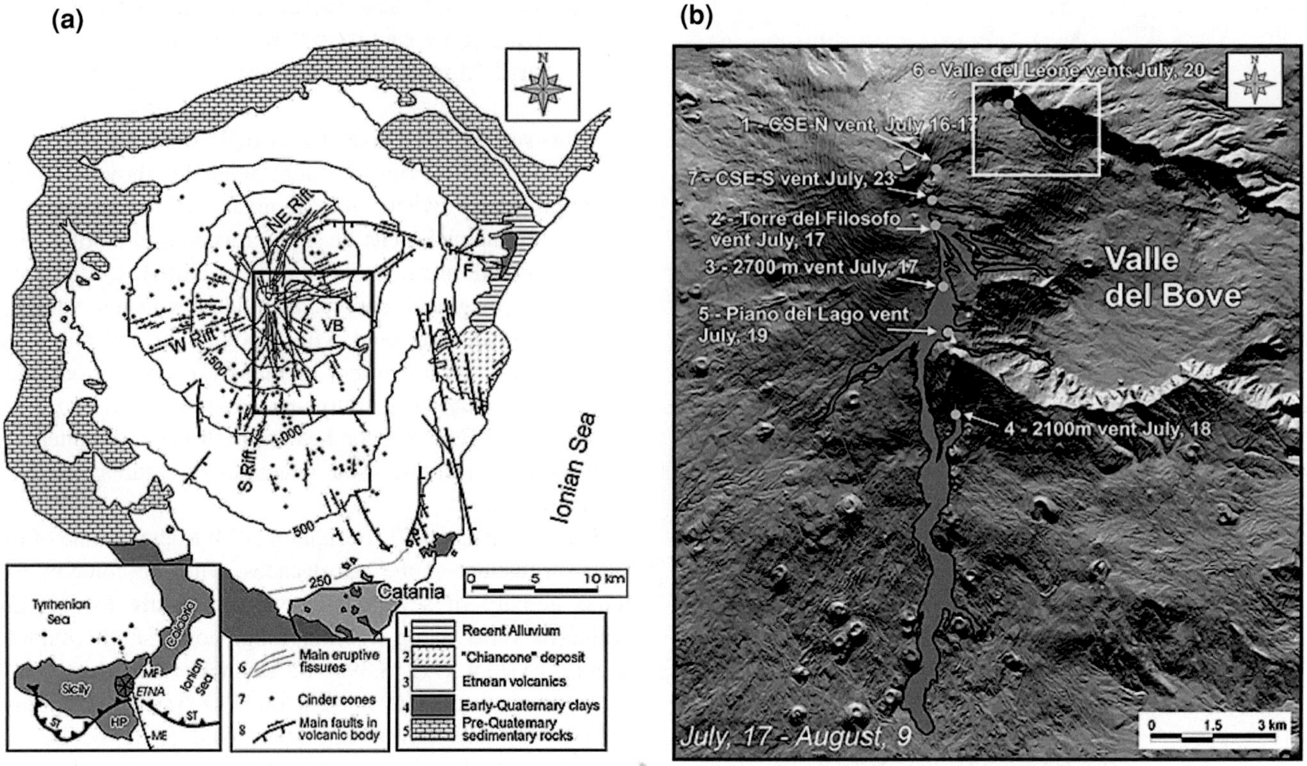

Fig. 8.39 Setting and location of vents of the July–August 2001 Valle del Leone eruption, Mt Etna. **a** Geological sketch map of Mt Etna showing the context of the July–August 2001 Valle del Leone vents and other vents; VB = Valle de Bove (Modified from Lanzafame et al. 2003). **b** Location and timing of the main vents superimposed on a DEM of the summit area of Mt Etna. The white box highlights the Valle del Leone area

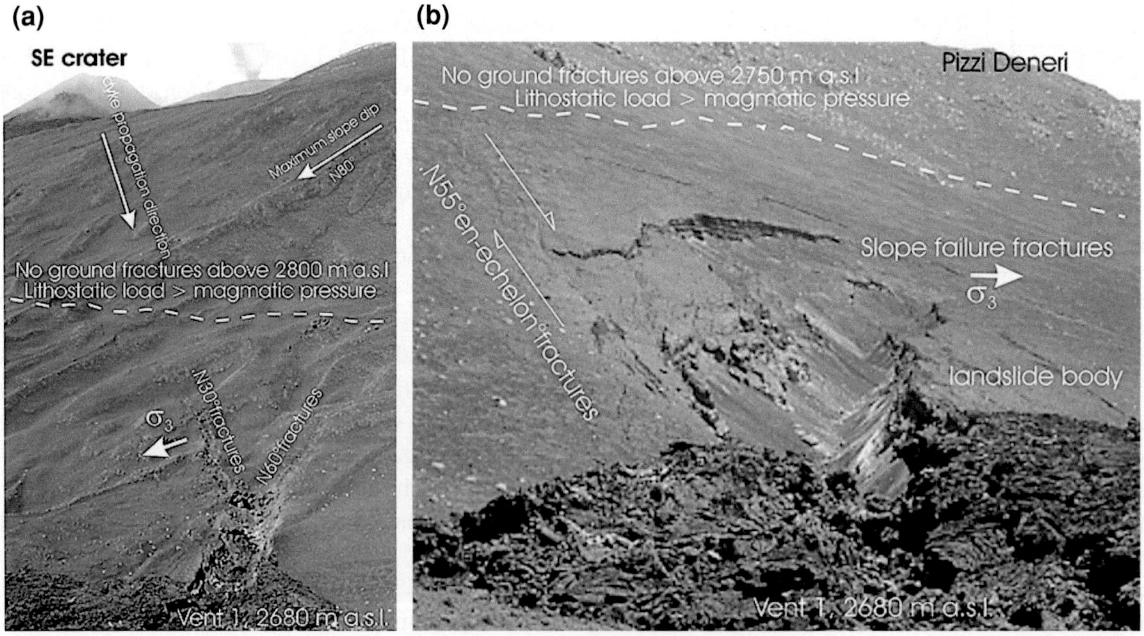

Fig. 8.40 Fracture system associated with the July–August 2001 Valle del Leone eruption, Mt Etna. **a** View from vent 1 towards the south, that is, the right flank of Valle del Leone, towards the Southeastern crater (CSE). Note that the N 30° fractures are perfectly aligned with the CSE crater and form a 50° angle with the maximum tip of topography. The propagation of the dyke was aided by $\sigma = \sigma_1$, which is a σ_3 induced by the lack of lateral confinement in the direction of maximum slope. Note also the absence of fractures above about 2,800 m a.s.l. where lithostatic load was high enough to constrain brittle deformation. **b** View from Vent 1 towards the north, that is, the left flank of the Valle del Leone. Note the N 55° left lateral, en echelon array of fractures caused by a local dextral shear due to the pull exerted by the topography in the direction parallel to the slope. Having failed to climb Pizzi Deneri rise, fractures stopped at 2,750 m a.s.l., and rotated to trigger the landslide

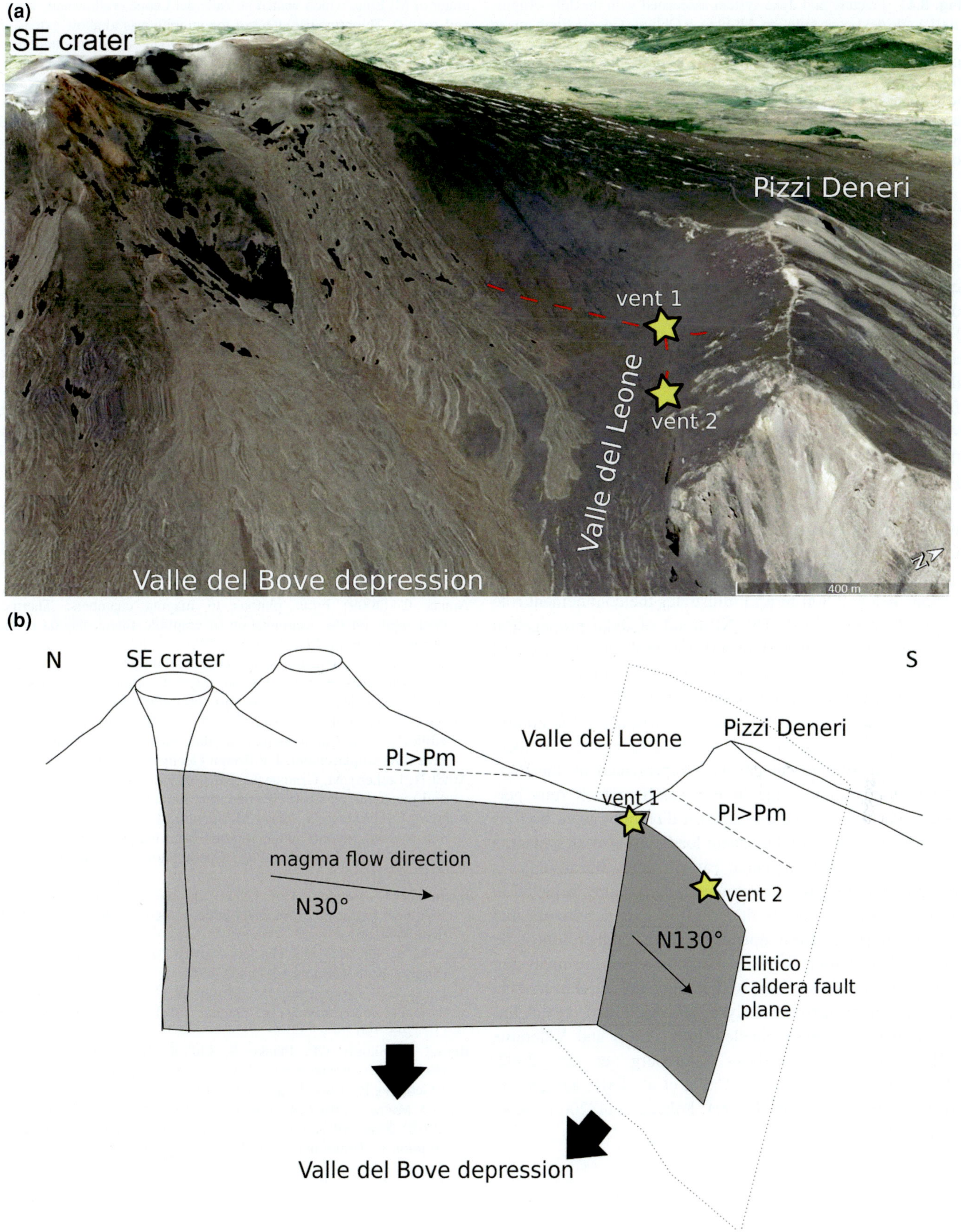
(a)
SE crater
Pizzi Deneri
vent 1
vent 2
Valle del Leone
Valle del Bove depression
N
400 m
(b)
N
SE crater
S
Valle del Leone
Pizzi Deneri
Pl>Pm
vent 1
Pl>Pm
magma flow direction
N30°
vent 2
N130°
Ellitico
caldera fault
plane
Valle del Bove depression

◄ **Fig. 8.41** Fracture and dyke system associated with the July–August 2001 Valle del Leone eruption, Mt Etna. **a** Oblique Google Earth image of Valle Del Leone, showing in the dashed red lines the trajectories of the fracture field developed (and dykes) from the Southeastern crater (top left) to Valle del Leone on 20 July 2001, the right-angle deviation, which coincides with the Ellitico caldera fault plane, and the location of the 2 vents in Valle del Leone (*Source* Google Earth). **b** Schematic 3D perspective diagram of (**a**) showing the inferred subsurface geometries of the dykes injected on 20 July 20 along the northern slope of the SE crater of Mt Etna, which vented at Valle del Leone (yellow star, vent 1 and vent 2). The principal effect of topography was to allow extension towards the Valle del Bove depression. The dyke orientation is perpendicular rather than parallel to the slope maximum dip and reflects the lateral attenuation of the magmatic pressure P_m over which tensile stresses induced by unconfined topography and structural discontinuities progressively play a greater role. Pl is the lithostatic load exerted by topography, where Pl > Pm conditions for fracturing are not developed

During the afternoon of 20 July, propagation of the fracture system resumed. After failing to cross-cut the steep caldera rim, where N55°-trending fractures attenuate to disappear at the elevation of 2,750 m a.s.l. (Pizzi Deneri in Fig. 8.40b), the fracture system bent abruptly to the SE by about 90°, following the bottom of the NW-trending Valle del Leone. This coincides with the older Ellitico caldera wall fault (Fig. 8.41a, b) down to an elevation of 2,600 m a.s.l., where mild eruptive lava fountaining built a 5 m high hornito and fed a lava flow that was active for the next 10 days (vent 2 in Fig. 8.41a, b).

The final inferred geometry of the Valle del Leone dyke is illustrated in Fig. 8.41b. The radial dyke departed from the CSE as a result of the magmatic pressure exerted on the CSE conduit wall. The surfacing of ground deformation at elevations below 2,800 m a.s.l. illustrates the confinement role of the lithostatic load. The NE-trend of dyke propagation was also aided by the pull towards the east–southeast exerted by the topography, i.e. perpendicular to the mass deficit of the Valle del Bove sector collapse valley (Fig. 8.41b). The sharp change of direction towards the SE, along the bottom of the Valle del Leone, is due to a combination of local factors. The Ellitico caldera wall (represented by the Pizzi Deneri topographic ridge in Fig. 8.41b), which runs perpendicular to the dyke propagation direction, acted both as an obstacle, due to its lithostatic load, and also as a magma pathway along the tangential caldera fault. Because P_m is magmastatic, magma can intrude along fractures oriented in any direction as long as there are no external factors that prevent fractures from opening such as higher lithostatic load and/or tectonic stresses. For this reason, in analysing dyke intrusions it is critical to determine the sense of magma flow, either by textural analysis of vesicle and crystal lineation (Coward 1980; Shelley 1985; Harp and Valentine 2018), or AMS analyses (Aubourg et al. 2002; Herrero-Bervera et al. 2001; Callot et al. 2001; Knight and Walker 1988; Poland et al. 2004; Polteau et al. 2008a; Roni et al. 2014).

References

Acocella V, Cifelli F, Funiciello R (2000) Analogue models of collapse calderas and resurgent domes. J Volcanol Geotherm Res 104(1–4):81–96

Allan AS, Morgan DJ, Wilson CJ, Millet MA (2013) From mush to eruption in centuries: assembly of the super-sized Oruanui magma body. Contribs Mineral Petrol 166(1):143–164

Alparone S, Andronico D, Giammanco S, Lodato L (2004) A multidisciplinary approach to detect active pathways for magma migration and eruption at Mt. Etna (Sicily, Italy) before the 2001 and 2002–2003 eruptions. J Volcanol Geotherm Res 136(1):121–140

Anderson EM (1937a) Cone-sheets and ring-dykes: The dynamical explanation. Bull Volcanol 1(1):35–40

Anderson EM (1937b) The dynamics of the formation of cone-sheets, ring-dykes, and caldron-subsidences. Proc Roy Soc Edinburgh 56:128–157

Annen C (2009) From plutons to magma chambers: Thermal constraints on the accumulation of eruptible silicic magma in the upper crust. Earth Planet Sci Lett 284(3–4):409–416

Annen C (2011) Implications of incremental emplacement of magma bodies for magma differentiation, thermal aureole dimensions and plutonism–volcanism relationships. Tectonophysics 500(1–4):3–10

Annen C, Lénat JF, Provost A (2001) The long-term growth of volcanic edifices: numerical modelling of the role of dyke intrusion and lava-flow emplacement. J Volcanol Geotherm Res 105(4):263–289

Arndt N, Lesher CM, Czamanske GK (2005) Mantle-derived magmas and magmatic Ni-Cu-(PGE) deposits. Econ Geol 100:5–24

Aubourg C, Giordano G, Mattei M, Speranza F (2002) Magma flow in sub-aqueous rhyolitic dikes inferred from magnetic fabric analysis (Ponza Island, W. Italy). Phys Chem Earth Parts A/B/C 27(25–31):1263–1272

Bachmann O, Bergantz GW (2004) On the origin of crystal-poor rhyolites: extracted from batholithic crystal mushes. J Petrol 45 (8):1565–1582

Behncke B, Neri M (2003) The July–August 2001 eruption of Mt. Etna (Sicily). Bull Volcanol 65(7):461–476

Bergantz GW, Schleicher JM, Burgisser A (2015) Open-system dynamics and mixing in magma mushes. Nature Geosci 8 (10):793–796

Berndt C, Skogly OP, Planke S, Eldholm O, Mjelde R (2000) High-velocity breakup-related sills in the Vøring Basin, off Norway. J Geophys Res: Solid Earth 105(B12):28443–28454

Biggs J, Mothes P, Ruiz M, Amelung F, Dixon TH, Baker S, Hong SH (2010) Stratovolcano growth by co-eruptive intrusion: the 2008 eruption of Tungurahua Ecuador. Geophys Res Lett 37(21)

Billi A, Acocella V, Funiciello R, Giordano G, Lanzafame G, Neri M (2003) Mechanisms for ground-surface fracturing and incipient slope failure associated with the 2001 eruption of Mt. Etna, Italy: analysis of ephemeral field data. J Volcanol Geotherm Res 122(3–4):281–294

Billings MP (1943) Ring dykes and their origin. Trans NY Acad Sci Ser.II v 5, 6,p. 131

Bischoff A, Nicol A, Barrier A, Wang H (2019a) Paleogeography and volcanic morphology reconstruction of a buried monogenetic volcanic field (part 2). Bull Volcanol 81(9):57

Bischoff A, Nicol A, Cole J, Gravley D (2019b) Stratigraphy of architectural elements of a buried monogenetic volcanic system. Open Geosci 11(1):581–616

Bischoff AP, Nicol A, Beggs M (2017) Stratigraphy of architectural elements in a buried volcanic system and implications for hydrocarbon exploration. Interpretation 5(3):141–159

Blažić L, Moreau J (2017) Discovery of Lower Cretaceous hydrothermal vent complexes in a late rifting setting, southern North Sea: insights from 3D imaging. J Geol Soc Lond 174(2):233–241

Bowen NL (1915) The later stages of the evolution of the igneous rocks. J Geol 23(S8):1–91

Breitkreuz C, Rocchi S (eds) (2018) Physical geology of shallow magmatic systems: dykes, sills and laccoliths. Advances in volcanology. IAVCEI-Springer, 409p

Breitkreuz C, Ehling BC, Pastrik N (2018) The subvolcanic units of the Late Paleozoic Halle Volcanic Complex, Germany: geometry, internal textures and emplacement mode. In: Breitkreuz C, Rocchi S (eds) (2018) Physical geology of shallow magmatic systems: dykes, sills and laccoliths. Advances in volcanology. IAVCEI-Springer, pp 295–307

Buchan KL, Goutier J, Hamilton MA, Ernst RE, Matthews WA (2007) Paleomagnetism, U-Pb geochronology, and geochemistry of Lac Esprit and other dyke swarms, James Bay area, Quebec, and implications for Paleoproterozoic deformation of the Superior Province. Canad J Earth Sci 44(5):643–664

Burchardt S (ed) (2018) Volcanic and igneous plumbing systems: understanding magma transport, storage, and evolution in the Earth's crust. Elsevier, 341p

Burchardt S, Walter TR, Tuffen H (2018) Growth of a volcanic edifice through plumbing system processes—volcanic rift zones, magmatic sheet-intrusion swarms and long-lived conduits. In: Burchardt S (ed) Volcanic and igneous plumbing systems. Elsevier, pp 89–112

Callot JP, Geoffroy L, Aubourg C, Pozzi JP, Mege D (2001) Magma flow directions of shallow dykes from the East Greenland volcanic margin inferred from magnetic fabric studies. Tectonophysics 335 (3–4):313–329

Cañón-Tapia E (2008) How deep can be a dyke? J Volcanol Geotherm Res 171(3–4):215–228

Cartwright J, Møller Hansen D (2006) Magma transport through the crust via interconnected sill complexes. Geology 34(11):929–932

Cas RAF (1983) A review of the palaeogeography and tectonic development of the palaeozoic lachlan fold belt of southeastern Australia. Geol Soc Austr Spec Pub, Sydney, p 104

Cas RAF, Allen RL, Bull SW, Clifford BA, Wright JV (1990) Subaqueous, rhyolitic dome-top tuff cones: a model based on the Devonian Bunga Beds, southeastern Australia and a modern analogue. Bull Volcanol 52(3):159–174

Cas RAF, Edgar C, Allen RL, Bull S, Clifford BA, Giordano G, Wright JV (2001) Influence of magmatism and tectonics on sedimentation in an extensional lake basin: the Upper Devonian Bunga Beds, Boyd Volcanic Complex, south-eastern Australia. In: White JDL, Riggs NR (eds) Volcaniclastic sedimentation in lacustrine settings. Blackwell Science Ltd., Oxford, pp 83–90

Cas RAF, Simmons JM (2018) Why deep-water eruptions are so different from subaerial eruptions. Front Earth Sci 6:198

Cas RAF, Wolff JA, Marti J, Olin PH, Edgar CJ, Pittari A, Simmons JM (2022). Tenerife, a complex end member of basaltic oceanic island volcanoes, with explosive polygenetic phonolitic calderas, and phonolitic-basaltic stratovolcanoes. Earth-Sci Rev 230:103990

Cas RAF, Wright JV (1987) Volcanic successions: modern and ancient. Allen & Unwin, London, 528p

Cashman KV, Giordano G (2014) Calderas and magma reservoirs. J Volcanol Geotherm Res 288:28–45

Chadwick WW Jr, Dieterich JH (1995) Mechanical modeling of circumferential and radial dike intrusion on Galapagos volcanoes. J Volcanol Geotherm Res 66(1–4):37–52

Charlier B, Namur O, Latypov R, Tegner C (eds) (2015) Layered intrusions. Springer, 748p

Chevallier L, Woodford A (1999) Morpho-tectonics and mechanism of emplacement of the dolerite rings and sills of the western Karoo, South Africa. South Afr J Geol 102(1):43–54

Chu R, Helmberger DV, Sun D, Jackson JM, Zhu L (2010) Mushy magma beneath Yellowstone. Geophys Res Lett 37:L01306

Corry CE (1988) Laccoliths: mechanics of emplacement and growth. Geol Soc Am Spec Pap 220:116p

Corti G, Bonini M, Conticelli S, Innocenti F, Manetti P, Sokoutis D (2003) Analogue modelling of continental extension: a review focused on the relations between the patterns of deformation and the presence of magma. Earth Sci Rev 63(3–4):169–247

Coward MP (1980) The analysis of flow profiles in a basaltic dyke using strained vesicles. J Geol Soc Lond 137(5):605–615

Cox KG (1993) Continental magmatic underplating. Phil Trans Roy Soc Lond. Series A Phys Eng Sci 342(1663):155–166

Cruden AR, McCaffrey KJW, Bunger AP (2018) Geometric scaling of tabular igneous intrusions: implications for emplacement and growth. In: Breitkreuz C, S (eds) (2018) Physical geology of shallow magmatic systems: dykes, sills and laccoliths. Advances in volcanology. IAVCEI-Springer, pp 12–38

Cruden AR, Weinberg RF (2018) Mechanisms of magma transport and storage in the lower and middle crust—magma segregation, ascent and emplacement. In: Burchardt S (ed) Volcanic and igneous plumbing systems. Elsevier, pp 13–53

Curry A, Gaynor SP, Davies JHFL, Ovtcharova M, Simpson G, Caricchi L (2021) Timescales and thermal evolution of large silicic magma reservoirs during an ignimbrite flare-up: perspectives from zircon. Contrib Mineral Petrol 176(12): 1-27

Daly RA (1905) The classification of igneous intrusive bodies. J Geol 13(6):485–508

Delcam A, Poppe S, Detienne M, Paguican EM (2018) Destroying a volcanic edifice—interactions between edifice instabilities and the volcanic plumbing system. In: Burchardt S (ed) Volcanic and igneous plumbing systems. Elsevier, pp 231–257

Denlinger RP, Morgan JK (2014) Instability of Hawaiian volcanoes. In: Poland MP, Takahashi TJ, Landowski CM (eds) Characteristics of Hawaiian volcanoes. United States Geol Surv Prof Pap 1801, pp 149–17

Doyle MG, Krapez B, Barley ME (2001) Volcanic facies architecture of the Woongarra Large Igneous province. In: 4th international archaean symposium extended abstracts, pp 149–150

Doyle MG, McPhie J (2000) Facies architecture of a silicic intrusion-dominated volcanic centre at Highway-Reward, Queensland, Australia. J Volcanol Geotherm Res 99(1–4):79–96

Dragoni M, Lanza R, Tallarico A (1997) Magnetic anisotropy produced by magma flow: theoretical model and experimental data from Ferrar dolerite sills (Antarctica). Geophys J Int 128(1):230–240

Druitt TH, Costa F, Deloule E, Dungan M, Scaillet B (2012) Decadal to monthly timescales of magma transfer and reservoir growth at a caldera volcano. Nature 482(7383):77–80

Druitt TH, Sparks RSJ (1984) On the formation of calderas during ignimbrite eruptions. Nature 310(5979):679–681

Dvorak JJ, Okamura AT, English TT, Koyanagi RY, Nakata JS, Sako MK, Tanigawa WT, Yamashita KM (1986) Mechanical response of the south flank of kilauea volcano, hawaii, to intrusive events along the rift systems. Tectonophysics 124(3–4):193–209

Eide CH, Schofield N, Jerram DA, Howell JA (2017) Basin-scale architecture of deeply emplaced sill complexes: Jameson Land, East Greenland. J Geol Soc Lond 174(1):23–40

Ernst RE, Buchan KL (2001) The use of mafic dike swarms in identifying and locating mantle plumes. In: Ernst RE, Buchan KL (eds) Mantle plumes: their identification through time. Geological Society of America, Boulder, pp 247–265

Ernst RE, Head JW, Parfitt E, Grosfils E, Wilson L (1995) Giant radiating dyke swarms on Earth and Venus. Earth Sci Rev 39(1–2):1–58

Ferré EC, Bordarier C, Marsh JS (2002) Magma flow inferred from AMS fabrics in a layered mafic sill, Insizwa. South Africa. Tectonophysics 354(1–2):1–23

Ferrucci F, Rasá R, Gaudiosi G, Azzaro R, Imposa S (1993) Mt. Etna: a model for the 1989 eruption. J Volcanol Geotherm Res 56(1–2):35–56

Fiorentini ML, Rosengren N, Beresford SW, Grguric B, Barley ME (2007) Controls on the emplacement and genesis of the MKD5 and Sarah's Find Ni–Cu–PGE deposits, Mount Keith, Agnew-Wiluna Greenstone Belt, Western Australia. Mineral Deposita 42(8):847–877

Francis EH (1982) Magma and sediment-I Emplacement mechanism of late Carboniferous tholeiite sills in northern Britain: President's anniversary address 1981. J Geol Soc Lond 139(1):1–20

Francis EH, Hopgood AM (1970) Volcanism and the Ardross fault, Fife. Scott J Geol 6(2):162–185

Francis EH, Walker BH (1987) Emplacement of alkali-dolerite sills relative to extrusive volcanism and sedimentary basins in the Carboniferous of Fife, Scotland. Earth Envir Sci Trans Roy Soc Edinburgh 77(4):309–323

Freymueller JT, Murray JB, Rymer H, Locke CA (2015) Ground deformation, gravity, and magnetics. In: Sigurdsson H, Stix J, Houghton BF, McNutt SR, Rymer H (eds) Encyclopedia of volcanoes. Academic Press, San Diego, pp 1101–1124

Galetto F, Acocella V, Caricchi L (2017) Caldera resurgence driven by magma viscosity contrasts. Nat Commun 8(1):1–11

Galland O, Bertelsen HS, Eide CH, Guldstrand F, Haug ØT, Leanza HA, Mair K, Palma O, Planke S, Rabbel O, Rogers B (2018b) Storage and transport of magma in the layered crust—formation of sills and related flat-lying intrusions. In: Burchardt S (ed) Volcanic and igneous plumbing systems. Elsevier, pp 113–138. https://doi.org/10.1016/B978-0-12-809749-6.00005-4

Galland O, Holohan E, van Wyk de Vries B, Burchardt S (2018a) Laboratory modelling of volcano plumbing systems: a review. In: Breitkreuz C, Riocchi S (eds) Physical geology of shallow magmatic systems: dykes, sills and laccoliths. Advances in volcanology. IAVCEI-Springer, pp 147–214

Galland O, Spacapan JB, Rabbel O, Mair K, Soto FG, Eiken T, Schiuma M, Leanza HA (2019) Structure, emplacement mechanism and magma-flow significance of igneous fingers–implications for sill emplacement in sedimentary basins. J Struct Geol 124:120–135

Geoffroy L, Aubourg C, Callot J, Barrat J (2007) Mechanisms of crustal growth in large igneous provinces: the north Atlantic province as a case study. In: Foulger GR, Jurdy DM (eds) Plates, plumes, and planetary processes. Geol Soc Am Spec Pap 430, pp 747–774

Geoffroy L, Callot JP, Aubourg C, Moreira M (2002) Magnetic and plagioclase linear fabric discrepancy in dykes: a new way to define the flow vector using magnetic foliation. Terra Nova 14(3):183–190

Geshi N, Ruch J, Acocella V (2014) Evaluating volumes for magma chambers and magma withdrawn for caldera collapse. Earth Planet Sci Lett 396:107–115

Geyer A, Martí J (2008) The new worldwide collapse caldera database (CCDB): a tool for studying and understanding caldera processes. J Volcanol Geotherm Res 157(4):375–386

Gibson IL, Walker GPL (1963) Some composite rhyolite/basalt lavas and related composite dykes in eastern Iceland. Proc Geol Assoc 74 (3):301–318

Gifkins CC, McPhie J, Allen RL (2002) Pumiceous rhyolitic peperite in ancient submarine volcanic successions. J Volcanol Geotherm Res 114(1–2):181–203

Gilbert GK (1877) Report on the geology of Henry Mountains. In: Department of the interior, U.S. Geograph Geol Surv Rocky Mountain Region. Washington D.C. Govt Printing Office, 160p

Giordano G, Caricchi L (2022) Determining the State of Activity of Transcrustal Magmatic Systems and Their Volcanoes. Ann Rev Earth Planet Sci 50:231-259

Giordano G, Cas RAF (2001) Structure of the upper devonian boyd volcanic complex, south coast New South Wales: implications for the Devonian-Carboniferous evolution of the Lachlan Fold Belt. Aust J Earth Sci 48(1):49–61

Glazner AF, Bartley JM, Coleman DS, Gray W, Taylor RZ (2004) Are plutons assembled over millions of years by amalgamation from small magma chambers? GSA Today 14(4/5):4–12

Goto Y, McPhie J (1998) Endogenous growth of a Miocene submarine dacite cryptodome, Rebun Island, Hokkaido, Japan. J Volcanol Geotherm Res 84(3–4):273–286

Goto Y, Tomiya A (2019) Internal structures and growth style of a quaternary subaerial rhyodacite cryptodome at Ogariyama, Usu Volcano, Hokkaido, Japan. Front Earth Sci 7:66

Gray DR (1997) Tectonics of the southeastern Australian Lachlan Fold Belt: structural and thermal aspects. Geol Soc Lond Spec Pub 121 (1):149–177

Gregg PM, De Silva SL, Grosfils EB, Parmigiani JP (2012) Catastrophic caldera-forming eruptions: thermomechanics and implications for eruption triggering and maximum caldera dimensions on Earth. J Volcanol Geotherm Res 241–242:1–12

Grove C (2013) Submarine hydrothermal vent complexes in the Paleocene of the Faroe-Shetland Basin: insights from three-dimensional seismic and petrographical data. Geology 41(1):71–74

Gualda GA, Ghiorso MS (2013) The Bishop Tuff giant magma body: an alternative to the standard model. Contribs Mineral Petrol 166 (3):755–775

Gualda GA, Pamukcu AS, Ghiorso MS, Anderson Jr AT, Sutton SR, Rivers ML (2012) Timescales of quartz crystallization and the longevity of the Bishop giant magma body. PloS one 7(5)

Gudmundsson A (2002) Emplacement and arrest of sheets and dykes in central volcanoes. J Volcanol Geotherm Res 116(3–4):279–298

Gudmundsson A (2003) Surface stresses associated with arrested dykes in rift zones. Bull Volcanol 65(8):606–619

Gudmundsson A (2005) The effects of layering and local stresses in composite volcanoes on dyke emplacement and volcanic hazards. CR Geosci 337(13):1216–1222

Gudmundsson A (2012) Magma chambers: Formation, local stresses, excess pressures, and compartments. J Volcanol Geotherm Res 237:19–41

Gudmundsson A, Brenner SL (2001) How hydrofractures become arrested. Terra Nova 13(6):456–462

Gudmundsson A, Lecoeur N, Mohajeri N, Thordarson T (2014) Dike emplacement at Bardarbunga, Iceland, induces unusual stress changes, caldera deformation, and earthquakes. Bull Volcanol 76 (10):869

Gudmundsson A, Marinoni LB, Marti J (1999) Injection and arrest of dykes: implications for volcanic hazards. J Volcanol Geotherm Res 88(1–2):1–13

Hacker DB, Rowley PD, Biek RF (2018). Catastrophic collapse features in volcanic terrains: styles and links to subvolcanic magma systems. In: Breitkreuz C, Riocchi S (eds) Physical geology of shallow magmatic systems: dykes, sills and laccoliths. Advances in volcanology. IAVCEI-Springer, 215–248.

Hald N, Tegner C (2000) Composition and age of Tertiary sills and dykes, Jameson Land Basin, East Greenland: relation to regional flood volcanism. Lithos 54(3–4):207–233

Hansen DM (2006) The morphology of intrusion-related vent structures and their implications for constraining the timing of intrusive events along the NE Atlantic margin. J Geol Soc 163(5):789–800

Hansen DM, Cartwright J (2006a) The three-dimensional geometry and growth of forced folds above saucer-shaped igneous sills. J Struct Geol 28(8):1520–1535

Hansen DM, Cartwright J (2006b) Saucer-shaped sill with lobate morphology revealed by 3D seismic data: implications for resolving a shallow-level sill emplacement mechanism. J Geol Soc 163 (3):509–523

Hanson RE (1991) Quenching and hydroclastic disruption of andesitic to rhyolitic intrusions in a submarine island-arc sequence, northern Sierra Nevada, California. Geol Soc Am Bull 103(6):804–816

Hanson RE, Schweickert RA (1982) Chilling and brecciation of a Devonian rhyolite sill intruded into wet sediments, northern Sierra Nevada, California. J Geol 90(6):717–724

Hanson RE, Wilson TJ (1993) Large-scale rhyolite peperites (Jurassic, southern Chile). J Volc Geotherm Res 54(3–4):247–264

Harp AG, Valentine GA (2018) Emplacement controls for the basaltic-andesitic radial dikes of Summer Coon volcano and implications for flank vents at stratovolcanoes. Bull Volcanol 80 (2): 1-20

Helz RT (1987) Differentiation behavior of Kilauea Iki lava lake, Kilauea Volcano, Hawaii: an overview of past and current work. In: Mysen BO (ed) Magmatic processes: physicochemical principles. Geochem Soc Spec Pub, pp 241–258

Herrero-Bervera E, Walker GPL, Cañon-Tapia E, Garcia MO (2001) Magnetic fabric and inferred flow direction of dikes, conesheets and sill swarms, Isle of Skye, Scotland. J Volcanol Geotherm Res 106 (3–4):195–210

Hildreth W (1981) Gradients in silicic magma chambers: Implications for lithospheric magmatism. J Geophys Res Solid Earth 86 (B11):10153–10192

Hildreth W (2004) Volcanological perspectives on Long Valley, Mammoth Mountain, and Mono Craters: several contiguous but discrete systems. J Volcanol Geotherm Res 136(3–4):169–198

Hildreth W, Fierstein J, Calvert A (2017) Early postcaldera rhyolite and structural resurgence at Long Valley Caldera, California. J Volcanol Geotherm Res 335:1–34

Horsman E, Broda RJ, Gwyn NZ (2018) Progressive construction of laccolithic intrusive centers: Henry Mountains, Utah, U.S.A. In: Breitkreuz C, S (eds) Physical geology of shallow magmatic systems: dykes, sills and laccoliths. Advances in volcanology. IAVCEI-Springer, pp 327–347

Houlé MG, Gibson HL, Lesher CM, Davis PC, Cas RAF, Beresford SW, Arndt NT (2008) Komatiitic sills and multigenerational peperite at Dundonald Beach, Abitibi greenstone belt, Ontario: volcanic architecture and nickel sulfide distribution. Econ Geol 103 (6):1269–1284

Hürlimann M, Garcia-Piera JO, Ledesma A (2000) Causes and mobility of large volcanic landslides: application to Tenerife, Canary Islands. J Volcanol Geotherm Res 103(1–4):121–134

Hutton DH (2009) Insights into magmatism in volcanic margins: bridge structures and a new mechanism of basic sill emplacement–Theron Mountains, Antarctica. Petroleum Geosci 15(3):269–278

Infante-Paez L, Marfurt KJ (2017) Seismic expression and geomorphology of igneous bodies: a Taranaki Basin, New Zealand, case study. Interpretation 5(3):121–140

Jackson CA-L (2012) Seismic reflection imaging and controls on the preservation of ancient sill-fed magmatic vents. J Geol Soc 169 (5):503–506

Jackson CA, Schofield N, Golenkov B (2013) Geometry and controls on the development of igneous sill–related forced folds: A 2-D seismic reflection case study from offshore southern Australia. Geol Soc Am Bull 125(11–12):1874–1890

Jamtveit B, Svensen H, Podladchikov YY, Planke S (2004) Hydrothermal vent complexes associated with sill intrusions in sedimentary basins. In: Physical geology of high-level magmatic systems, vol 234. Geological Society, London, Special Publication, pp 233–241

Jaupart C (2000) Magma ascent at shallow levels. In: Sigurdsson H, Stix J, Houghton BF, McNutt SR, Rymer H (eds) Encyclopedia of volcanoes. Academic Press, San Diego, pp 237–245

Jerram DA, Bryan SE (2018) Plumbing systems of shallow level intrusive complexes. In: Breitkreuz C, Riocchi S (eds) Physical geology of shallow magmatic systems: dykes, sills and laccoliths. Advances in volcanology. IAVCEI-Springer, pp 39–60

Johnson DJ (1992) Dynamics of magma storage in the summit reservoir of Kilauea Volcano, Hawaii. J Geophys Res: Solid Earth 97 (B2):1807–1820

Johnson RB (1968) Geology of the igneous rocks of the Spanish Peaks region, Colorado. US Government Printing Office, Washington D. C., 47p

Johnson SE, Schmidt KL, Tate MC (2002) Ring complexes in the Peninsular Ranges Batholith, Mexico and the USA: magma plumbing systems in the middle and upper crust. Lithos 61(3-4): 187-208

Judd JW (1893) On composite dykes in Arran. Quart J Geol Soc Lond 49(1–4):536–563

Kavanagh JL (2018) Mechanisms of magma transport in the upper crust —dyking. In: Burchardt S (ed) Volcanic and igneous plumbing systems. Elsevier, pp 55–88

Kavanagh JL, Rogers BD, Boutelier D, Cruden AR (2017) Controls on sill and dyke-sill hybrid geometry and propagation in the crust: the role of fracture toughness. Tectonophysics 698:109–120

Kavanagh JL, Sparks RSJ (2011) Insights of dyke emplacement mechanics from detailed 3D dyke thickness datasets. J Geol Soc Lond 168(4):965–978

Keating GN, Valentine GA, Krier DJ, Perry FV (2008) Shallow plumbing systems for small-volume basaltic volcanoes. Bull Volcanol 70(5):563–582

Keays RR, Lightfoot PC, Hamlyn PR (2012) Sulfide saturation history of the Stillwater Complex, Montana: chemostratigraphic variation in platinum group elements. Miner Deposita 47(1–2):151–173

Kennedy B, Wilcock J, Stix J (2012) Caldera resurgence during magma replenishment and rejuvenation at Valles and Lake City calderas. Bull Volcanol 74(8):1833–1847

Kennedy BM, Holohan EP, Stix J, Gravley DM, Davidson JR, Cole JW, Burchardt S (2018) Volcanic and igneous plumbing systems of caldera volcanoes. In: Burchardt (ed) Volcanic and igneous plumbing systems. Elsevier, pp 259–284

Kereszturi G, Németh K (2012) Monogenetic basaltic volcanoes: genetic classification, growth, geomorphology and degradation. In: Németh K (ed) Updates in volcanology—new advances in understanding volcanic systems. InTech, pp 3–88. https://doi.org/10.5772/51387

Knight MD, Walker GPL (1988) Magma flow directions in dikes of the Koolau Complex, Oahu, determined from magnetic fabric studies. J Geophys Res: Solid Earth 93(B5):4301–4319

Knopf A (1936) Igneous geology of the Spanish Peaks region, Colorado. Bull Geol Soc Am 47(11):1727–1784

Kokelaar BP (1982) Fluidization of wet sediments during the emplacement and cooling of various igneous bodies. J Geol Soc 139(1):21–33

Kokelaar BP, Bevins RE, Roach RA (1985) Submarine silicic volcanism and associated sedimentary and tectonic processes, Ramsey Island, SW Wales. J Geol Soc Lond 142(4):591–613
Konstantinou KI, Nolet G, Morgan WJ, Allen RM, Pritchard MJ (2000) Seismic phenomena associated with the 1996 Vatnajökull eruption, central Iceland. J Volcanol Geotherm Res 102(1–2):169–187
Lanzafame G, Neri M, Acocella V, Billi A, Funiciello R, Giordano G (2003) Structural features of the July–August 2001 Mount Etna eruption: evidence for a complex magma supply system. J Geol Soc 160(4):531–544
Lesher CM, Lightfoot PC (2012) Thematic issue on Ni–Cu–PGE deposits. Miner Deposita 47(1–2)
Lightfoot PC, Hawkesworth CJ (1997) Flood basalts and magmatic Ni, Cu, and PGE sulphide mineralization: comparative geochemistry of the Noril'sk (Siberian Traps) and West Greenland sequences. In: Mahoney JJ, Coffin MF (eds) Large igneous provinces: continental, oceanic, and planetary flood volcanism. AGU Mono, pp 357–380
Lightfoot PC, Keays RR (2005) Siderophile and chalcophile metal variations in flood basalts from the Siberian trap, Noril'sk region: Implications for the origin of the Ni-Cu-PGE sulfide ores. Econ Geol 100(3):439–462
Lipman PW, Mullineaux DR (1981) The 1980 eruptions of Mount St. Helens, Washington. US Geol Surv, Washington, 844p
Lucci F, Carrasco-Núñez G, Rossetti F, Theye T, White JC, Urbani S, Azizi H, Asahara Y, Giordano G (2020) Anatomy of the magmatic plumbing system of Los Humeros Caldera (Mexico): implications for geothermal systems. Solid Earth 11(1):125–159
Magee C, Bastow ID, van Wyk de Vries B, Jackson CA, Hetherington R, Hagos M, Hoggett M (2017) Structure and dynamics of surface uplift induced by incremental sill emplacement. Geology 45 (5):431–434
Magee C, Jackson CA (2020) Seismic reflection data reveal the 3D structure of the newly discovered Exmouth dyke swarm, offshore NW Australia. Solid Earth Discuss. https://doi.org/10.5194/se-2019-201
Magee C, Jackson CA, Schofield N (2013) The influence of normal fault geometry on igneous sill emplacement and morphology. Geology 41(4):407–410
Magee C, Muirhead JD, Karvelas A, Holford SP, Jackson CAL, Bastow ID, Schofield N, Stevenson CTE, McLean C, McCarthy W (2016) Lateral magma flow in mafic sill complexes. Geosphere 12 (3):809–841
Magee C, Stevenson CT, O'Driscoll B, Petronis MS (2012b) Local and regional controls on the lateral emplacement of the Ben Hiant Dolerite intrusion, Ardnamurchan (NW Scotland). J Struct Geol 39:66–82
Magee C, Stevenson CTE, O'driscoll B, Schofield N, McDermott K (2012a) An alternative emplacement model for the classic Ardnamurchan cone sheet swarm, NW Scotland, involving lateral magma supply via regional dykes. J Struct Geol 43:73–91
Marsh B (2004) A magmatic mush column rosetta stone: the McMurdo Dry Valleys of Antarctica. EOS Trans Am Geophys Union 85 (47):497–502
Marsh BD (2000) Magma chambers. In: Sigurdsson H, Stix J, Houghton BF, McNutt SR, Rymer H (eds) Encyclopedia of volcanoes. Academic Press, San Diego, pp 191–206
Marsh BD (2002) On bimodal differentiation by solidification front instability in basaltic magmas, part 1: basic mechanics. Geochim Cosmochim Acta 66(12):2211–2229
Marsh BD (2015) Magma chambers. In: Sigurdsson H, Stix J, Houghton BF, McNutt SR, Rymer H (eds) Encyclopedia of volcanoes. Academic Press, San Diego, pp 185–202
Maughan LL, Christiansen EH, Best MG, Gromme CS, Deino AL, Tingey D G (2002) The Oligocene Lund Tuff, Great Basin, USA: a very large volume monotonous intermediate. J Volcanol Geoth Res 113(1-2): 129-157
Martí J, Hurlimann M, Ablay GJ, Gudmundsson A (1997) Vertical and lateral collapses on Tenerife (Canary Islands) and other volcanic ocean islands. Geology 25(10):879–882
Martí J, Geyer A, Folch A (2009) A genetic classification of collapse calderas based on field studies, and analogue and theoretical modelling. Volcanology: the Legacy of GPL Walker. IAVCEI-Geological Society of London, London, pp 249–266
Mathieu L, De Vries BVW (2009) Edifice and substrata deformation induced by intrusive complexes and gravitational loading in the Mull volcano (Scotland). Bull Volcanol 71(10):1133
McGuire WJ, Pullen AD (1989) Location and orientation of eruptive fissures and feederdykes at Mount Etna; influence of gravitational and regional tectonic stress regimes. J Volcanol Geotherm Res 38 (3–4):325–344
McPhie J (1993) The Tennant Creek porphyry revisited: a synsedimentary sill with peperite margins, Early Proterozoic, Northern Territory. Aust J Earth Sci 40(6):545–558
Medialdea T, Somoza L, González FJ, Vázquez JT, de Ignacio C, Sumino H, Sánchez-Guillamón O, Orihashi Y, León R, Palomino D (2017) Evidence of a modern deep water magmatic hydrothermal system in the Canary Basin (eastern central A tlantic O cean). Geochem Geophys, Geosyst 18(8):3138–3164
Minakami T, Ishikawa T, Yagi K (1951) The 1944 eruption of volcano usu in Hokkaido, Japan. Bull Volcanol 11(1):45–157
Morgan S (2018) Pascal's principle, a simple model to explain the emplacement of laccoliths and some Mid-crustal Plutons. In: Burchardt S (ed) Volcanic and igneous plumbing systems. Elsevier, pp 139–165
Morley CK (2018) 3-D seismic imaging of the plumbing system of the Kora Volcano, Taranaki Basin, New Zealand: the influence of syn-rift structure on shallow igneous intrusion architecture. Geosphere 14(6):2533–2584
Muirhead JD, Airoldi G, Rowland JV, White JDL (2012) Interconnected sills and inclined sheet intrusions control shallow magma transport in the Ferrar large igneous province, Antarctica. Geol Soc Am Bull 124(1–2):162–180
Muirhead JD, Airoldi G, White JDL, Rowland JV (2014) Cracking the lid: Sill-fed dikes are the likely feeders of flood basalt eruptions. Earth Planet Sci Lett 406:187–197
Muirhead JD, Van Eaton AR, Re G, White JDL, Ort MH (2016) Monogenetic volcanoes fed by interconnected dikes and sills in the Hopi Buttes volcanic field, Navajo Nation, USA. Bull Volcanol 78 (2):11
Nahlawi H, Chakrabarti S, Kodikara J (2004) A direct tensile strength testing method for unsaturated geomaterials. Geotech Test J 27(4):6
Nakamura K (1977) Volcanoes as possible indicators of tectonic stress orientation—principle and proposal. J Volcanol Geotherm Res 2 (1):1–16
Naldrett AJ (1997) Key factors in the genesis of Noril'sk, Sudbury, Jinchuan, Voisey's Bay and other world-class Ni-Cu-PGE deposits: implications for exploration. Aust J Earth Sci 44(3):283–315
Naldrett AJ (1999) World-class Ni-Cu-PGE deposits: key factors in their genesis. Miner Deposita 34(3):227–240
Naldrett AJ (2004) Magmatic sulfide deposits: geology, geochemistry and exploration. Springer, 727p
Naldrett AJ, Federenko VA, Lightfoot PC, Kunilov VI, Gorbachev NS, Doherty W, Johan Z (1995) Ni-Cu-PGE deposits of Noril'sk region, Siberia: Their formation in conduits for flood-basalt volcanism. Trans Instit Min Metall Appl Earth Sci B 104:B18-36
Németh K, Martin U (2007) Shallow sill and dyke complex in western Hungary as a possible feeding system of phreatomagmatic volcanoes in "soft-rock" environment. J Volcanol Geotherm Res 159(1–3):138–152

Omosanya KO, Eruteya OE, Siregar ES, Zieba KJ, Johansen SE, Alves TM, Waldmann ND (2018) Three-dimensional (3-D) seismic imaging of conduits and radial faults associated with hydrothermal vent complexes (Vøring Basin, Offshore Norway). Marine Geol 399:115–134

Omosanya KO, Johansen SE, Eruteya OE, Waldmann N (2017) Forced folding and complex overburden deformation associated with magmatic intrusion in the Vøring Basin, offshore Norway. Tectonophysics 706:14–34

Orth K, McPhie J (2003) Textures formed during emplacement and cooling of a Palaeoproterozoic, small-volume rhyolitic sill. J Volcanol Geotherm Res 128(4):341–362

Pamukçu AS, Wright KA, Gualda GA, Gravley D (2020) Magma residence and eruption at the Taupo Volcanic Center (Taupo Volcanic Zone, New Zealand): insights from rhyolite-MELTS geobarometry, diffusion chronometry, and crystal textures. Contrib Mineral Petrol 175:1–27

Petford N (2003) Rheology of granitic magmas during ascent and emplacement. Ann Rev Earth Planet Sci 31(1):399–427

Petford N, Lister JR, Kerr RC (1994) The ascent of felsic magmas in dykes. Lithos 32:161–168

Pinel V, Jaupart C (2003) Magma chamber behavior beneath a volcanic edifice. J Geophys Res Solid Earth: Solid Earth 108(B2)

Planke S, Rasmussen T, Rey SS, Myklebust R (2005) Seismic characteristics and distribution of volcanic intrusions and hydrothermal vent complexes in the Vøring and Møre basins. In: Geological society, London, petroleum geology conference series. Geol Soc Lond, pp 833–844

Poland MP, Fink JH, Tauxe L (2004) Patterns of magma flow in segmented silicic dikes at Summer Coon volcano, Colorado: AMS and thin section analysis. Earth Planet Sci Lett 219(1–2):155–169

Poland MP, Miklius A, Montgomery-Brown EK (2014) Magma supply, storage, and transport at shield-stage Hawaiian volcanoes. In: Poland MP, Takahashi TJ, Landowski CM (eds.) Characteristics of Hawaiian volcanoes. US Geol Surv Prof Pap 1801, pp 179–236

Pollard DD (1973) Derivation and evaluation of a mechanical model for sheet intrusions. Tectonophysics 19(3):233–269

Pollard DD, Muller OH, Dockstader DR (1975) The form and growth of fingered sheet intrusions. Geol Soc Am Bull 86(3):351–363

Polteau S, Ferré EC, Planke S, Neumann ER, Chevallier L (2008a) How are saucer-shaped sills emplaced? Constraints from the Golden Valley Sill, South Africa. J Geophys Res: Solid Earth 113(B12)

Polteau S, Mazzini A, Galland O, Planke S, Malthe-Sørenssen A (2008b) Saucer-shaped intrusions: Occurrences, emplacement and implications. Earth Planet Sci Lett 266(1–2):195–204

Poppe S, Galland O, de Winter NJ, Goderis S, Claeys P, Debaille V, Boulvais P, Kervyn M (2020) Structural and geochemical interactions between magma and sedimentary host rock: the Hovedøya case, Oslo Rift, Norway. Geochem Geophys Geosyst 21: e2019GC008685. https://doi.org/10.1029/2019GC008685

Porreca M, Cifelli F, Soriano C, Giordano G, Mattei M (2015) Magma flow within dykes in submarine hyaloclastite environments: an AMS study of the Miocene Cabo de Gata volcanic units. Geol Soc Lond, Spec Publ 396(1): 133–157

Ramsay JG, Huber MI (1987) The techniques of modern structural geology, vol 2: folds and fractures. Academic Press, San Diego

Rawlings DJ (1993) Mafic peperite from the Gold Creek Volcanics in the Middle Proterozoic McArthur Basin. Northern Territory. Aust J Earth Sci 40(2):109–113

Reynolds P, Schofield N, Brown RJ, Holford SP (2018) The architecture of submarine monogenetic volcanoes–insights from 3D seismic data. Basin Res 30(S1):437–451

Rickwood PC (1990) The anatomy of a dyke and the determination of propagation and magma flow directions. In: Parke AJ, Rickwood PC, Tucker DH (eds) Mafic Dykes and Emplacement Mechanisms. Balkema, Rotterdam, pp 81–100

Roberge J, Wallace PJ, Kent AJ (2013) Magmatic processes in the Bishop Tuff rhyolitic magma based on trace elements in melt inclusions and pumice matrix glass. Contribs Mineral Petrol 165 (2):237–257

Rocchi S, Breitkreuz C (2018) Physical geology of shallow-level magmatic systems - an introduction. In: Breitkreuz C, Rocchi S (eds) Physical geology of shallow magmatic systems: dykes, sills and laccoliths. Advances in volcanology, IAVCEI-Springer, pp 1-10

Roche O, Druitt TH, Merle O (2000) Experimental study of caldera formation. J Geophys Res Solid Earth 105(B1):395–416

Roni E, Westerman DS, Dini A, Stevenson C, S (2014) Feeding and growth of a dyke–laccolith system (Elba Island, Italy) from AMS and mineral fabric data. J Geol Soc Lond 171(3):413–424

Rosengren N, Grguric B, Beresford SW, Fiorentini M, Cas R (2007) Internal stratigraphic architecture of the komatiitic dunite-hosted MKD5 disseminated nickel sulfide deposit, Mount Keith Domain, Agnew-Wiluna greenstone belt, Western Australia. Mineral Deposita 42(8):821–845

Ross P-S, White JDL (2005) Unusually large clastic dykes formed by elutriation of a poorly sorted, coarser-grained source. J Geol Soc 162(4):579–582

Rutherford MJ, Gardner JE (2000) Rates of magma ascent. In: Sigurdsson H, Stix J, Houghton BF, McNutt SR, Rymer H (eds) Encyclopedia of volcanoes. Academic Press, San Diego, pp 207–217

Saumur BM, Cruden AR (2016) On the emplacement of the Voisey's Bay intrusion (Labrador, Canada). Geol Soc Am Bull 128(1–2):147–168

Schirnick C, van den Bogaard P, Schmincke H-U (1999) Cone sheet formation and intrusive growth of an oceanic island—The Miocene Tejeda complex on Gran Canaria (Canary Islands). Geology 27 (3):207–210

Schmiedel T, Breitkreuz C, Görz I, Ehling BC (2014) Geometry of laccolith margins: 2D and 3D models of the Late Paleozoic Halle Volcanic Complex (Germany). Int J Earth Sci 104(2):323–333

Schofield NJ, Brown DJ, Magee C, Stevenson CT (2012a) Sill morphology and comparison of brittle and non-brittle emplacement mechanisms. J Geol Soc Lond 169(2):127–141

Schofield N, Heaton L, Holford SP, Archer SG, Jackson CA, Jolley DW (2012b) Seismic imaging of 'broken bridges': linking seismic to outcrop-scale investigations of intrusive magma lobes. J Geol Soc Lond 169(4):421–426

Schofield N, Jerram DA, Holford S, Archer S, Mark N, Hartley A, Howell J, Muirhead D, Green P, Hutton D, Stevenson C (2018) Sills in sedimentary basins and petroleum systems. In: Breitkreuz C, S (eds) Physical geology of shallow magmatic systems. Advances in volcanology. IAVCEI-Springer, pp 273–294

Self S, Thordarson T, Keszthelyi L (1997) Emplacement of continental flood basalt lava flows. In: Mahoney JJ, Coffin MF (eds) Large Igneous Provinces: continental, oceanic, and planetary flood volcanism. AGU Mono 100, pp 381–410

Shelley D (1985) Determining paleo-flow directions from groundmass fabrics in the Lyttelton radial dykes, New Zealand. J Volcanol Geotherm Res 25(1–2):69–79

Smith RL, Bailey RA (1968) Resurgent cauldrons. Geol Soc Am Mem 116:613–662

Song X, Wang Y, Chen L (2011) Magmatic Ni-Cu-(PGE) deposits in magma plumbing systems: features, formation and exploration. Geosci Front 2(3):375–384

Spera FJ (1984) Carbon dioxide in petrogenesis III: role of volatiles in the ascent of alkaline magma with special reference to xenolith-bearing mafic lavas. Contrib Mineral Petrol 88(3):217–232

Srivastava RK, Ernst RE, Peng P (eds) (2019) Dyke swarms of the world: a modern perspective. Springer Nature Switzerland. https://doi.org/10.1007/978-981-13-1666-1

Staude S, Markl G (2019) Remnant lenses of komatiitic dykes in Kambalda (Western Australia): occurrences, textural variations, emplacement model, and implications for other komatiite provinces. Lithos 342:206–222

Stewart AL, McPhie J (2003) Internal structure and emplacement of an Upper Pliocene dacite cryptodome, Milos Island, Greece. J Volcanol Geotherm Res 124(1–2):129–148

Stone P, Millward D, Young B, Merritt JW, Clarke S M, McCormac M, Lawrence D J D (2010) British regional geology: Northern England. British Geological Survey, Keyworth, Nottingham, 293p

Suppe J (1985) Principles of structural geology. Prentice-Hall, New York

Svensen H, Corfu F, Polteau S, Hammer Ø, Planke S (2012) Rapid magma emplacement in the Karoo large igneous province. Earth Planet Sci Lett 325:1–9

Svensen H, Jamtveit B, Planke S, Chevallier L (2006) Structure and evolution of hydrothermal vent complexes in the Karoo Basin, South Africa. J Geol Soc Lond 163(4):671–682

Svensen H, Planke S, Chevallier L, Malthe-Sørenssen A, Corfu F, Jamtveit B (2007) Hydrothermal venting of greenhouse gases triggering Early Jurassic global warming. Earth Planet Sci Lett 256 (3–4):554–566

Svensen H, Planke S, Malthe-Sørenssen A, Jamtveit B, Myklebust R, Eidem TR, Rey SS (2004) Release of methane from a volcanic basin as a mechanism for initial Eocene global warming. Nature 429 (6991):542–545

Tait S, Jaupart C, Vergniolle S (1989) Pressure, gas content and eruption periodicity of a shallow, crystallising magma chamber. Earth Planet Sci Lett 92(1): 107-123

Thomson K, Hutton D (2004) Geometry and growth of sill complexes: insights using 3D seismic from the North Rockall Trough. Bull Volcanol 66(4):364–375

Thomson K, Schofield N (2008) Lithological and structural controls on the emplacement and morphology of sills in sedimentary basins. Geol Soc Lond Spec Pub 302(1):31–44

Tibaldi A (2015) Structure of volcano plumbing systems: A review of multi-parametric effects. J Volcanol Geotherm Res 298:85–135

Tibaldi A, Pasquarè AF, Rust D (2011) New insights into the cone sheet structure of the Cuillin Complex, Isle of Skye, Scotland. J Geol Soc Lond 168(3):689–704

Tilling RI, Dvorak JJ (1993) Anatomy of a basaltic volcano. Nature 363 (6425):125–133

Tilling RI, Topinka L, Swanson DA (1984) Eruptions of Mount St. Helens: past, present and future. US Geological Survey, Washington

Trendall AF (1995) The Woongarra Rhyolite—a giant lava-like felsic sheet in the Hamersley Basin of Western Australia. Geol Surv Western Aust Rept 42:70 p

Troise C (2001) Stress changes associated with volcanic sources: constraints on Kilauea rift dynamics. J Volcanol Geotherm Res 109 (1–3):191–203

Valentine GA, Gregg TKP (2008) Continental basaltic volcanoes—processes and problems. J Volcanol Geotherm Res 177(4):857–873

van Wyk de Vries B, Davies T (2015) Landslides, debris avalanches, and volcanic gravitational deformation. In: Sigurdsson H, Houghton B, McNutt SR, Rymer H, Stix J (eds) The encyclopedia of volcanoes. Academic Press Elsevier, pp 665–685

Verma SP (1985) On the magma chamber characteristics as inferred from surface geology and geochemistry: examples from Mexican geothermal areas. Phys Earth Planet Int 41(2–3):207–214

Voight B, Janda RJ, Glicken H, Douglass PM (1983) Nature and mechanics of the Mount St. Helens rockslide-avalanche of 18 May 1980. Geotechnique 33(3):243–273

Wadge G (1976) Deformation of Mount Etna, 1971–1974. J Volcanol Geotherm Res 1(3):237–263

Walker GPL (1988) Three Hawaiian calderas: an origin through loading by shallow intrusions? J Geophys Res: Solid Earth 93 (B12):14773–14784

Walker GPL (1989) Gravitational (density) controls on volcanism, magma chambers and intrusions. Aust J Earth Sci 36(2):149–165

Walker GPL (1992) "Coherent intrusion complexes" in large basaltic volcanoes—a new structural model. J Volcanol Geotherm Res 50 (1–2):41–54

Walker GPL (1999) Volcanic rift zones and their intrusion swarms. J Volcanol Geotherm Res 94(1–4):21–34

Walker BH, Francis EH (1987) High-level emplacement of an olivine–dolerite sill into Namurian sediments near Cardenden, Fife. Earth Envir Sci Trans Roy Soc Edinburgh 77(4):295–307

Westerman D, S, Breitkreuz C, Stevenson C, Wilson P (2018a) Structures related to the emplacement of shallow-level intrusions. In: Breitkreuz C, Rocchi S (eds) Physical geology of shallow magmatic systems. Advances in volcanology. IAVCEI-Springer, pp 83–118

Westerman DS, Rocchi S, Dini A, Farina F, Roni E (2018b) Rise and fall of a multi-sheet intrusive complex, Elba Island, Italy. In: Breitkreuz C, Rocchi S (eds) Physical geology of shallow magmatic systems. Advances in volcanology. IAVCEI-Springer, pp 309–325

Willcock MAW, Bargossi GM, Weinberg RF, Gasparotto G, Cas RAF, Giordano G, Marocchi M (2015) A complex magma reservoir system for a large volume intra-to extra-caldera ignimbrite: Mineralogical and chemical architecture of the VEI8, Permian Ora ignimbrite (Italy). J Volcanol Geotherm Res 306:17–40

Zavala K, Leitch AM, Fisher GW (2011) Silicic segregations of the Ferrar Dolerite sills, Antarctica. J Petrol 52(10):1927–1964

Part C

Explosive Eruptions, Processes and Pyroclastic Deposits

Aims, Scope, and Structure

Explosive eruptions are among the most dangerous, energetic, and destructive natural phenomena known. They range from small, short phreatic and hydrothermal explosions to larger and longer magmatic and phreatomagmatic explosive eruptions that produce sustained, buoyant eruption columns, to sustained magmatic and phreatomagmatic super-eruptions that are capable of affecting global climate, food chains, and life. In Part C, we review the range of known explosive eruption phenomena, the spectrum of explosive eruption styles in subaerial and subaqueous environments, their dispersal processes, and their deposit characteristics.

In Chap. 9, we consider the range of volcanic explosive eruption styles, from the smallest and least intensive, to the largest and most intensive. Pyroclastic fall(out) deposits form when material has been explosively ejected from a vent and falls back to Earth through the atmosphere and/or a body of water. Explosive eruptions can also produce a spectrum of pyroclastic density currents (pdcs) which we describe in Chap. 10. Pdcs are potentially the most destructive of all volcanic phenomena, due to their high velocities, the large distances that some can travel, the large volume of volcanic debris they can carry and their high temperature. In Chap. 11, we consider the intriguing and provocative topic of what happens when pyroclastic density currents enter the sea or large lakes from subaerial vents, or are erupted from subaqueous vents. We provide an in-depth review and assessment of processes, experimental and numerical modelling, and deposit characteristics, including examples described in the literature.

Explosive Eruption Styles, Columns, and Pyroclastic Fallout Deposits

9

Subaerial and Subaqueous, Dynamics, Deposit Characteristics and Classification

Guido Giordano, Ray Cas, and John V. Wright

R. Cas et al., *Volcanology*, Springer Textbooks in Earth Sciences, Geography and Environment,
https://doi.org/10.1007/978-3-319-66613-6_9

Plate 9.1 **a** Typical strombolian eruption activity, captured in time lapse photography at night that highlights the parabolic paths of spatter fragments, Stromboli volcano, Italy (Photo by J. Alean and M. Fulle, taken 15 October 2005; *Source* Stromboli OnLine: https://www.swisseduc.ch/stromboli/ with permission R. Carniel). **b** Painting of the explosive fissure vent eruption at Mt Tarawera, New Zealand, in the early hours of 10 June 1886, created in 1888 by landscape artist Charles Blomfield based on eyewitness accounts (*Source* Wikimedia). The eruption vented a column of scoria and ash now thought to be $\sim$30 km high, while phreatomagmatic and hydrothermal explosions at Lake Rotomahana caused base surges that destroyed several Maōri villages and obliterated the famous Pink and White Terraces. By 6.00 am, the main phase of the eruption was over. **c** $\sim$10 km high eruption column from Kliuchevskoi volcano, Kamchatka, Russia, seen erupting on 30 September 1994, developing a strongly assymmetric eruption plume dispersing fallout deposits as it propagates downwind (*Source* NASA). **d** Spectacular $\sim$9 km high explosive eruption column from the 21 April 1990 eruption of Redoubt volcano, Alaska, USA. Most of the column formed as a buoyant ash cloud derived from pyroclastic flows flowing away from the viewer. A small steam-rich white plume is rising directly from the summit vent. (Photo by R. E. Clucas; *Source* United States Geological Survey)

Abstract

In this chapter, we consider the range of volcanic explosive eruption styles, from the smallest and least intense to the largest and most intense that produce pyroclastic fallout deposits. At the small-scale end of the spectrum of subaerial explosive eruptions, we include strombolian *sensu stricto*, halema'ma'uan and small-scale vulcanian and hydrothermal explosions. At the next level, there are small-scale hawai'ian magma/fire fountaining and larger scale vulcanian and hydrothermal explosions, followed by large-scale hawai'ian magma/fire fountaining, large vulcanian, and violent strombolian eruptions. We consider the eruptions that produce basaltic scoria through to rhyolitic pumice cones to be micro-plinian, at the small end of the spectrum of sub-plinian, plinian, and ultra-plinian explosive styles. We suggest that violent strombolian eruption style, which also produces cones, represents a transition between open-system degassing and closed-system degassing processes, and also that all the magmatic explosive eruption styles can transition into phreatomagmatic equivalents, irrespective of the scale, as demonstrated by the 2022 Hunga Tonga-Hunga Ha'apai eruption. The dynamics of the explosive eruption columns for all these styles are discussed, as are characteristic erupted masses and mass eruption rates. The characteristics of the deposits from each eruption style are described and comprehensively illustrated, particularly the field facies characteristics. Close to the vent, some proximal fallout deposits are welded, or pass into agglutinated spatter. We also review what is known about pyroclastic fallout processes in aqueous environments, and highlight how eruption dynamics, columns, and dispersal processes are different from subaerial processes. Several new proposed subaqueous eruption styles have recently emerged from recent research, including yalian, poseidic, neptunian, tangaroan-havre, and now hunga tongan. Although not enough is presently known about the fragmentation processes, intensity, and dispersal extent of these subaqueous eruption styles, it is important that they be acknowledged as a basis for driving future research. Finally, we consider how fit for purpose existing classification schemes for pyroclastic fallout deposits are. We suggest modifications to these, including recalibration of the logarithmic VEI, Eruption Magnitude and Explosive Intensity scales that have been applied for several decades, to accommodate smaller scale events that could previously not be included, but can be lethal in their impact, and therefore, need to be included in all schemes that reflect eruption scale and intensity.

9.1 Introduction: Pyroclastic Fallout-Forming Eruptions and Deposits —Definitions

A pyroclastic fall(out) deposit is formed after pyroclasts have been explosively ejected from a vent and fall back to Earth through the atmosphere and/or a body of water. Fallout deposits can be produced from a variety of explosive eruption styles in subaerial environments, including from magma or fire fountaining (Plate 9.1a, b), ballistic blasts, those producing sustained, buoyant, eruption columns (Plate 9.1 c), and ash fallout from buoyant ash clouds originating from pyroclastic density currents (Plate 9.1d; Fig. 9.1). Deposits include magma/fire fountaining spatter agglutinate, ballistic blast agglomerates, air-settled pumice and scoria fallout deposits, distal co-ignimbrite ash deposits, and water-settled ash and pumice deposits (Fig. 9.1). In subaqueous settings, the same origins for fallout deposits may occur in near-vent and relatively shallow water settings, but in addition, subaqueous effusive eruptions can also produce convective buoyant plumes of water with suspended ash and even pumice that has spalled from the surfaces of erupting lavas through quench fragmentation.

Based on the energy released by an eruption and on the particle size and densities, pyroclasts fall back to Earth under the influence of gravity and may experience different trajectories. Some, such as dense, heavy ballistic blocks and bombs (Chap. 4) are ejected explosively around the vent on parabolic ballistic trajectories through explosive momentum (e.g. Plate 9.1a; Fig. 9.2a). On the other hand, lighter and low-density vesiculated pyroclasts such as pumice, scoria, small lithic fragments, and ash are carried upwards in a buoyant eruption column of tephra and gas into the atmosphere (or subaqueously, a body of water) by the convection and turbulence of hot buoyant volcanic gas (Plate 9.1c, d; Fig. 9.2b). Pyroclasts fallout from the column onto the Earth's surface forming a pyroclastic fallout deposit dispersed around and away from the vent.

The geometry and size of a fallout deposit reflects the eruption intensity, duration, the grain size and densities of pyroclasts at vent, their physical properties (e.g. their terminal fall velocity; see Sect. 9.4), and, for the eruptions that produce a buoyant eruption column, the velocity and direction of atmospheric winds (Walker 1973; Settle 1978; Wilson et al. 1978; Suzuki 1983; Carey and Sparks 1986; Cas and Wright 1987; Wilson and Walker 1987; Armienti et al. 1988; Woods 1988; Pyle 1989; Carey and Sigurdsson 1989; Bursik and Woods 1991; Bursik et al. 1992; Sparks et al. 1992; Dobran et al. 1993; Ernst et al. 1996; Sparks et al. 1997; Bonadonna et al. 1998; Kaminski and Jaupart 1998; Koyaguchi and Ohno 2001; Bonadonna and Phillips 2003;

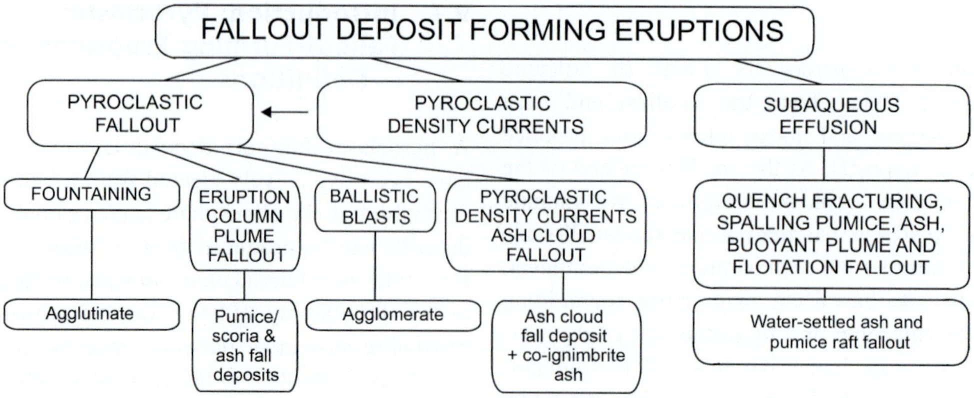

Fig. 9.1 Pyroclastic fallout process spectrum, depicting eruption styles and fallout deposit outcomes from different origins, focusing on subaerial settings

Fig. 9.2 End-member eruption columns. **a** Strombolian spray launching large ballistics, depicted by parabolic trajectories in time lapse photography, with subordinate ash particles which are easily blown downwind away by wind forming the orange cloud moving to the left. Stromboli volcano, Italy, 2005. (Photo by Tullio Ricci). **b** Buoyant Plinian eruption column ejecting a large amount of fine ash that is easily entrained in turbulent eddies formed as soon as gas mixes with air. The eruption column is ~15 km high. 18 May 1980, Mt. St. Helens. (*Source* United States Geological Survey)

Pfeiffer et al. 2005; Suzuki et al. 2005; Connor and Connor 2006; Carazzo et al. 2008; Koyaguchi et al. 2009; Mastin et al. 2009; Dellino et al. 2010; Woodhouse et al. 2013; Carey and Bursik 2015; Clarke et al. 2015; Cioni et al. 2015; Costa et al. 2016a, b; Trolese et al. 2019; Bonadonna et al. 2022).

Other fine-grained pyroclastic fallout deposits are generated in part from ash elutriated out of moving pyroclastic flows, forming buoyant trailing ash clouds that produce ash-cloud-derived fallout deposits. Some examples of this type of pyroclastic fall deposit can be more voluminous and may be more widely dispersed than those of ash fallout deposits derived directly from eruption columns (Sparks and Huang 1980; Cas and Wright 1987; Woods and Wohletz 1991; Eychenne et al. 2012; Engwell et al. 2014; Smith et al. 2016; see Sect. 9.2.3).

Fallout deposits show mantle bedding; that is, they locally maintain a uniform thickness while draping all but the steepest topography (Fig. 9.3a, b). Although pyroclastic deposits are generally poorly sorted compared with many sedimentary deposits, pyroclastic fallout deposits are relatively well-sorted compared to other types of pyroclastic deposits (cf. pyroclastic flows and surges; Chap. 10), with

Fig. 9.3 Examples of fallout deposits with draping geometry of fallout layers mantling paleotopography. **a** Several separate trachytic pumice fall deposits from the bedded sequence that mantles an erosional topography at Fogo volcano (São Miguel, Azores, Portugal). Hills and valleys are cut into older sub-horizontal pumice fall sequences. **b** Scoria fall deposits, with undulating stratification, Villarrica volcano, southern Chile

sorting values normally <2.0 (Figs. 9.4 and 9.5a, b), because of aerodynamic fractionation and hydraulic sorting during transport as they settle through the atmosphere and/or a body of water. An exception to this is in the near-vent (tens m to <4 km) fallout deposits which may also contain outsized, often dense ballistic blocks and bombs that were not transported by convective forces in the eruption column, but by explosive momentum and are anomalously large compared with convectively transported and hydraulically settled fragments. Fallout deposits are commonly internally massive, but sometimes they show diffuse planar internal stratification or lamination marked by grain size variations caused by variations in eruption column behaviour or changing wind direction or strength during the course of the eruption. However, they never have cross stratification or bedforms showing erosion or truncation of the underlying layers. Near the vent, air-fall deposits usually build up and form more or less wide cones or wedging sheets, and in some relatively low explosive intensity eruption styles, they are the principal component of spatter, scoria, and pumice

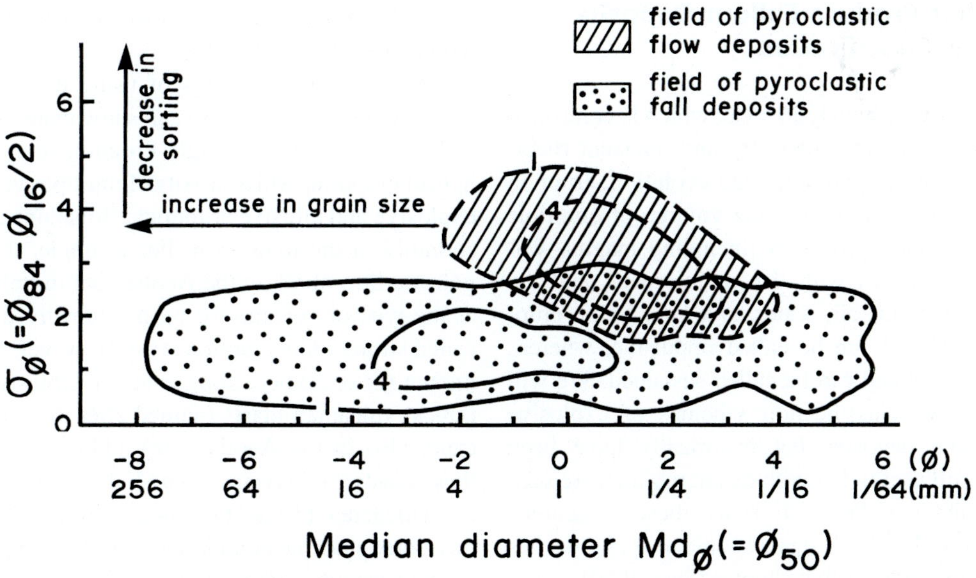

Fig. 9.4 Md_ϕ/σ_ϕ plot often used to show grain size characteristics of unconsolidated modern pyroclastic deposits. The grain size distribution of a sample is first determined by mechanical sieving. Cumulative frequency curves of the distribution are then drawn on arithmetic probability paper and the two Inman (1952) parameters of median diameter (Md_ϕ) and graphical standard deviation (σ_ϕ), which is a measure of sorting, are derived. Solid lines labelled 1% and 4% are contours for the field of pyroclastic fallout deposits and within these 99% and 96%, respectively, of sieve analyses of fall deposits occur (based on over 1300 analyses). Broken lines are similar contours for the field of pyroclastic flow analyses (based on about 800 analyses). (From Cas and Wright 1987, After Walker 1971, Walker 1980). For the practical details of grain size analysis of modern pyroclastic deposits, see Cas and Wright (1987), and for an updated and expanded discussion of the application to fallout deposits, see Sect. 9.6

Fig. 9.5 Typical, well-sorted fallout deposits. **a** Basaltic scoria lapilli fallout deposits, 35 ka Tower Hill volcanic complex, Newer Volcanics Province, Victoria, southeastern Australia. **b** Pumice lapilli fallout deposit, from the Hekla 4 eruption, Iceland. Note the very angular shape of the white pumice clasts and the subordinate presence of grey lithic clasts, which are smaller than the maximum pumice clast size, as a result of efficient aerodynamic, and hydraulic sorting

cones. Some proximal deposits are welded, or pass into agglutinated spatter. Carbonised wood is rare but, when found, is usually restricted to near-vent deposits. More widespread fallout deposits are sheet-like, or fan-like, thinning progressively with distance from the source.

9.2 An Introduction to Explosive Eruption Styles that Produce Fallout Deposits and Their Classification

Explosive eruptions vary hugely in style, which is controlled by volatile contents, magma chemistry and transient rheology, conduit and vent morphologies and evolution, eruption magnitude and intensity, duration, the volume of eruptible magma(s), fragmentation processes (including magma–water interaction), interaction with the environment, and the type of ambient medium (air, water, and ice). Some eruptions are short-lived and can be called impulsive, whereas others are longer-lived and can be called sustained. We will briefly introduce two classification schemes of explosive eruption styles and deposits that historically have been widely used, are still useful, but with recent research are seen to have some limitations. We will discuss these limitations, discuss other approaches, and suggest some new approaches in the concluding section of this chapter (Sect. 9.10).

There have been several attempts to describe the huge range of variables that determine eruption style, to capture the fundamental observables from which a classification scheme of eruption styles and their deposits can be constructed, and to create a common terminology for eruption styles for volcanologists. Terms such as hawai'ian, strombolian, vulcanian, and plinian have become widely used and entrenched in the literature, especially during the last century. These were originally based on type localities where a particular style of eruption is common; for example, Hawai'i(an) for fire fountains, Stromboli(an) for small to moderate spatter explosions, Vulcano(ian) for impulsive blasts, and based on the earliest, description of a plinian eruption, that of the 79 AD eruption at Vesuvius described by Pliny the Younger (see Sect. 9.2.2.3).

The main shortcomings of using those volcano-specific terms are that each volcano can show many different eruption styles, even during a single eruption (e.g. 2021 La Palma hybrid eruption, which involved multiple vents with varying explosive and effusive eruptions; Bonadonna et al. 2022) and certainly in the long term. For example, the marine stratovolcano Stromboli, in the Aeolian arc in Italy, is best known as the type of volcano for weak, short-lived fountain sprays called strombolian eruption style. However, Stromboli is also known for episodic, strong, sudden explosive blasts that are typical vulcanian blasts (named after the type volcano, Vulcano, also in the Aeolian arc), although some have called these blasts "paroxysms", which they are, but basically, they are vulcanian blasts (Giordano and De Astis 2021). Furthermore, placing boundaries between eruption styles may not be very satisfactory because they are artificial, depend on the parameters chosen to define them, and don't take account of varying intensities in all eruption styles.

The most widely used classification scheme of fallout deposits and their eruption styles, used for nearly 50 years, was proposed by Walker (1973) (Cas and Wright 1987;

Fig. 9.6a, adapted to include our modifications described below), known as the *Walker Dispersal–Fragmentation (D–F) classification scheme*. Walker's (1973) scheme applies only to fallout deposits derived from buoyant eruption columns rising above a vent and does not include fallout deposits originating from the buoyant ash clouds rising above pyroclastic density currents. Walker's approach was based on the characteristics of the fallout deposits measured in the field using areal extent, or *dispersal area*, *D*, and grain size characteristics thought to represent the *degree of fragmentation, F,* of the ejected pyroclasts. The empirical measure of D used is the area enclosed by the 0.01 T_{max} isopach (where T_{max} is the maximum thickness of the deposit; Fig. 9.6c). The empirical measure of F chosen is the percentage of a deposit finer than 1 mm at the point on the axis of dispersal of the fallout deposit where it is crossed by the 0.1 Tmax isopach; this can only be determined from sieve analysis of a sample collected in the field either at this point or, more practically, obtained graphically from sieve analyses of a few samples collected near the dispersal axis (Fig. 9.6c). D and F can only be determined from mapping the areal extent and thickness of a deposit and collecting samples at multiple localities in the field, which are then sieved in the laboratory. The classification is thus based on properties of the deposits and not from the characteristics of the eruptions, as was generally the practice previously.

Walker's (1973) classification is genetic and was designed to represent a spectrum of increasing dispersal and fragmentation, from small-scale magma "fire" fountains to the spectrum of larger scale Plinian-style eruptions. Phreatomagmatic eruptions were grouped separately, and two types were described: surtseyan and phreatoplinian (Fig. 9.6a). These two types have extremely high degrees of fragmentation, and are, respectively, generally associated with basic to intermediate magmas, and silicic magmas, although air-fall type and composition cannot be considered in mutually exclusive terms. Vulcanian air-fall deposits generated by short-lived, sudden explosions from closed vents were also defined in Walker's scheme.

Walker (1973) interpreted the classification parameters in terms of D reflecting *eruption column height* which controls dispersal, although wind strength and changing wind direction also control dispersal and F reflecting *explosiveness* which controls the degree of fragmentation of the ejected pyroclasts (Cas and Wright 1987; Fig. 9.6b). However, in the light of recent research, Walker's (1973) scheme has been shown to have some shortcomings, which will be discussed in Sect. 9.10.

A different approach was followed by Tsuya (1955) who classified *eruption magnitude* and intensity according to erupted volume and deposit area, which Newhall and Self (1982) developed by defining more specific deposit volume limits, and relating these to eruption column heights and eruption styles into the also widely used *Volcanic Explosivity Index (VEI)* scale (Table 9.1). This is essentially based on individual proxies of the energetics of different eruption styles, suggested to be, in decreasing order of reliability: volume of ejecta, column height, duration, and level of atmospheric injection. Constantinescu et al. (2021) suggested that for large-scale explosive eruptions that produce eruption columns with substantial and radially spreading umbrella clouds, the radius of the umbrella cloud should also be included as a criterion for assessing the VEI ranking (Table 9.1). They propose that tephra fallout dispersal patterns for such eruptions are more affected by the radius of the umbrella cloud than by eruption column height.

Although the VEI scale was introduced to quantify and provide the scale or magnitude of mostly explosive eruptions (Newhall and Self 1982), the approach has historically been to assess the scale of deposits from an entire eruption. More recent, detailed studies of deposits or eruption sequences are leading to subdivision of eruption sequences into distinctive phases or stages of an eruption (e.g. Houghton et al. 2014; Bonadonna et al. 2022), allowing VEI to be assigned to different phases of an overall single eruption. Bonadonna et al. (2022) dissected the 2021 La Palma eruption sequence in the Canary Islands, into several distinct phases, each of which they ranked at VEI 2, although the total eruption sequence was ranked at VEI 3, which reflects the larger scale impact of the eruption and the area that was impacted by volcanic hazards. In such cases, the larger scale, first order total deposit assessment of VEI provides an appraisal of the hazard magnitude and total impacted area, whereas the VEI of specific phases allows appraisal of impact area of specific eruption phases and styles during the eruption.

The VEI scale is not linear, but logarithmic, and therefore each level is an order of magnitude different from the next lower and higher levels. Its shortcomings are at the lower intensity-small volume end of the spectrum, because instead of VEI 0 corresponding to zero or on a log scale 1 (i.e. 10^0) m^3, VEI 0 was assigned a volume of $<10^4$ m^3 (Table 9.1). Houghton et al. (2013) proposed an extension of the VEI scale at the lowest intensity and volume end of the scale down to negative -6 VEI values, to address small volume deposits and fallout events, using a logarithmic scale. We feel that negative volcanic explosivity values are awkward to explain to lay people and the media given that during volcanic crises, our communication has to be simple and unambiguous. We therefore propose a recalibration of the VEI scale into a *modified or recalibrated VEI* scale* in Table 9.1. Our modified VEI* scale begins with VEI* 0, representing a minimum deposit volume of 1 m^3, with increasing volumes and VEI* values following a logical logarithmic progression upwards (Table 9.1). As discussed fully in Sect. 9.10 in a more comprehensive assessment of classification schemes for fallout deposits and their

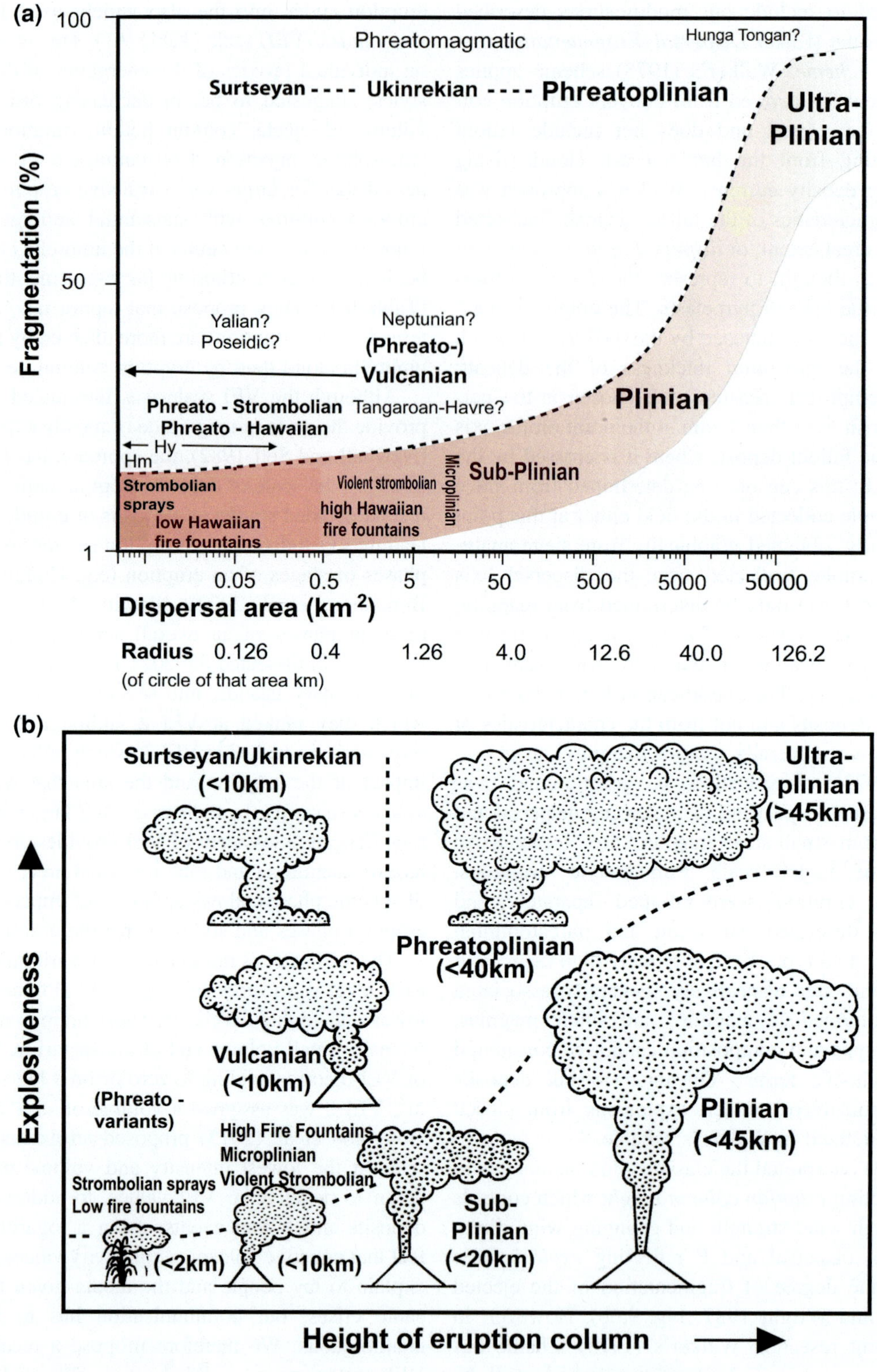

Fig. 9.6 Classification of fallout deposits. **a** Dispersal-Fragmentation (D-F) plot used to characterise different types of pyroclastic fallout deposits. D and F are defined in the text. The arrow depicts the possible wide range of dispersal of vulcanian deposits, controlled by the intensity of the blast. Hm is the very limited, almost insignificant, field of halema'uma'uan deposits as described by Orr et al. (2013) and Houghton et al. (2013). Hy represents hydrothermal explosions which range from very small mud explosions to very large blasts. Also shown, highly speculatively, are the possible positions of newly defined submarine explosive eruption styles, yalian, neptunian, poseidic, tangaroan, and hunga tongan (see Sects. 9.9.4.4–9.9.4.6). (Modified from Cas and Wright 1987, After Walker 1973). **b** Cartoon explaining

(continued)

D-F plot in terms of eruption column height and "explosiveness" for common subaerial explosive eruption styles (Modified from Cas and Wright 1987). **c** Representation of method used by Walker (1973) to calculate the two parameters D and F (defined in the text) used for classification, showing an elongated area of dispersal for a fallout deposit formed when a strong wind was blowing during an eruption and the dotted axis of dispersal. Contours are lines of equal thickness of the deposit (isopachs) showing the deposit becoming thinner away from the vent along the axis of dispersal and perpendicular to it. The maximum thickness isopach is 6.5 m, and the 0.1 and 0.01 of maximum thickness contours area shown in dashes. Crosses represent sample collection points. **d** Weight % of samples with grain size <1 mm, collected on or near the axis of dispersal plotted against distance from the vent, showing a clear increase in the percentage of fine ash with increasing distance. (**c** and **d** from Cas and Wright 1987)

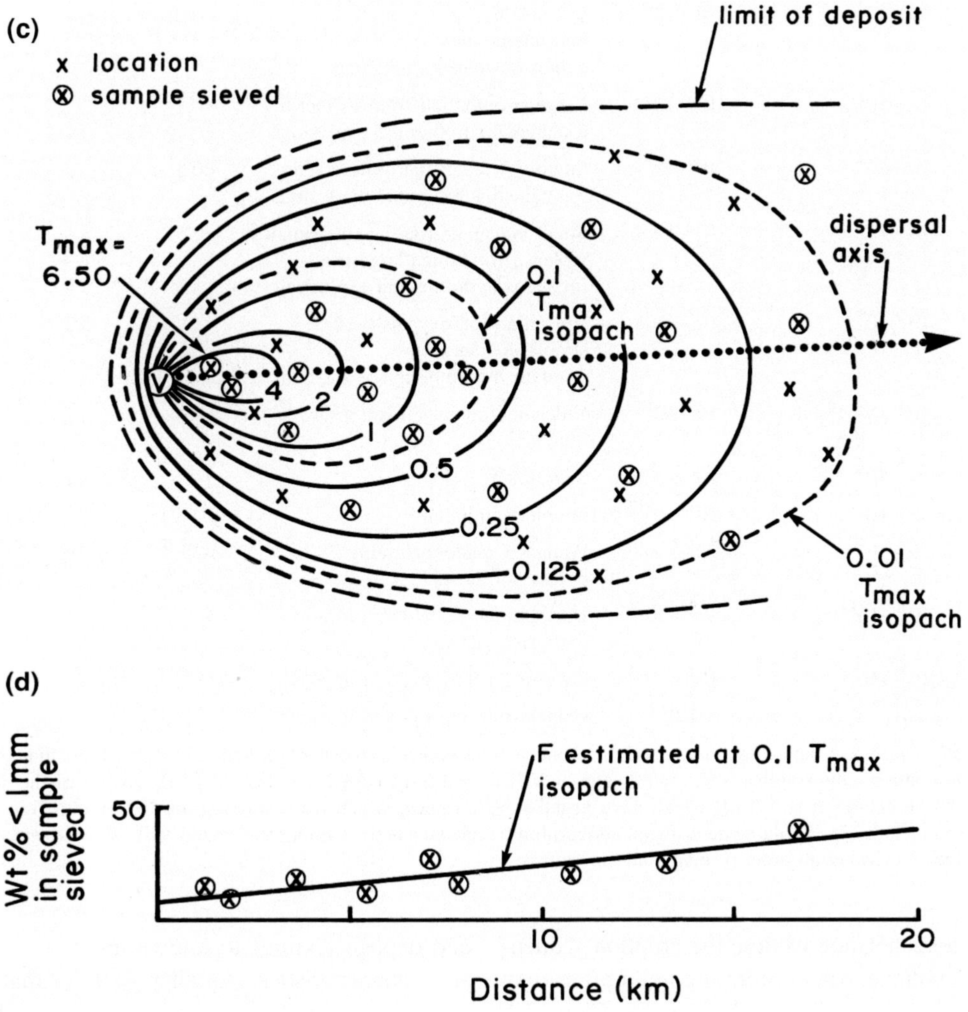

Fig. 9.6 (continued)

eruptions, the modified VEI* scale recalibrates the old VEI values by adding +4.

Crosweller et al. (2012) proposed that eruption magnitude can be linked to the *erupted mass*. Both VEI and Magnitude schemes are not related to any particular deposit type, so VEI and Magnitude may refer equally to effusive and explosive eruptions, and among the explosive eruptions disregard the eruption style, so that fallout deposits and ignimbrites are lumped together in terms of volume or mass and are not distinguished. The strength of the VEI and Magnitude classifications is in going beyond some of the limitations of Walker's (1973) scheme, as fall deposits may not be the dominant products of eruptions and consequently, there are many cases for which it may not be appropriate to

Table 9.1 Volcanic explosivity index (VEI) and new modified VEI* classification (Modified after Newhall and Self 1982; Constantinescu et al. 2021)

Old VEI	VEI* log_{10}dep vol (m^3)	Tephra volume of individual eruption/ explosion (m^3)	Mass (kg)	Common eruption style	Eruption column height (km)	Umbrella cloud radius (km)#	Stratospheric injection
[−4]	**0**	10^0–10^1	10^3–10^4	Strombolian sprays/very small hydrothermal explosions/halema'uma'uan	<0.01	–	None
[−3]	**1**	10^1–10^2	10^4–10^5	Strombolian sprays/small hydrothermal explosions/halema'uma'uan	<0.05	–	None
[−2]	**2**	10^2–10^3	10^5–10^6	Large strombolian sprays/small fire fountains/hydrothermal explosions	<0.1	–	None
[−1]	**3**	10^3–10^4	10^6–10^7	Small vulcanian (paroxysms)/ fire fountains/hydrothermal explosions	<0.5	–	None
0	**4**	10^4–10^5	10^7–10^8	Small vulcanian (paroxysms)/small violent strombolian/large fire fountains/hydrothermal explosions	<0.5	–	None
1	**5**	10^5–10^6	10^8–10^9	Vulcanian (paroxysms)/violent strombolian/large fire fountains/hydrothermal explosions	<1	<10	None
2	**6**	10^6–10^7	10^9–10^{10}	Vulcanian/micro-plinian/fissure fountains	1–5	<10	None
3	**7**	10^7–10^8	10^{10}–10^{11}	Subplinian	3–15	<10	Possible
4	**8**	10^8–10^9	10^{11}–10^{12}	Subplinian/plinian	10–25	10–100	Definite
5	**9**	10^9–10^{10}	10^{12}–10^{13}	Plinian ± caldera-forming pyroclastic flows	>25	100–200	Significant
6	**10**	10^{10}–10^{11}	10^{13}–10^{14}	Ultra-plinian ± caldera-forming pyroclastic flows	>40	200–500	Significant
7	**11**	10^{11}–10^{12}	10^{14}–10^{15}	Caldera-forming pyroclastic flows	1–25	500–1000	Significant
8	**12**	$>10^{12}$	$>10^{15}$	Caldera-forming pyroclastic flows	1–25	>1000	Significant

Note that the VEI scale is *not* linear but is logarithmic and therefore each VEI level is an order of magnitude more or less than the preceding or next level. In the first five columns, the original VEI scale began at VEI 0. We have added the above VEI 0, several of the smaller volume VEI categories proposed by Houghton et al. (2013) which have negative VEI values, which we don't support. In columns 6 to 8, we advocate our modified VEI* scale to include small volume deposits on a logarithmic scale (see text), with a recalibrated VEI* 0 representing a deposit of 1 m^3, and higher VEI* values increasing an order of magnitude in volume

use fall deposits nomenclature to describe eruption sequences (e.g. the large volume, caldera-forming ignimbrites may not even have an associated fallout deposit; see Chap. 10, Sect. 10.11). The main shortcomings of this latter group of classification schemes is the impossibility of capturing actual eruption styles, since there is no necessary link between magnitude, i.e. the total mass erupted, and intensity of the eruption, which is much more related to the mass eruption rates, the efficiency of the fragmentation process, and the efficiency of the thermal and kinetic exchange between the erupted mixture and the ambient environment (atmosphere or hydrosphere).

Moreover, all of the above classification approaches can only be applied to fallout deposits with relatively well-preserved original extent. For ancient deposits, more qualitative assessments are required that match visible clast and deposit textures to known eruption styles that produce such characteristics. Another issue is that some sustained explosive eruptions that produce collapse calderas (Chap. 16) and huge volume ignimbrites (hundreds to thousands of km^3 in volume; Chap. 10) from fissure vent-fed, continuously collapsing pyroclastic fountains, may produce no fallout deposits (e.g. the 2.08 Ma, >800 km^3 [>630 km^3 DRE] Cerro Galan ignimbrite eruption, northwestern Argentina, Cas et al. 2011a, b; Folkes et al. 2011; the 274 Ma, >1390 km^3 DRE, Ora ignimbrite, northern Italy, Willcock et al. 2013), or only small volumes of fallout deposits. Therefore, they don't rank on the Walker (1973) scale even though they have produced the most voluminous pyroclastic deposits of any explosive eruption style. By volume of erupted ignimbrite pyroclastic deposits alone, they would be categorised as VEI 7 to 8 (VEI* 11–12), but their eruption fountains were probably relatively low (<1 km?), and certainly not injecting into the stratosphere (cf. Table 9.1), although their co-ignimbrite ash clouds may instead be substantial (e.g. Woods and Wohletz 1991).

We will now briefly review the main explosive eruption styles, organised in terms of their main qualitative observational differences. These will be discussed, broadly following Walker's (1973) philosophy, by increasing the dimension of the explosive eruption column or plume, although other options could be followed, for example ordering by magnitude, or by closed vent versus open vent conditions. Afterwards, we will review the main conduit and plume processes that are useful for understanding the dispersal of fallout deposits, and finally we will describe different types of fallout deposits and relate them to eruption styles. In the following discussion of each eruption style, the listing of VEI explosivity indices for each style includes both the original VEI scheme value plus our modified or recalibrated VEI* value, as in Table 9.1, which is essentially the old VEI value + 4 (see Sect. 9.10 for a discussion of the rationale for this recalibration).

9.2.1 Small Explosive Eruptions (VEI −4 to 1, or VEI* 0–5)

Small-scale explosive eruptions include a wide variety of eruption styles that have proven to be very difficult to classify. These include eruptions of basaltic or mafic magmas, usually associated with, or evolving to, effusive activity. These also include small-magnitude eruptions from more viscous magmas such as andesite to rhyolite. All types are characterised by more or less pronounced unsteady and pulsatory behaviour, often resulting in rapid transitions from one style to another. Furthermore, the buoyant plumes formed during these types of explosive activity, if any, are relatively dilute, and weakly sustained, meaning that they are highly affected by winds at any level above the vent. These features generally result in locally dispersed deposits which are highly variable over time in terms of grain size and area of dispersal.

In particular, there has been some confusion about the distinction between strombolian *sensu stricto*, hawai'ian magma "fire" fountaining and low explosive intensity eruptions that build scoria and pumice cones. As discussed below, we define strombolian *sensu stricto* explosive eruptions to be discrete, spaced, short-lived (seconds to tens of seconds) spatter spray eruptions that build small spatter cones around the vent, whereas hawai'ian magma fountaining eruptions are sustained magma fountaining eruptions, lasting at least minutes and sometimes many weeks, that can build larger spatter cones as well as commonly feeding lava flows. Sustained explosive eruptions of largely vesiculated scoria and pumice, which produce low but buoyant eruption columns, often build scoria and pumice cones around the vent, and disperse scoria and pumice fallout deposits away from the vent are here called micro-plinian eruptions. Houghton et al. (2016) have quantified the differences in eruption dynamics between strombolian and hawai'ian eruption styles, using parameters such as eruption duration, erupted mass, and mass eruption rate (Fig. 9.7a), as discussed below. They clearly demonstrated that strombolian *sensu stricto* eruptions are shorter lived, erupt smaller masses of magma at lower magma mass eruption rates than hawai'ian magma "fire" fountain eruptions.

9.2.1.1 Strombolian *sensu stricto* Eruptions: Short-Lived, Time-Spaced Fire Fountain Spatter Sprays (VEI <0, or VEI* 0–3)

Several volcanoes, including the type volcano Stromboli in Italy, generally erupting mafic, low-silica composition magmas, are characterised by periods of open vent conditions where explosive eruptions consist of a series of discrete time transient explosive sprays of spatter separated by seconds to several hours, called *strombolian (sensu stricto) explosive eruptions* (e.g. Plate 9.1a; Fig. 9.7a–e). Explosion durations usually last from a few seconds to tens of seconds (Houghton and Gonnermann 2008). The events produce instantaneous or rapidly sequential, spaced sprays of largely fluidal spatter, scoria, bombs, and minor ash (Plate 9.1ba; Figs. 9.2a and 9.7a–e).

Unfortunately, in the past, the term "strombolian" was also applied to more sustained eruptions that produce scoria cones and dispersed scoria fallout deposits (Walker 1973; Self et al. 2014), which were then more appropriately called micro-plinian by Francis et al. (1990; also see Martin and Németh 2006 for a discussion of the distinction). The discussion by Walker (1973) on the meaning of "strombolian" is a little confusing. On the one hand, he associated hawai'ian and strombolian styles and deposits because they both produce spatter. But then he distinguishes strombolian because it produces more scoria, can produce scoria cones, and can have dispersal areas of >5 km^2. All of these are inconsistent with the strombolian *sensu stricto* eruption style and deposits, which are characteristic of the eruption style most of the time on Stromboli volcano in recent history. We recommend using the term strombolian (*sensu stricto*) for the spaced spatter spray eruption style described above, which produces spatter cones and ramparts, and as discussed below, using micro-plinian (and in some cases violent strombolian) for the sustained small column eruption style that produces scoria cones and dispersed scoria fallout deposits.

Strombolian *sensu stricto* eruption column or fountain/spray heights are generally from a few tens of m to <200 m (e.g. Barberi et al. 1993). The total amount of material ejected during single explosions is as little as

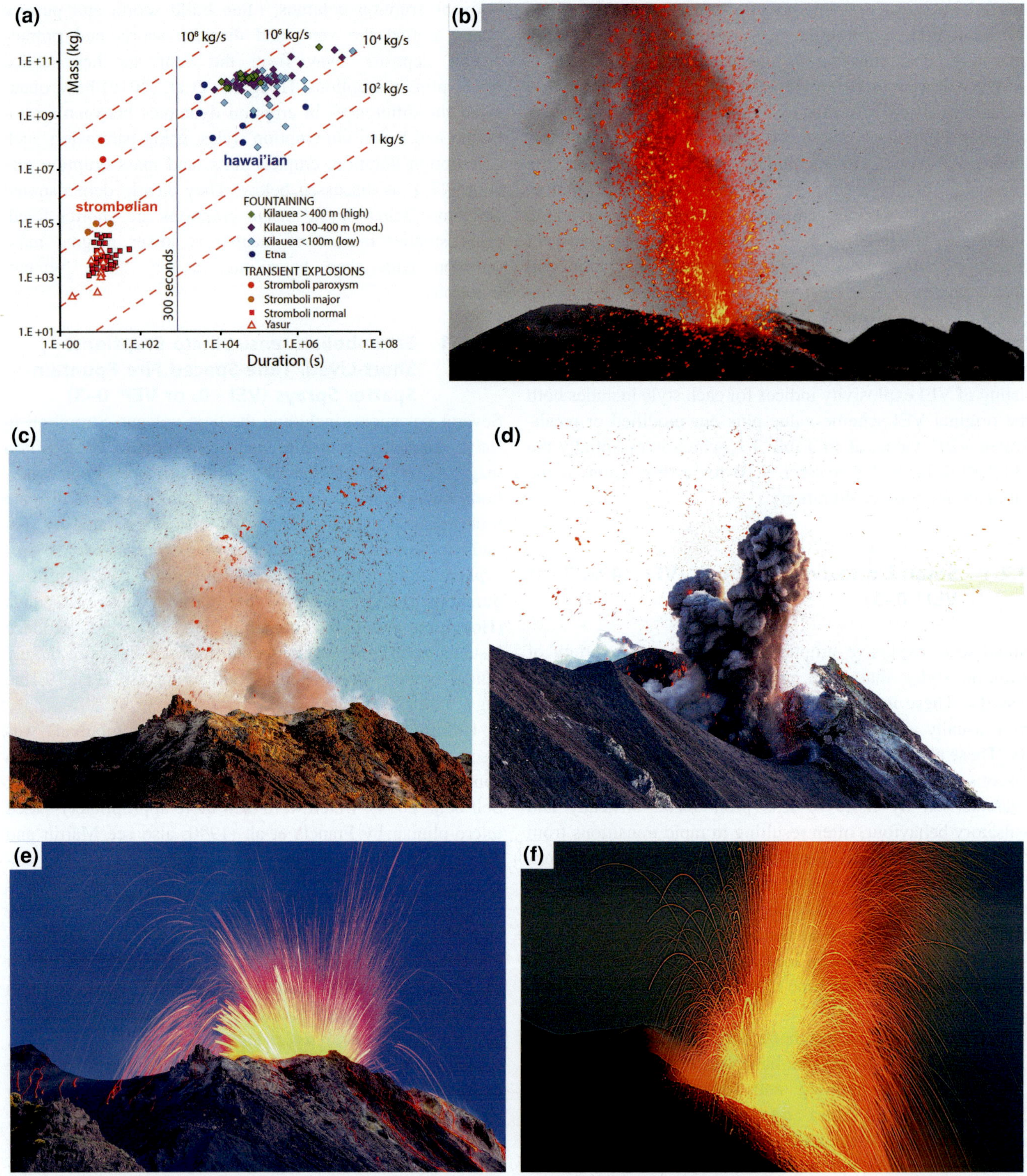
(a)
(b)
(c)
(d)
(e)
(f)
Mass (kg)
Duration (s)
10^8 kg/s
10^6 kg/s
10^4 kg/s
10^2 kg/s
1 kg/s
300 seconds
hawai'ian
strombolian
FOUNTAINING
Kilauea > 400 m (high)
Kilauea 100-400 m (mod.)
Kilauea <100m (low)
Etna
TRANSIENT EXPLOSIONS
Stromboli paroxysm
Stromboli major
Stromboli normal
Yasur
1.E+01
1.E+03
1.E+05
1.E+07
1.E+09
1.E+11
1.E+00
1.E+02
1.E+04
1.E+06
1.E+08

◀ **Fig. 9.7** Strombolian and hawai'ian magma ("fire") fountaining eruption styles. **a** Distinction between discrete strombolian *sensu stricto* spatter spray and sustained hawai'ian magma "fire" fountaining eruption styles based on eruption vent duration, magma mass erupted, and magma mass eruption rates (Modified from Houghton et al. 2016). The two outliers at the top of the strombolian field with the high mass and mass eruption rates are strombolian paroxysms, which we consider to be vulcanian explosive events. **b** Individual, short-lived, dilute daytime strombolian spatter spray explosion at Stromboli volcano, Italy on 26 June 2014; each spatter clast being ejected as a ballistic clast. **c** Higher concentration of spatter clasts during daytime strombolian spatter spray at Stromboli volcano, Italy. **d** Strombolian magma spray and explosion. **e** Daytime time lapse image of strombolian magma sprays, showing parabolic trajectories of spatter clasts. (Photos **c**, **d**, and **e** by Matteo Trolese). **f** Time lapse photo at night showing the cumulative effect of multiple strombolian spatter spray explosions at Stromboli volcano, clearly defining the parabolic path of individual spatter clasts (Photo by M. Fulle, taken 12–6-2012; *Source* Stromboli OnLine: https://www.swisseduc.ch/stromboli/ with permission R. Carniel). **g** Drifting ash cloud after a day time strombolian spatter spray event, Stromboli, 21 June 2014. **h** Low, continuous hawai'ian style fire fountain, Pu'u O'o volcano, Hawai'i, 1983. This fire fountain is feeding a lava flow. (Photo by J. Griggs; *Source* United States Geological Survey). **i** Fissure fire fountaining, with multiple focused discharge points, during the Bárdarbunga-Holuhraun eruption in Iceland that fed a rapidly expanding lava flow (Photo by Gro Birkfeldt Møller Pedersen). **j** Close-up of the fountain from one discharge point (Photo by Thórdís Högnadóttir)

Fig. 9.7 (continued)

10^{-2}–10^{1} m^3, which places this kind of phenomenon at the low end member of the explosive spectrum (VEI ~0.1/ VEI* ~3), at average mass eruption rates of 10^0–10^4 kg s^{-1} (Chouet et al. 1974; Ripepe et al. 1993; Cole et al. 2005). Measured exit velocities range from 3 to 100 m s^{-1} (Patrick et al. 2007), although the use of high-speed cameras revealed that the initial high-velocity bursts out of the vent may reach 400 m s^{-1} (Taddeucci et al. 2012). Patrick et al. (2007) imaged 344 strombolian eruptions on Stromboli volcano from 2001–2004, noting their short-lived nature with each eruption lasting between 6 and 41 seconds, with an average duration of 15 seconds.

Spatter bombs and scoria lapilli are ejected on ballistic, largely parabolic trajectories (Plate 9.1a; Figs. 9.2a and 9.7b) and largely build small and low spatter and scoria cones or ramparts around the vent. Minor ash fractions may be entrained in short-lived weak convective plumes that can drift away with winds and rarely reach higher than a few km (Fig. 9.2a; Patrick et al. 2007). In order to classify the range in strombolian *sensu stricto* explosion types and eruption styles Patrick et al. (2007) suggested three styles, naming *Type 1* the explosion dominated by only ballistic spatter bombs and scoria lapilli driven by gas thrust, *Type 2a* those where ash and bombs are ejected contemporaneously with less gas thrust, and *Type 2b* where only ash is ejected, involving buoyant convective rise of ash. In the latter case, the ash fraction may be abundant, resulting from either efficient fragmentation of a degassed and stiffened shallow plug, or just from the recycling of ash from previous explosions.

This type of mild and impulsive explosive activity represents the present-day normal activity at Stromboli (Barberi et al. 1993; Rosi et al. 2000; Harris and Ripepe 2007), from which the term *strombolian* activity originated.

Phases of such short-lived, time-spaced fire fountain sprays may last for centuries at volcanoes such as Stromboli (Rosi et al. 2000), or for shorter periods of years, such as at Arenal volcano, Costa Rica (Cole et al. 2005), and Yasur volcano, Vanuatu (Simons et al. 2020). It is common, however, that such prolonged periods of relatively steady open vent activity are punctuated by rapid transitions to effusive episodes or more intense short-lived vulcanian-style explosions, such is the case of paroxysms at Stromboli (see Sect. 9.2.1.3), when mass eruption rates may instantaneously reach 10^5–10^6 kg s^{-1} (Rosi et al. 2006; Giordano and De Astis 2021).

9.2.1.2 Hawai'ian Magma or Fire Fountaining (VEI −3 to 1, or VEI* 2–5)

In *(hawai'ian) magma or fire fountaining* activity, the eruption column is essentially a sustained, "fire" fountain of magma (Fig. 9.7g–i), formed when jets of disrupting magma or spatter are released, almost continuously in some cases, through the vent (Macdonald 1972; Taddeucci et al. 2015). Fire fountains are episodic events that are associated typically with eruptions of low-viscosity magmas (basalts, phonolites). Episodes may last hours to several days and may originate from linear fissures, such as commonly occur in Iceland and Hawai'i (Fig. 9.7i, and Fig. 5.13a; Duffield et al. 1982; Macdonald and Eaton 1984; Riel et al. 2015; Pedersen et al. 2017), and then concentrate over time into one or more, point sources. Magma or fire fountaining may also characterise periods of activity at central vents such as in the case at Mt Etna, Italy, and Pu'u O'o, Hawai'i (Fig. 9.7g, and Fig. 5.13b; Behncke et al. 2006; http://www.youtube.com/watch?v=5zMSLpRjsQI). Some of the most spectacular recent examples of sustained magma fountaining were the 2014–2015 Bárdarbunga-Holuhraun eruption in Iceland that fed a far flowing lava flow field (Fig. 9.7h, i), and the recent series of en echelon fissures that opened up on the Leilani Estate in the southeast rift zone, Kilauea volcano, Hawai'i, in May 2018 (see Fig. 5.13c).

Fountain heights from central vents are generally >200 m high (Houghton and Gonnermann 2008), and in such cases magma would be ejected at velocities of a few tens of m s^{-1} (Wilson and Head 1981). Fountains from linear fissures may show lower heights of tens of m (Duffield et al. 1982). Average mass eruption rates, both for central and linear eruptions, are in the order of 10^5–10^6 kg s^{-1} (Houghton and Gonnermann 2008), although fountains wax and wane over time scales of minutes (Gouhier et al. 2012). Short-lived fountain events probably rank as ~VEI 1/VEI* 5, but this is complicated because the products are not always just tephra deposits, but can include voluminous lavas.

The predominant products of these fire fountains are large spatter pieces, which fall back around the vent area, sometimes forming agglutinated spatter cones, and/or reconstituted lava that flows away from the vent (e.g. Fig. 5.13b, Pu'u O'o, Hawai'i, 1983; Bárbardunga, Iceland, 2014–2015, Pedersen et al. 2017). Poorly developed convective plumes above magma fountains may take the smallest ash-sized particles derived from lava sprays up to heights of a few hundred m, but all coarser fragments will already have fallen out ballistically from the column.

Fountaining is usually contemporaneous with, or transitions to effusive activity (Calvari and Pinkerton 2004). Continuous fire fountains usually feed lava flows (Fig. 5.13 b; Mangan et al. 2014; Pedersen et al. 2017) both from the base of the fountain where no or little fragmentation occurs, and also by agglutination of spatter clasts on landing, which then contribute mass and heat to the lava flow. Type localities for this kind of activity are Hawai'i, from where the term hawai'ian style became universally associated with fire fountaining, and Iceland. During the Bárdarbunga-Holuhraun fissure eruption in Iceland from 28 August 2014 to 27 February 2015, about 1.4 km^3 of lava was erupted emplacing an 85 km^2 lava field (Pedersen et al. 2017) and causing subsidence of about 50 m at the nearby caldera (Riel et al. 2015).

Fig. 9.8 Eruption styles of small explosive eruptions (VEI 0–3). **a** Initial small-scale vulcanian explosion at Sakurajima volcano, Japan, on 22 July 2013. Note the dense dark slug, and ballistics being ejected from the base of the rising slug on the right side. **b** Continued rise of the slug, formation of a weak ash plume at the top, and granular flows down the slope triggered by high concentration axisymmetric shower of ballistics and lobes of debris laden material from the margins of the column. **c** Drifting ash plume from the top of the eruption column, and buoyant uprise of an ash cloud from the top of the granular flows. (Photos **a**, **b** and **c** by T. Blaikie). **d** Weak drifting eruption plume from a short vulcanian explosion at Sakurajima volcano, Japan, 23 November 2009 (Photo by K. Berlin; *Source* Wikimedia, through Creative Commons Attribution-Share Alike 2.0 Generic licence). **e** Intense vulcanian explosion of Ngauruhoe, New Zealand, at 18.10 h on 19 February 1975. Collapse of dense interior of erupted slug to form pyroclastictic flows with ash plume rising above summit at + 45 s. (From Nairn and Self 1978; Photo by Graham Hancocks). **f** Sudden vulcanian blast (paroxysm) at Stromboli volcano, Italy, on 3 July 2019, which has all the hallmarks of a vulcanian explosion (Photo by S. Rinauro; From Giordano and De Astis 2021). **g** Buoyant eruption column that developed after the initial vulcanian blast observed in (**f**) (Photo by G. Scognamiglio; From Giordano and De Astis 2021)

(a)
Ballistics
(b)
Granular flow
(c)
Ash cloud
(d)
(e)

Fig. 9.8 (continued)

9.2.1.3 Weak Eruption Columns from Pulsatory or Episodic Explosive Eruptions (VEI 2–3, or VEI*6–7)

Many moderately explosive eruptions involve the generation of weak eruption columns. "Weak" *versus* "strong" reflects the competition between the vertical velocity in the column, which largely relates to the mass flux into the column, and the velocity of winds (Sparks et al. 1997; Bonadonna et al. 2005; Woodhouse et al. 2013). Weak columns are columns where the wind velocities are larger than the column uprise velocities and the column "leans over" as it is pushed in a downwind direction. These conditions are more likely, under normal wind velocities, for eruptions characterised by relatively low mass eruption rates and a pulsatory behaviour, i.e. not strongly sustained. Such columns are narrow and bent over by winds, rarely rising above the troposphere, i.e. usually lower than 10 km in height (e.g. Figs. 9.7c and 9.9c). Small, moderately sustained plumes can be established continuously or discontinuously for periods of days to weeks, and nowadays, with increasing population density closer to and around volcanoes, this kind of activity has become a major hazard even if the actual magnitude (total mass erupted) of the eruptions is relatively small (VEI 2–3/VEI* 6–7). Examples of such columns are the 1995–1996 eruptions at Mt Ruapehu volcano, New Zealand, the 2001–2 and 2002–3 eruptions at Etna, Italy (Fig. 9.9b, c), the 2005 eruptions at Augustine volcano (Alaska), and the ongoing eruptive phase at Sakurajima (Japan).

9.2.1.4 Vulcanian Activity: Violent, Paroxysmal, Vent Clearing/Opening, Short Duration Cannon-Like Explosions/Ballistic Blasts (VEI <1–3 or VEI* <5–8)

Vulcanian activity has been observed and described in a large number of historic eruptions since the 1888–90 eruptions of Vulcano in the Aeolian Islands, Italy, which are the "type example" of this activity (Mercalli 1907; Macdonald 1972). The term vulcanian explosion is usually given to sudden, impulsive gas-driven explosive blast eruptions (Fig. 9.8a–e; Clarke et al. 2015) that are common at andesitic–dacitic stratovolcanoes and viscous lava domes (e.g. the frequent explosive phases at Soufrière Hills volcano, Montserrat, Lesser Antilles, often began with a vulcanian explosion; see Fig. 10.4a), Sakurajima, Japan (Fig. 9.8a–d); Augustine volcano, Alaska, USA; Sumeru volcano, Java, Indonesia), but may also involve less evolved compositions, such as the paroxysmal phases at Stromboli volcano, Italy, in 2003 (Rosi et al. 2006), 2007 (Pistolesi et al. 2011) and 2019 (Giordano and De Astis 2021), and the 1975 eruption at Ngauruhoe volcano, New Zealand (Fig. 9.8e; Nairn 1976; Self et al. 1979; also see Fig. 10.4b).

Vulcanian eruptions *sensu stricto* are highly unsteady, transient events, marked by sudden explosions, although they may evolve into other more sustained eruptions styles, as discussed below. Where associated with viscous, dome-forming magmas, vulcanian eruptions occur at low magma ascent rates. Explosions involve the explosive destruction of a viscous plug or cap of magma, dome, or solidified rock obstructing a gas-rich conduit, in which pressure builds up to 10 MPa, and sudden failure and decompression occur (Clarke et al. 2009, 2015). The plug is most likely cooled, congealed magma that has risen after the previous explosion, and formed a very viscous mass of magma or a solid crust in the conduit, or it could be an older plug. The pressure rise may be due to the exsolution of magmatic gas in the conduit beneath the plug or cap, perhaps fed by injections of new hot magma from depth in the conduit into cooled magma high in the conduit, or to the heating and partial vaporisation of groundwater, but not due to violent mixing with external water, as in a phreatomagmatic eruption. Gas pressure builds up until the overlying cap fails, in tension or shear, leading to a vertically directed blast(s) of mostly dense clasts, including a significant component of accessory lithic clasts, which clears the vent to some degree. For rocks at temperatures up to 950 °C, pressures of a few MPa (or > tensile strength of the cap rock) are expected (Wilson 1980a). Because of the pressures involved, this cap rock is unlikely to be simply a layer of unconsolidated clasts.

Similarly, sudden, short-lived phreatic (e.g. Ontake volcano, Japan in 2014) and hydrothermal explosions (e.g. White/Whaakari Island in New Zealand in 2016, 2019) also represent the sudden release of pressure through cap rock, producing similar phenomena, although the source mechanism is not largely magmatic gas, but superheated external water, and the erupted components often contain little or no juvenile magmatic debris, although this can vary (e.g. 2019 White Island/Whakaari event).

A recently discovered phenomenon at Kilauea volcano during 2008–2018, involving the collapse of the vent walls causing rock falls to impact the surface crust of the lava lake in Halema'uma'u crater, also caused short vulcanian-like explosions (Orr et al. 2013). The sudden impact of the rock fall on the lava lake crust caused fracturing of the lava crust and decompression and vesiculation of the magma under the surface crust, leading to the explosive ejection of abundant lithic debris from the rock fall as well as juvenile vesiculated scoria. Nonetheless, the volume of erupted deposits was extremely small from 9 to 300 m^3, with erupted masses up to 10^5 kg, and mass discharge rates from 10^2 to 10^3 kg s^{-1} (Houghton et al. 2013). Houghton et al. (2013) have called this style of explosive eruption "*halema'uma'uan*". It is highly likely that this phenomenon has occurred on other volcanoes where vision into the depth vents has not been available. Some of the paroxysms on Stromboli could have such origins. For example, before the 2007 "paroxysmal" explosive eruption of Stromboli, there was a detectable collapse of the summit area of Stromboli associated with a marked drop in the magma level (Pistolesi et al. 2011). Destabilisation of the vent area could have led to intra-vent rock fall as a partial at least trigger for the subsequent explosive eruption.

Exit velocities during vulcanian explosive eruptions may reach 400 m s^{-1} (Self et al. 1979). At Soufrière Hills volcano, Montserrat, videos imaged exit velocities in the range of 40–140 m s^{-1} (Druitt et al. 2002). When initial velocities are supersonic (>343 m s^{-1}), a shock wave would also be propagated (Nairn 1976). Magma volumes erupted are typically <0.1 km^3 (Clarke et al. 2009). Activity during vulcanian eruptions proceeds as one or a number of discrete cannon-like explosions that last for seconds and occur at intervals from seconds to tens of minutes to hours. Such short-lived explosions produce a series of small eruption columns, usually <5 km high, with plumes strung out downwind of the volcano (Rose et al. 1978; Stix et al. 1997; Druitt et al. 2002; Wright et al. 2006). Eruption activity at Sakurajima volcano in Japan is currently characterised by regular discrete explosions (e.g. Iguchi et al. 2013; Fig. 9.8a–d). Depending on the intensity and sustained duration, vulcanian eruptions could range from VEI < 1–3 (VEI* < 5–8).

Discrete vulcanian blasts *per se* do not produce sustained, high eruption columns, but low weak plumes that drift with the wind (Fig. 9.8d). However, the sudden pressure released at the onset of vulcanian explosions triggers a decompression wave that propagates into the conduit and allows gas exsolution and magmatic fragmentation, which *may* then feed a more sustained plinian style phase of eruption. In such cases, higher eruption columns have reached heights of 10 to 15 km (Nairn and Self 1978; Druitt et al. 2002) with much larger areas of dispersal than with short discrete vulcanian explosions (*sensu stricto*).

Proximal vulcanian pyroclastic fall deposits are typically thinly bedded, and made of large ballistic blocks and fragments deposited around or into the vent. Finer-grained fractions have a much wider dispersal. Commonly, a large part of the material ejected with each explosion is not juvenile but includes a large fraction of country rocks as accessory lithics. Vulcanian-type explosions that do not eject any juvenile clasts are usually phreatic or hydrothermal explosions (e.g. Marini et al., 1993), as was the case with the deadly 2014 phreatic blast event at Ontake volcano (Maeno et al. 2016; Maeda et al. 2017; http://www.volcano.si.edu/volcano.cfm?vn=283040), and the 2019 hydrothermal explosion at Whaakari/White Island (Gibson 2020).

Small-volume pyroclastic flows are also frequently generated during vulcanian eruptions when large amounts of ejecta fall back around the vent (Fig. 9.8b–d). During the 1975 vulcanian explosion at Ngauruhoe volcano, New Zealand (Fig. 9.8e), a slug of material was ejected at an estimated initial velocity of ~400 m s^{-1}, and partial collapse of the column occurred at nearly 500 m above the crater rim to generate pyroclastic flows (Fig. 10.4b; also see Sect. 10.10.2). This event ejected 2×10^8 kg of rock, although only half of this was juvenile clast material. Half of the total volume collapsed to form the pyroclastic flow (Self et al. 1979).

The climactic phases of the 1995–2015 eruptions at Soufrière Hills volcano, Montserrat, involved initial vulcanian explosions (see Fig. 10.4a) that evolved into more stable eruption columns that produced associated pumiceous pyroclastic flows, as described in Druitt et al. (2002). However, not all vulcanian eruptions produce pyroclastic flows (e.g. Irazú volcano, Costa Rica, 1963–5; Sakurajima volcano, Japan (Fig. 9.8c), ongoing; Sumeru volcano, Indonesia, ongoing).

Self et al. (1979) and Wilson (1980a) estimated the maximum velocities of ejected debris in vulcanian explosions as a function of the pressure beneath the retaining plug at the time it fails. Their calculations indicate that initial velocities up to 200 m s^{-1} are readily explicable by magmatic gas contents (up to a few wt%), and pressures from a few to >10 MPa are consistent with rock strengths. However, for initial velocities above 300 m s^{-1}, the influence of external water must be postulated. Even if pressures of a hundred megapascals or more (which is far in excess of rock strengths) are assumed, the unreasonable magmatic water content of >10 wt% would be required. Heated groundwater is probably a significant feature in such explosions, but not necessarily an essential feature as proposed by Schmincke (1977).

Paroxysmal Vulcanian Explosive Events on Stromboli Volcano, Italy

A brief discussion is in order to explain why we include the paroxysmal explosions at Stromboli in the vulcanian eruption style. Very different to what has been just described for classic vulcanian explosions, Stromboli is an open conduit volcano, continuously erupting, where no viscous plug/dome exists within the conduit to justify the build-up of a gas overpressure below it. In fact, paroxysms at Stromboli have occurred either during effusive phases (e.g. the 5 April 2003 and the 15 March 2007, Calvari et al., 2011; 3 July and the 28 August 2019, Giordano and De Astis 2021) or punctuating ordinary strombolian activity (e.g. the 11 September 1930, Rittmann 1931, and the 3 July 2019, Giordano and De Astis 2021).

Paroxysms on Stromboli volcano are vulcanian in style (Fig. 9.8f) because they are short-lived (few minutes), vent clearing, cannon-like blasts, frequently associated with pyroclastic flows, and phenomenologically and instrumentally they are identical to vulcanian eruptions (see data in Ripepe and Harris 2008). Yet, the overpressured blasts are caused by the, still poorly understood, sudden rise of gas-rich, small batches of magma from a deep reservoir, inferred from petrological data to be at ca 10 km depth (200–300 MPa, Métrich et al. 2005, 2021; Francalanci et al. 2008; Pichavant et al. 2022). The high ascent velocity, up to 2 m s^{-1}, prevents the decoupling of the gas from the deep magma batch, which displaces the stiffer and degassed magma resident within the open conduit, and triggers an overpressured explosion just at the vent level that is able to deeply erode the summit crater. Accordingly, fragmentation depths calculated from geophysical data are very shallow (80–150 m; Ripepe and Harris 2008).

It is commonly agreed that the paroxysms result from the interaction of rapidly rising, phenocryst-poor, gas-rich fluidal magma with upper conduit, phenocryst-rich, variably degassed, viscous magma (e.g. Métrich et al. 2021; Pichavant et al. 2022). Whereas Métrich et al. (2021) propose that the intensity of paroxysms is related to subtle variations in magma compositions and crystallisation rates, Pichavant et al. (2022) propose that the ascent rate of the poorly crystallised deep magma and its volatile content are the principal determinants of the origins of paroxysms and their intensity. Pichavant et al. (2022) propose a model for paroxysms involving rapid ascent of magma (<2 m s-[1]; rising from depths of 7 km in a matter of a few hours) and decompression of deep low phenocryst magma, leading to volatile supersaturation and closed-system degassing and vesiculation at depths of 1.5 to 2 km below the volcano top. This is followed by interaction of the low phenocryst magma as it rises into the pre-existing high-phenocryst magma "reservoir" triggering degassing, cooling, and microlite crystallisation of the low-phenocryst magma for about 20 h. Finally, the overpressurised magma continues to rise through the conduit leading to inflation of the summit (Ripepe et al. 2021) and explosive degassing of the magma. This leads to ejection of highly vesiculated crystal-poor "golden pumice" pyroclasts, crystal-rich, less vesiculated pyroclasts, mixed magma pyroclasts, and accessory vent wall lithic ballistic ejecta.

An array of instrumentation around the summit of Stromboli volcano has allowed details of the vulcanian paroxysmal eruptions that occurred in 2003 (Rosi et al., 2001), 2007 (Pistolesi et al. 2011), and 2019 (Giordano and De Astis 2021; Ripepe et al. 2021) to be determined. The 2003 event lasted 8 min and consisted of 4 phases. Phase 1 precursory event was marked by heightened gas release and lasted 30 s. It was followed by Phase 2, the climax of the eruption, which was marked by an initial dark ash plume which rose to 4 km height and produced a fallout deposit, followed by multiple ballistic jets that hurled ballistics up to 2 km from the vent, 1400 m above the vent and were ejected at 170 m s^{-1}. The mass of fallout deposit was 1.1 to 1.4×10^8 kg, and the discharge rate was 2.8 to 3.6×10^6 kg s^{-1}. Phase 3, 45 s after the start of the eruption, consisted of a lateral and vertical ash plume that produced buoyant ash clouds. Phase 4 commenced 3 min after the start of the eruption and consisted of multiple small explosions that produced ash plumes up 600 m high. The climax of the eruption ranks as perhaps a VEI 1/VEI* 5 eruption based on the mass erupted, driven by a gas overpressure of 2.3 MPa (Rosi et al. 2006).

In the 2007 event, there was precursor seismic tremor and inflation, around the vent for 90 s before a major ballistic blast that lasted 30 s, followed by a second 12-s long explosion that produced a 3 km high buoyant plume. The total mass of tephra ejected was ~ 2.5×10^7 kg, as well as $1.9–2.3 \times 10^5$ kg of gas. The mass discharge rate during the second explosion was 1.9–2.3 10^6 kg s^{-1}. The gas overpressure was 7.9 MPa, the gas-ash exit velocity was 210 ± 40 m s^{-1}, and the ballistic ejection (muzzle) velocity was 100–155 m s^{-1}. There was then a 2-min-long period of gas release before a final scoria flow-forming event that lasted 90 s (Pistolesi et al. 2011). The total volume of tephra erupted as fallout was $1.7–2 \times 10^4$ m^3, which ranks the event as a VEI 0 or perhaps VEI 0.4 (~VEI* 3).

The 3 July and the 28 August 2019 blasts (Fig. 9.8f, g) were remarkably similar (Giordano and De Astis 2021). The only precursory indicator was an inflation of the edifice detected only a few minutes before the explosions. The 3 July jet expanded spherically at an initial mixture exit velocity of 210–220 m s^{-1}, calculated from ballistic exit velocities of 120–160 m s^{-1}, before feeding a vertical jet and laterally directed pyroclastic flows. Juvenile pyroclasts include a mixture of highly vesicular and aphyric pumice related to the deep magma batch and highly crystalline and poorly vesicular scoria representing the degassed magma resident in the conduit. Paroxysmal deposits are very thin, with little potential for preservation in the rock record, except for the largest events. These deposits form thick blankets dominated by variably agglutinated ballistic juvenile spatter and lithic blocks around the summit area (only 30 cm thick for the 3 July 2019 event, which triggered slides of glowing avalanches along the volcano flanks); farther away, the short-lived vertical eruption column fed the fallout of pumice and scoria. The 3 July 2019 event had a 8.4 km high column above the vent (9.1 km above sea level; Fig. 9.8g) and while the seismic duration of the explosive event was 5 min, the fallout lasted about 30 min on the island, producing only a 1–2 cm thick lapilli fall deposit. The calculated mass discharge rates for the July 3 and August 28 explosions are constrained at 1.1×10^6 kg s^{-1} and 3.6×10^5 kg s^{-1} respectively (Giordano and De Astis 2021).

Some authors in the literature have described the paroxysmal events at Stromboli as violent strombolian eruptions, but we disagree with such nomenclature as violent strombolian events (see next Sect. 9.2.1.5) are very closely spaced, almost sustained strombolian eruptions that are transitional into micro-plinian events. Violent strombolian eruptions are prolonged (days, months) scoria cone-building, during which the frequency of explosions and both the instantaneous and the average mass eruption rates are sustained for prolonged periods. Furthermore, no examples of violent strombolian eruptions are described associated with pyroclastic flows. We therefore suggest that the paroxysmal events of low-viscosity magmas which are impulsive and vent blasting should be classified as vulcanian in style, even if associated with open conduit volcanoes (Giordano and De Astis 2021), distinct from the cone-building, longer lasting, violent-strombolian and from high magma fountain eruption styles.

9.2.1.5 Violent Strombolian Explosive Eruptions (VEI 2–3 or VEI* 6–7)

Explosive eruptions from low-viscosity magmas can produce significant amounts of fine-grained material, as demonstrated during the 1943–44 period of the 9-year-long eruption of Parícutin volcano in Mexico (Foshag and Gonzales 1956; Pioli et al. 2008). Moderately sustained activity characterised the first phases of activity at Parícutin monogenetic volcano, which started on 20 February 1943 (Fig. 9.9a; http://www.youtube.com/watch?v=IqhF7wZDvWA). In just 6 days, a scoria cone 167 m high was built, and explosions were visually counted at up to 18 per minute feeding a pulsating eruption plume estimated to be about 5 km high (Foshag and Gonzales 1956). More dramatic at Parícutin was the period between 18 March and 9 June 1943, when the style of activity changed and much more ash was erupted feeding a moderately sustained convecting plume up to 6 km high (Pioli et al. 2008), which was blown downwind by winds. The eruption style was pulsing, due to the closely spaced explosions, and the magma mass eruption rate was ~10^4–10^5 kg s^{-1}. Ash was deposited as far away as 250 km from the volcano producing a continuous tephra cover, with lapilli-sized clasts transported for several km from the vent (Foshag and Gonzales 1956; Pioli et al. 2008). Walker (1973) described some scoria fall and cone-building deposits with unusually high values of fragmentation and dispersal as "*violent strombolian*" (MacDonald 1972; Valentine and Gregg 2008), emphasising the pulsatory nature of the explosions, even though the eruption style was not like strombolian *sensu stricto*. The small to moderate size eruption column was more like a small plinian eruption column, but the eruption style was markedly pulsatory, suggesting perhaps a transition between strombolian and micro-plinian.

Parícutin also experienced phases of vulcanian explosive activity, and the last 6 years were dominated by effusive lava-forming activity. Violent strombolian eruptions mostly classify as VEI 2 to 3, or as VEI* 6–7 on the recalibrated VEI* scale (Fig. 9.6a; Table 9.1).

Etna volcano is also often characterised by intense and prolonged magma/fire fountaining accompanied by the production of ash, but simultaneously produces buoyant, convective plumes, as was the case during the 2001–2002 and 2002–2003 eruptions (Fig. 9.9b, c; Calvari and Pinkerton 2004). In this type of activity, a more widely dispersed scoria fall deposit can attain a dispersal area much larger than the proximal scoria cone. In 2002, that eruption

Fig. 9.9 **a** Violent strombolian eruption column produced during the 1943 eruption of Parícutin volcano, Mexico, and the scoria cone and lava flow it produced (Photo by K. Segerstrom; *Source* United States Geological Survey). **b** Violent micro-plinian (violent strombolian) style at Torre del Filosofo locality during the 2002–2003 Etna eruption, Sicily, Italy. Discrete but closely spaced blasts fed the weak column. Note the small emerging dark plume disconnected from the main paler column, and a second active vent in the background, as in (**c**). **c** Three simultaneously active vents and eruption columns dispersing a downwind (complex?) fallout deposit via downwind plumes at different altitudes in the atmosphere at Etna on 30 October 2002 (*Source* NASA Johnson Space Centre). **d** Micro-plinian eruption column, 1973 Heimaey eruption, Iceland, that built the Eldfell scoria cone and dispersed scoria downwind. The weak eruption column can be seen to be bent over by winds to the right. (Photo by George Carman)

also produced simultaneous fallout from three simultaneously active, but separated vents, complicating the understanding and interpretation of such complex fallout deposits.

However, the establishment of moderately sustained plumes, although weak, may also be seen as a transitional regime that would lead, with increasing frequency and intensity of the explosions, towards the plinian style (see next section), which Francis et al. (1990) called *micro-plinian* because of its similarity in style to plinian eruptions, albeit on a much smaller scale.

To differentiate between violent strombolian and micro-plinian styles for modern eruptions, we suggest using "violent strombolian" to indicate those eruptions that form columns generally <8 km high, i.e. that cannot inject into the stratosphere at any latitude, and are marked by a rapid pulsing explosive style, reflected in diffusely stratified fallout deposits. Slightly taller columns, between ~ 8 and 13 km, that are sustained and stable, are transitional to the plinian family as they may or may not involve a stratospheric injection depending on the latitude, and can be called micro-plinian. They produce massive fallout deposits with minimal stratification, although some vertical grain size variations may occur. This difference is broadly represented in Fig. 9.6a, where violent strombolian eruptions form the upper end of the field limited by the 5 km^2 dispersal (and maybe slightly across), whereas micro-plinian eruptions also fall into the 5 km^2 field, but may spread into the lower end of the sub-plinian field. Even so, it seems that apart from column height, violent strombolian eruptions are marked by a sustained, but more pulsatory, explosive eruption style, producing cone-forming stratified deposits with significant fine ash, whereas micro-plinian eruption style is more sustained and produces more massive deposits (e.g. coarse scoria cone-forming and more widely dispersed massive deposits).

9.2.2 Plinian Eruptions: Small to Large Sustained Explosive Eruptions (VEI 2–8 or VEI* 6–12)

The spectrum of plinian style eruptions describes eruptions or eruption phases characterised by high eruption rates ($>10^6$ kg s^{-1}; Cioni et al. 2015), which feed mostly sustained eruption columns for periods of hours to days (e.g. Fig. 9.2b, c). These have been described in increasing order of magnitude/intensity as sub-plinian, plinian, and ultra-plinian. We add at the small end of the spectrum, micro-plinian (Francis et al. 1990), because of the similarity of style (sustained buoyant column and continuous fallout), even though column heights are lower than the rest of the plinian spectrum. The high mass discharge rates usually promote the formation of strong plumes for plinian and ultra-plinian eruptions, although micro-plinian and sub-plinian eruption plumes may be affected by strong winds, hence are classified as weak, as was the case of the 1995 eruption of Ruapehu volcano, New Zealand (Bonadonna et al. 2005). Plinian columns usually rise through the troposphere into the stratosphere, with column heights >8 km, and consequently, large deposit dispersal areas of >5 km^2 up to >5000 km^2 (Walker 1973).

9.2.2.1 Micro-Plinian Eruptions: Transition into Highly Explosive Eruptions (VEI 2–3 or VEI* 6–7)

This eruption style was originally called strombolian by Walker (1973) in his classification of fallout-forming eruptions and deposits, to describe events that produced scoria cones and dispersed scoria fallout deposits, even though the eruption style was nothing like the typical spaced sprays of spatter that characterise strombolian *sensu stricto* eruption style (Sect. 9.2.1.1). Recognising this misleading association, Francis et al. (1990) proposed using micro-plinian for small sustained eruption column forming eruptions that produced scoria cones, which we agree with (Fig. 9.6a). Walker (1973) also recognised that micro-plinian (his strombolian) scoria cones were different to violent strombolian scoria cones.

The scoria cone-forming phase of the basaltic 1973 Heimaey eruption (Iceland), which is often cited as a typical example of this style of eruption, began with lava fountains rising 50–100 m from up to 20 vents along the length of a 1.5 km long fissure (Thorarinsson et al. 1973; Self et al. 1974). Later, activity became centralised, and more explosive activity took place from three vents at the northern end of the fissure, building a scoria cone 200 m high, from which lava also flowed (http://www.youtube.com/watch?v=VX9e9ei6v3s). Individual explosions could be described as discrete fire fountain spray events. However, this was different from the activity described in Sect. 9.2.1.1 because they occurred in rapid succession (e.g. every 1–2 s), providing enough material and heat to maintain a weak buoyant eruption column, driven by convection, which reached heights of >5 km (e.g. Blackburn et al. 1976).

The later phase of the 1973 eruption on Heimaey transformed into a more sustained explosive phase that produced a buoyant eruption column and a coarse, near-vent cone of massive scoria and massive scoria fall deposits of limited dispersal downwind over the town (Fig. 9.9d). Self et al. (1974) and Walker (1973) described this as strombolian, even though the eruption style was nothing like strombolian *sensu stricto*, as described above and was more energetic and sustained. Francis et al. (1990) called such small sustained scoria cone-forming eruptions "*micro-plinian*" because of their similarity of style to larger scale plinian eruptions. Although the term micro-plinian would imply, according to some authors, the possible stratospheric injection which is a characteristic of plinian eruptions (Bonadonna et al. 2016), this should not be a necessary criterion for defining an eruption style or a deposit type, particularly for older deposits. The altitude of the tropopause (the border between the troposphere and the stratosphere) varies enormously from between about 8 km at the poles, 10–13 km at intermediate latitudes and up to 18 km at the equator, and in the long term also varies according to global climate change.

Similar small to moderately explosive micro-plinian activity may also be associated with weak explosive felsic magma eruptions, which can form pumice cones, but they are less common (e.g. the 776 AD pumice cone eruption at Mt Pilato, Lipari, Italy; Davì et al. 2011). Eruption columns from micro-plinian eruptions can also be weak, and easily deflected by strong winds (Fig. 9.9d).

9.2.2.2 Sub-Plinian Style: Unsteady Sustained Eruption Columns (VEI 3–4 or VEI* 7–8)

Sub-plinian eruptions involve magma mass eruption rates in the order of 10^6 kg s^{-1}, and eruption columns usually < 20 km in height (Fig. 9.2b). Calculations from the ejection of large ballistic lithic clasts indicate that in some eruptions muzzle velocities are as high as in some plinian events (>400 m s^{-1}), although the lower range is 100 m s^{-1} (Wilson 1978). Deposit dispersal areas range from 5 to 500 km^2, and, according to the Pyle (1989) classification, the thickness half-distance, b_t, is 0.5–4 km, and the clast half-distance, b_c, is 1–3 km.

Recent sub-plinian eruptions have been observed and described at Ruapehu in 1995 (Bonadonna et al. 2005), at Chaiten, Chile, in 2008 (Alfano et al. 2011, 2012), Puyehue

Fig. 9.10 Plinian-style eruption columns. **a** Sub-plinian eruption column during the 2015 eruption of Calbuco volcano, Chile, which rose to >15 km high (Photo courtesy of Angelo Castruccio; From Castruccio et al. 2016). **b** Sub-plinian eruption column from the 1993 eruption of Lascar volcano, Chile, which grew to 23 km high, showing a well-defined, spreading, umbrella region (Photo by J. Viramonte). **c** Plinian eruption column from the 1991 eruption of Mt Pinatubo in the Philippines, which grew to >40 km high, showing a more evenly spreading form with height (*Source* United States Geological Survey). **d** Contemporaneous drawing of the October 1822 eruption column from Vesuvius volcano, Italy, by pioneering volcanologist George J. P. Scrope (*Source* Wikimedia)

Cordón Caulle, Chile, in 2011 (see Fig. 9.28b–d; Pistolesi et al. 2015), and Calbuco, Chile, in 2015 (Castruccio et al. 2016; Fig. 9.10a), and Lascar volcano, Chile, in 1993 (Fig. 9.10b). Sub-plinian eruptions are characterised by sustained activity that continuously feeds the rising column. At the same time, significant pulsations in eruption intensity are reflected in variations over time in plume height, density, and degree of interaction with the atmosphere and winds, which in turn are reflected in the dispersal patterns and development of an internal stratigraphy in the associated deposits (see Sect. 9.7.4.3). Sub-plinian fallout deposits, like all plinian deposits, are dispersed, well-sorted deposits dominated by pumice clasts, and share many deposit characteristics with their larger scale Plinian cousins.

9.2.2.3 Plinian Style: Eruption Columns from Steady, Large-Scale, Sustained Eruptions (VEI 5–6 or VEI* 9–10)

"A cloud, from which mountain was uncertain, at this distance (but it was found afterwards to come from Mount Vesuvius), was ascending, the appearance of which I cannot give you a more exact description of than by likening it to that of a pine-tree, for it shot up to a great height in the form of a very tall trunk, which spread itself out at the top into a sort of branches; occasioned, I imagine, either by a sudden gust of air that impelled it, the force of which decreased as it advanced upwards, or the cloud itself, being pressed back again by its own weight, expanded in the manner I have mentioned; it appeared sometimes bright and sometimes dark and spotted, according as it was either more or less impregnated with earth and cinders".

This description of the 79 AD eruption at Vesuvius volcano, Italy, by Pliny the Younger (Letters. LXV. To Tacitus. The Harvard Classics. 1909–14), contains the first observational written account of a VEI 5/VEI* 9 eruption column, hence the name "plinian". Plinian eruption columns originate from quasi-steady sustained eruptions, characterised by magma mass eruption rates in the order of 10^6–10^8 kg s^{-1}, and eruption columns that are usually 20–35 km in height. Muzzle (or vent exit) velocities are usually > 400 m s^{-1} (Wilson 1978; Wilson and Parfitt 2008).

Plinian eruption columns (Fig. 9.2b; also see Figs. 9.10 and 9.28 of sub-plinian columns which show similar features) are classified as strong, when they are only weakly affected by winds during their ascent, due to the high eruption rates. Eruption columns are, as described by Pliny the Younger, buoyant, quite vertical, and upward expanding (Figs. 9.2b and 9.10b, c), up to the level where lateral spreading becomes dominant (level of neutral buoyancy; see Sect. 9.5; Fig. 9.10a–d). Plinian column heights involve an injection through the troposphere, and lateral spreading of the higher parts of the column in the stratosphere (Fig. 9.10a, b), often dominated by stratospheric wind jet streams, which leads to clearly defined downwind plumes (see Fig. 9.28c–f), and large dispersal areas. Deposit dispersal areas range from 500 to 5000 km^2, and according to the Pyle (1989) classification, the thickness half-distance (b_t) is 2–10 km, and the clast half-distance (b_c) is 3–8 km. Plinian eruptions are typically at VEI 5–6/VEI* 9–10 levels.

Plinian scale eruptions have been observed and described at Mt St Helens in 1980 (Fig. 9.2b), El Chichon, Mexico, in 1982, and Pinatubo, the Philippines, in 1991 (Fig. 9.10b). Other well-documented examples are the eruptions of Hekla, Iceland, in 1947 (Thorarinsson 1954, 1968), the 1932 eruption of Quizapu in the Chilean Andes (Larsson 1937), and Santá Maria, Guatemala, in 1902 (Rose 1972; Williams and Self 1983). Quite a number of plinian eruptions have occurred in historic times, and the deposits of the following examples have received detailed attention: the eruptions of Askja, Iceland, in 1875 (Sparks et al. 1981; Carey et al. 2008a, b, 2009, 2010), Tarumai, Japan, in 1667 (Suzuki et al. 1973), 1563 Fogo eruption, São Miguel, the Azores (Walker and Croasdale 1971), 4.6 ka Fogo A eruption, São Miguel, the Azores (Walker and Croasdale 1971; Pensa et al. 2015), and the 1912 Valley of Ten Thousand Smokes eruption, Novarupta-Katmai, Alaska, USA (Hildreth 1987; Fierstein et al. 1997; Houghton et al. 2004a, b; Hildreth and Fierstein 2012). Better known because of their archaeological significance are the Vesuvius 79 AD eruption, Italy (Gurioli et al. 2008; Cioni et al. 2000), and the Late Bronze Age Minoan 1600 BC (3.6 ka) eruption of Santorini, Greece (Bond and Sparks 1976; Druitt 2014; although the exact date of this eruption has been debated for many years; see Kutschera 2020). Historical records have complemented some of these studies, and have provided evidence of the duration of these events.

All of the historical plinian eruptions mentioned above seem to have taken place from central vents, and most of the older deposits studied have been mapped to point source vents, which are generally assumed to be located above cylindrical-shaped conduits. There is no doubt that some of these vents are aligned along linear fissures or ring fractures, but activity from different vents in many cases can be shown to be separated by long intervals, recognised for instance, by soils between their different fall deposits. However, detailed mapping by Nairn (1981) in the Okataina rhyolitic centre in New Zealand has shown that many plinian fall deposits and associated ignimbrites were erupted during simultaneous or sequential activity from multiple vents along fissures. These eruptions were often spread along fissures more than 10 km long but, as in basaltic fissure eruptions, activity seems to have been restricted to definite points. Vent types for large explosive silicic eruptions, during which plinian activity may lead to ignimbrite formation, are discussed in Chap. 10.

9.2.2.4 Ultra-Plinian Style (VEI 7–8 or VEI* 11–12)

The ultra-plinian style was introduced by Walker (1980), essentially to account for the dispersal of the fallout deposits of the 232 AD Taupo eruption in New Zealand. These deposits show an exceptionally large dispersal area in the order of several tens of thousands of km^2 (Fig. 9.6) and were suggested to have been derived from an eruption column as high as >50 km, driven by magma mass discharge rates of 10^8–10^{10} kg s^{-1} (Alfano et al. 2012; Cioni et al. 2015), representing the highest and most intense end member of the Plinian family (Wilson et al. 1978). No such activity has ever been witnessed in historical times, although the 15th January 2022 eruption of Hunga Tonga-Hunga Ha'apai volcano in Tonga, southwest Pacific, did produce an eruption column >50 km high (Proud et al. 2022), which we discuss later in this chapter as ultra-phreatoplinian. Recent work on the 232 AD Taupo pumice fall deposit suggests that the very large dispersal area is an artefact of changing wind direction during the eruption, questioning the very existence of an ultra-plinian style (Houghton et al. 2014).

However, it is somewhat semantic as to whether or not the classification should be based on the dispersal of individual phases and wind shifts during *a single essentially ongoing eruption*, or on the overall characteristics of the entire fallout deposit from the whole fallout stage of the ongoing eruption. We favour using the overall dispersal area because there is insufficient, detailed dispersal data available on all the individual phases from most fallout eruption sequences, and many sustained eruptions experience wind shifts and changing dispersal patterns. So, unless there is clear evidence for a significant time break within a fallout sequence, overall the various layers are the product of one eruption, not multiple eruptions. We can of course consider the variations in eruption parameters during a single eruption, such as column height and mass discharge rate. Other deposits also have dispersal characteristics that may be interpreted as ultra-plinian, such as the 1257 AD Samalas eruption from Indonesia (Vidal et al. 2015; see Sect. 9.7.4.4). Ultra-plinian eruptions are VEI 7–8/VEI* 11–12 scale eruptions.

9.2.2.5 Caldera-Forming Eruption Style

Walker's (1973) classification of explosive eruption styles was proposed based on fallout deposit characteristics and applies to eruptions from point source vents. This classification does not capture the fissure-fed pyroclastic fountain collapse style associated with caldera-forming eruptions that can produce voluminous pyroclastic flows and ignimbrites, as discussed in Chap. 10. Giordano and Cas (2021) proposed that pyroclastic flows formed by the partial or total collapse of eruption columns from point source vents can be included in Walker's (1973) classification to capture the magnitude of such eruptions on the VEI scale. These usually have some associated fallout deposits. By contrast, pyroclastic flows that form during caldera-forming eruptions from linear fissure vents originate from very complex and different column collapse dynamics, at much higher mass eruption rates (up to 10^{11} kg s^{-1}), which should be distinguished from point source fallout-forming eruptions as a distinctive style named "caldera-forming". In some cases, the eruptions that produce calderas begin from a point source vent, before evolving into a fissure vent and caldera collapse phase. The early point source eruption phase, which is essentially a plinian style eruption phase, can produce voluminous and widespread fallout deposits (e.g. Bandelier eruptions, New Mexico; Self et al. 1986, 1996). However, in other cases, caldera collapse begins almost immediately after the eruption begins, and no fallout deposits form, just ignimbrites (e.g. the 2.1 Ma Cerro Galan caldera-forming eruption, northwest Argentina; Folkes et al. 2011; Cas et al. 2011a, b). However, the ignimbrite-forming eruptions associated with caldera collapse can produce voluminous and widespread co-ignimbrite ash fallout deposits, as discussed next, although these are not classified in the Walker (1973) scheme, although they can be considered under the VEI scheme using their erupted volumes. In Chap. 10, we will discuss the details of this new caldera-forming eruption style.

9.2.3 Fallout from Ash Clouds Accompanying Pyroclastic Flows

During pyroclastic flow-forming eruptions, significant amounts of the explosively ejected ash-size magma particles may be elutriated out of the body of the flow into the overriding ash cloud (see Chap. 10; Figs. 10.10, 10.11; Sparks and Huang 1980; Woods and Wohletz 1991; Eychenne et al. 2012; Engwell et al. 2014; Martì et al. 2016), and will not be incorporated in the resulting pyroclastic flow deposit. The ash rises above the pyroclastic flows as a buoyant, trailing turbulent ash cloud (Fig. 9.11a; Chap. 10), which is taken to great heights by huge convective plumes. Because the ash particles are very fine-grained (nearly all <1 mm), there may be nearly 100% efficiency of conversion of heat to convective energy to drive the plumes. Atmospheric winds disperse the fine ash according to wind patterns, and it can fall out forming a widely dispersed co-ignimbrite ash-cloud fallout deposit.

Fine vitric ash is most easily lofted and separated from other components into the gas-rich, buoyant ash cloud, leading to concentration or enrichment of other components, including dense crystal fragments and lithic fragments, in the body of the pyroclastic flow and its deposits. Hay (1959) first showed that enrichment of crystals took place in a small basaltic andesite pyroclastic flow deposit from the 1902 eruption of Soufrière, St Vincent, and he attributed this to the selective loss of vitric ash. Lipman (1967) found a similar enrichment in crystals in a pumiceous rhyolitic ignimbrite erupted from Aso caldera, Japan. Walker (1972) and

Fig. 9.11 **a** A pyroclastic flow that formed during the eruption at Mt Unzen, Japan, on 24 June 1993, flowing from left to right, showing a 3D billowing ash cloud rising from the flow, part of which is lofting into the atmosphere (yellow arrow) (*Source* Volcanological Society of Japan). **b** Surtseyan phreatomagmatic explosion during the 2009 eruption of Hunga Tonga–Hunga Ha'apai volcano, Tonga, southwest Pacific Ocean. This shows white vapour-rich and dark solids-rich parts of multiple tephra jets with characteristic "cock's tail" shapes in the explosion plume. The pointy end of each jet traces the trajectory of ballistic blocks and bombs. (Photo taken 18–3-2009; *Source* Dana Stephenson/Getty Images, and Global Volcanism Program https://doi.org/10.5479/si.GVP.BGVN200902-243040). **c** Phreatomagmatic eruption column of eruption at Ukinrek maars in Alaska in 1977, resulting from magma rising through high-level permafrost in the subsurface (*Source* United States Geological Survey). **d** Pyroclastic flow entering the sea, Montserrat (Photo courtesy of S. Loughlin, British Geological Survey). **e** Entry of lava into the sea on 27 June 2018, Hawai'i, sending a plume of steam and ash into the atmosphere (*Source* United States Geological Survey). *Inset* telephoto image shows dark fragments of molten and semi-solid lava being blasted upward and outward during lava–seawater interaction (Photo by M. Patrick; *Source* United States Geological Survey)

Fig. 9.11 (continued)

Sparks and Walker (1977) demonstrated that the enrichment of crystals is a typical feature of ignimbrites and must be accounted for by substantial volumetric losses of the fine vitric component, as much as 50% of the original magma, which is deposited in associated air-fall ash deposits.

Both column- and flow-related fractionation processes appear to have the capacity to remove huge volumes of glassy ash into a buoyant ash cloud. Calculation of ash loss and conversely crystal concentration is done by comparing the proportion of crystals in pumice clasts, which represent whole or bulk samples of the erupting magma, with that in the body or matrix of the ignimbrite. This is calculated on a weight percentage basis to establish dense rock equivalents for crystals and glass, and to overcome the problem of vesicularity or porosity of vesiculated pyroclasts. Walker (1972) and Sparks and Walker (1977) showed that the matrix of ignimbrites, especially of outflow ignimbrites, is consistently enriched in crystals relative to whole pumice fragments. Intra-caldera ignimbrites show less ash loss, indicating derivation from lower eruption columns, and less efficient ash fractionation into ash clouds in intra-caldera pyroclastic flows. Cas and Wright (1987) describe the method used to calculate crystal concentration, and conversely ash loss factors.

The 1912 Valley of Ten Thousand Smokes eruption at Novarupta-Katmai produced a plinian column that was simultaneously buoyant and collapsing (see Sect. 9.5.2; Fierstein et al. 1997; Hildreth and Fierstein 2012). This indicates that the contribution of ash to the widely dispersed buoyant elements of the eruption that result in fallout deposits (eruption column, buoyant pyroclastic flow ash cloud) can be both vent-derived (column), as well as from the gas and particle fractions elutriated from the ground-hugging pyroclastic flows into the buoyant ash cloud.

9.2.4 Phreatomagmatic Explosive Eruption Columns (Surtseyan, Ukinrekian, and Phreatoplinian)

Some eruptions become explosive or change their eruption style to more or less explosive when rising magma interacts explosively and efficiently with external water or ice bodies producing distinctive phreatomagmatic explosive eruptions, eruption columns, and fallout deposits (Chap. 4; Fig. 4.18; Walker 1973; Cas and Wright 1987; White and Houghton 2000; Houghton et al. 2015). Such situations and settings include interactions with:

- *seawater* (e.g. the 1952–53 eruption of Myojinsho, Japan, Fig. 9.80, Fiske et al. 1957–58 eruption at Capelinhos volcano, Azores, Cole, et al. 2001, Fig. 10.100a–d; the 1963 eruption at Surtsey volcano, Iceland, Plate 4d, Fig. 4.18a, Thorarinsson et al. 1964); the 2009 and 2014–2015 surtseyan eruptions of Hunga Tonga–Hunga Ha'a-pai, Fig. 9.11b, Colombier et al. 2018), and more recently the spectacular phreatoplinian (?) eruption on 15 January 2022,
- *lakes* (e.g. the 1966 and 2020 eruptions at Taal volcano, Philippines, Moore 1967; the 2007 eruption at Ruapehu volcano, New Zealand, Kilgour et al. 2010),
- *melted ice- or snow caps* (e.g. in Iceland under the Vat-najokull ice cap in 1996 at Gjalp, Gudmundsson et al. 1997; in 2004 at Grimsvötn volcano, Iceland, Jude-Eaton

et al. 2012; at Eyjafjallajökull volcano, Iceland, in 2010, Gudmundsson et al. 2012, Magnusson et al. 2012, Fig. 4.18b; in Antarctica, Smellie and Hole 1997),

- *shallow groundwaters and permafrost* (e.g. 1973 Ukinrek maars, Alaska, Fig. 9.11c, and Fig. 14.18c, Self et al. 1980; Kienle et al. 1980; White and Ross 2011; Suona, Miyakejima volcano, Japan, Geshi et al. 2011; Mt Gambier, southeastern Australia, Van Otterloo et al. 2013), and
- *deep ground- and hydrothermal waters*, which are mostly known from the rock record (e.g. Nysiros volcano, Greece, Fig. 4.18f, Marini et al. 1993; Lago Albano maar, Colli Albani volcano, Italy, Giordano et al. 2002; 1315 AD eruption, Tarawera, New Zealand, Nairn et al. 2005; Solfatara maar, Campi Flegrei, volcano, Italy, Isaia et al. 2015; Chaps. 16, 18).

In addition, "rootless" phreatic and phreatomagmatic explosions can occur away from eruptive vents when:

- pyroclastic flows
  - enter the sea (e.g. Fig. 9.11d, Plate 11b, c, e, Montserrat, Lesser Antilles, 1995–2013, Edmonds and Herd 2005, Chap. 11; 2019 Stromboli, Giordano and De Astis 2021, Chap. 11); or
  - flow down river valleys or over water-saturated sediments (e.g. Toutle River, 1980 Mt St Helens, USA, Moyer and Swanson 1987, Fig. 4.18e).
- lavas flow
  - into the sea (e.g. southern coast of Hawai'i 1982–2018, Fig. 9.11e, Fig. 4.18d, and Fig. 16.24c, d; Mattox and Mangan 1997); or
  - over wetlands or marshy ground (e.g. 1783–84 Laki lava flow, Iceland, Hamilton et al. 2010; Fig. 16.24a, b).

Based on both the historical and the rock records, phreatomagmatic explosive eruptions are more likely at low, or relatively low, eruption rates, associated with basaltic compositions and small to moderate pulsatory eruptions, but may also occur associated with erupting felsic magmas (Austin-Erickson et al. 2008) up to plinian scale and intensities (Walker 1981a, b; Wilson 2001).

Surtseyan is the common term to describe explosive eruptions which involve explosive interaction between erupting magma with surface water bodies (oceans, lakes) or glacial ice. *Ukinrekian* is used here (Fig. 9.6) to describe explosive interaction between magma and groundwater aquifers or permafrost. Both surtseyan and ukinrekian explosive eruptions are usually pulsatory, marked by a rapid succession of explosions driven by rapid recharge of water into the vent or conduit after each phreatomagmatic explosion, which then triggers the next explosion (Fig. 9.11b, and Fig. 4.18a). These eruption styles generate a continuous succession of base surges, ballistic tephra jets, and simultaneous fallout around the vent (Fig. 9.11b, and Fig. 4.18a; see Chap. 4, Sect. 4.6.9; Chap. 10, Sect. 10.14). *Phreatoplinian* is used to describe explosive magma–water interaction at a plinian scale, usually involving magma and caldera lake interaction (Self and Sparks 1978; Walker 1981a, b), but could also originate from magma–seawater explosive interaction, as witnessed during the 15 January 2022 eruption of the marine stratovolcano Hunga Tonga–Hunga Ha'apai, Tonga.

Currently, there are no named eruption styles that specifically relate to magma–glacier interactions, and perhaps this should be considered, as discussed below. We know that subglacial eruptions begin in a confined cavern under the ice as magma rising into the base of a glacier causes melting to begin, creating a presumably mixed water steam-filled subglacial cavern into which the eruption begins (Edwards et al. 2015). Most eruptions are initially effusive, and the erupting lavas are subject to quench fragmentation. The same ambient glaciostatic-hydrostatic pressure constraints apply to such eruptions as in relatively shallow marine and lacustrine settings, whereby high glacio- and hydrostatic pressures will suppress volatile exsolution and explosive expansion of magmatic volatiles and superheated subglacial waters (Chaps 3, 4). At some depth however, where these ambient pressures are low enough (water depth, glacial thicknesses <1 km; Zimanowski and Büttner 2003) explosive phreatomagmatic interactions (Chap. 4) can begin, initially perhaps subglacially in the cavern and then as the eruption plume melts its way through the ice to the surface of the ice, or collapse of the overlying ice block occurs, and a meltwater pond forms, interactions become more like in an open water lake surtseyan eruption. It is unclear if there are any discernible features, other than the context within a preserved, ice-confined tuya edifice (Edwards et al. 2015; Chap. 16), in the initial subglacial explosive deposits that would render distinctive definable features to the deposits distinguishing them as uniquely *glacio-surtseyan,* as distinct from normal, open water surtseyan deposits.

Where widespread dispersal of ash results from extended explosive interaction between rising magma and glacial masses-meltwater ponds associated with large shield edifices (e.g. Eyjafjallajökull volcano eruption, Iceland, in 2010; VEI 3/VEI* 7; Gudmundsson et al. 2012; Magnússon et al. 2012), the term *glacio-surtseyan* or *glacio-phreatoplinian* may be appropriate. Defining this as a unique, definable new eruption style and deposit type distinct from phreatoplinian deposits originating from explosive interaction in a large caldera lake or ocean depends on defining distinctive characteristics of such deposits, compared with existing phreatoplinian deposits. The dispersal of these deposits is certainly more widespread than surtseyan deposits. The April–May 2010 glacio-surtseyan explosive eruption of Eyjafjallajökull volcano, apart from producing pulsing cock's tail ballistic jets out of the vent, also produced a buoyant ash-rich eruption column <10 km high that dispersed 0.27 km^3 of ash, including over much of Europe, even though it is classified as only a VEI 3/VEI* 7 intensity explosive eruption (Gudmundson et al. 2012).

Another sub-class of phreatoplinian eruptions has been introduced by the 15 January 2022 explosive eruption of Hunga Tonga–Hunga Ha'apai marine stratovolcano, Tonga, southwest Pacific Ocean. The eruption, which involved at least initial explosive magma–seawater interaction from a vent below sea level and was so spectacularly captured in satellite imagery, produced an eruption column between 40 and 57 km high (Proud et al. 2022) that dispersed fallout ash deposits radially for >100 km over the Tonga Islands, and also a stratospheric downwind plume of very fine ash and SO_2 westwards over Australia and beyond (see Sect. 9.9.4.7 below for further discussion).

Eruption columns originating from magma–water interaction usually show contrasting colour domains: white where the steam due to vaporisation of water is dominant, and dark where finger jets of fragmented magma and rock debris (having a characteristic "cock's tail" shape) dominate (Fig. 9.11b). The energy of the column is largely driven by the transfer of heat from the magma to the externally superheated water, which becomes energetically buoyant and drives convection in the column. In addition to fragmentation of magma and country rock due to explosive superheating of external water, fragmentation is also caused by cooling or quenching generated thermal shock of magma during magma–water interaction (see Chap. 4). For this reason, deposits associated with explosive magma–water interaction are characterised by high degrees of fragmentation (Chap. 4; Colombier et al. 2018; Fig. 9.6) and contain significant fine ash. Even close to the vent, fine ash is usually abundant in the deposits because of the clumping of fine wet ash and deposition as aggregates with high settling velocities. The effects of the abundance of ash and water in the eruption column can be contrasting and will be discussed in Sect. 9.5.3.

In addition, phreatomagmatic explosions can also be generated away from vents, where lava flows into or within water bodies in relatively shallow water depths. This is the cause of littoral explosions (e.g. Hawai'i; Fig. 9.11e, Figs. 4.18d, 5.28d–f, 16.24c, d; Mattox and Mangan 1997), and of the large fields of pseudo-craters or rootless cones generated where lava flows into shallow water bodies or even marshes (e.g. Iceland, Hawai'i; Fig. 5.28c, 16.24a, b; Hamilton et al. 2010).

Transport and depositional processes during phreatomagmatic explosive eruptions, especially pulsing surtseyan and ukinrekian eruptions, are very complex because simultaneously pyroclasts are falling out from the column, ballistic tephra jets are being generated, and base surges are propagating laterally. Not surprisingly, deposits can have hybrid characteristics of fallout and surge (Fig. 9.51d). Cas (1989) called these surge-modified fall deposits and fall-modified surge deposits, depending on the principal signature, i.e. fallout versus surge depositional structures and textures.

Cas et al. (2011a, b; 2017) highlighted the frequent transitions between magmatic and phreatomagmatic deposits in the Late Tertiary–Quaternary New Volcanic continental basaltic intraplate volcanic province (Chap. 16). They proposed using the term *phreato-vulcanian* for some deposits representing that transition, which van Otterloo et al. (2013) then applied to specific deposits at Mt Gambier volcanic complex. Latrutrie and Ross (2020), following this reasoning, recognised transitional deposits reflecting transitions from phreatomagmatic to both strombolian and hawai'ian eruption styles (i.e. "drying" of conduit and vent systems in both cases) preserved at the Hopi Buttes Volcanic Field, Northern Arizona, which they called *phreato-strombolian* and *phreato-hawai'ian*. In other words, the boundaries between distinct magmatic and phreatomagmatic eruptions styles are not clear-cut, and transitions need to be recognised, as suggested in Fig. 9.6.

9.3 Conduit processes that Influence Eruption Styles

Explosive eruption styles are largely dominated by the style of degassing during magma ascent and its influence on the physical state of the vesiculated magma (or magmatic foam) at the time of fragmentation. In Chap. 4 (Sect. 4.5), we discussed the mechanisms that promote volatile saturation, supersaturation, and eventually bubble nucleation and growth during ascent, as well as mechanisms of magma fragmentation. The flow behaviour of two-phase gas–liquid fluids has long been studied in chemical engineering (e.g. Wallis 1969; Hewitt 1982; Weisman 1983; Brennen 2005; Mudde 2005) and has subsequently been adapted to understanding conduit processes in volcanic gas–magma systems (Mader 1998; Gonnerman and Manga 2013).

Once gas bubbles form and grow as they decompress during the ascent of the magma in the conduit, they become buoyant, as their density is lower than that of liquid magma. Magma viscosity then controls the growth rate and tendency of gas bubbles to rise buoyantly, or not, through the magma, as well as the ascent rate of the magma. In this regard, an important concept that is usually not fully addressed is that volatiles exsolve as supercritical fluids, and it is only at shallow depths in the conduit that they decompress below their critical pressure, transforming initially into dense and then low-density vapours, which causes substantial increases in their molar volume (Chaps. 3, 4).

Based on the relative rise speed of bubbles and melt, two end-member systems of magma degassing during ascent can be defined, which control the eruption style, how sustained the explosive activity can be, and the extent of magma withdrawal from the reservoir. These two end members are called closed-system degassing and the open-system degassing (Chap. 4).

Open systems are defined as those in which degassed volatiles are able to either partly or totally escape from magma during magma ascent and before the eruption, constituting the process of outgassing (Saar and Manga 1999; Rust and Cashman 2004; Cashman 2004 and references therein; Pioli et al. 2012). This is the general case during low magma ascent rates, especially in low-viscosity magma, and less so for high-viscosity magma. In low-viscosity magmas, outgassing occurs largely through gas bubble escape (Houghton and Gonnermann 2008), and, for viscous magmas, through permeable bubble networks (Klug and Cashman 1996) and fractures (Gonnermann and Manga 2013; Gonnermann et al. 2012). Low-viscosity magmas allow total or partial outgassing during relatively fast ascent rates, as they are highly deformable and bubbles grow easily, coalesce, and become more buoyant, rising faster through the melt. Total outgassing prior to eruption will lead to effusive styles, whereas partial outgassing will allow the poorly to moderately vesicular magma to fragment approaching the surface through weak and short-lived explosivity.

Conduit flows in which bubbles rise at a higher velocity than magma are called "separated flows", and the gas can escape readily from the magma (see Gonnermann and Manga 2013 for a comprehensive review). Based on the relative volumetric flux of gas versus magma, different two-phase flow regimes can be defined (Fig. 9.12; Wallis 1969; Hewitt 1982; Weisman 1983; Mader 1998; Brennen 2005; Mudde 2005; Gonnerman and Manga 2013). Bubbly flows are those in which the ratio of volumetric flux of gas relative to magma is <1 and bubbles are dispersed throughout and carried by the continuous magma phase (Fig. 9.12). As the volumetric gas flux increases and the ratio approaches 1, bubbles coalesce and form a series of slugs (Fig. 9.12) which can rise faster than the magma and have diameters and spacing controlled primarily by the conduit dimensions. At much higher gas volumetric fluxes, slugs coalesce to form irregular slugs in a stage called churn flow, which is turbulent. The flow may transition to annular flow, in which a thin annular ring of magma is dragged by a fast-rising central column of gas and suspended liquid droplets (Fig. 9.12). Once the liquid phase has completely fragmented, the conduit is completely occupied by a continuous gas phase with suspended liquid magma particles, known as mist or disperse flow in chemical engineering and gas-particulate flow in volcanology (Fig. 9.12).

An important outcome of separated conduit flows in low-viscosity magmas is that as volatiles escape in the conduit, they cannot exert a significant drag on magma, even for short-lived, transient eruption phases, such as magma or fire fountains and strombolian sprays. As a result, magma ascent rate, eruption duration, and final erupted volume are ultimately controlled bottom-up, by the decay of the initial excess pressure in the magma reservoir that initiated magma ascent. The extreme case is that of open conduit volcanoes (e.g. Stromboli, Villarrica, Erta Ale, Erebus, and Kilauea), where magmastatic pressure (i.e. the magma density ρ_m, times the acceleration of gravity g, times the length of the conduit h) in the filled conduit normally compensates the total pressure in the magma reservoir (see Chap. 8). It is only magma recharge from below that induces overpressure in the reservoir and drives occasional eruptions, while degassing and outgassing occur throughout. When the magmastatic pressure in the magma in the conduit equals or exceeds the magma pressure in the reservoir, the eruption will terminate and magma will be drawn down, back into the conduit.

Closed systems are defined as those in which outgassing is minimal during ascent (Cashman 2004, and references therein), and all bubbles that form remain and grow within the magma. This is the general case for high magma ascent rates in relatively low-viscosity magma, and is common in high-viscosity magma. Closed-system degassing in low-viscosity magmas can only occur at very high magma ascent rates, so that bubble rise speed remains equal to the magma velocity, and this is usually attained only during short-lived transient events at the very beginning of high magma/fire fountain eruptions, which quickly transition to open degassing. In high-viscosity magmas, bubble mobility is negligible, and so, unless permeable outgassing is allowed through bubble coalescence and magma fracturing, bubbles nucleate and grow within the magma, forming highly vesiculated magma or foam, so lowering its bulk density and making it increasingly buoyant in the conduit. Usually, but not always, the gas overpressure will be entirely released through explosive magma fragmentation (Chap. 4), resulting in explosive outgassing. In high-viscosity magmas, the volumetric expansion of the vesiculating magma upwards in the conduit, combined with the bulk buoyancy of bubble-rich magma, maintains the upward flow and leads to an increase in the ascent rate. In contrast, in the case of low-viscosity magmas it is the high magma ascent rate, driven by overpressure in the magma reservoir, that minimises the time for bubble escape, so that the bubbles remain coupled with, and inside the rapidly ascending magma.

A very important outcome of closed-system degassing in viscous magmas is that the vertically directed volumetric expansion and enhanced buoyancy produce high mass fluxes up the conduit to the level of the fragmentation surface, which is typical of plinian explosive eruptions. Furthermore, the presence of the buoyant magmatic foam high in the conduit is a self-sustaining mechanism during decompression, as long as volatile saturated magma is fed up the conduit from the magma reservoir. This then controls, top-down, the duration and therefore the final volume of the eruption. The eruptible magma in the reservoir will in fact

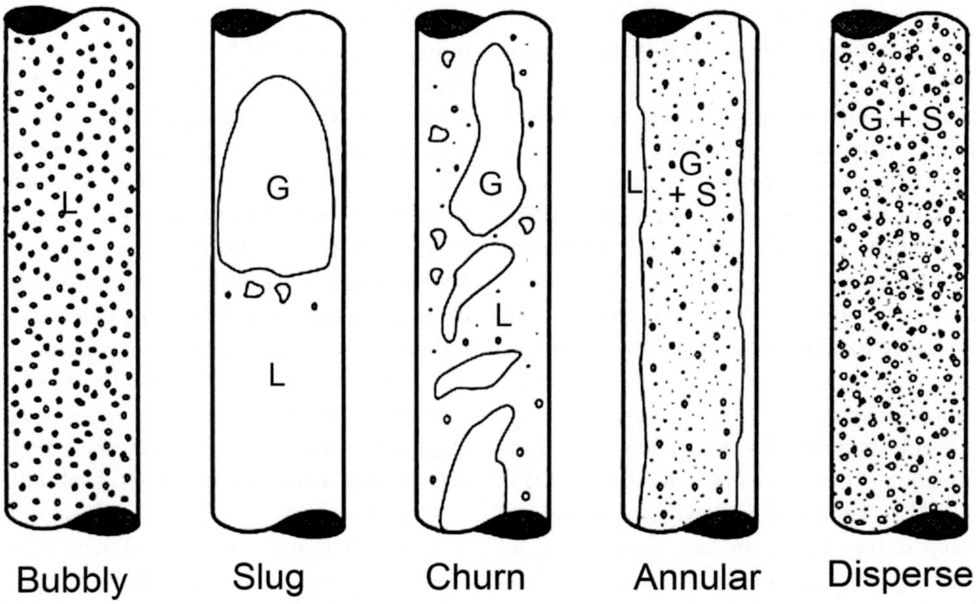

Fig. 9.12 Examples of two-phase conduit flow regimes from the chemical engineering literature based on the relative upward velocity of the gas (G) versus the liquid (L = magma) phase. Bubbly flows dominate in viscous systems where gas bubbles are suspended in a continuous liquid (magma) phase and cannot rise faster relative to the liquid, and so are carried by it. In low-viscosity systems, bubbles can rise at higher velocity than magma and are defined as separated flows (large gas domains in the magma). At increasing fluxes of the gas phase, the regime changes to slug flows, where large and spaced coalesced bubbles form, to churn flow to annular flows, where the liquid is excavated inside by the fast-rising central gas with an annulus of liquid magma which is fragmented into liquid/solid (S) particles near the conduit margins, to mist or disperse flow, or gas-particulate flow, in which liquid/solid (S) particles are fully suspended in a continuous gas phase in the conduit. (From Brennen 2005, After Weisman 1983)

continue to be withdrawn from the reservoir for as long as the conduit remains filled and magma foaming occurs at the top of the conduit. Furthermore, as pressure in the reservoir decreases, the volatile saturation level may migrate downwards into the reservoir, where foaming may occur allowing magma withdrawal well below lithostatic pressure. This depends on the ongoing availability of magmatic volatiles. In this type of eruption, it will be possible to reach underpressures in the reservoir, possibly triggering caldera collapses (see Chaps. 8 and 16).

9.3.1 Eruption Styles Resulting from Open Degassing Systems

Developed open systems, where pre- and syn-eruptive degassing is fully allowed, result in effusive eruptions of both low-viscosity and viscous magmas (Chaps. 5 and 6). However, some explosive eruptions can occur from open systems, and these originate from a range of conduit flow and degassing regimes where the competition between syn-eruptive degassing and magma ascent rates controls the ultimate eruption style.

9.3.1.1 Strombolian *sensu stricto* and Violent Strombolian Eruptions

Strombolian *sensu stricto* eruptions (see Sect. 9.2.1.1; i.e. short-lived, time-spaced, fire fountain sprays; Taddeucci et al. 2015; Plate 9.1a; Fig. 9.7a–e) have been interpreted in terms of very slow magma rise rates (<0.01 m s^{-1}; Parfitt 2004) in low-viscosity systems. In such cases, gas bubbles can grow, coalesce, and ascend more rapidly, forming gas slugs up to several m in diameter (Fig. 9.12), which burst near the surface, driven by the excessive pressure within each bubble (Fig. 9.13). The ejecta exit velocity from the vent at Stromboli is ~50 to 100 m s^{-1}. Mass eruption rates can vary from 10^0 to 10^3 kg s^{-1} (Taddeucci et al. 2015). The mechanism of formation of gas slugs in strombolian eruptions (reflected in their intermittent eruption style) is still debated, whether formed simply by coalescence in a separated flow regime as described above (e.g. Wilson and Head 1981) or associated with non-vertical conduit geometries, where bubbly foam can accumulate and eventually collapse to form slugs (e.g. Vergniolle and Jaupart 1986).

When each gas slug bursts at the magma surface in an open vent, it blasts out pyroclasts of the fragmented upper magma skin of the bursting bubble (Fig. 9.7a–c; Blackburn

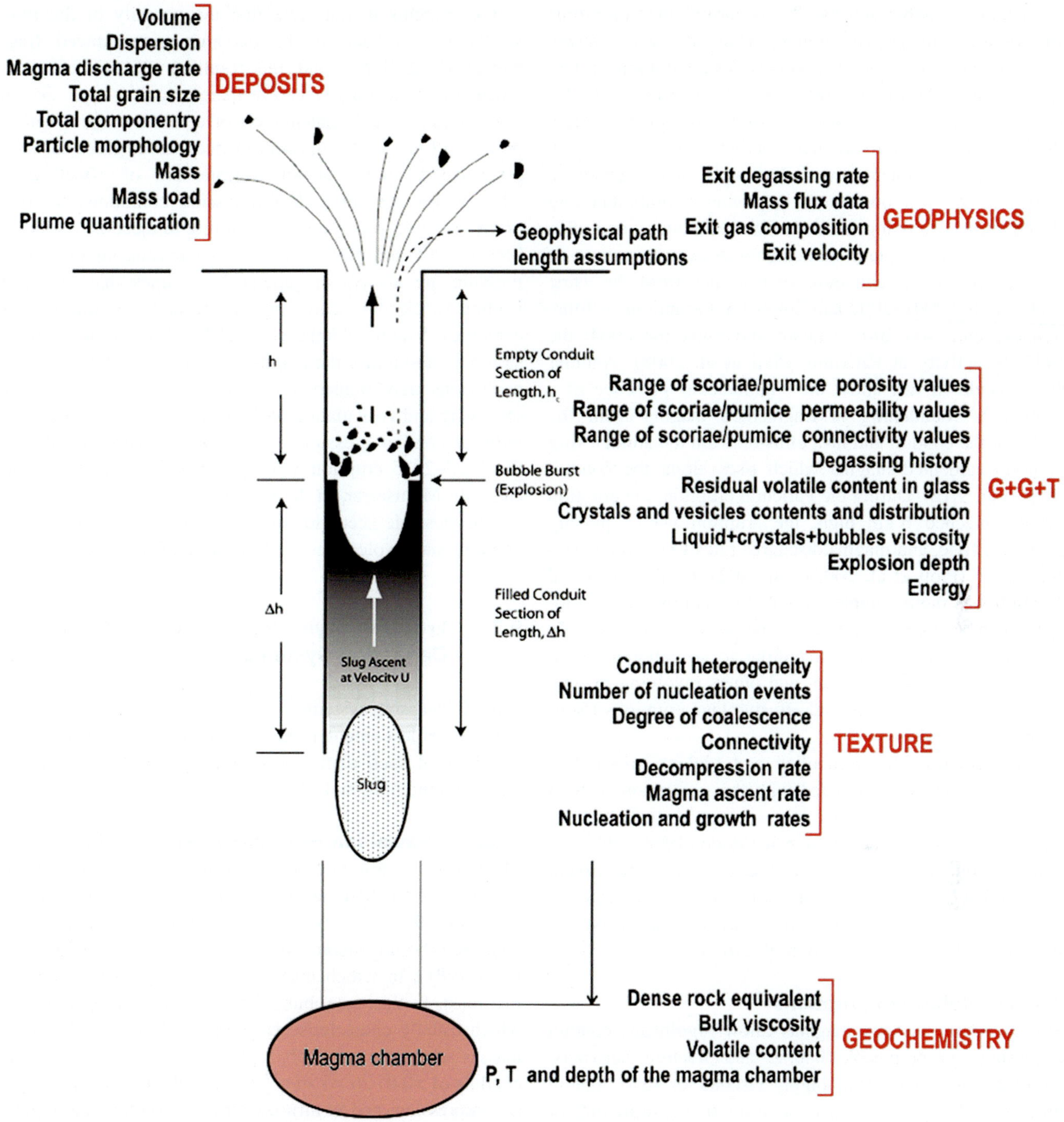

Fig. 9.13 Simple diagram depicting the rise, expansion, and bursting of gas bubble-slugs in an open conduit with an upper cooling surface producing intermittent strombolian sprays. G+G+T indicates the concurrence of geophysical, geochemical and textural approaches. (From Gurioli et al. 2015)

et al. 1976; Wilson 1980). While degassing in strombolian activity *sensu stricto* is continuous, the magma in the shallow conduit becomes stiffer because of volatile loss and associated crystallisation. This process of increasing magma viscosity reduces permeability to gas percolation and becomes more pronounced the longer the time interval between subsequent explosions, which may last from minutes to hours or more. The development of a "weak plug" made of degassed and stiffer and less permeable shallow resident magma has been indicated as a major control on driving the build up of the mild volatile overpressure that drives strombolian explosions (e.g. Thivet et al. 2020; Oppenheimer et al. 2020). This mechanism may also account for the slabby lava blocks found in some deposits (Houghton and Hackett 1984).

At higher magma and gas fluxes, strombolian eruptions transition to violent strombolian, where the very closely spaced explosions (every 1–2 s) may sustain columns a few km in height (Pioli et al. 2008; Sect. 9.2.1.5; Fig. 9.9a, b). This kind of activity is thought to be associated with a closely spaced slug-to-annular conduit flows (Fig. 9.12), suggesting that violent strombolian style may represent a transition between open vent degassing strombolian and closed vent degassing plinian style eruptions, and also explaining the frequent contemporaneous association of violent strombolian and effusive activities from the same vent (e.g. the 2001–2002 and 2002–2003 eruptions at Etna, Behncke and Neri 2003; Calvari and Pinkerton 2004; the 1943–46 activity at Parícutin, Pioli et al. 2008). Annular flows would in fact allow the simultaneous presence of a continuous fountaining of fragmented magma from the centre of the conduit and lava effusion from the annular ring of non-fragmented magma which rises along the conduit walls (Fig. 9.12). During the Parícutin eruption, the eruption column reached 6 km high, the eruption was a pulsing explosive style, and magma discharge rates were about 10^4–10^5 kg s^{-1} (Pioli et al. 2008). The 2021 La Palma hybrid eruption was more complex, involving multiple vents associatedwith a fissure system, some erupting explosively involving changes from ash puffing to strombolian *sensu stricto* to violent strombolian, and others involving magma fountaining feeding lavas and effusive lava eruptions (Bonadonna et al. 2022).

Fragmentation in strombolian and violent strombolian eruptions is largely associated with the generation of fluid instabilities in the low-viscosity magma as gas bubbles rapidly rise and expand through it (Mader 1998), although high ash content, even in near-vent deposits in some violent strombolian deposits, suggests some degree of external water–magma explosive interaction may also be involved in some cases (e.g. van Otterloo et al. 2013).

9.3.1.2 Vulcanian Eruptions

Slowly rising, viscous magmas allow significant conduit degassing via permeable bubble and fracture networks. However, once a shallow, highly viscous, or solid conduit plug or a dome is formed as a result of the stiffening of magma as it cools and degases, conditions may arise for the transition to local closed degassing. This means that even magmas in which pre-eruptive open degassing is significant may fail due to an overpressure (up to $\sim$10 MPa) developed at shallow levels by the residual exsolving gas in non-permeable (or poorly permeable) domains (Fig. 9.14), leading to sudden explosive events, called vulcanian explosive eruptions (Clarke et al. 2015; Fig. 9.8). Blasting of viscous materials results in brittle fragmentation, and pyroclasts vary from dense to vesicular, with ubiquitous evidence of post-fragmentation degassing (e.g. breadcrusted bombs).

Experiments demonstrate that the porosity of the plug substantially influences the decompression-induced fragmentation efficiency and the resulting exit velocities of vulcanian explosions (Alatorre-Ibargüengoitia et al. 2011). For example, the fragmentation of a dense magma with a porosity of just 17% will result in ejection velocities between 50 and 150 m s^{-1} at an overpressure of 10–20 MPa, whereas for the same decompression conditions, a 61% porous magma will fragment resulting in ejection velocities between 200 and 400 m s^{-1}. Blasts releasing internal overpressure are always in supersonic regimes that trigger a leading shock wave commonly heard and seen during vulcanian explosions (Clarke et al. 2015). Sudden decompression may destabilise the magmatic system, so that vulcanian explosions may transition into a plinian phase driven by exsolving and vesiculation of the magma under the suddenly reduced conduit pressures, as frequently witnessed during the 1995–2015 eruption activity at the Soufrière Hills volcano, on Montserrat, in the Caribbean.

The possible scenario for vulcanian paroxysms in open conduit, mafic volcanoes has been described in Sect. 9.2.1.4.

9.3.2 Eruption Styles Resulting from Closed Degassing Systems

Closed degassing systems always trigger explosive eruptions if the volatile content is high enough. There are two end members of such explosive eruptions, (hawai'ian) magma or fire fountaining and plinian eruptions.

9.3.2.1 Hawai'ian Magma/Fire Fountaining

Magma or fire fountaining eruptions of low-viscosity magmas (usually called hawai'ian magma/fire fountain eruptions; Fig. 9.7g–i; Taddeucci et al. 2015) are modelled as dispersed, homogeneous conduit flow regimes (Wilson and Head 1981), in which magma rise rates compete with gas rise rates. In this case, bubble coalescence and gas escape, which usually characterise eruptions of low-viscosity magmas, is limited by relatively high magma rise velocity in the conduit of >0.01 to <1 m s^{-1} (Parfitt 2004), allowing rapid decompression, and in turn favouring exsolution and bubble growth very near to the ground surface. The fountain exit velocity from the vent is $\sim$100 m s^{-1} (Parfitt, 2004). Mass eruption rates can vary from 10^2 to 10^6 kg s^{-1} (Taddeucci et al. 2015). Fragmentation in magma fountains is largely dynamic (see Fig. 4.10), i.e. due to the shear and tearing exerted by the expanding bubbles on the low-viscosity melt which drives liquid instabilities and failure (Namiki and Manga 2008). A very large range of pyroclasts originate from magma/fire fountains, from variably vesicular magma spatter clasts to small achneliths, Pele's hair, and even reticulite (Figs. 4.19, 4.20, Chap. 4).

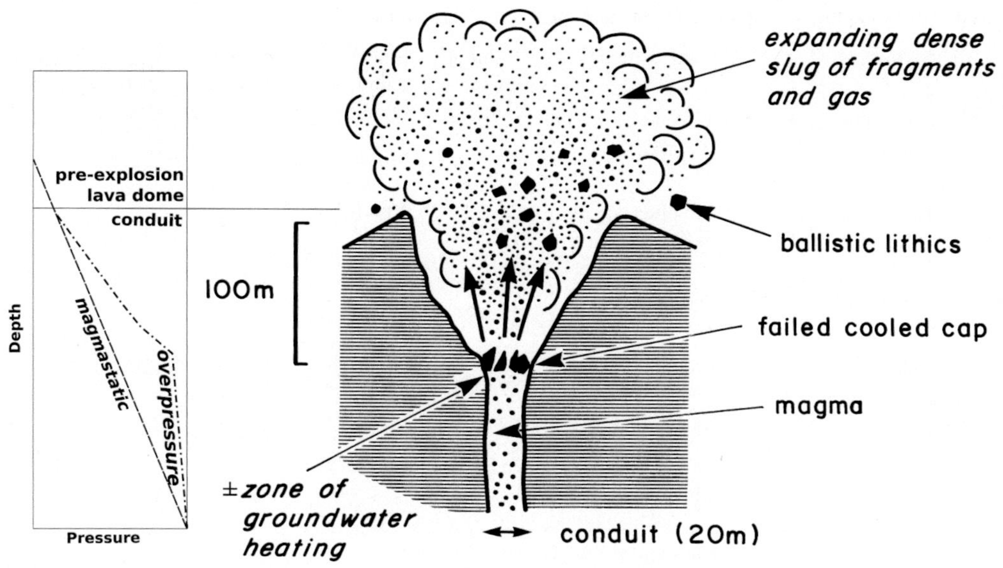

Fig. 9.14 Schematic cross-section through the crater of a stratovolcano at the time of a vulcanian explosion. Expanding "cauliflower"-shaped slug rapidly decelerates during this gas thrust phase (From Cas and Wright, After Self et al. 1979). On left, idealised pressure gradient in a vulcanian conduit

9.3.2.2 Plinian Eruptions

Plinian eruptions (including sub-plinian and ultra-plinian; Figs. 9.2b, 9.10) are largely associated with the explosive eruption of felsic magmas, during which conduit outgassing is largely prevented by magma viscosity and high magma rise velocities, although some mafic, plinian intensity and scale eruptions also occur. Calculations for the pre-vesiculation magma ascent rate for the 18 May 1980 Mt. St. Helens dacitic plinian eruption, based on the duration of the eruption, known erupted volume, and depth of the magma chamber at 6–7 km, suggest magma rise rates between 2 m s^{-1} (Carey and Sigurdsson 1985) and 15 m s^{-1} (Sparks et al. 1994). For the 2008 sub-plinian eruption of Chaitén volcano, Chile, Castro and Dingwell (2009) calculated an average conduit magma ascent velocity for vesicle-free rhyolite magma (i.e. pre-volatile exsolution and vesiculation) from a depth of 5 km of ~0.5 m s^{-1}, based on decompression experiments of samples with known mineralogy. Plinian eruptions are essentially relatively steady, high-energy events in which a continuous, turbulent flow of fragmented magma and gas is released through a conduit to the atmosphere (Cioni et al. 2015).

In Chap. 4, we discussed how the fragmentation of magma occurs during this type of eruption. Gas bubbles in rising silicic magma nucleate and grow until the volume occupied by bubbles has increased (by pressure decrease and gas exsolution), usually to a critical value of about 60–80%.

The growth of immobile or weakly mobile bubbles (i.e. dispersed conduit flow regime) feeds back into a progressive increase in magma buoyancy and rise velocity, which eventually triggers strain rate-controlled fragmentation of the vesiculated foam, which could be brittle at high strain rates, or plastic at low strain rates (Sparks 1978; Papale 1999). Fragmentation in this case is largely static, i.e. related to the downward propagation of a decompression wave as fragmented magma is progressively removed from the top of the magmatic column (see Fig. 4.11a, b). Theoretically, for perfectly closed degassing conditions, magma should completely fragment to ash, allowing the most efficient transformation of the thermal energy stored in the magma. However, in real eruptions, permeable bubble networks always form to some extent, so that the final grain size distribution and eruption dynamics are largely controlled by the competition between syn-eruptive permeable outgassing versus closed-system degassing (Rust and Cashman 2011). Shearing of the magma, either through friction with the conduit walls, or internally at high magma rise velocities, leads to different fragmentation processes at a larger scale.

Lower values of pyroclast vesicularity may be possible in the case of low-viscosity magmas or partly open degassing conditions at rapid decompression rates (Cashman 2004; Spieler et al. 2004; Vinkler et al. 2012).

The maximum velocity of the mixture as it leaves the vent is a function of gas pressure at the fragmentation level, which is the depth in the conduit at which magma fragmentation is taking place (Figs. 4.7a, 4.11a, b). The theoretical models of Wilson (1980) predict maximum plinian eruption vent exit velocities of 700 m s^{-1}, which agree with maximum velocities deduced from the sizes of the largest ballistic clasts ejected in these eruptions. Such exit velocities indicate that the volumetric discharge rates of magma can be as high as 10^6 m^3 s^{-1} (dense magma equivalent), which are

Table 9.2 Some properties of explosive eruption columns and their source eruptions

Eruption/fall deposit[9.1]	Style	Composition	Column height (km)	Mass eruption rate (kg s^{-1})	Volume (km^3)	Volume DRE (km^3)	Duration (hrs)	Principal source
Shikotsu 3.2 ka, Japan	Plinian	Rhyolite			100	24		1
Towada A-G (915 AD—11 ka), Japan	Plinian A, C, E; sub-plinian B, D, F, G	Rhyolite dacite	25–32 12–23	10^6–10^8				2
Mt St Helens 1980, USA	Plinian	Dacite	19	10^7	1.1	0.2	9	1, 2, 3
Novarupta-Katmai 1912, USA	Plinian	Rhyolite	26	10^7–10^8		13		4
Lower Bandelier 1.6 Ma, USA	Plinian	Rhyolite			100	24		1
Upper Bandelier 1.26 Ma, USA	Plinian	Rhyolite			70	17		1
La Primavera B 95 ka, Mexico	Plinian	Rhyolite			50	14		1
Upper Toluca 10.5 ka, Mexico	Plinian	Andesite			9	4	~20–30	1
Waimihia 3.3 ka, New Zealand	Plinian	Rhyolite (mixed)			29	7.1		1, 3
Taupo 232 AD[2]U2, 3, 4, 5, 6, New Zealand	Plinian–sub-plinian–ultra-plinian	Rhyolite	55	10^6–10^{10}	23	5.1		2, 4
Tarawera 1886, New Zealand	Plinian	Basalt	28	10^8				2
Vesuvius 79 AD, EU 1, 2, 3 (max), 4	Plinian	Phonolite	15, 26, 29, 23	10^7–10^9	6	1.4	~24	1, 2, 3
Agnano–Monte Spina 4.5 ka, Campi Flegrei, Italy	Sub-plinian–plinian	Trachyte	5, 23, 28	10^6–10^7		1.2		4
Askja 1875 B, C, D, Iceland	Sub-plinian B–plinian C, D	Rhyolite (mixed)	8, 23, 28	10^6–10^8	1	0.3	6.5	1, 2, 5
Hekla 1947, Iceland	Sub-plinian	Andesite	28	10^6	0.4	0.1	1	1
Minoan 1600 BC, Santorini, Greece	Plinian	Rhyolite	36	10^8			~20–40	1
Pinatubo 15 June 1991, Philippines,	Plinian	Dacite	<40	<10^9	~4	~2	9	5
Santá Maria 1902, Guatemala	Plinian	Dacite (mixed)	34	10^8	20	8.5		1, 3
Chaitén 2008, Chile	Sub-plinian	Rhyolite	19	10^7	0.5	~0.2	2–3 (6 May peak)	2, 6
Eyjafjallajökull 2010, Iceland	Glacio-surtseyan	Basalt	<10	5–10 × 10^5	0.27	~0.01	39	9
Cordón Caulle 2011, Chile	Sub-plinian	Rhyolite	9–12	10^6–10^7	~1	~0.3	72; max discharge period	7
Kilauea Iki 1959, Hawaii	Hawaiian	Basalt	0.6	10^5			36 × 24	2
Stromboli 2019, Italy	Vulcanian	Basaltic–shoshonitic	8.4	10^5–10^6	10^{-4}	10^{-4}	5 min	8
Stromboli 2002, Italy	Strombolian	Basaltic–shoshonitic	0.1–0.3	2–1700			Seconds	2

(continued)

Table 9.2 (continued)

Eruption/fall deposit[9.1]	Style	Composition	Column height (km)	Mass eruption rate (kg s^{-1})	Volume (km^3)	Volume DRE (km^3)	Duration (hrs)	Principal source
Hunga Tonga–Hunga Ha'apia 15 January 2022	Ultra-Phreatoplinian?	Andesitic-basaltic andesitic	~40–57	10^7–10^{10}	?	7?	2?	10

[9.1]*Sources of data* 1 Cas and Wright 1987; 2 Alfano et al. 2012; 3 Cioni et al. 2015; 4 Houghton et al. 2010; 5 Alfano et al. 2011; 6 Pistolesi et al. 2015; 7 Castruccio et al. 2016; 8 Giordano and De Astis 2021; 9 Gudmundsson et al. 2012; 10 Proud et al. 2022; Cronin et al. 2023 (oral presentation at IAVCEI conference) [2]Units of the Taupo 232 AD eruption named elsewhere in the text and figures: 2 Hatepe plinian pumice; 3 Hatepe phreatoplinian ash; 4 Rotongaio phreatoplinian ash; 5 Taupo ultra-plinian pumice; 6 Taupo ignimbrite

substantially greater than observed in historic eruptions (Table 9.2). The exit velocity of the gas–pyroclast mixture from the 18 May 1980 eruption of Mt St Helens was ~330 m s^{-1} (Carey and Sigurdsson 1985), and the volumetric uprise rate was 10^4 m^3 s^{-1} (Sparks et al. 1994). Mass eruption rates for sub-plinian eruptions are typically ~10^6 kg s^{-1}, for plinian eruptions ~10^6–10^8 kg s^{-1}, and for ultra-plinian eruptions up to ~10^{10} kg s^{-1} (Alfano et al. 2012; Cioni et al. 2015; Table 9.2).

A continuous gas blast can probably not be sustained for a long time, and from the available data, typical eruption durations range from hours to days (Table 9.2). The 1563 Fogo eruption in the Azores lasted up to about two days overall, but its plinian eruption phase was interrupted during this time, and the deposits contain interbedded small ignimbrite flow units and other layers. The duration of large ignimbrite-forming eruptions which sometimes follow an initial plinian phase can be much longer (Chap. 10).

As plinian eruptions involve the most intense, and often largest explosive events on Earth, their waxing and waning stages usually correspond to significant modifications of the conduit/vent geometries, as well as possible changes in the feeding system. As a plinian eruption proceeds, we can predict that generally, two things will happen with time:

- deeper, and more gas-depleted, levels of the magma chamber will be tapped, and
- widening of the vent by wall erosion will occur.

The effect of (a) is to reduce the gas velocity of the eruption column with time. The effect of (b) is to increase the mass eruption or discharge rate with time, and this will produce a column that grows in height (Fig. 9.15a). Both (a) and (b) cause the effective density of a plinian column to increase. This can continue until at some stage the density of the column becomes significantly greater than that of the atmosphere, at which time gravitational collapse of the column will occur to generate ignimbrite-forming pyroclastic flows. Models have shown that various combinations of magmatic gas content, gas velocity, and vent radius produce buoyant, convecting columns, whereas others produce collapsing columns (Sparks et al. 1978, Wilson 1980; Valentine and Wohletz 1989; Dobran et al. 1993; Koyaguchi and Woods 1996; Neri et al. 2003; Di Muro et al. 2004; Carazzo et al. 2008; Cioni et al. 2015; Trolese et al. 2019; Fig. 9.15b). From these, we can therefore predict when eruption column collapse will occur. Columns formed from magmas with high gas contents (>5 wt% water) are likely to be buoyant, whereas those with low gas contents (<1 wt% water) will form collapsing columns. In magmas with intermediate gas contents, collapse will occur when the vent radius exceeds a value defined in Fig. 9.15.

However, not all plinian eruptions continue to the collapse, or ignimbrite-forming stage, nor is collapse required to be total. There are many examples of early, intra-plinian partial collapses of eruption columns, which usually herald the eventual, main, climactic collapse phase (Fig. 9.16; e.g. the 4.6 ka Fogo A in the Azores, Pensa et al. 2015; the 4.5 ka Agnano–Monte Spina eruption at Campi Flegrei, Italy, De Vita et al. 1999 Romano et al. 2020; the 800 yr BP/~1280 AD eruption of Quilotoa, Ecuador, Di Muro et al. 2004, 2008). Furthermore, the 1912 eruption at Novarupta-Katmai (Alaska) shows that a sustained plinian column feeding fall deposits co-existed with contemporaneous collapses forming pyroclastic flows for the entire duration of the eruption (Hildreth and Fierstein 2012; Chap. 10).

There are also many examples of plinian fallout deposits without associated *major* ignimbrites (e.g. Askja 1875; Sparks et al. 1981, Carey et al. 2010; Santá Maria 1902; Williams and Self 1983; the 10.5 ka Upper Toluca plinian deposit, Bloomfield et al. 1977; and including the extensive, large volume 6.6 Ma Corte Blanco plinian deposit, northwestern Argentina, Tait 2004, Fig. 9.17).

9.4 Vent Evolution, Mixture Muzzle/Vent Exit Velocity, and Terminal Fall Velocity of Pyroclasts

Before we describe the dynamics of explosive eruption plumes and the different types of pyroclastic fall deposits, some parameters that are essential for understanding the transport and deposition of pyroclastic fall deposits need to

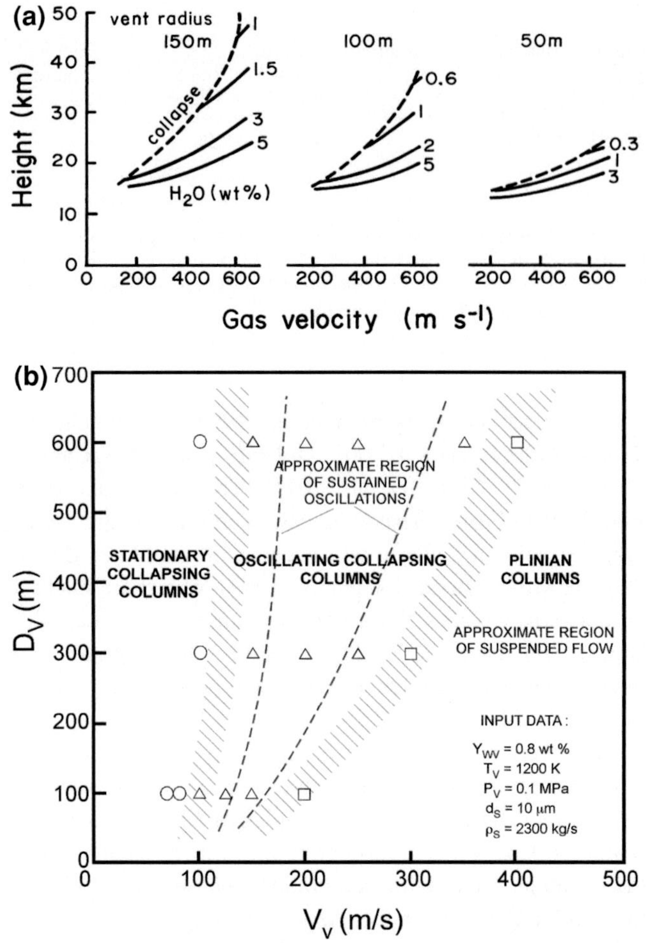

Fig. 9.15 Factors influencing buoyant versus collapsing explosive eruption columns. **a** Eruption column height as a function of the muzzle gas velocity for three different vent radii. In each case, curves are given for constant values of magmatic water content. Column collapse occurs for combinations of values to the left of the broken line. (From Cas and Wright 1987, After Wilson 1976). **b** Fields of collapsing, buoyant, and transitional plumes as modelled by varying muzzle gas velocity (V_V) and vent diameter (D_V). The cross-hatched regions identify the transition among stationary collapsing columns, oscillating columns, and plinian columns. Y_{WV} = water content, T_V = temperature, P_V = pressure, d_S = particle size, and ρ_S = density. (Fom Cioni et al. 2015, After Neri and Dobran 1994)

be discussed. These include the evolution of the vent, muzzle velocity, which is the exit velocity of the gas–pyroclast mixture from the vent, and the terminal fall velocity of pyroclasts.

9.4.1 Vents and Their Evolution

Vents are the surface expressions of conduits that allow the eruption of magma at the Earth's surface and are formed when magma conduits become open to the surface either as linear fissures or ~cylindrical pipes. For effusive or weakly explosive eruptions, vents are largely constructional in effect, which are formed by the piling up of erupted material around the vent, with these deposits usually dipping both outward and inward (Fig. 9.18a). In explosive eruptions of progressively stronger intensity, vent morphology becomes that of a crater, meaning that the constructional character is accompanied by the progressive enlargement and erosion of the vent area involving excavation of previously erupted deposits, often to pre-volcanic depths, exposing older erupted strata in the walls of the crater (Fig. 9.18b). Subsequent small eruptions could produce multiple small craters in the floor of the bigger crater produced by the last big explosive event (Fig. 9.18b). The formation and stabilisation of a crater during an eruptive phase is a dynamic process that relates to a series of factors, which include the depth of the fragmentation level, the mass eruption rate, the pressure of the eruptive mixture in the conduit and at vent, and the mechanical properties of the rocks forming the vent walls. In turn, the evolution over time of the vent shape feeds back into the mass eruption rates and ultimately in determining the eruption style.

In general, the pressure of a pyroclast + gas mixture in the conduit is related to the gas content at the time of fragmentation, which causes the mixture to accelerate upward. The transition from pressures inside the conduit to atmospheric pressure may occur smoothly or abruptly. This largely depends on the shape of the conduit. A pressure-balanced transition needs the progressive upward enlargement of the conduit (a shape known as de Laval nozzle) so that the conduit overpressure can maintain the upward acceleration up to the vent. However, if the conduit does not enlarge upward the mixture cannot accelerate to accommodate the overpressure from below, and becomes internally overpressurised. This condition is known as *choked flow*. The overpressure will be released once the mixture emerges into the atmosphere, with a radial blast often heard and seen at erupting volcanoes (Fig. 9.16).

As most vents are located in volcanoes where the shallowest rock layers are mechanically weak, choked flow will rapidly transition to pressure-balanced flow via the explosive erosion and excavation of a crater, the shape and depth of which allow a progressive decrease in the internal pressure of the pyroclastic mixture, its upward acceleration to maximum exit velocities, and consequent increase in mass eruption rate. By contrast, collapses of the crater wall may reduce the cross-sectional area of the vent and induce temporary transitions to overpressurised conditions. It is now clear that the evolution of the vent geometry during an explosive eruption plays an important role in its dynamics and the ultimate eruption styles (Koyaguchi et al. 2010). Some considerations on vent excavation energetics and controls can be found in Goto et al. (2001) and Valentine et al. (2012).

Fig. 9.16 Space station view of the 12 June 2009 eruption at Sarychev Peak, Kuril Islands. Partial collapses of the buoyant column can be observed at its base with different solids concentrations, as indicated by the different colours. The hole in the cloud cover is the result of the propagation of the initial shock wave. (*Source* NASA)

Fig. 9.17 The late Miocene (~6.6 Ma) Corte Blanco Tuff fallout deposit (middle 10 m high cliff), erupted from the Ramadas rhyolitic volcanic centre, northwestern Argentinian Andes, records an exceptional >36 km^3 (DRE) plinian eruption dominated by a buoyant column which peaked at heights of ~40 km a.s.l. (Tait 2004). With the presence of only minor thin, intra-plinian pyroclastic density current (ignimbrite) units, the bulk of the deposit chronicles a pumice fall from the beginning to the end of the eruption

9.4.2 Muzzle Velocity

The initial gas or *muzzle velocity* at the vent is one of the most useful physical parameters in comparing explosive pyroclastic fall-forming eruptions. During observed eruptions, this can be determined by measuring the fall times of ballistic blocks and bombs which are unaffected by the wind (Ripepe and Harris 2008), or by analysing film footage of

Fig. 9.18 Vents and their layered vent-wall successions, built up during multiple eruptions. **a** Exceptional exposure of the interior of a small scoria cone cut through by the AD 2000 caldera collapse vent at Miyakejima volcano, Miyakejima, Izu Islands, Japan. Note that the crater is fully constructional with deposits piled up and dipping inward and outward, and fed by a narrow dyke that has propagated through the substrate (centre left). **b** Crater of Colima volcano, Mexico, after vulcanian to plinian eruptions in April 2016, and superimposed by initial stages of lava dome-forming eruptions, represented by large blocks (Photo taken 24-5-2016)

eruptions (e.g. Taddeucci et al. 2012). In older deposits, one can measure the average maximum size of the largest fragments, and if the vent location is known these sizes can be used to estimate the minimum muzzle velocity based on values of particle radius, density, launch velocity, launch angle, and release height (Fagents and Wilson 1993; Mastin et al. 2009; Tsunematsu et al. 2016; Kilgour et al. 2019). Tsunematsu et al. (2016) calculated vent exit velocities of 145–185 m s^{-1} during the short-lived 2014 Ontake phreatic explosion in Japan, and Kilgour et al. (2019) calculated exit velocities of ballistics during the April 2016 night-time eruption of White/Whakaari Island volcano, New Zealand, of 50–60 m s^{-1}. See Sect. 9.6.2.3 for further discussion on using ballistic blocks and bombs to calculate vent exit muzzle velocities.

The distance that individual pyroclastic fragments are transported from the vent is controlled by many interacting factors, the most important being the heights to which particles are taken in the eruption column, the angle of ejection, the wind strength, and the terminal fall velocity of the particles.

9.4.3 Terminal Fall Velocity and Settling Velocity of Pyroclasts

When an object falls through the air, it accelerates until it reaches a constant or *terminal fall velocity*, which is the velocity at which the force of gravity and aerodynamic drag forces are in a state of balance. Particles with low terminal fall velocities (e.g. small dense grains, larger low-density vesiculated clasts) will travel downwind further for a given eruption column height and wind speed than larger particles with a higher terminal fall velocity.

The *settling velocity* of a particle in fluids of varying density and viscosity under laminar fluid conditions, which is often its terminal fall velocity, is best represented by Stokes' settling velocity equation:

$$v = \frac{g\,d^2(\rho_p - \rho_f)}{18\mu} \tag{9.1}$$

where v is the settling velocity, g is the acceleration due to gravity, d the clast diameter, ρ_p is the clast density, ρ_f is the fluid density, and μ the fluid viscosity.

Another general equation for the pyroclast terminal fall velocity w is

$$w = \left(\frac{\rho_p\,g\,d}{C_d\,\rho_f}\right)^{1/2} \tag{9.2}$$

where ρ_p is the density of the pyroclast of diameter d, characterised by a drag coefficient C_d when falling through a fluid of density ρ_f which may represent the mixture of gas and coupled small particles through which the pyroclast in consideration is falling (Dellino et al. 2005). The actual shape of the pyroclast also plays an important role in defining its terminal fall velocity (Dellino et al. 2005).

The drag coefficient can be represented by

$$C_D = \frac{2\,F_D}{\rho_f\,A\,u^2} \tag{9.3}$$

where F_D is the drag force acting on a particle moving through a fluid, ρ_f is the fluid density, A is the cross-sectional area of the particle, and u is the wind velocity (Alatorre-Ibargüengoitia and Delgado-Granados 2006).

Walker (1971) showed that for poly-component pyroclastic fall deposits (Fig. 9.19a) it is useful to plot histograms of weight percentages against terminal fall velocity (Fig. 9.19b), so grouping together all particles which fall at the same rate. When this is done, grain size histograms of pyroclastic fall samples become strongly unimodal, reflecting very efficient hydraulic sorting (Fig. 9.19b). Using simple size-based sorting parameters is

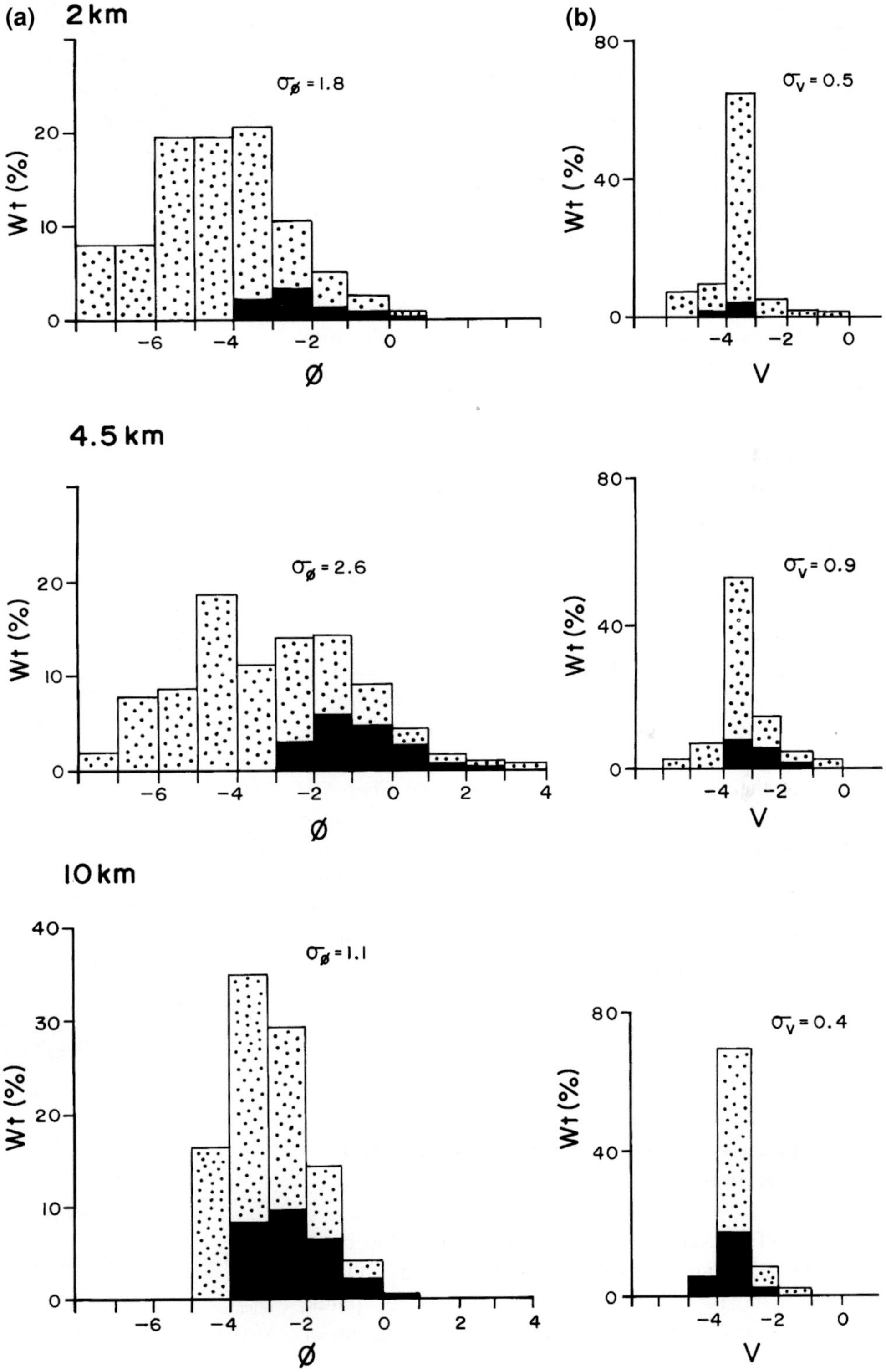

Fig. 9.19 Grain size characteristics of three samples of the Middle Pumice (144.6 ka) fallout deposit on Santorini volcano, Greece, taken at increasing distances from the probable vent (see also Fig. 9.68). **a** Histograms of the grain size distributions. Grain size distributions of air-fall deposits on a weight percentage basis are a function of the terminal fall velocity of ejecta which is controlled by both grain size and density. Less than 3 km from source, samples of the fall deposit contain >90 wt% pumice (stipple), and have unimodal histograms and a low σ_ϕ (sorting) value. The proportion of dense components (lithics and crystals; black) increases away from source. Between 3 and 5 km from source, this results in a bimodal grain size distribution, with a coarse mode due to pumice and a fine secondary mode due to the denser components, and an increase in σ_ϕ. At greater distances (>5 km), a decrease in the proportion of very coarse pumice clasts results in a restricted pumice size range with a mode closely corresponding to that of the dense components. The grain size distribution is unimodal and sorting improves markedly. **b** Histograms of grain size in weight percentages plotted against the terminal fall velocity of ejecta; V is defined as $-\log_2$ TV where TV is the terminal velocity in m s^{-1}. These group together all particles which fall at the same rate in the same class. By doing this, all the grain size histograms become strongly unimodal. (From Cas and Wright 1987)

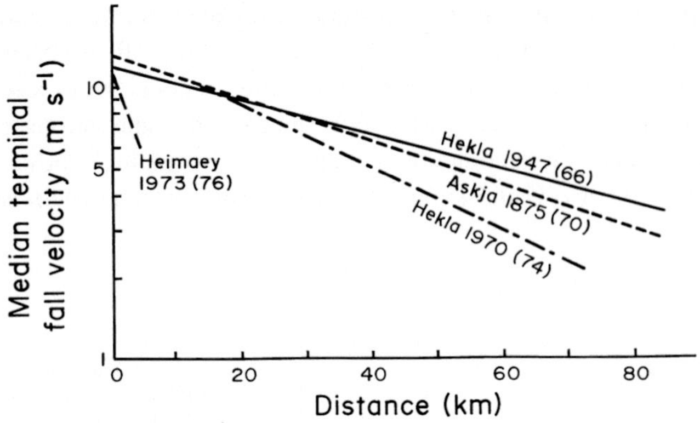

Fig. 9.20 Median terminal fall velocity plotted against distance from source for some pyroclastic fallout deposits. For each deposit, an indication of the wind speed (in km h^{-1}) is given in parentheses. (From Cas and Wright 1987, After Self et al. 1974)

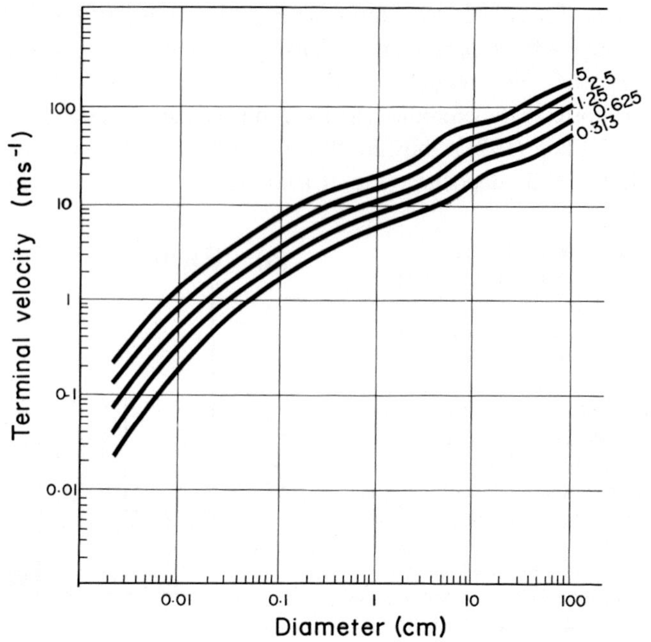

Fig. 9.21 Computed terminal fall velocities for cylindrical particles, which show good agreement with the behaviour of pyroclastic particles. The curves are for grains ranging in density from 0.313 and 5 g cm^{-3}. From Cas and Wright 1987, After Walker 1971)

not as useful for reflecting sorting efficiency because of the highly variable densities and shapes in most fallout deposits. Median terminal fall velocity in an air-fallout deposit gradually decreases with distance (Figs. 9.19, 9.20). The slope of the median terminal fall velocity-distance curve (Fig. 9.20) is controlled by eruption column height and wind speed. For the deposits shown in Fig. 9.20, the wind speed was approximately the same, and the slope of the line is therefore a function of eruption column height. Data on the terminal fall velocities of pyroclasts are not necessarily simple to measure, but sets of data are available in the literature (Figs. 9.21, 9.22; Walker 1971; Dellino et al. 2005).

As eruption columns are turbulent and characterised by significant variations of flow regimes, an appropriate way to express the sedimentation behaviour of particles as a function of their decreasing grain size and shape parameters is the use of their particle Reynolds number (see Chap. 2), which is a reflection of the fluid state around a solid particle settling through the fluid as a function of the diameter and shape of the particle (Fig. 9.23). This shows that proximal fall deposits are dominated by high Reynolds number particles (>500; generally large particles with high settling velocities) which distally decrease in favour of intermediate (0.4 < Re < 500) and further away to low particle Reynolds numbers (Re < 0.4; generally fine particles with low settling velocities) (Bonadonna et al. 1998).

9.5 Explosive Eruption Columns and Their Structure

The mass of gas, and juvenile and accessory lithic clasts (including ash) that are explosively ejected from and rise above a vent is called an eruption column or plume (Plate 9.1c; Figs. 9.2b, 9.10a-d). It is from such columns or

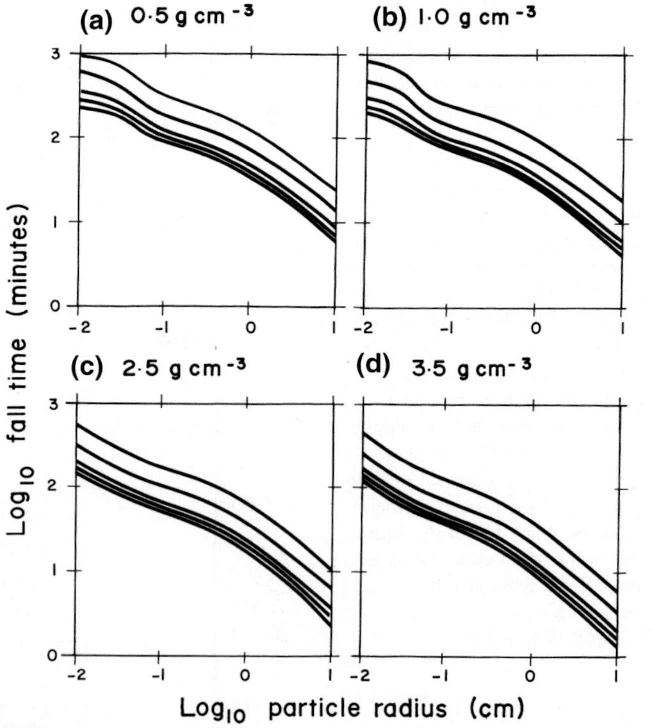

Fig. 9.22 Computed fall time of particles ranging in density from 0.5 and 5 g cm^{-3}. The curves (From bottom to top) are for particles released from heights of 5, 10, 20, 30, and 50 km. (From Cas and Wright 1987, After Wilson 1972)

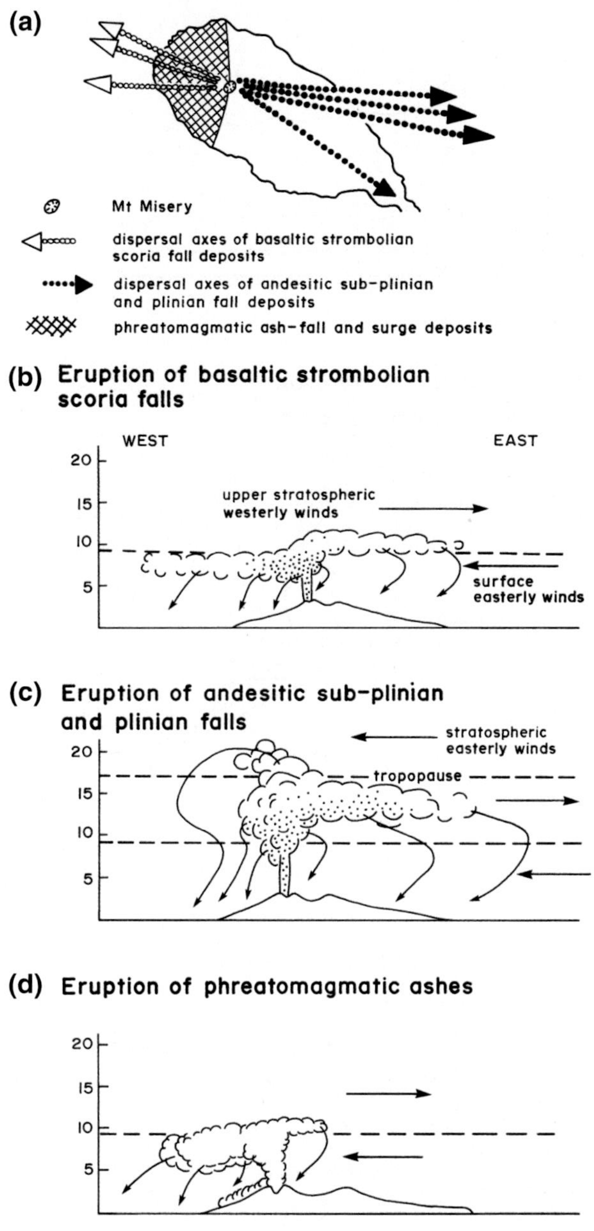

Fig. 9.23 The effect of a layered wind system and height of eruption columns on the dispersal of air-fall deposits erupted from Mt Misery (Liamuiga) volcano, St Kitts, Lesser Antilles. The lower tropospheric easterlies are today confined to altitudes of ~5–9 km depending on season. We have depicted the boundary during the wet summer season. The tropopause occurs at ~17–18 km. Trajectories are schematic transport paths of air-fall ejecta. (From Cas and Wright 1987, After Roobol et al. 1985)

plumes that pyroclasts fall back to Earth to form fallout deposits. Eruption columns can be short-lived ballistic jets with associated buoyant finer ash clouds or plumes (Fig. 9.11b), or magma/fire fountains or sprays of jetting magma (Plate a, b; Fig. 9.2a, Fig. 9.7a, b, d–f) or towering, buoyant, mushroom-shaped columns and plumes that can rise high into the troposphere and the stratosphere (Plate 9.1c; Fig. 9.2b, Fig. 9.10a–d). At a level in the atmosphere where the bulk density of the eruption column is equal to that of the atmosphere, the column begins to spread laterally in the atmosphere forming the mushroom or umbrella cloud which has a much greater diameter than the underlying column (Plate 9.1c; Fig. 9.10a–d). If there is little wind blowing at this level, the column spreads radially, dispersing the fallout deposits axi-symmetrically around the vent, with the deposits that fall to the ground being thickest and coarsest close to the vent (sometimes also forming a cone there), and becoming progressively thinner and finer in a concentric pattern away from the vent. However, as is commonly the case high in the atmosphere, if there is a strong wind blowing, the upper part of the eruption column or umbrella region is deflected downwind to form an asymmetrical downwind eruption plume or fan (e.g. Figs. 9.6c, 9.10a). Fallout deposits from such asymmetrical plumes will have a fan shape on the ground or on a map, called a *fan of dispersal*. Such deposits will be coarsest and thickest near the vent (sometimes also forming a cone there), and along the *axis of (the fan of) dispersal* (Fig. 9.6c).

Explosive eruption columns may take many forms (Figs. 9.2, 9.7, 9.8, 9.10, 9.11), and their energetics and dynamics have been discussed by many authors (Settle 1978; Wilson et al. 1978; Sparks and Wilson 1982; Sparks 1986; Woods 1988; Carey and Sparks 1986; Carey and Sigurdsson 1989; Bursik and Woods 1991; Dobran et al. 1993; Sparks et al. 1997; Bonadonna et al. 1998; Suzuki et al. 2005; Barsotti et al. 2008; Mastin et al. 2009; Dellino et al. 2010; Woodhouse et al. 2013; Clarke et al. 2015; Carey and Bursik 2015; Cioni et al. 2015; Costa et al. 2016a, b; Trolese et al. 2019; Aubry et al. 2023). The height reached by an eruption column, together with the atmospheric wind velocity profile, which may vary with height (e.g. Fig. 9.23), has for a long time been considered to control the dispersal of pyroclasts (Fig. 9.24a–c; e.g. Carey and Sparks 1986; Armienti et al. 1988; Bursik et al. 1992; Bursik 2001; Bonadonna et al. 1998; Bonadonna and Phillips 2003; Connor and Connor 2006; Costa et al. 2006; Byrne et al. 2007; Carazzo et al. 2008; Koyaguchi et al. 2009; Carey and Bursik 2015). However, it is not the height of the eruption column, but the height of the level of neutral buoyancy from which tephra are dispersed that is more important (Carey and Sparks 1986; Constantinescu et al. 2021). Observed eruption columns have attained heights between 2 and 57 km (Fig. 9.25; Fig. 9.26; Table 9.2; Wilson et al. 1978; Mastin et al. 2009; Woodhouse et al. 2013; Cioni et al. 2015; Proud et al. 2022).

Eruption column or plume height is a function of the rate of production of thermal energy at the vent (depending on its radius, mixture density, exit velocity) and the efficiency of conversion of thermal energy during the entrainment of cool atmospheric air, which in turn largely relates to grain size

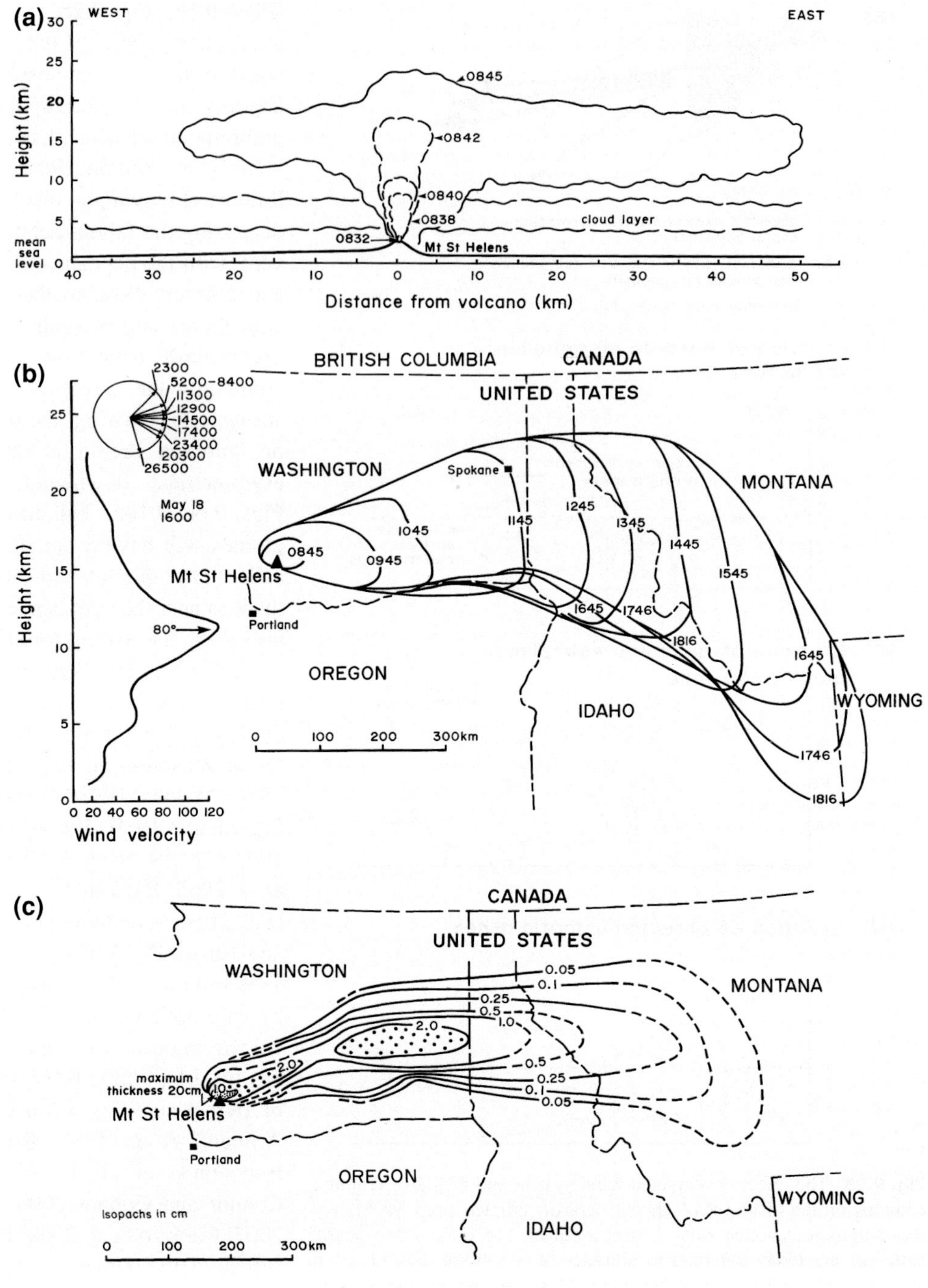

Fig. 9.24 Development of the eruption column, downwind plume, and dispersal of pyroclasts in the 18 May 1980 eruption of Mt St Helens volcano, USA. **a** East–west profile schematically showing early vertical growth and lateral expansion of the plume. **b** Isochron map showing maximum downwind extent of the edge of the plume carried by the fastest-moving wind layer (as observed on satellite photographs). On the left, an average wind speed profile measured at Spokane at 16.00 h is given. Circular wind diagram shows average directions to which wind was blowing at different altitudes and were again measured at Spokane at 16.00 h. **c** Isopachs of the 18 May pyroclastic fall deposit. Note the secondary thickening of the air-fall deposit 300 km downwind; the significance of this is discussed in Sect. 9.7.7. (From Cas and Wright 1987, After Sarna-Wojcicki et al. 1981)

distribution of pyroclasts (Wilson et al. 1978). The mass discharge rate and its steadiness or unsteadiness determine the extent to which the column is strong and whether it may become affected by the atmospheric wind profile (Sparks et al. 1997; Bonadonna et al. 2005; Woodhouse et al. 2013; Suzuki et al. 2016; Aubry et al. 2023). Column height, H_T, has been shown to have a relationship to magma volume eruption rate (Fig. 9.25, 9.26a; Wilson et al. 1978), erupted volume of magma in DRE (Fig. 9.26a; Mastin et al. 2009), and mass eruption or discharge rate (Fig. 9.26b, c, Mastin et al. 2009; Aubry et al. 2023). In addition, Trolese et al. (2019) found that for columns subject to partial column collapse, collapses significantly lower the eruption column height, expressed as a percentage of column mass that is involved in the collapse, Qc, for a given mass eruption rate (Fig. 9.26d). Their modelling showed that a linear decrease in column height occurred with an increasing percentage of collapsed mass. Their numerical data show up to a 45% reduction of plume height with respect to fully buoyant conditions, with a more marked effect at higher mass

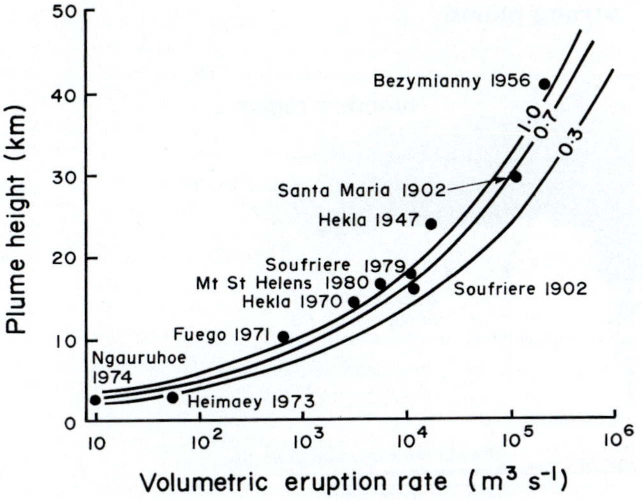

Fig. 9.25 Relationship between observed plume height and volumetric eruption rate. The theoretical curves for *F*-values of 1.0, 0.7, and 0.3 which reflect the efficiency of the conversion thermal energy to kinetic and potential energy. *F*-value of 1.0 occurs when the column is dominated by ash leading to optimum heat conversion, whereas *F*-value of 0.7 coincides with plinian columns with considerable ash fraction, and 0.3 coincides with low degrees of fragmentation, such as micro-plinian with abundant coarse pyroclasts (e.g. Heimaey 1973). (From Cas and Wright 1987, After Wilson et al. 1978)

eruption rates (10^9 kg s^{-1}; Fig. 9.26d). As a consequence, variation in grain size of pyroclastic fall deposits may relate to fluctuations of eruption column height due to partial collapses rather than fluctuations in mass eruption rate.

Strong columns are those in which the mass discharge rate is high and sustained, and are largely unaffected by winds, rising vertically to their level of neutral buoyancy (Fig. 9.27a). Weak columns are those in which the mass discharge rate is low and/or not sustained, and are bent over by winds (Fig. 9.27b).

Woodhouse et al. (2013) thoroughly reviewed existing models of eruption columns and concluded that the principal variables determining eruption column height are mass discharge or eruption rate, crosswind strength, and moisture content. Although there has been an assumption that eruption column height reflects magma discharge rate, Woodhouse et al. (2013) point out that the effects of strong crosswinds will reduce the eruption column height relative to a tranquil atmosphere, and could lead to underestimations of mass discharge rates. They also emphasise that moist or phreatomagmatic eruption columns could be higher than dry magmatic eruption columns because condensation of water vapour in the atmosphere will release latent heat and so

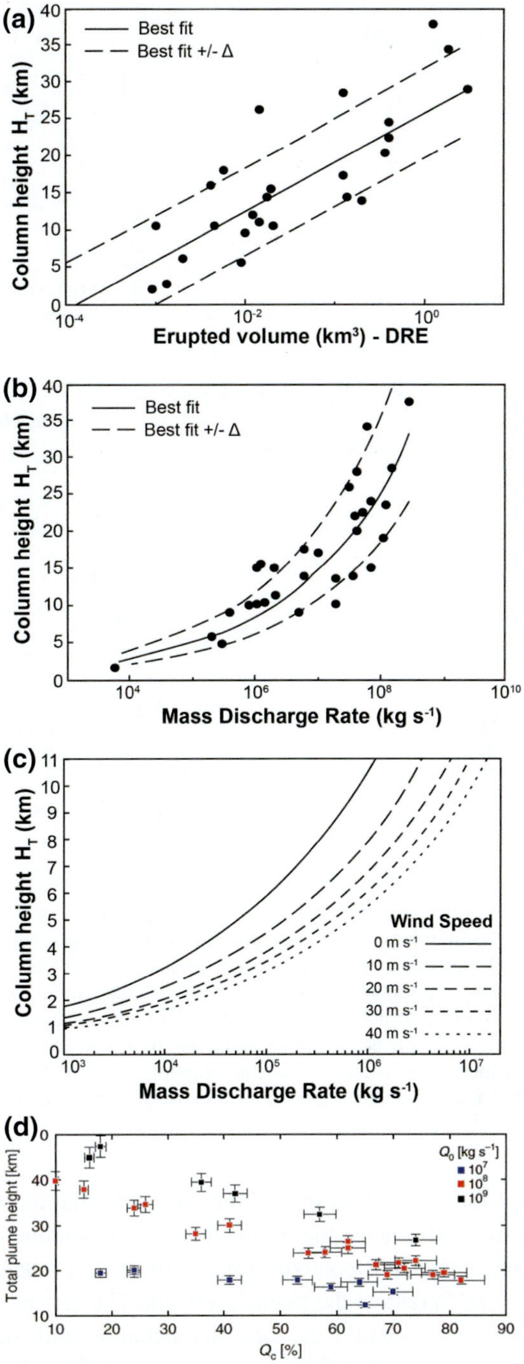

Fig. 9.26 Relationships between eruption column height, H_T, and **a** magma volume erupted, **b** magma mass eruption/discharge rate, and **c** the effects of crosswinds factored in (both **a** and **b** are from Mastin et al. 2009; **c** is from Woodhouse et al. 2013). **d** Relationship between column height and percentage of column mass that is involved in collapse, Qc, for a given mass eruption rate coded in different colours (From Trolese et al. 2019)

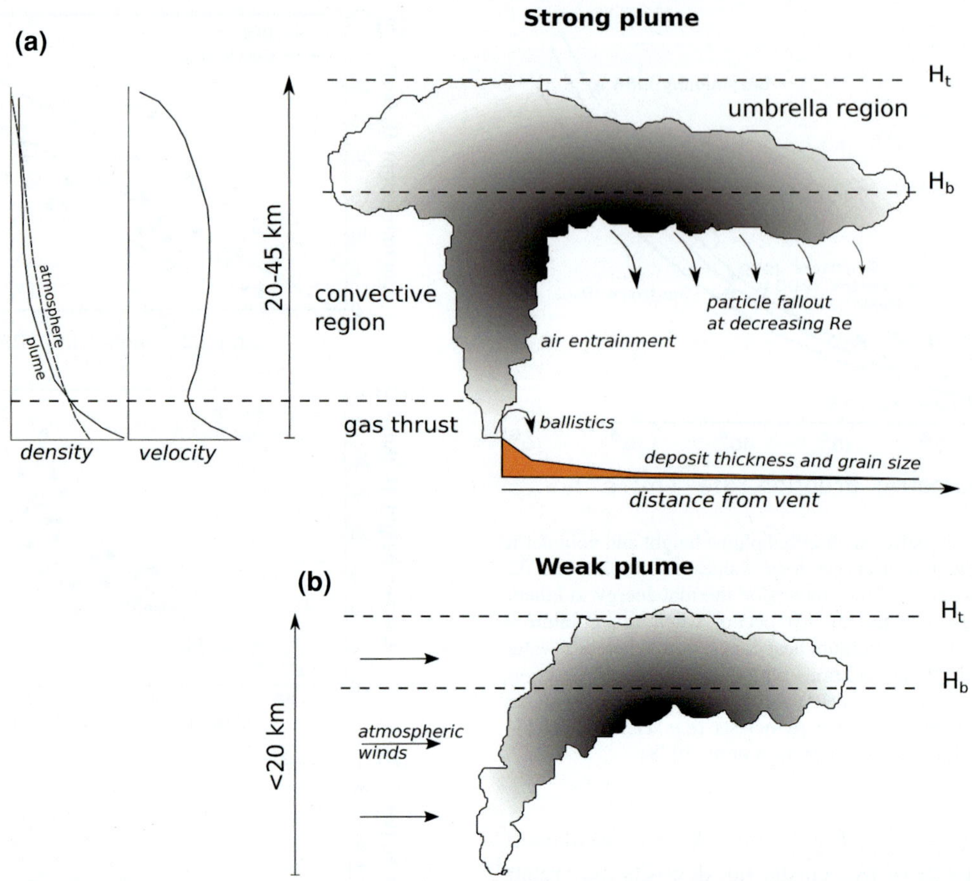

Fig. 9.27 Schematic representation of **a** strong and **b** weak eruption columns, noting how weak columns are deflected and bend in a downwind direction (Adapted from Bonadonna et al. 2015). **c** Schematic representation of a buoyant, plinian style explosive eruption column, depicting the three main dynamic zones above the vent—gas thrust, convective uprise, and umbrella zones, and the factors that contribute to dynamics of buoyant eruption columns (From Cioni et al. 2015). **d** Schematic depiction of a buoyant eruption column and umbrella region, with the source of fallout deposits in the umbrella region depicted as an areally extensive disc source with radius *R*. Release of tephra occurs from the whole area of the base of the umbrella region, not the centre, *C*, of umbrella. In all parts, H_t is the maximum height of the eruption column, and H_b is the level of neutral buoyancy of the eruption column, at which altitude it begins to spread radially, forming the umbrella region. (From Constantinescu et al. 2021)

enhance the thermal budget of the eruption column. Aubry et al. (2023) also noted the importance of the structure of the atmosphere and the potential impact of climate change in causing the height of the main layers and boundaries in the atmosphere to change as their temperature and density change, and how that then impacts on pyroclast dispersal processes.

The style of explosive activity is therefore important in controlling the character of the eruption column and includes a spectrum between fully maintained plumes and instantaneous explosions, as well as their possible partial collapse. The criterion to define the end members of this spectrum and its intervening continuum depends on the relative duration of the explosive event versus the time for the plume to rise. The duration of a discrete explosion is short and cannot support over time the rise of a plume; by contrast, a maintained plume forms when the supply rate at the vent is continuous during an eruption phase and longer than the time needed for the plume to rise (Wilson et al. 1978). Observed plume rise velocities are in the order of some tens of m s^{-1}, which corresponds to some minutes to reach maximum height depending on the scale of the eruption.

Discrete instantaneous explosions may produce transient plumes, whereas prolonged release of fragmented magma in a steady phase of an eruption forms a long-term, maintained buoyant plume. If discrete explosions occur in rapid succession (within seconds to a few minutes), a (more or less) maintained plume may also form.

Fig. 9.27 (continued)

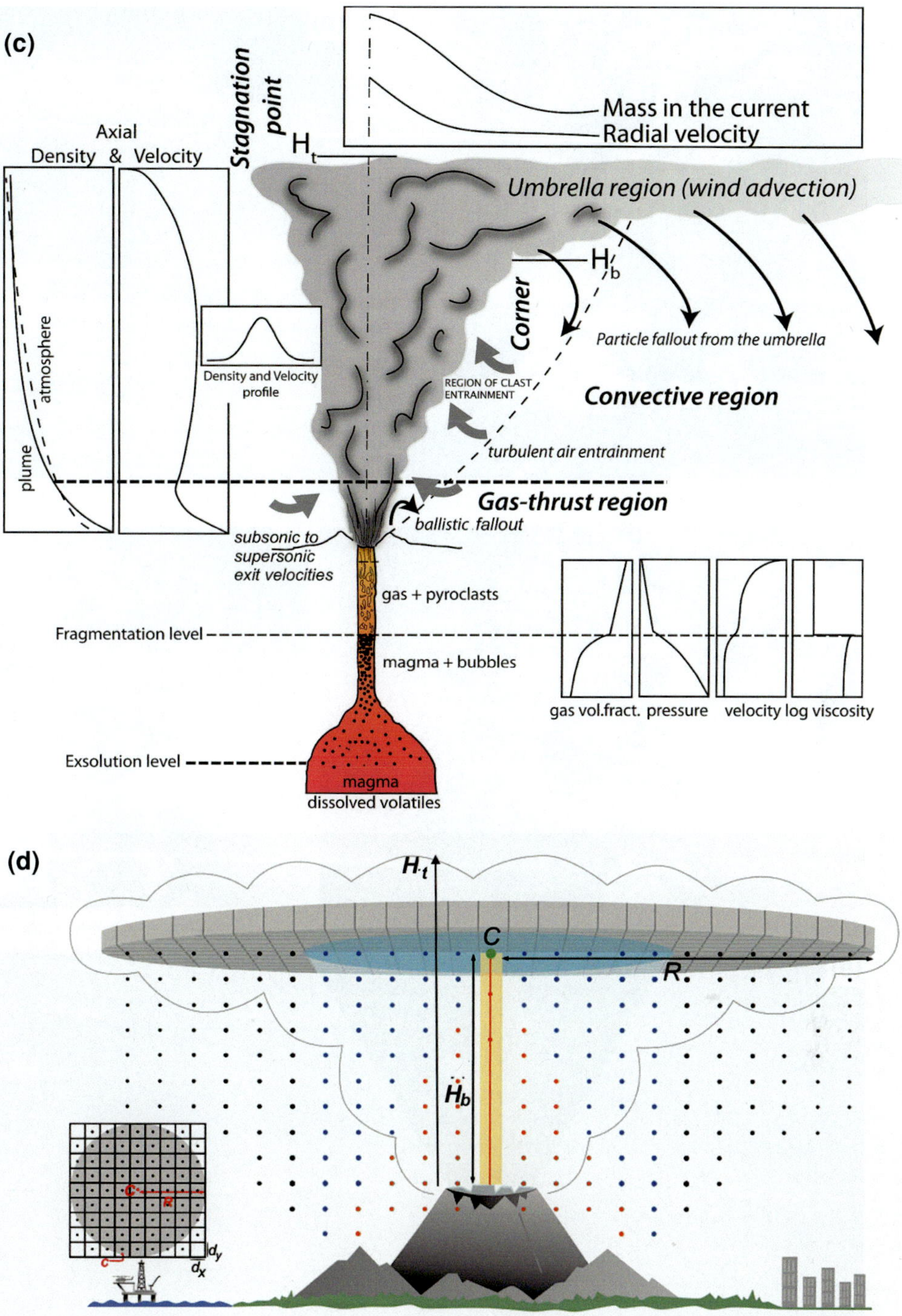

9.5.1 Sustained Plumes—Their Dynamic Architecture

Fully formed buoyant eruption columns (e.g. Plate 9.1c; Figs. 9.2b, 9.10a–d, 9.16, 9.28b) can be divided into three parts (Sparks et al. 1978; Sparks 1986; Wilson and Parfitt 2004; Cioni et al. 2015): (1) a basal gas thrust zone, (2) a convective uprise region, and (3) an upper umbrella cloud (Fig. 9.27a, c).

9.5.1.1 Gas Thrust Zone (or Inertial Region)

The gas thrust zone defines that part of the column in which uprise is essentially due to the initial explosive momentum of the pyroclast–gas mixture at the point of exit from the vent (Figs. 9.27a, c, 9.28a). This means that the exit velocity is fundamentally controlled by the exsolved volatile content and the mode of degassing, and could vary in velocity from subsonic to supersonic (Wilson and Walker 1987). The density of the pyroclastic mixture at the vent is always much

(a)
(b)
(c)
(d)
(e)

◀ **Fig. 9.28** Parts of a sustained eruption column (cf. Fig. 9.27c). **a** Gas thrust zone of the 18 May 2018, Mt. St. Helens, USA, plinian eruption column. The gas thrust zone is several hundred m high and driven by explosive momentum from the vent. At the top of gas thrust zone in the image, the column begins to expand as it loses momentum and thermal convection at the base of the convective uprise zone takes over to drive the vertical trajectory of the eruption column. (*Source* United States Geological Survey). **b** Eruption column from the 5 June 2011 eruption of Puyehue-Cordon Caulle volcano, Chile, showing the turbulent convective uprise zone, passing upwards into the radially expanding umbrella region of the column, which is also being blown downwind. The column is ~9 to 12 km high. (*Source* Chilean Airforce and Reuters). **c** Downwind eruption plume spreading from Puyehue-Cordon Caulle volcano on 4 June 2011 (*Source* NASA). **d** Downwind eruption plume spreading from Puyehue-Cordon Caulle volcano on 6 June 2011 (*Source* NASA). **e** Space Shuttle image showing ≥20 km high sub-plinian eruption plume being blown downwind from the 19 September 1994 eruption of Rabaul volcano, Papua New Guinea (*Source* NASA)

higher than the atmosphere and can thus be defined as being negatively buoyant. This means that the fate of this gas thrust jet should be to collapse back towards the ground as it rises and rapidly dissipates its kinetic energy. This fate may change depending essentially on the rate of air entrainment (Dellino et al. 2014; Suzuki and Koyaguchi 2015; Trolese et al. 2019). For dense hot jets (concentration of solid fraction at vent > 1%), the rate of entrainment is low and collapse is likely. For dilute hot jets (concentration of solid fraction at vent < 1%), the friction exerted by the atmosphere is substantial and, along with rapid deceleration, air is entrained, heated up, and expanded, lowering the bulk density of the gas–pyroclast mixture, and making it buoyant. For this reason, the shape of the inertial or gas thrust region is that of a narrow upturned cone, dark in colour due to the high initial concentration of pyroclastic material, and externally characterised by progressively larger eddies as a result of the upward propagating process of air entrainment and expansion. The coefficient of air entrainment is therefore a fundamental factor that influences the evolution of a volcanic plume (Dellino et al. 2014).

The top of the gas thrust region can be defined then as the region where the upward directed, initially negatively buoyant jet of ejecta and gas, driven by explosive momentum, progressively transforms (or not) into a state of buoyancy by the process of air entrainment.

Comparison between the total kinetic energy available for the range of possible mass eruption rates and exit velocities (u = up to 700 m s^{-1}) and the corresponding thermal energy demonstrates that in all highly explosive eruptions, the thermal energy released is completely dominant (>90%) over the initial kinetic energy released from decompression and expansion of the gas phase (Wilson et al. 1978).

The height of the initial gas thrust zone varies with the style of the activity. In most buoyant eruption columns (e.g. plinian group), the gas thrust zone makes up less than 10% of the total column height (Wilson et al. 1978). For discrete explosions (strombolian and vulcanian eruptions), this ranges from a few tens of m to a few hundred m in height (Blackburn et al. 1976; Self et al. 1979; Clarke et al. 2002; Allard 2010). For maintained eruption columns (plinian family), the range is from a few hundred m to a few km in some eruptions (1.5–4.5 km for initial gas velocities of 400–600 m s^{-1}; Wilson 1978; Sparks and Wilson 1976; Trolese et al. 2019).

Rapid deceleration of the gas thrust phase occurs between these heights, above which low density and light particles are incorporated (or not) into an eruption column driven by convection, whereas dense heavy pyroclasts fall out as ballistic clasts. Eruptions that involve crater enlargement are accompanied by an increase in mass discharge rate, which would dissipate the thrust momentum, and so collapses of the columns are more likely at increasing mass discharge rates.

9.5.1.2 The Convective Uprise Zone

The convective region of eruption columns (Figs. 9.27a, c, 9.28b) describes the region above the initial gas thrust zone, where the rise of the column is driven by thermal buoyancy, i.e. where air is entrained and heated by the release of thermal energy from juvenile particles, which drives upward convection of the gas–pyroclast mixture (Cioni et al. 2015). The turbulent eddies which form as a consequence of air ingestion, heating and expansion determine the classic cauliflower appearance and the cone or mushroom shape of the column.

The efficiency in heat transfer within the plume depends largely on grain size (Fig. 9.25). The time needed for the release of heat from particles depends on the properties of the material and the actual dimensions and shape of the particles (Thomas and Sparks 1992). Small particles cool much quicker than large particles. However, the residence time of a particle within the column, hence the time available for heat exchange, depends on the particle's terminal fall velocity and its Reynolds number (Woods and Bursik 1991; Bonadonna et al. 1998; Dellino et al. 2005), so that the contribution of each particle size to the incremental amount of heat released within the column depends on the competition between the vertical velocity component of the gas phase at each site compared with the terminal fall velocities of particles. Both longer residence time and heat release are hence favoured for fine particles (Wilson et al. 1978). For a given eruption, then, the total amount of ash-sized particles (which in turn depends on the efficiency of the fragmentation process) will largely determine the amount of heat that can be released at the time scale of the plume rise. For plinian eruptions, Sparks et al. (1979) calculated that the heat release is about 70% of the total thermal energy carried by the magma, based on total grain size distributions of deposits, the rest being lost largely due to the early fallout of large clasts.

In the convective uprise zone, buoyancy is dominant and the top is defined by the level of neutral buoyancy, H_b, where the column bulk density equals that of the surrounding atmosphere, and it begins to spread sideways (Figs. 9.27c, 9.28b; also Figs. 9.2b, 9.10a–d).

9.5.1.3 The Umbrella Cloud, Height, Radius, and Downwind Plume

The umbrella cloud is the part of the column that has reached the level of neutral buoyancy, H_b, and spreads radially or downwind, or both, to form an umbrella cloud and downwind plume (Plate 9.1c, d; Figs. 9.27a, c, 9.28b–e, 9.10a, b). The umbrella cloud extends from height H_b to height H_T (top of the umbrella region, Figs. 9.27a, c, 9.28b). H_T is a dynamic level depending on both the vertical momentum of the column acquired in the thermally convecting region, which defines the maximum height reached above H_b, and the mass-flow rate across H_b, which then sustains the laterally spreading umbrella.

A maintained, convecting, eruption column could reach heights of greater than 40 km during some large explosive (plinian) eruptions (e.g. 1993 Pinatubo eruption, Philippines; Wilson et al. 1978; Mastin et al. 2009; Suzuki and Koyaguchi 2009; Woodhouse et al. 2013; 15 January 2022 Hunga Tonga–Hunga Ha'apai eruption, Tonga, southwest Pacific Ocean; Proud et al. 2022). In contrast, in eruption columns that form from discrete explosions, convective recovery only takes columns to heights of a few km (VEI < 3), unless explosions occur in quick succession, in which case a weakly maintained plume develops.

Determining the height of the eruption column and umbrella region during modern explosive eruptions is not straightforward. Estimates of the heights of modern eruption columns can be undertaken visually viewed directly from ground level, or remotely using remote sensing techniques such as ground-based infrasound (e.g. Perrtu et al. 2020) or using satellite infrared imagery (e.g. Bachmeier 2020; Prata et al. 2020; Proud et al. 2022). Visual estimates can be very erroneous because of parallax and weather (clouds, rain) effects, and remote assessments contain inherent experimental errors. For example, estimates of the height of the 2020 Taal volcano, the Philippines, eruption column (Balangue-Tarriela et al. 2022) based on remote infrasound suggested a height of ~17 km (Perrtu et al. 2020), whereas satellite infrared assessment using data from the Himawari-8 weather satellite indicated a height of ~21 km (Bachmeier 2020; Balangue-Tarriela et al. 2022). The latter, however, was complicated by the masking effects of solid ice particles in the upper part of the eruption column (Bachmeier 2020). Micron size ice particles (<11 microns) also constituted the bulk of the uppermost part of the eruption column from the 2020 explosive eruption of Anak Krakatau, Indonesia, which was essentially ash-free (Prata et al. 2020). That eruption column was estimated to be ~16 km high using satellite infrared technology. Tupper et al. (2009) have also emphasised the difficulty in determining how much ash occurs in eruption columns due to the masking effects of water moisture, ice, and aerosol gases, at what height the ash is concentrated, and how to distinguish it from ice particles and aerosol gases.

For volcanologists, this raises a very important question in discussing eruption column/plume height as a control on tephra dispersal. The question is: what are we measuring when we consider eruption column height? In terms of eruption energy considerations, we are interested in maximum eruption column height, H_T, as summarised below, using a range of relationships between eruption parameters. However, in volcanology we also want to know the height from which the tephra fall, which is not necessarily the maximum height of the eruption column or plume, H_T. Carey and Sparks (1986) and Constantinescu et al. (2021) proposed that tephra dispersal is sourced from the base of the umbrella cloud (i.e. the height of the level of neutral buoyancy, H_b) which is therefore the effective release height of tephra from the column (Fig. 9.27d). The upper part of the umbrella region may be depleted in ash and more dominated by ice and aerosol gases, especially in humid tropical climatic zones, as validated for the 2020 Anak Krakatau eruption column (Prata et al. 2020).

In addition, Tupper et al. (2009), building on previous work, have indicated that explosive eruption columns in different climatic zones reach different altitudes given the same magma mass eruption rates, especially at MER at <10^7 kg s^{-1}. In humid tropical settings, atmospheric water vapour is entrained into the eruption column and enhances the convective strength and the height of the column, whereas in drier polar and temperate climatic zones, less atmospheric moisture is entrained, and convection strength and column height are lower. In both cases, the eruption column has difficulty breaching the tropopause, which varies according to latitude from 8–9 km above the poles to 17–18 km above the equator, but the height difference could be as much as 9 km. At MER of 10^8 kg s^{-1} and higher, eruption columns are energetic enough to inject into the stratosphere. Can this climate effect be detected in ancient fallout deposits, and has a lack of understanding of this effect led to misrepresentation of the eruption intensity through the VEI classification based on eruption column height assessment? Aubry et al. (2023) have also pointed out that as global climate changes the temperature, density and structure of the atmosphere changes and the heights of the main atmospheric layers and boundaries change. This impacts the dynamics of eruption columns and pyroclast dispersal, and so assessments of eruption column heights based on deposit dispersal patterns of old deposits could be erroneous if based on modern atmospheric structure, patterns and climate conditions.

An example of the difficulty in understanding the dynamic significance eruption column height relates to the spectacular explosive eruption of Hunga Tonga–Hunga Ha'apai volcano in Tonga, southwest Pacific Ocean, on 15 January 2022. The volcano, a marine stratovolcano that was barely emergent above sea level, had experienced low-intensity surtseyan explosive eruptions of andesitic magma in 2009, 2014–15, and from 20 December 2021 to 14 January 2022. It erupted with much greater intensity at ~4.00 pm on 15 January. New basaltic andesitic magma was erupted, and the eruption column grew spectacularly over an hour as captured by the Himawari-8 weather satellite. The eruption column height was initially estimated to have reached 25 km, which was then reassessed to ~40 km, about the same as the 1991 Pinatubo eruption column in the Philippines. NASA then proposed that the eruption column reached a height of ~55 km, well into the mesosphere (https://earthobservatory.nasa.gov/images/149474/tonga-volcano-plume-reached-the-mesosphere), which was refined to 57 km by Proud et al. (2022), using multiple satellite sensors to calculate plume height using the parallax effect from different geometries of the top of the plume produced by the different sensors. This maximum height is in the centre of the umbrella cloud, more or less above the vent area, in what is called the overshoot area, where momentum projects the column to its highest level. It seems that much of that upper part of the column consists of ice and aerosol gases, with little ash content (A. Prata, pers. comm.). Proud et al. (2022) suggest that infrequent satellite spot measurements and images during previous eruptions, such as the 1991 Pinatubo eruption, may have led to underestimation of the actual maximum plume height reached.

The problem then is how to compare estimates of eruption column height based on deposit characteristics and parameters (e.g. maximum average lithic clast size isopleth patterns; see Sect. 9.6.2.2 below) for prehistoric eruptions, with assessments of an eruption column height of modern eruption columns using satellite and other remote sensing technology? The two are not comparable. Based on the above considerations, the maximum height of eruption columns is largely irrelevant to ash dispersal processes because the top of the umbrella cloud most likely consists of ice particles and aerosol gases (e.g. Prata et al. 2020). As pointed out by Constantinescu et al. (2021) and discussed above, it is the height of the base of the umbrella cloud, H_b, that is effectively determined by column height assessments for prehistoric fallout deposits using isopleth patterns because that is the level in the convective part of the plume that pyroclasts fallout. In making comparisons with modern eruption column heights, the height of the base eruption cloud should be cited, as much as the maximum height. Determination of H_b from deposit dispersal characteristics is depicted by Eq. 9.20, but is also addressed by Carey and Sparks (1986), as depicted in Fig. 9.36a–d. This then may have implications for assessments of the Volcanic Explosivity Index (VEI) of an eruption if it is based largely on "eruption column height" rather than other parameters, because isopach, isopleth, and isomass patterns can be explained by a lower eruption column if the umbrella region has a large radius (Constantinescu et al. 2021).

Given the upward rate of dissipation of the initial momentum and the upward increasing rate of air entrainment, in turn accompanied by particle sedimentation and heat exchange, the shape of the forming umbrella will have different rates of upward enlargement. Wilson and Parfitt (2004) suggest that the spreading rate is 1:8 in the first quarter of the column dominated by the gas thrust region, increasing upward to 1:4 in the second quarter; 1:2 in the third quarter, and 1:1 (i.e. at 45°) near the umbrella region.

Early models have assumed that the bulk dynamics of volcanic plumes, including their maximum height, could be satisfactorily described by the dissipation of the thermal energy released when there is thermal equilibrium between the particles and the gas + air phase (pseudo-fluid approach, in Wilson 1986; Sparks et al. 1997). With this approach, the maximum height of an eruption column (H_t) is proportional to the release of thermal energy, and hence the mass eruption rate (Wilson et al. 1978; Settle 1978; Sparks et al. 1997).

For maintained eruption columns, the height can be predicted (Morton et al. 1956a, b; Wilson et al. 1978) from

$$H_T = 8.2\,Q^{1/4} \tag{9.4}$$

where H_T is the height of the column in m and Q is the steady rate of release of energy in watts. Aubry et al. (2023) parameterise this H_T relationship in terms of mass eruption rate (MER) based on a dataset of 130 eruptions which generate an empirical power law, $H_T = a\ MER^b$, in which the factor a and exponent b vary from eruption to eruption, but on average can be represented by $H_T = 0.345\ MER^{0.226}$. Q is related to the eruption conditions at the vent by

$$Q = \beta\, v\, \pi\, r^2 s\,(\theta - \theta_a)\,F \tag{9.5}$$

in which β, v, s, *and* θ are, respectively, the bulk density, velocity, specific heat, and temperature of the erupting fluid, θ_a is the temperature to which the eruption products ultimately cool (~270 K in most cases), r is the vent radius, and F is an efficiency factor (discussed below). The bulk density, β, is related to the density of the magmatic gas, ρ_g, the density of the pyroclasts, ρ_m, and the weight fractions of gas and pyroclasts, N and x_m:

$$\frac{1}{\beta} = \frac{x_m}{\rho_m} + \frac{N}{\rho_g} \tag{9.6}$$

If it is assumed that the predominant gas is water and that the erupting fluid is at atmospheric pressure, then for θ = 1200 K, ρ_g is 0.18 kg m^{-3}. The thermal properties of magma are dominated by the solid phase for gas contents of a few per cent by weight, and the value of s, the specific heat, is taken as 1.1 × 10^{-3} J kg^{-1} K^{-1}.

The maximum height of the eruption column, H_T, can also be expressed as a function of the volume discharge rate of magma (Sparks 1986; Fig. 9.25):

$$H_T = 5.773\,(1+n)^{-3/8}\,[\rho\,\phi\,s(\theta - \theta_a)]^{1/4} \tag{9.7}$$

where ϕ is the volume discharge rate in cubic metres per second, s is the specific heat, θ is the initial temperature of the erupting material, θ_a is the atmospheric temperature at sea level, ρ is the magma density, and n is the ratio of the vertical gradient of the absolute temperature to the lapse rate. Sparks et al. (1997) and Carey and Bursik (2015) simplified the relationship between column height, H_T, and volumetric discharge rate, ϕ:

$$H_T = 1.67\phi^{0.259} \tag{9.8}$$

compared with the proposal of Mastin et al. (2009) of $2.00\phi^{0.241}$.

Figure 9.25 depicts theoretical curves showing the relationships between maintained eruption column height and volumetric eruption or discharge rate of magma calculated from Eqs. 9.4–9.6, together with the heights of some observed eruption columns (Wilson et al. 1978; Table 9.2). The calculations coincide well with recorded column heights. The efficiency factor, F, measures the efficiency of conversion of heat to potential or kinetic energy, and curves with values of F = 1.0, 0.7 and 0.3 are used in Fig. 9.25. F, as previously described, is mainly controlled by the degree of fragmentation of the magma in the explosive event. Here, we are only considering magmatic eruptions, not the special case of explosions generated by magma–water interaction, which will be discussed below. In eruptions which generate a higher proportion of ash-sized ejecta, virtually all of the magmatic heat can be converted to mechanical energy, as the time required for the heat transfer is lower than the residence time in the column. Many plinian deposits have a substantial proportion of fine-grained particles, and Sparks and Wilson (1982) estimated at least 70% efficiency in the conversion of heat in selected plinian columns.

On the other hand, vulcanian eruptions produce a comparatively much higher proportion of coarse debris (because of a lower degree of fragmentation (Clarke et al. 2002; Alatorre-Ibargüengoitia et al. 2011), and columns are likely to be much less efficient in the conversion of heat. Consequently, observed eruption columns from these types of activity should fit a theoretical curve with a low F value, and this seems to be the case of the 1973 Heimaey eruption in Iceland (Fig. 9.25).

The maximum theoretical height expected for a stable, maintained eruption plume is about 55 km (Wilson et al. 1978). This corresponds to an initial gas velocity of 700 m s^{-1} (greater vent exit velocities are unlikely to occur on Earth; McGetchin and Ullrich 1973; Wilson 1978), which requires a volumetric eruption rate of 1.1 × 10^6 m^3 s^{-1}. Costa et al. (2018) discussed how super-eruptions may be fed by up to 2 orders or magnitude higher eruption rates, because they originate from caldera fissure vents rather than point source vents (see Chap. 10). Therefore, plume dynamics are very different, and maximum column heights for mass eruption rates of 10^{11} kg s^{-1} are modelled up to 60–70 km. Aubry et al. (2023) have also pointed out that eruption column height changes according to different climate zone settings at fixed mass eruption rates related to changing atmospheric temperature, humidity and density.

A standard atmosphere with a vertical decrease in temperature (environmental lapse rate) of 6.5 °C km^{-1} is used in the calculations of the theoretical curves in Fig. 9.25. However, substantial departures from the standard atmosphere can occur, and the scatter in the data from observed eruptions in Fig. 9.25 may partly reflect variations in vertical atmospheric temperature gradients (Settle 1978; Wilson et al. 1978).

More recent and more accurate models are now able to account for a series of other important factors, in addition to the mass eruption rate and the efficiency factor. These affect the ultimate dynamics of eruption plumes and consequently their maximum height, and include factors such as the thermal and mechanical disequilibrium between particles and gas (e.g. Bursik and Woods 1991), the aggregation of particles (e.g. Veitch and Woods 2002; Textor et al. 2006a, b), and multi-phase processes (e.g. Costa et al. 2006; Barsotti et al. 2008).

A very significant factor that affects the dynamics, the maximum height, and the spreading of plumes is the occurrence of crosswinds (Carey and Sparks 1986; Woodhouse et al. 2013). Equation 9.4 strictly applies to the vertical rise of an eruption column into a still atmosphere with no wind. This should be broadly applicable to most large explosive eruptions that feed strong columns, where upward velocities of a particle-rich plume are likely to be much greater over much of its height than the transverse wind velocity.

For strong winds and moderate to small eruption columns, the effect of wind on column height can be significant, and this is discussed in Settle (1978), Sparks et al. (1997), Bursik (2001), Bonadonna et al. (2005), and Woodhouse et al. (2013). In particular, strong winds, apart from tilting the eruption column and guiding the dispersal of pyroclasts, also promote a much-increased interaction of the column with the atmosphere, increasing the entrainment rate and contributing to the buoyancy of the column. Woodhouse et al. (2013) have proposed that wind strength can be factored into calculations of eruption column height as follows:

$$H_T = 0.318\,M^{0.253}\frac{1+1.373W_s}{1+4.266W_s+0.3527W_s} \qquad (9.9)$$

where M is the source mass flux in kg s^{-1}, and W_s is a dimensionless wind shear factor (Woodhouse et al. 2013).

To estimate the rise height of a plume generated from a discrete explosion (e.g. vulcanian or hydrothermal eruption), another equation must be used (Morton et al. 1956a, b):

$$H_T = 1.37\,Q^{1/4} \qquad (9.10)$$

This has the same form as Eq. 9.4, but Q is the total energy released in joules. Parfitt and Wilson (2008) have expressed this in terms of the total mass of solids and gas ejected, M_e, during a short explosive event as follows:

$$H_T = 0.042\,M_e^{1/4} \qquad (9.11)$$

Strong tropospheric and stratospheric winds can cause the umbrella region to the deflected in a downwind direction, causing the dispersal of gas, ash, and coarser pyroclasts to take on a fan shape (Fig. 9.28c-f) and leading to the dispersal of fallout deposits in an aymmetrical fan tens to hundreds of km downwind (see Figs. 9.31, 9.32, 9.33). The finest ash and gas can circumnavigate the globe.

In addition, the high plumes generated, particularly during plinian-type and ignimbrite-forming eruptions, must penetrate the level of the tropopause in the atmosphere (at heights of <6 at polar latitudes to 18 km in tropical latitudes, depending on latitude and season), and contribute fine ash and gaseous species to stratospheric dust veils. Some climatologists have therefore thought that volcanic eruptions might promote periods of climatic cooling. The topic is beyond the scope of this book, and critical reviews can be found by Rampino et al. (1979), Self and Rampino (1981), Simkin (1994), Briffa et al. (1998), De Silva and Zielinski (1998), Robock (2000), Oppenheimer (2003), Lowe et al. (2012), and Ernst and Youbi (2017).

There are numerous numerical modelling and computational approaches to modelling the dynamics of eruption plumes, the dispersal of tephra, and the impacts. Perhaps the most widely used program is *Tephra3D*, which is now in its eighth iteration (Folch et al. 2020; Prata et al. 2021), and since its initial development in 2003, it has been used for a range of purposes, including hazard assessment of tephra fallout, impact of volcanic ash clouds on aviation, determining eruption parameters, assessing the effects of super-eruptions on climate, environment, and humans, and forecasting ash dispersal and fallout patterns (Folch et al. 2020; Prata et al. 2021). An updated version of the Tephra3D code (FALL3D-8.1) includes improvements that allow the incorporation of forecasts of an ash or gas dispersal event from both a deterministic (incorporating previous factually constrained examples) and probabilistic (statistical predictions of various scenarios) points of view and the potential errors (Folch et al. 2022). This is a precursor to a forthcoming FALL3D-8.2 version. Another modelling approach proposed by Biass et al. (2019) uses a Matlab application called *TephraFits*. It utilises deposit dispersal properties using isopach, isomass, and isopleth maps, empirical methods of integration (exponential, power law, Weibull), and allows calculation of the deposit volume, mass, VEI and magnitude, and thickness decay pattern.

Constantinescu et al. (2021) have proposed that the emphasis on eruption column height in controlling tephra dispersal may mask the role of the radius of the umbrella cloud in controlling convective tephra dispersal processes and patterns. They propose that if the umbrella cloud expands substantially away from its centre, then the dispersal and fallout of the tephra is controlled by a disc source (i.e. the entire umbrella cloud), rather than a point source (Fig. 9.27d). This may mean that calculations of eruption column height based on deposit properties (e.g. average maximum lithic clast size) may lead to an overestimate of eruption column height, which could in turn lead to an overestimate of the volcanic explosivity index (VEI) where the principal parameter used to assess it is the eruption column height. This model only applies to sustained explosive eruptions that produce buoyant eruption columns and relatively symmetrical umbrella clouds. It does not apply to ballistic or other point source blasts and becomes difficult to apply in the case of fissure eruptions and eruptions affected by very strong winds that produce a very asymmetrical ash cloud or plume. Constantinescu et al. (2021) developed the Python application to model fallout deposition from umbrella clouds. The program and guidelines for its use are available at https://github.com/robertoon/umbrella_cloud_model.

The radius of the umbrella cloud, R, can be depicted by (Costa et al. 2013; Prata et al. 2020; Constantinescu et al. 2021)

$$R = \frac{[3\lambda N V^*]^{0.33}}{2\pi}\cdot t^{0.67} \qquad (9.12)$$

where λ is a constant (~0.2), N is the Brunt–Väisälä frequency (0.015), V^* is the volumetric flow rate (m^3 s^{-1}) in the umbrella cloud, and t is the time since the plume begins spreading at the level of neutral buoyancy. R is the average radius in the unevenly spreading umbrella cloud, and can also be represented as a function of the area, A, of the cloud (Prata et al. 2020):

$$R = (A/\pi)^{0.5} \tag{9.13}$$

The average radial flow velocity, v_R, in a radially spreading umbrella cloud with radius R is (Carey and Sparks 1986)

$$v_R = \frac{M^*}{2\pi R \rho_a (H_T - H_b)} \tag{9.14}$$

where M^* is the mass-flow rate of ejecta and supporting gas/air, H_T is the eruption column height, H_b is the level of neutral buoyancy, or the height of the base of the umbrella cloud, and ρ_a is the mean density of air between H_t and H_b.

9.5.2 Buoyant Plumes Derived from Pyroclastic Flows

The ash clouds which enclose and trail behind pyroclastic flows (see Chap. 10) can become buoyant, rise above moving pyroclastic flows, and form buoyant ash plumes well away from the vent (Plate 9.1d; Fig. 9.11a). Fine ash can be dispersed from these plumes as fallout ash deposits well beyond the limits of the pyroclastic flows and are called co-ignimbrite ash deposits. The physics of pyroclastic flow-generated ash-cloud plumes (also occasionally referred to as *phoenix plumes* in the literature) is essentially similar to that described above for vent-sourced explosive eruptions, except that they lack gas thrust zones. The major differences are (1) the source geometry, which instead of being a point source as for usual vent-derived eruptions, is the full surface area of the parent pyroclastic flow; and (2) the grain size, which is selectively made up of only those particle sizes that can be elutriated from the parent flow. Ash-cloud plumes develop immediately as convective plumes, in which the fine grain sizes carried allow an efficient exchange of heat with entrained air. Such plumes can interfere and eventually merge with the explosive plume generated at the vent, causing complex patterns of fine ash dispersal. The physics of ash-cloud plumes is discussed in Woods and Wohletz (1991), Calder et al. (1997), Herzog and Graf (2010), and Costa et al. (2018). An important recent advancement in understanding of co-ignimbrite fall origin and deposits has been achieved by numerical simulations, which show that pyroclastic flows at their maximum runout may still have absolute velocities of several tens of meters per second, turning to vertical as they literally lift-off. In the case of the numerical simulation of the caldera-forming Pozzolane Rosse ignimbrite (Colli Albani, Italy), Calabrò et al. (2022) showed that pyroclastic flows at their maximum runout, i.e. ~30 km from vent, can still have velocities of 60–200 m s^{-1}, and up tp 50–60% of the total mass, which can result in the formation of "rootless" plinian-size vertical columns rising directly from the flow front and able to entrain not just the smaller fine ash particles, but up to lapilli-sized clasts, matching observation of co-ignimbrite deposits found at more than 100 km distance across the high Apennine mountains.

Numerical modelling of the 74 ka Toba and 1815 Tambora co-ignimbrite plumes indicate maximum heights of 35 km and 23 km, respectively (Woods and Wohletz 1991), and much greater heights, up to 50–60 km, have been inferred to justify the large dispersal of the 39 ka Campanian Ignimbrite co-ignimbrite ash fall (Costa et al. 2018). However, ash-cloud plumes do not form uniformly along the flow, as the intensity of elutriation depends on changing flow velocity and ash concentration. Increased sedimentation at breaks-in-slope, for example, promotes elutriation of fine particles, so that a more intense rise of co-ignimbrite plumes may occur at significant distances from the vent (see Chap. 10, Fig. 10.14; Giordano and Dobran 1994).

9.5.3 Phreatomagmatic "Columns"/Tephra Jets

We discussed some of the physical controls on phreatomagmatic or hydrovolcanic explosions that generate surtseyan and phreatoplinian pyroclastic fall deposits in Chap. 4, Sect. 4.6.9. The major contrast between these types of eruptions and magmatic explosive eruptions is the very high degree of fragmentation of the magma. In magmatic explosive eruptions, fragmentation of the magma is driven by exsolution and explosive expansion of its volatiles. This produces the relatively coarse population of pyroclasts (pumice, scoria) observed in micro-plinian and plinian deposits. However, even in phreatomagmatic eruptions, it is likely that some vesiculation and disruption of the magma are caused by magmatic gases. Scanning electron microprobe (SEM) studies show that shards from surtseyan and phreatoplinian deposits, although having sharp, planar, fractured boundaries, can also contain vesicles, from low to high abundance, suggesting that magmatic fragmentation can play a role in the fragmentation process (Chap. 4; Self and Sparks 1978; Pardo et al. 2009; Mattsson 2010).

Eruption columns from this type of activity (Figs. 9.11c, 9.80b; Fig. 4.18b) are driven by rapid successions of hydrovolcanic explosions (Fig. 9.11b). Observations of the 1963 Surtsey eruption in Iceland suggest that these occur every few seconds to tens of seconds (Thorarinsson et al. 1964). Each explosion at Surtsey thrust out black "cock's tail" jets of ash and bombs (e.g. Plate 4d, Fig. 4.18a; cf.

Figs. 9.11b, 9.80a), which occasionally reached heights of about 1 km. Finer particles were then taken to greater heights by convection, occasionally as high as 9 km (Thorarinsson et al. 1964). During the 1979 Soufrière eruption, St Vincent, columns from phreatomagmatic explosions rose as high as 18 km (Sparks and Wilson 1982). As the steam–pyroclast mixture rises, condensation of the steam may occur in the column. This phase change requires a large change in volume, and a substantial increase in density. Although this may be partially compensated for by mixing in air from the side of the column, partial collapse of the column could occur. It is in this situation that *surtseyan tephra* or *debris jets* will form (Figs. 9.11b, 9.80a, and Plate 4d, 4.18a).

During phreatomagmatic eruptions, a great deal of heat, that in normal magmatic eruptions would be used to drive a convective plume, is used instead in the conversion of water to steam (the heat of vaporisation of water is 580 cal g^{-1} (1 cal ~ 4.18 J) at atmospheric pressure and 298 °K) (Wilson et al. 1978; Self and Sparks 1978; Cas and Giordano 2014, and references therein). The thermal energy used in vaporisation can only be recovered by condensation of the steam. Consequently, eruption column heights should be lower in a phreatomagmatic eruption than in a magmatic eruption with the same volumetric rate of discharge (Sparks and Wilson 1982; Koyaguchi and Woods 1996; Herzog et al. 1998). However, the much finer grained nature of the ejecta means that there will be more efficient transfer of thermal energy from the magma to eruption column, perhaps enhancing its height (Woodhouse et al. 2013). Hajimirza et al. (2022) demonstrated that eruption column height associated with the phreatomagmatic eruption phase of the 10th century Eldgjá fissure eruption in Iceland, was 17−18 km high compared with the magmatic explosive eruption column which was 11−17 km high. They attribute the higher phreatomagmatic eruption column to the entrainment of glacial meltwater above the subglacial vent into the eruption column and its vaporisation within the column.

The only potential observation of a phreatoplinian eruption is the spectacular 4.00 pm, 15 January 2022 eruption of Hunga Tonga–Hunga Ha‘apai volcano in Tonga, southwest Pacific Ocean, that was spectacularly captured by weather satellite imagery. The eruption column rose and expanded rapidly within minutes of the start of the eruption and it reached 57 km above the vent, rising through the stratosphere and into the mesosphere (Proud et al. 2022). The column produced a rapidly expanding umbrella cloud at least 60 km in diameter, before a downwind ash and SO_2 gas plume formed that spread westwards over Australia and beyond within hours. The sonic pressure wave was heard thousands of km away, and circumnavigated the globe 3 times, detected in atmospheric pressure increases each time it passed meteorological recording stations around the world. The 4.00 pm 15 January eruption of Hunga Tonga–Hunga Ha'apai succeeded earlier surtseyan phases of eruption in 2009, 2014–15, and from 20 December 2021 to 14 January 2022, marked by pulsing tephra jets and weak eruption plumes a few km high. The eruption indicated a continuous eruption of a large volume of magma at a high discharge rate and with a relatively high magmatic volatile content, as has been proposed for other, prehistoric, phreatoplinian eruptions (e.g. Orsi et al. 1992; Smith and Houghton 1995; De Rita et al. 2002).

9.6 Pyroclastic Fallout Deposits: Field Description and Reconstruction of Eruption Parameters

The description and interpretation of pyroclastic fallout deposits can be approached in a number of ways. However, before discussing classification schemes and their deposits, are there any simple divisions of fallout deposits we can use that still retain some genetic significance? Four types of pyroclastic fallout deposits can be distinguished on broad lithological and genetic grounds: (1) scoria fallout deposits, (2) pumice fallout deposits, (3) juvenile ash-fallout deposits, and (4) lithic-rich (ash) fallout deposits.

Scoria fallout deposits are composed largely of vesiculated basalt to basaltic andesite magma (Figs. 9.3b, 9.5a), or other types of low-viscosity magmas (e.g. phonolitic). These are the deposits that are usually characteristic of basaltic scoria cone-forming micro-plinian explosive activity (Fig. 9.18a) and hawai'ian and strombolian near-vent, spatter cone-forming (Plate 9.1a; Figs. 9.2a, 9.7d). The deposits can be very coarse-grained close to the vent in scoria cones, with a predominant grain size >64 mm, and contain large ballistic bombs, including irregularly shaped bombs and spatter fragments (Chap. 4). Away from the cones, scoria fall deposits are finer-grained and usually <5 m thick, thinning to <1 cm thick distally. In recent years, scoria fall deposits have also been described associated with more intense explosive eruptions up to plinian intensity (Williams 1983; Coltelli et al. 1998; Miyaji 2002; Perez et al. 2009; Costantini et al. 2009; Giordano et al. 2010; Miyaji et al. 2011; Vinkler et al. 2012).

Pumice fallout deposits (Figs. 9.3a, 9.5b, 9.17) are composed largely of clasts of vesiculated pumice, originating from high-viscosity magmas (andesite to rhyolite, phonolite, and trachyte). They form widely dispersed sheet-like deposits of the plinian family (Sect. 9.2.2). Deposits of one eruption are rarely >10 m thick, but very close to the vent deposits as thick as 25 m are known. At the vent, the predominant grain size may be >64 mm, and the deposits contain large lithic and pumice blocks and bombs.

However, pumice fall deposits are also rarely known to be associated with milder explosive activity forming pumice cones (e.g. Monte Pilato, Lipari; Davì et al. 2011).

Ash fallout deposits of predominantly juvenile ash fragments can be formed by a whole spectrum of pyroclastic processes. Pumice and scoria fall deposits have juvenile-dominated ash-fall deposits at their distal dispersal areas, and their character depends on downwind aeolian fractionation processes. Co-ignimbrite ash-fall deposits can be very extensive examples of ash fallout deposits. They may also contain accretionary lapilli caused by rain flushing, and would be difficult to distinguish from silicic phreatomagmatic (phreatoplinian) ash-fall deposits in the absence of field criteria and SEM imaging of the morphology of ash particles (Chap. 4). Phreatomagmatic eruptions characteristically form fine-grained deposits even close to the vent, and these often contain accessory fragments and accretionary lapilli (Chap. 4, Sect. 4.8.3). Air-fall ash deposits range in thickness from <1 mm to >1 m thick, even at more than 100 km away for co-ignimbrite ash-fall deposits and phreatoplinian deposits.

Lithic-rich (ash)fallout deposits are usually associated with small to mild explosive eruptions, are not very dispersed and rarely preserved in the rock record. Phreatic eruptions produce lithic-rich ash-fall deposits, and ballistic blocks may be very abundant around the vent. Lithic-rich ash-fall deposits can also form from ash clouds resulting from elutriation from dense clast pyroclastic flows (e.g. the eruption at Tunghuraua 2006; Eychenne et al. 2012). Vulcanian eruptions typically produce ash fallout deposits which may range from dense lithic-rich to scoriaceous types.

An alternative non-genetic approach to subdividing fallout deposits is to use lithological descriptions based on dominant grain size and component types, as shown in Chap. 14 (Tables 14.1, 14.2). For example, in this case most pumice fall deposits would be simply pumice lapilli deposits. Most of the coarser (>2 mm diameter clasts) near-vent equivalents of the deposits discussed above would then be called volcanic breccias. We will discuss the use of these terms in Chap. 14.

9.6.1 From Outcrop Scale Characteristics of Fallout Deposits to Attribution of a Specific Eruption Style

As discussed in previous sections, eruptions that produce pyroclastic fallout deposits vary greatly in magnitude (from VEI 0.1 to VEI > 5), conduit dynamics, style of fragmentation, mass discharge rate, plume dynamics (including the type and extent of interaction with the atmosphere), and dispersal of pyroclasts. All these variations are reflected in highly variable deposit characteristics including overall geometry and dispersal pattern, internal structures, componentry, and grain size. However, there are some common characteristics, at least for deposits of moderate to high explosivity, which are usually best preserved and exposed in medial reaches. At the outcrop scale, these include:

- tabular geometry and mantling the paleotopography (mantle bedding; Fig. 9.3a, b);
- massive to plane-parallel stratified internal structure (Figs. 9.3a, b, 9.5a, b);
- moderately to well-sorted grain size distributions (Fig. 9.5a, b), which further improve (Fig. 9.19), if aerodynamic properties of the particles (e.g. their terminal fall velocity; see Sect. 9.4) are taken into account (Fig. 9.20),
- clast-supported framework textures (Fig. 9.5a, b) in proximal and medial settings transforming laterally into distal fine ash deposits, although very proximal deposits could possibly be chaotic if they contain by numerous large ballistic blocks and bombs, and
- angular shape of the clasts (e.g. felsic pumice; Fig. 9.5a, b), unless pyroclasts have undergone post-fragmentation welding, agglutination, or rheomorphism (e.g. low-viscosity spatter and scoria); however, no signs of abrasion derived by particle–particle collisions during transport should be observed.

Regionally, with increasing distance from the vent, as represented by decreasing deposit thickness contours (isopachs) or grain size contours (isopleths) drawn on a map showing the deposit, changes include:

- overall exponential thinning of the deposit with distance from the vent (Figs. 9.30a, b, 9.31a, 9.32a, 9.33a),
- gradual fining of the overall grain size and improved sorting resulting ultimately in distal fine ash deposits (Figs. 9.19a, 9.31b, c, 9.32b–d, 9.33b, c),
- symmetrical radial dispersal around the vent at very proximal locations (e.g. a scoria cone; Fig. 9.18a) and/or when wind strengths are low (Fig. 9.30a), which
- for moderate to large explosive eruptions progressively transition to more or less asymmetrical, elongated, and complex dispersal map patterns at medial and distal reaches (Figs. 9.24a–c, 9.30b, 9.31a–c, 9.32a–d, 9.33a, b), depending on the strength of the column versus the wind strength profile(s) at the time.

Very proximal and very distal fall deposits can *sometimes* be sampled during, but mostly immediately after, an eruption, but are rarely preserved in the rock record. This is because at very proximal locations, on steep topography, they can be eroded or covered by younger deposits, and distally, where they are thin and fine, they may be rapidly eroded, washed out, or incorporated into the soil. Even deposits in the medial reaches may be difficult to preserve unless they fall in topographic lows, and/or are rapidly buried by other deposits, as is the case of many sub-plinian to plinian fall deposits covered by associated ignimbrites.

9.6.2 Measurable Properties of Fallout Deposits

Once a pyroclastic fallout deposit has been recognised in the field, several outcrops have been correlated, and data and samples have been collected, a series of important analyses of the data can be undertaken, and extrapolations made at the scale of the whole deposit, in order to quantify the main eruptive parameters and the eruption style. These are based on the overall downwind and lateral changes in the character of fall deposits away from the vent that include:

- decrease in thickness,
- decrease in largest five pumice and lithic clasts or maximum grain size,
- decrease in median grain size (increasing $M_{d\phi}$ values),
- increase in sorting (decreasing σ_{ϕ}, values),
- changes in component population,
- decrease in median terminal fall velocity, and
- changing mass per unit area.

In addition, the structure and the compositional characteristics of the deposit and the textures of the juvenile component may provide important information on the conduit dynamics and the evolution of the eruption.

We will now briefly summarise a possible workflow scheme which takes field data and uses it to interpret the following eruption and dispersal dynamics:

(1) magnitude of the eruption,
(2) conduit decompression rates,
(3) mixture exit velocities,
(4) height of the eruption column,
(5) average mass eruption rate,
(6) dispersal areas and the distribution of the mass load,
(7) fragmentation style and total grain size distribution, and
(8) temperature of emplacement and the related thermal structure of the plume.

Classifications of eruption styles from deposits are related to the reconstruction of all or part of the above parameters. However, the reliability of possible reconstructions of the eruptive parameters is strictly related to the amount of field data available, which of course will be less and less as we deal with progressively older deposits. In many cases from the prehistoric volcanic record, all that is left are isolated outcrops, which nonetheless preserve important specific characteristics, such as the geometry of the deposit at outcrop scale, pyroclast shapes/morphology, their texture, and deposit aggregate characteristics, including grain size, sorting, depositional structures and textures, and componentry. Documenting the outcrop scale characteristics of such poorly preserved fallout deposits is however very important, because it at least allows a qualitative assessment of the general eruption style, and may allow an assessment of the possible deposits and eruption magnitude.

9.6.2.1 Determining Source Vent Location, Dispersal Area, and Volume: Thickness (Isopach), Maximum Grain Size (Isopleth), and Isomass Maps

The areal extent and thickness variations of a fallout deposit can be described by constructing an *isopach map* of the changing thickness of deposits with distance from the vent using isopach contours (an isopach is a line that shows everywhere that deposit has a specific thickness, such as 50, 20, 5 cm; Figs. 9.30, 9.31, 9.32, 9.33). An isopach map is constructed by plotting on a map the location of every site where the thickness has been measured, the thickness value, then drawing multiple contour lines that depict different key thickness values of the deposit, which can be done manually (preferably), either on a hardcopy map or on a digital map on a computer. However, in using computing, care needs to be taken in defining boundary conditions, and ensure that interpolations made by the computer don't override real, but unusual patterns. In this regard, it is important that the person who has done the fieldwork and collected the field thickness data constructs the isopach map. Although programmed functions or apps on computers can save a lot of time, the results need to be checked or interrogated by the human brain to ensure that computer logic interpretations don't gloss over and mask important, real variations in the data.

The accuracy of the interpolation between contours relies on the accuracy of the thickness measurements at outcrops and on the number of measured thickness data points or outcrops over the dispersal area (e.g. Engwell et al. 2013). Much care should be given when measuring the thickness in identifying any possible syn-depositional erosion or reworking, such as may occur on very steep slopes, or other erosional processes that may alter the original local thickness, such as erosion by overriding or contemporaneous pyroclastic flows or surges (see Chap. 10). The maximum thickness of a pyroclastic fall deposit is measured in cm or m at each outcrop, and should only be measured where the original bottom and top of the deposit are preserved.

Isopach maps represent the basis for determining:

- the dispersal pattern, which can be related to the type of eruption and wind strength and directions, which can change during an eruption,
- the vent position, and
- the volume and the mass of the deposit.

Construction of an isopach map entails mapping out the extent of a deposit, sometimes over large areas. However, by mapping we do not mean continuously tracing on foot lithological boundaries between deposits, as these are usually so complex that no attempt is made to draw them exactly. Thus, a map of a relatively recent pyroclastic fall deposit generally shows its *inferred* original distribution, and not its present outcrop pattern, because between data points the deposit could be partially or completely eroded. Indeed, non-welded pyroclastic deposits may be ephemeral. In addition, outcrops may be so rapidly overgrown and badly weathered that they cannot be measured. For example, new road-cuts in tropical climates are sometimes completely overgrown within four or five years. However, where this has happened, a flat-blade spade can be used to dig down into a deposit and produce artificial outcrops (pits, short trenches) of the entire deposit thickness, or an original outcrop that is overgrown with vegetation can be carefully cleaned with a spade to produce a nice clean exposure. Pits and trenches should be filled in again after use to avoid injuries to people and animals, and where outcrops are cleaned this should be done responsibly, to minimise leaving loose debris, for example alongside roadside cuttings, which could become a hazard for road traffic.

Because pyroclastic fall deposits can change markedly laterally, it is preferable to use reasonably closely spaced data points. Although the spacing ultimately depends on the type of eruption, for the large plinian fall deposits localities within 2 km of each other if possible are favoured, and definitely <5 km spacing. Within these distances, correlation is more certain, and internal changes can be carefully documented. This is very important in correlating deposits between localities and in understanding the eruption and its stratigraphy.

The *bulk volume of air-fall deposits* can be calculated from isopach maps in various ways. A common method involves measuring the area enclosed by each isopach and then plot area against thickness. The thinning of fall deposits with distance follows a power law (Rose et al. 1973) or exponential decay pattern (Pyle 1989), as a result of different settling behaviours of particles (Rose 1993; Bonadonna et al. 1998). Therefore, the deposit thinning pattern can be approximated by one or more straight line segments if plotted on a logarithm (ln) of thickness versus the square root of the area enclosed by each isopach (Fig. 9.34; Pyle 1989; Fierstein and Natheson 1992).

Usually at least two, and up to five, straight line segments can be identified on such a diagram (). The most proximal, and very rarely preserved or visible segment, is the steepest (segment -1 in Fig. 9.34) and is associated with sedimentation from the margins of the gas thrust region and the lower part of the convective uprise region (Figs. 9.27c, 9.28a); this segment represents the thickest portion of the deposit, which rapidly thins and halves in only tens to a few hundreds of m from vent. The second segment (segment 0 in Fig. 9.34) represents the sedimentation from the margins of the upper convective region and extends to 1-10 s km from the vent (Figs. 9.27c, 9.28b). The inflection point between segment 0 and the successive and more gentle segment(s) in Fig. 9.34 roughly reflects the transition from sedimentation from the convective uprise zone in the plume to sedimentation is dominated by fallout from the umbrella cloud (Figs. 9.10a, c, 9.27c, 9.28b). Segmentation beyond this point (segments 1, 2, 3, Fig. 9.34) reflects the fallout of particles characterised by progressively lower Reynolds numbers (i.e. increasingly smaller and lighter with distance, Bonadonna et al. 1998), which in turn reflect the progressive lateral decrease of turbulence in the umbrella and downwind plume region.

The calculation of *deposit volume, V,* derives from the integration of the variation of thickness (T_x) from maximum (T_0 at vent) to zero (at infinite distance), across the isopachs of area A_x, resolving to the following equation (Pyle 1989):

$$V = \int_0^\infty \frac{[\pi r^2 \cdot k T_0 \exp(-k r) \cdot dr]}{\alpha} \tag{9.15}$$

where: *r* is the radial distance from the vent; T_0 is the intercept of the segment describing the thickness decay with distance on the lnT axis and corresponds to the extrapolated deposit maximum thickness at vent; *k* indicates the slope on the ln T *versus* $\sqrt{}$Area plot and is = ln (2)/b_t, where b_t is the thickness half-distance i.e. the distance along the dispersal axis at which the maximum thickness halves, and α indicates the average eccentricity *e* of the ellipsoidal isopachs, as the ratio of the minimum (Y)/maximum (X) axes of the ellipse, defined as $Y/X = (1 - e^2)^{0.5}$.

However, a strong uncertainty in the final calculations derives from extrapolation of the deposit decay pattern to both proximal and distal areas, where usually data are poor or unavailable (Sulpizio 2005), and from sedimentation processes perhaps being affected by particle–particle interaction and particle aggregation.

Based on the type of fitting of data on ln T *versus* $\sqrt{A}$, where A is the area of the isopach of thickness T, the following equations for the calculation of fallout volume have been proposed:

(a) Exponential fitting valid for elliptical isopachs (Pyle 1989):

$$V = \frac{2\pi T_0}{\alpha m^2} = \frac{13.08\,T_0\,b_t^2}{\alpha} \qquad (9.16)$$

and

$$V = 13.08\,T_0\,b_t^2 + \delta \qquad (9.17)$$

where, as above, T_0 is the extrapolated maximum thickness at vent, b_t is the thickness half-distance and α refers to the eccentricity of the ellipsoidal isopachs; $m = k\,\pi^{1/2} = \ln(2)/b_t$ and δ applies if more than one straight line segment appears on the plot, taking into account the distance R at which the break in slope of the segments occur and the ratio of the slope before (k) and after (k_1).

(b) Exponential fitting corrected for non-ellipticity of isopachs (Fierstein and Natheson 1992; a spreadsheet calculator is available online at Natheson and Fierstein 2015):

$$V = \frac{2T_0}{k^2} + 2T_0\left[\frac{(k_1 A_{ip}^{0.5} + 1)}{k_1^2} - \frac{(k A_{ip}^{0.5} + 1)}{k^2}\right] exp\left(-k A_{ip}^{0.5}\right) \qquad (9.18)$$

where the first component describes the thickness decay of the first near-vent segment of the plot and the second component represents the second segment that starts at area A_{ip} and is characterised by a slope k_1.

(c) Weibull function proposed to minimise errors derived by either exponential or power law fitting at proximal and distal areas (Bonadonna and Costa, 2012, 2013). A spreadsheet calculator is now available online to compare exponential, power law, and Weibull distributions for tephra volume calculations from isopach data (Daggit et al. 2014). Janeba et al. (2016) have calculated the volumes of 6 sub-plinian to plinian fallout deposits erupted from Hekla volcano, Iceland, using exponential thinning, power law, and Weibull function approaches, and found they produce variations in volume estimates, and each has advantages and disadvantages, depending on circumstances.

In addition, Legros (2000) proposed a useful approximation for calculating fallout deposit volumes, for fall deposits that are poorly exposed and their extent poorly constrained in the rock record. If one isopach contour can be constructed from even just a few thickness measurements at several locations, the thickness exponential decay thinning pattern can be approximated by constructing just one isopach contour for thickness T, which encloses an area of A. The minimum volume can then be calculated by applying the following formula:

$$V = 3.69\,T\,A \qquad (9.19)$$

Other empirical methods based on the semi-log plots of thickness versus square root of isopach area for extrapolating distal volumes from poorly preserved deposits are given in Sulpizio (2005).

The problem of extrapolation of un-preserved distal volumes was also approached in a completely different way by Walker (1980, 1981b, c), based on the study of the enrichment in free crystal concentration in the bulk deposit relative to phenocryst content in pumice clasts. This is based on the assumption that the glass/crystal ratio in crushed pumice (i.e. vesicle free) is representative of that in the erupting magma, and that fine vitric ash is selectively and more efficiently transported downwind in plumes than larger dense crystal fragments. Crystal fragments should thus fallout early, whereas fine vitric ash remains lofted and carried downwind, so that bulk deposits at proximal and medial distances become enriched in crystals to various degree, whereas distal fallout deposits are conversely enriched in fine vitric ash. Once the total bulk volume erupted had been estimated from the proportion of free crystals relative to the magmatic ratio as represented in pumice clasts, a straight-line extrapolation could be made at the low thickness end on an area plot to a selected limiting thickness value, so giving the same volume as would be calculated by integration of the area curve. It was found that for various deposits extrapolations to the same lower limiting thickness as little as 1 μm had nearly identical slopes, suggesting that total erupted volumes of other plinian deposits can be conveniently estimated on an area plot by extrapolation parallel to this slope, using the same limiting thickness of 1 μm.

Isopach maps can be transformed into *isomass maps* (kg m^{-2}) if the bulk mass of the deposit is measured at various distances from the vent, in order to account for bulk density variations due to lateral changes in grain size and componentry (e.g. Scollo et al. 2007; Fig. 9.35). Isomass maps depict contours that define areas where the mass of the fallout deposit is the same, obtained by determining the bulk density of a sample at a locality, which is converted into mass per unit area, knowing the thickness of the deposit at each locality. Data points record deposit mass in kg m^{-2} at each locality.

Many of the preceding procedures to determine deposit and eruption parameters can now be addressed using modelling software. For example, Biass et al. (2019) have provided a guide to using isopach, isomass, and isopleth maps to calculate deposit parameters including deposit volume, mass, rates of thickness decay, and eruption parameters such as VEI and magnitude, using a Matlab function called *TephraFits*.

9.6.2.2 Calculating Eruption Column Height and Mass Eruption Rate Using Construction of Maximum Average Lithic and Pumice Clast Isopleth Maps

Measuring the average maximum juvenile and lithic clast size (i.e. the population of the largest clasts that fell at a given distance from the vent) is an important field technique, which involves measuring, at numerous localities, the long axis (or the average of the three main axes) of several (we suggest 5) of the largest clasts in a deposit (Carey and Sparks 1986). In some detailed sections, the variation in grain size between different layers of one deposit is measured. Usually, the sizes of the five largest clasts per square metre are then averaged, and this would closely approximate the coarsest one percentile of the clast population. However, the sample area may change depending on the coarsest grain size present at each specific outcrop (i.e. larger area for larger clasts, smaller area for thin beds). Different sampling and measuring strategies can be adopted depending on the state of the outcrop (e.g. lithified vs. unconsolidated), and a review of methods and suggested recommendations can be found in Bonadonna et al. (2013). Houghton and Carey (2015) found that variations in measuring and averaging strategy (for example averaging the three axes versus just the maximum axis of the five largest clasts) produce very little change in the reconstruction of eruption parameters, such as eruption column height (see below). We suggest that for common lapilli-sized fall deposits, the measurement in situ of the maximum diameter of the five largest clasts in an area of 1 m^2 is appropriate both for consolidated outcrops where clasts cannot be extracted, and unconsolidated deposits. However, care should be taken in consolidated deposits as the visible maximum axis may not be the actual maximum dimension in 3D, and could lead to underestimations. In such cases, measurements should be cited as the apparent maximum average clast size.

In addition, in near-vent areas, anomalously large clasts should be ignored because they are likely to be of ballistic origin and would distort the analysis and interpretation of the data. The purpose of measuring the average maximum lithic and pumice clast sizes is to analyse the behaviour and dimensions of the convective eruption column using clasts that are transported by convection in the column only. Ballistic clasts are therefore inappropriate. The same comment applies when collecting samples for standard grain size analyses such as sieving (i.e. omit obviously outsize ballistic clasts from the sieve sample), unless it is the patterns of ballistic clasts and their dispersal that are being investigated, in which case the focus should be just on those.

Care should also be adopted when measuring the maximum clast size of fragile pumice clasts, which may have broken at the time of landing on the ground surface. Breakage on impact should result in the jigsaw-fit texture of broken pumice clasts. However, for this reason, it is preferable to measure the maximum lithic clast size instead of pumice, also because the type of information in terms of transportation process and plume dynamics given by the two populations is essentially the same, since the two clast populations should be aerodynamically equivalent at each site, with their settling controlled by the competition of their terminal fall velocity and the upward directed component of the fluid velocity in the plume (see Sect. 9.4). This is of course valid only beyond the limit of the impact of ballistic clasts, where particles only fall from the upper margin of the convective region and the umbrella region. As stated above, clasts considered to be ballistic in origin should not be measured, except where their ballistic behaviour is being specifically investigated.

Average maximum pumice (or scoria) and lithic clast sizes can be plotted up as *isopleth maps* (Figs. 9.31b, c, 9.32b, c, 9.33b, c), where an isopleth is defined as a contour line along which the average maximum clast size is the same. As with isopach maps, maximum-size isopleth maps with multiple isopleths representing declining clast sizes away from the vent are important in locating the source vent of pyroclastic fall deposits, and for comparing their dispersal patterns in order to characterise the type of eruption. Carey and Sparks (1986) calculated the averaged vertical velocity for convective plumes between 6 and 43 km in height, and for both no wind and crosswind conditions. They argue that the more asymmetrical and narrow the isopach and isopleth dispersal patterns, the stronger the crosswind. They then computed the calculated velocity field in the column with the terminal fall velocities of particles to derive a set of theoretical isopleth area *versus* maximum clast (pumice/lithic) size plots (Fig. 9.36a–d). These plots can be conveniently used to compare deposits for which isopleths have been reconstructed, in order to derive the height of the umbrella cloud, i.e. the level of neutral buoyancy, H_b, and maximum height, H_T, of the eruption column, respectively (Fig. 9.36; see Sect. 9.5.1.3), where usually $H_b/H_T = 0.7$ (Sparks 1986). For large-scale eruptions, the thickness of the umbrella region is several km. Determining the height of the base of the umbrella cloud, H_b, or the level of neutral buoyancy is important because that is essentially the release height of convectively transported pyroclasts that form fallout deposits from the eruption column of the plume (Carey and Sparks 1986; Constantinescu et al. 2021). That is, pyroclasts are not released from the top of the eruption column which is several km higher than H_b.

An important aspect of the Carey and Sparks (1986) calculations is that strong eruption columns (i.e. vertical, plinian scale columns where the vertical velocity largely exceeds that of local crosswinds; Fig. 9.27a) produce isopleths that are circular or weakly elongated, even in the presence of crosswinds. This is because, at the level of neutral buoyancy, the high rate of material injected into, and laterally spreading in the umbrella cloud, attenuates the effect of winds on lateral transport. By contrast, weak plumes (Fig. 9.27b), characterised by a lower mass discharge rate ($\leq 10^4$ kg s^{-1}), are easily bent over by winds at various levels, and resulting isopleths become strongly elongated (e.g. Fig. 9.31b). This results in an enlargement of the area of single isopleths, which is strongly asymmetrical downwind. For this reason, Carey and Sparks (1986) suggested that for highly elongated isopleths, the clast dimension in the crosswind range should be considered in estimating the plume height.

Wind profiles change each day at each location, however between latitudes of 30° and 60° in both the Northern and Southern Hemispheres, circum-terrestrial jet streams are permanently present just below the tropopause, at an average elevation of 11 km, blowing from east to west. For this reason, volcanic plumes reaching these altitudes are always deflected laterally towards the west. At tropical latitudes, it is more common to have no-wind conditions so that roughly concentric, circular isopach, and isopleth patterns are possible (Fig. 9.30a; e.g. Fogo A in the Azores, Walker and Croasdale 1980, Pensa et al. 2015; Pululagua in Ecuador, Papale and Rosi 1993).

For any deposit, on a plot of average maximum clast size against distance from the vent, a line drawn along the top of the resulting scatter will show the maximum range of fragments of a given size (Figs. 9.31e, 9.33c). When maximum lithic or denser juvenile sizes are plotted, this line usually shows a sharp inflection a few kilometres from the vent, and this is thought to reflect the distance range of ballistic fragments. Maximum pumice clast sizes usually do not show a similar pronounced inflection (Fig. 9.31d), because, as stated earlier, larger pumice bombs tend to break on impact with the ground surface and owing to their low density even the largest clasts are affected by the wind to some extent. Pyle (1989) represented the grain size variations of fall deposits on a semi-log plot of the average maximum clast size versus the square root of the isopleth area (Figs. 9.29b, 9.37a, b). Similar to thickness contours, the maximum grain size of both pumice and lithics also decays exponentially with distance, so the half-distance, b_c, i.e. the distance at which the maximum size at proximal locations halves, was defined to indicate the slope of the best fit line. This slope is different for different eruptions as it reflects the plume dynamics, but is the same for a specific eruption for clasts of different densities. However, fitting curves will be offset as denser clasts will have smaller isopleth areas for a given grain size (Fig. 9.29b). The column neutral buoyancy height, H_b, was then derived from b_c according to (Fig. 9.37c; Pyle 1989):

$$-b_c \sim \frac{0.41\,H_b}{(H_b^{0.5} - 7.3)} \tag{9.20}$$

In this case, this relationship is also valid for strong columns unaffected by crosswinds, as b_c will increase in the presence of strong winds (Fig. 9.36b).

Bonadonna and Costa (2013) proposed that eruption column height can be determined most accurately from grain size characteristics of the deposits such as trends in fining of the average maximum lithic clast size with distance from the

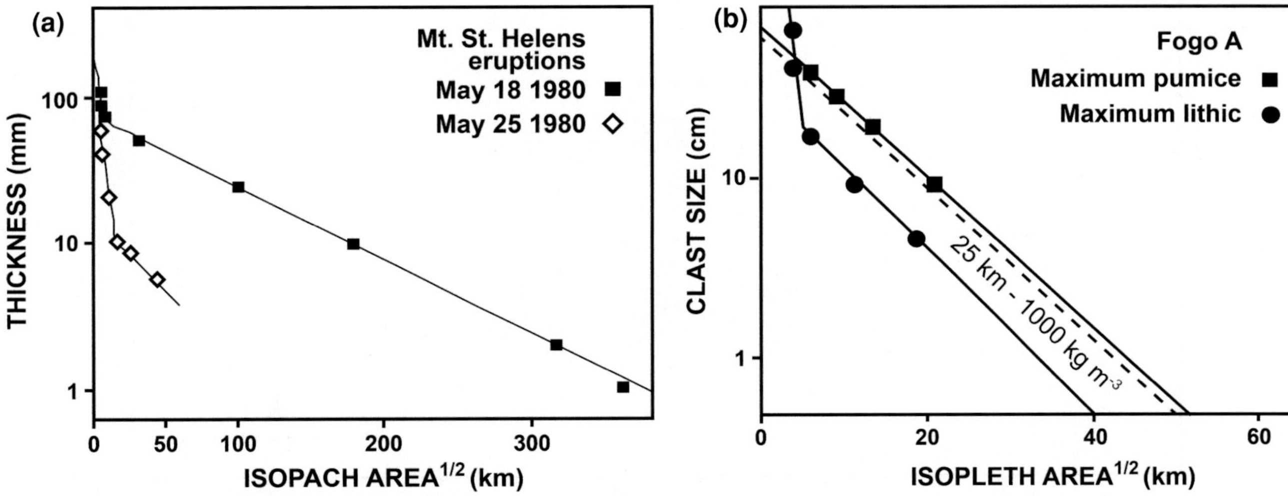

Fig. 9.29 Variations of thickness and grain size of fallout deposits across their dispersal from proximal to distal areas represented against the square root of the isopleth area. **a** exponential thinning (May 1980, Mt. St. Helens, USA); **b** exponential fining of the grain size (4.6 ka Fogo A, São Miguel, Azores, Portugal) (Adapted from Pyle 1989)

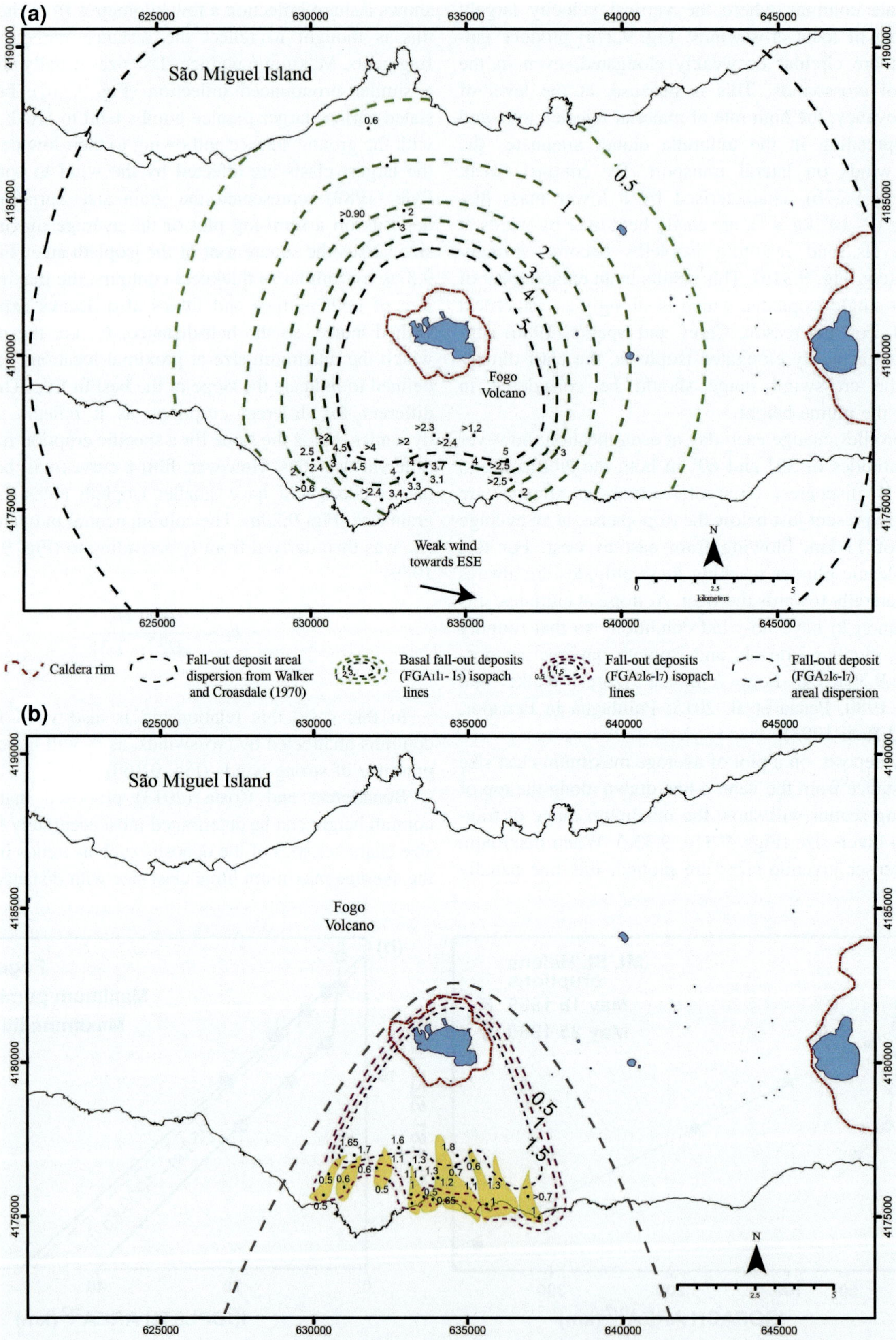

Fig. 9.30 Isopach maps of the 4.6 ka Fogo A pumice fallout deposits in cm, São Miguel, Azores, Portugal. **a** Prior to and **b** after the intra-plinian partial collapse phase. Fogo A erupted from the volcano Água de Pau with a caldera now occupied by Lago do Fogo. (From Pensa et al. 2015)

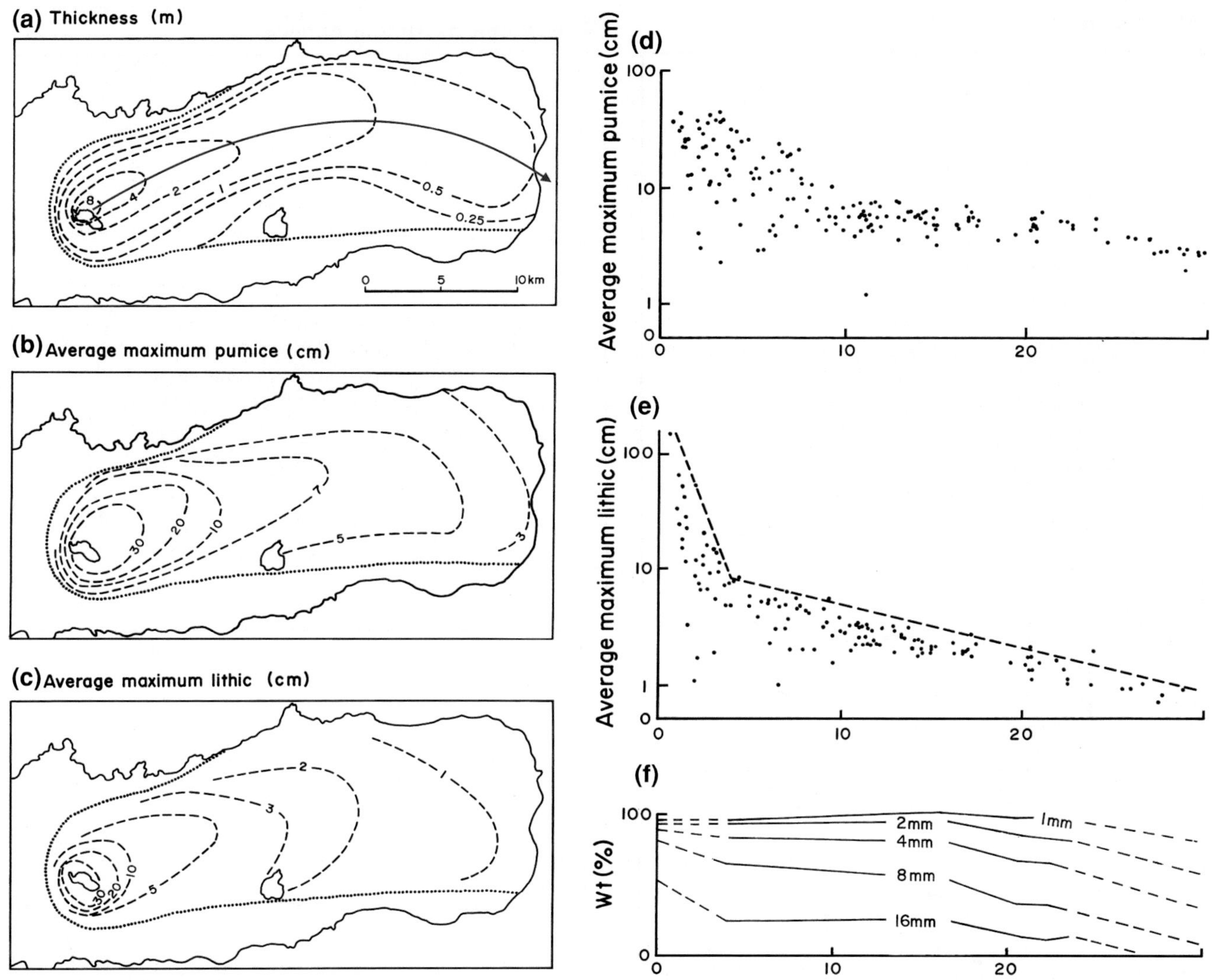

Fig. 9.31 Dispersal, thickness, and grain size characteristics of the 1563 Fogo plinian pumice fallout deposit, São Miguel, Azores, Portugal. The vent for the 1563 eruption of Água de Pau volcano was on the northwest flank of Lago do Fogo caldera. **a** Isopach map, with the axis of dispersal shown in a blue arrow. **b** and **c** Maximum size isopleth maps using the average diameter of the three largest clasts for pumice and lithics, respectively. **d–f** Maximum pumice, lithic, and total grain size variation with distance from vent. (From Cas and Wright 1987, After Walker and Croasdale 1971)

vent using Weibull probability distribution assessments. Based on analysis of a large database, they proposed that eruption column height, H_T, could be depicted as

$$H_T = 5.01\lambda_{ML}^{0.55} \quad (9.21)$$

where λ_{ML} represents the characteristic decay length scale of fining of average maximum lithic clast size.

If the maximum height of the eruption column can be calculated from lithic isopleth data (e.g. Fig. 9.36) using the method of Carey and Sparks (1986), then the volumetric eruption rate can be calculated using Eqs. 9.7, 9.8, or interpolated from Fig. 9.25 (Wilson et al. 1978) or Fig. 9.26a (Mastin et al. 2009). Alternatively, observations of column height versus magma volume eruption rate from modern observed eruptions (e.g. Fig. 9.26b; Mastin et al. 2009), even allowing for crosswinds (Fig. 9.26c; Woodhouse et al. 2013), can be used to assess mass eruption rates of past eruptions. The deposit volume and mass can be calculated from areal extent, the isopach data, and the determination of the deposit bulk density in the laboratory. The bulk volume or mass of accessory lithic clasts can be subtracted after the fraction of lithic clasts as part of the whole deposit has been determined from componentry determinations, or even using visual estimate charts, leaving the volume and mass of the residual magmatic components. Using vesicularity or density data of juvenile pyroclasts and the density of magma of the appropriate composition (e.g.

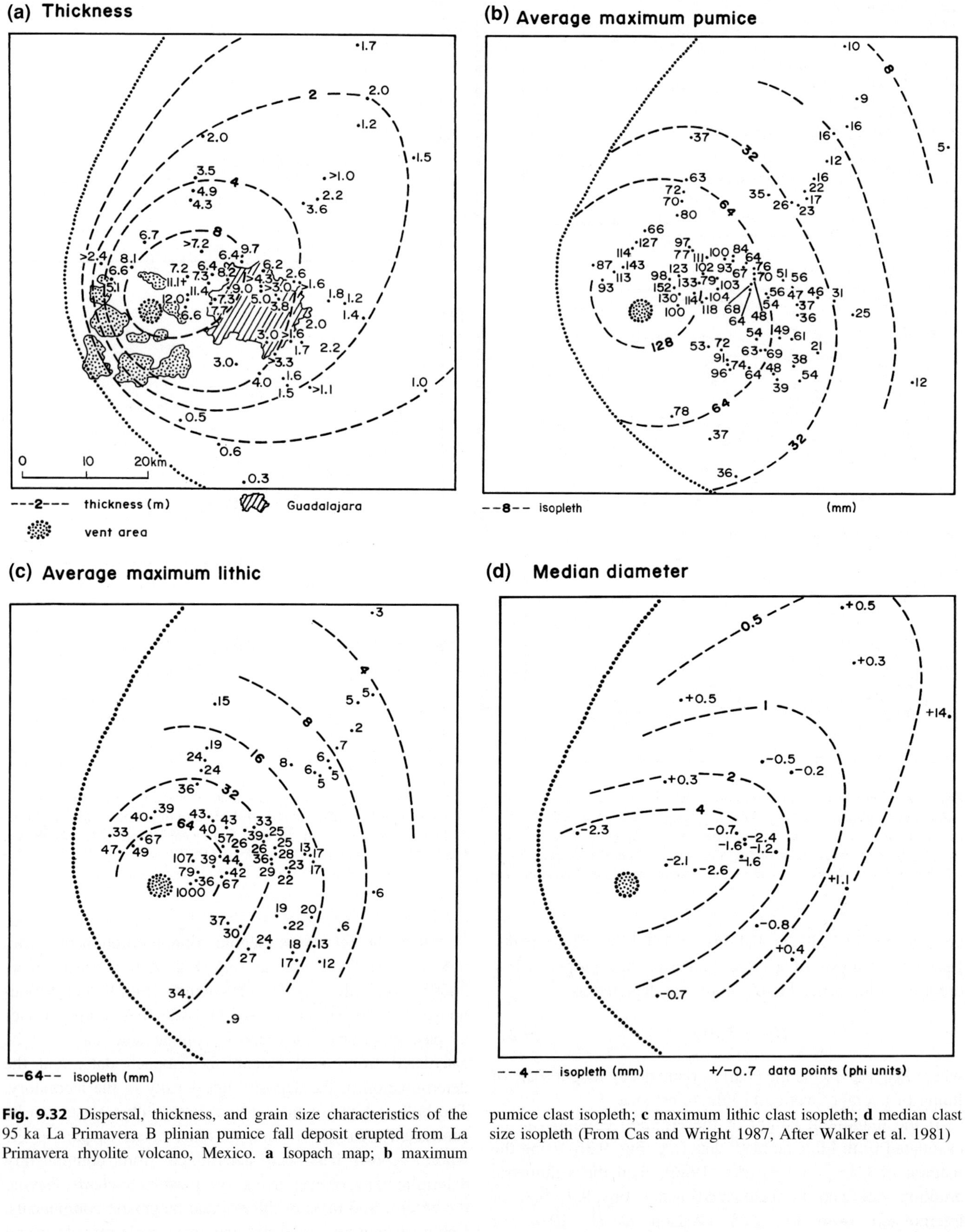

Fig. 9.32 Dispersal, thickness, and grain size characteristics of the 95 ka La Primavera B plinian pumice fall deposit erupted from La Primavera rhyolite volcano, Mexico. **a** Isopach map; **b** maximum pumice clast isopleth; **c** maximum lithic clast isopleth; **d** median clast size isopleth (From Cas and Wright 1987, After Walker et al. 1981)

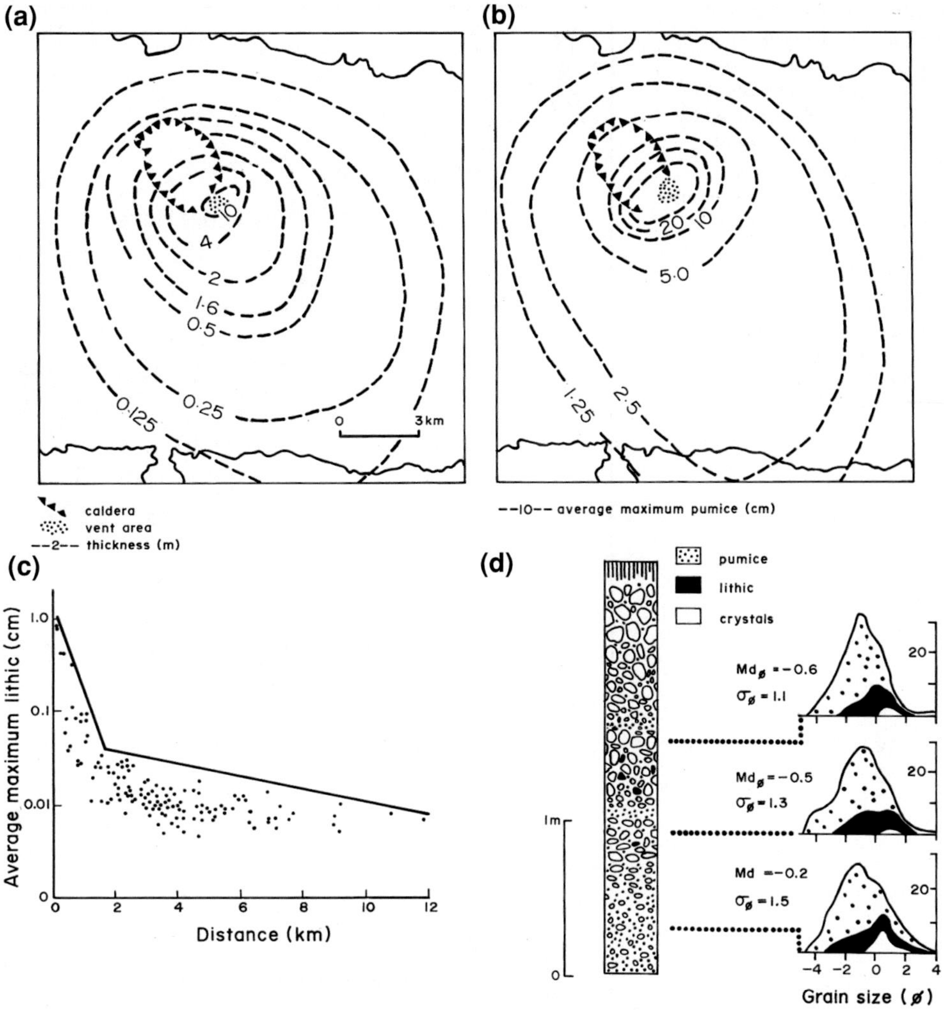

Fig. 9.33 Sub-plinian pumice fallout deposit from Terceira, Azores, Portugal. **a** Isopach map giving thickness in m. **b** Average maximum diameter of three largest pumice fragments in cm. **c** Average maximum diameter of three largest lithic fragments against distance from source. **d** Sect. 1.8 km SE of source on dispersal axis. Frequency curves of weight percentage against grain size show proportions of pumice, crystals, and lithics for three sieved samples, together with vertical variation in Md_{ϕ} and σ_{ϕ}. (From Cas and Wright 1987, After Self 1976)

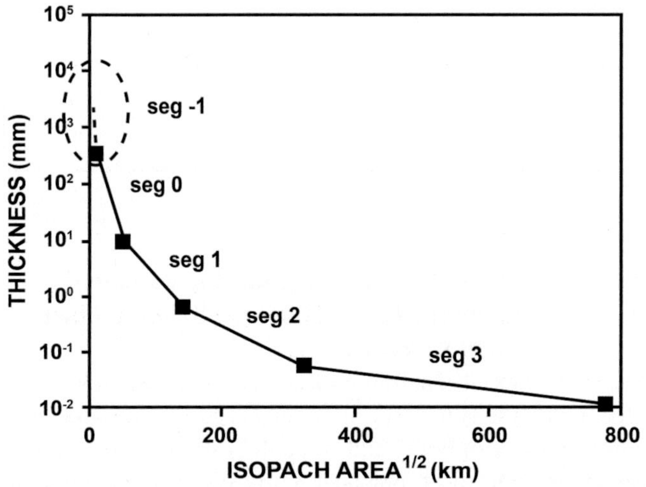

Fig. 9.34 Generalised diagram of log of thickness (T) versus square root of the isopach areas (A) showing possible segmentation reflecting different transport and sedimentation processes. Close to vent, the thickness is dominated by ballistic clasts, whereas laterally the changes in slope mostly depend on settling from the umbrella of particles characterised by decreasing values of Reynolds number. (Adapted from Houghton and Carey 2015)

rhyolite–dacite ~2,400 kg m^{-3}, andesite ~2,550 kg m^{-3}, and basalt ~2,700 kg m^{-3}), the erupted dense magma equivalent mass can then be calculated. Knowing both the magma discharge rate and the magma mass erupted, then the average duration of the eruption can be calculated.

However, fallout deposits are usually complex and so isopleth maps need to be defined for specific stratigraphic levels if possible, so that variations in intensity during an eruption can be assessed (e.g. 472 AD Pollena eruption; Sulpizio et al. 2005; the 4.6 ka Fogo A eruption, Pensa et al. 2015; the 232 AD Taupo eruption, Houghton et al. 2014; the 1257 AD eruption of Samalas, Vidal et al. 2015). Alternatively, if, as commonly happens, the largest maximum sizes are measured as representative of an entire fall succession, then calculations of the column height and mass discharge rate would refer to an average that slightly underestimates eruptive peak conditions.

Finally, it is important here to stress that even for deposits for which a very limited number of outcrops are available, it is possible to retrieve some important information by making

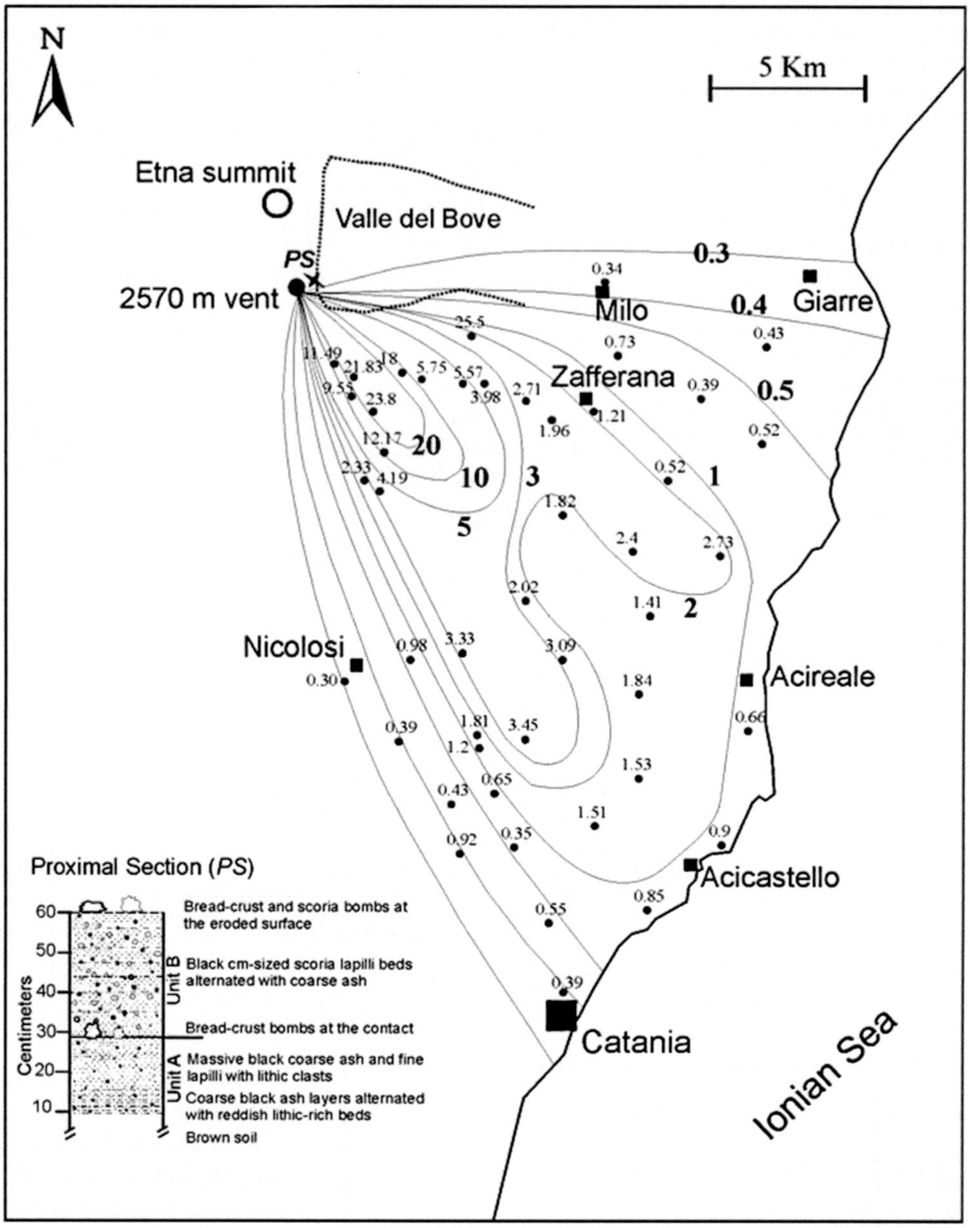

Fig. 9.35 Isomass map of the 2001 Etna fall deposits, showing the contours that define areas where the mass of the fallout deposit is the same, obtained by calculating the total mass of tephra per unit area at each locality. Data points and contours record deposit mass in kg m^{-2}. Isomass contours decrease in value away from the vent, as thickness, grain size, and componentry change with distance from vent. (From Scollo et al. 2007)

conservative assumptions based on clast dimension and distance from the vent. This is the case of the fall deposits of the most recent Holocene, plinian eruption at Mt Melbourne, Antarctica. There, using just two outcrops, one at the vent and the other in the ice cap 30 km to the northeast, Giordano et al. (2012) assumed conservatively that the two were along the axis of maximum dispersal, and using crosswind conditions 30 m s^{-1}, they determined a minimum column height of 14 km. At polar latitudes, the troposphere–stratosphere boundary is at just 5 km, and this implies significant stratospheric injection by the eruption column.

9.6.2.3 Calculation of Exit (Muzzle) Velocity: Maximum Grain Size and Ballistic Clast Travel Distance

Measurements of maximum clast size of ballistic clasts are used to analyse some aspects of the energetics of pyroclastic fall-forming eruptions and reliable estimates of muzzle velocities. For most practical purposes, this is going to involve only lithics much greater than 20 cm in diameter, which are so heavy that they follow largely parabolic ballistic trajectories and are unaffected by wind drift or turbulent convection in the eruption column. The distance at

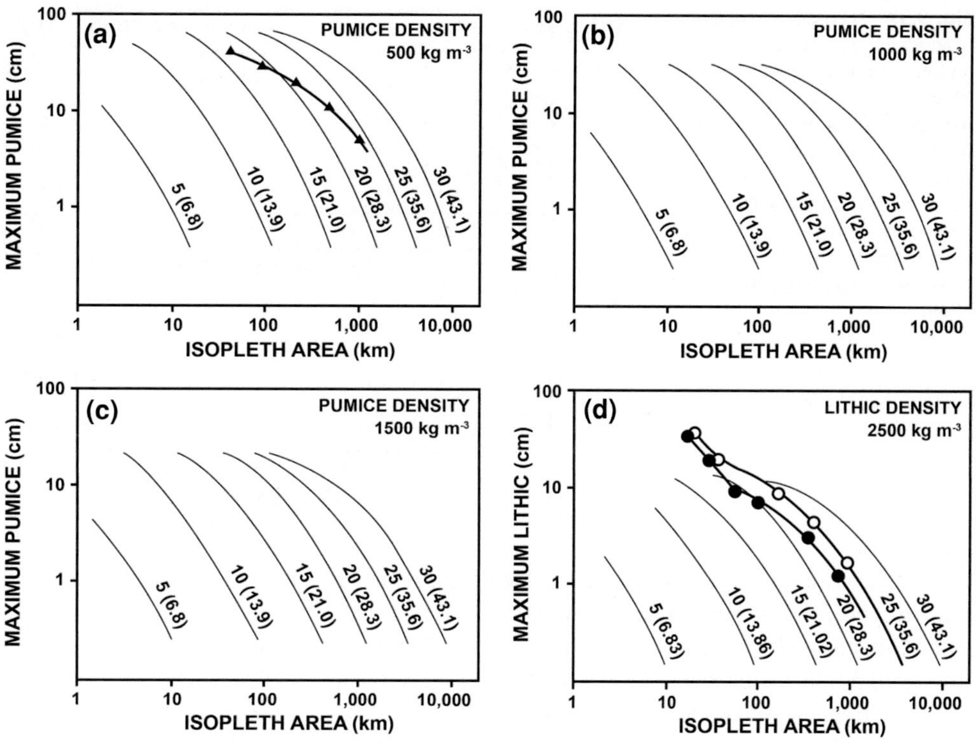

Fig. 9.36 Theoretical plots of isopleth area–clast size for eruption column heights ranging 7–43 km, with pumice densities of **a** 500 kg m^{-3}, **b** 1000 kg m^{-3}, **c** 1500 kg m^{-3}, and lithic density of **d** 2500 kg m^{-3}. Values in parentheses along curves correspond to H_b and H_t (in km). (From Carey and Sparks 1986)

which ballistics fall from the vent (that is, their range) can be used to estimate initial gas thrust velocities from the vent, or the muzzle velocities of the ballistics from the vent. Ballistic clasts can reach maximum distances of 4–5 km from the vent in the case of plinian eruptions. Wilson (1972) presented tables of calculated ranges for individual ballistic clasts of varying radii and density, launched at speeds between 10 and 1000 m s^{-1} and various angular elevations. Mastin et al. (2009) released a user-friendly computer program ("*Eject*" http://pubs.usgs.gov/of/2001/0045/) that allows calculation of the distance of landing from the vent for variable sets of parameters of the analysed clast (density, shape, drag properties) and eruption conditions (exit velocity, angle of ejection, atmospheric conditions, and volcano shape). Field measurement may also include the angle of impact and crater size and coupling with theoretical analysis facilitating the reconstruction of the relevant eruptive parameters (Alatorre-Ibarguengoita et al. 2006; Maeno et al., 2011; Beard et al. 2014; Fitzgerald et al. 2014; Tsunematsu et al. 2016; Biass et al. 2016; Kilgour et al. 2019).

For plinian deposits, the muzzle velocity can be conveniently calculated from measurements of maximum clast size using the equation of Wilson (1972, 1978):

$$u_0^2 = \frac{8 g\, r_0\, \rho_p}{3\, C_D\, \rho_0} \tag{9.22}$$

where u_O is the velocity of the gas (or muzzle velocity), C_D is the drag coefficient (~ 1 for plinian velocities), ρ_O is the effective density of the volcanic gas in the vent (0.25 kg m^{-3} for dusty gas), g is the acceleration due to gravity, r_O is the radius of the average maximum clast at vent, and ρ_p is its density (generally taken as 2,500 kg m^{-3} for lithics). Because of inaccessibility, it is usually impossible to measure r_O, but it can be estimated by plotting a graph of the product $r_O\ \rho_p$ of the largest clasts against distance from the vent, and extrapolation to the zero range (Wilson 1978). Applying Eq. 9.22 gives the maximum muzzle velocity during the eruption, and the average velocity is taken as half this value.

The transit of an object through a fluid is retarded by a drag force. This will apply to ballistic clasts as they leave a vent, and as they travel downwards, as well as to pumice clasts rising buoyantly as well as subsequently settling. Even rising eruption columns and thermal gas plumes will experience significant viscous drag as they rise. Given the low density and viscosity of air, the drag force is low, compared with the passage of clasts through water.

The viscous drag force, F_D, acting on a particle moving through a fluid is defined as

$$F_D = 0.5 \rho_p v^2 C_D A \tag{9.23}$$

where ρ_p is the density of the particle, v is the relative velocity of the particle relative to the fluid (which is the actual velocity in still air or water), C_D is the fluid drag coefficient, and A is the cross-sectional area of the particle. The drag coefficient has been defined in Eq. 9.3 above, and

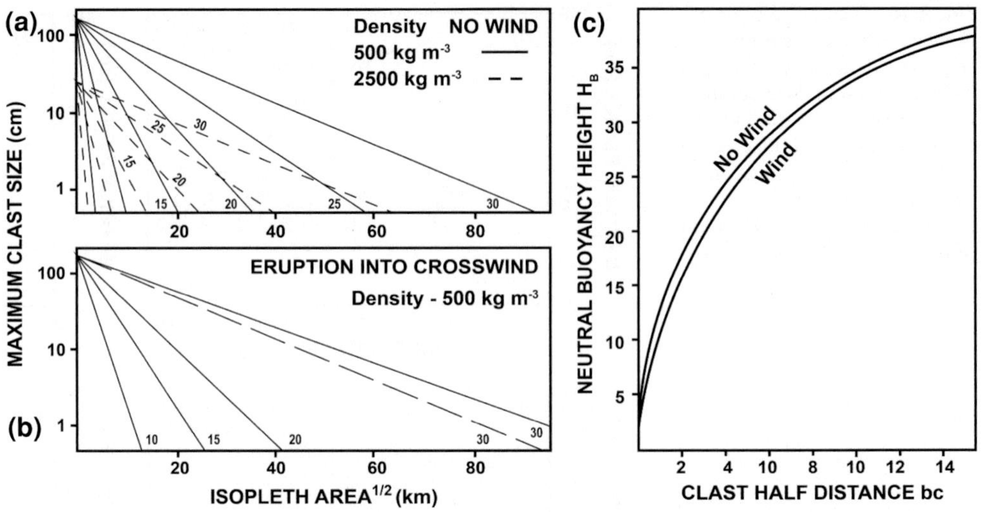

Fig. 9.37 Theoretical isopleths from Carey and Sparks (1986) replotted by Pyle (1989) on a ln (thickness) versus area$^{1/2}$ diagram for cases with **a** no wind, and **b** crosswind (maximum wind speed of 30 m s^{-1} at tropopause); numbers refer to predicted H_b; **c** relationship between maximum level of neutral buoyancy H_b and clast half-distance b_c (From Pyle 1989)

it varies according to particle shape, surface area, and particle Reynolds number. Alatorre-Ibargüengoitia and Delgado-Granados (2006) determined that C_D varied from as low as 0.62 for semi-round porous clasts (pumiceous) to 1.01 for irregular rough clasts. The drag coefficient can also be expressed as a function of the Reynolds number, since both are dimensionless:

$$C_D = F_D(Re) \tag{9.24}$$

where Re is the particle Reynolds number, defined as

$$Re = \frac{\rho_p d v}{\mu} \tag{9.25}$$

where ρ_p is the particle density, d is the particle diameter, and μ is the kinematic fluid viscosity. However, the exact nature of that function varies according to the characteristics of the particle (shape, density), and has to be determined experimentally in each case. For very low Reynolds numbers (<1, laminar fluid state), $C_D = 24/Re$.

In reality, although some ballistic clasts do behave unimpeded as individual ballistics, many (20% +) collide in transit, so that their paths are modified after the collisions (Vanderkluysen et al. 2012; Fitzgerald et al. 2014; Tsunematsu et al. 2014; Kilgour et al. 2019). Allowing for clast collisions and fluid drag effects, modelled ballistic particle exit velocity (muzzle velocity) can be represented as follows (Alatorre-Ibargüengoitia and Delgado-Granados 2006; Fitzgerald et al. 2014):

$$dv/dt = \frac{A\, C_d\, \rho_a (v-u)|v-u|}{2m} - g \tag{9.26}$$

where v is particle velocity, u is wind or gas velocity, t is time, A is the surface area of the particle, C_d is the drag coefficient, ρ_a is air density, m is particle mass, and g is the gravitational acceleration. Field data on the distribution of ballistic blocks around a source crater, distance from the source, diameter, density, and mass of the ballistic clasts, and where possible angle of impact, have been developed into a ballistic trajectory model by Tsunematsu et al. (2014) and Fitzgerald et al. (2014) that also accounts for some collisions of ballistics (or "ballistics behaving badly"—Vanderkluysen et al. 2012). Fitzgerald et al. (2014) mapped the distribution of ballistic blocks ejected during the 2012 hydrothermal blast on Tongariro volcano, New Zealand and calculated the exit velocity of ballistics to be 200 m s^{-1}, with a gas flow velocity of 150 m s^{-1}. Tsunematsu et al. (2014) calculated ballistic clast velocities for the 2014 Ontake volcano blast in Japan to be between 145 and 185 m s^{-1}, with drag coefficients varying between 0.62 and 1.01. Kilgour et al. (2019) generated 3D projections of ballistic trajectories (Fig. 9.38) and calculated the maximum velocities for ballistic blocks during the 2016 night-time hydrothermal explosion on White Island/Whakaari volcano, New Zealand, to have been 65 m s^{-1}. Giordano and De Astis (2021) calculated the maximum exit velocity of ballistics to be 160 m s^{-1} for the July 3 Stromboli 2019 vulcanian eruption by measuring distance, average diameter and density of equant lava blocks, using the free software E-ject (Mastin et al. 2009).

9.6.2.4 Definition of Fragmentation Style: Whole-Deposit Grain Size

Total grain size distribution (TGSD) of fall deposits is a very important parameter because it reflects the initial size distribution in the eruption column. The basis for the evaluation of the TGSD is the retrieval of areally well-distributed grain size data that cover the entire preserved dispersal of the studied fallout deposit.

Mechanical or granulometric analyses (sieving) of samples laboriously collected in the field are used as the main source of data when examining the grain size variations in non-welded and unconsolidated pyroclastic deposits.

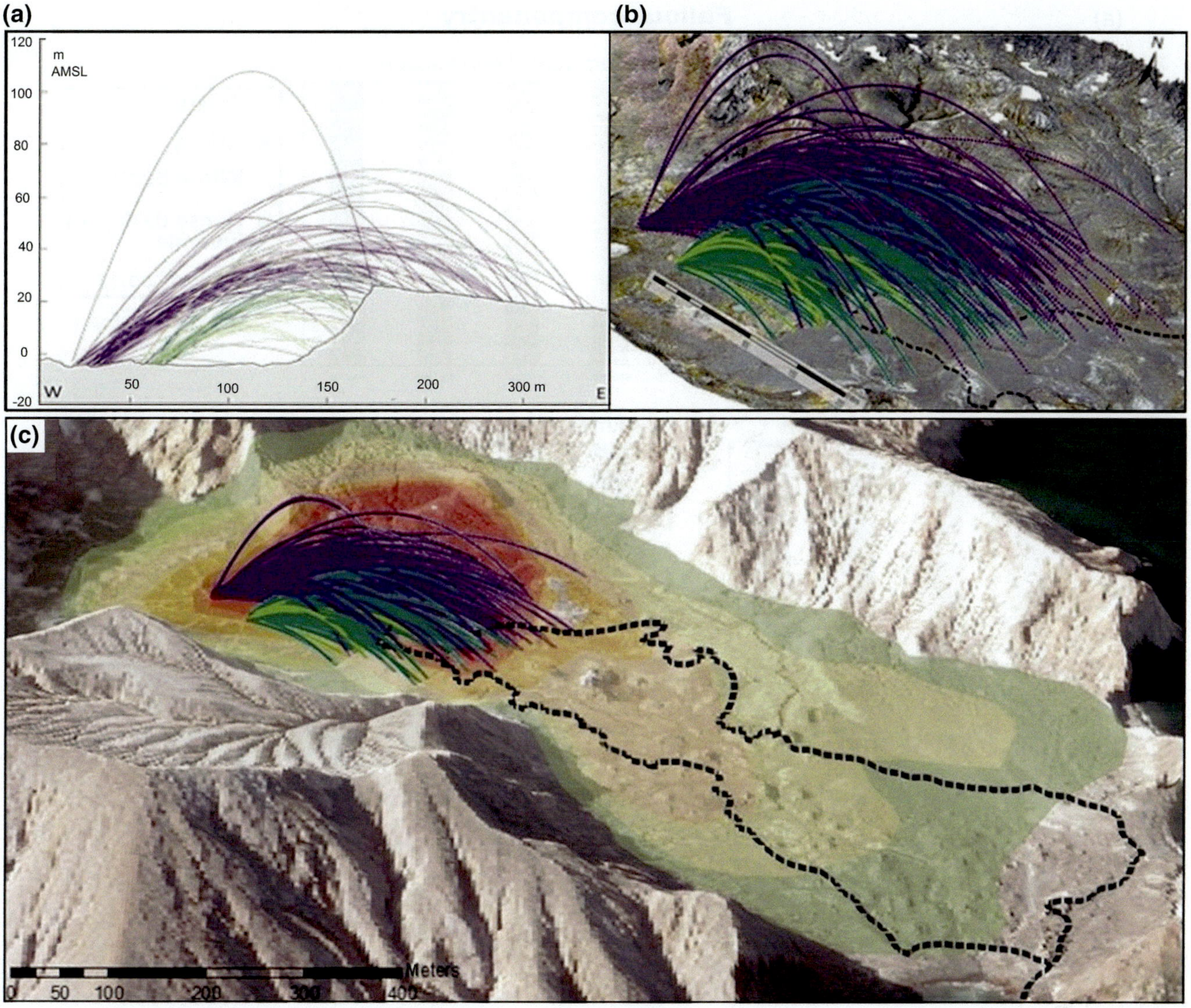

Fig. 9.38 Ballistic trajectory models for ballistics ejected during the 2016 hydrothermal explosive eruption of Whakaari/White Island volcano, New Zealand. **a** 2D trajectory model of three main colour-coded explosive pulses. **b** 3D model for the same ballistic sets. **c** As for **b** showing a crater-wide perspective, and on the floor of the crater, shaded in pastel colours reflecting thickness changes, the distribution of the thin pyroclastic surge deposit that was also formed during the eruption. Dashed line is the tourist pathway which was to become significant in the tragic hydrothermal explosion of 9 December 2019. (From Kilgour et al. 2019)

Generally, the methods described by Walker (1971) are followed. Analyses are made with a set of sieves with mesh sizes spaced at one-phi (Φ) intervals (where $\Phi = -\log_2 d$, d being the grain size in millimetres) and ranging in size from −5 to 4 phi (32 to 1/16 mm). The sieving is done manually, rather than on an automated shaker, to minimise the breakage of fragile pumice clasts. Obviously, the sample size needed to give a representative sieve analysis of a deposit becomes larger with increasing maximum grain size, and is also larger if the sorting in a deposit is apparently poor. For coarse plinian fall deposits within a few km of the vent, samples between 0.5 and 2.0 kg should be collected. In some very poorly sorted deposits, larger clasts can be measured (for instance, by linear traverses) and then much smaller samples of just the matrix are taken. For fine-grained deposits only containing ash-sized particles (<2 mm), samples weighing a few gm to a few tens of gm may be adequate. Laser diffraction analysis can provide very accurate data for very fine-grained samples or the finest ash sizes, but these provide volumetric distributions that must be converted to weight % equivalent sizes to be comparable with grain size curves obtained with mechanical sieving procedures (Rose and Durant 2009). Grain size data are presented as frequency histograms of grain sizes (Figs. 9.19, 9.39b), or as frequency–grain size curves (Fig. 9.40a, b) or as cumulative frequency plots (Fig. 9.40d).

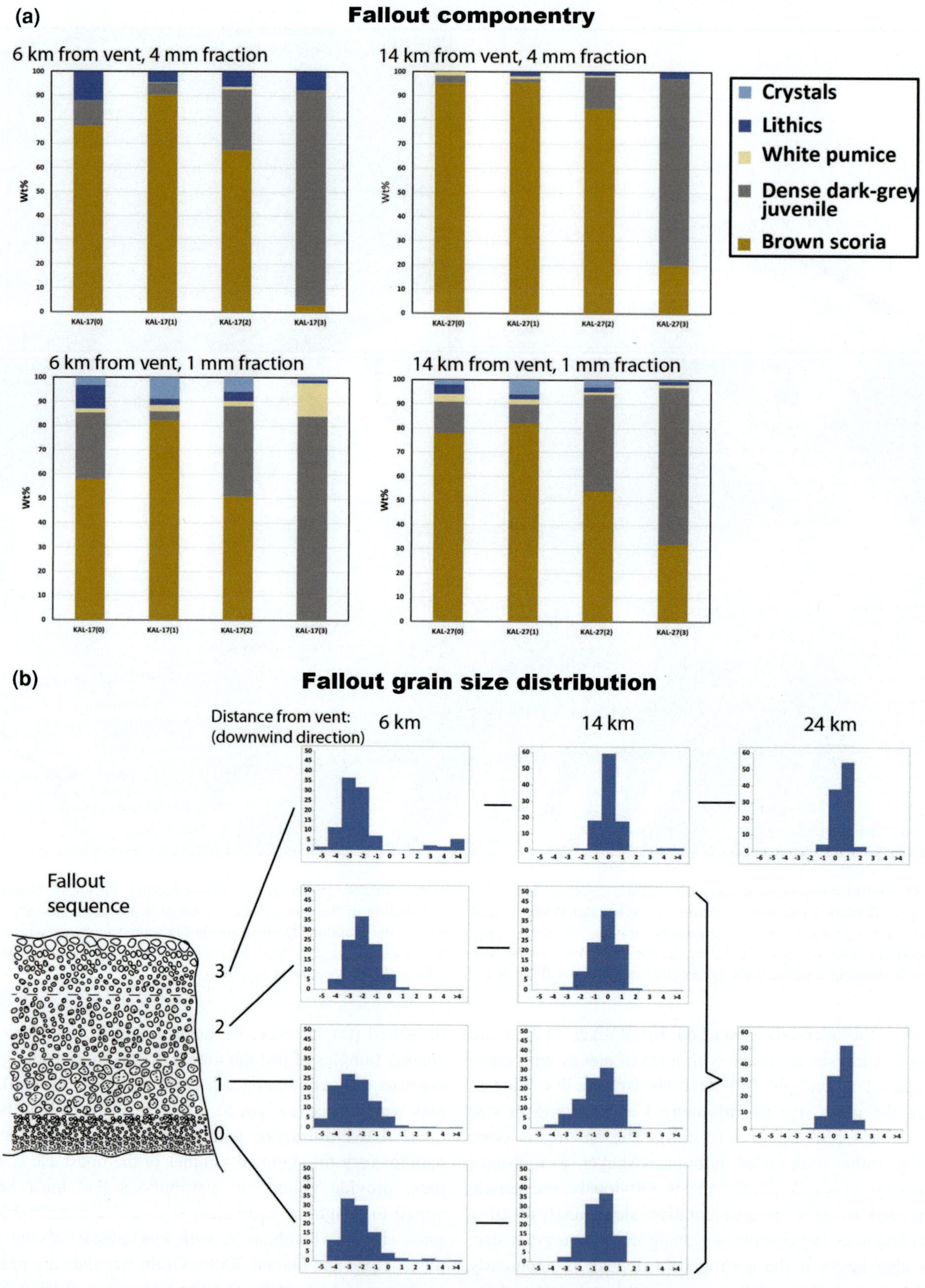

Fig. 9.39 **a** Componentry and **b** grain size downwind variations with distance from vent, from the 22–23 April 2015 fallout deposit from the eruption from Calbuco volcano, Chile (see Fig. 9.10a). Note the fining and improved sorting of the deposits in a downwind direction, the latter represented by a narrow histogram profile. (From Castruccio et al. 2016)

The relative proportions of the different components in a pyroclastic deposit reflect its mode of formation, and details of the transport processes. To determine the proportions of different components (e.g. pumice, crystal fragments, lithic fragments, vitric ash particles), these need to be counted. Different techniques are used to separate pumice, crystals, and lithics in the different size classes. The larger size classes (>4 mm) are hand-picked, but forceps are used for the fine end of this size range. Quite often with the >16 mm size classes this is done in the field, if field sieving and weighing can be carried out. For the size classes 2 mm to 0.5 mm, hand-picking is carried out under a binocular microscope using forceps or a camel-hair brush. It is important to separate out and accurately weigh on an analytical balance as much material as is needed to produce a statistically sound result (100 clasts). In the finest size classes, grains are usually counted either under a binocular microscope (0.25 and 0.125 mm) or under a petrological microscope (0.063 and <0.063 mm). Results are then presented in the frequency histograms to show the proportion of each component per grain size class (Figs. 9.19, 9.39a). Histograms can then be presented on the same diagram showing their variation with distance (Figs. 9.19, 9.39a), and then the contribution of each component to the total grain size must be weighted according to the representativeness over the total volume.

Downwind variations in grain size can also be plotted as multiple grain size frequency curves at different distances from the vent on the same diagram (Fig. 9.40a, b), which clearly depict downwind fining with distance. In addition, various methods have been employed to integrate over the whole-deposit volume the total grain size distribution (TGSD) measured at individual sites, but usually they involve dividing the isopach map up into different segments. Grain size data for the segments are averaged and then weighted according to the volume of each sub-unit (e.g. Murrow et al. 1980). The most effective way of segmentation and evaluation of the contribution of distal deposits for well-preserved young deposits proved to be the Voronoi tassellation, which places the grain size and thickness of individual outcrops at the centre of the relative Voronoi polygon (Fig. 9.40c). Details of this method of segmentation and grain size integration are given in Bonadonna and Houghton (2005), and also applied by Alfano et al. (2016) to the 2008–2013 Chaitén eruption deposits in Chile.

It is important here to recall that reliability of TGSD determinations relies on the existence of a large data set and accurate volume estimations, for which large uncertainties relate to the usual lack of both very proximal and distal data, as discussed in Sect. 9.6.2.1. A spreadsheet for the calculation of the TGSD is available in Biass and Bonadonna (2014). TGSD is fundamental for modelling plume dynamics (e.g. Girault et al. 2014), and an evaluation of the reliability of TGSD reconstructions from field data is provided in Costa et al. (2016b).

Early applications of TGSD calculations focused on using just the juvenile vitric component (pumice, vitric ash), which then allows calculation of the erupted magma volume and dense rock equivalent but leads to underestimates of total deposit volume because lithic and crystal components are excluded. This has been addressed by Mele et al. (2020), who undertook TGSD analyses of all three components from two deposits from Campi Flegrei caldera, Italy, including distal fine ash samples, and then also integrated them to generate a whole-of-deposit TGSD. They produced significantly improved and higher deposit volume calculations, and also found that this provided significant insights into fragmentation, dispersal, and settling processes. In particular, their analyses showed that accessory lithic components decreased in size to medial distances, and then became an insignificant component more distally, so reflecting a source rock textural and strength control on the fragmentation intensity of country rock. Crystal fragments were uniform in size and were most abundant in medial distances, and then were also insignificant in more distal localities, again indicating strength and textural control on the degree of fragmentation of magmatic crystals. Distal ashes were dominated by fine vitric ash particles, the size of which reflected the original vesicle size, with ash particle sizes being at least half the size of parent vesicles (Mele et al. 2020).

Whole-deposit cumulative frequency grain size curves for some deposits of highly explosive eruptions are given in Fig. 9.40d. What is surprising is the high proportion of fines found in some plinian deposits, including the Taupo ultra-plinian deposit. These only seem to differ from the phreatomagmatic deposits in having a tail of coarser grain sizes. Walker (1981a) suggested that the markedly different appearances of plinian and phreatoplinian deposits in the field reflect a different depositional process rather than the degree of fragmentation. However, details of the mechanisms of magma fragmentation and how they operate during explosive eruptions are still relatively poorly understood, and the clarification of grain size relationships and pyroclast morphology requires more work (see Chap. 4; e.g. Rust and Cashman 2011; Kueppers et al. 2006; Huebsch et al. 2023).

For lithified deposits, such as young sintered deposits, young variably welded deposits, and older diagenetically lithified deposits from the rock record, a different approach is required (Jutzeler et al. 2014, 2016). The preceding approach is only valid for non-lithified deposits that can be sieved. However, many deposits are variably lithified and cannot be disaggregated without damaging clasts and so distorting the grain size distribution. Various approaches have been used to undertake grain size analysis of lithified clastic rocks, and then produce a factually accurate grain size

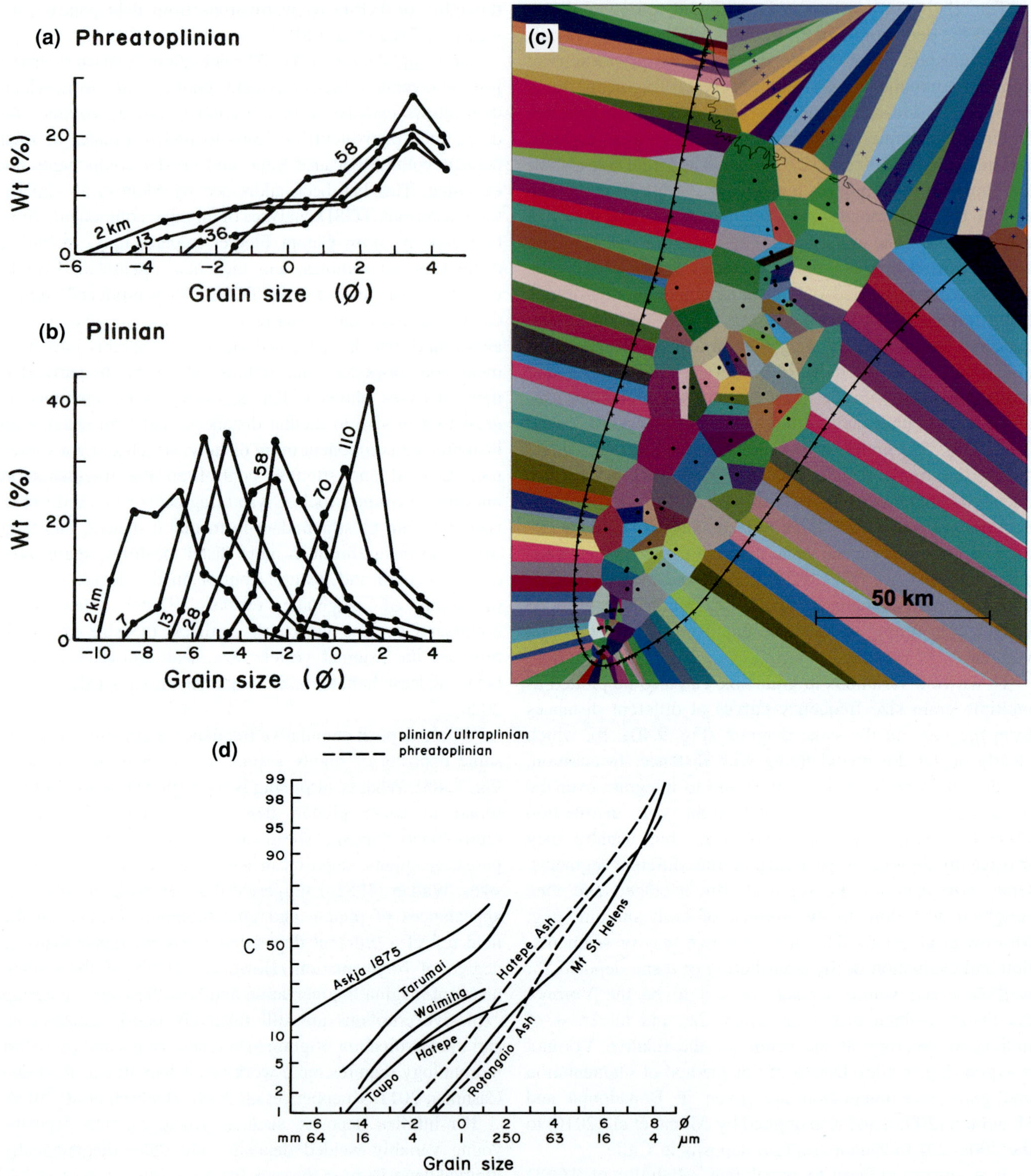

Fig. 9.40 Grain size data from the 1875 Askja fallout deposits, Iceland, showing contrasts between the **a** phreatoplinian (layer C) and **b** plinian (layer D) deposits. Frequency–grain size curves are for samples collected at various distances downwind. (From Cas and Wright 1987, After Sparks et al. 1981). **c** Voronoi tassellation applied to the 1995 Ruapehu sub-plinian deposit for the areal weighting of grain size data (black dots indicate sampling sites and the black line the areal extent of the tephra deposit) in the calculation of the total grain size distribution (From Bonadonna and Houghton 2005). **d** Total grain size distributions (TGSD) of some pyroclastic fall deposits formed by highly explosive silicic eruptions (see Table 9.2). C is weight percentage coarser than the size stated on the x-axis. (From Cas and Wright 1987, After Walker 1981d, Carey and Sigurdsson 1982; also see Cashman and Scheu 2015)

distribution, from which statistical data can be derived. The latest developments focus on applying functional stereology to 2D rock slab surfaces, high-resolution photos, and thin sections of lithified volcaniclastic deposits of any origin (fallout, ignimbrite, sediment, hyaloclastite, etc.; Jutzeler et al. 2014, 2016). In addition to grain size distribution, the method used involves documentation of the componentry, the grain size, and the shape parameters of each component.

Jutzeler et al. (2014) outline a method that involves using a polished or lacquered flat 2D surface of a slab of rock, or a thin section, or a high-resolution photo of either, or of an outcrop. Using image manipulation software such as Photoshop or ImageJ, single clasts are manually outlined, and colour-coded according to component type. Then, using image analysis software *AnalySIS,* clast parameters, including area, perimeter, maximum and intermediate diameters, aspect ratio, and shape factor, can be determined. From these, volumes of clasts can be extrapolated, and then by determining or assigning densities to clast types, weight % of components and by grain size can be assigned. The practical cutoff for grain size definition is 2 mm, perhaps as low as 1 mm in some cases, so this is a limitation. The fine grain size population cannot be determined in detail. However, if a fining grain size pattern or trend can be determined from the coarser grain sizes, the weight % in the finest grain size classes or bins can be extrapolated. Then, grain size statistical parameters can be determined.

9.6.2.5 Determining Conduit Decompression Rates: Density, Porosity, and Bubble Number Density of Juvenile Clasts

Pyroclastic fallout deposits can provide important information about conduit processes. Much of this information is stored in the vesiculation patterns of its juvenile fragments. In Chap. 4, we discussed the conduit conditions under which fragmentation occurs, and the different types and sizes of pyroclasts.

The density and porosity of juvenile clasts can be used as proxies for the bulk clast vesicularity and can be measured easily for selected grain sizes using the method of Houghton and Wilson (1989) and Shea et al. (2010). Density measurements of juvenile clasts are usually done in the size classes between 16 and 32 mm, by first weighing each clast in air, then wrapping each clast with a piece of water-proof paraffin film of known mass in water, then immersing the wrapped pumice clast in water, and measuring the mass of the immersed clast in water (Fig. 9.41). The bulk pumice density is then calculated using

$$\rho_{bulk} = \frac{m_{air}}{[m_{air} - (m_{water} - m_{filmwater})]} \tag{9.27}$$

Using the known dense magma density (dense rock equivalent, DRE), the bulk vesicularity of the pumice clasts can be determined, using

$$Vesicularity = \frac{(\rho_{DRE} - \rho_{bulk})}{\rho_{DRE}} \tag{9.28}$$

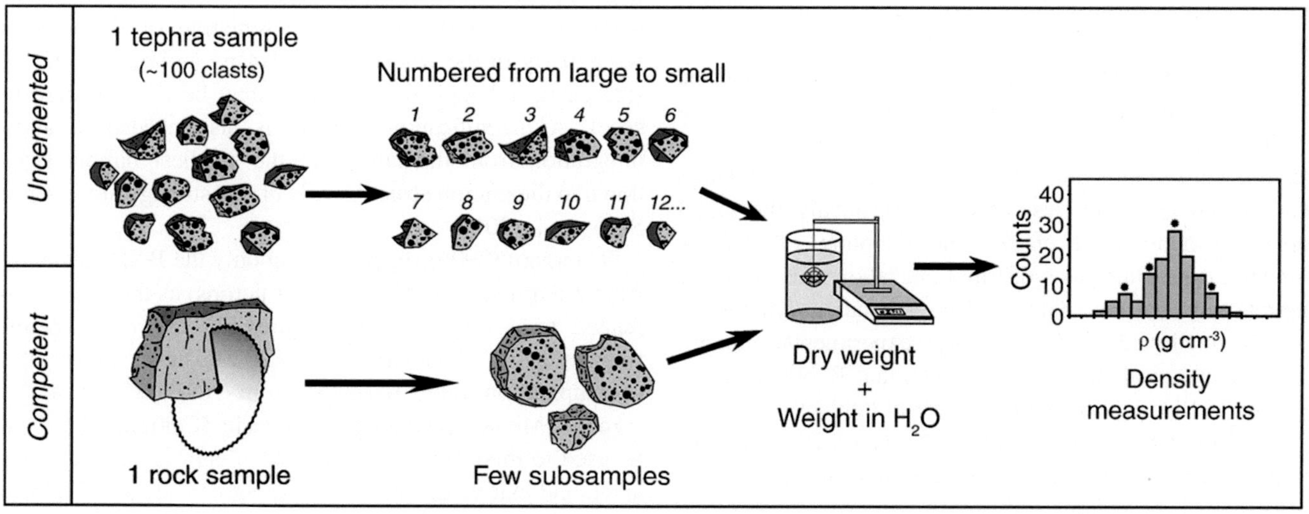

Fig. 9.41 Schematic diagram illustrating the method for measuring clast density in the laboratory, including pumice and scoria clasts (From Shea et al. 2010)

The number of clasts measured to produce statistically representative results is usually taken at 100 for a given grain size. Density and vesicularity data are then plotted as histograms, so that the statistical mode or the modes and the standard deviation can be determined. All of the above data entry, calculations, and plotting can be easily done using a programmed spreadsheet.

Usually, felsic plinian deposits show Gaussian or normal distributions of densities with relatively narrow dispersions, as a result of relatively steady conditions of conduit and fragmentation processes in viscous magmas. Such narrow ranges of density correspond to common pumice vesicularities of 60–85% by volume (Houghton and Wilson 1989). By contrast, sub-plinian or weaker explosive eruptions are characterised by non-sustained, pulsatory behaviour and much more variable conduit and fragmentation conditions that result in a usually much wider range of densities and possibly more than one mode. This variability may increase for explosive eruptions of low-viscosity magmas, where post-fragmentation relaxation of the glass may induce various degrees of rheomorphism and resorption of bubbles. Ranges of densities in such cases are wide and correspond to vesicularities of 10–70% by volume (e.g. Vinkler et al. 2012).

Total porosity is usually measured with He-pycnometers. This method allows measurement of the volume of the solid and both the volume of the connected and non-connected voids (after crushing of the material) and eventually the porosity. The degree of bubble connection in lapilli-sized pumice clasts is very important and is usually very high for plinian eruptions (up to 98%). Rust and Cashman (2011) propose that grain size distribution ultimately reflects the permeability of the fragmenting foam, where non-connected domains of bubbles will explode by gas overpressure, generating fine ash, whereas at increasing degrees of interconnection permeability allows degassing, and fragmentation will become less and less efficient, producing the larger lapilli to bomb-sized clasts. The coupling of density, porosity, and grain size distributions allows us to make inferences about the conduit and fragmentation dynamics.

At a smaller scale of observation, experimental, numerical, and analytical studies have shown that Bubble Number Density (BND; defined as the number of bubbles or vesicles per cm^3 or m^3; also see Chap. 4, Sect. 4.8.4) of juvenile clasts depends on decompression rates and on the physical properties of the magma (diffusivity, viscosity, surface tension, volatile content; Toramaru 2006, and references therein). Existing data sets show that BNDs vary several orders of magnitude with increasing explosive intensity from 10^4 cm^{-3} in strombolian eruptions, up to more than 10^9 cm^{-3} in plinian eruptions (Fig. 9.42; e.g. Campagnola et al. 2016; Simmons et al. 2017a, b; Romano et al. 2020).

In particular, Toramaru (2006) proposed that if homogeneous nucleation occurs (i.e. bubble nucleation at high degrees of supersaturation in an aphyric magma; see Chap. 4), then for a given magma composition with a determined BND, the decompression rate dP/dt can be calculated according to the following formula:

$$dP/dt = a_0\, D\, \sigma_t^2\, P_{\mathrm{w}}^{-0.33} T^{-0.5} N_V^{0.67} \quad (9.29)$$

where a_0 is a constant (= 1×10^{15}), D is the diffusivity of water in a silicate magma (m^2 s^{-1}; see Chap. 4, Eqs. 4.11–4.13), σ_t is surface tension (N m^{-1}), P_W is saturation pressure (MPa), T is temperature (in °K), and N_V is the number density of vesicles in the size range 1–10 μm, which are considered to be the last formed vesicle population resulting from a rise in the conduit during the final stages of decompression. Given that only a fraction of the volatile content of the magma contributes to the final small vesicle population, Simmons et al. (2017a, b) corrected the initial water content using the vesicle content of > 10 μm vesicles and applying the equilibrium models of Gardner et al. (1999) and Shea et al. (2010).

Toramaru (2006; Fig. 9.43) found that decompression rates for 14 plinian eruptions were in the order of 10^6 to 10^8 Pa s^{-1} and are proportional to the column height ($dP/dt \propto H_T^4$; Fig. 9.43) and the mass discharge rate ($dP/dt \propto Q$).

The calculation of BND for an eruption sequence is made using image analysis of vesicular juvenile clasts (Fig. 9.44; Shea et al. 2010) and requires the selection of a meaningful pyroclast grain size class on which the image analysis is performed, in order to be representative of the average vesicularity of the whole deposit (many authors use the 3ϕ size class, although this selection may be decided case by case). A magnification strategy is also needed to allow integration from the largest to the smallest bubble sizes, down to the micron scale. Details of this strategy are given in Shea et al. (2010).

Toramaru (2006) suggests using only the BND calculated from micron-scale bubbles (1–10 microns) as they should be related to a very rapid decompression event just before fragmentation. Romano et al. (2020) have shown that decompression rates retrieved from BND of the 4.5 ka Agnano–Monte Spina eruption deposits (Campi Flegrei) can be used to model conduit dynamics, depth of fragmentation level, and exit velocities that are in agreement with eruption column heights derived from field data. Furthermore, the authors documented the systematic decrease of BND at the transitions between fall and flow deposits, from 10^8 to 10^7 cm^{-3} respectively, indicating that transitions between fully buoyant and collapsing phases were controlled by the deceleration of magma ascent. By contrast, Simmons et al. (2017a, b) found that BND in pumices increased

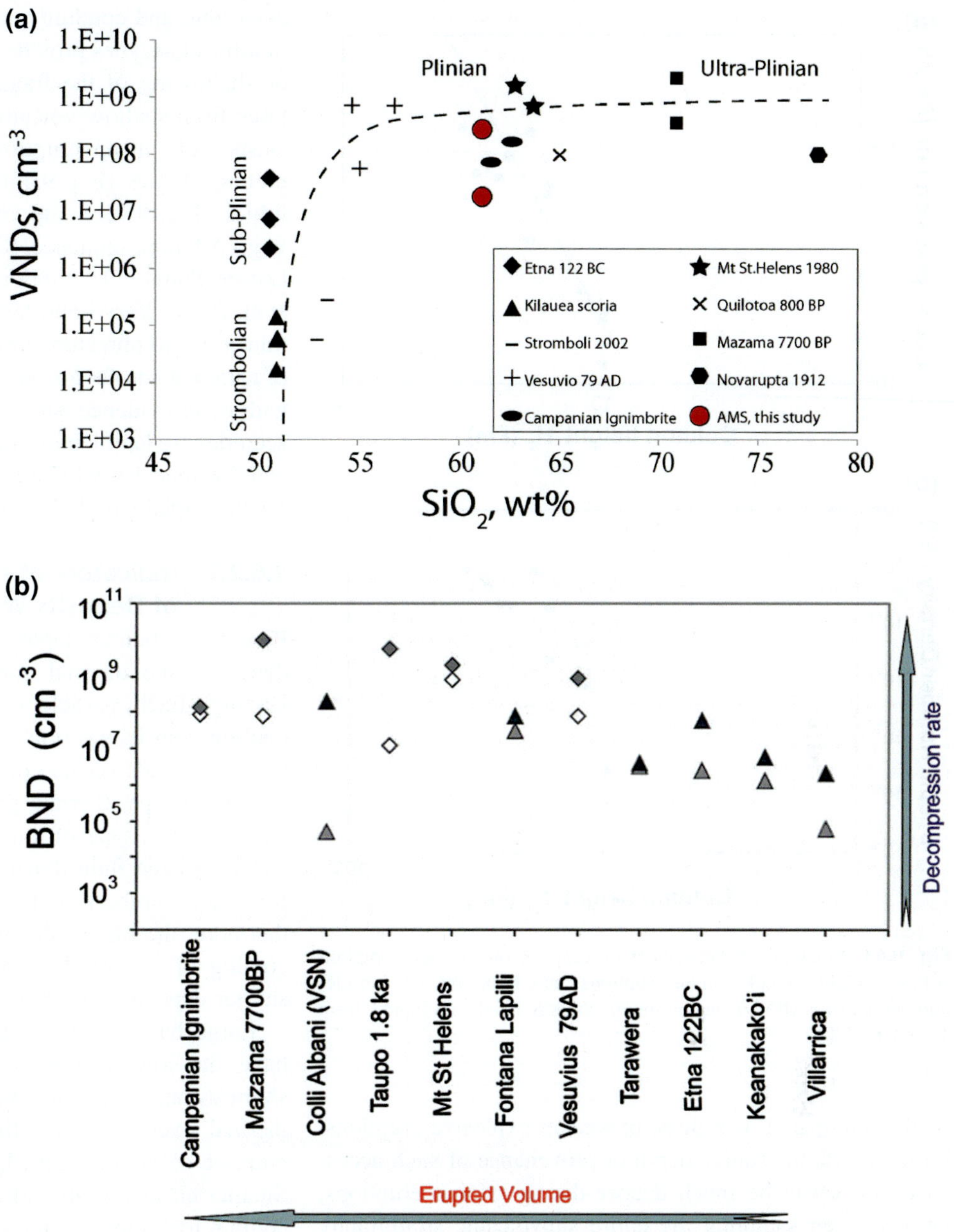

Fig. 9.42 **a** Comparison of bubble (or vesicle) number densities (B or/VND) for explosive eruptions with a range of magma compositions indicated by bulk SiO_2 content. **b** Relative relationships between BND, decompression rate and erupted volume for several eruptions. (*Data source references* Adams et al. 2006; Gurioli et al. 2008; Houghton and Gonnermann 2008; Houghton et al. 2010; Klug and Cashman 1994; Klug et al. 2002; Lautze and Houghton 2005; Mangan and Cashman 1996; Mastin et al. 2004; Polacci et al. 2003; Rosi et al. 2006; Sable et al. 2006, 2009; Shea et al. 2010; Signorelli et al. 1999; Vinkler et al. 2012; Campagnola et al. 2016). (From Campagnola 2014)

significantly between plinian fallout and initial ignimbrite-forming stages, and the final lag breccia and caldera collapse stage during the 184 ka Lower Pumice 1 eruption on Santorini. BNDs increased from 1.3×10^9 to 2.1×10^9 cm^{-3} to 4.5×10^9 to 1.1×10^{10} cm^{-3} corresponding to increased magma decompression rates from 7–11 MPa s^{-1} to 15–28 MPa s^{-1}, and magma discharge rates, from $\sim 10^7$ kg s^{-1} to a very high $\sim 10^{8.5}$ kg s^{-1}.

9.6.2.6 Lithic Fragment Types and Abundance

Accessory lithic clasts are an important part of many fall deposits (Fig. 9.45). As described in detail in Chap. 4 (Sects. 4.7, 4.8.1, 4.8.6), accessory lithic clasts result from the explosive erosion of the vent and the conduit walls. Besides their use in constructing isopleth maps, their type, abundance, and distribution within a fallout sequence provide very important information on the evolution of the eruption. For example, very shallow fragmentation levels will produce little conduit erosion and few if any lithics, as is usually the case with low magma/fire fountaining and scoria cone-building eruptions. As volatile content increases, the fragmentation level in the conduit deepens and exit velocities increase, leading, as one of the consequences, to an increased tendency to explosively erode the conduit and vent walls to accommodate the internal pressure (see Sect. 9.3). This usually results in more abundant lithic clasts in deposits of the plinian

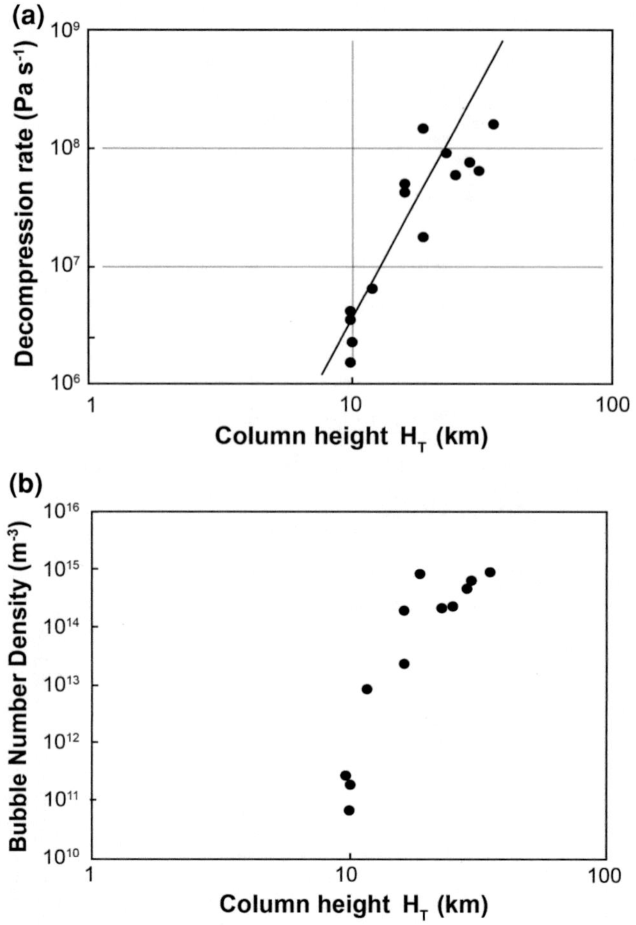

Fig. 9.43 Relationship between **a** decompression rate and eruption column height for 14 plinian eruptions, and **b** bubble (or vesicle) number density (BND) and eruption column height (Adapted from Toramaru 2006)

family, compared with those of weaker explosive eruptions. Furthermore, the source depth or provenance of such accessory lithics can be much deeper during plinian eruptions, perhaps even sampling the entire subvolcanic stratigraphy down to the magma chamber walls. Massaro et al. (2018) have used the lithic abundances and types in the 1995 BC (3.9 ka) Pomici di Avellino plinian fall deposit (Vesuvius, Italy), to reconstruct the evolution of the geometry of the conduit and to model feedbacks on magma ascent dynamics, demonstrating that conduits change shape over the course of an eruption and are more likely dykes rather than cylinders.

By contrast, scoria cones usually include lithic clasts derived from the immediate subsurface (either volcanic or pre-volcanic in origin) in the order of 0 to a few hundreds of m below the ground surface (Valentine 2012). Accessory lithic clast abundance is usually low in fall deposits: $\sim< 1\%$ in scoria cones and up to $\sim 10\%$ in plinian deposits; larger abundances may be associated with episodes of phreatomagmatism and/or episodes of vent opening, vent enlarging, and conduit/vent clogging and clearing. Changes in lithic clast types provide an opportunity to trace deepening or shallowing of the fragmentation level, including transitions from shallow volcanic edifice-derived accessory lithic clasts early in an eruption (Fig. 9.45a) to deeper, crustal plutonic lithics (Fig. 9.45b), or metasedimentary basement lithics (Fig. 9.45c) late in an eruption; for example, the Fogo A fallout sequence, São Miguel (Pensa et al. 2015), and Lower Pumice 1 eruption sequence, Santorini (Simmons et al. 2016). Very importantly, lithic clasts may help identify transitions to phreatomagmatic styles, where suddenly clasts of a known aquifer appear in a fall succession, in association with other evidence, such as sudden fining of the grain size, accretionary lapilli, base surge deposits (e.g. Vesuvius 79 AD and Avellino 1995 BC pumice fallout deposits; Barberi et al. 1989; Sulpizio et al. 2010).

9.6.2.7 Indicators of the Thermal State of Deposits When Deposited

Pyroclastic fallout deposits show a spectrum of field evidence of the thermal state of pyroclasts at the time of landing. Each particle experiences a variable degree of cooling after its ejection from the vent into the atmosphere, which depends on its composition and initial temperature, size and shape, thermal properties (thermal diffusivity and conductivity), type of ambient medium (hot dusty gas, air, water), related fluid dynamics (essentially expressed by the Reynolds number), and the duration of its trajectory. Furthermore, the rate of deposition is important, as the rate of cooling of a clast that has just landed may be longer or shorter depending on how quickly it is buried.

Large ballistic clasts (usually bomb-sized >20 cm) will have the lowest residence time in the column and the shortest trajectory in the atmosphere before landing, so their thermal exchange with the atmosphere and gases in the eruption column is limited, and cooling will also be limited. Smaller clasts (lapilli and ash size) will be entrained in the column for as long as their terminal fall velocity is less than the net upward-directed component of the local fluid velocity in the column. Hence, the residence time will be progressively longer for progressively smaller particles, and as a result thermal exchange will be more efficient, leading to cooling and contributing to the energy balance of the rising column. Furthermore, their fall trajectory in air after leaving the column or the umbrella region will be long, allowing significant cooling in air (Thomas and Sparks 1992). This general thermal behaviour of clasts results in laterally, thermally zoned fall deposits.

For example, basaltic pyroclastic fall-forming eruptions often seem to produce deposits showing a complete range of thermal facies, from variably agglutinated or welded spatter and scoria deposits in proximal coarse facies, including transitions to clastogenic lavas, to medial and distal

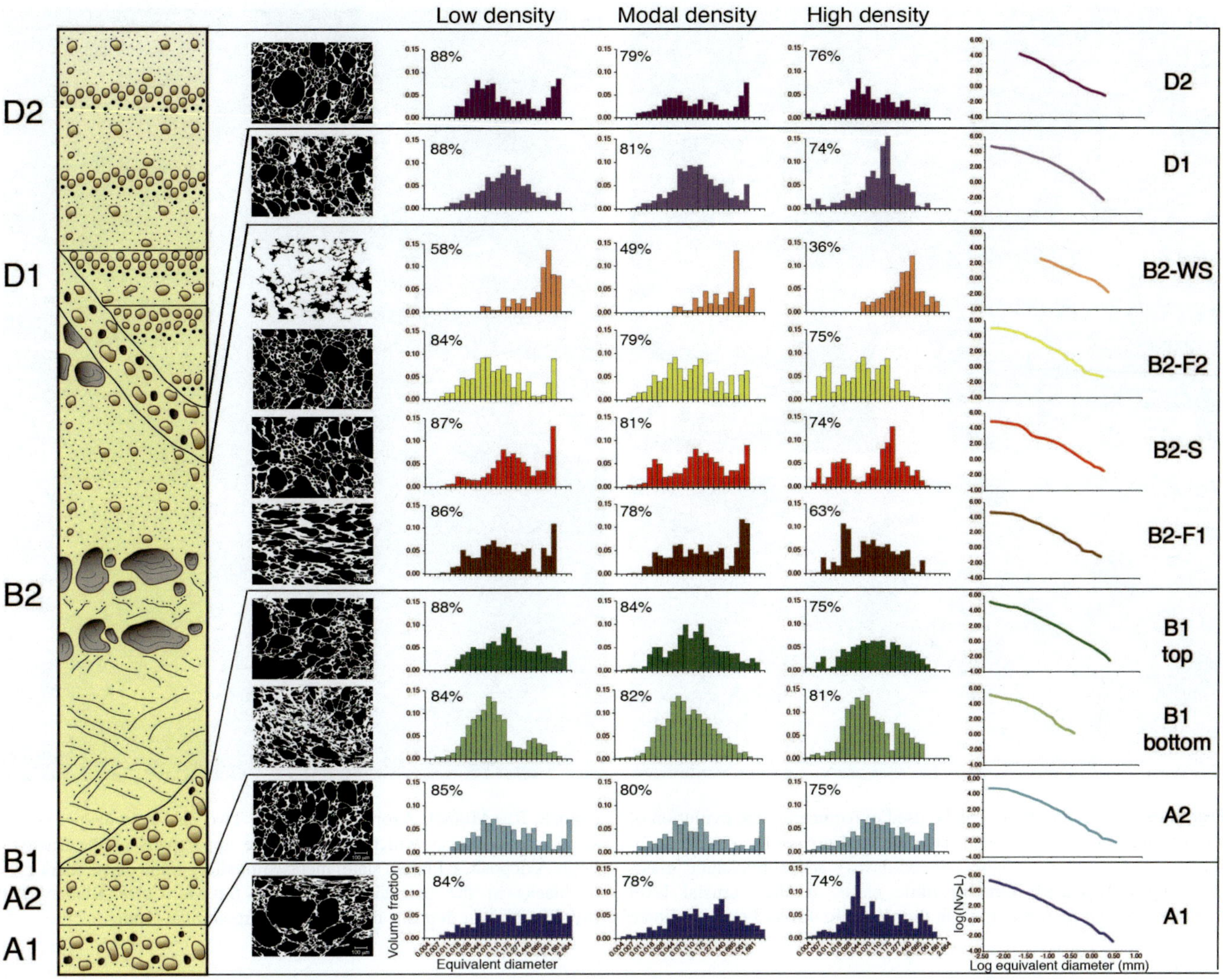

Fig. 9.44 Binarised scanning electron microscope images at 160 × magnification for the 4.5 ka Agnano–Monte Spina plinian pumice fallout deposits at Campi Flegrei volcano, Italy. A1, B1, and D1 are fallout deposits; A2, B2, and D2 are pyroclastic density current deposits. Histograms indicate the volume bubble/vesicle distributions for low-, modal-, and high-density selected pumices. The log–log diagram indicates the cumulative volume distribution. (Adapted from Campagnola 2014, and Romano et al. 2020)

non-welded scoria fall deposits. In contrast, pumice fall deposits are generally non-welded, although relatively rare proximal welded facies can be present (see Sect. 9.7.8). Lack of welding in felsic pumice deposits may not necessarily indicate a cold state of emplacement. Changes from pinkish to reddish to dark colours of pumice clasts have been reported as evidence of the thermal oxidation of iron (Walker 1981b).

Post-fragmentation plastic deformation of clasts in fact depends on the relaxation time of the magma compared with the time scale of cooling. For high-viscosity felsic magma, the time scale of relaxation of the magma is long, so clasts usually cannot deform at the time scale of pyroclastic fall cooling rates.

Thomas and Sparks (1992) analysed the heat loss for ideal spherical pumice clasts of various sizes and initial temperatures (Fig. 9.46), by considering the forced convection associated with heat removal exerted by the surrounding turbulent fluid (dusty gas or air) and by radiation, according to the following equations:

$$F_c = h_c(T_s - T_a) \tag{9.30}$$

$$F_r = \gamma(T_s^4 - T_a^4) \tag{9.31}$$

where F_c is the convective heat flux per unit area, F_r is the radiative heat flux per unit area, h_c is the convective heat transfer coefficient, γ is the black body emissivity, T_s the surface temperature, and T_a is the ambient temperature.

Fig. 9.45 Lithic clast types can be used to reconstruct the evolution of the erosion of the vent and conduit encompassing a variety of shallow-derived (e.g. variably altered lavas) and subvolcanic clasts including deep-seated intrusive lithic clasts. **a** High crustal level trachyte lithic clasts (dark grey) in the trachytic 4.6 ka Fogo A plinian deposit, São Miguel, Azores, Portugal. **b** Two pale grey deeper crustal syenitic lithic clasts which appear in the upper parts of the trachytic Fogo A deposit. **c** Platey, slate, metasedimentary lithic clasts (dark grey to black) in the extensive rhyolitic ~6.6 Ma Corte Blanco Tuff rhyolitic plinian deposit, northwestern Argentinian Andes

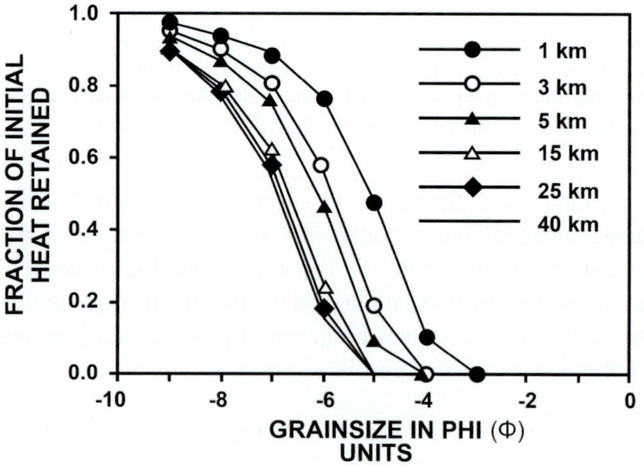

Fig. 9.46 Fraction of calculated heat retained (relative to initial magmatic temperature) by a spherical pumice particle after a fall from columns of various heights, for particles of various sizes (Modified from Thomas and Sparks 1992)

The time scale of heat loss was then compared with a column model (Bursik et al. 1992) for plume heights varying between 3 and 40 km, with the assumption that heat loss is entirely due to the free fall of the particle in air. The main conclusions by Thomas and Sparks (1992) were (1) cooling is largely due to convective heat loss; (2) initial total grain size (TGSD) distribution is important as clasts smaller than 1.6 cm (−4 Φ) cool so rapidly that they would land cold for any given condition; (3) the cooling of clasts larger than 1.6 cm and up to 25.6 cm (−8 Φ) is instead controlled by the height of fall and initial temperature (Fig. 9.47); and, (4) clasts larger than 6 cm would experience a fast cooling of their rim, potentially forming a chilled margin and bread-crusts if degassing of the interior is allowed.

Clearly, the calculation of the forced convection term is quite complicated as the fluid turbulence, particle concentration, and temperature also change through the different regions of the column, and they are different from the ambient air through which the clast eventually falls once it exits from the column margin.

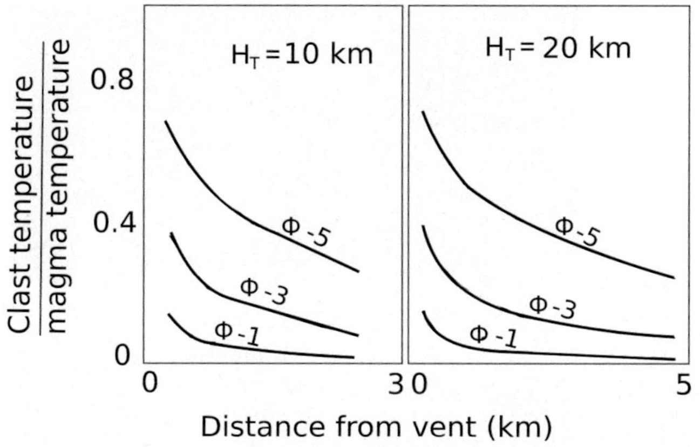

Fig. 9.47 Calculated temperature on landing of pyroclastic fallout clasts of different sizes (in phi sizes) with distance from vent, for columns of **a** 10 km height, and **b** 20 km height. Φ −5 is equivalent to 32 mm diameter, Φ −3 to 8 mm, and Φ −1 to 2 mm (Modified from Thomas and Sparks 1992)

Away from the proximal zone, where possible welding and pyroclast deformation may be observed (at the largest in the range of 1–2 km from the vent), the thermal state of pyroclastic fall deposits can be determined from the presence of charcoal and by measurement of the partial Thermal Remnant Magnetisation (TRM) of both juvenile clasts and accessory lithics. These methods are mostly applied to pyroclastic flow deposits (e.g. Caricchi et al. 2014; Pensa et al. 2015; Giordano et al. 2018), but may be equally used for fall deposits and will be described in detail in Chap. 10, Sect. 10.8.14.4.

9.7 Pyroclastic Fallout Deposits

We will discuss the deposit characteristics of fallout deposits in the same order as the source eruption styles were discussed in Sect. 9.2.

9.7.1 Strombolian *sensu stricto* and Hawai'ian Magma/Fire Fountain Deposits

These styles both involve fire fountaining of basaltic or other low-viscosity magmas. At one end of the spectrum, discrete, small-spaced fire fountain sprays (Sects. 9.2.1.1, 9.3.1.1) are called *strombolian sensu stricto* eruptions (Plate 9.1a; Figs. 9.2a, 9.7a, b) and at the other end, sustained magma fountaining events are called *hawai'ian magma/fire fountain* eruptions (Sects. 9.2.1.2, 9.3.2.1; Fig. 9.7d–f). The principal products of both are spatter, and lesser scoria. During sustained magma/fire fountaining, lavas may also form as spatter accumulating around the vent, often as a spatter cone, fuses and remobilise as a lava flow that flows away from the vent (e.g. 1983–2015 Pu'u O'o eruptions, Hawai'I; 2018 Leilani Estates eruption, Hawai'i; 2014–2015 Bárdabunga-Holuhraun eruption in Iceland).

During strombolian *sensu stricto* activity (i.e. short-lived, time-spaced magma/fire fountain sprays), the intervals between explosions are so long that a maintained column and convective plume cannot form, and the spatter and scoria fall deposit are restricted to the maximum distance of ballistic ejecta, in the order of tens to a few hundreds of m from the vent. Such is the case with the current activity at Stromboli volcano, Italy, excluding the episodic paroxysms or vulcanian events (Plate 9.1a; Figs. 9.2a, 9.7a, b), that are slowly building a cluster of cones that together may build up deposits to a few tens to hundreds of m in height and a maximum 1–1.5 km across (Plate 9.1a).

Hawai'ian magma fountaining activity produces a much higher proportion of lava spatter and agglutinate deposits much more rapidly, due to episodes of continuous magma fountaining (e.g. Fig. 9.7g, Pu'u O'o cone, Hawai'i; Fig. 9.7 h, I, Bárdabunga-Holuhraun, Iceland), compared with more explosive violent strombolian and micro-plinian deposits. Such deposits are similar to those built during strombolian *sensu stricto* fountain spray activity. Consequently, the formation of spatter cones and ramparts at the vent (Fig. 9.48a), made up of variably agglutinated spatter and scoria deposits (Fig. 9.48b, c), and through the fusion of spatter, lava flows flowing away from the vent (Fig. 9.7 g, i; Fig. 5.13a–c), is likely. Spectacular magma/fire fountaining fissure vents are associated with episodes of very intense effusive activity, as was the case during the 2 years long, 14 km^3, Laki eruption in Iceland in 1783–84 (Fig. 9.48a), and the 2014–2015 Bárdabunga-Holuhraun eruption in Iceland (book cover; Fig. 9.7 h, i).

The downwind fall deposits are fine-grained and composed almost entirely of scoria and are volumetrically small. Deposits usually contain achneliths, pele's tears, and pele's hairs, which are juvenile fragments with smooth, glassy surfaces formed from solidified, very fluidal lava spray (Chap. 4, Fig. 4.20a–c). Although a wide variety of shapes are possible (Chap. 4, Fig. 4.20a–c), the most extreme form is stretched-out, hair-like filaments of basaltic glass known as pele's hair (Fig. 4.20b, c; Duffield et al. 1977). Achneliths are especially common in hawai'ian spatter and scoria fall deposits. Hawai'ian deposits are usually free of accessory lithic clasts as a result of very shallow fragmentation levels.

Eruption column heights and muzzle exit velocities during hawai'ian and strombolian activity are usually low. Consequently, hawai'ian spatter and scoria fall deposits usually have a limited dispersal of less than 0.05 km^2 (Fig. 9.6a; Walker 1973) and a low degree of fragmentation, as do strombolian *sensu stricto* deposits (Fig. 9.6a). However, some episodes of lava fountaining, especially in the very early phases of hawai'ian activity can reach heights of

Fig. 9.48 Features of near-vent spatter deposits. **a** The spectacular fissure of the 1783–84 hawai'ian eruption at Laki, Iceland. Point sources along the fissure fed the building of low spatter cones and ramparts made of agglutinated spatter and scoria. The grey areas around represent the early and most intense and gas-rich phases of the fountaining that sprayed lapilli and ash high into the atmosphere producing thin deposits around the fissure. **b** Close-up view of an agglutinated spatter deposit from Iceland. **c** Partially agglutinated spatter deposits from a spatter cone at the ~35 ka Tower Hill volcanic complex, Newer Volcanics Province, Victoria, southeastern Australia

several hundreds of m (e.g. Kilauea Iki in 1959, Table 9.2), in which case, although observed activity would be considered typically hawai'ian, the resulting deposits are much more widely dispersed (Fig. 9.6a). Deposits associated with peak discharge may also contain reticulite pyroclasts (Chap. 4, Fig. 4.20d), characterised by up to 90% vesicularity (Mangan and Cashman 1996; Rust and Cashman 2011). The 1959 Kilauea Iki lava fountains reached heights of 600 m, and the downwind scoria fall deposit has a dispersal of about 0.7 km^2 (Richter et al. 1970). By dispersal area alone it could be classified also as violent strombolian; however, this scoria fall deposit is composed almost entirely of achneliths or fragments of them, and these should be a very large component of scoria fall deposits, even of those of wide dispersal, formed by magma fountaining. Many eruptions will also vary in observed style from hawai'ian to strombolian to micro-plinian (e.g. 1973 eruption on Heimaey, Iceland; Fig. 9.9d), and even to plinian (e.g. 1886 eruption of Tarawera, New Zealand; Plate 9.1b) and vice versa.

9.7.2 Deposits from Vulcanian and Other Impulsive Eruptions (Phreatic, Hydrothermal, and Halema'uma'uan)

Vulcanian activity occurs under closed or blocked conduit conditions (and open degassing conditions), and typifies the sudden violent opening phases of many explosive eruptions, when a rigid plug, or viscous domain of magma in the conduit, or a dome, or country rock is blasted by cannon-like explosions to open the vent (see Sect. 9.2.1.4). This may then (or not) lead to decompression of the subjacent magma, and ongoing magmatic exsolution-driven explosive eruption phase, which is different to the initial vulcanian blast phase. Magma compositions for vulcanian deposits are usually intermediate (basaltic-andesite, andesite, dacite), and vulcanian deposits are common products of andesite and basaltic-andesite stratovolcanoes.

Phreatic explosions, hydrothermal explosions, rock fall impact (halema'uma'uan eruptions), and other impulsive eruptions have a similar sudden blast-like expression, with some similarities and some minor differences in the deposits. As discussed in Sect. 9.2.1.4, we also include among the vulcanian eruptions the paroxysms at open conduit basaltic volcanoes, such as Stromboli (e.g. Giordano and De Astis 2021; Pichavant et al. 2022).

Pyroclastic fallout deposits from individual vulcanian eruptions are thin, small volume (<0.1 km^3; Fig. 9.49a), stratified lapilli-ash deposits (Fig. 9.50a, b), which contain large juvenile ballistic bombs and accessory blocks near the vent, sometimes with breadcrusted surfaces (Figs. 9.49b, 9.50c) and radially jointed interiors (Fig. 9.50d). At Soufrière Hills volcano, Montserrat, ballistics have been found up to 1.7 km from the vent (Druitt et al. 2002).

Vulcanian, phreatic, hydrothermal and halema'uma'un fallout beds usually contain abundant very high-level crust-derived accessory lithic clasts and ash, which are dominant, especially in the opening phases of eruptive cycles, as well as varying proportions of juvenile clasts. Ballistic clasts can be m-size in diameter, can be ejected hundreds of m above the vent, and when they land create impact craters that have areas at least twice that of the clasts. The ballistic clasts commonly shatter and fracture on impact, leading to a local halo of such shattered debris around the main impact crater (e.g. Pistolesi et al. 2011). In addition to ballistic ejecta, the deposits include variably bedded fallout and pyroclastic surge and flow deposits (Fig. 9.50a, b; Chap. 10). Large magmatic ballistic clasts often develop surface breadcrust fracture networks (Fig. 9.49b, Fig. 9.50 b), resulting from ongoing exsolution and vesiculation of the clast interior, and radial cooling joints (Fig. 9.50d) resulting from rapid cooling after ejection from the vent. Smaller juvenile pyroclasts can be poorly to highly vesiculated.

Hydrothermal explosive deposits are dominated by hydrothermally altered clasts (Chap. 4, Fig. 4.31 g), but these also occur in vulcanian, phreatic, and halema'uma'un deposits. However, they are usually so thin and fine-grained that they are soon eroded by wind and water, and have a very poor preservation potential in the rock record, especially on steep-sided stratovolcanoes.

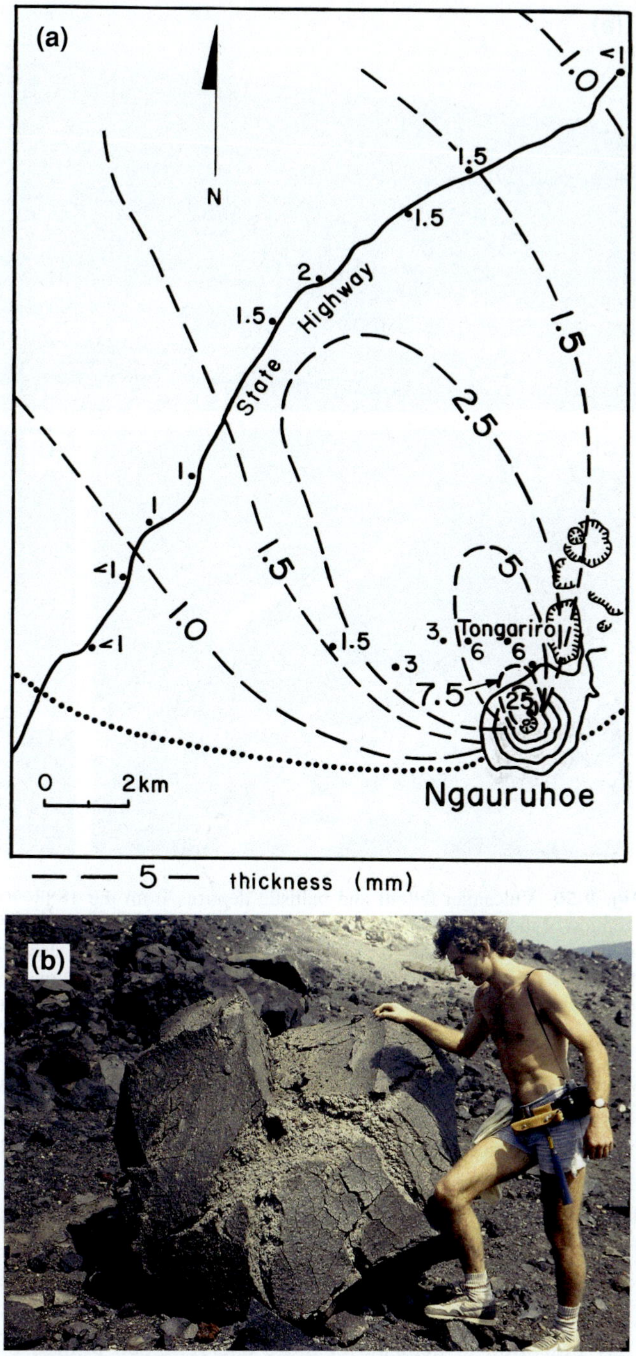

Fig. 9.49 **a** Isopach map of the vulcanian pyroclastic fallout deposit erupted from Ngauruhoe volcano, New Zealand, during 28–29 March 1974. Note thicknesses are in mm. (From Cas and Wright 1987, After Self et al. 1974). **b** Large ballistic breadcrust bomb erupted from Anak Krakatau, Indonesia. During the mid-July eruption, 1978, bombs (average diameter 1 m), lapilli, and ash were ejected every 5–15 min, covering the area within about 700 m of the crater

Vulcanian plumes are weak, not sustained, and not very high (usually < 10 km but up to a maximum of 15 km; such high columns reflect the transition to magmatic exsolution-driven activity), so they are usually strongly affected by winds (Fig. 9.8c, d), producing highly elongated and narrow dispersal patterns, reaching some km away from the vent (Fig. 9.49a). When eruption phases continue for a few years, like the current (2015) case at Sakurajima

Fig. 9.50 Vulcanian fallout and ballistic deposits from the 1888–90 eruption of the Fossa Cone, Vulcano, Italy. **a** Thinly stratified sequence of deposits mantling the outer slope of the crater. **b** Thin, steeply dipping, stratified, slope-draping vulcanian deposits (fall, surge) on the lower, outer slope of the Fossa Cone. **c** Large ballistic block with breadcrust texture, perched on the steep inner crater slope of the Fossa Crater. **d** Ballistic block with internal radial cooling joints on the rim of the steaming Fossa Crater (left). The volcanic islands of Lipari (middle ground) and Salina (twin peaks, background) are visible to the north

volcano, Japan, bedded sequences can be built up around and near the vent (Fig. 9.50a, b), but these are not likely to be of great thickness; for example, less than 2 m of ash-fall deposits accumulated just 800 m downwind of the vent on Irazú volcano, Costa Rica, from the multiple eruptions between March 1963 and mid-1964. Vulcanian fall successions may be associated with poorly dispersed surge deposits (e.g. Vulcano island, De Astis et al. 1997, 2013), but also may transition to deposits of more energetic sub-plinian or plinian phases after the vent has been opened and if explosive magmatic vesiculation occurs, including pyroclastic flow- and surge-forming stages (Fig. 9.8c; Ngauruhoe volcano New Zealand in 1975, Nairn 1976, Self et al. 1979; Fig. 10.4b; 1995–1999 phase of the eruption of Soufrière Hills, Monteserrat; Druitt et al. 2002).

Pistolesi et al. (2011) described the deposits of the 2007, paroxysmal, 5-min-long vulcanian eruption from Stromboli volcano, which consisted of an initial deposit of variably hydrothermally altered ballistic blocks of older lava and spatter agglutinate, and an associated lithic-rich, pyroclastic density current deposit, both resulting from an initial 30 s-long ballistics explosive pulse. This was followed by an ash fallout dominated by accessory lithic debris, resulting from a second explosive pulse lasting 12 s that produced a weak buoyant plume that rose to about 3 km above the vent, and also produced a small, lithic-rich pyroclastic density deposit. The final stage produced a small scoria flow deposit, dominated by vesicular golden scoria.

Giordano and De Astis (2021) described the paroxysmal vulcanian explosive eruptions of Stromboli on 3 July and 28 August 2019, and their deposits. Ballistic exit velocities were $\lesssim 160$ m s^{-1}, plume heights were ~ 8.4 and 6.4 km, respectively, and mass eruption rates were 1.1×10^6 kg s^{-1} and 3.6×10^5 kg s^{-1}, respectively. Both explosions were dominated by the rapid rise of a gas-rich batch of magma, commonly called “golden” or low porphyritic (LP) because of its production of highly vesicular, yellow, aphyric pumice, as well as black porphyritic scoria. The eruptions lasted for only about 5 min each. Although both eruptions produced vertical eruption columns, they also produced directed lateral blasts that produced small-volume pyroclastic flows that reached and flowed for several hundred m over the sea (Giordano and De Astis 2021; Chaps. 10, 11). The fallout deposits include ballistic spatter bombs from the 3 July eruption which cover a continuous area of $\sim$0.35 km^2 around the crater. The maximum equivalent diameter of the measured spatter is 200 cm, where the cover is commonly made up of the superposition and welding of two to four large spatter clasts 30–40 cm thick. Large ballistic lava blocks litter the summit region in association with spatter clasts, and landed up to 1300 m from the vent (in the sea). A lapilli fallout deposit of $\sim$1.5 cm thickness, consisting of golden pumice, dark porphyritic scoria, and accessory lava fragments occurs downwind around the coastal village of Ginostra, at elevations <250 m (Giordano and De Astis 2021).

Deposits from the 28 August eruption are dominated by sparse, shattered, accessory, dense lava ballistic blocks up to 86 cm in diameter with impact craters above 500 m elevation, and more vesiculated, smaller, juvenile bombs below 500 m elevations. A continuous lapilli and ash fallout deposit occurs on the northern slope below 300 m elevation, and also consists of golden pumice, dark porphyritic scoria, and accessory lava fragments, but this fallout deposit is <2 mm thick at the coast at Stromboli harbour (Giordano and De Astis 2021).

Phreatic/hydrothermal eruptions are somewhat similar to vulcanian in their impulsive style and produce similar near-vent, thin deposits of ballistic ejecta, fallout, and pyroclastic density current deposits that consist entirely or mostly of variably hydrothermally altered, and dense accessory lithic clasts. The absence or minor abundance of a magmatic component largely reduces the buoyancy of the related weak plumes, so the dispersal areas of the deposits are much smaller (Marini et al. 1993; Maeno et al. 2011, 2016; Kilgour et al. 2019). The very small volume Halema'uma'uan 2008–2018 deposits from Hawai'i consist of accessory basaltic lithic clasts and variably vesiculated juvenile clasts.

9.7.3 Violent Strombolian Deposits

During intense *violent strombolian* activity, as in the case of the 1943–45 eruption of Parícutin , Mexico (e.g. Pioli et al. 2008), some phases may be characterised by very closely spaced explosions capable of maintaining a weak plume that also builds cones of scoria, ash, and bombs (Fig. 9.9a). Deposits are typically stratified, and with cm- to m-thick layers, moderately sorted, lithic-poor, scoria lapilli and ash fall deposits (Fig. 9.51a–d) may have dispersal areas up to several km^2, and in some cases (e.g. Parícutin) tens of km^2, well beyond the base of the cone, which is usually less than 1.5 km in diameter (e.g. Valentine et al. 2005; Pioli et al. 2008). Violent strombolian deposits are typically diffusely stratified, indicating pulsatory explosive activity, relatively fine-grained close to the vent, and contain significant coarse ash in the matrix (Pioli et al. 2008; van Otterloo et al. 2013), which distinguish them from micro-plinian scoria cones and their deposits. Successive layers vary in grain size and fine ash content. Such characteristics are consistent with perhaps some magma–water explosive interaction, which is certainly the case at the Mt Gambier volcanic complex, in South Australia (van Otterloo et al. 2013). For the Parícutin deposits, Pioli et al. (2008) attribute the high ash content to relatively high magma viscosity, enhanced by groundmass crystallisation and high volatile contents. Nonetheless, the mass eruption rates were relatively modest at $\sim 10^4$–10^5 kg s^{-1}.

9.7.4 Deposits from the Plinian Spectrum of Eruptions

Plinian pyroclastic fall deposits are a common product of highly explosive eruptions of high-viscosity magmas (Sect. 9.2.2.2), but also originate from some eruptions of mafic magmas. The plinian family of deposits ranges from micro-plinian (described below) at the small dispersal and volume end of the scale upwards to sub-plinian to plinian, and the largest, ultra-plinian. We will describe comprehensively the characteristics of plinian deposits, including many generic characteristics that are common to other members of the family, and then briefly sub- and ultra-plinian which vary mostly in dispersal area and volume.

9.7.4.1 Deposits from Micro-Plinian Activity

Micro-plinian eruptions build scoria and pumice cones (Fig. 9.52a) and weakly dispersed scoria/pumice and ash deposits beyond (Fig. 9.6a). Micro-plinian deposits are massive and coarse close to the vent, and better sorted than violent strombolian deposits. They typically build scoria and spatter cones (Fig. 9.52a), resulting from fallout around the vent from a weak, but sustained, buoyant convective eruption column that may be active for hours to a few days, and may produce fallout deposits extending beyond the limits of the scoria cone (Fig. 9.9a–c). Such eruptions are transitional into sub-plinian. In the past, some authors have called these eruptions and deposits "strombolian", which we consider to be confusing relative to the actual eruption style that consistently occurs at Stromboli volcano, i.e. spaced, discrete magma/fire fountain sprays. We recommend that "strombolian" be used for fountain spray deposits, and micro-plinian for scoria and pumice cone-forming eruptions with buoyant

Fig. 9.51 Violent micro-plinian (violent strombolian) deposits. **a** Bedded sequence with uppermost bomb bed, just west of the cone, which can be seen in background, Parícutin volcano, Mexico. **b** Fine planar to undulating layered sequence 5 km north of the cone at Parícutin. Note the low angle trough-like erosion surface in the middle, above which layering changes from undulating to planar again, perhaps suggesting surge transport and deposition. (Photos **a** and **b** from Pioli et al. 2008). **c** Diffusely stratified, ash-rich deposits of violent micro-plinian eruption phases of the Alumbrera scoria cone (in the background) covered by late-stage lava, southern Puna Plateau, Argentina. **d** Diffusely stratified, ash-rich violent strombolian fall and surge that form part of a 60 m thick/high tuff cone. Note the slight undulations suggesting these are surge-modified fallout deposits. Mt Gambier violent strombolian tuff cone, Newer Volcanics Province, South Australia

eruption columns, as advocated by Francis et al. (1990), and applied to deposits from Etna during 2000 by Behncke et al. (2006).

Other well-documented historic eruptions producing scoria cones are the 1970 Hekla eruption, Iceland, cited by Walker (1973), the 1973 Heimaey eruption, Iceland (Self et al. 1974), and the scoria fall deposit erupted with the formation of Sunset Crater (1065 AD) in the San Francisco Volcanic Field, Arizona, USA (Amos et al. 1981). Besides producing a very widely dispersed scoria fall deposit (dense rock equivalent, DRE = 0.30 km^3), the Sunset Crater eruption also built a scoria cone 300 m high (DRE = 0.15 km^3). The gas thrust part of the column in the Sunset Crater eruption may have reached heights of several hundred m. Such energetic basaltic eruption columns (which in some examples are also known to have attained plinian proportions; see Sect. 9.7.4) may have formed through a combination of a relatively high magma discharge rate and high initial magmatic gas contents.

Much rarer are felsic pumice cones emplaced from micro-plinian activity, including cones that formed during the most recent activity of Lipari island, Italy (Dellino and La Volpe 1995; Davì et al. 2011).

Proximal deposits of scoria cones often consist of thick (tens to hundreds of m; Figs. 9.52a, b, 9.53a, c), massive to rather poorly bedded, very coarse-grained and sometimes red (oxidised) scoria with up to m-sized ballistic bombs and blocks (Fig. 9.52b). A variety of bombs and blocks may be found, including large scoriaceous fragments, less well-vesiculated spindle and cowpat shapes, sometimes bombs with breadcrusted surfaces, and dense lava blocks and slabs (Chap. 4, Fig. 4.19a–h). Large accessory lithics of country rock are usually uncommon and are usually very shallow derived, sourced from several m to a maximum of a

Fig. 9.52 **a** Alumbrera basaltic scoria and spatter cone and associated surrounding lava field, southern Puna Plateau, Argentina (After Báez et al. 2017). **b** Massive to diffusely stratified scoria with dispersed ballistic blocks and bombs of the cone facies of the Mt Leura volcanic complex, Newer Volcanics Province, Victoria, southeastern Australia. **c** Massive to bedded, micro-plinian scoria fallout outside of the cone, Mt Leura volcanic complex. The scoria deposits are mantling phreatomagmatic base surge deposits (pale colour) which underlie and are also interbedded with the scoria, heralding a transition from initial phreatomagmatic to magmatic volcanism in the volcanic centre. **d** Well-sorted, vesicular, micro-plinian scoria fallout deposit, Villarrica volcano, southern Chile. **e** Reverse-graded scoria grain flow deposits, forming the slopes of a scoria cone near Oia, Santorini, Greece

few hundreds of m in depth below the surface (Valentine 2012). Petrologically important mantle-derived nodules may occur in bombs, as "cored" bombs with a rind of lava around them or dispersed in larger bombs (Fig. 9.54a, b). Deep-derived (mantle or intrusive) xenoliths will show evidence of thermal equilibration with the magma, such as a rounded form, which indicates magmatic resorption, as well as adhesion of a magmatic coating, together forming cored bombs (Fig. 9.54a, b). By contrast, shallow accessory lithics usually do not show any melt-coating.

Beyond the cone, the deposits of micro-plinian eruptions are thinner, and massive to planar stratified (Fig. 9.52c). They are finer than the cone deposits (Fig. 9.52c–e), and are dominated by well-sorted, vesicular scoria clasts (Fig. 9.52 d). Within the cones, many of the observed beds are not simply air-fall layers but include mass-flow deposits formed by grain flow (Chap. 12), avalanching, and rolling of scoria down the steep, unstable, angle of repose slopes as the cone built up. Such beds are laterally discontinuous, irregular in geometry, and marked by reverse grading (Fig. 9.52e).

Bomb impact sag structures are not a common feature in scoria deposits. This is because ballistic bombs land in a thick, accumulating bed of coarse, loosely packed, unstratified scoria, which accommodates impact by breaking and/or reorganising of scoria clasts (cf. cohesive, wet, surtseyan base surge deposits, where bomb sags are common because

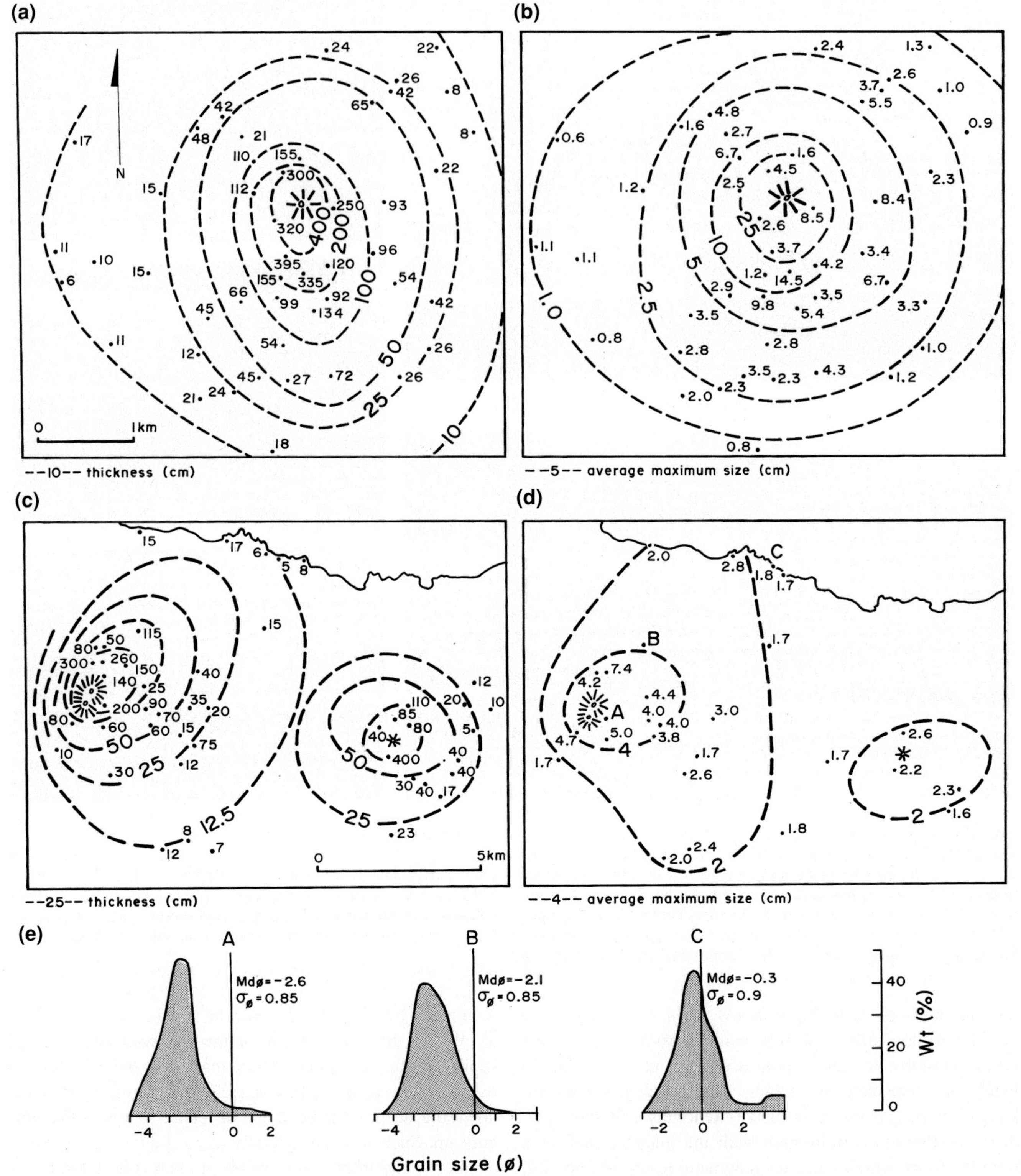

Fig. 9.53 Thickness and grain size characteristics of some micro-plinian, scoria cone-forming pyroclastic fallout deposits in the Azores, Portugal. Isopleth maps show the average diameter of the three largest scoria clasts. **a** and **b** Scoria fall deposit from the Galiarte cone, Terceira (From Cas and Wright 1987, After Self 1976). **c** and **d** Scoria fall deposits from Serra Gorda (west) and Cone 301 (east) on São Miguel. **e** Grain size distribution curves for the Serra Gorda deposit at the three locations in (**d**). (From Cas and Wright 1987, After Booth et al. 1978)

Fig. 9.54 Mantle xenoliths in near-vent, micro-plinian scoria cone deposits. **a** Peridotite mantle xenoliths in the remains of a spindle bomb, with a thin coating of basalt. Note the rounded, resorbed form of the xenolith. **b** Small, sub-rounded peridotite xenoliths in a basalt spindle bomb. Both from the Mt Noorat volcanic complex, Newer Volcanics Province, Victoria, southeastern Australia

of the finer grain sizes and the often wet, plastic state of the ash pile into which bombs fall). Layers of agglutinated lava spatter and scoria can be conspicuous when the activity transitions to a more sustained "hawai'ian" style. In this case, complete fusing or welding-together of the clasts may occur, and this is one way in which (clastogenic) lavas may be generated (Chap. 5). Rapid accumulation of spatter and scoria through fallout from low eruption columns or fountains is needed to produce such agglutinated, welded layers and clastogenic lavas (see Sect. 9.8 on welded air-fall deposits). In fact, many scoria cones are associated with lava flows that are either clastogenic in origin erupted as distinct through contemporaneous fire fountaining, or effusive phases (e.g. Fig. 9.52a; Báez et al. 2017). Micro-plinian scoria cone-building activity may also transition from and to phreatomagmatic activity and deposits (e.g. Fig. 9.52c). Occasionally vulcanian activity can also occur.

9.7.4.2 Plinian Fallout Deposits

Plinian pumice fallout deposits are not only a common eruptive product of all large rhyolite volcanoes, but are also frequently found as products of andesitic, dacitic, phonolitic, and trachytic explosive volcanoes (Fig. 9.55a-c, f–h). Classic historic examples of Plinian eruptions and their deposits are mostly felsic in composition and include the 79 AD Vesuvius eruption deposits (Lirer et al. 1973; Carey and Sigurdsson 1987; Cioni et al. 1995), the Minoan eruption deposits of 1600 BC, Santorini (Bond and Sparks 1976; Druitt et al. 1999, 2019; Druitt 2014), 1875 Askja eruption deposits, Iceland (Carey et al. 2009, 2010), 1980 Mt St Helens, USA (Sarna-Wojcicki et al. 1981; Waitt and Dzurisin 1981; Carey and Sigurdsson 1985), and the 1991 Pinatubo, Philippines deposits (Paladio-Melosantos et al. 1996; Rosi et al. 2001). However, basaltic/mafic scoria fall deposits which have plinian dispersal patterns are also known (Fig. 9.55d, e; Williams and Self 1983; Walker et al. 1984; Houghton et al. 2004a, b; Sable et al. 2006; Giordano et al. 2010; Miyaji et al. 2011). The characteristics of plinian pumice fall deposits and their eruptions are fairly well-defined, and the extensive literature about these has been reviewed by Cioni et al. (2015).

General Characteristics

Plinian pumice deposits are the most impressive types of pyroclastic fall deposits found in the field (Figs. 9.3a, 9.17, 9.55a–h). The deposits of individual eruptions may attain thicknesses near the vent of 25 m (e.g. the 172 ka Lower Pumice 2, near Thera, Santorini; Simmons et al. 2017b). However, they are usually thinner, but still as much as 10 m in proximal to medial areas (Fig. 9.55a–e), to cm in distal and off-dispersal axis areas (e.g. Simmons et al. 2016). Near-vent deposits are generally massive to crudely stratified (Fig. 9.55a–e), and can be very coarse, containing large pumice fragments of several cm in diameter, and some tens of cm (Fig. 9.55f), as well as metre-sized ballistic lithic blocks some of which may shatter on impact (Fig. 9.55g). Large pumice bombs are rarely found, because in most cases they shatter on impact, and disaggregate (cf. Fig. 9.55g). Away from the vent, and lateral to the dispersal axis, deposits become thinner, finer-grained, and are often diffusely stratified (Fig. 9.55h), Field mapping, documentation, and analysis of these changes are essential in making volcanological interpretations.

When mapped out, plinian pyroclastic fallout deposits are extensive, fan-shaped, sheet-like deposits, although strong winds in the upper troposphere and stratosphere cause the umbrella region to be distorted into a downwind fan of dispersal (Figs. 9.10a, 9.28c–f), producing a fan shape for the deposit dispersal area (Figs. 9.24c, 9.30b, 9.31a, 9.32a).

(a)
(b)
(c)
LP 2
Plin
LP 1 plinian fallout
(d)
(e)
(f)

◀ **Fig. 9.55** Characteristics of plinian fallout deposits. **a** Massive to stratified 1.6 Ma Lower Bandelier pumice fallout deposit, 30 km downwind from vent. The top is finer and distinctly planar stratified, and the pinkish, ash-rich unit above is the succeeding Lower Bandelier ignimbrite. **b** Very thick 0.6 Ma Granadilla pumice fallout deposit, 15 km from vent along the dispersal axis, Tenerife. **c** Thick, massive 184 ka Lower Pumice 1 plinian fallout deposit (base of cliff) with a layered uppermost part, overlain by lithic lag breccias and ignimbrite of the LP1, then another white plinian fallout deposit of the 172 ka LP2 eruptions sequence, and the LP2 lithic lag breccias. ~6 km from vent, Santorini, Greece. **d** At vent, thick stratified 10 June 1886 red–grey basaltic plinian deposit mantling the top of the rhyolitic 700 BP Kaharoa eruption lava domes and stratified pyroclastic deposits, exposed in the wall of the 1886 fissure vent, Mt Tarawera lava dome complex, Okataina Volcanic Complex, New Zealand. Note the white rhyolite ballistics blocks in the dark basaltic deposits. **e** Diffusely stratified, well-sorted, basaltic, plinian 1707 AD Hoei fallout deposit, Mt Fuji, Japan. **f** Outsize pumice block within the fines-poor, 8–12 ka, trachytic Ribeira Cha plinian fallout deposit, São Miguel, Azores. **g** Large ballistic impact sag structure within the Lower Pumice 1 plinian fallout deposit, ~4 km from vent, associated with eruption of the overlying lithic lag breccia. Australian volcanologist for scale. Akrotiri, Santorini. The red ballistic block has shattered on impact. **h** Stratified 0.31 Ma phonolitic, Fasnia fallout deposit, Tenerife. Some layers have distinct bases marked by a concentration of dark fine lithic fragments, others are more diffusely distributed

Fig. 9.55 (continued)

They have a large dispersal area, with D > 500 km^2 (Fig. 9.6a). Ballistic lithic blocks may be ejected up to 4–5 km from the vent and their sizes near the vent imply muzzle velocities of 400–600 m s^{-1} (Wilson 1978). These indicate very high rates of magma discharge occur, which in turn leads to very high eruption columns, with evidence suggesting that column heights of 20 to 40 km should be common during this type of activity, and some in the ultra-plinian field up to 50 km or more (see Sect. 9.5).

Proximal deposits of plinian eruptions within 5–10 km from the vent originate from the margins of the upper convective region up to the umbrella region. Medial deposits, which are typically lapilli-sized, may extend several tens of kilometres away from the vent, whereas distal ash-sized deposits occur up to several hundreds of km from the vent, both resulting from fallout from the bottom of the laterally spreading umbrella region and the downwind plume, respectively. Studies in which the total grain size populations of deposits have been determined (Fig. 9.39) suggest that the volume of fine ash produced during plinian eruptions is much more than was previously supposed (Cioni et al. 2015). Thus, the importance of widespread distal ash-fall deposits of plinian origin may have been underestimated.

From textural studies of juvenile clasts, Bubble Number Densities of several plinian deposits are found in the range of 10^8–10^{10} cm^{-3} (Fig. 9.42) and correspond to decompression rates in the order of 10^6 to 10^8 Pa s^{-1} (Toramaru 2006; Fig. 9.43a). Similar figures are also found for some mafic plinian eruptions (Fig. 9.42), suggesting common conduit processes.

Internal Deposit Characteristics and Lateral Changes

Many deposits at first sight are fairly homogeneous, at least in their lower parts (Fig. 9.55a–c; e.g. Avellino 1995 BC and

79 AD pumice fall deposits at Vesuvius), and this is thought to reflect continual fallout from a downwind plume steadily fed by a continuous gas jet. For this reason, such homogeneous deposits or parts thereof are also known as simple plinian deposits (Cioni et al. 2015). They are predominantly composed of hydraulically well-sorted, juvenile material: pumice, glass shards and, when the magma is porphyritic, free crystal fragments, as well as varying proportions of accessory lithic clasts (Fig. 9.56a–g). However, significant departures from homogeneity can occur in detail, including variations in componentry, grading, and layering.

Hydraulic Sorting and Lithic Clast Patterns

Plinian fallout deposits are typically hydraulically well-sorted, showing a clear difference between the diameter of large, low-density pumice clasts and small, dense lithic clasts (Fig. 9.56a, b). Both have been subject to convective transport in the column, which has efficiently sorted pyroclasts hydraulically. As a result, large, low-density pumice clasts and dense smaller lithic clasts fall out together because they have the same terminal fall velocity (Fig. 9.56a, b). An exception to this rule is in proximal vent settings where anomalously large ballistic clasts can also occur in such deposits. These must be disregarded when collecting samples for granulometric studies because they would distort the results for sorting and grain size characteristics. Similarly, many fallout deposits have a lithic clast-rich base (Fig. 9.56c). This results from initial blast triggered opening of the vent, releasing dense lithic debris into the column first, sourced from the substrate through which the vent was opened, and which begin to fall out before later released low-density pumice clasts.

Size and Compositional Grading

Reverse grading of pumice clasts has been described from a number of deposits (e.g. Lirer et al. 1973; Bond and Sparks 1976; Bloomfield et al. 1977; Rosi et al. 1999, Sulpizio et al. 2010). This is usually more likely to be found in deposits at distances of several to a few tens of km from the vent (Fig. 9.56b, d, e), and outside these limits may be only slightly developed, or may not occur at all. Nearer the vent (<5 km), deposits are often so coarse-grained that it is very difficult to detect any grading, and more distally (>50 km) deposits may be too fine-grained.

Wilson (1980) suggest that the major cause of reverse grading in plinian fall deposits is vent-widening by vent wall rock erosion during the eruption. As an eruption continues and the vent widens, the mass discharge rate increases and, because of the increase in mass and energy flux, increased convective velocities will cause the eruption column to rise, reflecting a waxing phase of the eruption. Particles of a given size will be taken to progressively greater heights in the column before being released and will then be transported downwind from the vent during the eruption. The increased proportion of larger clasts being transported downwind and deposited with time will build up a reversely graded deposit. Conversely, upward fining or normal size grading also occurs, and reflects the waning of the eruption intensity and eruption column height. Shifts in the wind direction or speed could also have this effect, as the position of the axis of dispersal (which is the maximum energy dispersal line) shifts. In stratified deposits, the stratification is often marked by subtle changes in pumice clast size, which could also be due to changing wind directions, or wind gusting and changes in velocity. However, local reverse grading could also be found in fallout deposit sequences, resulting from local deposition in water, or on very steep slopes followed by secondary mass (grain) flow (e.g. Duffield et al. 1979). Such deposits will have local irregularities in geometry, showing lensing and pinching and swelling (Fig. 9.56f).

Widening of the vent, together with an increased rate of vent erosion during the eruption, also explains why many plinian fall deposits show a vertical increase in the proportion of accessory lithics. An estimate of the lithic content therefore indicates the amount of vent wall erosion and the size of conduit. For example, the 4.6 ka Fogo A plinian pumice deposit on Sáo Miguel contains 0.09 km^3 of lithic fragments (14 wt%; Pensa et al. 2015), and if the magma source was at a depth of 5 km this would be equivalent to a hypothetical cored-out cylindrical conduit of diameter 78 m. However, conduit geometries are usually far from perfect cylinders, and Massaro et al. (2018) used the lithic amount and types to model the evolution of the magma feeding system during the 1995 BC Avellino pumice eruption at Vesuvius. Erosion is likely to be more important near the surface, where rocks are weak and less consolidated, and flaring of the vent is therefore probably likely (Wilson 1980).

Normal grading of pumice fall deposits is also relatively common (Fig. 9.56g) and, conversely, must reflect the waning stage of a column or a significant shift in wind direction, gusting, or velocity, and of the plume dispersal axis, so that at the one location the maximum clast size decreases over time. Settling of progressively finer grained particles also depends on the time lapse related to their settling velocity versus plume + wind drift. This means that usually fine particles do not settle at proximal and medial locations as they remain easily entrained in the eruption column and downwind plume, and are drifted to distal locations, so normal grading is rarely seen. However, at tropical latitudes, where cases of fallout deposits associated with conditions of no or little wind are not uncommon, normal grading has been described, such as at Cotopaxi, Ecuador (Barberi et al. 1995).

(a)
(b)
(c)
(d)
(e)
(f)
(g)

◀ **Fig. 9.56** Features of plinian fallout deposits and resedimented pumice deposits. **a** Hydraulically well-sorted plinian fallout deposit, with a clear difference in the diameter of low-density large pumice clasts and small dense lithic clasts, which have the same terminal fall velocity. Te Teko Pumice, Okataina volcanic centre, New Zealand. **b** Similar to (**a**), 172 ka Lower Pumice 2, Santorini, Greece. Note also the reverse grading of pumice clast sizes at the base. **c** Dark, fine-grained lithic fragment-rich layer at the base of the 144.6 ka Middle Pumice plinian fallout deposit, Santorini, representing the vent opening phase of the eruption (also see Figs. 9.67, 9.68). **d** Reverse size and compositional grading in the 3.9 ka Avellino fallout deposit, Vesuvius volcano, Italy. This shows compositional layering from white phonolitic Avellino plinian fall phase EU2 white, into EU3 grey tephri-phonolitic pumice upwards. The upwards coarsening reflects the waxing stage of eruption column growth, i.e. increasing in height. **e** Reverse size grading at the base of the 4.6 ka Fogo A pumice fallout deposit, São Miguel, Azores, Portugal, again reflecting a waxing stage of the eruption. **f** Reverse size grading in several stacked grain-flow deposits, resulting from resedimentation on the steep paleoslope that the primary fallout deposit was deposited on. Note the grain-flow deposit layers have irregular pinching and swelling or lensing geometry. **g** Two pumice fallout deposits showing both vertical normal and reverse size grading, interbedded with more poorly sorted and finer grained pyroclastic density current deposits. Precursor sequence of the 184 ka Lower Pumice 1 succession, Santorini. Both represent short-lived events and eruption columns

Layering/stratification

Internal layering, or stratification, also occurs in plinian fall deposits. It is usually best developed towards the vent and in distal locations and is commonly and significantly found towards the top of some deposits (Fig. 9.55a, c). Further away from the vent, such stratification may disappear, but then reappears in distal ash and fine lapilli deposits (e.g. Fig. 9.55h). Stratification varies from a crude internal layering to distinctly mappable fall units (e.g. Sulpizio et al. 2005, 2010; Miyaji et al. 2011; Pensa et al. 2015), although dividing such deposits up into discrete fall units can be difficult, unless there are sharp contacts or erosional breaks.

There are a number of causes for the development of stratification. Many eruptions are probably not truly steady and continuous throughout their duration, but pulsate irregularly, and this seems be common at the low-energy sub-plinian end member of the plinian family. Slight fluctuations in muzzle velocity and discharge rate will cause particles of a given size and terminal fall velocity to be released from different heights in the eruption column and plume with time. A pulsating column such as this would produce faint layering in a fall deposit. Wind shifts also cause stratification, defined by grain size changes, and these effects may be marked in distal ash deposits of plinian deposits.

More significant changes in activity are probably needed to cause distinct fall unit breaks. Explosive activity could temporarily cease, caused, for instance, by blockage of the conduit by the collapse of the vent walls. During such breaks, fine ash may settle out from the previous column as it dissipates, forming a thin, discrete fine-grained fall unit overlying the deposit of the maintained column. When activity recurs, the next fall unit may at first be rich in lithic fragments, as lithic fragments that blocked the vent are reamed out, and then show reverse grading of pumice clast sizes, as the eruption column re-establishes itself and increases in height and its ability to disperse pyroclasts further from the vent. Smaller collapses of the vent wall may just cause an increase in lithic clast content taken into the column, and these would be recognised as a layer of lithics, perhaps within a fall unit. There are also cases, like the 1707 Hoei eruption at Mt Fuji, where the historical record is very accurate and shows that the plinian eruption underwent six stop–start cycles over the 15 days of the eruption (Miyaji et al. 2011). However, the deposit today, even though it shows internal layering, does not indicate any evidence of pauses (Fig. 9.55e).

Critical for distinguishing a layering, which is caused by fluctuations during a steady phase of the eruption perhaps hours long, and layering that reflects much longer periods of eruption (days, weeks) and a stop–start behaviour of the column, is the detailed mapping and correlation of different sub-units. The reconstruction of different dispersal patterns, and changes in textural characteristics can reveal the multi-stage nature of stratified fall deposits (e.g. Houghton et al. 2014; Pensa et al. 2015). Cioni et al. (2015) propose that the former will be a simple stratified plinian fall deposit whereas the latter will be a complex or multiple (sub-plinian) deposit sequence.

All of the above mechanisms involve changes at the vent affecting the behaviour of the eruption column. However, stratification can also be generated away from the vent by particle aggregation, which could prematurely bring down fine ash from the plume in localised areas while the eruption continues (see next paragraph). Such fine-grained fall units will have isopach contours which do not close on the vent. Their distribution and thickness are related to the processes that favour aggregation, including the distribution and intensity of rain showers, and they may contain accretionary lapilli (Walker 1981d; Carey and Sigurdsson 1982).

In the above, we have only discussed the generation of stratification by the "random" processes which could occur at any time during a plinian event. The stratification and fall units found at the top of some deposits may, however, be related to more significant changes that are developing with time, as the plinian column becomes overloaded and unstable (e.g. Trolese et al. 2019). Sometimes, pyroclastic flow and surge deposits are found interbedded between separate fall units, particularly in proximal areas and where confined in paleovalleys, such is the case of the 4.6 ka Fogo A sequence in the Azores (e.g. Pensa et al. 2015), the 800 yr BP eruptive cycle of Quilota, Ecuador (Di Muro et al. 2004, 2008), and the 2015 Calbuco eruption, Chile (Romero et al. 2016). The deposits of these density current flows can be traced distally and laterally into fine ash-fall and surge deposits (Fig. 10.6). In many cases, plinian deposits are overlain by ignimbrites (Fig. 9.55a, c; Figs. 10.6b, 10.39a), and it seems that there is a continuum from a buoyant plinian eruption column stage to a pyroclastic flow-forming or collapsing column stage (Chap. 10). The stratification at the top of many plinian deposits may reflect instabilities in the column before wholesale, climactic collapse occurs and generates pyroclastic flows and ignimbrite deposits. For these reasons, a lot of care must be taken when studying fall sequences in the rock record, where the intervening flow deposits may not be exposed, and column instabilities and collapses may only be preserved as thin ash layers or stratification/layering of the fall deposits.

Where intra-plinian ignimbrites occur interstratified with pumice fallout deposits above and below (e.g. Fig. 10.6c), it is possible to decipher if the eruption column remained buoyant during the pyroclastic flow-forming phase (collapse of only part of the column) and continued to produce fallout at the same time as partial column collapse pyroclastic flows, or collapsed completely and was then regenerated by the growth of a new eruption column, indicating a time break was possible. In the former case, the grain size at the base of the fallout unit overlying the ignimbrite is essentially the same as the grain size at the top of the fallout unit beneath the ignimbrite. In the case of complete column collapse between two fallout-forming phases, the base of the fallout unit above the interstratified ignimbrite will the fine-grained, and may contain a concentration of lithic clasts, indicating vent opening, and then above the base, the fallout unit becomes coarser grained (i.e. it is reverse size graded), indicating a progressive increase in the height and energy of the newly formed eruption column.

Detailed chemical stratigraphy for the 1912 Novarupta-Katmai eruption sequence showed compositional homogeneity in the fallout deposits below and above ignimbrites, indicating that fallout deposits formed contemporaneously with pyroclastic flow deposits, and that the column was simultaneously sustained and collapsing (Hildreth and Fierstein 2012; see Chap. 10).

A number of plinian fall deposits are known to be compositionally zoned (Fig. 9.56d), or to contain mixed pumice clasts (Fig. 9.57a, b). Compositionally zoned plinian pumice fall deposits show an upward vertical increase in the proportion of a more basic juvenile component (e.g. the 1995 BC Avellino, Fig. 9.56d, and 79 AD pumice fall deposits at Vesuvius, Cioni et al. 1995, Sulpizio et al. 2010; 4.6 ka Fogo A deposit, Azores, Fig. 9.57a, b). Rarer examples are known where this type of zonation is reversed (e.g. Cape Riva unit, Santorini; Druitt 1985). The boundary between zones can be gradational or very abrupt. This is commonly

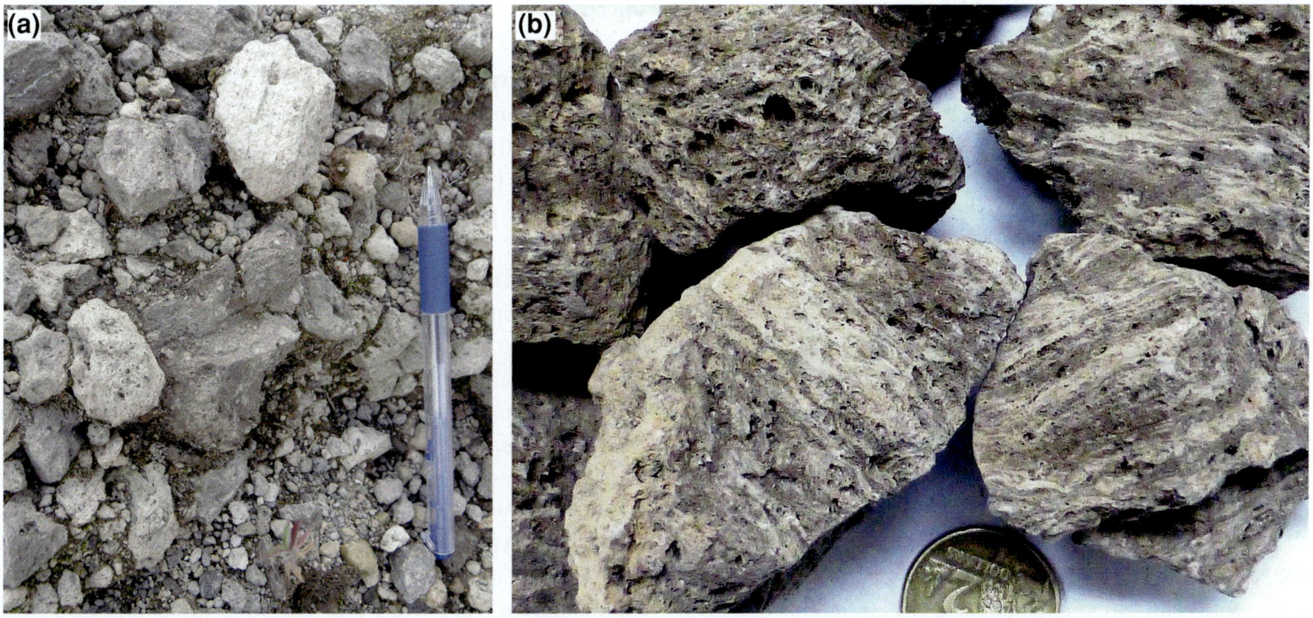

Fig. 9.57 White, dark, and mingled pumice from the top of Phase 2 fallout deposits of the 4.6 ka Fogo A fall deposit, São Miguel, Azores, Portugal. **a** White and dark pumice clasts in outcrop. White pumices are trachytic in composition and dark pumices more mafic trachy-dacite in composition. **b** Banded, compositionally mingled pumice clasts occur towards the top of the sequence

not marked by grain size differences, showing that the discharge of magma was steady, although the composition of the magma was changing. In some cases, there is an almost complete change in magma types, in others, there is just a slight change in the proportion of the two types. Banded-pumice clasts are common in some deposits and indicate mechanical mingling of the magma types before and during the eruption (e.g. Fig. 9.57b). Some deposits, although not distinctly zoned, can have a significant proportion of mingled pumice (e.g. the Fogo A deposits, Pensa et al. 2015, or the 1912 Novarupta-Katmai fall deposits, Hildreth and Fierstein 2012). Aspects of the role of magma mixing and mingling in triggering explosive volcanism were discussed in Sect. 4.5.7, Chap. 4.

9.7.4.3 Deposits from Sub-Plinian Eruptions

Sub-plinian deposits are pumice fall deposits which resemble plinian deposits in the field, as described above, especially near the vent, but when mapped out they have a smaller dispersal area and volume than plinian deposits (Fig. 9.58a; e.g. Self 1976; Cioni et al. 2003; Pardo et al. 2012; Vona et al. 2020). Walker (1973) set arbitrary D limits for them between 5 and 500 km^2 (Fig. 9.6a), and this implies column heights <20 km. However, the lower D values certainly overlap with violent strombolian dispersals and again it is important here to recall that there is a continuum or spectrum of processes and deposits that cannot be fully and artificially separated.

Sub-plinian eruptions are scaled-down plinian eruptions, and their mechanisms and dynamics can be treated as essentially the same (see Sects. 9.2.2.1, 9.3, 9.4). Large lithic clasts indicate that in some eruptions muzzle velocities are as high as in some plinian events (>400 m s^{-1}), although the lower range is 100 m s^{-1} (Wilson 1976). The mass discharge rate is likely to be lower for sub-plinian events, and this is the main factor controlling eruption column height and dispersal (Table 9.2). In particular, the lower mass eruption rates reduce the mass supply to the column, and ultimately the heat budget for its convection and rise, resulting in lower column heights, pulsatory and unsteady behaviour, and a lesser resistance to the effect of local winds, even at low elevations. Furthermore, the intimate unsteadiness of the eruption makes it more prone to episodes of vent or conduit closure and opening, resulting in stop–start behaviour, often over periods of activity longer than for plinian eruptions.

The resulting sub-plinian deposits are therefore commonly internally layered or distinctly stratified (Fig. 9.58b, c, f), with sub-units characterised by different dispersal patterns, over narrow and elongated areas (Fig. 9.58a). In medial areas, they form relatively thin, but well-sorted deposits (Fig. 9.58g). Many fallout deposits form during an early explosive phase accompanying the effusion of a small rhyolite dome or coulee (e.g. the 2011 Puyehue-Cordon Caulle eruption; Schipper et al. 2013; Pistolesi et al. 2015).

Some well-documented examples of sub-plinian fallout deposits are from Terceira in the Azores (Fig. 9.33; Self 1976), Vesuvius (e.g. the 472 AD Pollena eruption; Sulpizio et al. 2005; Vona et al. 2020; Fig. 9.58a–e), several eruptions of Mt Ruapehu, New Zealand (Pardo et al. 2012), 2008 Chaitén eruption, Chile (Alfano et al. 2011), and 2011 Cordón Caulle eruption, Chile (Pistolesi et al. 2015). All are well-stratified (Fig. 9.58b, c, f) which suggests that there were large fluctuations in the gas exit velocity, and hence mass discharge rate. This would also inhibit the development of a fully maintained convective plume, which would therefore not attain the heights associated with plinian columns. Sub-plinian eruptions can lead to the generation of ignimbrite-forming pyroclastic flows similar to the larger plinian ones. This is shown in the eruption of Krakatau in 1883, from which the pumice fall deposit which preceded an ignimbrite is sub-plinian rather than plinian in its characteristics (Self and Rampino 1981).

A number of basaltic or near-basaltic scoria fall deposits are now also known to be sub-plinian or even plinian in their dispersal characteristics. Examples include Pleistocene eruptions from Colli Albani, Italy (Giordano et al. 2010), Late Pleistocene eruptions of Masaya, Ecuador (Costantini et al. 2009), 3.9 ka eruption of Etna, Italy (Coltelli et al. 2005), 1886 Tarawera eruption, New Zealand (Walker et al. 1984; Houghton et al. 2004a, b; Plate 9.1b; Fig. 9.55d), and the 1707 Hoei eruption, Fuji, Japan (Miyaji et al. 2011; Fig. 9.55e).

9.7.4.4 Ultra-Plinian Deposits

"*Ultra-plinian*" was introduced as a separate deposit type by Walker (1980) to describe the most widely dispersed plinian fall deposits (Fig. 9.6), and specifically the Taupo pumice, which forms part of the eruption sequence of the 232 AD eruption of Lake Taupo caldera, New Zealand (Sect. 9.2.2.3). The products of this eruption will be discussed in more detail in Chap. 10.

The Taupo pumice is currently the most widely dispersed pyroclastic fall deposit known. The average largest pumice and lithic clast sizes are dispersed significantly further than for plinian deposits (Fig. 9.59a, b). The Taupo pumice only has a maximum measured thickness of 1.8 m which, by comparison with most near-vent plinian deposits, is relatively thin. Another feature is that the deposit is very enriched in free crystals. This results from the high degree of fragmentation, and from the loss of a large proportion of vitric material by column fractionation processes into a downwind, distal ash plume. Data from crystal concentration studies on the Taupo pumice show that 80% of the erupted

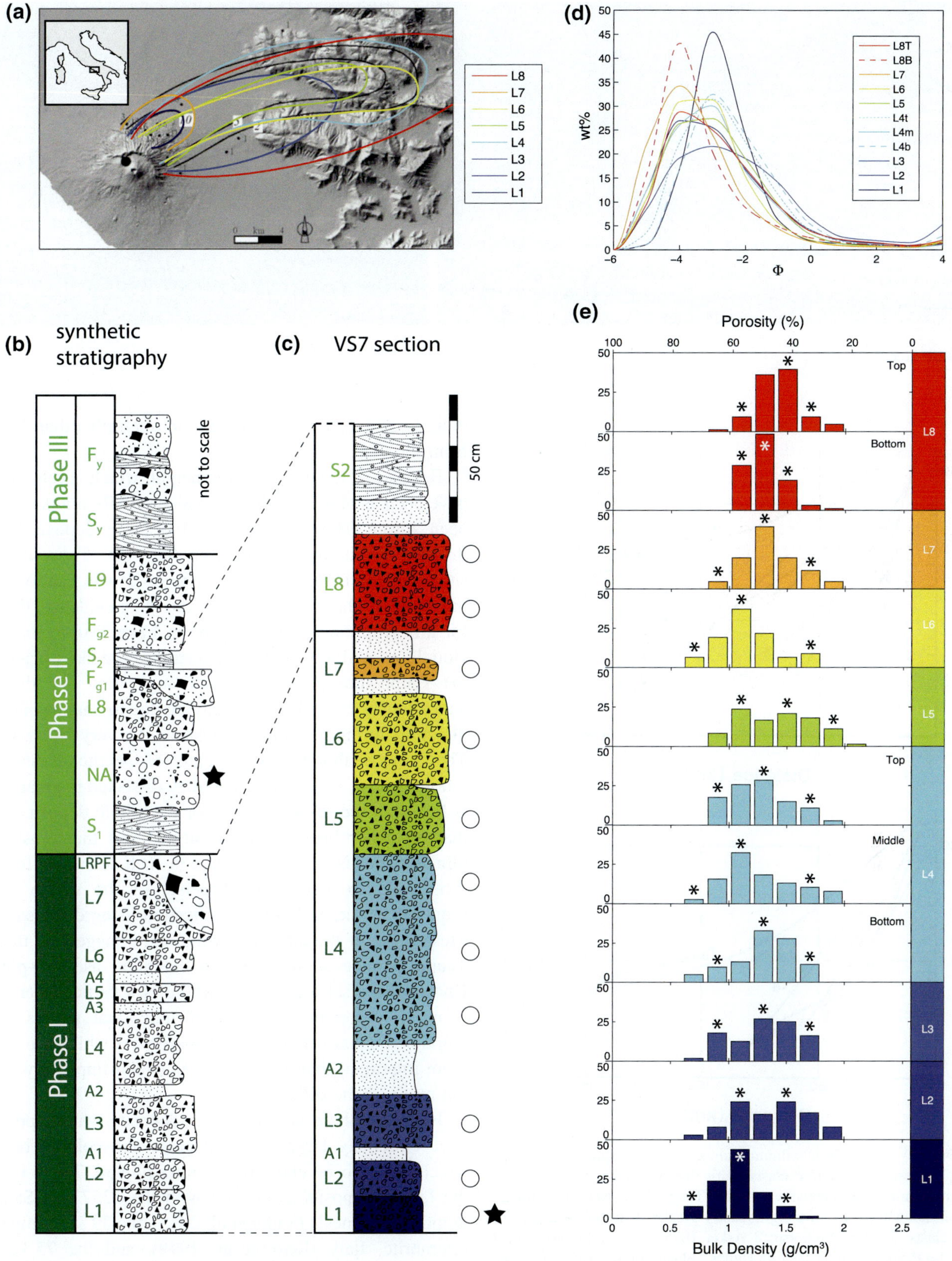

Fig. 9.58 Features of sub-plinian fallout deposits. **a–e** Pollena 472 AD sub-plinian deposits, Vesuvius volcano, Italy (From Vona et al. 2020). **a** Isopach maps of the various sub-units represented by overlapping the 10 cm isopach lines for each sub-unit. **b** Generalised stratigraphic sequence, showing the stratified nature of the succession. **c** Stratigraphic sequence at locality VS7 showing **d** grain size, **e** porosity distributions. **f** Stratified facies of the 1944 Vesuvius eruption, at the crater rim. **g** Massive pumice fallout deposit (white layer) of the 1887 Cotopaxi eruption, Ecuador, at medial distances. The layer is ~0.6 m thick and well-sorted

Fig. 9.58 (continued)

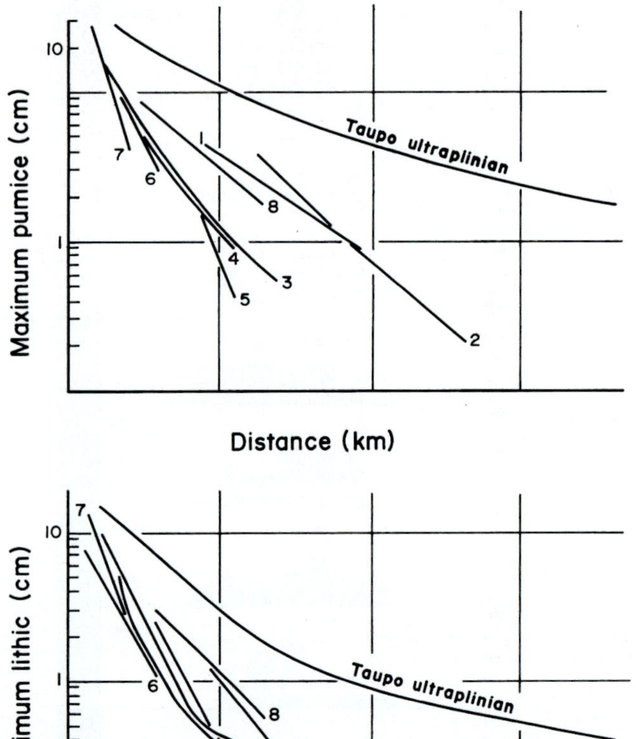

Fig. 9.59 Variation in maximum pumice and lithic diameter (average of the largest three or five clasts) with distance from vent for some plinian pumice fall deposits and the Taupo 232 AD ultra-plinian deposit, showing it to have significantly greater dispersal distances of large clasts. 1 32 ka Shikotsu; 2 Askja 1875; 3 Vesuvius AD 79; 4 95 ka La Primavera B; 5 10.5 ka Upper Toluca; 6 Fogo 1563; 7 4.6 ka Fogo A; 8 1.6 Ma Lower Bandelier. (From Cas and Wright 1987, After Walker 1980, and Self et al. 1986)

volume, mainly vitric ash, fell out to sea further than 220 km from the vent.

Because of the great column height, which is estimated to have been as high as >50 km (Walker 1980), but could have been lower (Houghton et al. 2014), the deposit is also thickest 20 km downwind from the vent. Thus, in these situations the isopach map must be used with caution as an indication of the vent position. From other evidence, the vent for this eruption is known to be in Lake Taupo. Recently, Houghton et al. (2014) have questioned the classification of the Taupo ultra-plinian deposits, suggesting that the very anomalously large dispersal may be due to changes in wind direction during the eruption, instead of a very high column. They propose that the dispersal area of specific phases of the fallout deposit sequence, which they dissected in great detail, is less than the dispersal area for the whole deposit. Given that no major breaks were identified and that most other plinian deposits are not dissected in such detail, we feel that it is valid to assess the dispersal of the whole fallout sequence as the basis for classification using dispersal area, irrespective of the effects of wind shift, suggesting that the Taupo pumice should still be classified as ultra-plinian. Known dispersal area for the whole of a particular eruption is more important than that of individual phases of an eruption for hazard assessment purposes, which we feel is more important than minimising potential impact by focusing more on the different parts.

However, there are also other deposits which appear to have extremely large dispersal areas and column heights, which can also be classified as ultra-plinian. These include the fallout deposits associated with the 1257 AD Samalas eruption, Indonesia (Vidal et al. 2015), the 39 ka Campanian ignimbrite, Italy (Rosi et al. 1999), and the 75 ka Los Chocoyos eruption of Lake Atitlán caldera, Guatemala (Kutterolf et al. 2016; Cisneros de León et al. 2021), and the

1.26 Ma Upper Bandelier Tuff eruption of Valles caldera, New Mexico, USA (Self et al. 2021; see Fig. 10.22d). However, Smith et al. (2016) and Costa et al. (2018) argue that the most widely dispersed ash is co-ignimbrite ash derived from the phase of the eruption when ring fissure vents were able to feed mass fluxes two orders of magnitude larger (up to 10^{11} kg s^{-1}) than those that can be attained by the largest single-vent plinian eruptions (10^9 kg s^{-1}). Another candidate is the ~6.6 Ma Corte Blanco Tuff (>36 km^3 DRE) which represents deposition from a large, relatively stable plinian style eruption column erupted from a point source vent, which is estimated to have peaked at heights as high as 40 km a.s.l., with only minor intra-plinian pyroclastic flows (Fig. 9.17; Tait 2004). Thick, fine-grained (garnet-bearing) ash correlatives are found throughout the sub-Andes of northwest Argentina and Bolivia, indicating the extremely large dispersal area of the eruption (Viramonte et al. 1994; Coira et al. 2018). This ranks as the largest plinian fallout deposit-forming eruption currently documented in the central Andes, and perhaps one of the largest recorded on Earth.

9.7.5 Fallout Deposits from Eruptions Affected by Magma–Water Interaction

Surtseyan and *phreatoplinian* are the terms commonly used to describe pyroclastic fall deposits resulting from eruptions which have taken place in the sea or a lake, or by contact with groundwater or glacial ice (Sect. 9.2.4). Such eruptions are most generally called phreatomagmatic or hydrovolcanic (Sect. 4.6.9, Chap. 4). Both types of deposits have extreme fragmentation, and this results from the magma–water interaction. Surtseyan pyroclastic fall deposits have moderate dispersal, while phreatoplinian deposits can be extremely widely dispersed (Fig. 9.6a).

There is a tendency to associate specific magma compositions with each of these two types of deposit, based on the compositions of the type examples used to define the terms originally (basaltic for surtseyan, rhyolitic for phreatoplinian; see below). However, given the right conditions, the eruption of any magma type may produce the dispersal and fragmentation characteristics defining these two types of deposits. It may thus be possible to find rhyolitic surtseyan deposits (Austin-Erikson et al. 2008), and andesitic, trachytic, and even basaltic and foiditic phreatoplinian deposits (e.g. De Rita et al. 2002).

In addition, we add a distinction between small-scale phreatomagmatic eruptions occurring as a result of magma interaction with surface bodies of water (surtseyan) and those with subsurface aquifers or permafrost, which we call ukinrekian, after the 1977 eruptions of the Ukinrek maars in Alaska (Kienle et al. 1980; Self et al. 1980; Ort et al. 2018; see Sect. 9.2.4 above).

9.7.5.1 Surtseyan and Ukinrekian Fallout Deposits

Phreatomagmatic activity usually produces maars, tuff rings, and tuff cones (Chap. 16, Sect. 16.3), which are more common in (but not exclusive to) basaltic volcanic fields. These constructional forms are largely built up from the deposits of base surges (Chap. 10, Sect. 10.14.2), ballistic ejecta, and thin ash-fall beds that occur downwind (Fig. 9.60a–d). Several examples of this type of eruption have been witnessed in historic times (e.g. 1960–64 eruption of Surtsey, Iceland, Thorarinsson et al. 1964; 1965 Taal, Philippines, Moore et al. 1966; and more recently in 2020, 2021; Ukinrek, Alaska, Fig. 9.11c, Kienle et al. 1980; Self et al. 1980; Ort et al. 2018; 1979 eruption of Soufrière, St Vincent, Shepherd and Sigurdsson 1982; 2005 eruption at Vanuatu, Nemeth et al. 2006; 2007 eruption at Ruapehu, New Zealand, Kilgour et al. 2010; 2008, 2014–15 Hunga Tonga–Hunga Ha'apai, Tonga eruption, Colombier et al. 2018; Fig. 9.11b; and again from 2021–2022, including surtseyan activity from 20 December 2021 to 14 January 2022, followed by the spectacular 15 January phreatoplinian eruption). Eruptions through crater or caldera lakes have also produced phreatomagmatic air-fall deposits and are also included under our broad definition. Glacio-surtseyan explosive activity was witnessed during the high-impact 2010 eruption of Eyjafjallajökull volcano, Iceland (Gudmindsson et al. 2012; Magnússon et al. 2012).

The principal difference between surtseyan and ukinrekian styles of eruption is that during *surtseyan* eruptions, magma erupts into a surface body of water, which is explosively superheated, whereas with *ukinrekian* eruptions, it is subsurface groundwater or permafrost that is superheated explosively by magma rising through it. Surtseyan deposits will be dominated by juvenile clasts, and minimal accessory components, because the explosive fragmentation surface sits at the sea or lake floor, where erupting magma interfaces with the body of water. In contrast, ukinrekian deposits will contain a significant proportion of crustal accessory lithic debris as well as juvenile clasts because the explosive fragmentation surface sits at the level of the subsurface aquifer driving the explosive magma–groundwater interaction. Most accessory xenoliths will therefore consist of the aquifer host rock, as well as those derived from strata located above the aquifer, and through which the conduit is explosively excavated. In many cases, the subsurface stratigraphy can be reconstructed from the xenoliths in the deposits. In addition, if the conduit is progressively excavated downwards driving the fragmentation surface deeper, this migration can be reflected in the changing xenolith

(a)
(b)
(c)
(d)
(e)
(f)
(g)

◂ **Fig. 9.60** Phreatomagmatic fallout deposits. **a** Quaternary, planar stratified basaltic ash fall deposits (lower, thin, dark, mantle-bedded deposits) from a maar located along the shoreline of Miyakejima island, Japan. **b** Detail of (**a**), showing abundant accretionary lapilli. Slight thickening–thinning of some strata may suggest surge-modified fallout influence. **c** Late Devonian, accretionary lapilli-rich, basaltic fall deposit, sourced from a nearby basaltic tuff cone. Boyd Volcanic Complex, Eden, southern New South Wales, Australia. **d** Planar laminated, rhyolitic lapilli and ash beds from the remnant tuff ring, Ramadas volcanic centre, ~6.6 Ma Corte Blanco Tuff, northwestern Argentinian Andes. **e** Planar laminated ash and fine lapilli beds, distal part of maar rim sequence, 35 ka basaltic Tower Hill volcanic complex, Victoria, Newer Volcanics Province, southeastern Australia. **f** Planar stratified, 25.4 ka bedded phreatoplinian ash, with some thin interbedded, coarser, thin ignimbrite units, Oruanui Formation, east of Taupo caldera, Taupo, New Zealand. **g** Very fine, wet, distal, laminated phreatoplinian ash at the base of the 0.22 Ma La Caleta member, Tajao, Tenerife. Note the soft state deformation, induced by the impact of lithic clasts from the base of the overlying magmatic plinian fallout deposit

population preserved through the stratigraphy of the pyroclastic deposits that accumulate around the vent.

The principal fallout component of surtseyan deposits are ballistic blocks transported by ballistic tephra (cock's tail) jets, and distal ash fallout, which may be preserved in lacustrine successions, but are likely to be reworked and resedimented in marine settings. In ukinrekian successions, tephra jet deposits will also occur, and co-surge ash fallout deposits may be deposited between base surge deposits in the constructional volcanic edifice sequences. Surtseyan eruptions build shallow water tuff cones and tuff rings, whereas ukinrekian eruptions are responsible for building subaerial maar volcanoes, tuff rings, and occasionally tuff cones.

The downwind air-fall deposits are thin, fine-grained ashes and fine lapilli-ashes, and internally they may be well-laminated (Fig. 9.60a–e) because phreatomagmatic activity occurs as a succession of discrete explosions. They often contain accretionary lapilli (Fig. 9.60b, c), and SEM photographs of the ash particles show very angular, broken surfaces (Chap. 4) due to the magma–water interaction (Heiken and Wohletz 1991; Büttner et al. 1999; Mattsson 2010). For air-fall deposits, they can be poorly sorted (Fig. 9.60b). Studies of the 1979 Soufrière, St Vincent, air-fall ash layer have shown that thickness and grain size decrease with distance, and bimodal sorting may be an important feature (Sigurdsson 1982, Brazier et al. 1982; Fig. 9.61). Bimodality and poor sorting are attributed to the premature fallout of aggregated wet or damp ash clumps in the eruption column or downwind plume at the same settling velocity as coherent juvenile and lithic lapilli (see Sect. 9.7.7). Accreted ash could occur as accretionary lapilli, or as unstructured ash clumps which may be difficult to distinguish as discrete clasts rather than fine ash matrix. During the 1979 Soufrière activity, such ash clumps were observed to fall and then disintegrate on impact. Nearer to the vent, this type of air-fall deposit is found interbedded with base surge deposits. Other evidence of such ashes being wet may be cavities in the ash, which confers to the ash deposit the appearance of a vesiculated tuff. These are however most likely formed as pore spaces between coalesced, wet accretionary lapilli when deposited (Rosi 1992). However, in near-vent situations transport and depositional processes are complex because of simultaneous fallout, surge, and ballistic transport and deposition. Deposits often have hybrid characteristics, such as apparent planar stratification, but subtle lensing or pinching and swelling, and very low-angle truncations. Cas (1989) called deposits with such hybrid characteristics *surge-modified fall deposits* and *fall-modified surge deposits*, depending on dominating characteristics (also see Chap. 10, Sect. 10.14.5.1).

Surtseyan eruption fall and surge sequences are known from modern oceans, lakes, and englacial settings, as well as the rock record (e.g. Verwoerd and Chevallier 1987; Cas et al. 1989; Sohn and Chough 1992; Moorhouse et al. 2015). They are usually also associated with intra-sequence debris flow and turbidite deposits resulting from syn-eruptive remobilisation of the wet tephra (e.g. Giordano et al. 2002). Ukinrekian deposits are widely described in the abundant literature on modern maar and tuff ring volcanoes and their eruption sequences (e.g. Sheridan and Updike 1975; Sohn and Chough 1992; White 1991; De Benedetti et al. 2008; van Otterloo et al. 2013; Jordan et al. 2013). However, in these accounts, the main focus is on the near-vent edifice deposits, which are dominated by base surge and ballistic tephra jet deposits, whereas distinctive, unambiguous, subaerial phreatomagmatic fallout deposits have rarely been described and identified. The long-term preservation potential of both deposit types is poor due to the effects of both marine erosion (surtseyan) and subaerial erosion (ukinrekian).

9.7.5.2 Phreatoplinian Deposits

The term "*phreatoplinian*" was introduced by Self and Sparks (1978) for the silicic, plinian scale analogue of surtseyan deposits, and they described several examples, documenting in detail widespread ash layers from the 25.4 ka Oruanui Formation (later redefined and described by Self 1983 as the Wairakei Formation; see also Wilson 2001), New Zealand, and from a phreatomagmatic phase of the 1875 Askja eruption (Self and Sparks 1978; Sparks et al. 1981; Carey et al. 2009, 2010; Figs. 9.62, 9.63). Walker (1981a) documented the Rotongaio ash and the Hatepe ash, both of which were formed during phreatomagmatic episodes during

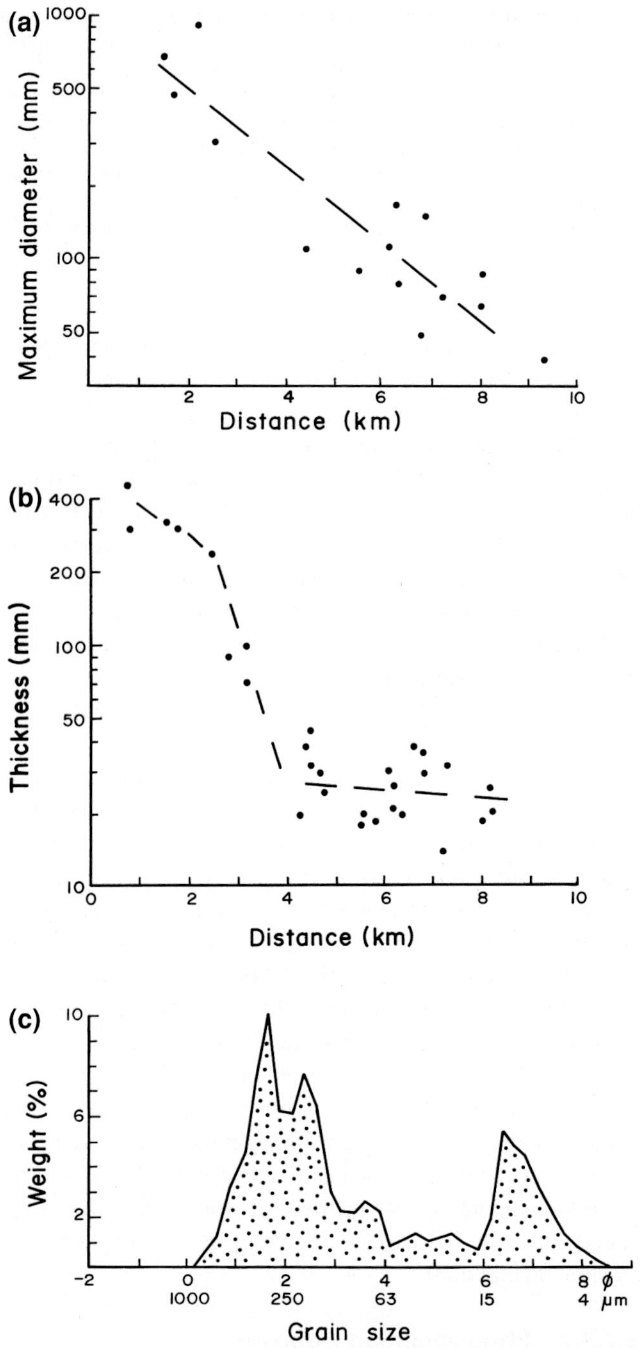

Fig. 9.61 Characteristics of the phreatomagmatic ash fallout deposit erupted in 1979 from Soufrière, St Vincent, Lesser Antilles. Variation in **a** maximum dense clast diameter and **b** thickness, both as a function of distance from vent. **c** Bimodal size distribution in a sample collected from the lower part of the deposit. (From Cas and Wright 1987, After Sigurdsson 1982)

the Taupo 232 AD eruption (Fig. 9.64). Other documented examples of phreatoplinian layers are from the Santorini Minoan 1600 BC eruption (Bond and Sparks 1976); the 79 AD and Avellino 1995 BC eruptions, Vesuvius (Rolandi et al. 1993; Cioni et al. 2003); 15 ka Neapolitan Yellow Tuff at Campi Flegrei (Orsi et al. 1992); 0.6 Ma Granadilla pumice member on Tenerife (Bryan et al. 2012); 161 ka Plateau Tuff, Kos, Greece (Allen and Cas 1998); Middle Pleistocene Pisolitic Tuffs, Colli Albani (De Rita et al. 2002); 4.6 ka Fogo A, Azores (Pensa et al. 2015). In practice, just as there are variations in eruption intensity and dispersal area across the plinian spectrum of eruptions, there will be variations in scale between surtseyan/ukinrekian and phreatoplinian eruptions and deposits.

All of these phreatoplinian fallout deposits were produced during phreatomagmatic phases of much larger eruptions which involved several different styles of activity. Collectively, these include plinian, base surge, and ignimbrite-forming activity. Most of the examples involved the eruption of rhyolitic magma through caldera lakes. However, the cases of Vesuvius, Campi Flegrei, and Colli Albani demonstrate that phreatoplinian activity may occur for almost any composition. At Colli Albani, where phreatoplinian activity also involved interaction of rising magma with a caldera lake, at least four caldera-forming eruptions, collectively known as the Pisolitic Tuffs, are all characterised by tephri-phonolitic to K-foiditic compositions (De Rita et al. 2002; Giordano et al. 2010). Critical in the assessment of the type of water involved is both the lithic content and the stratigraphic position of the phreatoplinian layer. For example, in the Fogo A fall deposit, three thin ash layers (a1, a2, and a3) open the sequence and are lithic-free, suggesting an interaction with a caldera lake (Pensa et al. 2015). In contrast, in the case of the Vesuvius 79 AD and Avellino 1995 BC eruptions, phreatoplinian ash layers and ensuing pyroclastic flow sequences are associated with the appearance of lithics, including Mesozoic limestones, derived from deep portions of the plumbing system, suggesting that water from a deep aquifer had gained access to the magma chamber or conduit (Barberi et al. 1989; Sulpizio et al. 2010).

Phreatoplinian fallout deposits have an extremely wide dispersal (Figs. 9.6a, 9.62, 9.63) indicating a high eruption column (of plinian proportions), yet all of the other characteristics of the deposits indicate that they have a phreatomagmatic origin. Deposits are very fine-grained, even close to the source, and accretionary lapilli are common. They may be finely laminated, and the laminae may show signs of soft state deformation (Fig. 9.60g) and near the vent they are often associated with base surge deposits (see Chap. 10). Deposits are poorly sorted for pyroclastic falls, especially considering their median grain size, and their size distributions are often bimodal or strongly negatively skewed, or both, indicating that they have an extended coarse tail (i.e. dominant fine ash fraction; cf. Figure 9.61c). In contrast, plinian deposits are positively skewed. SEM studies show that typical phreatomagmatic blocky shards are typical (see Chap. 4), although the phreatoplinian deposits of

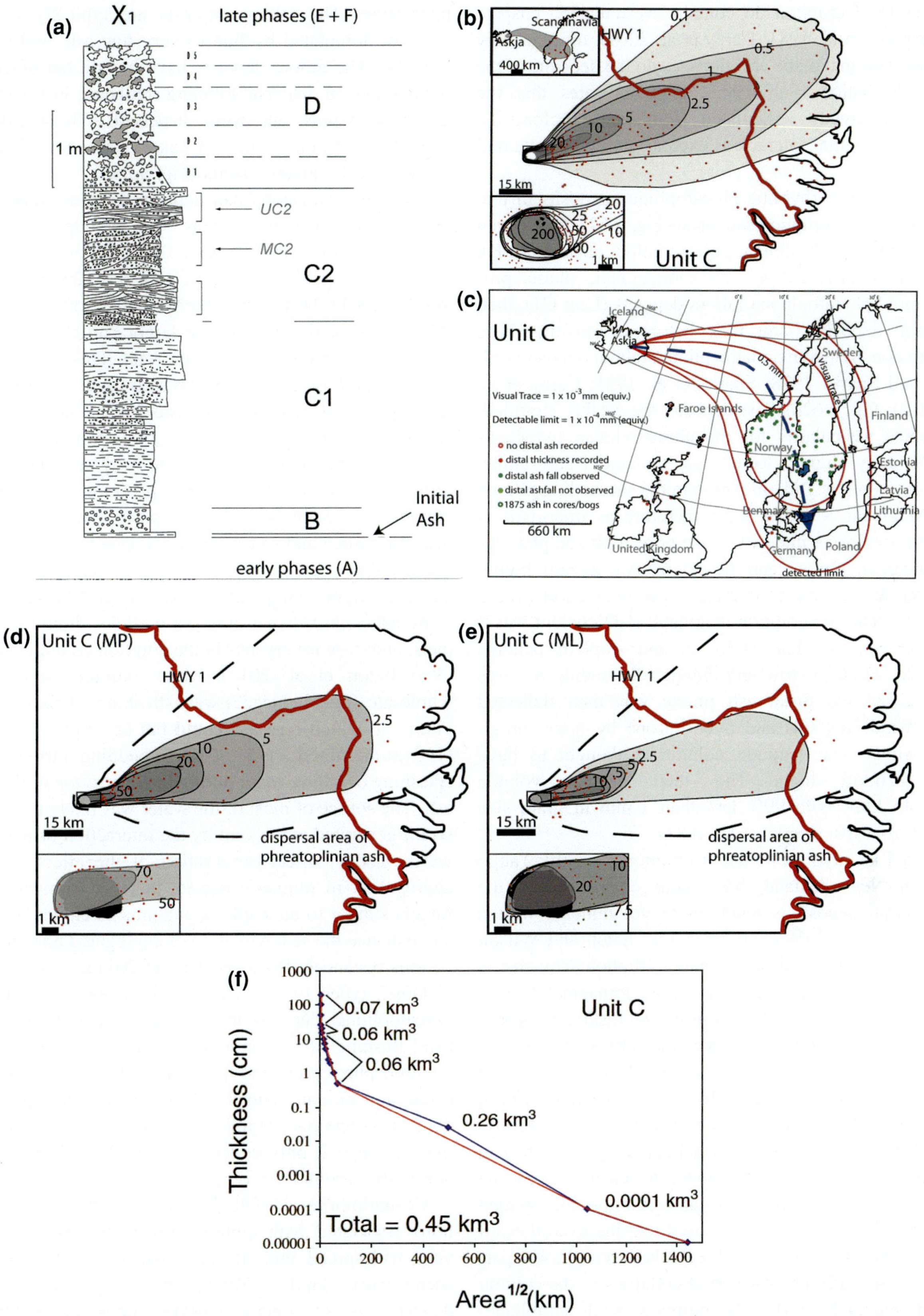

Fig. 9.62 Characteristics of the rhyolitic Askja 1875 phreatoplinian deposits, Iceland. **a** Context within the overall eruption sequence, where unit B is a pumice fallout deposit, C1 is a largely phreatoplinian fallout deposit, C2 consists of pdc deposits, and D represents a magmatic fallout event, including non-welded and welded pumice fallout deposits. **b** Dispersal pattern for C1 phreatoplinian deposits, in Iceland, and **c** showing the distal very fine ash deposited in Europe. **d** Pumice and **e** lithic clast isopleth patterns, and **f** thinning trend with distance from vent for unit C, plotted as area$^{1/2}$ of isopachs defined by multiple straight-line segments that more or less define a power law fit. (Adapted from Carey et al. 2010)

the Askja 1875 eruption do contain "vesiculated" cuspate shards (Sparks et al. 1981; Carey et al. 2009), more usually associated with magmatic eruptions, as do the deposits of the Oruanui Formation (Self 1983). This indicates that the magma had already vesiculated significantly before the vesiculated magma interacted explosively with external water.

The Askja 1875 rhyolitic phreatoplinian deposits in Iceland are part of an eruption sequence (Fig. 9.62a) that began with a magmatic, sub-plinian, pumice fallout phase (unit B), and then transformed into a phreatomagmatic phase, producing initial phreatoplinian fallout deposits (Unit C1), then base surge deposits (Unit C2), before culminating in a welded to non-welded magmatic plinian fallout deposit (Unit D; Self and Sparks 1978; Sparks et al. 1981; Carey et al. 2009, 2010; Fig. 9.62a). The deposits are weakly stratified. The pumice vesicularity for the phreatoplinian pyroclasts is ~83%, only slightly lower than the initial sub-plinian pumices and only slightly higher than the succeeding plinian phase. Magmatic volatiles were therefore as important in driving the explosive intensity of the phreatoplinian phase as in the magmatic phases, and the magma was already highly vesiculated at the time of magma–water interaction (Carey et al. 2009). The pre-eruption magmatic H_2O content was 3 wt%. The isopach (Fig. 9.62b, c) and isopleth patterns (Fig. 9.62d, e) are narrow and elongated, trending towards the ENE, but the distal ash plume was then deflected southeastwards towards and over Europe by upper atmospheric winds. The eruption column is thought to have reached ~26 km high. The total deposit volume was ~0.45 km^3, with 99% less than 1 mm in grain size (Carey et al. 2010; Sparks et al. 1981).

The 25.4 ka rhyolitic Oruanui eruption from the Taupo caldera in New Zealand, was more complex, and the phreatoplinian deposit is much more voluminous, with a volume of 430 km^3 (Wilson 2001; Van Eaton and Wilson 2013). Close to the vent, the Oruanui eruption sequence is overwhelmingly represented by phreatomagmatic ignimbrites (Fig. 9.63a, b), whereas in medial distances there is an intercalation of thin ignimbrites and phreatoplinian ash fallout deposits (Fig. 9.60f), and distally the deposits are all fallout (Fig. 9.63a). Originally documented by Self (1983), who subdivided the sequence into 6 units, Wilson (2001) extended the previous work and defined 10 phases/units in the composite stratigraphy. In medial to distal settings, the sequence is well-stratified, with numerous normally graded fallout units. Wilson (2001) proposed that the overall eruption was very stop–start, with breaks between phases, particularly in the early phases. Partial collapses of the column occurred frequently during the eruption, while simultaneously relatively dry pumiceous and wet phreatomagmatic fallout deposits, some with abundant accretionary lapilli, were being deposited. This suggests varying water–magma interaction. The fallout deposits are typically negatively skewed, dominated by fine to very fine ash, and some are bimodal. The fallout deposits have a wide fan of dispersal, with an axis of dispersal trending southeast, just north of the Chatham Islands, and were dispersed at least 1,000 kms from the vent in the centre of Lake Taupo, New Zealand, to beyond the Chatham Islands (Fig. 9.63c).

Laterally, phreatoplinian deposits become imperceptibly thin and fine over wide areas, with perhaps a power law thinning pattern being the best, although not a perfect fit for the pattern (e.g. Figure 9.62f; Carey et al. 2010; Fig. 9.63d; Wilson 2001). Downwind, there is little sorting of the size distribution, and it is only the larger particles that are lost with increasing distance from the vent. This type of grading is coarse-tail grading, and contrasts with the distribution grading typical of plinian deposits where, laterally, sorting affects the whole size distribution. Downwind sorting of fines must be inhibited in some way and involves the bringing down of different-sized particles in aggregates or clumps to explain why these deposits are so fine-grained near the source and so poorly sorted (see Sect. 9.7.7). These could fall as accretionary lapilli, but perhaps in many cases as unstructured clumps of wet ash, formed through electrostatic adhesion between fine ash particles, liquid water droplets, and even ice crystals in the eruption column and plume (Van Eaton et al. 2012). More extreme would be the wholesale (rain induced?) water-flushing of the downwind plume, in which case ash could fall as "mud-rain". Walker (1981a) described a type of micro-bedding attributed to the splashing of falling water during the deposition of the Hatepe ash. The source of most of the water was likely to have been steam-generated at the vent by the interaction of magma and water. Today, direct observation of the effect of particle aggregation in plumes suggests that the formation of ash fingers similar to atmospheric mammatus clouds does occur and enhance the sedimentation of aggregated fine particles at various distances (Rose and Durant 2011).

Other types of pyroclastic fall deposits can resemble phreatoplinian deposits in the field. One is formed by the local flushing of a downwind plinian or sub-plinian plume by atmospheric rain, and this type we have mentioned while discussing plinian deposits. Such an ash-fall deposit could contain accretionary lapilli, but an isopach map would show that the deposit only covered a very limited area, and not necessarily around the vent.

Co-ignimbrite ash-fall deposits (see the following section) associated with ignimbrite-forming eruptions can be very widespread and, if rain-flushed, could also contain accretionary lapilli. Texturally, co-ignimbrite ash-fall deposits should contain delicate shards and bubble wall fragments, suggesting fragmentation by exsolving magmatic gases, as opposed to blocky shards typical of phreatomagmatic fragmentation. It has now also been recognised that

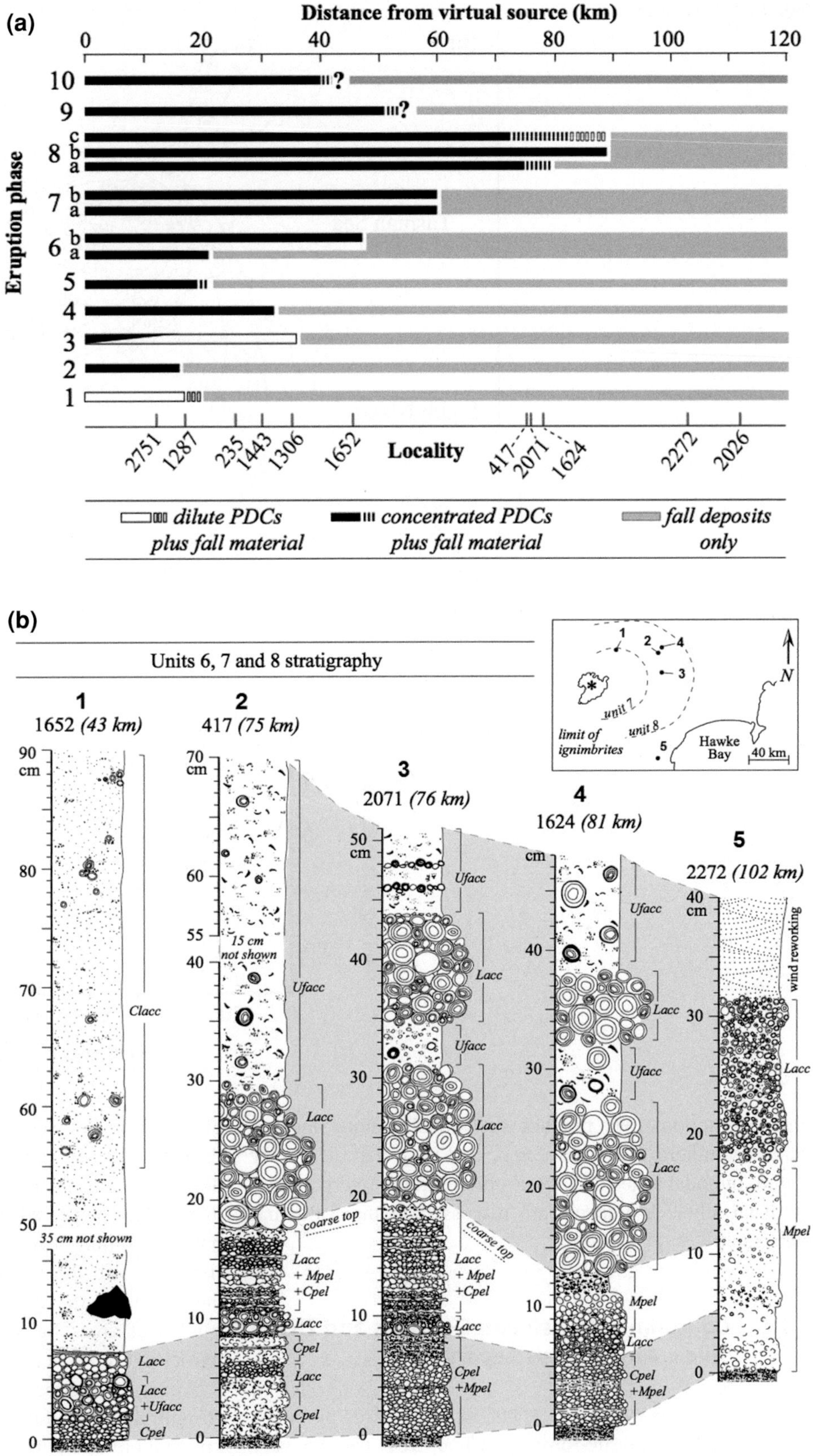

Fig. 9.63 Characteristics of the 25.4 ka rhyolitic Oruanui phreatoplinian deposits, Taupo caldera, New Zealand. **a** Schematic representation of the unit stratigraphy and lateral facies relationships and changes, between the fallout and ignimbrite deposits. **b** Lateral facies changes and correlations for part of the eruption sequence, showing the lateral change from predominantly ignimbrites near vent to fallout deposits with abundant accretionary lapilli away from vent. (**a** and **b** from Van Eaton and Wilson 2013). **c** Isopach map for the phreatoplinian fallout deposits. **d** Thickness thinning trend on a thickness (m) versus square root of isopach area diagram, defined by two straight line segments, the steeper representing the more proximal "wet" deposit area, and the distal area representing deposition of drier ash. (**c** and **d** from Wilson 2001)

Fig. 9.63 (continued)

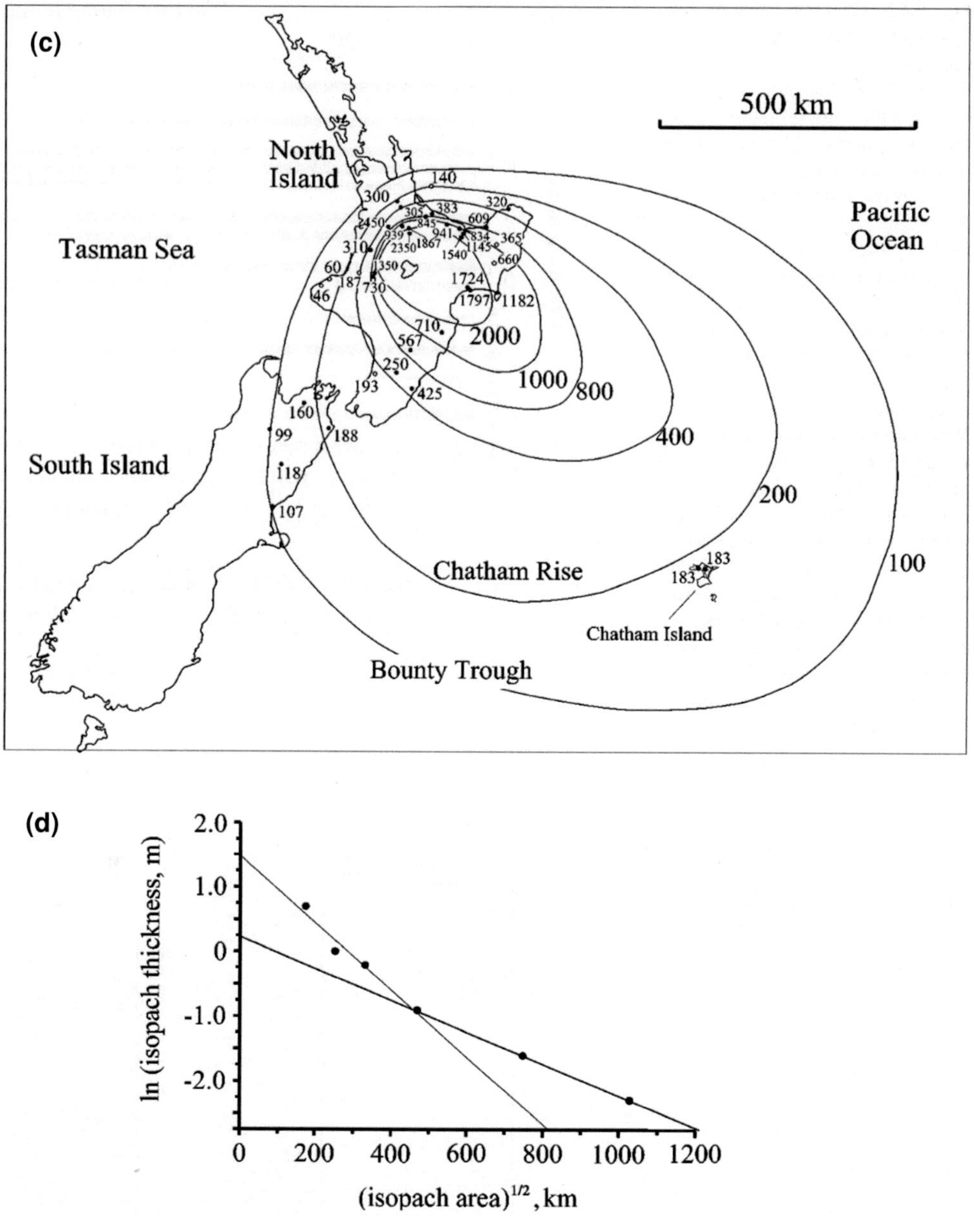

plinian eruptions may produce substantially more fine ash than has hitherto been suspected, and may deposit poorly sorted, bimodal ashes distally. Problems in the interpretation of distal silicic ash layers are discussed in the following paragraph.

9.7.6 Co-Ignimbrite Ash Fall Deposits and Distal Silicic Air-Fall Ash Layers

Large-magnitude explosive events, in many cases pyroclastic flow and ignimbrite-forming (Chap. 10), can be very complex, and this makes understanding their distal (>> 150 km from the vent) ash layers difficult. In the past, distal ash layers were generally assumed to be distal air-fall deposits derived from the buoyant eruption column and deposited from the downwind plume. Studies then indicated that widely dispersed ash deposits can be derived from the buoyant trailing ash clouds of pyroclastic flows, which can rise high into the atmosphere (e.g. Fig. 9.11a). Winds can then disperse the fine ash, depositing it as "*co-ignimbrite ash*" over areas far greater than the areas covered by the pyroclastic flow deposits.

Many widespread ashes have now been interpreted as co-ignimbrite ash-fall deposits. These can occur hundreds of km or more away from the source ignimbrites, as thin, fine, planar, variably laminated ash layers cm in thickness (e.g. Ninkovich et al. 1978; Sparks and Huang 1980; Woods and

Wohletz 1991; Machida and Arai 1983; Rose and Durant 2009; Perrotta and Scarpati 2003; Engwell et al. 2013; Smith et al. 2016; see also Chap. 10). Smith et al. (2016) undertook a combined stratigraphic and geochronological and geochemical study of the deposits resulting from the 39 ka Campanian Ignimbrite and caldera-forming eruption from Campi Flegrei caldera, Naples, Italy. They concluded that the widespread distal ash, which is found over much of eastern Europe and in the Mediterranean Ocean and the Black Sea, was sourced from the co-ignimbrite ash cloud from the late pyroclastic flows and not from the eruption column. Walker (1980) suggested that some widespread ash layers could also be formed by littoral explosions when hot pumice flows entered the sea.

Co-pyroclastic flow ash-fall particles can be an important component of small or moderate-size fall deposits through contemporaneous fallout deposition from both the eruption column and associated pyroclastic flows, producing deposits with bimodal grain size patterns (Sect. 9.2.3; Watanabe et al. 1999; Eychenne et al. 2012, 2015). Such combined deposition is expected to be relevant at volcano islands, as it has been recently witnessed during the 3 July and 28 August 2019 eruptions at Stromboli, where pumice lapilli fall from the buoyant eruption column occurred contemporaneously to the settling of wet ash from the ash cloud generated at the entry point of pyroclastic flows into the sea (Giordano and De Astis 2021; also see Chap. 11). A similar mechanism has been envisaged by Giordano et al. (2008) and Morgavi et al. (2019) to explain the accretionary lapilli-rich fallout layers that blanket the surge deposits of the Secche di Lazzaro succession at Stromboli volcano, associated with large-scale magma–seawater interaction that occurred at 5 ka due to edifice collapse into the sea.

The types of air-fall ash deposits which result from deposition out of the trailing ash cloud of pyroclastic density currents are variously termed layer 3 deposits (Sparks et al. 1973), co-ignimbrite ash-fall deposits (this is our preferred term; Sparks and Walker 1977), and vitric air-fall ash deposits (Wright 1980). These deposits can be very extensive and can have volumes which are comparable with those of ignimbrites (Sigurdsson and Carey 1989; Pyle et al. 2006). They are often interbedded as thin, fine, distal ash deposits in fluvial–alluvial deposits well away from the source volcano (e.g. Fig. 9.64), which in many cases is unknown. If the volume of vitric-enriched co-ignimbrite ash is known, then, from the assessment of the crystal concentration level in the parent ignimbrite, the total volume of the ignimbrite can also be computed (Walker 1972; Scarpati et al. 2014).

Fig. 9.64 Subaerial co-ignimbrite ash fallout deposits. **a** Co-ignimbrite ash fall layer (white) forming an extensive deposit interbedded in Miocene clastic sequences in the northwestern Argentinian Andes. **b** Distal co-ignimbrite ash fall deposits (thin white layers), San Pedro de Atacama, Chile

It must be said though that reliable and accurate calculations of the volume of co-ignimbrite fall deposits, the volume of the ignimbrite, and the ignimbrite crystal enrichment or concentration (i.e. the crystal content of juvenile clasts versus ignimbrite matrix) are all very difficult to obtain accurately, so much care should be taken in undertaking the measurements and computations, placing a lot of care in assessing their statistical significance.

9.7.7 Secondary Thickening and Bimodality: Ash Aggregation

The understanding of aggregation behaviour of fine ash particles into clumps of ash in the eruption column, umbrella region, and downwind plume is one of the current challenges in modern volcanology, as ash aggregation causes premature settling of fine particles at any site, as well as affecting the plume dynamics as a whole. Ash aggregation is also a major issue in understanding sedimentation processes in pyroclastic density currents as discussed in Chap. 10.

Aggregation is a poorly understood process and may be related to the condensation of vapour and adhesion of water droplets and ash particles in the plume, which could be important if water is involved in the eruption (>5% by weight in the plume), or to electrostatic attraction of ash particles, which are electrostatically charged (Carey and Sigurdsson 1982; Textor et al. 2006a, b; Taddeucci et al. 2011; Brown et al. 2012; Van Eaton et al. 2012; Van Eaton and Wilson 2013; Mueller et al. 2016, 2017).

Electrostatic forces are weak and only affect small particles <0.063 mm, whereas wet aggregation due to surface tension forces between moisture droplets and fine ash particles may involve larger particle sizes, usually <0.25 mm, but also larger. There is a full range of aggregates, from chaotic snowflake type that are largely destroyed on landing to rounded but unstructured pelletal lapilli, to more structured, concentric accretionary lapilli (Fig. 9.60b; see Chap. 4). Their formation is thought to reflect the presence of turbulent vertical updrafts of ash particles in the plume, which cross regions of different humidity, including dry zones where electrostatic forces dominate, allowing increasing aggregation until eventually the aggregate weight overcomes the upward force exerted by the plume upward velocity. The blending of widespread co-ignimbrite or co-surge plumes with the umbrella region is thought to provide the environment for the formation of well-structured accretionary lapilli (Brown et al. 2010; Van Eaton and Wilson 2013).

However, aggregates are not always preserved. In fact, how delicate accretionary lapilli can be preserved on landing is quite poorly understood, and often how they survive the fall into and incorporation within high-concentration pyroclastic flow deposits is also unclear (e.g. Pisolitic Tuffs at Colli Albani volcano, Italy, De Rita et al. 2002; Giordano et al. 2010; Oruanui Formation, New Zealand, Wilson 2001; see Chap. 10). In the most common cases, ash aggregation may be identified by secondary thickening and grain size bimodality of fine grained ash fallout deposits away from the vent.

The 1980 Mt St Helens fall deposit surprisingly showed secondary thickening beyond the 1 cm isopach (Fig. 9.24c; Sarna-Wojcicki et al. 1981; Eychenne et al. 2015). First, the air-fall deposit thinned exponentially to 1.0 cm thickness at 180 km, but then increased to 4.0 cm thickness at 250–300 km before thinning once again at greater distances. Carey and Sigurdsson (1982) attributed this to the premature fallout of aggregated particles in the downwind eruption plume. Sorem (1982) described ash clusters carefully collected while settling, and showed SEM photographs of these. These ashes are poorly sorted and bimodal. Durant et al. (2009) and Eychenne et al. (2015) propose that ice nucleating on ash particles in the atmosphere at medial distances has an effect in causing aggregation, forming clumps of ash called "hydrometeors", which then fell at settling velocities higher than that of the individual ash particles, and led to "hydrometeor-enhanced tephra sedimentation" fallout (Durant et al. 2009) and thickening in medial distances (>250 km) along the dispersal area.

Recent observation of fallout from active plumes has shown the frequent occurrence of gravitationally controlled, downward collapsing tephra fingers from the plume (Fig. 9.65) that form through the aggregation of fine particles even close to vent and convective instabilities at the base of eruption plumes (Rose and Durant 2009, 2011; Eychenne et al. 2015). This process may account for the bimodality of grain size shown by some fall deposits even at proximal locations.

Secondary thickening had previously been described in the air-fall deposit erupted from Quizapu in 1932 (Larsson 1937), but Brazier et al. (1982) recognised that secondary thickening and bimodality may be common features of distal ash deposits. They described several North American examples; some are the distal deposits of ignimbrite-forming eruptions, but others were from solely plinian eruptions.

Fig. 9.65 Fallout sedimentation from a prolonged (1 h) vulcanian eruption at Sakurajima volcano, Japan, in 2013. Particle fallout occurs preferentially in some locations of the downwind plume, defining downward-directed tephra fingers

Branney (1991) even proposed an extreme case of a phreatoplinian eruption from the rock record, where large-scale aggregation in the plume may have triggered collapses of discrete portions of the umbrella region to form "rootless" pyroclastic flows.

Bimodality has been used as a criterion to distinguish co-ignimbrite ashes from precursor plinian ashes in the deposits of multi-phase ignimbrite-forming eruptions (Sparks and Huang 1980). Eychenne et al. (2012) found that the distal basaltic fallout deposit from the 2006 Tungurahua eruption in Ecuador consists of bimodal components from the sub-plinian downwind plume, and fine ash from the contemporaneous buoyant ash cloud of pyroclastic flows. Similarly, Eychenne et al. (2015) found that some of the distal ash from the 18 May 1980 Mt St Helens eruption, USA, consists of coarse juvenile components from the plinian eruption column and plume, and suspended fine ash of largely accessory fragments from the initial blast. However, as has just been described, bimodality may also be a common feature of distal plinian deposits due to particle aggregation processes, as also suggested by Eychenne et al. (2012) for the Mt St Helens fallout deposit.

It is therefore apparent that, in distal situations, clearly distinguishing phreatoplinian, plinian, and co-ignimbrite ash-fall deposits is very difficult. By grain size characteristics alone, this may be impossible. Proximal characteristics of the deposits, and the eruption sequence, together with detailed grain size determinations, including deconvolving the grain size population into distinctive grain size classes which are then subject to SEM morphological analysis of the ash particles and their componentry (e.g. Eychenne et al. 2015), will provide the best approach. In addition, if secondary thickening is a common feature of distal plinian deposits, it therefore has important implications for calculating their volumes, and volumes of juvenile components (Eychenne et al. 2015). Conventional ways of estimating the volumes of plinian deposits are based on log-thickness *versus* square root of area plots. These plots give straight-line or exponential relationships in areas near the source (<150 km from the vent). However, secondary thickening in distal deposits suggests that this method is therefore unsuitable where dispersal of ash occurs over distances greater than 150 km from the vent (cf. Froggatt 1982; Sulpizio 2005).

9.8 Welded Fallout Deposits

As will be discussed more fully in Chap. 10 on pyroclastic flow deposits (see Sect. 10.5), welding involves the sintering together of hot, glassy fragments, irrespective of shape and size, under the influence of a compactional, lithostatic load or the collapse of very plastic pyroclasts that are deposited well above the glass transition temperature. During welding, the glassy fragments deform plastically, producing a bedding parallel fabric of flattened, elongated large pumice, scoria, or spatter fragments (fiamme) and glass shards, known as eutaxitic texture (see Fig. 9.66f).

Welded lapilli tuffs are common in the geological record and are generally called welded ignimbrites or welded ash-flow tuffs, implying deposition from a pyroclastic flow (Chap. 10). However, well-sorted welded air-fall lapillistones are a relatively common volcanic product, and are known to occur on several modern volcanoes, covering nearly the full range of magma compositions (Fig. 9.66). They have also been recognised in ancient volcanic successions (Suthren and Furnes 1980).

Sparks and Wright (1979) first recognised welded fall deposits on Santorini and in Iceland and discussed the differences in textures between welded ignimbrites and welded fallout deposits. The best documented examples so far are from Santorini volcano and Askja volcano, Iceland (Sparks and Wright 1979; Carey et al. 2008a, b; Boyce and Gertissier 2012; Figs. 9.66). These welded lapillistones can be distinguished from welded rocks formed from pyroclastic flows (ignimbrites) by their geometric form, textures, and field relations to non-welded counterparts. Their features indicate syn- to post-emplacement compaction and welding over a wide area and cannot simply be ascribed to the agglutination of spatter lumps on impact. We would restrict the use of the term "agglutination" to describe the process of deformation and sintering of air-fall pyroclasts, largely fluidal spatter clasts, when they impact on an accumulating bed near the vent. Flattening is thus caused by the momentum of the falling fluidal pyroclasts (Thomas and Sparks, 1992; Capaccioni and Cuccoli 2005; Sumner et al. 2005). This is distinct from welding, in a strict sense, which occurs under the influence of high eruption and deposit accumulation rate, so ensuring that pyroclasts remain hot and plastic, and load pressure imposed by already accumulated hot plastic tephra on the underlying, still hot, part of the deposit. Agglutination requires the eruption of quite fluid magma and, as we have discussed, is therefore a more important process in basaltic eruptions (Sects. 9.7.2–3). However, extensive welded scoria fall deposits are known, and agglutination could have been an important process near the vent during these eruptions (e.g. Figs. 9.66 h, 9.55d, e).

Welding has also been documented in "fused, or sintered" tuffs, where lava has induced welding of underlying air-fall layers (Christiansen and Lipman 1966; Schmincke 1967), but these are not formed by primary processes like the examples described here.

(a)
1875
(b)
(c)
(d)
(e)

◀ **Fig. 9.66** Features of welded fallout deposits. **a–d** The Askja 1875 unit D welded rhyolitic fallout deposit, along the northern margin of the Lake Öskjuvatn caldera, Iceland. **a** Topography draping nature of variably welded unit D. **b** Transition from non-welded black to grey basal zone passing gradationally upwards into moderately densely welded top in medial settings. Note the flattened black juvenile clasts near the top. **c** Profile through the proximal welded deposit. Black fragments are spatter clasts. The fine (red = oxidised) and grey matrix is welded. **d** Eutaxitic texture (lower left) deformed by ballistic lithic bomb. (All photos **a–d** by Rebecca Carey). **e** View of the black to pale 144.6 ka Middle Pumice welded dacitic fallout deposit (*aka* Thera welded tuff; above the stone-retaining walls), close to vent near the steps to Thera harbour, Santorini caldera, Greece. The base is buff coloured and non-welded to tack welded, and grades up into the black, diffusely stratified, obsidian, welded fall facies. Directly above, the welded fall facies are capped by a thick sequence of co-ignimbrite lag breccias (brown; unit MPB, Fig. 9.67a), recording collapse of the preceding plinian eruption column. **f** Welded phonolitic fallout deposit (or "eutaxite") at the base of 0.6 Ma Granadilla member (?), Guajara Peak, Las Cañadas caldera, Tenerife. This shows a clast-supported texture, defined by closely packed flattened pumice clasts that are now obsidian and define fiamme, and less well-flattened pumice clasts (buff coloured) that are interpreted to have fallen out from a higher, cooler part of the eruption column than the flattened clasts. **g** The andesitic Pinnacle Ridge tuff member is a sub-plinian, welded, coarse-grained lapilli fallout deposit up to 25 m thick mantling the slopes of Ruapehu volcano, New Zealand. Stratigraphically, it belongs to the most recent (post-glacial) cone-building episode of the volcano ≤ 15 ka. (Photo by Colin Wilson). **h** <4.5 ka black and red welded mafic scoria fallout deposit (3 m thick) displaying mantle bedding, northern slopes of Cotopaxi volcano, Ecuador (From Vezzoli et al. 2017, Fig. 11a)

Fig. 9.66 (continued)

9.8.1 Characteristics and Examples

Discussion of the more general aspects of welding textures and processes will be left until we describe welded ignimbrites in Chap. 10, Sect. 10.8.14.1. However, in the welded air-fall examples known, welding occurs outwards to a distance of 1–3 km from the probable vent position (Thomas and Sparks 1992; Carey et al. 2008a; Boyce and Gertisser 2012). Ignimbrites, on the other hand, can be welded at distances of 50 km or more from the source. The basic criteria used to recognise welded fallout deposits and distinguish them from welded pyroclastic flow deposits are:

- mantle bedding and deposition on steep slopes (>20°; Fig. 9.66a, g, h),
- internal stratification (Fig. 9.66b, c, e) and distinguishable fall or textural units perhaps reflected by rapid fluctuations in the compaction profile (Fig. 9.69),
- clast-supported texture, producing a welded lapillistone with a fiamme texture (Figs. 9.66b–d, f and 9.67c), and
- gradational vertical (Fig. 9.66b, c and 9.67d) and lateral transitions along strike in the same depositional unit from welded into non-welded pumice deposits with depositional and grain size characteristics of air-fall deposits (Fig. 9.67a–f).

The "Thera welded tuff" or Middle Pumice (144.6 ka) welded fallout deposit on Santorini in proximal settings forms a distinctive black, glassy, dacitic layer as much as 7 m thick (Fig. 9.66e, Fig. 9.67a–f), which in hand specimen has a well-developed eutaxitic texture of flattened lapilli-sized clasts. It is therefore strictly not actually a tuff, but a welded lapillistone. The welded part of the deposit is exposed for more than 3 km along the caldera wall and must have originally covered at least 1 km^2 (Fig. 9.67; Boyce and Gertisser 2012). Near the vent (Thera harbour steps), the deposit changes progressively from a non-welded lapillistone at the base upwards into a diffusely stratified, densely welded obsidian (zone B, Figs. 9.66e, 9.67a, e). Where thickest and most densely welded, the deposit is internally stratified and contains some conspicuous layers and lenses of coarse pink–orange (oxidised) uncollapsed pumice (zone A, Fig. 9.67a–d). Many of these layers are not laterally continuous and are thought to have formed by the rapid local accumulation of large pumice bombs, which cooled below the glass transition temperature high in the eruption column. These produce anomalous deviations in the compaction profile, and one is shown in Fig. 9.69a. Laterally, the welded lapillistone passes into a thick, coarse, non-welded lapilli pumice fallout deposit, forming the Middle Pumice deposit (zone D, Figs. 9.67a, f) which, near the welded lapillistone, is thermally darkened and black in colour (zone C, Fig. 9.67a). This layer mantles topography, and isopach (Fig. 9.68b) and maximum-clast size isopleth maps (Fig. 9.68c, d) are typical of plinian air-fallout deposits. These maps indicate that the vent was near the place where the welded deposit is thickest (Figs. 9.67a, b and 9.68a). The welded fallout deposit is directly overlain by a thick sequence of co-ignimbrite lag breccias (Chap. 10) resulting from the ensuing collapse of the plinian eruption column.

The Askja 1875 welded fallout deposits were formed during the same explosive rhyolitic eruption of this Icelandic volcano discussed previously in Sect. 9.7.4. The eruption produced a number of distinctive pyroclastic layers, and two of these pass into welded rocks (layers D2 and D4) on the northern rim of Öskjuvatn caldera (Fig. 9.66a–d). This welded deposit has a maximum thickness of 4 m and covers a minimum area of 1.6 km^2. It mantles topography and can be found on slopes as steep as 30° (Fig. 9.66a). It is internally stratified, and there are marked fluctuations in the compaction profile (Fig. 9.69b; Carey et al. 2008a). Welded air-fall deposits are deposited in layers, and can show vertical variations in grain size, sorting, components and, more importantly, accumulation rate. This can result in more irregular compaction profiles (especially relative to thickness) than with ignimbrites.

Other well-documented examples of welded fallout deposits include the Ruru Pass pantelleritic deposits on Mayor Island, New Zealand (Houghton et al. 1985), and the andesitic Pinnacle Ridge tuff member, Ruapehu volcano, New Zealand (Fig. 9.66g; Hackett and Houghton 1985). The Green Tuff of Pantelleria has been interpreted as fallout by some (Sparks and Wright 1979; Wright 1980; Wolff and Wright 1981) although more recent studies demonstrate thata basal non-welded fallout pumice is overlain by variably welded and rheomorphic ignimbrite (Orsi and Sheridan 1984; Scarani et al. 2023).

9.8.2 Conditions of Formation

The critical problem posed by welded fallout deposits is to determine under what conditions airborne ejecta can remain sufficiently hot during flight to weld and compact after deposition. The basic controls are viscosity, grain size, eruption discharge, deposit accumulation rate, and cooling rates. Low-viscosity magmas (up to 10^5 Pas) such as peralkaline magmas are prone to forming proximal welded fall deposits (Wolff and Wright 1981; Sumner et al. 2005; Di Genova et al. 2013). Field and experimental evidence suggests that grain size and cooling rate is the critical factor (Thomas and Sparks 1992; Capaccioni and Cuccoli 2005). Both the Middle Pumice and Askja welded fallout deposits are thickest and most densely welded near the vent, where

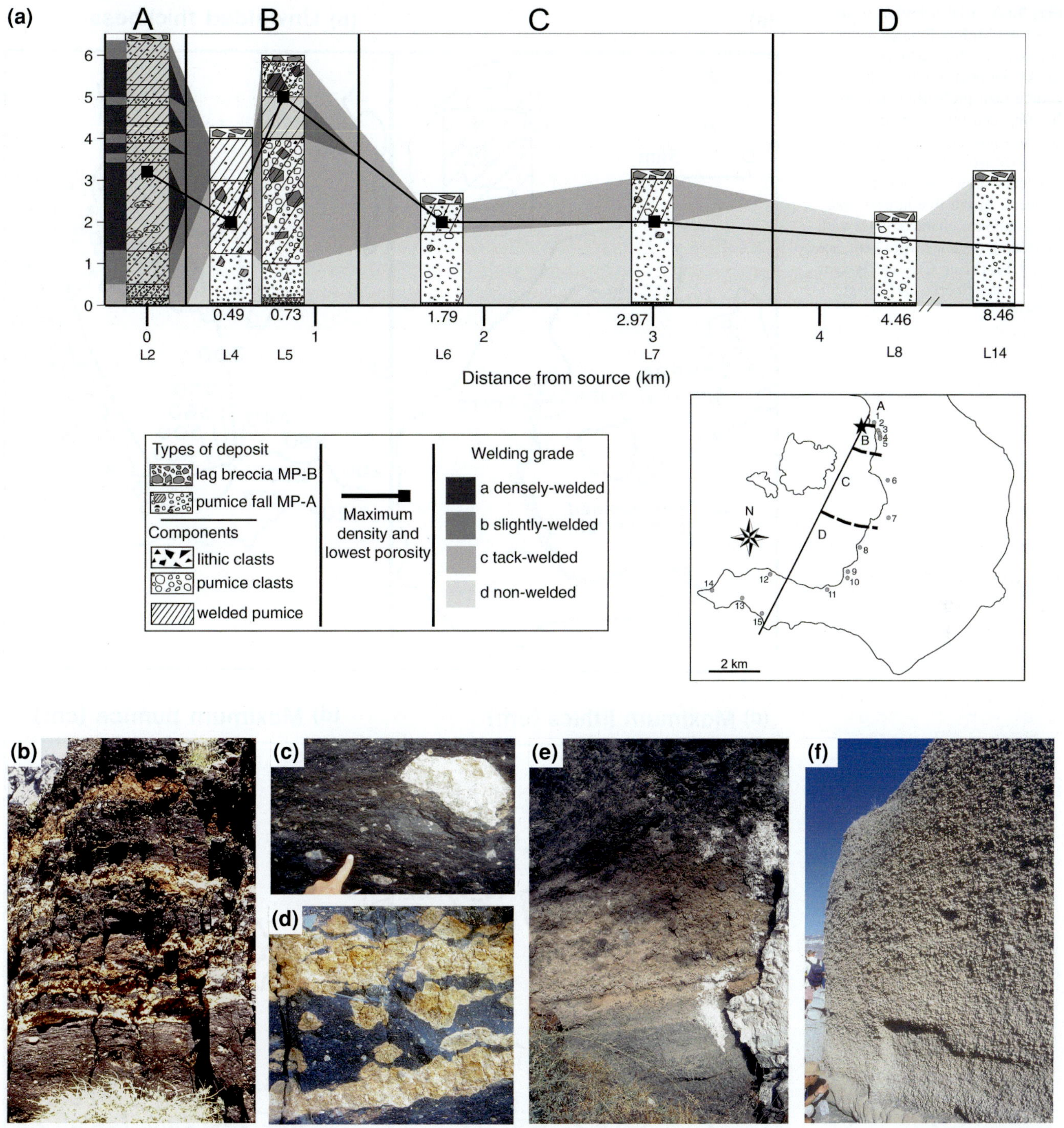

Fig. 9.67 **a** Lateral variations in welding and physical properties of the 144.6 ka Middle Pumice welded fallout deposit, in the caldera wall of Santorini, Greece. MPA = Middle Pumice plinian fallout; MPB = overlying co-ignimbrite lag breccias. (From Boyce and Gertisser 2012). This depicts a fully welded (with variations in intensity) profile near vent (**b–e**), to a non-welded profile away from vent (**f**). **b-d** Most densely welded tuff at Thera harbour (zone A in a); **b** Near-complete profile through densely welded zone (see Fig. 9.69a), which shows well-developed internal stratification composed of semi-continuous pink-orange coloured pumiceous layers; **c** Close-up view of well-developed internal eutaxitic welding foliation, with faintly visible fiamme and xenoliths. **d** Densely welded zone, showing black (eutaxitic) obsidian glass with pale coloured patches of clusters of non-welded pumice clasts, which cooled below the glass transition temperature high in the eruption column before deposition. **e** Base of the Middle Pumice showing transition from tack welded (buff coloured) at the base up into welded (black) fallout deposits (zone B). **f** Distal, completely non-welded, Middle Pumice fallout deposit, 8 km from the vent (zone D)

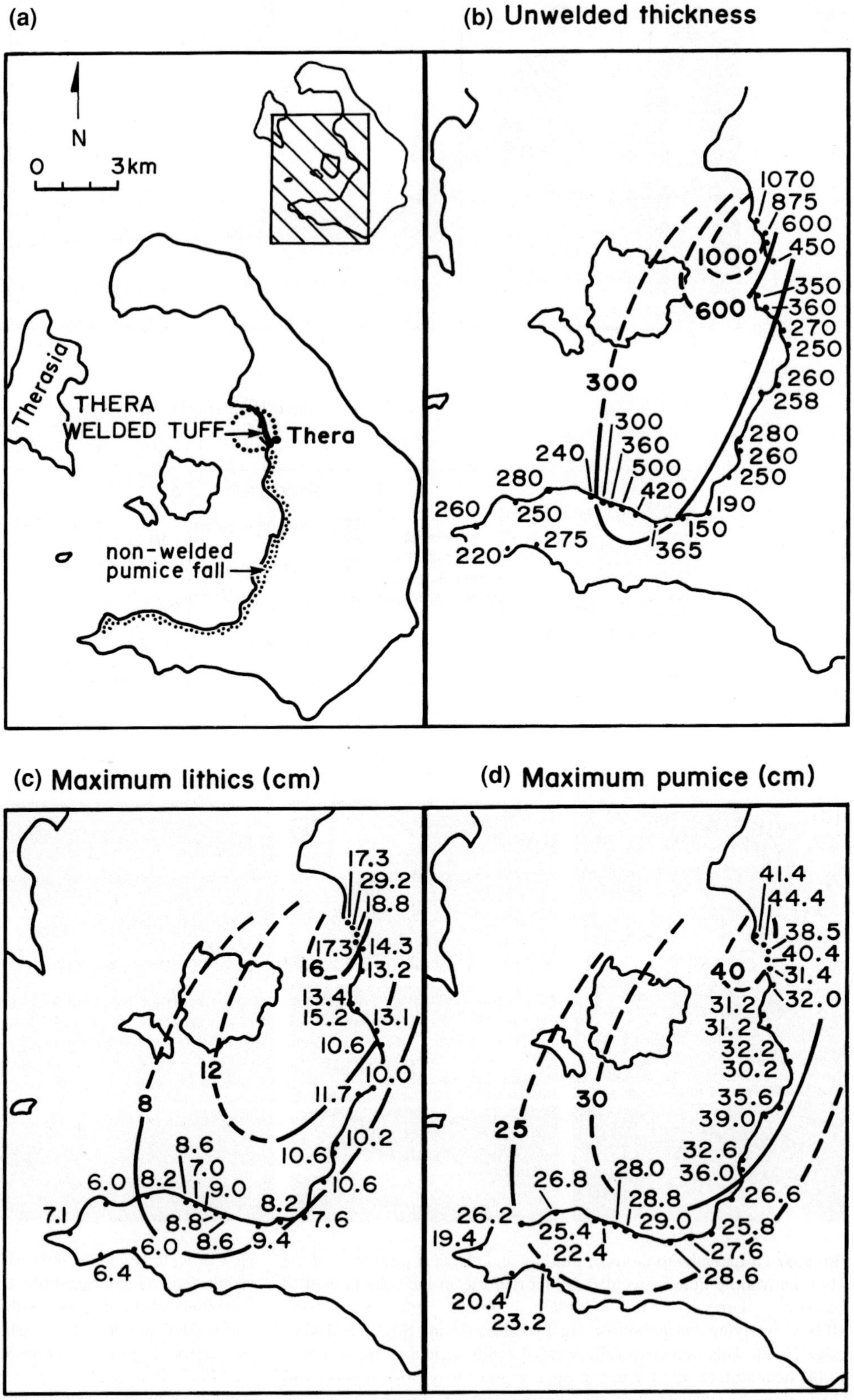

Fig. 9.68 **a** Distribution of the 144.6 ka Middle Pumice fallout deposit, Santorini. Dotted circle marks the area in which the source vent probably occurred. **b** Unwelded thickness. **c** and **b** Maximum lithic and pumice clast size isopleth pattern. Isopleths show the average diameter of the five largest casts. For grain size characteristics with increasing distance from vent, see Fig. 9.19. (From Cas and Wright 1987, After Sparks and Wright 1979)

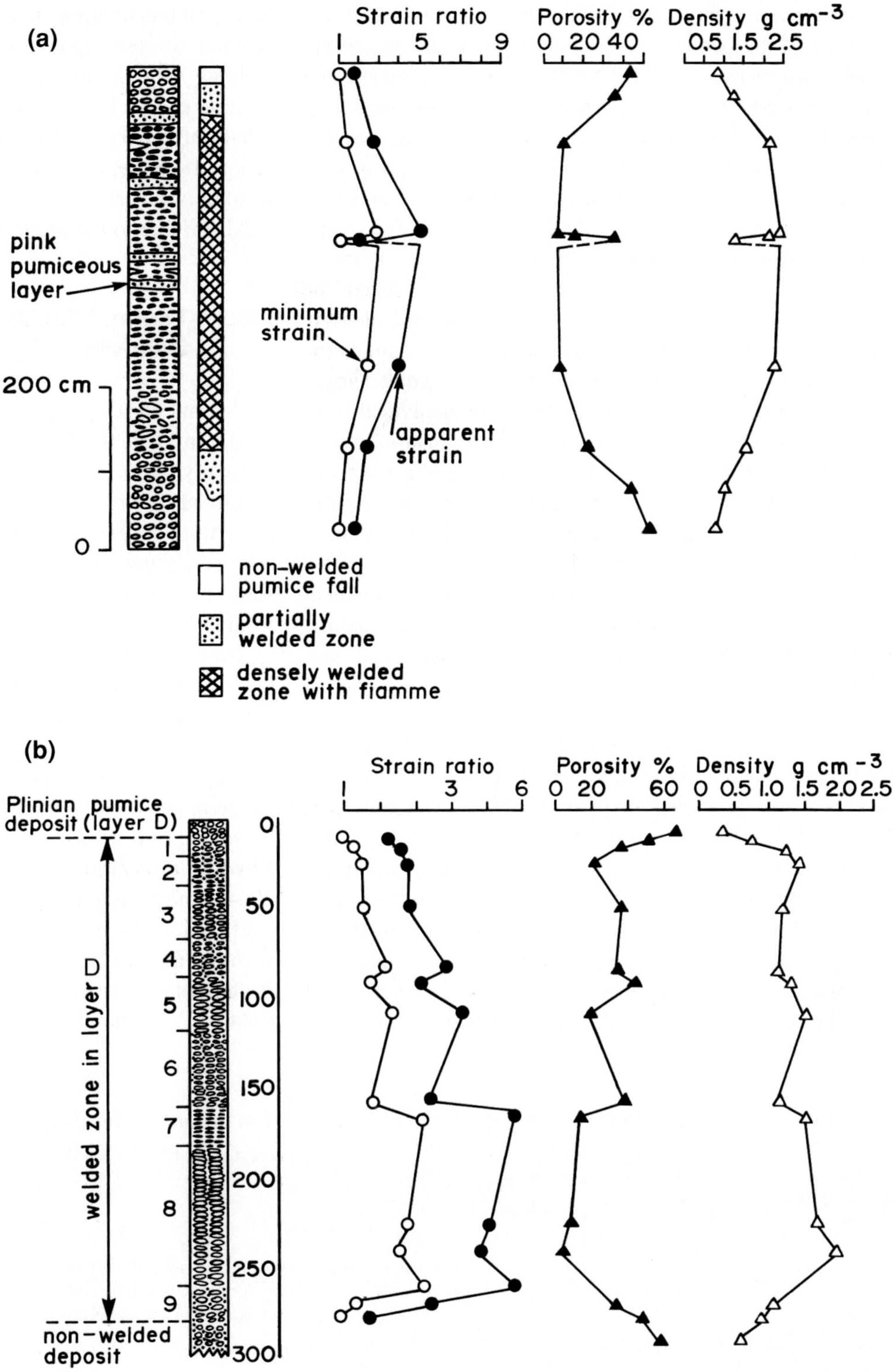

Fig. 9.69 Vertical profiles and physical properties of welded fallout deposits. **a** Compaction and lithological profile through the welded Middle Pumice fallout deposit (Thera welded tuff) where thickest at Thera harbour, Santorini. The minimum strain ratio assumes all ellipsoidal pumice clasts landed with their long axes parallel to bedding. **b** Compaction and lithological profile through the zone of densest welding at the thickest part of the 1875 Askja C welded fallout deposit, Iceland. (From Cas and Wright 1987, After Sparks and Wright 1979)

the grain size is large and cooling low due to shorter trajectories of the particles (cf. welded ignimbrites which are again most densely welded where they are thickest, but where they are ponded in a depression this could be a long distance from the vent). Accumulation rate was initially thought to play an important role in welding, by preventing radiative and convective cooling of deposited fragments, and by reducing the bulk heat loss of the deposit. The Askja 1875 eruption is sufficiently well-documented to allow an estimate of the accumulation rates which produced the welded tuff (Sparks and Wright 1979). For the densely welded zone, this was calculated as 20 cm min^{-1}, and at the boundary between incipiently welded and non-welded zones it was between 2 and 4 cm min^{-1}. However, analytical work demonstrates that the contribution of accumulation rate to welding is almost irrelevant compared to the effects of viscosity, grain size, and cooling rate (Thomas and Sparks 1992). Both the Thera and Askja welded air-fall tuffs are also mixed pumice deposits (Sparks and Wright 1979). Superheating the silicic component by the mafic component may have been an additional factor promoting welding. However, the importance of this is difficult to judge, because many pumice fall deposits show evidence of magma-mixing but are not welded.

Carey et al. (2008a) proposed that Askja 1875 welded fallout deposits (unit D) could be subdivided into 5 divisions (D1–D5), which varied in their degree of welding from welded (D2, D4) to non-welded (D1, D3, D5). D2 and D4 originated from phases of the Askja eruption when the source eruption column was fountain-like and from 1 to 4 km high (cf. 26 km high for the phreatoplinian phase). This prevented cooling of pyroclasts and controlled their accumulation rate, which was the key factor in determining whether or not welding would occur, in that the discharge and fountain/column heights could fluctuate rapidly to account for the rapid vertical variations in the degree of welding (Fig. 9.69b). Welding due to the influence of hot, discrete spatter bombs impacting into and supplying heat to a halo of surrounding pyroclasts is also documented in the Askja 1875 deposits (Carey et al. 2008b). This style of welding was termed “local welding” in contrast to “regional welding” due to the progressive accumulation of pyroclasts.

9.8.3 Thermal Facies Model for Welded Fallout Deposits

A thermal facies model for pyroclastic fallout deposits is shown in Table 9.3. The lateral changes indicated also occur vertically, with most of the welded examples that we have described displaying a range of these facies in vertical sections through deposits. What is evident from this type of analysis is the overlap between pyroclastic processes and some rheomorphic lava-forming processes in very densely welded fallout deposits.

However, considerable uncertainty must be attached to the interpretation of clastogenic lava facies forming from welded fallout deposits. Although welded/agglutinated facies are a common facies found in the deposits of basaltic eruptions, they are less likely to be found in those of more silicic magmas. Rhyolitic agglutinates *sensu stricto* are rare, however, but have been described in the Yamakogawa Rhyolite in central Kyushu, Japan (Furukawa and Kamata 2004), and in northern Chile (Self et al. 2008). More generally, these features in high-silica rocks are only going to be found in the fluidal ejecta derived from peralkaline magmas. Pantelleritic clastogenic lavas are found to erupt from the Cala di Tramontana volcanic centre in Pantelleria

Table 9.3 Thermal facies model for pyroclastic fall deposits (Modified from Cas and Wright 1987)

	(1) Clastogenic lava	(2) Rheomorphic welded tuff	(3) Densely welded tuff	(4) Partially welded tuff	(5) Non-welded pumice deposit
	Increasing distance from the vent → Decreasing accumulation rate → Decreasing temperature →				
Major processes	Agglutination rheomorphism	Load–pressure welding; rheomorphism	Load–pressure welding	Load–pressure welding	
Minor processes	Load-pressure welding	Agglutination			
Characteristic texture or structure	Massive lava-like rock; relict clast outlines	Linear fabric; flow-folds	Planar fabric or eutaxitic texture; near-zero porosity in zone of densest welding	Poorly developed eutaxitic texture; relict primary porosity	Loose pumice

(Stevenson and Wilson 1997), and occur on Mayor Island, New Zealand (Stevenson et al. 1993; Gottsmann and Dingwell 2002). Nonetheless, rheomorphic facies seem to be rare among silicic fallout deposits. Given the low-viscosity of kimberlite magmas, it has also been proposed that some kimberlite rock types might be welded (agglutinated) pyroclastic rocks (Brown et al. 2008; Chap. 18).

9.9 Eruptions and Eruption Columns that Produce Subaqueous, Water-Settled Fallout Deposits, and Their Characteristics

Pyroclastic fallout does not just occur on land. Eruption columns and downwind plumes erupted from vents on continents, oceanic volcanic islands, as well as buoyant ash plumes derived from subaerial pyroclastic flows (co-ignimbrite ash clouds), often disperse their pyroclast load over the oceans, and often thousands of km downwind from the source volcano, as deep-sea ash deposits. Distal ashes found on the seafloor may be very difficult to trace back to the source volcano or to an original eruption style. Fallout can also occur in lakes. And, as discussed below not all deep-sea "ash" deposits are pyroclastic in origin; some originate from effusive subaqueous eruptions, with spalling and dispersal of fine ash-sized particles originating by quench fragmentation of the surfaces of propagating hot sea floor lavas and dispersal by hot water plumes rising from the lava surfaces.

Pyroclasts falling from subaerial eruption plumes into aqueous environments are initially subject to the same transport and sorting regime through the atmosphere as for subaerial fallout deposits. However, in addition they are then subject to settling through water, and to any wave, wind, or aqueous currents that may affect pyroclasts once they fall into water. The physical behaviour of pyroclasts settling on the seafloor is significantly different to subaerial fallout regimes because of the effects of the denser water mass through which they are settling.

In addition, fallout deposits can be produced in aqueous environments directly from shallow water vents or deeper water vents. Although there are some very general similarities between subaqueous and subaerial explosive eruption styles, in detail, every subaqueous explosive eruption is significantly different to subaerial counterparts due to the way the water mass interacts dynamically with erupting magma, and secondly in the way that the water mass influences the dispersal processes (Cas and Simmons 2018). In Chap. 3, we concluded that the properties of the water mass significantly impact eruption processes, explosive intensity of subaqueous explosive eruptions, particularly deep ones, dispersal and depositional processes, and deposit characteristics.

Our understanding of subaqueous fallout deposits and processes is significantly limited by our inability to directly observe the processes. In situations where vents are submerged, direct observation of shallow water explosive phenomena is only possible from a safe distance from boats or the air, or in rare cases from "manned" submersibles or remotely operated submersible vehicles. Where a vent is submerged at depths of tens, to hundreds to thousands of m, there are limitations in what is known about subaqueous eruption processes and subaqueous eruption columns. Occasional small eruptions have been imaged (Fig. 9.70a, b) from submersibles or remotely operated vehicles (ROVs; e.g. Jason, Medea; e.g. Chadwick et al. 2008; Resing et al. 2011; Deardorff et al. 2011). Surface expression of such submarine eruptions is most commonly represented by discoloured, ash-laden plumes of water, floating, hot vesiculated clasts, and sometimes very extensive floating mats or rafts of pumice, which are dispersed by oceanic currents. These include the visible 2006 Home Reef and 2014–2015 Hunga Tonga–Hunga Ha'apai eruptions from shallow water vents in the Tonga volcanic arc, southwest Pacific Ocean (Bryan et al. 2012; Colombier et al. 2018), and the non-visible 2012 Havre submarine eruption from vents hundreds of m deep, further to the south in the Kermadec arc (Jutzeler et al. 2014; Carey et al. 2018; Manga et al. 2018; Sect. 9.9.4.6). As a result of the limitations of not being able to view subaqueous eruption columns directly, much of our understanding of their physics is theoretical and speculative, without all processes and physical constraints being well-understood (Cas and Simmons 2018).

However, we can say, based on simple physics, that submarine eruption columns, particularly deep-water ones, do not behave at all like subaerial ones. There are several important reasons for this, which we will discuss below.

Similarly, understanding the characteristics of subaqueous fallout deposits is much less advanced than for subaerial fallout deposits, because of the obvious problem of limited accessibility to modern subaqueous environments and deposit sequences, and relatively limited understanding of the way pyroclasts are dispersed and their dispersal patterns in water. However, recent research on modern submarine volcanoes is beginning to provide significant advances, and, in addition, uplifted ancient subaqueous volcanic successions also provide an important opportunity for learning about subaqueous fallout deposits.

9.9.1 Behaviour of Submarine Eruption Columns

Direct observations of submarine eruptions are rare, but observed phenomena include small-scale "explosive" events and effusive pillow lava-forming events at the West Mata

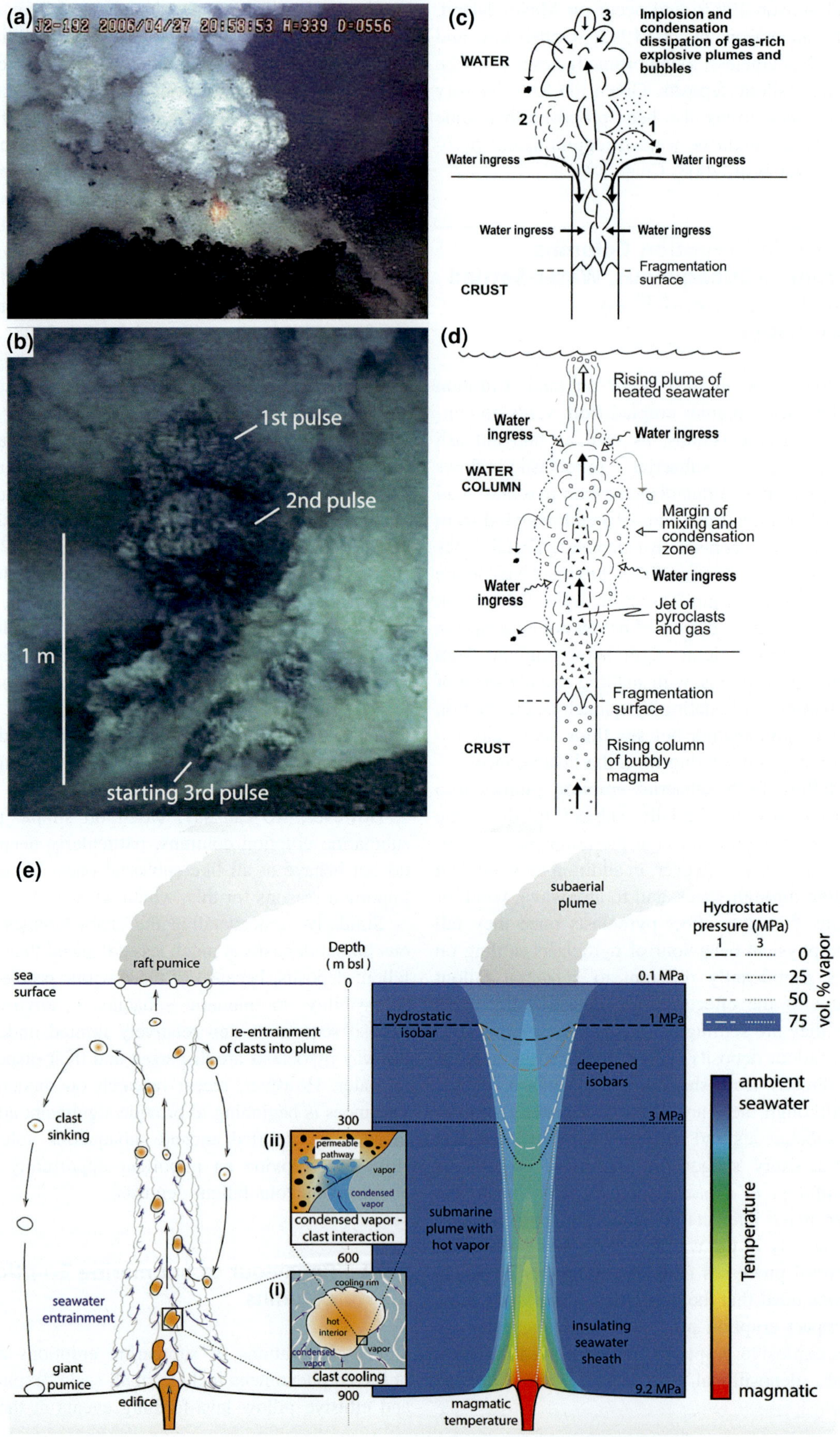
(a)
J2-192 2006/04/27 20:53:53 H=339 D=0556
(b)
1st pulse
2nd pulse
1 m
starting 3rd pulse
(c)
Implosion and condensation dissipation of gas-rich explosive plumes and bubbles
WATER
3
2
1
Water ingress
Water ingress
Water ingress
Water ingress
Fragmentation surface
CRUST
(d)
Rising plume of heated seawater
Water ingress
Water ingress
WATER COLUMN
Margin of mixing and condensation zone
Water ingress
Water ingress
Jet of pyroclasts and gas
Fragmentation surface
CRUST
Rising column of bubbly magma
(e)
sea surface
raft pumice
re-entrainment of clasts into plume
clast sinking
seawater entrainment
giant pumice
edifice
Depth (m bsl)
0
300
600
900
(ii)
permeable pathway
vapor
condensed vapor
condensed vapor - clast interaction
(i)
cooling rim
hot interior
condensed vapor
vapor
clast cooling
subaerial plume
0.1 MPa
hydrostatic isobar
1 MPa
deepened isobars
3 MPa
submarine plume with hot vapor
insulating seawater sheath
9.2 MPa
magmatic temperature
Hydrostatic pressure (MPa)
1
3
0
25
50
75
vol. % vapor
ambient seawater
Temperature
magmatic

◀ **Fig. 9.70** Imaged and modelled subaqueous explosive eruption columns. **a** Blast of small ejecta, with a core of incandescent lava or fire fountain, in 2006, Brimbank Crater, NW Rota 1 volcano, Marianas arc, at a water depth of ~550 m (From Chadwick et al. 2008). **b** Pulsing succession of explosions, Brimbank Crater, NW Rota 1 volcano, Marianas arc (From Deardorff et al. 2011). **c** Schematic representation of small pulsing (3 explosive pulses) eruption columns, showing decay of the column through cooling, condensation, and hydrostatic pressure-induced collapse or implosion. **d** Schematic representation of the behaviour of a higher and more sustained, moderate intensity, subaqueous explosive eruption columns, showing transition from a rising column of bubbly magma, fragmenting into a jet of pyroclasts, hot gas and hot water, which above the vent is subject to cooling and condensation of the gas, and ingress of cold water from the margins driven by the hydrostatic pressure. Columns forming in deep water are likely to dissipate before reaching the surface, but still generate a buoyant plume of hot water. **e** Proposed eruption column/plume model for the 2012 submarine eruption of the submarine Havre volcano, Kermadec arc, north of New Zealand (see Chap. 6). Clasts cool within a vapour-rich plume over time scales of seconds to minutes, enabling ongoing degassing during ascent to shallow depths. Segregation of raft and giant pumice occurs, here, according to influence of clast size; clasts that become water-logged sink and may become re-entrained in the plume. Insets show (i) interaction of hot condensed vapour with the exterior of a cooling clast, where quench fractures are beginning to develop on the upper clast rim, and (ii) the diffusion of water through permeable bubble pathways in a hot clast, resulting in vesicle-edge $H2O\mathit{t}$ enrichment and clast rehydration. The process of entraining seawater into a vapour-rich plume is more efficient at depth (blue arrows). The potential for deviation from hydrostatic pressure conditions within the plume is illustrated on the right by calculated depths of isobars at 1 and 3 MPa for varying vapour contents within the plume (0, 25, 50, and 75% by volume), assuming a simple vapour with constant density 10 kgm^{-3}. **f** Schematic representation of the pressure regime in a thermal plume of heated seawater above a vent, relative to that in the surrounding water mass, depicted by the changing degree of depression downwards of isobars. The plume cools upwards by mixing in of cold seawater, and becomes denser. (Parts **e**, **f**, and caption from Mitchell et al. 2018). **g** Hypothetical profile through a heated seawater plume above a subaqueous vent, showing depressed density profiles across the plume, the effect of which decreases upwards as dense cold water is mixed into the plume. **h** Hypothetical configuration of expanding gas bubble size and increasing abundance upwards (relative size of gas bubbles), and isobar depression up through subaqueous eruption columns as buoyant bubbles rise and decompress upwards; most likely for relatively shallow plumes (<500 m). **i** Complex pattern of bubble size and abundance, and isobar depression patterns as a result of initial bubble decompression and expansion up to the middle of the column, then cooling, condensation, and water ingress into a subaqueous eruption column upwards. There are low levels of isobar depression at the vent and at the top of the column, and a peak in the middle. **j** Schematic representation of a relatively shallow water (tens to hundreds of m) explosive eruption column. The rising buoyant column is subject to condensation of gases and mixing in of ambient water. Large pumice clasts become water-logged, contributing to collapse of the column, forming a hot mass flow of pyroclastic debris, steam, and water on the seafloor. Small clasts are entrained in the column, large clasts are dispersed by a combination of ballistic fallout, and buoyant flotation if they are vesiculated. (From Kano et al. 1996). **k** Schematic representation of a buoyant ash-bearing hot water plume above a vent, reaching its level of neutral buoyancy and then spreading laterally in the water mass as a density current from which fallout ash can be deposited (Modified from Newland et al. 2022)

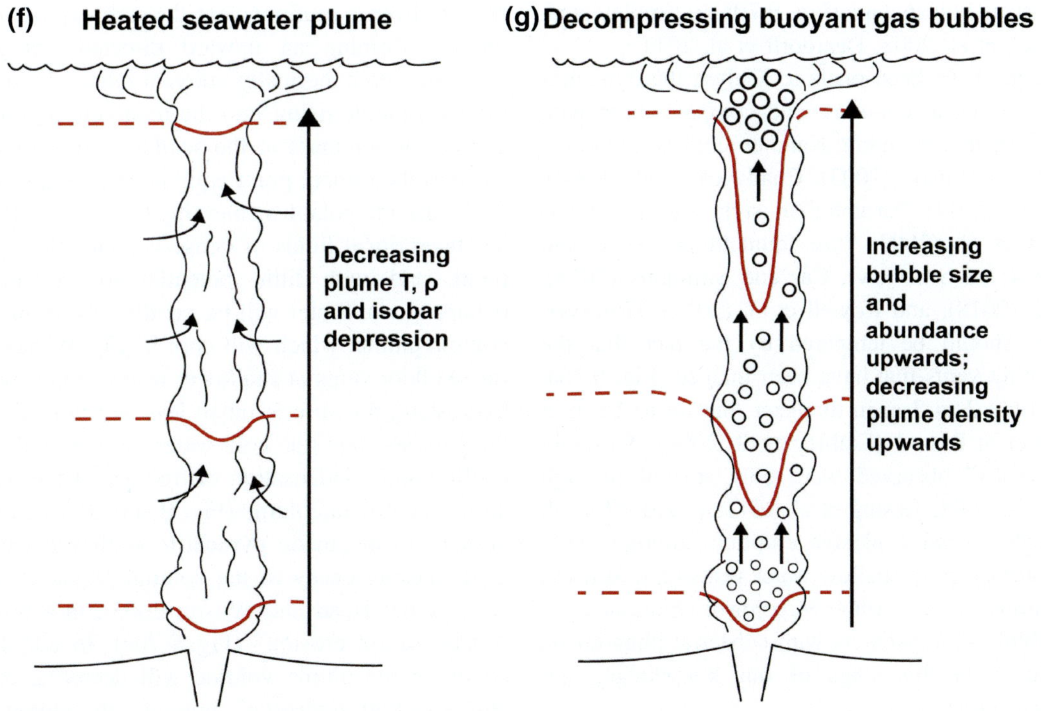

Fig. 9.70 (continued)

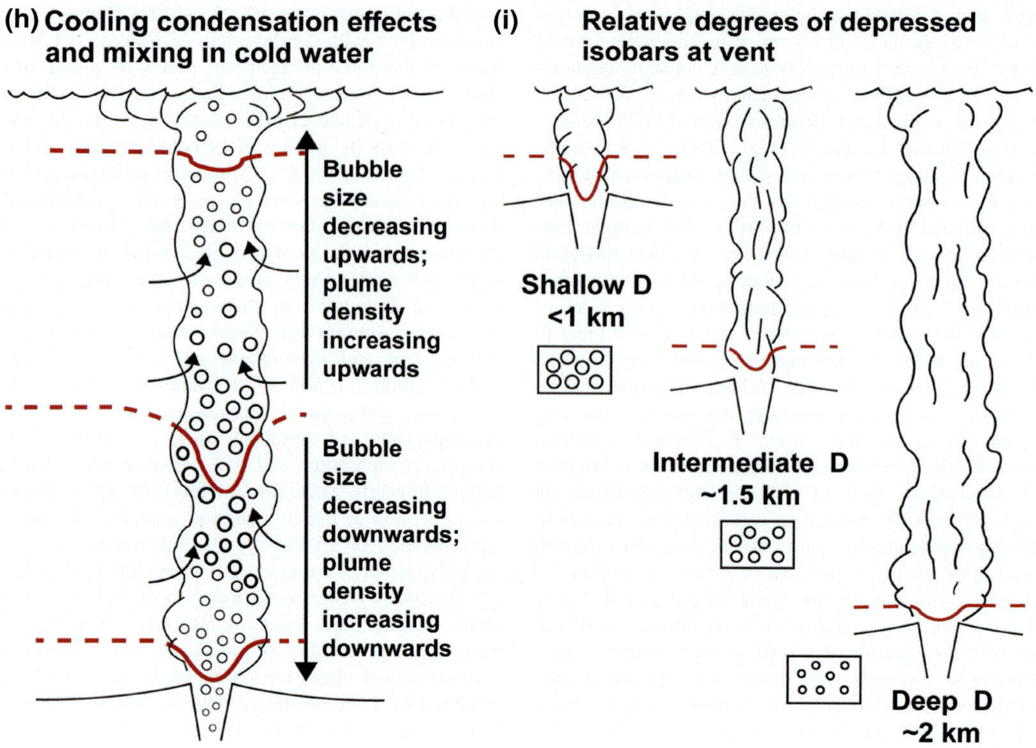

Fig. 9.70 (continued)

volcano in 2009 in the Lau Basin (Chadwick et al. 2008; Resing et al. 2011) and the 2004–2005 N.W. Rota 1 eruption in the Mariana volcanic arc (Fig. 9.70a, b; Embley et al. 2006; Chadwick et al. 2008; Deardorff et al. 2011).

Some attempts have been made to analyse the dynamics of subaqueous eruption columns, and compare them with subaerial counterparts, including Kano et al. (1996; "plinian style"), Head and Wilson (2003), Chadwick et al. (2008), Deardorff et al. (2011; "strombolian style and fire fountaining"), Cas et al. (2003; "fire fountaining"), Cas and Giordano (2014; an overview), Cas and Simmons (2018), Mitchell et al. (2018), and Rowell et al. (2022). However, such analyses should be tempered by the fact that the observed erupting vents that have been imaged (Fig. 9.70a, b) are small, only 2 to 3 m in diameter, but up to 15 m at Brimbank crater at Rota 1 (Embley et al. 2006). The only eruption "columns" observed were only several m high (Chadwick et al. 2008; Resing et al. 2011; Deardorff et al. 2011), although several explosive eruption columns breached the sea surface from shallow water vents (e.g. Surtsey, Iceland, Thorarinsson et al. 1964; Miyojinsho, Marianas arc, Fiske et al. 1998). Comparisons with subaerial phenomena are thus tenuous at this stage of our knowledge, and potentially misleading.

In general terms, eruption columns driven dominantly by exsolving magmatic volatiles are likely to be buoyant as they emerge from a seafloor vent through explosive momentum initially forming an upward directed gas jet and then evolving into a thermally buoyant relatively low-density hot to warm plume rising into the relatively high-density, cold, ambient water mass at the seafloor. At water depths at or close to the critical point equivalent pressure, for say water ($\sim$2 km), the volatiles emerging from a seafloor vent would be supercritical fluids or dense vapours close to the critical point, and with little potential to expand explosively (Chaps. 3, 4). Water will be rapidly mixed into the eruption column/plume, which will cool it (Fig. 9.70a–d). However, for sea floor vents at shallower water depths, where ambient hydrostatic pressure is much less and volatiles can exist in the gaseous state, the cold water mass will also cause rapid cooling and condensation of free gas at the margins of the gaseous eruption column (Fig. 9.70c, d), as well as of gas in vesicles in magmatic pyroclasts as they emerge above the vent, causing decay of the column (Kano et al. 1996; Cas et al. 2003; Head and Wilson 2003), in a process we call "condensation erosion" (Fig. 9.70c). In addition, as gases condense the plume volume will decrease, and under the effects of high hydrostatic pressure, the plume will collapse

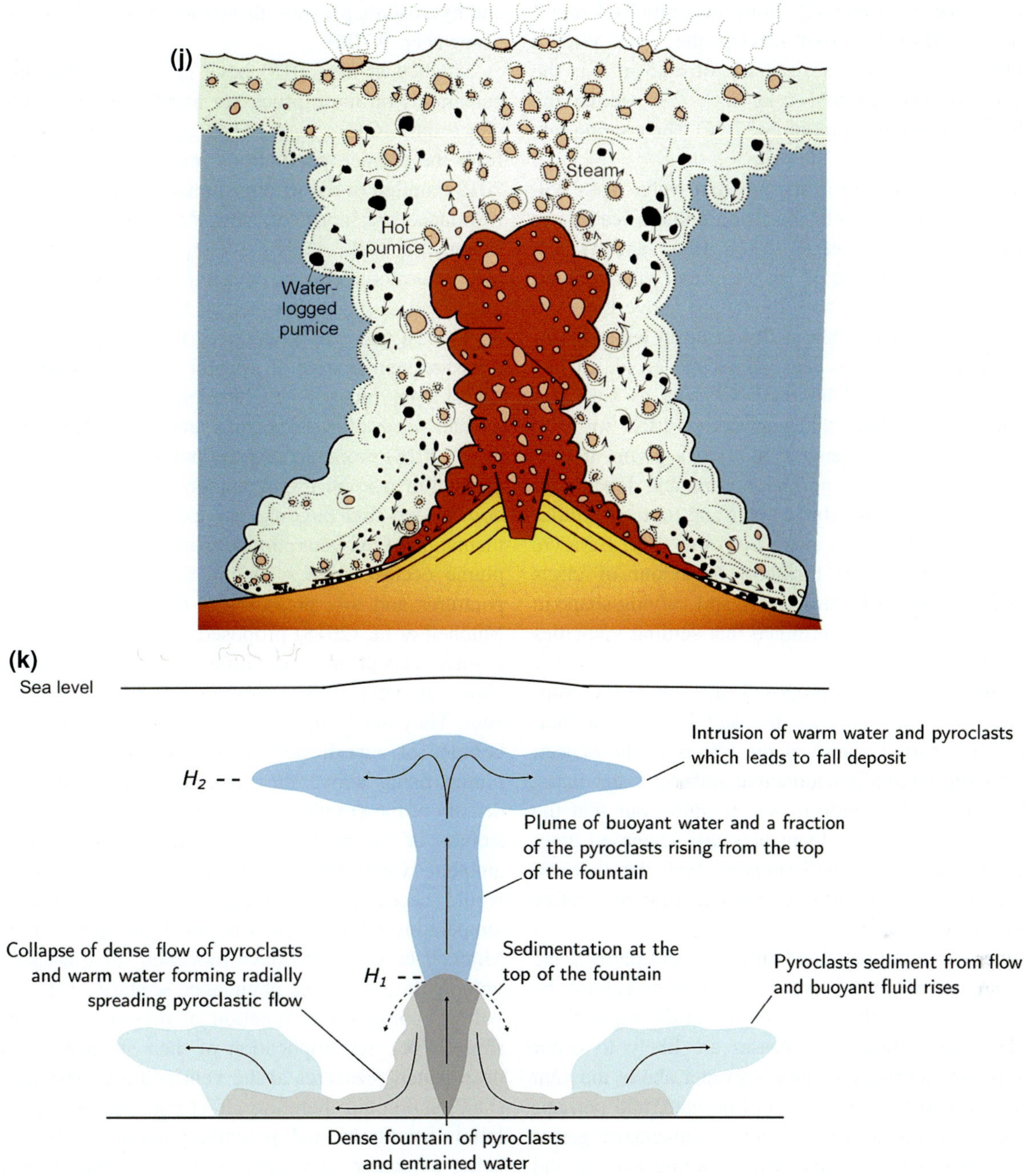

Fig. 9.70 (continued)

or implode, and ingress of water will occur (Fig. 9.70c). Deardorff et al. (2011) assessed the dynamics of cooling of small explosive plumes only several metres high from the 2006 pulsing, explosive events at the NW Rota 1 volcano, Marianas arc (Fig. 9.70b, c), at a water depth of ~550 m, with a magma eruption temperature at the vent of >700 °C, using the relationship:

$$v = \frac{\sqrt{(8r\,g\,\Delta\rho)}}{3\rho_{sw}\,C_D} \tag{9.32}$$

where v is the plume rise velocity, r is the plume radius, g is the acceleration due to gravity, $\Delta\rho$ is the density contrast between the seawater mass and the plume, ρ_{sw} is the density of seawater, and C_D is the drag coefficient of a sphere. The plume rise velocity decreased from an initial <4 m s^{-1} to <0.5 m s^{-1} within 1.5 m of leaving the vent. With a plume radius of 1 m and a rise rate of <0.5 m s^{-1}, the density contrast dropped to ~5 kg m^{-3}, 2 m above the vent, with a plume temperature of ~27 °C. The cooling rate was therefore >200 °C s^{-1}.

The hydrostatic pressure in the ambient water mass will drive the ingress of water into the rising column leading to mixing and cooling (Cas et al. 2003; Head and Wilson 2003). Nonetheless, a plume of hot water and steam should rise, and entrain pyroclasts, to some height above the vent, providing a source for the fallout of ash up to tens to hundreds of metres from the vent (Barreyre et al. 2011). Seawater heated to several hundred °C is up to 10% less dense (~900 kg m^{-3}) than cold seawater (~1,027 kg m^{-3}) (Safarov et al. 2009; Sharqawy et al. 2010), giving it considerable buoyancy in the normal ocean, and leading to thermal plumes of seawater above vents and rootless plumes above propagating hot lavas away from the vent. Barreyre et al. (2011) modelled the transport behaviour of clasts ranging from <1 mm to 12 mm in size and varying shape in such thermal plumes and determined that settling velocities vary according to shape.

Head and Wilson (2003) proposed that submarine eruption columns will be less dynamic and lower than their subaerial counterparts because of the effects of the denser, colder, and more thermally conductive ambient water mass. Deardorff et al. (2011) confirm this, pointing out that the combined effects of high ambient pressure, increased drag, and more efficient cooling will suppress both the size and intensity of subaqueous explosive eruption columns relative to subaerial ones.

For higher and more sustained eruptions and plumes, the processes are more speculative because they cannot be observed in total from the seafloor vent to the tops of the plumes. The same physical processes are likely to occur (Fig. 9.70d). As an eruption starts, seawater above the vent will be heated, and form a plume of hot water and perhaps steam rising upwards above the vent. As magmatic gas is released and seawater in contact with erupting magma and hot ejecta is boiled to steam, the bulk density of the column will decrease, making it more buoyant, and causing it to rise more quickly. The plume could be gas-dominated, or water-dominated (Mitchell et al. 2018), or more likely, this will change in the column from bottom to top, and from the margins inwards during an eruption and during the ascent of the column as condensation erosion takes effect (Fig. 9.70d), unless the vent is in shallow water. In that case, if the eruption is intense, a major subaerial eruption column may form above sea level, having many of the characteristics of normal subaerial eruption columns, as discussed earlier in this chapter. So, gas condensation, condensation erosion, and hydrostatic pressure-driven water ingress are all likely to occur (Fig. 9.70d).

Mitchell et al. (2018) have proposed some quantitative constraints on these processes using water speciation abundance data ($OH^{-}_{magmatic}$, $H_2O_{magmatic}$, $H_2O_{total\ incl.\ seawater}$) from the pumice populations produced during the 17 July 2012 eruption of the wholly submarine Havre volcano, in the Kermadec arc, north of New Zealand (Carey et al. 2018; Manga et al. 2018). They used these data as paleo-thermometers, to determine the quenching temperatures, and as paleo-barometers, to determine at what water depth clast quenching occurred. The eruption occurred from a fissure vent system at a water depth of ~900 m, on the rim of a previously formed caldera, the floor of which sits at a water depth of ~1,500 m. Carey et al. (2018) and Manga et al. (2018) proposed that the pumice populations were the product of quench fragmentation of erupting vesicular lava domes, and that ongoing gas exsolution and quench fragmentation of buoyant pumice rising in an eruption column or plume occurred. To reconcile the combined quench temperatures and low pressures at the moment of quenching, Mitchell et al. (2018) proposed that the quenching of most pumice except in situ dome carapace samples occurred above the vent, in the eruption column that rose above the vent. They also proposed that if sufficient gas bubbles (and bubbles of superheated seawater) existed in the eruption plume rising above the vent, this would lower the bulk density of the eruption plume, leading to a depression of the isobars in the plume downwards relative to those in the ambient water mass around the plume (Fig. 9.70e). This would account for some of the pressure quench temperature-inferred water depth implications of their data. Mitchell et al. (2018) have modelled three eruption plume volumetric gas bubble abundance scenarios (25, 50, and 75 vol % bubbles) as a function of pressure and temperature (Fig. 9.70e). An implication of their suggestions is that in deep water, pressures at the vent could be less than hydrostatic, and could therefore affect constraining volatile exsolution conditions and potentially eruption styles, perhaps being explosive rather than effusive as would be the case under high normal hydrostatic pressure regimes.

Although this is a valuable attempt to quantify the state of the eruption plume, there are many complex variables not taken account of. Mitchell et al. (2018) use simple static state gas bubble abundances and densities, not changing dynamic gas bubble abundances and densities as pressure and temperature conditions change within the plume vertically, laterally, and temporally. Their argument is based on speculative gas bubble abundances in the eruption column, not based on quantitative modelling of likely gas release rates from the erupting vesiculating magma. In particular, one of the free gas bubble abundances proposed 75% is highly unlikely at depths of 900 m. This estimate may be based on the fact that the vesicularity of the Havre pumice is very high ($\sim$70%) suggesting that this volume of free gas bubbles is released into the water mass at the same time as the pumice is being released. This could only happen if all bubble walls in the erupting pumice were bursting at the time of the eruption or in the column, which should then produce more ash glass shards than pumice clasts, for which there is no evidence. The concept could perhaps work where vents lie at shallow water depths of only a few hundred metres or less, under conditions of much lower hydrostatic pressure, almost simulating subaerial conditions with high gas overpressures.

Dynamic factors in any eruption plume include increasing pressures downwards in at least part of the plume, perhaps all of it, and in the water mass as a whole, the unknown effects of bubble cooling and condensation and volume reduction with time spent in the plume, and the effects of the higher hydrostatic pressure and density of the cold enclosing water mass in forcing ingress of dense water into the plume. Some aspects have been represented schematically by Mitchell et al. (2018), including declining plume temperature upwards, although the basis for the inferred temperatures is unclear (Fig. 9.70e). The specific volume of individual gas bubbles will be lowest at depth because of pressure constraints (Fig. 3.7), and theoretically will increase upwards as buoyant rising bubbles decompress, assuming they retain magmatic temperature and don't cool and condense, in which case they will contract in volume, causing a decrease in the bulk volumetric gas abundance levels upwards, and an increase in plume density.

Mitchell et al. (2018) have also assumed a density of steam at 10 kg m^{-3} in their modelling, but this will be highly variable as a function of ambient pressure and effects of cooling and condensation. And, as long as hydrostatic pressure in the water mass surrounding the eruption plume is higher than the pressure in the plume, dense cold water will be driven sideways into the plume. Isobar patterns can thus be drawn depicting many different scenarios, allowing for the variables just outlined. In Fig. 9.70f–i, speculative isobars and gas bubble sizes are depicted for (f) a heated seawater column; in (g) a column in which gas bubbles released from the vent at depth under pressure decompress and expand as they rise, leading to lower column densities upwards and progressively *greater* depression of the isobars *upwards*; in (h) speculative effect of cooling, condensation, and water ingress on bubble size/abundance and isobar depression particularly at the top of column (also see Fig. 9.70d), depicting low levels of gas bubble size and volumetric abundance, and isobar depression at the base with a peak in the middle, and upwards decrease in bubbles size, abundance and column temperature upwards; and (i) depicting the size and relative abundance of bubbles, and the degree to which isobars are depressed at vents at varying water depths, reflecting the increasing effects of hydrostatic pressure on gas bubble specific volumes.

Regarding the implication that pressures at the vent could be less than hydrostatic, and could therefore potentially affect eruption styles, this could occur in some cases where the seafloor is shallower than the critical pressure equivalent water depth. For all eruptions, the vent conditions before an eruption will likely be close to hydrostatic pressure. At water depths approaching water critical pressure equivalent depths, exsolved volatiles will more and more approach supercritical fluid state and density, which has a very low expansivity (Pioro and Mokry 2011) and would suppress the likelihood of explosive eruptions. The effect will also depend on the rate of gas bubble discharge from the vent, the rate of condensation, and ingress of water due to hydrostatic pressure. As discussed in Chaps. 3 and 4, explosive expansion of exsolved volatiles is most likely at pressures <10 MPa (=<1 km water depth). Another important consideration is that the behaviour of exsolving magmatic water is subject to freshwater critical pressure regime constraints rather than saline seawater critical pressure regime constraints. Mitchell et al. (2018) have not modelled any of these factors.

Kano et al. (1996) also indicated that if subaqueous eruption columns become inundated by seawater through condensation and water ingress, and overloaded with dense solids to the point that their bulk density exceeds that of the surrounding water mass, they can collapse and produce seafloor mass flows of pyroclastic debris, in addition to fallout of ballistic blocks and water-saturated pumiceous debris (Fig. 9.70j).

That this appears not to have happened in the case of the 2012 submarine eruption of Havre volcano, Kermadec arc, discussed above, suggests that the rate at which pumice clasts were discharged and the rate at which they became water-logged were insufficient to generate a sufficient volume and mass of pumice clasts denser than water that would render the eruption plume denser than the ambient water mass, which would lead to plume collapse and generation of mass-flow density currents. This is consistent with a slow release of vesiculated pumice clasts from the quench fragmented and spalling from the pumiceous carapace of lava domes (Carey et al. 2018; Manga et al. 2018). This in turn is consistent with the absence of any reported lithic clasts in any of the various pumice deposits, which would, if present, suggest explosive eruptions.

Rowell et al. (2022), in assessing the impact of subaqueous explosive eruptions seeming to suppress the release of climate-changing SO_2 to the atmosphere, have analysed and quantified aspects of the subaqueous eruption from vents in varying subaqueous depths. They consider the nature of resultant eruption columns as a function of magma mass eruption rate, ambient hydrostatic pressure as a function of water depth, mixing of pyroclasts, and ambient water and heat transfer processes. Rowell et al. (2022) propose that entrainment, thermal and physical mixing of external water into an erupting gas–pyroclast jet, affects the mass, momentum, and particle and enthalpy changes that control eruption column rise. Water ingested and mixed into the column and the efficiency of heat transfer between hot pyroclasts and mixed in water control dynamical parameters, such as bulk column temperature, density, velocity, and column radius (Rowell et al. 2022).

As proposed by Cas and Simmons (2018), Rowell et al. (2022) consider that increasing hydrostatic pressure as water depth increases suppresses bubble nucleation, bubble growth, magma decompression, and ascent rates, potentially stopping magmatic explosive fragmentation where hydrostatic pressure and water depth are high enough. As a gaseous explosive jet emerges above a vent on the seafloor and experiences decompression, there is a jet height, known as a decompression length, L_d, in which there is little or no interaction between the jet and the external water mass (Rowell et al. 2022). L_d, which is equivalent to 10 times the vent diameter, is determined by mass eruption rate (MER), gas pressure in the jet, and ambient hydrostatic pressure exerted on the periphery of the gas jet, which is a function of water depth. If a vent is at a water depth $< L_d$, an eruption column can rise through the sea surface and rise into the atmosphere, essentially behaving as a "dry" (i.e. no entrained external water) magmatic subaerial eruption column.

If the vent water depth exceeds L_d, then at a water depth above L_d, there is a transition from jet phase to turbulent phase flow which enhances mixing in of external water (e.g. Fig. 9.70d), which then affects the thermal dynamics of the column, as it rises through buoyancy rather than explosive momentum in the jet flow zone below L_d. Such an eruption column can breach the sea surface and rise above sea level, the height reached being dependent on vent water depth and hydrostatic pressure, the mass fraction of entrained external water in the column, and MER. Eruption columns may reach the stratosphere if the vent water depth is shallow and MER is high, with column height decreasing as vent water depth increases and MER decreases (Rowell et al. 2022). When the entrained water mass fraction is >30–40 wt% of the column, column collapse occurs. At water depths of hundreds of m or more, eruption jets and columns become overwhelmed with mixed in external water, reducing the thermal energy of the column and its ability to rise into the atmosphere, although steam plumes may do so (Rowell et al. 2022).

Cahalan and Dufek (2021) investigated the behaviour of explosive eruption columns at water depths of 200, 500, and 1,000 m, at gas-steam mass eruption rates (MER) of 0–9×10^6 kg s^{-1}, and at eruption temperatures ranging from 600 to 1,400 °C through numerical simulations. Their first simulations involved the discharge of a fluid gas jet or eruption column into the water mass free of suspended particles. They considered the role of steam and condensation of water in the eruption columns in determining column behaviour and the factors that determine if an eruption column will breach the sea surface. They found that steam MER is a greater determinant of column behaviour than vent width. At water depths between 500 and 1,000 m where the hydrostatic pressure is 5 to 10 MPa respectively, when steam cools and condenses to water droplets, its volume decreases 50x to 100x respectively, the density of the column increases, and its buoyancy and upward momentum decrease significantly (Cahalan and Dufek 2021). Column diameter decreases upwards as condensation takes effect. According to Cahalan and Dufek (2021), eruption columns originating from eruptions at a water depth of 1,000 m could not breach the sea surface (e.g. Fig. 9.70c), and from eruptions at 500 m water depth, breaches only occurred at MER of $>\sim 5 \times 10^6$ kg s^{-1}, and then only when eruption temperatures at vent were $> \sim 1200$ °C. Breaches of the sea surface by the eruption columns occurred commonly from vents at 200 m water depth at the modelled MER.

Cahalan and Dufek (2021) also simulated the effects of varying MER and water depths on sea surface displacement heights above the vent and surface water temperature changes, using a vent eruption temperature of 1,000 °C. They found that the sea surface was displaced upwards to greater heights with increasing MER. The effects were greatest and similar for eruptions from vents at 200 and 500 m water depth (<100 m vertical displacement at highest MER) but significantly less for 1,000 m depth eruptions, only a few m at 4×10^6 kg s^{-1} MER, but increasing to ~ 50 m at 8×10^6 kg s^{-1}. Similarly, sea surface temperatures increased with MER, rises were greatest and similar from eruptions at 200 and 500 m depths (<80 °C at the highest MER of 8×10^6 kg s^{-1}), and less from 1,000 m depth eruptions (<40 °C at the highest MER of 8×10^6 kg s^{-1}).

Cahalan and Dufek (2021) then simulated the behaviour of submarine eruption columns or jets with suspended solid particles to investigate how they affect eruption column/jet stability and heat transfer. They investigated the effects of particle size, density, solids volume fraction, eruption temperature, and MER at a water depth of 500 m. Results indicate that the presence of solids in the columns stabilised the columns irrespective of the volume present. At low MER increasing particle concentration and eruption temperature increased the column/jet height due to thermal input, but at higher MER addition of particles caused a decrease in column heights, and perhaps column collapse once a threshold loading is reached. They also speculate that particles that are entrained in a steam phase above the vent and are decelerating may begin to fall out into the adjacent warm water mixing zone but could then become entrained upwards again by the denser, warm buoyant water.

Newland et al. (2022) focused largely on the dynamics of deep submarine explosive eruptions. They use the concept of an eruption fountain although they don't clarify if they distinguish deep-water magma fire fountains from gas-charged explosive jet and eruption columns, which is an important distinction (Chap. 3). They also seem to assume that explosive eruptions can occur at any water depth, which as discussed in Chaps. 3 and 4 cannot occur deeper than the critical point pressure of water water depth equivalent of about 2200 m due to exsolved magmatic and superheated external water being a supercritical fluid with limited explosive expansion potential.

Newland et al. (2022) propose that with the onset of turbulence in the rising eruption column (see Rowell et al. 2022, as discussed above), mixing in and entrainment of ambient water, which becomes heated, causes the lower part of an eruption fountain to become denser than the ambient water mass and collapse, producing density currents of pyroclastic debris and hot water (Fig. 9.70k). However, as described above, heated water at the top of the fountain is buoyant relative to the cold ambient water mass and therefore rises (e.g. Fig. 9.70d), entraining some of the low-density (e.g. vesiculated pumiceous debris) and finer ash-size particles. Under varying conditions, this rising hot water plume may either begin to spread naturally into the ambient water mass at its neutral density level, or perhaps even rise to the sea surface. Newland et al. (2022) suggest that such a model can explain many dispersed ash deposits on the deep-seafloor, such as that produced during the 2012 Havre volcano eruption in the Kermadec arc (e.g. Murch et al. 2019).

In their modelling, Newland et al. (2022) assume a magma temperature of 1400 °C which is extremely high and consistent with ultramafic magmas rather than mafic or silicic ones. Their study evaluates the effects of heat transfer and mixing of an initial eruption column into the ambient water mass above the vent by considering the latent heat of magma, its heat capacity, the enthalpy (total heat content) of water, and the flux of both magma and water involved during mixing, on the density of the magma clast–water mixture or plume which drives its dynamics in rising upwards. Because the study focuses on deep-water eruptions, an assumption is made that the vesicularity of pyroclasts is low enough so that pyroclast density is greater than that of water, which is also not always the case, but is adequate for the study.

According to Newland et al. (2022), the temperature of the mixture of pyroclasts and water is related to the latent heat of the magma, L, the heat capacity of the magma, C_{pm}, magma mass flux, Q_m, water magma flux, Q_w, and the enthalpy of the water $h_w(T,P)$ at mixture temperature T and pressure P:

$$[L + C_{pm}(T_0 - T)]Q_m = h_w(T,P)\,Q_w \tag{9.33}$$

The bulk density of the rising column, ρ_c, which determines its buoyancy and upward velocity after the initial explosive thrust dissipates can be expressed by:

$$\rho_c = \frac{(Q_m + Q_w)}{[(Q_m/\rho_m) + (Q_w/\rho_w)]} \tag{9.34}$$

where ρ_m is the density of the magmatic clasts and ρ_w is the water density (Newland et al. 2022).

Bulk column density can also be expressed by:

$$\rho_c = [(\phi_p/\rho_p) + (1 - \phi_p)/\rho_w]^{-1} \tag{9.35}$$

where ϕ_p is the mass fraction of pyroclasts, with density of ρ_p.

It should be noted here that although the focus of the discussion here is on eruption columns or plumes produced by submarine explosions that transport pyroclasts, buoyant hot water plumes transporting quench fragmented debris spalled off erupting submarine lavas also behave similarly.

Calculations for an eruption at 1,000 m water depth and MER of 10^6 kg s^{-1} indicate that 20 m above the vent the water mixed into the rising column causes it to become dense and decelerate to zero speed 100 m above the vent (Newland et al. 2022). Analogue experiments that Newland et al. (2022) carried out show, not surprisingly, that a part of the column collapses and forms a sedimenting bottom density current. However, from the top of the temporarily stalled plume, which has lost significant solid load during partial collapse, hot buoyant water rises forming a secondary warm water plume (e.g. Fig. 9.70f, h, i, k) that can entrain fine suspended solids or pyroclasts and rise to its level of neutral buoyancy, where it spreads laterally and flows radially as a density current within the water mass (Fig. 9.70k).

The radially spreading flow carries suspended solids as long as its velocity exceeds the settling velocity of the suspended solids. Solid particles fallout progressively with the highest settling velocity particles falling out first and then slower and slower settling velocity particles falling out successively, forming a submarine fallout deposit (Newland et al. 2022). Newland et al. (2022) have calculated lateral distance of dispersal as a function of grain settling velocity and MER, which are compared with data for ash dispersal from deep-water eruptions at the Gakkel Ridge in the Arctic Ocean in 1999–2001 (Pontbriand et al. 2012) and the 2012 eruption of Havre volcano in the Kermadec arc (Murch et al. 2019).

Finally, subaqueous magma/fire fountaining eruption columns can occur down to water depths of at least several km, based on evidence of submarine spatter deposits on the modern deep-seafloor and in the geological rock record. The behaviour of such magma/fire fountains is likely to be driven largely by the buoyant uprise of a fluidal magma jet in the oceanic crust, rather than by a particulate gas flow. A significant difference with subaerial fire fountains, however, will lie in the way that the erupting fountain interacts with the ambient water body upon eruption above the seafloor, which will be discussed below, in Sect. 9.9.3.2.

9.9.2 Clast Dispersal Processes: Eruption Column Versus Oceanic Currents and Waves

The behaviour of pyroclasts in a subaqueous eruption column will depend on their density and concentration levels, whether or not they are entrained in a gaseous, steam-rich fluid, or higher density water, and how quickly, if at all, they become water-logged (Whitham and Sparks 1986; Barreyre et al. 2011; Fauria et al. 2017). Hot magmatic clasts may be quench fragmented by the water in the condensing eruption column, leading to secondary, syn-eruptive fragmentation processes, before fallout (Cas et al. 2003; Carey et al. 2018; Manga et al. 2018), particularly if relatively dense. However, Patel et al. (2013) found that pumice, reheated to 600 °C, was not subject to significant fragmentation when quenched by water. However, could pumice that has already cooled from its eruption temperature have been "hardened" or thermally cured making it less prone to quench fragmentation when reheated than clasts of magma being suddenly immersed in cold water? At low particle concentrations, pyroclasts could behave individually according to the balance of their density, settling velocity, and the shear stress of thermal convective currents trying to carry them upwards. Many pyroclasts will immediately fallout around the vent from the rising eruption plume according to their settling velocities through water, as discussed below in Sect. 9.9.3. Hot pumice fragments with densities <1000 kg m^{-3} will initially be naturally buoyant, but may become water-logged rapidly (Whitham and Sparks 1986; Fauria et al. 2017), so acquiring a bulk density greater than water, and begin to fall out together with other denser clasts around the vent. In many cases, high concentrations of dense or water-logged clasts will lead to syn-eruptive collapse of the subaqueous eruption column, producing a largely water-supported, syn-eruptive mass flow of pyroclastic debris along the seafloor away from the vent (Fig. 9.70j; Kano et al. 1996; Walker et al. 2008, 2019). Post-eruptively, pyroclastic and other volcaniclastic deposits on steep submarine volcano slopes may be remobilised by gravitational slides and slumps, and transform into sea floor turbidity currents (Walker et al. 2019; Chap. 12). Both syn- and post-eruptive turbidity currents would produce their own (ash) turbidite deposits, but in addition, are likely to produce lofted trailing ash clouds of fine ash particles that could rise and mix into the overlying water column as secondary ash plumes. From these secondary ash plumes fine ash fallout deposits, not derived directly from the vent or the above vent eruption plume, would be dispersed over the sea floor to form deep sea ash layers (Walker et al. 2019).

Walker et al. (2008) and Newland et al. (2022) recognised that as thermal submarine plumes rise above erupting vents significant parts of the rising plumes reach their level of neutral buoyancy in the water mass and may begin to spread laterally as a density current within the water mass and well above the sea floor. Material entrained in such currents may then fallout if their settling velocity exceeds the current velocity. However, in addition, many submarine eruptions lead to the formation of huge floating mats or rafts of pumice debris (Fig. 9.71a; e.g. Bryan et al. 2012; Jutzeler et al. 2014), indicating that much pumice can remain buoyant enough to rise to the sea surface. Pumice rafts can be very voluminous, and can be dispersed thousands of km by oceanic currents and waves, which can generate irregular drift patterns, that last for days, weeks, and months, sometimes even years after the end of the source eruption, as was the case after the 2012 submarine eruption of Havre volcano in the Kermadec arc, southwest Pacific Ocean (Fig. 9.71b; Jutzeler et al. 2014). The size of the long-distance transported clasts varies enormously, from boulder to lapilli to coarse ash size. If floating pumice clasts of any size become water-logged they will settle to the sea floor, leaving a trail of clasts from the pumice raft as its drifts. However, there is no systematic size decrease in the downcurrent deposit, which is very different from the dispersal of pyroclasts in a vertical eruption column and downwind plume resulting from a subaerial explosive eruption, which become finer with increasing distance. Fallout from subaerial plumes stops almost immediately after the end of the explosive eruption, with only the finest ash remaining suspended in the upper atmosphere for months after. In contrast, rafted floating pumice from submarine eruptions may continue to be dispersed for years after the eruption has stopped.

Pumice clasts in the floating mat of pumice are abraded as they rub against each other and become rounded while floating. During this process, much secondary fine ash-size debris may be created through abrasion, which settles to the seafloor as an ash layer that is unrelated to the initial explosive eruption (Jutzeler et al. 2014).

Above newly erupting subaqueous lavas, thermal plumes of heated water will also be buoyant (cf. Barreyre et al. 2011), rising above the lava because heated seawater is at up to several hundred °C and is up to 10% less dense than the surrounding cold seawater (Safarov et al. 2009; Sharqawy et al. 2010). The hot, glassy surfaces of such lava are cooling rapidly, contracting, becoming thermally stressed, and

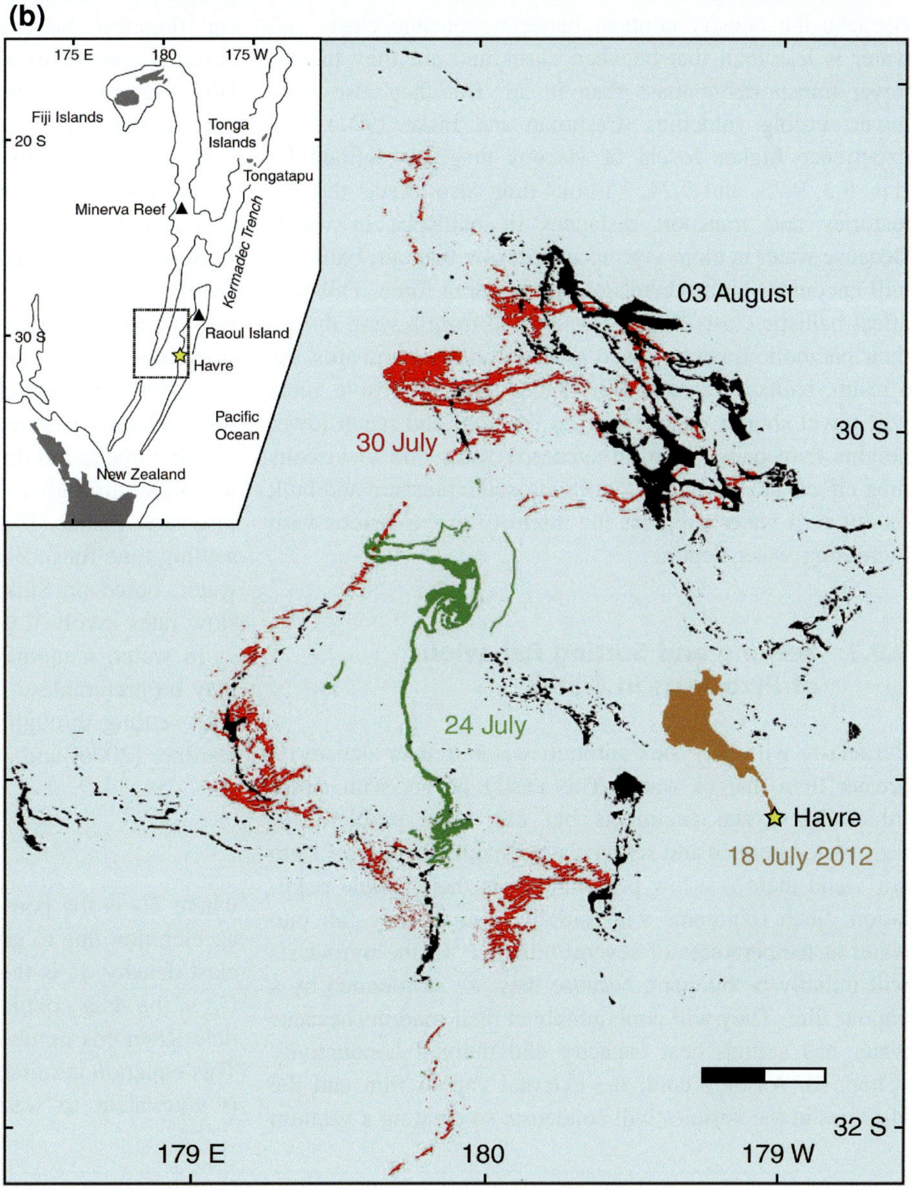

Fig. 9.71 **a** Raft of pumice clasts cm thick drifting in the southwest Pacific Ocean after the Home Reef volcano eruption during 7–16? August 2006, Tonga volcanic arc, southwest Pacific (*Source* NOAA; photo taken by Fredrik on the yacht Maiken, 17–8-2006. Thanks, Fred!). **b** Drift pattern of the pumice raft resulting from the 18 July 2012 submarine eruption of the submarine Havre volcano in the Kermadec volcanic arc over the months following the eruption (From Jutzeler et al. 2014)

spalling or "popping" a lot of fine glassy debris that is thermally exfoliating from the lava surface as a result of quench cooling fragmentation (White et al. 2015). Deposits of ash associated with new submarine lava-forming eruptions, such as the 2015 eruption from Axial Seamount volcano on the East Pacific Rise at water depths of >1,500 m, have traditionally been interpreted as pyroclastic in origin, which is highly unlikely given the water depth and the limiting effects of hydrostatic pressure (Chap. 4). We suggest that they result from fine quench fragmented spalled ash particles being entrained in thermal plumes above the lava. If the settling velocity of such debris is less than the rise velocity of the thermal plumes, that glassy debris will be dispersed upwards and may then fallout, producing layers or pockets of quench-derived, hyaloclastic fallout "ash" that is not of explosive origin (cf. Barreyre et al. 2011).

9.9.2.1 Ballistic Transport in Subaqueous Settings

Because the density contrast between volcanic clasts and water is less than that between clasts and air, they have a lower transportable mass than in air. But they also have lower settling velocities (Cashman and Fiske 1991), and experience higher levels of viscous drag, as defined by Eqs. 9.3, 9.23, and 9.24. Viscous drag also affects the trajectories and transport distances of ballistics in water. Because water is more viscous and denser than air, ballistics will encounter higher levels of viscous drag force. This will affect ballistic clasts as they are ejected from a vent, during their parabolic trajectory, and as they travel downwards. As a result, ballistics ejected from subaqueous explosive vents will travel shorter distances from the vent and reach lower heights than ballistics from subaerial vents due to viscous drag effects alone, especially if hydrostatic pressure and bulk modulus of water suppress the intensity of explosions with increasing water depth.

9.9.3 Settling and Sorting Behaviour of Pyroclasts in Water

Pyroclasts will only sink through water if their density is greater than that of water. This easily occurs with dense lithic and crystal fragments but can be a problem for vesiculated pumice and scoria clasts if their density is <1 gm cm^{-3} and there is a low permeability in their vesicle population. Such fragments will initially float. If they fall into water at temperatures of several hundred °C, the pyroclasts will initially be buoyant, because they are surrounded by a vapour film. They will cool quickly at their margins because water has a high heat capacity and thermal conductivity (Chap. 3). As they cool, the external vapour film and the H_2O gas in the vesicles will condense, so creating a vacuum in the vesicles which will draw water into the outermost vesicles (Whitham and Sparks 1986). If the permeability of the pumice clasts is high enough, the pumice clasts may become very rapidly water-saturated, so that their bulk density, including that of the ingested water, is greater than the host water body, and the pumice will then sink according to Stokes' Settling Law. Alternatively, if the permeability is low, the pumice clasts may float at least temporarily until they become water-logged and settle, or be transported by surface currents, and be rafted thousands of km from where they entered the sea (Fig. 9.71b; e.g. Manville et al. 1998; Bryan et al. 2012; Jutzeler et al. 2014; Fauria et al. 2017).

Even cold or poorly permeable pumice may become sufficiently water-logged with time and settle. Manville et al. (1998) propose that the time it takes for cold pumice to become water-saturated is proportional to the square of the clast radius. Fauria et al. (2017) have demonstrated that some pumice clasts may never become water-logged due to trapped bubbles of gas or air in the vesicle network. They can therefore be transported thousands of kilometres by currents and waves as part of pumice rafts or individually. However, glass shards derived from a highly vesiculated magma, and which contain no fully enclosed vesicles, will have the bulk density of dense glass, and will settle through water much more quickly than much larger and heavier, vesiculated pumice clasts (Jutzeler et al. 2014).

Settling of any solid through a static fluid (= laminar state) is controlled by Stokes' Settling Law, as defined in Eq. 9.1 and modified in Eqs. 9.36 and 9.37 to include consideration of water drag forces. If particle behaviour in air is compared with particle settling behaviour in water, the settling velocity in water is slower (Cashman and Fiske 1991), and the time to fall through water is significantly greater than for fallout through air because water has a higher density and viscosity. Fisher (1965) and Carey (1997) summarised the settling time for individual dense ash-sized particles through water, based on Stokes' Settling Law, which highlight the slow rates involved (Fig. 9.72a).

In water, a modified version of the Stokes relationship may be preferable to account for the drag forces associated with settling through water, as suggested by Burgisser and Gardner (2006) and Jutzeler et al. (2014)

$$U_T = \frac{4g\,d^2(\rho_p - \rho_f)}{3\mu\,C_D\,Re} \qquad (9.36)$$

where U_T is the particle terminal settling velocity, g is the acceleration due to gravity, d is the clast diameter, ρ_p is the clast density, ρ_f is the fluid density, μ is the fluid viscosity, C_D is the drag coefficient, and Re is the dimensionless particle Reynolds number. Other parameters are as in Eq. 9.1. This equation is suitable for particles with a $Re > 10^3$, which is equivalent to water-logged pumice clasts >5 mm, and

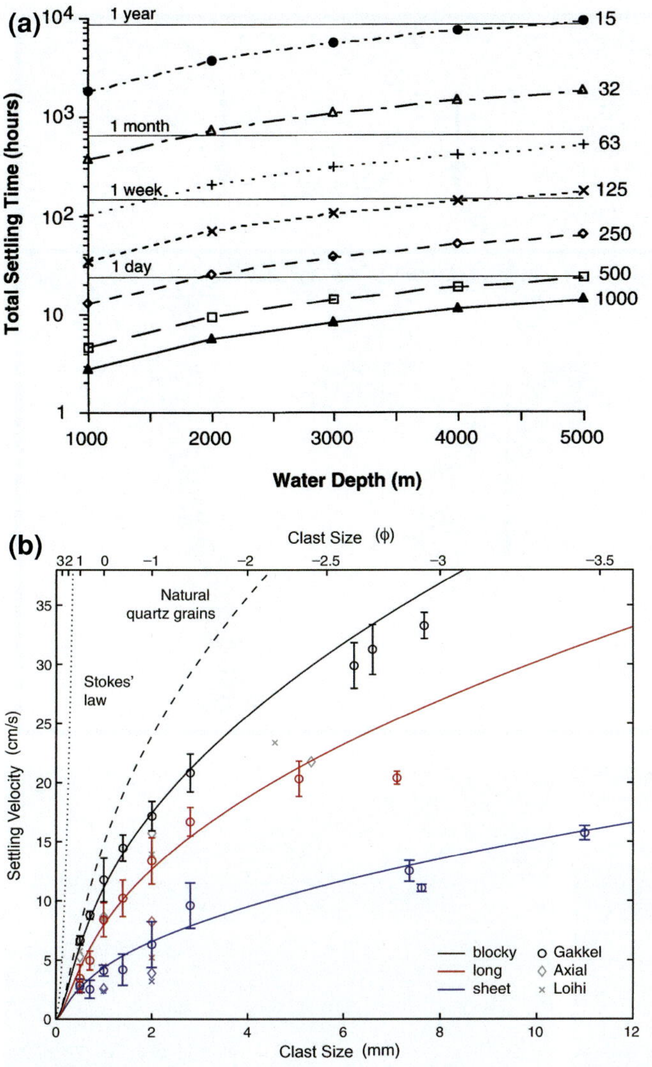

Fig. 9.72 **a** Settling times for ash particles of different sizes through water to different water depths (From Carey 1997). **b** Settling velocities of particles of different shapes, showing deviations from the Stokes Law settling behaviour (From Barreyre et al. 2011)

dense fragments (lithics, crystals) >3 mm (Jutzeler et al. 2014) C_D for particles settling through water is ~0.45.

Barreyre et al. (2011) assessed the effects of shape on settling velocities for particles ranging from <1 mm to 12 mm and different shapes including spheres, blocky, long, and platey, demonstrating marked variations in settling or terminal fall velocities, and differences relative to Stokes Law settling behaviour (Fig. 9.72b). The relationship they used was

$$w = \frac{R g d^2}{C_1 v + (0.75 C_2 R g d^3)^{0.5}} \quad (9.37)$$

where w is the settling velocity, R is the submerged specific gravity for clasts in water, g is the acceleration due to gravity, d is the particle diameter, C_1 is derived from Stokes law and is 18 for spheres, and C_2 is the asymptotic drag coefficient limit, which is 0.4 for spheres and increases as particles diverge in shape from spheres (Barreyre et al. 2011).

In addition, Carey (1997) recognised that as ash particles fall into water, they slow down, and are kept momentarily at the surface due to the surface tension of the water. The surface tension force of water is 72.8 mN m^{-1}, and so theoretically, if we represent the surface tension force as F = mg, water could support a particle of about 7 gm in mass, given that g = 980 cm s^{-2}, and if the appropriate shape and a tranquil interface scenario exist. Pumice, some lithic fragments, and especially ash particles could remain supported by surface tension at the surface of the water mass for some time. Turbulence at the surface (e.g. waves) could allow particles to break through the surface tension membrane and begin to sink. Ash particle concentration builds up in a thin surface layer due to their slow settling velocities in water.

Particles may settle through water independently, according to the above relationships (Eqs. 9.1, 9.36, and 9.37), or, as Carey (1997) and Manville and Wilson (2004) demonstrated in experiments, and Fiske et al. (1998) inferred from deposits, if a critical particle concentration is reached, domains of the water supported ash and pumice begin to sink gravitationally, forming vertical, downward directed ash-laden inverse plumes or density currents, because their bulk density exceeds that of water (Fig. 9.73a). Such currents deliver ash layers to the seafloor (Fig. 9.73b) at several orders of magnitude rates faster than through individual particle settling according to Stokes' Law.

However, the cause of this phenomenon is more than just density contrast between one water layer and another, otherwise such plumes would always form, because any liquid with even a few suspended, denser solids has a bulk density greater than that of the liquid alone. There are two possible, more specific causes that come into play. First, if enough particles in close proximity to each other are sinking, the cumulative shear that they all exert on the fluid surrounding them causes entrainment of fluid and coalescence of flowing liquid and sinking particles, so forming an aggregated, coupled liquid–solid density current flow that moves downward as an inverse plume under the force of gravity, relative to the stationary, lower density water mass (Fig. 9.73a). Secondly, vitric ash particles are electrostatically charged, as are water molecules. A sinking vitric ash particle probably has an envelope of attached water molecules, and if an ash-water aggregate moves closer to another,

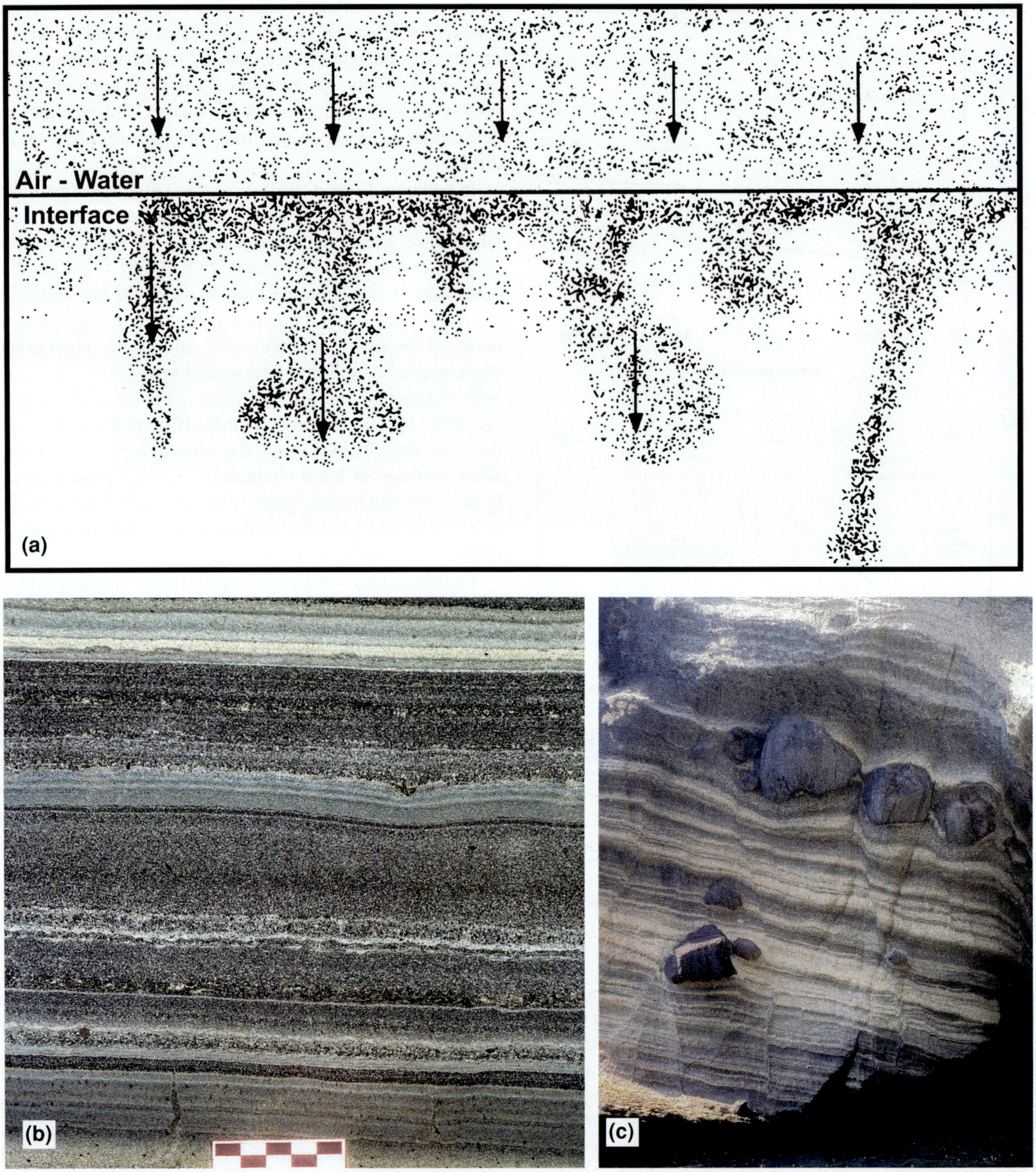

Fig. 9.73 **a** Downward, vertical density currents or inverse plumes of ash from an ash-loaded surface layer at the air–water interface and critical ash flux levels. This implies that deposition of distal ash fall layers in deep-sea sediments may be dominantly controlled by diffuse vertical gravity currents, as opposed to passive settling of individual particles through the water column. A process that could reduce the residence time of fine ash in the ocean and diminish the role of ocean currents in influencing the distribution patterns of individual tephra layers. (From Carey 1997). **b** Sequence of thin, variably graded water-settled ash beds from the Ordovician Borrowdale Volcanic Group, Lake District, UK. Scale bar is 5 cm. Did these form through vertical particle fallout, suggesting each layer represents an explosive event, or through a succession of vertical, inverse-density currents that then spread over the seafloor? **c** Planar stratified water-settled pumiceous fallout deposits that are variably graded, and impacted by lithic clast ballistic blocks. Did these originate from discrete explosions and free particle settling, or from small volume, inverse-density currents settling to the seafloor from the water column(?). Miocene, Koura Bay, Izu Peninsula, Japan

they may interact electrostatically, and so begin to form a dynamic larger aggregated mass of water and ash particles that under gravity acquire downwards directed momentum. This poses an intriguing question. Are fine variably graded ash layers found on the seafloor the result of a particle by particle settling according to Stokes' Settling Law, or do they represent small density currents that have reached the seafloor and spread laterally as dilute turbidity currents that then produce a thin graded bed or lamination (e.g. Fig. 9.73b, c)? As in most cases in nature, some layers form by the former, whereas some will form by the latter. The key controls will be the particle flux to the water surface, and the strength and turbulence of surface waves and currents, which will retard the settling of fine ash particles for some time.

Wiesner et al. (1995) sampled ash from the 1991 Pinatubo explosive eruption deposited in submarine sediment traps moored at depths of 1190 and 3730 m above the seafloor (4270 m water depth) in the South China Sea, at a distance of 580 km west of the volcano. They calculated the subaerial transport time to the depositional site, the sedimentation rate required in the time available, and determined that subaerial ash fallout to the sea surface required ash aggregation and fallout from the ash plume. Then submarine sedimentation to the sediment traps and the seafloor also required sedimentation by ash aggregates in the submarine environment, at velocities of 0.0217 ms^{-1} (1875 m/day), which is 2 to 3 orders of magnitude higher than that of individual ash particles (shards, pumice, mineral grains) in the size range of 11 to 88 μm. Although Carey (1997) disputed this, preferring the downward directed density current for fine ash sedimentation, as outlined above, both mechanisms can operate.

Cashman and Fiske (1991) also analysed the settling behaviour of pyroclasts in water, and compared the hydraulic sorting characteristics of water-settled fallout pumice deposits with subaerial fallout deposits. They demonstrated that the terminal fall velocities of dense lithic clasts of similar size and density are an order of magnitude lower in water than in air, and for pumice clasts, at least two orders of magnitude slower (Fig. 9.74a), assuming that all pumice clasts are water-saturated, which in nature they are not. They also proposed that in subaqueous pyroclastic fallout deposits, the ratio of the diameter of pumice clasts to dense lithic clasts which have settled at the same velocity and are therefore in hydraulic equivalence is between 10:1 and 5:1, whereas for subaerial deposits, the ratio is between 3:1 and 2:1. This reflects the more effective influence that water has over air as a hydraulic sorting agent, because of its greater density and viscosity, which create longer settling times. Although the Cashman and Fiske (1991) analysis may be correct for very short-lived, single explosion-generated fallout deposits during which all pyroclasts are released into the water mass simultaneously, it is not valid for long-lived fallout events in water during which there is a continuous rain of pyroclasts released at different times into the water mass, and it doesn't take account of the variable time for different pumice clasts to become water-logged before they settle.

For long-lived events, grain size or grain size ratios are not necessarily a good indicator of efficient hydraulic sorting in subaqueous settings. The density of "pumice" clasts can vary widely from significantly $<<1000$ kg m^{-3} to <2000 kg m^{-3}, and can change as pumice clasts become variably water-logged over an extended time (Manville et al. 1998; Fauria et al. 2017). Another factor is the duration of a fallout event and whether or not a similar size population of pumice and lithic clasts is delivered to the water surface for the duration of the event, or whether they change in response to changing intensity of the explosive eruption and the height of the eruption column. A long-lived fallout event is likely to produce deposits with highly variable grain sizes, with some large clasts with high settling velocities that enter the water column late, being deposited before or simultaneously as smaller clasts that begin to settle much earlier.

The rate and the degree to which pumice clasts become water-logged is a key determinant of settling velocity, and depends on the temperature, permeability, and size of the pumice clasts at the time they come into contact with water (Whitham and Sparks 1986; Manville et al. 1998). The settling velocity of individual pumice clasts can therefore vary, depending on their degree of water logging as they settle. In cases where very large pumice clasts are only partially water-saturated because of low permeability and have a very low settling velocity for such a large clast, they could have the same settling velocity as much, much smaller grains (e.g. Fig. 9.74b). Deposit grain sizes could thus be extremely variable, particularly in long-lived fallout events, even though the deposits were hydraulically well-sorted at the time of deposition. Since much pumice that is erupted or deposited into water remains buoyant and can be rafted thousands of km from the point of entry of the pumice into the sea (e.g. Bryan et al. 2012; Jutzeler et al. 2014), preserved subaqueous fallout deposits do not represent the size population that is erupted, and they can be very different to subaerial fallout deposits.

Grain size similarity or diversity is therefore clearly not a good indicator of how well a subaqueous fallout deposit is hydraulically sorted. Jutzeler et al. (2014) introduced the concept of a hydraulic sorting ratio (HSR), which is a measure of the fluid dynamic behaviour of dense clasts (lithics, crystals) relative to that of water-saturated pumice clasts as they settle through water, expressed as:

Fig. 9.74 **a** Terminal settling velocities of pyroclasts of different grain sizes and densities in air and water according to Stokes' Settling Law, showing orders of magnitude slower settling velocities in water than in air (From Cashman and Fiske 1991). **b** Deposit of large pumice clasts (cream colour) interleaving with diffusely, horizontally laminated, quartz and plagioclase crystal fragment-rich ash (dark), near the base of a subaqueous pumice and tuff cone sequence, Late Devonian Bunga Beds outlier, Boyd Volcanic Complex, Aragunnu Bay, southern New South Wales, Australia. The large pumice clasts are interpreted to have had the same settling velocity as the crystal fragments due to their partially water-logged state. Many of the pumice clasts are tube pumice, and they are visibly moderately crystal-rich. (See Cas et al. 1990 for discussion of the origin and setting)

$$\frac{(\rho_{dense}-\rho_{fluid})}{\rho_{fluid}} \cdot \left[\frac{(\rho_{pumice}-\rho_{fluid})}{\rho_{fluid}}\right]^{-1} \quad (9.38)$$

When the HSR ratio is 1, a deposit is hydraulically well-sorted, consistent with unhindered particle free fall. When the HSR ratio is <1, pumice clasts are too small to be in hydraulic equivalence with dense fragments, and when HSR > 1, pumice clasts are too large to be in hydraulic equivalence. Both indicate alternative mechanisms rather than simple particle-free settling or fallout, and probably involving hindered settling conditions associated with mass flow or tractional sedimentation.

The essential parameters that determine if a subaqueous fallout deposit is hydraulically well-sorted or not are the water depth, *d*, the duration of a fallout supply event, t_e, the time that it takes for a dense clast to settle to the seafloor, t_{DC}, determined by its terminal settling velocity in water, V_{wDC}, and water depth, and the time that it takes for a low-density pumice pyroclast to settle, t_P, which is determined by its terminal settling velocity in water, V_{wP}, and the water depth. If

$$t_e \leq t_P - t_{DC} = \frac{d}{V_{wP}} - \frac{d}{V_{wDC}} \quad (9.39)$$

then a fallout deposit will be hydraulically well-sorted as long as the deposit is formed by particle-free fall, and not by aggregate depositional processes. If

$$t_e \geq t_P - t_{DC} \quad (9.40)$$

a fallout deposit may not be hydraulically well-sorted because dense clasts released late in the event could overtake or be deposited at the same time as pumice clasts released at the beginning or early in the event, and there could be changes in grain size and clast density during the event.

9.9.4 Subaqueous Explosive Eruption Styles, and Deposit Characteristics

Cas and Giordano (2014), White et al. (2015), and Cas and Simmons (2018) have most recently reviewed what is currently known about aspects of subaqueous explosive eruptions. There are significant differences compared with subaerial counterparts largely due to the effects of the host water mass. This includes its ability to suppress the intensity of subaqueous explosive eruptions through hydrostatic pressure and bulk modulus (of water) effects, its cooling, quenching effects on hot pyroclasts, its density which slows settling velocities of pyroclasts, and its ability to raft away huge volumes of erupted low-density pumiceous material for thousands of km from the vent (Cas and Simmons 2018; also see Chap. 4). Nonetheless, there may be some similarities to some subaerial eruption styles, as discussed below.

9.9.4.1 Submarine Magma/Fire Fountaining Eruptions and Deposits

Many examples of submarine magma/fire fountaining spatter deposits, of a variety of magma compositions, are recorded in the literature from both the modern seafloor and ancient submarine rock successions (e.g. Smith and Batiza 1989; Mueller and White 1992; Cas et al. 1996, 2003; Simpson and McPhie 2001; Davis and Clague 2003; Fujibayashi and Sakai 2003). Known occurrences on the modern seafloor occur at depths down to at least 2,000 m (Smith and Batiza 1989; Davis and Clague 2003; Cas and Giordano 2014). Head and Wilson (2003) suggest that at water depths of several kilometres, submarine magma/fire fountains of basaltic magmas are driven mostly by CO_2. Although they assume that magma fragmentation in a magma/fire fountain can only occur when the level of vesiculation is 75% and that it involves gas bubble bursting, vesicle expansion at explosive rates will *not* occur at those depths, given that both CO_2 and H_2O will be compressed supercritical fluids, not gases. The spatter clasts described by Cas et al. (2003) at Ryugazaki, Japan, are very poorly vesicular, indicating that bubble expansion and vesiculation to levels of 75% did not cause fragmentation of the magma in the fountain. However, bubbles of exsolved supercritical CO_2 and H_2O would lower the density of the magma, and given the low viscosity of submarine basalt magmas, the fire fountains could be driven purely by buoyancy, exiting the vent as a fluidal, weakly vesiculated (<20%), low-viscosity magma jet (Cas et al. 2003; Cas and Giordano 2014).

As proposed by Cas et al. (2003), above the vent the magma/fire fountain would expand and flare. Fragmentation of the fire fountain magma then results from a combination of shearing due to differential velocity between the margins and centre of the fountain, extension as the fountain expands sideways above the vent, and quench fragmentation of the fountain magma and spatter clasts due to instantaneous cooling by the water mass, especially at the periphery of the fountain (Fig. 9.75a). With high to moderate height fountains, the larger surface area of the fountain, and the higher degree of expansion out of the vent will ensure more interaction with the water mass, leading to higher degrees of quench fragmentation of the magma and ejected spatter clasts (Fig. 9.75a, b). With lower fountains, there will be a high proportion of large spatter clasts to the matrix because there has been less interaction with the water mass (Fig. 9.75c).

The lower part of the Ryugazaki succession contains variable proportions of fluidal spatter clasts, but is dominated

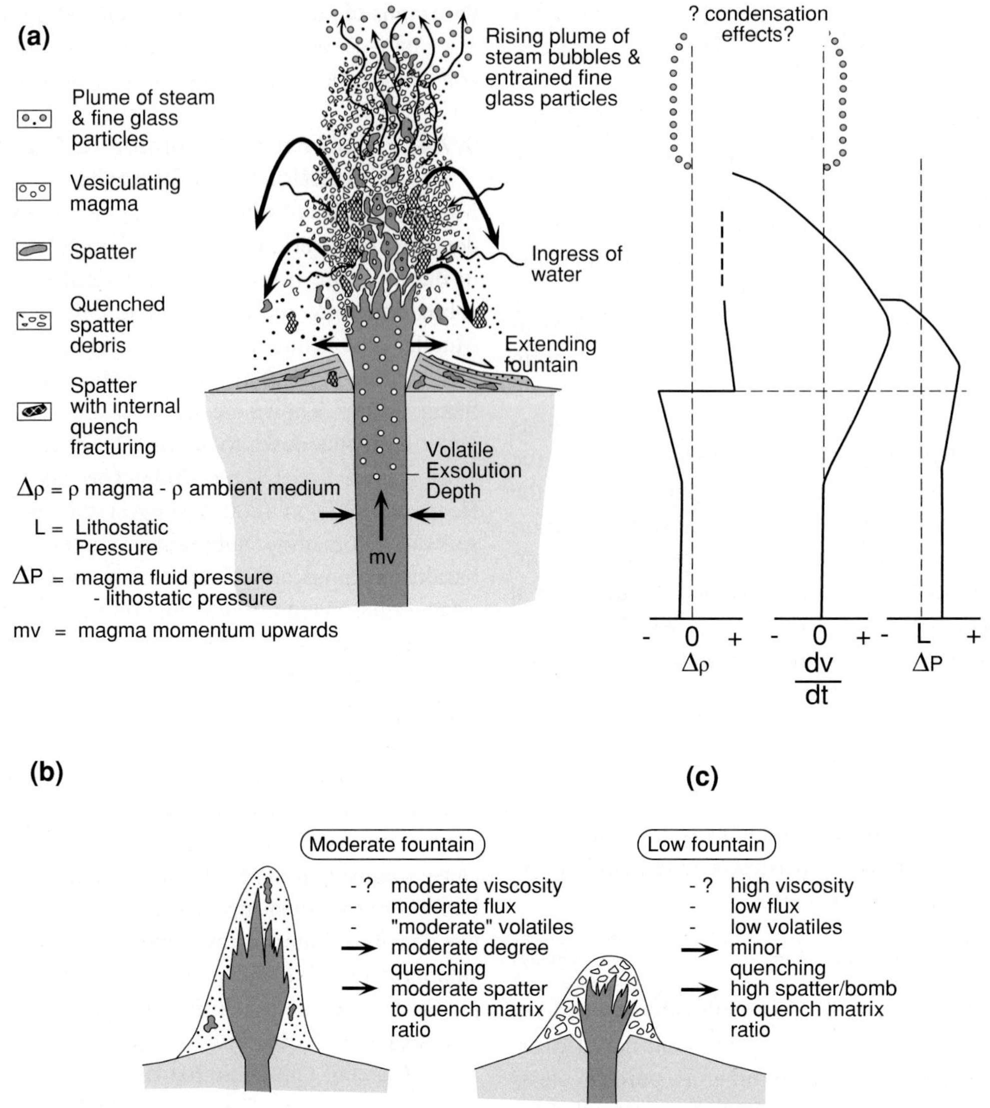

Fig. 9.75 Characteristics of submarine fire fountains. **a** Model for submarine magma/fire fountain and fragmentation processes, highlighting the expanding flaring fountain of magma above the vent and the cooling–quenching interaction with the ambient water body around the margins of the fountain. **b** High fountain with greater degree of quench fragmentation compared with (**c**), showing low degrees of quench fragmentation in a low fountain. **d** Multiple reverse-graded spatter clast layers, with abundant finely granulated matrix resulting from quench fragmentation. Miocene Green Tuff Belt, Ryugazaki, Hokkaido, Japan (and similarly for remaining images). **e** Planar stratified lapilli-tuff consisting of dispersed spatter fragments in a coarse ash-size matrix of blocky granular glassy fragments resulting from quench fragmentation of a high fountain. **f** Close-up view of spatter clasts with low vesicularity in an abundant matrix of blocky quench fragmented granules in stratified submarine fire fountain deposits. Note the network of internal fractures inside the spatter clasts, the chilled black glassy rim of the central spatter clast, and the truncated margin of the spatter clast at the right, resulting from quench fragmentation of a chilled spatter clast. **g** Uppermost, dark layer of clast-rich, spheroidal bomb agglomerate. Note the spheroidal jointing and blocky nature of matrix fragments, consistent with cooling-related exfoliation jointing and quench fragmentation. **h** Uppermost, dark layer of clast-rich, spheroidal bomb agglomerate connected to its feeder fissure dyke, with irregular margins, which cross-cuts the lower bedded spatter sequence. Note the person circled for scale. (From Cas et al. 2003)

by an abundant finer matrix of blocky glassy granules, resulting from quench fragmentation of the magma and spatter fragments in a high submarine fire fountain (Cas et al. 2003, Fig. 9.75a, d-f). Some of the spatter clasts are truncated by quench fracture surfaces and contain an internal network of quench fractures (Fig. 9.75f). The upper part of the succession is dominated by spheroidal bomb clasts, with a minor matrix of granular, blocky glassy fragments, forming an agglomerate erupted from a low fountain (Fig. 9.75g). The deposit is connected to its feeder dyke (Fig. 9.75h), and intrusive lobes of coherent magma mingle through the agglomerate in places (Cas et al. 2003).

Fig. 9.75 (continued)

Simpson and McPhie (2001) documented a thick Cambro-Ordovician succession of submarine, andesitic spatter clast breccias, interpreted as magma/fire fountain deposits, interbedded with coherent andesites (lavas, sills, dykes), and pumice breccias. Like the Ryugazaki fire fountain succession, the matrix of the spatter breccias consists of granular spatter debris, which Simpson and McPhie (2001) also interpreted as resulting from quench fragmentation of spatter clasts at the margins of, and in the fire fountain. The succession was interpreted as a submarine spatter and lava cone (Simpson and McPhie 2001). They note that submarine magma/fire fountain deposits are unlikely to show signs of agglutination (cf. subaerial magma/fire fountain deposits).

Silicic submarine magma/fire fountain deposits are rare, but Mueller and White (1991) recorded a convincing Archean example in the Abitibi Greenstone Belt, Quebec, Canada. Amoeboid, contorted spatter clasts are part of a massive to diffusely stratified depositional unit that is associated with a turbidite facies sequence of the same composition. The authors interpreted the association to result from a magma/fire fountain of fluidal rhyolite that collapsed and fed a hot, steam-rich, seafloor density current of pyroclastic debris. Mueller and White (1992) also recognised the likely interaction between the magma/fire fountain and the ambient water mass as the origin of the finer matrix lapilli-tuff breccia. Examples of Archean mafic spatter, fire fountain

Fig. 9.76 Mafic fire fountain deposits at the Potter Mine in the Archean Abitibi Greenstone Belt, Canada. **a** Thick massive deposit, with several large irregular, fluidal spatter clasts (e.g. cluster that the coin sits on), dispersed in a matrix of finer, blocky, spatter debris. **b** Several highly fluidal spatter clasts

deposits also occur in the Archean Abitibi belt, including well-preserved deposits at the Potter Mine (Fig. 9.76a, b). The effects of water depth and high hydrostatic pressure, which limits H_2O exsolution together with high lithostatic pressure, both have the effect of lowering magma viscosity making erupting magma more fluidal at the time of eruption.

9.9.4.2 Deep Submarine Equivalents of "Strombolian/Vulcanian" Style Eruptions?

Chadwick et al. (2008) and Deardorff et al. (2011) proposed that the weakly explosive submarine eruption of NW Rota I volcano in the Marianas arc in 2006 at water depths of 550 m, which was imaged from the ROV *Jason II* submersible, was similar in style to strombolian explosions from subaerial volcanoes. The eruption varied from lava effusion to gas streaming to explosive bursts of ballistics, producing weak buoyant plumes rising 10's of metres above the vent (Figs. 9.77a, 9.70a, b). Juvenile pyroclasts and recycled debris were deposited close to the vent. The vesicularity of juvenile pyroclasts varied from 0 to 38%, suggesting that small vesicles were not the cause of the explosions. Fragmentation probably resulted from both explosions driven by spaced, very large gas slugs during slug flow conditions (Fig. 9.77b; Chadwick et al. 2008), some component of explosive water–magma interaction in the conduit, perhaps hydrothermal explosive activity, and by cooling–quenching granulation. These complications, which are different to those associated with subaerial strombolian eruptions and described above, make it difficult to accept a description of the eruptions style as strombolian or vulcanian. Clague et al. (2009, and numerous earlier publications) and Davis and Clague (2003) have also documented the deposits of other submarine deposits on the modern ocean floor that they called strombolian.

Chadwick et al. (2008) also imaged a laterally flowing, turbulent, seafloor density current generated by a blast, analogous to base surges associated with subaerial phreatomagmatic explosions (Fig. 9.77c, d).

Jutzeler et al. (2016) summarised the characteristics of an uplifted and dissected (by coastal erosion), Miocene, submarine, basaltic-andesite scoria and spatter cone in the Nishiizu member of the Dogashima Formation, exposed on the Izu Peninsula, southeastern Japan. The 50 m thick succession is largely planar stratified with massive to diffusely layered beds up to 1 m thick, consisting of scoria, with vesicularity ranging from 10 to 35%, and spatter bombs. The deposits are generally moderately sorted, dominated by coarse ash and lapilli, and are fines depleted (>0.5 mm). The original eruption water depth is unknown. Jutzeler et al. (2016) suggest that the relatively fine grain size was in part caused by quench fragmentation of spatter clasts in the column above the vent and during fallout. There is little evidence of downslope resedimentation, suggesting an in situ fallout origin around a vent that allowed a substantial cone to form. Vesicles are generally small (0.01–3 mm, occasionally <10 mm), but vesicle size and density data do not allow distinction between vesicle–magma coupled (e.g. hawai'ian, plinian) or uncoupled (e.g. strombolian) styles of eruption. Bedding indicates a pulsing style of eruption, which could be due to the influence of some degree of explosive magma–water phreatomagmatic interaction. It is

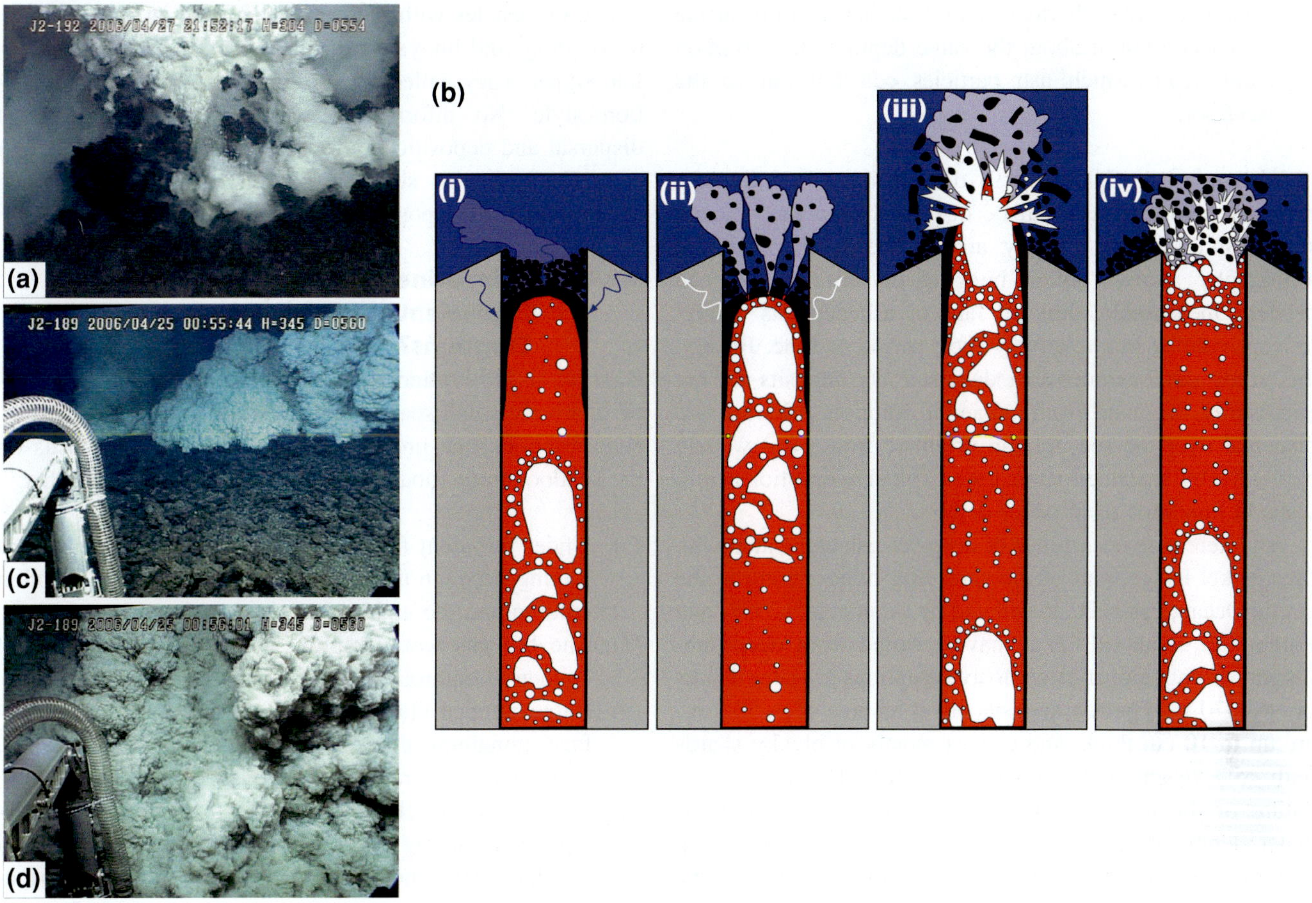

Fig. 9.77 Features of the submarine explosive eruption of NW Rota 1 volcano in the Marianas arc, at water depths of ~550 m. **a** Ballistic blast from Brimbank Crater, ejecting relatively dense bombs and blocks of basalt. **b** Proposed gas slug model for the subaqueous explosions from the NW Rota 1 submarine volcano. Red = magma; white = magmatic gases; black = solidified lava; blue = seawater. The model envisages four stages: (i) between explosions, seawater cools the top of the magma column, forming a solidified cap; (ii) when the next gas-rich pocket reaches the cap, it immediately begins to escape; (iii) the gas pocket forces the lava cap upward in the vent where explosions destroy it; (iv) with the vent uncapped, the eruptive burst proceeds at a higher level until all the gas in the pocket is vented and activity abruptly ceases, returning to the initial state. **c** Laterally flowing, turbulent, seafloor density current generated by a blast; elapsed times following blast are 19 and 36 s. (From Chadwick et al. 2008)

not appropriate to describe this succession as being strombolian in origin. Although much of the succession may have originated from discrete strombolian-like fountain sprays, the involvement of water in perhaps both explosive activity and through quench fragmentation make a subaerial magmatic explosive analogue inappropriate, unless phreato-strombolian is used, as suggested in Fig. 9.6a.

Generally, very little is known about the dispersal patterns and distances of modern submarine examples, other than the coarse breccias localised around the vent. However, in a fascinating study, Walker et al. (2008) documented the dispersal of ash and anhydrite-laden plumes away from and around the NW Rota 1 volcano vent producing fine ash deposits on the seafloor. They concluded that ash particles could be dispersed in three ways:

(a) Particulate, density gravity currents, mostly turbidity currents, could have originated directly from the vent during explosions (cf. Fig. 9.77c, d), or from remobilisation of original ash fall deposits on the slopes of the volcano that then carry ash and hydrothermal anhydrite particles down the slope and distribute thin, fine ash deposits onto the seafloor, as far as 18 km from the vent, but fine ash was also detected in near-bottom nepheloid layers tens of km from the vent.

(b) Lofted ash plumes separated from density gravity currents at depths where their density is that of the water mass and spread laterally as a mesopycnal density current within the density-stratified water mass (cf. Walker et al. 2011; Newland et al. 2022). Fallout of ash particles could then occur from this suspended layer.

(c) Magmatic-hydrothermal plumes spread laterally in the water column at about the same depth as the eruption vent, from which ash particles can fall out to the seafloor.

Walker et al. (2008, 2019) demonstrated that through these mechanisms, ash can be dispersed onto the volcano slopes and seafloor radially around the volcano. Where topography allows, bottom-hugging density currents can produce measurable, directed fans of ash deposits of significant volume in a relatively short period of time. Finally, this study demonstrates that deep-sea ash deposits do not necessarily originate from subaerial eruption columns and dispersal over the sea but can originate from even weakly explosive by sustained submarine explosive eruptions from vents at hundreds of m water depth.

An interesting occurrence of deep-sea ash associated with the Gakkel Ridge, part of the spreading ridge system in the Arctic Ocean, has been considered by Sohn et al. (2008) and Pontbriand et al. (2012) to have resulted from deep submarine strombolian style explosive eruptions at water depths of ~3,870 m. The ash deposits cover an area of ~50 km^2, are up to 10 cm thick, and consist mostly of blocky shards with <5% vesicles (Pontbriand et al. 2012). The blanket-like nature of the ash deposit suggests fallout origin, but the water depth, being greater than the critical pressure point depth equivalent for both water and carbon dioxide, and the extremely low vesicle abundance mitigate against an explosive origin driven by either magmatic or phreatomagmatic processes. Perhaps the ash particles were formed by quench fragmentation and/or spalling of fine glassy fragments from the surface of lava crust during submarine lava-forming eruption(s) or from quenching of a sustained seafloor magma fountain, and then lofted by warm buoyant water plumes rising above the vent area and the surface of the propagating seafloor lavas, forming hyaloclastite ash fallout deposits.

9.9.4.3 "Poseidic" Explosive Eruptions

Schipper et al. (2010) described pyroclasts from a 60 m high cone at water depths of ~1,000 m at the submarine Loihi seamount south of Hawaii. The pyroclasts are weakly to moderately vesiculated (~40%) basaltic pyroclasts, with simple vesicle populations indicating closed-system magmatic exsolution of volatiles in the sub-seafloor conduit. However, blocky ash particles with curviplanar surfaces suggest a hydromagmatic fragmentation origin (thermohydraulic or phreatomagmatic explosions and/or quench fragmentation of an already vesiculated magma). Schipper et al. (2010) distinguished this style of the submarine explosive eruption from strombolian styles because the simple vesicle population indicates coupled, closed-system uprise of vesicles with magma, rather than an open-system, decoupling. and buoyant uprise of gas slugs as in strombolian styles. They called this eruption style "poseidic" eruption style. No information was provided on pyroclast dispersal and deposition but given the growth of a seafloor cone at depth, this suggests relatively local dispersal and accumulation of deposits around the vent.

9.9.4.4 Submarine "Yalian" Eruptions: Equivalents of Felsic Surtseyan Eruptions?

Kano et al. (1996) and Kano (2003) analysed the dynamics of relatively small-scale, shallow water subaqueous explosive eruptions that produce significant pumice deposits on the seafloor from subaqueous vents. They proposed that:

(a) Under turbulent flow conditions, out of the vent rapid mixing between the hot, erupting gas-charged eruption column and the ambient water mass would occur.
(b) The hot gas would rapidly condense (which we call condensation erosion), into a buoyant hot water plume.
(c) Large pumice clasts would rise buoyantly to the surface, remaining there until they become water-logged, and sink, or are rafted away by surface currents.
(d) Small pumice clasts would rapidly become water-logged according to the mechanism of Whitham and Sparks (1986) and begin to fall out of the transformational eruption column (gas to water supported), together with dense lithic and crystal debris.
(e) If the mass flux of falling material is high enough, the collapsing mass could transform into an initially hot, gas-rich mass flow of pyroclasts, but due to mixing and condensation effects, these would rapidly transform into water-supported mass flows (debris flows, turbidity currents) flowing away from the volcano (Kano et al. 1996).

The Kano et al. (1996) model is based on a study of the Shinjima Pumice, which erupted subaqueously in the Aira caldera several thousand years ago, and was then uplifted by the emplacement of a young cryptodome. Kano et al. (1996) identified a massive to diffusely stratified, fines-depleted, blocky pumice clast breccia facies. The matrix contains bubble wall glass shards with brittle fracture terminations, suggesting the cooling effects of water. The deposit was interpreted to be a subaqueous density current deposit, triggered by high rates of fallout from the eruption column. Although they suggest that a significant fraction of the erupting pumice would have risen buoyantly and been dispersed by surface currents, they did not describe fallout deposits. However, their model for a subaqueous explosive eruption column was adopted in many subsequent studies, not only for understanding subaqueous pumiceous mass flows but also for fallout deposits.

The ~24 ka, rhyolitic Yali pumice cone, in the Hellenic arc in the southeastern Aegean Sea, Greece, described by Allen and McPhie (2000), contains a range of facies, including subaqueous fallout deposits. However, as Kano et al. (1996) and Kano (2003) pointed out, many other processes also occurred simultaneously (Allen and McPhie 2000). The sequence is >150 m thick, was erupted in water depths of ~200 m based on fossil evidence, and was subsequently uplifted. The succession is bedded throughout with beds ≤ 3 m thick (Fig. 9.78a). The principal component is white pumice, with a vesicularity range of 56–86%. Vesicles are generally elongated to tube vesicles, and from microns to several mm in diameter, but most clasts are finely vesicular. Pumice clast sizes range from 1.5 m to granule size (2–4 mm), indicating the deposits are fines depleted.

The principal facies are massive, coarse cobble to boulder size pumice block breccia beds up to 3 m thick that are clast-supported and lack a finer matrix. Pumice clasts are typically prismatic or polyhedral in shape, and their surfaces are marked by closely spaced cooling contraction joint sets (Fig. 9.78b), indicative of cooling–quenching contraction. A second facies, dominated by pebble size lapilli pumice clasts that are typically polyhedral in shape (Fig. 9.78c), occurs in well-sorted, diffusely layered beds from cm to 0.6 m thick that can form sequences up to 10 m thick (Fig. 9.78d). Although most beds are not graded, some are reverse graded, beds are wedge-shaped, and some low-angle bedding truncations are present. There is also a less well-sorted mixed clast facies. Deposition probably occurred through both suspension settling for tabular beds to resedimentation for wedging and truncation beds.

Allen and McPhie (2000) interpreted the coarse cobble to boulder-size pumice clast facies to have formed from hot, buoyant, floating clasts that were released passively from the surface of erupting vesicular lavas or domes, subject to quench fragmentation, and then ingested water into vesicles, became water-logged and settled from the water column onto the slopes of the pumice cone and beyond. Large pumice blocks would have taken longer to become water-logged and would have floated further before settling. The bedded pebble size pumice breccia facies shows evidence of lateral transport by subaqueous grain flows (massive to diffusely stratified, wedging beds, truncations). These originated from magmatic/phreatomagmatic explosive phases, during which hot, small pumice clasts quickly ingested water, and settled, forming an unstable steep slope (>10°) as a near-vent pumice cone began to build, with regular slope failure generating downslope grain flows. The mixed facies was interpreted to result from simultaneous large buoyant clast settling and explosive eruption-related grain flow.

During further research by Allen et al. (pers comm) on the characteristics of the pumice in the deposits of the explosive phases of the eruption of the Yali pumice cone, distinct populations could be identified. Most pumice clasts have a vesicularity range of 45–75%. Type 1 pumice, from the finer facies, have a unimodal vesicularity range of 55–60%, the largest (>50 μm) have complex vesicle shapes, walls >100 μm, and a low vesicle number density of N^{m}_{v} of 7.0×10^{7} cm^{-3} to 1.6×10^{8} cm^{-3}. Type 2 pumice clasts, from the coarser facies, have unimodal vesicularity range of 65–70%, a higher N^{m}_{v} of 1.7–3.4 $\times 10^{8}$ cm^{-3}, and vesicles are round to sub-round, including a population that is <50 μm. Allen et al. attribute the smaller vesicles in small Type 1 pumice to be due to some degree of phreatomagmatic explosive fragmentation, whereas the larger and more vesicular Type 2 pumices to have been fragmented by magmatic exsolution and explosive vesiculation. They find that the vesicle number density for Yali pumices is comparable to weakly explosive subaerial explosive eruption styles (e.g. sub-plinian, violent strombolian) which is not surprising given that the eruption centre erupted in water depths of 200 m, at a very low initial ambient hydrostatic pressure of ~2 MPa, but decreasing rapidly as the water column above the vent became steam bubble saturated, and a cone grew to above sea level.

Allen et al. (in prep) have called this style of shallow marine explosive activity "*yalian*", which they consider to be a less intense explosive style than neptunian, as discussed below. Given the context, however (<200 m water depth), then such shallow water felsic explosive eruptions may be comparable to shallow water mafic surtseyan explosive eruptions, although perhaps the degree of fragmentation is not as high.

Similar rhyolitic pumice cones were formed several times during the eruption of Tuluman volcano in the northern Bismarck Sea, Papua New Guinea, in 1953–1957 (Reynolds et al. 1980). These resulted from initial shallow marine eruptions producing large pumice rafts followed by a phase of explosive activity that produced a pumice cone that emerged above sea level, which in some cases was followed by an effusive lava-forming phase. In other cases, collapse of the cones or parts thereof occurred, presumably producing resedimented submarine pumice deposits. Details of the deposits are not available.

Cas et al. (1990) described two tuff and pumice cone successions preserved in the Late Devonian Bunga Beds outlier of the Boyd Volcanic Complex on the south coast of New South Wales, Australia. They considered them to have formed in the relatively shallow water below the wave base, resulting from the emergence of rhyolite domes onto the basin floor, triggering both phreatomagmatic and magmatic explosive activity from the dome tops. The initial deposits were interpreted to be hydraulically well-sorted crystal-rich

(a)
(b)
(c)
(d)

◀ **Fig. 9.78** Features of the ~24 ka now uplifted, submarine Yali pumice cone, Yali Island, southeast Aegean Sea, Greece. **a** View of a section through part of the thick diffusely stratified pumice cone succession, showing an interval of the cobble-boulder size pumice breccia in the middle between finer pebble size pumice breccia layers above and below. **b** Large polyhedral pumice clast with cooling contraction fracture network on the surface. **c** Close-up of fines depleted lapilli and block pumice breccia, made of polyhedral to blocky pumice clasts. Note the large clasts are internally fractured, probably due to quench cooling contraction, and many of these will have disaggregated to produce many of the smaller clasts. **d** Diffuse stratification in lapilli pumice breccia facies, marked by clast size variations, and wedging geometries of beds, indicating some component of lateral transport, not just vertical fallout

fallout and near-vent lateral surge deposits (Fig. 9.79a), followed by planar stratified variably fine pumice and ash fallout deposits (Fig. 9.74b, Fig. 9.79a, b) that contributed to the growth of dome top pyroclastic cones.

9.9.4.5 Submarine "Neptunian" Eruptions: Equivalents of Plinian Eruptions?

In modern environments, many inferences about the eruption and depositional processes also have to be made because of limited accessibility to subaqueous environments. Fiske et al. (1998) described visible processes resulting from the 1952–1953 shallow water eruption of Myojinsho submarine volcano in the Izu–Ogasawara arc south of Japan and made inferences about dispersal processes. Myojinsho is a cone-like feature which straddles the rim of a nearly circular submarine caldera ~ 10 km in diameter. The 12.5-month-long activity was dominated by the eruption of multiple subaqueous dacite lava domes that often grew above sea level and were destroyed by numerous (possibly as many as 1000), major phreatomagmatic ballistic explosions (Fig. 9.80a). Dense clasts were seen to fall back into the sea, then disappear beneath the water and settle. During a number of eruptions, significant volumes of floating pumice were seen, and cloudy surface plumes, carried away by currents, were interpreted to represent suspended ash-sized material and pumice (Fig. 9.80b, c). Fiske et al. (1998) considered that much of the suspended pumice and ash would have formed vertical density currents or plumes that descended into deep water from a clast-rich surface layer on or near the sea surface. These helped to build the subaqueous angle of the repose slopes of the edifice. Deposition therefore involved both individual clast settling or fallout or dense components as well as density current sedimentation (Fig. 9.80c). The full extent of the deposits is unknown.

Exploration of the Kermadec arc, north of New Zealand, led to the discovery of several submarine volcanoes with abundant pumice deposits (e.g. I.C. Wright and Gamble 1999; Wright et al. 2003). At Healy submarine caldera, Wright and Gamble (1999) and Wright et al. (2003) interpreted a large pumice deposit to have been erupted explosively from water depths of 500–1,100 m. Deposition of the pumice is considered to have involved ingestion of water through cooling of hot pumice (Whitham and Sparks 1986), which was considered to have been important in enhancing sedimentation of the pumice.

Allen and McPhie (2009) proposed that such subaqueous, pumice-forming eruptions could be called "*neptunian*". They considered that the hydrostatic pressure and the cooling effects of ambient ocean waters have significant effects on the eruption style, "pyroclast" characteristics, and dispersal processes, as previously discussed by Kano et al. (1996). They proposed that pyroclasts are dispersed by a combination of flotational buoyancy, especially large pumiceous blocks with hot, gas-rich interiors, fallout (see below), and collapse of columns due to the cooling effects of water on pyroclasts and condensation of their internal gases, making them dense and prone to collapse. Deposits include large, rafted, and settled pumiceous blocks, submarine fallout deposits, and hot to cold syn-eruptive pumiceous mass-flow deposits. Fragmentation mechanisms can vary from spalled blocks from erupting submarine domes to explosively fragmented and quench fragmented pumiceous debris. The explosive component is driven by both exsolving, magmatic volatiles and explosive superheating of seawater.

Subaerial plinian eruptions and deposits are defined by column heights, dispersal patterns and areas, and grain size characteristics. There are no such data available for subaqueously erupted pumice deposits and given the differences between subaerial and subaqueous eruption columns, environments and dispersal processes, it is inappropriate to call submarine eruptions "plinian".

9.9.4.6 "Tangaroan-Havre" Eruptions: Submarine Autoclastic Pumice Fallout Deposits from Non-Explosive Submarine Eruptions

Kano (2003) and Allen and McPhie (2009) proposed that during submarine lava dome-forming eruptions, a highly vesiculated pumiceous carapace is likely to liberate large, buoyant spalled pumice clasts that float to the surface, experience water ingestion into vesicles leading to some large clasts sinking and settling, whereas others undergo quench fragmentation that could generate a smaller pumice clast population (Fig. 9.81a). Although some minor phreatic explosive activity may accompany these processes (Fig. 9.81b), the majority of the pumice is generated through non-explosive vesiculation, and then physical spalling and/or quench fragmentation.

Rotella et al. (2013) proposed that widespread pumice deposits found on the seafloor around and up to 25 km from

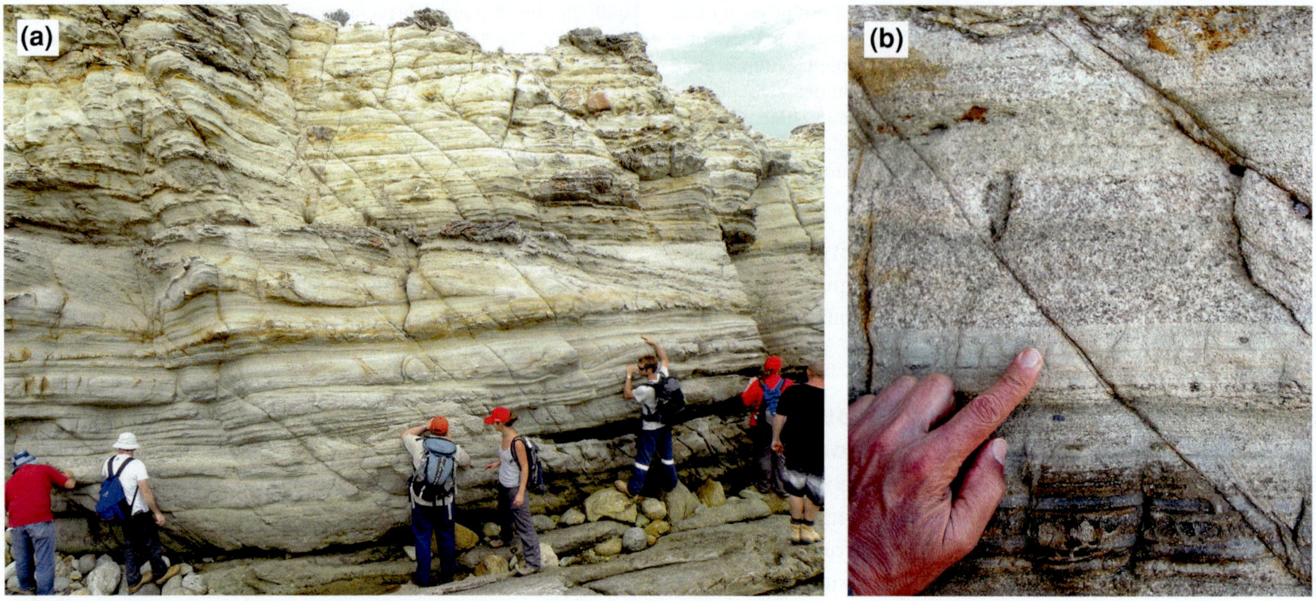

Fig. 9.79 **a** Lower planar, diffusely bedded, and low-angle cross-stratified crystal-rich and pumiceous subaqueous fallout deposits, overlain by finer grained, planar laminated pumice and ash fallout deposits, from a Late Devonian tuff and pumice cone succession, Bunga Beds outlier, Boyd Volcanic Complex, Aragunnu Bay, southern New South Wales, Australia (also see Fig. 9.74b). **b** Close-up of the upper thinly planar stratified pumice and ash subaqueous fallout deposits from the same locality as in (**a**)

Macauley volcano in the Kermadec arc originated from the non-explosive spalling and rafting of highly vesiculated pumice from erupting vesiculating magma from a submarine vent. They called this style of eruption activity "*tangaroan*" (named after their research ship, the *Tangaroa*, which is also the Maori name for the god of the sea). As hot spalled pumice rafts detached and floated away from the vent, they were subject to varying degrees of quench fragmentation into smaller clasts, dispersal by currents and fallout. The high levels of vesiculation (up to 90%) are the result of slow steady, non-explosive exsolution and vesicle growth at low gas overpressures relative to the ambient hydrostatic pressure (Cas and Giordano 2014; Cas and Simmons 2018).

The 2012 submarine eruption of Havre volcano, Kermadec arc, north of New Zealand, produced the largest known historic silicic submarine eruption (Carey et al. 2018; Manga et al. 2018; Mitchell et al. 2018). The eruption products were rhyolite lava domes erupted from vents at water depths of ~900 m (Chap. 6), from the ring fracture on the southern rim of the pre-existing caldera, and a range of pumice deposits, including giant pumice blocks up to 6 m in diameter (Figs. 6.24, 16.74d, e), which were deposited on the caldera floor at a depth of 1,500 m and around the volcano, as well as pumice dispersed thousands of km by a floating raft of pumice that reached the eastern seaboard of Australia (Jutzeler et al. 2014). The eruption has been interpreted to have been a non-explosive eruption that produced multiple lava domes with pumiceous carapaces from which pumice clasts up to many metres in diameter were spalled and detached from the lava surfaces during quench fragmentation while hot and gas-charged (Carey et al. 2018; Manga et al. 2018; Chap. 6). The buoyant pumice clasts then rose buoyantly in the water mass above the vent and became variably water-logged in transit (Manga et al. 2018; Mitchell et al. 2018). Some fell out of the water mass as submarine fallout inside and outside the caldera, producing a deposit of giant and smaller pumice clasts (Fig. 9.81c). The deposit is therefore an *autoclastic fallout deposit*, not a pyroclastic fallout deposit. Some large clasts show breadcrust-like, cooling contraction fracture networks (Fig. 9.81d), suggesting that large hot pumice blocks probably broke up to produce smaller pumice clasts. Some clasts floated to the sea surface and were then dispersed by oceanic currents (Fig. 9.71b), also producing a trail of water-settled, water-logged pumice and ash-sized particles abraded in the raft by frictional rubbing of large pumice clasts (Jutzeler et al. 2014). In addition, Murch et al. (2019) proposed that there was a magmatic-phreatomagmatic explosive phase of activity that produced an ash deposit with fluidal ash shards.

9.9.4.7 The 15 January 2022 Eruption of Hunga Tonga–Hunga Ha'apai Volcano, Tonga, Southwest Pacific Ocean

The spectacular explosive eruption of Hunga Tonga–Hunga Ha'apai (HTHH) volcano on 15 January 2022 was captured on satellite imagery, but even after over a year after the eruption, at the time of writing much remains unknown about it. The volcano is an almost totally submerged or

Fig. 9.80 Shallow submarine eruption of Myojinsho volcano in 1952–53, Izu–Bonin arc, south of Japan. **a** Major phreatomagmatic dome destroying explosion on 23 September 1952. Note "cock's tail" shape of jets of blocks and ash accompanying explosions; eruption column is some 600 m wide and 400 m high. (Photo by The Asahi Shimbun, taken at 13.12 JST; *Source* Fiske et al. 1998). The suddenness of these powerful explosions proved to be fatal to a research vessel that sailed over the vent the following day with the loss of its crew of 31 (including the nine scientists studying the eruption. **b** Numerous explosions from the submerged vent occurred during 25 September–10 October as eruption columns repeatedly rose into the air. Arrows follow in direction of downcurrent seawater plumes (paler tones) of suspended pyroclastic particles. (Photo by The Asahi Shimbun, taken 27-9-1952). **c** Depositional model for the Myojinsho eruption. Fallout of pyroclasts with high terminal velocities feed submarine gravity flows; subaerial fallout of pyroclasts with lower terminal velocities form particle hyper-concentrations in near-surface layer. Gravitational instability begins to form inverted plume-like vertical density currents, which later descend the volcano's slopes. (After Fiske et al. 1998)

submarine arc stratovolcano in the Tonga volcanic arc, located north of New Zealand. It was built onto the arc oceanic crust at a depth of ~ 2 km by largely submarine eruptions, building to sea level (Fig. 9.82a). Historically, the summit of the volcano has been marked by two emergent islands, Hunga Tonga and Hunga Ha'apai, from which the oldest volcanic deposits are dated ~1,040–1,180 AD (Cronin et al. 2017; Brenna et al. 2022). These islands are located on the northwestern rim of a submarine caldera that was produced during a prehistoric eruption (Fig. 9.82a).

Known historical eruptions (Global Volcanism Program, Smithsonian Institution, https://volcano.si.edu/volcano.cfm?vn=243040) occurred in 1912, 1937, 1988, 2009 (Vaughan and Webley 2010), and 2014–15 (Fig. 9.82b; Cronin et al. 2017). After more than 6 years, surtseyan eruption activity began again on 20 December 2021, continuing until the highly explosive 15 January 2022 eruption (Fig. 9.82c).

The eruptive activity of HTHH in 2009, 2014–15 (Fig. 9.82b), and from 19-12-2021 to 14-1-2022 (Fig. 9.82c) was low-intensity phreatomagmatic, surtseyan explosive

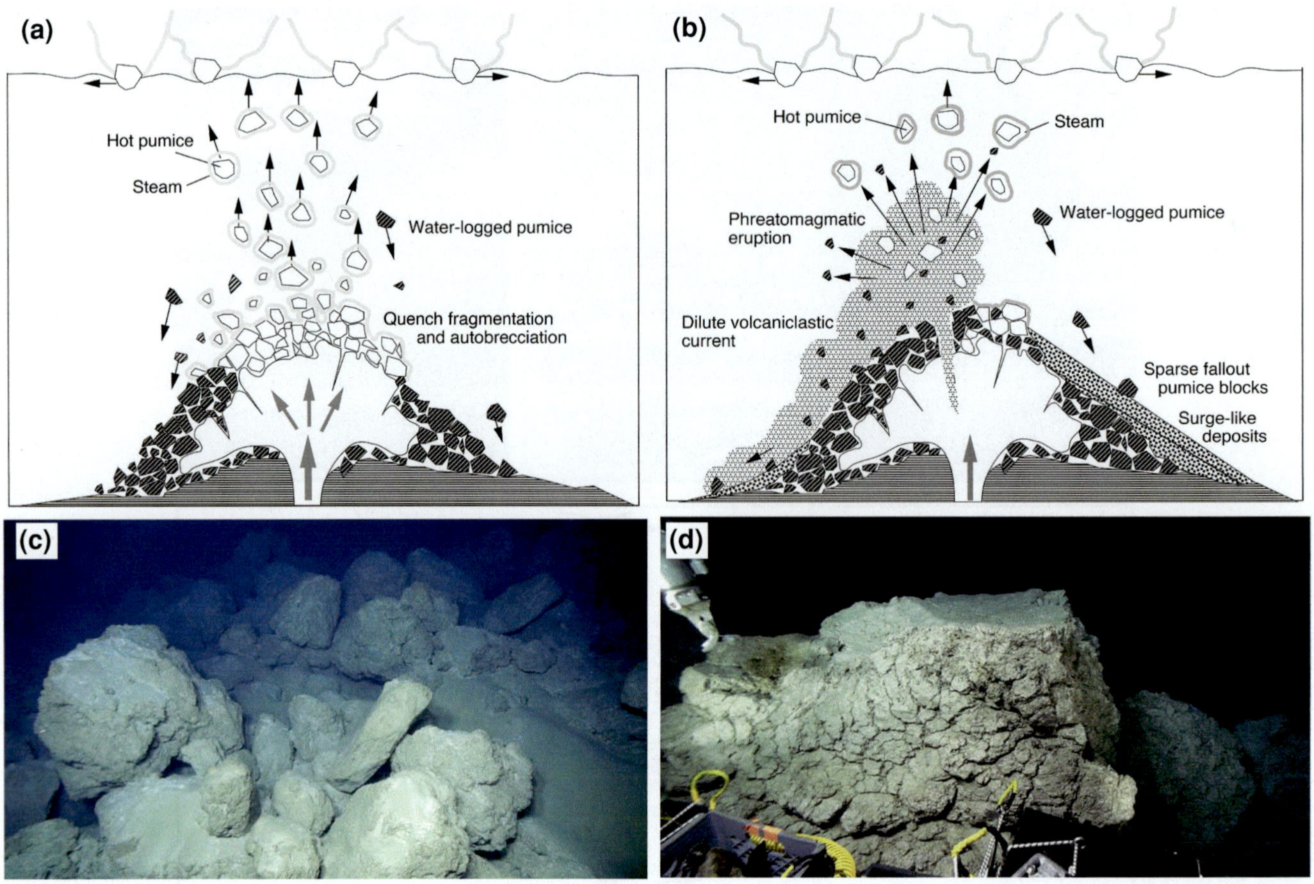

Fig. 9.81 Generation of pumiceous deposits through largely non-explosive spalling of the quench fragmented pumiceous carapace of an erupting subaqueous lava dome. **a** Passive spalling and floatation of pumice blocks. Many become water-logged, subject to quench fragmentation, and the debris "falls" to the bottom forming pumice deposits. **b** As in (**a**) but accompanied by minor phreatic steam explosions, sinking and mass-flow transport, and sedimentation. **c** Stacked giant 1 m pumice hyaloclastic blocks on the caldera floor of Havre submarine volcano, Kermadec arc, southwest Pacific, at 1500 m water depth, resulting from fallout from suspension after limited buoyancy and dispersal. **d** 2 m diameter giant pumice block from the deposit in (**c**), showing curviplanar bounding surfaces and a breadcrust surface texture, both resulting from quench cooling and fragmentation of an erupting highly vesiculated rhyolite magma at depths of ~900 m. (**c** and **d** from Carey et al. 2018)

activity, erupting from vents at or just below sea level causing a pulsing eruption with tephra jets of rock debris and ash being ejected from the shallow water vent in rapid succession (Fig. 9.82b). Small columns of ash and steam rose several km into the atmosphere above the vent (Fig. 9.82c; Vaughan and Webley 2010; Cronin et al. 2017). The 2009 and 2014–15 eruptions built small tuff cones above sea level around the vents (e.g. Fig. 9.82d), and these were variably eroded by ocean waves and rain in succeeding years. The 2014–2015 cone merged with the two old islands (Fig. 9.82d; Cronin et al. 2017; Brenna et al. 2022). The relatively low intensity of the surtseyan explosive activity from 19-12-2021 to 14-1-2022 was punctuated by two major explosive events on the 19 December 2021 and 13 January 2022. The 19 December 2021 eruption produced an eruption column and umbrella region up to 15 +/- 2 km high, covered an area of 21 x 10^3 km^2 and had a lateral volumetric flow rate of ~3.7 x 10^9 m^3 s^{-1} (Gupta et al. 2022). The 13 January 2022 major explosive event produced an eruption column and umbrella region up to 18 +/- 2 km high, covered an area of 30 x 10^3 km^2 and had a lateral volumetric flow rate of ~5.0 x 10^9 m^3 s^{-1} (Gupta et al. 2022). Both of the umbrella regions were ice rich, indicating the phreatomagmatic input of the water; the ash content was indeterminable (Gupta et al. 2022).

Unraveling and understanding the chronology of events of the major 15 January 2022 explosive eruption remains a moving target, to put it mildly! There have been several atempts at doing this (e.g. Gupta et al. 2022; Purkis et al. 2023). There is consensus that the sudden intense explosive eruption began at ~4.00 pm (UTC) or slightly before (Gupta et al. 2022; Purkis et al. 2023). According to Gupta et al. (2022), using analysis of satellite imagery, the initial umbrella region had a cloud top at ~31 km and an

Fig. 9.82 Features of Hunga Tonga–Hunga Ha'apai volcano, Tonga, and its recent eruptions. **a** Simple map and bathymetry showing the main features of the summit area of the volcano, including the caldera, the rim, the emergent islands Hunga Tonga and Hunga Ha'apai, and the locations of the 2009 (see Fig. 9.11b), and 2014–15 eruptions and tuff cones that formed (*Source* Cronin, 2022). **b** Image showing the surtseyan style of the 2014–15 eruption, originating from a vent in the newly formed tuff cone. See (**a**) for location. (Photo taken 14–1-2015; *Source* Global Volcanism Program, Smithsonian Institution, USA, courtesy of the New Zealand High Commission, Tonga). **c** Surtseyan explosive eruption from a vent between Hunga Tonga and Hunga Ha'apai islands, the day before the major explosive eruption of 15 January 2022 (*Source* Tonga Geological Services). **d** View of the 2014–15 tuff cone that has linked the pre-existing islands (Hunga Tonga, Hunga Ha'apai), partially eroded by waves and rain runoff (Satellite image visualisation, March 2018; *Source* Cindy Starr, NASA)

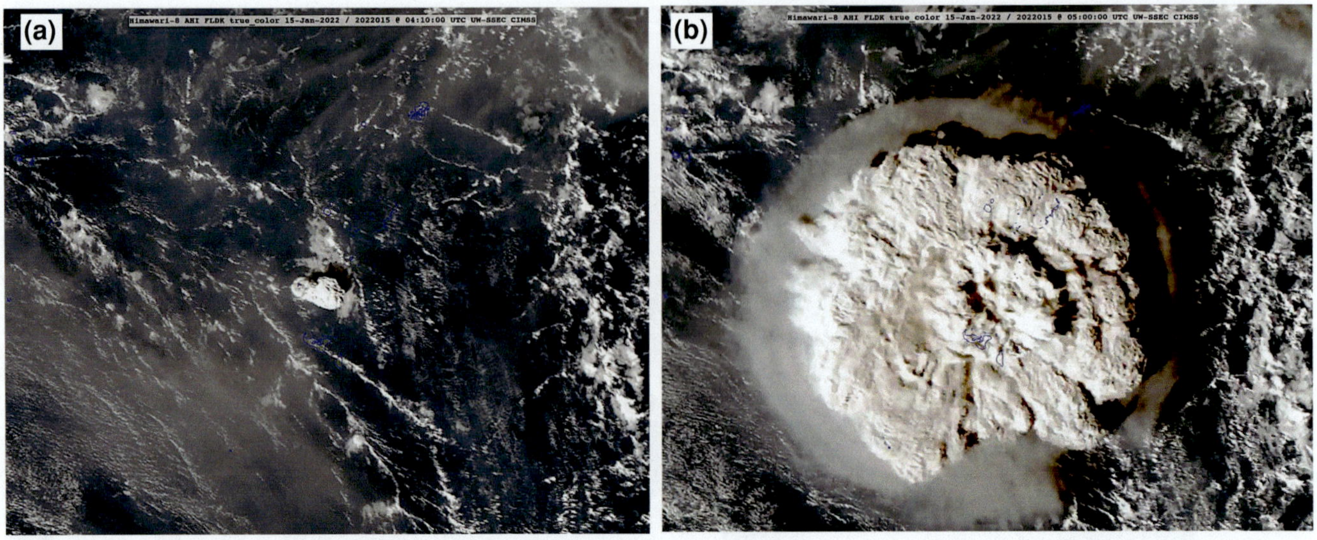

Fig. 9.83 **a** Satellite view of the 15 January 2022 eruption plume, Hunga Tonga–Hunga Ha'apai volcano, Tonga, at 4.10 pm (UTC), a few minutes after the start of the eruption. **b** 5.00 pm, 45 min after the peak of the highly explosive eruption. Note that there are two visible parts of the plume, a lower grey, radially spreading cloud at ~18 km altitude and an upper, still rising, and radially spreading cloud with lobate structure driven by gravity waves, and at an altitude of at least 25 km (A. Prata, pers. comm.). The lower grey cloud is probably dominated by fine ice particles formed through sublimation of water moisture in the eruption column, much of it derived from the initial at least phreatomagmatic blast, and carrying some very fine ash (A. Prata, pers. comm.), whereas the upper cloud could be more driven by magmatic gases (our interpretation). (*Source* Himawari-8 weather satellite, Japan Meteorological Agency)

overshoot height of 55–58 km, which Proud et al. (2022) refined to 57 km, well into the mesosphere. At 5.30 UTC a second umbrella region became visible near the tropopause at a height of ~17 km. The upper umbrella region covered an area of 170 x 10^3 km^2, reached a diameter of 500 km within 90 minutes (Fig. 9.83a, b; A. Van Eaton, pers. comm.) and had a volumetric flow rate of 5.0 x 10^{11} m^3 s^{-1}, orders of magnitude greater than the 19 December and 13 January explosive events (Gupta et al. 2022). According to Gupta et al. (2022) there was another explosive event at 8.42 UTC that produced another overshoot. Both the upper and lower umbrella clouds were again water/ice rich with an indeterminate ash content. Purkis et al. (2023) used atmospheric pressure disturbances, tsunami arrival times around Tonga and ear and eye witness accounts to develop their chronology. They identified 5 blasts, including two relatively weak ones at ~4.00 UTC, followed by 3 progressively more intense ones at 4.06, 4.18 and 4.56 UTC. Using pressure-distance formulae and known tsunamic run-up heights at different islands at different times Purkis et al. (2023) were able to compute the power of the 3 later explosions as 0.5, 4 an 15 Mt (equivalent dynamite), and proposed that the explosive eruption was a VEI 6.3 intensity eruption, the strongest since the 1883 Krakatau explosive eruption. In comparison, the 1991 Pinatubo, Philippines, eruption column reached 40 km in height, and also reached a radius of 250 km, but that took nearly 200 min after the eruption started (Fig. 9.84a). The HTHH explosive eruption has also been evaluated as the most intense since the 1883 eruption of Krakatau, based on the propagation of seismic-acoustic surface Lamb waves and atmospheric pressure waves (Matoza et al. 2022).

The most intense or peak explosive phase of the eruption lasted for 2 to 3 h, based on the frequency of lightning strikes (A Van Eaton, pers. comm.). It produced >400,000 lightning strikes (Vaisala Global Lightning Detection Network (GLD360) and then waned for a further 8 h, indicating an eruption duration of 11 h.

The explosive eruption destroyed much of the previously merged island, it dispersed pumice and ash widely over the Tongan islands, it generated atmospheric shock waves (Fig. 9.84b; M Barlow, University of Massachusetts Lowell) that circumnavigated the globe at least 3 times, and it was associated with submarine pyroclastic density currents (that travelled at least 80 km from the volcano) and caldera collapse (Seabrook et al. 2023), and it produced tsunamis that spread in all directions across the Pacific Ocean, causing minor to moderate damage in many countries, although only a limited number of fatalities (at least 4 in Tonga and 2 people drowned in Peru). The atmospheric waves even triggered anomalous independent sea-level rises or small tsunamis in the Mediterranean and Caribbean oceans, rather than through the direct passage of tsunamis from the Pacific Ocean to the other oceans (Matoza et al. 2022).

It is clear that something very different occurred at HTHH volcano on 15 January 2022 compared with the

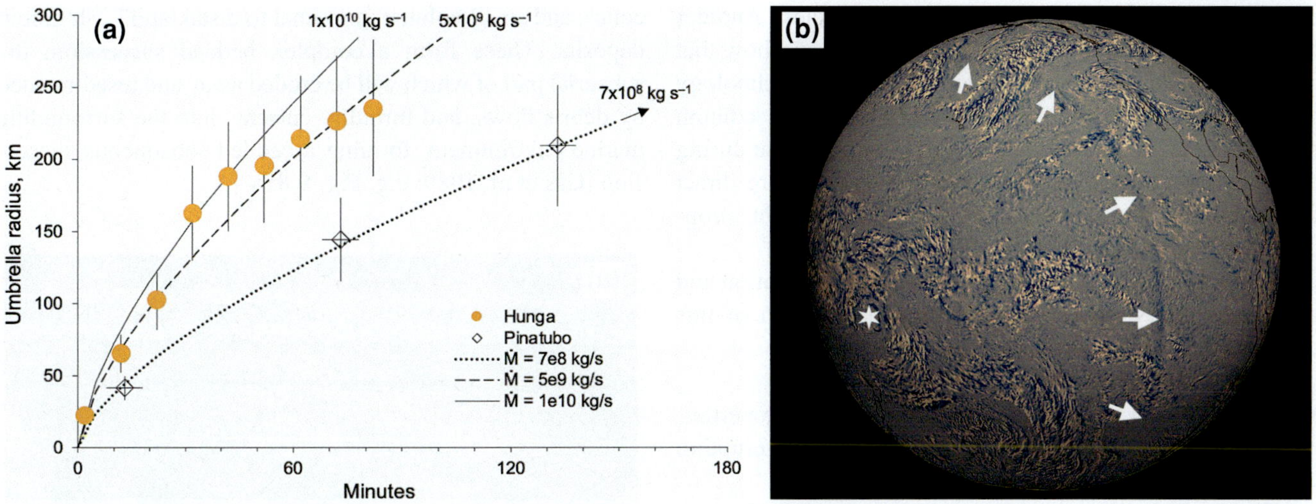

Fig. 9.84 **a** Plot comparing initial expansion rate of umbrella clouds from Hunga Tonga–Hunga Ha'apai volcano, Tonga, on 15 January 2022 and Mt Pinatubo volcano, Philippines, on 15 June 1991 (see Fig. 9.10c). Mass eruption rates derived from these measurements (annotated curves) suggest that the climactic eruption of HTHH was approximately an order of magnitude more intense than Pinatubo. However, the Pinatubo eruption continued at high intensity for several hours longer (not shown). Measurements were derived from cloud areas using 10-min GOES-17 satellite images for HTHH (Van Eaton, unpublished data), and hourly GMS satellite images for Pinatubo, digitised from Holasek et al. (1996) by Mastin and Van Eaton (2020). Vertical error bars give an uncertainty of 20% on the umbrella dimensions. X-axis shows minutes after the inferred start of umbrella expansion, which is 4:15 pm (UTC) for HTHH and 5:28 pm (UTC) for Pinatubo. (*Source* courtesy of A. Van Eaton, United States Geological Survey). **b** Atmospheric shock wave captured by Mathew Barlow, University of Massachusetts Lowell using NOAA's GOES-West satellite infrared radiance data (band 13)

low-intensity surtseyan activity in the previous decade. This sudden much more explosive event suggests that a significantly larger volume of new magma with perhaps a significantly higher dissolved magmatic gas content rose quickly in the vent of the volcano. Preliminary geochemical analyses of pyroclasts confirm that the magma composition was andesite to basaltic andesite with ~66–52 wt% SiO_2, compared with the andesitic composition of the surtseyan phase of volcanism (S. Barker et al., Victoria University Wellington, online seminar presentation to the Geological Society of Australia, 24-2-2022). Although some phreatomagmatic explosive energy must have contributed to the eruption, given that the vent was probably submerged, it is not clear that the energy released by the explosive eruption was all of phreatomagmatic origin. The vent was probably quite shallow (several tens of metres?), so hydrostatic pressure would not have exerted any influence in suppressing explosive intensity, nor would it have suppressed vesiculation of rising and erupting magma. Juvenile pyroclasts apparently have a variable vesicularity, with some very vesicular pumice (~70% + ?; S. Barker er al., as above) indicating that magma volatile content was high.

At present, the eruption style can be classified as phreatoplinian, based on likely magma–seawater phreatomagmatic explosive interaction, and using the eruption column height of 57 km high overshoot, it would be a phreato-ultra-plinian eruption. Based on an umbrella cloud radius of 250 km, the eruption would also be classified as phreato-ultra-plinian (Table 9.1; Constantinescu et al. 2021). Unfortunately, other parameters and data that are usually used to classify eruption style (erupted volume, area of dispersal, grain size, and degree of fragmentation; Chap. 9) are not yet available, and may not be for a long time, given that much of deposit will have been deposited in the ocean.

The VEI/VEI* ranking (Volcanic Explosivity Index) is also debatable, because it is usually based on several parameters, including eruption column height and importantly deposit volume (Newhall and Self 1982), but now also umbrella cloud radius. The ranking appears to be VEI6 or VEI*10 using eruption column height and umbrella cloud radius. However, even now there is a discussion on whether the 15 January column height was 40 km, similar in height to the 1991 Pinatubo eruption column height, or 57 km, but how that is defined, and the deposit volume, is unclear. So far the only volume estimate is at >6 km^3 (Seabrook et al. 2023) based on the bathymetry of the caldera surveyed after the eruption, but this of course is an indirect measure. We caution over the column height estimate, because it is not clear if "eruption column height" should be defined on the basis of the height to which ash is projected upwards and dispersed laterally from that height, and/or the height to which gas carrying only minor very fine ash rises at the top of the plume. In terms of the dispersal of deposits, it is only the ash component that is important, but gas may be more buoyant and rise higher than solid ash particles. Another factor is the resolution of satellite technology, and how that has changed over the decades. Does the satellite technology that has detected the 15 January HTHH eruption column behaviour and height have the same resolution as that during the Pinatubo eruption or is it better? Therefore, are direct comparisons between assessments of column height, properties, and behaviour made in 1991 and 2022 valid?

Finally, there is much still to learn about this eruption and the causes of its effects (also see Chap. 9). Much of this understanding will depend on:

- collecting and undertaken textural (vesicle properties, abundance) and geochemical analysis (pre-eruption volatile content from glass inclusions),
- new detailed deep-sea bathymetry to assess whether the size and shape of the caldera have changed as a result of the eruption to assess if the eruption was caldera-forming,
- using new deep-sea bathymetry to assess if the generation of the tsunamis was due to caldera collapse and seafloor displacement or due to a submarine landslide on the submarine slopes of the volcanic edifice or inner caldera wall, causing displacement of seafloor. If neither of these alternatives proves to be correct, does this indicate that the tsunamis were generated by explosive displacement of the water mass around the vent, and/or by shock waves propagating through the water mass, as they did through the atmosphere? and
- assessing how much of the explosive eruption was driven by phreatomagmatic interaction between the newly inject magma and ocean water, and how much by magmatic volatiles, and whether or not the driving parameters change during the eruption.

9.9.4.8 Surtseyan Fallout Deposits

When volcanic vents are located in shallow marine shelf or lacustrine water depths, explosive events are marked by pulsatory, explosive phreatomagmatic magma–water interaction, called surtseyan, that is dominated by ballistic blasts, tephra jets, and base surges that build shallow water to emergent tuff cones (Fig. 9.11b; see Sect. 9.7.5.1 for discussion of processes). This eruption phenomenon has been well-reviewed by White and Houghton (2000). During the iconic eruption of the monogenetic Surtsey tephra cone volcano offshore from Iceland during 1962–1964, which began in water depths of just 140 m, the pulsing, phreatomagmatic explosive style built a tephra cone above sea level in just a few days (Thorarinsson et al. 1964; see Plate 4d, Fig. 4.18a). Fallout deposits are largely near-vent, directed, coarse tephra jet fans that fall on the subaerial and nearby submarine slopes of the volcanic edifice, ballistics blocks, and bombs, interbedded with base surge deposits in the subaerial parts of cones, and small volume, proximal to distal lapilli-ash fallout deposits. These form a complex bedded succession, the subaerial part of which will be eroded away and resedimented by debris flows, and turbidity currents into the surrounding marine environment, forming a bedded subaqueous succession (Cas et al. 1989; e.g. Fig. 9.85a–c).

Fig. 9.85 Possible water-settled surtseyan fallout deposits. **a** Planar stratified lapilli-tuff deposits, Eocene–Oligocene Oamaru tuff cone complex, South Island, New Zealand. **b** Planar stratified, water-settled fallout(?) deposits, also Oamaru tuff cone succession. Trough cross-bedded tuff in the middle also indicates some bottom current reworking or lateral transport for that bed. **c** Planar bedded lapilli-tuff deposits, Eocene–Oligocene Bridge Point submarine tuff cone, between Dunedin and Oamaru, South Island, New Zealand. The massive poorly sorted nature of the beds probably indicates a mass-flow depositional origin

Other historic surtseyan eruptions that built cones above water include the eruption of Capelhinos in 1957 (see Fig. 10.100a, d), and the 2009, 2014–15 Hunga Tonga–Hunga Ha'apai, eruptions (see Fig. 16.12d).

9.9.5 Possible Origins of Pumice in Subaqueous Settings

The foregoing overview of subaqueous eruption styles and deposits has highlighted that pumice deposits found in subaqueous settings and sequences are not necessarily in situ pyroclastic fallout deposits, nor does the depth at which subaqueous pumice deposits are found represent the depth at which eruption occurred. There are multiple possible origins of pumice deposits in subaqueous settings, including:

(a) Fallout from subaerially erupted and downwind transported eruption plumes, then settling through the water column to the seafloor.
(b) Explosive eruptions from relatively deep-water vents ($<\sim 1{,}000$ m). Subaqueous eruptions can produce significant volumes of pyroclastic pumice that settles to the basin floor through water-settled fallout deposits, vertical, downward directed density currents descending from the near-surface, lateral flotation and transport by surface currents, and collapsing columns that produce basin floor, hot to cooling density currents that transform from gas-supported to water supported (e.g. Kano 2003; Kano et al. 1996).
(c) Entry of pumiceous pyroclastic flows into water, and passage over the water, producing fallout from the bases of these overflows through the water column (Cas and Wright 1987; Jutzeler et al. 2017).
(d) Entry of pumiceous pyroclastic flows into water and transformation into water-supported mass flows that flow along the seafloor as pumiceous turbidity currents, depositing variably graded and hydraulically sorted pumiceous deposits (Cas and Wright 1987; Cas and Wright 1987).
(e) In addition, subaqueous lavas, even in deep water, can vesiculate subaqueously to very high levels (>70%; cf. pumice carapaces of subaerial rhyolite lavas). These pumiceous carapaces can be subject to in situ autobrecciation and quench fragmentation, producing in situ, deep water, hyaloclastite, and autobreccia pumice deposits (e.g. Binns 2003; Allen and McPhie 2000; Rotella et al. 2013). Binns (2003) documented, imaged, and sampled such deposits at water depths of 700 to 950 m in the Woodlark Basin, Papua New Guinea.
(f) Such pumice blocks could also spall off erupting vesicular lavas and domes, producing large buoyant blocks, which may become water-saturated and settle to the seafloor, and/or can be subject to further quench fragmentation producing abundant smaller pumice that falls to the seafloor around the eruption site. These deposits are autoclastic fallout deposits, not pyroclastic. That is, apart from minor phreatic explosions, the pumice was *NOT* formed by explosive eruption activity.
(g) Buoyant pumice from such submarine effusive eruptions can rise to the ocean surface and be dispersed by pumice rafts thousands of km away by surface currents, leaving a trail of seafloor pumice fallout deposits that are not pyroclastic in origin, but autoclastic (Bryan et al. 2012; Jutzeler et al. 2014).
(h) Pumice deposits can also be resedimented from shallower water settings into deeper water by subaqueous mass flows (e.g. Fiske et al. 1998).

The origin of each subaqueous pumice deposit must therefore be evaluated on merit, rather than assumptions being made that all subaqueous pumice deposits are the products of in situ, deep-water, explosive activity.

9.9.6 Deep-Sea Ash Deposits: Links to Co-Ignimbrite Ash Plumes

Visible ash layers preserved in sediment cores have often been found hundreds to thousands of km from their potential sources. Many of these layers are the distal air-fall equivalents of large silicic explosive eruptions that have taken place from both island and continental volcanoes. Although such fine ash layers are difficult to map and correlate, detailed geochemistry of glass ash particles and geochronology has made this possible (e.g. 39 ka Campanian eruption ash; Smith et al. 2016). Widespread volcanic ash layers in deep-sea sediment cores can be used for stratigraphic purposes (e.g. Worzel 1959; Ninkovich and Heezen 1965; Ninkovich et al. 1966; Hays and Ninkovich 1970; Siani et al. 2004), assessing the extent and volumes of fallout deposits from major explosive eruptions (e.g. Huang et al. 1975, 1979; Carey and Sigurdsson 1978, 1980; Watkins & Huang 1977; Watkins et al. 1978; Ninkovich et al. 1978; Rose and Chesner 1987; Smith et al. 2016), considering their volcanological significance (e.g. Carey and Sigurdsson 1978; Ledbetter and Sparks 1979; Sparks and Huang 1980; Smith et al. 2016), impact on climate and humans (Lowe et al. 2012, 2015), and understanding volcaniclastic sedimentation processes in marine basins (e.g. Sigurdsson et al. 1980; Wiesner et al. 1995; Carey 1997). In the geological record, many of the stratigraphically important bentonite layers and "tonsteins" may be co-ignimbrite ashes, and they also provide an important record of major ancient silicic explosive eruptions.

In this section, we concentrate strictly on ash layers that result from the fallout of airborne ash. However, as discussed in the foregoing sections, substantial accumulations of ash can be generated from subaqueous explosive and non-explosive eruptions, and in addition can be resedimented as turbidites from shallow water primary pyroclastic deposit sources into deep-water basins. These may be difficult to distinguish from primary fallout deposits. Understanding dispersal patterns, geochemistry, geochronology, and sedimentological characteristics of deep-sea ashes is important to distinguish the two.

Although many of the volcanic ashes found in deep-sea cores are considered to be co-ignimbrite ash layers (e.g. Figs. 9.86 and 9.87), there are few definitive studies that specifically prove this. However, Smith et al. (2016) distinguished the more proximal plinian fallout deposit from the more widely dispersed co-ignimbrite ash deposits from the 39 ka Campanian eruption from Campi Flegrei caldera in Naples, Italy, using geochemistry and ash morphology. They and Marti et al. (2016) demonstrated that the widely dispersed ash is specifically derived from the co-ignimbrite ash cloud associated with the final and largest, caldera-forming ignimbrite, and is not a distal ash from the early plinian eruption column.

The largest known deep-sea ash deposit is that associated with the 74 ka Youngest Toba Tuff (YTT; Ninkovich et al. 1978), which erupted from the huge Lake Toba caldera, Sumatra, Indonesia (Knight et al. 1986; Chesner et al. 1991; Chesner 2012; Fig. 9.86). It has been found in many deep-sea cores in the northeast Indian Ocean (Fig. 9.86a). Close to the source, cores contain a coarse basal layer below the co-ignimbrite ash (e.g. core RC17-145, Fig. 9.86b), which peters out with distance. The YTT eruption is not known to have had a plinian eruption phase associated with it (Chesner 2012). During the eruption, 2,800 km^3 of magma erupted, 2,000km^3 of which was deposited as ignimbrite on Sumatra, and 800 km^3 of which was dispersed as distal co-ignimbrite ash to the northwest (Ninkovitch et al. 1978; Rose and Chesner 1987). Although most of this was deposited in the Bay of Bengal in the northeast Indian Ocean, deposits are also preserved on-land in Malaysia and India (Fig. 9.86a; Rose and Chesner 1987; Chesner et al. 1991).

The 39 ka Campanian eruption from Campi Flegrei caldera in Naples produced an initial plinian fallout deposit, four ignimbrites, and an ash layer distributed over the eastern Mediterranean and on-land over Eastern Europe, extending up to 2,500 km to the east of the vent (Sparks and Huang 1980; Smith et al. 2016). The deep-sea ash layer is clearly divisible into a coarse lower unit separated by a sharp boundary from a fine, graded upper unit (Fig. 9.87). In core RC9-191 (Fig. 9.87), the lower unit only shows a coarse mode, whereas the normally graded top possesses a coarse and a finer mode. The normally graded unit in core V10-67 (Fig. 9.87) has a bimodal size distribution throughout most of its thickness. Bimodality disappears downwind, because the coarser mode decreases in diameter with distance from source, but the finer mode shows no lateral variation. The basal coarse unit is interpreted to be from the documented plinian phase, whereas the fine-grained upper unit is the co-ignimbrite ash (Smith et al. 2016). The higher convecting plume above the collapsing column may also contribute to

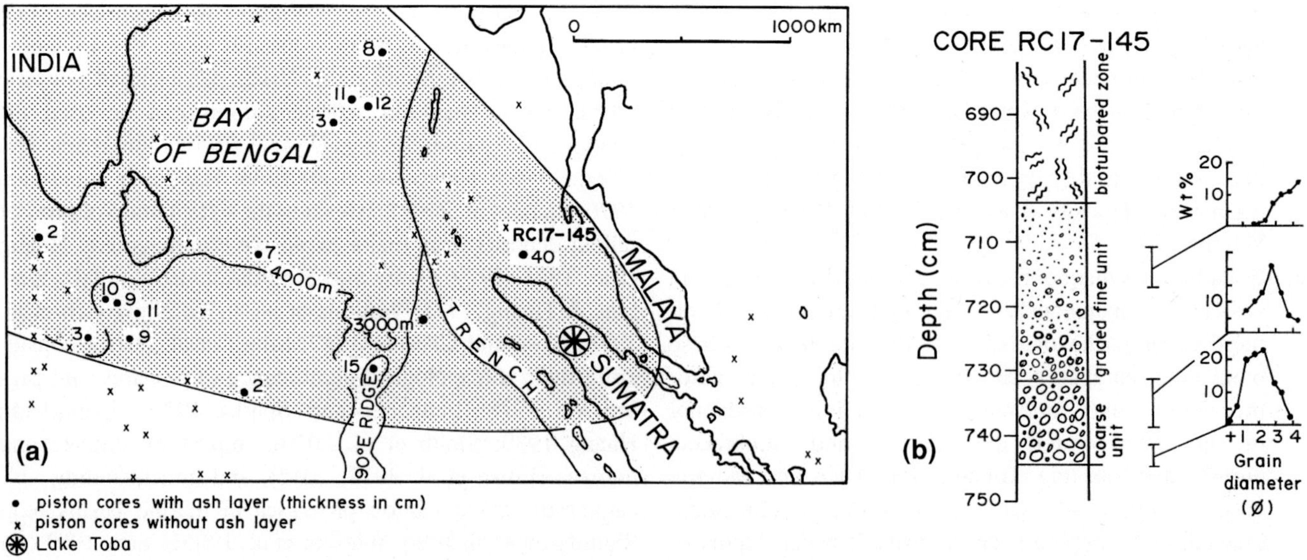

Fig. 9.86 **a** Map showing the distribution and thickness of the Toba ash layer in deep-sea cores. This is a product of an eruption from Lake Toba, Sumatra, in 74 ka, which produced a 2000 km^3 ignimbrite, and is the largest magnitude eruption documented in the Quaternary. **b** Section showing the ash layer in the most proximal core. The lower, coarser unit is interpreted as a distal plinian ash, and the upper graded fine unit is the co-ignimbrite ash. (From Cas and Wright 1987, After Ninkovich et al. 1978)

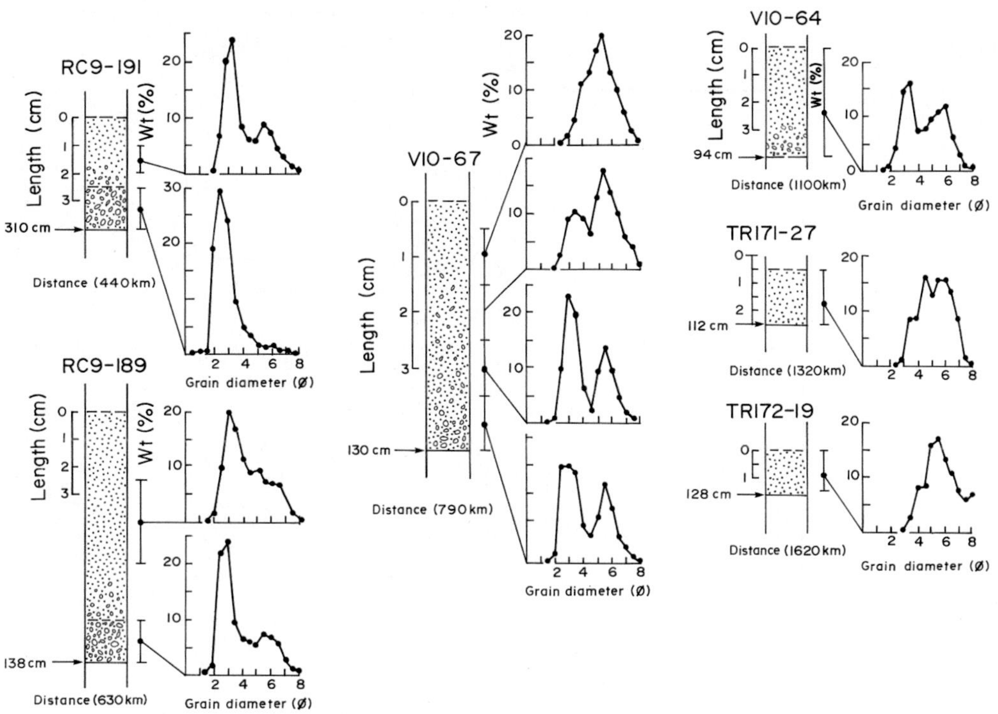

Fig. 9.87 Sections through the Campanian (Y-5) ash layer (39 ka) in six deep-sea cores from the eastern Mediterranean Sea at increasing distance from the Bay of Naples. This is correlated with the Campanian ignimbrite distributed over a large area in the Naples region, and erupted from Campi Flegrei caldera, Italy. (From Cas and Wright 1987, After Sparks and Huang 1980)

the coarse mode in this layer. Observations on the size grading of crystals in deep ash layers are also important and can be used to estimate eruption durations (see below).

Both volcanic and marine processes will control the accumulation of any deep-sea air-fall ash layer. Their interpretation and internal stratigraphy can be as complex as the subaerial products of the eruptions. More distally, the problem of distinguishing the different types of silicic air-fall ash, and depositional processes also becomes more complex.

The principal primary volcanic processes that influence deep marine ash deposits are (1) whether the ash has one input source, or two (eruption column-plume fallout, and co-ignimbrite ash-cloud fallout; cf. Eychenne et al. 2012, 2015); (2) subaerial atmospheric transport and sorting; and (3) whether or not particle-free-fall is unhindered through the water column (Fisher 1965; Cashman and Fiske 1991; Carey 1997), or whether they aggregate and settle (Wiesner et al. 1995), or form downward directed plumes, and then laterally flowing, seafloor density currents (Carey 1997).

In addition, several marine processes may affect the characteristics of deep-sea ash deposits:

- ocean currents,
- secondary resedimentation,
- bioturbation, and
- compaction.

9.9.6.1 Ocean Current Effects on Ash Dispersal and Sedimentation

Although wind transport is the dominant ash dispersal mechanism, the strength and direction of ocean currents will have an important effect on the distribution of the ash in the marine environment. Ninkovich and Shackleton (1975) attributed the W-shaped pattern of the Worzel L ash layer (230 ka) to the effect of the strong equatorial undercurrent, or Cromwell current, moving eastwards across the Pacific at a depth between 100 and 300 m, and a speed of 100–150 cm s^{-1} (Fig. 9.88). Besides surface and near-surface currents, bottom currents can cause reworking and redistribution, and in terminology applied to clastic sediments, such layers would be called contourites (Stow and Lovell 1979) (Fig. 9.89).

In contrast, Wiesner et al. (1995) summarised the distribution of the distal ash from the 1991 eruption from Pinatubo in the Philippines, which was dispersed to the WSW from the volcano. The deep-sea ash deposit at depths >4,000 m defines a fan of dispersal which matches the wind direction at the time of the eruption. They concluded that there was no evidence of the effects of surface currents in influencing the submarine distribution pattern of the ash. They calculated transport times and depositional times using sediment traps moored in the ocean above the seafloor and determined that ash fallout from the atmosphere

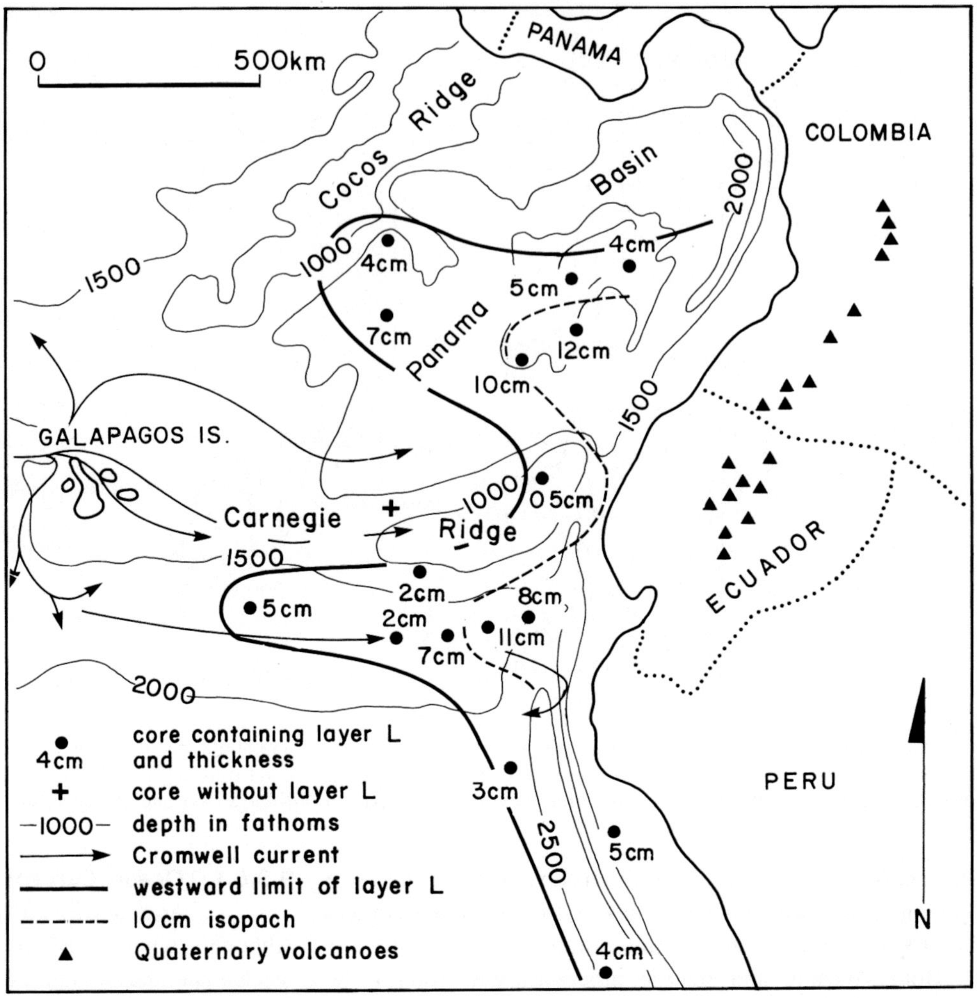

Fig. 9.88 Distribution patterns of the Worzel L ash layer (230 ka), probably erupted from a source in northern South America (From Cas and Wright 1987, After Ninkovich and Shackleton 1975)

occurred through ash aggregation, not individual ash particle fallout. They also determined that ash sedimentation in the submarine environment must have occurred at settling velocities 2 to 3 orders of magnitude greater than that for individual ash particles, requiring sedimentation by ash aggregates, which have significantly higher terminal settling velocities compared with individual very fine ash particles.

9.9.6.2 Secondary Resedimentation

Ash which is deposited on land in near-shore settings or shallow water settings, or which settles out on top of submarine topographic highs will be susceptible to secondary subaqueous resedimentation, generating turbidity currents (e.g. Walker et al. 2019). The consequence of such processes is an increase in ash-layer thickness in basins, and a thinning or removal of ash layers from highs and steep slopes. In the basins, a succession of ash turbidites may be deposited by repeated slumping during or after an eruption. Figure 9.89 shows several units within the Minoan 1600 BC ash layer, cored in a small basin southeast of Santorini in the Aegean Sea. The two units labelled D are both interpreted as turbidites, as they have erosional basal boundaries, show normal grading, and contain abundant carbonate, clay, and some rounded terrigenous quartz grains mixed with the ash (Watkins et al. 1978). Crystal-poor, fine-grained air-fall ash derived from the phreatomagmatic phase in the eruption, or co-ignimbrite ash, is probably accumulated on submarine highs and then slumped into the small deep-water basin where this core is located. Close inspection of Fig. 9.88 also suggests that slumping could at least be partly responsible for the distribution pattern of the Worzel L ash layer, given its proximity to the South American continental margin and submarine ridges.

9.9.6.3 Effects of Bioturbation

Bioturbation of ash layers mixes the ash with other sediments and can reduce visible ash layer thicknesses. In some cores through the Minoan ash layer, Watkins et al. (1978) found that over 50% of the original ash had been dispersed upward by this process.

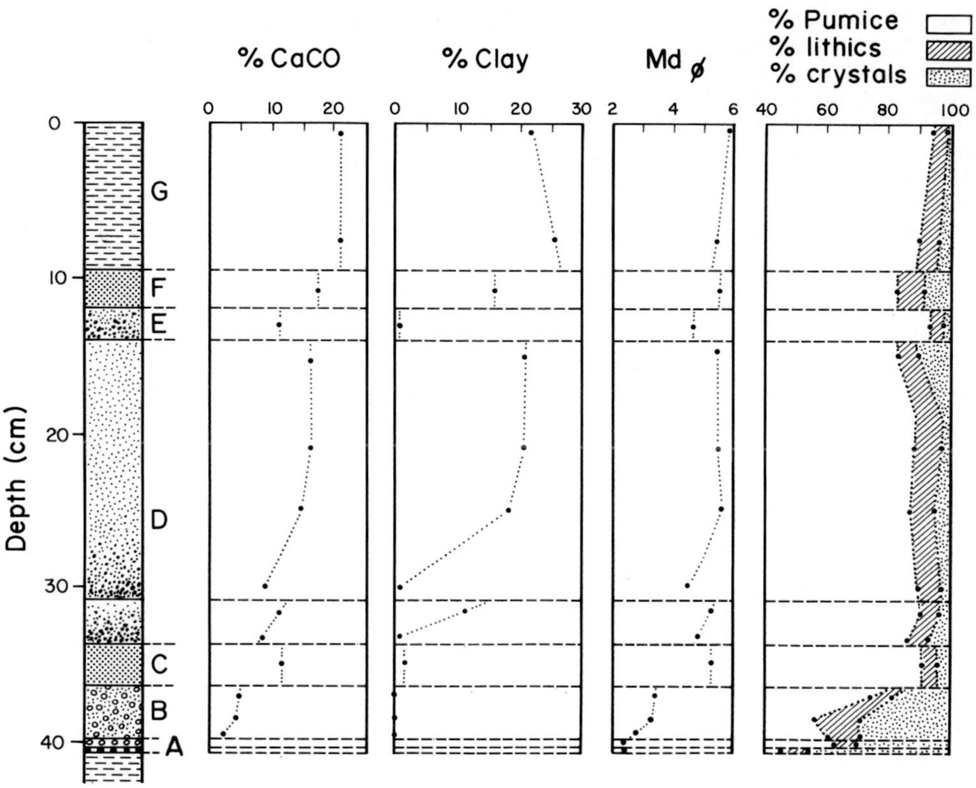

Fig. 9.89 Section showing lithological variations in the Minoan 1600 BC (3.6 ka) ash layer in core TR 172–9, taken approximately 170 km southeast of Santorini, Aegean Sea. From bottom to top: units A and B are interpreted as the plinian fallout; C, as a fine ash fall (co-ignimbrite?); D, as two ash turbidites resulting from resedimentation; E and F as laminated ash layers winnowed by bottom currents; G as pelagic muds admixed with ash. (From Cas and Wright 1987, After Watkins et al. 1978)

9.9.6.4 Post-Depositional Compaction

Post-depositional compaction can substantially reduce the thickness of an ash layer, due to an initial high interstitial water content between ash particles on deposition, which is expelled during compaction. A reduction factor of 50% is usually adopted when recalculating isopach contours to represent the thicknesses of freshly fallen ash (e.g. Watkins et al. 1978).

9.9.6.5 Using Deep-Sea Ash Layers to Estimate Eruption Duration

Ledbetter and Sparks (1979) suggested that deep-sea ash layers could also be used to estimate the duration of large-magnitude explosive eruptions from the vertical size-grading of feldspar phenocryst fragments near the base of deep-sea ash layers. The size grading is a function of the release time of the particles from the atmosphere, their settling velocity, the water depth at the site of the ash layer, and the duration of the eruption. This model predicts a zone at the base of the deep-sea ash layer where the size of the coarsest particle remains constant with height. Above this zone, the size of the coarsest particle decreases with height. The prominent break in the coarsest particle size-to-height curve represents the deposition of the last largest particle ejected at the end of the eruption. The finest particle deposited in conjunction with the last largest particle must have erupted at the beginning of the eruption. The difference in settling times between the largest and smallest particles at the critical level is equal to the duration of the eruption. Settling times can be calculated by knowing the water depth, and determining the terminal settling velocity of the coarsest and finest particles.

There are several assumptions that have to be made with calculations using the "*eruption duration model*". The most important is that size and density of the largest particles arriving at any downwind site remain the same throughout the eruption, and that the density of all particles is uniform, so that size variations reflect hydraulic sorting variations. For this to occur, the range of particle sizes ejected at the vent must not vary substantially, the eruption column height must not change substantially, and the wind direction and dispersal direction must not vary. If these conditions are not met, the vertical size distribution of particles would not show the pattern predicted in the model, and more complex patterns of grading would result. It is also necessary that the ash layer be recognised as a primary air-fall deposit. The normal size grading of an ash turbidite may approximate the size grading predicted by the model for the base of a primary ash-fall layer.

Another complication now, that was not envisaged when this model was first proposed, is that graded ash layers could also be the deposits of downward directed or inverse surface plumes, as proposed by Carey (1997). Such layers settle to the seafloor at orders of magnitude of higher velocities and

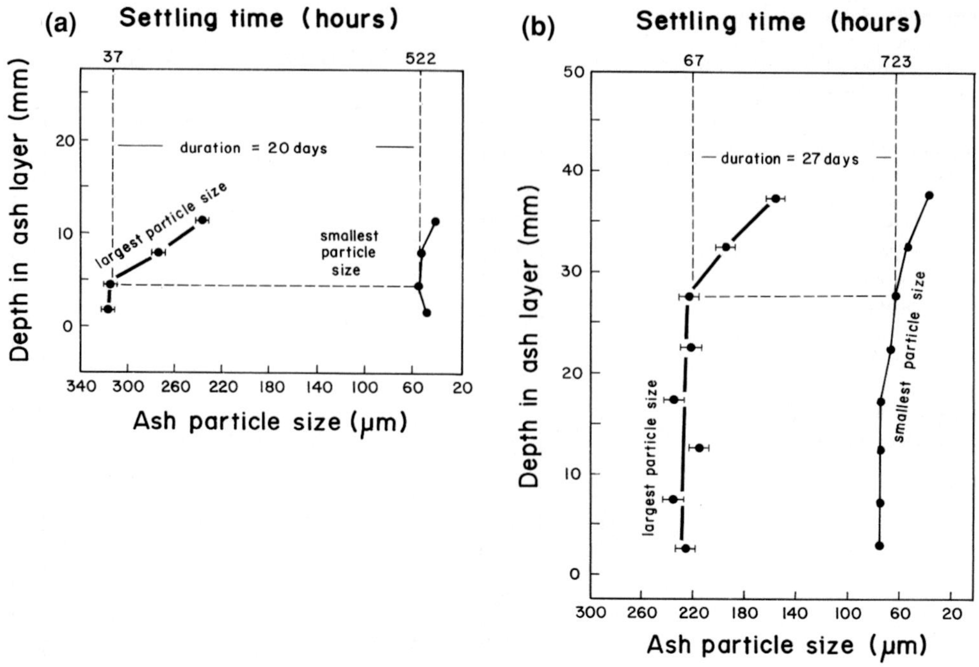

Fig. 9.90 Duration model applied to the Worzel D ash layer in two cores from the eastern equatorial Pacific. **a** Core TR 163–7 in water 3435 m deep. **b** Core TR 163–10 in water 3500 m deep. Worzel D is equivalent to the Los Chocoyos Ash erupted at 75 ka from Lake Atitlán caldera, Guatemala, which comprises an underlying plinian fallout deposit and overlying ignimbrite. (From Cas and Wright 1987, After Ledbetter and Sparks 1979)

shorter time periods than free particle settling rates. The eruption duration model applied to such deposits would provide estimates much longer than the actual duration. It is therefore important to search for other distinguishing sedimentary features.

Results from applying the Ledbetter and Sparks (1979) eruption duration model to the Worzel D ash layer are shown in Fig. 9.90. This layer is the distal counterpart of the Los Chocoyos Ash now radioisotopically dated 75 ka, which is the product of a major rhyolitic ignimbrite- and caldera-forming eruption associated with Lake Atitlán in Guatemala, as well as being the largest volume Quaternary eruption in the Central American region (Drexler et al. 1980; Cisneros de León et al. 2021). The estimated duration of this eruption was 20–27 days. The model has also been tested on the 74 ka Toba deep-sea ash layer (Fig. 9.86) and indicates that the duration of the eruption was about 10 days. Knowing the approximate volume of the total products in these two eruptions, it is possible to estimate the volumetric eruption rate of magma (Cas and Wright 1987, Fig. 8.34). This provides a means of comparing large-magnitude explosive eruptions with relatively small-scale ones. For more recent theory and analysis of the duration of the Worzel D ash layer eruption, see Iriyama et al. (2018).

Deep-sea ash layers are potentially the best record of age, location, duration, and magnitude of explosive volcanism, provided that individually they can be accurately correlated with known on-land source volcanoes and specific eruption events. However, this depends on several factors. First, it is essential to have a sound understanding of the stratigraphy and age of the sedimentary succession within which the ash layer occurs, including sound biostratigraphic controls, which are still often the major age control on the age of the ash layer (e.g. Ninkovich and Shackleton 1975; Ledbetter 1985). Secondly, sound geochemical major, trace, and isotopic analytical data must be generated to accurately fingerprint the ash layer composition to compare with potential source compositions (e.g. Ledbetter 1985; Engwell et al. 2014; Scudder et al. 2016). Thirdly, if possible, geochronology of crystal separates from deep-sea ashes (if they can be found and separated), such as high-resolution $^{40}Ar/^{39}Ar$ dating, is invaluable for narrowing the possible correlations (e.g. Ton-That et al. 2001). The study of ash layers in the ocean basins remains a powerful tool for further resolving eruption histories.

9.10 Assessing Classification Schemes of Pyroclastic Fallout Deposits and Fallout-Forming Eruptions, Their Weaknesses, and Possible Ways Forward

Classification is an artefact of the need by humans to recognise order for the purpose of communication. The problem is that nature doesn't understand this logic, and it will throw many variables at us to complicate this process and our appetite for order. Most volcanologists recognise that classification erects artificial boundaries in a continuum, but we like to pigeon-hole eruptions and their deposits into specific end members. However, as is becoming increasingly obvious, the nature of any individual eruption can change during the event, confronting us with several parts of that continuum in the one event, making it artificial to classify an eruption as one style or another, rather than subdividing it into several phases, each with a different eruption style (e.g. initial vulcanian phase, followed by evolving plinian phase). That doesn't negate the value of classification to facilitate scientific communication. We just have to be realistic about what classification can achieve and be flexible with classification, rather than feeling compelled to squeeze a whole event into one style when clearly the eruption has evolved from one style to another.

For example, for some modern eruptions, eruption style changes have been witnessed and recorded by multiple instruments in real time, as well as the corresponding deposit responses being documented. Characteristics of the eruption can be quantified almost immediately, including establishing a well-constrained chronology of events or phases and their duration, muzzle velocities, eruption column height, mass eruption rate, eruption column properties, including its temperature, mass, density, and internal structure. Then by remote sensing mapping and rapid sampling of the deposit after the eruption, the instrumentation-based interpretations can be verified with ground truthing data, and the deposits and the changing eruption phases can be classified. For such real-time eruptions, the hazards have been observed as well, and don't need to be evaluated deductively.

Classifications can thus be undertaken that are based on quantifiable parameters relating to the eruption and its changing phases. We know now that many eruptions begin with a short, impulsive, vent opening, vulcanian blast, and once the bottle has been uncorked, and the magmatic exsolution genie has been set free, the eruption quickly transitions into a sustained buoyant plume-forming eruption, driven by steady volatile exsolution. The whole eruption should not be categorised as one style, but as two distinct phases and styles, with each phase being driven by different dynamics that produce different deposit types and characteristics.

For other eruption sequences, for which the eruption was not witnessed, but the deposits are still largely intact, allowing quantification of deposit characteristics (e.g. area, thickness and grain size parameters, and changes therein), the process of defining dynamic eruption parameters is more long-winded, but can be deduced if enough detailed field mapping, sample collection, and sample analysis are undertaken. There is no shortcut, no magic black box that does these, and so comprehensive, dusty, sweaty fieldwork and data collection are essential. *No pain, no gain!* Classification can then be undertaken that accommodates the state of knowledge about the deposits (whole of deposit characteristics, changes through the eruption deposit sequence, and pyroclast characteristics) that can then be tied to eruption phenomena (cf. Walker 1973; Newhall and Self 1982; Bonadonna and Costa 2013; Pyle 2015; Constantinescu et al. 2021). It is only then that the hazards and their magnitude that accompanied that past eruption can be deduced. Collecting data about such unwitnessed eruptions and their deposits and evaluating eruption processes, and magnitude on any volcano, is just as important for future hazard assessment on that volcano as documenting real-time eruptions, because it provides a better long-term understanding of the range of behaviour of the volcano and its hazards.

For older eruption sequences, deposits may be incompletely preserved, they may be variably lithified and altered, making a more qualitative, descriptive approach to documenting deposits, and interpreting the eruption origins, the most practical way of understanding the eruption. This also requires extensive fieldwork, involving mapping, stratigraphic correlation, sample collection, and analysis. This is as important to understanding the history and evolution of a volcano and its hazards as the first two categories. Such sequences also need a classification platform to allow reconstruction of eruption events and their dynamics and hazards. This will invariably focus more on outcrop scale deposit characteristics and pyroclast characteristics.

As volcanologists working on both modern erupting volcanoes, as well as variably preserved ancient sequences, we need to address all of the above scenarios in our attempts to classify fallout (and other) deposits. For ancient poorly preserved deposits, the best approach is a descriptive to progressively genetic approach, as advocated in Chap. 14, that can then be related to other known deposits and their interpreted eruption origins. For the real-time deposits that were witnessed forming, and for the non-witnessed, but areally well-preserved deposits, the following assessments of classification schemes are relevant.

9.10.1 Walker's (1973) Dispersal-Fragmentation (D-F) Classification Scheme and Its Limitations

In Sect. 9.2, we introduced Walker's (1973) widely used classification for subaerial fallout deposits and briefly introduced also its main limitations which motivated our revision of the scheme, as shown in Fig. 9.6a. Walker's original quantitative scheme relies on accurate mapping of the distribution of fall deposits and detailed granulometric analysis to determine two parameters: dispersal (D), assumed to scale with the eruption column height and the mass eruption rate, and fragmentation index or degree of fragmentation of the deposit (F), assumed to scale with the explosive intensity of the eruption. As a reminder, the empirical measure of D used is the area enclosed by the $0.01T_{max}$ isopach (where T_{max} is the maximum thickness of the deposit; Fig. 9.6c). The empirical measure of F chosen is the percentage of a deposit finer than 1 mm at the point on the axis of dispersal where it is crossed by the $0.1T_{max}$ isopach; this can only be determined from sieve analysis of a sample collected either at this point or, more practically, obtained from sieve analyses of a few samples collected near the dispersal axis (Fig. 9.6c).

The main problems and limitations of Walker's (1973) approach are in the limitations of actually being able to collect the required field data for classification. D, for example, needs the definition of the maximum thickness of the deposit which is rarely measurable; often it's a closest best estimate based on deposit thinning trends. Although Walker (1973) interpreted D to reflect the eruption column height, Houghton et al. (2014) pointed out that shifts in wind direction during an eruption can increase the dispersal area, perhaps giving a misleading impression of maximum eruption column height, and an erroneous classification of the deposit. However, it should not matter what factors affect the total dispersal area of a deposit, which is paramount in assessing the hazards in an impact area. For example, if the total area of dispersal is deemed to be greater than a normal Plinian deposit due to wind effects, then even if its column height is plinian, the deposit dispersal is validly ultra-plinian. F needs to be defined at a specific locality along the dispersal axis which may not necessarily be available. In addition, older deposits in the rock record are usually consolidated or cemented and lithified, so that grain size analysis cannot be conducted easily or at all. Furthermore, there are problems on the arbitrary definition of boundaries between different eruption styles, especially at the low-intensity end of the classification, where Walker's diagram does not adequately capture the spectrum of hawai'ian, strombolian, violent strombolian, micro-plinian, and vulcanian eruption styles and deposits, as well as others discussed below, especially for the low explosive intensity, small volume, and area fields (Bonadonna et al. 2016). For example:

(a) Vulcanian explosions can range from minuscule to very large, and hence the dispersal can vary from almost zero to many km^2.
(b) Strombolian *sensu stricto* eruptions are single short events of very small scale that produce spatter cones over extended periods and many individual eruption events, not scoria cones and dispersed fallout deposits. They overlap only at the very small dispersal area with some very weak hawai'ian fountain events.
(c) What was classified as strombolian activity by Walker (1973) actually refers to scoria cone-forming activity resulting from a low plinian-style (or violent strombolian-style) eruption column, and producing a large proportion of scoria, which Francis et al. (1990) called micro-plinian, with which we agree (Fig. 9.6a).
(d) Most hawai'ian fountaining events are sustained and produce greater dispersal areas than strombolian *sensu stricto* events.
(e) All the main magmatic explosive eruption styles can be transitional into phreatomagmatic eruptions styles if the rising magma encounters external water, which then also contributes to the explosive dynamics and fragmentation intensity.
(f) Hence, hybrid eruption styles such as phreato-vulcanian, phreato-strombolian, phreato-hawai'ian need to be added to the phreatoplinian designation of Walker (1973) (Fig. 9.6a).
(g) Hydrothermal explosive eruptions were not included in Walker's scheme, yet they can be devastating (e.g. White /Whaakari Island eruption, New Zealand on 9 December 2019). Again, these are small-scale events producing ballistic and fallout deposits of only local extent (Fig. 9.6a).
(h) The extremely small-scale, impulsive Halema'uma'u explosions of 2008–2018 (e.g. Orr et al. 2013), triggered by rock fall impact onto the crust of the lava lake in the crater, are not accommodated. The deposits are amongst the smallest in volume, mass, and dispersal recorded (Houghton et al. 2013; Fig. 9.6a).
(i) Apart from the plinian spectrum, there is little scope in the Walker (1973) classification for variations in intensity and therefore dispersal area among individual eruption styles (e.g. vulcanian and hawai'ian) because rigid boundaries were set on dispersal area. We feel these rigid boundaries should be removed (Fig. 9.6a), although the original dispersal areas remain useful as guides of general dispersal areas for particular eruption styles.
(j) Apart from surtseyan, other subaqueous eruption styles are not accommodated, leading to some attempts to shove subaerial eruption styles under water and pretend that everything is the same. It is not.

In addition, the meaning of F is not as clear as was suggested above. High F-values are not just the result of high degrees of fragmentation during magmatic explosive eruptions, but also reflect intense fragmentation during "wet" phreatomagmatic eruptions, and eruption plumes in which premature deposition of fines is promoted by moist particle aggregation, and fallout of aggregates of ash or clumps, rather than the individual ash particles, as is assumed in the classification.

Also, the fields for phreatomagmatic ash-fall deposits, which are simply divided into surtseyan and phreatoplinian, are not clearly defined, and it is becoming clear that many deposits represent transitions between magmatic and phreatomagmatic styles, because they show characteristics of both. As a result, new terms such as phreato-vulcanian (van Otterloo et al. 2013), phreato-hawai'ian, and phreato-strombolian (Latrutrie and Ross 2020) have been proposed. Surtseyan eruptions and deposits are distinctive in setting from deposits resulting from subsurface magma–aquifer interaction, which we call ukinrekian (Fig. 9.6a). The as yet poorly understood eruption of Hunga Tonga–Hunga Ha'apai volcano, Tonga, southwest Pacific Ocean, on 15 January 2022 may warrant being distinguished from land-based phreatoplinian eruptions as "hunga tongan (ultra-)phreatoplinian" (Fig. 6a). This highlights that eruption styles can be ephemeral and are not just simply classifiable to very succinctly defined end-member boxes in a classification scheme. We have therefore amended the Walker (1973) classification scheme to reflect transitions between fields rather than hard boundaries (Fig. 9.6a).

Because many fallout deposits are the cumulative product of *several changes in dynamics and style* during the same eruption event, actual classifications should refer to individual layers from each eruption phase (or style) that show distinctive features (e.g. grain size, componentry, and dispersal). If dispersal and fragmentation data is not available for each layer from an eruption sequence, this cannot be done, but at least this provides an approach for future research. Furthermore, in the rock record, the poor preservation of fallout deposits, especially original areal extent, usually prevents precise determination of D and F (see Sect. 9.7.9).

Significant problems also occur when attempts are made to classify small to moderate intensity subaerial eruptions and their small volume deposits (e.g. Bonadonna and Costa 2013; Bonadonna et al. 2016).

9.10.2 Area Plots Related to Deposit Characteristics (Thickness, Grain Size)

Walker (1980, 1981c) also compared and classified plinian style fallout deposits using plots of area enclosed by/versus thickness contours (Fig. 9.91a), or area enclosed by various grain size parameter isopleths (median grain size, maximum lithic, or pumice clast size; Fig. 9.91b) on logarithmic scales, which demonstrate an exponential thinning and grain size pattern away from the vent. Pyle (1989) refined this by plotting log of thickness versus square root of areas enclosed by thickness contours (isopachs) for 23 eruptions, but he did not formalise this as a classification scheme. He also quantified rates of deposit thinning, represented by the thickness half-distance, b_t (i.e. the distance away from the vent at which the maximum thickness at vent is halved and then halves again), and the maximum clast size decay rate, b_c (i.e. the distance at which the average maximum clast size halves with respect to most proximal locations, and then again). Pyle (1989) suggested that the grain size characteristics of fallout deposits could be depicted by the "half-distance" ratio, b_c/b_t, and that a classification scheme of fallout deposits could be constructed by plotting b_c/b_t versus b_t (Fig. 9.92a, b). Such plots can then also be contoured for column height (Fig. 9.92a, b), and used for relatively accurate deposit volume calculations (Eqs. 9.16, 9.17).

In Fig. 9.92a, b, b_t is a proxy of the dispersal and the b_c/b_t ratio is a proxy of the influence of the total grain size distribution (TGSD) on column height. This approach is based on the correlation between the clast half-distance b_c with the column height (see Fig. 9.37b), and on the observation that for a given column height the dispersal varies a lot with b_c/b_t, varying between 0.5 and 3. This variation captures the influence of the initial TGSD on the effective dynamics of the eruption column, and specifically of the amount of particles that remain entrained in any given parcel of the column. A coarse-grained TGSD (positively skewed) would not allow a significant development of a plume and a quite rapid decline of the maximum lithic size with distance. This would result in high values of b_c/b_t. By contrast, if TGSD is negatively skewed (rich in fines) then the deposit would thin slowly and have a high b_t value. According to Pyle, the most energetic eruptions are characterised by large b_c (i.e. a high eruption column) and small b_c/b_t (i.e. a large dispersal).

9.10.3 The Volcanic Explosivity Index (VEI)

In Sect. 9.2, we introduced a widely and long used classification scheme for explosive eruption styles, the Volcanic Explosivity Index (VEI; Newhall and Self 1982) that attempts to reflect the explosive intensity of an eruption based on the deposit volume, related to eruption column height (Table 9.1). The scale for the VEI is logarithmic.

However, as already discussed, the Volcanic Explosivity Index (VEI) has limitations. First, as it was defined, linked to pyroclastic fallout deposit volumes, the base-level explosivity was 0, assigned a deposit volume of 10^4 m^3 (rather than 0), with activity described as "non-explosive, gentle, effusive", which makes it difficult to classify smaller volume deposits that are explosive in origin. However, Houghton et al. (2013) proposed that the VEI scale should be extended at the low-intensity-small deposit volume end of the scale, to take into account small-scale explosive events that on the current scale are subsumed. This proposal focused on the realisation that rock falls onto the surface of lava lakes could trigger short impulsive explosive eruptions, based on events at Halema'uma'u crater in Hawai'i in 2008. The very small volume (10^0 to 10^2 m^3) of the deposits from such small explosive events introduces negative VEI values on the scale introduced by Houghton et al. (2013). Houghton et al. (2013) include at their low-end member, volumes as low as

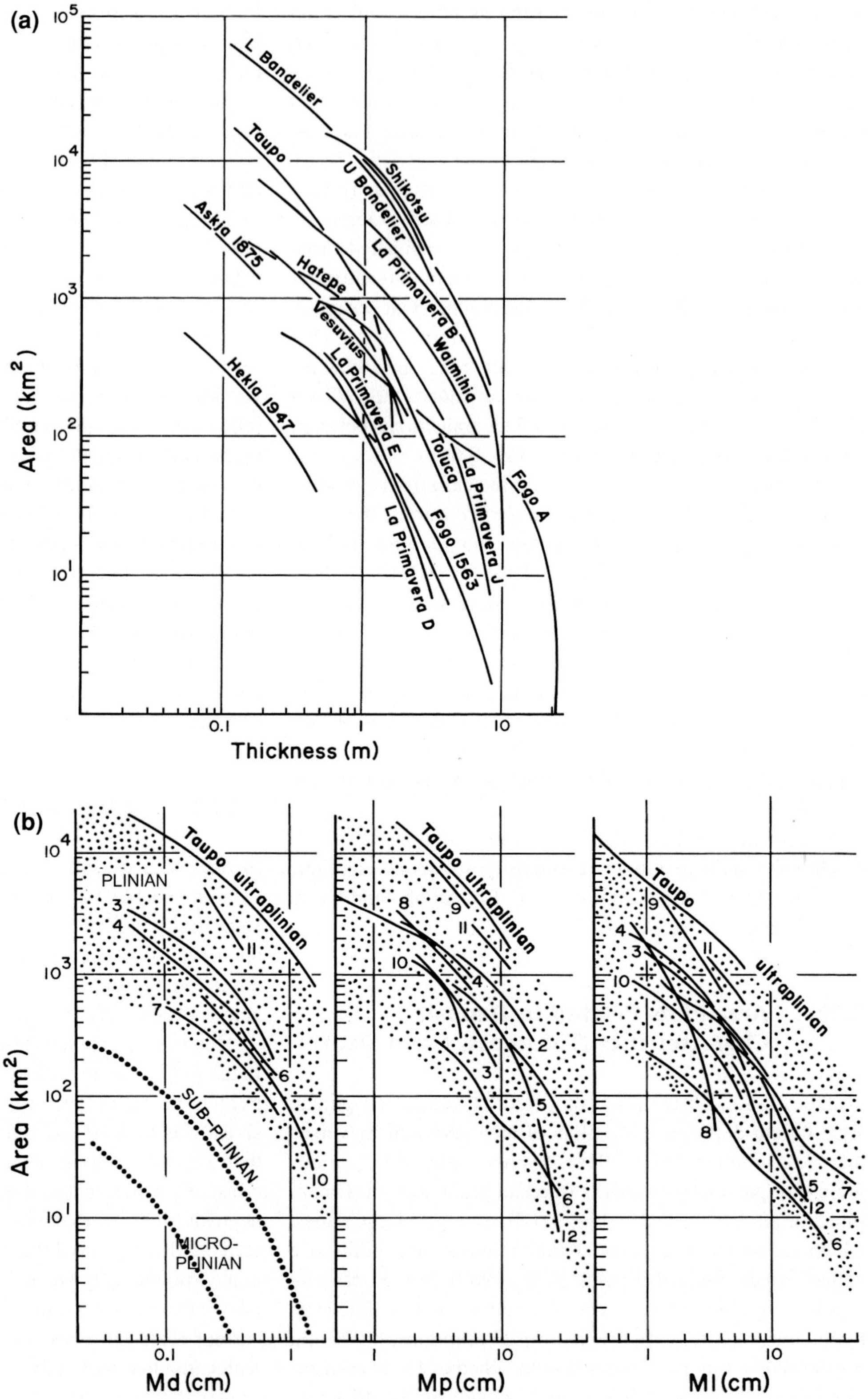

Fig. 9.91 Area plots used to classify plinian type deposits. **a** Plot of the area enclosed by each isopach against thickness for some plinian fall deposits and the Taupo ultra-plinian deposit. **b** Plot of the area enclosed by isopleths of median grain size (Md), maximum pumice diameter (Mp) and maximum lithic diameter (Ml). Stipple is field of plinian deposits. 1 32 ka Shikotsu; 2 Askja 1875; 3 Vesuvius 79 AD; 4 95 ka La Primavera B; 5 Upper Toluca; 6 Fogo 1563; 7 Fogo A; 8 1.6 Ma Lower Bandelier; 9 1.26 Ma Upper Bandelier; 10 La Primavera J; 11 3.3 ka Waimihia; 12 Minoan 1600 BC. (From Cas and Wright 1987, After Walker 1980, 1981b, and Self et al. 1986)

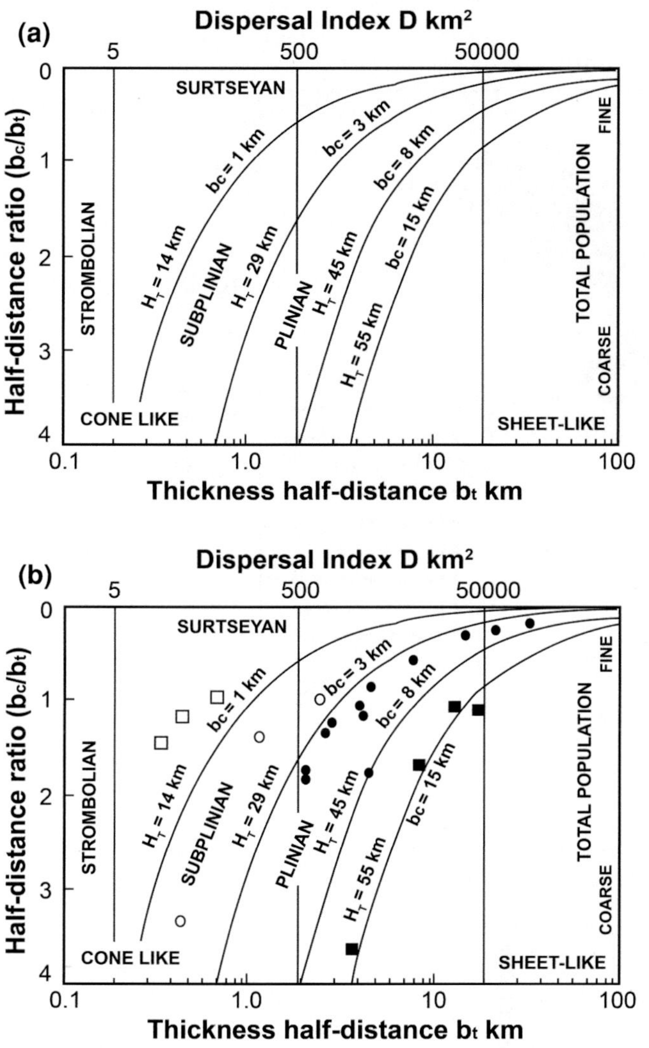

Fig. 9.92 Pyle (1989) classification of eruptions and deposits based on the fallout deposit thickness half distance as a proxy for the dispersal, and the ratio between the clast and the thickness half distances as a proxy of the influence of the total grain size distribution on column height. **a** The essential classification; **b** showing data points underpinning the classification. See Pyle (1989) for specifics of the deposits used. (From Pyle's 1989)

0.01 m^3 (25 kg; VEI -6) and 0.1 m^3 (250 kg), which represent events that can only rarely be recorded and documented, are inconsequential in most cases unless someone is standing right next to the explosion vent, and will not be preserved in the rock record. It also seems counter-intuitive to have a negative value on a scale of explosivity, which is difficult to explain to the general public and media, and is an important consideration during volcanic crises.

Has the VEI outlived its usefulness, as currently defined? To remain relevant as a scheme, we suggest that its volume scale needs to be redefined or recalibrated, beginning with 10^0 m^3 equating to VEI 0, then 10^1 m^3 defining VEI 1, etc. as discussed further in Sect. 9.10.5 and outlined in Table 9.1. There is also the issue that deposit volume does not necessarily reflect eruption explosiveness, because it is not necessarily related to the volume eruption rate. A particular volume of deposit could be produced quickly through a short, intense phase of high eruption rate, or over a considerable time during an extended eruption of lower intensity (explosivity).

9.10.4 Process-Focused Classifications

Although the VEI classification scheme of Newhall and Self (1982) has a process implication by relating deposit volume to an inferred eruption column height, it did not directly relate either to quantitative eruption dynamics, such as actual explosiveness or mass eruption rate. A clear distinction should be made between classifications of deposits and classifications of observed eruption styles and eruption parameters, because, although a link of course exists, the proxies used to classify may be substantially different.

Pyle (1989, 2015) proposed two new indices reflecting processes, that is, the eruption Intensity Index (EI), the proxy for which is Mass Eruption Rate (MER, kg s^{-1}), and eruption Magnitude Index (MI), the proxy for which is Mass Erupted (ME, kg). The indices are applicable to both explosive eruptions and effusive eruptions and can be plotted against each other (Fig. 9.93a). In both indices, "Mass" refers to dense magma mass equivalent erupted, not deposit mass, and both indices are logarithmic functions, defined by Pyle (1989, 2015) as

$$Intensity\ Index,\ EI = (log_{10} mass\ eruption\ rate) + 3 \quad (9.38)$$

$$Magnitude\ Index,\ MI = (log_{10} mass\ erupted) - 7 \quad (9.39)$$

It is unclear why the integers 3 and 7 were included in the index formulae other than to provide a better comparison with VEI values and to facilitate graphical representation, which we will discuss in Sect. 9.10.5. The scheme proposed by Pyle (1989, 2015) does not classify or distinguish between eruption styles, which Bonadonna et al. (2022) used to good effect in comparing the 2021 La Palma hybrid eruption with other eruptions and different eruption styles.

Bonadonna and Costa (2013) applied a probability approach to extrapolating trends in deposit characteristics (area, thickness, grain size) into eruption dynamic parameters by applying Weibull distributions to the previously used characteristics. They found that thickness change patterns correlate well with deposit volume (hence eruption magnitude) and grain size parameter patterns (e.g. median grain size, average maximum lithic size) correlate well with eruption column height, hence eruption intensity or magma discharge rate. From this, they proposed a classification into

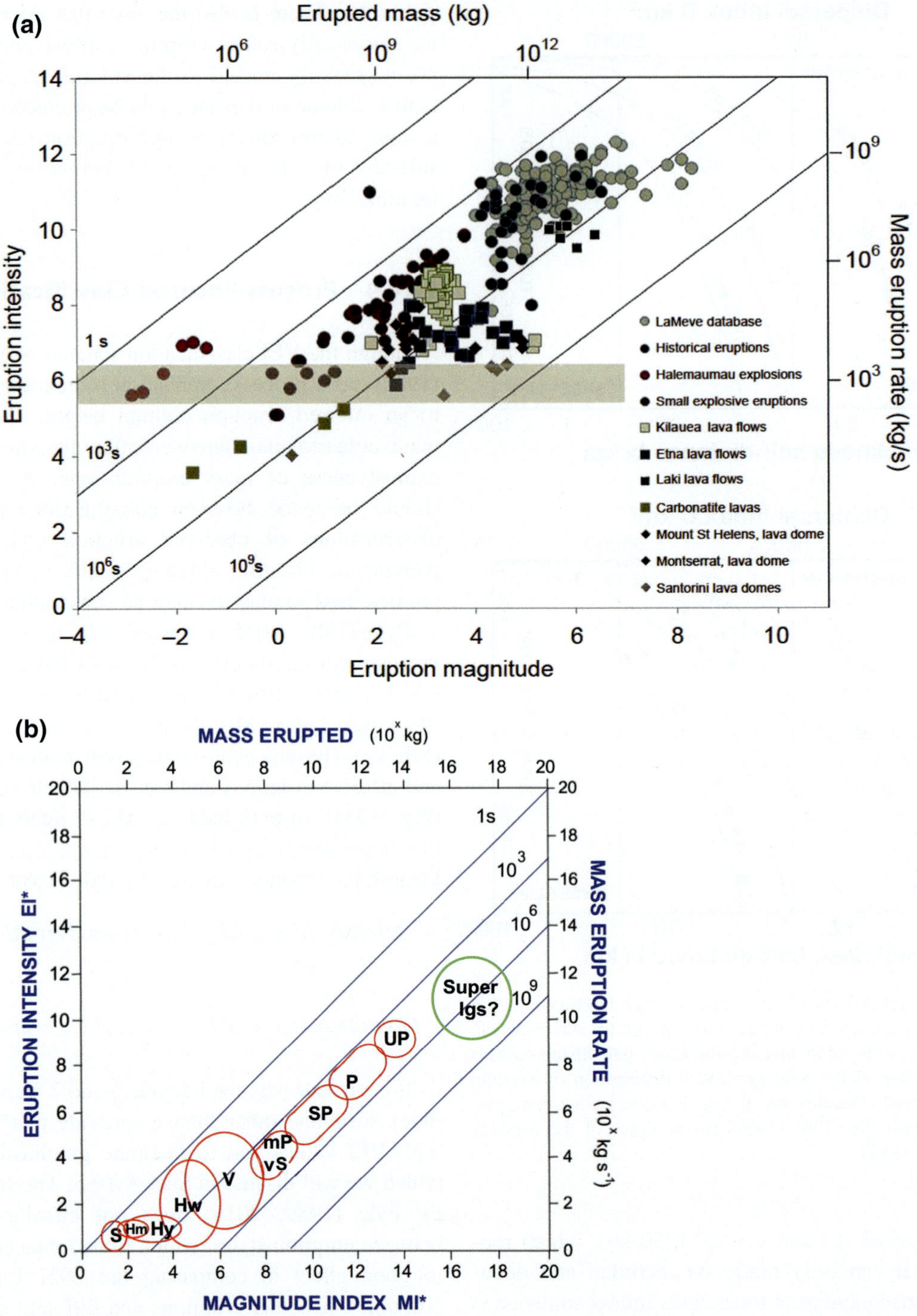

Fig. 9.93 **a** Plot of Eruption Intensity Index (EI), a log function, the proxy for which is Mass Eruption Rate (MER) against Eruption Magnitude Index (MI), also a log function, the proxy for which is Mass Erupted (ME), for both pyroclastic and effusive eruptions. Both EI and MI are defined in the text. Also shown as oblique contours are eruption duration time lines, in seconds, from 1 to 10^3 s (=~20 min), 10^6 s (≥1 day). 10^9 s (=3 years). (From Pyle 2015). **b** A proposed plot of modified Eruption Intensity, EI*, against modified Eruption Magnitude Index, MI*, as defined in the text. The scale for all parameters begins with 0; all scales are logarithmic functions to the base 10. Approximate fields for the main subaerial fallout-forming styles and deposits are shown. S = strombolian *sensu stricto*; Hm = halema'uma'uan; Hw = hawai'ian fire fountaining; Hy = hydrothermal; V = vulcanian; mP = micro-plinian; vS = violent strombolian; SP = sub-plinian; P = plinian; UP = ultra-plinian. MER estimates and possible durations of peak activity from Table 9.2 and other sources. For comparison, the approximate field for caldera-forming super- or giant ignimbrites, many of which don't have major eruption column sourced primary fallout deposits, is shown in green to emphasise the extreme mass of the deposits and the inferred extremely high MER compared with the largest known fallout-forming eruption styles. Based on data in Pyle (2000) and Mason et al. (2004)

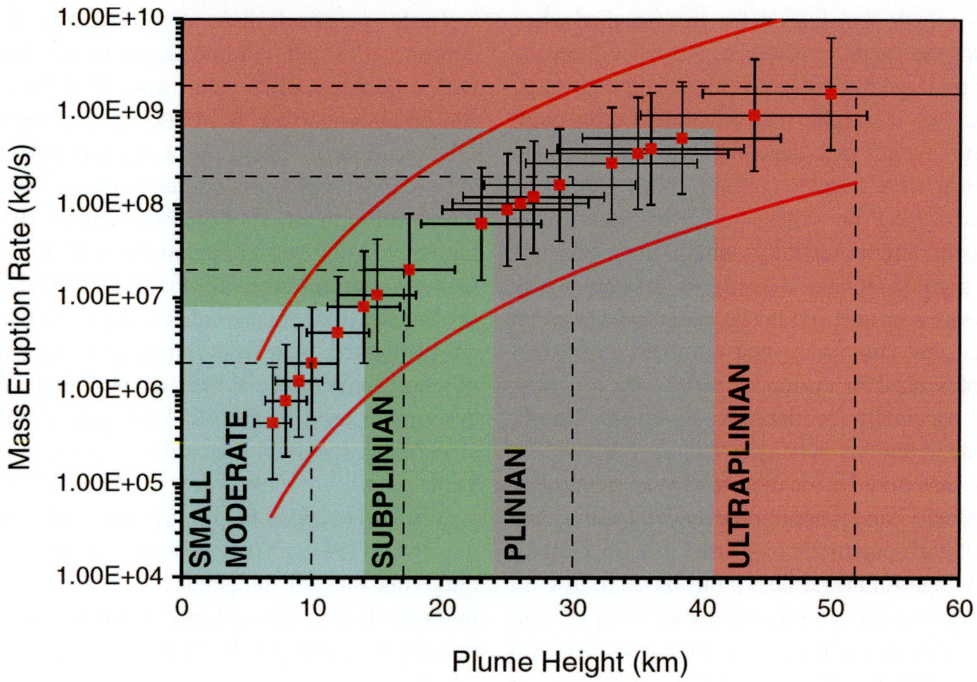

Fig. 9.94 Classification of highly explosive eruption styles using a plot of $\log_{10}$MER (mass eruption rate, kg s^{-1}) against $\log_{10}H_T$ (eruption column height, km), which, in using empirical data sets, discriminates ultra-plinian, plinian, sub-plinian from weaker explosive eruption styles, including strombolian, hawai'ian, vulcanian, violent strombolian, but does not discriminate the weaker styles, nor does it include phreatomagmatic eruption styles (From Bonadonna and Costa 2013)

small-moderate, sub-plinian, plinian, and ultra-plinian deposits by plotting calculated mass eruption rate against calculated (or observed) plume height (Fig. 9.94), both of which reflect the intensity of explosiveness.

Gonnermann and Manga (2013) related magma discharge rate and magma ascent rate to magma viscosity plotting both on logarithmic scales, to distinguish mafic hawai'ian, strombolian, and effusive eruptions from felsic sub-plinian explosive and effusive dome-forming eruptions (Fig. 9.95).

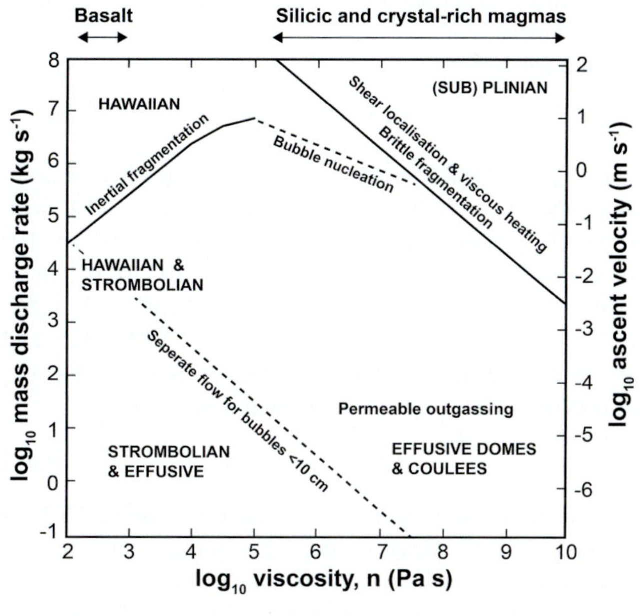

Fig. 9.95 Classification of eruption styles based on eruption characteristics (magma mass discharge rate, magma ascent velocity in the conduit) and magma properties (composition, viscosity) (After Gonnermann and Manga 2013)

9.10.5 Some Proposals for Eruption Indices as a Way Forward

Having critically assessed existing classification schemes for fallout deposits and their eruptions, we have some suggestions for both deposit characteristics based and process-focused schemes to make them more logically constrained, easier to use, and easier to compare.

In an ideal world, it would make sense for all indices to have a simple base value of 1 for the parameter that the classification is based on (e.g. deposit volume, mass, and eruption rate). Unfortunately, this was not done with any of the principal classifications or indices used when they were defined. There is a historical reason for this, in that when they were devised, the focus in research and hazard assessment was largely on large explosive eruptions and their large volume deposits. As a result, the schemes began with moderate intensity events and deposits, but low-intensity events and deposits were largely ignored. With habitation and tourism encroaching on volcanoes, even small-scale events and deposits are potentially lethal, and thus are becoming more significant and a focus of research. Any event that represents a hazard should be encompassed in all schemes or scales that summarise and compare explosiveness, but the scales used in current schemes are too contracted to do this, especially for low-intensity events. Rather than discarding the previous schemes that have served us well, we suggest that they be recalibrated to accommodate small volume, intensity, and magnitude events by setting the minimum value of each parameter measured to 1 (m^3, kg, kg s^{-1}). By using a logarithmic scale to the base 10 for all indices, we can devise more comprehensive, simple, and useable systems. Although some people may disagree with this approach, sometimes such revisions are necessary to resurrect old valuable schemes, that are no longer fully fit for purpose, into useable schemes that are fit for today, and for future use. We therefore make suggestions as to how this can be done.

This would require the VEI scale to be expanded and redefined or recalibrated on a logarithmic scale to the base 10 of deposit volume, V, defining a simple modified or recalibrated VEI*:

$$\mathrm{VEI}^* = \log_{10}\mathrm{V} \tag{9.40}$$

Using this $\log_{10}$ function, a minimum deposit volume in the range of 1 (10^0) to 10 (10^1) m^3 becomes VEI* = 0. Sequentially then, higher volumes such as 10^1 to 10^2 m^3 = VEI* 1, 10^2 to 10^3 m^3 = VEI* 2, 10^3 to 10^4 m^3 = VEI* 3, 10^4 to 10^5 m^3 = VEI* 4 etc. .... 10^{12} m^3 = VEI* 12, and beyond if needed. This requires a simple correction of + 4 to all previous VEI (original) numbered assessments:

$$\mathrm{VEI}^* = \mathrm{VEI} + 4 \tag{9.41}$$

In addition, four smaller categories preceding VEI* 4 (formerly VEI 0), based on deposit volumes < 10^4 m^3 (the previous minimum volume classified), are now included (see Table 9.1 for comparison between VEI and VEI* scales). This expansion allows the classic strombolian *sensu stricto* eruption style to be included, whereas it was excluded in the previous VEI scheme because the volume of most strombolian explosion deposits are <10^4 m^3. It also now allows the deposits of new and other old explosive eruption styles to be included (e.g. halema'uma'uan, hydrothermal).

In this proposal, there are no negative VEI* values for a deposit, all small volume deposits can be accommodated with a positive VEI* value, and even decimal point values can be assigned (e.g. a 14,000 m^3 deposit becomes a VEI* 4.15 eruption). There is a logical direct link between $\log_{10}$(deposit volume) and the modified VEI* number in that the power number, x, of deposit volume, 10^x m^3, is the VEI* number. In the proposed modified VEI* scheme, the confusing descriptor terms for the old VEI 0 ("non-explosive … gentle, effusive" for *pyroclastic* deposit volume of <10^4 m^3) are eliminated. New descriptors are defined for the new scale of explosivity that as best as possible correlate with the descriptors used in the old VEI scale of Newhall and Self (1982) but inevitably includes some new descriptors (see Table 9.1).

We also propose to simplify and make more logical both the Pyle (1989, 2015) Eruption Intensity Index, EI, by removing the + 3 integer in Eq. 9.38, so that the modified Intensity Index, EI*, the proxy for which is mass eruption rate (kg s^{-1}; Fig. 9.93b), becomes simply

$$\mathit{Intensity\ Index,\ EI}^* = (\mathit{log_{10} mass\ eruption\ rate}) \tag{9.42}$$

For the Magnitude Index, MI, the proxy for which is the deposit mass (kg) we suggest removing the—7 integer in Eq. 9.39, so that the modified Magnitude Index, MI*, becomes simply

$$\mathit{Magnitude\ Index,\ MI}^* = (\mathit{log_{10} mass\ erupted}) \tag{9.43}$$

Again, these indices then have a logical minimum base (Fig. 9.93b). For example, for a 1 kg s^{-1} MER, the EI* is 0, for a 10^3 kg s^{-1} MER, the EI* is 3, and for a 10^6 kg s^{-1} MER, the EI* is 6. Similarly, for a 1 kg deposit mass, the eruption magnitude index, MI* is 0, for a 10 kg deposit mass, the MI* is 1, for a 10^2 kg deposit mass, the MI* is 2, for a 10^3 kg deposit mass, the MI* is 3, for a 10^4 kg deposit mass, the MI* is 4, etc. With the eruption duration contours also shown in Fig. 9.93b, fields for different eruption styles can be approximated, taking into account typical mass eruption rates, masses erupted, and durations for each typical eruption style (Fig. 9.93b).

9.10.6 Finally ....

Although trying to construct a single classification scheme that addresses all aspects of subaerial fallout deposits and their eruption styles and characteristics is a holy grail, it is an unrealistic and counter-productive goal, and becomes more so the more we learn. In reality to benefit the most from our knowledge of all the different facets of fallout-forming eruptions and their deposits, we need to develop multiple classification schemes that are each fit for a specific purpose. Some will account for large-scale field deposit characteristics, such as areal extent, thickness changes, grain size characteristics and trends (Walker 1973), or deposit volumes (Newhall and Self 1982). Others will account for deposit characteristics at the outcrop scale, which is important for those working in the rock record, and some will address the dynamics of the eruptions, eruption columns, and downwind plumes (e.g. Pyle 1989; Pyle et al. 2006; Bonadonna and Costa (2013). They are all useful in characterising particular eruption styles and their deposits.

While a single, unique, and accepted classification for all subaerial fallout deposits is not available (Bonadonna et al. 2016), there is a general consensus in the retention and use of the specific nomenclature for the basic eruption styles and deposits that we have adopted in this chapter of strombolian, hawai'ian, sub-plinian, plinian, ultra-plinian, vulcanian, phreatic, surtseyan, ukinrekian, and phreatoplinian deposits (e.g. Fig. 9.6a), which is largely taken from Walker's (1973) classification. We have also noted the existence of hybrid eruption styles between truly magmatic and phreatomagmatic end members, the names of which should reflect such hybrid styles (e.g. phreato-vulcanian; Fig. 9.6a). However, individual deposit or rock units resulting from a single eruption may consist of sub-units resulting from different phases of an eruption characterised by changing styles of eruption. Only detailed fieldwork mapping of the deposits and their stratigraphic variations would lead to identification of all phases of an eruption. These include the onset, the climax, the waning, and closing stages, sometimes involving multiple cycles during a single event, and also possibly encompassing more or less prolonged pauses.

Regarding a classification scheme for subaqueous fallout deposits and subaqueous eruptions, there is no quantitative one available. It is not sensible to try to use the subaerial eruption style and deposit names, let alone the classification scheme, and apply them to subaqueous deposits because the quantitative criteria and data that underpin classification schemes of subaerial fallout deposits are just not available because of limitations of access to explosive eruptions, their deposits, and mapping their extent on the modern seafloor. Fragmentation, dispersal, and sedimentation processes are also very different and more complex due to the major influence of the subaqueous environment (Cas and Simmons 2018). At this stage, it is better just to describe what is known about the deposits and interpret as best as possible the eruption, fragmentation, dispersal, and sedimentation processes, all of which are going to be more complex than for subaerial eruptions. Some new eruption style names have been suggested (yalian, neptunian, tangaroan, and hunga tongan), reflective of particular volcanoes or events, as was the case with subaerial events in the early years. These can be usefully applied, deposit characteristics quantified, and a classification scheme progressively developed. We have *speculatively* included those submarine eruption styles on the Walker classification diagram (Fig. 9.6a), even though we don't have D and F fields and laboratory data.

References

Adams NK, Houghton BF, Hildreth W (2006) Abrupt transitions during sustained explosive eruptions: examples from the 1912 eruption of Novarupta, Alaska. Bull Volcanol 69(2):189–206

Alatorre-Ibargüengoitia MA, Delgado-Granados H (2006) Experimental determination of drag coefficient for volcanic materials: calibration and application of a model to Popocatépetl volcano (Mexico) ballistic projectiles. Geophys Res Letts 33(11)

Alatorre-Ibargüengoitia MA, Delgado-Granados H, Farraz-Montes IA (2006) Hazard zoning for ballistic impact during volcanic explosions at Volcán de Fuego de Colima (Mexico). Geol Soc Am Spec Pap 402:209–216

Alatorre-Ibargüengoitia MA, Scheu B, Dingwell DB (2011) Influence of the fragmentation process on the dynamics of Vulcanian eruptions: an experimental approach. Earth Planet Sci Lett 302 (1):51–59

Alfano F, Bonadonna C, Gurioli L (2012) Insights into eruption dynamics from textural analysis: the case of the May, 2008, Chaitén Eruption. Bull Volcanol 74(9):2095–2108

Alfano F, Bonadonna C, Volentik ACM, Connor CB, Watt SFL, Pyle DM, Connor LJ (2011) Tephra stratigraphy and eruptive volume of the May, 2008, Chaitén eruption, Chile. Bull Volcanol 73 (5):613–630

Alfano F, Bonadonna C, Watt S, Connor C, Volentik A, Pyle DM (2016) Reconstruction of total grain size distribution of the climactic phase of a long-lasting eruption: the example of the 2008–2013 Chaitén eruption. Bull Volcanol 78(7):46

Allard P (2010) A CO_2-rich gas trigger of explosive paroxysms at Stromboli basaltic volcano, Italy. J Volcanol Geotherm Res 189(3–4):363–374

Allen SR, Carey RJ, McPhie J (in prep) Yalian eruptions: subaqueous, moderate-low intensity pumice cone eruptions

Allen SR, Cas RAF (1998) Rhyolitic fallout and pyroclastic density current deposits from a phreatoplinian eruption in the eastern Aegean Sea, Greece. J Volcanol Geotherm Res 86(1):219–251

Allen SR, McPhie J (2000) Water-settling and resedimentation of submarine rhyolitic pumice at Yali, eastern Aegean, Greece. J Volcanol Geotherm Res 95(1–4):285–307

Allen SR, McPhie J (2009) Products of neptunian eruptions. Geology 37(7):639–642

Amos RC, Self S, Crowe B (1981) Pyroclastic activity of Sunset Crater: evidence for a large magnitude, high dispersal strombolian eruption. Eos 62:1085

Armienti P, Macedonio G, Pareschi MT (1988) A numerical model for simulation of tephra transport and deposition: applications to May 18, 1980, Mount St. Helens eruption. J Geophys Res: Solid Earth 93 (B6):6463–6476

Aubry TJ, Engwell SL, Bonadonna C, Mastin LG, Carazzo G, Van Eaton A, Jessop D, Grainger RG, Scollo S, Taylor IA, Jellinek M (2023, in press) New insights into the relationship between mass eruption rate and volcanic column height based on the IVESPA dataset. Authorea Preprints. https://doi.org/10.22541/essoar.167214588.86276236/v1

Austin-Erickson A, Büttner R, Dellino P, Ort MH, Zimanowski B (2008) Phreatomagmatic explosions of rhyolitic magma: experimental and field evidence. J Geophys Res: Solid Earth 113(B11)

Bachmeier S (2020) Eruption of the Taal Volcano in the Philippines CIMSS Satellite Blog. https://cimss.ssec.wisc.edu/satellite-blog/archives/35406

Báez W, Nuñez GC, Giordano G, Viramonte JG, Chiodi A (2017) Polycyclic scoria cones of the Antofagasta de la Sierra basin, Southern Puna plateau, Argentina. Geol Soc Lond Spec Pub 446 (1):311–336

Balangue-Tarriela MIR, Lagmay AMF, Sarmiento DM, Vasquez J, Baldago MC, Ybañez R, Trinidad JR, Thivet S, Gurioli L, de Vries BVW, Aurelio M, Rafael DJ, Bermas A, Escudero JA (2022) Analysis of the 2020 Taal Volcano tephra fall deposits from crowdsourced information and field data. Bull volcanol 84(3):35

Barberi F, Cioni R, Rosi M, Santacroce R, Sbrana A, Vecci R (1989) Magmatic and phreatomagmatic phases in explosive eruptions of Vesuvius as deduced by grain-size and component analysis of the pyroclastic deposits. J Volcanol Geotherm Res 38(3–4):287–307

Barberi F, Coltelli M, Frullani A, Rosi M, Almeida E (1995) Chronology and dispersal characteristics of recently (last 5000 years) erupted tephra of Cotopaxi (Ecuador): implications for long-term eruptive forecasting. J Volcanol Geotherm Res 69 (3):217–239

Barberi F, Rosi M, Sodi A (1993) Volcanic hazard assessment at Stromboli based on review of historical data. Acta Vulcanol 3: 173–187

Barreyre T, Soule SA, Sohn RA (2011) Dispersal of volcaniclasts during deep-sea eruptions: settling velocities and entrainment in buoyant seawater plumes. J Volcanol Geotherm Res 205(3):84–93

Barsotti S, Neri A, Scire JS (2008) The VOL-CALPUFF model for atmospheric ash dispersal: 1. Approach and physical formulation. J Geophys Res: Solid Earth 113(B3)

Behncke B, Neri M (2003) The July–August 2001 eruption of Mt. Etna (Sicily). Bull Volcanol 65(7):461–476

Behncke B, Neri M, Pecora E, Zanon V (2006) The exceptional activity and growth of the Southeast Crater, Mount Etna (Italy), between 1996 and 2001. Bull Volcanol 69(2):149–173

Biass S, Bonadonna C (2014) TOTGS: total grainsize distribution of tephra fallout. URL https://vhub.org/resources/3297

Biass S, Bonadonna C, Houghton BF (2019) A step-by-step evaluation of empirical methods to quantify eruption source parameters from tephra-fall deposits. J Appl Volcanol 8(1):1–6

Biass S, Falcone J-L, Bonadonna C, Di Traglia F, Pistolesi M, Rosi M, Lestuzzi P (2016) Great balls of fire: a probabilistic approach to quantify the hazard related to ballistics—a case study at La Fossa volcano, Vulcano Island, Italy. J Volcanol Geotherm Res 325:1–14

Binns RA (2003) Deep marine pumice from the Woodlark and Manus Basins, Papua New Guinea. Explosive subaqueous volcanism. American Geophysical Union, Washington, D.C., pp 329–343

Blackburn EA, Wilson L, Sparks RSJ (1976) Mechanisms and dynamics of strombolian activity. J Geol Soc 132(4):429–440

Bloomfield K, Rubio GS, Wilson L (1977) Plinian eruptions of Nevado de Toluca volcano, Mexico. Geol Rund 66:120–146

Bonadonna C, Cioni R, Costa A, Druitt T, Phillips JC, Pioli L, Andronico D, Harris A, Scollo S, Bachmann O, Bagheri G, Biass S, Brogi F, Cashman KV, Dominguez L, Durig T, Galland O, Giordano G, Gundmundsson M, Hort M, Hoskuldsson A, Houghton B, Komorowski JC, Kuppers U, G. L, Le Pennec JL, Macedonio G, Manga M, Manzella I, de' Michieli Vittrui E, Neri A, Pistolesi M, Polacci M, Riperpe M, Rossi E, Scheu B, Sulpizio R, Tripoli B, Valade S, Valentine GA, Vidal CM, Wallenstein N (2016) MeMoVolc consensual document: dynamics and classification of volcanic explosive eruptions. Bull Volcanol 78(84)

Bonadonna C, Cioni R, Pistolesi M, Connor CB, Scollo S, Pioli L, Rosi M (2013) Determination of the largest clast sizes of tephra deposits for the characterization of explosive eruptions: a study of the IAVCEI commission on tephra hazard modelling. Bull Volcanol 75(1):1–15

Bonadonna C, Connor CB, Houghton BF, Connor L, Byrne M, Laing A, Hincks TK (2005) Probabilistic modeling of tephra dispersal: Hazard assessment of a multiphase rhyolitic eruption at Tarawera, New Zealand. J Geophys Res: Solid Earth 110(B3)

Bonadonna C, Costa A (2012) Estimating the volume of tephra deposits: a new simple strategy. Geology G32769:32761

Bonadonna C, Costa A (2013) Plume height, volume, and classification of explosive volcanic eruptions based on the Weibull function. Bull Volcanol 75(8):1–19

Bonadonna C, Costa A, Folch A, Koyaguchi T (2015) Tephra dispersal and sedimentation. In: Sigurdsson H, Houghton B, McNutt SR, Rymer H, Stix J (eds) The Encyclopedia of volcanoes. Academic Press Elsevier, Cambridge, pp 587–597

Bonadonna C, Ernst GGJ, Sparks RSJ (1998) Thickness variations and volume estimates of tephra fall deposits: the importance of particle Reynolds number. J Volcanol Geotherm Res 81(3):173–187

Bonadonna C, Houghton BF (2005) Total grain-size distribution and volume of tephra-fall deposits. Bull Volcanol 67(5):441–456

Bonadonna C, Macedonio G, Sparks RSJ (2002) Numerical modelling of tephra fallout associated with dome collapses and Vulcanian explosions: application to hazard assessment on Montserrat. In: Druitt TH, Kokelaar BP (eds) The eruption of Soufrière Hills Volcano, Montserrat, From 1995 to 1999. Geological Society, London, Memoirs, pp 517–537

Bonadonna C, Phillips JC (2003) Sedimentation from strong volcanic plumes. J Geophys Res: Solid Earth 108(B7)

Bonadonna C, Pistolesi M, Biass S, Voloschina M, Romero J, Coppola D, Folch A, D'Auria L, Martin-Lorenzo A, Dominguez L, Pastore C, Reyes-Hardy MP, Rodriguez F (2022) Physical Characterization of Long-Lasting Hybrid Eruptions: The 2021 Tajogaite Eruption of Cumbre Vieja (La Palma, Canary Islands). J Geophys Res: Solid Earth 127(11):e2022JB025302

Bond A, Sparks RSJ (1976) The Minoan eruption of Santorini, Greece. J Geol Soc Lond 132:1–16

Booth B, Croasdale R, Walker GPL (1978) A quantitative study of five thousand years of volcanism on Sao Miguel, Azores. Phil Trans Royal Soc London. Series A, Math Phys Sci 288(1352):271–319

Boyce JA, Gertisser R (2012) Variations in welding characteristics within the Plinian air-fall deposit of the Middle Pumice eruption, Santorini, Greece. J Volcanol Geotherm Res 221:71–82

Branney MJ (1991) Eruption and depositional facies of the Whorneyside Tuff Formation, English Lake District: an exceptionally large-magnitude phreatoplinian eruption. Geol Soc Am Bull 103 (7):886–897

Brazier S, Davis AN, Sigurdsson H, Sparks RSJ (1982) Fall-out and deposition of volcanic ash during the 1979 explosive eruption of the Soufriere of St. Vincent. J Volcanol Geother Res 14(3–4):335–359

Breard ECP, Lube G, Cronin SJ, Fitzgerald R, Kennedy B, Scheu B, Montanaro C, White JDL, Tost M, Procter JN (2014) Using the spatial distribution and lithology of ballistic blocks to interpret eruption sequence and dynamics: August 6 2012 Upper Te Maari eruption, New Zealand. J Volcanol Geotherm Res 286:373–386

Brenna M, Cronin S, Smith I, Pontesilli A, Tost M, Barker S, Tongaonevai S (2022) Post-caldera volcanism reveals shallow priming of an intra-ocean arc andesitic caldera: Hunga volcano, Tonga. SW Pacific. Lithos 412–413:106614

Brennen CE (2005) Fundamentals of multiphase flow. Cambridge Univ Press, p 345

Briffa KR, Jones PD, Schweingruber FH, Osborn TJ (1998) Influence of volcanic eruptions on Northern Hemisphere summer temperature over the past 600 years. Nature 393(6684):450–455

Brown RJ, Buse B, Sparks RSJ, Field M (2008) On the welding of pyroclasts from very low-viscosity magmas: examples from kimberlite volcanoes. The J Geol 116(4):354–374
Brown RJ, Branney MJ, Maher C, Dávila-Harris P (2010) Origin of accretionary lapilli within ground-hugging density currents: evidence from pyroclastic couplets on Tenerife. Geol Soc Am Bull 122 (1–2):305–320
Brown RJ, Bonadonna C, Durant AJ (2012) A review of volcanic ash aggregation. Phys Chem Earth Parts a/b/c 45:65–78
Brown SK, Crosweller HS, Sparks RSJ, Cottrell E, Deligne NI, Guerrero NO, Hobbs L, Kiyosugi K, Loughlin SC, Siebert L (2014) Characterisation of the quaternary eruption record: analysis of the large magnitude explosive volcanic eruptions (LaMEVE) database. J Appl Volcanol 3(1):1–22
Bryan SE, Cook AG, Evans JP, Hebden K, Hurrey L, Colls P, Jell JS, Weatherley D, Firn J (2012) Rapid, long-distance dispersal by pumice rafting. PLoS ONE 7(7):e40583
Burgisser A, Gardner JE (2006) Using hydraulic equivalences to discriminate transport processes of volcanic flows. Geology 34 (3):157–160
Bursik M (2001) Effect of wind on the rise height of volcanic plumes. Geophys Res Lett 28(18):3621–3624
Bursik MI, Sparks RSJ, Gilbert JS, Carey SN (1992) Sedimentation of tephra by volcanic plumes: I. Theory and its comparison with a study of the Fogo A plinian deposit, Sao Miguel (Azores). Bull Volcanol 54(4):329–344
Bursik MI, Woods AW (1991) Buoyant, superbuoyant and collapsing eruption columns. J Volcanol Geotherm Res 45(3):347–350
Büttner R, Dellino P, Zimanowski B (1999) Identifying magma–water interaction from the surface features of ash particles. Nature 401 (6754):688–690
Byrne MA, Laing AG, Connor C (2007) Predicting tephra dispersion with a mesoscale atmospheric model and a particle fall model: Application to Cerro Negro volcano. J Appl Meteorol Clim 46 (2):121–135
Cahalan RC, Dufek J (2021) Explosive submarine eruptions: The role of condensable gas jets in underwater eruptions. J Geophys Res: Solid Earth 126(2):e2020JB020969
Calabrò L, Esposti Ongaro T, Giordano G, de'Michieli Vitturi M (2022) Reconstructing pyroclastic currents' source and flow parameters from deposit characteristics and numerical modeling: The Pozzolane Rosse ignimbrite case study (Colli Albani, Italy). J Geophys Res: Solid Earth 127(5):e2021JB023637
Calder ES, Sparks RSJ, Woods AW (1997) Dynamics of co-ignimbrite plumes generated from pyroclastic flows of Mount St. Helens (7 August 1980). Bull Volcanol 58(6):432–440
Calvari S, Pinkerton H (2004) Birth, growth and morphologic evolution of the 'Laghetto'cinder cone during the 2001 Etna eruption. J Volcanol Geotherm Res 132(2):225–239
Calvari S, Spampinato L, Bonaccorso A, Oppenheimer C, Rivalta E, Boschi E (2011) Lava effusion—a slow fuse for paroxysms at Stromboli volcano? Earth Planet Sci Letts 301(1–2):317–323
Campagnola S (2014) Large-scale plinian eruptions of the Colli Albani and the Campi Flegrei volcanoes: insights from textural and rheological studies. PhD Thesis, Roma Tre University
Campagnola S, Vona A, Romano C, Giordano G (2016) Crystallization kinetics and rheology of leucite-bearing tephriphonolite magmas from the Colli Albani volcano (Italy). Chem Geol 424:12–29
Capaccioni B, Cuccoli F (2005) Spatter and welded air fall deposits generated by fire-fountaining eruptions: cooling of pyroclasts during transport and deposition. J Volcanol Geotherm Res 145 (3):263–280
Carazzo G, Kaminski E, Tait S (2008) On the dynamics of volcanic columns: A comparison of field data with a new model of negatively buoyant jets. J Volcanol Geotherm Res 178(1):94–103
Carey RJ, Houghton BF, Thordarson T (2008a) Contrasting styles of welding observed in the proximal Askja 1875 eruption deposits I: Regional welding. J Volcanol Geotherm Res 171(1–2):1–19
Carey RJ, Houghton BF, Thordarson T (2008b) Contrasting styles of welding observed in the proximal Askja 1875 eruption deposits II: Local welding. J Volcanol Geotherm Res 171(1):20–44
Carey RJ, Houghton BF, Thordarson T (2009) Abrupt shifts between wet and dry phases of the 1875 eruption of Askja Volcano: microscopic evidence for macroscopic dynamics. J Volcanol Geotherm Res 184(3):256–270
Carey RJ, Houghton BF, Thordarson T (2010) Tephra dispersal and eruption dynamics of wet and dry phases of the 1875 eruption of Askja Volcano, Iceland. Bull Volcanol 72(3):259–278
Carey RJ, Soule SA, Manga M, White JDL, McPhie J, Wysoczanski R, Jutzeler M, Tani K, Yoerger D, Fornari DJ, Caratori-Tontini F, Houghton B, Mitchell S, Ikegami F, Conway C, Murch A, Fauria K, Jones M, Cahalan R, McKenzie W (2018) The largest deep-ocean silicic volcanic eruption of the past century. Sci Adv 4(1)
Carey S (1997) Influence of convective sedimentation on the formation of widespread tephra fall layers in the deep sea. Geology 25(9): 839–842
Carey S, Bursik M (2015) Volcanic plumes. In: Sigurdsson H, Houghton B, McNutt SR, Rymer H, Stix J (eds) The encyclopedia of volcanoes. Academic Press Elsevier, Cambridge, pp 571–585
Carey S, Sigurdsson H (1985) The May 18, 1980 eruption of Mount St. Helens: 2. Modeling of dynamics of the Plinian phase. J Geophys Res: Solid Earth 90(B4):2948–2958
Carey S, Sigurdsson H (1987) Temporal variations in column height and magma discharge rate during the 79 AD eruption of Vesuvius. Geol Soc Am Bull 99(2):303–314
Carey S, Sigurdsson H (1989) The intensity of plinian eruptions. Bull Volcanol 51(1):28–40
Carey S, Sparks RSJ (1986) Quantitative models of the fallout and dispersal of tephra from volcanic eruption columns. Bull Volcanol 48(2):109–125
Carey SN, Sigurdsson H (1980) The Roseau Ash: deep-sea tephra deposits from a major eruption on Dominica, Lesser Antilles Arc. J Volcanol Geotherm Res 7(1):67–86
Carey SN, Sigurdsson H (1978) Deep-sea evidence for distribution of tephra from the mixed magma eruption of the Soufriere on St. Vincent, 1902: Ash turbidites and air fall. Geology 6(5):271–274
Carey SN, Sigurdsson H (1982) Influence of particle aggregation on deposition of distal tephra from the May 18, 1980, eruption of Mount St. Helens volcano. J Geophys Res: Solid Earth 87 (B8):7061–7072
Caricchi C, Vona A, Corrado S, Giordano G, Romano C (2014) 79 AD Vesuvius PDC deposits' temperatures inferred from optical analysis on woods charred in-situ in the Villa dei Papiri at Herculaneum (Italy). J Volcanol Geotherm Res 289:14–25
Cas RAF (1989) Physical volcanology. In: Johnson RW, Knutson J, Taylor SR (eds) Intra-plate Volcanism in Eastern Australia and New Zealand. Cambridge University Press, Cambridge, pp 55–87
Cas RAF, Allen RL, Bull SW, Clifford BA, Wright JV (1990) Subaqueous, rhyolitic dome-top tuff cones: a model based on the Devonian Bunga Beds, southeastern Australia and a modern analogue. Bull Volcanol 52(3):159–174
Cas RAF, Blaikie T, Boyce J, Hayman P, Jordan S, Piganis F, Prata G, van Otterloo J (2011a) VF01: factors that influence varying eruption styles (from magmatic to phreato-magmatic) in intraplate continental basaltic volcanic provinces: The Newer Volcanics Province of south-eastern Australia. IUGG 2011a Fieldtrip guide, 128p
Cas RAF, Giordano G (2014) Submarine volcanism: a review of the constraints, processes and products, and relevance to the Cabo de Gata volcanic succession. Italian J Geosci 133(3):362–377

Cas RAF, Landis CA, Fordyce RE (1989) A monogenetic, surtla-type, Surtseyan volcano from the Eocene-Oligocene Waiareka-Deborah volcanics, Otago, New Zealand: a model. Bull Volcanol 51(4): 281–298

Cas RAF, Moore CL, Scutter CR, Yamagishi H (1996) Fragmentation processes and water depth significance of Miocene submarine fire fountain deposits, Ryugazaki, Hokkaido, Japan. Western Pacific geophysics meeting. Eos 77:W126

Cas RAF, Simmons JM (2018) Why deep-water eruptions are so different from subaerial eruptions. Front Earth Sci 6:198

Cas RAF, Van Otterloo J, Blaikie TN, Van den Hove J (2017) The dynamics of a very large intra-plate continental basaltic volcanic province, the Newer Volcanics Province, SE Australia, and implications for other provinces. Geol Soc Lond Spec Publ 446: SP446–448

Cas RAF, Wright JV (1987) Volcanic successions: modern and ancient. Allen and Unwin, 529p

Cas RAF, Wright HM, Folkes CB, Lesti C, Porreca M, Giordano G, Viramonte JG (2011b) The flow dynamics of an extremely large volume pyroclastic flow, the 2.08-Ma Cerro Galán Ignimbrite, NW Argentina, and comparison with other flow types. Bull Volcanol 73 (10):1583–1609

Cas RAF, Yamagishi H, Moore L, Scutter CR (2003) Miocene submarine fire fountain deposits, Ryugazaki Headland, Oshoro Peninsula, Hokkaido, Japan: implications for submarine fountain dynamics and fragmentation processes. In: White JDL, Smellie JL, Clague DA (eds) Explosive subaqueous volcanism. AGU Mono, pp 299–316

Cashman KV (2004) Volatile controls on magma ascent and eruption. Washington DC Am Geophys Union Geophys Mono Series 150:109–124

Cashman KV, Fiske RS (1991) Fallout of pyroclastic debris from submarine volcanic eruptions. Science 253(5017):275–280

Cashman KV, Scheu B (2015) Magmatic fragmentation. In: Sigurdsson H, Houghton B, McNutt SR, Rymer H, Stix J (eds) The encyclopedia of volcanoes. Academic Press Elsevier, Cambridge, pp 459–471

Castro JM, Dingwell DB (2009) Rapid ascent of rhyolitic magma at Chaitén volcano, Chile. Nature 461(7265):780–783

Castruccio A, Clavero J, Segura A, Samaniego P, Roche O, Le Pennec JL, Droguett B (2016) Eruptive parameters and dynamics of the April 2015 sub-Plinian eruptions of Calbuco volcano (southern Chile). Bull Volcanol 78(9):62

Chadwick WW, Cashman KV, Embley RW, Matsumoto H, Dziak RP, De Ronde CEJ, Lau TK, Deardorff ND, Merle SG (2008) Direct video and hydrophone observations of submarine explosive eruptions at NW Rota-1 volcano, Mariana arc. J Geophys Res: Solid Earth 113(B8)

Chadwick WW, Paduan JB, Clague DA, Dreyer BM, Merle SG, Bobbitt AM, Caress DW, Philip BT, Kelley DS, Nooner SL (2016) Voluminous eruption from a zoned magma body after an increase in supply rate at Axial Seamount. Geophys Res Lett 43(23):12,063–012,070

Chesner CA (2012) The Toba caldera complex. Quat Int 258:5–18

Chesner CA, Rose WI, Deino AL, Drake RE, Westgate JA (1991) Eruptive history of Earth's largest Quaternary caldera (Toba, Indonesia) clarified. Geology 19(3):200–203

Chouet B, Hamisevicz N, McGetchin TR (1974) Photoballistics of volcanic jet activity at Stromboli, Italy. J Geophys Res 79 (32):4961–4976

Christiansen RL, Lipman PW (1966) Emplacement and thermal history of a rhyolite lava flow near Fortymile Canyon, southern Nevada. Geol Soc Am Bull 77(7):671–684

Cioni R, Civetta L, Marianelli P, Metrich N, Santacroce R, Sbrana A (1995) Compositional layering and syn-eruptive mixing of a periodically refilled shallow magma chamber: the AD 79 Plinian eruption of Vesuvius. J Petrol 36(3):739–776

Cioni R, Marianelli P, Santacroce R, Sbrana A (2000) Plinian and subplinian eruptions. In: Sigurdsson H, Stix J, Houghton BF, McNutt SR, Rymer H (eds) Encyclopedia of volcanoes. Academic Press Elsevier, Cambridge, pp 477–494

Cioni R, Longo A, Macedonio G, Santacroce R, Sbrana A, Sulpizio R, Andronico D (2003) Assessing pyroclastic fall hazard through field data and numerical simulations: example from Vesuvius. J Geophys Res: Solid Earth 108(B2)

Cioni R, Pistolesi M, Rosi M (2015) Plinian and subplinian eruptions. Chapter 29 In: Sigurdsson H, Houghton B, McNutt SR, Rymer H, Stix J (eds) The encyclopedia of volcanoes. Academic Press Elsevier, Cambridge, 520–534

Cisneros de León A, Schindlbeck-Belo JC, Kutterolf S, Danišík M, Schmitt AK, Freundt A, Pérez W, Harvey JC, Wang KL, Lee HY (2021) A history of violence: magma incubation, timing and tephra distribution of the Los Chocoyos supereruption (Atitlán Caldera, Guatemala). J Quat Sci 36(2):169–179

Clague DA, Paduan JB, Davis AS (2009) Widespread strombolian eruptions of mid-ocean ridge basalt. J Volcanol Geotherm Res 180 (2):171–188

Clarke AB, Esposti Ongaro T, Belousov A (2015) Vulcanian eruptions. In: Sigurdsson H, Houghton B, McNutt SR, Rymer H, Stix J (eds) The encyclopedia of volcanoes. Academic Press Elsevier, Cambridge pp 506–518

Clarke AB, Phillips JC, Chojnicki KN (2009) An investigation of Vulcanian eruption dynamics using laboratory analogue experiments and scaling analysis. Stud Volcanol 2:155–166

Clarke AB, Voight B, Neri A, Macedonio G (2002) Transient dynamics of vulcanian explosions and column collapse. Nature 415 (6874):897–901

Coira B, Kay SM, Viramonte JG, Kay RW, Galli C (2018) Origin of late Miocene peraluminous Mn-rich garnet-bearing rhyolitic ashes in the Andean Foreland (Northern Argentina). J Volcanol Geotherm Res 364:20–34

Cole PD, Guest JE, Duncan AM, Pacheco J-M (2001) Capelinhos 1957–1958, Faial, Azores: deposits formed by an emergent surtseyan eruption. Bull Volcanol 63(2–3):204–220

Cole PD, Fernandez E, Duarte E, Duncan AM (2005) Explosive activity and generation mechanisms of pyroclastic flows at Arenal volcano, Costa Rica between 1987 and 2001. Bull Volcanol 67 (8):695–716

Colombier M, Scheu B, Wadsworth FB, Cronin S, Vasseur J, Dobson KJ, Hess KU, Tost M, Yilmaz TI, Cimarelli C, Brenna M (2018) Vesiculation and quenching during surtseyan eruptions at Hunga Tonga-Hunga Ha'apai Volcano, Tonga. J Geophys Res: Solid Earth 123(5):3762–3779

Coltelli M, Del Carlo P, Pompilio M, Vezzoli L (2005) Explosive eruption of a picrite: the 3930 BP subplinian eruption of Etna volcano (Italy). Geophys Res Lett 32(23)

Coltelli M, Del Carlo P, Vezzoli L (1998) Discovery of a plinian basaltic eruption of roman age at Etna volcano, Italy. Geology 26 (12):1095–1098

Connor LJ, Connor CB (2006) Inversion is the key to dispersion: understanding eruption dynamics by inverting tephra fallout. In: Coles SG, Connor CB, Connor LJ (eds) Mader HM. Statistics in Volcanology, Geol Soc Lond, pp 231–242

Constantinescu R, Hopulele-Gligor A, Connor CB, Bonadonna C, Connor LJ, Lindsay JM, Charbonnier S, Volentik A (2021) The radius of the umbrella cloud helps characterize large explosive volcanic eruptions. Comms Earth Environ 2(1):1–8

Costa A, Folch A, Macedonio G (2013) Density-driven transport in the umbrella region of volcanic clouds: implications for tephra dispersion models. Geophys Res Letts 40:4823–4827

Costa A, Macedonio G, Folch A (2006) A three-dimensional Eulerian model for transport and deposition of volcanic ashes. Earth Planet Sci Lett 241(3):634–647

Costa A, Pioli L, Bonadonna C (2016a) Assessing tephra total grain-size distribution: insights from field data analysis. Earth Planet Sci Letts 443:90–107

Costa A, Suzuki YJ, Cerminara M, Devenish BJ, Ongaro TE, Herzog M, Van Eaton AR, Denby LC, Bursik M, Vitturi MD, Engwell S (2016b) Results of the eruptive column model inter-comparison study. J Volcanol Geotherm Res 326:2–5

Costa A, Suzuki YJ, Koyaguchi T (2018) Understanding the plume dynamics of explosive super-eruptions. Nat Comm 9(1):1–6

Costantini L, Bonadonna C, Houghton BF, Wehrmann H (2009) New physical characterization of the Fontana Lapilli basaltic Plinian eruption, Nicaragua. Bull Volcanol 71(3):337–355

Cronin S (2022) Why the volcanic eruption in Tonga was so violent, and what to expect next. The Conversation January 16, 2022. https://mail.google.com/mail/u/0/#search/Shane+Cronin/FMfcgzGmtXCRxPtdgDXpmkstLfQGmjNv

Cronin SJ, Brenna M, Smith IE, Barker S, Tost M, Ford M (2017) New volcanic island unveils explosive past. Eos 98:18–23

Crosweller HS, Arora B, Brown SK, Cottrell E, Deligne NI, Guerrero NO, Hobbs L, Kiyosugi K, Loughlin SC, Lowndes J, Nayembil M (2012) Global database on large magnitude explosive volcanic eruptions (LaMEVE). J Appl Volcanol 1(1):4

Daggitt ML, Mather TA, Pyle DM, Page S (2014) AshCalc–a new tool for the comparison of the exponential, power-law and Weibull models of tephra deposition. J Appl Volcanol 3(1):1–8

Davì M, De Rosa R, Donato P, Sulpizio R (2011) The Lami pyroclastic succession (Lipari, Aeolian Islands): a clue for unravelling the eruptive dynamics of the Monte Pilato rhyolitic pumice cone. J Volcanol Geotherm Res 201(1):285–300

Davis AS, Clague DA (2003) Hyaloclastite from Miocene seamounts offshore central California: compositions, eruption styles, and depositional processes. In: L WJD, L SJ, A CD (eds) Explosive subaqueous volcanism. AGU Mono 140, pp 129–142

De Astis G, La Volpe L, Peccerillo A, Civetta L (1997) Volcanological and petrological evolution of Vulcano island (Aeolian Arc, southern Tyrrhenian Sea). J Geophys Res: Solid Earth 102(B4):8021–8050

De Astis G, Lucchi F, Dellino P, La Volpe L, Tranne CA, Frezzotti ML, Peccerillo A (2013) Chapter 11 Geology, volcanic history and petrology of Vulcano (central Aeolian archipelago). Geol Soc Lond Mem 37(1):281–349

De Benedetti AA, Funiciello R, Giordano G, Diano G, Caprilli E, Paterne M (2008) Volcanology, history and myths of the Lake Albano maar (Colli Albani volcano, Italy). J Volcanol Geotherm Res 176(3):387–406

De Rita D, Giordano G, Esposito A, Fabbri M, Rodani S (2002) Large volume phreatomagmatic ignimbrites from the Colli Albani volcano (Middle Pleistocene, Italy). J Volcanol Geotherm Res 118(1):77–98

De Silva SL, Zielinski GA (1998) Global influence of the AD 1600 eruption of Huaynaputina, Peru. Nature 393(6684):455–458

De Vita S, Orsi G, Civetta L, Carandente A, D'Antonio M, Deino AL, Di Cesare T, Di Vito MA, Fisher RV, Isaia R (1999) The Agnano-Monte Spina eruption (4100 years BP) in the restless Campi Flegrei caldera (Italy). J Volcanol Geotherm Res 91(2):269–301

Deardorff ND, Cashman KV, Chadwick WW (2011) Observations of eruptive plume dynamics and pyroclastic deposits from submarine explosive eruptions at NW Rota-1, Mariana arc. J Volcanol Geotherm Res 202(1):47–59

Dellino P, Dioguardi F, Mele D, D'Addabbo M, Zimanowski B, Buettner R, Doronzo DM, Sonder I, Sulpizio R, Duerig T (2014) Volcanic jets, plumes, and collapsing fountains: evidence from large-scale experiments, with particular emphasis on the entrainment rate. Bull Volcanol 76(6):1–18

Dellino P, Dioguardi F, Zimanowski B, Büttner R, Mele D, La Volpe L, Sulpizio R, Doronzo DM, Sonder I, Bonasia R (2010) Conduit flow experiments help constraining the regime of explosive eruptions. J Geophys Res: Solid Earth 115(B4)

Dellino P, La Volpe L (1995) Fragmentation versus transportation mechanisms in the pyroclastic sequence of Monte Pilato-Rocche Rosse (Lipari, Italy). J Volcanol Geother Res 64(3–4):211–231

Dellino P, Mele D, Bonasia R, Braia G, La Volpe L, Sulpizio R (2005) The analysis of the influence of pumice shape on its terminal velocity. Geophys Res Lett 32(21)

Di Genova D, Romano C, Hess K-U, Vona A, Poe BT, Giordano D, Dingwell DB, Behrens H (2013) The rheology of peralkaline rhyolites from Pantelleria Island. J Geophys Res 249:201–216

Di Muro A, Neri A, Rosi M (2004) Contemporaneous convective and collapsing eruptive dynamics: the transitional regime of explosive eruptions. Geophys Res Lett 31(10)

Di Muro A, Rosi M, Aguilera E, Barbieri R, Massa G, Mundula F, Pieri F (2008) Transport and sedimentation dynamics of transitional explosive eruption columns: the example of the 800 BP Quilotoa Plinian eruption (Ecuador). J Volcanol Geotherm Res 174(4): 307–324

Dobran F, Neri A, Macedonio G (1993) Numerical simulation of collapsing volcanic columns. J Volcanol Geotherm Res 98 (B3):4231–4259

Drexler JW, Rose WI, Sparks RSJ, Ledbetter MT (1980) The Los Chocoyos Ash, Guatemala: a major stratigraphic marker in Middle America and in three ocean basins. Quat Res 13(3):327–345

Druitt TH (1985) Vent evolution and lag breccia formation during the Cape Riva eruption of Santorini, Greece. J Geol 93(4):439–454

Druitt TH (2014) New insights into the initiation and venting of the Bronze-Age eruption of Santorini (Greece), from component analysis. Bull Volcanol 76(2):794

Druitt TH, Edwards L, Mellors RM, Pyle DM, Sparks RSJ, Lanphere M, Davies M, Barreirio B (1999) Santorini volcano. Geol Soc Lond Mem 19

Druitt TH, McCoy FW, Vougioukalakis GE (2019) The Late bronze age eruption of Santorini volcano and its impact on the ancient mediterranean world. Elements 15(3):185–190

Druitt TH, Young SR, Baptie B, Bonadonna C, Calder ES, Clarke AB, Cole PD, Harford CL, Herd RA, Luckett R (2002) Episodes of cyclic Vulcanian explosive activity with fountain collapse at Soufrière Hills Volcano, Montserrat. Geol Soc Lond Mem 21:281–306

Duffield WA, Bacon CR, Roquemore GR (1979) Origin of reverse-graded bedding in air-fall pumice, Coso Range, California. J Volcanol Geother Res 5(1–2):35–48

Duffield WA, Christiansen RL, Koyanagi RY, Peterson DW (1982) Storage, migration, and eruption of magma at Kilauea volcano, Hawaii, 1971–1972. J Volcanol Geother Res 13(3–4):273–307

Duffield WA, Gibson EK, Heiken G (1977) Some characteristics of Pele's hair. J Res US Geol Surv 5(1):93-101

Durant AJ, Rose WI, Sarna-Wojcicki AM, Carey S, Volentik AC (2009) Hydrometeor-enhanced tephra sedimentation from the 18 May 1980 Mount St. Helens (USA) volcanic cloud. J Geophys Res 114:B03204

Edmonds M, Herd RA (2005) Inland-directed base surge generated by the explosive interaction of pyroclastic flows and seawater at Soufriere Hills volcano, Monsterrat. Geology 33(4):245–248

Edwards BR, Gudmundsson MT, Russell JK (2015) Glaciovolcanism. In: Sigurdsson H, Houghton B, Mcnutt SR, Rymer H, Stix J (eds.) The encyclopedia of volcanoes. Academic Press Elsevier, pp 377–394

Embley RW, Chadwick Jr WW, Baker ET, Butterfield DA, Resing JA, De Ronde CE, Tunnicliffe V, Lupton JE, Juniper SK, Rubin KH, Stern RJ (2006) Long-term eruptive activity at a submarine arc volcano. Nature 441(7092):494–497

Engwell SL, Sparks RSJ, Aspinall WP (2013) Quantifying uncertainties in the measurement of tephra fall thickness. J Appl Volcanol 2(1):1

Engwell SL, Sparks RSJ, Carey S (2014) Physical characteristics of tephra layers in the deep sea realm: the Campanian Ignimbrite eruption. Geol Soc Lond Spec Pub 398(1):47–64

Ernst GGJ, Sparks RSJ, Carey SN, Bursik MI (1996) Sedimentation from turbulent jets and plumes. J Geophys Res: Solid Earth 101 (B3):5575–5589

Ernst RE, Youbi N (2017) How Large Igneous Provinces affect global climate, sometimes cause mass extinctions, and represent natural markers in the geological record. Palaeogeog Palaeoclimat Palaeoecol 478:30–52

Eychenne J, Le Pennec J-L, Troncoso L, Gouhier M, Nedelec J-M (2012) Causes and consequences of bimodal grain-size distribution of tephra fall deposited during the August 2006 Tungurahua eruption (Ecuador). Bull Volcanol 74(1):187–205

Eychenne J, Cashman KV, Rust A, Durant A (2015) Impact of the lateral blast on the spatial pattern and grain size characteristics of the 18 May 1980 Mount St. Helens fallout deposit. J Geophys Res: Solid Earth 120(9):6018–6038

Fagents SA, Wilson L (1993) Explosive volcanic eruptions—VII. The ranges of pyroclasts ejected in transient volcanic explosions. Geophys J Int 113(2):359–370

Fauria KE, Manga M, Wei Z (2017) Trapped bubbles keep pumice afloat and gas diffusion makes pumice sink. Earth Planet Sci Lett 460:50–59

Fierstein J, Houghton BF, Wilson CJ, Hildreth W (1997) Complexities of plinian fall deposition at vent: an example from the 1912 Novarupta eruption (Alaska). J Volcanol Geotherm Res 76(3–4):215–227

Fierstein J, Nathenson M (1992) Another look at the calculation of fallout tephra volumes. Bull Volcanol 54(2):156–167

Fisher RV (1965) Settling velocity of glass shards. In: Deep sea research and oceanographic abstracts. Elsevier, pp 345IN7347–7346IN8353

Fiske RS, Cashman KV, Shibata A, Watanabe K (1998) Tephra dispersal from Myojinsho, Japan, during its shallow submarine eruption of 1952–1953. Bull Volcanol 59(4):262–275

Fitzgerald RH, Tsunematsu K, Kennedy BM, Breard ECP, Lube G, Wilson TM, Jolly AD, Pawson J, Rosenberg MD, Cronin SJ (2014) The application of a calibrated 3D ballistic trajectory model to ballistic hazard assessments at Upper Te Maari, Tongariro. J Volcanol Geotherm Res 286:248–262

Folch A, Mingari L, Gutierrez N, Hanzich M, Macedonio G, Costa A (2020) FALL3D-8.0: a computational model for atmospheric transport and deposition of particles, aerosols and radionuclides. Part I: model physics and numerics. Geosci Model Developmt 13:1431–1458

Folch A, Mingari LA, Prata A (2022) Ensemble-based forecast of volcanic clouds using FALL3D-8.1. Front Earth Sci:1304. https://doi.org/10.3389/feart.2021.741841

Folkes CB, Wright HM, Cas RA, de Silva SL, Lesti C, Viramonte JG (2011) A re-appraisal of the stratigraphy and volcanology of the Cerro Galán volcanic system, NW Argentina. Bull Volcanol 73 (10):1427–1454

Foshag WP, Gonzales GR (1956) Birth and development of Paricutin volcano, Mexico. US Geol Surv Bull 965-D

Francalanci L, Bertagnini A, Métrich N, Renzulli A, Vannucci R, Landi P, Del Moro S, Menna M, Petrone CM, Nardini I (2008) Mineralogical, geochemical, and isotopic characteristics of the ejecta from the 5 April 2003 paroxysm at Stromboli, Italy: Inferences on the pre-eruptive magma dynamics. Am Geophys Union Geophys Monog Ser 2008:331–345

Francis PW, Glaze LS, Pieri D, Oppenheimer CMM, Rothery DA (1990) Eruption terms. Nature 346:519

Froggatt PC (1982) Review of methods of estimating rhyolitic tephra volumes; applications to the Taupo Volcanic Zone, New Zealand. J Volcanol Geotherm Res 14(3–4):301–318

Fujibayashi N, Sakai U (2003) Vesiculation and eruption processes of submarine effusive and explosive rocks from the Middle Miocene Ogi Basalt, Sado Island, Japan. In: White JDL, Smellie JL, Clague DA (eds) Explosive subaqueous volcanism. AGU Mono 140, pp 259–272

Furukawa K, Kamata H (2004) Eruption and emplacement of the Yamakogawa Rhyolite in central Kyushu, Japan: a model for emplacement of rhyolitic spatter. Earth Planets Space 56(5): 517–524

Gardner JE, Hilton M, Carroll MR (1999) Experimental constraints on degassing of magma: isothermal bubble growth during continuous decompression from high pressure. Earth Planet Sci Lett 168(1–2):201–218

Gertisser R, Preece K, Keller J (2009) The Plinian Lower Pumice 2 eruption, Santorini, Greece: magma evolution and volatile behaviour. J Volcanol Geotherm Res 186(3–4):387–406

Geshi N, Németh K, Oikawa T (2011) Growth of phreatomagmatic explosion craters: a model inferred from Suoana crater in Miyakejima Volcano, Japan. J Volcanol Geotherm Res 201(1):30–38

Gibson P (2020) New Zealand Whakaari volcano eruption and occupational hygiene responses. Occup Health Sthn Africa 26 (1):29–29

Giordano G, Benedetti AADE, Diana A, Diano G, Esposito A, Fabbri M, Gaudioso F, Marasco F, Mazzini I, Miceli M (2010) Stratigraphy, volcano tectonics and evolution of the Colli Albani volcanic field. Geol Soc Lond Spec Pub IAVCEI 3:43–98

Giordano G, Cas RAF (2021) Classification of ignimbrites and their eruptions. Earth-Sci Rev 220:103697

Giordano G, De Astis G (2021) The summer 2019 basaltic Vulcanian eruptions (paroxysms) of Stromboli. Bull Volcanol 83(1):1–27. https://doi.org/10.1007/s00445-020-01423-2

Giordano G, De Rita D, Cas RAF, Rodani S (2002) Valley pond and ignimbrite veneer deposits in the small-volume phreatomagmatic 'Peperino Albano'basic ignimbrite, Lago Albano maar, Colli Albani volcano, Italy: influence of topography. J Volcanol Geotherm Res 118(1–2):131–144

Giordano G, Dobran F (1994) Computer simulations of the Tuscolano Artemisio's second pyroclastic flow unit (Alban Hills, Latium, Italy). J Volcanol Geother Res 61(1–2):69–94

Giordano G, Lucci F, Phillips D, Cozzupoli D, Runci V (2012) Stratigraphy, geochronology and evolution of the Mt. Melbourne volcanic field (North Victoria Land, Antarctica). Bull Volcanol 74 (9):1985–2005

Giordano G, Porreca M, Musacchio P, Mattei M (2008) The Holocene Secche di Lazzaro phreatomagmatic succession (Stromboli, Italy): evidence of pyroclastic density current origin deduced by facies analysis and AMS flow directions. Bull Volcanol 70(10):1221–1236

Giordano G, Zanella E, Trolese M, Baffioni C, Vona A, Caricchi C, De Benedetti AA, Corrado S, Romano C, Sulpizio R, Geshi N (2018) Thermal interactions of the AD79 Vesuvius pyroclastic density currents and their deposits at Villa dei Papiri (Herculaneum archaeological site, Italy). Earth Planet Sci Lett 490:180–192

Girault F, Carazzo G, Tait S, Ferrucci F, Kaminski É (2014) The effect of total grain-size distribution on the dynamics of turbulent volcanic plumes. Earth Planet Sci Lett 394:124–134

Gonnermann HM, Manga M (2013) Dynamics of magma ascent in the volcanic conduit. In: Rust A, Gregg TKP, Lopes R (eds) Modeling Volcanic processes: the physics and mathematics of volcanism. Cambridge Univ Press, pp 55–84

Goto A, Taniguchi H, Yoshida M, Ohba T, Oshima H (2001) Effects of explosion energy and depth to the formation of blast wave and crater: field explosion experiment for the understanding of volcanic explosion. Geophys Res Lett 28(22):4287–4290

Gottsmann J, Dingwell DB (2002) The thermal history of a spatter-fed lava flow: the 8-ka pantellerite flow of Mayor Island, New Zealand. Bull Volcanol 64:410–422

Gouhier M, Harris A, Calvari S, Labazuy P, Guéhenneux Y, Donnadieu F, Valade S (2012) Lava discharge during Etna's January 2011 fire fountain tracked using MSG-SEVIRI. Bull Volcanol 74(4):787–793

Gudmundsson MT, Sigmundsson F, Björnsson H (1997) Ice–volcano interaction of the 1996 Gjálp subglacial eruption, Vatnajökull, Iceland. Nature 389(6654):954–957

Gudmundsson MT, Thordarson T, Höskuldsson Á, Larsen G, Björnsson H, Prata FJ, Oddsson B, Magnússon E, Högnadóttir T, Petersen GN, Hayward CL, Stevenson JA, Jónsdóttir I (2012) Ash generation and distribution from the April-May 2010 eruption of Eyjafjallajökull, Iceland. Sci Repts 2(1):572

Gupta AK, Bennartz R, Fauria KE, Mittal T (2022) Eruption chronology of the December 2021 to January 2022 Hunga Tonga-Hunga Ha'apai eruption sequence. Comm Earth Environ 3(1):314

Gurioli L, Andronico D, Bachèlery P, Balcone-Boissard H, Battaglia J, Boudon G, Burgisser A, Burton MR, Cashman K, Cichy S, Cioni R, Di Muro A, Dominguez L, D'Oriano C, Druitt T, Harris AJL, Hort M, Kelfoun K, Komorowski JC, Kueppers U, Le Pennec JL, Menand T, Paris R, Pioli L, Pistolesi M, Polacci M, Pompilio M, Ripepe M, Roche O, Rose-Koga E, Rust A, Schiavi F, Scharff L, Sulpizio R, Taddeucci J, Thordarson T (2015) MeMoVolc consensual document: a review of cross-disciplinary approaches to characterizing small explosive magmatic eruptions. Bull Volcanol 77:1–33

Gurioli L, Harris AJL, Houghton BF, Polacci M, Ripepe M (2008) Textural and geophysical characterization of explosive basaltic activity at Villarrica volcano. J Geophys Res: Solid Earth 113(B8)

Hackett WR, Houghton BF (1985) Pinnacle ridge member, whakapapa formation: a welded airfall deposit from Ruapehu volcano, Taupo Volcanic Zone. NZ Geol Surv Rec 8:24–29

Hajimirza S, Jones TJ, Moreland WM, Gonnermann HM, Thordarson T (2022) Quantifying the water-to-melt mass ratio and its impact on eruption plumes during explosive hydromagmatic eruptions. Geoche Geophys Geosyst 23(5):e2021GC010160

Hamilton CW, Thordarson T, Fagents SA (2010) Explosive lava–water interactions I: architecture and emplacement chronology of volcanic rootless cone groups in the 1783–1784 Laki lava flow, Iceland. Bull Volcanol 72(4):449–467

Harris A, Ripepe M (2007) Temperature and dynamics of degassing at Stromboli. J Geophys Res: Solid Earth 112(B3)

Hay RL (1959) Formation of the crystal-rich glowing avalanche deposits of St. Vincent, BWI. J Geol: 540–562

Hays JD, Ninkovich D (1970) North Pacific deep-sea ash chronology and age of present Aleutian underthrusting. Geol Soc Am Mem 126:263–290

Head III JW, Wilson L (2003) Deep submarine pyroclastic eruptions: theory and predicted landforms and deposits. J Volcanol Geother Res 121(3–4):155–193

Heiken G, Wohletz K (1991) Fragmentation processes in explosive volcanic eruptions. In: Fisher RV, Smith GA (eds) Sedimentation in volcanic settings, vol 45. SEPM Special Publication SEPM Soc Econ Pal; Mineral, pp 19–26

Herzog M, Graf H-F, Textor C, Oberhuber JM (1998) The effect of phase changes of water on the development of volcanic plumes. J Volcanol Geotherm Res 87(1):55–74

Herzog M, Graf HF (2010) Applying the three-dimensional model ATHAM to volcanic plumes: dynamic of large co-ignimbrite eruptions and associated injection heights for volcanic gases. Geophys Res Lett 37(19)

Hewitt GF (1982) Chapter 2 of handbook of multiphase systems. McGraw-Hill, New York

Hildreth W (1987) New perspectives on the eruption of 1912 in the valley of ten thousand smokes, Katmai National Park, Alaska. Bull Volcanol 49(5):680–693

Hildreth W, Fierstein J (2012) The Novarupta-Katmai eruption of 1912: largest eruption of the twentieth century: centennial perspectives. US Geol Surv Prof Pap 1791:259p

Holasek RE, Self S, Woods AW (1996) Satellite observations and interpretation of the 1991 Mount Pinatubo eruption plumes. J Geophys Res 101:27635–27656

Houghton B, Carey RJ (2015) Pyroclastic fall deposits. In: Sigurdsson H, Houghton B, McNutt SR, Rymer H, Stix J (eds) The encyclopedia of volcanoes. Academic Press Elsevier, Cambridge, pp 599–616

Houghton BF, Carey RJ, Cashman KV, Wilson CJN, Hobden BJ, Hammer JE (2010) Diverse patterns of ascent, degassing, and eruption of rhyolite magma during the 1.8 ka Taupo eruption, New Zealand: Evidence from clast vesicularity. J Volcanol Geotherm Res 195(1):31–47

Houghton BF, Carey RJ, Rosenberg MD (2014) The 1800a Taupo eruption: "III wind" blows the ultraplinian type event down to Plinian. Geology 42(5):459–461

Houghton BF, Gonnermann HM (2008) Basaltic explosive volcanism: constraints from deposits and models. Chem Erde-Geochem 68 (2):117–140

Houghton BF, Hackett WR (1984) Strombolian and phreatomagmatic deposits of Ohakune Craters, Ruapehu, New Zealand: a complex interaction between external water and rising basaltic magma. J Volcanol Geother Res 21(3–4):207–231

Houghton BF, Swanson DA, Rausch J, Carey RJ, Fagents SA, Orr TR (2013) Pushing the Volcanic Explosivity Index to its limit and beyond: constraints from exceptionally weak explosive eruptions at Kīlauea in 2008. Geology 41(6):627–630

Houghton BF, Taddeucci J, Andronico D, Gonnermann HM, Pistolesi M, Patrick MR, Orr TR, Swanson DA, Edmonds M, Gaudin D, Carey RJ, Scarlato P (2016) Stronger or longer: discriminating between Hawaiian and Strombolian eruption styles. Geology 44(2):163–166

Houghton BF, Wilson CJN (1989) A vesicularity index for pyroclastic deposits. Bull Volcanol 51:451–462

Houghton BF, Wilson CJN, Del Carlo P, Coltelli M, Sable JE, Carey R (2004a) The influence of conduit processes on changes in style of basaltic Plinian eruptions: Tarawera 1886 and Etna 122 BC. J Volcanol Geotherm Res 137(1):1–14

Houghton BF, Wilson CJN, Fierstein J, Hildreth W (2004b) Complex proximal deposition during the Plinian eruptions of 1912 at Novarupta, Alaska. Bull Volcanol 66(2):95–133

Houghton BF, Wilson CJN, Smith RT, Gilbert JS (2000) Phreatoplinian eruptions. In: Sigurdsson H, Stix J, Houghton BF, McNutt SR, Rymer H (eds) Encyclopedia of volcanoes. Academic Press Elsevier, Cambridge, pp 513–525

Houghton B, White JD, Van Eaton AR (2015) Phreatomagmatic and related eruption styles. Chapter 30 In: Sigurdsson H, Houghton B, McNutt SR, Rymer H, Stix J (eds) The encyclopedia of volcanoes. Academic Press Elsevier, Cambridge, pp 537–552

Houghton BF, Wilson CJN, Weaver SD (1985) The Ruru Pass tephra, a peralkaline welded airfall tuff from Mayor Island. NZ Geol Surv Rec 8:30–36

Huang TC, Watkins ND, Shaw DM (1975) Atmospherically transported volcanic glass in deep-sea sediments: volcanism in sub-Antarctic latitudes of the South Pacific during Late Pliocene and Pleistocene time. Geol Soc Am Bull 86(9):1305–1315

Huebsch M, Kueppers U, Carazzo G, Cimarelli C, Hess KU, Hornby A, Lejeune AM, Michaud-Dubuy A, Perugini D, Weller D, Heap MJ, Dingwell D (2023) Fragmentation behavior of young pyroclasts from Mt. Pelée, Martinique. Bull Volcanol 85(11):1–9

Iguchi M, Tameguri T, Ohta Y, Ueki S, Nakao S (2013) Characteristics of volcanic activity at Sakurajima volcano's Showa crater during the period 2006 to 2011. Bull Volcanol Soc Japan 58:115–135

Iriyama Y, Toramaru A, Yamamoto T (2018) Theory for deducing volcanic activity from size distributions in plinian pyroclastic fall deposits. J Geophys Res: Solid Earth 123(3):2199–2213

Isaia R, Vitale S, Di Giuseppe MG, Iannuzzi E, Tramparulo FD, Troiano A (2015) Stratigraphy, structure, and volcano-tectonic evolution of Solfatara maar-diatreme (Campi Flegrei, Italy). Geol Soc Am Bull 127(9–10):1485–1504

Janebo MH, Thordarson T, Houghton BF, Bonadonna C, Larsen G, Carey RJ (2016) Dispersal of key subplinian–Plinian tephras from Hekla volcano, Iceland: implications for eruption source parameters. Bull Volcanol 78(10):66

Jordan SC, Cas RAF, Hayman PC (2013) The origin of a large (> 3 km) maar volcano by coalescence of multiple shallow craters: Lake Purrumbete maar, southeastern Australia. J Volcanol Geother Res 254:5–22

Jude-Eton TC, Thordarson T, Gudmundsson MT, Oddsson B (2012) Dynamics, stratigraphy and proximal dispersal of supraglacial tephra during the ice-confined 2004 eruption at Grímsvötn Volcano, Iceland. Bull Volcanol 74(5):1057–1082

Jutzeler M, Marsh R, Carey RJ, White JDL, Talling PJ, Karlstrom L (2014) On the fate of pumice rafts formed during the 2012 Havre submarine eruption. Nat Commun 5

Jutzeler M, Manga M, White JDL, Talling PJ, Proussevitch AA, Watt SFL, Cassidy M, Taylor RN, Le Friant A, Ishizuka O (2017) Submarine deposits from pumiceous pyroclastic density currents traveling over water: an outstanding example from offshore Montserrat (IODP 340). Geol Soc Am Bull 129(3–4):392–414

Jutzeler M, White JDL, Proussevitch AA, Gordee SM (2016) Vesiculation and fragmentation history in a submarine scoria cone-forming eruption, an example from Nishiizu (Izu Peninsula, Japan). Bull Volcanol 78(2):1–25

Kaminski E, Jaupart C (1998) The size distribution of pyroclasts and the fragmentation sequence in explosive volcanic eruptions. J Geophys Res: Solid Earth 103(B12):29759–29779

Kano K (2003) Subaqueous pumice eruptions and their products: a review. In: White JD, Smellie JL, Clague DA (eds) Explosive subaqueous volcanism. AGU Mono 140:213–229

Kano K, Yamamoto T, Ono K (1996) Subaqueous eruption and emplacement of the Shinjima Pumice, Shinjima (Moeshima) Island, Kagoshima Bay, SW Japan. J Volcanol Geotherm Res 71(2):187–206

Kienle J, Kyle PR, Self S, Motyka RJ, Lorenz V (1980) Ukinrek Maars, Alaska, I. April 1977 eruption sequence, petrology and tectonic setting. J Volcanol Geotherm Res 7(1):11–37

Kilgour G, Gates S, Kennedy B, Farquhar A, McSporran A, Asher C (2019) Phreatic eruption dynamics derived from deposit analysis: a case study from a small, phreatic eruption from Whakāri/White Island, New Zealand. Earth, Planets, Space 71(1):36

Kilgour G, Manville V, Della Pasqua F, Graettinger AH, Hodgson KA, Jolly GE (2010) The 25 September 2007 eruption of Mount Ruapehu, New Zealand: directed ballistics, surtseyan jets, and ice-slurry lahars. J Volcanol Geotherm Res 191(1):1–14

Klug C, Cashman KV (1994) Vesiculation of May 18, 1980, Mount St. Helens magma. Geology 22(5):468–472

Klug C, Cashman KV (1996) Permeability development in vesiculating magmas: implications for fragmentation. Bull Volcanol 58(2–3):87–100

Klug C, Cashman KV, Bacon CR (2002) Structure and physical characteristics of pumice from the climactic eruption of Mount Mazama (Crater Lake), Oregon. Bull Volcanol 64(7):486–501

Kokelaar BP (1983) The mechanism of Surtseyan volcanism. J Geol Soc 140(6):939–944

Koyaguchi T, Ochiai K, Suzuki YJ (2009) The effect of intensity of turbulence in umbrella cloud on tephra dispersion during explosive volcanic eruptions: Experimental and numerical approaches. J Volcanol Geotherm Res 186(1–2):68–78

Koyaguchi T, Ohno M (2001) Reconstruction of eruption column dynamics on the basis of grain size of tephra fall deposits: 1. Methods. J Geophys Res: Solid Earth 106(B4): 6499–6512

Koyaguchi T, Suzuki YJ, Kozono T (2010) Effects of the crater on eruption column dynamics. J Geophys Res: Solid Earth 115(B7)

Koyaguchi T, Woods AW (1996) On the formation of eruption columns following explosive mixing of magma and surface-water. J Geophys Res: Solid Earth 101(B3):5561–5574

Kueppers U, Scheu B, Spieler O, Dingwell DB (2006) Fragmentation efficiency of explosive volcanic eruptions: A study of experimentally generated pyroclasts. J Volcanol Geother Res 153(1–2):125–135

Kutschera W (2020) On the enigma of dating the Minoan eruption of Santorini. Proc Nat Acad Sci 117(16):8677–8679

Kutterolf S, Schindlbeck JC, Anselmetti FS, Ariztegui D, Brenner M, Curtis J, Schmid D, Hodell DA, Mueller A, Pérez L, Pérez W (2016) A 400-ka tephrochronological framework for Central America from Lake Petén Itzá (Guatemala) sediments. Quat Sci Rev 150:200–220

Larsson W (1937) Vulkanische Asche vom Ausbruch des Chilenischen Vulkans Quizapú (1932) in Argentina gesammelt. Bull Geol Inst Uppsala 26:27–52

Latutrie B, Ross PS (2020) Phreatomagmatic vs magmatic eruptive styles in maar-diatremes: a case study at Twin Peaks, Hopi Buttes volcanic field, Navajo Nation, Arizona. Bull Volcanol 82(3):1–25

Lautze NC, Houghton BF (2005) Physical mingling of magma and complex eruption dynamics in the shallow conduit at Stromboli volcano, Italy. Geology 33(5):425–428

Ledbetter MT (1985) Tephrochronology of marine tephra adjacent to Central America. Geol Soc Am Bull 96(1):77–82

Ledbetter MT, Sparks RSJ (1979) Duration of large-magnitude explosive eruptions deduced from graded bedding in deep-sea ash layers. Geology 7(5):240–244

Legros F (2000) Minimum volume of a tephra fallout deposit estimated from a single isopach. J Volcanol Geotherm Res 96(1):25–32

Leonard GS, Begg JG, Wilson CJN (2010) Geology of the Rotorua area: scale 1:250,000. Lower Hutt. In: Inst Geol & Nuc Sci Ltd.

Lipman PW (1967) Mineral and chemical variations within an ash-flow sheet from Aso caldera, southwestern Japan. Contrib Mineral Petrol 16(4):300–327

Lirer L, Pescatore T, Booth B, Walker GPL (1973) Two plinian pumice-fall deposits from Somma-Vesuvius, Italy. Geol Soc Am Bull 84(3):759–772

Lowe J, Barton N, Blockley S, Ramsey CB, Cullen VL, Davies W, Gamble C, Grant K, Hardiman M, Housley R (2012) Volcanic ash layers illuminate the resilience of Neanderthals and early modern humans to natural hazards. Proc Natl Acad Sci 109(34):13532–13537

MacDonald GA (1972) Volcanoes. A discussion of volcanoes, volcanic products, and volcanic phenomena. Prentice-Hall, 510 pp

Macdonald GA, Eaton JP (1984) Hawaiian volcanoes during US Prof Pap 1955, 1964p

Machida H, Arai F (1983) Extensive ash falls in and around the Sea of Japan from large late Quaternary eruptions. J Volcanol Geotherm Res 18(1):151–164

Mader HM (1998) Conduit flow and fragmentation. Geol Soc Lond Spec Pub 145(1):51–71

Maeda Y, Kato A, Yamanaka Y (2017) Modeling the dynamics of a phreatic eruption based on a tilt observation: Barrier breakage leading to the 2014 eruption of Mount Ontake, Japan. J Geophys Res: Solid Earth 122(2):1007–1024

Maeno F, Nakada S, Nagai M, Kozono T (2013) Ballistic ejecta and eruption condition of the vulcanian explosion of Shinmoedake volcano, Kyushu, Japan on 1 February, 2011. Earth Planets Space 65(6):609–621

Maeno F, Nakada S, Oikawa T, Yoshimoto M, Komori J, Ishizuka Y, Takeshita Y, Shimano T, Kaneko T, Nagai M (2016) Reconstruction of a phreatic eruption on 27 September 2014 at Ontake volcano, central Japan, based on proximal pyroclastic density current and fallout deposits. Earth, Planets, Space 68(1):82

Magnússon E, Gudmundsson MT, Roberts MJ, Sigurðsson G, Höskuldsson F, Oddsson B (2012) Ice-volcano interactions during the 2010 Eyjafjallajökull eruption, as revealed by airborne imaging radar. J Geophys Res: Solid Earth 117(B7)

Manga M, Fauria KE, Lin C, Mitchell SJ, Jones M, Conway CE, Degruyter W, Hosseini B, Carey R, Cahalan R (2018) The pumice raft-forming 2012 Havre submarine eruption was effusive. Earth Planet Sci Lett 489:49–58

Mangan MT, Cashman KV (1996) The structure of basaltic scoria and reticulite and inferences for vesiculation, foam formation, and fragmentation in lava fountains. J Volcanol Geotherm Res 73(1): 1–18

Mangan MT, Cashman KV, Swanson DA (2014) The dynamics of Hawaiian-style eruptions: a century of study: Chapter 8. In: Poland MP, Takahashi TJ, Landowski CM (eds) Characteristics of Hawaiian volcanoes. US Geol Sur, Washington

Manville V, Wilson CJN (2004) Vertical density currents: a review of their potential role in the deposition and interpretation of deep-sea ash layers. J Geol Soc 161(6):947–958

Manville V, White JDL, Houghton BF, Wilson CJN (1998) The saturation behaviour of pumice and some sedimentological implications. Sediment Geol 119(1):5–16

Marini L, Principe C, Chiodini G, Cioni R, Fytikas M, Marinelli G (1993) Hydrothermal eruptions of Nisyros (Dodecanese, Greece). Past events and present hazard. J Volcanol Geotherm Res 56(1–2):71–94

Marti A, Folch A, Costa A, Engwell S (2016) Reconstructing the plinian and co-ignimbrite sources of large volcanic eruptions: a novel approach for the Campanian Ignimbrite. Scientific reports 6

Martin U, Németh K (2006) How Strombolian is a "Strombolian" scoria cone? Some irregularities in scoria cone architecture from the Transmexican Volcanic Belt, near Volcán Ceboruco, (Mexico) and Al Haruj (Libya). J Volcanol Geotherm Res 155(1–2):104–118

Mason BG, Pyle DM, Oppenheimer C (2004) The size and frequency of the largest explosive eruptions on Earth. Bull Volcanol 66 (8):735–748

Massaro S, Costa A, Sulpizio R (2018) Evolution of the magma feeding system during a Plinian eruption: The case of Pomici di Avellino eruption of Somma-Vesuvius, Italy. Earth Planet Sci Letts 482:545–555

Mastin LG (2001) A simple calculator of ballistic trajectories for blocks ejected during volcanic eruptions. In: US Geol Surv Open-File Report 2001–45

Mastin LG, Christiansen RL, Thornber CR, Lowenstern JB, Beeson M (2004) What makes hydromagmatic eruptions violent? Some insights from the Keanakāko'i Ash, Kīlauea Volcano, Hawai'i. J Volcanol Geotherm Res 137(1):15–31

Mastin LG, Guffanti M, Servranckx R, Webley P, Barsotti S, Dean K, Durant A, Ewert JW, Neri A, Rose WI (2009) A multidisciplinary effort to assign realistic source parameters to models of volcanic ash-cloud transport and dispersion during eruptions. J Volcanol Geotherm Res 186(1):10–21

Mastin LG, Van Eaton AR (2020) Comparing simulations of umbrella-cloud growth and ash transport with observations from Pinatubo, Kelud, and Calbuco Volcanoes. Atmosphere 11:1038. https://doi.org/10.3390/atmos11101038

Matoza RS, Fee D, Assink JD, Iezzi AM, Green DN, Kim K, Toney L, Lecocq T, Krishnamoorthy S, Lalande JM, Nishida K et al (2022) Atmospheric waves and global seismoacoustic observations of the January 2022 Hunga eruption. Science, Tonga. https://doi.org/10.1126/science.abo7063(2022)

Mattox TN, Mangan MT (1997) Littoral hydrovolcanic explosions: a case study of lava–seawater interaction at Kilauea Volcano. J Volcanol Geotherm Res 75(1):1–17

Mattsson HB (2010) Textural variation in juvenile pyroclasts from an emergent, Surtseyan-type, volcanic eruption: the Capelas tuff cone, São Miguel (Azores). J Volcanol Geotherm Res 189(1):81–91

McGetchin TR, Ullrich GW (1973) Xenoliths in maars and diatremes with inferences for the Moon, Mars, and Venus. J Geophys Res 78 (11):1833–1853

Mele D, Costa A, Dellino P, Sulpizio R, Dioguardi F, Isaia R, Macedonio G (2020) Total grain size distribution of components of fallout deposits and implications for magma fragmentation mechanisms: examples from Campi Flegrei caldera (Italy). Bull Volcanol 82(4):1–12

Métrich N, Bertagnini A, Landi P, Rosi M, Belhadj O (2005) Triggering mechanism at the origin of paroxysms at Stromboli (Aeolian Archipelago, Italy): the 5 April 2003 eruption. Geophys Res Letts 32(10)

Métrich N, Bertagnini A, Pistolesi M (2021) Paroxysms at Stromboli volcano (Italy): source, genesis and dynamics. Front Earth Sci 9:593339. https://doi.org/10.3389/feart.2021.593339

Mitchell SJ, McIntosh IM, Houghton BF, Carey RJ, Shea T (2018) Dynamics of a powerful deep submarine eruption recorded in H2O contents and speciation in rhyolitic glass: the 2012 Havre eruption. Earth Planet Sci Lett 494:135–147

Miyaji N (2002) The 1707 eruption of Fuji volcano and its tephra. Glob Environ Res 6(2):37–40

Miyaji N, Kan'no A, Kanamaru T, Mannen K (2011) High-resolution reconstruction of the Hoei eruption (AD 1707) of Fuji volcano, Japan. J Volcanol Geotherm Res 207(3):113–129

Moore JG, Nakamura K, Alcaraz A (1966) The 1965 eruption of Taal Volcano. Science 151(3713):955–960

Moorhouse BL, White JDL, Scott JM (2015) Cape Wanbrow: a stack of surtseyan-style volcanoes built over millions of years in the Waiareka-Deborah volcanic field, New Zealand. J Volcanol Geotherm Res 298:27–46

Morgavi D, Valentini L, Porreca M, Zucchini A, Di Michele A, Ielpo M, Costa A, Rossi S, Perugini D (2019) Volcanic ash aggregation enhanced by seawater interaction: the case of the Secche di Lazzaro phreatomagmatic deposit (Stromboli). Ann Geophys 61(6 sup):673

Morton BR, Taylor G, Turner JS (1956a) Turbulent gravitational convection from maintained and instantaneous sources. Proc Royal Soc Lond A Math Phys Eng Sci 234(1196):1–23

Morton BR, Taylor G, Turner JS (1956b) Turbulent gravitational convection from maintained and instantaneous sources. In: Proc

Royal Soc Lond A Math Phys Eng Sci. The Royal Society, London pp 1–23

Moyer TC, Swanson DA (1987) Secondary hydroeruptions in pyroclastic-flow deposits: examples from Mount St. Helens. J Volcanol Geotherm Res 32(4):299–319

Mudde RF (2005) Gravity-driven bubbly flows. Ann Rev Fluid Mech 37:393–423

Mueller W, White JDL (1992) Felsic fire-fountaining beneath Archean seas: pyroclastic deposits of the 2730 Ma Hunter Mine Group, Quebec, Canada. J Volcanol Geotherm Res 54(1):117–134

Mueller SB, Kueppers U, Ayris PM, Jacob M, Dingwell DB (2016) Experimental volcanic ash aggregation: Internal structuring of accretionary lapilli and the role of liquid bonding. Earth Planet Sci Lett 433:232–240

Murch AP, White JD, Carey RJ (2019) Characteristics and deposit stratigraphy of submarine-erupted silicic ash, Havre volcano, Kermadec Arc, New Zealand. Front Earth Sci 7:1

Murrow PJ, Rose WI, Self S (1980) Determination of the total grain size distribution in a vulcanian eruption column, and its implications to stratospheric aerosol perturbation. Geophys Res Lett 7(11): 893–896

Nairn IA (1976) Atmospheric shock waves and condensation clouds from Ngauruhoe explosive eruptions. Nature 259:190–192

Nairn IA (1981) Some studies of the geology, volcanic history, and geothermal resources of the Okataina Volcanic Centre, Taupo Volcanic Zone, New Zealand. PhD Thesis, Victoria University

Nairn IA, Hedenquist JW, Villamor P, Berryman KR, Shane PA (2005) The ~ AD1315 Tarawera and Waiotapu eruptions, New Zealand: contemporaneous rhyolite and hydrothermal eruptions driven by an arrested basalt dike system? Bull Volcanol 67(2):186–193

Nairn IA, Self S (1978) Explosive eruptions and pyroclastic avalanches from Ngauruhoe in February 1975. J Volcanol Geother Res 3(1–2):39–60

Namiki A, Manga M (2008) Transition between fragmentation and permeable outgassing of low viscosity magmas. J Volcanol Geotherm Res 169(1):48–60

Nathenson M, Fierstein J (2015) Spread sheet to calculate tephra volume for exponential thinning. https://vhub.org/resources/3716

Németh K, Cronin SJ, Charley DT, Harrison MJ, Garae E (2006) Exploding lakes in Vanuatu-"Surtseyan-style" eruptions witnessed on Ambae Island. Episodes 29(2):87–93

Neri A, Dobran F (1994) Influence of eruption parameters on the thermofluid dynamics of collapsing volcanic columns. J Geophys Res: Solid Earth. 99(B6):11833–11857

Neri A, Esposti Ongaro T, Macedonio G, Gidaspow D (2003) Multiparticle simulation of collapsing volcanic columns and pyroclastic flow. J Geophys Res: Solid Earth 108(B4)

Newhall CG, Self S (1982) The volcanic explosivity index (VEI) an estimate of explosive magnitude for historical volcanism. J Geophys Res 87(C2):1231–1238

Newland EL, Mingotti N, Woods AW (2022) Dynamics of deep-submarine volcanic eruptions. Sci Rept 12(1):3276

Ninkovich D, Heezen BC (1965) Santorini tephra. In: Whittand WF, Bradshaw R (eds) Submarine geology and geophysics. Proceedings of the 17th symposium colston research society. Butterworths, London, pp 413–453

Ninkovich D, Opdyke N, Heezen BC, Foster JH (1966) Paleomagnetic stratigraphy, rates of deposition and tephrachronology in North Pacific deep-sea sediments. Earth Planet Sci Lett 1(6):476–492

Ninkovich D, Shackleton NJ (1975) Distribution, stratigraphic position and age of ash layer "L", in the Panama Basin region. Earth Planet Sci Lett 27(1):20–34

Ninkovich D, Sparks RSJ, Ledbetter MT (1978) The exceptional magnitude and intensity of the Toba eruption, Sumatra: an example of the use of deep-sea tephra layers as a geological tool. Bull Volcanol 41(3):286–298

Oppenheimer C (2003) Climatic, environmental and human consequences of the largest known historic eruption: Tambora volcano (Indonesia) 1815. Prog Phys Geogr 27(2):230–259

Oppenheimer J, Capponi A, Cashman KV, Lane SJ, Rust AC, James MR (2020) Analogue experiments on the rise of large bubbles through a solids-rich suspension: a "weak plug" model for Strombolian eruptions. Earth Planet Sci Lett 531:115931

Orr TR, Thelen WA, Patrick MR, Swanson DA, Wilson DC (2013) Explosive eruptions triggered by rockfalls at Kīlauea volcano, Hawai'i. Geology 41(2):207–210

Orsi G, D'Antonio M, de Vita S, Gallo G (1992) The Neapolitan Yellow Tuff, a large-magnitude trachytic phreatoplinian eruption: eruptive dynamics, magma withdrawal and caldera collapse. J Volcanol Geotherm Res 53(1):275–287

Orsi G, Sheridan MF (1984) The Green Tuff of Pantelleria: rheoignimbrite or rheomorphic fall? Bull Volcanol 47(3):611–626

Ort MH, Lefebvre NS, Neal CA, McConnell VS, Wohletz KH (2018) Linking the Ukinrek 1977 maar-eruption observations to the tephra deposits: new insights into maar depositional processes. J Volcanol Geotherm Res 360:36–60

Paladio-Melosantos MLO, Solidum RU, Scott WE, Quiambao RB, Umbal JV, Rodolfo KS, Tubianosa BS, Delos Reyes PJ, Alonso RA, Ruelo HB (1996) Tephra falls of the 1991 eruptions of Mount Pinatubo. In: Newhall CG, Punongbayan R (eds) Fire and mud: eruptions and lahars of Mount Pinatubo, Philippines. University of Washington Press, Seattle, pp 513–536

Papale P (1999) Strain-induced magma fragmentation in explosive eruptions. Nature 397(6718):425–428

Papale P, Rosi M (1993) A case of no-wind plinian fallout at Pululagua caldera (Ecuador): implications for models of clast dispersal. Bull Volcanol 55(7):523–535

Pardo N, Cronin SS, Németh K, Brenna M, Schipper CI, Breard E, White JDL, Procter JN, Stewart B, Agustín-Flores J (2014) Perils in distinguishing phreatic from phreatomagmatic ash; insights into the eruption mechanisms of the 6 August 2012 Mt. Tongariro eruption, New Zealand. J Volcanol Geotherm Res 286:397–414

Pardo N, Macias JL, Giordano G, Cianfarra P, Avellán DR, Bellatreccia F (2009) The ~ 1245 yr BP Asososca maar eruption: the youngest event along the Nejapa-Miraflores volcanic fault, Western Managua, Nicaragua. J Volcanol Geotherm Res 184(3):292–312

Pardo N, Cronin SJ, Palmer AS, Németh K (2012) Reconstructing the largest explosive eruptions of Mt. Ruapehu, New Zealand: lithostratigraphic tools to understand subplinian–plinian eruptions at andesitic volcanoes. Bull Volcanol 74(3):617–640

Parfitt EA (2004) A discussion of the mechanisms of explosive basaltic eruptions. J Volcanol Geotherm Res 134(1):77–107

Patrick MR, Harris AJL, Ripepe M, Dehn J, Rothery DA, Calvari S (2007) Strombolian explosive styles and source conditions: insights from thermal (FLIR) video. Bull Volcanol 69(7):769–784

Patel A, Manga M, Carey RJ, Degruyter W (2013) Effects of thermal quenching on mechanical properties of pyroclasts. J Volcanol Geother Res 258:24–30

Pedersen GB, Höskuldsson A, Dürig T, Thordarson T, Jonsdottir I, Riishuus MS, Óskarsson BV, Dumont S, Magnússon E, Gudmundsson MT, Sigmundsson F et al (2017) Lava field evolution and emplacement dynamics of the 2014–2015 basaltic fissure eruption at Holuhraun, Iceland. J Volcanol Geotherm Res 340:155–169

Pensa A, Cas R, Giordano G, Porreca M, Wallenstein N (2015) Transition from steady to unsteady Plinian eruption column: the VEI 5, 4.6 ka Fogo A Plinian eruption, São Miguel, Azores. J Volcanol Geotherm Res 305:1–8

Perez W, Freundt A, Kutterolf S, Schmincke H-U (2009) The Masaya Triple Layer: a 2100 year old basaltic multi-episodic Plinian

eruption from the Masaya Caldera Complex (Nicaragua). J Volcanol Geotherm Res 179(3):191–205
Perrotta A, Scarpati C (2003) Volume partition between the plinian and co-ignimbrite air fall deposits of the Campanian Ignimbrite eruption. Mineral Petrol 79(1–2):67–78
Perttu A, Taisne B, De Angelis S, Assink JD, Tailpied D, Williams RA (2020) Estimates of plume height from infrasound for regional volcano monitoring. J Volcanol Geotherm Res 402:106997
Pfeiffer T, Costa A, Macedonio G (2005) A model for the numerical simulation of tephra fall deposits. J Volcanol Geotherm Res 140 (4):273–294
Pichavant M, Di Carlo I, Pompilio M, Le Gall N (2022) Timescales and mechanisms of paroxysm initiation at Stromboli volcano, Aeolian Islands, Italy. Bull Volcanol 84(4):1–26
Pioli L, Bonadonna C, Azzopardi BJ, Phillips JC, Ripepe M (2012) Experimental constraints on the outgassing dynamics of basaltic magmas. J Geophys Res: Solid Earth 117(B3)
Pioli L, Erlund E, Johnson E, Cashman KV, Wallace P, Rosi M, Granados HD (2008) Explosive dynamics of violent Strombolian eruptions: the eruption of Parícutin Volcano 1943–1952 (Mexico). Earth Planet Sci Lett 271(1):359–368
Pioro I, Mokry S (2011) Thermophysical properties at critical and supercritical conditions. Heat Transfer-Theoretical Analysis, Experimental Investigations and Industrial Systems, 22, 654
Pistolesi M, Cioni R, Bonadonna C, Elissondo M, Baumann V, Bertagnini A, Chiari L, Gonzales R, Rosi M, Francalanci L (2015) Complex dynamics of small-moderate volcanic events: the example of the 2011 rhyolitic Cordón Caulle eruption, Chile. Bull Volcanol 77(1):1–24
Pistolesi M, Delle Donne D, Pioli L, Rosi M, Ripepe M (2011) The 15 March 2007 explosive crisis at Stromboli volcano, Italy: assessing physical parameters through a multidisciplinary approach. J Geophys Res: Solid Earth 116(B12)
Polacci M, Pioli L, Rosi M (2003) The Plinian phase of the Campanian Ignimbrite eruption (Phlegrean Fields, Italy): evidence from density measurements and textural characterization of pumice. Bull Volcanol 65(6):418–432
Pontbriand CW, Soule SA, Sohn RA, Humphris SE, Kunz C, Singh H, Nakamura KI, Jakobsson M, Shank T (2012) Effusive and explosive volcanism on the ultraslow-spreading Gakkel Ridge, 85 E. Geochem Geophys Geosyst 13(10):1–22
Prata AT, Folch A, Prata AJ, Biondi R, Brenot H, Cimarelli C, Corradini S, Lapierre J, Costa A (2020) Anak Krakatau triggers volcanic freezer in the upper troposphere. Sci Repts 10(1):1–3
Prata AT, Mingari L, Folch A, Macedonio G, Costa A (2021) FALL3D-8.0: a computational model for atmospheric transport and deposition of particles, aerosols and radionuclides–Part 2: model validation. Geosci Model Develop14(1):409–436
Proud SR, Prata AT, Schmauß S (2022) The January 2022 eruption of Hunga Tonga-Hunga Ha'apai volcano reached the mesosphere. Science 378(6619):554–557
Purkis SJ, Ward SN, Fitzpatrick NM, Garvin JB, Slayback D, Cronin SJ, Palaseanu-Lovejoy M, Dempsey A (2023) The 2022 Hunga-Tonga megatsunami: Near-field simulation of a once-in-a-century event. Science Adv 9(15):eadf5493
Pyle DM (1989) The thickness, volume and grainsize of tephra fall deposits. Bull Volcanol 51(1):1–15
Pyle DM (2000) Sizes of volcanic eruptions. In: The encyclopedia of volcanoes (pp. 263–270). Academic Press
Pyle DM (2015) Sizes of volcanic eruptions. In: Sigurdsson H, Houghton B, McNutt SR, Rymer H, Stix J (eds) The encyclopedia of volcanoes. Academic Press, pp 257–264
Pyle DM, Ricketts GD, Margari V, van Andel TH, Sinitsyn AA, Praslov ND, Lisitsyn S (2006) Wide dispersal and deposition of distal tephra during the Pleistocene 'Campanian Ignimbrite/Y5'eruption, Italy. Quat Sci Rev 25(21):2713–2728
Resing JA, Rubin KH, Embley RW, Lupton JE, Baker ET, Dziak RP, Baumberger T, Lilley MD, Huber JA, Shank TM, Butterfield DA et al (2011) Active submarine eruption of boninite in the northeastern Lau Basin. Nat Geosci 4(11):799–806
Reynolds MA, Best JG, Johnson RW (1980) 1953–57 eruption of Tuluman Volcano: rhyolitic volcanic activity in the Northern Bismarck Sea. Geol Surv Papua New Guinea Mem 7
Richter DH, Eaton JP, Murata KJ, Ault WU, Krivoy HL (1970) Chronological narrative of the 1959–60 eruption of Kilauea volcano, Hawaii. US Geol Surv Prof Pap No. 537
Riel B, Milillo P, Simons M, Lundgren P, Kanamori H, Samsonov S (2015) The collapse of Bárðarbunga caldera, Iceland. Geophys J Int 202(1):446–
Ripepe M, Harris AJ 2008) Dynamics of the 5 April 2003 explosive paroxysm observed at Stromboli by a near-vent thermal, seismic and infrasonic array. Geophysl Res Letts 35(7)
Ripepe M, Lacanna G, Pistolesi M, Silengo MC, Aiuppa A, Laiolo M, Massimetti F, Innocenti L, Della Schiava M, Bitetto M, La Monica FP, Nishimura T, Rosi M, Mangione D, Ricciardi A, Genco R, Coppola D, Marchetti E, Delle Donne D (2021) Ground deformation reveals the scale-invariant conduit dynamics driving explosive basaltic eruptions. Nat Commun 12:1683. https://doi.org/10.1038/s41467-021-21722-2
Ripepe M, Rossi M, Saccorotti G (1993) Image processing of explosive activity at Stromboli. J Volcanol Geotherm Res 54(3–4):335–351
Rittmann A (1931) Der ausbruch des Stromboli am 11 September 1930. Zeit Vulkanol 14:47–77
Robock A (2000) Volcanic eruptions and climate. Rev Geophys 38 (2):191–219
Rolandi G, Mastrolorenzo G, Barrella AM, Borrelli A (1993) The Avellino plinian eruption of Somma-Vesuvius (3760 yBP): the progressive evolution from magmatic to hydromagmatic style. J Volcanol Geotherm Res 58(1):67–88
Romano C, Vona A, Campagnola S, Giordano G, Arienzo I, Isaia R (2020) Modelling and physico-chemical constraints to the 4.5 ka Agnano-Monte Spina Plinian eruption (Campi Flegrei, Italy). Chem Geol 532:119301
Romero JE, Morgavi D, Arzilli F, Daga R, Caselli A, Reckziegel F, Viramonte J, Díaz-Alvarado J, Polacci M, Burton M, Perugini D (2016) Eruption dynamics of the 22–23 April 2015 Calbuco Volcano (Southern Chile): analyses of tephra fall deposits. J Volcanol Geotherm Res 317:15–29
Roobol MJ, Smith AL, Wright JV (1985) Dispersal and characteristics of pyroclastic fall deposits from Mt. Misery volcano, West Indies. Geol Rundsch 74(2):321–335
Rose WI (1993) Comment on 'another look at the calculation of fallout tephra volumes' by Judy Fierstein and Manuel Nathenson. Bull Volcanol 55(5):372–374
Rose WI, Anderson AT, Woodruff LG, Bonis SB (1978) The October 1974 basaltic tephra from Fuego volcano: description and history of the magma body. J Volcanol Geotherm Res 4(1–2):3–53
Rose WI, Chesner CA (1987) Dispersal of ash in the great Toba eruption, 75 ka. Geology 15(10):913–917
Rose WI, Durant AJ (2009) Fine ash content of explosive eruptions. J Volcanol Geotherm Res 186(1):32–39
Rose WI, Durant AJ (2011) Fate of volcanic ash: aggregation and fallout. Geology 39(9):895–896
Rose WI Jr (1972) Notes on the 1902 eruption of Santa Maria volcano, Guatemala. Bull Volcanol 36(1):29–45
Rose WI Jr, Bonis S, Stoiber RE, Keller M, Bickford T (1973) Studies of volcanic ash from two recent Central American eruptions. Bull Volcanol 37(3):338–364

Rosi M, Bertagnini A, Landi P (2000) Onset of the persistent activity at Stromboli volcano (Italy). Bull Volcanol 62(4–5):294–300

Rosi M, Bertagnini A, Harris AJL, Pioli L, Pistolesi M, Ripepe M (2006) A case history of paroxysmal explosion at Stromboli: timing and dynamics of the April 5, 2003 event. Earth Planet Sci Lett 243 (3):594–606

Rosi M, Paladio-Melosantos M, Di Muro A, Leoni R, Bacolcol T (2001) Fall vs flow activity during the 1991 climactic eruption of Pinatubo Volcano (Philippines). Bull Volcanol 62(8):549–566

Rosi M, Vezzoli L, Castelmenzano A, Grieco G (1999) Plinian pumice fall deposit of the Campanian Ignimbrite eruption (Phlegraean Fields, Italy). J Volcanol Geotherm Res 91(2):179–198

Rotella MD, Wilson CJN, Barker S, Wright IC (2013) Highly vesicular pumice generated by buoyant detachment of magma in subaqueous volcanism. Nat Geosci 6(2):129–132

Rowell CR, Jellinek AM, Hajimirza S, Aubry TJ (2022) External surface water influence on explosive eruption dynamics, with implications for stratospheric sulfur delivery and volcano-climate feedback. Front Earth Sci 10:788294

Rust AC, Cashman KV (2004) Permeability of vesicular silicic magma: inertial and hysteresis effects. Earth Planet Sci Lett 228(1–2): 93–107

Rust AC, Cashman KV (2011) Permeability controls on expansion and size distributions of pyroclasts. J Geophys Res: Solid Earth 116 (B11)

Saar MO, Manga M (1999) Permeability-porosity relationship in vesicular basalts. Geophys Res Lett 26(1):111–114

Sable JE, Houghton BF, Wilson CJN, Carey RJ (2006) Complex proximal sedimentation from Plinian plumes: the example of Tarawera 1886. Bull Volcanol 69(1):89–103

Sable JE, Houghton BF, Wilson CJN, Carey RJ (2009) Eruption mechanisms during the climax of the Tarawera 1886 basaltic Plinian eruption inferred from microtextural characteristics of the deposits. Stud Volcanol: the Legacy of George Walker (special Publications of IAVCEI) 2:129–154

Safarov J, Millero F, Feistel R, Heintz A, Hassel E (2009) Thermodynamic properties of standard seawater: extensions to high temperatures and pressures. Ocean Sci 5(3):235–246

Sarna-Wojcicki AM, Shipley S, Waitt Jr RB, Dzurisin D, Wood SH (1981) Areal distribution, thickness, mass, volume, and grain size of air-fall ash from the six major eruptions of 1980. In: Lipman PW, Mullineaux DR (eds) The 1980 eruptions of Mount St. Helens, Washington. US Geol Surv Prof Pap 1250, pp 577–600

Scarani A, Faranda CF, Vona A, Speranza F, Giordano G, Rotolo SG, Romano C (2023) Timescale of emplacement and rheomorphism of the Green Tuff ignimbrite (Pantelleria, Italy). J Geophys Research: Solid Earth 128(7):e2022JB026257

Scarpati C, Sparice D, Perrotta A (2014) A crystal concentration method for calculating ignimbrite volume from distal ash-fall deposits and a reappraisal of the magnitude of the Campanian Ignimbrite. J Volcanol Geotherm Res 280:67–75

Schipper CI, Castro JM, Tuffen H, James MR, How P (2013) Shallow vent architecture during hybrid explosive–effusive activity at Cordón Caulle (Chile, 2011–12): evidence from direct observations and pyroclast textures. J Volcanol Geotherm Res 262:25–37

Schipper CI, White JDL, Houghton BF, Shimizu N, Stewart RB (2010) "Poseidic" explosive eruptions at Loihi Seamount. Hawaii. Geology 38(4):291–294

Schmincke H-U (1967) Fused tuff and peperites in south-central Washington. Geol Soc Am Bull 78(3):319–330

Schmincke H-U (1977) Phreatomagmatische Phasen in quartären Vulkanen der Osteifel. Geol Jahrb (a) 39:3–45

Scollo S, Del Carlo P, Coltelli M (2007) Tephra fallout of 2001 Etna flank eruption: analysis of the deposit and plume dispersion. J Volcanol Geotherm Res 160(1):147–164

Scudder RP, Murray RW, Schindlbeck JC, Kutterolf S, Hauff F, Underwood MB, Gwizd S, Lauzon R, McKinley CC (2016) Geochemical approaches to the quantification of dispersed volcanic ash in marine sediment. Prog Earth Planet Sci 3(1):1

Seabrook S, Mackay K, Watson S, Clare M, Hunt J, Yeo I, Lane E, Clark M, Wysoczanski R, Rowden A, Hoffmann L, Armstrong E, Williams M (2023) Pyroclastic density currents explain far-reaching and diverse seafloor impacts of the 2022 Hunga Tonga Hunga Ha'apai eruption. Research Square Springer Nature, 20p (in press)

Self S (1976) The recent volcanology of Terceira, Azores. J Geol Soc 132(6):645–666

Self S, de Silva SL, Cortés JA (2008) Enigmatic clastogenic rhyolitic volcanism: The Corral de Coquena spatter ring, North Chile. J Volcanol Geotherm Res 177(4):812–821

Self S, Goff F, Gardner JN, Wright JV, Kite WM (1986) Explosive rhyolitic volcanism in the Jemez Mountains: vent locations, caldera development and relation to regional structure. J Geophys Res 91 (B2):1779–1798

Self S, Heiken G, Sykes ML, Wohletz K, Fisher RV, Dethier DP (1996) Part I: Field guide to the Bandelier Tuff and Valles caldera. Field excursions to the Jemez Mountains, New Mexico. New Mexico Bur Mines Min Res Bull 134:7–55

Self S, Kienle J, Huot J-P (1980) Ukinrek Maars, Alaska, II. Deposits and formation of the 1977 craters. J Volcanol Geotherm Res 7 (1):39–65

Self S, Rampino MR (1981) The 1883 eruption of Krakatau. Nature 294:699–704

Self S, Schmidt A, Mather TA (2014) Emplacement characteristics, time scales, and volcanic gas release rates of continental flood basalt eruptions on Earth. Geol Soc Am Spec Pap 505:319–337

Self S, Sparks RSJ (1978) Characteristics of widespread pyroclastic deposits formed by the interaction of silicic magma and water. Bull Volcanol 41(3):196–212

Self S, Sparks RSJ, Booth B, Walker GPL (1974) The 1973 Heimaey strombolian scoria deposit, Iceland. Geol Mag 111(6):539–548

Self S, Wilson L, Nairn IA (1979) Vulcanian eruption mechanisms. Nature 277(5696):440–443

Self S, Wolff JA, Wright JV (2021) Tsankawi pumice fall unit B, a very widespread and powerfully emplaced plinian deposit. In: AGU fall meeting, New Orleans. https://agu.confex.com/agu/fm21/meetingapp.cgi/Paper/911876

Settle M (1978) Volcanic eruption clouds and the thermal power output of explosive eruptions. J Volcanol Geotherm Res 3(3):309–324

Sharqawy MH, Lienhard JH, Zubair SM (2010) Thermophysical properties of seawater: a review of existing correlations and data. Desalin Water Treat 16(1–3):354–380

Shea T, Houghton BF, Gurioli L, Cashman KV, Hammer JE, Hobden BJ (2010) Textural studies of vesicles in volcanic rocks: an integrated methodology. J Volcanol Geotherm Res 190(3–4):271–289

Shepherd JB, Sigurdsson H (1982) Mechanism of the 1979 explosive eruption of Soufriere volcano, St. Vincent. J Volcanol Geotherm Res 13(1):119–130

Sheridan MF, Updike RG (1975) Sugarloaf Mountain Tephra—A Pleistocene rhyolitic deposit of base-surge origin in northern Arizona. Geol Soc Am Bull 86(4):571–581

Signorelli S, Vaggelli G, Francalanci L, Rosi M (1999) Origin of magmas feeding the Plinian phase of the Campanian Ignimbrite eruption, Phlegrean Fields (Italy): constraints based on matrix-glass and glass-inclusion compositions. J Volcanol Geotherm Res 91 (2):199–220

Sigurdsson H (1982) Tephra from the 1979 Soufriere explosive eruption. Science 216(4550):1106–1108

Sigurdsson H, Carey S (1989) Plinian and co-ignimbrite tephra fall from the. Bull Volcanol 51(4):243–270

Sigurdsson H, Sparks RSJ, Carey SN, Huang TC (1980) Volcanogenic sedimentation in the Lesser Antilles arc. J Geol:523–540

Simkin T (1994) Distant effects of volcanism-how big and how often. Science 264(5161):913–914

Simmons JM, Carey RJ, Cas RAF, Druitt TH (2017a) High magma decompression rates at the peak of a violent caldera-forming eruption (Lower Pumice 1 eruption, Santorini, Greece). Bull Volcanol 79(6):42

Simmons JM, Cas RA, Druitt TH, Carey RJ (2017b) The initiation and development of a caldera-forming Plinian eruption (172 ka Lower Pumice 2 eruption, Santorini, Greece). J Volcanol Geotherm Res 341:332–350

Simmons JM, Cas RAF, Druitt TH, Folkes CB (2016) Complex variations during a caldera-forming Plinian eruption, including precursor deposits, thick pumice fallout, co-ignimbrite breccias and climactic lag breccias: the 184ka Lower Pumice 1 eruption sequence, Santorini, Greece. J Volcanol Geotherm Res

Simons BC, Jolly AD, Eccles JD, Cronin SJ (2020) Spatiotemporal relationships between two closely-spaced Strombolian-style vents, Yasur, Vanuatu. Geophys Res Lett https://doi.org/10.1029/2019GL085687

Simpson K, McPhie J (2001) Fluidal-clast breccia generated by submarine fire fountaining, Trooper Creek Formation, Queensland, Australia. J Volcanol Geotherm Res 109(4):339–355

Smellie JL, Hole MJ (1997) Products and processes in Pliocene-Recent, subaqueous to emergent volcanism in the Antarctic Peninsula: examples of englacial Surtseyan volcano construction. Bull Volcanol 58(8):628–646

Smith RT, Houghton BF (1995) Vent migration and changing eruptive style during the 1800a Taupo eruption: new evidence from the Hatepe and Rotongaio phreatoplinian ashes. Bull Volcanol 57: 432–439

Smith TL, Batiza R (1989) New field and laboratory evidence for the origin of hyaloclastite flows on seamount summits. Bull Volcanol 51(2):96–114

Smith VC, Isaia R, Engwell SL, Albert PG (2016) Tephra dispersal during the Campanian Ignimbrite (Italy) eruption: implications for ultra-distal ash transport during the large caldera-forming eruption. Bull Volcanol 78:1–15

Sohn YK, Chough SK (1992) The Ilchulbong tuff cone, Cheju Island, South Korea. Sedimentology 39(4):523–544

Sohn RA, Willis C, Humphris S, Shank TM, Singh H, Edmonds HN, Kunz C, Hedman U, Helmke E, Jakuba M, Liljebladh B (2008) Explosive volcanism on the ultraslow-spreading Gakkel ridge, Arctic Ocean. Nature 453(7199):1236–1238

Sorem RK (1982) Volcanic ash clusters: tephra rafts and scavengers. J Volcanol Geotherm Res 13(1):63–71

Soriano C, Galindo I, Martí J, Wolff J (2006) Conduit-vent structures and related proximal deposits in the Las Cañadas caldera, Tenerife, Canary Islands. Bull Volcanol 69(2):217–231

Sparks RSJ (1978) The dynamics of bubble formation and growth in magmas: a review and analysis. J Volcanol Geotherm Res 3:1–37

Sparks RSJ (1986) The dimensions and dynamics of volcanic eruption columns. Bull Volcanol 48(1):3–15

Sparks RSJ, Barclay J, Jaupart C, Mader HM, Phillips JC (1994) Physical aspects of magmatic degassing I. Experimental and theoretical constraints on vesiculation. In: Carroll MR, Holloway JR (eds) Volatiles in magmas, pp 413–443

Sparks RSJ, Bursik MI, Ablay GJ, Thomas RME, Carey SN (1992) Sedimentation of tephra by volcanic plumes. Part 2: controls on thickness and grain-size variations of tephra fall deposits. Bull Volcanol 54(8):685–695

Sparks RSJ, Bursik MI, Carey SN, Gilbert J, Glaze LS, Sigurdsson H, Woods AW (1997) Volcanic plumes. John Wiley, New York

Sparks RSJ, Huang TC (1980) The volcanological significance of deep-sea ash layers associated with ignimbrites. Geol Mag 117 (05):425–436

Sparks RSJ, Walker GPL (1977) The significance of vitric-enriched air-fall ashes associated with crystal-enriched ignimbrites. J Volcanol Geotherm Res 2(4):329–341

Sparks RSJ, Wilson L (1976) A model for the formation of ignimbrite by gravitational column collapse. J Geol Soc 132(4):441–451

Sparks RSJ, Wilson L (1982) Explosive volcanic eruptions—V. Observations of plume dynamics during the 1979 Soufrière eruption, St. Vincent. Geophys J Int 69(2):551–570

Sparks RSJ, Wilson L, Hulme G (1978) Theoretical modeling of the generation, movement, and emplacement of pyroclastic flows by column collapse. J Geophys Res: Solid Earth 83(B4):1727–1739

Sparks RSJ, Wilson L, Sigurdsson H (1981) The pyroclastic deposits of the 1875 eruption of Askja, Iceland. Phil Trans Math Phys Eng Sci 299(1447):241–273

Sparks RSJ, Wright JV (1979) Welded airfall tuffs. In: Chapin, CE, Elston WE (eds) Ash-flow tuffs. Geol Soc Amer Spec Pap 180, pp 155–166

Spieler O, Kennedy B, Kueppers U, Dingwell DB, Scheu B, Taddeucci J (2004) The fragmentation threshold of pyroclastic rocks. Earth Planet Sci Lett 226(1):139–148

Stevenson RJ, Briggs RM, Hodder AP (1993) Emplacement history of a low-viscosity, fountain-fed pantelleritic lava flow. J Volcanol Geotherm Res 57(1–2):39–56

Stevenson RJ, Wilson L (1997) Physical volcanology and eruption dynamics of peralkaline agglutinates from Pantelleria. J Volcanol Geotherm Res 79(1–2):97–122

Stix J, Torres RC, Narváez L, Raigosa JA, Gómez D, Castonguay R (1997) A model of vulcanian eruptions at Galeras volcano, Colombia. J Volcanol Geotherm Res 77(1):285–303

Stow DAV, Lovell JPB (1979) Contourites: their recognition in modern and ancient sediments. Earth-Sci Rev 14(3):251–291

Sulpizio R (2005) Three empirical methods for the calculation of distal volume of tephra-fall deposits. J Volcanol Geotherm Res 145 (3):315–336

Sulpizio R, Cioni R, Di Vito MA, Mele D, Bonasia R, Dellino P (2010) The Pomici di Avellino eruption of Somma-Vesuvius (3.9 ka BP). Part I: stratigraphy, compositional variability and eruptive dynamics. Bull Volcanol 72(5):539–558

Sulpizio R, Mele D, Dellino P, La Volpe L (2005) A complex, Subplinian-type eruption from low-viscosity, phonolitic to tephri-phonolitic magma: the AD 472 (Pollena) eruption of Somma-Vesuvius, Italy. Bull Volcanol 67(8):743–767

Sumner JM, Blake S, Matela RJ, Wolff JA (2005) Spatter. J Volcanol Geotherm Res 142(1):49–65

Suthren RJ, Furnes H (1980) Origin of some bedded welded tuffs. Bull Volcanol 43(1):61–71

Suzuki T (1983) A theoretical model for dispersion of tephra. Arc Volcanism: Phys Tectonics 95:113

Suzuki T, Katsui Y, Nakamura T (1973) Size distribution of the Tarumai Ta-b pumice-fall deposit. Bull Volcanol Soc Japan 18:57–64

Suzuki YJ, Costa A, Cerminara M, Ongaro TE, Herzog M, Van Eaton AR, Denby LC (2016) Inter-comparison of three-dimensional models of volcanic plumes. J Volcanol Geotherm Res 326:26–42

Suzuki YJ, Koyaguchi T (2009) A three-dimensional numerical simulation of spreading umbrella clouds. J Geophys Res: Solid Earth 114(B3)

Suzuki YJ, Koyaguchi T (2015) Effects of wind on entrainment efficiency in volcanic plumes. J Geophys Res: Solid Earth 120(9): 6122–6140

Suzuki YJ, Koyaguchi T, Ogawa M, Hachisu I (2005) A numerical study of turbulent mixing in eruption clouds using a three-dimensional fluid dynamics model. J Geophys Res: Solid Earth 110(B8)

Taddeucci J, Edmonds M, Houghton B, James MR, Vergniolle S (2015) Hawaiian and strombolian eruptions. In: Sigurdsson H, Houghton B, McNutt SR, Rymer H, Stix J (eds) The encyclopedia of volcanoes. Academic Press Elsevier, Cambridge, pp 485–503

Taddeucci J, Scarlato P, Capponi A, Del Bello E, Cimarelli C, Palladino DM, Kueppers U (2012) High-speed imaging of strombolian explosions: the ejection velocity of pyroclasts. Geophys Res Lett 39(2)

Taddeucci J, Scarlato P, Montanaro C, Cimarelli C, Del Bello E, Freda C, Andronico D, Gudmundsson MT, Dingwell DB (2011) Aggregation-dominated ash settling from the Eyjafjallajökull volcanic cloud illuminated by field and laboratory high-speed imaging. Geology 39(9):891–894

Tait, MA (2004) Dynamics of a large volume plinian eruption: dispersal of the late Miocene Corte Blanco Tuff, Ramadas Volcanic Centre, Andes Mountains, Salta, Argentina. PhD Thesis, Monash University, 254p

Textor C, Graf H-F, Herzog M, Oberhuber JM, Rose WI, Ernst GGJ (2006a) Volcanic particle aggregation in explosive eruption columns. Part I: Parameterization of the microphysics of hydrometeors and ash. J Volcanol Geotherm Res 150(4):359–377

Textor C, Graf H-F, Herzog M, Oberhuber JM, Rose WI, Ernst GGJ (2006b) Volcanic particle aggregation in explosive eruption columns. Part II: Numerical experiments. J Volcanol Geotherm Res 150(4):378–394

Thivet S, Gurioli L, Di Muro A, Derrien A, Ferrazzini V, Gouhier M, Coppola D, Galle B, Arellano S (2020) Evidences of plug pressurization enhancing magma fragmentation during the September 2016 basaltic eruption at Piton de la Fournaise (La Réunion Island, France). Geochem Geophys Geosyst 21(2):e2019GC008611

Thomas RME, Sparks RSJ (1992) Cooling of tephra during fallout from eruption columns. Bull Volcanol 54(7):542–553

Thorarinsson S (1954) The tephra-fall from Hekla on March 29th 1947. HF Leiftur

Thorarinsson S (1968) On the rate of lava-and tephra production and the upward migration of magma in four Icelandic eruptions. Geol Rundsch 57(3):705–718

Thorarinsson S, Einarsson T, Sigvaldason G, Elisson G (1964) The submarine eruption off the Vestmann Islands 1963–64. Bull Volcanol 27(1):435–445

Thorarinsson S, Steinthorsson S, Einarsson T, Kristmannsdóttir H, Óskarsson N (1973) The eruption on Heimaey, Iceland. Nature 241 (5389):372–375

Ton-That T, Singer B, Paterne M (2001) 40Ar/39Ar dating of latest Pleistocene (41 ka) marine tephra in the Mediterranean Sea: implications for global climate records. Earth Planet Sci Letts 184 (3–4):645–658

Toramaru A (2006) BND (bubble number density) decompression rate meter for explosive volcanic eruptions. J Volcanol Geotherm Res 154(3–4):303–316

Trolese M, Cerminara M, Ongaro TE, Giordano G (2019) The footprint of column collapse regimes on pyroclastic flow temperatures and plume heights. Nat Commun 10(1):1–10

Tsunematsu K, Chopard B, Falcone J, Bonadonna C (2014) A numerical model of ballistic transport with collisions in a volcanic setting. Comput Geosci 63:62–69

Tsunematsu K, Ishimine Y, Kaneko T, Yoshimoto M, Fujii T, Yamaoka K (2016) Estimation of ballistic block landing energy during 2014 Mount Ontake eruption. Earth Planets Space 68:88

Tsuya H (1955) Geological and Petrological Studies of Volcano, Fuji, V.: 5. On the 1707 eruption of Volcano Fuji. Bull EQ Res Inst (tokyo) 33:341–382

Tupper A, Textor C, Herzog M, Graf HF, Richards MS (2009) Tall clouds from small eruptions: the sensitivity of eruption height and fine ash content to tropospheric instability. Nat Hazards 51(2): 375–401

Valentine GA (2012) Shallow plumbing systems for small-volume basaltic volcanoes, 2: evidence from crustal xenoliths at scoria cones and maars. J Volcanol Geotherm Res 223:47–63

Valentine GA, Gregg TKP (2008) Continental basaltic volcanoes—processes and problems. J Volcanol Geotherm Res 177(4):857–873

Valentine GA, Groves KR (1996) Entrainment of country rock during basaltic eruptions of the Lucero volcanic field, New Mexico. J Geol 71–90

Valentine GA, Krier D, Perry FV, Heiken G (2005) Scoria cone construction mechanisms, Lathrop Wells volcano, southern Nevada, USA. Geology 33(8):629–632

Valentine GA, White JDL, Ross P-S, Amin J, Taddeucci J, Sonder I, Johnson PJ (2012) Experimental craters formed by single and multiple buried explosions and implications for volcanic craters with emphasis on maars. Geophys Res Lett 39(20)

Valentine GA, Wohletz KH (1989) Numerical models of Plinian eruption columns and pyroclastic flows. J Geophys Res: Solid Earth 94(B2):1867–1887

Van Eaton AR, Muirhead JD, Wilson CJ, Cimarelli C (2012) Growth of volcanic ash aggregates in the presence of liquid water and ice: an experimental approach. Bull Volcanol 74(9):1963–1984

Van Eaton AR, Wilson CJN (2013) The nature, origins and distribution of ash aggregates in a large-scale wet eruption deposit: Oruanui, New Zealand. J Volcanol Geotherm Res 250:129–154

van Otterloo J, Cas RAF, Sheard MJ (2013) Eruption processes and deposit characteristics at the monogenetic Mt. Gambier maar complex, SE Australia: implications for alternating magmatic and phreatomagmatic activity. Bull Volcanol 75:737–758

Vanderkluysen L, Harris AJ, Kelfoun K, Bonadonna C, Ripepe M (2012) Bombs behaving badly: unexpected trajectories and cooling of volcanic projectiles. Bull Volcanol 74(8):1849–1858

Vaughan RG, Webley PW (2010) Satellite observations of a surtseyan eruption: Hunga Ha'apai, Tonga. J Volcanol Geotherm Res 198(1–2):177–186

Veitch G, Woods AW (2002) Particle recycling in volcanic plumes. Bull Volcanol 64(1):31–39

Vergniolle S, Jaupart C (1986) Separated two-phase flow and basaltic eruptions. J Geophys Res: Solid Earth 91(B12):12842–12860

Verwoerd WJ, Chevallier L (1987) Contrasting types of surtseyan tuff cones on Marion and Prince Edward islands, southwest Indian Ocean. Bull Volcanol 49(1):399–413

Vezzoli L, Apuani T, Corazzato C, Uttini A (2017) Geological and geotechnical characterization of the debris avalanche and pyroclastic deposits of Cotopaxi Volcano (Ecuador). A contribute to instability-related hazard studies. J Volcanol Geotherm Res 332:51–70

Vidal CM, Komorowski J-C, Métrich N, Pratomo I, Kartadinata N, Prambada O, Michel A, Carazzo G, Lavigne F, Rodysill J (2015) Dynamics of the major plinian eruption of Samalas in 1257 AD (Lombok, Indonesia). Bull Volcanol 77(9):1–24

Vinkler AP, Cashman K, Giordano G, Groppelli G (2012) Evolution of the mafic Villa Senni caldera-forming eruption at Colli Albani volcano, Italy, indicated by textural analysis of juvenile fragments. J Volcanol Geother Res 235:37–54

Viramonte J, Reynolds JH, Del Papa C, Disalvo A (1994) The Corte Blanco garnetiferous tuff: a distinctive late Miocene marker bed in northwestern Argentina applied to magnetic polarity stratigraphy in the Río Yacones, Salta Province. Earth Planet Sci Lett 121(3–4):519–531

Vona A, Romano C, Giordano G, Sulpizio R (2020) Linking magma texture, rheology and eruptive style during the 472 AD Pollena Subplinian eruption (Somma-Vesuvius). Lithos 370:105658

Waitt R, Dzurisin D (1981) Proximal air-fall deposits from the May 18 eruption-stratigraphy and field sedimentology. In: Lipman PW, Mullineaux DR (eds) The 1980 eruptions of Mount St. Helens, Washington. US Geol Surv Prof Pap 1250, pp 601–616

Walker GPL (1971) Grain-size characteristics of pyroclastic deposits. J Geol 79(6):696–714
Walker GPL (1972) Crystal concentration in ignimbrites. Contribs Mineral Petrol 36(2):135–146
Walker GPL (1973) Explosive volcanic eruptions—a new classification scheme. Geol Rundsch 62(2):431–446
Walker GPL (1980) The Taupo pumice: product of the most powerful known (ultraplinian) eruption? J Volcanol Geotherm Res 8(1): 69–94
Walker GPL (1981a) Characteristics of two phreatoplinian ashes, and their water-flushed origin. J Volcanol Geotherm Res 9(4):395–407
Walker GPL (1981b) New Zealand case histories of pyroclastic studies. In: Self S, Sparks RSJ (eds) Tephra studies. Springer, Netherlands, Dordrecht, pp 317–330
Walker GPL (1981c) The Waimihia and Hatepe plinian deposits from the rhyolitic Taupo Volcanic Centre. New Zeal J Geol Geophys 24 (3):305–324
Walker GPL (1981d) Generation and dispersal of fine ash and dust by volcanic eruptions. J Volcanol Geotherm Res 11(1):81–92
Walker GPL, Croasdale R (1971) Two Plinian-type eruptions in the Azores. J Geol Soc 127(1):17–55
Walker GPL, Self S, Wilson L (1984) Tarawera 1886, New Zealand—a basaltic plinian fissure eruption. J Volcanol Geotherm Res 21(1–2):61–78
Walker GPL, Wright JV, Clough BJ, Booth, B (1981) Pyroclastic geology of the rhyolitic volcano of La Primavera, Mexico. Geol Rundsch 70:1100–1118
Walker SL, Baker ET, Lupton JE, Resing JA (2019) Patterns of fine ash dispersal related to volcanic activity at west mata volcano, NE Lau Basin. Front Marine Sci 6:593
Walker SL, Baker ET, Resing JA, Chadwick WW, Lebon GT, Lupton JE, Merle SG (2008) Eruption-fed particle plumes and volcaniclastic deposits at a submarine volcano: NW Rota-1, Mariana Arc. J Geophys Res: Solid Earth 113(B8)
Wallis GB (1969) One dimensional two-phase flow. McGraw-Hill
Watanabe K, Ono K, Sakaguchi K, Takada A, Hoshizumi H (1999) Co-ignimbrite ash-fall deposits of the 1991 eruptions of Fugen-dake, Unzen Volcano, Japan. J Volcanol Geotherm Res 89 (1):95–112
Watkins ND, Huang TC (1977) Tephras in abyssal sediments east of the North Island, New Zealand: chronology, paleowind velocity, and paleoexplosivity. New Zeal J Geol Geophys 20(1):179–198
Watkins ND, Sparks RSJ, Sigurdsson H, Huang TC, Federman A, Carey S, Ninkovich D (1978) Volume and extent of the Minoan tephra from Santorini Volcano: new evidence from deep-sea sediment cores. Nature 271:122–126
Weisman J (1983) Two-phase flow patterns. Chapter 15. In: Cheremisinoff NP, Gupta R (eds) Handbook of fluids in motion. Ann Arbor Science Publishers
White JDL (1991) Maar-diatreme phreatomagmatism at Hopi Buttes, Navajo Nation (Arizona), USA. Bull Volcanol 53(4):239–258
White JDL, Houghton B (2000) Surtseyan and related phreatomagmatic eruptions. In: Sigurdsson H, Stix J, Houghton BF, McNutt SR, Rymer H (eds) Encyclopedia of volcanoes. Academic, San Diego, pp 495–513
White JDL, Ross P-S (2011) Maar-diatreme volcanoes: a review. J Volcanol Geotherm Res 201(1):1–29
White JDL, Schipper CA, Kano K (2015) Submarine explosive eruptions. In: Sigurdsson H, Houghton B, McNutt SR, Rymer H, Stix J (eds) Encyclopedia of volcanoes. Academic Press Elsevier, Cambridge, pp 553–569
Whitham AG, Sparks RSJ (1986) Pumice. Bull Volcanol 48(4): 209–223
Wiesner MG, Wang Y, Zheng L (1995) Fallout of volcanic ash to the deep South China Sea induced by the 1991 eruption of Mount Pinatubo (Philippines). Geology 23(10):885–888
Willcock MAW, Cas RAF, Giordano G, Morelli C (2013) The eruption, pyroclastic flow behaviour, and caldera in-filling processes of the extremely large volume (> 1290 km^3), intra-to extra-caldera, Permian Ora (Ignimbrite) Formation, Southern Alps, Italy. J Volcanol Geother Res 265:102–126
Williams SN (1983) Plinian airfall deposits of basaltic composition. Geology 11(4):211–214
Williams SN, Self S (1983) The october 1902 plinian eruption of Santa María volcano, Guatemala. J Volcanol Geother Res 16(1–2):33–56
Wilson CJ (2001) The 26.5 ka Oruanui eruption, New Zealand: an introduction and overview. J Volcanol Geotherm Res 112(1–4):133–174
Wilson L (1972) Explosive volcanic eruptions-II The atmospheric trajectories of pyroclasts. Geophys J Int 30(4):381–392
Wilson L (1978) Energetics of the Minoan eruption. In: Doumas C (ed) Thera and the aegean world I: geosciences. The Thera Foundation, London, pp 221–228
Wilson L (1980) Relationships between pressure, volatile content and ejecta velocity in three types of volcanic explosion. J Volcanol Geotherm Res 8(2–4):297–313
Wilson L, Head JW (1981) Ascent and eruption of basaltic magma on the Earth and Moon. J Geophys Res: Solid Earth 86(B4):2971–3001
Wilson L, Parfitt EA (2008) Fundamentals of physical volcanology. In: Blackwell, Oxford, p 256
Wilson L, Sparks RSJ, Huang TC, Watkins ND (1978) The control of volcanic column heights by eruption energetics and dynamics. J Geophys Res: Solid Earth 83(B4):1829–1836
Wilson L, Walker GPL (1987) Explosive volcanic eruptions-VI. Ejecta dispersal in plinian eruptions: the control of eruption conditions and atmospheric properties. Geophys J Int 89(2):657–679
Wolff JA, Wright JV (1981) Rheomorphism of welded tuffs. J Volcanol Geother Res 10(1–3):13–34
Woodhouse MJ, Hogg AJ, Phillips JC, Sparks RSJ (2013) Interaction between volcanic plumes and wind during the 2010 Eyjafjallajökull eruption, Iceland. J Geophys Res: Solid Earth 118(1):92–109
Woods AW (1988) The fluid dynamics and thermodynamics of eruption columns. Bull Volcanol 50(3):169–193
Woods AW, Bursik MI (1991) Particle fallout, thermal disequilibrium and volcanic plumes. Bull Volcanol 53(7):559–570
Woods AW, Wohletz K (1991) Dimensions and dynamics of co-ignimbrite eruption columns. Nature 350(6315):225–227
Worzel JL (1959) Extensive deep sea sub-bottom reflections identified as white ash. Proc Natl Acad Sci USA 45(3):349
Wright HM, Roberts JJ, Cashman KV (2006) Permeability of anisotropic tube pumice: model calculations and measurements. Geophys Res Lett 33(17)
Wright IC, Gamble JA (1999) Southern Kermadec submarine caldera arc volcanoes (SW Pacific): caldera formation by effusive and pyroclastic eruption. Mar Geol 161(2):207–227
Wright IC, Gamble JA, Shane PA (2003) Submarine silicic volcanism of the Healy caldera, southern Kermadec arc (SW Pacific): I–volcanology and eruption mechanisms. Bull Volcanol 65(1):15–29
Wright JV (1980) Stratigraphy and geology of the welded air-fall tuffs of Pantelleria, Italy. Geol Rundsch 69(1):263–291
Zimanowski B, Büttner R (2003) Phreatomagmatic Explosions in Subaqueous Volcanism. In: White JDL, Smellie JL, Clague DA (eds) Explosive subaqueous volcanism. AGU, pp 51–60

Subaerial Pyroclastic Density Currents (Pyroclastic Flows and Surges)

10

Origins, Types, and Deposit Characteristics

Guido Giordano, Ray Cas, and John V. Wright

R. Cas et al., *Volcanology*, Springer Textbooks in Earth Sciences, Geography and Environment,
https://doi.org/10.1007/978-3-319-66613-6_10

Plate 10.1 Two recent examples but very different types of pyroclastic density current. **a** Subaerial, dense to dilute pumice and ash flow erupted from Mt Pinatubo, Philippines, 15 June 1991, with blue pick-up in full retreat. The passengers survived! (Photo by Alberto Garcia, acknowledged as one of the great images of the twentieth century; *Source* Redux Pictures). **b** In contrast, dilute base surges radiating away from phreatomagmatic Surtseyan explosions taking place at the shallow marine vent of Anak Krakatau volcano, Indonesia on 23 December 2018 (*Source* video frame from YouTube, by YtNature https://bit.ly/YtNature)

Abstract

Explosive eruptions can produce a spectrum of pyroclastic density currents (pdcs) which we describe in this chapter. Pdcs are potentially the most destructive of all volcanic phenomena, due to their high velocities, the large distances that some can travel, the large volume of volcanic debris they can carry, and their high temperature. Pyroclastic flows include pumice/scoria and ash flows, blast flows, and block and ash flows. Pumice/scoria and ash flows are characterised by variably vesiculated pumice and scoria and finer ash and result from partial or wholesale gravitational collapse of initially buoyant explosive eruption columns or directly from a boil-over fountain of the pyroclastic mixture directly from the vent (s). In some cases, column collapses are preceded or accompanied by pyroclastic fallout, but some extremely large volume pyroclastic flow systems, especially those derived from very large, crystal-rich magma reservoirs that source major caldera collapse events and explosive super-eruptions, seem to originate from ring fractures rather than point source vents and immediate eruption column or fountain collapse, as soon as the eruption begins. Pumice/scoria and ash flows can flow from <1 to >100 km from source, and range in volume from <1 to >1000 km^3. Deposits, called ignimbrites, range from felsic to mafic in composition, and they are massive to diffusely stratified, non-welded to largely welded. They are generally poorly sorted, and include variable proportions of accessory lithic clasts, including significant breccia horizons. Surface area and volume of ignimbrites are related and reflect variable mass eruption rates and eruption durations, which can be used to distinguish between column collapse and caldera collapse associated ignimbrites. Lateral blast flows originate from sector collapse of parts of high-relief stratovolcanoes, or the margins of high-relief lava domes, which are very polymictic, and can flow several tens of km. Block and ash flows are relatively small in volume, monomictic, lack accessory lithics, originate from lava dome collapse and have limited flow distances (<15 km). Pyroclastic surges range from vent-centric base surges derived from phreatomagmatic-phreatic-hydrothermal explosions with limited flow distances, to marginal, co-pyroclastic flow surges that can continue to propagate for the length of pyroclastic flows. Although we focus on deposit facies characteristics to interpret the flow dynamics of pdcs, post-depositional effects are also considered, particularly welding, which is a common, although not universal characteristic of ignimbrites, but absent or rare in block and ash flow deposits, and absent in blast flow and base surge deposits.

10.1 Introduction

In this chapter, we examine the origin, the transportation, and depositional processes associated with pyroclastic flows and surges (Plate 10.1a, b), and summarise the characteristics of their deposits. Pyroclastic flows and surges are end-member types of a spectrum of (more or less) hot, gravity influenced, ground-hugging, particle-laden, gas-rich, pyroclastic density currents (or pdcs). They are gravity controlled, density and velocity stratified, and variably turbulent. Although some researchers now prefer to use the term pdc all the time, we use it only as an umbrella term, and continue to focus on the end-member types because they have different flow behaviour and produce different deposit types that are usually distinguishable in the field.

Pyroclastic flows (or *dense to dilute pyroclastic density currents*) are hot, variably turbulent, gravity driven, mass flows of gas, and volcanic debris (Plate 10.1a). They grade vertically and laterally in terms of density and velocity. Their interior carries the largest part of the volcanic debris and is called body of the flow, usually formed by a basal and high particle concentration flow zone (or undercurrent, usually at densities of 100–1000 kg m^{-3}) grading upward more or less abruptly into a more dilute and turbulent flow zone (usually at densities of 10–100 kg m^{-3}). Different bulk flow behaviours arise from the combination of the density profile and the velocity profile within the body: i) dominantly frictional flows are those where the basal high concentration zone is faster than the dilute zone, the flow front is relatively thin and the bulk flow propagation is controlled by particle interactions and gas pore pressure occuring within the undercurrent; ii) dominantly expanded and turbulent flows are those where the more dilute zone is faster and overrides the undercurrent, the flow front is relatively thick and bulk flow propagation is controlled by the level of turbulence, the rate of particle settling to the basal undercurrent and air entrainment. Needless to say, the density and velocity profiles inside the body of the same pyroclastic flow event may change laterally and over time. The body in turn transitions upward and laterally into a marginal, progressively lower particle concentration, ash-cloud zone at the front, sides and trailing above and behind the body. The ash clouds of pyroclastic flows have much lower particle concentrations than the interior, with densities as low as <10 kg m^{-3}. They are highly turbulent, and are called ash-cloud surges when they are still denser than atmosphere and retain a lateral component of motion. Three surge types based on context within a pyroclastic flow system can be identified: frontal ash-cloud or ground surges, trailing ash-cloud surges, and lateral ash-cloud surges. These marginal surges are continuously present during a pyroclastic flow event, enclosing the interior or body of the flow from which they continuously receive mass, momentum and thermal energy, and are active to the distal reaches of the pyroclastic flow (perhaps even 100 km or more). However, some ash cloud surges may detach from the body of the pyroclastic flow at topographic barriers or hydraulic jumps, and then flow independently for relatively short distances (hundreds of metres to few kilometres), but with an ability to overpass topographic highs. The ash cloud zone inevitably becomes buoyant upward and laterally as air entrainment reduces its density below that of the atmosphere, lofting and rising as a buoyant ash plume (plume density < than air density at ~ 1 kg m^{-3}) above and at the front of the laterally spreading pyroclastic flow.

Pyroclastic surges are characterised by high levels of turbulence, low particle concentrations, and low sedimentation rates, defining them as *dilute pyroclastic density currents*. Some pyroclastic surges can be initiated in their own right as low-concentration flow systems at the vent, which we call impulsive surges (e.g. phreatomagmatic base surges; Plate 10.1b). Impulsive vent generated surges, such as base surges, are short-lived and flow only limited distances from the vent (up to a few km) before they literally run out of gas, deposit their coarse particle load, and stop. The hot, ash clouds of both pyroclastic flows and surges may become buoyant and rise into the atmosphere, from where fine ash is dispersed or falls out as a fallout ash layer, perhaps even long distances from the active flow front.

These definitions have been used to conceptualise discrete end-member types of flows, and flow regimes, mostly based on observations of modern pdc events, but also on inferences made from the characteristics of different deposit types. These were then used to infer a dichotomy of flow processes (pyroclastic flows and surges).

A rich scientific debate over the last forty years on the flow and depositional behaviour of pdcs has broadened and modified the perspective and demonstrated that the variations in deposit or facies characteristics depend mostly on mass flux, particle concentration, temperature, turbulence levels, particle interaction, and sedimentation rate in the flow system (e.g. Fisher 1966, 1979; Sparks 1976; Sparks et al. 1978; Wright and Walker 1983; Wright 1981; Cas and Wright 1987; Palladino and Valentine 1995; Branney and Kokelaar 1992, 1997, 2002; Bursik and Woods 1996; Dade and Huppert 1996, 1997; Wilson 1997; Druitt 1998; Giordano 1998a; Freundt 1999; Burgisser and Bergantz 2002; Dellino et al. 2008; Cas et al. 2011; Andrews and Manga 2012; Doronzo 2012; Sulpizio et al. 2014; Brown and Andrews 2015; Dufek et al. 2015, 2016; Roche et al. 2016; Breard et al. 2016; Wright et al. 2016; Giordano and Doronzo 2017; Brand et al. 2014, 2023; Breard and Lube 2017; Lube et al. 2015, 2019; Palladino and Giordano 2019; Trolese et al. 2019; Shimizu et al. 2019, 2021; Báez et al. 2020a, b; Smith et al. 2020; Giordano and Cas 2021; Roche et al. 2021, 2022; Calabrò et al. 2022; Doronzo et al. 2022). The properties or characteristics of pyroclastic density currents can vary and change vertically and laterally, and with time. Higher particle concentration develops in pdcs near the base, away from the vent, beyond the column collapse deflation zone, in the so-called depositional or flow boundary layer, whereas lower particle concentrations persist in the enclosing ash cloud. Pdcs are therefore ephemeral, 4D (dimensional) flow systems, as a result of transient (i.e. time-dependent) processes associated with changing eruptive conditions and flow and depositional regimes along flow paths and in response to topography and changes in topography.

These variations in flow behaviour or flow architecture of pdcs are reflected in usually very complex, laterally and vertically changing facies architecture of the resultant deposits (e.g. Fisher 1979; Wright et al. 1981; Giordano 1998a; Branney and Kokelaar 2002; Cas et al. 2011). There are surge deposits (low angle, cross-stratified ash deposits) beneath, above, and lateral to massive pyroclastic flow body deposits in the rock record (e.g. Sparks et al. 1973; Fisher 1979; Wright et al. 1981; Branney and Kokelaar 2002; Cas et al. 2011) that testify to a flow architecture of a frontal/leading, lateral and trailing ash-cloud surge that encloses the body of the flow at any time during waxing and waning of the mass flux, which may result also in stratified surge facies to be present within massive pyroclastic flow body deposits emplaced during a continuous depositional process. Although, in recent years, some researchers have emphasised that deposits do not necessarily reflect the flow system, this may only be the case with the early phase of flow (e.g. Valentine 2020). However, the deposits give us the best indication of the state of the flow during the time interval that deposition is occurring along the flow path. We should therefore use the deposit characteristics to understand transportation and depositional processes. In particular, we stress that the understanding of flow dynamics from deposits can only be obtained by reconstructing and interpreting holistically the deposit in terms of lithofacies architecture, thickness variations, grain size distributions, dispersal area, volume and emplacement temperatures (Giordano and Cas 2021). For example, although the lithofacies architecture relates to processes at the flow boundary layer which directly reflect conditions of the depositional system at emplacement (e.g. Cas et al. 2011; Roche et al. 2016), the overall ignimbrite dispersal across topography, its lateral thickness, grain size and emplacement temperature variations embed the essential information on the transport system (Calabrò et al. 2022; Silleni et al. 2024). Interestingly, although the thermal state of pyroclastic flow deposits varies hugely from welded ignimbrites to totally unconsolidated deposits (e.g. Walker 1983), only recently quantitative studies have highlighted the importance of the temperature of emplacement as a key to understanding initiation, transportational and depositional processes (e.g. Lesti et al. 2011; Trolese et al. 2017, 2018, 2019, 2024; Giordano et al. 2018; Pensa et al. 2019).

The study and interpretation of deposits can therefore reveal the evolution over time of the eruption that fed the pdc, and of the role of changing transportational and depositional mechanisms and regimes. In addition, it is important to emphasise that real deposit characteristics are essential to benchmark laboratory experimental and numerical simulations of pyroclastic density current flow and depositional behaviour (Calabrò et al. 2022). Without relating these simulations to real deposit clast population properties and overall deposit characteristics, numerical and analogue experiments do not accurately simulate true conditions.

Eruptions that feed pdcs range from very small volume (<<1 km^3) to very large volume (>1000 km^3). Consequently, the associated flow events range from short-lived (minutes) to sustained over significant intervals of time (hours--days), with implications for the relative development of what we define as flow or surge, as well as depositional units in the resulting deposits.

Remarkably, the huge spectrum of pdc types has never been included in the classification of eruption styles (hawai'ian to plinian), which are only based on pyroclastic fall deposit dispersal patterns and grain size characteristics (see Chap. 9), so we will introduce in this chapter (Sect. 10.11) our proposal for the classification of pdc types and deposits, and related eruptions (Giordano and Cas 2021).

Herein the emphasis is on depositional geometry, facies, and their characteristics as the basis for interpreting the flow dynamics of pdcs. Therefore, we consider the concepts and definitions of pyroclastic flows and pyroclastic surges as a framework for considering the processes that are reflected in the sedimentological facies characteristics of the deposits (i.e. those that reflect high sedimentation rates from flows and low sedimentation rates from surges, respectively). We will show they are useful ones, and we therefore retain them. However, the two end-member flow types are not mutually exclusive and share lateral spatial and temporal changes and relationships (Giordano 1998a; Branney and Kokelaar 2002; Burgisser and Bergantz 2002; Druitt et al. 2002; Doronzo 2012; Doronzo and Dellino 2014; Dufek et al. 2015; Dufek 2016). In terms of bulk behaviour, pyroclastic flow deposits are better controlled by the underlying topography than surge deposits because pyroclastic flows develop denser undercurrents as a result of high mass fluxes and particle concentrations (Cole et al. 2002; Druitt et al. 2002; Giordano et al. 2002; Pittari et al. 2006, Cas et al. 2011; Doronzo and Dellino 2014; Brand et al. 2016, 2017, 2023; Báez et al. 2020a, b).

We will differentiate processes that form surges that are intimately associated with the evolution of sustained pyroclastic flows over time and influenced by topography, from those pyroclastic surges that originate as dilute, impulsive currents and remain dominantly so over time and space. The ash-cloud surge that forms the front of a pyroclastic flow can be called the frontal ash-cloud surge (also previously called the ground surge), the ash-cloud surge trailing above and behind the pyroclastic flow is called the trailing ash-cloud surge, and the ash-cloud surge that spreads laterally along the lateral margins of the pyroclastic flow is called the lateral ash-cloud surge.

The differentiation between sustained pyroclastic flows with their associated surges (Sects. 10.7, 10.14.1.2–10.14.1.3) and impulsive or individual pyroclastic surges, such as base surges (Sect. 10.14.1.1), is also very important because it distinguishes the depositional behaviour in different types of pdc, which is very important where the interpretation of deposits is used to build hazard maps. In addition, the anatomy of pdcs is examined in detail, and consideration is given to the characteristic depositional processes associated with the head, body, and ash cloud of pyroclastic flows, and of impulsive surges, as well as the deposits produced.

It must be emphasised however that sedimentary lithofacies in pyroclastic density current deposits are scale invariant, so their full spectrum from massive to stratified reflects closely the depositional conditions that can be found in ignimbrites and surge deposits of any dimension and composition but do not appear related to any specific eruption and transportation style, so we remark that in order to understand the flow behaviour, the lithofacies architecture of pdc deposits should always be accompanied by an accurate reconstruction of the spatial thickness, grain size and emplacement temperature variations.

In our experience, a useful distinction to discriminate pdc flow types is between:

(a) Dominantly frictional flows in which the bulk flow behaviour (i.e. macroscopic flow velocity, invasion area, topographic confinement and maximum runout) is controlled by particle interactions and gas pore pressure (fluidization) within the relatively thin (metres to tens of metres in thickness) and high concentration (usually 100-1000 kg m^{-3}) basal parts, with relatively minor internal stratification, which may flow in an overall laminar fashion. In this kind of flow these dense undercurrents form the bulk of the flow body, are the fastest and drive the bulk flow propagation, while the overlying low concentration current transforms into the trailing ash cloud, which may locally detach and move independently, especially where flows are small relative to topography roughness (Fisher 1995). Frictional flows are typical of small volume flows along rough and steep stratovolcano valleys (e.g. block and ash flows), where valleys favour the channelisation and the rapid development of a dense and gas fluidised undercurrent controlled by topography, which eventually reaches the maximum runout and stops by particle interlocking (e.g. Roobol et al. 1987; Fuji and Nakada 1999; Giordano et al. 2002; Druitt et al. 2002; Pensa et al. 2019). Dominantly frictional flows also invariably characterise the basal undercurrent of larger volume and sustained pyroclastic flows. However in such cases the literature distinguishes between those where the bulk flow behaviour is controlled by that of the high concentration frictional basal undercurrent (e.g. Cas et al. 2011; Roche et al. 2016), from those where the basal undercurrent is continuously fed/aggraded by the overriding more dilute body flow zone, as described below (e.g. Fisher et al. 1993; Baer et al. 1997; Ort et al. 2003; Silleni et al. 2020, 2024; Calabrò et al. 2022);
(b) Dominantly expanded, thick and stratified flows where the bulk flow behaviour is controlled by the more dilute part of the flow body which overrides and continuosly feeds the basal undercurrent, being faster, frictionless and dominated by gas turbulence. In this kind of flows the macroscopic flow velocity, runout and buoyancy (ranging from negative while flows move along the ground, to positive at final runout and flow lift-off) are controlled by the particle sedimentation rate and air entrainment. The maximum extent of the resulting ignimbrite coincides (more or less) with the maximum runout of the parent flow at the point of flow lift-off (e.g. Bursik and Woods 1996; Andrews and Manga 2011; Calabrò et al. 2022). These are thought to be typical of ignimbrites associated with large volume flows, mostly caldera-related (and axisymmetric), thick with respect to topography (hundreds of metres to kilometres in thickness) and/or travelling on near-flat ground (with respect to flow), where density stratification and interaction with topography do not induce significant detachment of undercurrents (e.g. Wilson 2001; Giordano and Doronzo 2017; Shimizu et al. 2019; Calabrò et al. 2022). It is important to underline that the basal undercurrents formed at the base of these expanded and stratified flows invariably experience frictional behaviour at the time of deposition, so that the resulting deposit lithofacies are usually not dissimilar to those derived from wholesale frictional flows (Silleni et al. 2024). This is why we emphasise how important it is to reconstruct the entire lithofacies architecture, geometry and attitude to topography of an ignimbrite before addressing its parent flow properties.

10.2 Characteristics of Modern Pyroclastic Flows and Associated Surges

Pyroclastic flows (Plate 10.1a) are potentially the most destructive of all volcanic phenomena (Fig. 10.1), due to their high velocities, the high dynamic pressure and destructive force they can apply to infrastructure and forests, the large distances that some flow types are capable of travelling, the large volume of volcanic debris they can carry and their high temperature.

Serious loss of life has been caused by several historic pyroclastic flows (Table 10.1; Witham 2005). Small historic flows have been observed to move from 1 or 2 km, up to about 20 km from vent at speeds from 20 m s^{-1} to as high as 60 m s^{-1} (>70 to >200 km hr^{-1}; 1968 eruption at Mayon volcano, Philippines, Moore and Melson 1969; 1974 and 2018 eruptions at Fuego volcano, Guatemala, Davies et al. 1978a; post-May 18, 1980 flows at Mt St Helens, USA, Rowley et al. 1981; 1996–1998 events at Soufrière Hills, Montserrat, Cole et al. 2002; 2015 eruption of Colima, Mexico, Pensa et al. 2018, 2019; 2019 eruptions at Stromboli, Italy (Giordano and De Astis 2021).

Pdcs, including pyroclastic flows, primarily move along topographic lows because of the excess bulk density induced by their particle load, relative to atmospheric density; hence they are density currents. However, some currents can also flow up and over topography. Theoretical analysis based on measurements of the heights of mountains climbed and

Fig. 10.1 Examples of towns destroyed during historic pyroclastic flow events with major loss of life. **a** Today's town of St Pierre on the Caribbean island of Martinique, Lesser Antilles, with Mt Pelée in the background (Photo by Reg Barker). **b** St Pierre after its destruction in 1902 by pyroclastic flows generated by the explosive collapse of an andesitic lava dome from the summit of Mt Pelée. The description of this eruption by Lacroix (1904) set a milestone in modern volcanology. (Photo by Angelo Heilprin). **c** Part of the excavated, ancient Roman city of Herculaneum (foreground) that was destroyed and buried under the pyroclastic flow deposits of the 79 AD eruption of Vesuvius (background), and which also buried Pompei. The modern city of Napoli is built on the eroded 79 AD deposits in the middle ground

overpassed by pyroclastic flows suggests that average speeds of >100 m s^{-1} (>300 km hr^{-1}) are not uncommon (e.g. the 22 ka Ito pyroclastic flow deposit, Yokoyama 1974; the 39 ka Campanian ignimbrite, Fisher et al. 1993; the 75 ka Los Chocoyos ignimbrite, Cisneros de León et al. 2021; the 232 AD Taupo ignimbrite, Wilson and Walker 1985). However, large volume pyroclastic flows are not uniquely associated with high flow velocities, as some deposits have been interpreted to reflect very slow emplacement velocities (e.g. the 2.08 Ma Cerro Galán ignimbrite, ~10 km hr^{-1}, Cas et al. 2011; the 18.8 Ma Peach Spring Tuff, 18–72 km hr^{-1}, Roche et al. 2016; the 73 ka Campo de la Piedra Pomez ignimbrite, ~ 17–61 km hr^{-1}, Báez et al. 2020b) . Furthermore, the large volume of volcanic debris emplaced implies that these flows lasted much longer than the time required to cover the run-out distance, resulting in deposits

Table 10.1 Some historical eruptions that have caused the greatest damage and number of fatalities

Year	Start date	Country	Volcano name	Volcano type (GVP)	Fatalities	VEI	Main phenomena
2018	22/12/2018	Indonesia	Anak Krakatau	Stratovolcano	437	3	*da, ts, ph, pf*
2018	03/06/2018	Guatemala	Fuego	Stratovolcano	>200	3?	*pf*
1995	18/07/1995	Montserrat	Soufrière Hills	Stratovolcano	19	3	*ph, pf, ld, mf*
1991	09/06/1991	Philippines	Pinatubo	Stratovolcano	640^	6	*pf, ld, ph, cc, mf*
1991	03/06/1991	Japan	Unzen	Dome	43		*ld, pf*
1986	21/08/1986	Cameroon	Lake Nyos	Maar	1746^		*cl*
1985	13/11/1985	Colombia	Nevado Del Ruiz	Stratovolcano	23,080*	3	*pf, ph, mf, ts*
1982	28/03/1982	Mexico	El Chichón	Stratovolcano	2000	5	*ph, pf, mf*
1980	27/03/1980	USA	Mt St Helens	Stratovolcano	57	4–5	*ph, da, pf, ld*
1963	17/03/1963	Indonesia	Agung	Stratovolcano	1584^	5	*pf, lf, mf*
1951	21/01/1951	Papua New Guinea	Lamington	Stratovolcano	3000^	4	*pf, ld, da, mf*
1937	29/05/1937	Papua New Guinea	Rabaul (Tavurvur)	Pyroclastic shield	507*	4	*pf, ph, mf, ts*
1931	13/12/1931	Indonesia	Merapi	Stratovolcano	1300^	3	*pf, ld, mf*
1930	18/12/1930	Indonesia	Merapi	Stratovolcano	1369^	3	*pf, ld, mf*
1929	02/11/1929	Guatemala	Santa María	Stratovolcano	5000^	3	*pf, ld, ph, da, mf*
1919	19/05/1919	Indonesia	Kelut	Stratovolcano	5110*	4	*pf, mf, cl*
1911	30/01/1911	Philippines	Taal	Stratovolcano	1335^	3	*pf, ph, ts*
1909	29/08/1909	Indonesia	Semeru	Stratovolcano	5500^	2	*pf*
1906	06/04/1906	Italy	Vesuvius	Somma volcano	700^	4	*pf, lf, cc, mf*
1902	24/10/1902	Guatemala	Santa María	Stratovolcano	1500*	6	*ph, mf*
1902	08/05/1902	Martinique	Pelée	Stratovolcano	30,000^	4	*pf, ld, ph, mf, ts, cl*
1902	07/05/1902	St Vincent	Soufrière	Stratovolcano	1680*	4	*pf, ph, mf, ts, cl*
1887	26/06/1887	Ecuador	Cotopaxi	Stratovolcano	1000	4	*pf, mf*
1883	6/06/1883	Indonesia	Krakatoa	Stratovolcano	36,417	6	*pf, cc, ts*
1845		Colombia	Nevado Del Ruiz	Stratovolcano	1000		*mf*
1822		Indonesia	Gulanggung	Stratovolcano	4000	5	*mf*
1815	05/04/1815	Indonesia	Tambora	Stratovolcano	>100,000	7	*pf, mf, cc, ts*
1814	01/02/1814	Philippines	Mayon	Stratovolcano	1335	4	*pf, mf*
1792	21/05/1792	Japan	Unzen	Dome	15,000	3	*da, ts*
1631	16/12/1631	Italy	Vesuvius	Stratovolcano	3360	4	*pf, mf*
1600		Peru	Huaynaputina	Stratovolcano	>1500	6	*pf, mf, cc*
1640		Ecuador	Tungurahua	Stratovolcano	5000	3	*pf*
1586		Indonesia	Kelut	Stratovolcano	10,000	4	*mf*
1257		Indonesia	Samalas (Rinjani)	Stratovolcano		7	*ph, pf, cc*
79	24/08/79	Italy	Vesuvius	Stratovolcano	18,000	5	*pf, mf, cc*
1600BC		Greece	Santorini	Multiplex		6–7	*ph, mf, pf, cc*

Records from 1900 to 1991 are from Whitman 2005 (*Sources* *NOAA; ^EM-DAT); other eruptions compiled from various sources listed in Wikipedia. *pf* pyroclastic flow; *ld* lava dome; *lf* lava flow; *ph* phreatic/phreatomagmatic explosion; *mf* lahar; *ts* tsunami; *cl* crater lake explosion; *cc* caldera collapse; *da* debris avalanche that always show some sort of vertical compositional zoning or textural variation, reflecting their aggradation over time, either progressively or incrementally.

Although modern pdcs have flowed up to 20 km, large volume Quaternary pumice and ash pyroclastic flow deposits or ignimbrites (see below) suggest that larger flows have travelled distances of >100 km from the vent. The 1 Ma Kidnappers ignimbrite pyroclastic flow sourced from the Taupo Volcanic Zone in New Zealand is known to have flowed >200 km from the source Mangakino caldera (Wilson et al. 1995), and the pyroclastic flow that deposited the Peach Springs Tuff ignimbrite flowed >170 km from the source Silver Creek Caldera in Nevada, USA (Roche et al. 2016). Bryan et al. (2010) have even suggested that the pdcs that deposited extremely large volume rheomorphic ignimbrites associated with large igneous provinces, such as flood basalt provinces, may have flowed >600 km, although the mechanisms and flow state are not understood. Can pyroclastic flows flow such a distance, or do they deposit closer to vent, weld and then flow as (tube-fed) rheomorphic, lava-like flows? Here's a research opportunity for someone!

10.3 Different Types of Pyroclastic Flows and Their Origins

Pyroclastic flows and surges are generated by a number of different mechanisms (Figs. 10.2, 10.3 and 10.4). From understanding of observed modern eruptions, these can be split initially into six main types, which include some sub-types:

(1) Partial collapse of short-lived explosive eruption columns from point source vents (e.g. vulcanian, sub-plinian; Figs. 10.3a and 10.4a, b).
(2) Partial to total collapse of a sustained explosive eruption columns from point source vents: partial collapses occur while the plume is still buoyant and fallout continues; total collapses are climactic events that usually herald the termination of the eruption (e.g. sub-plinian, plinian, phreato-plinian; partial collapses are called intra-plinian collapse; Figs. 10.3b and 10.4c).
(3) Continuous collapse of low "boiling over" types of eruption columns or pyroclastic fountains (Figs. 10.3c and 10.4d, e).
(4) Continuous high mass flux "boiling over" pyroclastic fountains from ring fractures during caldera collapses (Fig. 10.3c).
(5) Instantaneous lateral blasts from stratovolcanoes and lava domes, associated with landslide sector collapses (Figs. 10.3d and 10.4f).
(6) Short-lived, small volume lava dome or lava-flow collapse, through purely gravitational collapse, or involving collapse triggered by an explosion from within the dome (Figs. 10.3e and 10.4g, h).

These different origins produce discrete *pyroclastic flow types* (Figs. 10.2 and 10.4) characterised by different mass fluxes (i.e. the amount of material collapsing and feeding the pyroclastic flow), durations, initial temperature, concentration, grain size distribution, and velocity, resulting in different levels of flow turbulence, mobility, and different sedimentation patterns (e.g. Bursik and Woods 1996; Dufek 2016; Shimizu et al. 2019; Trolese et al. 2019).

Based on their componentry, the different types of pyroclastic flows and related deposits can be distinguished as follows:

(a) *Pumice and ash flows*: Characterised by variably vesiculated juvenile pyroclasts of felsic composition, and usually resulting from collapse of variably buoyant

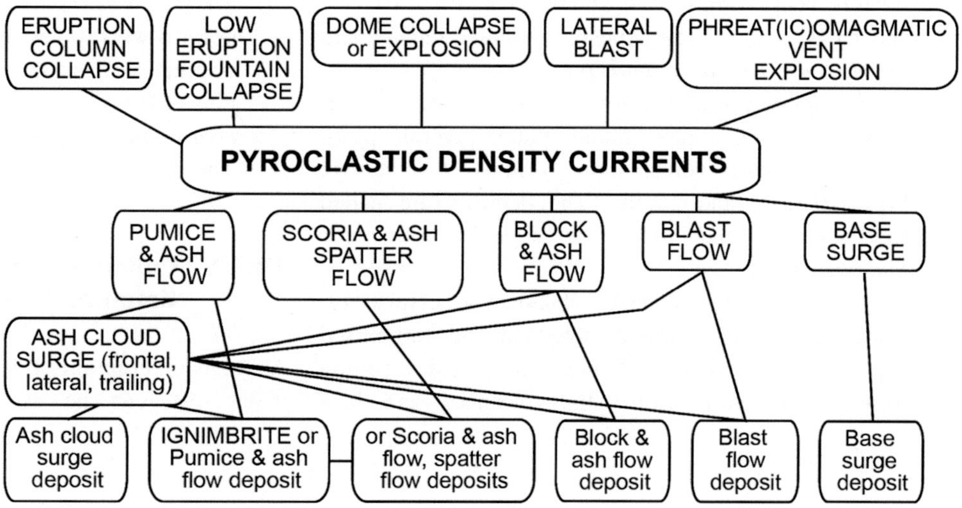

Fig. 10.2 Process and deposit linkage spectrum diagram for all pyroclastic density currents as discussed in the text

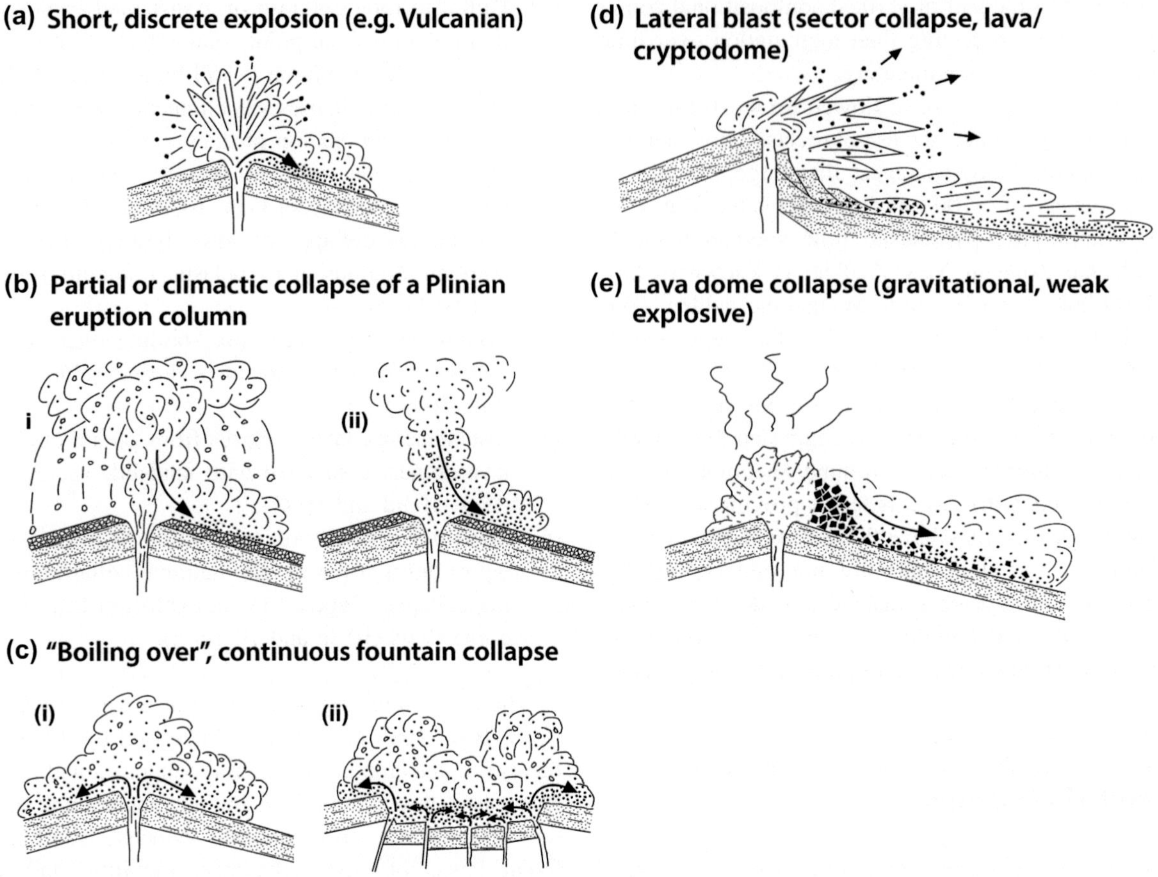

Fig. 10.3 Mechanisms by which different pyroclastic flow types are generated **a** Short Vulcanian blast and immediate collapse to form a small volume pyroclastic flow. **b** Collapse of buoyant sub-plinian to plinian eruption columns; (i) partial collapse producing an intra-plinian pyroclastic flow, while fallout deposits continue to form, and (ii) climactic, end of eruption complete collapse of the column. Flow types are pumice/scoria and ash flows. **c** Continuous low column or pyroclastic fountain collapse; (i) single vent, forming pumice/scoria and ash flows, and (ii) from ring fractures, forming sustained high mass flux pumice/scoria and ash flows. **d** Lateral blast from a cryptodome or lava dome, forming a blast flow. **e** Gravitational collapse of the margins of a steep sided, high aspect ratio lava dome forming a block and ash flow

explosive eruption columns during partial or complete column collapse, and collapsing low pyroclastic fountains, including caldera-forming fountains from ring fracture vents (origins 1–4 above; Figs. 10.2, 10.3a–c and 10.4a, c, d, e). The deposits are called *pumice and ash flow deposits* or *ignimbrites*.

(b) *Scoria, spatter, and ash flows*: Characterised by variably vesiculated and/or fluidal spatter juvenile pyroclasts of mafic composition (origins 1–4 above; Figs. 10.3a–c, and 10.4b). The deposits are called *scoria/spatter and ash flow deposits,* or *mafic ignimbrites*.

(c) *Blast flows*: Characterised by abundant accessory volcanic lithic clasts, variably shredded and carbonised vegetation, and minor juvenile clasts, and resulting from lateral blasts from stratovolcanoes and lava domes, often associated with landslide sector collapse (origin 5, sometimes 1 above; Figs. 10.3d and 10.4f). The deposits are called *blast flow deposits*.

(d) *Block and ash flows*: Dominated almost exclusively by non- to weakly vesicular juvenile lava dome clasts and originating from the gravitational collapse of the margins of lava domes, and/or explosive collapse of parts of lava domes (origin 6 above; Figs. 10.3e and 10.4g–i). The deposits are called *block and ash flow deposits*.

10.3.1 Collapse of Short-Lived Eruption Columns from Point Source Vents

A range of eruption styles, from vulcanian to sub-plinian (Chap. 9), can be associated with the generation of

Fig. 10.4 Examples of eruptions producing different types of pyroclastic flows. **a** Vulcanian blast, Soufrière Hills volcano, Montserrat, collapsing instantaneously to form small volume pyroclastic flows (Photo by B. Voight; *Source* National Science Foundation). **b** Collapsing, low vulcanian blast column, 1975 eruption of Ngauruhoe volcano, New Zealand, forming scoria, spatter, and ash flows (Photo by Ian Nairn). **c** Collapsing buoyant eruption column, 7 August 1980, Mt St Helens, forming small volume pumice and ash flows (Photo by P. Lipman; *Source* United States Geological Survey). **d** Boiling over, low collapsing fountain, and **e** its evolution ($\sim$20 s later), showing development of a pumice and ash flow, 22 July 1980, Mt St Helens (Photos by H. Glicken; *Source* Cas and Wright 1987). **f** Lateral and upward directed hot blast on 18 May 1980, Mt St Helens, triggered by a cold landslide resulting from collapse of the north slope of the volcano (detached blocks visible in lower part of image). The lateral blast then produced a dynamic flow of gas, rock debris, and minor juvenile

(continued)

cryptodome debris called a blast flow that devastated the landscape to the north (Photo by G. Rosenquist; *Source* United States Geological Survey, Modified from Pallister et al. 2017). **g** Lava blocks detaching gravitationally from the margin of the summit lava dome complex (rock face height is about 10 m), triggering **h** the initial block and ash flow of 3 June 1991, Mt Unzen, Japan (*Source* video frames from YouTube, by C. Gomez, University of Canterbury, New Zealand). **i** Firefighters frantically escaping the final run out of devastating ash-cloud surge associated with block and ash flow which claimed 43 lives on 3 June 1991, Mt Unzen, Japan (also see Figs. 10.12c and 10.98c) (Photo by Ikitsugo Manako)

Fig. 10.4 (continued)

impulsive, small volume, short-lived pyroclastic flow events involving collapse of short-lived, non-buoyant to temporarily buoyant explosive eruption bursts or columns from point source vents (Figs. 10.3a and 10.4a, b). In these cases, the effective density of the vertical debris-laden eruption column, or parts of it, is greater than that of the atmosphere, and gravitational collapse occurs almost immediately after the explosion, generating a pyroclastic flow.

Such small volume pyroclastic flows can be associated with almost any magma composition and generate deposits that, based on componentry, range from scoria and ash flow deposits for the mafic types to pumice and ash flow deposits for the more felsic types. The observed historic examples of this type have produced small volume pyroclastic flow deposits (<<1 km^3). Observations suggest that small collapses occur when a short-lived, over-pressurised explosion (e.g. vulcanian) ejects a dense slug of pyroclastic fragments to an altitude of a few hundred m, part of which (or all of it) then collapses back to the ground and is deflected laterally as a short-lived pyroclastic flow.

Observed historic eruptions during which this type of pyroclastic flow formed include Mt Lamington, Papua New Guinea (1951; Taylor 1958), Mayon, Philippines (1968; Moore and Melson 1969), Fuego, Guatemala (1974; Davies et al. 1978a), Ngauruhoe, New Zealand (1975; Fig. 10.4b; Nairn and Self 1978), Soufrière Hills, Montserrat (1997; Calder et al. 2002; Cole et al. 2002; Loughlin et al. 2002; Druitt et al. 2002), and Stromboli (2019; see Fig. 10.89, Giordano and De Astis 2021). All produced scoria and ash flows, except the Montserrat events that formed andesitic pumice and ash flows, surges and fallout deposits. These types of pyroclastic flows have also been called nuées ardentes and pyroclastic avalanches (Nairn and Self 1978).

Druitt et al. (2002) provide a valuable account of a series of 88 vulcanian explosive eruptions and events during 1997 at the Soufrière Hills volcano, Montserrat. Explosions originated from the growing lava dome and its feeder conduit and occurred at spacings between 2.5 and 63 h. Explosions were marked by an initial intense phase followed by a waning phase lasting from 1 to 3 h. Each explosive burst collapsed within 20 s, generating surges that flowed up to 1–2 km, and pumiceous pyroclastic flows that flowed up to 6 km down all major valleys around the lava dome. In addition, some eruptions generated buoyant columns that rose 3–15 km, producing dispersed fallout deposits. Each explosion ejected an average of 3×10^5 m^3 of magma, at exit velocities of 40–140 m s^{-1}, causing ballistic blocks to be ejected up to 1.7 km from the vent. The ejecta are dominated by angular, vesicular andesite pumice with up to 55% vesicles, indicating that the conduit was occupied by a viscous foam plug at the time of eruption, that was presumably retarding gas escape from deeper in the conduit.

Similar eruptive parameters were associated with the small pyroclastic flows produced by the 3 July and 28 August 2019 vulcanian eruptions at Stromboli volcano, Italy (Giordano and De Astis 2021; see Fig. 10.89; Chap. 11, Sect. 11.6.1.1). Pyroclastic flow-forming explosions had ballistic exit velocities of 110–120 m s^{-1} and gas velocities of 210–220 m s^{-1}, with bouyant plumes up to 8.4 km in height and mass eruption rates up to 1.1×10^6 kg s^{-1}. Pyroclastic flows were directional along breaches in the crater rim and rushed along the 35° inclined slope of the volcano at an average speed of ~50 m s^{-1}, but with an instantaneous velocity of up to 100 m s^{-1}. On entry into the sea, a sudden change in slope, rapid entrainment of seawater, and sedimentation, slowed the flows down, and they were only able to reach a few hundreds of m off-shore before ash-cloud lofting occurred (see Fig. 11.9b). Light, buoyant pumice formed extensive pumice rafts, which floated for days around the island.

10.3.2 Partial Collapse (Intra-plinian) to Total Collapse from Sustained Sub-plinian to Plinian Eruption Columns from Point Source Vents

The ignimbrite deposits of pyroclastic flows resulting from the collapse of sub-plinian to plinian eruption columns (Chap. 9) erupted from single point source vents (Figs. 10.3b and 10.4c) are frequently <1 km^3 in volume, and rarely larger (≤ 10 km^3). Much larger pyroclastic flow deposits on Earth (≥ 10 km^3 up to >1000 km^3), result instead from continuous fountain collapse from ring fault fissure vents associated with caldera collapses, as discussed in Sect. 10.3.3. Only the eruption of a few small to intermediate-volume pyroclastic flows has been witnessed and described in modern times, and even so, there is little observational information for these.

Well-studied examples of small volume, pumice, and ash flow ignimbrite deposits from single vents include those from Mt St Helens in 1980 (Fig. 10.4c; Rowley et al. 1981; Brand et al. 2014, 2016, 2017, 2023), Lascar in 1993, northern Chile (Deruelle et al. 1995; Sparks et al. 1997), and Calbuco in 2015, southern Chile (Romero et al. 2016). None of the Mt St Helens pumice flows travelled very far, and all the deposits are minor in volume (<<1 km^3). The generally known examples of larger volume events are the eruption and pyroclastic flow deposits from the 79 AD eruption of Vesuvius, Italy (Cioni et al. 1992; Carey and Sigurdsson 1987; Shea et al. 2011; Giordano et al. 2018), the Valley of Ten Thousand Smokes ignimbrite erupted at Novarupta-Katmai, Alaska, in 1912 (Hildreth and Fierstein 2012), those formed during the eruptions of Komagatake, Japan, in 1929 (Aramaki and Yamasaki 1963), and those from Mt Pinatubo, Luzon, Philippines, in 1991 (Plate 10.1a; Scott et al. 1996). Two notable, and larger, ignimbrite-forming eruptions occurred in the nineteenth century: Krakatau, west of Java, in 1883 (van Bemmelen 1949; Self and Rampino 1981; Self 1992; Chap. 11), and Tambora on the island of Sumbawa, also in Indonesia, in 1815 (Self et al. 1984, 2004). However, these were formed during caldera-forming events, which most likely evolved into the continuous collapse of "boiling over" pyroclastic fountains described in the following section.

Sparks and Wilson (1976) and Sparks et al. (1978) presented a theoretical model for the formation of ignimbrites based on the continuous gravitational collapse of plinian eruption columns erupting from single vents (Figs. 10.3b and 10.5a, b). This model helps to explain many features of ignimbrites. The mass flux involved in the partial to total collapses of point source vent-derived plinian eruption columns, the degree of air ingestion, as well as the heights (up to a few km) from where collapse occurs, account for the thermal and kinetic properties of the forming pyroclastic currents, and may explain the widely variable distributions, degrees of topographic control and temperatures of emplacement of these types of single-vent-derived ignimbrites (Trolese et al. 2019). The column collapse model is also appealing because it explains why many ignimbrites are underlain by plinian pyroclastic fallout deposits (e.g. the 0.6 Ma Granadilla eruption sequence, Tenerife, Bryan et al. 2000; 184 and 172 ka Lower Pumice 1 and 2 eruption sequences, Santorini, Simmons et al. 2016; Gertisser et al. 2009; the 75 ka Los Chocoyos eruption from Lake Atitlán caldera, Guatamala; the 4.6 ka Fogo A eruption sequence, São Miguel; Pensa et al. 2015a; Fig. 10.6a–c; Chap. 9; the 800 yr BP (~ 1280 AD) Quilota eruptive cycle, Ecuador; Di Muro et al. 2004, 2008).

Furthermore, detailed stratigraphic reconstruction of the June 6–8, 1912 eruption at Novarupta-Katmai, Alaska, demonstrates that the formation of ignimbrites can be contemporaneous with continuous fallout, suggesting that a plinian column can generate both fallout and ignimbrite deposits simultaneously by partial collapse (Fig. 10.3b; Fierstein and Hildreth 1992; Houghton et al. 2004; Hildreth and Fierstein 2012). At Mt Pinatubo, Philippines, stratigraphic relations are consistent with the broadly synchronous emplacement of pyroclastic flow and pumice fallout deposits during much of the climactic eruption on 15 June 1991 (Scott et al. 1996). Scott et al. (1996) infer that pumice and ash flows were generated by either quasi-continuous, low-level collapse of portions of a sustained plinian column or by repeated brief collapse of the entire column over a 3–5 h period. Other observations at Pinatubo suggest if the column was alternating between collapse and convective rise, periods of collapse must have been brief (minutes?). Further examples include the 1815 eruption of Tambora, Indonesia (Self et al. 1984; Sigurdsson and Carey 1989), the 800 yr BP eruption of Quilota,

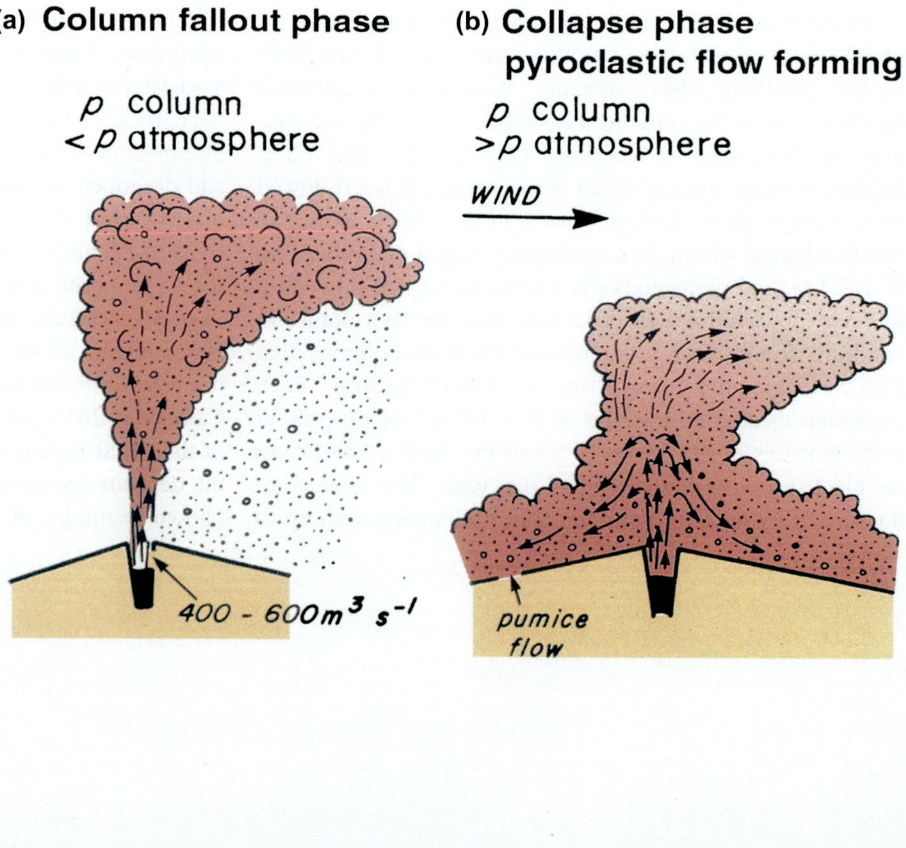

Fig. 10.5 Simple, classical plinian column collapse model. **a** Buoyant eruption column that produces plinian fallout deposits. **b** Collapse of the eruption column to produce pyroclastic flows. This is typical of the generation of relatively small volume pyroclastic flows from central vents from stratovolcanoes. (From Cas and Wright 1987)

Fig. 10.6 Examples of eruption sequences resulting from collapse of buoyant sub-plinian to plinian eruption columns to produce initial pumice and ash fallout deposits from a buoyant column and pyroclastic flow deposits or ignimbrite from a collapsing column. **a** 0.22 Ma Caleta eruption sequence, consisting of a fine, white, basal phreatomagmatic ash fallout and surge unit, overlain by a very well sorted, plinian pumice fallout deposit, overlain by the poorly sorted Caleta ignimbrite. Tajao, Tenerife, Canary Islands, Spain. Note the fine-grained base of the ignimbrite, which is reverse graded up into the body. **b** The 0.6 Ma Granadilla ignimbrite (massive unit at top of road-cutting) overlying its preceding 9 m thick diffusely stratified plinian pumice fallout deposit, which mantles irregular paleotopography, east of San Isidro, Tenerife. **c** Two intra-plinian ignimbrites interbedded with on-going plinian fallout deposits, and overlain by the final climactic ignimbrite, 4.6 ka Fogo A eruption sequence, São Miguel, Azores, Portugal. Note the Fogo A plinian fall deposits mantle an irregular paleovalley and ridge topography, cut into two older fallout deposit sequences, whereas the ignimbrites have a valley-pond geometry, wedging out against the paleo-ridge. The ignimbrites pass laterally into very thin lateral ash-cloud surge deposits that are interstratified with the fallout deposits. (Photo by A. Pensa)

Ecuador (Di Muro et al. 2004, 2008), the 79 AD eruption of Vesuvius, Italy (Carey and Sigurdsson 1987; Cioni et al. 1992), the 4.6 ka Fogo A eruption sequence, Azores (Walker and Croasdale 1971; Pensa et al. 2015b), the 0.28 Ma Poris eruption, Tenerife (Edgar et al. 2002) to name but a few of many.

Small volume pyroclastic flows can be generated as a result of short-lived, partial collapse of sustained eruption columns during sub-plinian and plinian eruptions (Fig. 10.3b). The resulting deposits are found as variably thick, massive, and poorly sorted, to cross-stratified depositional units interbedded with characteristically well-sorted fall deposits and are usually named "intra-plinian ignimbrites" (Fig. 10.4a). Intra-plinian ignimbrite deposits are generally small in volume and suggest short-lived, partial collapses of eruption columns, although deposits of much larger volumes are also described (Wright 1981; Houghton et al. 2004; Pensa et al. 2015b). The 95 ka Rio Caliente intra-plinian ignimbrite is a multi-flow unit deposit formed during a major Pleistocene explosive rhyolitic eruption of La Primavera volcano, Mexico (Wright 1981). The eruption sequence of the ignimbrite is complex and it occurs between lower and upper plinian fallout deposits. Air-fall layers, pyroclastic surge, mudflow, and fluviatile reworked pumice deposits also occur interbedded between ignimbrite flow units. Near-vent facies of the ignimbrite includes co-ignimbrite lag breccias (Sect. 10.8.10.3), which locate a central vent but one which could not have been associated with a well-defined edifice. The eruption sequence shows the reinstatement of plinian activity several times during the eruption after initial column collapse occurred.

This, together with erosional breaks and evidence that solidified fragments of already welded ignimbrite were re-ejected, all suggest the eruption lasted a relatively significant time period. Nearly 90 km^3 of tephra were erupted in total. The associated plinian pumice fall is one of the largest known having a volume of 50 km^3 and the ignimbrite, plus co-ignimbrite ash-fall deposits, have a compound volume of nearly 40 km^3.

Partial or total collapse of eruption columns can be caused by a variety of factors, including: (1) rapid increase in eruption mass discharge rate, usually associated with vent widening (Wilson et al. 1980); this is because the increase in width of the eruptive jet makes air entrainment less efficient and restricted to its outer region, resulting in the persistence, at the point of decay of vertical momentum, of a more or less pronounced inner region denser than the atmosphere which collapses back toward the surface forming pyroclastic flows (e.g. Trolese et al. 2019, and references therein); (2) vent widening is usually accompanied by the ejection of a large volume of dense lithic clasts explosively excavated from the vent into the column (e.g. Shea et al. 2011; Simmons et al. 2016), caused by the increase in discharge and/or by a short-lived vent blockage due to vent/conduit-wall collapse, followed by explosive vent clearing; (3) tapping and ejection of denser magma (volatile-poor, more crystallised, and/or more mafic) into the eruption column (Wilson et al. 1980; Balcone-Boissard et al. 2011); or (4) pulsatory phreatomagmatic eruption styles, which may involve eruption of denser, less vesicular pyroclasts (Di Muro et al. 2004; Van Eaton et al. 2012).

10.3.3 Continuous Collapse of Low, "Boiling over" Eruption Columns or Pyroclastic Fountains from Point Source Vents and Caldera Ring Fissure Vents

Most ignimbrites preserved in the rock record are more likely related to caldera collapses and range in volume from a few km^3 to >1000 km^3 (Giordano and Cas 2021; Roche et al. 2021). As early as 1960, Smith (1960a) suggested that such ignimbrites could be formed by an eruption column collapse mechanism, but on a very large scale. However, there is little to no consideration in the existing literature of the specific role that the transition from single point source vent to ring fault fissure conduits may have on the dynamics of the pyroclastic fountains. Thus, the model by Sparks and Wilson (1976) and Sparks et al. (1978) described in the previous section for the collapse of a point source vent-derived plinian eruption column is usually applied "*as-it-is*" to caldera-related ignimbrites. Although this practice may seem to be justified by the common presence of plinian fallout deposits beneath caldera-forming ignimbrites, there are cases where early plinian fallout deposits originate from a point source vent, whereas, the ensuing ignimbrites originate from ring fissure vents heralding the start of caldera collapse (e.g. the 1.6 and 1.26 Ma Lower and Upper Bandelier Tuff eruptions, Valles caldera, New Mexico, USA; Self et al. 1986, 1996; see Fig. 10.22d, e).

Nonetheless, many other ignimbrites, especially very large volume ignimbrites, are not underlain by any precursor fallout deposit, indicating the absence of a buoyant eruption column, even right from the start of the eruption (e.g. de Silva et al. 2006; Cas et al. 2011; Willcock et al. 2013). Numerical modelling also suggests that continuous fountaining is a viable mechanism for the generation of large volume ignimbrites (Giordano and Dobran 1994; Bursik and Woods 1996; Costa et al. 2018).

Later in this chapter (Sect. 10.11) we will discuss and demonstrate that caldera-related pyroclastic flows and associated ignimbrites erupted from caldera ring faults (Fig. 10.3) should not be envisaged and modelled just like the collapse of a previously buoyant single vent plinian column (see Jessop et al. 2016). Instead, they are, from their very beginning (i.e. from the formation of ring fault conduits), fed by sustained fountaining at mass fluxes much higher (>10^9 kg s^{-1} and up to 10^{11} kg s^{-1}) than those related to collapses of point source vent plinian columns ($\leq 10^9$ kg s^{-1}). Consequently, the presence or not of a plinian fall deposit at the base of caldera-related ignimbrites is in our opinion totally unrelated to the understanding of the dynamics of the ring-fissure-fed pyroclastic flows (e.g. Costa et al. 2018), and only tells whether or not during that specific eruption an early, pre-collapse point source vent existed, feeding a buoyant plinian column, with no requirement for an earlier buoyant column phase to occur (e.g. Cas et al. 2011; Willcock et al. 2013). At the same time, caldera subsidence does not necessarily take place totally in one event but may well be stepwise or incremental. Ring faults may open and close and shift position during the course of an eruption, therefore allowing the opening of single vents and short-lived plinian columns at different locations and times, before eventually evolving into fissure-fed pyroclastic fountains as ring faults propagate. Some caldera-related ignimbrites have been described as interlayered with thin plinian fallout deposits that may have such significance (e.g. the 25.4 ka Oruanui ignimbrite, New Zealand, Wilson 2001; the 0.76 Ma Bishop Tuff, USA, Hildreth and Wilson 2007).

The continuous high-intensity fountaining or boiling-over eruption style, directly from caldera ring fractures, even from the very beginning of an eruption, is therefore our preferred eruption mechanism for caldera-related pyroclastic flows and ignimbrite deposits (e.g. Bacon 1983). However, caldera-forming eruptions are rare and the few that have occurred in historical times are at the low end of the magnitude spectrum, and were either not witnessed (e.g. ~680 AD Rabaul caldera, New Britain, Papua New Guinea, McKee et al. 2015; ~1000 AD Changbaishan caldera, China-North Korea border, Sun et al. 2014), or not well constrained (e.g. 1257 AD Samalas (Rinjani) caldera, Lombok, Lavigne et al. 2013; Vidal et al. 2015; 1815 Tambora, Sumbawa, Stothers 1984; Self et al. 1984, 2004; Sigurdsson and Carey 1989; 1883 Krakatau, Verbeek 1885; Self and Rampino 1981; Self 1992).

Smaller volume boil-over flows from point source vents have also been described. The climactic eruption of Mt Pinatubo on 15 June 1991 was observed by geologists (although under poor viewing conditions due to the arrival of Typhoon Yunya), and culminated in formation of a small caldera, 2.5 km in diameter (Scott et al. 1996). This is inferred to have been created as the summit dome foundered, which disrupted the vent forcing the eruption through growing fractures and faults in the old summit dome. The onset of this is marked stratigraphically by the abrupt appearance of the lithic-rich facies (Sect. 10.8.10.3). In contrast to the radial distribution of pumiceous and ash flow deposits, lithic debris from the disintegrating summit was preferentially funnelled through the deep canyons of major drainages that headed on the former summit dome (Scott et al. 1996). Reduced height of pyroclastic flow formation in the eruption column, or perhaps a transition to low pyroclastic fountains issuing from numerous vents in the foundering caldera, may have localised the lithic-rich facies.

Rare, low intensity boiling over columns or fountains have instead been witnessed from single point source vents (Figs. 10.3c and 10.4d, e). Eye-witness accounts recorded by Wolf (1878) of the eruption of Cotopaxi volcano, Ecuador, in 1877 suggest that small volume pyroclastic flows may have originated directly out of the vent, and formed without the collapse of an eruption column, or from a column so dense that it only rose a small height above the vent and instantaneously collapsed and "bubbled over". Observations of the Mt St Helens 1980 eruption also suggest that many of the small volume pumiceous pyroclastic flows (e.g. Fig. 10.4c) were not generated by collapse of a high eruption column, but from low columns or fountains (Rowley et al. 1981). Many pumice flows seemed to spread out from bulbous inflated masses of pyroclasts as they upwelled a small height above the vent. The activity on 22 July 1980 at Mt St Helens, illustrated this particularly well, showing the development and instantaneous collapse of a fountain about 500 m high (Fig. 10.4d, e). Descriptions such as a "pot boiling over" were given (Rowley et al. 1981), and there are obvious similarities to the eyewitness descriptions given of the Cotopaxi eruption in 1877.

Mechanisms of continuous low-fountaining or boil-over directly from point-source vents or from caldera ring fractures from the beginning of an eruption may therefore be very important in the generation of some small, but especially for large volume pyroclastic flows (Figs. 10.3c and 10.4d, e; de Silva et al. 2006; Cas et al. 2011; Willcock et al. 2013).

10.3.4 Lateral Blasts (Forming Blast Flows)

Blast flows are rare events on Earth and the first witnessed example was the blast that was triggered by the lateral, sector, or landslide failure of the flanks of Mt St Helens volcano on May 18 1980 (Voight et al. 1981; Figs. 10.3d and 10.4f). The effect of the blast flow on the forest in its path is well known (see Fig. 1.15e). Sector collapse of the flank of the volcano occurred as a consequence of a prolonged cryptodome intrusion within the edifice (see Chap. 6). The giant sector collapse and landslide (Chap. 12), triggered by a magnitude 5.2 earthquake at 8.32 PDT, released the confining pressure on a rising dacite cryptodome, which was intruded high into the north flank of the volcano (Christiansen and Peterson 1981). The sector collapse landslide transformed into a dynamic rock debris avalanche, and was quickly followed by a laterally directed blast (Fig. 10.4f, taken after about 30 s), the direction of which was influenced by the instantaneous formation and topography of the amphitheatre-shaped landslide scar on the northern slope of the volcano. Explosions were generated by flashing of superheated groundwater, snow, and ice, as well as release of magmatic gases when the cryptodome and its hydrothermal system were depressurised and exposed by removal of the landslide mass.

The blast flow was relatively "hot and dry", consisting almost entirely of older volcanic rock debris from the collapsing edifice, shredded vegetation from the pre-existing forest that the blast flow destroyed, and relatively minor juvenile material derived from the cryptodome (<1%; Voight et al. 1981). At the time of emplacement much of the deposit was as hot (or cool) as 100 °C (Hoblitt et al. 1981). The flow has been variously called a pyroclastic surge and pyroclastic flow, but because it was triggered by a lateral volcanic explosion, its very dynamic nature and flow dynamics (Hoblitt et al. 1981; Kieffer 1981; Esposti Ongaro et al. 2011; Sect. 10.12), we prefer the term blast (pyroclastic) flow, as first introduced by Kieffer (1981). The eruption from an explosive disintegation of a cryptodome is completely different from gravitational collapse of lava domes that generate block and ash flows (Sects. 10.3.5 and 10.13). The initial events at Mt St Helens on 18 May also then triggered a nine-hour-long dacitic plinian eruption that produced pumice flows and ignimbrite deposits (e.g. Rowley et al. 1981; Brand et al. 2014, 2017; Mackaman-Lofland et al. 2014; Pallister et al. 2017).

Several historic eruptions have involved a sector collapse followed by a lateral blast, similar to the one observed at Mt St Helens (Belousov et al. 2007). The eruption of Bezymianny in 1956 (Gorshkov 1959; Belousov 1996; Belousov et al. 2007) produced a directed blast and pyroclastic flows as well as a very high eruption column dispersing air-fall ash but is sometimes given as an example of explosive dome collapse (e.g. Wright et al. 1980; Voight et al. 2002). The "agglomerate flow" of Gorshkov (1959) may have been a similar volcaniclastic debris avalanche flow to the Mt St Helens rockslide avalanche, containing a large proportion of non-juvenile material, judging from the horseshoe-shaped amphitheatre that was left after the eruption. However, the deposits also had a substantial amount of juvenile material and were originally interpreted as pyroclastic flow deposits.

Other eruptions, at Bandai-san, Japan in 1888 (Nakamura 1978), and Sheveluch, in Kamchatka in 1964 (Gorshkov and Dubik 1970), seem to have produced similar chaotic deposits, but made entirely or almost entirely from non-juvenile fragments. Therefore, there could be a broad spectrum of deposits produced by collapses of this kind, ranging from recognisable explosively generated block- and ash-flow deposits to blast flow deposits with no (or very little) juvenile material, to volcaniclastic debris avalanche deposits, as well as other types of pyroclastic flow deposit. Distinguishing such deposits, which lack a high proportion of juvenile fragments, from those formed by gravitational debris avalanche and debris flows is very difficult, even in the late Quaternary record. Criteria by which to identify debris avalanche deposits have been discussed by Ui (1983) and Siebert (1984).

Blast flows can also be generated by lateral blasts from lava dome complexes. Examples include the devastating eruption of Mt Pelée, on Martínique on 8 May 1902 (Lacroix 1904), and the sector collapse and lateral blast from the Soufrière Hills volcano, Montserrat, on 26 December 1997 (Voight et al. 2002; Ritchie et al. 2002; Sparks et al. 2002). The nature of the highly destructive and deadly pyroclastic density currents that destroyed the town of St Pierre (Fig. 10.1b) during the catastrophic 8 May 1902 eruption of Mt Pelée has been debated in the literature. The accounts of Lacroix (1904) indicate the eruption which generated block and ash flows was formed by a directed low angle explosion from some point near the base of the dome which was occupying the summit crater. Although the idea of a directed blast persists in the literature, later studies suggested that St Pierre was destroyed by an overriding ash-cloud surge (Roobol and Smith 1975; Fisher et al. 1980; Fisher and Heiken 1982; see Fig. 10.31a, b). The concept of an incredibly destructive initial flow pulse preceding the main part of a pyroclastic flow during the 1902 eruption was revived by Walker and McBroome (1983). This event killed almost 30,000 people, almost the entire population of St Pierre, and remains one of the most deadly, explosive events in historic times.

10.3.5 Lava Dome or Lava-Flow Collapse (Forming Block and Ash Flows and Bomb and Ash Flows)

This mechanism typically operates on lava dome complexes at the summit of steep-sided andesitic–dacitic volcanic cones, or during the eruption of silicic domes not related to major edifices. Fragmental flows of broken lava, called *block and ash flows*, are generated when the margins of an unstable, actively growing lava dome or lava-flow collapse passively (i.e. non-explosively) by gravitational collapse (Figs. 10.3e and 10.4g, h). These events are very short-lived at source (seconds-minutes) and involve a limited volume of volcanic debris, usually in the order of 10^5–10^7 m^3 (e.g. the collapses of the dacitic lava dome complex at Mt Unzen, Japan from 1991 to 1995, Nakada and Fujii 1993; Nakada et al. 1999; those from the andesitic dome during the 1995–2013 eruption at Soufrière Hills, Montserrat, Calder et al. 2002; and those from Colima summit dome in 2015, Pensa et al. 2018, 2019).

In addition, block and ash flows could be generated by explosions from within the dome, some of which can become explosively directed blasts (Fig. 10.3d), such as the 5 November 2010 event at Mt Merapi, Indonesia (Trolese et al. 2018). However, pressure release within a dome due to an initial gravitational collapse could lead to explosive collapse so, in some cases, both processes may occur (cf. Chap. 16; Chap. 6, Sect. 6.2.5.3). Explosions could also be triggered by explosive interaction between the growing dome with groundwater or ice. Such an eruption could therefore be considered to be phreatomagmatic. This also leads to the possibility that phreatic explosions could generate pdcs containing no juvenile fragments (e.g. Sheridan 1980). Therefore, different processes may have occurred at about the same time, and identifying the relative importance of each is the challenge.

Other terms used in the past for block and ash flows are hot avalanches and nuées ardentes (e.g. Lacroix 1904; Bardintzeff 1984; Cas and Wright 1987; Abdurachman et al. 2000). There is perhaps a semantic issue, for flows which are triggered by passive gravitational collapse of lava dome margins, without any explosive trigger. *Sensu stricto*, they are hot or cold rock falls and avalanches, whereas the term pyroclastic flow denotes an explosive origin. However, it is often impossible to tell whether or not explosive input was involved. Block and ash flows are small volume pyroclastic flows, dominated by dense lava dome juvenile clasts. Even the deposits of many separate flows or flow units accumulated during the same eruption typically have volumes <<1 km^3. Some recent examples indicate that block-and-ash flows associated with the highest peaks in dome extrusion rates may generate far-reaching and high energy pdcs, which are much more mobile than common block-and-ash flows (e.g. 2010 Mt Merapi, Komorowski et al. 2013; Trolese et al. 2018; 2015 Colima, Capra et al. 2016; Pensa et al. 2018).

Examples of historic eruptions during which block and ash flows were generated either by passive gravitational collapse of a dome or by explosively triggered lava dome or lava-flow collapse, include the eruptions of Mt Pelée, Martínique in 1902 and 1929–32 (Lacroix 1904; Perret 1935; Fisher and Heiken 1982), Mt Merapi, Indonesia in 1930, 1942–3, 1994, 2006, 2010 (Voight et al. 2000; Charbonnier and Gertisser 2008; Komorowski et al. 2013; Cronin et al. 2013; Trolese et al. 2018), the eruptions Hibok-Hibok, Philippines in 1951 (Macdonald and Alcaraz 1956), Mt Lamington, Papua New Guinea in 1951 (Taylor 1958) Santiaguito, Guatamala in 1967 and 1973 (Stoiber and Rose 1969; Rose et al. 1976, 1977), Colima, Mexico in 1991, 1999, 2015 (Saucedo et al. 2004; Sarocchi et al. 2011; Capra et al. 2016; Pensa et al. 2018, 2019), Mt Unzen, Japan in 1991–1995 (Sato et al. 1992; Yamamoto et al. 1993; Nakada and Fujii 1993; Nakada et al. 1999; Uehara et al. 2015), Soufrière Hills, Montserrat in 1995–2013 (Calder et al. 2002; Cole et al. 2002).

The most recent disaster from block and ash flows occurred on June 3, 2018 at Fuego volcano (Guatemala), when several hundreds of people were killed. Detailed description of the deposits along with the absence of a growing lava dome in the crater at the time of the event suggest that the material mobilised was taphra accumulated by small explosive eruptions in the preceding days and temporarily perched at the summit, which then collapsed (Risica et al. 2022). Another form of non-explosive mechanism of formation of pyroclastic density currents may also occur at basaltic volcanoes due to the gravitational destabilisation of low viscosity lava flow fronts or by the collapse of agglutinated hot material at the crater rim. This kind of rather unexpected event at mafic volcanoes has occurred repeatedly in the past years at Etna (Behncke 2009; Andronico et al 2018) and at Stromboli (Di Roberto et al. 2014; Di Traglia et al 2023), and their deposits are made of scoriaceous bombs and ash, so that although their gravitational origin can be broadly associated with that forming block and ash flows, descriptively they are bomb and ash flow deposits.

10.4 The Standard Eruption Column Collapse Model and Eruption Sequence in Forming Ignimbrites

Most of the eye witnessed mechanisms generating pyroclastic density currents have been from the collapse of relatively small volume vulcanian, sub-plinian, and plinian buoyant columns from point source vents. The largest was probably that of the 1991 eruption of Mt Pinatubo in the Philippines (8–10 km^3), which was well monitored even though the climax occurred under poor viewing conditions. These often, although not always (as in the case of Mt Pinatubo; Sect. 10.3.2), indicate a common sequence of activity:

- an early plinian phase producing a sustained, buoyant eruption column, and a resulting pumice fall deposit (Chap. 9; Fig. 10.5a), followed by
- a short-lived to sustained collapse of the eruption column, producing a pyroclastic flow phase, and deposition of ignimbrite and co-ignimbrite surge deposits (Fig. 10.5b).

The resultant stratigraphy thus consists of a basal plinian pumice fallout deposit, overlain by the ignimbrite (Fig. 10.6a, b). This sequence is thought to represent either the progressive tapping of deeper and less gas-rich, denser levels of the magma reservoir, and/or collapse of the conduit walls, causing increased discharge of dense juvenile clasts, and accessory lithic fragments, respectively. Both lead to overloading of the eruption column by dense solids, so that it is no longer buoyant, causing it to collapse, forming pyroclastic flows. Theoretical modelling of the origin of such pyroclastic flows has shown that the change from plinian to ignimbrite-forming activity varies with vent radius, gas velocity, H_2O content, and magma discharge rate (Wilson 1976; Sparks and Wilson 1976; Wilson et al. 1978, 1980; Chap. 9). These models suggest that the height of eruption columns, when collapsing above the vent, is between 0.6 and 9 km for the range of values of the controlling parameters considered.

However, more recent numerical models show that such figures at the higher end are totally unrealistic as collapses can occur only from within the gas thrust jet region of the column (i.e. before it becomes buoyant; see Chap. 9), which, even for plinian intensity, is confined to below 4–5 km in height at the most (Trolese et al. 2019). In addition, as the jet rises, dilution of the pyroclastic mixture occurs through air entrainment, progressively eroding the inner and denser part of the jet with height. As a consequence, the higher the zone of collapse, the more dilute will the resulting pyroclastic flow be. Partial collapses (<50% of the mass flux) may occur at various heights within the jet region (up to 5 km height), but substantial to total fountaining (>50% of the mass flux) is only attained at much lower heights (<2 km), as increasing the radius of the vent lowers the exit velocity of the jet (Fig. 10.7a; Trolese et al. 2019). These figures are important as it is still largely believed that high energy pdcs, such as those related to the well-known 232 AD Taupo ignimbrite (see Sect. 10.9.1.1), derive their mobility from the height of the collapse, when in fact, this belief is totally unrealistic and should be abandoned.

When eruption column collapse occurs, the pyroclastic mixture flows away from the volcano as a density current with high velocity. Initial velocities of the flows range from 60 to 300 m s^{-1}. Flows from large eruptions may still have velocities of 100 m s^{-1} at distances of 50 km from the source. The models show that initially the mixture is likely to be a highly turbulent, rapidly moving density current with a low particle concentration.

Numerical modelling of collapsing plinian columns is based essentially on these concepts (Sparks et al. 1978; Dobran et al. 1993; Neri et al. 2003; Suzuki et al. 2016; Trolese et al. 2019). When partial or total collapse occurs depends on the combination of magma–water content, vent radius and "muzzle" or exit velocity of the gas-solids mixture from the vent. Widening or flaring of the vent during the eruption will drive the column towards collapse conditions, as magma mass discharge rate increases, leading to reduction in the surface area available for air entrainment, and eventually to column overload.

An important concept recently discussed by Trolese et al. (2019) is the collapsing mass flux as the main control on the thermodynamics and mobility of pyroclastic flows. The percentage of the erupting mass flux that collapses and feeds pyroclastic flows is controlled by the amount of air entrainment before collapse. Air entrainment controls, for any given mass eruption rate, the partitioning between the mass flux that collapses and the mass flux that becomes buoyant. As air entrainment increases with increasing height, the absolute maximum height of collapse is inversely related to the percentage of collapsing mass flux (Fig. 10.7a). Hence the mobility of the resulting flow decreases as the collapse height increases, because the percentage of the collapsing mixture is less than the percentage of the mass flux that becomes buoyant (Fig. 10.7a, b). Actually, numerical simulations show that at the highest collapse elevations, the amount of collapsing mass is so low that many times it does not even reach the topography as air entrainment occurs during collapse further decreasing the density and promoting recycling of the partial collapse into the buoyant plume. By contrast, at lower collapse elevations and lower air entrainment levels, the pyroclastic mixture may collapse at nearly the total mass eruption flux available and at temperatures close to magmatic (Fig. 10.7b, c). We stress, therefore, that the most important initial parameter for understanding pyroclastic flow dynamics is the collapsing mass flux, which also controls the initial pyroclastic flow temperature (Fig. 10.7c; see Sect. 10.8.14), and not the column collapse height. An important outcome of increasing the percentage of mass that collapses is that the mass that is partitioned into the buoyant eruption column is reduced and consequently the total height of the plume also decreases during episodes of partial and total collapse (Fig. 10.7b; cf Chap. 9).

Scenarios of three model ignimbrite-forming eruptions (Fig. 10.8) show the variations in column height and gas velocity that take place as the conduit is eroded and gas content decreases during the course of an eruption. In the first case (Fig. 10.8a), vent radius increases while exsolved water content remains constant at 3.45 wt%. The eruption column height grows steadily as the mass eruption rate increases with the enlargement of the conduit, until sudden collapse occurs at a critical value and an ignimbrite-forming fountain is established. Eruption velocity in the vent slowly increases throughout the eruption. In the second case (Fig. 10.8b), vent radius increases and exsolved gas content decreases. Variation in height of the eruption column is similar to that found in the first case, though fountain height decreases after the collapse event. Eruption velocity decreases throughout the eruption, due to the falling gas content. In the third case (Fig. 10.8c), vent radius remains fixed and gas content decreases during the eruption. There is a much smaller variation in plinian column height than in the first two cases. Again, fountain height gradually decreases after the collapse event.

In Fig. 10.9a, combinations of the controlling parameters are plotted and fields of convecting plinian and collapsing columns are shown. From knowledge of the stratigraphy of a deposit, and by estimating values of the controlling parameters, it is possible to track the progress and changes in energetics of an eruption to a first-order approximation. Three examples are shown in Fig. 10.9b, c, d. Eruptions that terminate before column collapse conditions are reached will only produce plinian fall deposits (Fig. 10.9b; Chap. 9).

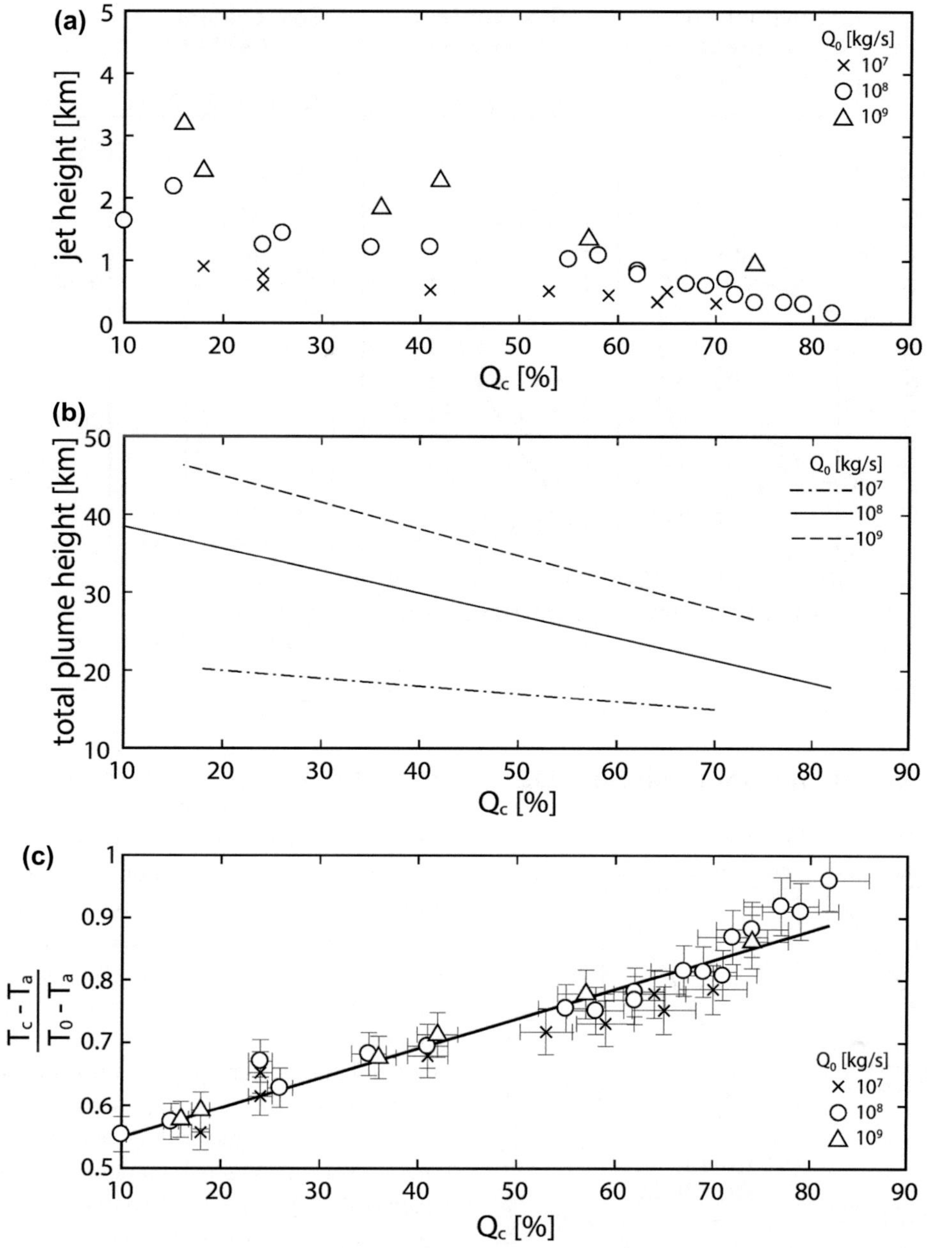

Fig. 10.7 Results of numerical simulations of partial to near total collapse of sustained plinian columns with mass eruption rates, Q_O, ranging from 10^7 to 10^9 kg s^{-1}, at varying gas contents from 0.5 to 6 wt %. **a** Jet height in km at collapse as a function of the average percentage of mass collapsed, Qc. Note that all values of collapse heights are below 3.5 km and that total or near total collapses (Qc > 60%) occur at jet heights below 1 km. **b** Total plume height in km as a function of the average percentage of mass collapsed, Qc; note that at increasing % of collapse the proportion of pyroclastic mixture that remains buoyant decreases and so does, as a consequence, the total height of the column that could be observed from distance. **c** Temperature at collapse, Tc, relative to the air temperature, Ta, with respect to the initial magmatic temperature, T_O, as a function of the average percentage of mass collapsed, Qc; in this case, irrespective of the mass eruption rate, there is a linear decrease in temperature at decreasing % of collapse, where the initial temperature can decrease by up to 50%, indicating that the regime of collapse has a strong control on the temperature of the resulting pyroclastic flows and ultimately on the temperature of emplacement, and thus on the related hazard. In essence, welded ignimbrites can only result from total or near total collapses of the columns. (Modified from Trolese et al. 2019)

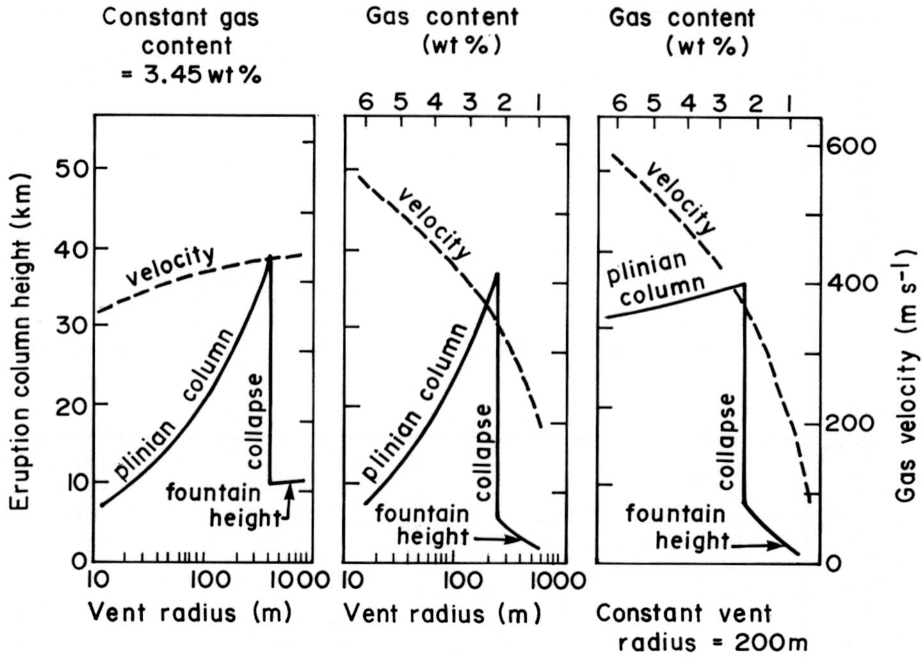

Fig. 10.8 Scenarios of three model ignimbrite-forming eruptions (see text for discussion) (From Cas and Wright 1987, After Wilson et al. 1980)

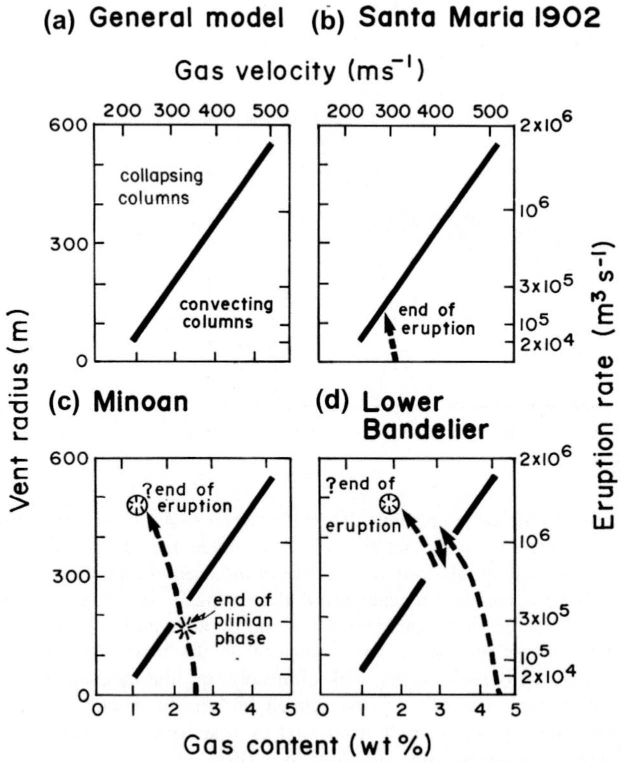

Fig. 10.9 **a** General model showing plot of vent radius, gas content, eruption rate and gas velocity, relating these parameters to convecting plinian, and collapsing ignimbrite-forming eruption columns (After Wilson et al. 1980). **b–d** Changing conditions during three eruptions (After Williams and Self 1983; Wilson 1980; Self et al. 1986, 1996). (From Cas and Wright 1987)

Williams and Self (1983) suggested that the Santá Maria 1902 eruption in Guatemala resulted from release of volatile-rich magma at the top of a volatile stratified magma chamber. As the eruption proceeded, denser, less volatile-rich magma was encountered and structural instabilities caused collapse of the conduit walls, abruptly shutting off the magma supply. The remaining magma was unable to generate sufficient volatile pressure to clear the conduit, and the plinian phase ended; it was 20 years later when volatile-poor magma reached the surface as lava.

The 1600 BC (3.6 ka) Minoan eruption (Santorini, Greece) was caldera-forming and the single vent plinian column was maintained until caldera collapse conditions developed (Fig. 10.9c), although a phase of phreatomagmatic activity between the plinian and pyroclastic flow-forming phases may have complicated the transition between the two (Druitt 2014). As the 1.6 Ma Lower Bandelier plinian eruption column (Valles caldera, New Mexico; Self et al. 1986, 1996) neared the point of collapse (Fig. 10.9d), slight fluctuations in gas velocity and mass discharge rate caused instabilities in the column and small volume intra-plinian pyroclastic flow units to form. Instabilities in eruption columns nearing collapse conditions would account for the stratification found at the tops of many plinian deposits (Chap. 9). The following catastrophic column collapse is instead more likely related to the opening of caldera ring fractures. The transition from single vent plinian eruption columns to ring fault eruption columns involves the enlargement of the vent area and therefore, even though the geometry changes from axisymmetric (circular vent) to linear, the net effect is the increase of the mass eruption rate, the reduction of the exit velocity for given gas contents, the reduction of air entrainment, hence leading to collapse (cf. Jessop et al. 2016).

Some deposits show much more complex eruption sequences than indicated above, and ignimbrite flow units are interstratified with plinian fall units (Chap. 9; e.g. the 1991 eruption of Pinatubo, Philippines, Scott et al. 1996; 4.6 ka Fogo A sequence, São Miguel, Pensa et al. 2015b; Fig. 10.6c; 95 ka Rio Caliente ignimbrite, Mexico, Wright 1981), or are found laterally equivalent to continuous fall deposition as in the case of the 1912 Valley of Ten Thousands Smokes ignimbrite at Novarupta in Alaska (Houghton et al. 2004; Hildreth and Fierstein 2012), and the 1815 eruption of Tambora, Sumbawa (Sigurdsson and Carey 1989). Ignimbrites that are interstratified with contemporaneous plinian fallout deposits are called intra-plinian ignimbrites (Sect. 10.3.2). In terms of plume dynamics, these ignimbrites indicate that the plinian column only promoted partial collapses while remaining buoyant. During partial collapses, a significant part of the mass flux is partitioned into the collapsing flow and this should result in fluctuation of the column height, as the buoyant mass flux component is temporarily reduced (cf. Chap. 9).

Very complicated eruption sequences can also be produced if external water-magma interaction is involved and collapse of phreatoplinian eruption columns takes place (Sheridan et al. 1981; Self 1983; Wilson 2001; Edgar et al. 2002; De Rita et al. 2002). Fluctuating magmatic and phreatomagmatic explosive influences during an extended plinian eruption can produce remarkably complex sequences of interstratified fallout deposits, intra-plinian ignimbrites, and surge deposits (e.g. 0.24 Ma Poris Member, Tenerife; Edgar et al. 2002).

However, not all ignimbrite-forming eruptions have an earlier plinian phase. This is particularly the case for very large volume, caldera-forming eruptions, such as the large volume, caldera-forming, 2.08 Ma Cerro Galán ignimbrite eruption, Argentina (Cas et al. 2011; Folkes et al. 2011a, b), and the Permian Ora ignimbrite eruption, Italy (Willcock et al. 2013, 2015a). This suggests that these very large ignimbrites formed by a rather different eruption mechanism than that proposed for the well-described small volume ignimbrites associated with collapse of single vent plinian columns. Right from the start, these large volume caldera-forming eruptions must have involved a continuous fountaining of the gas-pyroclast mixture, issuing directly from caldera ring fracture vents, and directly feeding a long lived, high mass flux pyroclastic flow event, with the appearance of "a pot boiling over" (Fig. 10.3c; Jessop et al. 2016). However, observations at Mt St Helens have shown that even during some small eruptions, a high eruption column is not an essential prerequisite for generating a pyroclastic flow (e.g. Cas and Wright 1987, Fig. 5.12). In such cases the venting eruptive mixture may be so dense, or the magma discharge rate is so high that a well-defined, maintained eruption column does not form.

10.5 Anatomy of Pyroclastic Density Currents

Irrespective of their size, duration, geometry, and componentry, pyroclastic density currents have the common external appearance of a billowing, generally dark, highly turbulent cloud of ash and gas that expands both laterally and vertically (Figs. 10.4 and 10.10).

Since the description of the pyroclastic flows during the 1980 eruption at Mt St Helens (Rowley et al. 1981; Brand et al. 2014, 2016, 2023), many pyroclastic density currents have been observed and well described, such as from the 1991–1995 crisis at Mt Unzen, Japan (Yamamoto et al. 1993; Nakada et al. 1999), the 1991 eruption at Colima, Mexico (Saucedo et al. 2004), the 1991 eruption of Mt Pinatubo, Philippines (Scott et al. 1996), and the 1995–2013 eruption at Soufrière Hills, Montserrat (Cole et al. 2002; Loughlin et al. 2002). Understanding of the anatomy of pyroclastic density currents and their dynamics has also increased substantially with the aid of analogue and numerical modelling (Roche et al. 2010, 2011, 2016; Dobran et al. 1993; Giordano and Dobran, 1994; Neri et al. 2003; Dartevelle et al. 2004; Esposti Ongaro et al. 2008; Doronzo et al. 2011; Sulpizio et al. 2014; Lube et al. 2015; Breard and Lube 2017; Smith et al. 2020; Gueugneau et al. 2021; Brosch et al. 2022; Calabrò et al. 2022).

Based on morphology, three main morphological and dynamic parts of pyroclastic flows can be recognised in longitudinal and cross-sectional views (Figs. 10.10 and 10.11):

- the flow front or head,
- the body, and
- the enclosing ash cloud.

The *flow front or head* is the ground-hugging, leading "front-edge" of pyroclastic density currents that actively forces the current to penetrate and displace the atmosphere and is the first part of the flow where entrainment of air, vegetation, and ambient water/snow occurs. The flow front

Fig. 10.10 Anatomy of pyroclastic flows: head or front, body (internally stratified in density and velocity) and ash cloud. **a** Pyroclastic flows generated during the 1975 eruption at Ngauruhoe volcano, New Zealand (Photo by Herb Spannagl; *Source* Williams 2001). A = ash-cloud or tail; B = body; F = head or frontal ash cloud. Note the darker colour for the body and pale colour for the tail, indicative of different particle concentrations, and also the larger turbulent billows in the trailing and frontal ash cloud compared with the body. **b** Pyroclastic flow on 24 June 1993 during the eruption at Mt Unzen (Japan). Note that the interior of the flow is entirely and three-dimensionally enveloped by the billowing ash-cloud, part of which is lofting above and trailing the laterally moving flow. (*Source* Volcanological Society of Japan). **c** Essential elements of a pyroclastic density current in 2D longitudinal section

is visibly highly turbulent (Figs. 10.10a, b and 10.11a), usually consists of multiple, vigorously billowing lobes and intervening clefts, and represents the front of the ash-cloud surge that encloses the entire pyroclastic flow (Figs. 10.10b and 10.11a–c), and spreads laterally from the sides of the flow, and trails behind and above it (Fig. 10.10a, b). The flow state of the flow front is the most ephemeral in the pdc. Although it develops at the initiation of the flow, and is sustained during flow advancement, the turbulence levels, and the number and shape of pulsing flow lobes and clefts, constantly change. Once the front reaches its maximum runout, which usually takes minutes to tens of minutes based on observed flow velocities but could even last for hours for very large flow systems, the flow front stalls, and the turbulent frontal ash cloud begins to rise buoyantly. The flow system behind the flow front may still be transporting material to the runout limits, but a forward propagating flow front will not be visible unless the flow front re-activates during waxing pulses fed by increased discharge from the vent into the flow system.

The flow front is followed by the "*body*" of the current, which is enclosed in the ash cloud and has less interaction with the atmosphere (Figs. 10.10a, c and 10.11b, c). The body carries the largest part of the volcanic debris, is denser than the atmosphere but is internally stratified in terms of concentration and velocity, moves along topographic lows (Fig. 10.12), and is rarely visible, usually being completely enveloped by the frontal, lateral and trailing ash cloud (Figs. 10.10 and 10.11). The extent to which the body is controlled by topography depends on its internal particle concentration, velocity profiles (Fig. 10.13) and the flow mass flux. In sustained flows, the body stabilises and feeds the aggradation of the bulk of the deposit over time (Fig. 10.13), which is also dependent on variations in mass flux supply rates at vent (Doronzo 2012). The body grades vertically and laterally, and more or less abruptly, into the ash cloud (Figs. 10.10, 10.11, 10.12 and 10.19). Because the body is not visible, there is considerable debate on its internal stratification and about how turbulent it is (e.g. Cas et al. 2011). The internal turbulence of the flow body varies as a function of the flow velocity and particle concentration profiles, which control the flow bulk density, bulk viscosity, and flow thickness (Fig. 10.13). This implies that the body may experience low levels of turbulence at its base where the particle concentration is high, becoming more and more turbulent as it progressively grades to less dense regions upward. Although usually the body is represented as one flow region (Figs 10.10, 10.11, 10.12), due to its internal stratification, an accurate description of the body should include both the dense basal current that feeds the deposit and part of the overriding more dilute current that feeds both the dense basal current and the ash cloud surge (Figs 10.13 and 10.19). The bulk flow behaviour, i.e. the visible flow propagation velocity, the degree of topographic confinement, the maxiumum runout, depend on whether the leading part of the body is the basal high concentration, essentially laminar undercurrent (Fig. 10.13, concentration profile c and velocity profile d) or the more dilute and

Fig. 10.11 Morphological anatomy of pyroclastic flows. **a** Frontal view of spreading pyroclastic flow lobes, during the 24 June 1993, eruption at Mt Unzen, Japan. Note the multiple flow lobes (L), separated by clefts in the flow front. The small white boxes are farm buildings, which were destroyed. (*Source* Volcanological Society of Japan). **b** Essential elements of a pyroclastic density current in frontal cross-section, and **c** in plan view. **d** Aerial view of multiple levees and depositional lobes from 1980 Mt St Helens eruptions, USA (*Source* United States Geological Survey). **e** Ground view of coarse, pumiceous flow front of depositional lobe, with Mt St Helens collapse scar in background (Photo by D. Swanson, taken 30-5-1980; *Source* United States Geological Survey)

turbulent overriding part of the body flow (Fig. 10.13, concentration profile a and velocity profile b). In the former case, pyroclastic flows will reach their maximum runout due to freezing of the undercurrent; in the latter case, pyroclastic flows will reach their maximum runout due to dissipation of the negative buoyancy (induced by sedimentation and air entrainment). The flow may still have high velocity when it literally lifts off forming huge and vigorous co-ignimbrite plumes, which may reach heights of plinian columns (Fig. 10.19; Calabrò et al. 2022). In addition the two parts of the body at topographic barriers and at the distal end may detach and move independently (e.g. Shimizu et al. 2019, Silleni et al. 2024). Although some pyroclastic flows are visibly high velocity flows, and it has traditionally been assumed that flow velocities have to be high to maintain flow mobility, some research indicates that high particle concentrations of fine ash traps gas, keeps the flow fluidised, and allows long distance flow (Roche et al. 2006, 2022; Druitt et al. 2007), even at low velocities (Cas et al. 2011; Roche et al. 2016; Báez et al. 2020b). However, it is important to take into account that the velocities recorded in the deposits are more directly related to the high concentration basal undercurrent and not necessarily to the more dilute part of the body (Fig. 10.19). In consideration of the vertical and lateral stratification in terms of density and velocity, flows will certainly experience significant transitions downcurrent and across topography (Figs. 10.14 and 10.19). It is important to distinguish between regions where the body is essentially

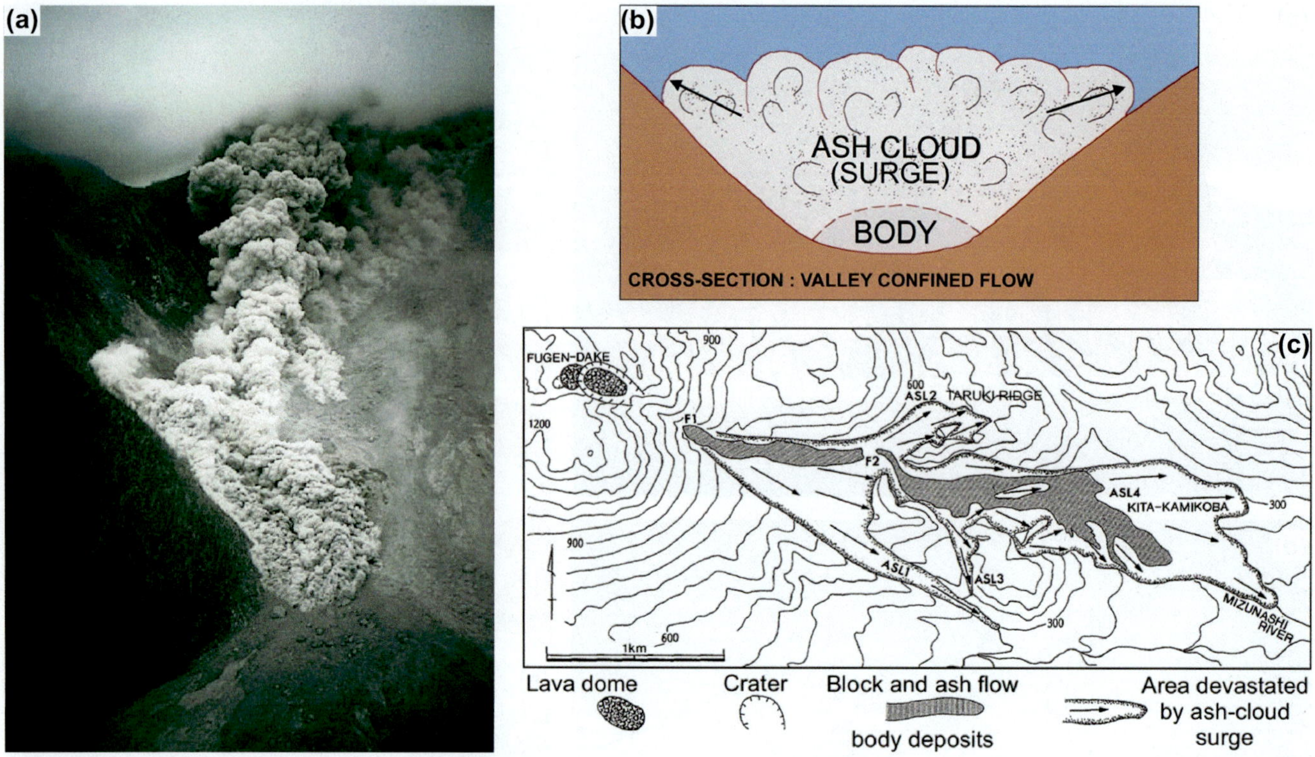

Fig. 10.12 Effects of paleotopography on the flow behaviour of pyroclastic flows. **a** Valley confinement of a pyroclastic flow in the Tar River Valley from the Soufrière Hills lava dome, Montserrat. Note the pale ash cloud climbing the valley wall at a bend, whereas the body is deflected around the bend. (*Source* Montserrat Volcano Observatory). **b** Schematic representation of the expansion of the flow sideways up confining valley walls. **c** Map showing the distribution of the block and ash flow body deposit and detached ash-cloud surge deposits from the 3 June 1991 flow event at Fugen-Dake, Mt Unzen, Japan (see Fig. 10.4g–i). The surges (arrows) detached from the body of the flow in the valley axis and surmounted higher topography on the valley walls and adjacent ridges, as well as flowing further down valley. The surge that surmounted Taruki Ridge climbed 200 m, and the surge that inundated Kita-Kamikoba killing 43 people, including volcanologists Harry Glicken, and Katia and Maurice Krafft. (From Yamamoto et al. 1993)

dilute (thick), turbulent, and faster than its high concentration undercurrent (velocity profile b in Fig. 10.13 and t1–t3 in Fig. 10.19), and regions where the body deflates and is essentially the high concentration (thin) undercurrent that leads the flow propagation (velocity profile d in Fig. 10.13 and tf in Fig. 10.19). The former body regions have been successfully modelled as dilute inertial flows at average densities of 10–100 kg m^{-3} (e.g. Bursik and Woods 1996; Andrews and Manga 2012; Shimizu et al. 2019; Brosch et al. 2022; Calabrò et al. 2022), whereas for the latter, at densities of 100–1000 kg m^{-3}, granular frictional models have been shown to be appropriate (e.g. Kelfoun et al. 2009; Smith et al. 2020; Gueugneau et al. 2021).

The "*ash cloud*" is the highly dynamic enveloping dilute part of a current, carrying progressively finer and lighter particles upward. Conceptually, the ash cloud consists of an inner portion close to the higher concentration part of the flow (the body) which is still denser than air (>1 kg m^{-3}) and moves laterally (surge), grading upward, at the sides and at the front in a vertically buoyant (density <1 kg m^{-3}) and trailing ash cloud (Figs 10.13 and 10.19). The ash cloud is visibly turbulent everywhere (Figs. 10.10, 10.11 and 10.12). This ash cloud is characterised by large eddies of progressively paler colour upward, as air is entrained and the concentration of particles decreases (Figs. 10.10a, b, 10.11a and 10.12). Upwards, as the ash cloud becomes progressively less dense than the atmosphere, it becomes buoyant and lofts above the laterally moving flow (Fig. 10.10b). This buoyant ash cloud, also known as an "ash plume" or "phoenix cloud", is capable of transporting and depositing large volumes of (co-ignimbrite) fallout ash large distances beyond the limits of the pdc deposits (e.g. the 74 ka Toba co-ignimbrite ash, Ninkovich et al. 1978; Rose and Chenser 1987; 39 ka Campanian co-ignimbrite ash, Smith et al. 2016; the 75 ka Los Chocoyos Ash, Cisneros de León et al. 2021; Chap. 9, see Sects. 9.7.6 and 9.9.6).

Depending on topography, pdcs may be channelled or confined in a valley (Fig. 10.12a–c) or expand on flat land (Fig. 10.11a). In the latter case, the frontal expansion of the flow generates instabilities and the formation of lobes with different flow velocities. Even in narrow, steep-sided valleys, dilute ash clouds may spread laterally up, onto topographic highs (Fig. 10.12b), and may even detach from the more topographically confined, high-concentration undercurrents flowing along the valley axis (Fisher 1995; Ogburn et al. 2014). This occurred several times during block and ash pyroclastic events at Mt Unzen, Japan, in 1991 (Fig. 10.12c; Yamamoto et al. 1993; Fujii and Nakada 1999), and is documented from ~2000 yr BP pumiceous pyroclastic flow deposits erupted from Mt Pelée, accounts of which describe zones of topographic lift-off and buoyant plume development in northern Martinique (Wright et al. 2016; see Fig. 10.31c).

However, it is also important to stress that the flow size with respect to topographic roughness and the type of density stratification inside the flow both play an important role in defining the degree and style of interaction with topography. Small directional flows, which quickly develop a high concentration basal undercurrent sharply grading into the overlying ash cloud, are more prone to show a strong channelisation of the dense undercurrent along rough topographies and detachment from the overriding ash cloud, which may travel independently across topographic highs (Figs. 10.12a, c, 10.31 and 10.32). By contrast, large and thick flows with smoother internal density profiles are less affected by small topographies and may spread radially unless topography becomes of comparable size (Figs. 10.36 and 10.62; cf Andrews and Manga 2012; Giordano and Doronzo 2017; Calabrò et al. 2022). Detached ash cloud surges may be very hazardous even though their dilute nature is rarely associated with high dynamic pressures or with the emplacement of more than a few decimetres of fine pyroclastic material. Historical casualties, injuries and damage from ash cloud surges mostly derive from their high temperature (Baxter et al. 2017), which is initially the same as the parent high concentration flow (Pensa et al. 2019). Examples of their deadly consequences are 79AD Vesuvius eruption (Mastrolorenzo et al. 2010; Pensa et al. 2023), the June 3, 1991 event at Mt Unzen, Japan (Yamamoto et al. 1993), and the 2018 Fuego eruption, Guatemala (Risica et al. 2022).

10.6 Factors Influencing the Behaviour of Pyroclastic Density Currents

Pyroclastic density currents are density and velocity stratified (Fig. 10.13) and subject to flow transformations along their flow paths during the flow event. Understanding of all the contributing and controlling factors of these phenomena is still to be achieved.

The main factors that control the flow behaviour of pdcs and their mobility are the eruption style and origin of the pdc, the collapsing mass flux that feeds the pdc and its duration, the collapse height, the temperature and volatile content of the juvenile clasts, the initial total grain size and density distribution of the clast population, the evolution along flow of the particle concentration and velocity profiles, the degree of turbulence and coupling of fine ash with the gas phase, the degree of in-flow ash aggregation, the amount of entrainment of air, ambient vegetation and water and their impact on the lateral and vertical flow temperature gradients, the topography, and the duration of the flow event in comparison with the duration of the collapsing event at source. All of the above affect the kind of particle support mechanisms (Cas and Wright 1987; Branney and Kokelaar 2002; Sulpizio et al. 2014; Dufek 2016), the sedimentation rate at different times and locations along the flow path (Bursik and Woods 1996; Shimizu et al. 2019; Calabrò et al. 2022), and all can change during an eruption, affecting the flow mobility (i.e. runout; Shimizu et al. 2019; Giordano and Cas 2021; Roche et al. 2021, 2022; Brosch et al. 2022). The lack of direct observation of the interior of pdcs has so far prevented the definition of the unique roles and interplay of these factors.

Analogue modelling helps in understanding the sedimentation processes (e.g. Roche et al. 2008, 2011, 2016; Dufek et al. 2009; Andrews and Manga 2012; Rowley et al. 2014; Sulpizio et al. 2014; Lube et al. 2015; Breard and Lube 2017; Smith et al. 2020; Brosch et al. 2022), although due to technical limitations, models deal with cold or only slightly warm, and usually water or air supported analogues. There are also problems in scaling the physical parameters to those in the much larger scale, real world (Andrews and Manga 2012; Sulpizio et al. 2014). Numerical modelling has allowed significant advances in the understanding of the bulk behaviour of pyroclastic density currents, including temperature and the role of coupling of gas and particles (e.g. Valentine and Wohletz 1989; Dobran et al. 1993; Burgisser and Bergantz 2002; Neri et al. 2003; Dartvelle et al. 2004; Esposti Ongaro et al. 2008, 2020a, b; Doronzo et al. 2010, 2011; Roche 2015; Dufek et al. 2015; Dufek 2016; Shimizu et al. 2019, 2021; Roche et al. 2021; Calabrò et al. 2022), but computational limitations have so far prevented their use in fully describing what happens to particles at the point of sedimentation and the polydisperse nature of pdcs, largely because most numerical models cannot handle the complete grain size spectrum carried by pyroclastic density currents, and all the variables involved in determining flow dynamics.

The consequence of the density and velocity stratification of pdcs (Druitt 1998; Branney and Kokelaar 2002; Burgisser and Bergantz 2002; Dufek 2016) is that these parameters are characterised by gradients, which induce variations in rheology and shear within the pdc itself. Sedimentation is controlled by the amount of debris that the current cannot fully support anymore, which moves towards the base, where particles experience variable degrees of shear and

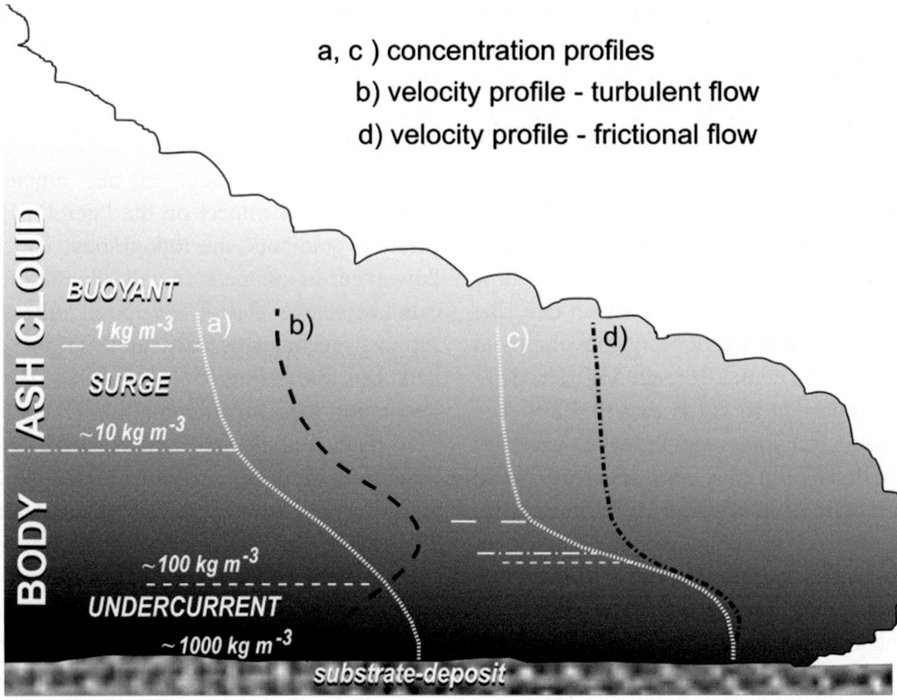

Fig. 10.13 Schematic representation of the vertical variations in horizontal flow velocity and particle concentration in a pyroclastic density current. Flow density boundaries that separate the body from the ash cloud, as well as their internal subdivisions (body undercurrent, ash cloud surge and buoyant ash cloud) are indicative. The concentration profiles (a, c) decrease continuously upwards, and the shape of the velocity profile may vary between end-members that define different flow types. Where the more dilute part of the body moves faster than the rest of the flow (b), the bulk flow propagation is controlled by its turbulent behaviour. Where the basal high concentration undercurrent moves faster than the rest of the flow (d), the bulk flow propagation is controlled by its frictional behaviour. In both cases the velocity profile is broadly parabolic, suggesting that the flow velocity is highest above ground level, and implying the flow front must roll over. The proposed concentration and velocity gradients are qualitative and vary among flows and within flows with time and space (see Fig. 10.14). Conceptually important is that regions at high concentrations and low velocities will have low Reynolds numbers therefore more likely to experience bulk laminar flow behaviour, whereas at low concentrations and/or high velocities turbulence will dominate. This implies that gradations will occur within pdcs as they evolve in time and space. The flow behaviour near the vent will be markedly different to flow behaviour at the distal advancing flow margins.

processes that delay their incorporation into the actual deposit. The region where deposition occurs is named the *depositional boundary layer* (Fisher 1966, 1995; Valentine 1987) or lower flow boundary layer (Branney and Kokelaar 2002), and it may define either a sharp or variably gradational rheological interface with the overridding and moving current (Fig. 10.14). Depending on the flow state, the depositional boundary layer may be only cm to tens of cm thick in a low particle concentration, highly turbulent flow (e.g. near vent), or many m thick in a high particle concentration granular flow zone in medial to distal reaches. The nature of this gradation may vary from proximal to distal parts of the flow system. Coupling between the stratified flow and the depositional system results in the vertical aggradation of the pdc deposit. By contrast, if and when the high-concentration basal layer decouples from the overriding low-concentration ash cloud, it may move laterally for some distances and deposit independently (Figs. 10.12a, c and 10.31).

A key concept is that the particle load in pdcs is polydisperse in terms of grain size (particles usually range from microns of the fine ash fraction, to decimeters and metres for blocks and bombs), density (pumice density is usually <1000 kg m^{-3} and lithic clast density is usually >2500 kg m^{-3}), particle temperature and rheology (behaviour ranges from brittle for lithics and pumice below the glass transition temperature to ductile for spatter clasts, pumice and ash above the glass transition temperature, that deform during and after transport) and aerodynamic properties (shapes vary from equant, to oblate, to prolate and from smooth to highly irregular). Furthermore, based on the fragility and shape of particles, the amount of particle–particle interaction within the flow usually produces abrasion and spalling of clasts, causing changes in the particle-size distribution during flow (Dufek and Manga 2008; Manga et al. 2011; Keuppers et al. 2012). This begins in the conduit flow stages, involving milling of large clasts (e.g. Campbell et al. 2013), and continues in the pyroclastic flow beyond the deflation zone.

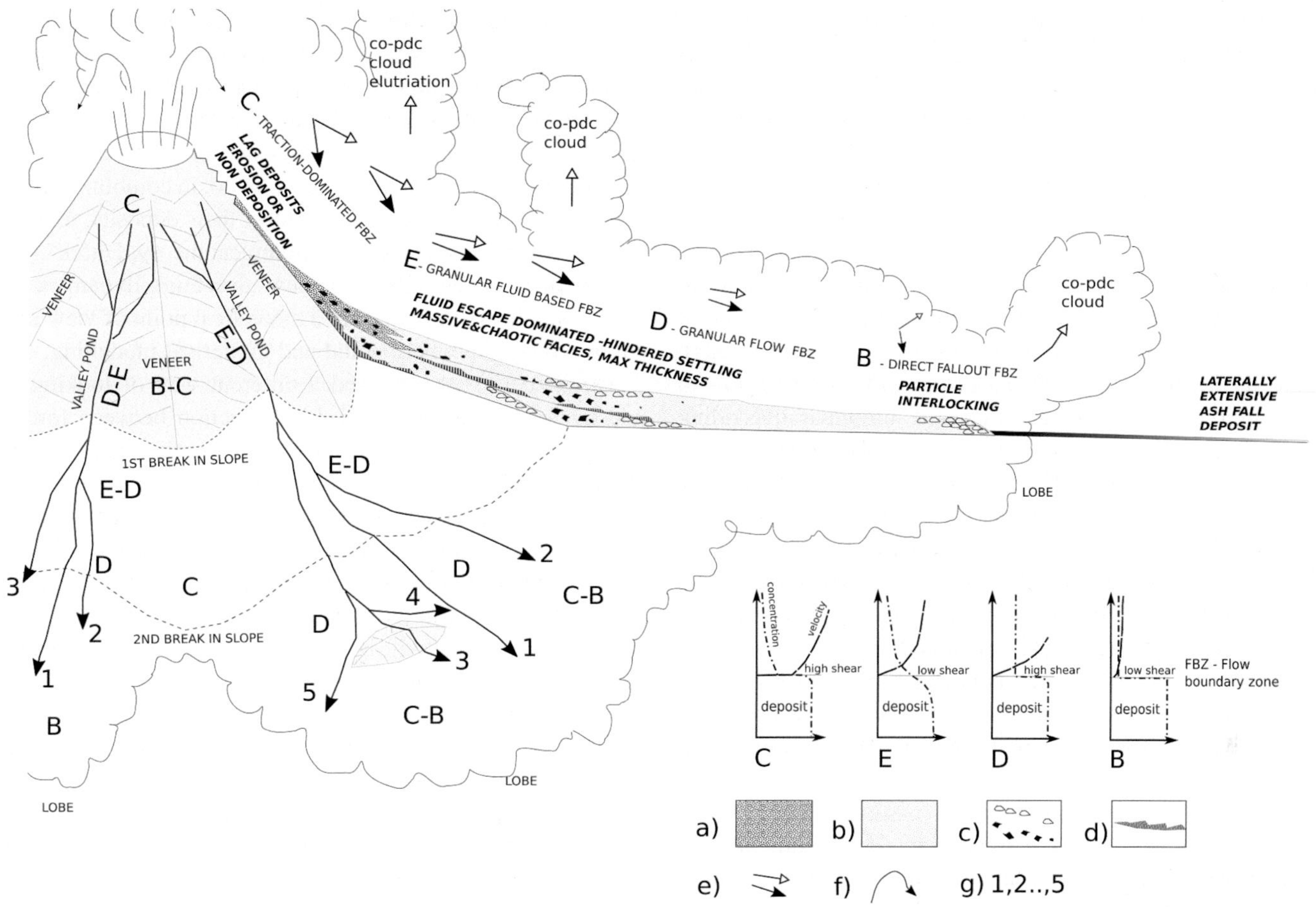

Fig. 10.14 Schematic model (not to scale) summarising the main possible depositional scenarios along and across a pyroclastic flow, including concepts from Branney and Kokelaar (2002). Waxing and waning at source may occur several times over sustained flow events and result in complex vertical and lateral facies architectures. Legend: B, C, D, E are vertical profiles of velocity and particle concentration in pyroclastic flows across the flow boundary zone; **a**) lithic-rich facies; **b**) standard ignimbrite facies from massive to stratified; **c**) lithic (black) and pumice (white) concentration zones; **d**) ground breccia layers; **e**) particle (black) and gas (white) velocity vectors; **f**) pyroclastic flow paths; **g**) possible order of depositional events as a response to the interplay between eruption dynamics, interaction with topography and deposition

In experiments, Dufek and Manga (2008) found that collision (above the bedload) and frictional interaction (in the bedload) between pumice clasts produced fine ash that is mostly 10–100 μm in size. Such ash has gas-trapping capacity, that enhances flow mobility (Roche et al. 2006; Druitt et al. 2007). This implies that by varying the current bulk properties of velocity, particle size and particle concentration, particle support mechanisms will change in time and location along the flow path for different particle populations.

Another key aspect of this changing behaviour is the manner in which the dynamic fluid of the flow changes. In low particle concentration flows, in which solids are widely dispersed in gas, the gas is the dynamic fluid that controls the behaviour of the solids by virtue of its velocity and level of turbulence. Once particle concentration increases, the finest particle size couples with the gas and together they constitute the dynamic fluid that then controls the behaviour of larger particles. At high particle concentrations the ash and gas form a dense, fluidised granular flow that effectively forms the dynamic fluid that controls the behaviour of the coarsest particles. As the bulk density of the dynamic fluid increases, turbulence levels in the flow decrease, more particle-to-particle interaction occurs, and the developing granular dispersion applies more force and buoyancy to larger clasts, causing them to continue moving. This basal high-concentration part of the flow behaves as a granular flow, in which turbulence is significantly suppressed, grain-to-grain momentum transfer and fluidisation contribute more to flow mobility than turbulence, and friction becomes progressively more important (Roche et al. 2010, 2011; Dufek 2016).

Increasing particle concentration, especially of fine ash, reduces the permeability of a flow, slowing down the rate of gas loss. This is particularly the case with very fine ash, which in high concentrations, traps the hot gas, hinders its natural tendency to rise and escape, and thus provides upwards directed fluidisation support to the ash particles, hindering their settling or sedimentation. The concentrated dispersion of gas and fine ash then provides support and hinders settling of coarser sizes. The rheology and fluid dynamic behaviour of the flow system varies in vertical profile from top (low particle concentration pdc or surge) to base (high particle concentration granular flow) at any point along the flow profile, from proximal to distal and through time, as the flow system changes from waxing phase flow to waning phase flow. In addition, the turbulent overriding surge of a pdc may detach from the basal granular flow, producing two discrete flow pulses (i.e. the basal dense flow and the ash cloud surge; Figs. 10.31 and 10.32), with different flow dynamics, from the one pdc event.

A useful approach to such complexity can be found in the sedimentological literature dedicated to aqueous sediment gravity flows. Lowe (1982) identified three main particle populations in terms of their behaviour in aqueous granular flows: particles ranging from clay to medium-sand size, i.e. the equivalent of fine ash, are those that can be maintained in suspension by the turbulence of the fluid medium alone, independent of their concentration; particles ranging from coarse-sand to small-pebble gravel size, i.e. the equivalent of coarse ash- to lapilli-size, are not fully suspended within dilute flows, whereas in concentrated suspensions with a wide particle size dispersion, these coarser grains can be supported by the combined effects of turbulence, hindered settling, and buoyant lift provided by the interstitial mixture of fluid and finer-grained sediment. The grain size population comprised from pebble and cobble-sized clasts (lapilli to block sizes) can be transported by saltation, rolling and sliding, aided by the effects of buoyant lift and dynamic pressure.

Sparks et al. (1978) undertook the first significant numerical modelling of the physics of flow behaviour of pyroclastic flows, and their analysis has been the basis for many subsequent reappraisals. Their model was of an initial turbulent flow after eruption column collapse, followed by in-flow grain segregation, leading to a high particle concentration flow state. Cas and Wright (1987) followed this approach, and provided an assessment of the flow dynamics of pyroclastic flows and grain support mechanisms, focusing on the importance of fluidisation, hindered settling, grainflow and dispersive pressure in the lower, high particle concentration part of pyroclastic flows, and increased levels of turbulence upwards in the transition to the ash cloud, and in high velocity flows. Branney and Kokelaar (2002) summarised the various mechanisms of particle support and segregation but focused on the role of turbulence and the depositional processes at the basal depositional boundary layer, involving progressive aggradation. Burgisser and Bergantz (2002) undertook an analysis the particle behaviour of pyroclasts in response to the turbulent eddy structure in pdcs, in particular as a function of their settling behaviour (Stokes Number) and turbulence induced eddy velocities and properties. Sulpizio et al. (2014) take a holistic view, as we do, in combining what we know from observations of pdcs, their deposits, experiments, simulations, and sedimentological theory. Dufek et al. (2015) and Dufek (2016) review the fluid dynamics of pyroclastic flows mostly from a theoretical point of view and based mostly on experimental and numerical modelling.

Doronzo (2012) proposed a different way of looking at pdc dynamics and proposed the distinction between forced convection and inertial flows, based on the main forces acting on and within a pdc which contribute to their mobility. Forced pdcs (or parts thereof) are supported by high mass flux either from a high collapsing mass discharge rate at vent or due to channellisation in paleovalleys along flow and owe their mobility to an efficient transport of mass, energy, and momentum. By contrast, inertial flows (or parts thereof) are those where the flow dissipates mass, energy, and momentum without any external forcing agent. In essence, forced flows occur where and when the internal pressure is supported and favour the lateral transport of the flow properties (mass, energy, and momentum), whereas inertial flows occur where the internal pressure drops favouring deposition (Doronzo et al. 2022). Flow transformation between forced and inertial conditions can and does occur laterally across topography, vertically within flow and over time during the same event (cf. "accumulative" versus" depletive" flows in Branney and Kokelaar 2002).

We now briefly analyse the main grain-support mechanisms and present a perspective on the processes in pyroclastic flows, within a 4D framework.

10.6.1 Fluid Turbulence

The fluid in a pdc is dominantly hot gas. Gas expansion and mixing with air generate turbulence, as the low viscosity and high velocities of gas easily determine Reynolds numbers >500–1000 (see Chap. 2, Eq. 2.7), which defines transition to a turbulent flow state, especially in the low-density portions of the pdc, such as in the dilute part of the body, at the flow front and in the ash cloud. Fluid turbulence in pdcs is also generated during eruption column collapse and then deflection laterally, by virtue of high velocities, because Re is directly proportional to velocity. However, fine ash particles tend to be fully supported by fluid turbulence and couple with gas (Valentine and Wohletz 1989; Burgisser and Bergantz 2002; Doronzo and Dellino 2011; Sulpizio et al. 2014), because their settling velocities are lower or comparable to the upward component of the gas eddy velocity

(Allen 1984) and the drag forces of gas acting on the suspended ash particles (Burgisser and Bergantz 2002; Sulpizio et al. 2014; Dufek et al. 2015).

Whether or not particular particles are entrained in suspension or are subject to tractional entrainment or granular flow, can be assessed using the Rouse Number, R_n (Sparks et al. 1978; Valentine 1987; Sulpizio et al. 2014):

$$R_n = \frac{v_T}{k\, v_{shf}} \qquad (10.1)$$

where v_T is the particle terminal settling velocity, k is the von Karman's constant ($\sim$0.4), and v_{shf} is the flow shear velocity. Flow shear velocity is a function of the drag coefficient or coefficient of friction of the ground surface and the flow velocity (Sparks et al. 1978). When R_n is low ($\sim$<0.5), small particles are efficiently transported in suspension by fluid turbulence, when R_n = 0.5–2.5, particles saltate, whereas at R_n > 2.5 particles slide or roll (Middleton and Southard 1984; Sulpizio et al. 2014). Small particles can easily be transported by turbulent suspension, unless particle concentration becomes high. Sparks et al. (1978; Fig. 10.15) demonstrated that under stable flow, only very high velocity flows (>100 m s^{-1}) are capable of suspending pumice and lithic fragments a few cm in diameter in a turbulent gas. Larger clasts would experience saltation or bed rolling/sliding, but the capacity to carry large pumice and lithic clasts in suspension increases with flow velocity and flow density (Fig. 10.15).

Burgisser and Bergantz (2002) analysed the behaviour of particles in pdcs relative to the turbulent eddy structure of the currents. In pyroclastic flows with a segregated lower high particle concentration flow system and an upper fully turbulent flow system, the size and velocity of turbulent eddies depends on particle concentration. The stability of the density gradient within such a stratified flow system occurs at Gradient Richardson Number of >0.25, which relates density (gravitational) factors to the destabilising velocity (shear) factors (Dufek 2016).

The behaviour of particles in a turbulent eddy in a pdc can be understood in terms of the Stokes Number, S_T, and a stability factor, Σ_T. The Stokes Number measures the coupling between the gas and particles and the particle response time to forcing by gas turbulence (i.e. whether the particle will remain entrained in the fluid or will decouple and settle out). The Stokes Number, S_T, is defined (Burgisser and Bergantz 2002) as

$$S_T = \frac{t_v}{f}\frac{\Delta U}{\delta} = \frac{1}{f}\frac{\Delta\rho\, d^2}{18\mu}\frac{\Delta U}{\delta} \qquad (10.2)$$

where t_v is the particle response time, ΔU is the eddy velocity, f is the drag factor function of the particle Reynolds

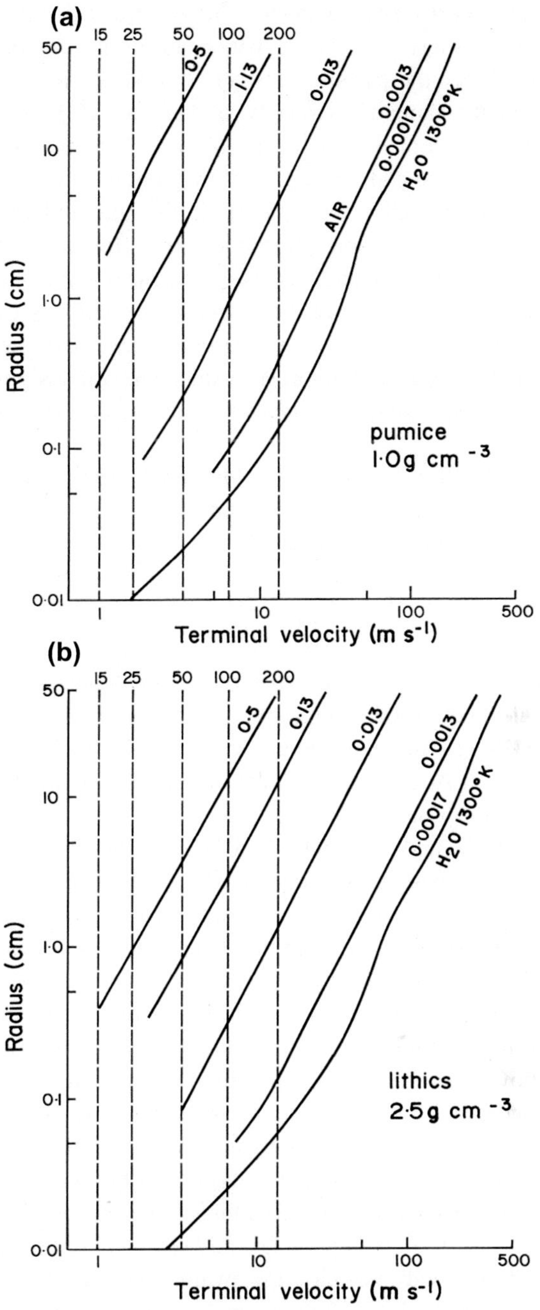

Fig. 10.15 Terminal velocity grain size curves are given for **a** pumice clasts and **b** lithic clasts in fluids of several different densities. The curves for the densities 0.0013 and 0.00017 g cm^{-3} are for air at room temperature and steam at 1300°K, respectively. The gaseous medium in a pyroclastic flow is assumed to have a density between that of cold air (Q = 0.0013 g cm^{-3}) and that of hot steam. The curve for H_2O at 1300 °K is an exact curve using the solutions of Walker et al. (1971), whereas the other solutions plot an approximate equation (Reynolds Number of >500). The broken vertical lines represent the shearing stress velocities of flows with velocities of 15, 25, 50, 100, and 200 m s^{-1} and a friction factor of 0.01. The intersection of any terminal velocity curve with any value of shearing stress velocity defines the size of the largest particle that can be suspended by a flow of that density and velocity. (From Cas and Wright 1987, After Sparks et al. 1978)

number, δ is the eddy diameter, $\Delta\rho$ is the density contrast between particles and gas (which in dilute regimes is approximately the particle density), d is the particle diameter, μ is the dynamic viscosity of the gas. If S_T is <<1, particles are coupled with the gas, if it is ~ 1, particles reside at the edge of an eddy, and may or may not drop out of it. If S_T is >>1, particles decouple from the gas and are not affected by its motion (Burgisser and Bergantz 2002). Importantly, as dynamic viscosity increases (e.g. a dense gas-fine ash fluid), S_T decreases and larger particles can be suspended. Alternatively, if particle size and density increase, then it is difficult to suspend such particles through turbulent suspension, and they segregate to the base of the flow.

The stability factor, Σ_T, is the ratio of terminal fall velocity of particles, U_T, to the eddy velocity, ΔU, which tries to keep particles suspended:

$$\Sigma_T = \frac{U_T}{\Delta U} \tag{10.3}$$

When Σ_T is >>1, particle behaviour is dominated by gravitational forces and they drop out or sediment from the eddy, where Σ_T is <<1, they remain supported by the eddy and remain in it. At $\Sigma_T \sim 1$, particles are metastable in the eddy and may drop out.

Burgisser and Bergantz (2002) describe five particle behaviour types as a function of S_T and Σ_T, and in Fig. 10.16, these are linked to eddy velocity and particle grain sizes. In what they call Fall behaviour, Σ_T is >>1, and $S_T > 1$, particles sediment from the eddy (non-equilibrium particles). They may however still be moved in saltation or bed rolling/sliding along the ground by the shear exerted by the flow. At the other end of the spectrum, in the Homogenous Transport zone, the particles (mostly equilibrium particles) are suspended by turbulence under conditions of $\Sigma_T << 1$, and $S_T << 1$. This is typical of the surge zone of a pdc. In between they describe three zones. In an "Unroll" zone ($\Sigma_T \sim 1$, $S_T > 1$) particles only sediment out if subject to maximum eddy velocity downwards, whereby both inertial and gravitational forces cause sedimentation. In the Margins zone, ($\Sigma_T < 1$, $S_T \sim 1$), particles are at the periphery of eddies and may or not drop out. In the Turbulent Sedimentation zone ($\Sigma_T \sim 1$, $S_T < 1$), particle sedimentation can occur easily, or be delayed depending on eddy spatial distribution.

All of the above assumes minimal particle interaction, which is most relevant to dilute or low particle concentration pdcs, such as surges or very high velocity pyroclastic flows (e.g. blast flows). Burgisser and Bergantz (2002), however, point out that increasing particle concentration and interaction towards the base suppresses turbulence, and eddy size. They propose a density-dilution parameter, D_D, which is a function of the degree of particle interaction:

$$D_D = \frac{t_c f}{t_v} = \frac{3\mu f}{\rho\, v_{\mathrm{rms}}\, d} \tag{10.4}$$

where t_c is the time between particle interaction, v_{rms} is the root mean square of the particle velocity, and other parameters are as defined above (Burgisser and Bergantz 2002). Where $D_D < 1$ (dense flow), particles don't have time to respond to gas dynamics before the next collision, and the flow is dominated by particle collisions and granular flow (= the basal flow zone of pdcs; Fig. 10.17). Where $D_D > 1$, (dilute flow), there are minimal particle collisions, and pdcs are wholly turbulent dilute surges, or they can segregate into basal granular flow and upper dilute flow zones (Fig. 10.17).

The process of coupling of gas and fine ash modifies the nature of the fluid, which in pdcs has to be treated as a mixture of gas + fine ash particles (dusty gas), i.e. gas + those particles that are fully suspended by the turbulent velocity of the gas, and move kinematically in unison with it (equilibrium particles). The settling velocities of other particles which are not coupled with the gas and fine ash (non-equilibrium particles) are affected by the density of the coupled gas-ash fluid mixture. The density and the viscosity of this mixture are higher than those of the gas alone and therefore the time and length scales of turbulence decrease at increasing particle concentration, whereas the buoyancy effects on slightly larger particles increase.

Fluid turbulence will therefore be the principal grain support mechanism in low particle concentration, dilute pdcs (surges), at the flow fronts of pyroclastic flows, and in the ash-cloud surges at the front, sides, top and trailing pyroclastic flows. It will be the principal grain-support process in the early, proximal parts of all pyroclastic flows, even in the basal parts, because irrespective of the mechanism of initiation of the pyroclastic flow (column collapse, blast, dome collapse) all flows will initially be high velocity, turbulent dispersions. However, beyond the deflation zone, as sedimentation occurs and pdcs quickly lose gas, a basal higher particle concentration granular flow zone develops. Importantly, the amount of air entrainment is an important control of the level of turbulence, from flow initiation throughout its extent and duration, especially for the front and outer zones of the flow. Given the low settling velocities of fine ash particles, the question is how do such particles, with low settling velocities, end up in the basal high particle concentration granular flow zone? It is likely that in-flow aggregation of ash through capillary liquid bonding and electrostatic forces (Schumacher and Schmincke 1995), forms ash aggregates that have higher settling velocities and settle from the fully turbulent upper flow zone into the basal

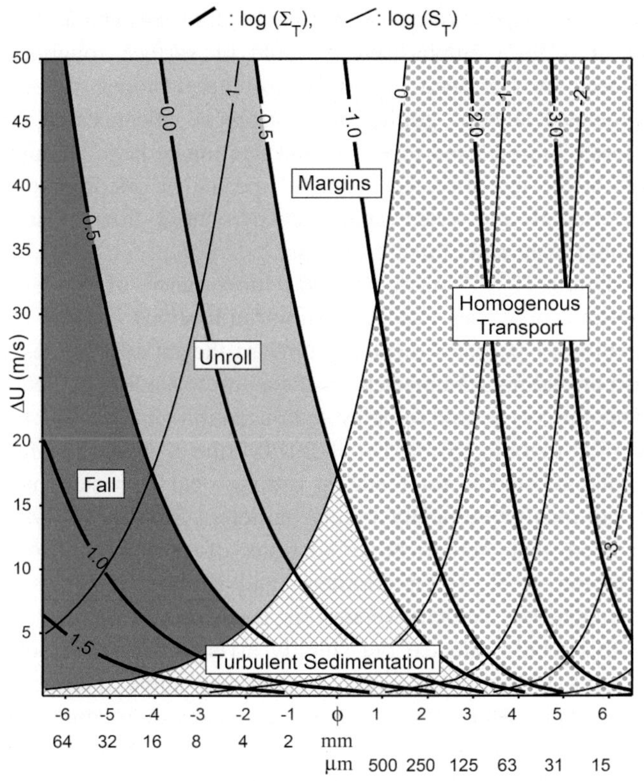

Fig. 10.16 Pumice clast (1000 kg m^{-3}) behaviour in pdcs as a function of eddy velocity. Heavy lines depict the stability factor (Σ_T), whereas light lines depict Stokes Number (S_T). See text for description of the transportational and depositional regimes depicted. (From Burgisser and Bergantz 2002)

flow zone. Furthermore, the permeability of the basal high-concentration granular flow decreases, causing the fine ash and dusty gas medium to be trapped during deposition. In any case, the level and importance of turbulence will decline and other grain-support mechanisms discussed below will progressively become more important in the basal granular flow zone. The gradationally enclosing ash cloud however will remain turbulent throughout the flow event, right to the distal margins, allowing ash aggregation to occur throughout.

10.6.2 Hindered Settling (Sedimentation-Fluidisation), Fluidisation, and Granular Temperature

The heavier particle populations (either larger or denser, and including aggregated ash clumps) characterised by terminal fall velocities higher than the upward fluid velocity in turbulent eddies are not fully supported by fluid turbulence and will progressively tend to settle and move towards the bottom of the current, where concentration increases within the basal part of the body of the flow, or underflow (Fig. 10.19). As the competence of the flow varies over time and from proximal to distal reaches, the grain size and density threshold values of the settling particles change. Fluidisation of the high concentration underflows has been long proposed as a major mechanism for particle support and flow mobility

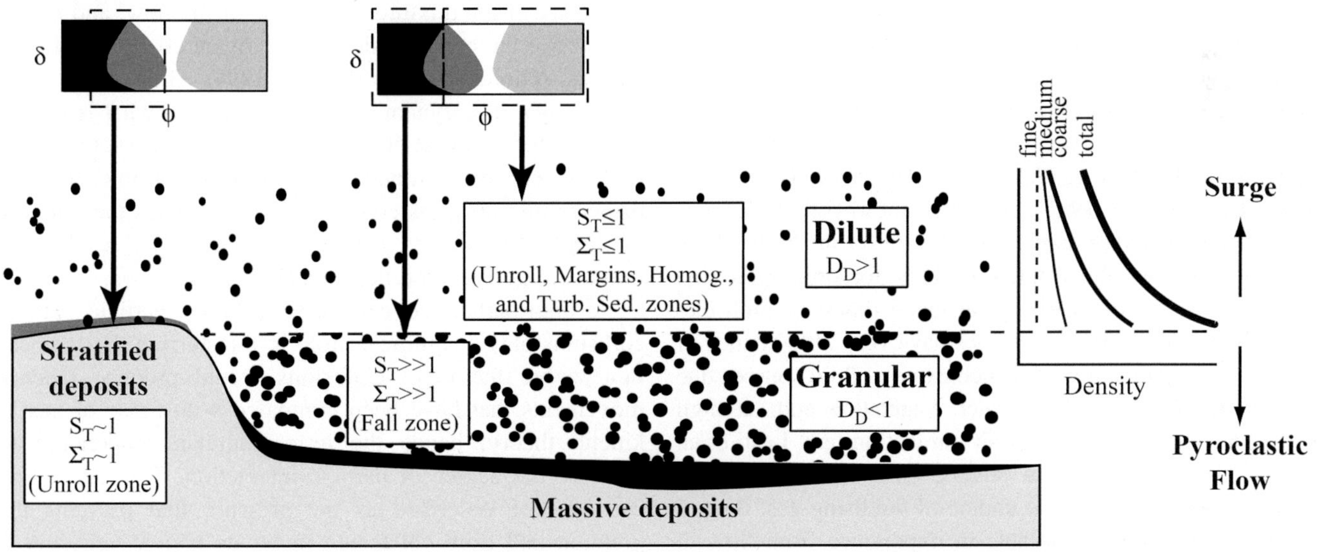

Fig. 10.17 Schematic cross-section of a density stratified pdc that is partly valley confined. This depicts an upper dilute flow zone (ash-cloud surge) and a lower higher particle concentration granular flow zone (pyroclastic flow body), and the limiting physical parameters, Stokes Number, S_T, Stability factor, Σ_T, and density-dilution parameter, D_D. (From Burgisser and Bergantz 2002)

of pyroclastic flows (see Cas and Wright 1987 for a summary). Fluidisation, as an industrial process, occurs when an upward stream of gas (or liquid) passes at a certain critical fluid velocity (the minimum fluidisation velocity, U_{mf}) through a bed of cohesionless particulate solids, and the drag force exerted across the bed by the fluid is equal to the buoyant weight of the bed (Wilson 1980). In this state, the bed loses coherence, and takes on a fluid-like character. Sparks (1976) showed theoretically that pyroclastic flows can only be semi-fluidised. In polydispersive systems (grain populations with a wide range of grain sizes or densities, or both), such as in pyroclastic underflows, before the largest particles become fluidised the gas flow velocity exceeds the terminal fall velocity (U_t) of the smallest sizes. Such particles are entrained by the gas and may be carried out and lost from the system if the particle concentration is low enough, or through gas escape pathways. This process is called elutriation and is important industrially and in the transport of pyroclastic flows.

However, away from the deflation zone, as gas is lost and a basal high particle concentration granular underflow develops, fluid turbulence tends to be suppressed by increasing particle concentration and plays only a minor role. However, at high particle concentrations, particularly of fine ash, gas escape is hindered, and if the upward directed drag force of the gas is high enough to buoyantly support ash particles, then the granular layer develops a pore pressure and can remain in a fluidised and mobile state (Roche 2012; Aravena et al. 2021). At the top of the granular flow layer however, both hot gas and fine ash will be transferred into the ash cloud.

Several different sources of gas as a fluidising agent have been suggested in pyroclastic flows apart from exsolving magmatic gas, such as the ingestion of air at the flow front, and sides, as well as vegetation, surface water, or ice over which flows move (e.g. Walker et al. 1980a, b; Wright et al. 1984; Roobol et al. 1987). However, these volatile sources are ephemeral, and probably not relevant for the body of the flow that follows the front, especially for prolonged sustained flows. Needless to say, in many desert or high elevation localities, pyroclastic flows have very little chance of interacting significantly with such external fluid sources due to the absence of surface water, vegetation and, at high elevations, also the effects of air rarefaction (e.g. Lesti et al. 2011). Gas exsolved from pumice clasts within the flow (Sparks 1978) is a possible source of fluidising gas, though ephemeral and highly variable in importance from flow to flow. Chedeville and Roche (2015) undertook experiments that indicate that air trapped in substrate irregularities, no matter how small, over which a pyroclastic flow moves, also provides a significant source of fluidising gas for the body of a granular pyroclastic flow. Wright et al. (1984) and Roobol et al. (1987) emphasised the role of surface roughness caused by the steep (up to 40°) irregular slopes, densely vegetated sinuous gullies of the tropical volcanoes of the Lesser Antilles, and ingestion and ignition of large amounts of lush vegetation as a major means of incorporating external air and water vapour into pyroclastic flows in such topographic and climatic settings.

Druitt (1995) recognised the importance of hindered settling processes (sedimentation-fluidisation), where the settling of particles displaces upward the mixture of gas + ash providing a counter-effect of support to settling particles. This process is able to delay sedimentation and promote the current mobility (Breard et al. 2019; Lube et al. 2019). Most analogue models for hindered settling deal with stationary conditions of vertically falling particles. Giordano (1998a) noticed that numerical simulations of pyroclastic flows (Dobran et al. 1993; Giordano and Dobran 1994) show instantaneous velocity vectors for particles and gas that are both laterally directed but slightly diverging, with velocity vectors for particles being at a higher angle relative to topographic slope (Fig. 10.14). This tendency is more pronounced at specific sites, such as at breaks in slope (hydraulic jumps), where the sudden reduction of flow velocity promotes particle settling and at the same time a stronger net upward flux of fluid (Burgisser and Bergantz 2002), which in a dynamic environment, is equivalent to what is described in static experiments. This effect is seen in numerical simulations by both an increase of the divergence between particles' and gas' instantaneous velocity vectors, by an increase of density of the mixture at the base of the flow, and by the rise, above the break in slope, of buoyant co-ignimbrite ash plumes (Giordano and Dobran 1994; Giordano 1998a; Fig. 10.14). In a dynamic flow environment what is important then for the onset of hindered settling processes is a high particle concentration and high sedimentation rate. In polydisperse mixtures this condition also facilitates the trapping of significant volumes of fine-grained particles, resulting in massive, poorly sorted deposits (Fig. 10.14).

At high particle concentrations (>30%), particle–particle collisions play a significant role in transferring the momentum within the flow. Collisions would produce random movements that have been compared with those of gas in kinetic theory, hence the term granular temperature to describe the degree of particle interaction. This interaction can also be described as the pressure that prevents the granular bed from collapsing under its own weight and is called dispersive pressure (Bagnold 1956; Lowe 1982). As downward movements are not favoured because the substrate is a no-through barrier, the dispersive pressure that results from particle–particle interaction is upward and laterally directed, and this will be very relevant as a support mechanism at the very base of high concentration granular underflows, but even within high concentration flows above the base. Currents that are fully supported by dispersive pressure are called grain flows or granular flows. The high particle concentration parts of pyroclastic flows can be described as "modified grain flows or granular flows" as they always show a combination of processes and the presence of interstitial fluid (dusty gas) (Branney and Kokelaar 2002).

Fluidisation through sustained gas pore pressure, particle–particle collision, and dispersive pressure will therefore be major grain-support mechanisms in high particle concentration, basal granular flows that will develop away from the column collapse deflation zone and extend to the distal domains of the flow. These mechanisms are of only minor importance in surges, and only where basal high particle concentration zones develop (e.g. coarse, massive proximal base surge deposits).

10.6.3 Vibration-Induced Dynamical Weakening

Valverde and Soria-Hoyo (2015) addressed the issue of how far-flowing pyroclastic flows can remain fluidised and mobile within their high concentration undercurrent. They undertook rotating drum experiments of granular powders, in which the powders were subject to shear during rotation as well as having an imposed regular vibration. The experiments simulate the vibrational energy that is imparted to high particle concentration pyroclastic flows by interaction with substrate surface roughness. They found that the vibrations generated fluid-particle oscillations that create gas pore pressure, that contribute to reduction of internal dynamical friction and to the dynamic viscosity of the flow. They called this effect "dynamical weakening". Oscillatory flow creates a gas-solids drag, even without net gas flow, that retards settling of particles, and creates an effective particle volume fraction in the flow in the range of 0.3–0.5. Theoretically, no other fluidising gas source is required through such a mechanism, but in pyroclastic flows, such a gas (magmatic, and ingested external air and gas) exists, and will also provide some fluidisation support. Vibration-induced oscillatory flow can become another important grain support mechanism in the dynamics of pyroclastic flows, particularly when other mechanisms begin to wane in their influence, in medial to distal flow settings.

10.6.4 Clast Buoyancy

Buoyancy is described by Archimedes Principle, by which an object immersed in a fluid will experience an upward directed force that arises from the tendency of the fluid to oppose its displacements, with an intensity that depends on the weight of the volume of displaced fluid (V_{disp}) of a given density (ρ_{fl}). This buoyancy force, B, can be represented as

$$B = \rho_{fl} \, V_{disp} \, g \qquad (10.5)$$

and indicates that buoyancy becomes more relevant as a particle support mechanism as the density of the pdc increases. Many pyroclastic flow deposits show variable degrees of coarse tail grading of large dense lithic and low-density pumice clasts, and in many instances reverse and lateral gradings, which can be due to the effect of buoyancy in a high particle concentration flow (Sparks 1976).

10.6.5 Support on an Interface During Traction: Sliding, Rolling, and Saltation

In dilute pyroclastic density currents, most of the clasts are transported at the interface by traction, between the substrate/deposit and the overriding and shearing current. This kind of transportation depends on flow shear and particle characteristics, mainly their weight (size and density) and drag coefficient (shape), and can produce sliding and rolling, where the drag is continuous, or saltation, if the particles are re-entrained within the current for a while (Dufek and Bergantz 2007; Dufek et al. 2009; Douillet et al. 2014).

Transport on an interface typically produces bedforms associated with surge deposits (e.g. Valentine 1987; Dellino et al. 2004). Large and heavy particles may also experience such transportation in high-concentration pyroclastic flows depending on shear velocity, dynamic pressure, and aided by buoyancy. Roche (2015) has demonstrated that both dilute and high particle concentration pdcs can capture and uplift dense accidental lithic clasts from the substrate, due to the development of a pressure gradient and uplift force between the substrate and the interior of the flow (Chap. 12).

10.6.6 Internal Friction, Strength, and Particle Interlocking

As a pdc decelerates and deposits, mechanisms of particle support change progressively. The final stages that precede deposition at the base and at the front of a pyroclastic flow are dominated by the effects of friction between particles (Roche et al. 2010, 2011). Many frontal lobes and lateral levees of pyroclastic flow deposits have steep margins with high angles of repose (Figs. 10.11d, e, 10.18a, b and 10.21) that have been suggested as indicating emplacement by frictional or *en masse* freezing of the flow once particles have nearly reached their maximum packing density, and

Fig. 10.18 High relief (~ 1.5 m), steep-sided depositional lobes at the termination of recently deposited non-welded ignimbrites. **a** Pumice and ash flow deposits from the 1993 eruption of Lascar volcano, Chile. **b** Aerial view of lobate form of ignimbrite from the 22 July 1980, Mt St Helens eruption, USA (also see Fig. 10.11d, e) (*Source* United States Geological Survey)

deposition is dominated by particle friction (Pittari et al. 2005; Charbonnier and Gertisser 2012; Jessop et al. 2012). This process is helped by the buoyancy of large pumice clasts that are forced to the front, upwards, and around the margins of the main flow lobes, producing highly permeable domains where the interstitial fluid (dusty gas) is easily lost, preventing maintenance of most of the particle support mechanisms described above.

The occurrence of large and dense clasts in pyroclastic flow deposits well away from source and embedded within ash and pumice matrix has previously also been explained by the onset of yield strength in high-concentration granular flows or flow zones, which would then resemble debris flows of sedimentary environments (Wilson and Head 1981; Chap. 12). This is due to very high particle concentrations that develop in the basal parts and during the later stages of a flow event and deposition. Wilson and Head (1981) modelled the 1980 Mt St Helens pumice and ash flows that produced the lobate deposits as Bingham rheology flows with yield strength. They estimated a flow viscosity of about 4 Pa s just prior to cessation of flow (cf. water = 0.001 Pa s).

Such flows exert a significant dynamic pressure, which allows them to mobilise clasts on the substrate (Pittari et al. 2005). Roche (2015) has also shown that large clasts can be uplifted into the flow due to development of a pressure gradient from the substrate to the interior of a flow.

In some cases the presence of moisture may produce cohesion in the later stages of flow during phreatomagmatic eruptions, and where pyroclastic flows entrain significant amounts of water from the environment (e.g. where pdcs flow into river channels) pyroclastic flows have been observed to transform along flow into water-supported lahars or debris flows (e.g. Mt Unzen eruption, Japan, 1991–1995; Uehara et al. 2015; Chap. 12).

10.6.7 In-flow Ash Aggregation: An Inconvenient Truth in Most Models

In most previous assessments of the flow dynamics of pdcs, all particles have been treated as cohesionless, which is in fact incorrect. In moist flows (e.g. phreatomagmatic flows), capillary bonding or liquid bonding (Schumacher and Schmincke 1995; Mueller et al. 2016) causes aggregation of fine ash particles, producing larger transportable ash aggregates. In hot flows, thermal particle agglutination may occur and also produce transportable aggregates of "sticky" ash (Freundt 1998). Even in dry flows, where water moisture is insignificant, electrostatic attraction and aggregation of fine ash particles (<32 μm, and even <63 μm; Del Bello et al. 2015), and even coarser fragments (Schumacher and Schmincke 1995) occurs, again producing larger transportable ash aggregates. Such larger transportable aggregates affect fluid dynamic reconstructions and concepts of sedimentation processes and physics but have been largely overlooked. Such in-flow ash aggregation explains very well why even high velocity pdcs retain so much fine ash, rather than it being elutriated out of the flow, and also the poorly sorted nature of pyroclastic flow deposits. Views of pdc flow behaviour are therefore still relatively simplistic, and there is still much to be learned about the role of in-flow ash aggregation on flow and depositional behaviour, and deposit characteristics.

10.7 Flow Transport and Depositional Models: 4D Ephemeral Flow Model—Vertical, Lateral, and Temporal Variations in Flow Properties and Behaviour

It is proposed that pyroclastic flows, irrespective of their size and duration, are highly ephemeral, 4D, density currents, capable of changing their flow state in 4D space. Every flow is different, and every flow changes its flow state in 4D space. Single pulse triggered pyroclastic flows may behave differently to continuously fed pyroclastic flows, at least at different time scales. So again, there is still much to learn about pyroclastic flow transport and depositional dynamics.

Within our proposed 4D ephemeral model, all pdc flow models previously proposed find a collocation that is useful in describing specific parts of pyroclastic flow deposition and emplacement. Some pyroclastic flows are "hares", at least initially, but many become "tortoises" in distal settings (and/or everywhere during waning flow stages), explaining other features such as levees, and high relief terminal lobe geometries, which cannot be explained by high velocity, highly turbulent flow states. For example, Fisher et al. (1993), Giordano and Dobran (1994), Dade and Huppert (1996), Calabrò et al. (2022) proposed the emplacement of pyroclastic flows from a dilute current able to "drain" the particle load along topographic traps leading to the detachment of underflows from the overriding ash cloud, which can occur during the early stages of flow in proximal areas; for example, as described for the 25 June and 26 December 1997 surge-derived pyroclastic flows at Montserrat (Sparks et al. 2002; Druitt et al. 2002). More or less sustained deposition by progressive aggradation of a flow deposit from stratified density currents proposed by Branney and Kokelaar (1992, 2002), and Bursik and Woods (1996) is the likely mechanism that accounts for the majority of deposition in proximal areas and for the formation of high concentration undercurrents in medial and distal areas. Progressive aggradation involves a continuous transfer or settling of particles out of suspension from the upper turbulent part of the body flow into the lower higher particle concentration granular flow part towards a basal depositional boundary layer, leading to the progressive thickening of the deposit as the flow passes over it. In some cases, progressive aggradation is also the dominant depositional mechanism at distal reaches, especially for low particle concentration pyroclastic flows (Fig. 10.19).

As progressive aggradation occurs, most pyroclastic flows develop high particle concentration underflows mostly at medial to distal reaches (Fig. 10.19). This segregation process occurs during flow and must be kept distinct from deposition that occurs when parts or the entire high particle concentration underflow come to rest. Recent numerical models have captured such partitioning by the coupling of two-layer models, with an upper dilute layer that feeds a basal high-concentration layer in turn feeding deposition (Shimizu et al. 2019, 2021). The basal, poorly expanded, semi-fluidised, basal granular flow, with relatively high bulk viscosity, flows in an overall laminar flow shear fashion (Fig. 10.19). Deposition can keep occurring by progressive aggradation from the base of the high particle concentration zone, at high sedimentation rates, as likely happens at medial reaches of the flow, or by freezing en masse and shear-friction induced increments or pulses. This also preserves a vertical organisation reflecting the final stage of flow, as likely occurs in distal reaches and at final runout. In the latter case, the lower, high particle concentration zone of the current may have flowed some distance as a high particle concentration granular mass flow under laminar flow conditions, maintained by high internal gas pressure in the gas that is trapped between fine ash particles in the matrix of the flow, until the flow literally runs out of gas! (e.g. Sparks et al. 1973, 1978; Sparks 1976; Wright and Walker 1981; Wilson 1984; Francis 1993; Palladino and Valentine 1995; Roche et al. 2016; Wright et al. 2016; Giordano and Doronzo 2017). Deposition in such degassed granular flows occurs largely by frictional freezing and shear. In all cases much of the dilute upper ash cloud part of pdcs can detach from the lower part of the flow and flow up and over topography or flow beyond the limits of the lower high particle concentration part of the flow (e.g. Fig. 10.12c), and/or become buoyant and loft into the atmosphere as a buoyant co-pdc ash-cloud plume (Figs. 10.10b, 10.14 and 10.19; Wright et al. 2016; Calabrò et al. 2022), from which widely dispersed fine ash fallout can occur (Chap. 9).

Figures 10.14 and 10.19 illustrate a generalised conceptual model for the main downflow transitions experienced by pdcs. To understand the facies and geometries of pdc deposits, it is important to understand not only how concentration, velocity, and temperature are vertically stratified, but most importantly how they transition laterally. If the internal flow pressure is maintained, transportation is favoured and deposition occurs mostly by vertical aggradation, both in dilute and high-concentration flows; by contrast if the flow pressure quickly drops, deposition occurs by direct fallout in dilute systems and by en masse freezing in high concentration flows. In any of the above cases, the variations in sedimentation rate across the flow will control the variations in final thicknesses and facies (Giordano and Doronzo 2017).

Of course, transitions may occur anytime and anywhere during and along a pyroclastic flow path, given appropriate conditions, and involving 4D changes in mass flux, velocity and degree of interaction with the ambient atmosphere and ground surface.

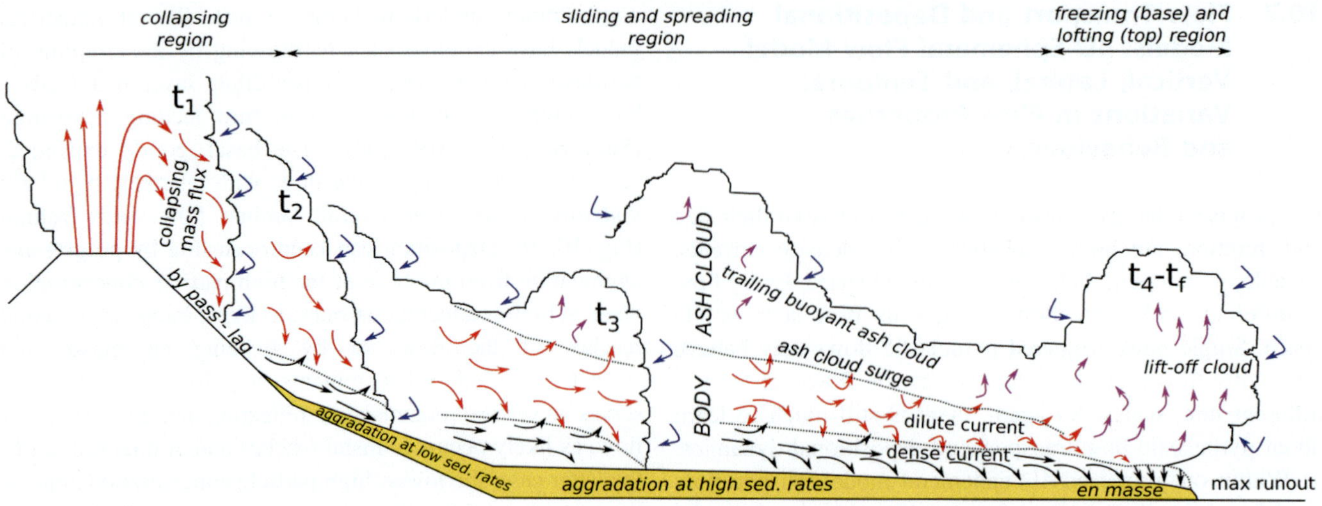

Fig. 10.19 Conceptual model for the evolution over time (t_1–t_4) and at steady state (t_4–t_f) of the transportational-depositional systems within a generalised pyroclastic flow; cf. with Fig. 10.14 to cross-correlate with expected deposit facies. The flow in the collapse zone is the thickest, dilute and fully turbulent (red arrows), internally unstructured and initially (at t_1) experiences most of the frontal air entrainment (blue arrows). As the flow front moves ahead into the sliding and spreading region (at t_2 to t_3), the dilute and turbulent flow (red arrows) internally segregates a lower dense high concentration, self-fluidised zone or current, with progressive suppression of the internal turbulence, and both transporting (black arrows parallel to topography) and depositing (black arrows pointing at an angle to the ground). Exchange of material occurs continuously across the poorly defined boundary between upper dilute (ca. ≤ 100 kg m^{-3}) and lower dense zones (ca. ≥ 100 kg m^{-3}; black and red arrows crossing the dashed line). The upper part of the dilute zone which retains a lateral motion is the ash cloud surge and elutriates upward fine ash feeding a trailing buoyant ash cloud (purple arrows), especially where sedimentation is enhanced, such as at breaks in slope. Depending on the balance of material fed from the dilute zone into the dense zone and material deposited from the base of the dense part of the flow, the head of the pyroclastic flow can be driven by the basal dense zone (e.g. at t_2) or by the overriding dilute zone (e.g. at t_3). As the flow reduces laterally its density due to sedimentation and air entrainment, the dilute body zone may merge with the ash cloud and literally lift off at maximum runout (at t_4). The detachment of the lofting cloud (vertical red arrows) from the basal dense zone may occur at velocities that are still tens of m s^{-1} in the dilute flow where up to 50% or more of the total mass flux is still concentrated, which may give rise to co-ignimbrite columns as high as plinian columns, able to entrain not just the fine ash fraction but up to lapilli-sized clasts. At reaching the maximum runout distance (at t_4) and for as long as a steady collapsing mass flux feeds the pyroclastic flow (at t_f) the rate of deposit aggradation at the base of the dense current may become so rapid that it ultimately freezes *en masse* by particle interlocking, in many cases after flowing as a high particle concentration, granular mass flow for some distance under bulk laminar flow conditions, mobilised by gas trapped between fine matrix ash particles (e.g. Fig. 10.18)

Deposits from pdcs may therefore show complex lateral and vertical facies associations that relate to the interplay between the transport-depositional system and topography (Fig. 10.14; e.g. Sparks 1976; Wilson 1980; Wright et al. 1981; Fisher and Heiken 1982; Roobol et al. 1987; Fisher 1990, 1995; Valentine 1987; Valentine et al. 1992; Druitt 1992, 1995; Palladino and Valentine 1995, Giordano 1998a; Giordano et al. 2002; Branney and Kokelaar 2002; Dellino et al. 2004; Pittari et al. 2006; Cas et al. 2011; Doronzo et al. 2010, Doronzo 2012; Wright et al. 2016; Giordano and Doronzo 2017; Palladino and Giordano 2019; Báez et al. 2020a, b; Smith et al. 2020; Pimentel et al. 2021). Pyroclastic density currents are highly energetic, geologically short-lived events, and deposition generally occurs over a very large range of transportational and depositional regimes, and sedimentation rates, compared with most sedimentary environments such as fluvial or eolian environments, and their surface process regimes.

The partitioning between the sedimentation rate and the transport rate at different depositional sites along the flow path is controlled essentially by (Fig. 10.19; see Doronzo et al. 2022):

- variations at source with time of the mass feeding rate (mass flux) to the pyroclastic density current, which influences the vertical facies association of a pdc deposit,
- distance from the eruption point, which influences the lateral facies variations of a pdc deposit, and
- topography, which influences both the vertical and lateral facies association of a pdc deposit).

In particular, the valleys parallel to flow direction may act as a lateral confinement to the spreading of a pdc, favouring channelling and forcing of the flow (Fig. 10.12a, b). For example, the case of the 1912 pdcs at Novarupta (Hildreth and Fierstein 2012), or of the 1980 pdcs at Mt St Helens

(Rowley et al. 1981; Hoblitt et al. 1981; Brand et al. 2014, 2016, 2017, 2023), the 1991 pdcs at Mt Pinatubo (Scott et al. 1996), the 1995–1999 pdcs at Montserrat (Cole et al. 2002), and the 2015 pdcs at Colima volcano, Mexico (Pensa et al. 2018, 2019). In contrast, valleys or ridges at an angle to flow direction, as well as breaks in slope may act as loci or traps for sedimentation because they represent hydraulic jumps, which cause changes in the immediate flow conditions (Fig. 10.14; e.g. Roobol et al. 1987; Valentine et al. 1992; Giordano and Dobran 1994; Fisher et al. 1993; Fisher 1995; Giordano 1998a; Giordano et al. 2002; Druitt et al. 2002; Pittari et al. 2006; Wright et al. 2016; Giordano and Doronzo 2017; Silleni et al. 2020, 2024; Trolese et al. 2024). Furthermore, topography changes as deposition fills it in, so that successive pulses of the pdc may experience different paleotopographic conditions (e.g. Watkins et al. 2002).

Although considerable work has been done on the significance of the facies associations for the 2D downcurrent evolution of pdcs, the 3D internal architecture of pdc deposits and the lateral facies changes may reveal fundamental aspects of the emplacement history, spatial and temporal changes in flow behaviour, and the time evolution of the eruption. Reconstructing the internal facies architecture therefore allows the development of 4D models for pdcs, from the onset to the end of the flow event.

10.7.1 Depositional Model: Fore-Stepping and Back-Stepping Stacking Patterns, and Depositional Time Lines

De Rita et al. (1998) recognised that the internal vertical and lateral facies associations of pdc successions record the progradational (waxing) and retrogradational (waning) phases of deposition during an eruption. They named such internal architecture "fore-stepping-back-stepping stacking pattern" (F-B), analogous with nomenclature given to sedimentary rock successions (e.g. Mutti et al. 1994).

The F-B architecture relates to waxing and waning of the transport energy. Fore-stepping is associated with increasing transport energy with time, which produces progradation of depositional units (Figs. 10.20a and 10.1). Progradational units show an *off-lap* geometry with the underlying substrate and may be associated with vertical aggradation or not. This defines the extent, if any, of by-pass areas, i.e. proximal areas where pdcs are erosional or not, or are only poorly depositional, and deposits lack or have limited thickness (Fig. 10.20b; e.g. Giordano 1998a; Brown and Branney 2004). Back-stepping is instead related to waning of transport energy with time and progressive upslope deposition associated with an *on-lap* geometry. Back-stepping is always associated with vertical aggradation of the succession (Fig. 10.20a2, b). Fore-stepping and back-stepping are

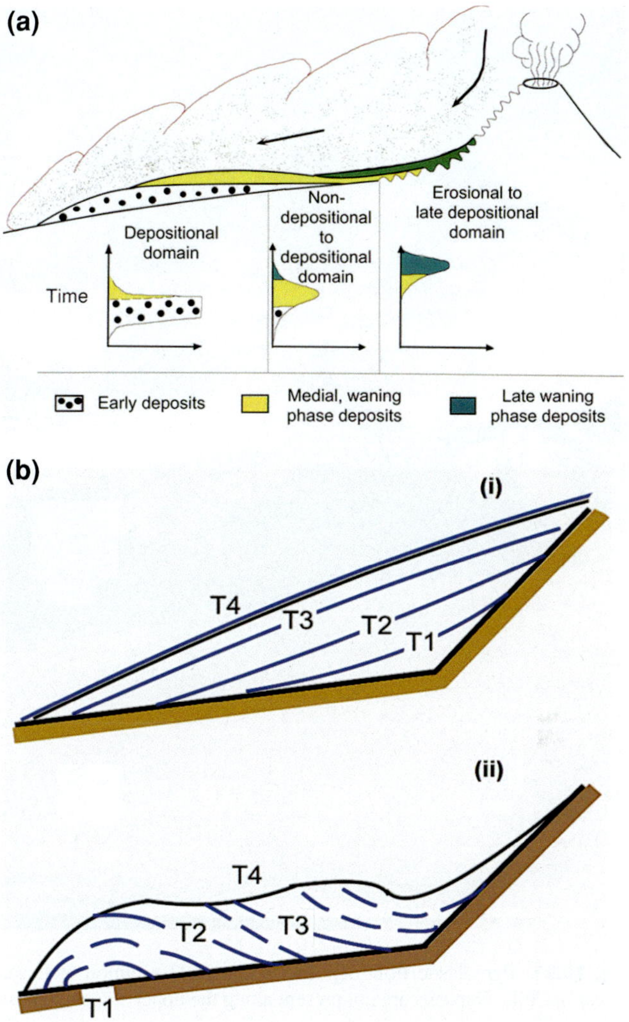

Fig. 10.20 Representation of fore-stepping (prograding) and back-stepping (retrograding) depositional behaviour in pyroclastic flow deposits (Adapted from Giordano 1998a). **a** Schematic representation of non-depositional and depositional stages in a pyroclastic flow that in proximal high slope areas is initially erosional, to non-depositional and by-passing. In medial areas, beyond the break in slope, the flow is initially non-depositional to weakly depositional. In distal areas, as the flow front velocity wanes, the flow becomes depositional, and then the principal depocentre back-steps towards the volcano. **b** Time line-facies architecture from two pdc scenarios. (i) Aggrading and prograding (fore-stepping) deposition from a relatively low capacity, low velocity flow at the base in slope of a stratovolcano, fills and smooths topography; time lines are parallel to the deposit surface, and the time vector is almost perpendicular to base and top of deposit, migrating in a downflow direction. (ii) Deposition from en masse flow pulses from an initially high capacity and high velocity flow creates topography, and then as the flow wanes, the depocentre backsteps; time planes dip back towards the volcano and the time vector migrates back towards volcano. *Note* In a large enough flow scenario (ii) may evolve from (i) up-slope

typically recognised as 2D, along-flow structures, but the 3D geometry is equally important (Báez et al. 2020a, b). Typically, mass eruption rates associated with eruptions, either explosive or effusive, are far larger than the ability of the

Fig. 10.21 Pyroclastic flow deposits of the 1993 eruption at Lascar volcano, Chile. Deposits are not present along the upper slopes (by-pass zone) and become thicker after breaks-in-slope. Between the 1st and the 2nd breaks-in-slope the deposit is a tabular low aspect ratio sheet, whereas after the 2nd breaks-in-slope it forms an apron made of superposed and laterally juxtaposed, and progressively, back-stepping, high aspect ratio depositional lobes. Depositional lobes are numbered in order of emplacement, based on superposition relationships

environment to distribute this excess mass on the Earth's surface. This process is responsible for the typical fan geometry of lava fields and pyroclastic flow deposits, similar in shape to alluvial fans at the break in slopes or at deltas. The typical resulting geometry of volcanic successions is similar to the compensation cycles in sedimentary environments (Mutti and Sonnino 1981; De Rita et al. 1998). Compensation cycles are constructed by the lateral juxtaposition of lobes or lenses due to meandering of pyroclastic flows or lava flows around areas of deposition (Fig. 10.21; e.g. LaBerge et al. 2009).

An excellent example of F-B geometry along flow direction and compensation cycles geometry perpendicular to flow direction is provided by the 1993 deposits from Lascar volcano in Chile (Fig. 10.21; Sparks et al. 1997; Calder et al. 2000; Jessop et al. 2012). The eruption produced a succession of pyroclastic density currents (or a sustained pdc) that flowed down the slopes of the stratovolcano. The upper reaches are very steep and pdcs left virtually no deposit. This area is therefore a by-pass area. At the first major break in slope, where inclination abruptly changes from over 25° to less than 10°, deposits are thin and evenly distributed. The smooth top surface of the deposits, where no pumice levees can be observed, suggests that steady and uniform deposition occurred along that reach. Further downslope, where topography widens, the total thickness of the deposit gradually increases to form a low aspect ratio sheet (Sect. 10.8.4.3).

Several, sub-parallel longitudinal pumice levees, as well as erosional channels, can be seen at the top of the deposit, suggesting that with time, the locus of sedimentation shifted laterally to form an anastomosed arrangement of compensation cycles. Further downslope two perfectly preserved pyroclastic fans can be observed (Fig. 10.21) at a second major break in slope. Deposition at this site was promoted by deceleration of the pyroclastic flows, rich in coarse pumice clasts that form multiple lobes with steep levees dominated by large pumice and lithic clasts (Fig. 10.18).

The levees formed through frictional freezing due to particle interlocking resulting from the coarse, permeable nature of the margins which cannot be supported by fluidisation (Jessop et al. 2012), whereas the finer, more ash-rich interior channelised material continued to flow as fluidised plugs (Sect. 10.6.5; Fig. 10.18).

The fans preserve the compensation cycle geometry very well, related to meandering of pdcs due to the rapid, pulse-like aggradation, or incremental aggradation of successive high aspect ratio lobes that induce shifting of the locus of sedimentation laterally on relatively low slopes. In this case, the geometry is relatively simple with lobes deposited in back-stepping arrangement (numbered 1–6, Fig. 10.21). This must relate to the progressive reduction of transportation energy of the pyroclastic flows. However, it should be clear that long-lasting eruptions may experience several phases of both waxing and waning of mass flux and of flow competence, and therefore deposits may show repeated F-B architecture within the same eruption unit.

Although an F-B architecture in ignimbrites is probably commonly present in some parts of the deposit area, it may not always be easy to recognise in the field. One of the main reasons is that amalgamation processes may obliterate internal evidence of incremental deposition or aggradation by flow pulses or lobes, especially at intermediate reaches where sustained pyroclastic flows are most likely well-developed and steadily feed deposition over significant spatial and temporal extents. Giordano (1998a) described evidence of proximal stratified deposits, which laterally fade out and amalgamate into a single depositional unit in intermediate settings, while Branney and Kokelaar (2002) similarly describe several depositional features "fading" laterally into more uniform massive and chaotic lithofacies. Similarly, lateral gradation in distal settings from a uniform massive and chaotic lithofacies to a stratified stack of sub-units has also been described (Cas et al. 2011; Wright et al. 1981). Large volume ignimbrites in the rock record probably do not preserve very proximal and distal outcrops due to post-depositional erosion, not to mention that they are generally sourced from low-relief caldera complexes, rather than high-relief stratovolcanoes, and their volume commonly buries the underlying topography.

The recognition and reconstruction of the complex internal F-B architecture of an ignimbrite unit if present is fundamental as it implies that deposition is essentially diachronous and defines the way isochron time-lines cut through the deposit (Giordano 1998a; Branney and Kokelaar 2002; Báez et al. 2020a, b). For example, in a purely aggradational deposit, after initial deposition begins away from the vent, depositional time-lines will be more or less parallel to the underlying topography and the tops and bottoms of the deposit, whereas in a granular or plug-flow depositional time-lines would change from the proximal region, where aggradation occurs, to becoming more or less perpendicular to topography downcurrent, i.e. sub-vertical (Fig. 10.20b). Consequently, evidence of possible changes of eruptive characteristics, such as a change in chemistry of the magma, should be found, within these two end-members, up-stratigraphy or up-slope respectively. Our 4D model suggests that individual ignimbrites may show complex internal time-line patterns that may change their arrangement at different localities depending on local flow dynamics and interaction with topography (Fig. 10.14).

Lateral and vertical compositional zoning (in terms of componentry, juvenile clast texture, and chemistry) are therefore important for the reconstruction of the internal architecture of ignimbrites and will be discussed later (Sect. 10.8.13).

10.8 Ignimbrites: The Deposits of Pumice and Ash and Scoria, Spatter and Ash Flows

Ignimbrites are the deposits of *pumice and ash, and scoria, spatter and ash* pyroclastic flows. Since the early works by Marshall (1935), Martin (1959), Smith (1960a, b), Ross and Smith (1961), Fisher (1966) and Sparks et al. (1973), the term "ignimbrite" has been used in many different ways in the literature, to describe welded tuffs, ash-dominated tuffs, and felsic pumice and ash flow deposits. *Ignimbrite* is redefined here as the rock or deposit formed from pumice and ash through to scoria and ash pyroclastic flows, irrespective of the degree of welding or volume or chemistry. We therefore now include, inside this broad "ignimbrite" umbrella, compositional distinctions that are in use to distinguish pumice and ash flow deposit for the felsic types, and scoria and ash flow deposits for the mafic types. Ignimbrites therefore can vary in magma composition (felsic, intermediate, mafic), explosive eruption mechanism (magmatic, phreatomagmatic), the deposit thickness and volume, the state of welding (non-welded, welded), and the lithification state (unconsolidated, lithified). The common factors are similar flow behaviour and depositional origins, as well as originating from some level of eruption column or fountain collapse that generates a pyroclastic flow.

Ignimbrites vary from unconsolidated ash deposits texturally similar to mud flow deposits, to vapour-phase lithified sillar deposits, to hard, densely welded lapilli tuffs which may be difficult to distinguish from some lava flows. They dominantly show a poorly sorted texture, and the grain size varies from almost entirely sub-millimetre ash particles to clasts over a metre in diameter immersed in a fine ash matrix. Such variations may also occur laterally and vertically across one ignimbrite unit.

In size, ignimbrites range over at least six orders of magnitude of volume and include the largest single eruptive units known, covering tens of thousands of km^2 and having volumes of more than 1000 km^3. Even though eruptions that generate such large volume units have never been witnessed, they must be the most cataclysmic of all geological phenomena. Even small- to moderate-volume, historic ignimbrite-forming eruptions (Table 10.1) are awesome: e.g. the 1600 BC eruption of Santorini, which has been linked with the rapid decline of the Late Bronze Age Minoan civilisation on Crete and adjacent islands; the 79 AD eruption of Vesuvius, in which the towns of Pompeii and Herculaneum were destroyed (Fig. 10.1c); Tambora which in 1815 caused the deaths of more than 10,000 people on Sumbawa directly as a result of the eruption, and perhaps as many as a further 100,000 as a result of tsunamis and an ensuing famine which affected the remainder of the island and nearby Lombok and Bali; the lesser known but even larger eruption of Samalas (Rinjani) on Lombok in 1257 AD, that led to the destruction of the island Kingdom and its capital Pamatan which remains "lost" entombed in ignimbrite, and is believed to have caused widespread crop failures, famine and pestilence in Mediaeval Europe and elsewhere (killing as many as 20,000 people in London alone); the eruption of Krakatau in 1883 which set in motion tsunamis killing more than 30,000 people on neighbouring islands; Mt St Helens in 1980 which cost an estimated 1.1 billion USD in property damage (or $3.5 billion at today's prices); and El Chichón in Mexico in 1982, after which 2000 people were missing.

We now describe the characteristics of ignimbrites, from large-scale regional features and relationship to topography, to outcrop scale mesoscopic textures and depositional structures, to microscopic features. Other useful reviews are those of Branney and Kokelaar (2002) and Brown and Andrews (2015).

10.8.1 Occurrence, Composition, and Size of Ignimbrites

Ignimbrites are common volcanic products, being found in all volcano-tectonic settings: oceanic islands (e.g. Iceland, Azores, and Canary Islands), island arcs (e.g. Lesser Antilles), microcontinental arcs (e.g. New Zealand, Italy), continental margin arcs (e.g. Andes) and continental interiors (e.g. Western United States). Rhyolite, dacite, and andesite are the most common compositions, but trachytic and phonolitic ignimbrites are also common. In addition, mafic ignimbrites, with compositions ranging from basaltic to foiditic, have progressively been more widely recognised. Many of the most voluminous ignimbrites are rhyolitic and dacitic, related to calderas, and some are compositionally zoned, indicating in some cases the tapping of large-zoned magma chambers (Smith 1979; Hildreth 1979). Ignimbrites can have alkaline compositions (e.g. in the Azores, Tenerife, Philippines, and Italy). Pantelleritic ignimbrites, the name of which comes from the island of Pantelleria (Sicily), have been recorded from a number of areas, e.g. the East African Rift, Gran Canaria, and the Western United States. Ignimbrites have been recognised in geological formations of all ages.

In size, ignimbrites range from the order of 10^7 m^3 (VEI 4/VEI* 8; see Chap. 9 for explanation) to >10^{12} m^3 (>1000 km^3; >VEI 8/VEI* 12) in volume. The largest are restricted in their occurrence to continental margins and interiors, and large islands. They typically result from eruptions of silicic calc-alkaline magmas and tend to form extensive sheets or shields (Chap. 16), such as in the ignimbrite flare-up of the Altiplano Puna in Central Andes (De Silva and Gosnold 2007). The smallest are found in all settings and are a common product of stratovolcanoes, although not restricted to this type of centre (Chap. 16). Even maar volcanoes can produce ignimbrites (Fig. 10.22c; Freundt and Schmincke 1985a, b;

Fig. 10.22 Varying areal extent of some ignimbrites and accompanying fallout deposits. **a** Map showing the distribution of the pumice and ash flow deposits from the various eruptions of Mt St Helens during 1980 (From Rowley et al. 1981). **b** Topography controlled 1912 Valley of Ten Thousand Smokes ignimbrite (shown in yellow), Katmai, Alaska. The ignimbrite was erupted from the Novarupta crater, marked by the dome N, but the magma was sourced from under the Katmai caldera just off the map to the east). The superposed isopach maps relate to the fallout deposits associated with the different phases of the eruption (C-H). (From Houghton et al. 2004). **c** Radial distribution of the 355 ka mafic Villa Senni Formation, consisting of the Pozzolanelle ignimbrite (pink) overlying the Tufo Lionato ignimbrite, around the Colli Albani caldera, near Rome, Italy. The dashed white lines indicate isopach contours in cm of the basal fall deposit. In green, the distribution of the <23 ka, small volume Peperino Albano phreatomagmatic ignimbrite from the Lago Albano maar, the youngest centre of the Colli Albani caldera. (Modified from Giordano et al. 2010). **d** Simplified map of Valles caldera and the surrounding Bandelier Tuff ignimbrite shield, Jemez Mountains, New Mexico, USA. Below: distribution of the Lower (1.6 Ma) and Upper (1.26 Ma) Bandelier ignimbrites and outline of caldera ring fractures. Eruption of the Upper Bandelier ignimbrite initiated the collapse of the Valles caldera, nested within the Toledo caldera; note, the semi-circular depression further to the northeast is thought to be an older volcanic feature. RP = Redondo Peak resurgent dome of uplifted (>1000 m) intra-caldera ignimbrite. Above: isopach contours showing distributions of main plinian fallout units associated with the Lower (in m) and Upper (in cm) Bandelier ignimbrites; the preserved upwind part of Upper Bandelier fallout unit B shows exceptional crystal concentration consistent with an eruption column of ultra-plinian scale (Self et al. 2021). (Adapted from Self et al. 1996). **e** Landsat 7 natural-colour image of same area showing the topographic margin of the Valles caldera, its arcuate line of Cerro Toledo lava domes that define the structural ring fault mainly to the north, the central resurgent dome or block, and the surrounding Bandelier ignimbrite shield forming the Pajarito and Plateau (see Figs. 10.26b, c and 10.28b) and Jemez Plateau (see Fig. 10.65c). (*Source* NASA Earth Observatory). Also see Fig. 10.61 for preserved areal extent of the very large volume 2.08 Ma Cerro Galán ignimbrite, northwestern Argentina

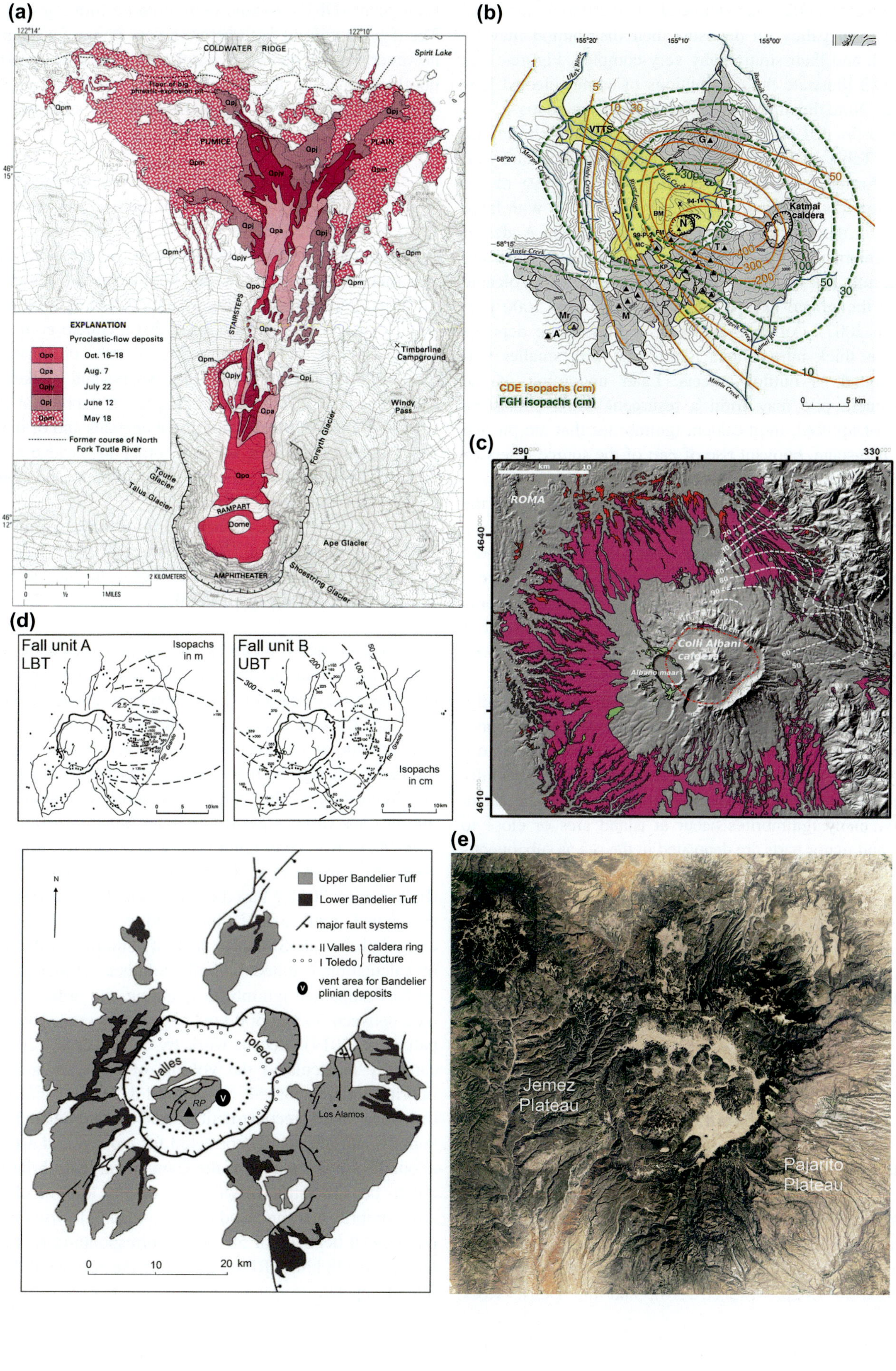
(a)
COLDWATER RIDGE
Spirit Lake
PUMICE
PLAIN
STAIRSTEPS
EXPLANATION
Pyroclastic-flow deposits
Qpo Oct. 16–18
Qpa Aug. 7
Qpjy July 22
Qpj June 12
Qpm May 18
Former course of North Fork Toutle River
Timberline Campground
Windy Pass
Forsyth Glacier
Toutle Glacier
Talus Glacier
RAMPART
Dome
Ape Glacier
AMPHITHEATER
Shoestring Glacier
2 KILOMETERS
1 MILES
(b)
VTTS
Katmai caldera
CDE isopachs (cm)
FGH isopachs (cm)
5 km
(c)
ROMA
Colli Albani caldera
Albano maar
(d)
Fall unit A
LBT
Isopachs in m
Fall unit B
UBT
Isopachs in cm
Rio Grande
Upper Bandelier Tuff
Lower Bandelier Tuff
major fault systems
II Valles
I Toledo
caldera ring fracture
vent area for Bandelier plinian deposits
Valles
Toledo
RP
Los Alamos
0 10 20 km
(e)
Jemez Plateau
Pajarito Plateau

Giordano et al. 2002; van Otterloo and Cas 2016). Ignimbrites tend to form valley fill deposits, their distribution may be localised, and their stratigraphy very complex. Figures 1.5a and 10.22 illustrate the distributions of some selected ignimbrites. Note the differences in scales between maps.

The estimated volumes of a number of ignimbrites are listed in Table 10.2, which emphasises those associated with caldera-forming eruption styles (Sect. 10.11). Most of the largest ignimbrites described are those associated with large calderas in the Western United States, and in the Andes of South America, and in many cases, these consist of densely welded tuff. The Fish Canyon Tuff in the San Juan Volcanic Field is thought to have a minimum volume of 5000 km^3 (Lipman 2007). An enormous part of these large deposits occurs as thick intra-caldera ignimbrite, with smaller volumes found as outflow sheets. Later updoming of the intra-caldera pile may form a resurgent dome consisting mostly of uplifted, intra-caldera ignimbrites that are pushed up by late stage, buoyant rise of part of the source reservoir magma (e.g. Figs. 10.22d, e and 10.61; Chap. 16).

There are many problems with volume estimates in the literature (see discussion in the next section, Sect. 10.8.1.1). First, data from which estimates have been made vary from detailed maps and volcanological studies to simply gross estimates. Secondly, a significant or major part of some ignimbrites is welded tuff, but the proportion of welded to non-welded material is generally not given; therefore, no allowance is made for compaction, although this is an important factor when making volume comparisons. Thirdly, published studies have tended to describe the larger, more spectacular ignimbrites. In addition, the volume lost into a widely dispersed, co-ignimbrite ash-fall deposit is generally not included in these estimates or is only roughly calculated. Finally, many ignimbrites occur at island arcs or close to coasts and major parts are deposited in the sea as subaqueous "pyroclastic flow" deposits and ash turbidites (Chap. 11). Today, a summary of caldera-forming ignimbrite volumes can be searched in the Collapse Caldera Database (http://www.gvb-csic.es/CCDB.htm; Geyer and Martí 2008), and in the LaMEVE (Large Magnitude Explosive Volcanic Eruptions) database (Crossweller et al. 2012; http://www.bgs.ac.uk/vogripa/view/controller.cfc?method=lameve).

10.8.1.1 Methods and Problems in Calculation of Ignimbrite Volumes and Other Relevant Parameters (VEI, Magnitude, DRE)

The calculation of the bulk volume of ignimbrites is the basis for defining some of the most useful comparative descriptors, such as the Volcanic Exposivity Index (VEI; Newhall and Self 1982, or VEI* as discussed in Chap. 9, Table 9.1), the Magnitude, (M, Pyle, 2000, or M* as discussed in Chap. 9, Sect. 9.10), and the Dense Rock Equivalent (DRE) volume of ignimbrite-forming eruptions. The deposit volume is therefore a very important parameter; however, its calculation is a quite difficult task and is very poorly defined in the literature (e.g. Mason et al. 2004). The main sources of problems can be summarised as follows:

(a) Degree of erosion, hence state of outcrop preservation, often reflecting the age of the unit (e.g. Wilson 1991; Cook et al. 2016).
(b) Accuracy in the reconstruction of detailed isopach maps, which include the knowledge of pre-eruption topography, usually largely buried, and definition of the isopach 0 m (e.g. Best et al. 2013; Silleni et al. 2020).
(c) Definition of the amount of ash elutriated into co-ignimbrite ash clouds (e.g. Sparks and Walker 1977; Cook et al. 2016; Martí et al. 2016; Calabrò et al. 2022).
(d) Definition of the density of the deposits and of the lithic content at the unit scale, in order to retrieve a Dense Rock Equivalent volume (e.g. Streck and Grunder 1995; Cook et al. 2016).
(e) Calculation of the volume of intra-caldera ignimbrite fill (e.g. Giordano et al. 2010; Best et al. 2013; Cook et al. 2016), commonly taken as equal to the volume of outflow, but not often verifiable.

Based on such difficulties, the literature has usually been quite vague about the methods used to produce the volumes proposed (see discussions in Mason et al. 2004; Crossweller et al. 2012; Cook et al. 2016; Silleni et al. 2020). Shortcuts have commonly been taken such as "*overall area × average thickness*", though both are usually very poorly defined, if at all. Interestingly, the volume of some well-studied ignimbrites has varied significantly over time. For example, the 2.08 Ma Cerro Galán ignimbrite was originally estimated at 1000 km^3 (Sparks et al. 1985), and later revised at 630 km^3 (Folkes et al. 2011a) based on more careful consideration of the actual total extent, intra-caldera fill and elutriated co-ignimbrite ash. Similarly, lack of definition of the method of calculation resulted in bulk volumes estimated for the 39 ka Campanian ignimbrite spanning one order of magnitude between 50 and 500 km^3 (e.g. Fisher et al. 1993; Scarpati et al. 2014). Recent work by Silleni et al. (2020) has carefully recalculated the volume of the Campanian ignimbrite; this is based on a well-defined isopach map, described in its reconstruction method and related uncertainties, and afterwards corrected for erosion and ash elutriation. The bulk volume of the Campanian ignimbrite turned out to be 144 km^3, to which up to 365 km^3 of elutriated co-ignimbrite ash can be added for a total bulk volume of the pdc-related deposits of 561 km^3, corresponding to 225 km^3 DRE, and 248 km^3 DRE by including the volume of the basal fallout deposit.

Table 10.2 Characteristics of selected ignimbrites and their source calderas (Modified from Giordano and Cas 2021)

Ignimbrite	Caldera volcano	Age	Caldera diameter (km)	Total DRE volume (km^3)	Composition[1]	Crystals[2]	Outflow area (km^2)	Outflow bulk volume (km^3)	Average thickness (m)	Equivalent radius (m)	Aspect ratio	Max runout (km)	Temperature emplacement	No. cooling units
Atana	La Pacana	4.1 Ma	60 × 35	1600	RD	CR	6750	240	35.6	46,364.7	3.83×10^{-4}	30	Unwelded-welded	1
Bishop Tuff	Long Valley	0.76 Ma	17 × 32	650	R	CP	2200	200	90.9	26,469.5	1.72×10^{-3}	40	Welded	8
Bishop Tuff, cooling unit	"	"	"	"	"	"	2200	25	11.4	26,469.5	2.15×10^{-4}		"	1
Campanian	Campi Flegrei	39 ka	15 × 12	> 150	T	CP	12,000	146	12.2	61,819.5	9.84×10^{-5}	80	>580 °C	1
Cerro Galán	Cerro Galan	2.08 Ma	35 × 20	> 630	RD	CR	7200	324	45.0	47,885.2	4.70×10^{-4}	74	>580 °C	1
Ito	Aira	22 ka	20 × 20	143	R	CP	15,000	249.9	16.7	69,116.4	1.21×10^{-4}	80	Unwelded-welded	1
Kidnappers	Mangakino	1.0 Ma	40 × 30	1200	R	CP	45,000	450	10.0	119,713.0	4.18×10^{-5}	>200	Unwelded	1
Lund Tuff	White Rock	29.2 Ma	40 × 40	4400	R	CR	12,000	2300	191.7	61,819.5	1.55×10^{-3}	80	Welded	1
Mamaku	Rotorua	225 ka	20 × 16	146	D-R	CP	3200	225	70.3	31,923.5	1.10×10^{-3}	> 30	Unwelded-welded	1
Oruanui, unit10	Taupo	25.4 ka	28 × 16	530	R	CP	6358	200	31.5	44,998.2	3.50×10^{-4}	40	Unwelded	1
Peach Spring Tuff	Silver Creek	18.8 Ma	25 × 25	1300	R	CP	32,000	650	20.3	100,950.9	1.01×10^{-4}	>170	Welded	1
Pinatubo	Pinatubo	1991	2.5 × 2.5	4.5	D	CR	326	5.5	16.9	10,189.3	8.28×10^{-4}	16	Unwelded	1
Pompeii	Vesuvius	79 AD	5 × 5	1.5	Ph-T	CP	150	1.5	10.0	6911.6	7.23×10^{-4}	15	<440 °C	1
Pozzolane Rosse	Colli Albani	46 ka	10 × 10	63	Te	CP	2500	59	23.6	28,216.6	4.18×10^{-4}	33	>580 °C	1
Taupo	Taupo	232 AD	28 × 16?	35	R	CP	15,000	15	1.0	69,116.4	7.23×10^{-6}	80	200–400 °C	1
Tor de Cenci Fm	Colli Albani	560 ka	10 × 10	22	F	CP	2000	20	10.0	25,237.7	1.98×10^{-4}	32	200–400 °C	1
Villa Senni Fm, compound	Colli Albani	355 ka	10 × 10	28	PhTe	CP-CR	2000	30	15.0	25,237.7	2.97×10^{-4}	29	>580 °C	2
Villa Senni Fm, Pozzolanelle	"	"	"	"	PhTe	CR	2000	20	10.0	25,237.7	1.98×10^{-4}	29	>580 °C	1
Villa Senni Fm, Tufo Lionato	"	"	"	"	PhTe	CP	2000	10	5.0	25,237.7	9.91×10^{-5}	25	>580 °C	1

(continued)

Table 10.2 (continued)

Ignimbrite	Caldera volcano	Age	Caldera diameter (km)	Total DRE volume (km^3)	Composition[1]	Crystals[2]	Outflow area (km^2)	Outflow bulk volume (km^3)	Average thickness (m)	Equivalent radius (m)	Aspect ratio	Max runout (km)	Temperature emplacement	No. cooling units
Valley of Ten Thousand Smokes	Katmai	1912	3 × 2	13.5	R-D-A	CP	180	11	61.1	7571.3	4.04×10^{-3}	22	Welded	1
VTTS, depositional unit	”	”	”	”	”	”	180	1.22	6.8	7571.3	4.48×10^{-4}		Welded	”
Toba Tuff	Toba	74 ka	100 × 35	2800	RD	CR	20,000	1700	85.0	79,808.7	5.33×10^{-4}	>120	Unwelded-welded	1

[1] R = rhyolite, D = dacite, RD = rhyodacite, T = trachyte, Ph = phonolite, Te = tephrite, Ph = Te phonotephrite, F = foidite
[2] CR = crystal-rich, CP = crystal-poor

Wilson (1991) suggested that erosion has a much larger effect on the calculation of the total areal extent of an ignimbrite with respect to its volume, which is usually, for the largest part, contained in the valley pond/topography filling-building facies. Based on this consideration we favour the estimates that are based on careful reconstructions of the ignimbrite isopach maps (e.g. Wilson 1991; Streck and Grunder 1995; Giordano et al. 2010; Best et al. 2013), in order to make clear the raw dataset, upon which any further correction is made. In fact, the evaluation of the amount of ignimbrite eroded, as well as the maximum original extent of the sheet, must be clearly related to the actual outcrops (Cook et al. 2016; Cavazos and Carrasco-Núñez 2020).

Furthermore, crystal concentration studies (e.g. Sparks and Walker 1977; Scarpati et al. 2014) and careful reconstruction of distal co-ignimbrite ash deposits (e.g. Eychenne et al. 2012; Cook et al. 2016; Marti et al. 2016; Smith et al. 2016) are essential to constrain the amount of ash elutriated that must be added to the ignimbrite volume, an amount that numerical simulations and available field reconstructions, especially for intermediate to large volume ignimbrites, constrain to ⋝50% of the total erupted mass (e.g. Calabrò et al. 2022). In addition, assessment of the volume of intra-caldera ignimbrite is important, which largely depends on the timing of the collapse (see discussion in Sect. 16.7.3), and is dependent on availability of either subsurface data or exhumation/erosion of the intra-caldera fill deposits (e.g. Willcock et al. 2013; Cook et al. 2016). Best et al. (2013) proposed four different models to compute the intra-caldera volume that encompass the most conservative assignment to the caldera area of the thickness at caldera rim, to various estimations of the ratio between intra-caldera and extra-caldera volumes, based on timing and geometry of collapse (piston vs. trapdoor vs. subsidence) during the course of the eruption.

In summary, we suggest that it is both timely to revise existing volume estimates, and to provide more rigorous and replicable calculations for the volume of ignimbrites. Best practice is the reconstruction of detailed isopach maps from which quantitative data can be extracted in terms of thickness variations, volume, and areal extent of the ignimbrites (Giordano and Cas 2021; see also Sect. 10.11), whereas simplistic and poorly constrained averaging of areas and thickness, or unsupported "guesstimations" of the ash elutriation and intra-caldera volumes, should be avoided.

10.8.2 Source Vents

Several types of source vents for ignimbrites have been proposed (Fig. 10.22). Observed small volume, ignimbrite-forming eruptions have been erupted from central vents. The most voluminous caldera-forming ignimbrites are generally associated with continuous ring fissure or linear volcano-tectonic fissure vent eruptions.

10.8.2.1 Central or Point Source Vents

A number of Quaternary ignimbrites have been related to central or point source vents (Figs. 10.5 and 10.22a, b) by several methods, including direct observation of historic and modern eruptions, mapping their distribution, mapping the distribution of co-eruptive pumice fall deposits (Chap. 9), and identifying proximal facies (e.g. from the distribution of co-ignimbrite breccias; Sect. 10.8.10.3). These include small volume ignimbrites from stratovolcanoes (e.g. notably at Mt St Helens in 1980, Fig. 10.22a; Soufrière Hills volcano, Montserrat, Cole et al. 2002), as well early ignimbrites that precede caldera formation. Examples of the latter include the 8–10 km^3 eruption of Mt Pinatubo, Philippines in 1991; the 5 km^3 79 AD Vesuvius eruption and the 10 km^3, 151 ka Sutri Formation at Vico volcano, Italy. At Mt Pinatubo precursor activity produced minor fall, flow and surge activity from a central vent, which by June 15 had developed into a sustained plinian column from which pumice and ash flows ($\sim$5.5 km^3) were more or less continuously emplaced (Plate 10.1a; Scott et al. 1996). This culminated in foundering of the summit dome and the eruption of lithic-rich debris that chocked the conduit, shut down the eruption and resulted in the formation of a small "funnel-type" caldera (see Fig. 16.57f). The 79 AD Vesuvius eruption commenced with a plinian column from a central vent (Sigurdsson et al. 1985; Cioni et al. 1992; Gurioli et al. 2005), which was later followed by ignimbrites that both preceded and accompanied the caldera collapse. The Sutri Formation starts with a fall deposit and early ignimbrites, which have been interpreted, based on dispersal, to have originated from a central vent, but are then followed by caldera collapse and generation of ignimbrites from a ring fracture vent near the end of the eruption, represented by a large volume scoria, spatter and ash ignimbrite, and contemporaneous lithic breccias (Fig. 10.23; Bear et al. 2009a, b).

Similarly, the 248 km^3 (DRE) Campanian ignimbrite erupted 39 ka from Campi Flegrei caldera, Italy (Silleni et al. 2020), is characterised by an early fall deposit and small volume ignimbrites from a central vent, followed by the bulk of the Campanian ignimbrite, which is dominated proximally by co-ignimbrite breccias rich of deep seated lithics, interpreted to indicate the timing of the caldera collapse and transition to a ring fracture vent system (Rosi et al. 1996). If this is correct, then caldera collapse, forming the Campi Flegrei present day caldera, had to happen near the end of the Campanian eruption. Likewise, the 183.5 ka Lower Pumice 1 eruption on Santorini commenced with central vent eruption phases, producing plinian fallout and pumice and ash ignimbrite deposits, and then climactic

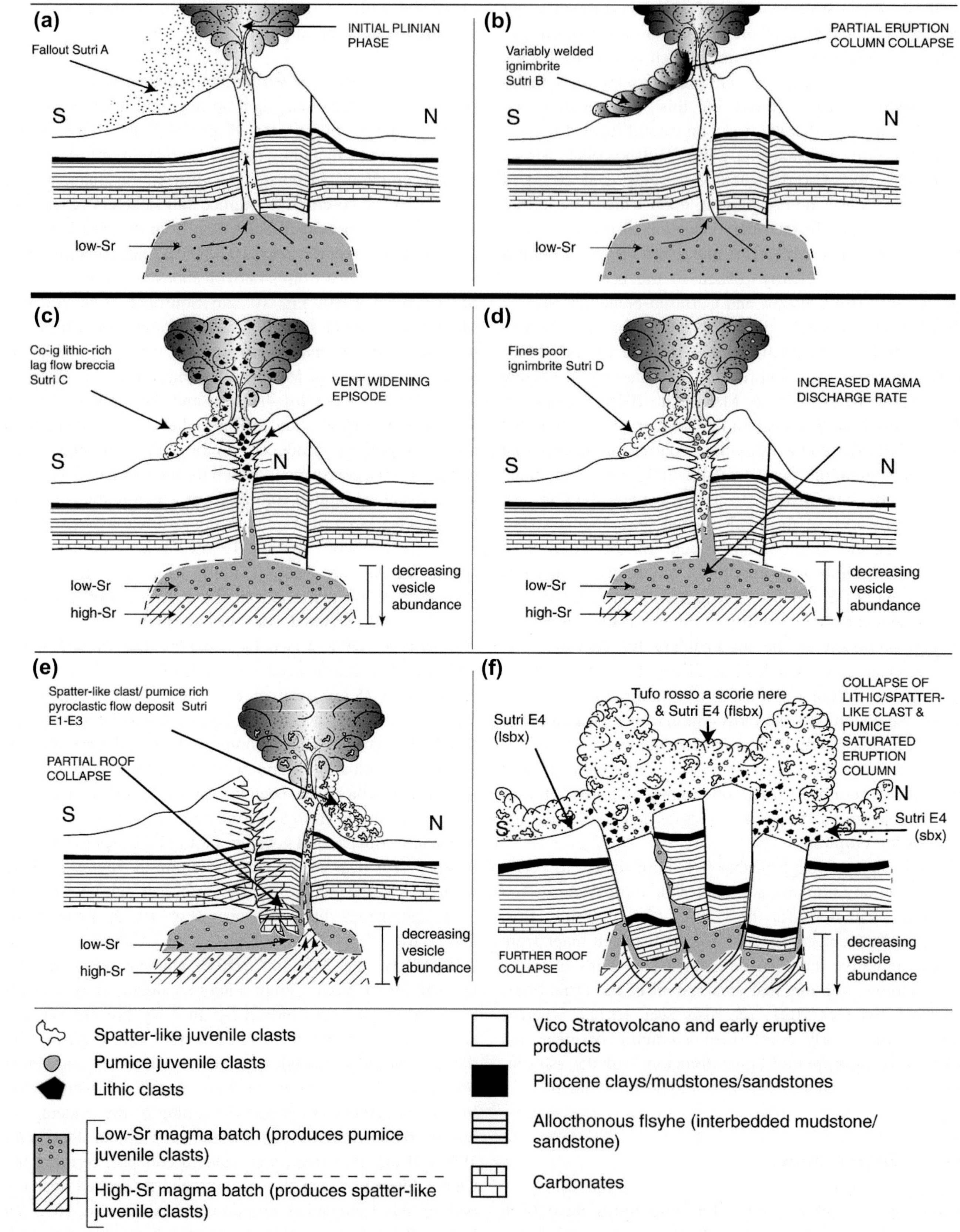

Fig. 10.23 a–f Eruption stages during the 151 ka Sutri eruption from Vico caldera, Roman Magmatic Province, central Italy. Stages A–D show central vent eruption, producing a short-lived plinian fallout deposit and several small volume ignimbrites, resulting from column collapse. Stage E represents the development of radial fissures and onset of caldera collapse, shortly followed by Stage F with caldera collapse and formation of a ring fracture fissure vent system. (From Bear et al. 2009a)

caldera collapse, represented by lithic-rich ignimbrite and lag breccia deposits (Simmons et al. 2016). During the 1883 eruption at Krakatau, caldera collapse seems to have taken place very late in the eruption, which commenced from a central vent (Self and Rampino 1981). The start of the ultra-plinian and ignimbrite-forming, 35 km^3 (DRE), Taupo AD 232 eruption has been related to a central vent (Sect. 10.9.1.1). Sudden collapse of the vent area is thought to have greatly widened the vent, and to have triggered the eruption of the violent Taupo ignimbrite. The 30 km^3 (DRE) Rio Caliente intra-plinian ignimbrite, Mexico, appears mainly, if not entirely, to have erupted from a central vent (Wright 1981) and, although La Primavera rhyolite volcano lacks a well-defined caldera, post-eruption subsidence is recognised by a circular area of (now uplifted) lake sediments $\sim$10 km in diameter.

Some central vent eruptions have apparently been unrelated to any pre-existing volcano. Just as Paricutin was "born in a cornfield" in Mexico in 1943, Wright and Walker (1981) suggested the Quaternary (<0.65 Ma) Acatlán ignimbrite erupted similarly erupted from a vent on flat ground based on the presence of compositionally zoned co-ignimbrite lag breccias (Sect. 10.8.10.3), although this could be part of a larger caldera system. The Valley of Ten Thousand Smokes ignimbrite in Alaska in 1912 was erupted from a new vent at Novarupta, located 10 km away from the summit of Mt Katmai volcano from beneath which magma was tapped by lateral withdrawal through a sill, causing syn-eruptive collapse of the 4 km wide $\times$ 1 km deep Katmai caldera, 10 km away from the eruption vent (Hildreth and Fierstein 2012; Fig. 10.22b)!

10.8.2.2 Ring Fissure or Linear Fracture Fissure Vents During Caldera Collapse

Ignimbrites associated with caldera formation have long been thought to be erupted along a continuous, more or less circular, ring fracture or fissure vent system, along which caldera collapse occurs during the eruption (Smith and Bailey 1968; Fig. 10.22c–e). The ring fissure vent model has been popular in the Western United States since the caldera collapse model was proposed by Williams (1942). This was based on the formation of the Mt Mazama or Crater Lake caldera, and it explains the radial distribution of thin outflow sheets associated with very much thicker intra-caldera ignimbrites, which are thought to have ponded within the depression formed by a continually subsiding caldera floor block (Fig. 10.24).

Detailed field studies of the 7.7 ka Mt Mazama pumice and ash flow deposits indicate that eruption from around the ring fracture must have occurred while caldera collapse was taking place (Bacon 1983; Druitt and Bacon 1986). The eruption sequence can also be divided into two stages, as for the previous examples cited in Sect. 10.8.2.1: single vent and ring fissure vent phases (Fig. 10.24a, b). The initial plinian fallout phase was erupted from a single vent. Collapse of this column generated a series of pumice and ash flows deposited to the north and east of Mt Mazama. Approximately 30 km^3 of magma were expelled before collapse of the roof of the magma chamber occurred, and the inception of caldera formation ended the single vent phase. Later pumice and ash flows are believed to have been erupted from multiple vents around the ring fracture as caldera collapse continued (Fig. 10.24c). The composition of lithic clasts within proximal co-ignimbrite breccias (Sect. 10.8.10) varies around the rim and certain types can be correlated to probable vent locations. During the ring vent phase, another 20 km^3 of magma may have been erupted. Such a two-stage model for caldera-related ignimbrite-forming eruptions has been adopted for many, both felsic (e.g. Santorini, Druitt and Sparks 1984; Simmons et al. 2016; Valles, Self et al. 1986, 1996, Fig. 10.22d, e), and mafic ignimbrites (Vico, Bear et al. 2009a; Colli Albani, Vinkler et al. 2012, Fig. 10.22c; also see Sects. 10.10.3 and 10.10.4).

Large volume (VEI8/VEI*12; Chap. 9), rhyodacitic, crystal-rich ignimbrites (the "monotonous intermediate" of Hildreth 1981; Maughan et al. 2002) usually lack a pre-ignimbrite plinian deposit, which suggests that the caldera collapse commenced immediately at the beginning of the eruption, without the existence of an initial single vent phase (de Silva et al. 2006; Willcock et al. 2013, 2015a). At Cerro Galán, for example, eruption from around the ring fissure was perhaps the important mechanism throughout. This would make an eruption column unstable and collapse, because of the very mass eruption rate (Fig. 10.5). Francis et al. (1983) and Folkes et al. (2011a, b) indicated that very large mass eruption rates would be produced by sinking of the caldera floor block into the magma reservoir along outward dipping ring fractures (Fig. 10.25). Vertical or inward-dipping ring fractures, as suggested in some models of ring fissure eruptions, would inhibit very high eruption rates (Roche et al. 2000; Geyer et al. 2006; Acocella 2007), and would require significant explosive excavation of the vent to facilitate caldera collapse, generating large volumes of co-ignimbrite lithic breccias. The thickness of intra-caldera ignimbrite exposed by the Cerro Galán resurgent dome is 1200 m, and there is no doubt that caldera collapse must have been concurrent with eruption. Outward dipping ring faults allow vents to open by extension as the magma chamber roof block subsides, facilitating increased magma discharge rates, and minimal explosive erosion of vent wall country rock (Cas et al. 2011). In contrast, vertical

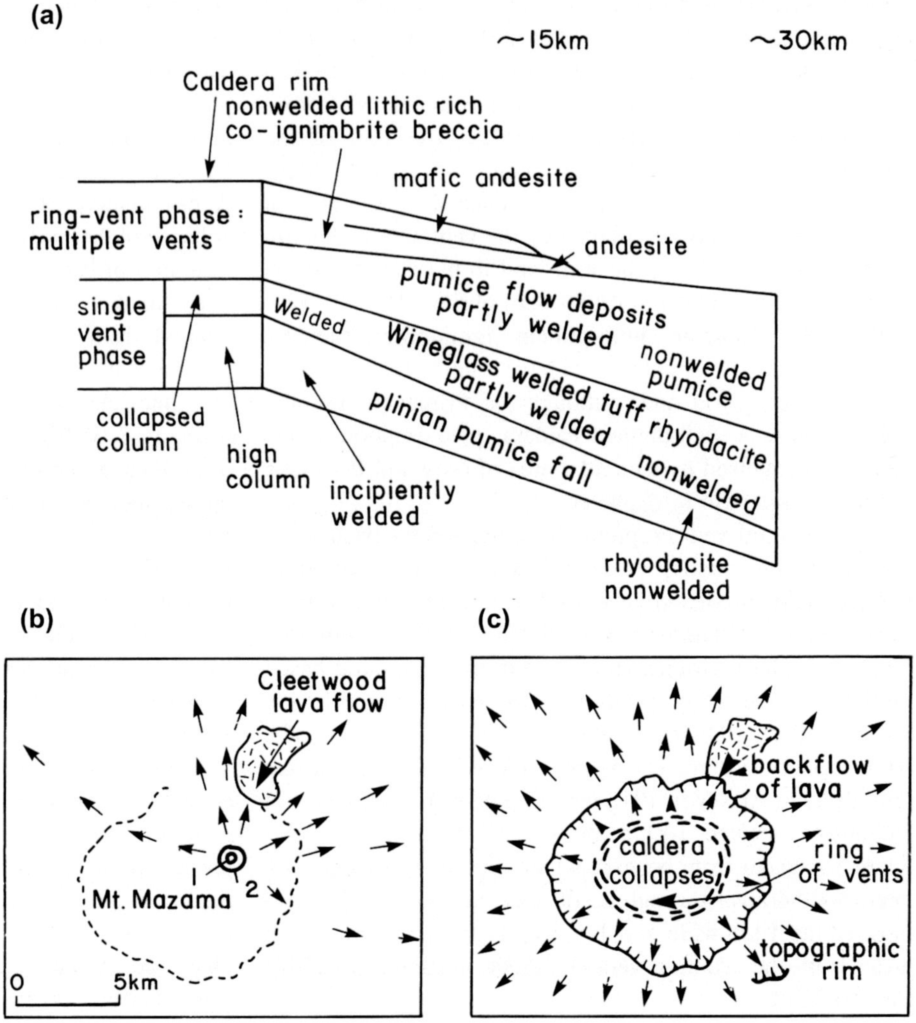

Fig. 10.24 Summary of the 7.7 ka Mt Mazama ignimbrite-forming eruption, Crater Lake caldera, Oregon, USA. **a** Stratigraphic relations and vent type for the products. Note also the last erupted pumice flows were mafic andesites in contrast to the earlier rhyodacites, and spectacular compositional zonation is shown. **b** Single vent phase of the eruption; 1 indicates vent for plinian eruption and 2 the enlarged vent during eruption of the pumice flows that deposited the Wineglass welded tuff. Arrows show direction of movement of pumice flows, which were confined to topographic low areas. **c** Ring vent phase during which pumice flows surmounted most topographic barriers. The ring fissure is shown inboard of the present topographic rim which would indicate scarp retreat by land sliding and slumping of the caldera wall during or after caldera-collapse. (From Cas and Wright 1987, After Bacon 1983)

to inward dipping ring faults hinder caldera collapse and enhance explosive erosion of the vent walls, by which they widen. Resulting ignimbrite deposits are therefore likely to be lithic-rich, whereas with outward dipping faults lithic clast content will be relatively low. Lithic clast content can therefore be used as a proxy for how much vent wall country rock has been excavated by explosive erosion and may reflect the geometry of the ring fracture system associated with calderas. In their study of the Permian Ora ignimbrite, northern Italy, Willcock et al. (2013; see Fig. 10.85) demonstrated that co-ignimbrite breccias which underlie the 1.3 km thick crystal-rich, intra-caldera welded ignimbrite around ring fractures, herald the initiation of caldera collapse from the onset of the eruption.

In the Sierra Madre Occidental volcanic belt in Mexico, huge volume ignimbrites were erupted from rectilinear fault fissure vents that bound many calderas (Aguirre-Diaz 2014; Aguirre-Díaz and Labarthe-Hernández 2003). The fissure vent systems are controlled by major rift or graben faults, usually through several vents aligned along normal faults which form the caldera walls. Such graben calderas occur in areas of regional extension (Chaps. 16 and 17).

However, the vent type and eruption mechanisms for the largest ignimbrites will not always be clear. Models of central vent and fissure vent eruptions must be tested by the analysis of eruption sequences and facies, and some of the answers are likely to be found by further examination of ancient volcanic terrains where the roots of large centres and calderas are exposed. Vent type may change as an eruption proceeds from a central vent to ring or linear fractures (e.g. Bacon 1983; Druitt and Sparks 1984; Rosi et al. 1996; Bear et al. 2009a, b; Vinkler et al. 2012; Simmons et al. 2016). Even if eruption is initiated along a fissure, or if one or more fissures develop due to later caldera collapse, localisation to multiple vents or a central vent along the fissure may be a natural progression. Note that in the theoretical

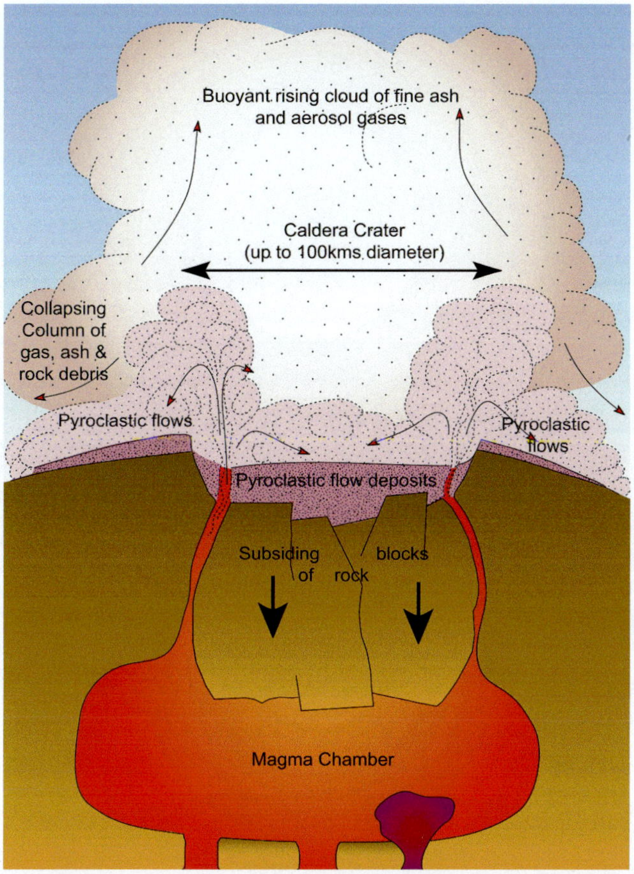

Fig. 10.25 Caldera collapse model, facilitated by outward dipping ring faults bounding the subsiding roof block. The ring faults become fissure vents, opening by extension as the roof block subsides into the magma reservoir. This then allows for rapid increase in mass eruption rate of the ensuing ignimbrite, which infills the subsiding caldera with a very thick intra-caldera ignimbrite, relative to the much thinner extra-caldera or outflow ignimbrite

column-collapse models (Figs. 10.5, 10.6, 10.7, 10.8 and 10.9) a cylindrical vent is assumed, but the results can vary significantly if the geometry of the conduit is not axisymmetric as in the case of a dyke or ring-fissures (e.g. Jessop et al. 2016).

An important concept is that calderas collapse with a variety of styles that control the timing and kinematics of ring fault nucleation and slip (Chap. 16). Analogue models by Roche and Druitt (2001) and Geyer et al. (2006) indicate that shallow and sill-like (oblate) reservoirs promote the coherent collapse of a piston and development of annular ring fractures. In contrast, magma withdrawal from deep and vertically elongated (prolate) reservoirs induces the progressive upward fracturing of the roof until complex sets of ring faults eventually reach the surface. These contrasting collapse styles have been named coherent or piston-like, and chaotic collapses respectively (Chap. 16, see Fig. 16.60). Coherent collapses are more common for large volume ignimbrites >100 km^3 (Gregg et al. 2012; Cashman and Giordano 2014), whereas chaotic collapses are more commonly associated with smaller volume ignimbrites. Such different collapse modes certainly have an influence on the time evolution of the eruption. In this regard, Giordano and Cas (2021; Sect. 10.11, see Fig. 10.93) noted that large and very large volume ignimbrites usually show relatively simple internal lithofacies organisation, high temperature of emplacement and evidence of relatively slow velocities (e.g. Cas et al. 2011; Roche et al. 2016), indicative of sustained high-intensity mass fluxes throughout the life of the pyroclastic flow event, which is reconcilable with coherent, wholesale caldera collapse. By contrast, smaller volume ignimbrites show much larger internal variabilities as well as evidence of eruption point migration during the course of the eruption (e.g. Wilson 2001; Hildreth and Wilson 2007; Pittari et al. 2008), and even stop-start behaviours (e.g. 355 ka Villa Senni eruption, Vinkler et al. 2012; 4.1 ka Agnano-Monte Spina eruption, de Vita et al. 1999; Romano et al. 2020; Novarupta-Katmai 1912 eruption, Hildreth and Fierstein 2012), which is likely in response of the complex process of segmentation and migration of ring fractures in chaotic caldera collapses.

10.8.3 Settings: Extra-Vent (Outflow) and Intra-Caldera

The setting where ignimbrites are emplaced is important both for understanding the transportational and depositional processes, but also, in ancient rock successions, for locating the source areas, which is very useful in ore deposit exploration (Chap. 18). We will separate the discussion of setting into those associated with calderas, and those related to point source volcanoes such as stratovolcanoes, lava domes, and large maar volcanoes. Caldera-forming ignimbrites can be found dominantly within two settings: the outflow or extra-caldera/extra-vent, and the intra-caldera settings (Figs. 10.22, 10.23, 10.24 and 10.25; Chap. 16). Outflow ignimbrite deposits are usually 1 to >100 m thick and can form extensive sheets outside and around caldera margins, covering usually 10^3–10^4 km^2. In multiple eruption caldera complexes, the superposition of several outflow ignimbrites builds ignimbrite shields, typically with low slopes (2–5°; Fig. 10.26), which progressively bury pre-existing topography (Fig. 10.27; Chap. 16). Around stratovolcanoes outflow ignimbrites also contribute to the building of the cone sequence and the surrounding ring-plains. Outflow ignimbrite successions typically appear as a vertically

Fig. 10.26 Gently sloping, extra-caldera ignimbrite shields. **a** Ignimbrite shield gently sloping away from the Cerro Galán caldera, northwestern Argentina (see Fig. 10.61 for plan view). This shield is built up by several rhyodacitic ignimbrites, the top one and most extensive is the 2.08 Ma, 630 km^3 Cerro Galán ignimbrite (see Fig. 10.28a). **b** Oblique aerial view of the Bandelier ignimbrite shield of the Valles caldera, Pajarito Plateau, Jemez Mountains, New Mexico, USA (looking northwards to Los Alamos; see Fig. 10.22d, e for plan and satellite views). This shield comprises two major rhyolitic ignimbrites erupted at 1.6 and 1.26 Ma, which are further subdivided into several flow units and cooling units (top orange-coloured layers). (Photo by P. Fisher; *Source* Wikimedia through Creative Commons Attribution 4.0 International licence). **c** Canyon exposure through three eroded Bandelier Tuff ignimbrite units (1–3) making up the shield. In this case: unit 1 = 1.6 Ma Lower Bandelier ignimbrite unwelded to incipiently welded in the middle; units 2, 3 = two cooling units of the 1.26 Ma Upper Bandelier ignimbrite. Looking south down Cochiti Canyon, southern Pajarito Plateau (see also Fig. 10.28b)

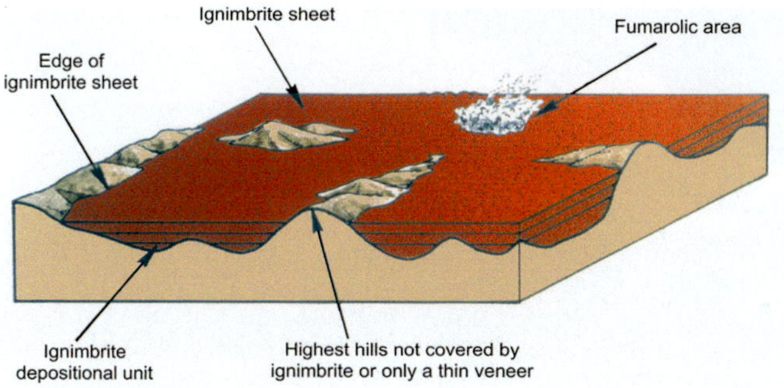

Fig. 10.27 Schematic diagram showing the topography in-filling, burial and on-lap relationships of an ignimbrite shield succession with bedrock

accumulated sequence of tabular sheets, separated by paleosoils and/or reworked deposits (Fig. 10.28).

The usual lack of, or minor occurrence of intervening primary volcanic deposits in outflow settings must be carefully assessed, and results either from volcanic quiescence and erosional effects or, more likely, limited volcanic activity and restricted dispersal from the vent area between major ignimbrite-forming events, as for example, is currently the case at Campi Flegrei, Bay of Naples (Orsi et al. 2004). Also, at the Colli Albani volcano, Italy, the western part of the ignimbrite shield consists almost entirely of 7 major ignimbrites separated by paleosoils and emplaced with an average recurrence time of 40 kyr (Fig. 10.28b). However, these stratigraphic breaks do not correlate with periods of volcanic quiescence, because to the east (i.e. downwind) the same ignimbrites are interbedded with numerous small volume fallout deposits from sub-plinian activity sourced from the Colli Albani caldera, as well as evidence of effusive activity having occurred (Giordano et al. 2010).

With stratovolcanoes the extra-vent setting has a rougher topography, characterised by higher slope angles (10–30°) and deeply incised erosional valleys and canyons which often act to confine the pyroclastic flows. Steep slopes result in substantial erosional processes after ignimbite emplacement, so that ignimbrites are usually highly dissected and ignimbrite successions show very complex architectures that may not be easy to reconstruct, especially in older successions (Chaps. 1, 15, and 16).

The intra-caldera setting contrasts markedly with the outflow environment (Chap. 16). As caldera collapse usually occurs during the eruption, intra-caldera ignimbrites can have thicknesses varying from 100 s to >1000 m (Fig. 10.29), and they are usually welded to varying degrees (Sects. 10.8.14.1 and 10.9.3). They are confined to the caldera area, which usually ranges between 100 s and rarely >1000 km^2, a much more limited area than outflow settings. However, thick intra-caldera ignimbrites usually also have thinner outflow counterparts. Although the limited extent makes intra-caldera ignimbrites more rarely observable in the rock-record, their preservation potential is very high, both because of their thickness and welding, and because they were emplaced in a depositional trap. They are then often easily and quickly covered by sedimentary and other volcanic deposits and preserved.

Small volume ignimbrites that are not caldera-forming can occur in both extra-vent settings, where they are commonly sourced from volcanoes with a point source vent, such as stratovolcanoes, domes, and some large maar volcanoes, as well as being erupted from point source vents within calderas. The deposits are usually more lobate, have a limited lateral extent, generally 1–10 km from, the vent, and have an asymmetrical distribution.

10.8.4 Geometry of Ignimbrites

The original geometry of ignimbrites reflects the transportational and depositional processes and the interaction of pyroclastic flows with topography (Giordano and Doronzo 2017; Palladino and Giordano 2019). In the first instance, the final geometry of an ignimbrite is controlled by the volume of material carried by the body of the flow, relative to the accommodation volume of the topography, secondly by the velocity and momentum of the flow, and thirdly by the duration of the flow event. Small volume, slow pyroclastic flows produce valley confined ignimbrites. Large volume ignimbrites bury topography and produce sheet-like ignimbrites. As already discussed, recently erupted ignimbrites all have lobate, relatively high relief levees, and lobate margins (Figs. 10.11d, e, 10.18 and 10.21), reflecting frictional locking in a granular flow or plug flow state at the end of the flow event. We now analyse the significance of the final geometry of pyroclastic flow deposits in relation to topography and distance from source.

10.8.4.1 Maximum Distance from Vent and Uphill Reach—Indicators of Pyroclastic Flow Mobility and Mass Flux

Pyroclastic flow mobility and behaviour within topography are very important for volcanic hazard studies and because they reflect transportational processes. Large pyroclastic flows can travel for distances of tens of km or more, and their mobility has been considered spectacular because of their ability, in some cases, to surmount topographic barriers. For example, the Ito pyroclastic flow (Yokoyama 1974) must have surmounted a 600 m high mountain pass 60 km from source, and the Campanian ignimbrite (Fisher et al. 1993; Silleni et al. 2020) crossed a 685–1000 m high ridge of the Sorrento Peninsula 20 km away from source (Chap. 11, see

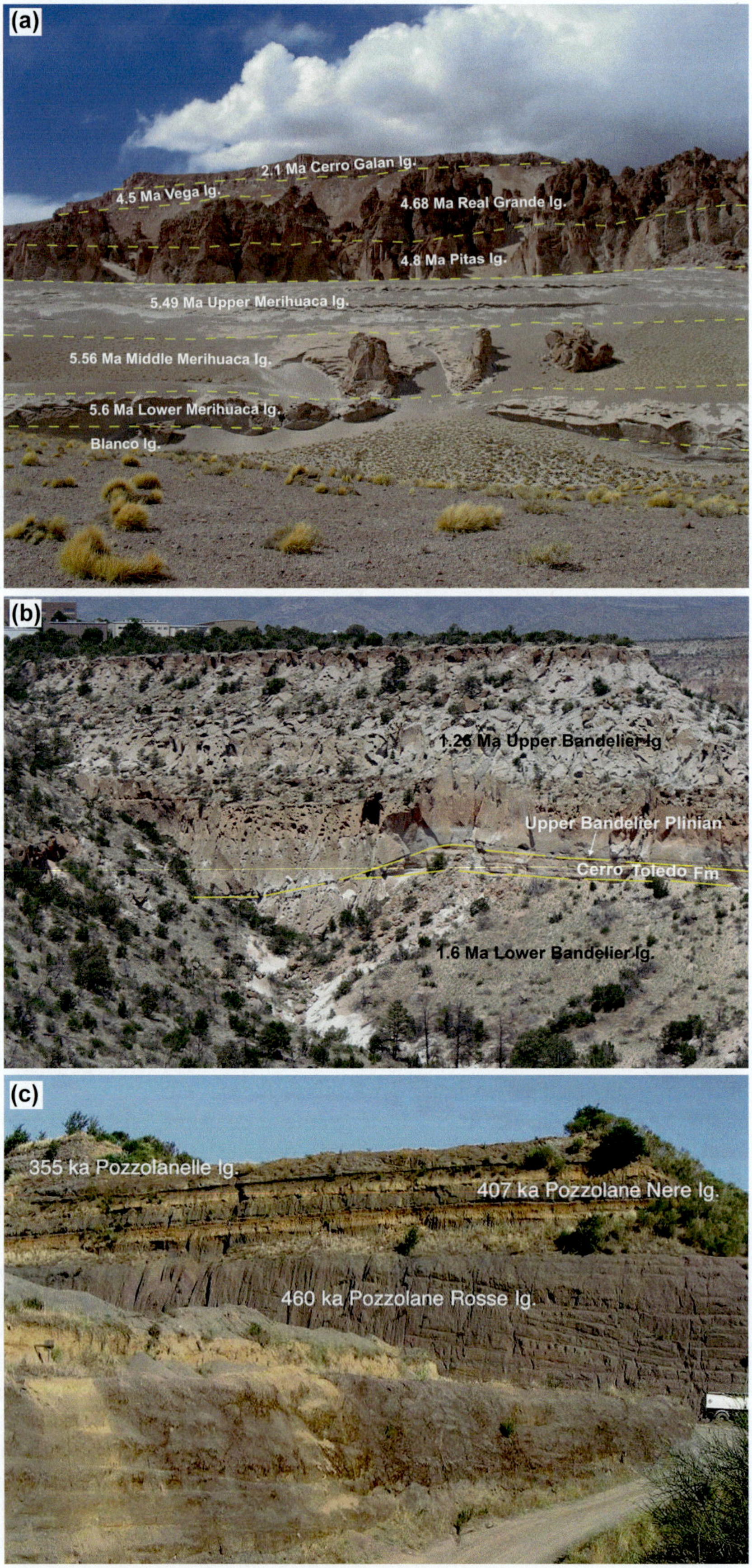

Fig. 10.28 Internal outflow ignimbrite shield successions. **a** Rhyodacitic ignimbrite shield stratigraphy of the Cerro Galán caldera, 10 km from the caldera margin, at Las Pitas, northwestern Argentina, showing stacked, multiple, large volume non-welded to welded ignimbrites, all erupted from the caldera. **b** Closer view of the internal succession of the rhyolitic Valles caldera ignimbrite shield (see Fig. 10.26b, c); Los Alamos airport, northern Pajarito Plateau, Jemez Mountains. (Photo by John Wolff). This shows the Lower Bandelier (white, slope-forming) and Upper Bandelier ignimbrites (grey to salmon pink, cliff-forming), and intervening Cerro Toledo Formation (stratified sub-plinian fallout deposits, erupted from Valles caldera rhyolite domes between 1.6 and 1.26 Ma). Note down cutting of a paleovalley through the Cerro Toledo fallout deposits and Lower Bandelier ignimbrite prior to infilling by Upper Bandelier ignimbrite. Remnant lenses of the Upper Bandelier plinian fallout deposit also mantle the slope cut into the Cerro Toledo Formation, which were largely removed by succeeding pdcs. Lastly, several zones of welding induration can be seen in the Upper Bandelier ignimbrite; the lower pale grey to salmon-coloured part is vitric, above that the tuff is largely recrystallised (see Fig. 10.65c). **c** Part of the tephri-phonolitic outflow ignimbrite shield succession of the Colli Albani caldera, near Rome, Italy. There are very well-developed paleosoils between the ignimbrites

Fig. 10.29 Exposures through very thick, intra-caldera ignimbrite successions. **a** 800 m thick exposure through part of the 28 Ma, dacitic La Garita caldera, San Juan Volcanic Field, Colorado, USA, which is the source of the largest ignimbrite (>5,000 km^3) known on Earth, the Fish Canyon Tuff. **b** Part of the >1200 m thick, 1300 km^3, 274 Ma rhyolitic Ora intra-caldera ignimbrite, northern Italy (also see Figs. 10.85 and 10.86)

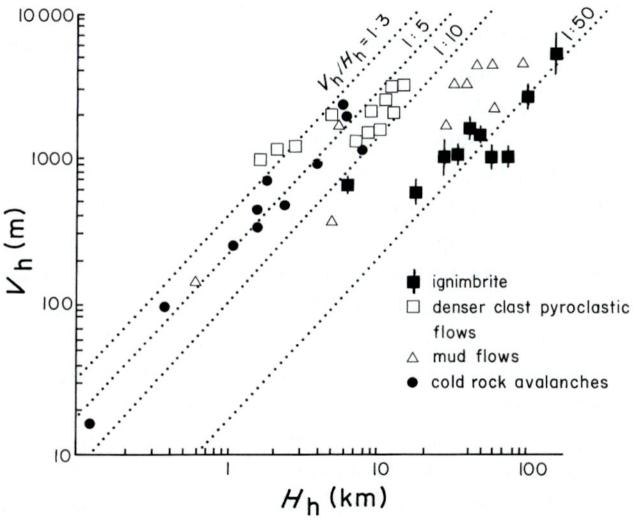

Fig. 10.30 Relationship between the vertical height dropped (V_h) and the horizontal distance travelled (H_h) for types of pyroclastic flow and debris flow. V_h for ignimbrites is usually taken as the reconstructed height of the volcano before caldera collapse, and bars indicate the uncertainty. The data indicate that ignimbrites are apparently more mobile than other types of debris flow. (From Cas and Wright 1987, After Sparks 1976)

Fig. 11.22a–d). Pumiceous density currents erupted 2,000 yr BP from Mt Pelée, on Martinique, passed over topography 500 m high 10 km from vent, and changed character beyond that point (Wright et al. 2016; Fig. 10.31c).

In Table 10.2, the maximum distance that ignimbrites are preserved from the source vent is also given for a number of ignimbrites. The data demonstrate the ability of pumice flows to travel large distances, often over gently sloping ground, but sometimes over significant topographic barriers hundreds of m high. Flow distances vary from <1 km for small volume ignimbrites, to >70 km for the 650 km^3 Peach Springs Tuff in Nevada (Roche et al. 2016), and >200 km for the 450 km^3 Kidnappers outflow ignimbrite in New Zealand (Wilson et al. 1995). For both of these, large volume, continuous eruption, and favourable topography are instrumental in long distance flow.

As discussed previously, there are two fundamentally different views on how pyroclastic flows move. Some workers accounted for this mobility by suggesting that pyroclastic flows are highly expanded, and thicker than the heights of mountains they are able to surmount (Yokoyama 1974; Fisher et al. 1993; Dade and Huppert 1996; Bursik and Woods 1996; Legros and Kelfoun 2000; Andrews and Manga 2011; Calabrò et al. 2022). Other workers have proposed that given sufficient momentum, a high concentration flow could surmount such barriers (Miller and Smith 1977; Sparks et al. 1978). However, as already expressed elswhere in this chapter, this dichotomy may not always be real, but reflect a spectrum of flow types. In many cases flow stratification may account for the existence of both a large scale expanded and dilute flow able to surmount high mountain regions, and at each location feeding the lower highly concentrated and fluidised undercurrent that may even move independently under its own momentum (e.g. Fisher et al. 1993). Irrespective of the flow type, some other authors have suggested that mobility of pyroclastic flows results dominantly from the heights of the eruption column collapse or height descended during flow (e.g. down stratovolcanoes), similar to the concept applied to explain mobility of cold rock avalanches (Fig. 10.30; also see Chap. 12) that are known to travel uphill (e.g. the Saidmarreh landslide in Iran climbed over a 600 m barrier, Kent 1966; the May 18 Mt St Helens debris avalanche surmounted a ridge 300 m high).

The "energy line" concept was introduced by Hsu (1975) for rock avalanche runout, and for pyroclastic flows by Sheridan (1979). The method links the maximum distance travelled, the maximum heights climbed and height of the column collapse by implicitly assuming the flow as a uniform high-concentration frictional flow.

However, as discussed in Sects. 10.6 and 10.7, pyroclastic flows are very different from cold rock avalanches at the first order because the majority of pdcs, especially with increasing volumes, are highly sustained by the collapsing mass flux at vent for minutes to hours, which is inherently a huge difference with respect to the impulsive nature of rock avalanches that collapse in seconds (e.g. the May 18 Mt St Helens debris avalanche). Furthermore, pyroclastic flows are not homogeneous, but are inherently density stratified (e.g. Breard et al. 2016) and undergo flow transformation along flow and over time as they interact with topography (e.g. Giordano 1998a; Wright et al. 2016). Under some conditions, the overridding dilute and turbulent part of the body and the ash cloud surge (Fig. 10.19) can even decouple from and travel independently from the basal high concentration underflow (Figs. 10.12c and 10.31; Fisher et al. 1980; Fisher 1990; Yamamoto et al. 1993; Druitt et al. 2002; Fujii and Nakada 1999; Wright et al. 2016; Sect. 10.14).

Over rough terrain, flow transformations, decoupling and divergent flow directions, including backflows, are especially likely (Figs. 10.31c and 10.32; Giordano and Dobran 1994; Ort et al. 2003; Porreca et al. 2003; Wright et al. 2016; Smith et al. 2020). In addition, numerical simulations show that sustained column collapse (fountaining) feeding ignimbrites cannot occur from heights much greater than 1–2 km above the vent, because as the jet region of the column extends upward, the progressive amount of air entrainement reduces its density allowing only partial collapses (Trolese et al. 2019). This improved understanding of how column collapses work rules out the height of collapse as a major factor controlling pyroclastic flow mobility, which must instead be related to the mass flux of collapse (Shimizu et al. 2019; Giordano and Cas 2021; Roche et al. 2021). Importantly, this is not equal to the mass eruption rate at vent, but represents a percentage of it, which varies between 10 and 90% (Trolese et al. 2019). As a consequence, the use of the energy line applied to ignimbrites should be adandoned, although it will be relevant to many impulsive lava dome collapse events that generate block and ash flows.

Flow mobility is therefore likely to be a function of the nature of the terrain, mass and momentum fluxes, and processes of self-fluidisation (Sect. 10.6), with relative roles changing during flow and between flows. In some cases, momentum will be dominant, in others self-fluidisation will be dominant. Numerical simulations by Bursik and Woods (1996), Dufek (2016), Shimizu et al. (2019), and Calabrò et al. (2022) suggest that the maximum runout of pdcs is mainly related to mass flux, and we will discuss this aspect in more detail in Sect. 10.11 on classification of ignimbrites (Giordano and Cas 2021; Roche et al. 2021).

Other main factors that affect flow mobility are its permeability, which is related to the concentration and particle grain size distribution, and the degree of coupling between particles and gas. Numerical simulations with single-sized particles illustrate that at increasing particle diameter the flow permeability increases and gas escapes more easily, limiting flow runout (Giordano and Dobran 1994; Roche et al. 2006; Fig. 10.33). By decreasing the particle size the coupling between gas and particles increases and the flow, with identical eruptive parameters (mass eruption rate, exit velocity, height of column collapse), would travel farther than a coarser-grained flow (Fig. 10.33) The estimation of the Total Grain Size Distribution (TGSD) for interpreting and modeling of pdc processes and mobility is therefore essential, although rarely computed for real ignimbrites (e.g., Pozzolane Rosse ignimbrite in Calabrò et al. 2022), whereas it is now done routinely for pyroclastic fall deposits. Similarly, the importance of the grain size distribution on processes of self-lubrication and pyroclastic flow runout are highlighted by large-scale analogue models (Breard et al. 2019; Lube et al. 2019; Brosch et al. 2022; Dellino et al. 2019). Given the very complex nature of pyroclastic flows, accurate flow models that also take into account the effects of the TGSD and mountainous terrain are required for accurate mobility predictions (e.g. Doronzo et al. 2010; Calabrò et al. 2022).

Nevertheless, with a simple but useful and easy first approach, measurements of the heights climbed by a

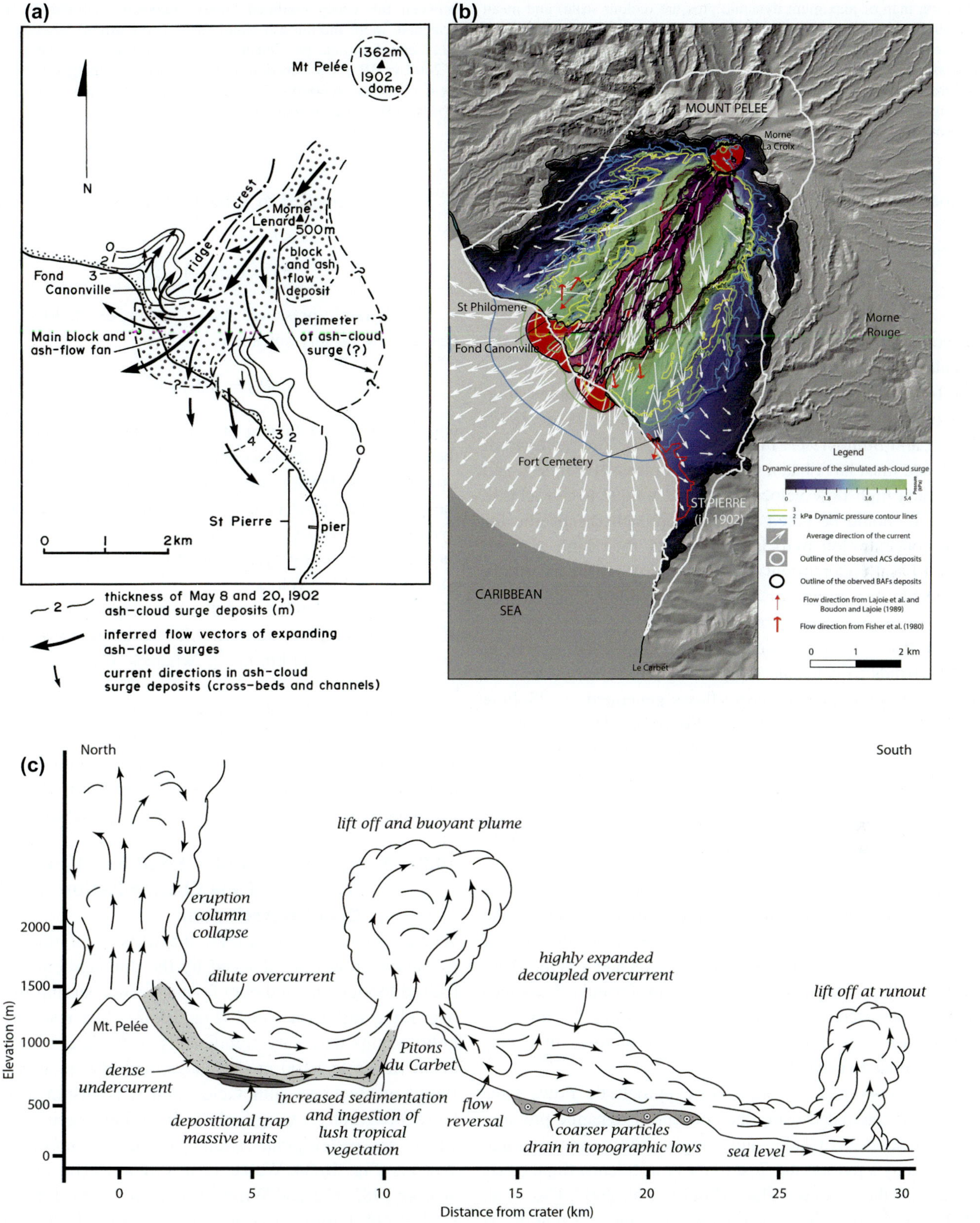

Fig. 10.31 Extent of ash-cloud surge deposits relative to valley pond body deposits. **a** Distribution of the block and ash flow and associated ash-cloud surge deposits from the 8 and 20 May 1902 eruptions of Mt Pelée, Martinique, and their inferred flow directions. The main block and ash-flow lobe fills the Rivière Blanche valley. Notice how at Fond Canonville ash-cloud surges moved around the ridge and then in an opposite direction to the main flow. (From Cas and Wright 1987, After Fisher et al. 1980). **b** Numerical simulation of the 8 May event,

(continued)

showing map of maximum dynamic pressure (colour scale) and mean direction of ash-cloud surge (white arrows whose length corresponds to the velocity). Purple to red colours correspond to the simulated block and ash flow deposit ($\geq$0–30 m thick). Also see that the model does not match the backward direction measured at Fort Cannonville. (From Gueugneau et al. 2020, through Creative Commons Attribution 4.0 International licence). **c** Schematic representation of the relationship between the valley confined "body" deposits (ignimbrite) and the detached, more mobile ash-cloud surge (overcurrent) associated with pumiceous pyroclastic density current eruptions at Mt Pelée at ~2000 yr BP. Flow direction of denser material segregating in valleys form dilute overcurrents is indicated by arrowheads toward reader (westward). (From Wright et al. 2016)

pyroclastic flow can be used to estimate minimum paleoflow velocities from the simple potential energy to kinetic energy relationship:

$$v = \sqrt{(gh)} \tag{10.6}$$

where h is the height climbed, v is the velocity and g is the acceleration due to gravity. Absolute-minimum velocities of 60–160 m s^{-1} are inferred from several ignimbrites (e.g. Yokoyama 1974; Sparks 1976; Francis and Baker 1977; Miller and Smith 1977; Fisher et al. 1993). The Saidmarreh landslide must have had a velocity of at least 100 m s^{-1}, and thus such high flow velocities are not unique to pyroclastic flows. It should be noted that this approach is specific for high concentration pyroclastic flows or their high concentration basal undercurrent, whereas it may not be as simply applicable for surges or for the more diluted and turbulent part of pyroclastic flow body.

Druitt et al. (2002) used the maximum uphill runout at valley bends to calculate the velocity of the surge-derived, high concentration pyroclastic flows generated on 25 June and 26 December 1997 at Soufrière Hills volcano, Montserrat. They balanced the centrifugal forces and gravity forces acting at bends using the following formula:

$$\tan\theta = \frac{v^2}{R\, g \cos\alpha} \tag{10.7}$$

where θ is the inclination of the line connecting the height of the flow/deposit across the two flanks of the valley at the bend, v is the pyroclastic flow velocity, R is the curvature of the bend, g is the gravitational acceleration and α is the slope of the valley along-flow. Based on this formula, velocities were estimated between 6 and 10 m s^{-1} for the high-concentration pyroclastic flows and up to 22 m s^{-1} for the associated surge clouds.

Flow velocities for pre-historic pyroclastic flows have also been calculated using the concept of dynamic pressure (Valentine 1998; Clarke and Voight 2000) and the force required by a pyroclastic flow to move accidental lithic clasts at rest on the ground surface (Pittari et al. 2007; Cas et al. 2011), as follows. The force required to move a boulder at rest on a surface is

$$F_1 = u_s\, m\, g \tag{10.8}$$

The force applied to a boulder at rest by a pyroclastic flow is

$$F_2 = 0.5\, \rho_{fl}\, v^2\, C_D\, A \tag{10.9}$$

The dynamic pressure, P_D, or power, exerted by a pyroclastic flow on any surface is

$$P_D = 0.5\, \rho_{fl}\, v^2 \tag{10.10}$$

where u_s is the coefficient of static friction, m is the mass of the boulder, g is the acceleration due to gravity, ρ_{fl} is the assumed bulk density of the pumiceous pyroclastic flow, v is the pyroclastic flow velocity, C_D is the coefficient of aerodynamic drag (=0.8), and A is the surface area of the clast facing the oncoming pyroclastic flow force. These equations can be solved for velocity.

Pittari et al. (2007) found that lithic-rich pyroclastic flows of the Abrigo ignimbrite (Tenerife) travelled at velocities up to 87 m s^{-1} (>300 km hr^{-1}), whereas Cas et al. (2011) calculated that the very large volume Cerro Galán pyroclastic flow travelled at <3 m s^{-1} (~10 km hr^{-1}) and Báez et al. (2020b) calculated velocities comprised between 4.7 and 17 m s^{-1} (~ 17-61 km hr^{-1}). Again, it should be noted that these formulas and calculations apply for determining the velocity of the basal undercurrent of pyroclastic flows.

10.8.4.2 Ignimbrite Valley Pond and Veneer Deposits—Flow Channelisation and Ash Cloud Detachment

Where channelled in valleys, the density stratification of pyroclastic flows is enhanced by the drainage effect of the incised topography and may favour the decoupling from the overriding turbulent cloud (Fig. 10.32). The resulting ignimbrite units may vary significantly in thickness from <0.1 m on ridges to >100 m in valleys, and also in lithofacies. Based on thickness, lithofacies, and relationship to underlying topography, two ignimbrite facies have been identified, which are *valley pond ignimbrite* and *ignimbrite veneer deposits* (Fig. 10.34; Walker et al. 1980a, 1981a; Wilson and Walker 1985; Cas and Wright 1987; Scott et al. 1996; Bryan et al. 1998; Giordano et al. 2002; Cole et al. 2002; Druitt et al. 2002; Pittari et al. 2006). Although originally conceived to be characteristic of high-velocity pyroclastic flows such as the 232 AD Taupo ignimbrite, and the 1400 yr BP

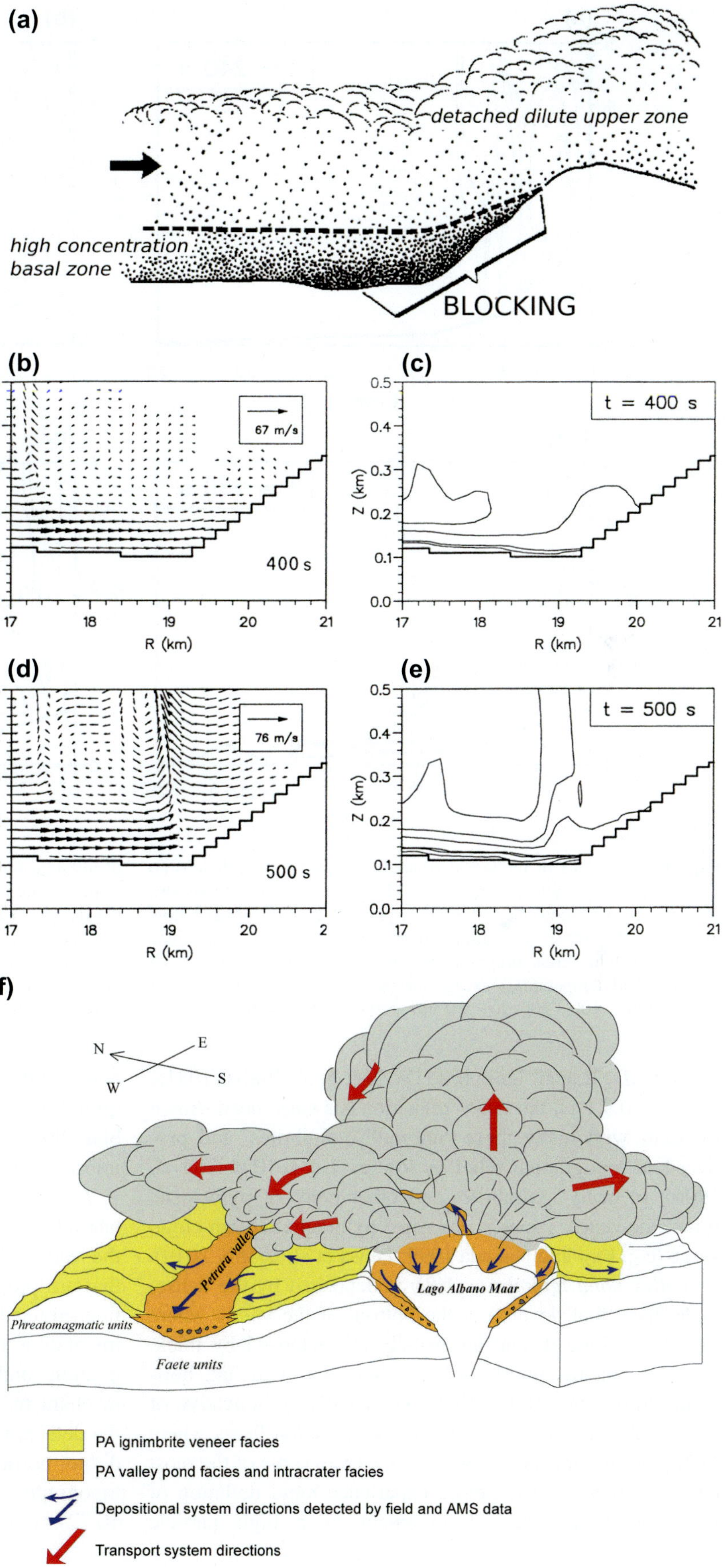

Fig. 10.32 **a** Blockage of basal high particle concentration part of a pdc at transverse topographies, and detachment of the more mobile ash-cloud surge part of the flow system (From Valentine 1987). **b–e** Snapshots of results of numerical simulation of a pdc attempting to cross an obstacle 250 m high at 19 km from vent, shown at 400 and 500 s (From Giordano and Dobran 1994). **b** and **d** Particle velocity vectors (the upper box shows maximum velocity in the frame); **c** and **e** Mixture density contours (the contour levels of mixture density plots, beginning from the outer region, correspond to 0.8, 1.2, 10, 20, 50, 100, 200, 300, and 500 kg m^{-3}). Note that at 400 s velocity is significantly slowed, but vectors are upslope directed, while at 500 s a backflow is generated which promotes deposition at break in slope and the rise of a phoenix co-ignimbrite plume. **f** Schematic reconstruction of flow directions and undercurrent directions controlled by paleotopography, as deduced from anisotropy of magnetic susceptibility (AMS) data in the Peperino Albano ignimbrite (Modified from Porreca et al. 2003; see Fig. 10.22c for volcanic setting)

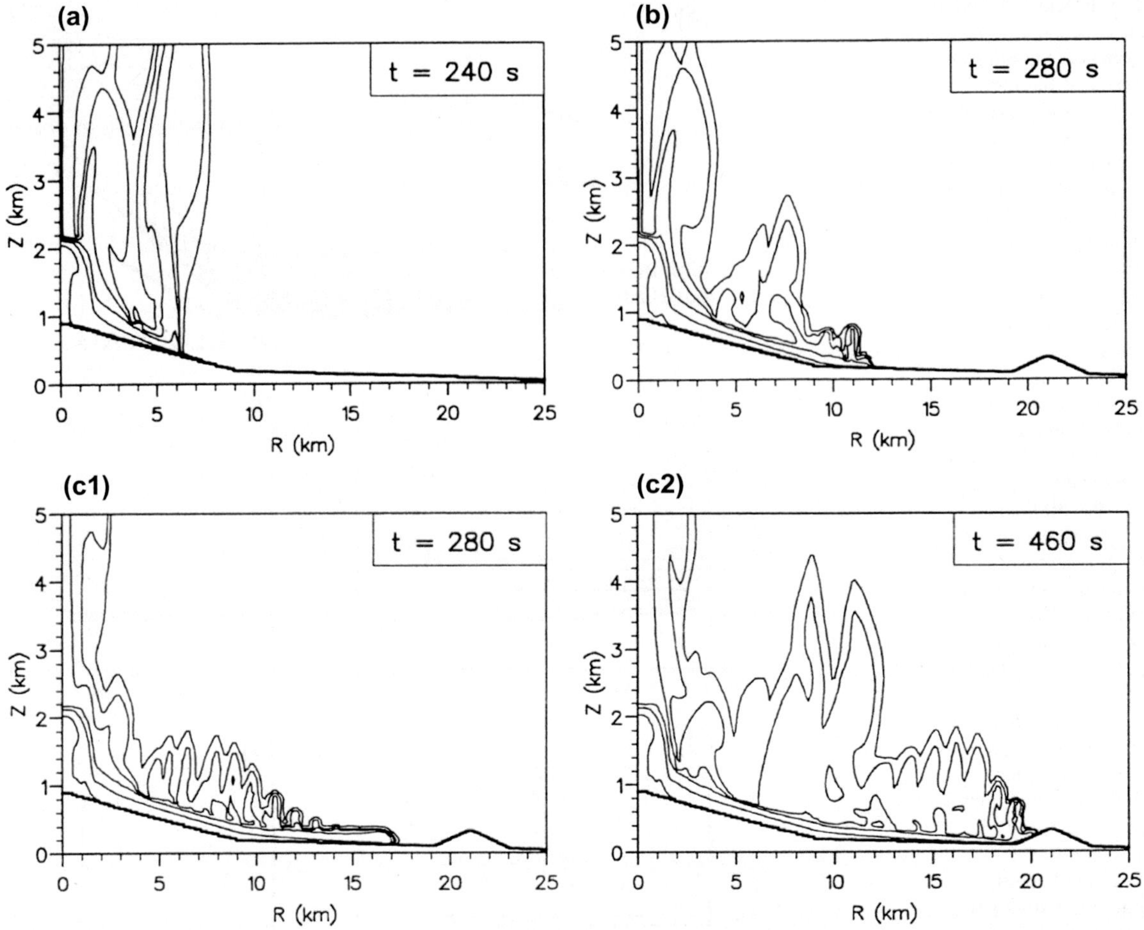

Fig. 10.33 Results of numerical simulation showing the relationship between grain-size and pdc mobility. The numerical simulations (**a**), (**b**), and (**c**1, **c**2) show the influence of grain size on pyroclastic flow mobility. The three simulations have identical initial conditions (collapse height, mass eruption rate, mixture density at vent, exit velocity, and temperature) except for the particle grain size which ranges from 100, 200 to 500 μm respectively, with associated decreasing densities of 2200, 2000, 1700 kg m^{-3}. The striking effect of increasing particle size (only one grain size simulated) is to decrease the degree of coupling with gas, so that the related pyroclastic flows become more permeable to gas escape, which promotes faster deposition. Note the huge and fast vertical gas escape and ash-cloud buoyancy in (**a**) and (**b**) with respect to (**c**), and the lower degree of flow mobility. (From Giordano and Dobran 1994)

(~680 AD) Rabaul ignimbrite (Walker et al. 1980a, 1981a, c; Sect. 10.9.1.1), the facies transition has since been shown to occur with many more "normal" ignimbrites, and produced in topographic relief as low as 1.5 m (Pittari et al. 2006). In many ignimbrites and other types of pyroclastic flows, the veneer facies is deposited by the mobile ash-cloud surge (e.g. Fuji and Nakada 1999; Pensa et al. 2015b, 2019).

Valley pond ignimbrite (VPI) is deposited by the denser underflows that develop at the bottom of the valley where they flow along it, and are usually characterised by thick, massive, chaotic, poorly sorted lithofacies, with flat, horizontal tops (Figs. 10.6c, 10.34 and 10.35a–f), indicative of sedimentation from high particle concentration flows where turbulence is suppressed (Sect. 10.6). Even some of the most energetic flows can therefore experience rapid deflation of the flow system and development of a high particle concentration underflow in valleys (e.g. the 232 AD Taupo ignimbrite, Wilson and Walker 1985; the 1980 Mt St Helens blast flow, Voight et al. 1981; the 39.8 ka Campanian Ignimbrite, Silleni et al. 2024).

Ignimbrite veneer deposits (IVD) can be traced thinning laterally upslope from the valley ponded facies and draping the valley slopes and ridges, usually displaying lithofacies that are progressively more stratified to cross-stratified and finer-grained (Figs. 10.34 and 10.35a, b, g), indicating sedimentation from dilute, pulsing, turbulent currents where traction and direct fallout are important sedimentation mechanisms (Sect. 10.6). IVD can therefore form both from the detachment of ash cloud surges but also where the more dilute portions of a stratified flow body make it across topographic highs (e.g. Fisher 1995; Figs 10.13, 10.19, 10.32). It could be expected that being more dilute the

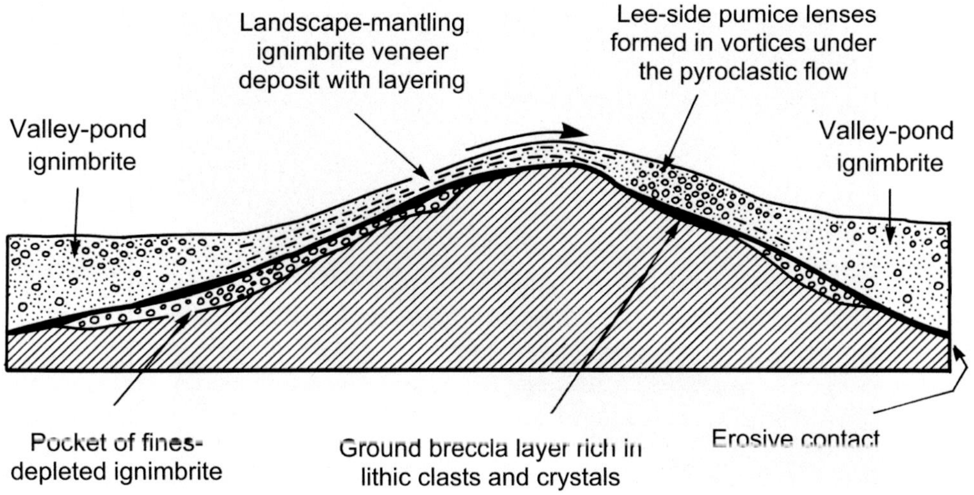

Fig. 10.34 Distinction between thick, coarse, massive valley pond ignimbrite (VPI) and thin, finer-grained, stratified ignimbrite veneer deposit (IVD) facies as well as other facies of the 232 AD Taupo ignimbrite, New Zealand, as discussed in the text (From Cas and Wright 1987, After Walker 1981b)

former are short-lived, localised and characterised by relatively low dynamic pressure, carrying and depositing almost only ash (e.g. the September 15, 1991 event at Unzen; Fuji and Nakada 1999; Figs 10.12c; 10.99e), whereas the latter can be much more energetic and widespread, carrying and depositing cross stratified coarse debris (e.g. the Taupo AD 232 IVD, Figs 10.34, 10.81a; see Sect. 10.9.1). It is important to mention that IVD are thin and have a much lower preservation potential in the rock record with respect to valley pond deposits, as they can easily be eroded away. Depending on the thickness and grain size of the IVD, and the climate, erosion can occur in just a few seasons. This has led to a general understudy and underestimation of the hazard posed by the detachment of dilute parts of the body and of ash cloud surges which can move unpredictably across rough topographies, and, even where ash is the dominant particle size (i.e. in ash cloud surges) and low concentration prevents the development of severe impacts from dynamic pressure, their temperature can be deadly (e.g. 2018 eruption of Fuego, Guatemala; Risica et al. 2022). For example, detailed stratigraphic and paleotemperature investigations (see Sect. 10.8.14) show that the victims of the 79AD Vesuvius eruption found in the archeological excavation of the ancient Roman town of Herculaneum (Fig. 10.1c) were killed by the very high temperature (> 500°C) of an ash cloud surge detached from nearby valley confined pyroclastic flows, and only later buried by the following incoming flows (Pensa et al. 2023).

Whereas some ignimbrites contain both facies, with transition across topography (Figs. 10.6c, 10.34 and 10.35a, b), others may preserve thin ash-cloud surge deposits on topographic ridges, and others appear to have only valley pond facies (e.g. Cerro Galán ignimbrite; Cas et al. 2011), with a flat top abutting against the valley walls, suggesting that mechanisms of transport and deposition may vary substantially, from highly energetic, diluted and at least partially turbulent, to more dominantly high concentration and laminar. In many such cases, original ash-cloud surge deposits may have been eroded from ridges since the eruption.

10.8.4.3 Aspect Ratio and Topological Aspect Ratio

The *aspect ratio of ignimbrites*, as for lavas (see Chap. 2, Sect. 2.6.2), is a useful and synthetic measure that describes the geometry and distribution of the deposit. This can be defined as the ratio of the average thickness (T) to the horizontal extent (D), i.e. T/D, where D is taken as the diameter of a circle with a surface area equal to that of the deposit. Walker (1983) discussed the distinction between high aspect ratio ignimbrites (HARI) and low aspect ratio ignimbrites (LARI), where the T/D ratio is 10^{-2} to 10^{-3} and 10^{-4} to 10^{-5} respectively. The proposed distinction was related to the intensity of the flow event in terms of discharge rate feeding the flow and its mobility. At one end of the spectrum are the HARI, usually small volume, valley confined pyroclastic flow deposits. At the other end of the spectrum, are the LARI that are very widespread and relatively thin ignimbrites, such as the 232 AD Taupo ignimbrite, representing a single flow that travelled radially outwards in all directions, almost regardless of topography. It climbed hills up to 1500 m above the vent in Lake Taupo, and is interpreted to have erupted with an extremely high magma discharge rate and travelled with an extremely high velocity (Sect. 10.9.1; Walker et al. 1980a, 1981c; Wilson and Walker 1985).

Although still useful and generically used in the existing literature to distinguish topography confined from topography mantling ignimbrites, the quantitative assessment of aspect ratio of ignimbrites (Table 10.2) has progressively waned. The main difficulties in accurately calculating T and D for ignimbrites are similar to those listed above for the calculation of bulk volumes (Sect. 10.8.1.1). In order to

(a)
(b)
partly welded with fiammé
sillar
non-welded
IVD
paleovalley wall
1
m
0
(c)
(d)
(e)
Paleovalley slope
(f)
(g)

◀ **Fig. 10.35** Valley pond ignimbrite and ignimbrite veneer deposit facies. **a** Two valley ponded, massive, poorly sorted depositional units from the 1.26 Ma Upper Bandelier ignimbrite, infilling topography and with finer-grained, diffusely stratified veneer facies draping-up the paleoslope formed by canyon wall incised into epiclastic volcanic gravels and conglomerates. Cochiti Canyon, Pajarito Plateau, New Mexico, USA. **b** Field sketch of relations in (**a**) showing gradual infilling of paleovalley by depositional units of the ignimbrite; note pumice concentrations and their alignment at the base and margins of each unit (From Self et al. 1996). **c** Two valley ponded ignimbrite facies in cross-section ponded in a low relief paleotopography in a Holocene succession of pyroclastic deposits in the Okataina Volcanic Centre, New Zealand. The thin ignimbrite units pass rapidly into very thin veneer ash layers that drape a low paleo-ridge. **d** Valley-ponded 232 AD Taupo ignimbrite in the Ngaruroro Valley, Kaimanawa Mountains, New Zealand. The pyroclastic flow had to surmount topography 500 m above the valley floor to reach this site. The terrace represents the original top of the deposit, much of which has been eroded. **e** Transition from thick, columnar jointed valley pond (left) to thinner but still massive ridge drape on a bedrock paleo-ridge (right) in the large volume, 2.08 Ma Cerro Galán Ignimbrite, Curuto, Paycqui, northwestern Argentina. The scree represents poorly exposed Ordovician metasedimentary bedrock. People circled for scale. (From Cas et al. 2011). **f** Erosional remnant of massive, poorly sorted, small volume, valley ponded Holocene Secche di Lazzaro phreatomagmatic ignimbrite, Punta Lena, Stromboli, Italy. The thinly stratified wavy deposits under the ignimbrite are a succession of base surge deposits. (From Giordano et al. 2008a, b). **g** Surge-like, fine-grained, ignimbrite veneer deposits of the 232 AD Taupo ignimbrite

calculate these parameters, detailed isopach maps should be reconstructed, but many studies indicate very poorly constrained data for "average thickness" and "areal extent" of ignimbrite units. Furthermore, in the rock record, erosion of distal outcrops may severely affect the evaluation of the maximum areal extent (e.g. Wilson 1991). More important is that the aspect ratio should be applied to individual depositional units which are not always easy to reconstruct unless very detailed lithostratigraphic studies are performed (e.g. De Rita et al. 1998). The typical high aspect ratio Rio Caliente ignimbrite (Wright 1981), for example, was made by the piling up of multiple flow unit deposits; hence a cumulative aspect ratio tells little about the dynamics of individual pyroclastic flows that formed that stack.

Nevertheless, aspect ratio is useful in comparing deposits that have comparable patterns of dispersal. However, ignimbrite dispersal patterns can be very different, as they can be found dispersed radially in continuous fans around vents/calderas (e.g. the Cerro Galán ignimbrite, Cas et al. 2011; Fig. 10.61) or have very directional dispersal patterns, confined within narrow elongate valleys (e.g. the Valley of Ten Thousand Smokes ignimbrite, Hildreth and Fierstein 2012; Fig. 10.22b). This makes it questionable if the use of a single value to compare ignimbrite geometries is useful. The combined effects of paleotopography and laterally directed *vs* uniformly spreading pyroclastic flows are reflected in the final geometry of the ignimbrite in a non-unique way.

In Table 10.3, we compare several ignimbrites of similar aspect ratio (10^{-4}) with very different volumes and field characteristics, which show that the aspect ratio may not capture fundamental differences in terms of the related flow behaviour and mobility. For example, the 63 km^3, single unit, tephritic Pozzolane Rosse ignimbrite (460 ka, Colli Albani volcano, Italy) is found across topographic barriers over 400 m high at 25 km from vent (Giordano and Dobran 1994; Calabrò et al. 2022), although it shares the same aspect ratio as the 630 km^3, single unit, rhyodacitic Cerro Galán ignimbrite, which in contrast had no ability to surmount any of the topographic highs in the surrounding landscape (Cas et al. 2011). This simple comparison highlights that the geometry (aspect ratio) of ignimbrites does not necessarily reflect the kinetic energy of pdcs or their emplacement dynamics; see also comparison of the ~25 ka Lajes and Angra igimbrites, Terceira, Azores (Pimentel et al. 2021).

So, whereas aspect ratio as a single parameter applied to ignimbrites, seems to have several limitations, deposit thickness variations with distance and topography are proxies at each site for the competition between sedimentation and transport ability, integrated over the entire time scale of the flow event. With this approach, Giordano and Doronzo (2017) reappraised the concept of aspect ratio for ignimbrites by defining a "*topological aspect ratio*" (*ARt*) as the ratio of the local deposit thickness (*Ht*) to the distance between that site and the maximum runout distance (*Lt*). *Lt* represents the "runout ahead" of each location and varies between a maximum value at the most proximal outcrop (i.e. where the "runout ahead" is a maximum) and 0 at the maximum runout distance. This approach implies the definition of flow paths along which *Ht* and *Lt* can be calculated from an isopach map (Fig. 10.36a). Even in the absence of certainties on the maximum travel distance, the distance of the farthest outcrop (e.g. for ancient ignimbrites) can be used for the calculation of *Lt* as the reference for all localities along the flow path. The topological aspect ratio is therefore calculated at each locality where *Ht* is the proxy for the tendency for the flow to deposit, whereas *Lt* is the proxy for the tendency for the flow to continue transporting the pyroclastic material.

Although *ARt* in itself is a site-specific value, its spatial gradient $\Delta ARt/\Delta x$ along the selected flow path allows visualisation of different downcurrent trends (Fig. 10.36b). For example, an ignimbrite sheet with constant or increasing thickness shows a positive gradient of *ARt* (i.e. where *ARt* increases with distance). This means that the pyroclastic flow, while depositing at each site, was still able to continue transporting more pyroclastic material forwards (e.g. from proximal to medial localities in paths 1–3 and proximal localities in path 4 in Fig. 10.36b). These conditions

Table 10.3 Features of four low aspect ratio ignimbrites

Ignimbrite	Cerro Galan	Villa Senni	Pozzolane Rosse	Peperino Albano
Age (Ma)	2.08	0.355	0.460	0.002
Chemistry	Rhyodacite	Tephri-phonolite to tephrite	Tephrite	K-foidite
Areal distribution	Radial	Radial	Radial	Directional
Extracaldera bulk (preserved) volume (km^3)	324 (97.2)	30 (23)	59 (35)	<1
Dispersal (preserved) area (km^2)	7200 (2160)	2000 (1600)	2500 (1600)	75
Max preserved distance (km)	74	29	33	8.5
Average thickness H (m)	45	15	24	4
Average radius L (km)	48	25	28	5
Aspect ratio H/L	4.70×10^{-4}	2.97×10^{-4}	4.18×10^{-4}	8×10^{-4}
Basal fallout	No	Sub-plinian scoria	Sub-plinian scoria	Sub-plinian pumice
Max. height climbed (at distance)	–	300 m (20 km)	400 m (20 km)	
T°C	<780 to >620	>580	>580	240–350
% <1/16 mm	>20%	<4%	<10%	>30%

Note while sharing similar low aspect ratios (10^{-4}) these ignimbrites show very different field characteristics. The mafic, intermediate volume, Villa Senni, and Pozzolane Rosse ignimbrites were emplaced across significant topographic barriers (300–400 m at 20 km distance), and the small volume Peperino Albano shows valley pond and veneer facies, whereas the large volume, felsic Cerro Galan ignimbrite, although radially dispersed for up to 100 km from vent, is a purely valley pond facies with no evidence of parent flows having climbed any topographic obstacle

correspond to a dominantly forced regime, in which, at any given locality, both transportation and sedimentation are affected by external drivers. These could be a prevailing high mass flux, such as occurs in proximal regions and in high mass flux-fed pyroclastic flows, or due to a local increase of mass flux due to confining topography. The interpretation of an ignimbrite sheet with constant or increasing thickness is that the internal pressure is efficiently transferred inside the flow, allowing for the sustained lateral transport along with steady deposition at each site, where deposits are dominantly massive and chaotic, although internal organisation and cross stratification are possible where velocity flux is high.

On the other hand, a negative gradient (i.e. where *ARt* decreases with distance) implies that the deposit thins faster than the lateral decrease of the potential runout (e.g. medial to distal localities in path 4, Fig. 10.36b). This suggests that sedimentation is fast as internal inertia becomes dominant, affecting both transportation and sedimentation. These conditions are typical of pyroclastic surges or across topographic barriers where internal flow pressure drops quickly promoting fast sedimentation. Intermediate conditions imply a flattening of the *ARt* gradient and indicate a sort of balance between forced and inertial regimes (distal localities in paths 1–3, Fig. 10.36b).

With a similar approach, Palladino and Giordano (2019) introduced a competence ratio, to show how the maximum lithic size and maximum pumice size vary with distance.

10.8.4.4 Thickness Decay Patterns

The geometry of fallout deposits, especially dispersal area (Walker 1973) and thickness decay patterns with distance (Pyle 1989), is widely used to interpret the explosive eruption and dispersal mechanisms (Chap. 9). By contrast there has been very little use of isopachs and thickness decay patterns for ignimbrites, mostly because, unlike for fall deposits, there is a very limited number of studies which reconstruct accurately isopachs and dispersal areas.

In a pioneering study, Wilson (1991) suggested that for the 25.4 ka Oruanui ignimbrite there is an exponential relationship between thickness and cumulative isopach area. Though intriguing, such a relationship was not generalised by the author, as the Oruanui ignimbrite is deposited on a rough terrain and isopachs certainly do not follow a simple concentric pattern like most fall deposits do. Wilson (1991) then cautiously warned about the possible fortuity of such a relationship. Much more recently Silleni et al. (2020) have instead shown for a few more ignimbrites for which isopach maps are available, that the exponential relationship between

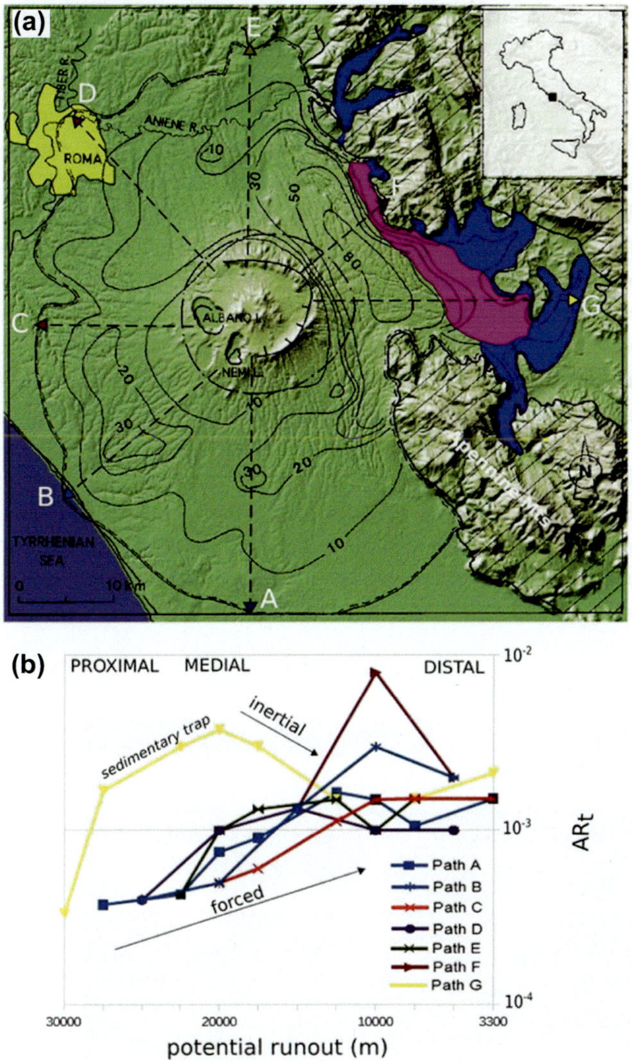

Fig. 10.36 Method for the calculation of the topological aspect ratio *ARt* and its spatial variations applied to the 460 ka Pozzolane Rosse ignimbrite, Colli Albani caldera, central Italy. *ARt* represents the local competition between the tendency to sediment (thickness *Ht*) versus the ability to further transport (distance from maximum runout *Lt*). **a** Isopach map of the Pozzolane Rosse, superposed onto a DEM. Arrows indicate directions along which the topological aspect ratio *ARt*, defined as the ratio between the local ignimbrite thickness Ht and the distance Lt of that site from the farthest outcrop, is calculated at regular pace along flow paths (A to G in clockwise order from south direction) starting from the inferred vent (at caldera centre). The pink area indicates where the ignimbrite is stratified (inertial behaviour) downflow of major deposition (thickening to 80 m in cross-flow valley) and the blue where it is channelised in paleovalleys (forced behaviour). **b** The spatial gradient of *ARt* along paths A to G illustrates different trends: in open flat plains (paths A–E) the Pozzolane Rosse ignimbrite shows rather uniform thickness and a mild increasing trend of *ARt*, characteristic of transportational regime forced by a high mass flux; at cross-flow sedimentary traps (path G) the thickness increases suddenly and then thins, producing a decreasing trend of *ARt*, characteristic of inertial transportational regime (see text for explanation). (From Giordano and Doronzo 2017)

thickness and isopach area is a general feature and must relate to the emplacement processes, just like for fall deposits. This common characteristic thickness decay pattern has been shown to be consistent with depth-averaged numerical models of turbulent, large-scale pyroclastic currents (Bursik and Woods 1996; Calabrò et al. 2022). In general, an exponential thickess decay can be associated with a deposit progressively aggraded by a flow which loses material along its flow path, so internal time-lines are expected to be parallel to the base (Fig. 10.20bi) and also grain-size should decrease laterally (Doronzo et al. 2022). By contrast, a flow (or part thereof) that transports material wholesale and freezes *en masse* is expected to accumulate mostly at its distal end, forming a debris fan, with complex internal time-lines (Fig. 10.20bii) and no obvious lateral grading of grain sizes. Calabrò et al. (2022) have succesfully numerically modeled the thickness decay with distance, the mobility across topography and maximum runout, the lateral grain size distribution and deposit temperature of the Pozzolane Rosse ignimbrite, constraining initial flow conditions in terms of mass flow rates feeding the flow (6-7 x 10^{10} kg s^{-1}) and average initial flow density (25-70 kg m^{-3}), demonstrating that measurable field indicators can be used to model quantitatively the initial flow parameters and track their downcurrent evolution.

Later in Sect. 10.11, where we discuss the classification of ignimbrites, we show a log thickness versus cumulative isopleth area for a larger ignimbrite dataset (see Fig. 10.96). This confirms the relationship between log thickness and cumulative isopach area and suggests that such quantities are strong descriptors for a wide range of ignimbrites, however, different in terms of volume, areal extent, and relationship with topography.

10.8.5 Conformable Versus Erosive Base

A remarkable feature of most ignimbrites is the lack of erosional interaction with their substrate. Commonly, ignimbrites overlie a paleosoil or delicate and unconsolidated fallout deposits made of pumice and ash, with little or no evidence of scouring and entrainment of clasts (e.g. Fig. 10.6a). There are even cases at the base of the Villa Senni ignimbrite (Sect. 10.10.1) from the Colli Albani volcano near Rome, where in situ fossil grass flattened on the paleosoil surface is preserved. The Cerro Galán Ignimbrite (Sect. 10.9.2) also shows little evidence of erosional interaction with the paleosoil and alluvial substrate (Cas et al. 2011).

What does this tell us? Was the flow velocity at ground level so low as to not be capable of exerting an erosional shear force on the substrate? Was the flow velocity above ground level higher, and so the flow front was rolling over the substrate, like the tracks of a bulldozer, but without doing any bulldozing? Was the internal gas pressure in the flow directed downwards at the base, keeping the substrate in place? Or was it a combination of these?

Nevertheless, the contact between the base of an ignimbrite and its substrate can also show evidence of substrate erosion. Given their extreme dynamics, it is intuitive to consider that pyroclastic flows would always be able to scour and shear the substrate. During the climactic eruption of Mt Pinatubo in 1991 pumice and ash flows removed virtually all vegetation and much soil and unconsolidated colluvium from its steep, proximal slopes in an area of almost 30 km^2 (Scott et al. 1996). The passage of energetic flows etched and fluted canyon walls (some as high as 200 m). Clearly, at least parts of some proximal flows were expanded in order to affect areas high above canyon floors, and all were sufficiently mobile to pass through the area and leave only small patches of pyroclastic flow deposits in protected areas. Scott et al. (1996) suggest that much of this proximal area was in the deflation zone, the zone in which pdcs were rapidly descending from the eruption column and segregating into density-stratified flows; if so, the deflation zone extended 3–4 km out from the vent. The highly energetic blast flow observed at Mt St Helens in 1980 (Sect. 10.12), scoured erosional furrows up to 9 km from source (Kieffer and Sturtevant 1988). The base of the 232 AD Taupo ignimbrite, in some proximal areas scours older substrate (Fig. 10.37a; also see Sect. 10.9.1.1, Fig. 10.80a, b).

The 161 ka Kos Plateau Tuff ignimbrite on the island of Kos, Greece was capable of eroding substrate fossiliferous mudstone intraclasts up to 2 m in length locally, and contains a layer of chaotic breccia up to 6 m thick sandwiched between Plio-Pleistocene sedimentary units below and ignimbrite above (Allen and McPhie 2001; Chap. 11, see

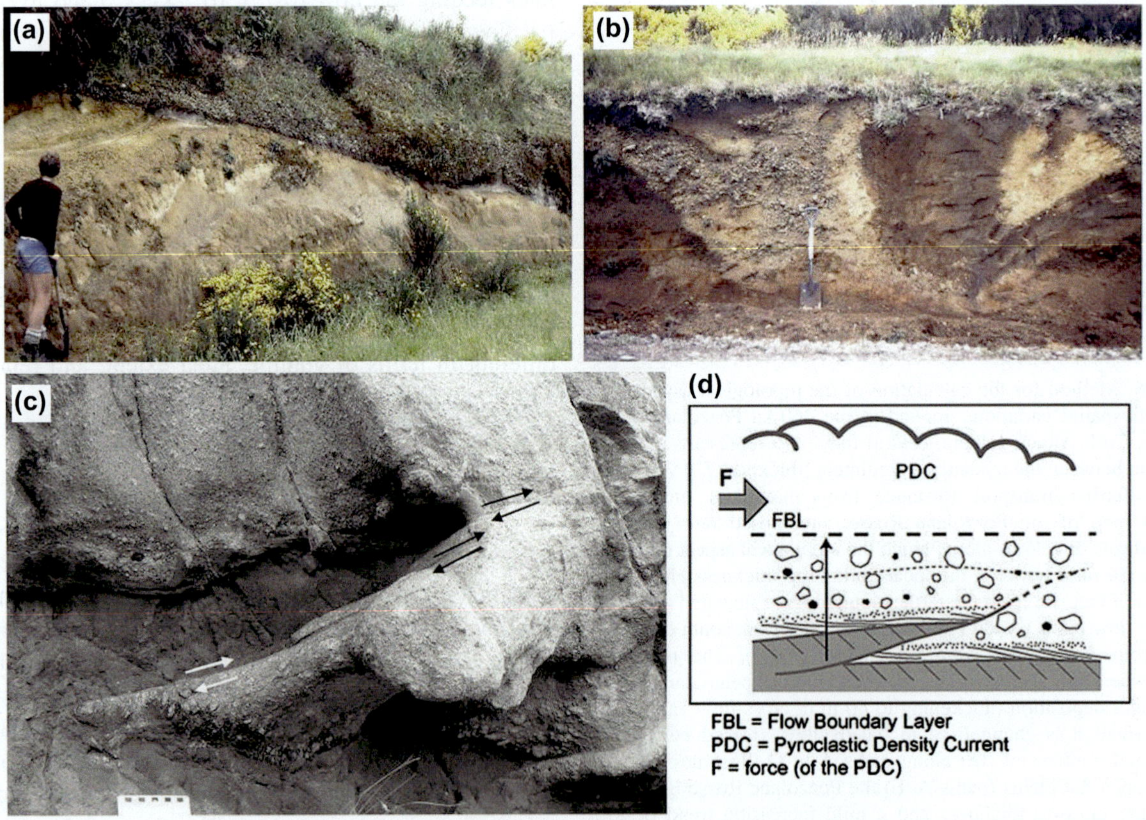

Fig. 10.37 Examples of erosional capacity of pyroclastic flows. **a**, **b** The 232 AD Taupo ignimbrite, New Zealand. **a** Highly discordant basal contact with its substrate in proximal areas. **b** Uprooted tree "soil-flap" (dark brown) 35 km from vent; spade rests on soil mixed with ignimbrite. Flow is left to right. **c** Shearing at the base of the non-welded 1.3 Peperino Tipico Monte Cimino ignimbrite, central Italy, at the contact with substrate mudstones. Several sub-parallel thrust planes can be seen, the main ones marked by movement vectors. **d** Interpretation of (**c**) showing the relationship between the active pyroclastic density current, flow/depositional boundary layer, and the force exerted by the flow through the deposit and the substrate to cause frictional coupling and thrusting between the deposit and the substrate mudstones. (**c** and **d** from LaBerge et al. 2006)

Fig. 11.19g, i, j). Allen and McPhie (2001) suggest this formed synchronously with the passage of a highly energetic pyroclastic flow where it traversed wet, unconsolidated sediment. Shear-induced liquefaction, together with possible seismic shaking associated with the eruption, possibly caused failure. This failed portion continued to move downcurrent and disintegrate by lateral shear, probably for no more than a few tens to hundreds of m, until it was deposited en masse. Even smaller volume and less energetic ignimbrites, such as the 1993 pumice and ash flows from Lascar volcano in Chile (Sparks et al. 1997), and 2–3 ka scoria and ash flows from Mt Misery (Liamuiga) volcano on St Kitts, Lesser Antilles (Roobol et al. 1987), may also show local scouring and erosion at their base, particularly on steep volcano slopes. At a paleovalley bend the 1.3 Ma Peperino Tipico from Cimino volcano, central Italy, shows local development of reverse faults cutting both the soil underlying the ignimbrite and the lower ignimbrite deposits that are dragged upward by the rest of the main ignimbrite (Fig. 10.37c; LaBerge et al. 2006). These features have been interpreted as the result of temporary coupling of a shearing pyroclastic flow with its substrate with a rapid downward migration of the force of the flow into both the previously emplaced ignimbrite and the underlying paleosoil, inducing thrusting between deposit and substrate (Fig. 10.37d).

10.8.6 Some General Deposit Characteristics of Ignimbrites

In preceding sections, we have begun to identify some of the very complex lithological variations and facies associations (both laterally and vertically) that ignimbrites may present to the geologist in the field (e.g. Figs. 10.14, 10.34 and 10.35). Nonetheless, for the most part, ignimbrites are typically poorly sorted, thick (up to tens of m), massive and internally chaotic deposits (Fig. 10.38a, b). They contain variable amounts of ash, angular to roughly rounded pumice and/or scoria lapilli and bombs, and lithic clasts, occasionally up to >1 m in diameter. However, these massive deposits, which form the main ignimbrite body deposit, may grade laterally and vertically into stratified depositional units or facies (Fig. 10.38c, d; see Figs. 10.41b, 10.46 and 10.47), as well as into welded deposits or cooling units (Fig. 10.38e; also see Figs. 10.26c and 10.28a, b; Sect. 10.8.14.1). They may also show stratified basal deposits that form distinctive facies layers (see Fig. 10.40) from the overlying body deposit(s).

In general, ignimbrites are distinguished from other types of volcaniclastic mass flow deposits (mudflows, debris flows, lahars, rock avalanches; Chap 12) in that they:

- possess a relatively uniform juvenile clast composition with a similarly composed vitric ash matrix and crystal content (although this may change vertically and laterally),
- in many cases contain a significant vent-derived (accessory), lithic component which can form significant horizons of lithic breccias especially in proximal locations,
- show systematic vertical and lateral clast size and abundance grading of larger lithic and (to a lesser extent) juvenile, vesicular clasts,
- may show compositional zoning, both mesoscopically (pumice vs. scoria component), microscopically and geochemically,
- contain evidence of a high emplacement temperature, including:
  - hot blocks,
  - salmon pink colour,
  - welded to non-welded character,
  - columnar jointing,
  - vapour-phase crystallisation (sillar) associated with steam fumaroles,
  - presence of carbonised wood, or
  - thermal remanent magnetisation (TRM) of lithic clasts,
- occasionally contain accretionary lapilli, and
- often (but not universally) develop gas segregation or gas escape pipes and structures.

Most ignimbrites are composed of more than one depositional unit (Fig. 10.38b, c d). However, it is apparent multiple flow depositional units may arise in a multitude of ways, and criteria to distinguish them are imprecise. They may form from deposition by separate pyroclastic flows (flow units *sensu stricto*) or flow pulses during the same eruption and flow event, which implies a stop-start or pulsing pattern of deposition. Changes in eruption dynamics at vent and inherent unsteadiness of the pyroclastic flow may cause pulsing within the same sustained flow system resulting in separate depositional units and lithofacies, although they have aggraded continuously (Sect. 10.7; some authors refer to these as pulse units). At vent a pyroclastic flow could rapidly divide into several subflows affecting different sectors of particularly, deeply-ravined stratovolcanoes. As a pyroclastic flow advances, it could also further split into more subflows or flow lobes (as observed at Mt St Helens in 1980; Rowley et al. 1981; Figs. 10.11d and 10.18b), each represented in the field by a discrete flow depositional unit (Cas et al. 2011). Topographic obstacles may also cause the flow to split, as well as induce backflows where obstacles cannot be surmounted. In the rock record it may be difficult to distinguish multiple depositional units

Fig. 10.38 Some general aspects of ignimbrite depositional units. **a** Cliff (~20 m high) of massive (single? depositional unit), 1883 Krakatau ignimbrite, Rakata Island, Krakatau caldera, Indonesia. **b** Typical massive depositional ignimbrite units 800 yr BP (~1280 AD) Quilotoa Ignimbrite, Ecuador. **c** 161 ka Kos Plateau Tuff, Kos, Greece, showing subtle, but defined multiple ignimbrite depositional units. Person circled for scale. **d** Cliffs (up to 25 m) of 1600 BC Minoan ignimbrite, Akrotiri, Santorini, Greece, showing thick, massive basal flow unit (12 m thick; cliff face to right), overlain by succession of thin to diffusely stratified depositional units (main cliffs in background with small erosional benches). **e** Succession of depositional units preserved in densely welded ignimbrite, outflow facies of the 28 Ma Fish Canyon Tuff, San Juan Volcanic Field, USA. Massive units (Layer 2b) are separated by finer-grained layers and partings (Layers 1 and 2a; see discussion Sect. 10.8.7 regarding possible origins). **f** Example of longer duration depositional boundaries between ignimbrites with complex, intervening erosional surfaces, soils and other volcanic and epiclastic deposits, southern coast of Terceira, Azores, Portugal. Time period indicated is about 60 kyr; distance between sections ~1 km. LAI = Lajes-Angra ignimbrite (~25 ka); LMI = Linhares-Matela ignimbrite (~35 ka); VFI = Vila Nova-Fanal ignimbrite (~40 ka); CCI = Caldeira-Castelinho ignimbrite (~70 ka); GVI = Grota do Vale ignimbrite (~86 ka). Note depositional boundaries within ignimbrites are also indicated. (Adapted from Gertisser et al. 2010)

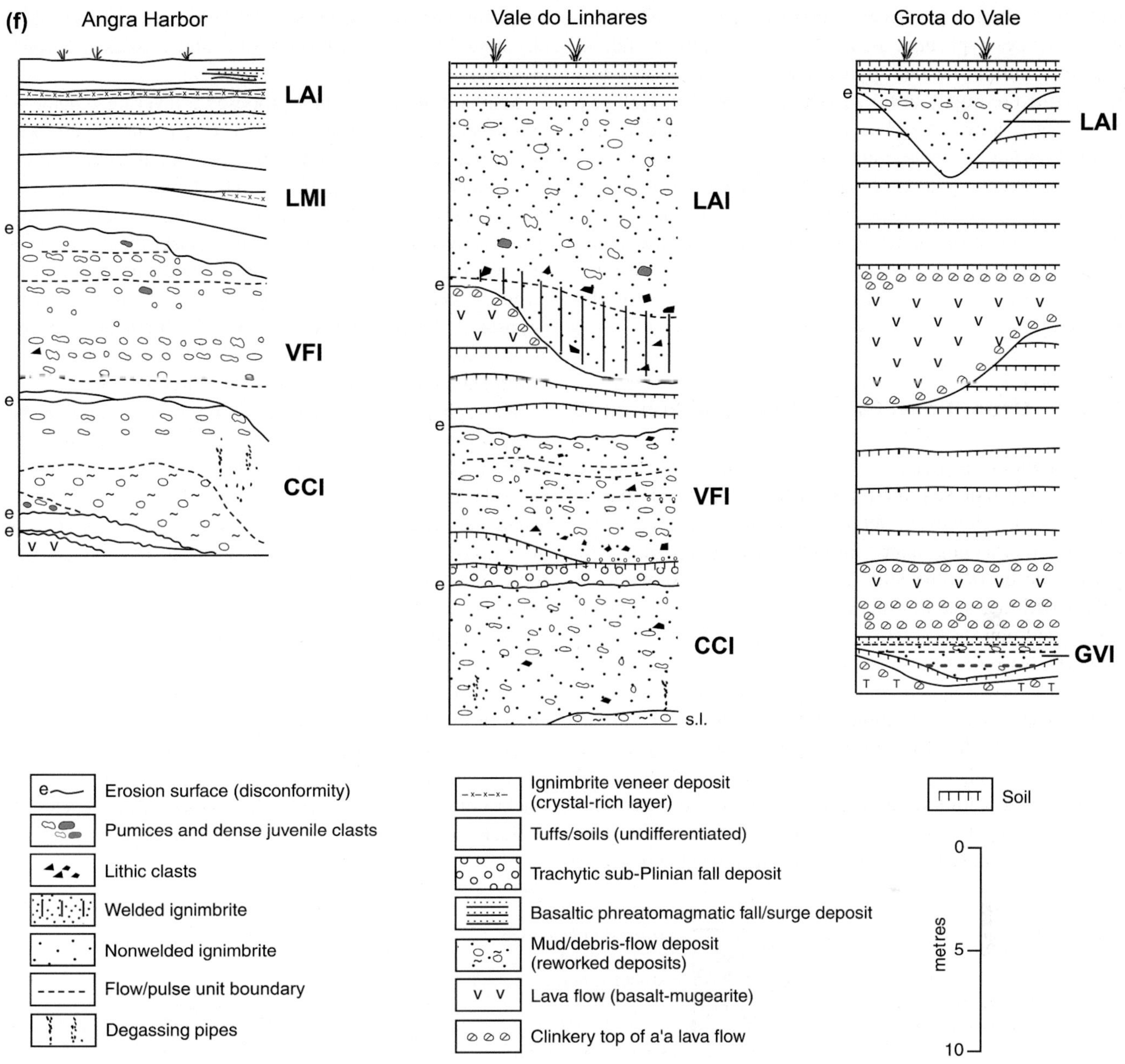

Fig. 10.38 (continued)

that arise from the above scenario from those that may arise from a succession of discrete, spaced, eruption and pyroclastic flow events during a long-lasting eruptive event, such as in the 1995–2013 eruption at Montserrat (Druitt and Kokelaar 2002), or the 1980 eruption at Mt St Helens (Rowley et al. 1981; Brand et al. 2017; Pallister et al. 2017; Fig. 10.22a); i.e. where time breaks between events are not too long. In fact, time breaks during eruption events are probably more common than usually recognised in the rock record (e.g. Aramaki 1984; Wilson and Walker 1985; Giordano 1998b; Adams et al. 2001; Kaneko et al. 2007; Romano et al. 2020).

Where time breaks are of longer duration, widespread intervening erosional surfaces, paleosoils, or reworked deposits may be present, together with other volcanic products (Fig. 10.38f). This results in more complex stratigraphic relations with ignimbrites often filling in valleys cut into older ignimbrites. Repeated infilling and excavation of the same valley is commonly observed with ignimbrites where, because of compaction effects, fluvial erosion often restores a similar pre-existing drainage pattern. The ignimbrites are typically associated with pumice fallout deposits, current-bedded pyroclastic surge deposits, and fine-grained (co-ignimbrite) air ash fallout beds, as well as a complicated array of facies, both lithic- and pumice-rich, formed during the same eruptions.

In the remaining parts of Sect. 10.8, we will document in more detail ignimbrite depositional models and facies, and how flow processes affect fabrics and the component distribution of the final deposit, as well as how other features listed above affect the final lithological character of the ignimbrite.

10.8.7 Depositional Facies Model and Sub-units

It is commonly recognised that the deposits of pyroclastic flows do not represent the physical flow state of the parent flow system. However, deposit characteristics are fundamental in reflecting many important dynamic aspects of pyroclastic flows, especially during deposition. As we have subdivided the anatomy of pyroclastic flows into the head, body, and tail (Sect. 10.5), consequently there must be deposits that reflect those parts. Deposits from the head of a pyroclastic flow can be found from proximal right through to distal areas in many ignimbrites, not just at the distal margin. These deposits therefore must reflect the flow state of the flow during its history, not just in the final stages of flow and deposition. In addition, the nature of the basal contact with substrate along the entire flow path, also reflects the flow state of the pyroclastic flow during the flow event at different times and varying distances from the source vent. Trying to disconnect flow state from deposit characteristics is therefore incorrect.

Sparks et al. (1973) first proposed a layering scheme for pyroclastic flow deposit units. The separate layers were interpreted as reflecting different depositional regimes within a pyroclastic flow (Fig. 10.39). The lowest layer, *layer 1,* was thought to be the deposit of a dilute pyroclastic surge which moved in advance of the pyroclastic flow, called the ground surge or here, the frontal ash-cloud surge, and/or the turbulent head of the flow. *Layer 2* was thought to be the deposit of the pyroclastic flow body (a high particle concentration underflow), and *layer 3* the deposit of the overriding ash cloud, called the trailing ash-cloud surge.

This scheme however, though simple and intuitive, has some limitations in describing the complex nature of pyroclastic flow deposits, particularly in process terms, as layer 1 deposits are often not observed, and layer 3 deposits are usually eroded away. Layer 2 deposits, which often form most if not all of a preserved ignimbrite succession, can also show internally significant lithofacies variations due to the duration of the event and changing flow dynamics or changes in the nature of components being released at the vent, which may be easily confused with processes occurring at the head and tail, without implying any stop-start behaviour of the flow process. That is, a single flow can produce a very complex association of facies that change vertically and laterally in the deposit. Furthermore, each layer is not an isochron line, deposited everywhere at same time; each layer is diachronous.

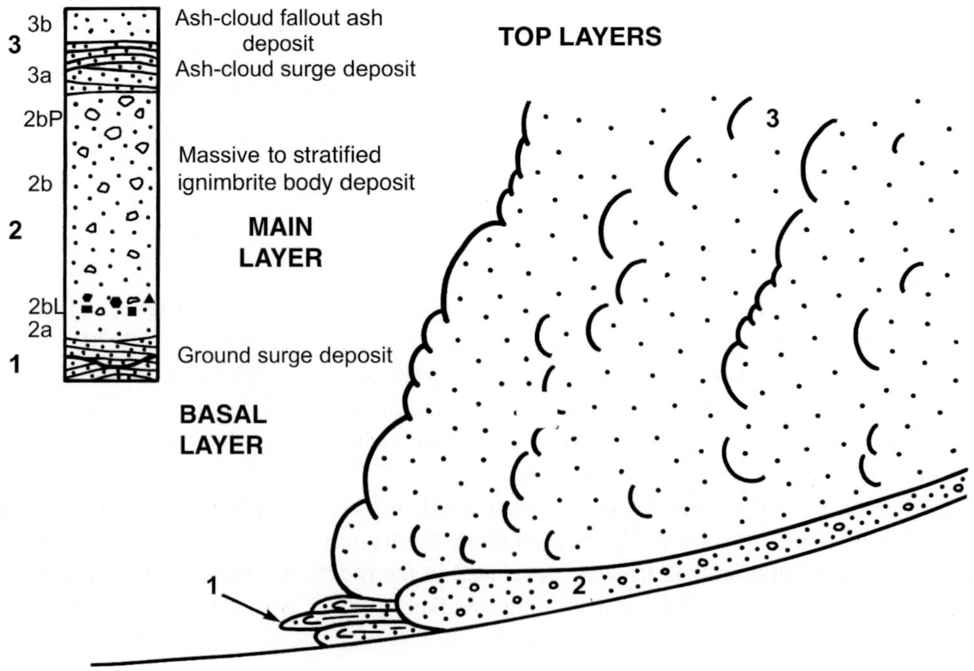

Fig. 10.39 Simple, generalised ignimbrite facies model as first proposed by Sparks et al. (1973). Although this idealised model is still valid, care must be taken when interpreting basal layers and facies divisions within the main ignimbrite, as several changing, transportation, depositional and time-dependent processes laterally and vertically may affect the emplacement of pyroclastic flows. See text for discussion of the origins of each facies. (From Cas and Wright 1987)

Furthermore, in their model, Sparks et al. (1973) suggested that layer 2 deposits would represent the flow that as a whole was halted en masse by freezing when sources of fluidisation ran out. Although this may reflect the state of a flow at its distal margin, this is not the general case in proximal and even medial reaches of a flow. We therefore redefine layer 2 deposits as the main part of ignimbrites, whether constituted by individual (*simple Layer 2*) or multiple depositional units or pulses (*compound Layer 2*), which may record a very complex depositional history (Sect. 10.7). This is because the internal facies architecture of pyroclastic flow deposits depends on the sedimentation rate occurring at different locations along flow path and over time, and is mostly controlled by the relative coupling between the fluid (dusty gas) phase and particles along the flow path (Giordano 1998a; Branney and Kokelaar 2002; Doronzo 2012), which can change. Depicting pyroclastic flows in terms of simple vertical profiles or gradients may be accurate only at a single point along the flow path, but at different points along the flow path that profile varies. Also, flows cannot be depicted by uniform flow conditions or flow state throughout the extent of the flow and for the duration of the flow event because they are such complex, ephemeral 4D systems (Sect. 10.7).

We utilise a descriptive field-based subdivision of the deposit architecture, with no genetic implications about the constituent layers. We describe the *basal contact*, *basal layers, main ignimbrite body*, whether consisting of a single depositional unit or if it is compound, and *top layers*. We also assign the original basic three-fold layer/deposit distinction of Sparks et al. (1973) of Layers 1, 2, and 3, with the caution that their origin should be ascertained case by case. In addition, we add a Layer 0, which is the contact of the base of the ignimbrite and its substrate, the nature of which is very important for understanding flow dynamics along the flow path.

10.8.7.1 Basal Contact [Layer 0]

The basal contact can tell us much about the flow and its behaviour (Sect. 10.8.5, Fig. 10.37). It is as important to study and document the nature of this contact, as the deposits, because it records the behaviour of the flow front along its flow path, whereas the deposits in different localities, from proximal to distal, record the evolution of the flow in time and space. The contact on outcrop (mesoscopic) scale can appear passive (i.e. non-erosive), slightly erosive, sheared, moderately or highly erosive (discordant, channelised). On a regional scale, all such contact variations may be present reflecting different behaviour of the flow front in space (especially topographic control) and time. Where a basal contact is discordant with the substrate, it is important to assess if that discordant surface was produced by pre-eruption landscape erosion, or by erosion by the flow (e.g. substrate clasts in the base of the ignimbrite.

10.8.7.2 Basal Deposits [Layer 1]

Both simple and compound Layer 2 ignimbrites show basal deposits that form distinctive facies layers that are different from the overlying deposit associated with the main part or succeeding depositional (flow) units of the ignimbrite. However, not all ignimbrites develop Layer 1 deposits or preserve them, these basal deposits are usually thin, in the order of cm or up to 1 m or so. There are three types of Layer 1 or basal deposits, and which ones develop depends on the flow state of the pyroclastic flow, especially at the flow front. The three layer 1 deposit types are called "ground or frontal ash-cloud surge deposits", "ground layers or breccias", and "jetted-fines-depleted deposits". Most Layer 1 deposits represent the short-lived transient passage of the head of the pyroclastic flows as it propagates at the atmosphere/topography boundary. The head stagnates once the flow reaches its maximum runout and becomes more or less stationary and stable for the rest of the duration of the flow event and does not interact with the undisturbed environment ahead of the stationary flow front.

Some ignimbrites, such as the thick 1.26 Ma Upper Bandelier Tuff outflow ignimbrite or even thin outflow ignimbrites show well developed up to m-thick, basal cross-stratified deposits interpreted as emplaced from a ground surge or frontal ash-cloud that preceded the body of the pyroclastic flow (Fig. 10.40a–d). On the other hand, ground layers or breccias (see Sect. 10.8.10.3) are thin, discontinuous, fines-depleted, coarse-grained and consisting largely of lithic clasts and crystal fragments, that are thought to have been deposited from the very highly turbulent head of a pyroclastic flow, where only the heavier fraction is deposited, all the rest being fully fluidised by the gas velocity and elutriated into the ash cloud. The occurrence of fines-depleted coarse pumice lenses at the very base of the Taupo ignimbrite succession has also been interpreted as a vigorous flow-head process, where these low-density clasts may be "jetted" ahead of the flow front due to super-heating and almost explosive expansion of ingested air and volatilised vegetation (Wilson and Walker 1985; Cas and Wright 1987). In some other ignimbrites pumice "dune-like" bedforms, discontinuous lenses or lobes have been described in this position (see Figs. 10.48 and 10.57f; Cas and Wright 1987; Pittari et al. 2005). Where ground breccias and jetted deposits occur, frontal ash-cloud surge deposits are unlikely to occur, because the former requires higher velocity flow conditions than the latter.

However, all these types of basal layers may or may not include deposits that actually originate at the head of a pyroclastic flow. Valentine and Fisher (1986) showed that basal layers may equally originate from turbulent boundary layers developed at the base of a pyroclastic flow, due to local interaction with a rough topography **or** vegetation/water ingestion. The emplacement of these deposits will depend on

Fig. 10.40 Basal layer deposits (Layer 1) and associations with Layer 2 deposits. **a** Canyon wall exposure of profile through the 60 m thick, 1.26 Ma Upper Bandelier Tuff outflow eruption sequence, Pajarito Plateau, New Mexico, USA. The lowest unit (above dashed orange line) is the massive stratified plinian fall deposit; this overlies a succession of sub-plinian fall deposits erupted from Valles Caldera (Cerro Toledo rhyolite domes) between 1.6 and 1.26 Ma. The plinian fall is overlain by a basal Layer 1, dune bedded, diffusely stratified ground surge deposit, and then the very thick, massive, main ignimbrite body deposit (Layer 2), which is tens of m thick. **b** At the other extreme of ignimbrite unit thickness, a succession of very thin ignimbrite depositional units at Tajao, Tenerife, the lowest of which has a well developed, low angle, cross-stratified ground surge layer (Layer 1) overlain by two thin, <50 cm thick Layer 2 ignimbrite depositional units, which show typical poor sorting, normal coarse-tail grading of large, dense, dark lithic clasts, and reverse coarse-tail grading of large white, pumice clasts. Tajao, Tenerife. **c** Cross-stratified and planar stratified ground surge deposit at the base of the Ramadas intra-plinian ignimbrite overlying a paleosoil, and passing into coarse, massive Layer 2 of the ignimbrite. ~6.6 Ma Corte Blanco Tuff, San Antonio de Los Cobres, northwestern Argentina. **d** Low angle and dune bedded Layer 1 ground surge deposit, between two massive, poorly sorted ignimbrite Layer 2 deposits. The horizon is interpreted as a ground or frontal ash-cloud surge deposit of the upper ignimbrite because of their similar pink thermal colouration, indicating their hotter temperature of emplacement. 1.26 Ma Upper Bandelier ignimbrite. **e** Cross-stratified surge facies deposited on 18 May 1980 by the pumiceous pyroclastic flows descending along the very proximal steep slope of Mt St Helens, USA

local flow and substrate roughness conditions (e.g. Fig. 10.40e), thus explaining their usually discontinuous occurrence and independence of distance from vent. Branney and Kokelaar (2002) also suggest that a highly turbulent head may not develop at all for some pyroclastic flows, hence the need to take into consideration the temporal evolution of depositional behaviour from a pyroclastic flow along its path. In this context some fines-depleted lenses of rough rounded pumice, usually interpreted as "jetted-deposits", could be the early deposition of pumice frontal or lateral levees during a fore-stepping or back-stepping phase of deposition, as well as during lateral shifting of the locus of sedimentation (Pittari et al. 2006; Pallister et al. 2017; Sects. 10.6 and 10.7; Figs. 10.11d, e, 10.14 and 10.20).

As pdcs form near vent, especially in the cases of column collapse and lateral gas expansion (Sect. 10.3), particle concentration is likely to be low (<30%) and velocity and gas turbulence high. Sparks et al. (2002) inferred velocities up to 120 m s^{-1} for the dilute proximal density current of the 26 December 1997 eruption at Soufrière Hills volcano, Montserrat. These conditions favour sedimentation of large and dense clasts from direct fallout from turbulent suspension, and bed-load traction, producing *lag breccia deposits* and/or surge deposits, with most of the particle load by-passing proximal areas. The resulting deposits, if any, are thin, coarse-grained, fines-depleted and clast supported. Their structure may be massive to stratified, depending on the relative importance of processes of direct fallout from the turbulent and dilute flow and of the shear exerted by overridding current at the basal flow boundary layer. Proximal lag breccia deposits may be laterally traced into basal deposits to the main ignimbrite (ground layers, ground breccias, ground surges). However, some care should be taken with their interpretation, as these basal deposits may be covered by Layer 2 deposits much later in the eruption as a result of back-stepping processes leading to incremental sedimentation stages (Fig. 10.20a). In such cases, the basal layer would bear no relationships with the phase of pyroclastic flow that emplaced the overlying ignimbrite. The basal layer would be indicative of the transition from waxing to waning flow, and the main ignimbrite (Layer 2), the waning stage of flow.

10.8.7.3 Main Ignimbrite Deposit [Layer 2]

The main ignimbrite layer (layer 2 in the Sparks et al. 1973 ignimbrite model; Fig. 10.39) is deposited by the basal part of the body of pyroclastic flows, i.e. the high concentration basal undercurrent. This usually develops fully at a certain distance from vent as a result of the tendency of the particle load to concentrate towards the base, counter-acted by several supporting mechanisms described in Sect. 10.6. An important concept that must be taken into account is that, different from the initial definition of Layer 2 by Sparks et al. (1973), we herein imply that the basal undercurrent of the flow may vary from very short lived to sustained over many hours, accounting for the progressive aggradation of ignimbrites from relatively unsteady to steady conditions. We also emphasize, again differently from intial definitions of Layer 2, that the overriding dilute part of the flow body continuously interacts with the basal high concentration undercurrent (Figs 10.13 and 10.19), exchanging mass, momentum and heat at variable rates along flow (e.g. Shimizu et al. 2019), so that the final geometry (runout and thickness variations along flow), and facies architecture of the main ignimbrite deposit (grain size distribution and sedimentary structures of Layer 2) depend on their interplay (Figs 10.13, 10.14, 10.19).

Our redefined Layer 2 facies form mostly at breaks in slope of stratovolcanoes, along incised paleovalleys or simply on low slopes, as the competence of the flow declines, and the increasing sedimentation rate induces the rise of the basal flow-depositional boundary layer (Fig. 10.14), either progressively, or incrementally, in *en masse* steps (Fig. 10.20b). These conditions are favourable for the formation of high concentration underflows, where suppression of turbulence and processes of hindered settling, fluidisation and particle–particle interactions become dominant. Druitt et al. (2002) described the lateral transformation of dilute flows (surges) into high concentration pyroclastic flows that even detached and flowed independently, confined in valleys, during the 25 June and 26 December 1997 dome collapses at Soufrière Hills volcano. They estimated that the sedimentation rate needed to generate a pyroclastic flow from a pyroclastic surge is in the order of few mm s^{-1}, in agreement with figures derived from theoretical studies (Druitt 1998; Freundt 1999).

The deposits of the main part or body of the ignimbrite are generally poorly sorted and matrix supported, dominated by massive to layered facies (Figs. 10.35, 10.38, 10.40a–d and 10.41a–f), and are much thicker in medial areas and sometimes in distal areas, compared with near vent, steeper slope proximal areas (Figs. 10.14, 10.19, 10.21). The general texture consists of lapilli to bomb-sized pumice (or other juvenile clasts) and lithic clasts dispersed in a fine, vitric ash matrix (Fig. 10.41). Massive deposits are favoured where velocity and density gradients do not allow high shear to act at deposition (Fig. 10.14 case E), whereas diffuse layering may be developed where significant shear acts at the flow boundary layer (Fig. 10.14 cases C and D). More regular bedding and bedforms are evidence of step-wise aggradation (Pollock et al. 2019), and Smith et al. (2020) showed that bedding can also be developed within the basal undercurrent at critical flow conditions (for example at hydraulic jumps at specific topographic conditions), in the form of dunes and antidunes.

A remarkable feature of many ignimbrites (but certainly not all!) is the presence of a lower, fine-grained, reversely graded division at the base of Layer 2, which grades upward into a more homogeneous, and coarser facies (Figs. 10.6a and 10.41d–f). This lower division has usually been referred to as the fine-grained basal layer or Layer 2a (Sparks et al. 1973; Cas and Wright 1987) and is generally less than 1 m thick. It should not be confused with our expanded usage of basal layers for Layer 1 deposits described above. By comparison with sedimentary debris flows, this reversely graded layer 2 has been interpreted to form as a boundary layer as a result of the dispersive pressure and shearing in the base of the flow, which lifts large clasts, both dense lithics and light pumice, up to a neutral buoyancy level (Johnson 1970; Sparks 1976; Roche 2015); here lithic clasts may become concentrated to form a distinctive Layer 2bL lithic concentration zone (Figs. 10.6a, 10.39, 10.41e, f and 10.43a; Sparks et al. 1973). Alternative interpretations suggest that reverse grading in Layer 2a may reflect incremental deposition from progressively waxing flow conditions (Branney and Kokelaar 2002; e.g. Fig. 10.38e). In addition, large pumice clasts may be concentrated near the top of Layer 2, forming Layer 2bP pumice concentration zone (Figs. 10.39, 10.40b and 10.57a–d, f). These zones are discussed further in Sect. 10.8.10.

The complete facies architecture of the main ignimbrite body deposit reflects spatial and temporal variations in the mass flux variations at vent, as well as during flow along the flow length due to interaction with topography (Sect. 10.7), together with changes in flow properties (particle concentration, density) and grain support mechanisms during flow. At Roccamonfina stratovolcano in Italy, Giordano (1998a) noticed that, just downslope of the main-break in slope, which represents a hydraulic jump (i.e. a location where a sudden change in topography affects the flow behaviour), deposits contain the maximum lithic clast size (Sect. 10.8.9) and are relatively fines-depleted. Similarly, at Mt Misery, St Kitts, Roobol et al. (1987) described the accumulation of lithic breccias concentrated at the break in slope where flows moved from gullies across hydraulic jumps onto the outer coastal flanks of the volcano. Due to flow expansion and increased turbulence at this break in slope, sedimentation of larger and denser clasts is enhanced, and conversely, the upward flow of gas causes fines elutriation and the rise of co-ignimbrite ash clouds (Figs. 10.14 and 10.19). Further decelerations and reduction of transport energy promote either a complete decoupling and freezing, resulting in high relief lobes (Fig. 10.18), or else, at lower sediment load, traction carpet sedimentation (Giordano 1998a).

10.8.7.4 Top Layers [Layer 3]

Although usually eroded away, the final stages of deposition of either a single pulse or a sustained pyroclastic flow are dominated by traction and direct fallout from the fine-grained, low-concentration, turbulent trailing and lateral ash cloud or tail. These deposits, called Layer 3, are usually thin, ash rich, diffusely stratified to low angle cross-stratified with low profile surge dune bedforms (Fig. 10.42a–e). Fine ash may remain lofted and rise as a phoenix cloud and may be dispersed as co-ignimbrite ashes well away from the area of deposition of the main ignimbrite (Fig. 10.42f; Chap. 9), but may also mantle the ignimbrite as a very fine, thin, massive to faintly laminated ash layer only mm to a few cm thick. In the ignimbrite facies model such facies are called Layer 3a (ash-cloud surge) and 3b (ash-cloud fallout). Thin, distal lateral and trailing ash-cloud surge and fall deposits may be preserved if rapidly buried, such as within contemporaneous fall deposits, where ash-cloud surge deposits may represent lateral or distal deposits associated with intra-plinian pyroclastic flows (Figs. 10.6c and 10.42e), or under other successive pyroclastic flow deposits (Fig. 10.42b, d). While common Layer 3 deposits derived from ash-clouds are invariably ash-dominated, it is important to underline as discussed earlier (Sect. 10.5), that recent numerical simulations and field data on co-ignimbrite fall deposits suggest that some expanded pyroclastic flows may literally lift off at their maximum runout still retaining high velocities that turn upwards at speeds of several tens to more than 100 m s^{-1}, capable of generating rootless "ash" columns where the grain-sizes entrained may not be limited to elutriated fine-ash, but can include lapilli-sized particles (Calabrò et al. 2022).

10.8.8 Poor Sorting and Crystal-Enrichment

A main and common characteristic of ignimbrites is their poor sorting (Walker 1971, 1983; Cas and Wright 1987; Figs. 10.40, 10.41, 10.42, 10.43 and 10.44), due to the co-existence of abundant ash-sized particles, lapilli- to bomb-sized juvenile clasts and lithic clasts, within most ignimbrite lithofacies (Fig. 10.43a, b). Although much ash is elutriated from flows (Fig. 10.43c) the abundance of fine ash matrix in the deposits of the main ignimbrite deposit (Layer 2) indicates that a dominant high particle concentration granular flow regime exists at the base of almost all pyroclastic flows, even blast flows, with high particle concentration suppressing turbulence, and limiting the ability of hot buoyant gas to elutriate fine ash (Sect. 10.6). A role in trapping of ash into the high concentration undercurrent could also be played by the downward pressure exerted by the overriding flow (Figs 10.13 and 10.19). In contrast, sedimentary turbidites

Fig. 10.41 Typical massive to diffusely layered, and poorly sorted facies of Layer 2, the main ignimbrite body deposit. **a** 2.08 Ma Cerro Galán rhyodacitic ignimbrite, Tacuil, Argentina, showing matrix support of platey pinkish granite lithic clasts, which define a sub-horizontal fabric. Pumice fragments are white, and the matrix contains abundant biotite flakes, which are visible as black speckles. **b** Diffusely stratified, lithic-rich facies of the rhyolitic 1600 BC Minoan ignimbrite, Monolithos, Santorini. Scale 20 cm. Graphic log and grain size data from this depositional unit are given in Fig. 10.47b, c. **c** Poorly sorted, fine ash dominant, non-welded ignimbrite, 1.6 Ma Lower Bandelier ignimbrite, New Mexico, USA. Many of the pumice fragments are marked by stretched tube vesicle fabric, indicating high (continued)

shear stresses and magma conduit velocities. **d** Base of pumice-rich, fines-depleted dacitic 35 ka Choiseul pumice flow deposit, Choiseul, Saint Lucia, Lesser Antilles. Note prominent Layer 2a (right) above paleosoil (dashed orange line). **e** 0.28 Ma phonolitic Poris ignimbrite, Tenerife, Canary Islands. This shows a fine-grained Layer 2a at the base which is reversely graded up into the main Layer 2 that is very lithic-rich representing a Layer 2bL lithic concentration layer. **f** Basal part of the thickest flow unit of the Minoan ignimbrite, showing a prominent reversely graded Layer 2a and 2bL lithic concentration zone deposited on top of white Minoan mudflow unit, Akrotiri, Santorini (see Fig. 10.38d). Scale 20 cm. Graphic log and grain size data from this depositional unit are given in Fig. 10.43a

usually have well-developed vertical size-grading, because they are expanded and highly turbulent, unless they result from high particle concentration flows. Highly turbulent, fully expanded Saharan dust storms do not produce thick massive deposits that fill topography. They produce thin, mantling, fallout-like, loess-like deposits. A remarkable feature is that poor sorting usually characterises the massive and chaotic lithofacies as well as more stratified divisions within ignimbrite units, indicating that poorly stratified facies result from shearing in granular flows, not upper flow regime turbulent flow as in sedimentary deposits.

Nonetheless, ignimbrites become overall finer-grained and better sorted away from vent (Fig. 10.43c) by sedimentation of large lithic clasts in transit, and through abrasion and breakdown of large, fragile pumice clasts, by which more, fine vitric ash matrix is generated in the flow (Dufek and Manga 2008; see Fig. 10.48).

The gas flow within pyroclastic flows, however, to various extents, affects the grading and sorting in the finer-grained fractions. A method has been developed to analyse this type of gas grading, and the details are given in Wright and Walker (1981) and Wright (1981). The method basically assumes that permeable dilute flows will lose a lot of fine-grained ash (into the overriding turbulent cloud), while less-permeable high-concentration (under-)flows will lose relatively smaller amounts. A qualitative assessment can be made in the field because, if the magma was porphyritic, highly expanded, turbulent flows should have lost a high proportion of fine, variably cuspate, and even vesicular vitric ash through elutriation into the ash cloud. Therefore, away from the vent, the matrix of the deposits should be more enriched with dense crystal fragments (Fig. 10.43c), compared with the proportion of crystals in large pumice clasts, or fiamme. In the laboratory, it is possible to quantify the crystal content of crushed juvenile clasts and that of the ash matrix. The ratio is called the crystal-enrichment factor. It can also be used conversely, to calculate the fraction of ash lost or elutriated into buoyant co-ignimbrite clouds (Fig. 10.43c), and then calculate more correctly, the total erupted volume of deposits, and from that, the erupted dense magma erupted volume. However, the reliability of crystal concentration studies depends on the statistical significance of the determinations of juvenile clast pheonocryst content versus free crystal content in the ash matrix. Furthermore, we must remember that volume calculations are usually heavily affected by uncertainties and errors (see Sect. 10.8.1.1), so a lot of care must be taken when assessing both the crystal-enrichment factor and the ash fraction elutriated.

Some calculated crystal concentration factors are up to 10:1, implying a loss of fine vitric ash equal to the preserved volume of the ignimbrites (Walker 1972). However, the degree of crystal enrichment and vitric ash loss occurs in both buoyant explosive eruption columns as well as during passage of pyroclastic flows (Walker 1972). Calculating crystal enrichment and vitric ash loss is therefore very important in understanding eruption and pyroclastic flow dynamics and eruption volumes during explosive eruptions (e.g. Scarpati et al. 2014; Marti et al. 2016).

10.8.9 Massive and Chaotic Facies Versus Stratified Facies, and Types of Internal Contacts

The most common lithofacies in ignimbrites is a massive and chaotic lapilli tuff, dominated by ash-sized matrix (usually >70%), with dispersed lapilli-sized pumice and/or scoria and lithic clasts (e.g. Figs. 10.40 and 10.41). The ash matrix consists of glass shards, crystal fragments, and fine lithic material. Fine ash (<63 μm) is abundant. Grain size distributions are highly polydisperse, skewed, and polymodal (cf. the usually unimodal distributions in fall deposits; Chap. 9). Polymodality is largely accounted for by variable component distributions, that reflect different densities and settling behaviours of different components (Fig. 10.45). This figure also shows the effect of gas streaming through the ash matrix on deposition from a flow to form gas segregation pipes, which are enriched in lithic and crystal components (see Sect. 10.8.16).

Massive and chaotic lapilli tuff facies may show gradational or rapid transitions, both vertically and laterally, to facies characterised by a better internal organisation, which range from faint layering defined by weak grain size variations or concentrations of lithics and pumice clasts (see next section), to well-stratified facies, characterised by discrete stratigraphic contacts and alternations of coarser and finer-grained layers or beds, usually with a low angle cross stratification (Fig. 10.46a–e). Massive and chaotic facies are generally thick (1 to >100 m; Figs. 10.38 and 10.40a). Well-stratified divisions are generally thinner, relatively fines-depleted, and better sorted than the massive and chaotic facies (Figs. 10.40b and 10.46a–e).

Fig. 10.42 Co-ignimbrite ash-cloud surge and fall deposits (Layer 3). **a–c** Distribution of ash-cloud surge deposit in the 1.26 Ma Upper Bandelier ignimbrite, Pajarito Plateau, New Mexico, USA. **a** Averaged maximum thickness in cm; note isopachs increase with distance, beginning ~15 km from Valles caldera rim and become thicker and more continuous away from source (From Fisher 1979). **b** Distal facies association showing thin, Layer 3a ash-cloud surge lens interbedded with very pumice-rich, reversely graded Layer 2 deposits; deposits are indurated by vapour-phase crystallisation (see Sect. 10.8.14.3) (From Cas and Wright 1987). **c** Schematic cross-section along transport direction indicating measured variation in height and wavelength, and internal stratification of dune bedforms, and interpretation (below) showing relative lateral thickness changes of the pyroclastic flow versus its overriding ash-cloud surge developing above it (From Self et al. 1996). **d** Low angle cross-stratified and dune bedded ash-cloud surge deposit, between two massive, poorly sorted ignimbrite Layer 2 deposits. The horizon is interpreted as the trailing ash-cloud surge deposit (Layer 3a) of the lower flow deposit because of the gradational relationship between the two. Tocomar, northwestern Argentina. **e** Lateral or trailing ash-cloud surge deposit (Layer 3a) of an intra-plinian ignimbrite, preserved between two plinian pumice fallout horizons, all part of the 308 ka Fasnia eruption sequence, Tenerife, Canary Islands, Spain. **f** Distal plinian and co-ignimbrite ash-fall deposit (Layer 3b) from the 1.6 Ma Lower Bandelier eruption, 550 km ESE of Valles caldera, at Floydada, Texas, USA (From Cas and Wright 1987)

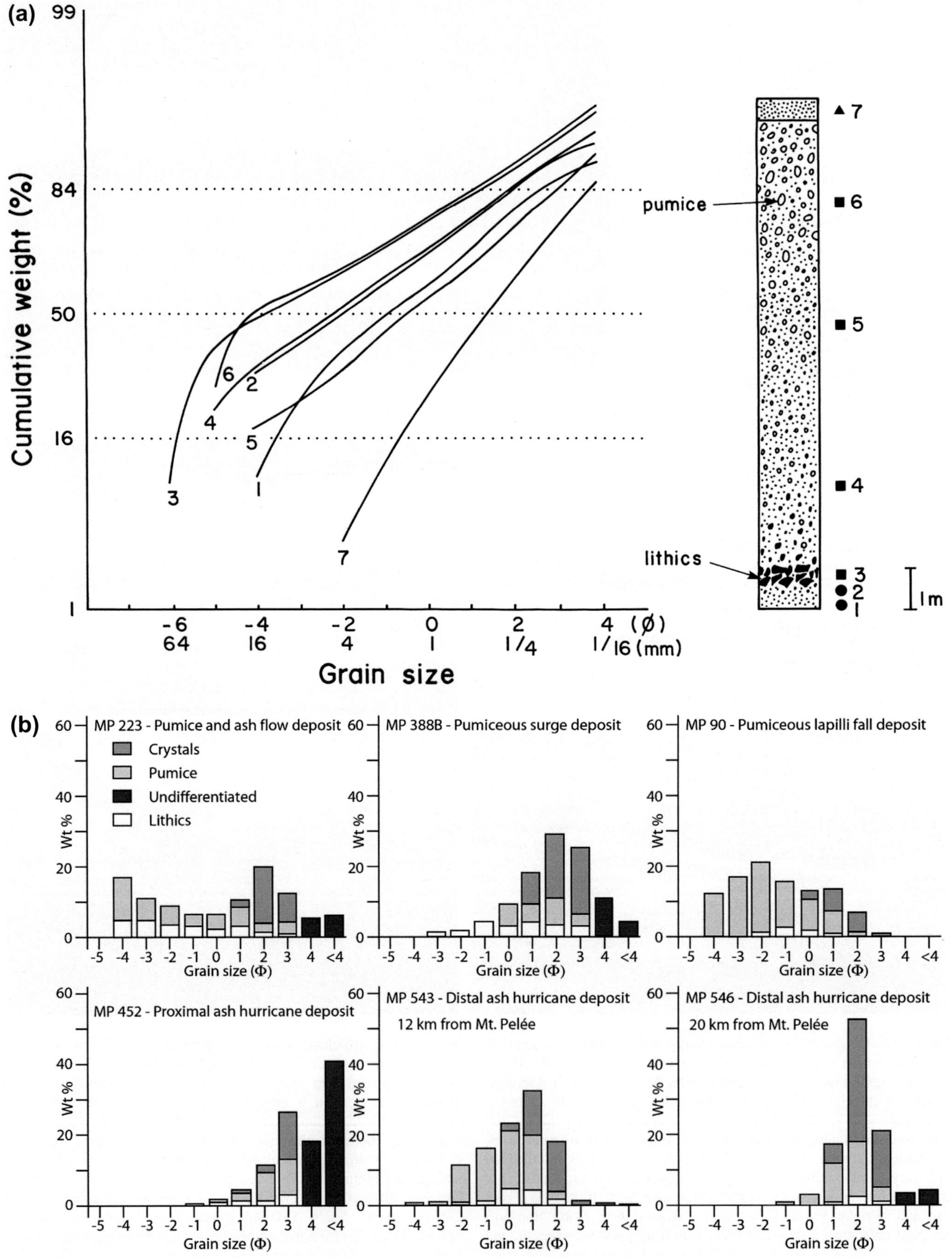
(a)
Cumulative weight (%)
99
84
50
16
1
-6 64
-4 16
-2 4
0 1
2 1/4
4 (Ø) 1/16 (mm)
Grain size
pumice
lithics
1m
1 2 3 4 5 6 7
(b)
MP 223 - Pumice and ash flow deposit
Crystals
Pumice
Undifferentiated
Lithics
MP 388B - Pumiceous surge deposit
MP 90 - Pumiceous lapilli fall deposit
MP 452 - Proximal ash hurricane deposit
MP 543 - Distal ash hurricane deposit
12 km from Mt. Pelée
MP 546 - Distal ash hurricane deposit
20 km from Mt. Pelée
Wt %
Grain size (Φ)
-5 -4 -3 -2 -1 0 1 2 3 4 <4
0 20 40 60

◀ **Fig. 10.43** **a** Cumulative frequency grain size curves for samples of the thickest flow unit of the 1600 BC Minoan ignimbrite, Akrotiri, Santorini (Figs. 10.38d and 10.41f). Closed circles = samples from Layer 2a; closed squares = Layer 2b; closed triangle is a co-ignimbrite ash-fall deposit (Layer 3a). The relatively gentle slope of the cumulative frequency curves indicates poor sorting of the deposits, while their convergence towards the finer fractions indicates the matrix essentially remains the same throughout the flow unit. Grain size grading therefore only affects the coarse part of the distribution, which is called coarse-tail grading. (From Cas and Wright 1987). **b**, **c** Grain size characteristics of ~2,000 yr BP pumice and ash flow deposits erupted from Mt Pelée, Martinique. **b** Grain size and component abundances in different types of deposit. Ash hurricane deposit translates as the topography-mantling IVD facies of a low aspect ratio ignimbrite (see Fig. 10.34). **c** Various granulometric trends with distance from the crater of Mt Pelée for three distal low aspect ratio ignimbrite deposits (Units I, II and III). (From Wright et al. 2016)

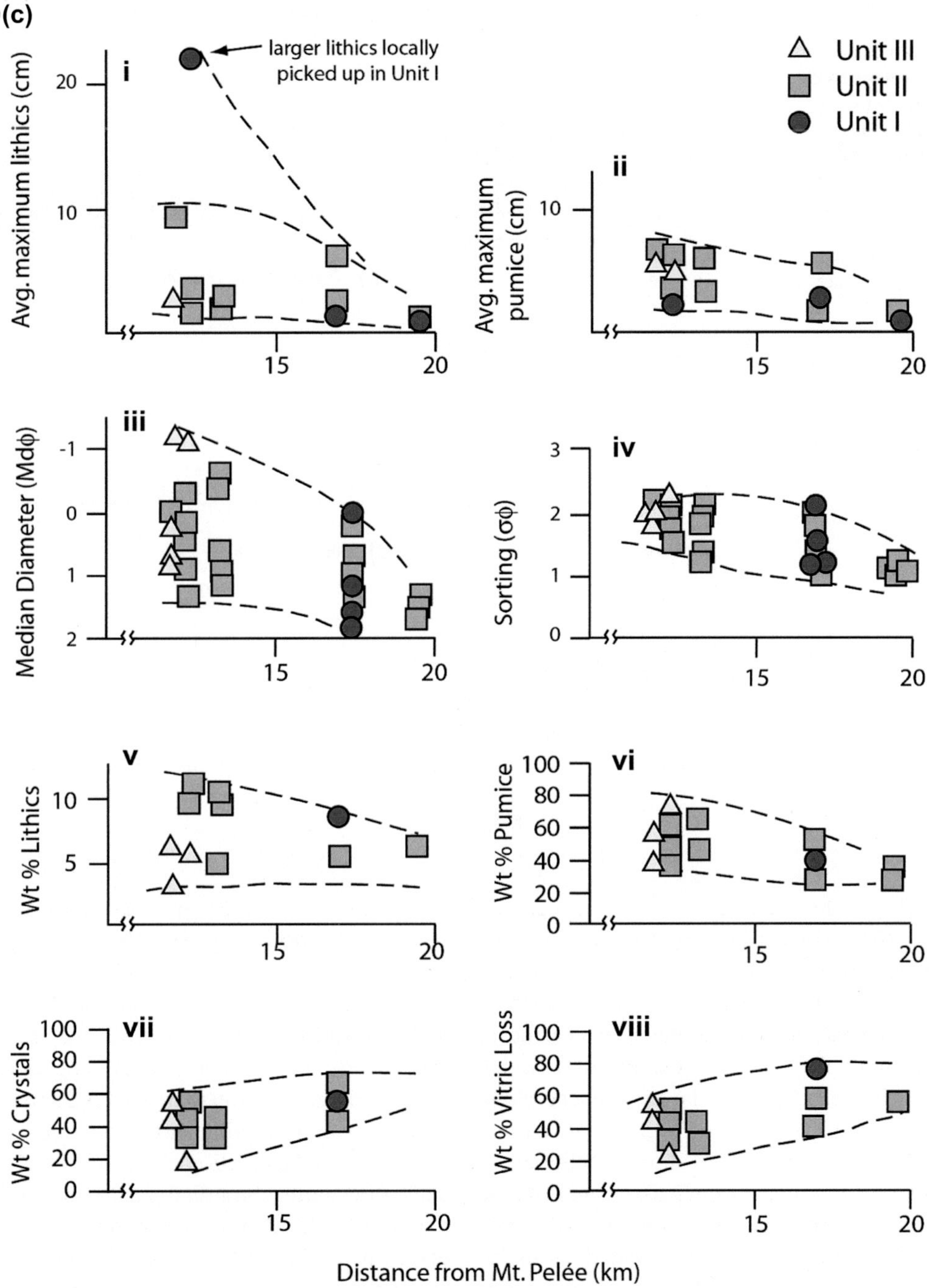

Fig. 10.43 (continued)

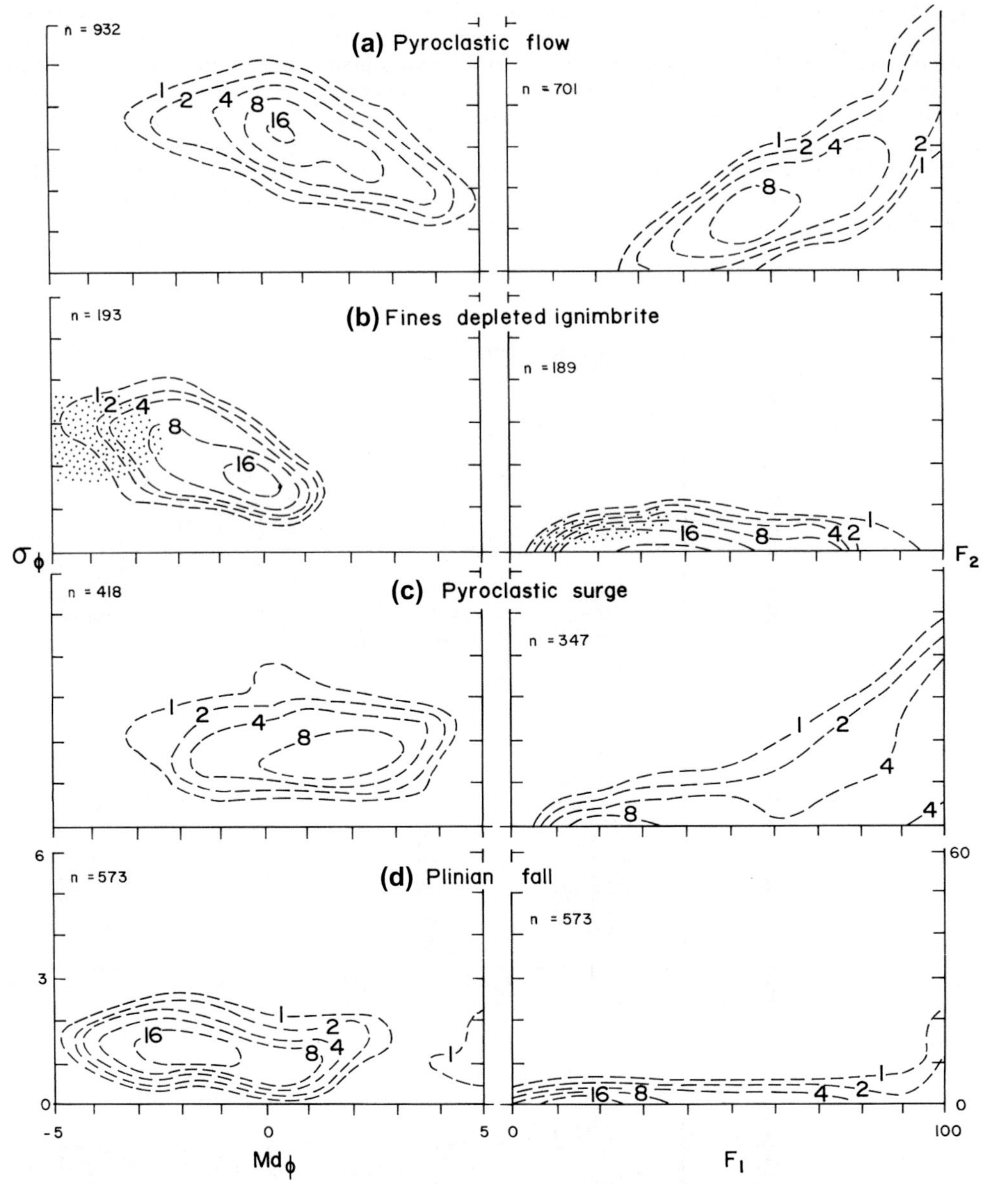

Fig. 10.44 Plots of Md_ϕ (median phi grain size) versus σ_ϕ (phi sorting parameter) (left columns), and of weight percentage finer than 1/16 mm (F_2), versus weight percentage finer than 1 mm (F_1) for pyroclastic flows and three kinds of fines-depleted pyroclastic deposits. In (**b**) various kinds of fines-depleted facies associated with ignimbrite are included: ground layers/breccias, fines-depleted ignimbrite, and gas segregation pipes. The dotted area contains co-ignimbrite lag breccias not included in the contoured plot. Note that fields in (**b**), (**c**), and (**d**) overlap, showing that grain size parameters alone are not unique criteria for determining origin. (From Cas and Wright 1987, After Walker 1983)

Based on varying sedimentation rates, flow shear and gradational *versus* sharp contact between the flow and the deposit, Branney and Kokelaar (2002) described four different types of basal flow boundary zone, dominated respectively, and at increasing concentration, by suspension fallout, traction, granular flow and fluid-escape, all of which result in different lithofacies and may be experienced along flow and over time during a single pyroclastic density current event (Fig. 10.14). Many ignimbrites show repetitions of both vertical and lateral transitions from massive and chaotic to stratified facies. Lateral transitions may occur across paleovalleys and along flow as a result of variations of the transport system and are gradational. Vertical variations record temporal changes of the pyroclastic flow dynamics, both in a sustained event and in multiple events that emplace compound units. Vertical transitions can be gradational or sharp (Fig. 10.46). Sharp transitions usually show very gentle undulations (Fig. 10.46b) and suggest that impulsive erosion stages can punctuate the aggradation of ignimbrite units, perhaps implying a stop-start style of transportation and deposition. The occurrence of thin stratified facies interbedded within thick massive and chaotic divisions has been used as a criterion to define distinctive flow depositional units. However, caution should be used considering the significance and the duration of the time break involved between such flow units. Stratified facies may in fact result from variations of the depositional regime in a sustained current, or else a lateral shift of the locus of sedimentation or indeed by a significant change of eruption dynamics at source. The latter case should be discernable by extensive lateral correlation of the stratified division accompanied by some measurable change in componentry and/or texture of the juvenile clasts.

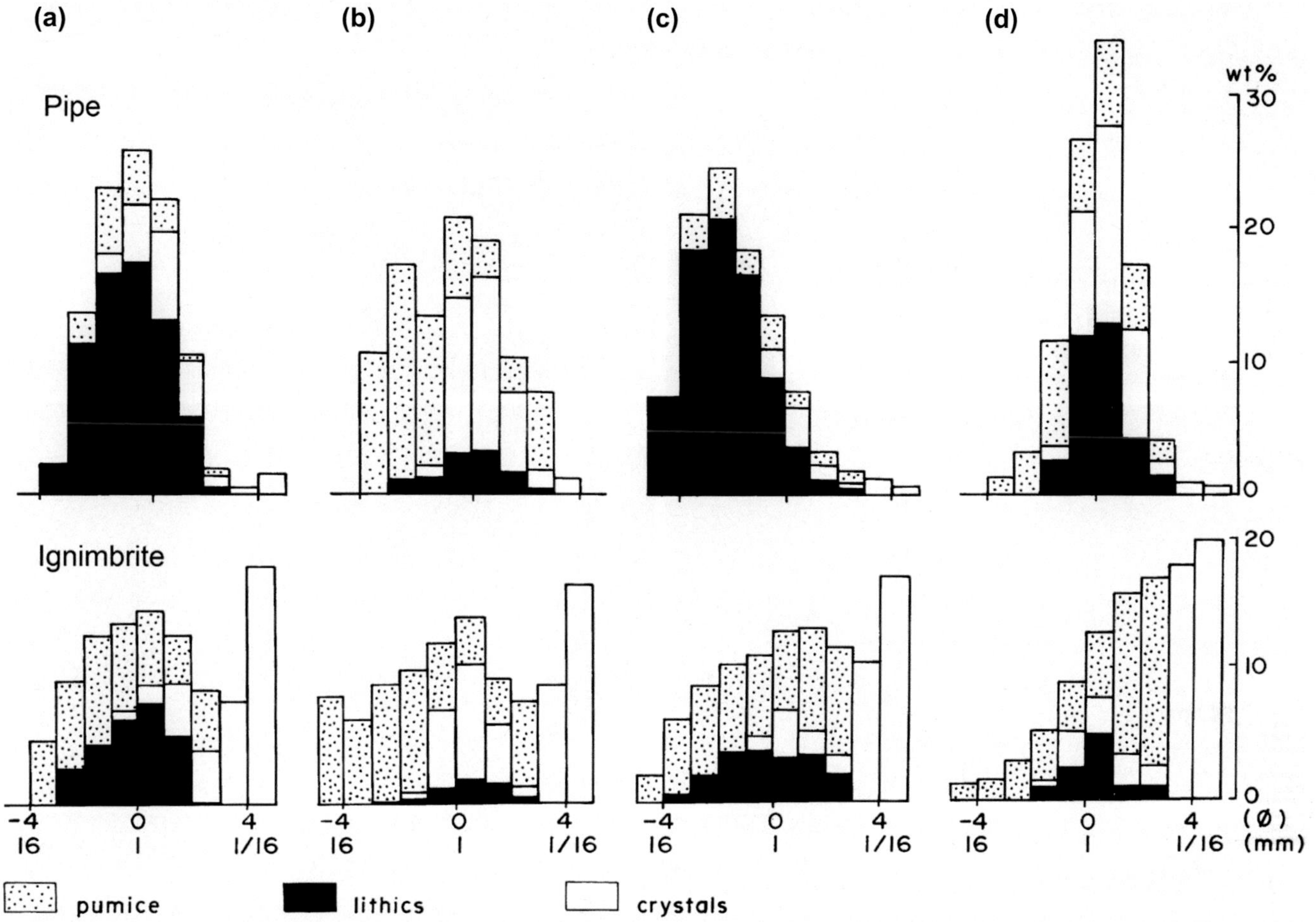

Fig. 10.45 Grain size histograms and componentry of samples of ignimbrite (bottom row) and gas segregation pipes (top row) which they contain. **a** Vesuvius AD 79, Italy. **b** 25 ka Angra ignimbrite, Terceira, Azores, Portugal. **c** 75 ka Los Chocoyos ignimbrite, Lake Atitlán caldera, Guatemala. **d** 39 ka Campanian ignimbrite, Italy. Deposits show varying degrees of grain size polymodality and skewness, and the pipes varying degrees of lithic and crystal enrichment. (From Cas and Wright, After Walker 1971)

10.8.10 Vertical and Lateral Grading of Lithics and Pumice, and Facies Within Ignimbrites

The main portion of an ignimbrite (Layer 2) may show various types of crude grading, both in local outcrops, especially of the larger clasts (Figs. 10.39, 10.40b and 10.41e, f). When grading processes affect only the coarse part of the grain size distribution, this is called coarse-tail grading (cf. continuous grading found in classical turbidites; Chap. 12; Allen 1984). Cumulative frequency grain size curves of samples collected within the same depositional unit converge towards the finer-grained fractions (Fig. 10.43a), indicating that this part of the distribution or "matrix" essentially remains uniform throughout the depositing flow. Grading is therefore a function of grain size and is largely controlled by the hindered settling velocities of particles.

The "textbook" ignimbrite vertical facies profile (Fig. 10.39) shows normal coarse-tail grading of larger, dense lithic clasts at the bottom (Layer 2bL), and reverse coarse-tail grading of larger pumice clasts at the top (Layer 2bP), forming well-defined concentration zones at the top and bottom, respectively (e.g. Figs. 10.40b and 10.41e, f; also see Figs. 10.6a and 10.57). These textbook coarse tail grading patterns can be interpreted in two ways in terms of flow dynamics, either they represent: (1) effective hydraulic sorting within a high particle concentration, short-lived flow (dense lithic fragments settle, light pumice fragments are buoyant in the ash-rich flow matrix; e.g. Sparks 1976; Palladino and Valentine 1995), or (2) the vertical variation in componentry in the deposit reflects changes in componentry released from the vent into the flow, which is progressively aggrading a deposit (Branney and Kokelaar 2002; Sect. 10.7). Reverse coarse-tail grading of lithics and normal coarse-tail grading of pumice is more rarely found, and an absence of grading of any type is also common (Cas and Wright 1987). Reverse coarse-tail grading of lithic clasts can only be interpreted in terms of progressive aggradation

Fig. 10.46 Variations in the internal organisation of the main ignimbrite body deposits or Layer 2. **a** Upwards transition from planar, diffusely bedded ignimbrite to massive ignimbrite, reflecting increasing mass flux, steadier flow, and higher particle concentration. The lower depositional unit is overlain by the lithic-rich base of the overlying depositional unit, which is still part of the same eruption. 172 ka Lower Pumice 2, Akrotiri, Santorini, Greece. **b** Abrupt upward transition from massive, poorly sorted ignimbrite to undulating stratified facies to massive facies, reflecting suddenly changing flow conditions. 190 ka Abrigo ignimbrite, Los Abrigos, Tenerife, Canary Islands, Spain. **c** Coarse tail, normally graded and diffusely stratified ignimbrite, San Isidro, Tenerife. Person circled for scale. **d** Planar to dune cross-stratified ignimbrite, 1600 BC Minoan ignimbrite, Santorini. Person circled for scale. **e** River section through the 1980 Pumice Plain at Mt St Helens showing diffusely stratified to cross-stratified ignimbrite, suggesting transport and deposition under highly unsteady or pulsating conditions (Photo by K. Allstadt; *Source* United States Geological Survey)

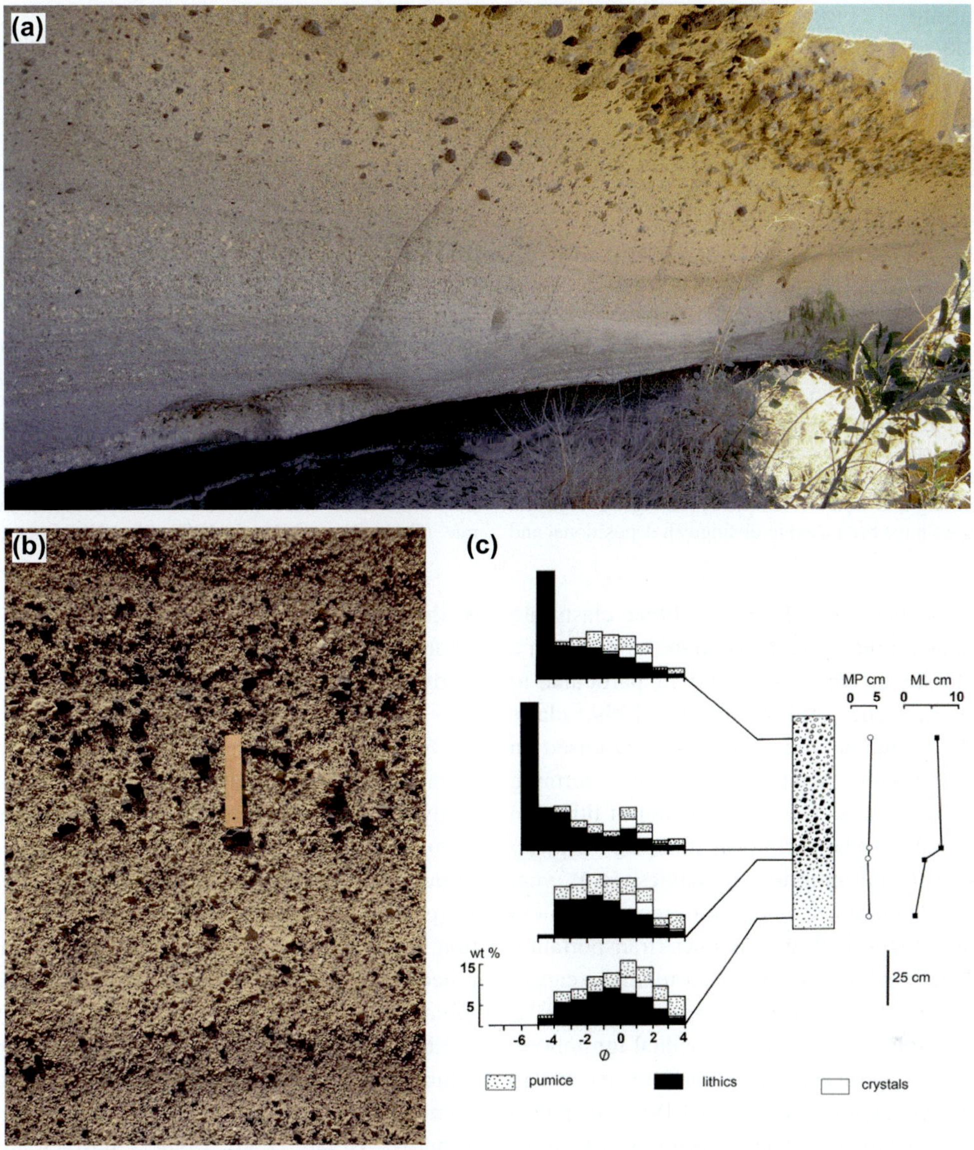

Fig. 10.47 Reverse coarse-tail grading of lithic clasts in ignimbrite depositional units as a result of progressive aggradation. **a** The phonolitic 220 ka Caleta ignimbrite, Tajao, Tenerife, Canary Islands, Spain. More massive reversely graded lithic-rich layer occurs above a diffusely stratified base to the deposit. Thickness is ~4 m. **b**, **c** Reverse coarse-tail grading of larger lithics and grain size componentry in thin depositional unit within the rhyolitic 1600 BC Minoan ignimbrite, Monolithos, Santorini, Greece. This forms part of diffusely stratified package of more lithic-rich units towards the top of the ignimbrite (central unit with scale in Fig. 10.41b), consistent with the progressive aggradation from a sustained, pulsating flow (Sects. 10.8.9 and 10.8.10). Laterally such diffusely stratified facies appear to thicken into topographic depressions where the bulk of pyroclastic flows was channelised. Grain size studies indicate only minor elutriation of fine-grained ash, implying low gas flow rates and deposition from a dense, boundary layer. (From Wright 1979)

(Fig. 10.47a, b, c) because large dense lithic clasts would sink to the bottom of a high particle concentration fluidised "plug" flow. The repetitious occurrence of lithic and pumice concentration zones has been often used to define flow units. However, although this can certainly be the case especially in thick depositional units, it must be taken into account that the pulsatory nature of eruptions and of pyroclastic flows may produce repetitions of concentration zones as a result of influxes of lithics and pumices from a sustained, but pulsing eruption into a sustained or pulsing flow. Understanding causes in different cases needs the detailed reconstruction of the internal facies architecture of the ignimbrite, and the assessment of the lateral persistence of what is defined as a flow and/or depositional unit.

Compared with other components, lithic clasts, whether they are cognate, accessory or accidental (cf. Chap. 4, Sect. 4.8.3), are very useful to study, as they are less affected than pumice by processes of breakage along the flow path, so that the size of the lithic clast in the deposit can be assumed to be approximately equal to its size at vent (excluding friable

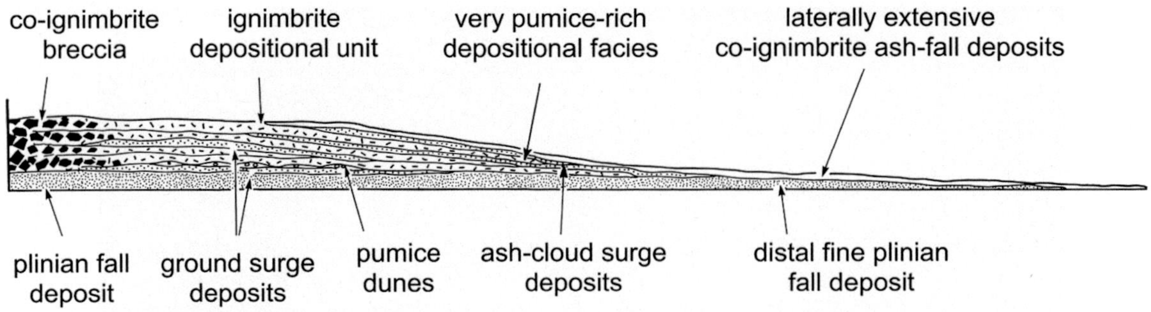

Fig. 10.48 Legacy generalised facies model for the proximal to distal internal architecture of an ignimbrite based on studies of Upper and Lower Bandelier Tuff ignimbrites, Jemez Mountains, New Mexico, USA. Overall, close to source the ignimbrite may be coarsest, and loses its largest lithic component in a downflow direction. Distal flow facies are represented by thin, reversely graded pumice-rich flow units, which may represent the splitting into separate flow lobes (cf. Figs. 10.11d and 10.18b). Stratified facies have been used to distinguish depositional and flow unit boundaries, especially ground or ash-cloud surge deposits. Jetted deposits are represented by pumice dunes. However, it is now recognised that such boundaries may also reflect variations in the depositional regime over time and space during continuous aggradation. A correct interpretation should be made based on a 3D reconstruction, discriminating those transitions that reflect emplacement of different flow units, from those that reflect local variations in the sedimentation rate. (From Cas and Wright 1987, After Wright et al. 1981)

lithic fragments). Furthermore, the dense lithic clasts are more susceptible than pumice to changes in the tendency of a pyroclastic flow to begin depositing, and in particular to changes in flow competence and/or capacity. Lithic clasts may be scattered throughout the deposit or organised in lenses and concentration zones, sometimes forming well-defined breccia horizons. The final position of lithics in the ignimbrite depends on the reciprocal interplay of many processes that reflect the eruption dynamics (e.g. vent dynamics, vent wall lithology and stratigraphy, column dynamics) and the pyroclastic flow dynamics (transportation dynamics, transportational and depositional regimes, capacity), all varying greatly from one ignimbrite to another, as well as during the emplacement of an individual ignimbrite.

Lateral variations in deposit characteristics from proximal to distal settings can be significant (Fig. 10.48) and reflect downcurrent changes in the competence of the pyroclastic flow, velocity, concentration, and grain size (Calabrò et al. 2022; Doronzo et al. 2022). Such changes can be summarised in isopleth maps or variograms from measurement in the field of the maximum and/or average dimensions of lithic and pumice clasts at each locality and at comparable stratigraphic levels within depositional units (see Figs. 10.54, 10.55 and 10.56). These may be difficult to construct unless careful regional mapping and correlation between outcrops is undertaken, which is time consuming, but both informative and rewarding. Sitting in a lab isn't going to give you that information!

10.8.10.1 Lithic Clast Types

Recognition of the type of lithic clasts contained in an ignimbrite and their grading is important as they give a first insight into:

- stratigraphy of the volcano,
- the level of and changes in the magma fragmentation level in the conduit during the eruption,
- the timing and magnitude of vent/caldera collapse events, and
- dynamics of the transport system.

Accessory lithic clasts are derived from and entrained from the walls of the vent during conduit-flow, and especially at and above fragmentation level (Dufek et al. 2012; Campbell et al. 2013). Vertical variations of lithic types through an ignimbrite sequence can give information about shifts of the magma fragmentation level with time. Derivation from the upper levels of the conduit/vent wall is reflected by a dominance of relatively unaltered volcanic lithic clasts, representing episodes of vent widening and shallow conduit-erosion (Fig. 10.49a). Deeper crustal level lithic clasts could include variably altered plutonic (e.g. syenitic clasts in ignimbrites from the Azores, Fig. 10.49b; Tenerife, Bryan et al. 1998; Pittari et al. 2008; and the Roman Magmatic Province, De Benedetti et al. 2010), or (meta-) sedimentary basement clasts (e.g. from the Cerro Galán Ignimbrite, Fig. 10.49c). These may or may not be exposed in the surrounding volcanic hinterland and may reflect both the deepening of the fragmentation level and/or episodes of destabilisation of the conduit as well as of formation of caldera ring faults (e.g. Suzuki-Kamata et al. 1993; Rosi et al. 1996; Bear et al. 2009a, b).

Mechanisms that allow for erosion of lithic clasts from country rock walls include pneumatic fracturing of country rock by debris laden highly pressured gas or steam, and by deep-level explosions that propagate shock waves and fracture country rock (e.g. Heiken et al. 1988). During waning discharge phases of an eruption, decreasing pressure may allow gravitational collapse of previously pneumatically/tectonically fractured rock into the conduit. During peak magma discharge stages of the eruption, such lithics were prevented from collapsing by high conduit gas pressures (e.g. Papale and Dobran 1994).

Fig. 10.49 Lithic clast types in ignimbrites. **a** Shallow derived accessory lithics of dominantly basaltic lava fragments (dark) from episodes of vent/upper conduit erosion and widening, Sete Cidades volcano, São Miguel, Azores, Portugal. **b** Deep seated syenitic fragments (white and grey clasts) and reddened hydrothermally altered clast, 4.6 ka Fogo A ignimbrite, São Miguel. **c** Clusters of locally picked-up accidental lithics of slate from the nearby metasedimentary bedrock, 2.08 Ma Cerro Galán ignimbrite, northwestern Argentina. Note that clusters of clasts indicate minimal mixing of the clasts in the flow

The occurrence of *accidental* lithic clasts in ignimbrites is generally sporadic, and sometimes they are difficult to distinguish from accessory lithics. Accidental lithic clasts can be incorporated into pyroclastic flows from the ground surface even tens of km from vent, and indicate that pyroclastic flows can locally develop the ability to rip-up clasts from the substrate (Fig. 10.49c; e.g. Suzuki-Kamata 1988; Buesch 1992; Bryan et al. 1998; Cas et al. 2011; Roche 2015; Roche et al. 2016; Wright et al. 2016). Accidental lithics are usually found toward the base of the ignimbrite, because once the substrate is covered with the early blanket of pyroclastic deposits, it is less likely that the subsequent parts of the pyroclastic flow will "see" the bedrock. However, in a rough topography, successive pyroclastic flows can incorporate accidental lithics from valley walls and ridge slopes. Topography may therefore be important by itself in determining the abundance of accidental lithics in ignimbrites and the stratigraphic height at which they appear (e.g. Suzuki-Kamata and Kamata 1990; Buesch 1992; Bryan et al. 1998).

Accidental lithic clasts also include intraclasts, used in the sense utilised in sedimentary successions, i.e. irregular rip-up clasts of pre-existing semi-cohesive substrate sediment incorporated into a density current by its erosional capacity. Intraclasts in ignimbrites can include normal, irregular semi-cohesive intraclasts of mudstone (e.g. Allen and Cas 1998; Allen and McPhie 2001), or of aggregates of previously deposited ignimbrite or fallout deposits (De Rita et al. 1998; Allen and Cas 1998; Allen and McPhie 2001). Intraclasts of pumice or particles of ash can be eroded from underlying unconsolidated pyroclastic deposits but may be difficult to distinguish from the juvenile components of the host ignimbrite, unless they are compositionally or texturally distinctive.

Another mechanism for incorporating accidental clasts into ignimbrites, not related to the erosional capability of a pyroclastic flow, is large-scale caldera wall, gravitational sector or landslide collapse events during major caldera-forming eruptions. These produce intra-caldera megabreccias interstratified and gradational with huge volume intra-caldera ignimbrites (Lipman 1976) that commonly can be tied back to the stratigraphy in the exposed host caldera wall. Clasts range from tens of m down to cm in diameter.

10.8.10.2 Shapes of Lithic Clasts

Lithic clasts in ignimbrites are generally angular. However, they may also show some degree of rounding, especially the largest clasts (Fig. 10.50; e.g. Allen and Cas 1998). There are several possible causes of rounding of lithic clasts, as follows:

(a) Conduit and vent milling, i.e. rounding occurring in the conduit and vent (Wilson et al. 1980; Campbell et al. 2013). This process is likely to occur for large dense clasts that could not be ejected from the vent because the drag force exerted by the upward flowing pyroclastic mixture is unable to exceed their gravitational settling forces and velocity. This could happen if either or both the exit velocity and density of the gas-pyroclast mixture are too low, causing the large clasts to be reworked in the conduit and abraded during collisions with other lithics or with the conduit walls, until finally they are ejected out of the vent at increased gas flow velocities.

(b) Rounding occurring during pyroclastic flow transport. Dense lithic clasts are less susceptible to abrasional rounding compared with fragile pumice, especially where being transported in isolation as lone lithic clasts in a matrix fluid of gas, ash, and pumice, which would buffer incipient collisions between dispersed lithic clasts. However, during phases of flow involving high lithic clast concentration transport in restricted parts of flows, significant clast to clast collisions, abrasion, and clast rounding could occur. Although pyroclastic flow transport durations are usually very short, which is not normally conducive to abrasional rounding, the velocities of pyroclastic currents (tens to hundreds of km hr^{-1}) could generate high enough clast momentum to ensure rapid abrasion especially of large clasts, by clast collisions where and when lithic clast concentrations are high in a flow.

(c) Rounding of accidental lithics by normal surface processes prior to incorporation in the pyroclastic flow. Lithic clasts rounded by tractional surface processes, such as river transport, rapidly become rounded and smoothed. By contrast short term, episodic clast collisions and abrasion during pyroclastic flow may cause rough rounding but not smoothing. Based on smoothing of their surface, some lithic clasts in the Kos Plateau Tuff ignimbrite were interpreted as accidental clasts (Allen and Cas 1998).

10.8.10.3 Types of Co-ignimbrite Lithic Breccias and Lithic Concentrations

There are several types of lithic breccias and lithic clasts concentrations which can inform the dynamics of the eruption producing the ignimbrite, as well as flow dynamic processes.

Lag Breccias

Co-ignimbrite lag-fall breccias (Wright and Walker 1977, 1981), or now termed more generally *co-ignimbrite lag breccias* (Wright 1981; Druitt and Sparks 1982; Walker 1985; Yamagata 1991; Cole et al. 1993), are coarse (lithic clasts cm to >2 m diameter), thick (up to 50 m) lithic breccias which occur in proximal-to-vent locations (Fig. 10.50a–l). These are inferred to form at an eruption climax, particularly at the onset of caldera collapse, both through processes of direct fallout and by segregation of lithic clasts during flow from a highly expanded flow at proximal localities (Fig. 10.51). Lag breccias are often fines-depleted, clast supported and are usually diffusely to

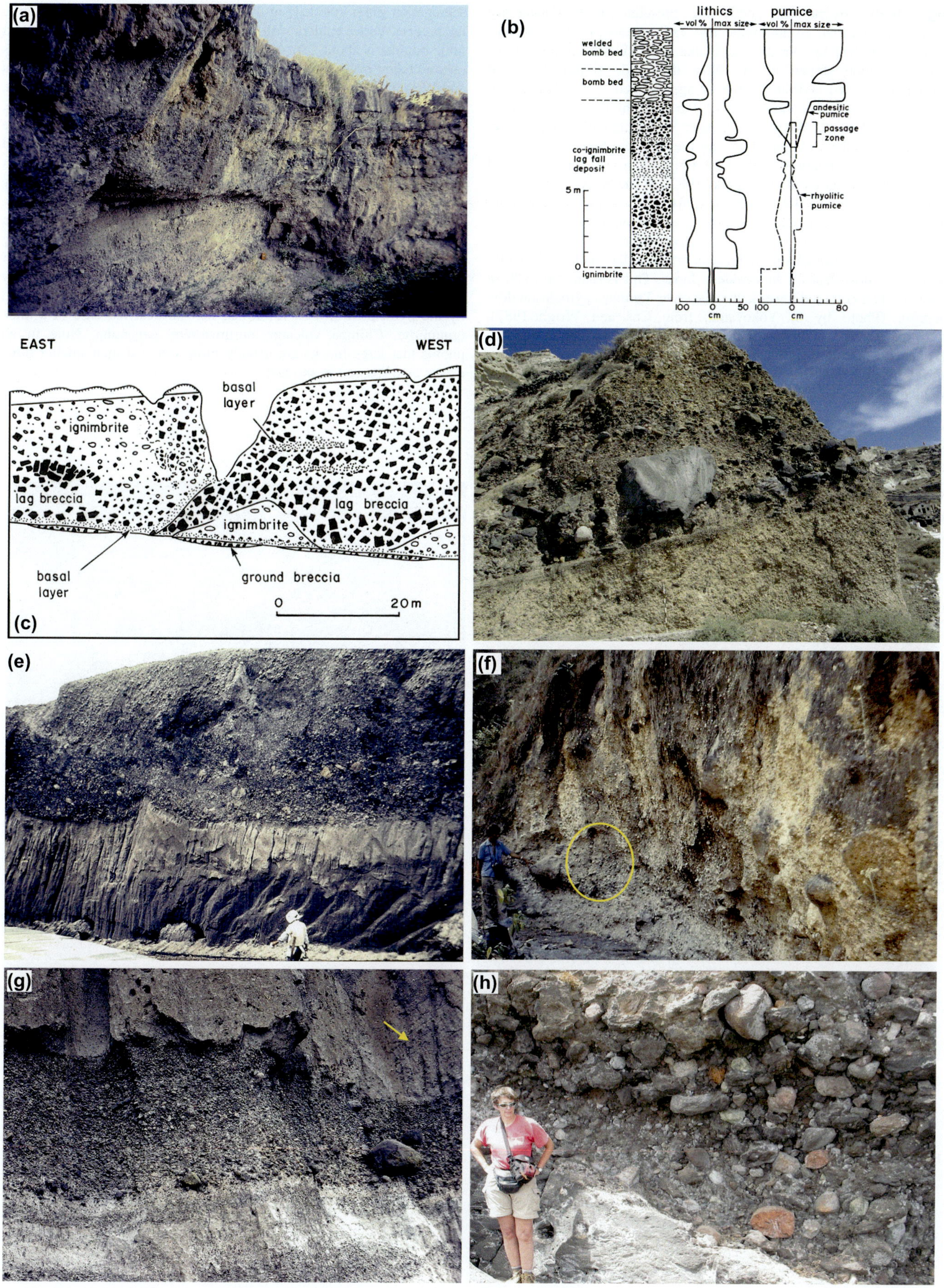
(a)
(b)
lithics
pumice
vol %
max size
welded bomb bed
bomb bed
andesitic pumice
passage zone
co-ignimbrite lag fall deposit
rhyolitic pumice
5 m
0
ignimbrite
100
0
50
100
0
80
cm
cm
EAST
WEST
(d)
basal layer
ignimbrite
lag breccia
lag breccia
ignimbrite
basal layer
ground breccia
0
20 m
(c)
(e)
(f)
(g)
(h)

◂ **Fig. 10.50** Co-ignimbrite lithic lag breccias. **a**, **b** Quaternary (<0.65 Ma) Acatlán co-ignimbrite lag "fall" deposit, central Mexico, as first described by Wright and Walker (1977). **b** Measured stratigraphic section shown in (**a**). Note the crude stratification and prominent welded andesitic bomb or agglutinated spatter bed at the top of the deposit. Location is ~ 1.5 km from the most likely source vent. (From Cas and Wright 1987, After Wright and Walker 1977). **c**, **d** 22 ka Cape Riva co-ignimbrite lag breccia, Ioa, Santorini, Greece. **c** Field sketch of relations in the Santorini caldera wall (From Cas and Wright 1987, After Druitt and Sparks 1982). **d** The breccia with a very large clast sits above the fine-grained base of the ignimbrite, without disturbance, indicating that the breccia is not an aggregate of ballistics, but was transported laterally by the pyroclastic flow gliding over the substrate. **e** The Kamewarizaka breccia, proximal facies of the 22 ka Ito pyroclastic flow deposit, Aira caldera, Japan. This is found up to 30 m thick. Lighter coloured lower half is the older Tsumaya pyroclastic flow deposit. (Photo by S. Yokoyama; from Cas and Wright 1987). **f** Co-ignimbrite lag breccia of the 95 ka Rio Caliente ignimbrite, La Primavera rhyolite volcano, Mexico. Geologist's hand is on sub-rounded boulder of welded ignimbrite identical lithologically to welded Rio Caliente ignimbrite. Note vertical lithic-rich gas segregation pipes (circled). **g** 1.6 Ma Lower Bandelier co-ignimbrite lag breccia, Valles caldera wall, Cat Mesa, New Mexico, USA. Largest lithic boulder is about 2 m. Note again lithic-rich gas segregation pipes (arrow) rising from top of breccia into overlying ignimbrite. (From Cas and Wright 1987). **h** The 39 ka Campanian co-ignimbrite lag breccia, Campi Flegrei, Italy. **i**, **j** 172 ka Lower Pumice 2 eruption sequence, Akrotiri, Santorini. **i** An horizon of lag breccia enclosed within pumice and ash ignimbrite. **j** Fine lag breccia matrix, showing clast-supported texture, sparse pumice clasts, and interstitial fine ash matrix. Most clasts are volcanic, but one is metasedimentary basement (circled). **k** Diffusely stratified lag breccia of the 184 ka Lower Pumice 1 eruption sequence, sitting above the prominent 2 m thick LP1 plinian pumice fallout deposit, Akrotiri, Santorini. The largest blocks are >1 m. The lag breccia sits above fine, thin (~40 cm) ignimbrite, at the level of Layer 2bL. The spectacular impact sag was caused by a red hydrothermally altered ballistic block that shattered on impact, and was not transported by the pyroclastic flow. **l** Lithic lag breccia near the base of a thick ignimbrite, Chimpa volcano, northwestern Argentina. Note in all photos that large blocks are usually more rounded than smaller lithic clasts, as discussed in the text

Fig. 10.50 (continued)

crudely stratified (Fig. 10.50a, b, k). They show lateral and vertical, sharp or gradational transition and interfingering with "normal" pumice and ash rich ignimbrite (Fig. 10.50c, g, j; Wright 1981; Druitt and Sparks 1982; Walker 1985; Gertisser et al. 2009; Simmons et al. 2016). These deposits can contain large gas segregation pipes (Sect. 10.8.16; Fig. 10.50f, g) and irregular lithic-enriched lenses and pods (e.g. Wright 1981; Druitt and Sparks 1982; Cas and Wright 1987). Lag breccias (as well as caldera wall collapse mega-breccias; Lipman 1976) in intra-caldera ignimbrites, also occur as breccia horizons within ignimbrites.

Numerical simulations however question the origin of proximal breccias as true lag breccias and suggest that the coarse fraction at impact immediately forms an undercurrent able to flow laterally (e.g. Valentine 2020). For example, Druitt and Sparks (1982) interpreted the 22 ka Cape Riva lag breccia on Santorini (Fig. 10.50c, d) to have undergone some degree of lateral transport, and segregation of the lag breccias to have resulted from initially expanded flows which formed under high pressures at the base of the collapsing column in the deflation zone (Fig. 10.51). Many co-ignimbrite lag breccias have been documented, and are often interpreted to indicate the timing of final caldera collapse, representing break-up of the foundering crustal roof block(s) of the magma chamber, and/or the country rock along which ring faults propagate, especially when the breccias contain a significant amount of deep seated accessory xenoliths (e.g. Cape Riva, Santorini, Druitt and Sparks 1982; Mt Mazama ignimbrite, Crater Lake, USA, Bacon 1983, Druitt and Bacon 1986; Lower Bandelier Tuff, USA, Self et al. 1986; Shikotsu pumice flow deposit, Japan, Yamagata 1991; Campanian ignimbrite, Italy, Rosi et al. 1996; Poris Member, Tenerife, Edgar et al. 2002; Sutri Formation, Italy, Bear et al. 2009a; Villa Senni Formation, Italy, Giordano et al. 2006, Vinkler et al. 2012; Lower Pumice 1, Santorini, Simmons et al. 2016; Lower Pumice 2, Santorini, Gertisser et al. 2009). Interfingering of such facies with normal ignimbrite (Druitt and Sparks 1982; Fierstein and Hildreth 1992) is probably related to pulsing in eruption energy and discharge rate. Many of the lithics in the exposed lag breccias of the 95 ka Rio Caliente ignimbrite, Mexico, appear to be fragments of welded Rio Caliente ignimbrite, indicating considerable vent widening, or migration, during the eruption (Fig. 10.50f; Wright 1981). Co-ignimbrite lag breccias are however not confined to larger-scale eruptions and volcanoes, including the seminal Acatlán example in Mexico (Fig. 10.50a, b), which was erupted on a fairly flat terrain and whose most likely conduit is now plugged by a rhyolite dome. They may also occur associated with the eruption of small volume pyroclastic flows (<1 km^3) from smaller volcanic centres (e.g. Laacher See volcano, Germany; Freundt and Schmincke 1985b).

The lateral distribution of co-ignimbrite lag breccias is marked by an inflection in the rate of fall off of lithic clast size (Wright and Walker 1977; Wright 1981; Walker 1985; see Fig. 10.54c). In exceptionally violent eruptions a lag breccia can spread out to 20 km from source (e.g. 232 AD Taupo eruption, New Zealand, Walker 1985, see Fig. 10.79b; Tosu breccia, Japan, Suzuki-Kamata and Kamata 1990; 161 ka Kos Plateau Tuff, Greece, Allen and Cas 1998), although these are more likely to be ground breccias, as discussed next.

Ground Breccias

Lithic breccias that occur beneath the base the main Layer 2 ignimbrite are known as basal breccias or "ground breccias" (Fig. 10.52). Co-ignimbrite ground breccias are a type of "basal layer", i.e. occurring at the base of an ignimbrite without any genetic implication (Sect. 10.8.6). They are enriched in dense clasts, and commonly depleted in fine ash relative to the overlying pumiceous ignimbrite facies (Fig. 10.52b, c). The ground layer breccia facies were first recognised at the base of the Taupo ignimbrite and interpreted to have formed in the highly fluidised head of a high velocity pyroclastic flow (Walker et al. 1981b; Cas and Wright 1987; see Figs. 10.79c and 10.81a, d). Ground breccias are sharply overlain by, or grade relatively abruptly both upward and laterally into normal pumiceous ignimbrite facies (Fig. 10.52b). Ground breccias can occur in proximal settings, where they may be several m thick, but also frequently occur well away from "proximal areas", usually as discontinuous lenses, that are cm to tens of cm thick, and generally in sharp contact with the overlying ignimbrite. Branney and Kokelaar (2002) argue that ground breccias may not be solely the result of flow head processes, but may also reflect the temporal evolution of a pyroclastic flow, involving an initial highly energetic blast-like flow pulse from the vent.

Other Intra-ignimbrite Breccias

Many ignimbrites contain simple *basal lithic concentration zones* (2bL) in Layer 2b (Sparks et al. 1973; Sparks 1976), or at different stratigraphic heights forming complex multi-storey arrangements (e.g. Roobol et al. 1987; Buesch 1992; Cole et al. 1993; De Rita et al. 1998; Bear et al. 2009a, b). *Intra-ignimbrite breccias and lithic concentration zones* are usually matrix supported, enclosed in normal fine-grained ignimbrite. Lithic breccias can be traced laterally into horizons only cm thick or consisting of a trail of scattered lithic clasts. Contacts with the hosting ignimbrite range from gradational to sharp, planar to undulose due to loading or erosion (Fig. 10.53). Lithic breccia geometry also varies from tabular

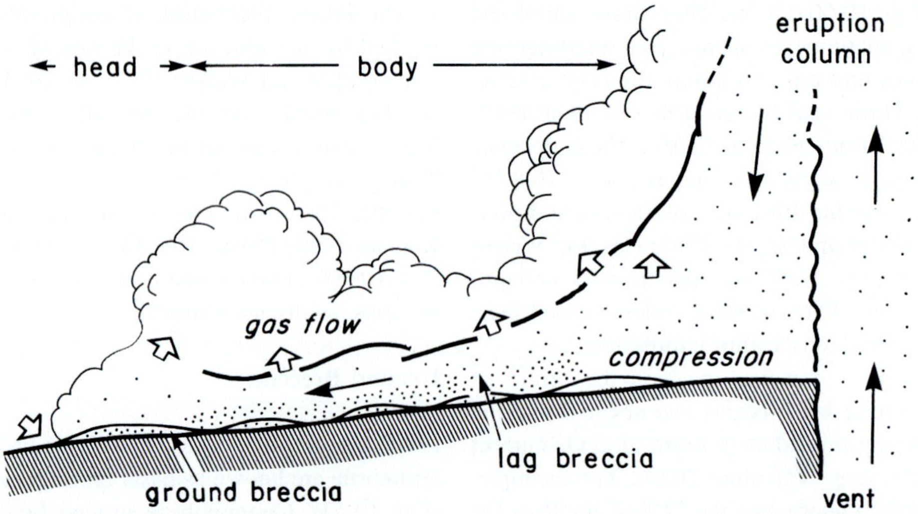

Fig. 10.51 Model for the proximal segregation of lithic clasts in pumice and ash flows generated by eruption column-collapse. Compression of the particle-gas mixture at the base of the collapsing column generates high pore pressures within a high particle concentration flow body. Decompression of the gas phase as the flow moves away laterally causes strong fluidisation with the body, and the segregation of lithics to generate the lag breccia. Ingestion of air at the flow-head causes strong fluidisation, and the sedimentation of the ground breccia. Open arrows schematically depict the passage of gas through the flow system. (From Cas and Wright 1987, After Druitt and Sparks 1982)

Fig. 10.52 Features of ground breccias. **a** Spectacular ground breccia to the 161 ka Kos Plateau Tuff (KPT), ignimbrite E (the climactic ignimbrite), at an outcrop on the island's axial basement ridge 200 m above sea level, Kos, southeastern Aegean Sea, Greece. The pyroclastic flow had to climb to this altitude from sea level. The fine-grained unit under the breccia is an earlier ignimbrite unit. **b** The matrix of the KPT is fines depleted, contains pumice and some ash matrix. Same locality as (**a**). **c** Another outcrop of the KPT ignimbrite E ground breccia, south coast of Kos. **d** Lithic ground breccia in the 18 May 1980 Mt St Helens, USA pyroclastic flow deposits, formed just downstream of the first main break in slope of the volcano edifice

Fig. 10.53 Other intra-ignimbrite lithic-rich horizons. **a** Trail of lithics within the 161 ka Kos Plateau Tuff ignimbrite succession, Sunny Beach, Kos, Greece. **b** Lithic load casts in thin ignimbrites in the 190 ka Abrigo ignimbrite, Tenerife, Canary Islands, Spain. **c** Intra-ignimbrite lithic lens, 4.6 ka Fogo A ignimbrite, São Miguel, Azores Portugal. **d** Lithic lenses sheared by the overriding pyroclastic flow producing bedforms with regular wavelength in the 18 May 1980 Mt St Helens, USA pyroclastic flow deposits

to lensoidal. Internally, lithic breccias may be massive and ungraded, or inverse to normal graded. Elongated clasts may show some degrees of imbrication. Degassing structures may also be associated with the lithic horizons (e.g. Roobol et al. 1987; Buesch 1992; Bear et al. 2009b).

The concentration of lithic clasts toward the base of an ignimbrite (Layer 2bL) has been interpreted largely as a result of coarse-tail density settling during flow in a high particle concentration flow (Fig. 10.39; Sparks 1976), or inherited from the distribution of lithics in the eruption column before collapse, during which a rough vertical zonation develops, according to terminal fall velocity of solid particles (e.g. Valentine and Wohletz 1989). This is then translated into a lateral zonation in the pyroclastic flow as the column collapses and a laterally moving flow is initiated (Fig. 10.51). However, frequent multi-storey arrangements of lithic horizons point toward the dominant effect of the temporal evolution of the eruption that fed the pyroclastic flow in terms of fluxing or even late ejection from the vent, and availability of lithic clasts that are then fed into the pyroclastic flow system. For example, during the climactic eruption of Mt Pinatubo on 15 June 1991, the sudden and apparently short-lived production of abundant lithic clasts late in the emplacement of pumice and ash flow deposits suggests that vent conditions changed markedly as caldera collapse occurred near the end of the eruption (Scott et al. 1996). Low lithic contents in the underlying pumice-fall and pumiceous and ash flow deposits indicate that little vent erosion occurred during most of the eruption. Scott et al. (1996) infer that a large amount of lithic debris was created as the summit dome foundered, so entraining large amounts of dome rock. Some of the dense lithic material settled through the flows near the vent, forming lag breccia deposits (≤ 5 m thick), while some travelled farther to form lithic-rich pyroclastic flow deposits, which occur at or near the top of the pumice and ash flow deposits and mark the end of the climactic phase of the eruption.

The stratigraphic position of breccias and their componentry (shallow vs. deep origin of lithics) can help to identify processes such as vent wall collapse and widening, and caldera collapses. At the onset of an eruption, it is likely that conduit opening and clearing processes result in availability of large volumes of shallow derived lithic debris (e.g. Kieffer and Sturtevant 1984). At steady mass eruption rates, this process evolves to an equilibrium stage where volcanic jets are nearly pressure balanced (Kieffer and Sturtevant 1984) and lithic clasts become less available. However, departures from a steady discharge from the vent can promote renewal of vent erosion and vent wall collapse, particularly if conduit pressures fall and increase cyclically. Conduit wall erosion and collapse also determines the amount of lithics available. Numerical simulations suggest that erosion prevails when the pyroclast-gas mixture pressure inside the conduit exceeds the lithostatic pressure of the surrounding rocks. At lower mixture pressures the conduit walls may collapse and lithic debris may become available during subsequent pulses of the eruption energy (e.g. Dobran et al. 1993). Deep crustal fracturing during caldera collapses is a further source for availability of abundant deep-seated lithics.

10.8.10.4 Lateral Grading of Lithic Clast Sizes and Abundance

Many ignimbrites are known to show a decrease in maximum size of lithic clasts with distance from source (Fig. 10.54a–c). The distance that accessory and cognate lithic clasts (i.e. vent-derived lithics) are transported away from vent in pyroclastic flows will reflect the competence and/or the capacity of the host pyroclastic flows to transport dense lithics, and their flow dynamics (Branney and Kokelaar 2002). While the vertical superposition of breccias and lithic concentration zones, as well as the internal vertical grading of lithics in an ignimbrite, can be interpreted in a number of ways by the combination of eruption and transport controlling factors (Sparks 1976; Palladino and Valentine 1995; Branney and Kokelaar 2002), the lateral grading can be safely interpreted as being due to the dynamics of the transportation system (Giordano 1998a; Palladino and Giordano 2019). An absence of lateral grading would suggest a significant competence of the parent pyroclastic flow, such as may be the case for high particle concentration Bingham-rheology flows, characterised by internal high yield strength. By contrast, most if not all known ignimbrites show breccias and concentration zones of vent-derived lithics relatively close to vent (<20 km), and contouring of maximum vent-derived lithic clast sizes commonly produces an away from vent fining isopleth pattern (Figs. 10.54, 10.55 and 10.56; e.g. Kuno et al. 1964; Sparks 1975; Wright and Walker 1977; Bailey and Carr 1994; Streck and Grunder 1995; Giordano 1998a; Milner et al. 2003; Scarpati et al. 2015; Palladino and Giordano 2019). These demonstrate that lateral lithic distribution is largely affected by competence variations due to distance from vent and by interaction of pyroclastic flows with topography, where channelling in valleys or spreading over diverging ridges would respectively increase or decrease the flow competence.

On stratovolcanoes, the main breaks in slope of the edifice may exert a fundamental influence on the internal dynamics of pyroclastic flows (e.g. Freundt and Schmincke 1985b; Robool et al. 1987; Giordano and Dobran 1994; Giordano 1998a; Bryan et al. 1998). Ignimbrites are usually absent along the upper and steep slopes of stratovolcanos due to a combination of post-depositional erosion and initial non-deposition (i.e. by-passing because of steep slope; e.g. Sparks et al. 1997). Quaternary ignimbrites from Roccamonfina stratovolcano and from Tenerife have been reported in onlap relationship with the topography (Giordano 1998a; Bryan et al. 1998), verifying that pyroclastic flows on steep slopes can be non-depositional and fully competent to transport the sediment load, whereas on lower slopes they are prograding and aggrading. At Roccamonfina volcano, ignimbrites thicken downslope of the area of the main break of the edifice slope. At break in slope locations ignimbrites are thickest, massive and contain the maximum lithic clast sizes, and may overlie a scoured substrate (Fig. 10.56). Furthermore, they are relatively fines-depleted. The association, at the break in slope, of an erosional base to the ignimbrite, maximum lithic clast size, maximum thickness, and massive facies, have been interpreted to be a result of loss of flow capacity. This in turn leads to an increase in sedimentation rate, which results in the development of a hindered settling zone (Giordano 1998a). The fines-depletion is the result of gas segregation from the flow resulting from increased turbulence and flow expansion due to impact of the flow with the lower slope (Druitt 1995; Fig. 10.14).

10.8.10.5 Pumice Concentration Zones

Pumice concentration zones (PCZ) are frequently observed in ignimbrites, generally at the top of distinct depositional units (Layer 2bP; Figs. 10.39, 10.40b and 10.57a–d, f). Pumice clasts range from lapilli- to bomb-sized and may be matrix-supported or fines-depleted and are more common in medial and distal reaches. Clasts are generally rough rounded to well rounded, indicating that inter-particle friction occurred (Manga et al. 2011). PCZs have been interpreted to form as a result of buoyancy and shear, experienced by light and large pumice clasts which tend to roll and bounce and eventually concentrate at the sides, top, and front of pyroclastic flow deposits (Fig. 10.14). Rarely are PCZs found at the base of ignimbrites (e.g. Fig. 10.57e), and there they have been interpreted as a result of dynamic, almost explosive, flow head processes (e.g. jetted-deposits of Wilson and

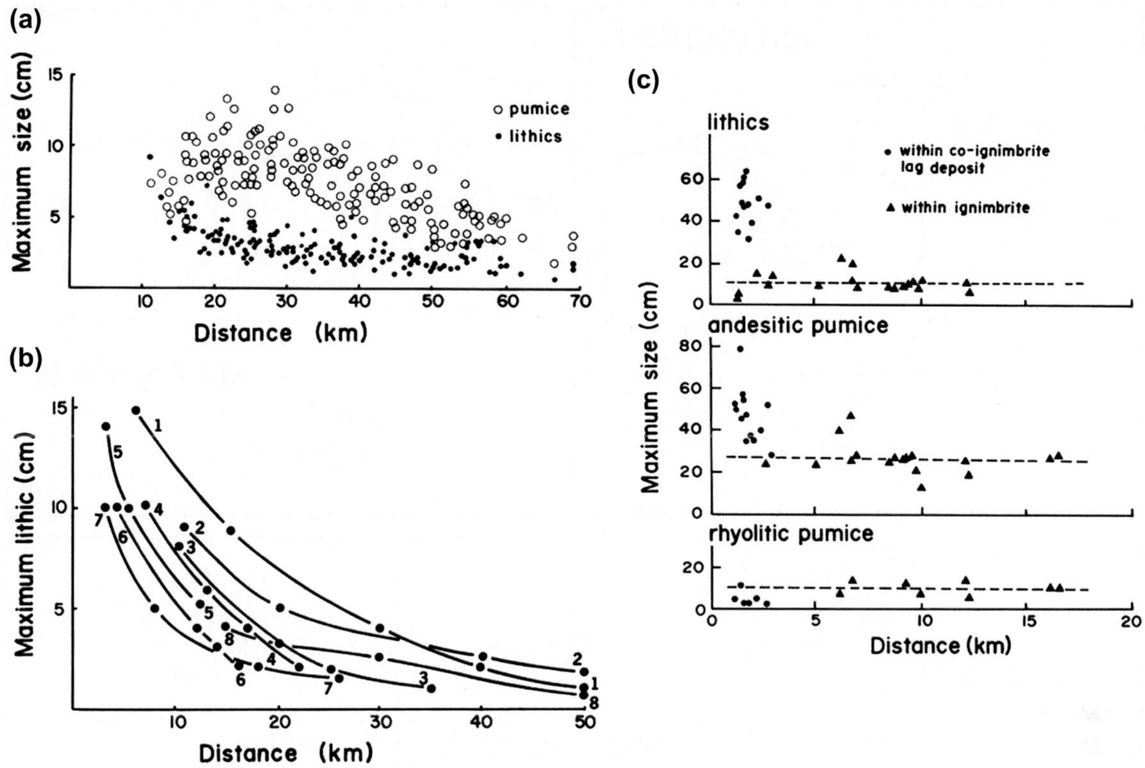

Fig. 10.54 Relationship of the average maximum lithic clast size to distance from source for several Quaternary ignimbrites (Layer 2 deposits). **a** Lateral variation in average maximum lithic and pumice size for the 22 ka Ito pyroclastic flow deposit, southern Kyushu, Japan. **b** Lateral variation in average maximum lithic size for 1 13 ka Towada pumice flow deposit; 2 Ito pyroclastic flow deposit; 3 32 ka Shikotsu pumice flow deposit; 4 ignimbrite C, Vulsini; 5 21 ka Lajes-Angra ignimbrite, Terceira; 6 ignimbrite F ($\sim$170 ka Onano ignimbrite), Vulsini; 7 ignimbrite E, Vulsini; 8 120 ka Kutcharo pumice flow deposit. **c** Average maximum size of the five largest clasts for the main, compositionally zoned, depositional unit of the Quaternary (<0.65 Ma) Acatlán ignimbrite and co-ignimbrite lag breccia, central Mexico. (From Cas and Wright 1987, After Yokoyama 1974, Sparks 1975, and Wright and Walker 1977)

Walker 1985; Cas and Wright 1987; Fig. 10.48), or alternatively as early emplacement of frontal or lateral pumice-rich flow lobes (Figs. 10.11d, e and 10.18), that are later overlain by more "standard" ignimbrite facies (e.g. Branney and Kokelaar 2002; Pittari et al. 2005; Pallister et al. 2017).

Although in some ignimbrites larger pumices may show a decrease in average maximum size from source (e.g. Fig. 10.54a), in many others no simple relationship is found, or maximum size may even increase. Palladino and Valentine (1995) and Palladino and Giordano (2019) suggest that an increase in pumice size with distance, together with development of PCZ are the result of flows that are largely in transportation mode in proximal regions, where they leave thin and coarse deposits, whereas they are highly depositional in distal regions, where they form stacks of pumice-rich lobes (Figs. 10.11d, e, 10.18, 10.21 and 10.42b).

10.8.11 Horizontal Fabric, Clast Imbrication, and Other Paleocurrent Indicators

Many ignimbrite lithofacies may show subtle evidence of an internal organisation, such as horizontal fabric defined by alignment of pumice, lithic clasts, and even mineral grains (e.g. Fig. 10.41a). In some cases, *clast imbrication* (i.e. en echelon organisation) of more or less flat and elongate lithic and pumice clasts is developed (Fig. 10.58), and in some cases, even platey minerals such as biotite (Cas et al. 2011). Imbrication is generally defined by an upstream inclination of the long or intermediate axis of the clast with respect to bedding or to the base of the ignimbrite. Clasts orientate themselves to such a position to minimise resistance to the surrounding or overpassing flow (Postma et al. 1988). Clast alignment and imbrication developed in high-concentration flows indicate the occurrence of shear at the time of deposition (Branney and Kokelaar 2002).

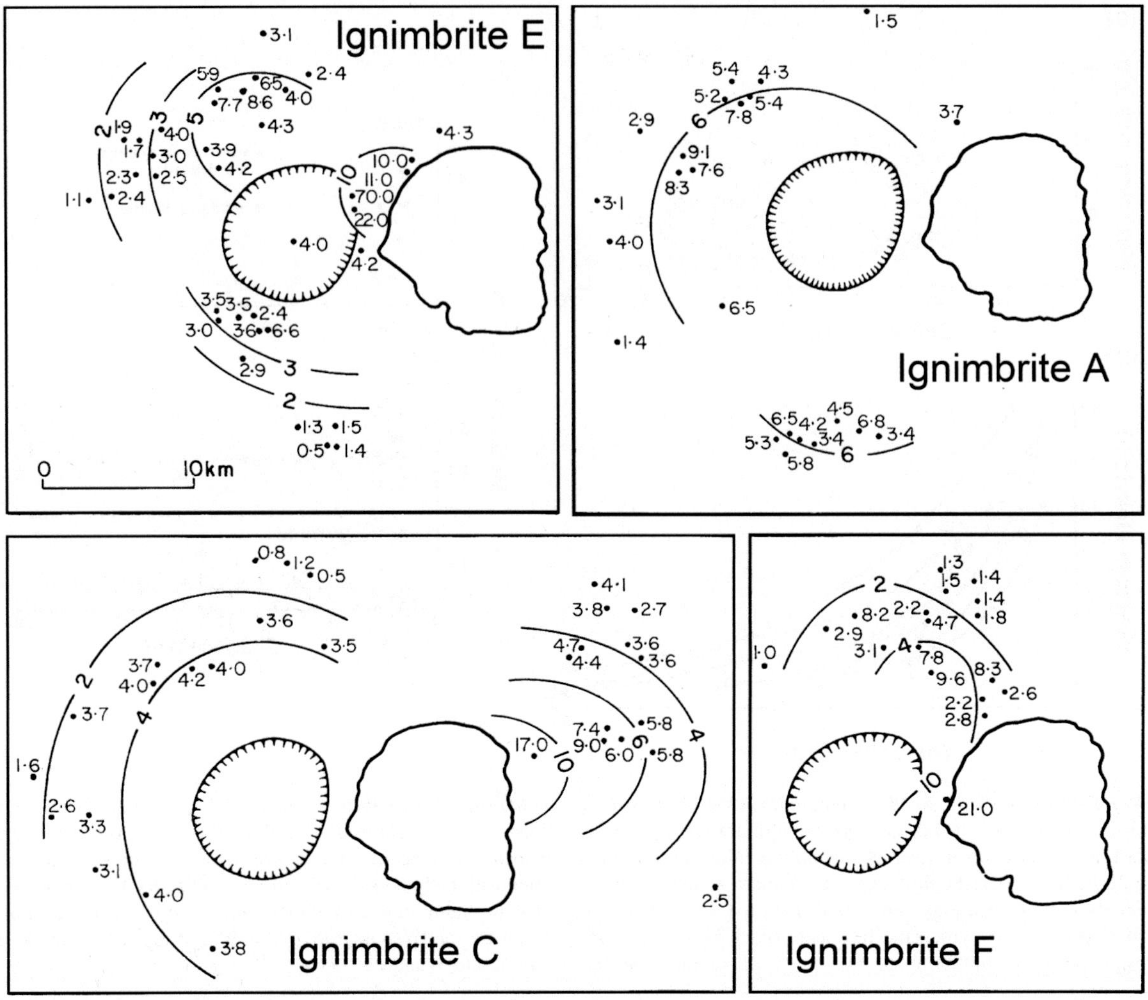

Fig. 10.55 Isopleth maps showing the average diameter of the three largest lithic clasts in four ignimbrites erupted between ~300 and 170 ka from the Vulsini Volcanic District, central Italy. Average size decreases away from Latera caldera (left with hachures) for ignimbrites A, E, and F, and Bolsena caldera (right) for ignimbrite C. (From Cas and Wright 1987, After Sparks 1975)

Clast imbrication due to the alignment of pumice or spatter clasts is also commonly observed in high grade to rheomorphic ignimbrites (e.g. Schmincke and Swanson 1967; Wolff and Wright 1981; Branney and Kokelaar 1992; Sect. 10.8.14.1). Branney and Kokelaar (1992) infer that syn-depositional agglutination and (simple) shear deformation are important processes in the development of imbricated, highly attenuated fiamme in high-grade ignimbrite sheets (cf. Wolff and Tuberville 1994).

Flow direction has been determined for a number of ignimbrites by measuring the preferred orientation of elongate crystals and lithic fragments (Suzuki and Ui 1982), tree logs (Froggatt et al. 1981), pumice, lithic clast, and crystal fragment imbrication (Mimura 1984), as well as sole structures (groove marks) on the base of ignimbrite deposits (Pittari and Cas 2004), welding fabrics (Elston and Smith 1970; Rhodes and Smith 1972; Kamata and Mimura 1983), and magnetic fabrics (see next section). Suzuki and Ui (1982) have shown that flow orientation was strongly controlled by the pre-existing topography for the Ata pyroclastic flow deposit. Flow lineations oriented radially away from source were only found in samples collected from the surface of the ignimbrite sheet. Samples collected from the bottoms of valley-fill deposits had lineations parallel to the axis of the valleys. Samples collected from valley walls tended to be parallel to the slope of the valley. Froggatt et al. (1981) also noted that in the Taupo ignimbrite the orientation of tree logs in areas far from the source was parallel to valley axes, even though these could be perpendicular to the general outflow direction.

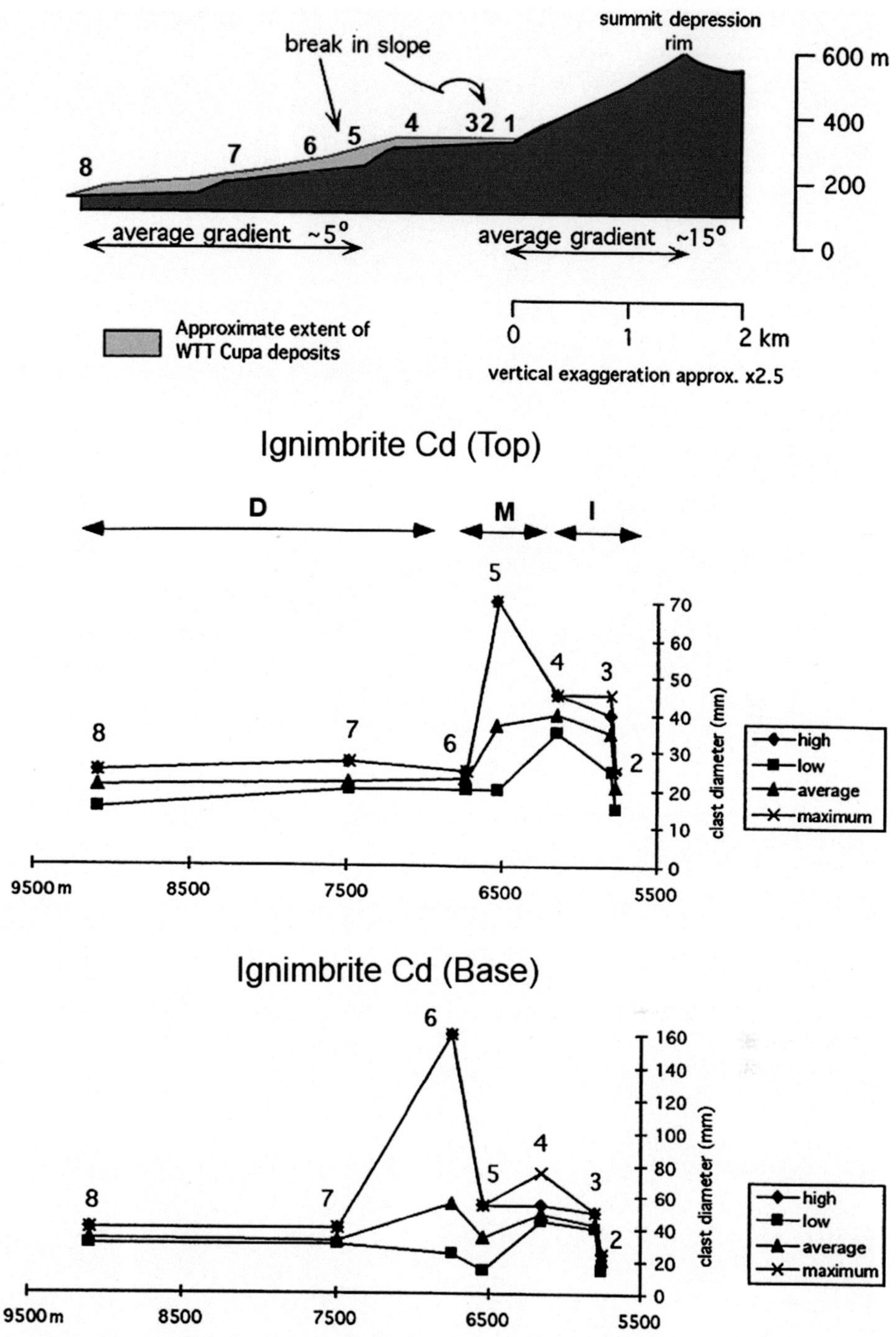

Fig. 10.56 Lateral grading of lithics in the 300 ka WTT ("White Trachytic Tuff") Cupa ignimbrite (unit Cd), Roccamonfina volcano, southern Italy. This shows (top) cross-section and deposit thickness across the volcano break in slope, and (below) variations of lithic clasts size (maximum, minimum, and average dimension of largest 5 for each locality 2–8; unit Cd is not present at locality 1) for the top and base of the same depositional unit. Note that lithic clast sizes increase downslope (I) to a maximum (M), then laterally rapidly decline to values that are then more or less maintained (D). The M site is also where the unit is thickest and fines-depleted, suggesting that together with increased deposition, the upward flux of gas and fine particles promoted the development of co-ignimbrite ash clouds (cf. Fig. 10.14). (From Giordano 1998a)

10.8.12 Anisotropy of Magnetic Susceptibility (AMS) Fabrics

Anisotropy of magnetic susceptibility (AMS) is an assessment of the presence or absence of a preferred orientation of the magnetic fabric of tiny ferro- and para-magnetic mineral grains within a deposit, in this case in the matrix of pyroclastic flow deposits (Fig. 10.59). The ash matrix of ignimbrites often preserves an AMS fabric. The direct comparison between the AMS and petrofabric verifies that AMS reflects the microscopic physical fabric of magnetic minerals, hence alignment or imbrication preserved by the ash components, usually not visible at the mesoscopic scale (Plenier et al. 2005). AMS studies show that imbrication of magnetic particles is developed in many ignimbrites, regardless of the degree of welding, and is a pre-compaction, syn-depositional feature (Wolff and Turbeville 1994). In ignimbrites, the AMS method was originally applied to infer the source location of a deposit and determine flow direction (Ellwood 1982; Hillhouse and Wells 1991; Incoronato et al.

Fig. 10.57 Different types of pumice concentration in ignimbrites. Different types of pumice concentration in ignimbrites. **a** Pumice concentration zone (Layer 2bP) at the top of a thin pumice and ash ignimbrite, with a 5 cm basal frontal ash-cloud surge deposit, and a Layer 2a. 1993 Lascar eruption sequence, Chile. **b** Coarse-tail reverse graded pumice concentration zone at top of 2 m thick ignimbrite depositional unit, 600 ka Abades Ignimbrite, Los Abrigos, Tenerife Canary Islands, Spain. **c** Pumice concentration zones (PCZs) at the top of two ignimbrite depositional units, 4.7 Ma Real Grande ignimbrite, Las Pitas, northwestern Argentina. People circled for scale. **d** Succession of thin, trachytic ignimbrite depositional units, each with fine-grained bases and PCZs marking their tops. Pumice clasts are rough rounded. 8–12 ka Ribeira Chà, eruption sequence, Agua de Pau, São Miguel, Azores, Portugal. **e** Pumice "dunes" deposited between the base of the 0.76 Ma Bishop Tuff ignimbrite (salmon pink colouration) and stratified plinian fallout deposit, Long Valley, California, USA. **f** Well-developed PZCs on the tops of two sequential depositional units of the 1600 BC Minoan ignimbrite, Santorini. Grain size studies indicate these were emplaced with high gas flow rates, in keeping with the extreme sorting of pumice and fines depletion

Fig. 10.58 Well-defined imbrication of platey lithic clasts in the ~650 ka Arico ignimbrite, Poris, Tenerife, Canary Islands, Spain. Flow direction was from left to right

1983; Knight et al. 1986; MacDonald and Palmer 1990; Ort 1993; Ort et al. 2003, 1999; Seaman et al. 1991; Le Pennec 2000). Later applications demonstrated that AMS can also retrieve information about transport and depositional processes (Figs. 10.60 and 10.61; Fisher et al. 1993; Le Pennec et al. 1998; Porreca et al. 2003; Gurioli et al. 2007; Giordano et al. 2008a, b; LaBerge et al. 2009; Cas et al. 2011; Willcock et al. 2015a), including relationships between flow directions and topographic obstacles in the flow path, degree of turbulence of the flow, and coupling between the transport and the depositional system (Baer et al. 1997; LaBerge et al. 2009).

The frequent preservation of a sub-horizontal AMS fabric defined by minute magnetic minerals is remarkable, defining a sub-horizontal alignment of some of the smallest particles in a pyroclastic flow. Such a pervasive fabric is interpreted to represent high shear stress at the flow boundary and a broad laminar shear flow state during deposition. The depth of this laminar flow zone is unknown and could vary from a thin basal "boundary layer" in highly expanded, high velocity pyroclastic flows or the proximal, initial parts of flows, to a thick laminar flow zone encompassing the entire body of the flow, kept mobile by trapped gas, in waning "highly depositional" low velocity granular flow (e.g. at distal settings; Cas et al. 2011).

10.8.13 Compositional Zoning

Many ignimbrite sheets show progressive vertical changes in mineral and chemical composition, representing the inverted sequence of magma erupted from zoned chambers (e.g. the 7.7 ka Mt Mazama ignimbrite, Crater Lake caldera; Fig. 10.62). Wright and Walker (1977, 1981) described

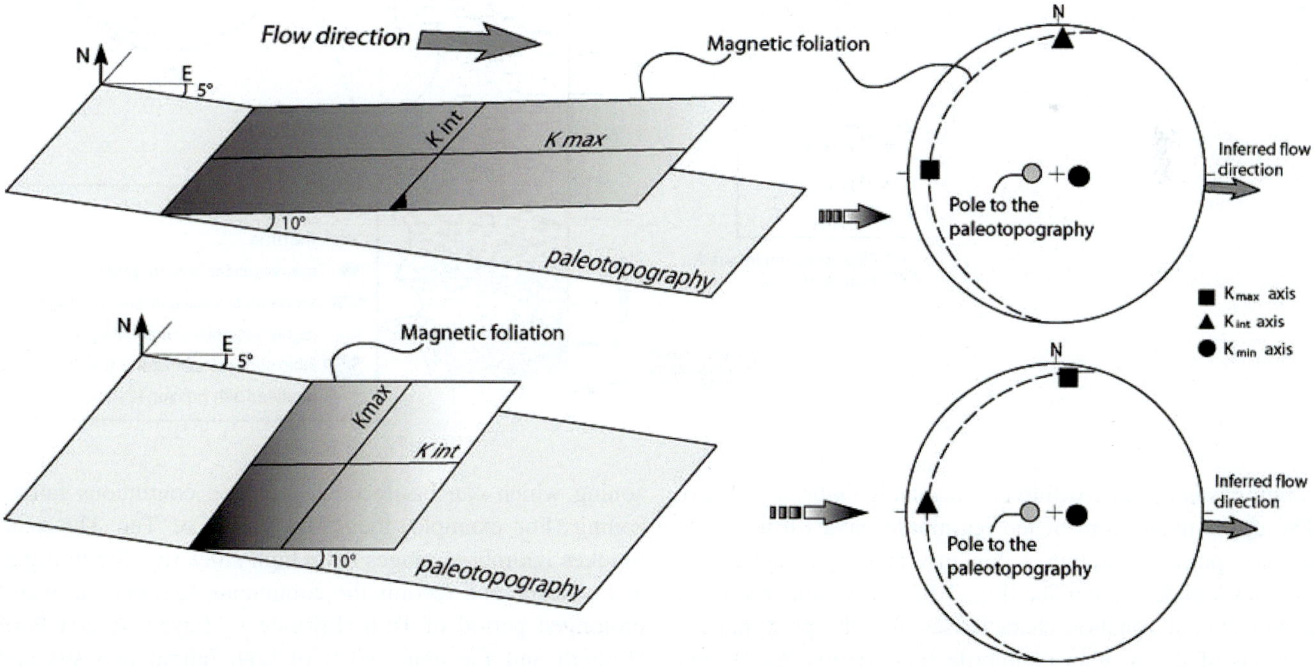

Fig. 10.59 Schematic representation of AMS fabric showing the low angle imbrication of the magnetic fabric orientation plane that contains the K_{max} and K_{min} magnetic component vectors, produced by alignment of the magnetic mineral grains relative to the paleotopographic depositional surface. Magnetic foliation can be imbricated with K_{max} axis oriented parallel (upper case) or orthogonal to the flow direction (lower case). Corresponding stereoplots of AMS axes show the same orientation of magnetic foliation and inferred flow direction in both cases. AMS is sensitive indicator of fabric elements that yield a primary flow azimuth irrespective of welding. (From Giordano et al. 2008b)

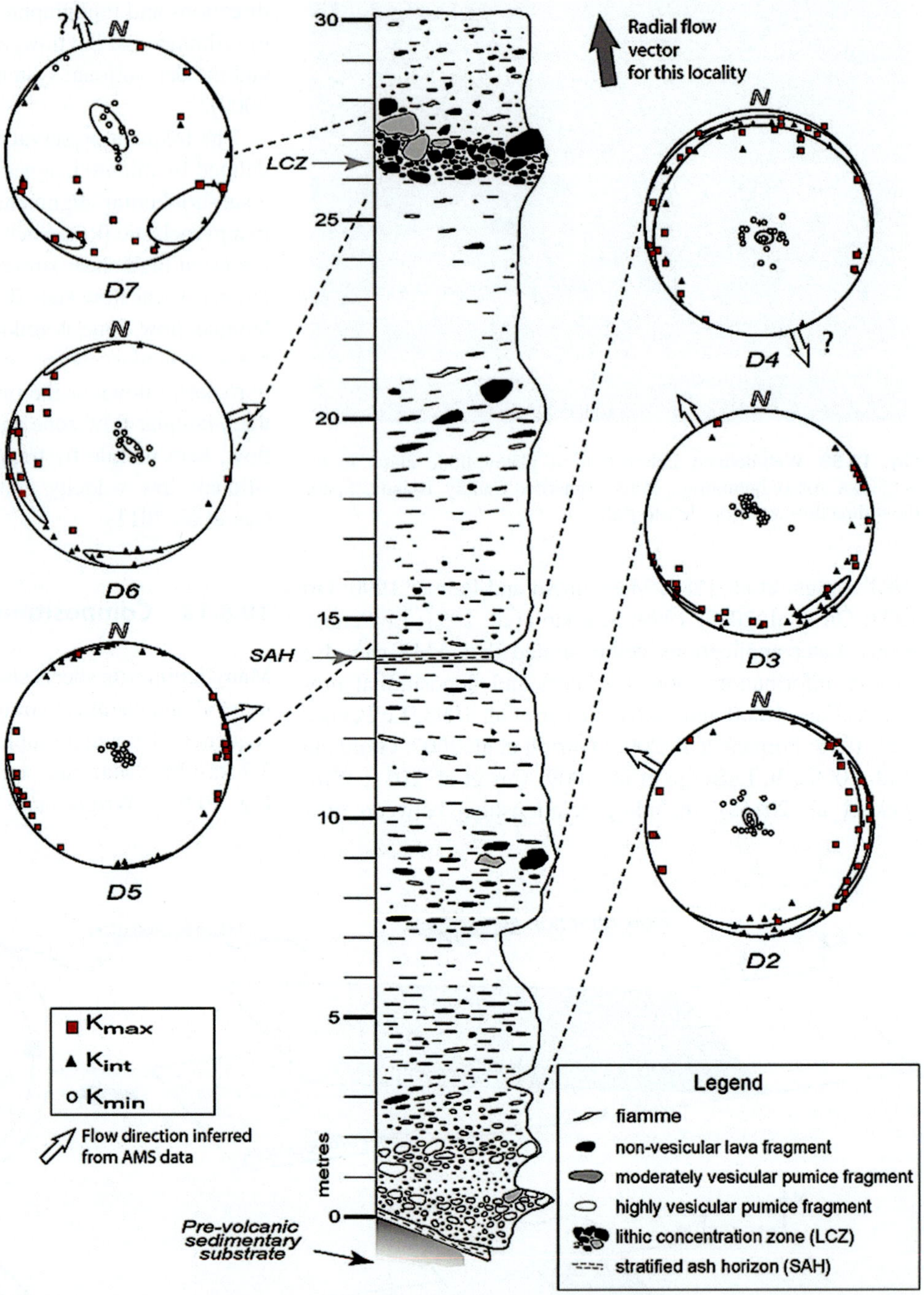

Fig. 10.60 AMS flow orientations through Pleistocene Monte Cimino ignimbrites, Cimini dome complex, central Italy. This shows that the flow direction has meandered through the flow event, and also in this case that deposition occurred progressively through aggradation. (From LaBerge et al. 2008)

compositional zoning within the Acatlán ignimbrite, Mexico. The upper main 30 m of the ignimbrite passes from white rhyolite pumice, through a thin mix zone (passage zone), into black andesitic pumice (Fig. 10.54c). A similar vertical compositional zonation characterises also the proximal lag breccias of the Acatlán ignimbrite (Fig. 10.50a, b). Wright and Walker (1977) suggested that the Acatlán ignimbrite records the lateral preservation of the vertical compositional zonation developed at proximal locations in a plug flow. However, several other ignimbrites show compositional zoning which can be traced laterally to continuous fallout zoning. For example, the 1912 Valley of Ten Thousand Smokes ignimbrite ranges from high-silica rhyolite to dacite and andesite and records the continuous aggradation over a prolonged period of 16 h (Episode I—Layers A and B of Hildreth and Fierstein 2012) of both fallout deposits and ignimbrite with similar compositional zoning. This indicates that vertical zoning is much more likely to record the temporal variations of erupted magma (e.g. Mt Mazama ignimbrite, Bacon and Druitt 1988; Druitt and Bacon 1989;

Fig. 10.61 Map of the preserved extent of 2.08 Ma Cerro Galán ignimbrite, northwestern Argentina, showing directions of pyroclastic flow (arrows) defined by AMS fabrics, which overall define a radial paleoflow pattern around the Cerro Galán Caldera. White arrows represent samples from the base of the ignimbrite; light grey arrows correspond to middle heights; dark grey arrows indicate samples from the top of the ignimbrite. In some places, the pyroclastic flow meandered as it flowed. Numbers are locality numbers. Note also the well-defined, current topographic rim of the caldera (white line), the structurally interpreted caldera shape (dashed line) and the topographically well-defined resurgent dome in the centre. (From Cas et al. 2011; AMS fabrics measured by C. Lesti)

Bishop Tuff, Hildreth and Wilson 2007), hence supporting progressive or incremental aggradation of the deposits rather than a plug flow-emplacement (Branney and Kokelaar 1997).

Common reverse compositional zoning, i.e. the upward transition within an ignimbrite to more mafic products, underpin models of magma withdrawal from zoned and density stratified magma chambers (e.g. Spera et al. 1986; Druitt and Bacon 1986). However, many ignimbrites show quite complex zoning patterns, including early eruption of mafic magmas (e.g. the Cape Riva eruption, Santorini; Druitt et al. 1999), or mingling and mixing of different magma types (e.g. Valley of Ten Thousand Smokes ignimbrite, Alaska; Hildreth and Fierstein 2012; the mafic P1 ignimbrite, Gran Canaria; Freundt and Schmincke 1995a, b), and double zonation (e.g. Zaragoza ignimbrite, Mexico; Carrasco-Núñez and Branney 2005), suggesting that perhaps the standard model for some magma chambers is too simplistic and that more complex magma chamber, or reservoir, architecture is required (e.g. Cashman and Giordano 2014; see discussions in Chap. 8, Sect. 8.4.3, and Chap. 16, Sect. 16.7.3).

10.8.14 Emplacement Temperature

Pyroclastic flow processes and deposits record the exchange of the initial thermal energy involved in explosive, ignimbrite forming eruptions with the environment. A complete understanding of thermal exchange paths from pre-eruption magmatic temperatures, through fragmentation processes (±magma–water interaction), conduit processes, eruption,

Fig. 10.62 7.7 ka Mt Mazama ignimbrite, showing spectacular vertical compositional zonation from rhyodacite at the base to andesite at the top. Note also the spectacular columns and spires that rise from the upper part of the canyon walls described as fossil fumaroles by Williams (1942; see Sect. 10.8.14.3). The Pinnacles, Crater Lake, Oregon, USA

air ingestion through the plume and pyroclastic flow processes and post-depositional cooling is still to be fully achieved.

Eruption temperatures span several hundreds of degrees, from 1200 °C down to 650 °C, depending on melt chemistry and volatile content (Chap. 2), or lower in the case of phreatomagmatism (Chap. 4, Sect. 4.6). Eruption styles determine the grain size distribution of juvenile clasts, which control the column dynamics (cf. Chap. 9) and the rate of heat loss from each particle.

A significant amount of heat can be dispersed through eruption column processes, including, the height of the column and amount of air entrainment, as well as the height from which column collapse occurs. Trolese et al. (2019) show that the initial temperature of pyroclastic flows strongly depends on the style of eruption column collapse. Bulk flow temperature may drop by up to 55% with respect to magmatic temperature at vent, due to the pre-collapse degree of air ingestion in the eruption column (Fig. 10.7). They suggested that the welding "*window*" for ignimbrites, i.e. for deposition above the glass transition temperature (T_g; Chaps. 2 and 4), relates to the percentage of mass collapsing from the column into the flow which should be above 50%, which is more likely for continuous boil-over eruptions from ring fractures during caldera collapses (cf. Jessop et al. 2016). Welding, as a consequence, is much more common in large volume caldera-forming ignimbrites with respect to small volume ignimbrites derived from point-sourced partial to total column collapses (Trolese et al. 2019; Table 10.2).

Direct thermocouple measurements of pdc temperatures undertaken immediately after the several eye-witnessed events of column collapse occurred in 1980 at Mt St Helens (18 May, 25 May, 12 June, 22 July, 7 August, and 17 October 1980) confirm that the style of column collapse is the main control over the pdc temperature. In particular, the less the air entrainment, then the higher the pdc temperature (Banks and Hoblitt 1981, 1996).

Once formed, other ways for the flow to locally exchange heat and lose temperature are the rapid entrainment of air on steep slopes of stratovolcanoes (e.g. Doronzo et al. 2012; Trolese et al. 2018, 2024), and the ingestion of ambient water, vegetation and accidental lithics (e.g. McClelland et al. 2004; Zanella et al. 2007; Giordano et al. 2018). Similarly, shallow derived and "cold" accessory lithic clasts can significantly lower the bulk temperature of pyroclastic flows and act as a heat sink during and after deposition (Banks and Hoblitt 1996; Martí et al. 1991). Sedimentation processes and rates, and the final thickness of the deposit also bear a significant influence on the thermal history of the deposit (Doronzo et al. 2016). The complex combination of the processes outlined above causes a huge variability in the temperature of emplacement of ignimbrites, even within a single eruption sequence (e.g. the 4.6 ka Fogo A eruption; Pensa et al. 2015c), in turn influencing significant syn- and post-depositional processes. Accordingly, ignimbrites range from completely welded to partially welded to non-welded (Table 10.2; next section). Ignimbrites may also undergo vapour-phase crystallisation as a result of their emplacement temperature and percolation of hot gases and be lithified and develop columnar jointing even though they are not "strictly" welded. Such rocks are sometimes termed "sillar" (Sect. 10.8.14.3).

Many studies on the temperature of emplacement of pdc deposits ranging from large volume ignimbrites to small block-and-ash flow deposits show that pyroclastic flow deposits conserve heat from proximal to distal settings (e.g. Banks and Hoblitt 1996; Lesti et al. 2011; Benage et al. 2014; Trolese et al. 2017, 2018; Pensa et al. 2018, 2019). This suggests that air mixing along a flow is limited to the outer turbulent ash clouds that envelope the pyroclastic currents, only weakly affecting its high concentration interior (e.g. Benage et al. 2016). Doronzo et al. (2016) propose that particle concentration controls the temperature of the multi-phase flow of the high concentration pdc interior, and that the sedimentation rate controls the heat transfer both into the deposit and into the flow. Hence the deposit temperature may provide an indication of the actual flow dynamics at the time of deposition, provided that most of the particles are thermally in equilibrium with the flow. This is more likely in fine ash-dominated flows, with the predominant small size of the particles allowing continuous and efficient heat transfer from the particles to the gas phase as it mixes with entrained air and cools (e.g. Thomas and Sparks 1992). By contrast, in block and ash flows, the dominant large clasts will retain most of the heat irrespective of the actual flow process and will produce a deposit temperature

Fig. 10.63 Mesoscopic and microscopic welding textures in ignimbrites. **a** Sketch of cut blocks of the 0.76 Ma Bishop Tuff, California, USA, showing flattening of pumice clasts (fiamme) with decreasing porosity, and increasing compactional strain (R_s) (From Cas and Wright 1987, After Ragan and Sheridan 1972). **b** Outcrop showing transition from the densely welded interior (massive black bottom zone with horizontal fiamme fabric) to the weakly welded top, with equant, dark, phonolitic pumice clasts visible ~0.65 Ma Arico ignimbrite, Tenerife, Canary Islands, Spain. **c** Slabbed surface of weakly welded Arico ignimbrite showing variably compacted, black phonolitic scoria and spatter clasts, some still preserving vesicles. **d** Outcrop showing eutaxitic texture defined by plastically flattened pumice fragments (dark fiamme) which form a planar welding foliation, 4 Ma Atana ignimbrite, northern Chile. **e** Outcrop of moderately welded ignimbrite erupted from Platoro caldera, part of ~29 Ma Treasure Mountain Tuff, San Juan Volcanic Field, Colorado, USA. **f** Slabbed surface of moderately densely welded, lithic-rich, Early Devonian, Snowy River ignimbrite, eastern Victoria, Australia. **g** Outcrop of very densely welded ignimbrite showing strongly attenuated fiamme (parataxitic texture), La Piedad, Central Mexico. **h** Thin section of moderately densely welded, 230 ka Kaingaroa ignimbrite, Taupo Volcanic Zone, New Zealand, showing weakly plastically deformed bubble wall shards in the matrix, to strongly deformed shards around rigid crystal fragments; ppl, field of view ~2 mm across. **i** Thin section of densely welded 7 Ma Rattlesnake Tuff, eastern Oregon, USA, showing strong alignment of shards and collapsed bubble walls; ppl, field of view ~3 mm. **j** Extremely densely welded ignimbrite, showing relict shard welding textures and foliation that verges on being rheomorphic flow layering, as well as spectacular perlitic cracks. Basal vitrophyre zone, 274 Ma intra-caldera Ora ignimbrite, Bolzano, northern Italy; also see Fig. 10.66. (Photomicrograph by M. Willcock; ppl, field of view ~2 mm)

that may be out of thermal equilibrium (e.g. Pensa et al. 2018, 2019).

A very important outcome of recent thermal studies is that the upward heat flux from the basal high-concentration flow can maintain the overriding ash-cloud surge at temperatures as high as the undercurrent. At Colima, Pensa et al. (2019) estimated from charcoalised bark of trees standing out of the 2015 block-and-ash flow deposit (see Fig. 10.99k) that the ash cloud had temperatures of more than 400 °C. In that case the ash cloud did not detach from the undercurrent, but other cases are known where at valley bends, the ash-cloud surge detached and moved independently across topography. Due to their low concentration and ash-grain size, ash-cloud surges rapidly wane and cool. Nevertheless, their associated hazard is poorly understood and much more related to their temperature impact with respect to their

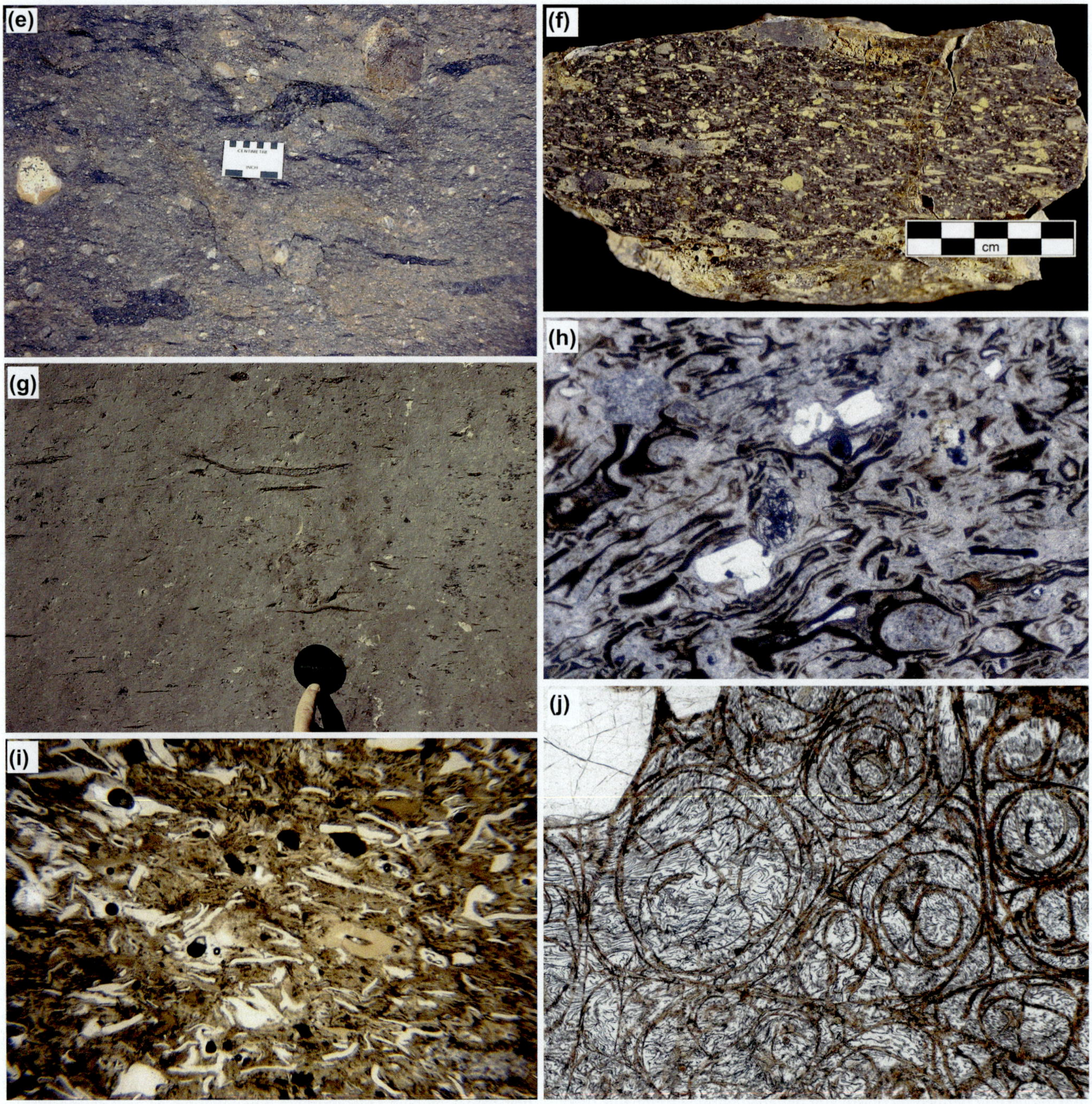

Fig. 10.63 (continued)

dynamic pressure. A sad and deadly example of such thermal impact of ash cloud surges is the 3 June 1991 event at Mt Unzen (Japan), when 43 people were killed including three volcanologists (Yamamoto et al. 1993; Fujii and Nakada 1999, see Figs. 10.4i and 10.12c). Other examples include the 2018 Fuego (Guatemala; Risica et al. 2022) pyroclastic flows and possibly November 2010, high energy flow at Mt Merapi (Indonesia; Trolese et al. 2018). The recent unique finding of a vitrified human brain at Herculaneum below the pyroclastic flow deposits of the 79 AD eruption of Vesuvius (Petrone et al. 2020a, b; see Fig. 10.74c) also indicates that the body of the victim must have been first hit by a detached hot and dilute ash cloud surge (Pensa et al. 2023), which quickly vanished leaving just a few centimetres of ash on ground, allowing a very high cooling rate able to vitrify the human brain, and this before the later burial of the entire town by much thicker ignimbrites.

10.8.14.1 Welding in Ignimbrites

Welding is the sintering together of hot juvenile plastic or "glassy" fragments ranging in size from bomb- to micron-sized glass shards (e.g. Smith 1960a, b; Ross and Smith 1961; Smith and Bailey 1966; Cas and Wright 1987; Branney and Kokelaar 1992; Freundt 1998; Sparks et al. 1999; Quane and Russell 2005; Pioli and Rossi 2005; Willcock and Cas 2014). Welding is recognisable at mesoscopic outcrop scale through the presence of plastically flattened pumice clasts, called "*fiamme*" that define a welding foliation which can be used (with some care) as a measure of the dip and strike of the ignimbrite (Fig. 10.63a–g; also see Fig. 10.82d). Although many ignimbrites are completely non-welded (e.g. Figs. 10.35c, d, g, 10.38a–d, 10.41, 10.52, 10.53 and 10.57a, b, f), others may be partially welded, involving larger juvenile fragments and ash matrix in the middle being welded, whereas the top and bottom remain non-welded because they readily lose heat to the atmosphere and substrate (see Figs. 10.64 and 10.65), and some are almost pervasively welded from top to bottom, involving all the ash fraction.

Controls on Welding

Welding occurs if juvenile pyroclastic fragments are still above the glass transition temperature (T_g; Giordano et al. 2005; Russell and Quane 2005; Chaps. 2 and 4) when they are transported and deposited. They can then deform plastically and sinter or fuse together. As indicated above, plastically deformed lapilli- to bomb-sized fragments form stretched and flattened glassy juvenile clasts (usually originally highly vesiculated pumice fragments) or fiamme (Fig. 10.63a, d–g), and/or spatter clasts, which may partially preserve, or not, some evidence of their original vesicularity (Fig. 10.63b, c). At the microscopic scale, tiny glass shards are plastically deformed and annealed during welding, to varying degrees, from weakly welded to intensely welded (Fig. 10.63h–j). Most intense welding produces a dense, obsidian-like, vitrophyric rock comprising completely annealed glassy fragments (Fig. 10.63j; also see Figs. 10.66 and 10.67). Both macroscopic and microscopic welding fabrics are called *eutaxitic texture* and can be preserved in both very young and very old ignimbrites (Fig. 10.63).

The most important controls on welding are as follows:

- juvenile pyroclast rheology (dependent on eruption temperature, composition, vesicularity, and water content),
- eruption and transport styles,
- deposition rate,
- lithostatic load (dependent on the thickness of the deposit), and
- lithic content in a deposit.

If magma viscosity is low, due to composition and/or high temperature, pyroclasts may be prone to weld or agglutinate. If eruption columns are high before they collapse, they can ingest cold atmospheric air, leading to cooling of the column. Similarly, turbulent pyroclastic density currents are likely to ingest cold air at their margins, leading to internal cooling. Rapid rates of deposition will inhibit loss of heat by pyroclasts, and significant deposit thickness, although not essential, may cause compaction and flattening of pyroclasts. Marti et al. (1991) demonstrated that lithic clasts act as heat sinks in a pyroclastic system and deposit, so causing juvenile clasts such as pumice and vitric ash to cool above the glass transition temperature, T_g. For example, ignimbrites on Tenerife contain significant volumes of lithic lasts, which may have contributed to the few welded ignimbrites there.

The parameter that controls the rheology and the ability of pyroclasts to flow, sinter and agglutinate is T_g, the glass transition temperature (or temperature interval) which can be defined kinetically or thermodynamically. Kinetically, T_g defines the temperature boundary above which juvenile pyroclasts will deform viscously, whereas below which, they will behave elastically when subjected to an applied stress (Giordano et al. 2005 2008a, b; Chaps. 2 and 4), which is usually the lithostatic load exerted by a deposit, or the shear stress during deposition. T_g varies significantly for different melt compositions, but also depends on cooling rate, crystal and bubble content, H_2O content and applied stress rate. Even for a given composition, T_g varies according to the cooling rate. However, for all compositions, T_g occurs when viscosity reaches a threshold of 10^{12-13} Pa s during cooling.

Experimental and theoretical studies indicate that minimum welding temperatures range from about 600–750 °C for rhyolitic compositions, depending on load pressure and H_2O content of the glass (Friedman et al. 1963; Bierwirth 1982; Bindeman and Valley 2003; Grunder et al. 2005; Giordano et al. 2005; Quane and Russell 2005; Russell and Quane 2005; Fig. 10.64). Welding experiments conducted on rhyolitic ash (<2 mm) from the Upper Bandelier ignimbrite (Bierwirth 1982) showed for an ignimbrite 100 m thick emplaced at 850 °C, dense welding (<20% porosity) would occur within one week (Fig. 10.64a). For even a 1 km thick intra-caldera ignimbrite at 650 °C, dense welding would only be achieved after one year, assuming that crystallisation of the groundmass did not prevent compaction. Welding temperatures were shown to be lowered by the effect of H_2O content in the glass, with 0.5 wt% water lowering welding temperatures by $\sim$60 °C compared with anhydrous conditions (Fig. 10.64b).

The results of reverse conductive thermal modelling of drill core samples from a 20 m section of incipiently to welded Upper Bandelier ignimbrite are shown in

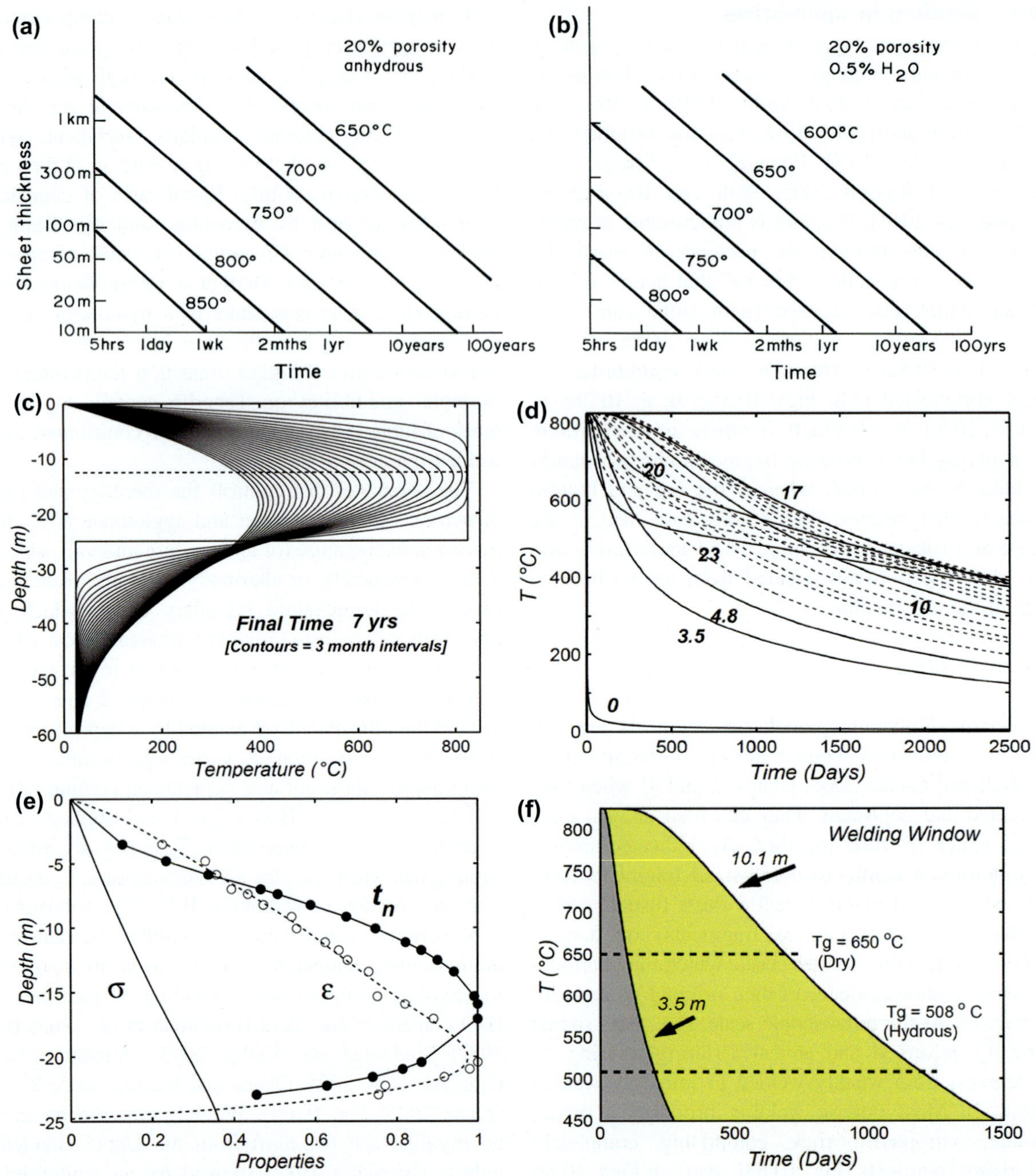

Fig. 10.64 Results of experimental and theoretical studies of welding conducted on the 1.26 Ma Upper Bandelier ignimbrite, New Mexico, USA, New Mexico, USA. Experimentally derived model of time to produce dense welding for **a** anhydrous and **b** with 0.5 wt % H_2O for various ignimbrite thicknesses and emplacement temperatures. The model is least reliable for thin flows and long times, due to the effects of cooling. Sheet thickness was calculated for porous ash (1 bar = 7 m). (From Cas and Wright 1987, After Bierwirth 1982). **c**–**f** Simplified conductive cooling model derived from 20 drill core samples collected at ~1 m intervals from which various physical parameters were measured, including density, porosity, and strain. **c** Temperature distributions for a 25 m thick deposit are shown during first 7 years of cooling at 3-month intervals. **d** Model cooling curves (T°C vs. time) for specific sample depths; labels are sample depths associated with heavy lines. **e** Summary of relationships between stratigraphic load (σ, MPa), strain recorded by individual samples (Ɛ), and residence time (tn) at T°C > T_g (glass transition temperature). **f** Model cooling curves predicted for samples at depths 3.5 and 10.1 m. The area under each curve between the original emplacement temperature (Te) and T_g is a measure of the residence time at temperatures consistent with welding (i.e. welding window). (From Russell and Quane 2005)

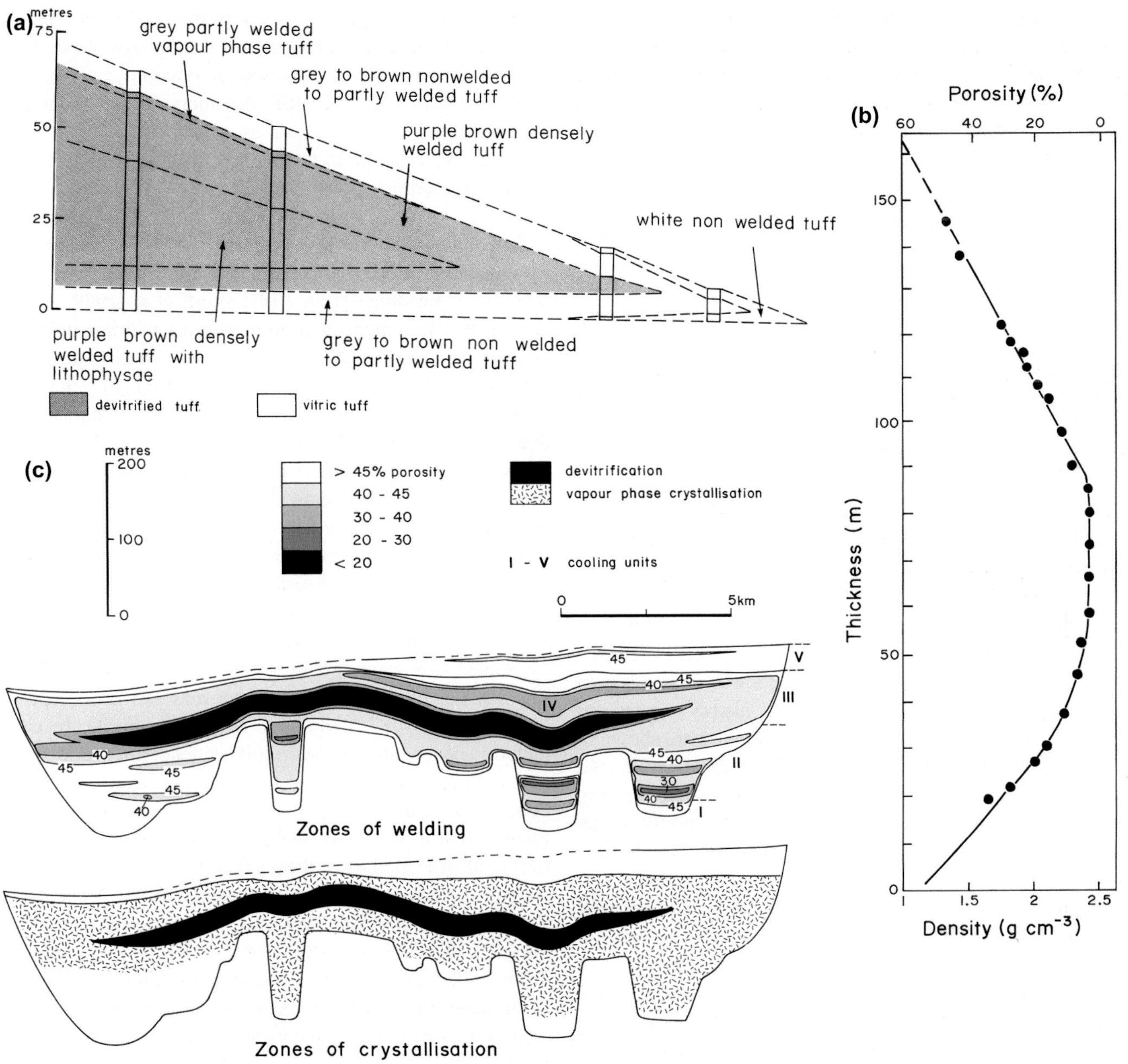

Fig. 10.65 Cooling units and zonal patterns in ignimbrites. **a** Schematic cross-section based on four measured sections of the ~12 Ma Yucca Mountain Tuff, Nevada, USA, showing simple pattern of welding and devitrification zones in an ignimbrite; length of section illustrated is 6 km (From Cas and Wright 1987, After Lipman and Christiansen 1964). **b** Simple ignimbrite cooling unit as exemplified by the 0.76 Ma Bishop Tuff, California, showing density-porosity profile (From Cas and Wright 1987, After Ragan and Sheridan 1972). **c** Cross-section through the 1.26 Ma Upper Bandelier Tuff, southern Jemez Plateau, New Mexico (see Fig. 10.22d, e), showing complex welding and crystallisation zones. The ignimbrite is a compound cooling unit and shows an overall upward increase in the degree of welding (porosity reduction) in cooling units I–III. Recognisable depositional units are much thinner in units IV and V, and nearer the source they pass into densely welded tuff continuing the trend towards higher temperature of emplacement of successive pumice flows. Note the topography that the ignimbrite fills in is cut into older ignimbrites (including the 1.6 Ma Lower Bandelier Tuff), volcanics and basement as old as Precambrian; cross-section is approximately normal to movement direction of the pumice flows, towards the reader. (From Cas and Wright 1987, After Smith and Bailey 1966)

Fig. 10.64b–d (Russell and Quane 2005). Samples were taken at approximately 1 m intervals and used as proxies for strain. The observed strain ($\mathcal{E}_T$) is ascribed to a combination of a time-dependent viscous compaction ($\mathcal{E}_v$) and a time-independent mechanical compaction ($\mathcal{E}_m$) described by

$$\varepsilon_T = \frac{(1-\phi_o)}{\alpha}\ln\left\{1+\frac{\alpha\,\sigma\,\Delta t}{\eta_o(1-\phi_o)}\exp\frac{\alpha\,\phi_o}{(1-\phi_o)}\right\}+\frac{\sigma}{E\,(1-\phi_o)} \tag{10.11}$$

where ϕ_o is the original porosity, σ is load, and η_o and E are the viscosity and Young's modulus of the deposit at zero porosity, respectively. The quantity α is an experimentally determined parameter used to relate the viscosity and porosity of porous aggregates. Simple conductive heat transfer models were used to generate cooling curves for individual samples (Russell and Quane 2005).

In Fig. 10.64c, transient temperature distribution is shown for a total time of 7 yr by plotting the temperature distribution in the ignimbrite and the underlying ground at increments of 3 months. Within three years the hottest portion of the body has cooled below 0.5 of the initial temperature (T_e). Natural convection on the upper boundary surface, relative to the base causes the asymmetric temperature distribution. The model shows the lower contact is held near 400 °C for about 3 yr. Cooling history curves for depths corresponding to each of the 20 samples are shown in Fig. 10.64d. These predict the time each sample would take to cool from Te to some critical temperature where welding and compaction cease (i.e. T_g). The total time and the actual path taken are dependent on sample depth. In Fig. 10.64e, relationships are shown between stratigraphic load (σ), strain recorded by individual samples ($\mathcal{E}$), and residence time (tn) at T (°C) > T_g. Stratigraphic load increases nearly linearly with depth whereas residence at temperatures > T_g peaks two-thirds of the way down the section, and maximum strain is found 80–90% down the section. The position of peak values of these properties therefore need not be coincident. Figure 10.64f illustrates the concept of residence time spent in the welding "window", i.e. the relative time each sample resides at conditions suitable for welding. The value of T_g chosen (hydrous versus anhydrous) causes up to an order of magnitude variation in this parameter.

Timing of Welding and Formation of Welded Textures

It is now apparent, given the large range of T_g for a given magma type, welding and the formation of welded textures of an ignimbrite can occur during different stages of its evolution:

- pre-depositionally, such as within the conduit or eruption column before it collapses, and during pyroclastic flow through a process of in-flow agglutination,
- syn-depositionally, at the flow-depositional boundary layer, or
- following deposition, under the effects of gravity and/or deposit load.

There are a number of examples of vents or conduits that have been described that contain densely welded ignimbritic rocks (e.g. Reedman et al. 1987; Wolff et al. 1989; Sparks et al. 1999). Fragments with low viscosity within a turbulent conduit flow, eruption column, as well as during fire fountaining (cf. Chap. 9), may sinter and agglutinate randomly, depending on casual collisions between hot fragments of ductile magma. Such fragments may be highly irregular in shape or may acquire some aerodynamic shape, which may be preserved if cooled below T_g during their trajectory in air. Alternatively, if they are deposited still above T_g they could further deform and flatten under their own mass and fluidity, and/or through the load exerted by overlying deposits, such as in the case of some welded fall deposits (Chap. 9). This kind of welding can be developed in proximal lag deposits associated with some ignimbrites (Wright and Walker 1977; Fig. 10.50a, b), or in scoria/spatter and ash flow deposits (see Fig. 10.89c). A number of co-ignimbrite breccias include spatter agglutinates, with evidence of lateral transport (e.g. spatter breccias of the Sutri Formation, Vico caldera, Bear et al. 2009a, b; Campanian ignimbrite, Campi Flegri, Rosi et al. 1996; welded spatter agglutinates, Santorini caldera, Mellors and Sparks 1991; Onano Formation, Vulsini volcano, Palladino and Simei 2005; Villa Senni Formation, Colli Albani volcano, Vinkler et al. 2012; see Sect. 10.10, and Figs. 10.87, 10.88, 10.89, 10.90 and 10.91).

Juvenile fragments may continue to deform, sinter and agglutinate once incorporated into the pyroclastic flow, as well as within the dilute and turbulent upper part, well above the basal flow boundary layer. In this case, agglutination of particles would increase their aggregate settling velocity and therefore promote deposition (Freundt 1998; Branney and Kokelaar 2002). Occurrence of out-sized highly irregular, deformed spatter or fiamme clasts randomly oriented within ignimbrites is evidence of cooling of those fragments below their T_g before deposition within a deposit that then underwent welding.

Fragments that are deposited above their T_g can sinter and agglutinate during deposition of pyroclastic flows, under the shear exerted by the moving, high particle concentration basal flow (Schmincke and Swanson 1967; Chapin and Lowell 1979; Branney and Kokelaar 1992; Andrews and

Branney 2011; Lavellée et al. 2015). Freundt (1998) demonstrated experimentally that sticky particles agglutinate within fluidised beds and classified the kind of syn-depositional welding which occurs within pyroclastic flows as "*aggregation welding*". The main characteristic produced by welding during deposition is flattening of previously vesiculated pumice clasts. However, in some high-grade ignimbrites, the flattened pumices are seen to be stretched caused by a shear-induced stretching and a fabric that indicates a directional component. This is termed a rheomorphic fabric (see below). In sections parallel to the plane of foliation of stretched fiamme they show a well-defined lineation, and evidence of deformation by simple shear (cf. Chap. 6, Sect. 6.2.4; see Fig. 10.67). Evidence of syn-depositional welding is the quite common occurrence of thin welded horizons associated with compositional layering within ignimbrites (Chapin and Lowell 1979; Fisher and Schmincke 1984; Sumner and Branney 2002). Welded ignimbrites may also preserve a directional AMS (Sect. 10.8.12), which matches with imbrication of fiamme or lithic fragments (e.g. Pioli et al. 2008; Willcock et al. 2015a).

The third stage where welding can occur is within the deposit, depending on the temperature of emplacement, the physical state of the "glass" particles at the time of deposition, and the lithostatic load, i.e. the final thickness of the overlying deposit (load-welding; Freundt 1998; Riehle et al. 1995). Theoretically, a load is not required to cause welding if pyroclasts are still fluidal and well above T_g at the time of deposition; they simply mould into the form of neighbouring clasts and deform under their own plasticity. For fragments close to T_g at the time of deposition, a deposit load enhances the development of welding. Load or compaction welding in felsic ignimbrites as first proposed by Smith (1960a, b) is what historically has been regarded as welding in the volcanological literature.

Welding Zonation

Characteristically, when load-compaction welding approaches completion, three zones of welding intensity are produced within ignimbrite deposits: dense welding in the interior, where heat loss after deposition is minimal, partial welding marginal to the central welded zone, and non-welding at the top and bottom of the deposit, where heat loss after deposition is most rapid to the atmosphere above and the substrate below (Figs. 10.64c and 10.65; Cas and Wright 1987). In ignimbrites that show such a welding pattern the welding must have occurred post-depositionally, and not as a result in-transport agglutination, and not as a result of in-transport agglutination.

Smith (1960b) classified ignimbrites that showed such simple welding variations as *simple cooling units* (Fig. 10.65a, b). Riehle (1973) found that these variations could be predicted from theoretical analysis of the cooling of a sheet emplaced with uniform temperature. Many ignimbrites show a succession of several zones, of variable thickness, from densely welded to non-welded (Fig. 10.65c), which were classified as *compound cooling units*. However, as described above, it is now recognised some of these types of alternation may be due to syn-depositional agglutination welding, rather than load-compaction (Branney and Koelaar 1992). Some care must be taken therefore when describing cooling units, especially thinner deposits, as the alternation of welded and non-welded zones in an ignimbrite may not refer to welding that occurred during cooling, but due to the continuous aggradation of an ignimbrite with changing compositional characteristics (grain size, chemical composition, lithic content, volatile content, etc.).

In the welded zones, plastically flattened fragments, including both fiamme and matrix glass shards, define a planar foliation (or eutaxitic texture; Fig. 10.63). Many workers have recorded systematic changes in bulk density and porosity through welded ignimbrites (Fig. 10.65b, c). Typically, bulk density is at a maximum, whereas porosity is at minimum in zones of dense welding, and conversely in non-welded zones. Ragan and Sheridan (1972) showed in the Bishop Tuff that these features are related to the degree of flattening and stretching of the pumice clasts, which can be considered a measure of the compactional strain (Fig. 10.64e). During pure vertical compactional welding flattened pumice clasts are like discs without a preferred orientation (Ragan and Sheridan 1972; Fig. 10.63a), whereas where flattened pumice clasts are all elongated in a preferred direction lateral shear and stretching induced by post-depositional rheomorphic (see below) flow has also contributed to flattening. Strain ratio can be measured in welded tuffs by applying techniques developed by structural geologists to measure tectonic strain (Ragan and Sheridan 1972; Sheridan and Ragan 1976; Sparks and Wright 1979; Wolff and Wright 1981; Quane and Russell 2005).

Depending on the porosity and permeability of the ignimbrite, both the amount of gas trapped in, and the low percolation rate would also promote some degree of water resorption within the glass, which may lower T_g and increase the chances of post-depositional welding (Sparks et al. 1999; see Fig. 10.64a, b, f). Compaction also decreases porosity and permeability of the deposit favouring water vapour retention, and resorption into glass.

Welding is often associated with distinctive colour changes, which are due to different thermal oxidation states of iron as well devitrification and chemical changes (e.g. K and Fe element concentrations; Koralay and Kadioglu 2008). In densely welded zones columnar cooling joints are often well developed (Sect. 10.8.14.2). Spherulites and lithophysae may also be present (see Chap. 13, Sect. 13.2 for

Fig. 10.66 Basal vitrophyre to the 274 Ma intra-caldera Ora Ignimbrite, Bolzano, northern Italy. The ignimbrite and vitrophyre are very crystal rich, but the matrix in between, which microscopically has extremely intensely welded glass shard textures (Figs. 10.63 and 10.86c), is still fresh, black obsidian glass, making it perhaps the oldest glass vitrophyre preserved

a comprehensive discussion of their origin). Where densely welded ignimbrite has a glassy obsidian appearance, it is called a *vitrophyre* (Figs. 10.66 and 10.63j). Vitrophyre horizons occur in the zone of most intense welding (see Fig. 10.69). However, they are most likely to occur towards the base of aggradational ignimbrites (e.g. Pioli and Rossi 2005), or very thick, intra-caldera ignimbrites. In older ignimbrites, such vitrophyre zones may have been devitrified, but can be rarely preserved even in Paleozoic ignimbrites (e.g. Willcock et al. 2013; Willcock and Cas 2014; Fig. 10.66). Densely welded vitrophyric ignimbrite may also experience the formation of perlitic cracks, resulting from cooling contraction conchoidal fracturing in isotropic glass. Perlite cracks have a sub-concentric pattern (e.g. Fig. 10.63j), and once formed can allow vapour-phase fluids and meteoric water to percolate along them, that then initiate hydration and alteration of the glass (von Aulock et al. 2013).

Rheomorphism

Where welding affects the ash matrix to produce a continuous fluidal, plastic mass that remains above T_g, secondary mass flowage of the welded deposit is possible, especially down slopes, and was termed *rheomorphism* (e.g. Wolff and Wright 1981). During rheomorphism, fiamme may become stretched, defining a down-dip flow lineation (Fig. 10.67a, c). This is sometimes termed a *parataxitic* texture.

However, as indicated earlier, it has now been widely recognised that similar flow textures can be produced as syn-depositional structures due to agglutination of particles within the lower part of an aggrading pyroclastic flow (Fig. 10.67b; Freundt 1998; Branney and Kokelaar 2002; Andrews and Branney 2011; Lavallée et al. 2015). Differentiating the two processes can only be made on a case by case basis, and while welding and deformation of pyroclasts may commence within the moving flow, the process may continue by the influence of gravity on the mass of the non-particulate flow following deposition (e.g. Fig. 10.67d). One such detailed study is that of the 15.8 Ma Nuraxi high-grade rhyolitic ignimbrite, Sardinia, Italy (Pioli and Rosi 2005). This shows that where deposited in fairly flat areas the welded tuff displays regular and consistent, laminar flow features which likely formed at the depositional boundary layer (Fig. 10.67c). However, where deposited on an irregular topography, rheomorphic structures are more complex in terms of style of deformation and scale and include several phases characterised by different fold geometries and styles. While deformation may have started during deposition, it was still occurring after by secondary spreading of the welding mass driven mainly by gravity.

Where the extent of welding and rheomorphism is extreme, as in the case of the Nuraxi ignimbrite, the original clastic texture may be lost, the ignimbrite may appear to be lava-like, and it can be easily mistaken for a coherent lava (Fig. 10.68a, b). The original ignimbrite deposit has in fact behaved and flowed like a lava after deposition, developing a flow banding like shear layering, flow folds, and even undergoing autobrecciation (e.g. Bonnischen and Kauffman 1987; Henry and Wolff 1992). Flow folds (upright, recumbent, sheath) can be well developed (Fig. 10.68c, d; Wolff and Wright 1981; Bonnischen and Kauffman 1987; Henry and Wolff 1992; Pioli and Rosi 2005; Branney et al. 2004; Andrews and Branney 2011; Brown and Bell 2013). Post-depositional diapiric structures formed by the buoyant rise of partially welded lithofacies into rheomorphic host rocks may also be found (Mundula et al. 2013).

Distinctive features may help in recognising the ignimbrite origin, such as the presence of angular, broken crystal fragments in what was the ash matrix, abundant lithic clasts, the low aspect ratio of the unit compared with viscous lava counterparts, and transition into less welded zones, where the original vitriclastic or pyroclastic textures are still visible (cf. Chap. 6; Bonnischen and Kauffman 1987; Henry and Wolff 1992). Many (but not all) examples of rheomorphic welded ignimbrites are of peralkaline rhyolite composition, having unusually low magma viscosities compared with calc-alkaline rhyolites.

Welding Classification Schemes

Based on the textures and physical characteristics of the ignimbrite, several classification schemes of the welding

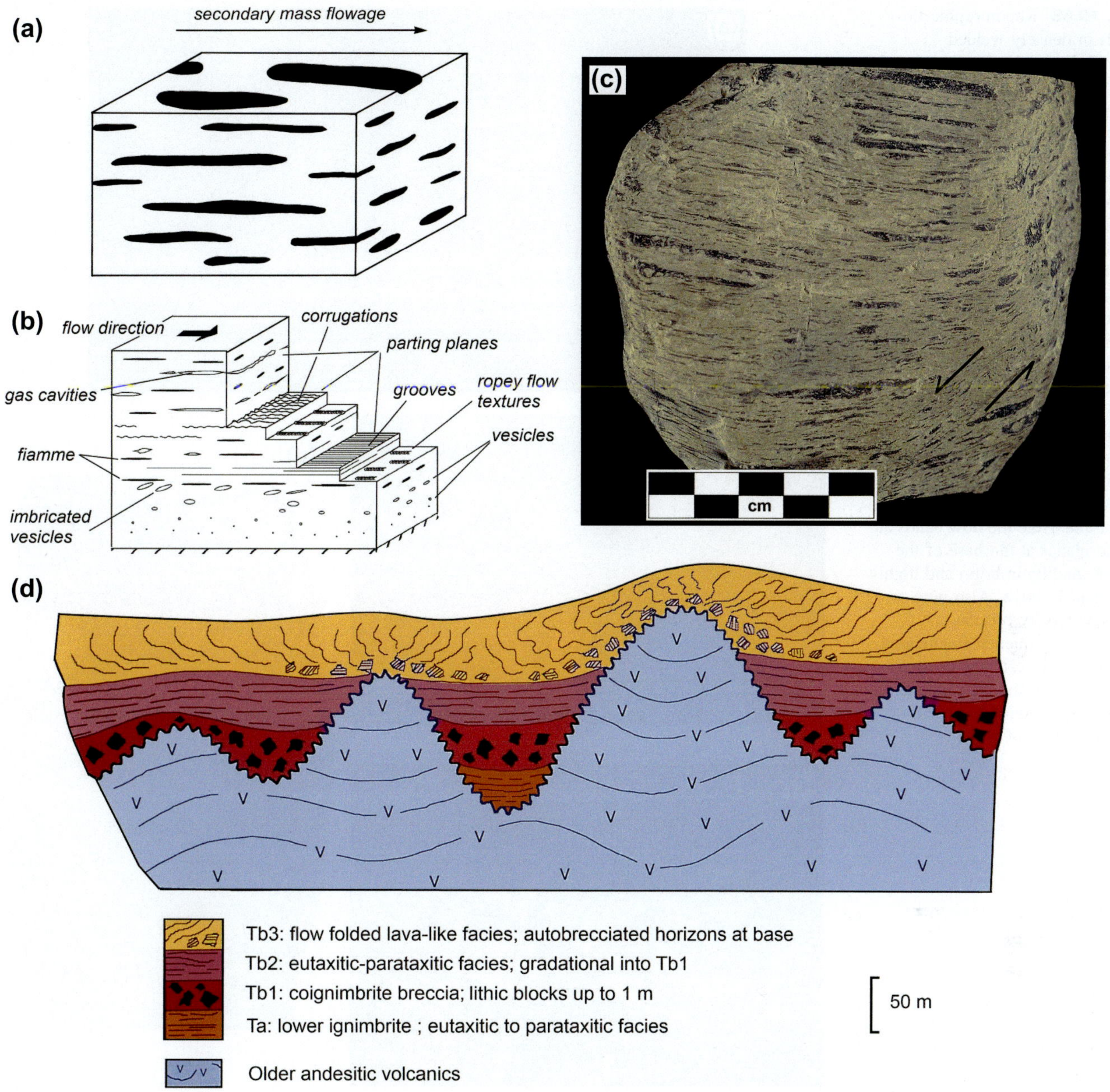

Fig. 10.67 Rheomorphism of densely welded ignimbrite. **a** Block diagram depicting flattened fiamme, which have also been stretched as a result of post-depositional or secondary rheomorphic flow of the ignimbrite during welding, defining a down-dip slope lineation. Note the different aspect of fiamme on the two faces of the block. (From Cas and Wright 1987). **b** Schematic representation of syn-depositional rheomorphic flow structures in the basal part of the 15.8 rhyolitic Nuraxi ignimbrite, Sardinia, Italy. Structures indicating flow direction include elongated vesicles, parting planes with stretched gas cavities, parallel ridges and grooves, ropey surfaces, and elongation of fiamme. The basal part of the ignimbrite (2–3 m) displays the highest concentration of syn-depositional flow structures, but the very base of the deposit (10–20 cm thick) is typically massive. The main body of the tuff is prominently foliated and lineated and has parataxitic texture, except where deposited on irregular topography where large-scale flow folds develop by secondary mass flowage. (Adapted from Pioli and Rosi 2005). **c** Stream polished boulder of rheomorphic, Late Devonian rhyolitic ignimbrite, Yalwal region, southern New South Wales, southeastern Australia. Note the syn-compaction, syn-rheomorphic flow ductile shear zone cross-cutting the welding foliation at 45°, with a sense of motion of left side down, right side up. It is not known which is the correct way up of this sample since it was collected as loose float in a riverbed. **d** Schematic reconstruction showing rheomorphic flow structures developed on irregular topography. Permian trachytic Mt Toussaint ignimbrite, Lower Bowen Volcanics, Collinsville, Queensland, Australia. The uppermost unit (Tb3) consists of flow banded and flow folded lava-like facies. The fold structures indicate downslope secondary flowage of several-m-thick layers of the ignimbrite along slopes normal to deeply incised valleys cut into older andesitic volcanic rocks. Note the presence of autobrecciated horizons suggesting local cooling below the T_g, and brittle failure during flowage

Fig. 10.68 Rheomorphic flow folds in densely welded ignimbrites. **a** Highly attenuated fiamme (parataxitic texture) showing minor flow folding. Stream-polished block, Permian trachytic Mt Toussaint ignimbrite, Lower Bowen Volcanics, Collinsville, Queensland, Australia. **b** Flow folds visible in outcrop in Carboniferous rheomorphic welded ignimbrite, Campelles, Pyrenees, Spain. **c** Recumbent lava-like fold in outcrop, Mt Toussaint ignimbrite. **d** Slabbed base of a rhyolitic rheomorphic ignimbrite, showing flow folds of variable geometry, amplitude and wavelength, 1070 Ma Kathleen ignimbrite, Pussy Cat Group, West Musgrave Ranges, Central Australia. Note the flow bands are more planar at the base of the block (and ignimbrite) and highly attenuated above. Also note jig-saw blocky breccia, representing quench fragmented hyaloclastite resulting from the interaction of the base of the rheomorphic ignimbrite with a shallow body of water

intensity have been proposed (Fig. 10.69a). Field-based textural classifications are quite simple and range from incipient welding for those ignimbrite divisions that show only sintering and stretching or flattening of large particles, to moderate, to dense welding, where the process is pervasive and involves also the fine ash matrix (e.g. Smith 1960b). This range also corresponds to the scheme that assesses welding intensity in terms of grades of welding, ranging from low-grade (non-welded), to intermediate grade (non-rheomorphic welded ignimbrites) to high-grade ignimbrites (rheomorphic welded ignimbrites; Walker 1983; Branney and Kokelaar 1992).

Some welding classification schemes also involve measurement of physical parameters, such as the density, the porosity, degree of stretching of pumice/fiamme, compressive strength, strain, and evaluation of the micro-texture of the ignimbrite (Fig. 10.69a; Quane and Russell 2005; Mundula et al. 2009). The most comprehensive scheme to date is that by Quane and Russell (2005; Fig. 10.69a, b), who propose a welding ranking scale, ranging from I (non-welded, unconsolidated) to VI (very densely welded, obsidian vitrophyre; Fig. 10.69c). This is based on textural and physical measurements on 100 drill core samples collected from a single cooling unit of the Upper Bandelier

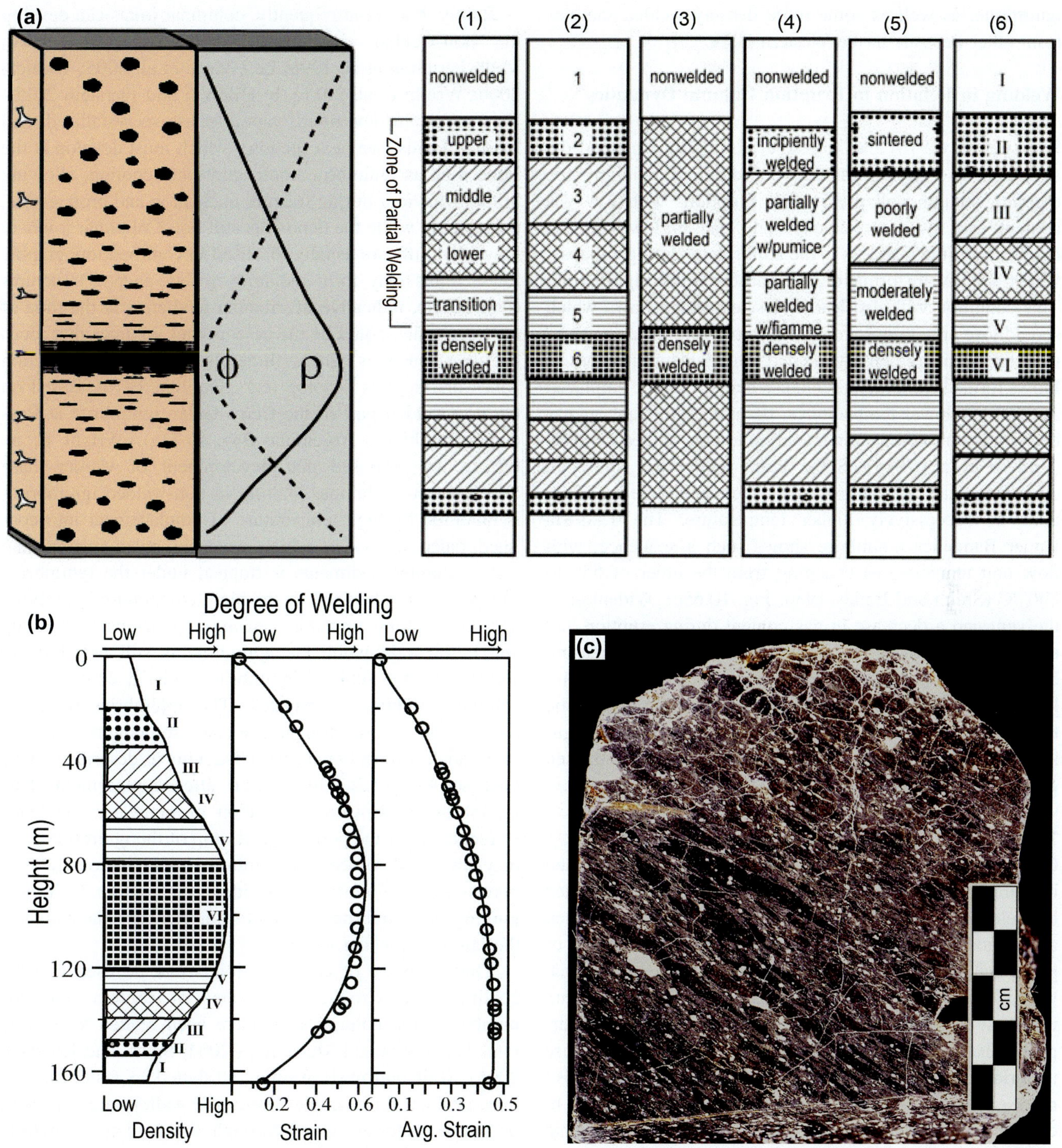

Fig. 10.69 Welding hierarchy for ignimbrites. **a** Various schemes for classifying zones of equal welding intensity in a single ignimbrite cooling unit based on changes in physical properties and textures. Columns on left represent textural variations in ash shards (open) and pumice lapilli (solid) as well as typical variations in density (σ) and porosity (Φ). (1) Smith (1960b; petrographic observations Bandelier and Battleship Rock Tuffs); (2) Smith and Bailey (1966; porosity Upper Bandelier Tuff); (3) Sheridan and Ragan (1976; density Bishop Tuff) and Peterson (1979; fiamme elongation Apache Leaf Tuff); (4) Streck and Grunder (1995; density Rattlesnake Tuff); (5) Wilson and Hildreth (2003; density Bishop Tuff); (6) Quane and Russell (2005; textural and physical measurements mainly Upper Bandelier Tuff). **b** Ranking scheme for welding of Quane and Russell (2005) applied to Bishop Tuff. (**a**, **b** after Quane and Russell 2005). **c** Example of rank VI (most densely welded) showing eutaxitic texture with obsidian-like fiamme in which no former traces of vesicles can be observed. In thin section, this would show an obsidian-like vitrophyre with thoroughly collapsed and annealed shard textures

ignimbrite, as well as some more densely welded samples from other deposits in the Western USA.

Welding in Relation to Eruption Column Dynamics

Sparks et al. (1978) and Trolese et al. (2019) explained the variations in welding intensity of different units by the eruption column collapse model. Eruptions with low gas content and low gas velocity will lead to low collapse heights and little heat loss to the atmosphere during collapse, which will favour formation of a densely welded ignimbrite. This condition will also lead to less-expanded flows which should lose only minor amounts of vitric ash, as seems to be the case for many densely welded intra-caldera ignimbrites. Flows formed from higher collapse heights will be emplaced at relatively lower temperatures, because significant air has been mixed into the column, leading to non-welded deposits (Trolese et al. 2019). Some large ignimbrites show an upward increase in welding due to the emplacement of flow units of successively higher temperature. The 1.26 Ma Upper Bandelier ignimbrite shows such a sequence, with flow unit temperatures changing from the order of 550 to 800 °C (Smith and Bailey 1966; Fig. 10.65c). Widening of the vent and a decrease in gas content during eruption and column collapse height could both explain increasing emplacement temperature. Welding intensity may also vary laterally as a result of progressive cooling of the pyroclastic flow and results in a lateral welding zonation in the ignimbrite deposit (Smith 1960b; Lipman and Christiansen 1964; Wright et al. 2011a, b; Fig. 10.65a, c).

10.8.14.2 Columnar Joints

Columnar joints commonly form in high-grade welded ignimbrites during cooling, as the ignimbrite is transformed by welding in a largely solid, hot coherent rock body which accommodates volumetric contraction with development of polygonal joints (Fig. 10.70). This process is identical to that which occurs in lavas and intrusions (Chaps. 5, 6 and 8), and involves cooling contractional fractures propagating perpendicular to the cooling surfaces that propagate in response to various factors. Initially the top and bottom of the deposit represent cooling surfaces. As these cool and attempt to contract, multiple cooling cells form, bounded by polygonal sets of fractures, which are tensional in origin. Columns are often smaller at the margins, reflecting high cooling rates, and then merge inwards to form larger diameter columns inwards, where the cooling rate is slower. More complicated curving and radiating joint patterns have been attributed to the local effects of degassing (Sheridan 1970; see below) or, by analogy with flood basalt lavas, where surface water gains access into the interior of the cooling ignimbrite (Long and Wood 1986; Sect. 5.5.4).

Rarely, but not infrequently, columnar joints can develop in non-welded ignimbrites (e.g. Sheridan 1970; Vatin-Perignon et al. 1996; Le Pennec et al. 2005; Wohletz 2006; Wright et al. 2011a, b; Hildreth and Fierstein 2012). This occurs in devitrified, vapour-phase recrystallised, and lithified facies (see next section), which must develop at the same time as ignimbrite compaction and cooling, allowing brittle behaviour during fracture nucleation and propagation, presumably while the deposit is still hot. Columnar joints in unwelded tuffs are usually localised in confined topographic lows. Joints may form radiating, rosette-shaped structures (Fig. 10.71), indicative of irregular isotherms at the time of cooling, which could be the case around volumes that focus heat-loss, such as where fumarolic gas escape/migration zones or fractures develop (e.g. Sheridan 1970). Based on the non-welded part of the Cerro Galán ignimbrite at Paycuqui, northwest Argentina (Fig. 10.71a), Wright et al. (2011a, b) proposed that development of vapour-phase alteration and columnar jointing in non-welded ignimbrites is favoured by high temperature of emplacement into confined paleo-valleys in which water in paleo-streams and water-saturated sediments is trapped under the ignimbrite. This water is volatilised by the high temperature ignimbrite, and then streams upwards, causing vapour-phase alteration and lithification while the ignimbrite is still hot, but not welded. As the lithified ignimbrite cools, it contracts and columnar jointing propagates. The ignimbrite preserves non-welded texture, with pumice clasts showing no evidence of thermal, compactional (load) flattening as described above (also see Fig. 13.2c). In addition, after emplacement of an ignimbrite, downward percolation of rainwater can leach elements out of the porous, glassy top of the ignimbrite, and may also lead to secondary mineral precipitation in open pore spaces and lithification inside the ignimbrite. Columnar jointing in ignimbrites is therefore not always indicative of the ignimbrite being welded.

The geometry of columnar joints can be used to infer cooling rates. Cross-sectional area of columns and the number of the column sides have been related to cooling rates (Toramaru and Matsumoto 2004; Sporli and Rowland 2006), with narrower spacings attributed to faster cooling rates attained for example where paleovalleys are narrower, or where water is fluxed through the cooling ignimbrite (Wohletz 2006; Wright et al. 2011a, b). Wright et al. (2011a, b) also found that there was a hierarchical order of joints, from first order, straight long joints, to shorter, more closely spaced, second-order joints that propagate from and initially perpendicular to the first-order joints, and then third-order joints propagate perpendicular to second order joints, etc.

Recently regional-scale joint patterns have been described from ignimbrites in Bolivia and Chile (Tibaldi and Bonali 2018). In plan view, these show circular to semi-circular

Fig. 10.70 Columnar jointing in densely welded ignimbrites. **a** 4.2 Ma Huichapan Tuff, near Amealco, central Mexico, with columns becoming wider and curved upwards (From Cas and Wright 1987; for detailed description of this impressive locality, see Aguirre-Díaz and López-Martínez 2009). **b** 0.76 Ma Bishop Tuff, Long Valley, California, USA, showing radiating columns plunging at 45°. Such fan-shaped columns have been attributed to the local effects of degassing or meteoric infiltration. **c** Late Devonian Boyd Volcanic Complex, Eden, New South Wales, Australia. The axes of the columns plunge 60°, but in this case indicating that primary bedding has been tilted 30° by tectonic deformation

geometries and intersections that have "X" and "Y" patterns. In section view, they are from sub-vertical to vertical and do not affect the underlying deposits. This joint architecture is interpreted as resulting from differential contraction after emplacement of the ignimbrite deposit above a complex topography, although the prevailing tectonic stress regime may act to nucleate faults. The authors suggest the pattern is probably common, but recognisable only in plan view and before tectonic deformation obscures and overprints it.

10.8.14.3 Vapour-Phase Crystallisation

Vapour-phase crystallisation or alteration results from the percolation of hot gases and vapours through ignimbrites during cooling which cause chemical reaction with the glassy ash, and formation of secondary minerals, causing it to solidify (Fig. 10.72a). The most important gas sources are probably diffusion from juvenile vitric particles, the gas trapped during deposition, and heated ground and/or surface water (Sheridan 1970; Sparks et al. 1999; Keating 2005). In

Fig. 10.71 Columnar jointing in non-welded ignimbrites, commonly showing radiating clusters of columns. **a** 2.08 Ma Cerro Galán ignimbrite, Paycuqui, northwestern Argentina. Also see Fig. 10.82a. The depositional setting is a paleovalley. **b** 460 ka Pozzolane Rosse ignimbrite, Colli Albani caldera, central Italy. Height of cliff ~10 m. **c** 39 ka Campanian ignimbrite, Campi Flegrei caldera, Italy

particular, the process of devitrification (i.e. micro-crystallisation) of glass during cooling starts immediately after deposition and mobilises volatiles that do not enter within the dry crystal framework and redistribute them in the ignimbrite deposit (Sheridan 1970; Vaniman 2006; Stimac et al. 2008). In felsic compositions, devitrification-crystallisation is dominated by feldspars (up to 90%) and silica phases (tridymite, cristobalite, quartz) and the volatiles liberated also are not just water but are enriched in halogens and trace elements that, by percolating through the ignimbrite, contribute to form a variety of vapour-phase minerals that cement the deposit. The main products of vapour-phase crystallisation are tridymite, cristobalite and alkali feldspar which form at >450–650 °C and occur as drusy infills of matrix micro-porosity and pumice cavities, so forming a cement and reducing pore space. Other mineral phases include iron oxides and minor metal sulphides and chlorides, and less commonly scapolite (marialite), biotite, amphibole, and clinopyroxene (Chap. 13, see Fig. 13.2d, e and g–i; Sheridan 1970; Stimac et al. 2008; Caporuscio et al. 2012).

The extent of vapour-phase crystallisation depends on the permeability and thickness of the deposit and may be

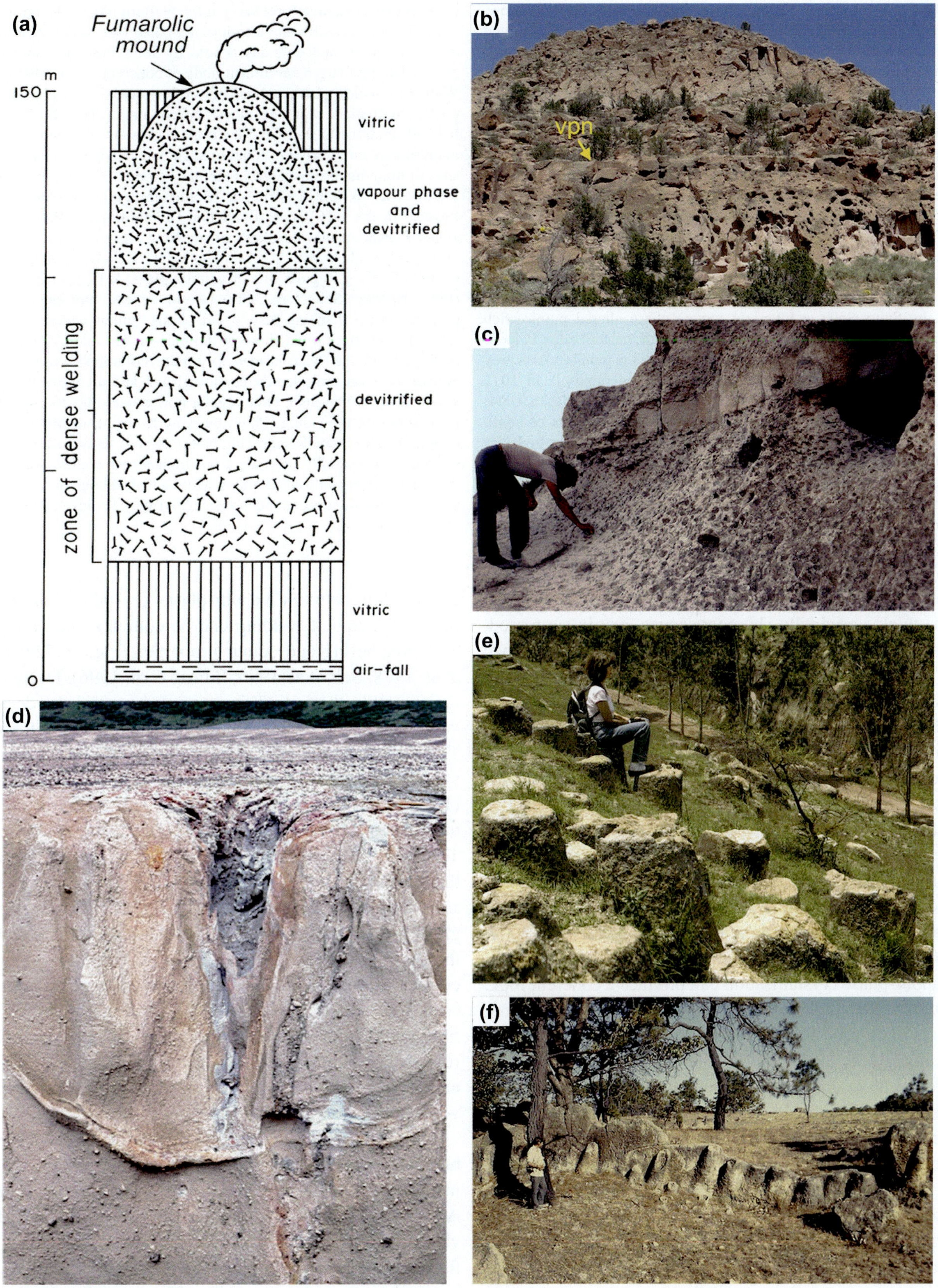
(a)
Fumarolic mound
m
150
0
vitric
vapour phase and devitrified
zone of dense welding
devitrified
vitric
air-fall
(b)
vpn
(c)
(d)
(e)
(f)

◀ **Fig. 10.72** Features of vapour-phase crystallisation and fumarolic activity. **a** Zones of high temperature vapour-phase crystallisation and devitrification, 0.76 ka Bishop Tuff, California, USA. Fumarolic mounds project from top of vapour-phase zone through non-welded ignimbrite. (From Cas and Wright 1987, After Sheridan 1970). **b** Complex interrelationship between vapour-phase crystallisation and degree of welding, base of 1.26 Ma Upper Bandelier Tuff, Ancho Canyon, Pajarito Plateau, New Mexico, USA. The onset of vapour-phase crystallisation coincides with change from pink to orange colouration, and decrease of large etched out pumice fragments and excavations by ancestral Pueblo people (called "*cavates*") within the non-welded zone. The "vapour phase notch" (vpn) marks the internal contact between the glassy lower portion and the devitrified central portion of the ignimbrite. Above the notch welding intensity increases (shown by columnar jointing), masking the effects of vapour-phase crystallisation. Height of cliff section is ~60 m. (Photo by Steve Self). **c** Intensely vapour-phase altered pumice-rich ignimbrite, distal facies of Upper Bandelier Tuff, eastern edge Pajarito Plateau (see Fig. 10.42a, b). Alteration has turned the pumice clasts grey and led many to be etched out. **d** Funnel-shaped fumarole in 1912 non-welded ignimbrite, Valley of Ten Thousand Smokes, Alaska, USA. Orifice flares to ~2 m wide at surface. Note concentrations of lapilli along irregular cracks, from which fine ash was elutriated by high gas flux. (From Hildreth and Fierstein 2012). **e–h** Low temperature "steam" pipes and columns, locally known as "fairy stools", 95 ka Rio Caliente ignimbrite, La Primavera rhyolite volcano, Mexico. Pipes and related zeolite (clinoptilolite-heulandite) cemented features are found exposed over an extensive area (>60 km^2) in the western part of the ignimbrite's outflow sheet, possibly the location of a paleo-lake. **e** Part of a rounded cluster of pipes and columns (Photo by John Pint). **f** Annular pipes grouped along margin curvilinear body of coherent tuff or sillar (George Walker for scale). **g** Large tabular body of massive sillar, with some partly coalesced shallow plunging pipes and columns (Photo by John Pint). **h** Field relations in the Rio Salado area. Clusters of pipes are radially distributed around massive zones of sillar, which appear to have been the sites of more intense or elevated thermal activity. (From Cas and Wright 1987, After Wright 1979). **i–k** Mordenite ± clinoptilolite cemented features in the 1.26 Upper Bandelier Tuff ignimbrite, Paliza Canyon, Jemez Plateau, south of Valles caldera, New Mexico, USA. **i** Schematic cross-section illustrating field relations at Paliza Canyon (Image courtesy of Steve Self). **j** Regular line of annular pipes; **k** Wall of coherent tuff (Photo by Steve Self). **l–m** Unusual evenly spaced, mordenite cemented columns in the 0.76 Ma Bishop Tuff, Long Valley caldera, California, USA. **l** Exposure on the shore of the man-made Lake Crowley, where unaltered tuff has been eroded to expose columned caves. Columns extend from wave-eroded caves through tuff and up above the outcrop which is some 8 m high. Note regular horizontal joints thought to have formed during unloading (Photo by Amy Kuwata). **m** Model showing cooling history of the Bishop Tuff. Note (left) low temperature mordenite columns are spatially separate from higher temperature welding and vapour-phase crystallisation zones. Mordenite is thought to precipitate in down-welling fingers of condensed steam (right). (From Randolph-Flagg et al. 2017)

pervasive where most of the vapour-phase is retained, such as in thick, originally homogeneous, matrix-supported, non-welded ignimbrites (e.g. Smith and Bailey 1966; Stimac et al. 1996; Fig. 10.65c). Highly permeable deposits such as pumice lapilli fall beds, or thin ignimbrite units usually do not show vapour-phase crystallisation as they are unable to retain volatiles, although exceptions are known (e.g. in pumice fall deposits from the 1912 Valley of Ten Thousand Smokes eruption, Alaska; Hildreth and Fierstein 2012). Furthermore, in order for vapour-phase crystallisation to occur, the water–rock ratio must not exceed certain values, such as in the case of many phreatomagmatic rocks or tuffs deposited in water, in which case alteration to clay and zeolites is more common (e.g. Hay 1966; Braithwaite 2003; Mormone and Piochi 2020; also see below).

Vapour-phase crystallisation can produce a coherent rock. The term *sillar* (a Peruvian word first applied by Fenner 1948) is often used for such rocks but has also been applied to incipiently welded tuffs in which ash grains are barely sintered at point contacts and show no other deformation of juvenile clasts. Identical rocks texturally in outcrop and hand specimen can also be produced by precipitation of low temperature zeolite minerals, which have also been called sillar irrespective of secondary mineralogy or temperature of formation.

Seminal studies of the 1.26 and 1.6 Ma Bandelier Tuff ignimbrites, New Mexico (Smith 1960b; Smith and Bailey 1966) found vapour-phase crystallisation occurs in specific zones associated with more densely welded ignimbrites, typically above the base and commonly towards the top forming the outer envelope around the welded sheet (Figs. 10.65c and 10.72a, b; see also Keating 2005). The relationship between vapour-phase crystallisation and the onset of incipient welding can be complex and controlling both weathering and colour properties (Fig. 10.72b; Crowe et al. 1978; Self et al. 1996; Stimac et al. 1996). Depositional facies within the ignimbrite may also control the degree of vapour-phase crystallisation; e.g. in the Upper Bandelier Tuff pumice-rich distal facies of the ignimbrite are intensely vapour-phase altered (Figs. 10.72c and 10.42b). Sheridan (1970) described fumarolic mounds and ridges of sillar on top of the 0.76 Ma Bishop Tuff, California (Fig. 10.72a). These are characterised by their greater induration relative to the surrounding ash or tuff, and stand 0.5–15 m above the surrounding terrain. Fumarole mounds were formed in greatest numbers where devitrification-crystallisation within the sheet was intense (Fig. 10.72a). The ridges are straight or curved features up to 600 m long, and appear to follow early conjugate joint sets which controlled the distribution of fumarolic mounds and ridges. The mineralogy of the fumarolic mounds is similar to that of the vapour-phase zone and indicates high temperatures indicative of transport of hot volcanic gases through these pathways.

Dramatic fumarolic activity was famously observed by botanist Robert F. Griggs who in 1916 explored the aftermath of the 6–8 June 1912 Novarupta-Katmai eruption, Alaska. Pumice and ash flows deposited in the Ukak River valley buried snow fields and glacial streams which were flashed into steam forming thousands of fumaroles, and hence the name "Valley of Ten Thousand Smokes". Gases emerged along fissures within the ignimbrite, with measured temperatures ≥ 350 °C recorded in 1917. Fumarolic activity continued for a number of years as subsequent rain and snow melt seeped into the ignimbrite, but by the 1930s most of the

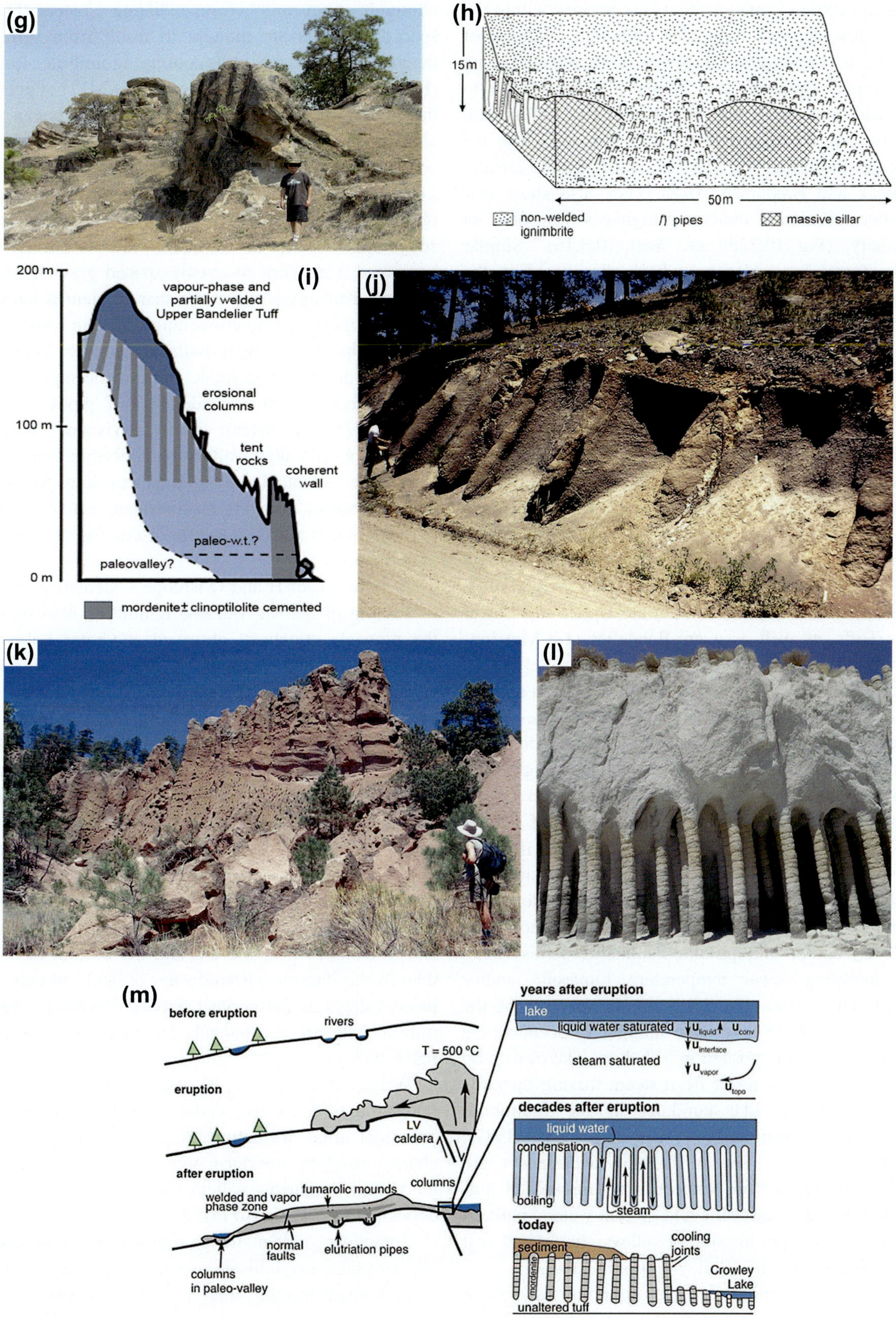

Fig. 10.72 (continued)

fumaroles had cooled as the residual heat trapped within the pyroclastic flow deposits dissipated, although evidence of ongoing heat loss continues (Hogeweg et al. 2005). Today, better cemented chimney- and funnel-like forms (up to several m wide and deep), and varicoloured zones of mineral precipitation and alteration mark extinct fumaroles near the surface of the ignimbrite (Fig. 10.72d; Hildreth and Fierstein 2012; Hudak and Bindeman 2018). They also show evidence of fines-elutriation indicating high gas flow rates, at least initially (Fig. 10.72d; cf. Sect. 10.8.16). Similar funnel-like forms have also now been discovered in the Upper Bandelier Tuff ignimbrite, which are characterised by high temperature vapour-phase mineral assemblages and evidence of fines-elutriation (Caporusico et al. 2012).

A group of enigmatic, low temperature paleo-fumarolic or -geothermal features, including peculiar "steam" pipes or columns, have also been recognised in several ignimbrites. Williams (1942) first described tubular spire or "fumarolic pinnacles" from the 7.7 ka Mt Mazama ignimbrite, Crater Lake (Fig. 10.62). Since then somewhat similar features, although not identical, have found within and near the top of the 95 ka Rio Caliente ignimbrite, La Primavera rhyolite volcano (Fig. 10.72e–h; Wright 1979; Cas and Wright 1987; Figs. 8.46b, c and 8.47), and in both the Upper and Lower Bandelier Tuffs, as well as a pre-Bandelier (1.8 Ma) ignimbrite, around Valles caldera (Fig. 10.72i, k; Self et al. 1996, 2022; Bailey and Self 2010). Pipes vary from annular structures with indurated rims and non-welded cores (lacking evidence of elutriation) to coherent circular columns, generally ≤ 1 m diameter and up to several m high; they may also occur with more massive zones (walls) up to 10 s of m high, and tabular and curvilinear bodies of similar coherent tuff (or sillar), which do not obviously fit into any zonal welding scheme. These features are more resistant to erosion than the surrounding non-welded ignimbrite owing to the precipitation of zeolite minerals, principally mordenite (100–130 °C) and/or clinoptilolite-heulandite series (125–230 °C), implying lower temperature fumarolic and/or geothermal activity (see Fig. 13.2j, k). In the case of the Rio Caliente ignimbrite they were thought to occur where the outflow sheet was deposited in a shallow lake or marshy ground and assumed to result from steam fluxing upwards through selected pipe-like conduits in the ignimbrite post-depositionally (Wright 1979; Cas and Wright 1987). Around Valles caldera, several isolated occurrences are attributed to the interaction between hot, non-welded ignimbrite and circulating meteoric and ground waters, perhaps associated with the rapid incision of valleys, that boiled as it seeped into hot ignimbrite (Bailey and Self 2010).

Since then, even more unusual columns have been described in the Bishop Tuff (Fig. 10.77k; Randolph-Flagg et al. 2017). These generally form regular spaced columns although, more rarely, form radiating clusters. The regular spaced columns are thought to result from surface water percolating into the still cooling ignimbrite from above (Fig. 10.72l). Randolph-Flagg et al. (2017) suggest that as the ignimbrite was emplaced, it instantly began to slump along paleo-drainages. Into these low-lying regions, water pooled forming ponds or lakes, including in and near the Long Valley caldera, which then percolated into the porous ignimbrite causing it to heat and boil. Numerical simulations suggest instabilities at the liquid water and steam interface generated a network of evenly spaced gravitational fingers composed of upwelling steam that condenses before reaching the surface, and down-welling water (in which mordenite was precipitated) which boils at its base (Fig. 10.72m). A similar downwelling model has now been proposed by the same authors for the Bandelier Tuff pipes and columns (Fig. 10.72i, k), where it is envisaged slumping of over-steepened valley walls exposed friable, warm, unaltered ignimbrite that enabled ponded meteoritic water to permeate into the deposit, allowing localised convective chimney cells to form (Self et al. 2022; see for discussion of current models of formation).

A more recent H and O isotope systematics investigation of cooling of the Mt Mazama ignimbrite also suggests that the pinnacle structures observed at Crater Lake represent late-stage cooling features in the ignimbrite rather than vigorous fumaroles (Hudak and Bindeman 2018). This model is somewhat different from that of Randolph-Flagg et al. (2017) because it invokes groundwater rather than surface water to hydrate the pinnacles. At Crater Lake groundwater from a high-elevation source is necessary to reconcile the isotopic data. Emergent springs feeding streams could then percolate down through the upper portion of the ignimbrite in the incipient channels of the streams where the pinnacles are now found. Water sources from above and below help to explain the greater depletions in H and O isotopes near the top and the bottom of the deposit than in the interior. Groundwater is likely to concentrate in paleo-valleys at the contact between bedrock and the ignimbrite and stream channels are likely to reoccupy these paleo-valleys.

Wright et al. (2011a, b) also linked vapour-phase crystallisation of the Cerro Galán ignimbrite and concomitant columnar joint formation to a paleo-topographic setting characterised by interface with external groundwater from the bottom of narrow valleys into which the ignimbrite was deposited (see Fig. 10.71a).

Some low-grade mafic ignimbrites show characteristics of vapour-phase crystallisation, such as pervasive lithification and columnar jointing (Fig. 10.71b, c; also see Fig. 10.77j), but produce very distinctive mineral assemblages compared to silicic ignimbrites. Several ignimbrites in central and southern Italy which range in composition from tephrite to phonolite and trachyte (e.g. the Campanian ignimbrite, Campi Flegrei; and in the Roman Magmatic Province, the Pozzolane Rosse ignimbrite, Colli Albani, and the Tufo Rosso a Scorie Nere, Vico caldera), are pervasively zeolitised (chabazite, phillipsite as main mineral phases; Langella et al. 2002; de'Gennaro et al. 1999, 2000; Giampaolo et al. 2006; Bear et al. 2009c; Campbell et al. 2019; see Fig. 13.2f, l). These minerals form at relatively low temperatures (<450 °C), and combined with the mafic composition of the glass, this prevents crystallisation of mineral phases typical of felsic systems at higher temperatures. Zeolitisation of the Vesuvius AD 79 ignimbrite has also occurred, with alteration "plumes" and haloes of zeolitised ash and large m-scale steam or gas pipes formed where pyroclastic flows were deposited in a shoreline environment, and generated by the convective, upward percolation of vapourised seawater (Giordano et al. 2018; Chap. 11, see Plate 11.1f, Fig. 11.15).

Lastly we note, none of the variety of steam, gas, or fumarolic pipes and columns described above should be confused with "*tent rocks*" (or hoodoos), that are features commonly observed in some eroded and dissected ignimbrite terrains (e.g. around Valles caldera, where they are sometimes found together at the same location, Fig. 10.72i; Cappadocia, Turkey, Sarikaya et al. 2015). Tent rocks are cone-shaped structures often capped by a boulder, or a remnant of welded or indurated ignimbrite, which retard erosion of the softer deposit below them. When the caps finally drop, the soft necks are quickly destroyed. They can be used to determine bedrock erosion rates and quantify landscape evolution (Sarikaya et al. 2015).

10.8.14.4 Thermal/ThermoRemanent Magnetisation of Lithic Clasts

Lithic clasts, both accessory and accidental, but also juvenile clasts, can be used to determine the emplacement temperature of ignimbrites, by the determination of their partial thermoremanence or *thermal/thermoremanent magnetisation* (pTRM) with paleomagnetic techniques (Aramaki and Akimoto 1957; Wright 1978; Hoblitt and Kellog 1979; McClelland and Druitt 1989; Mandeville et al. 1994; Ort et al. 1999; Bardot 2000; Cioni et al. 2004; Porreca et al. 2006, 2008; Sulpizio et al. 2008; Zanella et al. 2007; Paterson et al. 2010; Fontana et al. 2011; Lesti et al. 2011; Pensa et al. 2015a, c, 2018; Uehara et al. 2015; Trolese et al. 2017, 2024; Giordano et al. 2018). The best lithic types for such analysis are accessory and accidental lava lithic clasts, which have an original magnetisation acquired in situ during cooling below the Curie temperature, at the time when the lava was emplaced, and before the explosive or erosional event had extracted the clast from its source and incorporated it into the pyroclastic flow.

The principle is that as magnetic minerals cool below their Curie temperature in a solidifying lava (generally between 630–580 °C depending on the type of magnetic minerals), they acquire magnetisation and magnetic components that are parallel to the Earth's magnetic field at the time of emplacement. This magnetic component can be measured in the laboratory as a vector and, if for example we drill and measure lava cores from one lava unit (i.e. one cooling unit), these will all have exactly the same magnetic signature such as inclination, declination, and intensity of magnetisation, irrespective of where within that lava unit they were drilled. But, when lava clasts are extracted from a lava unit and then incorporated within a pyroclastic flow, they tumble around randomly, so their magnetic vectors, originally parallel when they where in situ, if measured, would be completely randomly oriented. However, these clasts are also subjected to the thermal regime of the pyroclastic flow during transport and deposition within the ignimbrite (Fig. 10.73a) Assuming that lithic clasts were cold at the time of incorporation into the pyroclastic flow (which may not necessarily be the case for lithics extracted from within the conduit or for cognate lithics), they will be heated up to the pyroclastic flow temperature for the duration of the transportational event and then they will become embedded in the pyroclastic flow deposit, wherein they have generally enough time to equilibrate with the deposit temperature and eventually cool to ambient temperature. Depending on the deposit temperature, during this second thermal stage, the original magnetisation component of each lava clast will be either: (1) completely, or (2) partially demagnetised. The first case applies to ignimbrites with emplacement temperatures above the Curie temperature of the magnetic minerals of the lava lithics (or other clast types if sampled) (Fig. 10.73b), whereas the second applies to ignimbrite emplacement temperatures below that Curie temperature (Fig. 10.73c).

After total or partial demagnetisation, a new magnetisation is acquired during the cooling of the ignimbrite, with a vector oriented parallel to the Earth's magnetic field at the time of ignimbrite emplacement. If the ignimbrite was emplaced above the Curie temperature, then lithic clasts would show only one new, high temperature magnetic component parallel to the Earth's magnetic field at the time of ignimbrite emplacement (Fig. 10.73b). If the ignimbrite was instead emplaced at a temperature below the Curie temperature, then the lithic clasts will have a two-component magnetisation: an original high temperature component with a random orientation in different lithic clasts and a low temperature component that is parallel in all lithics and

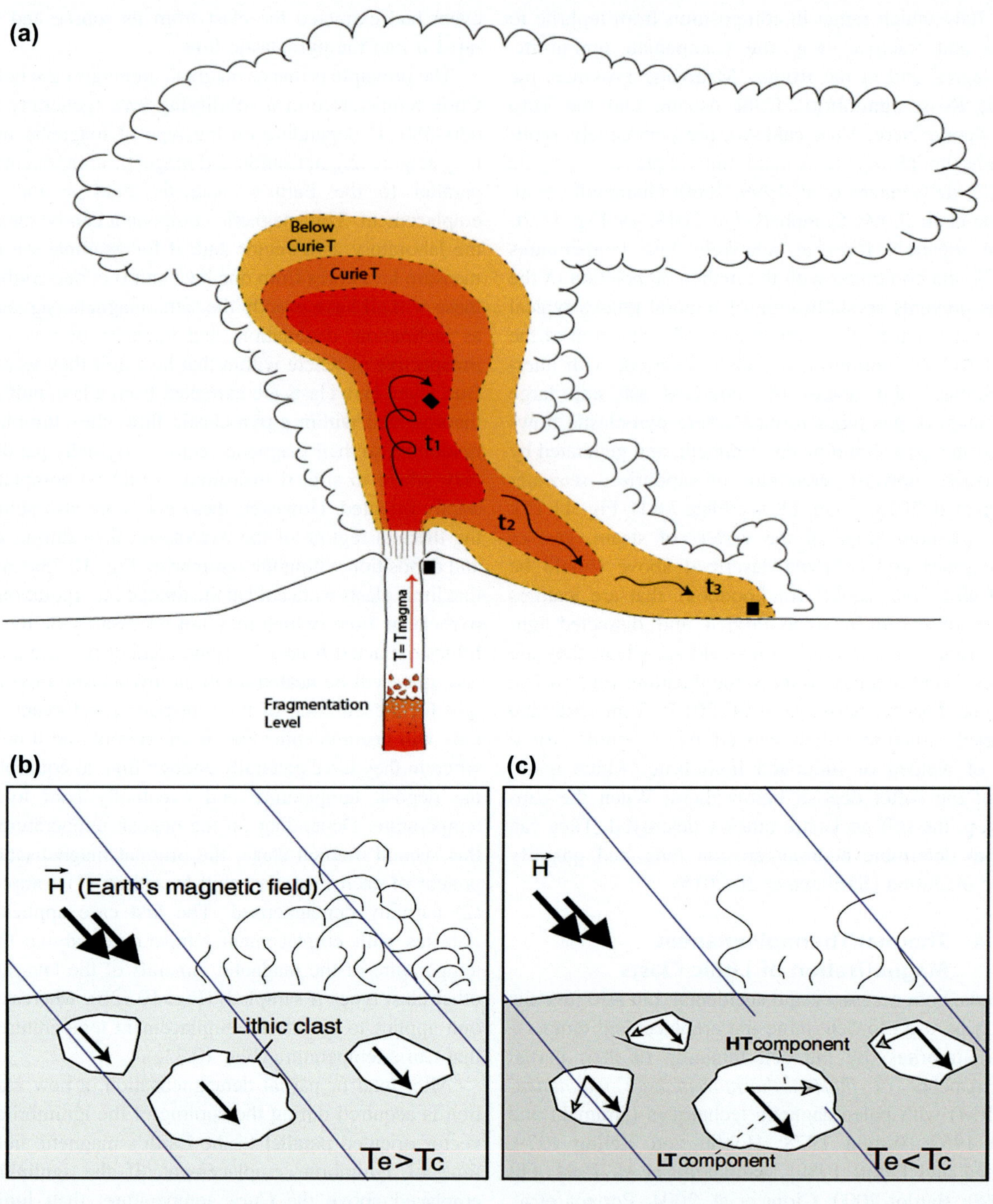

Fig. 10.73 Principles of acquisition of Thermal Remanent Magnetisation (TRM) magnetic components in lithic clasts in ignimbrites; see text for explanation. **a** If accessory lithics (or juvenile lithic clasts) ejected from the vent (black square) are entrained in the eruption column and then in any resulting pyroclastic flow, they are heated up. If small enough (1–2 cm) they may reach equilibrium with the pyroclastic flow temperature at emplacement. If small enough (1–2 cm) the lithics may reach equilibrium with the pyroclastic flow temperature at emplacement. That temperature may be above or below the Curie temperature for magnetic minerals in the lithic clast depending on the temperature structure of the pyroclastic flow. **b** If the emplacement temperature (Te) exceeds the Curie Temperature (Tc) of magnetic minerals in the lithic clasts the orientation of the Earth's magnetic lines of force (blue lines) and their strength, H, at that location, and magnetic polarity (black arrows) will be set within the clasts as TRM as they cool below the Tc. Clasts that already had a primary TRM from the time they originally formed, will be demagnetized and the TRM will be reset. (Modified from Pensa et al. 2015c). **c** In an ignimbrite that was deposited at temperatures below the Tc of magnetic minerals in the clasts, any pre-existing, high temperature TRM magnetic component in lithic clasts inherited from the original source rock, represented by heavy black arrows, will be preserved. However, upon deposition, the magnetic component in all clasts inherits a secondary, but weaker TRM (light arrows), that is oriented parallel to the Earth's magnetic field at the time of deposition, so creating two magnetic components. (From Lesti et al. 2011)

parallel to the Earth's magnetic field at the time of the ignimbrite emplacement (Fig. 10.73c). The emplacement temperature (T_e) of the lithic clasts can, therefore, be determined by progressive thermal demagnetisation (PTD) of the components present within the clast as well as their orientation in space. This technique is very valuable in non-welded deposits where other thermal indicators are absent.

Temperatures of emplacement have been found to be quite variable for different ignimbrites. Large volume ignimbrites such as the Cerro Galán ignimbrite are often emplaced at temperatures above the Curie temperature (>580°) throughout (Lesti et al. 2011), and is explained by very high mass eruption rates and a concomitant very low eruption column fountaining (no fall layer at the base of the ignimbrite), with very little interaction with the atmosphere and the environment (Cas et al. 2011). Smaller volume ignimbrites, originating by collapse of high columns, such as those of the historical plinian eruption at Vesuvius, the 4.6 ka Fogo A eruption, the May 18, 1980 Mt St Helens eruption show instead much lower temperatures, generally <400–500 °C (Cioni et al. 2004; Zanella et al. 2007, 2008; Pensa et al. 2015c; Giordano et al. 2018; Trolese et al. 2024), indicating that both the mass eruption rates and the amount of air-entrainment during the column phase and subsequent collapse probably played a significant role in the final temperature of the deposit. This is consistent with high emplacement temperatures retrieved from deposits from observed boil-over eruptions, such as 22 July 1980 ignimbrite at Mt St Helens (Banks and Hoblitt 1981, 1996; Paterson et al. 2010). Also, the amount of shallow derived accessory and accidental lithics entrained play a significant role in cooling the deposit (Martí et al. 1991) and have an effect on the emplacement temperature because lithic clasts act as a heat sink, absorbing heat from the pyroclastic flow and deposit (Pensa et al. 2015a). Giordano et al. (2018) showed that at localities where the 79 AD Vesuvius pyroclastic flows were able to knock down edifices and mix in wall debris, temperatures locally dropped by up to 200 °C with respect to the ignimbrite not contaminated by wall debris.

Variations of several hundreds of degrees have been found in the Minoan ignimbrite (200 to >580 °C; McClelland and Druitt 1989). For the Taupo ignimbrite, McClelland et al. (2004) found an increase in temperature with distance, suggesting that proximal sites were affected by the presence of an early caldera lake, indicating the local environment of emplacement may also play a significant role in the final emplacement temperature. Alternatively, cooling of the early part of the flow occurred because of its highly expanded, turbulent state, whereas later parts of the flow were less expanded, flowed farther, and lost less heat. Pyroclastic flow deposits that originate from phreatomagmatic phases usually show lower temperatures (<250 °C; Porreca et al. 2006, 2008; van Otterloo and Cas 2016; Trolese et al. 2017). However, Trolese et al. (2017) were able to constrain to just a few tens of degrees the downcurrent cooling of the VEI 6/VEI* 10 Colli Albani ignimbrites, suggesting that the pyroclastic currents, at least in their basal part that directly relates to the deposit, conserve thermal state, whereby most of their thermal characteristics are acquired in the very early collapse phases (Trolese et al. 2019).

Pensa et al. (2015a, c) determined the emplacement temperature of three ignimbrites from the 4.6 ka plinian eruption of Fogo volcano in the Azores, two resulting from partial collapses of the plinian eruption column, producing small volume intra-plinian ignimbrites interbedded with plinian fallout deposits, and the third, the climactic ignimbrite resulting from total collapse of the eruption column that ended the eruption. The first intra-plinian ignimbrite was emplaced at 350–400 °C based on pTRM analysis of accessory lithic clasts, the second intra-plianian ignimbrite, which is welded in places, was emplaced at ~600 °C, whereas the third climactic ignimbrite, gave a pTRM emplacement temperature of 250–370 °C (Pensa et al. 2015a, c). The differences in emplacement temperatures were attributed to different collapse heights of the column, and different abundances of juvenile and accessory lithic clasts for each pyroclastic flow event (Pensa et al. 2015a).

Whereas most TRM studies focus on analysing accessory and accidental clasts, it is worth noting here that Uehara et al. (2015) used the same approach on juvenile lava dome clasts from the block and ash flow and contemporaneous debris flow deposit sequence from the 1991–1995 eruption of Mt Unzen, Japan, providing another TRM approach to determining pyroclastic flow emplacement temperature. The approach and results are summarised in Sect. 10.13 below.

10.8.14.5 Reflectance of Charcoal Plant Fragments

Low-grade ignimbrites may preserve carbonised wood (Fig. 10.74a, b) which may be used as evidence of a hot state of emplacement (Scott and Glasspool 2005; Ascough et al. 2010; Caricchi et al. 2014; Pensa et al. 2015c, 2018, 2019, 2023; Trolese et al. 2018; Petrone et al. 2020a; Schito et al. 2022). Charcoal can be studied in polished section, because charcoal is highly reflective. Experimental work demonstrated that optical reflectance (R%) of the charcoal increases with the temperature of emplacement of pyroclastic flow deposits (e.g. Fig. 10.74b; Scott and Glasspool 2005). For a long time, coal has been studied by reflectance microscopy of polished blocks under oil to assess their diagenetic maturity or rank. Coal petrologists recognise fundamental particles called macerals that include inertinite: these are now widely believed to represent fossil plant cellular material (Scott et al. 2008).

Scott et al. (2008) found that the temperature range determined using charcoal reflectance (R%) for block and ash flow deposits at Soufrière Hills volcano was 325–525

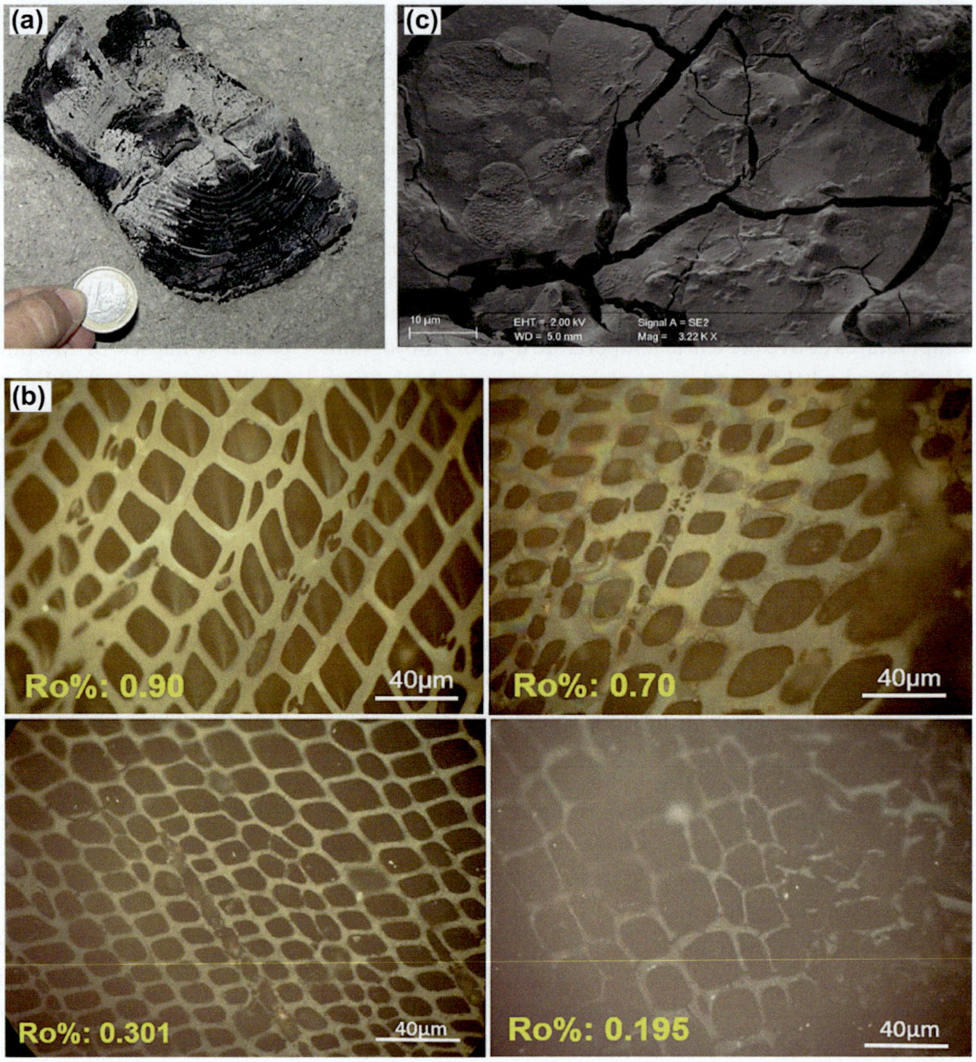

Fig. 10.74 Examples of organic materials found in ignimbrites. **a** Piece of carbonised tree trunk in non-welded 79 AD Vesuvius pyroclastic flow deposits, Herculaneoum, Italy. **b** Reflected light photomicrographs of four different polished thin sections from charcoal samples taken in the Villa dei Papiri archeological site in Herculaneum, buried under the 79 AD Vesuvius deposits. Samples have reflectance values R% varying from 0.9 and 0.195, which correspond to a temperature between 380 and 240 °C, respectively. (Modified from Caricchi et al. 2014). **c** SEM image of the exceptional finding of a vitrified human brain at Herculaneum, showing neural cells perfectly preserved. This indicates a complex process of sudden heat exposure during an early ash-cloud surge event of the 79 AD eruption of Mt Vesuvius and a time break before the arrival in town of the ensuing pyroclastic flow, allowing the fast cooling and vitrification of the body. (From Petrone et al. 2020b)

°C, compared with direct deposit temperature measurement of 365–640 °C. Hudspith et al. (2010) found a good agreement in trends between emplacement temperature values obtained for the 232 AD Taupo ignimbrite by this method and values obtained with TRM analyses. Caricchi et al. (2014) showed that charcoal embedded in the 79 AD Vesuvius ignimbrites provides reflectance data that indicate average deposit temperature around 380 °C (Fig. 10.74b), well in agreement with those retrieved from pTRM data. The specific case of the Herculaneum archeological site shows also the importance of thermal interactions between pdcs with buildings at a very local scale (Giordano et al. 2018).

Pensa et al. (2015a) calibrated the two methods, by collecting lithic clasts for TRM and charcoal fragments from the same stratigraphic level and localities for two ignimbrites from the 4.6 ka Fogo A eruption sequence. They showed that pTRM analysis of lithics and reflectance of charcoal give very similar emplacement temperature values in each of the two ignimbrites, which each had different emplacement temperatures (pTRM: 430 °C, R%: 380–460 °C; and pTRM: 300–350 °C, R%: 330–350 °C, respectively). A third ignimbrite, which is welded in places, gave a pTRM emplacement temperature of ~600 °C, but contains no charcoal (Pensa et al. 2015a), perhaps because at such temperatures, all carbonaceous material is destroyed. Based on excellent correspondence between pTRM and R% temperature estimates, Pensa et al. (2015c) confirmed that charcoal reflectance can be used as an independent method to evaluate the temperature of emplacement of pyroclastic flow deposits. Schito et al. (2022) also calibrated the signal of Raman spectroscopy on charcoal, which provides reliable measures of the paleotemperature.

An additional benefit of charcoal is that it can preserve evidence of peak temperatures, even for short thermal events, as the process of charcoalification is not retrograde. For example, at Colima volcano, Pensa et al. (2019) found temperatures as high as 500 °C in the charcoal of the standing, partly buried tree trunks which were affected by the 11 May 2015, 50 min-long, block-and-ash flow event, compared with the lower temperatures (300 - 480°C) of the charcoal embedded within the deposits just below. This was interpreted as the effect of an early inflated flow pulse at peak temperature, whereas the deposit below was emplaced diachronously from the wake of the flow at a lower temperature. In other cases, the charcoal reflectance shows bimodal or polymodal distributions such as at Merapi (2010 eruption; Trolese et al. 2018) and Herculaneum (Vesuvius 79 AD eruption; Petrone et al. 2020a), suggesting that the highest values may record the peak thermal event whereas the lower values of the equilibration with later cooler burial. In the latter case of Herculaneum, the reconstruction of both the detailed stratigraphy of the 79 AD Vesuvius eruption and the associated temperatures registered by polymodal reflectance of the entombed charcoal, allowed Pensa et al. (2023) to recognise that the first flow event hitting the town was a very hot ash cloud locally detached from pyroclastic flows at temperatures much higher than 500°C, which left a deposit of ash on ground only a few cm thick (deposit S1). It was however, able to instantaneously kill every person in town (Petrone et al. 2020a, b). This early ash cloud event was later followed by other pyroclastic flows which eventually buried the entire town but at lower temperatures (Caricchi et al. 2014).

Although limited to deposits with preserved carbonised wood, the charcoal reflectance method is very promising because of the high quality of the data it can produce about the thermal state of pdc deposits, but it is still under-utilised.

10.8.15 Accretionary Lapilli in Some Ignimbrites

Accretionary lapilli are lapilli-sized, spherical pellets of ash commonly exhibiting a concentric internal layering structure (see Chap. 4, Sect. 4.8.4). They have been described from pyroclastic fall (Chap. 9, Sect. 9.7.5), surge and flow deposits, and can even be reworked. They are believed to form by the accretion of fine ash around some nucleus, either a water droplet or solid particle, or both (Moore and Peck 1962; Walker 1981a; Gilbert and Lane 1994; Schumacher and Schmincke 1995; Mueller et al. 2016). The variety called cored or armoured lapilli have a recognisable lithic core and a thick (sometimes 1–2 cm) shell of unstructured coarse ash. It is important to stress that accretionary lapilli are not indicative solely of pyroclastic fall deposits, as often seems to have been assumed by workers in ancient successions.

Accretionary lapilli are found in young ignimbrites (Fig. 10.75a, b), as well as older, lithified ones (Fig. 10.75c, d). Several ignimbrites, with evidence of an origin through large scale phreatomagmatic explosive activity, contain accretionary lapilli and ash aggregates of various types that have been interpreted to have formed within the pyroclastic flow; e.g. the 25.4 ka Oruanui ignimbrite, New Zealand (Fig. 10.75a; Wilson 2001; Van Eaton and Wilson 2013; Sect. 10.9.2); the Pisolitic Tuffs succession at Colli Albani, Italy (De Rita et al. 2002; Giordano et al. 2010); ignimbrites of the Bandas del Sur, Tenerife (Brown et al. 2010). They are also certainly common in flow or surge deposits from small volume steam-rich phreatomagmatic eruptions (Self and Sparks 1978; e.g. the 5 ka Secche di Lazzaro succession, Stromboli; Giordano et al. 2008a, b), or phreatomagmatic phases of larger magmatic eruptions (Fig. 10.75b). Furthermore, they can form well away from vents in pyroclastic flows and their trailing ash clouds (Giordano et al. 2008a, b; Brown et al. 2010; Figs. 10.75d and 10.76), as well as in secondary eruption columns generated when pyroclastic flows interact explosively with surface water into which they flow (Morgavi et al. 2019; Chap. 11). As accretionary lapilli may have some mechanical cohesion and strength immediately after formation, it is certainly possible that some occurrences in ignimbrites may derive from entrainment within the pyroclastic flow of accretionary lapilli that have fallen out of the co-ignimbrite ash cloud, or directly from the buoyant column (Van Eaton and Wilson 2013). This is probably the case where accretionary lapilli are found concentrated at the top of depositional units, grading to accretionary lapilli-bearing ash-cloud (surge) deposits.

Accretionary lapilli can also be found concentrated in gas segregation pipes (next section) in phreatomagmatic ignimbrites (Fig. 10.77e, f). If they are already present in an ignimbrite deposit at the time of deposition, they could become concentrated, through elutriation of fine ash up through the gas pipes, leaving lags of accretionary lapilli in the pipes (e.g. the 25.4 ka Oruanui ignimbrite, New Zealand, Fig. 10.77e; Devonian Boyd Volcanic Complex, New South Wales, Australia, Fig. 10.77f), as also happens to dense lithic and crystal fragments described below. They could also form in situ within an ignimbrite through gases streaming up through pyroclastic flow deposits (Walker 1971).

10.8.16 Gas Segregation Pipes and Structures

Gas segregation pipes (aka gas escape or elutriation pipes) in ignimbrites were first recognised by Walker (1971, 1972), who described them as “fossil fumaroles”. Walker demonstrated that they were enriched in heavy components (crystals and lithic fragments; Fig. 10.77, and see Fig. 10.45), although they can also contain abundant accretionary lapilli (Fig. 10.77e, f). He further suggested that the pipes formed by gas streaming through the ash matrix of a pyroclastic flow after deposition, elutriating fine ash from the ignimbrite matrix up and out of the ignimbrite, so forming fines-depleted pipe-like structures. Since Walker’s studies, gas segregation pipes have been commonly found in both

Fig. 10.75 Accretionary lapilli in ignimbrites. **a** Abundant accretionary lapilli in the phreatomagmatic 25.4 ka Oruanui ignimbrite, New Zealand. Note the spherical form, and internal concentric layering of ash. Accretionary lapilli are up to 1 cm in diameter. **b** Accretionary lapilli in dilute pdc deposits associated with phreatomagmatic first phase (Unit B) of the 161 ka Kos Plateau Tuff ignimbrite eruption sequence, Kos, Greece. **c** Accretionary lapilli dispersed through the ash matrix of an Early Devonian ignimbrite, Snowy River Volcanics, Victoria, Australia. Note the fine concentric ash layer rimming most accretionary lapilli. **d** Example of accretionary lapilli (finger) formed within the ash-cloud of a dilute pdc showing characteristic couplet of layers (as explained in Fig. 10.76), Early Devonian, Snowy River Volcanics

young and old ignimbrite deposits (Fig. 10.77). In older successions, they can be especially important for distinguishing pyroclastic flow deposits from epiclastic debris-flow deposits (e.g. Duyverman and Roobol 1981), although some care should be taken as mass-flow deposits may preserve somewhat similar water escape pipes (Postma 1983; Chap. 11). They have also been produced during fluidisation experiments (Wilson 1980; Druitt 1995; Roche et al. 2001a).

Gas segregation structures vary from only several cm to generally ≤ 2 m long and a few grain diameters to several cm wide and are characteristically depleted in fines and enriched in crystals and lithics (Cas and Wright 1987; Pacheco-Hoyos et al. 2020; Figs. 10.77 and 10.45). Much larger segregation pipes may occur in coarse, near-vent co-ignimbrite breccia facies (see Fig. 10.50f, g). In cross-section they may have irregular shapes, reflecting complex geometries in three dimensions; pod-like structures may also be found (cf. experiments of Wilson 1980; see Pacheco-Hoyos et al. 2020 for detailed description and classification). Such gas segregation structures commonly occupy the upper half of a depositional unit, and are now thought to form where the rate of aggradation of the deposit (or *deposition-triggered aggradational segregation*) is high enough to trap significant amounts of gas between particles, which then migrates upward following deposition (Giordano and Dobran 1994; Pacheco-Hoyos et al. 2020). For this reason, truncation of gas segregation pipes could be a good indicator of the occurrence of short time breaks between the emplacement of successive depositional units (Fig. 10.77a).

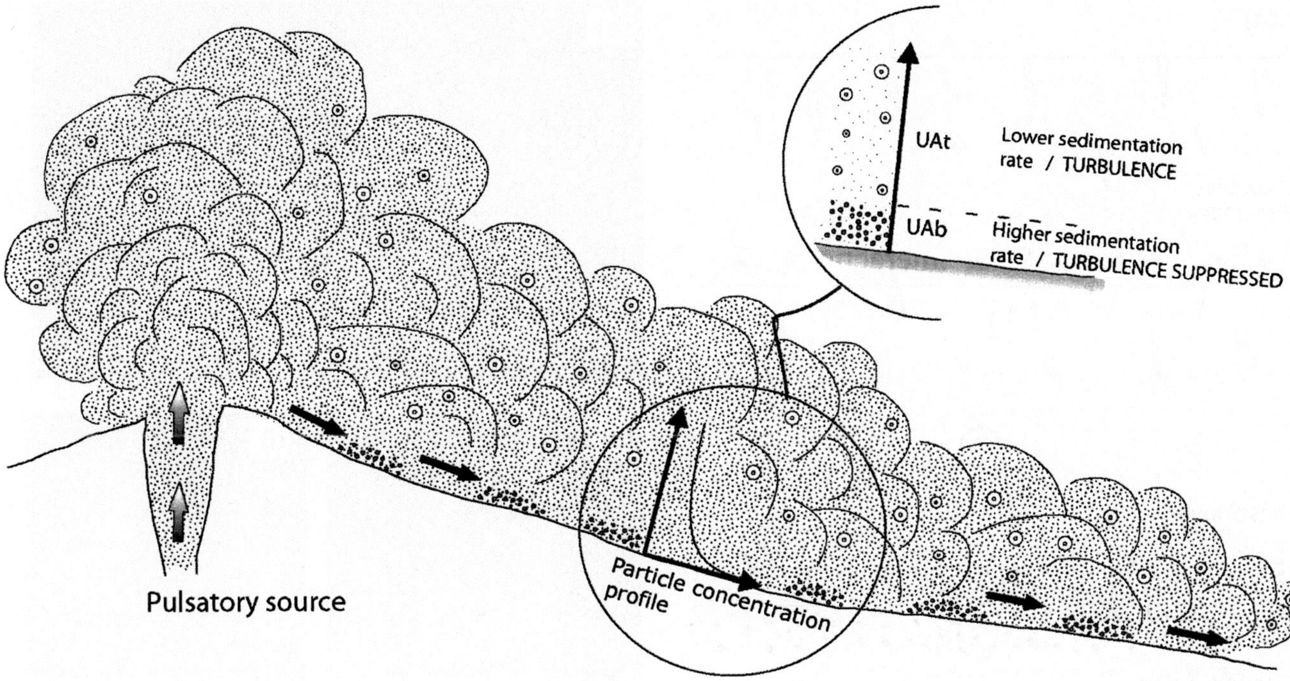

Fig. 10.76 Model for formation of accretionary lapilli within the ash cloud of a pyroclastic flow. Individual beds comprise a distinctive couplet of layers: basal layer (UAb) and top accretionary lapilli-rich layer (UAt). UAb layers were emplaced from an early, basal, higher concentration zone, characterised by a higher sedimentation rate, which would suppress turbulence. UAt is generated in a more dilute traction zone, where ash can aggregate to form accretionary lapilli during the turbulent transport. (From Giordano et al. 2008b)

However, these may also simply indicate times where aggradational segregation was reduced, only to be later reactivated during the same advancing flow, which may account for more complex pipe shapes and anastomosing geometries, and truncations (Pacheco-Hoyos et al. 2020). Post-emplacement upward gas flow could also occur for some period of time (Fig. 10.77a, h). In a study of the lower and upper portions of the Huichapan Ignimbrite (Mexico), Pacheco-Hoyos et al. (2020) concluded that deposition-triggered pipe formation is enhanced by poor sorting, low shearing, and presence of large clasts (>32 mm). Pipe-free ignimbrites likely derive from relatively better-sorted dispersions with few coarse clasts, although more importantly, strong shearing during deposition may be the main process that diminishes pipe formation by enhancing more diffuse gas escape.

Additionally, gas segregation pipes may also be found at the bases of ignimbrites due to volatilisation of substrate ground water onto which the hot deposits are emplaced (Cas and Wright 1987; Fig. 10.77a). Good examples of this type of pipe are found in the Minoan ignimbrite, where flow units are interbedded with torrent deposits and heated ground water was the gas source (Fig. 10.77g; Bond and Sparks 1976), and the 151 ka Sutri Formation from Vico volcano, Italy (Bear et al. 2009c). Gas segregation pipes can also form when vegetation is entrained into pyroclastic flows and is combusted, forming gas pipes above burnt vegetation especially large logs (Fig. 10.77h, i). Gas segregation features may be overprinted by vapour-phase crystallisation (Fig. 10.77j), and/or incorporated within larger fumarolic structures (Fig. 10.72d; Caporuscio et al. 2012).

Small gas segregation pipes have only rarely been recorded from pyroclastic surge deposits. Examples are found in ash-cloud surge deposits in the Upper Bandelier Tuff (Fisher 1979; Self et al. 1996). These cut the internal lamination of the surge deposits and were formed by post-emplacement gas flow from the underlying parent pyroclastic flow unit.

10.9 Pumice and Ash Flow Felsic Ignimbrite Deposits

Pumice and ash flows are the typical pyroclastic flows associated with the explosive eruption of felsic magmas and deposition of ignimbrites. There are two general depositional settings for ignimbrites with respect to the vent: outside of the vent, commonly called *outflow ignimbrites*, and inside the vent, generally referring to ignimbrite deposits ponded within calderas, hence called *intra-caldera ignimbrites*.

In the following sections, we refer to a number of case studies to compare and contrast some of the range and variations in depositional facies and eruption and transport dynamics. Firstly, we summarise two end-member examples

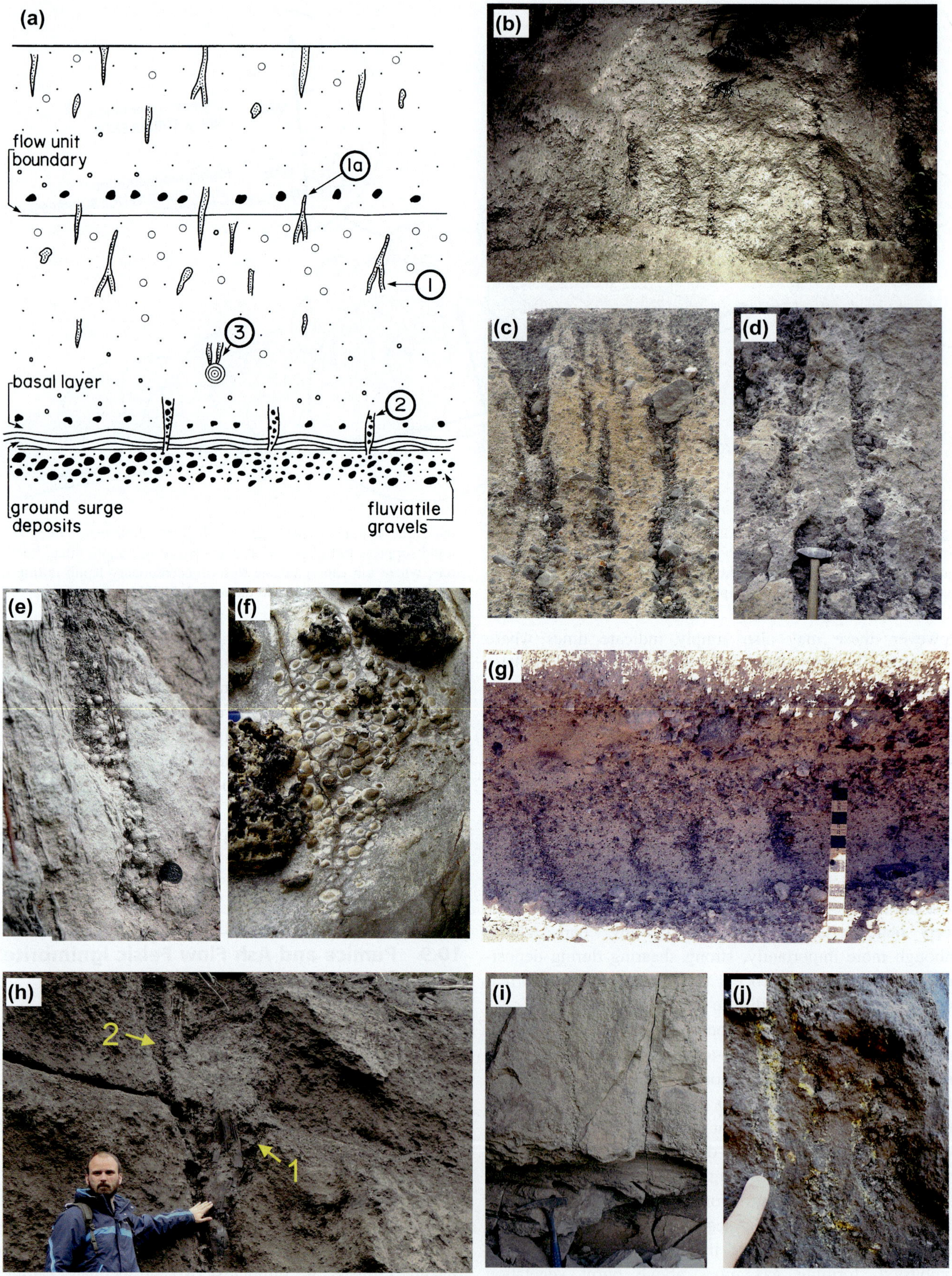
(a)
flow unit boundary
1a
1
3
basal layer
2
ground surge deposits
fluviatile gravels
(b)
(c)
(d)
(e)
(f)
(g)
(h)
2
1
(i)
(j)

◀ **Fig. 10.77** Occurrence of gas segregation or escape pipes in ignimbrites. **a** Types of gas segregation structure. 1 Pipes and pods formed entirely by intraflow gas sources during aggradational segregation and/or further post-depositional upward gas flow (located in the upper parts of depositional units); 1a formed by post-emplacement upward gas-flow (cross-cutting depositional units); 2 formed from volatilised groundwater; 3 formed above burnt vegetation and carbonised logs. (From Cas and Wright 1987). **b** Typical lithic-enriched, fines-depleted, vertical gas segregation pipes in pumiceous 1400 yr BP Rabaul ignimbrite, Rabaul, New Britain. **c** Swarm of very coarse-grained gas segregation pipes in the 1600 BC Minoan ignimbrite, Oia, Santorini, Greece. The pipes are filled with lithic clasts, resulting from the elutriation of ash as gas streamed through the pipes, leaving a lag of lithic debris, which could not be ejected by the gas stream. The height of the outcrop is 2 m. **d** Gas segregation pipes also filled with coarse lithic fragments, 1.26 Upper Bandelier ignimbrite, Pajarito Plateau, New Mexico, USA. **e** Abundant accretionary lapilli in gas segregation pipe, phreatomagmatic 25.4 ka Oruanui Formation ignimbrite, New Zealand. **f** Accretionary lapilli concentrated in a gas segregation pipe in a phreatomagmatic ignimbrite, Late Devonian Boyd Volcanic Complex, Eden, New South Wales, Australia. **g** Small, lithic enriched gas segregation pipes formed at base of ignimbrite flow unit deposited on top of fluviatile flash flood deposits, Minoan ignimbrite, Monolithos, Santorini. **h** Coarse, gas segregation pipes issued from a carbonised tree trunk within the intra-plinian ignimbrite of the 2015 Calbuco eruption, Chile. The tree trunk is standing within the lower depositional unit of the ignimbrite and was hewed down by the following pdc. Note that the right branch of the gas pipe (1) stops at the boundary between the lower and the upper flow units, separated by a thin pumice fallout layer which testifies a short time break in pdc deposition; by contrast the left branch (2) crosses the two flow units as the trunk must have kept carbonising after emplacement of the upper unit and gas must have "chosen" that as its preferential way for outgassing. (Photo by Daniele Morgavi). **i** Irregular gas segregation pipes rising off carbonised log (hammer), 561 ka Tor de' Ceni phreatomagmatic ignimbrite, Colli Albani volcano, central Italy. **j** Coarse gas segregation pipes displaying post-emplacement vapour-phase alteration in the 157 ka Ronciglione ignimbrite, Vico caldera, central Italy

(the 232 AD Taupo eruption, of Lake Taupo, New Zealand, and the 2.08 Ma Cerro Galán ignimbrite, Argentina; Table 10.2) illustrating normal pyroclastic flows versus violent pyroclastic flows found in outflow settings. Next, we consider large volume rhyolitic phreatomagmatic ignimbrite-forming eruptions, principally the 25.4 ka Oruanui eruption also erupted from the Taupo Volcanic Zone, New Zealand. Here we demonstrate that phreatomagmatic explosive activity need not be a small-scale phenomenon and that large volumes of phreatomagmatic ignimbrite can be produced. Lastly, we describe the large volume (>1390 km^3 DRE) Permian (274 Ma) Ora Formation (northern Italy), where the extremely thick intra-caldera ignimbrite is very well exposed and preserved along the entire extent of the source 42×40 km^2 caldera.

10.9.1 Outflow Ignimbrites: The "Tortoise and the Hare"

Studies over many years have identified two end-member pyroclastic flows and their ignimbrite deposits: "normal" pyroclastic flows, and "violent" pyroclastic flows. Normal pyroclastic flows are relatively slow moving from at least medial and into distal settings, and their deposits are largely topography confined. By contrast, "violent" pyroclastic flows are fast flowing, at least in proximal to medial settings, and can surmount significant topography, leaving deposits on ridges as well as in valleys, and are marked by distinctive flow head deposits, resulting from very energetic processes. We have coined these two end-member pyroclastic flow types, "tortoises" and "hares". Tortoise-like flows produce a range of volumes of deposits, from small, to extremely large (super-eruption scale), whereas hare-like pyroclastic flow deposits are in the small to medium scale volume range.

The 2.08 Ma Cerro Galán ignimbrite (Argentina) is a typical large volume, monotonous intermediate composition ignimbrite, which in spite of its >630 km^3 DRE volume shows field characteristics of very passive emplacement, so we take it to represent the deposit of a "tortoise-like" pyroclastic flow. The Taupo ignimbrite (New Zealand), emplaced during the 232 AD eruption of Lake Taupo caldera, is a medium volume (35 km^3 DRE), crystal-poor rhyolitic ignimbrite associated with probably the most powerful and fastest pyroclastic flows documented, based on extent, relationship of deposits to topography, and distinctive facies types. It is the type "hare-like" pyroclastic flow, which are able to surmount hills and mountains hundreds of m high, and to deposit a very low aspect ratio deposit found both ponded in valleys as well as draping highs forming veneer deposits. We will describe the 232 AD Taupo ignimbrite first because of its historical precedence in terms of research.

10.9.1.1 "Hare-Like" Outflow ignimbrites—The Great Taupo 232 AD Eruption

The studies by Walker (Walker 1980, 1985; Walker et al. 1980a, b, 1981b, c), Wilson (Wilson and Walker 1981, 1983, 1985), Houghton (Smith and Houghton 1995; Houghton et al. 2010, 2014), and co-workers on the products of this ignimbrite-forming eruption have greatly stimulated volcanology, yielding new insights into explosive volcanism and extending the range and scale of known volcanic phenomena. The eruption produced a great variety of pyroclastic products, including perhaps the most powerful plinian and most violent ignimbrite-forming events yet documented. The youth and excellent preservation of the deposits enabled the eruption sequence to be examined and reconstructed in such detail.

The 232 AD Taupo Eruption Sequence

The products of the 232 AD Taupo eruption (Fig. 10.78a, b) are dispersed over a large part of the North Island of New Zealand. All the deposits are rhyolitic, with a mean SiO_2 content of 74%, low crystal content (2–3.5 wt%), and a restricted pre-eruption magmatic H_2O content between 3.6 and 4.3 wt%. The eruption sequence (Wilson and Walker 1985; Houghton et al. 2010; Fig. 10.78a) consists of: an initial minor phreatomagmatic ash (Unit 1), followed by a well-sorted, typical plinian pumice-fallout deposit, the Hatepe pumice (Unit 2), two very fine-grained, stratified phreatoplinian ashes, the Hatepe ash (Unit 3) and the Rotongaio ash (Unit 4), which have small coalesced accretionary lapilli, and which are separated from each other in proximal areas by an erosional break, then a very widely dispersed "ultra-plinian" pumice-fallout deposit, the Taupo pumice (Unit 5; see discussion Chap. 9, Sect. 9.7.4.4), followed by the climactic Taupo ignimbrite (Unit 6), and finally a lava dome (Unit 7). The total erupted volume was about 100 km^3 (35 km^3 dense rock equivalent), of which 3.5 km^3 (DRE) constitutes the early eruption phase deposits, 5.8 km^3 (DRE) is found in the Taupo ultra-plinian pumice fall deposit, 12.1 km^3 (DRE) in the Taupo ignimbrite, another 6–8 km^3 (DRE) in a co-ignimbrite ash fall, and the rest represents a speculative intra-caldera volume.

The total thickness of air-fall deposits amounts to over 5 m near the vent, nearly all of the material having been blown to the east of the vent by strong south-west to westerly winds. Airfall deposits more than 10 cm thick were deposited over an area of >30,000 km^2, while an area of about 20,000 km^2 was devastated by the Taupo ignimbrite-forming flow (Fig. 10.79; Wilson and Walker 1985). Close to vent the Taupo ignimbrite eroded earlier fallout phases and soil substrate from the ground surface, and uprooted mature trees over 30 km away (e.g. Figs. 10.37a, b and 10.78). In addition, mud flows and floods reached far beyond the limits affected by primary eruption products. The duration of the

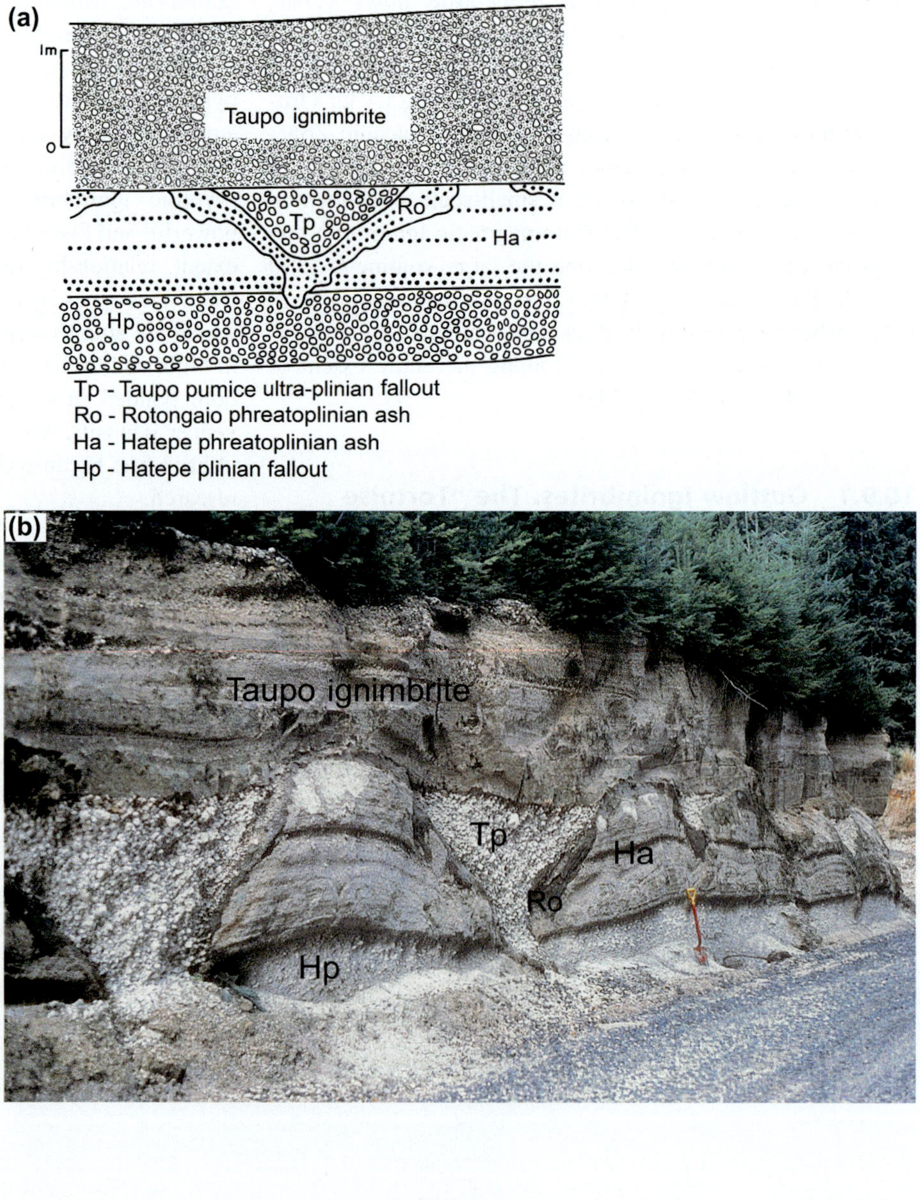

Fig. 10.78 Stratigraphy of the 232 AD Taupo eruption, Taupo Volcanic Zone, New Zealand. **a** Simplified sketch showing stratigraphic relations, based on sequence of aggradational (depositional) and degradational (erosional) events deduced from the outcrop shown in (**b**). **b** Outcrop reveals prominent infilled erosion gullies between the Hatape pumice and Rotongaio ash, and strongly discordant erosion surface between the Taupo ignimbrite and Taupo ultra-plinian pumice. Outcrop is 20 km ESE of the vent; see text for further discussion. (From Cas and Wright 1987; Photo by Colin Wilson)

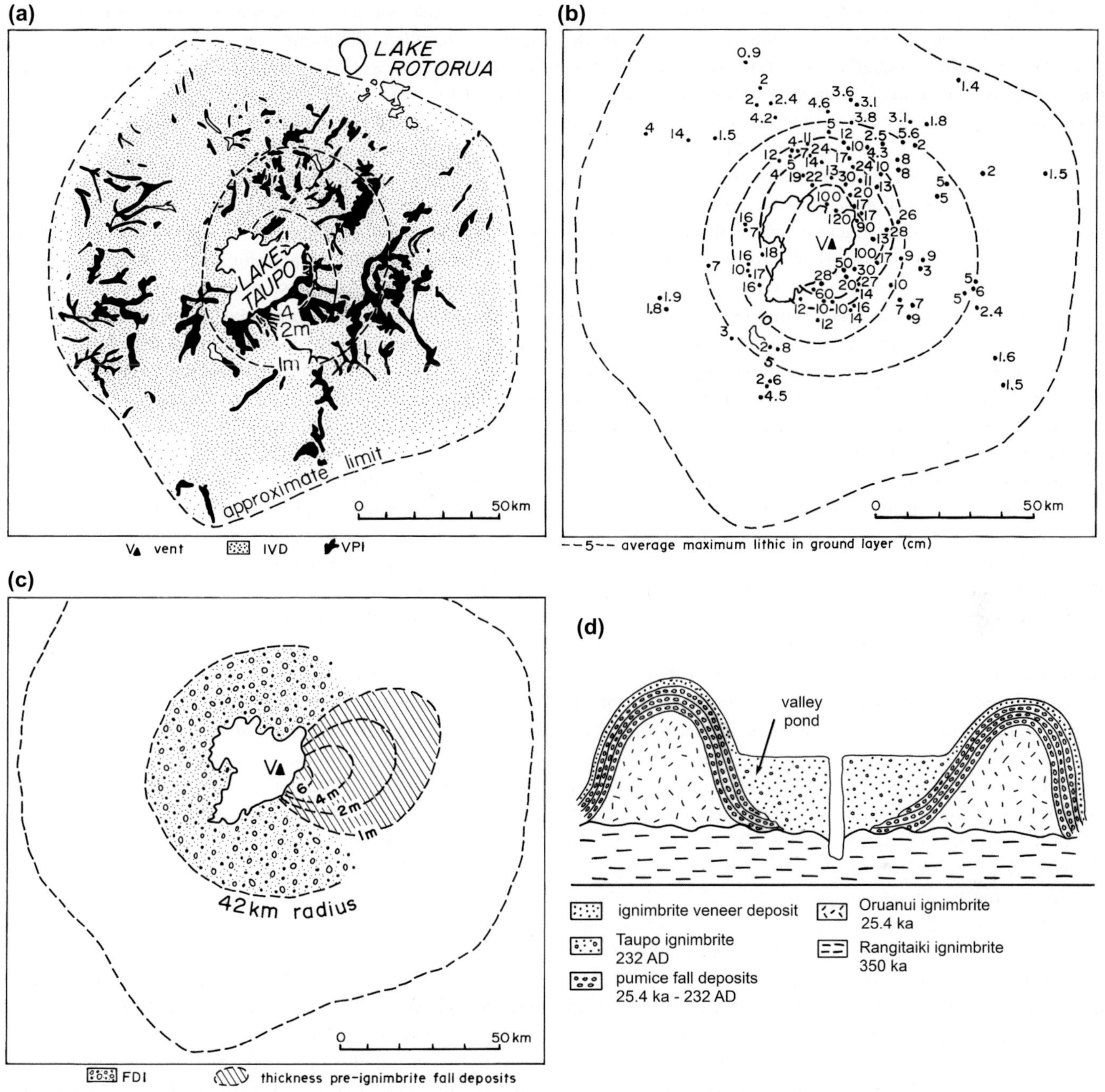

Fig. 10.79 Areal extent and relations of the 232 AD Taupo ignimbrite. **a** Map of the Taupo ignimbrite showing its radial distribution around the source Taupo caldera, irrespective of topography, and distinguishing the ignimbrite veneer deposit (IVD) found on topographic ridges (stipple), from valley-ponded ignimbrite (black). Isopachs show the average thickness of IVD. **b** Isopleth map of the average maximum diameter of the three largest lithic clasts in the Taupo ignimbrite ground layer. **c** Distribution of fines-depleted ignimbrite and pre-ignimbrite fall deposits excluding the Taupo pumice. **d** Schematic geological relations east of Lake Taupo. Valley pond of Taupo ignimbrite is up to about 40 m thick. The deep valley shown is floored by the welded Rangitaiki ignmbrite (Whakamaru Group), while both the Oruanui and Taupo ignimbrites are completely non-welded. Note the distinctive erosion surface created on top of the Oruanui ignmbrite which was influenced by climatic conditions created by the Last Galacial Maximum until ~ 17 ka (Wilson and Manneville 2004). (From Cas and Wright 1987, After Walker et al. 1980b, 1981a, b)

entire eruption is unclear. The locally developed erosional scoured contact between the Hatepe ash and Rotongaio ash (Fig. 10.78) is not thought to represent a significant time break, because elsewhere the two are concordant (Smith and Houghton 1995). Walker et al. (1981a) believed that this gullying represented a brief episode during the eruption when a large part of the Lake Taupo was erupted as a giant "volcanogene waterspout" (see Cas and Wright 1987, Fig. 8.50

c). If this is correct, the whole eruption sequence, except for the lava dome, could represent as little as a day or several days. The lava dome (including giant floating pumice blocks; Chap. 16) may have been erupted over weeks, months, or even tens of years after the explosive event. If the Taupo eruption were repeated, it would have catastrophic effects in New Zealand, given the modern demographics.

From various lines of evidence, the vent position migrated along a NE trending fracture fissure during the whole eruption, from south for the Hatepe phases, north for the Rotongaio phase and central for the Taupo phases, at or near the Horomatangi Reefs in Lake Taupo (Wilson and Walker 1981; Smith and Houghton 1995; Houghton et al. 2010; Cole et al. 1998). Horomatangi Reefs may be the remnants of a lava dome. Houghton et al. (2010) studied juvenile clast vesicle characteristics and showed that vesicularity was high for the Hatepe phases (mean ~75%) low for the Rotongaio phase (~45%) and extremely high for the Taupo phase, rising from 79% for the initial fallout to ~80% for the ignimbrite phase. The pumice clasts in the Hatepe (Units 2, 3) and Taupo phases (Units 5, 6) show very high bubble number densities, small vesicles, and little evidence of bubble coalescence, implying very rapid ascent. The Rotongaio pumice clasts (Unit 4) have low bubble number densities, widely spaced bubbles, which often display shearing, collapse, and coalescence textures; the interpretation is initial rapid ascent and then a slower ascent involving upper crustal storage before eruption. The final effusive dome (Unit 7) has moderate bubble number densities, large complex coalesced vesicles, suggesting steady ascent. The pre-eruption magma volatile content was between 3.6 and 4.3 wt% H_2O, but magma discharge rates varied from 10^5 to 10^6 kg s^{-1} for the phreatomagmatic phases (Units 1, 3, 4) to 10^7 for the Hatepe plinian, to 10^8 for the Taupo ultra-plinian, to 10^{10} kg s^{-1} for the Taupo ignimbrite, and finally 10^3 kg s^{-1} for the final effusive phase (Houghton et al. 2010). The sequence of phases can be interpreted in terms of varying magma ascent and discharge rates, relocation of vent, varying degrees of magma–water interaction (Talbot et al. 1994; Houghton et al. 2010), perhaps late-stage tumescence of the caldera, so exposing a dry-land vent, and subsequent detumescence submerging the vent (also see Cas and Wright 1987, Fig. 8.50).

Taupo Ignimbrite

The abrupt switch from the Taupo ultra-plinian fall to Taupo ignimbrite flow-forming activity is thought to have been caused by a drastic increase in mass eruption or discharge rate from 10^8 to 10^{10} kg s^{-1}, as indicated by textural evidence of pumice clasts (Houghton et al. 2010), and cannot simply have been due to the collapse of the very high ultra-plinian column. Wilson and Walker (1985) suggest that the high eruption rate during the ultra-plinian phase drained the upper part of the magma chamber, triggering the caldera collapse and the development of a 10 km long fissure vent. This greatly enlarged the vent, suddenly increasing the mass discharge rate beyond that which would allow for the maintenance of a stable convective column (cf. Fig. 10.9). The ignimbrite-producing flow then formed by column collapse. The lithic content of the pre-ignimbrite airfall deposits (1 km^3) implies that vent widening by explosive erosion was important and would eventually have led to column collapse if the eruption had followed a normal course of events. Indeed, occasional partial collapses of the ultra-plinian column did occur, generating a number of early ignimbrite units found interbedded with the Taupo pumice near the source (see Fig. 10.81a). A sudden increase in the volume (2 km^3) and sizes of lithics in the Taupo ignimbrite support the idea that there was a drastic change in eruption conditions (Cole et al. 1998).

Wilson and Walker (1981) suggested that the parent flow of the Taupo ignimbrite erupted over only 400 s, in batches of material which gradually coalesced so that from about 40 km outwards the flow was a single wave of material. During most of its passage, the flow was interpreted to consist of a head (strongly fluidised by ingested air) which generated Layer 1 deposits (ground breccia, fines-depleted jetted ignimbrite; there are no conventional frontal or ground surge deposits), and a body plus tail which generated Layer 2 deposits. The flow initially moved at very high velocities over a deeply dissected topography incised into deposits of the 25.4 ka Oruanui Formation (Fig. 10.79a, d), and over a locally mountainous terrain at speeds probably exceeding 250–300 m s^{-1} near the vent, and remained high (locally >100 m s^{-1}) to the outer limits of the flow. The flow-head was highly erosive and locally in proximal areas, almost unbelievably, scalped and overturned strata of the early fallout phases over which it passed (Fig. 10.80a, b). A variety of processes acting in response to fluidisation or flow kinetics operated within the flow to produce the great variety of facies.

An outstanding feature of the Taupo ignimbrite is that it was deposited radially up to 60 km from vent and all elevations, covering an area >20,000 km^2 (Fig. 10.79a–c). The ignimbrite occurs in two contrasting forms, each having quite different relationships to the pre-existing land surface (Figs. 10.34 and 10.79a, d). One is a landscape mantling *ignimbrite veneer deposit* (IVD; Fig. 10.35g), while the other fills in depressions as "conventional" ignimbrites do and is called *valley pond ignimbrite* (VPI; Figs. 10.35d and 10.81a). The IVD is stratified and occasionally shows bedforms (Fig. 10.35g) and pumice concentration lenses on the lee side of ridges (Figs. 10.34 and 10.81c). All of the evidence suggests this was left behind as a "trail-marker" on high topography by the rapidly moving Taupo flow. The Taupo VPI facies is fairly normal, massive, poorly sorted, matrix-supported ignimbrite (Fig. 10.81b).

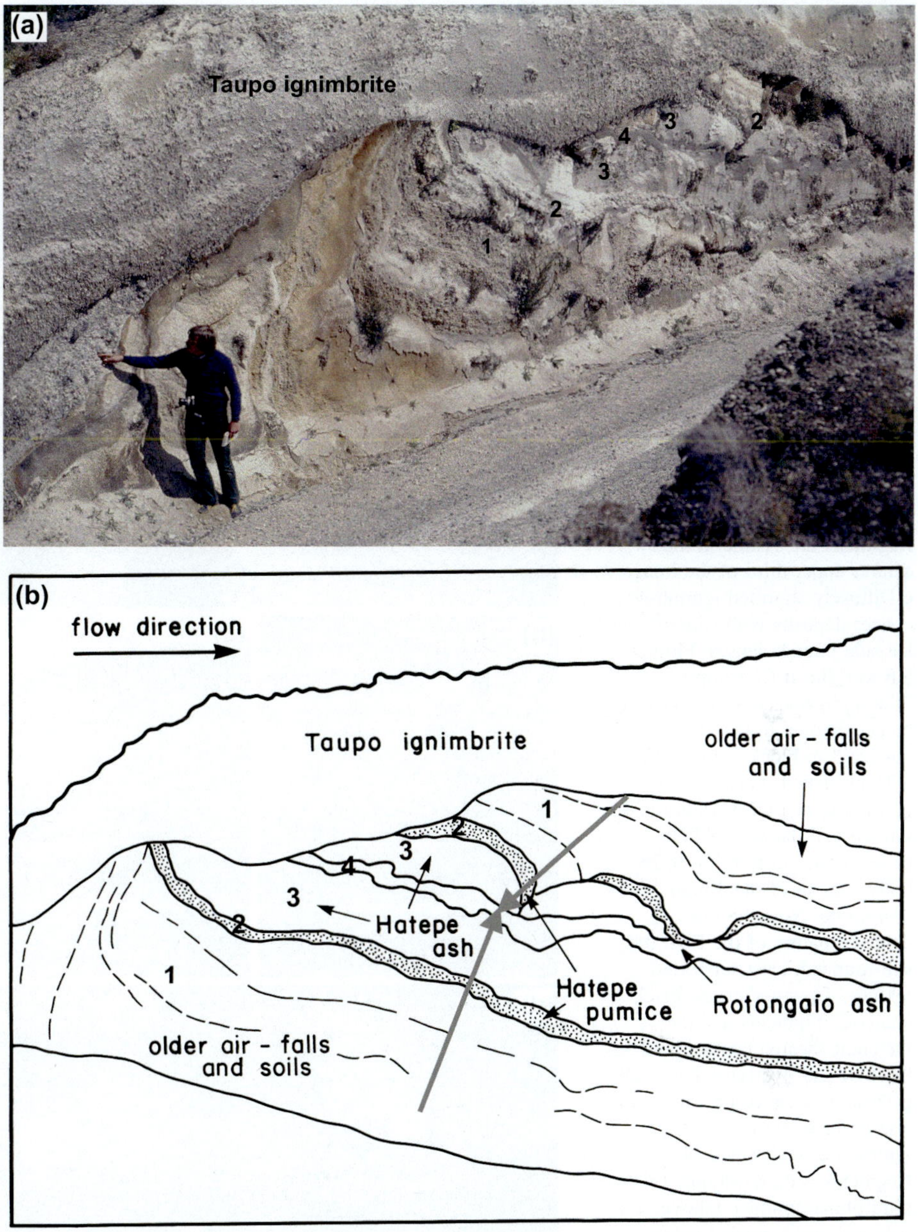

Fig. 10.80 Erosional capacity of the Taupo 232 AD pyroclastic flow. **a** The base of the 232 AD Taupo ignimbrite truncates bedding in the early Taupo eruption deposits that have been recumbently folded over by the base of the flow, when the flow was in a highly erosional, waxing flow stage. The massive deposit overlying the erosional discordance was deposited during the waning flow stage. 1 = older fallout deposits and soils; 2 = Hatepe pumice; 3 = Hatepe ash; 4 = Rotongaio ash. **b** Interpretative sketch of the outcrop shown in (**a**), from a slightly different perspective. Location is 15 km east of the vent. The Taupo ultra-plinian pumice is absent, presumably eroded prior to the overfolding event. Grey arrows show stratigraphic younging directions. (Modified from Cas and Wright 1987)

In addition, the Taupo ignimbrite has two distinctive Layer 1 facies developed, including a lithic-rich ground breccia, and a jetted, fines-depleted, ignimbrite facies. Ground breccias in the Taupo ignimbrite (Fig. 10.81a, d) extend far from the vent (Fig. 10.79b). They have been interpreted as generated in the extremely fluidised head of the high velocity Taupo flow (Walker et al. 1980b, 1981b). The >10 cm average maximum lithic size isopleth extends 35 km approximately radially around the vent (Fig. 10.79b).

The Layer 1 fines-depleted ignimbrite (FDI; Fig. 10.81e, f) is very distinctive in the field, and is characterised by large rounded to subrounded framework pumice clasts, with few fines and abundant charcoal fragments (Fig. 10.81f). The FDI is restricted to an area where vegetation was not covered by earlier co-eruptive fall deposits (Fig. 10.79c). The FDI was interpreted to be material jetted out from the flow head (Wilson and Walker 1985), although other interpretations are possible (Sect. 10.8.10.3). The fines-depleted facies extend 42 km from the vent. At its outer limits, the Taupo ignimbrite is represented by distal facies which comprises a thin, charcoal-bearing ash and fine-grained lapilli deposit, often soilified (Fig. 10.81g).

To reconcile the end-member characteristics of the Taupo ignimbrite (highly energetic flow head and ignimbrite veneer deposits, together with normal, massive, poorly sorted valley pond ignimbrite), we propose that the flow front or head of the Taupo pyroclastic flow behaved like a blast flow (Sect. 10.12) capable of flowing over topography, producing the ground breccia, fines-depleted ignimbrite, and the ignimbrite veneer deposits. This was followed by the trailing, deflating part of

Fig. 10.81 Some facies of the 232 AD Taupo ignimbrite. **a** Valley pond ignimbrite facies (VPI) passing into ignimbrite veneer deposit (IVD), overlying very coarse ground breccia (GB). This is at 8 km east of the source vent, where the Taupo ignimbrite erodes a local, near-vent, precursor fine-grained pdc deposit (white layer below orange dashed line). Flow is left to right. (Also see Fig. 10.35c for images of VPI, and 10.35f, 10.37 for IVD facies). **b** Normal, Taupo ignimbrite valley ponded facies, showing typical poor sorting and matrix support texture. Note that pumice clasts are sub-angular to sub-rounded. Irregular charcoal fragments are visible in the centre, upper third of the image. **c** Diffusely stratified ignimbrite veneer deposits with (ridge) lee-side pumice lenses. Flow is left to right. **d** Close-up of lithic-rich, fines depleted ground breccia shown in (**a**); the large outsize 3 m block is mudstone. **e** Thick, jetted, fines depleted ignimbrite or FDI. Locality is 30 km west of the vent. **f** FDI beneath the ash rich Layer 2b body deposit of the Taupo ignimbrite. Note the fines depleted nature of the FDI, the sub-rounded, clast-supported texture of pumice clasts, black charcoal fragments and the incipient sheared contact between the FDI and overlying Layer 2b. **g** Distant facies comprising thin, fine-grained (soilified) Taupo ignimbrite (with charcoal fragments = c), overlying Taupo ultra-plinian pumice (above paleosoil, with spade blade). Locality is 50 km east of the vent

the pyroclastic flow, progressively forming a more standard, high concentration valley-confined undercurrent, moving as a relatively slow moving fluidised granular flow down valley axes. In addition, during deflation there should have been a transitional regime where the internal stratification of the flow became progressively better developed with time and distance. Wilson (1985) suggests deflation from a highly turbulent flow began from about 13 km from the caldera. This suggests that the Taupo pyroclastic flow was both a hare (initially) and then a more normal tortoise (later). Even hares can transform from initial turbulent flow systems to segregated basal granular flow, as proposed generically by Sparks et al. (1978) for pyroclastic flows.

The origin of the initial "violent" blast-like phase, which is a rare occurrence, but not unique to the Taupo ignimbrite (e.g. the Campanian ignimbrite, Campi Flegrei caldera; Fisher et al. 1993), remains a major conundrum that is still unresolved. As discussed in Sects. 10.3, 10.4, and 10.8.14, a great height of collapse cannot be invoked as a cause for an excess in the pyroclastic flow mass flux (see Fig. 10.7a), and in the case of the Taupo eruption (as for the Campi Flegrei caldera) not even the height of the volcano, which is a large caldera depression, can support the simple transformation of potential energy onto kinetic energy to justify the violence of the pyroclastic flow. One possibility is that excess volatiles, for example, released from a suddenly depressurised geothermal system, may become involved in the early caldera collapse phase and drive massive overpressure in the pyroclastic flow system. Or, extremely rapid magma ascent, driven perhaps by early, caldera-forming, magma chamber roof collapse, caused delay of volatile exsolution until very late in the ascent, and then instantaneous explosive exsolution at near ground surface level occurred, driving a lateral blast.

10.9.1.2 "Tortoise-Like" Outflow Ignimbrites—The 2.08 Ma Cerro Galán Ignimbrite

By contrast, the 2.08 Ma Cerro Galán ignimbrite (CGI) is very different from the Taupo ignimbrite. The Cerro Galán ignimbrite was erupted from the 25 km × 30 km Cerro Galán caldera in the Andes of northwestern Argentina (Fig. 10.61; Francis et al. 1983; Sparks et al. 1985; Cas et al. 2011; Folkes et al. 2011a). The caldera has experienced 8 ignimbrite-forming eruptions over at least the last 6 million years. Each eruption was progressively larger, culminating in the 650 km^3 DRE Cerro Galán eruption at 2.08 Ma (Folkes et al. 2011a). The eruption products are totally different to those from the Taupo eruption. There are no fallout deposits, indicating that there was no buoyant plinian eruption column, right from the start and throughout the eruption. There are no co-ignimbrite breccias or ignimbrite veneer deposits. There is only a very voluminous massive ignimbrite, which is currently preserved radially around the caldera up to 90 km from the original caldera margin (Fig. 10.61). The ignimbrite forms a sloping shield around the caldera in proximal to medial distances (Fig. 10.26a). In distal areas it has been strongly eroded but is preserved in paleovalleys, where relationships indicate that it was an almost totally valley ponded deposit (Cas et al. 2011; Figs. 10.61 and 10.35d).

The CGI has a rhyodacitic composition (SiO_2: 68–72 wt %; high K calc-alkaline); it is very crystal rich (40–55% DRE) and the magma had a pre-eruption magmatic H_2O content ranging from 2 to 6.3 wt % (Folkes et al. 2011b; Wright et al. 2011b). It is marked by two main pumice clast populations, a dominant white pumice with abundant large vesicles, and a less abundant, but ubiquitous grey pumice with abundant small vesicles. The grey pumice represents a hotter, deeper crustal magma (800–840 °C, 3–5 kbar) that was ultimately the source of the cooler shallower magma that produced the white pumice (770–810 °C, 1.5–2.5 kbar; Folkes et al. 2011b). Thermobarometry calculations using samples from older ignimbrites erupted from Cerro Galán caldera, and the Al in amphibole thermobarometer suggests crystallisation depths of 6–10 km for the white pumice magma and 14–18 km for the grey pumice magma (Folkes et al. 2011b). The grey pumice magma was a recharge magma that was injected into the white pumice magma reservoir leading to magma mingling, which probably triggered the CGI eruption through addition of magma volume, heat and perhaps volatiles to the shallower chamber (Wright et al. 2011b; Folkes et al. 2011b).

The ignimbrite most commonly consists of one depositional unit (Fig. 10.82a), up to 80 m thick. However, in several distal locations where the flow interacted with steep valley and ridge topography and split into multiple flow lobes of pulses, it consists of multiple stacked depositional units each up to 20 m thick (Fig. 10.82b); these are sometimes separated by thin ground surge deposits (Fig. 10.82c). Internally all depositional units are massive, poorly sorted, and unbedded, although there are subtle variations in pumice clast content and size, and there is a pervasive sub-horizontal fabric defined by alignment of the long axes of elongate pumice clasts, slabby lithic fragments and even biotite flakes only millimetres long (Figs. 10.41a and 10.49c).

Where there are exposures of the contact between the ignimbrite and the paleo-valley bedrock, all the fabrics in the ignimbrite are sub-horizontal, even where the valley walls have steep slopes. In distal settings, where the ignimbrite consists of multiple depositional units the boundaries of sub-units are sub-horizontal and onlap the paleo-topographic slope (Fig. 10.82b). These onlap relationships indicate that the pyroclastic flow was not a high velocity flow. Nowhere does layering in the ignimbrite drape paleo-slopes, and even where the ignimbrite has buried paleo-topographic highs the deposits are massive, and do not look like stratified ignimbrite veneer deposits (Cas et al. 2011). In addition, AMS

Fig. 10.82 Some deposit characteristics of the 2.08 Ma Cerro Galán Ignimbrite (CGI). **a** Massive, single depositional unit of the CGI, in an open valley-ponded setting (see sloping paleovalley wall under the ignimbrite in the foreground. Ojo de Curuto, near Paycuqui, northwestern Argentina. People circled for scale. **b** Photo of cactus (with some outcrop for scale!), showing multiple thin depositional units where the distal pyroclastic flow interacted with a high relief valley and ridge topography. Unit boundaries are horizontal, whereas the dashed yellow line shows the slope of the paleovalley bedrock, indicating that sub-horizontal depositional units are onlapping paleotopography. The basal unit has wavy stratification at its base, representing a ground surge to that unit. Outcrop height is 10 m. Tacuil, northwestern Argentina. **c** Ground surge deposit between fine-grained, massive, distal depositional units. Tacuil, northwestern Argentina. **d** Eutaxitic fiamme texture in crystal rich welded ignimbrite, defined by porphyritic fiamme. Los Patos River, Maderes mine camp, northwestern Argentina

fabrics indicate that the pyroclastic flow travelled along and parallel to paleo-valley axes, not over topography (Lesti et al. in prep.).

The deposit is largely non-welded, but to the east up to 20 km from the caldera, and immediately to the south, the CGI is moderately to densely welded (Lesti et al. 2011; Fig. 10.83a), and commonly has eutaxitic and fiamme textures there (Fig. 10.82d). The CGI was emplaced almost everywhere above 620 °C, with a minimum emplacement temperature of 580 °C (Fig. 10.83a). The maximum temperature drop was 200 °C from the fragmentation temperature to depositional temperature, indicating significant heat

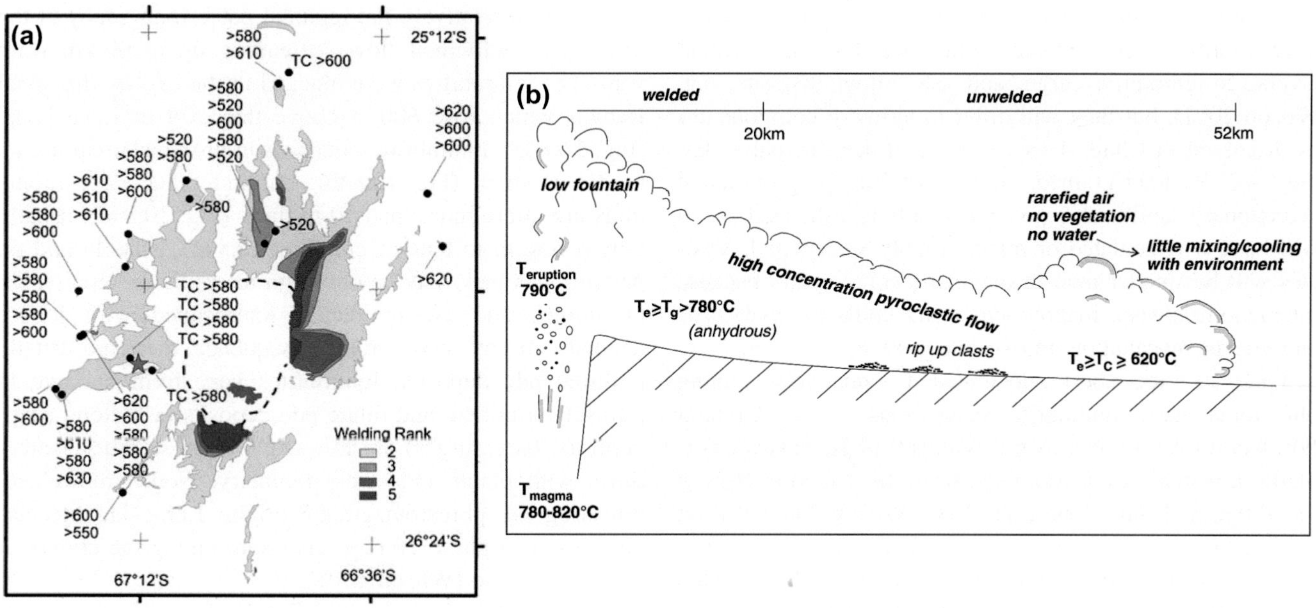

Fig. 10.83 **a** Areal pattern of unblocking temperatures for lithic clasts analysed using TRM step heating processes for the 2.08 Ma Cerro Galán and older Toconquis Group ignimbrites erupted from the Cerro Galán caldera. Also shown is the distribution of intensity of welding or welding rank using the Quane and Russell (2005) scheme. The ignimbrite is mostly non-welded, except close to the caldera to the north, east and south. **b** Thermal model for the Cerro Galán Ignimbrite eruption and pyroclastic flow, showing the high level of heat retention, or low degree of cooling. (From Lesti et al. 2011)

retention during the eruption and during pyroclastic flow (Lesti et al. 2011). Heat retention was facilitated by a low eruption column or fountain, limiting cooling and mixing of cold atmospheric air into the column, as well as into the flow (Fig. 10.83b; Cas et al. 2011). This is consistent with a flow that is not very turbulent.

As a result of these observations, Cas et al. (2011) considered that from the beginning of the eruption the CGI eruption column was like a constantly collapsing fountain, which never produced fallout deposits and constantly fed pyroclastic debris into the radially spreading pyroclastic flow system, probably from fissure vents defining the collapsing caldera. Although the flow was probably turbulent near vent, it very quickly deflated into a high particle concentration granular flow with suppressed turbulence, fluidised by gas trapped in the high particle concentration ash matrix. Calculations using dynamic pressure levels suggest that the flow moved at velocities <3 m s^{-1}, or $\sim$10 km hr^{-1}, and the sub-horizontal fabric indicates that deposition occurred under a laminar shear flow regime (Cas et al. 2011). This explains the retention of heat, allowing the ignimbrite to be welded over a large area. The Cerro Galán pyroclastic flow was a tortoise flow, not a hare!

Why is the Cerro Galán eruption sequence so different from the Taupo sequence? Both originated from high silica magmas, with generally similar volatile contents, so they might be expected to have erupted similarly. They erupted very differently, with the Taupo eruption beginning with eruptions from several point source vents, before caldera collapse and opening of a ring fissure vent system during the Taupo ignimbrite phase. The CGI eruption appears to have started immediately from a caldera-forming fissure vent system. The timing and style of caldera collapse is therefore likely to have controlled the eruption style and eruption sequence (see Sect. 10.11).

10.9.2 Phreatomagmatic Ignimbrites

So far, we have considered only ignimbrites erupted through magmatic explosive eruptions, marked by highly vesiculated pumice and shards. Can silicic phreatomagmatic explosive activity produce voluminous ignimbrites? The answer is clearly yes, represented by both young and old examples, including the large volume 25.4 ka Oruanui Formation in New Zealand (Self 1983; Wilson 2001; Wilson et al. 2006; Van Eaton and Wilson 2013), the intermediate volume, 560 ka Tor de Cenci ignimbrite, Colli Albani volcano, Italy (De Rita et al. 2002; Giordano et al. 2010), and the Late Carboniferous Cana Creek Tuff of the New England region, New South Wales, Australia (McPhie 1986). The principal requirement is a large source of external water, which could be a large crater lake (e.g. caldera lake), ground water, or glacial into which silicic magma erupts. The characteristics of phreatomagmatic ignimbrites are similar to magmatic ignimbrites as discussed in this current chapter (i.e. massive to diffusely stratified, poorly sorted, valley ponded facies, finer stratified veneer facies, lithic lag breccias, ground layers, co-ignimbrite surge, and ash fallout deposits; e.g. Wilson 2001), but they will differ in terms of componentry as discussed in Chap. 4 (Sects. 4.6 and 4.9). In particular, they will be finer-grained, contain varying proportions of accretionary lapilli, and juvenile lapilli to ash-sized clasts will be less vesiculated or more variably vesiculated. Vesicles will be smaller than in magmatic counterparts because interaction between magma and water chills the pyroclasts and arrests vesiculation and vesicle growth.

One of the best documented, relatively young phreatomagmatic ignimbrite successions is the Oruanui Formation (also known as the Wairakei or Kawakawa Formation), which was also erupted from the Taupo caldera in the Taupo Volcanic Zone (Self 1983; Wilson 2001; Wilson et al. 2006; Van Eaton and Wilson 2013). It is the largest known phreatomagmatic explosive eruption yet documented with a total volume of deposits of 1170 km^3 (=530 km^3 DRE; Wilson 2001). The composition is rhyolitic (average SiO_2: 74–76 wt%), sourced from a long-lived, polygenetic regional silicic source, with evidence of input from mafic and intermediate sources as well (Wilson et al. 2006). The specific Oruanui magma body was generated and erupted within 40 kyr (Wilson et al. 2006), and the eruption produced a rectilinear caldera, nested within the older Whakamaru caldera. The source vents were located within the confines of the modern Taupo caldera, but also within the confines of a very extensive lake ponded within the Taupo rift zone, Lake Huka (Fig. 10.84a). There was thus abundant surface water to interact with the large volume of erupting magma.

The eruption sequence or stratigraphy consists of interleaved fallout and ignimbrite deposits (Fig. 10.84b, c), which Self (1983) subdivided initially into 6 phases or members, whereas Wilson (2001) identified ten phases (Fig. 10.84c). Wilson (2001) interpreted the eruption to have been spasmodic, with time breaks of up to weeks or months between some eruption phases. The eruption culminated and climaxed during phases 7–10 which produced the most voluminous fallout and ignimbrite deposits, and farthest flowing pyroclastic flow deposits (90 km, phase 8; Fig. 10.84c). Ignimbrites were produced during every phase, reflecting repeated eruption column collapses and in some cases partial collapses.

The fallout deposits were widely dispersed, largely to the ESE, with some deposits preserved on Chatham Island 800 km from the vent. Fallout deposits are usually fine-grained, stratified, planar and regionally extensive, and commonly have accretionary lapilli. These characteristics are consistent with phreatoplinian deposits (Chap. 9).

The Oruanui pyroclastic flow deposits (phreatomagmatic ignimbrites) have a largely radial distribution (Fig. 10.84a), suggesting a relatively flat regional paleo-topography at the time, with individual flows extending up to 90 km from vent. The ignimbrites were blocked to the east by the Axial Ranges which rose 600 m above the level of Lake Huka (Fig. 10.84a). Ignimbrite sequences in proximal areas are up to 200 m thick (Fig. 10.84b), but individual depositional units are rarely more than 10 m thick (Fig. 10.84d), usually only a few m to tens of cm (Fig. 10.84e). Distally and on paleo-ridges they may be only mm to cm thick, represented by surges and veneer facies, interbedded with fallout deposits. In any phase eruption sequence, there are usually multiple pdc deposits. Ignimbrites vary from massive to diffusely stratified, and dilute pdc deposits (ash-cloud surge deposits, base surges) are low angle cross-stratified, sometimes with pinch and swell geometry. None are welded, reflecting the phreatomagmatic origin. Lithic lag breccias only occur in phase 10 deposits, representing the climax of caldera collapse (Wilson 2001).

A common feature of both fallout and ignimbrite deposits is the presence of large accretionary lapilli, up to cm in diameter (e.g. Figs. 10.75a and 10.77e; Self 1983; Wilson 2001; Van Eaton and Wilson 2013). These have variable origins, from formation in the eruption column, to rain-flushing through down-wind ash plumes, to formation in co-ignimbrite ash clouds. They vary from well defined and lithified, to diffuse ash aggregates (Van Eaton and Wilson 2013).

Like the younger Oruanui Formation, the less well preserved, Late Carboniferous, rhyolitic Cana Creek Tuff sequence of the New England region of northeastern New South Wales, Australia, is also marked by an association of interpreted phreatoplinian fallout ashes with accretionary lapilli, and multiple ignimbrite units, some also bearing accretionary lapilli (McPhie 1986). McPhie (1986) estimated the original erupted volume to have been >100 km^3 (DRE). She interpreted the eruption to have also been pulsating, driven by varying degrees of explosive magma: water interaction at a source caldera centre. This not only caused fluctuations in the behaviour of the eruption column, producing ash fallout deposits, as well as multiple ignimbrite units, but also produced voluminous syn- to post-eruptive fluvial debris flow and sheet-flood deposits of juvenile volcanic debris. However, the magnitude and stratigraphic complexity is much less than documented for the Oruanui sequence, suggesting a shorter-lived event.

In addition, the 10 km^3, 560 ka Tor de Cenci ignimbrite shares very similar facies to those described above for the calc-alkaline Oruanui Formation and the Cana Creek Tuff, even though characterised by a K-foiditic composition (Fig. 10.84f, g; De Rita et al. 2002; Giordano et al. 2010). Furthermore, detailed study of the temperature of emplacement of the Tor de Cenci ignimbrite shows temperatures between 400 and 200 °C from proximal to distal, with a

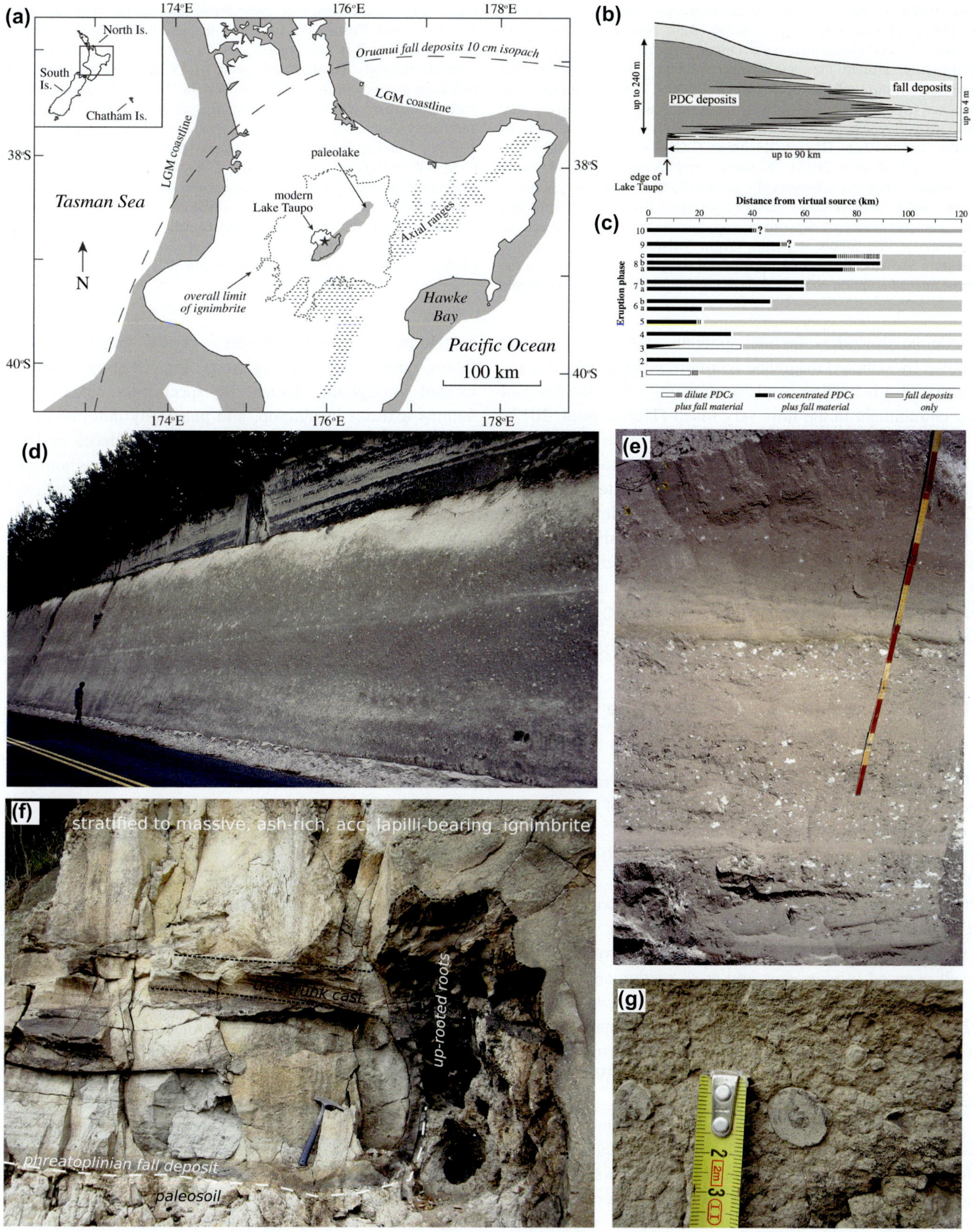
(a)
174°E
176°E
178°E
North Is.
South Is.
Chatham Is.
Oruanui fall deposits 10 cm isopach
LGM coastline
LGM coastline
38°S
38°S
paleolake
modern Lake Taupo
Tasman Sea
Axial ranges
N
overall limit of ignimbrite
Hawke Bay
Pacific Ocean
100 km
40°S
40°S
(b)
up to 240 m
PDC deposits
fall deposits
up to 4 m
up to 90 km
edge of Lake Taupo
(c)
Distance from virtual source (km)
Eruption phase
dilute PDCs plus fall material
concentrated PDCs plus fall material
fall deposits only
(d)
(e)
(f)
stratified to massive, ash-rich, acc. lapilli-bearing ignimbrite
tree trunk cast
up-rooted roots
phreatoplinian fall deposit
paleosoil
(g)

◀ **Fig. 10.84** Features of some phreatomagmatic ignimbrites. **a–e** 25.4 ka Oruanui Formation, Taupo Volcanic Zone, New Zealand. **a** Extent of the 25.4 ka phreatomagmatic Oruanui Formation showing radial distribution around the paleo-Taupo Caldera. The star shows the average location of multiple eruption points. The ignimbrite was blocked from flowing eastwards by the Axial Ranges (presently ≥ 750 m). Also shown is the NNE trending regionally extensive paleo-Lake Huka, and the Last Glacial Maximum (LGM) coastline. (From Van Eaton and Wilson 2013). **b** Facies architecture of the Oruanui Formation, showing schematically the relationship between ignimbrite facies and ash fallout facies relative to distance from source vents, and the contemporaneous deposition and lateral facies equivalence of ignimbrites and fallout deposits. (From Wilson 2001). **c** Schematic diagram showing the increasing reach of the pdcs, culminating in the most intense eruption phase 8, and the contemporaneous deposition and lateral facies equivalence of ignimbrites and fallout deposits. (From Van Eaton and Wilson 2013). **d** Thick massive flow deposit of Oruanui ignimbrite, which is fine-grained, poorly sorted, and texturally homogeneous through the thickness seen (~ 15 m; horizontal lines are bulldozer scrapings). (From Cas and Wright 1987). **e** Succession of thin phreatomagmatic pdcs of the Oruanui ignimbrite with white pumice and barely visible, darker accretionary lapilli (also see Figs. 10.75a and 10.77a). **f** and **g** 561 ka Tor de' Cenci phreatomagmatic ignimbrite, Colli Albani volcano, central Italy. **f** Above the basal thin phreatoplinian fall deposit, the ash-rich, massive to stratified ignimbrite shows casts of knocked down tree trunks which are oriented parallel to flow direction. Flow direction from right to left. **g** Abundant and large accretionary lapilli are present both scattered through the matrix (as in photo), as well as in concentration structures

moderate downflow thermal gradient, comparable to that shown by standard magmatic ignimbrites, indicating that the increased ash fraction available does not affect the level of air entrainment in the flow very much (Trolese et al. 2017).

Rare phreatomagmatic ignimbrites have also been described from some maar volcanoes (Freundt and Schmincke 1985a, b; Giordano et al. 2002; van Otterloo and Cas 2016), associated with more normal maar rim base surge deposits. These include some that are characterised by valley pond and veneer deposit facies (e.g. Peperino Albano; Giordano et al. 2002; Fig. 10.22c), as well as containing co-ignimbrite lithic breccias (e.g. Laacher See; Freundt and Schmincke 1985b).

10.9.3 Intra-caldera Ignimbrites: The Permian Ora Ignimbrite

There were very few detailed studies of the facies characteristics, facies architecture, and assessment of the way that intra-caldera ignimbrites fill-in calderas, until the comprehensive studies of Willcock et al. (2013, 2015a, b), and Willcock and Cas (2014) who described the internal stratigraphy and facies architecture of the well-preserved Permian Ora rhyolite ignimbrite and caldera in northern-most Italy (Fig. 10.85a, b). The Ora system is very well exposed in plan view and cross-section and is the last of a succession of nested calderas erupted between 280 and 274 Ma. It consists of two coalesced, contemporaneous caldera depressions, which measure 42 $\times$ 40 km, and have a surface area of 1500 km^2 (Morelli et al. 2012; Willcock et al. 2013).

The Ora ignimbrite has a preserved volume of >1290 km^3, which is DRE because all the ignimbrites are welded. Inside the caldera, the intra-caldera ignimbrite succession is up to 1350 m thick, and the less well-preserved outflow succession ranges from 10 up to 230 m thick (Figs. 10.85b and 10.86a). The ignimbrites are very crystal rich (cf. many of the large volume caldera successions in the Andes), containing on average 40–50% crystal fragments, $\sim 2\%$ lithics, and $\sim 20\%$ fiamme (e.g. Fig. 10.86b). They are welded throughout, defined by spectacular eutaxitic and fiamme textures in outcrop, and plastically deformed shard textures in thin section (Fig. 10.86b, c; Willcock and Cas 2014). The ignimbrite is columnar jointed throughout (e.g. Fig. 10.86a). There is no initial plinian fallout deposit.

On face value, the ignimbrite appears to be very monotonous. However, through detailed graphic logging, correlation of sections and detailed component point counting and geochemistry, Willcock et al. (2013, 2015b) were able to demonstrate that the intra-caldera fill ignimbrite succession has a stratigraphy (Fig. 10.85b), defined by a lithic lag breccia at the base, a remarkable nearly pristine vitrophyre near the base of the ignimbrite succession (Fig. 10.67; perhaps the oldest known), and then vertical variations in crystal content and size (27–55%), biotite content (7–15%), and fiamme abundance and size. The diffuse stratigraphy is sub-horizontal (Fig. 10.86a) and is matched by a sub-horizontal AMS paleomagnetic fabric, that was used to identify intra- and extra-caldera flow direction patterns (Willcock et al. 2015a). However, geochemistry and mineral content, especially biotite content, indicate that the southern caldera began to erupt first, with the least evolved magma, then the northern caldera eruption chimed in, and overall, the most evolved magma was erupted last.

Most importantly, the presence of a definable, albeit subtle stratigraphy, demonstrates that the caldera system began to collapse from the beginning of the eruption, and was filled incrementally, by a nearly continuous succession of pyroclastic flow pulses, but not by a single, steady, uniform supply of pyroclastic debris. Definable AMS fabrics also demonstrates that pyroclastic flow pulses flowed laterally within the caldera, eventually surmounting the caldera rim into the outflow region (Willcock et al. 2015a). The lack of an initial fallout deposit indicates that, as for the Cerro Galán caldera-forming eruption, there was no initial buoyant eruption column, suggesting again a continuously collapsing eruption fountain, that was fissure fed, as the caldera appears

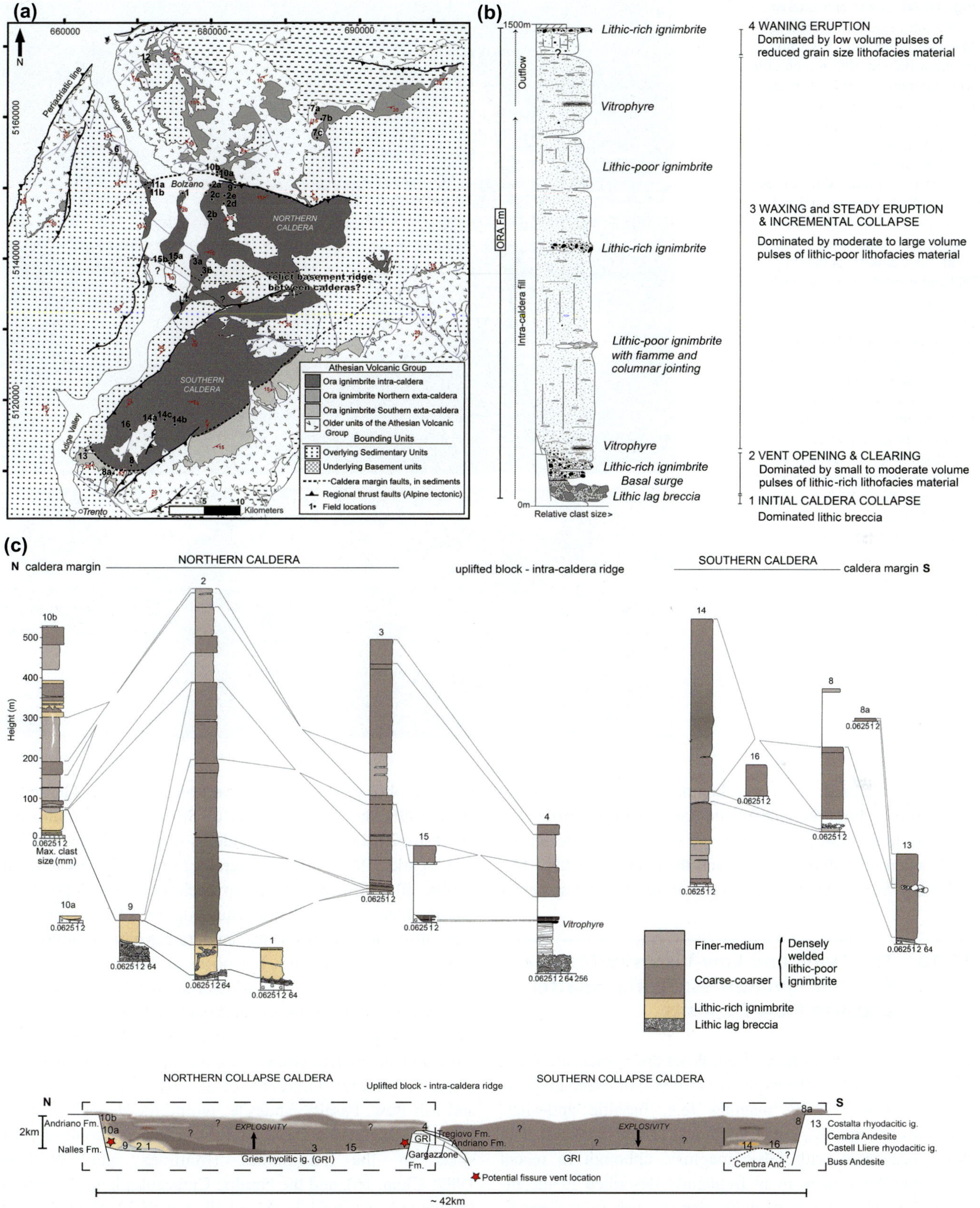

Fig. 10.85 Regional characteristics of the Permian (274 Ma) Ora ignimbrite, Bolzano, northern Italy. **a** Geological map showing the areal extent of the intra- and preserved extra-caldera Ora ignimbrite, its relationship to the older Athesian volcanic group (also mostly caldera sequences), sediments and basement. Numbers refer to measured section locations, some of which are shown in (**c**). **b** Generalised stratigraphic log through the Ora Formation showing internal lithofacies and summary of the four main phases of the eruption. **c** Stratigraphic and facies architecture depicted in a north (left) to south (right) correlation panel, and interpretive cross-section (bottom). (Adapted from Willcock et al. 2013; all original figures digitised by M. Willcock)

Fig. 10.86 Some characteristics of the intra-caldera Permian (274 Ma) Ora Ignimbrite, Bolzano, northern Italy. **a** 500 m high cliff, which is part of the 1350 m thick Ora intra-caldera fill. Note the pervasive columnar jointing. The white dashed lines define some of the sub-units shown in Fig. 10.85b. (From Willcock and Cas 2014). **b** Very crystal rich nature of the Ora Ignimbrite matrix, with large clear volcanic quartz and pinkish potassium feldspar crystal fragments visible. Also shown is a wispy flattened pumice fragment, or fiamme, which has a noticeably lower crystal content than the matrix. **c** Photomicrograph of very densely welded Ora Ignimbrite from the intra-caldera setting, showing extreme plastic deformation of shards between abundant volcanic quartz crystal fragments (From Willcock and Cas 2014)

to have collapsed from the beginning of the eruption. Interestingly, both the Cerro Galán and Ora ignimbrite systems are lithic clast poor, which we will discuss later in the book when describing calderas and their origins (Chap. 16). (Also see Sect. 10.8.2.2 for discussion of the vent system.)

10.10 The Mafic and Low-Viscosity Magma Sourced Ignimbrites: Scoria, Spatter, and Ash-Flow Deposits

Scoria and ash, or spatter and ash flows are pyroclastic flows that may result from the explosive eruption of mafic magmas or other low-viscosity magmas (e.g. basaltic andesites, andesites, phonolites, some trachytes). These are rarer than those associated with felsic magmas, although in recent years they have been more frequently described in the literature. Small volume scoria/spatter and ash flow deposits (or mafic ignimbrites) are relatively common at stratovolcanoes (e.g. the 1975 eruption of Ngauruhoe; Figs. 10.4b and 10.10a; Nairn and Self 1978; Mt Misery/Liamuiga, St Kitts, Roobol et al. 1987; Lascar volcano, Chile, Gardeweg et al. 1998; the 2019 eruption of Stromboli, Giordano and De Astis 2021), and some calderas (e.g. Santorini, Mellors and Sparks 1991; Vico caldera, central Italy, Bear et al. 2009a, b). However, mafic pyroclastic flow deposits of intermediate volume ignimbrites (up to VEI 7/VEI* 11), may also be produced by caldera-forming eruptions at explosive mafic calderas, but no examples are known of > 100 km^3 DRE (Cashman and Giordano 2014). Examples of intermediate volume, fissure and caldera-forming scoria, and ash flow deposits include those from the K-alkaline Colli Albani mafic caldera volcano, on the outskirts of Rome, Italy (described below; Giordano et al. 2010), extensive scoria and ash flow, basaltic-andesite, ignimbrite deposits erupted from the Laguna de Bay and Taal volcano-tectonic depressions in the Philippines (Martinez and Williams 1999; Chap. 16), and the basaltic Caracautin ignimbrite from Llaima volcano in Chile (Marshall et al. 2022; Valdivia et al. 2022). A further example is the compositionally intriguing, basaltic, welded P1 ignimbrite ($\sim$45 km^3 DRE) emplaced during the 14.1 Ma caldera-forming eruption of Tejeda caldera, Gran Canaria (Freundt and Schmincke 1995a, b). This represents the first voluminous eruption of highly differentiated as well as mafic magmas on top of the basaltic Miocene shield volcano (Chap. 16).

10.10.1 General Characteristics of Scoria, Spatter, and Ash Ignimbrites

In general, mafic ignimbrites have many similar facies characteristics as felsic ignimbrites, including all the major textural zones described above. They can be cm to tens of m thick, massive to diffusely stratified, poorly sorted, matrix supported. They can have the same lithic breccia textural-genetic types (Sect. 10.8.10). Small volume mafic scoria/spatter and ash flows are commonly fines-poor (e.g. Mt Misery/Liamuiga, Roobol et al. 1987), whereas larger volume mafic ignimbrites are more ash rich, though usually more coarse than fine ash, and can flow tens of km from vent (e.g. Colli Albani ignimbrites, Giordano et al. 2006, 2010; Llaima volcano ignimbrites, Chile, Marshall et al. 2022; Valdivia et al. 2022).

Some mafic ignimbrites have remarkable spatter-rich and/or "cauliflower" bomb horizons (Fig. 10.87; Mellors and Sparks 1991; Gardeweg et al. 1998). Spatter agglutinates may occur in mafic ignimbrites as an equivalent of pumice concentration zones (Fig. 10.87c). However, spatter agglutinates are reported in a number of ignimbrites ranging in composition and size, often concomitant with lithic breccias (Wright and Walker 1977, Figs. 10.50a, b and 10.54c); Mellors and Sparks 1991, Fig. 10.87; Rosi et al. 1996; Valentine et al. 2000; Palladino and Simei 2005; Allen 2004; Bear et al. 2009b, Fig. 10.90; Vinkler et al. 2012, Fig. 10.91). In some cases these are associated with caldera collapse, and Vinkler et al. (2012) interpreted such spatter agglutinates as resulting from the eruption of partially degassed magma that intruded up to shallow levels of the developing ring faults before the main phase of caldera collapse, and before the collapse eventually triggered their eruption. On Gran Canaria, compositional zonation of the P1 ignimbrite is the result of vertically changing proportions of four component magmas (rhyolite-trachyte-trachyandesite-basalt), which were intensely mixed during eruption (Freundt and Schmincke 1995a, b). The larger volume of 26 km^3 of erupted basalt compared with only 17 km^3 of the evolved magmas suggest that the basaltic magma was erupted and emplaced in a fashion commonly only attributed to felsic magmas because it utilised the felsic P1 magma chamber and its ring-fissure conduits. The intensity of welding, especially at the base of basaltic ignimbrite, and the presence of spherical particles and of mantled and composite particles formed by accretion and coalescence in a viscous state imply that the flow was a suspension of hot (1160 °C) magma droplets (Freundt and Schmincke 1995a).

10.10.2 Small Volume, Vulcanian Scoria, Spatter and Ash Flow Deposits: 1975 Eruption of Ngauruhoe Stratovolcano, New Zealand and 2019 Eruption of Stromboli, Italy

A 7-h long eruption of Ngauruhoe volcano in New Zealand on 19 February 1975 was marked by an initial gas streaming phase and 9 vulcanian explosions, during which ash fallout and multiple scoria/spatter and ash flow events occurred (Nairn and Self 1978; Figs. 10.4b and 10.10a). The flows were small volume ($<10^6$ m^3), and formed steep sided, lobate deposits up to 1.5 m high (Fig. 10.88a, b), that have well-defined levees and channels, and are coarse-grained, and fine ash depleted (Fig. 10.88c, d), compared with felsic ignimbrite counterparts. The flows travelled less than 2.5 km from the vent, and only 0.6 km beyond the base of the volcano, or break in slope, which represents a change from a slope angle of 20° to 6°. They were described as inertial avalanches because of the limited flow distance, the lobate morphology, and the lack of evidence of fluidisation as a flow mechanism (Nairn and Self 1978). Juvenile clasts are andesitic spatter with ropey and cauliflower textures, irregular scoria with moderate vesicularity, as well as outsize blocks up to 5 m in length, many with breadcrust texture (Fig. 10.88c, d).

On 3 July and 28 August 2019, two vulcanian explosions occurred at the basaltic Stromboli volcano, Italy, feeding for about 2.5 min directional pyroclastic flows that rushed down the 35° inclined slope of the Sciara del Foco (Fig. 10.89a, b), at an average velocity of 50 m s^{-1} and maximum instant velocity of 100 m s^{-1} (Giordano and De Astis 2021). The pyroclastic flows slowed down at the entry point into the sea and travelled for about 500–800 m over the sea, with rapid inflation of their flow fronts, and eventually lofting once most of the pyroclastic material had been deposited into the water mass (see Chaps. 9 and 11). The pyroclastic flows showed very thin and fast-flowing flow fronts that preceded the development of the ash cloud, suggesting that they moved as high concentration granular flows. Pyroclastic material included spatter clasts to scoria to highly vesicular juvenile clasts, shoshonitic basalt in composition, and with densities ranging from 1500 to 500 kg m^{-3}. In proximal areas, a continuous bomb bed of agglutinated ballistic spatter was deposited (Fig. 10.89c).

Fig. 10.87 Mafic ignimbrites of the 54 ka Upper Scoriae 2 eruption sequence, Santorini, Greece. The eruption began with a fallout phase from a vent near present-day Nea Kameni. Scoria and ash flows then swept down the flanks of the Skaros lava shield volcano (which occupied much of the present-day caldera). These transported lithic blocks and fluidal spatter rags up to 3 m in diameter that settled rapidly from the flows to form lag deposits (Mellors and Sparks 1991). These lag deposits typically are 10-20 m thick. **a** Prominent horizon (~20 m thick) of welded spatter agglutinate in the caldera wall north of Thera. **b** Spatter bomb-rich scoria flow deposit with highly welded (pink) interior laid down close to the summit of the Skaros shield. **c** Spatter bomb-rich concentration zone at the top of a scoria and ash flow deposit. Note the imbrication of bombs and their fluidal form and ropey surface features

10.10.3 Moderate Volume Plinian and Small Caldera-Forming Scoria, Spatter and Ash Flow Deposits: Vico Caldera, Central Italy

Vico caldera is a phonolitic lava shield and explosive caldera volcano, in central Italy, 50 km north of Rome. It has an active eruption history from 420–95 ka, and is probably extinct, or experiencing a long repose period. Its eruption history is dominated early by effusive lava and lesser explosive activity (>260 ka), followed by a major phase of explosive ignimbrite and fallout-forming eruptions, that culminated in an 8 km diameter caldera-forming event at 151 ka, the Sutri eruption (Bear et al. 2009a, b, c). From 151–95 ka, the volcano has experienced smaller scale explosive and lava-forming eruptions.

The Sutri Formation represents a complex phonolitic eruption sequence, including an initial small volume fallout deposit, up to 75 cm thick, dispersed to the south from a point source vent, and then a succession of ignimbrites (units Sutri B, C, D), with a variety of co-ignimbrite breccias (Bear et al. 2009b), that culminated in the climactic, most

Fig. 10.88 Features of small volume andesitic scoria and ash flow deposits from the 19 February 1975 vulcanian eruption of Mt Ngauruhoe, New Zealand. **a** Dark, lobate form of the flow deposits, some with levees visible. **b** High relief, 1–1.5 m thick, lobate deposits with steep sides. **c** Cauliflower bombs and coarse scoria, on the fines depleted surface of one lobate deposit. Much of the original fine ash of the matrix has probably been ablated by wind erosion. **d** 2 m long juvenile block with breadcrust texture, on surface of fines-depleted lobate deposit

voluminous and widespread ignimbrite (Tufo Rosso e Scorie Nere Ignimbrite, or unit Sutri E; Fig. 10.90a, f), and caldera collapse. The ignimbrites are typically pumiceous to spatter clast bearing, and include an initial very thin, welded ignimbrite, up to 1 m thick (Fig. 10.90b), and then a succession of non-welded ignimbrites with lithic-rich lag breccias (Fig. 10.90c), and some spectacular mega-spatter clast lag breccias found up to 11 km from vent, with spatter clasts <2.5 m long (Fig. 10.90d, e). Ignimbrites are typically poorly sorted, matrix-supported (Fig. 10.88), and vary from 40 cm (unit Sutri B) to m (units Sutri C, D, E) to >60 m thick (unit Sutri E). The Sutri E scoria/spatter and ash ignimbrite extends up to 25 km from the caldera rim, is extensively zeolitised, and the eruption sequence has a minimum volume of 5 km^3 DRE.

The variation in the characteristics and dispersal patterns of the ignimbrites and co-ignimbrite breccias indicate that multiple vents sourced different units, probably located along an incipient ring fracture or piecemeal block fault margins (Fig. 10.23a–e), typical of chaotic caldera collapses (see Chap. 16, Sect. 16.7.3; Roche et al. 2000). The final climactic ignimbrite, Sutri E, which has radial distribution around the volcano, indicates that the ring fracture vent system had completely unzipped and caldera collapse was fully involved in its eruption (Fig. 10.23f). Although there is an initial buoyant eruption column, the rest of the eruption was marked by episodic, to continuous collapsing column, or a fountaining style of eruption, represented by spectacular spatter breccias (Fig. 10.90d, e), and the climactic ignimbrite (unit Sutri E), with abundant spatter clasts (Fig. 10.90f).

10.10.4 Outflow Mafic Ignimbrites Associated with Caldera Formation: 355 ka Villa Senni Ignimbrites, Colli Albani Volcano, Italy

The Colli Albani caldera-stratovolcano-maar volcanic complex is located only 25 km ESE of Rome, Italy (Giordano et al. 2010). It has an eruption history spanning the last 600 kyr (see Fig. 15.13). From 600 to 355 ka it experienced seven large volume, ignimbrite and caldera-forming eruptions, the last of which was the Villa Senni eruption. Remarkably, all these 10–100 km^3, caldera-forming

Fig. 10.89 Scoria and ash flow generated during the vulcanian eruption (paroxysm) of 3 July 2019 at Stromboli volcano, Italy. The magma composition is shoshonitic basalt. **a** At 14.46.35 h, rushing down the 35° inclined slope of the Sciara del Fuoco at an average speed of 50 m s^{-1}; and **b** at 14.46.43 h, slowing and inflating at the entry point into the sea. In both images splash circles can be seen where large ballistic blocks from the vent or out of the flow are impacting the sea surface. (**a** and **b** are reproduced by courtesy of LGS, Laboratory of Experimental Geophysics, Università di Firenze, Italy; http://lgs.geo.unifi.it/). **c** At the summit, ballistic spatter clasts formed a continuous cover (~0.35 km^2). The maximum equivalent diameter of the measured spatter is 200 cm, where the cover is commonly made up of the superposition and welding of two to four large spatter clasts. (From Giordano and de Astis 2021)

eruptions range in composition from K-foidites to tephrites to phono-tephrites, making the Colli Albani volcano probably the most explosive mafic volcano on Earth, in spite of the very low viscosity of its magmas (e.g. Campagnola et al. 2016). From 350 to 20, ka it then went through a strato-volcano building phase and then a monogenetic eruption phase (Giordano et al. 2006). The Villa Senni Formation (VSN) was the last caldera-forming eruption at Colli Albani volcano, 355 ka ago (Watkins et al. 2002; Giordano et al. 2006, 2010; Vinkler et al. 2012; see also Fig. 15.14).

The VSN begins with an eastward dispersed plinian basal fallout deposit (VSN_0) followed by two low aspect ratio and radially distributed ignimbrites erupted in rapid succession: the "Tufo Lionato" and "Pozzolanelle" ignimbrites (Figs. 10.91a and 10.22c). The early phase of the eruption produced a sustained column that emplaced dominantly fall deposits. The very basal deposit is a parallel planar to faintly cross-stratified, fine- to coarse-ash surge deposit, consistently a few cm thick, and extending more than 20 km from the caldera. This is overlain by plane parallel stratified, broadly reverse graded, scoria lapilli fallout deposits up to 140 cm thick at the most proximal exposure, along the northern caldera wall. The juvenile material is made of poorly vesicular scoria which is aphyric to sub-aphyric (minor leucite). Blocky shards and outsized lithic clasts suggest that magma–water interaction occurred during the early phases of the eruption, probably from a pre-existing caldera lake (Vinkler et al. 2012). The upper part of the fall

Fig. 10.90 Features of the 151 ka Sutri eruption mafic ignimbrites, Vico caldera, Roman Province, central Italy. **a** Eruption sequence stratigraphy, showing the eruption phases and variations in ignimbrite facies through the eruption (From Bear et al. 2009b). **b** Thin welded ignimbrite, unit Sutri B, with eutaxitic texture defined by flattened, aligned pumice fiamme. Poggio del Cavaliere, Ronciglione. **c** Lag breccia at the base of ignimbrite unit C, exposed on the current topographic caldera rim at Poggio del Cavaliere, Ronciglione. **d** Fines depleted, spatter clast breccia (unit Sutri E4sbx), showing irregular spatter rags (grey) up to 2.5 m long separated by coarse matrix, at Canepina, 7 km from the rim of Vico caldera. **e** Sketch of the same outcrop as in (**c**) showing geometry of the spatter breccia unit and relationships to other units (From Bear et al. 2009b). **f** Massive outcrop of the zeolitised, climactic ignimbrite (unit E5; Tufo Rosso e Scorie Nere), showing irregular black spatter and scoria clasts dispersed in the fine, orange, zeolitised matrix. Vertical striations are quarrying gouges. Cava di Favricola

Fig. 10.91 Features of the 355 ka Villa Senni ignimbrites, Colli Albani volcano, Roman Province, central Italy. **a** The stratigraphy of the Villa Senni eruption sequence (see text), is highlighted by a basal fallout deposit (VSN0), followed by the Tufo Lionato ignimbrite (VSN1), then co-ignimbrite breccias (VSN2b) at the base of the Pozzolanelle ignimbrite (VSN2), in turn overlain by final fall deposits (VSN3) (After Vinkler et al. 2012). **b** Outcrop of the Villa Senni ignimbrites, showing their massive, sheet-like character, overlying the 460 ka Pozzolane Nere ignimbrite. PNR = Pozzolane Nere; VSN_1 = Tufo Lionato; VSN_2 = Pozzolanelle ignimbrite. **c** Typical zeolitised orange brown, poorly sorted, massive, matrix-supported orange-red Tufo Lionato, with a spectacular twisted spatter fragment. Note the chilled rim and the highly vesicular inner core. **d** Outcrop of the Villa Senni ignimbrites, 7 km from vent, showing the Tufo Lionato (VSN_1), overlain by co-ignimbrite breccia (VSN_{2b}) of the Pozzolanelle ignimbrite (VSN_2) above. **e** Typical massive, poorly sorted, matrix-supported Pozzolanelle ignimbrite, showing a typical poorly-vesiculated, dark scoria clast with abundant, visible, white, large leucite crystals. **f** Fines depleted, scoria and spatter rich matrix of the Pozzolanelle ignimbrite. Note the bread-crusted and twisted scoria clasts

deposit is interbedded with thin intra-plinian ignimbrites that herald the main collapse event associated with the overlying Tufo Lionato ignimbrite.

The VSN outflow ignimbrites cover more than 2000 km^2 (Fig. 10.22c) with a minimum bulk volume of 30 km^3 (18 km^3 DRE). The total volume is almost 50 km^3 if an inferred intra-caldera volume is also included (Giordano et al. 2010). Based on estimated erupted volume, which does not include ash loss, intra-caldera volume, and distal erosion, the minimum subsidence at the 8×8 km^2 structural caldera is $\sim$280 m. The chemical compositions range from tephri-phonolite (Tufo Lionato ignimbrite) to phono-tephrite, tephrite and also K-foidites (Pozzolanelle ignimbrite), which define a broad chemical zonation toward more mafic compositions (44–46 wt% SiO_2, 6–9 wt% K_2O; Freda et al. 1997; Boari et al. 2009; Conticelli et al. 2010; Vinkler et al. 2012).

10.10.4.1 Tufo Lionato Ignimbrite (VSN_1)

The Tufo Lionato ignimbrite is a distinctive yellow to red tuff (i.e. the colour of a lion; Fig. 10.91c) radially distributed around the volcano as far as 35 km from vent (Fig. 10.22c). The ignimbrite thickness ranges from 2 to 10 m (Fig. 10.91a, b) and fills pre-existing paleovalleys, where gas pipes are observed. The estimated bulk volume is 10 km^3. The unit is a vapour-phase altered sillar, lithified by a remarkably uniform zeolite alteration (Fig. 10.91c; Giampaolo et al. 2006). To the east of the volcano, where the Apennine mountains were an obstacle to pyroclastic flows, the Tufo Lionato occurs up to 450 m in elevation and was channelled for some 10 km along the Aniene valley, where flows had to surmount a saddle at least 250 m high at 20 km from vent in order to get across the mountain range (Watkins et al. 2002).

The Tufo Lionato ignimbrite is mostly massive and chaotic (Fig. 10.91a–c). The base locally shows a thin lithic-rich, fines-poor ground layer. At distal locations and in paleovalleys, the unit may show some layering and development of pumice concentration zones. The VSN_1 ignimbrite is matrix supported (70–90% of ash). Based on their shape and texture, juvenile clasts can be classified as *scoria* (80–90%) and *spatter* (Vinkler et al. 2012). Spatter clasts show fluidal morphologies, such as twisted, flattened, elongated (Fig. 10.91c), with longer axes usually not horizontally aligned, but not due to welding. Juvenile clasts are poorly to moderately vesicular (20–60%), and aphyric to poorly porphyritic (clinopyroxene + leucite + phlogopite). Lithic clasts are mostly shallow derived (lavas), although the proportion of deep-seated lithics (pre-volcanic basement, intrusives) increases slightly toward the top, suggesting that the Tufo Lionato was largely erupted before the main caldera collapse event.

Watkins et al. (2002) recognised three distinct lithofacies based on gradual up-sequence changes in the components of the deposit and characteristics of the juvenile clasts. The lower yellow facies is characterised by a dominance of scoria clasts. The transition from the yellow to the overlying red facies is marked by the appearance of the spatter clasts combined with a slight increase in the grain size and enrichment in lithics. The transition into the uppermost brown facies corresponds with an increase in the vesicularity of the juvenile clasts, increasing free crystal content, and an increase in deep seated xenoliths. These characteristics indicate that the Tufo Lionato was largely emplaced by progressive aggradation and that the internal stratigraphy records temporal variations in the nature of erupted material and eruption dynamics.

10.10.4.2 Pozzolanelle Ignimbrite (VSN_2)

The Pozzolanelle ignimbrite covers essentially the same area as the Tufo Lionato ignimbrite (Fig. 10.22c). Thickness is on average >20 m and up to 80 m in proximal areas (Fig. 10.91a, b), and the total bulk volume is 20 km^3. The basal contact with the underlying Tufo Lionato ignimbrite is generally quite sharp (Fig. 10.91b, d). The Pozzolanelle is a dark grey to dark red, un-lithified ignimbrite, dominated by massive and chaotic facies throughout (Fig. 10.91b, d, e). The matrix (60–90%) is made of fine scoria lapilli, coarse-ash shards, and crystal fragments of cm-sized leucite (up to 30% of the upper deposit at proximal locations), clinopyroxene and biotite. The matrix is often fines-depleted (Fig. 10.91f). The framework clasts are largely made of dark reddish to black, aphyric to highly porphyritic (up to 35% clinopyroxene-biotite) scoria clasts (Fig. 10.91e). Characteristically the groundmass glass of scoria varies from aphyric to highly crystalline (Vinkler et al. 2012). Scoria clasts (Fig. 10.91e) are vesicular to poorly vesicular (30–60%), denser along flattened rims, rounded or flattened to spatter. Accessory lithics include a large proportion of deep-seated metamorphic and intrusive lithologies. The characteristic absence of fine ash in the ignimbrite matrix, which plots in the fines-depleted ignimbrite field of Walker (1983; cf. Giordano and Dobran 1994), is explained as a result of fragmentation of poorly vesicular magma and not the result of elutriation of fine ash during flow. Nonetheless, coarse-grained gas segregation pipes generally containing cm-sized scoria lapilli and lithics are frequently observed toward the top.

At proximal locations the Pozzolanelle ignimbrite is underlain by lapilli to block-sized (maximum diameter 2 m), fines-depleted, co-ignimbrite lag and ground breccias, up to tens of metres in thickness (Fig. 10.91d). These breccias show gradational to sharp upper/lateral transition to the Pozzolanelle ignimbrite and are found as far as 8 km from the caldera rim. The breccias are dominated by juvenile clasts (60–90%), with up to 30% of volcanic, basement-derived, and intrusive accessory lithics. Breccias associated with spatter bombs, with foiditic composition, not observed elsewhere in the VSN, are exposed in the northern-northwestern sector of the volcano (Conticelli et al. 2010; Vinkler et al. 2012), indicating that ring fractures tapped different magma types at different locations along a ring fracture fissure vent. The co-existence of spatter clasts and deep-seated lithics in the co-ignimbrite breccias indicates a complex eruptive mechanism involving both a relatively shallow fragmentation level along with the extraction of deep-seated xenoliths.

At several locations in the south and in the east, the top of the Pozzolanelle ignimbrite is conformably covered by a succession of dm- to m-thick, planar bedded scoria lapilli beds, which are probably fallout deposits from plinian to sub-plinian plumes erupted in the aftermath of the eruption.

10.10.4.3 Textures and Eruption History

The explosive eruption of large volumes of low-viscosity mafic magmas is problematic. Juvenile clasts are very poorly vesicular compared with felsic counterparts (Vinkler et al. 2012), averaging around 40%. By contrast, bubble size distribution and number densities have been found to be higher than for other known mafic eruptions and almost at values typical of standard plinian eruptions (Figs. 10.92), suggesting that high decompression rates played a significant role. At the same time the groundmass is always

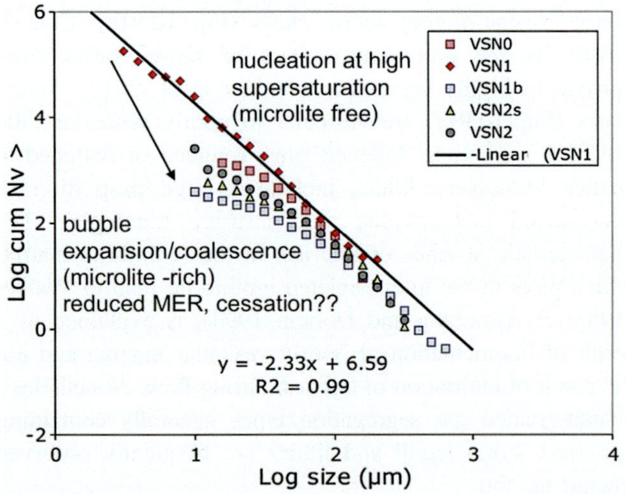

Fig. 10.92 Bubble size and number and density data for Villa Senni ignimbrites, Colli Albani volcano. There is no evidence of significant post-fragmentation bubble expansion, indicating pre-fragmentation nucleation of bubbles at low supersaturation and a growth event in low-viscosity magma at depth (stagnation at transition VSN1 to VSN1b). The absence of very large bubbles suggests permeability limited bubble growth ± early fragmentation and T < T_g. (After Vinkler et al. 2012)

extensively crystallised, with the exception of the very early phases of the eruption, in the basal fallout and in the lower part of the lower Tufo Lionato ignimbrite.

Vinkler et al. (2012) interpreted the eruption to have started with a sustained column from a central conduit, which emplaced the lower Plinian fallout deposit. This phase led to the early development of an inner caldera collapse, and the first development of a nearly steadily collapsing column issuing from ring fractures, from which the Tufo Lionato ignimbrite was emplaced radially. At the top of the Tufo Lionato, i.e. after the eruption of almost 10 km^3 of magma, bubble number density of juvenile clasts drastically decreases (Fig. 10.92), suggesting a slowing in the eruption rate. During the temporary waning of the eruption, which could have been a brief pause at the end of the Tufo Lionato phase, magma started to intrude the developing outer caldera ring fractures, producing both temporary magma ponding and degassing at shallow level. This magma was then erupted as fire fountaining spatter along with the main co-ignimbrite breccia at the base of the overlying Pozzolanelle ignimbrite, which signals the main caldera collapse event during the Villa Senni eruption (Vinkler et al. 2012). The extensive occurrence of leucite microlites in the groundmass of scoria clasts in the Pozzolanelle ignimbrite indicates the onset of bulk decompression and degassing of the magma reservoir during a waning phase or temporary halt of the eruption at the end of the Tufo Lionato eruption, leading to the main phase of caldera collapse. A change in viscosity due to extensive crystallisation is the likely explanation for the sustained explosive phase associated with the eruption of the syn-collapse Pozzolanelle ignimbrite.

10.11 Classification of Ignimbrites

The term "*ignimbrite*" encompasses a huge variety of deposits associated with pyroclastic flows, irrespective of their sizes (which span >5 orders of magnitude, from <0.01 km^3 to >1000 km^3 as bulk volume), magma chemistries (which span from mafic to high silica rhyolites), juvenile clast textures (which range from crystal poor to crystal rich types, from poorly vesicular scoria to highly vesicular pumice), geometries (from high to low aspect ratios), emplacement temperature (from non-welded, to vapour-phase indurated, to welded deposits), and related eruption styles and rates (which range from impulsive collapse of plinian/vulcanian columns from a central vent, to sustained boil-over from caldera ring fractures).

It is remarkable that while a nomenclature and classification for fallout deposits exists (Chap. 9), which describes the spectrum of phenomena associated with increasing size and intensity of the related eruptions, ignimbrite-forming eruptions have never been similarly classified. This is in our opinion a major deficiency of modern volcanology, which has prevented so far, the development of standardised methods to measure comparable descriptors of ignimbrites; see for example the discussion in Sect. 10.8.1.1, about the disparate and very poorly defined methods for defining the volume of ignimbrites. In spite of their very substantial presence in the Earth's volcanic rock record, much larger than fallout deposits, so far, the terminology for ignimbrite types does not correspond to or reflect eruption styles, as is the case for fallout deposits!

Walker (1983) noticed that macroscopically ignimbrites vary in terms of their geometry and relationship to topography, and their temperature of emplacement. He therefore proposed the high to low aspect ratio trend (see Sect. 10.8.4.3) and low- to high-grade trend, i.e. from unwelded to welded ignimbrites (see Sect. 10.8.14). However, these trends were not related to each other and did not represent a classification scheme.

As a matter of fact, attempting a classification for ignimbrites is not an easy task. A meaningful classification should be based on observables, that could be relatively easy to retrieve from deposits of both modern and ancient successions and that reflect fundamental processes associated with eruption and pyroclastic flow dynamics. However, most of the research of recent years has focused on sedimentological aspects and mostly on the distinction between dilute and concentrated currents and related deposit lithofacies. While, as we have extensively described in previous sections, such distinctions certainly help in understanding the lithofacies architecture of ignimbrites, they are scale-invariant, i.e. pdc of any scale and type invariably develop at some places and at sometimes, both dilute and concentrated parts and related deposits, though in varying proportions. This has prevented so far, the use of lithofacies associations to classify the immense spectrum of pyroclastic density current types and associated eruptions.

In our opinion, the classification of ignimbrites must rely on the comparison of bulk property descriptors of the deposit, as is done for fall deposits (Chap. 9), and not on their internal complexities. In this regard, numerical simulations have long suggested that the geometry of ignimbrites (runout and area of dispersal, volume, thickness decay patterns, relationship to topography) is, as for fallout deposits, an excellent tracer of variable sedimentation patterns associated with variable flow types and eruption styles (Giordano and Dobran 1994; Bursik and Woods 1996; Dade and Huppert 1996; Dufek 2016; Shimuzu et al. 2019; Esposti Ongaro et al. 2020a, b; Calabrò et al. 2022).

Giordano and Cas (2021) have recently proposed a classification scheme based on the dispersal area and the bulk deposit volume of outflow ignimbrites as macro-descriptors, which are well distributed on a log–log diagram and show a defined relationship (Fig. 10.93). The ignimbrite bulk volume is a proxy for the magnitude of the eruption (Sect. 10.8.1.1). The dispersal area is a proxy for the way that volume was distributed along topography, and therefore relates to the competition between transport and sedimentation along the flow path of pyroclastic flows (Giordano and Doronzo 2017), and specifically is related to the mass flux (Bursik and Woods 1996; Dufek 2016; Shimizu et al. 2019; Fig. 10.94a). The classification scheme

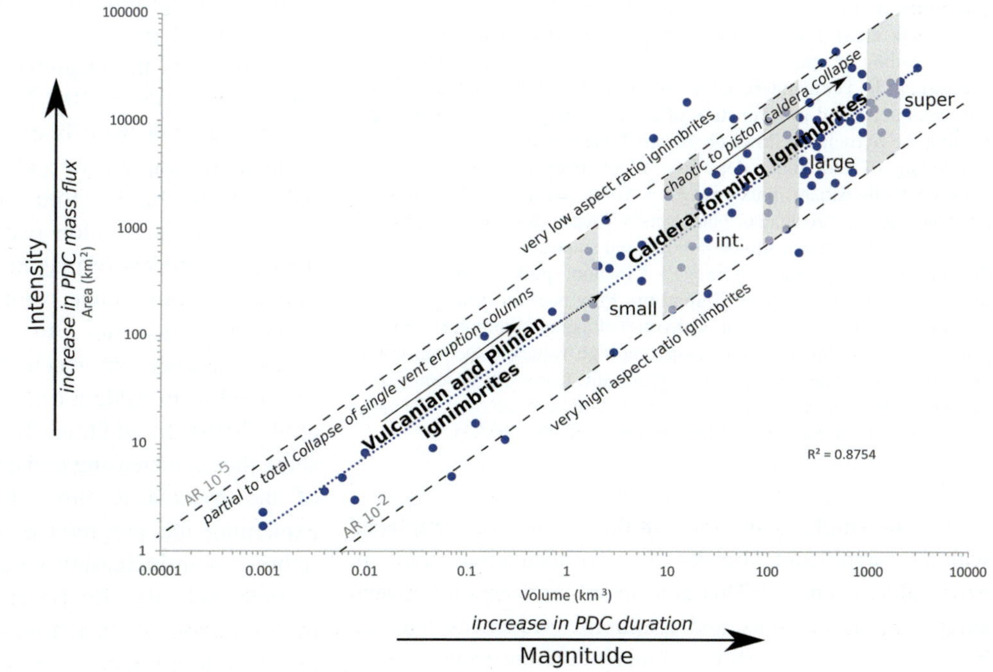

Fig. 10.93 Classification of ignimbrite types, plotting their area against volume; see text for explanation (From Giordano and Cas 2021)

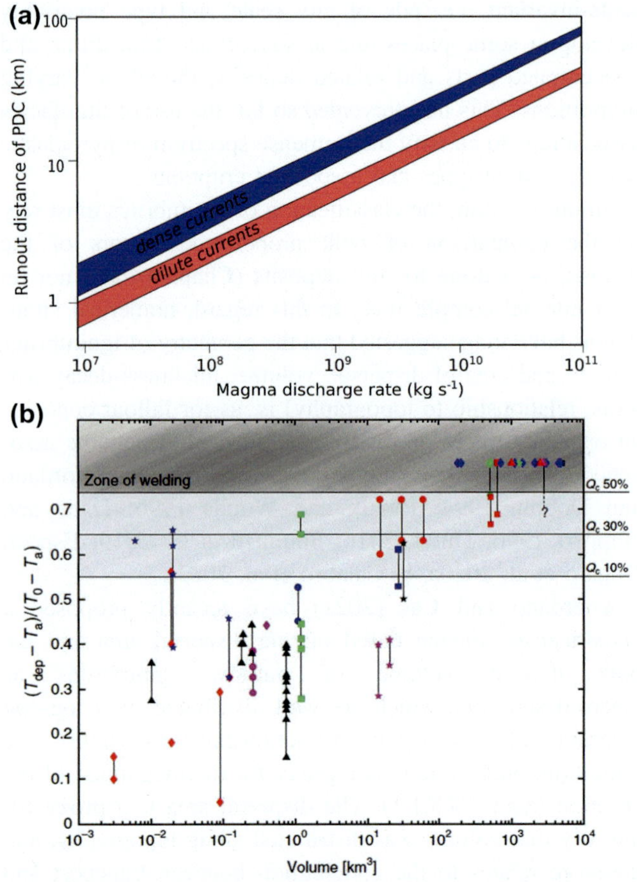

Fig. 10.94 Relationships between mass eruption rate (MER), runout distance, temperature of emplacement, and volume of ignimbrites. **a** Relationship between magma mass discharge or eruption rate and runout distance of pyroclastic density currents as derived from numerical simulations (From Shimizu et al. 2019). The results show a direct control of mass flux in the flow on the final runout. High concentration flows (blue area) attain farther reach with respect to dilute flows (pink area). Internal variations are controlled by the ratio of the rate of feeding of mass to the basal zone of the flow versus the rate of deposition. **b** Temperature of emplacement versus volume of ignimbrites interpreted as a function of the percentage of collapsing mass feeding the pyroclastic flows *Qc* (From Trolese et al. 2019, and see for full listing of individual deposits). Black triangles = Vesuvius; red circles = Colli Albani; pink triangle = Toba; red diamond = Montserrat; blue square = Krakatau; black star = Taupo; pink star = Santorini; green star = Ora; green square = Fogo; red square = Cerro Galán; pink diamond = Lascar; blue circle = El Chichon; pink circle = Colima; blue diamond = Great Basin Ignimbrite Province; red triangle = Yellowstone; black square = Fish Canyon Tuff; green diamond = Bishop Tuff; blue star = Mt St Helens. Note that high values of Qc imply high mass discharge rates feeding the pyroclastic flows, so by comparing (**a**) and (**b**) we can deduce a logical connection between mass discharge rate, runout, temperature, and final volume (cf. Fig. 10.93)

applies to single ignimbrite outflow units, i.e. emplaced without significant breaks in sedimentation and in extra-caldera settings. This is to avoid comparison between single ignimbrite units and ignimbrite sequences that are built up by the superposition of many different eruption units (e.g. the Racer Canyon Tuff, Nevada, which is made up of 12 different cooling units; Best et al. 2013).

Some workers have suggested that, based on the progressive aggradation theory for ignimbrites, there should be in principle no relationship between final volume and dispersal area (Branney and Kokelaar 2002, p. 118). However, Fig. 10.93 shows very clearly that this is not the case and that ignimbrites show a remarkable log–log relationship between total dispersal area and bulk volume ($R^2 = 0.8$). An equivalent power law relationship between final volume and runout of ignimbrites (which is the radius of a circle with the equivalent dispersal area) is shown by numerical simulations in Dufek (2016).

Understanding the processes behind these geometrical relationships is essential to be able to classify ignimbrites. In theory, the final volume should not directly affect the total dispersal, as in principle the maximum lateral extent which is controlled by the peak mass flux (Shimizu et al. 2019), could be attained at any time during the duration of a sustained flow event and deposit thickness can by piled up irregularly during a pulsatory event. This is especially true for large volume ignimbrites, where the total volume is certainly not erupted in a single instant. Giordano and Cas (2021) suggest that the only logical way to relate the areal extent of a single outflow ignimbrite to its bulk volume is that both the runout and the volume are, at a first order, proxies for mass flux, the final volume being attained across the entire flow time duration. To understand this, we must recall that ignimbrites (different from fall deposits, for example, which mostly relate to eruptions from single vents) derive from a huge variety of vent source configurations, that span from single vents of variable diameters associated with vulcanian, sub-plinian and plinian columns (Sect. 10.8.2.1), to caldera ring faults of progressively increasing length and variable geometries, as the magnitude of the caldera-forming eruption increases (Sect. 10.8.2.2).

So, although there is a physical limitation to the maximum mass eruption rate at single vents, controlled essentially by the Earth's force of gravity, along linear fissure vents the mass eruption rate feeding pyroclastic flows can increase by orders of magnitude. The largest known plinian eruptions from single vents are characterised by mass eruption rates up to 10^9 kg s^{-1}. Caldera-forming super-eruptions can reach 10^{11} kg s^{-1} and maybe more, sustained from caldera ring fissure vent systems (e.g. Costa et al. 2018). In addition, Trolese et al. (2019; Fig. 10.94b) show that at increasing collapsing mass flux, the temperature of the pyroclastic flows becomes progressively higher, explaining the progressive tendency to have welded ignimbrites with increasing volumes.

Independently, Roche et al. (2021) demonstrated that mass eruption rate increased significantly during the collapse phase of an ignimbrite-forming eruption compared with the preceding fallout deposit forming buoyant column phase, and that runout distance is proportional to mass eruption rate. This is comparable to the findings of Giordano and Cas (2021) relating mass eruption rate to ignimbrite area and volume.

These factors allow consideration of Walker's (1983) intuitions about geometry and emplacement temperature as the best observables to classify ignimbrites. On these grounds, it is possible to place a boundary at around 1–2 km^3 (although these values can vary slightly) in Fig. 10.93, which subdivides the small ignimbrites related to collapse of columns issuing from single vents, from those that are related to collapses of calderas. Small ignimbrites should be named according to the eruption style they are associated with. For example, ignimbrites with associated fallout deposits indicate column collapse origins, and are therefore vulcanian, sub-plinian, and plinian ignimbrites (Table 10.4). Larger ignimbrites relate to a distinctive eruption style dominated by multiple sources and fissure vent eruptions at progressively increasing mass fluxes, that is different to those presently identified for fallout deposits forming eruptions such as hawai'ian to plinian. We propose the term "*caldera-forming eruptions*" (Table 10.4); hence, the related ignimbrites should be named "caldera-forming ignimbrites".

It must be noted that although many caldera-forming ignimbrites, especially at the smaller end of the volume spectrum, are commonly underlain by plinian fall deposits, these are the result of a pre-caldera, single vent eruption phases, the eruption parameters of which (e.g. the mass eruption rate feeding the plinian column) should not be confused with those characterising the caldera collapse phase when pyroclastic flows are fed by caldera ring fault fissure vents. Detailed works on caldera-forming sequences (e.g. the 7.7 ka Mt Mazama eruption, Bacon 1983; the 25.4 ka Oruanui ignimbrite, Wilson 2001, Fig. 10.84c; the 4.6 ka Fogo A eruption sequence, Pensa et al. 2015a) have shown that initial plinian phases of the eruption can be punctuated by early column collapses that fed what we now would call intra-plinian ignimbrites according to current definitions, whereas many plinian ignimbrites occur above a fall deposit, with no need to be only interbedded within. These can be followed later in the eruption by the caldera-forming ignimbrite(s). This also clarifies and provides a proper nomenclature for all those caldera-forming ignimbrites that have no basal, underlying plinian fallout deposits (e.g. the 2.08 Ma Cerro Galán ignimbrite, Cas et al. 2011; the 274 Ma Ora ignimbrite, Willcock et al. 2013), as it should be clear by now that there is no requirement for a caldera-forming eruption to be preceded by phases of any point source eruption column forming specific eruption style.

Giordano and Cas (2021) also proposed a distinction within the caldera-forming ignimbrites family, into small (up to 10 km^3), intermediate (up to 100 km^3), large (up to 1000 km^3) and super (>1000 km^3) (Fig. 10.93). These distinctions, though based on subjective limits, encompass the important transition between chaotic and coherent caldera collapse styles proposed in the modelling literature (Roche and Druitt 2001; Gregg et al. 2012; see Chap. 16, Sect. 16.7.3). Chaotic caldera collapses correspond to piecemeal calderas, where activation of multiple intersecting caldera magma reservoir roof block faults of varying orientations follows complex spatial and temporal patterns, as opposed to coherent collapses which correspond to single piston calderas, where a well-defined, outward dipping ring fault system encircles the collapsing caldera block (Fig. 10.95). Analogue and numerical modelling suggest that chaotic collapses are more likely associated with small to intermediate volume caldera-forming ignimbrites, whereas piston collapses are more likely associated with large volume caldera-forming ignimbrites related to maximum magma discharge rates (Roche et al. 2001b; Geyer et al. 2006; Gregg et al. 2012). The consequences of these end-member types of caldera collapses on caldera-forming ignimbrites are significant. Chaotic collapses should involve a shift of vent locations over time and contemporaneous multiple vent sourcing, changes in their geometric configurations and mass eruption rates, as fractures propagate and differential sinking of caldera blocks occur. By contrast, coherent magma chamber roof block collapses should involve a much more continuous process of magma boil-over eruption style at very high mass eruption rates from a relatively steady vent geometric configuration associated with the continuous ring fault along which the coherent caldera floor subsides.

Distinguishing caldera-forming ignimbrites by volume allows capture of other important variables in the eruption processes associated with different styles of caldera collapse. For example, the degree of internal stratigraphic complexity of ignimbrites decreases with increasing volumes and this can be explained by the chaotic versus coherent collapse style. At the same time, Fig. 10.93, which is contoured for values of aspect ratios that allow distinction between topography mantling ($<10^{-5}$, e.g. Taupo ignimbrite), topography filling ($>10^{-2}$, e.g. 1991 Pinatubo ignimbrite) and topography burying/building ($>10^{-5}$ - $< 10^{-2}$) geometries, shows that above 1000 km^3 the range of aspect ratios drastically narrows to values of 10^{-3}–10^{-4}. This suggests that at increasing mass fluxes and flow durations associated with piston collapses, ignimbrites develop topography burying geometries and restricted flow types, similar to the "tortoise" end member described for the Cerro Galán ignimbrite in Sect. 10.9.1.2. Furthermore, unlike smaller volume caldera-forming ignimbrites which show a wide spectrum of emplacement temperatures (Sect. 10.8.14.1), large volume caldera-forming ignimbrites are almost ubiquitously welded,

Table 10.4 Characteristics of eruption styles combining caldera-forming eruptions and their related ignimbrite types (From Giordano and Cas 2021)

Eruption style	Feeding	Intensity ($kg\ s^{-1}$)	Vent geometry	Column type	Column height (km)	PDC	Ignimbrite type	Ignimbrite volume (km^3)	Composition
Strombolian s.s	Long term intermittent	$<10^4$	Single point	Jet	<0.5		N/A		Mafic
Hawaiian	Sustained (short lived)	$<10^6$	Fissure/single point	Jet ± weak convective	<5		N/A		Mafic
Violent Strombolian	Short term intermittent	$<10^6$	Single point	Weak convective	<10		N/A		Mafic-intermediate
Vulcanian	Intermittent	$<10^7$	Single point	Weak convective	<15	Possible	Vulcanian ignimbrites	$<10^{-1}$	Mafic-intermediate-felsic
Sub-Plinian	Sustained	10^6–10^7	Single point	Weak to strong convective	15–20	Possible	Sub-plinian ignimbrites	$\leq 10^{-1}$	Mafic-intermediate-felsic
Plinian (and Ultra-)	Sustained	10^7–10^9	Single point	Strong convective	20–35 (50?)	Frequent	Plinian ignimbrites	$\leq 10^0$	Mafic-intermediate-felsic
Caldera-forming	Sustained	10^9–10^{11}	Segmented ring faults (chaotic collapse)	Collapsing fountain + co-ignimbritic	Up to 80*	Dominant	Small to intermediate caldera-forming ignimbrites	10^0–10^2	Mafic-intermediate-felsic
			Segmented to continuous ring faults			Dominant	Large caldera-forming ignimbrites	10^2–10^3	Intermediate-felsic
			Continuous ring faults (piston collapse)			Dominant	Super caldera-forming ignimbrites	$>10^3$	Intermediate-felsic

* Plume heights up to 80 km for caldera-forming eruptions are numerical results from Costa et al. (2018)

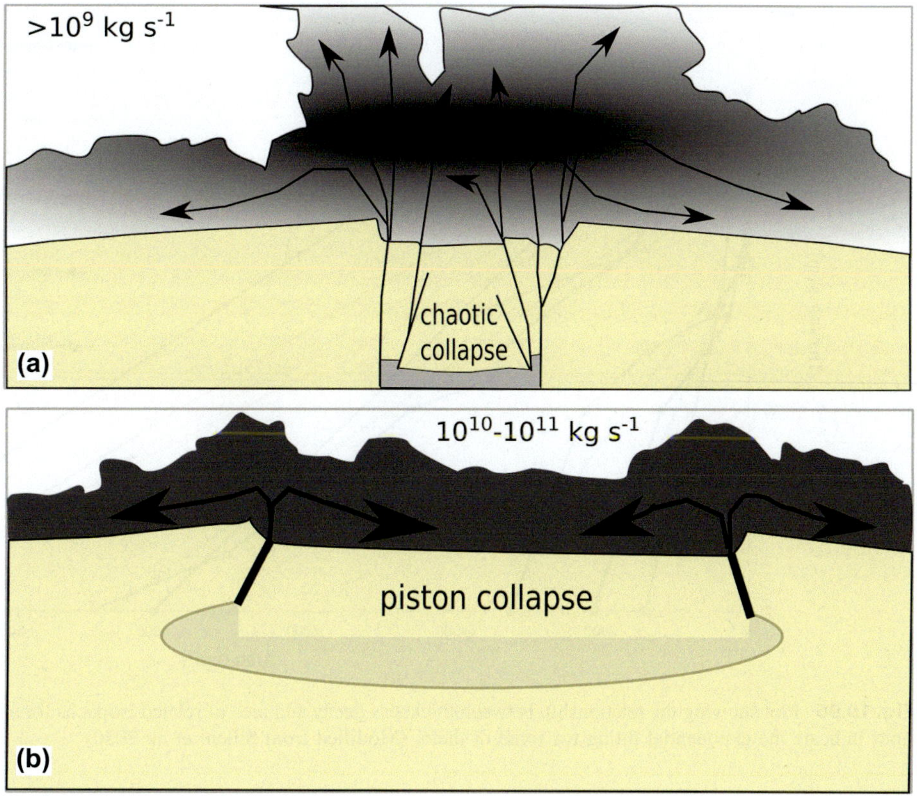

Fig. 10.95 Potential control of caldera collapse mechanism on the source configuration and the mass flux feeding the outward moving pyroclastic flows. **a** Chaotic caldera collapses promote complex spatial and temporal patterns of caldera fault unzipping, multiple sources. **b** Piston caldera collapses are characterised by well-defined ring faults and coherent subsidence of the caldera floor. Arrows and caldera faults thickness are qualitatively proportional to mass flux as well as the grey scale in the pyroclastic flows. (From Giordano and Cas 2021)

again in agreement with sustained boil-over fountaining at high mass fluxes which prevents significant entrainment of air (Sparks et al. 1978; Jessop et al. 2016; Trolese et al. 2019).

Interestingly, a similar narrowing of the aspect ratio variability is observed for the single-vent-derived ignimbrites at the higher end of their volume spectrum (1–2 km^3), where aspect ratios also narrow around values of 10^{-3}-10^{-4} (Fig. 10.93). This evidence can be interpreted in terms of the transition from very small volume, impulsive partial collapses from variably sustained columns, which produce a very wide range of pyroclastic flow types reflected in the wide variability of aspect ratios, to larger volume and sustained collapses that favour a more steady high mass flux feeding of the resulting pyroclastic flows, and restrict the flow types to forced convection (Giordano and Doronzo 2017).

To summarise, Fig. 10.93 shows that for any given mass flux feeding the pyroclastic flow, which controls the maximum runout and dispersal area, the duration of the eruption will increase the final volume of the ignimbrite and progressively increase its aspect ratios. By contrast, for any given final volume, the increase in mass flux enlarges the area covered and decreases the aspect ratio of the associated ignimbrite.

The new classification scheme for ignimbrites proposed here focuses on the use of macro-descriptors for ignimbrites which can be properly compiled through good field work, measuring standard parameters such as the thickness of ignimbrites in outcrops, their areal distribution, their lithofacies architecture and the evidence for thermal state at emplacement. These measures provide the basis for identifying the boundaries of individual cooling/depositional units, constructing isopach maps, determining maximum areal extent and bulk volume, and thus allowing linking the physical descriptors to eruption processes and caldera collapse processes. Although our scheme shows that the internal facies complexities of individual cooling/depositional units are important for reconstructing the time–space evolution of a flow they, nevertheless, should not overshadow the significance of the reconstruction of the overall morphology of ignimbrites, which has not yet received the attention needed by the volcanological community.

A major challenge for the future in physical field-volcanology of ignimbrites is to reach much more standardised ways of measuring such parameters, to reduce uncertainties and to make measurements reproducible, such as has been done for the description of fallout deposits (e.g. Bonadonna et al. 2016). For example, the exponential trends of log thickness variations versus isopach areas of individual units suggest that lateral decay patterns of thickness should be reappraised (e.g. Wilson 1991; Giordano and Doronzo 2017; Silleni et al. 2020, 2024; Sect. 10.8.4.4), as well as grain size decay patterns (i.e. from isopleth maps; Palladino

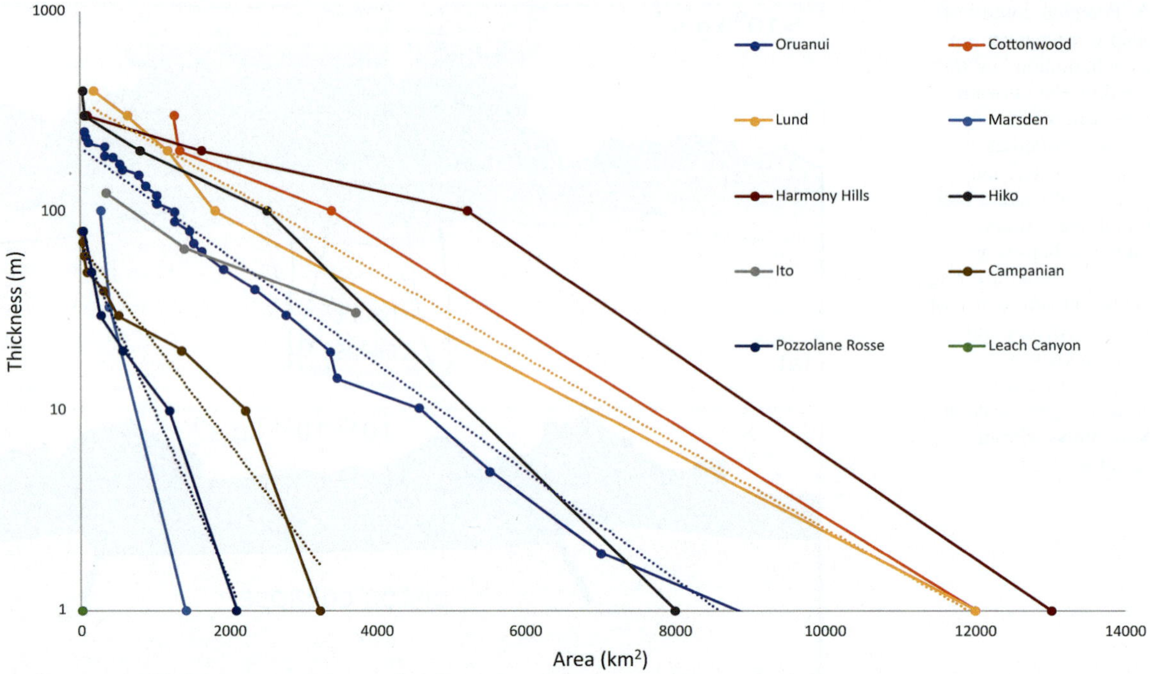

Fig. 10.96 Plot showing the relationship between thickness decay and area of related isopachs for a number of well-described ignimbrites. Dotted lines indicate the exponential fitting for some of them. (Modified from Silleni et al. 2020)

and Giordano 2019; Doronzo et al. 2022), which may provide other ways to compare and classify ignimbrites. In Fig. 10.96, we show a log thickness versus cumulative isopleth area for several well-studied ignimbrites. Considering the large uncertainties in the definition of the outer isopach extent (arbitrarily set at 1 m instead of 0 m in order to be used in a log scale), caldera-forming ignimbrites seem to show well-defined exponential fits with R^2 always >0.9. By analogy with similar exponential trends for fall deposits, we speculate that variations of the best-fit exponential relate to eruption and dispersal processes. However, a full understanding needs more investigation, such for the presence of breaks in slope along the curve, which may relate to along flow changes in depositional patterns (for example across topography; cf. Wilson 1991). Furthermore, the exponential fitting in Fig. 10.96 is in good agreement with sedimentation models (Bursik and Woods 1996; Andrews and Manga 2012). Calabrò et al. (2022) have recently demonstrated that thickness variation and exponential decay patterns reflect the initial flow density, and that the grain size distribution, emplacement temperature, and maximum runout can constrain the initial mass flow rate and total grain size distribution.

10.12 Blast Flows and Their Deposits

Since the 1980 eruption at Mt St Helens, which was initiated by a landslide and debris avalanche that then triggered a lateral *blast flow* (terminology of Kieffer 1981; Voight et al. 1981; Fig. 10.4f), recognition of horseshoe shaped depressions cutting the top of stratovolcanoes and their associated deposits of debris avalanches with hummocky topographies at the base of volcanoes (Fig. 10.97a; Chaps. 12 and 16), have been recognised worldwide as representing the results of sector collapse (e.g. Socompa volcano, Francis et al. 1985; Ollague volcano, Tibaldi et al. 2006; Roccamonfina volcano, De Rita and Giordano 1996; Bezymianny in 1956, Gorshkov 1959; Belousov et al. 2007; Bandai-san, Japan in 1888, Nakamura 1978; Sheveluch, in Kamchatka in 1964, Gorshkov and Dubik 1970; Soufrière Hills volcano on Montserrat in 1997, Voight et al. 2002). By analogy with the 1980 Mt St Helens 1980 event, sector collapse and debris avalanches represent a potential footprint for associated lateral blast flow events and deposits. Lateral blast flow deposits, however, are small in volume (<<1 km^3), and easily eroded.

(a)
(b)
(c)
(d)
(e)
D
C
B
A
(f)
Windy Ridge (6.6 km)
MSHI Campsite (13.3 km)
Layer B
Older tephra deposits
May 18, 1980
Lower pumice-rich fallout
Basal lithic-rich fallout
Accretionary lapilli unit
Layer D
Pyroclastic surge unit
Layer C
Layer A
Layer A_0
10 cm
(g)
(h)
O
H
P
D

◀ **Fig. 10.97** The blast flow deposit from the 18 May 1980 eruption of Mt St Helens, USA. **a** View from the pumice plain of the northward open horseshoe-shaped depression formed after the sector collapse that triggered the lateral blast (Photo by M. Trolese). **b** Valley ponded blast flow deposit, showing a flat top with abundant rafted tree trunks. (From Hoblitt et al. 1981; photo taken in Smith Creek Canyon soon after the eruption). **c** Carpet of blow-down tree trunks, aligned in the flow direction, as well as in situ, splintered tree stumps with splinters also pointing in the flow direction. **d** The distal damage zone at ~15 km from Mt St Helens, visible in the background; note the patchy occurrence of blowdown and still standing trees but stripped of bark. **e** Trench cutting showing gradation from a lithic and wood debris rich base (layer A) into a massive, poorly sorted lapilli-ash (layer B), into a low angle, trough cross-stratified ash (layer C). Layer D, which is mostly eroded from the outcrop is a very fine ash layer. **f** Stratigraphic columns for two sections of the blast deposit measured at Windy Ridge and Mt St Helens Institute Campsite; distance from vent is indicated in km. See text for discussion of details of layers A-D. (Adapted from Pallister et al. 2017). **g** Coarse grain size of the basal part of the valley ponded blast flow deposit northeast of the volcano. The clast population includes porphyritic dacite (grey), obsidian (black), pumiceous dacite (light), and hydrothermally altered clasts (brown). **h** Close-up texture of the blast flow deposit, showing the varied componentry and coarse matrix. O = obsidian; D = dense dacite; P = pumiceous dacite; H = hydrothermally altered lithic

Several blast flow events and their deposits have been documented, including the May 18 1980 Mt St Helens event (Kieffer 1981; Voight et al. 1981; Walker and McBroome 1983; Fisher 1990; Belousov et al. 2007; Esposti Ongaro et al. 2011; Gardner et al. 2018), Bezymianny (Gorshkov 1959; Belousov et al. 2007) and Soufrière Hills (Sparks et al. 2002; Voight et al. 2002; Belousov et al. 2007). The flows were triggered by powerful blasts that were directed laterally and exerted enormous force on trees and infrastructure. We therefore call this pyroclastic density current type, *blast flows*, and their deposits, blast flow deposits. These types of flows are initiated by violent blasts that result from instantaneous decompression of a high-level volatile bearing magma body (and/or hydrothermal system), when sector collapse of a large part of a volcanic edifice occurs.

There has also been debate about the details of the initial state of the blast flows. Kieffer (1981) proposed a direct lateral blast with an initial supersonic gas jet phase that generated shock waves that destroyed the forests in the proximal "blast" zone (Figs. 1.5a and 1.15e), and then a following subsonic phase that produced largely valley-controlled pyroclastic flows in distal areas (Figs. 1.5a and 10.97b). Belousov et al. (2007) suggest that the initial phase was a directed ballistic "burst" that collapsed to produce energetic flows that generated sufficient dynamic pressure to destroy the forests but were not necessarily supersonic. Either way, the flows were directed laterally, and here we will rather focus on the behaviour of the flows and their deposit characteristics.

The most significant characteristics of blast flows are their mobility and force, or at least the force of the initial flow pulse. Blast flows are able to flow over topography hundreds of m high, and in the cases of Mt St Helens and Bezymianny, flow distances were up to 30 km from source (e.g. Fig. 1.5a). The Mt St Helens flows famously destroyed forests up to 30 km from the crater, aligning tree trunks parallel to flow paths (Figs. 1.15e and 10.97c, d), irrespective of topography within 15 km, and then being influenced by topography for the remaining 15 km, travelling along valleys (Kieffer 1981). Kieffer (1981) calculated flow velocities of >300 m s^{-1} (>1000 km hr^{-1}) in proximal areas, based on calculations of the Mach flow number (>>1). Esposti Ongaro et al. (2011, 2012) calculated proximal flow velocities of 140 m s^{-1} (~500 km hr^{-1}) and dynamic pressures of >100 kPa in proximal areas, and flow velocities of 70 m s^{-1} and dynamic pressures of 0.5–10 kPa in distal areas, based on numerical modelling. Gardner et al. (2018), based on evidence for distal blast lift-off from the ground, suggested the velocity of the distal end of the flow was still between 40 and 115 m s^{-1}. Dynamic pressures required to cause severe damage to the forests in the proximal area at Mt St Helens is 20 kPa, but trees can be destroyed at pressures of ~2 kPa (Valentine 1998; Esposti Ongaro et al. 2011, 2012). For comparison, Wilson (1985) estimated the flow velocity of the 232 AD Taupo pyroclastic flow also to have been ~300 m s^{-1}. The difference is that the Taupo flow event involved collapse of a vertical buoyant explosive eruption column, whereas the blast flow events considered originated from lateral blasts from volcanic edifices.

The deposits of the 1980 Mt St Helens blast flow event have been very closely studied (Hoblitt et al. 1981; Moore and Sisson 1981; Waitt 1981; Walker and McBroome 1983; Fisher 1990; Druitt 1992; Pallister et al. 2017). They have a fan-like dispersal to the north, extending up to 30 km from the crater (Fig. 1.5a). As indicated above, up to 15 km from the crater, trees were blown down and aligned (Fig. 1.15e and 10.97c, d), forming a flow lineation that tracks the path of the flow and its pulses (Kieffer 1981; Fisher 1990; Gardner et al. 2018). Although the deposits are commonly <1 m thick, they can be up to 10 m thick in valleys (Fig. 10.97b), and distally only cm thick. Away from the proximal area, the deposits fill valleys and have flat tops, just like normal ignimbrite valley pond deposits (Fig. 10.97b), and on adjacent slopes they are like ignimbrite veneer deposits (Hoblitt et al. 1981). Although there are many detailed internal and lateral variations in the deposits, recorded in the foregoing papers, there is a general stratigraphy that is consistent in these papers (Fig. 10.97e). The deposits were emplaced at between 100 and 300 °C. In addition, the deposits of blast flows at Mt St Helens, Bezymianny and Soufrière Hills have some general similarities, which Belousov et al. (2007) summarise, choosing a simple A, B, C stratigraphic subdivision, in which our units B and C, that will be described below, are combined in their unit B.

The base of the Mt St Helens blast flow deposit, which we call *Layer A_0*, is a layer of blow-down tree-trunks and timber debris (Fig. 10.97e, f), variably rooted in, or mingled with soil. *Layer A* is a coarse gravelly or sandy textured layer with grain size and thickness decreasing outward from the source, and significant fines depletion is evident. The basal part is rich in vegetation debris, accessory edifice and accidental lithic debris, substrate regolith rip-up material and juvenile clast material, which passes up into a fines-depleted lapilli clast-rich zone (Fig. 10.97e, f). *Layer B* consists of a massive, homogenous more ash rich lapilli-ash which has the aspect of a more normal pyroclastic flow deposit (Fig. 10.97e–g), especially in valley settings, where it also contains gas segregation pipes (Hoblitt et al. 1981; Walker and McBroome 1983). The componentry is diverse, consisting of a mixture of debris from the pre-existing edifice (dacite, andesite, hydrothermally altered lithic clasts, obsidian), as well as juvenile debris from the cryptodome (dense dacite, pumiceous dacite) (Fig. 10.97g, h), and varying proportions of shredded vegetation and soil. *Layer C* is cm to tens of cm thick and marked by surge dune forms and low angle cross stratification (Fig. 10.97e, f). *Layer D* is very fine, contains accretionary lapilli, is rarely more than several cm thick and is agreed by all authors to be an ash fallout of dilute trailing ash clouds derived from the blast flow.

The relationship between layers A and B is generally abrupt, but locally their interface is gradational (Moore and Sisson 1981; Waitt 1981; Fig. 10.97f). The total thickness of this assemblage of layers rarely exceeds 1 m, except in hollows or valleys where it can be several m thick. Walker and McBroome (1983) demonstrated that the basal facies lie in the fines-depleted field and the massive Layer B unit lies in the normal ignimbrite field on plots of sorting vs. median granisize, and plots of F_2 (the weight percentage finer than 1/16 mm) versus F_1 (the weight percentage finer than 1 mm). Walker and McBroome (1983) and Walker (1983) interpret the Mt St Helens deposits to be the products of very violent, low aspect ratio pyroclastic flows. They interpreted Layer A as the fines-depleted ground layer deposited from the fluidised head. Layer B seems very similar to valley pond ignimbrite, Layer C is like an ash-cloud surge deposit, and also occurs on gently sloping valley walls, as a thin veneer deposit (Walker and McBroome 1983). Distally the deposits are surge-like (Druitt 1992).

The flow dynamics have been widely discussed (see cited references above). Our preferred interpretation is that of Hoblitt et al. (1981). The Mt St Helens blast flow clearly changed its character dramatically in a short period of time and over a relatively short distance from an initial laterally directed, sonic to supersonic blast the would have generated shock waves, and generated a horizontal jet of gas, air, vegetation debris, accidental lithics and some juvenile clast debris that did not deposit anything, created high dynamic pressures, and damaged the forest for up to 35 km to the north (Layer A_0; *phase 1*). Contemporaneously, the flow system behind the blast front quickly deflated and formed a trailing, high velocity flow of pyroclasts that generated high enough dynamic pressures to continue destroying forests and began to deposit fines-depleted deposits in proximal areas (<15 km; Layer A; *phase 2*). Contemporaneously, the flow system trailing behind the second phase of the flow deflated even more and transformed into a more normal, higher particle concentration flow phase that was influenced by topography and produced valley ponded and veneer deposits in the outer zone (Layer B), and surge deposits (Layer C) in distal most areas (*phase 3*).

This model is consistent with the 4D, evolution of pyroclastic flow dynamics proposed previously in Sect. 10.7. It is also consistent with our suggestions for the evolution of the 232 AD Taupo flow in New Zealand, even though the initiation of the two flow systems and the componentry was different (Sect. 10.9.1.1). In both cases we envisage a flow system that evolved very rapidly both temporally and spatially, changing its character in time and space during the 400 s event. Most other models of pyroclastic flows take a single, simple snapshot view of the whole flow event, characterising it as one type of flow or another that does not change in transit.

It is important to stress especially from a volcanic hazard point of view (Chap. 16), that the deposits of blast flows, representing one of the most violent, energetic, and forceful volcanic phenomena, have an extremely poor preservation potential in the rock record because they are thin, non-welded and easily eroded.

10.13 Block and Ash Flows and Their Deposits

Block and ash flows (BAFs) originate from the relatively small, finite source volumes of erupting lava domes or their margins (Sect. 10.3.1). They are therefore generally small in volume (<<1 km^3), and short-lived flows, flowing only a few km from source, and rarely more than 10 km, especially when generated by passive gravitational collapse of lava dome margins (e.g. Mt Unzen, Japan, 1991–1995, Nakada et al. 1999; Merapi, Indonesia, 2006, Charbonnier et al. 2008; Colima, Mexico, 1998, 2015, Saucedo et al. 2004; Capra et al. 2016). When explosive destruction of lava domes is involved, the flows can be more dynamic, flowing up to 15 km from source (e.g. 2010 Merapi flows, Komorowski et al. 2013; Charbonnier et al. 2013; Cronin et al.

2013; Trolese et al. 2018). In addition to the small volume, limiting factors on flow distance of BAFs are the dense nature of lava dome clasts, and the coarse and permeable nature of the ash matrix, which rapidly allows exsolved gas to dissipate from the flows. The fragmentation mechanism producing clasts in BAFs is largely non-explosive or weakly explosive, resulting initially from gravitational collapse of large blocks from the steep lava dome or flow margins, followed by impact shattering of hot, brittle, glassy lava dome blocks, as they tumble downslope and impact with the substrate and other blocks.

Most of the ash matrix results from in-flow comminution of large blocks by such impact processes. In addition, von Aulock et al. (2013) demonstrated that during cooling, obsidian lava domes can undergo high temperature crack formation associated with spherulite growth (at ~450 °C) and perlitic cracking (at ~400 °C). Such cracking not only weakens the dome rock, making it more prone to collapse gravitationally, but also contributes to rapid comminution of dome collapse blocks as they tumble downslope. The ash fraction can be limited in abundance and is generally coarse, especially in comparison with pumice and ash flow ignimbrite deposits. Block and ash flows are therefore dominantly inertial or granular flows, dominated by frictional interaction and collisions between clasts and with the substrate (Sarocchi et al. 2011; Charbonnier et al. 2013), lacking many of the flow and grain support mechanisms of pumice and ash flows, except in the volumetrically small ash cloud that develops downslope.

Examples of recent eruptions that generated well-studied block and ash flows and deposits include Merapi in 1930, 1942–3, 1994, 2006, 2010 (Voight et al. 2000; Charbonnier and Gertisser 2008; Komorowski et al. 2013; Cronin et al. 2013; Trolese et al. 2018), Colima in 1991, 1998, 1999, 2015 (Saucedo et al. 2004; Sarocchi et al. 2011; Capra et al. 2016; Pensa et al. 2018, 2019), Mt Unzen in 1991–1995 (Sato et al. 1992; Yamamoto et al. 1993; Nakada and Fujii 1993; Nakada et al. 1999; Uehara et al. 2015), and Soufrière Hills, Montserrat in 1995–2013 (Calder et al. 2002; Cole et al. 2002).

These examples highlight two end-member settings for generation of block and ash flows. In the first, BAFs are generated episodically, due to slow or episodic growth of small lava domes, which once collapsed, cease growing or take time to regenerate (e.g. Colima, Merapi). At the other end of the spectrum, dome growth is more or less continuous over an extended period of time over months to years, and lava dome collapses occur frequently, almost continuously (e.g. Mt Unzen 1991–1995). In the former case, BAF deposits form discrete, valley confined units in the landscape around the volcanic edifice (Fig. 10.98a) and in older volcanic stratigraphy, are interbedded with and separated by other volcanic products, reworked volcanic sediments, or incipient soils. The deposits form valley ponded and overbank deposits (e.g. Merapi, Colima; Fig. 10.98a, b; e.g. Sarocchi et al. 2011; Charbonnier and Gertisser 2008; Pensa et al. 2018; Trolese et al. 2018). In the Unzen apron case, the deposits formed a spreading stratified apron or fans (Fig. 10.98c) tens to hundreds of m thick, preserving many of the nearly 10,000 dome collapse and BAF events that occurred from 1991 to 1995 (Nakada et al. 1999; Uehara et al. 2015). In both cases, the deposits can be reworked during and immediately following weather events that produce debris flows and hyperconcentrated flood flow deposits (Nakada et al. 1999; Sarocchi et al. 2011; Charbonnier et al. 2013; Uehara et al. 2015). At Soufrière Hills volcano, another lava dome dominated the eruption that extended for more than 15 years from 1995, also building significant debris aprons in the main valleys. Here the deposits are more diverse (block and ash flows, pumice and ash flows resulting from vulcanian eruptions, rare blast flows, debris avalanche deposits), and eruption events were more spaced than at Mt Unzen (Druitt and Kokelaar 2002).

The deposits of block and ash flows range from largely valley-ponded body deposits, with some thinner overbank deposits (Fig. 10.99a, b), and veneer surge deposits (Fig. 10.99c, m). The sequences of multiple event, apron deposits (e.g. Mt Unzen), are stratified through build-up of multiple flow deposits (Fig. 10.99d, f, j). They also show transitions from coarse, bouldery valley pond facies to finer-grained, thinner valley slope or veneer deposits (e.g. Fig. 10.99e).

The body deposits of BAFs are massive, very poorly sorted, have a relatively coarse ash matrix (Fig. 10.99d, f, g–k; cf. pumice and ash flow deposits), and contain large, generally non-vesicular to poorly vesicular, cognate lithic blocks, including outsize blocks >10 m in diameter (Fig. 10.99h, i). Many blocks preserve radial cooling joints and columns, which show they were emplaced as hot blocks (Fig. 10.99i), as well as impact shatter fractures. Clasts range from angular and slabby to more spheroidal in form and are bounded by planar to curvi-planar surfaces resulting from impact shattering of hot glassy blocks. If corners or edges are rounded, it's a rough rounding resulting from abrasive chipping of corners during transport, rather than prolonged tractional smoothing by fluvial processes. Some blocks may show weak to strong breadcrust textures, resulting from post-fragmentation volatile exsolution from the exterior of clasts, and expansion and cracking of the chilled clast surfaces. Clasts are all of the same composition, with few, if any, accessory lithic clasts and few accidental lithic clasts; that is, deposits are almost monolithologic. Individual flow units may show reverse, coarse tail grading in many examples (Fig. 10.99j). They may contain gas segregation pipes (Sect. 10.8.16), although these are not very common in BAF deposits because of the permeable

Fig. 10.98 Large scale geometry of block and ash flows and relationship to topography. **a** 2015 valley confined block and ash flow deposits from Colima volcano, Mexico, showing the linear geometry of the deposits transitioning into a depositional fan on the lower depositional slope and lower topographic relief beyond the break in slope of the volcano. **b** Overbank deposits spilling onto the valley margins from the entrenched, valley-confined flow system in the middle reaches of the flow and depositional system, 2015 Colima volcano eruption sequence. **c** Developing apron of multiple block and ash flow events that is filling and burying major valleys around the Fugen-dake summit dome complex from the 1991–1995 eruption sequence, Mt Unzen, Japan. Note the active high aspect ratio, source, summit lava dome; while at the base of slope, the block and ash flow apron transitions into a debris flow and hyperconcentrated flood flow depositional system on the coastal plain (foreground). (*Source* Volcanological Society of Japan)

coarse matrix. BAFs are hot enough to carbonise vegetation (Pensa et al. 2018, 2019; Trolese et al. 2018; Fig. 10.99k), and devolatilisation during carbonisation of wood debris in hot BAF deposits after deposition may generate degassing pipes above carbonised logs.

Surface features of deposits include the presence of levees, steep flow fronts, and the presence of large surface blocks, all of which indicate a high concentration during transportation. Block and ash flows also create frontal and enclosing ash-cloud surges, which produce ground surge deposits beneath the body deposits (Fig. 10.99l), and on top of them in valley settings, whereas on adjacent ridges only thin, overbank pebbly body deposit lobes are deposited (Fig. 10.99b, m), as well as thin, fine, lateral surge veneer deposits (Fig. 10.99m).

BAFs formed by explosive disintegration events of lava domes can be much more variable in flow behaviour because some flows are generated by explosive activity, often sharp vulcanian blasts involving low eruption column collapse, and some from dome gravitational collapse. Some are dense clast dominated, and valley confined, others have clasts that can be highly vesiculated to dense, and are more dynamic, expansive and flow and spread over topography (2010 Merapi eruption, Komorowski et al. 2013; Cronin et al. 2013; Trolese et al. 2018).

In the prolonged, Unzen-like eruptions, primary BAF apron deposits (Fig. 10.98c), can be remobilised as debris flows and hyperconcentrated flood flows (called lahars by some) during rain events that could occur during or in between eruptions. During the Mt Unzen dome eruption in 1991–1995, there were numerous instances of primary block and ash flows transforming during flow into hot, steaming, watery debris flows and hyperconcentrated flood flows when BAFs flowed into rivers, and mixed with water, or through heavy rain remobilising still hot BAF deposits on the steep, upper slopes of the volcano. Such remobilised deposits are very difficult to distinguish from the primary deposits because the componentry is the same, and textures can be very similar because the duration of the resedimentation event is short, but many had longer run-out than the BAFs (Uehara et al. 2015). Recently the gravitational collapse of already deposited hot material accumulated along steep volcano slopes (e.g. 2018 Fuego event, Risica et al. 2022) has been recognized as another type of non-explosive PDC type, similar to BAFs but unrelated to dome collapses.

10.13.1 Emplacement Temperatures of Block and Ash Flows

The emplacement temperatures of BAF deposits vary greatly depending on (1) the temperature(s) of the portion of dome collapsed, which may vary from cold in the outer carapace to magmatic temperature at the core (Uehara et al. 2015); (2) the efficiency of air entrainment which depends from the interaction with topography (Pensa et al. 2018, 2019; Trolese et al. 2018); and (3) the grain size distribution that controls the efficiency of thermal exchange between juvenile particles and entrained air, more efficient for ash-sized particles than for block-sized clasts.

Because of the uncertainty in distinguishing primary BAF deposits from resedimented deposits, Uehara et al. (2015) employed thermal remanent magnetisation (pTRM) to determine the temperature of emplacement of both (see Sect. 10.8.14.4) by analysis of dome lava clasts collected from BAF deposits from the 1991–1995 eruption of Mt Unzen in Japan. Whereas most TRM studies focus on analysing the magnetic components of accessory and accidental clasts, Uehara et al. (2015) used the same approach on juvenile lava dome clasts from the block and ash flow and contemporaneous debris flow deposit sequence at Unzen. Using pTRM analyses, they could distinguish between hot block and ash flow deposits, deposited above the Curie temperature, and cold, debris flow deposits, but there was ambiguity in distinguishing warm block and ash flow deposits, derived from collapse of the partially cooled margins of lava domes, and warm debris flows. The latter could have formed either by transformation of primary block and ash flows in transit into warm debris flows by mixing with river water, or by remobilisation of primary, still hot block and ash flow deposits into warm debris flows, for example during heavy monsoon rains.

Uehara et al. (2015) found that deposits with a single magnetic component were emplaced at >580 °C, and were interpreted to represent high temperature primary block and ash flow deposits. Deposits with two magnetic components, the lowest temperature one being parallel to the Earth's magnetic field, were interpreted as warm block and ash flows emplaced at temperatures from 100 to 520 °C and derived from the cooled margins of a lava dome. Deposits with multiple magnetic components were interpreted to be cold to warm syn-eruptive debris flow deposits, mobilised from primary block and ash flow deposits contemporaneously with their deposition. Readers interested in the methodology are referred to Uehara et al. (2015).

Pensa et al. (2018) demonstrated that BAF deposits from the 2015 dome eruption of Colima volcano, Mexico, were emplaced at temperatures between 170 and 385 °C, using both pTRM analysis of dome clasts and optical reflectance analysis of carbonised vegetation from the deposits. Using only optical reflectance analysis of carbonised vegetation from BAF deposits from the 2010 dome eruption of Merapi volcano, Indonesia, Trolese et al. (2018) proposed emplacement temperatures ranging from 240 to 320 °C. Charbonnier and Gertisser (2008) used data from digital thermometer measurements inside an inundated bunker at Merapi to estimate the temperature of the body deposits of >400 °C, just hours after the eruption, and thermal effects on plastic objects in surge deposits to estimate temperatures of >100 °C. Interestingly, such energetic flows show a relatively homogenised temperature in the ash matrix but much lower than expected, between 450 °C and <250 °C, in agreement with estimations from injuries and melting of plastic objects (Jenkins et al. 2013; Baxter et al. 2017). Trolese et al. (2018) noticed that for energetic events at Merapi in 2010, similarly to most other such cases worldwide, the largest thermal drop occurs very early along the upper reaches of the volcano (1-2 km from source), whereas once the BAF is formed and confined within valleys the subsequent drop in temperature is very minor (up to >10-15 km from source). Based on this evidence they subdivided the thermal behaviour of energetic BAFs into domains: the very proximal dissipative domain from the laterally more extensive conservative domain, and suggested that energetic events involve significant initial expansion and air entrainment along the steep upper slopes of the stratovolcanoes, where also most of the ash-fraction forms allowing fast thermal exchange with the entrained air (see also Doronzo et al. 2012 for the role of steep

Fig. 10.99 Features of block and ash flow deposits. **a** Valley ponded block and ash flow deposits (right) and overbank deposits (left); 2015 block and ash flow deposits, Colima volcano, Mexico. **b** Coarse, bouldery but thin overbank lobes that spilled over onto soil (brown domain) at the margin of the valley confined block and ash flow deposits (distant right); 2015 block and ash flow deposits, Colima volcano. **c** Fine, undulating stratified veneer surge deposits, 2015 block and ash flow deposits, Colima volcano. **d** Thick, stratified, coarse block and ash flow apron deposits, representing multiple dome collapse and block and ash flow events, incised through erosion in the Akamatsu valley, Mt Unzen, Shimabaru, Kyushu, Japan. **e** Transition from bouldery valley ponded facies to finer veneer facies deposits, Kita-Kamikoba, Mt Unzen. (*Source* Volcanological Society of Japan). **f** Poorly sorted, massive to diffusely stratified, monolithologic, valley ponded, block and ash flow deposits, Akamatsu valley, Mt Unzen. The texture varies from matrix-supported to locally clast-supported, with a

(continued)

coarse ash matrix. **g** Coarse ash and lapilli matrix of Mt Unzen block and ash flow deposits, dominated by lava dome clasts. **h** Outsize dome blocks in poorly sorted, still hot, steaming, valley ponded block and ash flow deposit, Kita-Kamikoba, Mt Unzen (*Source* Volcanological Society of Japan). **i** Outsize dome block with radial cooling joints, in poorly sorted, block and ash flow deposit, Akamatsu valley, Mt Unzen. **j** Stacked, reverse graded, block, and ash flow depositional units, Mt Unzen. **k** Burnt and truncated tree trunks within valley pond facies, 2015 block and ash flow deposits, Colima volcano. Flow from left to right as shown by splintered tree trunk in the centre. **l** Fine, diffusely stratified, ground surge deposit at the base of a very coarse, bouldery, block and ash flow deposit, overlying paleosoil with abundant carbonaceous material (dark); Akamatsu valley, Mt Unzen. The depositional slope is about 20°. **m** Veneer lateral surge deposits, and fine-grained, overbank and fine-grained black and ash flow deposit. Taruki Plateau, Mt Unzen. **n** Outcrop of weakly welded block and ash flow deposit from Mt Meagher volcano, Cascade Volcanic Arc, British Columbia, Canada

Fig. 10.99 (continued)

topography), whereas once the high concentration BAF is formed, air entrainment becomes almost irrelevant.

10.13.2 Welding in Block and Ash Flow Deposits

Welding is rare in block-and-ash flow deposits, but one case (2360 yr BP) from Mt Meager in the northern Cascades arc in Canada has been recorded (Fig. 10.99n; Michol et al. 2008). Heap et al. (2014) state the welding must have occurred while lava dome blocks were above their glass transition temperature, T_g, which, depending on dissolved magmatic water content, they estimate was between 550 and 700 °C. They also specified that blocks must remain above their T_g long enough to experience viscous deformation (up to ~ 30 days), which can only occur at high mass flux rates in the flow and accumulation of a deposit in proximal settings that is ~ 100 m thick. This is significantly thicker than most BAF deposits and is only likely to occur if deposits are ponded in very steep topography with deep valleys or canyons, which was the case with the Mt Meager BAF deposits (Heap et al. 2014).

10.13.3 Flow Dynamics in Block and Ash Flows

Block and ash flows have significant potential energy from collapse to depositional site, at the base of slope. This translates into kinetic energy. Being dominated by dense clasts, flows are strongly gravity, and topography controlled, except for the associated surges which can be mobile, and even detach from the dense body of the flow (e.g. Mt Unzen, Yamamoto et al. 1993; Fujii and Nakada 1999; Figs. 10.4i and 10.12c; Mt Pelée 1902, Fisher et al. 1980; Fig. 10.31a, b). As cohesionless granular flows, grain support is maintained in large part by grain-to-grain collisions and momentum transfer. Turbulence arises from high flow velocities (<100 km hr^{-1}) on steep slopes, and perhaps ingestion and volatilisation of vegetation, and generates associated frontal and enclosing ash-cloud surges and deposits. The turbulent and low-density ash-cloud surges can flow onto and over topography leaving surge deposits. Once detached from the source dome outsize blocks mostly move under their own momentum in a bounding or saltating manner on proximal steep slopes, directed into valleys by topography. On lower slopes they can become roughly coupled with the dense granular flow, and they are then moved as bedload, rolling and bouncing downslope, and often shattering due to impacts with the substrate and other clasts. Because of the coarseness, block and ash flows rapidly lose gas, deflate and begin to deposit by frictional, laminar shear flow locking, which limits flow distances to just a few km from source (e.g. Mt Unzen 1991–1995; Merapi 2006, Colima 1998, 1999, 2015), unless initiation also involved volatile driven explosive activity (e.g. Merapi 2010).

10.14 Low Particle Concentration or Dilute Density Currents: Pyroclastic Surges and Their Deposits

Pyroclastic surges are regarded as highly turbulent, highly expanded, low particle concentration, or dilute pdcs (Burgisser and Bergantz 2002; see Sect. 10.6). Wohletz and Sheridan (1979) described surges as time-transient, unsteady flows of tephra that occur as a pulse or series of pulses in which the kinetic energy rapidly decays. Surges are complex three-phase systems, being mixtures of solids, gases, and water vapour, the proportions varying from one surge to another, and even during the flow of individual surges (Allen 1984). Hot "dry" (i.e. dominated by gas and solids) surges are associated with magmatic explosive events, whereas lower temperature, "wet" (i.e. mixtures of gas, steam, and solids) surges are commonly associated with phreatomagmatic, phreatic and hydrothermal explosive events. However, volumetrically the proportion of solids (and thus their concentration) is subordinate to gas and liquid phases in most surges, and therefore much less than in pyroclastic flows (also see Walker 1983; Burgisser and Bergantz 2002). Surges can result from eruptions of any magma type and can arise from both magmatic and phreatomagmatic eruptions (Chap. 4).

During the initial flow stages in pyroclastic surges, solid particles are widely dispersed within the fluid phase(s) and are supported and driven through turbulence in the gaseous phase. The solids thus behave in a particulate fashion during transportation, and are free to sort themselves hydraulically, according to the internal eddy dynamics and settling velocities of the particles. In comparison, in viscous mass flows particle interaction and particle–fluid cohesion hinder sorting to a large degree. Similarly, as the wetness of surges increases, particle freedom decreases due to adhesion processes involving clumping of fine ash due to liquid bonding effects (Schumacher and Schmincke 1995; Mueller et al. 2016).

Surges have densities higher than the ambient atmospheric density, so their passage will be largely gravity-controlled, and they generally follow topographic lows (Moore 1967; Waters and Fisher 1971). However, because of their largely turbulent and low-density nature, stemming from an initial high velocity explosive thrust and/or high gas content relative to solids (especially fines, e.g. Walker 1983), they can also climb significant topographic highs and mantle them with a veneer of ash (Nairn 1979; Fujii and Nakada 1999). Ash-cloud surges can detach from their host pyroclastic flows and flow separately over topography as dilute pdcs, causing significant damage to infrastructure and igniting buildings (e.g. Figs. 10.1b and 10.31a, b; Fujii and Nakada 1999; Gueugneau et al. 2019).

Given the low concentration and highly expanded nature of surge flows, their runout distances are generally rather limited to a maximum of a few km from source, especially in the case of base surges associated with impulsive and short-lived phreatomagmatic explosions which may be found within 1–5 km from their parent maars of tuff cones (see Chap. 16, Sect. 16.3). However, when surges stem from a pyroclastic flow they can be generated at any stage in the flow path, at the front, laterally and above the flow, so the deposits of surges associated with ignimbrites and BAF deposits can be found much further from source, up to several tens of km.

10.14.1 Types and Origin of Surges

As discussed in Sect. 10.1, pyroclastic surges can be subdivided into three basic types:

- base surges,
- ground surges, and
- ash-cloud surges.

Each type has different origins and may therefore contain widely different proportions of the three main phases (solid, gas, and liquid). This may in turn lead to surges with varying physical properties, transportational and depositional modes, and facies characteristics of their deposits. The latter two surge types are genetically associated with pyroclastic flow dynamics and deposits

10.14.1.1 Base Surges

Base surges (Moore 1967) originate from the base of an explosive phreatomagmatic eruption column and its vent as a radially spreading, outward moving, ground-hugging, turbulent flow of fluid and ash (Plate 10.1b; Fig. 10.100). They develop from phreatomagmatic blasts in the vent, which eject dilute mixtures of solids, gases, and steam as debris jets. These expand rapidly on eruption, collapse, and then spread radially, or flow along directed paths as turbulent surges. Other phreatomagmatic surges originate from minor column collapse events (Fig. 10.100c). Base surges will generally be steam-rich and wet, unless they are extremely hot when erupted and so contain little condensed water; these are called dry base surges. Base surge is also used to describe similar blasts produced at phreatic and hydrothermal vents and craters, where either heated ground water or hydrothermal fluids initiate explosive activity (Chaps. 16 and 18).

Phreatomagmatic surge-forming eruptions are usually composed of multiple base surges, resulting from successions of pulsing explosions, with negligible time breaks. Base surge-forming eruptions may therefore lead to deposits which are stacks of surge deposited beds that build maar volcano rims, tuff rings, and tuff cones. Successive surges may erode and rework the deposits of preceding surges to varying degrees.

10.14.1.2 Ground or Frontal Ash-Cloud Surges of Pyroclastic Flows

Ground surges or frontal ash-cloud surges form the turbulent flow fronts of all pyroclastic flows (Figs. 10.10, 10.11a and 10.19). In the deposit record, their role is identified where their deposits (Layer 1) directly underlie pyroclastic flow body deposits (Layer 2; Sect. 10.8.7; Figs. 10.39, 10.40a–d and 10.99l). They may also form from weak directed blasts at vent, and from small partial collapses of eruption columns.

The origin of frontal or ground surges from the front of pyroclastic flows is likely to be the result of ingestion of cold air or vegetation over which the flow is moving, into clefts at the flow front, and are then instantaneously superheated and expand (Fisher 1979; Wilson and Walker 1982), forming expanding gaseous cells in the flow front (see Fig. 10.11a). In some cases, they could have enough momentum and excessive kinetic energy to be ejected forward, out of the head of the moving pyroclastic flow as a cell of highly turbulent gas and vapour with low particle concentration, which moves forward as a surge. The surge deposit is then immediately overridden and buried under the basal facies of the body of the pyroclastic flow (Figs. 10.39 and 10.40a–d; Sparks and Walker 1973; Sparks et al. 1973; Wright et al. 1981; Scarpati et al. 2015).

10.14.1.3 Trailing and Lateral Ash-Cloud Surges of Pyroclastic Flows

Also forming parts of all pyroclastic flows are trailing and lateral ash-cloud surges, which enclose the body of the pyroclastic flow above, behind, and laterally (Figs. 10.10a–c, 10.11b, c and 10.19). These form as gas and fine ash are elutriated and stream out of the head and body of the flow under the influence of fluidisation processes (Fisher 1979; Sect. 10.6.2), forming the billowing ash cloud which trails behind and lateral to a pyroclastic flow, which give rise to surge deposits (Layer 3a; Fig. 10.42a–d) that directly overlie pyroclastic flow body deposits (Sect. 10.8.7).

These ash clouds are continuous with the frontal ash-cloud surge, and the energy in this type of surge is also derived entirely from the parent pyroclastic flow. The surge acquires its momentum and kinetic energy from the pyroclastic flow out of which it originates. The energy of the ash cloud surge and its potential ability to detach and travel laterally, independent of the body, largely depends on the

Fig. 10.100 Base surge eruptions. **a** Map showing early surtseyan phases of the 1957–58 eruption of Capelhinos, which added an ~2.5 km^2 tephra and lava promontory to the western end of Faial Island, Azores, Portugal. Pink circle is location of lighthouse. (Adapted from Zhao et al. 2019). **b** Radially spreading base surge at the base of an eruption column during the first few days of the eruption that commenced on 28 September 1957 (*Source* Portuguese Historical Museum). **c** At Capelhinos in early October, where steam has blown downwind to expose a dense debris-laden central column collapsing to feed a base sure; lighthouse is circled for scale (Photo provided by Victor Forjaz). **d** Radially spreading white base surge initiated at the base of dark ballistic blast plume with cock's tail jets from "cinder island", which by mid-November had constructed a tephra isthmus that had joined with the mainland (*Source* Portuguese Historical Museum). **e** Dramatic close-up of collapsing blast plume from Kavichi shallow (~25 m b.s.l.) submarine volcano, Solomon Islands, southwest Pacific. At Kavichi volcanism is frequently active, but rarely observed. The most recent eruption took place during 2016, which included phreatomagmatic explosions as illustrated here (Photo by Alex DeCiccio; *Source* Wikipedia through Creative Commons Attribution-Share Alike 4.0 International licence). **f** Base surge spreading radially from the base of collapsing ballistic blast plume, Hunga Tonga–Hunga Ha'apai eruption of 2009, Tonga, southwest Pacific (*Source* Dana Stephenson/Getty Images, and Global Volcanism Program https://doi.org/10.5479/si.GVP.BGVN200902-243040). In all examples, note the multiple turbulent cells

internal stratification of the body of the parent pyroclastic flow. In particular, the energy of the ash cloud surge does not arise from the basal high particle concentration, but rather from the overriding more dilute and turbulent part of the body (Fig. 10.19), which, depending on the degree of coupling between the two layers, may transfer momentum and heat to the ash-cloud. The turbulence and momentum of the ash cloud surge have two sources: (1) thermal convective circulation as the hot gas from the body rises and expands (Pensa et al. 2019), and (2) the overall high velocity at the coupled boundary between the dilute part of the body and the ash cloud surge (Fig. 10.19), which contributes to high Reynolds Numbers, as discussed previously (Sect. 10.6.1). The velocity of the turbulent eddies determines the support mechanism of the solids and their maximum particle size (Andrews and Manga 2012).

Trailing and lateral ash-clouds may eventually become buoyant lofting above the pyroclastic flow or may be able to detach and flow independently of the body of the pyroclastic flow (e.g. 1912 Novarupta eruption, Alaska, Hildreth and Fierstein 2012; 1991 Mt Unzen eruption, Yamamoto et al. 1993). The latter may happen for example at valley turns, where these surges may surmount hillslopes in very different directions with respect to the parent high concentration, valley confined flow, thus posing a serious and unexpected hazard. Pensa et al. (2019) were able to retrieve the temperature of the ash-cloud overriding the 2015 block and ash flow at Colima volcano by studying the charcoal on the trunks of standing trees with height and determined temperatures as high as 500 °C. Such a high temperature indicates that in case of detachment, even at low ash particle concentrations, these ash clouds can be deadly, such as occurred in 1991 at Mt Unzen (see Figs. 10.4i and 10.12c).

10.14.2 Transportation and Grain-Support Processes in Surges

The transportation of pyroclastic detritus in surges is dominated by turbulent eddies in the supporting gaseous phase (Fig. 10.100). This is based on direct observations of base surges (e.g. Moore 1967; Waters and Fisher 1971; Kienle et al. 1980; Self et al. 1981; Ort et al. 2018), and on the depositional structures observed in the deposits of all types of surges (e.g. Fisher and Waters 1970; Schmincke et al. 1973; Waters and Fisher 1971; Cas and Wright 1987; Sohn and Chough 1989; Cole et al. 2001). Observed surges are dominated by turbulent billowing clouds of gas, steam, and ash (e.g. Figs. 10.10, 10.11a, 10.100 and 10.101a). On this basis, at least the peripheral parts of surges are highly turbulent. However, much of the unobservable inner, lower parts, or the "body" of the surge, can also be considered to be largely turbulent by virtue of the tractional sedimentary structures such as (anti)dunes (sometimes called sand waves), cross stratification and horizontal lamination commonly found in their deposits (Fig. 10.101a, b; Moore 1967; Fisher and Waters 1970; Waters and Fisher 1971; Schmincke et al. 1973; Wohletz and Sheridan 1979; Allen 1984; Cas and Wright 1987; Sohn and Chough 1989; Cole et al. 2001; Dellino et al. 2004; Douillet et al. 2013b). All of these structures can only be produced by grain-by-grain tractional transport (rather than *en masse*) indicating low grain concentrations in the transporting medium.

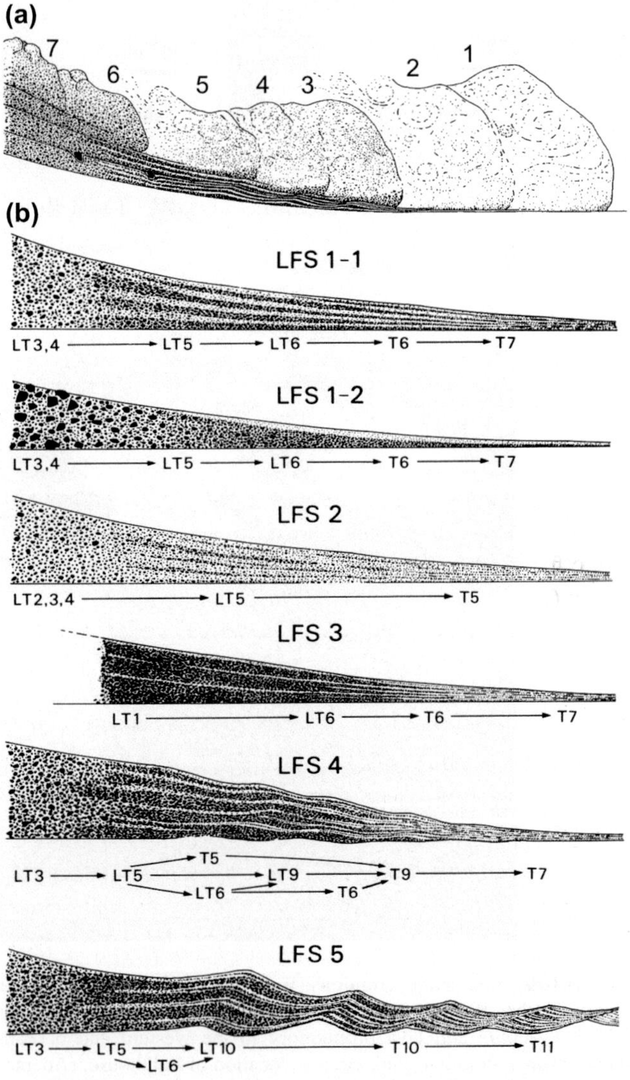

Fig. 10.101 a Schematic representation of base surge transport and resulting deposit characteristics downstream. Numbers represent successive pulses or turbulent cells in a single base surge event that may each produce a single layer or cross-bed in the deposit bedset of that surge. **b** Most common lateral facies sequence changes in deposits, from proximal (left) to distal (right), based on statistical lateral facies sequence analysis of deposits of Suwulbong tuff ring, Cheju Island, South Korea. LT(number) refers to different lapilli-tuff facies, and T (number) refers to different tuff facies. (From Sohn and Chough 1989)

Many studies have addressed the flow dynamics of pyroclastic surges and low-density pdcs (e.g. Wohletz and Sheridan 1979; Allen 1984; Cas and Wright 1987; Valentine 1987; Dellino and La Volpe 2000; Burgisser and Bergantz 2002; Andrews and Manga 2012). The energy for base surge initiation and flow is derived initially from the thermal energy of the rising magma (Fig. 10.102). On contact with ground water or surface water, this thermal energy is converted into mechanical explosive energy, as the heat from the magma is used to superheat and explosively expand the external water source (Fig. 10.102; Chap. 4). The explosive energy is translated into momentum and kinetic energy. Potential energy is also created during eruption, especially where base surges are derived from minor column collapse. This potential energy, in turn, translates into kinetic energy. In addition, during at least the early stages of flow, extra thermal energy is liberated from juvenile fragments, and latent heat is released when condensation of gaseous water to liquid water droplets occurs (Fig. 10.102). The thermal energy produces turbulent convective circulation, and the large initial explosive thrust ensures high lateral transport velocities, which are also responsible for turbulent flow conditions. In the Reynolds Number criterion for turbulence (Eq. 2.9, Re = U D ρ/η), the low viscosity (η) of the driving fluid (gas) and the high velocity (U) of the surge both contribute to high Reynolds Numbers. Both the convective and velocity-induced turbulence entrain and support grains (see below). The energy for ash-cloud surges associated with the margins of pyroclastic flows derives from the host pyroclastic flows, including their thermal, potential, and kinetic energy (Fig. 10.102).

The uplift force of turbulent fluid eddies is the main grain-support process during pyroclastic surge flow events (Burgisser and Bergantz 2002). Suspended grain-support will be maintained as long as the drag force of these uplifting turbulent eddies exceeds the settling velocities of grains at the appropriate Reynolds Number for the turbulence level (Allen 1984; Burgisser and Bergantz 2002; Dellino et al. 2004; Andrews and Manga 2012). The energy that generates turbulence for grain support is of two types: (1) thermal convective turbulence, due to the heat of the gaseous fluids in the surge, which want to expand, are buoyant and want to rise, and velocity-induced turbulence in a medium of low viscosity; and (2) high lateral velocity. Turbulence due to fluidisation-induced streaming of external air is considered to be minor because of the low particle concentrations.

These conditions hold for an initial highly turbulent state and for small grain sizes. However, surge deposits are known to contain fragments several centimetres or more in diameter, which were not transported and deposited as ballistic blocks. They must have been at least entrained and moved by surges as well. The actual competence of surges to transport different sizes in true suspension has been modelled by Burgisser and Bergantz (2002; Fig. 10.16). However, it is clear that during incipient sedimentation, if not before, a large proportion of the solids are saltated (bounced), rolled, and dragged over the substrate as bed load under the influence of a large lateral shear velocity (Fig. 10.101a; Chap. 12). More of the clast population will experience this mode of transport as the surge loses kinetic energy and begins to deflate. Some surge deposits in proximal areas are marked by a massive, poorly sorted facies, indicating extremely high sedimentation rates and high particle concentrations in the lower part of the surge at the time of deposition (Sohn and Chough 1989; Fig. 10.101a, b). Such facies appear to represent transition into a flow regime similar to pyroclastic flows.

In a single surge, some transport will take place through turbulent suspension and some through shear-induced traction and grain collisions. There is therefore very likely to be a "grain carpet" layer at the base of most surges, out of which the bulk of sedimentation will occur, as well as a more dilute, turbulent, trailing ash cloud. The grain-layer population is dragged along by the shear stress operating at the base of the flow from the lateral flow of gas and suspended solids. Individual grains will also propel neighbours by colliding with them and transferring momentum, which constitutes a form of inertial dispersive pressure. Base surges originate as a vent blast and propel a poorly sorted grain population from which a significant coarse fraction will tend

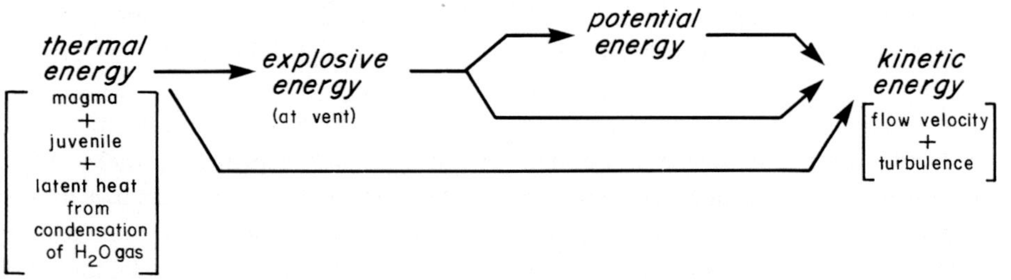

Fig. 10.102 The complex energy chain involved in the initiation and flow of dry and wet pyroclastic surges. This chain, with variations, can also be applied to the various types of pyroclastic flows, and is thus a useful representation of energy transformations in all pyroclastic density currents. (From Cas and Wright 1987)

to drop out quickly. As the surge loses energy and deflates, more solids, particularly the coarser fractions (including moist ash clumps) will cease being transported in a suspension mode and will become entrained in a traction mode of transport in the depositional boundary layer. The change from high concentration, shear-induced granular flow conditions to low grain concentration traction should be reflected by transition vertically and laterally from massive or diffusely layered beds to dune bedforms, cross stratification, and discrete lamination (Fig. 10.101a, b; Wohletz and Sheridan 1979; Sohn and Chough 1989).

In wet phreatomagmatic surges such as base surges, this simple analysis is complicated by the process of clumping or adhesion of particles in the very moist atmosphere of the moving surge (Allen 1984; Mueller et al. 2016). Clumping and adhesion refer to the aggregation of moist grains, especially fine ash, with moisture droplets through "liquid bonding" and electro-static forces (Mueller et al. 2016), so increasing the effective dynamic transported grain size. For continued suspension transport of these clumps, the upward component of the fluid drag force must therefore also exceed the settling velocity of the ash clumps. As the moisture content increases, and cohesive clumps of ash develop, transport becomes pseudo-viscous near the base of the surge. It is still not wholly viscous, because the high shear stress operating probably continually reconstitutes wet clumps and smears them out. That is, the moisture-laden, driving gaseous fluid is distinct from the cohesive clumps it is propelling, since they have not totally amalgamated into a cohesive, viscous fluid. In this sense, wet surge transport is quite different from other types of particulate tractional transport or mass-flow transport because it is a complex polyphase system. It has not been mathematically modelled. Evidence of this complex phase transportation by wet surges is found in accounts of observed base surges and their deposits. Plastering of muddy ash on the "upwind side" of trees and buildings was documented at Taal (Moore 1967). That these surges were still turbulent is attested by the "sandblasting" effects in abrading and stripping of bark and foliage, and in the near-blast zone, by the total destruction of vegetation (Moore 1967).

The potential sources of the fluids providing grain-support in surges include magmatic volatiles (both those initially present and those that diffuse from juvenile fragments within the surge), external water incorporated during phreatomagmatic eruptions and heated to steam, and volatiles derived from devolatilisation of vegetation through and over which surges travel. Significant mixing of air into the billowing top of the surge is also likely to occur (Allen 1984; Andrews and Manga 2012), which will provide turbulent thermal buoyancy support for the finest grain sizes transported by the surge. Mixing with the ambient medium occurs along the top of the surge due to the growth of Kelvin–Helmholtz instability waves (billows) on the top of the head (Allen 1984). The upper part of the head is diluted by this, and becomes less dense and therefore slower, whereas the head continues to propagate at a steady velocity, until it becomes buoyant (Andrews and Manga 2012).

Andrews and Manga (2012) also address the nature and scale of turbulent structures or cells/eddies in surges, which represent the essential dynamic element of surges, and is important given that they propose that each laminae is the product of a single eddy. Dilute pdcs are dominated by transient turbulent cells or eddies, some of which extend the full height, h, at the front of the current, although others form smaller cells near the bases. Regularly spaced large eddies may also form along the top surface of the tail due to Kelvin–Helmholtz instabilities at the interface between the current and the atmosphere, producing flame-like plumes. The largest eddies generally have lengths (parallel to flow) of h, and heights of 0.25–0.5 h. Smaller eddies may form parts of larger eddies or form independently. Large cells have turbulence life spans of 3 s in the experiments and smaller ones, <1 s.

During flow propagation the current segregates into a lower density, slower moving, upper part, and a denser faster moving lower part, which has a thickness of 0.3–0.5 of the original flow height, and in which eddies propagate at the velocity of the current front (Andrews and Manga 2012). At the termination of a surge flow, a buoyant ash cloud rises and forms a plume. Particles remain suspended, but the flow front ceases to propagate forward. Turbulent eddies may begin to propagate back upstream as the rising plume causes rollback of the flow front. The upstream propagation velocity of eddies is about the same as the initial downstream propagation velocities, $\sim 0.1\ \mathrm{m\ s^{-1}}$ in the experiments (Andrews and Manga 2012).

10.14.3 Depositional Processes in Surges

Any transporting medium begins to deposit its load when horizontal shear forces and turbulent eddy forces that are needed to maintain discrete suspension grain support or tractional entrainment drop below the threshold level at which gravitational and frictional forces become dominant. Surges, although initiated by an "explosive" thrust, once in motion, are gravity-controlled density currents (Allen 1984). If upward directed eddy velocities fall below the settling velocity of the suspended particles, which varies according to their size and density, a change from suspended transport to surface traction shear transport will commence. We have discussed the numerical modelling of Burgisser and Bergantz (2002) who defined the particle behaviour in relation to Stokes Number and a stability factor Σ_T, which is the ratio of the particle settling velocity of the turbulent eddy velocity

(Sect. 10.6). This allows individual particle behaviour and the conditions under which they sediment to be considered (Fig. 10.16).

Convective turbulence will be dissipated as the surge cools in transit, perhaps by mixing cold air at the front and top, by depositional loss of hot particle load, and through loss of latent heat during condensation of steam to water. As the velocity declines, the sediment carrying capacity of the surge will also decrease, and the surge as a density current will be dissipated (Allen 1984).

Given the low density of surges, the coefficient of friction between the surge and its substrate will be low while the surge is inflated and carrying a large suspended sediment population. However, when the surge begins to deflate significant grain fallout will produce a concentrated basal tractional grain layer. There will be a higher coefficient of friction between this grain layer and the substrate, permitting sedimentation and further slowing of the surge.

Frictional retardation of a surge can also occur by mixing of air at the head and top of the surge. Allen (1984) suggested that the dissipation rate of a density surge is

$$Dissipation\ rate = h_s/L \tag{10.12}$$

where h_s, the thickness of the body of the surge, is equal to 0.01H, H being the height of the head, and L, the streamwise length of the mixing zone, is equal to about 2–3 H. The dissipation rate is in the order of 0.01–0.001 (Allen 1984).

For relatively dry surges, grain transport will pass from suspended and surface traction modes of transport to increasing degrees of shear tractional transport. The rate at which grain transfer from the suspended to the traction mode occurs will control whether relatively low grain concentration tractional transport and sedimentation occurs or high shear adhesion to the substrate, producing corresponding structures, such as low angle dune bedforms, cross stratification and well-defined lamination, or whether high grain concentration shear-induced grain-layer flow occurs, with attendant high degrees of particle interaction, deposition by freezing of the grain mass and deposition of a massive bed.

Although there has been a general assumption that sedimentation from surges is progressive through fallout as surges dissipate, Andrews and Manga (2012) have demonstrated that sedimentation occurs in waves. In experiments using heated talc powder density currents, they noted that sedimentation occurred through propagating waves of sediment associated with short-lived ephemeral, migrating turbulent eddies that represent depositional-erosional pulses. They propose that single laminae in the deposits of dilute pdcs (surges) represent the deposit of a single turbulent eddy, resolving a long-term uncertainty about surge sedimentation processes. The sedimentation time for each lamination is therefore limited to the duration of each eddy (<3 s), and the extent of each lamination is about the length of the eddy or 2–3 times, given it migrates laterally. They propose that a surge deposit 0.5 m thick, consisting of 50 × 1 cm thick laminae, would be deposited in 300 s (5 min). The sediment waves tracked the flow head, were large near the front, smaller behind, and were regular in more proximal areas. When thermal buoyancy and lift-off of ash plumes occurred and the flow waned, sedimentation became chaotic, sedimentation waves stacked and even tracked backwards as the plume lift-off zone regressed backwards.

Dellino et al. (2008) proposed that the typical association of coarse basal layers draped by ash layers found in surge deposits can be used to determine the velocity and concentration of the surge flow, by considering that the clasts in the basal part are entrained by shear stress, while the ash blanketing the top is settled directly from the turbulent suspension. They applied the Turbulent Boundary Layer Theory and used the grain size characteristics of the surge deposit to develop a probabilistic quantitative assessment of the surge velocity and concentration as the basis for calculating the local dynamic pressure (which is 0.5 ρ U^2, where ρ is the surge density and U is the surge velocity). This approach and relevant software are described in Dioguardi and Dellino (2014).

In wet surges the condensation of steam will also be an important influence in causing sedimentation, particularly if there is a large fine ash fraction. As fine ash and water droplets begin to adhere to each other, producing clumps (e.g. Mueller et al. 2016), the effective transported grain size will be increased by this. Condensation should therefore lead to an increased rate of sedimentation compared with the pre-condensation rate of sedimentation, because the greater effective transported grain size may exceed the prevailing competency of the surge, and because of the adhesion effects between the substrate and moist particles in the surge. The rate at which condensation takes place will therefore be important and will depend on the moisture content of the erupting mass and the temperature of the surge. If that is only slightly above the boiling/condensation point of water (100 °C), then condensation and ash clumping should be significant.

If surges do segregate into a lower concentrated granular flow layer and a more dilute suspension cloud, then some deposits should contain a separate "body" layer and an associated co-surge ash fallout layer. The presence of topography may induce increased sedimentation rates and the development of high concentration underflows resulting in massive, chaotic facies that can be valley-ponded (e.g. Giordano et al. 2002).

10.14.4 Characteristics of Surge Deposits

The facies or deposit characteristics of surges can be readily discussed in terms of some of the essential facies parameters introduced in Chap. 1, including

- geometry,
- grain size,
- sorting,
- clast shape and vesicularity, and
- depositional structures.

In the following we address the first four parameters, while describing in more detail depositional structures found in surge deposits in the next section (Sect. 10.14.5).

10.14.4.1 Geometry

The geometry of surge deposits will depend on the type of surge, the topographic control, and post-depositional erosion. Base surges form one of the principal facies of tuff rings and maar volcano rims (Chap. 16, Sect. 16.3), where they form deposit sequences composed of multiple layers representing multiple surge events, from a prolonged phreatomagmatic eruption event. Such sequences may be tens of m thick around the vent (e.g. Crowe and Fisher 1973; Schmincke et al. 1973; Sheridan and Updike 1975; Fisher 1977; Sohn and Chough 1989; De Benedetti et al. 2008) but will thin and wedge out hundreds of m to a km from the vent. As such, a thick succession (several tens of m) of multiple base surge deposits in the rock record is indicative of proximity to the vent.

The deposit from a single base surge is a thin, sheet-like, fan-shaped deposit, with minor variations in thickness being controlled by pre-depositional topography and the morphology of depositional bedforms (e.g. dune forms). Overall they thin away from the vent (Chough and Sohn 1990; Fig. 10.101a, b). Individual layers may be up to 1 m thick near the vent, although usually tens of cm or less, and thin distally to only mm to several cm thick. Base surges usually do not flow further than 10 km from the vent, and usually no further than one to several km. Successive base surge deposits may be separated by thin co-surge ash-cloud ash fallout deposits up to several cm thick, derived from a trailing ash cloud (cf. co-ignimbrite ash-fall deposits). Sequences of base surge deposits may contain intercalated, thin, near-vent pyroclastic flow deposits (e.g. Fisher et al. 1983; van Otterloo and Cas 2016) and laterally directed, ballistic jet fan layers (e.g. Graettinger et al. 2015).

Ground or frontal ash-cloud surge deposits (Layer 1), originating from the head of a pyroclastic flow, will usually form a deposit no more than 1 m thick, but usually only cm to tens of cm thick (Fig. 10.39), and will be directly overlain by the body deposit (Layer 2) of the parent pyroclastic flow (Fig. 10.40a–c). Ground surges can form or persist to any distance, even tens of km or more from vent, but should always be accompanied by a body deposit of the pyroclastic flow.

Trailing and lateral ash-cloud surge deposits (Layer 3), originating from the trailing and lateral margins of pyroclastic flows, if preserved, will occur as a thin (cm up to 1 m thick) sheet of fine ash, mantling the body deposit (Layer 2) of the host pyroclastic flow (e.g. Fisher 1979), but can also be deposited laterally beyond the limits of the body deposit. However, the preservation potential of ash-cloud surge deposits is low because of the effects of post-depositional erosion, unless quickly buried beneath further eruptive products (e.g. another ignimbrite, Fig. 10.42b, d; or contemporaneous fallout deposits from the same plinian eruption sequence, Fig. 10.42e).

10.14.4.2 Grain Size

The grain size of surge deposits (Fig. 10.103) reflects the degree of fragmentation at the time of eruption, the intensity of in-flow abrasion, and the competency of surges to carry particular grain sizes (Chap. 1). The coarseness and variance in the grain size is thought to be a reflection of the levels of turbulence and velocity (Walker 1983, 1984). As Crowe and Fisher (1973) pointed out, because base surge-forming eruptions are highly pulsatory and of variable explosive strength, there may be considerable changes in the flow-power and in the grain size characteristics of successive surges, and even within single surges (Fig. 10.104a). Near to the vent ballistic fragments and air-fall materials may also be incorporated into surge deposits (e.g. Fig. 1.15b). However, both base surge and ground surge deposits, because of their high energy state and because their sources at the vent and the heads of pyroclastic flows may contain considerable coarse debris, are capable of carrying, and do carry, significant amounts of large lapilli-size clasts (Fig. 10.104a). However, ash-cloud surges, being the products of continued elutriation processes in the head and body of pyroclastic flows, will be low energy systems and their deposits will be fine-grained (Fig. 10.42a), rarely containing lapilli, and then probably only highly vesicular, low-density ones. Wet phreatomagmatic surge deposits will be very fine-grained, because of the efficiency of fragmentation at the vent, not necessarily because they are distal fine ashes, and may contain accretionary lapilli (Fig. 10.104b). Dry surges are coarser, and fines-depleted relative to pyroclastic flow deposits (Walker 1984).

10.14.4.3 Sorting

The sorting characteristics of the different surge deposits will also be variable (Fig. 10.103), and will be controlled by the

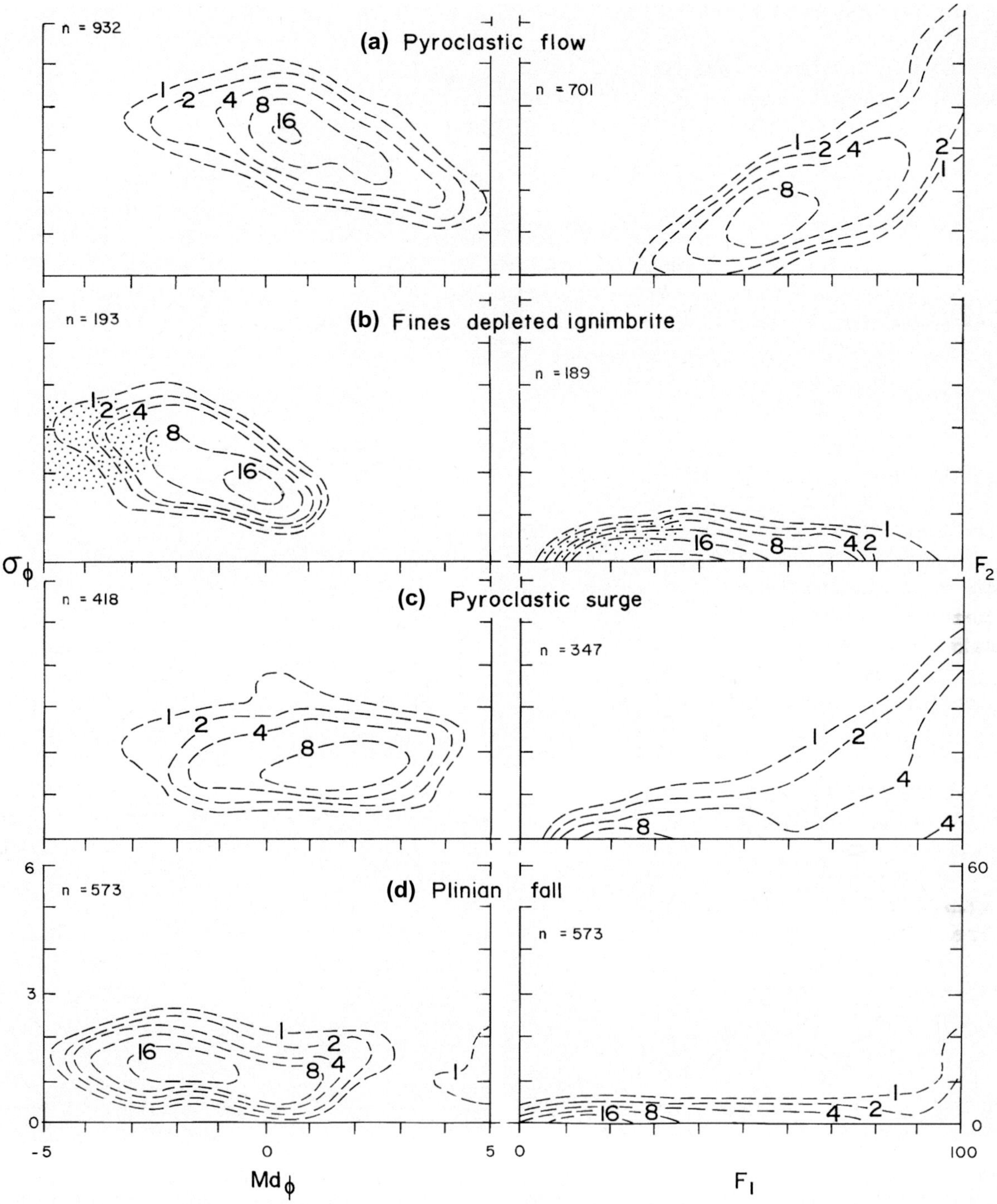

Fig. 10.103 Grain size and sorting characteristics of pyroclastic flow deposit types, fallout deposits, and surge deposits This is summarised by plots of Md_{ϕ}/σ_{ϕ} (column 1) and of weight percentage finer than 1/16 mm (F_2), versus weight percentage finer than 1 mm (F_1) (column 2) for pyroclastic flows **a**, three kinds of fines-depleted pyroclastic flow deposits **b**, surges **c** and plinian fallout deposits **d**. In (**b**) various kinds of fines-depleted facies associated with ignimbrites are included: ground layers, fines-depleted ignimbrite, and gas segregation pipes. The dotted area contains co-ignimbrite breccias not included in the contoured plot. The data show that pyroclastic flow deposits are poorly sorted, surge deposits are moderately sorted deposits, and fallout deposits are well sorted. Note that fields in (**b**), (**c**), and (**d**) overlap, showing that grain size parameters alone are not good criteria of origin. (From Cas and Wright 1987, After Walker 1983)

Fig. 10.104 Pre-depositional structures in base surge deposits. **a** Large channels infilled by stratified basaltic base surge ash deposits, 4.5 ka Mt Schank volcano, Newer Volcanics Province, South Australia. The underlying dark breccia unit is the top of a fire fountain fed clastogenic lava. The channels probably present a combination of original lava flow morphology and surge erosion. The base of the surge deposits contains chunks of the a'a clinker, although not huge amounts, suggesting much was transported downstream. Flow outwards from quarry face. Sadly, this outcrop no longer exists. **b** Close-up of right-hand channel in (**a**); bedding on-laps the channel margin, and on the left side part of the channel undercuts the lava breccia. On the right-hand side a small channel is filled in. **c** Trough cross-bed set scouring underlying strata and indicating base surge flow perpendicular to the outcrop, out of the outcrop. Balos maar-scoria cone complex, Santorini, Greece. **d** Large U-shaped channel at El Golfo tuff cone, Lanzarote, Canary Islands, Spain. This forms part of a laterally extensive erosive surface with a U- and V-shaped rill morphology, and is thought to have formed by surface runoff from surges after the deposition of their pyroclastic material (Pedrazzi et al. 2013). Flow is into page. **e** More complex U-shaped erosional channel, Koko Crater, Oahu, Hawai'i. According to Fisher (1977), V-shaped bases and stratigraphy of deposits suggest that these were fluviatile erosional gullies reshaped and re-emphasised by younger base surges erupted from the Hanauma Bay crater complex to the south. (From Cas and Wright 1987; Dick Fisher for scale)

degree of fragmentation, the density variations in the debris being carried, the turbulence levels, adhesion processes and the time available for sorting. As discussed in Chap. 1, sorting is a hydraulic/pneumatic process, and in populations containing grains with a variety of size, density and shape, variation in grain size will not necessarily reflect poor hydraulic sorting. Nevertheless, given the origins of the three surge types, it is likely that ash-cloud surge deposits will be better sorted than base surge or ground surge deposits. In addition, because of the relatively low particle concentration in surges, the sorting will be significantly better than in most pyroclastic flow deposits, but less so than in fall deposits (Fig. 10.103). However, field observations show that major grain size variations occur between the layers in a single surge deposit, especially base surge deposits, as well as in a downstream direction, suggesting a highly pulsatory mode of flow and grain transport within a single surge. Secondly, sorting is generally still much poorer than in most normal sedimentary systems transporting equivalent grain sizes, even in turbidites, the nearest sedimentary analogue, because generally siliciclastic sediments consist of components with similar densities. Volcanic sediments and pyroclastic deposits will not be as well size-sorted because of the wide range in density of the components, especially grains with variable vesicle content.

10.14.4.4 Shape and Vesicularity of Juvenile Pyroclasts

Shape and vesicularity of grain populations in the deposits of the three surge types will be more a reflection of the mode of fragmentation of the magma during eruption than of flow processes. Hence, base surge deposits, being of phreatomagmatic origin, may be dominated by poorly to variably vesicular, blocky fragments where the erupting magma is poorly vesiculated (Chap. 4), as will the deposits of ground and ash-cloud surges whose host pyroclastic flows have been associated with phreatomagmatic eruptions (e.g. Self 1983). However, ground and ash-cloud surge deposits derived from eruptions driven by magmatic explosions (Chap. 4) will contain abundant vesicular fragments, although concentration processes during the initiation of ground surges may concentrate denser lithics and crystals in the surges. Even some base surge deposits may contain at least some highly vesiculated pyroclasts if magma vesiculation was already advanced at the time of magma–water interaction, or if some pyroclasts have been recycled from the crater walls from deposits from earlier magmatic explosive eruptions.

Abrasional rounding during base surge transport events is unlikely to occur given their low particle concentration, and the short-lived nature of the flow event. However, clasts that have been milled and recycled from the vent, may have developed some degree of rough rounding. Ground and ash-cloud surge deposits may however contain some rounded clasts, with some degree of rounding being derived from transit of those clasts through the body of the pyroclastic flow, before transfer into associated marginal surges.

10.14.5 Depositional Structures in Surge Deposits

Depositional structures are diverse in surge deposits, and have been recognised as a response to varying flow and physical conditions ever since surges were recognised as a dilute, low particle concentration pyroclastic transportational and depositional agent (e.g. Moore 1967; Fisher and Waters 1970; Waters and Fisher 1971; Crowe and Fisher 1973; Schmincke et al. 1973; Sheridan and Updike 1975; Fisher 1977, 1979; Wohletz and Sheridan 1979; Allen 1984; Fisher et al. 1983; Sigurdsson et al. 1987; Sohn and Chough 1989; Chough and Sohn 1990; Cole et al. 2001; Dellino et al. 2004; Gençalioğlu-Kuşcu et al. 2007; Vazquez and Ort 2006; Ort and Carrasco-Núñez 2009; Pardo et al. 2009; Andrews and Manga 2012; Jordan et al. 2013; van Otterloo et al. 2013; Pedrazzi et al. 2013; Douillet et al. 2013a, b). Many of these authors have also recognised lateral changes in the nature of bedforms or associated internal depositional structures, or both, with distance from the vent, as a response to changing flow conditions as dissipation of surge energy occurs away from vent (e.g. Sohn and Chough 1989; Fig. 10.101a, b). We will first discuss base surge depositional structures.

10.14.5.1 Base Surge Deposits

As with normal sedimentary structures, depositional structures in base surge deposits can be classified as

- pre-depositional,
- syn-depositional, and
- post-depositional.

Note we use this classification in the sense that it applies to the nature of the immediate depositional surface or relative timing, which may be pre- or syn-eruptive.

Pre-depositional Structures

Pre-depositional structures include erosional gullies and U- or V-shaped channels carved out of the depositional surface and infilled by base surge deposits (Fig. 10.104). The origin of these depressions may be normal surface erosional (or aggradational) processes, producing channels before the eruptive event occurs (e.g. Fig. 10.104a, b). However, erosion caused by the turbulence of base surges is a common feature of many maar, tuff ring and tuff cone successions.

(e.g. Moore 1967; Fisher 1977; Verwoerd and Chevallier 1987; Pedrazzi et al. 2013). In relatively simple cases (e.g. Fig. 10.104c), it is possible to infer that a two-stage process filled these channels: firstly, erosion by a waxing passing current generated the U- or V-shaped geometries of the channels, which secondly, were completely filled in by the pyroclastic debris transported by the waning phase of the same surge or by later surges. In other more complexes cases, surface runoff from surges following deposition of their pyroclastic load (e.g. Fig. 10.104d; Pedrazzi et al. 2013), or some combination of fluviatile erosion and base surge activity (e.g. Fig. 10.104e; Fisher 1977) have been invoked. Considering the setting, with large volumes of condensing water vapour available, erosion surfaces due to rainfall events should be expected within base surge successions due to short response periods during phreatomagmatic eruptions (Nemeth and Cronin 2007).

Syn-depositional Structures

Syn-depositional features include a variety of accretion structures and current bedforms formed during base surge transport (Fig. 10.105). Base surge often shows evidence of being wet and "sticky" when deposited. Accretionary and cored lapilli are common (Fig. 10.105a, b), and deposits may be plastered and adhere to the "upwind-side" of vertical or near vertical surfaces such as basement outcrops, sea cliffs, trees, and vegetation (e.g. Fig. 10.105c).

Current bedforms are ubiquitous and include wave-like (or dune-like) structures called (anti)dune forms or (anti)-dunes, and their internal, low angle cross stratification (Fig. 10.105d–g), massive beds without structure and variously crudely stratified to planar stratified or laminated beds (Fig. 10.105d, h–j). Wohletz and Sheridan (1979) documented lateral facies changes including changes in syn-depositional structures relative to distance from the vent, based on studies of many volcanic centres in the Western USA. They proposed a transition from proximal dune bedforms (sandwave in their terminology) to dominantly massive, structureless beds at medial distances and planar beds in distal reaches, which they related to the changing flow conditions within a surge with time and distance. However, the documented grain size characteristics for these facies (Wohletz 1983; Sheridan and Wohletz 1983) indicate that all are dominated by ash-sized material which could affect the types of depositional bedforms developed.

In contrast, Sohn and Chough (1989) and Chough and Sohn (1990) propose that the array of facies is much more diverse and complex, and downflow transitions may vary depending on the initial grain size and moisture content of different surge flow systems. In general, they propose changes from coarse massive, block-lapilli-tuff near vent to diffuse planar stratification in coarse lapilli-tuff, to dune bedforms in finer lapilli-tuff and distally into laminated tuff (Fig. 10.101). Such transitions are a reflection of declining flow competency (ability to carry large clasts), capacity, velocity, and turbulence levels.

In our experience, surge deposits are very complex and their vertical and lateral facies architecture change from case to case due to the complex interplay between the inherently unsteady, pulsatory nature of the phreatomagmatic eruptions, the highly turbulent nature of the surge, variations in surge physical properties (temperature, moisture content, degree of condensation, intensity of fragmentation, grain size variation, cohesion and adhesion effects, degree of explosive thrust, velocity), and the variable topography onto which they flow.

The terms dune form, dune, and antidune have all been applied to the dune-like structures that characterise dune-form facies for both ash and lapilli surge deposits. Allen (1984) points out that "dune", and "antidune" have been used almost invariably synonymously with aqueous ripples and dunes, which form under conditions of unidirectional, aqueous current flow, involving cohesionless grain aggregates. Asymmetrical sedimentary aqueous ripples and dunes, with downcurrent dipping, angle of repose cross-beds form under lower flow regime or velocity conditions, whereas more symmetrical, low profile anti-dunes with low angle cross-beds dipping upstream, downstream, or draping over the dune, form under conditions of high velocity and high bed shear stress, called upper flow regime (Chap. 12). On these criteria, gaseous surge dune forms are anti-dunes. However, interpreting gaseous surge bedforms in terms of similar flow conditions and regimes as aqueous systems is incorrect because the physical conditions involved in the formation of pyroclastic dune forms and aqueous dunes and antidunes are quite different.

Surge dune-like forms may be symmetrical or asymmetrical, but always have low angle slopes and cross-beds. Asymmetry may be either upflow or downflow. Dune-form bedding set thickness decreases approximately logarithmically away from the vent (Wohletz and Sheridan 1979) and average wavelength and amplitude also decrease away from the vent (Fig. 10.106; Moore 1967; Waters and Fisher 1971; Wohletz and Sheridan 1979; Gençalioğlu-Kuşcu et al. 2007). Wavelength and wave height are logarithmically proportional to each other (Allen 1984). Gençalioğlu-Kuşcu et al. (2007) have summarised the amplitude and wavelength of historic and modern (Holocene-Pleistocene) base surge dune forms from Taal (Luzon), Laacher See (Germany), Ubehebe (California) volcanoes, and Cora maar (Turkey). These generally show amplitudes of 30–100 cm, with wavelengths from 1 to 10 m. At Ubehebe crater, where deposits are finer-grained, wavelength ranges from 30 to

Fig. 10.105 Syn-depositional features of base surge deposits. **a** Accretionary lapilli in basaltic, ash base surge deposits interbedded with planar fine lapilli layers. Basaltic 35 ka Tower Hill maar, Newer Volcanics Province, Victoria, southeastern Australia. **b** Cored lapilli forming small dune-like feature. Note imbrication of cored lapilli; flow is from right to left. Basaltic Koko Crater, Oahu, Hawai'i. **c** Mafic base surge ash plastered against the almost vertical side of basement reef limestone, Koko Crater (From Cas and Wright 1987). **d** Alternating coarse and fine, massive to diffusely bedded, planar bedded and dune bedded, low angle cross-stratified, rhyolitic base surge deposit sequence in the Tocomar tuff ring succession, northwestern Argentina. Gradational changes in depositional structures and grain size indicate that the generation of multiple base surges was a nearly continuous process, reflecting fluctuating surge energy levels, and pulsing explosions varying in eruption intensity at the vent. Flow from right to left. **e** Typical progressive, climbing base surge (anti)dunes, with low profiles, and low angle cross-stratification. Cross-beds generally dip, and thin downstream, but also continue and drape over the crest of the next dune form. Note the regular alternation of coarse- and fine-grained crossbeds, indicating pulsing current behaviour. Basaltic Lake Purrumbete maar, Newer Volcanics Province, Victoria. Flow from right to left. **f** Climbing, progressive antidunes, and planar to wavy stratification in the basaltic <300 ka Cora maar rim succession, Cappadocia Volcanic Province, Turkey. Flow from left to right. **g** Asymmetrical antidune forms in basaltic ash deposits, Tower Hill maar. Note the small wavelength and amplitude in these ash base surge deposits (coin for scale). Flow from right to left. **h** Proximal, planar to undulating, crudely stratified base surge deposits. Note semi-continuous boundaries between different depositional layers, consisting of poorly sorted lapilli-rich layers, composed of juvenile lava and poorly vesicular scoria, alternating with lapilli-poor layers with a yellow (palagonitised) ash matrix. These facies can be interpreted as the result of rapid deposition from a high concentration boundary layer developed at the base of an unsteady flow as it propagated forward from vent. Basaltic El Golfo tuff cone (Pleistocene), Lanzarote, Canary Islands, Spain. Flow is oblique from left to right, approximately into page. **i** Planar stratified basaltic base surge deposits with variations in grain size, and layer geometry varying from laterally continuous to subtle lensing. Note the very low profile antidune form in the middle. Tower Hill maar. **j** Planar bedded base surge deposits, abutting and draping large substrate boulders, underlying a thin ignimbrite deposit, which is overlain by a debris flow deposit. Holocene Secche di Lazzaro phreatomagmatic ignimbrite, Stromboli, Italy

Fig. 10.105 (continued)

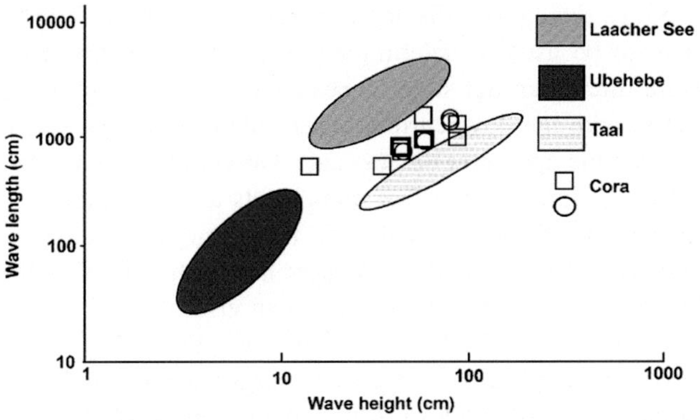

Fig. 10.106 Relationship between base surge antidune wavelength and amplitude (wave height) for four well-studied maar phreatomagmatic volcanoes. At Cora (Cappadocia Volcanic Province, Turkey), circles are for dune bedsets low in the maar rim succession, boxes are for dune bedsets high in the maar rim sequence. At Ubehebe Crater (California, USA) deposits are fine-grained, and have shorter wavelength and amplitude than other deposits. (From Gençalioğlu-Kuşcu et al. 2007)

300 cm and amplitude is ≤ 10 cm (Fig. 10.106), consistent with observations elsewhere in comparing lapilli ash bedforms with ash bedforms (Fig. 10.105d–g).

Allen (1984) related the documented variations in bedforms and cross stratification relationships in base surge deposits to sedimentation rate, surge temperature, and moisture content in a scheme that classifies base surge (anti)dune bedforms and their internal cross stratification, and relates these to depositional rate, temperature, and moisture content (Fig. 10.107). In this scheme, Allen (1984) recognised three categories of surge dune bedform: progressive, stationary, and regressive, which are also found in other studies (Cole 1991; Schmincke et al. 1973; Gençalioğlu-Kuşcu et al. 2007). In progressive bedforms (types A1, A2, and B; Fig. 10.107), the crests of the dune forms migrate in the direction of surge flow. Where the sedimentation rate is high, the internal arrangement of cross-strata resembles normal sedimentary climbing ripples and cross stratification (type A1). Progressive dune forms are

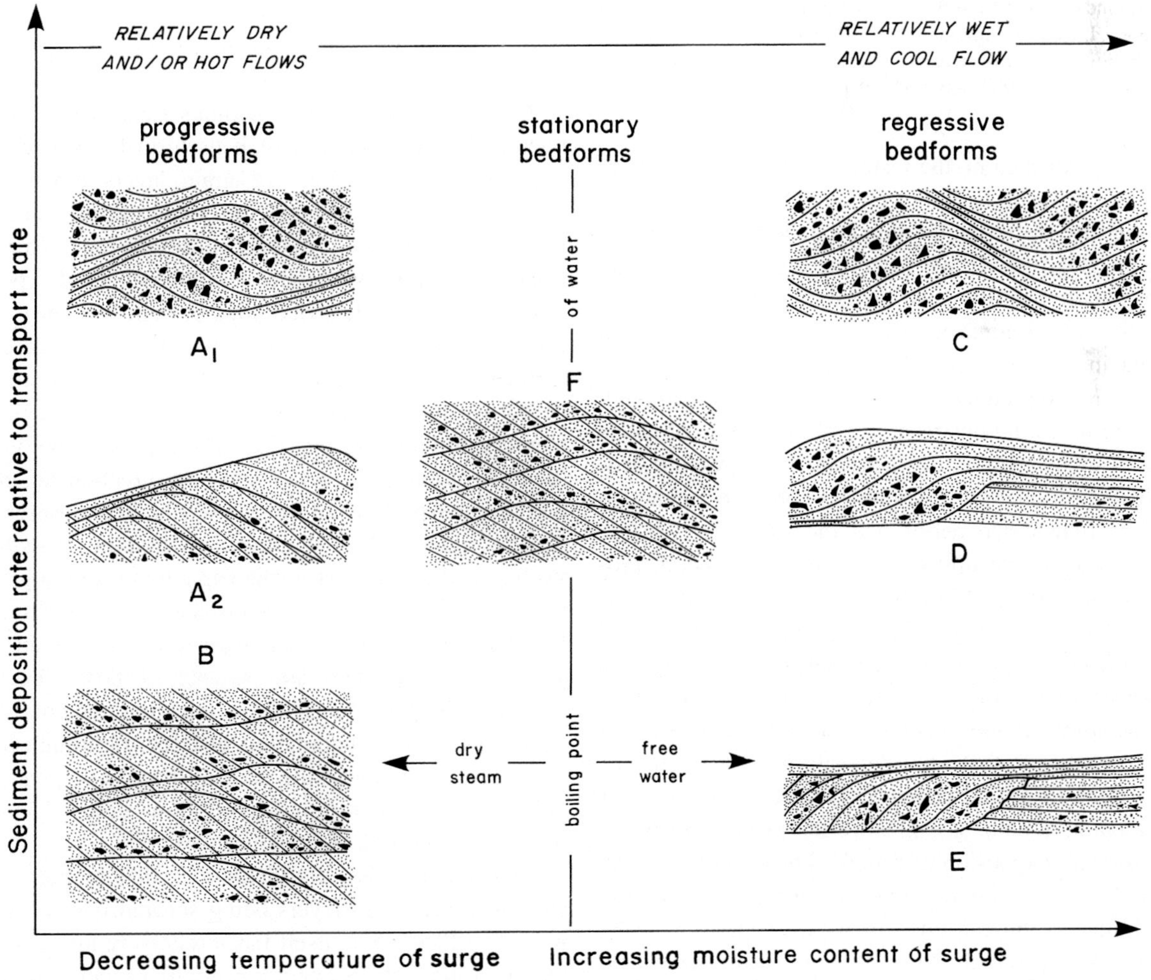

Fig. 10.107 Classification of base surge (anti)dune bedforms and internal cross-stratification related to depositional rate (relative to transport rate; vertical axis) and surge temperature and moisture content (horizontal axis) (From Cas and Wright 1987, After Allen 1984)

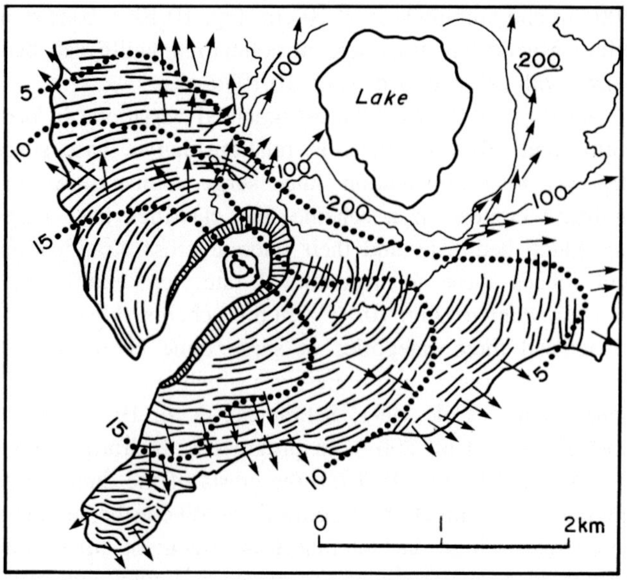

Fig. 10.108 Map showing traces and orientations of base surge antidunes around the vent from the 1966 phreatomagmatic eruption of Taal volcano, Luzon, Philippines; also shown are the interpreted directions of flow and the decreasing wavelength away from the vent. (From Cas and Wright 1987, After Moore 1967)

considered by Allen to be the result of surges that are relatively dry or hot, or both. Stationary dune forms (type F) are distinguished by crests which migrate neither upstream nor downstream. Regressive dune forms are distinguished by an upstream migration of the crest of the dune form, as represented in the crestal point of successive sigmoidally-shaped cross-strata in a single set. Allen suggests that regressive dune forms are the products of wet or cool surges. Although the crests are retreating upstream, the through-flow of solids within individual layers is downstream. In regressive bedforms, the depositional process involves cohesive interaction between the transported debris and the substrate. Coarser fragments occur on the upstream side of the crest (cf. progressive dune forms). "Climbing" effects, in this case upstream, result from both high sedimentation rates and adhesion-plastering due to wetness. Allen (1984) suggests that stationary dune forms (Fig. 10.107) result from surges with temperatures at the condensation or boiling point of water. He points out that only deposits with stationary or regressive dune forms are known to contain accretionary or armoured lapilli, or both, as well as vesicles which are due to post-depositional expansion of trapped hot volatiles in wet, impermeable ashes. Small amplitude, irregular adhesion ripples, similar to those formed in wind-swept snowfields, are another, less common, form of dune forms associated with surge deposits (Allen 1984).

Surge dune forms display sufficient regularity in form for them to be used in determining surge transport directions. More often than not surge dunes have an element of asymmetry, with the downcurrent or lee-side being slightly steeper than the upcurrent or stoss side. Observations on modern volcanoes have shown that dune crests are continuous and are orientated perpendicular to the surge flow direction (Fig. 10.108; Moore 1967; Waters and Fisher 1979; Allen 1984). Internal cross stratification can also be used to infer flow directions (Fig. 10.105d–g; also see Fig. 1.15g).

Cross-stratification angles in surge dunes are highly variable (Fig. 10.105d–g). Although they approach the angle of repose in some cases, in most instances they are considerably less than the angle of repose, suggesting that the origin of the cross stratification is not analogous to low flow regime, aqueous, dune formation. In addition, surge dune cross-beds often drape dune crests, thinning over them, and can be traced from the upcurrent to downcurrent side of the dune. They tend to be thinner on the upcurrent side and dip at a lower angle than on the down current side where they are slightly thicker and more steeply dipping. Cross-beds can show significant grain size variations from one cross-bed to the next, reflecting energy pulsing, and this is never the case with sedimentary cross-beds.

In aqueous sedimentary settings steep, angle of repose, tractional (e.g. fluvial) cross-beds form by passive, inertial gravitational avalanching of grain layers down angle of repose (>25°), downstream face of the dune, under conditions of subcritical aqueous flow (Froude Number <1; Chap. 12). Cross-bed angles less than angle of repose indicate deposition under conditions of high bed shear stresses, usually indicating high flow velocities at the depositional surface, which is most commonly the case in surge dune bedforms, making them more like high flow regime sedimentary antidunes (Chap. 12). The aspect of the outcrop of the cross stratification relative to flow direction can give very different perspectives of cross-bed sets. In exposures parallel to flow, dune forms, dipping, and draping cross-beds are visible, and are closest to their maximum dip angle. While they can be used to determine flow direction (e.g. Fig. 10.105d–g), in 2D outcrops there can be an error of perhaps 30° in such assessments of flow direction. In exposures perpendicular to flow direction, the cross-bed sets may appear to be undulating bed sets or trough-like sets, with the orientation of the axes of troughs perpendicular or oblique to the outcrop, indicating flow direction into or out of the outcrop (e.g. Fig. 10.104c–e).

Low angle cross stratification is due to high velocity, current-driven grain layers being sheared over an irregular, waved surface which itself has a relatively low relief due to high bed shear stress operating. Andrews and Manga (2012) propose that such forms could form as sediment waves as large flow eddies migrate and dissipate, not from sustained uni-directional flow. They propose that each lamination is the deposit of a single, short-lived turbulent eddy or pulse, which is supported by the clear variation in grain size between successive cross-beds in most lapilli-ash surge antidune bedsets (e.g. Fig. 10.104h). The implication of this is that high bed shear forces, through high velocity flow, were operating (Dellino et al. 2008). In the pyroclastic base surges particulate matter is driven by a gas or gaseous fluid which has a density and viscosity orders of magnitude less than that of water, and the steam rich nature of the flow, with ash clumping and adhesion occurring means that the flow system is neither cohesionless nor cohesive. Simple analogies with aqueous flow regime concepts should be avoided.

In the case of dry surges, which are simple, two-phase gas–solid systems, the processes may be analogous to aeolian processes, albeit at hurricane strength velocities (Doronzo et al. 2016). With dry surge dune forms, normal asymptotic, angle of repose cross-strata of avalanche origin may occur at low velocities (cf. aeolian sand dunes) or lower profile ripples may form at high velocities. However, perhaps surprisingly, there is no established flow regime concept or known order of bedforms for wind blown sediments. In addition, the competency of surges in transporting coarse debris is well beyond the limits of normal surface winds, except hurricanes and tornadoes, and low angle cross stratification is common.

Massive base surge facies (e.g. Fig. 10.105d) represent high particle concentration parts of surges and result from high sedimentation rates. In base surge deposits at Suwolbong tuff ring in Korea (Fig. 10.101), Sohn and Chough (1989) found that poorly sorted massive block-lapilli-ash facies occurred closest to vent, passing downstream into diffusely stratified (block)lapilli-ash then into lapilli-ash antidunes and then distally, laminated fine lapilli-ash and ash facies. The massive facies represent a high particle concentration, granular flow, and depositional regime. Massive ash facies are considered by Wohletz and Sheridan (1979) to be transitional between dune form and distal planar bedded facies. They describe beds of massive facies as being frequently lens-shaped and occurring on the lee-side of dune forms. They are ungraded, and generally unstructured internally except for occasional planar to wave-like diffuse layering or pebble trains. Wohletz and Sheridan (1979) consider massive beds to be deposited by a deflating, highly concentrated stage of surge flow, perhaps involving some fluidisation. The massive character is indicative of high particle concentrations and very rapid rates of deposition, probably involving en masse freezing of the grain population. The diffuse layering probably reflects internal shearing at the time of deposition in the highly concentrated grain aggregate. However, almost no surge deposits contain gas segregation pipes.

Planar bedded facies (e.g. Fig. 10.105d, h–j) consists of layers mm to several cm thick which are planar, slightly wavy or even locally lensoidal, and can be dominated by lapilli, or alternations of lapilli and ash layers, or ash layers. Just as in the cross-stratified sets of dune forms, major grain size variations are common from layer to layer and between different bed sets. These can form crudely stratified to planar stratified or laminated beds dominated by lapilli, lapilli and ash, or ash. Layers may be semi-continuous and taper (Fig. 10.105h), but others are laterally continuous (Fig. 10.105d, i). The grain size variation between layers again suggests a highly pulsatory depositional process. Planar bedded facies may be difficult to distinguish from bedded fall deposits, but lack of good sorting, and evidence of lateral transport in the form of subtle scours, cross stratification, low angle truncations, subtle lensing and pinching and swelling is often identifiable in planar stratified surge bedsets (Fig. 10.105d, h–j). In addition, in near vent settings debris is falling into propagating base surges, so that some planar stratified deposits are actually hybrid deposits, which depending on the dominant characteristics can be named *fall modified base surge deposits*, or *base surge modified fall deposits* (Cas 1989; see Figs. 9.51d and 16.17c).

Post-depositional Structures

Post-depositional structures commonly associated with base surge deposits include ballistic bomb impact sags which are well preserved in wet, unconsolidated surge deposits which deform plastically (Fig. 10.109a; also see Fig. 1.15b). Flame structures at the base of surges have also been described (Crowe and Fisher 1973), and soft state deformation under the shearing influence of the succeeding surge, or due to plastic slumping downslope while in a wet, cohesive state is common (Fig. 10.109b, c). Faulting and low angle sliding, in particular as a result of crater widening, can result in complex intra-eruptive unconformities and soft state deformation in younger, draping base surge deposits (Fig. 10.109d, e).

10.14.5.2 Frontal Ash-Cloud or Ground Surge, and Trailing/Lateral Ash-Cloud Surge Deposits—Dilute PDC's

Ground, trailing, and lateral ash-cloud surge deposits will be most readily recognised when associated with a massive Layer 2 body deposit. Frontal ash-cloud or ground surge deposits occur as thin deposits (Layer 1; Fig. 10.40) below the body deposit of a pyroclastic flow, whereas trailing ash-cloud surge deposits (Layer 3a) occur above body deposits (Sect. 10.8.7; Fig. 10.42b, d). However, they may also occur lateral to the body deposit, as the deposits of lateral surges, perhaps interbedded with contemporaneous intra-plinian fallout deposits (Fig. 10.42e), or above other

Fig. 10.109 Post-depositional structures in base surge deposits. **a** Ballistic impact sag structure, showing soft state deformation of wet, phreatomagmatic surge deposits. Ballistic trajectory from left to right, and surge flow direction from left to right based on dune geometry and cross-bedding orientation. Basaltic 35 ka Tower Hill maar, Newer Volcanics Province, Victoria, southeastern Australia. **b** Contorted and slumped layer interbedded with planar stratified basaltic base surge deposits. Koko Crater, Oahu, Hawai'i (From Cas and Wright 1987). **c** Soft state slump folds in rhyolitic base surge deposits, Ramadas maar-tuff ring, ~6.6 Ma Corte Blanco Tuff, San Antonio de Los Cobres, northwestern Argentina. **d** Post-depositional but syn-eruptive normal fault scarp cutting older base surge deposits, and draped by steeply dipping, onlapping younger sequence. Exposure along the present crater rim of the basaltic Lake Purrumbete maar, Newer Volcanics Province, Victoria. Flow is right to left up the scarp (arrow). **e** Detail of inset in (**d**), showing soft state deformation of younger eruptive sequence adhered to fault scarp. These show small-scale folds with sinistral movement sense, consistent with the flow direction suggesting deformation under the shearing influence of succeeding surges

deposits in older sequences. They are usually finer-grained than the body deposits, diffusely bedded, dune bedded with low angle cross stratification, and plane laminated (Figs. 10.40 and 10.42; e.g. Fisher 1979). Exactly the same flow dynamic issues apply as to base surges, except that they are derivative from pyroclastic flows and must be interpreted within the context of the flow dynamics of the pyroclastic flow system (e.g. Fisher 1979; Douillet et al. 2013a, b).

Spectacular dune bedform fields were produced from pdcs during eruptions of El Chichón volcano in Mexico in 1982 (Sigurdsson et al. 1987). The pdcs that produced the dune fields originated from the collapse of plinian explosive eruption columns and were not base surges. They were small volume (<1 km^3) pumiceous flows and given that dunes occur from near the vent to distal margins 8 km from vent, the pdcs were dilute, low concentration flows. Dune forms

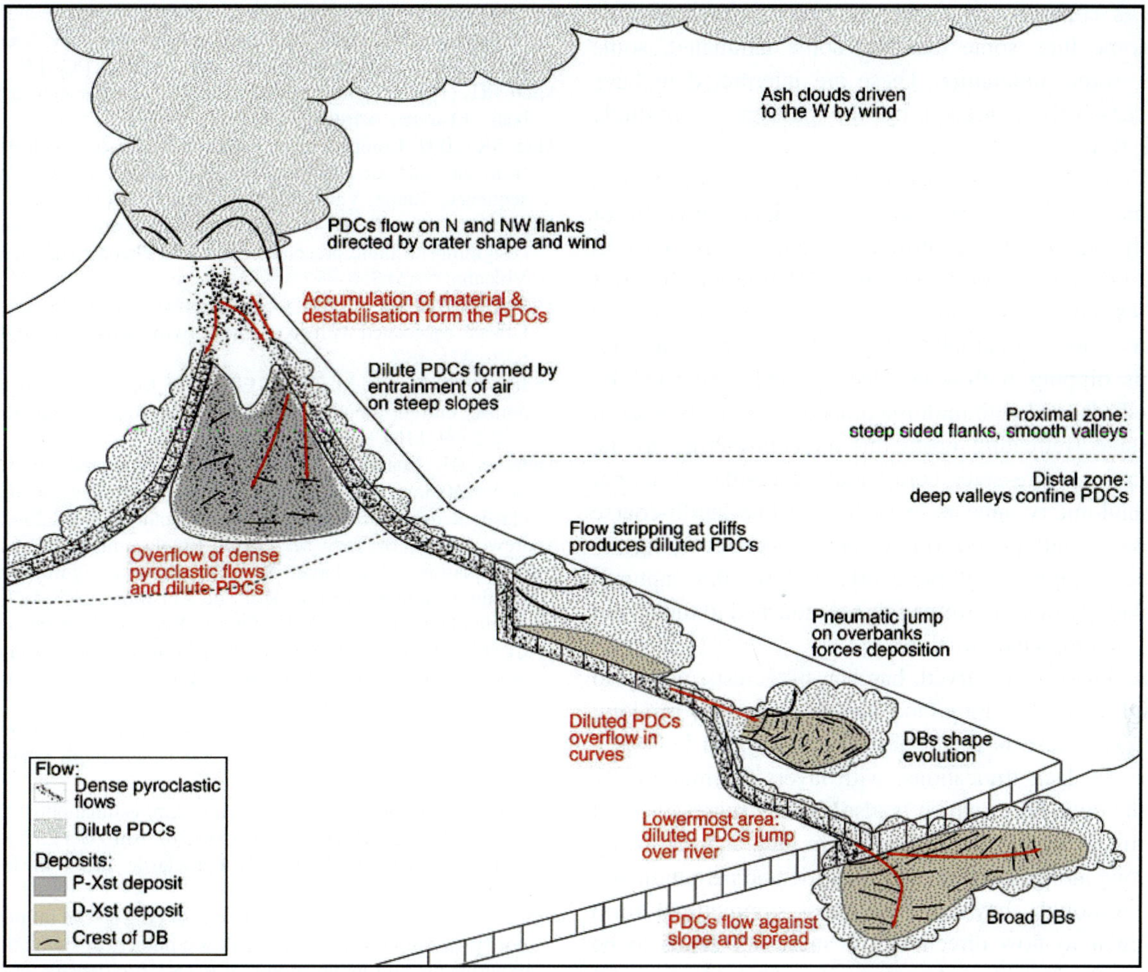

Fig. 10.110 Model for formation and deposition of cross-stratified (dune bedded) dilute pdc deposits from the August 2006 eruption of Tungurahua volcano, Ecuador. P-Xst = proximal cross-stratified; D-Xst = distal cross-stratified; DB = dune bedforms. (From Douillet et al. 2013b)

range from almost imperceptible undulations near the vent to more definable forms from 3 km from the vent to distal margins, ranging from 0.2 to 4 m high, wavelength ranges from 1 to 40 m, and ridge crests are up to 90 m long. Both dune height and wavelength decrease with distance from vent. These dune forms are the largest regular sets of dune forms known from either pyroclastic flow or base surge systems. Dune forms are generally asymmetrical, with shorter and steeper upcurrent or stoss sides more gently dipping and longer downcurrent or lee sides. Cross-beds range from 1 to cm thick and mostly dip in a down current direction at angles of <20°. Beds are mostly truncated by the stoss side and are lensoidal to planar. Grain size varies significantly from one cross-bed to the next, indicating a pulsating transport and depositional regime. Clasts are up to 10 cm down to ash, and the deposits plot in the fines-depleted ignimbrite field of Walker (1983; Fig. 10.103). The flows were interpreted as high velocity, highly turbulent flows.

Douillet et al. (2013a, b) have provided a thoughtful account of the surge dune form deposits associated with the pyroclastic flow deposits from the August 2006 eruptions of Tungurahua volcano, Ecuador. Pdcs were triggered by gravitational collapse of thick fallout and ballistic deposits on the steep upper slopes of the stratovolcano. Strictly, they are secondary flows of pyroclastic debris, not primary pdcs, although they did occur contemporaneously with the eruption. Massive, poorly sorted body deposits were deposited in steep upper slope valleys, but where these impacted on valley cliffs, they generated dilute pdcs, that flowed into proximal overbank settings as well as continuing further down slope where they encountered further hydraulic/pneumatic topographic jumps (cliffs), generating more dilute pdcs (Fig. 10.110).

The proximal overbank deposits are marked by longitudinal mounds, or elongate dune bed forms. They are up to 1.8 m high and up to 17 m long, but measure up to 27 m along the crest, but they can be smaller. Their internal structure is complex, consisting of multiple layers, some coarse, some fine, some massive, some laminated, some lensoidal, some undulating. These are interpreted to have been successively aggraded by the passage of multiple overbank flows.

Distal dune bedforms consist of three types: transverse, lunate, and 2D dune forms. Transverse dunes have linear crests perpendicular to the flow direction, are up to 1.6 m high and up to 17 m long, and up to 19 m along the crest trend. They have steeper, shorter upstream or stoss sides and gentler lee sides. Internally they are cross-bedded with cross-beds dipping both ways, leeside beds truncated by stoss side beds, and with multiple draped dune crests stacked in regressive (retreating upstream) all contributing to the building of a regressive, compound, dune-like structure. Their granulometry appears to be largely fine lapilli-coarse ash. These are interpreted to have been successively aggraded by the passage of multiple overbank flows, that captured the morphology of early formed dunes, attached to them, and aggraded a compound structure.

Lunate dunes have curved, barchan-like crests, are up to 1.5 m high and 12 m long, and measure up to 24 m along the crest. They also have steeper stoss sides than lee sides, and internally lack truncations, with layers asymmetrically draping the dune crest, which is also slightly regressive. 2D dune forms are up to 1.1 m high, 9.5 m long and can be up to 38 m long along the crest, which is oriented perpendicular to flow. Although they have the most symmetrical form perpendicular to flow direction, the stoss side tends to be longer than the lee side. Internal structure is not known. Of the distal dune forms, the transverse tend to be most proximal, the lunate more distal and the 2D dune bedforms, the most distal. All are interpreted to result from relatively slow moving currents, with each layer deposited by a pulse (or sediment wave of Andrews and Manga 2012), within an ongoing pdc event.

References

Abdurachman EK, Bourdier JL, Voight B (2000) Nuées ardentes of 22 November 1994 at Merapi volcano, Java, Indonesia. J Volcanol Geotherm Res 100(1–4):345–361

Acocella V (2007) Understanding caldera structure and development: an overview of analogue models compared to natural calderas. Earth Sci Rev 85(3–4):125–160

Adams NK, de Silva SL, Self S, Salas G, Schubring S, Permenter JL, Arbesman K (2001) The physical volcanology of the 1600 eruption of Huaynaputina, southern Peru. Bull Volcanol 62(8):493–518

Aguirre-Diaz G (2014) Calderas and ignimbrites of the central sector of the Mexican Volcanic Belt. Elsevier, p 300

Aguirre-Díaz GJ, Labarthe-Hernández G (2003) Fissure ignimbrites: fissure-source origin for voluminous ignimbrites of the Sierra Madre Occidental and its relationship with Basin and Range faulting. Geology 31(9):773–776

Aguirre-Díaz GJ, López-Martínez M (2009) Geologic evolution of the Donguinyó-Huichapan caldera complex, central Mexican Volcanic Belt, Mexico. J Volcanol Geoth Res 179(1–2):133–148

Allen JRL (1984) Sedimentary structures: their character and physical basis. Elsevier, Amsterdam, p 592

Allen SR (2004) Complex spatter- and pumice-rich pyroclastic deposits from an andesitic caldera-forming eruption: the Siwi pyroclastic sequence, Tanna, Vanuatu. Bull Volcanol 67(1):27–41

Allen SR, Cas RAF (1998) Lateral variations within coarse co-ignimbrite lithic breccias of the Kos Plateau Tuff, Greece. Bull Volcanol 59(5):356–377

Allen SR, McPhie J (2001) Syn-eruptive chaotic breccia on Kos, Greece, associated with an energetic pyroclastic flow. Bull Volcanol 63(6):421–432

Andrews BJ, Manga M (2011) Effects of topography on pyroclastic density current runout and formation of coignimbrites. Geology 39 (12):1099–1102

Andrews BJ, Manga M (2012) Experimental study of turbulence, sedimentation, and coignimbrite mass partitioning in dilute pyroclastic density currents. J Volcanol Geotherm Res 225–226:30–44

Andrews GDM, Branney MJ (2011) Emplacement and rheomorphic deformation of a large, lava-like rhyolitic ignimbrite: Grey's Landing, southern Idaho. Geol Soc Am Bull 123(3–4):725–743

Andronico D, Di Roberto A, De Beni E, Behncke B, Bertagnini A, Del Carlo P, Pompilio M (2018) Pyroclastic density currents at Etna volcano, Italy: The 11 February 2014 case study. J Volcanol Geother Res 357:92–105

Aramaki S (1984) Formation of the Aira Caldera, southern Kyushu, ~22,000 years ago. J Geophys Res: Solid Earth 89(B10):8485–8501

Aramaki S, Akimoto SI (1957) Temperature estimation of pyroclastic deposits by natural remanent magnetism. Am J Sci 255(9):619–627

Aramaki S, Yamasaki M (1963) Pyroclastic flows in Japan. Bull Volcanol 26(1):89–99

Aravena A, Chupin L, Dubois T, Roche O (2021) The influence of gas pore pressure in dense granular flows: numerical simulations versus experiments and implications for pyroclastic density currents. Bull Volcanol 83(11):1–20

Ascough PL, Bird MI, Scott AC, Collinson ME, Cohen-Ofri I, Snape CE, Le Manquais K (2010) Charcoal reflectance measurements: implications for structural characterization and assessment of diagenetic alteration. J Archaeol Sc 37(7):1590–1599

Bacon CR (1983) Eruptive history of Mount Mazama and Crater Lake Caldera, Cascade Range, U.S.A. J Volcanol Geotherm Res 18 (1):57–115

Bacon CR, Druitt TH (1988) Compositional evolution of the zoned calcalkaline magma chamber of Mount Mazama, Crater Lake, Oregon. Contrib Mineral Petrol 98(2):224–256

Baer EM, Fisher RV, Fuller M, Valentine GA (1997) Turbulent transport and deposition of the Ito pyroclastic flow: determinations using anisotropy of magnetic susceptibility. J Geophys Res 102 (B10):22565–22586

Báez W, Bustos E, Chiodi A, Reckziegel F, Arnosio M, de Silva S, Peña-Monné JL (2020a) Eruptive style and flow dynamics of the pyroclastic density currents related to the Holocene Cerro Blanco eruption (Southern Puna plateau, Argentina). J South Am Earth Sci 98:102482

Báez W, de Silva S, Chiodi A, Bustos E, Giordano G, Arnosio M, Suzaño N, Viramonte JG, Norini G, Groppelli G (2020b) Pulsating flow dynamics of sustained, forced pyroclastic density currents: insights from a facies analysis of the Campo de la Piedra Pómez ignimbrite, southern Puna, Argentina. Bull Volcanol 82:1–32

Bagnold RA (1956) The flow of cohesionless grains in fluids. Philos Trans Math Phys Eng Sci 249(964):235–297

Bailey JE, Self S (2010) The properties and formation of erosional pipe-shaped structures in ignimbrites around the Valles Caldera. Geol Soc Am Abs Prog 42(5):51

Bailey RA, Carr RG (1994) Physical geology and eruptive history of the Matahina Ignimbrite, Taupo volcanic zone, North Island, New Zealand. N Z J Geol Geophys 37(3):319–344

Balcone-Boissard H, Boudon G, Villemant B (2011) Textural and geochemical constraints on eruptive style of the 79 AD eruption at Vesuvius. Bull Volcanol 73(3):279–294

Banks NG, Hoblitt RP (1981) Summary of temperature studies of 1980 deposits. In: Lipman PW, Mullineaux DR (eds) The 1980 eruptions of Mount St. Helens, Washington. US Geol Surv Prof Pap 1250:295–313

Banks NG, Hoblitt RP (1996) Direct temperature measurements of deposits, Mount St. Helens, Washington, 1980–1981. US Geol Surv Prof Pap 1387

Bardintzeff JM (1984) Merapi volcano (Java, Indonesia) and Merapi-type Nuée ardente. Bull Volcanol 47(3):433–446

Bardot L (2000) Emplacement temperature determinations of proximal pyroclastic deposits on Santorini, Greece, and their implications. Bull Volcanol 61(7):450–467

Baxter PJ, Jenkins S, Seswandhana R, Komorowski JC, Dunn K, Purser D, Voight B, Shelley I (2017) Human survival in volcanic eruptions: thermal injuries in pyroclastic surges, their causes, prognosis and emergency management. Burns 43(5):1051–1069

Bear AN, Cas RAF, Giordano G (2009a) The implications of spatter, pumice and lithic clast rich proximal co-ignimbrite lag breccias on the dynamics of caldera forming eruptions: the 151 ka Sutri eruption, Vico Volcano, Central Italy. J Volcanol Geotherm Res 181(1–2):1–24

Bear AN, Cas RAF, Giordano G (2009b) Variations in eruptive style and depositional processes associated with explosive, phonolitic composition, caldera-forming eruptions: the 151 ka Sutri eruption, Vico Caldera, central Italy. J Volcanol Geoth Res 184(3):225–255

Bear AN, Giordano G, Giampaolo C, Cas RAF (2009c) Volcanological constraints on the post-emplacement zeolitisation of ignimbrites and geoarchaeological implications for Etruscan tomb construction (6th–3rd century B.C) in the Tufo Rosso a Scorie Nere, Vico Caldera, Central Italy. J Volcanol Geothem Res 183(3–4):183–200

Behncke B (2009) Hazards from pyroclastic density currents at Mt. Etna (Italy). J Volcanol Geother Res 180(2–4):148–160

Belousov A (1996) Deposits of the 30 March 1956 directed blast at Bezymianny volcano, Kamchatka, Russia. Bull Volcanol 57:649–662

Belousov A, Voight B, Belousova M (2007) Directed blasts and blast-generated pyroclastic density currents: a comparison of the Bezymianny 1956, Mount St Helens 1980, and Soufrière Hills, Montserrat 1997 eruptions and deposits. Bull Volcanol 69(7):701–740

Benage MC, Dufek J, Degruyter W, Geist D, Harpp K, Rader E (2014) Tying textures of breadcrust bombs to their transport regime and cooling history. J Volcanol Geotherm Res 274:92–107

Benage MC, Dufek J, Mothes PA (2016) Quantifying entrainment in pyroclastic density currents from the Tungurahua eruption, Ecuador: Integrating field proxies with numerical simulations. Geophys Res Lett 43(13):6932–6941

Best MG, Christiansen EH, Gromme S (2013) Introduction: the 36–18 Ma southern Great Basin, USA, ignimbrite province and flareup: Swarms of subduction-related supervolcanoes. Geosphere 9 (2):260–274

Bierwirth PN (1982) Experimental welding of volcanic ash. Thesis, Monash University

Bindeman IN, Valley JW (2003) Rapid generation of both high- and low-δ18O, large-volume silicic magmas at the Timber Mountain/Oasis Valley caldera complex, Nevada. Geol Soc Am Bull 115(5):581–595

Boari E, Avanzinelli R, Melluso L, Giordano G, Mattei M, De Benedetti AA, Morra V, Conticelli S (2009) Isotope geochemistry (Sr–Nd–Pb) and petrogenesis of leucite-bearing volcanic rocks from "Colli Albani" volcano, Roman Magmatic Province, Central Italy: inferences on volcano evolution and magma genesis. Bull Volcanol 71(9):977–1005

Bonadonna C, Cioni R, Costa A, Druitt T, Phillips J, Pioli L, Andronico D, Harris A, Scollo S, Bachmann O, Bagheri G, Wallenstein N (2016) MeMoVolc report on classification and dynamics of volcanic explosive eruptions. Bull Volcanol 78:1–12

Bond A, Sparks RSJ (1976) The Minoan eruption of Santorini, Greece. J Geol Soc 132(1):1–16

Bonnichsen B, Kauffman DF (1987) Physical features of rhyolite lava flows in the Snake River Plain volcanic province, southwestern Idaho. Geol Soc Am Spec Pap 212:119–145

Braithwaite RL (2003) Geological and mineralogical characterization of zeolites in lacustrine tuffs, Ngakuru, Taupo Volcanic Zone, New Zealand. Clays and Clay Miner 51(6):589–598

Brand BD, Bendaña S, Self S, Pollock N (2016) Topographic controls on pyroclastic density current dynamics: insight from 18 May 1980 deposits at Mount St. Helens, Washington (USA). J Volcanol Geotherm Res 321:1–17

Brand BD, Mackaman-Lofland C, Pollock NM, Bendaña S, Dawson B, Wichgers P (2014) Dynamics of pyroclastic density currents: Conditions that promote substrate erosion and self-channelization—Mount St Helens, Washington (USA). J Volcanol Geotherm Res 276:189–214

Brand BD, Pollock N, Sarocchi D, Dufek J, Clynne MA (2017) Field-trip guide for exploring pyroclastic density current deposits from the May 18, 1980, eruption of Mount St. Helens, Washington. US Geol Surv Sci Invest Rept 2017-5022-C:34 pp. https://doi.org/10.3133/sir20175022C

Brand BD, Pollock N, Vallance JW, Ongaro TE, Roche O, Trolese M, Giordano G, Mashall A, Criswell CW (2023) Advances in our understanding of pyroclastic current behavior from the 1980 eruption sequence of Mount St. Helens volcano (Washington), USA. Bull Volcanol 85(4):24

Branney MJ, Barry TL, Godchaux M (2004) Sheathfolds in rheomorphic ignimbrites. Bull Volcanol 66(6):485–491

Branney MJ, Kokelaar P (1992) A reappraisal of ignimbrite emplacement: progressive aggradation and changes from particulate to non-particulate flow during emplacement of high-grade ignimbrite. Bull Volcanol 54(6):504–520

Branney MJ, Kokelaar P (1997) Giant bed from a sustained catastrophic density current flowing over topography: Acatlán ignimbrite. Mexico Geol 25(2):115–118

Branney MJ, Kokelaar P (2002) Pyroclastic density currents and the sedimentation of ignimbrites. Geol Soc Lond Mem 27:145 p

Breard EC, Jones JR, Fullard L, Lube G, Davies C, Dufek J (2019) The permeability of volcanic mixtures—implications for pyroclastic currents. J Geophys Res: Solid Earth 124(2):1343–1360

Breard EC, Lube G (2017) Inside pyroclastic density currents–uncovering the enigmatic flow structure and transport behaviour in large-scale experiments. Earth Planet Sci Lett 458:22–36

Breard EC, Lube G, Jones JR, Dufek J, Cronin SJ, Valentine GA, Moebis A (2016) Coupling of turbulent and non-turbulent flow regimes within pyroclastic density currents. Nat Geosci 9(10):767–771

Brosch E, Lube G, Esposti Ongaro T, Cerminara M, Breard EC, Meiburg E (2022) Characteristics and controls of the runout behaviour of non-Boussinesq particle-laden gravity currents–a large-scale experimental investigation of dilute pyroclastic density currents. J VolcanoL Geotherm Res 432:107697

Brown DJ, Bell BR (2013) The emplacement of a large, chemically zoned, rheomorphic, lava-like ignimbrite: the Sgurr of Eigg Pitchstone, NW Scotland. J Geol Soc 170(5):753–767
Brown RJ, Andrews G (2015) Chapter 36 Pyroclastic density currents. In: Sigurdsson H, Houghton B, McNutt SR, Rymer H, Stix J (eds) The encyclopedia of volcanoes. Academic Press, Elsevier, pp 631–648
Brown RJ, Branney MJ (2004) Bypassing and diachronous deposition from density currents: evidence from a giant regressive bed form in the Poris ignimbrite, Tenerife, Canary Islands. Geology 32(5):445–448
Brown RJ, Branney MJ, Maher C, Dávila-Harris P (2010) Origin of accretionary lapilli within ground-hugging density currents: evidence from pyroclastic couplets on Tenerife. Geol Soc Am Bull 122 (1–2):305–320
Bryan SE, Cas RAF, Martí J (1998) Lithic breccias in intermediate volume phonolitic ignimbrites, Tenerife (Canary Islands): constraints on pyroclastic flow depositional processes. J Volcanol Geotherm Res 81(3–4):269–296
Bryan SE, Cas RAF, Martí J (2000) The 0.57 Ma plinian eruption of the Granadilla Member, Tenerife (Canary Islands): an example of complexity in eruption dynamics and evolution. Geol Mag 103(1–4):209–238
Bryan SE, Peate IU, Peate DW, Self S, DlA J, Mawby MR, Marsh JS, Miller JA (2010) The largest volcanic eruptions on Earth. Earth Sci Rev 102(3–4):207–229
Buesch DC (1992) Incorporation and redistribution of locally derived lithic fragments within a pyroclastic flow. Geol Soc Am Bull 104 (9):1193–1207
Burgisser A, Bergantz GW (2002) Reconciling pyroclastic flow and surge: the multiphase physics of pyroclastic density currents. Earth Planet Sci Lett 202(2):405–418
Bursik IM, Woods WA (1996) The dynamics and thermodynamics of large ash flows. Bull Volcanol 58(2):175–193
Calabrò L, Esposti Ongaro T, Giordano G, de Michieli Vitturi M (2022) Reconstructing pyroclastic currents' source and flow parameters from deposit characteristics and numerical modelling: the Pozzolane Rosse Ignimbrite case study (Colli Albani, Italy). J Geophys Res: Solid Earth 127:e2021JB023637
Calder ES, Luckett R, Sparks RSJ, Voight B (2002) Mechanisms of lava dome instability and generation of rockfalls and pyroclastic flows at Soufrière Hills Volcano, Montserrat. Geol Soc Lond Mem 21(1):173–190
Calder ES, Sparks RSJ, Gardeweg MC (2000) Erosion, transport and segregation of pumice and lithic clasts in pyroclastic flows inferred from ignimbrite at Lascar Volcano, Chile. J Volcanol Geotherm Res 104(1–4):201–235
Campagnola S, Vona A, Romano C, Giordano G (2016) Crystallization kinetics and rheology of leucite-bearing tephriphonolite magmas from the Colli Albani volcano (Italy). Chem Geol 424:12–29
Campbell LS, Giordano G, Stock MJ, Langella A, Bish DL, Gatta GD (2019) Reconstruction of residual melts from the zeolitized explosive products of alkaline-mafic volcanoes. Am Min: J Earth Planet Mater 104(1):2–16
Campbell ME, Russell JK, Porritt LA (2013) Thermomechanical milling of accessory lithics in volcanic conduits. Earth Planet Sci Lett 377–378:276–286
Caporuscio FA, Gardner JN, Schultz-Fellenz ES, Kelley RE (2012) Fumarolic pipes in the Tshirege Member of the Bandelier Tuff on the Pajarito Plateau, Jemez Mountains, New Mexico. Bull Volcanol 74(5):1023–1037
Capra L, Macías JL, Cortés A, Dávila N, Saucedo R, Osorio-Ocampo S, Arce JL, Gavilanes-Ruiz JC, Corona-Chávez P, García-Sánchez L, Sosa-Ceballos G, Vázquez R (2016) Preliminary report on the July 10–11, 2015 eruption at Volcán de Colima: pyroclastic density currents with exceptional runouts and volume. J Volcanol Geotherm Res 310:39–49
Carey S, Sigurdsson H (1987) Temporal variations in column height and magma discharge rate during the 79 AD eruption of Vesuvius. Geol Soc Am Bull 99(2):303–314
Caricchi C, Vona A, Corrado S, Giordano G, Romano C (2014) 79 AD Vesuvius PDC deposits' temperatures inferred from optical analysis on woods charred in-situ in the Villa dei Papiri at Herculaneum (Italy). J Volcanol Geotherm Res 289:14–25
Carrasco-Núñez G, Branney MJ (2005) Progressive assembly of a massive layer of ignimbrite with a normal-to-reverse compositional zoning: the Zaragoza ignimbrite of central Mexico. Bull Volcanol 68(1):3
Cas RAF (1989) Physical volcanology. In: Johnson RW, Knutson J, Taylor SR (eds) Intra-plate volcanism in Eastern Australia and New Zealand. Cambridge University Press, Cambridge, pp 55–87
Cas RAF, Wright JV (1987) Volcanic successions: modern and ancient. Allen and Unwin, London, p 528
Cas RAF, Wright HMN, Folkes CB, Lesti C, Porreca M, Giordano G, Viramonte JG (2011) The flow dynamics of an extremely large volume pyroclastic flow, the 2.08-Ma Cerro Galán Ignimbrite, NW Argentina, and comparison with other flow types. Bull Volcanol 73 (10):1583–1609
Cashman KV, Giordano G (2014) Calderas and magma reservoirs. J Volcanol Geotherm Res 288:28–45
Cavazos JA, Carrasco-Núñez G (2020) Anatomy of the Xáltipan ignimbrite at Los Humeros Volcanic Complex; the largest eruption of the Trans-Mexican Volcanic Belt. J Volcanol Geotherm Res 392:106755
Chapin CE, Lowell GR (1979) Primary and secondary flow structures in ash-flow tuffs of the Gribbles Run paleovalley, central Colorado. Geol Soc Am Spec Pap 180:137–154
Charbonnier SJ, Gertisser R (2008) Field observations and surface characteristics of pristine block-and-ash flow deposits from the 2006 eruption of Merapi Volcano, Java, Indonesia. J Volcanol Geotherm Res 177(4):971–982
Charbonnier SJ, Gertisser R (2012) Evaluation of geophysical mass flow models using the 2006 block-and-ash flows of Merapi Volcano, Java, Indonesia: towards a short-term hazard assessment tool. J Volcanol Geotherm Res 231–232:87–108
Charbonnier SJ, Germa A, Connor CB, Gertisser R, Preece K, Komorowski JC, Lavigne F, Dixon T, Connor L (2013) Evaluation of the impact of the 2010 pyroclastic density currents at Merapi volcano from high-resolution satellite imagery, field investigations and numerical simulations. J Volcanol Geotherm Res 261:295–315
Chédeville C, Roche O (2015) Influence of slope angle on pore pressure generation and kinematics of pyroclastic flows: insights from laboratory experiments. Bull Volcanol 77(11):1–13
Chough SK, Sohn YK (1990) Depositional mechanics and sequences of base surges, Songaksan tuff ring, Cheju Island, Korea. Sedimentology 37(6):1115–1135
Christiansen RL, Peterson DW (1981) The 1980 eruptions of Mount St. Helens. Chronology of the 1980 eruptive activity. In: Lipman PW, Mullineaux DR (eds) The 1980 eruptions of Mount St. Helens, Washington. United States Geol Surv Prof Pap pp 17–67
Cioni R, Gurioli L, Lanza R, Zanella E (2004) Temperatures of the A. D. 79 pyroclastic density current deposits (Vesuvius, Italy). J Geophys Res: Solid Earth 109(B2)
Cioni R, Sbrana A, Vecci R (1992) Morphologic features of juvenile pyroclasts from magmatic and phreatomagmatic deposits of Vesuvius. J Volcanol Geotherm Res 51(1):61–78
Cisneros de León A, Schindlbeck-Belo JC, Kutterolf S, Danielo M, Schmitt AK, Freundt A, Pérez W, Harvey JC, Wang KL, Lee HY (2021) A history of violence: magma incubation, timing and tephra

distribution of the Los Chocoyos supereruption (AtitlistCaldera, Guatemala). J Quat Sci 36(2):169–179
Clarke AB, Voight B (2000) Pyroclastic current dynamic pressure from aerodynamics of tree or pole blow-down. J Volcanol Geotherm Res 100(1–4):395–412
Cole JW, Brown SJA, Burt RM, Beresford SW, Wilson CJN (1998) Lithic types in ignimbrites as a guide to the evolution of a caldera complex, Taupo volcanic centre, New Zealand. J Volcanol Geotherm Res 80(3–4):217–237
Cole PD, Calder ES, Sparks RSJ, Clarke AB, Druitt TH, Young SR, Herd RA, Harford CL, Norton GE (2002) Deposits from dome-collapse and fountain-collapse pyroclastic flows at Soufrière Hills Volcano, Montserrat. Geol Soc Lond Mem 21(1):231–262
Cole PD, Guest JE, Duncan AM (1993) The emplacement of intermediate volume ignimbrites: a case study from Roccamonfina Volcano, Southern Italy. Bull Volcanol 55(7):467–480
Cole PD, Guest JE, Duncan AM, Pacheco JM (2001) Capelinhos 1957–1958, Faial, Azores: deposits formed by an emergent surtseyan eruption. Bull Volcanol 63(2):204–220
Conticelli S, Boari E, Avanzinelli R, De Benedetti AA, Giordano G, Mattei M, Melluso L, Morra V (2010) Geochemistry, isotopes and mineral chemistry of the Colli Albani volcanic rocks: constrains on magma genesis and evolution. In: Funiciello R, Giordano G (eds) The Colli Albani Volcano. IAVCEI Spec Publ 3 Geol Soc Lond 104–140
Cook GW, Wolff JA, Self S (2016) Estimating the eruptive volume of a large pyroclastic body: the Otowi Member of the Bandelier Tuff, Valles caldera, New Mexico. Bull Volcanol 78:1–11
Costa A, Suzuki Y, Koyaguchi T (2018) Understanding the plume dynamics of explosive super-eruptions. Nat Commun 9(1):654
Cronin SJ, Lube G, Dayudi DS, Sumarti S, Subrandiyo S (2013) Insights into the October–November 2010 Gunung Merapi eruption (Central Java, Indonesia) from the stratigraphy, volume and characteristics of its pyroclastic deposits. J Volcanol Geotherm Res 261:244–259
Crosweller HS, Arora B, Brown SK, Cottrell E, Deligne NI, Guerrero NO, Hobbs L, Kiyosugi K, Loughlin SC, Lowndes J, Nayembil M, Siebert L, Sparks RSJ, Takarada S, Venzke E (2012) Global database on large magnitude explosive volcanic eruptions (LaMEVE). J Appl Volcanol 1(1):1–13
Crowe BM, Fisher RV (1973) Sedimentary structures in base-surge deposits with special reference to cross-bedding, Ubehebe Craters, Death Valley, California. Geol Soc Am Bull 84(2):663–682
Crowe BM, Linn GW, Heiken G, Bevier ML (1978) Stratigraphy of the Bandelier Tuff in the Pajarito Plateau. Applications to waste management. Los Alamos Sci Lab, New Mexico. Inform Rept LA-7225-MS:57 pp
Dade WB, Huppert HE (1996) Emplacement of the Taupo ignimbrite by a dilute turbulent flow. Nature 381(6582):509–512
Dade WB, Huppert HE (1997) Reply to C. J. N Wilson's comment on "Emplacement of the Taupo ignimbrite by a dilute turbulent flow." Nature 385:307–308
Dartevelle S, Rose WI, Stix J, Kelfoun K, Vallance JW (2004) Numerical modeling of geophysical granular flows: 2. Computer simulations of plinian clouds and pyroclastic flows and surges. Geochem Geophys 5(8)
Davies DK, Quearry MW, Bonis SB (1978) Glowing avalanches from the 1974 eruption of the volcano Fuego, Guatemala. Geol Soc Am Bull 89(3):369–384
De Benedetti AA, Caprilli E, Rossetti F, Giordano G (2010) Metamorphic, metasomatic and intrusive xenoliths of the Colli Albani volcano and their significance for the reconstruction of the volcano plumbing system. In: Funiciello R, Giordano G (eds) The Colli Albani Volcano. Geol Soc Lond IAVCEI Spec Publ 3:153–176
De Benedetti AA, Funiciello R, Giordano G, Diano G, Caprilli E, Paterne M (2008) Volcanology, history and myths of the Lake Albano maar (Colli Albani volcano, Italy). J Volcanol Geotherm Res 176(3):387–406
de'Gennaro M, Cappelletti P, Langella A, Perrotta A, Scarpati C (2000) Genesis of zeolites in the Neapolitan Yellow Tuff: geological, volcanological and mineralogical evidence. Contrib Mineral Petrol 139(1):17–35
de'Gennaro M, Incoronato A, Mastrolorenzo G, Adabbo M, Spina G (1999) Depositional mechanisms and alteration processes in different types of pyroclastic deposits from Campi Flegrei volcanic field (Southern Italy). J Volcanol Geoth Res 91(2–4):303–320
Del Bello E, Taddeucci J, Scarlato P, Giacalone E, Cesaroni C (2015) Experimental investigation of the aggregation-disaggregation of colliding volcanic ash particles in turbulent, low-humidity suspensions. Geophys Res Lett 42(4):1068–1075
Dellino P, Dioguardi F, Doronzo DM, Mele D (2019) The rate of sedimentation from turbulent suspension: an experimental model with application to pyroclastic density currents and discussion on the grain-size dependence of flow runout. Sedimentology 66 (1):129–145
Dellino P, Isaia R, La Volpe L, Orsi G (2004) Interaction between particles transported by fallout and surge in the deposits of the Agnano-Monte Spina eruption (Campi Flegrei, Southern Italy). J Volcanol Geotherm Res 133(1–4):193–210
Dellino P, La Volpe L (2000) Structures and grain size distribution in surge deposits as a tool for modelling the dynamics of dilute pyroclastic density currents at La Fossa di Vulcano (Aeolian Islands, Italy). J Volcanol Geoth Res 96(1–2):57–78
Dellino P, Mele D, Sulpizio R, La Volpe L, Braia G (2008) A method for the calculation of the impact parameters of dilute pyroclastic density currents based on deposit particle characteristics. J Geophys Res: Solid Earth 113(B7)
Deruelle B, Medina ET, Figueroa OA, Maragano MC, Viramonte JG (1995) The recent eruption of Lascar volcano (Atacama-Chile, April 1993): petrological and volcanological relationships. Comptes rendus Académie des sciences. Série 2. Sciences de la terre et des planètes 321(5):377–384
De Rita D, Giordano G (1996) Volcanological and structural evolution of Roccamonfina volcano (Italy): origin of the summit caldera. Geol Soc Lond Spec Pub 110(1):209–224
De Rita D, Giordano G, Milli S (1998) Forestepping-backstepping stacking pattern of volcaniclastic successions: roccamonfina volcano, Italy. J Volcanol Geotherm Res 80(1–2):155–178
De Rita D, Giordano G, Esposito A, Fabbri M, Rodani S (2002) Large volume phreatomagmatic ignimbrites from the Colli Albani volcano (Middle Pleistocene, Italy). J Volcanol Geother Res 118(1–2):77–98
De Silva SL, Gosnold WD (2007) Episodic construction of batholiths: insights from the spatiotemporal development of an ignimbrite flare-up. J Volcanol Geotherm Res 167(1–4):320–335
De Silva S, Zandt G, Trumbull R, Viramonte JG, Salas G, Jiménez N (2006) Large ignimbrite eruptions and volcano-tectonic depressions in the Central Andes: a thermomechanical perspective. Geol Soc Lond Spec Pub 269(1):47–63
de Vita S, Orsi G, Civetta L, Carandente A, D'Antonio M, Deino A, di Cesare T, Di Vito MA, Fisher RV, Isaia R, Marotta E, Necco A, Ort M, Pappalardo L, Piochi M, Southon J (1999) The Agnano–Monte Spina eruption (4100 years BP) in the restless Campi Flegrei caldera (Italy). J Volcanol Geotherm Res 91:269–301
Dioguardi F, Dellino P (2014) PYFLOW: a computer code for the calculation of the impact parameters of dilute pyroclastic density currents (DPDC) based on field data. Comput Geosci 66:200–210

Di Muro A, Neri A, Rosi M (2004) Contemporaneous convective and collapsing eruptive dynamics: the transitional regime of explosive eruptions. Geophys Res Lett 31(10)

Di Muro A, Rosi M, Aguilera E, Barbieri R, Massa G, Mundula F, Pieri F (2008) Transport and sedimentation dynamics of transitional explosive eruption columns: the example of the 800 BP Quilotoa Plinian eruption (Ecuador). J Volcanol Geotherm Res 174(4):307–324

Di Roberto A, Bertagnini A, Pompilio M, Bisson M (2014) Pyroclastic density currents at Stromboli volcano (Aeolian Islands, Italy): a case study of the 1930 eruption. Bull Volcanol 76:1–14

Di Traglia F, Borselli L, Nolesini T, Casagli N (2023) Crater-rim collapse at Stromboli volcano: understanding the mechanisms leading from the failure of hot rocks to the development of glowing avalanches. Nat Haz 115(3):2051–2068

Dobran F, Neri A, Macedonio G (1993) Numerical simulation of collapsing volcanic columns. J Geophys Res: Solid Earth 98 (B3):4231–4259

Doronzo DM (2012) Two new end members of pyroclastic density currents: Forced convection-dominated and inertia-dominated. J Volcanol Geotherm Res 219–220:87–91

Doronzo DM, Dellino P (2011) Interaction between pyroclastic density currents and buildings: numerical simulation and first experiments. Earth Planet Sci Lett 310(3–4):286–292

Doronzo DM, Dellino P (2014) Pyroclastic density currents and local topography as seen with the conveyer model. J Volcanol Geotherm Res 278:25–39

Doronzo DM, de Tullio MD, Dellino P, Pascazio G (2011) Numerical simulation of pyroclastic density currents using locally refined Cartesian grids. Comput Fluids 44(1):56–67

Doronzo DM, Giordano G, Palladino DM (2022) Energy facies: a global view of pyroclastic currents from vent to deposit. Terra Nova 34:1–11

Doronzo DM, Martí J, Dellino P, Giordano G, Sulpizio R (2016) Dust storms, volcanic ash hurricanes, and turbidity currents: physical similarities and differences with emphasis on flow temperature. Arab J Geosci 9(4):1–9

Doronzo DM, Martí J, Sulpizio R, Dellino P (2012) Aerodynamics of stratovolcanoes during multiphase processes. J Geophys Res: Solid Earth, 117(B1)

Doronzo DM, Valentine GA, Dellino P, de Tullio MD (2010) Numerical analysis of the effect of topography on deposition from dilute pyroclastic density currents. Earth Planet Sci Lett 300(1–2):164–173

Douillet GA, Pacheco DA, Kueppers U, Letort J, Tsang-Hin-Sun È, Bustillos J, Hall M, Ramón P, Dingwell DB (2013a) Dune bedforms produced by dilute pyroclastic density currents from the August 2006 eruption of Tungurahua volcano, Ecuador. Bull Volcanol 75(11):1–20

Douillet GA, Rasmussen KR, Kueppers U, Lo Castro D, Merrison JP, Iversen JJ, Dingwell DB (2014) Saltation threshold for pyroclasts at various bedslopes: wind tunnel measurements. J Volcanol Geotherm Res 278–279:14–24

Douillet GA, Tsang-Hin-Sun È, Kueppers U, Letort J, Pacheco DA, Goldstein F, Von Aulock F, Lavallée Y, Hanson JB, Bustillos J, Robin C, Ramón P, Hall M, Dingwell DB (2013b) Sedimentology and geomorphology of the deposits from the August 2006 pyroclastic density currents at Tungurahua volcano, Ecuador. Bull Volcanol 75(11):1–21

Druitt TH (1992) Emplacement of the 18 May 1980 lateral blast deposit ENE of Mount St. Helens, Washington. Bull Volcanol 54(7):554–572

Druitt TH (1995) Settling behaviour of concentrated dispersions and some volcanological applications. J Volcanol Geoth Res 65(1–2):27–39

Druitt TH (1998) Pyroclastic density currents. Geol Soc Lond Spec Pub 145(1):145–182

Druitt TH (2014) New insights into the initiation and venting of the Bronze-Age eruption of Santorini (Greece), from component analysis. Bull Volcanol 76(2):1–21

Druitt TH, Avard G, Bruni G, Lettieri P, Maez F (2007) Gas retention in fine-grained pyroclastic flow materials at high temperatures. Bull Volcanol 69(8):881–901

Druitt TH, Bacon CR (1986) Lithic breccia and ignimbrite erupted during the collapse of Crater Lake Caldera, Oregon. J Volcanol Geotherm Res 29(1):1–32

Druitt TH, Calder ES, Cole PD, Hoblitt RP, Loughlin SC, Norton GE, Ritchie LJ, Sparks RSJ, Voight B (2002) Small-volume, highly mobile pyroclastic flows formed by rapid sedimentation from pyroclastic surges at Soufrière Hills Volcano, Montserrat: an important volcanic hazard. Geol Soc Lond Mem 21(1):263–279

Druitt TH, Edwards L, Mellors RM, Pyle DM, Sparks RSJ, Lanphere M, Davies M, Barriero M (1999) Santorini Volcano. Geol Soc London Memoir 165

Druitt TH, Kokelaar BP (2002) The eruption of Soufriere Hills Volcano, Montserrat, from 1995 to 1999. Geol Soc Lond Memoir 645

Druitt TH, Sparks RSJ (1982) A proximal ignimbrite breccia facies on Santorini, Greece. J Volcanol Geotherm Res 13(1):147–171

Druitt TH, Sparks RSJ (1984) On the formation of calderas during ignimbrite eruptions. Nature 310(5979):679–681

Dufek J (2016) The fluid mechanics of pyroclastic density currents. Annu Rev Fluid Mech 48(1):459–485

Dufek J, Bergantz GW (2007) Suspended load and bed-load transport of particle-laden gravity currents: the role of particle–bed interaction. Theor Comp Fluid Dyn 21(2):119–145

Dufek J, Esposti Ongaro T, Roche O (2015) Pyroclastic density currents: processes and models. In: Sigurdsson H, Houghton B, McNutt SR, Rymer H, Stix J (eds) The encyclopedia of volcanoes. Academic Press, Elsevier, pp 617–629

Dufek J, Manga M (2008) In situ production of ash in pyroclastic flows. J Geophys Res: Solid Earth 113(B9)

Dufek J, Manga M, Patel A (2012) Granular disruption during explosive volcanic eruptions. Nat Geoscisci 5(8):561–564

Dufek J, Wexler J, Manga M (2009) Transport capacity of pyroclastic density currents: experiments and models of substrate-flow interaction. J Geophys Res: Solid Earth 114(B11)

Duyverman HJ, Roobol MJ (1981) Gas pipes in Eocambrian volcanic breccias. Geol Mag 118:265–270

Edgar CJ, Wolff JA, Nichols HJ, Cas RAF, Martí J (2002) A complex Quaternary ignimbrite-forming phonolitic eruption: the Poris Member of the Diego Hernández Formation (Tenerife, Canary Islands). J Volcanol Geotherm Res 118(1–2):99–130

Ellwood BB (1982) Estimates of flow direction for calc-alkaline welded tuffs and paleomagnetic data reliability from anisotropy of magnetic susceptibility measurements: Central San Juan Mountains, southwest Colorado. Earth Planet Sci Lett 59(2):303–314

Elston WE, Smith EI (1970) Determination of flow direction of rhyolitic ash-flow tuffs from fluidal textures. Geol Soc Am Bull 81 (11):3393–3406

Esposti Ongaro T, Cerminara M, Charbonnier SJ, Lube G, Valentine GA (2020a) A framework for validation and benchmarking of pyroclastic current models. Bull Volcanol 82(6):1–7

Esposti Ongaro T, Clarke AB, Voight B, Neri A, Widiwijayanti C (2012) Multiphase flow dynamics of pyroclastic density currents during the May 18 1980 lateral blast of Mount St. Helens. J Geophys Res: Solid Earth 117:B06208

Esposti Ongaro T, Komorowski JC, Legendre Y, Neri A (2020b) Modelling pyroclastic density currents from a subplinian eruption at

La Soufrière de Guadeloupe (West Indies, France). Bull Volcanol 82(12):1–26

Esposti Ongaro T, Neri A, Menconi G, Vitturi MDM, Marianelli P, Cavazzoni C, Baxter PJ (2008) Transient 3D numerical simulations of column collapse and pyroclastic density current scenarios at Vesuvius. J Volcanol Geothem Res 178(3):378–396

Esposti Ongaro T, Widiwijayanti C, Clarke AB, Voight B, Neri A (2011) Multiphase-flow numerical modeling of the 18 May 1980 lateral blast at Mount St. Helens, USA. Geology 39:535–538

Eychenne J, Le Pennec JL, Troncoso L, Gouhier M, Nedelec JM (2012) Causes and consequences of bimodal grain-size distribution of tephra fall deposited during the August 2006 Tungurahua eruption (Ecuador). Bull Volcanol 74(1):187–205

Fenner CN (1948) Incandescent tuff flows in southern Peru. Geol Soc Am Bull 59(9):879–893

Fierstein J, Hildreth W (1992) The plinian eruptions of 1912 at Novarupta, Katmai National Park, Alaska. Bull Volcanol 54 (8):646–684

Fisher RV (1966) Mechanism of deposition from pyroclastic flows. Am J Sci 264(5):350–363

Fisher RV (1977) Erosion by volcanic base-surge density currents: U-shaped channels. Geol Soc Am Bull 88(9):1287–1297

Fisher RV (1979) Models for pyroclastic surges and pyroclastic flows. J Volcanol Geothem Res 6(3):305–318

Fisher RV (1990) Transport and deposition of a pyroclastic surge across an area of high relief: the 18 May 1980 eruption of Mount St. Helens, Washington. Geol Soc Am Bull 102(8):1038–1054

Fisher RV (1995) Decoupling of pyroclastic currents: hazards assessments. J Volcanol Geotherm Res 66(1):257–263

Fisher RV, Heiken G (1982) Mt. Pelée, Martinique: May 8 and 20, 1902, pyroclastic flows and surges. J Volcanol Geotherm Res 13 (3):339–371

Fisher RV, Orsi G, Ort M, Heiken G (1993) Mobility of a large-volume pyroclastic flow—emplacement of the Campanian ignimbrite, Italy. J Volcanol Geoth Res 56(3):205–220

Fisher RV, Schmincke H-U (1984) Pyroclastic rocks. Springer, Berlin, p 472

Fisher RV, Schmincke H-U, Van Bogaard P (1983) Explosive Volcanism Origin and emplacement of a pyroclastic flow and surge unit at Laacher See, Germany. J Volcanol Geotherm Res 17(1):375–392

Fisher RV, Smith AL, Roobol MJ (1980) Destruction of St. Pierre, Martinique, by ash-cloud surges, May 8 and 20, 1902. Geology 8 (10):472–476

Fisher RV, Waters AC (1970) Base surge bed forms in maar volcanoes. Am J Sci 268(2):157–180

Folkes CB, de Silva SL, Wright HM, Cas RAF (2011a) Geochemical homogeneity of a long-lived, large silicic system; evidence from the Cerro Galán caldera, NW Argentina. Bull Volcanol 73(10):1455–1486

Folkes CB, Wright HM, Cas RAF, de Silva SL, Lesti C, Viramonte JG (2011b) A re-appraisal of the stratigraphy and volcanology of the Cerro Galán volcanic system, NW Argentina. Bull Volcanol 73 (10):1427–1454

Fontana G, Mac Niocaill C, Brown RJ, Sparks RSJ, Field M (2011) Emplacement temperatures of pyroclastic and volcaniclastic deposits in kimberlite pipes in southern Africa. Bull Volcanol 73 (8):1063–1083

Francis PW (1993) Volcanoes. A planetary perspective. Clarendon Press, Oxford

Francis PW, Baker MCW (1977) Mobility of pyroclastic flows. Nature 270(5633):164–165

Francis PW, Gardeweg M, Ramirez CF, Rothery DA (1985) Catastrophic debris avalanche deposit of Socompa volcano, northern Chile. Geology 13(9):600–603

Francis PW, O'Callaghan L, Kretzschmar GA, Thorpe RS, Sparks RSJ, Page RN, De Barrio RE, Gillou G, Gonzalez OE (1983) The Cerro Galan ignimbrite. Nature 301:51–53

Freda C, Gaeta M, Palladino DM, Trigila R (1997) The Villa Senni Eruption (Alban Hills, central Italy): the role of H_2O and CO_2 on the magma chamber evolution and on the eruptive scenario. J Volcanol Geotherm Res 78(1–2):103–120

Freundt A (1998) The formation of high-grade ignimbrites, I: experiments on high- and low-concentration transport systems containing sticky particles. Bull Volcanol 59(6):414–435

Freundt A (1999) Formation of high-grade ignimbrites Part II. A pyroclastic suspension current model with implications also for low-grade ignimbrites. Bull Volcanol 60(7):545–567

Freundt A, Schmincke H-U (1985a) Hierarchy of facies of pyroclastic flow deposits generated by Laacher See–type eruptions. Geology 13 (4):278–281

Freundt A, Schmincke H-U (1985b) Lithic-enriched segregation bodies in pyroclastic flow deposits of Laacher See Volcano (East Eifel, Germany). J Volcanol Geotherm Res 25(3):193–224

Freundt A, Schmincke H-U (1995a) Eruption and emplacement of a basaltic welded ignimbrite during caldera formation on Gran Canaria. Bull Volcanol 56(8):640–659

Freundt A, Schmincke HU (1995b) Petrogenesis of rhyolite-trachyte-basalt composite ignimbrite P1, Gran Canada, Canary Islands. J Geophys Res: Solid Earth 100(B1):455–474

Friedman I, Long W, Smith RL (1963) Viscosity and water content of rhyolite glass. J Geophys Res 68(24):6523–6535

Froggatt PC, Wilson CJN, Walker GPL (1981) Orientation of logs in the Taupo Ignimbrite as an indicator of flow direction and vent position. Geology 9(3):109–111

Fujii T, Nakada S (1999) The 15 September 1991 pyroclastic flows at Unzen Volcano (Japan): a flow model for associated ash-cloud surges. J Volcanol Geoth Res 89(1–4):159–172

Gardeweg MC, Sparks RS, Matthews SJ (1998) Evolution of Lascar volcano, northern Chile. J Geol Soc Lond 155(1):89–104

Gardner JE, Nazworth C, Helper MA, Andrews BJ (2018) Inferring the nature of pyroclastic density currents from tree damage: the 18 May 1980 blast surge of Mount St. Helens, USA. Geology. https://doi.org/10.1130/G45353.1

Gençalioğlu-Kuşcu G, Atilla C, Cas RAF, Kuşcu İ (2007) Base surge deposits, eruption history, and depositional processes of a wet phreatomagmatic volcano in Central Anatolia (Cora Maar). J Volcanol Geotherm Res 159(1–3):198–209

Gertisser R, Preece K, Keller J (2009) The Plinian Lower Pumice 2 eruption, Santorini, Greece: magma evolution and volatile behaviour. J Volcanol Geotherm Res 186(3–4):387–406

Gertisser R, Self S, Gaspar JL, Kelley SP, Pimentel A, Eikenberg J, Barry TL, Pacheco JM, Queiroz G, and Vespa M (2010) Ignimbrite stratigraphy and chronology on Terceira Island, Azores In: Groppelli G and Viereck-Goette L (eds) Stratigraphy and geology of volcanic areas. Geol Soc Amer Specl Pap 464:133–154. https://doi.org/10.1130/2010.2464(07)

Geyer A, Folch A, Martí J (2006) Relationship between caldera collapse and magma chamber withdrawal: an experimental approach. J Volcanol Geotherm Res 157(4):375–386

Geyer A, Martí J (2008) The new worldwide collapse caldera database (CCDB): a tool for studying and understanding caldera processes. J Volcanol Geotherm Res 175(3):334–354

Giampaolo C, Lo Mastro S, De Rita D, Giordano G (2006) Lateral and vertical zeolite grade variations in the Tufo Lionato ignimbrite unit (Colli Albani, Roma, central Italy). In: Bowman RS, Delop SE (eds) Zeolite '06'-7th International conference on the occurrence, properties and utilisation of natural zeolites, Socorro, New Mexico, USA

Gilbert JS, Lane SJ (1994) The origin of accretionary lapilli. Bull Volcanol 56(5):398–411

Giordano D, Nichols ARL, Dingwell DB (2005) Glass transition temperatures of natural hydrous melts: a relationship with shear viscosity and implications for the welding process. J Volcanol Geotherm Res 142(1–2):105–118

Giordano D, Russell JK, Dingwell DB (2008a) Viscosity of magmatic liquids: a model. Earth Planet Sci Lett 271(1–4):123–134

Giordano G (1998a) The effect of paleotopography on lithic distribution and facies associations of small volume ignimbrites: the WTT Cupa (Roccamonfina volcano, Italy). J Volcanol Geotherm Res 87(1–4):255–273

Giordano G (1998b) Facies characteristics and magma–water interaction of the White Trachytic Tuffs (Roccamonfina Volcano, southern Italy). Bull Volcanol 60(1):10–26

Giordano G, Cas RAF (2021) Classification of ignimbrites and related pyroclastic flows. Earth Sci Rev 220:103697. https://doi.org/10.1016/j.earscirev.2021.103697

Giordano G, De Astis G (2020) The summer 2019 basaltic Vulcanian eruptions (paroxysms) of Stromboli. Bull Volcanol. https://doi.org/10.1007/s00445-020-01423-2

Giordano G, De Astis G (2021) The summer 2019 basaltic Vulcanian eruptions (paroxysms) of Stromboli. Bull Volcanol 83(1):1–27

Giordano G, De Benedetti A, Diana A, Diano G, Esposito A, Fabbri M, Gaudioso F, Marasco F, Mazzini I, Miceli M (2010) Stratigraphy, volcano tectonics and evolution of the Colli Albani volcanic field. In: Funiciello R, Giordano G (eds) The Colli Albani Volcano. Geol Soc Lond IAVCEI Spec Publ 3:43–97

Giordano G, De Benedetti AA, Diana A, Diano G, Gaudioso F, Marasco F, Miceli M, Mollo S, Cas RAF, Funiciello R (2006) The Colli Albani mafic caldera (Roma, Italy): stratigraphy, structure and petrology. J Volcanol Geotherm Res 155(1–2):49–80

Giordano G, De Rita D, Cas R, Rodani S (2002) Valley pond and ignimbrite veneer deposits in the small-volume phreatomagmatic 'Peperino Albano' basic ignimbrite, Lago Albano maar, Colli Albani volcano, Italy: influence of topography. J Volcanol Geotherm Res 118(1–2):131–144

Giordano G, Dobran F (1994) Computer simulations of the Tuscolano Artemisio's second pyroclastic flow unit (Alban Hills, Latium, Italy). J Volcanol Geotherm Res 61(1):69–94

Giordano G, Doronzo D (2017) Sedimentation and mobility of PDCs: a reappraisal of ignimbrites' aspect ratio. Sci Rep 7:4444

Giordano G, Porreca M, Musacchio P, Mattei M (2008b) The Holocene Secche di Lazzaro phreatomagmatic succession (Stromboli, Italy): evidence of pyroclastic density current origin deduced by facies analysis and AMS flow directions. Bull Volcanol 70(10):1221–1236

Giordano G, Zanella E, Trolese M, Baffioni C, Vona A, Caricchi C, De Benedetti AA, Corrado S, Romano C, Sulpizio R, Geshi N (2018) Thermal interactions of the AD79 Vesuvius pyroclastic density currents and their deposits at Villa dei Papiri (Herculaneum archaeological site, Italy). Earth Planet Sci Lett 490:180–192

Gorshkov GS (1959) Gigantic eruption of the volcano bezymianny. Bull Volcanol 20(1):77–109

Gorshkov GS, Dubik YM (1970) Gigantic directed blast at Shiveluch volcano (Kamchatka). Bull Volcanol 34(1):261–288

Graettinger AH, Valentine GA, Sonder I (2015) Circum-crater variability of deposits from discrete, laterally and vertically migrating volcanic explosions: experimental evidence and field implications. J Volcanol Geotherm Res 308:61–69

Gregg PM, de Silva SL, Grosfils EB, Parmigiani JP (2012) Catastrophic caldera-forming eruptions: thermomechanics and implications for eruption triggering and maximum caldera dimensions on Earth. J Volcanol Geotherm Res 241–242:1–12

Grunder AL, Laporte D, Druitt TH (2005) Experimental and textural investigation of welding: effects of compaction, sintering, and vapor-phase crystallization in the rhyolitic Rattlesnake Tuff. J Volcanol Geotherm Res 142(1–2):89–104

Gueugneau V, Charbonnier S, Esposti Ongaro T, de' Michieli Vitturi M, Peruzzetto M, Mangeney A, BouchutF, Patra A, Kelfoun K (2021) Synthetic benchmarking of concentrated pyroclastic current models. Bull Volcanol 83:1–32

Gueugneau V, Kelfoun K, Druitt T (2019) Investigation of surge-derived pyroclastic flow formation by numerical modelling of the 25 June 1997 dome collapse at Soufrière Hills Volcano, Montserrat. Bull Volcanol 81:1–14

Gurioli L, Houghton BF, Cashman KV, Cioni R (2005) Complex changes in eruption dynamics during the 79 AD eruption of Vesuvius. Bull Volcanol 67(2):144–159

Gurioli L, Zanella E, Pareschi MT, Lanza R (2007) Influences of urban fabric on pyroclastic density currents at Pompeii (Italy): 1. Flow direction and deposition. J Geophys Res: Solid Earth 112(B5)

Hay RL (1966) Zeolites and zeolitic reactions in sedimentary rocks. Geol Soc Am Spec Pap 85:131p

Heap MJ, Kolzenburg S, Russell JK, Campbell ME, Welles J, Farquharson JI, Ryan A (2014) Conditions and timescales for welding block-and-ash flow deposits. J Volcanol Geotherm Res 289:202–209

Heiken G, Wohletz K, Eichelberger J (1988) Fracture fillings and intrusive pyroclasts, Inyo Domes, California. J Geophys Res: Solid Earth 93(B5):4335–4350

Henry CD, Wolff JA (1992) Distinguishing strongly rheomorphic tuffs from extensive silicic lavas. Bull Volcanol 54(3):171–186

Hildreth W (1979) The Bishop Tuff: evidence for the origin of compositional zonation in silicic magma chambers. Geol Soc Am Spec Pap 180:43–76

Hildreth W (1981) Gradients in silicic magma chambers: implications for lithospheric magmatism. J Geophys Res: Solid Earth 86 (B11):10153–10192

Hildreth W, Fierstein J (2012) The Novarupta-Katmai eruption of 1912—largest eruption of the twentieth century; centennial perspectives. US Geol Surv Prof Pap 1791

Hildreth W, Wilson CJN (2007) Compositional zoning of the Bishop Tuff. J Petrol 48(5):951–999

Hillhouse JW, Wells RE (1991) Magnetic fabric, flow directions, and source area of the Lower Miocene Peach Springs Tuff in Arizona, California, and Nevada. J Geophys Res: Solid Earth 96(B7):12443–12460

Hoblitt RP, Kellogg KS (1979) Emplacement temperatures of unsorted and unstratified deposits of volcanic rock debris as determined by paleomagnetic techniques. Geol Soc Am Bull 90(7):633–642

Hoblitt RP, Miller CD, Vallance JW (1981) Origin and stratigraphy of the deposit produced by the May 18 directed blast. In: Lipman PW, Mullineaux DR (eds) The 1980 eruptions of Mount St. Helens, Washington. US Geol Surv Prof Pap 1250:401–419

Hogeweg N, Keith TE, Colvard EM, Ingebritsen SE (2005) Ongoing hydrothermal heat loss from the 1912 ash-flow sheet, Valley of Ten Thousand Smokes, Alaska. J Volcanol Geotherm Res 143(4):279–291

Houghton BF, Carey RJ, Cashman KV, Wilson CJN, Hobden BJ, Hammer JE (2010) Diverse patterns of ascent, degassing, and eruption of rhyolite magma during the 1.8 ka Taupo eruption, New Zealand: evidence from clast vesicularity. J Volcanol Geotherm Res 195(1):31–47

Houghton BF, Carey RJ, Rosenberg MD (2014) The 1800a Taupo eruption: "Ill wind" blows the ultraplinian type event down to Plinian. Geology

Houghton BF, Wilson CJN, Fierstein J, Hildreth W (2004) Complex proximal deposition during the Plinian eruptions of 1912 at Novarupta, Alaska. Bull Volcanol 66(2):95–133

Hsu KJ (1975) Catastrophic debris streams (sturzstroms) generated by rockfalls. Geol Soc Am Bull 86(1):129–140

Hudak MR, Bindeman IN (2018) Conditions of pinnacle formation and glass hydration in cooling ignimbrite sheets from H and O isotope systematics at Crater Lake and the Valley of Ten Thousand Smokes. Earth Planet Sci Lett 500:56–66

Hudspith VA, Scott AC, Wilson CJN, Collinson ME (2010) Charring of woods by volcanic processes: an example from the Taupo ignimbrite, New Zealand. Palaeogeogr Palaeoclim Palaeoecol 291 (1–2):40–51

Incoronato A, Addison FT, Tarling DH, Nardi G, Pescatore T (1983) Magnetic fabric investigations of pyroclastic deposits from Phlegrean Fields, southern Italy. Nature 306(5942):461–463

Jenkins S, Komorowski JC, Baxter PJ, Spence R, Picquout A, Lavigne F (2013) The Merapi 2010 eruption: an interdisciplinary impact assessment methodology for studying pyroclastic density current dynamics. J Volcanol Geotherm Res 261:316–329

Jessop DE, Gilchrist J, Jellinek AM, Roche O (2016) Are eruptions from linear fissures and caldera ring dykes more likely to produce pyroclastic flows? Earth Planet Sci Lett 454:142–153

Jessop DE, Kelfoun K, Labazuy P, Mangeney A, Roche O, Tillier JL, Trouillet M, Thibault G (2012) LiDAR derived morphology of the 1993 Lascar pyroclastic flow deposits, and implication for flow dynamics and rheology. J Volcanol Geotherm Res 245–246:81–97

Johnson AM (1970) Physical processes in geology. Freeman, Cooper, San Francisco, p 577p

Jordan SC, Cas RAF, Hayman PC (2013) The origin of a large (>3 km) maar volcano by coalescence of multiple shallow craters: Lake Purrumbete maar, southeastern Australia. J Volcanol Geotherm Res 254:5–22

Kamata H, Mimura K (1983) Flow directions inferred from imbrication in the Handa pyroclastic flow deposit in Japan. Bull Volcanol 46 (3):277–282

Kaneko K, Kamata H, Koyaguchi T, Yoshikawa M, Furukawa K (2007) Repeated large-scale eruptions from a single compositionally stratified magma chamber: an example from Aso volcano, Southwest Japan. J Volcanol Geotherm Res 167(1–4):160–180

Keating GN (2005) The role of water in cooling ignimbrites. J Volcanol Geotherm Res 142(1–2):145–171

Kelfoun K, Samaniego P, Palacios P, Barba D (2009) Testing the suitability of frictional behaviour for pyroclastic flow simulation by comparison with a well-constrained eruption at Tungurahua volcano (Ecuador). Bull Volcanol 71:1057–1075

Kent PE (1966) The transport mechanism in catastrophic rock falls. J Geol 74(1):79–83

Kieffer SW (1981) Fluid dynamics of the May 18 blast at Mount St. Helens. In: Lipman PW, Mullineaux DR (eds) The 1980 eruptions of Mount St. Helens, Washington. US Geol Surv Prof Pap 1250:379–401

Kieffer SW, Sturtevant B (1984) Laboratory studies of volcanic jets. J Geophys Res: Solid Earth 89(B10):8253–8268

Kieffer SW, Sturtevant B (1988) Erosional furrows formed during the lateral blast at Mount St. Helens, May 18, 1980. J Geophys Res: Solid Earth 93(B12):14793–14816

Kienle J, Kyle PR, Self S, Motyka RJ, Lorenz V (1980) Ukinrek Maars, Alaska, I. April 1977 eruption sequence, petrology and tectonic setting. J Volcanol Geother Res 7(1–2):11–37

Knight MD, Walker GPL, Ellwood BB, Diehl JF (1986) Stratigraphy, paleomagnetism, and magnetic fabric of the Toba Tuffs: constraints on the sources and eruptive styles. J Geophys Res: Solid Earth 91 (B10):10355–10382

Komorowski JC, Jenkins S, Baxter PJ, Picquout A, Lavigne F, Charbonnier SJ, Gertisser R, Preece K, Cholik N, Budi-Santoso A, Surono, (2013) Paroxysmal dome explosion during the Merapi 2010 eruption: processes and facies relationships of associated high-energy pyroclastic density currents. J Volcanol Geotherm Res 261:260–294

Koralay T, Kadioglu YK (2008) Reasons of different colors in the ignimbrite lithology: micro-XRF and confocal Raman spectrometry method. Spectrochimica Acta Pt a: Mol Biomol Spectrosc 69 (3):947–955

Kueppers U, Putz C, Spieler O, Dingwell DB (2012) Abrasion in pyroclastic density currents: insights from tumbling experiments. Phys Chem Earth Parts a/b/c 45–46:33–39

Kuno H, Ishikawa T, Katsui Y, Yago Y, Yamasaki M, Taneda S (1964) Sorting of pumice and lithic fragments as a key eruptive and emplace¬ment mechanisms. J Geol Geogr 35:223–238

Lacroix A (1904) La Montagne Pelee et ses eruptions. Masson, Paris

LaBerge RD, Giordano G, Cas RAF, Ailleres L (2006) Syn-depositional substrate deformation produced by the shear force of a pyroclastic density current: an example from the Pleistocene ignimbrite at Monte Cimino, northern Lazio, Italy. J Volcanol Geotherm Res 158(3–4):307–320

LaBerge RD, Porreca M, Mattei M, Giordano G, Cas RAF (2009) Meandering flow of a pyroclastic density current documented by the anisotropy of magnetic susceptibility (AMS) in the quartz latite ignimbrite of the Pleistocene Monte Cimino volcanic centre (central Italy). Tectonophysics 466(1–2):64–78

Langella A, De Simone P, Calcaterra P, de'Gennaro M (2002) Evidence of the relationship occurring between zeolitisation and lithification in the yellow facies of the Campanian ignimbrite (southern Italy). Stud Surf Sci Catal 142:1775–1782

Lavallée Y, Wadsworth FB, Vasseur J, Russell JK, Andrews GD, Hess KU, von Aulock FW, Kendrick JE, Tuffen H, Biggin AJ, Dingwell DB (2015) Eruption and emplacement timescales of ignimbrite super-eruptions from thermo-kinetics of glass shards. Front Earth Sci 3:2

Lavigne F, Degeai JP, Komorowski JC, Guillet S, Robert V, Lahitte P, Oppenheimer C, Stoffel M, Vidal CM, Pratomo I, Wassmer P (2013) Source of the great AD 1257 mystery eruption unveiled, Samalas volcano, Rinjani Volcanic Complex, Indonesia. Proc Nat Acad Sci 110(42):16742–16747

Le Pennec J-L (2000) Identifying ash flow sources with directional data: an application to the Kizilkaya ignimbrite, central Anatolia. J Geophys Res: Solid Earth 105(B12):28427–28441

Le Pennec J-L, Chen Y, Diot H, Froger J-C, Gourgaud A (1998) Interpretation of anisotropy of magnetic susceptibility fabric of ignimbrites in terms of kinematic and sedimentological mechanisms: an Anatolian case-study. Earth Planet Sci Lett 157(1):105–127

Le Pennec J-L, Temel A, Froger JL, Sen S, Gourgaud A, Bourdier JL (2005) Stratigraphy and age of the Cappadocia ignimbrites, Turkey: reconciling field constraints with paleontologic, radiochronologic, geochemical and paleomagnetic data. J Volcanol Geotherm Res 141 (1–2):45–64

Legros F, Kelfoun K (2000) On the ability of pyroclastic flows to scale topographic obstacles. J Volcanol Geoth Res 98(1–4):235–241

Lesti C, Porreca M, Giordano G, Mattei M, Cas RAF, Wright HMN, Folkes CB, Viramonte JG (2011) High-temperature emplacement of the Cerro Galán and Toconquis Group ignimbrites (Puna plateau, NW Argentina) determined by TRM analyses. Bull Volcanol 73 (10):1535–1565

Lipman PW (1976) Caldera-collapse breccias in the western San Juan Mountains. Colorado. Geol Soc Am Bull 87(10):1397–1410

Lipman PW (2007) Incremental assembly and prolonged consolidation of Cordilleran magma chambers: evidence from the Southern Rocky Mountain volcanic field. Geosphere 3(1):42–70

Lipman P, Christiansen R (1964) In the Piapi Canyon Formation, Southern Nevada. Geol Surv Res 1964: Chapter B, 74

Long PE, Wood BJ (1986) Structures, textures, and cooling histories of Columbia River basalt flows. Geol Soc Amer Bull 97(9):1144–1155

Loughlin SC, Calder ES, Clarke A, Cole PD, Luckett R, Mangan MT, Pyle DM, Sparks RSJ, Voight B, Watts RB (2002) Pyroclastic flows and surges generated by the 25 June 1997 dome collapse, Soufrière Hills Volcano, Montserrat. Geol Soc Lond Memoirs 21(1):191–209

Lowe DR (1982) Sediment gravity flows; II, Depositional models with special reference to the deposits of high-density turbidity currents. J Sed Res 52(1):279–297

Lube G, Breard ECP, Cronin SJ, Jones J (2015) Synthesizing large-scale pyroclastic flows: experimental design, scaling, and first results from PELE. J Geophys Res: Solid Earth 120(3):1487–1502

Lube G, Breard EC, Jones J, Fullard L, Dufek J, Cronin SJ, Wang T (2019) Generation of air lubrication within pyroclastic density currents. Nat Geosci 12(5):381–386

Macdonald GA, Alcaraz A (1956) Nuées ardentes of the 1948–1953 eruption of Hibok-Hibok. Bull Volcanol 18(1):169–178

MacDonald WD, Palmer HC (1990) Flow directions in ash-flow tuffs: a comparison of geological and magnetic susceptibility measurements, Tshirege member (upper Bandelier Tuff), Valles caldera, New Mexico, USA. Bull Volcanol 53(1):45–59

Mackaman-Lofland C, Brand BD, Taddeucci J, Wohletz K (2014) Sequential fragmentation/transport theory, pyroclast size–density relationships, and the emplacement dynamics of pyroclastic density currents—a case study on the Mt. St. Helens (USA) 1980 eruption. J Volcanol Geotherm Res 275:1–13

Mandeville CW, Carey S, Sigurdsson H, King J (1994) Paleomagnetic evidence for high-temperature emplacement of the 1883 subaqueous pyroclastic flows from Krakatau Volcano, Indonesia. J Geophys Res: Solid Earth 99(B5):9487–9504

Manga M, Patel A, Dufek J (2011) Rounding of pumice clasts during transport: field measurements and laboratory studies. Bull Volcanol 73(3):321–333

Marshall AA, Brand BD, Martínez V, Bowers JM, Walker M, Wanless VD, Andrews BJ, Manga M, Valdivia P, Giordano G (2022) The mafic Curacautín ignimbrite of Llaima volcano, Chile. J Volcanol Geother Res 421:107418

Marshall P (1935) Acid rocks of the Taupo-Rotorua volcanic district. Trans Roy Soc NZ 64:323–366

Martí A, Folch A, Costa A, Engwell S (2016) Reconstructing the plinian and co-ignimbrite sources of large volcanic eruptions: a novel approach for the Campanian Ignimbrite. Sci Rep 6:21220

Martí J, Diez-Gil JL, Ortiz R (1991) Conduction model for the thermal influence of lithic clasts in mixtures of hot gases and ejecta. J Geophys Res: Solid Earth 96(B13):21879–21885

Martin RC (1959) Some field and petrographic features of American and New Zealand ignimbrites. N Z J Geol Geophys 2(2):394–411

Martinez MML, Williams SN (1999) Basaltic andesite to andesite scoria pyroclastic flow deposits from Taal caldera, Philippines. J Geol Soc Philipp 54(1&2):1–18

Mason BG, Pyle DM, Oppenheimer C (2004) The size and frequency of the largest explosive eruptions on Earth. Bull Volcanol 66 (8):735–748

Mastrolorenzo G, Petrone P, Pappalardo L, Guarino FM (2010) Lethal thermal impact at periphery of pyroclastic surges: evidences at Pompeii. PLoS ONE 5

Maughan LL, Christiansen EH, Best MG, Grommé CS, Deino AL, Tingey DG (2002) The Oligocene Lund Tuff, Great Basin, USA: a very large volume monotonous intermediate. J Volcanol Geotherm Res 113(1–2):129–157

McClelland E, Wilson CJN, Bardot L (2004) Palaeotemperature determinations for the 1.8-ka Taupo ignimbrite, New Zealand, and implications for the emplacement history of a high-velocity pyroclastic flow. Bull Volcanol 66(6):492–513

McClelland EA, Druitt TH (1989) Palaeomagnetic estimates of emplacement temperatures of pyroclastic deposits on Santorini, Greece. Bull Volcanol 51(1):16–27

McKee CO, Baillie MG, Reimer PJ (2015) A revised age of AD 667–699 for the latest major eruption at Rabaul. Bull Volcanol 77(7):65

McPhie J (1986) Primary and redeposited facies from a large-magnitude, rhyolitic, phreatomagmatic eruption: Cana Creek Tuff, Late Carboniferous, Australia. J Volcanol Geother Res 28(3–4):319–350

Mellors RA, Sparks RSJ (1991) Spatter-rich pyroclastic flow deposits on Santorini, Greece. Bull Volcanol 53(5):327–342

Michol KA, Russell JK, Andrews GDM (2008) Welded block and ash flow deposits from Mount Meager, British Columbia, Canada. J Volcanol Geotherm Res 169(3–4):121–144

Middleton GV, Southard JB (1984) Mechanics of sediment movement. SEPM Press

Miller TP, Smith RL (1977) Spectacular mobility of ash flows around Aniakchak and Fisher calderas, Alaska. Geology 5(3):173–176

Milner DM, Cole JW, Wood CP (2003) Mamaku Ignimbrite: a caldera-forming ignimbrite erupted from a compositionally zoned magma chamber in Taupo Volcanic Zone, New Zealand. J Volcanol Geotherm Res 122(3–4):243–264

Mimura K (1984) Imbrication, flow direction and possible source areas of the pumice-flow tuffs near Bend, Oregon, U.S.A. J Volcanol Geoth Res 21(1):45–60

Moore JG (1967) Base surge in recent volcanic eruptions. Bull Volcanol 30(1):337–363

Moore JG, Melson WG (1969) Nuées Ardentes of the 1968 eruption of Mayon Volcano. Philippines. Bull Volcanol 33(2):600–620

Moore JG, Peck DL (1962) Accretionary Lapilli in volcanic rocks of the Western Continental United States. J Geol 70(2):182–193

Moore JG, Sisson TW (1981) Deposits and effects of the May 18 pyroclastic surge. In: Lipman PW, Mullineaux DR (eds) The 1980 eruptions of Mount St. Helens, Washington. US Geol Surv Prof Pap 1250:421–438

Morelli C, Marocchi M, Moretti A, Bargossi GM, Gasparotto G, Waele BD, Klötzli U, Mair V (2012) Volcanic stratigraphy and radiometric age constraints at the northern margin of a mega-caldera system: Athesian Volcanic Group (Southern Alps, Italy). GeoActa 11:1–43

Morgavi D, Valentini L, Porreca M, Zucchini A, Di Michele A, Ielpo M, Perugini D (2019) Volcanic ash aggregation enhanced by seawater interaction: the case of the Secche di Lazzaro phreatomagmatic deposit (Stromboli). Ann Geophys 61(6 Sup):673

Mormone A, Piochi M (2020) Mineralogy, geochemistry and genesis of zeolites in Cenozoic pyroclastic flows from the Asuni area (Central Sardinia, Italy). Minerals 10(3):268

Mueller SB, Kueppers U, Ayris PM, Jacob M, Dingwell DB (2016) Experimental volcanic ash aggregation: internal structuring of accretionary lapilli and the role of liquid bonding. Earth Planet Sci Lett 433:232–240

Mundula F, Cioni R, Mulas M (2013) Rheomorphic diapirs in densely welded ignimbrites: the Serra di Paringianu ignimbrite of Sardinia, Italy. J Volcanol Geotherm Res 258:12–23

Mundula F, Cioni R, Rizzo R (2009) A simplified scheme for the description of textural features in welded Ignimbrites: the example of San Pietro Island (Sardinia, Italy). Boll Soc Geol It 128(3):615–627

Mutti E, Davoli G, Mora S, Papani L (1994) Internal stacking patterns of ancient turbidite systems from collisional basins. In: Submarine fans and turbidite systems: Gulf Coast Section SEPM proceedings, 15th annual research conference, pp 257–268

Mutti E, Sonnino M (1981) Compensation cycles: a diagnostic feature of turbidite sandstone lobes. Int. Ass. Sed. 2nd European meeting, Bologna, Abstracts, pp 120–123

Nairn IA (1979) Rotomahana—Waimangu eruption, 1886: base surge and basalt magma. NZ J Geol Geophys 22(3):363–378

Nairn IA, Self S (1978) Explosive eruptions and pyroclastic avalanches from Ngauruhoe in February 1975. J Volcanol Geotherm Res 3(1–2):39–60

Nakada S, Fujii T (1993) Preliminary report on the activity at Unzen Volcano (Japan), November 1990–November 1991: dacite lava domes and pyroclastic flows. J Volcanol Geoth Res 54(3):319–333

Nakada S, Shimizu H, Ohta K (1999) Overview of the 1990–1995 eruption at Unzen Volcano. J Volcanol Geotherm Res 89(1):1–22

Nakamura Y (1978) Geology and petrology of Bandai and Nekoma volcanoes. Sci Rep Tohoku UnivSer 3(14):67–119

Németh K, Cronin SJ (2007) Syn-and post-eruptive erosion, gully formation, and morphological evolution of a tephra ring in tropical climate erupted in 1913 in West Ambrym, Vanuatu. Geomorphology 86(1–2):115–130

Neri A, Esposti Ongaro T, Macedonio G, Gidaspow D (2003) Multiparticle simulation of collapsing volcanic columns and pyroclastic flow. J Geophys Res: Solid Earth 108(B4)

Newhall CG, Self S (1982) The volcanic explosivity index (VEI) an estimate of explosive magnitude for historical volcanism. J Geophys Res: Oceans 87(C2):1231–1238

Ninkovich D, Sparks RSJ, Ledbetter MT (1978) The exceptional magnitude and intensity of the Toba eruption, Sumatra: an example of the use of deep-sea tephra layers as a geological tool. Bull Volcanol 41:286–298

Ogburn SE, Calder ES, Cole PD, Stinton AJ (2014) The effect of topography on ash-cloud surge generation and propagation. Geol Soc Lond Mem 39:179–194

Orsi G, Di Vito MA, Isaia R (2004) Volcanic hazard assessment at the restless Campi Flegrei caldera. Bull Volcanol 66(6):514–530

Ort MH (1993) Eruptive processes and caldera formation in a nested downsag collapse caldera: Cerro Panizos, central Andes Mountains. J Volcanol Geotherm Res 56(3):221–252

Ort MH, Carrasco-Núñez G (2009) Lateral vent migration during phreatomagmatic and magmatic eruptions at Tecuitlapa Maar, east-central Mexico. J Volcanol Geotherm Res 181(1–2):67–77

Ort MH, Lefebvre NS, Neal CA, McConnell VS, Wohletz KH (2018) Linking the Ukinrek 1977 maar-eruption observations to the tephra deposits: new insights into maar depositional processes. J Volcanol Geother Res 360:36–60

Ort MH, Orsi G, Pappalardo L, Fisher RV (2003) Anisotropy of magnetic susceptibility studies of depositional processes in the Campanian Ignimbrite, Italy. Bull Volcanol 65(1):55–72

Ort MH, Rosi M, Anderson CD (1999) Correlation of deposits and vent locations of the proximal Campanian Ignimbrite deposits, Campi Flegrei, Italy, based on natural remanent magnetization and anisotropy of magnetic susceptibility characteristics. J Volcanol Geotherm Res 91(2):167–178

Pacheco-Hoyos JG, Aguirre-Díaz GJ, Dávila-Harris P (2020) Elutriation pipes in ignimbrites: an analysis of concepts based on the Huichapan Ignimbrite, Mexico. J Volcanol Geotherm Res 403:107026

Palladino DM, Giordano G (2019) On the mobility of pyroclastic currents in light of deposit thickness and clast size trends. J Volcanol Geotherm Res 384:64–74

Palladino DM, Simei S (2005) Eruptive dynamics and caldera collapse during the Onano eruption, Vulsini, Italy. Bull Volcanol 67(5):423–440

Palladino DM, Valentine GA (1995) Coarse-tail vertical and lateral grading in pyroclastic flow deposits of the Latera Volcanic Complex (Vulsini, central Italy): origin and implications for flow dynamics. J Volcanol Geotherm Res 69(3):343–364

Pallister JS, Clynne MA, Wright HM, Van Eaton AR, Vallance JW, Sherrod DR, Kokelaar BP (2017) Field-trip guide to Mount St. Helens, Washington-An overview of the eruptive history and petrology, tephra deposits, 1980 pyroclastic density current deposits, and the crater. USGS Scientific Investigations Rept (2017–5022-D)

Papale P, Dobran F (1994) Magma flow along the volcanic conduit during the Plinian and pyroclastic flow phases of the May 18, 1980, Mount St. Helens eruption. J Geophys Res: Solid Earth 99 (B3):4355–4373

Pardo N, Macias JL, Giordano G, Cianfarra P, Avellán DR, Bellatreccia F (2009) The ~1245 yr BP Asososca maar eruption: the youngest event along the Nejapa-Miraflores volcanic fault, Western Managua, Nicaragua. J Volcanol Geotherm Res 184(3–4):292–312

Paterson GA, Roberts AP, Mac Niocaill C, Muxworthy AR, Gurioli L, Viramonté JG, Navarro C, Weider S (2010) Paleomagnetic determination of emplacement temperatures of pyroclastic deposits: an under-utilized tool. Bull Volcanol 72(3):309–330

Pedrazzi D, Martí J, Geyer A (2013) Stratigraphy, sedimentology and eruptive mechanisms in the tuff cone of El Golfo (Lanzarote, Canary Islands). Bull Volcanol 75(7):1–7

Pensa A, Capra L, Giordano G (2019) Ash clouds temperature estimation. Implication on dilute and concentrated PDCs coupling and topography confinement. Sci Rep 9(1):1–14

Pensa A, Capra L, Giordano G, Corrado S (2018) Emplacement temperature estimation of the 2015 dome collapse of Volcán de Colima as key proxy for flow dynamics of confined and unconfined pyroclastic density currents. J Volcanol Geotherm Res 357:321–338

Pensa A, Cas RAF, Giordano G, Porreca M, Wallenstein N (2015a) Transition from steady to unsteady Plinian eruption column: the VEI 5, 4.6 ka Fogo A Plinian eruption, São Miguel, Azores. J Volcanol Geotherm Res 305:1–18

Pensa A, Giordano G, Cas RAF, Porreca M (2015b) Thermal state and implications for eruptive styles of the intra-Plinian and climactic ignimbrites of the 4.6 ka Fogo A eruption sequence, São Miguel, Azores. Bull Volcanol 77(11):1–19

Pensa A, Giordano G, Corrado S, Petrone PP (2023) A new hazard scenario at Vesuvius: deadly thermal impact of detached ash cloud surges in 79CE at Herculaneum. Sci Rep 13:5622. https://doi.org/10.1038/s41598-023-32623-3

Pensa A, Porreca M, Corrado S, Giordano G, Cas RAF (2015c) Calibrating the pTRM and charcoal reflectance (Ro%) methods to determine the emplacement temperature of ignimbrites: Fogo A sequence, São Miguel, Azores, Portugal, as a case study. Bull Volcanol 77(3):1–19

Perret FA (1935) The eruption of Mt. Pelée, 1929–1932. Carnegie Institution of Washington

Petrone P, Giordano G, Vezzoli E, Pensa A, Castaldo G, Graziano V, Sirano F, Capasso E, Quaremba G, Vona A, Miano MG (2020b) Preservation of neurons in an AD 79 vitrified human brain. PLoS ONE 15(10):e0240017

Petrone P, Pucci P, Niola M, Baxter PJ, Fontanarosa C, Giordano G, Graziano V, Sirano F, Amoresano A (2020a) Heat-induced brain vitrification from the Vesuvius eruption in 79CE. New Engl J Med 382(4):383–384

Pimentel A, Self S, Pacheco JM, Jeffery AJ, Gertisser R (2021) Eruption style, emplacement dynamics and geometry of peralkaline ignimbrites: insights from the Lajes-Angra Ignimbrite Formation, Terceira Island, Azores. Front Earth Sci 9:503

Pioli L, Lanza R, Ort M, Rosi M (2008) Magnetic fabric, welding texture and strain fabric in the Nuraxi Tuff, Sardinia, Italy. Bull Volcanol 70(9):1123–1137

Pioli L, Rosi M (2005) Rheomorphic structures in a high-grade ignimbrite: the Nuraxi tuff, Sulcis volcanic district (SW Sardinia, Italy). J Volcanol Geotherm Res 142(1):11–28

Pittari A, Cas RAF (2004) Sole marks at the base of the late Pleistocene Abrigo Ignimbrite, Tenerife: implications for transport and depositional processes at the base of pyroclastic flows. Bull Volcanol 66 (4):356–363

Pittari A, Cas RAF, Edgar CJ, Nichols HJ, Wolff JA, Martí J (2006) The influence of palaeotopography on facies architecture and pyroclastic flow processes of a lithic-rich ignimbrite in a high gradient setting: the Abrigo Ignimbrite, Tenerife, Canary Islands. J Volcanol Geotherm Res 152(3–4):273–315

Pittari A, Cas RAF, Martí J (2005) The occurrence and origin of prominent massive, pumice-rich ignimbrite lobes within the Late Pleistocene Abrigo Ignimbrite, Tenerife, Canary Islands. J Volcanol Geoth Res 139(3):271–293

Pittari A, Cas RAF, Monaghan JJ, Martí J (2007) Instantaneous dynamic pressure effects on the behaviour of lithic boulders in pyroclastic flows: the Abrigo Ignimbrite, Tenerife, Canary Islands. Bull Volcanol 69(3):265–279

Pittari A, Cas RAF, Wolff JA, Nichols HJ, Larson PB, Martí J (2008) The use of lithic clast distributions in pyroclastic deposits to understand pre-and syn-caldera collapse processes: a case study of the Abrigo Ignimbrite, Tenerife, Canary Islands. Dev Volcanol 10:97–142

Plenier G, Camps P, Henry B, Ildefonse B (2005) Determination of flow directions by combining AMS and thin-section analyses: implications for Oligocene volcanism in the Kerguelen Archipelago (southern Indian Ocean). Geophys J Int 160(1):63–78

Pollock NM, Brand BD, Rowley PJ, Sarocchi D, Sulpizio R (2019) Inferring pyroclastic density current flow conditions using syn-depositional sedimentary structures. Bull Volcanol 81:1–16

Porreca M, Giordano G, Mattei M, Musacchio P (2006) Evidence of two Holocene phreatomagmatic eruptions at Stromboli volcano (Aeolian Islands) from paleomagnetic data. Geophys Res Lett 33 (21)

Porreca M, Mattei M, Giordano G, De Rita D, Funiciello R (2003) Magnetic fabric and implications for pyroclastic flow and lahar emplacement, Albano maar, Italy. J Geophys Res: Solid Earth 108 (B5)

Porreca M, Mattei M, MacNiocaill C, Giordano G, McClelland E, Funiciello R (2008) Paleomagnetic evidence for low-temperature emplacement of the phreatomagmatic Peperino Albano ignimbrite (Colli Albani volcano, Central Italy). Bull Volcanol 70(7):877–893

Postma G (1983) Water escape structures in the context of a depositional model of a mass flow dominated conglomeratic fan-delta (Abrioja Formation, Pliocene, Almeria Basin, SE Spain). Sedimentology 30(1):91–103

Postma G, Nemec W, Kleinspehn KL (1988) Large floating clasts in turbidites: a mechanism for their emplacement. Sed Geol 58(1):47–61

Prata G (2012) Understanding cyclicity in the deposits of maar volcanoes, Tower Hill Volcanic Complex. MSc, Monash University, Victoria

Pyle DM (1989) The thickness, volume and grainsize of tephra fall deposits. Bull Volcanol 51(1):1–15

Quane SL, Russell JK (2005) Ranking welding intensity in pyroclastic deposits. Bull Volcanol 67(2):129–143

Ragan DM, Sheridan MF (1972) Compaction of the Bishop tuff, California. Geol Soc Am Bull 83(1):95–106

Randolph-Flagg N, Breen S, Hernandez A, Manga M, Self S (2017) Evenly spaced columns in the Bishop Tuff (California, USA) as relicts of hydrothermal cooling. Geology 45(11):1015–1018

Reedman AJ, Park KH, Merriman RJ, Kim SE (1987) Welded tuff infilling a volcanic vent at Weolseong, Republic of Korea. Bull Volcanol 49:541–546

Rhodes RC, Smith EI (1972) Distribution and directional fabric of ash-flow sheets in the northwestern Mogollon Plateau, New Mexico. Geol Soc Am Bull 83(6):1863–1868

Riehle JR (1973) Calculated compaction profiles of rhyolitic ash-flow tuffs. Geol Soc Am Bull 84(7):2193–2216

Riehle JR, Miller TF, Bailey RA (1995) Cooling, degassing and compaction of rhyolitic ash flow tuffs: a computational model. Bull Volcanol 57(5):319–336

Risica G, Rosi M, Pistolesi M, Speranza F, Branney MJ (2022) Deposit-derived block-and-ash flows: the hazard posed by perched temporary tephra accumulations on volcanoes; 2018 Fuego Disaster, Guatemala. J Geophys Res: Solid Earth 127(6):e2021JB023699

Ritchie LJ, Cole PD, Sparks RSJ (2002) Sedimentology of deposits from the pyroclastic density current of 26 December 1997 at Soufrière Hills Volcano, Montserrat. Geol Soc Lond Memoirs 21 (1):435–456

Roche O (2012) Depositional processes and gas pore pressure in pyroclastic flows: an experimental perspective. Bull Volcanol 74:1807–1820

Roche O (2015) Nature and velocity of pyroclastic density currents inferred from models of entrainment of substrate lithic clasts. Earth Planet Sci Lett 418:115–125

Roche O, Attali M, Mangeney A, Lucas A (2011) On the run-out distance of geophysical gravitational flows: Insight from fluidized granular collapse experiments. Earth Planet Sci Lett 311(3):375–385

Roche O, Azzaoui N, Guillin A (2021) Discharge rate of explosive volcanic eruption controls runout distance of pyroclastic density currents. Earth Planet Sci Lett 568. https://doi.org/10.1016/j.epsl.2021.117017

Roche O, Buesch DC, Valentine GA (2016) Slow-moving and far-travelled dense pyroclastic flows during the Peach Spring super-eruption. Nat Commun 7

Roche O, De Vries BVW, Druitt TH (2001b) Sub-surface structures and collapse mechanisms of summit pit craters. J Volcanol Geotherm Res 105(1):1–18

Roche O, Druitt TH (2001) Onset of caldera collapse during ignimbrite eruptions. Earth Planet Sci Lett 191(3–4):191–202

Roche O, Druitt TH, Cas RAF (2001a) Experimental aqueous fluidization of ignimbrite. J Volcanol Geotherm Res 112(1–4):267–280

Roche O, Druitt TH, Merle O (2000) Experimental study of caldera formation. J Geophys Res: Solid Earth 105(B1):395–416

Roche O, Gilbertson MA, Phillips JC, Sparks RSJ (2006) The influence of particle size on the flow of initially fluidised powders. Pow Techn 166(3):167–174

Roche O, Henry CD, Azzaoui N, Guillin A (2022) Long-runout pyroclastic density currents: analysis and implications. Geology 50 (10):1172–1176

Roche O, Montserrat S, Niño Y, Tamburrino A (2008) Experimental observations of water-like behavior of initially fluidized, dam break granular flows and their relevance for the propagation of ash-rich pyroclastic flows. J Geophys Res: Solid Earth 113(B12)

Roche O, Montserrat S, Niño Y, Tamburrino A (2010) Pore fluid pressure and internal kinematics of gravitational laboratory air-particle flows: insights into the emplacement dynamics of pyroclastic flows. J Geophys Res: Solid Earth 115(B9)

Romano C, Vona A, Campagnola S, Giordano G, Arienzo I, Isaia R (2020) Modelling and physico-chemical constraints to the 4.5 ka Agnano-Monte Spina Plinian eruption (Campi Flegrei, Italy). Chem Geol 532 119301

Romero JE, Morgavi D, Arzilli F, Daga R, Caselli A, Reckziegel F, Perugini D (2016) Eruption dynamics of the 22–23 April 2015 Calbuco Volcano (Southern Chile): analyses of tephra fall deposits. J Volcanol Geotherm Res 317:15–29

Roobol MJ, Smith AL (1975). A comparison of the recent eruptions of Mt. Pelée, Martinique and Soufrière, St. Vincent. Bull Volcanol 39 (2):214–240

Roobol MJ, Smith AL, Wright JV (1987) Lithic breccias in pyroclastic flow deposits on St. Kitts, West Indies. Bull Volcanol 49(5):694–707

Rose WI, Chesner CA (1987) Dispersal of ash in the great Toba eruption, 75 ka. Geology 15(10):913–917

Rose WI, Grant NK, Hahn GA, Lange IM, Powell JL, Easter J, Degraff JM (1977) The evolution of Santa Maria volcano, Guatemala. J Geol 85(1):63–87

Rose WI Jr, Pearson T, Bonis S (1976) Nuee ardente eruption from the foot of a dacite lava flow, Santiaguito Volcano, Guatemala. Bull Volcanol 40(1):23–38

Rosi M, Vezzoli L, Aleotti P, De Censi M (1996) Interaction between caldera collapse and eruptive dynamics during the Campanian Ignimbrite eruption, Phlegraean Fields, Italy. Bull Volcanol 57 (7):541–554

Ross CS, Smith RL (1961) Ash-flow tuffs: their origin, geologic relations, and identification. United States Government Printing Office

Rowley PD, Kuntz MA, McLeod NS (1981) Pyroclastic flow deposit. In: Lipman PW, Mullineaux DR (eds) The 1980 eruption of Mount St. Helens, Washington. US Geol Surv Prof Pap 489–512

Rowley PJ, Roche O, Druitt TH, Cas RAF (2014) Experimental study of dense pyroclastic density currents using sustained, gas-fluidized granular flows. Bull Volcanol 76(9):1–13

Sarikaya MA, Çiner A, Zreda M (2015) Fairy chimney erosion rates on Cappadocia ignimbrites, Turkey: insights from cosmogenic nuclides. Geomorphology 234:182–191

Sarocchi D, Sulpizio R, Macías JL, Saucedo R (2011) The 17 July 1999 block-and-ash flow (BAF) at Colima Volcano: new insights on volcanic granular flows from textural analysis. J Volcanol Geotherm Res 204(1):40–56

Sato H, Fujii T, Nakada S (1992) Crumbling of dacite dome lava and generation of pyroclastic flows at Unzen volcano. Nature 360:664–666

Saucedo R, Macías JL, Bursik M (2004) Pyroclastic flow deposits of the 1991 eruption of Volcán de Colima, Mexico. Bull Volcanol 66 (4):291–306

Scarpati C, Sparice D, Perrotta A (2014) A crystal concentration method for calculating ignimbrite volume from distal ash-fall deposits and a reappraisal of the magnitude of the Campanian Ignimbrite. J Volcanol Geotherm Res 280:67–75

Scarpati C, Sparice D, Perrotta A (2015) The ground layer of the Campanian Ignimbrite: an example of deposition from a dilute pyroclastic density current. Bull Volcanol 77:1–10

Schito A, Pensa A, Romano C, Corrado S, Vona A, Trolese M, Morgavi D, Giordano G (2022) Calibrating carbonization temperatures of woodfragments embedded within pyroclastic density currents through Raman spectroscopy. Minerals 12(2):203

Schmincke H-U, Fisher RV, Waters AC (1973) Antidune and chute and pool structures in the base surge deposits of the Laacher See area, Germany. Sedimentology 20(4):553–574

Schmincke H-U, Swanson DA (1967) Laminar viscous flowage structures in ash-flow tuffs from Gran Canaria, Canary Islands. J Geol 641–644

Schumacher R, Schmincke H-U (1995) Models for the origin of accretionary lapilli. Bull Volcanol 56(8):626–639

Scott AC, Glasspool IJ (2005) Charcoal reflectance as a proxy for the emplacement temperature of pyroclastic flow deposits. Geology 33 (7):589–592

Scott AC, Sparks RSJ, Bull ID, Knicker H, Evershed RP (2008) Temperature proxy data and their significance for the understanding of pyroclastic density currents. Geology 36(2):143–146

Scott WE, Hoblitt RP, Torres RC, Self S, Martinez MML, Nillos TJ (1996) Pyroclastic flows of the June 15, 1991, climactic eruption of Mount Pinatubo. In: Newhall CG, Punongbayan R (eds) Fire and mud: eruptions and lahars of Mount Pinatubo. University of Washington Press, Philippines, pp 545–570

Seaman SJ, McIntosh WC, Geissman JW, Williams ML, Elston WE (1991) Magnetic fabrics of the Bloodgood Canyon and Shelley Peak Tuffs, southwestern New Mexico: implications for emplacement and alteration processes. Bull Volcanol 53(6):460–476

Self S (1983) Large-scale phreatomagmatic silicic volcanism: a case study from New Zealand. J Volcanol Geotherm Res 17(1–4):433–469

Self S (1992) Krakatau revisited: the course of events and interpretation of the 1883 eruption. GeoJournal 28(2):109–121

Self S, Gertisser R, Thordarson T, Rampino MR, Wolff JA (2004) Magma volume, volatile emissions, and stratospheric aerosols from the 1815 eruption of Tambora. Geophys Res Lett 31(20)

Self S, Goff F, Gardner JN, Wright JV, Kite WM (1986) Explosive rhyolitic volcanism in the Jemez Mountains: vent locations, caldera development and relation to regional structure. J Geophys Res 91 (B2):1779–1798

Self S, Heiken G, Sykes ML, Wohletz K, Fisher RV, Dethier DP (1996) Part I: Field guide to the Bandelier Tuff and Valles caldera. Field excursions to the Jemez Mountains, New Mexico. New Mexico Bur Mines Min Res Bull 134:7–55

Self S, Rampino MR (1981) The 1883 eruption of Krakatau. Nature 294(5843):699–704

Self S, Rampino MR, Barbera JJ (1981) The possible effects of large 19th and 20th century volcanic eruptions on zonal and hemispheric surface temperatures. J Volcanol Geotherm Res 11(1):41–60

Self S, Rampino MR, Newton MS, Wolff JA (1984) Volcanological study of the great Tambora eruption of 1815. Geology 12(11):659–663

Self S, Randolph-Flagg N, Bailey JE, Manga M (2022) Exposed columns in the Valles Caldera ignimbrites as records of hydrothermal cooling, Jemez Mountains, New Mexico, USA. J Volcanol Geotherm Res 426:107536

Self S, Wolff JA, Wright JV (2021) Tsankawi pumice fall unit B, a very widespread and powerfully emplaced plinian deposit. In: AGU Fall meeting, New Orleans. https://agu.confex.com/agu/fm21/meetingapp.cgi/Paper/911876

Shea T, Gurioli L, Houghton BF, Cioni R, Cashman KV (2011) Column collapse and generation of pyroclastic density currents during the AD 79 eruption of Vesuvius: the role of pyroclast density. Geology 39(7):695–698

Sheridan MF (1970) Fuarmolic Mounds and Ridges of the Bishop Tuff, California. Geol Soc Am Bull 81(3):851–868

Sheridan MF (1979) Emplacement of pyroclastic flows: a review. Geol Soc Am Spec Pap 180:125–136

Sheridan MF (1980) Pyroclastic block flow from the September, 1976, eruption of La Soufrière volcano, Guadeloupe. Bull Volcanol 43 (2):397–402

Sheridan MF, Barberi F, Rosi M, Santacroce R (1981) A model for Plinian eruptions of Vesuvius. Nature 289:282–285

Sheridan MF, Ragan DM (1976) Compaction of ash-flow tuffs. In: Chilingarian GV, Wolf KH (eds) Compaction of coarse-grained sediments. Develop. in Sedimentol. Elsevier, Amsterdam, pp 677–713

Sheridan MF, Updike RG (1975) Sugarloaf Mountain Tephra—a Pleistocene rhyolitic deposit of base-surge origin in northern Arizona. Geol Soc Am Bull 86(4):571–581

Shimizu HA, Koyaguchi T, Suzuki YJ (2019) The run-out distance of large-scale pyroclastic density currents: a two-layer depth-averaged model. J Volcanol Geotherm Res 381:168–184

Shimizu HA, Koyaguchi T, Suzuki YJ, Brosch E, Lube G, Cerminara M (2021) Validation of a two-layer depth-averaged model by comparison with an experimental dilute stratified pyroclastic density current. Bull Volcanol 83(11):1–12

Siebert L (1984) Large volcanic debris avalanches: characteristics of source areas, deposits, and associated eruptions. J Volcanol Geotherm Res 22(3–4):163–197

Sigurdsson H, Carey S (1989) Plinian and co-ignimbrite tephra fall from the 1815 eruption of Tambora volcano. Bull Volcanol 51 (4):243–270

Sigurdsson H, Carey S, Cornell W, Pescatore T (1985) The eruption of Vesuvius in AD 79. Nat Geogr Res 1(3):332–387

Sigurdsson H, Carey SN, Fisher RV (1987) The 1982 eruptions of El Chichon volcano, Mexico (3): physical properties of pyroclastic surges. Bull Volcanol 49(2):467–488

Silleni A, Giordano G, Isaia R, Ort MH (2020) The magnitude of the 39.8 ka Campanian ignimbrite eruption, Italy: method, uncertainties and errors. Front Earth Sci 8:543399

Silleni A, Giordano G, Ort MH, Isaia R (2024) The transport and deposition of the 39.8 ka Campanian Ignimbrite large scale pyroclastic density currents (Italy). Geol Soc Am Bull. https://doi.org/10.1130/B37500.1

Simmons JM, Cas RAF, Druitt TH, Folkes CB (2016) Complex variations during a caldera-forming Plinian eruption, including precursor deposits, thick pumice fallout, co-ignimbrite breccias and climactic lag breccias: the 184 ka Lower Pumice 1 eruption sequence, Santorini, Greece. J Volcanol Geotherm Res 324:200–219

Smith G, Rowley P, Williams R, Giordano G, Trolese M, Silleni A, Parsons D, Capon S (2020) A bedform phase diagram for dense granular currents. Nat Comm 11(1):1–1

Smith RL (1960a) Ash flows. Geol Soc Am Bull 71(6):795–841

Smith RL (1960b) Zones and zonal variations in welded ash flows. US Geol Surv Prof Pap 35FF:149–159

Smith RL (1979) Ash-flow magmatism. In: Chapin CE, Elston WE (eds) Ash flow tuffs. Geol Soc Am Spec Pap 180:5–27

Smith RL, Bailey RA (1966) The Bandelier Tuff: a study of ash-flow eruption cycles from zoned magma chambers. Bull Volcanol 29 (1):83–104

Smith RL, Bailey RA (1968) Resurgent cauldrons. In: Coats RR, Hay RL, Anderson CA (eds) Studies in volcanology (Howell Williams volume). Geol Soc Am Memoir 116:153–210

Smith RT, Houghton BF (1995) Vent migration and changing eruptive style during the 1800a Taupo eruption: new evidence from the Hatepe and Rotongaio phreatoplinian ashes. Bull Volcanol 57 (6):432–439

Smith VC, Isaia R, Engwell SL, Albert PG (2016) Tephra dispersal during the Campanian Ignimbrite (Italy) eruption: implications for ultra-distal ash transport during the large caldera-forming eruption. Bull Volcanol 78(6):1–15

Sohn YK, Chough SK (1989) Depositional processes of the Suwolbong tuff ring, Cheju Island (Korea). Sedimentology 36(5):837–855

Sparks RSJ (1975) Stratigraphy and geology of the ignimbrites of Vulsini Volcano, Central Italy. Geol Rundsch 64(1):497–523

Sparks RSJ (1976) Grain size variations in ignimbrites and implications for the transport of pyroclastic flows. Sedimentology 23(2):147–188

Sparks RSJ (1978) Gas release rates from pyroclastic flows: a assessment of the role of fluidisation in their emplacement. Bull Volcanol 41(1):1–9

Sparks RSJ, Barclay J, Calder ESH, R. A., Luckett R, Norton GE, Pollard L, Robertson RA, Ritchie L, Voight B, Young SR, Woods AW (2002) Generation of a debris avalanche and violent pyroclastic density current: the Boxing Day eruption of 26 December 1997 at the Soufrière Hills volcano, Montserrat. In: Druitt TH, Kokelaar BP (eds) The eruption of Soufrière Hills volcano, Montserrat, from 1995–1999. Geol Soc Lond Memoir 409–434

Sparks RSJ, Gardeweg MC, Calder ES, Matthews SJ (1997) Erosion by pyroclastic flows on Lascar Volcano, Chile. Bull Volcanol 58 (7):557–565

Sparks RSJ, Self S, Walker GPL (1973) Products of ignimbrite eruptions. Geology 1(3):115–118

Sparks RSJ, Tait SR, Yanev Y (1999) Dense welding caused by volatile resorption. J Geol Soc 156(2):217–225

Sparks RSJ, Voight B (2002) Small volume, highly mobile pyroclastic flows formed by rapid sedimentation from pyroclastic surges at Soufriere Hills Volcano, Montserrat: an important volcanic hazard. In: Druitt TH, Kokelaar BP (eds) The eruption of Soufriere Hills Volcano, Montserrat, from 1995 to 1999. Geol Soc Lond Memoir 263–279

Sparks RSJ, Walker GPL (1973) The ground surge deposit: a third type of pyroclastic rock. Nature 241(107):62–64

Sparks RSJ, Walker GPL (1977) The significance of vitric-enriched air-fall ashes associated with crystal-enriched ignimbrites. J Volcanol Geother Res 2(4):329–341

Sparks RSJ, Wilson L (1976) A model for the formation of ignimbrite by gravitational column collapse. J Geol Soc 132(4):441–451

Sparks RSJ, Wilson L, Hulme G (1978) Theoretical modeling of the generation, movement, and emplacement of pyroclastic flows by column collapse. J Geophys Res: Solid Earth 83(B4):1727–1739

Sparks RSJ, Wright JV (1979) Welded air-fall tuffs. In: Chapin, Elston (eds) Ash flow tuffs. Geol Soc Am Spec Pap 155–166

Spera FJ, Yuen DA, Greer JC, Sewell G (1986) Dynamics of magma withdrawal from stratified magma chambers. Geology 14(9):723–726

Spörli KB, Rowland JV (2006) 'Column on column'structures as indicators of lava/ice interaction, Ruapehu andesite volcano, New Zealand. J Volcanol Geotherm Res 157(4):294–310

Stimac J, Hickmott D, Abell R, Larocque ACL, Broxton D, Gardner JE, Chipera S, Wolff JA, Gauerke E (1996) Redistribution of Pb and other volatile trace metals during eruption, devitrification, and vapor-phase crystallization of the Bandelier Tuff, New Mexico. J Volcanol Geotherm Res 73(3):245–266

Stimac J, Nordquist G, Suminar A, Sirad-Azwar L (2008) An overview of the Awibengkok geothermal system, Indonesia. Geothermics 37 (3):300–331

Stoiber RE, Rose WI Jr (1969) Recent volcanic and fumarolic activity at Santiaguito Volcano, Guatemala. Bull Volcanol 33(2):475–502

Stothers RB (1984) The great Tambora eruption in 1815 and its aftermath. Science 224(4654):1191–1198

Streck MJ, Grunder AL (1995) Crystallization and welding variations in a widespread ignimbrite sheet; the Rattlesnake Tuff, eastern Oregon, USA. Bull Volcanol 57(3):151–169

Sulpizio R, Dellino P, Doronzo DM, Sarocchi D (2014) Pyroclastic density currents: state of the art and perspectives. J Volcanol Geotherm Res 283:36–65

Sulpizio R, Zanella E, Macías JL (2008) Deposition temperature of some PDC deposits from the 1982 eruption of El Chichon volcano

(Chiapas, Mexico) inferred from rock-magnetic data. J Volcanol Geotherm Res 175(4):494–500
Sumner JM, Branney MJ (2002) The emplacement history of a remarkable heterogeneous, chemically zoned, rheomorphic and locally lava-like ignimbrite:'TL'on Gran Canaria. J Volcanol Geotherm Res 115(1):109–138
Sun C, You H, Liu J, Li X, Gao J, Chen S (2014) Distribution, geochemistry and age of the Millennium eruptives of Changbaishan volcano, Northeast China—a review. Front Earth Sci 8(2):216–230
Suzuki K, Ui T (1982) Grain orientation and depositional ramps as flow direction indicators of a large-scale pyroclastic flow deposit in Japan. Geology 10(8):429–432
Suzuki YJ, Costa A, Cerminara M, Ongaro TE, Herzog M, Van Eaton AR, Denby LC (2016) Inter-comparison of three-dimensional models of volcanic plumes. J Volcanol Geother Res 326:26–42
Suzuki-Kamata K (1988) The ground layer of Ata pyroclastic flow deposit, southwestern Japan—evidence for the capture of lithic fragments. Bull Volcanol 50(2):119–129
Suzuki-Kamata K, Kamata H (1990) The proximal facies of the Tosu pyroclastic-flow deposit erupted from Aso caldera, Japan. Bull Volcanol 52(5):325–333
Suzuki-Kamata K, Kamata H, Bacon CR (1993) Evolution of the caldera-forming eruption at Crater Lake, Oregon, indicated by component analysis of lithic fragments. J Geophys Res: Solid Earth 98(B8):14059–14074
Talbot JP, Self S, Wilson CJN (1994) Dilute gravity current and rain-flushed ash deposits in the 1.8 ka Hatepe Plinian deposit, Taupo, New Zealand. Bull Volcanol 56(6–7):538–551
Taylor GAM (1958) The 1951 eruption of Mount Lamington, Papua. Australian Dept of National Developments Canberra, Bureau of Mineral Resources. Geology Geophys Bull (38)
Thomas RME, Sparks RSJ (1992) Cooling of tephra during fallout from eruption columns. Bull Volcanol 54:542–553
Tibaldi A, Bistacchi A, Pasquare FA, Vezzoli L (2006) Extensional tectonics and volcano lateral collapses: insights from Ollagüe volcano (Chile-Bolivia) and analogue modelling. Terra Nova 18 (4):282–289
Tibaldi A, Bonali FL (2018) A model to explain joint patterns found in ignimbrite deposits. Bull Volcanol 80(3):1–5
Toramaru A, Matsumoto T (2004) Columnar joint morphology and cooling rate: a starch-water mixture experiment. J Geophys Res: Solid Earth 109(B2)
Trolese M, Cerminara M, Giordano G, Brand BD, Pollock N, Cifelli F, Mattei M (2024) Thermal transient PDC behavior induced by topographic drops: a test case at Mt. St. Helens, USA. Earth Planet Sci Lett 627:118441
Trolese M, Cerminara M, Ongaro TE, Giordano G (2019) The footprint of column collapse regimes on pyroclastic flow temperatures and plume heights. Nat Comm 10(1):1–10
Trolese M, Giordano G, Cifelli F, Winkler A, Mattei M (2017) Forced transport of thermal energy in magmatic and phreatomagmatic large volume ignimbrites: paleomagnetic evidence from the Colli Albani volcano, Italy. Earth Planet Sci Lett 478:179–191
Trolese M, Giordano G, Komorowski JC, Jenkins SF, Baxter PJ, Cholik N, Corrado S (2018) Very rapid cooling of the energetic pyroclastic density currents associated with the 5 November 2010 Merapi eruption (Indonesia). J Volcanol Geotherm Res 358:1–12
Uehara D, Cas RAF, Folkes CB, Takarada S, Oda H, Porreca M (2015) Using thermal remanent magnetisation (TRM) to distinguish block and ash flow and debris flow deposits, and to estimate their emplacement temperature: 1991–1995 lava dome eruption at Mt. Unzen Volcano, Japan. J Volcanol Geotherm Res 303:92–111
Ui T (1983) Volcanic dry avalanche deposits—identification and comparison with nonvolcanic debris stream deposits. J Volcanol Geotherm Res 18(1–4):135–150
Valdivia P, Marshall AA, Brand BD, Manga M, Huber C (2022) Mafic explosive volcanism at Llaima Volcano: 3D X-ray microtomography reconstruction of pyroclasts to constrain shallow conduit processes. Bull Volcanol 84:1–18
Valentine GA (1987) Stratified flow in pyroclastic surges. Bull Volcanol 49(4):616–630
Valentine GA (1998) Damage to structures by pyroclastic flows and surges, inferred from nuclear weapons effects. J Volcanol Geotherm Res 87(1):117–140
Valentine GA (2020) Initiation of dilute and concentrated pyroclastic currents from collapsing mixtures and origin of their proximal deposits. Bull Volcanol 82(2):20
Valentine GA, Fisher RV (1986) Origin of layer 1 deposits in ignimbrites. Geology 14(2):146–148
Valentine GA, Wohletz KH (1989) Numerical models of Plinian eruption columns and pyroclastic flows. J Geophys Res: Solid Earth 94(B2):1867–1887
Valentine GA, Wohletz KH, Kieffer SW (1992) Effects of topography on facies and compositional zonation in caldera-related ignimbrites. Geol Soc Am Bull 104(2):154–165
Valentine GA, Perry FV, WoldeGabriel G (2000) Field characteristics of deposits from spatter-rich pyroclastic density currents at Summer Coon volcano, Colorado. J Volcanol Geotherm Res 104(1):187–199
Valverde JM, Soria-Hoyo C (2015) Vibration-induced dynamical weakening of pyroclastic flows: insights from rotating drum experiments. J Geophys Res: Solid Earth 120(9):6182–6190
Van Bemmelen RW (1949) The geology of Indonesia and adjacent archipelagoes, general geology, vol IA. Martinus Nijhoff, The Hague, pp 25–28
Van Eaton AR, Herzog M, Wilson CJN, McGregor J (2012) Ascent dynamics of large phreatomagmatic eruption clouds: the role of microphysics. J Geophys Res: Solid Earth 117(B3)
Van Eaton AR, Wilson CJN (2013) The nature, origins and distribution of ash aggregates in a large-scale wet eruption deposit: Oruanui, New Zealand. J Volcanol Geotherm Res 250:129–154
van Otterloo J, Cas RAF (2016) Low-temperature emplacement of phreatomagmatic pyroclastic flow deposits at the monogenetic Mt Gambier Volcanic Complex, South Australia, and their relevance for understanding some deposits in diatremes. J Geol Soc 2015–2122
van Otterloo J, Cas RAF, Sheard MJ (2013) Eruption processes and deposit characteristics at the monogenetic Mt. Gambier Volcanic Complex, SE Australia: implications for alternating magmatic and phreatomagmatic activity. Bull Volcanol 75(8):737–758
Vaniman D (2006) Tuff mineralogy. In: Heiken G (ed) Tuffs-their properties, uses, hydrology, and resources. Geol Soc Am Spec Pap 11–15
Vatin-Pérignon N, Poupeau G, Oliver RA, La Venu A, Labrin F, Keller F, Bellot-Gurlet L (1996) Trace and rare-earth element characteristics of acidic tuffs from Southern Peru and Northern Bolivia and a fission-track age for the Sillar of Arequipa. J South Am Earth Sci 9(1):91–109
Vazquez JA, Ort MH (2006) Facies variation of eruption units produced by the passage of single pyroclastic surge currents, Hopi Buttes volcanic field, USA. J Volcanol Geotherm Res 154(3–4):222–236
Verbeek RDM (1885) Krakatau. Landsrukkerij, Batavia, 546 pp
Verwoerd WJ, Chevallier L (1987) Contrasting types of surtseyan tuff cones on Marion and Prince Edward islands, southwest Indian Ocean. Bull Volcanol 49(1):399–413
Vidal CM, Komorowski JC, Métrich N, Pratomo I, Kartadinata N, Prambada O, Michel A, Carazzo G, Lavigne F, Rodysill J, Fontijn K (2015) Dynamics of the major plinian eruption of Samalas in 1257 AD (Lombok, Indonesia). Bull Volcanol 77:1–24

Vinkler AP, Cashman KV, Giordano G, Groppelli G (2012) Evolution of the mafic Villa Senni caldera-forming eruption at Colli Albani volcano, Italy, indicated by textural analysis of juvenile fragments. J Volcanol Geotherm Res 235:37–54

Voight B, Constantine EK, Siswowidjoyo S, Torley R (2000) Historical eruptions of Merapi volcano, central Java, Indonesia, 1768–1998. J Volcanol Geoth Res 100(1):69–138

Voight B, Glicken H, J JR, Douglass PM (1981) Catastrophic rockslide avalanche of May 18. In: Lipman PW, Mullineaux DR (eds) The 1980 eruptions of Mount St. Helens, Washington. US Geol Surv Prof Pap 347–377

Voight B, Komorowski J-C, Norton G, Belousov A, Belousova M, Boudon G, Francis P, Franz W, Sparks RSJ, Young S (2002) The 1997 Boxing Day sector collapse and debris avalanche, Soufrière Hills Volcano, Montserrat. In: Druitt T, Kokelaar BP (eds) The eruption of Soufrière Hills volcano, Montserrat, from 1995–1999. Geol Soc London Memoir 363–407

Von Aulock FW, Nichols ARL, Kennedy BM, Oze C (2013) Timescales of texture development in a cooling lava dome. Geochim Cosm Acta 114:72–80

Waitt RB (1981) Devastating pyroclastic density flow and attendant air fall of May 18—stratigraphy and sedimentology of deposits. In: Lipman PW, Mullineaux DR (eds) The 1980 eruptions of Mount St. Helens, Washington. US Geol Surv Prof Pap 439–458

Walker GPL (1971) Grain-size characteristics of pyroclastic deposits. J Geol:696–714

Walker GPL (1972) Crystal concentration in ignimbrites. Contr Min Petrol 36(2):135–146

Walker GPL (1973) Explosive volcanic eruptions—a new classification scheme. Geol Rund 62(2):431–446

Walker GPL (1980) The Taupo pumice: product of the most powerful known (ultraplinian) eruption? J Volcanol Geoth Res 8(1):69–94

Walker GPL (1981a) Characteristics of two phreatoplinian ashes, and their water-flushed origin. J Volcanol Geotherm Res 9(4):395–407

Walker GPL (1981b) New Zealand case histories of pyroclastic studies. In: Self S, Sparks RSJ (eds) Tephra studies. D Reidel, Dordrecht, pp 317-330

Walker GPL (1983) Ignimbrite types and ignimbrite problems. J Volcanol Geotherm Res 17(1):65–88

Walker GPL (1984) Characteristics of dune-bedded pyroclastic surge bedsets. J Volcanol Geother Res 20(3–4):281–296

Walker GPL (1985) Origin of coarse lithic breccias near ignimbrite source vents. J Volcanol Geoth Res 25(1–2):157–171

Walker GPL, Croasdale R (1971) Two Plinian-type eruptions in the Azores. J Geol Soc 127(1):17–55

Walker GP, Heming RF, Sprod TJ, Walker HR (1981a) Latest major eruptions of Rabaul volcano. In: Johnson RW (ed) Cooke-Ravian volume of volcanological papers. Geol Surv Papua New Guinea Mem 10:181–193

Walker GPL, Heming RF, Wilson CJN (1980a) Low-aspect ratio ignimbrites. Nature 283:286–287

Walker GPL, McBroome LA (1983) Mount St. Helens 1980 and Mount Pelée 1902—flow or surge? Geology 11(10):571–574

Walker GPL, Self S, Froggatt PC (1981b) The ground layer of the Taupo ignimbrite: a striking example of sedimentation from a pyroclastic flow. J Volcanol Geotherm Res 10(1):1–11

Walker GPL, Wilson CJN, Froggatt PC (1980b) Fines-depleted ignimbrite in New Zealand—the product of a turbulent pyroclastic flow. Geology 8(5):245–249

Walker GPL, Wilson CJN, Froggatt PC (1981c) An ignimbrite veneer deposit: the trail-marker of a pyroclastic flow. J Volcanol Geotherm Res 9(4):409–421

Waters AC, Fisher RV (1971) Base surges and their deposits: Capelinhos and Taal volcanoes. J Geophys Res 76(23):5596–5614

Watkins SD, Giordano G, Cas RAF, De Rita D (2002) Emplacement processes of the mafic Villa Senni Eruption Unit (VSEU) ignimbrite succession, Colli Albani volcano, Italy. J Volcanol Geothem Res 118(1):173–203

Willcock M, Mattei M, Hasalova P, Giordano G, Cas RAF, Morrelli C (2015a) Flow behaviour in the intra-caldera setting: an AMS study of the large (>1290 km^3) Permian Ora ignimbrite. In: Ort M, Porreca M, Geissman JW (eds) The use of paleomagnetism and rock magnetism to understand volcanic processes. Geol Soc London Spec Publ 396(1):1–11

Willcock MAW, Bargossi GM, Weinberg RF, Gasparotto G, Cas RAF, Giordano G, Marocchi M (2015b) A complex magma reservoir system for a large volume intra-to extra-caldera ignimbrite: mineralogical and chemical architecture of the VEI8, Permian Ora ignimbrite (Italy). J Volcanol Geotherm Res 306:17–40

Willcock MAW, Cas RAF (2014) Primary welding and crystallisation textures preserved in the intra-caldera ignimbrites of the Permian Ora Formation, northern Italy: implications for deposit thermal state and cooling history. Bull Volcanol 76(6):1–16

Willcock MAW, Cas RAF, Giordano G, Morelli C (2013) The eruption, pyroclastic flow behaviour, and caldera in-filling processes of the extremely large volume (>1290 km^3), intra- to extra-caldera, Permian Ora (Ignimbrite) Formation, Southern Alps, Italy. J Volcanol Geotherm Res 265:102–126

Williams H (1942) The geology of Crater Lake National Park, Oregon: with a reconnaissance of the Cascade Range Southward to Mount Shasta. Carnegie Institution Washington Publ 540

Williams K (2001) Volcanoes of the South Wind: A field guide to the volcanoes and landscape of Tongariro National Park. Tongariro Natural History Society (NZ)

Williams SN, Self S (1983) The October 1902 plinian eruption of Santa Maria volcano, Guatemala. J Volcanol Geotherm Res 16(1–2):33–56

Wilson CJ, Blake S, Charlier BLA, Sutton AN (2006) The 26· 5 ka Oruanui eruption, Taupo volcano, New Zealand: development, characteristics and evacuation of a large rhyolitic magma body. J Petrol 47(1):35–69

Wilson CJN (1980) The role of fluidization in the emplacement of pyroclastic claws: an experimental approach. J Volcanol Geotherm Res 8(2–4):231–249

Wilson CJN (1984) The role of fluidization in the emplacement of pyroclastic flows, 2: experimental results and their interpretation. J Volcanol Geotherm Res 20(1–2):55–84

Wilson CJN (1991) Ignimbrite morphology and the effects of erosion: a New Zealand case study. Bull Volcanol 53:635–644

Wilson CJN (1997) Comment on "Emplacement of the Taupo ignimbrite by a dilute turbulent flow" by B.W. Dade and H.E. Huppert. Nature 385:306–307

Wilson CJN (2001) The 26.5 ka Oruanui eruption, New Zealand: an introduction and overview. J Volcanol Geotherm Res 112(1):133–174

Wilson CJN, Houghton BF, Kamp PJJ, McWilliams MO (1995) An exceptionally widespread ignimbrite with implications for pyroclastic flow emplacement. Nature 378(6557):605–607

Wilson CJN, Walker GPL (1981) Violence in pyroclastic flow eruptions. In: Self S, Sparks RSJ (eds) Tephra studies. D. Reidel, Dordrecht, pp 441–448

Wilson CJN, Walker GPL (1982) Ignimbrite depositional facies: the anatomy of a pyroclastic flow. J Geol Soc 139(5):581–592

Wilson CJN, Walker GPL (1985) The Taupo eruption, New Zealand I. General aspects. Philos Trans R Soc Lond A: Math Phys Eng Sci 314(1529):199–228

Wilson L (1976) Explosive volcanic eruptions—III. Plinian eruption columns. Geophys J Int 45(3):543–556

Wilson L, Head JW (1981) Morphology and rheology of pyroclastic flows and their deposits, and guidelines for future observations. In: Lipman PW, Mullineaux DR (eds) The 1980 eruptions of Mount St. Helens, Washington. US Geol Surv Prof Pap 513–524

Wilson L, Sparks RSJ, Huang TC, Watkins ND (1978) The control of volcanic column heights by eruption energetics and dynamics. J Geophys Res: Solid Earth 83(B4):1829–1836

Wilson L, Sparks RSJ, Walker GPL (1980) Explosive volcanic eruptions —IV. The control of magma properties and conduit geometry on eruption column behaviour. Geophys J Int 63(1):117–148

Witham CS (2005) Volcanic disasters and incidents: a new database. J Volcanol Geothem Res 148(3):191–233

Wohletz K (2006) Fractures in welded tuff. In: Heiken G (ed) Tuffs-their properties, uses, hydrology, and resources. Geol Soc Am Spec Pap 17–31

Wohletz KH (1983) Mechanisms of hydrovolcanic pyroclast formation: grain-size, scanning electron microscopy, and experimental studies. J Volcanol Geoth Res 17(1):31–63

Wohletz KH, Sheridan MF (1979) A model of pyroclastic surge. In: Chapin CE, Elston W (eds) Ash flow tuffs. Geol Soc Am Spec Pap 177–194

Wolf T (1878) Der Cotopaxi und seine letzte Eruption am 26 Juni 1877. Neues Jb Miner Geol Palaeont 113–167

Wolff JA, Ellwood BB, Sachs SD (1989) Anisotropy of magnetic susceptibility in welded tuffs: application to a welded-tuff dyke in the Tertiary Trans-Pecos Texas volcanic province, USA. Bull Volcanol 51(4):299–310

Wolff JA, Turbeville BN (1994) Comment on 'A reappraisal of ignimbrite emplacement: progressive aggradation and particulate to non-particulate flow transitions during emplacement of high-grade ignimbrite' by Branney, M. J., and Kokelaar, P., 1994. Bull Volcanol 56:134–137

Wolff JA, Wright JV (1981) Rheomorphism of welded tuffs. J Volcanol Geotherm Res 10(1):13–34

Wright HMN, Folkes CB, Cas RAF, Cashman KV (2011a) Heterogeneous pumice populations in the 2.08-Ma Cerro Galán Ignimbrite: implications for magma recharge and ascent preceding a large-volume silicic eruption. Bull Volcanol 73(10):1513–1533

Wright HMN, Lesti C, Cas RAF, Porreca M, Viramonte JG, Folkes CB, Giordano G (2011b) Columnar jointing in vapor-phase-altered, non-welded Cerro Galán Ignimbrite, Paycuqui, Argentina. Bull Volcanol 73(10):1567–1582

Wright JV (1978) Remanent magnetism of poorly sorted deposits from the Minoan eruption of Santorini. Bull Volcanol 41(2):131–135

Wright JV (1979) Formation, transport and deposition of ignimbrites and welded tuffs. PhD, Imperial College, University of London, 451 pp

Wright JV (1981) The Rio Caliente ignimbrite: analysis of a compound intraplinian ignimbrite from a major late Quaternary Mexican eruption. Bull Volcanol 44(2):189–212

Wright JV, Roobol MJ, Smith AL, Sparks RS, Brazier SA, Rose WI, Sigurdsson H (1984) Late Quaternary explosive silicic volcanism on St Lucia, West Indies. Geol Mag 121(1):1–15

Wright JV, Self S, Fisher RV (1981) Towards a facies model for ignimbrite-forming eruptions. In: Self S, Sparks RSJ (eds) Tephra studies. D. Reidel, Dordrecht, pp 433–439

Wright JV, Smith AL, Roobol MJ, Mattioli GS, Fryxell JE (2016) Distal ash hurricane (pyroclastic density current) deposits from a ca. 2000 yr BP Plinian-style eruption of Mount Pelée, Martinique: distribution, grain-size characteristics, and implications for future hazard. Geol Soc Am Bull 128(5–6):777–791

Wright JV, Smith AL, Self S (1980) A working terminology of pyroclastic deposits. J Volcanol Geotherm Res 8:315–316

Wright JV, Walker GPL (1977) The ignimbrite source problem: significance of a co-ignimbrite lag-fall deposit. Geology 5(12):729–732

Wright JV, Walker GPL (1981) Eruption, transport and deposition of ignimbrite: a case study from Mexico. J Volcanol Geotherm Res 9 (2):111–131

Yamagata K (1991) Formation of lithic breccia and vent evolution during the 32 ka eruption of Shikotsu caldera, Japan. Geog Repts Tokyo Met Univ 26:227–240

Yamamoto T, Takarada S, Suto S (1993) Pyroclastic flows from the 1991 eruption of Unzen volcano, Japan. Bull Volcanol 55(3):166–175

Yokoyama S (1974) Flow and emplacement mechanism of the Ito pyroclastic flow from Aira caldera, Japan. Tokyo Kyoiku Daigaku Sci Rep Sec C 12:17–62

Zanella E, Gurioli L, Lanza R, Sulpizio R, Bontempi M (2008) Deposition temperature of the AD 472 Pollena pyroclastic density current deposits, Somma-Vesuvius, Italy. Bull Volcanol 70 (10):1237–1248

Zanella E, Gurioli L, Pareschi MT, Lanza R (2007) Influences of urban fabric on pyroclastic density currents at Pompeii (Italy): 2. Temperature of the deposits and hazard implications. J Geophys Res: Solid Earth 112(B5)

Subaqueous Flows of Pyroclastic Debris and Their Deposits

11

Origins, and Distinguishing Hot Gas-Supported Flows and Deposits from Water-Supported Ones

Ray Cas, John V. Wright, and Guido Giordano

R. Cas et al., *Volcanology*, Springer Textbooks in Earth Sciences, Geography and Environment,
https://doi.org/10.1007/978-3-319-66613-6_11

(a)
(b)
(c)
(d)
(e)
(f)
EU4pf
Degassing pipe
(Zeolitisation)
EU3pf
EU2/3pf

Abstract

In this chapter, we consider the intriguing and provocative topic of what happens when pyroclastic density currents enter the sea or large lakes from subaerial vents, or are erupted from subaqueous vents. Does dynamic interaction with the ambient water mass occur? Do the flows mix in ambient water, and transform from hot, gas-supported pyroclastic (mass) flows into water-supported mass flows (e.g. syn-eruptive debris flows and megaturbidity currents)? Can we recognise differences in the characteristics of the deposits of subaqueous mass-flow density currents of pyroclastic debris, and subaerial pyroclastic flow deposits? What do we call the deposits? Initially, it is important to consider the terminology that has developed historically to describe such subaqueous deposits. By analogy with subaerial pyroclastic flows, use of the term "pyroclastic flow" should be restricted to demonstrably hot, gas-rich mass flows of pyroclastic debris. Using this definition, very few examples of subaqueous pyroclastic deposits have been described with both evidence for hot emplacement and of having been wholly submerged. From the results of more recent experimental and numerical modelling studies, together with observations of some contemporary eruptions, we conclude that, when pyroclastic flows enter water, they are generally disrupted explosively and/or ingest water and transform into water-supported mass flows, although some higher mass and volume flux flows may instantaneously displace the shoreline seawards and deposit near shore ignimbrite. Based on a multitude of likely scenarios in which pyroclastic flows may enter subaqueous environments (either through eruption from subaerial vents and flow of pyroclastic flows into a body of water or through eruption from subaqueous vents), we then provide an in-depth review and assessment of examples described in the literature. There is no evidence to suggest that welding in wholly subaqueous environments is common, although, we highlight several situations where this potentially could occur.

11.1 Introduction

The concept of subaqueous pyroclastic flows originates from Lacroix's (1904) observations of the 1902 eruptions of Mt Pelée on the Caribbean island of Martinique, in the central Lesser Antilles volcanic arc. During this activity, many pyroclastic flows continued out to sea (Plate 11.1a; Cas and Wright 1987, 1991), destroying and setting fire to ships anchored offshore, while their subaqueous lateral equivalents broke underwater telegraph cables nearly 20 km from the coast (Lacroix 1904; also see Fiske 1963 for the early history of the concept). More recently, pyroclastic flows entered the sea numerous times in the period from 1995 to 2010 during the 1995–2013 eruption of the Soufrière Hills volcano, on Montserrat in the northern Lesser Antilles (Plate 11.1b), during which significant pyroclastic fan deltas formed at the coastline at the mouths of valleys that acted as conduits for the pyroclastic flows (Plate 11.1c). Other less well-documented examples occurred during the 1986 and 2006 explosive eruptions of Augustine volcano in the Aleutian arc, south-central coastal Alaska, USA (Plate 11.1d). The most recent examples include small-volume pyroclastic flows which entered the sea at Stromboli volcano, Italy, during two paroxysmal events that occurred on 3 July and 28 August 2019 (Plate 11.1e). Explosive interaction between pyroclastic flows entering the sea was also witnessed during the explosive eruption of Anak Krakatau on 23 December 2018 (Plate 10.1b).

Studies in the 1960s and 1970s of welded ignimbrites within Ordovician, shallow marine-deltaic successions in the Murrisk region, western Ireland (Stanton 1960; Dewey 1963), and in the shallow marine Ordovician successions of

◄ **Plate 11.1** Historic pyroclastic flows entering the sea and their deposits. **a** Towering ash cloud surge 4,000 m above pyroclastic flow travelling down Rivière Blanche and out to sea, during an eruption in December 1902, Mt Pelée, Martinique, Lesser Antilles (Photo by Angelo Heilprin; *Source* Wikimedia). **b** Pyroclastic flow erupted during a series of lava dome collapses in mid-July to early September 1996 from the Soufrière Hills volcano entering the sea via the Tar River on the southeast coast of Montserrat, Lesser Antilles. Note the dark plume around the front and margins of the flow, where some of it is propagating underwater. (Photo by Simon Young; *Source* British Geological Survey). **c** Prograding pyroclastic fan-delta made up of still hot pyroclastic flow deposits from the Soufrière Hills volcano on reaching the sea at the mouth of the Tar River valley. This has caused the shoreline to migrate seawards or regress on this steeply sloping shoreline zone. (Photo by Simon Young; *Source* British Geological Survey). **d** Augustine volcano, Alaska, showing the light-coloured, lobate pyroclastic flow deposits from eruptions in 1986 and 2003. Some pyroclastic flows clearly entered the sea as shown by the truncated deposits at coastal cliffs, especially on the right. (Photo by C. Read, Alaska Volcano Observatory, 2006; *Source* United States Geological Survey). **e** 3 July 2019 pyroclastic flow entering the sea after flowing down the 30° slope of the Sciara del Fuoco on the northern flank of Stromboli volcano, Italy. The flow consists of two flow pulses, visible as two lobes propagating over the sea, and originating from a sudden (paroxysmal) vulcanian blast near the summit. (Photo courtesy of LGS, Laboratory of Experimental Geophysics, University of Firenze). **f** Excavations at the Villa dei Papiri, Herculaneum showing pyroclastic flow deposits of the 24–25 August 79 AD eruption of Vesuvius deposited here at the paleoshoreline, Bay of Naples, Italy. The yellow dashed lines identify zones or plumes of pervasive zeolite alteration thought to be induced by convection of external seawater during cooling of the pyroclastic flows (chronicled EU2/3pf-EU4pf; described later in text). (Photo courtesy of Superintendence for Archaeology of Herculaneum)

North Wales (Francis and Howells 1973), together with the interpretations of Tertiary subaqueous pyroclastic eruptions and pyroclastic flow deposits in Washington state (Fiske 1963) and Japan (Fiske and Matsuda 1964), elevated research interest in understanding the processes, and in interpreting the deposits. Fiske (1963) and Fiske and Matsuda (1964) focused on one end member scenario, the wholly subaqueous eruption and deposition by pyroclastic flows, whereas Francis and Howells (1973) addressed the other end member situation, involving subaerial eruption and then passage of pyroclastic flows into the sea at an Ordovician continental shoreline and shelf. The entry of pyroclastic density currents (flows, surges; Chap. 10) into water is not just confined to marine settings, but may also occur where volcanism occurs within or in proximity to large freshwater lakes, especially caldera lakes.

What happens to pyroclastic flows on entry into water and their deposits has now been debated for several decades (e.g. Plate11.1f, Fig. 11.1; Cas and Wright 1987, 1991; Stix 1991). Critically important to the discussion are, firstly, can they maintain their integrity as hot, gas-supported pyroclastic flows underwater, or do they transform into syn-eruptive, warm to cold, water-supported mass flows of pyroclastic debris through the ingestion of ambient water into the initial hot, gas-rich flows, or can both occur? Secondly, what role does the thermal interaction between hot flows and the water mass play in these processes, which is likely to be most vigorous where the pyroclastic flows are hottest, when they first interact with water in shoreline locations, resulting in phreatic steam explosions or "littoral blasts" (Fig. 11.1b), or more passive interactions with seawater (Plate 11.1f)? Thirdly, if the deposits of such initial events are subsequently remobilised by subaqueous mass-flow processes, can these resedimented mass-flow deposits be distinguished from primary syn-eruptive deposits, and can these be uniquely identified as post-eruptive mass-flow deposits when their componentry may be identical to primary pyroclastic flow deposits? Because such flows cannot be directly witnessed, much of the focus on this topic has been on understanding and interpreting the textural characteristics of deposits preserved in the geological rock record, and sampling deposits in the modern oceans through drill coring (e.g. Mandeville et al. 1994; Schindlbeck et al. 2013).

Cas and Wright (1987, 1991) and Stix (1991) provided comprehensive reviews of this topic, at the time focusing largely on the assessment of the evidence about the paleoenvironmental settings of known uplifted ancient deposits, and interpreting their documented textural characteristics for evidence of a hot state of emplacement. In addition, White (2000) presented an overview of processes and a range of scenarios associated with subaqueous eruptions that produce density currents. Since then, further studies of deposits, and experimental simulations and numerical modelling, have been undertaken to develop a better understanding of the flow behaviour of pyroclastic flows in subaqueous environments. In this chapter, we give an updated summary of what is now known about the entry of pyroclastic flows into subaqueous environments, and their deposits.

11.2 Terminology—Use of "Subaqueous Pyroclastic Flow"

One of the first issues to appreciate in understanding subaqueous pyroclastic flows and their deposits is that the term "pyroclastic flow" has been frequently used, particularly in the older literature, too loosely. Implicitly, it has been used in ways that do not accurately differentiate primary, hot gas-rich pyroclastic flows from mass flows which may have resulted from the transformation of gas-supported flows into water-supported ones, or secondary mass flows carrying redeposited pyroclastic debris. Cas and Wright (1991) considered the terminological distinctions between the terms subaqueous pyroclastic flow (deposit) and mass flow (deposit) of pyroclastic debris, and traced the history of terminology; to which interested readers are referred for further discussion. Previously, in Chap. 10, Sect. 10.1, we defined subaerial pyroclastic flows (or dense to dilute pyroclastic density currents) as *"hot, variably turbulent, gravity driven, mass-flows of gas and volcanic debris. They grade vertically and laterally from a basal and interior, high particle concentration flow zone (the body of the pyroclastic flow) into a marginal, low particle concentration, ash-cloud surge flow zone at the front, sides and trailing above and behind the body*". A fundamental aspect of this definition is that the erupted volcanic gas is the fluid that determines the dynamics of the pyroclastic flow. In subaqueous settings, if ambient water becomes mixed into a pyroclastic flow, the flow essentially becomes an aqueous mass flow of pyroclastic debris. In such cases, it may be difficult to determine if the deposit was formed syn-eruptively or post-eruptively (i.e. epiclastically; Chap. 14).

By analogy with subaerial pyroclastic flows, we therefore suggest, the term "pyroclastic flow" should be restricted to *demonstrably* hot, gas-rich mass flows of pyroclastic debris. Several crucial features of pyroclastic flows distinguish them from other types of gravity-controlled mass flows or density currents (Chap. 10):

(a) They are hot; many were deposited at temperatures of hundreds of °C, enabling some deposits to weld in a zonal pattern, whereas others were emplaced as "cool" as ~100–200 °C (Quane and Russell 2005; Trolese et al. 2019).

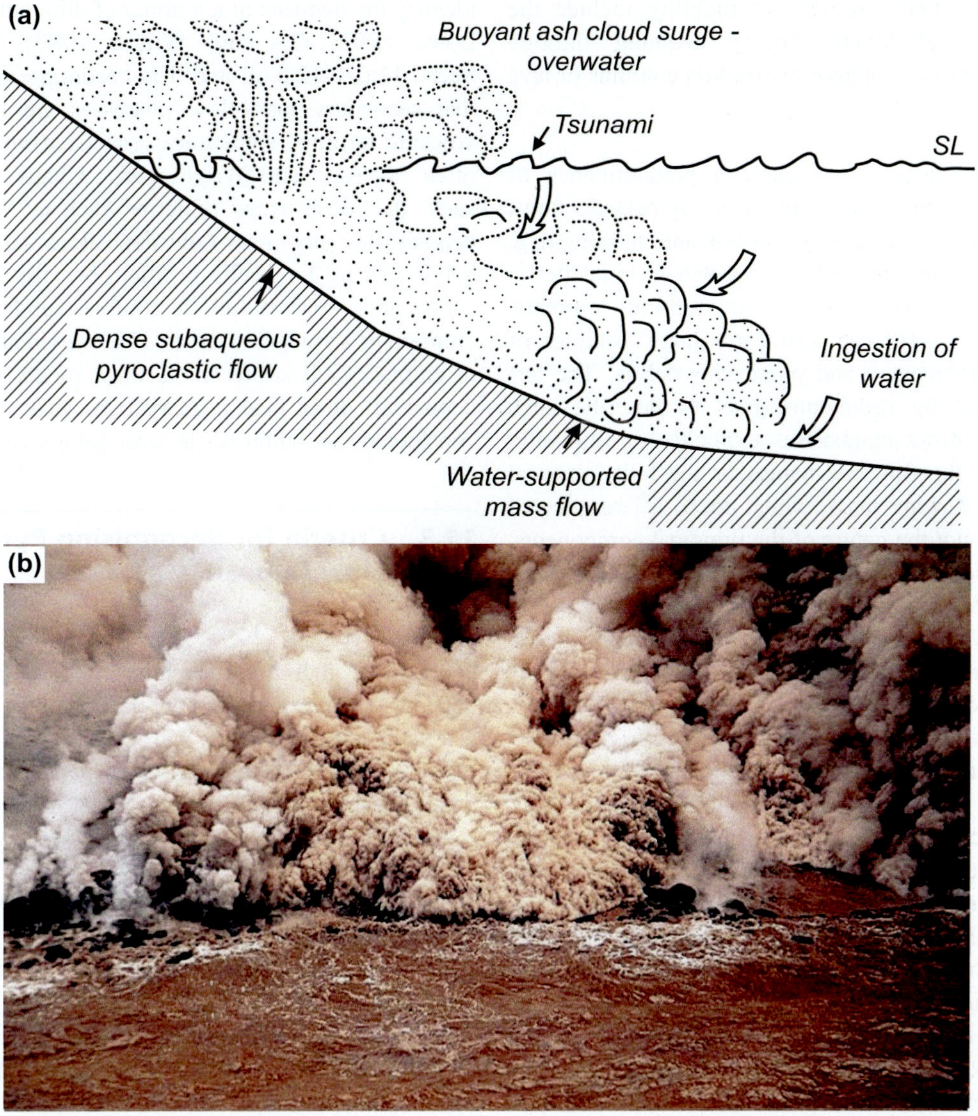

Fig. 11.1 Interactions between subaerial pyroclastic flows entering the sea. **a** Schematic illustration showing a generalised interaction scenario where a hot, pyroclastic flow generated on land enters the sea or a lake at a high relief shoreline typical of island arc volcanoes. This shows a flow dense enough to dive underwater and transform into water-supported debris flows and turbidity currents, while a dilute ash cloud surge continues to flow over the surface of the sea. The littoral zone is potentially an area of vigorous pyroclastic flow-water interaction where the flow is hottest and first interacts with water producing violent phreatic steam explosions. Another closely associated phenomenon is the generation of tsunami waves. This situation is analogous to the extensively observed mid-July–early September 1996 phases of the eruption of Soufrière Hills volcano, Montserrat (Plate 11.1b) and the entry of pyroclastic flows into the sea at Stromboli volcano, Italy, in 2019 (Plate 11.1e). (After Cas and Wright 1991). **b** White phreatic explosions and steam plumes at the front of a pyroclastic flow entering the littoral zone from the Tar River valley on 1 September 1996, Soufrière Hills volcano (Photo by Simon Young; *Source* British Geological Survey)

(b) They are gas-rich; sources of gas include exsolving magmatic gases derived directly from the vent during an explosive eruption and from juvenile clasts (Sparks 1976), vaporised external water incorporated during phreatomagmatic water–magma explosive interaction, and air mixed into the initial eruption column (Wilson and Walker 1987), or ingested into the head of the moving pyroclastic flow (Wilson and Walker 1982).

(c) They may result directly from explosive eruption or lava dome collapse.

(d) The solids component consists of a substantial proportion of vesiculated juvenile magmatic pyroclastic fragments (pumice or scoria), or dense, lava dome clasts.

(e) They are variably gas-fluidised density currents in which interstitial gas plays a fundamental part in the dynamics of flow (Sparks 1976; Wilson 1980, 1984;

Druitt 1998). Other sources of mobility include the momentum and kinetic energy acquired through explosive thrust or collapse of eruption columns or lava domes.

Although pyroclastic flows carry a predominance of pyroclastic debris, this is not unique to pyroclastic flows alone. Secondary mass-flow types in volcanic terrains (e.g. avalanches, slides, slumps, debris flows/lahars, mud flows, hyper-concentrated flows, slurry flows, turbidity currents, etc.; Chaps. 12, 15–16) can also carry large volumes of pyroclastically fragmented and pumiceous debris. The terminology applied by sedimentologists to describe such secondary mass flows emphasises explicitly or implicitly basic aspects of the flow mechanics or the bulk physical character of the flowing mass, which are to a large degree interrelated, *and* not the nature of the principal components, unless this is added to the mechanism name. Nevertheless, differences in textural and architectural organisation of components (the facies characteristics) in mass-flow deposits allow the interpretation of different flow mechanisms in the geological record.

Taking a semantic point of view, Cas and Wright (1991) also pointed out that, in the term "pyroclastic flow", the word "*flow*" is a noun meaning a fluid-like movement, process, or mechanism through which matter is moved, whereas the word "*pyroclastic*" is an adjective qualifying the noun "flow"—i.e. qualifying a flow process or mechanism. It does not qualify or describe the nature of the debris being carried which would be grammatically incorrect, although, this has been done commonly in the past. To make the distinction between primary gas-supported processes and deposits of the Roseau Tuff ignimbrite on Dominica (Lesser Antilles), from the offshore deposited, equivalent subaqueous mass-flow deposits, Carey and Sigurdsson (1980) proposed the term "subaqueous pyroclastic debris flow". In this term, the adjective "*pyroclastic*" qualifies the nature of the debris being transported in a water-supported debris flow. In the term "pyroclastic flow", the adjective "pyroclastic" should indicate that the flow is somehow uniquely dynamically "pyroclastic" in nature, i.e. gas-supported. This further implies a proven derivation directly from a contemporaneous explosive eruption or lava dome collapse event (Cas and Wright 1991).

Mass flows of pyroclastic debris may be contemporaneous with explosive eruptions (e.g. 18 May 1980, Mt St Helens eruption, Chap 12, Sect. 12.2.3), or postdate them, and their deposits may be difficult to distinguish from the deposits of true gas-rich pyroclastic flows, especially non-welded ones. However, the sedimentology of mass-flow processes and deposits is now sufficiently advanced to identify the deposits of a number of different flow types (e.g. Lowe 1982; Cas and Wright 1987; Chap. 12, e.g. Sects. 12.6.4–12.6.6), and it is becoming increasingly clear that secondary mass-flow processes are important in all volcanic terrains, sometimes associated with eruptions but often not (Cas and Wright 1987; Chaps. 12, 15–16). As such, the loose usage of terms such as "pyroclastic flow deposit" for any mass-flow deposit containing pyroclastic debris can no longer be justified, because it confuses concepts, and, most of all, it confuses understanding of an eruption or a volcanic system. Unless evidence for a hot state of emplacement is preserved, other modes of transport and deposition other than primary pyroclastic flow have to be considered, and appropriate terminology should be used.

11.3 Criteria for Recognising Pyroclastic Flow Deposits

Cas and Wright (1987, 1991) identified three mutually dependent criteria for identifying true pyroclastic flow deposits: (1) presence of pyroclastic debris; (2) facies characteristics indicating a pyroclastic (i.e. gas-supported) mass-flow mode of transport and deposition; and (3) evidence for a hot state of emplacement of the deposits.

11.3.1 The Presence of Explosively Fragmented Debris

Explosive eruptions (Chap. 4, Sects. 4.5, 4.6) produce a wide range of pyroclast types with varying morphologies and textural features (Chap. 4, Sect. 4.8; Macdonald 1972; Fisher and Schmincke 1984; Heiken and Wohletz 1985; Zimanowski et al. 2015; Cas and Wright 1987). Diagnostic juvenile clasts/grain types should be present and include (hot) radially jointed blocks and bread-crust textured or shaped bombs, variably vesiculated glassy debris, ranging from highly vesicular pumice, scoria, and cuspate shards, to the less vesicular, blocky glassy debris and accretionary lapilli that may be produced by phreatomagmatic explosive eruptions (Chap. 4, Sect. 4.8). Blocky shards may be difficult to distinguish from debris produced by quench fragmentation, lava dome gravitational collapse, and surface process erosion. Substantial accessory and accidental lithic clasts may also be present (Chap. 4, Sect. 4.8). However, not all pumice found in subaqueous settings is necessarily pyroclastic. Autoclastic pumice, produced by quench fragmentation of the highly vesiculated carapace of subaqueously erupted domes and lavas, can also occur in subaqueous settings (e.g. Manga et al. 2018).

11.3.2 Facies Characteristics of Pyroclastic Flow Deposits

The characteristics of the deposits of the three main types of pyroclastic flows (pumice-flow deposits or ignimbrites, scoria and ash-flow deposits, and block and ash-flow deposits) contain diverse facies which have been documented in a great number of publications (e.g. Sparks 1976; Wright et al. 1980a, b, 1981; Wilson and Walker 1982; Fisher and Schmincke 1984; Freundt and Schmincke 1986; Cas and Wright 1987; Roobol et al. 1987; Branney and Kokelaar 2002). While it is beyond the scope of this chapter to discuss these again in any detail (see Chap. 10 for an extensive summary), the key criteria are very poor hydraulic sorting and lack of continuous grading, with clasts of variable density (low-density pumice, high-density lithic clasts) and size (block, bomb, lapilli, ash, and fine ash) all mixed together. If hot enough, such poorly sorted deposits, which are massive to diffusely stratified, may also show evidence of welding textures, gas-escape pipes, or other indications of a hot state of emplacement (Chap. 10, Sect. 10.8.14).

11.3.3 Evidence for a Hot State of Emplacement

Several criteria are known for determining whether a deposit was emplaced in a hot state (Chap. 10, Sect. 10.8.14). These largely depend on the aggregate properties of the deposit, and not on the presence of particular types of clast or clast morphologies (such as jointed or bread-crust blocks or bombs) which may be subsequently redeposited with little or no abrasion. These criteria are:

- welding textures and associated devitrification features,
- clearly defined columnar jointing,
- gas segregation pipes,
- steam or fumarolic pipes and columns,
- syn-depositional thermo-remanent magnetisation (TRM),
- a red thermal oxidation colouration effect, and
- carbonised wood and vegetation.

11.3.3.1 Welding Textures

The only unequivocal evidence for welding is at the microscopic scale where syn-depositional plastic deformation of clearly defined cuspate shards and pumice vesicles can be recognised (Cas and Wright 1987, 1991; Busby et al. 2003; Chaps. 10, 13; Sect. 10.8.14.1). At the mesoscopic or outcrop scale, an apparent eutaxitic texture is by itself, without convincing evidence at the microscopic scale, often misleading because of pseudo-eutaxitic textures (e.g. apparent fiamme). These can be produced depositionally through the alignment of flat, platey pumice during transport and deposition (e.g. Niem 1977), diagenetically through the effects of post-depositional compaction (e.g. Niem 1977; Branney and Sparks 1990), and by syn- to post-depositional devitrification, hydrothermal alteration, metamorphism and/or deformation (e.g. Allen 1988; Chap. 13). However, even where alteration has destroyed original microscopic vitriclastic textures in an ignimbrite, devitrification features such as spherulites indicate that the original rock was a dense glass, either welded pyroclastic or coherent lava or intrusion (Chaps. 10, 13).

11.3.3.2 Clearly Defined Columnar Jointing

Well-developed, polygonal, columnar jointing is indicative of cooling in hot coherent igneous bodies such as lavas, intrusions, and welded and sintered ignimbrites (e.g. Chap. 10, Sect. 10.8.14.2). Ignimbrites are distinguished from the former by a microscopic welded vitriclastic texture and/or field facies characteristics (Chap. 10). However, a crude blocky jointing, resembling poorly developed columnar jointing, may occur in vapour-phase altered non-welded ignimbrites (Wright et al. 2011; Chap. 10, Sects. 10.8.14.2, 10.8.14.3), but similar jointing may be developed in any relatively massive rock mass as a result of mild tectonic movements. Independent evidence of a hot state of emplacement is required.

11.3.3.3 Gas Segregation Pipes

Gas segregation, gas escape, or elutriation pipes are structures that are produced in pyroclastic flow deposits as a consequence of hot gas streaming or fluidisation processes operating within the pyroclastic flow mass during deposition, or immediately thereafter in the deposit (Cas and Wright 1987; Pacheco-Hoyos et al. 2020; Chap. 10, Sect. 10.8.16). They are generally up to about 50 cm in vertical length and a few cm wide; but quite often, they are only several cm long and a few grain diameters wide. They are characteristically depleted in fines and enriched in crystals and lithics. Much larger gas segregation pipes >2 m long may occur in coarse-grained, near-vent ignimbrite facies. Water-escape pipes may occur in mud flows, hyper-concentrated flood flow, debris flows, and turbidites (e.g. Lowe 1975), but these are commonly poorly defined and have associated "dish" structures (Chap. 12, Sect. 12.6.5) that are absent from hot gas-rich pyroclastic flow deposits. One exception to the rule is described from the coastal facies of the 320 ka Matahina ignimbrite in the northern Taupo Zone, New Zealand (Bailey and Carr 1994). However, the Matahina dish-like features, which are spatially associated with tongues and diapirs of pumiceous debris, appear to be orders of magnitude larger and more complex than typical dish-and-pillar structures found in epiclastic sediments (see Sect. 11.6.1.2; Fig. 11.14).

11.3.3.4 Steam or Fumarolic Pipes and Columns

These features are significantly different in scale and character from "gas segregation pipes", and although they may form immediately on deposition in many cases they likely represent more protracted vapour phase or paleofumarolic activity (Chap. 10, Sect. 10.8.14.3). Sheridan (1970) described fumarolic mounds and ridges of incipiently welded tuff (or *sillar;* Fenner 1948) above the 0.76 Ma Bishop Tuff, California, which are characterised by high-temperature (>450–650 °C) vapour-phase mineral pore-filling cements (tridymite, cristobalite, and alkali feldspar). Wright (1979) described somewhat similar features as well as unusual "*steam pipes*" within the 95 ka Rio Caliente ignimbrite of La Primavera volcano, Mexico. Pipes vary from annular structures with indurated rims and non-welded cores (lacking evidence of elutriation) to coherent circular columns generally ≤ 1 m diameter and up to several m high (see Fig. 10.72e, f; Cas and Wright 1987, Figs. 8.46b, c, 8.47). In this case, however, cements consist of zeolite minerals (clinoptololite–heulandite) implying low-temperature (125–230 °C) fumarolic or geothermal activity, which might be expected if the ignimbrite was deposited in a shallow lake or on marshy ground. Comparable low-temperature (mordenite-cemented; 100–130 °C) pipes and columns have now been discovered in the Bandelier Tuff and older ignimbrites (1.26–1.8 Ma), New Mexico (Fig. 10.72 g; Self et al. 1996; Bailey and Self 2010), and Bishop Tuff (Fig. 10.77 h, i; Randolph-Flagg et al. 2017). These are attributed to the interaction between hot, non-welded ignimbrite and circulating meteoric and groundwaters perhaps associated with the rapid incision of valleys into the cooling ignimbrite). Randolph-Flagg et al. (2017) suggest that as the Bishop Tuff ignimbrite was emplaced, it instantaneously began to slump along paleo-drainages. Into these low-lying regions, where water pooled forming ponds or lakes, which then percolated into the porous ignimbrite causing it to heat and boil.

More recently, alteration "plumes" and haloes of zeolitised ash and large m-scale gas pipes have been described from within non-welded ignimbrite deposited at Herculaneum during the 79 AD eruption of Vesuvius (Giordano et al. 2018; see Plate 11.1f, Fig. 11.15). Notably, these occur at the paleoshoreline and are interpreted to be generated by the convective, upward percolation of vaporised seawater.

11.3.3.5 Syn-Depositional Thermo-Remanent Magnetisation

The thermo-remanent magnetisation (TRM) method has been applied to establishing a hot state of emplacement for young, subaerial deposits of volcanic debris (e.g. Aramaki and Akimoto 1957; Wright 1978; Hoblitt and Kellog 1979; Kent et al. 1981; Mandeville et al. 1994; Pensa et al. 2015a, b; Uehara et al. 2015; Giordano et al. 2018; Pérez-Rodríguez et al. 2019 Chap. 10, Sect. 10.8.14.4), as well as to a number of deposits interpreted as subaqueous pyroclastic flow deposits (e.g. Yamazaki et al. 1973; Kato et al. 1971; Wright and Mutti 1981; Mandeville et al. 1994). It depends on a fragmental deposit as a whole and/or its clasts acquiring a uniform syn-depositional TRM as it cools below the Curie temperature of magnetic minerals (which is $\sim$540 °C for magnetite), or the maximum value of the blocking temperatures (Tb spectrum) after deposition. However, some syn-eruptive mud flow ("lahar") deposits have also been shown by this method to have been emplaced at high temperatures (Lock 1978).

11.3.3.6 Red Thermal Oxidation Colouration

Young pyroclastic deposits often have a diffuse pinkish red or "salmon-pink" colour, resulting from the thermal oxidation of the outer porous margins during cooling (Sparks et al. 1980). In older deposits, such a colouration may be difficult to detect as a result of later weathering and diagenetic effects.

11.3.3.7 Carbonised Wood and Vegetation

Pyroclastic flows on forested tropical and temperate volcanoes may incorporate vegetation which ignites because of the high temperatures of the flows. Carbonised wood is not only useful for ^{14}C dating the deposits but, using optical reflectance techniques, can also be used to estimate emplacement temperatures in young pyroclastic deposits (Chap. 10, Sect. 10.8.14.5; e.g. Maury 1973; Caricchi et al. 2014; Pensa et al. 2015a, b; Trolese et al. 2018, 2019). Nevertheless, carbonised logs and debris can also be redeposited by mass-flow processes, and so they cannot be used to assess the temperature of emplacement of those resedimented deposits. Similarly, in subaqueous deposits, carbonised wood cannot be used to assess the emplacement temperature of the host deposits because these could have been cooled by mixing with ambient water by the time of deposition in the subaqueous environment, whereas the carbonisation occurred in a subaerial environment, not in the subaqueous deposit.

In conclusion, unless all three criteria for recognising true pyroclastic flow (*sensu stricto*) deposits can be established, the possibility should be considered that deposits of juvenile volcanic debris may have been transported by secondary gravity and water-supported mass flows, formed either contemporaneous with, or even post-dating explosive eruptions.

11.4 Criteria for Recognising '*Subaqueous*' Pyroclastic Flow Deposits

The next question is: *should/can these criteria be applied to suspected subaqueous counterparts?* Given that many authors believe that hot gas-rich pyroclastic flows can enter water and continue to flow unmodified, we propose that it is important the same criteria be used. If, and only if, evidence exists for a *hot state of emplacement*, as discussed in the previous section, then the term "subaqueous pyroclastic flow" deposit may be applicable. However, if the characteristics of the subaqueous deposits are consistent with deposition by water-supported mass-flow processes (Chap. 12, Sects. 12.6.4–12.6.6), then the appropriate genetic mass-flow terminology should be used (e.g. volcaniclastic debris flow, vitric or ash turbidite, etc.), even if the deposit contains or is composed largely of pyroclastic debris. For a final note on terminology, see Sect. 11.7.

This approach is therefore consistent with that now used in subaerial settings where subaerial primary pyroclastic phenomena (e.g. pyroclastic flows, blasts, and surges) are distinguished from secondary surface sedimentary transport processes that have remobilised pyroclastic debris (e.g. lahars, debris flows, hyper-concentrated flows; Smith 1986; Chap. 12), even when primary pyroclastic transport processes transform syn-eruptively into water-supported phenomena, such as debris flows and hyper-concentrated flows (Smith 1986; also discussed in Chap. 12, Sects. 12.6.4–12.6.6). The resultant deposits are named genetically to reflect the type of transportational and depositional agents, not the nature of the debris being carried.

The criteria required to be able to apply the term "subaqueous pyroclastic flow (deposit)" are now considered.

11.4.1 Evidence for a Wholly Subaqueous Depositional Setting

In addition to establishing the evidence for a hot pyroclastic flow mode of deposition, *it is necessary to justify the interpretation of the depositional environment*, especially for ancient geological examples (Cas and Wright 1991; Chap. 12).

Firstly, was the water depth truly deep enough for the deposits to have been emplaced in a wholly submerged state? In this regard, much of the older literature on ancient "subaqueous pyroclastic flow" deposits lack adequate documentation of the sedimentology of ambient host sedimentary facies; especially depositional structures and use of fossils for the paleoenvironmental reconstruction, or authoritative studies on these aspects have not been cited. The spatial *and* temporal relationships between the pyroclastic deposits and the ambient sedimentary facies have to be established. For example, even though marine fossils may occur in sedimentary strata beneath and above the pyroclastic deposits, this does not necessarily indicate the pyroclastic deposits were emplaced beneath water. If the fossils and sediments indicate shallow water, and the pyroclastic deposits are very thick (e.g. ignimbrites tens of m thick), then much, if not all of the deposits could have been emplaced essentially subaerially. This is because the emplacement processes (e.g. large volume, high mass flux pyroclastic flows) could have caused the shoreline to be pushed back instantaneously, i.e. regressed, with some time passing before transgression occurred and marine conditions were restored above the deposit. Alternatively, relatively rapid uplift or falls in sea level may be induced volcano-tectonically or glacio-eustatically on the scale of several thousand years (Chap. 15), which could be very difficult to detect in ancient volcanic successions.

In contrast, if absolutely clear sedimentological and paleoecological evidence exists for deep-water conditions, then the context is clear, but that evidence still has to be presented.

11.4.2 Evidence for a Hot State of Emplacement

The criteria cited above for recognising a hot state of emplacement in subaerial deposits of pyroclastic debris, can be evaluated to determine their applicability in subaqueous settings. Thermal oxidation of substrate sediments is unlikely to occur in a wholly subaqueous deposit, although deposits that are emplaced in shallow water, with their tops emergent above water level, may experience post-emplacement weathering and oxidation of the top in a cold state that could be mistaken for thermal oxidation. The presence of charcoal material in subaqueous deposits is not useable because these will have been derived from subaerial environments and then transported into and under water. Any reflectance-derived temperature determinations will relate to what happened in the subaerial environment, not in the subaqueous setting.

Other criteria for recognising a hot state of emplacement, including microscopic evidence of plastic, thermally induced deformation of glass shards and vesicle collapse in pumice clasts, columnar jointing, and TRM, are still valid. Gas segregation pipes are potentially more equivocal, because, in subaqueous settings, water-escape pipes can also form as a result of very rapid rates of deposition of cold sediment in water that look somewhat like gas escape pipes. However, vesicles in densely welded deposits (sometimes termed *vesicular welded tuffs*) may be good paleoenvironmental indicators of the deposition of ignimbrites in shallow marine

(Wright and Coward 1977; Reedman et al. 1987; McArthur et al. 1998; Riggs and McConnell 2017), or lacustrine (Yamada 1973; Kano 1990, 1991) settings. In ancient, metamorphosed deposits these may be recognised by abundant siliceous nodules (amygdales) which represent gas cavities that have since been infilled by drusy quartz (see Fig. 11.12h, i) or *lithophysae*, while, in some younger basins, depending on their diagenetic history, they are known to be filled by hydrocarbons and are (perhaps to the surprise of many volcanologists) significant petroleum reservoirs (see Chap. 18, Sect. 18.9.4.2, Fig. 18.101).

Welding in subaqueously emplaced deposits requires the juvenile pumice and shard pyroclasts to remain plastic during and after deposition, in turn requiring them to remain above the glass transition temperature (T_g; Chap. 2) during eruption, transport, and deposition, which can only occur if transported by volcanic gas at magmatic temperatures. When water is ingested into a transport system pyroclasts will rapidly cool below the T_g and solidify, preventing welding from occurring in the resultant deposits. Gas retention is most likely to occur close to the vent and under conditions of very high mass flux in the transport system, which retards the ingress of external water driven by high hydrostatic pressure underwater. Sparks et al. (1980) proposed a theoretical model whereby the conversion of liquid water to steam at the margins of a hot gas-supported pyroclastic flow entering a body of water is converted into steam, which is then driven into the body of the subaqueous pyroclastic flow by the hydrostatic pressure if water depths are high. Under high pressure, steam (H_2O) is more soluble in silicate melts and glass than under low pressure, and since increased dissolved H_2O in a silicate melt reduces the viscosity of the melt (Chap. 2), Sparks et al. (1980) proposed that if enough steam can be absorbed into the juvenile pyroclasts in the interior of the subaqueous pyroclastic flow, the resulting deposit may weld.

We agree with this theoretical assessment, however, for this process to occur the "Goldilocks and the three bears principle" needs to be invoked. In some cases, pyroclasts will be hot enough and if isolated from external water they may weld upon deposition. Alternatively, pyroclast temperatures may not be hot enough to spontaneously weld, but if just enough steam can infiltrate the deposit interior, it may weld. These two scenarios are most likely to occur near the vent, but do not necessarily occur with each subaqueous eruption. In other cases, pyroclast temperatures are not high enough to cause welding, and if not enough steam invades the interior, the system may remain hot, but not hot enough to cause welding. In many cases, too much water may be mixed into flows through various mechanisms (small volume and low mass flux flows, explosive interaction and mixing between hot flow and ambient water, high levels of flow turbulence, and long-lived flows) leading to water-supported mass flows. Whether or not the right amount occurs will depend on many factors including the temperature of the flow mass and pyroclasts (have they already cooled significantly through air ingestion?), the mass flux of pyroclasts and gas, the flow state (high levels of turbulence will lead to large volumes of liquid water being mixed, causing flow transformation into a water-supported mass flow, as will explosive interaction at the shoreline between the hot pyroclastic flow and the water mass, and the rate of ingress of steam into a subaqueous flow and its deposit.

There are good examples of microscopic welding textures preserved in pyroclastic deposits that appear to have been erupted and deposited subaqueously (e.g. Kokelaar and Busby 1992), but whether or not this is related to access of externally generated steam into the deposits or due to the inherently high emplacement temperature is unclear.

11.5 Factors Affecting Pyroclastic Flows on Entry into Water—Experimental and Numerical Modelling Studies

Many factors affect the way in which pyroclastic flows enter bodies of water (sea, lakes), and the way in which they interact with the sea and behave subsequently. These include:

- density and componentry of flows relative to the ambient water mass,
- water depth, subaerial and subaqueous topographies, and angle of entry into the water mass,
- mass and volume flux of flows into water,
- temperature of the flows and thermal interaction with a body of water,
- velocity of the flows, and
- mixing mechanisms of ambient water into the flow.

11.5.1 Density and Componentry of Flows Relative to the Ambient Water Mass - Experimental Simulations

The bulk density of a pyroclastic flow is controlled by its componentry and proportions of solid components to hot gas, which may vary from the base and interior to the top and lateral margins. Pumice and ash flows in which the pumice clasts have a vesicularity of 60% (density of 960 kg m^{-3}) or less will have a bulk density less than pure water (1,000 kg m^{-3}), and will thus be buoyant in water. Dense clast pyroclastic flows such as block and ash flows, will not be buoyant in water. Even pyroclastic flows in which the average clast density is slightly greater than that of water will also be buoyant in water, as the dispersed nature of solids in hot gas in such flows gives them a bulk density less than that of water.

Monaghan et al. (1999) simulated the entry of pyroclastic density currents (pdc's) of varying density into water in simple tank experiments. Red-dyed, saline currents of varying density were released down a ramp from a lock at one end of the tank. The tank was filled with two layers of water: an upper, clear fresh one, simulating the atmosphere, and a lower blue-dyed saline one simulating the sea. Monaghan et al. (1999) showed that density currents less dense than the bottom blue saline layer flowed along the interface between the clear fresh and the blue saline layers and barely generated a "bow wave" (or simulated tsunami; Fig. 11.2a). Currents denser than the basal blue saline layer flowed into the basal layer and then along the bottom of the tank while generating a substantial bow wave (Fig. 11.2c, d). Density currents with a density about the same as the basal layer, also flowed along the interface between the two tank layers, but were not as buoyant as in the first case, and generated only a small bow wave (Fig. 11.2b). It was clear from these experiments that there is some immiscibility between currents entering a standing body of water at velocity due to their density and/or chemical composition (varying contents of dissolved NaCl), and/or velocity or momentum.

McLeod et al. (1999) conducted a series of more intricate analogue experiments to simulate the entry of pyroclastic flows and a range of other volcanic mass flows (lahars, debris flows, sediment-bearing floods, and jökulhlaups) into the sea. Using a 2.42 m long, 0.175 m wide tank with a depth of 22 cm, filled with fresh water and a ramp inclined at 20°, experimental currents, some with suspended solid particles, were released from a lock at the top of the ramp. Three sets of experiments were conducted, where a range of currents of differing density were fed down the slope into the tank of water: These involved either: (1) saline solution of differing density, (2) simple suspensions of silicon carbide (SiC) in fresh water composed of varying proportions of suspended solid grains, or (3) complex suspensions of SiC and polypropylene particles in methanol to simulate a range of clast densities and particle sizes, and low-density gas in hot pyroclastic flows. For more details of the experimental set-up and various mixtures to produce the range of density currents the reader is referred to McLeod et al. (1999).

In the saline experiments, currents with densities of 1,050, 1,100, 1,150, and 1,200 kg m^{-3} were released into the freshwater tank. The lowest density current generated a bow wave or tsunami at the "shoreline", causing the current to stall, then generated an anti-clockwise circulating, turbulent mixing zone as the water displaced by the tsunami flowed back onshore, onto the ramp. The mixing zone or cell at the shoreline allowed water and air to be mixed into the current. The mixing cell then collapsed gravitationally forming an offshore directed, turbulent, bottom-density current, which mixed in more water at its upper contact (Fig. 11.3a). The higher density saline currents entered the fresh water with progressively less stalling at the shoreline as each current density increased and developed directly into a subaqueous flow (Fig. 11.3b). These generated tsunami waves and bottom density currents with each denser current moving faster along the bottom than less dense currents.

In the experiments with suspended silicon carbide grains (i.e. particulate currents with current bulk densities of 1,100, 1,200, 1,300, and 1,400 kg m^{-3}), the same patterns occurred as in the dense saline current experiments, involving tsunami generation, shoreline mixing zone, small overflow and dominant underflow (Fig. 11.3c). Tsunami generation, decreasing levels and duration of mixing at the shoreline, and increasing bottom density current velocities were proportional to (increasing) current densities. That is, bulk density determined the nature of the interaction, regardless of the constitution of the currents. However, these particulate currents also experienced sedimentation, which then caused the reduction of current density along the tank during each flow event, and due to loss of momentum decreasing flow velocity. Deposited sediment concentration/thickness was greatest at the base of the ramp where the slope changed (hydraulic jump) and the current slowed or stalled, and then thinned down the tank as the current velocity waned further. In addition, there were some peaks in sediment thickness offshore, where the underflows stalled as a result of surface water wave disturbance, which could also be related to water depth, although this was not investigated.

For the most complex currents with methanol as the interstitial fluid, two size populations of silicon carbide (mean grain sizes of 110 μm and 15 μm), representing denser lithics and crystal fragments and ash, respectively, and polypropylene particles (mean grain size 250 μm) simulating larger pumice, were added in various proportions. The current densities on entry into water ranged from 944 to 1116 kg m^{-3}. All currents entering the water mass generated an offshore-directed bow wave or tsunami, and a nearshore turbulent mixing zone. The mixing cloud appeared to be stratified, with smaller and lower density particles high in the cloud, and heavier and higher density particles nearer the base, reflecting stratification in the subaerial flow prior to entry into water. The lower, denser part of the mixing cloud collapsed to form a dense underflow along the bottom of the tank, and a lower density upper part spread as a buoyant flow at the surface of the water body (Fig. 11.3d). In general, the buoyant upper flow moved more slowly and not as far as the dense underflow, polypropylene fragments floated to the surface forming simulated pumice rafts, and plumes of fine SiC ash descended from the base of the buoyant flow into the underlying water mass, so allowing the buoyant current

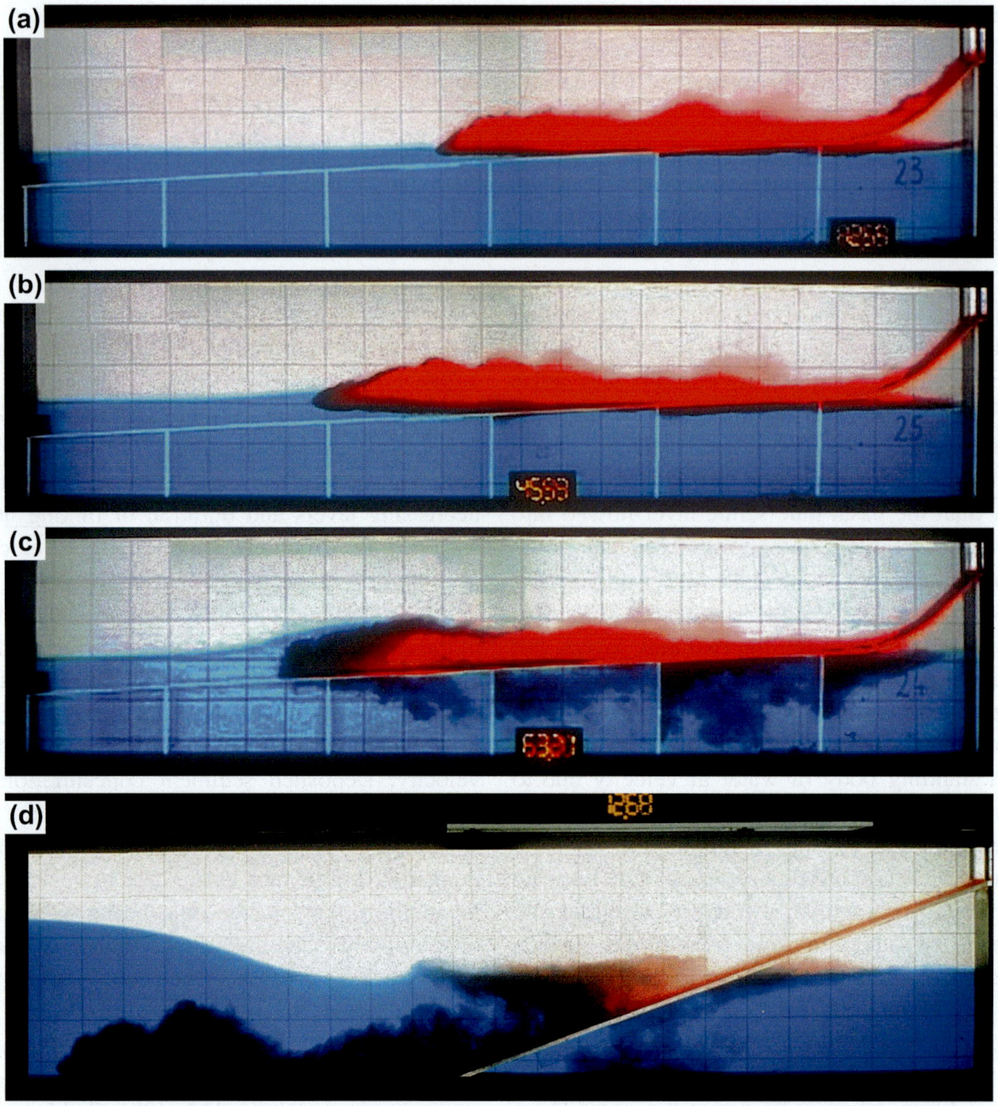

Fig. 11.2 Experiments simulating pyroclastic density currents (pdc's) entering the sea. The pdc's are dyed red and contain variable concentrations of dissolved salt. They are released from a lock to the right of the tank and flow down a ramp with a slope angle of 15° into the tank with two layers of water, a clear, fresh-water layer at the top simulating the atmosphere, and a blue-dyed, saline layer, simulating the ocean, with a density of 1,030 kg m^{-3}. **a** Red current has a density of 1,018 kg m^{-3}, which is more than the clear layer, but less than the blue layer, and so it flows along the interface, generating only a tiny bow wave. **b** Red current density is 1,029 kg m^{-3}, about the same as the underlying blue layer. The red current is not as buoyant as in (**a**). It flows along the interface, generating a moderate bow wave or tsunami, and its base is more immersed in the underlying blue layer than in (**a**). **c** Red current density is 1,097 kg m^{-3}, much denser than the blue saline layer. It enters the blue layer without stalling and develops into a density underflow. The displacement of the blue water mass by the volume of the red current has generated a major bow wave or simulated tsunami, as discussed by Monaghan et al. (1999). **d** As in (**c**) but the ramp slope is 20°. The current generates a very large bow wave/tsunami, but in addition, at its top, a dilute secondary current is being stripped off the top of the main descending current and begins to flow along the interface because of its buoyancy. All currents show mixing with the lower blue layer at the fronts and bases, shown by the dark grey-blue margins, and with the upper clear layer, represented by the lighter-coloured irregular plumes along the top of the currents mixing into the clear layers. In (**c**) the dark downward-directed plumes represent leakage of the dense experimental current along the edges of the ramp and should be ignored. In each successive experiment, the shoreline (where the ramp intersects the boundary between the clear and blue water layers) is successively more displaced to the left because of the increasing density of the currents, which successively displace more water, and the very low ramp slope. (Experiments conducted by A. Kos, J. Monaghan, R. Cas at Monash University)

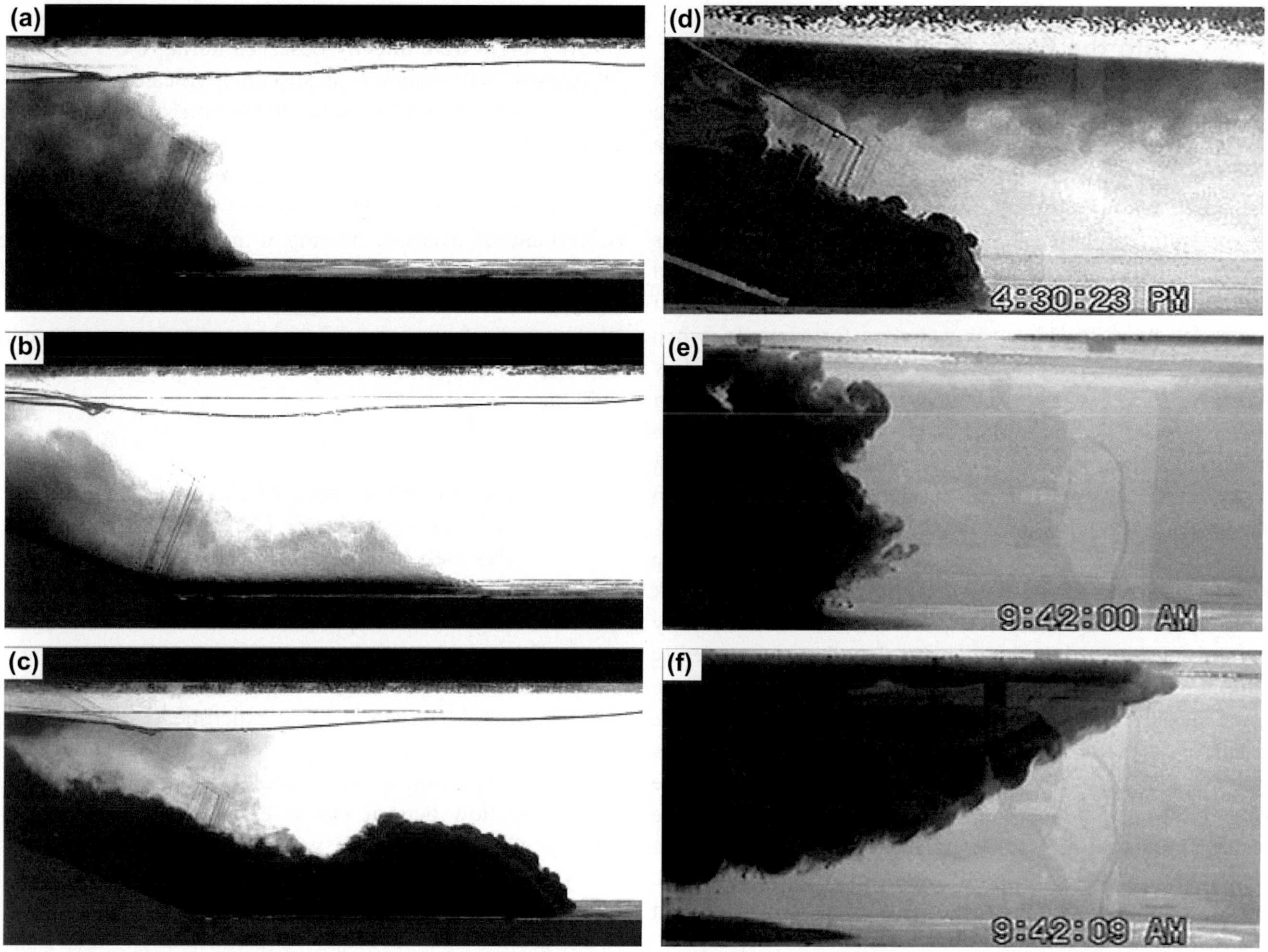

Fig. 11.3 Photographs of simulated experimental pyroclastic density current entering the sea involving saline solution and SiC ± water flows (left), and complex particulate flows (right); water depths are 22 cm (Modified from McLeod et al. 1999). **a** Collapse of a saline turbulent cloud. **b** Underwater saline gravity current. **c**. Underwater SiC ± water gravity current. **d** Concurrent buoyant and dense flows; the front of the buoyant flow is out of view to the right of the picture. **e** A dense flow that has stopped moving forward at the base of ramp (active sedimentation) and has lift-off occurring along the entire flow length; the flow front is at ~1.5 m horizontally from the water–ramp intersection, which is ~0.9 m from the base of the ramp. **f** A secondary buoyant flow

to become even more buoyant and so, prolonging its life. The dense underflow slowed down as it deposited dense clast SiC fragments (Fig. 11.3e); its density decreased and distally it formed a buoyant plume that rose to the surface forming a secondary buoyant flow (Fig. 11.3f). With the lowest density currents, no underflow formed, all material being fed into a buoyant overflow.

Taken together the main conclusions of these two experimental studies indicate, on entering seawater: (1) denser subaerial pdc's will usually directly split into a dense underflow on the seabed, and a low-density buoyant flow moving over the surface; (2) subaerial pdc's with bulk densities less than or similar to seawater will travel along the surface as a turbulent cloud, depositing much of the load in nearshore areas, and (3) surface bow waves or tsunamis will commonly accompany such events (also see Bougouin et al. 2020). However, it is important to stress, these analogue pdc's were all cold-water experiments, not at the hundreds of °C of true pyroclastic flows, so no account of the effects of thermal energy transfer from flows to the sea at the point of entry and beyond is possible. Thermal interaction with the water mass enhances mixing with it (see Sect. 11.5.4, describing "hot" experiments; Freundt 2003), and hot pumice clasts can absorb water quickly (Whitham and Sparks 1986; Fauria et al. 2017), causing their bulk density to increase and affecting their hydraulic behaviour. However, the experiments simulating the behaviour of pdc's flows entering water and their flow behaviour based on their bulk density attributes are valid. Another limitation of tank experiments relates to scaling experimental observations to real-world phenomena.

11.5.2 Topography or Angle of Entry into the Water Mass

Cas and Wright (1991) proposed that the slope of the topography and thus the angle of entry or incidence of a subaerial flow flowing into water may significantly affect the behaviour of pyroclastic flows entering a water mass with momentum, particularly considering the phase differences between interstitial gas and a water mass. They considered that the surface tension of the water surface would make it easier for flows to skip over water, as for example, when throwing flat skipping stones onto a water surface at a low angle, or as very dense, spherical or cylindrical, bouncing ("dam busting") bombs did when released at low angles with momentum from low-flying (at an altitude of 18 m, and an air speed of 390 km/h) aircraft during Operation Chastise in the Second World War. In additional experiments undertaken by Monaghan et al. (1999), the effects of changing the slope of the entry ramp for saline density currents into a tank were also investigated. In experiments in which density currents entered at low entry angles they flowed seamlessly and more readily over water (Fig. 11.4a), while flows of the same density entering at high angles, burrowed into the water mass in mixing zones before rebounding and then flowing laterally over the simulated water surface (Fig. 11.4b). Again we emphasise, as with all cold state analogue experiments as described above, there are some limitations in applying results to natural hot pyroclastic flows entering water, especially regarding the effects of thermal energy transfer.

Recently, pyroclastic flows entering the sea at a steep (~30°) angle were directly observed during the 3 July and 28 August 2019 eruptions at Stromboli volcano, Italy (Plate 11.1e; Giordano and De Astis 2021). These illustrated that flows became readily inflated by water ingestion and vaporisation, leading to the partitioning of an underwater current (seemingly carrying the largest part of the pyroclastic debris) and an overflow flowing at the sea surface for some hundreds of m before lofting off (Plate11.1e). As a result of the pyroclastic debris entering the sea, a tsunami wave was triggered which was up to 2 m in amplitude at source and then rapidly attenuated to a few tens of cm (http://lgs.geo.unifi.it/).

11.5.3 The Possible Role of Mass and Volume Flux of Pyroclastic Flows Entering the Sea in Displacing the Shoreline

Cas and Wright (1991) speculated that large volume pyroclastic flows with high mass and volume flux entering a shoreline zone with a low gradient should be able to displace the water mass (as seen in some experiments in Fig. 11.2), pushing the shoreline seaward, effectively causing it to regress instantaneously. This could produce a very thick pyroclastic flow deposit that is essentially deposited subaerially without interaction with the sea in what was formerly a shelf environment because the sea at the shoreline has been displaced seawards. Legros and Druitt (2000) investigated this concept experimentally, and also

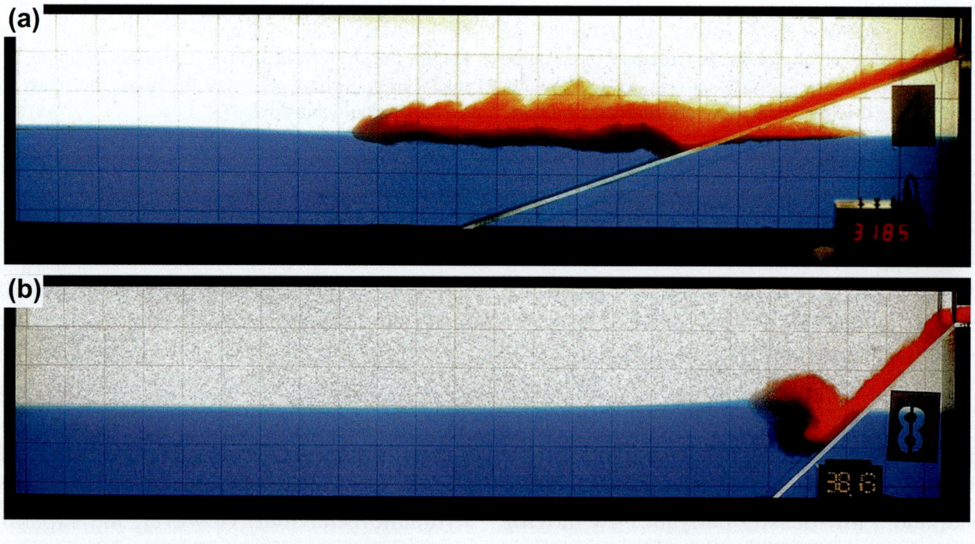

Fig. 11.4 Same experimental set-up as in Fig. 11.2. **a** Intermediate density red current (1,030 kg m^{-3}) descending a ramp at 20° inclination, and transforming into an interflow with only minor disturbance and interaction with underlying blue layer, but partially immersed in it. **b** Similar density current descending at 45°, plunging into and then rebounding to the interface. (Experiments conducted by A. Kos, J. Monaghan, R. Cas at Monash University)

numerically, given the impossibility of scaling the processes in an experimental tank to natural situations involving pyroclastic flows with volumes of >10 km^3 flowing into the sea.

Their experimental set-up involved entry of a variably dense (due to varying dissolved NaCl), analogue pdc from a lock at the side a fan-shaped tank (with a projected fan apex angle of 5.9°), that allowed a sustained current to be fed into the tank. The density current was released into the tank from the lock at the surface level of a dense saline layer that partially filled the tank, which then spread slightly radially into the tank. There was no upper aqueous layer and no ramp (cf. Monaghan et al. 1999), so the density currents entered the saline water mass horizontally, simulating a gently sloping coastal plain and low angle of incidence of the current into the sea. The main, and not surprising findings of the experiments were twofold: (1) where the current density was less than or equal to that of the tank saline layer, it spread along the top of the saline layer but displaced the saline layer completely "seaward" for a distance R before thinning and flowing over it (cf. Fig. 11.2a, b); and (2) for a current denser than the tank saline layer, it also displaced the tank saline layer a distance R, before descending beneath it and transforming into a thinner underflow (cf. Fig. 11.2c, d). R in the two scenarios is not the same, with R being greater for dense currents (Legros and Druitt 2000).

Legros and Druitt (2000) also modelled numerically the entry of pyroclastic flows into the sea. For *density currents with a density greater than the sea/saline tank water layer*, they adopted the Bernoulli equation for an inviscid turbulent fluid, in which the hydrostatic pressure must be equal to the dynamic pressure, that is

$$(\rho_c - \rho_w)gh = 0.5\,\rho_w U^2 \tag{11.1}$$

The density current velocity is then

$$U = \sqrt{2[(\rho_c - \rho_w)/\rho_w]gh} \tag{11.2}$$

where U is the velocity, ρ_c is the current density, ρ_w is the seawater/tank water density (1020 kg m^{-3}), g is the acceleration due to gravity, and h is the pdc thickness. $g(\rho_c - \rho_w)/\rho_w$ is the reduced gravity, g', of the density current in water. For density currents which lose energy through turbulent drag (e.g. at the upper margins),

$$U = Fr\sqrt{g'h} \tag{11.3}$$

in which Fr is the Froude number, with a value of 1 ± (Legros and Druitt 2000).

For a radially spreading density current, the flow thickness, h, can be expressed as a function of the volumetric flux, Q:

$$h = Q/2\pi rU \tag{11.4}$$

where r is the radial distance from source. Where $h > H$, the entire depth of the seawater/tank water at any point along the flow path, the sea is completely displaced, but at the point where h < H, the density current goes beneath the water surface and flows as a normal gravity current underflow. The maximum distance from the pre-existing shoreline that the shoreline is displaced, R, by the flow entering the shoreline, is the location where h = H, where R is given by

$$\mathrm{R} = \mathrm{Q}/2\pi Fr\sqrt{g'H^3} \tag{11.5}$$

For a submarine slope of θ, H = R tan θ, and

$$\mathrm{R} = (Q/2\pi Fr)^{0.4}\left(1/g'\tan^3\theta\right)^{0.2} \tag{11.6}$$

Using this equation indicates that the shoreline displacement distance increases as the density of the current approaches that of the sea, as it becomes more buoyant due to the low-density contrast.

For *density currents with a density less than the sea/tank water*, there is a complication in that part of the density current will be partially immersed in seawater, the thickness of which is designated h_w, and much of it, with a thickness of h_a, will be emergent in the atmosphere, which has a density of ρ_a (Legros and Druitt 2000). As a result, in Eq. 11.1, the total dynamic pressure will be the sum of the dynamic pressures of the parts of the flow in water and in air, i.e.

$$(\rho_c - \rho_a)g\,h_a + (\rho_w - \rho_c)g\,h_w = 0.5\left[\rho_a(h_a/h)U^2 + \rho_w(h_w/h)U^2\right] \tag{11.7}$$

The current velocity can then be reconfigured to (Legros and Druitt 2000)

$$U = \sqrt{\frac{2gh}{[\rho_a/(\rho_c - \rho_a)] + [\rho_w/(\rho_w - \rho_c)]}} \tag{11.8}$$

This relationship can be reduced to a simpler form for currents in which $\rho_w \sim \rho_c \gg \rho_a$ as follows:

$$U = \sqrt{2[(\rho_w - \rho_c)/\rho_w]gh} \tag{11.9}$$

and for currents in which $\rho_w \gg \rho_c \sim \rho_a$ by

$$U = \sqrt{2[(\rho_c - \rho_a)/\rho_c]gh} \tag{11.10}$$

The thickness of the flow that is immersed, h_w, which causes the seawater to be displaced is

$$h_w = h[(\rho_c - \rho_a)/(\rho_w - \rho_c)] \tag{11.11}$$

The distance that the shoreline is displaced is given by

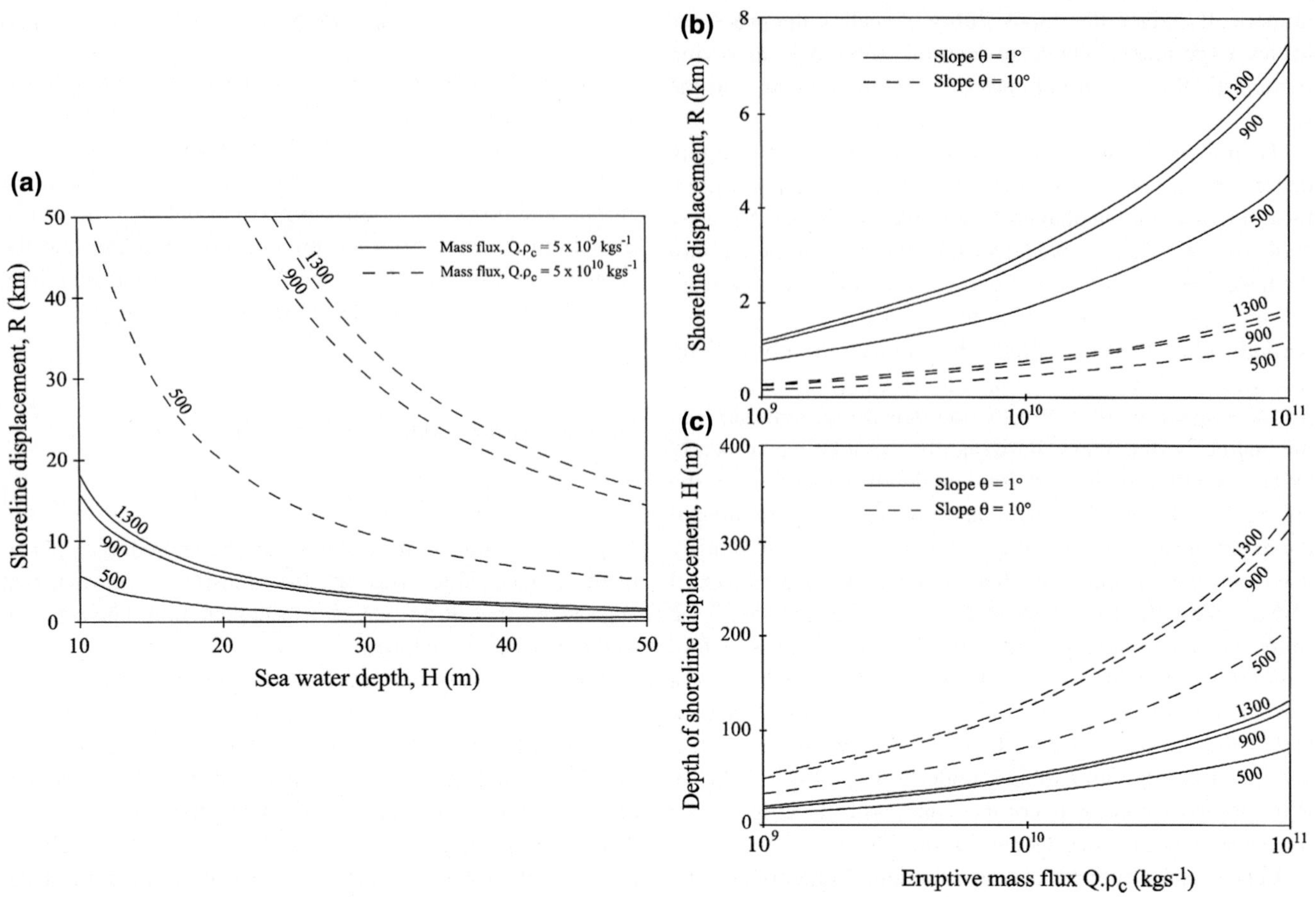

Fig. 11.5 Results of numerical simulations of the entry of pyroclastic flows into water. **a** Calculated relationships between different pyroclastic flow densities (numbers on curves in kg m^{-3}), mass flux (solid and dashed curves in kg s^{-1}), shoreline displacement distance (km), and seawater depth (m) for scenarios where pyroclastic flows enter the sea where the coastal plain and shelf slopes are very minor (up to a few degrees). **b** Calculated relationships between pyroclastic flow density, seafloor slope, shoreline displacement distance, and eruptive mass flux. **c** Corresponding calculated water depths. (From Legros and Druitt 2000)

$$R = [Q/2\pi Fr] \times \sqrt{[1/gH^3]\left[\rho_a/(\rho_c-\rho_a)+(\rho_w/(\rho_w-\rho_c)][(\rho_c-\rho_a)/(\rho_w-\rho_a)]^3\right.} \quad (11.12)$$

On a slope of θ, R becomes

$$R = [Q/2\pi Fr]^{0.4} \times \left\{[1/g\tan^3\theta]\left[\rho_a/(\rho_c-\rho_a)+(\rho_w/(\rho_w-\rho_c)][(\rho_c-\rho_a)/(\rho_w-\rho_a)]^3\right.\right\}^{0.2} \quad (11.13)$$

When the eruption begins from a subaerial vent on a volcanic island with a radius of R_o, and the distance of shoreline displacement is designated R_D, R in Eqs. 11.5, 11.6, and 11.12 is replaced by $[(R_o + R_D)^2 R_D^3]^{0.2}$ (Legros and Druitt 2000).

Key caveats to the numerical simulations are that the currents have to be sustained, and at a high eruption or discharge rate (Legros and Druitt 2000). Using the above relationships, Legros and Druitt (2000) calculated multiple scenarios to illustrate distances of shoreline displacement as functions of varying flow densities and water depths (Fig. 11.5a), and seafloor slope and eruption mass flux (Fig. 11.5b, c), using estimated discharge rates from major eruptions that are caldera-forming ($\sim 10^9$ kg s^{-1}). The results show that dense flows and flows with high mass flux displace shorelines further than low density and lower mass flux flows, and all flows displace the shoreline more so where water depths are shallow (Fig. 11.5a). Furthermore, denser flows displace the shoreline further on lower slopes than steeper slopes and further than low-density flows (Fig. 11.5b, c).

Lastly, here, Legros and Druitt (2000) also considered whether or not the deposition of pyroclastic flows erupted during the 1883 eruption of Krakatau, west of Java, could have resulted from them displacing the shoreline. These deposits are currently preserved within a 15-km radius of the volcano in the offshore shallow submarine environment of the Sunda Straits (Self and Rampino 1981; Francis and Self

1983; Mandeville et al. 1994, 1996). Prior to the eruption, the offshore shelf was several tens of m deep for up 10–15 km offshore to the north. This area was inundated during the eruption, even creating temporary islands of ignimbrite and pumice banks (Verbeek 1884; Self and Rampino 1981). Since the eruption, these non-welded deposits have been eroded by marine currents. Coring and TRM analyses have determined that the temperature of emplacement was >350– ~550 °C (Mandeville et al. 1994). The average discharge rate during the eruption was 10^9 kg s^{-1}, but was probably higher during peak discharge. Using these constraints, Legros and Druitt (2000) calculated that the flow front could have displaced the sea, so allowing deposition of hot pyroclastic flows under essentially subaerial conditions, not under water.

11.5.4 Temperature of the Flows and Thermal Interaction with a Body of Water

There have been very few quantitative studies of the thermal interaction between pyroclastic flows entering an ambient body of water and that water. Howells et al. (1985) speculated that pyroclastic flows entering the sea and flowing underwater may be insulated from the overlying liquid water mass by a vapour film of steam (Leidenfrost effect), so allowing them to retain heat and thus being able to weld. This explanation was proposed to explain welding features within a succession of Ordovician ignimbrites described from North Wales (see Fig. 15.15). These ignimbrites are up to several tens of m thick, often densely welded and sometimes rheomorphic, and interbedded with and can be traced from subaerial environmental lithofacies assemblages to offshore shelf paleoenvironments. However, the process was not quantified, and paleo-ecological affinities of invertebrate fossils and sedimentological evidence in the shelf sediments above and below the ignimbrites indicate maximum water depths of only tens of m (Brenchley and Pickerill 1980; Cas and Wright 1987; Sect. 11.6.1.2). This suggests that these large-volume pyroclastic flows would have displaced the shoreline, and been emplaced in essentially subaerial conditions as proposed by Cas and Wright (1987, 1991) and is consistent with the modelling of Legros and Druitt (2000) discussed above. Busby (1984a, b, 1986), Kokelaar and Busby (1992) and Kessel and Busby (2003) further considered such a Leidenfrost steam insulation effect to be important in explaining welding in deep-water submarine pyroclastic deposits, including submarine caldera filling ignimbrites.

Allen and Cas (2001) undertook simple calculations to determine how much cooling a pumiceous pyroclastic flow would undergo when flowing over the sea, such as during the 161 ka Kos Plateau Tuff (KPT) eruption, Greece. They based this on the principle of conservation of thermal energy:

$$T_f = T_p - \left[m_w c_w \left(T_{100} - T_w + m_w L + m_v c_v \left(T_f - T_{100}\right)\right)\right] / m_p c_p \quad (11.14)$$

where T_f is the final equilibrium temperature of the flow, T_p is the initial temperature of the flow, T_w is the initial water temperature, T_{100} is the temperature of water heated to boiling point, 100 °C, m_w is the mass of water ingested into that part of the flow flowing over water, L is the latent heat of vaporisation of water (2.3 × 10^6 J kg^{-1} K^{-1}), c_w is the heat capacity of water (4228 J kg^{-1} K^{-1}), c_v is the heat capacity of steam (>2 J kg^{-1} K^{-1} at >100 °C), and c_p is the heat capacity of the ignimbrite, taken as that of pumice (1200 J kg^{-1} K^{-1}).

However, the problem with such a multi-variable relationship is that there is often not a unique solution, especially for old deposits, because some assumptions have to be made about the values for some of the variables. Multiple scenarios using different values for the variables is the only approach that can be presented, without a unique solution being possible. Nevertheless, results showed that the KPT flows were able to flow over the sea for at least 35 km from the source to the south. Allen and Cas (2001) found that the amount of cooling during flow over water depended on the mass of water that was ingested into the flow and vaporised as steam within the flow, the mass of the pyroclastic flow, and its heat capacity. Cooling of tens to several hundred °C was possible for some of the KPT flows. Some of the pyroclastic flows erupted from Krakatau in 1883 flowed 45–60 km over the sea, but were still hot enough to set fire to coastal villages and forests on Sumatra when they reached there. The Campanian Ignimbrite flowed over the Bay of Naples for over 20 km to reach the Sorrento Peninsula, where TRM data show that the deposit was emplaced above the magnetite Curie temperature, i.e. above 580 °C (Ort et al. 2015).

In a significant advance to the analogue modelling approach, Freundt (2003) undertook an extensive series of experiments and calculations to assess the effects of the entry of "real" hot granular flows of pyroclastic debris into water. Freundt (2003) used a 3 m long tank which was 48 cm deep and 36 cm wide, with a ramp at one end having a slope of 26°. Poorly sorted granular flows of pumice and ash, with <4 mm grain size, weighing from 2.5 to 7 kg were heated to temperatures of from 61 to 403 °C and released from a chute sloping at 36° onto the ramp. The granular flow masses had a loosely packed density of 960 kg m^{-3} before release, and flowed into water depths of 4–19 cm at fluxes from 900 to 2,200 g s^{-1} and flow velocities of 1.5–2.5 m s^{-1}. Thermocouples, pressure sensors and video cameras recorded the flow events. Freundt (2003) also discussed scaling issues and limitations between his experiments and real-world pyroclastic density currents entering bodies of water, to which interested readers are referred.

In the experiments, a basal granular flow, dominated by saltating grains, and a very fine ash cloud developed in the chute during subaerial initial flow. Freundt (2003) distinguished the nature of the interaction between relatively "cool" (<150 °C) and hot (>250 °C) experimental pyroclastic flows and the water in the tank, noting that flows of intermediate temperature showed transitional behaviour. For the cool flows, a series of tsunami waves, a bottom-flowing turbidity current, and ash-cloud surges that flowed over the water formed from the point of entry into water. For the hot flows (Fig. 11.6), more voluminous ash-cloud surges formed (Fig. 11.6b, c, f), a tsunami wave formed at the point of entry of the hot flows into the water (Fig. 11.6a, b), where intense mixing and phreatic explosions also occurred, forming an ash plume and ash-cloud surges as well as a laterally flowing mass of material along the surface. These generated more explosions away from the "shoreline" (Fig. 11.6b, c, d, e), producing fallout of ash and pumice, as well as downward sinking plumes of sediment (Fig. 11.6c–f) from the surface mixing zone, which then transformed into bottom turbidity currents (Fig. 11.6g). An underflow also formed from the shoreline mixing zone.

The cool flow deposits were quite well sorted and normally graded, and were thickest at the base of the ramp, where deposits from slumps on the ramp accumulated. Although the hot deposits were normally graded and more poorly sorted, they were also marked by local mounds on the surface, resulting from sedimentation from downward sinking plumes from the water surface, superimposed on the tank bottom turbidity current deposit. Hot flow deposits extended further into the tank than cool flow deposits. Pumice rafts were common after the cool flow experiments, whereas almost no pumice rafts formed with the hot flow experiments. A turbidity current underflow formed further from the "shoreline" in the hot flow experiments than in the cooler flow experiments, related to deceleration rates of the flows. Freundt (2003) also noted that steam explosions did not occur in experimental flows where ash particle temperatures were <200 °C; this is because at those temperatures insufficient volumes of steam are formed from water interstitial to ash particles to coalesce and expand explosively. He also pointed out that rapid heat transfer could occur where there was a large population of fine grains/ash, but this then required a mechanism of mixing in enough water to interact with the surfaces of those ash particles.

Allen et al. (2012) undertook similar experiments, as part of a study to determine the origin of the regionally extensive submarine Znp tephra deposits in the Niigata Basin, off the northwest coast of Honshu, central Japan. They documented in more detail the characteristics of the deposits from hot experimental pumice-lithic and ash flows with a flow density of $\sim 1{,}010$ kg m^{-3} into a similar water tank as used by Freundt (2003). They also found that for hot flows phreatic explosions occurred at the shoreline, generating an ash-cloud surge over-flow, and a mixing zone between the experimental pyroclastic flow and the water, from which a bottom turbulent turbidity current formed. Initial flow temperatures ranged from 243 to 310 °C. Overall, the deposits showed a concentration of dense clasts in proximal areas, particularly adjacent to the break in slope at the foot of the ramp (a hydraulic jump), and then increasing pumice clast concentration was noted into medial distances and increasing ash content into distal settings. Deposit thickness declined from proximal to distal, sorting improved in a down-flow direction, and grain size decreased. In some experiments, pumice formed an upper layer above the dense clast zone in proximal areas and thickened into medial areas where sometimes it showed reverse grading.

To end here, we also refer the reader to Chap. 12 (Sect. 12.6.5) regarding the dynamics of subaqueous granular mass flows, which are relevant as soon as pyroclastic flows entering water ingest water and transform into water-supported mass flows.

11.5.5 Velocity of the Flows

Not a great deal can be said about the velocity behaviour of real pyroclastic flows entering water and, while some observations can be made from the experiments described above, there are scaling problems in translating these to the dimensions of large, natural pyroclastic flows entering the sea. Freundt (2003; see above) showed that experimental ash-cloud surges travelled faster over water than the turbidity current underflows, and that the hot flows travelled faster than the cool flows (Fig. 11.6h). Flow velocity is important because it determines the level of flow turbulence and therefore affects the amount of mixing of the flows with the ambient water mass at the front, top, margins, and for over-flows travelling over water, at the base. However, flow velocity can be counteracted by high levels of particle concentration/flow density in determining levels of turbulence.

11.5.6 Mixing of Ambient Water into Flows

As discussed in Sects. 11.5.1–11.5.4, analogue simulations of pdc's entering a body of water have shown that zones of mixing occur at the shoreline entry points, and the degree of mixing is dependent on the properties of the experimental currents and the slope and topography of the entry points

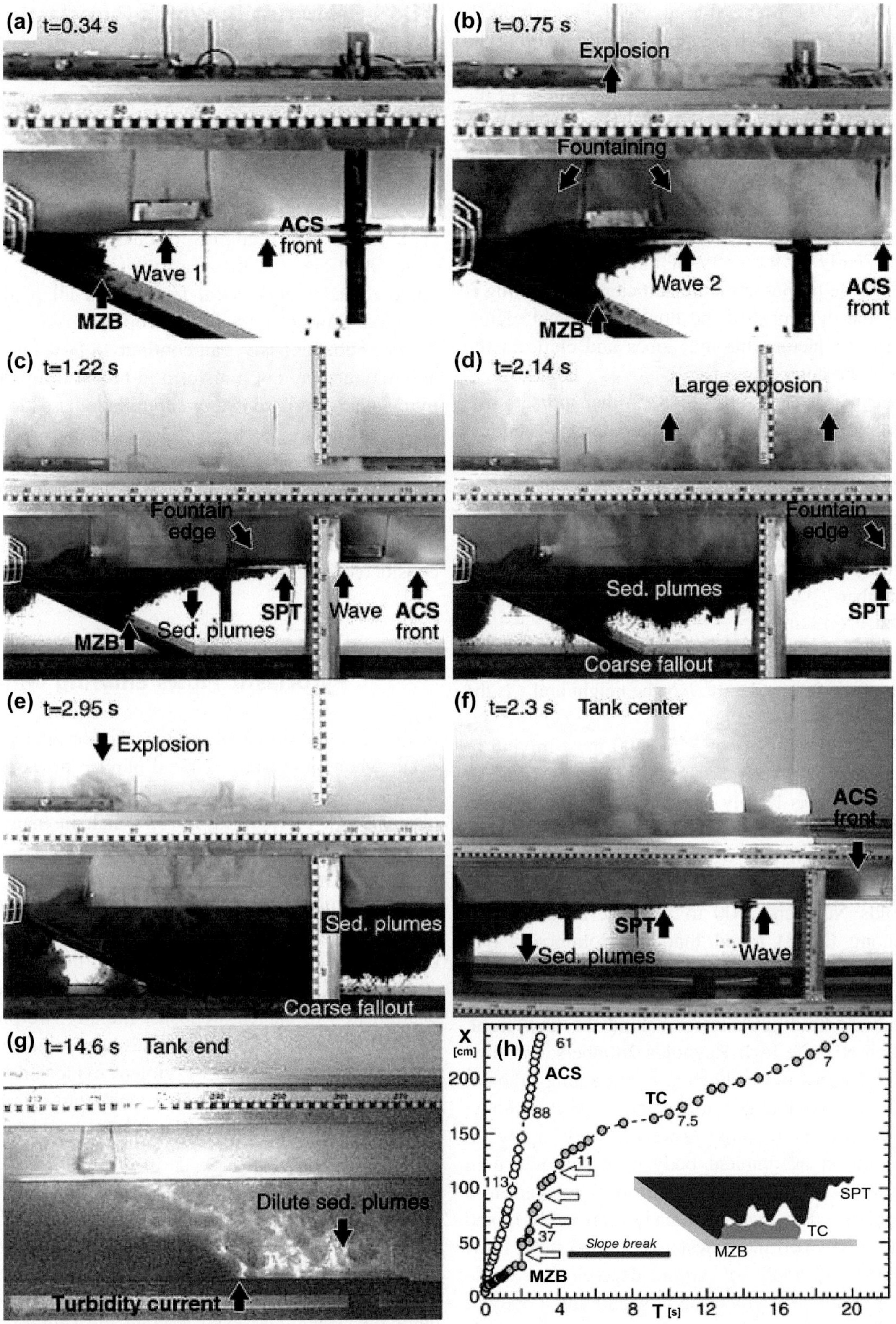

Fig. 11.6 Video prints of experimental flows of hot-ash (350 °C) flowing into water showing the first few seconds of entry (**a**–**f**), and at the end of the run in g the distal turbidity current advancing through sinking ACS (ash cloud surge) derived sediment plumes. Note larger scales in **a** and **b**, and how sites of distinct explosions vary between **b**, **d**, and **e**, the one farther out in d being the most vigorous. SPT marks the tip of the underwater sediment-plume region, and MZB is the basal front of the mixing zone. Coarse fallout from sediment plumes is seen in **d**, **e**. Position of flow fronts of ACS and turbidity current (TC) over time is shown in **h**; X is Distance from shoreline in cm, and T is Time in seconds; white arrows indicate sediment-plume additions to turbidity current as illustrated in the cartoon inset. (From Freundt 2003, experimental run 15; video of this run is available as Electronic Supplementary Material to the original publication)

(Monaghan et al. 1999; McLeod et al. 1999; Freundt 2003; Figs. 11.1 and 11.2). Monaghan et al. (1999) noted that the degree of mixing increased as the entry angle of experimental (pyroclastic) density current (i.e. ramp angle in the experimental tank) increased, supporting the earlier findings of Britter and Linden (1980). They also noted that most mixing occurred in the head of the density current. In a more detailed analysis, Bhaganagar (2017) has shown that for dense currents that enter a lower density medium most mixing occurs early in the entry of interaction of experimental currents. He proposed that the efficiency of mixing is controlled by the dynamics of the flow head (level of turbulence being one factor), through lobes and clefts in the flow front, and the entry angle.

Balasubramanian and Zhong (2018) further indicate that degree of mixing is influenced by the density contrast between the density current and the ambient fluid, the height of the density current flow front, and the slope of the terrain controlling the entry angle of the density current into ambient fluid. However, they found that the degree of mixing or entrainment of ambient water was affected by the Reynolds Number:

$$Re = Uh/v \tag{11.15}$$

where U is the flow front velocity, h is the height and v is the kinematic fluid viscosity.

At low Reynolds Numbers of 845–2700 most mixing or entrainment of ambient water occurred at the flow front. In this regime, the flow is largely shearing under the ambient fluid, and the interface between the two is marked by low-amplitude undulations or waves called Holmboe waves, which entrain 9 times less ambient fluid than the flow front. From Reynolds Numbers 2700 to >12,000, Balasubramanian and Zhong (2018) found that the Holmboe waves transformed into large amplitude Kelvin–Helmholtz instabilities with significant vortex structure that progressively increased ambient fluid entrainment behind the flow front. However, even at these high Reynolds Numbers, the flow front was still 2 times more effective at entraining ambient fluid than the vortexes that developed above the flow body.

These concepts of mixing, however, only apply to aqueous currents in an ambient body of water, which in terms of the discussions here are strictly only relevant to a gaseous pyroclastic flow that has already thoroughly mixed in water and transformed into a water-supported mass flow. The constraints may apply to varying degrees to a dense particulate gas-supported pyroclastic flow entering a body of water if density contrast is the main factor. However, little research has been done to assess the mixing processes of fluid systems with fluids that are in different phases (gaseous flow in water), let alone quantify them.

11.6 Scenarios in Which Pyroclastic Flows Enter Subaqueous Environments

Cas and Wright (1987, 1991) identified multiple, specific scenarios during which pyroclastic density currents interact with and enter a body of water, either through eruption from subaerial vents and flow of pyroclastic flows into a body of water, or through eruption from subaqueous vents (Fig. 11.7). The scenarios discussed below represent the main situations, but there are variations in all of these that could reflect unique local features. In all of the scenarios depicted, flows of pyroclastic debris entering water, or being erupted subaqueously, can continue to flow into deep water, and may or may not transform in transit from gas-supported pyroclastic flows to water-supported mass flows of pyroclastic debris, as affected by the factors discussed above. The examples presented below represent some of the best-documented examples in which both the deposit characteristics and the evidence for the paleoenvironment of deposition are well documented. Other examples have been considered in Cas and Wright (1991).

11.6.1 Subaerially Erupted Pumiceous Pyroclastic Flows Entering Water

In this first and principal section, we consider scenarios that occur when subaerially erupted pumice flows enter the sea. Broadly, we are therefore considering lower density pyroclastic flows ($\pm 1{,}000$ kg m^{-3}), which in many cases may enter the sea on a low topography shoreline at a low angle of incidence ($\sim$<30°). However, no large-scale ignimbrite-forming eruption has ever been witnessed, never mind observed entering the sea. We therefore have to apply both observations of modern pyroclastic flows and analogue experiments as described above, as well as the results of examination of ancient deposits. Initially, we present a scenario in which the pumice flow has been almost destroyed as a gas particulate system by violent explosions on entering the sea, which we believe is a valid starting point for discussion, or generic model (Fig. 11.7a).

11.6.1.1 Explosive Disintegration and Mixing at, or Near the Shoreline—A Generic Model

Many pyroclastic flows, especially small volume ones, will interact explosively with the water mass, generating phreatic explosions at the shoreline (or "littoral blasts") where they enter the water, and even beyond the shoreline where hot surge overflows may be formed (Fig. 11.7a; Dufek et al. 2007). In some cases, these explosions will only affect parts of the flows; in others, they could terminate the passage of

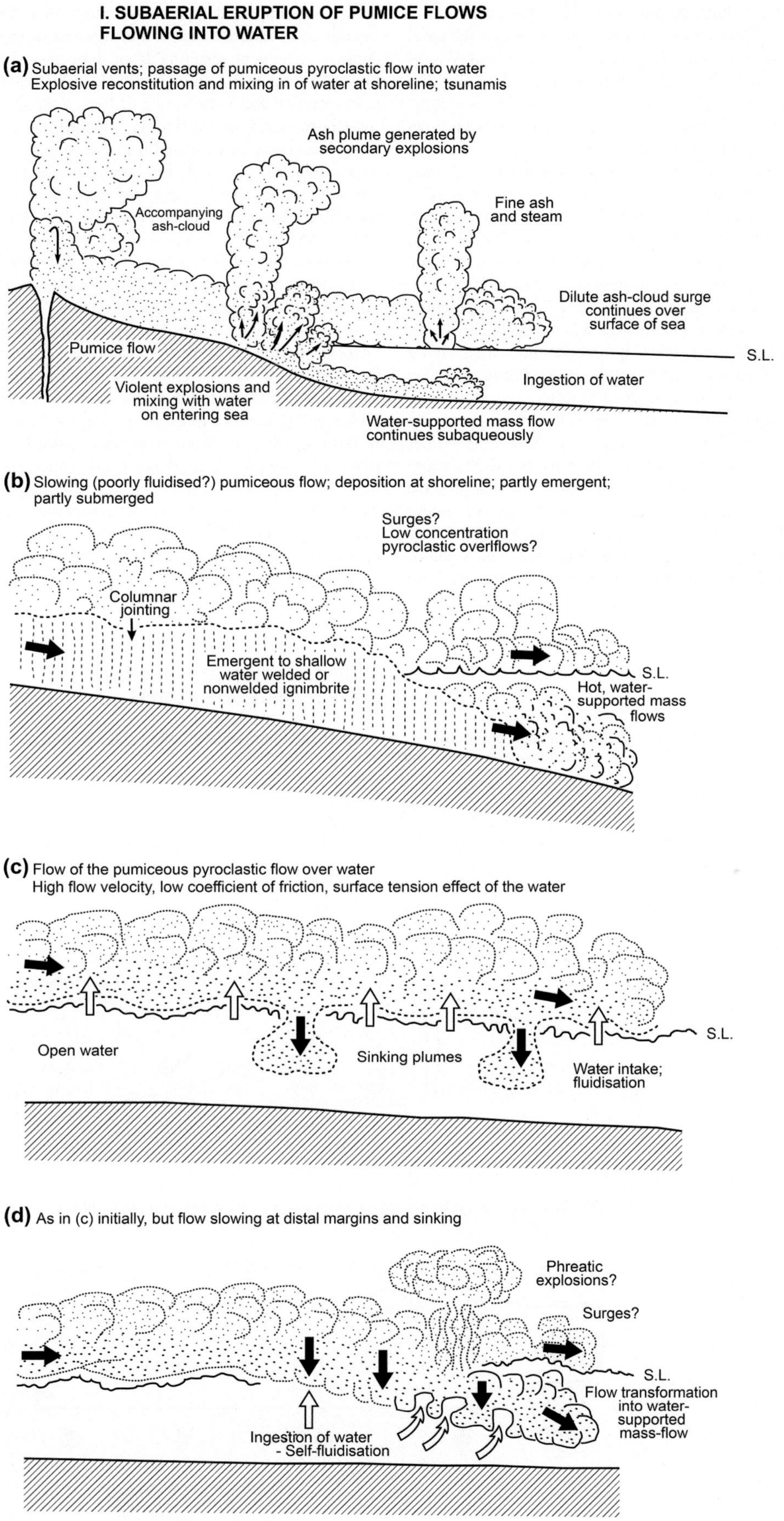
I. SUBAERIAL ERUPTION OF PUMICE FLOWS
FLOWING INTO WATER
(a) Subaerial vents; passage of pumiceous pyroclastic flow into water
Explosive reconstitution and mixing in of water at shoreline; tsunamis
Ash plume generated by secondary explosions
Accompanying ash-cloud
Fine ash and steam
Dilute ash-cloud surge continues over surface of sea
Pumice flow
S.L.
Ingestion of water
Violent explosions and mixing with water on entering sea
Water-supported mass flow continues subaqueously
(b) Slowing (poorly fluidised?) pumiceous flow; deposition at shoreline; partly emergent; partly submerged
Surges?
Low concentration pyroclastic overlflows?
Columnar jointing
Emergent to shallow water welded or nonwelded ignimbrite
S.L.
Hot, water-supported mass flows
(c) Flow of the pumiceous pyroclastic flow over water
High flow velocity, low coefficient of friction, surface tension effect of the water
S.L.
Open water
Sinking plumes
Water intake; fluidisation
(d) As in (c) initially, but flow slowing at distal margins and sinking
Phreatic explosions?
Surges?
S.L.
Flow transformation into water-supported mass-flow
Ingestion of water
- Self-fluidisation

◀ **Fig. 11.7** Several of the multiple scenarios in which pyroclastic flows can interact with a body of water when they flow into it from a subaerial eruption, or are erupted underwater (Modified from Cas and Wright 1987, and Cas and Wright 1991). **a** General scenario for the entry of a pumiceous pyroclastic flow into water at a low gradient shoreline, causing phreatic explosions, an ash plume, and a surge-like overflow, that in turn may experience further phreatic explosive disruption. At the shoreline a mixing zone between the flow and the water mass forms, enhanced by phreatic explosions, and leads to the formation of a water-supported, subaqueous mass flow (debris flow, turbidity current) that produces a hydraulically well-sorted, graded bedded deposit. Tsunamis can also form. **b** Large volume pumice and ash pyroclastic flow entering a low gradient shoreline at high mass and volume flux. Such flows are capable of instantaneously pushing back the shoreline well out to sea, as they generate a tsunami bow wave (cf. Fig. 11.2c, d), leading to a subaerial platform sheet ignimbrite deposited under subaerial conditions in what was a shelf setting. Such ignimbrites can be hot enough to weld. Offshore, in deeper water, the flow may transform into a water-supported subaqueous mass flow through mixing at the flow front. **c** High-velocity, low-density flow over water facilitated by surface tension. Basal parts of the flow and dense clasts can drop out, forming downward sinking plumes of debris that produce seafloor deposits with turbidite characteristics. **d** As in (**c**), under decelerating flow conditions, leading to large mass sinking of the flow material, self-fluidisation as it sinks, and transformation into seafloor mass flows. **e** Eruption column originating from a submarine vent and collapsing from subaerial levels or from a subaqueous level to form an initial gas-supported pyroclastic flow that may transform through mixing in of external water into a water-supported mass flow. Some deposits near vent resulting from deposition from the initial gaseous flow part may be hot enough to weld. **f** Subaqueous, eruption at a high mass eruption and volume rate of large volume, gas-supported flows of pyroclastic debris in a subsiding, subaqueous caldera depression. These are at least partially insulated from the ambient water mass by escaping over-pressured gas and low thermal conductivity across the gas envelope at the top of the erupting mass. Rapid eruption and deposition rates hinder ingress of water, ensure heat retention and allow welding in some cases, especially near vent(s). **g** Eruption and passage of dense clast (block and ash) pyroclastic flows, such as block and ash flows into water. Phreatic shoreline explosions, surge overflows and buoyant ash plumes may form, and because the flow is dense, most of it mixes and flows into and under water, transforming into subaqueous, water-supported mass flows. Differing scenarios exist depending on the gradient of the shoreline. The shallower the gradient, the greater surface area for interaction at the shoreline leads to more vigorous and prolonged phreatic explosive activity at and beyond the shoreline. **h** Wholly subaqueous eruption and passage of dense clast flows of pyroclastic debris, which, depending on mass and volume eruption rate, may be gaseous near vent but transform into syn-eruptive, water-supported mass flows of pyroclastic debris away from vent. **i** Post-eruption resedimentation of pyroclastic debris derived from cold, previously deposited pyroclastic flow deposits, by subaqueous water-supported mass flows

Fig. 11.7 (continued)

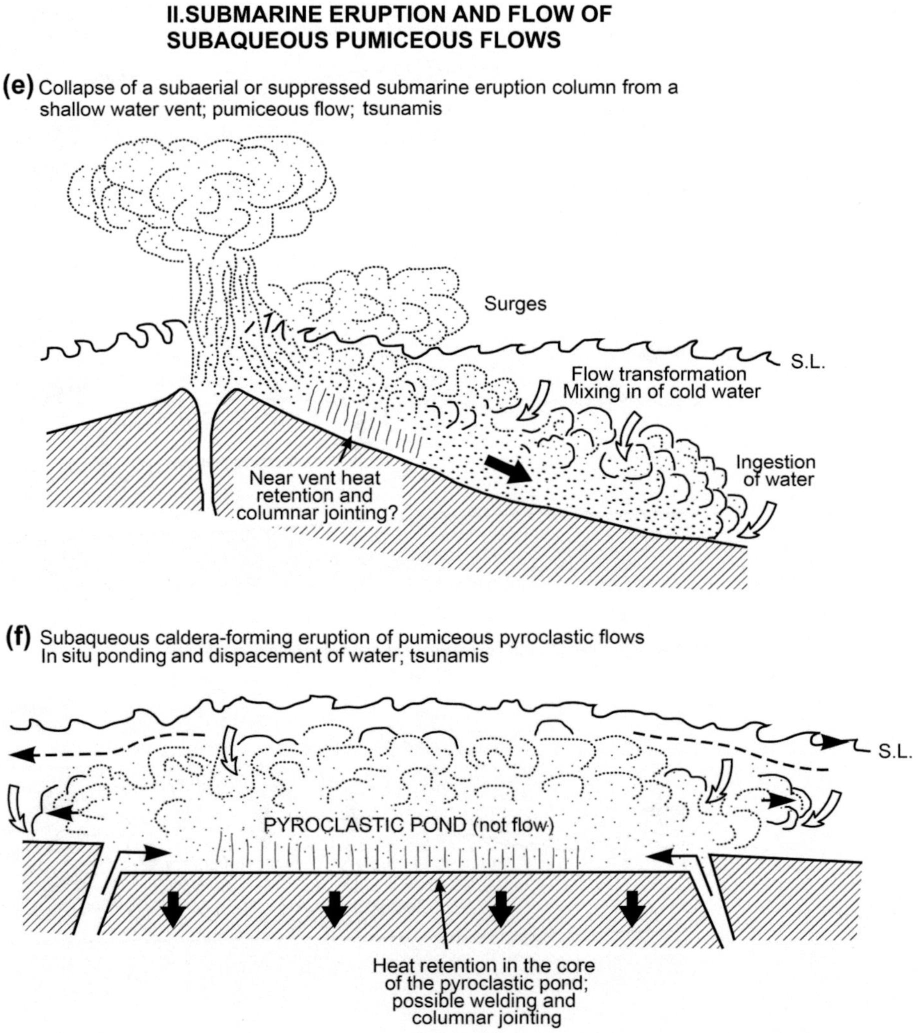

Fig. 11.7 (continued)

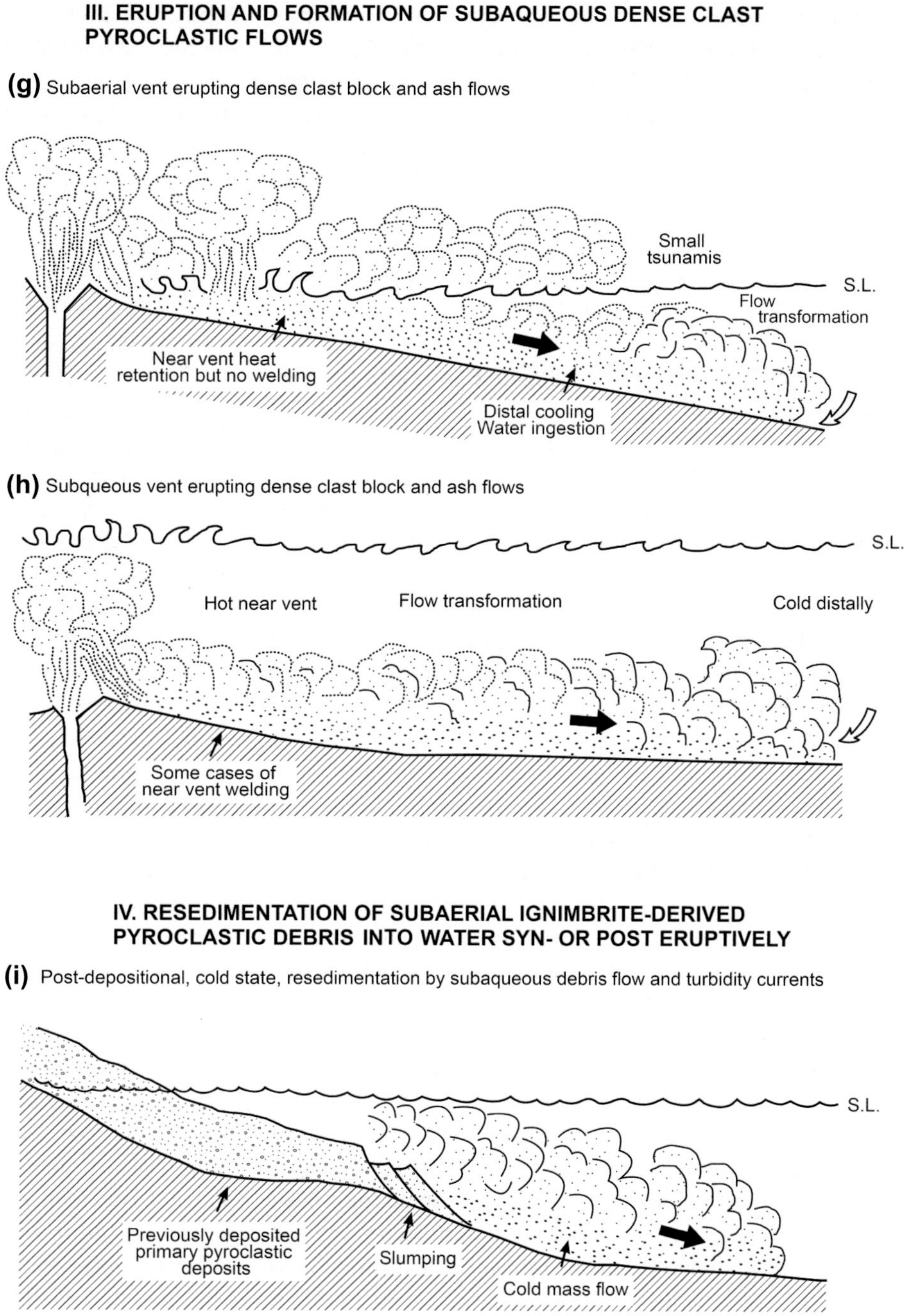

the flows, and/or generate secondary "rootless" phreatic base surges capable of flowing radially back onto land or out to sea (e.g. 12–13 July 2003 eruption of Soufrière Hills volcano, Montserrat; Dufek et al. 2007; cf. Edmonds and Herd 2005; Edmonds et al. 2006). This helps explain land-ward directed surges observed after pyroclastic flows entered the sea during the eruptions of Mt Pelée 1902 (Fisher and Heiken 1982), and deduced from anisotropy of magnetic anisotropy (AMS) studies of the ~5 ka Secche di Lazzaro succession at Stromboli (Giordano et al. 2008). Dufek et al. (2007) indicate that such explosions are enhanced when pumice fragments 1–5 mm in diameter are abundant; because of the improved efficiency of steam generation at such grain sizes which optimises the surface area for interaction between hot pyroclasts and external water. Such phreatic explosions can also generate buoyant subaerial ash plumes that disperse ash and accretionary lapilli away from the shoreline (Walker 1979; Edmonds et al. 2006; Herd et al. 2005). During the 3 July 2019 pyroclastic flow at Stromboli, ash "rain" occurred following the lofting of the ash cloud which formed as the flow entered the sea and inflated due to rapid mixing with water (see Fig. 11.9b). Phreatic explosions are likely to generate mixing of the pyroclastic flow mass and the water mass, allowing transformation of at least part of the flow system into a subaqueous, water-supported mass flow (Fig. 11.7a; Freundt 2003).

12–13 July 2003 Entry of Pyroclastic Flows into the Sea, Montserrat

Although the pyroclastic flows which reached the shoreline and interacted with the sea during the 1995–2013 eruption of the Soufrière Hills volcano (Plate 11.1b, c; Fig. 11.1b) were mostly dense clast block and ash flows, they nonetheless provide a very good eyewitness account of all of the phenomena described above. Specifically, the 12–13 July 2003 lava dome collapse event illustrates very well the range of phenomena when pyroclastic flows enter water and interact explosively with it (e.g. Edmonds and Herd 2005; Edmonds et al. 2006). Numerous pyroclastic flows entered the ocean during the collapse, depositing approximately 90% of the total material into the submarine environment (Trofimovs et al. 2006, 2008). The pyroclastic flows caused major phreatic explosive activity when they entered the sea. During peak collapse conditions, as the main flow penetrated the air-ocean interface, phreatic explosions generated a radially flowing, pyroclastic surge that flowed both seawards, and landwards, travelling as much as 4 km inland and up to 300 m in elevation. The entry of the flow also produced a tsunami(s) that was up to several m high and travelled at least 65 km to Guadeloupe (Edmonds et al. 2006; Herd et al. 2005). At the shoreline, the phreatically disrupting pyroclastic flow transformed into a large, water-supported, coarse debris or granular flow transforming into a deep-water turbidity current underflow that produced a hydraulically well-sorted and graded bedded, syn-eruptive, turbidite deposit offshore (see Sect. 11.6.3.1 and Fig. 11.42; Trofimovs et al. 2006, 2008).

A succession of such small-volume pyroclastic flow events at Soufrière Hills volcano gradually built a prograding bedded pyroclastic fan delta, which extended out from the mouth of the Tar River valley (Fig. 11.8).

3 July and 28 August 2019 Entry of Pyroclastic flows into the Sea, Stromboli, Italy

Between 3 July and 28 August 2019 , Stromboli experienced its most intense explosive activity of recent years, with two vulcanian paroxysmal explosions that occurred with only very minor precursory phenomena (e.g. Giordano and De Astis 2021; http://lgs.geo.unifi.it/). Both explosions were dominated by the rapid rise of a gas-rich batch of magma, commonly called "golden" or LP (low porphyritic) because of its production of highly vesicular, yellow, aphyric pumice (Di Carlo et al. 2006). This explosively displaced the column of magma filling the persistently open conduit of the volcano, resulting in Vulcanian explosions. The explosive events were similarly very short-lived, each lasting about 5 min. Both explosions were directed, displacing part of the summit cones seaward and producing both vertical and lateral jets of gas and pyroclastic debris into the atmosphere contemporaneously. The vertical jets gave rise to 4–5 km tall buoyant plumes (Fig. 11.9a), while the lateral blasts transformed into laterally directed pyroclastic flows (Plate 11.1e, Fig. 11.9a, b) that rushed down the Sciara del Fuoco sector collapse scar on the volcano's northern flank, reaching the sea in about 20–30 s, at an average speed of some 50 m s^{-1}.

Pyroclastic flows entered the sea at a high angle of about 30° giving rise to a tsunami wave which was measured up to 2 m in total amplitude at the pressure-buoy directly in line of the flow on 3 July (http://lgs.geo.unifi.it/). Fortunately, the fragmented nature of the volcanic debris entering the sea and its relatively small volume caused the rapid attenuation of the tsunami wave. Thus, the popular beaches of the island, just a few km away from the site of the tsunami generation, were reached within 2–3 min by waves of only 20–30 cm which largely passed unnoticed by the population. At the shoreline, pyroclastic flows encountered a drastic break in slope and were visually affected and spread laterally, generating a buoyant ash cloud as the flows spread over water. The flows moved $\sim$800 m over the sea surface and rapidly waned. An interesting observation is that the vertically rising buoyant ash cloud detached from the flow (as photographed in the 28 August event; Fig. 11.9b), indicating that the basal part of the flow rapidly lost the ability to transfer mass (ash) and heat to the ash clouds, suggesting that most of the debris was rapidly lost below sea level, transforming and travelling as a submarine mass flow. However, the buoyant ash cloud was able to produce fallout deposits of pumice and ash (Fig. 11.9c).

Fig. 11.8 View from east of the Tar River mouth pyroclastic fan-delta, Monserrat, built from repeated pyroclastic flows entering the sea after flow down the Tar River valley, causing a depositional regression as the shoreline migrates seaward. The summit of the active Soufrière Hills volcano is in cloud on the skyline. (Photo by Simon Young; *Source* British Geological Survey)

11.6.1.2 Pumiceous Flows Reaching the Shoreline (in a Waning Flow State) Depositing at, and Beyond the Shoreline

In this scenario, pumiceous pyroclastic flows are erupted from a subaerial vent or volcanic centre, and the subaerial pyroclastic flows wane and transform into a rapidly depositing system at the shoreline (Fig. 11.7b). If the flow system is small in volume, the scenario will largely be as already depicted in Fig. 11.7a, and discussed above. If the flow system is voluminous (km^3) and of high mass and volume flux (i.e. ignimbrite-forming), the deposit may be thick and continuous from the subaerial (is)land margin into the offshore area. If voluminous enough, ignimbrite deposits tens of m thick will cause the shoreline to be displaced seawards, regressing instantaneously (Fig. 11.7b), and generating a major tsunami(s). This happened during the

Fig. 11.9 Pyroclastic flows entering the sea at Stromboli volcano, Italy, during two eruptions in 2019. **a** Vertical eruption column rising a few seconds after the onset of the explosion at 14:46 UTC at Stromboli volcano, 3 July 2019. Note the associated directional pyroclastic flow ash cloud developing to the right. Inset shows the view from opposite direction of pyroclastic density current flowing down the Sciara del Fuoco at 14.46.35. The circular turbulent eddies in the sea represent ballistics launched in the initial explosion from the summit crater falling directly into water; these were up to 2 m in diameter and reached more than 100 m off the coast. (Photo courtesy of LGS, Laboratory of Experimental Geophysics, University of Firenze). **b** 28 August 2019 eruption at Stromboli. Note the lofting ash cloud of the lateral pyroclastic flow fully detached from the flow as pyroclastic debris rapidly sank into the sea. (*Source* Il Tempo.it). **c** Pumice lapilli deposited at the Ginostra Village harbor on a fishing boat. The late "rain" of ash splashing observed on the boat is related to the ash cloud generated at the time of pyroclastic flow entering the sea on 3 July

passage of pyroclastic flows into the sea during the 1815 eruption of Tambora volcano, on the island of Lombok, Indonesia (Self et al. 1984), and the 1883 eruption of Krakatau described earlier (Self and Rampino 1981; Simkin and Fiske 1983; Mandeville et al. 1994, 1996; Legros and Druitt 2000). The shoreline was displaced seawards up to 0.5–1 km after deposition of the Vesuvius 79 AD pyroclastic flows, making the once harbour cities of Pompeii and Herculaneum now stranded far inland, far from the seashore (Vogel and Marker 2010).

Only after subaerial and subsequent shallow marine erosion of the top of the ignimbrite deposit, and/or crustal subsidence and/or absolute sea level rise, will the shoreline and marine environments transgress back over the eroded top of the pyroclastic flow deposit into the previous offshore area, so producing marine deposits above the erosional unconformity at the top of the ignimbrite. In this scenario, the pyroclastic flow will have effectively deposited most of its bulk load under subaerial conditions due to instantaneous evacuation of the sea, even though in the paleo-offshore area the deposit is underlain and overlain by shallow marine sediments. The ignimbrite may have become welded if hot enough, largely free of any interaction with the ambient water, although superheating of the water-saturated substrate sediments could trigger phreatic explosions through the deposit. This could cause liquefaction and upward injection of the substrate into the base of the ignimbrite, and lead to fluxing of steam causing alteration (e.g. formation of various crystalline, spherulitic and vesicular textures through the welding profile; McArthur et al. 1998), diapiric structures which deform the welding fabric, or rootless vents (Wright and Coward 1977; Kokelaar and Koniger 2000; Lipman 2018), or, in the case of non-welded ignimbrite, zeolitised steam pipes and alteration plumes and haloes (Cas and Wright 1987; Giordano et al. 2018).

"Shallow Water" Ordovician Ignimbrites of the Northern Margin of the Welsh Basin, North Wales

We envisage that many of the welded Ordovician ignimbrites deposited in North Wales, during the evolution of the Lower Paleozoic Welsh Basin (Francis and Howells 1973; Howells et al. 1979, 1986; Reedman et al. 1987; Kokelaar and Koniger 2000; Chap.15), were deposited in such a context, with the hot, large volume pyroclastic flows instantaneously displacing the shoreline seawards (Cas and Wright 1987, 1991; Reedman et al. 1987; Orton 1991; McArthur et al. 1998; cf. Legros and Druitt 2000). The presence of shallow marine sediments both below and above the ignimbrites led some authors to suggest that the ignimbrites were emplaced subaqueously (Francis and Howells 1973; Howells et al. 1979; Fritz and Howells 1991; Kokelaar and Koniger 2000). However, using the paleo-ecological affinities of the shallow marine invertebrate fossils of the associated sedimentary successions, Brenchley and Pickerill (1980) estimated shelf water depths in the northern Welsh Basin of several tens of m. Howells et al. (1985) indicated that the ambient environments were tidal to sub-tidal, and Cas and Wright (1987) calculated water depths to be <10 m based on the amplitude of cross-beds in sedimentary rocks between the major ignimbrites. These data indicate that it is highly likely that the tens of m thick ignimbrites displaced the shoreline seawards and were deposited under essentially subaerial conditions.

The Garth Tuff, which is the lowest and most extensive member of the Capel Curig Volcanic Formation, is preserved over a strike length of $\sim$40 km, spanning multiple environments from subaerial alluvial fan in the north to shoreline to offshore shelf in the south, based on the facies characteristics of sediments beneath and above the ignimbrite (Fig. 11.10a; Francis and Howells 1973; Howells et al. 1979, 1985; Orton 1991; Fritz and Howells 1991). The Garth Tuff outcrops as a massive, thick welded ignimbrite that varies from 12 to 150 m thick (Fig. 11.10b; McArthur et al. 1998). It commonly has an irregular base of ignimbrite load casts sagging into underlying sediments that have been mobilised as flame structures or diapiric features around the load casts, with many m of relief in places (Figs. 11.10c, d; Francis and Howells 1973). In the distal-most marine setting in the south, it consists of a trail of isolated pods of ignimbrite that are disconnected from the main and continuous sheet to the north (Fig. 11.10a; Howells et al. 1985). In subaerial to shoreline settings, the Garth Tuff is underlain by trough cross-bedded fluvial braidplain sandstones (Fig. 11.10e), and in offshore settings by variably fossiliferous, in places wave rippled (Fig. 11.10f), cross-bedded and hummocky stratified sandstones which occur in local sedimentary packages representing deltaic distributary channels and channel mouth bars. These are separated by extensive areas of variably fossiliferous mudstones with thin Bouma sequence sandstone and siltstone beds and occasional hummocky cross-stratification, representing deltaic inter-distributary bays (Orton 1991; McArthur et al. 1998). The top of the ignimbrite is variably eroded, and is overlain by tractionally reworked volcanic sediments (Fig. 11.10g).

The Garth Tuff is strongly welded, often to the base, based on eutaxitic shard textures, and shows evidence of rheomorphic deformation of the welding foliation (McArthur et al. 1998; Chap. 10). In places the welded base has experienced quench fragmentation and preserves peperitic breccia textures with jigsaw-fit textures (Fig. 11.10h). The top of the Garth Tuff is extensively eroded, with much of the non-welded top removed, forming an irregular top and preserving the non-welded top only locally (McArthur et al. 1998). These features indicate that although the base was deposited on wet subaerial braidplain sediment to the north, and offshore deltaic sediments to the south (Orton 1991; McArthur et al. 1998), the top of the ignimbrite was

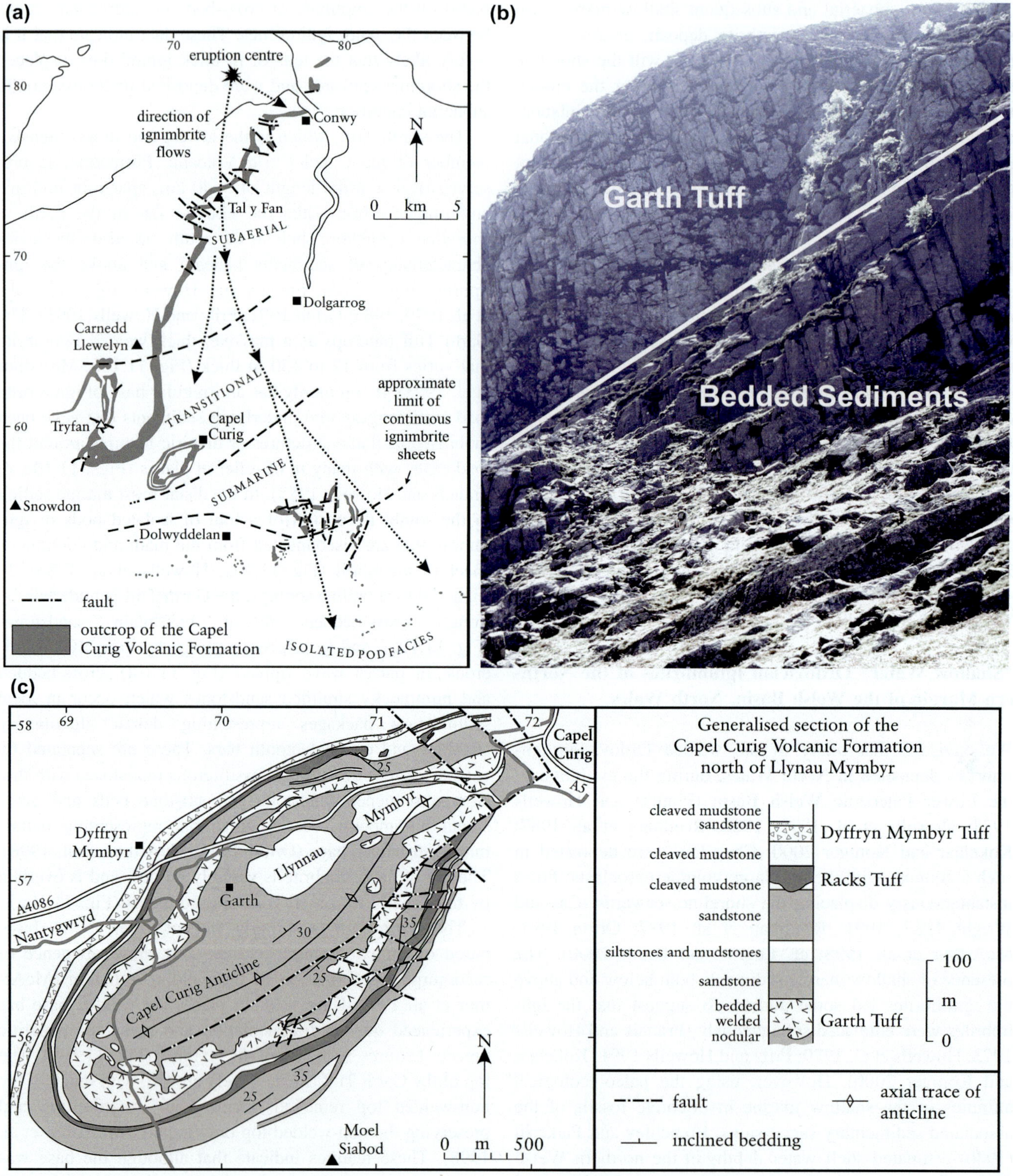

Fig. 11.10 Features of Ordovician Garth Tuff ignimbrite, Capel Curig Formation, which was deposited in the shallow, southeasterly sloping braid plain to deltaic shoreline zone, North Wales. **a** Distribution of the preserved remains of the Capel Curig Volcanic Formation, which is dominated by the lowermost Garth Tuff and its distal disconnected pods (*Source* Stephenson et al. 2000, After Howells et al. 1985). **b** Massive Garth Tuff ignimbrite tens of metres thick in cliff in background, overlying bedded sediments in foreground, Gallt-yr-Ogof. **c** Geological map of the Capel Curig Volcanic Formation in the Capel Curig Anticline, North Wales. Note the highly irregular base of the Garth Tuff due to large-scale ignimbrite load casts and sedimentary flame structures resulting from thermal and loading fluidisation of the water-saturated sediments under the Garth Tuff during its emplacement. (*Source* Stephenson et al. 2000, After Francis and Howells 1973).

(continued)

d Schematic cross-section of the Garth Tuff on the southeastern limb of the Capel Curig Anticline, highlighting the irregularity of the base where it has sagged down into underlying sediments at the time of emplacement (From Cas and Wright 1987, After Francis and Howells 1973). **e** Trough cross-bedding in braid plain sandstones below the Garth Tuff, Capel Curig Anticline. Outcrop is ~1 m high. **f** Symmetrical wave ripples beneath the Garth Tuff, Cwm Clorad Isaf, west of Capel Curig. **g** Large-scale cross-bedding in reworked volcanic sediments (at top of photo) above the Garth Tuff, Plas-y-Brenin, North Wales. **h** Peperite hyaloclastite breccia at the base of the Garth Tuff, with angular, blocky clasts of welded ignimbrite (light) separated by black mudstone matrix. Some clasts preserve jigsaw-fit relationships. **i** Isolated Carreg Alltrem pod of the Garth Tuff, at Dolwyddelan, North Wales. The cliff is ~30 m high, and consists of welded ignimbrite surrounded by mudstones and fine sandstones

Fig. 11.10 (continued)

emergent and subject to subaerial erosion after emplacement, before subsidence and/or transgression allowed marine environments to encroach again. It is likely, given the thickness of the ignimbrite (<150 m) and the shallow deltaic environment that the large-volume pyroclastic flow instantaneously displaced the shoreline, and the sea, seawards on the gently sloping shoreline-deltaic surface.

The disconnected offshore ignimbrite pods are up to 30 m thick, up to 200 m in diameter (Fig. 11.10i), separated by up to 1500 m from each other, and are also largely welded (Howells et al. 1985). They were interpreted as forming by the disintegration of the pyroclastic flow in an offshore submarine environment, with pods of the ignimbrite foundering down into the subjacent sediments through heat-induced fluidisation of the pore water in the underlying sediments, presumably like the large load casts at the base of the continuous sheet to the north (Howells et al. 1985). However, this does not explain how the pods became so disconnected by such large distances. A more feasible explanation was suggested by McArthur et al. (1998), who proposed that the marine paleoenvironment was shallower than previously envisaged, and that the whole ignimbrite was emergent when deposited. They also proposed that most of this originally continuous sheet where the pods occur was eroded away by subaerial erosion, scalping the sheet from above the subsided load casts, which are now preserved as isolated pods. It is very difficult to envisage how else the disconnected pods formed, and a modern analogue from the west Java coastline abutting the shallow Sunda Strait shelf is illustrated in Fig. 11.11. Orton (1991) proposed that regression due to regional uplift was already occurring before the Garth Tuff erupted and emplaced. There was thus a seaward migration of paleoshoreline environments (i.e. a regression) from before to after the emplacement of the Garth Tuff (ignimbrite) based on sedimentary facies characteristics.

The stratigraphically younger, Pitts Head Tuff ignimbrite (PHT) forms the base of the Snowdon Volcanic Group (Reedman et al. 1987). This was similarly deposited in a subaerial to shelf setting, but by this time the regional topographic slope of $\sim 2^{\circ}$ was from subaerial alluvial fan and braidplain in the southwest, to shoreface and shallow marine in the northeast (Fig. 11.12a; Reedman et al. 1987). The PHT also includes a major body of welded ignimbrite, ≥ 700 m thick which is interpreted as an intra-caldera fill and source vent (Reedman et al. 1987). The lowest and most extensive of two outflow ignimbrite sheets which comprise the PHT is up to 90 m thick, pervasively welded (Fig. 11.12b, c), commonly to the base, and shows abundant evidence of rheomorphism (Chap. 10), including flow folding of the welding foliation and autobrecciation at its base (Reedman et al. 1987; Kokelaar and Koniger 2000). Sediments under the ignimbrite where it flowed into water are clearly subaqueous, preserving beach, shoreface and pro-delta facies which are variably fossiliferous, and indicate water depths of <25 m (Brenchley and Pickerill 1980; Wright and Coward 1977; Reedman et al. 1987). Beach face planar bedding, symmetrical wave ripples, hummocky and swaley cross-stratification as well as angle of repose cross-bedding have all been recorded (Fig. 11.12d, e; Wright and Coward 1977; Reedman et al. 1987). The top of the ignimbrite has been variably eroded and reworked by shallow marine tractional currents (Fig. 11.12f, g; Wright and Coward 1977; Reedman et al. 1987; Cas and Wright 1991).

The PHT ignimbrite also has an irregular base where it flowed into water, with diapiric structures along its base, interpreted as rootless vents and fluidisation structures resulting from superheating of substrate, water-saturated sediments (Fig. 11.12h; Wright and Coward 1977; Reedman et al. 1987; Kokelaar and Koniger 2000). Although siliceous nodules or lithophysae and spherulites occur in many areas and throughout the PHT, they are particularly common near the base where the ignimbrite flowed over shoreface to shelf sediments (Fig. 11.12h, i; Wright and Coward 1977; Reedman et al. 1987). The nodules which are generally <1–5 cm in diameter are thought to represent original gas vesicles infilled with drusy quartz (i.e. amygdales), some of which occur in bubble trains extending parallel to diapiric structures. These formed as a result of heated vaporised, substrate waters rising from under the ignimbrite and streaming up through it leading to significant vapour phase alteration. More detailed analysis by Kokelaar and Koniger (2000) has suggested a more complicated picture, where wet seafloor sediments buried by the ignimbrite were heated and locally fluidised by steam, and hot ignimbrite with variable thicknesses of sedimentary substrate detached and slid downslope. Directions of sliding and the order of piling-up of slide sheets are shown by hot-state (rheomorphic) deformation fabrics and the geometric relations of detachment surfaces. Extensional disruption of the ignimbrite is marked by breaks in the sheet via which fluidised sediments were mobilised, locally to form rootless vents.

The lower ignimbrite of the PHT is likely to have been deposited by a pyroclastic flow that overwhelmed the shallow shelf environment, and given the great thickness (<150 m) of the ignimbrite and the shallow water depths, making it highly

Fig. 11.11 Erosional remnants of Plio-Pleistocene welded ignimbrite forming disconnected metre- to tens of metre-sized pods partly enclosed within onlapping shallow marine sands on the margin of the Sunda Shelf. Anyer Beach, approximately 50 km east of Krakatau, west coast of Java, Indonesia

unlikely that the ignimbrite was deposited wholly submerged in the shallow marine area. The pyroclastic flow clearly displaced the shoreline, causing instantaneous regression (Fig. 11.12a; Cas and Wright 1991; Reedman et al. 1987). The ignimbrite thus appears to have been deposited subaerially on the evacuated shelf, except perhaps in the most distal offshore settings. Eventually, through erosion of the top and regional subsidence, the sea transgressed again so that overlying sediments are fossiliferous (Orton 1987). Prograding pyroclastic fan-deltas on volcanic islands such as Montserrat (Fig. 11.8) provide a snapshot of the regressions that emplacement of pyroclastic flow deposits at shorelines can cause, although it is noted that the shoreline slope on such islands is much steeper than envisaged here, and so the regressions are less marked at these islands than where shoreline zone slopes are low. In addition, the volumes of the Montserrat pdc's were orders of magnitude smaller than the Welsh Basin ignimbrites.

Ordovician Ignimbrites of the Mweelrea Formation, Murrisk, South Mayo Trough, Ireland

Ignimbrites within the Ordovician Mweelrea Group of the South Mayo Trough in the Murrisk region, western Ireland, were originally interpreted by Stanton (1960) and Dewey (1963) to have been emplaced and welded in a shallow marine-deltaic setting. However, Pudsey (1984) favoured a largely subaerial fan-delta setting, transitional distally into a shelf setting, based on detailed sedimentological analysis. She described the ignimbrites as massive, non-graded and not welded, commonly having reworked tops overlain by conglomerate, consistent with subaerial emplacement and subsequent erosion.

More recent analysis by Riggs and McConnell (2017) provides an interesting case study of shoreline disruption during an ignimbrite displacement event and its effect on sedimentation, and how the shoreline succession responds and recovers. Sedimentary facies record the progradation of alluvial facies northwestward across shallow- and marginal-marine Mweelrea "passage beds" to marine Glenummera Formation facies (Fig. 11.13a). Early in Mweelrea Formation sedimentation, the first of a series of several, ignimbrites (individually <20 m thick) was emplaced, causing a disruption in this progradation. The preserved thickness of this lowermost ignimbrite is up to about 12 m, and the sedimentary response varied depending on the environment that was affected. In areas represented by fluvial facies, emplacement of the ignimbrite caused overlying sediment to become richer in feldspar as well as coarser grained with more abundant heavy-mineral laminations, but the depositional style was unchanged. The ignimbrite overlies decreasing thickness of fluvial sandstone westward, where it then directly overlies shoreline and shallow-marine passage beds. Here, sandstone that overlies the ignimbrite is fluvial, indicating that emplacement of the ignimbrite brought about an abrupt progradation (Fig. 11.13b). At the westernmost outcrops of the Mweelrea Formation, the ignimbrite lies in contact with the underlying marine, graptolite-bearing Glenummera Formation. Marine

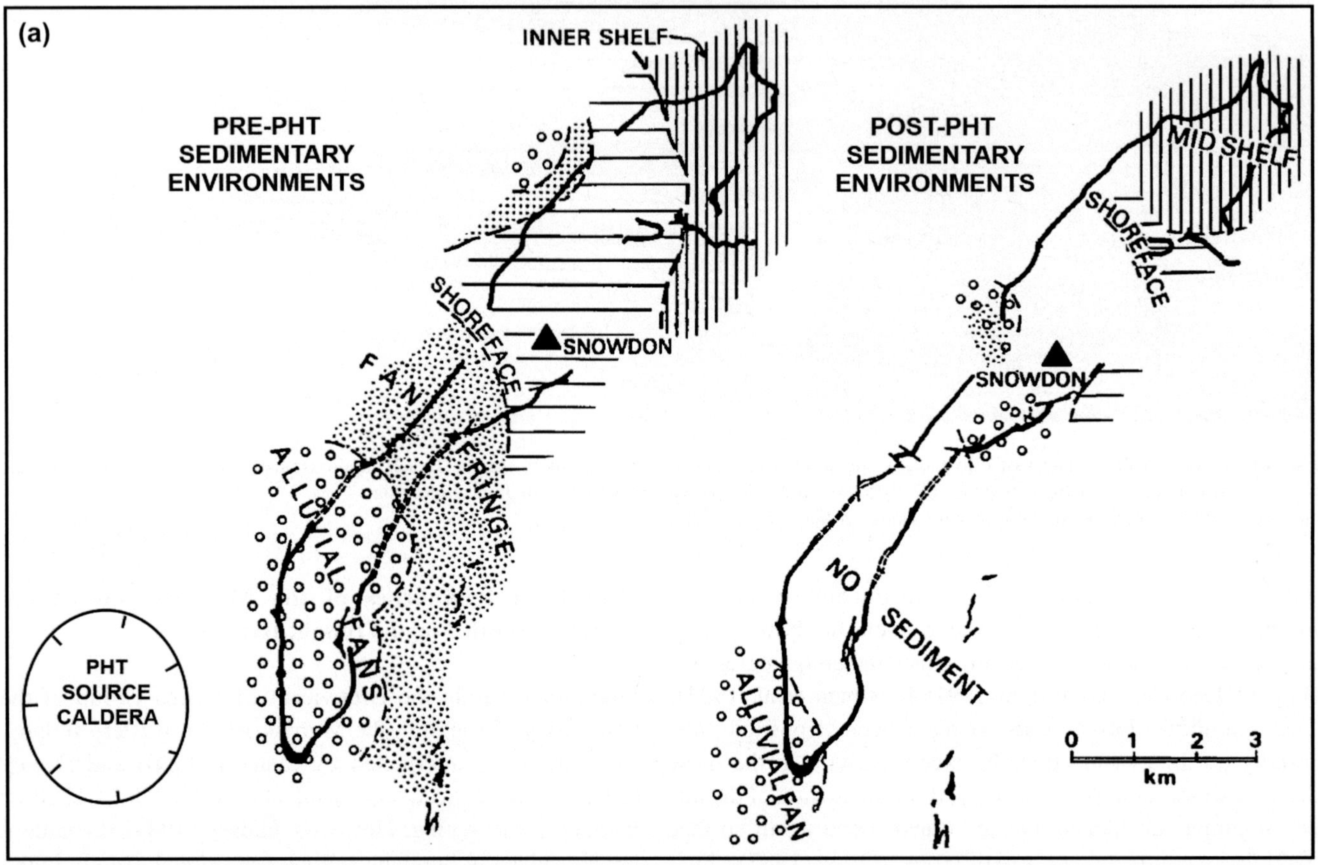

Fig. 11.12 Features of Ordovician Pitts Head Tuff (PHT) ignimbrite, base of the Snowdon Volcanic Formation, which was deposited in the shallow, northeasterly sloping shoreline zone, North Wales. **a** Changes in the depositional environment before and after the emplacement of the PHT, showing that the shoreline regressed to the northeast through the displacement of paleoenvironments. (From Reedman et al. 1987). **b** Columnar joined densely welded PHT, north of Llanberis Pass. **c** Fiamme eutaxitic texture of the welded PHT. **d** Swaley and hummocky cross-stratification beneath the PHT, Pen-y-Pass. **e** Angle of repose cross-bedding below the PHT, Pen-y-Pass. **f** Eroded top of the PHT, showing planar stratified volcanic sediments (arrow) above massive, cleaved (non-welded) ignimbrite, Pen-y-Pass. **g** Shallow water, trough cross-bedded, reworked sediment above the PHT, Pen-y-Pass. **h** Schematic longitudinal section through the PHT, showing the transition from subaerial to marine setting in the sediments below and above the ignimbrite, and the clear displacement of the shoreline before and after emplacement. This does not, however, necessarily reflect wholly subaqueous emplacement of the ignimbrite, most of which could have been emplaced subaerially by instantaneous displacement of the shoreline seaward at the time of entry of the pyroclastic flow. (From Reedman et al. 1987). **i** Zone of siliceous nodules (amygdales, lithophysae, and spherulites) near the base of the PHT, Pen-y-Pass

Fig. 11.12 (continued)

sedimentation continued after emplacement of the ignimbrite, although a coarse conglomerate horizon immediately overlying the ignimbrite together with the development of a delta suggests that sediment input was vastly increased.

As to the degree of welding, the ignimbrites in general consist of flattened red pumice clasts, lithic lapilli, and broken feldspar and quartz crystals in a purple matrix. Microscopic examination of this matrix shows preservation of relict vitric shards, some showing delicate, undeformed cuspate and bubble wall junctions, confirming that the ignimbrites are poorly welded to completely unwelded (Riggs and McConnell 2017). The lowermost ignimbrite is also characterised by the development of quartz nodules (<10 cm diameter) which, as in North Wales, are similarly ascribed to vapour streaming from the ignimbrite/sediment interface (Riggs and McConnell 2017).

Fig. 11.12 (continued)

Miocene Koura Formation, Shimane Peninsula, Japan

Kano (1990, 1991) documented a ~20 m thick, poorly sorted pumice-lithic clast deposit with wood fragments and armoured lapilli in the early Miocene Koura Formation on the Shimane Peninsula, southwestern Honshu, Japan. The deposit is underlain by plane to cross-bedded sandstones interpreted as beach and foreshore deposits, whereas immediately above there are planar-bedded turbidites. The pyroclastic deposit has a poorly sorted 18 m thick basal zone with normal coarse-tail grading of lithic clasts, and crude reverse coarse-tail grading of pumice clasts, passing gradationally into a planar to low-angle cross-stratified pumice and ash interval, overlain by a thin, locally preserved, laminated, fine ash horizon that is 25 cm thick. The top appears to be eroded, cut by channels. Gas or steam escape pipes up to 10 m long cut through the deposit. This profile is identical to a normal subaerial ignimbrite, ash-cloud surge and fallout sequence (Chap. 10). Kano (1990, 1991) interpreted the deposit to be an ignimbrite resulting from the passage of a subaerial pyroclastic flow into the shoreline and shallow water zone of a lake and to have displaced water so that it was emplaced under subaerial conditions.

Miocene Carpenter Ridge Ignimbrite in the Lacustrine Moat, La Garita Caldera, Colorado, USA

Lipman (2018) reported on the passage of the Miocene Carpenter Ridge Tuff ignimbrite pyroclastic flow into the lacustrine moat of the La Garita caldera in Colorado. At this

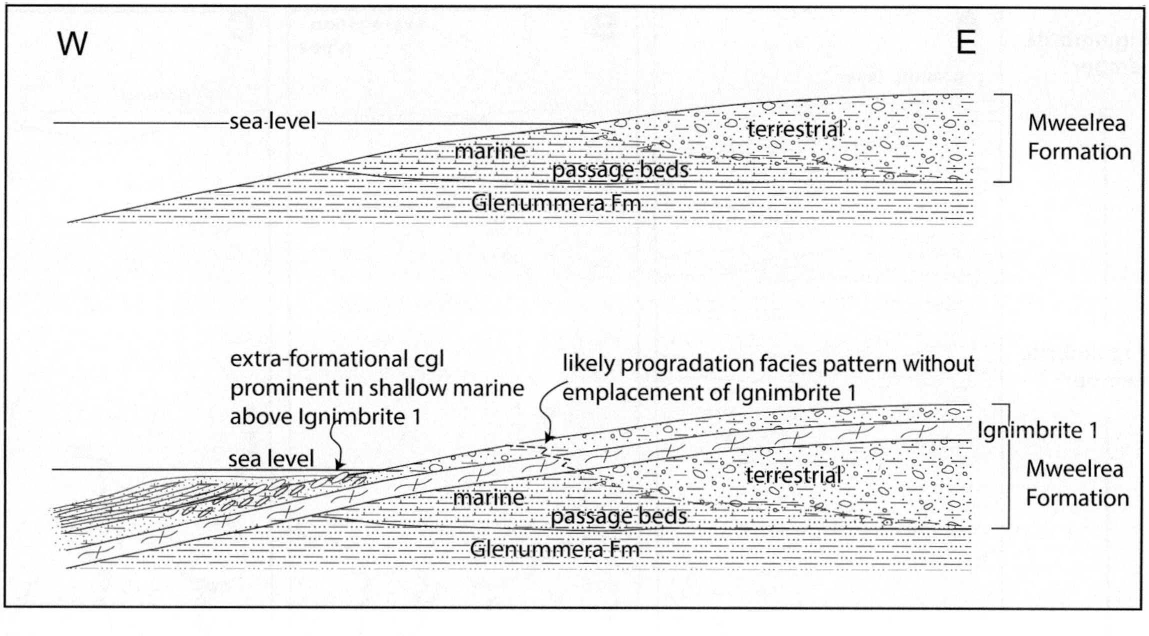

Fig. 11.13 Sedimentary response to ignimbrite emplacement across a fluvial to shallow marine transition, Ordovician Mweelrea to Glenummera Formations, South Mayo Trough, western Ireland. Above: facies prior to emplacement of lowermost ignimbrite. Below: facies after emplacement of ignimbrite showing the abrupt westward transition of facies; compare with dashed line to signify likely continuation of progradation without disruption. (From Riggs and McConnell 2017)

location in the Carpenter Ridge Tuff ignimbrite, sub-vertical zones of non-welded tuff tens to hundreds of m wide, surround "knobs" of highly welded, devitrified ignimbrite tens to hundreds of m in diameter with abundant spherulites and lithophysae (cf. Fig. 11.12 h, i). He interpreted the non-welded zones around the knobs to have formed through the fluidisation and vertical gas escape in wide zones from the non-welded base of the ignimbrite. This occurred where the source pyroclastic flow entered the moat, causing volatilisation of moat water or groundwater, driving fluidised diapirs of the basal non-welded ignimbrite up through the welded core of the ignimbrite (Lipman 2018).

Matahina Ignimbrite (320 ka), Taupo Volcanic Zone, New Zealand

The Matahina ignimbrite is a large 320 ka rhyolite, compound cooling unit ignimbrite with an outflow volume of 120 km^3 located in the northern part of the Taupo Zone, New Zealand. The ignimbrite was erupted from the Okataina Volcanic Centre (Chap. 16). It flowed mainly east and north, reaching the Bay of Plenty where the lower ignimbrite member is represented by distal subaqueous facies characterised by abundant liquefaction convolution features (Bailey and Carr 1994; Fig. 11.14). Along the Bay of Plenty coast, the Matahina ignimbrite, including its basal plinian fallout deposit member, rests directly on a thick marine and estuarine sequence consisting of poorly consolidated, massive-thinly bedded, fossiliferous, locally ripple-marked, shallow-water volcaniclastic sediments. In most of this coastal area, the bed immediately beneath the Matahina succession is a brown, 2–3 m thick, massive, fine sand, commonly showing bioturbation and worm or mollusc tubes. The environment suggested by this and the underlying sediments is a low-energy, shallow, estuarine bay or lagoon not unlike those observed along the present-day coastline, which typically has up to 15 m water depth with a 3 m tidal range (Bailey and Carr 1994).

The basal air-fall units show evidence for deposition and settling in shallow water and include texturally distinct "doublets" with coarse lower and upper finer units, as well as cross-bed lenses and rip-up clasts (Fig. 11.14). The lower ignimbrite is a single flow unit, 7–10 m thick and entirely nonwelded. The overlying middle ignimbrite member also represents a single flow unit, 10–45 m thick, with sections >20 m incipiently to partly welded and crudely columnar jointed. The two members are readily distinguishable by colour: the lower, greyish-white throughout, and the middle, salmon pink along its non-welded base. Both flow units contain gas segregation pipes which are most abundant in the lower part of the middle member and rise mainly from its basal contact. Pipes are less common in

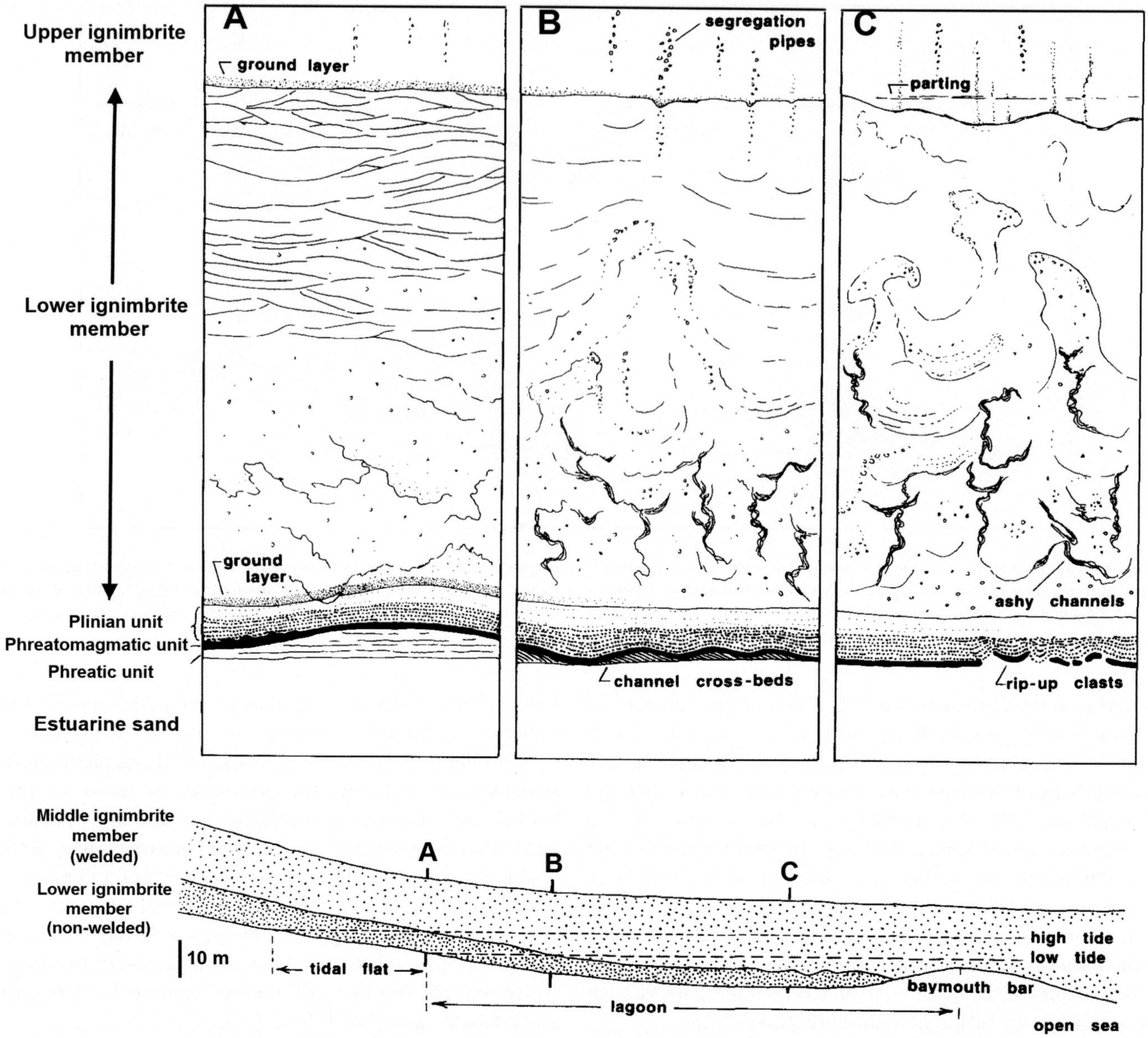

Fig. 11.14 Schematic relations and features in the subaqueously emplaced lower pyroclastic flow member of the Matahina ignimbrite, Bay of Plenty coast, North Island, New Zealand (From Bailey and Carr 1994). Inferred water depth increases from left to right. A = intertidal zone: typical of outcrops 1–2 km inland from coast; B = mid-tidal zone; and C = subtidal zone. In parts of the flow that were totally submerged liquefaction convolutions or fluid escape structures occupy its entire thickness. These include tongues and diapirs (defined by pumice trains and ill-defined, finer-grained marginal zones), contorted ashy channels, and subtle dish-like features that rise nearly to the top of the flow unit, but are best developed in the lower half. In shallower water exposures, diapirs are less well developed, more widely spaced, and occupy only the lower half of the flow unit. Ashy channels occur only in the lower third; the upper third shows faint, discontinuous, subhorizontal bedding or flow layering, which appears bowed around and upward over the diapirs. In the intertidal areas, convolutions and diapirs are ill-defined, and ashy channels are very thin, extremely fine-grained, and generally ptygmatic. Subhorizontal bedding in the upper half of the unit is defined by lenticular packets that resemble through cross-lamination but show little internal structure. These lenses generally decrease in length and thickness upward, suggesting a gradual decline in velocity of the depositing medium, either water-saturated ash or ash-laden water, or possibly they relate to syn-eruptive redeposition by tidal or alluvial processes. In all exposures, regardless of apparent water depth, there is a tendency for the flow to fine-upward, the upper part being relatively ash-rich and containing few and only small pumice lapilli, while coarser lapilli and blocks tend to be more abundant in the diapirs and the lower part of the unit

the lower member and occur mainly in its upper part (Fig. 11.14); some pass continuously upward through the lower/middle member contact and merge with pipes in the middle member, suggesting that both were hot and degassing simultaneously. Thus, the two members were emplaced in rapid succession.

Liquefaction convolutions are restricted to the lower ignimbrite unit, and not found in the subjacent air-fall and estuarine deposits. They include a variety of fluid escape features, ranging from ill-defined tongues and diapirs to subtle dish-like structures (Fig. 11.14). Within and around these structures, the pumiceous matrix is crudely segregated into coarse and fine domains, some of which show flow structure. Locally, pumice lapilli and blocks, as well as lithics, are entrained in irregular swirls or segregated in pockets. Intimately associated with these features are smaller scale, tortuous, wispy, anastomosing "channels' of fine ash (which in 3D are both pipe- and sheet-like) with sinuous or braided internal bedding or flow structure. Many of these exhibit ptygmatic and drag-fold-like forms, suggesting that, while differential flow was occurring within the channels, they themselves were being deformed by mass movement of the enclosing, coarser, less well-sorted matrix. Bailey and Carr (1994) suggest the array of structures was produced by vigorous fluid churning of the pyroclastic flow while in a water- and/or vapour-saturated state, either during or shortly after emplacement, and that the fines within the channels were winnowed from the mass of the flow and concentrated by fluid fluxes. The convolutions are best developed in the most distal, deeper water exposures. They change systematically shoreward, decreasing in height and complexity, and are displaced progressively downward by crude, discontinuous, subhorizontal lenticular bedding or flow layering in the upper part of the flow (Fig. 11.14).

Although most of the liquefaction features in the lower ignimbrite member may have resulted directly from interaction of the hot pyroclastic flow with sea water during or immediately after subaqueous deposition, Bailey and Carr (1994) speculate that contemporaneous seismic activity may have triggered disruption of the gas-charged but still intact flow, facilitating ingress of water to the flow interior, initiating pyroclast–water interaction. Bailey and Carr (1994) report no evidence of littoral zone craters or phreatic deposits, suggesting less violent escape of hot water, steam, and/or residual magmatic volatiles did occur during and after liquefaction. Subsequent eruption of the much thicker, larger volume, middle member ignimbrite overwhelmed and buried the entire lagoonal sequence (Fig. 11.14), forcing the shoreline to be displaced an unknown distance seaward with the deposit and remaining sufficiently hot to weld.

24–25 August 79 AD Eruption of Vesuvius, Bay of Naples, Italy

As indicated above, pyroclastic density currents during the 79 AD eruption displaced the shoreline between Pompeii and Herculaneum by some 0.5–1 km seaward. In an innovative thermal study, Giordano et al. (2018) investigated pyroclastic flow deposits that progressively destroyed and buried a large ancient Roman construction, the Villa dei Papiri, which was built on the shore just outside Herculaneum. The chronology of the 79 AD eruption of Vesuvius is now well understood based on both the eyewitness accounts of Pliny the younger and modern forensic volcanological analysis (e.g. Sigurdsson et al. 1982; Carey and Sigurdsson 1987; Cioni et al. 1992; Gurioli et al. 2002). Herculaneum did not experience the initial pumice fallout which commenced at about 1 pm on 24 August when a buoyant column rose up to 30–33 km high (Chap. 9), being located off the fallout dispersal axis. The stratigraphic succession records the first pyroclastic flow deposits (E2/3pf; EU3pf) which were associated with the partial collapses of the column during the transition between phonolitic EU2 and phono-tephritic EU3 plinian phases, and later pyroclastic flow deposits (EU4) of the caldera-forming phase which took place on 25 August (Fig. 11.15; Chap. 10).

Giordano et al. (2018) measured the TRM of lithic clasts embedded in the ash matrix of pyroclastic flow deposit massive and chaotic facies and compared the results with temperatures retrieved from the reflectance of charcoal fragments collected at the very same sites by Caricchi et al. (2014). The reconstruction of the thermal interactions at Villa dei Papiri shows that the first incoming pyroclastic flow had the highest degree of mixing with both debris from the unroofing of the edifice and partial collapses, along with the interaction with the sea facing the villa. The pyroclastic flow was able to unroof the villa and partially destroy sections of it, but most of the main structure remained in place. At the same time, the flow crossed the shoreline just in front of the villa, and entered the sea depositing and pushing the coastline seaward. The subaqueously deposited ignimbrite is associated with zeolitised alteration plumes and some large m-scale steam or gas pipes (Fig. 11.15a, b; Plate 11.1f). Deposit temperatures are highly heterogeneous, with the lowest (gas pipe = 180–200 °C) reflecting rapid post-emplacement heat transfer due to convective mixing

Fig. 11.15 Features of the 79 AD pyroclastic flows, Vesuvius volcano, Italy. **a** View seawards from the Villa dei Papiri excavations at Herculaneum showing pyroclastic density current deposit stratigraphy of the 79 AD eruption, and location of thermal sampling sites (VP). Yellow dashed lines identify zones or plumes of pervasive zeolitic alteration likely induced by convection of external seawater during cooling; compare sample VP6 from zeolite facies overprinting early EU2/3pf, with later EU4pf (VP9) pyroclastic flow which was not influenced by seawater. White numbers indicate elevations in m a.s.l. (Photo and elevation survey courtesy of Superintendence for Archaeology of Herculaneum). **b** Detail of the large gas pipe affecting EU2/3pf at sampling site VP8. **c** Cartoon illustrating the different thermal interactions of the incoming pyroclastic density current with the environment either by mixing with water or building debris, which promote rapid changes of temperature of the deposit in different though very close settings. The temperature at the base of the incoming pyroclastic flow (or below the depositional boundary layer, dbl) is everywhere ~350 °C, similar to that of the deposit unless mixing with external cooling agents occurs. (From Giordano et al. 2018)

with seawater/vapour and building debris (Fig. 11.15c). Time scales of such processes are <1 h, i.e. the time scale needed for 1–3 cm sized lithics to equilibrate with the enclosing deposits (Giordano et al. 2018).

After a short pause associated with the restoration of the EU3 buoyant plume, a second major flow (EU3pf) reached the Villa dei Papiri, partly burying the still-standing main edifice. EU3pf, however, did not cause major collapses to the structure, nor did it interact with seawater, previously buried by the EU2/3pf deposits. Deposit temperatures in EU3pf average ~350 °C, reflecting the average temperature of the parent pyroclastic flow near its depositional base and more concentrated part of the flow close to the depositional boundary layer (Fig. 11.15c; Chap. 10). The following flow, EU4pf, reached the Villa after the transition from the buoyant EU3 phase and the beginning of caldera collapse, and has a deposit temperature of up to 400 °C (Fig. 11.15a).

27 August 1883 Eruption of Krakatau Volcano, Indonesia, and Entry of Pyroclastic Flows Into/Under the Sea

The 1883 eruption of Krakatau volcano, west of Java, remains one of the largest (~9km^3 DRE), the most powerful and famous explosive eruptions of recent historic times (Fig. 11.16). Through the combined effects of pyroclastic flows and eruption-induced tsunamis it is thought to have caused >30,000 fatalities. In the period from May to August

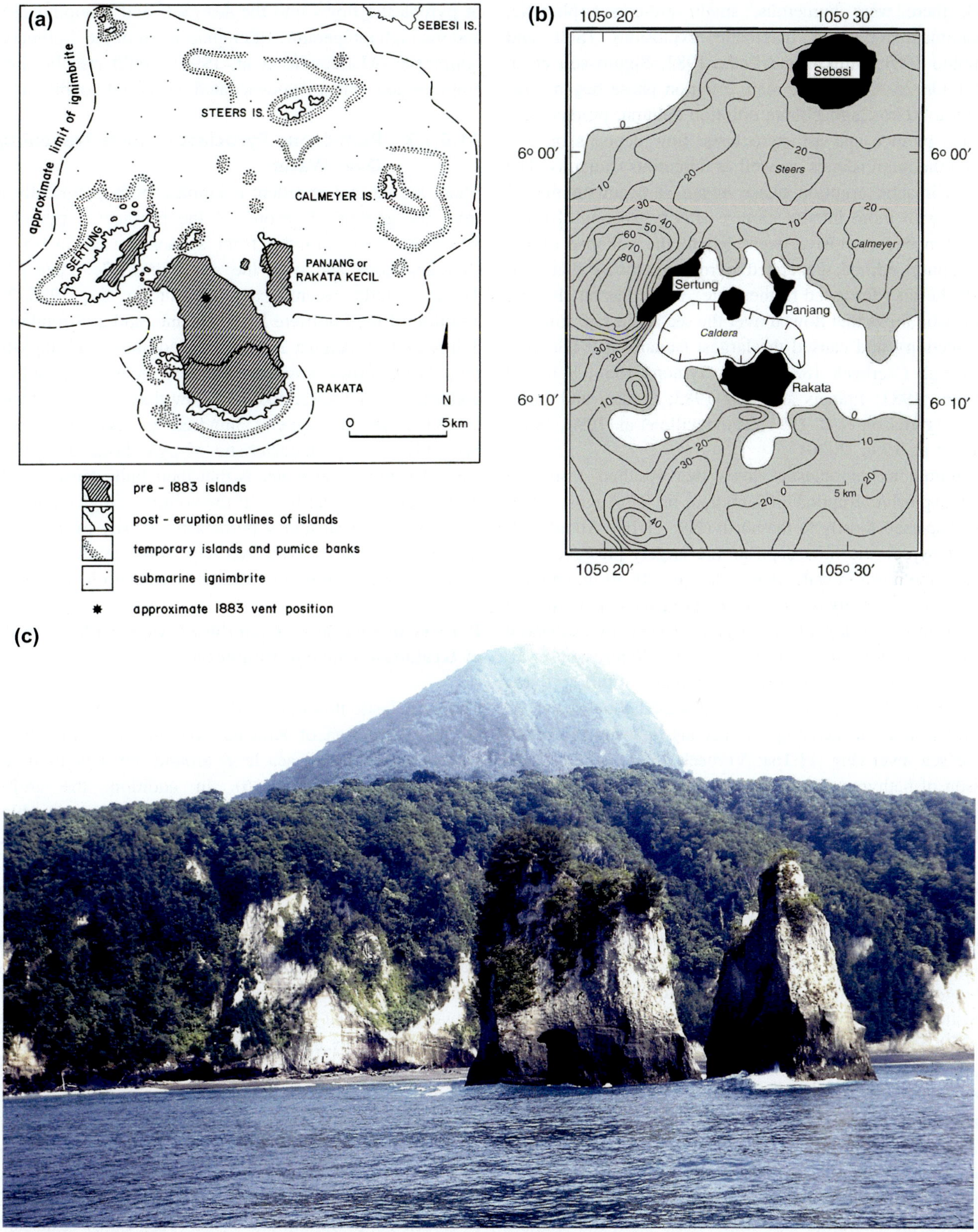

Fig. 11.16 Features of the 1883 Krakatau eruption and deposit distribution. **a** Distribution of 1883 Krakatau ignimbrite and new ignimbrite islands (Calmeyer, Steers) and pumice banks formed in the shallow waters of the shelf north of Krakatau (From Cas and Wright 1987, After Self and Rampino 1981). **b** Isopach map constructed from bathymetric changes of the submarine ignimbrite around Krakatau (From Carey et al. 2000). **c** Proximal ignimbrite deposits of the 1883 eruption on the southern side of Rakata (photo taken a century later in 1986)

1883, there were numerous, small, pre-cursor phreatic, phreatomagmatic, and magmatic explosions (Self and Rampino 1981; Simkin and Fiske 1983; Sigurdsson et al. 1991). On 26 August, the main eruption phase began, producing a 20 km high plinian eruption column, pumice fall-out and pumice and ash pyroclastic flow deposits on and around the main islands, as well as numerous tsunamis. The main, climactic eruption phase began at 6 am on the 27 August with more intense climax activity at about 10 am. This climactic explosive event led to the collapse of the pre-eruption edifice, generated pyroclastic flows that "entered" the sea (discussed further here), as well as far-flowing pyroclastic flows that flowed over the sea at least 80 km (see next section), and caused the largest tsunami that was 30–40 m high (Verbeek 1884, 1886; Symons 1888; Self and Rampino 1981; Francis and Self 1983; Simkin and Fiske 1983; Sigurdsson et al. 1991; Mandeville et al. 1994, 1996; Carey et al. 1996, 2000).

Voluminous pyroclastic flows that resulted from the major explosive eruption of Krakatau on 27 August 1883, were dispersed mostly to the north (Fig. 11.16). About 13.6 km^3 of pyroclastic flow deposits are largely preserved offshore to the north in water depths of only 20–30 m, and this water depth is largely due to post-eruptive erosional removal of ~20 m of the top of the original deposit by (subaerial and then) shallow marine ocean currents (Sigurdsson et al. 1991). Immediately after the eruption, much of this area was not navigable by ship and where there was shallow seawater, several islands of steaming pumiceous deposits protruded above sea level (Fig. 11.16a; Verbeek 1884, 1886). Comparison of bathymetric charts before and after the eruption, indicate significant shallowing of the sea around Krakatau, with pyroclastic flow deposits beneath the sea ranging from 10 to 90 m thick (Fig. 11.16b; Verbeek 1884; Sigurdsson et al. 1991; Carey et al. 1996; Mandeville et al. 1996). The offshore deposits are texturally (poorly sorted) and compositionally identical to the subaerial ignimbrites preserved on Krakatau (Fig. 11.16c). TRM studies of cored samples of the *subaqueously preserved* deposits (but not necessarily deposited subaqueously) indicate that they were emplaced at >350–550 °C (Mandeville et al. 1994, 1996).

As discussed earlier (Sect. 11.5.3), Legros and Druitt (2000) undertook numerical modelling of the entry of the Krakatau pyroclastic flows into the sea using eruption parameters (e.g. mass flux of 10^9 kg s^{-1}) to constrain the modelling, and proposed that if the discharge was high enough and sustained, the flows could have instantaneously displaced the shoreline and water mass from the shallow shelf to the north of Krakatau, before sweeping back in after ignimbrite deposition. The tsunamis generated during the eruption largely appear to have been generated by the entry of pyroclastic flows into the sea (Self and Rampino 1981), and the high temperature of emplacement of the "submarine" ignimbrite (Mandeville et al. 1994), could also be due to emplacement under sea-evacuated, subaerial conditions.

11.6.1.3 Pumiceous Pyroclastic Flows Travelling Over Water

There are several excellent examples, both historically and from the geological record, of the passage of pyroclastic flows that were erupted from subaerial vents across the shoreline zone and over the sea (Fig. 11.7c; Plate 11.1). These include recent examples during the 1995–2012 eruption of the Soufrière Hills volcano, and at Stromboli on 3 July and 28 August 2019 described above, during which pyroclastic flows entered the sea, generating both an underflow and a surge-like overflow that were able to flow over the sea for several hundred m. As discussed in Sect. 11.5, a key characteristic of pyroclastic density currents that flow over water is that their bulk density be less than that of the water mass. If the volume is high enough and the mass flux is sustained pyroclastic density overflows can flow tens of km over the sea (e.g. some pyroclastic flows from the 1883 eruption of Krakatau volcano in Indonesia).

Pyroclastic Overflows from the 27 August 1883 Eruption of Krakatau Volcano, Indonesia

The pyroclastic flows and resultant ignimbrite deposits from the 1883 eruption of Krakatau volcano are currently preserved largely below sea level around the remains of Krakatau volcano (Fig. 11.16). In addition, the eruption produced pyroclastic density currents which were able to flow over the sea in a fan-like pattern to the north of Krakatau and reach land masses 40–80 km from their source (Fig. 11.17a; Carey et al. 1996, 2000). The pyroclastic flows were still hot enough to set fire to coastal villages and tropical forests on islands in transit to, and on Sumatra after traversing tens of km over the sea (Verbeek 1884; Simkin and Fiske 1983; Carey et al. 1996, 2000). Forests on islands between Krakatau were completely destroyed, with trees knocked down by the force of the pyroclastic flows, and set on fire. People on Sumatra were severely burned. The pyroclastic deposits commonly contain charcoal debris, testifying to flow temperatures of > 300 °C at the source of the fires. In addition, tsunamis reaching up to 30 m in run-up heights on landfalls in coastal settings were also catastrophic, accounting for most of the death toll of 36,000 (Verbeek 1886; Simkin and Fiske 1983; Carey et al. 2000). Eyewitness accounts from observers on ships up to 80 km from Krakatau reported the sudden arrival of ash and pumice-laden "hurricane force winds" (i.e. pyroclastic

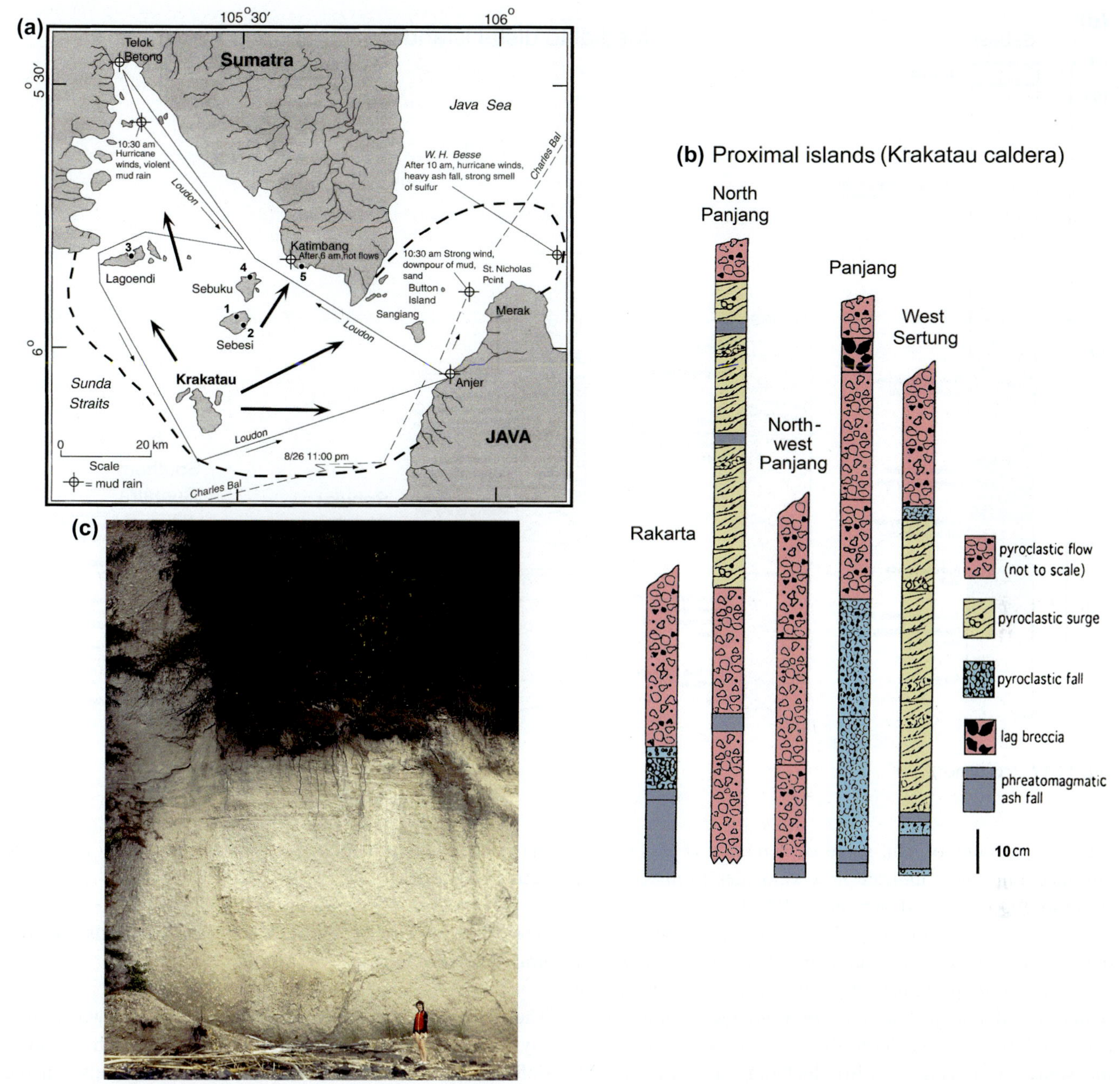

Fig. 11.17 Features of the Krakatau pyroclastic flow system that flowed over the sea resulting from the climactic phase of the eruption at 10 am on 27 August 1883. **a** Dispersal pattern of flows, which also notes eyewitness accounts from ship tracks (the *Loudon, W. H. Besse,* and *Charles Bal*) in distal localities. Dashed line encloses inferred area of flow and surge run out. (Modified from Carey et al. 1996). **b** Logged sections of deposits from proximal sites on the islands forming the Krakatau caldera (see Fig. 11.16 for locations). These deposits are generally the coarsest and also include the highest proportion of cross-stratified surge deposits. Note, pyroclastic flow deposits are not plotted according to same 10 cm scale at bottom right. (Modified from Mandeville et al. 1996). **c** Proximal flow and surge (stratified) deposits on north end of Panjang (*aka* Rakata Kecil). **d** Sections from medial-to-distal islands from about 10–35 km from source (see Fig. 11.17a for locations). Those in medial positions (Sects. 1, 2) are dominated by finer typical massive, poorly sorted pyroclastic flow deposits, and some unusual fines depleted coarse, rounded pumice deposits. Sections from distal settings (Sects. 3–5), including one from southernmost Sumatra, are the thinnest and finer than in proximal and medial positions. (Adapted from Carey et al. 1996)

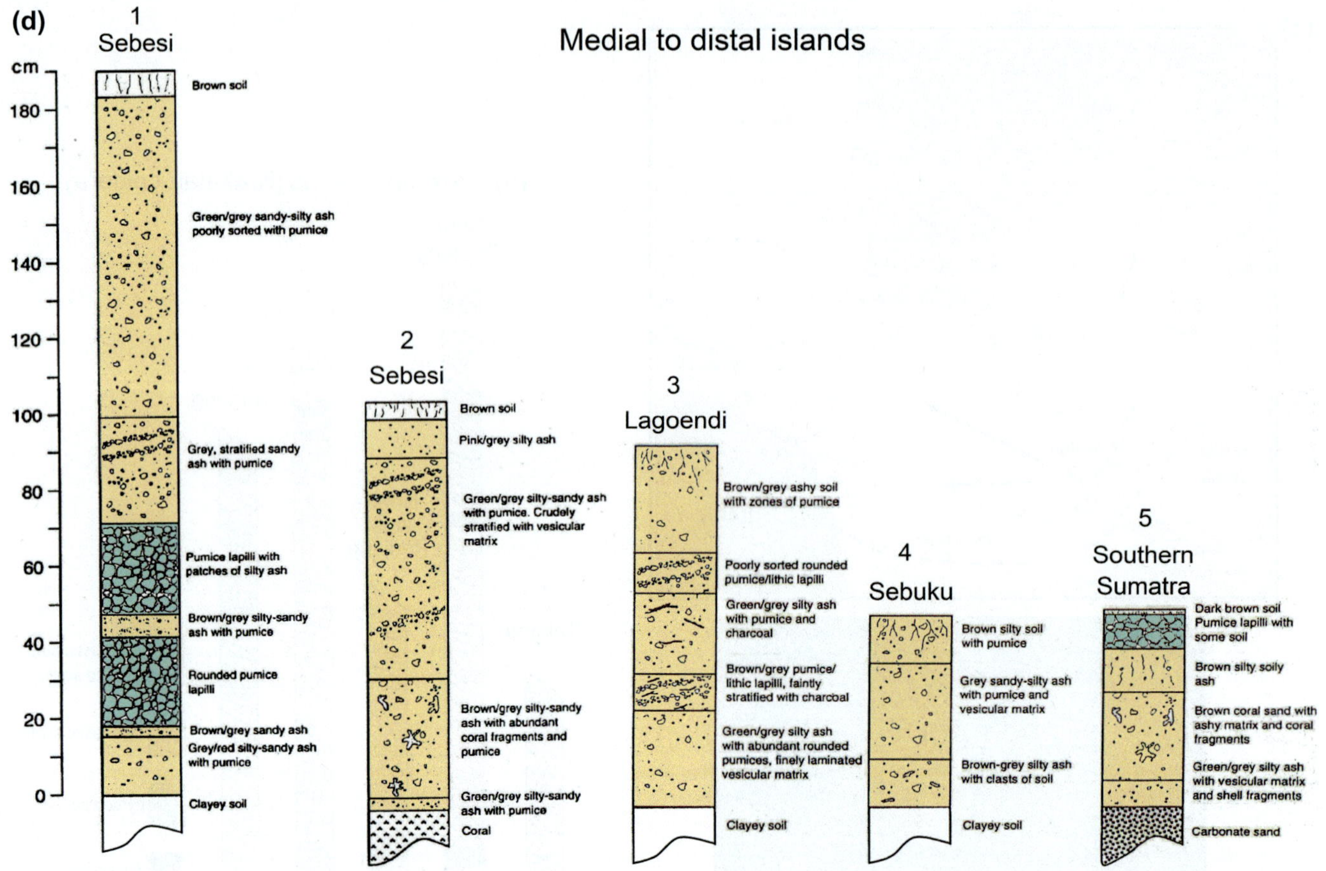

Fig. 11.17 (continued)

surges), and rains of ash, some of it muddy (Fig. 11.17a). Distance–time plots indicate flow velocities from Krakatau of up to 300 km hr^{-1} (Carey et al. 1996).

Carey et al. (1996) logged sections of the pyroclastic density current deposits found on islands to the north of Krakatau, from proximal sites on islands making up the Krakatau caldera, just a few km from the eruption vent (Fig. 11.17b, c), to medial to distal islands 10–35 km from the source vent (Fig. 11.17d), including a section at the southern end of Sumatra, being the most distal. These sections show an overall fining and thinning of the deposits from proximal to distal locations. At all sections the deposits are marked by multiple, thin, non-welded depositional units, indicating a pulsatory flow system. Deposits range from massive to planar and dune cross-stratified deposits, although the latter becomes less prominent with distance from source (Fig. 11.17b–d). The deposits are poorly sorted and mostly plotted in the pyroclastic flow deposit field of Walker (1983; Chap. 10) using median grain size versus sorting plots (Carey et al. 1996). The references to rains of "muddy" ash suggest a phreatomagmatic influence, which Self and Rampino (1981) and Carey et al. (1996) attribute to the interaction of pyroclastic flows entering and flowing over the sea, rather than major phreatomagmatic explosive activity at vent.

Kos Plateau Tuff (161 ka), Dodecanese Islands, Greece and Western Turkey

The ~90 km^3, 161 ka Kos Plateau Tuff (KPT) was probably erupted from a caldera, the location of which is poorly defined, but occurred in a submarine setting between the volcanic islands of Kos and Nisyros, in the southeastern part of the Hellenic volcanic arc, Aegean Sea, Greece (Fig. 11.18a; Keller et al. 1990; Cas and Wright 1987, 1991; Allen and Cas 1998a, b, 2001; Allen et al. 1999; Allen 2001; Allen and McPhie 2001; Pe-Piper et al. 2005; Nomikou et al. 2013). The studies of Allen and others have demonstrated that at the time of the KPT eruption sea-level was 60–80 m lower. Allowing for lower sea level, and tectonic effects, the area to the north of the present island of Kos was largely subaerial, whereas the area to the south, west, and east of the inferred caldera was a marine environment over which the KPT pyroclastic flows travelled (Fig. 11.18b; Allen and Cas 2001). Allen logged sections through the preserved deposits of the KPT on Kos, the surrounding islands and western

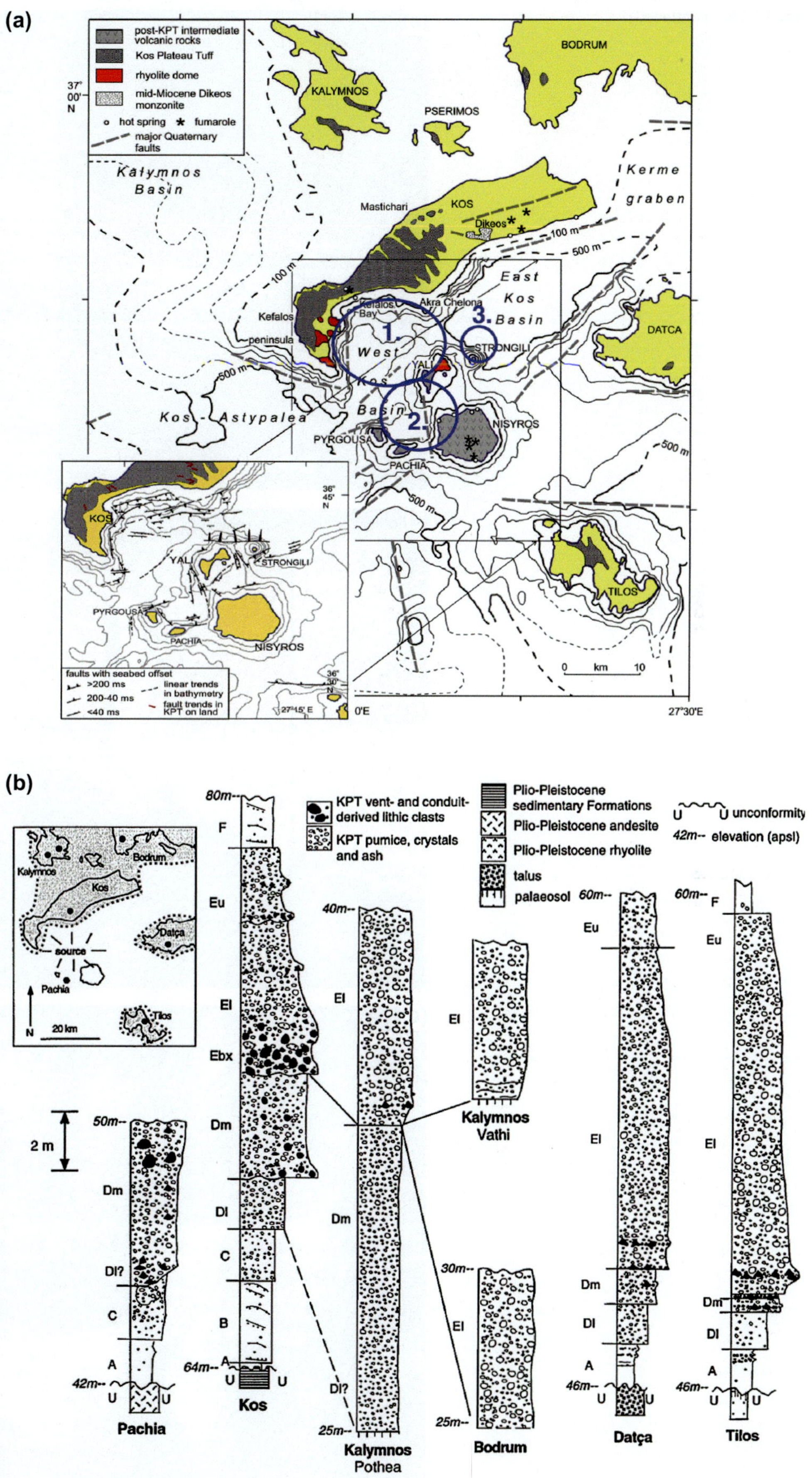

Fig. 11.18 Location and sections through the 161 ka Kos Plateau Tuff (KPT), southeastern Aegean Sea. **a** Map showing the location of outcrops of the KPT (grey shade) on the island of Kos, surrounding Dodecanese islands, Greece, and the Bodrum and Datça peninsulas of western Turkey. Blue circles indicate proposed locations of the offshore caldera; 1: Kos–Yali caldera inferred by Cas and Wright (1987), Allen (2001); 2: Pe-Piper et al. (2005); 3: Nomikou et al. (2013). Inset (bottom left) shows major sea-floor faults identified from seismic sections. (Modified from Pe-Piper et al. 2005; distribution of KPT is taken from Allen et al. 1999). **b** Stratigraphy of the KPT on Kos and surrounding islands and land areas. The distribution of land areas at the time of the KPT eruption to the north of Kos is shaded grey in the inset. (From Allen and Cas 2001). **c** View of the eroded plateau that the KPT forms in central Kos at ~100 m elevation. The large ~4 m vent-derived boulder is an erosional relict of the ground breccia formed by the climactic ignimbrite unit E, and occurs about 15 km from vent. **d** View of the KPT sequence (A to E) at Petrol Valley, Kos. here the basal units (A-D) are ponded in a paleovalley and onlap a paleo-ridge, and the top unit E forms an undulating plateau. **e** Oyster bank within unconsolidated Plio-Pleistocene shallow marine sediments directly below the KPT on Kos

Fig. 11.18 (continued)

coastal areas of Turkey (Fig. 11.18a, b), demonstrating significant variations in the stratigraphy and facies regionally (Allen 2001; Allen and Cas 2001). The thickest, most stratigraphically complete, and coarsest section occurs on Kos, north of the inferred vent area (Fig. 11.18b), where in places the KPT forms a 30 m thick, plateau-like blanket on the island (Fig. 11.18c, d). On Kos, the KPT unconformably infills erosional paleovalleys and generally overlies an incipiently developed paleosoil developed above unconsolidated Plio-Pleistocene shallow marine fossiliferous sediments, with faunas very similar to modern ones from the northern coast of the island (Fig. 11.18e; Waltschew 1997). It is unclear if the area between Kos and the inferred vent area was emergent land, sea or a combination of both at the time of the KPT eruption. However, apart from this southern coastal zone, most of Kos appears to have been an emergent landmass (Fig. 11.18b).

Allen et al. (2001) considered that the erupted volume of the KPT was ~ 60 km^3 DRE, and that prior to caldera formation, the precursor volcanic edifice was an andesitic edifice, based on the abundance of andesite lithic clasts ($\sim 3km^3$) in the ignimbrite succession. The KPT deposits are preserved up to 60 km from the inferred source, but are thought to have extended a greater distance, especially the phreatomagmatic and co-ignimbrite fallout ashes. Allen (2001) and Allen and Cas (2001) subdivided the KPT into 6 units (A–F), most of which consist of multiple depositional units, indicating a pulsing eruption and transport system.

The first phase of the eruption was phreatomagmatic (Unit A = surge and fallout ashes, Fig. 11.19a; Unit B cross-stratified, surge-like flow lapilli-ash deposit, Fig. 11.19b), indicating access of water to the vent. Phase 2 ignimbrites Units C (Fig. 11.19b–d) and D (Fig. 11.19c–e) are magmatic, and indicate a waxing eruption, producing successively larger volume ignimbrites, with increasing lithic clast abundances and sizes (Fig. 11.19b, d, e). Phase 3 Unit E ignimbrite (Fig. 11.19c, f, g) was also magmatic and represents the climax of the eruption, producing a high-energy, low aspect ratio ignimbrite (LARI) package; Unit D may also have been a LARI. The final, waning phase 4, which was possibly phreatomagmatic, produced small volume pyroclastic flow and surge and fallout deposits (Unit F) that are only preserved on Kos and Tilos (Fig. 11.19h), and a fallout ash. Ignimbrites D and E have impressive co-ignimbrite ground lithic breccias on Kos, 15 or more km from vent (Fig. 11.19d–g; Allen and Cas 1998a), but elsewhere, these are much less abundant and finer. Allen and Cas (1998a) assessed the flow dynamics based on distribution, abundance, and size of lithic clasts, and determined that the E flow(s), with the coarsest lithic breccias in the deposits (Fig. 11.19b, f), was initially in a highly erosive waxing phase up to 16 km from vent, and then beyond that in more distal settings it was a waning flow system. This deposited significant ground breccias with some vent-derived clasts >4 m in diameter (Fig. 11.18c).

On the northwestern shore of Kos, ignimbrite E was capable of eroding substrate fossiliferous mudstone intraclasts up to 2 m in length (Fig. 11.19g). In this same area unusual outcrops of a chaotic breccia are exposed within a narrow (1 km across) area directly along the coast (Allen and McPhie 2001). This layer of chaotic breccia is up to 6 m thick and is sandwiched between Plio-Pleistocene sedimentary units below and ignimbrite E above (Fig. 11.19g, I, j). The lower parts of the chaotic breccia comprise distinct domains of unconsolidated sediment or locally lower KPT (A-D) units that are deformed and/or mingled with pumiceous ignimbrite. The upper part is dominated by a matrix of mingled-to-pervasively mixed ignimbrite and sediment that contains sediment domains as large as 2–10 m in diameter. Allen and McPhie (2001) suggest this was formed synchronously with the passage of a highly energetic pyroclastic flow (ignimbrite E) where it traversed wet, unconsolidated sediment. Shear-induced liquefaction, together with possible seismic shaking associated with the eruption, possibly caused failure. This failed portion continued to move downcurrent and disintegrate by lateral shear, probably for no more than a few tens to hundreds of m, until it was deposited en masse.

Allen (2001) and Allen and Cas (2001) demonstrated that the D and E ignimbrites which were deposited on distant land masses (Tilos island, Daçta Peninsula, Turkey; Fig. 11.18a, b) from flows over water, especially to the east, were finer and contained fewer and significantly smaller vent derived accessory lithic clasts compared with flow over land deposits. They attribute this to lithic clasts dropping out of the base of flows over water and sinking, whereas where the flows travelled wholly on land to the north lithic clasts were saltated along the solid ground as bedload. Even so, the high energy E pyroclastic flows were able to transport dense 20 cm diameter lithic clasts over water. A schematic representation of the flow behaviour of the KPT climactic Unit E pyroclastic flows is shown in Fig. 11.20. Based on the interpretation of new seismic data, Pe-Piper et al. (2005) presented an alternative view, that the KPT was deposited on a subaerial coastal plain or at the most very shallow marine and thus it is unlikely that pyroclastic flows crossed large stretches of deeper water towards Tilos.

From a theoretical perspective, Dufek and Bergantz (2007a, b) conducted numerical simulations to test the capacity of the KPT D and E pyroclastic flows to carry large lithic clasts over water to the south of the vent, with what they called a "leaky" base, compared with over-the-land

Fig. 11.19 Features of the 161 ka Kos Plateau Tuff (KPT) on Kos, Greece. **a** The basal phreatomagmatic unit A, showing surge cross-stratification above the substrate paleosoil, and upper planar lamination from fallout. This unit locally contains accretionary lapilli. It is overlain by the coarser, poorly sorted Unit B. Portobello, central southern coast of Kos, the closest outcrop to the inferred vent, 11–12 km away. **b** Diffusely stratified, low-angle cross-stratified Unit B, at Plaka, central Kos. The unit was deposited by a low-concentration pyroclastic density current at ~15 km from vent. The coarse, massive, overlying unit with a basal lithic concentration zone is Unit C ignimbrite. **c** Outcrop of most of the KPT sequence except Unit E, Portobello, Kos, dominated by massive to diffusely stratified Unit C

(continued)

and D ignimbrite units. Units A and B form the very thin basal interval, Unit C forms the lower white half of the outcrop, while slightly tan Unit D forms the upper half; person circled for scale. **d** Outcrop of Units B, C, and D ignimbrites with a coarse ground layer breccia at the base of Unit D, Golden Beach area, northern Kos. **e** Close-up view of Unit D ground layer breccia, overlying Unit C ignimbrite, Golden Beach area, which is dominated by variably altered andesitic vent-derived boulders. **f** Ground breccia of Unit E, "Goat farm" locality in central Kos, at ~120 m above sea level. Most lithic clasts are vent-derived, having been transported ~17 km, probably over both sea and land and then up the central ridge of Kos. Maximum lithic clast size is >1.5 m diameter. **g** Unit E ground breccia with an irregular eroded basal contact and containing, elongated variably fossiliferous mudstone intra-clasts eroded from the nearby substrate, southwest of Mastichari. **h** Massive to planar to low-angle cross-stratified Unit F, near Plaka. **i** Chaotic zone along the contact between unit E ignimbrite and underlying Plio-Pleistocene sediments, beach section, west-southwest of Mastichari. From bottom to top this shows surface lithified sediments, a zone showing the most shear (including deformed worm burrows), rafts of soft sediment and breccia, then the base of unit E with mingling. Note irregular flame structures composed of sandy matrix intruding into the base of the ignimbrite. The gradational boundaries and highly irregular shapes of the different domains can be attributed to shear-induced mechanical mixing during flowage (see text). Height of outcrop is ~6 m. **j** Detail of flame structure near (**i**), containing a jumbled mixture of soft sediment and entrained angular fragments of early lithified, mainly aeolian dune (carbonate-cemented) coarse sandstone. Note the absence of vent-derived lithic clasts

Fig. 11.19 (continued)

flows to the north of the vent, with a saltation lower boundary. They found that if the eruptive volume flux was $\sim 2 \times 10^6$ m^3 s^{-1}, assuming a fissure ring fracture vent to accommodate the high volume eruption rate, then the D pyroclastic flows could reach the island of Tilos (35 km from vent), over the sea and transport the small lithic clast sizes (5–10 cm) found there, assuming a leaky flow base. However, the more energetic E pyroclastic flows could not transport their coarse >20 cm lithic clasts across the sea to Tilos. This implies that the E flows had a saltating base, formed after the D event. This could have resulted either by the filling of the ocean to sea level with ignimbrite D deposits or by an extensive and thick raft of floating pumice on the sea surface, derived from the earlier phases of the eruption, allowing the E flows to skip across the sea as if they had a saltation base. Alternatively, eruption flux must have been significantly higher than calculated during the climactic Unit E stage of the eruption.

Koya Ignimbrite (7.3 ka), Kyushu, Japan

The 7.3 ka VEI 7 eruption of Kikai caldera, located 40–50 km south and offshore of the Kyushu coastline, produced a widespread plinian pumice fallout unit (Unit A), and the climactic low aspect ratio Koya ignimbrite (Unit C). The pyroclastic flow travelled radially over the sea, reaching islands 50 km to the south of the caldera, as well as many locations on Kyushu to the north reaching 80 km from the caldera (Fig. 11.21; Maeno et al. 2006). On islands that form the emergent caldera margin, the early part of the ignimbrite and caldera-forming eruption is also represented by a stratified succession of fallout and variably welded, surge-like

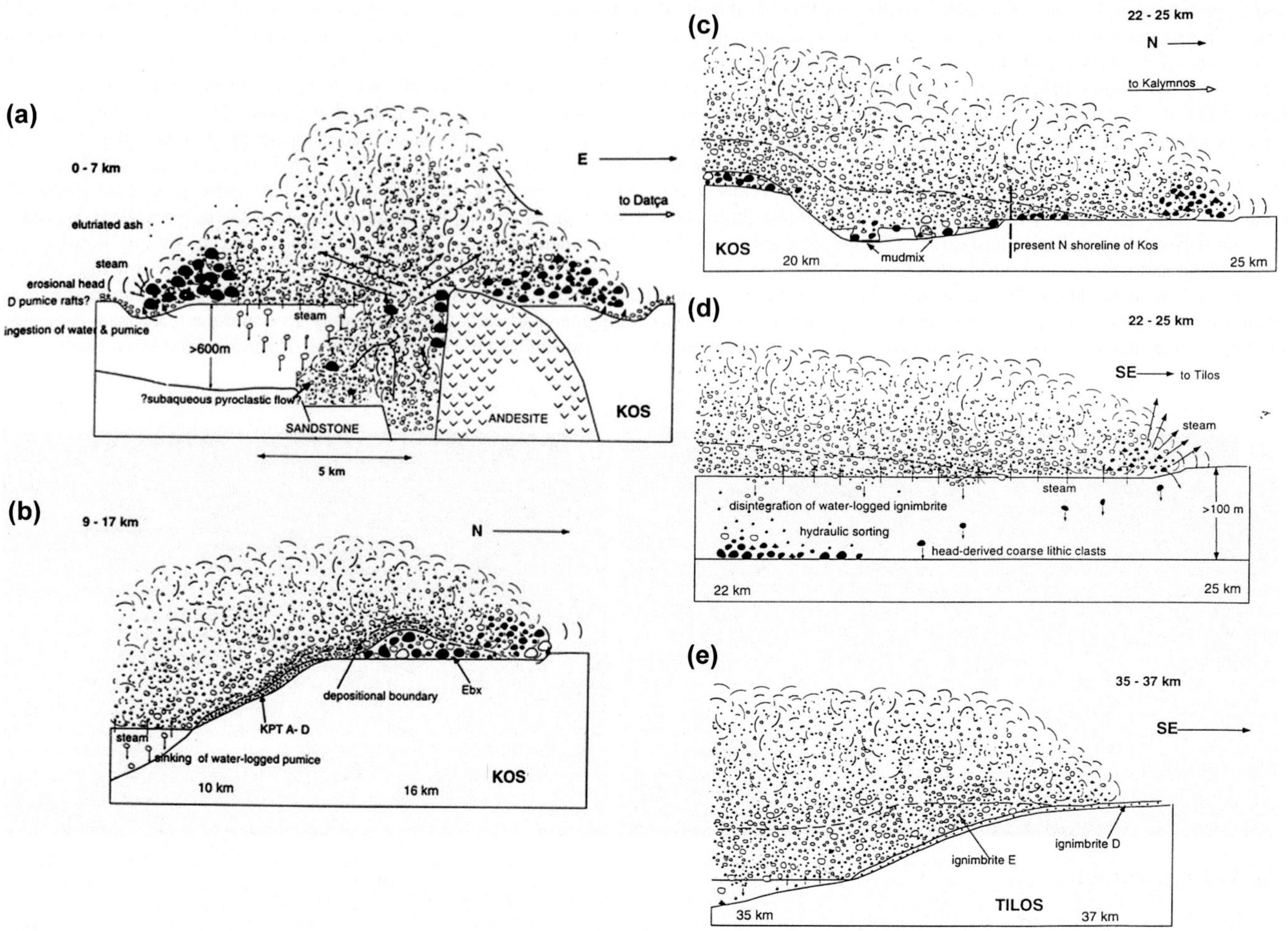

Fig. 11.20 Schematic representation of the flow behavior of the Kos Plateau Tuff Unit E ignimbrite, signifying the climactic, largest volume and most energetic event of the eruption. **a** At vent, an andesitic volcanic edifice through which rhyolitic magma was erupting explosively and incorporating large volumes of andesitic lithic clasts into the collapsing eruption column and flow system. **b** Flows from the vent reaching the southern coastline of the island of Kos and depositing coarse lithic-rich facies against and on the central ridge which is currently up to 120 m above sea level and would have been higher 161 ka ago. Ebx = ground layer breccia. **c** The flow system flowing north of Kos, and eroding large intra-clasts of fossiliferous mudstone into its base (Ebx), as well as chaotic breccia formed by shear-induced liquefaction and mechanical mixing during flowage possibly over wet, unconsolidated ground. **d** Part of the flow system flowing over water towards Tilos, with large lithic clasts dropping out of the "leaky" base and sinking through the water mass. **e** The lithic clast depleted flow reaching Tilos after flowing 30 km over the sea. (From Allen 1998)

deposits (Unit B). These signify multiple partial column collapse events during a magmatic phase of explosive eruption, and also include a 20 m thick, very lithic-rich, proximal facies of the Koya ignimbrite coinciding with the climactic caldera collapse phase (Maeno and Taniguchi 2007, 2009). The widespread Koya Unit C ignimbrite has a lithic and crystal-rich ground breccia layer overlain by a poorly sorted pumice-rich ignimbrite facies almost everywhere away from the caldera rim (Maeno et al. 2006, Maeno and Taniguchi 2007). It has a maximum preserved thickness of ~5 m away from the caldera rim, and is a LARI-type ignimbrite (Chap. 10). Using high resolution seismic, Shimizu et al. (2024) mapped the submarine equivalent of the Koya ignimbrite radially for 40 km in water depths up to 200 m. The unit thins from 30 to 2 m, has a volume of ~71 km^3, and was interpreted as a syn-eruptive water-transformed (mega)turbidite deposit.

In a similar example, the older ~25 ka Ito pyroclastic flow, erupted from Aira caldera to the north, appears to have flowed partially over the sea (Yokoyama 1974; Baer et al. 1997), although there is little information on whether there are differences between the over water and over land deposits.

Campanian Ignimbrite (39 ka), Campi Flegrei, Bay of Naples, Italy

The 39 ka VEI 7 Campanian ignimbrite eruption began with a very widely dispersed plinian fallout forming eruption

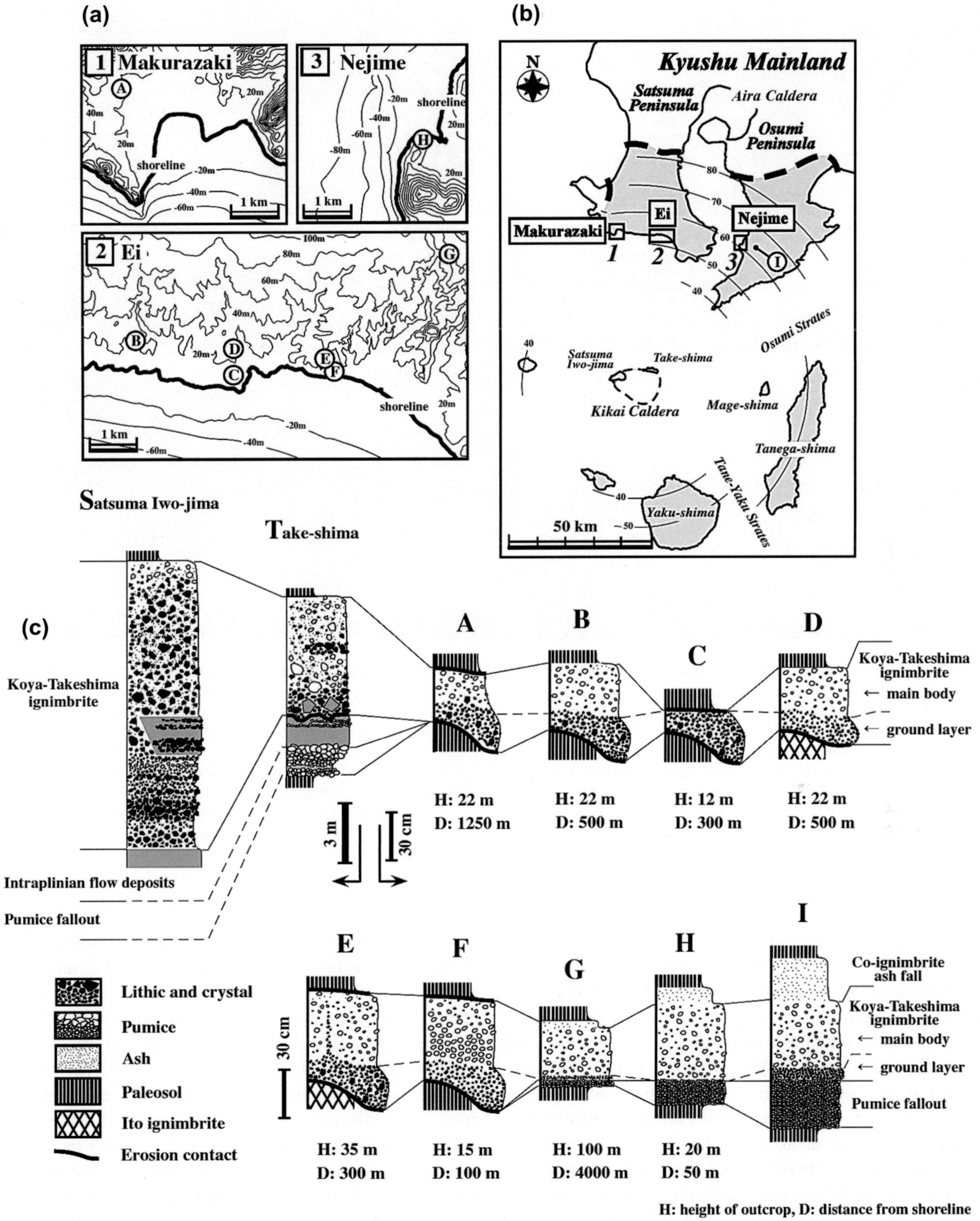

Fig. 11.21 Features of the 7.3 ka Koya ignimbrite, Kyushu, Japan. **a** Locations of measured sections of the Koya Ignimbrite along the Kyushu coastline. **b** Location of Kikai caldera, preserved area of outcrops of the Koya ignimbrite (pale grey shade), and distance from caldera of the measured sections. **c** Measured sections of the Koya ignimbrite from proximal (Satsuma Iwo-Jima, Takeshima) to distal locations 60 km from vent (Section H, I). (From Maeno et al. 2006)

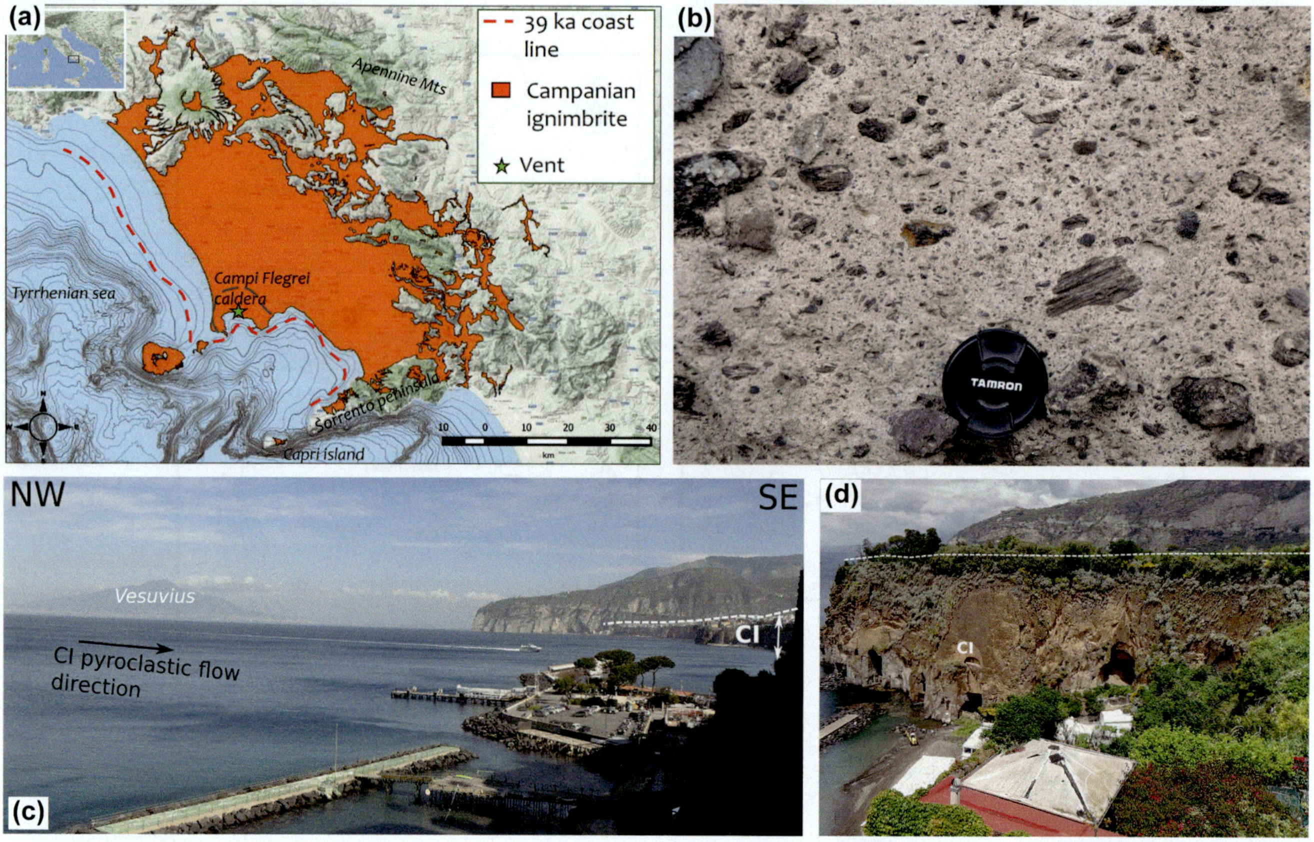

Fig. 11.22 Features of the 39 ka Campanian ignimbrite, Campi Flegrei caldera, Naples, Italy. **a** Areal extent of the onshore Campanian ignimbrite, Bay of Naples, Italy. The paleo-coastline at the time of the eruption 39 kyr ago was 65 m below present sea level and is indicated with the dashed red line. The deposits on the Sorrento Peninsula to the southeast of the Campi Flegrei caldera must relate to pyroclastic currents able to flow over the sea for more than 20 km. These largely maintained their internal dynamic properties as the deposits were emplaced hot (>580 °C), and show classic valley ponds, ash-rich massive and chaotic facies. (From Silleni et al. 2020). **b** Typical fine ash-matrix supported pumice and lithic lapilli texture of the Campanian ignimbrite on the Sorrento Peninsula, deposited after pyroclastic currents travelled over the sea a distance of some 20 km. The facies is virtually identical to the internal massive and chaotic texture shown elsewhere by the ignimbrite. **c** Panoramic view of the Bay of Sorrento southeast of Naples. This was largely submerged when the Campanian ignimbrite was emplaced, which forms the terrace where Sorrento town is built. The Campi Flegrei caldera was the vent for the Campanian ignimbrite (CI) pyroclastic flows and is located to the northwest, i.e. to the left of Vesuvius in the photograph. **d** Campanian ignimbrite in Sorrento town

phase (Scarpati and Perrotta 2016; Smith et al. 2016), followed by the column collapse, pyroclastic flow and ignimbrite-forming phase (Fisher et al. 1993; Civetta et al. 1997; Fedele et al. 2008; Scarpati et al. 2015; Silleni et al. 2020). The ignimbrite has a low aspect ratio (between 1 and 6×10^{-4}, Silleni et al. 2020), is spread over 7,500 km^2, climbed ridges >1000 m in elevation and reached up to 80 km from vent, which was located inside the Campi Flegrei caldera (Fisher et al. 1993; Fig. 11.22a). The fallout deposit volume was 20 km^3 or 5.3 km^3 DRE (Scarpati and Perrotta 2016), whereas the ignimbrite sequence has been recently carefully recalculated to a bulk volume of $\sim$300 km^3, or $\sim$120 km^3 DRE (Silleni et al. 2020). The deposits are trachytic, and compositionally zoned, indicating eruption of the most evolved magma first, with the most mafic erupted last (Civetta et al. 1997; Fedele et al. 2016; Forni et al. 2016). The main Campanian ignimbrite is a grey ash-rich and black pumice lapilli tuff and is distributed radially from the vent (Fig. 11.22a,b). Half of the ignimbrite was deposited offshore, as the caldera is adjacent to the coast. Towards the southeast, the pyroclastic flows must have crossed a stretch of sea deeper than 100 m for about 20 km based on the reconstructed coastline at the time of its emplacement, before reaching the Sorrento Peninsula.

Seismic surveys in the Bay of Naples indicate the presence of up to 40 m of Campanian ignimbrite in the seafloor stratigraphy (Milia and Torrente 2007; Milia 2010; Chap. 15, see Fig. 15.23). Beyond this stretch of open sea, the Campanian ignimbrite is found on the Sorrento Peninsula, where it forms an up to 40 m high cliff where Sorrento town is built (Fig. 11.22c, d), and it also occurs farther across the ridge of the Peninsula. The ignimbrite deposited on the Sorrento Peninsula is not significantly different from that resulting from flow only over land. That is, it is typical, poorly sorted, fine ash matrix-supported massive ignimbrite (Fig. 11.22b), indicating that there was little or no interaction between the pyroclastic flow and the sea. Furthermore, the temperature of the deposit inferred from TRM analysis of lithic clasts shows values >580 °C (i.e. above the Curie temperature for magnetite; Ort et al. 2015). Silleni et al. (2020) interpreted the low aspect ratio of the ignimbrite, the ability of the flow to climb over high ridges and the little interaction with seawater as indicating a highly transportational flow able to maintain the internal pressure across its whole extent, due to high mass flux within the flow, associated with very high velocity and/or high mass flux eruption column/fountain collapse rates (cf. Giordano and Doronzo 2017).

11.6.1.4 Pumiceous Pyroclastic Flows Flowing Over Water, and then Sinking or Depositing from the Base

Although this process must happen when some pyroclastic flows travel over water and wane (Fig. 11.7d), there are few subaqueous deposits that have been interpreted through this process based on the characteristics of the deposits (Dufek and Bergantz 2007a, b; Jutzeler et al. 2017). As discussed in the previous section, Dufek and Bergantz (2007a, b) considered that it was possible that the limited lithic clast content in ignimbrite unit D in outcrops of the Kos Plateau Tuff on the island of Tilos, Greece (Allen and Cas 2001), could be explained by larger lithic clasts dropping out from the base and sinking to the seafloor as the pyroclastic flow travelled over water. This must have happened from the base of pyroclastic flows as they flowed from the submarine caldera southeastwards to Tilos, while the bulk of the flow over the sea reached Tilos and deposited a lithic clast poor ignimbrite deposit there (Fig. 11.20d, e).

Jutzeler et al. (2017) describe a ~4 Ma, 1.2 m thick succession of pumice and ash deposits cored 35 km offshore from Montserrat in water depths of ~800 m at site U1396 during IODP expedition 340; also see Sect. 11.6.3.1. The deposits are stratified, poorly sorted, lack obvious grading, and are interpreted to have resulted from a combination of individual particle (pumice, lithic fragments, ash, and planktonic foraminifera) settling from the base of over-passing pyroclastic flows flowing over the sea (Fig. 11.23a), as well as sinking plumes of aggregates of pyroclasts released from the base of the over-passing pyroclastic flows (Fig. 11.23b), similar to a model proposed by Carey (1997) for the formation of some subaqueous fallout deposits through water. The deposits are underlain and overlain by deep-water, fossiliferous hemipelagic mudstone. In addition, some of the stratification results from the sinking plumes of pyroclasts being deflected sideways as density currents impacting the seafloor (Fig. 11.23b; Jutzeler et al. 2017).

For particles that fall freely from the base of a pyroclastic overflow, the duration of an event and the water depth will determine how well the seafloor deposit is hydraulically sorted (Fig. 11.23a). For a short-lived event in moderately to deep water, the deposit will be well sorted, with the highest settling velocity (HSV) grains at the base grading upwards to slowest settling velocity (LSV) grains at the top of the deposit, which will usually be the finest grained, unless there are large buoyant pumice clasts in the population. For a prolonged event and/or if water depth is shallow, the bottom (early releases HSV particles dominant) and the top of the deposit (early and late released LSV particles dominant) may be well sorted but the middle will be poorly sorted because the LSV grains released from the base of the flow early in the flow event will be depositing at the same time as HSV grains released late in the flow event (Fig. 11.23a).

11.6.1.5 Subaerially or Shallow Water Erupted Pumiceous Pyroclastic Flows Flowing into Water and Transforming into Subaqueous Volcaniclastic Megaturbidity Currents

As we have emphasised, there are many observed instances of pyroclastic flows entering the sea, and many may be transformed into submarine mass flow underflows (e.g. Fig. 11.1). Many deposit a graded bedded deposit tens of cm to m thick with turbidite characteristics. In this next scenario, we consider deposits that are orders of magnitude larger, tens of m thick, massive to stratified near the top, grading into ash-size grain sizes at the top of the deposits, unless truncated by the base of an overlying, amalgamated depositional unit, and lacking evidence for heat retention (e.g. welding). Such a facies has the scale and volumetric magnitude of subaerial large-volume ignimbrites, often those that result from caldera-forming eruptions. However, they are better hydraulically sorted, with more, high settling velocity clasts (dense lithics, crystal fragments, and/or large pumice clasts) near the base, grading overall to more low settling velocity, usually finer grained clasts, near the top (fine vitric and crystal ash, vesicular pumice; Cas and Wright 1987, 1991; Stix 1991). We call these hydraulically well-sorted, very large volume, monomictic deposits, *volcaniclastic megaturbidite facies or deposits,* because they have the textural characteristics of terrigenous turbidite deposits (Chap. 12) but are an order to several orders of magnitude larger in volume.

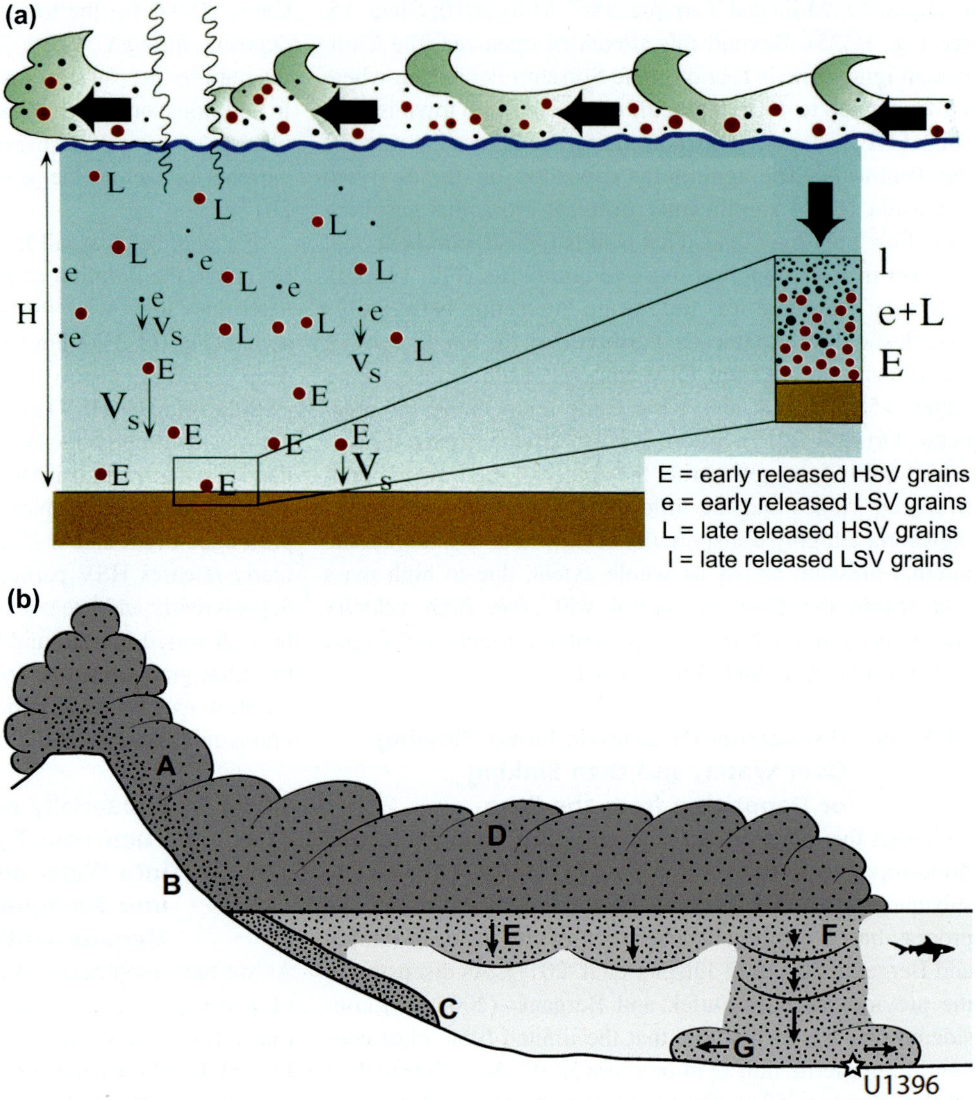

Fig. 11.23 Schematic representation of two depositional mechanisms from the base of a pyroclastic overflow travelling over the sea. **a** Cartoon depicting the release of high settling velocity (HSV) and low settling velocity (LSV) particles released early (E, e) and late (L, l) during a pyroclastic flow event, and the resultant internal sorting and grading resulting from a prolonged flow event where water is moderately deep to deep. **b** Depiction of sedimentation scenario (A–G) from the ~4 Ma Montserrat event (IODP 340, site U1396), as discussed by Jutzeler et al. (2017). This shows subaerial and submarine underflows (A–C), the development a shoreline-derived overflow (D), and inverted plumes of pyroclasts sinking from the base of the overflow (E and F). On impact with the seafloor, the plume spreads radially triggering a seafloor hugging density current (G), contributing to stratification in the total deposit. (From Jutzeler et al. 2017)

Some might argue that they could result from post-eruptive remobilisation of newly deposited pyroclastic ignimbrite deposits from the shoreline zone, by perhaps a storm or seismic or tsunami trigger, forming a normal turbidity current. Although this may be the case with some at the thinner, small volume end of the spectrum (beds <1 m thick), in checking the sedimentology literature, it seems that what sedimentologists call *megaturbidites* or *seismo-turbidites*, triggered by storm, seismic and/or tsunami events, are >1 m thick (Ricci Lucchi and Valmori 1980), usually a few metres to 10 m thick, but are rarely more than 20-25 m thick (Ricci Lucchi and Valmori 1980; Cita et al. 1984; Mutti et al. 1984; Rosell and Wiezorak 1989; Polonia et al. 2013).

Therefore, given the monomictic character of what we call volcaniclastic megaturbidite deposits (i.e. they appear to have been sourced from a contemporaneous explosively erupting volcanic source and are dominated by juvenile pyroclasts and a monomictic population of angular volcanic crystal fragments) and the anomalously thick and inferred voluminous nature of the deposits relative to sedimentary megaturbidite deposits, the most likely origin of these deposits is that they result from a contemporaneous large volume, explosive eruption from a subaerial or relatively shallow marine vent producing pyroclastic flows (Chap.10). These then transform from gas-supported pyroclastic flows into water-supported mega-turbidity currents or mass-flows upon entry into, and interaction with, a large body of water. These are represented diagrammatically by a larger scale, higher flux version of Fig. 11.7a, whereby a continuous “megaturbidy current” underflow is fed for a sustained period of time. Several examples are now discussed.

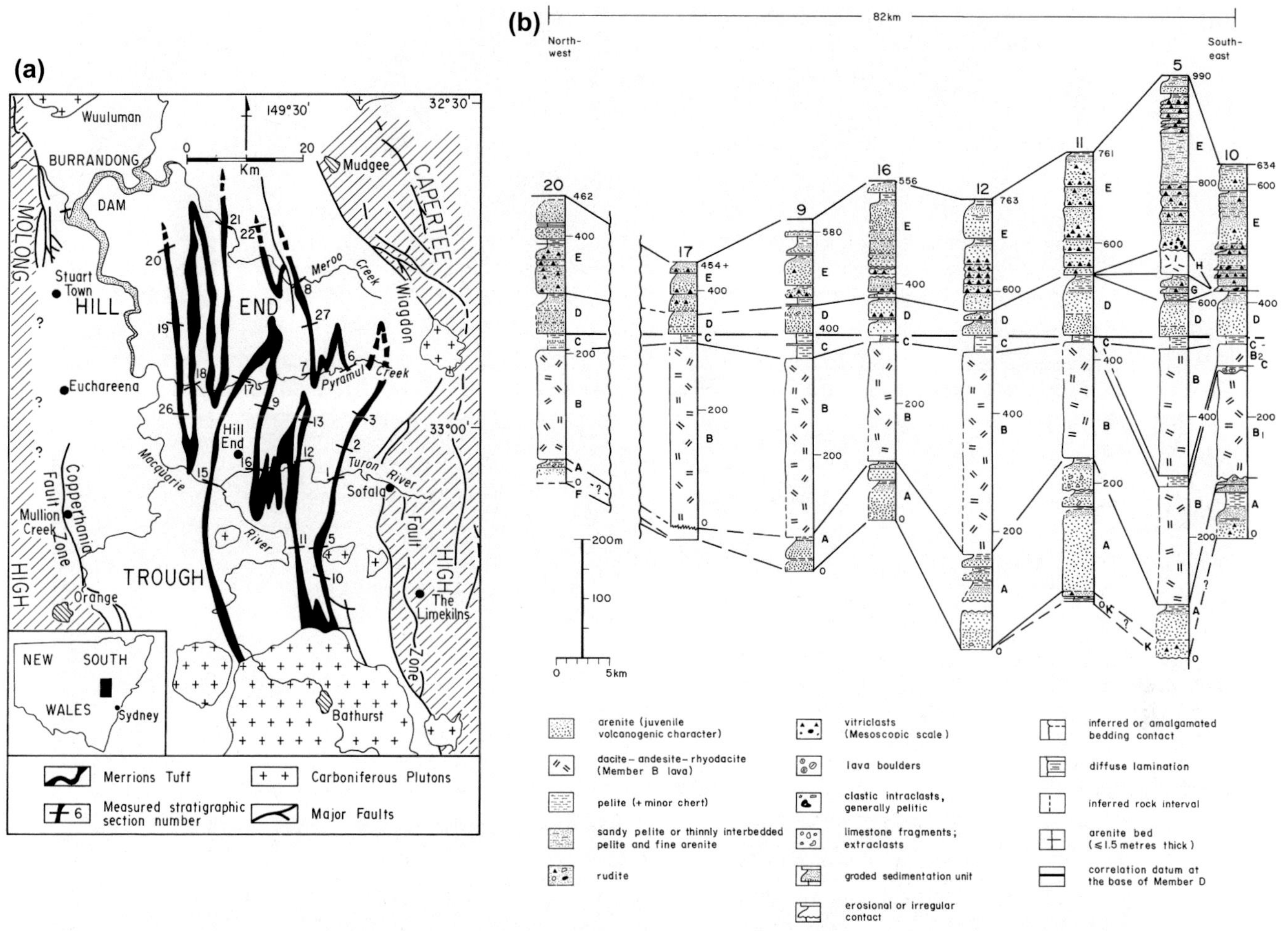

Fig. 11.24 Setting and characteristics of the Lower Devonian Merrions (Tuff) Formation, central New South Wales, Australia. **a** Regional geological setting of the Merrions Formation (black) in the Siluro-Devonian Hill End Trough, showing its extensive nature and the location of measured stratigraphic sections. **b** Selected measured sections through the Merrions Formation, showing the extremely thick, variously graded or amalgamated, crystal-rich, volcaniclastic sedimentation units, intercalated with very thick submarine dacite lavas, and pelitic intervals. Note how laterally extensive individual sedimentation units are. All sedimentation units are plagioclase crystal-rich, with minor volcanic quartz fragments, except those of Unit D, which contains a distinctive assemblage of sub-equal amounts of volcanic quartz, plagioclase, and potassium feldspar crystal fragments (see Fig. 11.25a), making an excellent internal stratigraphic marker, and indicating derivation from multiple magma batches. (From Cas 1977)

Lower Devonian Merrions (Tuff) Formation, Central New South Wales, Australia

The Merrions Formation is a regionally extensive, prominent stratigraphic and structural marker within in a >5,000 m thick Siluro-Devonian turbidite deep water basin-fill succession of the Hill End Trough (Fig. 11.24a), which was one of many Siluro-Devonian basins developed in the Paleozoic Lachlan Orogen of southeastern Australia (Packham 1968; Cas 1983a, b; Chap. 17). The Hill End Trough succession is metamorphosed to greenschist facies, but generally preserves original or pseudomorphed components and many textures. Lithologically the Hill End Trough succession is dominated by graded bedded terrigenous turbidite sandstones and hemipelagic mudstones, clearly indicating a sub-wave base, deep-water depositional setting. The Merrions Formation is up to 1 km thick and consists of a succession of crystal-rich, dacitic to rhyodacitic, volcaniclastic mega-graded depositional units (megaturbidites) intercalated with volumetrically equivalent far-flowing dacite/rhyodacite lavas and hemipelagic mudstones and thin turbidites (Fig. 11.24b; Cas 1977, 1979, 1983b). The formation has a preserved, palinspastically unfolded areal extent of >3,400 km^2, and some individual depositional units are as, or almost as extensive. Prior to deformation, uplift and erosion, their extent would have been even greater.

The volcaniclastic megaturbidites are composed of whole and broken, euhedral to anhedral and variably resorbed crystal fragments of volcanic quartz, albitised plagioclase, orthoclase (probably after sanidine; Fig. 11.25a) and minor

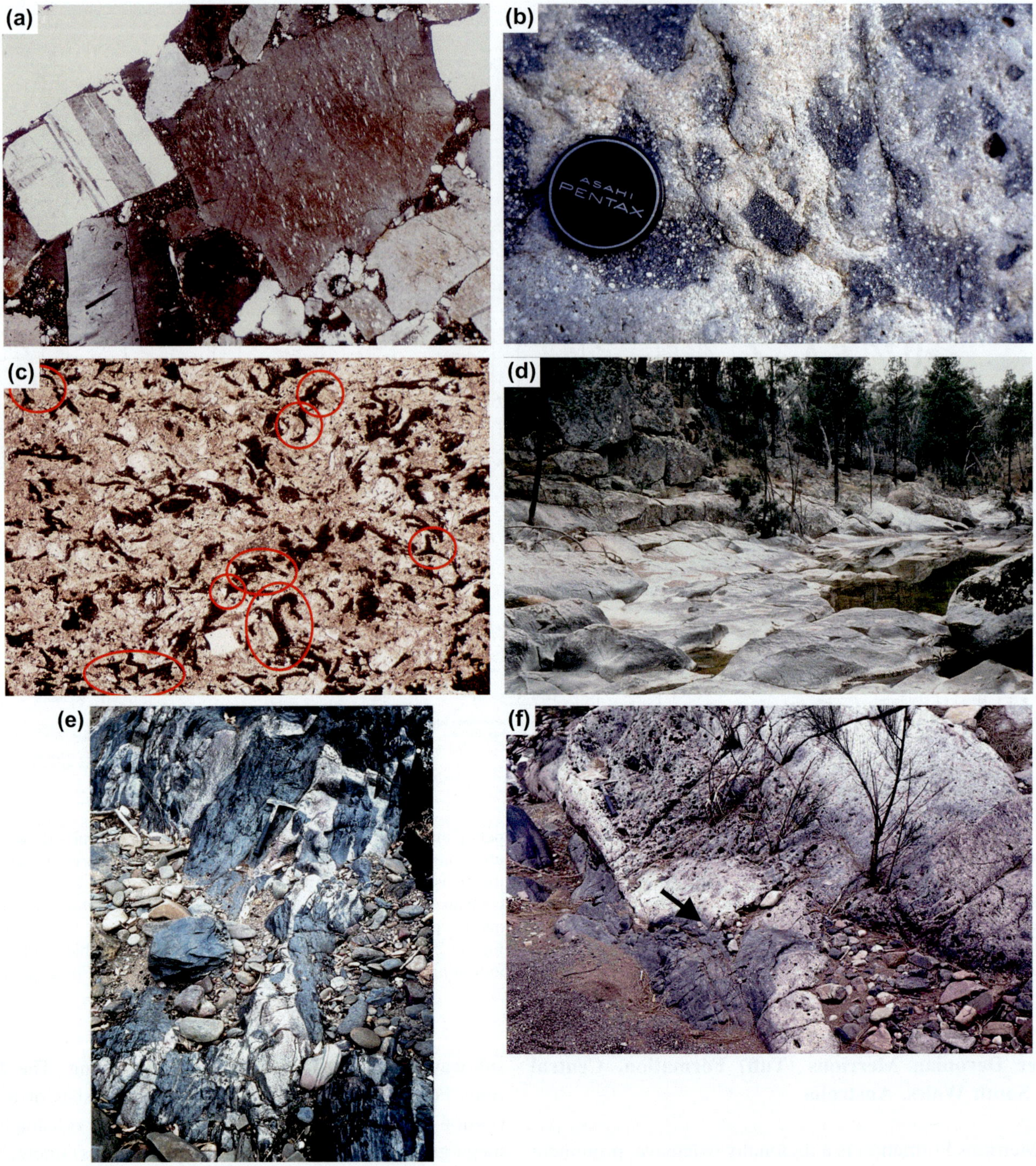

Fig. 11.25 Features of the Lower Devonian Merrions (Tuff) Formation crystal-rich volcaniclastic sedimentation units, Hill End Trough, central New South Wales. **a** Photomicrograph of angular to euhedral crystal fragments of volcanic quartz, plagioclase, and potassium feldspar. Note how little fine matrix there is between the crystal fragments, indicating a very high level of hydraulic sorting. Width of field of view ~3 mm, x-nicols. **b** Outcrop view of irregular to blocky porphyritic clasts of dacite, with ~65% groundmass that has been chloritically altered, dispersed in a crystal-rich matrix in very thick volcaniclastic sedimentation units. Turon River, Turondale. **c** Photomicrograph of greenschist facies metamorphosed, silicified, fine-grained top of graded volcaniclastic mass-flow sedimentation units, preserving pseudomorphed bubble wall and Y-shaped shards, some of which are circled in red. Width of field of view ~2 mm, ppl. Downwards the sedimentation unit passes into a coarse, crystal-rich facies, as in (**a**). **d** Massive outcrop of volcaniclastic megaturbidite deposit, Berraga Creek. **e** Irregular intra-clast breccia at the base of a megaturbidite unit, with dark mudstone intra-clasts separated by crystal-rich sandstone matrix. Turon River, Turondale. **f** Sharp base of megaturbidite depositional unit overlying dark hemipelagic deep-water mudstone, with a large flame structure (arrow) developed along the base. Turon River, Turondale

angular to irregular vitriclasts (porphyritic clasts of dacite and rhyodacite in which the original glassy groundmass has been chloritically altered during regional metamorphism; Fig. 11.25b). In addition, the fine-grained graded tops of depositional units contain pseudomorphed cuspate glass shards, replaced by epidote (Fig. 11.25c). The average crystal fragment content of the volcaniclastic units is 64.2% (Fig. 11.25a), much higher than the average 37.6% phenocryst content of the associated lavas and glassy vitriclasts (Fig. 11.25b; Cas 1983b), and in extreme instances, crystal contents exceed 80% (due to diagenetic compaction). The volcaniclastics are therefore largely closed-framework grain aggregates with a subordinate fine matrix component, except at their tops. Apart from some intra-basinal mudstone intra-clasts (Fig. 11.25e), the whole grain or component assemblage is monomictic (i.e. derived from a single volcanic source), and given the abundance of euhedral to angular volcanic crystals, the fine grain size and textural immaturity, can only have been derived from large volume pyroclastic sources. This assessment formed the basis for several attempts to date the Merrions Formation, using Rb/Sr whole rock (Cas et al. 1976) and U/Pb zircon crystal SHRIMP techniques (Jagodzinski and Black 1999; Black and Jagodzinski 2003; Packham 2003). U/Pb zircon SHRIMP geochronology suggested a likely age range of ~411 to 407 Ma for emplacement of both the multiple volcaniclastic megaturbidites and lavas of the Merrions Formation.

The multiple sedimentation units are medium- to coarse-grained, and range from 1 to tens of metres thick; some are over 100 m thick (Figs. 11.24b, 11.25d). In some cases, they become finer upwards, grading from a coarse, grain-supported crystal-rich base (Fig. 11.25a, b) into fine siliceous pelitic tops, that contain well-preserved, microscopic cuspate shard pseudomorphs (Fig. 11.25c). Sedimentation units are largely massive or internally structureless, occasionally diffusely stratified, with sharp planar to erosive bases (Fig. 11.25e, f). They are normally graded, or graded only at their tops, they may be coarse tail-graded, or, where they are overlain in amalgamated contact with the coarse base of the next sedimentation unit, they may be non-graded (Figs. 11.24b). Where graded, they show clear changes in grain size and componentry upwards (Fig. 11.26a, b).

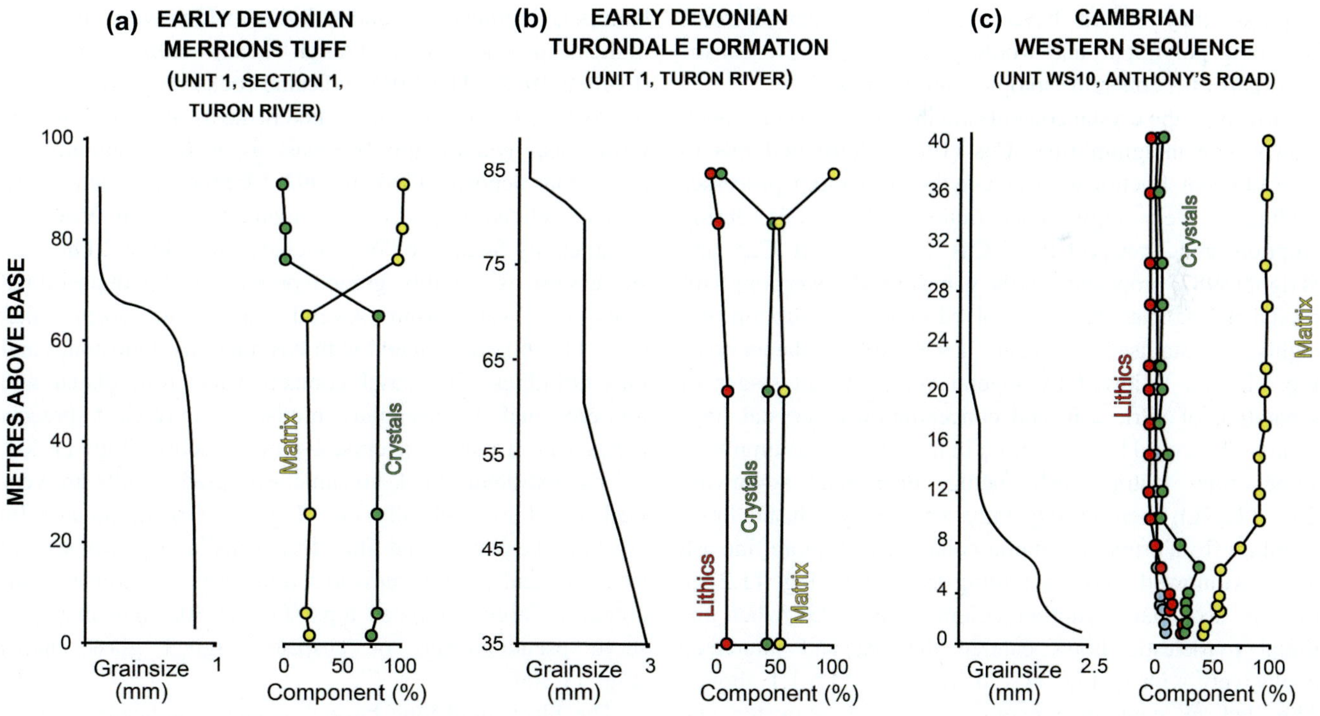

Fig. 11.26 Grain size and componentry profile through several Paleozoic volcaniclastic megaturbidite deposits from southeastern Australia, based on thin section point counting. **a** The basal 90 m thick, single volcaniclastic megaturbidite depositional unit of the ~410 Ma Merrions (Tuff) Formation, Turon River, Turondale, central New South Wales. Note the abrupt grain size grading and change from coarse crystal-rich base to fine vitric and crystal fragment ash-rich top from 60 to 70 m above the base. **b** Basal, single megaturbidite unit, again showing abrupt size and componentry grading at 80 m above the base, the ~419 Ma, Turondale Formation, Turon River, Turondale, central New South Wales **c** Megaturbidite depositional unit WS10, Cambrian Western Volcano-sedimentary Sequence, Anthony's Road, western Tasmania, showing a less significant coarse base, and a more significant graded, fine vitric ash top (also see Fig. 11.29b)

Although the fragmentation mode of the monomictic grain assemblage was pyroclastic, as evidenced by cuspate shards and huge volumes of angular volcanic crystal fragments, the deposits show no evidence of having been emplaced by hot, gas phase pyroclastic transportation processes such as pyroclastic flows (they lack welding, columnar jointing, gas escape pipes). They are also much better hydraulically sorted than pyroclastic flow deposits, and although some subaerial crystal-rich pyroclastic fallout deposits are well sorted (Chap. 9), the subaqueous Merrions Formation depositional units are orders of magnitude thicker than crystal-rich fallout deposits. Given their thickness, sorting and grading, the sedimentation units are akin to very thick, voluminous, high particle concentration turbidity current deposits (Cas 1979; Chap. 12); i.e. they are crystal-rich volcaniclastic megaturbidites.

They are texturally juvenile and have not been reworked by sedimentary surface processes, but they have been transported and deposited by water-supported mass-flow processes (Cas 1979, 1983b). Their juvenile character and the fact that they were used together with the lavas to define a tight whole rock Rb/Sr isochron (Cas et al. 1976) suggests that they were deposited contemporaneously with eruption. If they had been reworked and then redeposited long after eruption, they would have mixed with older detritus, becoming polymictic and would probably not have constituted part of a coherent isotopic suite of rocks.

However, the crystal contents are in many instances much higher than in ignimbrites. Cas (1983b) attributed this to very efficient fractionation and hydraulic sorting processes leading to very high concentrations of crystals during eruption and transportation. Cas (1983b) and Cas and Wright (1987) proposed that the genesis of the Merrions Tuff crystal-rich volcaniclastics involved explosive eruption of a highly crystallised magma ($\sim$30–40% phenocrysts; Fig. 11.27a), followed by several very efficient stages of separation of vitric ash, and concentration of crystal fragments. This could have occurred in the eruption column with some fine (vitric) ash being dispersed downwind (Fig. 11.27b), then subsequently also in pyroclastic flows resulting from eruption column collapse, with more fine ash being winnowed into the trailing ash cloud (Fig. 11.27c), then in secondary eruption columns generated when the initial pyroclastic flows entered the sea and interacted explosively with it (Fig. 11.27d; see Sect. 11.6.1.1 above). This led to contemporaneous mixing of seawater and transformation into submarine water-supported mass-flow turbidity currents, during which more fine ash was elutriated from the head and body of such seafloor flows into the trailing fine sediment cloud (Fig. 11.27e, f). As a result, very well hydraulically sorted megabeds were deposited.

Given that there are many sedimentation units and that these occur in compositionally discrete stratigraphic packages (Cas 1977), with interstratified background basinal sediments (hemipelagic pelites, terrigenous greywacke turbidites), it appears that such a process was repeated many times, sometimes in quick succession, other times with time breaks (Jagodzinski and Black 1999; Black and Jagodzinski 2003). In addition, the stratigraphic record of the Hill End Trough preserves other similar major explosive events preserved as submarine volcaniclastic megaturbidite deposits (e.g. the $\sim$417 Ma Turondale Formation, Fig. 11.28a; Packham 1968, 2003; Jagodzinski and Black 1999; and, part of the Kowmung Volcaniclastics, Fig. 11.28b, Cas et al. 1981).

Cambrian Western Volcano-Sedimentary Sequence, Mt Read Volcanics, Western Tasmania, Australia

The Cambrian Mt Read volcanic belt (MRV) of western Tasmania, Australia, is well known as the host to numerous metallic ore deposits, especially volcanic-hosted massive sulphide (VMS) deposits (Fig. 11.29a; see Chap. 18 for a detailed discussion of VMS deposits and their volcanic setting). The MRV rocks are intercalated with deep-water hemipelagic mudstones and turbidites which were emplaced in the deep-water Dundas Trough volcano-sedimentary basin (Corbett 1992). The MRV volcanic rocks range from rhyolite to basalts, and include coherent lavas and sub-volcanic intrusions, hyaloclastite breccias, as well as voluminous pumiceous deposits (Corbett 1992; McPhie and Allen 1992).

The relatively poorly documented Western Volcano-Sedimentary Sequence (WS), includes an at least 2 km thick succession of variably graded pumice-crystal depositional units that vary from several metres to 80 m thick (Fig. 11.29b), intercalated with basinal black mudstones and thin turbidites. The basal contacts vary from planar and non-erosional to irregular, mudstone intra-clast breccia zones with a matrix of crystal–pumice debris (Fig. 11.30a, b). The extremely thick volcaniclastic graded units are very well-sorted hydraulically (see Fig. 11.26c), dominated by crystal, tube pumice and lithic fragments in the lower several to 10 m, and grading upwards into very fine-grained, silicified, variably laminated tops (Fig. 11.30c) that may preserve pseudomorphs of cuspate original glass shards (Fig. 11.30d).

The black mudstone basinal sequence indicates a quiet, deep-water, anoxic paleoenvironment. Although the clast population in the thick volcaniclastic units is overwhelmingly juvenile, angular volcanic debris, and almost certainly pyroclastic given the large volume of pumice, cuspate glass shards and angular volcanic crystal (phenocryst) fragments,

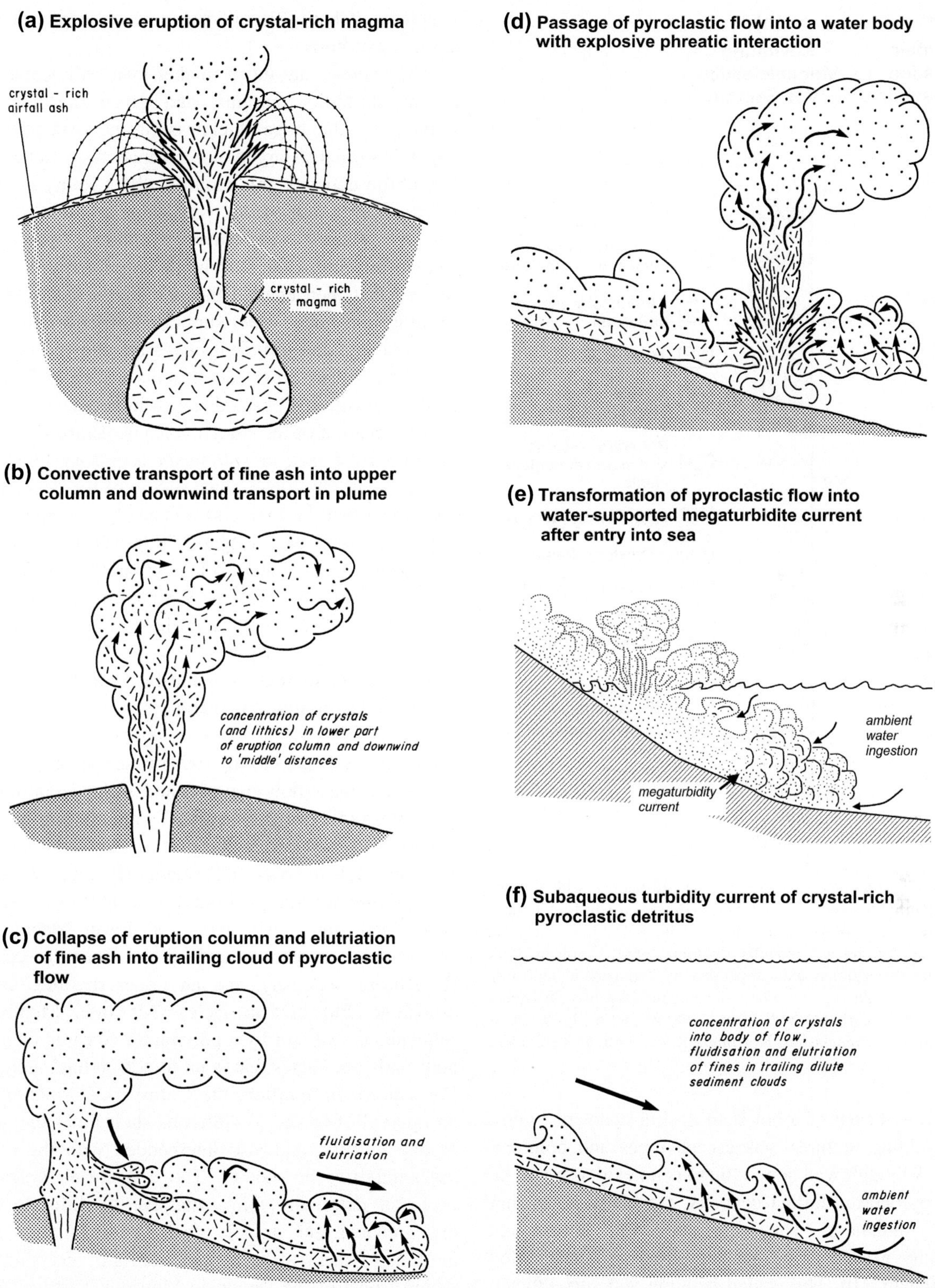

Fig. 11.27 Stages in the formation of crystal-rich domains at the bases of subaqueous megaturbidite deposits, including: **a** the explosive eruption of crystal-rich magma; **b** winnowing of fine vitric ash from an explosive eruption plume into the downwind plume; **c** further winnowing of vitric ash into the trailing ash cloud of pyroclastic flows resulting from column collapse; **d** further segregation of fine ash from coarse crystal debris in secondary explosion plumes as subaerial pyroclastic flows enter a body of water, interacting explosively with it. **e** and **f** Further winnowing of fine vitric ash into the trailing turbulent ash sediment cloud of subaqueous water-supported mass flows that form as gas-supported pyroclastic flows ingest water and transform into water-supported megaturbidity currents. (Modified from Cas and Wright 1987)

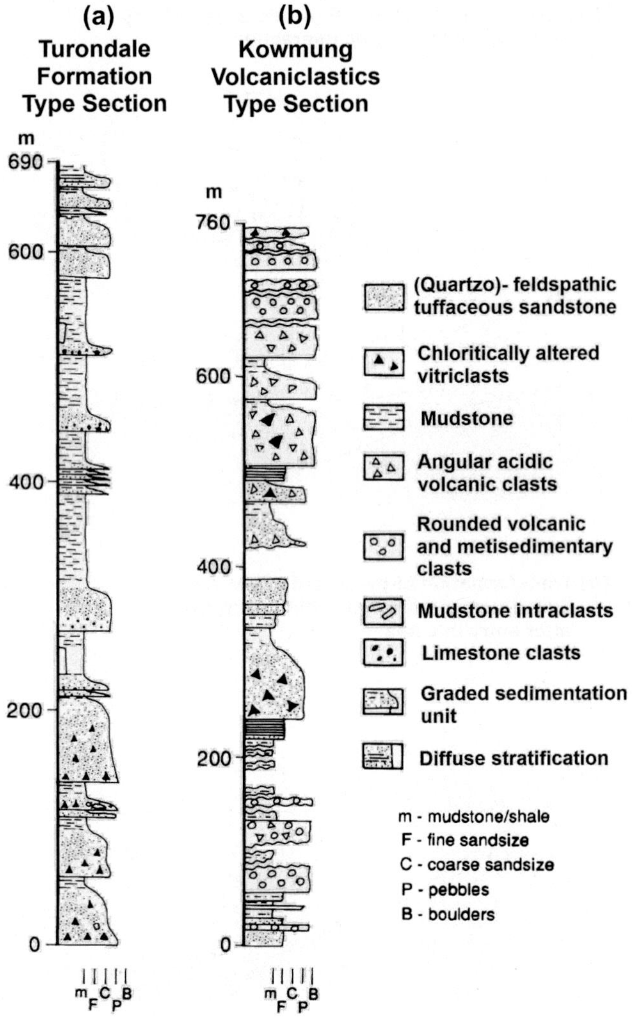

Fig. 11.28 Graphic logs through two other formations with volcaniclastic megaturbidite deposits in the Siluro-Devonian Hill End Trough, central New South Wales, Australia. **a** The ~419 Ma Turondale Formation at Turondale, showing multiple graded volcaniclastic megaturbidite depositional units separated by hemipelagic mudstone and thin, graded bedded greywacke sandstone turbidite beds. **b** Graded volcaniclastic megaturbidite beds from ~240 to 580 m, Early Devonian Kowmung Volcaniclastics, Kowmung River, north of Yerranderie

there is no evidence of a hot state of emplacement (with a lack of welding textures, gas escape pipes and columnar jointing). Although well sorted, there is no evidence that the units are pyroclastic fallout deposits, especially given their enormous thickness. They are too well sorted and hydraulically graded to be gas-supported pyroclastic flow deposits (Chap. 10). The graded bedding and very efficient hydraulic sorting indicate efficient transport and sorting by turbulent, water-supported mass flow, specifically turbidity currents, again producing in the volcaniclastic megaturbidite deposits. The absence of shallow water tractional, or storm hummocky and swaley cross-stratification sedimentary structures (Chap. 12) indicate a water depth greater than storm wave base.

The deposits are much thicker than sedimentary seismo- or megaturbidity currents, and given the huge volume implied by the huge thickness of the depositional units, which is similar to large volume subaerial pyroclastic flow or ignimbrite deposits, this requires direct sourcing of the WS volcaniclastic units from very large volume subaerial or shallow marine explosive eruptions, perhaps even caldera-forming scale. During the eruption, pyroclastic flows formed that then entered the sea and mixed with it, transforming into huge, water-supported megaturbidity currents. An expanded and turbulent flow state allowed very efficient hydraulic sorting when the currents waned and transitioned into a depositional phase distal from source.

The location of the source volcanic centre(s) is unknown, but given the lack of significant accessory lithics and the overall relatively fine-grained nature of the deposits, it does not appear that the source(s) was nearby, despite the implied large volume of each megaturbidite bed. Given the numerous depositional units, often separated by hemipelagic mudstones (Fig. 11.29b), the WS records an extended period of highly explosive activity in the region.

Cambrian Comstock Formation, Mt Read Volcanics, Western Tasmania, Australia

White and McPhie (1996, 1997) described crystal-rich volcaniclastic mass-flow deposits and welded ignimbrite from the Comstock Formation of the Tyndall Group, which is the uppermost lithostratigraphic unit of the Cambrian Mt Read Volcanic Belt in western Tasmania (Fig. 11.29). The Comstock Formation ranges from 100 to 800 m thick, and was subdivided into two parts, a lower dacitic-andesitic crystal-rich sandstone and lithic clast breccia facies (Lynchford Member), and an upper rhyolitic crystal-rich sandstone and breccia facies (Mt Julia Member). This sequence is overlain by a polymictic volcanic conglomerate and sandstone facies that is up to 500 m thick (Zig-Zag Hill Formation). In addition, the Comstock Formation contains an interval of in situ fossiliferous shelf limestone at its base in one locality, and also interbedded with the crystal-rich volcaniclastics in another. The crystal-rich volcaniclastics occur in extremely thick, massive to diffusely stratified, amalgamated and variably graded sedimentation units that are tens of m to >100 m thick, and are very similar to the facies aspect of the Devonian Merrions Tuff (cf. Fig. 11.24 b), described above in this section. They are interbedded with laminated mudstones and grade into them in some places. The dominant population of crystal fragments consists mostly of angular juvenile plagioclase and volcanic quartz, which make up 35–65 modal % of the facies.

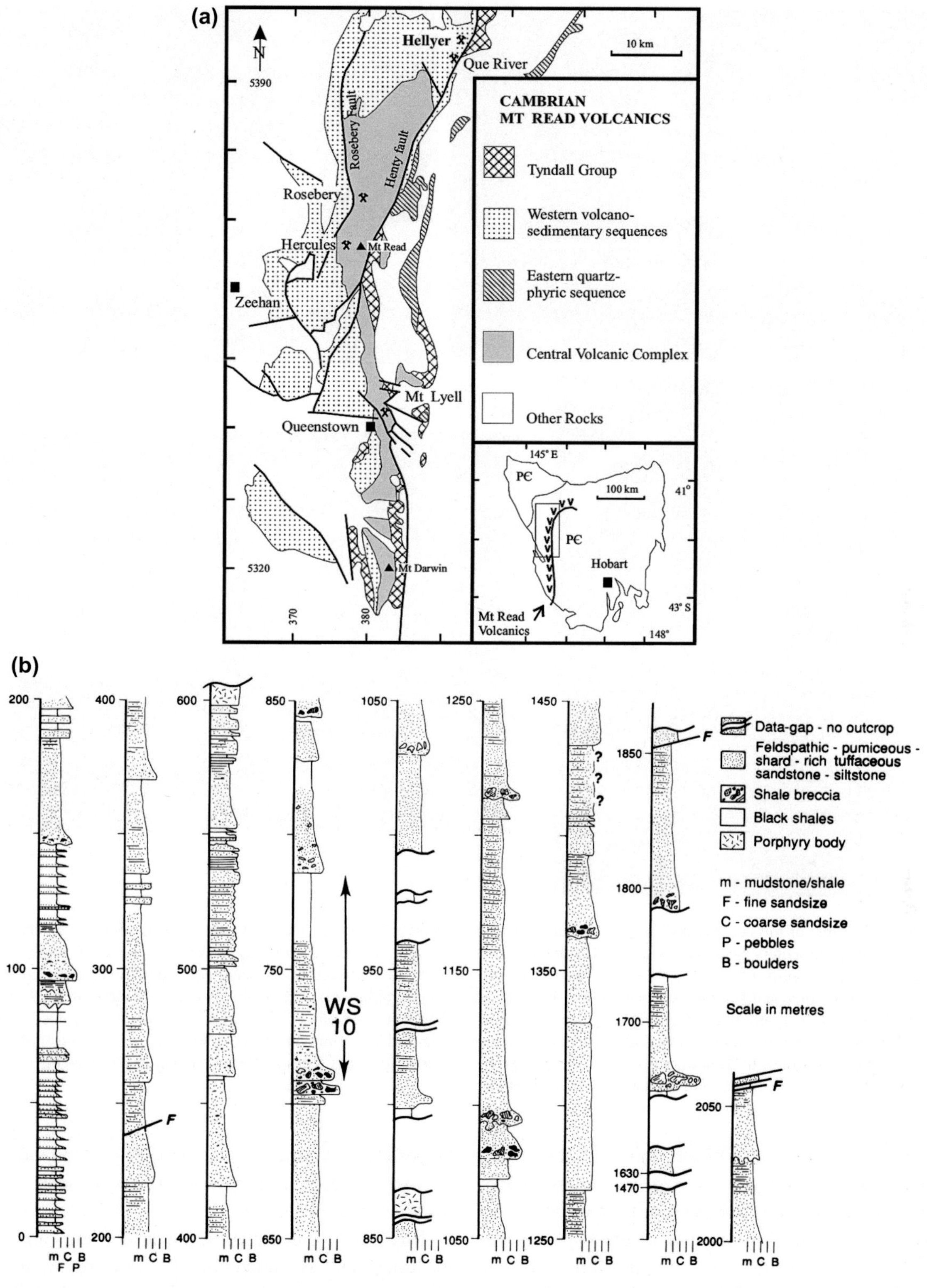

Fig. 11.29 Setting and features of the Western Volcano-Sedimentary Sequence, western Tasmania, Australia. **a** Location and regional geological setting of the Cambrian Mt Read Volcanic Belt in western Tasmania, showing its main stratigraphic subdivisions and major mines. The main volcanic unit (Central Volcanic Zone) passes westwards into the Western Volcano-Sedimentary Sequence (WS) which is a mixed volcanic–sedimentary succession. Both were emplaced in the deep-water Dundas Trough. (From Schardt et al. 2002). **b** Graphic log through the at least 2 km thick WS exposed in cuttings along the Anthony's Road between the Hercules Mine and Queenstown, showing multiple, tens of m thick, graded, volcaniclastic megaturbidite depositional units interbedded with basinal black mudstones and thin sandstone turbidites. Note the position of megaturbidite depositional unit WS10 illustrated in Fig. 11.26c. (Based on unpublished thesis of D. Hutton, Monash University)

Fig. 11.30 Some features of the volcaniclastic megaturbidite units of the Western Volcano-Sedimentary Sequence, western Tasmania, Australia. **a** Irregular black mudstone intraclasts (dark) with "wispy" margins in a matrix of pumice and crystal fragments (grey) at the base of the 70 m thick depositional unit WS10 (see Fig. 11.29b), testifying to the erosional capacity of the original megaturbidity current to scour the basinal black mud substrate. **b** Photomicrograph of the matrix of unit WS10, showing sericite altered tube pumice clasts (foliated texture, centre of field of view, cloudy sericite altered feldspar, and minor volcanic quartz fragments). Width of field of view ~4 mm, ppl. **c** Silicified, laminated fine-grained graded top of a graded volcaniclastic megaturbidite unit. **d** Relict pseudomorphed original cuspate glass shards preserved in the graded top of a graded volcaniclastic megaturbidite unit. Width of field of view ~2 mm, ppl. All outcrops and samples are from Anthony's Road exposures (Fig. 11.29)

As for the Merrions Tuff, White and McPhie (1996, 1997) interpreted the origin of the extremely thick, crystal-rich depositional units to have resulted from the entry of pyroclastic flows erupted from subaerial explosive centres (location unknown) into the sea. At the shoreline, the pyroclastic flows transformed into water-supported mass flows, probably associated with shoreline phreatic explosive activity. Fine ash was winnowed out of the flows both at the shoreline as well as during subaqueous mass flow. Deposition occurred onto a shelf to offshore sloping, sub-wave base deeper water environment (White and McPhie 1996, 1997).

Rhyolitic welded ignimbrite also occurs in situ in two locations and as allochthonous clasts up to several m in diameter in other places in the upper Mt Julia Member (White and McPhie 1996, 1997). The ignimbrite also contains angular volcanic quartz and plagioclase in a welded matrix of eutaxitically deformed glass shards. The ignimbrite is up to 60 m thick and extends at least 600 m along strike at one locality. Unfortunately, the actual lower contact with the underlying host volcaniclastic sediments is not exposed, but the top is abruptly overlain by coarse, polymicitic, but largely volcaniclastic mass-flow conglomerate of the Zig-Zag Hill Formation. White and McPhie (1996, 1997) proposed that the rapid accumulation of the deposits of water-transformed flows of pyroclastic debris in the offshore environment probably led to rapid seafloor shallowing, allowing some high mass-flux pyroclastic flows to enter relatively shallow water, allowing heat retention and welding in the deposits, some of which may have had emergent tops or submerged tops. Whether or not the pyroclastic flows were able to push back the shoreline momentarily, as suggested above for Ordovician shelf-emplaced ignimbrites in North Wales, is uncertain.

Oligocene Ohanapecosh Formation, Mt Rainier National Park, Washington, USA

The Ohanapecosh Formation in Washington State is historically significant in the discussion on pyroclastic flows in water because it was one of the first volcaniclastic sequences described and interpreted as originating from subaqueous pyroclastic flows (Fiske 1963). Fiske recognised variably graded, very thick (tens of m), volcaniclastic "slurry" deposits and thin (<3 m) volcaniclastic turbidite deposits with andesite to dacite lithic clasts, pumice clasts, and shards. He interpreted the thick facies as subaqueous pyroclastic flow deposits sourced from subaqueous vents in a lacustrine environment (only plant fossils are known), even though he noted there was no evidence of welding or columnar jointing. Some of the thin beds were interpreted as water-settled fallout deposits and resedimented ash deposits by Fiske (1963; Cas and Wright 1991). Minor accretionary lapilli occurs in some thin mudstone and sandstone beds, and represent some water-magma explosive activity, probably at the shoreline where pyroclastic flows entered a water body.

Jutzeler et al. (2014a) reassessed the characteristics and origins of the Ohanapecosh Formation volcaniclastic facies, and defined 16 major facies. They noted that the compositional range of depositional facies was from intermediate to mafic, and that hydraulic sorting was well developed in the very thick depositional units (Fig. 11.31a). This includes changes from dense, clast-supported breccias, sometimes conglomerates, at the base, to more pumiceous and fine crystal- and shard-rich ash towards the tops of depositional units (Fig. 11.31b). In some units, the average maximum pumice clast size increased upwards defining a reverse, coarse-tail grading, whereas dense lithic clasts define a normal coarse-tail grading. Pumice clasts ("fiamme") are often elongate dand aligned parallel to bedding (Fig. 11.31 b), which Jutzeler et al. (2014a) interpreted as being due to depositional alignment and post-depositional diagenetic compaction, rather than hot state thermal welding.

Jutzeler et al. (2014a) interpreted the sequence to have resulted from several source vents, most likely located subaerially, based on the common occurrence of rounded to subrounded dense lithic clasts, which require subaerial to shoreline tractional environments to abrade. The abundance of pumice (30%), shard and crystal-bearing matrix ($\sim$50%) also indicate a low ambient pressure explosive origin, such as a subaerial vent (Fig. 11.32). The componentry is consistent with generation of some of the thickest units as resulting directly from subaerial pyroclastic flows entering water and mixing and transforming into syn-eruptive ("eruption-fed") water-supported mass flows (Fig. 11.32a), which Fiske (1963) called slurries and Jutzeler et al. (2014a) consider to be megaturbidity currents. Water-supported turbidity current transport allowed efficient hydraulic sorting and grading to develop. In addition, Jutzeler et al. (2014a) interpreted some of the thinner beds to have resulted from post-eruptive resedimentation from shoreline or shallow water storage aprons (Fig. 11.32b).

Oligocene Dali Ash, Rhodes, Greece

The Dali Ash on the island of Rhodes in the southeastern Aegean Sea, has also been important in the development of concepts regarding subaqueous pyroclastic flow deposits. It was first described by Mutti (1965) as a 5 m thick massive unit with abundant glass shards, which he considered to be welded. It occurs in association with overlying Bouma cycle ash turbidites containing plant detritus, deep-water marls and distal ambient turbidites (Fig. 11.33a, b), with abundant planktonic foraminifera testifying to a sub-wave base deep-water setting, and for a number of years was considered an important example of a deep-water welding.

The massive unit (or megaturbidite) is composed of >90% juvenile volcanic detritus, chiefly glass shards and fragmented euhedral quartz and feldspar crystals and is essentially monomictic, and rhyolitic in composition (Fig. 11.33c, d; Wright and Mutti 1981). By comparison, it is finer grained and much better hydraulically sorted than other pyroclastic flow deposits. This includes a thin, planar laminated, crystal-rich basal layer interpreted to have formed in a water-inflated flow head, which allowed an early stage of traction plus fallout from a more dilute dispersion, followed by deposition of the overlying more concentrated body of the flow. Wright and Mutti (1981) reconsidered the thermal characteristics, failing to find textural or TRM evidence for welding or a hot state of deposition. Finding foraminifera mixed into the deposit (Fig. 11.33d) they suggested that it was a fully water-supported, cold, density current. Wright and Mutti (1981) carefully considered many of the problems in interpreting submarine pyroclastic flows and their deposits.

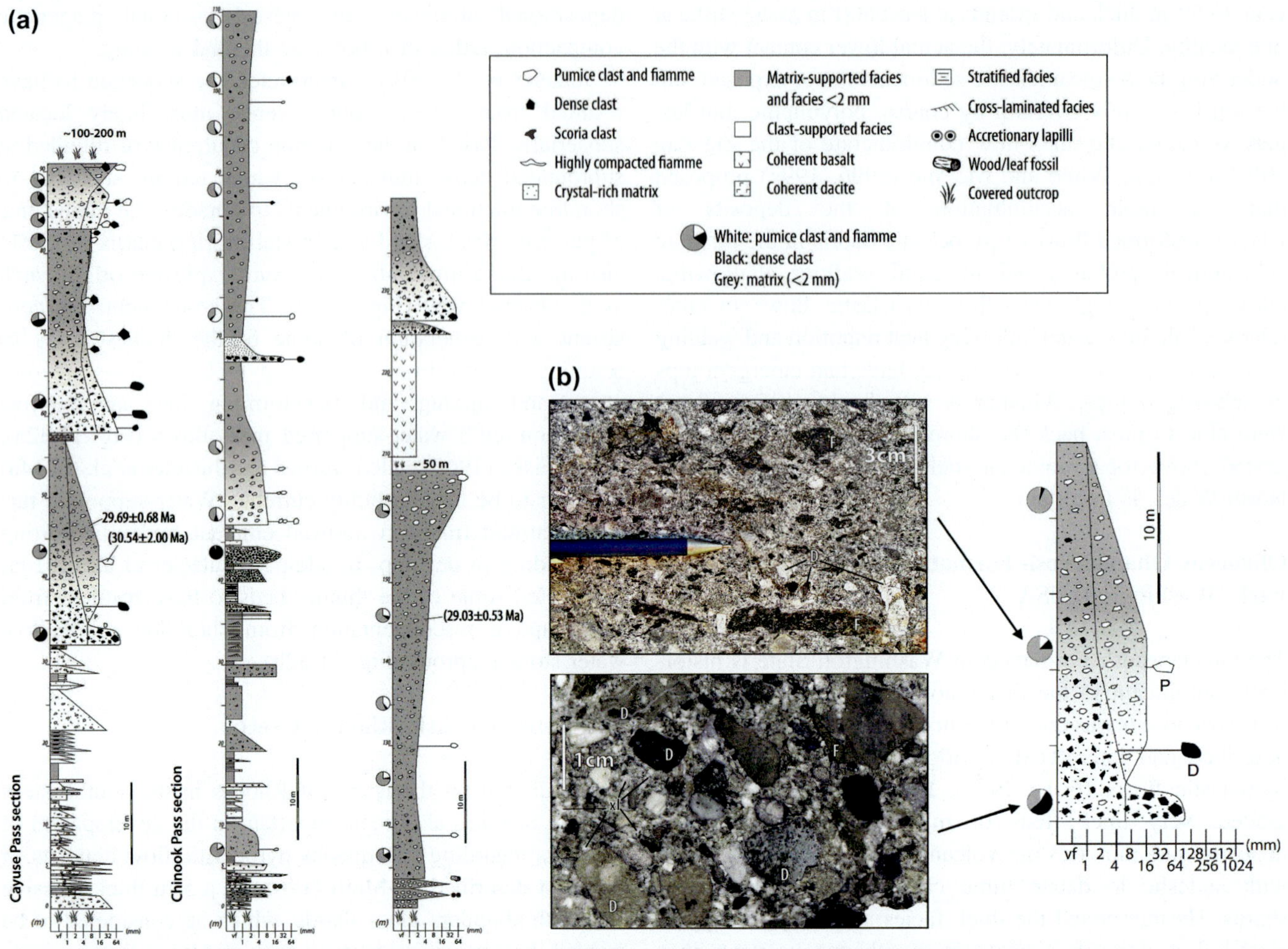

Fig. 11.31 Features of the Oligocene Ohanapecosh Formation, Mt Rainier National Park, USA. **a** Graphic log of facies features, showing the extremely thick, mega-graded depositional units, associated with thin-graded bedded turbidite facies beds. **b** Graphic log of bed 40, which is the radiometrically dated unit in the middle of the left-hand column. This shows the hydraulic sorting of dense, variably rounded lithic clasts (D), feldspar crystal fragments (xl), and rare "fiamme" (F) near the base, and flattened, aligned black pumice lapilli (F) in the middle supported by a grey ash matrix. The two grain size profiles through the depositional unit depict the mean clast size for dense lithic clasts (D, left) and pumice clasts (P, right). (From Jutzeler et al. 2014a)

As a result, the Dali Ash was re-interpreted as the product of a high particle concentration, water-supported, turbidity current mass flow that was the lateral equivalent of a subaerial pyroclastic flow, with the thin turbidites being subsequent slump deposits from a possible shoreline zone or tuff cone (Wright and Mutti 1981).

Miocene Shikoku Basin Syn-eruptive Turbidites, South of Japan

The ocean drilling programs (DSDP, ODP, and IODP) have not only drilled in many volcanic maritime settings, frequently intersecting volcanogenic sediments in aprons around volcanoes, but also hundreds of km away from source volcanoes in deep marine sedimentary basins. In most cases, these represent resedimented volcanic sediments derived through epiclastic processes that have eroded or remobilised volcanic sediment from subaerial source volcanoes or old volcanic terrains. This sediment has then been fed onto submarine volcaniclastic aprons by submarine mass-flow transport, mostly by turbidity currents (Chap. 12). In other cases, however, the deposits appear to be the distant offshore equivalents of pyroclastic flows that entered the sea from subaerial explosive eruptions, and then transformed by ingestion of ambient water into water-supported volcaniclastic mass flows, with some transporting their sediment hundreds of km offshore from their sources.

Schindlbeck et al. (2013) and Kutterolf et al. (2014) have summarised the characteristics of four volcaniclastic deposits ("tuffaceous sandstones") intersected in deep sea drilling in holes IODP 322, and C0011A and B in the Nankai Trough, 200 km south of Honshu, drilled in a water depth of about

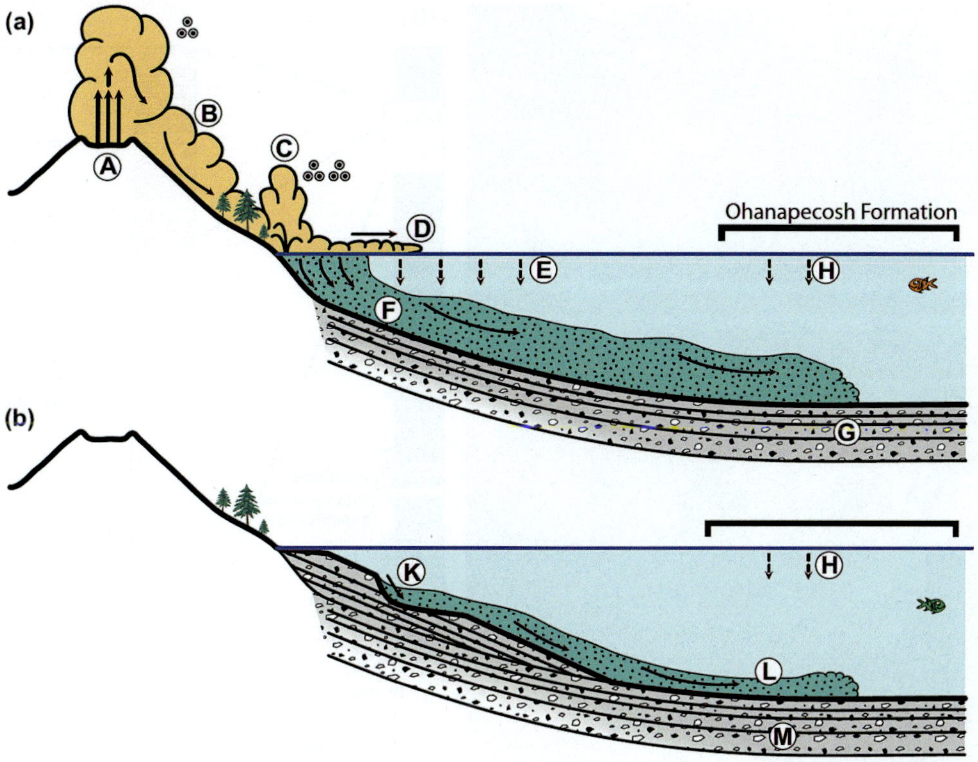

Fig. 11.32 Interpreted origin of the Ohanapecosh Formation. **a** Syn-eruptive, eruption-fed volcaniclastic megaturbidite deposits showing the transformation of subaerial pyroclastic flows (A, B) into water-supported megaturbidity currents as the pyroclastic flows enter the water body (C). Dilute pyroclastic density current flows over the water body (D). Additionally, the generation of some accretionary lapilli at vent and at the shoreline is shown. Pumice clasts from dilute pyroclastic density currents may stay buoyant and create a pumice raft (E). Dense part of the pyroclastic flow enters water and transforms into a subaqueous volcaniclastic density current (F) that deposits very thin to extremely thick, tabular beds on the basin floor (G). Background sedimentation (H) produces fine-grained thin beds. **b** Post-eruptive mass flow resedimentation of shore zone stored deposits (K) from (**a**) into deep water by water-supported turbidity currents (L) that form tabular beds on the basin floor (M). Background sedimentation (H) produces interlayers of fine-grained thin beds. (Modified from Jutzeler et al. 2014a)

5,000 m. Three of the deposits were interpreted to have been derived from major rhyolitic explosive eruptions on Honshu to the north, at least 150 km to perhaps as much as 350 km away, whereas the fourth has an island arc geochemical signature, interpreted to have been derived from a shallow marine explosive eruption from the Izu Bonin arc, about 200 km to the east.

The key characteristics of the volcaniclastic deposits are: (1) they are thicker (1 to 7 m thick) than other "ambient" turbidites (usually < 30 cm thick in the logged drill holes), suggesting source events were large in volume, and (2) they are monomictic and dominated by juvenile pyroclastic debris (angular pumice, glass shards, angular crystal fragments, and in one case even cored lapilli). The latter is consistent with rhyolitic pyroclastic sources, whereas the other thin turbidites are polymictic, and contain variably rounded debris. The deposits of pyroclastic debris are generally massive to diffusely stratified and show well-developed density grading of components, with dense lithic, pumice and crystal fragments concentrated near the base and low-density pumiceous and shard material, and sedimentary intra-clasts near the top. Schindlbeck et al. (2013) and Kutterolf et al. (2014) propose that all four deposits resulted from syn-eruptive transformation of gas-supported pyroclastic flows as they flowed into the sea, continuing into very deep water as highly mobile, far-flowing turbidity currents.

These examples indicate that the submarine water-supported turbidity currents derived from gas-supported pyroclastic flows can be more mobile and flow further than their gas-supported pyroclastic flow counterparts, and that hundreds of km of flow on the seafloor are possible given sufficient volume supply and favourable seafloor topography. Given that submarine terrigenous turbidity currents can flow well over 1,000 km (e.g. the >1,500 km 1928 Grand Banks event; terrigenous turbidites ultimately sourced from the Himalaya found >2,000 km offshore on the submarine

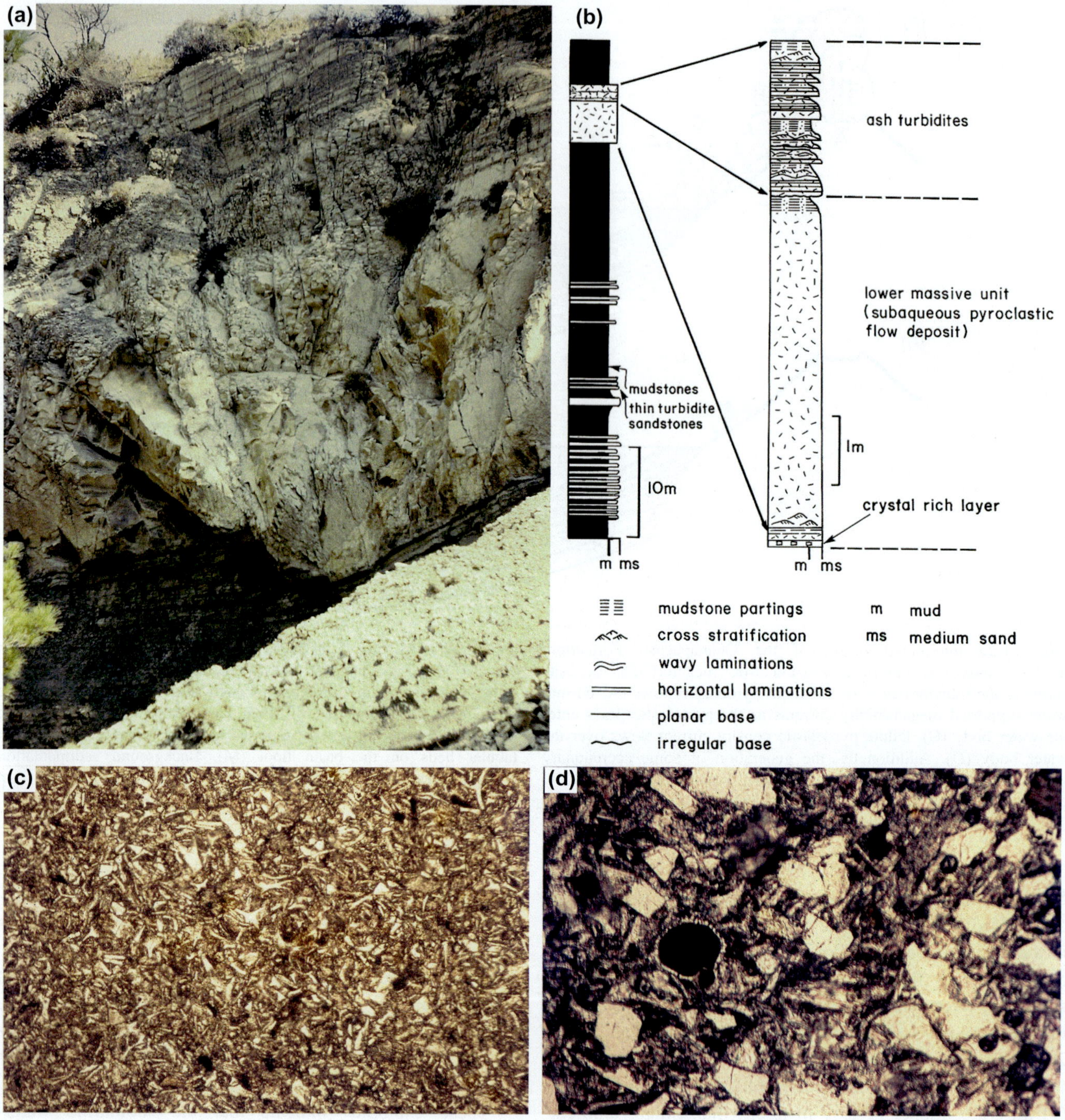

Fig. 11.33 Characteristics of the Oligocene Dali Ash, island of Rhodes, southeastern Aegean Sea, Greece. **a** Outcrop of the massive 5 m thick unit, with overlying thin ash turbidite beds. **b** Graphic log of Dali Ash lower massive unit (megaturbidite), overlain by thin ash turbidite beds, in a succession of deep-water basinal mudstones and thin distal turbidites. **c** Photomicrograph showing non-welded, delicate glass shards and bubble wall fragments from within the middle of the lower massive (megaturbidite) unit. Magnification 10x, ppl. **d** Close-up of fragmented quartz and feldspar crystals and entrained foraminifera test, middle of the lower massive unit. Magnification 25x, ppl. (**a**, **b** From Cas and Wright 1987, After Wright and Mutti 1981)

Bengal Fan), this would appear also possible for pyroclastic flow initiated submarine turbidity currents.

Offshore Bedform Field, Macauley Volcano, Kermadec Arc, North of New Zealand

Pope et al. (2018) have documented fields of high amplitude, long wavelength dune-like waveforms or bedforms on the submarine slopes of Macauley volcano, which is located in the Kermadec oceanic arc, north of New Zealand, using high-resolution bathymetry and seismic profiles. Macauley volcano is a marine stratovolcano which has a submerged summit caldera, with only Macauley Island perched on the eastern rim of the caldera being above sea level.

The fields of bedforms occur at water depths of 2000 m–3000 m, they extend radially about 20 km from the submarine caldera rim to the southwest and north and have a combined estimated volume of 27.5 km^3. The bedforms decrease in amplitude from 140 to 10 m downslope, and in wavelength from 1500 to 250 m. The bedforms show internal layering which is mostly upstream and upslope dipping, although a topset of layers drape the morphological form. The lee side of the dipping layers (down current) appears to be truncated.

Pope et al. (2018) interpret the deposits to have resulted from syn-eruption density currents derived from a subaerial eruption, and the bedforms to be "cyclic steps" (Cartigny et al. 2011) rather than antidunes. The formation of cyclic steps apparently involves subcritical (depositional; see Chap. 12) upstream or stoss side flow conditions and supercritical (erosional) downstream or lee side flow conditions, indicating that the crests of the bedforms acted as hydraulic jumps. However, the actual nature of the density current and its flow dynamics have not been described. Seismic sections through the bedform field show that the crests of the bedforms more or less coincide with subjacent ridges on a buried unconformity surface (perhaps the surface of a hummocky slope landslide deposit?), and so it is unclear whether or not the bedforms could form on a flat depositional surface or whether irregularities on the depositional surface are required for the bedforms to form in phase with the substrate irregularities.

The only known significant pyroclastic sequence known from on-land exposures on Macauley Island is the 5.7 ka Sandy Bay Tephra, but Pope et al. (2018) comment that dredged samples of submarine pumice are not similar to Sandy Bay Tephra clasts, which is also noted by Rotella et al. (2013). Rotella et al. (2013) dredged samples from the seafloor slope around Macauley volcano, including two localities on the surface of the southwestern bedform field. They interpreted the pumice to originate from passive detachment of the highly vesicular carapace of a subaqueously erupting rhyolite magma on the basis of different vesicle morphology and number densities compared with Sandy Bay pumice clasts. There is also no age data on the submarine pumice clasts, or componentry or grain size data to relate the submarine pumice to any known on-land sequence, and so ascribing the bedform deposit to a syn-eruptive density current origin is speculative, lacking convincing evidence.

11.6.2 Subaqueous Eruption of Pumiceous Flows

Wholly subaqueous vents could also produce flows of pumice and ash debris if they are shallow enough. We will summarise two scenarios, the first small-volume eruptions from point source vents, and the second, large volume eruptions from fissure vents that could include fracture vents associated with caldera formation and collapse. White (2000) also presents an overview of this scenario.

11.6.2.1 Relatively Small Volume Pumiceous Flows Erupting from Shallow to Deep-Water Subaqueous Vents

In this scenario, explosive eruptions occur from a shallow to moderately deep-water subaqueous vent driven by exsolving magmatic volatiles and/or explosive phreatomagmatic interaction with external water (Fig. 11.7e; also cf. Plate 10.1b during the 23 December 2018 eruption triggered by earlier flank collapse of the submarine cone of Anak Krakatau; Williams et al. 2019; Chap. 12). Such a concept was first considered by Fiske and Matsuda (1964) to account for the deposits of the Miocene Waidara Tuff in Japan, and has subsequently been considered by Cas and Wright (1987, 1991), Kokelaar and Busby (1992), Kano et al. (1996), White (2000), Head and Wilson (2003), Allen and McPhie (2001, 2009), Allen et al. (2020) and Jutzeler et al. (2014b). Such eruptions could be purely explosive or also involve initial eruption of a lava dome that transforms into an explosive eruption that feeds a subaqueous flow of pyroclasts, which may be variably gas-supported or water-supported, and which has consequences for the final deposit textures, as discussed in several case studies.

Miocene Waidara Tuff Member, Tokiwa Formation, Fossa Magna Basin, Central Honshu, Japan

The Miocene Waidara Tuff D described by Fiske and Matsuda (1964) represents one of the first documented accounts of a submarine explosive eruption and transformation of the erupting dacite pumice, dense clasts, crystal fragments and shards into a syn-eruptive water-supported mass flow like a turbidity current. The host succession occurs in a deep-water, rift basin called the Fossa Magna in Central

Honshu that includes mudstones containing open ocean planktonic foraminifera and deep-water (150–500 m) benthonic foraminifera. The Waidara Tuff consists of several packages of massive lapilli-tuff beds up to 50 m thick followed by about 200 thinner, finer lapilli-tuff and tuff beds each less than 1 m thick, of which Unit D is one (Fig. 11.34a). Fiske and Matsuda (1964) highlight that the thick basal unit is coarser than the thinner units above, which themselves become progressively finer grained upwards, and, how well hydraulically and size-sorted all depositional units are compared with subaerial ignimbrites. Dense non-vesicular dacite clasts are concentrated at the base of depositional units, whereas less dense and finer grained pumiceous debris, crystals and shards are concentrated at the top (=density grading). The package defines a vertical double grading represented by overall upwards fining, as well as fining of each bed upwards (cf. Dali Ash, Fig. 11.33). Beds have sharp bases and sole structures, and some are erosional. Neither the thick massive bed nor thinner beds show any evidence of emplacement in a hot state.

As summarised by Cas and Wright (1991), Fiske and Matsuda (1964) interpreted the succession of upward-fining, thick-to-thin beds to have resulted from a single subaqueous explosive eruption, in which the basal thick unit was fed directly from column collapse of an active subaqueous eruption column as a warm to cold, water-supported mass flow of pyroclastic debris (Fig. 11.34b, stages 1 and 2). The overlying thinner units were interpreted to represent the waning of the eruption, and the decline of the eruption column, with successive slumping of small volumes of pyroclastic debris off the near-vent flanks of the centre into the adjacent deeper water (Fig. 11.34b, stage 3). All beds have characteristics of high-concentration turbidity currents (Chap. 12). There is no evidence that the mechanism of emplacement was by a hot gas-supported pyroclastic flow.

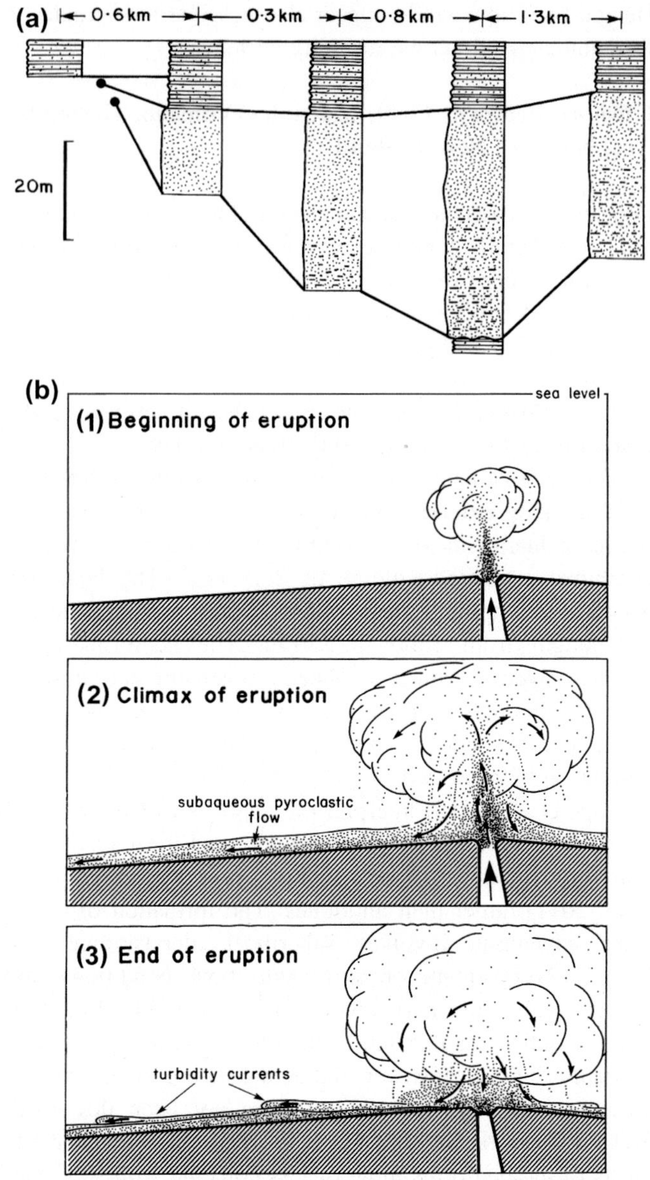

Fig. 11.34 Facies architecture and origin of the submarine, Miocene Waidara Tuff Unit D, Japan. **a** Facies architecture of Unit D, showing a basal massive depositional unit overlain by a succession of thin depositional units which are individually size graded and overall become finer upwards, giving the succession a doubly graded aspect. **b** The interpreted origin of the Waidara Tuff D eruption. This shows stage 1, initial onset of submarine explosive eruption; stage 2, collapse of the submarine eruption column, directly feeding a mass-flow of pyroclastic debris which becomes rapidly water-supported; and finally, stage 3, a succession of small volume slump generated high particle concentration turbidity currents, producing the upward fining, doubly graded succession of thin beds. (From Cas and Wright 1987, After Fiske and Matsuda 1964)

Shinjima Pumice (11 ka), Shinjima Island, Aira Caldera, Kyushu, Japan

The submarine emplaced Shinjima Pumice (Holocene) was exposed in 1780 when part of the previously submerged floor of the Aira Caldera was uplifted, presumably as a result of shallow intrusion of magma into the sedimentary and volcanic fill of the caldera (Kano et al. 1996). The Shinjima Pumice is underlain by tuffaceous sandstone to mudstone turbidites with marine molluscs indicating water depths of 100–200 m, whereas molluscs in strata above indicate water depths of 50–100 m (Kano et al. 1996). The 40 m thick Shinjima Pumice deposit consists of highly vesiculated rhyolite pumice, crystal fragments, glass shards, and rare clasts of obsidian and lithified black mudstone. The unit is massive to diffusely stratified, and is organised into multiple depositional units 1–10 m thick, each with sharp to diffuse boundaries and with reverse coarse-tail size grading of pumice clasts. The deposit is well-sorted, bimodal with the dominant clast sizes being fine lapilli and coarse ash, and it is depleted in fine ash. Pumice clasts are blocky to elongate, vesicles are mostly tube vesicles, and large pumice blocks and lapilli are highly fractured, with a polygonal fracture network with the morphology of cooling contraction cracks. Shards typically have curvi-planar surfaces, suggestive of magma-water interaction (Kano et al. 1996).

Kano et al. (1996) interpreted the Shinjima Pumice to have been erupted explosively from a relatively shallow subaqueous vent, since it is only known from the subaqueous setting of the Aira Caldera. Explosive magmatic vesiculation and phreatomagmatic interaction between the erupting, vesiculating magma and external water are likely to have been drivers of the explosive eruption. The eruption column was like a collapsing fountain, with an initial gas and steam-charged column with suspended pyroclasts collapsing into a density current that quickly ingested water and transformed into a water-supported mass flow (Fig. 11.35; Chap. 9). Interaction between the ejected pyroclasts and external water is likely to have caused cooling contraction quench fragmentation of the pyroclasts producing polygonal crack networks and blocky pyroclasts. Many pumice clasts, as they cooled, would have allowed water to permeate into the vesicles, so causing them to become water-logged and denser. Smaller pumice clasts would have become dense more quickly than large clasts, so settling out of the density current before the larger clasts, resulting in inverse coarse-tail grading.

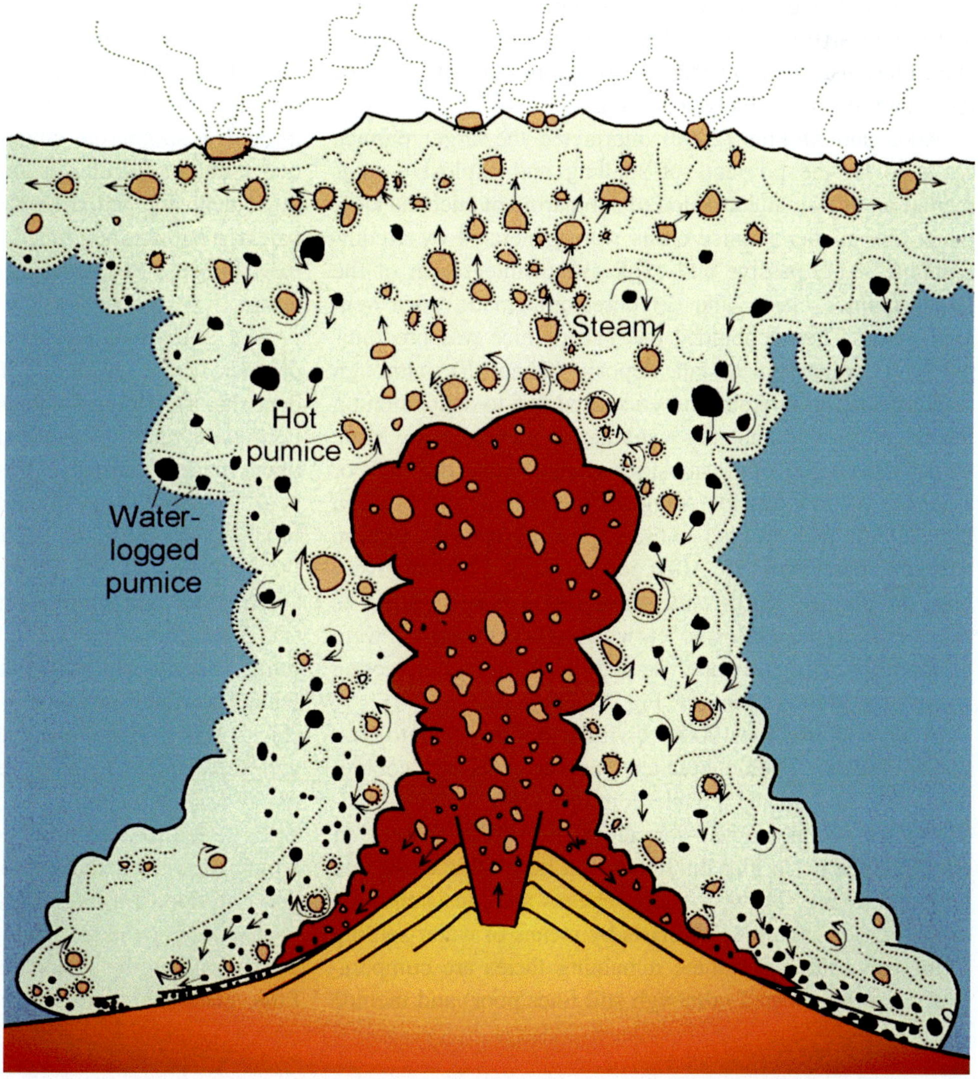

Fig. 11.35 Eruption model for the subaqueous 11 ka Shinjima Pumice, Aira caldera, Kyushu, Japan. This shows a subaqueous collapsing pyroclastic fountain or column, transforming from hot and gas rich-(red) to a water-supported mass flow of pyroclastic debris laterally away from the base of the column, and a mixing zone enclosing the core of the column (pale yellow). Dotted envelopes around hot pumice clasts (orange) signify steam envelopes. Some clasts are buoyant, but water-logged pumice clasts (black) settle in the column and in the mass flow. (Adapted from Kano et al. 1996)

Yali Pumice Deposit (~31 ka), Kos–Yali Çaldera, Dodecanese Islands, Greece

A strikingly similar deposit to the Shinjima Pumice in terms of facies characteristics and potential eruptive processes is the Yali pumice deposit, Greece (Chap. 9). The Yali pumice deposit is a thick (>150 m), Pleistocene succession of submarine, rhyolitic pumice that has been uplifted and exposed on the small island of Yali, which Cas and Wright (1987) and Allen and Cas (2001) place on the southeastern margin of the inferred Kos–Yali caldera (see Fig. 11.18a).

The Yali deposit comprises moderately to well-sorted, multiple cm-to-m thick beds of loose pumice clasts that are fines-depleted, and include only rare dense juvenile and basement lithic clasts (Allen and McPhie 2000; Fig. 11.36a–c). The dominant pumice type is a white aphyric rhyolite, with vesicularity ranging from 36 to 86 volume % and typically strongly elongated (tube pumice). Larger pumice blocks and boulders have quenched margins and internal polyhedral joints (Fig. 11.36d), whereas the smaller pumice clasts are polyhedral, angular to subrounded, blocky pebbles and granules (Fig. 11.36e). These smaller clasts lack quenched margins, but have angular to sharp, curvi-planar surfaces and are pervasively fractured.

Allen and McPhie (2000) interpreted the larger pumice clasts to be the products of spalling and explosive fragmentation of a small extrusion of submarine pumiceous lava, while the smaller pumice clasts were generated by a combination of (1) passive and explosive disintegration of the larger pumice clasts, and (2) phreatomagmatic open vent explosions. They concluded the Yali pumice was predominately a water-settled fall deposit (Chap. 9), although resedimentation in grain flows occurred due to deposition on the steep unstable flanks of a submarine pumice cone. More recent data based on vesicularity analysis, and the lack of evidence of an extrusion, suggests the magma ascended in the conduit and erupted entirely in a pyroclastic fashion within an open vent (SR Allen pers. comm.). The negatively buoyant pumice lapilli and minor dense clasts settled as showers from the explosive jet, whereas coarser blocks were temporarily buoyant and settled from the convecting water-supported column (cf. Fig. 11.35).

Allen and McPhie (2000) divided the deposit into four major facies: cobble–boulder, pebble, mixed cobble-dominant, and mixed pebble-dominant. The cobble–boulder facies occurs in well-sorted, massive, tabular beds of large pumice up to 3 m thick that lack both lithic clasts and fine matrix (Fig. 11.36c). These features are consistent with deposition of large pumice clasts by means of water-settling from suspension. The three remaining facies are comparatively less well-sorted (although still fines poor) and include rare lithic clasts. These exhibit massive, or internally diffusely stratified 0.1–2 m thick beds of coarser grained pumice clasts, which onlap or are separated by wedge-shaped, thinner (<0.1 m) beds of finer grained (fine pebble to granule) pumice clasts (Fig. 11.36a, b, e). The grain size and sorting of the pebble facies resemble those of fallout, but the bedforms suggest resedimentation by submarine gravity flows (modified grain flows) of water-logged, cohesionless pumice. The mixed facies probably resulted from downslope resedimentation of unstable pebble and cobble–boulder facies simultaneously with the settling of large pumice clasts from suspension. Syn- to post-eruptive resedimented grain flow facies are significant in the most southwestern part of the islet, which is interpreted as the most distal sector of the uplifted submarine pumice cone (Fig. 11.36a; Allen and McPhie 2000; SR Allen pers. comm.).

Miocene Dogashima Formation, Shirahama Group, Izu Peninsula, Japan

Cashman and Fiske (1991) first suggested that hydraulically very well sorted pumice deposits near the town of Dogashima on the Izu Peninsula, Japan, were the result of fallout through a submarine environment from an explosive eruption column. Jutzeler et al. (2014b) later demonstrated that those deposits, called Dogashima Formation Member 2, are part of a submarine graded mass-flow deposit resulting from the explosive disintegration of an erupting submarine lava dome.

The Shirahama Group consists of submarine basaltic to dacitic lavas, hyaloclastite breccias, water-settled fall deposits, a variety of volcaniclastic mass-flow deposits, and it contains open marine planktonic foraminifera, suggesting open marine sub-wave base depositional setting (Jutzeler et al. 2014b). Dogashima Formation Member 2 consists of a basal, monomictic, dense andesite lithic clast breccia (unit D2-2 in, Fig. 11.37a–c), overlain gradationally by progressively finer and more stratified andesitic pumice breccias (D2-3 to D2-7; Fig. 11.37a, b, d, e). The base of the andesite lithic clast breccia cross-cuts substrate and defines a 600 m wide erosional channel scoured 15 m into substrate (Fig. 11.37a). Jutzeler et al. (2014b) interpreted the D2-2 andesite breccia to have resulted from initial syn-eruptive explosive destruction of an erupting seafloor lava dome. This collapse led to decompression of the erupting dome magma which then triggered explosive vesiculation and formation and eruption of a large volume of pumice.

The D2-3 facies overlying the D2-2 andesite lithic breccia consist of highly vesiculated pumice and smaller andesite lithic clasts, with little ash matrix, forming a hydraulically

(a)
30
4 c-b
20
3 mc, mp, p
2 mc
1 mp
10 m
asl
60
7 mc, mcp, p
50
6 mp, p
40
5 c-b, mc, mp
pebble
cobble boulder
f m c f c
2 mm 64 256
Yali fallout
10
9
70
8 mc, c-b
(b)
(c)
(d)
LEVEL MATIC

◀ **Fig. 11.36** Features of the ~31 ka Yali pumice deposit, an uplifted submarine pumice cone forming the southwestern sector of the island of Yali, on the inferred margin of the Kos–Yali caldera, southeastern Aegean Sea, Greece. (See Fig. 11.18a for location). **a** Graphic log illustrating more distal deposits of the cone which show a significant proportion of resedimented mass (grain) flow deposits. Here, it comprises eight units that are mostly less than 10 m thick. Unit 1 = mp: dominated by mixed pebble-dominant facies which is weakly planar stratified. Unit 2 = mc: dominated by mixed cobble-dominant facies which is also weakly planar stratified. Units 3, 7 = mc, mp, p: interbedded mixed facies and pebble facies. Unit 4 = c–b: dominated by cobble–boulder facies. Unit 5 = c–b, mc, mp: interbedded cobble–boulder and mixed facies. Unit 6 = mp, p: interbedded mixed pebble-dominant facies and pebble facies. Unit 8 = mc, c–b: interbedded mixed cobble-dominant and cobble–boulder facies. The submarine pumice cone deposit is then overlain by a thin, well-sorted pumice pebble bed (unit 9), talus breccia of blocks of carbonate-cemented pumice (unit 10), which are in turn overlain by limestone, bioclastic sandstone, and a younger subaerial Yali fallout deposit. (From Allen and McPhie 2000). **b** View to more medial pumice cone deposits. Prominent middle unit (~20 m thick) comprises interbedded mixed cobble-dominant facies with finer intervals of pebble facies occurring in wedge-shaped beds. **c** Cobble-boulder facies. **d** Pumice boulder showing quenched margins and polyhedral jointing patterns. **e** Planar stratified to wedge-shaped medium beds of pumice pebbles with sharp upper and lower contacts. Most beds are dominantly ungraded but some have finer grained tops; note occasional larger cobble-sized fragments

Fig. 11.36 (continued)

very well-sorted clast-supported pumice and lithic clast breccia (Fig. 11.37d). The upward-directed column of erupting pumice and lithic clasts collapsed almost immediately forming a water-supported mass flow carrying still hot andesite clasts that produced units D2-2 and D2-3 (Fig. 11.37a–d; Jutzeler et al. 2014b). The variably water-logged large pumice clasts and the small dense andesite lithic clasts have the same settling velocity in water (Cashman and Fiske 1991). Planar to cross-stratified units D2-4, D2-6 and D2-8 (Fig. 11.37a, b) consist of finer pumice breccia in which clasts are rounded to subrounded and were interpreted by Jutzeler et al. (2014b) to have resulted from reworking and resedimentation from shallower water and by tractional currents in the submarine channel defined by D2-2 (Fig. 11.37a). D2-5 was also considered to be a small syn-eruption fed pumice breccia mass-flow unit and D2-7 (Fig. 11.37e), a syn-eruptive suspension fallout deposit, accompanied by some lateral current movement producing low angle truncations and pinch and swell bedding (Fig. 11.37e).

Carboniferous and Cretaceous Submarine Welded Tuffs, in the Vosges, France, and Anatolia, Turkey

Schneider et al. (1992) documented two ancient examples of welded tuffs interbedded with marine strata in the Vosges region of France (Carboniferous), and Anatolia region in Turkey (Cretaceous). The Carboniferous rhyolitic deposit described is up to 15 m thick, graded, with mudstone intra-clasts in the base and a welded zone only about 1 m thick which is marked by fiamme (Fig. 11.38a). It is hosted by sub-wave base fossiliferous mudstones and siltstones. The Cretaceous rhyodacitic deposit is about 30 m thick, massive and slightly graded, and apparently welded in the

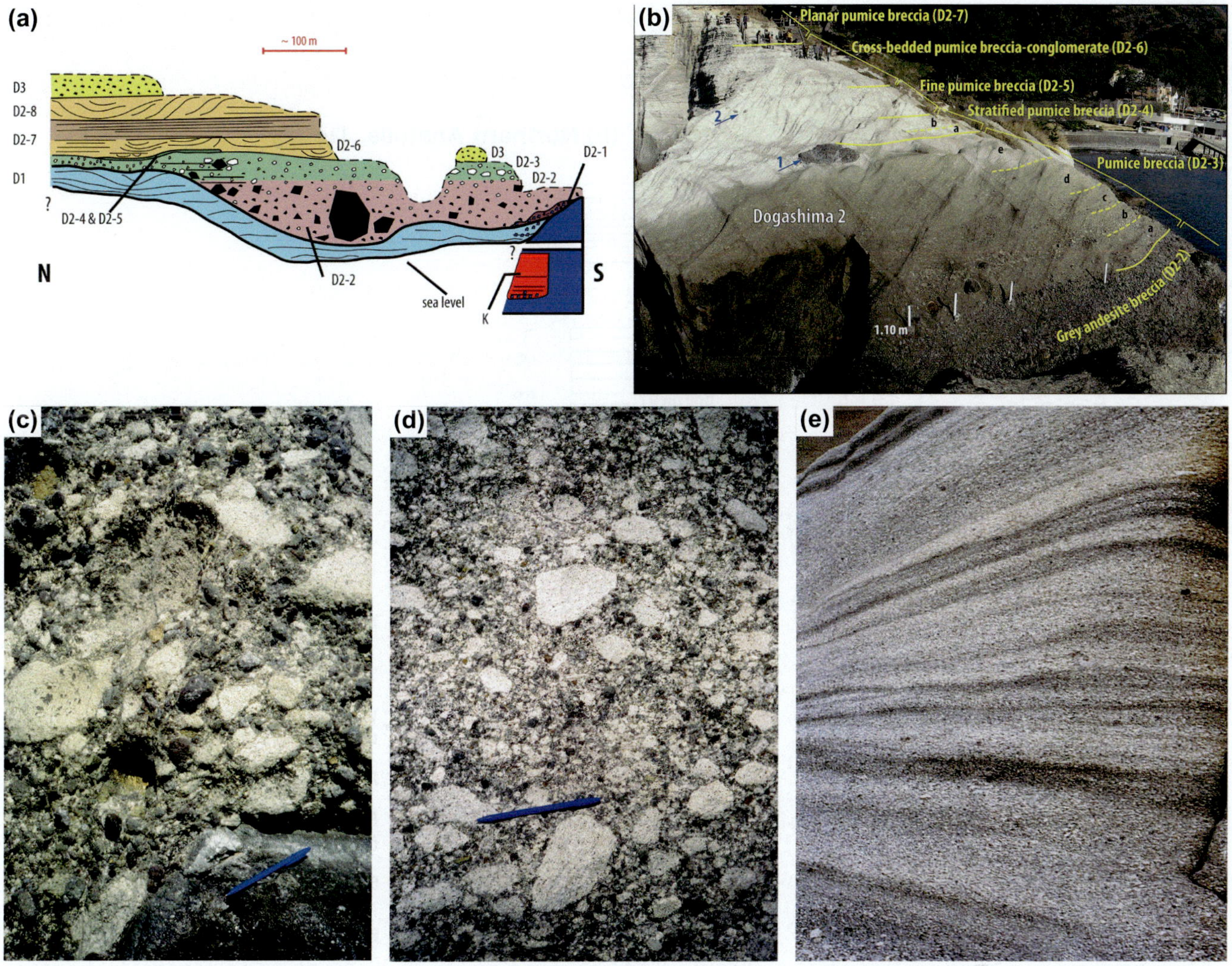

Fig. 11.37 Features of the Miocene Dogashima Formation Member 2, Izu Peninsula, Japan. **a** Interpretative coastal cross-section of Member 2, showing the basal channel form nature of the deposit and the transition from basal dense andesite clast (black polygonal shapes) breccia upwards into various pumice clast (white shapes) breccia units (green, yellow, brown). The association of D2-2 and D2-3 represents the syn-eruptive water-supported mass flow breccia deposit. (From Jutzeler et al. 2014b). **b** Almost complete vertical profile through Member 2 at Matsuzaku Harbour, Dogashima. The very dark unit at sea level is the initial D2-2 largely andesite lithic clast breccia, which is overlain by the progressively lighter coloured, largely andesite pumice breccia units (D2-3–D2-7) discussed in the text. Lettered sub-units (**a**–**e**) are graded. Grey clasts numbered 1 and 2 are dense, outsize andesite lithic clasts. (From Jutzeler et al. 2014b). **c**–**e** Selected lithologies observed at Matsuzaku Harbour, Dogashima. **c** Transition from D2-2 andesite lithic breccia to D2-3 andesite pumice breccia, showing mixing of grey dense andesite and white andesite pumice clasts. **d** Hydraulically well-sorted pumice (white) and small dense andesite (dark) lithic clasts in the massive D2-3 pumice breccia. Note how little fine matrix there is between clasts. **e** Planer to wavy bedded fine pumice breccia of unit D2-7, interpreted as resulting from suspension fallout influenced by lateral current flow, producing low angle pinch and swell and truncations. Height of outcrop ~1.5 m

upper 20 m, marked by fiamme texture (Fig. 11.38b). It occurs in marine fossiliferous marls and pelagic micritic limestones. Variably altered microscopic textures include apparent fluidal eutaxitic matrix textures (Schneider et al. 1992).

These two deposits require eruption, transport, and emplacement of small volumes of pyroclasts while they remain above the glass transition temperature. This requires exclusion of contact with external cold liquid water, which would cause immediate chilling of pyroclasts through the glass transition temperature, so preventing welding. This in turn requires a hot gas-rich transport system (eruption column, seafloor gas-supported pyroclastic mass-flow system) that remains isolated from the enclosing water mass for long enough to deposit and weld before hydrostatic pressure, gas condensation and turbulence at the flow margins cause physical mixing of external water into the interior of the flow. This is most likely to happen in settings close to vents

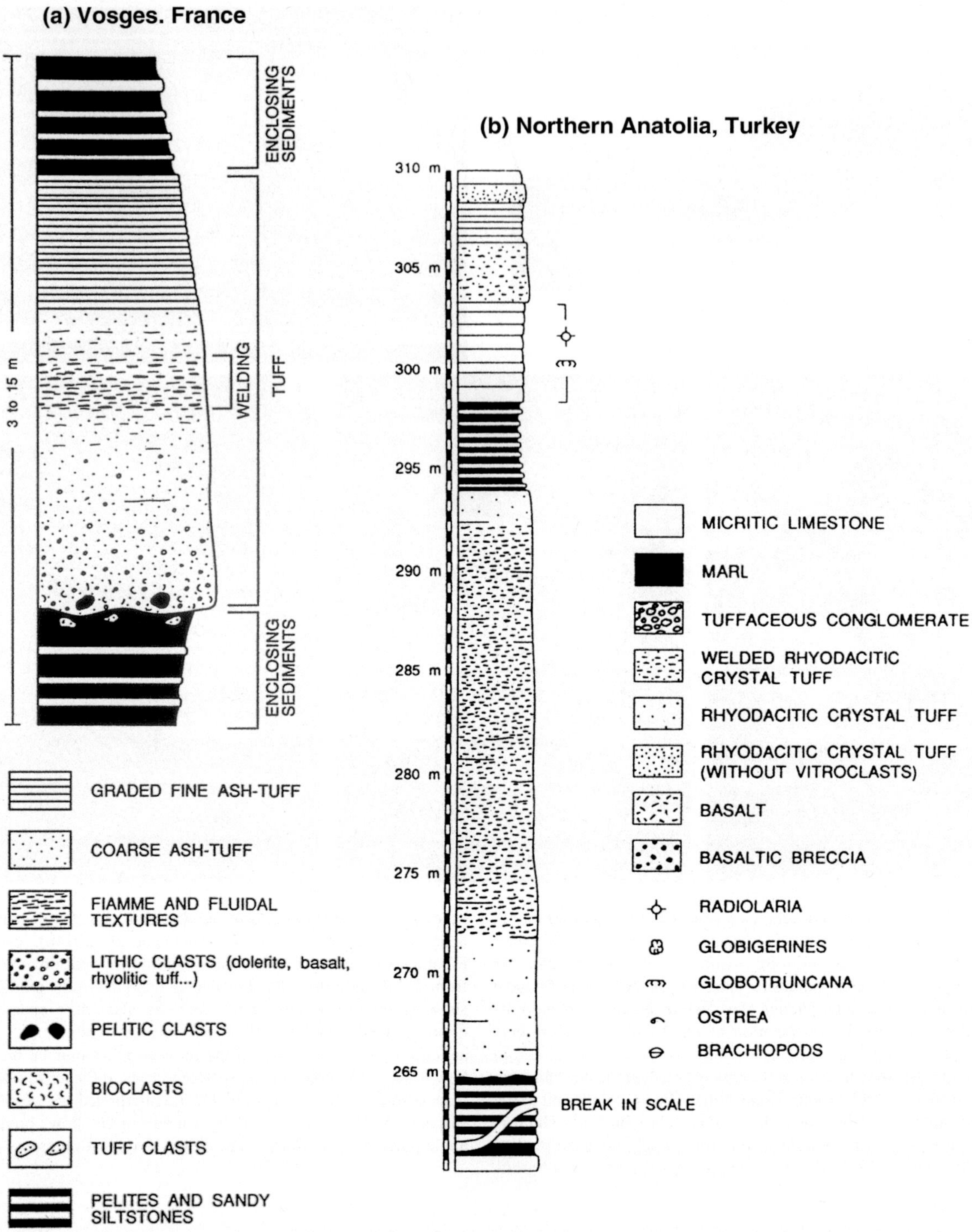

Fig. 11.38 Near vent subaqueous welded deposits. **a** Idealised summary graphic log of the Carboniferous (Visean) submarine welded tuff unit in the southern Vosges Mountains, France, showing the graded unit with a thin welding zone within a fossiliferous mudstone succession. **b** Graphic log the Upper Cretaceous submarine welded tuff unit in the Göllüveren Yayla area of Northern Anatolia, Turkey, showing the 30 m thick graded unit with a 20 m thick apparent welding zone within a fossiliferous mudstone succession. (From Schneider et al. 1992)

with high eruption mass flux, which is the setting proposed for the Carboniferous example by Schneider et al. (1992).

Ordovician Metal Man Tuff, County Waterford, Ireland

As described by Fritz and Stillman (1996), the Ordovician rhyolitic Metal Man Tuff is up to 30 m thick and distinctly graded. It comprises a basal breccia containing mudstone intraclasts and rhyolite clasts, grading up into a 15 m thick matrix-supported pumice fiamme breccia with both a macroscopic and microscopic eutaxitic texture (plastically deformed shards). The deposit becomes columnar jointed upwards, and passes into a 3 m thick zone of fine laminated tuff. Perlite texture has also been noted. The base of the unit is irregular with load casts into underlying mudstones. The underlying sediments are fine-grained fossiliferous muddy limestone and fossiliferous terrigenous black mudstone, whereas the overlying sediments are black tuffaceous mudstones. Brachiopods in mudstones indicate a shelf setting, and the mudstones indicate a sub-wave base hemipelagic origin.

As for the examples described above by Schneider et al. (1992), the location of the source vent, and whether it was subaerial or submarine is unknown, but Fritz and Stillman (1996) propose that the vent was nearby to allow transportation and deposition by a submarine gas-supported pyroclastic flow without ingestion of significant external water, so allowing welding to occur during deposition.

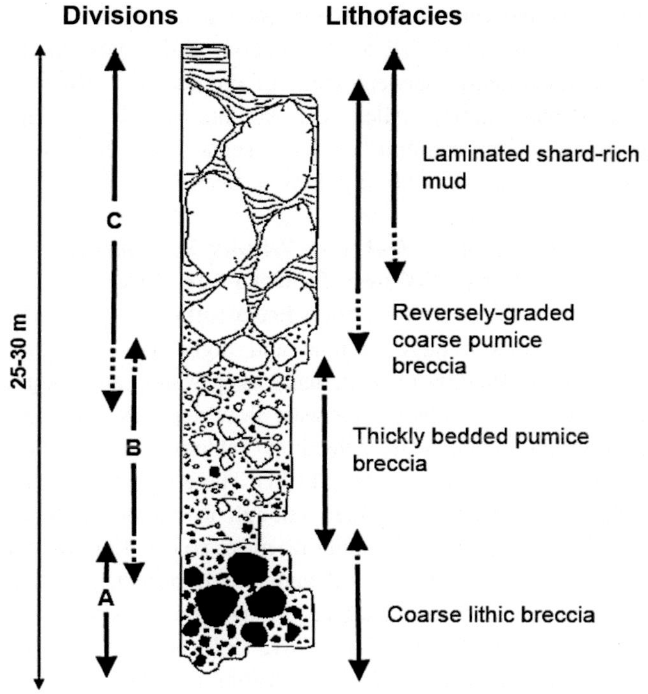

Fig. 11.39 Schematic summary graphic log of the depositional facies of the uplifted, submarine, Upper Pliocene Filakopi Pumice Breccia from the island of Milos, southern Aegean Sea, Greece. This shows three main depositional divisions: (A) basal poorly sorted, andesite and dacite clast breccia, which is reverse to normal coarse-tail graded, overlain by (B) poorly sorted massive pumice breccia, and finally (C) reverse coarse-tail graded, outsize pumice clast breccia with laminated shard rich matrix in the upper zone. (From Stewart and McPhie 2004)

Pliocene Filakopi Pumice Breccia, Milos, Greece

Stewart and McPhie (2003) described the Upper Pliocene rhyolitic Filakopi Pumice Breccia (FPB) on the island of Milos in the central Hellenic volcanic arc, Aegean Sea. The FPB outcrops between fossiliferous and bioturbated mudstones, in which bivalves indicate sub-wave base water depths of several hundred m. The FPB is <25–30 m thick, and consists of three textural and component divisions (Fig. 11.39). The basal division (A) is up to 8 m thick and is a lithic breccia with clasts of andesite and dacite up to 2.6 m in diameter. Its base is concordant to highly erosional with the substrate mudstone and pumice deposits, and includes substrate clasts. Its top is gradational and diffusely stratified. Division A changes from reverse graded to normally graded upwards. The second division (B) is up to 20 m thick, and is a very thickly bedded, poorly sorted pumice breccia with dispersed rhyolite obsidian, as well as andesite and dacite clasts, with diffusely defined, often wedge-shaped beds. Pumice clasts are up to 1 m in diameter. The uppermost division (C) is also up to 20 m thick and is a reversely graded, clast-supported pumice breccia. The matrix changes upwards from fine pumice lapilli at the base to fine shard-dominated ash upwards, which is thinly bedded between the large pumice clasts. Pumice clasts show quenched margins and interiors represented by closely spaced normal joints at the margins and polygonal joint patterns internally. Pumice clasts in both divisions B and C have vesicularity levels of up to 80%.

Stewart and McPhie (2004) interpret the sequence to have resulted from relatively shallow (several hundred m) explosive excavation of a vent through a pre-existing andesite-dacite substrate. The explosive eruption was driven by magmatic exsolution and vesiculation of a rhyolitic magma. The initial slug of explosively ejected substrate lithics rapidly collapsed forming a dense clast mass flow, which then entrained several pulses of vesiculated rhyolitic pumice debris into the mass-flow system. Many of the pumice clasts were quenched by the water mass and so the mass flow was probably water and steam supported. Large buoyant pumice clasts formed a buoyant plume and perhaps a sea-surface floating raft of pumice. As this buoyant pumice became water-logged (cf. Whitham and Sparks 1986), it began to sink with the largest clasts sinking more slowly than the smaller clasts, hence producing reverse coarse grading. Simultaneously, finer ash and pumice may have been forming vertical density currents or plumes (cf. Carey 1997) which descended to the seafloor and deposited the fine-bedded matrix between the large pumice clasts (Stewart and McPhie 2004). Allen and McPhie (2009) cite this eruption as an example of their subaqueous Neptunian explosive style of eruption (Chap. 9, Sect. 9.9.4.5).

11.6.2.2 Shallow Water to Wholly Submarine, Large Volume, Caldera-Forming Explosive Pumice Eruptions

In this scenario, large volume and high mass flux rate, eruptions in shallow to moderately deep water can produce thick, variably welded, pyroclastic flow deposits, including ignimbrites, ponded in subsiding calderas, and outflow sheets in marine settings (Fig. 11.7f). The eruption mass and volume flux have to be high enough to exclude external water from infiltrating the erupting mass, so ensuring that the transport system remains hot and gas-supported, and thus preventing pyroclast quenching through the glass transition temperature (Cas and Wright 1987, 1991). However, a significant problem in ancient successions is determining the absolute water depth *of the eruption.* This is because, firstly, fossils and sedimentary structures in host sedimentary successions are rarely very specific in terms of depth indications. In addition, the host eruptions that potentially produce subaqueous pyroclastic flows occur at the pre-existing bottom water depths *prior* to downfaulting during caldera collapse, which may be hundreds or more m shallower than the depths to which the floor of the caldera has subsequently subsided to, and could even have been subaerial (e.g. Johnson et al. 2015).

The most recent example of this scenario was the 15th January 2022 eruption of Hunga Tonga-Hunga Ha'apai marine stratovolcano, in the Tonga archipelago. Relatively little is yet known about this extremely intense explosive eruption. The summit of the volcano was a pre-existing caldera that was completely submerged, except for two small islands on its rim, Hunga Tonga and Hunga Ha'apai (Cronin 2022). After the eruption, detailed submarine bathymetry surveying by NIWA, New Zealand, reported by Mackay (2023), showed that the 2022 eruption produced an 850 m deep, 5 km diameter caldera involving the eruption of 9.5 km3 of magma and rock. The surveys of the seafloor around the volcano showed that there were thick deposits of debris that extend at least 80 km on the seafloor from the volcano. This material must have been transported by submarine mass flows, with some flows breaking several seafloor communication cables in the process. The nature of these deposits (pyroclastic debris? rock avalanche debris? remobilized outer volcano slope deposits?) and their thermal state at the time of deposition is as yet unknown. It is also unknown how they formed—was it by collapse of either the subaerial part of the 57 km high eruption column, or from its subaqueous base, or by gravitational collapse of part of the preexisting edifice, is as yet unknown.

Based on analysis of the chronology of events using several sources, Purkis et al. (2023) proposed that 3 significant tsunamis were generated during the eruption that all coincided with major explosions, the largest and most destructive tsunami coinciding with the highest intensity explosion at 4.56 UTC. That tsunami caused significant damage across the Tongan islands and involved a run-up onto cliffs as high as 45 m on some islands. Purkis et al. (2023) suggested that it seems that the tsunamis were triggered by the explosions but recognised that other factors could also contribute. Factors that could contribute include water displacement by the shock wave of the explosions pushing the sea surface up and triggering lateral tsunami flow, caldera and/or partial edifice collapse (landslide, debris avalanche) and displacement of water by the large volume submarine mass flows that were generated during the eruption. We now discuss examples shallow water to submarine caldera forming eruptions from the geological record that have been well documented.

Subaerial to Shallow Water Ordovician Lower Rhyolitic Tuff Formation, North Wales

In addition to the Pitts Head Tuff Formation (described in Sect. 11.6.1.2), the overlying Lower Rhyolitic Tuff Formation (LRTF) of the Ordovician Snowdon Volcanics Group in North Wales was also erupted and deposited in a subaerial to shelf setting, forming a caldera. The LRTF is dominated by ≥ 500 m of welded and non-welded rhyolitic ignimbrite that is ponded in an asymmetrical subsidence structure interpreted as a submarine trapdoor caldera (Fig. 11.40; Reedman et al. 1985; Howells et al. 1986). The caldera was some 15 km in diameter (Fig. 11.40a), and the volume of intra-caldera ignimbrite has been estimated to be >30 km^3 (Howells et al. 1986). The intra-caldera ignimbrite is massive to diffusely stratified (Fig. 11.41a) and is underlain by sedimentary clast and volcanic clast megabreccias thought to be associated with early stages of caldera collapse. The ignimbrite is welded at the base, mostly in the southwest but is otherwise non-welded. Shard and pumice textures are well preserved although widespread recrystallisation of the ignimbrite matrix has occurred (Howells et al. 1986).

In addition to the caldera fill succession, there is a variably bedded (<1.5 m thick beds) outflow sheet of ignimbrite (Fig. 11.41b–e) that is <70 m thick, extends for 25 km from the caldera margin to the northeast (Fig. 11.40a) and could represent as much as 20 km^3 in volume. The outflow sheet also preserves shard textures, but is not welded and locally contains mudstone intra-clasts and brachiopod and trilobite fossil fragments, indicating emplacement in a submarine environment where these occur (Howells et al. 1986),

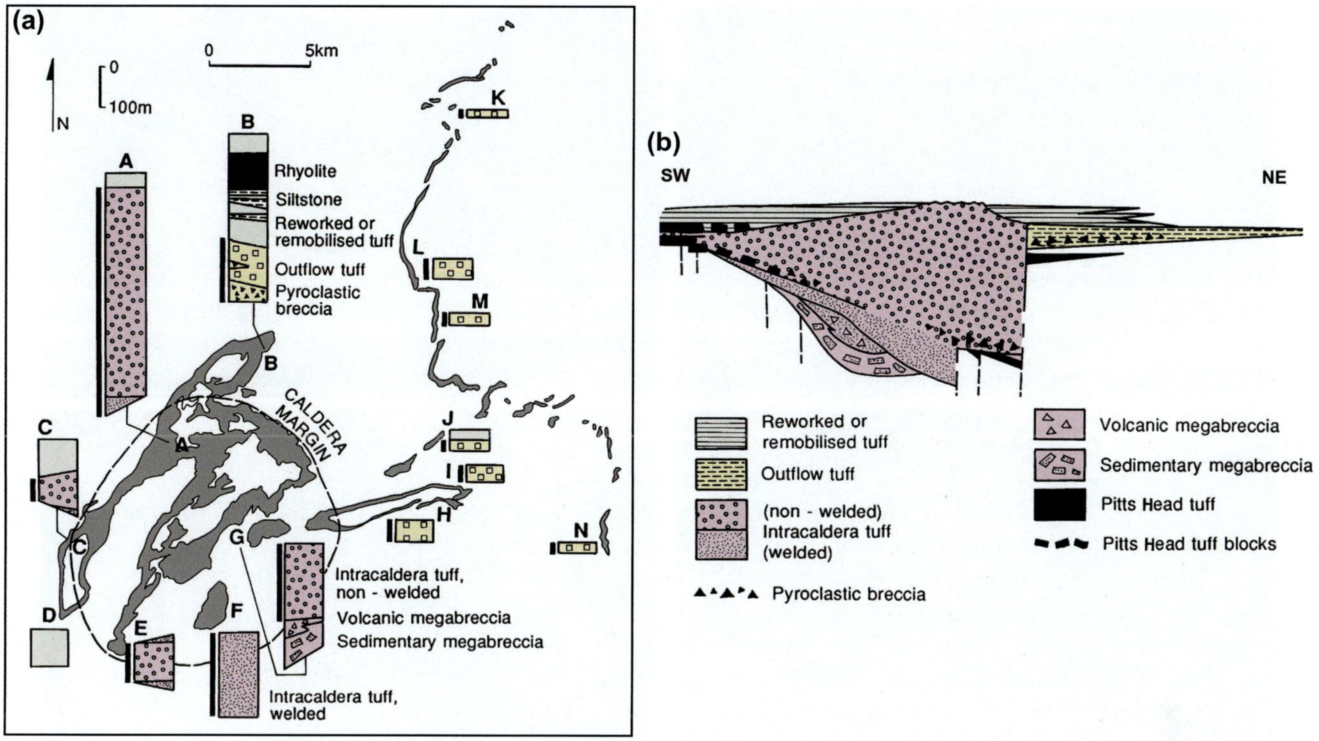

Fig. 11.40 Extent and architecture of the Ordovician Lower Rhyolitic Tuff Formation (LRTF), North Wales. **a** Map and stratigraphic sections for the LRTF, showing the distribution of the nearly circular caldera and the sheet-like outflow ignimbrites and associated lithofacies. The LRTF is >500 m thick in the northwest of the asymmetrically subsided trapdoor caldera, and less than 100 m thick in the outflow sheet. It is most intensely welded in the south around section F. Vertical bars to the left of the sections indicate the thickness of the primary ignimbrite present. **b** Schematic cross-section through the LRTF caldera and outflow ignimbrite. Asymmetrical trapdoor collapse was associated with the formation of sedimentary and volcanic megabreccias consisting of substrate clasts. The base of the LRTF is unconformable to disconformable on substrate, and much of its top is reworked and also marked by an unconformity. (After Cas 1992; Modified from Howells et al. 1986)

as well as mixing in of substrate sediment and water. The top 1–6 m of the outflow sheet is finer grained than the main part, interpreted to represent settling of fine vitric ash suspended in the water column after the eruption and depositional event (Howells et al. 1986).

Substrate sedimentary rocks include cross-bedded fossiliferous sandstones and mudstones interpreted to represent shallow marine settings affected by strong tractional current and storm currents, and basaltic volcanic rocks that include pillow lava and associated hyaloclastite breccia (Howells et al. 1986). Tuffaceous turbidite beds also occur to the northeast (Orton 1987). Orton (1987) proposed a pre-eruption paleogeography of a subaerial horst block under most of the caldera transitioning to shoreface and offshore to the northeast. However, over most of its extent, the LRTF lies disconformably to strongly unconformably on its substrate (Howells et al. 1986), indicating the removal of significant intervals of the pre-existing marine sedimentary substrate and significant parts of the older welded Pitts Head Tuff. This indicates an extended period of significant erosion, probably under subaerial conditions, prior to the emplacement of the LRTF, indicating pre-eruption regional updoming from shallow marine settings to above sea level. A likely cause of this was intrusion of a large volume of magma into shallow crustal levels prior to eruption causing regional tumescence (Orton 1987).

According to Howells et al. (1986), the strata overlying the LRTF include the largely basaltic Bedded Pyroclastic Formation above the thick caldera fill succession, consisting of volcaniclastic sediments, pyroclastic deposits and pillow lavas, that are conformable to unconformable on the top of the LRTF (Fig. 11.40b), and deeper water black mudstones and turbiditic siltstones and sandstones to the east above the outflow sheet. However, above the very thick welded caldera fill succession, there are conglomerates with welded ignimbrite clasts derived from the LRTF and reworked tuffaceous sediments with tractional depositional structures (Fig. 11.41f; Howells et al. 1986). The conglomerates indicate clearly that at least some of the LRTF caldera was exposed to subaerial to shoreline environments, reworking and erosional regimes (cf. seismic reflection profiles of modern intra-caldera fill at Santorini volcano, central Hellenic arc; Johnston et al. 2015). Conformably above the outflow sheet to the northeast, there are thinly bedded,

Fig. 11.41 Features of the Ordovician Lower Rhyolitic Tuff Formation, North Wales. **a** Part of massive to diffusely stratified intra-caldera LRTF ignimbrite, Mt Snowdon; near location A, Fig. 11.40a. Thickness in field of view ~50–60 m. **b** Dipping medial LRTF outflow sheet overlying sediments and Pitts Head Tuff (PHT) ignimbrite, Idwal Valley; near location B, Fig. 11.40a. **c** Crudely bedded outflow LRTF north of Capel Curig township; near location I, Fig. 11.40a. Unit thickness ~40 m. **d** Contact between two beds (arrow) in the LRTF outflow ignimbrite north of Capel Curig, from outcrop in (**c**). **e** Massive base passing upwards into multiple thin beds, eastern limb of Idwal Syncline, Idwal Valley, from location in (**b**). **f** Trough cross-bedded volcaniclastic sandstone above the LRTF, Pont y Gromlech, Llanberis Pass; midway between locations A and B, Fig. 11.40a. **g** Thinly bedded, graded sandstone–siltstone turbidites and/or water-settled fallout deposits above the LRTF north of Capel Curig. Graded beds are between 1 and 4 cm thick; location as in (**c**)

fine-graded tuffaceous deposits (e.g. Fig. 11.41g), representing tuffaceous turbidites and even water-settled fallout deposits, indicating likely deposition of the outflow sheet in an offshore, sub-wave base shelf setting.

Although the caldera succession is underlain and overlain by shallow marine sedimentary strata, unconformities and disconformities beneath and above the caldera succession, as well as conglomerates above it indicate that the caldera eruption probably occurred subaerially, above sea level, after regional magma intrusion induced regional tumescence. There then appears to have been a period of downfaulting during caldera collapse, detumescence and marine transgression after the eruption leading to erosion and reworking of the top of the LRTF. A significant part of the outflow sheet to the northeast may have flowed into water and been emplaced in shallow water to sub-wave base water depths. However, there are insufficient details about the internal facies architecture of the outflow sheet to determine if it was emplaced by a hot gas-supported mass-flow system or a water-supported mass-flow system. Inclusions of substrate sediment and fossil fragments in the outflow sheet are suggestive of mixing in of water and transformation into a water-supported mass flow.

Shallow to Deep-water Jurassic Vandever Mountain Ash Flow Tuff, Mineral King Caldera Complex, Sierra Nevada, California, USA

Busby-Spera (1984a, b, 1985, 1986) and Kokelaar and Busby (1992) described the geology of the Jurassic Vandever Mountain Ash Flow Tuff (VMT) preserved in the Mineral King Caldera Complex roof pendant and caldera complex in a Cretaceous granite batholith in the Sierra Nevada mountains in California. The succession is metamorphosed to upper greenschist facies. The underlying sediments are thin-bedded calcareous, pelitic hemipelagic rocks, turbidites and megabreccias, whereas the overlying strata are black pyritic shales, thin turbidite sandstones and debris flow deposits. The VMT is ~480 m thick, and, apart from the upper 30 m, it is a massive poorly sorted ignimbrite that looks similar texturally to subaerial ignimbrites. The upper 30 m is bedded and at the top begins to have turbidite characteristics. Kokelaar and Busby (1992) demonstrated that back-scatter polished thin sections preserved plastic shard and pumice welding textures in the matrix of the ignimbrite.

Busby (1984a, b) proposed that the VMT, as well as three other very thick rhyolitic ignimbrite units in the Mineral King pendant, formed through caldera-forming eruptions in shallow to moderately deep water, at estimated water depths ranging from 10 to ~500 m. Busby (1984a, b), and subsequently Kokelaar and Busby (1992) proposed a collapsing fountaining style of eruption column. Busby (1984a, b) proposed that high hydrostatic pressure suppressed formation of a buoyant eruption column, so explaining the absence of identifiable fallout deposits. She also proposed that the bulk of the erupting pyroclastic mass was emplaced by laminar flow, with turbulence at the margins, in the syn-eruptive subsidence depocentre. With a high mass eruption rate in a ponded subsiding caldera depocentre, the erupting flow mass would not have interacted with the ambient water mass, so retaining heat and enabling submarine welding in this proximal to vent setting. Given that caldera subsidence of several hundred m occurred, the explosive eruption would have commenced in relatively shallow water, but below wave base.

Deep-Water Ordovician Bald Mountain Caldera, Maine, USA

Busby et al. (2003) and Kessel and Busby (2003) described the geology of the subaqueous Ordovician rhyolitic Bald Mountain Caldera, Maine. The basal stratigraphy consists of ~3,000 m of massive to pillow basalt lavas and intrusions and associated basalt breccias interpreted as hyaloclastite. Interbedded very thin carbonaceous mudstones indicate an anoxic, sub-wave base water depth. The basalt sequence is overlain by ~500 m of massive, blocky, poorly vesicular, pumice clast-rich depositional units up to 150 m thick, interbedded with graded pumice units up to 2 m thick, and some basalt lavas and associated hyaloclastite breccia. Busby et al. (2003) interpret the thick pumice units to be ignimbrites produced by subaqueous phreatomagmatic explosive activity, with the blocky pumice clasts resulting from quench fragmentation of larger pumice clasts erupted into a subaqueous environment. This sequence also hosts VMS deposits about one third up from the base, which subdivide the succession into a nearly aphyric footwall succession and a more crystal-rich hanging wall succession, and also have different geochemistry. The VMS horizon therefore appears to represent a significant boundary, and perhaps a hiatus in volcanism, however brief. The 150–200 m intervals of footwall ignimbrites lack fine ash matrix (cf. subaerial ignimbrites) and are dominated by lapilli-size clasts. Busby et al. (2003) attribute to this to the effects of hydrostatic pressure suppressing the intensity of vesiculation and explosive fragmentation, and leading to a collapsing fountain-like eruption of pumiceous lapilli debris feeding submarine flows. Welding textures are only found in a vertical vent structure, but not in the pumiceous depositional units.

The ~300 m thick hanging wall pumiceous sequence occurs in depositional units up to 60 m thick, some of which are graded, and in which the pumice clasts are not blocky. There are also intercalated basalt lavas, and poly-lithological volcanic breccias with angular clasts up to 1 m in diameter, interpreted as caldera wall collapse breccias. Rhyolite lavas and intrusions also occur and the hanging wall pumice succession is overlain by up to 1,500 m of coherent rhyolite lavas, intrusions into mudstones producing peperitic margins and a succession of variably graded rhyolitic lithic lapilli-tuff beds interpreted to have resulted from subaqueous lava dome collapse events producing mass-flows including turbidites (Busby et al. 2003).

Busby et al. (2003) interpret the Bald Mountain succession to have resulted from a piecemeal caldera collapse event in an arc rift setting. They propose that the water depth of the explosive eruption was $\sim$1,400 m, based on fluid inclusion paleo-barometry data from the sulphide deposits. However, as discussed further in Chaps. 16 and 18, if the sulphide mineralisation formed after some caldera collapse associated with the eruption of the footwall succession, the water depth at which the sulphide formed could be several hundred m greater than that at which the explosive caldera-forming eruption began, allowing for caldera collapse subsidence height changes.

11.6.3 Eruption and Formation of Subaqueous Dense Clast Pyroclastic Flows

As documented in this chapter, both observations of recent historic eruptions and experiments indicate denser clast pyroclastic flows could enter bodies of water after subaerial lava dome collapse events that produce block and ash flows, and transform into submarine and, if the water depth is deep enough, lacustrine turbidity currents (Fig. 11.7g). However, they may also form from subaqueous lava dome collapse (Fig. 11.7h), as described earlier during the initial part of the Dogashima Formation eruption (Fig. 11.37; Jutzeler et al. 2014b). Other examples include, the late Miocene-early Pliocene Upper Donzurobo Formation and Miocene Takashibiyama Formation scoria flows in Japan (Kato et al. 1971; Yamazaki et al. 1973; Kano et al. 1994), and an unusual occurrence associated with the subaqueous fountaining-style eruption of poorly vesiculated, fluidal felsic clasts described from the Archean Hunter Mine Sequence, Canada (Mueller and White 1992).

11.6.3.1 Subaerial Dense Clast Pyroclastic Flows Entering Water

We have previously mentioned a number of well-described examples documenting this scenario including the seminal observations of Lacroix (1904) during the 1902 eruption of Mt Pelée on Martinique, and dome collapse events during the 1995–2013 eruption of the Soufrière Hills volcano Montserrat. Between 1999 and 2009, numerous lava dome collapse events occurred generating larger and smaller volume pyroclastic flows, which entered the sea at the mouth of the Tar River valley (Fig. 11.8). In this period more than 380 $\times$ 10^6 m^3 of material entered the sea (Le Friant et al. 2010, and built a composite submarine fan deposit offshore Tar Valley (Fig. 11.42a). In this section, we highlight two events on Montserrat during this period, firstly, again those associated with activity on 12–13 July 2003 and secondly with the dome collapse on 20 May 2006.

12–13 July 2003 and 20 May 2006 Submarine Deposits, Monserrat

The general scenario of the 12–13 July 2003 lava dome collapse, and particularly the observed phenomena associated with entry of the pyroclastic flows into the sea and their explosive disintegration and mixing at the shoreline, has already been described and discussed in some detail in Sect. 11.6.1.1. In this section, attention is focused on the description of the submarine deposits. The pyroclastic flows were dense clast, block, and ash flows resulting from the collapse of 210 $\times$ 10^6 m^3 of the lava dome rock dominated by variably vesiculated lava dome clasts. During peak collapse conditions, as the main flow entered the sea, phreatic explosions were observed and a surge cloud decoupled from the pyroclastic flow to travel 2 to 3 km over the ocean surface before settling. At the shoreline, the phreatically disrupting main flow body transformed into a large, water-supported, turbidity current underflow which produced a hydraulically well-sorted and graded bedded, syn-eruptive, turbidite deposit offshore from Montserrat (Fig. 11.42b, c; Trofimovs et al. 2006, 2008). The offshore deposit contains no evidence of having been hot at the time of deposition (Trofimovs et al. 2006, 2008). The nearshore facies is massive, coarse, fines-depleted, boulder-rich deposit that is normal size graded only at the top (Fig. 11.42c). The offshore deposit is thin, sand size, normal size graded (Fig. 11.42c) and contains the typical Bouma sequence a–d divisions that characterise water transported turbidity current deposits (Chap. 12; Trofimovs et al. 2006, 2008). Trofimovs et al. (2006, 2008) proposed that hot, subaerial, gas-supported block and ash flows transformed into submarine water-supported granular and turbidity current flow through mixing in of water at the shoreline during the shoreline phreatic explosions.

The dome collapse event on 20 May 2006 was similar to the 2003 event, although a smaller volume of material was involved (115 $\times$ 10^6 m^3). Block and ash pyroclastic flows reached the sea, phreatic explosions occurred at the shoreline, surges were generated that flowed up to 3 km over the sea and also onto land and small tsunamis were triggered (Trofimovs et al. 2012). Most of the collapsing mass was deposited in the sea. The seafloor bathymetry was well documented after the 2003 collapse event and then again after the 2006 event, providing accurate information about

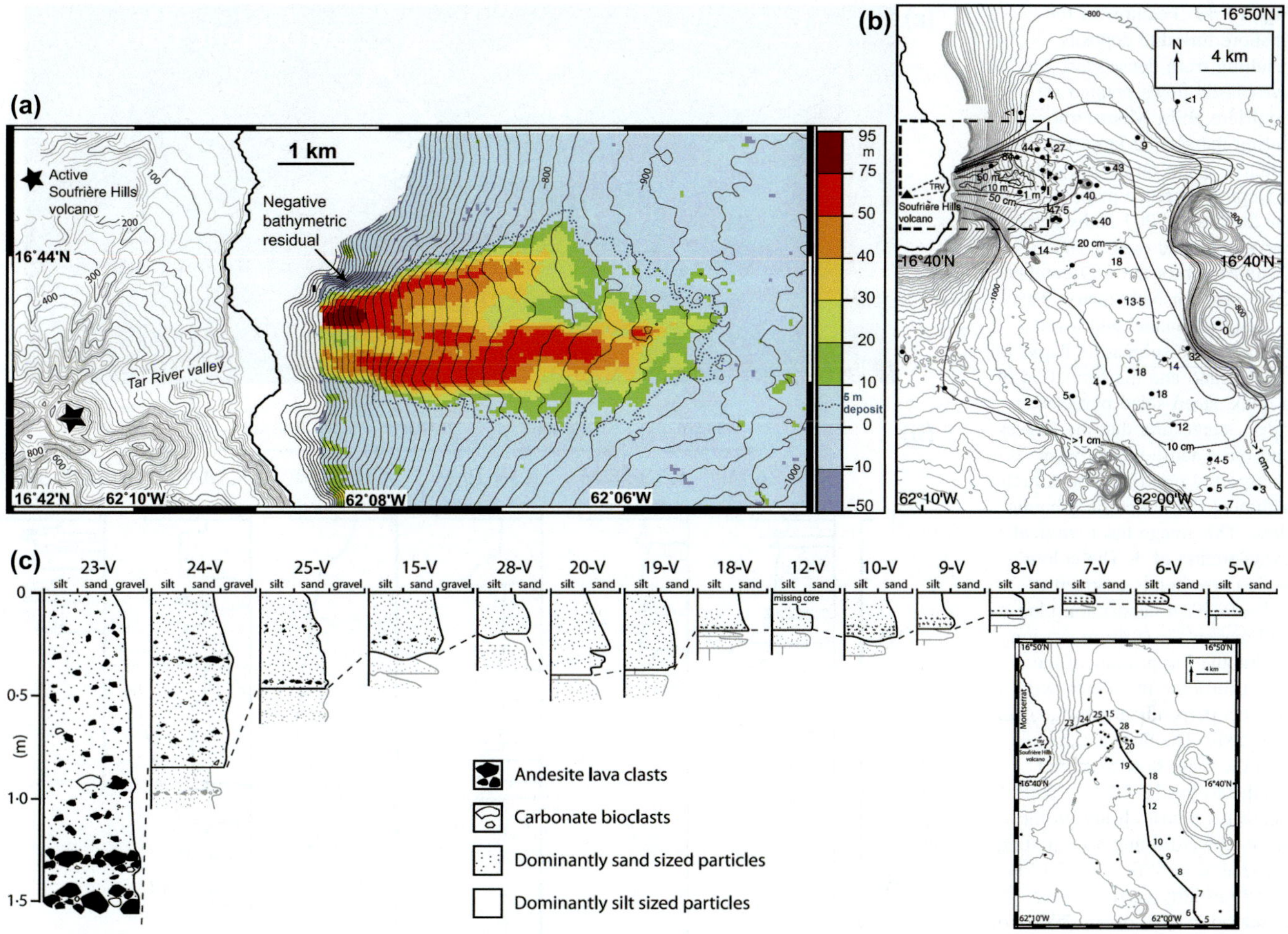

Fig. 11.42 Features of the offshore turbidite deposits resulting from the transition of dense clast subaerial block and ash flows into submarine water-supported turbidity currents, Soufrière Hills volcano, Montserrat, building both a shoreline fan-delta (see Fig. 11.8) and an offshore composite submarine fan. **a** Detailed map of the 1999–2009 composite submarine fan deposit offshore Tar Valley. Colours indicate depth difference between two swath bathymetric surveys in 1999 and 2009. Black contour lines show the 1999 bathymetry with a 25 m contour interval. (From Le Friant et al. 2010). **b** Isopach map of the thickness of a volcanic turbidite deposit resulting from the entry of lava pyroclastic flows into the sea on 12–13 July 2003, which was the largest dome collapse event during the entire Soufrière Hills eruption. During peak collapse conditions, the main flow was accompanied by shoreline phreatic explosions and transformation into a water-supported turbidity current offshore. Black box shows the approximate area of detailed survey in (**a**). (From Trofimovs et al. 2008). **c** Down-flow graphic log profiles of the cored turbidite in (**b**), showing marked changes in thickness and grain size from proximal (left-hand graphic logs) to distal (right-hand logs) settings over a distance of more than 60 km on the seafloor. Almost all log profiles show upward-fining size grading. (From Trofimovs et al. 2008)

the thickness of the 2006 offshore deposits. In addition, the 2006 deposits were extensively cored.

The 2006 submarine deposits are thickest just offshore from the mouth of the Tar River Valley and pyroclastic fan delta (Fig. 11.8). Here they formed a depositional ridge or lobe up to 54 m high and 3.5 km long beyond the mouth of a levèed channel on the fan delta and 7 km from the shoreline (Fig. 11.43a). At an offshore break in slope from 11° to <7°, at a depth of ~700–800 m, the deposit becomes thinner and spreads as a fan. In this deeper water, the 2006 deposit is only a maximum of 50 cm thick but extends more than 40 km on the deepening seafloor. The deposit consists of variably altered andesite lithic debris organised into multiple thin massive to graded, sandy, Bouma a division depositional units, some passing upwards into laminated b intervals, with some grading upwards into silt (cf. Fig. 11.43b). Depositional units are commonly amalgamated, and the multiple units are thought to have resulted from reflections of the main flow by submarine topography. Trofimovs et al. (2012) interpret the near-shore ridge to have been deposited by a high concentration granular flow (debris flow), and the more distal deposits by turbidity currents. Some of the distal turbidity currents were able to climb submarine topography 200 m high, some 24 km from

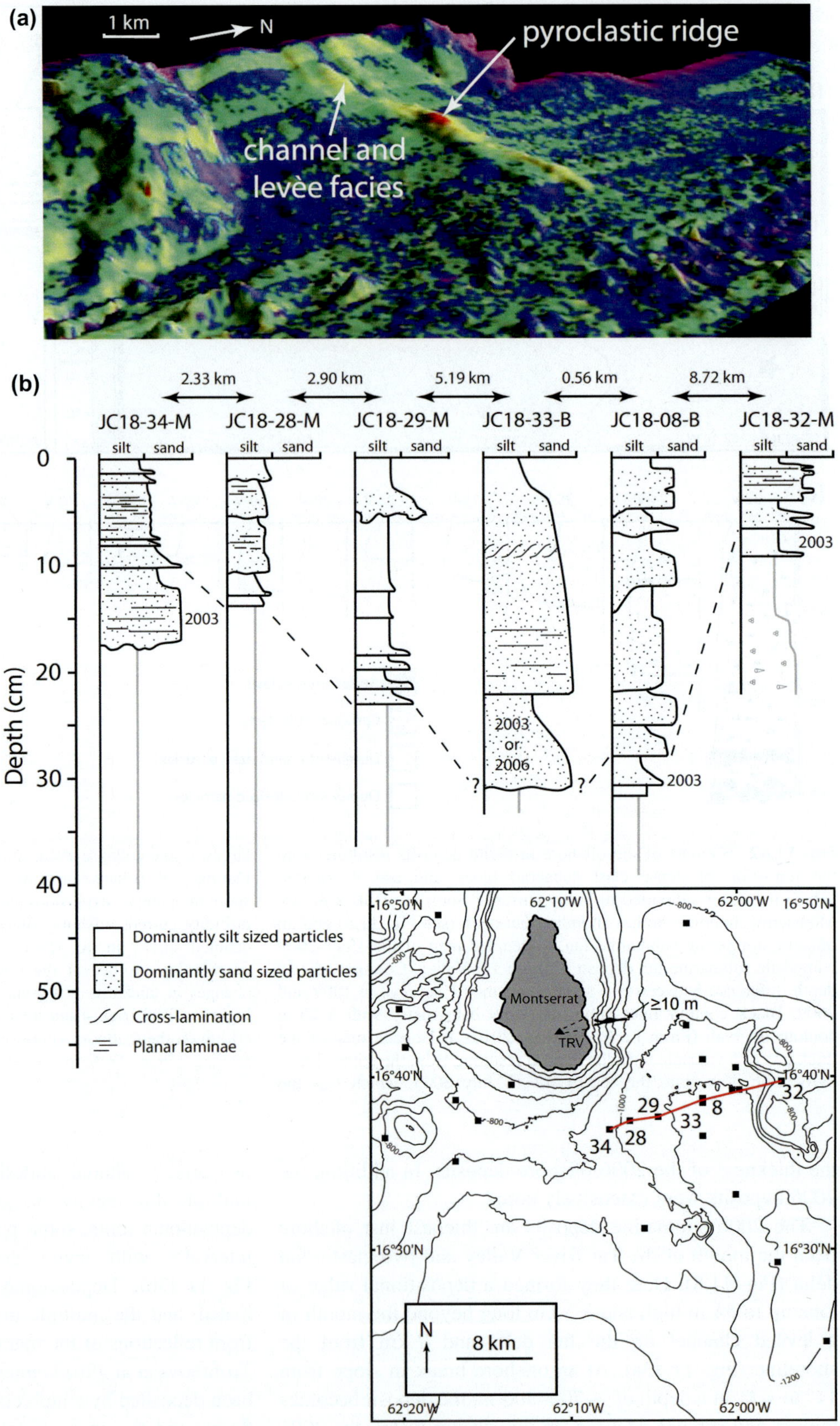

Fig. 11.43 Features of the offshore turbidite deposits resulting from dense clast pyroclastic flows erupted at Soufrière Hills, Montserrat, during the second largest dome collapse event on 20 May 2006. **a** Depth difference map between the 2005 and 2007 bathymetric surveys draped over a 3D visualisation of the 8 December 2007 seafloor bathymetry. This highlights channel and levee facies on the submarine flanks of the volcano and the formation of a thick, steep-sided pyroclastic ridge, suggesting that the largest and most dense blocks were transported proximally as a high sediment concentration granular flow. This image has a vertical exaggeration of 3. The colour scale corresponds to green to red = 0–50 m deposition, blue and magenta = 0–40 m erosion. **b** Graphic log profiles of the cored turbidite in an east–west transect perpendicular to the main flow axis. The deposits thicken into the main flow axis, and similar to the 12–13 July 2003 deposits, almost all log profiles show upward-fining size grading. At several localities, the 20 May distal turbidity currents were observed to have run up 200 m of topography and eroded up to 20 cm of underlying substrate. The high energy of the May 2006 submarine flows resulted in material being transported further than the larger dome collapse of July 2003. The imaged proximal deposits and pyroclastic ridge ($\geq$ 10 m) are shown offshore from the base of the Tar River Valley. (From Trofimovs et al. 2012)

Montserrat, and erode 20 cm of pre-existing substrate sediments. The proposed transformation appears to have occurred syn-eruptively as a result of the subaerial dome collapse event and the passage of a block and ash flow into water.

11.6.3.2 Subaqueous Eruption and Flow of Dense Clast Flows

Cas and Wright (1991) cited the deposits of the Miocene Donzurobo Formation in Japan as an example of this scenario. White (2000) has reviewed a wide variety of examples from shallow-water lacustrine basaltic phreatomagmatic centres to deep-water Archean fire fountain-style eruptions and deposits which greatly expand the range of this type of phenomenon and products, and their preservation in the geological record.

Late Miocene-Early Pliocene Upper Donzurobo Formation, Osaka, Honshu, Japan

Kato et al. (1971) and Yamazaki et al. (1973) described a bedded succession of mass-flow deposits of angular, glassy rhyolite clasts, crystal fragments and ash of the same material exposed in the Upper Donzurobo Formation, Nijo-san area, Osaka, Japan. Only 5% of the debris is silt size or finer. Depositional units are from 2 to 4.5 m thick, mostly massive, and not size graded, although some units are normal size graded at their tops. Some units contain a crude basal cross-bedding. Finer, bedded, and cross-laminated `tuffs' occur intercalated in the succession, and "cold" debris flow deposits occur at the top of the succession, based on their TRM properties (Kato et al. 1971).

Very few details of the depositional setting are provided, except that some fine tuffaceous beds are graded, show slump structures and have basal load casts, suggesting a subaqueous, sub-wave base setting of unknown water depth, interpreted by Kato et el. (1971) as a relatively shallow lake setting. Kato et al. (1971) and Yamazaki et al. (1973) proposed that their units A, B, C, and E were emplaced in a hot state, with a clast Curie temperature at ~ 500 °C. In spite of these inferred temperatures, the deposits were not welded, or sintered, and could be sieved. Although magnetisation data for the clasts in individual units show a relatively uniform orientation, there is nonetheless a scatter of up to 50° $\sim$ inclination within individual units. Their units G, I and J show a wide scatter of TRM orientations and were interpreted as cold debris flow deposits.

These deposits appear to be proximal, near-vent deposits. Yamada (1984) suggests relatively shallow depths of perhaps several tens of m, and a flow distance of less than 3 km, although the basis for these estimates is not clear. Given that dense clasts predominate, and that the silt size fraction constitutes < 5% of the deposits, the original transporting phenomena appear to have been either block and ash flows or debris avalanches of some sort, perhaps derived from nearby dome collapses into a shallow body of water.

Miocene Takashibiyama Formation, Shimane Peninsula, Honshu, Japan

Kano et al. (1994) described submarine "scoria" deposits emplaced in an outer shelf to bathyal setting based on benthic foraminifera in associated mudstones in a sequence exposed in the Takashibiyama Formation on the Shimane Peninsula, in southwestern Honshu. The deposit is 5–7 m thick, has an irregular lithic breccia base, is relatively poorly sorted, and consists of non- to poorly vesiculated mafic lapilli and coarse ash fragments, which have cracked outer margins and curvi-planar surfaces indicative of quench cooling fracturing. Vesicles and gas-escape pipes occur in the matrix, and limited TRM polarity direction data indicate a likely hot state of emplacement (Kano et al. 1994). The location of the vent is unknown, but Kano et al. (1994) propose a nearby submarine vent and short-distance transport by a mass flow of "scoria" debris. This they interpret to have been hot and gas-supported near vent, where matrix vesicles and gas pipes occur, and to have transformed into a water-supported mass flow distally, where these features are absent from the deposit, over a distance of 500 m.

Archean Hunter Mine Group, Quebec, Canada

Mueller and White (1992) and White (2000) have described a submarine bedded sequence of fluidal rhyolitic spatter clast breccias associated with finer graded tuffaceous deposits and laminated iron formation in the Hunter Mine sequence of the Abitibi Greenstone Belt (2730 Ma). The iron formation indicates a sub-wave base depositional setting. The eruption sequence begins with a basal 5–20 m thick, fluidal clast-rich breccia with abundant intact fluidal spatter clasts with clast-supported texture (Fig. 11.44a). The fluidal clasts vary in vesicularity from 0 to 60%. They were interpreted to represent deposition from a gas-insulated fire fountain around the seafloor vent with some lateral flow as a hot, steam-rich laminar state granular mass flow (Fig. 11.44b; Mueller and White 1992). The top of this facies is marked by an increase in the abundance of sub-equant blocky clasts, interpreted to represent the effects of quench fragmentation of such clasts at the margin of the fire fountain. This passes gradationally into a poorly sorted, largely matrix-supported blocky clast breccia with some spatter clasts, with a coarse granular matrix that is 10–70 m thick, diffusely layered and is interpreted to represent ingestion of water into the margins of the fire fountain, which collapsed and transformed into a succession of coarse, water-supported granular mass-flow pulses (Fig. 11.44a, c). The top of the sequence is represented by a 5 m thick, fine-grained, sand-sized sequence of

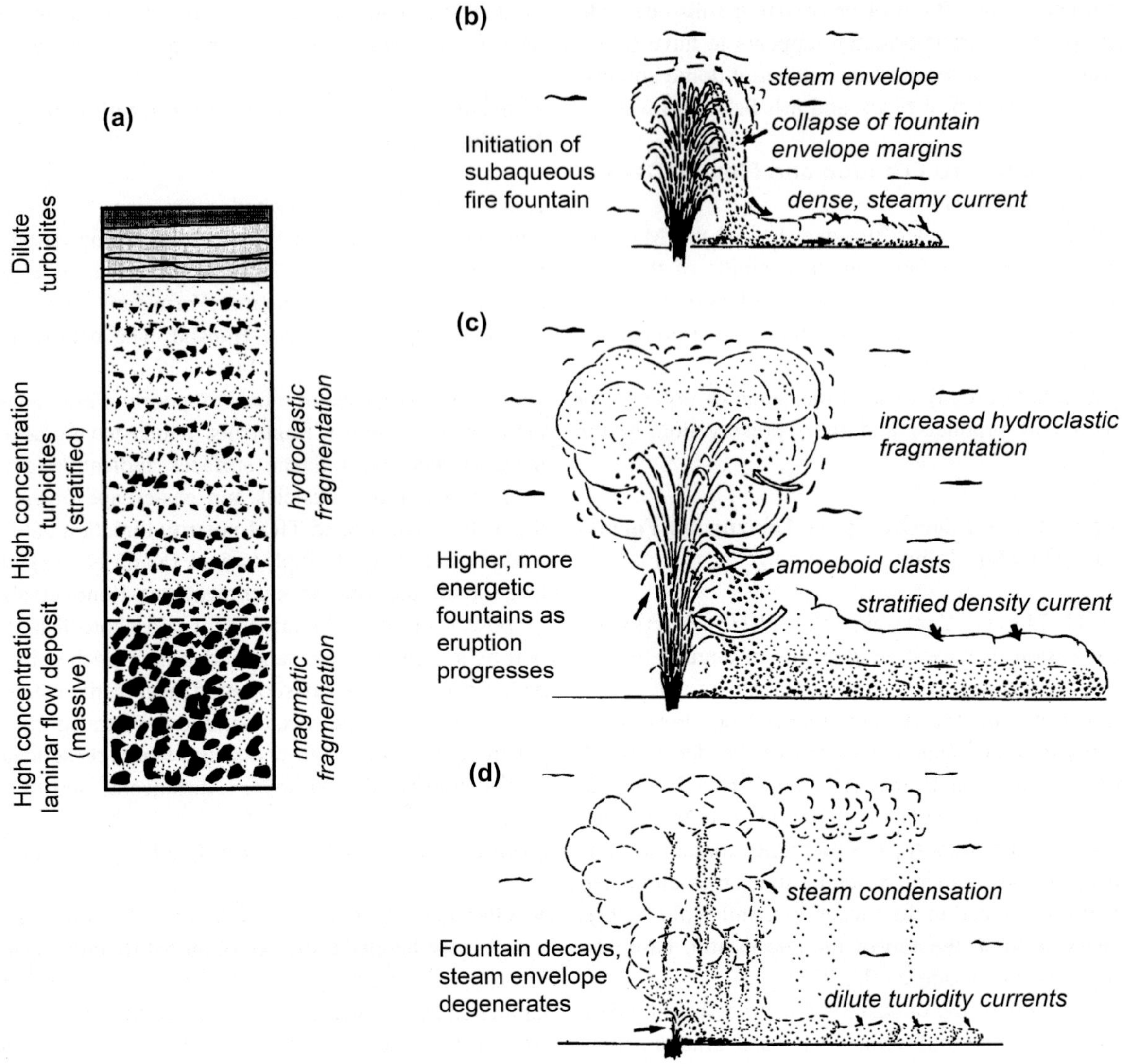

Fig. 11.44 Eruption style and felsic deposit facies association for the submarine fire fountain eruption that produced the Archean Hunter Mine sequence, Abitibi Greenstone Belt, Quebec, Canada. **a** Stratigraphy of an eruption sequence, showing a high particle concentration granular mass flow deposit consisting of a basal, massive felsic magmatic spatter breccia facies. This was generated from **b** an initial submarine fire fountain of highly fluidal magma. This is overlain by a massive to diffusely stratified, matrix-supported finer breccia facies, resulting from **c** ingress of water into the fire fountain causing quench fragmentation of the magmatic clasts in the fountain or column, its collapse and lateral transport and deposition by high particle concentration turbidity currents. This is overlain by thin, finer grained, variably graded turbidite beds, representing **d** the waning stage of the eruption and continued pulsing collapse of the eruption column and generation of small-volume turbidity currents. (Modified from White 2000)

thin, graded beds with shallow basal scours into underlying strata, and some with low angle dune cross stratification interpreted as aqueous turbidites representing waning of the fire fountain and the eruption (Fig. 11.44a, d; Mueller and White 1992; White 2000). The final deposit in an eruption sequence is a fine tuffaceous siltstone that represents water settling fallout of very fine suspended ash from the water mass after the eruption had ceased (Fig. 11.44a).

11.6.4 Resedimentation of Subaerial Ignimbrite-Derived Pyroclastic Debris into Deep Water

In almost all cases where pyroclastic flows emplace deposits in the shoreline to offshore region, gravitational, erosional (e.g. storms) and seismic triggers will remobilise some of the the deposits as mass flows into deeper water settings

(Fig. 11.7i). This can occur contemporaneously with initial pyroclastic flow deposition, shortly after or long after. This variable timing allows for a range of no to increasing degrees of sedimentary reworking before final mass-flow resedimentation. Deposits may therefore preserve the componentry of the original pyroclastic deposits, or show varying degrees of modification, including rounding and abrasion of components and mixing in of multiple components from other sources (e.g. Freundt and Schmincke 1998; Schmincke and Sumita 1998).

Schmincke and Sumita (1998) and Freundt and Schmincke (1998) reported on the characteristics of the offshore volcaniclastic apron around the volcanic island of Gran Canaria, Canary Islands. This was based on ODP drilling in water depths >3,000 m and at distances of 60 to >80 km from the source volcanic vents on Gran Canaria (Chap. 16). The deposits encountered are well-bedded, and massive to graded, and show well-developed density grading and turbidite characteristics. They identified syn-eruptive, syn-ignimbrite-forming layers as being thick (>50 cm), dominated by angular monomictic assemblages of pumice, crystal fragments and shards. On the other hand, layers resulting from post-eruptive reworking and resedimentation were < 30 cm thick, and polymictic, involving mixing of clast populations from more than one source; these showed some evidence of rounding of clasts, resulting from reworking prior to resedimentation.

An ancient example from the geological rock record of post-eruptive resedimented volcaniclastic subaqueous mass-flow deposits, likely to have been sourced from variably reworked subaerial to shoreline ignimbrite deposits, occur in the Archean Yilgarn Craton of Western Australia, north of Kalgoorlie, at a locality called Kanowna Dam. There, a submarine volcaniclastic fan consists of graded bedded turbidites, each bed ranging from cm to 2 m in thickness (Fig. 11.45a). The facies is dominated by angular fragments of volcanic quartz and feldspar, with localised concentrations of variably rounded porphyritic dacite clasts (Fig. 11.45b, c), which testify to reworking in subaerial and/or shoreline tractional zones before being resedimented into sub-wave base deep water by turbidity currents. The deposits are massive and generally size graded, with locally spectacular load casts and flame structures at the bases of beds (Fig. 11.45d). Channel scours and intra-formational slump horizons also occur in a sandstone-dominated succession that is over 80 m thick, and appears to have been a channelised part of a submarine volcaniclastic submarine fan (Krapez and Hand 2008). This outcrop forms part of the regionally extensive Black Flag Beds or Group, that consists of black mudstones, volcaniclastic megaturbidites of felsic composition, and local occurrences of mafic and felsic lavas and autoclastic deposits, as well as resedimented volcaniclastic sediments (e.g. Hand et al. 2002; Krapez and Hand 2008). The volcaniclastic megaturbidites reflect the passage of subaerially or shallow submarine explosively erupted pyroclastic flows into water and transformation into water-supported mass flows (see Sect. 11.6.1.4). The Kanowna Dam exposure represents surface sedimentary process reworking and resedimentation of debris eroded from subaerial ignimbrites into deep water by submarine mass flows (Fig. 11.7i), although no subaerial ignimbrites are now preserved anywhere in the Yilgarn Craton.

Allen and Freundt (2006) undertook an experimental investigation of the transport processes and deposits resulting from the resedimentation of cold, unconsolidated ignimbrite into water using a flume tank. Moderate collapse events generated high-density turbidity currents, trapping pumice in the flow, causing them to saturate. These low-density pumice clasts were easily remobilised by wave activity and passing currents and accumulated on the gentle slope at the bottom of the resedimented deposit. Turbidity currents produced hydraulically better-sorted deposits than the original ignimbrite, segregating dense lithic and crystal fragments into the base of the deposit and lower density pumice and fine ash into the upper part of deposit. Large collapse events formed slumps, producing poorly sorted mounds similar in texture to the original starting material. As the matrix of the ignimbrite sample became saturated with water, moderate and large collapse events generated debris flows and slurries that deposited massive, poorly sorted deposits.

11.7 A Final Note on Terminology

As addressed in detail in Chap. 14, in the first instance, deposit terminology should be descriptive, not genetic, highlighting key components, textural make-up and depositional structures. If clear unequivocal evidence exists in deposits for emplacement in a subaqueous setting by a hot, gas-supported mass flow (e.g. clear microscopic evidence of fluidal, welded glass shard textures), the term subaqueous pyroclastic flow (deposit) can be justifiably used. If evidence for deposition in a hot state is lacking, but evidence exists for syn-eruptive generation and transformation into a water-supported mass flow (e.g. efficient vertical hydraulic sorting profile such as well-developed hydraulic sorting of components, grain size grading), a term such as syn-eruptive, water transformed, pyroclastic debris flow or pyroclast-rich (mega) turbidity current (turbidite) deposit is more appropriate. If there is clear evidence of some degree of reworking or hiatus between original deposition by pyroclastic flows and resedimentation of variably reworked pyroclastic debris by subaqueous mass flows (e.g. abundant rounded clasts, polymictic clast assemblage), then terms such as volcaniclastic debris flow (deposit), or volcaniclastic turbidity current (deposit) should be applied, even if the

Fig. 11.45 Features of reworked and resedimented volcaniclastic sediments resulting from erosion and remobilisation of subaerial and shoreline zone deposited ignimbrites, Archean Black Flag Group, Kanowna Dam, Boorara Domain, Kalgoorlie Terrane, Yilgarn Craton, Western Australia. **a** Graphic log through 40 m of a submarine volcaniclastic fan dominated by massive, graded turbidites consisting of dacitic composition and derivation. Pale grey beds are pebbly, dark grey beds are largely sandstone grading into siltstone. (From Krapež and Hand 2008). **b** Platform exposure of tabular bedded volcaniclastic turbidites. **c** Bedded pebble conglomerate and sandstone turbidites showing scour structures and lensing of the pebble-rich layers. **d** Tabular graded turbidite with irregular basal load casts and delicate flame structures of underlying mudstone (light colour) injecting upwards into the sandstone beds

original clast population was pyroclastic. The key determinants are the interpreted mechanism of final deposition (hot gas-supported *versus* cold to warm water-supported) and whether or not evidence exists for the final transport and depositional mechanism being syn-eruptive or post-eruptive in environments that can be demonstrated to have been clearly subaqueous.

11.8 Summary

From the foregoing assessment of subaqueous flows of pyroclastic debris and their deposits, as well as those of Cas and Wright (1987, 1991), several generalisations can be made:

(a) In modern oceanic settings, the environment of deposition of volcaniclastic deposits is usually well constrained. In ancient volcanic successions, great care is required to fully understand the depositional setting of the volcaniclastic deposits, involving not only assessment of the paleoenvironmental affinity of the host sediments, including the possible water depths, but also the contact stratigraphic relationships (conformable, unconformable) before final interpretation is made.

(b) Hot, gas-supported pyroclastic flows that enter the sea from subaerial vents and eruptions will be subject to some degree of explosive interaction at the shoreline, including perhaps disintegration.

(c) Through explosive interaction and further physical mixing offshore after entering water such flows, or significant parts of them, ingest large volumes of ambient water and transform into water-supported, subaqueous mass flows.

(d) Large-volume and high mass-flux pyroclastic flows entering the sea can displace the sea seawards, generating tsunamis, and allowing the pyroclastic flows to deposit thick, hot ignimbrites that are capable of welding in evacuated shelf settings because they don't mix in water, which has been "bulldozed" offshore. Such entry points may build prograding, partially subaerial pyroclastic fans if the deposits are thick enough. Subsequently, transgression of the sea covers the ignimbrite(s) by a marine environment again.

(e) Some large-volume pyroclastic flows can travel tens of km offshore above the sea surface transporting mass and heat with little interaction with water. Resulting ignimbrites deposited on-land on other land masses after travelling over water, may be virtually identical as facies and temperature of emplacement to those emplaced by fully on-shore pyroclastic flows. However, large lithic clasts may have dropped out of the base of the flows as they travelled over water.

(f) All pyroclastic flows entering the sea will produce some offshore aqueous mass flow underflows (e.g. turbidity currents), as well as low-density gas to steam-rich overflows. Pumiceous pyroclastic flows produce more overflows, whereas high-density pyroclastic flows produce a greater percentage of aqueous underflows such as debris flows and turbidity currents. Some syn-eruptive turbidity currents can flow hundreds of km offshore into very deep submarine basins, as shown by deep sea drilling.

(g) Some subaqueously erupted, gas-supported pyroclastic flows may be generated and maintain their integrity away from the vent. However, these usually only appear to occur in some proximal settings for small volume eruptions (e.g. associated with low collapsing fountaining eruptions), or if the mass discharge rate is high, so excluding ambient water from infiltrating into the erupting and flowing mass on the seafloor (e.g. submarine caldera-forming scale explosive eruptions). While some deposits may remain hot enough to weld, more commonly such flows will likely ingest water near the vent and transform into water-supported mass flows.

(h) Subaqueous, water-supported mass flows of pyroclastic debris are nearly always better hydraulically sorted than gas-supported pyroclastic flow deposits, and compared with subaerial ignimbrites. The bases of water-supported mass-flow deposits will be rich in dense clasts such as lithic clasts and coarse crystal fragments, as well as some large pumice clasts, forming a clast-supported texture depleted in fine ash. Upwards, maximum and average clast size will usually decrease, forming a size grading into fine ash sizes often dominated by shards at the top, although in some cases, concentrations of buoyant large pumice clasts may also occur at the top of the deposit.

(i) Following variable degrees of shoreline reworking, resedimentation of volcaniclastic sediment derived from erosion of subaerial to shoreline zone deposited pyroclastic flow deposits may occur. This is deposited as offshore submarine fan-deltas and submarine fans, dominated by volcaniclastic turbidity currents, debris flows, and slump and slide deposits that build submarine volcaniclastic aprons around volcanic islands and offshore from continental margins.

(j) Finally, the terminology to be used for the subaqueous deposits of pyroclastic debris should be applied with caution, otherwise, misleading inferences can be made about the eruption, transport and depositional processes, and their timing.

References

Allen RL (1988) False pyroclastic textures in altered silicic lavas, with implications for volcanic-associated mineralisation. Econ Geol 83:1424–1446

Allen SR (1998) Volcanology of the Kos Plateau Tuff, Greece: the product of an explosive eruption in an archipelago. PhD Thesis, Monash University

Allen SR (2001) Reconstruction of a major caldera-forming eruption from pyroclastic deposit characteristics: Kos Plateau Tuff, eastern Aegean Sea J Volcanol Geotherm Res 105(1–2):141–162

Allen SR, Carey R, McPhie (2020) New insights into the eruption of submarine pumice cones: Yali revisited. Geology, in prep

Allen SR, Cas RAF (1998a) Lateral variations within coarse co-ignimbrite lithic breccias of the Kos Plateau Tuff, Greece. Bull Volcanol 59(5):356–377

Allen SR, Cas RAF (1998b) Rhyolitic fallout and pyroclastic density current deposits from a phreatoplinian eruption in the eastern Aegean Sea, Greece. J Volcanol Geotherm Res 86(1–4):219–251

Allen SR, Cas RAF (2001) Transport of pyroclastic flows across the sea during the explosive, rhyolitic eruption of the Kos Plateau Tuff, Greece. Bull Volcanol 62:441–456

Allen SR, Freundt A (2006) Resedimentation of cold pumiceous ignimbrite into water: facies transformations simulated in flume experiments. Sedimentology 53(4):717–734

Allen SR, Freundt A, Kurokawa K (2012) Characteristics of submarine pumice-rich density current deposits sourced from turbulent mixing of subaerial pyroclastic flows at the shoreline: field and experimental assessment. Bull Volcanol 74(3):657–675

Allen SR, McPhie J (2000) Water-settling and resedimentation of submarine rhyolitic pumice at Yali, eastern Aegean, Greece. J VolcanolGeother Res 95(1-4):285-307

Allen SR, McPhie J (2001) Syn-eruptive chaotic breccia on Kos, Greece, associated with an energetic pyroclastic flow. Bull Volcanol 63(6):421–132

Allen SR, McPhie J (2009) Products of neptunian eruptions. Geology 37:639–642. https://doi.org/10.1130/G30007A.1

Allen SR, Stadlbauer E, Keller J (1999) Stratigraphy of the Kos Plateau Tuff: product of a major quaternary explosive rhyolitic eruption in the eastern Aegean, Greece. Int J Earth Sci 88(1):132–156

Aramaki S, Akimoto SI (1957) Temperature estimation of pyroclastic deposits by natural remanent magnetism. Am J Sci 225:619–627

Baer EM, Fisher RV, Fuller M, Valentine G (1997) Turbulent transport and deposition of the Ito pyroclastic flow: determinations using anisotropy of magnetic susceptibility. J Geophys Res: Solid Earth. 102(B10):22565–22586

Bailey JE, Self S (2010) The properties and formation of erosional pipe-shaped structures in ignimbrites around the Valles Caldera. Geol Soc Am Abs Prog 42(5):51

Bailey RA, Carr RG (1994) Physical geology and eruptive history of the Matahina ignimbrite, Taupo volcanic zone, North Island, New Zealand. NZ J Geol Geophys 37(3):319–344

Balasubramanian S, Zhong Q (2018) Entrainment and mixing in lock-exchange gravity currents using simultaneous velocity-density measurements. Phys Fluids 30(5):056601

Bhaganagar K (2017) Role of head of turbulent 3-D density currents in mixing during slumping regime. Phys Fluids 29(2):020703

Black LP, Jagodzinski EA (2003) Importance of establishing sources of uncertainty for the derivation of reliable SHRIMP ages. Aust J Earth Sci 50(4):503–512

Branney MJ, Kokelaar P (2002) Pyroclastic density currents and the sedimentation of ignimbrites. Geol Soc Lond Mem 27:145

Branney MJ, Sparks RS (1990) Fiamme formed by diagenesis and burial-compaction in soils and subaqeuous sediments. J Geol Soc Lond 147(6):919–922

Brenchley PJ, Pickerill RK (1980) Shallow subtidal sediments of Soudleyan (Caradoc) age in the Berwyn Hills, North Wales, and their palaeogeographic context. Proc Geol Assoc 91:177–194

Britter RE, Linden PF (1980) The motion of the front of a gravity current travelling down an incline. J Fluid Mech 99(3):531–543

Busby CJ, Kessel LO, Schulz KJ, Foose MP, Slack JF (2003) Volcanic setting of the Ordovician Bald Mountain massive sulfide deposit, northern Maine. Econ Geol Monog 11:219–244

Busby-Spera CJ (1984a) The lower Mesozoic continental margin and marine intra-arc sedimentation at Mineral King, California. In: Bachman S, Crouch J (eds) Tectonics and sedimentation along the California margin. Soc Econ Pal Mineral Soc Pacific Sect 135–156

Busby-Spera CJ (1984b) Large volume rhyolite ash-flow eruptions and submarine caldera collapse in the lower Mesozoic Sierra Nevada, California. J Geophys Res 89(B10):8417–8427

Busby-Spera CJ (1985) A sand rich submarine fan in the lower Mesozoic Mineral King Caldera complex, Sierra Nevada, California. J Sed Petrol 55:376–391

Busby-Spera CJ (1986) Depositional features of rhyolitic and andesitic volcaniclastic rocks of the Mineral King submarine caldera complex, Sierra Nevada, California. J Volcanol Geotherm Res 27:43–76

Carey S (1997) Influence of convective sedimentation on the formation of widespread tephra fall layers in the deep sea. Geology 25(9):839–842

Carey S, Sigurdsson H (1980) The Roseau Ash: deep-sea tephra deposits from a major eruption on Dominica, Lesser Antilles arc. J Volcanol Geotherm Res 7:67–86

Carey S, Sigurdsson H (1987) Temporal variations in column height and magma discharge rate during the 79 AD eruption of Vesuvius. Geol Soc Am Bull 99(2):303–314

Carey S, Sigurdsson H, Mandeville C, Bronto S (1996) Pyroclastic flows and surges over water: an example from the 1883 Krakatau eruption. Bull Volcanol 57(7):493–511

Carey S, Sigurdsson H, Mandeville C, Bronto S (2000) Volcanic hazards from pyroclastic flow discharge into the sea: examples from the 1883 eruption of Krakatau, Indonesia. In McCoy F, Heiken G (eds) Volcanic hazards and disasters in human antiquity. Geol Soc Am Spec Pap 345:1–14

Caricchi C, Vona A, Corrado S, Giordano G, Romano C (2014) 79 AD Vesuvius PDC deposits' temperatures inferred from optical analysis on woods charred in-situ in the Villa dei Papiri at Herculaneum (Italy). J Volcanol Geotherm Res 289:14–25

Cartigny MJ, Postma G, Van den Berg JH, Mastbergen DR (2011) A comparative study of sediment waves and cyclic steps based on geometries, internal structures and numerical modeling. Marine Geol 280(1–4):40–56

Cas RAF (1977) Basin characteristics of the early Devonian part of the hill end trough, New South Wales, based on a stratigraphic analysis of the Merrions Tuff. J Geol Soc Aust 24(7–8):381–401

Cas RAF (1979) Mass-flow arenites from a Palaeozoic interarc basin, New South Wales, Australia: mode and environment of emplacement. J Sed Petrol 49:29–44

Cas RAF (1983a) A review of the paleogeography and tectonic development of the Lachlan Fold Belt of southeastern Australia. Geol Soc Aust Spec Publ 10:104p

Cas RAF (1983b) Submarine 'crystal-tuffs': their origin using a Lower Devonian example from southeastern Australia. Geol Mag 120:471–486

Cas RAF, Flood RH, Shaw SE (1976) Hill end trough: new radiometric ages. Search 7:205–207

Cas RAF, Powell CM, Fergusson CL, Jones JG, Roots WD, Fergusson J (1981) The lower devonian kowmung volcaniclastics: a deep-water succession of mass-flow origin, northeastern Lachlan Fold Belt, NSW. J Geol Soc Aust 28(3–4):271–288

Cas RAF, Wright JV (1987) Volcanic successions: modern and ancient. Allen and Unwin, London, p 528

Cas RAF, Wright JV (1991) Subaqueous pyroclastic flows and ignimbrites: an assessment. Bull Volcanol 53(5):357–380

Cashman KV, Fiske RS (1991) Fallout of pyroclastic debris from submarine volcanic eruptions. Science 253:275–280

Cioni R, Marianelli P, Sbrana A (1992) Dynamics of the AD 79 eruption: stratigraphic, sedimentologic and geochemical data on the successions of the Somma-Vesuvius southern sector. Acta Vulcanol 2:109–123

Cita MB, Beghi C, Camerlenghi A, Kastens KA, McCoy FW, Nosetto A, Parisi E, Scolari F, Tomadin L (1984) Turbidites and megaturbidites from the Herodotus abyssal plain (eastern Mediterranean) unrelated to seismic events. Marine Geol 55(1–2):79–101

Civetta L, Orsi G, Pappalardo L, Fisher RV, Heiken G, Ort M (1997) Geochemical zoning, mingling, eruptive dynamics and depositional processes—the Campanian Ignimbrite, Campi Flegrei caldera, Italy. J Volcanol Geotherm Res 75(3–4):183–219

Corbett KD (1992) Stratigraphic-volcanic setting of massive sulfide deposits in the Cambrian Mount Read Volcanics, Tasmania. Econ Geol 87(3):564–586

Cronin S (2022) Why the volcanic eruption in Tonga was so violent, and what to expect next. The Conversation January 16, 2022.https://mail.google.com/mail/u/0/#search/Shane+Cronin/FMfcgzGmtXCRxPtdgDXpmkstLfQGmjNv

Dewey JF (1963) The lower Palaeozoic stratigraphy of central Murrisk, County Mayo, Ireland, and the evolution of the South Mayo Trough. Quart J Geol Soc Lond 119:313–343

Di Carlo I, Pichavant M, Rotolo S, Scaillet B (2006) Experimental crystallization of a high-K arc basalt: the golden pumice, Stromboli volcano (Italy). J Petrol 47(7):1317–1343

Dufek J, Bergantz GW (2007a) Dynamics and deposits generated by the Kos Plateau Tuff eruption: controls of basal particle loss on pyroclastic flow transport. Geochem Geophys Geosyst 8(12)

Dufek J, Bergantz GW (2007b) Suspended load and bed-load transport of particle-laden gravity currents: the role of particle–bed interaction. Theor Comput Fluid Dyn 21(2):119–145

Dufek J, Manga M, Staedter M (2007) Littoral blasts: Pumice-water heat transfer and the conditions for steam explosions when pyroclastic flows enter the ocean. J Geophys Res: Solid Earth 112 (B11)

Druitt TH (1998) Pyroclastic density currents. In Gilbert JS, Sparks RSJ (eds) The physics of explosive volcanic eruptions. Geol Soc Lond Spec Publ 145(1):145–182

Edmonds M, Herd RA (2005) Inland-directed base surge generated by the explosive interaction of pyroclastic flows and seawater at Soufriere Hills volcano, Montserrat. Geology 33(4):245–248

Edmonds M, Herd RA, Strutt MH (2006) Tephra deposits associated with a large lava dome collapse, Soufrière Hills Volcano, Montserrat, 12–15 July 2003. J Volcanol Geotherm Res 153(3–4):313–330

Fauria KE, Manga M, Wei Z (2017) Trapped bubbles keep pumice afloat and gas diffusion makes pumice sink. Earth Planet Sci Letts 460:50–59

Fedele L, Scarpati C, Lanphere M, Melluso L, Morra V, Perrotta A, Ricci G (2008) The Breccia Museo formation, Campi Flegrei, southern Italy: geochronology, chemostratigraphy and relationship with the Campanian Ignimbrite eruption. Bull Volcanol 70 (10):1189–1219

Fedele L, Scarpati C, Sparice D, Perrotta A, Laiena F (2016) A chemostratigraphic study of the Campanian Ignimbrite eruption (Campi Flegrei, Italy): insights on magma chamber withdrawal and deposit accumulation as revealed by compositionally zoned stratigraphic and facies framework. J Volcanol Geotherm Res 324:105–117

Fenner CN (1948) Incandescent tuff flows in southern Peru. Geol Soc Am Bull 59(9):879–893

Fisher RV, Heiken G (1982) Mt. Pelée, Martinique: May 8 and 20, 1902, pyroclastic flows and surges. J Volcanol Geotherm Res 13(3–4):339–371

Fisher RV, Orsi G, Ort M, Heiken G (1993) Mobility of a large-volume pyroclastic flow—emplacement of the Campanian ignimbrite, Italy. J Volcanol Geotherm Res 56(3):205–220

Fisher RV, Schmincke HU (1984) Pyroclastic rocks. Springer, Berlin, New York, Heidelberg, p 472

Fiske RS (1963) Subaqueous pyroclastic flows in the Ohanapecosh Formation, Washington. Geol Soc Am Bull 74:391–406

Fiske RS, Matsuda T (1964) Submarine equivalents of ash flows in the Tokiwa Formation, Japan. Am J Sci 262:76–106

Forni F, Bachmann O, Mollo S, De Astis G, Gelman SE, Ellis BS (2016) The origin of a zoned ignimbrite: Insights into the Campanian Ignimbrite magma chamber (Campi Flegrei, Italy). Earth Planet Sci Letts 449:259–271

Francis EH, Howells MF (1973) Transgressive welded ash-flow tuffs among the Ordovician sediments of NE Snowdonia. J Geol Soc Lond 129:621–641

Francis P, Self S (1983) The eruption of Krakatau. Sci Am 249(5):172–187

Freundt A (2003) Entrance of hot pyroclastic flows into the sea: experimental observations. Bull Volcanol 65(2–3):144–164

Freundt A, Schmincke H-U (1986) Emplacement of small volume pyroclastic flows at Laacher See (East Eiffel, Germany) Bull Volcanol 48:39–60

Freundt A, Schmincke H-U (1998) Emplacement of ash layers related to high-grade ignimbrite P1 in the sea around Gran Canaria. In: Weaver PPE, Schmincke H-U, Firth JV, Duffield WA (eds.) Proc ODP Sci Res 157:201–218

Fritz WJ, Howells MF (1991) A shallow marine volcaniclastic facies model: an example from sedimentary rocks bounding the subaqueously welded Ordovician Garth Tuff, North Wales, UK. Sed Geol 74(1–4):217–240

Fritz WJ, Stillman CJ (1996) A subaqueous welded tuff from the Ordovician of County Waterford, Ireland. J Volcanol Geotherm Res 70(1–2):91–106

Giordano G, De Astis G (2021) The summer 2019 basaltic Vulcanian eruptions (paroxysms) of Stromboli. Bull Volcanol 83(1):1–27. https://doi.org/10.1007/s00445-020-01423-2

Giordano G, Doronzo DM (2017) Sedimentation and mobility of PDCs: a reappraisal of ignimbrites' aspect ratio. Sci Repts 7 (1):4444

Giordano G, Porreca M, Musacchio P, Mattei M (2008) The Holocene Secche di Lazzaro phreatomagmatic succession (Stromboli, Italy): evidence of pyroclastic density current origin deduced by facies analysis and AMS flow directions. Bull Volcanol 70(10):1221–1236

Giordano G, Zanella E, Trolese M, Baffioni C, Vona A, Caricchi C, De Benedetti AA, Corrado S, Romano C, Sulpizio R, Geshi N (2018) Thermal interactions of the AD79 Vesuvius pyroclastic density currents and their deposits at Villa dei Papiri (Herculaneum archaeological site, Italy). Earth Planet Sci Letts 490:180–192

Gurioli L, Cioni R, Sbrana A, Zanella E (2002) Transport and deposition of pyroclastic density currents over an inhabited area: the deposits of the AD 79 eruption of Vesuvius at Herculaneum, Italy. Sedimentology 49(5):929–953

Hand JL, Cas RAF, Ong L, Brown SJ, Krapez B, Barley ME (2002) Syn-and post-eruptive volcaniclastic sedimentation in late archaean subaqueous depositional systems of the black flag group, Eight Mile Dam, Kalgoorlie, Western Australia. In: Altermann W, Corcoran PL (eds) Precambrian sedimentary environments: a modern approach to ancient depositional systems. Int Assoc Sed Spec Publ 18:235–258

Head JW, Wilson L (2003) Deep submarine pyroclastic eruptions: theory and predicted landforms and deposits. J Volcanol Geotherm Res 121:155–193

Heiken G, Wohletz KH (1985) Volcanic ash. University of California Press, p 246

Herd RA, Edmonds M, Bass VA (2005) Catastrophic lava dome failure at Soufrière Hills volcano, Montserrat, 12–13 July 2003. J Volcanol Geotherm Res 148(3–4):234–252

Hoblitt RP, Kellogg KS (1979) Emplacement temperatures of unsorted and unstratified deposits of volcanic rock debris as determined by palaeomagnetic techniques. Geol Soc Am Bull 90(1):633–642

Howells MF, Campbell DG, Reedman AJ (1985) Isolated pods of subaqueous welded ash-flow tuff: a distal facies of the Capel Curig Volcanic Formation (Ordovician), North Wales. Geol Mag 122:175–180

Howells MF, Reedman AJ, Campbell SDG (1986) The submarine eruption and emplacement of the lower Rhyolitic Tuff Formation (Ordovician), North Wales. J Geol Soc Lond 143:411–423

Howells MF, Leveridge BE, Addison R, Evans CDR, Nutt MJC (1979) The capel curig volcanic formation, Snowdonia, North Wales; variation in ash-flow tuffs related to emplacement environment. In: Harris AL, Holland CH, Leake BE (eds) The caledonides of the British Isles reviewed. Geol Soc Lond Spec Publ 8:611–618

Jagodzinski EA, Black LP (1999) U-Pb dating of silicic lavas, sills and syneruptive resedimented volcaniclastic deposits of the Lower Devonian Crudine Group, Hill End Trough, New South Wales. Aust J Earth Sci 46(5):749–764

Johnston EN, Sparks RS, Nomikou P, Livanos I, Carey S, Phillips JC, Sigurdsson H (2015) Stratigraphic relations of Santorini's intra-caldera fill and implications for the rate of post-caldera volcanism. J Geol Soc Lond 172(3):323–335

Jutzeler M, Manga M, White JD, Talling PJ, Proussevitch AA, Watt SF, Cassidy M, Taylor RN, Le Friant A, Ishizuka O (2017) Submarine deposits from pumiceous pyroclastic density currents traveling over water: an outstanding example from offshore Montserrat (IODP 340). Geol Soc Am Bull 129(3–4):392–414

Jutzeler M, McPhie J, Allen SR (2014) Facies architecture of a continental, below-wave-base volcaniclastic basin: the Ohanapecosh formation, Ancestral Cascades arc (Washington, USA). Geol Soc Am Bull 126(3–4):352–376

Jutzeler M, McPhie J, Allen SR (2014) Submarine eruption-fed and resedimented pumice-rich facies: the Dogashima Formation (Izu Peninsula, Japan). Bull Volcanol 76(10):867

Kano K (1990) An ash-flow tuff emplaced in shallow water, Early Miocene Koura formation southwest Japan. J Volcanol Geotherm Res 40:1–9

Kano K (1991) Volcaniclastic sedimentation in a shallow-water marginal basin: the Early Miocene Koura formation SW Japan. Sed Geol 74(1–4):309–321

Kano K, Orton GJ, Kano T (1994) A hot Miocene subaqueous scoria-flow deposit in the Shimane Peninsula, SW Japan. J Volcanol Geotherm Res 60(1):1–14

Kano K, Yamamoto T, Ono K (1996) Subaqueous eruption and emplacement of the Shinjima Pumice, Shinjima (Moeshima) Island, Kagoshima Bay, SW Japan. J Volcanol Geotherm Res 71(2–4):187–206

Kato I, Muroi I, Yamazaki T, Abe M (1971) Subaqueous pyroclastic flow deposits in the Upper Donzurobo Formation, Nijosan district, Osaka, Japan. J Geol Soc Japan 77:193–206

Keller J, Rehren TH, Stadlbauer E (1990) Explosive volcanism in the Hellenic Arc: a summary and review. In: Hardy DA, Keller J, Galanopoulos VP, Flemming NC, Druitt TH (eds) Thera and the Aegean world III, vol 2. Earth sciences. Proceedings of third international congress on Santorini, Greece 3–9 September 1989, pp 13–26

Kent DV, Ninkovich D, Pescatore T, Sparks SR (1981) Palaeomagnetic determination of emplacement temperature of Vesuvius AD 79 pyroclastic deposits. Nature 290(5805):393

Kessel LG, Busby CJ (2003) Analysis of VHMS-hosting ignimbrites erupted at bathyal water depths (Ordovican Bald Mountain sequence, northern Maine). In: White JD, Smellie JL, Clague DA (eds) Explosive subaqueous volcanism. American Geophys Union Monog 140:361–379

Kokelaar P, Busby C (1992) Subaqueous explosive eruption and welding of pyroclastic deposits. Science 257(5067):196–201

Kokelaar P, Königer S (2000) Marine emplacement of welded ignimbrite: the Ordovician Pitts Head Tuff, North Wales. J Geol Soc Lond 157(3):517–536

Krapež B, Hand JL (2008) Late Archaean deep-marine volcaniclastic sedimentation in an arc-related basin: the Kalgoorlie Sequence of the Eastern Goldfields Superterrane, Yilgarn Craton, Western Australia. PreCamb Res 161(1–2):89–113

Kutterolf S, Schindlbeck JC, Scudder RP, Murray RW, Pickering KT, Freundt A, Labanieh S, Heydolph K, Saito S, Naruse H, Underwood MB, Wu H (2014) Large volume submarine ignimbrites in the Shikoku Basin: an example for explosive volcanism in the Western Pacific during the Late Miocene. Geochem Geophys Geosyst 15 (5):1837–1851

Lacroix A (1904) La Montagne Pelée et ses eruptions. Masson et Cie, Paris, p 653

Le Friant A, Deplus C, Boudon G, Feuillet N, Trofimovs J, Komorowski JC, Sparks RSJ, Talling P, Loughlin S, Palmer M, Ryan G (2010) Eruption of Soufrière Hills (1995–2009) from an offshore perspective: insights from repeated swath bathymetry surveys. Geophys Res Letts 37(19):L11307. https://doi.org/10.1029/2010GL043580

Lock BE (1978) Ultrahigh-temperature volcanic mudflows amongst the Drakensberg volcanic rocks: new criteria for their recognition. Trans Geol Soc Sth Africa 81:55–59

Legros F, Druitt TH (2000) On the emplacement of ignimbrite in shallow-marine environments. J Volcanol Geotherm Res 95(1–4):9–22

Lipman PW (2018) When ignimbrite meets water: Megascale gas-escape structures formed during welding. Geology 47(1):63–66

Lowe DR (1975) Water escape structures in coarse grained sediments. Sedimentology 22:157–204

Lowe DR (1982) Sediment gravity flows: II. Depositional models with special reference to the deposits of high density turbidity currents. J Sed Petrol 52:279–297

Macdonald GA (1972) Volcanoes. Prentice-Hall, New Jersey, p 510

Mackay K (2023) https://www.abc.net.au/news/2022-11-21/sea-floor-map-reveals-huge-cavity-left-behind-by-tonga-volcano/101674708

Maeno F, Imamura F, Taniguchi H (2006) Numerical simulation of tsunamis generated by caldera collapse during the 7.3 ka Kikai eruption, Kyushu, Japan. *Earth, planets and space* 58(8):1013–1024

Maeno F, Taniguchi H (2007) Spatio-temporal evolution of a marine caldera-forming eruption, generating a low-aspect ratio pyroclastic flow, 7.3 ka, Kikai caldera, Japan: Implication from near-vent eruptive deposits. J Volcanol Geotherm Res 167(1–4):212–238

Maeno F, Taniguchi H (2009) Sedimentation and welding processes of dilute pyroclastic density currents and fallout during a large-scale silicic eruption, Kikai caldera, Japan. Sed Geol 220(3–4):227–242

Mandeville CW, Carey S, Sigurdsson H (1996) Sedimentology of the Krakatau 1883 submarine pyroclastic deposits. Bull Volcanol 57 (7):512–529

Mandeville CW, Carey S, Sigurdsson H, King J (1994) Paleomagnetic evidence for high-temperature emplacement of the 1883 subaqueous pyroclastic flows from Krakatau Volcano, Indonesia. J Geophys Res: Solid Earth 99(B5):9487–9504

Manga M, Fauria KE, Lin C, Mitchell SJ, Jones M, Conway CE, Degruyter W, Hosseini B, Carey R, Cahalan R (2018) The pumice raft-forming 2012 Havre submarine eruption was effusive. Earth Planet Sci Lett 489:49–58

Maury R (1973) La matière organique des bois fossiles, indicatrices des conditions thermiques de raise in place des br6ches volcaniques. CR Acad Sci Paris 276, Ser D:917–920

McArthur AN, Cas RAF, Orton GJ (1998) Distribution and significance of crystalline, perlitic and vesicular textures in the Ordovician Garth Tuff (Wales). Bull Volcanol 60(4):260–285

McLeod P, Carey S, Sparks RSJ (1999) Behaviour of particle-laden flows into the ocean: experimental simulation and geological implications. Sedimentology 46:523–536

McPhie J, Allen RL (1992) Facies architecture of mineralized submarine volcanic sequences; Cambrian Mount Read Volcanics, western Tasmania. Econ Geol 87(3):587–596

Monaghan JJ, Cas RA, Kos AM, Hallworth M (1999) Gravity currents descending a ramp in a stratified tank. J Fluid Mech 379:39–69

Milia A (2010) The stratigraphic signature of volcanism off Campi Flegrei (Bay of Naples, Italy). In: Groppelli G, Viereck-Goette L (eds) Stratigraphy and geology of volcanic areas. Geol Soc Am Spec Pap 464:155–170

Milia A, Torrente MM (2007) The influence of paleogeographic setting and crustal subsidence on the architecture of ignimbrites in the Bay of Naples (Italy). Earth Planet Sci Letts 263(3–4):192–206

Mueller W, White JDL (1992) Felsic fire-fountaining beneath Archean seas: pyroclastic deposits of the 2730 Ma Hunter Mine Group, Quebec, Canada. J Volcanol Geotherm Res 54(1–2):117–134

Mutti E (1965) Submarine flood tuffs (ignimbrites) associated with turbidites in Oligocene deposits of Rhodes Island (Greece). Sedimentology 5(4):265–288

Mutti E, Lucchi FR, Seguret M, Zanzucchi G (1984) Seismoturbidites: a new group of resedimented deposits. Marine Geol 55(1–2): 103–116

Niem AR (1977) Mississippian pyroclastic flow and ash-fall deposits in the deep-marine Ouachita flysch basin, Oklahoma and Arkansas. Geol Soc Am Bull 88:49–61

Nomikou P, Bell KL, Papanikolaou D, Livanos I, Fero Martin J (2013) Exploring the Avyssos-Yali-Strogyli submarine volcanic complex at the eastern edge of the Aegean Volcanic Arc. Zeitschrift für Geomorphologie 57(3):125–137

Ort, MH, Giordano G, Zanella E, Isaia R (2015) Incorporation of air into the Campanian Ignimbrite pyroclastic density current. In: AGU Fall Meeting Abstracts

Orton G (1987) Discussion on the submarine eruption and emplacement of the Lower Rhyolitic Tuff Formation (Ordovician), North Wales. J Geol Soc Lond 144(3):523–524

Orton GJ (1991) Emergence of subaqueous depositional environments in advance of a major ignimbrite eruption, Capel Curig Volcanic Formation, Ordovician, North Wales: an example of regional volcanotectonic uplift? Sed Geol 74:251–286

Pacheco-Hoyos JG, Aguirre-Díaz GJ, Dávila-Harris P (2020) Elutriation pipes in ignimbrites: an analysis of concepts based on the Huichapan Ignimbrite, Mexico. J Volcanol Geotherm Res 403:107026

Packham GH (1968) The lower and middle Palaeozoic stratigraphy and sedimentary tectonics of the Sofala-Hill End–Euchareena region, NSW. Proc Linn Soc NSW 93(1):111–163

Packham GH (2003) U-Pb SHRIMP ages of volcanic zircons from the Merrions and Turondale formations, New South Wales, and the Early Devonian time-scale: a biostratigraphic and sedimentological assessment. Aust J Earth Sci 50(2):169–179

Pensa A, Porreca M, Corrado S, Giordano G, Cas RAF (2015) Calibrating the pTRM and charcoal reflectance (Ro%) methods to determine the emplacement temperature of ignimbrites: Fogo A sequence, São Miguel, Azores, Portugal, as a case study. Bull Volcanol 77(3):18

Pensa A, Giordano G, Cas RA, Porreca M (2015b) Thermal state and implications for eruptive styles of the intra-Plinian and climactic ignimbrites of the 4.6 ka Fogo A eruption sequence, São Miguel, Azores. Bull Volcanol 77(11):99

Pe-Piper G, Piper DJ, Perissoratis C (2005) Neotectonics and the Kos Plateau Tuff eruption of 161 ka, South Aegean arc. J Volcanol Geotherm Res 139(3–4):315–338

Pérez-Rodríguez N, Morales J, Goguitchaichvili A, García-Tenorio F (2019) A comprehensive paleomagnetic study from the last Plinian eruptions of Popocatepetl volcano: absolute chronology of lavas and estimation of emplacement temperatures of PDCs. Earth Planets Space 71(1):80

Polonia A, Bonatti E, Camerlenghi A, Lucchi RG, Panieri G, Gasperini L (2013) Mediterranean megaturbidite triggered by the AD 365 Crete earthquake and tsunami. Sci Repts 15(3):1285

Pope EL, Jutzeler M, Cartigny MJ, Shreeve J, Talling PJ, Wright IC, Wysoczanski RJ (2018) Origin of spectacular fields of submarine sediment waves around volcanic islands. Earth Planet Sci Letts 493:12–24

Pudsey CJ (1984) Fluvial to marine transition in the Ordovician of Ireland–A humid-region fan-delta? Geol J 19(2):143–172

Purkis SJ, Ward SN, Fitzpatrick NM, Garvin JB, Slayback D, Cronin SJ, Palaseanu-Lovejoy M, Dempsey A (2023) The 2022 Hunga-Tonga megatsunami: Near-field simulation of a once-in-a-century event. Science Adv 9(15):eadf5493.

Quane SL, Russell JK (2005) Ranking welding intensity in pyroclastic deposits. Bull Volcanol 67(2):129–143

Randolph-Flagg N, Breen S, Hernandez A, Manga M, Self S (2017) Evenly spaced columns in the Bishop Tuff (California, USA) as relicts of hydrothermal cooling. Geology 45(11):1015–1018

Reedman AJ, Colman TB, Campbell SD, Howells MF (1985) Volcanogenic mineralization related to the Snowdon Volcanic Group (Ordovician), Gwynedd, North Wales. J Geol Soc Lond 142 (5):875–888

Reedman AJ, Howells MF, Orton G, Campbell SDG (1987) The Pitts Head Tuff Formation: a subaerial to submarine welded ash-flow tuff of Ordovician age, North Wales. Geol Mag 124:427–439

Ricci Lucchi F, Valmori E (1980) Basin-wide turbidites in a Miocene, over-supplied deep-sea plain: a geometrical analysis. Sedimentology 27(3):241–270

Riggs N, McConnell B (2017) Sedimentary response to ignimbrite emplacement across a fluvial–shallow-marine transition: Ordovician Mweelrea Formation, South Mayo Trough, Western Ireland. Geol J 52(4):667–680

Roobol MJ, Smith AL, Wright JV (1987) Lithic breccias in pyroclastic flow deposits on St Kitts, West Indies. Bull Volcanol 49:694–707

Rosell J, Wieczorek J (1989) Main features of megaturbidites in the Eocene of southern Pyrenees. Ann Soc Geol Pol 59(1–2):3–16

Rotella MD, Wilson CJ, Barker SJ, Wright IC (2013) Highly vesicular pumice generated by buoyant detachment of magma in subaqueous volcanism. Nat Geosci 6(2):129–132

Scarpati C, Perrotta A (2016) Stratigraphy and physical parameters of the Plinian phase of the Campanian Ignimbrite eruption. Geol Soc Am Bull 128(7–8):1147–1159

Scarpati C, Sparice D, Perrotta A (2015) Facies variation in the Campanian Ignimbrite. Rend Online Soc Geol Italia 33:83–87

Schardt C, Cooke DR, Gemmell JB, Large RR (2002) Geochemical modeling of the zoned footwall alteration pipe, Hellyer volcanic-hosted massive sulfide deposit, Western Tasmania, Australia. Econ Geol 96(5):1037–1054

Schindlbeck JC, Kutterolf S, Freundt A, Scudder RP, Pickering KT, Murray RW (2013) Emplacement processes of submarine volcaniclastic deposits (IODP Site C0011, Nankai Trough). Marine Geol 343:115–124

Schmincke H-U, Sumita M (1998) Tephra event stratigraphy and emplacement of tephra layers, Mogan and Fataga stratigraphic intervals. Part II: origin and emplacement of volcaniclastic layers. In: Weaver PPE, Schmincke H-U, Firth JV, Duffield WA (eds) Proc ODP Sci Res 157:267–291

Schneider JL, Fourquin C, Paicheler JC (1992) Two examples of subaqueously welded ash-flow tuffs: the Visean of southern Vosges (France) and the Upper Cretaceous of northern Anatolia (Turkey). J Volcanol Geotherm Res 49(3–4):365–383

Self S, Heiken G, Sykes ML, Wohletz K, Fisher RV, Dethier DP (1996) Part I: Field guide to the Bandelier Tuff and Valles caldera. Field

excursions to the Jemez Mountains, New Mexico. New Mexico Bur Mines Min Res Bull 134:7–55
Self S, Rampino MR (1981) The 1883 eruption of Krakatau. Nature 294(5843):699–704
Self S, Rampino MR, Newton MS, Wolff JA (1984) Volcanological study of the great Tambora eruption of 1815. Geology 12(11):659–663
Sheridan MF (1970) Fumarolic mounds and ridges of the Bishop Tuff, California. Geol Soc Amer Bull 81(3):851-868
Shimizu S, Nakaoka R, Seama N, Suzuki-Kamata K, Kaneko K, Kiyosugi K, Iwamaru H, Sano M, Matsuno T, Sugioka H, Tatsumi Y (2024) Submarine pyroclastic deposits from 7.3 ka caldera-forming Kikai-Akahoya eruption. J Volcanol Geotherm Research 448:10801
Sigurdsson H, Carey S, Mandeville C (1991) Submarine pyroclastic flows of the 1883 eruption of Krakatau volcano. Nat Geogr Res 7 (3):310–327
Sigurdsson H, Cashdollar S, Sparks RSJ (1982) The eruption of Vesuvius in AD 79: reconstruction from historical and volcanological evidence. Amer J Archeol 1:39–51
Silleni A, Giordano G, Isaia R, Ort MH (2020) The magnitude of the 39.8 ka Campanian Ignimbrite eruption, Italy: method, uncertainties and errors. Front Earth Sci 8: 543399.
Simkin T, Fiske RS (1983) Krakatau 1883—The volcanic eruption and its effects. Smithsonian Institution Press, Washington DC, p 464
Smith GA (1986) Coarse-grained nonmarine volcaniclastic sediment: Terminology and depositional process. Geol Soc Am Bull 97 (1):1–10
Smith VC, Isaia R, Engwell SL, Albert PG (2016) Tephra dispersal during the Campanian Ignimbrite (Italy) eruption: implications for ultra-distal ash transport during the large caldera-forming eruption. Bull Volcanol 78(6):45
Sparks RSJ (1976) Grain size in ignimbrites and implications for the transport of pyroclastic flows. Sedimentology 23:147–188
Sparks RSJ, Sigurdsson H, Carey SN (1980) The entrance of pyroclastic flows into the sea, II. Theoretical considerations on subaqueous emplacement and welding. J Volcanol Geotherm Res 7 (1–2):97–105
Stanton WI (1960) The lower Palaeozoic rocks of southwest Murrisk, Ireland. Quart Jl Geol Soc Lond 116:269–296
Stephenson D, Bevins RE, Millward D, Highton I, Parsons I, Stone P, Wadsworth WJ (2000) Caledonian Igneous Rocks of Great Britain. Geol Conserv Rev Ser 17, Peterborough, p 648
Stewart AL, McPhie J (2004) An Upper Pliocene coarse pumice breccia generated by a shallow submarine explosive eruption, Milos, Greece. Bull Volcanol 66(1):15–28
Stix J (1991) Subaqueous, intermediate to silicic-composition explosive volcanism: a review. Earth Sci Rev 31(1):21–53
Symons GJ (ed.) (1888) The eruption of Krakatoa and subsequent phenomena. Rept Krakatoa Committee of the Royal Society, Trubner, London
Trofimovs J, Amy L, Boudon G, Deplus C, Doyle E, Fournier N, Hart MB, Komorowski JC, Le Friant A, Lock EJ, Pudsey C (2006) Submarine pyroclastic deposits formed at the Soufrière Hills volcano, Montserrat (1995–2003): what happens when pyroclastic flows enter the ocean? Geology 34(7):549–552
Trofimovs J, Foster C, Sparks RSJ, Loughlin S, Le Friant A, Deplus C, Porritt L, Christopher T, Luckett R, Talling PJ, Palmer MR (2012) Submarine pyroclastic deposits formed during the 20th May 2006 dome collapse of the Soufrière Hills Volcano, Montserrat. Bull Volcanol 74(2):391–405
Trofimovs J, Sparks RSJ, Talling PJ (2008) Anatomy of a submarine pyroclastic flow and associated turbidity current: July 2003 dome collapse, Soufriere Hills volcano, Montserrat, West Indies. Sedimentology 55(3):617–634
Trolese M, Cerminara M, Ongaro TE, Giordano G (2019) The footprint of column collapse regimes on pyroclastic flow temperatures and plume heights. Nat Commun 10(1):2476
Trolese M, Giordano G, Komorowski JC, Jenkins SF, Baxter PJ, Cholik N, Raditya P, Corrado S (2018) Very rapid cooling of the energetic pyroclastic density currents associated with the 5 November 2010 Merapi eruption (Indonesia). J Volcanol Geotherm Res 358:1–12
Uehara D, Cas RAF, Folkes C, Takarada S, Oda H, Porreca M (2015) Using thermal remanent magnetisation (TRM) to distinguish block and ash flow and debris flow deposits, and to estimate their emplacement temperature: 1991–1995 lava dome eruption at Mt. Unzen Volcano, Japan. J Volcanol Geotherm Res 303:92–111
Verbeek RDM (1884) The Krakatoa eruption. Nature 30:10–15
Verbeek RDM (1886) Krakatau. lmprimerie de l'Etat, Batavia
Vogel S, Märker M (2010) Reconstructing the Roman topography and environmental features of the Sarno River Plain (Italy) before the AD 79 eruption of Somma-Vesuvius. Geomorphology 115(1–2):67–77
Walker GPL (1979) A volcanic ash generated by explosions where ignimbrite entered the sea. Nature 281(5733):642
Walker GPL (1983) Ignimbrite types and ignimbrite problems. J Volcanol Geotherm Res 17:65–88
Waltschew A (1997) Plio-/pleistozäne und rezente mikro- und molluskenfaunen von Kos (Griechenland, Ägäis). Natur und Mensch, Naturhistorische Gesellschaft Nürnberg e.V:52–64
White JD (2000) Subaqueous eruption-fed density currents and their deposits. Precamb Res 101(2–4):87–109
White MJ, McPhie J (1996) Stratigraphy and palaeovolcanology of the Cambrian Tyndall Group, Mt Read Volcanics, western Tasmania. Aust J Earth Sci 43(2):147–159
White MJ, McPhie J (1997) A submarine welded ignimbrite-crystal-rich sandstone facies association in the Cambrian Tyndall Group, western Tasmania, Australia. J Volcanol Geother Res 76 (3-4): 277-295
Whitham AG, Sparks RSJ (1986) Pumice. Bull Volcanol 48(4):209–223
Williams R, Rowley P, Garthwaite MC (2019) Reconstructing the Anak Krakatau flank collapse that caused the December 2018 Indonesian tsunami. Geology 47(10):97–76
Wilson CJN (1980) The role of fluidisation in the emplacement of pyroclastic flows: an experimental approach. J Volcanol Geotherm Res 8:231–249
Wilson CJN (1984) The role of fluidisation in the emplacement of pyroclastic flows, 2: experimental results and their interpretation. J Volcanol Geotherm Res 20:55–84
Wilson CJN, Walker GPL (1982) Ignimbrite depositional facies: the anatomy of a pyroclastic flow. J Geol Soc Lond 139:581–592
Wilson L, Walker GPL (1987) Explosive volcanic eruptions-VI. Ejecta dispersal in plinian eruptions: the control of eruption conditions and atmospheric properties. Geophys J Internat. 89(2):657-679
Wright HMN, Lesti C, Cas RAF, Porreca M, Viramonte JG, Folkes CB, Giordano G (2011) Columnar jointing in vapor-phase-altered, non-welded Cerro Galán Ignimbrite, Paycuqui, Argentina. Bull Volcanol 73(10):1567–1582
Wright JV (1978) Remanent magnetism of poorly sorted deposits from the Minoan eruption of Santorini. Bull Volcanol 41:131–135
Wright JV (1979) Formation, transport and deposition of ignimbrites and welded tuffs. PhD thesis, Imperial College, University of London, p 451

Wright JV, Coward MP (1977) Rootless vents in welded ash-flow tuffs from northern Snowdonia, North Wales, indicating deposition in a shallow water environment. Geol Mag 114(2):133–140

Wright JV, Mutti E (1981) The Dali Ash, Island of Rhodes, Greece: a problem in interpreting submarine volcanigenic sediments. Bull Volcanol 44:153–167

Wright JV, Self S, Fisher RV (1980a) A working terminology of pyroclastic deposits. J Volcanol Geotherm Res 8:315–336

Wright JV, Self S, Fisher RV (1981) Towards a facies model for ignimbrite eruptions. In: Self S, Sparks RSJ (eds) Tephra studies. Reidel, Dordrecht, pp 433–439

Wright JV, Smith AL, Self S (1980b) A working terminology of pyroclastic deposits. J Volcanol Geotherm Res 8:315–336

Yamada E (1973) Subaqueous pumice flows in the Onikobe Caldera, Miyagi Prefecture, Japan. J Geol Soc Japan 79:585–597

Yamada E (1984) Subaqueous pyroclastic flows: their development and their deposits. In: Kokelaar BP, Howells MF (eds) Marginal basin geology: volcanic and associated sedimentary and tectonic processes in modern and ancient marginal basins. Geol Soc Lond Spec Publ. pp, 29–35

Yamazaki T, Kato I, Muroi I, Abe M (1973) Textural analysis and flow mechanisms of the Donzurobo subaqueous pyroclastic flow deposits. Bull Volcanol 37:231–244

Yokoyama S (1974) Mode of movement and emplacement of Ito pyroclastic flow from Aira Caldera, Japan. Sci Rept Tokyo Kyoiku Daigaku 12:17–62

Zimanowski B, Büttner R, Dellino P, White DL, Wohletz KH (2015) Magma-water interaction and phreatomagmatic fragmentation. In: Houghton B, McNutt SR, Rymer H, Stix J (eds) Sigurdsson H. The Encyclopedia of Volcanoes Elsevier, Amsterdam, pp 473–484

GPSR Compliance

The European Union's (EU) General Product Safety Regulation (GPSR) is a set of rules that requires consumer products to be safe and our obligations to ensure this.

If you have any concerns about our products, you can contact us on ProductSafety@springernature.com

In case Publisher is established outside the EU, the EU authorized representative is:

Springer Nature Customer Service Center GmbH
Europaplatz 3
69115 Heidelberg, Germany

Batch number: 10756791

Printed by Printforce, the Netherlands